COMPUTER METHODS AND ADVANCES IN GEOMECHANICS
VOLUME 1

PROCEEDINGS OF THE TENTH INTERNATIONAL CONFERENCE
ON COMPUTER METHODS AND ADVANCES IN GEOMECHANICS
TUCSON/ARIZONA/USA/7 – 12 JANUARY 2001

Computer Methods and Advances in Geomechanics

Edited by

Chandra S. Desai, Tribikram Kundu, Satya Harpalani, Dinshaw Contractor & John Kemeny

The University of Arizona, Tucson, Arizona, USA

VOLUME 1

Keynote papers; 1 *Computer methods;* 2 *Neural networks, expert systems, reliability, optimization;* 3 *Constitutive models: Soils and rocks;* 4 *Constitutive models: Interfaces and joints;* 5 *Microcracking, localization, fracture, Nano-micro-macro correlation;* 6 *Applications, validation, construction processes;* 7 *Geoenvironmental engineering and coupled processes.*

A.A. BALKEMA/ROTTERDAM/BROOKFIELD/2001

The texts of the various papers in this volume were set individually by typists under the supervision of each of the authors concerned.

Published by
A.A. Balkema, P.O. Box 1675, 3000 BR Rotterdam, Netherlands
Fax: +31.10.413.5947; E-mail balkema@balkema.nl; Internet www.balkema.nl
A.A. Balkema Publishers, 2252 Ridge Road, Brookfield, VT 05036-9704, USA
Fax: 802.276.3837; E-mail: info@ashgate.com

For the complete set of two volumes, ISBN 90 5809 183 X
For Volume 1, ISBN 90 5809 184 8
For Volume 2, ISBN 90 5809 185 6

Printed in the Netherlands

Computer Methods and Advances in Geomechanics, Desai et al. (eds) © 2001 Balkema, Rotterdam, ISBN 90 5809 183 X

Table of contents

* Theme papers

2 *Neural networks, expert systems, reliability, optimization*

3 *Constitutive models: Soils and rocks*

4 *Constitutive models: Interfaces and joints*

5 Microcracking, localization, fracture, Nano-micro-macro correlation

6 *Applications, validation, construction processes*

Computer Methods and Advances in Geomechanics, Desai et al. (eds) © 2001 Balkema, Rotterdam, ISBN 90 5809 183 X

Preface

The Tenth International Conference on Computer Methods and Advances in Geomechanics was held in Tucson, Arizona, USA during 7 – 12 January 2001. The previous conferences in the series were held at Vicksburg, USA 1972; Blacksburg, USA 1976; Aachen, Germany 1979; Edmonton, Canada 1982; Nagoya, Japan 1985; Innsbruck, Austria 1988; Cairns, Australia 1991; Morgantown, USA 1994 and Wuhan, China 1997.

Since their inception, the conferences have grown continuously and significantly, in size and scope; the latter entails computer methods as well as other emerging and important topics. About 320 papers accepted for the tenth conference involve a wide range of topics: Computer methods; Neural networks, expert systems, reliability, optimization; Constitutive models: Soils and rocks; Constitutive models: Interfaces and joints; Microcracking, localization, fracture, Nano-micro-macro correlation; Applications, validation, construction processes; Geoenvironmental engineering and coupled processes; Laboratory and field testing, Parameter determination; Dynamics, earthquake analysis, liquefaction; Soil/Rock-structure interaction; Ground improvement and geosynthetics; Petroleum geomechanics, hydraulic fracturing, geological simulation; Pavement geomechanics; Foundations, footings, piles; Slopes, dams, walls, landslides; Tunnels, underground works, mining; Flow, seepage, consolidation; Education.

The tenth conference, held at the dawn of the new millennium, heralds new and innovative activities in a number of areas of geomechanics. The theme of the conference, *Fundamentals through Applications*, addresses the important issue of where the developments stand today and what the future will be in the application of research developments for the solution of complex problems of modern technology.

The tenth conference program is designed to encourage critical discussions so as to lead to an assessment of past developments and future application and research needs. The outcome is expected to help engineers and researchers to shoulder important responsibilities toward the use of the computer methods, constitutive models, laboratory testing and field validations for solution of a wide range of geotechnical problems, and to pursue advanced research for the understanding and definition of the issues that remain to be addressed.

We thank authors of the keynote, theme and contributed papers for timely submission and participation in the conference, and the members of the Organizing Committees for their support and guidance. We also express our appreciation to the technical and financial cosponsors, including the Department of Energy, USA, and various industrial organizations. Thanks are also due to the members of the staff including the Engineering Professional Development, University of Arizona (U of A), Mrs Rachèlè Logan of the Department of Civil Engineering and Engineering Mechanics, U of A, and Mr Jose Francisco Perez Avila for assistance for the web page.

Editors

Keynote papers

Computer Methods and Advances in Geomechanics, Desai et al. (eds) © 2001 Balkema, Rotterdam, ISBN 90 5809 183 X

Modeling the behavior of rockfill

E.E.Alonso

Department of Geotechnical Engineering and Geosciences, Universitat Politecnica de Catalunya, Barcelona, Spain

L.A.Oldecop

Earthquake Engineering Institute (IDIA), Universidad Nacional de San Juan, Argentina

ABSTRACT: Field experience reveals that rockfill behaviour is strongly dependent of water action. Tests in specially designed oedometer cells show that relative humidity of the air filling the void space is the controlling variable to introduce the effect of water. Fracture processes of rockfill particles provide also the necessary background to understand bulk behaviour. A strain hardening plasticity model, which incorporates these fundamental frameworks, has been developed. It explains main features of observed rockfill behaviour and its predictions have been satisfactorily compared with experimental results. Coupled flow deformation problems have also been tackled in the paper. A double porosity formulation seems consistent with the physics of the problem. Once the coupled model is formulated through the appropriate balance equations. Some simulations have been performed. They show the capabilities of the formulation developed and shed light on long term deformation mechanisms.

1 INTRODUCTION

Rockfill is a common construction material in engineering works (embankments, zoned and faced dams) which has received considerable attention from basic as well as applied geotechnical research. Rockfill shows unique characteristics of behaviour. Most structures built with crushed rock experience continuous deformations along many years after the completion of their construction. Moreover, if rockfill is impounded (as it occurs in the upstream shell of zoned dams, for example), significant additional deformations develop. This phenomenon is known as collapse and it implies an increment of strain at constant confining stress. In a few cases it was observed that collapse can take place even in non impounded rockfills, for example due to wetting by rainfall (Bauman, 1960; Naylor et al, 1997, Soriano & Sánchez, 1999, Sowers et al, 1965)

Additional difficulties arise from the fact that both, load induced and collapse strains, do not occur as a unique event, but rather as a time-dependent process (Nobari & Duncan, 1972). This fact leads to practical problems during the life of the structure, since operation has to be frequently interrupted in order to carry out remedial works. Post-construction time-dependent deformations can lead to different kinds of related problems: loss of freeboard in dams (Cetin et al, 2000), increase of seepage hazard due to cracking in clay core dams (Cetin et al, 2000), slab cracking in concrete faced dams (Bauman, 1960) and damage to roads, rails and buildings supported by a deforming embankment (Soriano & Sánchez, 1999). Post-construction deformation of the supporting embankment is particularly critical for high velocity railways, since the allowable deformations of rails for safe operation is very small (Rein et al, 1992; Soriano & Sánchez, 1999)

Experimental evidence (Marsal, 1973; Nobari & Duncan, 1972) supports the idea that the collapse behaviour of rockfill is related to particle breakage and the loss of rock strength due to water presence. Laboratory testing also shows that the features of rockfill which have influence on its mechanical behaviour are the rock strength and durability, the grain size distribution and the particle shape. Some rocks may be very sensitive to water action, determining high levels of post-construction strains, when the rockfill is subjected to impoundment or even to the sole environmental action. Ataturk Dam, a 170 meters high zoned rockfill dam with central clay core, experienced extremely severe post-construction settlements (up to 7 meters in ten years). This was caused by the deficient behaviour of the rock material used in the construction of the rockfill shells, a vesicular basalt, which was rapidly weathered in contact with the reservoir and rainwater. Such large deformations affected the overall safety of the dam due to the loss of most of its freeboard and also due to longitudinal cracking which reached the impervious core (Cetin et al, 2000). Although in most structures the collapse phenomenon is not as intense as in Ataturk Dam and hence, it does not imply a serious hazard, this case shows how severe can be the consequences that may arise from disregarding the influence of water in rockfill behaviour.

Finite element analysis is currently used in the design of large rockfill dams. Constitutive equations used in FE models are rather simplistic: Kulhawy's hyperbolic model (Nobari & Duncan, 1972; Veiga Pinto, 1983; Soriano & Sánchez, 1999) and K-G model (Naylor et al, 1986; 1997). None of these constitutive models can take into account the dependence of rockfill mechanical behaviour on the water action in a comprehensive manner. This is usually done by means of numerical artifices. Nobari & Duncan (1972) proposed such a procedure based on the "double oedometer" concept, usually applied to the computation of collapse strains in soils. The method considers two sets of material parameters for the constitutive model (the hyperbolic model in this case). The first set describes the material behaviour at its initial water content and the second set corresponds to the final water content (usually saturated). At a certain point of the analysis, the material properties are changed and the resulting strains are computed. Naylor et al (1986) extended this method introducing the possibility of using any constitutive model (including plasticity models). This type of approach has the shortcoming that the intermediate behaviour between the initial and final states remains unknown. Therefore, it is unable to reproduce gradual wetting processes, like those which occur in a natural environment, involving water addition by rainfall, water transport (in liquid and vapour states) within the rockfill structure and water loss due to evaporation.

The aim of this paper is to present a new mathematical tool for the modelling of rockfill structures, which takes into account the influence of water in the behaviour of the material. It makes use of an elasto-plastic constitutive equation, specially formulated for rockfill, on the basis of fracture propagation concepts and experimental results obtained from mechanical tests conducted in an environment with relative humidity control. The constitutive equation is integrated into a general FE computational code for thermohydromechanical analysis. This allows to simulate the transfer of liquid water and vapour, which occur within a rockfill structure in a natural environment.

2 A CONSTITUTIVE MODEL FOR ROCKFILL

Two distinct features of rockfill behaviour: time dependent strains and collapse strains, can be explained by a unique phenomenon acting at a micromechanical level. Since it seems that particle breakage is always involved in rockfill deformation,

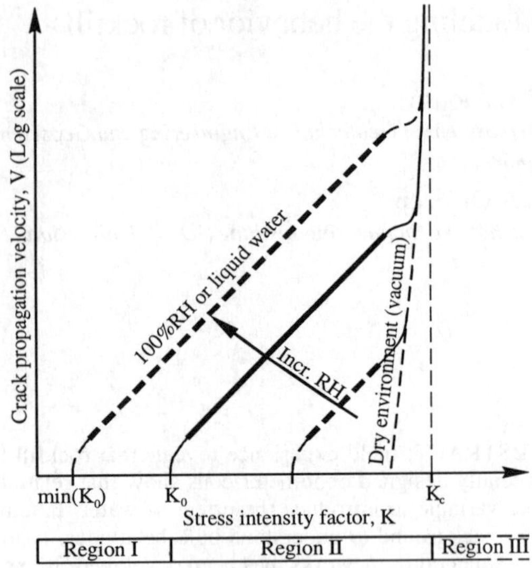

Figure 1. Schematic subcritical crack growth curves and conceptual model (Alonso & Oldecop, 2000).

it could be expected that fracture propagation phenomena determine those particular features.

Subcritical crack growth mechanism was proposed by Alonso & Oldecop (2000) to explain why both time dependent strains and collapse strains, simultaneously depend on the stress state and the water action. A rockfill element can be viewed as composed by a certain number of rock particles, each one containing a large number of micro and macro-cracks. Those cracks will propagate due to the interaction of load and water effects, a phenomenon which is known as environmentally assisted subcritical crack propagation (Broek, 1986). Each time one of those cracks fails, a particle breakage occurs and hence an increment of strain occurs.

Figure 1 shows the typical subcritical crack growth curves as can be obtained from fracture mechanic tests on rock specimens (Atkinson & Meredith, 1987) . The crack velocity propagation is a function of the applied loads and the water action. Load effects is taken into account by means of the stress intensity factor, K (Broek, 1986). Water action is conveniently measured by means of the relative humidity measured in the testing environment, i.e. the air surrounding the specimen. In the same figure, a conceptual model for rockfill deformation is presented. The stress intensity factor axis is divided in three regions. Cracks lying in region I ($K<K_0$) do not propagate at all. Hence in a steady state, when the rockfill is not undergoing any deformation, all cracks contained in the rockfill particles lie in region I. If a load increment is then applied, cracks will move along the K axis, following the particle geometry, the particular position of each crack and the crack length. Some cracks will fall in region III

where the stress intensity factor is greater than the fracture toughness, K_c. Those cracks will fail instantaneously producing the rockfill instantaneous component of strain, as observed in experiments (Sowers et al, 1965). Some other cracks will fall in region II and will grow, with a finite value of the propagation velocity, until they fail. These cracks determine the rockfill time–dependent component of strain.

If the relative humidity is increased, Figure 1 shows that cracks lying in region II will increase their propagation velocity. Moreover, some cracks from region I will move towards region II, due to the reduction in the limit value K_0 as the relative humidity increases. This will produce an additional increment of strain, which is not related to an increment in load, i.e. a collapse deformation.

The proposed conceptual model explains the characteristic behaviour of rockfill and its dependence on loads and water action. According to this model one may conclude that the key parameter which controls the influence of water in the mechanical behaviour of rockfill is the relative humidity measured in the air surrounding the rock particles (provided the gas phase is in thermodynamical equilibrium with the rock particles)

The previous reasoning led to the idea that carrying out mechanical tests of rockfill under controlled relative humidity conditions could provide a better knowledge of the influence of water in the rockfill compressibility properties. Large-scale one-dimensional compression tests were carried out on a rockfill type material, a crushed slate of Cambric origin. A relative humidity control system was added to a Rowe type oedometer (Oldecop & Alonso, 2000). Some experimental results, obtained with this equipment, are shown in Figure 2 and Figure 3. Figure 2.a shows the paths followed by the tests performed in the total suction-vertical stress space. Figure 2.b shows the measured vertical strain vs the applied vertical stress. Figure 3 shows a plot of the measured collapse strain in test 4 (with RH control) vs total suction.

Total suction, ψ, is used here as a parameter to measure the water action on rockfill behaviour, in an identical way as could be done with the relative humidity. This can be justified by assuming isothermal conditions and that thermodynamic equilibrium between gas and liquid phases exists within the specimen. Under these conditions, relative humitity and total suction are bi-univocally related by the psichrometric relationship (Coussy, 1995), and hence they can be used as interchangeable parameters.

The previously presented experimental results clearly show that the mechanical behaviour of rockfill depends on the total suction (or alternatively on

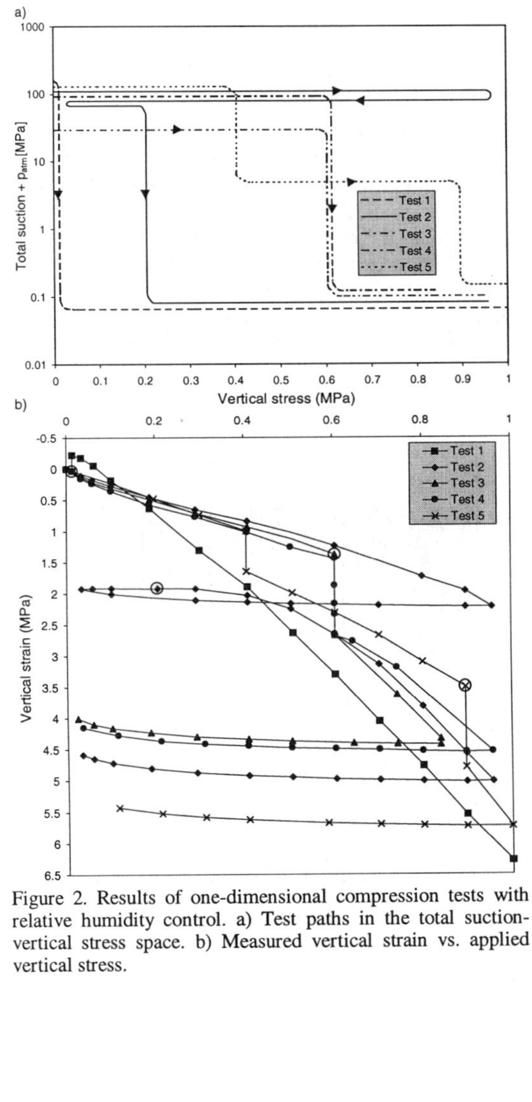

Figure 2. Results of one-dimensional compression tests with relative humidity control. a) Test paths in the total suction-vertical stress space. b) Measured vertical strain vs. applied vertical stress.

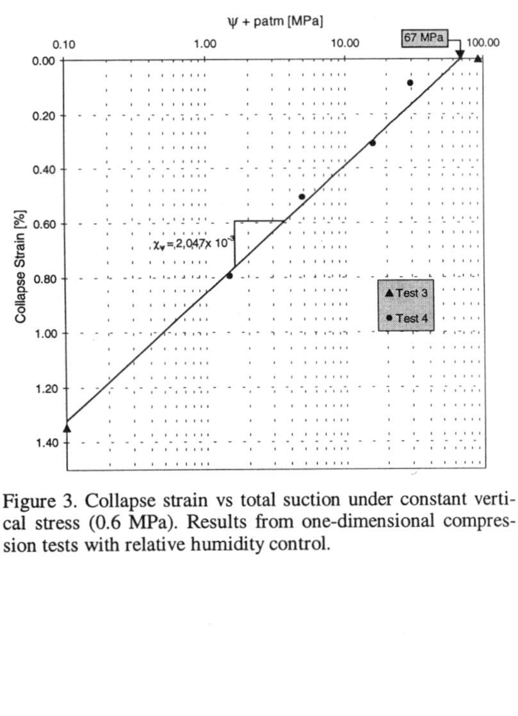

Figure 3. Collapse strain vs total suction under constant vertical stress (0.6 MPa). Results from one-dimensional compression tests with relative humidity control.

the RH). Normal compression lines in Figure 2 show that the material compressibility increases as total suction decreases. Measured data of Figure 3 show a strong correlation between the measured collapse strain and total suction. On the basis of this experimental evidence, a new elasto-plastic constitutive model for rockfill compression was developed. The influence of water in the mechanical behaviour is introduced by means of the total suction. The formulation is briefly presented here. A comprehensive description of the model is given in Oldecop & Alonso (2000).

The constitutive model is formulated in three steps. In a first step, only isotropic volumetric behaviour is considered and water influence is introduced. A yield surface is obtained in the mean stress-total suction space. In a second step, the volumetric yield surface is combined with some other triaxial formulation in order to include the effect of deviatoric stresses. Finally, the formulation is extended to general three-dimensional stress states by combining the triaxial yield surface with the Von Mises failure criteria.

As follows from the experimental results presented in Figure 2.b, linear normal compression lines are assumed for the material at different total suction values. The compressibility index in normal compression is a function of the total suction for mean stresses in excess of a threshold value, p_y. If the mean stress, p, is lower than p_y, then the normal compressibility index is a constant. Parameter p_y is interpreted as the stress value which marks the onset of particle breakage. Strains for $p<p_y$ are due to the rearrangement of the granular structure (i.e., referring to Figure 1, when $p<p_y$, every crack contained in the rock particles lies in region I). The incremental strain-stress relationships in normal compression are:

$$\text{for } p < p_y \quad d\varepsilon = \lambda^i\, dp \tag{1}$$

$$\text{for } p \geq p_y \quad d\varepsilon = [\lambda^i+\lambda^d(\psi)]\, dp \tag{2}$$

where ε is the total volumetric strain, λ^i is the normal compressibility index for $p<p_y$ and $(\lambda^i + \lambda^d)$ the slope of the normal compression line when $p \geq p_y$.

Also a linear strain-stress relationship is assumed for unloading-reloading paths. The elastic strain increments within unloading-reloading (URL) paths are given by:

$$d\varepsilon^e = \kappa\, d\sigma \tag{3}$$

where the slope κ is assumed to be independent of the water action. Test 1, in Figure 2b, shows a moderate swelling behaviour when the specimen is flooded under low stress. This is assumed to be a reversible (elastic) strain. A linear relationship is assumed to exist between the swelling strain and the logarithm of ψ:

$$d\varepsilon^\psi = \kappa_\psi \frac{d\psi}{(\psi + p_{atm})} \tag{4}$$

where p_{atm} is the atmospheric pressure and κ_ψ is the expansion/compression index due to suction changes, which is assumed to be independent of the stress level for simplicity.

The yield surface in the p-ψ space is given by:

$$F(p, \psi) = p_0(\lambda^i+\lambda^d(\psi)-\kappa)-p_y\lambda^d(\psi)-p_0^*(\lambda^i - \kappa) = 0 \tag{5}$$

where p_0 is the yield stress and p_0^* is the hardening parameter. Equation (5) is the yield surface for $p>p_y$. If the stress state does not exceed the threshold value p_y, the yield surface simply reads:

$$F(p,\psi) = p_0 - p_0^* = 0 \tag{6}$$

A strain hardening law is proposed:

$$dp_0^* = d\varepsilon^p / (\lambda^i - \kappa) \tag{7}$$

where ε^p is the volumetric plastic strain.

The compressibility coefficient, $\lambda^d(\psi)$, is related to changes in total suction by:

$$\lambda^d(\psi) = \lambda^d_0 - \alpha_\psi \operatorname{Ln}\left(\frac{\psi + p_{atm}}{p_{atm}}\right) \text{ and } \lambda^d \geq 0 \tag{8}$$

where λ^d_0 and α_ψ are model parameters. λ^d_0 is the compressibility index for the saturated material. If the material is dried (ψ increases) the compressibility decreases until a "very dry state" is reached. In such a state, the compressibility index is λ^i (since $\lambda^d = 0$). For the experimental results presented in Figure 3, $\psi \geq 67$ MPa defines the very dry state. It should be noted that the proposed hardening law (eq. 7) is identical to the stress-strain relationship, which arises from the model in the very dry state.

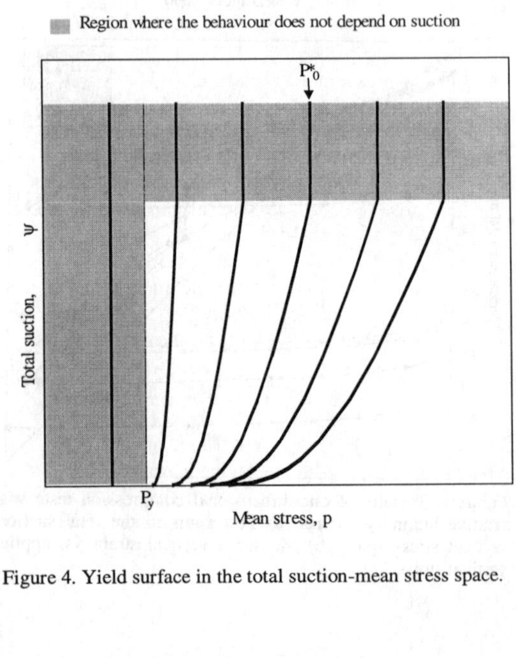

Figure 4. Yield surface in the total suction-mean stress space.

Introducing expression (8), for $\lambda^d(w)$ in eq. (5), the yield surface in the p-ψ plane is obtained. A plot of this yield surface for different values of the hardening parameter, p_o^*, is shown in Figure 4. The regions (p,ψ) in which the rockfill behaviour does not depend on total suction are indicated in grey. They correspond to stress levels below the threshold stress p_y and to very dry states ($\psi \geq 67$ MPa).

The proposed model can be easily extended to triaxial stress states by combining it with the modified Cam-Clay model, which is used here for simplicity:

$$F(p, q) = q^2 - M^2 p (p_0 - p) = 0 \qquad (9)$$

where q is the deviator stress, p_0 is the yield stress in isotropic compression and M is the slope of the critical state line (a model parameter). Any other hardening model could, however, be used in the deviatoric plane. Rearranging expression (5):

$$p_0 = [p_0^* (\lambda^i - \kappa) + p_y \lambda^d]/(\lambda^i + \lambda^d - \kappa) \qquad (10)$$

Equations (9) and (10) together with expression (8) define the yield surface in the p-ψ-q space. In order to integrate the proposed constitutive model in a FE computer code, it has to be extended to general three-dimensional stress states. Using the Von Mises failure criterion, the generalised version of eq. (9) reads (Gens & Potts, 1988):

$$F(p, J) = 3 J^2 - M^2 p (p - p_0) \qquad (11)$$

where J^2 is the second invariant of the deviatoric stress tensor and θ is the Lode's angle. The modified Cam-Clay model usually overestimates the coefficient of earth pressure at rest. In order to avoid this problem, the original associated flow rule is changed to non-associated by introducing a plastic potential of the form (Ohmaki, 1982):

$$G(p, J) = 3 J^2 - \alpha M^2 p (p - p_0) \qquad (12)$$

where α is a model parameter.

Summarising the proposed model elements, the yield surface is given by eqs. (11), (10) and (8), the plastic potential by eqs. (12), (10) and (8), the hardening law by eq. (7) and the behaviour in the elastic range by eqs. (3) and (4). Model parameters are: p_y, λ^i, λ^d_0, α_ψ, κ, κ_ψ, M and α.

The one-dimensional compression tests shown in Figure 2 were simulated with the model developed. A comparison between model performance and testing results for tests no. 5 is shown in Figure 5. The following rockfill parameters were used in this simulation: $\lambda^i = 2.3 \times 10^{-2}$ MPa^{-1}; $\lambda^d_o = 5.1 \times 10^{-2}$ MPa^{-1}; $p_y = 0.4$ MPa; $\kappa = 0.26 \times 10^{-2}$ MPa^{-1} $\kappa_\psi = 0.04 \times 10^{-2}$ MPa^{-1} $\alpha_\psi = 0.78 \times 10^{-2}$ MPa^{-1}. The measured partial collapse under a vertical stress $\sigma_v = 0.4$ MPa is not well captured in this particular test but the overall comparison is quite satisfactory.

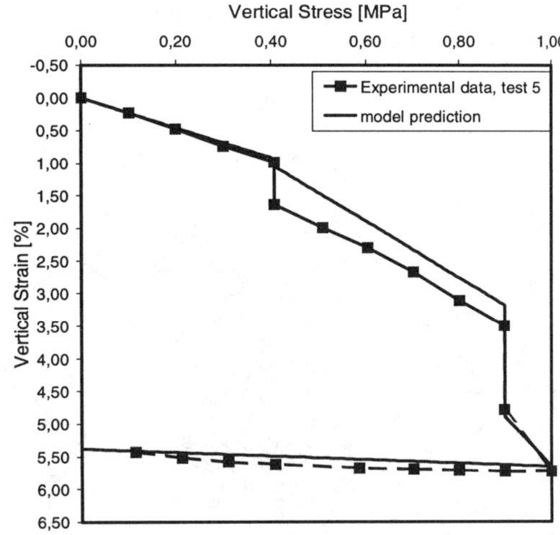

Figure 5. Relative humidity controlled oedometer test no. 5. Comparison between model and experimental results.

3 FLOW PROCESSES IN A ROCKFILL

The very high permeability of a rock fill makes it extremely unlikely in practice a condition of saturated flow. In a zoned earth and rockfill dam the upstream shoulder becomes inundated when the water reservoir level is increased. The downstream shoulder is however typically subjected to rainfall infiltration (and evapotranspiration) from and towards the atmosphere.

Similar hydraulic boundary conditions are expected in rockfill embankments used in highway or railway construction. The large open voids of the rockfill remain therefore essentially filled with air and water vapor during the lifetime of these structures. As illustrated in Figure 6, liquid water flow is likely to develop as a film transport mechanism around the large rock blocks. If the rock blocks are not fully saturated water may also penetrate into the particles through the connected porosity and also along cracks. Specially relevant for the purposes of rockfill deformation are the cracks which develop at the particle contacts.

A simplified framework for this complex transport mechanism could be established by postulating two energy states of the rockfill water: the energy associated to the state of the water in open voids and the water energy inside the rock blocks. The first one will control the flow of free water. This free water will be retained by capillary actions at the grain-to-grain contacts and also by rock surface effects. The fundamental mechanisms of water transfer under these circumstances have been discussed, for instance, in Gili (1988). In the present work a generalized Darcy type of equation for nonsaturated material will be used, instead. It is a simple formulation in which the rate of water transfer is controlled by the bulk degree of saturation through a relative permeability term.

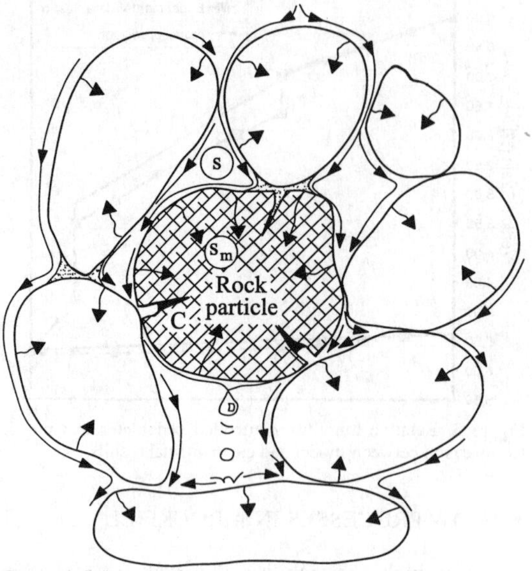

Figure 6. Scheme to show water transfer in a rockfill.

The local loss of water towards (or from) the rock blocks will be made equivalent to a sink (source) term, which depends on the difference of water energy (or total suction) in the open voids and the rock matrix (s and s_m respectively; see also Figure 6). A simple linear formulation for this sink term is therefore:

$$q = -\frac{dw^m}{dt} = -\rho_w \alpha'(s - s_m) \quad (1)$$

where q is the flow rate interchanged between the open voids and the rock particles; w^m is the mass of microstructural water (water in the rock particles) and α' is a phenomenological transfer coefficient. If ϕ_m is the porosity of the rock particles, the rate of change of the microstructural mass of water may be also defined as:

$$\frac{dw^m}{dt} = \rho\,(1-\phi)\,\phi_m \frac{dS_l^m}{dt} \quad (2)$$

where S_l^m is the microstructral degree of saturation and φ is the bulk (or macrostructural) porosity of the rockfill. Computational results are presented later in terms of the transfer parameter $\alpha = \alpha'/\phi_m$ (MPa s)$^{-1}$.

Water retention properties are also required to complete the formulation. They were determined for the Cambric slate. The following expression may, for instance, define the matrix degree of saturation of the rock particles:

$$S_l^m = \exp\left(-s_m / p_o\right) \quad (3)$$

where p_o is a reference suction (model parameter).

The coupled flow-deformation problem of the rockfill may be defined through the equations of water mass balance and mechanical equilibrium.

3.1 *Water mass balance*

$$\frac{\partial}{\partial t}\left(\theta_l^w S_l \phi + \theta_g^w S_g \phi\right) + \nabla \cdot \left(\mathbf{j}_l^w + \mathbf{j}_g^w\right) = f^w = q \quad (4)$$

In this expression φ is the rockfill porosity, θ_l^w and θ_g^w are the masses of water per unit volume of liquid and gas respectively; $S_g = 1 - S_l$ and j_l^w and j_g^w are the flow rates (both diffusive and advective of water in liquid and gas phases. The term f^w describes the external supply of water and, in our case it will be represented by the local sink term q as given in equation (1).

3.2 *Stress equilibrium*

It may be expressed in terms of total stress components, σ, and body forces, b, as:

$$\nabla \cdot \boldsymbol{\sigma} + \mathbf{b} = 0 \quad (5)$$

Internal capillary stress components are assumed to be negligible and therefore formulations used for unsaturated soils (Alonso et al. 1988) are not required here.

The set of equations (1), (4) and (5) may be solved provided a number of constitutive relations and equilibrium restrictions are included (Olivella et al. 1994, 1996). For the open void structure of the rockfill the following information must be provided: the generalized Darcy's law for liquid flow transfer, a Fick's law for vapor diffusion transfer and a water retention relationship between bulk degree of saturation and water suction. The local flow (sink/source term) requires the specification of a water retention relationship (such as equation (3)) and the local transfer coefficient α (equation (1)).

Rock particles do not deform in the model developed. Rockfill deformation, including particle breakage, is represented by the overall macroscopic constitutive equations developed in a previous section. They may be conveniently described as:

$$d\sigma = \mathbf{D}_{ep}\left(d\varepsilon - d\varepsilon_o(s_m)\right) \quad (6)$$

where $\mathbf{D}_{ep}$ is the elastoplastic constitutive matrix and $\mathbf{d\varepsilon_o}$ are the strains associated with changes in the particles water content or, its equivalent, the microstructural suction s_m. Expressions for $\mathbf{D}_{ep}$ and $\mathbf{d\varepsilon_o}(s_m)$ may be derived on the basis of the constitutive relations presented before.

The formulation outlined above has been implemented into an existing numerical tool for thermo-hydro-mechanical analysis of unsaturated/saturated porous media (CODE_BRIGHT). Details of the theoretical basis and numerical formulation of this code may be found in Olivella et al. (1994, 1996). Macrostructural suction is selected as the independent suction variable. At every time increment, s_m may be determined locally through equations (1) and (2).

4 APPLICATION OF THE MODEL

Test no. 2 has been selected to compare the predictions of the coupled flow-deformation model with measured behavior. The stress path of the test was presented in Figure 2. The test sample was simulated by 10 elements of high (macro) permeability, which guarantees a fast transmission of boundary suctions to the whole sample. Figure 7 shows the stress and suction values imposed in the test no. 2 along time. This sequence reproduces the actual testing data. The sample is first loaded, unloaded and reloaded at constant suction (s = 92 MPa). Relative humidity is then increased until saturation and a final loading-reloading cycle is then imposed at zero suction. The preceeding set of constitutive model parameters were used in the simulation.

Macro and micro suctions are initially assumed to be in equilibrium. The change in macrovoid suction initiates a transient mechanism between open voids and rock particles, which may be seen in Figure 7 when macro and micro suction have different values. The time evolution of micro suction inside the rock particles is controlled by the α parameter. Three values have been used in three alternative simulations of test results. In all cases the mechanical response of the model is governed by the microstructural suction. When α reaches a high value ($\alpha = 10^{-5}$ (MPa s)$^{-1}$) in case c (Figure 7) local transfer of water is fast and micro and macro suctions evolve in parallel adopting the same values.

Calculated and measured mechanical response of the sample is presented in Figure 8 for the same 3 values of α selected for the analysis. Local water transfer affect in a significant way the results. When the suction is first reduced (wetting) at a constant p stress (0.2 MPa) the value of α controls the velocity of microstructural suction reduction. If the reduction is slow, the reaction of the rockfill (small swelling) is also delayed and no instantaneous deformation is calculated. However, when α is large enough, the reduction in microstructural suction follows closely the boundary change and a small instantaneous rebound is computed. Evidence of the delayed local transfer of water is also seen when computed deformations at the highest value of applied stress are compared. When α is small (Figure 8a) there is still a difference between micro and macro suctions. The first one tends slowly towards equilibrium and this process induces time-delayed deformations at constant p stress. This phenomenon is progressively reduced as α increases.

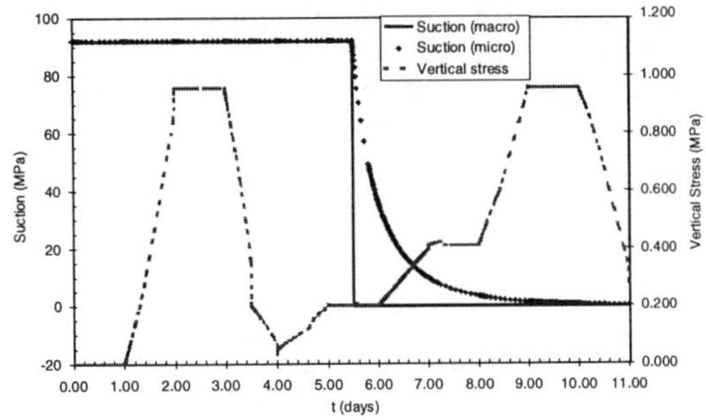

(a) $\alpha = 8x10^{-8}$ (MPa s)$^{-1}$

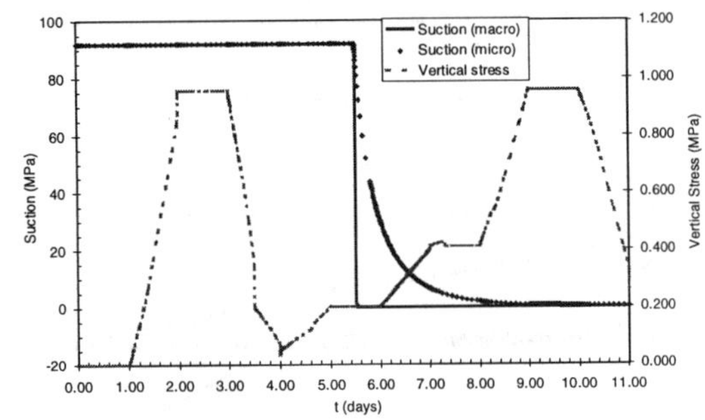

(b) $\alpha = 1x10^{-7}$ (MPa s)$^{-1}$

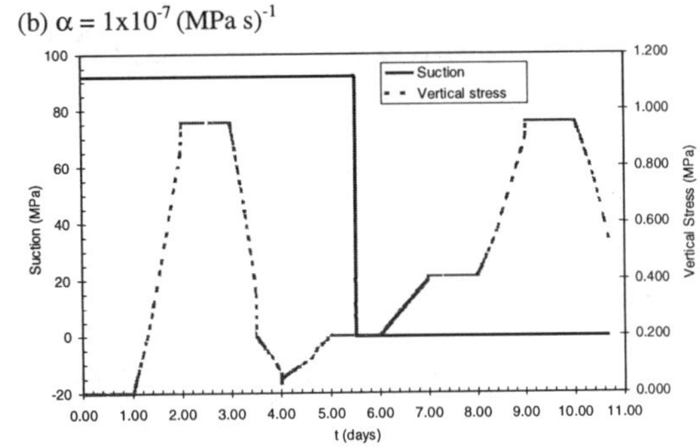

(c) $\alpha = 1x10^{-5}$ (MPa s)$^{-1}$

Figure 7. Suction and vertical stress histories on test no. 2. Microstructural suction computed for three values of α (a, b, c).

This example points out the capabilities of the double structure framework. Note that it introduces a "secondary" time effect linked to the transfer of water inside the rock particles. This time effect is not the only one explaining the delayed deformation effects in rockfill. In fact tests at constant (overall) suction (presumably in equilibrium with microstructural suction) show a finite crack propagation velocity, which results in delayed deformations. This additional time effect has not explicitly taken into account in the model developed. The dependence of crack propagation velocity on microstructural suction seems consistent with the postulated mechanism of subcritical crack propagation. In fact, it is the potential energy of the water partially filling the active cracks that actually controls its propagation. Microstructural suction, as defined above is probably a convenient average approximation to the suction of water in the vicinity of crack tips.

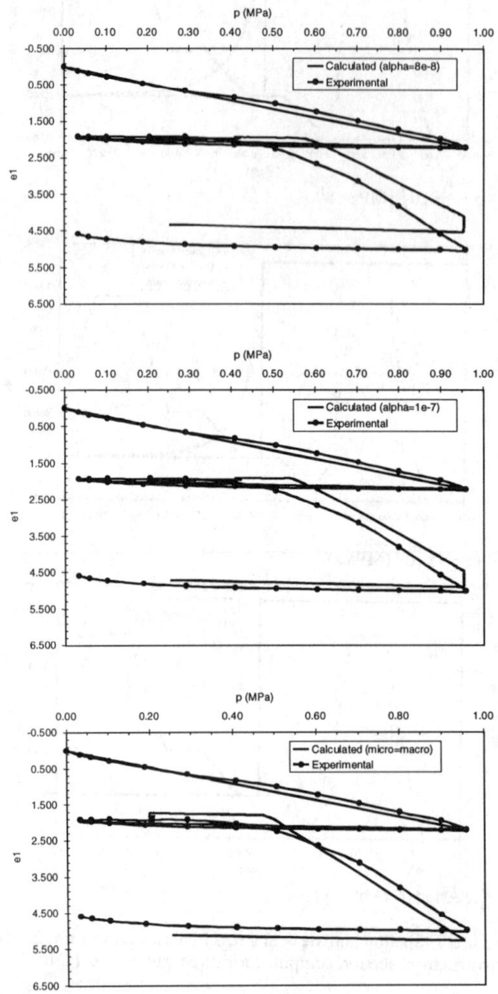

Figure 8. Results of the simulation of test no. 2 for three values of parameter α.

5 CONCLUSIONS

A subcritical crack propagation mechanism provides a convenient framework to develop a macroscopic constitutive model for rockfill materials. The model includes the effect of relative humidity changes on rockfill behavior. Laboratory tests as well as field observations demonstrate that humidity changes have a relevant effect on rockfill behavior.

The macroscopic model has been cast in terms of hardening plasticity. It is believed to be the first one to incorporate in a consistent manner water effects on rockfill behavior. The model was first developed for isotropic stress states and later extended to triaxial and general stress conditions. In its simplest form, for isotropic stress conditions, the model is described by six parameters, which may be ideally derived from compressibility tests, in which relative humidity is controlled.

Some parallelism between the behavior of moderately active unsaturated soils and rockfill may be found. The elastoplastic model developed by Alonso, Gens and Josa (1990) for unsaturated soils has also formal similarities with the model presented here. However, major differences remain and they have been collected in Table 1 to underline the fundamental differences between rockfill and unsaturated soil behavior.

Table 1. Comparison between unsaturated soil and rockfill behavior.

Unsaturated rockfill	Unsaturated soil
Collapse is associated with particle breakage and a subsequent rearrangement of structure	Collapse is associated with particle rearrangement
Particle toughness is a fundamental property	Particle strength does not affect overall behaviour
The effect of suction is to control particle breakage velocity	The effect of suction is to "prestress" soil structure
Particle toughness is included in the model through a single parameter (confining stress, p_y) For $p<p_y$ no time delayed deformations exist (no collapse)	There is no equivalent parameter
TOTAL suction controls water induced effects	MATRIC suction controls water induced effects
Time delayed deformations (and hence collapse) is inhibited for very dry states	There is no equivalent concept
Yield stress for the very dry state is conveniently chosen as a hardening parameter	Yield stress of the saturated soil is conveniently chosen as the hardening parameter
Elastoplastic strains (instantaneous and delayed) are linearly related to confinig stress for the relevant range of stresses in practice	Elastoplastic stains are linearly related to log (confining stress)
In a compacted material yield starts at zero stress	A compacted material is characterized by an initial yield locus

A further development to the mechanical model comes from the consideration of water transfer mechanisms in rockfill materials. Since wetting or drying the rock particles has a very distinct effect on overall behavior, a procedure to model the local water interchanges has been set out. Particle breakage has been related to the suction or water energy prevailing inside rock particles. The method developed includes a secondary time effect whose relevant features have been explored in the paper. The coupled flow-deformation model presented provides a comprehensive tool for the analysis of rockfill structures subjected to environmental changes as well as mechanical loading.

REFERENCES

Alonso, E.E., F. Batlle, A. Gens & A. Lloret 1988. Consolidation analysis of partially saturated siols. Application to earthdam construction. *Proceedings of the 6th Int. Conf. on Num. Meth. in Geomech.* Innsbruck: 1303-1308.

Alonso, E.E., A. Gens & A. Josa 1990. A constitutive model for partially saturated soils. *Géotechnique*. 40: 405-430.

Alonso, E.E. & L.A. Oldecop 2000. Fundamentals of rockfill collapse. *1st Asian Conference on Unsat. Soils. Singapore.*

Atkinson, B. K. & Meredith, P. G. (1987). The theory of subcritical crack growth with applications to minerals and rocks. *Fracture Mechanics of Rock,* B. K. Atkinson, ed. London: Academic Press Inc., 111-166.

Bauman, P. 1960. Rockfill dams: Cogswell and San Gabriel Dams. *Transactions of the ASCE,* 125 (2): 29-57

Broek, D. 1986. *Elementary engineering fracture mechanics.* Dordrecht: Martinus Nijhoff Publishers.

Coussy, O. 1995. *Mechanics of porous continua.* John Wiley & Sons Ltd. Chichester

Cetin, H., Laman, M. & Ertunç, A. 2000. Settlement and slaking problems in the world's fourth largest rock-fill dam, the Ataturk Dam in Turkey. *Engineering Geology.* 56: 225-242.

Gens, A. & D. M. Potts 1988. Critical state models in computational geomechanics. *Engineering Computations Journal*: 178-197

Gili, J. 1988. *Modelo microestructural para medios granulares no saturados.* Tesis Doctoral. Barcelona: UPC.

Marsal, R. J. 1973. Mechanical properties of rockfill. *Embankment Dam Engineering. Casagrande Volume.* Hirschfeld, R. C. & Poulos, S. J., eds. John Wiley & Sons.

Naylor, D. J., Maranha das Neves, E., Mattar, D. y Veiga Pinto, A. A 1986. Prediction of construction performance of Beliche Dam. *Géotechnique,* 36 (2): 221-233.

Naylor, D. J., Maranha, J. R., Maranha das Neves, E. & Veiga Pinto, A. A. 1997. A back-analysis of Beliche Dam. *Géotechnique* 48 (2): 221-233.

Nobari, E. S. & Duncan, J. M. 1972. *Effect of reservoir filling on stresses and movements in earth and rockfill dams.* Department of Civil Engineering, Report No. TE-72-1, University of California.

Oldecop, L.A. & E.E. Alonso 2000. A model for rockfill compressibility. *Géotechnique*. In press.

Olivella, S., J. Carrera, A. Gens & E.E. Alonso 1994. Nonisothermal multiphase flow of brine and gas through saline media. *Transport in Porous Media.* 15:271-293.

Olivella, S., A. Gens & A. Josa 1996. Numerical formulation for a simulator (CODE_BRIGHT) for the coupled analysis of saline media. *Engineering Computations.* 13, 7: 87-112.

Ohmaki, S. 1982. Stress-strain behaviour of anisotropically, normally consolidated cohesive soil. *Proceedings 1st Int. Symp. Num. Methods on Geomechanics. Zurich: 250-269.*

Rein, J.A., Soriano, A., Pradera E. 1992. Estabilidad de los terraplenes del ferrocarril de alta velocidad del NAFA. *III Simposio Nacional sobre Taludes y Laderas Inestables*, La Coruña: 509-520.

Soriano, A. & Sánchez, F. J. 1999 Settlements of railroad high embankments. *Proc. XII European Conf. on Soil Mech. and Geotech. Eng.*, Netherlands.

Sowers, G. F., Williams, R. C. & Wallace, T. S. 1965. Compressibility of broken rock and settlement of rockfills. *Proc. 6th ICSMFE,* 2, Montreal: 561-565.

Veiga Pinto, A.A. 1983. *Previsao do comportamento de enrocamento.* PhD Thesis. Lisboa: Laboratorio Nacional de Engenheria Civil.

Computer Methods and Advances in Geomechanics, Desai et al. (eds) © 2001 Balkema, Rotterdam, ISBN 90 5809 183 X

Geotechnical issues in deep and ultra deep waters

C.P. Aubeny, D.J. Murff & J.M. Roesset
Texas A&M University, College Station, Tex., USA

ABSTRACT: The oil industry's success in finding hydrocarbons in deep water has lead to a number of important economic and technical challenges involving, among others, new drilling procedures, new structural concepts and configurations, new types of foundations and new materials. In this paper we discuss some of the problems that have been encountered in the area of geotechnical engineering, with special emphasis on deep and ultra-deep waters. After a brief review of the historical developments and a short description of typical soil profiles in the sea bottom for different parts of the world, the paper concentrates on three main topics in need of additional integrated theoretical and experimental research: the design of reliable anchors that can resist significant vertical loads (or loads with an important vertical component) as imposed by vertical tethers or taut mooring lines; the characterization of the subsea soil conditions and their properties; and the identification of geotechnical hazards such as over pressured sands, hydrate formations, and slope instability.

1 INTRODUCTION

The oil industry has now ventured into very deep water off the continental shelves in search of large hydrocarbon resources and in doing so has had to face a number of important economic and technical challenges. The transition from 1,500 feet of water to 3,000 feet occurred over a ten-year period during which intensive research and development efforts were carried out. As these efforts progressed, recognition of the relative merits of the various structure and foundation concepts under consideration gradually evolved. This work took place concurrently by widely dispersed, international teams engaged in the design, fabrication, and installation of the platforms. The traditional steel jackets of the Gulf of Mexico or concrete gravity structures of the North Sea have been replaced by compliant structural systems consisting of large floating bodies attached to the seabed by vertical tethers or mooring lines (either catenary or taut moorings). The transition from 3,000 to 5,000 or even 6,000 feet, which has very recently taken place, has seemingly involved more incremental innovations. For the challenge now facing industry, the transition from 6,000 to 10,000 feet of water or more, another major research and development effort on a global scale may be needed.

The contributions of the geotechnical engineering community to the development and evolution of offshore engineering, the problems encountered in shallow and deep waters, and the issues and research needs created by the move into ultra-deep waters were discussed by Lacasse (1999) in an excellent review paper presented at the 31st Offshore Technology Conference (OTC). In this paper she emphasized the need for collaboration between geologists, geophysicists and geotechnical engineers in addressing these issues. She also underscored the importance of both in situ and laboratory testing. We agree wholeheartedly with these two points but we also feel that there is above all a need for well coordinated and balanced experimental (field and laboratory) and theoretical (analytical, numerical and computational) work. New theoretical formulations that have not been properly validated by experimental observations are of little practical value, and advanced computations that do not reflect the real material behavior in situ are useless. On the other hand, the use of empirical approximations based on limited observations to predict the performance of real foundations at a vastly different scale, without a solid theoretical basis, is far from satisfactory.

In this paper we discuss some of the problems that have been encountered in the area of geotechnical engineering, requiring new applications of analysis methodologies and new experimental data. Further, we discuss some recent examples of progress in research directed at solving these problems using new innovative concepts that is being carried out at different research institutions. Two general problems unique to deepwater development are highlighted. The first of these is the design of reliable anchors that can resist lateral and vertical (uplift) loads, can be accurately positioned and can be quickly and efficiently installed. The second of these involves geotechnical hazards that are associated with the deepwater environment. For example, severe drilling problems have occurred with relatively shallow (<2000 ft. below mud line), high-pressured sands frequently encountered in deep-water environments. Sands of this type have caused numerous borehole instabilities requiring termination of drilling and relocation at great expense. A main consideration in both areas is our ability to determine with acceptable accuracy the properties of the soil and its in situ conditions in water depths of several thousand feet. Finally we speculate on new challenges that our next step into deep water is likely to reveal.

2 HISTORICAL DEVELOPMENT

Lee (1982) presented an interesting review of the history of offshore drilling and development at the 1982 Conference on the Behavior of Offshore Structures (BOSS Conference). A very brief summary of that study and an updating to account for recent progress will be included here for completeness.

Over-water drilling and development of hydrocarbon resources began in Baku, in the Caspian Sea, in 1877, and around ten years later off the coast of California where piers constructed from the beach near Santa Barbara were used to support the drilling rigs and facilities. This activity was extended to coastal and inland waterways and lakes, especially in the marshes of southern Louisiana and Lake Maracaibo, over the next 50 years. In 1947 the first offshore platform out of the sight of land was installed in the Gulf of Mexico off the coast of Louisiana in 20 feet of water. Subsequently, water depths for exploration and production activities gradually increased to about 500 feet by the mid-seventies. Virtually all of the platforms up to this time were steel jacket platforms supported by steel pipe piles. Toward the end of this period activities significantly increased outside the Gulf of Mexico, particularly in the Middle East, Australia, and the North Sea.

Over the next fifteen years water depth limits continued to increase to about 1500 feet. The main workhorse remained the steel jacket platform supported on steel pipe piles but new innovations became more prominent. In the North Sea, the use of giant, concrete gravity platforms founded on mat foundations became common. The hard, flat glaciated seafloor and the availability of inland deepwater construction sites were especially well suited for such structures. Further, the use of subsea production systems, well clusters, and satellite wells became common. During this period however, it became obvious that relatively rigid, fixed bottom structures were being pushed to their limits. It became more and more difficult and expensive to hold the natural periods well below the dominant sea state periods. The use of compliant systems with periods well in excess of the dominant sea state periods afforded a solution to this dilemma.

In the past ten years or so, the compliant system has become the preferred solution for virtually all deepwater development (with the exception of subsea systems that typically have limited wells and are often used in conjunction with or for augmentation of development.) Several compliant concepts are being used at the time of this writing including the Tension Leg Platform (TLP), the Floating Production System (FPS), the Spar, and the Compliant Piled Tower (CPT). Selection of a specific concept depends on water depth, environmental conditions, and functional requirements among other variables. The CPT obviously must be founded on piles but the other three systems have more foundation options. Piles were initially used to anchor the first few compliant systems and drag embedment anchors have been selected on a much more limited scale. However, a recent innovation called the suction caisson has quickly proved to have significant advantages over the other options, especially in very deep water. The suction caisson will be discussed in more detail in subsequent sections.

3 TYPICAL SOIL PROFILES

Seafloor soil conditions are a major consideration in the design of foundations for offshore platforms. Three examples of widely differing offshore foundation conditions are presented below: (1) the weak compressible Gulf of Mexico sediments which historically have required pile foundations for permanent structures, (2) the more competent glaciated, North Sea soils which are capable of supporting massive gravity structures on shallow foundations, and (3) collapsible calcareous soils off of Australia which require special foundation design measures; e.g. drilled and grouted piles.

3.1 *Gulf of Mexico*

Typical soil conditions on the upper continental slope of the Gulf of Mexico (GOM) include: (1) high plasticity clays with liquid limits in the range LL=65-100 and plastic limits on the order of 25, (2) under to normally consolidated clays with water contents ranging from more than 100% at the mudline to about 35% at depths greater than 150 ft, (3) undrained shear strength which is very low at the mudline and which increases roughly linearly with depth, and (4) high compressibility with reported compression indices (Bryant and Delflache, 1971) in the range 0.53-1.54. Figure 1 illustrates a typical range of soil strengths. While measured undrained strength can vary widely according to the method of measurement (e.g., miniature vane, field vane, unconfined compression, UU triaxial compression) all methods show the above-mentioned trend of increasing strength with depth. Young et al (1983) report sensitivities (the ratio of undisturbed to remolded strength) as high as 6.

Soil conditions on the upper continental shelf are relatively uniform to water depths of about 500 ft. Beyond this depth conditions can become more complex, largely due to movement of the underlying salt. Campbell (1997) discusses a number of hazards associated with these more complex regions including steep and potentially unstable slopes, recently active faults, landslide areas, over-pressured sands (shallow water flows,) and gas hydrates. These hazards will be discussed in more detail subsequently.

3.2 *North Sea*

In many south and central North Sea locations, soils are primarily stiff clays and sands; however, normally consolidated clay profiles do occur in deeper waters. Figure 2a (Lunne et al, 1985; Gullfaks "A") illustrates an overconsolidated clay profile in 440 ft of water, comprised of glaciomarine clays of medium plasticity (LL=41-50) with water contents near the plastic limit. The soil is moderately to highly overconsolidated (OCR=3-15) with correspondingly high undrained shear strengths in the range 3,000-6,000 psf. In contrast, a normally to lightly overconsolidated (OCR=1-2) clay profile occurs at Gullfaks C in 720 ft of water (Figure 2b), with undrained shear strengths in the upper 30 ft on the order of 400-800 psf. This is more typical of sediments in the Norwegian Trench. Finally, Figure 2c (Sleipner Area 4) shows a profile comprised of 70 ft of fine dense sand overlying a moderately overconsolidated (OCR=1-3) stiff glaciomarine clay, with undrained shear strengths generally in the range 2,000-4,000 psf.

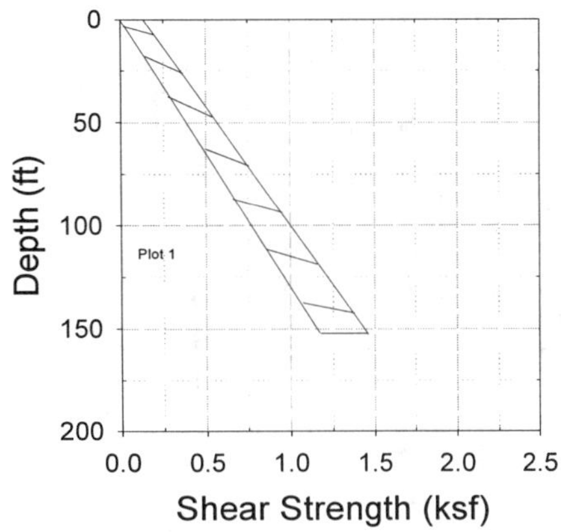

Figure 1. Undrained Shear Strength Profile in Gulf of Mexico

3.3 *Australia*

Calcareous soils occur extensively in offshore areas around Australia as well as other parts around the world including Brazil, India, Indonesia, and Saudi Arabia. Calcareous soils can comprise a variety of soil types, but those of most concern are carbonate silts and sands of biological origin characterized by high void ratio (greater than unity), high compressibility, and variable cementation. These soils can occur to depths exceeding 300 ft. While calcareous soils can have drained strength properties equal to or exceeding those of silica soils, strengths are typically mobilized at large strain levels; hence, foundation designs are often controlled by compressibility considerations. The extreme contractive behavior of calcareous soils during compression and shear also makes them susceptible to weakening under cyclic loading conditions.

Shaft friction of piles in calcareous soils can be extremely low, with measured capacities well below (on the order of 20% or less) those predicted by conventional methods. In fact early recognition of geotechnical problems associated with calcareous soils occurred when piles free fell during installation. Skin friction is much lower than for piles in silica soils, generally less than 600 psf (Gunasena et al, 1995.) The low mobilized friction appears to be primarily a consequence of the low lateral stresses along the pile shaft rather than a low soil-pile interface friction angle (Murff, 1987.) Drilled and grouted pile installations have been employed to increase shaft resistance to levels comparable to silica soils. Skin friction on the inner pile boundary is

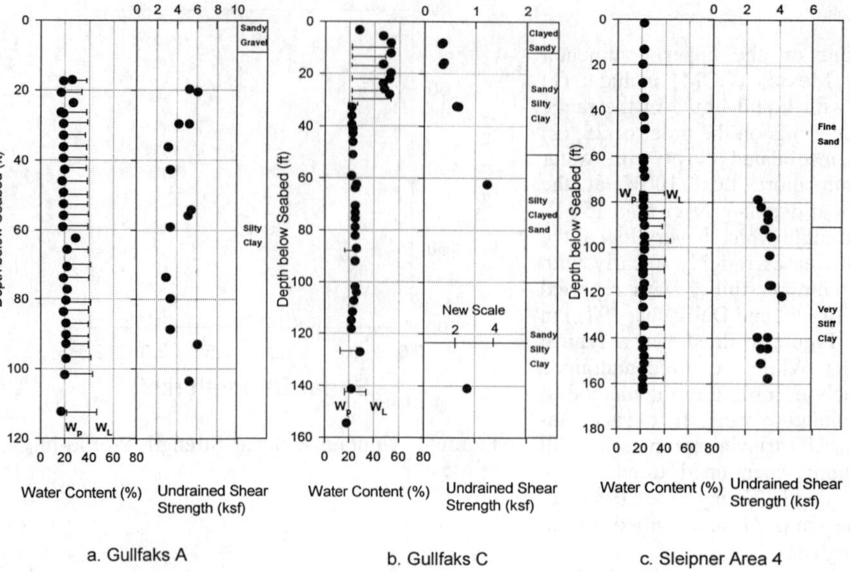

Figure 2. Typical North Sea Soil Profiles (after Lunne et al, 1985)

apparently high enough for plugging to occur during driving; nevertheless, end bearing capacity of piles in sands is typically about half of that in silica soils, which may necessitate the use of belled piles. Experience with small shallow foundations, spud cans for jackup exploration rigs for example, similarly shows somewhat lower bearing capacity in calcareous soils relative to silica soils. Larger gravity platforms supported by mat or independent suction caisson foundations are also considered feasible, provided that design pressures are maintained below collapse pressures (Roland and Hoeg, 1993.)

4 CONVENTIONAL PLATFORMS

Conventional offshore structures for oil and gas production are braced steel jacket platforms that are pinned to the seabed. In special cases, massive gravity platforms that rely on self-weight to maintain stability are used. Jacket structures are typically supported on pile foundations, although, as will be discussed subsequently, newer design alternatives such as suction caisson foundations may be used. Gravity structures are founded on shallow mat foundations designed on principles of classical shallow foundation theory. Mud mats are shallow foundations that provide temporary support for jacket platforms during installation.

4.1 *Pile Foundations*

Waves, winds, and currents impose large lateral loads on jacket structures. Piles must therefore have sufficient axial capacity to resist overturning moments and sufficient lateral capacity to resist lateral forces. Accepted practice (e.g., API, 1993) is to model the pile as an elastic beam-column with the soil continuum modeled as discrete, uncoupled, nonlinear springs; i.e., lateral and axial resistance modeled by 'p-y' curves and 't-z' curves, respectively. Guidelines for constructing p-y and t-z curves have been developed empirically (e.g., Matlock, 1970; Reese, et al, 1975) from load test data corresponding to certain types of soil conditions. The resulting pile-spring model is used to simulate a compliant foundation in the structural analysis of the jacket as well as to design the pile itself. In offshore practice driven open-ended tubular piles are used with diameters ranging up to 96 inches or more. Although test data are limited for piles of this size and capacity - the capacities of the largest available test piles are an order of magnitude less than that for a large offshore structure - estimates of axial pile capacity in offshore practice parallels that of land-based practice. For example, shaft resistance in cohesive soils is estimated from a total stress 'α-method' approach while in sands it is estimated from an effective stress friction formulation limited by an upper bound of mobilized friction. Actual behavior of piles is affected by factors such as the possible development of a soil plug inside the pile, remolding of the soil during installation, thixotropy and reconsolidation, rate effects and cyclic loading. Standard design procedures do not explicitly address these effects; hence, the practice is largely empirical. Wave equation analyses of piles during driving are needed for estimates of penetration rates, driving stresses, and hammer selection. These analyses are also used in a reverse sense to interpret driving records and estimate in-place pile capacity. Early wave equation models used idealized spring-mass systems for spatial discretization, while more recent versions employ finite difference and finite element formulations.

As we have discussed, most offshore platforms are founded on piles. Offshore pile design practice has evolved over its fifty-year history to a relatively mature state. However, the very large extrapolations that are made from the empirical database leave a lot of room for improvement. This remains a very important area where research breakthroughs could pay huge dividends

4.2 *Mud Mats*

Mud mats are basically shallow foundations designed to temporarily support jacket platforms on the seabed during installation, with pile foundations later installed to provide permanent support. Mud mats must be designed to support (1) vertical loading due to the buoyant weight of the jacket, (2) eccentric loads due to piles hanging off inside the legs of the jacket, and (3) lateral loads due to wave impact and accidental installation loads. Affixing shear plates to the mud mat may enhance lateral load capacity. Due to the low undrained shear strengths of typical GOM soils near the mudline, large areas may be required to limit bearing pressures and avoid excessive penetration. Mud mats are typically designed based on classical bearing capacity theory with empirical modifications for load inclination and eccentricity (API, 1993.) While advances in these design procedures are not among the most urgent needs, improvements in the prediction of penetration of shallow foundations on soft, non-homogeneous soil under complex loading would be a welcome addition to the art.

4.3 *Gravity Structures*

Gravity structures achieve their stability against environmental loads by virtue of their large buoyant weight and large diameter base. The use of these structures is most widespread in the central North Sea where conditions are very favorable to their unique features. The first major offshore gravity structure, the Ekofisk Tank, was installed in 1973 in the Norwegian sector. A number of these structures were installed in the 1970's and 1980's but their use recently has been minimal. Much of the central North Sea is 300ft to 600ft deep with a flat seabed consisting of very hard clays and dense sands due to glaciation. Because of their very large size (base diameters often in excess of 300ft), these structures can support many wells (often more than 40), high production rates, and the enormous deck loads requisite for such a facility. In addition, the deep fjords along the coast of Norway and Scotland have made ideal fabrication sites.

The typical gravity structure is fabricated of reinforced concrete. It has a large base composed of vertical cylinders and is perhaps 300ft in diameter and 150ft high. Three or four concrete columns extend from the base to support the deck structure. The base usually has cookie cutter type skirts which penetrate a few meters into the seabed forming a shear connection to inhibit sliding. The foundation can usually be classified as a shallow mat subjected to vertical and horizontal load and overturning moment. The critical failure mechanism is usually sliding. To date these structures have performed very well. Compared to steel structures though, their use has been sparse and limited to special applications.

5 NEW CONCEPTS

In water depths greater than about 1,500 feet it becomes increasingly difficult to keep the natural period below the dominant frequency of the sea wave spectrum. As a result costs tend to rise dramatically. These issues have led to the development of compliant structural systems ranging from slender vertical towers to floating structures attached to the sea floor by mooring lines.

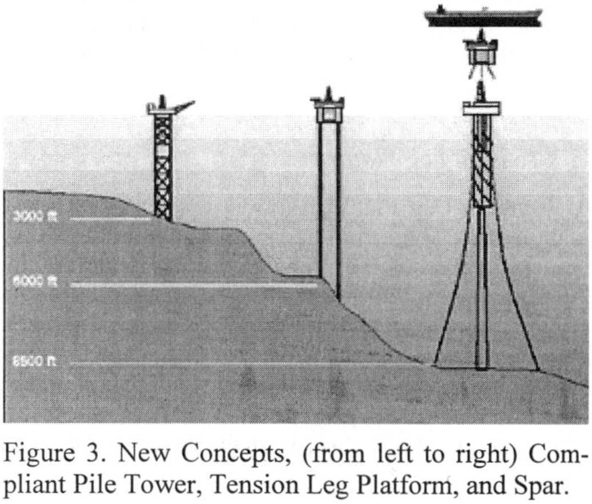

Figure 3. New Concepts, (from left to right) Compliant Pile Tower, Tension Leg Platform, and Spar.

5.1 *Compliant Pile Tower*

The compliant piled tower (CPT) is a slender steel space frame, bottom-founded structure of approximately constant cross section. The tower has significant axial strength to support the deepwater production activities but is very flexible in bending so that the natural period is quite long relative to a conventional bottom founded structure. To add to this flexibility, the piles are extended upward and joined to the structure well above the seabed. The stiffnesses of the very long pile extensions are relatively small giving rise to a very flexible rotational restraint at the base of the tower.

Conventional wisdom is that the CPT is cost competitive in moderately deep waters from perhaps 1500 to 2500 ft. water depths but it seems unlikely that it can be used in the very deep waters currently being developed. To date, Texaco's Petronius Tower, installed in 2000 in 1,600 ft. of water, is the only CPT that has been installed. As yet, it is too soon to know how successful this concept will be.

5.2 *Floating Production Systems*

Most of the solutions for deep and ultra-deep water consist of large diameter floating bodies anchored to the sea bottom. Among these, semi-submersible FPS's (Floating Production Systems) and tanker based FPSO's (Floating Production Storage and Offloading systems) have been around for some time but have now started to be considered as viable alternatives for deep and ultra-deep water. FPSO's consisting of either converted tankers or brand new vessels have become very popular in many parts of the world such as offshore Brazil (where many, if not most, of the deep water innovations are being tried), the North Sea, and the West Coast of Africa. They are not allowed yet in the Gulf of Mexico but there is considerable pressure from industry to have them recognized and a number of important studies to predict their behavior under extreme environmental conditions, such as hurricanes, and to evaluate their risk relative to that of other solutions are now in progress.

5.3 *Tension Leg Platforms*

A Tension Leg Platform (TLP) consists of a semi-submersible vessel moored by vertical tendons connected to the seafloor. The excess buoyancy of the structure keeps the tendons in tension even under the worst storm loading conditions. Conoco constructed the first TLP in the British sector of the North Sea in the early 1980's in 500 ft of water. This structure was selected for relatively shallow water as a test of the concept prior to use in deep waters. Since then several deep water TLP's (>1500 ft.) have been installed. The foundations for TLP's can consist of large diameter pipe piles that provide resistance to uplift through skin friction, or suction caissons, which will be described subsequently. Loading on the foundation consists of a tensile mooring force and cyclic loading due to wave loading on the superstructure. Loading is predominantly vertical with lateral forces estimated at less than 10% of vertical forces. A critical loading condition most likely occurs when a change in mean sea level due to a storm surge occurs in conjunction with large wave loading. TLP's can provide a feasible platform alternative to water depths up to about 1,500 to at least 3,500 ft.

5.4 *Spar*

The Spar platform is an old idea, which has recently emerged as a popular deepwater alternative for 2,000 to 6,000 ft of water because of various advantages over other solutions. It is essentially a deep-draft, truncated cylinder, with a moonpool that supports a platform by means of excess buoyancy. The hull is a steel cylinder, which encloses soft tanks in the bottom and hard tanks in the top part to provide the required buoyancy. Variable ballast allows adjustments to be made to maintain a constant draft when the topside loads change. The spar is tied to the sea bottom by vertical tethers or more commonly by catenary or taut mooring lines. Drilling and production risers run down a centerwell, which shields them from wave and current loading. As a result the spar is very stable, relatively insensitive to deck loads and easy to transport. All natural periods of moored spars are much longer than that of waves; therefore, their linear dynamic response to wave action is very small (the structure is very flexible in all six degrees of freedom as a rigid body). There are

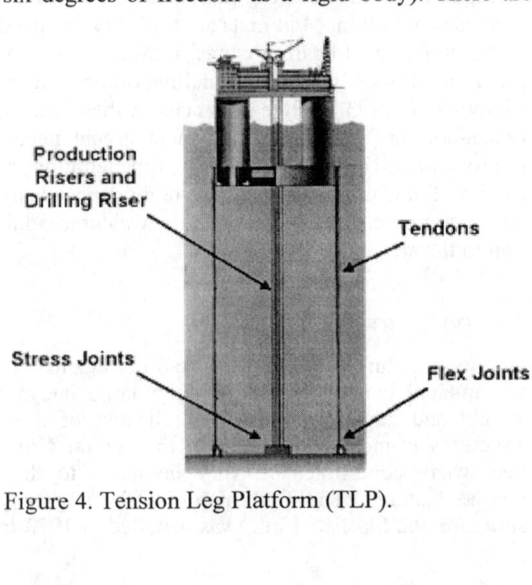

Figure 4. Tension Leg Platform (TLP).

however nonlinear effects, due primarily to difference frequencies, which can cause significant responses in the natural frequencies of the platform, and which must be taken into account. A substantial number of experimental studies in wave basins and numerical studies (nonlinear dynamic analyses in the time and frequency domains) have been carried out over the last five years.

6 DEEP WATER ANCHORS

Floating systems often require anchors designed to resist uplift rather than compression-dominated conventional foundations like those used for jacket and gravity platforms. In addition to static mooring forces, these anchors must resist cyclic loads generated by the combined action of currents, waves, and wind on the floating structure. Conventional piles have been used as the means of anchorage for TLP's and other floating structures. However, the increased complexity of pile installation at greater water depths and their relative inefficiency in resisting lateral loads has led the offshore industry to consider alternative methods of anchoring floating structures to the seafloor. Anchors being considered for these applications include suction caissons, drag embedment anchors, and suction embedment plate anchors (SEPLA's.)

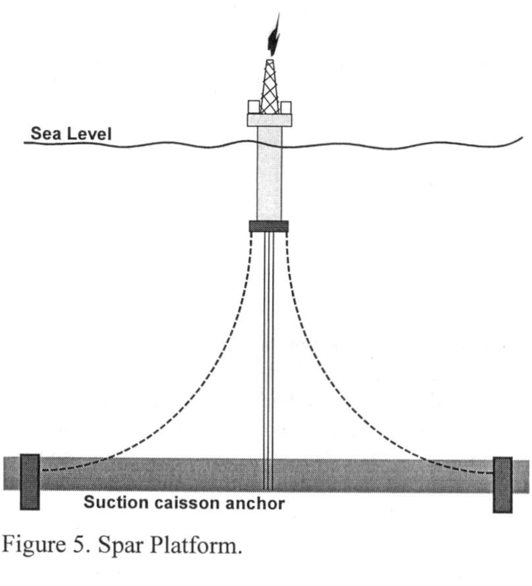

Figure 5. Spar Platform.

The nature of the anchor load will depend on type of platform and mooring system. TLP's are anchored by vertical tendons and hence subject the anchors to mainly vertical uplift loading. Catenary moorings apply essentially horizontal loads on anchors, while taut moorings are inclined up to about 30° to 35° from horizontal. Much of the experience to date has been with catenary moorings; however, in ultra deep waters catenary moorings require excessively large offsets, resulting in an increased likelihood of encroaching upon adjacent tracts, crossing mooring lines from adjacent facilities, or crossing undersea pipelines. Hence, taut mooring systems provide a very attractive means of mooring floating platforms in ultra deep waters.

6.1 *Suction Caissons*

Suction caissons are large cylindrical foundation elements installed by pressure differential, herein referred to as "suction," after partial penetration due to dead weight. Among their major advantages, suction caissons can be installed in a relatively short time without heavy installation equipment (underwater hammers); hence, they can often provide an attractive alternative to piles. They can be installed as either single or multi-cell units and can be used either as foundations for fixed structures or as anchors for floating structures. As fixed foundations they can be suitable for gravity structures, jackets, jackups, and subsea systems. As anchors they can be used for floating platforms including TLP's. Suction caissons are a viable alternative for soil conditions ranging from soft normally consolidated clays to dense sands and stiff clays. They also have potential in calcareous soils, although experience in this area is limited. Optimal aspect ratios (depth/diameter) are generally smaller in competent soils (1.5 to 3.0 in stronger clays; <1.0 in dense sands) and larger in soft soils (as much as 5 or more.)

When subjected primarily to vertical uplift, e.g., a TLP application, suction caissons resist pullout through suction and skin friction. The suction provides resistance to uplift by mobilizing a 'reverse end bearing' at the bottom of the caisson. This reverse end bearing mechanism has been confirmed by centrifuge testing (Clukey and Morrison, 1993); however, measurements indicate that the theoretical maximum suction is not fully mobilized, with measured 'suction efficiencies' ranging from roughly 55 to 90 percent, depending on the loading conditions. This reduction in capacity may be related to the fact that the soil beneath the caisson is loaded in extension, which typically mobilizes substantially lower undrained shear strength than compression. Redistribution of pore pressures may also be a factor, particularly under slow or cyclic loading conditions.

When tethered to catenary or taut moorings, suction caissons will be subjected to horizontal or inclined loads, respectively. Under these conditions, both theoretical and model studies (Keaveny et al,

1994) show that locating the load attachment point below the top of the caisson to minimize rotation will optimize the load capacity. Under predominantly horizontal loading, suction caissons behave in a manner similar to short laterally loaded piles. In this case a collapse mechanism comprised of a failure wedge near the free surface (e.g. mudline) and flow-around zone (Randolph and Houlsby, 1984) at depth appears to govern the lateral load capacity (Murff and Hamilton, 1993.) The effects of load inclination on collapse mechanisms and load capacity are not well understood at present and both analytical and experimental investigations are ongoing in this area.

The offshore oil industry is currently designing and using suction caissons to anchor all types of floating exploration and production platforms. However, there is a paucity of performance data on suction caisson behavior and our current level of understanding is limited. This is especially true for long, slender caissons subjected to inclined, cyclic loads as required for deepwater applications in soft cohesive sediments. Because of the uncertainty involved, conservative assumptions and relatively large factors of safety are currently needed to ensure adequate performance of these foundation systems. The prediction of both installation performance and load carrying capacity are both areas of intense interest.

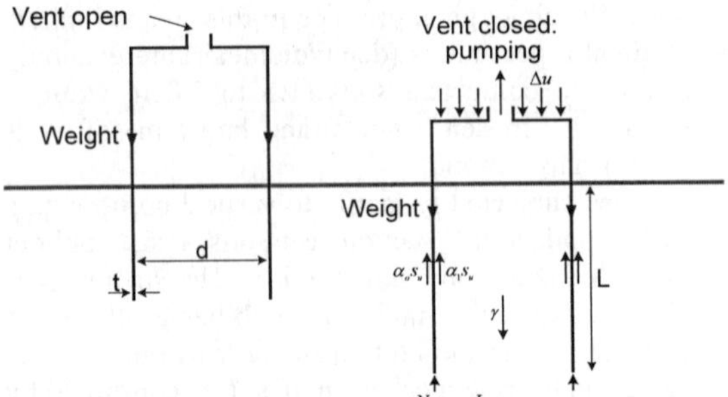

Figure 6. Suction Caisson.

6.2 *Drag Embedment Anchors*

Drag anchors (Dunnevant, 1991) are a commonly used foundation element in mooring systems. A drag anchor is basically a bearing plate inserted into the seafloor soil by dragging with wire rope or chain. Pullout resistance is provided by the bearing resistance of the plate as well as friction along the embedded portion of the mooring line. While industry has been using drag embedment anchors (with catenary mooring lines) for temporary moorings for many years, the practice is highly empirical, with installations being carried out by trial and error. Limited data from field trials and anchor retrieval operations indicate that some of the newer drag anchors can have substantial vertical pullout capacity, raising the possibility that drag anchors may serve as foundations for taut line mooring systems.

Two basic issues arise in connection with drag embedment anchors: (1) the trajectory of the anchor during installation and (2) the holding capacity of the anchor. The trajectory issue is particularly critical because, in contrast to other types of anchor systems, the physical location of the anchor is subject to considerable uncertainty even for uniform seafloor soil conditions. Since the optimal anchor geometry (fluke angle) for maximum penetration depends on soil type, layered soil profiles (e.g., typical North Sea profiles) can add to the uncertainty of anchor penetration depth. Measurement systems for monitoring anchor trajectory have been developed to mitigate some of the uncertainty of drag anchor penetration.

Limit equilibrium analyses have been applied to the problems of anchor trajectory and holding capacity (e.g., Neubecker and Randolph, 1996). While providing valuable insights into anchor behavior, these solutions are usually calibrated to centrifuge or field test data; hence, they cannot be reliably extrapolated to other site conditions. Further, much of the analytical work in this area has been directed toward horizontal loading conditions associated with catenary systems, with little effort directed towards inclined loading. Prediction and design methods must therefore be significantly improved if vertically loaded drag embedment anchors are going to be used for permanent facilities with confidence and routinely accepted by certification and permitting agencies.

6.3 *SEPLA's*

A recent concept for taut anchors is the Suction Embedded Plate Anchor (SEPLA.) A SEPLA is a taut embedded anchor installed by (1) inserting a vertically oriented plate anchor attached to a suction caisson, (2) withdrawing the suction caisson, and (3) rotating the plate into an inclined position. This installation method allows for more accurate positioning of the anchor, thereby avoiding many of the uncertainties associated with drag-embedment anchors. Since the suction caisson is used only as an insertion tool in this design, it can be re-used. Hence, the SEPLA anchor can potentially achieve significant cost savings over suction caisson anchors. Aside from some limited model and field trials (Dove and Rorass, 2000), very little performance data exists and their behavior is not well understood. Analytical and experimental model studies are needed to address a number of basic issues regarding plate anchor behavior such as the effects of plate geometry, embedment depth, anchor installation disturbance, consolidation and pore pressure redistribution, load orientation, and cyclic load behavior.

6.4 *Current Research Work*

Suction caissons, suction piles, bucket foundations or skirted foundations, all names commonly used for similar but slightly different types of foundations, have been the subject of considerable interest and research since the early 1980's (Senpere and Auvergne, OTC 1982). A number of experimental studies with scaled models in the laboratory (Steensen-Bach, 1992, S.N. Rao et al, 1997; El Gharbawy, 1998), real caissons in the ocean (Tjelta et al, 1986), and models in centrifuges (Clukey & Morrison, 1993) have been conducted, as well as theoretical and numerical work (Sukumaran & McCarron, 1999; Chicata, 2000). A proper understanding and simulation of the behavior of these systems requires accounting for the installation procedure, the build up and dissipation of pore pressures, the partial consolidation of the soil as a function of time and the true nonlinear behavior of the soil, as a two phase medium, for both drained and undrained loading conditions. The complexity of this type of analysis is such that it is rarely undertaken. Yet sophisticated finite element analyses with thousands of degrees of freedom have been often performed for the foundattions of the North Sea platforms.

Chicata (2000) conducted full simulations of the self-weight initial penetration of the caisson, followed by the suction induced penetration, partial consolidation and undrained or drained pull out tests, and compared his predictions with the experimental data for two model caissons tested by El Gharbawy (1998), with reasonable success although not a perfect agreement. He used a finite element formulation with a boundary surface plasticity model for the soil, treated as a two-phase medium based on mixture theory. To study the interaction between the caisson and the porous medium a frictional contact interface was included and the coupling equations were written in a way to extract easily the equivalent forces due to the effective stresses. The model allowed computation of the friction forces on the inside and outside surfaces of the caisson and prediction of whether a plug would form on the inside. The results of this study pointed out on one hand the difficulty in obtaining all the parameters needed to define the nonlinear constitutive soil model from the available experimental data. Of particular importance was the value of the friction coefficient between the caissons and the soil, which had not been measured. The numerical predictions could reproduce the self weight penetration values within 15% for one case and within 4% for the other (the longer caisson); the suction penetration was predicted within 3% for the first case and within 18% for the second; the predicted undrained pull out capacity was 75% of the experimental value for the shorter caisson and 90% for the other, while the drained capacity was overestimated by 17% for the former and by 40% for the latter. Considering the

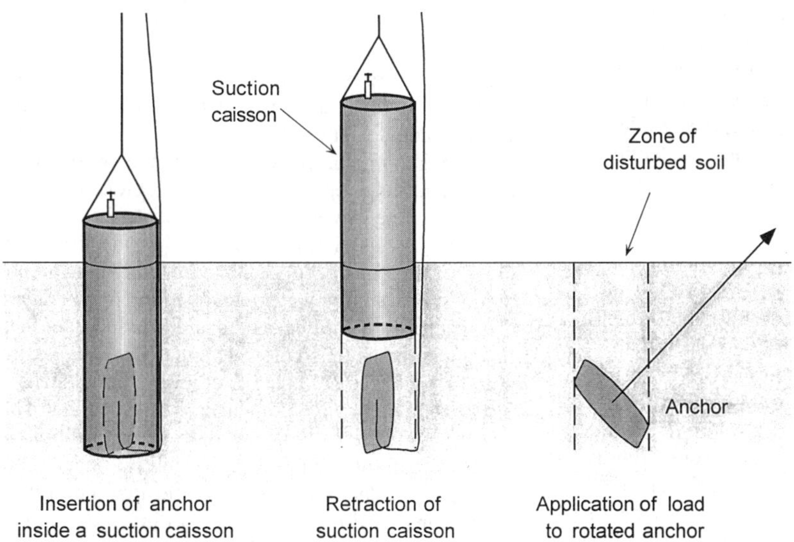

Figure 7. Suction Embedded Plate Anchor.

uncertainties in the values of the soil parameters this type of agreement seems reasonable and promising for most cases. In addition, Chicata was able to derive some important conclusions on the relative importance of the suction force, the friction on the exterior surface of the caisson and the inside friction on the pullout capacity for the drained and undrained tests. He was also able to predict the formation of a plug for the undrained pullout test. For the drained case, plug formation depended on the rate of pullout (fully drained or only partially drained conditions.) These types of studies are badly needed and can provide an appropriate understanding of the behavior of these foundations (or anchors) under different conditions when properly validated with experimental work.

7 DEEPWATER HAZARDS

7.1 *Shallow Water Flows*

Shallow water flows (SWF) refer to flow of water into a borehole from an over-pressured formation at shallow depths. They are quite common in GOM waters exceeding 2,000 ft at depths below the mudline of 300 to 2,000 ft (Medley, 1998). Typically, during a drilling operation conductor pipes of varying dimensions are drilled down to target depth and the annulus between the pipe exterior and sediment is cemented or grouted. When shallow water flows are encountered during this operation the high pore pressures can destabilize the hole and eventually vent to the surface through the annulus, usually carrying large quantities of sand. The high pressures can also wash out grout placed in the annulus before it sets up. Potential economic losses associated with shallow water flows in the tens or hundreds of million dollars have motivated the development of a number of strategies for addressing the problem, including improved prediction and detection of SWF's, permeability reduction by injecting a viscous fluid, and unconventional casing installation methods (Medley, 1998.)

The latter approach entails driving rather than setting conductors to the desired depth. Driving the conductor through SWF zones will avoid creating a weakened annulus around the conductor thereby minimizing the potential for high pore pressures to vent to the surface. Strictly speaking, it is not necessary to drive the conductor for its entire depth; rather the conductor need only be driven through the over-pressured sand zone as well as some distance above and below the over-pressured zone to provide a seal layer to contain the shallow water flow. Still, driving a relatively slender conductor (20 to 30 inches in diameter) at such depths presents a formidable challenge.

Developing an effective strategy to counter SWF's requires that the depth of SWF layers be known in advance of conductor installation. A number of tools have been applied to address this problem including the piezocone to delineate stratigraphy and measure excess pore pressures, the small diameter piezoprobe, the hydraulic fracture tool to measure fracture gradient, temperature probes, cameras to monitor water flow and drilling returns, and open-hole petrophysical logging of a borehole (Dutt et al, 1997.)

To date there has been no solution that seems to generally apply to the SWF problem. Furthermore the cause of the excess pore pressure is not well understood. To improve our present understanding of this problem it would be essential, as in all cases involving geohazards, to perform coordinated studies involving geologists, geophysicists and geotechnical engineers. The three main steps of the research should be:

- Detection of the over-pressured sands, including their depths below the mudline and their thicknesses from basic geological studies complemented by geophysical exploration methods.
- Estimation of the pore pressures in the sands from geological dating of the deposits coupled with large strain numerical consolidation studies and possibly from the ratio of P and S wave velocities.
- Estimation of the hydraulic fracture pressures for the materials involved through basic engineering mechanics studies using assumed or, where possible, measured sediment properties.

Once the pore pressures in the sands and the hydraulic fracture pressures are known, an appropriate drilling program can be developed adapted to the window between these two values.

From geological considerations it is possible to define and map regions where the geological attributes favor the development and existence of overpressured sands. These maps provide the first step. Once an area has been identified has having the potential for shallow water flows more detailed site characterization studies, as those discussed later, can be conducted using shallow seismic core data, available or specially collected multibeam data, shallow 3D seismic surveys, and measured or calculated soil properties. If these studies were to be carried out for a number of sites one would be able to accumulate an important database, which would facilitate future predictions. Drilling experience within the area can also provide additional and valuable information.

Once the geological processes have been identified, numerical models can be developed to simulate the role that each one of these processes may have in the development of excess pore water pressures. There are a number of possible mechanisms that should be explored such as rapid sediment deposition followed by slow and incomplete consolidation of the underlying sediments, faulting and intrusive processes, diapirism, or lateral transport of sediments on slopes due to slope failures with overrunning deposits. All these processes can be modeled using numerical methods. The formulation must include however a nonlinear constitutive soil model, a two phase medium with undrained or partially drained conditions, and perhaps even a three-phase medium to account for the presence of gas as well as large strains and displacements.

The last step is the prediction of the hydraulic fracture pressures for the sand layers, requiring a coordinated laboratory and numerical study of the mechanical behavior of these materials.

7.2 *Gas Hydrates*

Gas hydrates (Paull, 1997) are solid ice-like compounds of water and low molecular weight gases (mostly methane.) They occur under conditions of high pressure, low temperature, and high gas concentrations that commonly exist under conditions of rapidly accumulating sediment. Temperature increase due to drilling and production operations may melt gas hydrates leading to reduced shear strength and thaw settlement, with consequent reduction in foundation capacity. Additional possible problems can include corrosion of subsea structures, seabed erosion, and requisite avoidance of seafloor chemosynthetic communities that tend to occur at these sites. The two main problems are first the identification and location of hydrate deposits at a site, and next the prediction of the possible effects of these deposits if they were to be subjected to high temperatures as a function of time Gas hydrates can be detected in principle from seismic reflection data; a prominent reflector called the bottom simulating reflector (BSR) is believed to represent the base of the gas hydrate stability. However, other elements present in the soil can cause however similar reflections making difficult to recognize whether the BSR is really due to hydrates. The difficulty in locating hydrates increases when one is not dealing with a relatively large monolithic formation but rather with dispersed nodules. The use of shear waves instead of the more traditional P waves, as discussed later, would seem to offer an increased opportunity to detect changes in stiffness associated with the presence of gas hydrates. This is an area that requires more research, but this has to be primarily verification work in the field in places with known existence or lack of existence of hydrates. These types of studies are very expensive, requiring specialized equipment as well as boat time. Less commonly gas hydrates may be directly sampled by piston or pressure core samples at a more localized level. The prediction of the possible effects of melting of the hydrates due to drilling operations or to the heat from the oil in the risers lends itself to numerical analysis when the material properties and the boundary and initial conditions can be reliably established.

8 SOIL CHARACTERIZATION

As pointed out by Lacasse (1999) "Advanced calculations shall not be a substitute for high quality site investigations. Important subsurface characteristics will go undetected, ignored or oversimplified. Using a range of values will not necessarily guarantee a correct result." It is clear that there is little point in using sophisticated and expensive analysis procedures if one does not know the characteristics of the soil or its mechanical properties, and that more emphasis has to be placed on the correct determination of these properties in situ. On the other hand this does not justify the use of incorrect or sloppy methods of analysis that will introduce additional systematic errors into what is already a poor prediction.

If the collaboration between geologists, geophysicists and geotechnical engineers is important for all aspects of geotechnical offshore engineering, it is essential for the characterization of the sub-soil and the identification of geohazards. The need for a coordinated effort among these different specialists to solve the problem of shallow water flows was already discussed above. When soil properties are difficult to obtain accurately, information on the geological history of the region, even if it is of a qualitative nature, can be of great help in selecting reasonable ranges of values. Determination of these properties at the site for low levels of strain, over an extended region rather than at a single point, lends itself to the use of geophysical methods.

Accurate determination of the soil properties and of all the parameters needed to calibrate and define nonlinear constitutive models is always a daunting task, even on land. It is extremely difficult on the sea bottom and increasingly so for deep and ultra-deep waters. For analysis we need both stiffness properties in a general sense (that is to say the parameters that define the relations between stresses and strains for different states and levels of strains) and the resistances or capacities (ultimate values, limit envelopes, etc.) We need these properties for the soil as it is at the site with the existing state of stresses and its own history of deposition process, loading and unloading. Laboratory tests can provide us with information on the behavior of the material and its mechanical properties under a wide range of strain and loading paths for drained and undrained conditions. This information is essential to construct an appropriate constitutive model but it does not provide the initial conditions corresponding to the state of the soil in situ. It is extremely difficult to obtain high quality samples in deep water, where the state of stress changes dramatically when extracting the sample. In situ tests are therefore necessary to complement the laboratory information. Some of these tests only provide information on ultimate soil resistance under a particular type of loading. Others, based primarily on geophysical procedures, can provide the elastic moduli of the material corresponding to very low levels of strain. Among these, the Spectral Analysis of Surface Waves Method (SASW) has been extensively used on land both to determine the soil properties and their variation with depth and to detect relatively shallow cavities or inclusions. In principle, the same methodology can be used under water considering interface waves rather than generalized surface waves. The advantages and potential of this method were presented by Stokoe (1999) at the 31st OTC. The analysis methodology to perform the inversion and to estimate the shear wave velocity of the soil layers as a function of depth from the recorded data is essentially the same as that used for land applications with the addition of a layer of water (Roesset, 1988; 1989, Roesset et al, 1990). The identification of pockets of hydrates or inclusions with different stiffness from that of the surrounding medium can again be done using similar techniques to those employed on land (Nogueira et al, 1994). While there is still room for improvement in the analysis procedures to make the inversion more reliable and more economical, the main need now is test these procedures in the real ocean environment on a large scale.

It is important to note that even well established and commonly used in situ and laboratory procedures can in many cases benefit from more rigorous analyses for interpreting the data and determining accurate values of the desired properties. It is not uncommon for instance to assume that there is a uniform distribution of stresses in a test when this is not really the case, to ignore the true boundary conditions imposed by the equipment on the soil sample, or to consider that the response of a finite size specimen that could be modeled with a number of finite elements represents the properties of the material at one point. The justification for using these simplifying assumptions becomes more tenuous as it becomes feasible to perform more realistic simulations of each test.

9 CONCLUSIONS

The push to develop deepwater hydrocarbon resources presents an array of geomechanics problems in need of solutions. We hasten to add that there is still much room for improvement of the more traditional offshore geotechnical problems as well. In the previous pages we have attempted to summarize some of the historical developments, recent innovations, and the ensuing new geotechnical challenges. We have also discussed briefly some of the research that is being conducted involving both experimental and theoretical work. In the following we summarize some of the perceived research needs in relation to anchors and shallow water flows.

Suction anchors: While suction anchors subjected to vertical loads have been successfully designed and installed on numerous projects, the mechanisms of pullout resistance are still imperfectly understood, particularly for long, slender caissons. Improved understanding of issues such as setup effects, skin friction on the inner and outer surfaces of the anchor, and soil anisotropy effects, would be beneficial in optimizing designs and allowing designs to be used with a greater degree of confidence. For horizontally loaded suction anchors, the P-ultimate framework proposed by Murff and Hamilton (1993) appears to provide a suitable overall approach to the undrained capacity problem for relatively simple conditions; however, it does not address many of the real complexities. For example, suction anchor behavior under cyclic loading in stratified soil profiles, or with partial drainage, are issues in need of further attention. As floating platforms tethered to taut moorings are a promising solution in deep and ultra-deep waters, there will likely be a critical need for developing reliable methods for estimating load capacity under inclined loading conditions.

Drag anchors: Unlike other anchor systems, the embedment depth of drag anchors cannot be closely controlled; hence, reliable predictive methods are needed for (1) estimating anchor trajectories for differing types of anchor geometries and soil conditions, including soft clays, sands, and stratified profiles, and (2) estimating anchor pullout capacities for a known embedment depth. A variety of factors that can effect anchor capacity require investigation, including embedment depth, geometry, soil type, rate of loading, and load inclination. Most efforts to date have been directed toward anchor capacity for horizontal loads corresponding to catenary mooring systems. As noted above, deepwater applications are likely to require taut moorings having a significant component of vertical load; hence, methods for predicting the capacity of anchors subjected to inclined loads are needed.

SEPLA's: The position of a SEPLA is closely controlled; hence, predicting embedment depth is not an issue. Otherwise, a SEPLA is simply an embedded plate, and the needed research for predicting pullout capacity overlaps much of that which is needed for estimating the pullout capacity of a drag embedment anchor. However, an issue unique to SEPLA's that may be significant is the disturbance effect created by the insertion and retraction of the suction caisson. Since this can create an annulus of weakened soil extending directly from the anchor to the surface, it has the potential for reducing pullout capacity under loading conditions having a large vertical component.

Shallow Water Flows: An improved understanding of the causes of overpressured sands can lead to improved methods of predicting their occurrence. We must be able on one hand to anticipate the possible existence of these sands at the site where drilling is planned and to compute the window between the hydraulic fracture pressure of the material and the existing pore water pressures in order to plan an appropriate drilling program. Shallow water flows have already resulted in the need to abandon drilling operations at a site with a considerable economic loss and are therefore real hazards.

Addressing the broad range of issues described above can be pursued by an array of activities that can include the following:

- Review of existing analysis and design methodologies, their limitations, and the range of soil and site conditions over which they apply.
- Development and application of analytical and numerical models, including constitutive models, for improved understanding of soil response, failure mechanisms, and the influence of problem variables (e.g., anchor geometry, soil strength.)
- Experimental studies, including single gravity and centrifuge tests, for validating and calibrating analytical and numerical models.
- Development of simplified analytical models (e.g. plastic limit analysis) to provide a basis for practical design methodologies.
- Collection and evaluation of actual field performance data.

We would like to emphasize again the need to coordinate these different activities in a balanced way and to involve geologists, geophysicists and geotechnical engineers in close collaboration, rather than proceeding in a piecemeal fashion looking independently at the different parts of the puzzle.

REFERENCES

American Petroleum Institute. 1993. *Recommended Practice for Planing, Designing, and Constructing Fixed Offshore Platforms.*

Bryant, W.R. and Delflache, A.P 1971. Geotechnical charts of the deep water portion of the Gulf of Mexico. *Proc. Offshore Technology Conference, Dallas.*

Campbell, K.J. 1997. Fast-Track Development: The Evolving Role of 3D Seismic data in deepwater Hazards Assessment and Site Investigation. *Proc. Offshore Technology Conference, Houston.*

Christensen, N.H. and F. Haahr (1992) " A Computer Program to Analyze Suction Effects" *Proc 24th OTC pp. 331-338 (OTC 6845), Houston.*

Clukey, E.C. and Morrison, M.J. 1993. A centrifuge and analytical study to evaluate suction caissons for TLP applications in the Gulf of Mexico. *Design and Performance of Deep Foundations: Piles and Piers in Soil and Soft Rock.*

Dove, P. and Rorass, H. 2000. Trial results from deepwater anchors, polyester moorings. *Offshore.* March.

Dunnevant, T.W. 1991. *Short Course on Design of Floating Production Systems, Oct. 14-25.* Univ. Texas Austin.

Dutt, R.N., Hamilton, T.K., Pelletier, J.H., and Doyle, E.H. 1997. Recent advances in deepwater Gulf of Mexico Geotechnical Investigations. *Proc. Offshore Technology Conference, Houston.*

El-Gharbawy, S. and Olson, R.E. 1998.The pullout capacity of suction caisson foundations for tension leg platforms. *Eighth Intl. Offshore and Polar Engrg Conf, Montreal.*

Gunasena, U., Joer, H.A., and Randolf, M.F. 1995. Design approach for grouted driven piles in calcareous soil. *Proc. Offshore Technology Conference, Houston.*

Keaveny, J.M., Hansen, S.B., Madshus, C. and Dyvik, R. 1994. Horizontal capacity of large scale model anchors. *Proc. XIII Intl. Conf. on Soil Mech. and Found. Engrg, New Delhi.*

Lacasse, S. 1999. Ninth OTRC Honors Lecture: Geotechnical contributions to offshore development. *Proc. Offshore Technology Conference, Houston.*

Lee. 1982. *Behavior of Offshore Structures.*

Lunne, T., Christoffersen, H.P., and Tjelta, T.I. 1985. Engineering use of piezocone data in North Sea clays. *11th Intl. Conf. Soil Mech. and Found. Engrg., Vol. 2:* 907-912.

Matlock, H. 1970. Correlations for design of laterally loaded piles in soft clay. *Proc. Offshore Technology Conference, Houston.*

Medley, G.H. 1998. Shallow Water Flow Technology Update. *Proc. Offshore Technology Conference, Houston.*

Murff, J.D. 1987. Pile capacity in calcareous sands: state of the art. *J. Geotech. Engrg., Vol. 113, No. 5, May:* 490-507.

Murff, J.D. and Hamilton, J.M. 1993. P-Ultimate for undrained analysis of laterally loaded piles. *J. Geotech. Engrg., Vol. 119, No. 1, January:* 91-107.

Nogueira, A. C., J. M. Roesset, and J. L. Tassolas. 1994. Effect of cavities on the propagation of surface waves. *Proc., 8th IACMAG Conference, West Virginia.*

Neubecker, S.R. and Randolph, M.F. 1996. The static equilibrium of drag anchors in sand. *Can. Geotech. J. 33:* 574-583.

Paull, C.K. 1997. Drilling for gas hydrates: Ocean Drilling Program Leg 164. *Proc. Offshore Technology Conference, Houston.*

Quiros, G.W., Young, A.G., Pelletier, J.H., and Chan, J. H-C. 1984. Shear strength interpretation of Gulf of Mexico clays. *Texas Civil Engineer, Houston, March:* 14-20.

Rao, S. N., R. Ravi and C. Ganapathy (1997) Pullout behavior of model suction anchors in soft marine clays. *Proc. 7th ISOPE Conference pp. 740-744 Honolulu.*

Randolph, M.F. and Houlsby, G.T. 1984. The limiting pressure on a circular pile loaded laterally in cohesive soil. *Geotechnique, 34, No. 4:* 613-623.

Reese, L.C., Cox, w.R., and Koop, F.D. 1975. Field testing and analysis of laterally loaded piles in stiff clay. *Proc. Offshore Technology Conference, Houston.*

Roesset, J. M. 1988. Soil water structure interaction. *Proc. Intl. Symp. on Modelling Soil Water Structure Iteraction SOWAS '88, Delft, Holland.*

Roesset, J. M. 1989. Non-destructive testing of soils and pavements - The SASW Method. *Intl. Conf. on Analytical and Experimental Methods in Mechanics, Capri.*

Roesset J.M, S.G.Wright and M. Sedighi-Manesh. 1990. Analytical investigation of seismic surface waves in the seafloor. *Proc. NATO Symp. on Shear Waves in Marine Sediments, La Spezia, Italy.*

Roland, B. and Hoeg, K. 1993. Design and construction of gravity platforms on calcareous sediments. *Offshore Australia 1993 Conference.*

Senpere, D. and Auvergne, G.A. 1982. Suction anchor piles – a proven alternative to driving or drilling. *Proc. Offshore Technology Conference, Houston.*

Steensen-Bach, J.O. 1992. Recent model tests with suction piles in clay and sand. *Proc. Offshore Technology Conference, Houston.*

Stokoe, K. H., and B. L. Rosenblad (1999) "Offshore Geotechnical Investigations with Shear Waves." 31st Offshore Technology Conference (OTC 10823).

Sukumaran, B. and McCarron, W. 1999. Total and effective stress analysis of suction caissons for Gulf of Mexico conditions. *Proc. of the OTRC '99 Conf., Austin.*

Tjelta, T.I., Guttormsen, T.R., and Hermstad, J. 1986. Large-scale penetration test at a deepwater site. *Proc. Offshore Technology Conference, Houston.*

Vasquez Chicata, L. F. G. 2000. Estimation of pile capacity considering installation effects. *Ph. D. Dissertation The University of Texas at Austin.*

Young, A.G., Quiros, G.W., Ehlers, C.J. Effects of offshore sampling and testing on undrained soil shear strength. *Offshore Technology Conference, Houston.*

Computer Methods and Advances in Geomechanics, Desai et al. (eds) © 2001 Balkema, Rotterdam, ISBN 90 5809 183 X

Practical use of numerical methods

J.T.Christian
Waban, Mass., USA

A.Urzua
Prototype Engineering, Winchester, Mass., USA

ABSTRACT: Numerical methods have, over the past few decades, moved from a peripheral area of research to a major component of geotechnical analysis and design. Today geotechnical practitioners have access to a broad spectrum of computer software, ranging from specifically geotechnical software running on microcomputers to large general-purpose analytical software running on super computers to software for particular geotechnical applications to spreadsheet add-ons. Each of these has its place and its advantages and disadvantages. However, the geotechnical engineering firm is still faced with problems of how to maintain the necessary complement of software and personnel to take advantage of these capabilities and how to pay for it. In fact, each project creates its own demands for analysis, and it is common to find that the existing capabilities do not exactly match the needs of a new project. An example of the seismic analysis of an existing hydraulic fill dam demonstrates how many software packages, including six finite-element codes, are combined to perform a complicated analysis, developing intermediate results to guide further efforts. Finally, some thoughts are presented about areas to which the numerical analysis community might profitably direct its attention.

1 INTRODUCTION

Numerical methods have come a long way over the last several decades. When the senior author started his career, the numerical methods taught in most graduate courses were based on Marchant calculators, computers were awkward and inaccessible, and the major figures in the geotechnical community were skeptical of the utility of numerical methods – if not frankly hostile. In the 1960s the development of finite element software suited to geotechnical problems, combined with expansion in the power and availability of computers, led to rapid advances in the application of numerical methods to geotechnical problems. Of course, finite difference methods, which had been available for many years, and boundary element methods, which were developed shortly after finite element methods became popular, also participated in these developments. It is also worth noting that these methods have long histories, including significant implementations before digital computers became widely available. For example, Courant (1943) described what was essentially a complete implementation of constant strain triangles to solve Laplace's equation for the warping function of a structural element under torsion.

Many of the major early advances were done with resources that seem paltry by today's standards. For example, the RAM of the IBM 7094 available to researchers at MIT in the early 1960s was 64K words, each 36 bits long, which corresponds to between 256K and 288K bytes in today's measurement. At the same time the 7094 at the University of California at Berkeley had only 32K words, but this is the machine on which most of the seminal work on finite elements was done. Anyone working in the 1970s remembers the frustration of working on IBM 360s and 370s with 500K bytes of memory and the joy with which later virtual machines were greeted.

In addition to the difficulties of fitting large problems onto small machines, early work necessarily dealt with issues of convergence, stability, singularity, and accuracy. One of the most difficult tasks facing the developer of numerical methods was to convince new students as well as potential users and clients that these issues are important and that numerical results can be wrong. In the opinion of the authors this problem continues today. It is not unusual at to find finite element or finite difference software applied to problems with no regard for possible errors.

This conference includes descriptions of current advances on all fronts in geomechanics. Numerical methods have solved – or at least made tractable – many problems that were once thought to be impenetrable. At the same time the power of today's computers has made possible sophisticated displays of results that could barely be imagined by the early developers. However, in addition to the ever-present matters of convergence, stability, singularity, and accuracy, one set of issues remains as important today as it was at the beginning. How do the developments find their way into the main stream of practice? How does the practicing engineer maintain the capacity to use this technology? How does he or she adapt available software and expertise to the new project in hand? Following some experiential observations, the paper describes a particular case in which available resources had to be applied to a major analysis and then concludes with some thoughts on current needs.

2 CURRENT STATE OF PRACTICE FOR GEOTECHNICAL ENGINEERS

In 1998, the ASCE convention in Boston included a panel discussion on the topic, "What has the Finite Element Method Done for (or to) Geotechnical Engineering?" Write-ups of the four presentations were published by the Boston Society of Civil Engineers Section of ASCE (Christian 1999, Duncan 1999, Whittle 1999, and Marr 1999). They present four views on the current state of practice, which are more remarkable for their agreement than disagreement. In particular, all four papers pay attention to what can be done with available software and some of the problems involved in applying it to a particular problem.

It is clear from the above papers, that software available to the practicing geotechnical engineer falls into one of three broad categories:

a) Software written specifically for geotechnical applications but with user-friendly interfaces. Examples include such finite element codes as CRISP, PLAXIS, and SIGMAW; finite difference codes such as FLAC; pile analysis codes; and various slope stability codes. Each of these has strengths and weaknesses; it is beyond the present scope to go into these. However, a general rule is that those features that were considered when the code was first written tend to work better than those that were added later because of user demand. It is always easier – and therefore more reliable – to include a feature that was included in the original formulation, even if it was not implemented in the first version of the code.

b) Large programs such as ABACUS, ADINA, and ANSYS that were written originally for general engineering analysis but can accept special modules for geotechnical features. The most common of these additional modules have been routines to incorporate non-linear constitutive models appropriate for soils and rocks. These programs often have their own user interfaces or are easily matched with such front and back ends as FEMAP, IDEAS, or PATRAN.

c) Special purpose programs that have never been made user friendly. Most are written in FORTRAN 77, or even an earlier dialect, and require that the input be prepared in a particular format. The output is usually rudimentary. Much of the software for geotechnical earthquake engineering falls into this category. Since the demand for the software is not high enough to warrant the investment necessary to upgrade their interfaces, they are not likely to be changed soon. The user must often be quite familiar with the inner workings of the programs to make them perform well.

The trend over the past ten or fifteen years has been to move software from category c to category a. That modern graduates are accustomed to attractive software has added pressure for the change. Unfortunately, the trend has also removed the user further from the workings of the software, so engineers are less familiar with what their software is doing than they used to be. The situation is analogous to what has happened to the automobile: modern cars are easier to drive and more reliable, but few people even pretend to know what is going on under the hood.

How then does the modern small geotechnical firm take advantage of what is available? A rough estimate is that for something in the neighborhood of $10,000 the firm can acquire hardware and software to get one of the sets of geotechnical finite element software running. However, this is only the initial part of the cost. Personnel must be trained to use the software effectively, or the firm must hire persons already familiar with it. Furthermore, the software must be exercised regularly lest the users loose their skills. Few things are as frustrating as climbing a learning curve the second time. The software also has to be maintained as hardware evolves and as the programmers introduce new features. All of this costs money, most of it chargeable to overhead. One result is that practical work is usually done with off-the-shelf software.

3 A PRACTICAL CASE STUDY

Recently the authors participated in the study of the behavior of a large (nearly 300 feet high) rock- and earthfill dam for several levels of seismic shaking. The central portion of the dam had been built by the hydraulic fill method. Others performed the seismic hazard studies, the field investigation, and the laboratory investigations, and various commitments had already been made about the types of analyses that would be performed. Time did not permit developing special-purpose software to carry out the full analysis; available software had to be used.

Because the analysis is still under review at the time of writing, it is not possible to identify the project or to describe the results in detail. The authors, together with other participants, intend to prepare a full technical paper on the analysis at the first opportunity. In the mean time, the following summary of the activities provides a background for some of the managerial issues.

The analysis consisted of the following steps:

a) A review of the results of the field and laboratory studies to establish a first estimate of the dynamic properties. The most useful results were identification of the boundaries of the various zones; values of unit weight, plasticity index, and void ratio; and cross-hole measurements of shear-wave velocities over a very limited depth of the dam. By comparing estimates of small strain shear modulus computed from the Hardin & Black (1968) relations with the measured shear-wave velocity values, it was possible to generate a general estimate of the pattern of shear modulus throughout the cross-section.

b) The commercial SIGMA/W system (Geo-Slope 1999b) was used to generate a finite element mesh and, by gravity turn-on, to compute the initial total stresses.

c) The commercial SEEP/W system (Geo-Slope 1999a) was used, together with results from the field and laboratory program, to compute the steady-state flow regime. The same finite element mesh was used.

d) SEEP/W can plot the equipotentials, but it cannot plot the flow net. A proprietary program that can plot the flow net using Christian's (1980) technique provided the flow net and a useful check on the SEEP/W results. The flow net was also helpful in presentations to clients and discussions with other engineers.

e) Pore pressures computed from the seepage analyses were subtracted from the total stresses to obtain initial effective stresses. Since the finite element meshes were the same in all cases, the subtraction was simply a matter of manipulation of the output files.

f) From the effective stresses and the other results of the field and laboratory studies the dynamic shear modulus pattern was computed. Again a small program that did the calculation element by element and prepared the output in a format compatible with the dynamic finite element programs did this.

g) Dynamic analyses were done with the same finite element mesh input to two programs. First, QUAD4M (Hudson *et a*l. 1994) solved for the dynamic response by step-by-step analysis in the time domain. Next, a microcomputer version of FLUSH (Lysmer *et al.* 1975) did the same analysis in the frequency domain. Because the two methods of analysis truncate the input motion differently and handle damping and boundary conditions differently, the results are somewhat different, and both sets of results were carried forward for the rest of the analysis. In most cases, the QUAD4M results indicated larger accelerations and stresses.

h) To account for three-dimensional effects, the estimated cyclic stresses and strains and peak accelerations were corrected on the basis of the results of Makdisi *et al.* (1982). Since both the dynamic programs provide estimates of the fundamental period of the dam and the spectral shapes of the input motion were available, this was simply done element by element.

i) For each element the most current thinking on liquefaction analysis, as presented by Youd *et al.* (2000), gave factors of safety against liquefaction for each element. These were then contoured, using the SURFER program, to give a visual presentation of the liquefaction hazard.

j) A further concern was the settlement of the top of the dam. Tokimatsu & Seed (1981) provide a method for estimating volumetric strain of unsaturated sands during earthquakes, and Ishihara & Yoshimine (1992) give similar results for the densification following total or partial liquefaction of saturated soils. Both results are expressed graphically, so analytical versions were developed and used, element by element, to estimate the strain of the saturated and unsaturated portions of the dam.

k) The permanent strains were calculated for each element independently, so it was necessary to redistribute these strains to restore connectivity between elements. The easiest way to do this was to use the same mesh in the PCFEAP program (NISEE 1998) and to treat the volumetric strain as the result of a temperature change in a section with varying coefficients of thermal expansion. Again, a small program prepared the appropriate input element by element.

l) Results involving deformed meshes were plotted using the capabilities of ADOBE Photo Shop. SURFER plotted contours of factors of safety, stresses, and so on. The settlement of the top surface was plotted in Power Point.

The result of this analysis was a clear, comprehensive description of what the dam would be expected to do during several levels of earthquake shaking. Graphical output was essential, especially in communicating results to other engineers and to the client. However, graphical output programs require that the user understand how the program works and be prepared to devote some time to getting the presentation right. It has been well said that modern graphical programs make it very easy to prepare very bad plots; this problem must be recognized and fought.

Six different pre-existing finite element programs were used, and the communication between them relied on the use of one mesh for all analyses. Small special programs working one element at a time easily modified the format and calculated the intermediate results. Although this seems less efficient than using one special program that encompassed the entire analysis, the reverse was the case. The analysis went off quite smoothly.

One unexpected benefit of dividing the analysis among different programs working on the same mesh was that it is easy to inspect intermediate results. All too often one sees the input and the final output of a complicated analysis but does not know what is going on inside the process. In this case, as in most cases, knowing the intermediate results made it possible to alter the course of the analysis and to reduce the number of blind alleys pursued.

4 FINAL COMMENTS

It is tempting to prognosticate. The past history of predictions like Bill Gates' that 640K RAM ought to be enough for anyone or Arthur D. Little's that five 1950s vintage Xerox copiers would satisfy the national demand gives one pause. Therefore, we will not venture into that thicket. However, there are some observations about current issues that we do offer as possible fodder for discussion.

The profession has come a long way since the days when analysts had to carry large boxes of cards about under their arms. Computers are more powerful by orders of magnitude; user-friendly software is widely available; graphical input and output – both on screen and in plots – is a staple of modern computing. However, some issues remain. Users still need to know what their software is doing and need to be sensitive to issues of convergence, singularity, stability, and accuracy.

Constitutive relations for geological materials have evolved beyond the dreams of the earliest workers. Unfortunately, so has the complexity of their expression. If modern constitutive relations are to be widely used in practice, the developers need to do two things. First, they need to simplify the descriptions and reduce the number of parameters. Of course, a full description of soil behavior does require a large number of parameters, but guidance is needed on what parameters are most important and which can be approximated with little damage. Second, they need to a much better job of clarifying what is similar and what is different between different models. There is some broad consensus on many aspects of constitutive relations, and no point is served by using differences in notation to hide this fact. The user needs to know what distinguishes one model from the next.

The most difficult problems in geotechnical analysis almost always involve interaction between different materials. Dynamic soil-structure interaction, behavior of piles, design of tunnel linings, and stability of reinforced soil walls are examples. These are composite structures, and they are difficult to analyze. A great deal of effort has gone into trying to understand such systems, but much more remains to be done. This is a challenge to the next generation of researchers.

Relatively less attention has been paid to the reliability of computational models. How well do the models represent reality? How sensitive are they to changes in parameters? As reliability – particularly probabilistic reliability – comes into more practical use, this question becomes more important. Knowledge of errors – how large they are, where they come from, and what to do about them – needs to be emphasized.

Finally, we would echo Whittle's (1999) comment that future developments will require progress on three fronts. First, improvements are needed in "measurement, interpretation, and representation of material properties in finite element analyses," particularly for aspects of behavior and special soils that remain poorly understood. Second, integration of numerical methods with stochastic approaches is inevitable but requires a great deal of work if the promise of these methods is to be realized in practice. Third, substantial educational efforts are needed to train geotechnical engineers in the most effective use of numerical and analytical methods.

5 REFERENCES

Christian, J. T. 1980. Flow Nets from Finite Element Data. *International Journal for Numerical and Analytical Methods in Geomechanics* 4(2): 191-196.

Christian, J. T. 1999. What has the Finite Element Method Done for (or to) Geotechnical Engineering? *Civil Engineering Practice, Journal of The Boston Society of Civil Engineers Section.* 14(2): 73-74.

Courant, R. 1943. Variational methods for the solution of problems of equilibrium and vibration,. *Bulletin of the American Mathematical Society* 49: 1-23.

Duncan, J. M. 1999. Applying the Finite Element Method to Practical Use in Geotechnical Engineering. *Civil Engineering Practice, Journal of The Boston Society of Civil Engineers Section.* 14(2): 75-80.

Geo-Slope International Ltd. 1999a. *SEEP/W*, version 4.23 documentation. Calgary. Alberta, Canada.

Geo-Slope International Ltd. 1999b. *SIGMA/W*, version 4.21 documentation. Calgary, Alberta, Canada.

Hardin, B. O., & Black, W. L. 1968. Vibration Modulus of Normally Consolidated Clay. *Journal of the Soil Mechanics and Foundations Division, ASCE.* 94(SM2): 353-369.

Hudson, M., Idriss, I. M., & Beikae, M. 1994. *User's Manual for QUAD-4M.* Center for Geotechnical Modeling, University of California, Davis.

Ishihara, K., & Yoshimine, M. 1992. Evaluation of Settlements in Sand Deposits Following Liquefaction during Earthquakes. *Soils and Foundations.* 32: 173-188.

Lysmer, J., Udaka, T., Tsai, C. F., & Seed, H. B. 1975. *FLUSH: A Computer Program for Approximate 3-D Analysis of Soil-Structure Interaction Problems*, Report EERC 75-30. Earthquake Engineering Research Center, University of California, Berkeley.

Makdisi, F. I., Kagawa, T., & Seed, H. B. 1982. Seismic Response of Earth Dams in Triangular Canyons. *Journal of the Geotechnical Engineering Division, ASCE.* 108(GT10): 1328-1337.

Marr, W. A. 1999. Effective Uses of Finite Element Analysis in Geotechnical Engineering. *Civil Engineering Practice, Journal of The Boston Society of Civil Engineers Section.* 14(2): 89-98.

National Information Services for Earthquake Engineering NISEE 1998. CDROM Software Library, University of California, Berkeley.

Tokimatsu, K., & Seed, H. B. 1981. Evaluation of Settlements in Sand Due to Earthquake Shaking

Journal of Geotechnical Engineering, ASCE. 113(8): 861-878.

Whittle, A. J. 1999. The Role of Finite Element Methods in Geotechnical Engineering. *Civil Engineering Practice, Journal of The Boston Society of Civil Engineers Section.* 14(2): 81-88.

Youd, T. L., Idriss, I. M., & 19 others 2000. Liquefaction Resistance of Soil. paper submitted to *Journal of Geotechnical and Geoenvironmental Engineering, ASCE.*

Computer Methods and Advances in Geomechanics, Desai et al. (eds) © 2001 Balkema, Rotterdam, ISBN 90 5809 183 X

Prediction of the lateral capacity of suction caissons

W. Deng, J. P. Carter & H. Taiebat
Department of Civil Engineering, The University of Sydney, N.S.W., Australia

ABSTRACT: The ultimate capacity under inclined loading, including purely horizontal loading, is a critical issue in the design of suction caissons, and accurate methods of predicting the capacity under this form of loading are required in order to produce reliable designs. Theoretical investigations using an economical 3-D finite element method have been carried out of suction caissons subjected to horizontal loading. Simplified solutions, which take into account the influence of the aspect ratio of the caisson, the load application point, and plan area of the suction caisson, and the strength parameters of the soil are developed from the results of the finite element analysis. The simplified equations are examined in the light of results from 35 individual tests from 5 previous independent experimental studies. The application of the theoretical solutions to the design of suction caissons is also discussed.

1 INTRODUCTION

1.1 *Description of the problem*

Resembling an inverted bucket in shape, suction caisson foundations have proved to be competitive alternatives to more traditional foundation solutions such as piles and drag anchors in various types of soils and for a wide range of fixed and floating platforms. Despite the wide use of suction caisson foundations, design is largely conducted on a case by case basis, as no generic solutions are available to assist geotechnical engineers.

In the research described in this paper, an extensive investigation has been carried out of a suction caisson subjected to horizontal loading using a semi-analytical 3-D finite element modelling approach. This research deals specifically with the typical cases illustrated in Figure 1. The soil is assumed to be either normally consolidated or of uniform strength. The suction caisson is assumed to be cylindrical with an overall diameter d and an embedded length L. It is assumed that a purely horizontal force, H_D is applied at the caisson wall by a chain attached to a lug located at a depth D.

1.2 *Previous research on lateral capacity of piles*

The related plane strain problem of a laterally loaded rigid pile embedded in a cohesive soil under undrained condition has been studied by several researchers. Broms (1964) approximated the ultimate lateral load using the theory of plasticity. He indicated that for plane strain conditions the ultimate lateral resistance or limiting pressure, expressed as the ultimate lateral load divided by the projected area of the pile shaft, is approximately $9.14s_u$ for a smooth circular pile, and $11.42s_u$ for a rough square pile, where s_u is the undrained shear strength of the soil. Poulos & Davis (1980) studied this problem for piles with rectangular cross sections using the classical theory of plasticity and limit analysis. The ultimate lateral resistance for a square smooth pile was found to be $8.28s_u$, while for a square rough pile the value was quoted as $11.14s_u$. Randolph & Houlsby (1984) used the classical theory of plasticity and developed exact solutions for the limiting lateral resistance of a circular pile in cohesive soil.

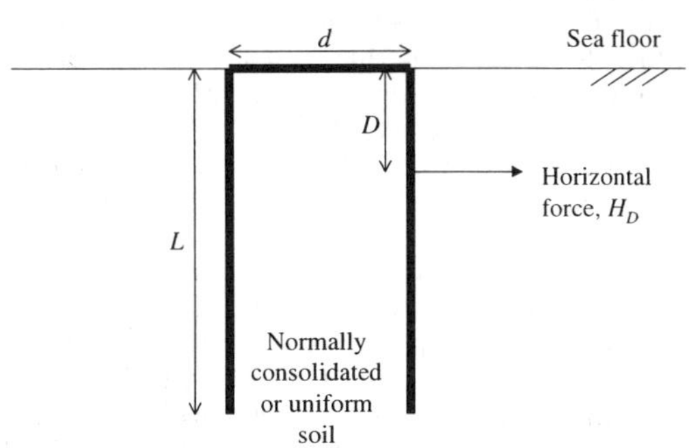

Figure 1 Study case

For a perfectly rough pile, the ultimate lateral pressure was shown to be approximately $11.94s_u$.

2 FINITE ELEMENT STUDY

A series of finite element analyses has been performed to investigate the effects of various parameters on the lateral capacity of caisson foundations. The three-dimensional nature of the problem requires a 3-D finite element analysis. However, application of conventional finite element codes usually requires extensive computational time and effort, especially in analyses that include non-linear and inelastic material behaviour.

A semi-analytical finite element approach (Zienkiewicz & Taylor, 1989) has been adopted in formulation of an efficient method of analysis. Application of this method to problems with axi-symmetric geometry and non-symmetric loading has shown an effective reduction in computational time. This method has been used for analysis of the cases considered in this study.

2.1 *Semi-analytical finite element study*

In many physical problems, such as the problem of a laterally loaded caisson, the loading is three-dimensional but the geometry and material properties do not vary in one coordinate direction, i.e., the circumferential direction. In such cases it is possible to solve a set of algebraic equations arising from a substitute problem, not involving that particular coordinate, and to synthesize the true answer from a series of simplified solutions. In general, this type of analysis has been termed the 'semi-analytical method' in finite element analysis.

A combination of the two dimensional finite element process and a Fourier series in the third dimension was first developed in the context of linear analysis by Wilson (1965). This method was applied to the determination of stresses and displacements within complex structures of revolution subjected to non-symmetric thermal and mechanical loads. This method has been modified and extended further to include non-linear analysis as well as analysis of consolidation problems (Meissner, 1976; Winniki & Zienkiewicz, 1979; Booker & Small, 1979; Carter & Booker, 1983; Runesson & Booker, 1982 and 1983; Lai & Booker, 1991; Taiebat & Carter, 2000).

Both continuous and discrete Fourier series can be used in the semi-analytical method. The continuous Fourier series approach is very useful for elastic analyses of problems with relatively simple loading conditions. However, for analyses which incorporate elasto-plasticity there are difficulties associated with the calculation of the values of harmonic forces as well as summing the large number of Fourier

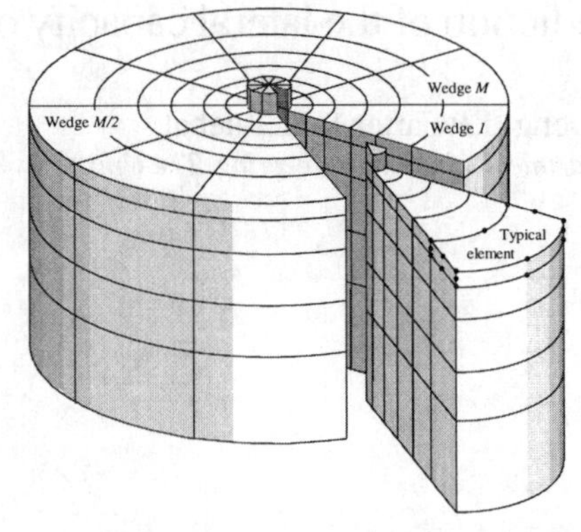

Figure 2 Typical arrangement of wedges and elements

terms. In contrast to the continuous Fourier series method, the discrete Fourier series terminates after a finite number of terms. In addition, the integration of pseudo forces occurring in plasticity presents no extra difficulty in the discrete Fourier series method.

The discrete Fourier series approach was therefore seen as preferable for problems involving material non-linearity, and so it was adopted in the formulation of the semi-analytical finite element method used in this study.

This approach is based on the assumption that the field quantities such as displacements can be represented by their discrete Fourier transforms. In this method advantage is taken of the axi-symmetric nature of the problem geometry. Therefore, only one wedge from a cylinder of the soil-foundation material requires modelling (Figure 2). Instead of solving a very large number of algebraic equations for the full three-dimensional problem, a smaller number of equations arising from a substitute problem are solved (Taiebat & Carter, 2000). Further efficiencies can be obtained by taking advantages of the fact that the geometry of each wedge has reflective symmetry about its bisecting plane, so that the standard 20 noded hexahedral element changes to its equivalent 12 noded half element in the semi-analytical method (Figure 3), which results in further reduction in computer storage and computation time.

The semi-analytical approach adopted in this study reduces the computation time to less than 5

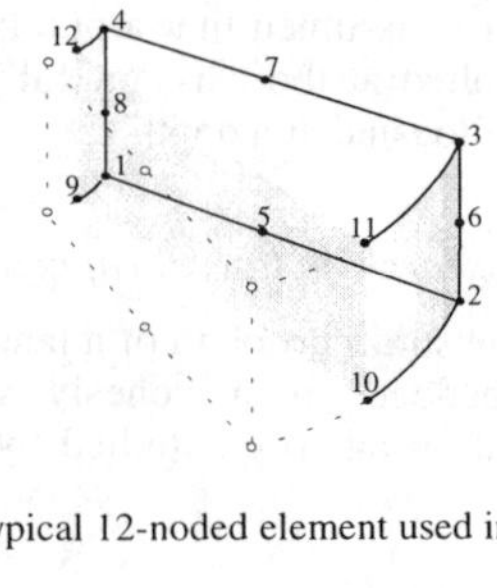

Figure 3 Typical 12-noded element used in modal analysis

percent of the time required for a standard three-dimensional finite element solution. The method has been encoded in the finite element program AFENA (Carter & Balaam, 1995).

In the study presented here, all analyses were performed using 12-noded isoparametric finite elements (Figure 3) with a reduced integration scheme. The number of wedges in the cylinder that models the geometry has been adopted as 12 in all analyses.

2.2 *Soil model*

Investigations of the properties of seabed materials from environments such as the outer North-West Shelf of Australia, suggest that in some locations the soil is normally consolidated or lightly over-consolidated. In this case the strength and stiffness would normally increase approximately linearly with depth. In other offshore environments different soil profiles can also exist, *e.g.*, they may consist of cohesive soils with approximately uniform strength. The uniform strength case is likely to be more important in relatively shallow water, where the deposits may have been influenced by lower sea levels in the past, and also in cases where complex soil profiles are most conveniently idealised as a uniform material with some representative value of strength. For example, Larsen (1989) reported a series of model tests conducted in a meltwater deposit from the Nivå tileworks, which is a homogeneous deposit preconsolidated by the pressure of overlying ice. Uniform strength conditions are probably encountered less frequently in practice than deposits where strength increases with depth, but nevertheless solutions for the uniform strength case are of some significance.

An elastic, perfectly plastic soil model was adopted in the finite element study and the analysis was conducted assuming undrained conditions adopting a total stress approach. In both cases described above, the Tresca yield criterion was used to describe the shear strength of the seabed soils and in general the undrained shear strength, s_u can be expressed as

$$s_u = s_{uo} + kz \qquad (1)$$

where, s_{uo} is the undrained strength at the surface of the soil layer, and k is its gradient with the depth, z, below the seabed. Typical values of 0 to 30 kPa and 0 to 3 kPa/m were adopted for s_{uo} and k respectively in the analyses. For uniform soils, $k = 0$. It is assumed that the soil is isotropic and linear elastic until the onset of plastic yielding. In all cases, the shear modulus was assumed to be 333 kPa at the surface of the soil layer and varying proportionally with depth at a gradient of 66.7 kPa/m. However, as there is a free surface in this problem, the deformations are relatively unconfined, so that the ultimate capacity will be relatively insensitive to the precise details of the modulus profile adopted in the analysis.

2.3 *Parametric study*

The caisson is in the shape of an upturned bucket with vertical cylindrical sides. In all cases it has been assumed that the caisson has been installed without causing any disturbance to the soil. The caisson was modelled as an isotropic elastic material using solid elements with a Young's modulus of 9×10^9 kPa and Poisson's ratio of 0.49. The soil inside the caisson was modelled as the same material as the soil outside the caisson.

Four different values of the aspect ratio have been considered in this study, viz. L/d = 1.0, 1.5, 2.0 and 2.5. The ratio of lug depth to the length of the caisson, D/L, was taken as: 0, 0.22, 0.33, 0.44, 0.53, 0.67, 0.73, 0.89 and 1.0.

In all cases considered in this research, the interface condition between the soil and the caisson has been considered as perfectly rough.

3 SIMPLIFIED SOLUTIONS

It is worthwhile investigating whether it is possible to develop a simplified method for the estimation of the ultimate value of the lateral loads, which quite clearly will be a function of aspect ratio of the caisson (L/d), the normalised lug depth (D/L), some measure of the undrained strength of the soil (s_u), and the base area of the caisson ($A = \pi d^2/4$) or its projected area in a vertical plane (Ld).

3.1 *Normally consolidated soil*

In this case the ultimate horizontal load that may be applied to a suction caisson, H_D, can be expressed in the form

$$H_D = N_h A s_{u(2/3L)} \qquad (2)$$

where, $s_{u(2/3L)}$ denotes the undrained shear strength at a depth equivalent to 2/3 of the caisson length. N_h can be defined as the lateral capacity factor. By fitting a curve to the finite element predictions of the ultimate lateral capacity it was found that N_h can be described by an expression of the form

$$N_h = \frac{\alpha}{\sqrt[4]{\left(\frac{\alpha}{6.3} - \frac{D}{L}\right)^4 + \left(\beta \frac{D}{L}\right)^4}}. \qquad (3)$$

The dimensionless parameters α and β are related to the aspect ratio, L/d, and they can be expressed as follows

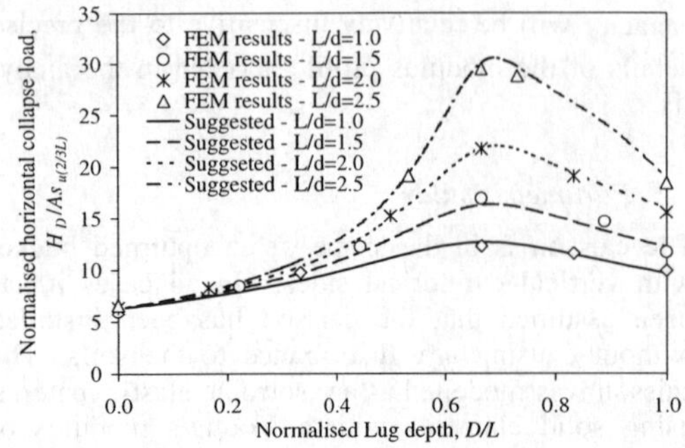

Figure 4 Lateral capacity ratio - normally consolidated soils

$$\alpha = 7.02\left(\frac{L}{d}\right)^{-0.3785} \tag{4}$$

$$\beta = 1.58\, e^{-0.775\left(\frac{L}{d}\right)}. \tag{5}$$

The finite element predictions and the curves described by equations (3) to (5) are presented in Figure 4, from which it is clear that the equations fit the finite element predictions to sufficient accuracy for most practical purposes.

3.2 *Uniform soil*

In this case the ultimate horizontal load that may be applied to a suction caisson, H_D, was assumed to have the following form

$$H_D = N_h L d s_u . \tag{6}$$

Note that the form of equation (6) differs from that of equation (2). In equation (6), it has been more convenient to express the ultimate load in terms of the projected area of the caisson, rather than its cross sectional area.

For uniform soil, the relationship between the lateral capacity factor, N_h, and the aspect ratio of the caisson, L/d, the load application point, D/L is shown on Figure 5. It is interesting to find that for the same value of D/L, the lateral capacity is independent of the aspect ratio of the caisson, L/d. By curve fitting it was found that N_h can be described by an expression of the form

$$N_h = \frac{3.6}{\sqrt{\left(0.75 - \left(\frac{D}{L}\right)\right)^2 + \left(0.45\left(\frac{D}{L}\right)\right)^2}} . \tag{7}$$

The curve described by equation (7) is also presented in Figure 5, where it appears that equation (7) fits the finite element predictions well.

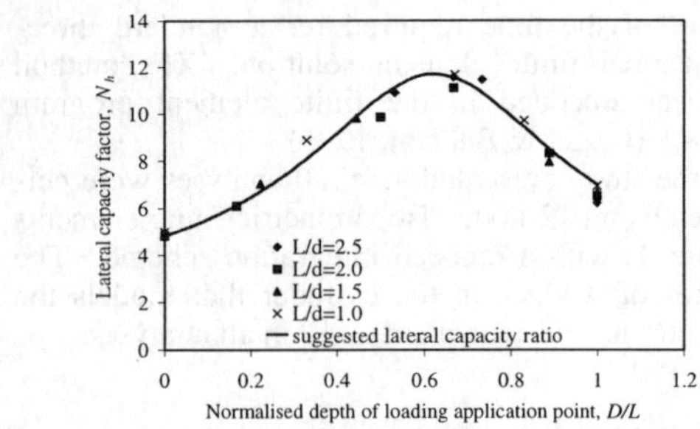

Figure 5 Lateral capacity ratio – uniform soils

It is noted that the optimal load application point is roughly at $D/L = 0.63$. When the load is applied at this point, the largest horizontal capacity is mobilised, *i.e.*, $H_D/(Lds_u) = 11.7$. It is interesting to note that this value of the normalised capacity agrees well with the plane strain solutions for laterally loaded piles derived by Randolph & Houlsby (1984).

4 COMPARISONS WITH EXPERIMENTAL RESULTS

Several experimental studies have been carried out of suction caissons subject to lateral loading, for cases where the subsoil is effectively undrained. Larsen (1989) reported 4 lateral load tests being carried out in a reconstituted natural clay and in kaolin, however only one result was actually quoted in the literature. Keaveny *et al.* (1994) reported the results of five large (1.4 m × 0.7 m × 1.4 m deep) model-scale tests performed in the field at Lysaker, Oslo, Norway. Three static tests were reported and in these tests the load was applied at 10° to the horizontal, which may be well approximated as a lateral load case. . Two centrifuge tests, at 120g, investigating the response of suction caissons to catenary loading were carried out by House (2000). The prototype size of the caisson was 4.8 m in diameter and 14.4 m in length. The loading point was 9.6 m below the caisson top. The soil used for this series of tests was normally consolidated kaolin, prepared by self-weight consolidation of the clay slurry on a fixed beam geotechnical centrifuge. Watson *et al.* (2000) reported 2 lateral load tests on caissons with skirt lengths of 40% and 50% of the foundation diameter. These tests were conducted in kaolin clay and calcareous silt. The lateral capacity of the caisson in calcareous silt continued to increase beyond the initial yield point. However, this type of behaviour was not observed in the tests in kaolin clay.

Figure 6 shows the theoretical solutions for L/d = 0.4, 0.5 and 2.85 given by expressions (3) to (5), together the test results from House (2000) where the aspect ratio was L/d = 2.85, and the results from

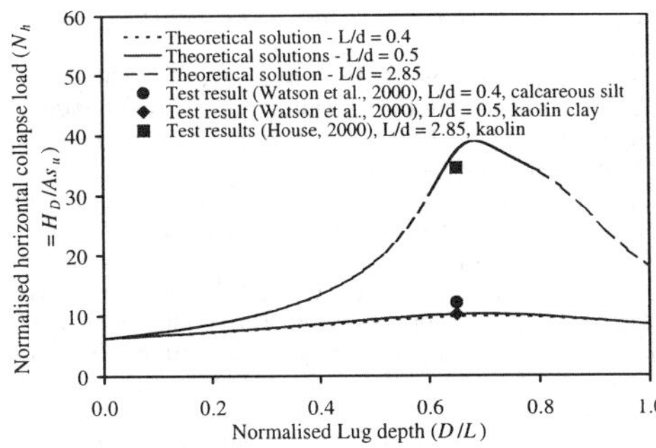

Figure 6 Comparison with tests results – normally consolidated soils

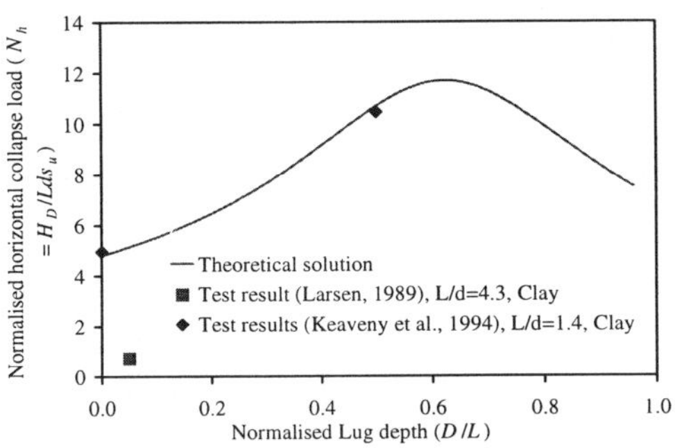

Figure 7 Comparison with test results – uniform soils

Watson *et al.* (2000) where the aspect ratio was 0.4 for calcareous silt and 0.5 for kaolin clay. It may be seen from this figure that the theoretical solution proposed for the lateral capacity agrees well with the data for these tests in clay.

Figure 7 shows the theoretical solutions given by expressions (6) and (7), together with the test results from Keaveny *et al.* (1994), where the aspect ratio was $L/d = 1.4$, and the results from Larsen (1989) for the test in which $L/d = 4.3$. In Figure 7, the results of test 1 of Keaveny *et al.* (1994) was scaled to reflect of the effect of attaching a mudline liner, which enforced a two-sided failure mechanism.

It can be seen from both Figure 6 and Figure 7 that the theoretical predictions agree with all the test results except two. One of the two exceptions is the test result obtained by Watson *et al.* (2000) for calcareous silt with an aspect ratio of 0.4. However, it is noted that the dilation of the calcareous silt had a significant impact on this test result, and it was also observed that the lateral capacity of the caisson in calcareous silt continued to increase beyond the initial yield point, so that the ultimate capacity was difficult to determine. Another exception is the result obtained by Larsen (1989). The aspect ratio (L/d)

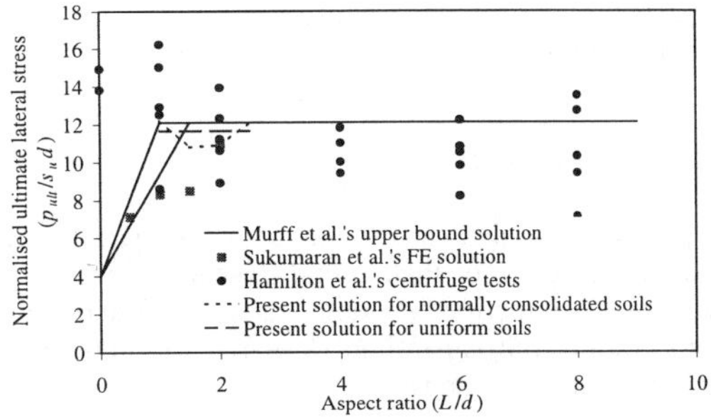

Figure 8 Comparison with the upper bond solutions and centrifuge test results

for this test result is 4.3, which is outside the range of the application of expressions (1) and (2).

Hamilton *et al.* (1991) also carried out experimental investigations of the ultimate soil resistance on laterally loaded piles. The tests were designed to determine ultimate soil resistance values for depths ranging from the mudline ($L/d = 0$) to over 10 diameters ($L/d = 10$) below the mudline. A total of 27 tests were performed in soils with a shear strength of 9 kPa at the surface, increasing at a rate of 1.4 kPa/m. A comparison between the present theoretical solutions and the tests results is presented in Figure 8, together with the upper bound solutions obtained by Murff *et al.* (1993). It is clear from this figure that the solutions developed in this research generally agree reasonably well with the experimental results over the limited range of values of L/d for which they were designed (0.5 to 2.5). Murff *et al.*'s approximate solution is also a reasonable fit to the experimental data over the full range of L/d from 0 to 10.

From the comparisons with all the test results considered here it is clear that the present solutions give very good estimates for the ultimate lateral capacity of caissons for cases where the soil is cohesive and non-dilating and for caisson aspect ratios ranging from 0.5 to 2.5.

5 CONCLUSIONS

This paper has described the results of a numerical study of the lateral capacity of suction caissons. Consideration has been given to the influence of the aspect ratio of the caisson (L/d), the load application point (D/L), plan area of the suction caisson (A) and the strength parameters of the soil. The simplified solutions developed in this research were also examined in the light of results from 35 individual tests from 5 previous independent experimental studies. It is concluded that the theoretical solutions developed in this paper may be used with some confidence for estimation of lateral capacity of suction caissons in cohesive and non-dilating soils, for caisson aspect ratios ranging from 0.5 to 2.5.

6 ACKNOWLEDGEMENT

The work described in this paper forms part of the research program of the Special Research Centre for Offshore Foundation Systems, established and supported under the Australian Research Council's Research Centres Program.

REFERENCES

Booker, J. R. & Small, J. C. 1979. Finite element analysis of the consolidation of layered soils, *Proc. 3rd Int. Conf. in Australia on finite element methods*, Sydney. 485-500.

Broms, B. B. 1964. Lateral Resistance of Piles in Cohesiveness Soils. *Journal of the Soil Mechanics and Foundations Division, ASCE*, Vol. 90, SM2, 27-63.

Carter, J. P. & Booker J. R. 1983. Consolidation of axi-symmetric bodies subjected to non-axi-symmetric loading, *Int. J. Num. Anal. Meth. Geom.*, 7, 273-281.

Carter, J.P. & Balaam, N.P. 1995. *AFENA – A Finite Element Numerical Algorithm – Users' Manual*, Centre for Geotechnical Research, The University of Sydney.

Hamilton, J. M., Phillips, R., Dunnavant, T. W. .& Murff, J. D. 1991. Centrifuge study of laterally loaded pile behaviour in clay. *Proceedings International Conference Centrifuge 1991*, ISSMFE, 345-361

House, A. 2000. The Response of Suction Caissons to Catenary Loading. *Proceedings of 4th Australia New Zealand Young Geotechnical Professionals Conference*, Perth, Australia, House & Watson (eds), 119-124.

Keaveny, J. V., S. B. Hansen, C. Madshus, & R. Dyvik, R. 1994. Horizontal Capacity of Large-scale Model Anchors. *Proceedings 13th International Conference On Soil Mechanics and Foundation Engineering*, New Delhi, Vol 2, 677-680.

Lai, J. Y.& Booker, J. R. 1991. Application of Discrete Fourier series to the finite element stress analysis of axi-symmetric solids, *Int. J. Num. Methods in Engineering,* 31, 619-47

Larsen, P. 1989. Suction Anchors as an Anchoring System for Floating Offshore Constructions. *Proceedings, Offshore Technology Conference, OTC 6209*. Houston, USA. 535-540.

Meissner, H. E. 1976. Laterally loaded pile in cohesionless soil, *Proc. 2nd Conference in Numerical Methods in Geomechanics*, Virginia, Desai, C. S.(ed.). 1353-1365.

Murff J. D. & Hamilton J. M. 1993. *P*-Ultimate for Undrained Analysis of Laterally Loaded Piles. *Journal of Geotechnical Engineering, ASCE*, Vol. 119(1). 91-107.

Poulos H. G. & Davis E. H. 1980. *Pile foundation analysis and design*, John Wiley and Sons, New York, USA

Randolph, M.F. & Houlsby, G.T. 1984. The Limiting Pressure On a Circular Pile Loaded Laterally in Cohesive Soil. *Geotechnique*, 34(4), 613-623

Runesson, K. & Booker, J. R. 1983. Finite element analysis of elastic-plastic layered soil using discrete Fourier series expansion, *Int. J. Num. Meth. Engng*, Vol 19. 473-478

Runesson, K. & Booker, J. R. 1982. Efficient finite element analysis of 3-D consolidation, *Proc. 4th Int. Conference in Numerical Methods in Geomechanics*, Edmonton. 359-364.

Taiebat, H. A. & Carter, J. P. 2000. A Semi-Analytical Finite Element Method for Three-Dimensional Consolidation Analysis, *Computers and Geotechnics*, Accepted for publication.

Watson, P. G., Randolph, M. F. & Bransby, M. F. 2000. Combined Lateral and Vertical Loading of Caisson Foundations. *Proceedings, Offshore Technology Conference, OTC 12195*. Houston, USA. 797-808

Wilson, E. L. 1965. Structural analysis of axi-symmetric solids, *Journal of American Institute of Aeronautic and Astronautics*, 3, 2269-2274.

Winnicki, L. A. & Zienkiewicz, O. C. 1979. Plastic (or visco-plastic) behaviour of axi-symmetric bodies subjected to non-symmetric loading, semi-analytical finite element solution, *Int. Journal for Numerical Methods in Engineering*, 14, 1399-1412.

Zienkiewicz, O. C. & Taylor, R. L. 1989. *The Finite Element Method*, 4th Edition, McGraw-Hill, New York.

Computer Methods and Advances in Geomechanics, Desai et al. (eds) © 2001 Balkema, Rotterdam, ISBN 90 5809 183 X

Problem solving by interactive use of numerical models and field measurements

M. Doležalová, I. Hladík, V. Zemanová & J. Danko
Dolexpert-Geotechnika, Prague, Czech Republic

ABSTRACT: To predict true engineering performance of geotechnical structures, the link between numerical modeling and field measurements should not be restricted to pure prediction and verification or to a single back analysis procedure. These are only particular stages, which should be followed by other stages like calibration of numerical models according to field measurements and updating the measurement system according to numerical solutions. This is an interactive and iterative process of input and feedback between numerical modeling and monitoring, which allows modification and improvement of both the model and the monitoring system. Such a process contributes to determination of proper input data at investigation and design stages and allows safety check of the structure according to the monitoring results during construction and operation stages. This is demonstrated in the paper by case histories of large civil engineering structures.

1 INTRODUCTION

20th century can be noted not only for construction of large civil engineering structures with their impact on the environment, but also for development of tools to predict and control their behavior and to prevent the unfavorable effects.

Due to computerization, an enormous development of measuring techniques and data processing on one side, and numerical methods for solving boundary value problems of engineering mechanics on the other occurred. Both field measurements and numerical modeling solely represent powerful tools for engineering design and operational control. However, their use in an interactive way has an additional, synergic potential, which facilitates solution of complex geotechnical problems and design of unique civil engineering structures.

This is an interactive, multi-stage approach with repeated input and feedback between numerical modeling and monitoring, whilst prediction, validation and back analysis procedures represent its particular steps. Typically, different numerical methods of continuum and discontinuum mechanics, several linked up 2D and 3D numerical models and various types of measurements can be applied to solve the problem. This iterative process allows improvement of both the numerical model and the measurement system, which yields a firm basis for decision making at different design and construction stages of civil engineering structures.

To illustrate that, three case histories are presented in the paper.

The first work is a part of an extensive site characterization program for High Level Waste (HLW) disposal in Hungary. Full-scale in situ tests performed in an underground rock laboratory (URL) were modeled and interpreted by numerical models. A series of linking up 2D and 3D, uncoupled and coupled computational models were used and constitutive models and parameters well describing the host rock mechanical response were obtained.

A closure system of a unique structure, an unlined underground gas storage built at a depth of 900 m in competent granodiorite (Haje, Czech Republic) is described in the second case history. The closure system is created by a pressurized water chamber, which is confined by four shotcrete plugs fixed into the surrounding rock mass and exposed to one-side gas/water pressure up to 13 MPa. To solve this rock-structure interaction problem, the above interactive design procedure including full-scale in situ tests, numerical models and monitoring was applied.

In the third case history, long-term control of mining operation via numerical models and monitoring is described. The Jiri Open-Pit Coal Mine is located in Western Bohemian spa region with artesian thermal springs. Due to removal of the overburden, unloading occurs to the protective rock barrier forming the mine bottom and confining the underlying pressurized aquifer. To prevent hydraulic fracturing, the pressure head has been reduced by flowing wells. To estimate the feasibility of the mining advance at minimum pressure head reduction, i.e. at minimum environmental impact, numerical models calibrated according to field measurements and a monitoring system updated according to numerical solutions have been applied since 1976 up to the present time. Selected stages of this long-term interactive procedure and the current situation are outlined in the paper.

2 NUMERICAL CODES

A modified version of the original CRISP FEM Code (Britto et al. 1987) called CRISP-PATH was used for the most of solutions listed below. The modified version has been developed since 1993, extending the facilities of CRISP to several new areas of application-especially to solution of larger 3D problems and to appropriate constitutive modeling:

- *new constitutive models*, namely path-dependent nonlinear constitutive model (Doležalová 1991), multisurface elasto-plastic model, multilaminate elasto-viscoplastic model, elasto-plastic model with hardening- softening;
- *efficient solvers of arising linear and nonlinear systems of equations*, namely linear iterative solvers (Hladík et al. 1997) and nonlinear solvers of Newton type (Hladík 1998).

For solving specific problems of jointed rock, the 2D Universal Distinct Element Code (ITASCA 1993) was applied.

3 ROCK MASS CHARACTERIZATION VIA NUMERICAL MODELING OF CONVERGENCE MEASUREMENTS

Since about 40% of the electricity of Hungary is produced by Paks Nuclear Power Plant, to find an appropriate geological formation for final disposal of HLW is a vital problem for the country.

An extensive site investigation program including more than 100 different test types was performed for characterization of Boda Claystone Formation (BCF), which is a sedimentary rock formation of Middle Permian age located in the Western Mecsek Mts. As a part of the program, full-scale in situ tests (convergence measurements, borehole water pressure tests, rock mass temperature and thermal gradient monitoring), which were modeled and interpreted later by numerical methods (Doležalová et al. 1998, 1999), were performed in an exploratory tunnel serving as an URL at a depth of 1100m.

3.1 *Convergence measurements*

Convergence measurements record the deformation of the tunnel wall caused by stress release due to excavation. Radial displacements are measured in directions determined by convergence pins installed in the vicinity of the tunnel face. An arrangement used for BCF according to the proposal of Canadian AECL is shown in Fig.1a. The length of pins was 1.2 m (crown) and 1.8 m (bottom) and their distance (A-A′) was measured after each of at least 7 blasts. The initial face distance was between 0.7-1.0 m.

The convergence measurements were carried out altogether in 12 sections in the Alfa-1 tunnel advancing dip direction and exploring the BCF under variable geological conditions. Along with regular measurement results used for estimation of in situ stress field ($K_0 = 1.0$), frequent change of the direction of maximum convergence in near-by profiles was registered, which could hardly be caused by

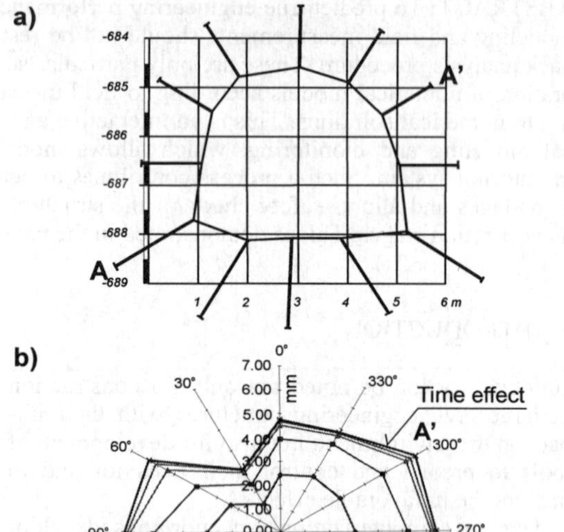

Figure 1. Convergence measurements in BCF Exploratory Tunnel (a), results in tectonically disturbed profile K3 (b) and undisturbed profile K8 (c)

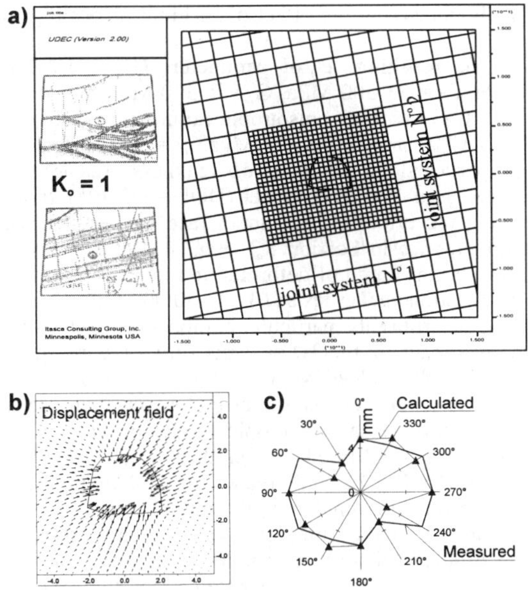

Figure 2. UDEC model of Section K3 (a), displacement vectors (b) and comparison of calculated and measured convergence (c)

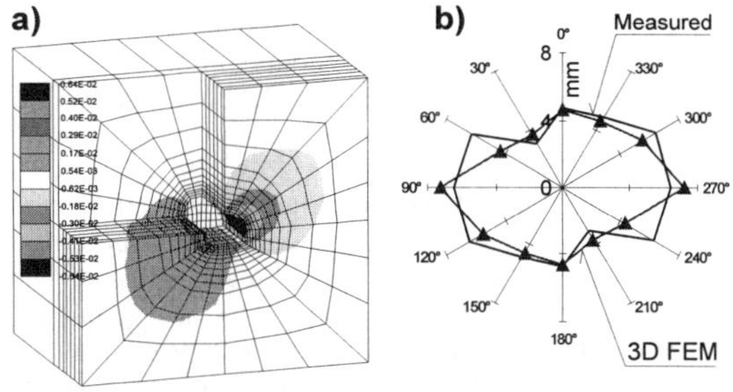

Figure 3. Transversally isotropic 3D FEM model of Section K3: horizontal displacements (a) and comparison of the measured and calculated convergence (b)

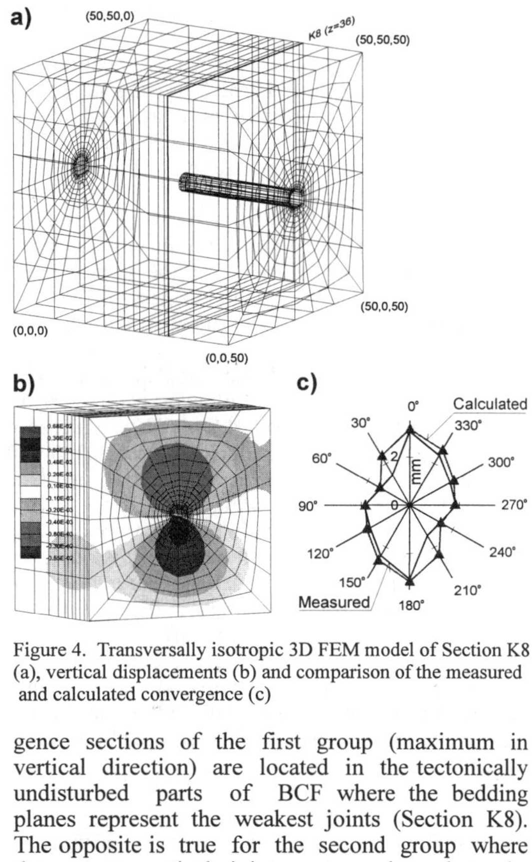

Figure 4. Transversally isotropic 3D FEM model of Section K8 (a), vertical displacements (b) and comparison of the measured and calculated convergence (c)

such a variability of primary stress state. Two groups were distinguished: maximum convergence in near-to vertical direction (Section K8 in Fig.1c) and maximum convergence in near-to horizontal direction (Section K3 in Fig.1b). Segment A-A′ here represents the sum of radial displacements of points A and A′ in Fig.1a.

Section K8 belongs to the so-called Complete Instrument Array, where along with convergence pins, radial extensometers were installed by AECL as well. By comparing these independent measurement results, the accuracy of convergence measurements was verified with satisfactory results.

3.2 *Numerical modeling*

The basic idea was that convergence measurements carried out under variable geological conditions give information not only on the in situ stress field, but on the rock mass structure as well. Thus, assuming for in situ stress field $K_0 = 1.0$, they can be used for identification of constitutive models and parameters describing the rock mass mechanical behavior.

The analysis was made in following steps:

1) The influence of the geological conditions on the measurement results was analyzed. The direction of maximum convergence showed to be dependent on the orientation of the weakest joint system.

2) The joint systems characteristic of tectonically undisturbed (bedding planes) and disturbed parts of BCF (bedding planes and near-to-vertical joints) were specified. It was found that the convergence sections of the first group (maximum in vertical direction) are located in the tectonically undisturbed parts of BCF where the bedding planes represent the weakest joints (Section K8). The opposite is true for the second group where the near-to-vertical joint system showed to be the weakest one (Section K3).

3) A 2D distinct element (UDEC) parametric study for explicit modeling of joints was performed and the joint properties fitting the measurement results determined by a trial and error procedure.

4) A transversally isotropic constitutive model for BCF was introduced with moduli derived from the properties of jointed rock. This model was applied in continuum analyses using finite elements.

5) The value of the independent shear modulus was refined by a 3D FEM parametric study. Based

on these results the effect of the tunnel face advance on the convergence measurements and other in situ tests were modeled using large 3D FEM models.

3.2.1 *2D UDEC model*

If a 2D model simulates convergence measurements, an assumption should be made concerning the initial convergence v_0 occurring prior installation of the convergence pins. In this study $v_0 \approx v_c$ (v_c is convergence due to creep, v due to excavation) was assumed for 2D models and an approximation of the measured sum $v_c + v \approx v_0 + v$ was aimed at.

Mohr-Coulomb elastic-perfectly-plastic constitutive model was applied for blocks and its version with zero cohesion and tensile strength after failure for joints. The material properties corresponding to convergence measurement results in Section K3 (Fig.1b) were determined in the following steps:

1) Use of the properties of intact rock determined by laboratory and field tests for blocks (Somosvári 1997) and homogenization of the region by increasing the stiffness of all joints.

2) Determination of the deformation modulus and joint stiffness fitting the minimum convergence.

3) Keeping this stiffness for the joint system No1 (Fig.2a), reduction of the stiffness of the joint system No2 in order to model the maximum convergence.

According to this procedure and the tests, BCF can be characterized by the following parameters: $E_r = 40\ 000$ MPa, $\nu = 0.20$, $\varphi = 42°$, c = 17.5 MPa, $\sigma_t = 6.5$ MPa, $\gamma = 27$ kN/m^3 for blocks, $k_n = k_t = 300\ 000$ MPa/m, $\varphi = 42°$, c = 8 MPa, $\sigma_t = 3$ MPa for joints No1 and $k_n = k_t = 30\ 000$ MPa, $\varphi = 42°$, c = 8 MPa and $\sigma_t = 3$ MPa for joints No2. Here E, ν, φ, c, σ_t, k_n, k_t denote the deformation modulus, Poisson ratio, friction angle, cohesion, tensile strength, unit weight, normal stiffness and shear stiffness.

The calculated displacements are shown in Fig. 2b, the comparison of convergence patterns in Fig. 2c. The model proved to be universal, as by changing the direction of the weakest joint system, it was possible to model convergence measured in different profiles.

3.2.2 *2D and 3D FEM models*

High stiffness in one direction and low stiffness in the perpendicular one characterizes the deformational response of BCF discovered by UDEC. Since such a behavior can be described by transversally isotropic constitutive model with five constants as well, its elastic-perfectly-plastic version was applied in continuum analyses using 2D and 3D FEM.

The deformation moduli were calculated using approximate formulas derived for single set of uniformly spaced (s = 0.5 m) joints: $1/E = 1/E_r + 1/(k_n s)$, the Poisson ratios were estimated by empirical formulas given in (Wittke 1990) and the independent shear modulus was refined by a 3D FEM parametric study, analyzing the effect of bedding planes dip as well.

The resulting set of parameters (E_1 = 10 900 MPa, E_2 = 31 500 MPa, G_2 = 5 500 MPa, ν_1= 0.20, ν_2 = 0.17) with an anisotropic failure envelope (c = 17.5 MPa, $\varphi = 42°$, $c^j = 8$ MPa, $\varphi^j = 42°$; j denotes properties along the planes of weakness) was applied for a large 3D FEM model (DOF = 42 495) simulating the effect of the tunnel face advance on the convergence in Section K3 (Fig.3). It can be noted that similar range of deformation moduli was found by inflation packer tests (10 000-30 000 MPa).

Concerning the particular components of the total convergence v_t, the 3D analysis fitting the measurement results yielded: $v_0+v+v_c = (0.4+0.3+0.3)v_t$.

This conclusion, as well as the universal character of the transversally isotropic model developed for BCF was confirmed by another large 3D FEM model (DOF = 72 260) created for modeling of coupled mechanical-hydraulic-thermal response of BCF according to the complex measurement results in Section K8 (Fig.4a). Although a quite different convergence pattern was registered in this profile (Fig.1c), a good fit was obtained (Fig. 4c) using almost the same set of deformational parameters, but with changed directions: E_1 = 31 500 MPa, E_2 = 12 500 MPa, G_2 = 5 500 MPa, ν_1= 0.17, ν_2 = 0.20. Along with the deformational parameters, the permeability of the Excavation Disturbed Zone was determined as well and a relatively high value (k = 10^{-8} m/sec) was obtained.

4 INTERACTIVE DESIGN PROCEDURE OF A PRESSURE PLUG

Since 1998 the underground natural gas storage Pribram-Haje built at a depth of 900 m in competent granodiorite in the Czech Republic is under operation. The unique storage facility consists of a net-

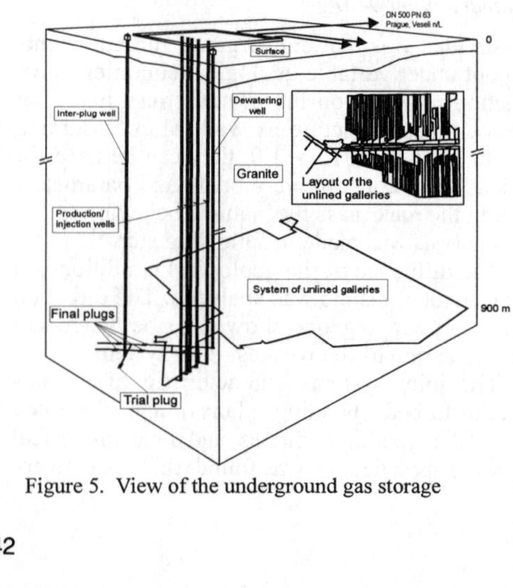

Figure 5. View of the underground gas storage

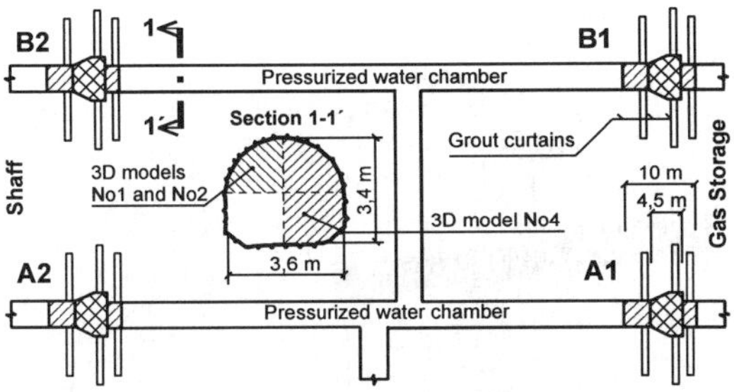

Figure 6. Closure system of the Haje unlined underground gas storage with four pressure plugs

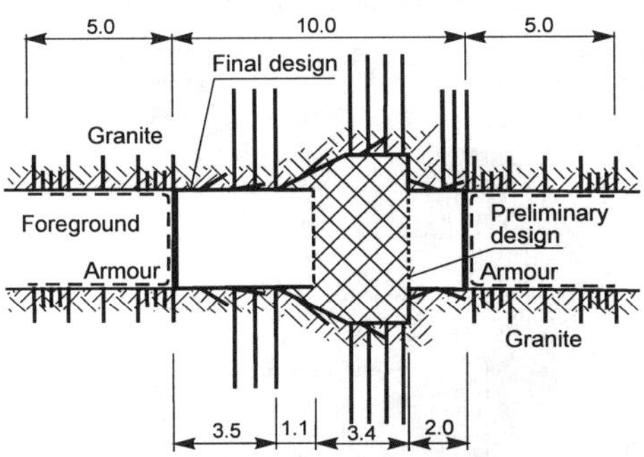

Figure 7. Steel fibre reinforced pressure plug fixed in granodiorite and sealed by grout curtains and contact grouting

work of unlined tunnels with a total length of about 45 km (Fig.5) and a closure system, which is formed by a pressurized water chamber confined by four shotcrete plugs built in tunnels and fixed in to the surrounding rock mass (Fig.6). The steel fibre reinforced shotcrete plugs exposed to one-side gas/water pressure up to 13 MPa are the most vulnerable and therefore the most important elements of the facility.

To produce a stable and gas/waterproof plug/rock mass complex, a number of problems had to be solved. The optimum shape and length of the plugs and the bearing capacity of the plug/rock mass complex had to be determined, the deformation and leakage of the whole complex under working and limit conditions predicted and the feasibility of the solution had to be verified via large-scale in situ tests and experimental operation of the gas storage.

To solve these problems, a step-by-step approach using large-scale in situ tests, monitoring and numerical modeling in an interactive way was shown to be useful (Fig.8). By comparing the calculation and measurement results, the numerical models could be calibrated on one side and the design of the plug, its construction technology and the layout of measurements could be improved on the other (Doležalová & Horáček 2000).

In this procedure two stages can be distinguished: the feasibility analysis and the final design.

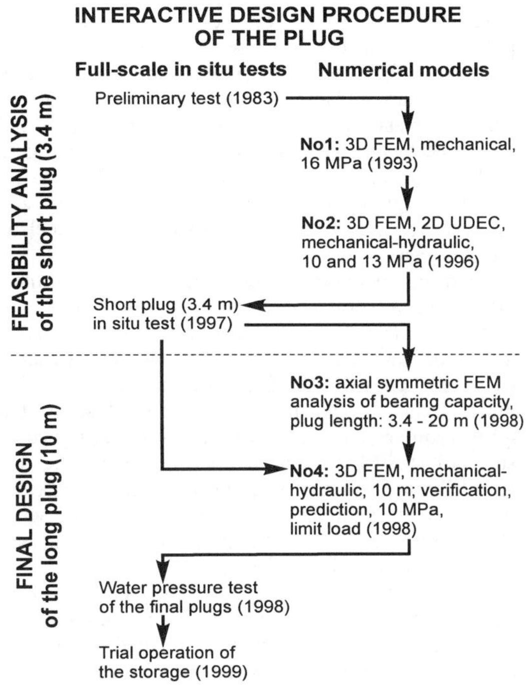

Figure 8. Link of in situ tests and numerical models in the design procedure of the plug

4.1 *Feasibility analysis of the plug*

Aiming at a cost-saving design, the performance of a short plug with a length of about 3.4 m and maximum cross section of 5.6 x 5.4 m was investigated first. The design of this plug, which later formed the central part of the final plug (Fig.7), was based on a preliminary field test performed as early as in 1983. Axial displacement of the plug of about 0.63 mm under a pressure of 6.8 MPa was measured and the mechanical stability of the shotcrete plug/ granodiorite complex demonstrated.

To analyze the performance of this complex under pressure increasing up to 16.5 MPa, a 3D FEM

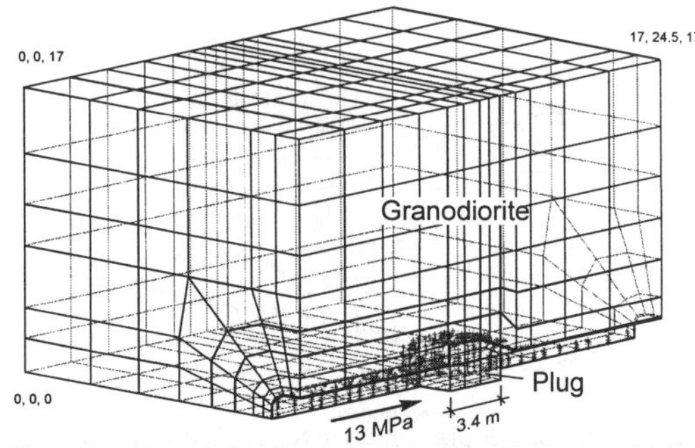

Figure 9. 3D model of the short plug, plastification at 13 MPa

model (DOF = 13 500) representing a quarter of the plug and the surrounding rock mass was elaborated (Figs.6 and 9). The primary stress state was given by the overburden weight and K_0 = 1.4. Elastic-perfectly-plastic constitutive model was applied and the properties of basic materials were as follows: E_r = 25 000 MPa, ν = 0.2, φ = 48°, c = 13 MPa, γ = 27 kN/m^3 for granodiorite and E = 27 000 MPa, ν = 0.16, φ = 55°, c = 2 MPa, γ = 24 kN/m^3 for steel fibre shotcrete.

First, an uncoupled mechanical problem simulating the excavation and construction sequence was solved and the model was calibrated using convergence measurements performed in the access tunnel (measured convergence 2-3 mm, calculated 2.7-3.9 mm). Axial displacement (0.57 mm) close to the measured one (0.63 mm) was calculated at the pressure of 6.8 MPa and axial displacements 1.3 and 1.7 mm for elastic and elasto-plastic plugs were predicted at the maximum pressure of 16.5 MPa. High shear strength mobilization along the plug/rock mass contact was found (Fig. 9).

To estimate the effect of the stress release due to excavation on the permeability of the rock mass, a UDEC parametric study was performed. Assuming different joint patterns with regular spacing, the joint opening and the equivalent coefficients of permeability of porous medium were computed. Increase of k from 10^{-10} m/sec to k = 10^{-6} m/sec was found.

Using the Biot consolidation theory, the above 3D FEM model, new data concerning the material of the plug (steel fibre reinforced shotcrete: γ = 23 kN/m^3, E = 36 000 MPa, ν = 0.16, c = 5.6 MPa, φ = 54°) and the above range of k, deformations (2 mm), stresses, pore pressures and leakage of the plug/granodiorite complex under pressure of 13 MPa were predicted for the field trial of the short plug. This full-scale in situ test was carried out in a drift near the closure system (Fig.5).

Mechanical behavior (axial displacement 1.3-1.8 mm at 13 MPa) was satisfactory, but problems with the watertightness of the whole system were found. Important conclusions concerning the shotcreting and grouting technology were made and a change of the plug length was proposed.

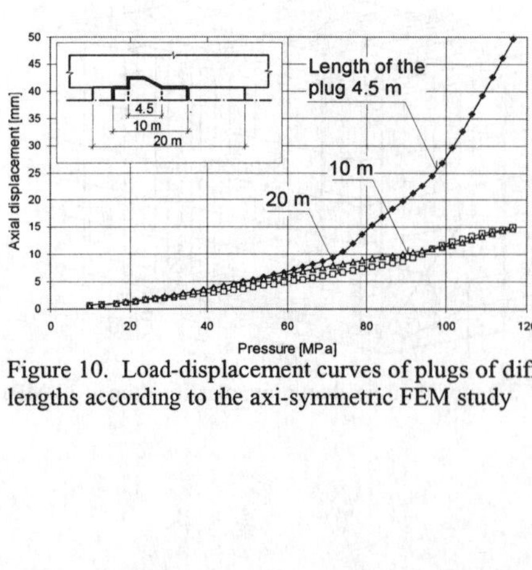

Figure 10. Load-displacement curves of plugs of different lengths according to the axi-symmetric FEM study

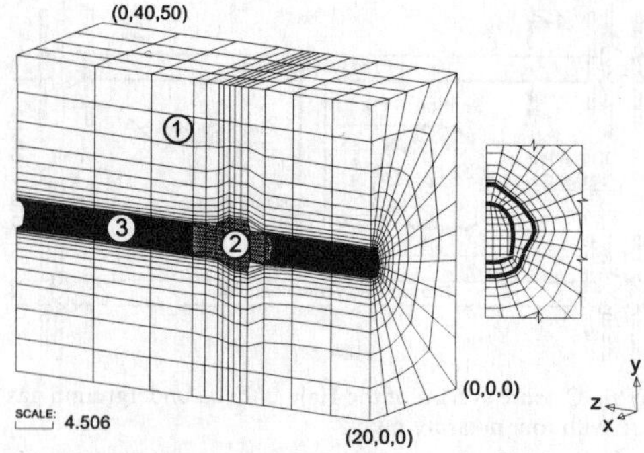

Figure 11. 3D model of the final plug (length 10 m)

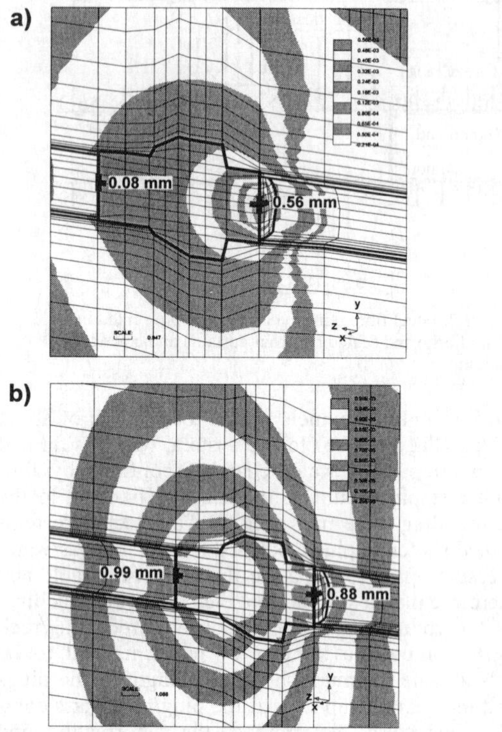

Figure 12. Axial displacement contour lines at the beginning (a) and at the end of consolidation (b) at a pressure of 10 MPa

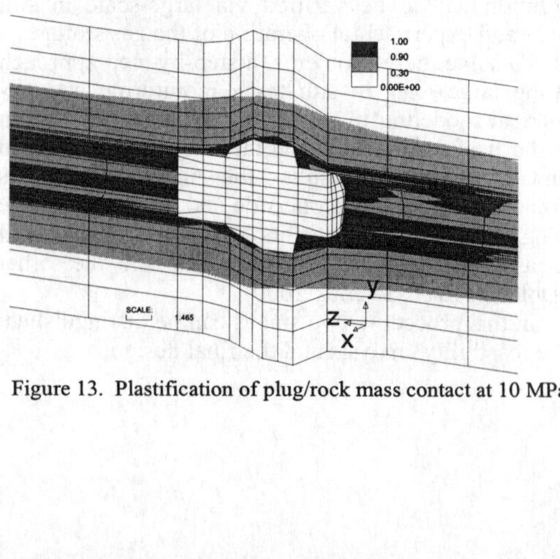

Figure 13. Plastification of plug/rock mass contact at 10 MPa

4.2 *Final design of the plug*

To support the final design, plugs of different lengths (3.5; 4.5; 8; 10; 20 m) have been analyzed by the aid of 3D and axial-symmetric FEM studies. The objective was to determine the bearing capacity of plugs of different lengths and to select the optimum size of the plug.

In Fig.10 the load-displacement curves of plugs of different lengths are compared. When calculating these curves, a proper account for dilatancy, i.e. application of the non-associated flow rule is necessary, since in the opposite case, an unrealistic volume change and bearing capacity of the rock mass can be calculated. Comparing the curves, it can be concluded that the optimum length is 10 m.

To predict the mechanical and hydraulic behavior of the final plug/granodiorite complex, a large coupled 3D FEM model considering also the irregular excavation shape of plugs was created (Fig.11).

This model solved three basic problems. First, the numerical model, i.e. the mechanical and hydraulic properties of the plug/rock mass complex, was verified by simulating the full-scale in situ test of the short plug described above. Second, the deformation and leakage of the 10 m long final plug during the water pressure test (10 MPa) were predicted. Third, the bearing capacity of the plug/rock mass complex was investigated by solving a series of mechanical-hydraulic problems.

Good agreement for axial plug displacement (measured between 1.3 mm and 1.8 mm, calculated at 1.5 mm) was obtained, but a difference for the leakage (measured 6.5 l/min, calculated at 3.1 l/min) was found by the verification procedure.

The predicted performance of the final plug and the surrounding rock mass under the testing pressure of 10 MPa is shown in Fig. 12. Axial displacements on the pressure/non-pressure sides at the beginning (0.55 / 0.08 mm in Fig.12a) and at the end of consolidation (0.88 / 0.99 mm in Fig.12b) were predicted. According to these results, saturation of the plug produces larger deformation on the non-pressure side than on the pressure one. This finding can indicate defect of the plug, therefore it was recommended to measure the axial displacements on both sides of the plugs.

The results of the water pressure test of the final plugs and their comparison with the FEM prediction is shown in Figs.14a,b. The above finding has been approved for the more permeable B2 plug (axial displacements 0.6 / 1.95 mm), while the opposite is true for the less permeable A2 plug (0.45 / 0.15 mm).

A plastification of the plug/rock mass contact (Fig.13), was calculated again, which approved the need of efficient grouting of the whole contact area including the adjacent rock (Fig.7). A staged contact grouting performed after each stage of the pressure increase during the water pressure test showed to be highly effective.

At present, the gas storage is under operation and the monitoring results show fairly positive trends.

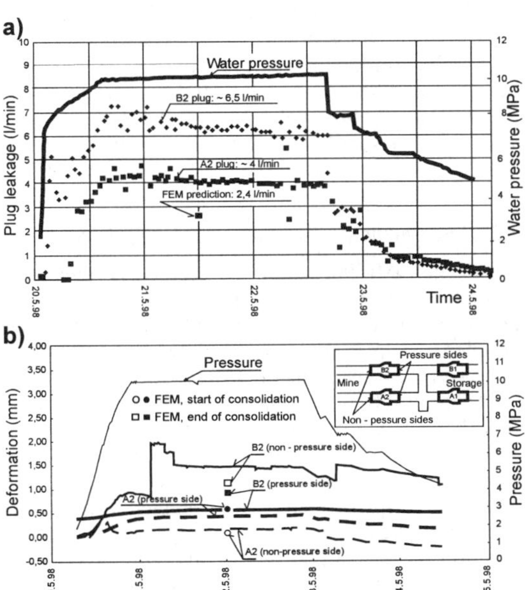

Figure 14. Water pressure test of the final plugs: comparison of the measured and calculated leakage (a) and deformation on the pressure sides and non pressure sides of the A2 and B2 plugs (b)

5 MINING ADVANCE CONTROL VIA NUMERICAL MODELING AND MONITORING

The case history of the Jiri Mine is noted for two features: it is a facility, where numerical modeling and monitoring have been used in an interactive way for more than 20 years and where the decision making of the Authorities has been based on the numerical models.

The Sokolovsko Coalfield is located in a famous spa region with artesian thermal springs. A cross section of Jiri Mine with characteristic geological and hydrogeological conditions of the region is shown in Fig. 15. The artesian water bearing stratum is formed by crystalline rocks and the pressure aquifer is confined by tuffitic claystone forming an impervious rock barrier. This stratum underlies the 40 m thick upper coal seam covered by stiff fissured clays. By removing the overburden (120-150 m), a substantial unloading occurs to the protective rock barrier.

The open-pit mining in these conditions is limited by hydraulic fracturing hazard along the confining boundary, i.e. along the rock barrier and pressure aquifer contact. To prevent this hazard, the pressure head has been reduced by flowing wells since 1989.

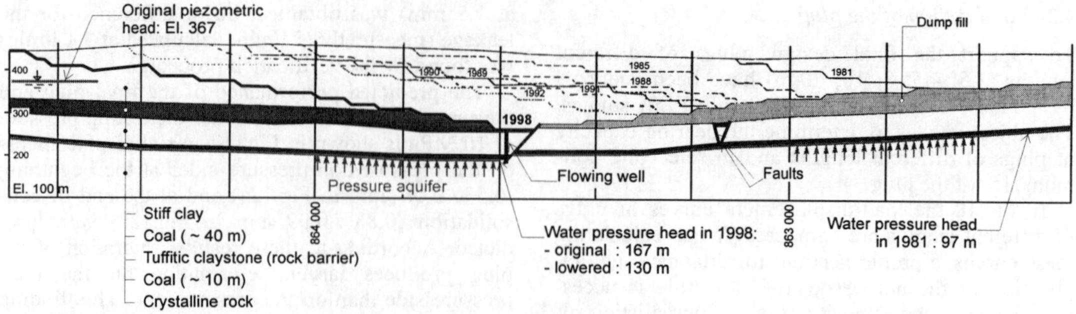

Figure 15. Cross section of Jiri Open-Pit Coal Mine loaded by artesian water pressure

To estimate the feasibility of the mining advance at minimum pressure head reduction, i.e. at minimum environmental impact, numerical models calibrated according to field measurements and monitoring system updated according to numerical solutions have been applied since 1976.

5.1 *Monitoring of the mine and model calibration*

The above interactive process was started by a pure prediction using FEM (Doležalová et al. 1977), which predicted the maximum heave of the unloaded mine bottom of about 0.80 m, instead of 7.0 m produced by early physical models from equivalent materials with incorrect unloading sequence.

To resolve the problem and get information on the feasibility of the mining advance, extensive monitoring of the mine bottom heave over a large area has been suggested and realized.

Bench marks at a distance of about 25-30 m were installed in dewatering and other adits along the coal seam bottom at a length of more than 3 km (Fig. 16). Precise leveling of the bench marks was provided four times a year and the results approved the above FEM prediction, i.e. that excavation of a 10 m thick layer induces heave no more than 0.03 - 0.05 m. Precise leveling carried out up to now, together with

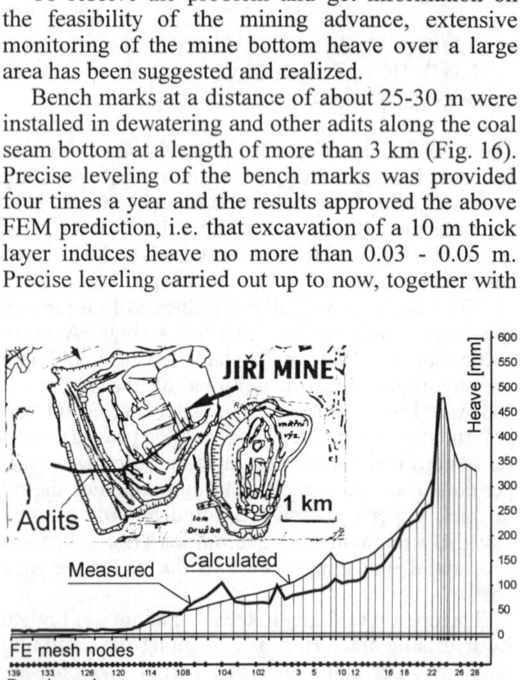

Figure 16. Mine bottom heave measured in adits and back analyzed by FEM

the piezometric head records registered by observational wells, form a valuable data basis for calibration of numerical models. In this way, structural changes of the rock barrier caused by unloading could be introduced in numerical models.

Non-linear constitutive model well describing the path dependence of deformational parameters below failure surface (Doležalová 1991) and its elasto-plastic version were used. The basic parameters like E_{in}, E_{unl}, c, φ were determined according to the field tests (Hudek et al. 1986), and only the exponents describing their change with the normal and shear stress level were calibrated according to the measured heave. The results of such a trial-and-error fitting procedure are shown in Fig. 16. The obtained set of data has been used without any change since 1991, and for tuffitic claystone are as follows: $\gamma = 22.7\ kN/m^3$, E_{in}=200 MPa, E_{unl}=1400 Mpa, $\nu_{in} = 0.25$, c = 0.27 MPa, φ = 28.3°, σ_t = 0.2 MPa.

Another important value calibrated according to the measured heave was the lateral coefficient at rest K_0. A relatively high value K_0=0.6-0.8 was obtained, which was a significant finding concerning the hydraulic fracturing hazard (Doležalová et al. 1986).

5.2 *Assessment of hydraulic fracturing hazard*

Up to the early 90s, the mine crossed an area with tectonic faults mostly perpendicular to the mining advance. These conditions could be analyzed by uncoupled and later coupled 2D FEM models simulating cross sections of the mine (Fig. 15).

According to the site investigations, impermeable rock barrier was assumed, the mining advance and the water pressure head reduction were modeled and, as a rule, a good match of the calculated and measured displacements was obtained. The corresponding stress state was evaluated via criteria quantifying the safety against hydraulic fracturing. Different components of the total stress tensor computed for the rock barrier along the confining boundary were related to the aquifer pressure, and the tensile strength of rock was alternatively included or excluded.

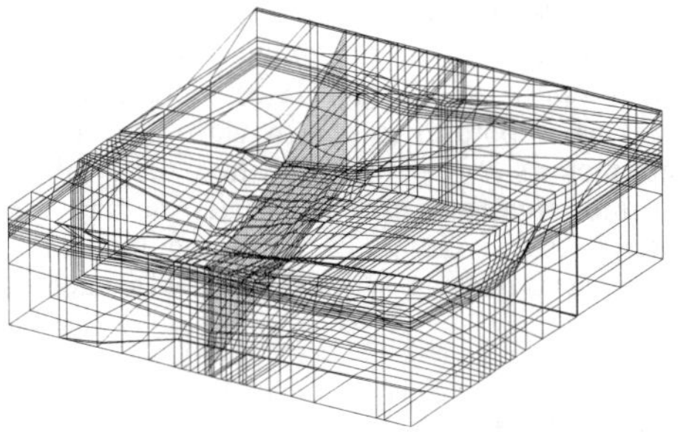

Figure 17. Longitudinal section modeled using UDEC

This analysis discovered the unfavorable influence of tectonic faults, the possible unbalance of the rock pressure and fluid pressure near them and that the hydraulic fracturing hazard is considerably larger in the horizontal direction than in the vertical one. Accordingly, the possible failure mode of the rock barrier is not cracking by bending, as predicted by early physical models. Instead, progressive hydraulic fracturing along the faults and seams and opening of flow paths resulting in reduction of impermeability of the rock barrier occur.

Such critical situations have already happened, because the water pressure reduction by flowing wells is limited regarding the limits of the withdrawal from the aquifer. The actual failure mode of the rock barrier was found to be, however, different from the predicted ones. Occasional rupture of casings of wells occurred at a depth of 20 - 30 m (six times so far) and only small springs and small heave of the mine bottom of about 0.035 m were observed. Except the first case when the damage of the well was detected later, quick packering and grouting of the ruptured boreholes prevented the pressure drop of the aquifer in all cases.

An attempt was made to model the first case (Doležalová & Zemanová 1997) making use of capabilities of UDEC to account for the structure of jointed rock, fracturing of joints, fluid flow through fractures, grouted bolts for modeling the grouted double casing of wells, etc. In order to cross the frequent parallel faults, a 2D model in longitudinal direction was formulated (Fig. 17, 18a). The effect of the slopes was compensated by strengthening the confining boundary and steady flow was assumed.

A number of computed quantities matched the observation (12 m pressure drop, well casing rupture at a depth of 20-25 m, gap of about 0.07-0.3 m) and the results shown in Figs. 18b,c,d helped to clarify a possible process of the borehole rupture. Due to excavation between 1993 and 1995, fractures along the planes of weakness occur and the rock and fluid pressure unbalance initiates fluid flow from the aquifer. If the fluid pressure in the horizontal fracture is larger than the overburden weight, rupture of the well (stiff element in the jointed rock) occurs.

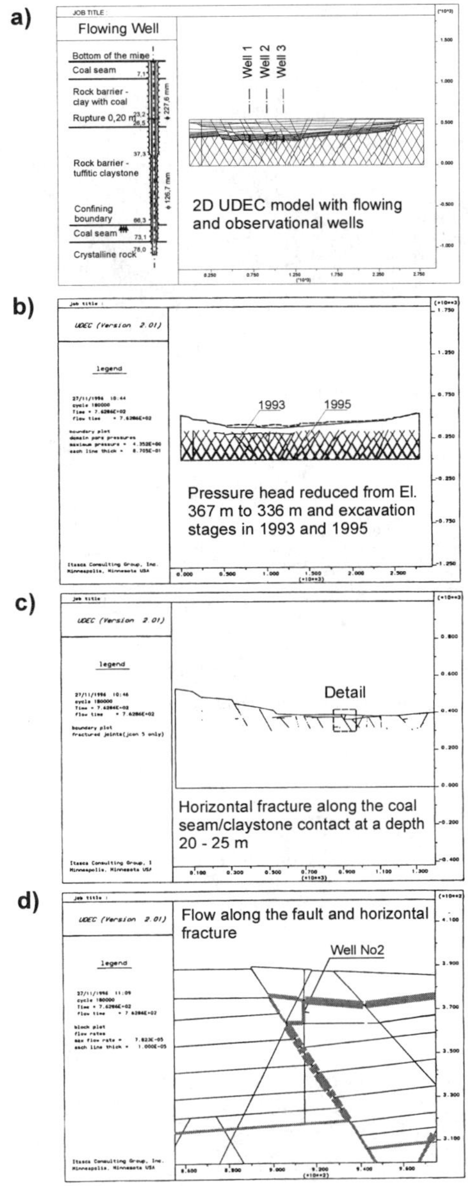

Figure 18. Results of the UDEC analysis of the borehole casing rupture

In spite of these findings, due to the simplifications, the modeled scenario differs from the reality as to the magnitude of the events and their course in time. Fortunately, the reality is much more favorable.

At present, only large, coupled 3D models well simulating the 3D geometry of the mine, mining advance and faults are applied. Permeability depending on shear and tensile failure was implemented in the code, and alternative models with impermeable and conditionally permeable rock barrier elaborated.

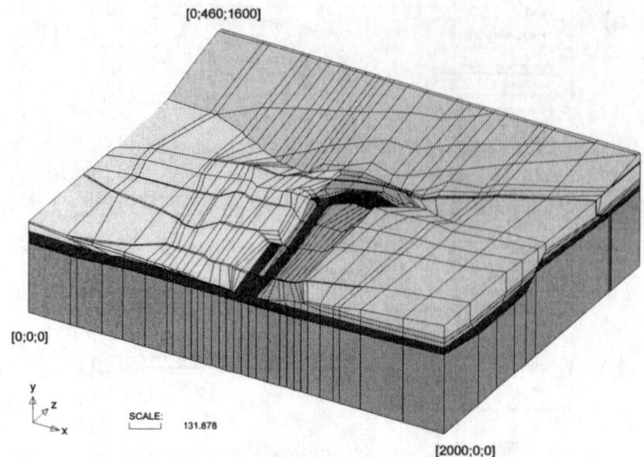

Figure 19. 3D FEM model of Jiri Mine

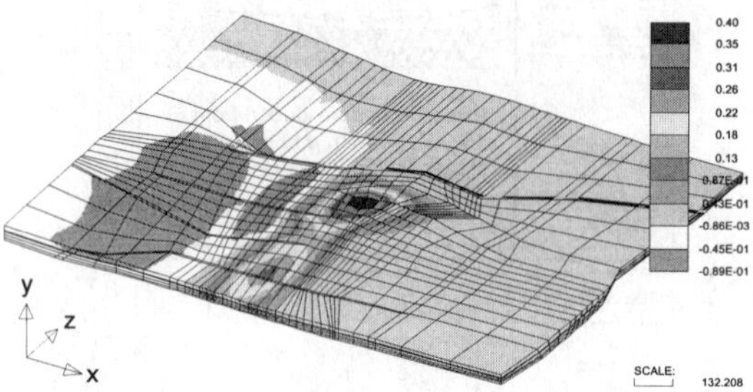

Figure 20. Heave due to mining advance in 2000

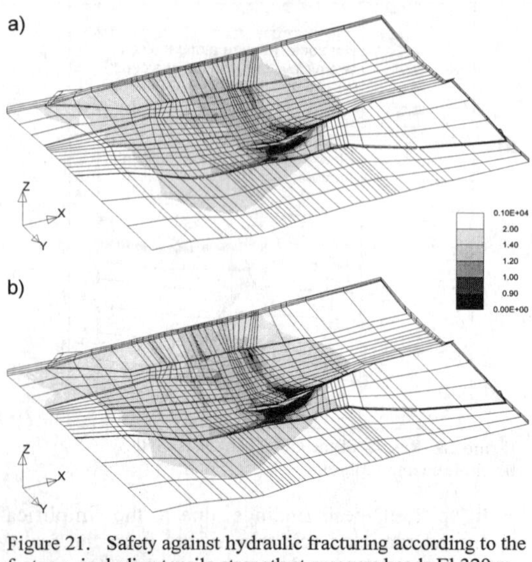

Figure 21. Safety against hydraulic fracturing according to the factor s_{4t} including tensile strength at pressure heads El.329 m (a) and El.332 m (b)

The last model (DOF = 150 027) elaborated for decision making of the Health Department and the Mining Authorities concerning the mining advance in 2000 is shown in Fig. 19. The model was validated by measurement results corresponding to the mining advance in 1996-1999 without any calibration. The model predicts relatively large heave in 2000 (Fig. 20) and possible opening of flow paths in the area of a new fault (Figs. 21a,b). Here a view of the rock barrier bottom is shown with contours of the safety factor $s_{4t} = (\sigma_3 + \sigma_t)/p_w$, where the symbols denote minor total principal stress, tensile strength and water pressure along the confining boundary. Water pressure head at El. 329 m was recommended.

6 CONCLUSION

The efficiency of interactive use of numerical models and field measurements at investigation, design, construction and operation of civil engineering structures is demonstrated in the paper. Discontinuum and continuum constitutive models and parameters of a host rock for HLW disposal were derived, design and construction of a unique closure system managed and long-term operation control of an open-pit mine in extremely difficult conditions carried out.

REFERENCES

Britto, A. M. & Gunn, M. J. 1987. *Critical state soil mechanics via finite elements.* Chichester: Ellis Horwood Limited.

Doležalová, M. 1986. Deep excavation loaded by artesian water pressure. *8. Danubian European Conference on Soil Mech. and Found. Eng., Vol.1.* Nürnberg: 247-256.

Doležalová, M. 1990. On interactive modeling and its use for open-pit mine stability analyses. *2nd European Specialty Conference on Numerical Methods in Geotechnical Engineering.* Santander: 679-692.

Doležalová, M. 1991. Numerical solution of rheological problems. Chapter 12 in Jaroslav Feda: *Creep of Soils and Related Phenomena.* Academia-Elsevier: 348-385.

Doležalová, M., Mikulášková, V., Škoch, V. 1977. Stability of a deep excavation bottom. *Proc. of 9th Int. Conf. ISSMFE.* Tokyo: 47-50.

Doležalová, M., Zemanová, V. 1997. FEM and DEM analysis of hydraulic fracturing hazard of an open-pit coal mine. *Workshop: Tests and Applications of Computing Codes on Geoenvironmental and Civil Engineering Problems.* Budapest: 115-159.

Doležalová, M., Zemanová, V., Kovács, L. 1998. Rock mass characterization via field measurements and numerical methods. *NUMGE98.* Udine (separate print).

Doležalová, M., Zemanová, V., Danko, J., Kovács, L. 1999. Geotechnical investigation and modelling - rock mass characterization via convergence measurements and numerical models. *Scientific Symposium on Rock Mechanics and Tunneling, Vol.2.* Zagreb (in print).

Doležalová, M., Horáček, M. 2000. Numerical modeling, testing and monitoring of a closure system under high pressure. *Ed.: T.R. Stacey, AITES-ITA 2000, World Tunnel Congress.* Durban: 123-131.

Hladík, I. 1998. *Efficient solution of linear and nonlinear systems of equations for elasticity and elasto-plasticity FEM analyses with unstructured irregular grids.* PhD Thesis, University of Innsbruck, Austria.

Hladík, I., Reed, M.B. & Swoboda, G. 1997. Robust preconditioners for linear elasticity FEM analyses. *Int. J. Num. Meth. Engrg.* 40: 2109-2127.

Hudek et al. 1986. Final report on geotechnical properties of coal seam Josef and volcanic claystone stratum according to in situ measurement, Technical Report, PUDIS, Prague (in Czech).

Somosvári, Zs. 1997. Laboratory and fields tests of BCF in Alfa-1 Tunnel, GEOCONSULT, Miskolc (in Hungarian).

UDEC (Universal Distinct Element Code), Version 2.0, ITASCA Consulting Group, Inc. Minneapolis, Minnesota, USA, 1993.

Wittke, W. 1990. *Rock Mechanics.* Springler-Verlag Berlin, Heidelberg.

Computer Methods and Advances in Geomechanics, Desai et al. (eds) © 2001 Balkema, Rotterdam, ISBN 90 5809 183 X

Gravity flow sewers: Modeling to capture limit states for buried thermoplastic pipes

I.D. Moore
Department of Civil and Environmental Engineering, University of Western Ontario, Canada

ABSTRACT: Buried thermoplastic pipes have stability and serviceability limit states that are influenced by a number of factors not captured by conventional linear or nonlinear two-dimensional buried pipe analyses. Theoretical advances have been required to support the development of rational design methods. Advances in material modeling have involved the development of both nonlinear viscoelastic and viscoplastic models. Explicit three-dimensional modeling of the pipe wall is required to provide estimates of the peak tensile stresses induced as different elements of the pipe wall interact. Three-dimensional modeling is also needed to obtain useful estimates of the compressions that cause local buckling. A number of specific calculation techniques are described, including those used to interpret pipe response in four different laboratory tests on pipe.

1 INTRODUCTION

Thermoplastic pipes are now used internationally to provide low-cost solutions for a variety of municipal engineering applications. While design methods for metal culvert and sewer products have generally formed the basis of design for thermoplastic pipes, there is a growing awareness that thermoplastic pipes have a number of unique characteristics. The performance limits for these structures need to be clearly understood to ensure optimal economy and reliability. This paper describes a number of numerical procedures that have been developed and used to investigate thermoplastic sewer pipes so that rational design methods can be developed for North American and other jurisdictions.

The new finite element techniques are described and illustrated with examples of their use to predict and interpret a number of different laboratory tests. These models are being used by the author and others to investigate the behavior of a variety of buried pipes and geosynthetics.

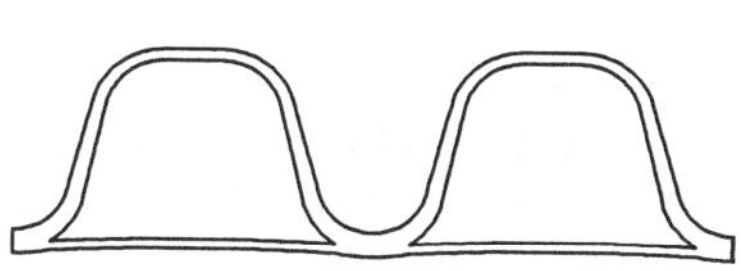

Figure 1. Wall geometry for typical buried thermoplastic pipe.

2 STATEMENT OF THE PROBLEM

2.1 *Profile geometry*

Figure 1 shows the profile of a typical thermoplastic sewer pipe, manufactured from High Density Polyethylene (HDPE), Polyvinylchloride (PVC) or Polypropylene (PP). This profile consists of a corrugated element to provide bending stiffness and an inner liner to improve hydraulic performance. Various other wall geometries are also in use, including ribbed and cellular profiles.

Analysis of these pipe structures has traditionally assumed two-dimensional plane strain response, where wall area per unit length A and second moment of area per unit length I are used to characterize the geometry of the pipe wall. Profile design has largely focussed on maximizing the second moment of area of the wall relative to the amount of polymer material, (I/A), so as to limit pipe deflections under earth loads. Research is underway to explore the implications of the wall geometry, e.g. Moore and Hu (1995), Moore (1995) and Moore and Laidlaw (1997), work that has been given further impetus by recent observations of both cracking and local buckling in a small number of installed structures.

2.2 *Time-dependent behavior*

Thermoplastics like HDPE and PP exhibit time and strain-rate dependent constitutive behavior. Conventional pipe analysis features the selection of an equivalent secant or tangent modulus and modeling of the thermoplastic as a linear elastic material.

The need for more detailed evaluations of behavior in the laboratory and the field, however, has prompted work to develop more sophisticated procedures, the objective being the reliable prediction of time-dependent load-rate dependent response.

2.3 *Pipe testing*

A number of test methods are being used to investigate and characterize the behavior of thermoplastic pipes. Firstly, empirical measurements of pipe resistance to two-point loading are used to quantify the stiffness and creep resistance of these structures. The pipe is compressed between two parallel plates, Figure 2, with loading under constant rate of deformation of 12mm/min used to define pipe stiffness in ASTM D2412, and time dependent deformations used to measure pipe creep in the British Standard Creep test (BS 4962).

Selig and Hashash (1989) clearly established that profiled thermoplastic pipes can experience significant decreases in circumference under the influence of overlying fill. A simple hoop compression test has therefore been developed to investigate the limit states in profiled thermoplastic pipe associated with this hoop compression, Di Francesco (1993), Laidlaw (1999). A typical test arrangement is shown in Figure 3, where the pipe is placed vertically within a steel pressure vessel and surrounded by an annulus of backfill soil. This is in turn contained within an air bladder placed at the inner surface of the steel vessel; inflation of the air bladder places the soil and the pipe within under radial compression. Measurement of decreases in the pipe diameter ΔD can be used to quantify the circumferential strains:

$$\varepsilon_\theta = \Delta D/D \qquad (1)$$

Pipes so tested may exhibit limit states like local buckling or tearing in the inner liner, Figure 4, and instrumentation such as strain gauges may be used to assess local bending within the pipe wall as a result of profile geometry or non-uniform soil bedding.

The hoop compression cell provides a simple axisymmetric test that is relatively easy to perform and evaluate. However, pipes placed horizontally in the field experience biaxial earth pressures that lead to ovaling deformations as well as hoop compression. Another pipe test system has therefore been developed to better approximate field conditions. The biaxial test cell, Figure 5, is essentially a stiff steel box developed to contain the pipe and a block of soil surrounding it. An air bladder acting on the soil surface simulates the overburden pressure associated with a deep embankment or landfill. The walls are lubricated to ensure the earth pressures reach the pipe and are not transferred to the walls in friction. Stiff walls lead to lateral earth pressures close to those associated with zero lateral strain (so called K_0 conditions).

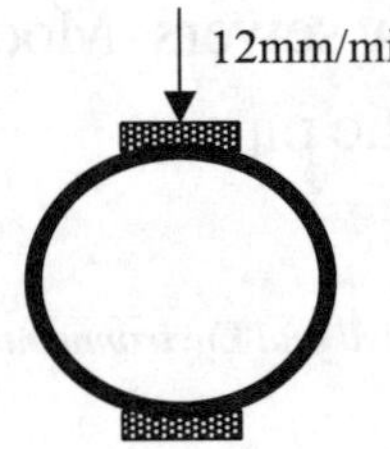

Figure 2. Pipe segment under parallel plate loading.

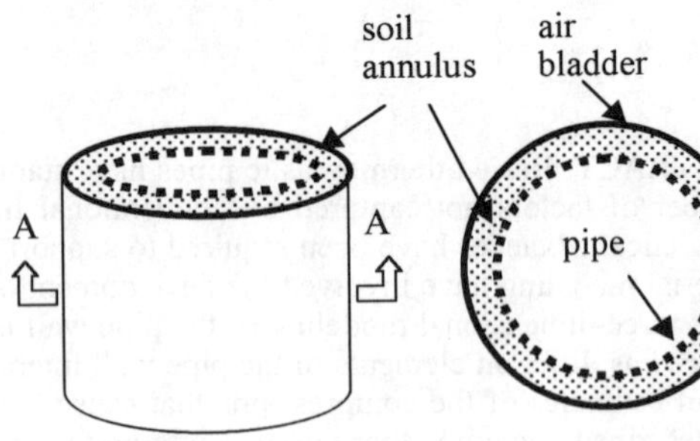

a. Perspective view b. Section A-A

Figure 3. Hoop compression cell.

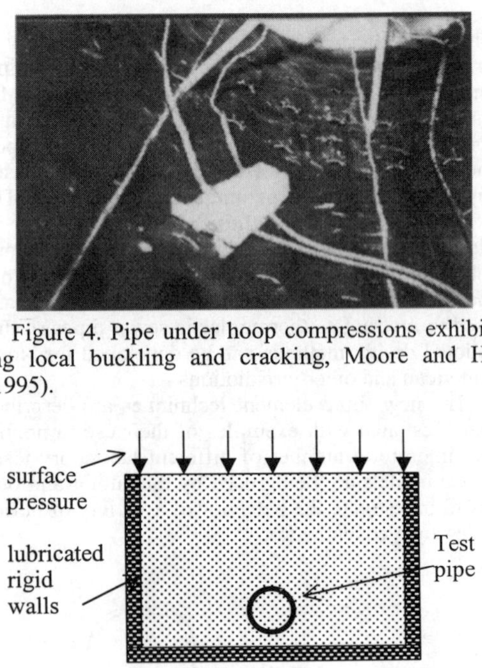

Figure 4. Pipe under hoop compressions exhibiting local buckling and cracking, Moore and Hu (1995).

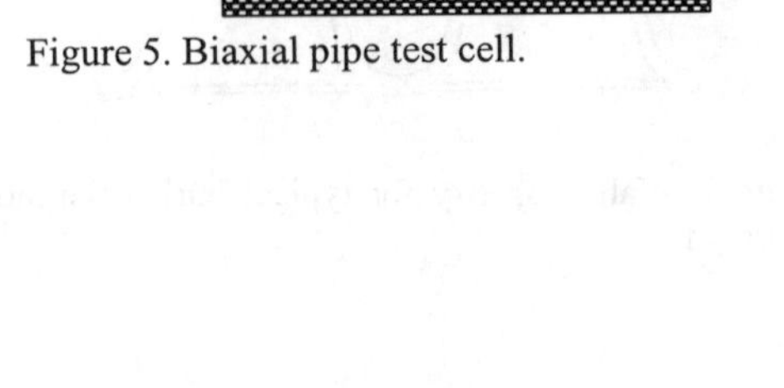

Figure 5. Biaxial pipe test cell.

3 TIME DEPENDENT MATERIAL MODELS

3.1 *Introduction*

Three different classes of constitutive model are being developed and used to characterize the time dependent material response of HDPE, PP and PVC:

- Linear viscoelastic (LVE);
- Nonlinear viscoelastic (NVE); and
- Viscoplastic (VP).

Each is described here, with details about the test data used to obtain material parameters and specific parameter values used in the example calculations described in subsequent sections.

3.2 *Linear viscoelastic modeling (HDPE, PP)*

Conventional linear rheology has been found very effective at small levels of material strain. A multi-Kelvin formulation is employed, featuring an independent elastic spring of modulus E_0 placed in series with a set of N Kelvin elements (elastic spring modulus E_l, dashpot viscosity η_l), Figure 6. Each Kelvin element captures response over one cycle of log time, so behavior can be modeled from seconds to years. Parameters are evaluated using long term creep or relaxation data.

For example, the PP material used in the lined corrugated pipe examined later in this paper was modeled with N=9 Kelvin elements, where E_0=1245MPa, E_1=6880MPa, η_1=1.45 MPa.days and ν=0.46. The number of independent parameters is reduced by making successive Kelvin elements have spring modulus 0.61 times that of the previous, and dashpot viscosity 12.46 times greater.

A number of HDPE materials have been examined by Moore and Hu (1996), all featuring similar behavior in the linear range: N=9, E_0=1120MPa, E_1=3615MPa, η_1=0.503 MPa.days and ν=0.46. Each successive Kelvin element has spring modulus 0.845 times that of the previous, and dashpot viscosity 10.0 times greater.

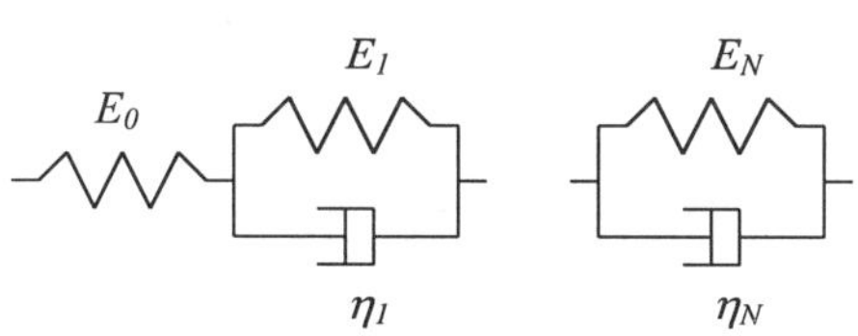

Figure 6. Multi-Kelvin rheological model.

3.3 *Nonlinear viscoelastic model (HDPE)*

The constitutive equations for the NVE model are based on a division of strain into elastic and viscoelastic parts:

$$\dot{\varepsilon} = \dot{\varepsilon}^e + \dot{\varepsilon}^v \tag{2}$$

where the elastic shear modulus G is used to define the elastic strains

$$\dot{\varepsilon}^e_{ij} = \dot{S}_{ij} / G \tag{3}$$

and the viscoelastic strains are defined using a series of six nonlinear Kelvin elements featuring modulus E_l, relaxation parameter τ_l and accumulated element strain ε^v_l

$$\dot{\varepsilon}^v_{ij} = \sum_{l=1}^{6} \left\{ \frac{3}{2} \frac{S_{ij}}{E_l \tau_l} - \frac{(\varepsilon^v_l)_{ij}}{\tau_l} \right\} \tag{4}$$

Assuming the material is incompressible (reasonable for HDPE with its high value of Poisson's ratio), shear modulus is given by $3G = E$. The deviatoric stress tensor S_{ij} is defined by

$$S_{ij} = \sigma_{ij} - \frac{1}{3}\sigma_{kk}\delta_{ij} \tag{5}$$

The material functions are expressed in terms of parameter ψ

$$E_l = (m)^{l-1} E_1 \tag{6}$$

$$\tau_l = (\mu)^{l-1} \tau_1 \tag{7}$$

$$E_0 = a_0 \exp(-a_1 \psi^3) \tag{8}$$

$$E_1 = b_0 + b_1 \exp(-\psi / b_2) \tag{9}$$

$$\tau_1 = c_0 \exp(-c_1 + c_2\psi + c_3\psi^2 - c_4\psi^3) \tag{10}$$

where

$$\psi = \sigma_e \exp(\varepsilon_e) \tag{11}$$

for equivalent or effective stress and strain quantities given by

$$\sigma_e = \sqrt{\frac{3S_{ij}S_{ij}}{2}} \tag{12}$$

$$\varepsilon_e = \sqrt{\frac{2\varepsilon_{ij}\varepsilon_{ij}}{3}} \tag{13}$$

Long term creep data is used to obtain the material parameters. Tests conducted at stress levels from 5.2 MPa to 25.2 MPa on samples cut from a plain HDPE pipe gave m = 0.635, $\mu = 10.0$, a_0 = 1400 MPa, a_1 = 4.65×10^{-5} MPa^{-3}, b_0 = 260 MPa, b_1 = 19545 MPa, b_2 = 8.3 MPa, c_0 = 8.64×10^2 s, c_1 = 10.5, c_2 = 0.218 MPa^{-1}, c_3 = 1.37×10^{-3} MPa^{-2}, c_4 = 1.66×10^{-4} MPa^{-3}. The model provides reasonable predictions up to strains of 15% (well beyond yield).

3.4 *Viscoplastic model (HDPE)*

Similarly, the multi-axial viscoplastic constitutive equations are formulated as follows:

$$\dot{\varepsilon}_{ij} = \dot{\varepsilon}_{ij}^{e} + \dot{\varepsilon}_{ij}^{I} \tag{14}$$

$$\dot{\varepsilon}_{ij}^{I} = \frac{3}{2} C \left(\frac{\sigma_e}{X} \right)^n \frac{S_{ij}}{\sigma_e} \tag{15}$$

$$X_0 / X = \alpha + \left\{ \beta / (\gamma + W^I) \right\}^{1/2} \tag{16}$$

$$\alpha = d_1 \exp(d_2 \dot{\varepsilon}_e^{I\,d_3}) \tag{17}$$

$$W^I = \int \sigma_{ij} d\varepsilon_{ij}{}^{I} \tag{18}$$

where X is a scalar state variable, W^I the inelastic work and $\dot{\varepsilon}_e^I$ the effective inelastic strain rate defined by

$$\dot{\varepsilon}_e^I = \sqrt{\frac{2 \dot{\varepsilon}_{ij}^I \dot{\varepsilon}_{ij}^I}{3}} \tag{19}$$

and X_0=1 MPa. Model parameters $C = 0.01 s^{-1}$, $n = 8.0$, $\gamma = 10^{-4}$ MPa.

Data has been obtained for samples cut from two pipe specimens (one plain and one corrugated) tested at six different constant engineering strain rates: 10^{-1} per second to 10^{-6} per second. The material constants are E = 1450 MPa, $\beta = 7.056 \times 10^{-5}$ MPa, $d_1 = 1.042 \times 10^{-3}$, $d_2 = 3.829$, $d_3 = 2.547 \times 10^{-2}$ for the plain HDPE pipe material (Zhang and Moore, 1997). For the corrugated HDPE pipe material classified as PE345420C according to ASTM, the material constants are determined using tensile constant strain rate data as E = 1350 MPa, $\beta = 7.744 \times 10^{-5}$ MPa, $d_1 = 1.055 \times 10^{-3}$, $d_2 = 3.829$, $d_3 = 2.25 \times 10^{-2}$ (Zhang and Moore, 1998).

3.5 *Incremental finite element solution*

For problems in which material response is nonlinear and load path dependent, an incremental approach is needed. Since the constitutive equations of both the NVE and VP models are formulated in the rate form, they are readily adapted to incremental finite element analysis. Zhang and Moore (1997b) described the finite element implementation, incorporating the two nonlinear models together with procedures based on the work of Carter et al (1978) to include the effects of geometrical nonlinearity.

4 AXISYMMETRIC AND HARMONIC SOLUTIONS

4.1 *Axisymmetric geometry and loading*

Analysis of pipe response in the hoop compression cell is undertaken using linear and nonlinear soil and thermoplastic models implemented within an axi-symmetric finite element formulation. The discretization focuses on the r-z plane allowing explicit modeling of the pipe wall as well as the soil annulus (section 5.3 features an example analysis).

4.2 *Biaxial earth pressures*

Harmonic three-dimensional analysis of a buried profile walled pipe can be undertaken using a two dimensional finite element mesh and Fourier series to characterize variations around the pipe circumference. Figures 7 and 8 show how the finite element mesh is used to represent one segment of a typical lined corrugated HDPE, PP or PVC pipe. Harmonic terms n = 0 and n = 2 are used to represent the normal and shear stress components of the geostatic earth pressures (vertical σ_v, horizontal σ_h):

$$\sigma = \sigma_0 + \sigma_2 \cos 2\theta \tag{20}$$

$$\tau = \tau_2 \sin 2\theta \tag{21}$$

where

$$\sigma_0 = (\sigma_v + \sigma_h) / 2 \tag{22}$$

$$\sigma_2 = -\tau_2 = (\sigma_v - \sigma_h) / 2 \tag{23}$$

Further explanation of this analysis is provided by Moore (1995), together with example calculations.

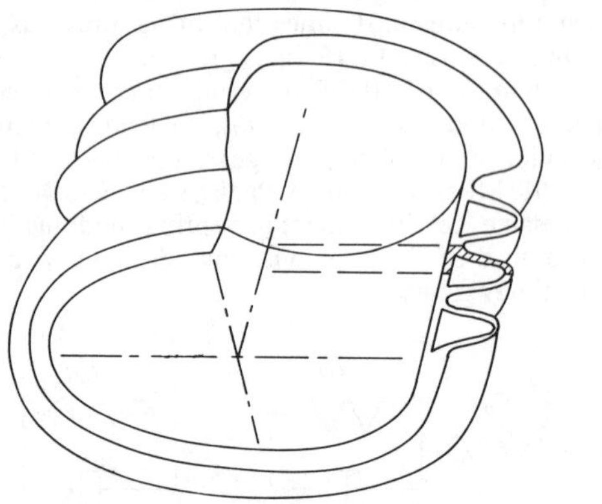

Figure 7. Discretized segment of lined corrugated pipe.

4.3 *Pipes under parallel plate loading*

Fourier series can also be used to represent the boundary conditions associated with pipes under parallel plate loading. The vertical force V is assumed to act over axial length L and circumferential distance δ, giving Fourier coefficients of radial pressure for harmonic n of:

$$F_n = 4 \, (V/\delta L) \sin n\delta / (n \pi) \tag{24}$$

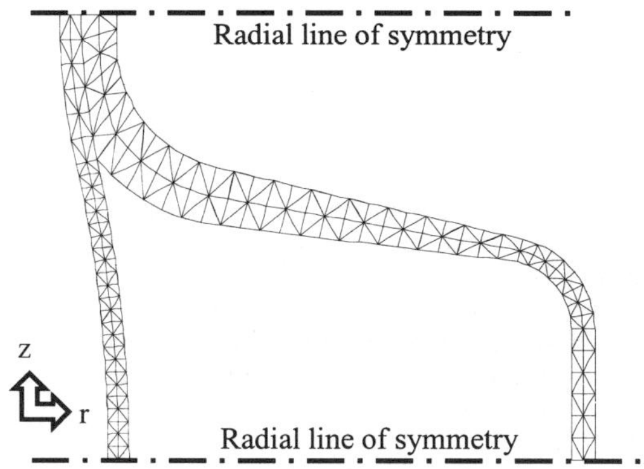

Figure 8. Mesh for segment of profiled polymer pipe.

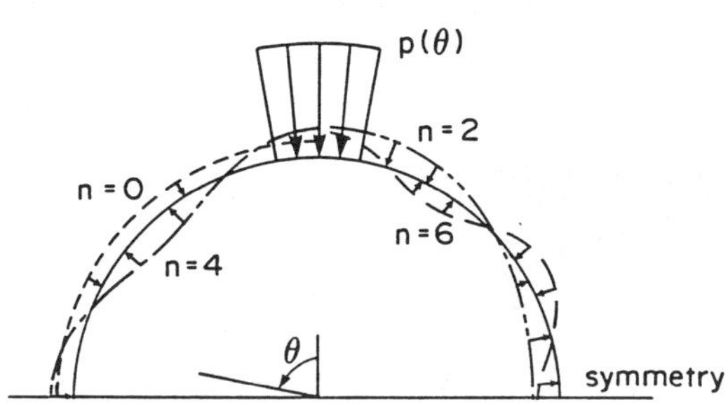

Figure 9. Harmonic loading for pipe under parallel plate load.

5 EXAMPLES

5.1 *Introduction*

Three sets of example problems are now examined featuring profiled HDPE and PP pipes.

5.2 *Polypropylene and polyethylene pipes under parallel plate loading*

A number of HDPE and PP pipes have been evaluated to predict pipe response under parallel plate loading. The objective of this work is to develop loads required to deform the pipe a certain percentage of pipe diameter (i.e. the pipe stiffness) or the extent to which pipe under fixed vertical load deforms over time. This ensures that the pipe is satisfactory prior to making the huge capital investments needed to fabricate the first new pipe.

5.2.1 *Polypropylene pipe, LVE analysis, BS 4962*

Figure 10 shows the mesh used to represent one half of a 300mm long segment of 450mm diameter PP pipe examined using the British Standard Creep Test. Both measured and predicted values of the vertical pipe diameter change with time are shown in Figure 11. Two predictions were made (prior to testing), one set ignoring and the other including the effects of geometrical nonlinearity (the change in stiffness as the pipe adopts an oval shape). Two sets of test measurements were then obtained.

The comparison of predictions with measurements were generally very successful, though measured deflections after 10 days exceeded those obtained with the analysis. The finite element analysis was also used to estimate the strains in the pipe. Strains were found sufficient to generate nonlinear response at certain locations in the pipe (the extreme fibers at the crown, the invert and the springlines).

Figure 10. Mesh for the 450mm diameter PP pipe.

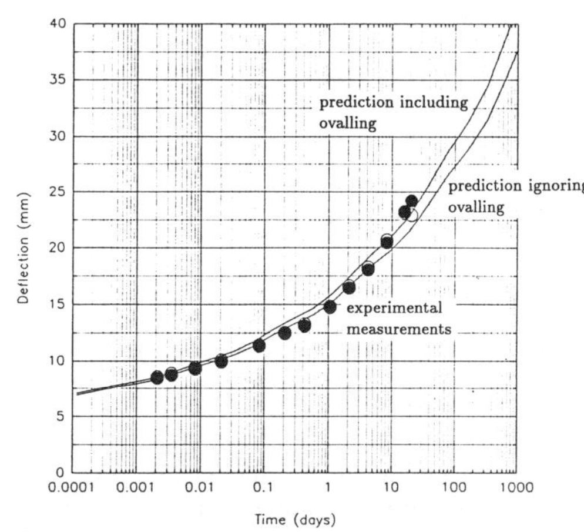

Figure 11. Deflection with time of PP pipe under parallel plate loading (BS creep test).

5.2.2 *Plane HDPE pipe, ASTM D2412.*

A short segment of 322mm diameter plain HDPE pipe with wall thickness of 31mm was tested under parallel plate loading to 20% vertical diameter decrease. Vertical diameter was decreased at a rate of 12mm per minute, the ASTM D2412 standard. Figures 12 and 13 show the deformed mesh at 20% vertical deflection and the measured and calculated response (load versus deflection) respectively. Each of the three material models was used to estimate the pipe response, incorporating considerations of geometrical nonlinearity (solutions denoted LVE+GN,

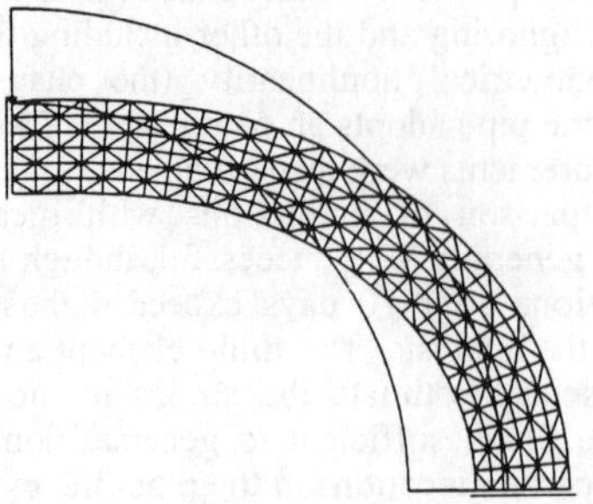

Figure 12. Deformed mesh for thick HDPE pipe loaded to 20% vertical diameter decrease.

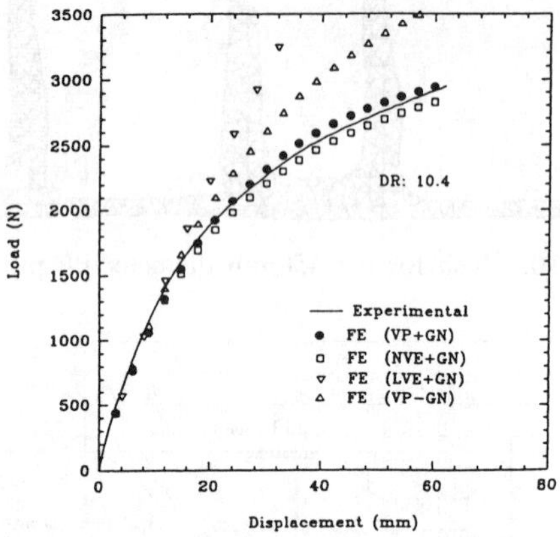

Figure 13. Response of the thick HDPE pipe.

NVE+GN and VP+GN). The viscoplastic model was also used to calculate the behavior without consideration of the effects of geometrical nonlinearity (VP-GN) so as to gauge the influence of changing shape.

Figure 13 clearly indicates that the two materially and geometrically nonlinear solutions have successfully captured the key features of this pipe loading test. The linear viscoelastic model appears to work successfully up to above 5% vertical diameter decrease. After that, the nonlinear material response geometrical nonlinearity also becomes significant after 5% deflection.

5.3 *Local stresses in buried corrugated pipe*

Laidlaw (1999) tested a 525mm diameter corrugated HDPE pipe in hoop compression to investigate local bending within the corrugated profile using measurements of surface strain at various locations on the corrugated profile. Finite element analysis was performed using the VP model and a conventional elastoplastic model for the soil (parameters based on data reported by Selig, 1990; modulus 30MPa, Poisson's ratio 0.2, angle of internal friction 38°).

The mesh shown in Figure 14 was used, composed of 6 noded triangular elements for both the pipe and the surrounding soil. Figure 15 shows the measured and calculated values of axial strain and Figure 16 shows the zones of shear failure predicted in the soil surrounding the pipe. Those areas where calculations indicate that the shear strength of the soil is fully mobilized result from the relatively low radial (hoop) stiffness of the corrugated HDPE pipe. This causes a lack of radial confinement, and principal stress ratio sufficient to induce shear failure. Success here in calculating strains like those observed in the test is indicative of the effectiveness of both the geometrical and material modeling. Calculated values of tensile stress are used to assess the pipe resistance to brittle fracture; compressive stresses are used to evaluate the susceptibility of the profile wall to local buckling.

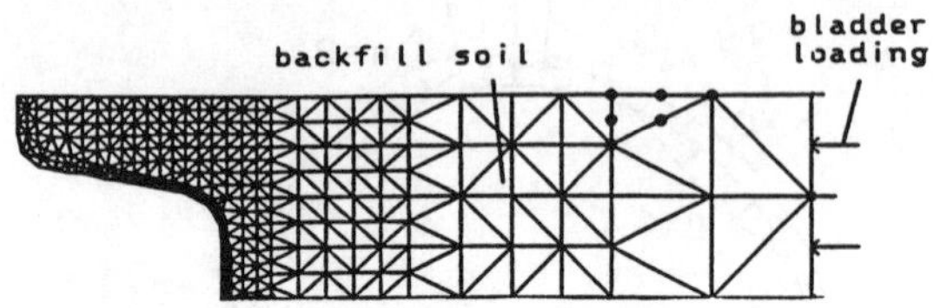

Figure 14. Finite element mesh for analysis of the hoop compression test.

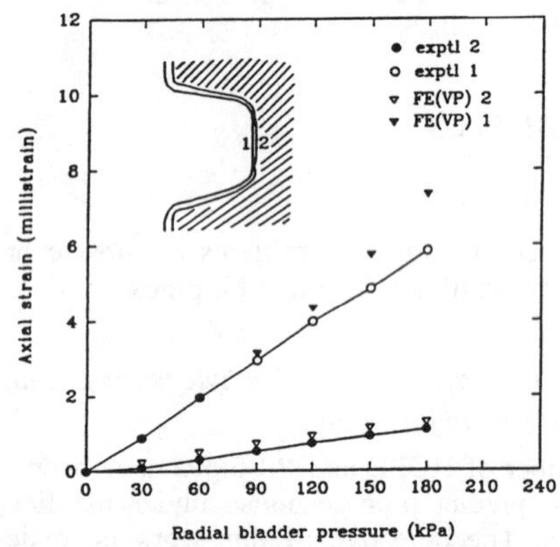

Figure 15. Axial strain on the inner and outer surfaces of the corrugated pipe wall.

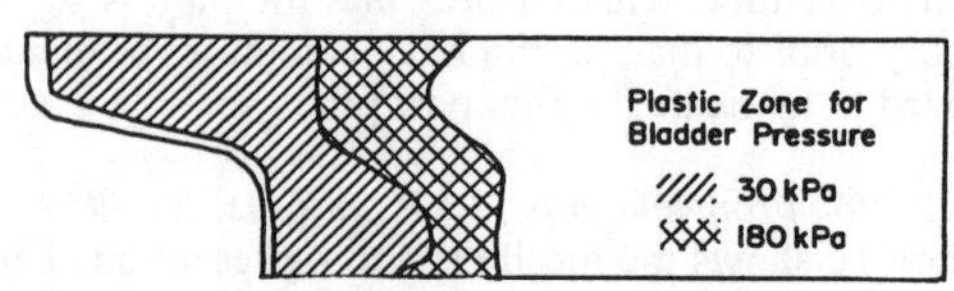

Figure 16. Zones of shear failure in the soil surrounding the corrugated HDPE pipe.

5.4 *Interpretation of strain readings for plane pipe under biaxial earth pressures*

The laboratory test facilities illustrated in Figure 5 are being used to investigate a variety of pipe burial conditions, including those associated with the plain perforated HDPE pipes commonly used to collect leachate at the base of municipal solid waste landfills. Studies have been conducted to determine the effect on stress and strain of the perforations in the pipe and the point loads applied by the coarse gravel backfill. Laboratory measurements include changes in vertical and horizontal pipe diameter, and strain gauge readings from rosettes placed on the inner and outer pipe surfaces. Of particular interest are the peak tensile stresses, since these can result in brittle fracture if sustained over a long period of time.

The use of a strain gauge rosette on the surface of the pipe enables the two principal strains tangential to the pipe surface to be calculated. Given that the normal stress is known to be zero, the constitutive behavior of the material can be used to calculate the two principal stresses. Simple linear elasticity has generally been used in the past to interpret stress from such strain gauge readings. Such calculations have used codified HDPE modulus values (e.g. the short term modulus of 760MPa specified by AASHTO, 1994) or have used a secant elastic modulus estimated from creep or relaxation data.

Only one particular set of strain gauge readings need be considered here to illustrate the impact of the constitutive models used to calculated stress from strain. The example strain gauge rosette featured major principal strains of 0.84% and 0.76% which developed at an almost constant rate over the period of the test (12 hours). This is modeled here as constant strain rates of 1.95×10^{-5} and 1.77×10^{-5} per second (variable strain rates can also be treated if they occur during the test).

Figure 17 shows predictions for the increase in the major principal stress over the 12 hour period of the test. The three values of major principal stress calculated at the conclusion of the test are 11.5MPa, 6.6MPa and 4.5MPa for the elastic, linear viscoelastic and viscoplastic constitutive models respectively.

Given that the hydrostatic design stress (long term tensile capacity) for this material is 5MPa, the three different values are significantly different.

6 SUMMARY AND CONCLUSIONS

Linear viscoelastic, nonlinear viscoelastic and viscoplastic models have been developed to characterize the time dependent response of thermoplastic pipes and geosynthetics. A number of different analyses of buried thermoplastic sewer pipes are being performed to provide guidance regarding pipe performance and design.

1. Accurate predictions can be obtained for pipe response under parallel plate loading; these allow pipe profile designers to evaluate pipe stiffness and creep characteristics prior to investing the considerable sums required to develop a new product line;
2. Assessments can be made of local bending stresses within the pipe wall; this three-dimensional bending is generally responsible for the peak tensile stresses; these finite element calculations are permitting assessment of resistance to crack development, or the selection of pipe resins to sustain the expected levels of tensile stress;
3. Predictions are now possible of the magnitude of compressive stresses in profile components like the inner liner and the side-wall of the corrugation; these are needed to evaluate the susceptibility to local buckling within the pipe wall, permitting burial limits to be established or designers to choose profile geometry that is adequate for a specified depth;
4. Better quality assessments can be made of stress based on strain measurements obtained during pipe tests; this is particularly valuable when the pipe burial conditions are complex, making analysis difficult or impossible; for example, stresses have been assessed in perforated leachate collection pipes buried within coarse stone backfill.

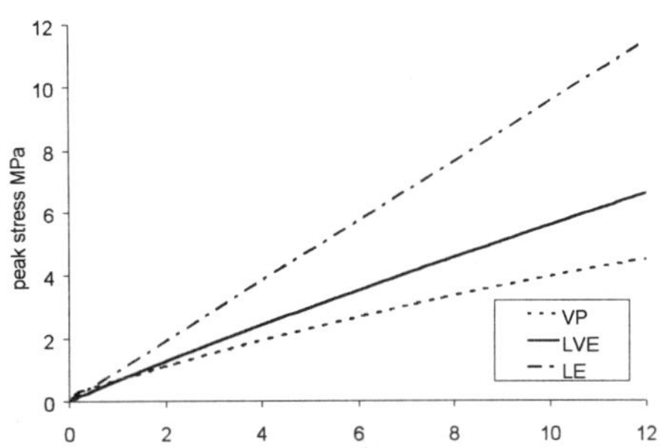

Figure 17. Stress development over the 12 hour duration of the buried pipe test; solutions for different models used to calculate stress from strain.

7 ACKNOWLEDGEMENTS

The work reported here was undertaken in collaboration with a number of individuals whose assistance is gratefully acknowledged: Dr Richard Brachman, Ms Frieda Hu, Dr Tim McGrath, Dr Kerry Rowe, Dr Ernie Selig, Mr Andrew Tognon, Dr Charles Zhang. Financial support for this work was provided by the

Natural Sciences and Engineering Research Council of Canada.

REFERENCES.

AASTHO 1994 LRFD Bridge Design Specifications, American Association of State Highway and Transportation Officials, Washington D.C.

Carter, J. P., Booker, J.R. and Davis, E.H. 1977 Finite deformation of an elasto-plastic soil, *Int. J. Numerical and Analytical Methods in Geomechanics*, 1, 25-43.

DiFrancesco, L. 1993 *Laboratory testing of high density polyethylene drainage pipes*, M.Sc. thesis, Department of Civil Engineering, University of Massachusetts, Amherst.

Laidlaw, T.C. 1999 *Ground support for profiled HDPE pipe*, MESc thesis, Department of Civil and Environmental Engineering, The University of Western Ontario.

Moore, I.D. 1994 Three dimensional time dependent models for buried HDPE pipe, In H.J. Siriwardane (Ed) Proceedings of the Eighth International Conference on Computer Methods and Advances in Geomechanics, 2, 1515-1520, A. A. Balkema, May, Morgantown, WV, USA.

Moore, I.D. 1995 Three-dimensional response of deeply buried profiled polyethylene pipe, *Transportation Research Record* 1514, 49-58.

Moore, I.D. and Hu, F. 1995 Response of profiled polyethylene pipe in hoop compression, *Transportation Research Record* 1514, 29-36.

Moore, I.D. and Hu, F. 1996 Linear viscoelastic modelling of profiled high density polyethylene pipe, *Canadian Journal of. Civil Engineering*, 23, 395-407.

Moore, I.D. and Laidlaw, T.C. 1997 Corrugation buckling in HDPE pipes - measurements and analysis. 76th Annual Meeting, Transportation Research Board, Washington, D.C., January 11-15, 1997. Preprint Paper 97-0565, 24 p.

Moore, I.D. and Zhang, C. 1998 Finite element analysis of thermoplastic pipe response, *Transportation Research Record*, Washington, DC, 1624, 225-230.

Selig, E.T. 1990 Soil properties for plastic pipe installations, Special Technical Publication 1093, Buried Plastic Pipe Technology, ASTM, 141-158.

Hashash, N.M.A. and Selig, E.T. 1990 Analysis of the performance of a buried high density polyethylene pipe, In Mitchell, Sargand and Hurd (Eds) *Structural Performance of Flexible Pipes*, Balkema, Rotterdam, 95-103.

Zhang, C. and Moore, I.D. 1997a Mechanical response of high density polyethylene: Part II - uniaxial constitutive modeling, *Polymer Engineering and Science*, 37, 2, 415-420.

Zhang, C. and Moore, I.D. 1997b Finite element modeling of inelastic deformation of ductile polymers, *Geosynthetics International*, 4 (2), 137-163.

Zhang, C. and Moore, I.D. 1998 Nonlinear finite element analysis for thermoplastic pipes, *Transportation Research Record*, 1624, 225-230.

Computer Methods and Advances in Geomechanics, Desai et al. (eds) © 2001 Balkema, Rotterdam, ISBN 90 5809 183 X

Prediction and reality in geotechnical practice

D.M. Potts & L. Zdravkovic
Department of Civil and Environmental Engineering, Imperial College, London, UK

ABSTRACT: One of the most important ingredients of any geotechnical analysis is the constitutive model used to represent soil behaviour. Many such models currently exist but very few can describe all facets of real soil behaviour based on parameters which can be obtained from standard site investigation tests. The more accurate the model is at reproducing real soil behaviour, the greater is the number of input parameters required to define it. This paper reviews the ability of certain types of constitutive model to reproduce real behaviour. This is done by considering field cases in which predictions are compared with field measurements.

1 INTRODUCTION

The use of numerical analysis, such as the finite element method, when applied to geotechnical problems, can be extremely complex. While in principle the method can be used to provide a solution to most of the problems that may arise, there are approximations which can lead to errors. These approximations can be classified into two groups. Firstly, there are approximations in the numerical method (i.e. finite element method) itself and secondly, there are approximations arising from the idealisations made by the user when reducing the real problem to a form which can be analysed (Potts & Zdravkovic, 1999).

Examples of the second group are the many potential errors which can be associated with a user's lack of 'in depth' understanding of the constitutive model employed to represent soil behaviour. There is a large number of constitutive models available. These range from very simple models (e.g. linear elastic) through to much more complex models which are rarely used in practice (e.g. MIT-E3 (Whittle, 1993)). When undertaking an analysis of any form it is important to consider which models are appropriate for the situation under consideration. It may be that under certain circumstances simple models are sufficient to obtain reasonable results, but for other types of problem a greater degree of sophistication is required.

This paper reviews several case histories in which the authors have been involved. All of these have involved finite element analyses using the computer code ICFEP[1] which has an extensive library of constitutive models (Potts & Zdravkovic, 1999). In the majority of cases the analysis were of the Class "A" type, being performed before the field measurements were made. Attention will be placed on the type of constitutive model employed and the ability of the analyses to predict the field behaviour.

2 LINEAR ELASTIC PLASTIC MODELS

In circumstances where there are no sensitive structures or installations close by, it may be possible to use very simple models to analyse deep excavations. For example Fernie *et al.* (1996) described the design and performance of the Eastbourne Wastewater Treatment Scheme at Langley Point, a 14m deep box very close to the English Channel. Figure 1 shows the soil profile and a section through the structure containing the Works.

To form the excavation, ground within the box was de-watered, which was only turned off after the base slab had been constructed, allowing pore water pressures under the slab to return to the levels that existed before construction started. Piles linked to the base slab restrained floatation in the long term.

Although movements remote from the excavation were not an issue, movements of the walls and structural forces in the walls and base slab were. Finite element analyses were therefore carried out. From the site investigation it was possible to

[1] Imperial College Finite Element Program, written by D.M. Potts.

determine the soil stratigraphy and estimate values of soil strength. However, it was only possible to obtain limited information on soil stiffness. Consequently a very simple linear elastic perfectly plastic (Mohr-Coulomb) constitutive model was employed to represent the soil. The structural members were assumed to be linear elastic.

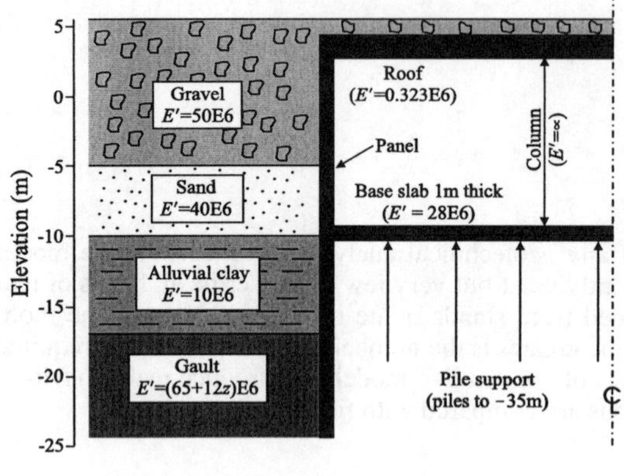

Figure 1: Cross-section for the Wastewater Works at Langley Point

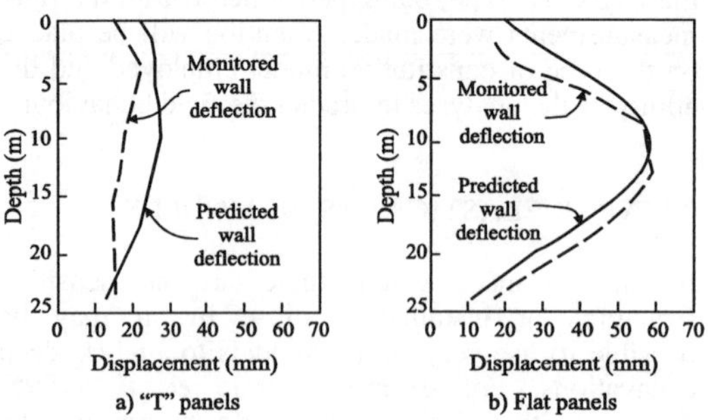

Figure 2: Predicted and observed wall displacements for two types of panels

Figure 2 compares predictions of wall displacement (made before construction) with the field observations for the situation when construction had been completed. It can be seen that the agreement between prediction and observation is good. Because of the restraint afforded by the piles beneath the base slab, the behaviour of the structure depended to a large extent on the changes in water pressure acting on it. There was therefore relatively little soil structure interaction. The comparatively simple soil model performed adequately and it is unlikely that the use of a more sophisticated model would have given predictions in much better agreement with the observations.

It is worth noting that the predictions of the soil movements remote form the box walls were not realistic. Another situation where the use of such simple models does not provide accurate predictions is the analysis of surface subsidence due to tunnelling. Figure 3 shows predictions and field measurements from the recently constructed Jubilee Line extension in London. The results presented are for the construction of the first running tunnel beneath St. James's Park (Nyren, 1998). Being a park, the ground movements were not influenced by surface structures. Soil conditions consist of London Clay, in which the tunnel is constructed, overlain by Thames Gravel and Made Ground. Soil parameters were obtained from a comprehensive site investigation.

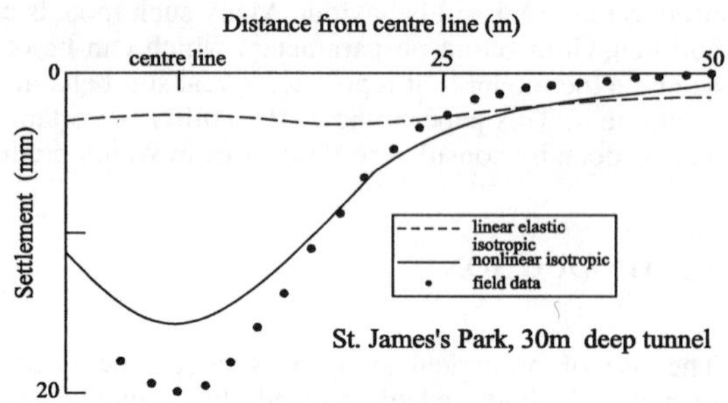

Figure 3: Effect of a pre-yield model on predicting settlement profile

Results based on a simple linear elastic plastic model of the same type as used for the excavation problem discussed above are shown by the dashed line. Clearly they are not in agreement with the field data. The predicted settlement profile is of the wrong shape, being too shallow and too wide when compared to the field data. Such results are typical of predictions of tunnel induced movements in stiff overconsolidated clays based on simple linear elastic perfectly plastic constitutive soil models (Potts & Zdravkovic, 2001).

3 SMALL STRAIN STIFFNESS

The reason why the simple linear elastic perfectly plastic model was unable to accurately predict the soil movements remote form the wall in the excavation problem, and failed totally to predict realistic subsidence troughs above tunnels in stiff clay, is that soil behaviour is nonlinear even at small strain levels. For example, Figure 4 shows the secant shear modulus (normalised by mean effective stress p') decay with shear strain observed in triaxial tests on London Clay. It is clar that even under small strain perturbations the soil behaviour is nonlinear.

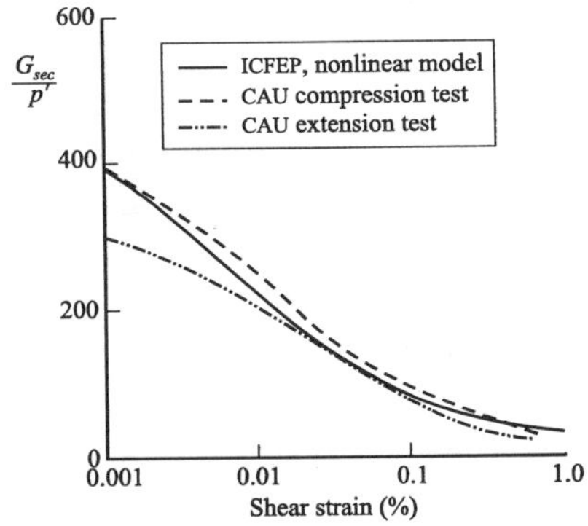

Figure 4: Small strain curves from triaxial tests

As the site investigation data for the Jubilee line extension included laboratory triaxial tests equipped with devices to accurately measure small strains, it is possible to include this behaviour into a constitutive model. The simplest way, but not necessarily the best way, is to account for the small strain behaviour by replacing the linear elastic parameters in the simple model by a nonlinear elastic model (see Addenbrooke *et al.*, 1997 and Potts & Zdravkovic, 2001).

Figure 4 shows that such a model is capable of capturing the behaviour observed in the laboratory. The result of using the model to predict the subsidence trough above the Jubilee Line extension tunnel at St. James's Park is shown in Figure 3. Clearly the predictions are in much better agreement with the observations than those based on the simpler linear elastic plastic model.

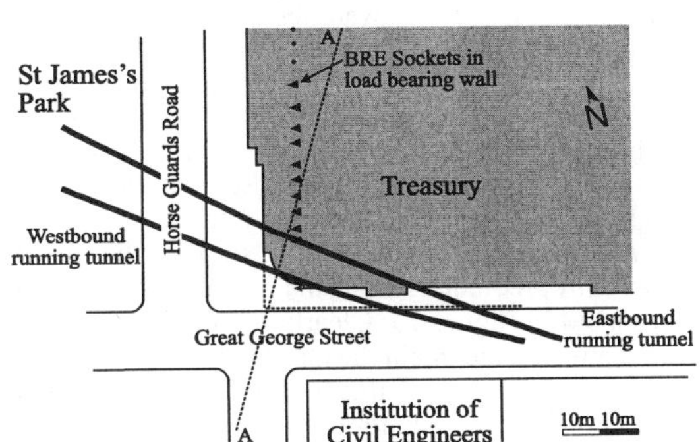

Figure 5: Plan of Treasury site

Adjacent to St. James's Park, and across the road from where the greenfield measurements shown in Figure 4 were obtained, is the Treasury building, see Figure 5. As can be seen, the running tunnels of the Jubilee line extension pass obliquely under the corner of the building. Finite element analyses, using the same nonlinear elastic plastic constitutive model, were performed to model the tunnel induced response of the building. Again these could be compared with field observations. Only a summarised description of the analyses and the results are included here, further details are given in Standing *et al.*, 1998.

The Treasury is a massive stone-clad brick-masonry structure, approximately 210m long and 100m wide, with four storeys above ground and two basement levels. The foundations consist of strips and pads connected by an unreinforced concrete slab founded in the Terrace Gravels which overlie London Clay. As part of the Jubilee Line extension project, two running tunnels were excavated under one corner of the building (shown in plan in Figure 5). The westbound tunnel was the first to be excavated. Following this there was a rest period before compensation grouting and excavation of the eastbound tunnel. Compensation grouting was implemented after driving the westbound tunnel and during construction of the eastbound tunnel. The level of the tubes-a-manchette (TAMs) used for this work was approximately 16m below ground level, see Figure 6, extending beneath the basement between the tunnels and the foundation slab.

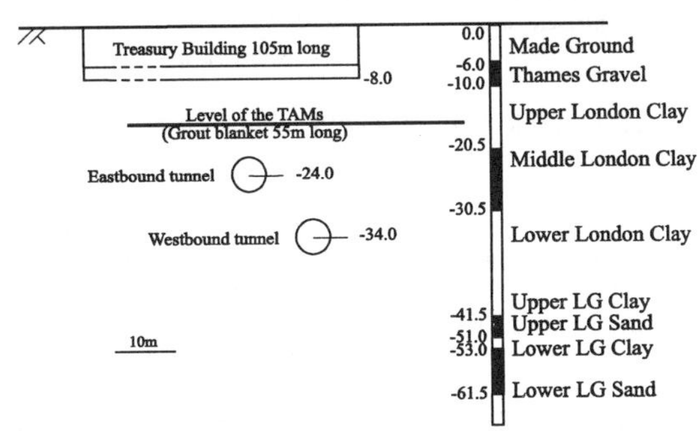

Figure 6: Cross-section through Treasury site

Figure 7 shows settlement troughs after excavation and construction of the westbound tunnel (i.e. the first tunnel to be constructed). Field measurements at ground surface from St. James's Park (greenfield) and at foundation level of the Treasury are presented. Comparison of observations from these two sites indicates the influence of the building. It reduces the maximum settlement, but increases the width of the settlement trough. The maximum settlement beneath the Treasury is beneath its corner, slightly offset from the tunnel centre-line. Also shown is the finite element prediction. The agreement between the prediction and field observations is excellent.

Figure 8 presents settlement troughs at foundation level for the Treasury, 18 weeks after construction of the eastbound tunnel. Both observation and finite element predictions are presented. The agreement between predictions and observation is again excellent.

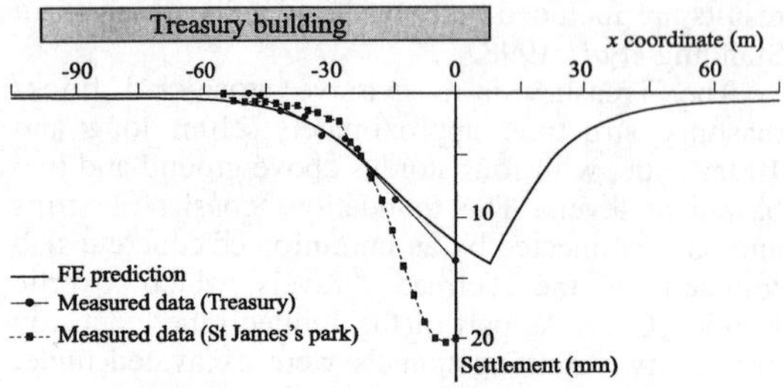

Figure 7: Settlement trough after WB tunnel construction

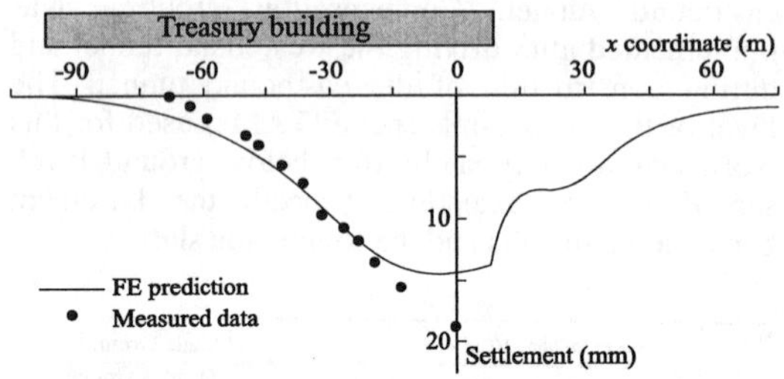

Figure 8: Settlement trough 18 weeks after EB tunnel construction

Another example which shows the merits of accounting for small strain soil behaviour concerns the installation of an oil platform. The installation in the northern North Sea of the Hutton tension leg platform (TLP) represented an important development in offshore engineering. The use of large driven piles as tension anchors was novel and a number of special studies were undertaken to help design the foundations and validate their performance in service (Jardine & Potts, 1988 and Jardine *et al.*, 1988).

One area of concern was the vertical movement of the anchor piles. As shown in Figure 9, the TLP foundations consisted of four clusters of eight piles, one cluster at each corner of the platform. The piles were 1.83m in diameter and 60m long. To monitor movements of the pile clusters 'settlement gauges' were installed on the sea bed. These gauges were of the manometer type and the position of one such gauge is indicated on Figure 9. As the gauges were installed to provide early warning of any likely problems with the anchor piles, it was necessary to estimate the movements that were likely to occur. Comparison of measured movements with these predictions would enable the performance of the piles to be assessed.

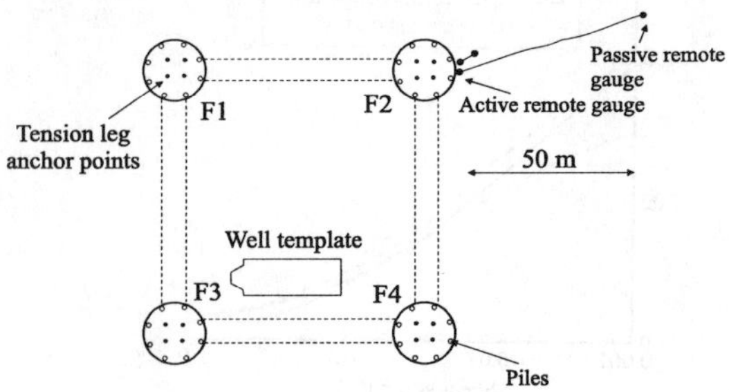

Figure 9: Layout of Hutton TLP platform

In conventional pile group design movements are calculated assuming the soil to be essentially linear elastic and predictions are usually conservative if they are greater in magnitude than those observed in the field. However, for the Hutton TLP situation conservative estimates of movement would need to be smaller in magnitude than those observed (i.e. it is necessary to underpredict the displacement at which the piles pull out). Selecting soil stiffness values based on experience with conventional pile design was therefore not appropriate

To obtain a more accurate estimate of soil stiffness, laboratory triaxial tests were performed in which the samples were fitted with local strain instrumentation. These provided accurate data on soil stiffness and in particular how it varied with strain level. Such data was then used to establish theoretical nonlinear elastic relationships for how the elastic shear and bulk moduli varied with both stress and strain level. These were then incorporated into a form of the modified Cam Clay model and used in a finite element analysis to predict the behaviour of the Hutton TLP foundations. It should be noted that a simple plasticity model, such as Mohr-Coulomb, was not appropriate, as the clay soils at Hutton had relatively low over consolidation ratios.

A comparison of results from some of the triaxial tests with predictions from the constitutive model is presented in Figure 10. Clearly the model is capable of capturing the main facets of the observed soil behaviour. During platform installation the tethers connecting the foundations to the buoyant deck were pre-tensioned. For each pile cluster there were four tethers and these were tensioned to give an upward force of approximately 3000 tonnes. During this process the movements of the anchor piles were observed. These movements are plotted against tether load for pile cluster F2 in Figure 11. Also shown on this Figure are predictions from the finite element analysis (made before the event) and from conventional pile group analysis based on isotropic elastic soil behaviour. Clearly, the latter predictions are not in agreement with the observations, predicting too large movements. In this respect they are not conservative. In comparison, the predictions from the finite element analyses, which take account of the small strain soil behaviour, are in excellent agreement with the observations. As noted by Jardine & Potts, 1988 the finite element analysis was able to accurately capture many other aspects of the observations. This example clearly indicates the importance of accounting for small strain soil behaviour.

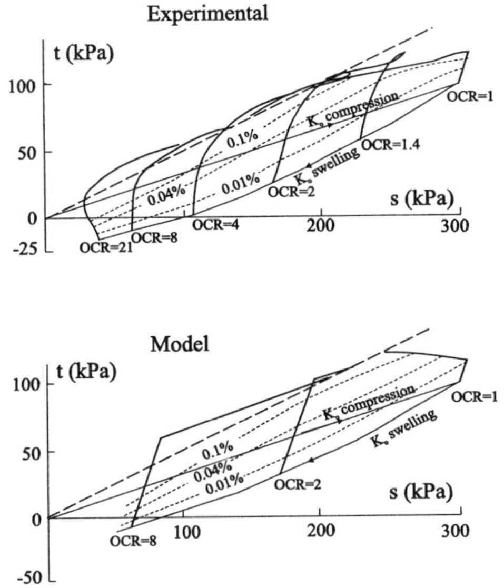

Figure 10: Experimental and simulated triaxial stress paths

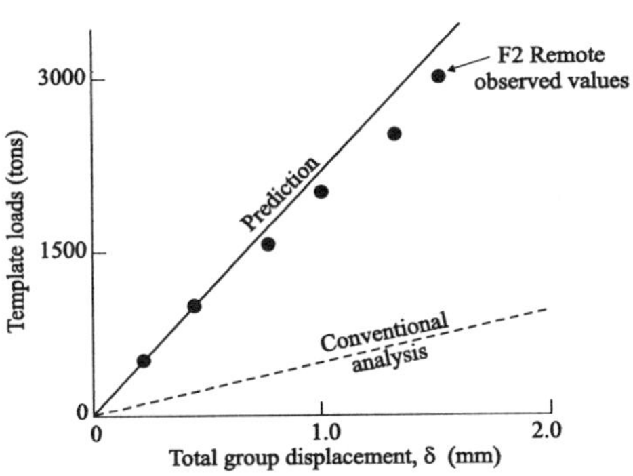

Figure 11: Observed and predicted displacement of Hutton TLP piles

4 CHOICE OF SOIL MODEL

In the previous section it has been shown that it is often necessary to account for the small strain behaviour of soil in any geotechnical analysis. In the examples presented this small strain behaviour has been accounted for using a nonlinear elastic formulation within a conventional plasticity model (i.e. either Mohr-Coulomb or modified Cam Clay). In this section the alternative of modelling the small strain behaviour in terms of an elasto-plastic formulation will be investigated. Again, results from a case history will be used.

Roadford dam is a 41m high rockfill dam that has recently been constructed in the United Kingdom. It has an upstream asphaltic concrete membrane and a typical cross section of the dam is shown in Figure 12. Details of the design of the dam and of the properties of the rockfill are given by Wilson & Evans, 1990. As indicated on Figure 12, instrumentation was installed in the dam to monitor its behaviour.

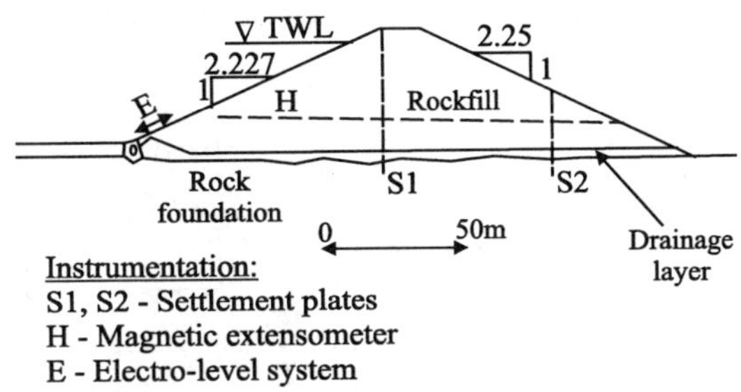

Figure 12: Cross-section through Roadford dam

Finite element analysis of the dam was performed using two different constitutive models (Potts & Zdravkovic, 2001). One of these models was a nonlinear elastic perfectly plastic (Mohr-Coulomb) model. The other was a complex work hardening/softening elasto-plastic model of the Lade type, having a double yield surface. In both cases the model parameters were determined from standard oedometer and triaxial laboratory tests. A comparison of the predictions from the constitutive models with this laboratory data is shown in Figures 13 and 14. While both models are able to accurately reproduce the oedometer behaviour and the shear stress - strain triaxial data, the nonlinear elastic perfectly plastic model has difficulty in reproducing the volumetric strain behaviour observed in the triaxial tests. Adjustments to the model parameters enabled a slightly better fit to be obtained, but this was at the expense of a less good fit to the shear stress - strain and oedometer data.

A comparison between the finite element predictions and the observations from the settlement gauges S1 and S2, see Figure 12, during construction of the dam are shown in Figure 15. Analyses using both soil models give reasonable agreement with the observations. It should be noted that even better agreement could be reached if account was taken of the fact that the top part of the dam was constructed at a higher relative density then assumed at the design stage. However, such agreement between the two analyses is not so good for horizontal movements, as can be seen in Figure 16 which compares predictions with observations from the horizontal gauge, H (see Figure 12). While the analyses using the complex Lade model gives horizontal movements in agreement with the observations, those from the analysis using the simpler model are inaccurate. The reason for the success of the complex model and the failure of the simpler model to predict horizontal displacements is that in the former both elastic and plastic components of strain are mobilised, whereas in the latter only elastic strains are developed during construction. It appears that while such a detail is not important for the prediction of vertical displacements, it is for accurate prediction of horizontal displacements.

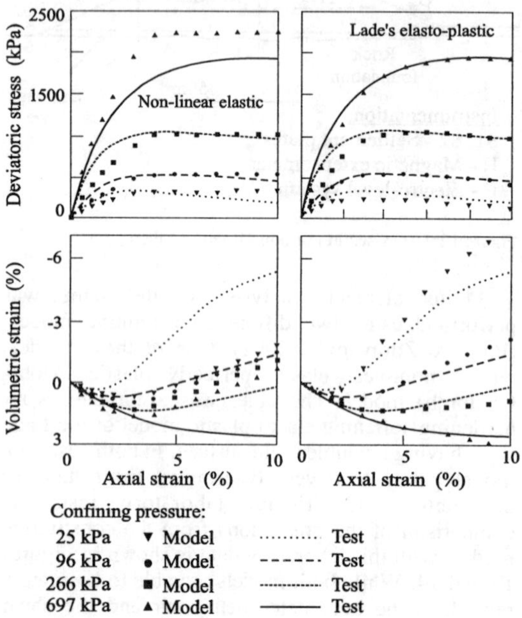

Figure 13: Simulation of triaxial tests

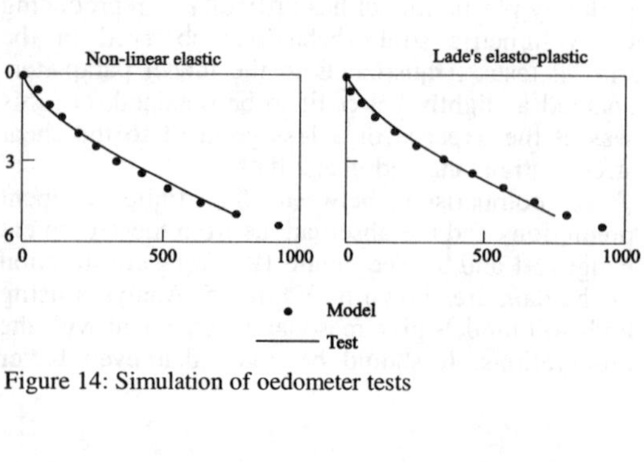

Figure 14: Simulation of oedometer tests

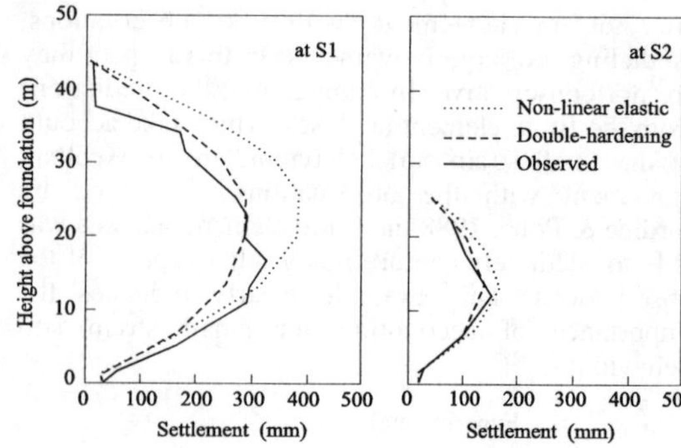

Figure 15: Comparison of predicted and observed settlements

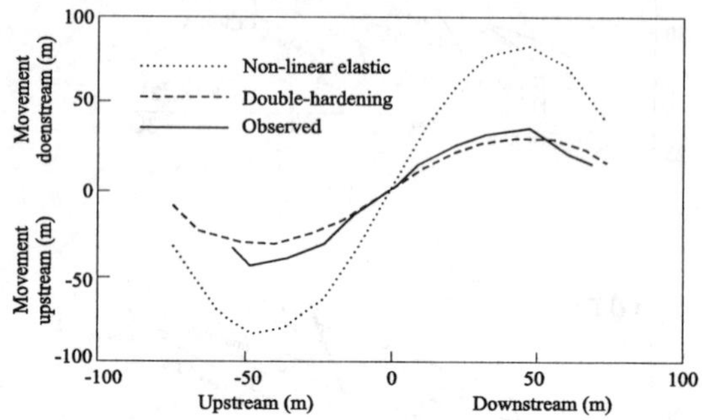

Figure 16: Comparison of predicted and observed horizontal displacements

It should be noted that for this dam accurate prediction of horizontal movements was important, especially during impounding, as it was the behaviour of the upstream membrane and its connection with the upstream toe inspection gallery that was of concern. The observation that analysis based on nonlinear elastic behaviour pre-yield have difficulties in predicting the measured horizontal movements was also reported by Potts *et al.*, 1990 when analysing the failure of Carsington dam.

5 ANISOTROPIC SOIL BEHAVIOUR

In all the analyses described above soil behaviour was assumed to be isotropic. This is a common assumption, even though most soils are likely to be anisotropic as a result of the way they were formed (i.e. one dimensional sedimentation). One of the reasons for making such an assumption is that it is difficult to investigate anisotropic behaviour with standard laboratory tests. There are only a few pieces of equipment, such as the hollow cylinder apparatus and the directional shear device, which allow simultaneous control of both the magnitude of the intermediate principal stress and the direction of the principal stresses. Such control is needed for a systematic investigation of both inherent and induced anisotropy. Recent research using such equipment has clearly shown that many soils are anisotropic in their behaviour.

To indicate the likely effects of soil strength anisotropy a case history involving the construction of an embankment on soft ground will be considered.

The Champlain Clay covers a large part of a densely populated region of Eastern Canada. Construction of the road infrastructure requires numerous overpasses, containing embankments approximately 5m in height. As the clay is only lightly overconsolidated, it is soft and weak and it is therefore difficult to design and construct embankments to the required height.

For this reason the geotechnical group of Laval University decided in 1972 to undertake a research programme involving the construction of test embankments on soft Champlain deposits. The objective of this project was to investigate the failure conditions of fills built on soft, sensitive foundations and also to study the magnitude and rate of settlement of such structures. Four embankments were constructed on instrumented clay foundations at the Saint-Alban site in the Province of Quebec: the first one, which is the subject of this study, was built until failure occurred; three others were built to a smaller height and with different side slopes in order to study settlement behaviour.

Test embankment A, described by La Rochelle *et al.*, 1974, was the subject of finite element analyses (Zdravkovic *et al.*, 2001), with the objective of investigating the effect of anisotropic soil behaviour on embankment height.

The geometry of this boundary value problem is presented in Figure 17. One side of the embankment has a slope of 2:1 (horizontal to vertical) inclination. The stability of this slope is enhanced by the presence of 1.5 m heigh berm, so that any failure is forced onto the other side of the embankment, where the slope is steeper, with 1.5:1 inclination. The predicted height of the embankment, based on conventional limit equilibrium analysis, was 4.6 m.

The construction sequence was such that the first 0.6m of embankment was lifted in a day, followed by 0.3m per day until a height of 1.5m was reached. After that two 0.3m layers were lifted per day, until failure occurred at about an embankment height of 4 m, which was less than the predicted 4.6m.

A detailed geotechnical investigation was carried out at the Saint-Alban site, with a substantial number of boreholes and both in-situ and laboratory experiments on Champlain Clay samples. Details of the results of these tests are given by Tavenas & Chapeau, 1973 and Sarrailh & Tavenas, 1972.

In general, the soil profile on the site has a weathered clay crust extending down to ~2.0 m below the ground surface, followed by a soft silty marine clay deposit down to ~13.7 m. Beneath this clay layer there is a layer of a dense, fine to medium sand, followed by bedrock. Ground water level is at 0.7 m below the ground surface. The embankment fill is granular.

For the finite element analyses, these foundation soil conditions were modelled using both the MIT-E3 soil model (to simulate anisotropic soil strength) and the modified Cam Clay soil model, MCC (to simulate isotropic soil strength).

The undrained strength variation with depth for different loading conditions, simulated by the MIT-E3 model, is presented in Figure 18. It can be seen from this figure that the predicted direct simple shear strength (DSS) matches, on average, the strength obtained from vane tests. Also, the simulated CIU strengths match closely the experimental CIU strengths. It can therefore be concluded that the observed anisotropic strength has been modelled very well with the MIT-E3 model. Also presented in Figure 18 are the profiles of plane strain compression (PSC) and plane strain extension (PSE) undrained strength.

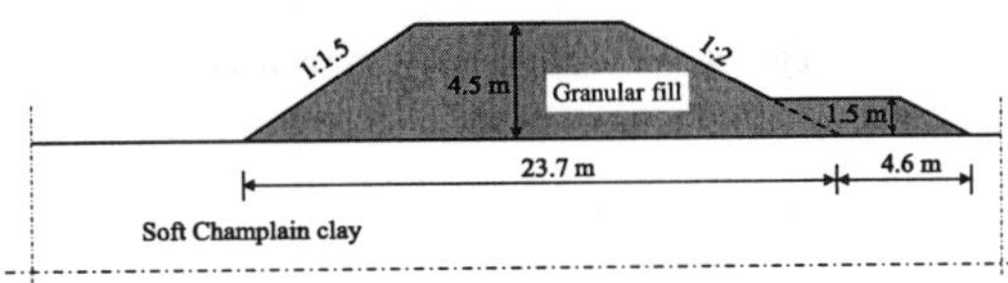

Figure 17: Geometry of test embankment at Saint-Alban

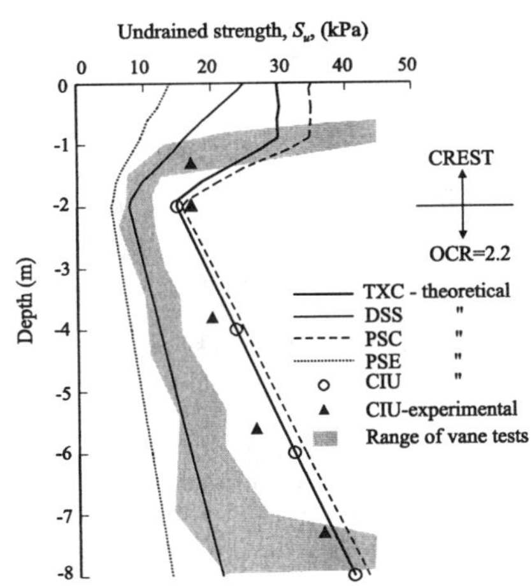

Figure 18: Undrained strength profiles at Saint-Alban

For the isotropic analyses, the undrained strength in triaxial compression was adopted and the modified Cam Clay model was used to simulate the triaxial compression strength profile (TXC), as predicted by the MIT-E3 model in Figure 18.

The same construction sequence and rate were simulated as those implemented on site, such that failure occurred under undrained conditions.

Figure 19 shows the development of horizontal displacement, u, of the embankment toe with the embankment height. This horizontal displacement is normalised by the maximum horizontal displacement, u_{max}, predicted just before the slope failed.

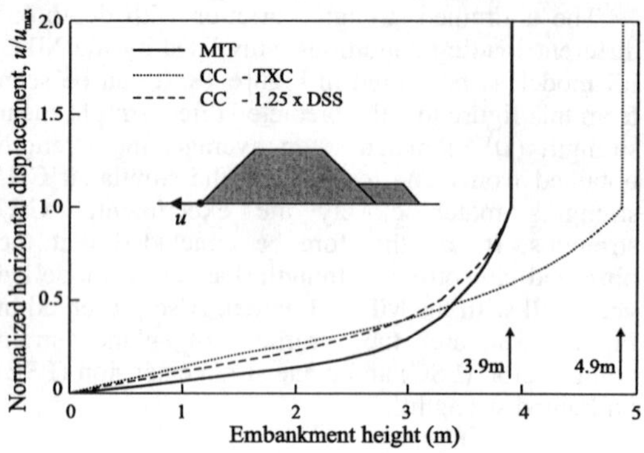

Figure 19: Failure of Saint-Alban embankment - original geometry

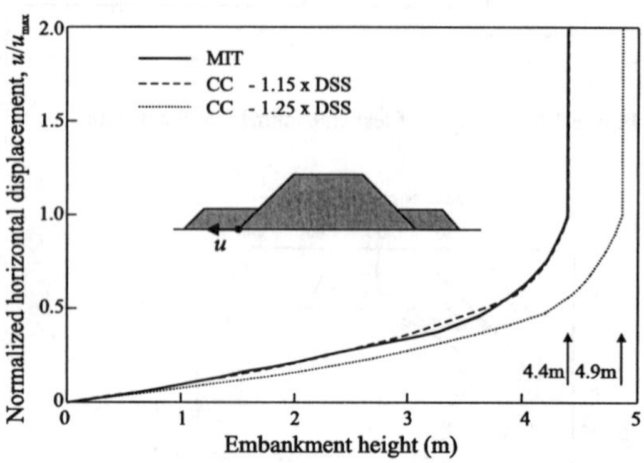

Figure 20: Failure of Saint-Alban embankment - modified geometry

The analysis performed using the MIT-E3 soil model, which simulated the anisotropic soil strength as depicted in Figure 18, predicted failure at an embankment height of 3.9m, which is very close to the 4m observed on site. For the same embankment, but now analysed with the MCC model with an isotropic undrained strength corresponding to the triaxial compression profile in Figure 18, the failure height is grossly overestimated at 4.9m. Back analysis using the MCC model to achieve a failure height of 3.9m with an isotropic strength resulted in the necessary strength being 1.25 times the DSS strength. Analysis with the isotropic strength being equal to the DSS profile, which is what some design practice suggests, is largely conservative, predicting failure at only 3.3 m embankment height.

A further set of analyses was performed, where the embankment geometry was altered to have berms on both sides of the embankment, see insert in Figure 20. Again both anisotropic and isotropic analyses were performed and the results are presented in Figure 20, as normalised horizontal displacement of the embankment toe versus embankment height.

An analysis using the MIT-E3 model predicted that, with the help of a berm, the embankment could now be erected to 4.4m. The same embankment was then analysed with the MCC model, with the isotropic strength back-calculated from the previous set of analyses (i.e. 1.25 times the DSS strength, see Figure 19). This analysis predicted an embankment height at failure of 4.9m and therefore overestimated that obtained by the analysis with the MIT-E3 model. To obtain the same failure height as the MIT-E3 analysis, the MCC analysis needed a strength of only 1.15 times the DSS strength.

Several conclusions can be drawn from this embankment study:

- Analysis which assumes isotropic soil behaviour and takes either the triaxial compression strength profile or the DSS strength profile will either overestimate or underestimate the embankment height, respectively;
- Taking account of soil anisotropy, observed through laboratory and field experiments, enables a more accurate prediction of embankment height;
- If an isotropic strength profile is back-calculated from the observed behaviour of a certain embankment geometry, it is not necessarily appropriate if the embankment geometry becomes slightly altered.

Clearly in the future, as more experimental data becomes available, and user friendly constitutive models are developed, it will be possible to model anisotropic behaviour in most field problems. At present the necessary laboratory test data is rarely available and even if it were, the determination of the model parameters is not straight forward.

6 CONCLUSIONS

The examples given in this paper demonstrate that care must be exercised when choosing a constitutive model for carrying out geotechnical analysis. In particular, it has been shown that the small strain behaviour and whether or not this involves purely elastic or a combination of elastic and plastic strains can have significant effects on the accuracy of any predictions.

There are many constitutive models available for the analysis of geotechnical problems, but there are no universally applicable models. Pre-failure models that perform well for some situations do not perform well for others, although some models have wider application than others. It is therefore important for the analyst to consider which issues and facets of soil behaviour are likely to dominate and then choose a model accordingly. In the case of the Saint-Alban embankment, anisotropy was important but for other problems perfectly reasonable predictions were made without considering it.

It is pointless using a model which will predict settlements accurately when lateral movements are an issue. Likewise, certain models may predict the behaviour of a structure reasonably well but they do not predict remote movements well and so potential damage of remote structures and/or services is underpredicted. Problems where stress changes occur on a boundary which is remote from the point of interest present a real challenge (e.g. tunnels) and it is generally necessary to use sophisticated constitutive models for the analysis of this class of problem.

Not withstanding the above difficulties it has been shown that, with experience and the appropriate software, it is possible to make accurate Class 'A' predictions (before the event).

ACKNOWLEDGEMENTS

The authors would like to thank Mr K. Higgins and Dr N. Kovacevic of the Geotechnical Consulting Group and Dr T. Addenbrooke from Imperial College for their contributions to some of the case histories discussed in this paper.

REFERENCES

Addenbrooke, T.I., Potts, D.M. & Puzrin, A.M. 1997. The influence of pre-failure soil stiffness on the numerical analysis of tunnel construction. *Geotechnique* 47, 3: 693-712.

Fernie, R., Kingston, P.J., St John, H.D., Higgins, K.G. & Potts, D.M. 1996. Case history of a "stepped box" excavation in soft ground at the sea front, Langley Point, Eastbourne. *Geotechnical Aspects of Underground Construction in Soft Clay*, R.J. Mair & R.N. Taylor (Eds): 123-130. Rotterdam: Balkema.

Jardine, R.J. & Potts, D.M. 1988. Hutton tension leg platform foundations: prediction of driven pile behaviour. *Geotechnique* 38, 2: 231-252.

Jardine, R.J., Hight D.W. & McIntosh W. 1988. Hutton tension leg platform foundations: measurement of pile axial load-displacement relations. *Geotechnique* 38, 2: 219-230.

La Rochelle, P., Trak, B., Tavenas, F.A. & Roy, M. 1974. Failure of a test embankment on a sensitive Champlain clay deposit. *Canadian Geotechnical Journal* 11: 142-164.

Nyren, R. 1998. Field measurements above twin tunnels in London Clay. *PhD thesis*, Imperial College, University of London

Potts, D.M., Dounias, G.T. & Vaughan P.R. 1990. Finite element analysis of progressive failure of Carsington embankment. *Geotechnique* 40, 1: 79-101.

Potts, D.M. & Zdravkovic, L. 1999. *Finite element analysis in geotechnical engineering: Theory*. London: Thomas Telford.

Potts, D.M. & Zdravkovic, L. 2001. *Finite element analysis in geotechnical engineering: Applications*. London: Thomas Telford.

Sarrailh, J. & Tavenas, F.A. 1972. Etude geotechnique preliminarie du site de Saint-Alban. *Internal report GCN-72-09-02 (MS-N2)*. University of Laval, Quebec, Canada.

Standing, J.R., Farina, M. & Potts, D.M. 1998. The prediction of tunnelling indiced building settlements - A case study. *Tunnels & Metropolises*, A. Negro & A.A. Ferreira (Eds): 1053-1058. Rotterdam: Balkema.

Tavenas, F.A. & Chapeau, C. (1973). Etude en laboratorie des proprietes geotechniques de l'argile de Saint-Alban. *Internal report GCN-73-04-03 (MS-N5)*. University of Laval, Quebec, Canada.

Whittle, A.W. 1993. Evaluation of a constitutive model for overconsolidated clays. *Geotechnique*, 43, 2:289-314.

Wilson, A.C. & Evans, J.D. 1990. The use of low grade rockfill at Roadford Dam. *The Embankment Dam*. London: Thomas Telford: 15-21.

Zdravkovic, L., Potts, D.M. & Hight, D.W. 2001. The effect of anisotropy on the behaviour of an embankment on soft clay. In preparation.

Computer Methods and Advances in Geomechanics, Desai et al. (eds) © 2001 Balkema, Rotterdam, ISBN 90 5809 183 X

Back analysis of strain localization occurring in the vicinity of geostructures

S. Sakurai
Hiroshima Institute of Technology, Japan

ABSTRACT: Field measurements play an important role in observational methods for the excavation of geostructures like tunnels, caverns, slopes etc. The observational method cannot function effectively unless the measurement results are interpreted properly. For achieving proper interpretation back analysis is a promising means. The role of back analysis is not only to assess the stability of the structures, but also to re-evaluate the adequacy of design parameters that have been used in the design stage. Moreover, it should be emphasized that the back analysis should predict an unexpected catastrophic failure, though the structures have been designed in such a way that the catastrophic failure have never occurred. In this paper a back analysis is described, which can predict strain localization occurring in the ground of the vicinity of the structures. It is obvious that strain localization may cause a catastrophic failure of the structures.

1. INTRODUCTION

It is well recognized that deformational behavior of geostructures like tunnels, caverns, slopes, retaining walls, etc. observed during construction often differs from the one predicted by numerical analysis, though a sophisticated computer program is used. This difference mainly comes from the difficulty in the determination of input data such as geological conditions, characteristics of joints, material properties, underground water conditions, etc., which are used in numerical analysis.

In order to overcome these difficulties, an observational method is often adopted. In the method, observation and field measurements are carried out during the construction and the adequacy of design and construction methods are assessed, and if necessary, they are modified in order to achieve rational design and construction procedures. The observational method should also assess the stability of the structures during construction. In particular an unexpected catastrophic failure must be predicted if it tends to occur, though the structures have been designed in such a way that the catastrophic failure would have never occurred. The observational method sounds very impressive, however, the method cannot function effectively unless the measurement results are interpreted adequately. One of the most powerful tools for interpreting field measurements is back analysis, which can assess the adequacy of modelling of rocks and soils, and the input data (design parameters) used in the design analysis, properly.

2. MODELING OF ROCKS AND SOILS

It is assumed that the concerned rocks are highly jointed, so that the continuum mechanics approach can be adopted even for rocks. In order to clarify the deformational mechanism of rocks and soils, triaxial compression tests and torsion tests were performed in laboratories. The test results demonstrate that Young's modulus never changes, while shears modulus decreases as an increase of shear strain (Sakurai and Akayuli, 1998a). Considering these results, Sakurai et al (1998b) proposed strain-induced damage and defined the anistropic damage parameter d as follows;

$$d = \frac{1}{2(1+\nu)} - m \tag{1}$$

where ν is Poisson's ratio
m is a parameter, the physical meaning of which is the ratio of shear modulus to Young's modulus.

The parameter m is a function of the maximum shear strain as shown in the following equation.

$$m = f(\gamma \max) \tag{2}$$

One of the experimental results is shown in Fig. 1

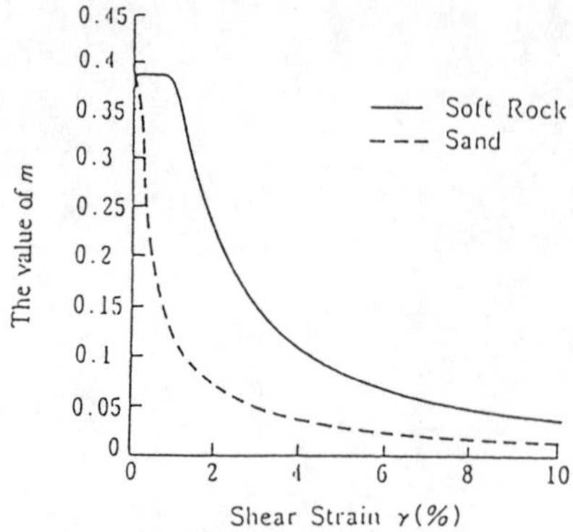

Figure 1. Relationship between m and γ

3. CONSTITUTIVE EQUATION

The following constitutive equation in plane strain condition is derived considering the above defined parameter *m*.

$$\{\sigma'\} = [D']\{\varepsilon'\} \tag{3}$$

where

$$[D'] = \frac{E}{1-\nu-2\nu^2}\begin{bmatrix} 1-\nu & \nu & 0 \\ \nu & 1-\nu & 0 \\ 0 & 0 & m(1-\nu-2\nu^2) \end{bmatrix} \tag{4}$$

Eq.(3) is valid in the local coordinate system. The $x'-axis$ is parallel to a potential sliding plane (see Fig. 2).

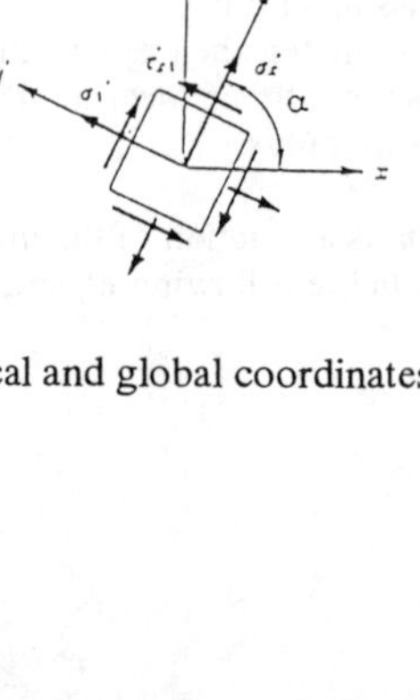

Figure 2. Local and global coordinates

Eq.(4) can be transformed to global coordinates as follows:

$$[D] = [T][D'][T]^T \tag{5}$$

where $[T]$ is a transformation matrix.

When the principal stress directions are known, the direction of a potential slip plane is determined, considering the Mohr-Coulomb's failure criterion. It should be noted that there exist, in general, two potential slip planes at the given point within a material under a stressed condition as shown in Fig. 3. It is seen from Fig. 3 that one is located at the angle of α_1 counted counter-clockwise from the direction of minimum principal stress (compression is positive), and the other is at the angle of α_2. The angle of α_1 and α_2 are determined for geomaterials which obey the Mohr-Coulomb yielding criterion as follows:

$$\alpha_1 = 45° + \frac{\phi}{2}, \quad \alpha_2 = 135° - \frac{\phi}{2} \tag{6}$$

where ϕ is the internal friction angle.

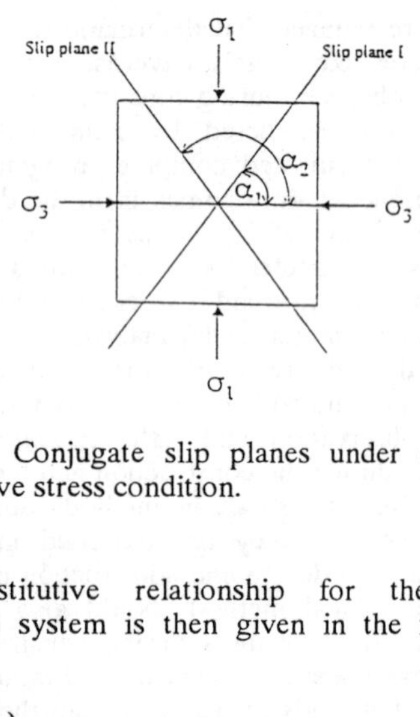

Figure 3. Conjugate slip planes under a triaxial compressive stress condition.

The constitutive relationship for the global coordinate system is then given in the following form:

$$\{\sigma\} = [D]\{\varepsilon\} \tag{7}$$

where

$$[D] = \left[\frac{1}{2}\left([D_1]^{-1} + [D_2]^{-1}\right)\right]^{-1} \tag{8}$$

$[D_1]$ and $[D_2]$ are the matrices defined for two conjugate potential slip planes, which can be calculated by Eq.(4) for slip planes I and II, respectively. If the parameter m is obtained as a function of shear strain, in addition to Young's modulus and Poisson's ratio, all the elements of the stress matrix can be determined.

It is noted that there are two ways for the cause of displacement in the ground. One is due to the change of stress caused by excavation. The other is due to the reduction of strength of soils and/or rocks. In other words, in the second case the displacements occur as the parameter of m decreases, even though there is no excavation. This type of decrease of the parameter m may be caused by loosening due to gravitation, weathering and/or creep. However, it should be noted that no matter what cause might be the increase of shear strain caused the reduction of the parameter m (see Fig. 4)

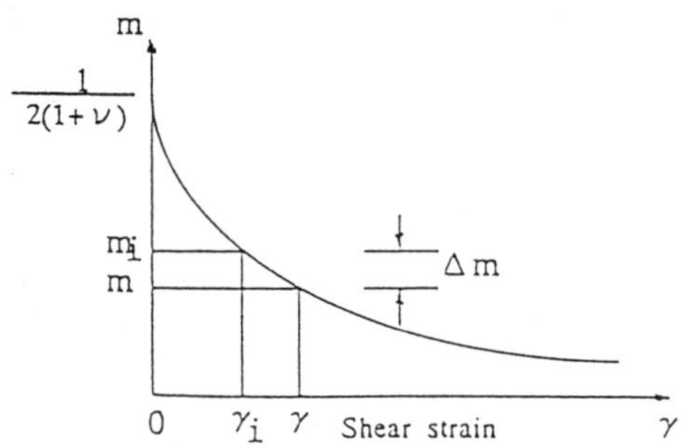

Figure 4. Relationship between increments of parameter m and shear strain.

In this study, the finite element method is adopted. When the parameter m decreases in a certain increment, say Δm (see Fig.4), the displacement increment $\Delta\delta$ is then related to Δm in the following equation:

$$[K]\{\Delta\delta\} = \{\Delta P\} \tag{9}$$

where $\{\Delta P\}$ is the external forces acting at each nodal point, which are the equivalent to the reduction of m values, and it can be represented by the following equation:

$$\{\Delta P\} = \int [B]^T [D][\Delta C]\{\sigma\} dV \tag{10}$$

$\lfloor K \rfloor$ is a stiffness matrix, $\lfloor B \rfloor$ is a matrix relating strain occurring in an element to displacements at nodal points of the element. $\lfloor D \rfloor$ is the elasticity matrix for isotropic elastic materials. $\{\sigma\}$ is stress in damaged regions where the anisotropic damage parameter d increases so that the parameter m decreases as a function of shear strain. $\lfloor \Delta C \rfloor$ is given for a two-dimensional case in the following form:

$$[\Delta C] = \begin{bmatrix} 0 & 0 & 0 \\ 0 & 0 & 0 \\ 0 & 0 & \dfrac{\Delta m}{m \bullet m_i} \end{bmatrix} \tag{11}$$

In back analysis m_i and m may be determined so as to minimize the following equation:

$$\varepsilon = \frac{\sum_{i=1}^{N} \left(u_i^{\,m} - u_i^{\,c}\right)^2}{\sum_{i=1}^{N} u_i^{\,m}} \tag{12}$$

where u_i^{m} and u_i^{c} are measured and computed displacements at measuring point *i*, respectively, and *N* is the total number of measuring points.

4. APPLICATION OF THE PROPOSED CONSTITUTIVE EQUATION TO ENGINEERING PRACTICE.

4.1 *Laboratory physical model test of tunnels*

Two parallel tunnels with a shallow depth were studied as an example problem. First, in order to investigate the mechanism of deformation occurring around the tunnels, physical model tests were carried out in a two dimensional fashion.

In the physical model tests thousands of aluminum pellets were piled up to simulate a highly jointed rock mass. Two parallel tunnels were set in the medium of piled aluminum pellets by applying air pressure on the inner surface of the tunnels, which was equivalent to overburden pressure. The excavation of the tunnels was simulated by reducing the air pressure gradually until a collapse occurred. Since the details of the tests have been published elsewhere (Sakurai and Akayuli, 1998a), only some of the results are shown here.

Fig. 5 indicates the maximum shear strain distribution, which was recorded just before the collapse occurred. It is seen from this figure that the maximum shear strain distribution tends to concentrate in the region above the tunnels, and shear bands start to be formed. This is known as a strain localization phenomenon.

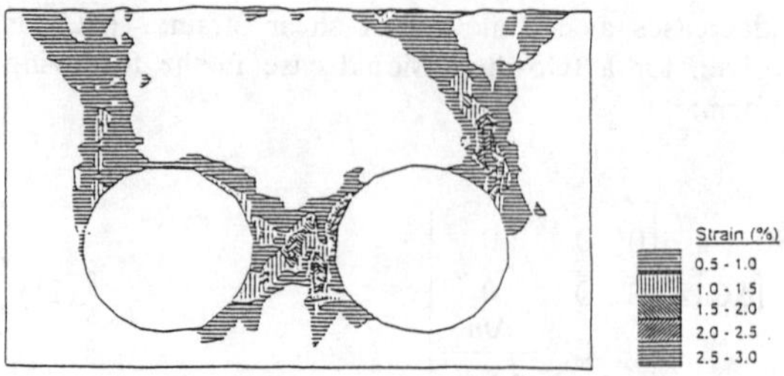

Figure 5. Maximum shear strain distribution around the two parallel tunnels.(Experiments)

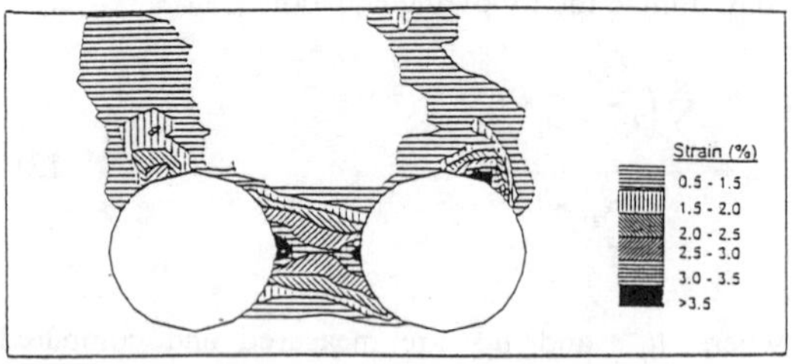

Figure 6. Maximum shear strain distribution around the two parallel tunnels. (Numerical analyses)

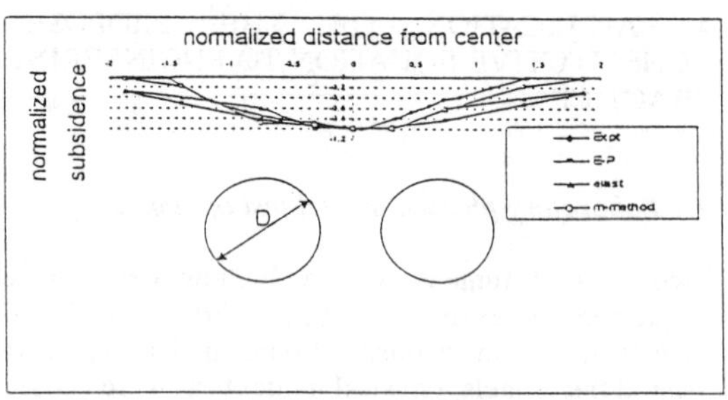

Figure 7. Comparison of numerical results for subsidence calculated by three different models, i.e. (1) elastic, (2) elasto-plastic, and (3) strain-induced damage.

The parameter *m* is back-calculated so as to obtain the maximum shear strain distribution similar to that of the physical model test. When the appropriate value of the parameter m is determined, finite element analysis is then carried out and the maximum shear strain distribution is obtained as shown in Fig. 6. Comparing Fig. 5 with Fig. 6, it is seen that there is a good agreement. The surface subsidence calculated by the strain-induced damage analysis is compared with the ones calculated by the conventional elastic and elasto-plastic analyses, as shown in Fig. 7. The experimental results are also plotted in this figure. It can be seen that the subsidence calculated by considering the strain-induced damage shows a better agreement with the experimental results.

Consequently, the proposed constitutive equation can be used to simulate the deformational behavior of highly jointed materials like aluminum pellets used for the physical model test. This means that it may also be applicable for analyzing the deformational behavior of geomaterials like highly jointed rocks and soils.

4.2 *Slope*

A high cut slope along a highway was monitored during excavation(see Fig. 8). The measurements were performed on the slope surface by surveying. Large displacements were measured after heavy rain fall. In order to stabilize the slope, the potential sliding plane must be first investigated. For this purpose the back analysis described in this paper was carried out, and the strain-induced damage parameter m was determined so as to minimize Eq.(12) and to obtain a close agreement between computed and measured surface displacements.

When all the values of m as well as Young's modulus(Poisson's ratio was assumed) were determined by the back analysis, displacement vectors were calculated and shown in Fig. 9. This figure clearly shows that a potential sliding plane has already taken place. Thus, taking into account the back-analyzed potential sliding plane, the length of anchoring was easily designed for stabilizing the slope. The comparison between the computed and measured displacements is shown in Fig. 10. It is obvious from the figure that a good agreement between the two were obtained.

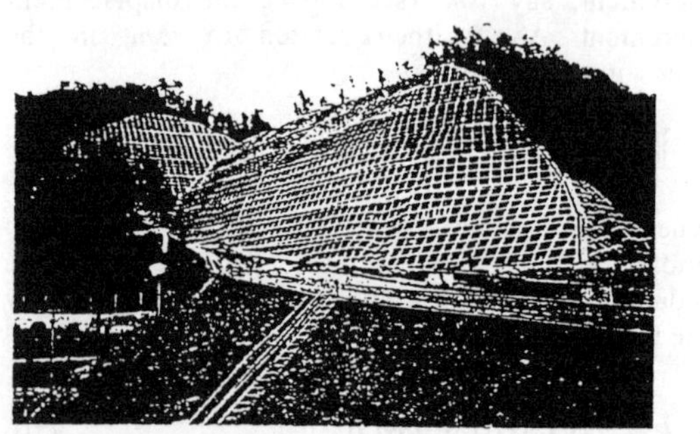

Figure 8. General view of slope

5. CONCLUSIONS

1. Field measurements are commonly carried out for achieving the rational design and construction procedures of the geostructures during

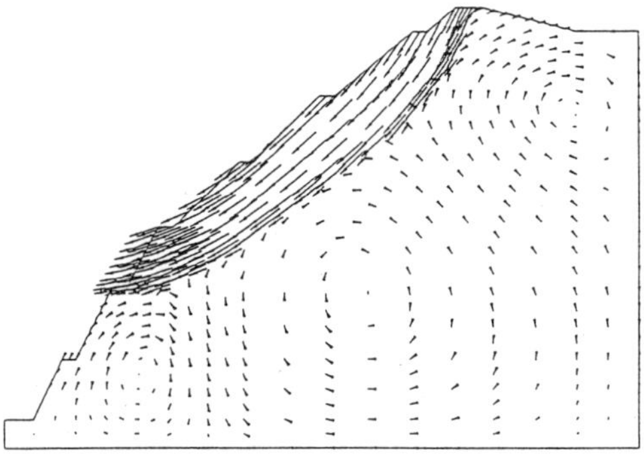

Figure 9. Displacement vectors back-analyzed from the measured surface displacements

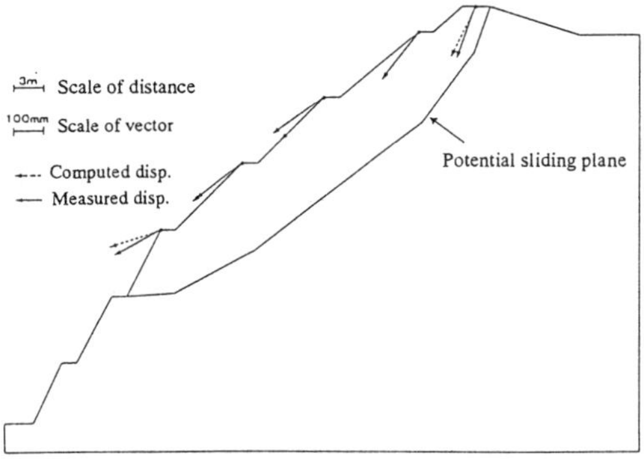

Figure 10. Comparison between calculated and measured displacements on the slope surface

construction. However, a crucial problem of field measurements is how to interpret their results. In order to interpret the field measurement results properly back analysis is a very promising means.
2. An unexpected catastrophic failure should be predicted at its early stage if it tends to occur. Thus, a back analyses should be capable to predict it. For this purpose a constitutive equation has been proposed on the basis of the concept of strain-induced damage of rocks and soils.
3. Laboratory experiments and some of the case studies are shown for demonstrating the applicability of back analysis based on the proposed constitutive equation. It is then concluded that the proposed strain-induced damage analysis can predict the catastrophic failure occurring in the ground in the vicinity of the structures.

ACKNOWLEDGEMENTS

The author would like to thank Mr. T. Nakayama, former graduate student of Kobe University, for his help in performing a back analysis determining a sliding plane of the slope.

REFERENCES

Sakurai, S. and Akayuli, C.F.A. 1998a, *Deformational analysis of geomaterials considering strain-induced damage, Proc. 4th European Conference on Numerical Methods in Geotechnical Engineering (NUMGE 98)*, Udine, 729-738

Sakurai, S. Hiraoka, A. and Hori, K. 1998b, *Strain-induced damage of rocks, Proc. 3nd. Int. Conference on Mechanics of Jointed and Faulted Rock(MJFR-3)*, Wien, 21-27

Sakurai, S. 1999, *Interpretation of the results of displacement measurements in geotechnical engineering projects, Proc. 5th Int. Sympo. on Field Measurements in Geomechanics-FMGM99*, Singapore, 13-18

Computer Methods and Advances in Geomechanics, Desai et al. (eds) © 2001 Balkema, Rotterdam, ISBN 90 5809 183 X

A damage mechanics model and its application to verification of an underground theater analysis

G. Swoboda
Department of Civil Engineering, University of Innsbruck, Austria

D. Zhao
Department of Civil Engineering, University of Innsbruck, Austria & Department of Civil Engineering, Lanzhou Railway Institute, People's Republic of China

Q. Yang
Department of Hydraulic Engineering, Tsinghua University, Beijing, People's Republic of China

ABSTRACT: This paper presents a damage mechanics method and its application to the prediction of damage of jointed rock mass . The anisotropic damage model is developed based on small-deformation behaviour during isothermal process. The damage variable is a second-order damage tensor which is assumed to describe the damage effect of jointed rock mass. Anisotropic damage is achieved by introducing a damage-dependent elasticity tensor, which is subject to the general principles of continuum mechanics: symmetry, positive definite, etc. The damage evolution law is based on the linear thermodynamic principle, and the related multiplier is determined by some consideration of endochronic theory. The elasticity tensor defined in the model is verified by employing a laboratory-used cement mortar model with different cases of joint distribution. The same model is analyzed by both the developed damage mechanics method and the normal FE method which uses joint elements to simulate jointed cracks. The results are compared and the elasticity behaviour of the jointed mass is concluded. Finally, an engineering example is presented as an application of the model.

1 INTRODUCTION

The design of structures related to jointed rock mass has made great progress in the theory of mechanical behaviour analysis. Much work has been performed in damage mechanics, either by the micromechanical approach or the phenomenological method. Micromechanical models have the distinct advantage of being able to capture structure details at the microscale and mesoscale and of permitting formulation of the kinetic equation for damage evolution based on the physical process involved (Krajcinovic 1983, Halm et al. 1996, Swoboda et al. 1999). Continuum Damage Mechanics(CDM), which employs some continuum variables to describe cracks and joints, provides another means of modelling jointed rock mass. Kawamoto et al. (1988) first suggested a CDM model for cracked rock mass. Thereafter, additional advances were presented for some cases of anisotropic damage analysis (Stumvoll et al. 1993, Chan et al. 1997).

This paper presents a CDM model for jointed rock mass and its application in engineering. The anisotropic damage model is introduced based on small-deformation behaviour during isothermal processes. The damage variable is a second-order damage tensor which is assumed to describe the internal structure of materials. Anisotropic damage is achieved by introducing a damage-dependent elasticity tensor which is subject to the general principles of continuum mechanics: symmetry, positive definite, etc. The model includes mechanisms, like damage, plasticity and dilatancy, in the process of static loading. Based on the physical observation that the plasticity flow restrains damage and the damage reinforces the plasticity flow, damage and plasticity are taken as two different dissipative mechanisms possessing different dissipative inequalities and loading surfaces. The damage evolution law is based on the linear thermodynamic principle. The related multiplier is determined by some consideration of endochronic theory. Due to limited space, the descriptions of anisotropic damage definition, damage elasticity and damage evolution law are emphasized.

Another predominant work of this paper particularly concentrates on damage elasticity verification for jointed crack systems, which are simulated in a cement mortar 3-D model. Different aspects influencing the mechanical response of jointed mass are considered, for example the variation of joint orientation, multi-set joint systems, etc. In each case the analysis uses both the damage model developed in this paper and a finite element model with a type of joint element to simulate the frictional slippage of the cracks. The results are compared and properties of the damage elasticity of the jointed mass are concluded.

An engineering example for an underground theater, a completely underground cavern structure existing in the geological environment of jointed rock mass, having dimensions of $31m \times 37m$ for the stage, a diameter of $40m$ for the auditorium, and a cavern height of $23m$, is presented. The proposed damage model is applied to predict the damage-induced mechanical behaviour. The results show that the model can predict not only the damage positions but also the case of damage evolution.

2 DAMAGE DEFINITION FOR JOINTED ROCK MASS

Cracks in geomaterials can be classified in three cases: (a) A large amount of cracks exist in the material, but the influence on mechanical behaviour from a single crack is negligible(Fig. 1a). The continuum mechanics theory is still suitable for this case. (b) The effect from a single crack is not negligible, and cracks are regularly distributed to form jointed rock mass(Fig. 1b,c). This is the case this paper deals with using continuum damage mechanics. (c) Figure 1d shows the faults in the rock mass, which is usually simulated by contact or a joint element in FE calculation.

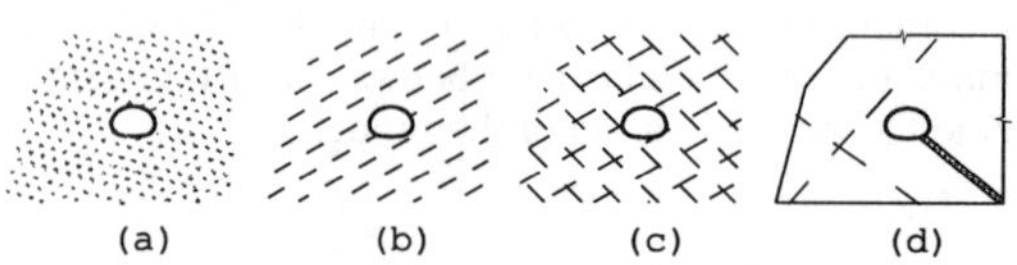

Figure 1: Tunnels in the rock mass with different discontinuities

2.1 *Damage definition*

Damage must be defined by a continuous homogeneous variable as an internal variable in continuum thermodynamics theory. In geology, each set of average distributed joints is usually described by three parameters (see Fig. 2), i.e., ω, the separation factor of one set of cracks, β, the angle between horizontal and the line of maximum dip of the joint plane and α, the angle between north and the horizontal projection of the line of maximum dip of the joint plane. To integrate these information, a second-order damage tensor is introduced as follows:

$$\Omega_{ij} = \omega n_i n_j \tag{1}$$

where $\mathbf{n} = [l, m, n]^T$ is the direction vector of damage tensor $\mathbf{\Omega}$ of the crack set. To position the joint plane in a rectangular coordinate system, an additional angle between x axis and north, γ is introduced (see Fig. 2). Thus, the spherical coordinates can be written as follows,

$$\left\{ \begin{array}{lcl} l &=& \sin\beta\cos[\pi - (\alpha+\gamma)] \\ m &=& \sin\beta\sin[\pi - (\alpha+\gamma)] \\ n &=& \cos\beta \end{array} \right. \tag{2}$$

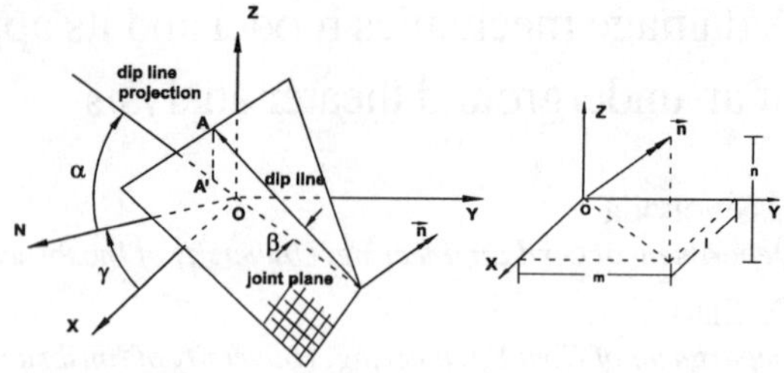

Figure 2: Description of spatial joint plane and spherical coordinate system

2.2 *Damage evaluation for multiple sets of joints*

For the damage state of multiple sets of joints, the global damage tensor can be obtained by employing the principle of energy equivalence, i.e., the energy dissipated by the global damage tensor is equivalent to the sum of the energy dissipated by the damage tensor related to each set of joins. Refer to the damage elasticity described in the latter section of this paper; the explicit formulation of calculating the global damage tensor is given in the following:

$$\mathbf{\Omega}_g = \mathbf{I} - \left[\sum_{n=1}^{N}(\mathbf{I} - \mathbf{\Omega}_n)^{-1} - (N-1)\mathbf{I}\right]^{-1} \tag{3}$$

2.3 *Active damage tensor*

In the jointed rock mass the damage is assumed to be tensile strain-oriented. To reflect the two cases of opening and closing of cracks as well as the different crack contact situations due to the surface roughness and even filling material in the cracks, an active damage tensor is defined on the basis of the principal strain as follows,

$$\tilde{\mathbf{\Omega}} = \mathbf{P}^+ : \mathbf{\Omega} \tag{4}$$

where $\mathbf{P}^+$ is a positive projection tensor(Ortiz 1985),

$$P^+_{ijkl} = Q^+_{ik} Q^+_{jl} \tag{5}$$

where

$$\mathbf{Q}^+ = \sum_{i=1}^{3} H(\varepsilon_\nu)\mathbf{p}^\nu\mathbf{p}^\nu \tag{6}$$

where $\mathbf{p}^\nu$ and ε_ν are the ν-th principal normal vector and value, respectively, of strain ε. $H(*)$ can be called a modified Heaviside function and is defined as follows,

$$H(\varepsilon_\nu) = \left\{ \begin{array}{ll} 1 & \text{if } \varepsilon_\nu > 0, \\ Hd & \text{otherwise.} \end{array} \right. \tag{7}$$

$Hd(0 \leq Hd < 1)$ is a material constant to reflect the properties of crack contact.

3 DAMAGE ELASTICITY

This section introduces the damage information on the elasticity of material to obtain the damage-dependent elasticity tensor **D=D(Ω)**. According to the general principles of continuum mechanics (Malvern 1969), the isotropic tensor function $D_{ijkl}(\Omega_{mn})$, which satisfies the symmetric condition, takes the form

$$\begin{aligned} D_{ijkl} \;=\; & A_1\delta_{ij}\delta_{kl} + A_2(\delta_{ik}\delta_{jl} + \delta_{il}\delta_{jk}) \\ & +A_3(\Omega_{ij}\delta_{kl} + \Omega_{kl}\delta_{ij}) + A_4(\Omega_{ik}\delta_{jl} \\ & +\Omega_{il}\delta_{jk} + \Omega_{jk}\delta_{il} + \Omega_{jl}\delta_{ik}) \\ & +A_5\Omega_{ij}\Omega_{kl} + A_6(\Omega_{ik}\Omega_{jl} + \Omega_{il}\Omega_{jk}) \\ & +A_7(\Theta_{ij}\delta_{kl} + \Theta_{kl}\delta_{ij}) + A_8(\Theta_{ik}\delta_{jl} \\ & +\Theta_{il}\delta_{jk} + \Theta_{jk}\delta_{il} + \Theta_{jl}\delta_{ik}) \\ & +A_9(\Theta_{ij}\Omega_{kl} + \Omega_{ij}\Theta_{kl}) + A_{10}(\Theta_{ik}\Omega_{jl} \\ & +\Omega_{ik}\Theta_{jl} + \Theta_{il}\Omega_{jk} + \Omega_{il}\Theta_{jk}) \\ & +A_{11}\Theta_{ij}\Theta_{kl} + A_{12}(\Theta_{ik}\Theta_{jl} + \Theta_{il}\Theta_{jk}) \end{aligned} \tag{8}$$

where $\Theta_{ij} = \Omega_{im}\Omega_{mj}$; $A_1, A_2, ..., A_{12}$ are functions of invariants of Ω_{ij} (Bazant 1983). To avoid the very complicated expression of $\boldsymbol{D}(\boldsymbol{\Omega})$ in practical use, an intermediate function ϕ_{ij} is introduced here to establish expression of the damage-dependent elasticity tensor,

$$\tilde{\boldsymbol{D}} = \boldsymbol{\phi} \cdot \boldsymbol{D}^0 \cdot \boldsymbol{\phi} \tag{9}$$

or

$$\tilde{D}_{ijkl} = \lambda\phi_{ij}\phi_{kl} + \mu(\phi_{ik}\phi_{jl} + \phi_{il}\phi_{jk}) \tag{10}$$

where $D^0_{ijkl} = \lambda\delta_{ij}\delta_{kl} + \mu(\delta_{ik}\delta_{jl} + \delta_{il}\delta_{jk})$ is the elasticity tensor of virgin material. According to the previous damage definition, it is clear that $\tilde{\boldsymbol{D}} = \boldsymbol{D}^0$ if $\boldsymbol{\Omega} = \mathbf{0}$ and $\tilde{\boldsymbol{D}} = \mathbf{0}$ if $\boldsymbol{\Omega} = \mathbf{I}$. Satisfying this condition, ϕ_{ij} is defined as,

$$\phi_{ij} = \delta_{ij} - m\tilde{\Omega}_{ij} - (1-m)\tilde{\Omega}_{im}\tilde{\Omega}_{mj}, \quad 0 \le m \le 1 \tag{11}$$

where m is assumed to be the only damage-related material constant and λ and μ are the Lame's constants of the virgin rock mass. The condition $0 \le m \le 1$ is needed to satisfy the positive definite requirement.

4 DAMAGE EVOLUTION LAW

In this section a damage evolution law is established in the framework of thermodynamics under isothermal process. It is assumed that damage evolution is completely determined by its conjugate force.

The potential function ϕ for damage evolution can be expressed as follows(Chow *et al.* 1992),

$$\phi^2 = \frac{1}{2}R_{ij}B_{ijkl}R_{kl}, \qquad B_{ijkl} = BI_{ijkl} \tag{12}$$

where R_{ij} is the conjugate force tensor, I_{ijkl} is the fourth-order unit tensor and B is a material constant governing the behavior of damage propagation. Based on the theory of thermodynamics, damage propagation is obtained from the relation:

$$\dot{\Omega}_{ij} = \dot{\lambda}\frac{\partial\phi}{\partial R_{ij}} = \dot{\lambda}\frac{B_{ijkl}}{2\phi}R_{kl} \tag{13}$$

where $\dot{\lambda}$ is a multiplier.

Introducing the Endochronic theory, which assumes that the rate of an internal variable is proportional to its conjugate force, one obtains the damage evolution equation in the intrinsic time space as

$$\frac{d\Omega}{dz} = C_{ijkl}R_{kl} \tag{14}$$

where C_{ijkl} is a constant coefficient tensor and dz is the infinitesimal increment of intrinsic time scale defined in elastic strain ε^e_{ij} space:

$$dz = \sqrt{de_{ij}de_{ij}}, \qquad e_{ij} = \varepsilon^e_{ij} - \frac{1}{3}\delta_{ij}\varepsilon^{kk}_e \tag{15}$$

From equations 13 and 14 the evolution law with intrinsic time can be written as follows,

$$\frac{d\Omega}{dz} = \dot{\lambda}\frac{B_{ijkl}}{2\phi}R_{kl} \tag{16}$$

Damage propagation is controlled through the threshold value of the damage potential ϕ^0, and the multiplier $\dot{\lambda}$ is defined as follows:

$$\dot{\lambda} = \begin{cases} 1 & \text{if } \phi > \phi^0, \quad \text{propagation;} \\ 0 & \text{if } \phi \le \phi^0, \quad \text{elastic.} \end{cases} \tag{17}$$

The equation calculating the increment of the damage tensor is obtained as

$$d\Omega = \begin{cases} \frac{B_{ijkl}}{2\phi}R_k dz & \text{if } \phi > \phi^0; \\ 0 & \text{if } \phi \le \phi^0. \end{cases} \tag{18}$$

The equation used to calculate the conjugate force under the given loading process is:

$$R_{ij} = P^+_{ijkl}\tilde{R}_{kl} = P^+_{ijkl}\left(\frac{1}{2}\varepsilon^e_{ij}\frac{\partial D_{ijkl}}{\partial\Omega_{mn}}\varepsilon^e_{kl}\right) \tag{19}$$

5 DAMAGE ELASTICITY VERIFICATION FOR A JOINTED-MASS MODEL

In this section the damage-dependent elasticity tensor is verified by employing a model with regularly distributed jointed cracks. The verification calculation is performed in two ways, namely with the finite element model using joint elements to simulate cracks and with the continuous model to analyze the material response based on the damage theory developed in this paper. Finally, the results obtained with these two models are compared with each other and with experimental results from a laboratory test performed under compression.

5.1 *verification model*

The model, which is designed to simulate the joint distribution of jointed rock mass, is geometrically described in Figure 3a. Figure 3b shows the relationship between the rectangular coordinate system of FE calculation and the geological information, α, β and γ, defining the spatial position of joints. γ is equal to zero in the case of this verification model. Figure 3c is the distribution of joints for the double-set case,

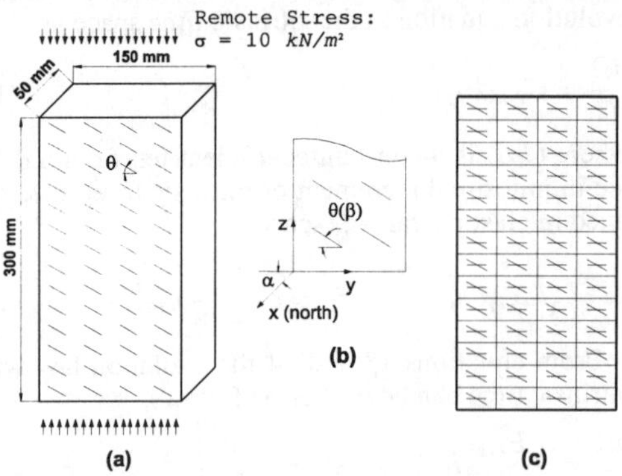

Figure 3: (a)Damage elasticity verification model, (b)relationship between geological coordinate and calculation coordinate systems, (c)distribution of joints in case of double sets

in which one set of joints remains horizontal and the other set rotates from 0° to 90°. All joint planes have the same separation factor, namely $\omega = 0.5$.

The problem type for the joint element model is plane stress. Damage calculation is based on a 3-dimensional model. Parallel calculations for both models are made under different joint orientations, e.g. $\theta = 0°, 15°, 30°, 45°, 60°, 75°, 90°$(Fig. 3).

5.2 *Mesh generation using cell method*

Because of the very tidy joint layout property and in order to save the time needed to generate a complicated mesh, a mesh-generation cell method was developed. Using this method a mesh for a regular cell is first generated. Next, the entire mesh is generated with the help of a cell mesh-generation program. Figure 4 shows the joint distribution for the case of $\theta = 30°$ and the cell mesh as well as the whole FE mesh.

5.3 *Material and load*

The material used for both the joint element model and the damage model is the same. It is rapid-hardening cement mortar with the following parameters:

Young's modulus E: 1000kN/m^2
Poisson's ratio ν: 0.1

The acting load is remote stress, $\sigma = 10kN/m^2$. The compression conditions are similar to those of the laboratory test.

Figures 6–7 give part of the damage verification results for the model of single-set joints. E is the normalized apparent Young's modulus of the damaged

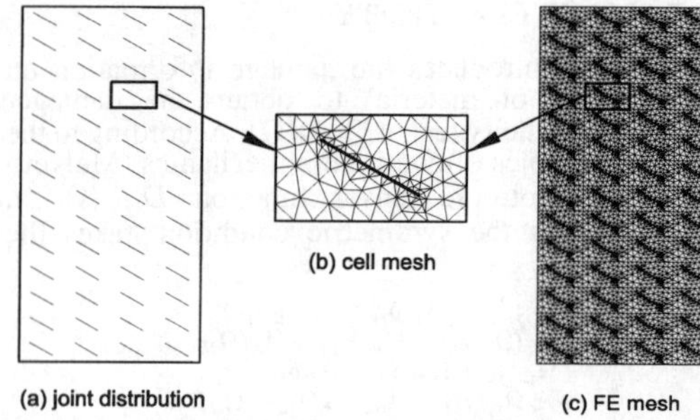

Figure 4: Joint element model

5.4 *Verification results*

Figure 5 shows deformations from the joint element model and the damage model.

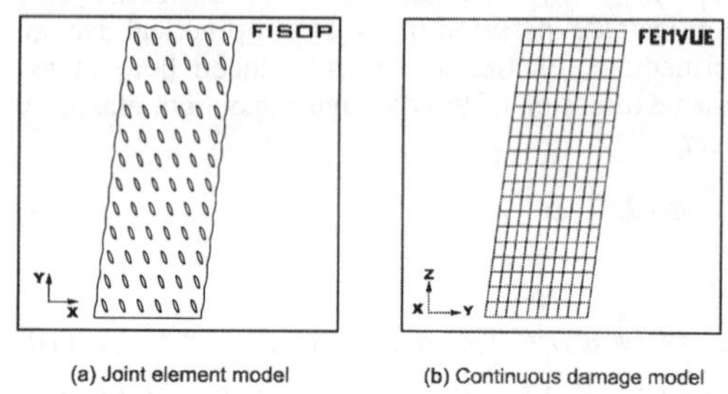

Figure 5: Comparison of displacements

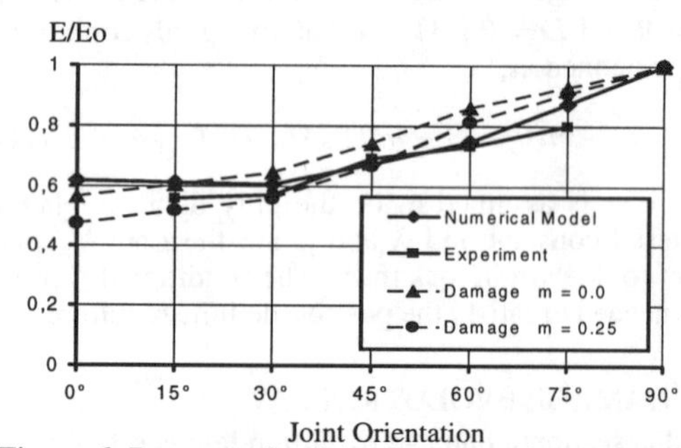

Figure 6: Damage elasticity (compression, Hd=1.0)

material, whereas E_o stands for the Young's modulus of the virgin cement mortar material. The results from the joint element model and the damage model show quite a consistent tendency of elasticity of damaged solid for different joint orientation(θ) relative to the load direction. To properly choose the material parameter *m*, the result of the damage calculation is close to that of the joint element model(see Figure 6). The damage results also show better consistency with the experiment result(Stumvoll 1991). Damage elasticity decreases while $m(0 \leq m \leq 1.0)$ increases.

Figure 7 shows that damage elasticity decreases while $Hd(0 \leq Hd \leq 1.0)$ increases. Hd, as described

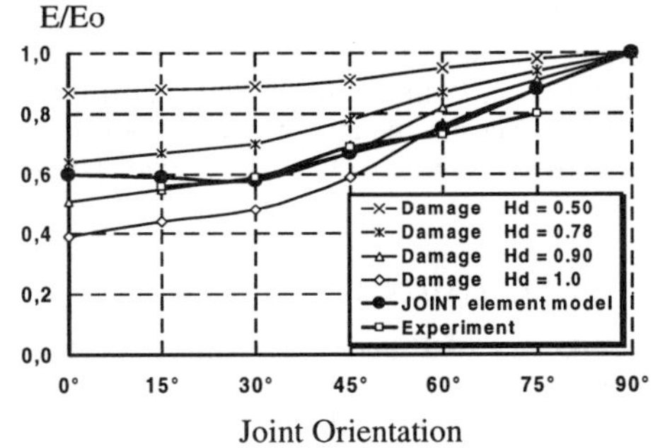

Figure 7: Damage elasticity (compression, m=0.5)

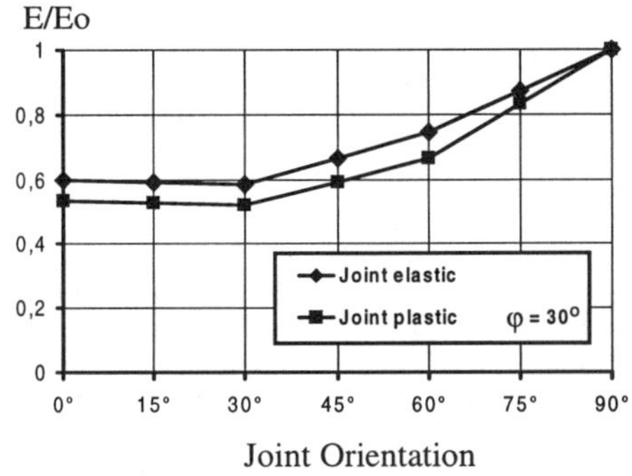

Figure 8: Impact of joint friction

in eqn. (7), is a material constant reflecting the contact effect of cracks. This means that the definition of Hd in the suggested damage model is reasonable.

In addition, the comparative research about the plastic behaviour of the material on the surface of joints is performed and the result is shown in Figure 8. It is seen that the difference in damage elasticity is about 7%.

Results of double-set joint models(Fig. 3c) are drawn in Figure 9. The damage elasticity is also seen to be quite well closed to the results of the joint element model in the case of $m = 0.5$.

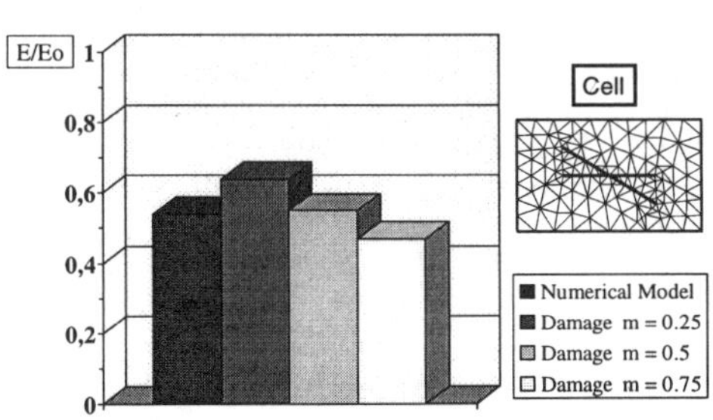

Figure 9: Damage elasticity for double joint system

6 DESIGN OF AN UNDERGROUND THEATER: ANALYSIS IN JOINTED ROCK

For the design of an underground theater for 900 persons in the heart of Linz structural analysis should consider the impact of the orientation of the joint system. The quality of the rock mass at the site of the theater excavation is very high, but a strong impact of the three main faults is expected. The layout of the total underground structure, including stage, auditorium, entrance hall, garage system and an existing tunnel system is shown in Figure 10. Due to the large layout of the entire structure three-dimensional models should be used to describe the impact of the underground excavation. The models should be based on

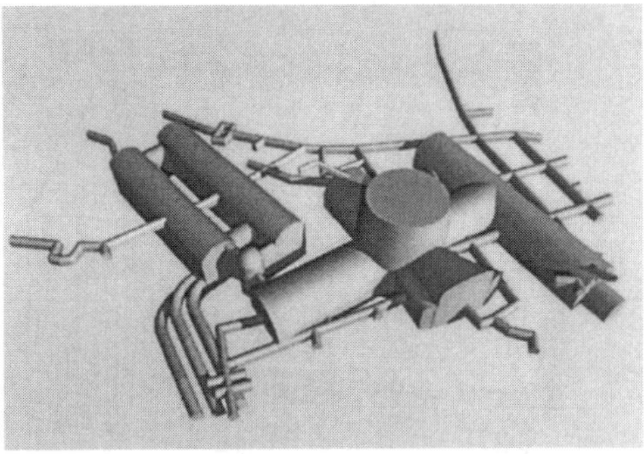

Figure 10: Layout of the underground theater

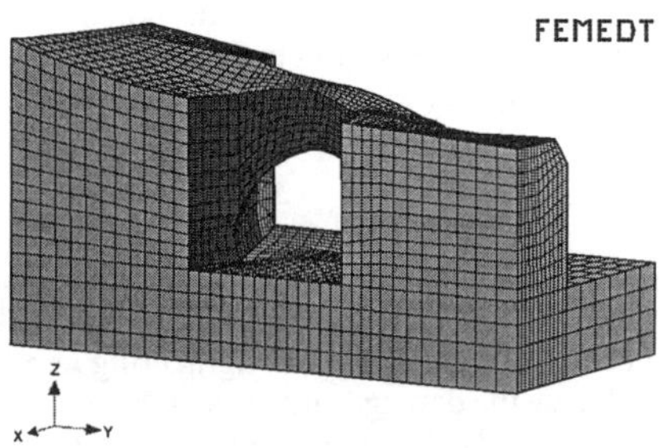

Figure 11: Finite element model for rock bridge

damage mechanics. For analysis of the "rock bridge" with a span of 40m above the auditorium the three-dimensional model shown in Figure 11 was used. It include modelling of the rock and support of the shotcrete lining. Excavation was simulated in four excavation steps. The global damage tensor is based on the orientation of the joint system and the orientation of the model, as shown in Figure 12. For interpretation of the design the plot of the damage tensor is of great importance. If the damage is zero, the joint system is closed. Any value above zero shows the amount

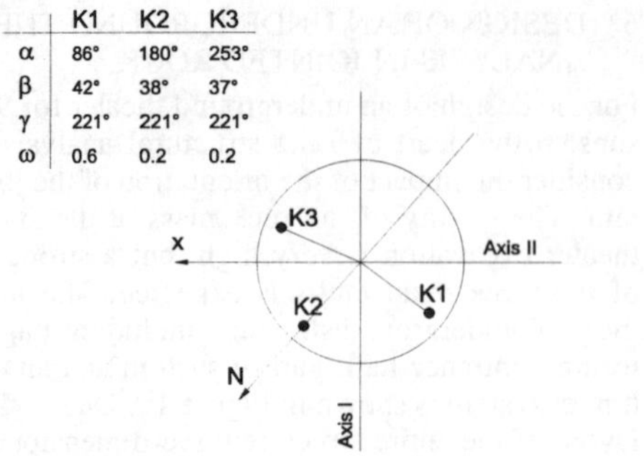

Figure 12: Geological information for 3 sets of joint system

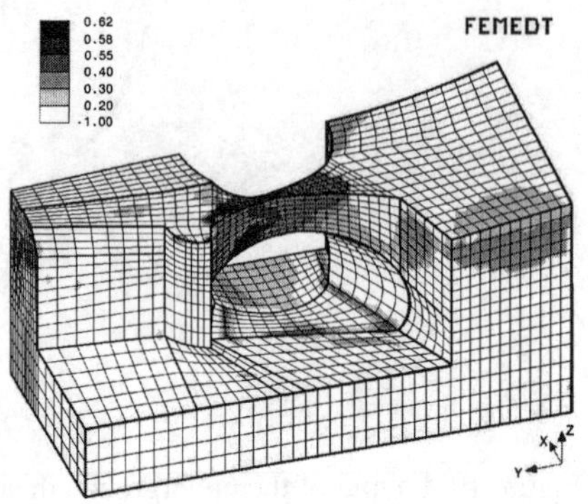

Figure 13: Damage in y direction

of opening in the joint system. From this result the layout of the bolt system can be designed. Figure 13 shows the damage in y direction. It is seen that most of the rock structure is completely undamaged.

7 CONCLUSIONS

The presented CDM model for jointed rock mass has been verified for various cases of joint distribution and successfully applied to an engineering application used to locate the damage and predict the damage evolution. A second-order damage tensor is defined and a method suggested with which the primary global damage tensor for multi sets of joints can be obtained. The geological information is used to determine the primary damage tensor of the jointed rock mass. A one-parameter damage elasticity tensor is defined as the function of the virgin material properties and the active damage tensor which is developed to consider the anisotropic properties of jointed rock mass. The effectiveness of the damage-dependent elasticity tensor is verified for a laboratory test model. The results show that the suggested model can describe the mechanical behaviour of jointed mass quite effectively. The numerical results are in agreement with the experiments. The damage evolution law is presented on the basis of thermodynamics and the endochronic theory. Application to the underground theater analysis has ensured that the model can predict not only the damage positions but also the case of damage evolution.

ACKNOWLEDGMENTS

The work reported here was supported by the Austrian National Science Foundation, Fonds zur Förderung der wissenschaftlichen Forschung, project No. S08004-TEC and the Austrian Industrial Research Promotion Fund, Forschungsförderungsfonds für die gewerbliche Wirtschaft, contract No. 800507 .

REFERENCES

Bazant, Z.R. 1983. Comment on orthotropic models for concrete and geomaterials. *Journal of Engng. Mech. ASCE.* 109:849-865.

Chow, C.L. & X.E. Chen 1992. An anisotropic model of damage mechanics based on the endochronic theory of plasticity. *Int. J. Fracture.* 55:115-130.

Chan, K.S., S.K. Bodner & A.E. Fossum 1997. A damage mechanics treatment of creep failure in rock salt. *Int. Journal of Dama. Mech.* 6:121-152.

Halm, D. & A. Dragon 1996. A model of anisotropic damage by ??microcrack?? growth; unilateral effect. *Int. J. Dama. Mech.* 5:384-402.

Kawamoto, T., Y. Ichikawa & T. Kyoya 1988. Deformation and fracturing behavior of discontinuous rock mass and damage mechanics theory. *Int. Journal of Numer. Anal. Meth. Geomech.* 12:1-30.

Krajcinovic, D. 1983. Constitutive equations for damaging materials. *J. Appl. Mech.* 50:355-360.

Malvern, L.E. 1969. *Introduction to the mechanics of a continuous medium.* Englewood Cliffs, NJ: Prentice-Hall.

Ortiz, M. 1985. A constitutive theory for the inelastic behaviour of concrete.*Mech. Mater.* 4:67-93.

Stumvoll, M. & G. Swoboda 1993. Deformation behaviour of ductile solids containing anisotropic damage. *J. Engng. Mech., ASCE.* 19:1321-1332

Stumvoll, M. 1991. Numerical model for discontinuous rock mass considering damage propagation under non-proportional and compressive loading. Dissertation. University of Innsbruck

Swoboda, G. & Q. Yang 1999. An energy-based damage model of geomaterials—II, deduction of damage evolution laws. *Int. J. Solids Structures.* 36:1735-1755.

1 Computer methods

Computer Methods and Advances in Geomechanics, Desai et al. (eds) © 2001 Balkema, Rotterdam, ISBN 90 5809 183 X

Developments and applications of discontinuum modelling to rock engineering

Giovanni Barla & Marco Barla
Department of Structural and Geotechnical Engineering, Politecnico di Torino, Italy

Masantonio Cravero
CNR, Torino, Italy

Corrado Fidelibus
Università di Taranto, Italy

ABSTRACT: The paper is to discuss the role of discontinuum modelling in rock engineering. This modelling procedure is more frequently used at present than in the past, when continuum modelling was nearly the only method adopted for design analysis. This trend is favoured by the advances in computational methods, which make it possible to use both 2D and 3D simulations of discontinua. Also important is the increased ability in the identification and quantification of design parameters for both the single element (intact rock and discontinuities) and system behaviour (jointed rock mass). Examples are taken from case studies in tunnelling and rock slopes, when the problem is to assess the stability conditions. Also considered is the role of discontinuum modelling in the study of flow through jointed rock masses.

1. INTRODUCTION

The use of computer-based numerical methods for the design analysis of rock engineering problems has been growing steadily over the past few decades. One reason for this is the gradual increase of interest of specialists in numerical methods toward the development of techniques and tools applicable to specific engineering problems in geomechanics. Another reason is the recognition by practitioners of the advantages which can be gained in dimensioning complex engineering structures by the adoption of these methods.

As the interest was moving toward an improved description of rock mass behaviour, the most effort was devoted initially to the developments and applications of methods for the analysis of continuum and equivalent continuum problems. While specialized equations and criteria were being derived and validated for the description of joint strength and deformability, discontinuum modelling was getting a growing attention in view of the solution of problems in practical situations, although continuum modelling continues to be to date the most frequently applied method. It is the aim of this paper to underline the advantages which can be gained by the use of discontinuum modelling in rock engineering practice.

2. TUNNEL ENGINEERING

The prediction of the rock mass response to the excavation of a tunnel is a complex engineering problem. The interest at the design analysis stage is to assess the stability conditions of the excavation in the "intrinsic state" (i.e. when no support/stabilization measures are installed) and when methods of tunnel excavation/construction and support are adopted.

The key of this process is the level of understanding achieved in describing the rock mass conditions (in terms of geological, geomechanical, in situ stress and hydrogeological parameters) and the ability to account for the fundamental components of rock mass behaviour, by using appropriate methods for the analysis of stresses and deformations in the rock mass around the tunnel and in its structural components (pre-support/pre-stabilization measures; primary and final support, etc.).

A number of methods are available for the stress analysis of tunnels, from the earliest closed-form solutions to the most recent numerical modelling methods. With the computational power today available at a reasonable cost, it is possible to solve increasingly sophisticated problems. In particular, with the advent of numerical methods, we have assisted to the development of techniques which have been conceived to model realistically the rock mass behaviour. This is quite different from the early start of geomechanics, when the methods of analysis and the solutions used were mostly taken from other engineering disciplines.

If we restrict our attention to numerical modelling methods and rigorous analysis of tunnelling problems, two procedures are applicable as follows:

- Equivalent continuum: the rock mass is treated as a continuum with equal in all directions input data for the strength and deformability properties, which define a given constitutive relation for the medium: elastic, elasto-plastic, etc.

- Discontinuum: the rock mass is represented as a discontinuum and most of the attention is devoted to the characterization of the rock elements and discontinuities (including major discontinuities and joint systems).

2.1 Continuum modelling

The use of continuum modelling in tunnel engineering makes it essential to simulate the rock mass response to excavation by introducing an equivalent continuum. The most common way to solve this problem, which seems to have gained wide acceptance, is to scale the intact rock properties down to the rock mass properties by using the empirically defined relations given by Hoek & Brown (1997). The next step is the adoption of the appropriate constitutive relations for the rock mass.

A number of computer-based numerical methods have been developed over the past few decades and provide the means for obtaining appropriate solutions to tunnel engineering problems in the framework of the equivalent continuum approach. Among the most frequently used methods are the finite element (FEM), the finite difference (FDM) and the boundary element (BEM) methods, as recently reviewed by Gioda and Swoboda (1999).

2.2 Discontinuum modelling

With the understanding that rock joints and discontinuities in a rock mass play a key role in the response of a tunnel to excavation (joints can create loose blocks near the tunnel profile and cause local instability; joints weaken the rock mass and enlarge the displacement zone caused by excavation; joints change the water flow in the vicinity of the excavation, Shen & Barton 1997), the use of discontinuum modelling has been gaining increasing attention in tunnel engineering.

As recently reviewed by Kawamoto & Aydan (1999), problems involving discontinuum modelling by numerical methods can be solved to a full extent by the distinct element method (DEM) and discontinuum deformation analysis (DDA). In particular, the first method as implemented in the UDEC and 3DEC codes is frequently being applied to tunnel engineering. However, it is to be recognized that discontinuum modelling in general is not being used for design analysis of tunnels as extensively as continuum methods and is considered to be a relatively "new" and "not yet proven" numerical method for design analysis.

2.3 Discussion

Accordingly, when investigating a particular problem at the design analysis stage in tunnel engineering the decision is to choose between continuum and discontinuum modelling of rock mass behaviour. This may be based on the analysis of the likely mechanisms (sliding along joints, opening of joints, block rotation and movement, etc.) which may influence the tunnel stability and the joint spacing relative to the size of the excavation. Consideration may be given to a suggested range of Q-values for which discontinuum modelling will be more appropriate than continuum modelling - Q $\cong 0.1 - 100$ (Barton 1998).

The use of numerical modelling in engineering practice, connected with the need to adopt modelling schemes (continuum versus discontinuum modelling) which are the most appropriate in order to analyse a given problem (provided that sufficient data are available), points out that "the modelling of the components, rock, rock joints and discontinuities is far more logical and relevant than present black box continuum models" (Barton 1999). The comparison shown in Figure 1 well demonstrates this point of view, especially if critical mechanisms of the physical problem under study are to be included in the analysis.

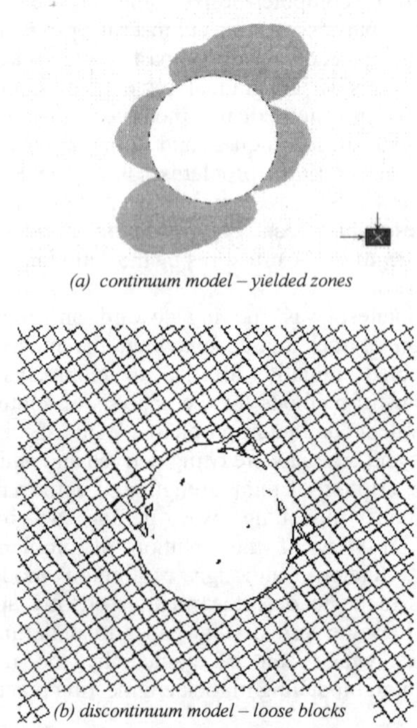

(a) continuum model – yielded zones

(b) discontinuum model – loose blocks

Figure 1. Comparison of (a) continuum ubiquitous joints and (b) discontinuum modelling results when analysing typical instability mechanisms around a TBM excavated tunnel in a weak rock mass.

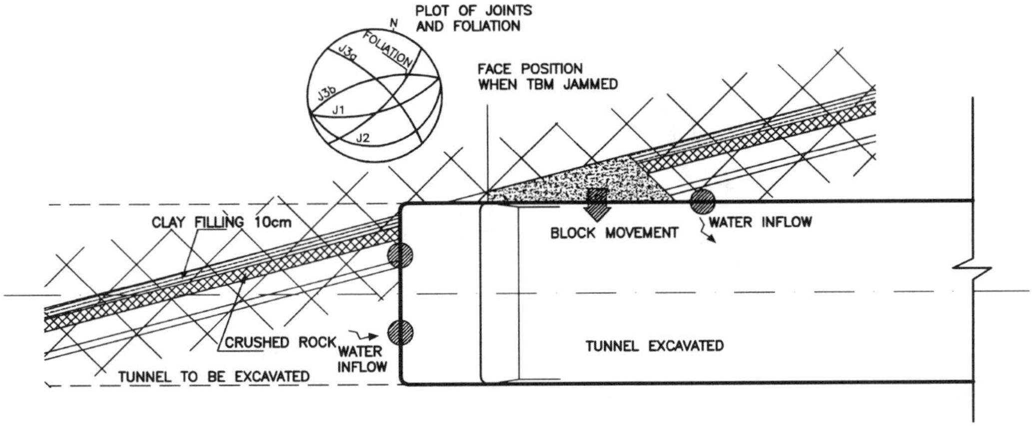

Figure 2. Sketch of the rock conditions at the face (plan view).

There is a relevant question which arises with reference to the use of discontinuum modelling in tunnel engineering and rock engineering, in general. This is in relation to the complexity of the discontinuum model to be used in order to be certain that the critical structural features have not been left out of the analysis (Hart 1993). From one side, the difficulty is in providing sufficient geological data; from the other side, the computer hardware requirements may exceed that available.

Based on the experience gained so far, a possible approach to this problem consists in using discontinuum modelling in connection with rock mass classification methods in a framework of cross-validation of the expected tunnel response to excavation. Once the specific model has been proven to be acceptable in a given rock mass condition, it may be improved and further validated during tunnel excavation. A set of guidelines, given as a function of Q values, can be used as a form of preliminary verification of a numerical model (Barton 1999).

2.4 Case example

2.4.1 Background information

The case study considers the crossing of a fault zone in a TBM excavated tunnel in the Susa Valley, near Torino (Italy). This tunnel (4.75 m diameter) is being excavated by an open TBM configuration through quartzitic micaschists under a cover which is to reach 800 m maximum. The sketch shown in Figure 2 gives a simplified illustration of the conditions at the tunnel face where the TBM became temporarily stuck as a consequence of overstressing and a 25 cm block movement of the right sidewall (Figure 3).

The quartzitic micaschists, which have been extensively tested both in the laboratory and in situ (Barla et al. 1999), are characterized by the presence of three to four joint systems including foliation. Based on geological mapping once the TBM could drill a few meters ahead of the section where it jammed, at least two sub-parallel discontinuities could be evidenced, the second of which (a fault with strike N66E and dip 83° to the S, which intercepts the tunnel axis) has a clay filling and gouge with aperture ranging from a few centimeters to more than a decimeter.

The rock mass conditions were estimated on a 7 m tunnel length, with RMR index equal to 31. According to a more complete Q-logging estimate due to Barton (1997), an extreme range of Q-values of about 0.007 ("exceptionally poor" - locally) to 0.3 ("poor") showed a weighted mean of all recordings of about 0.05. Water

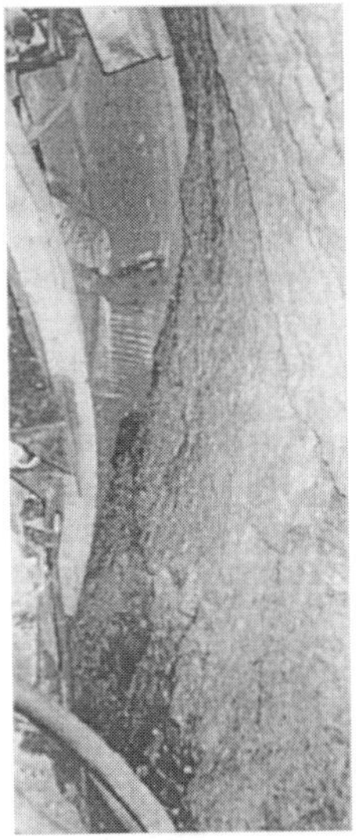

Figure 3. Photograph showing the head of TBM jammed with rock overstressing and block movement of the right sidewall.

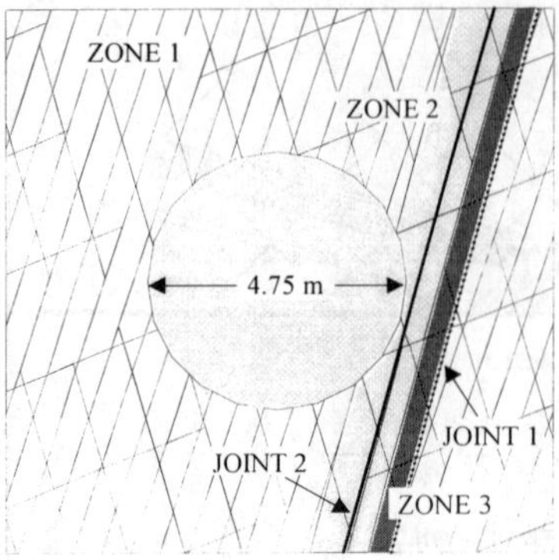

Figure 4. Detail of UDEC deterministic model showing regions with different material and joint properties.

is flowing through the joints at a temperature of 20°. No quantitative data have been recorded of the water pressure, which seems unlikely to be exceeding, with an overburden of 650 m, a maximum of 6 to 7 MPa (outside the tunnel drainage area).

2.4.2 *Discontinuum modelling*

With the understanding that discontinuities (major discontinuities and jointing) play a key role in the development of tunnel instability, the problem was investigated by adopting discontinuum modelling, with the 2D distinct element code, UDEC. The rock mass surrounding the tunnel was represented by two discontinuous models as follows.

2.4.3 *Deterministic model*

The rock mass was considered to be intercepted by three sets of joints, and foliation. Different spacing, degree of persistence and shear strength properties were introduced in the model so as to simulate the rock mass conditions away from the fault zone as illustrated in Figure 4. The joints were assumed to be Mohr-Coulomb joints, i.e. elasto-perfectly plastic joints. The blocks were treated as elasto-plastic material which follows Mohr-Coulomb criterion. The properties of rock blocks and joints are listed below:

Material Properties		Zone (Figures 4 and 5) 1	2	3
E_m	[GPa]	60	30	10
ν_m	[-]	0.25	0.35	0.35
c	[MPa]	34	6.0	2.8
ϕ	[°]	38	36	34

c = cohesion; ϕ = friction angle; E_m = Young's modulus; ν_m = Poisson's ratio.

Joint Properties		Zone (Figures 4 and 5) 1	2	3	Joint (Figures 4 and 5) 1 - 2
K_n	[GPa/m]	40	5×10^{-3}	10×10^{-3}	1.25×10^{-3}
K_s	[GPa/m]	4	5×10^{-4}	10×10^{-4}	1.25×10^{-4}
c	[MPa]	0.1	0	0	0
ϕ	[°]	33	22	22	22

K_n, K_s = normal and shear stiffness; c = joint cohesion; ϕ = joint friction angle.

2.4.4 *Discrete Feature Network (DFN) model*

A Discrete Feature Network (DFN) model was also created by using the procedures implemented in the FracMan code (Dershowitz et al. 1995). The results of geological mapping on the right wall of the tunnel formed the basis for defining the input data in terms of orientation, size and degree of fracturing for the joint sets J1, J2 and J3 a, b. These were superimposed with foliation and discrete features represented by the two sub-parallel discontinuities described above.

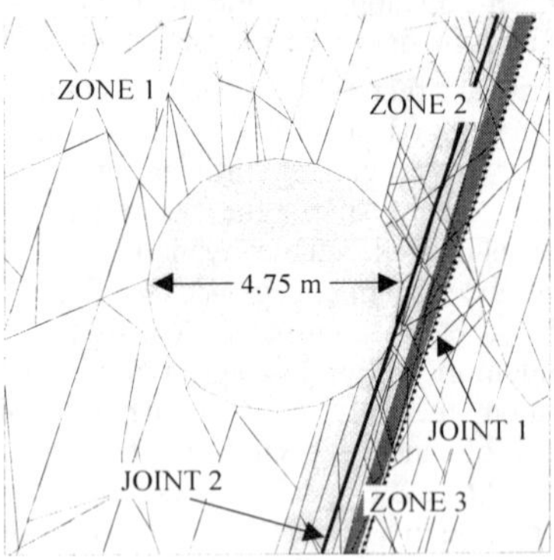

Figure 5. Detail of UDEC-DFN model showing regions with different material properties and joints.

The UDEC model was obtained by taking a cross section orthogonal to the tunnel axis through the 3D DFN volume. This network of 2D fractures is not directly amenable to UDEC analysis as the fractures need to be well connected and no isolated fractures can be handled by the code. At the same time, it was necessary to cross-validate the model against the rock mass conditions around the tunnel. This was carried out by increasing the block size in the original 2D network model both at the crown and on the left wall, to account for the generally better rock mass conditions in these zones.

A random process which cancels the fracture traces exceeding a given limit-length fixed was used in conjunction with the Set edge command which is available with the UDEC code. A typical model finally developed is shown in Figure 5. As for the deterministic model described above, the joints were assumed to behave as Mohr-Coulomb joints and the blocks were treated as elasto-plastic material. The input data are the same as listed above.

2.4.5 Results

The analysis were carried out for both the deterministic and discrete feature network models as shown in Figures 4 and 5 by simulating full face excavation without any support (i.e. in the intrinsic state). The initial state of stress in the model was taken to be as follows (Barla et al. 1999):

σ_1 = maximum principal stress = 21 MPa
σ_2 = minimum principal stress = 4.2 MPa
θ_1 = angle of σ_1 with respect to the vertical axis = 15°.

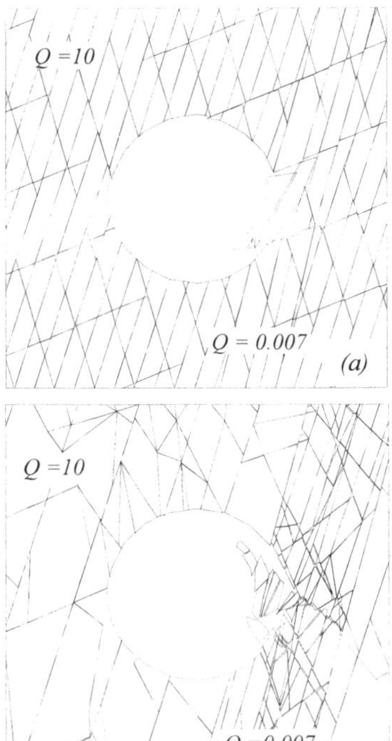

Figure 6. Block movements around the tunnel; (a) deterministic model, (b) DFN model.

The results obtained are plotted in Figure 6 a, b by showing the block movements around the tunnel. It is observed that both the deterministic and discrete feature network model capture well the overstressing conditions of the right wall of the tunnel. If one compares this pattern of behaviour with the actual conditions observed in the tunnel, the remarkable similarity is the block movement developing on the right wall, whereas stable conditions are experienced along the periphery away from the fault zone at the crown and left wall.

3. SLOPE ENGINEERING

Rock slope stability covers a vast area in geomechanics from geological, geomorphological and hydro-geological characterization to the assessment of strength-deformability properties of the rock mass and discontinuities at different scale. Of great relevance are both the understanding of external forces (i.e. groundwater influence, seismicity, external loading) and stabilization measures (i.e. drainage, reinforcement, anchoring, etc.) when applied in order to prevent and/or control instability from developing.

If the interest is centered on the identification of potential failure mechanisms, one is to look for modelling procedures which allow for the understanding of the in situ conditions and potential development of such conditions as function of the key parameters determining stability. However, modelling should not be created with the intention of simulating failure according to a preconcieved mechanism, although it need be recognized that the identification and analysis of failure modes (i.e. toppling failure, plane and wedge sliding failure, footwall buckling failure, etc.) still remains a very useful tool for the study of stability conditions of rock slopes (Hoek & Bray 1981, Sharp 1996).

By restricting to modern numerical modelling methods, one should recognize once again that he has the option to choose between equivalent continuum and discontinuum procedures, as already emphasized with reference to tunnel engineering. However, when dealing with rock slopes, the importance of discontinuities and description of their characteristics in controlling the relevant physical and mechanical processes points out the role of discontinuum modelling.

3.1 Continuum Modelling

The use of continuum modelling in slope engineering leads to the representation of the constitutive equations of the rock mass comprised of the discontinuities and rock material properties by using the empirical equations given by Hoek and Brown (1997) in order to obtain the rock mass properties of the equivalent continuum. However, great care is required, as clearly stated by the proponents of such equations: even rock masses which appear to be good candidates for the application of equivalent continuum modelling may be intercepted by discontinuities controlling stability conditions.

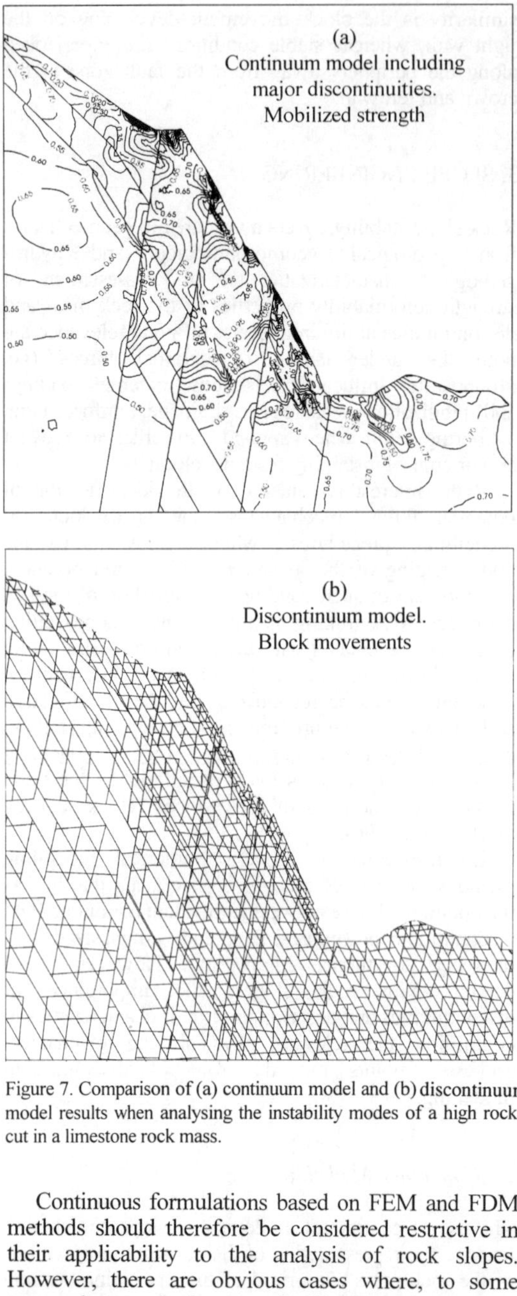

Figure 7. Comparison of (a) continuum model and (b) discontinuum model results when analysing the instability modes of a high rock cut in a limestone rock mass.

Continuous formulations based on FEM and FDM methods should therefore be considered restrictive in their applicability to the analysis of rock slopes. However, there are obvious cases where, to some extent, these methods are applicable, in particular when use is made of joint and interface elements which are available with them.

3.2 Discontinuum Modelling

The capability of discontinuum modelling to correctly simulate the behaviour of the rock mass, including the discontinuity response, and, to some extent, the interactive coupling of slope and hydraulic response, makes it ideal for design analysis of rock slopes, provided that adequate input data from field are obtained. Based on currently available numerical codes, this implies the use of discrete element codes such as UDEC.

As for the case of application in tunnel engineering, also for slopes the use of discontinuum modelling is complex in terms of the input data and experience gained in the analytical representation of physical processes. Perhaps, it is fair to say that to some extent one is forced into a position of implicit trust in the model (Sharp 1996). At the same time, there are obvious limitations as the analyses predict the stress-deformation responses of both the blocks and discontinuities, and are not able to study overall equilibrium and incorporate specific failure modes.

3.3 Discussion

The option of using continuum or discontinuum modelling of rock mass behaviour for design analysis of rock slopes is available. As already observed, the application of continuum representations in the case of rock slopes need to incorporate in the model the major discontinuities, in order to allow discrete displacements to occur along them.

If the input data are available, in terms of the representation of all salient geological structures, including rock joint systems, there is an obvious implication in using discontinuum modelling for a better representation of rock slope problems. The case of a very high cut shown in Figure 7, with an overall height of about 120 m, gives a clear indication of the advantages gained by using discontinuum modelling compared with continuum modelling, although incorporating the influence of major geological structures (Barla et al. 2000).

3.4 Case example

3.4.1 Background information

On November 5, 1994, during the severe rainfall event which hit the Piemonte and Liguria regions in northern Italy, a plane shear slope instability occurred near the town of San Raffaele Cimena (Torino), with dramatic consequences as shown in Figure 8. The slope movements occurred by translational sliding and resulted in collapse of houses and loss of lives (Barla et al. 1998).

The landslide covers a total area of 4000 m^2, with a volume of mobilized rock (sandy marl and weak sandstone layers) estimated to be in the range of 20000 m^3. The instability resulted in sliding of rock blocks with movement down-dip along the bedding. The dip of the bedding planes in the area ranges from 15° to 20°, with a dip direction of 250°-270°.

Figure 8. Aerial photograph showing the San Raffaele Cimena landslide site. Also shown is the trace of the E-W cross section given in Figure 9.

Plane shear instabilities at the site are essentially governed by the unfavourable orientation of the bedding planes, which occur nearly parallel to the slope surface, and the shear strength properties of these same planes. In the case under study the translational failure took place on one sliding plane down-dip along the beddings. This was clearly ascertained based on the investigations undertaken following the event which included detailed geological mapping and drilling of boreholes in the sliding area and in the near surrounding. As illustrated in Figure 9, the most likely plane of shear failure, dipping at 18° to the West, could be inferred.

Two additional factors could be shown to have an important influence on the plane shear slope instability:

1) undercutting at the toe of the slope, associated with the lateral release of the down-dipping bedding planes, due to a road cut running from the top down, in the landslide area;

2) the exceptionally heavy rainfall event which determined a significant rise of the groundwater table within the slope, also favoured by the inability of the drainage system of the house complex uphill to avoid water to flow down-slope without control.

Laboratory tests were conducted on rock material and discontinuities, including saw-cut surfaces and bedding planes. The samples for testing were obtained from cores and included fine grained sandy marl and weak sandstone (Barla et al. 1998). The direct shear tests on sawcut surfaces taken from sandy marl and weak sandstone gave the basic friction angle to be in the range from 30° to 33°, with the minimum value holding true for marl. A limited number of tests could be conducted on the bedding planes with the normal stress ranging between 0.1 and 1.0 MPa. These tests in two cases involved discontinuities which did not undergo any previous displacement, whereas in three cases these clearly pertained to the sliding plane (the tests were conducted on samples taken from rock blocks and cores near the sliding plane): the residual friction angle was found to be 22°.

3.4.2 *Discontinuum modelling*

With the main purpose to investigate the instability phenomenon a study was undertaken with the use of numerical modelling. The distinct element method (DEM) was adopted to simulate the weak discontinuum and computations were performed with UDEC. The

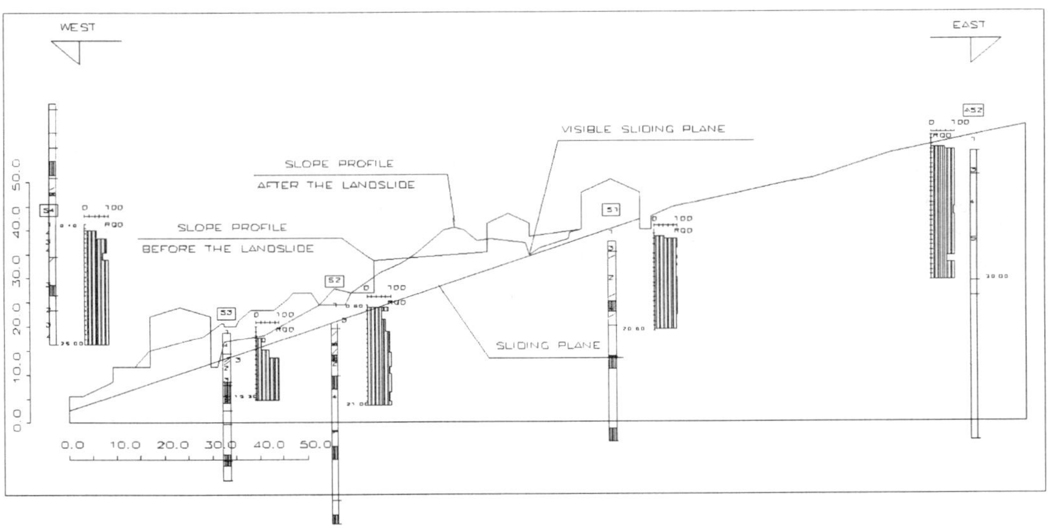

Figure 9. San Raffaele Cimena landslide: E-W cross section through the landslide area showing the plane of shear failure and the slope profiles prior and subsequent to the landslide.

E-W cross section of Figure 9 was taken as representative of the slope, with the purpose to simulate both the conditions prior and subsequent to the shear plane instability. As shown in Figure 10 a, the slope section was discretized into rectangular blocks subdivided by the bedding planes and subvertical joints.

The rock blocks were modelled as linear elastic and isotropic. The shear behaviour of the discontinuities (bedding planes and subvertical joints) was modelled by using the continuously yielding option which is available in the UDEC code in order to simulate a shear stress-shear strain curve of the softening type. To this end the Joint Initial Friction (JIF) and Joint Roughness (JIR) parameters were as shown in the table below, where also given are the other parameters used in the analyses.

Material Properties		Discontinuity Properties	Bedding planes	Subvertical Joints
E_m [GPa] =	2.0	ϕ_r [°] =	22	22
ν_m [-] =	0.25	JIF =	0.62	0.62
c [MPa] =	0.5	JIR [m] =	0.001	0.001
ϕ [°] =	40	K_n [GPa/m] =	8	8
		K_s [GPa/m] =	0.8	0.8

where: E_m = Young's modulus, ν_m = Poisson's ratio, c = cohesion of intact material, ϕ = friction angle of intact material, ϕ_r = joint residual friction angle, JIF = joint initial friction parameter, JIR = joint roughness parameter, K_n = normal stiffness along joint, K_s = shear stiffness along joint.

Modelling was performed by considering two different geometrical conditions for the slope:

(a) model with no cut in the toe zone;

(b) model with cut in the toe zone.

In each case the following computational steps were considered:

(1) consolidation under self-weight loading, to reach an initial equilibrated configuration;

(2) water head in the rock discontinuities;

(3) iterative analysis, to reach a final configuration.

The interest was centered on the influence of the water head in the rock dicontinuities by assuming this to change from 50, 70 per cent water head up to full saturation.

3.4.3 Results

The results obtained for cases (a) and (b) above are illustrated in Figures 10 and 11 by plotting the displacement vectors of rock blocks along the slope.

It is clear from Figure 10 that, with the water head in excess of 70 per cent above the sliding plane and with no cut in the toe zone, some form of instability develops only on the upper layers of the slope, with a maximum displacement down-slope of the order of

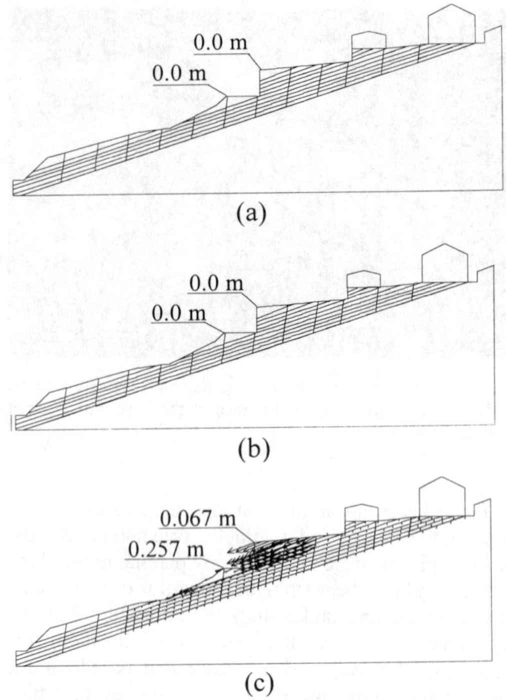

Figure 10. UDEC results showing block translational sliding along the slope. No cut in the toe zone; simulated conditions of: (a) 50 per cent, (b) 70 per cent, (c) 100 per cent water head.

0.257 m, when a full saturation condition is simulated. This trend of behaviour motivated the analysis of translational sliding by introducing the effect of undercutting at the toe zone, with simulated conditions of 70 per cent water head.

One may observe from Figure 11 a to c that in this case the instability phenomenon is initiated at the toe and develops up-slope progressively. As shearing occurs along the bedding planes, the subvertical joints open and block sliding takes place. This type of behaviour reproduces well the actual conditions in the field. It is relevant to remark that shearing penetrates at the depth of 5 m at the toe, as direct consequence of the slope cut and absence of passive reaction.

4. HYDRAULIC BEHAVIOUR OF JOINTED ROCK MASSES

The analysis of the hydraulic behaviour of jointed rock masses implies the study of fluid flow through joints, which requires an appropriate understanding and modelling of joint network characteristics. This understanding is linked directly, in hydrologic and transport problems, to geo-structural exploration at given scales and borehole testing for joint transmissivity, along with the interpretation and

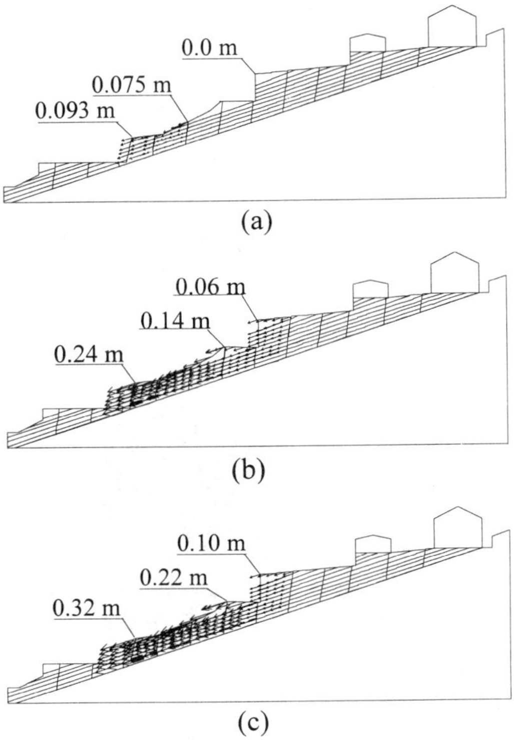

Figure 11. UDEC results showing progressive block translational sliding along the slope. Cut in the toe zone; simulated conditions of 70 per cent water head.

synthesis of the pertinent joint parameters. Joint modelling and flow analysis should be carried out by joint system models (Dershowitz and Einstein, 1988).

A joint system model should be able to reproduce the essential features of natural jointing, i.e. major disjunctive structures (faults, shear zones, ...) and medium to small size joints. Although the hydraulically effective branches of a joint system could be a small portion of the geometrical system itself, this results from a connectivity or a flow analysis performed on the whole joint network. Having established that the joint network is the fundamental precursor of any flow analysis, rules must be followed in order to construct a synthetic joint network (SJN). A class of joint network simulators is of the stochastic-geometric type, which requires the parameters of the joint location process and distributional forms for joint attitude, size, and transmissivity. Correlation among parameters and functional relations with depth can also be introduced (Dershowitz 1992).

The generated joint model statistically conforms to the characteristics of the actual joint network and reflects lack or evidence of spatial correlation. Other rules can be derived from the knowledge of the genetic processes and allow grouping of joints coming from a single deformation event. A structural trend can also be reproduced by iterating geometrical transformations giving raise to a fractal pattern (Hobbs 1976). Furthermore, jointing style, emerging from a genetic-mechanical process can be used as an indirect improvement for the stochastic-geometric modelling, in the sense that a hierarchical dependence between joints is allowed by generating primary and higher order joint sets (Barton 1995).

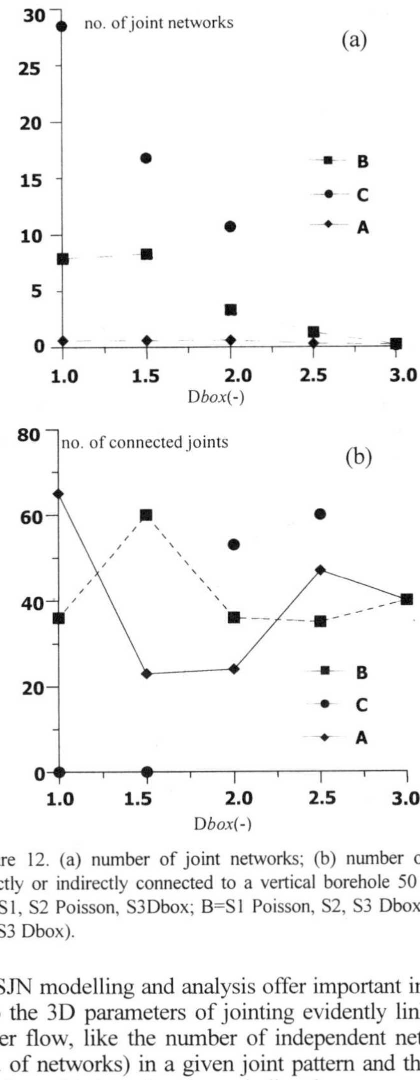

Figure 12. (a) number of joint networks; (b) number of joints directly or indirectly connected to a vertical borehole 50 m long (A=S1, S2 Poisson, S3Dbox; B=S1 Poisson, S2, S3 Dbox; C=S1, S2, S3 Dbox).

SJN modelling and analysis offer important insights into the 3D parameters of jointing evidently linked to water flow, like the number of independent networks (no. of networks) in a given joint pattern and the total number of joints directly or indirectly connected to a borehole (no. of connected joints). Figure 12 (obtained by means of some FracMan realizations of a joint system made of three strictly orthogonal joint sets: S1 horizontal, S2 trending NS, S3 WE) shows the trend of the two parameters above when the location process of joint centers is changed from pure Poisson to fractal box having different strength of clustering. All the other joint set parameters (joint radius=15 m, joint intensity 0.1 m^2/m^3 for each one set) are constant for all the realizations.

The synthetic joint network that is the output of stochastic-geometric simulators is only one of the possible realizations coming from populations of random variables. In this sense a SJN collects the intrinsic variability of the rock mass jointing and the answers from just one SJN cannot be conclusive about the hydraulic behavior of the joint system. The analyses should explore more than one rock mass picture, and this can be accomplished by a Monte Carlo procedure.

4.1 Hydrologic model

A SJN can be submitted to flow analysis provided the transmissivity parameter and storativity (when unsteady flow is considered) are given to each joint. The flow model is purely hydrological when the flow regime and the volumetric deformation of the jointed rock mass, accompanying the pressure variation of the fluid, are uncoupled. In this hypothesis the flow regime is described by the diffusion equation. In the general 3D case the SJN is very complex and even 100000 or more significant joints must be modelled when an aquifer has to be analyzed. The resulting computational task could be not practicable, especially when flow analyses are part of a stochastic simulation procedure involving several realizations of the SJN. These technical aspects are crucial and different solution schemes have been provided using FEM or BEM.

Apart from the relative merits of the different schemes, some forms of simplifying assumptions need be introduced when very large SJN are considered. The following three categories of approximations can be handled: the equivalent porous medium (EPM); the stochastic continuum medium (SCM); the equivalent pipe network (EPN). All these methods provide a consistent reduction of the computational effort. The application of the first method depends on the existence of a representative elementary volume (REV), based on one or more generated networks. The other two are in general applicable but introduce a certain degree of approximation.

4.2 Examples of SJN in flow and pollutant analyses

Two purely explanatory examples using the SJN concept for flow and pollutant analyses are described in the following. Both examples refer to a cubic domain (1km edge), confined by an impermeable formation except for the N & S sides, where a 10 m hydraulic head is applied (Figure 13). A 100 m deep well P is located in the middle of the domain, with a steady state discharge of 1 l/s. The SJN is made of 4 joint sets, generated according to the parameters given below:

	no. joints	ϑ_m	φ_m	ϑ_{st}	φ_{st}	L_1	L_2	T_m
1	200	0.	80.	1.	1.	700.	700.	1.e+06
2	400	45.	10.	5.	5.	50.	50.	1.e+05
3	400	135.	5.	5.	5.	50.	50.	1.e+05
4	400	45.	50.	5.	5.	50.	70.	5.e+05

φ_m, ϑ_m(°) average plunge, azimuth of the pole, φ_{st}, ϑ_{st} standard deviation; L_1 , L_2 , (m) average sides of the rectangular joints, T_m(m^2/s) average transmissivity

(a)

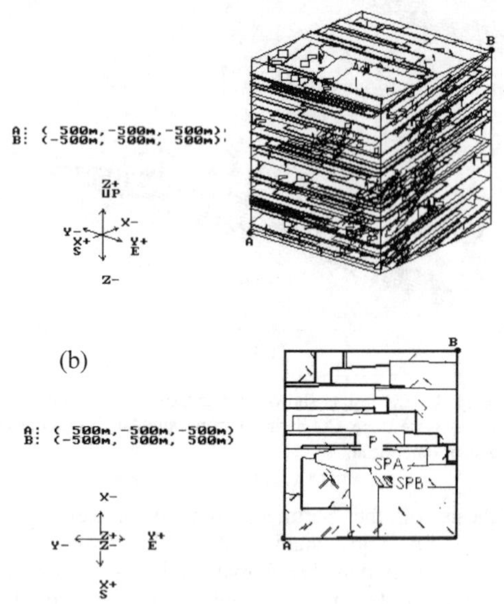

Figure 13. View of a SJN in the cubic domain (a); plan showing the well P and boreholes SPA and SPB (b).

Joint attitude (mean θ_m, φ_m) is bivariate normal. The joints have rectangular shape, with lognormally distributed edges L_1 , L_2 (standard deviation 1/10 of the mean size) and the location process of the 3 first sets is Poisson, being the fourth set fractal with a density of 1.5. Transmissivity also is lognormal with standard deviation 1/10 of the mean value.

4.2.1 Case example 1

This case example deals with the appraisal of the steady state heads observed in the central well P and in two 20m deep boreholes, located 100 m (SPA) and 200 m (SPB) apart from the well. A simulation consisting of 200 SJN realisations, carried out with MAFIC (Miller et al. 1995), allows the frequency histograms (Figure

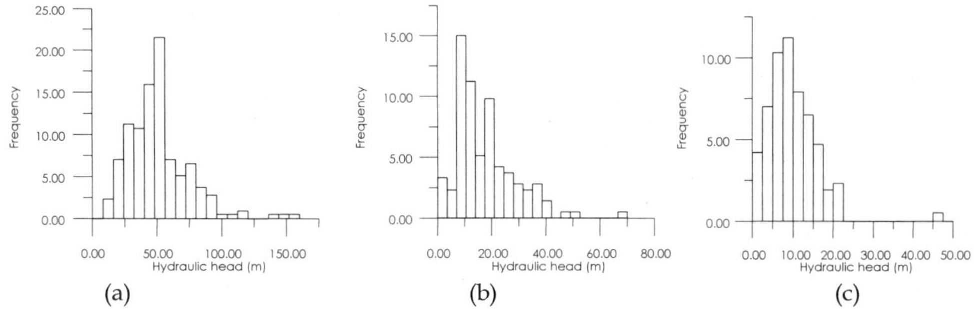

Figure 14 Variation of the hydraulic head (m) in comparison to the hydrostatic condition, for the well P (a) and the boreholes SPA (b) and SPB (c).

14) of the hydraulic head variations to be obtained in comparison to the hydrostatic condition. SPB is not connected to the well 43% of the time, SPA 35%, while the central well is not connected to the head boundaries 3% of the times, being the average head variations 49.4 m for the well, 11.5 m for SPA and 5.6 m for SPB.

The asymmetry shown by the histograms, where the classes below the mode show higher frequencies, can be explained by the presence of the fractal joint set. Namely in absence of this set one can reasonably expect an almost symmetric behaviour around the mean. On the contrary the fractal set produces joint clusters that, in most cases, act favourably improving the connectivity among joints and a greater rock mass permeability. In other words the isolated clusters have no effect on water percolation, whereas the connected clusters increase the percolation efficiency of the whole joint system: the net effect is the histogram asymmetry.

4.2.2 Case example 2

With the second example, the intent is to evaluate the pollution potential at the well when the borehole SPA releases, for 7 days, 0.01 l/s containing 10 g/l of pollutant. The pollution analysis (particle tracking or purely convective approach - Foxford et al. 1997) refers to a one month period of production at the well and only 15 cases from the fully connective SJN's have been chosen. The sample is small but this evaluation was made only for demonstration purposes. The flow pattern used is that of the steady state solution and the result of the pollutant concentrations at the well for the 15 models is given in Figure 15.

It is noted that the concentration values reflect the variability of the geometrical properties of the joint system, that delays or anticipates the convective flow of pollutant according to the joint connectivity. The network IB (the trend is not entirely represented) conveys 80 mg/l after 7 days to the well. This anomalous behaviour probably depends on the interposition among the pollutant source SPA and the production well of a cluster of conductive joints.

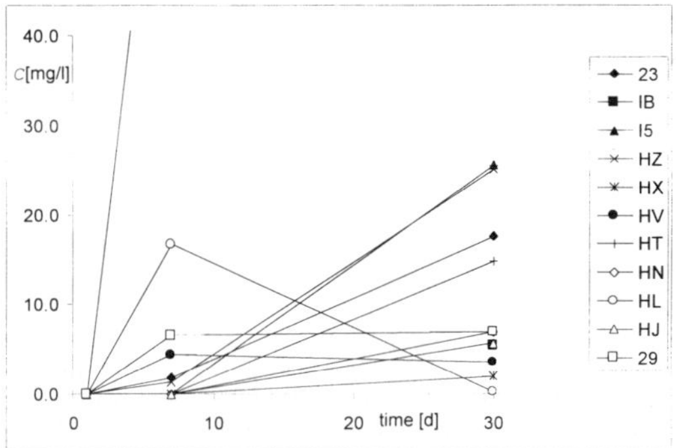

Figure 15. Pollutant concentration c (mg/l) versus time at the production well P.

Also to note is that the quality of the results depends on the reliability of the fracture data and that the model predicts a variability of the response that a porous-like medium conceptualization would not be able to show.

5. CONCLUDING REMARKS

This paper has endeavoured to discuss the importance of discontinuum modelling as a design analysis aid in rock engineering. Following a presentation of the issues faced when applying this modelling procedure to tunnel engineering, slope engineering and analysis of the hydraulic behaviour of jointed rock masses, case examples were briefly discussed. These were intended to show the merits of discontinuum modelling in rock engineering practice by describing the differences in the level of information which can be gained by the use of discontinuum representations of rock mass, in particular if compared with continuum modelling.

The use of a model which accounts for the essentially discontinuous nature of the rock mass in engineering design analysis has the great advantage that one need to consider the components of its real behaviour (rock blocks and discontinuities), including the coupling with ground water conditions. This is a consequence of the increased ability to describe the rock mass as a discontinuum from the geometrical, physical and mechanical points of view. It is not to say that equivalent continuum modelling should be ruled out: there are cases where there is no valid substitute to it, as for the analysis of overstressed and weak rock masses. Also, if used with judgement, continuum modelling with the inclusion of discrete features, such as major discontinuities, is essential for comparative calibration of discontinuum modelling results, in the understanding of the overall response of the engineering structure being analysed.

REFERENCES

Barla G., Barbero M., Castelletto M. & Maderni G. 1998. A case of plane shear slope instability during the severe rainfall event of 4-6 November 1994 in Piemonte (Italy). *2nd International Symposium on Hard Soils and Soft Rocks, Naples (Italy), 12-14 October 1998.*

Barla G., M. Barla & L. Repetto 1999. Continuum and discontinuum modeling for design analysis of tunnels. *9th Int. Congr. on Rock Mechanics, Paris (France), 25-28 August 1999.*

Barla G., M. Barla, U. Rabagliati, L. Repetto 2000. Reprofiling of a high rock cut slope following instability. *To be published.*

Barton, C. C. 1995. Fractal analysis of scaling and spatial clustering of fractures. *Fractals in the Earth Sciences*. C. C. Barton & P. R. La Pointe eds., Plenum Press.

Barton N. 1997. Pont Ventoux Hydropower Project. Tunnel F2 - Pont Ventoux. Assessment of geological conditions and required support for ch 2350-4000. Report to Nocon. 25 October 1997.

Barton N. 1998. Quantitative description of rock masses for the design of NMT reinforcement (Special Lecture 1*). Int. Conf. on Hydro Power Development In Himalayas, Shimla (India) 1998.*

Barton N. 1999. General report concerning some 20th century lessons and 21st century challenges in applied rock mechanics. *9th Int. Congr. on Rock Mechanics, Paris (France), 25-28 August 1999.*

Dershowitz W. S. 1992. Interpretation and synthesis of discrete fracture orientation, size, shape, spatial structure and hydrologic data by forward modelling. *ISRM Int. Conf. on Fractured and Jointed Rock Masses, Lake Tahoe (CA-USA), 1992.*

Dershowitz, W., G. Lee, J. Geier, T. Foxford, P. LaPointe & A. Thomas 1995. *Interactive discrete fracture data analysis, geometric modelling, and exploration simulation.* User documentation, version 2.5, Golder Associates Inc.

Dershowitz W.S. & H.H. Einstein 1998. Characterizing rock joint geometry with joint system models. *Rock Mechanics and Rock Engineering*, 21, 21-51.

Foxford, T., E. Sudicky, W. Dershowitz, D.A. Shuttle & T. Eiben 1997. *PaWorks, pathways analysis and solute transport.* User documentation, version 1.4. Golder Associates Inc.

Gioda G. & G. Swoboda 1999. Developments and applications of the numerical analysis of tunnels in continuous media. *Int. J. Numer. Anal. Meth. Geomech.*, 23, 1393-1405.

Hart R.D. 1993. An introduction to distinct element modeling for rock engineering. *Comprehensive Rock Engineering.* Volume 2. Analysis and Design Methods (Hudson J.A., editor). 245-263. Pergamon Press, Oxford (UK).

Hobbs B.E., W.D. Means & P.F. Williams 1976. *An outline of structural geology*. Wiley Int. Ed.

Hoek E. & J.W. Bray 1981. *Rock slope engineering.* 3rd Edn. Institute of Mining and Metallurgy. London.

Hoek E. & E.T. Brown 1997. Practical estimates of rock mass strength. *Int. J. Rock Mech. Min. Sci.* Vol. 34, 1156-1186.

Kawamoto T. & O. Aydan 1999. A review of numerical analysis of tunnels in discontinuous rock masses. *Int. J. Numer. Anal. Meth. Geomech.*, 23, 1371-1391.

Miller, I., G. Lee, W. Dershowitz & G. Sharp 1995. *MAFIC, matrix/fracture interaction code with solute transport.* User documentation, version b1.5, Golder Associates Inc.

Sharp J.C. 1996. Present understanding and predictive capability of the stability conditions for very high rock cuts. *Eurock '96, Turin (Italy).* Special lectures, Vol. 3.

Shen B. & N. Barton 1997. The disturbed zone around tunnels in jointed rock masses. *Int. J. Rock Mech. Min. Sci.* Vol. 34, pp. 117-125.

Computer Methods and Advances in Geomechanics, Desai et al. (eds) © 2001 Balkema, Rotterdam, ISBN 90 5809 183 X

Computational approaches to jointed rock mass mechanics

W.G. Pariseau
Department of Mining Engineering, University of Utah, USA

ABSTRACT: Finite element (FEM), distinct element (DEM) and discontinuous deformation analysis (DDA) are the three principal types of numerical approaches to mechanics of jointed rock masses. Special elements and interfaces were developed in FEM to accommodate the important role of joints in rock mass mechanics. Particle flow code (PFC) and numerical manifold method (NMM) are recent improvements to DEM and DDA, all of which have the advantage of allowing independent rigid body motion of blocks defined by intersecting joints and, thus, accommodating large displacements at joints. The main disadvantage of these approaches is the limited number of joints that can be handled in comparison with reality. The use of an equivalent properties formulation within the FEM approach is an alternative worth considering, provided elements need not be representative volumes. In this case, graded meshes may be used that insure numerical quality while accommodating "geology". Joint failure mechanisms are retained using influence functions.

1 INTRODUCTION

Natural rock masses generally contain many structural and material discontinuities such as joints, fractures, faults, bedding planes, dikes, sills and geologic contacts. Structural discontinuities, "joints" for brevity, make the rock mass more compliant and weaker than intact rock between the joints. Joints are not indefinitely continuous; intact rock bridges are generally present between joints, so rock masses are usually stiffer and stronger than joints alone. Changes in rock types across geologic contacts also complicate behavior of jointed rock masses.

Joints and intact rock have been studied extensively under laboratory conditions where a typical test specimen dimension is 10 cm. Detailed insight concerning the role of micro-structure in the determination of elastic moduli and strength of intact rock has been gained from such studies, especially from the study of microcracks in igneous rock types. Joint roughness has also received much scientific attention, especially with the surge of interest in fractal geometry. However, as a practical matter, grain-scale or sub-millimeter detail in laboratory testing is of no great importance to the design of excavations and foundations in rock masses.

Laboratory test specimens of intact rock and joints do contain many grains and are thus representative samples (relative to grain size). Hence, elastic moduli and strength are readily obtained from the slope and peak of laboratory compressive stress-strain curves of intact rock. Stiffness and strength of joints may be determined from direct shear tests. Both sets of properties are needed for analysis and design.

The size of a representative sample of a jointed rock mass is dictated, not by grain size, but by joint spacing, typically a meter or more. A representative volume element (RVE or REV) of a jointed rock mass is thus meters on edge and far too large for routine physical measurement. Results from a few large and costly experimental rock block tests that have been done have defied easy characterization in terms of elastic moduli and strengths.

This theme contribution is mainly personal opinion informed by years of experience in application of rock mechanics to mine engineering problems. No attempt at a detailed critique of the enormous literature extant is made. Such review is better left to expert individual practitioners of the various approaches available for jointed rock mass mechanics. Hopefully, this contribution will provide some stimulation to thought on where the state of the art is and what the real obstacles to improvement in rock mass engineering are. Some insight concerning an equivalent properties approach to jointed rock mass mechanics is presented. What to do about the "joints" remains a key issue in rock mass engineering.

2 APPROACHES

The main approaches to dealing with joints in large rock masses lead to numerical models that necessarily rely on the computer for implementation. The earliest and still quite popular numerical approach is the finite element method (FEM). Rock mechanics applications appeared in the 1960's, many by students of the prodigious E. L. Wilson at Cal-Berkeley whose thesis was on the finite element method (Wilson 1963). This most important advance, made possible with the advent of the digital computer, was followed by the innovative distinct element method (DEM), pioneered in the 1970's by Peter Cundall (1971). A third major development occurred in the late 1980's with the appearance of discontinuous deformation analysis (DDA), brought to the fore by Gen-Hua Shi (1988). Development focusing largely on computational efficiency in handling ever larger models has been on-going in all approaches. In this regard, increased computational speed, fast memory size and lower hardware costs tend to advance all approaches equally. For this reason, comparative advantages are generated mainly through software improvements.

2.1 *FEM*

In FEM, the most common way of taking effects of joints on rock mass behavior into account is by simply reducing the values of elastic moduli and strengths determined by laboratory testing of intact samples. Elastic moduli and strengths may be reduced by different "scale" factors, both in the range of zero to one. Numerical values of these scale factors for any particular site are based on engineering judgment during forward design and may be refined by back-analysis after excavation has begun. The same scaling of material properties is applicable to the boundary element method (BEM) that saw rapid development in the 1970's and the appearance of reference books in the early 1980's. (e.g., Banerjee and Butterfield 1981, Crouch and Starfield 1983, Telles 1983). Large faults may be represented directly in a FE or BE model, but generally one-to-one modeling is impractical because of the enormous number of joints present in any sizeable rock mass. Early applications of FEM to mine engineering include mine subsidence, slope stability analysis in surface mines and stope stability underground (e.g., Dahl 1969, Pariseau and Stout 1972, Pariseau and Kealy 1973). These early applications were limited to a few hundred elements; meshes of over 100,000 elements are now run on desktop PC's. Special elements may be used to model important interfaces, bedding plane contacts or major faults. However, use is restricted and problematic at interface intersections.

A major FEM advantage is the ready accommodation of water flow in "wet" masses where effective stress is critical to analysis. Although most rock mechanics analysis are based on a small strain assumption, large displacements and finite strains can also be accommodated. The much discussed phenomenon of localization (discontinuity in displacement gradient) can also be handled in FEM, although mesh sensitivity is problematic. Software fixes range from the obvious of putting small elements in the path of expected localization to adaptive meshing, while at the fundamental level, a characteristic length may be introduced into the constitutive equations, for example, by strain gradients. Alternatively, if localization is viewed as the appearance in time of discontinuities in displacement, rather than strain (Ode, 1960), then a different type of element interpolation function is needed (Simo and others, 1993).

2.2 *DEM*

In DEM, major joint sets are assumed to transect the entire region producing a "brick wall" model composed of numerous closely fitted individual blocks that may be deformable and have intact elastic moduli assigned to them. Finite block strains are avoided; overall deformation is mainly controlled by slip and separation along joints, that is, by block interaction. Joint stiffness and strength constrain these interactions. Accommodation of large displacement on joints is the principal advantage of the DEM. In this regard, discontinuous joints that may be present in reality are deleted in advance of any model calculation for the simple reason that joint extension is not generally allowed. Block breakage is also problematic for most DE computer programs. The approach is computationally demanding and, as a practical matter, is limited to relatively few rock blocks in comparison to reality. Consider three orthogonal joint sets spaced two meters apart and a tunnel 10 meters wide. As a rule of thumb, the analysis region should be ten times the size of the excavation, a 100 meters wide. A three-dimensional model for rock-support interaction with face advance would contain a minimum of 125,000 blocks. DEM models of more than 10,000 blocks are unusual. Thus, while the DE approach seemingly allows for direct use of laboratory rock and joint properties, sizeable jointed rock masses cannot be represented in detail. Large blocks defined by major joint sets will contain numerous lesser joints; the strength and moduli of these blocks may need scaling ("equivalent properties) as in the FE approach to jointed rock mass behavior. Localization does not appear to have entered DEM development. DEM is also a "dry" approach in the sense that coupled hydro-mechanical problems are not addressed; conventional DEM analyses are restricted to dry rock masses. This restriction is important in many practical cases, for example in wet surface mines where water is a major determinant of slope stability.

DEM is uniquely suited to analysis of disaggregated rock rubble behavior where blocks may come out of contact with adjacent blocks and no longer form a "brick wall", muck mechanics in mining parlance. Strictly speaking, this class of problems is not jointed rock mass mechanics, although the distinction is blurred in the analysis of avalanches at the surface and caving underground. Applications in the mining industry range from blast casting to ball milling (e.g. Ryu 1989, Acharya 1991). In the latter application, the "blocks" are various sized spheres of rock and steel. While the spherical steel balls are realistic, the spherical shape of particles is not, so there is room for improvement. Particle breakage is also an issue as is the absence of water. Although, development of particle clusters allows for different particle shapes and particle breakage in a "particle flow code" (PFC), the shapes obtained are far from realistic. Recent application to the mechanics of transporting broken ore and waste rock through orepasses and chutes in underground mining appears promising in this neglected area of mine engineering (Beus and others, 1998).

2.3 *DDA*

The DDA approach began as back-analysis of displacements in blocky rock masses using a least squares fit allowing for rigid body motion of the blocks (Shi and Goodman, 1985). The original motivation was back analysis of a large rock block test. The approach was subsequently developed into a new method of analysis of blocky rock systems (Shi, 1988). There are similarities between DDA and DEM because of the similarities of the blocky rock systems analyzed. However, DEM integrates Newton's second law of motion directly by a relaxation method in consideration of forces. DDA is more akin to FEM and is based on energy minimization and the solution of a system of stiffness equations with displacements as unknowns. Like DEM, DDA has constraints of no-tension and no-overlap between blocks that are deformable. Displacements within a block are usually interpolated using linear displacement functions, such blocks are analogous to constant strain finite elements. The number of degrees of freedom is larger than in FEM. Shi (1988) states that DDA is more efficient than DEM and that the computing "cost" of DDA is about five times FEM. The considerable benefit is greater physical realism through allowance for slip and separation of blocks defined by joints and is certainly worth the added cost. Ma (1999) presents a review of the past ten years of DDA developments.

A relatively recent innovation by Shi (1995) that builds on DDA is the numerical manifold method (NMM). This method depends on interpolation functions similar to FEM but these functions may be discontinuous, unlike FEM. Blocks defined by joints are deformable; large displacements and rotations of blocks are allowed. Two meshes are required by NMM. One defines physical boundaries to blocks that may separated by structural or material discontinuities. Meshing of individual blocks is not required. The other mesh is a mathematical one that overlies the physical extent of blocks. The mathematical mesh is similar to a conventional finite element mesh but has an important practical advantage because it does not need to conform to physical boundaries. This feature is a significant practical advantage because it allows for relatively simple input with automatic generation of the entire model. Similar to DEM and DDA, as the number of joints increases, the number of blocks becomes a limiting factor to practical application. As in DEM, block properties, block failure, localization and effective stress in wet ground are also questions to be addressed in DDA development.

3 NRVE ALTERNATIVE

Alternatives to representing joints as such in rock block models of large rock masses are few. Perhaps the only viable alternative is an equivalent properties approach. Such an approach may be an arbitrary scaling or a more methodical mechanics-based analysis, either forced by the enormous number of joints usually present in sizeable rock masses.

Most rational equivalent properties formulations are based on the assumption of a representative volume element (RVE). Assumption of an RVE limits application to excavations that are large relative to the RVE. The reason is that no element, cell or block may be smaller than the RVE. If joints are spaced two meters apart, then the RVE should be at least 20 meters on edge. If the excavation is 10 meters wide, then elements about one meter on edge are needed for reliable numerical results. Assignment of material properties to these small, sub-RVE size elements at the excavation wall is clearly problematic. If the excavation were 200 meters wide, then the assumption of an RVE could be useful. Excavations 200 meters wide are rare underground. In surface mines, typically, more than 200 meters deep, benches of about 20 meters high are common, so slope stability analysis using a single element per bench would be justifiable using RVE-equivalent properties. However, an additional joint set at a four meter spacing would be a problem.

A joint is actually a thin layer of material with properties that are distinct from the properties within the adjacent volumes, as shown in Fig. 1. The contacts with adjacent volumes are thus material discontinuities. Elastic stability at a material discontinuity requires continuity of tractions and displacements, while discontinuities in some stresses and some strains are allowed (e.g., Jaunzemis, 1967).

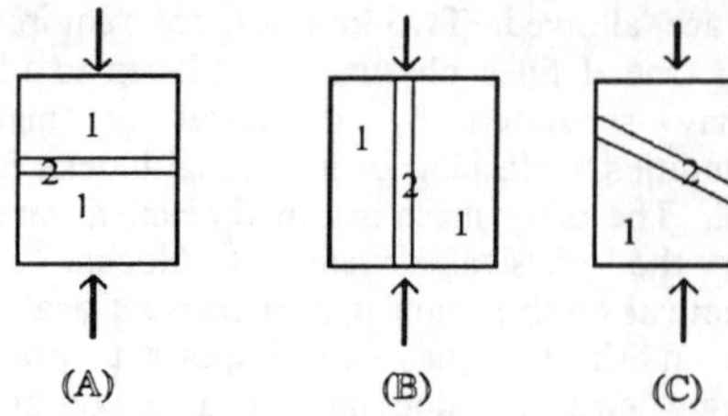

Figure 1. Single joints in an element: (A) flat, (B) vertical, (C) inclined

Under uniform applied stresses or displacements, the volume averages of stress or strain within a sample are equal to the applied stresses or surface strains regardless of sample size (e.g., Willis 1981). Thus, materials properties of jointed rock samples determined by simulating laboratory testing need not be based on the assumption of an REV. Such properties relate volume averages of the constitutive variables, e.g., solid stress-strain in Hooke's law (Pariseau and Moon 1988), fluid velocity-hydraulic gradient in Darcy's law (Pariseau 1991) in arbitrary sized, non-representative volume elements (NRVE properties approach). The advantage of the NRVE approach is the rigor associated with continuity requirements and the elimination of the assumption of an RVE. Small elements may be used near excavation walls to ensure reliable numerical results, while large elements away from the excavation walls may be used for economy. The disadvantage is that each element in a mesh may have different properties, a feature that increases computational cost.

3.1 *One dimension*

Consider the simple geometric arrangement of joint and intact rock shown in Fig. 1(a). Equilibrium in the direction of the applied force requires equality of corresponding *average* stresses. Thus $F=\int \sigma dV = V<\sigma> = V_1\sigma_1 = V_2\sigma_2$ where V is total volume, the angle brackets signify *overall* average, subscripted volumes correspond to regions occupied by intact rock and joint, respectively, and the subscripted stresses are *local* averages over the related volumes. Equivalent elastic properties relate overall averages of stress and strain. Thus, $\{<\epsilon>\}=[S^*]\{<\sigma>\}$ where {} is a 6x1 column matrix, [ ] is 6x6 square matrix and S_{ij}^* are equivalent elastic compliances. The question is then what are the S_{ij}^* for Fig. 1(a)? A direct solution considers the intact rock and joint as linear springs in series where the displacements add, while the forces are equal. Thus, $(1/E)^*=f_1(1/E_1)+f_2(1/E_2)$ where the f's are volume fractions and the E's are in the direction of the applied load. Thus, the overall compliance is a volume weighted average of intact rock and joint compliances. If $E_1=100E_2$, $f_1=0.93$ and $f_2=0.07$, then E* is about $0.13E_1$. Direct solution for the geometric arrangement in Fig. 1(b), assuming forces add while displacements are equal (springs in parallel model, $\int \epsilon dV = V<\epsilon> = V_1\epsilon_1 = V_2\epsilon_2$.) leads to the result $(E^*)=f_1(E_1)+f_2(E_2)$ which is a volume weighted average modulus that is about $0.93E_1$. The spread between these limits is almost an order of magnitude. A similar analysis leads to estimates of equivalent shear moduli G*. Nothing is said about Poisson's ratio. A direct, one-dimensional analysis of the inclined joint in Fig. 1(c) is problematic.

3.2 *Three dimensions*

A more rigorous, three dimensional analysis of the geometric arrangement in Fig. 1(c) is based on an energy equivalence and, again, the assumption of an RVE. Energy of the RVE is related to the integral $\int \{\epsilon\}^t\{\sigma\}dV = <\epsilon>^t<\sigma>V + \int \{\Delta\epsilon\}^t\{\Delta\sigma\}dV$ where the superscript-t means transpose, Δ is a deviation from an overall average, $\{\epsilon\}=<\epsilon>+\{\Delta\epsilon\}$, $\{\sigma\}=<\sigma>+\{\Delta\sigma\}$ and $\int\{\Delta\epsilon\}dV = \int\{\Delta\sigma\}dV = 0$. Assumption of uniform stress ($\{\Delta\sigma\}=0$), which satisfies equilibrium but violates compatibility of strain, leads to a lower bound on the equivalent (isotropic) elastic moduli. Assumption of uniform strain ($\{\Delta\epsilon\}=0$), which satisfies compatibility but violates equilibrium, leads to an upper bound. These bounds are also volume weighted averages and are the classic Reuss and Voight moduli estimates, respectively, (Hill 1952). Interestingly, these bounds are isotropic moduli even though the sample is obviously anisotropic. The RVE assumption allows for neglect of the integral on the right side of the energy expression, but clearly the uniform stress and uniform strain assumptions have the same effect on the integral of the deviation product $\int\{\Delta\epsilon\}^t\{\Delta\sigma\}dV$.

The NRVE analytical approximation to the equivalent elastic properties relationship is $\{<\epsilon>\}=[S^*]\{<\sigma>\}$ or in expanded form

$$\begin{Bmatrix} <\varepsilon_c> \\ <\varepsilon_e> \end{Bmatrix} = \begin{bmatrix} S^*_{cc} & S^*_{ce} \\ S^*_{ec} & S^*_{ee} \end{bmatrix} \begin{Bmatrix} <\sigma_c> \\ <\sigma_e> \end{Bmatrix} \quad \textbf{(1)}$$

Equivalent moduli [C*] and compliances [S*] are mutual inverses, so $\{<\sigma>\}=[C^*]\{<\epsilon>\}$. The vectors of stress and strain are partitioned in "compatible" and "equilibrium parts $\{\epsilon\}^t = (\epsilon_c \epsilon_e)$ and $\{\sigma\}^t = (\sigma_c \sigma_e)$

where subscripts "c" and "e" range from 0 to 6, while c+e=6. Such partitioning makes all combinations of the product $\{\Delta\epsilon\}^t\{\Delta\sigma\}=0$.

When the dimensional partitioning corresponds to the uniform strain assumption (c=6, e=0), then $[C^*]=<[S_{cc}]^{-1}>$ which are the volume weighted average moduli, that is, the Voight or upper bound moduli mentioned previously. When partitioning corresponds to the uniform stress assumption (c=0, e=6), then $[C^*]=<[S_{ee}]>^{-1}$ which is the inverse of the volume weighted average compliances, the Reuss or lower bound moduli. Thus, NRVE contains the Voight and Reuss bounds as special cases. There are five other orthogonal combinations of the vectors of stress and strain deviation in the NRVE approximation that also cause the deviation product to vanish. The other c-e combinations are 5-1, 4-2, 3-3, 2-4, 1-5. Reducing out of equilibrium stress lowers the upper bound, while increasing compatibility raises the lower bound. With reference to Fig. 1, with the z-direction normal to the joint, the equilibrium stresses (σ_e) are($<\tau_{yz}>$ $<\tau_{xz}>$ $<\sigma_{zz}>$), and the compatible strains (ϵ_c) are($<\epsilon_{xx}>$ $<\epsilon_{yy}>$ $<\gamma_{xy}>$) in the 3-3 case. Corresponding deviations are zero. The optimum is the 3-3 combination that, perhaps not too surprisingly, is the combination that also satisfies continuity requirements (compatibility and equilibrium) for elastic stability.

3.3 *Element example*

Figure 2 shows a generic brick element that is intersected by seven joints and the corresponding equivalent moduli for all combinations of compatible and equilibrium specification. The moduli corresponding to the 0-6 combination are at the 1-point on the x-axis and are the Reuss moduli. These moduli are lower bounds and are isotropic. The moduli at the extreme right at the 7-point on the x-axis correspond to the 6-0 combination; these moduli are the Voight upper bounds which are also isotropic. Increasing equilibrium from the left raises the lower bounds, while decreasing incompatibility from the right lowers the upper bounds and also introduces anisotropy as seen in Fig. 2. Point 4 on the x-axis corresponds to the best (3-3) combination that satisfies continuity requirements across joints.

The "true" equivalent compliances [S*] and moduli [C*] may be obtained from solutions to "appropriate boundary value problems" (Willis, 1981). Appropriate boundary value problems are simply laboratory-like numerical experiments using uniform applied stresses or surface strains. The results allow one to compute overall averages of stress and strain, $<\{\sigma\}>$ and $<\{\epsilon\}>$, without the assumption of an RVE. A detailed finite element analysis of the brick shown in Fig. 2. contains 391 nodes that define 139 polyhedrons formed by the thin

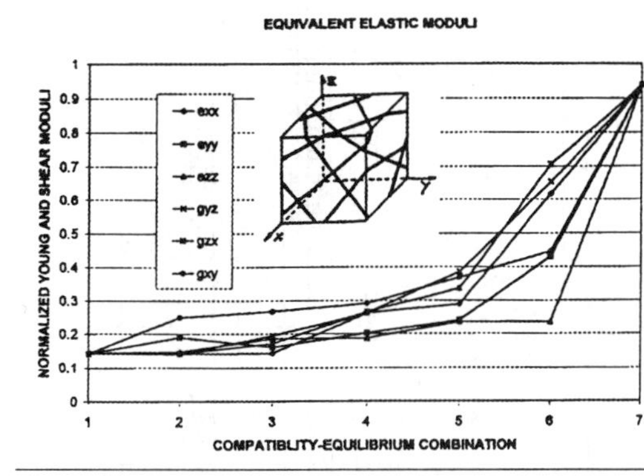

Figure 2. Elastic moduli for seven combinations of compatibility (c) and equilibrium (e) with c+e=6. X-axis: 1=0+6, 2=1+5, 3=2+4, 4=3+3, 5=4+2, 6=5+1, 7=6+0. Y-axis: Young's moduli divided by Young's modulus of intact rock, shear moduli divided by shear modulus of intact rock. Intact rock and joints are isotropic. Ratio of intact to joint moduli is 100. The joints occupy about 7% of the element volume.

joint layers. These polyhedrons have a variable number of corners, but are handled in a finite element code adapted for polyhedral elements that are divided into tetrahedrons. There are 1716 tetrahedrons in the brick element shown in Fig. 2. The reduction to one eight-node element indicates the potential economy of the NRVE approach to jointed rock mass mechanics (factor of over 1000). The NRVE moduli agree with the "true" moduli within 1% to 2%. Details of the elastic-plastic response are given by Pariseau (1996).

3.4 *Surface mine example*

Figure 3 shows one half of a finite element mesh used in this example from Pariseau (2000). Pit depth H is 366 m (1200 ft); slope angle is 45° bench height h is 24.4 m (80 ft). Bottom of the pit is 110 m (360 ft) wide when fully mined. These are realistic dimensions for many open pit mines. Elements at the slope face are 1-bench cubes, that is, 24.4 m on edge, while elements remote from the face are three times as large in section but have the same thickness. Overall mesh dimensions are about $5H$ by $10H$, so the outer boundaries are well away from the slope faces. These outer boundaries are fixed in the normal direction. There are 6132 nodes and 2970 elements in the mesh which is actually a slab (3D).

The full mesh is used because of the joints. Both joint sets dip to the right, the first at 60°, the second at 30°. Joints in the left pit wall thus dip sub-parallel to the slope; the 30° set dips into the face, while the 60° set dips below the face. Slip of joints into the pit would be expected here. However, both joint sets

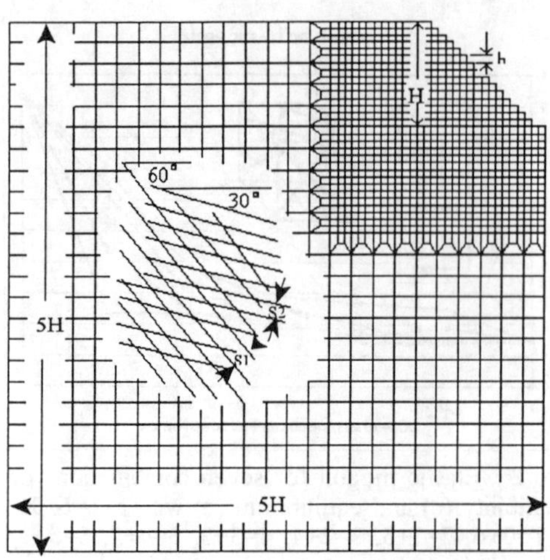

Figure 3. One-half of mesh used and joint pattern.

rock and joints is 100. Spacing of both joint sets is 6.1 m. The smaller elements contain about 11 joints each, while the large elements contain about 34 joints each. Joint volume averages less than 1% of element volume. There are 55,529 joints in the mesh; the number of joint segments is much greater because of joint intersections. No macroscopic element failures occur, as seen in Figure 4, although during initial gravity loading, about 13% of the joints yield.

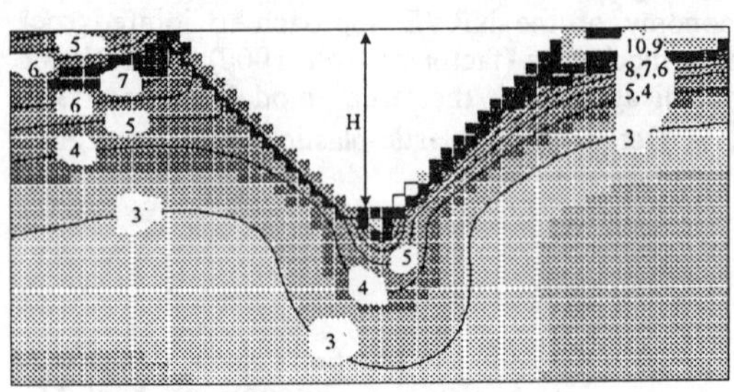

Figure 4. Safety factor contours near the excavation.

Excavation changes the joint yielding pattern as illustrated in Figure 5. Joints near the toe of the pit wall on the right side of the pit actually unload and improve toe safety, although some yielding on one or the other or both joint sets occurs behind the pit wall higher up. The left pit wall, where the joints dip into the pit, is threatened by yielding on both joint sets to a depth of about 75 m and by yielding on the steeply dipping (60°) joint set in a large region below and to the left of the slope toe.

3.5 *Underground mine example*

This example is a circular tunnel excavated to a radius of 2.74 m in a jointed rock mass containing

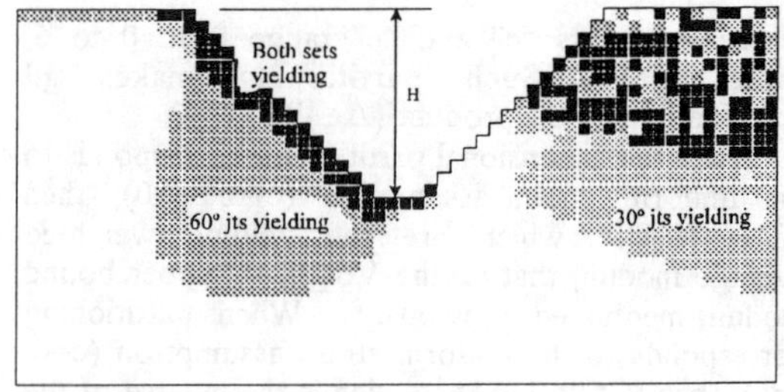

Figure 5. Regions of joint set yielding after excavation.

joints 0.005 m thick, spaced 0.55 m and dipping 60 degrees, as shown in Fig. 6 (Pariseau, 1998). The preexcavation stress state was generated by applying forces to the top and left boundaries of the mesh dip into the pit wall on the right-hand side of the pit. "Toppling" might be expected here because of the steeply dipping joint set, although the 30° set complicates this expectation.

Joints and intact rock are assumed to follow Drucker-Prager yield $(J^*_2)^{1/2} - I^*_1 = 1$ where J^*_2 and I^*_1 are anisotropic forms of the second invariant of deviatoric stress and the first stress invariant, respectively. Modulus and strength contrast between while keeping the bottom and right boundaries fixed in the normal direction. In the absence of joints, these forces would develop an in-plane hydrostatic stress state with the horizontal and vertical normal stresses equal to 20.7 MPa. There are more than 3,300 jointed elements in the mesh which contains 940 macro-elements. Almost all elements contain one joint; some of the larger elements contain more than a dozen joints. All joints enter the plastic domain during generation of the initial stress state; none of the intact rock between joints yields, however. Rock and joint moduli and strengths are in the ratio of 100 as shown. An incremental, iterative, tangent stiffness elastic-plastic model was used based on a parabolic yield function: $J^*_2 - I^*_1 = 1$.

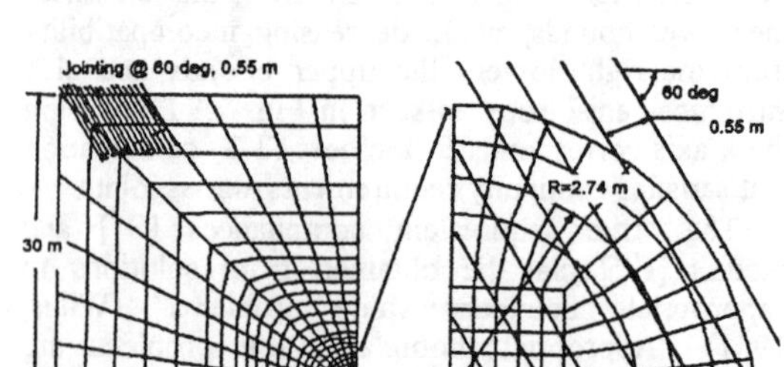

Fig. 6. Finite element mesh and rock mass jointing for the NRVE tunnel example. Quarter mesh - left. Excavation detail - right.

Results in the form of stress concentrations in the tunnel wall at the radius to element centers are shown in Fig. 7. The radial and circumferential stresses that would be present in the absence of joints are also shown (about 0.2 and 1.8, respectively). Effect of joints is significant, especially for those in tension (negative) near 0° and 180°.

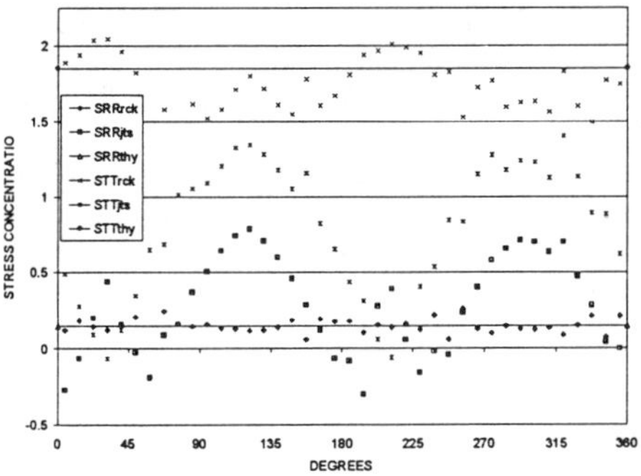

Fig. 7. Stress concentrations in the example tunnel wall. SRR=radial, STT=tangential stress.

4 CONCLUSION

The mechanics of jointed rock masses is dominated by "joints", structural discontinuities of many geologic types that are usually present in field-scale rock excavations of civil and mining engineering origin. Various numerical approaches to dealing with joints include finite element, distinct element and discontinuous deformation analysis (FEM, DEM, DDA). Special elements and interfaces are used in FEM, while direct representation through block formation is used in DEM and DDA. Particle flow code (PFC) and numerical manifold method (NMM) have advanced capabilities of DEM and DDA, respectively. In FEM, arbitrary scaling of elastic moduli and strength is a common method of dealing with joint effects. An alternative approach in FEM is the use of equivalent properties. Although conventional equivalent properties formulation are limited by the assumption of a representative volume element (REV), removal of this restriction of possible using a non-RVE (NRVE) approach.

Over ten years experience shows that comparisons of equivalent properties obtained from the NRVE analytical approximation with true properties obtained from boundary value problems show excellent agreement (e.g. Pariseau 1993, 1995, 1999). Several examples illustrate the considerable savings in computational effort possible using NRVE. Analytical averaging joint detail up to the element level allows numerical analyses to proceed. Influence functions allow for examination of sub-element detail and whether joint yielding occurs. More realistic stability evaluation of surface and underground excavations is thus possible using FEM.

Further improvement in application of all computational approaches to jointed rock mass mechanics would seem to require advances in fundamental aspects. For example, localization raises several basic questions regarding characterization of rock failure, not to mention behavior beyond the elastic limit but before the point of localization is reached (damage?). But is rock failure a matter of stability or strength, fracture or flow or both? How should extension of existing joints, development of new fractures and localization be described in the real world of jointed rock masses? Does localization actually occur in jointed rock masses? Clearly, there is a need for fundamental work to improve application in rock engineering.

ACKNOWLEDGMENT

This work was supported in part by the National Science Foundation under Grant CMS-9610026.

REFERENCES

Acharya, A. (1991) A discrete element approach to ball mill mechanics. M.S. thesis, Univ. of Utah, 103 pp.

Banerjee, P.K. and R. Butterfield (1981) *Boundary Element Methods in Engineering Science.* McGraw-Hill, London 452 pp.

Beus, M., S.R. Iverson and B.M. Stewart (1998) Design analysis of underground mine ore passes: current research approaches. *Proc. 100th Can. Inst. Min. Met. Pet.*, CIM, Montreal, 8 pp.

Crouch, S.L. and A.M. Starfield (1983) *Boundary Element Methods in Solid Mechanics.* G. Allen & Unwin, London, 322 pp.

Cundall, P.A. (1971) A computer model for simulating progressive, large scale movements in blocky rock systems. *Proc. Intl. Symp. Rock Fracture*. Nancy, France, paper II-8.

Dahl, D.H. (1969) A finite element model for anisotropic yielding in gravity loaded rock. Ph.D dissertation, Penn. State Univ., 155 pp.

Hill, R. (1952) The elastic behavior of a crystalline aggregate. *Proc. of the Physical Society (Sec. A)*, Vol. 65, pp 349-354.

Jaunzemis, W. (1967) *Continuum Mechanics.* The Macmillian Co., N.Y., pp165-166.

Ma, M.Y. (1999) Development of discontinuous deformation analysis; the first 10 years (1986-1996). *Proc. Third Intl. Conf. on Analysis of Discontinuous Deformation* (B. Amadei, ed). American Rock Mechanics Association, Alexandria, Virginia, pp 17-32.

Ode, H. (1960) Faulting as a velocity discontinuity in plastic deformation. *The Geological Society of America Memoir 79: Rock Deformation* (D.

Griggs and J. Handin: eds). Waverly Press, Baltimore, pp 293-321.

Pariseau, W.G. and K. Stout (1972) Open pit mine slope stability: the Berkeley Pit. *Proc. 13th U.S. Symp. Rock Mech.* ASCE, N.Y., pp 367-395.

Pariseau, W.G. and C.D. Kealy (1973) Support potential of hydraulic backfill. *Proc. 14th U.S. Symp. Rock Mech.* ASCE, N.Y., pp 501-526.

Pariseau, W.G. and H. Moon (1988). Elastic moduli of well-jointed rock masses. *Proc.6th Int. Conf. on Numerical Methods in Geomechanics.* Balkema, pp 815-822.

Pariseau, W.G. (1991). Estimation of permeability in well-jointed rock masses. *Proc. 7th Int. Conf. Computer Methods and Advances in Geomechanics.* Balkema, pp 1589-1594.

Pariseau, W.G. (1993). Equivalent properties of a jointed Biot material. *Int. J. Rock Mech. Min. Sci. & Geomech. Abstr.* Vol. 30, No. 7, pp 1151-1157.

Pariseau, W.G. (1995). Non-representative volume element modeling of equivalent jointed rock mass properties. *Proc. 2nd Int. Conf. Mechanics of Jointed and Faulted Rock.* Balkema, pp 563-568.

Pariseau, W.G. (1996). Equivalent strength of porous fractured rock. *Proc. 11th Conf. Engineering Mechanics.* ASCE, N.Y., pp 216-219.

Pariseau, W.G. (1998) An equivalent properties approach to geoplasticity. *Proc. 7th Intl. Symp. Plasticity and Its Current Applications.* Neat Press, Fulton, MD, pp 593-596.

Pariseau, W.G. (1999) An equivalent plasticity theory for jointed rock mass. *Intl. J. Rock Mech. Mining Sci.* Vol 36, No. 7, pp 907-918.

Pariseau, W.G. (2000) Equivalent properties approach to jointed viscoplastic rock slopes. To appear in: *Proc. 4th North American Rock Mechanics Symposium.* Seattle, July 29-Aug. 3, 2000.

Ryu, Chang-Ha. (1989) Numerical modeling of fragmentation in jointed geologic media. Ph.D. dissertation, Univ. of Utah, 201 pp.

Shi, Gen-Hua (1988) Discontinuous deformation analysis: a new numerical model for the statics and dynamics of block systems. Ph.D. dissertation, Univ. California - Berkeley. Univ. Microfilms, Ann Arbor, 377 pp.

Shi, Gen-Hua (1995) Numerical manifold method. *Proc. Intl. Conf. on Analysis of Discontinuous Deformation* (J. C. Li, C-Y Wang and J. Sheng, eds). Dept. Civil Engr., National Central Univ., Chungli, Taiwan, R.O.C, pp 187-222.

Shi, Gen-Hua and R.E. Goodman (1985) Two-dimensional discontinuous deformation analysis. *Intl. J. Numer. Analy. Meth. Geomech.* Vol. 9, No. 6, pp 541-556.

Simo, J.C., J. Oliver and F. Armero (1993) An analysis of strong discontinuities induced by strain-softening in rate-independent inelastic solids. *Computational Mechanics.* Vol. 12, pp 277-296.

Telles, J.C. F. (1983) *The Boundary Element Method Applied to Inelastic Problems.* Springer-Verlag, Berlin, 243 pp.

Willis, J.R. (1981) Variational and related methods for the overall properties of composites. *Advances in Applied Mechanics.* Vol. 21, pp 1-78.

Wilson, E.L. (1963) Finite element analysis of two-dimensional structures. Ph.D dissertation, Univ. of California, Berkeley. Univ. Microfilms, Ann Arbor, 72 pp.

Computer Methods and Advances in Geomechanics, Desai et al. (eds) © 2001 Balkema, Rotterdam, ISBN 90 5809 183 X

Far-field displacements of soil in scaled boundary finite-element method

John P. Wolf & Farid Moussaoui
Institute of Hydraulics and Energy, Department of Civil Engineering, Swiss Federal Institute of Technology, Lausanne, Switzerland

ABSTRACT: The scaled boundary finite-element method has been shown to be a powerful computational procedure to calculate the dynamic-stiffness matrix of the unbounded soil at the structure-soil and the displacements in the neighbouring soil. In this paper, the method is extended to determine the displacements also in the remaining part of the unbounded soil called the far-field. The radiation condition at infinity is satisfied exactly. Excellent agreement with analytical solutions is achieved. The extension to viscoelastic material with e.g. the constant hysteretic model is straightforward.

1 STATE OF THE ART

To analyse dynamic soil-structure interaction by the substructure method, the interaction force-displacement relationship in the degrees of freedom on the structure-soil interface of the unbounded soil is required (Figure 1).

In the frequency domain the amplitudes of the displacements $\{u(\omega)\}$ are related to those of the interaction forces $\{R(\omega)\}$ by the dynamic-stiffness matrix $[S^{\infty}(\omega)]$ (superscript ∞ for unbounded) as

$$\{R(\omega)\} = [S^{\infty}(\omega)]\{u(\omega)\} \tag{1}$$

In addition to $[S^{\infty}(\omega)]$, the displacements $\{u(\omega)\}$ in the unbounded soil must also be determined. The scaled boundary finite-element method, alias consistent infinitesimal finite-element cell method, is applied (Wolf and Song 1996; Song and Wolf 1997).

In this boundary-element method based on finite elements only the structure-soil interface is discretised with surface finite elements yielding a reduction of the spatial dimension by one. The structure-soil interface need not follow a plane, spherical or cylindrical contour, as is required in classical methods, but can consist of an irregular surface. The material properties may change arbitrarily in directions parallel to the structure-soil interface. No fundamental solution is necessary, and thus no singular integrals need to be evaluated, and general anisotropic material can be analysed. For an unbounded (semi-infinite or infinite) soil the radiation condition at infinity is satisfied exactly. No discretisation of free and fixed boundaries and interfaces between different materials is required as is apparent in Figure 1b (free surface denoted as side-face). No fictitious eigenfrequencies occur when analysing unbounded soil in frequency domain. This fundamental-solution-less boundary-element method is a semi-analytical procedure to solve partial differential equations in unbounded and bounded media with the polynomial approximation of finite elements in the two circumferential directions and an analytical solution in the radial direction.

This novel procedure is described in great detail in the book by Wolf and Song (1996), where partial differential equations of the hyperbolic type (e.g. wave propagation in time domain), the parabolic type (e.g. diffusion im time domain) and the elliptic type (statics and e.g. wave propagation in frequency domain) are addressed. It is called the consistent infinitesimal finite-element cell method reflecting the mechanically-motivated derivation. The same equations are derived starting from the governing partial differential equations using the Galerkin's weighted residual method in (Song and Wolf 1997). The reader is referred to (Wolf and Song 1996; Song and Wolf 1997; Song and Wolf 1998) for details of the derivations and a literature survey.

The method introduces a scaling centre O (Figure 1) from which the total structure-soil interface must be visible. The soil-structure interface (boundary) is discretised with (curved) surface finite elements. The free surface, i.e. the part of the boundary whose extension passes through the scaling centre, needs not to be discretised. Mathematically, the scaling corresponds to a transformation of the coordinates for each finite

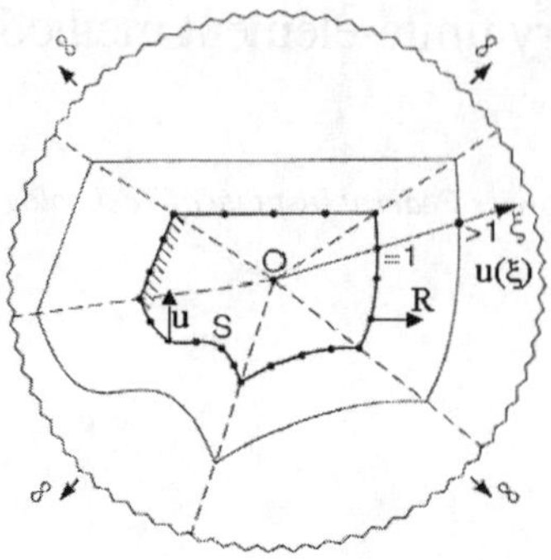

(a) unbounded soil

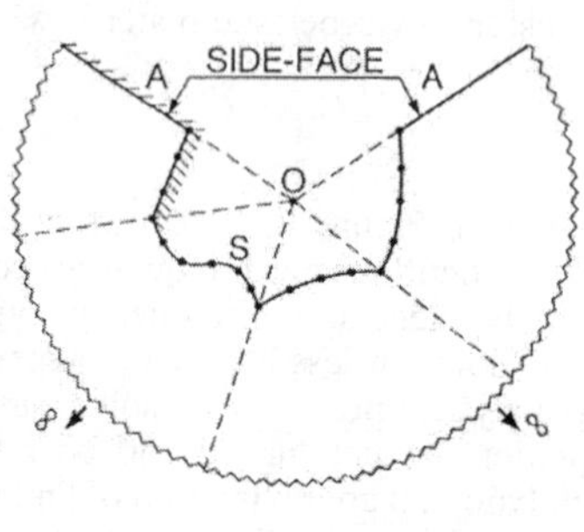

(b) unbounded soil with free surface

Figure 1: Interaction force-displacement relationship on discretised structure-soil interface of unbounded soil.

element, resulting in the two curvilinear local coordinates η, ζ in the circumferential directions on the boundary and the dimensionless radial coordinate ξ representing the scaling factor. $\xi = 0$ and $\xi = 1$ correspond to the centre O and the boundary (structure-soil interface), respectively. The unbounded soil is defined by $1 \leq \xi < \infty$.

After applying the scaled boundary transformation to the governing partial differential equations of elastodynamics, Galerkin's weighted residual method is formulated in the two circumferential directions η, ζ. For vanishing body loads the scaled boundary finite-element equation in displacement in the frequency domain follows as:

$$[E^0]\xi^2\{u(\xi)\},_{\xi\xi} + \\ ((s-1)[E^0] - [E^1] + [E^1]^T)\xi\{u(\xi)\},_{\xi} + \\ ((s-2)[E^1]^T - [E^2])\{u(\xi)\} + \omega^2[M^0]\xi^2\{u(\xi)\} = 0 \quad (2)$$

$\{u(\xi)\}$ are the displacements at ξ corresponding to the degrees of freedom in the nodes on the structure-soil interface, and s is the spatial dimension ($= 2$ or $=$ 3). The coefficient matrices $[E^0]$, $[E^1]$, $[E^2]$ and $[M^0]$ are calculated and assembled similarly as the static-stiffness and mass matrices of finite elements on the structure-soil interface.

Introducing $\bar{\xi} = \omega\,\xi$, equation (2) can be rewritten as:

$$[E^0]\bar{\xi}^2\{u(\bar{\xi})\},_{\bar{\xi}\bar{\xi}} + \\ ((s-1)[E^0] - [E^1] + [E^1]^T)\bar{\xi}\{u(\bar{\xi})\},_{\bar{\xi}} + \\ ((s-2)[E^1]^T - [E^2])\{u(\bar{\xi})\} + [M^0]\bar{\xi}^2\{u(\bar{\xi})\} = 0 \quad (3)$$

Equation (3) is a linear second-order ordinary differential equation. In the interaction force-displacement relationship, the dynamic-stiffness matrix $[S^\infty] = [S^\infty(\omega, \xi)]$ is related to the displacements as:

$$\xi^{2-s}[S^\infty]\{u(\bar{\xi})\} = -[E^0]\bar{\xi}\{u(\bar{\xi})\},_{\bar{\xi}} - [E^1]^T\{u(\bar{\xi})\} \quad (4)$$

Combining equations (4) and (3) leads to the scaled boundary finite-element equation in dynamic stiffness:

$$(\xi^{2-s}[S^\infty] + [E^1])[E^0]^{-1}(\xi^{2-s}[S^\infty] + [E^1]^T) \\ -(s-2)\xi^{2-s}[S^\infty] - \bar{\xi}(\xi^{2-s}[S^\infty]),_{\bar{\xi}} \\ -[E^2] + \bar{\xi}^2[M^0] = 0 \quad (5)$$

Equation (5) is a nonlinear matrix-valued ordinary differential equation of first order.

The high-frequency asymptotic expansion equals:

$$\xi^{2-s}[S^\infty] = \mathrm{i}\bar{\xi}[C_\infty] + [K_\infty] + \sum_{j=1}^{m} \frac{1}{(\mathrm{i}\bar{\xi})^j}[A_j] \quad (6)$$

The constant $[C_\infty]$ follows from an eigenvalue problem and is constructed as positive definite and the radiation condition is thus satisfied. The matrices $[K_\infty]$, $[A_j]$ follow from solving linear equations (Wolf and Song 1996).

As shown in (Song and Wolf 1998), the scaled boundary finite-element method in displacement can be solved analytically for a bounded domain. To determine the dynamic-stiffness matrix of the unbounded medium with respect to the boundary at $\xi_i =$ 1 (structure-soil interface) (Figure 2), another similar boundary at $\xi_e > \xi_i$ is introduced. The dynamic-stiffness matrix at ξ_e is calculated as the boundary condition using the asymptotic expansion (6) which satisfies the radiation condition. This permits the dynamic-stiffness matrix at the soil-structure interface and the displacements for $\xi_i = 1 \leq \xi \leq \xi_e$ to be obtained.

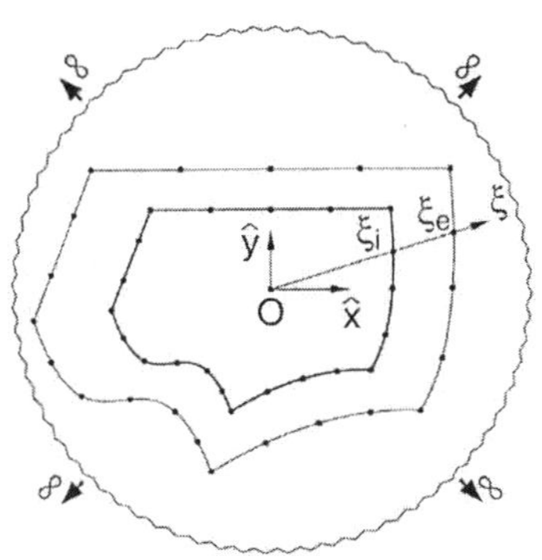

Figure 2: Unbounded soil on exterior of boundary corresponding to $\xi_i \leq \xi < \infty$.

As the solution is written as a power series (Gantmacher 1977), an infinite number of terms would be required for the unbounded soil with $\xi \rightarrow \infty$.

In the present paper, a numerical method is addressed to calculate the displacements $\{u(\xi)\}$ for $\xi > \xi_e$, which is denoted as the far-field of the unbounded soil. The extension to viscoelastic material with e.g. the constant hysteretic model is examined in the Appendix.

2 COMPUTATIONAL PROCEDURE

To be able to enforce the boundary conditions at infinity and at the structure-soil interface, the scaled boundary finite-element equation in displacement (equation (3)) is not solved directly but the two first-order differential equations (equations (4) and (5)) are addressed. The nonlinear first-order ordinary differential equation (5) is solved numerically starting from a very large $\bar{\xi}$ with the boundary condition determined from equation (6) for decreasing $\bar{\xi}$ (up to $\xi_i = 1$) yielding the dependent variables $\xi^{2-s}[S^\infty]$. Then the linear first-order ordinary differential equation (4) is solved numerically from $\xi_i = 1$ on the structure-soil interface with the known displacements corresponding to the definition of the dynamic-stiffness matrix for increasing $\bar{\xi}$ resulting in $\{u(\bar{\xi})\}$. The ordinary differential equations are thus solved numerically in the $\bar{\xi}$-direction. The transformation to the new coordinate system does, however, permit the radiation condition to be introduced rigorously for $\xi \rightarrow \infty$.

The numerical method used to compute $\xi^{2-s}[S^\infty]$ from equation (5) is a predictor/corrector scheme. In the low-frequency range, the error diminishes. Equation (5) is formulated as $[W],_{\bar{\xi}} = F(\bar{\xi}, [W])$ where $[W] = \xi^{2-s}[S^\infty]$ and F follows from equation (5).
The backward predictor/corrector scheme from step $n+1$ to step n is written as:

$$\left\{\begin{array}{lcl} [\tilde{W}] & = & [W]^{n+1} - \Delta\bar{\xi}F(\bar{\xi}^{n+1}, [W]^{n+1}) \\ [W]^n & = & [W]^{n+1} \\ & - & \frac{1}{2}\Delta\bar{\xi}\left(F(\bar{\xi}^{n+1}, [W]^{n+1}) + F(\bar{\xi}^n, [\tilde{W}])\right) \end{array}\right. \tag{7}$$

The linear first-order ordinary differential equation (4) is solved by an explicit Runge-Kutta $(4, 5)$ formula (Dormand and Prince 1980).

Thus, an efficient numerical procedure results to calculate very accurately the dynamic stiffness and displacements throughout the unbounded soil by solving numerically two first-order ordinary differential equations.

To avoid solving an eigenvalue problem to determine the coefficients in equation (6), only the term with $[C_\infty]$ is used in the asymptotic expansion. As in the high-frequency limit the response of the structure-soil interface is modelled as dashpots with coefficients equal to mass density times velocity corresponding to the direction, the coefficients of $[C_\infty]$ can be constructed directly.

In the far-field, the dynamic-stiffness matrix $[S^\infty]$ in equation (4) can be replaced by the high-frequency asymptotic expansion (equation (6)) by including the first two terms only, resulting in the following linear first-order ordinary differential equation

$$[E^0]\bar{\xi}\{u(\bar{\xi})\},_{\bar{\xi}} + ([E^1]^T + \mathrm{i}\bar{\xi}[C_\infty] + [K_\infty])\{u(\bar{\xi})\} = 0 \tag{8}$$

In this case, a series expansion (Gantmacher 1977) can be used to determine an approximate analytical solution. The boundary condition $\{u(\bar{\xi}_e)\}$ follows from the analytical solution of the domain between the two similar boundaries (Song and Wolf 1998). Thus the numerical solution of equation (5) is avoided.

3 EXAMPLES

3.1 Truncated semi-infinite wedge

The simplest case where all important features are preserved is addressed : the two-dimensional Helmholtz wave equation with the boundary discretised with a two-node line finite element. The properties of the scaled boundary finite-element method are illustrated using this example in (Wolf and Song 2000). The out-of-plane motion u of a shear plate in the form of a truncated semi-infinite wedge of shear modulus G and mass density ρ is addressed (Figure 3).

The sides 1-∞ and 2-∞ are fixed and free, respectively. The scaling centre is selected at O . As the extensions of the sides 1-∞ and 2-∞ pass through the scaling centre, no discretisation is required on these side-faces extending to infinity of the truncated semi-infinite wedge. The side 1-2, the structure-soil interface, is discretised with a two-node line finite element. The displacement varies linearly from 0 at node 1 to 1 at node 2. On the side 2-∞, the traction (flux) τ_n vanishes.

The following parameters are selected : $x_1 = r_0$, $y_1 = -r_0/\sqrt{5}$, $x_2 = r_0$, $y_2 = 0$. The dimensionless frequency $\omega\, r_0/c_s$ (with the shear-wave velocity $c_s = \sqrt{G/\rho}$) is chosen as 1. The displacements for $\xi \leq 8$ are determined.

As only one degree of freedom on the boundary in node 2 remains, all the equations are scalars. The exact solution satisfying the radiation condition is given by

$$u(\xi) = cH_\alpha^{(2)}\left(\frac{\omega \xi r_0}{c_s}\right) \tag{9}$$

where the length $r_0 = x_2$ as shown in Figure 3, $\alpha = \sqrt{E^2/E^0}$ and $H_\alpha^{(2)}$ is the Hankel function of second kind and of order α. The integration constant c follows from the boundary condition $u(\xi = 1) = 1$.

The high-frequency asymptotic expansion for S^∞, equation (6), is evaluated at $\bar{\xi} = 8$. $S^\infty(\bar{\xi})$ follows from equation (5). $u(\bar{\xi})$ is calculated from equation (4) with the boundary condition $u(\xi = 1) = 1$. The increment is selected as $\triangle\bar{\xi} = 0.2$. Figure 4a shows the excellent agreement between the computed solution and the exact one for $1 \leq \xi \leq 8$. Using the displacement at $\xi = 8$, equation (8) can be integrated numerically yielding good agreement for larger ξ. When the term A_1 of equation (6) is included in equation (8), the results shown in Figure 4b are obtained.

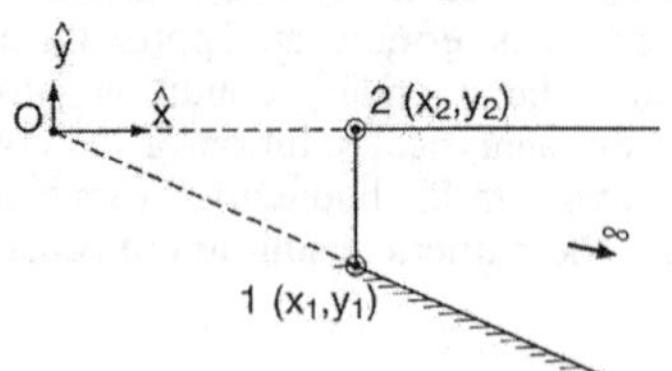

Figure 3: Truncated semi-infinite wedge.

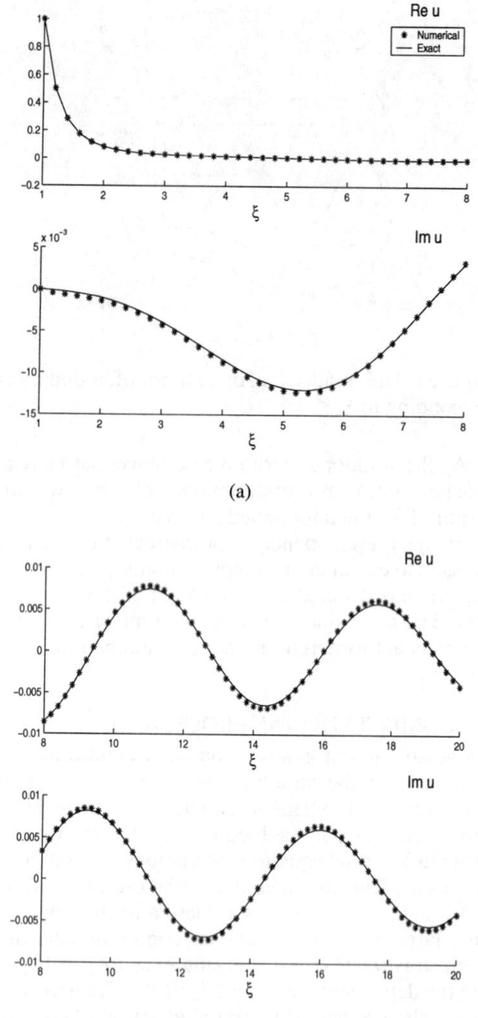

Figure 4: Real and imaginary parts of displacements for truncated semi-infinite wedge.

3.2 Circular cavity embedded in full-plane

As an example involving matrix operations, the out-of-plane motion of a circular cavity of radius r embedded in a full-plane of shear modulus G and mass density ρ is addressed. The out-of-plane displacement u and the shear stresses τ_{zr} and $\tau_{z\theta}$ in the polar coordinate system $\hat{r}$, θ are shown in Figure 5.

For a prescribed displacement in the circumferential direction on the structure-soil interface (the boundary) at $\xi = 1$ equal to the eigenfunction $\cos(n\theta)$ (n an integer), the displacements satisfying the radiation condition at infinity are given by:

$$u_n(\xi, \theta) = c_n H_n^{(2)}\left(\frac{\omega \xi r}{c_s}\right)\cos(n\theta) \tag{10}$$

c_n follows from the boundary condition at $\xi = 1$. The

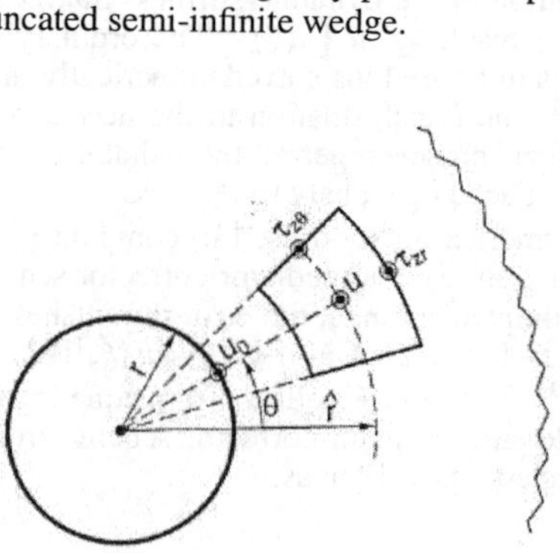

Figure 5: Circular cavity embedded in full-plane.

dimensionless frequency $\omega r/c_s$ is selected as 1 and the integer n is equal to 2.

In the scaled boundary finite-element method the boundary is discretised with 20 2-node line finite elements. The high-frequency asymptotic expansion for $[S^{\infty}(\bar{\xi})]$ (equation (6)) is evaluated at $\bar{\xi}=8$. Equation (5) is integrated with an increment $\triangle\bar{\xi}=0.1$ down to $\xi=1$ leading to only 70 steps. After enforcing the displacement boundary condition, equation (4) is integrated with the same increment. At $\xi=8$ the displacements on the circumference as of function of θ are plotted in Figure 6. Excellent agreement with the exact solution of equation (10) is obtained. Figure 7 shows the contours of the displacements for $1\leq\xi\leq 8$. These contours compare very well with those of the exact solution of equation (10) plotted in Figure 8.

The far-field displacements are calculated using equation (8), applying the boundary condition at $\xi=8$ determined from the integration of equation (4). Figure 9 compares the displacements at $\xi=15$ with the exact solution (equation (10)). Excellent agreement is obtained. Figure 10 presents the contours of the displacements in the region $1\leq\xi\leq 15$.

4 CONCLUSIONS

The scaled boundary finite-element method is extended to the analysis of the displacements in the far-field. The radiation condition at infinity is exactly satisfied.

The numerical procedure consists of two steps. In the first step, the scaled boundary finite-element equation in dynamic stiffness, a nonlinear first-order ordinary differential equation, is solved for decreasing radial coordinate down to the structure-soil interface. The high-frequency asymptotic expansion of the dynamic-stiffness matrix satisfying the radiation condition serves as the starting value (boundary condition). In the second step, the interaction force-displacement relationship involving the dynamic-stiffness matrix, a linear first-order ordinary differential equation, is integrated for increasing radial coordinate yielding the displacements. As an alternative, an analytical procedure can be used. The high-frequency asymptotic expansion of the dynamic-stiffness matrix is substituted in the interaction force-displacement relationship. Using a series expansion, the displacements in the far-field are obtained. The numerical procedure can be straightforwardly extended to viscoelastic material with the e.g. constant hysteretic model.

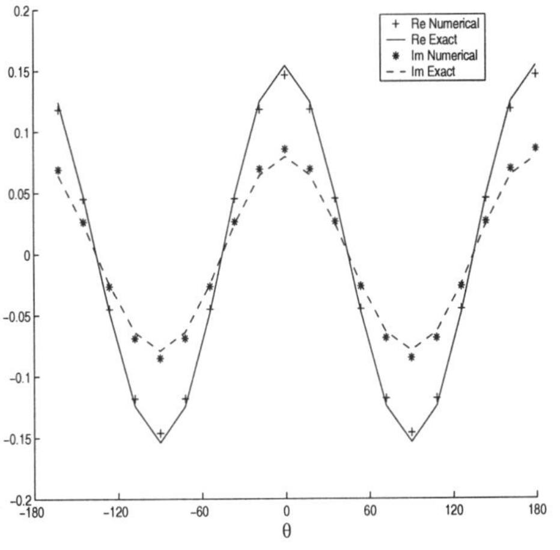

Figure 6: Displacements on circumference at truncation boundary $\xi=8$.

REFERENCES

Gantmacher, F. 1977. *The Theory of Matrices*, Volume 2. Chelsea, New York.

Song, C. and J. Wolf 1997. The scaled boundary finite-element method - alias consistent infinitesimal finite-element cell method - for elastodynamics. *Computer Methods in Applied Mechanics and Engineering 147*, 329–355.

Song, C. and J. Wolf 1998. The scaled boundary finite-element method: analytical solution in frequency domain. *Computer Methods in Applied Mechanics and Engineering 164*, 249–264.

Wolf, J. and C. Song 2000. The scaled boundary finite-element method - a primer: solution procedures. *Computers and Structures*. in press.

Wolf, J. and C. Song 1996. *Finite-Element Modelling of Unbounded Media*. Chichester, John Wiley and Sons.

Dormand, J. R. and P. J. Prince 1980. A family of embedded Runge-Kutta formulae. *Journal of Computational and Applied Mathematics 6*, 19–26.

A MATERIAL DAMPING

The equations (5) and (6) together with equation (4) to determine the displacements in the far-field apply to linear elastic material. They can straightforwardly be extended to viscoelastic material using the correspondence principle. For demonstration the constant hysteretic material model is addressed with the same damping coefficient for shear and dilatational waves. As an intermediate result, the dynamic-stiffness matrix at the structure-soil interface for the viscoelastic material of the unbounded soil is also obtained.

The correspondence principle states that the solution for hysteretic material follows from the elastic re-

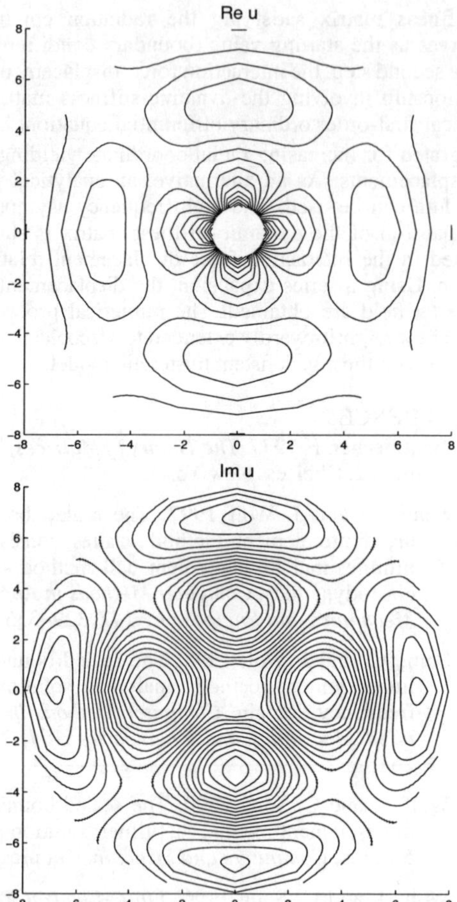

Figure 7: Contours of displacements for cavity problem determined with scaled boundary finite-element method.

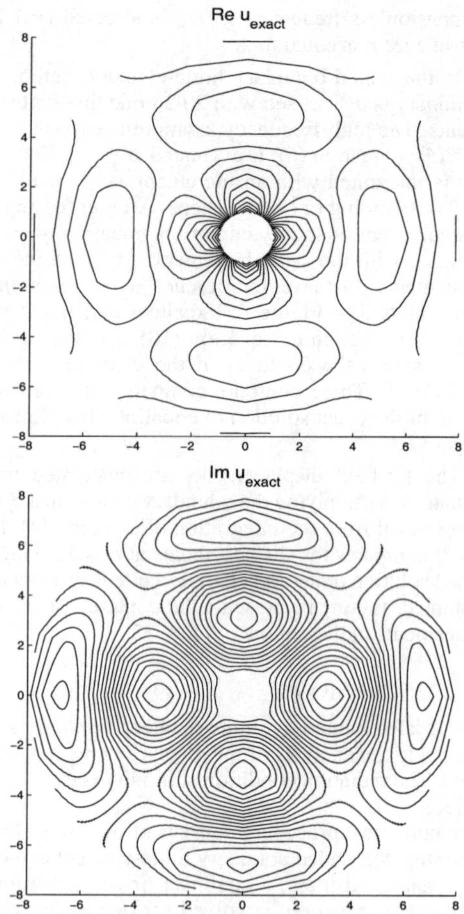

Figure 8: Contours of displacements for cavity problem determined with exact solution.

sults by replacing the real Lamé constants by the corresponding complex ones (multiplication by $1 + 2\,\mathrm{i}\,\zeta$ with the hysteretic damping ratio ζ). This yields

$$\begin{aligned} [E^{0*}] &= (1 + 2\,\mathrm{i}\,\zeta)[E^0] \\ [E^{1*}] &= (1 + 2\,\mathrm{i}\,\zeta)[E^1] \\ [E^{2*}] &= (1 + 2\,\mathrm{i}\,\zeta)[E^2] \end{aligned} \qquad (11)$$

The scaled boundary finite-element equation in dynamic stiffness, equation (5), equals

$$\begin{aligned} &(\xi^{2-s}[S^\infty] + [E^{1*}])[E^{0*}]^{-1}(\xi^{2-s}[S^\infty] + [E^{1*}]^T) \\ &-(s-2)\xi^{2-s}[S^\infty] - \bar{\xi}(\xi^{2-s}[S^\infty])_{,\bar{\xi}} \\ &-[E^{2*}] + \bar{\xi}^2[M^0] = 0 \end{aligned} \qquad (12)$$

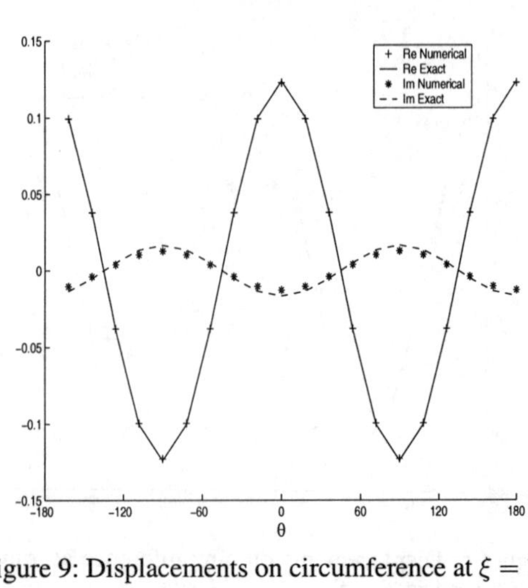

Figure 9: Displacements on circumference at $\xi = 15$.

with the dynamic-stiffness matrix of the hysteretic material $[S^{\infty}]$.

The high-frequency asymptotic expansion, equation (6), is written as

$$\xi^{2-s}[S^{\infty}] = \mathrm{i}\bar{\xi}[C^*_{\infty}] + [K^*_{\infty}] + \sum_{j=1}^{m} \frac{1}{(\mathrm{i}\bar{\xi})^j}[A^*_j] \quad (13)$$

where

$$\begin{aligned} [C^*_{\infty}] &= (1+2\,\mathrm{i}\,\zeta)^{1/2}[C_{\infty}] \\ [K^*_{\infty}] &= (1+2\,\mathrm{i}\,\zeta)[K_{\infty}] \\ [A^*_j] &= (1+2\,\mathrm{i}\,\zeta)^{3/2}[A_j] \end{aligned} \quad (14)$$

The interaction force-displacement relationship (equation (4)) reads

$$\xi^{2-s}[S^{\infty}]\{u(\bar{\xi})\} = -[E^{0*}]\bar{\xi}\{u(\bar{\xi})\}_{,\bar{\xi}} - [E^{1*}]^T\{u(\bar{\xi})\} \quad (15)$$

No modification occurs in the numerical integration procedures for equations (12) and (15). Even in the elastic case, the operations involve complex numbers. The boundary condition for equation (15) remains the same.

The truncated semi-infinite wedge and the circular cavity embedded in a full-plane are analysed with a hysteretic damping ratio $\zeta = 0.05$. The parameters of the definition of the problems and of the solution procedures specified in Section 3 remain unchanged. The analytical solutions of equations (9) and (10) still apply with $c^*_s = (1+2\,\mathrm{i}\,\zeta)^{1/2}c_s$.

The dynamic-stiffness coefficient of the truncated semi-infinite wedge is decomposed as

$$S^{\infty} = K^{\infty}(k(a) + \mathrm{i}\;a\;c(a)) \quad (16)$$

with the static-stiffness coefficient K^{∞}, the spring coefficient $k(a)$ and the damping coefficient $c(a)$. The dimensionless frequency equals $a = \omega\;\xi\;r_0/c_s$. The analytical solution for the truncated semi-infinite wedge equals

$$S^{\infty} = -E^{0*}a^*\frac{H^{(2)}_{\alpha-1}(a^*)}{H^{(2)}_{\alpha}(a^*)} - E^{1*} + E^{0*}\,\alpha \quad (17)$$

with $a^* = (1+2\;\mathrm{i}\;\zeta)^{-1/2}a$. Good agreement for the dynamic-stiffness coefficient of the numerical procedure with the exact solution is obtained in Figure 11. The same also applies for the displacements in Figure 12.

In Figure 13 the displacements on the circumference at $\xi = 8$ of the cavity embedded in a full-plane with hysteretic damping are compared with the exact result.

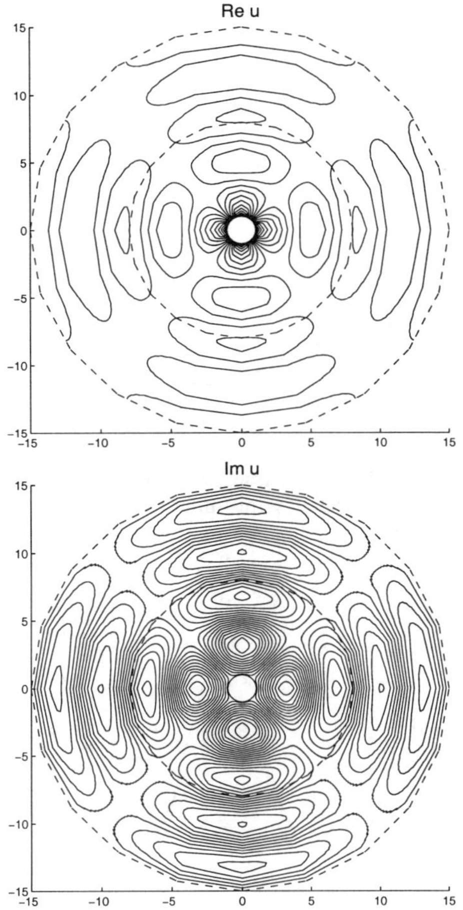

Figure 10: Contours of displacements for cavity problem including far-field.

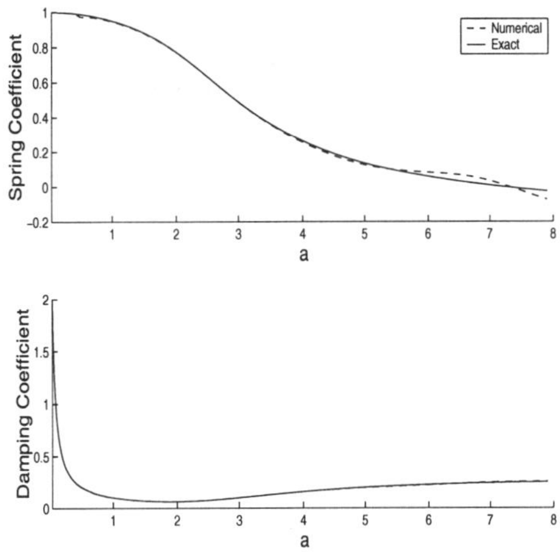

Figure 11: Dynamic-stiffness coefficient for truncated semi-infinite wedge with hysteretic damping.

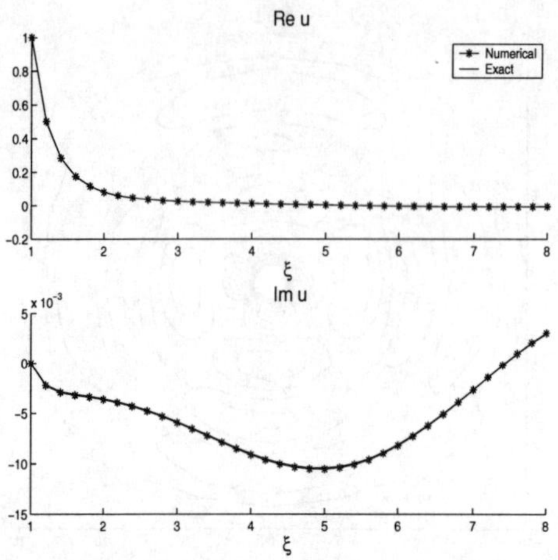

Figure 12: Real and imaginary parts of displacements for truncated semi-infinite wedge with hysteretic damping.

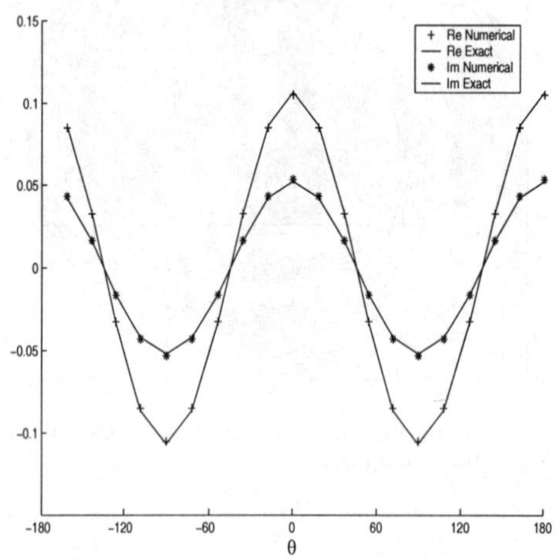

Figure 13: Displacements on circumference at truncation boundary $\xi = 8$ for cavity problem with hysteretic damping.

Computer Methods and Advances in Geomechanics, Desai et al. (eds) © 2001 Balkema, Rotterdam, ISBN 90 5809 183 X

Element-free method in solving discontinuous rock engineering problems

Weiyuan Zhou & Xiaodong Kou
Department of Hydraulic Engineering, Tsinghua University, Beijing, People's Republic of China

ABSTRACT: Element-free Method (EFM) which is based on Moving Least Squares Method, is a meshless method to solve complex boundary problems. It requires only nodes distributed inside and on the boundaries of the region under consideration, so no element information is needed for computation. This makes Element-free Method very convenient in application and especially superior in solving crack propagation problems. In this paper, the relevant theories of EFM and crack simulation used to trace crack propagation under complex stress conditions is presented. Numerical examples are given with very satisfying results.

1 INTRODUCTION

Since numerical simulation of crack propagation has been a very important problem all along in the field of engineering, many researchers have been engaged in it and have achieved great successes(Aoki et al. 1987, Gerstle et al. 1987, Koh et al. 1988, Wawrzynet & Ingraffea 1987). However there's not yet a method which can perfectly solve the problems of crack propagation under complex stress conditions such as compression induced crack simulation and arbitrarily-shaped crack propagation. Thus to study new methods in simulating crack propagation is of great significance.

Element-free Method(EFM), initiated by Belytschko (Belytschko et al. 1994, Lu et al. 1994), provides efficient approaches to treat crack propagation problems. It is a method to solve partial differential equations under complex boundary conditions. EFM is similar to Finite Element Method(FEM) in many aspects. The principal difference is that EFM requires only physical boundaries of the model to be taken into account and nodes in it, and no Element connectivity is needed. This makes EFM very convenient to trace crack propagation because nodes can been easily added along the crack line to simulate the crack and there is no need to re-mesh at the crack tips along with the crack growth. Moreover, to improve calculating accuracy of crack tip Stress Intensity Factor(SIF), dense pattern of nodes can be translated around crack tips through the fixed nodes which makes EFM more accurate and efficient.

EFM has been applied to crack propagation simulation by many researchers(Belytschko et al. 1994, Belytschko et al. 1995, Belytschko & Tabbara 1996). Their research mostly focused on tension induced cracks. Since cracks in practical environments are usually under complex conditions, more works need to be done in this field. In this paper, the authors submit new models on this subject.

2 GLOBAL BALANCE EQUATIONS OF EFM

2.1 *Moving Least Squares Method*

The mathematical basis of EFM is based on Moving Least Squares Method(MLSM). The details of MLSM see document (Lancaster & Salkauskas 1981).

Suppose there are n nodes, the displacement vector is:

$$\boldsymbol{u} = \left(\boldsymbol{u}_1^{\mathrm{T}}, \boldsymbol{u}_2^{\mathrm{T}}, \cdots, \boldsymbol{u}_n^{\mathrm{T}}\right)^{\mathrm{T}} \tag{1}$$

$$\boldsymbol{u}_i = \left(u_i,\ v_i\right)^{\mathrm{T}}, \quad i = 1,\ 2,\ \cdots,\ n \tag{2}$$

where u_i, v_i is displacement components of node i along x, y respectively.

Given a m-dimensional basis vector $\boldsymbol{p}(\boldsymbol{x})$ which is as follows in 2-dimension:

In linear order:

$$\boldsymbol{p}(\boldsymbol{x}) = \left(1,\ x,\ y\right)^{\mathrm{T}} \tag{3}$$

In quadratic order:

$$p(x) = \left(1,\ x,\ y,\ x^2,\ xy,\ y^2\right)^{\mathrm{T}} \tag{4}$$

In cubic order:

$$p(x) = \left(1, x, y, x^2, xy, y^2, x^3, x^2y, xy^2, y^3\right)^{\mathrm{T}} \tag{5}$$

The approximate displacements of an arbitrary point in the region are (Belytschko et al. 1994, Lu et al. 1994):

$$d(x) = (u, v)^{\mathrm{T}} = [N_1(x), N_2(x), \cdots, N_n(x)]u \tag{6}$$

$$N_i(x) = \begin{bmatrix} n_i(x) & 0 \\ 0 & n_i(x) \end{bmatrix}, \quad i = 1, 2, \cdots, n \tag{7}$$

where $n_i(x)$ is the shape function of node i obtained from MLSM:

$$n_i(x) = \sum_{j=1}^{m} p_j(x)\left[A^{-1}(x)B(x)\right]_{ji} \tag{8}$$

$$A(x) = \sum_{i=1}^{n} w_i(x)p(x_i)p^{\mathrm{T}}(x_i) \tag{9}$$

$$B(x) = \left[w_1(x)p(x_1), \cdots, w_n(x)p(x_n)\right] \tag{10}$$

where $w_i(x)$ is a weight function given as the following different from that of Belytschko:

$$w_i(r_i) = \begin{cases} \dfrac{r_{mi}^2}{r_i^2 + \varepsilon^2 r_{mi}^2}\left(1 - \dfrac{r_i^2}{r_{mi}^2}\right)^k & (r_i \le r_{mi}) \\ 0 & (r_i > r_{mi}) \end{cases} \tag{11}$$

where $r_i = \|x - x_i\|$ is the distance between x_i and x, r_{mi} is the influence radius of node i, ε is a minor value, and k is a positive integer.

The weight function of this form has great priority over that of Belytschko concerning approximation precision because it is k-1 order continuous.

k and ε should be selected carefully to improve precision. According to the authors' experiences, $k = 4$ and $\varepsilon = 1.0$ are good selections.

2.2 *Element Free Method*

The global balance equations of EFM can be derived from the theory of minimum potential energy. It should be mentioned that the boundary constraint is simulated by penalty function method:

$$Ku = f \tag{12}$$

$$K = \begin{bmatrix} K_{11} & K_{12} & \cdots & K_{1n} \\ \vdots & \vdots & \cdots & \vdots \\ K_{n1} & K_{n2} & \cdots & K_{nn} \end{bmatrix} \tag{13}$$

$$f = \left(f_1^{\mathrm{T}}, f_2^{\mathrm{T}}, \cdots, f_n^{\mathrm{T}}\right)^{\mathrm{T}} \tag{14}$$

$$K_{ij} = \int_{\Omega} B_i^{\mathrm{T}} D B_j \mathrm{d}\Omega + \int_{\Gamma_u} \alpha N_i^{\mathrm{T}} R N_j \mathrm{d}\Gamma \tag{15}$$

$$f_i = -\int_{\Omega} B_i^{\mathrm{T}} \sigma_0 \mathrm{d}\Omega + \int_{\Omega} N_i^{\mathrm{T}} \bar{f} \mathrm{d}\Omega + \int_{\Gamma_t} N_i^{\mathrm{T}} \bar{t} \mathrm{d}\Gamma + \int_{\Gamma_u} \alpha \bar{u}_\theta N_i^{\mathrm{T}} R r \mathrm{d}\Gamma \tag{16}$$

$$B_i(x) = \begin{bmatrix} \dfrac{\partial n_i(x)}{\partial x} & 0 \\ 0 & \dfrac{\partial n_i(x)}{\partial y} \\ \dfrac{\partial n_i(x)}{\partial y} & \dfrac{\partial n_i(x)}{\partial x} \end{bmatrix} \tag{17}$$

$$R = \begin{bmatrix} \cos^2\theta & \sin\theta\cos\theta \\ \sin\theta\cos\theta & \sin^2\theta \end{bmatrix}, \quad r = \begin{pmatrix} \cos\theta \\ \sin\theta \end{pmatrix},$$

$$\alpha = \frac{c(1-\mu)}{2(1+\mu)(1-2\mu)}E$$

where Ω is the calculated region. Γ_t is force boundary. Γ_u is constraint boundary. σ_0 is initial stress vector. $\bar{f}$ is body force vector. $\bar{t}$ is boundary force vector. u_θ is boundary displacement vector. θ is boundary constraint direction. α is penalty factor. $c = 10^3$-10^4, μ, E is poison ratio and elastic modulus of the material.

The integrals in the equations of EFM are calculated by Gauss quadrature performed on a rectangular cell structure independent of the nodes similar to the way by Belytschko (Belytschko et al. 1994).

3 SIMULATION OF CRACKS

Cracks can be treated as interfaces. Given an interface in region Ω as shown in Figure 1. The two faces of the interfaces are numbered as 1,2. The angle between local axis s and global axis x is θ. The normal and shear stiffness of the interface are k_n and k_s respectively. The interface has two effects:

(1) When using Moving Lest Squares Method, the influence region of a node never extends across the interface. As show in Figure 1, the influence nodes

of the Gauss point x_G are limited to the shade region of radius r_m, where r_m is the maximum of influence radii of nodes.

(2) The differences between displacements on face 1,2 produce new potential energy which contribute to the global balance equations.

The contribution of the interface to the equations can be obtained from theory of minimum potential energy:

$$\int_s \boldsymbol{N}_i^{(1)\mathrm{T}} \boldsymbol{A} \boldsymbol{N}_j^{(1)} \mathrm{d}s \rightarrow \boldsymbol{K}_{ij} \tag{18}$$

$$\int_s \boldsymbol{N}_i^{(2)\mathrm{T}} \boldsymbol{A} \boldsymbol{N}_j^{(2)} \mathrm{d}s \rightarrow \boldsymbol{K}_{ij} \tag{19}$$

$$-\int_s \boldsymbol{N}_i^{(1)\mathrm{T}} \boldsymbol{A} \boldsymbol{N}_j^{(2)} \mathrm{d}s \rightarrow \boldsymbol{K}_{ij} \tag{20}$$

$$-\int_s \boldsymbol{N}_i^{(2)\mathrm{T}} \boldsymbol{A} \boldsymbol{N}_j^{(1)} \mathrm{d}s \rightarrow \boldsymbol{K}_{ij} \tag{21}$$

where Superscripts 1,2 denote the two faces of the interface, and

$$\boldsymbol{A} = \boldsymbol{T}^{\mathrm{T}} \boldsymbol{P} \boldsymbol{T} \tag{22}$$

$$\boldsymbol{T} = \begin{bmatrix} \cos\theta & \sin\theta \\ -\sin\theta & \cos\theta \end{bmatrix}, \quad \boldsymbol{P} = \begin{bmatrix} k_s & 0 \\ 0 & k_n \end{bmatrix} \tag{23}$$

Mohr-Coulomb criteria is used to determine the states of the interface . If tensile failure occurs at a certain point on the interface, normal stiffness k_n at this point is set to zero. Otherwise, if shear failure occurs, Shear Stiffness k_s at this point is set to zero. So iterative calculation should be used in simulation of cracks.

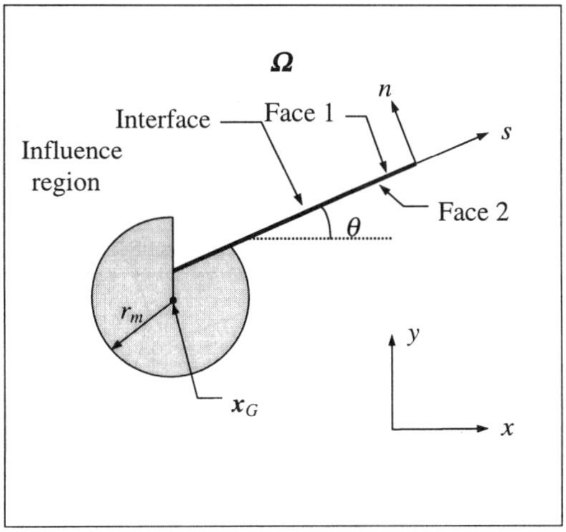

Figure 1. Simulation of a crack

4 CALCULATION OF CRACK TIP STRESS INTENSITY FACTOR (SIF)

Given a crack as shown in Figure 2. Take the body forces and surface forces of the crack into account. $\boldsymbol{\Gamma}$ and $\boldsymbol{\Gamma}'$ are positive in counterclockwise direction. $l_1 = A_1C_1$ and $l_2 = A_2C_2$. They are lines along the two crack surfaces, l_1 and l_2 are positive from left to right. Let

$$P_{1j} = w\delta_{1j} - \sigma_{jk}\frac{\partial u_k}{\partial x_1} \tag{24}$$

where w is the strain energy density. $\boldsymbol{\sigma}_{jk}$ is stress tensor, and u_k is displacement component.

Let $\boldsymbol{\Gamma}_1 = \boldsymbol{\Gamma} + l_1 - \boldsymbol{\Gamma}' - l_2$, a relationship can be derived from Green's formula:

$$J_1 = \int_{\Gamma_1} P_{1j} n_j d\boldsymbol{\Gamma} = \iint_{\Omega} f_k \frac{\partial u_k}{\partial x_1} d\boldsymbol{\Omega} \tag{25}$$

where n_j is the component of the outward unit normal of $\boldsymbol{\Gamma}_1$, and f_k is the body force density.

We have the next relationship:

$$l_1 = l_+ - l'_+, \quad l_2 = l_- - l'_-, \quad \boldsymbol{\Omega} = \boldsymbol{A} - \boldsymbol{A}' \tag{26}$$

where $l_+ = A_1O$, $l_- = A_2O$, $l'_+ = C_1O$, $l'_- = C_2O$, and $\boldsymbol{A}$ is the area surrounded by $\boldsymbol{\Gamma}$, $\boldsymbol{A}'$ is the area surrounded by $\boldsymbol{\Gamma}'$.

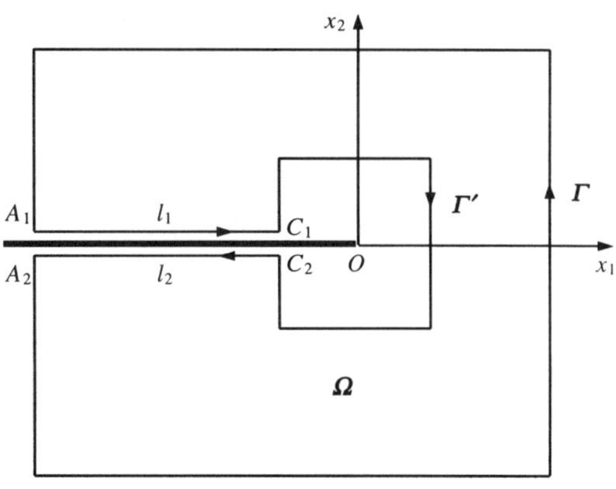

Figure 2. Calculation of SIF

Then an equation can be derived by substitution of equation (26) into (25):

$$J = J' \tag{27}$$

where:

$$J = \int_{\Gamma} P_{1j} n_j d\boldsymbol{\Gamma} + \int_{l_+ - l_-} P_{1j} n_j dl - \iint_A f_k \frac{\partial u_k}{\partial x_1} d\boldsymbol{A} \tag{28}$$

$$J' = \int_{\Gamma'} P_{1j} n_j d\boldsymbol{\Gamma} + \int_{l'_+ - l'_-} P_{1j} n_j dl - \iint_{A'} f_k \frac{\partial u_k}{\partial x_1} d\boldsymbol{A} \tag{29}$$

It can be proved that:

$$\lim_{\Gamma' \to 0} J' = \int_{\Gamma'} P_{1j} n_j d\boldsymbol{\Gamma} \tag{30}$$

The right term of above equation equals to the energy release rate G of the crack tip(Rice 1968), thus, it can be obtained from equation (27) that:

$$G = J = \int_{\Gamma} P_{1j} n_j d\boldsymbol{\Gamma} + \int_{l_+ - l_-} P_{1j} n_j dl - \iint_A f_k \frac{\partial u_k}{\partial x_1} d\boldsymbol{A} \tag{31}$$

The relationship between G and mode-I, II SIFs of a mixed-mode crack tip is as follows:

$$G = \alpha\left(K_{\mathrm{I}}^2 + K_{\mathrm{II}}^2\right) \tag{32}$$

$$\alpha = \begin{cases} \left(1-\mu^2\right)/E & \text{(Plain strain)} \\ 1/E & \text{(Plain stress)} \end{cases} \tag{33}$$

Now consider two independent elastic field 1,2. Suppose SIFs under these two fields are $K_{\mathrm{I}}^{(1)}, K_{\mathrm{II}}^{(1)}$ and $K_{\mathrm{I}}^{(2)}, K_{\mathrm{II}}^{(2)}$ respectively, and taking no account of body forces and surface forces for the crack under filed 2. Field 3 is the superposition of field 1,2, Thus:

$$\sigma_{ij}^{(3)} = \sigma_{ij}^{(1)} + \sigma_{ij}^{(2)}, \quad \varepsilon_{ij}^{(3)} = \varepsilon_{ij}^{(1)} + \varepsilon_{ij}^{(2)},$$

$$u_k^{(3)} = u_k^{(1)} + u_k^{(2)}, \quad f_k^{(3)} = f_k$$

Substitute above equations into equation (31), we can obtain:

$$G^{(3)} = G^{(1)} + G^{(2)} + \int_{\Gamma} P_{1j}^{(1,2)} n_j d\boldsymbol{\Gamma} + \int_{l_+ - l_-} P_{1j}^{(1,2)} n_j dl - \iint_A f_k \frac{\partial u_k^{(2)}}{\partial x_1} d\boldsymbol{A} \tag{34}$$

where:

$$P_{1j}^{(1,2)} = \varepsilon_{ij}^{(1)} \sigma_{ij}^{(2)} \delta_{1j} - \sigma_{jk}^{(1)} \frac{\partial u_k^{(2)}}{\partial x_1} - \sigma_{jk}^{(2)} \frac{\partial u_k^{(1)}}{\partial x_1} \tag{35}$$

The following expression can be derived from equation (32):

$$G^{(3)} = \alpha\left[\left(K_{\mathrm{I}}^{(1)} + K_{\mathrm{I}}^{(2)}\right)^2 + \left(K_{\mathrm{II}}^{(1)} + K_{\mathrm{II}}^{(2)}\right)^2\right] = G^{(1)} + G^{(2)} + 2\alpha\left(K_{\mathrm{I}}^{(1)} K_{\mathrm{I}}^{(2)} + K_{\mathrm{II}}^{(1)} K_{\mathrm{II}}^{(2)}\right) \tag{36}$$

Then the following equation can be found by comparison between equation (34) and (36):

$$2\alpha\left(K_{\mathrm{I}}^{(1)} K_{\mathrm{I}}^{(2)} + K_{\mathrm{II}}^{(1)} K_{\mathrm{II}}^{(2)}\right) = \int_{\Gamma} P_{1j}^{(1,2)} n_j d\boldsymbol{\Gamma} + \int_{l_+ - l_-} P_{1j}^{(1,2)} n_j dl - \iint_A f_k \frac{\partial u_k^{(2)}}{\partial x_1} d\boldsymbol{A} \tag{37}$$

Equation (37) can be used to determine $K_{\mathrm{I}}^{(1)}, K_{\mathrm{II}}^{(1)}$. Let $K_{\mathrm{I}}^{(2)} = 1$, $K_{\mathrm{II}}^{(2)} = 0$ to get $K_{\mathrm{I}}^{(1)}$, and let $K_{\mathrm{I}}^{(2)} = 0$, $K_{\mathrm{II}}^{(2)} = 1$ to get $K_{\mathrm{II}}^{(1)}$.

5 TRACING CRACK PROPAGATION

The process to trace crack propagation by EFM is shown in Figure 3.

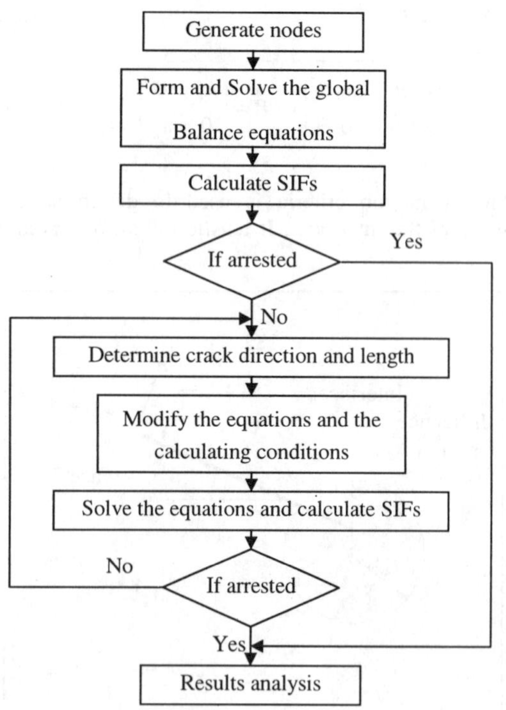

Figure 3. Flow chart to trace crack propagation by Element-free Method

In the current study, the maximum circumferential stress criteria has be used. The direction of crack growth measured from the current crack line is determined by

$$K_{\mathrm{I}} \sin\theta_0 + K_{\mathrm{II}}(3\cos\theta_0 - 1) = 0 \tag{38}$$

$$K_{\mathrm{I}} \cos\frac{\theta_0}{2}(3\cos\theta_0 - 1) - K_{\mathrm{II}} \sin\frac{\theta_0}{2}(9\cos\theta_0 + 5) > 0 \tag{39}$$

The critical condition of the crack tip is

$$K_\theta = \frac{1}{2}\cos\frac{\theta_0}{2}[K_{\mathrm{I}}(1+\cos\theta_0) - 3K_{\mathrm{II}}\sin\theta_0] = K_{\mathrm{IC}} \tag{40}$$

where K_{IC} is the fracture toughness. K_θ is called effective stress intensity factor.

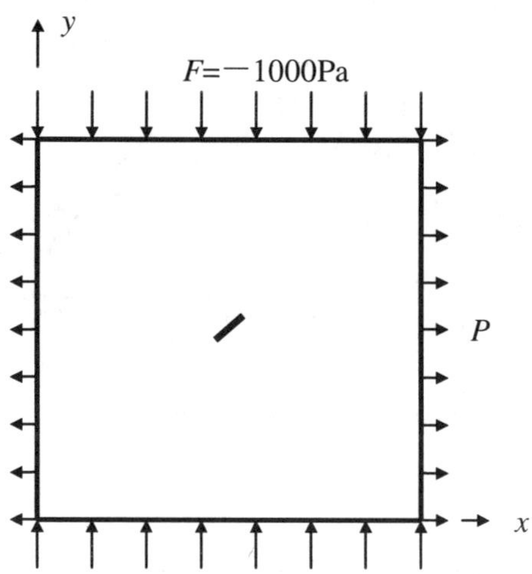

Figure 4. Plate containing a crack

6 EXAMPLE RESULTS

6.1 *Tracing compression-induced crack propagation*

Given a plain strain rectangular specimen containing a crack as shown in Figure 4. The side of the specimen is 10m long, and the coordinates of the two ends of the crack is (4.5m,4.5m), (5.5m,5.5m). Axial compressive stress *F*=-1000Pa and lateral stresses are applied to the specimen. We will study the crack propagation in three lateral stress cases of P=100Pa, P=0, P=-100Pa. Linear basis is employed for the Moving Least Squares Method in this example.

The crack length of the first step is set to 0.1m, then the following step length is selected in proportion to the effective stress intensity factor.

High dense and radially distributed nodes are translated around the crack tips to improve SIF accuracy. Toughness is set to zero in order to study the crack propagation pattern.

Figure 5a, b, c illustrate the crack path and nodes distribution after 30 steps tracing under lateral stresses of P=100Pa, P=0, P=-100Pa respectively. Figure 5d shows the relationship between effective stress intensity factor K_θ and crack growth length. It can be seen that crack propagation path will lead to the direction of the major principal compressive stress and the lateral stresses have significant effects on the stability of the crack. Under lateral compression or no lateral stress, the decreasing of K_θ along with crack growth length shows that the crack propagation is stable. Otherwise, under lateral tensile stress, the crack propagation will be unstable even if the stress is very small. This results coincide very well with experimental phenomenon(Horii & Nemat-Nasser 1985).

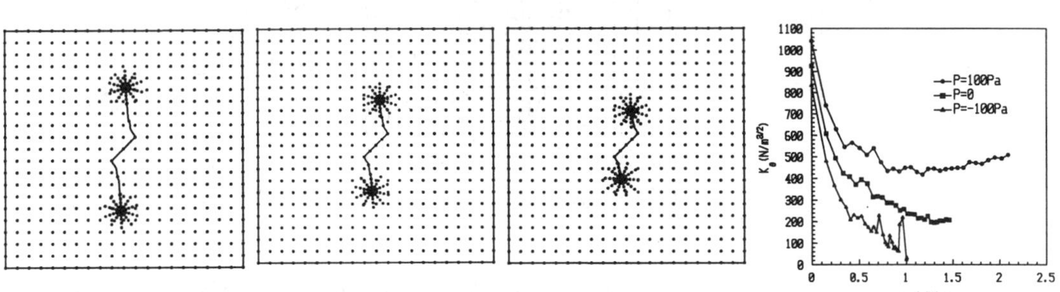

a. P=100Pa b. P=0 c. P=-100Pa d. K_θ-Length curve

Figure 5. Element-free method tracing crack propagation

6.2 *Simulation of multi-crack propagation*

Consider a plane strain 10m square rectangular plate with two cracks shown in Figure 6a.

The ends of the cracks are A, B and C, D. The plate is subjected to biaxial compression load of –1000Pa in y direction and –100Pa in x direction. In this example, we analyze the propagation the cracks under compression-shear load and linear basis is employed for the Moving Least Squares Method.

The step length is estimated in the same way as in the preceding example, but two points should be noted in case of the cracks coming close up.

(1). If the distance between two tips becomes less than the current step length, these two tips are joined by a straight crack line.

(2). If the distance between a tip and another crack line becomes less than the current step length, the crack tip is extended to the point of intersection between the tangent of the tip and that crack line.

Two types of crack position are considered:

(1). Coordinates of tips A, B, C, D are (4.0, 3.5), (5.0, 4.5), (5.0, 5.5), (6.0, 6.5) respectively.

(2). Coordinates of tips A, B, C, D are (4.25, 3.5), (5.25, 4.5), (4.75, 5.5), (5.75, 6.5) respectively.

Figures 6b-6f show the results of crack simulation for these two types of crack position.

The propagation of crack position 1 is studied firstly. Figure 6b shows the crack path and corresponding nodes distribution after 21 steps of calculating, and Figure 6c shows the relationship between effective stress intensity factor K_θ and crack growth length of tips A and B. It can be seen that for the crack position 1, two cracks propagate separately, no tips connection occurs. It is also found that the propagation of each tip is shown

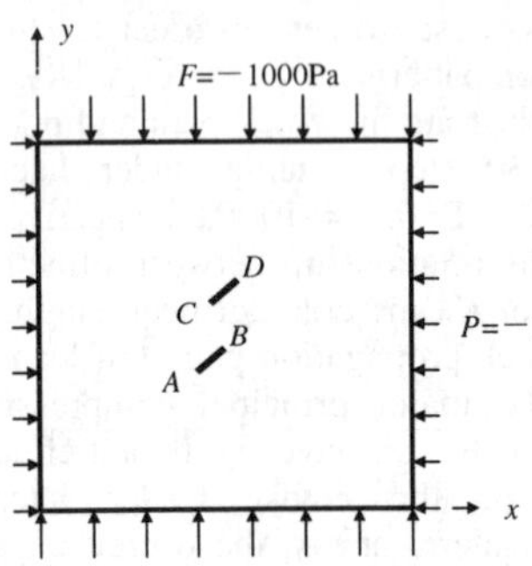

a. Plate with two cracks

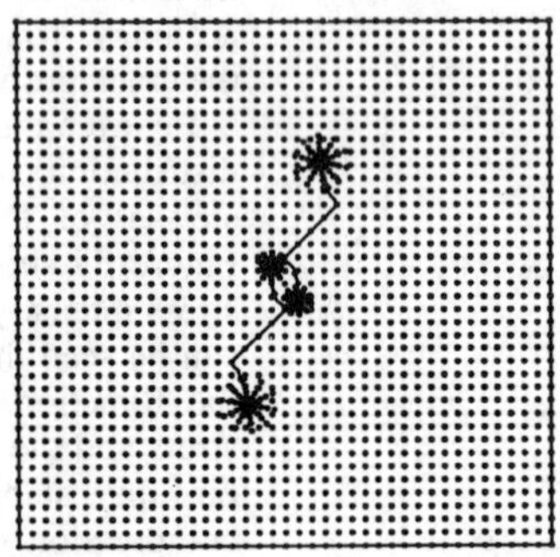

b. Crack path and nodes distribution (Step 21 for crack position 1)

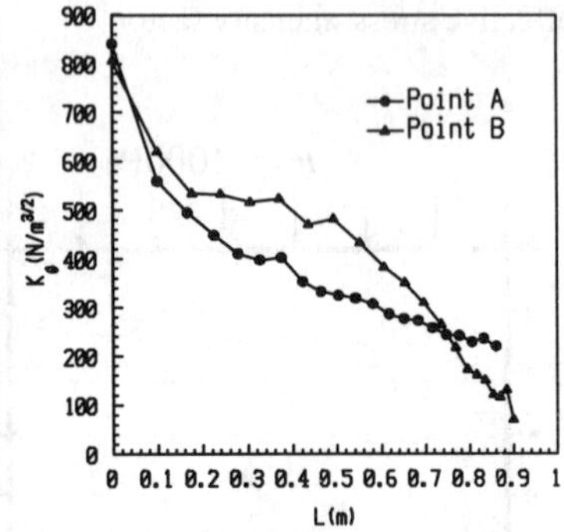

c. K_θ-Length curve (Step 21 for crack position 1)

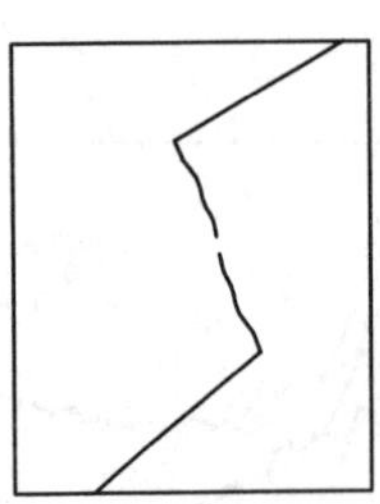

d. Amplified vicinity of tips (Step 10 for crack position 2)

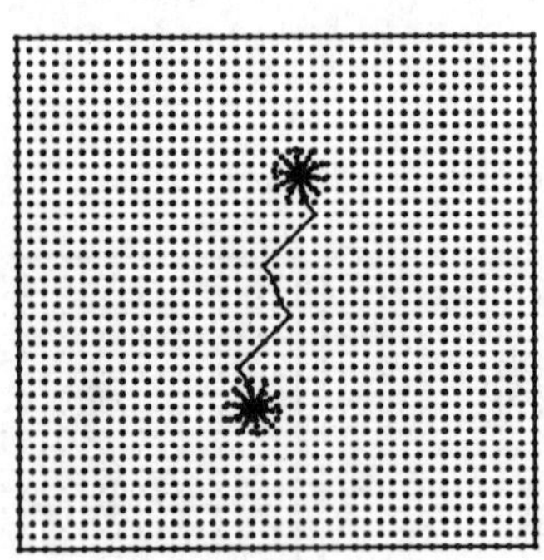

e. Crack path and nodes distribution (Step 22 for crack position 2)

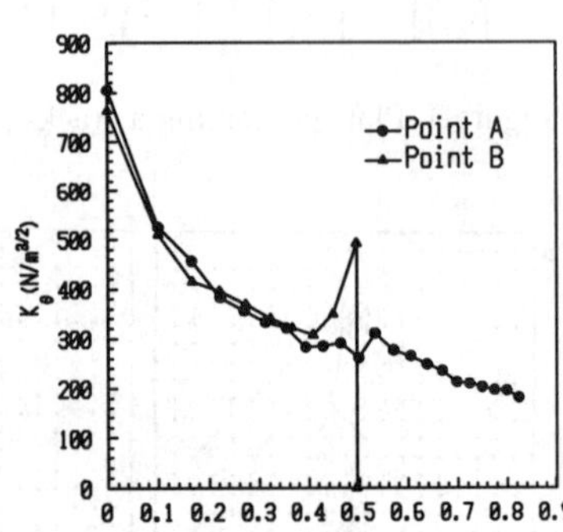

f. K_θ-Length curve (Step 22 for crack position 2)

Figure 6. Element-free method tracing multi-crack propagation in stable state by the continually decrease of K_θ along with crack growth. The path and characteristics coincide very well with the phenomenon shown in literature(Horii & Nemat-Nasser 1985).

Comparing K_θ of tip A with tip B, we find that K_θ of tip B begins with larger value than tip A at the early stage of propagation, but the situation is reversed at the later stage. The reason may be that, at the early stage, with the reduction of the distance between B and C, the stress intensity factor of tip B increases as a result of the increasing interaction between tip B and C, and later on the contrary, tip B goes farther and farther from tip C and approaches the initial line of the other crack. The sliding of this crack weakens the stress intensity factor of tip B.

Then the case of crack position 2 is studied. Figure 6d shows the amplified vicinity of tips B and C after 10 steps. It can be seen that tips B and C come in a distance smaller than the current step length. According to the measures proposed above concerning tips in small distance, if the fracture toughness of the material is small enough so that the cracking of tip B is not arrested before the 10th step, tips B and C will be joined instantly in the 11th step and the two cracks are looked as a single crack later on with just two tips A and D.

Figure 6e shows the crack path and nodes distribution after 22 steps of calculating, and Figure 6f shows the relationship between K_θ and crack growth length of tips A and B in the process. At the beginning, K_θ of tip B is close to that of tip A. But before the connecting of tip B with tip C, K_θ of tip B increases enormously because of the strong interaction between the crack tips. Then the two cracks are joined as a single crack. Tip B and C stops propagating and K_θ of tip B decreases to zero. The two tips A and D continue to propagate in the direction tending to the major principal compressive stress and they still experience a stable growth shown by the decrease of K_θ.

7 CONCLUSION

This paper has described the theories of EFM, studied the method to trace crack propagation under complex loads and demonstrated the performance of EFM. Numerical examples show the following advantages of EFM in tracing crack propagation.

1. Simplicity. EFM requires only nodes, no re-mesh is needed during the tracing process. So it is very convenient.

2. High accuracy and efficiency. High density and radially distributed nodes translated around crack tips makes EFM be able to obtain relatively accurate value of SIF with sparsely distributed nodes.

ACKNOWLEDGEMENT

This paper is supported by CNSFC-59909004.

REFERENCES

Aoki S., Kishimoto K. & Sakata M. 1987, Finite Element Computation of Dynamic Stress Intensity Factor for a Rapidly Propagating Crack Using J-integral. *Comput. Mech.*, 2: 54-62

Belytschko T., Lu Y.Y. & Gu L. 1994, Element-free Galerkin Methods. *Int. J. for Num. Methods in Engrg.*, 37: 229-256

Belytschko T., Lu Y.Y. & Gu L. 1995, Crack Propagation by Element-free Galerkin Method. *Engrg. Fract. Mech.*, 51(2): 295-315

Belytschko T., Gu L.. & Lu Y.Y. 1994, Fracture and Crack Growth by Element-free Galerkin Methods. *Modell. Simul. Mater. Sci. Engrg.*, 2: 519-534

Belytschko T. & Tabbara M. 1996, Dynamic Fracture Using Element-free Galerkin Methods. *Int. J. Num. Methods in Engrg.*, 39: 923-938

Gerstle W.H., Matha L. & Ingraffea A.R. 1987, Finite And Boundary Element Modelling of Crack Propagation in Two- and Three-dimensions. *Engrg. with Computers*, 2: 167-183

Horii H. & Nemat-Nasser S. 1985, Compression induced Microcrack Growth in Brittle Solids: Axial Splitting and Shear Failure. *J. of Geophysical Research*, 90(B4): 3105-3125

Koh H.M., Lee H.S. & Haber R.B. 1988, Dynamic Crack Propagation Analysis Using Eulerian-Lagrangian Kinematic Descriptions. *Comput. Mech.*, 3: 141-155

Lancaster P. and Salkauskas K. 1981, Surface generated by moving least squares methods. *Math. Comput.*, 37, 141-158.

Lu Y.Y., Belytschko T. and Gu L. A. 1994, New Implementation of the Element-Free Galerkin Method. *Comput. Methods Appl. Mech. Engrg.*, 113: 397-414

Rice J.R. 1968, A Path-independent Integral and Approximate Analysis of Strain Concentrations by Notches and Cracks. *ASME J. of Applied Mech.*, 35: 379-386

Wawrzynet P.A. and Ingraffea A.R. 1987, Interactive Finite Element Analysis of Fracture Processes and Integrated Approach. *Theor. Appl. Fracture Mech.*, 8: 137-150

Computer Methods and Advances in Geomechanics, Desai et al. (eds) © 2001 Balkema, Rotterdam, ISBN 90 5809 183 X

An adaptive approach to limit analysis on soil mechanics

I. D. S. Pontes Filho
Department of Civil Engineering, Federal University of Pernambuco, Recife, Brazil

L. A. Borges & N. Zouain
Department of Mechanical Engineering, Federal University of Rio de Janeiro, Brazil

ABSTRACT The main objective of this work is to presents an adaptive mesh refinement procedure for finite element models in limit analysis with application on soil mechanics. We use an a posteriori indicator based on the local directional interpolation error and a recovering scheme to compute second derivatives of the finite element solution. The proposed mesh adaptation process gives improved results in localizing regions of rapid or abrupt variations of variables, whose location is not know a priori. Limit analysis of the bearing capacity of the footing on the slope in plane strain is considered in the applications.

1 INTRODUCTION

The main objective of this work is to presents an adaptive mesh refinement procedure for limit analysis on soil mechanics. This procedure uses an a posteriori estimator of the local directional interpolation error and a recovering scheme to compute the first and second derivatives of the finite element solution (Borges et al, 1999; Borges et al, 1998).

We place particular emphasis on the anisotropic mesh adaptation process, generated by the proposed directional indicator. The goal of that approach is to achieve a mesh adaptive strategy accounting for mesh size refinement, as well as redefinition of the oriented element stretching. This way, the mesh turns aligned with the direction of maximum curvature of the function graph. This mesh adaptation procedure gives improved results in localizing regions of rapid or abrupt variations of the variables, whose location is not know a priori (Peiró, 1989; Almeida et al., 1998;). Limit analysis aims at determining the loads that will cause the phenomenon of incipient plastic colapse in a body made of an elastic ideally plastic material. This phenomenon is characterized by the development of continuously increasing strains under constant load. Localized plastic deformations or slip bands are present in many collapse situations. The accuracy in numerical solutions of limit analysis is seriously affected by local singularities arising from these slip bands.

In limit analysis, an a priori error estimate, in a sense local as provided by the standard error analysis in the finite element method, is not available. Those facts disclose the need of a mesh refinement indicator which can be extracted a posteriori from the numerical solution. In this paper, firstly, the theoretical framework of limit analysis is presented. Following, some issues of the adopted techniques for estimating the interpolation error, based on derivatives recovery schemes, are discussed. Finally, we presents a directional interpolation error estimator and adaptive mesh refinements for limit analysis applications.

2 LIMIT ANALYSIS

Under the consideration of proportional loading, the limit analysis problem consists in finding a load factor α such that the body undergoes plastic collapse when subject to the references loads F uniformly amplified by α. In turn, a system of loads produces plastic collapse if there exists a stress field in equilibrium with these loads, which is plastically admissible, and related, by the constitutive equations to, a plastic strain rate field kinematically admissible. Let us translate into mathematical notation the above definition. Limit analysis consists in finding a load factor α, a stress field T, a plastic strain rate field D^p and a velocity field v such that:

$$D^p = \mathcal{D}v \quad v \in V \tag{1}$$

$$T \in S(\alpha F) \tag{2}$$

$$T \in \partial X(D^p) \quad \Longleftrightarrow \quad D^p \in \partial Ind_P(T) \tag{3}$$

Equation (1) imposes that the collapse plastic strain rate is related to a kinematically admissible velocity field v by means of the tangent deformation operator $\mathcal{D}$. The symbol $S(\alpha F)$ in (2) denotes the set of all stress fields in equilibrium with the given system of forces αF, that is satisfying the Principle of Virtual Power:

$$\int_{\mathcal{B}} T\cdot\mathcal{D}v d\mathcal{B} = \int_{\mathcal{B}} \alpha b\cdot v d\mathcal{B} + \int_{\Gamma_\tau} \alpha\tau\cdot v d\Gamma \quad \forall v \in V \quad (4)$$

where b and τ are body and surface loads respectively, and Γ_τ the region of Γ where tractions are prescribed.

The constitutive relation (3) describes the behavior of an elastic ideally-plastic material. The symbol $\partial X(D^p)$ denotes the subdifferential of the plastic dissipation potential X. This subdifferential (Borges et al., 1996) is the set of all stress fields T complying with:

$$X(D^{p*})-X(D^p)\geq \int_{\mathcal{B}} T\cdot(D^{p*}-D^p)d\mathcal{B} \quad \forall D^{p*}\in W \quad (5)$$

For these non-hardening materials, the dissipation function X is related to the set of admissible stress fields P by

$$X(D^p) = \sup_{T^*\in P} \int_{\mathcal{B}} T^*\cdot D^p d\mathcal{B} \quad (6)$$

In the second part of (3), $Ind_P(T)$ defines the indicator function of the set P, which is 0 if $T\in P$ and $+\infty$ if $T\notin P$. The set $\partial Ind_P(T)$ is empty for $T\notin P$ and coincides with the cone of normals to P at T, denoted $C_P(T)$ and defined by:

$$C_P(T) = \{D^p\in W \mid (T^*-T)\cdot D^p \leq 0, \forall T^*\in P\} \quad (7)$$

The cone $C_P(T)$ is a generalization of the elementary concept of normals to hypersurfaces. It contains only the classical normal whenever smooth differential geometry applies. For singular points, on the boundary of P, this cone is bounded by all the normals defined in regular points of an infinitesimal neighborhood surrounding the considered vertex or edge (Pontes Filho et al., 1997). Therefore, (3) is written $D^p\in C_P(T)$ by additionally enforcing $T\in P$.

Often, the set P is defined, by means of a yield function f, as:

$$P = \{T\in W' \mid f(T)\leq 0 \quad x\in\mathcal{B}\} \quad (8)$$

In many cases f is a vector function whose components f_j, the plastic modes, are regular convex functions of T. Then, the inequality above is understood componentwise.

At any stress T of the boundary of P representing a non-singular point of all active plastic modes f_j, the flow law (3) is equivalent to

$$D^p = \nabla f(T)\dot{\lambda} \quad (9)$$

$$f(T)\dot{\lambda} = 0 \qquad f(T)\leq 0 \quad \dot{\lambda}\geq 0 \quad (10)$$

where $\nabla f(T)$ is the gradient of f, and $\dot{\lambda}$ the vector of plastic multipliers.

The discret limit analysis problem consists in finding a load factor $\alpha\in R$, a stress vector $T\in R^q$, a velocity vector $v\in R^n$ and a plastic multipliers vector $\dot{\lambda}\in R^m$, such that the system represented by a deformation matrix $B: R^n\to R^q$ and a convex function $f(T)\in R^m$, undergoes plastic collapse for some load being proportional to a given force vector $F\in R^n$. The discretized version of the limit analysis formulations leads to a finite dimensional problem that can be seen as a discret version of eqns (1-3), that is:

$$Bv - \nabla f(T)\dot{\lambda} = 0 \quad (11)$$

$$B^T T - \alpha F = 0 \quad (12)$$

$$F\cdot v = 1 \quad (13)$$

$$f(T)\dot{\lambda} = 0 \; f(T)\leq 0, \quad \dot{\lambda}\geq 0, \; j=1,...,m \quad (14)$$

A Newton-like algorithm for solving this discrete problems is described by Borges et al, (1996) and is not discussed in the present work.

We use a curved triangular mixed with six nodes intended fo the C^0 (quadratic interpolation of geometry and velocities), three nodes at vertices for the discontinuos linear interpolation of the deviatoric stress and, piecewise constant interpolation for the mean stress.

3 ESTIMATOR FOR THE INTERPOLATION ERROR

In the methods considered here, gradients and/or Hessians of the solutions obtained on a given mesh are smoothed and then used in the interpolation error estimator.

We present an anisotropic a posteriori error estimator for the diference between a given function u and a discrete function $u_h\in V_h$ which is a good approximation of u. That is (Almeida et al, 1998), there exists a constant C such that

$$\|u - u_h\|_{L^P} \leq C\|\mathcal{H}(u_h(x))(x - x_0).(x - x_0)\|_{L^P} \quad (15)$$

where $\mathcal{H}(u_h(x))$ denotes the recovered Hessian matrix obtained from the information given by the finite element solution u_h. This shows that the interpolation error at some point x is governed by the behaviour of the second order derivative in such point and the error depends on the direction $(x - x_0)$ and the recovered Hessian matrix value in this point $\mathcal{H}(u_h(x))$.

The above result suggests the use of (15) as a directional error estimator in the terminology used by Peiró (1989). Since the recovered Hessian matrix is not positive definite, it can not be taken as a metric tensor. As an alternative, Peiró introduced the metric tensor:

$$\mathbf{G} = \mathbf{Q\Lambda Q^T} \quad (16)$$

where Q is the matrix of eigenvectors of the recovered Hessian matrix and $\Lambda = diag(|\lambda_1|, |\lambda_2|)$, and $(|\lambda_i|), i = 1, 2)$, are the absolute value of the associated eigenvalue $(|\lambda_1| \leq |\lambda_2|)$. According to this definition, the metric tensor field G is at least positive semi-definite. In particular, a zero eigenvalue poses no difficulty since it leads to infinite mesh sizes that are inhibited by the element size limitation or by the bounds of the computational domain.

Following the above ideas it is proposed to use an error estimator that provide a measure of the second derivative contribution in each element (Venere, 1996). Considering a given finite element mesh $\Im$ of the domain Ω, the indicator value corresponding to each element $T \in \Im$ is defined by the following expression (Feijoó et al, 1997; Ainsworth and Oden, 1997; Zienckiewicz and Zhu, 1987).

$$\eta_T = \left(\int_T [G(u_h(x_0))(x - x_0).(x - x_0)]^p d\Omega\right)^{\frac{1}{p}} \quad (17)$$

where x_0 is the center of this element.

The global indicator η is given by:

$$\eta = \left(\sum_{T \in \Im} \eta_T^p\right)^{\frac{1}{p}} \quad (18)$$

Several approaches have been proposed in the framework of the finite element method, in order to recover first derivatives (Zienkiewicz and Zhu, 1992a; Zienkiewicz and Zhu, 1992b). The weighted average is one among these procedures and consists on turning the inter-elements discontinuos field ∇u_h into a continuos field $\nabla_R u_h$. This is made by employing the same element basis function used to construct the approxiamtion u_h to compute the field ∇u_h. Then, a weighted average of ∇u_h computed on the elements surrounding a node N is adopted as the value $\nabla_R u_h(\mathbf{X}_N)$ of the recovered gradient at this node, where $\mathbf{X}_N$ is the coordinate of node N. The weighted average is computed using weights given by the inverse of the distance between the node N and the points of superconvergence of the gradient (the center of the element in the case of linear triangles and the Gauss points near midside nodes in the case of quadratic triangles). The weights can also be defined by the area of the elements surrounding the node.

Second derivates can be recovered by using the same approaches used for the first derivative recuperation. In fact, taking $\nabla_R u_h$ as a new field, we can reapply the algorithm in order to find $\nabla_R(\nabla_R u_h)$. The symmetric part of the approximation is retained in order to ensure the symmetry of the recovered Hessian matrix.

In order to approximate functions presenting strong variations in the derivatives, the adapted mesh turns to be oriented by means of the stretching of its elements in the direction of maximum curvature of the function graph. Whenever this stretching is very large, it may cause poor precision when computing weighted averages. To overcome this situation, the original domain Ω_T is locally transformed into a standard unstretched domain Ω_T^*. For instance, when using mesh generator based on the advancing front technique (Peiró, 1989) this domain transformation is naturally chosen in accordance with the domain mapping which is part of the mesh generation algorithm.

Considering the neighborhood of an arbitrary node N, the mesh generation algorithm tends to create triangular elements which are equilateral when viewed in the transformed domain given by the following operator:

$$S(N) = \frac{1}{s(N) * h(N)}\mathbf{e_1} \otimes \mathbf{e_1} + \frac{1}{h(N)}\mathbf{e_2} \otimes \mathbf{e_2} \quad (19)$$

where $\mathbf{e}_i(N), i = 1, 2$ are the eingenvectors of the Hessian matrix $\mathcal{H}(u_h(N))$, the length $h(N)$ is the element size in the $\mathbf{e}_2$ direction, and $s \geq 1$ is the element stretching in the $\mathbf{e}_1$ direction of the generated mesh at node N. Those parameters are all dynamically defined along the mesh adaptation process. Finally, the mapping which transforms the stretched element into the standard (equilateral triangle) is written

$$x_M = S(N)(X_M - X_N) \quad (20)$$

where M is a generic point belonging to some element adjacent to the node N_i and the coordiantes X_N, x_N, X_M and x_M denotes the points N and M before and after the mapping respectively.

4 ADAPTIVE PROCEDURE

The present adaptive procedure takes into account the global error indicator, η, given by (17), for each triangulation $\Im_k$. Then, the objective is to find a new mesh $\Im_{k+1}$, with a given number of elements Nel. This new finite element mesh is generated trying to produce a uniform distribution of the interpolation error estimator over all elements.

Our remeshing algorithm is based on the *advancing front technique*. In this technique, the mesh generator tries to build equilateral triangles in the metric defined by the variable metric tensor **S** defined by (19). To evaluate the mapping parameters we proceed as follows

1. Compute η_T in each element and then the global quantity η.

2. Given a number of elements Nel in the new adapted mesh, the expected local indicator is given by

$$\eta^* = \frac{\eta}{\sqrt{Nel}} \tag{21}$$

3. The decreasing or increasing rate of the element size is estimated by

$$\beta_T = \left(\frac{\eta^*}{\eta_T}\right)^{\frac{1}{3}} \tag{22}$$

   From this rate distribution β_T, computed elementwise, nodal values are then obtained.

4. The size of the new element, to be generated at node N, is

$$h_{k+1}(N) = \beta(N)\; h_k(N). \tag{23}$$

   If necessary, the threshold values for the new element size are then enforced

$$\underline{\alpha} * h_k \leq h_{k+1} \leq \overline{\alpha} * h_k \leq L \tag{24}$$

   where L represents the characteristic length of the domain Ω. The two parameters above, $\underline{\alpha}$ and $\overline{\alpha}$, are used in order to ensure progressive mesh adaptation.

5. The stretching factor s at node N is defined by

$$s(N_k) = \sqrt{\frac{|\lambda_2|}{|\lambda_1|}} \tag{25}$$

   where $|\lambda_1| \leq |\lambda_2|$ are the absolute eigenvalue of the Hessian matrix $\mathbf{H}(u_h(N))$. This stretching factor must bounded in order to ensure that the new length of the element in the direction $\mathbf{e}_1$, sh_{k+1}, is no greater than the characteristic length L of the domain Ω, i.e., $s \leq L/h_{k+1}$.

6. The new size distribution, h_{k+1} is then uniformly scaled. Due to the limitation on the values of h and s the number of elements in the new adapted mesh may be different from the expected Nel. To enforce the required new number of elements, the elements size h is modified as follows:

$$h_{k+1} \leftarrow \sqrt{\frac{Nel_{new}}{Nel}}\; h_{k+1} \quad \text{with} \tag{26}$$

$$Nel_{new} = \frac{4}{\sqrt{3}} \int_\Omega \frac{2}{sh^2} d\Omega \tag{27}$$

The adaptive strategy described above is repeated until the interpolation error estimator in the mesh, $\Im_k$, becomes lower than a given admissible relative error $\overline{\gamma}$, that is until,

$$\frac{\eta_k}{\left(\|u_h\|_{L_2(\Omega)} + \eta_k\right)^{\frac{1}{2}}} \leq \overline{\gamma} \tag{28}$$

5 NUMERICAL APPLICATION

As application we consider the determination of the bearing capacity of the footing on the slope of angle 60^0, made up of an ideally plastic Mises material, under plane strain condition. This is a classical problem ($\alpha = 3.04$) of plastic strain localization in geomechanics (Finn, 1967).

In Figure 1 it is shown the evolution of the bearing capacity and the global error along the adaptive process.In three adaptation steps, we obtain a reduction of 13% in the collapse load ($\alpha = 3.05$).

In the same figure it is plotted the sequence of meshes obtained with the remeshing and the evolution of the region of plastic deformation along the process. Those results show that the proposed procedure is effective in capturing the localized plastic deformation, increasing the nodal points of the mesh only near the region where there are discontinuities, aligned and stretching the elements in the direction of the slip bands.

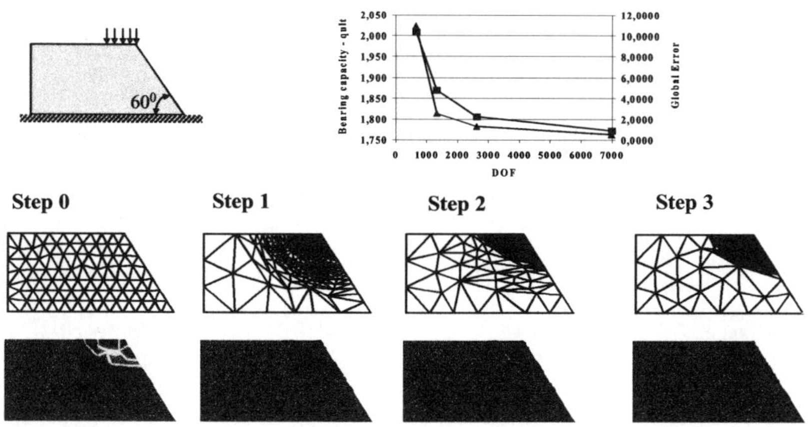

Plastic multiplier field

Figure 1 - Footing on the slope -- Error, bearing capacity,mesh and plastic deformation.

6 BIBLIOGRAPHY

Ainsworth, M. and Oden, J.T. (1997), A Posteriori Error Estimaton in Finite Element Analysis, Comput. Methods Appl. Mech. Engeneering, pg42-88.

Almeida, R.C.; Feijóo, R.A. and Galeão, A.C.N. and Padra, C. and Simões, R. (1998), Adaptive Finite Element Computational Fluid Dynamics Using an Anisotropic Error Estimator, Computational Mechanics - New Trends and Applications, Fourth World Congress on Computational Mechanics, Argentina.

Borges, L.; Feijóo, R.A.; Padra, C. and Zouain,N. (1998) A Directional Error Estimator For Adaptive Finite Element Analysis, Computational Mechanics - New Trends and Applications, Fourth World Congress on Computational Mechanics, Argentina.

Borges, L.A.; Zouain, N. and Huespe, A.E. (1996), A Nonlinear Optimization Procedure for Limit Analysis, European Journal of Mechanics A/Solids, vol.15, pp 487-512

Borges, L.; Feijóo, R.A. and Zouain,N. (1999), A directional error estimator for adaptive limit analysis, Mechanics Research Communications, vol. 26, n.5, pp 555-563.

Finn, D.L. (1967), Applications of Limit Plasticity in Soil Mechanics, J. Soil Mech. Foundations Division, ASCE, SM5, pp 101-119

Peiró, J. (1989), A Finite Element Procedure for the Solution of the Euler Equations on the Unstructured Meshes, Phd Thesis, Department of Civil Engineering, University of Swansea, UK.

Pontes Filho, I.D.S., Borges, L.A., Zouain, N., Lopes, F.R. (1997), An approach to limit analysis with coneshaped yield surfaces, Int. Journal for Numerical Methods in Engineering, vol. 40, 4011-4032 .

Venére, M.J. (1996), Técnicas Adaptativas para el Método de Elementos Finitos en Dos y Tres Dimensiones, Phd Thesis, Instituto Balseiro, Centro Atómico Bariloche, Argentina.

Zienkiewicz, O.C. and Zhu, J.Z. (1987),A simple error estimator and adaptive procedure for practical engineering analysis, Int. Journal for Numerical Methods in Engineering, vol24,pp 333-357.

Zienkiewicz, O. C. and Zhu, J. Z. (1992), The Superconvergent Patch Recovery and a posteriori error estimates. Part 2: Error estimates and adaptivity, Int. J. Numer. Methods Eng.,vol33, pp1365-1382.

Zienkiewicz, O. C. and Zhu, J. Z. (1992), Superconvergent Patch Recovery Techniques - Some Further Tests, Commun. Appl. Numer. Methods, vol 9, pp251-258.

Computer Methods and Advances in Geomechanics, Desai et al. (eds) © 2001 Balkema, Rotterdam, ISBN 90 5809 183 X

Object-oriented design of a finite element framework for geotechnical applications

P. Fritz
Division of Geotechnical Engineering, Swiss Federal Institute of Technology, Zürich, Switzerland

X. Zheng
Union Bank of Switzerland UBS, Zürich, Switzerland

ABSTRACT: Engineering software has conventionally been written in FORTRAN. This paper describes an alternative methodology for finite element programming based on object-oriented techniques. The objectives of the Finite Element framework presented are twofold: to provide scientists in the field of geotechnical engineering with the means to develop tailored applications for solving specific problems, and to develop an example application for geomechanics for a de facto proof of the applicability of the framework. The first aim requires extendability (with regard to, e.g., statical or physical modeling, algorithms and task management), understandability from the programmer's point of view, and maintainability. The second aim involves offering a basic set of the most frequently used features and a satisfying user interface, which should also comply with the object oriented paradigm.

The architectural design of the framework is based on a complete separation between concepts which concern the FE method itself, and items which concern managing an analysis, starting from the input, executing an analysis and assembling the results. Accordingly, the class architecture is based on two subsystems: one for the *project and resource management*, and the other for *FE modeling*. Each subsystem includes several main abstractions (class hierarchies) which are used to model different concepts, each for its own domain of competence.

The current contribution, which covers the design part of the framework, is the second of a set. The first one was published at the 4th European Conference on Numerical Methods in Geotechnical Engineering, Udine (Fritz & Zheng 1998). It illuminated the object-oriented analysis of the framework.

1 INTRODUCTION

> *As long as there were no machines, programming was no problem at all; when we had a few weak computers, programming became a mild problem, and now we have gigantic computers, programming has become an equally gigantic problem (Dijkstra, 1972).*

The life cycle of software viewed in the light of the object-oriented paradigm comprises

- Object-oriented requirement analysis (OOA)
- Object-oriented design (OOD)
- Object-oriented programming (OOP)
- Testing
- Maintenance

The analysis phase, where the external characteristics of the system which are visible to the user (programmer) and the key abstractions of the framework are defined, has been outlined by Fritz & Zheng (1998). The current report concentrates on the design phase, where the internal architecture of the system is established, classes are designed and their interrelationship and interaction are described.

Since OOP is still a relatively new concept, the understanding of how a good design may be characterized has varied substantially in the last years. Therefore, personal interpretation and preference play an important role, even taking into account the many guidelines published on the subject (e.g. Riel, 1996). As an illustration of this the following example may serve. It is not always obvious how far the capabilities of a class should go. An often used guideline of OOP is that the whole knowledge of a class should be locally embedded. E.g. an element should also know how to plot and print itself. In this project a different approach has been advocated: Plotting is considered to be an independent concept which is modeled accordingly outside of any element or node class. Whereas the first model will be simpler for smaller projects, the second model with its extensive separation of concerns will be adequate for more sophisticated frameworks.

2 ARCHITECTURE

Much effort has been invested in the FE framework IMAGINE (**I**ntegrated **M**odeling and **A**nalysis in **G**eotechnics by **fin**ite **E**lements) to achieve that it is lucid and manageable for programmers, adaptable to future requirements, and that its data organization guarantees data integrity and consistency, avoiding redundancies.

IMAGINEs architecture relies on completely separating everything which concerns the FE method from the items which concern managing an analysis. Accordingly, the class architecture is based on two subsystems: one for the *project and resource management*, and the other for *FE modeling* (Fig. 1). Following this figure, the top level classes of the architecture are briefly described.

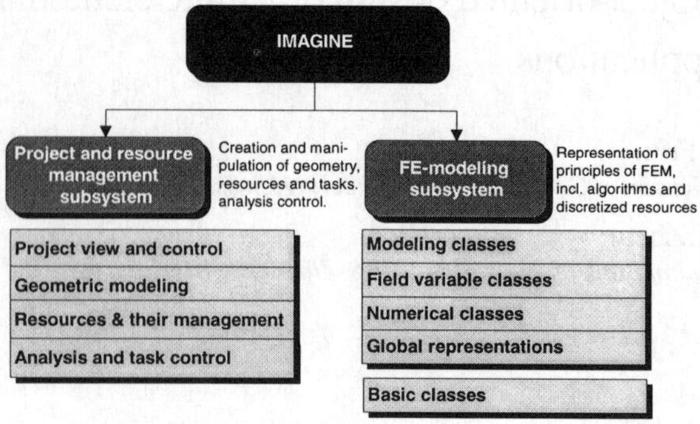

Fig. 1 General class architecture

3 PROJECT AND RESOURCE MANAGEMENT SUBSYSTEM

3.1 *Project view and control*

The top level class of the architecture is class **CFemProject** (Fig. 2). **CFemProject** is the abstraction of a whole FE analysis project: it "knows" everything about the geometry of the project, it has a resource database to store and query mathematical or engineering data like loads, constraints and material properties; and it contains a set of tasks which can be used to execute a complex computation which may depend on dynamically changing boundary conditions, e.g. a mesh refinement based on accuracy criteria. **CFemProject** class itself does not know anything about what kind of engineering project or physical system it represents, nor about the type of resources and tasks it contains. This is the responsibility of its derived specializations. Such specializations may be defined with regard to physical field concepts (e.g. stress analysis, seepage analysis, .etc.), or according to the modes of these fields (plane deformation, plane stress, etc.).

CFemProject abstracts the domain of the data of a project, together with its manipulations. In the design of IMAGINE, this domain is completely separated from the view of the data. In this context, view comprehends not only organizing the actual display of the data, but also receiving and forwarding all user interactions which could change the display either directly or by changing the underlying data. More information about the underlying *Document-View Concept* may be gained e.g. from Microsoft Visual C++ (1993).

In IMAGINE such a view will be called "workbench". To start with, three workbenches are defined: the **GGeoModelingWorkBench**, the **FemModelingWorkBench**, and the **FemTaskWorkbench.** The **GeoModelingWorkBench** is used to define and edit the geometry of the project by means of solid modeling. In the **FemModelingWorkBench** all the FE resources, including the mesh, may be defined and modified. The **FemTaskWorkbench** provides a graphics editor, in which the computational tasks may be defined and assembled in a kind of a flowchart.

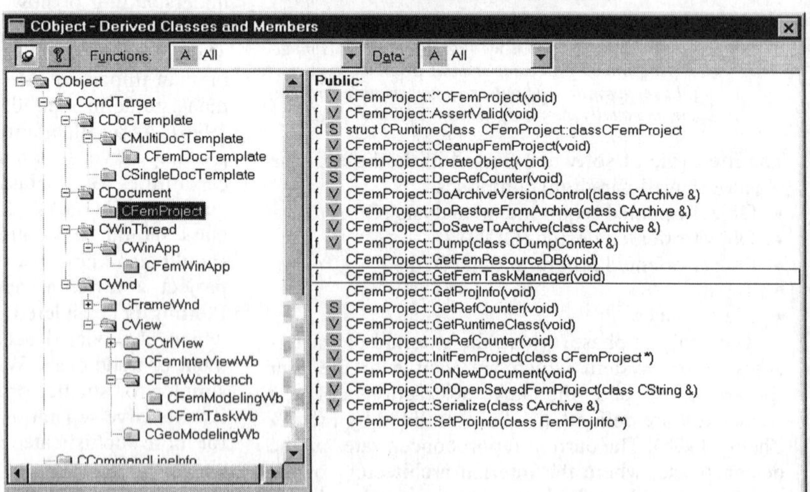

Fig. 2 Basic classes participating in the Document-View Concept

3.2 *Geometric modeling*

The geometric representation is modeled by two class hierarchies, **GeoDomain** and **GeoDomainRep** (Fig. 3). They are used to numerically represent project geometry. Although class **GeoDomain** appears more close to common concepts, as default, it does just provide two group of interfaces: one group for **CFemProject**, which will query the geometric properties when required, the other group for **GeoDomainRep**, where the geometric data are stored and managed. The most important features of **GeoDomainRep** are the capabilities of *querying*, *computing* and *spatial decomposition*. "Querying" refers to the capability of all geometric objects to be queried quickly on demand. "Computing" allows the determination of geometric quantities like volume, area, and center. The most demanding capability is the spatial decomposition. Here, spatial decomposition means the spatial subdivision into a regular mesh. This may also be referred as the process of automatic mesh generation. To be as general as possible for both 2D as well as 3D problems, geometry is expressed as a solid model. However, to our knowledge, up till now no algorithm for automatic FE mesh generation exists for the boundary model data structure. Therefore, although **GeoDomainRep** may read the geometry according to the Boundary Model approach, it is responsible to convert it to a form accessible for automatic mesh generation engines.

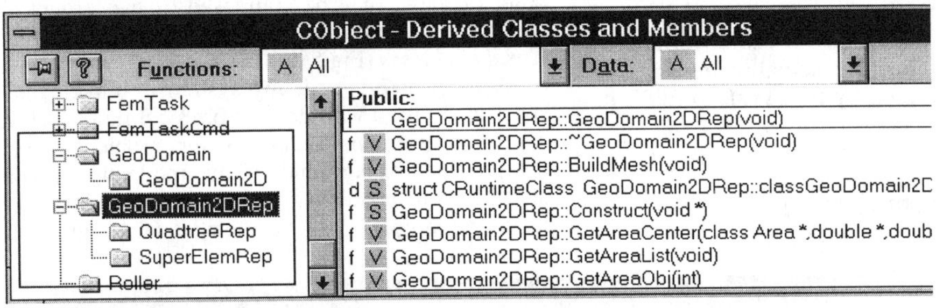

Fig. x3 Geometric modeling abstraction

3.3 *Resources and their management*

An FE run needs a variety of resources to completely specify a specific application problem. These may include material properties, loads, etc. Ultimately they must be known at the discretized finite element level. This aspect will be discussed in the context of the *FE modeling subsystem*. Here, in the context of the *Project and resource management subsystem*, these so called project resources are treated at a higher level. E.g. a zone may be defined where certain characteristics apply. This zone is completely independent of the FE mesh, which at this abstraction level is not known. Project resources are modeled by IMAGINE in the **FemResource** class hierarchy (Fig. 4), and include zones with specific type information, material laws and corresponding properties, loads in a broad sense, e.g. for a deformation analysis, it stands for forces, constraints, and coordinate system resources.

The geometry and all resources of the project may be defined completely independently of the FE mesh, and thus provide e.g. a solid foundation for a fully automatic adaptive analysis even with complex geometric profile and boundary conditions.

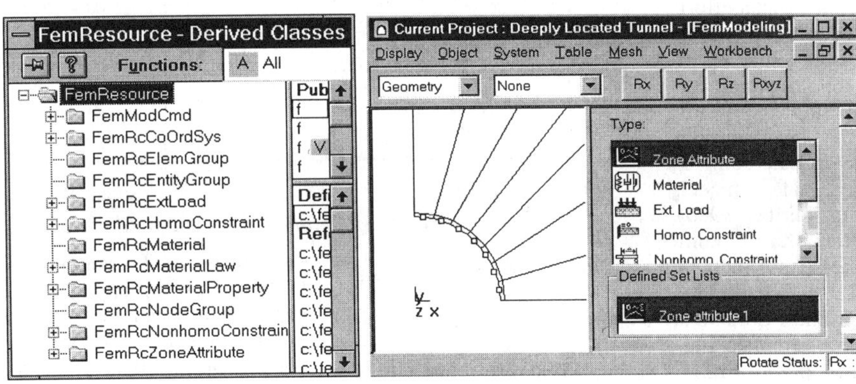

Fig. 4 Resource class hierarchy

3.4 *Analysis and task control*

Contrary to a conventional FE package, where an analysis is split into pre- and postprocessing and computation, IMAGINE abstracts the analysis and task control in an own, flexible class hierarchy.

FemAnalysis has the role of a control center at the engineering level: It sends messages to initialize and record the system state at its various stages, it communicates with other key components of the project like **FemStruct**, **FemSolver** and **ResourceManager**, and it fetches the required **FemTask** objects and dispatches them. **FemTask** contains the individual tasks which are modeled in their own class hierarchy. Examples are "elastic computation task", or "next layer task" of a cut-and-cover tunnel. **FemTaskCmd** contains the actual commands or events which may be used by **FemTask** to fulfill its duty.

4 FE MODELING SUBSYSTEM

4.1 *Introduction*

The *FE modeling subsystem* is the second leg of the IMAGINE framework. It is responsible for the abstractions of the general concepts of FEM. These mainly include the general principles, the data representations, and the geometric and topologic concerns of the discretization. Special efforts have been made to cover future extension requirements. Examples of this may be extensions to include complex material models (c.f. Fritz, 1983), structural complexities (joints, lining), and field coupling (e.g. seepage-deformation), etc.

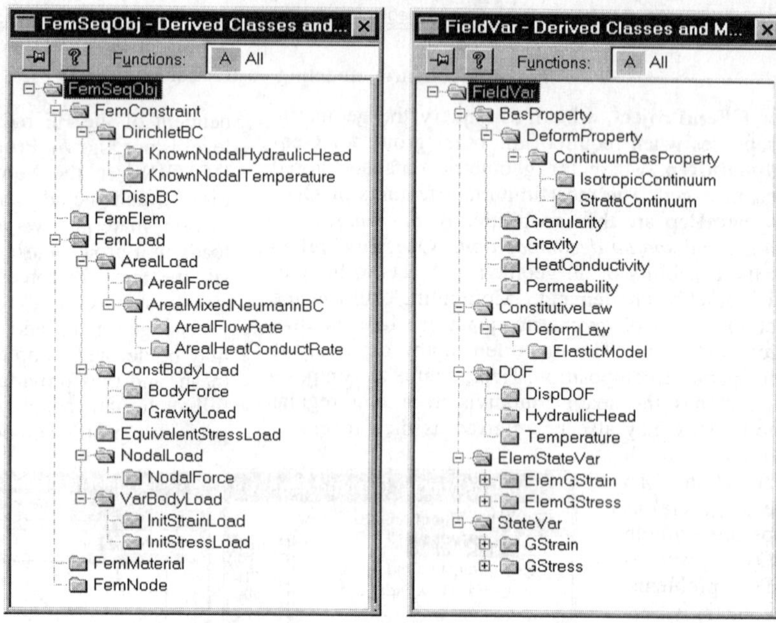

a) Modeling classes b) Field Variable classes

Fig. 5 *Modeling* classes and *field variable* classes

The *FE modeling subsystem* defines a set of generic architectural classes, each abstracting a specific part of FEM. A simple FE application could be built just with these classes. For more advanced applications subclasses may be derived which make available required extensions. However, most of the fundamental concerns, such as element analysis, local-global indexing, loading, associations between field variables, and storage management, etc., are covered by the basic classes and may be used directly.

4.2 *Modeling Classes and Field Variable Classes*

The physical side of FEM is governed by differential equations which are based on two fundamental elements: the definition domain (e.g. finite elements, supports), and the physical field variables which act in this domain. Consequently, the numerical model for the physical engineering problem, from an object point of view, may be composed of two groups of abstractions: *modeling classes* and analytic or *field variable classes* (Fig 5).

Usually an FE modeling system is implemented specifically for a well-defined physical field, such as a seepage problem, a deformation problem, where the physical background of the modeling concept and the corresponding field variables are clearly related and simple. However, it would be a great advantage if an FE framework could offer the possibility to develop applications for different physical engineering problems, e.g. stress analysis problems, groundwater flow problems, swelling problems and convection problems. With the conventional FE approach this would lead to a bunch of subclasses, each defined specifically for a certain field type, such as a Laplace Element, a Deformation Element, etc., just similar to the widely used structural elements, like truss and beam elements. This would be opposed to the object-oriented principle of localization of concerns: a *modeling class*, e.g. a finite element, should be separable from the physical field acting within it.

IMAGINE pursues a different course. It relies on two independent abstractions (Fig. 5), **FemSeqObj** is the root the class tree which contains all modeling concepts, and **FieldVar** contains all types of field variables.

The most important *modeling classes* of **FemSeqObj** are **FemElem**, **FemNode**, **FemMaterial**, **FemLoad** and **FemConstraint.** The responsibilities of the *modeling classes* are further split into two levels: the connectivity part and the analytical part. In IMAGINE, the connectivity part is considered to be the bones of the FE modeling concept, while the analytical part is viewed as the meat. At the architectural level, the *modeling classes* are mainly used to cover the connectivity concerns - the connections between bones - so that a generic frame can be built up. This is the reason for calling them "*modeling classes*". However, the analytical concerns of the *modeling classes* require both connectivity components and physical field variable components. The role of the base class **FemSeqObj** is to establish the connection via an ID between modeling components and their assembled structure of the whole FE system, which is abstracted in class **FemStruct**.

The physical field variable components in IMAGINE are modeled as a set of *field variable architectural classes* with the following main components: **DOF**, **StateVar**, **ElemStateVar**, **BasProperty** and **ConstitutiveLaw**. At this level, each of them only addresses the generic concept it represents. Subclasses may be developed to model effective components. For example, if the framework is limited to simple elastic deformation, then **DOF** is derived to hold displacements, **StateVar** stresses and strains, and **MaterialProperty** probably to hold the elastic modulus and Poisson's ratio.

4.3 *Element Concept*

On a first look, the introduction of an element data type (or class) is expected to be a straightforward matter. However, on a second look it is recognized, that the design of such an abstraction is not obvious and could lead to a variety of different approaches. At a first glance, it appears that element data types could be constructed according to the conventional "element type" concept. And effectively this approach could be used with certain element types like a truss, beams, or even simple shells. However, difficulties would appear integrating further extensions or configuration changes, such as general material models or new shapes, due to the lack of abstraction and separation of concern. Such an approach requires that for any possible combination of shape, material and field variable type, a new element data type has to be derived and many common operations like shape function derivatives have to be duplicated. This is obviously not the way to fulfill the promises of OOP techniques.

As a matter of fact, the responsibilities at the element level rely on a composit concept. It is a duty of a correct design to separate as far as possible the concepts that a generic element type is composed of. IMAGINE takes into account the following independent concepts: connectivity, geometric shape and shape functions, numerical integration, element field variables, material response, and coordinate system. In OO terminology, the abstraction of a generic element, **FemElem,** is configurable in terms of **CoOrdSys**, **FemShape**, **FemMaterial**, **ElemStateVar**, etc. The objects representing these capabilities will be created externally and bound to **FemElem** dynamically. Furthermore several **ElemStateVar** objects representing different field variable types are allowed to be linked.

4.4 *Node and Degrees of Freedom (DOF) concept*

The class **FemNode** (c.f. 4a) is used to model the node concept of a discretized finite element system. Similarly to **FemElem,** due to its extensive separation of concerns, it has currently no subclasses. It derives its power from a multitude of aggregate objects and member functions. As default, the properties of a **FemNode** include: coordinates, a list of **DOF** objects, connectivity information, the values of primary nodal field variables, and the values of generic nodal loads. The concept of the degrees of freedom at a node of a discretized finite element system is intimately related to the node concept. The **DOF** concept as introduced by IMAGINE should not be interpreted as the "number of degrees of freedom". **DOF** and its subclasses contain information about how many unknowns associated with a certain type of field variable (deformation, heat transfer, groundwater flow, etc.) apply in a **FemNode**, the state of these unknowns (free or constrained) and their indices in the discretized FE system.

4.5 *Load and Constraint Concept*

The class **FemLoad** and its subclasses are first hierarchically structured according to geometric characteristics: **NodalLoad** concerns concentrating loads. **ArealLoad** represents distributed loads over element borders. **BodyLoad** captures distributed loads along element boundaries. Since these features are common to all types of loads regardless of the physical backgrounds, they are also considered to be generic classes of the IMAGINE framework. Further, derived subclasses are forced to take physical significance into account. In the case of deformation problems, the subclasses may be **NodalForce**, **ArealForce**, and **GravityForce**.

The abstraction of the constraint concept is done similarly to that of **FemLoad**. The generic constraint data type, **FemConstraint,** is responsible for "where" to apply the constraints, the subclasses then handle the issues of to "what" and "how" they should be applied. As default, the constraints will be imposed on **DOF** objects.

4.6 *Material Concept*

In OOP, the material behavior forms an independent level of other FE classes: in a strict sense it neither fits in the *project management subsystem* nor in the *FE modeling subsystem*. The reason is that the concept of the material behavior is actually not a part of the principles of FEM, but rather of system states. In fact, the material concept could have been separated

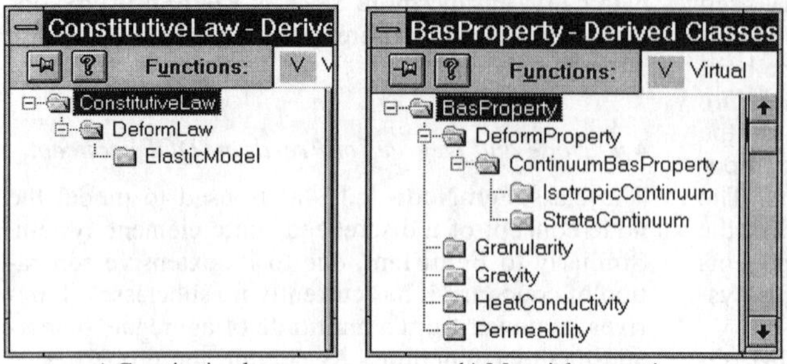

a) Constitutive law b) Material properties

Fig. 6 Material concept

from the *FE modeling subsystem* and integrated into a new subsystem. The main reason not to do that but to leave it in the *FE modeling subsystem* was to make it more convenient for other objects of the *FE modeling subsystem* to communicate with it.

The abstractions for the material lead to several class hierarchies in **IMAGINE**: a generic material data type **FemMaterial** (Fig. 5a) in the *modeling class* hierarchy, and a material constitutive law hierarchy **ConstitutiveLaw** and a material property hierarchy **BasProperty** in the field variables hierarchy as shown in Fig. 6. The class **FemMaterial** serves as a container for material properties and constitutive laws. Various types of material properties and laws can be defined and linked. According to the current design, **FemMaterial** is a kind of architectural class for **FE** modeling , while material **ConstitutiveLaw** and material **BasProperty** are kinds of architecture classes pertaining to field variables.

4.7 *Other Classes*

Numerical classes are primarily responsible for providing integration support for **FemElem.** In FEM, due to the mathematical complexity, element analysis (e.g. integration, derivation, etc.) is usually not done directly in the physical space of the element. Instead, some standard and regular element shapes are provided in a mapped space in form of standard shape templates. Each element template is defined by a set of shape functions, which may be defined here. Once an element shape is defined, it can be used for all physical field problems thus avoiding coding redundancy.

Global representations: a discretized FE system is a discretized representation of the original partial differential equations (PDE) for the overall (global) geometric definition domain. All the local objects are just parts of the global system and must be compiled to build up the global representation. The discretized FE representation in terms of local objects is abstracted in class **FemStruct.** The most important components of **FemStruct** are **FemElem**, **FemNode** and **FemMaterial**. The other important class hierarchy which is responsible for global concerns is **FemSolver**. It contains a hierarchy of various type of solvers, and all methods which depend on these types, including assembling the global stiffness matrix based on the local data of **FemStruct** and solving the system of equations.

Basic classes: often IMAGINE needs to manipulate, to operate on and to represent modeling objects like elements and nodes and field variable objects like stress and strain. For this purpose, a set of basic matrix and geometric data types has been designed to cover fundamental geometric and algebraic requirements.

5 CONCLUSIONS

The authors are aware that presenting the design of such an elaborated framework on just a few pages does not necessarily contribute to its understanding. Therefore it is recommended to read first the introduction about the object-oriented analysis part (Fritz & Zheng 1998), and, if more information is required, the comprehensive report of (Fritz & Zheng 2000).

REFERENCES

Dijkstra Edsger 1972: The humble programmer. *Communications of the ACM, V.15, no.10, Oct.*

Fritz P. 1983: RHEO-STAUB User's Manual. *Report Swiss Federal Institute of Technology, Zürich. [http://www.igt.ethz.ch/www/search/publication.asp?publication=1].*

Fritz P., Zheng Xiong 1998: Object-Oriented Analysis of a Finite Element Framework for Geotechnical Applications. *Proc. 4th European Conference on Numerical Methods in Geotechnical Engineering, Udine [http://www.igt.ethz.ch/www/search/publication.asp?publication=143].*

Fritz P., Zheng X. 2000: A Finite Element Framework for Geotechnical Applications based on Object-Oriented Programming. *Report Swiss Federal Institute of Technology,* Zürich. http://www.igt.ethz.ch/www/search/publication.asp?publication=145.

Microsoft Visual C++ 1993: Microsoft Visual C++, Programmers Guides. *Microsoft Corporation.*

Riel Arthur 1996: Object-Oriented Design Heuristics. Addison-Wesley, Reading MA.

Computer Methods and Advances in Geomechanics, Desai et al. (eds) © 2001 Balkema, Rotterdam, ISBN 90 5809 183 X

Nonlinear FEM analysis for dynamic response of saturated soil layers using a parallel algorithm

M. Kawamura
Department of Architecture and Civil Engineering, Toyohashi University of Technology, Japan

J. Tanjung
Department of Mechanical and Structural System Engineering, Toyohashi University of Technology, Japan

ABSTRACT: In this study a parallel 3D FEM formulation was developed to solve the nonlinear dynamic response of saturated soil layers and to overcome the limitation of a single processor computer due to consuming time and computer memory requirement. The Domain Decomposition Method (DDM) was introduced to separate the analysis of a whole domain into several non-overlapping subdomains through an interface problem. Each subdomain is addressed into an individual processor. The interface problem is iteratively solved by the Conjugate Gradient (CG) procedures with an interprocessor communication based on hypercube like paradigm. A good agreement of the comparison of the experimental with the numerical results shows the accuracy of this proposed algorithm. And also, as we expected, the proposed procedure has efficiently reduce the consuming time as well as the memory requirement.

1 INTRODUCTION

Analyses of seismic response of civil structures and soil layers are complicated in some aspects. First, saturated soil layers are complex materials, consisted of granular soil skeletons and pore water. The saturated soil behaves as a two-phase system. Second, the behaviors of the granular skeleton is often nonlinear and path dependent. The analyses are further complicated by presence of the inertia forces of the difference phase and coupling of them (Ishihara 1996). Therefore, a fast, accurate and efficient analysis tool, such as FEM, is required.

In this study, the governing equations are derived by idealizing the saturated soil as a porous material, i.e. the granular soil skeleton coupled with the pore water. Effect of material non-linearity is modeled according to the radial mapping type of the bounding surface plasticity theory within the framework of elastoplasticity and critical state of soil mechanics. The Galerkin 3D finite element approximation is used to solve these governing mathematical relations.

Since the elastoplastic responses are generally path dependent, it is more suitable to solve the equation of motion in an incremental procedure. The time integration scheme of Newmark method is used to define the response at certain discrete time. Calculation for non-linearity of the granular soil skeleton is carried out by an iterative solution of the modified Newton-Raphson algorithm.

Calculations of the above analytical procedures in a conventional serial machine require a long computing time as well as a large computer memory and often exceed the computer capacity. A parallel algorithm for multiprocessor machines is proposed to overcome the above limitation. In the parallel algorithm, an original problem is separated into several subdomains using the Domain Decomposition Method (DDM). The solution of the problem at first is conducted for solving an interface problem. An iterative solver of the Conjugate Gradient (CG) is used for this purpose. The overall solution is obtained by backward/forward substitution of the interface quantity into each subdomain.

Finally, an experimental data of the earth pressure problem of a saturated soil [Kawamura 1979] was used to evaluate the accuracy of the computation. The efficiency of the parallel algorithm was also investigated for checking the increase of the computing time as well as the computer memory requirement for different number of subdomains and processors.

2 FUNDAMENTAL EQUATIONS

2.1 *Governing Equations*

The equation of motion for porous material which is cooperated with the mass conservation, as written in the following equations, are used as a starting point of the problem (Zienkiewicz & Shiomi 1984).

$$\sigma_{ij,i} + \rho g - \rho \ddot{u}_i - \rho_f \ddot{w}_i = 0 \quad (1)$$

$$-\frac{\dot{w}_i}{k_{ij}} = p_{,i} - \rho_f g + \rho_f \left(\ddot{u}_i + \frac{\ddot{w}_i}{f} \right) \quad (2)$$

$$\alpha \dot{\varepsilon}_{ii} = -\dot{w}_{i,i} - \frac{1}{Q} \dot{p} \quad (3)$$

in which σ_{ij} are the components of the total stress tensor for a saturated material and tensile normal stresses are positive; p is the pore water pressure, compression is positive; ρ and ρ_f are the mass density of the mixture and the pore water, respectively; f is the porosity; g is the gravitational acceleration; k_{ij} is the permeability tensor; u_i and w_i are the displacement of the granular soil and the pore water, respectively. The superposed dot in Equations 1-2 implies time derivative, while in Equation 3 implies rate; $(\ )_{,i}$ denotes the first order derivative with respect to coordinate x_i. A constant α represents the contact area of the soil particles. The value of α is very close to unity for most of soil-like materials, α is taken as unity in Equation 3 and in all subsequent development. A constant Q denotes the compressibility due to compressibility of the granular soil and the pore water, and the value is approximated by the following equation.

$$\frac{1}{Q} = \frac{f}{K_f} + \frac{1-f}{K_s} \quad (4)$$

where K_f and K_s are the bulk module of the granular soil and the pore water, respectively.

2.2 *Constitutive Relations*

The granular soil and the pore water have individual properties, which follow the constitutive relations below.

$$\dot{\sigma}_{ij} = D_{ijkl} \dot{\varepsilon}_{kl} + Q \delta_{ij} (\delta_{kl} \dot{\varepsilon}_{kl} - \zeta) \quad (5)$$

$$\dot{p} = -Q (\delta_{ij} \dot{\varepsilon}_{ij} - \zeta) \quad (6)$$

in which δ_{ij} is the Kronecker's delta; superposed dot implies rate; a tensor ε_{ij} and ζ denote the strains of the granular soil and the volumetric strains of the pore water, respectively.

$$\dot{\varepsilon}_{ij} = \frac{1}{2} (u_{i,j} + u_{j,i}) \quad (7)$$

$$\zeta = -w_{i,i} \quad (8)$$

The four-rank tensor D_{ijkl} represents the components of elastoplastic tensor. The radial mapping type of the bounding surface plasticity theory within the framework of critical state of soil mechanics is used to construct the four-rank tensor. An explicit formulation of D_{ijkl} tensor is written as follows.

$$D_{ijkl} = G(\delta_{ki}\delta_{ij} + \delta_{kj}\delta_{li}) + \left(K - \frac{2}{3} G \right) \delta_{ij}\delta_{kl}$$
$$- h(L) \left[3KF_{,\bar{I}}\delta_{ij} + \frac{G}{J} F_{,\bar{J}} s_{ij} + \frac{\sqrt{3}G}{\cos 3\theta} \frac{F_{,\theta}}{\beta J} \left(\frac{s_{in}s_{nj}}{J^2} - \frac{3S^3 s_{ij}}{2J^4} - \frac{2\delta_{ij}}{3} \right) \right] \quad (9)$$
$$\left[3KF_{,\bar{I}}\delta_{kl} + \frac{G}{J} F_{,\bar{J}} s_{kl} + \frac{\sqrt{3}G}{\cos 3\theta} \frac{F_{,\theta}}{\beta J} \left(\frac{s_{kn}s_{nl}}{J^2} - \frac{3S^3 s_{kl}}{2J^4} - \frac{2\delta_{kl}}{3} \right) \right]$$

where I, J and S are the first, second and third invariant of the effective stress; θ is 'Lode' angle; overbar indicates the value on the bounding surface; β is similarity ratio, and β represents proximity of the actual stress point to the bounding; $h(L)$ is heavy step function; L is the loading index; F is the equation of the bounding surface. Comprehensive formulations of the bounding surface plasticity model are completely derived by Bardet (1986), Dafalias (1986), Dafalias et al. (1986), Wang et al. (1990), Crouch & Wolf (1994a).

3 PARALLEL FEM SOLUTION

Spatial discretization of the governing mathematical relation in Equations 1-6 were carried out using standard Galerkin finite element approximation based on Weighted Residual Method (WRM) (Zienkiewicz & Taylor 1991). Final semi discrete equation of motion is represented in the incremental form. The Newmark's method for $\gamma=0.50$ and $\beta=0.25$ was used to integrate the equation of motion in time domain.

DDM was applied to separate the analysis of whole domain into several non-overlapping subdomains through the interface problem. If the interface quantity is denoted as a traction vector $\boldsymbol{\lambda}$, the interface problem of equation of motion is written as follows (Kawamura & Tanjung 2000).

$$\sum_{s=1}^{Ns} \left[\boldsymbol{B}^{(s)} \overline{\boldsymbol{K}}_{Tn}^{(s)^{-1}} \boldsymbol{B}^{(s)^T} \right] \lambda_n = \sum_{s=1}^{Ns} \left[B^{(s)} \overline{\boldsymbol{K}}_{Tn}^{(s)^{-1}} \Delta \overline{\boldsymbol{R}}_n^{(s)} \right] \quad (10)$$

where

$$\overline{\boldsymbol{K}}_{Tn}^{(s)} = \frac{4}{\Delta t^2} \boldsymbol{M}^{(s)} + \frac{2}{\Delta t} \boldsymbol{C}^{(s)} + \boldsymbol{K}_{Tn}^{(s)} \quad (11)$$

$$\Delta \overline{\boldsymbol{R}}_n^{(s)} = \Delta \boldsymbol{R}_n^{(s)} + \boldsymbol{M}^{(s)} \left\{ \frac{4}{\Delta t} \dot{\boldsymbol{d}}_{n-1}^{(s)} + 2 \ddot{\boldsymbol{d}}_{n-1}^{(s)} \right\} + 2 \boldsymbol{C}^{(s)} \dot{\boldsymbol{d}}_{n-1}^{(s)} \quad (12)$$

in which $M^{(s)}$, $C^{(s)}$ and $K_T^{(s)}$ are the mass, damping and tangent stiffness matrices of each subdomain, respectively. The vector $\Delta R_n^{(s)}$ is the increment of excitation loading vector. The matrix $B^{(s)}$ is signed Boolean matrix, which localizes the subdomain quantity to subdomain interface. *Ns* is number of subdomains; *s* is subdomain number; *n* is time step number; Δt is the time step.

A feature of the parallel algorithm used in this study is that most of the computation processes, i.e. forming or evaluating the variables with superscript *(s)* in Equations 10-12, are performed in subdomain basis without interprocessor communication. The interprocessor communication is only needed when Equation 10 is solved using the iterative solver of Conjugate Gradient (CG) algorithms. However, as mentioned by Kawamura & Tanjung (2000), these computations are also amenable to parallel manner. The interprocessor communication was established by invoking the hypercube paradigm and PVM library (Geist et al. 1996).

After the interface quantity $\boldsymbol{\lambda}$ has been defined, the incremental displacement $\Delta d_n^{(s)}$ for each individual subdomain can be obtained by backward/forward substitution as follows.

$$\Delta \boldsymbol{d}_n^{(s)} = \overline{\boldsymbol{K}}_{Tn}^{(s)^{-1}} \left\{ \Delta \overline{\boldsymbol{R}}_n^{(s)} - \boldsymbol{B}^{(s)^T} \boldsymbol{\lambda}_n \right\} \tag{13}$$

When $\Delta d_n^{(s)}$ has been obtained, the states of stress using bounding surface approach and dynamic imbalance vector are calculated as the usual procedures in FEM. Since the responses are non-linear, the iterative procedure has to be applied until the imbalance vanishes. For this purpose, the modified Newton-Raphson algorithm was used.

The final procedure of this analysis is to define the acceleration, velocity and displacement responses at current time step configuration. These values are calculated according to the following equations.

$$\ddot{\boldsymbol{d}}_n^{(s)} = \frac{4}{\Delta t^2} \Delta \boldsymbol{d}_n^{(s)} - \frac{2}{\Delta t} \dot{\boldsymbol{d}}_{n-1}^{(s)} - \ddot{\boldsymbol{d}}_{n-1}^{(s)} \tag{14}$$

$$\dot{\boldsymbol{d}}_n^{(s)} = \dot{\boldsymbol{d}}_{n-1}^{(s)} + \frac{\Delta t}{2} \ddot{\boldsymbol{d}}_{n-1}^{(s)} + \frac{\Delta t}{2} \ddot{\boldsymbol{d}}_n^{(s)} \tag{15}$$

$$\boldsymbol{d}_n^{(s)} = \boldsymbol{d}_{n-1}^{(s)} + \Delta \boldsymbol{d}_n^{(s)} \tag{16}$$

4 APPLICATION ON THE EARTH PRESSURE PROBLEM

4.1 *FEM Model and Material Parameters*

As the model of FEM analysis, a shaking table test model was selected, as shown in Figure 1. The model with the dimensions of 2.00 m long, 1.00 m wide and 0.56 m high, was subjected to horizontal excitation load of a pure sinusoidal motion with the amplitude of acceleration of 300 gals and the frequency of 3 Hz. The rigid steel wall was allowed to rotate around the base of the wall. Saturated Toyoura sand was used as a backfill material. The interface between the rigid wall and the backfill was represented by a simple slip interface element. The interface element was assumed to be effective only in the sticking and slipping mode. The response of the wall was assumed in linear region only, while the responses of both of the backfill and the interface element were non-linear. The non-linearity of the backfill was modeled according to the bounding surface plasticity theory and the interface element follows Coulomb friction law.

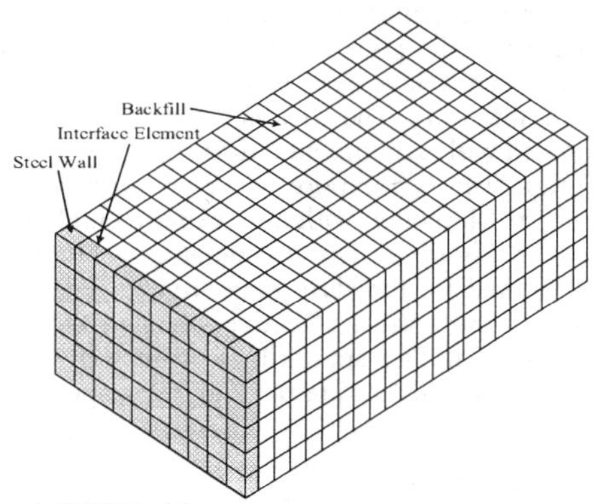

Figure 1. FEM Model

Table 1. Material Properties

Property	Symbol	Value	Unit
Steel wall			
Shear modulus	G	84.0	GPa
Poisson's ratio	ν	0.3	
Density	ρ	78.6	kN/m^3
Backfill of Toyoura sand			
Coeff. of earth pres. at rest	K_O	0.8	
Slope of iso. cons. line	λ	0.005	
Slope of elastic rebound line	κ	0.14	
Slope of critical state line	Nc	1.72	
	Ne/Nc	0.95	
Parameter of the ellipse 1	Rc	2.25	
	Re/Rc	1.0	
Parameter of the hyperbola	Ac	0.001	
	Ae/Ac	1.0	
Parameter of the ellipse 2	T	-0.1	
Projection center parameter	C	0.0	
Elastic nucleus parameter	s	1.0	
Hardening parameter	hc	2.0	
	he/hc	1.0	
Initial void ratio	e_{in}	0.64	
Poisson's ratio	ν	0.33	
Density of the saturates soil	ρ	20.0	kN/m^3
Density of the pore water	ρ_f	10.0	kN/m^3
Coefficient of permeability	k	1.2×10^{-4}	m/s
Bulk modulus of granular	K_s	3.70×1.0^3	GPa
Bulk modulus of pore water	K_f	2.08×10^2	GPa
Interface element			
Normal stiffness	C_{nn}	44×10^3	GPa
Shear stiffness	C_{ss}	22×10^{-2}	GPa

Due to limitation of the experimental data for the bounding surface plasticity model, material properties in this study was collected from the data obtained by Crouch & Wolf (1994b). The direct shear test result of the interface element mentioned by Hazarika & Matsuzawa (1997) was used as material property data for the interface element. Detail of the material parameters is listed in Table 1. The model was analyzed using eight processors in Cray SGI Origin/2000 machine for 1000 time steps and a constant time step of 0.001 second. Material damping ratio of the granular soil skeleton was assumed 5%.

4.2 *Computation Accuracy*

In order to evaluate the computation accuracy, the responses of the pore water pressure as well as the effective stress have been picked up at several nodes behind the wall. The analysis results furthermore compared with experimental data.

Comparison of the pore water pressure responses of the FEM model with the experimental data is shown in Figure 2. The computation accuracy was also checked for the earth pressure coefficient (K_{AE}) respect to the normal effective stress in the excitation direction. The approximate values of analysis and experimental data are 0.8 and 0.9, respectively.

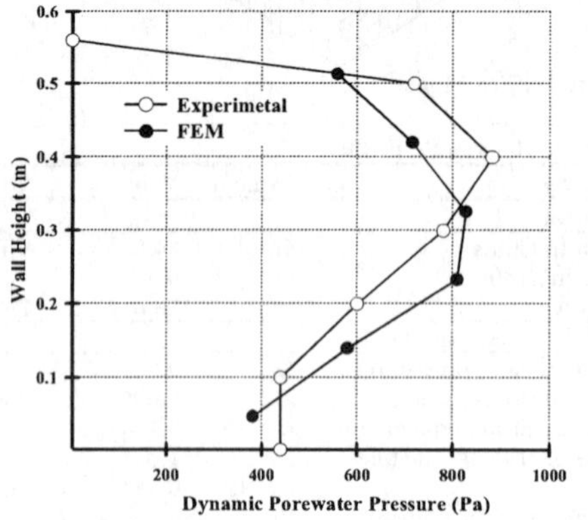

Figure 2. Comparison of Experiment and FEM Result

4.3 *Computation Efficiency*

The computation efficiency was evaluated by analyzing the constant size of subdomain problem with increasing the number of subdomains and processors. The size of each subdomain was kept constant as 392 elements, while the number of processors were increased from 1 to 16 corresponding with the number of subdomains. The total computing time, communication time as well as the computer memory requirement were used as the efficiency indicators. These indicators were recorded at the first convergence of the first time step. The results are shown in Table 2. Comparing with one subdomain (*Ns*=1), the computation time (Exc. Time) and the computer memory requirement (Mem. Req.) for 16 subdomains are increased 41.0 % and 2.1 %, respectively.

Table 2. Computation Performance

Ns	NEQ	Exc. Time (second)	Mem. Req. (MB)
1	3072	125.60	5.887
2	5760	129.51	5.894
4	10,800	144.52	5.910
8	20,880	167.73	5.941
16	40,368	177.14	6.010

5 CONCLUSIONS

The accurate and efficient parallel 3D FEM analysis for the non-linear dynamic responses of the saturated soil layers has been developed in this study. The accuracy of the computation was shown by the comparison of the FEM analysis with the experimental data. The efficiency of the computation is indicated by the capability to reduce the computation time and the computer memory requirement.

REFFRENCES

Bardet, J.P. 1986. Bounding Surface Plasticity Model for Sands. *J. Engg. Mech.* 112(11):1198-1217.

Crouch, R.S. & J.P. Wolf 1994a. Unified 3D Critical State Bounding Surface Plasticity Model for Soils incorporating Continuos Plasticity Loading Under Cyclic Path, Constitutive Relations. *Int. J. Num. Anal. Meth. Geomech.*18:735-758

Crouch, R.S. & J.P. Wolf 1994b. Unified 3D Critical State Bounding Surface Plasticity Model for Soils incorporating Continuos Plasticity Loading Under Cyclic Path, Calibration and Simulations. *Int. J. Num. Anal. Meth. Geomech.*18:759-784

Dafalias, Y.F. 1986. Bounding Surface Plasticity I, Mathematical Foundation and Hypoplasticity. *J. Engg. Mech.* 112(9):966-987

Dafalias, Y.F. et al. 1986. Bounding Surface Plasticity II, Application to Isotropic Cohesive Soils. *J. Engg. Mech.* 112(12):1263-1291

Geist, A.L. et al. 1996. *PVM, A User's Guide and Tutorial for Networked Parallel Computing.* Massachusetts: MIT Press.

Hazarika, H & H. Matsuzawa 1997. Coupled Shear Band Method and Its Application to the Seismic Earth Pressure Problem. *Soils. Found. JGS.* 37(3):65-77

Ishihara, K. 1996. *Soil Behavior in Earthquake Geotechnics.* Oxford: Clarendon Press.

Kawamura, M. 1979. *Study on Lateral Erth Pressures on Retaining Walls during Earthquake and Heavy Rainfalls (in Japanese).* Nagoya: Nagoya University.

Kawamura, M & J.Tanjung 2000. Parallel FEM Analysis for Dynamic Response of Saturated Soil Layer. *J. Struc. Engg. JSCE* 46A:703-712

Wang, Z.L. et al. 1990. Bounding Surface Hypoplasticity Model for Sand. *J. Engg. Mech.* 116(5):983-1001

Zienkiewicz, O.C. & T. Shiomi 1984. Dynamic Behavior of Saturated Porous Material, The Generalized Biot Formulation and Its Numerical Solution. *Int. J. Num. Anal. Meth. Geomech.* 8:71-96.

Zienkiewicz, O.C. & R.L 1991. Taylor. *The Finite Element Method.* London: Mc-Graw-Hill

Computer Methods and Advances in Geomechanics, Desai et al. (eds) © 2001 Balkema, Rotterdam, ISBN 90 5809 183 X

A combined discrete and continuum analysis of jointed rock masses

J.-S. Lin & C.-Y. Koo
Department of Civil and Environmental Engineering, University of Pittsburgh, Pa., USA

J.-C. Chern
Sinotech Engineering Consultants Incorporated, TaipeiTaiwan

ABSTRACT: A discrete-continuum model in the framework of the manifold method is presented here for analyzing jointed rock masses. The focus here is to demonstrate the capability of the combined method by studying rock masses that contain joints of two different geometric scales. When joint scales become very wide apart, it would not be feasible to model them with equal details within one analysis. A reasonable simplification can be achieved by preserving the dominant kinematic constrains. In this respect, the combination of discrete and continuum methodologies, such as the manifold method, may constitute an attractive alternative. This study models the primary set of joints as discontinuities that divide a rock mass into individual blocks. Within each block the secondary joint set is handled using an equivalent continuum constitutive model. Numerical experiments were conducted with emphasis on the variation of principal stress directions by the presences of discontinuities.

1 BACKGROUND

Mechanical behavior of jointed rock masses are often modeled either by a discrete or by a continuum approach depending upon the characteristics of the problem at hand. However, this may not always be appropriate. For instances, rock masses often contain more than one sets of joints and they may also have very different geometric scales which makes it infeasible to model them with equal details within one analysis. The immediate question in dealing with such a rock mass is, how much simplification can one make and still grasp the essential behavior of the rock mass. Good estimates of the displacement field provide the basis for evaluating failure modes and the stresses. It is essential that they be reasonably approximated. To achieve this goal, an acceptable simplification should at least preserve the gross kinematics of a rock mass.

In the following discussion, only two sets of joints are considered: one dominant or primary and the other secondary. The primary joint set is the one that has a wider spacing, and that it governs the overall kinematics of a rock mass. To preserve the rock mass kinematics, this joint set has to be modeled explicitly. This can be achieved via a discrete approach by modeling joints as physical discontinuities. These discontinuities cut a rock mass into individual discrete blocks.

The secondary joint set is also the source of some relative movements. But being confined by the primary joint set, this movement per se is not significant. It, however, has significant impact on the stress that a rock can withstand. The net effect of the secondary joint set may, therefore, be modeled satisfactorily using an equivalent continuum constitutive model.

The present study will employ a mixed discrete-continuum method, the manifold method (Shi, 1997; Lin, 1995), to illustrate how to modeled a jointed rock masses with two joint sets that have very different geometric scales.

2 MANIFOLD METHOD

In the manifold method, a problem is defined by a two-layer description. The first layer of the description is called a physical mesh. The geometric boundaries of a problem domain and all the discontinuities are defined in a physical mesh. The second layer of the description is called a mathematical mesh. A mathematical mesh is similar to that of a finite element mesh with a major difference: The manifold mathematical mesh does not have to conform to a problem geometry. The mathematical mesh can be a mesh of some regular pattern, or a combination of some of arbitrary figures. The mesh or shape size may be chosen according to the problem geometry, solution accuracy requirements, and the physical property zoning. In this study a regular triangular grid is employed as a mathematical mesh. The mathematical mesh has to be large enough to cover every point of the physical mesh.

Figure 1(a) depicts a jointed rock sample which is placed between two caps ready for a uniaxial test. This problem layout shows the physical boundaries and discontinuities encountered. Thus, this layout is served as a physical mesh in a manifold formulation. As for the rock itself, a discretization using a triangular mesh is selected. A mathematical mesh reflecting this consideration is depicted in Figure 1(b). As common in the manifold method, the two layers of meshes do not have to conform to each other. Putting these two meshes together gives a manifold description of a problem as illustrated in Figure 1(c).

At each node of the mathematical mesh, a polynomial can be used to describe the displacement field. Similar to a finite element, a cover is employed to describe the displacement of a region common to 3 nodes. But in addition, a weighing function is introduced to take care discontinuity. Each discontinuity encountered can have distinct properties. Surface of discontinuities may be separated or come to contact with each other just as in the discrete element. When surfaces are in contact, the normal force is obtained from no penetration requirement, while the shear force is determined from relative sliding movement together with Mohr-Coulomb failure criteria.

3 EQUIVALENT CONTINUUM MODELS

There are many equivalent continuum models for representing jointed rock masses. Analytical solution considering compliance and anisotropy induced by joints represent one approach (e.g., Gerrard, 1982; Amadei, 1982). Another approach requires numerical implementation but is much more versatile. Multi-laminated model is perhaps one of the most comprehensive model in this category. It was proposed by Zienkiewics and Pande (1977) to model the ronounced" anisotropy of rock due to the presence of two or three sets of weak planes. The normal and shear stresses on a typical weak plane can be found by Cauchy law, i.e.,

$$\begin{bmatrix} \sigma_n \\ \tau \end{bmatrix} = T_i \sigma$$

Each weak plane is associated with a yield function that is defined in terms of the stress state and stress history as follows,

$$F_i = F_i(\sigma_n, \tau, \varepsilon^p)$$

The weak plane may exhibit work-hardening or work-softening characteristics. A hardening parameter that is tied to a plastic strain measure can be used to model this effect. When there are many sets of weak planes, Zienkiewics and Pande suggested the intersection of the individual yield functions be used in representing the gross yield function of a rock element, i.e.,

$$F = F_1 \cap F_2 \ldots\cap F_k = 0$$

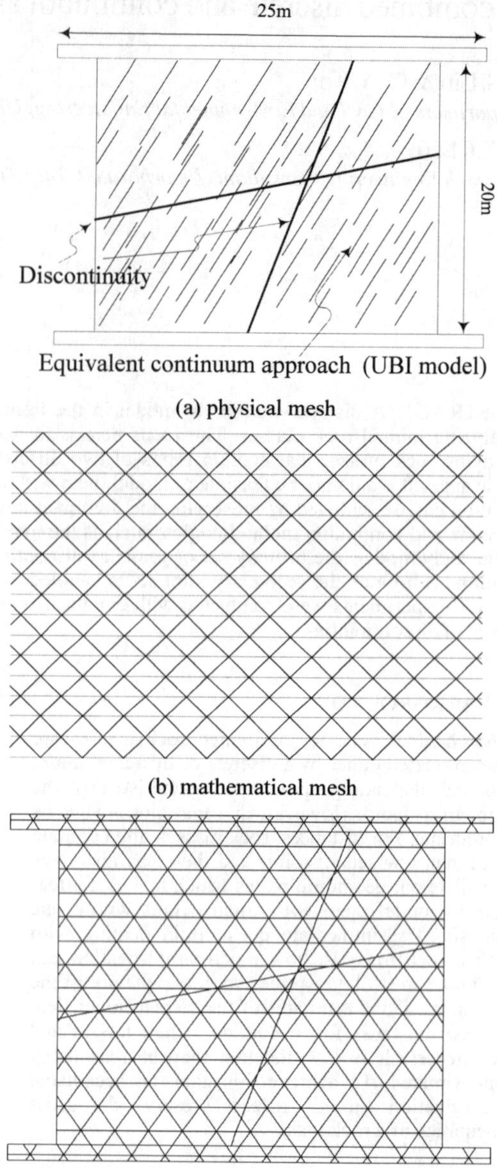

(a) physical mesh

(b) mathematical mesh

(c) manifold description

Figure 1. A manifold description of a rock sample under uniaxial test setup

One may also adopt an associated or a non-associated flow rule to model the plastic sliding and dilatancy of weak planes.

Ubiquitous joint model (Flac, 1993) is basically a simplified version of the multi-laminated model: one fixed weak plane direction is used. This study further assumes smooth weak plane surface. As a result, no dilatancy takes place at yielding. This simple version is adopted for the present study. Whenever the yield function of a weak plane is reached, the stress exceeds the yield strength is relaxed.

Only one set of weak plane within each continuum element is considered in this study. Each weak plane may also be assigned a residual strength once the yielding strain is exceeded.

4 COMPARISONS BETWEEN DISCRETE AND EQUIVALENT CONTINUUM MODELS

Plane stress numerical tests were conducted for rock samples that contain one single set of joint that delimits a sample into two blocks. Two sets of tests were conducted in order to compare the differences between discrete element and the equivalent continuum model results. First, each rock sample is modeled as a uniform continuum by adopting UBI model such that the joint of a fixed orientation can occur anywhere within the sample. Each sample is covered by 100 to 1000 triangular elements. Within each element, the displacement field is linear. All elements use the same set of parameters. Mathematical and physical meshes similar to Figure 1 are employed as depicted in Figure 2(a) except that there is no physical discontinuity. Initially, the rock is elastic until the yielding strength is reached, then UBI model is triggered. Several joint orientations, defined by an angle θ, are considered. They are 30, 50 and 60 degrees from horizontal, respectively. The strength parameters used are, $\phi_p = \phi_r = 30$ deg and $c_p = c_r = 30$ kN/m^2.

Parallel sets of tests using discrete elements were also carried out. Each rock sample in this case consists of two discrete blocks. They are separated by one single joint that has the same orientation as its counter part in the continuum model. In the discrete model, the rock is elastic. The displacement field within each block is linear.

Results from both sets of tests using same peak and residual strengths are shown in Figure 2(b) and 2(c). Both sets of results exhibit similar gross elasto-plastic stress-strain curve. The yielding strengths from both tests are about the same values. The main differences lie in the displacement. Since the two models employ vast different degrees of freedom, it is not expected to find significant differences in the elastic limit strain. In the discrete model, the failure is instantaneous. In the continuum model, the failure progresses also very fast, and the sample failed

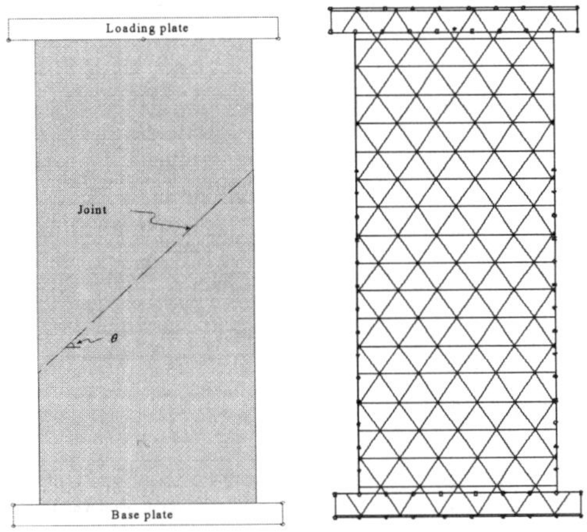

(a) Sketch of Discrete model (left) and equivalent continuum model (right)

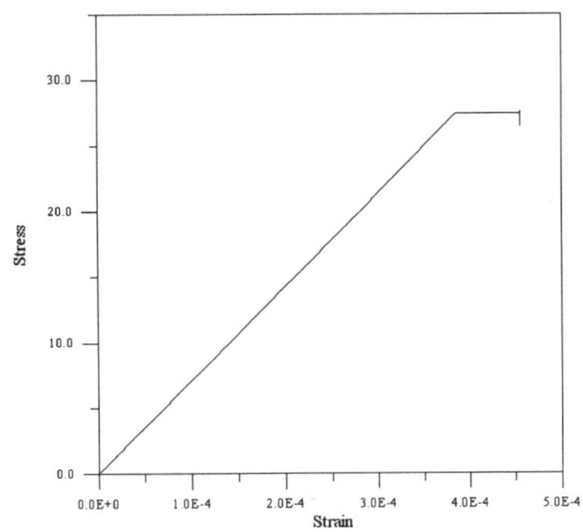

Results obtained by DDA

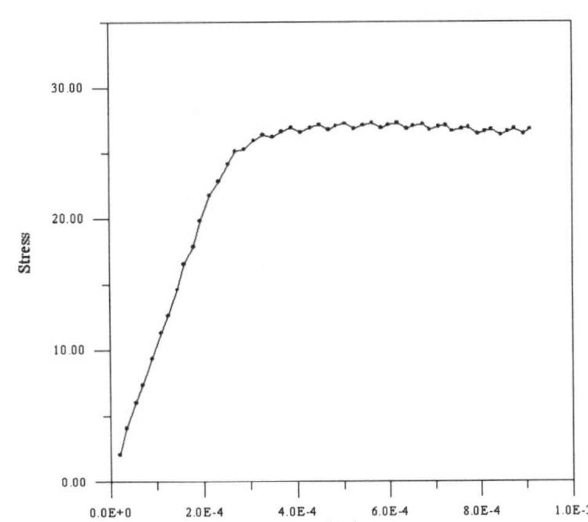

Results obtained by the Manifold method

(b) Joint angle θ=50^0

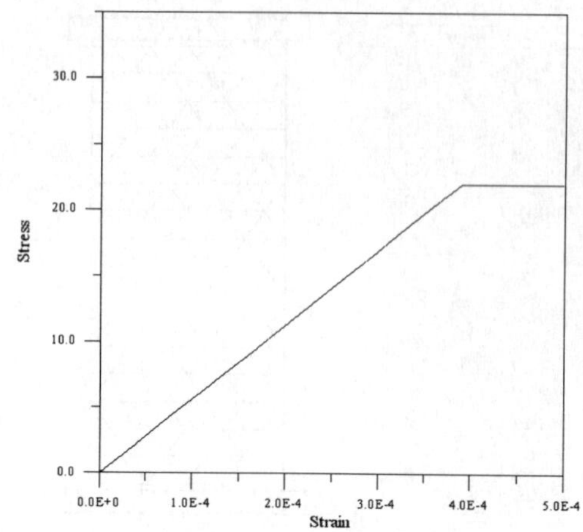

Results obtained by DDA

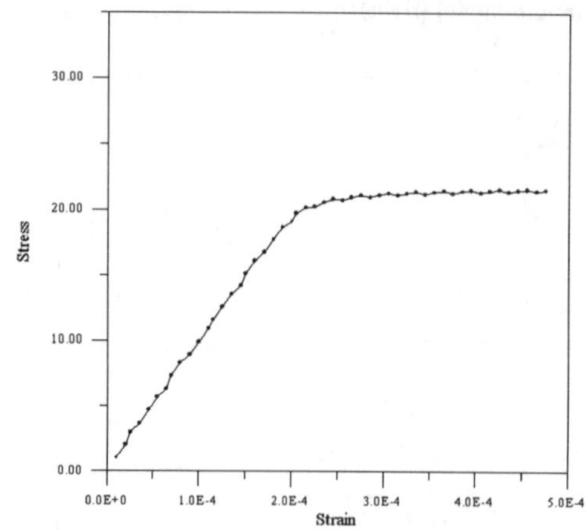

Results obtained by the Manifold method

(c) Joint angle θ=60^0

Fig. 2 Stress-strain curve obtained by DDA and the manifold method

within a few time steps but the elastic limit is greater.

The preceding results demonstrate that a continuum model, when incorporate a proper constitutive law for joints, can provide a reasonable basis for estimating the mechanical behavior of a jointed rock.

5 TESTS ON SAMPLES CONTAINING TWO SET OF JOINTS

Tests were also conducted on a sample contains two sets of joints of different scales. First of all, a single primary joint of various orientations were adopted. The secondary joint set was fixed at a θ of 60^0. The problem layout and the associated manifold description are depicted in Figure 3 (a).

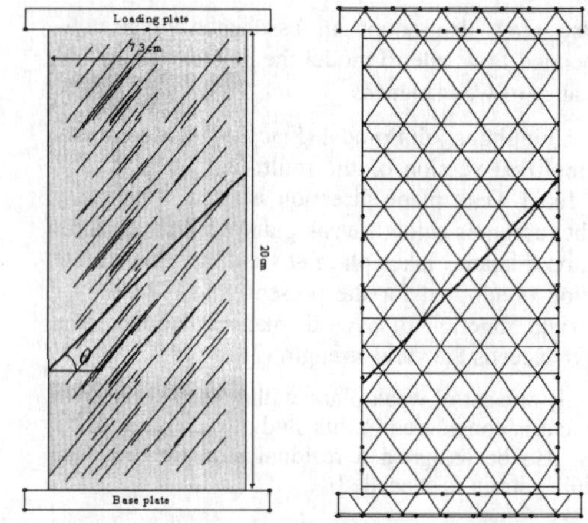

(a) Physical model layout (left) and the associated manifold description (right)

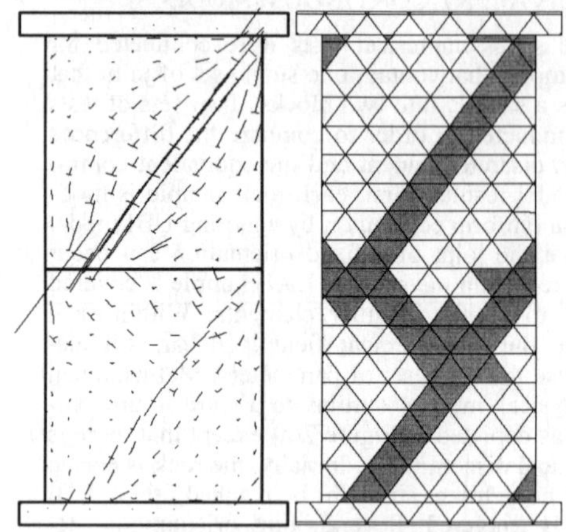

(b) Joint angle θ=0^0

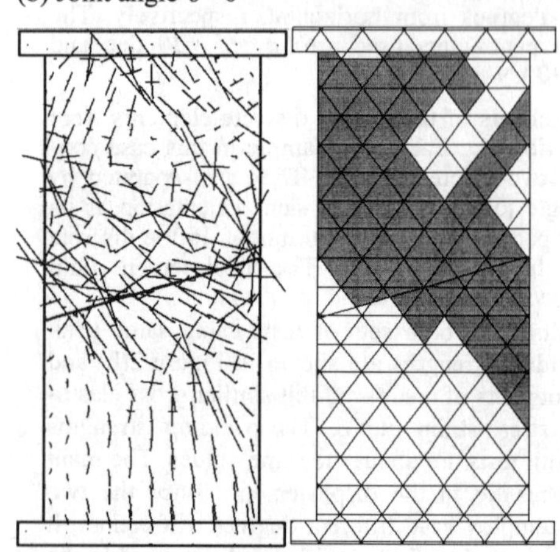

(b) Joint angle θ=20^0

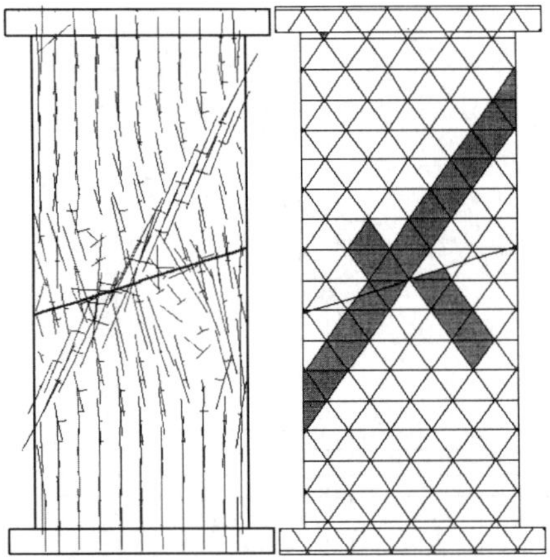

(c) Joint angle θ=20^0

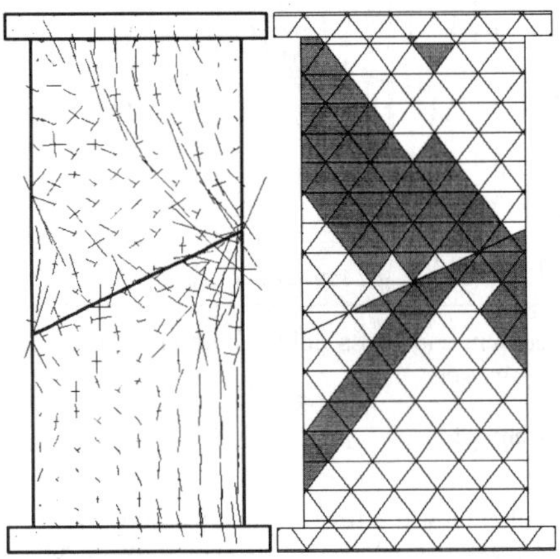

(d) Joint angle θ=30^0

Figure 3 Principle stress field of the specimen and the failure area for θ= 0^0, 20^0, 30^0 respectively.

When the primary joint set is horizontal, the sample failed along the secondary joint sets in each part delimited by the primary joint set as shown in Figure 3 (b). When the primary joint is inclined, the behavior become much more complicated. For the case that the primary joint is inclined at an angle of 20^0, the failure zone developed as well as the principal stress orientations within the secondary joint set, as depicted in Fig 3 (c), are affected by the frictional characteristics of the frictional properties of the primary joint set. If the friction on the primary joint is reduced to zero, the overall behavior of the sample, in this particular configuration, is as if the joint does not exist (Figure 3 (d)). When the primary joint is inclined at 30^0, similar behavior as that of 20^0 inclination is observed. The overall behavior observed is similar to that of creating a conjugate joint set within the continuum part.

6 PRIMARY JOINT SET CONSISTS OF TWO DISCONTINUITIES

A further study is carried out by adopting a more general primary joint set as depicted in Figure 4. The physical discontinuities used in this case do not undergo much relative movement under the applied load. Therefore, the load deformation curves are expected to be close to that contains only secondary joint set. Three cases were analyzed: (1) both primary and secondary joint sets are considered, (2) only secondary joint set is considered, and (3) only primary joint set is considered. The displacement vector and principle stresses are depicted in Fig. 4. The stress-strain curves obtained are presented in Fig. 5. These results demonstrate that a wide range of complex behavior can be modeled with various combination of discrete and equivalent continuum modeling techniques.

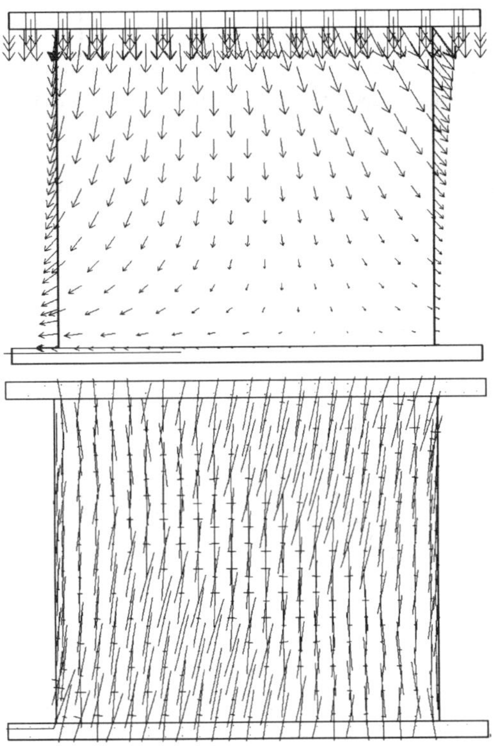

(a) displacement and stress fields: only secondary joint set

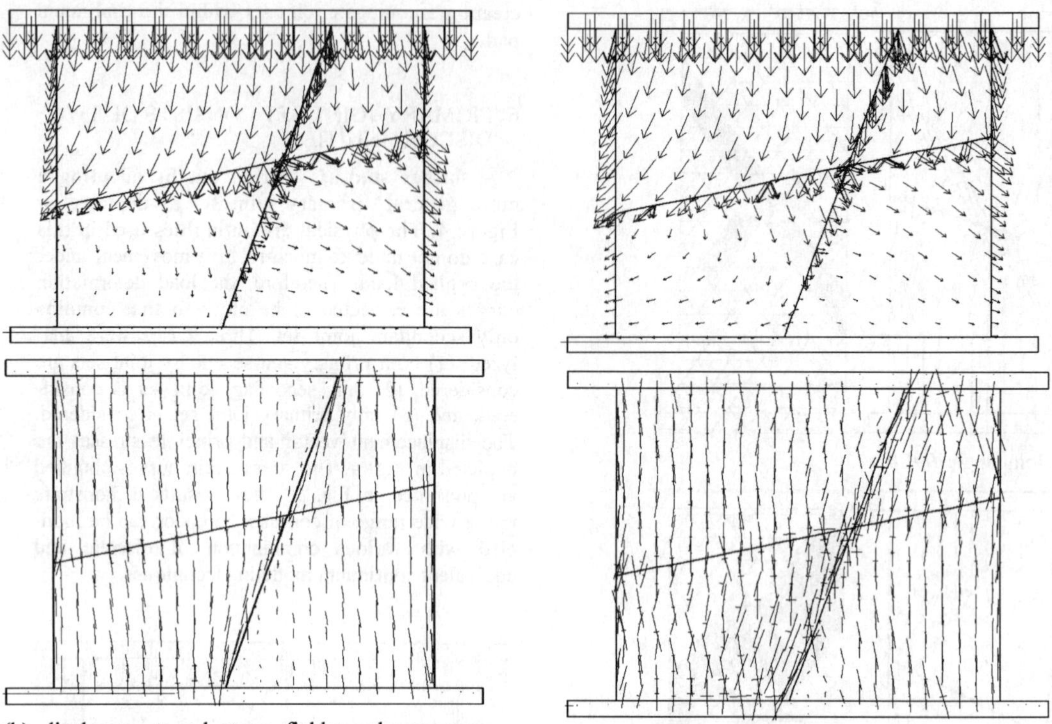

(b) displacement and stress fields: only promary joint set

(c) displacement and stress fields: both promary and secondary joint sets

Figure 4 Deformation vectors and principle stress distribution at the end of tests

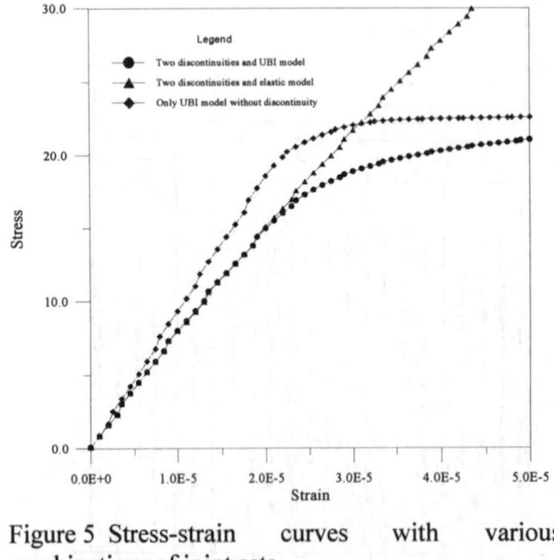

Figure 5 Stress-strain curves with various combinations of joint sets

7 CONCLUSIONS

This study shows the recent advance of discrete continuum interaction formulation, such as that based in manifold method, may provide a framework for enhancing the level of analysis involving jointed rock masses. Moreover, it also shows even simple model, such as UBI, couple with explicit discrete model has the potential for describing a very complex behavior. Work is currently underway in further exploring the effects of various combinations and their implications. The present work may readily be extended to assess the in-situ rock mass strength and stability.

8 REFERENCES

Amadei B. 1982. The influence of rock anisotropy on measurement of stresses in situ. Ph.D. thesis, University of California, Berkeley.

Curran, J. H. and Ofoegebu G. I. *Modeling Discontinuities in Numerical Analysis*. 443-468. Constitutive laws and numerical modeling.

Gerrard, C. M. 1982. Joint compliances as a basis for rock mass properties and the design of supports. *Int. J. Rock Mech. Min. Sci. & Geomech. Abstr*. 9: 285-305.

ITASCA. Flac user? manual.

Koo C-Y and Chern J-C, 1997. Modeling of progressive fracture in jointed rock by DDA method. Proceeding of Second International Conference on the Analysis of Discontinuous Deformation. pp. 186-200. Kyoto, Japan.

Lin, J-S. 1995. Continuous and Discontinuous Analysis using the Manifold Method. *Proceeding of the Working Forum on the Manifold of Material Analysis*. pp. 1-20. California.

Shi, G. H. 1997. Numerical Manifold Method. *Proceeding of Second International Conference on the Analysis of Discontinuous Deformation.* pp. 1-36. Kyoto, Japan.

Zienkiewics, O. C. and Pande, G. N. 1977. Time-dependent multi-laminated model of rocks? numerical study of deformation and failure of rock massses. *Int. J. of numer. anal. methods in geomechanics*. 1: 219-247.

Computer Methods and Advances in Geomechanics, Desai et al. (eds) © 2001 Balkema, Rotterdam, ISBN 90 5809 183 X

A new numerical method – Finite-cover element-free method

Maotian Luan, Rong Tian & Qing Yang
Department of Civil Engineering and State Key Laboratory of Coastal and Offshore Engineering, Dalian University of Technology, People's Republic of China

ABSTRACT: Presented in the paper is a so-called finite-cover element-free method (FCEFM) which is aimed to solve both continuous and discontinuous deformation problems without meshing. The method is mathematically based on the finite cover of manifold and the multiple weighted moving least square (MWMLS) approximation. The entire domain is overlaid by a series of overlapped mathematical covers. While cut by joints, interfaces of different media and physical boundaries, a mathematical cover may be divided into two or more completely disconnected parts that are defined as physical covers. A cover weight function defined on each physical cover in conjunction with MWMLS approximation is used to determine shape function of a point under consideration. Afterwards, discrete equations of the boundary-value problem with discontinuity can be established using variational principle. Through numerical analyses, it is shown that the proposed FCEFM that takes successfully advantages of both manifold and mesh-free methods together is theoretically rational and practically applicable.

1 INTRODUCTION

The actual deformation and failure process of geo-structures and materials is a complex progressive course involving initially elastic deformation, crack propagating, large-scale displacement and even movement of a discrete system.

By virtue of the finite cover technique of manifold, manifold method (MM) (Shi 1992) integrates conventional finite element methods (FEM), discontinuous deformation analysis (DDA) (Shi 1988) and analytical methods in a united mathematical framework and can deal with both continuous and discontinuous deformation problems such as contact and multi-body interaction. Element-free Galerkin method (EFGM) (Belytschko et al. 1994), on the basis of moving least square (MLS) method (Lancaster & Salkaushas 1981), is very effective in simulating crack propagation which is a key issue in modeling failure or/and damage behavior of structures or materials without meshing as required in MM. The kernel of MM is finite cover technique while mesh-free approximation (so-called mesh-free technique) mainly derived from MLS method is a key feature of EFGM. MM and EFGM have own advantages in handling discontinuous deformation problems. The main purpose of the paper is to explore the possibility to work out a new numerical method by combining the finite-cover technique and mesh-free concept together.

2 FUNDAMENTAL OF FCEFM

2.1 *Finite cover system and finite cover technique*

Suppose that the domain Ω of problem under consideration is bounded by its boundary Γ. A disk or ball respectively in two- or three-dimensional problem is denoted as C. The disk or ball which overlaps with the domain Ω is defined as a mathematical cover (MC) denoted as C^M. When cut by one or more physical lines, such as joints or cracks and interfaces of different media as well as physical boundaries, a MC may be divided into two or more parts which are defined as physical covers (PCs) denoted as C^P. For continuous deformation problems, a MC can only reproduce a PC and such a MC is called as a continuous cover. For discontinuous deformation problems, however, a MC may be partitioned into two or more PCs and such a MC is termed as a discontinuous cover. All PCs form a finite physical cover system of the domain that is a discretization pattern of the domain of problem. The technique of decomposing a domain

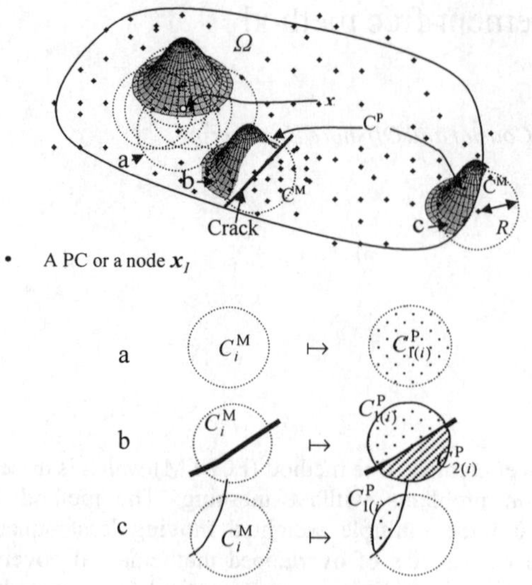

Figure 1. A finite PC system of the domain and the definitions of a MC, its PCs and cover weight function.

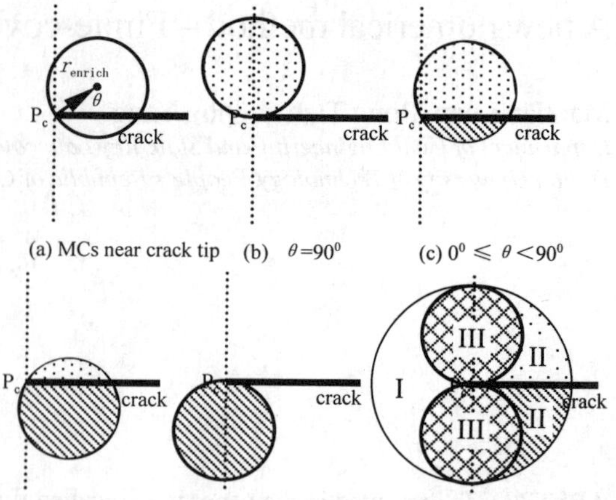

(a) MCs near crack tip (b) θ=90° (c) 0° ≤ θ <90°

(d) 270°< θ <360° (e) θ=270° (f) numerical model at a crack tip

Figure 2. Numerical model at a crack tip.

by a finite physical cover system is called as finite cover technique (FCT). Using FCT, discontinuous characteristics of materials can be captured easily and naturally.

A PC is alternatively called a node, denoted as $\{x_I\}$, $(I = 1, 2, \cdots, n^{\mathrm{P}})$. The degrees-of-freedom (DOF) of a node are defined on the center of its according MC. Hence for a MC, the DOFs of all PCs are defined on one point. The process of forming the cover system of problem considered is illustrated in Figure 1.

2.2 *A crack-tip numerical model (CTNM)*

For the crack-tip problem, an important aspect is how to distribute MCs so that smoothness of mesh-free approximations can be achieved and numerically implemented with a desired accuracy. The following procedure is proposed.

First, a circle in the neighborhood of crack tip is placed with the center at the crack tip P_c and a radius of r_{enrich}. As shown in Figure 2(a), a number of MCs are then distributed along the circumference of circle with different polar angle $\theta \in [0°, 360°)$. As observed from Figure 2(b)~(e), a MC with $\theta \in [0°, 90°)$ or $\theta \in (270°, 360°)$ is a discontinuous cover while a MC with $\theta \in [90°, 270°]$ is a continuous cover. By joining continuous and discontinuous covers together, the area near crack tip is divided into three sub-zones, i.e. zone I, zone II and zone III as displayed in Figure 2(f). Zone I is continuous because all MCs in this sub-area are continuous while zone II is discontinuous because it is covered only with discontinuous covers. Zone III between zones I and II is a transition zone since the number of continuous covers decreases and the number of discontinuous covers increases from zone I to zone II. Therefore continuous and discontinuous properties of materials near the crack tip can be simulated naturally by the FCT.

2.3 *Multiple weighted moving least-square (MWMLS) approximation*

On each PC, i.e. $\mathrm{C}_I^{\mathrm{P}}$, a cover weight function denoted by $\delta_I(x)$ $(x \in \mathrm{C}_I^{\mathrm{P}})$ is to be defined. According general definition of a weight function and considering continuity of a weight function, the following requirements on the weight function should be fulfilled

$$\begin{cases} \delta_I(x) = 1 \\ \delta_I(x) = 0 \quad (x \notin \operatorname{int} \partial \mathrm{C}_I^{\mathrm{P}}) \\ \delta_I(x) > 0 \quad (x \in \operatorname{int} \partial \mathrm{C}_I^{\mathrm{P}}) \\ \text{at least } C^1 \text{continuously differentiable} \end{cases} \tag{1}$$

Such a cover function $\delta_I(x)$ can be given by the following equation with a shape as portrayed in Figure 1

$$\delta_I(x) = \frac{1}{1 + \left(\dfrac{r(x)}{R}\right)^2} \left(1 - \left(\frac{r(x)}{R}\right)^2\right)^{2k} \tag{2}$$

where $r(x) = \|x - x_I\|$ is the distance from a sample point x to the node x_I, R is the radius of MC, $k \geq 1$.

$k=1$ is used in this study. The choice of radius of equally-spaced mathematical covers is studied as follows.

A necessary condition for a well-defined MLS approximation is that there are, at least m, e.g. n_k cover weight functions with non-zero value for each sample point $x \in \Omega$

$$n_k \ge m \text{ on } x \in \Omega \tag{3}$$

Equation (3) can be rewritten

$$\bar{n}_k = \beta m \text{ on } x \in \Omega \tag{4}$$

where $\bar{n}_k$ is the average value of n_k, standing for the average number of PCs in the PC set of any sample point x, $\beta > 1$.

Denote the area of the I-th PC as $S_{C_I^P}, (I = 1, 2, \cdots, n^P)$, the area of a MC as $S_{C^M} = \pi R^2$, and the area of the domain of problem as S_Ω, then Equation (4) can be represented in term of corresponding area

$$\sum_{I=1}^{n^P} S_{C_I^P} = \beta m S_\Omega \tag{5}$$

Notice that a PC is a part of its own MC

$$S_{C_I^P} = \alpha S_{C^M} = \alpha \pi R^2 \tag{6}$$

where $\alpha \in (0,1]$ is termed as overlaying ratio which defines the fraction of the area located in the domain of problem to the full area of the MC. When $\alpha = 1$, the MC locates completely inside of the domain. Otherwise, the MC is intersected into a number of PCs by physical lines.

Substituting Equation (6) into Equation (5) yields

$$\bar{\alpha} n^P \pi R^2 = \beta m S_\Omega \tag{7}$$

The relation between the number n^P of PCs and mathematical cover radius R can be deduced while βm is replaced by $\bar{n}_k$

$$R = \sqrt{\bar{n}_k S_\Omega / \bar{\alpha} n^P \pi} \tag{8}$$

Shown in Figure 3 is the average overlaying ratio $\bar{\alpha}$ varying with $\bar{n}_k$ for a square domain with equally-spaced MCs. For a linear basis functions, $\min\{n_k\} \ge 3$. $m=3$, $\bar{n}_k = 8$ and $\bar{\alpha} = 1$ are used here.

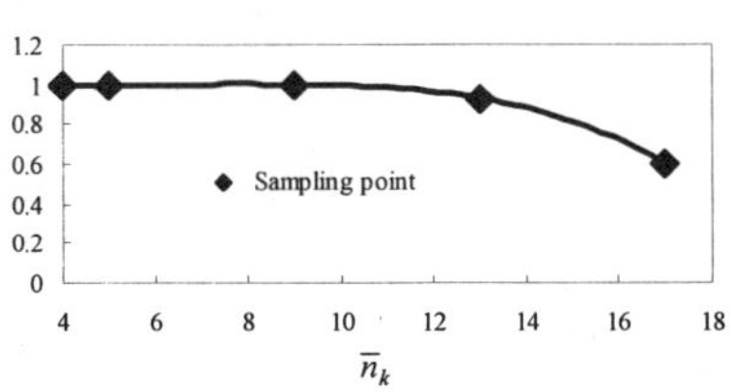

Figure 3. Relation of $\bar{\alpha}$ and $\bar{n}_k$.

An approximation $\bar{u}(x)$ of the global field function $u(x)$ at a given point x in the domain Ω is defined by

$$u(x) \approx \bar{u}(x) = \sum_{j=1}^{m} P_j(x) a_j(x) = P^T(x) a(x) \tag{9}$$

where $P(x)$ is the basis function and it is required to be complete. As usual, a monomials, e.g. linear basis, $P^T = [1, x, y]$, in the coordinates of $x^T = [x, y]$, can be chosen. $a(x)$ is an unknown coefficient vector, m is the number of the unknown coefficients. For the linear basis, $m = 3$.

In order to search the optimum approximation of the field variables by Equation (9), the MSL method is used. For a given point x, the weighted L_2 norm can be constructed by virtue of the known field variables u_I for all overlapped PCs covering the point x with a center of x_I

$$J(x) = \sum_{I}^{n_k} \delta_I(x) [u_I - u(x)]^2 \tag{10}$$

where n_k is the number of PCs covering point x.

According to the general procedure of MSL methods, the shape function $\varphi_I(x)$ of x can be obtained

$$\varphi_I(x) = P^T(x) A^{-1}(x) B_I(x) \tag{11}$$

where

$$A(x) = \sum_{I=1}^{n_k} \delta_I(x) P(x_I) P^T(x_I) \tag{12}$$

$$B(x) = (\delta_1(x) P(x_1),\ \delta_2(x) P(x_2),\ \cdots,\ \delta_{n_k}(x) P(x_{n_k})) \tag{13}$$

The matrix $A(x)$ is called the moment matrix, $B_I(x) = \delta_I(x) P(x_I)$.

The cover weight functions $\delta(x)$ are employed to weight the square distances at an arbitrary position x. It is noted that the weighting scheme in Equation (10) is different from the way used in MLS methods in that multiple different weight factors $\delta_I(x)$ taken as the value of the I-th cover weight function at x participated in weighting the square distance in $J(x)$. In order to distinguish from the conventional MLS methods, this approximation is defined as multiple weighted moving least-square (MWMLS) approximation. It is alternatively termed as multiple fixed least-square (MFLS) approximation by Onate (1996).

Then the discrete equations of elastostatic boundary-value problems can be established using variational principle. The essential boundary conditions are enforced with a penalty method. Due to the limited space, the details can be found elsewhere (Luan et al. 2000).

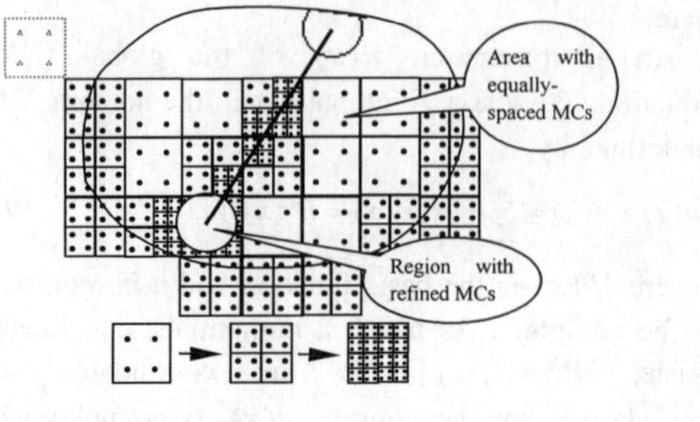

(a) Arrangement of Gauss points in the area covered by equally-spaced MCs

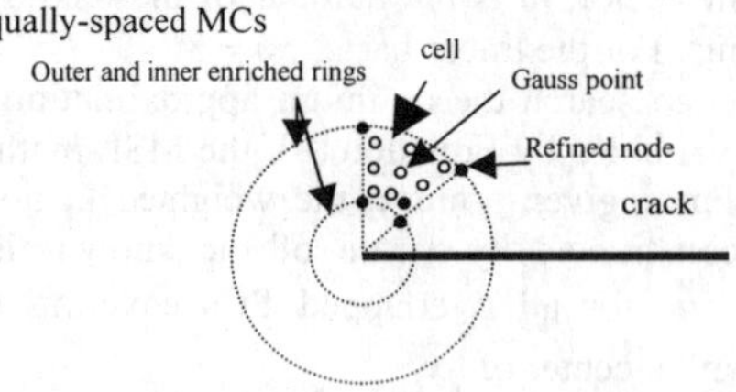

(b) Arrangement of Gauss points in area covered by refined MCs

Figure 4. Gauss point arrangement.

2.4 *Quadrature*

In computing the stiffness matrix and force vector in resulting equations, quadrature over the domain Ω is required. In order to keep the method mesh-free completely, the following Gaussian quadrature is used.

1. In the area with equally-spaced MCs, a so-called "moving window procedure" is used to arrange Gauss points. As shown in Figure 4a, a rectangle or square window scans from one end of the domain to another step by step. Within each step, Gauss points are located inside the domain until the domain is scanned with the window with no overlapping. When the window is intersected with physical lines, it will be divided into four or more sub-windows equally to density Gauss points if necessary.

2. For the regions such as crack tip where MC is need to be refined, Gauss points are arranged in the cells surrounded by outer enriched ring, inner enriched ring and nodes in corners as shown in Figure 4b.

3 NUMERICAL EXAMPLES

3.1 *Calculation of stress intensity factor (SIF)*

To verify the capability and reasonability of the proposed method in simulating the singularity at the crack tip, a uniformly tensioned square plate with a horizontal central inclined crack is analyzed as a benchmark problem.

The geometry and boundary conditions of the problem are shown in Figure 5. The crack length is $2a = 1$, applied vertical tension traction is $q=1$, $\alpha \in [0^\circ, 90^\circ]$ is the angle between the crack inclination and the loading direction. The Young's modulus is taken as $E = 10\,\text{MPa}$, and Poisson's ratio is $\nu = 0.3$. The exact values of the K_I- and K_{II}-type stress intensity factors (SIFs) can be calculated by $K_I = q\sin^2\alpha\sqrt{\pi a}$, $K_{II} = q\sin\alpha\cos\alpha\sqrt{\pi a}$.

The equally-spaced MCs are arranged with 0.25× 0.25 intervals, the radius of the MC is taken as 0.36 calculated by Equation (8). The 5×12 (5 rings, 12 MCs/ring) MCs are refined within the circle with a radius of $R=a$ at each crack tip based on the aforementioned CTNM. The 3 × 3 and 6 × 6 Gaussian quadratures are used in the region with equally-spaced MCs and refined regions respectively. For α=0°, 10°, 20°, 30°, 40°, 50°, 60°, 70°, 80°, 90°, the K_I- and K_{II}-type SIFs are assessed as given in Figure 7.

It is observed that the calculated results agree with exact solutions rather well.

In addition, shown in Figure 8 are the calculated vertical stress σ_y, horizontal stress σ_x, shear stress τ_{xy}, displacement fields and Von Mises stress

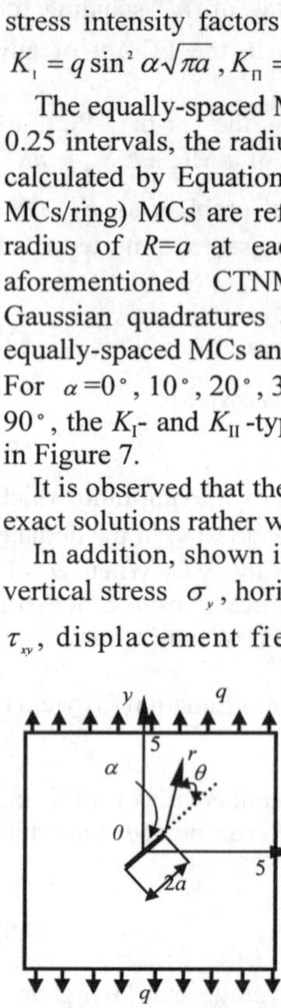

Figure 5. A crack tip problem.

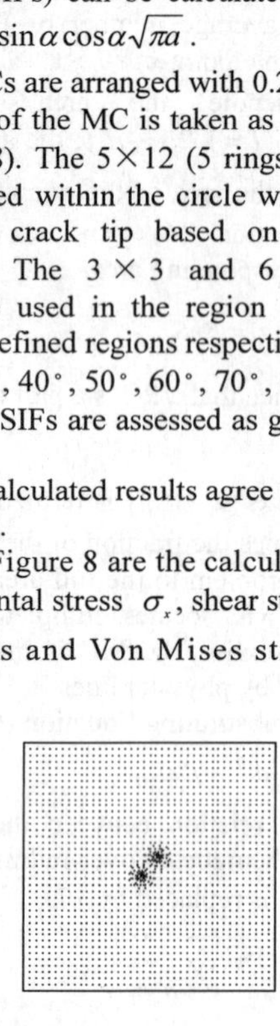

Figure 6. PC configuration.

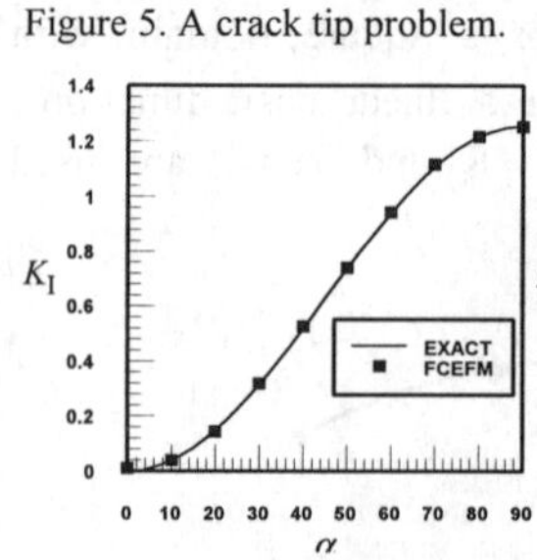

(a) $K_I \sim \alpha$ curve

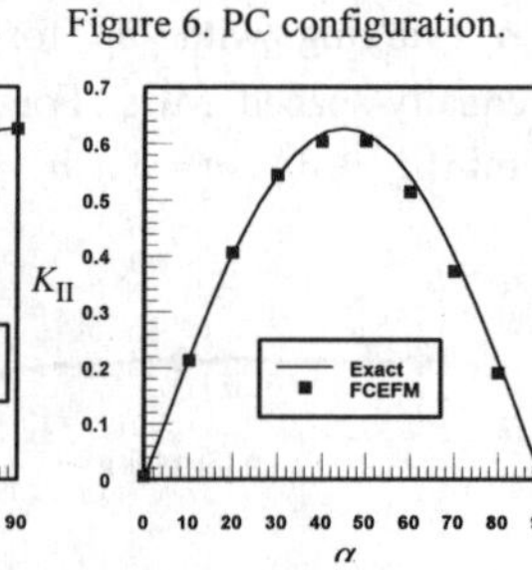

(b) $K_{II} \sim \alpha$ curve

Figure 7. Calculated stress intensity factors (SIFs).

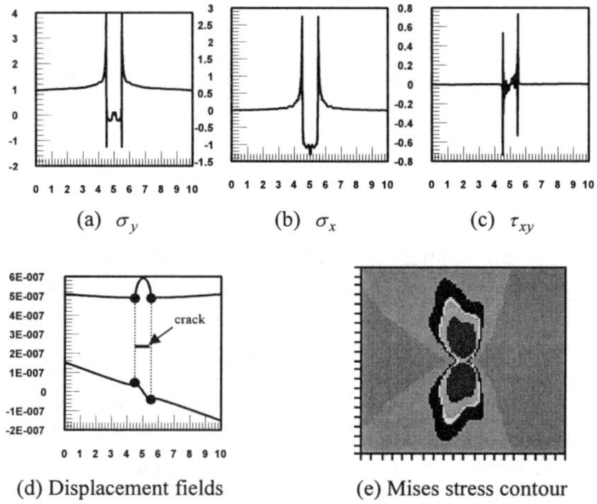

Figure 8. Simulation results at crack tips.

contour around crack tips for $\alpha = 90^{\circ}$.

In summary, the results presented here indicate that the new method can predict and reveal the displacement and stress fields around a crack tip reliably.

3.2 *Mixed-mode crack propagation analysis*

Shown in Figure 9 is a four-point single-notched shear beam which is one of well-known test problems for mixed-mode crack propagation first tested by Arrea & Ingraffea (1982) and then re-examined by Xie et al (1995) and among others. This problem is analyzed using the presented method based on linear elastic fracture mechanics theory.

The geometry, boundary conditions and material properties of the beam are given in Figure 9. The criterion of maximum circumferential stress

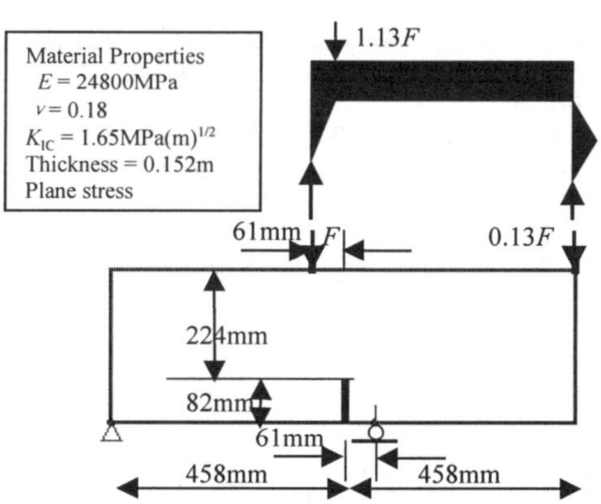

Figure 9. The four-point single notched beam and its material properties and boundary conditions.

$G(\theta)_{max}$ is used to analyze this mixed-mode crack propagation problem.

The proposed method is numerically implemented and can be used for automatic simulation of crack propagation with successive steps. At the beginning, a small initial size of increment of crack propagation is given by users. Then the load is applied step by step. Within each loading step, the effective SIF K_{sita}, is evaluated based on the $G(\theta)_{max}$ theory. Then K_{sita} is compared with fracture toughness K_{IC}. If K_{sita} exceeds K_{IC}, the applied load holds unchanged and the crack propagation is traced by invoking a stress releasing procedure to control local stability. The stress releasing procedure is mainly used to release unbalanced force by the crack growth under an invariable applied load. And this procedure is iteratively run and makes the choosing of load increment easy. If the crack is stable under current load or an equilibrium state can be achieved by the stress releasing procedure, the crack growth will stop at the current loading stage. Afterwards, a new incremental load is applied and a new cycle of propagation is started. If local stability does not exist, the crack will keep propagating under the current load until eventual damages occur. The peak of loading is then reached.

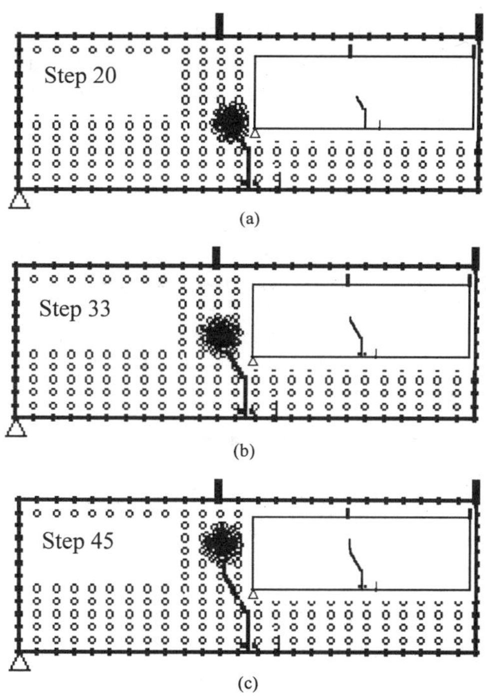

Figure 10. Deformed configurations of the model
(a) step=20; (b) step=33; and (c) step=45

As the crack propagates, a new crack-tip location is being updated by the new propagation direction computed using the $G(\theta)_{max}$ theory and the given increment of the crack propagation. Meanwhile, the cover system of the model is rebuilt for next step of propagation. Since the method is mesh-free as EFGM, updating of the model is easily fulfilled.

The load is imposed on the beam with increment of 20kN per step. Deformed configurations of the model using the $G(\theta)_{max}$ theory are shown in Figure 10 for different steps. And variations of K_I, K_{II} and K_{sita} with time step are displayed in Figure 11. It should be noted that during horizontal segments of the load curve, the crack keeps propagating and stress releasing procedures are invoked while the applied load stays unchanged.

It is found from Figure 11 that crack propagation before step 9 is dominated by shear action. Starting from step 9, the beam failure is controlled by tension cracking until complete damages occur. Therefore, shear cracking turns into tension cracking during a mixed-mode crack propagation and the unstable propagation of crack is caused eventually by the tension action.

The crack mouth sliding displacement (CMSD) varying with load is manifested in Figure 12. It can be observed that the calculated results are in good agreement with experimental results.

It can be concluded that the method developed in the paper can automatically and effectively deal with

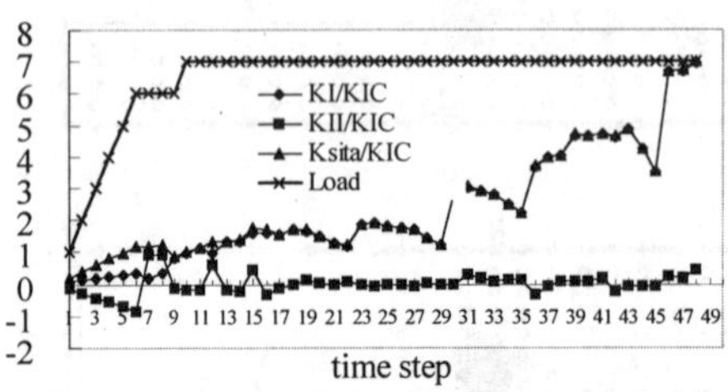

Figure 11. K_I/K_{IC}, K_{II}/K_{IC} and K_{sita}/K_{IC} versus time step.

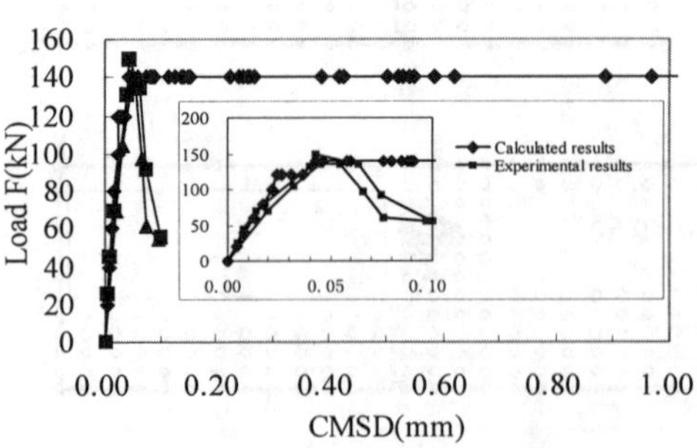

Figure 12. Load versus CMSD and comparisons between calculated results and experimental results.

the mixed-mode crack propagation problem, even for a nonlinear fracture problem, with use of a very small propagation increment size and a stress releasing procedure.

4 CONCLUSIONS

Through numerical examples, it is shown that the proposed FCEFM integrates both advantages of EFGM and MM together successfully in a natural manner. It is theoretically rational and practically applicable. Based on finite cover technique of manifold and MWMLS approximation, FCEFM has potential capability in simulating complex problems involving crack propagation, contact, moving boundaries of jointed or blocky systems and interaction of multiple bodies in a mathematically consistent manner without meshes or grids as required in MM.

ACKNOWLEDGEMENTS

The financial supports from the Trans-Century Training Programme Foundation for the Talents offered by the Ministry of Education of China and from the National Natural Science foundation through the grant No.59709008 are gratefully acknowledged.

REFERENCES

Arrea, A. & A.R. Ingraffea 1982. Mixed-mode crack propagation in mortar and concrete. *Report No.*81-83, Cornell University, Ithaca, N.Y.

Belytschko, T., Y.Y. Lu & L. Gu 1994. Element free Galerkin methods. *International Journal of Numerical Methods in Engineering*, 37(2):229-256.

Lancaster, P. & K. Salkaushas 1981. Surface generated by moving least square methods. *Mathematics and Computers*, 37:141-158.

Luan, Maotian, Rong Tian & Qing Yang 2000. Finite-cover element-free method and its applications. *Chinese Journal of Geotechnical Engineering* (to be published).

Onate, E., S. Idelsohn, O.C. Zienkiewicz, R.L. Taylor & C. Sacco 1996. A stabilized finite point method for analysis of fluid mechanics problems. *Computer Methods in Applied Mechanics and Engineering*, 139:315-346.

Shi, G. H. 1992. Manifold method. *Fractured and Jointed Rock Masses*, Lake Tahoe, California.

Shi, G. H. 1988. Discontinuous deformation analysis: a new numerical model for the statics and dynamics of block systems. *PhD. thesis*, University of California, Berkeley.

Xie, M., W.H. Gerstle & P. Rahulkumar 1995. Energy-based automatic mixed-mode crack-propagation modeling. *Journal of Engineering Mechanics, ASCE*, 121(8):914-923.

Computer Methods and Advances in Geomechanics, Desai et al. (eds) © 2001 Balkema, Rotterdam, ISBN 90 5809 183 X

Evaluation of implicit and explicit algorithms for the implementation of a material model including anisotropy and hysteretic nonlinearity

L.X. Luccioni
Geosyntec Consultants, Los Angeles, Calif., USA

J.M. Pestana
University of California, Berkeley, Calif., USA

ABSTRACT: The description of materials exhibiting anisotropy and nonlinear response in general boundary value problems require the numerical implementation of increasingly complex constitutive laws. Fully implicit algorithms schemes have been demonstrated to be very robust and efficient for a wide variety of model formulations, but they are difficult and impractical to apply except to very simple constitutive laws. Explicit algorithms, on the other hand, are easier to use but result in lower computational efficiency and require tight numerical controls to assure convergence. This paper compares the performance of implicit and explicit algorithms for the integration of a set of constitutive laws for lightly overconsolidated clays with includes anisotropic behavior and small strain hysteretic nonlinearity.

1 INTRODUCTION

If we consider that the practical objective of developing constitutive laws is to solve boundary value problems, it seems reasonable to think that the 'ultimate' test of constitutive relations is the 'boundary value problem' test and not the 'one element' test. Nevertheless, the solution of a boundary value problem is a function of the numerical algorithm used to integrate the differential expressions describing material response. In addition, uncertainties in the predicted response arise from uncertainty in materials properties, state variables (e.g., initial stresses and previous stress history), as well as from the inaccurate geometry or other simplifications. To address this apparent shortcoming, the 'boundary value problem' test can be subdivided into two separate components: (a) the numerical implementation evaluating the accuracy, stability and efficiency of the numerical algorithm, and (b) the overall model performance by comparing measured quantities such as displacements and stress fields with the ones in a "well controlled" physical experiment predicted by the proposed formulation. In this context, a constitutive law may be evaluated for its amenability and robustness for numerical implementation, thus addressing the problem of "realistic description" of material response separately. Once the numerical aspect has been investigated, simple boundary value problems where "analytical" solutions are available may be used to test the overall performance of the model. In this way, uncertainties that are irrelevant to the theoretical and numerical formulation are minimized.

This paper compares the performance of implicit and explicit algorithms to integrate a new set of constitutive laws developed to describe the response of lightly overconsolidated clays (Pestana & Luccioni, 1999). The finite element implementation is performed within FEAP, a multi purpose nonlinear finite element code developed at the University of California, Berkeley (Taylor, 1998). Numerical simulations of single element tests as well as a boundary value problem confirm the robustness, accuracy, and efficiency of the algorithms at the local and global levels (Luccioni & Pestana, 1999; Luccioni & al., 2000).

2 MODEL FORMULATION.

Constitutive models based on the Critical State Soil Mechanics framework (e.g., Wood, 1990) are perhaps the most widely used plasticity model for clays used to perform analyses of boundary value problems in geomechanics. Nevertheless, these models do not incorporate key elements of material response, such as anisotropic stress-strain-strength behavior or small strain nonlinearity. The numerical procedures, described later, are illustrated here by implementing a recently developed constitutive model for lightly overconsolidated clays. The model, referred to as BEAR-CLAY, is based on the theory of incrementally linearized plasticity which is extensively documented in the literature (e.g., Prévost, 1978). Model formulation includes three important components to describe the observed clay response: a) an elastoplastic framework for normally consolidated clays with a single anisotropic yield function with dependence on the third invariant of the stress tensor (cf., Fig. 1), b) equations describing the small strain non-linearity and hysteretic stress-strain response in unload-reload cycles, and c) non-associated flow and hardening rules to describe the evolution of anisotropic stress-strain properties.

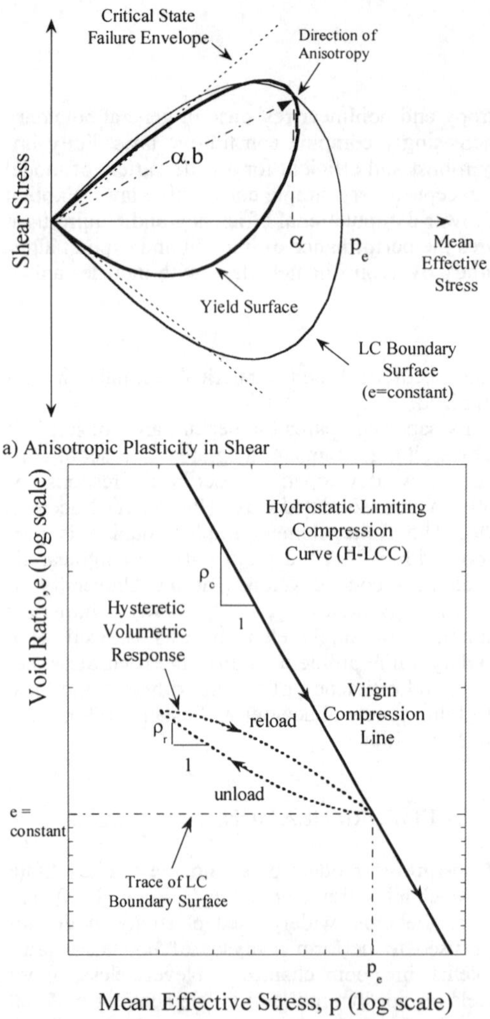

a) Anisotropic Plasticity in Shear

b) Compression Response

Figure 1: Proposed conceptual model for material response for lightly overconsolidated clays (Pestana & Luccioni, 1999).

Soils exhibit a high degree on nonlinearity in the "elastic" (i.e., recoverable strains) regime that is better described by a Perfectly Hysteretic Formulation. The main characteristic of this formulation is that the tangent stiffness decreases monotonically with continued straining following the last reversal state (i.e., G_{sec}/G_{max} as a function of shear strain, Fig. 2).

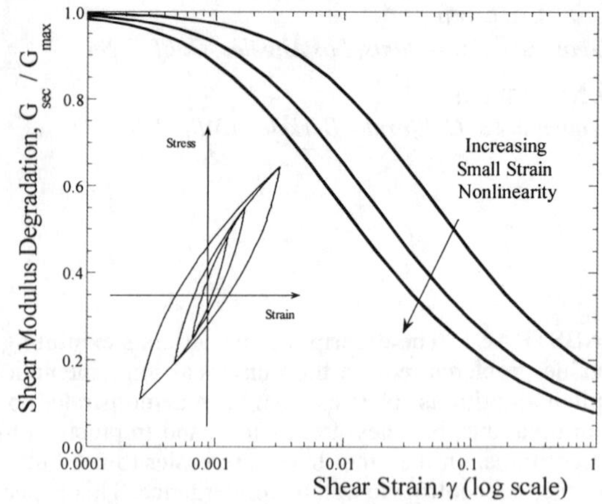

Figure 2: Modeling of small strain nonlinearity in shear.

The Perfectly Hysteretic model predicts fully recoverable strains in a closed unload-reload cycle, but dissipates energy according to the prescribed stiffness reduction law. The effect of small strain nonlinearity and anisotropy has been proven to be of significant importance in numerous soil-structure interaction problems such as the prediction of deformations around ground openings in soft clays (e.g., Luccioni et al., 1999). Since soil nonlinearity is strongly strain level dependent, the numerical integration algorithm must be able to maintain accuracy while, at the same time, be computationally efficient. Full details of model formulation are presented by Pestana and Luccioni (1999).

3 FINITE ELEMENT IMPLEMENTATION

Several numerical algorithms have been proposed for the integration of elastoplastic and elastic-viscoplastic constitutive relations, since these are found to be applicable to a large number of materials such as metals, concrete, soils and rocks (e.g., Simo and Ortiz, 1985, Lush et al., 1989). These algorithms fall into one of two categories: implicit and explicit algorithms. Recent advances in classical computational plasticity have established the superiority of fully implicit algorithms to solve boundary value problems using the finite element method for relatively simple plasticity models (e.g., Simo and Hughes, 1998). However, as the constitutive laws become more complex, the attractiveness of the fully implicit algorithm decreases significantly. First, the formulation requires the second Frechet derivative of the yield function with respect to the state variable tensors. Second, the solution of the local set for highly nonlinear yield functions is by no means trivial, as the local Newton scheme may not converge. Finally, an explicit expression for the algorithmic consistent tangent may not be available, hence at best, only a quasi-Newton method may be used and quadratic rate of convergence can not be expected. The following sections briefly describe the key element of a set of implicit and explicit algorithms used to integrate the BEAR-CLAY model while full details are available in the literature (Luccioni & Pestana, 1999, Luccioni & al., 2000)

3.1 *Implicit Algorithm*

3.1.1 *Local Stress Integration*

The problem of rate-independent plastic flow is described by a set of differential/algebraic equations and is integrated by using the implicit backward Cauchy-Euler method corresponding to the lowest hierarchy of Backward Differentiation Formula (BDF) method. The split operation procedure involves two steps: a) the integration of the elastic equations to obtain the elastic predictor (initial condition for the plastic equations) and b) the relaxation of the "elastically" predicted stresses onto an appropriately updated yield surface using the plastic corrector. In contrast to previous derivations (e.g., Borja, 1991), an iterative solution algorithm is required, even for the elastic regime, because of the complexity resulting from the hysteretic small strain nonlinearity. The resulting predictor and corrector steps are solved using the Newton-Raphson iterative method, which requires the derivation of the elastic and plastic Jacobian to obtain a quadratic rate of convergence.

3.1.2 *Global Solution Algorithm*

The finite element computer code FEAP is displacement driven, thus the unknowns are the nodal displacement, and the balance between external and internal forces is represented by a system of implicit nonlinear equations: $F(u) = 0$. $F(u)$ and u are vectors and the matrix of the first derivatives $\partial F_i/\partial u_j$ (the Jacobian matrix) is the tangent stiffness matrix for the system K_s.

Simo and Taylor (1985), have shown that the Newton-Raphson method in combination with the consistent algorithmic tangent leads to a quadratic rate of convergence. Although theoretically possible in some cases, the complexity inherent to more sophisticated models makes the derivation of the consistent tangent very cumbersome and impractical (Aravas, 1987) In this case, several alternatives are available. The simplest one uses the initial elastic tangent (initial stress method), but this method is not appropriate for pressure dependent geomaterials, which get stiffer as the confining pressure increase and may lead the algorithm to diverge altogether. Another variation consists on computing numerically the tangent stiffness instead. In this case, the convergence of the method is linear rather than quadratic; since the error (u_i-u_{exact}) is no longer squared at every step. Nevertheless, if the computed matrix K_{si} is close to the consistent tangent, then the factors multiplying the error at each iteration are reasonably small. The proposed global algorithm uses a quasi-Newton method with the Broyden-Fletcher-Goldfarb-Shano (BFGS) updates as described by Matthies and Strang (1979). The BFGS update has the following advantages: a) The updated matrix K_{si} satisfies the quasi-Newton equation $K_{si}(u_i-u_{i-1}) = F(u_i)-F(u_{i-1})$, b) It preserves symmetry and positiveness and c) the new search direction is computationally inexpensive.

The algorithmic consistent tangent derived for the non-linear elastic predictor is used as the starting tangent for the quasi-Newton scheme. This tangent is then updated accordingly using BFGS procedure and a line search algorithm is used to compute new solutions. For the following steps, the numerical tangent obtained by finite differentiation is used as the 'base' tangent to be updated by the BFGS scheme (Luccioni & al., 2000)

3.2 *Explicit Algorithm*

For classical explicit integration schemes the discrete constitutive equations are much simpler to formulate, but their accuracy depends significantly on the selected step size. Sloan (1987) first proposed the application of an automatic stepping with error control numerical algorithm for the prediction of collapse load using a relatively simple constitutive model. The use of automatic substepping and load stepping with error control algorithm overcomes the main limitation of explicit techniques, since the system adapts as the estimated error changes.

3.2.1 *Local Stress Integration*

Gear (1971) shows results indicating that fully explicit algorithms are, in general, unstable and do not converge. However, in this work we do dot enforce that yield conditions are satisfied exactly (i.e., $\varphi = 0$), but only in an approximated way, $|\varphi| < TOL$, where TOL is user specified. This concept may be viewed as a penalty regularization of the yield condition used for viscoplastic materials (Duvaut & Lion, 1972). The integration algorithm used here follows the work of Sloan (1987) which establishes a mechanism to control the error in the integration scheme. The return mapping algorithm is used to integrate the continuum rate equations using the first order accurate forward Euler scheme. The trial state is found based on converged values and the elastic predictor, while for "loading states" (i.e., plastic conditions) the plastic corrector step is invoked. For the most general case of "plastic" loading conditions, the system of nonlinear equations describing the state and memory parameters can be reformulated as the solution of a single scalar equation (i.e., the consistency equation) in terms of the plastic loading parameter, $\Delta\lambda$. The consistency equation is solved Newton's iterative method and as a result only this single scalar equation needs to be solved instead of the inversion of a 12x12 Jacobian for the fully implicit algorithm (in the case of BEAR-CLAY).

The error control scheme is based on local extrapolation and requires that the constitutive laws be re-integrated with a second order method. The modified Euler method uses the converged state achieved from the forward Euler as base values. The substep size is continually updated to insure that a measure of error for a given substep, R_{k+1}, is less than a specified tolerance, STOL :

$$R_{k+1} = \left\| {}^{2}\sigma_{k+1} - {}^{1}\sigma_{k+1} \right\| / \left\| {}^{2}\sigma_{k+1} + {}^{1}\sigma_{k+1} \right\| < \text{STOL}$$

where ${}^{1}\sigma_{k+1}$ and ${}^{2}\sigma_{k+1}$ are the stress tensors obtained with the first and second order forward integration algorithms. Figure 3 shows results of a parametric study for STOL for simulations of a Normally Consolidated (NC) clay in plane strain compression and extension undrained tests. The figure is plotted in terms of axial strain in log space in order to showcase the effect of small shear strain nonlinearity.

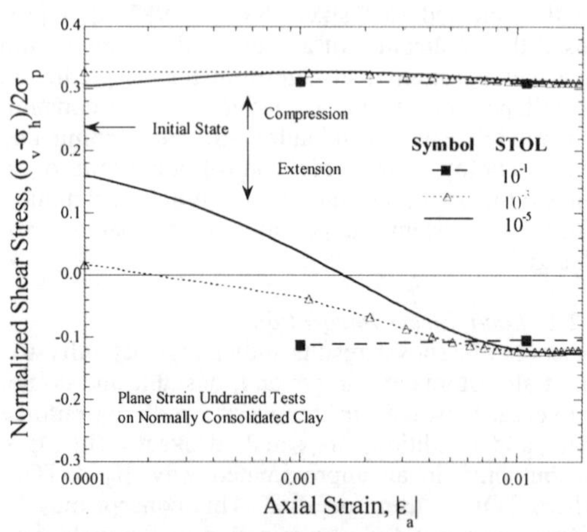

Figure 3: Parametric study on local algorithm for NC clay

The value of STOL is problem and constitutive model dependent. On one hand, if the value set for STOL is too large, the solution obtained may be completely inaccurate and thus not satisfactory. On the other hand, if STOL is chosen too small the problem may not converge. Luccioni investigated the range between 10^{-1} to 10^{-5} which was originally reported by Sloan (1987). While for the compression case a value of STOL of 10^{-3} seems to be sufficient, in extension a significant difference exist in the initial portion of the stress-strain curve (following a stress reversal state) while at larger strains the difference is negligible for most practical purposes.

3.2.2 *Global Solution Algorithm*

Abbo and Sloan first presented an incremental global algorithm based on an automatic load stepping scheme with error control. The formulation tries to minimize the 'drift' (from equilibrium) by calculating the residual forces at the end of each load increment and added these to the applied forces for the next increment.

Luccioni and Pestana (1999) implemented a slightly modified version of the Abbo and Sloan algorithm which allows the automatic computation of the coarse lad increment size, making the global algorithm entirely automatic from the Gauss point level to the coarse level. If more than two consecutive unsuccessful substeps are detected at any level, the current coarse level is divided by two, whereas if more than two consecutive coarse successful steps are detected, the current coarse level is multiplied by 1.2. These bounds are completely heuristic and are based on a significant number of simulations using the BEAR CLAY model, hence caution should be used in generalizing these numbers to other models or situations. This fact constitutes a drawback and a strength since it makes this technique flexible enough to accommodate very strong nonlinear behavior. In the solution scheme, the elastoplastic continuum tangent is assembled and used for the solution of the problem, In contrast to the implicit algorithm, the elastoplastic continuum tangent is not significantly different from the algorithmic "consistent tangent as step increments are much smaller.

Luccioni and Pestana evaluated the global explicit algorithm performance separately, for a different 'quality' of the Jacobian for the J_2 flow model with isotropic and kinematic hardening laws. A four-element patch test, shown in Figure 4, is used to compare the performance of the different algorithms. A summary of the material properties for the J_2 flow model is presented in Table 3. The surface load applied on the external boundary has a magnitude of 15 kN, which is higher that the value of the yield stress, $\sigma_Y = 10$ kN. The local stress integration for the J_2 flow model consists of the efficient implicit return mapping algorithm. Figure 4 illustrates two measures of computational effort, CPU time and number of coarse iterations, as a function of DTOL (i.e., residual force for a coarse step). As the tolerance DTOL is becoming looser, using the consistent tangent leads to a more accurate and efficient solution algorithm. These results demonstrate that in the context of the incremental technique, continuum tangent may be used as long as the user defined tolerance is kept strict enough (< 10^{-3} in this case).

Table 3. J2 Flow Model Properties Summary

Shear Modulus	10^5
Yield Stress	10
Yield Increment	10
Exponent Isotropic	300
Hardening Modulus	30

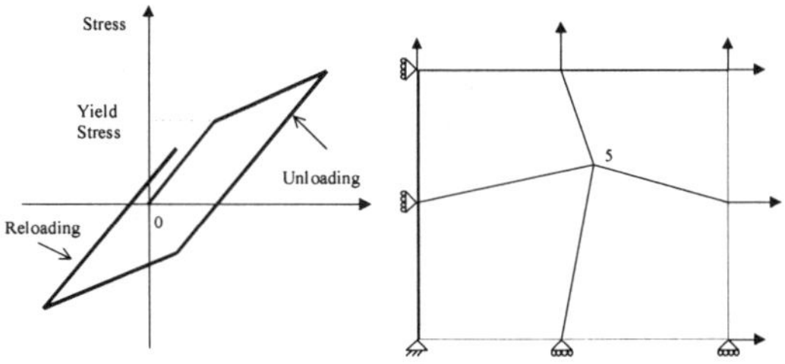

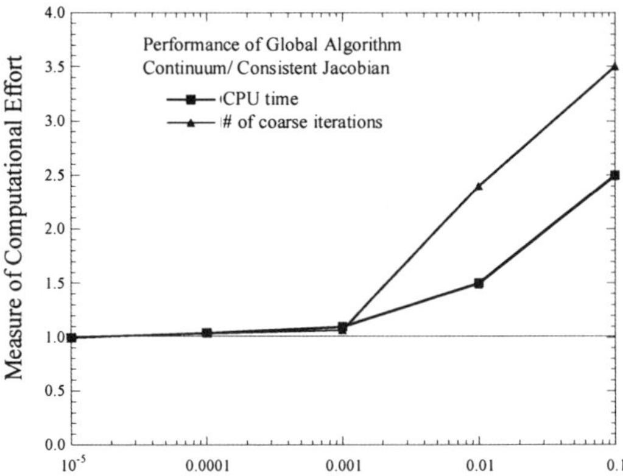

Figure 4: Performance of explicit global algorithm as a function of the quality of the Jacobian.

4 COMPARISON OF PERFORMANCE

This section describes numerical simulations used to validate the proposed implicit and explicit integration algorithms. Both local and global algorithms are coded into a series of Fortran90 subroutines and they are linked to the main nonlinear finite element code, FEAP. All computations were performed with double precision arithmetic on a 32-bit architecture DEC 3000 workstation at the University of California at Berkeley. In this example, we consider the plane strain problem of a vertically loaded flexible strip footing of width B on a uniform 'finite' deposit of normally consolidated clay. The initial geostatic stresses were generated within FEAP using a body force command to simulate the gravity load at each Gauss point and a prescribed value of K_0 of 0.53 describing the ratio of horizontal to vertical stress. The finite element mesh is composed of 200, four-noded quadrilateral 'soil' elements with a 2×2 Gaussian integration rule employed for each element. The soil profile is underlying by a rigid and rough bedrock. For the explicit algorithm, the load is applied using different global tolerance DTOL, whereas the local user tolerance is fixed to a value of 10^{-3}.

Figure 5 shows the load settlement curve at the centerline of the flexible foundation for the implicit and explicit algorithm. For loose tolerances, the implicit technique tends to underpredict the load for a given value of settlement and the explicit technique overpredicts the load for a given value of settlement. For the explicit algorithm, the results indicate that values of DTOL less than 10^{-3} are adequate whereas for the implicit a number of substeps larger than 5 is required. The load-settlement curve corresponding to DTOL equal to 10^{-4} compares very well with the one computed using the global implicit-iterative algorithm with 20 steps. The results from both procedures are in very good agreement and for most practical purposes identical. The cost for the two techniques is quite similar with 649 seconds CPU time for the 20 steps iterative solution and 780 seconds for the incremental solution with DTOL equal to 10^{-4}.

5 SUMMARY

Fully implicit stress integration algorithms are attractive from the computational point of view but they are typically very cumbersome, and in many cases impractical, to implement for complex nonlinear constitutive soil models. For these models, the choice of an explicit automatic substepping with error control algorithm represents a computationally efficient alternative. The paper has presented a brief summary of key elements of implicit and explicit algorithms with automatic error control for the integration of nonlinear elastoplastic constitutive laws. The global explicit technique is found to be computationally efficient and accurate using the continuum tangent as long as the user tolerance DTOL is tight enough. The numerical procedure is illustrated here by the implementing a recently developed constitutive model for lightly overconsolidated clays included anisotropic behavior with small strain nonlinearity in shear into a finite element computer code. The numerical algorithm proposed here can be directly extended to other models without conceptual changes and is not dependent on the selected set of constitutive expressions used to illustrate the procedure. For cases in which the algorithmic consistent Jacobian can not be determined, preliminary comparisons suggest that explicit and implicit algorithms have similar computational performances in terms of CPU time.

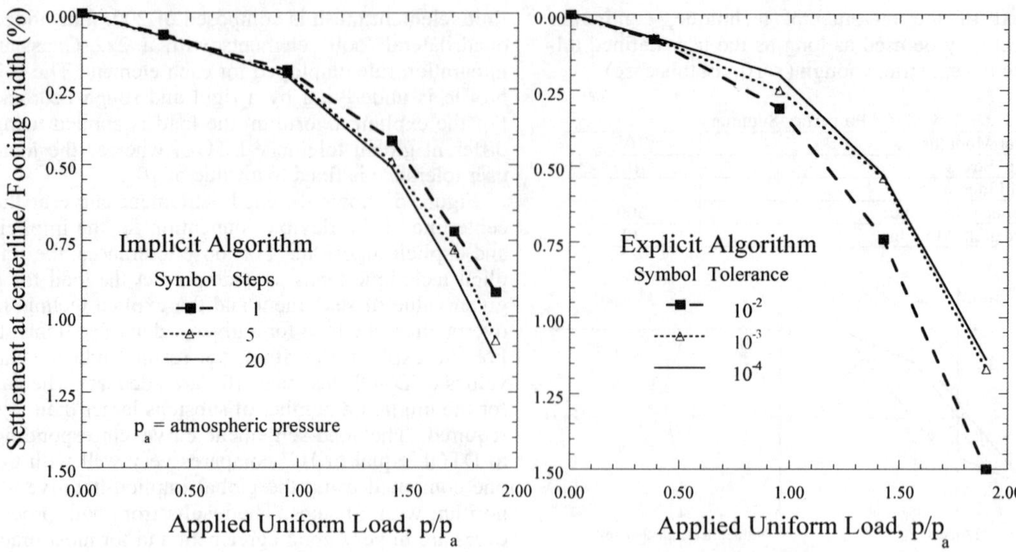

Figure 5: Comparison of predicted settlement curves using implicit and explicit integration algorithms.

6 ACKNOWLEDGMENTS

Support for this research was provided by the National Science Foundation (NSF) Grant No. CMS 9612136 with the University of California, Berkeley. Professors P. Papadopoulos, K. Miller and R. Taylor valuable input during this research study is gratefully acknowledged.

REFERENCES

Abbo, A.J. and Sloan W. (1996) “An Automatic Load Stepping Algorithm with Error Control" Int. J. for Numerical Meth Engng, 39, , 1737-1759

Aravas, N. (1987) “On the Numerical Integration of a Class of Pressure-Dependent Plasticity Models”, *Int. J. Numer. Meths. Engrg.*, 24, 1395-1416.

Borja R.I. (1991) “CAM-CLAY Plasticity, Part II: Implicit integration of Constitutive equations based on a nonlinear stress predictor”, *Comput. Meth. Appl. Mech. Engng*, 88, 225-240

Duvaut G., and Lion J.L.(1972) “Les Inequations en Mecanique et en Physique”, Dumod, Paris.

Gear, C.W. (1971) “Numerical Initial Value Problems in Ordinary Differential Equations” Prentice-Hall, Englewood Cliffs, N.J.

Luccioni, L. and Pestana, J.M. (1999) " Explicit Algorithms for the Implementation of a Nonlinear Elastoplastic Model for Lightly Overconsolidated Clays," UCB/GT/99-22. University of California, Berkeley.

Luccioni, L., Pestana, J.M. and Koutsoftas, D. “Modeling of Deformations around Deep Excavations in Soft Soils,” in Proc. of Geo-engineering for Underground Facilities, Illinois, June (1999)

Luccioni, X.L., Pestana, J.M. and Rodriguez-Marek, A. (2000) An Implicit Integration Algorithm for the Finite Element Implementation of a Nonlinear Anisotropic Material Model including Hysteretic Nonlinearity", *Computer Methods in Applied Mechanics and Engineering* (to appear)

Lush, A.M., Weber, G. and Anand, L. (1989) "An Implicit Time-Integration Procedure for a Set of Internal Variable Constitutive Equations for Isotropic Elasto-Viscoplasticity", *Int. J. of Plasticity*, 5 (3), 521-549.

Matthies, H. and Strang, G. (1979) “The Solution of Nonlinear Finite Element Equations” *Int. J. Numer. Meths. Engrg.*, 14, 1613-1626

Pestana, J.M. and Luccioni, X.L. (1999)"A Simplified Constitutive Model for Lightly Overconsolidated Clays,” *Report UCB/GT 99-23*, University of California, Berkeley.

Prevost J.H. (1978) “Plasticity Theory for Soil Stress-Strain Behavior” ASCE, *J. of Mech. Div.*, 104 (EM5),1177-1194

Simo, J.C., and Ortiz, M. (1985) "A Unified Approach to Finite Deformation Elastoplastic Analysis Based on the Use of Hyperelastic Constitutive Equations," *Comput. Meths. Appl. Mech. Engrg.*, 49, 221-245.

Simo, J.C. and Taylor R.L (1985) “Consistent Tangent Operator for Rate Independent Elasto-Plasticity”, *Comput. Meths. Appl. Mech. Engrg.*, 48, 79-116

Simo, J.C. and Hughes T.J.R. (1998) "Computational Inelasticity", *Interdisciplinary Applied Mathematics*, Springer.

Sloan, S.W. (1987) “Substepping Schemes for the Numerical Integration of Elastoplastic Stress-Strain Relations”, *Intl. J. For Num. Meth. In Engr.*, 24, 893-911.

Taylor, R.T. (1999) "FEAP manual," Dep. Civil & Env. Engng., University of California, Berkeley

Wood, D.M (1990) Soil Behavior and Critical State Soil Mechanics. Cambridge University Press, 462pp.

Computer Methods and Advances in Geomechanics, Desai et al. (eds) © 2001 Balkema, Rotterdam, ISBN 90 5809 183 X

EFGM analysis for saturated soil

A. Murakami
Graduate School of Natural Science and Technology, Okayama University, Japan

H. Kawabata
Shiga Prefectural Government, Japan

S. Aoyama
Graduate School of Agriculture, Kyoto University, Japan

ABSTRACT: A formulation of element-free Galerkin Method (EFGM) for saturated soil is presented. The equilibrium of stress and the continuity of pore water constitute a set of governing equations for two-phase media such as saturated soil. A weak form of the equation of stress equilibrium is derived and its discretization based on the EFGM strategy, e.g., the moving least square method (MLSM), is presented under the elastic constitutive equation. The continuity equation of pore water, on the other hand, is discretized in two ways: 1) the MLSM, 2) the difference scheme proposed by Akai and Tamura utilizing the framework of the background cell. Boundary conditions are treated with help of the penalty method. A special focus is placed upon the formulation including the automatic collocation of background cells based on nodal and boundary data. A numerical efficiency is demonstrated through several analyses. First, the accuracy of numerical computation is examined through one-dimensional consolidation problem by comparing with closed and usual finite element solutions. Numerical results for two-dimensional problems are then described and the feasibility of the current numerical scheme is emphasized. Strain localization phenomenon is pointed out as an example of further applications.

1 INTRODUCTION

The finite element method (FEM) has been a powerful tool for analyzing geotechnical problems. Recent developments in numerical skills for FE simulations allow us to solve the more complicated behavior of saturated soil in infinitesimal and even finite deformational fields. The dependency of numerical solutions on mesh arrangement and mesh type can often be confronted when dealing with the softening behavior of saturated soil within a localized deformation.

Cosserat continuum and strain gradient plasticity models are adopted as the remedy for such numerical difficulties. In an analysis which adopts such non-local models, the dilemma of appropriately determining the parameters in the models in order to obtain well-posed solutions is also encountered.

EFGM, developed by Belytschko *et al.* (Belytschko *et al.*, 1994), is expected to be an alternative way to obtain well-posed solutions for the above type of situation, because meshes are not used. Several researchers have reported that EFGM is advantageous in that the shear locking of the cantilever beam, the membrane shear, and shells in the analysis can be avoided.

The objective of this paper is to examine the formulation and the numerical accuracy of the EFGM strategy for saturated soil within an infinitesimal deformation. The arrangement of the paper is as follows. In the second section, the procedure of EFGM is outlined. In the third section, the formulation of EFGM for saturated soil is provided. In the last section, a numerical solution is examined based on different nodal arrangements for a one-dimensional consolidation problem and a comparison is made between FE and EFGM solutions under both elastic and elasto-plastic constitutive relations.

2 EFGM

The so-called meshless method was first proposed by Nayroles *et al.* as the diffuse element method (DEM) (Nayrloes *et al.*, 1992). Belytschko *et al.* further developed Nayroles' DEM to derive the shape function of EFGM by adopting MLSM and the appropriate weight function for interpolation and to integrate the stiffness by proposing the background cell structure. The main difference between FE and EFGM is that EFGM does not require the connectivity information of mesh and nodes. From the viewpoint of the numerical formulation, the shape functions of the methods are different, although FE and EFGM are the same up to the weak form. As seen in Figure 1, the shape functions in EFGM are derived from the interpolation of nodes which are included within a circle called a 'support' (see Figure 2).

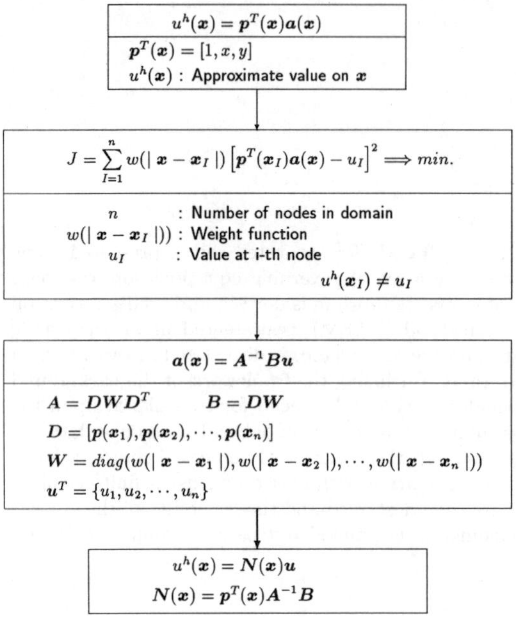

Figure 1. Shape function by MLSM (Belytschko *et al.*, 1994)

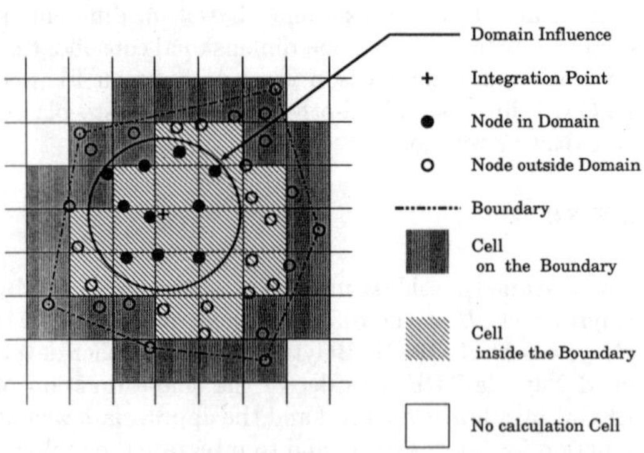

Figure 2. Notion of the 'support' (Belytschko *et al.*, 1994)

The stiffness is then evaluated at each integration points of the background cell instead of the element. The resultant stiffness equation for EFGM is different from that of FE, even though the arrangement of the meshes in FE and the background cells in EFGM are the same. Details of the derivation of the stiffness equation for two phase media are shown in the subsequent section.

3 NUMERICAL FORMULATION

3.1 Governing equation

A set of governing equations is given for analyzing the behavior of saturated soil.

Governing equations

$$\sigma_{ij} = \sigma'_{ij} + \delta_{ij}p_w \tag{1}$$

$$\dot{\sigma}_{ij,j} = 0 \tag{2}$$

$$\dot{\epsilon}_{ij} = (\dot{u}_{i,j} + \dot{u}_{j,i})/2 \tag{3}$$

$$\dot{\sigma}'_{ij} = D_{ijkl}\dot{\epsilon}_{kl} \tag{4}$$

$$v_i = -k_{ij}h_{,j} \tag{5}$$

$$h = p_w/\gamma_w + \Omega \tag{6}$$

$$\dot{\epsilon}_v = v_{i,i} \tag{7}$$

Boundary conditions

$$\dot{\sigma}_{ij}n_j = \hat{\dot{t}}_{si} \qquad (on\ \Gamma_\sigma) \tag{8}$$

$$\dot{u}_i = \hat{\dot{u}}_i \qquad (on\ \Gamma_u) \tag{9}$$

$$h = \hat{h} \qquad (on\ \Gamma_h) \tag{10}$$

$$-k_{ij}h_{,j} = \hat{v}_i \qquad (on\ \Gamma_v) \tag{11}$$

Initial conditions

$$\sigma'_{ij} = \sigma'_{ij}\,|_{t=0} \qquad (in\ V) \tag{12}$$

$$h = h\,|_{t=0} \qquad (in\ V) \tag{13}$$

where $\dot{\sigma}_{ij}$: stress rate, $\dot{\sigma}'_{ij}$: effective stress rate, $\dot{p}_w$: excess pore pressure rate, $\dot{\epsilon}_{ij}$: strain rate, D_{ijkl} : stress-strain relation, v_i : flow rate, k_{ij} : permeability, h : total head, γ_w : unit weight of water, Ω : head, $\dot{\epsilon}_v$: volumetric strain rate, n_j : unit normal of the boundary, Γ_σ : stress boundary, Γ_u : displacement boundary, Γ_h : head boundary, Γ_v : flow rate boundary, V : whole boundary.

Weak form of the equation for the stress equilibrium and the equation for the continuity of water are obtained in a similar manner to that of a usual FE. A numerical integration of the stiffness is carried out based on the shape functions appearing in Figure 1 at the fourth order integration points of each background cell. They are assembled into the stiffness equation. The boundary conditions are imposed by the penalty method, because the nodal quantity does not necessarily coincide with the variable in its physical meaning, e.g., displacement or head. The derived stiffness equations are summarized in Figure 3.

4 NUMERICAL RESULTS and DISCUSSIONS

A one-dimensional consolidation problem is solved by EFGM, and this solution and a theoretical one (Lambe

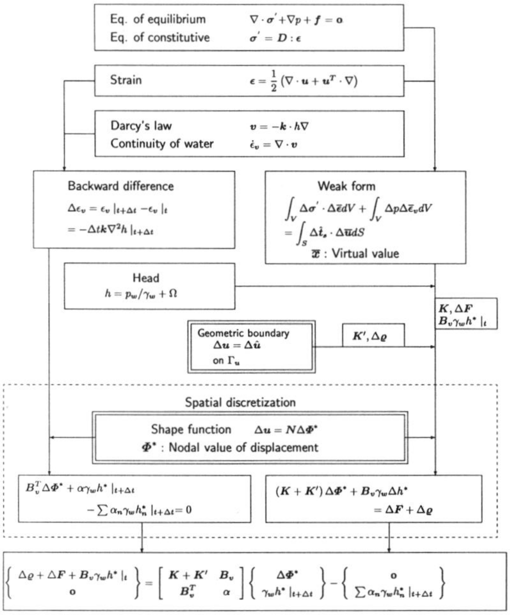

Figure 3. Derivation of stiffness equation (revised from Iizuka and Ohta, 1987)

and Whitman, 1969) are compared in order to examine the numerical accuracy of the solution procedure by EFGM under different patterns of nodal arrangements as seen in Figure 4. To gain a comparable impression, the results by the FE model are also included in the subsequent figures.

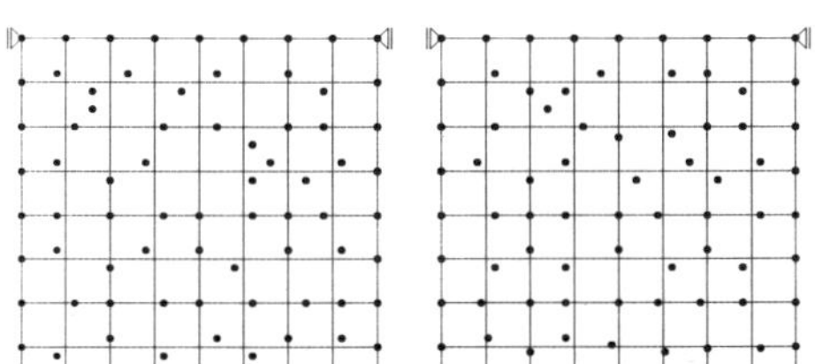

(a) Case 1 (b) Case 2

Figure 4. Collocation of nodal points

The material constants used in both numerical analyses reported here are as follows: Young's modulus is $E = 100 \times 98.1kPa$ and Poisson's ratio is $\nu = 0.25$, Permeability is $k = 1.0 \times 10^{-5} cm/sec$. The top surface subjected to the equally-distributed load as the traction boundary is also imposed to the drainage boundary. The coefficient of the Penalty method is 10^4. In Figure 4, 81 nodal points are scattered.

A numerical comparison is made under two different patterns of nodal arrangements in Figures 5 and 6. The agreement among the solutions is satisfactory for both figures. It can be seen that the EFGM strategy described herein provides highly accurate numerical solutions.

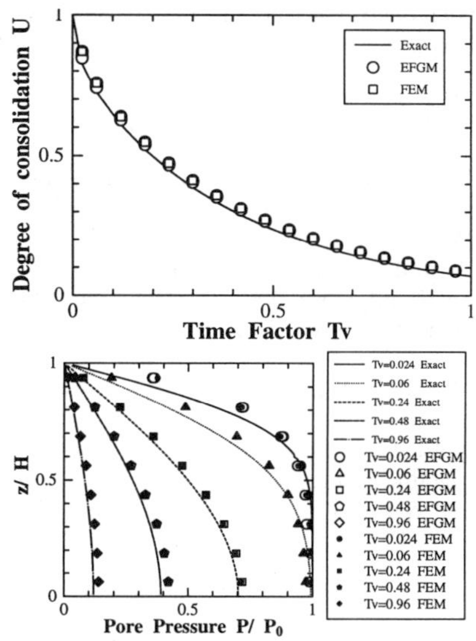

Figure 5. Numerical result (Case #1)

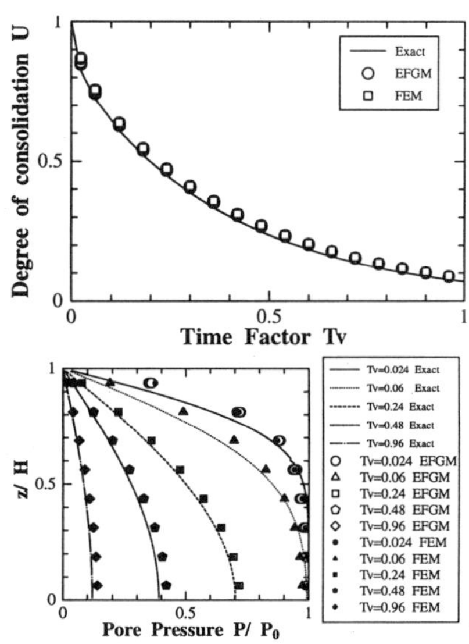

Figure 6. Numerical result (Case #2)

5 CONCLUSION

An EFGM based formulation for simulating the behavior of saturated soil has been developed, and its implementation has been presented. According to a comparison between the numerical solutions and the theoretical ones, the successful application of the numerical model to the analysis of a geotechnical problem has demonstrated that the EFGM strategy is also a useful tool.

REFERENCES

Belytschko, T., Y.Y. Lu and L. Gu 1994. Element-free Galerkin methods, *Int. J. Numer. Meth. Engng.*, 37: 229–256.

Iizuka, A. and H. Ohta 1987. A determination procedures of input parameters in elasto-viscoplastic finite element analysis, *Soils and Foundations*, 27(3): 71–87.

Lambe, T. and R. Whitman 1969. *Soil mechanics*, Wiley.

Nayroles, B., G. Touzot and P. Villon 1992. Generalizing the finite element method: Diffuse approximation and diffuse elements, *Comp. Mech.*, 10: 307–318.

Computer Methods and Advances in Geomechanics, Desai et al. (eds) © 2001 Balkema, Rotterdam, ISBN 90 5809 183 X

Higher order finite elements to reduce 2D numerical wave dispersion

J.F. Semblat
Engineering Modelling Department, Laboratoire Central des Ponts et Chaussées, Paris, France

J.J. Brioist
Centre d'Etudes Techniques Maritimes et Fluviales, Compiègne, France

ABSTRACT : The analysis of seismic wave propagation in soils by numerical means is not easy since one has to take into account both following constraints : the description of the infinite geometrical extent of the medium and the accuracy of the local computation of wave propagation. For finite difference or finite element methods, the second condition is known through numerical wave dispersion making the propagation characterictics depend on the features of the numerical model. This paper firstly recalls several results on 1D dispersion of numerical waves. Higher order finite elements are very efficient to reduce numerical dispersion. In two dimensions, the analysis of numerical dispersion is performed considering different types of element patterns (i.e. different 2D spatial arrangements). The accuracy of the different element types and the various 2D patterns are investigated to find out the case with the lowest numerical dispersion. These numerical investigations are compared by analyzing seismic wave propagation due to foundation vibration. The influence of numerical dispersion on the computed propagated seismic waves is studied.

1. NUMERICAL WAVE DISPERSION

The numerical analysis of seismic or mechanical wave propagation can be performed through various approaches : finite differences, finite elements, boundary elements, spectral methods (Bonnet 1999, Guéguen 2000, Komatitsch 1999, Semblat 2000a,b). All these methods have particular advantages and drawbacks for wave propagation analysis : numerical accuracy, description of infinite media, heterogeneous media, complex constitutive behaviour...

The finite element method is powerfull for the two latter but has some drawbacks concerning the accuracy of wave propagation computations. Both following types of numerical errors are generally involved : numerical dispersion making the wave velocity depend on the model characteristics and algorithmic damping for the wave amplitude (Hughes 1987). These errors are influenced by the time-integration scheme, the element type and geometry, the degree of interpolation, the wave type...

Considering the 1D scalar wave equation in the frequency domain (Aki 1980, Bourbié 1987), mechanical wave propagation in a linear viscoelastic medium can be described by the following equation :

$$\frac{\partial^2 u(x,\omega)}{\partial x^2}+\frac{\rho\omega^2}{E^*(\omega)}u(x,\omega)=0 \qquad (1)$$

where u is the displacement, ρ the mass density, ω the circular frequency and E^* the complex modulus.

The theoretical solution of equation 1 can be written using the complex wave number k^* including two terms : one for propagation and one for damping (Bourbié 1987, Semblat 1998). The solution then is :

$$u(x,\omega)=u(0,\omega).\exp\left(ik^*(\omega)x\right) \qquad (2)$$

A similar form of equation 2 can be considered for the numerical solution of equation 1. This solution is then expressed through the estimated wave number k_h as follows :

$$u_h(x,\omega)=u(0,\omega).\exp\left(ik_h(\omega)x\right) \qquad (3)$$

where h represents the element size.

For the numerical analysis of seismic wave propagation by the finite element method, one must take into account the accuracy of the estimation of the wave number. The numerical dispersion reveals the influence of the model characteristics on wave propagation modelling (Bamberger 1980, Ilhenburg 1995).

2. 1D ACCURACY OF HIGHER-ORDER FINITE ELEMENTS

Various theoretical researches on error estimation for the solution of Helmholtz equation have been proposed taking into account the degree of interpolation of the finite elements (Ihlenburg 1995, Stewart 1998). A specific numerical analysis has been proposed elsewhere (Semblat 2000) to investigate the efficiency of higher order finite elements towards 1D numerical wave dispersion. These results show that, for the same number of degree of freedom (DOF), the accuracy of wave propagation computations is increasing with the degree of the finite element interpolation. For the elements with the lowest degree, the wave velocity can be strongly overestimated.

For 2D and 3D analysis, one has to take into account several other issues : the influence of the element geometry, the effect of the spatial arrangement of the elements, the wave type as well as the order of the elements.

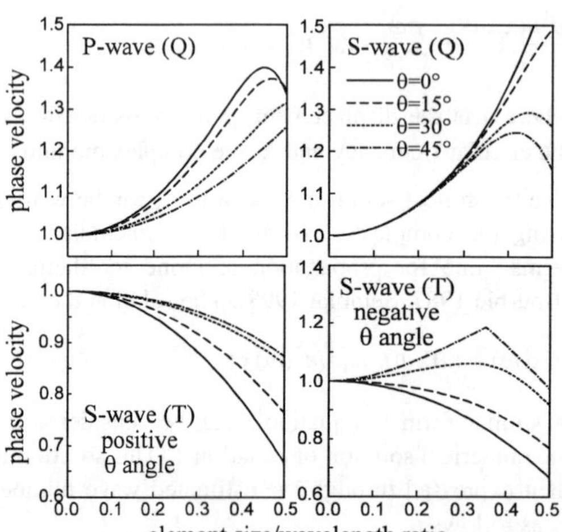

Figure 1 : Theoretical numerical dispersion for two types of element geometry : "Q" quadrilateral elements, "T" triangular elements (Bamberger, 1980).

3. 2D NUMERICAL DISPERSION

3.1 *Influence of the element geometry*

The theoretical researches made by Bamberger et al. (1980) are very interesting since they give an estimation of the numerical error on phase velocity of waves. As shown in figure 1, for a linear elastic medium, the wave velocity is not constant and depends on the element size/wavelength ratio. These curves correspond to quadrilateral elements (top) and triangular elements (bottom) for different wave incidences (angle θ). For a ratio of several tenth of the wavelength, the numerical error on velocity can reach values of several tens of percents. In figure 1, the influence on numerical dispersion of the wave type and of the angle of incidence is very large. It can be noticed that the element geometry has a strong effect on the accuracy of the numerical solution. However, for triangular finite elements, the influence of the 2D spatial arrangement has to be assessed.

3.2 *2D spatial arrangements*

Since we will analyse afterwards the accuracy of different finite element orders, we only consider two types of 2D spatial arrangements. These 2D element patterns are depicted in figure 2. The first one is a regular set of triangles called "half-square" pattern (a). The second one combines all possible orientations to get a much more isotropic pattern called "square-diamond" (b). In the following section we will analyse the efficiency of both 2D patterns towards numerical dispersion for a particular problem of seismic wave propagation.

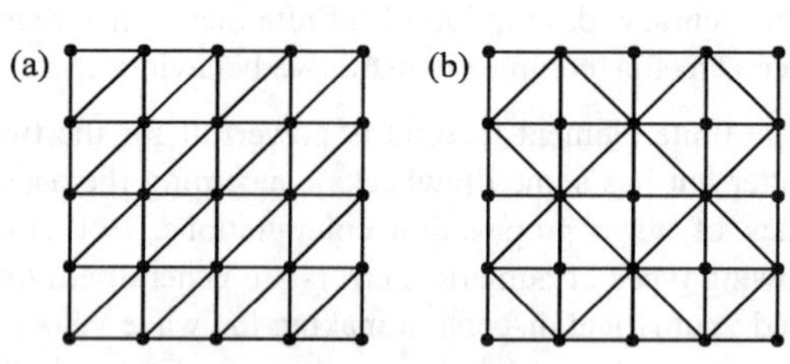

Figure 2 : Types of 2D element pattern considered : half-square (a) and square-diamond (b).

4. ANALYSIS FOR FOUNDATION VIBRATION

4.1 *Modelling of waves generated by foundation vibration*

One considers the particular problem of seismic waves generated by foundation vibration and propagating in a centrifuged soil mass. The complete description of the experiments is given in (Chazelas 2001). Figure 3 shows the foundation above the soil mass and displays the force vs time curve considered herein. The time step is Δt=0.5µs and the time integration algorithm is a Newmark non dissipative one. The element size is Δh=0.01m. The mechanical properties of the soil are the following : mass density ρ=1600kg/m3, Young's modulus E=100MPa and Poisson's ratio ν=0.24. The numerical analysis is performed with the FEM/BEM code CESAR-LCPC (Humbert 1989).

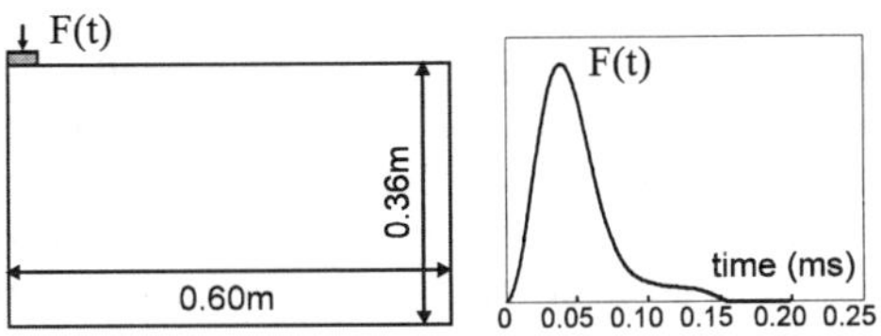

Figure 3 : Model for the analysis of waves generated by foundation vibration in the centrifuge (Chazelas 2001).

4.2 *Comparisons of 2D patterns efficiency*

The two element patterns depicted in figure 2 are compared towards numerical dispersion considering the propagation problem of figure 3. The horizontal displacement along the soil surface is displayed in figure 4. The square-diamond pattern (SD) appears to be more accurate than the half-square pattern (HS) since the numerical dispersion is stronger with the latter (velocity increase). The SD pattern being much more "isotropic" than the HS pattern, this result is still logical. We will now consider the square-diamond pattern to investigate the numerical accuracy at different element order.

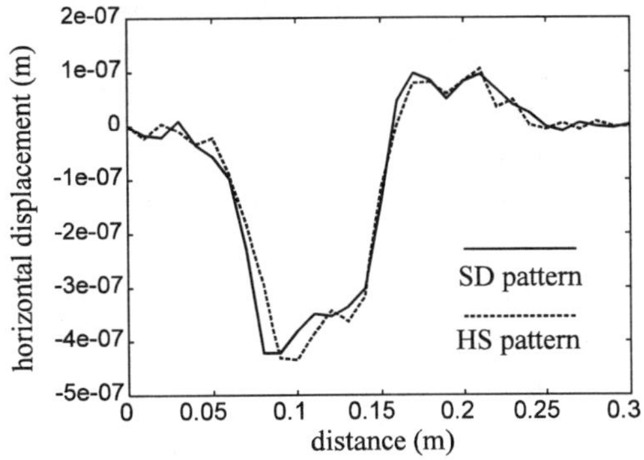

Figure 4 : Horizontal displacement along the surface at time t=0.75ms : comparison of both element patterns (SD : square-diamond, HS : half-square).

5. ORDER AND PATTERN VS DISPERSION

5.1 *Element order and constant DOF density*

To compare the efficiency of different finite element types (orders) towards 2D numerical dispersion, one considers three types of triangular elements : linear 3-node elements, quadratic 6-node elements and fourth order 15-node elements. The comparison is really interesting if the total number of degrees of freedom is the same with all element types. We then consider, with the square-diamond pattern, various element sizes to have a constant number of DOF.

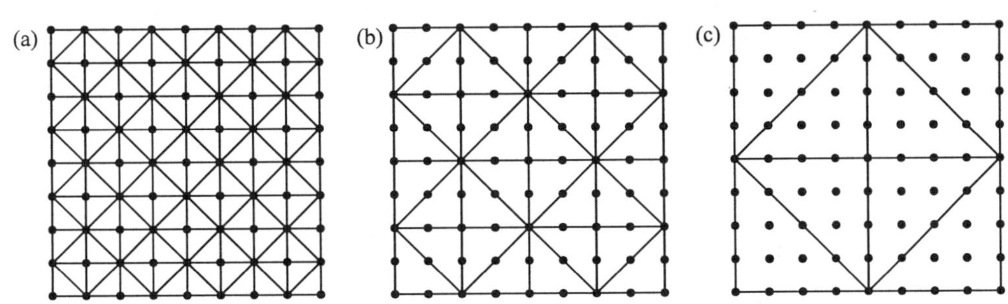

Figure 5 : Partial meshes with identical number of DOF for the comparison of different element orders : linear 3-node elements (a), quadratic 6-node elements (b) and higher-order 15-node elements (c).

As shown in figure 5, the element size being Δh=0.01m for the 3-node linear elements (a), it is chosen twice as large and four times larger for quadratic and higher order finite elements (resp.). The degree of freedom density appears to be the same with each element type (figure 5). For the complete model of figure 3, the total number of nodes for each of the three finite element mesh is 2257 and the corresponding number of elements are 4320 linear elements, 1080 quadratic elements and 270 fourth order finite elements. The wave field generated by foundation vibration is then computed with the three different models.

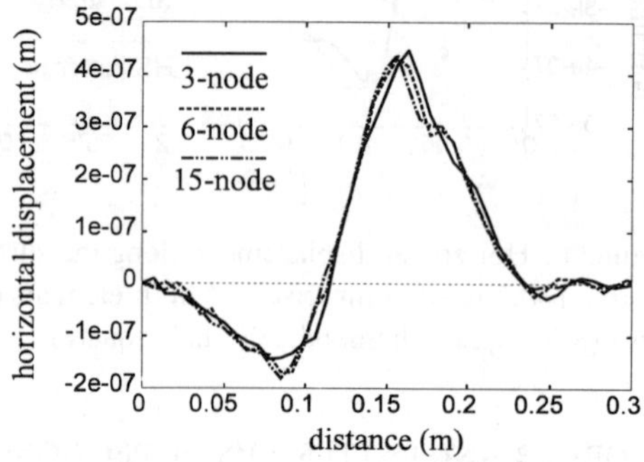

Figure 6 : Horizontal displacement along the diagonal at time t=0.75ms for different element orders.

5.2 *Lowest 2D numerical dispersion*

Using the three different models partly presented in figure 5 (square-diamond pattern), we compute the wave field generated under the foundation for the three different finite element orders considered (linear, quadratic and fourth order). We firstly analyse the horizontal displacement along the diagonal under the foundation at time t=0.75ms. As shown by Chazelas et al. (2001) with a non-constant velocity profile, large displacements due to Rayleigh waves appear along the diagonal. The computations presented here nevertheless involve a constant velocity profile to make a precise analysis of numerical dispersion.

Figure 6 displays three curves giving the horizontal displacement along the diagonal for the three element types (orders). Since the distance of propagation is short, the absolute delays between the three curves are small. The numerical error on wave velocity (numerical dispersion) make this delay increase with the propagation distance (e.g. Semblat 2000b). In figure 6 however, the velocity increase due to stronger numerical dispersion is larger for linear elements and even quadratic elements than for fourth order 15-node elements. For the square-diamond pattern, the efficiency of higher order finite elements towards 2D numerical dispersion is then better than for quadratic elements. For the same number of DOF, that is identical computational cost, quadratic 6-node elements are also more accurate than linear elements.

The vertical displacement along the diagonal is also plotted (figure 7) showing the same type of results for the numerical dispersion. These curves are enlarged (fig. 7, right) to have a better view on the numerical wave velocity variations. For identical node densities, it obviously appears to be larger for smaller degrees of interpolation.

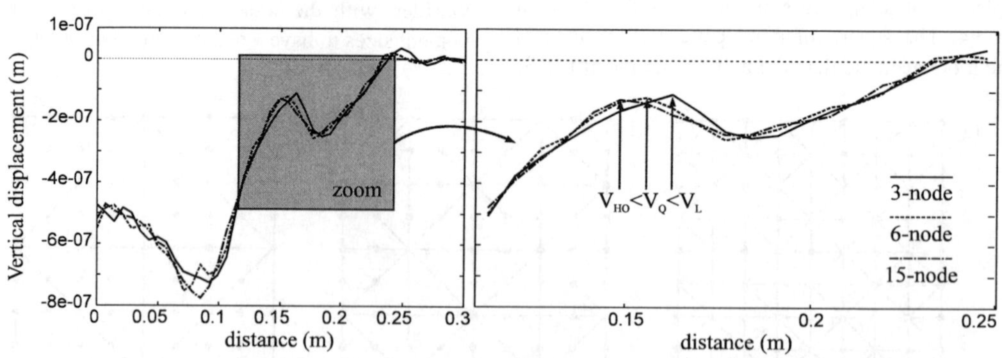

Figure 7 : Vertical displacement along the diagonal at time t=0.75ms for different element orders (left) and velocity variation in enlarged view (right).

5.3 *Influence of 2D numerical dispersion on wave characterization*

Since 2D numerical dispersion modifies propagation delays for both horizontal and vertical displacements, it can influence the characterization of waves if different types of seismic waves are involved. For instance, some numerical computations are proposed in (Chazelas 2001) to analyse different wave types (P, S, Rayleigh) in the centrifuge. We will then try to assess the differences between several finite element types for wave characterization.

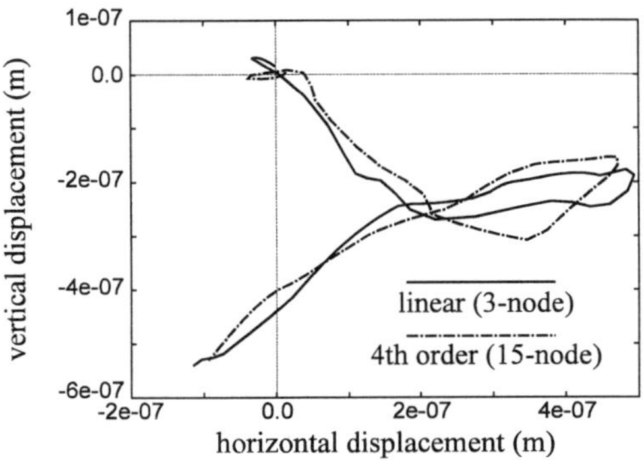

Figure 8 : Vertical vs horizontal displacement at point x=0.1m/y=-0.1m for linear and higher order elements.

Figure 8 displays vertical vs horizontal displacement at point x=0.1m/y=-0.1m with time as a parameter. Two different curves are given : one for the linear 3-node elements model and the other for the higher order 15-node elements one. The discrepancy between both curves is much larger than the slight velocity variation observed in figures 6 and 7. For the same number of DOF (figure 5), the particle motion displayed in figure 8 for linear elements is significantly modified when changing the element order. Its direction and the size of the loop are rather different. This discrepancy can be very important for the characterization of various seismic wave types. In the previous subsection, lower order finite elements were shown to overestimate the wave velocity. Since, for 2D propagation, this estimation is influenced by the angle of incidence (figure 1), the numerical dispersion has some stronger effects on 2D particle motion than on wave velocity itself. The efficiency of higher order finite elements towards numerical dispersion is then even better for particle motion estimation. It is fundamental for an accurate numerical characterization of various seismic wave types.

6. CONCLUSION

Numerical dispersion is depending on many parameters such as mesh refinement, time integration scheme, wave type... In this paper, we mainly focus on 2D numerical dispersion through the influence of element geometry, spatial arrangement and element order. The numerical modelling of wave propagation due to foundation vibration allows us to find the best spatial arrangement : "square-diamond" pattern that is the most "isotropic" one.

For a constant node density (i.e. the same computational cost), the efficiency of higher order finite elements, already shown previously in one-dimension (Semblat 2000b), also appears for the reduction of 2D numerical dispersion ("square-diamond" pattern). Their accuracy for particle motion estimation is even larger since numerical dispersion has some effects on velocities, but is also concerned in 2D by some particular issues such as the wave incidence and the wave type. The analysis of the different wave types generated and their behaviour towards 2D numerical dispersion for large propagation distances would be of great interest.

REFERENCES :

Aki K., Richards P.G. 1980. *Quantitative seismology*, San Francisco: Freeman ed.

Bamberger A., Chavent G., Lailly P. 1980. *Analysis of numerical schemes for linear elastodynamics equations* (in french), Report No.41, INRIA.

Bonnet M. 1999. *Boundary integral equation methods for solids and fluids*, Wiley.

Bourbié T., Coussy O., Zinszner B. 1987, *Acoustics of porous media*, Technip, Paris.

Chazelas J.L., Abraham O., Semblat J.F. 2001. Characterization of different seismic wave types generated by foundation vibration in the centrifuge, *4th Int. Conf. on Recent Advances in Geotechnical Earthq. Eng and Soil Dynamics*, San Diego.

Clough R.W., Penzien J. 1993. *Dynamics of structures*, McGraw-Hill.

Guéguen P., Bard P.Y., Semblat J.F. 2000, From soil-structure to site-city interaction, *12th World Conf. on Earthquake Eng.*, Auckland, New-Zealand.

Hughes T.J.R. 1987. *Linear static and dynamic finite element analysis*, Prentice-Hall, Englewood Cliffs, New Jersey.

Humbert P. 1989. CESAR-LCPC : A general finite element code (in french), *Bull. des Laboratoires des Ponts et Chaussées*, 160 : 112-115.

Ihlenburg F., Babuška I. 1995. Dispersion analysis and error estimation of Galerkin finite element methods for the Helmholtz equation, *Int. J. for Numerical Methods in Eng.*, 38 : 3745-3774.

Komatitsch, D., Vilotte, J.P., Vai, R., Castillo-Covarrubias, J.M., Sanchez-Sesma, F.J. 1999. The spectral element method for elastic wave equations - Application to 2D and 3D seismic problems, *Int. Journal for Numerical Methods in Eng.*, 45, pp.1139-1164.

Semblat J.F. 1997. Rheological interpretation of Rayleigh damping, *Journal of Sound and Vibration*, 206 (5) : 741-744.

Semblat J.F., Luong M.P. 1998. Wave propagation through soils in centrifuge testing, *Journal of Earthquake Engineering*, 2(1) : 147-172.

Semblat J.F., Duval A.M, Dangla P. 2000a. Analysis of seismic site effects : BEM and modal approach vs experiments, *12th World Conf. on Earthquake Eng.*, Auckland, New-Zealand.

Semblat J.F., Brioist J.J. 2000b. Efficiency of higher order finite elements for the analysis of seismic wave propagation, *Jal of Sound and Vibration*, 231(2) : 460-467.

Stewart J.R., Hughes T.J.R. 1998. A tutorial in elementary finite element error analysis: a systematic presentation of a priori and a posteriori error estimates, *Comput. Methods Appl. Mech. Eng.*, 158 : 1-22.

Computer Methods and Advances in Geomechanics, Desai et al. (eds) © *2001 Balkema, Rotterdam, ISBN 90 5809 183 X*

Node-centric displacement discontinuity method for plane elasticity problems

S.Vijaykumar
Rock Engineering Group, University of Toronto, Ont., Canada

J.H.Curran
Department of Civil Engineering, University of Toronto, Ont., Canada

T.E.Yacoub
Rocscience Incorporated, Toronto, Ont., Canada

ABSTRACT: In this paper, a powerful new displacement discontinuity (DD) formulation for plane linear elastic problems, which preserves inter-element continuity of tractions and displacements at nodes, is introduced. The continuous DD variation between elements is achieved by treating inter-element nodes as the points of specifications of unknown DD values, *i.e.*, the nodes, not the elements, play the central role. Thus, the most important sources of error in the displacement discontinuity method (DDM) implementation *i.e.*, inter-element displacement and traction incompatibility, are eliminated. On the other hand, certain conceptual and computational difficulties arise with respect to element integrations, by placing the nodes at the ends of the elements. These potential problems are addressed, though, by employing the ideas of invariant imbedding and continuation of singular and near-singular integrals, to develop a suitable integration ansatz. Several examples, which traditionally have been considered intractable by the conventional boundary element formulations, are presented.

1 INTRODUCTION

An underground mine consists of complex, three-dimensional arrangements of multiple, connected excavations. Typical boundary conditions are tractions free excavations surfaces and the stress distribution approaching the initial stress field outside of the zone of influence of the excavations. In order to accommodate the far field boundary conditions in mining stress analysis problems, a large rockmass volume must be modeled and as a result the ratio of excavation surface area to rockmass volume is low. Therefore, numerical methods based on volume discretization techniques (*e.g.*, finite element method) are computationally expensive.

Boundary element methods (BEM) are well suited for mining problems because only the boundaries of the excavations (*i.e.*, stopes, shafts *etc.*) need to be discretized (Brady & Bray 1978). For mining problems involving fractures, faults and thin slit-like structures such as seams, a version of BEM known as the displacement discontinuity method (DDM) is most appropriate (Crouch & Starfield 1983). With the traditional DD technique, the unknown parameters are represented at node(s) located inside of the element. This leaves the displacements and stresses between the elements discontinuous. Displacement discontinuities possess a higher-order mathematical singularity that the more familiar direct boundary element method or fictitious stress method. This higher-order singularity has been found to be a hindrance towards widespread applicability of the method. The higher-order singularity has a direct bearing on the singular behavior of the element integration (Wiles & Curran 1982).

Several examples are presented which show that the node centric DDM proves to be more efficient and robust that the traditional DDM.

2. NODE-CENTRIC DDM

Consider a homogenous, isotropic, linear elastic material subjected to a displacement discontinuity with density, d_k, at a point P (Fig. 1a). The stresses, σ_{ij}, and the displacements, u_i, at any point Q, in the domain can be represented as

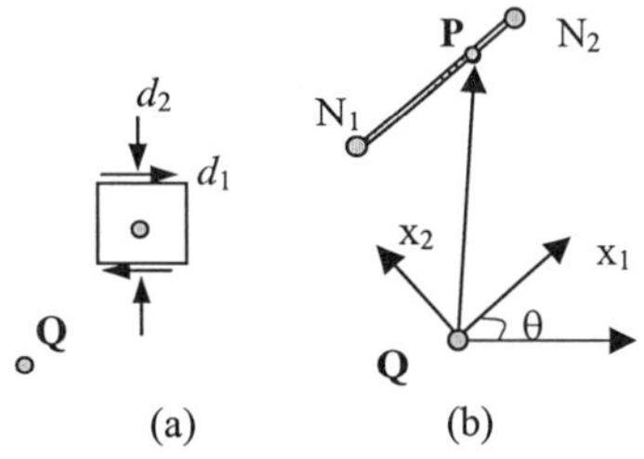

Figure 1 Notation.

$$\sigma_{ij}(Q) = G_{ijk}(P,Q)\, d_k(P) \tag{1}$$

$$u_i(Q) = H_{ik}(P,Q)\, d_k(P) \tag{2}$$

Throughout this paper repeated indices indicate the usual summation convention and *i, j, k* = 1, 2. The Green's functions, G_{ijk} and H_{ik} in eqns. (1) and (2) for a shear DD density, d_1, are given by

$$G_{111} = \frac{G}{\pi(\kappa+1)}\frac{1}{r^6}\left[12x_1^3x_2 - 4x_2^3x_1\right] \tag{3}$$

$$G_{221} = \frac{G}{\pi(\kappa+1)}\frac{1}{r^6}\left[12x_2^3x_1 - 4x_1^3x_2\right] \tag{4}$$

$$G_{121} = G_{211} = \frac{-G}{\pi(\kappa+1)}\frac{1}{r^6}\left[2x_1^4 + 2x_2^4 - 12x_1^2x_2^2\right] \tag{5}$$

$$H_{11} = \frac{-G}{2\pi(\kappa+1)}\frac{1}{r^2}\left[(\kappa-1)x_2 + 4\frac{x_1^2x_2}{r^2}\right] \tag{6}$$

$$H_{21} = \frac{-G}{2\pi(\kappa+1)}\frac{1}{r^2}\left[(\kappa-1)x_1 + 4\frac{x_2^2x_1}{r^2}\right] \tag{7}$$

where

$$\kappa = \begin{cases} 3-4\nu & \text{for plane strain} \\ \dfrac{3-\nu}{1+\nu} & \text{for plane stress} \end{cases}$$

G and ν are the shear modulus and Poisson's ratio, respectively and $r = \sqrt{x_1^2 + x_2^2}$. For a normal DD density, d_2, the Green's functions are

$$G_{112} = \frac{-G}{\pi(\kappa+1)}\frac{1}{r^6}\left[2x_1^4 + 2x_2^4 - 12x_1^2x_2^2\right] \tag{8}$$

$$G_{222} = \frac{-G}{\pi(\kappa+1)}\frac{1}{r^6}\left[2x_1^4 - 6x_2^4 + 12x_1^2x_2^2\right] \tag{9}$$

$$G_{122} = G_{212} = \frac{G}{\pi(\kappa+1)}\frac{1}{r^6}\left[12x_1x_2^3 - 4x_1^3x_2\right] \tag{10}$$

$$H_{12} = \frac{G}{2\pi(\kappa+1)}\frac{1}{r^2}\left[(\kappa-1)x_1 - 4\frac{x_2^2x_1}{r^2}\right] \tag{11}$$

$$H_{22} = \frac{-G}{2\pi(\kappa+1)}\frac{1}{r^2}\left[(\kappa-1)x_2 + 4\frac{x_2^3}{r^2}\right] \tag{12}$$

If we distribute displacement discontinuities of densities, d_k, over a number of curves, c_λ, then by superposition, the stresses and displacements at point *Q* become

$$\sigma_{ij}(Q) = \sum_{\lambda} \int_{c_\lambda} G_{ijk}(P,Q) d_k(P) dS(P) \tag{13}$$

$$u_i(Q) = \sum_{\lambda} \int_{c_\lambda} H_{ik}(P,Q) d_k(P) dS(P) \tag{14}$$

where the summation is performed over each of the curves, c_λ. If σ_{ij} or u_i are specified along c_λ, then eqns. (13) and (14) can be solved for the unknown DD densities, d_k. In practice it is necessary to discretize the curves, c_λ (*i.e.*, boundaries) into elements and assign a simple approximate functionality for d_k. For the case when *Q* approaches *P*, equations (13) and (14) become the standard indirect boundary element equations.

The main difference between the constant element and the node-centric formulation lies in the functionality of the DD distribution (Fig. 2). In the node-centric method, we assume a linear variation for the displacement discontinuity density, d_k, *i.e.*,

$$d_k(e) = D_k^{N_1} + \frac{x_1}{L}(D_k^{N_1} + D_k^{N_2}) \tag{15}$$

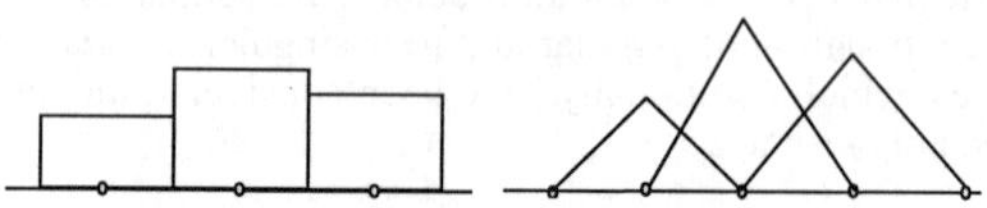

(a) Block function for a constant element — (b) Hat function for a node-centric element

Figure 2. Functionality of the distribution

where *e* denotes an element with length *L*, and N_1 and N_2 are the nodes located at the element ends. The integrals that need to be evaluated in equations (13) and (14) have the form

$$I_{ijk}^0(e) = \int G_{ijk}(x_1, x_2)\, dx_1 \tag{16}$$

$$I_{ijk}^1(e) = \int x_1 G_{ijk}(x_1, x_2)\, dx_1 \tag{17}$$

$$J_{ik}^0(e) = \int H_{ik}(x_1, x_2)\, dx_1 \tag{18}$$

$$J_{ik}^1(e) = \int x_1 H_{ik}(x_1, x_2)\, dx_1 \tag{19}$$

The integrations in equations (16)-(19) are done analytically and are tabulated in the appendix. Suppose the boundary of the problem to be considered is divided into *P* elements. It is necessary to find the values of the displacement discontinuities (normal and shear) which give the total stress and displacement components consistent with the boundary conditions. In general, if the boundary is represented by *P* elements, induced stresses and displacements at node *m* due to distributions of normal and shear displacement discontinuities (D_k) at node n can be written as

$$\sigma_{ij}^{m} = A_{ijk}^{mn} D_k^n \tag{20}$$

$$u_i^m = B_{ik}^{mn} D_k^n \tag{21}$$

where A_{ijk} and B_{ik} are the coefficient matrices. The influence coefficients for each node in the element are evaluated as

$$A_{ijk}^{mn} = \sum_e \alpha_{ijk}^{mn}(e) \tag{22}$$

where

$$\begin{bmatrix} \alpha_{ijk}^{mN_1}(e) \\ \alpha_{ijk}^{mN_2}(e) \end{bmatrix} = \frac{1}{x_1^{N_2} - x_1^{N_1}} \begin{bmatrix} x_1^{N_2} & -1 \\ -x_1^{N_1} & 1 \end{bmatrix} \begin{bmatrix} \bar{I}_{ijk}^{0}(e) \\ \bar{I}_{ijk}^{1}(e) \end{bmatrix} \tag{23}$$

$x_1^{N_1}$ and $x_1^{N_2}$ are the local coordinates of the two nodes of the element (Fig. 1) and $\bar{I}_{ijk}$ and $\bar{J}_{ik}$ are defined as

$$\bar{I}_{ijk}^{0}(e) = t_{il}(e) I_{mlk}^{0}(e)\, t_{mj}(e) \tag{24}$$

$$\bar{I}_{ijk}^{1}(e) = t_{il}(e) I_{mlk}^{1}(e)\, t_{mj}(e) \tag{25}$$

$$\bar{J}_{ik}^{0}(e) = t_{il}(e) J_{lk}^{0}(e) \tag{26}$$

$$\bar{J}_{ik}^{1}(e) = t_{il}(e) J_{lk}^{1}(e) \tag{27}$$

The displacement coefficient matrix, B_{ik}, is evaluated in the same manner of equation (22) by replacing I's by J's. The summation in equation (22) is performed over all the elements e which contain the node n, where n is either N_1 or N_2 of the element, based on the global node numbering system.

Equations (20) and (21) represent a set of linear algebraic equations which, after substitution of the appropriate boundary conditions, can be solved for the unknown displacement discontinuities D_k^n for each node.

2.1 *Integration ansatz*

A mathematically rigorous handling of singular integrals involves smooth weighting functions, *i.e.*, C^n functions on smooth boundaries and C^m on boundaries at infinity. The values of m and n depend on the order of the singularity. In the case of the DD method, the value of n and m should be at least 1. However when we use flat elements in discretizing the boundary, smooth boundaries cannot be maintained. Furthermore, even with a smooth boundary we have to deal with finite domains where the point of singularity is coincident with one of the end points.

Since we are using flat elements, the integrals are regular when the field point does not lie in the domain of the integration. The integrals can thus be obtained using standard integration. The integrations in equations (16-19) are done analytically and are tabulated in the appendix.

The procedure for dealing with the non-integrable singularity lying within the domain of integration is based on Hadamard's idea of the finite part integral (Hadamard 1923). Accordingly, the integration is performed formally and evaluated by ignoring the infinite terms. Further justification for using the finite part integral is given in (Vijayakumar & Cormack 1988) using the continuation approach.

Hence, the formal integrals given in the appendix can be used with proper interpretation and modification. When the field point lies on one of the nodes, there will be two neighboring elements which exhibit singular behavior. Following Hadamard's procedure, the integrals for the two elements are evaluated separately in the finite part sense. Thus when the element contains a singular point, we need to evaluate the integrals given in the appendix only at the non-singular end of the element.

In summary, we use the following integration ansatz:

1. If the singularity is at one of the end points, the integration is performed formally and evaluated at the opposite end, *i.e.* 'the non-singular end' of the domain.
2. When the domain is non-smooth, the normal direction at the node is assumed to be the mean of the normal directions of the elements which form the node.

In carrying out these integrations we note the special way the variation of the functions is constructed. Since this functionality (*i.e.*, the hat function) is dependent on the scale of the element, the integral should also possess the same scale dependency (or the scale invariance, so to speak). This means that when the element length is scaled by a factor λ, the value of the integral should scale by a factor λ^{α} where α is the order of the singularity (Rosen & Comack 1995). This imposes the condition that the $\log r$ term in the integration appears with proper scaling of the element length H as the term $\log(r/H)$. For the limiting case $r \rightarrow 0$, the evaluation of the integrals is done in Hadamard's finite part sense (*ibid*).

3. NUMERICAL APPLICATIONS

To illustrate the power of the new method, several examples are presented to test the validity of the node-centric DD formulation.

3.1 *Intersecting cracks*

Here, we consider two intersecting pressurized cracks in the form of a cross. This is an example of the class of physico-geometric situations which using the traditional DD method would need much finer discretization in the vicinity of the intersection because of the lack of inter-element continuity of the displacement discontinuities. Similar to the single crack problem, we discretize each crack into 10 elements. Since both cracks are subdivided uniformly, the centre node of one crack coincides with that of the other crack. We assign the unknown displacement discontinuities to each of the coinciding nodes independently. This is possible since in the DD method the cracks are modeled as a superposition of forces, instead of two separate boundaries. However, the stress state at the intersection point, which needs to be taken into consideration for the traction collocation of the external pressure, provides only one condition for the two intersecting nodes. As a consequence, we are left with only one pair of independent collocation equations at the intersecting nodes. This results in an under-determined system of equations. However, one can easily furnish an additional pair of equations by collocating tractions at any non-nodal boundary point such as the mid-point of an element at the nodes of coincidence. The conventional wisdom is that the best performance is obtained by collocating at the points at which the unknown parameters are assigned. By collocating at a point other than a node, we are not straying away from this convention since all the nodal positions have already been taken care of.

In order to test the robustness of the method, we carried out three sets of numerical experiments. The two cracks in the first model share nodes at the intersection (Fig. 3a). The center node of the horizontal crack is moved along the vertical crack a distance of 20% of the element length in the second model (Fig. 3b). The third model moves both center nodes by 20% of the element length along their respective cracks from the original position (Fig. 3c).

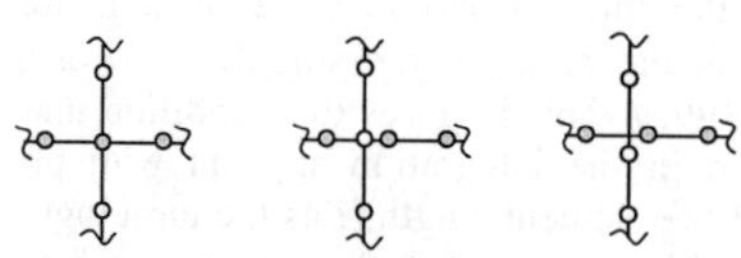

(a) (b) (c)

Figure 3. Different node configuration

Figure 4 depicts the DD values for the horizontal crack for several positions of the "center" node. The solution for a single crack is also provided as a convenient magnitude comparison. It is difficult to get an explicit analytical expression as the DD distribution of the cruciform crack in order to evaluate the accuracy of the node-centric method. However, from Stallybrass (1970) we can obtain the strain energy, W, of the system as

$$W = 1.4914\pi \frac{(\kappa + 1)}{8G} \tag{28}$$

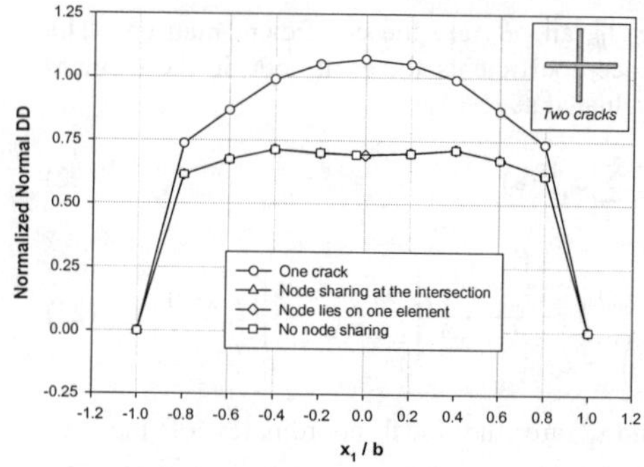

Figure 4. Normal DD variation along horizontal crack for different "center" node position

for a cruciform crack of unit half length subjected to uniform unit pressure. From our DD calculations using 10 elements, we get a factor value of 1.55 instead of 1.4914 for the strain energy, which is within 4.1% of the exact answer while the constant DD method gives the factor of 1.64 that is within about 9.8% of the exact answer.

As shown in the figure, the results of the node-centric formulation are very insensitive to the node configuration of the intersecting cracks.

3.2 *Uniformly pressurized circular hole*

This problem is analysed using 36 nodes uniformly placed along the boundary of a circle of radius, a. The tractions at the nodes are based on the tangent plane of the circular hole.

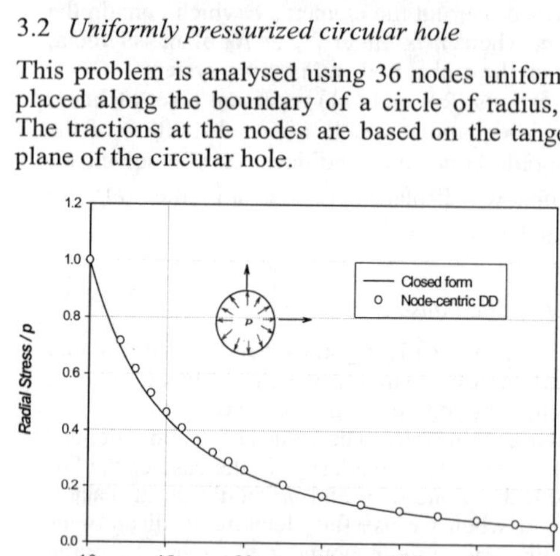

Figure 5. Radial stress variation for circular hole

Figure 5 compares the radial stress distribution outside the hole obtained from the node-centric DD method with the analytical solution (England 1971)

$$\sigma_{rr} = p\frac{a^2}{r^2} \tag{29}$$

3.3 *Circular hole with intersecting crack*

Having established the applicability of the node-centric DD method to external problems, we consider the more difficult problem of the intersection of a joint or fault with the boundary of an excavation. Again we need to deal with the issue of stress collocation when nodes coincide (node sharing) as discussed in section 2.1.

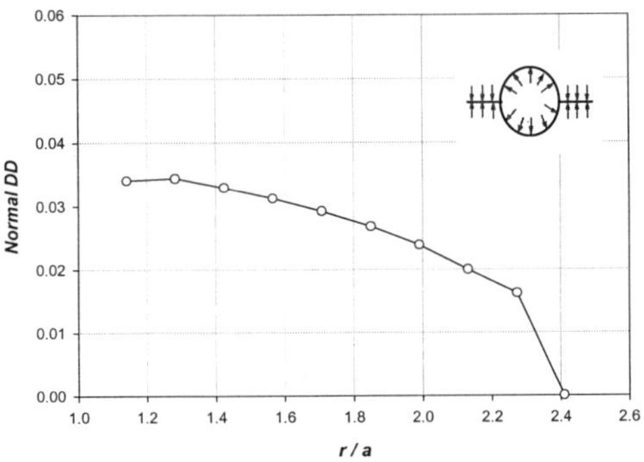

Figure 6. DD solution along the crack

Figure 6 illustrates the normal DD variation along the crack when the same uniform pressure is applied in the hole as well as in the crack. We emphasize here that with the node-centric DD method, the crack and the hole are discretized independently. It has common practice with the constant DD elements to discretize them in combination so that the effects of the infinite stresses at the element ends are minimized.

4. DISCUSSIONS

4.1 *Inter element continuity and node-sharing*

A significant breakthrough of the node-centric formulation is the handling of node-sharing and coincidence of nodes (node-sharing of higher types) in a systematic way. The issue of node sharing has been one of the stumbling blocks of the DD method. In the present context, it was the inability to formulate the DD method in a node-centric manner. In the node-centric formulation, assignment of quantities at the nodes ensures inter-element continuity of these quantities in a natural way. As shown in references (Crawford & Curran 1982, Shou and Crouch 1995), lack of inter-element continuity can introduce such large errors in the stress calculations as to render them useless for highly interacting elements. In contrast, the node-centric formulation is particularly attractive in this respect.

4.2 *Higher-order DDM*

A superficial perusal of the present work may mislead the reader into thinking that this is an attempt to use a higher-order DD variation than the constant DD element originally presented by Crouch (1976). This was the case with the early work by Crawford and Curran (1982) and by Shou and Crouch (1995). But all of the previous works did not, to quote Shou and Crouch, 'make the tractions and displacements continuous at the nodes between elements'. This is precisely the aim of the node-centric method. Regarding the order of the displacement discontinuity, we should emphasize here that, a linear variation along the element is the lowest order one can use.

4.3 *Element intersections*

In general purpose large scale applications, very complex physico-geometric situations can arise. Certainly for many practical problems discretizing the boundary into uniformly sized elements is not possible. The unbounded stresses due to a discontinuous DD at the node adversely affect the stress calculations in the surrounding elements. One may encounter complex situations arising from intersection of simpler geometric elements such as the intersection of cracks or cracks emanating from holes or passing through them. In such geometric situations, it is virtually impossible to avoid the unbounded stress regions in the calculations. In the case of intersecting cracks, for example, unless careful re-discretization is done, unbounded stresses along one crack would adversely affect the stress calculations along the other crack. The present node-centric DD formulation intrinsically avoids having unbounded stresses unless otherwise warranted by the geometry of the problem, *e.g.*, a corner. There would not be any need for a re-discretization. Hence the method should be robust for general purpose large scale applications.

5.CONCLUSIONS

We have shown the efficacy of node-centric formulation through several examples presented in this paper. Even though node sharing has been implemented in both the direct boundary method and fictitious stress technique, node sharing using the DD method has not been possible until now because of its higher-order singularity.

Through a proper integration ansatz, we have overcome the associated difficulties by imposing a continuous variation of the displacement discontinuities. Without a node-centric formulation, the ability of the DD method to solve practical geomechanics problems such as joints intersecting excavations is limited. In such problems, a more careful and finer discretization is usually required to overcome the infinite stresses generated at the element end. The node-centric formulation obviates this problem by imposing stress continuity.

REFERENCES

Brady B. H. G. & J. W. Bray 1978. The boundary element method for elastic analysis of tabular orebody extraction, assuming complete plane strain. *Int. J. Rock Mech. Min. Sci. & Geomech. Abstr.* 15: 29-37.

Crawford A. M. & J. H. Curran 1982. Higher order functional variation displacement discontinuity elements. *Int. J. Rock Mech. Min. Sci. & Geomech. Abst.*, 19:143-148.

Crouch S. L. 1976. Solution of plane elasticity problems by the displacement discontinuity method. *Int. J. Num. Meth. Engng.* 10:301-343.

Crouch S. L. & A. M. Starfield 1983. *Boundary Element Methods in Solid Mechanics.* London: George Allen & Unwin.

England A. H. 1971. *Complex Variable Methods in Elasticity*, Wiley-Interscience.

Hadamard, J. 1923. *Lectures on Cauchy's Problem in Linear partial Differential Equations*, New Haven CT: Yale University Press.

Rosen D. & D. E. Cormack 1995. The continuation approach: A general framework for the analysis and evaluation of singular and near-singular integrals. *SIAM J. App. Math.*, 55:723-762.

Shou K. J. & S. L. Crouch 1995. A higher order displacement discontinuity method for analysis of crack problems. *Int. J. Rock Mech. Min. Sci. & Geomech. Abst.* 32:49-55.

Stallybrass M. P. 1970. A Pressurized crack in the form of a cross', *Quart. Journ. Mech. and Applied Math.* 23(I).

Vijayakumar S. & D. E. Cormack 1988. An invariant imbedding method for singular integral evaluation on finite domains. *SIAM J. App. Math.* 48:1335-1349.

Wiles T. D. & J. H. Curran 1982. A general 3-D displacement discontinuity method. *Proc. 3rd Int. Conf. Num. Meth. Geomech.*, 103-111.

Appendix: Two-dimensional DD Green function integrals for constant and linear variations

	Constant	Linear
Shear DD	$I_{111}^{0} = \frac{G}{\pi(\kappa+1)}\left[\frac{3x_2}{r^2} - \frac{2x_2^3}{r^4}\right]$	$I_{111}^{1} = \frac{G}{\pi(\kappa+1)}\left[\frac{4x_1x_2}{r^2} - \frac{2x_1x_2^3}{r^4} - 2\tan^{-1}\left(\frac{x_1}{x_2}\right)\right]$
	$I_{221}^{0} = \frac{G}{\pi(\kappa+1)}\left[-\frac{x_2}{r^2} + \frac{2x_2^3}{r^4}\right]$	$I_{221}^{1} = \frac{G}{\pi(\kappa+1)}\left[-\frac{2x_1x_2}{r^2} + \frac{2x_1x_2^3}{r^4}\right]$
	$I_{121}^{0} = \frac{G}{\pi(\kappa+1)}\left[-\frac{x_1}{r^2} + \frac{2x_2^2x_1}{r^4}\right]$	$I_{121}^{1} = \frac{G}{\pi(\kappa+1)}\left[\frac{4x_2^2}{r^2} - \frac{2x_2^4}{r^4} + \ln\left(\frac{r}{H}\right)\right]$
	$J_{11}^{0} = \frac{1}{2\pi(\kappa+1)}\left[\frac{x_1x_2}{r^2} - \frac{(\kappa+1)}{2}\tan^{-1}\left(\frac{x_1}{x_2}\right)\right]$	$J_{11}^{1} = \frac{1}{2\pi(\kappa+1)}\left[-\frac{x_2^3}{r^2} - \frac{(\kappa+3)}{2}x_2\ln\left(\frac{r}{H}\right)\right]$
	$J_{21}^{0} = \frac{1}{2\pi(\kappa+1)}\left[\frac{x_2^2}{r^2} - \frac{(\kappa-1)}{2}\ln\left(\frac{r}{H}\right)\right]$	$J_{11}^{1} = \frac{1}{2\pi(\kappa+1)}\left[-\frac{x_2^3}{r^2} - \frac{(\kappa+3)}{2}x_2\ln\left(\frac{r}{H}\right)\right]$
Normal DD	$I_{112}^{0} = \frac{G}{\pi(\kappa+1)}\left[-\frac{x_1}{r^2} + \frac{2x_2^2x_1}{r^4}\right]$	$I_{112}^{1} = \frac{G}{\pi(\kappa+1)}\left[\frac{4x_2^2}{r^2} - \frac{2x_2^4}{r^4} + \ln\left(\frac{r}{H}\right)\right]$
	$I_{222}^{0} = \frac{G}{\pi(\kappa+1)}\left[-\frac{x_1}{r^2} - \frac{2x_2^2x_1}{r^4}\right]$	$I_{222}^{1} = \frac{G}{\pi(\kappa+1)}\left[-\frac{2x_2^2}{r^2} + \frac{2x_2^4}{r^4} + \ln\left(\frac{r}{H}\right)\right]$
	$I_{122}^{0} = \frac{G}{\pi(\kappa+1)}\left[-\frac{x_2}{r^2} + \frac{2x_2^3}{r^4}\right]$	$I_{122}^{1} = \frac{G}{\pi(\kappa+1)}\left[-\frac{2x_1x_2}{r^2} + \frac{2x_1x_2^3}{r^4}\right]$
	$J_{12}^{0} = \frac{1}{2\pi(\kappa+1)}\left[\frac{x_2^2}{r^2} + \frac{(\kappa-1)}{2}\ln\left(\frac{r}{H}\right)\right]$	$J_{22}^{0} = \frac{1}{2\pi(\kappa+1)}\left[\frac{x_1x_2^2}{r^2} + \frac{(\kappa-1)}{2}x_1 + \frac{(\kappa+1)}{2}x_2\tan^{-1}\left(\frac{x_1}{x_2}\right)\right]$
	$J_{22}^{0} = \frac{1}{2\pi(\kappa+1)}\left[\frac{x_1x_2}{r^2} + \frac{(\kappa+1)}{2}\tan^{-1}\left(\frac{x_1}{x_2}\right)\right]$	$J_{22}^{1} = \frac{1}{2\pi(\kappa+1)}\left[-\frac{x_2^3}{r^2} + \frac{(\kappa-1)}{2}x_2\ln\left(\frac{r}{H}\right)\right]$

Computer Methods and Advances in Geomechanics, Desai et al. (eds) © 2001 Balkema, Rotterdam, ISBN 90 5809 183 X

Landslide hazard assessment using Monte Carlo simulation based on GIS

Mowen Xie, Guoyun Zhou & Tetsuro Esaki
Institute of Environmental Systems, Kyushu University, Fukuoka, Japan

ABSTRACT: Based on GIS, this paper had presented the study of landslide hazard assessment of large mountain area by using the Monte-Carlo method to simulate the random distribution of geo-mechanical parameters. In the use of Monte-Carlo method, the probability of failure had been discussed with the change of parameters and distribution. With the example of Sasebo of Japan, we had gotten the distribution of probability of failure with this method.

1. INTRODUCTION

It has been paid attention to the failure of slope in the mountain area. The correct results based on the traditionally mechanical formula are relying on the detail mechanical parameters. But the expense to get detail parameters of large mountain area is very high. Generally, the gotten parameters of the rock are very limited at the initial stage of project such as city plan, environment assessment and best place for project. So how to assess the landslide hazard of large area is very important at the condition that only limited parameters had gotten. And then based on this assessment to detect the dangerous spots for future study or keep away from these spots.

Now we have developed many methods to assess the stability of slope, but the correct results of these methods are based on the correct geotechnical and mechanical parameters, and unfortunately, these parameters can not correctly got and we always get these by limited tests and experimental judge. In fact, the parameters of geological unit are the random variables that have certain random distributions. In this study, taking the parameters of rock-soil as variables, we had used the Monte-Carlo method to simulate the distribution of parameters.

For the powerful function of GIS, the technique of GIS has been playing more and more importance roles in the study of environment assessment and disaster forecast. In the landslide hazard assessment many practical methods had been developed too [1,2]. On the other hand, the error induced by the incorrect parameters is much bigger than the error random of parameters and lack of the suitable model, so it is realistic to select the relatively simple model for the study of landslide hazard assessment by GIS.

In this study, we still use the model of infinite slope that has been wildly used in the GIS model of landslide hazard assessment. This model is best suit for the feature of pixel of GIS-grid. At the same time in this study we has introduced the matrix of weigh value of influence of adjacent pixels to study the neighborhood analysis.

In this study, we had used the GIS software for windows Arc/View3.1[3] and Visual Basic 6.0.

2. THE MONTE-CARLO METHOD FOR THE STABILITY OF SLOPE

The safety of slope K_s is considered as the quotient of resistant force and slip force on the slip face. Because of the random of parameters, the K_s is hard to reflect the safety of slope. On the other hand, the different models may give out different K_s, and more the different importance and complex will demand the different standard of safety assessment. So giving the failure probability P_{ks} to illustrate the probability to get that K_s is more suitable to reflect the random of parameters of rock-soil.

2.1 *Monte-Carlo method*

The facts that will influence the safety of slope are the variables that possess the certain random distributions of parameters. If we had gotten the function of distribution of K_s, then the failure probability P_{ks} is:

$$P_{ks}(Ks<[Ks]=\int_0^{[Ks]} f(Ks)d(Ks) \qquad (2\text{-}1)$$

In this formula, $[K_s]$ is the design safety and the P_{ks} is the probability that the K_s are less then the design safety.

In the practical use because it is very hard to get the $f(Ks)$, it is always to use the Monte-Carlo method to calculate the P_{ks}. The method is that get the random variables of parameters based on the random distribution of parameters and then input into the formula of K_s, after many times calculations we can get the distribution function of $f(Ks)$, and can calculate the probability of P_{ks}.

2.2 *Random sample*

Based on the uniform distribution on [0,1] we can get the sample of the random variable which possesses a certain probability distribution. Set Z as random variable and its distribution is $F(u)$):

$$P\{Z<u\}=F(u) \qquad (2\text{-}2)$$

If $F(u)$ is continuous and strictly dull up, and then

$$X=F(Z) \qquad (2\text{-}3)$$

is uniform distribution on [0,1], and it is that X is random variable of uniform distribution on [0,1],

$$Z=F^{-1}(X) \qquad (2\text{-}4)$$

is random distribution that possesses the distribution of $F(u)$, and in it $F^{-1}(X)$ is the inverse function of $F(Z)$. If we had gotten samples $X_k=\{X_1,X_2,...\}$ of uniform distribution on [0,1], based on

$$Z_k=F^{-1}(X_k), k=1,2,... \qquad (2\text{-}5)$$

We can get the samples of distribution $F(Z)$.

In this study, we had considered four random variable distributions: Uniform distribution, negative exponential distribution, normal distribution and logarithmic normal distribution.

Random variable of uniform distribution on [0,1] are gotten by the method of LCG:

$$y_i = ay_{i-1}Mod(m)$$
$$r_i = y_i/m \qquad (2\text{-}6)$$

2.2.1 *Uniform distribution*

Its distribution function is

$$f(x)=\begin{cases}0 & x\notin[a,b]\\ \frac{1}{b-a} & x\in[a,b]\end{cases}$$

$$x=\frac{\quad}{2} \qquad (2\text{-}7)$$

Random sample is produced by

$$x_i = r_i(b-a)+a \qquad (2\text{-}8)$$

2.2.2 *Negative exponential distribution*

Its distribution is

$$f(x)=\begin{cases}0 & x\le 0\\ \lambda e^{-\lambda x} & x>0\end{cases}$$

$$\lambda=\frac{1}{\mu} \qquad (2\text{-}9)$$

Random sample is produced by

$$x_i=-E(x)\ln r_i \qquad (2\text{-}10)$$

$E(x)$ is the mathematical expectation of random variable x.

2.2.3 *Normal distribution*

Its distribution is

$$f(x)=\frac{1}{\sqrt{2\pi\sigma^2}}\exp[-\frac{1}{2}(\frac{x-\mu}{\sigma})^2] (2\text{-}11)$$

$$\mu=\frac{1}{n}\sum_{i=1}^{n}\ln x_i=\bar{x}$$

$$\sigma=\frac{1}{n}\sum_{i=1}^{n}(x_i-\bar{x})^2$$

Random sample is produced by

$$x_i=J=(\sum_{i=1}^{12}r_i-6.0)\sigma+\mu \qquad (2\text{-}12)$$

2.2.4 *Logarithmic normal distribution*

Its distribution function is

$$f(x)=\frac{1}{x\sqrt{2\pi\sigma^2}}\exp[\frac{-(\ln x-\mu)^2}{2\sigma^2}] (2\text{-}13)$$

$$\mu=\frac{1}{n}\sum_{i=1}^{n}\ln x_i \quad \sigma^2=\sqrt{\frac{\sum_{i=1}^{n}(\ln x_i-\mu)^2}{n}}$$

After normalized the function, its random sample is produced by the method of normal distribution.

$$\mu_1=\ln(\frac{\mu^2}{\sqrt{\mu^2+\sigma^2}}),\ \sigma_1=\sqrt{\ln((\frac{\sigma}{\mu})^2+1)}$$

$$x_i=\exp(J(\mu_1,\sigma_1)) \qquad (2\text{-}14)$$

2.3 *The formula of safety*

The calculation of safety is done by the formula of infinite slope (see Fig1):

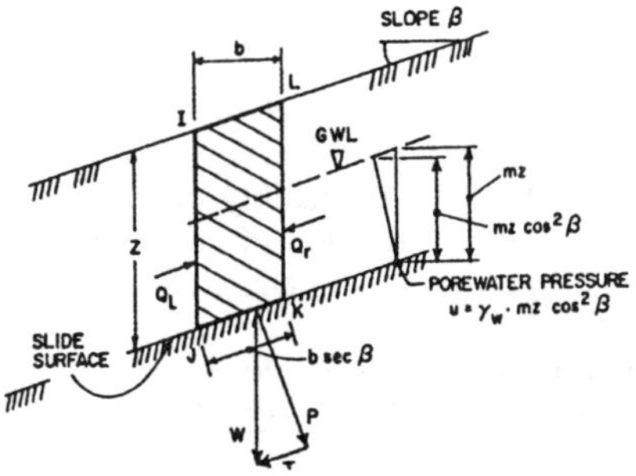

Fig.1 infinite slope assume

$$K_s = \frac{c' + (\gamma - m\gamma_w)Z\cos^2\beta\tan\Phi'}{\gamma Z\sin\beta\cos\beta} \qquad (2\text{-}15)$$

c' – Effective cohesion of slip face

Φ' – Effective angle shearing resistance of slip face

β – slope surface inclination

γ – Unit weight of slip block

γ_w – Unit weight of water

Z – Height of slip block

m – Proportion of underground water

2.4 *Calculating process and related problem discuss*

Monte-Carlo simulation is done as following flowchart:

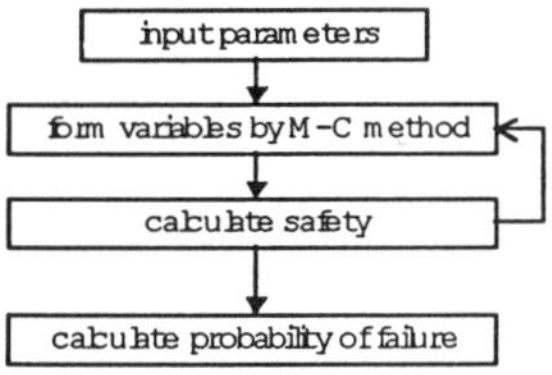

Fig2. Flowchart of Monte-Carlo simulation

The times of Monte-Carlo simulation is determined by precision. We had taken many times calculations between 50 and 50000 times, and form the point of precision demanded by the engineering, 200 times is enough.

In this study, we had selected c, Φ, γ and underground water h_w as random variables to take Monte-Carlo simulation. Generally speaking, the distribution of parameters are normal or logarithmic normal distribution If c, Φ, γ and h_w are normal or logarithmic normal distributions, the result of Ks is same distribution as parameters

From Fig 3 we can see that the probability of failure of slope is exponentially increasing with the increase of slope angle under condition of that other parameters are unchanging.

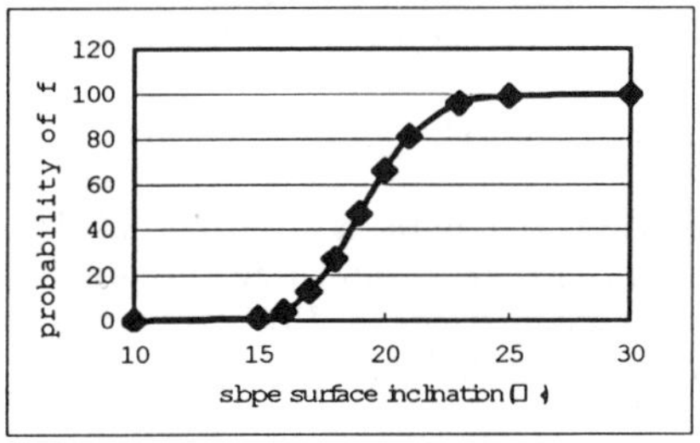

Fig.3 Slope and probability of failure

In the study of sensitive of parameters, we had study the effect of the change of parameters on the value of probability. Form Fig.4 we get that in this project, the effect of underground water is biggest, and the effect of c, γ is small. For different project, because of the relative value of each parameter, the sensitive of change is different.

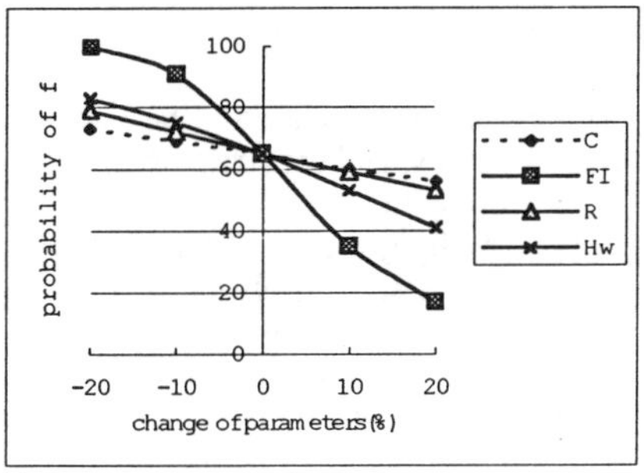

Fig.4 Sensitive of parameters to probability

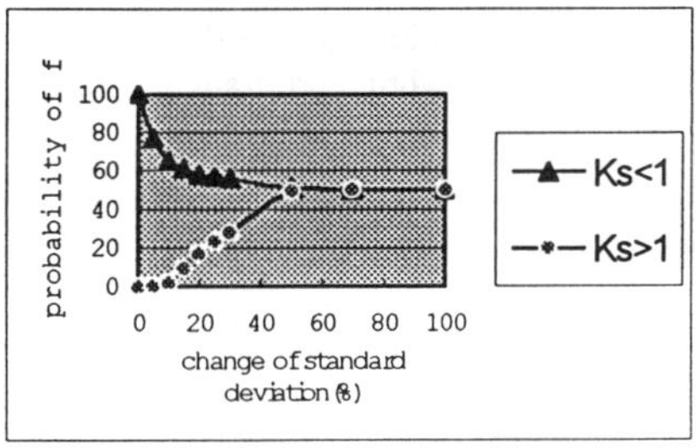

Fig.5 Standard deviation and probability

The influence of deviation of parameters on the probability is showed on Fig.5. With the increase of deviation, the probability of failure is coming to

50% under two conditions of safty Ks<1 and Ks>1 which based on the mean values

3. METHOD OF GIS AND THE HAZARD ASSESSMENT OF AREA OF SASEBO

3.1 *Basic data and program flowchart*

Sasebo is located in the northern part of Nagasaki of Japan. In this area , most often occurred landslide can be summarized as Hokusho type landslide, which is shown in Fig.6.

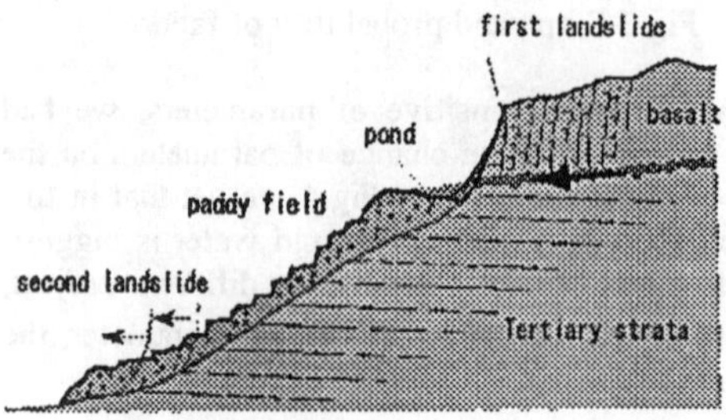

Fig.6 Hokusho type landslide model

In this study, it had selected the area of 1720x1540(m) for study, and the size of grid pixel is 10x10(m). The geographical parameters can be easily gotten in GIS by Arc/View, and the possible thickness of landslide is gotten by the Kriging interpolator from known points.

The outside forces such as raining, earthquake or man-made effect always course the failure of nature slope. And raining is always the main factor of failure. In this article, the raining condition had been transferred to the condition of underground water, and then gets the probability of landslide hazard under the possible outside force.□

The safety of slope is calculated by the formula of infinite slope, and the whole program is done by VB, the flowchart is:

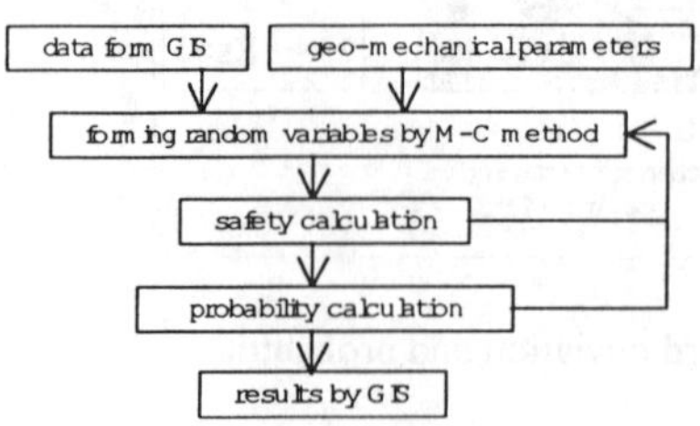

□□□ Fig.7 program flowchart

3.2 *Results*

By the program, we can get the safety and probability of failure of each pixel (in this study, the probability is set as the probability of safety less than 1). Fig.8 is the distribution of probability and table 1 is the dangerous classify and probability.

Table 1 probability and dangerous classify

Classify	Probability(%)
No-danger	<30
Low-danger	30-50
Middle-danger	50-70
Danger	70-90
High-danger	>90

Form Fig.8 we can see the dangerous spots and their probability of failure. At some spots of the probability of failure is bibber than 90%, the landslide hazard had token place. Form this danger distribution, we can concentrate ours spots for detail study or devoid the spots.

3.3 *Neighborhood analysis*

In the calculation of GIS, each pixel is assumed as an infinite slope and in fact that there is mutual influence between adjacent pixels. For this neighborhood analysis, we introduce the matrix of weigh value of influence of adjacent pixels:

$$U = \begin{bmatrix} w_{11}, w_{12}, \ldots, w_{1n} \\ \ldots\ldots\ldots\ldots\ldots \\ w_{n1}, w_{n2}, \ldots, w_{nn} \end{bmatrix} \tag{3-1}$$

in this, w_{ij} is the weigh value of influence on calculating pixel. So the probability of calculating pixel is:

$$P'_{ks} = \frac{1}{\sum_{j=1}^{n}\sum_{i=1}^{n} w_{ji}} \sum_{j=1}^{n}\sum_{i=1}^{n} w_{ji} P_{ks}(j,i) \tag{3-2}$$

and $P_{ks}(j,i)$ is probability of adjacent pixels, n is odd number.

The range (n) of influence of adjacent pixels is concerned with the pixel size and the size of possible landslide.

In this study, we use the decline value $a(a<1)$ of influence of adjacent pixel to deduce the w_{ij}. If $n=5$, the matrix of weigh value of influence of adjacent pixels is:

$$U = \begin{bmatrix} a^4,a^3,a^2,a^3,a^4 \\ a^3,a^2,a^1,a^2,a^3 \\ a^2,a^1,a^0,a^1,a^2 \\ a^3,a^2,a^1,a^2,a^3 \\ a^4,a^3,a^2,a^3,a^4 \end{bmatrix} \quad (3\text{-}3)$$

Set $a = 0.8$ and $n = 9$, the result of probability are shown on Fig.9 and we can see that the failure distribution is been smoothed. and more the show of random single failure pixel is disappear. In this study, we had taken the study of compare of different $a, n(a = 0.4 - 0.8, n = 5 - 9)$, and the range of failure is different with the change of a and n, but the danger degree and spots are unchanged

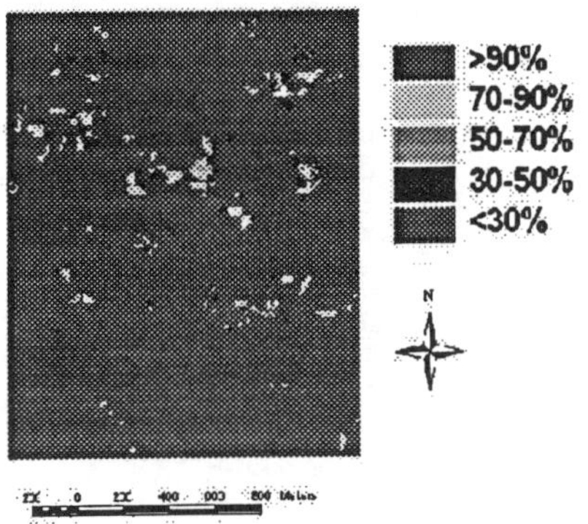

Fig.8 Probability of failure(By Monte-Carlo method)

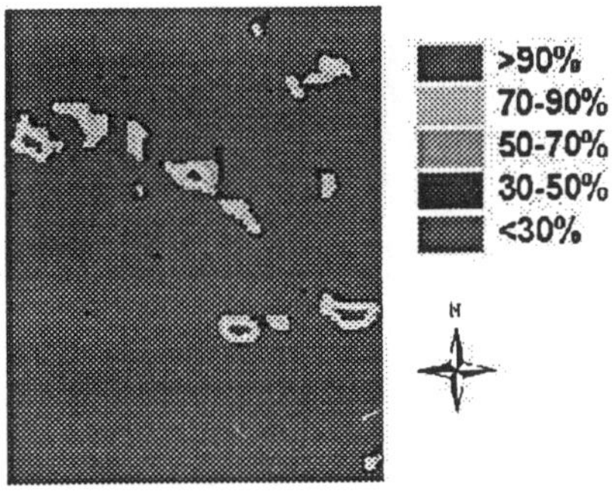

Fig.9 Probability of failure
(considering the influence of adjacent pixels)

4. CONCLUSION AND DISCUSSION

Form this paper, the following conclusions are reached:

1. For the assessment of landslide hazard at the initial stage of project, it is better to choice the relatively simple model for the calculation of safety because of the uncertain of parameters. Based on the random distribution of parameters, it is easy to get the probability of failure of Ks<[Ks].

2.GIS may play more and more important role in the study of landslide hazard assessment of large mountain area. Using the grid data, it is easy to calculate the attributes of each pixel and to show the picturesque results. In this paper, the distribution of safety and probability had gotten using the model of infinite slope. But in the future study, it is important that how to properly concern the pixel and landslide on the base of physical and mechanical concept.

3. The using of matrix of weigh value of influence of adjacent pixels can reduce the random and unharmonious

In a word, based on the powerful function of GIS, it is easy to assess the landslide hazard of large then take detail study of dangerous spots or avoid those spots

REFERENCE

C.J. Van Westen, 1998. GIS in landslide hazard assessment: a view, with example from the Colombia Andes, MOUNTAIN ENVIRONMENT & GEOGRAPHIC INFORMATION SYSTEMS, Edited by Martin F. Price and D. Lan Heywood, Taylor & Francis publisher

P. Aleotti, R. Chowdhury, 1999. landslide hazard assessment: summary review and new perspectives, Bull Engineering Geology Environment (1999) 58:21-44, Springer-Verlag

ESRI, 1996. AreView GIS, Using AreView GIS, Environmental Systems Research Institute, Inc.

Computer Methods and Advances in Geomechanics, Desai et al. (eds) © 2001 Balkema, Rotterdam, ISBN 90 5809 183 X

An object-oriented FE code for modeling the viscoplastic response of soft rocks

O.M.L.Yahya & M.Aubertin – *Département CGM, Ecole Polytechnique de Montréal, Qué., Canada*
M.Julien – *Golder Associates Ltd, Montréal, Qué., Canada*
R.Foerch – *Northwest Numerics Inc., Seattle, Wash., USA*
G.Cailletaud – *Centre des Matériaux de l'Ecole des Mines de Paris, Evry, France*

ABSTRACT: The response of soft rocks can be quite complex as it usually involves strong time and history dependencies. Over the years, various constitutive models have been formulated to describe their inelastic behavior. The authors have developed SUVIC, a model based on the unified theory of inelastic flow with internal state variables (ISV). It has been shown how the SUVIC model can be used to describe fairly complex mechanical responses observed during laboratory experiments, including creep, plasticity and relaxation.
In this paper, the authors recall the model features and introduce a comprehensive finite element code, named ZéBuLon, which was selected for the numerical implementation of SUVIC. This 3D object-oriented finite element program is well-suited for a variety of structural calculations including geomechanical applications. When used with SUVIC, ZéBuLon can provide valuable insight for a number of purposes such as the design of underground openings. Some calculations are presented and discussed to illustrate the general capabilities of the constitutive and numerical models.

1 INTRODUCTION

For the long-term prediction of the mechanical fields around man made excavations, one needs a reliable representation of the material behavior. In that regard, the limitations of classical creep laws with elementary hardening/softening rules have already been pointed out (Aubertin et al. 1999a,b).

The inelastic behavior of soft rocks, such as rocksalt and potash, is highly sensitive to history and rate effects (e.g., Cristescu and Hunsche 1997; Aubertin et al. 1998a). Because there are many deformation mechanisms acting locally at the microstructural level, the construction of an appropriate constitutive model to represent the inelastic behavior of these rocks can be quite a challenge. There has been many attempts over the years, and most have had limited success because they often neglected important features of the mechanical response. Nevertheless, progresses made over the last fifteen years or so, inspired by new concepts in materials sciences and favored by increased capabilities in computing tools (both software and hardware), now allow a more complete and elaborate treatment of soft rocks inelastic behavior. It is nowadays possible to use these new tools for structural calculations aimed at designing underground openings.

Various analyses conducted on underground excavations in potash and rocksalt deposits have highlighted the progressive wall and roof convergence (e.g., Mraz et al. 1993; Van Sambeek, 1998). Elaborate studies have also been performed for strategic projects such as storage of hydrocarbons or disposal of toxic and radioactive wastes.

In most applications, the analysis of underground structures in soft rocks involves numerical calculations. The degree of accuracy that can be expected from such calculations depends on numerous factors, among which the adequacy of the constitutive model appears as one of the most significant. As investigations involving elaborate loading conditions and extended periods of time have shown the limitations of oversimplified models, more rational and flexible models were developed over the past two decades to better respond to these specific needs. Some of these models can be found in Conference Proceedings (e.g., Ghoreychi et al. 1996; Aubertin and Hardy 1998).

Such a model is presented below. It is based on the fundamental assumption that there exists a critical surface, usually defined in stress space (known as the damage initiation threshold - DIT), below which there is little or no fracturing. A low yield strength within the constituent grains allows soft rocks to deform inelastically in a ductile (fully plastic) manner, as long as the deviatoric stress remains below the DIT (Aubertin et al. 1998a, 1998b).

In the ductile regime of inelastic flow, the behavior of soft rocks can be very similar to that of metals. The physical mechanisms that control inelastic behavior are mostly related to dislocations motion by glide, cross slip, or climb. These processes affect the microstructural state in various ways, which manifest themselves through the rate (or time) and history dependencies of material responses to external loading.

In the next section, the authors recall briefly the SUVIC model developed at the Ecole Polytechnique de Montréal. The following section deals with the numerical implementation of SUVIC into ZéBuLon, an object-oriented finite element (FE) thermomechanical code developed in France. Because of the enormous possibilities it offers, FE is the method of choice for such type of calculations. Sample calculation results on pillars and surrounding rock, representative of openings typically encountered in salt mines, are finally shown.

2 UNIFIED MODELING

Constitutive modeling of soft rock behavior in the ductile regime of inelastic flow requires a proper representation of the viscous nature and of memory effects imbedded within the constitutive equations. An interesting approach is to rely on internal state variables (ISV) associated to specific mechanical characteristics of the material. For each ISV, an evolution law can be developed to describe the competitive action of hardening and recovery processes. The ISV model presented below is an updated version of SUVIC (**S**train rate history dependent **U**nified **V**iscoplastic model with **I**nternal state variables for **C**rystalline materials), a model initially developed for rocksalt behavior in the context of rate-dependent plasticity (Aubertin et al. 1991). It uses three types of ISV to describe mixed (kinematic and isotropic) hardening : a tensorial back stress (B_{ij}), an isotropic yield strength (R) and a normalizing scalar drag stress (K). During transient inelastic flow, the internal state variables evolve and may eventually reach saturation before or at steady-state, when all rate variables (internal and external) become constant and codirectional. In the following, a new hyperbolic sine law is used to describe the relationship between the inelastic strain rate and deviatoric stress at steady-state (Aubertin et al. 1999a, Julien 1999, Yahya et al. 2000).

In this unified theory framework, material hardening and resulting mechanical state are entirely described at a given time by the current value of the external (observable) state variables (stress σ, strain rate $\dot{\varepsilon}$ and temperature T) and of the appropriately selected internal (hidden) state variables (here B_{ij}, R and K). This approach allows a good representation of inelastic responses associated to different types of behavior.

2.1 *Basic concepts*

The proposed constitutive equations, developed within the unified inelasticity theory framework, consider that only elastic and viscoplastic straining occurs. With the ISV modeling approach (e.g. Lemaître and Chaboche 1990), the material structural state can be described using the so-called internal stress σ_i. This internal stress generally opposes the action of the applied deviatoric stress σ. The model equations presented below include three distinct internal state variables with evolution laws. Two variables, R and K (the scalar yield strength and the drag stress respectively), account for isotropic effects, while B (the tensorial back stress) accounts for kinematic (or flow induced anisotropic) effects. The combination of *B* and *R* constitute the internal stress. Each of the three ISV evolves according to phenomenological expressions with competing components for strain hardening, strain induced dynamic recovery, and time/thermally induced static recovery.

2.2 *Set of equations*

For small perturbations under isothermal loading, the total strain rate tensor $\dot{\varepsilon}_{ij}$ is given as the sum of the elastic and inelastic strain rate tensors:

$$\dot{\varepsilon}_{ij} = \dot{\varepsilon}^{e}_{ij} + \dot{\varepsilon}^{i}_{ij} \tag{1}$$

The elastic strain rate tensor is expressed here by the generalized Hooke's law for isotropic materials. The inelastic component of the strain rate tensor is given by a single kinetic law:

$$\dot{\varepsilon}^{i}_{ij} = A\left\langle \frac{X_{ae} - R}{K} \right\rangle^{N} n_{ij} \tag{2}$$

with

$$X_{ae} = \left[\frac{3}{2}\left(S_{ij} - B_{ij}\right)\left(S_{ij} - B_{ij}\right)\right]^{\frac{1}{2}} \quad (2a)$$

and

$$n_{ij} = \frac{3}{2}\frac{S_{ij} - B_{ij}}{X_{ae}} \quad (2b)$$

$$S_{ij} = \sigma_{ij} - \delta_{ij}\ \sigma_m \quad (2c)$$

where σ_{ij} is the stress tensor; S_{ij} is the deviatoric stress tensor; σ_m is the mean stress, δ_{ij} is the Kronecker delta; n_{ij} is a directional tensor that gives the inelastic strain rate the same orientation as the deviatoric component of the active stress tensor. In these equations, the kinematic internal variable B is the sum of separately evolving constituents, B_s (for short range effect) and B_l (for long range effect):

$$B_{ij} = \sum_{\alpha = s,l} B_{ij,\alpha} = B_{ij,s} + B_{ij,l} \quad (3)$$

Two such components are commonly used to represent the kinematic backstress in unified models (e.g. Freed and Walker 1993, Krauz and Krausz. 1996). It is considered here that the short range back stress B_s grows much faster and saturates much sooner than its long range counterpart B_l.

Phenomenologically, each ISV evolves through a competitive process of the following type, given for a generic internal state variable Y ($\in B_{ij,s}$, $B_{ij,l}$, R, K):

$$\dot{Y} = f_h - f_d - f_s \quad (4)$$

where f_h, f_d and f_s are respectively the hardening term that accounts for strengthening (hardening) mechanisms, and the dynamic and static recovery terms that account for softening (recovery) mechanisms. The hardening and dynamic recovery terms both evolve with inelastic strain. The static recovery term, on the other hand, evolves with time and cumulated value of Y. The functional forms giving the evolution of each particular term depend on the type of internal state variable. They are given with full details in Aubertin et al. 1999a and Julien 1999.

Within the adopted modeling framework, it is considered that steady-state flow is independent of previous mechanical events, or memory effects. Therefore, the corresponding kinetic law can be formulated using only external variables. Because steady-state flow is associated, at the microscopic level, to dislocations climb, glide and/or cross slip mechanisms, the mathematical relationship can be represented by an hyperbolic sine function, so that one can write:

$$\dot{\varepsilon}_e^i = \dot{\varepsilon}_o \sinh^n\left(\frac{\sigma_e'}{\sigma_o}\right) \quad (5)$$

where σ_o, $\dot{\varepsilon}_o$ and n are material constants for steady-state flow. In the preceding equations, the well-known von Mises equivalencies are used for stress (σ_e ; its saturation value being σ_e') and inelastic strain rate ($\dot{\varepsilon}_e^i$).

The good correlation between equation (5) and measured values obtained on rocksalt samples submitted to creep and CSR tests has been confirmed by results presented by Julien (1999). The generally good agreement between the model complete set of equations and various laboratory test results have also been shown in Aubertin et al. (1999) and Yahya et al. (2000).

3 FE CALCULATIONS FOR UNDERGROUND STRUCTURES

The finite element program ZéBuLoN was selected for the numerical implementation of the SUVIC model. This code has been developed at the École Nationale Supérieure des Mines de Paris, France (Besson et al. 1997, 1998). Until recently, it had mostly been used for thermo-mechanical problems involving metallic and composite materials.

The current version of ZéBuLoN used here, written in C++, has been developed with modular programming to provide better control of data, and a high modularity to allow quick implementation of new algorithms using predefined structures. It provides a suitable computational framework for geomechanical problems with inclusion of non-linearities arising from material behavior, loading sequences and mesh refinement upon localization (if necessary).

For illustration purposes, plane strain calculations were performed on the system shown in Figure 1. It includes 3 levels of mine stopes with vertical pillars located at an equivalent depth between 130 and 230 m. The calculation time period was limited to approximately 2 years. Figure 2 shows the distribution of Mises equivalent stress at the end of this period (units MPa). Such calculations allow the assessment of a given design and the ensuing evolution of the overall structure behavior before its

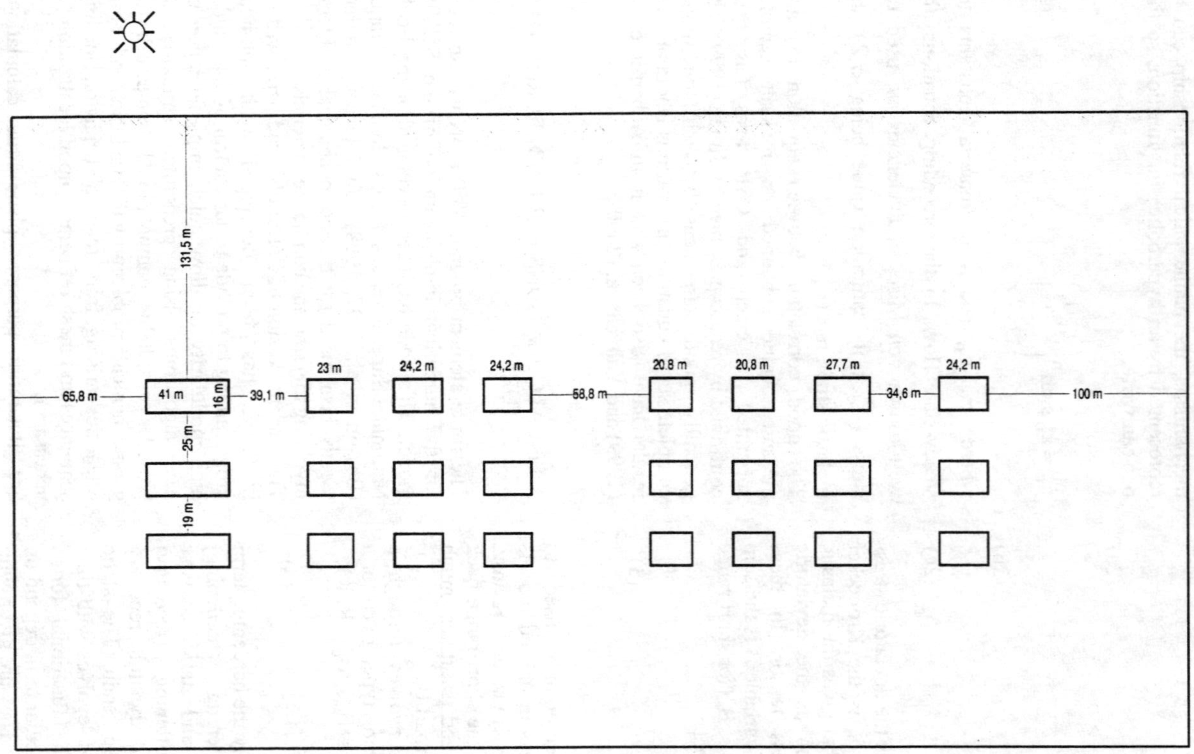

Figure 1. Multi-level room and pillar mine.

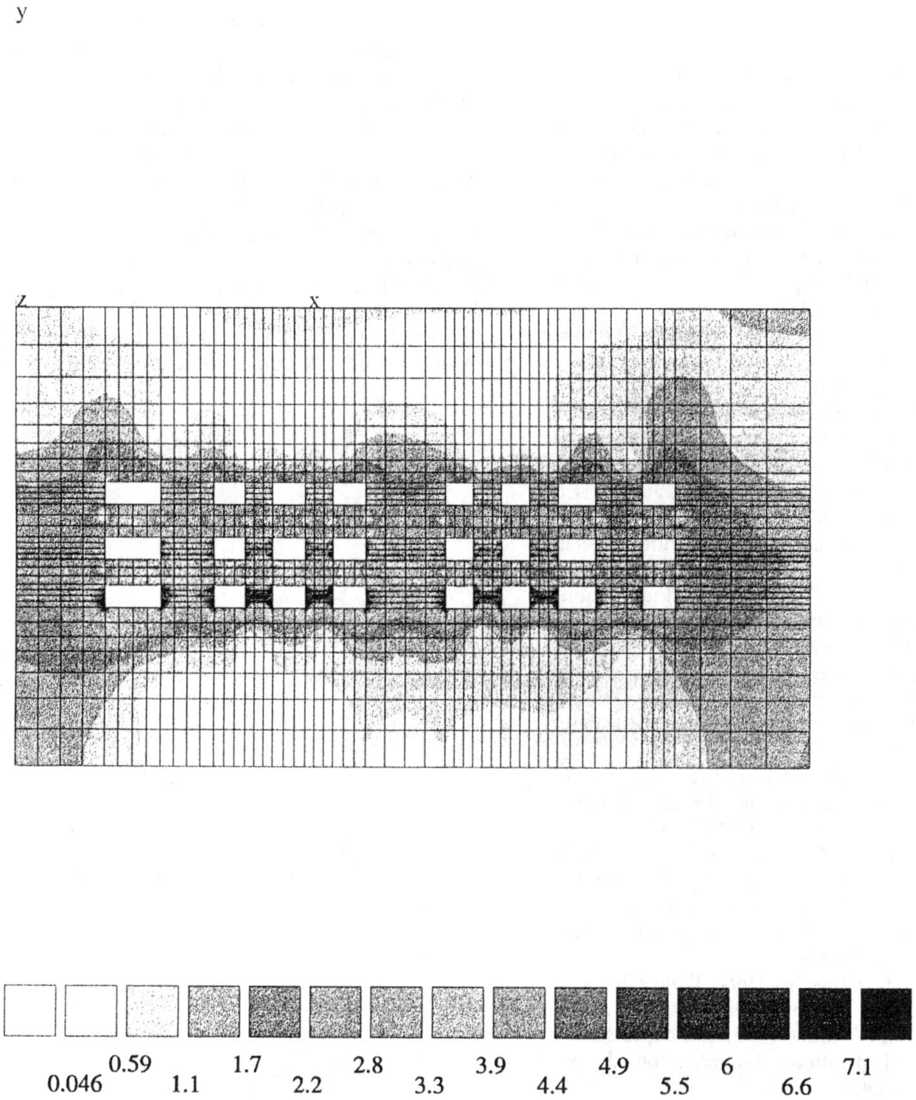

Figure 2. Distribution of Mises equivalent stress for a multi-level room and pillar mine.

construction in situ. The results shown here illustrate some of the constitutive and numerical capabilities of the model. Other structural 2D/3D calculations are presented in Julien (1999) who also give information on material parameters and calculation methods.

5 CONCLUSION

A modeling framework, based on the unified theory of inelastic flow using internal state variables (ISV), is presented. The latest version of the SUVIC model, developed in such context, is then introduced. The authors have implemented SUVIC into a comprehensive object-oriented finite element code, called ZéBuLon. This code is particularly well suited to perform analysis of underground openings, taking into account excavation sequence and material non linearity. In a sample calculation, the model is applied to a multilevel mine excavated in rocksalt.

REFERENCES

Aubertin, M. & H.R., Hardy Jr. (eds) 1998. The Mechanical Behavior of Salt. Proc. of 4th Conference, June 1996, Montréal, Canada, Trans Tech Publications, Clausthal-Zellerfeld, Germany.

Aubertin, M., D.E., Gill & B., Ladanyi 1991. A unified viscoplastic model for the inelastic flow of alkali halides. Mechanics of Materials 11:63-82.

Aubertin, M., M.R., Julien & L., Li 1998a. The semi-brittle behavior of low porosity rocks. Keynote lecture. Proc. NARMS'98, 3rd North American Roch Mechanics Symposium, Cancun, Mexico, Vol. 2, pp. 65-90.

Aubertin, M., M.R. Julien, S. Servant & D.E. Gill 1999b. A rate dependent model for the ductile behavior of salt rocks. Canadian Geotechnical Journal, 36:660-674.

Aubertin, M., , J. Sgaoula, S. Servant, M. Julien, D.E. Gill, & B. Ladanyi 1998b. An up-to-date version of SUVIC-D for modeling the behavior of salt. Proceedings of 4th Conference, June 1996, Montréal, Canada, Trans Tech Publications, Clausthal-Zellerfeld, Germany, pp. 205-220.

Aubertin, M., O.M.L. Yahya & M. Julien 1999a. Modeling mixed hardening of alkali halides with modified version of an internal state variables model. International Journal of Plasticity, 15:1067-1088.

Besson, J. & R. Foerch 1997. Large scale object-oriented finite element code design. Comput. Methods Appl. Mech. Eng., 142:165-187.

Besson, J., R. Leriche, R. Foerch. & G. Cailletaud 1998. Object-oriented Programming Applied to the finite element method – Part II. Application to material behaviors. Revue européenne des éléments finis, 7:567-588.

Cristescu, N.D., & U. Hunsche, 1997. Time effects in rock mechanics, John Wiley & Sons.

Freed, A. D., & K.P. Walker 1993. Viscoplasticity with creep and plasticity bounds. Int. J. of Plasticity, 9:213-242.

Ghoreychi, N., P. Bérest, H.R. Hardy & M. Langer (eds) 1996. Proc. 3rd Conference on the Mechanic Behavior of Salt, Palaiseau (1993), Trans Tech Publications, Clausthal-Zellerfeld, Germany.

Julien M.R. 1999. Une modélisation constitutive et numérique du comportement mécanique du sel gemme. Ph.D. thesis, Ecole Polytechnique de Montréal.

Krauz A.S. & K. Krausz 1996 (eds). Unified Constitutive laws of Plastic Deformation, Academic Press, San Diego.

Lemaître, J. & J.L. Chaboche, 1990. Mechanics of Solid Materials. Cambridge University Press, Cambridge.

Mraz, D., L. Rothenburg & M.B. Dusseault, 1993. Design of mines in salt and potash. Proc. 7 Int. Cong. on Rock Mechanics, Aachen, Germany, Balkema, Rotterdam, Vol. 3, pp. 1837-1845.

Van Sambeek, L.L. 1998. Salt pillar design equation. Proceedings, 4th Conference on the Mechanical Behavior of Salt, Montréal, Aubertin, M., & Hardy, H.R., (eds) Trans Tech Publications, Clausthal–Zellerfeld, Germany, pp. 495-508.

Yahya, O.M.L., M. Aubertin, & M. Julien, 2000. A unified representation of plasticity, creep and relaxation behavior of rocksalt. Int. J. of Rock Mech. and Mining Sciences. In press.

2 Neural networks, expert systems, reliability, optimization

Computer Methods and Advances in Geomechanics, Desai et al. (eds) © 2001 Balkema, Rotterdam, ISBN 90 5809 183 X

Expert system for evaluation of groundwater pollution potential from CAFO

K.N.M.Ahsanuzzaman, K.Strevett & M.Zaman
School of Civil Engineering and Environmental Science, University of Oklahoma, Norman, Okla., USA

ABSTRACT: A knowledge-based expert system to evaluate groundwater pollution potential due to leaching of nitrate and phosphate from land application of Concentrated Animal Feeding Operation (CAFO) waste is proposed in this study. The expert system is divided into four modules: surface loading, soil sorption, vadose zone transport and saturated zone transport. The human experts assigned rating values and relative weight to most significant parameters for each module. A two-stage weighted-average method is used for evaluation of pollution potential. In the first stage, rating values for each module are calculated by applying weighted-average technique on all module parameters. And in the second stage, pollution potential is evaluated from the calculated rating value of each module in first stage and the weighting value suggested for each module by the human experts. Finally, the expert system is validated against a site assessment done by a group of experts for a CAFO facility in Oklahoma.

1. INTRODUCTION

The waste generated from Concentrated Animal Feeding Operation (CAFO), often called manure, is commonly used as supplemental nutrient (e.g. nitrogen, phosphorus) source for plants and crops. The manure from most of the CAFOs except swine is usually applied to the cultivating land without any treatment. For swine CAFOs, the manure is stabilized by lagoon prior to land application. Even after stabilization, this lagoon effluent can still contain high concentrations of nitrogen and phosphorus. Inefficient application of these wastes could result in leaching of these nutrients to the groundwater and could become a potential health risk to the downstream well users.

Kellogg & Lander (1999) reported that the potential for groundwater pollution from manure application has increased over the years. In 1992, 114 counties in United States, as opposed to 28 counties in 1949, was in high risk of groundwater pollution since these counties produced excessive manure nutrient than the nutrient required for potential plant uptake and removal (Kellogg & Lander 1999). Thus, it has become a growing concern for the existing CAFOs to evaluate the risk involved from their management practices and to identify factors affecting the risk. For new CAFOs, it has become equally important to conduct site assessment with respect to groundwater pollution potential before building a new CAFO. Currently, there is no tool available that can specifically evaluate groundwater pollution potential from land application of manure (Chowdhury & Canter 1997-98). So far, effort has been given to develop expert systems for evaluating groundwater contamination from pesticide leaching (Arora & Mcternan 1993, Crowe & Mutch 1992, 1994). But none of those expert systems consider manure application. Although EPA (1997) developed a methodology to evaluate overall environmental risk from swine CAFOs, it lacks in addressing land application and consequent groundwater transport. This study focuses on developing a knowledge-based expert system that can evaluate the groundwater pollution potential at a downstream well from land application of manure. Expert systems has some advantages over numerical models: it is simple to use, limits the need for skilled individual to run numerical models, relatively less expensive than conducting numerical modeling and requires significantly less data. On the other hand, expert systems often have shortcomings in proving its reliability. Expert systems could be made more reliable by validating against field data, but such data in most cases is difficult and expensive to collect.

The expert system developed herein is designed to aid farmers in developing best management practices for land application of manure and selecting sites suitable for building new CAFOs. The proposed expert system could also be used as a preliminary assessment tool for the regulatory agencies in selecting most vulnerable sites from both new and existing CAFOs. The preliminary assessment results can then be used to identify the sites those need further extensive study.

2. DEVELOPMENT OF EXPERT SYSTEM

A typical expert system involves four stages: problem identification, knowledge acquisition, knowledge representation, and verifying and validation of the expert system (Chowdhury & Canter 1997-98). Problem identification involves defining the goals and then developing the structure of the system. The problem identified is manure (plus fertilizer) applied to agricultural land contain excess nutrient above that required for potential plant uptake. This excess nutrient could increase the groundwater pollution potential at a downstream well from transport through the vadose and saturated zone. Based on the problem definition, the expert system is divided into four modules: surface loading, sorption, vadose zone transport and saturated zone transport. The surface loading module is developed to address the source of nutrient, the sorption module addresses retardation of the nutrient transport through the subsurface soil layers, and the vadose and the saturated zone transport address the nutrient transport from the source to the sink (i.e. downstream well).

Knowledge acquisition is the process of extracting, organizing, and structuring the knowledge for the expert system (Chowdhury & Canter 1997-98). The sources of knowledge in the expert system are published literature, human experts, and relevant information databases. A two-stage weighted-average method is used for the evaluation of pollution potential. In the first stage, human experts selected a set of most important input parameters for each module. The users are asked to input those parameters by answering a set of questions. Based on different ranges of inputs a rating (on 1 to 5 scale, 5 being the most critical) is assigned to each parameter within a module. For modules having more than one rating parameters, weighted average rating is calculated by the expert system based on the rating of each parameter and relative weights (on 1 to 5 scale, 5 being the most critical) assigned to all the parameters. In the second stage, the overall weighted average rating (OWAR) is calculated from the rating value obtained from each module and the relative importance weight assigned to all four modules by the human experts. Finally, groundwater pollution potential of each nutrient (nitrate and phosphate) is assigned based on the calculated overall weighted average rating. It should be noted that human experts from a group discussion assigned all the ratings and relative weights used in the expert system. It should also be noted that in the expert system developed in this study, the pollution potential for each nutrient (nitrogen or phosphorus) is evaluated separately.

Rule-based knowledge representation, which is very common for environmental expert systems (Chowdhury & Canter 1997-98), is adopted in this study. Rule-based expert system is structured from a series of If-Then-Else questions. For example, If answer to a question is 'yes', then input the parameter value, else (i.e. question answer is 'no') ignore the parameter or quit from giving output. It should be noted that there are two types of questions used here: mandatory questions, and optional questions. Mandatory questions are those users should give input for the parameter value asked, otherwise the expert system will not give any output. Finally, optional questions are those user can answer 'no'. In which case, the expert system will ignore the parameter and the user will eventually get a lower level of output for groundwater pollution potential. For simplicity, mandatory questions are set for parameters those can be obtained from the literature.

The final stage of the expert system is verification of the code and then validating the expert system with respect to field data. The validation of this expert system was very difficult due to lack of relevant field data. For this study, the expert system is validated with respect to an extensive site assessment done by a group of engineers to get permit for a new swine CAFO.

3. SURFACE LOADING

Surface loading module evaluates the impact of the nutrient source. This is accomplished by evaluating the mass of nutrient applied (from manure, and fertilizer if used) and potential nutrient uptake by the harvested crop. First, the expert system calculates (eqn.1) the total recoverable waste (TRW) generated in the CAFO from annual livestock inventory and the data presented in Table 1. Common animals in Oklahoma (USDA-NRCS 1999) are selected in Table 1. Nitrogen (N_{man}) and phosphorus (P_{man}) concentrations in manure for the CAFO site is calculated from eqns. (2) and (3), respectively. Manure nitrogen (N) and phosphorus (P) contents presented in Table 1 are the ones available for plant uptake i.e. after all kinds of losses. Finally, total nitrogen (N_{app}) and phosphorus (P_{app}) applied per acre land in a cultivation season from both manure and fertilizer nitrogen (N_{fer}) and phosphorus (P_{fer}) is calculated from eqns. (4) and (5), respectively.

Potential crop nutrient uptake is estimated from commonly used nutrient application rates in the

United States (Table 2). In Table 2, N_{crop} and P_{crop} are the nitrogen and phosphorus application rates for different crops, respectively. The crops selected for the expert system are those common in Oklahoma. According to the latest USDA study, percent of land used for winter wheat cultivation is 73.5%, for soybeans 8.6%, for sorghum 7.4%, for corn 5.7% and for cotton 4.8% (USDA 2000). It should be noted that only average phosphorus application rates are reported by Lander & Moffitt (1996) in Table 2. For each crop, minimum and maximum P application rates are calculated from eqn. (6) and (7), respectively. In both equations (6 & 7), a division factor of 7.2 is used to account for the stoichiometric mass ratio of N:P uptake for plant biomass (Nultsch 1971). It should also be noted that there are no N or P application rates for sorghum reported in Table 2. The expert system will calculate average, maximum or minimum N or P application rates for sorghum from N or P application rates (for N 3.55 lb/bu and for P 0.36 lb/bu (Lander et al. 1998)) and average, maximum or minimum anticipated yield per acre land, respectively. The latter can be found at the State's Department of Agriculture. Alternatively, user's experience can be used.

Table 1. Manure nutrient estimation.

Animal types	Animals /AU†† *	Manure produce (tons/AU)*	Manure recovery factor*	N/ Animal (lb)†	P/ Animal (lb) †
Beef Cows	1.0	11.5	0.80	30. 4	29.7
Milk Cows	0.74	15.24	0.65	57.6	22.1
Heifers	1.82	12.05	0.65	7.83	4.73
Steers, Calves, Bulls	1.64	10.59	0.80	17.1	14.8
Breeding Hogs & Pigs	2.67	6.11	0.75	5.7	6.21
Other Hogs & Pigs	9.09	14.69	0.75	3.42	3.40
Broiler Chicken	455	14.97	1.0	0.883	0.257
Layer Chicken	250	11.45	1.0	1.23	0.456
Turkeys	67.0	8.18	0.8	2.97	1.16

*Lander et al. (1998), †Calculated, ††Animal Unit

$$\text{TRW (tons)} = \sum_{i=1}^{n} \{(\text{manure produce} \times \text{manure recovery factor}) / (\text{animals/AU}) \times \text{no. of animals})\} \quad (1)$$

Where, n = no. of animal types.

$$N_{man} = \sum_{i=1}^{n} (\text{N/animal}) \times \text{number of animals} / \text{TRW} \quad (2)$$

$$P_{man} = \sum_{i=1}^{n} (\text{P/animal}) \times \text{number of animals} / \text{TRW} \quad (3)$$

$$N_{app} = N_{man} \times \text{manure applied} + N_{fer} \quad (4)$$

$$P_{app} = P_{man} \times \text{manure applied} + P_{fer} \quad (5)$$

$$\text{Min } P_{crop} = (\text{Min - Avg})\, N_{crop} / 7.2 + \text{Avg } P_{crop} \quad (6)$$

$$\text{Max } P_{crop} = (\text{Max - Avg})\, N_{crop} / 7.2 + \text{Avg } P_{crop} \quad (7)$$

Table 2. Crop nutrient estimation.

Crops	Yield Unit (YU)	N_{crop}* (lb/acre) Avg	Min	Max	P_{crop} (lb/acre) Avg*	Min†	Max†
Corn, field	Bu	129	78	160	57	50	62
Cotton	Bale	88	66	131	48	45	54
Sorghum	Bu	-	-	-	-	-	-
Soybean	Bu	22	12	44	47	45	50
Winter Wheat	Bu	66	35	101	38	33	43

*Lander & Moffitt (1996); †Calculated

The parameters selected for rating are N_{app}, and N_{crop} for nitrogen and P_{app}, and P_{crop} for phosphorus transport. Beside minimum, average and maximum application rates, the experts for rating use one more application rate, named critical rate. Critical application rate is the additional nutrient application required over the maximum application rate (Table 2) to cause the unconfined aquifer nutrient concentration to exceed EPA standard. EPA water quality standards for nitrogen and phosphorus are 10 and 5 mg/L, respectively (USEPA 1999). These values along with saturated-zone void ratio equal to 0.4 are used to calculate the additional nitrogen and phosphorus required for critical application rate. The additional application rates found are 11 and 5.5 lb/acre per foot of soil layer for nitrogen and phosphorus, respectively. Using these numbers, eqns. (8) and (9) are developed for calculating the critical nutrient application rates. Rating values suggested for this module is shown in Table R1. It should be noted that all rating table titles in this article start with the abbreviation 'R', and all other tables do not start with any letter. User's input for this module are number of animal types, tons of manure and fertilizer (if any) applied, harvested crop type, and unconfined aquifer thickness.

Table R1. Rating for surface loading.

Condition	Rating
N_{app} or P_{app} < Min N_{crop} or Min P_{crop}	1
N_{app} or P_{app} < Avg N_{crop} or Avg P_{crop}	2
N_{app} or P_{app} < Max N_{crop} or Max P_{crop}	3
N_{app} or P_{app} < Critical N_{crop} or Critical P_{crop}	4
N_{app} or P_{app} > Critical N_{crop} or Critical P_{crop}	5

$$\text{Critical } N_{crop} = \text{Max } N_{crop} + 11 \times \text{aquifer thickness} \quad (8)$$

$$\text{Critical } P_{crop} = \text{Max } P_{crop} + 5.5 \times \text{aquifer thickness} \quad (9)$$

4. SORPTION

The sorption module evaluates the impact of nutrient adsorption on the soil surface. Sorption phenomenon causes the soil to adsorb solutes (e.g. nutrients) to its surface and consequently slow down the solute transport. This slowing down of solute transport due to sorption is called retardation. The higher the retardation, the longer the transport time and thus, the lower the pollution potential. Thus, the rating scale of 5 in sorption module indicates low retardation and 1 indicates high retardation.

4.1 *Nitrate sorption*

Factors affecting nitrate sorption are pH, AEC (Anion Exchange Capacity) of soil and concentration and ionic strength of the adsorbate chemical (Bellini et al. 1996; Qafoku et al. 2000a). Bellini et al. (1996) concluded after studying nitrate and chloride leaching through a soil column that retardation of anions is a direct function of AEC at native soil pH. To further validate the study done by Bellini et al. (1996), which was limited to analysis on only one soil, Qafoku et al. (2000a) conducted nitrate and chloride leaching for 16 different soils from all over the world. In addition to the column leaching, soil mineralogy was also studied for all those soils by Qafoku et al. (2000b). Qafoku et al. (2000a) concluded that nitrate leaching is significantly influenced by pH and concentration of leaching solution. Nitrate adsorption is linear with any concentration of the leaching solution at pH > 6.5. While for pH <6.5, nitrate adsorption is nonlinear for leaching solution concentration less than 70 mg/L and linear for concentration over 70 mg/L. Some of the results of nitrate retardation factors (RF) at different pH from Qafoku et al. (2000a) are presented in Table 3. The results presented in Table 3 is for AEC equal to 1.15 cmol/kg, bulk density equal to 1.13 g/ml, and moisture content equal to 0.53. Another finding by Qafoku et al. (2000a) was that AEC can be correlated with the retardation factor (RF) for all soils at their native pH and for leaching solution nitrate concentration of 70 mg/L (Eqn. 10). From Eqn. 10, AEC is 0.14 cmol/kg for RF equal to 1, which means that for AEC less than 0.14 cmol/kg there will be no sorption.

Table 3. RF with pH & Nitrate concentrations.

pH	RF at different nitrate concentrations (mg/L)			
	70	140	280	420
4.21	3.86	2.96	1.87	1.58
4.45	3.19	2.39	1.69	1.32
5.47	2.24	1.81	1.64	1.55
6.47	1.92	1.32	1.32	1.49

Source: Qafoku et al. (2000a)

$$AEC = -0.1674 + 0.3061 \times RF \quad (10)$$

Based on the above discussion, a rating table for evaluating the impact of nitrate sorption is developed for soil native pH, AEC of soil and nitrate concentration in groundwater (Tables R2). Relative weight of each parameter is also presented in Table R2. Rating values assigned for pH and nitrate concentration are based on the data presented in Table 3 and eqn. 10 is used for AEC. The five rating values in ascending order correspond to approximate RF values of 4, 3, 2.5, 2, and 1, respectively. The RF values for differentiating ratings are based on maximum and minimum RF values and a linear delineation between those limits (Table 3). Mandatory input parameter selected for nitrate sorption is pH. While inputs for AEC and nitrate concentrations could be ignored.

Table R2. Rating for nitrate sorption.

pH [4]*	AEC (cmol/kg) [5]	NO_3^- (mg/L) [3]	Rating
< 4.0	>1.15	< 70	1
4.0 – 4.5	0.75 – 1.15	70 – 150	2
4.5 - 5.5	0.5 - 0.75	150 – 250	3
5.5 – 6.5	0.14 - 0.5	250 – 400	4
>6.5	<0.14	> 400	5

*[ ] Relative weights of each parameter

4.2 *Phosphate sorption*

Factors influencing phosphate adsorption are soil pH (Barrow 1984, Naidu et al. 1990), aluminum and iron oxides in soil (Parfitt 1978, Borggaard 1983, Borggaard et al. 1990, van der Zee & van Riemsdijk 1986) calcium content (Naidu et al. 1990) and organic matter in soil (Borggaard et al. 1990). As discussed previously, pH has significant affect on anion sorption. Unlike nitrate sorption, phosphate (P) sorption does not always decrease with pH increase. Naidu et al. (1990) showed for four strongly acidic Fijian soils that phosphate sorption decreases with increasing pH up to pH of 6. For pH > 6, phosphate sorption begins to increase. Naidu et al. (1990) suggested that the increase in P sorption for pH > 6 was caused by formation of insoluble Ca-P compounds. Thus, presence of calcium in soil will retard phosphate leaching at pH > 6. Ratings for pH impact on P sorption (Table R3) were based on sorption versus pH plots for four different soils presented by Naidu et al. (1990).

Aluminum and iron oxides have significant affect on P adsorption (Borggaard 1983, van der Zee et al. 1986). Aluminum and iron oxides exchange hydroxyl groups on the oxide surface with phosphate during formation of surface complexes (Borggaard,

1983, van der Zee & van Riemsdijk 1986). Thus, the higher the aluminum and iron content in the soil, the higher the P sorption (Borggaard et al. 1990). Borggaard et al. (1990) and van der Zee & van Riemsdijk (1986) developed two different linear correlation equations for phosphate sorption with respect to aluminum and iron contents in soil. Since the correlation presented by Borggaard et al. (1990) was proved to be a better representation of P sorption, it was used in rating for Al and Fe (Table R3).

Borggaard et al. (1990) had shown that organic matter does not compete with phosphate for adsorption sites. However, organic matter affects phosphate adsorption indirectly by inhibiting aluminum oxide crystallization, which in turn results high P adsorption. Appelt et al. (1975) found no change in phosphate adsorption with addition of dissolved organic matter. On the contrary, Sibanda & Young (1986) found that organic matter in solution decreased P sorption on soils and on aluminum and iron oxides. Due to current controversy of the impact of organic content on phosphate sorption, organic content in soil is not selected as a rating parameter.

Bottani et al. (1993) conducted P sorption capacity of three different soils (named C2, R3 and LP). The authors considered a Langmuir model as the adsorption isotherm for phopsphate sorption. Results of the study and calculated RF values for three phosphate concentrations (5, 10 and 50 mg/L) are presented in Table 4. RF values for all the soils at equilibrium phosphate concentrations (Ce) of 5 and 10 mg/L are very high, while for Ce equal to 50 mg/L RF is less than 4. Among the factors affecting phosphate sorption mentioned earlier, Bottani et al. (1993) considered all but aluminum and iron oxides (Table 4). Soils R3 and LP with similar textural classification give similar RF values. Of the significant factors affecting phosphate sorption, pH and Ca content are different for these soils. Soil R3 has high pH and high Ca content, while soil LP has low pH and low Ca content. Thus, Ca content for soil R3, and pH for soil LP are the most contributing factors for phosphate sorption. Again, soil C2 and R3 has similar Ca content, but different pH. Thus, a high RF value for C2 could be due to low pH. Finally, C2 and LP have similar pH values but different Ca content. As Ca-P complexes form at pH > 6 (Naidu et al. 1990), the difference in RF values could be attributed to different Ca content in the soil. Based on the RF values for different Ca content and for different phosphate concentrations in Table 4, rating for Ca and phosphate concentration is done (Table R3). Relative weights for each parameter are also presented in Table R3. Since pH is the only data most commonly found in literature, it is selected as the mandatory input parameter.

Table 4. Typical values for phosphate sorption.

Soil	C2			R3			LP		
Clay (%)*	40.6			17			18		
Silt (%)*	40.2			60.5			61		
Sand (%)*	19.2			22.5			21		
pH (1:1)*	5.9			7.4			5.8		
Organic mater (%)*	5.1			6.2			3.41		
TKN (%)*	0.23			0.34			0.21		
Extractable Phosphate (mg/L)*	23			11.2			16.4		
Ca (mmol/kg)*	118			107			41		
K (l/mg)*	0.102			1.271			1.259		
b (mg/kg)*	274.9			261.4			258.3		
Specific Surface Area (m^2/gm)*	31.73			28.74			10.55		
Ce (mg/L)†	5	10	50	5	10	50	5	10	50
RF†	41	22	3.4	21	7	1.3	21	7	1.3

* Bottani et al. 1993; †Calculated

Table R3. Rating for phosphate.

pH [4]*	Ca (mmol/kg) [4]	Al + Fe (mmol/kg) [5]	PO_4^{-3} (mg/L) [3]	Rating
< 4.5 & > 6.5	> 100	> 100	> 50	1
4.5 – 5.0	50 – 100	50 – 100	20 – 50	2
6.0 – 6.5	20 – 50	20 – 50	10 – 20	3
5.0 - 5.5	10 – 20	10 – 20	5 – 10	4
5.5 - 6.0	< 10	< 10	< 5	5

*[ ] Relative weights of each parameter

From the literature review, it has been found that phosphate transport could be retarded 30 to 40 times as compared to 1 to 4 times for nitrate due to sorption depending on soil and solute characteristics (Table 3 & 4). Thus, it could be concluded that phosphate transport would be much slower than that of nitrate. Since sorption for nitrate and phosphate are significantly different, no relation has been considered while rating for sorption. That is why one should not compare sorption ratings for nitrates and phosphates. Furthermore, it should be noted that unlike nitrate sorption, phosphate sorption studies lack in reporting data for retardation factors. Therefore, the expert system might have some shortcomings in representing P sorption.

5. VADOSE ZONE TRANSPORT

The vadose zone transport module evaluates the impact of the most important parameters related to solute transport through vadose zone. One-dimensional downward transport through vadose zone is considered based on the assumption that manure is applied uniformly on a flat land and lateral dispersion is minimum due to low concentration gradient in that direction. In downward direction, the maximum possible infiltration velocity equal to the saturation permeability (K_s) could be achieved at saturation. Since during irrigation, water is applied to the land, soil moisture content could reach very close to the saturation point. As at near saturation point, soil suction potential reaches zero, it could be assumed that maximum possible soil moisture velocity during irrigation is equal to the unsaturated permeability (K_u) at 0.4 moisture content, which is the moisture content near saturation for most of the soils (Table 5). Based on the assumptions, K_u is calculated for all USDA soil classes using eqns. 11 through 13 (Table 5). Considering K_u as the soil moisture velocity, distance travel in 90 (D_{90}) and 365 days (D_{365}) are calculated and presented in Table5. Duration of either 90 or 365 days is chosen because the nutrient application would follow a cycle of either a minimum of one cultivation season (90 days) or a maximum of one year (365 days). Using D_{90} and D_{365}, rating values for soil types are developed (Table R4). Soil types having very close D_{90} or D_{365} values are assigned same rating value. Since K_s is the basic soil parameter used for soil type rating, users will be asked to input this value first. In case the user cannot input K_s, the expert system will require input of vadose zone soil type.

$$\text{Effective water content, } S_e = \frac{0.4 - thr}{ths - thr} \tag{11}$$

$$\text{Relative permeability, } K_r = S_e^{0.5}\,[1 - (1 - S_e^{1/m})^m]^2 \tag{12}$$

$$K_u = K_r \times K_s \tag{13}$$

The other input parameters for the rating are water table depth and annual rainfall. The ratings and relative weights for water table depth and rainfall are shown in Table R4. Water table depth is the distance the nutrients need to travel to reach groundwater. The maximum water table depth for rating is chosen on three considerations: (1) the higher the water table depth, the longer it will take to reach the groundwater, (2) for nitrate transport, longer travel time in the vadose zone might allow the nitrate to denitrified into gaseous nitrogen, and (3) only three soils could flow more than 25 foot in a year (Table 5). The upper and lower rating limits for annual rainfall are based on the maximum and minimum of 30-years-average rainfall in Oklahoma, respectively. Both water table depth and annual rainfall are mandatory input parameters.

Table R4. Rating for vadose zone transport.

Soil Type [5]*	Depth (ft) [4]	Rainfall (inch) [3]	Rating
Silty Clay, Clay, Silty Clay Loam	> 25	< 20	1
Sandy Clay, Silt	15 - 25	20 - 30	2
Silt Loam, Clay Loam	10 - 15	30 - 40	3
Sandy Clay Loam, Loam	5 - 10	40 – 50	4
Sand, Loamy Sand, Sandy Loam	< 5	> 50	5

*[ ] Relative weights of each parameter

6. SATURATED ZONE TRANSPORT

The saturated zone transport module evaluates impact by considering solute transport through the groundwater from the CAFO to the nearest downstream water well. Factors selected for evaluating the saturated zone transport module are soil permeability (K_s), porosity (n), hydraulic gradient (i), distance of water well (d), and planning period (t). All of these input parameters are optional for the user. If the user cannot input one of the parameters, an input of the saturated zone soil type must be provided and the expert system will input the necessary parameter(s) from Table 6 for that soil type. It should be noted that the hydraulic gradient used in Table 6 is the average for the Central Oklahoma Aquifer (Christenson 1992).

The rating for soil types or permeability is done in a similar way with the vadose zone transport.

Table 5. Approximate water velocity in vadose zone for different soil types.

Soil Type	thr *	ths *	n *	m *	Se†	Kr†	Ks* (ft/d)	Ku = KrKs (ft/d)!!	D_{90}† (ft)	D_{365} † (ft)
Sand	0.045	0.43	2.68	0.627	0.792	2.40E-01	23.39	5.62E+00	5.06E+02	2.05E+03
Loamy Sand	0.057	0.41	2.28	0.561	0.830	2.35E-01	11.49	2.70E+00	2.43E+02	9.87E+02
Sandy Loam	0.065	0.41	1.89	0.471	0.826	1.48E-01	3.48	5.16E-01	4.64E+01	1.88E+02
Sandy Clay Loam	0.1	0.39	1.48	0.324	0.862	7.15E-02	1.03	7.37E-02	6.63E+00	2.69E+01
Loam	0.078	0.43	1.56	0.359	0.773	4.00E-02	0.819	3.28E-02	2.95E+00	1.20E+01
Silt Loam	0.067	0.45	1.41	0.291	0.739	1.22E-02	0.354	4.31E-03	3.88E-01	1.57E+00
Clay Loam	0.095	0.41	1.31	0.237	0.810	1.24E-02	0.205	2.53E-03	2.28E-01	9.23E-01
Silt	0.034	0.46	1.37	0.270	0.742	9.11E-03	0.197	1.79E-03	1.61E-01	6.54E-01
Clay	0.068	0.38	1.09	0.083	0.904	7.63E-04	0.157	1.20E-04	1.08E-02	4.39E-02
Sandy Clay	0.1	0.38	1.23	0.187	0.893	1.78E-02	0.094	1.68E-03	1.51E-01	6.13E-01
Silty Clay Loam	0.089	0.43	1.23	0.187	0.765	2.18E-03	0.055	1.20E-04	1.08E-02	4.38E-02
Silty Clay	0.07	0.36	1.09	0.083	0.966	6.91E-03	0.016	1.09E-04	9.79E-03	3.97E-02

* Carsel & Parrish (1988), †Calculated

Where the groundwater velocity (V) is calculated from eqn. (14), and distance travel is calculated for 10 (D_{10}) and 20 (D_{20}) years (Table 6). Rating for soil type is done in same way as the vadose zone transport module (Table R5). Rating for distance of water well and time period is done arbitrarily, except the fact that no rating value is assigned for distance less than 300 ft. The selection of 300 ft is based on current Oklahoma regulation that states that no manure shall be land applied within 300 feet of an existing public or private drinking water well (USDA 1998). The ratings and relative weights assigned to each parameter are shown in Table R5.

$$V = K_s \times i / n \quad (14)$$

Table 6. Groundwater velocities for USDA soils

Soil Type	Ks (ft/d)*	n*	i††	V† (ft/d)	D_{10}† (ft)	D_{20}† (ft)
Sand	23.386	0.43	0.01	5.44E-01	1985	3970
Loamy Sand	11.490	0.41	0.01	2.80E-01	1023	2046
Sandy Loam	3.481	0.41	0.01	8.49E-02	310	620
Sandy Clay Loam	1.031	0.39	0.01	2.64E-02	97	193
Loam	0.819	0.43	0.01	1.90E-02	70	139
Silt Loam	0.354	0.45	0.01	7.87E-03	29	57
Clay Loam	0.205	0.41	0.01	4.99E-03	18	36
Silt	0.197	0.46	0.01	4.28E-03	16	31
Clay	0.157	0.38	0.01	4.14E-03	15	30
Sandy Clay	0.094	0.38	0.01	2.49E-03	9	18
Silty Clay Loam	0.055	0.39	0.01	1.41E-03	5	10
Silty Clay	0.016	0.36	0.01	4.37E-04	2	3

*Carsel & Parrish (1988), †Calculated, ††Christenson (1992)

Table R5. Rating for soil type

Soil Type [5]*	Distance (ft) [4]	Planning period (yrs) [3]	Rating
Silty Clay, Silty Clay Loam	> 2000	< 5	1
Sandy Clay, Clay	1500 - 2000	5 - 10	2
Silt Loam, Clay Loam, Silt	1000 - 1500	10 - 15	3
Sandy Clay Loam, Loam	500 - 1000	15 – 20	4
Sand, Loamy Sand, Sandy Loam	300 - 500	> 20	5

*[ ] Relative weights of each parameter

Finally, the expert system will calculate the overall weighted average rating (OWAR) from the ratings obtained for each module and from the relative weights assigned to each module (Table R6). Then the expert system will use OWAR to get the groundwater pollution potential from Table R7. It should be noted that phosphate pollution potential is from very low to high, while nitrate pollution potential is from low to very high. This is because phosphate has very high sorption potential and is likely to have much less pollution potential than nitrate.

Table R6. Relative weigh of each module.

Module	Weigh
Surface loading	5
Sorption	2
Vadose zone transport	3
Saturated zone transport	3

Table R7. Pollution potential.

OWAR		Pollution
Nitarte	Phosphate	Potential
-	1	Very Low (VL)
1	2	Low (L)
2	3	Medium (M)
3	4	Moderately High (MH)
4	5	High (H)
5	-	Very High (VH)

7. VALIDATION OF THE EXPERT SYSTEM

The expert system proposed in this study was validated against a site assessment done by a group of experts for a swine CAFO in Oklahoma. The farm was prohibited by the court from land application of waste on a 450-acre land. After an extensive evaluation by the experts, appointed by the court, the farm was recommended to purchase an additional 600-acre land to get the permit.

The farm was raising 27,000 sows annually and was growing corn and wheat. Since, wheat require less nutrient than corn (Table 2), the expert system evaluation was conducted for wheat. The input parameters for the expert system were obtained from soil exploration, soil and water quality analysis and from hydrogeologic atlas (Morton & Goemaat 1973). A summarized output of the expert system output is listed in Table 7 (see Appendix for intermediate outputs). It is found that before purchasing additional land (original), the farm had high N and medium to moderately high P pollution potential. While after having 1050 acres of land for manure application (permit requirements), the pollution potential for N improved to moderately high to high. Thus, the expert system output is in compliance with the decision made by the group of experts.

Table 7. Validation of the Expert System

	OWAR		Pollution Potential	
	N	P	N	P
Original	4	3.7	H	M to MH
Permit Requirements	3.4	3.7	MH to H	M to MH

8. CONCLUSION

The expert system developed in this study is very simple to use and require less monetary investment in collecting data. Furthermore, the expert system would be a useful preliminary site assessment and site-screening tool for both new and existing CAFO sites. It would also be very useful for the farmers in employing better management practice for land application of manure. However, the expert system needs to be validated against additional field data in order to increase reliability.

9. REFERENCE

Appelt, H., N. T. Coleman & P.F. Pratt 1975. Interaction between organic compounds, minerals & ions in volcanic-ash-derived soils. II. Effects of organic compounds on the adsorption of phosphate. *Soil. Sci. Soc. Am. Proceedings* 39:628-630.

Arora, P. A. & W. F. Mcternan 1994. An expert system to determine the probability of pesticide leaching. *Agri. Water mgmt*. 25:57-70.

Bellini G., M.E. Sumner, D. E. Radcliffe & N. P. Qafoku 1996. Anion transport through columns of highly weathered acid soils: Adsorption and retardation. *Soil Sci. Soc. Am. J.* 60:132-137.

Borggaard, O. K. 1983. Influence of iron oxides on phosphate adsorption by soil. *J. Soil. Sci.* 34:333-341.

Borggaard, O. K., S. S. Jorgensen, J.P. Moberg & B. Raben-Lange 1990. Influence of organic matter on phosphate adsorption by aluminum and iron oxides in sandy soils. *J. Soil. Sci.* 41:443-449.

Bottani, E. J., L.E. Cascarini de torre, E. A. Fertitta, S. Lopez & N. O. Barbaro 1993. Adsorption of gases and phosphate ions on some Argentinean soil samples. *Soil Sci.* 155(1):190-196.

Carsel, R. F. & R. S. Parrish (1988). Developing joint probability distributions of soil water retention characteristics. *Water Resour. Res*. 24:755-769.

Chowdhury, A. K. M. M. & L. W. Canter 1997-98. Expert systems for groundwater management. *J. Env. Sys.* 26(1):89-109.

Christenson, S.C., R. B. Morton & B. A. Mesander 1992. Hydrogeologic maps of the Central Oklahoma Aquifer. *Hydrogeologic atlas HA-724*. U. S. Geologic Survey.

Crowe, A. S. and J. P. Mutch 1992. EXPRESS: An expert systems for assessing the fate of pesticides in the subsurface. *Env. Monitoring and Assessment*. 23:19-43.

Crowe, A. S. and J. P. Mutch 1994. An expert systems approach for assessing the potential for pesticide contamination of groundwater. *Ground Water*. 32(3):487-498.

EPA 1997. Cumulative Risk Index Analysis (CRIA) (Swine concentrated animal feeding operations). Version 6.0. *USEPA*, Region 6, Dallas, TX.

Kellogg, R. L. & C. H. Lander 1999. Trends in the Potential for Nutrient Loading from Confined Livestock Operations. Poster Presentation for *"The State of North America's Private Land," a Conference held on January*, at Chicago, Illinois. USDA-NRCS. http://www.nhq.nrcs.usda.gov/land/pubs/ntrend.html

Lander C. H., D. Moffitt & K. Alt 1998. Nutrients available from livestock manure relative to growth requirements. Resource Assessment and Strategic Planning. Working Paper 98-1. *USDA-NRCS.* http://www.nhq.nrcs.usda.gov/land/pubs/nlweb.html

Lander, C. H. & D. Moffitt 1996. Nutrient use in cropland agriculture (commercial fertilizer and manure): nitrogen and phosphorus. Working paper no. 14. *USDA-NRCS.* http://www.nhq.nrcs.usda.gov/RCA_PAPERS/WP14/wp14 text.html

Morton, R.B. & Goemaat, R. L. 1973. Reconnaissance of water resources of Beaver County, Oklahoma. *Hydrogeologic Atlas HA-450*. USGS

Naidu, R., Syers, J. K., Tillman, R. W. & Kirkman, J. H. 1990. Effect of liming on phosphate sorption by acid soils. *J. Soil Sci.* 41:165-175.

Nultsch, W. 1971. *General Botany*. Academic Press, NY. pp.439

Parfitt, R. L. 1978. Anion adsorption by soils and soil materials. *Advances in Agronomy*. 30:1-50.

Qafoku , N. P., M. E. Sumner & D.E. Radcliffe 2000a. Anion transport in columns of variable charge subsoils: Nitrate and Chloride. *J. Environ. Qual*. 29:484-493.

Qafoku , N. P., M. E. Sumner & L.T. West 2000b. Mineralogy and chemistry of some variable charge subsoils. *Commun. Soil Sci. Plant Anal.* 31(7&8):1051-1070.

Sibanda, H.M. & S. D. Young 1986. Competitive adsorption of humus acids and phosphate on oethite, gibbsite and two tropical soils. *J. Soil Sci.* 37:197-204

USDA-NASS *crop acreage, yield and production graphics* 2000, http://www.usda.gov/nass/aggraphs/crops.htm

USDA-NRCS 1999. *Resource assessment Division.* http://www.nhq.nrcs.usda.gov/land/lgif

USDA 1998. *An act relating to agriculture*. Passed by Oklahoma legislature.

USEPA 1999. *'National recommended water quality criterion'*, EPA 822-Z-99-001.

Van der Zee, S. E. A. T. M. & van Riemsdijk, W. H. 1986. Sorption kinetics and transport of phosphate in sandy soil. *Geoderma* 38:293-309.

10. APPENDIX

Surface loading: Aquifer thickness = 200 ft, Land area = 450 acres, Crop = Wheat, N_{app} = 205 lb/ton, Critical N_{crop} = 2300 lb/ton, Critical P_{crop} = 1160 lb/ton. Rating = 4 for both N and P.
Sorption: pH = 6.7, Ca = 21 mmol/kg, NO_3^- = 3.75 mg/l. Overall N sorption rating = 3.7 and P sorption rating = 2.
Vadose zone transport: K_s = 150 ft/d, rainfall = 24 inch, water table depth > 45 ft. Rating = 3.
Saturated zone transport: K_s = 150 ft/d, n = 0.3, i = 0.005, V = 2.5 ft/d, nearest well < 500 ft (assumed), Planning period 20 years (assumed). Rating = 5.

Computer Methods and Advances in Geomechanics, Desai et al. (eds) © 2001 Balkema, Rotterdam, ISBN 90 5809 183 X

Intelligent identification of deformation behaviors of rockmass at permanent shiplock zone of Three Gorges Project

Xia-Ting Feng & Shiwei Bai
Institute of Rock and Soil Mechanics, Chinese Academy of Sciences, Wuhan, People's Republic of China

Zhiqiang Zhang
College of Resources and Civil Engineering, Northeastern University, Shenyang, People's Republic of China

ABSTRACT: Considering complicated behaviors of rockmass during excavation or in long term, three intelligent methods, with combination of neural networks, genetic algorithms and numerical analysis, were proposed. One is to use the model built using information monitored at upper strata to estimate deformation behaviors produced by next excavation of rock mass at lower level. Others are to use the model built using historical information to estimate deformation behaviors of the system in future by intelligent time series analysis or intelligent finite element analysis with neural network constitutive model. The complicated stress-strain relationship and historical-future relationship of information are represented by different neural networks and learned using genetic algorithms in global optimum range. Applications to permanent shiplcok of Three Gorges Project indicate that effectiveness of these intelligent methods.

1 INTRODUCTION

Permanent shiplock is one of main buildings of Three Gorges Project in China. It is a double five-stage shiplock, one of the largest artificial navigation architecture excavated in rockmass in the world. The permanent shiplock is constructed in rock masses with axial direction of 111° and total length 1617m. The single shiplock room is 280m in length, 34m in width, and 5m in depth for storing water. The shiplock room is excavated in granites and has high and steep slopes at both sides. The slope angle varies with weathered degree of granites. For example, it is 1:1 at full weathered granite zone, 1:0.5 at moderately weathered zone, and 1:0.3 at slightly weathered and fresh rockmass. The sidewall of the shiplock room is straight with height of 40~50m.The deepest excavation for slope is about 170m. Therefore, the permanent shiplock is located at this deep slot with both high and steep slope and wall. The slope at head section of the third shiplock room is the highest at permanent shiplock region. Geological structures are the most complex and there is a fault F_{215} in this area. It forms a key region to be taken more attention. Any excess deformation would result in invalidation of the building. Therefore, prediction of deformation of the key region of shiplock during construction and function period and its control becomes one of the most important tasks in permanent shiplock (Zhang & Zhou 1999, Zhu & Zhang 1998).

Considering advantages of neural networks (Rumelhart et al. 1986, Kohonen 1988, Grossberg 1988, Holland 1975, 1992, Kartam et al 1997, Feng et al. 1999, 2000) and genetic algorithms (Goldberg 1989, Goldberg et al. 1992, Feng et al. 1999, 2000), this paper proposed an intelligent methodology for the reasonable estimating deformation behaviors of rock masses at permanent shiplock.

2 ESTIMATION OF DEFORMATION BEHAVIORS OF THE ISOLATED ROCKMASS AND THE STRAIGHT WALL OF SHIPLOCK DUE TO EXCAVATION

2.1 *Intelligent methodology*

Deformation data monitored at the excavated strata at upper level is used to recognize model and its parameters. Thus, deformation behaviors of strata at lower level due to excavation can be estimated from this model and parameters. And then newly monitored deformation data can be feedback to input to improve the recognized model and parameters. Doing like this, deformation behaviors of strata at much lower level due to excavation can be estimated from this improved model and parameters. Figure 1 shows this intelligent methodology.

2.2 *Neural network constitutive model*

In this intelligent prediction methodology, the core problem is how to acquire a reasonable nonlinear stress-strain relationship. Considering the highly nonlinear stress–strain relationship, the following neural network constitutive model can be built according to stress path dependency (Figure 2).

$$\begin{aligned} &\mathrm{NN}: R^n \to R^m \\ &\Delta\sigma_j = \mathrm{NN}(\Delta\varepsilon_j, \sigma_{j-1}, \varepsilon_{j-1}) \end{aligned} \tag{1}$$

or

$$\begin{aligned} &\mathrm{NN}: R^n \to R^m \\ &\Delta\sigma_j = \mathrm{NN}(\Delta\varepsilon_j, \sigma_{j-1}, \varepsilon_{j-1}, \sigma_{j-2}, \varepsilon_{j-2}) \end{aligned} \tag{2}$$

or

$$
\begin{aligned}
&\mathrm{NN}: R^{n} \rightarrow R^{m} \\
&\Delta\sigma_{j} = \mathrm{NN}(\Delta\varepsilon_{j}, \sigma_{j-1}, \varepsilon_{j-1}, \cdots, \sigma_{j-k}, \varepsilon_{j-k})
\end{aligned} \tag{3}
$$

The neural network model is directly obtained from load-deformation data of granite at lab experiment. How can the rich stress-strain information that is hidden in this global structural response data be extracted and used to obtain nonlinear material stress-strain model? One way is to use numerical analysis to obtain stress-stain information at any inside point of the material using global response information, for example load vs. deflection data, from a structural test. A fundamental premise of this method, and its most crucial feature, is that the structural test generates complex spatial patterns of stress and strain, and is therefore rich in information. Perhaps all that has been recorded experimentally is the deflection at a particular location in a complex structure at a series of different structural load levels. To reach this, numerical analysis (e.g., the finite element analysis) can be used as a bridge.

During the learning process, the topology of the network and its connection weights are encoded in a binary string. The network may be fully connected. Not only nodes on neighboring layers but also nodes on different layers are all connected. Granularity bits define size of connection. The binary string will be evolved using genetic operations such as reproduction, crossover, and mutation to obtain global optimum model.

And then the learned neural network model can be embedded to a finite element analysis to perform estimation of deformation behaviors of rock mass.

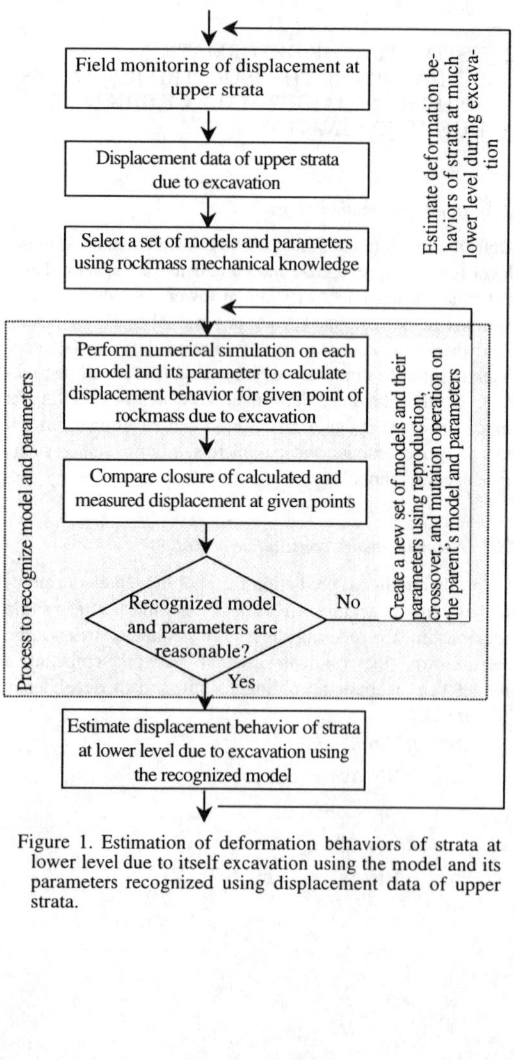

Figure 1. Estimation of deformation behaviors of strata at lower level due to itself excavation using the model and its parameters recognized using displacement data of upper strata.

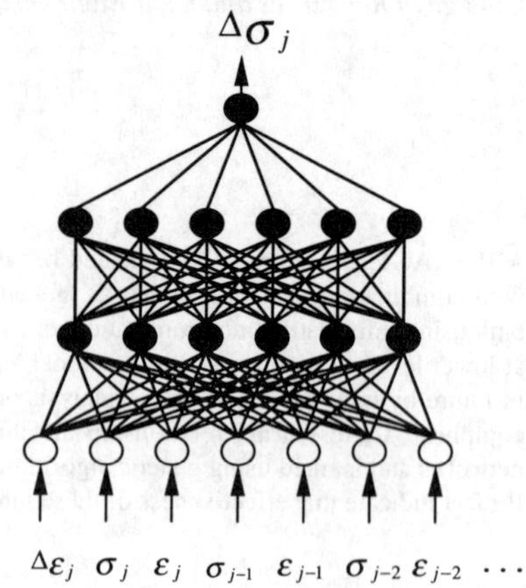

Figure 2. A neural network model to represent nonlinear stress -strain relationship

2.3 *Engineering geological and excavation zones*

According to weathered degree, rock masses are divided into four zones such as heavy weathered zone, moderately weathered zone, slightly weathered zone, and fresh zone. Considering effect of damage and unloading due to excavation, rock masses near the slope and wall of shiplock are damaged. And it forms an unloading deformation zone between the damaged zone and undistributed zone (Figures 3 and 4).

Table 1. Input rock mechanical parameters to neural analysis for deformation behavior estimation

	A	B	C	D	E	F_{215}
E (GPa)	7.51	9.68	18.95	32.1	0.3	3.0
$tg\phi$	1.3	1.4	1.5	1.7	0.7	1.0
C (MPa)	1.0	1.0	1.6	2.0	0.1	0.5
μ	0.24	0.22	0.23	0.23	0.32	0.35
R_t (MPa)	1.0	1.0	1.2	1.5	0.2	0.5
γ (kN/m^3)	26.5	27	27	27	25	26.5

A: Moderately weathered zone, B: Damaged zone, C: Unloading deformation zone, D: Slightly weathered or fresh zone, E: Heavy weathered zone.

2.4 *Input parameters*

The geo-stress is set to be $\sigma_x = 4.79 + 0.01168 \times H, \sigma_y =$

1.599+ 0.03039 $\times H, \sigma_z = 4.7152+0.01027 \times H$, $\tau_{xy} = 0$, $\tau_{yz} = 0$, and $\tau_{zx} = 0.0$ for fresh zone, and $\sigma_x = \mu\gamma H, \sigma_y = \gamma H$, $\sigma_z = \mu H/(1-\mu)$, $\tau_{xy} = 0$, $\tau_{yz} = 0$, and $\tau_{zx} = 0.0$ for the heavy weathered rock mass. Where γ is unit weight of rock mass ($\times 10^6$N/m^3). The other input parameters to the numerical model are shown in Table 1.

2.5 *Estimated deformation behaviors of top surface of the isolated rockmass and the straight wall of shiplock room*

Accumulated displacement increment from excavation step 8 to step 12 that was monitored from May 1997 to April 1999 at TP/BM26GP02, TP/BM27GP02, TP/BM28GP02, TP/BM29GP02, TP/BM10GP01, and TP/BM11GP01 are used for recognition of the constitutive model. Displacements of TP/BM71GP01 installed at the top of stand wall at north slope and TP/BM98GP02 installed at the top of stand

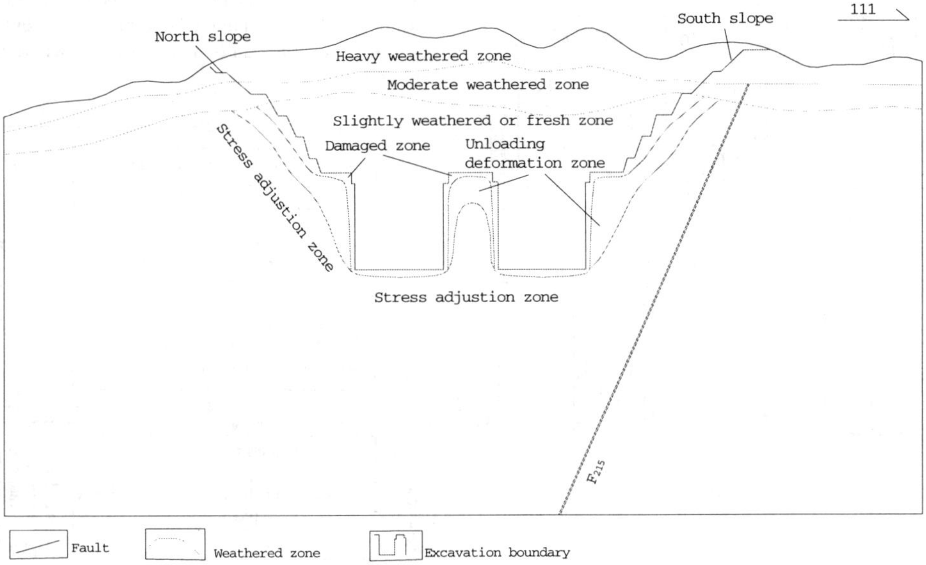

Figure 3. Engineering geological zones, scale:1:500

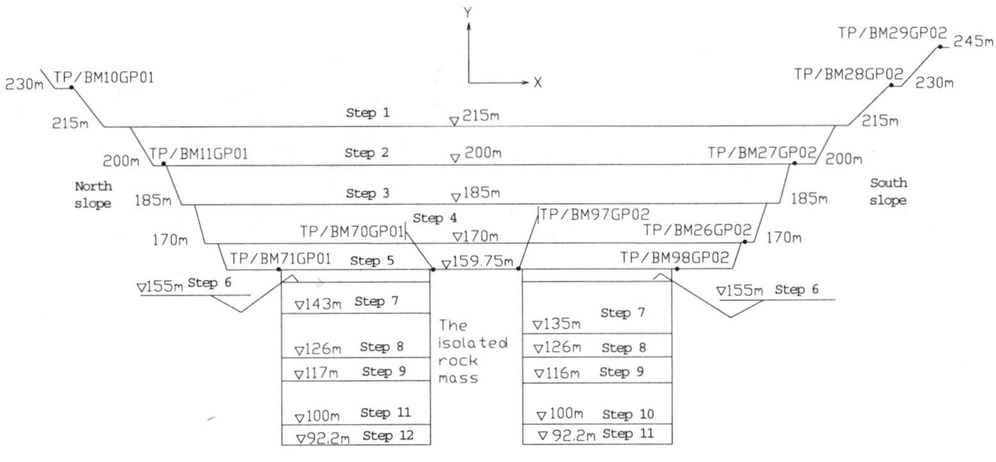

Figure 4. Excavation zones and arrangement of displacement monitoring points

wall at south slope are estimated as new cases. With application of model recognized, accumulative displacement increment at x-direction (i.e., vertical to axial of shiplock) from excavation Step 8 to Step 12 (i.e., at the period of January 1998 to April 1999 at TP/BM71GP01 and TP/BM98GP02 are calculated. Their calculated values are about 26 mm and 24mm, respectively, compared with measurement of 23.4 mm and 18 mm (as shown in Figure 5). Error is incurred from deformation complexity of stand wall. It is good agreement with the measurement.

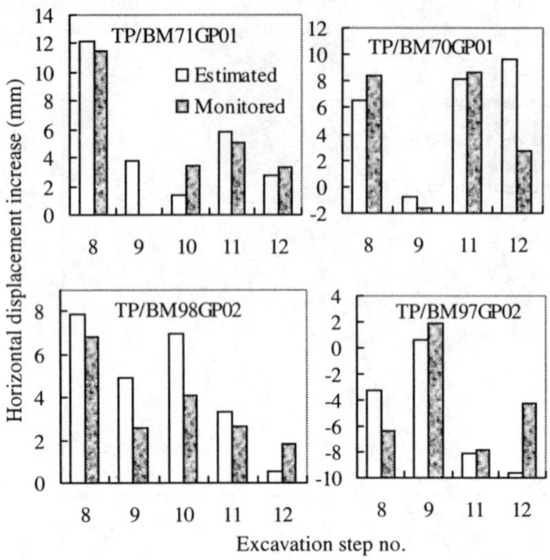

Figure 5. Comparison of estimated horizontal deformation of the top surface of the isolated rockmass and top of the straight wall induced by each step of excavation of shiplock room with the monitoring values. They aren't used for recognition of parameters

3 ESTIMATION OF DEFORMATION BEHAVIORS OF THE ISOLATED ROCKMASS AND THE STRAIGHT WALL OF SHIPLOCK AT FUNCTION PREIOD

3.1 *Intelligent methodology*

Similar to above, deformation data monitored at the current stage is used to obtain model and its parameters. Thus, deformation behaviors of rockmass in future can be estimated from this model and parameters. And then newly deformation data can be feedback to input to improve the recognized model and parameters. Furthermore, deformation behaviors of strata in newly future can be estimated from this improved model and parameters. Figure 6 shows this intelligent methodology.

3.2 *Nonlinear dynamic modeling methodology*

With combination of neural network and genetic algorithm, a new method is proposed to perform nonlinear estimation of permanent ship lock rock mass at Three Gorges Project. Deformation of the shiplock rockmass is considered as a nonlinear dynamic system. The nonlinear relationship between current deformation characteristics and its history is learned on a multiplayer feed ward neural network (Figure 7) using measured data.

$$x_{i+p} = \mathrm{NN}(p, h_1, \cdots, h_l, 1)\left(x_i, x_{i+1}, \cdots, x_{i+(p-1)}\right)$$

The number of historical points, p, is recognized automatically. Topology of the network and its connection weights are evolved using the proposed new generation of genetic algorithm. Here we use a multi-lag extrapolation procedure for evolving eigenvalues in future. Both the latest measured eigenvalues and the previous predicted values are taken as the input data to predict future values. In turn, consequent predictions can be made. The results are shown in Figure 8.

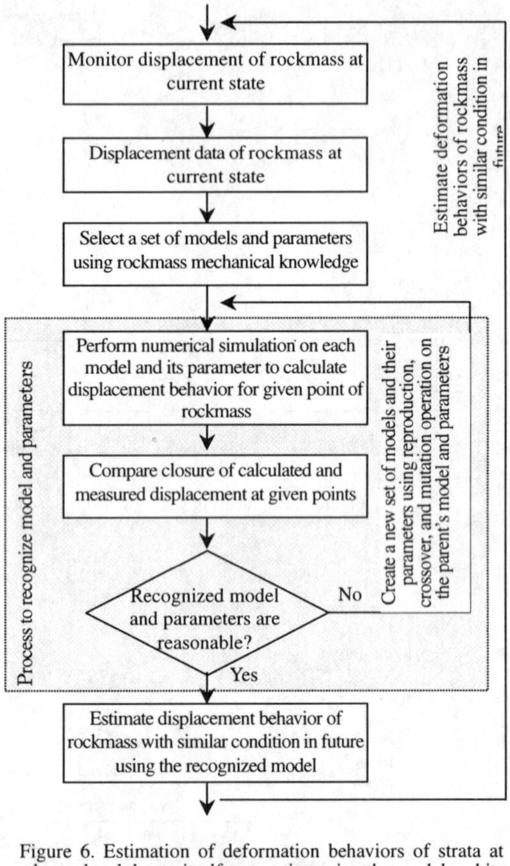

Figure 6. Estimation of deformation behaviors of strata at lower level due to itself excavation using the model and its parameters recognized using displacement data of upper strata.

4 CONCLUSIONS

The following conclusions can be drawn from above:

1) Deformation monitored at previous excavation steps implied abundant information related to that induced by

oncoming excavation. Therefore, it is possible to use the model built using information monitored at upper strata to estimate deformation behaviors produced by next excavation of rock mass at lower level.

2) Historical deformation implied abundant information related to the future of the system. Therefore, it is possible to use the model built using historical information to estimate incoming deformation behaviors of the system.

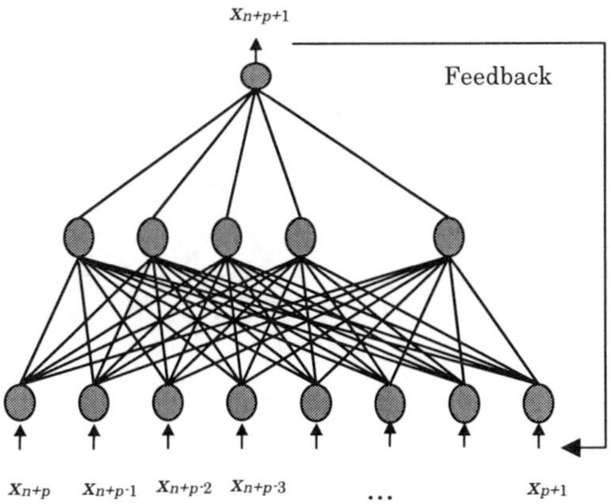

Figure 7. Neural network of nonlinear dynamic system for identifying deformation of shiplock rockmass

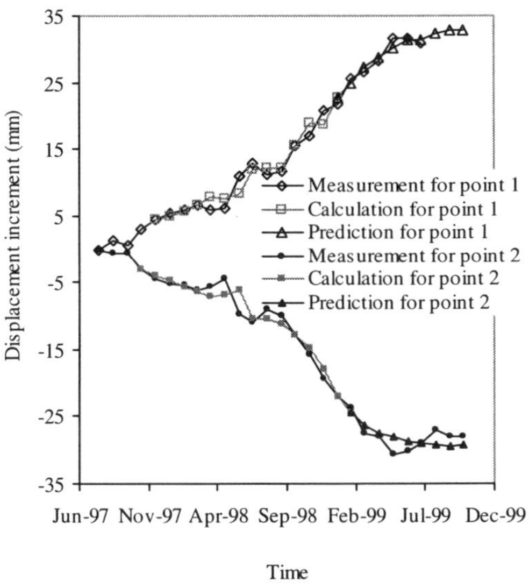

Figure 8. Predicted horizontal deformation at top of north sidewall (point 1: TP/BM70GP01) and the south sidewall (point 2: TP/BM97GP02) of the isolated rockmass at the head section of the third shiplock room

3) Stress-strain relationship at complicated geological and engineering condition is often highly nonlinear, particularly has high stress path dependency. As a high efficacious nonlinear representation and learning, neural networks are good for describing the stress-strain relationship by learning historical information or information obtained at similar geological and engineering conditions.

4) Engineering excavation induces damage of rock masses. As progress of excavation, damaged zone is extended gradually and unloading deformation will be occurred at some zones. Therefore, deformation behaviors of the damaged and unloading deformation zones shall be focally considered. The damage evolution process shall be reflected in numerical modeling.

5) Considering points mentioned above, the proposed intelligent methodology is effective to estimate reasonably deformation behaviors of complicated rock masses like permanent shiplock region of Three Gorges Project.

ACKNOWLEDGEMENT

Financial supports by National Science Foundation of China and National Key Basic Research Development Program are greatly acknowledged.

REFERENCES

Kohonen T. 1998. An introduction to neutral computing, *Neural Networks*, 1:3-16.

Grossberg S. 1988. Nonlinear neural Networks: principles, mechanisms, and architectures, *Neural Networks*, 1: 17-61.

Feng Xia-Ting Yang Chengxiang 1999. Intelligent rock mechanics. Part II: Intelligent recognition of model and parameters, *Chinese Journal of Rock Mechanics and Engineering*, 18(3): 350~353

Feng Xia-Ting, Zhang Zhiqiang 1999. Evolutionary neural network method of displacement back analysis, *Chinese Journal of Rock Mechanics and Engineering*, 18 (5):529~533

Feng, Xia-Ting, Zhang,Zhiqiang and Sheng, Qian Identifying rockmass mechanical parameters of Three Gorges Project permanent shiplock using intelligent displacement back analysis method, *International Journal Rock Mechanics and Mining Science.* in press.

Goldberg DE. *Genetic algorithms in search, optimization, and machine learning*, Addison-Wesley, Reading MA. 1989; 77, 106-122.

Goldberg, DE. Deb K. and Clark JH. 1992. Genetic algorithms, noise, and the sizing of populations, *Complex Systems*, 6: 19, 333-362.

Holland, JH.1975. *Adaptation in natural and artificial systems*, University of Michigan Press.

Holland JH. *Adaptation in natural and artificial systems*, MIT Press, Cambridge, MA, 1992.

Kartam, N., Flood, I. Garret, JH. Jr. 1997. *Artificial Neural networks for Civil Engineers: Fundamentals and Applications*, American Society for Civil Engineers, 216.

Rumelhart, DE., Hinton, GE. And Williams, R. J.1986 'Learning internal representations by err propagation', in: Rumulhart DE., and McClelland JL. (eds), *Parallel Distribution Processing*, MIT Press, Cambridge, MA.

Zhang, Youtian, Zhou Weiyuan 1999. *Deformation and stability of rock slopes*, Beijing: China Hydroelectric Press.

Zhu, W. and Zhang, Y. 1998. Effect of Reinforcing the High Jointed Slopes of Three Gorges Flight Lock, *Rock Mechanics and Rock Engineering*, 31(1) : 63-77.

Computer Methods and Advances in Geomechanics, Desai et al. (eds) © 2001 Balkema, Rotterdam, ISBN 90 5809 183 X

Displacement prediction for soil nailed structure based on ANN

L.Gao, K.H.Xie, H.W.Ying & D.J.Zhang
Department of Civil Engineering, Zhejiang University, Hangzhou, People's Republic of China

ABSTRACT: Soil nailing has become an important excavation support system for its good performance and cost- effectiveness. It is complicated to predict deformation of soil nailed support system during excavating. The Artificial Neural Network (ANN) is developed very quickly these years, which can be applied in diverse applications such as complex non-linear function mapping, pattern recognition, image processing and so on, and has been widely used in many fields, including geotechnical engineering. In this paper, the artificial neural network is applied for deformation prediction for soil nailed support system in deep excavation. The time series neural networks-based model for predicting deformation is presented and used in an engineering project. The results predicted by the model and those observed in the field are compared. It is shown that the artificial neural network-based method is effective in predicting the displacement of soil nailed structure during excavating.

1 INTRODUCTION

As one of the in situ ground reinforcement technique, soil nailing has been established for over 20 years and has been popular in many countries for it's good performance and cost-effectiveness. Soil nailing is a practical and proven technique used in retaining excavations and stabilizing slopes by reinforcing the ground in situ with relatively small, closely spaced, fully bonded inclusions, usually steel bars, which can improve native soil's mechanical characteristics and create a uniform composite mass where soil are locked together. The face of excavation or slope is locally stabilized by sprayed concrete or precast concrete panel (Schlosser 1982, Bruce et al. 1987, Gassler 1990, Chen 1998).

Several factors have contributed to the popularity of soil nailing. One of the important features is that it is convenient to adjust design parameters during excavation to make design more rational and to safeguard the stability of the soil nailed structure. Field monitoring and prediction that provides actual and forecasting information relating to the stability of the structure is thus necessary. During constructing the soil nailed structure, it is very simple to measure the horizontal displacement around the excavation. Therefore, the horizontal displacement is usually measured in order to monitor the stabilization of the support system during excavating. However, it is very complicated to predict the horizontal displacement because it is affected by many factors and the relationships among the horizontal displacement and these factors are highly nonlinear.

These years, Artificial Neural Network (ANN) is developed very quickly. The ANN is basically a massive parallel computational model that imitates the function of the human brain. Analogous to a biological model that consists of a large number of interconnected neurons, the ANN comprises a large number of similarly connected computational elements called artificial neurons. The ANN has been used as an alternative to the traditional mathematical models to simulate complex patterns. It is able to recall relationships between data by learning from train data and forming neural connection during the learning process. It can also be represented as a black-box where input data is transformed one or more times.

The ability of the ANN to approximate arbitrary highly non-linear functions and learn through examples has been broadly applied in diverse applications such as complex non-linear function mapping, pattern recognition, image processing and so on. This also lends it to many useful applications in civil engineering. In particular, the applications of ANN in geotechnical engineering has recently experienced rapid growth (Ghaboussi 1991, 1992, 1997, Wong 1992, Goh 1994, 1995, 1996, Chang et al. 1995, Ellis et al. 1995, Basheer et al. 1996, Lee et al. 1996, Teh et al. 1997, Kiefa et al. 1997, Zhu et al. 1998, Shi et al. 1998, Sun et al. 1998, Mathew, et al. 1999).

2 THE ANN MODEL

There exists a variety of network architectures and learning algorithms, among which, the feed-forward network trained by the error back propagation algorithm (BP network) is used widely. Figure 1 shows a typical BP network which consists of an input layer, an output layer, and one or more hidden layers. In this paper the BP network is employed. A detailed description of the BP network is beyond the scope of this paper, which has been introduced by numerous papers and books (Rumelhart et al. 1986, Karayiannis & Venetsanopoulos 1993, Flood & Kartam 1994).

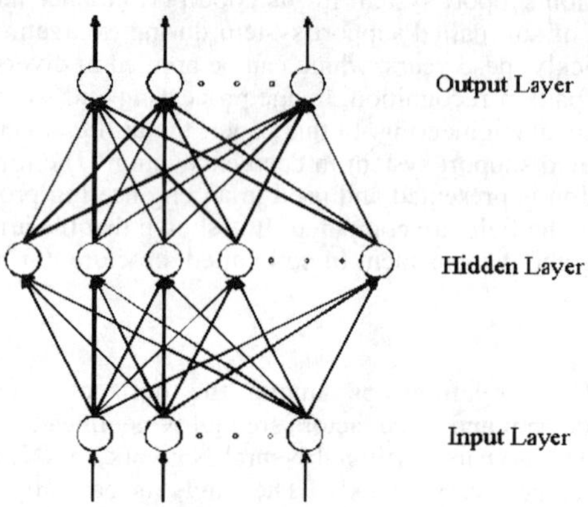

Figure 1. BP Network architecture

As mentioned above, the horizontal displacement of soil nailed structure is influenced by many factors such as soil properties, depth of excavation, nail spacing, nail length, and so on, which can be analyzed by using finite element method (Shen et al. 1981, Tabrizi 1996, Chen & Cui 1998). However, these factors will not be concerned in this paper. This paper focuses on only the use of BP network in predicting the horizontal displacement of soil nailed structure during excavating. Five sequential measured horizontal displacements are used as inputs, and the output represents the predicted displacement.

Designing a BP network structure not only includes determining the number of inputs and output variables, but also includes determining the number of hidden layers and the number of neurons in each hidden layer, and the latter is more difficult. Generally, there is no direct way for determining the number of hidden layers and the relevant neurons. The most appropriate number of hidden layers and the relevant neurons can provide the most accurate prediction. So, the number of hidden layers and the relevant neurons are determined by training and testing.

3 TRAINING AND TESTING OF THE BP NETWORK FOR PREDICTING THE DISPLACEMENT OF SOIL NAILED STRUCTURE

The measured data of a soil nailed structure are used as training patterns (Table 1) and testing patterns (Table 2). The structure is 5.4 meters high with a slope of 3H:10V. The nail length is 9 meters with

Table 1. Training patterns for the ANN

Training patterns	Input values					Output values
	I_1	I_2	I_3	I_4	I_5	O_1
1	1.5	1.8	5.4	6.3	5.3	6.5
2	1.8	5.4	6.3	5.3	6.5	6.3
3	5.4	6.3	5.3	6.5	6.3	6.0
4	6.3	5.3	6.5	6.3	6.0	9.5
5	5.3	6.5	6.3	6.0	9.5	8.9
6	6.5	6.3	6.0	9.5	8.9	11.4
7	6.3	6.0	9.5	8.9	11.4	11.8
8	6.0	9.5	8.9	11.4	11.8	15.7
9	9.5	8.9	11.4	11.8	15.7	19.2
10	8.9	11.4	11.8	15.7	19.2	20.2
11	11.4	11.8	15.7	19.2	20.2	25.4
12	11.8	15.7	19.2	20.2	25.4	27.2
13	15.7	19.2	20.2	25.4	27.2	30.6
14	19.2	20.2	25.4	27.2	30.6	33.3
15	20.2	25.4	27.2	30.6	33.3	39.4
16	25.4	27.2	30.6	33.3	39.4	41.2
17	27.2	30.6	33.3	39.4	41.2	42.9
18	30.6	33.3	39.4	41.2	42.9	44.2
19	33.3	39.4	41.2	42.9	44.2	45.3
20	39.4	41.2	42.9	44.2	45.3	45.4
21	41.2	42.9	44.2	45.3	45.4	47.7
22	42.9	44.2	45.3	45.4	47.7	48.6
23	44.2	45.3	45.4	47.7	48.6	50.3
24	45.3	45.4	47.7	48.6	50.3	50.1
25	45.4	47.7	48.6	50.3	50.1	49.8
26	47.7	48.6	50.3	50.1	49.8	51.2
27	48.6	50.3	50.1	49.8	51.2	52.3
28	50.3	50.1	49.8	51.2	52.3	53.3
29	50.1	49.8	51.2	52.3	53.3	53.9
30	49.8	51.2	52.3	53.3	53.9	53.9

1.5 meters of vertical spacing and 1.1 meters of horizontal spacing. The retaining structure is constructed in soft ground.

Table 2. Testing patterns for the ANN

Testing patterns	Input values					Output values
	I_1	I_2	I_3	I_4	I_5	O_1
1	51.2	52.3	53.3	53.9	53.9	54.3
2	52.3	53.3	53.9	53.9	54.3	54.4
3	53.3	53.9	53.9	54.3	54.4	54.4
4	53.9	53.9	54.3	54.4	54.4	58.0
5	53.9	54.3	54.4	54.4	58.0	59.1
6	54.3	54.4	54.4	58.0	59.1	58.2
7	54.4	54.4	58.0	59.1	58.2	57.8
8	54.4	58.0	59.1	58.2	57.8	59.1
9	58.0	59.1	58.2	57.8	59.1	58.5
10	59.1	58.2	57.8	59.1	58.5	57.6
11	58.2	57.8	59.1	58.5	57.6	57.4
12	57.8	59.1	58.5	57.6	57.4	57.6
13	59.1	58.5	57.6	57.4	57.6	58.8
14	58.5	57.6	57.4	57.6	58.8	55.3
15	57.6	57.4	57.6	58.8	55.3	61.8

After trying a different number of hidden layers and the relevant neurons, one hidden layer with 18 neurons is used. The sigmoid function shown in Equation (1) is selected as the activation function to transfer input variables to the hidden layer, and the same function is adopted as the neuron transfer function between the hidden layer and the output layer.

$$f(x) = 1/(1+e^{-x}) \tag{1}$$

Because of the form of the sigmoid function, the network outputs change within the range of 0 to 1. If the pattern values are in a different range, they are normalized generally before training the BP network. Since the network outputs can never reach 0 or 1, the values such as 0.1 and 0.9 is used to represent the smallest and largest pattern values. The pattern values can be normalized by using the following equation:

$$\bar{x} = 0.1 + \frac{0.8(x - x_{\min})}{(x - x_{\max})} \tag{2}$$

where, $\bar{x}$ = normalized values; x = the pattern value; $x_{\min}$ and $x_{\max}$ = the minimum and the maximum pattern value.

The training will stop when all patterns have an error that is less or equal to the target error. The root mean squared error (RMS) is used as the target error. The RMS error is defined by the following equation:

$$RMS = \frac{\sum_{i=1}^{N(p)}\sum_{j=1}^{N(o)}(t(i,j)-o(i,j))}{N(p)} \tag{3}$$

where, $N(p)$ = the total number of training patterns; $N(o)$ = the output number of the BP network; $t(i,j)$ = the BP network ith output for pattern j; $o(i,j)$ = the output training pattern j for output i.

The target RMS error is set as 0.01. The training result is illustrated in Fig. 2. The finial RMS error is 0.0046 when training end.

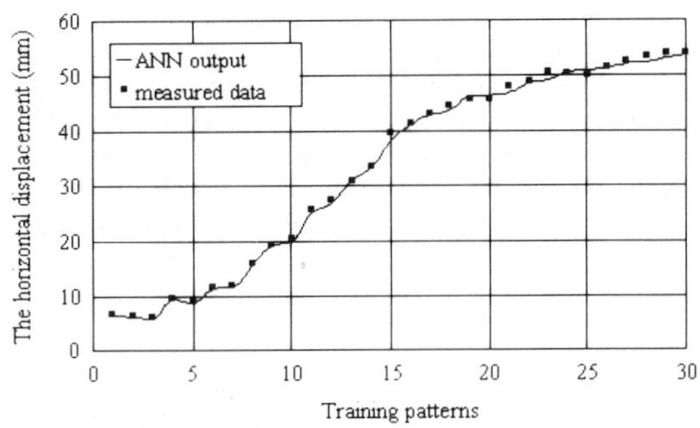

Figure 2. The result of training

The trained BP network is used for prediction. The testing result is illustrated in Figure 3. The absolute minimum error and the absolute maximum error are 0.2% and 9.3%, respectively. The result shows that the prediction of the network agree the

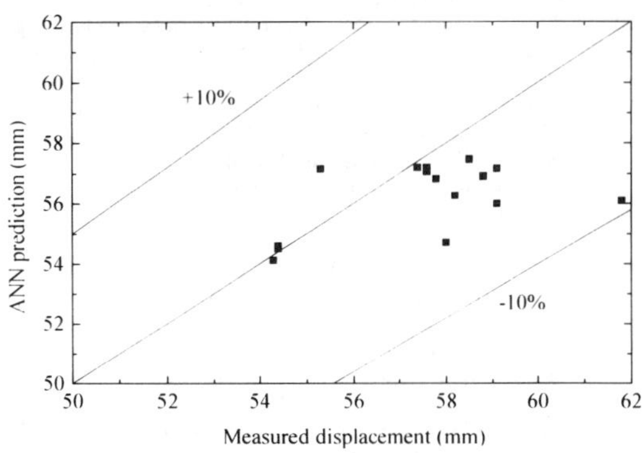

Figure 3. The result of prediction

measured data well and thus demonstrates that the BP network can be employed to predict effectively the horizontal displacement of soil nailed structure.

4 CONCLUSIONS

The artificial neural network simulates the function of the human brain. It can establish complex relationship between input and output patterns by learning from training set. The ANN has gained popular applications in geotechnical engineering.

In this paper, the use of BP network in predicting the horizontal displacement of soil nailed structure is presented. Five sequential measured horizontal displacements from a soil nailed retaining project are used as the inputs, and the output represents the predicted displacement. After comparing the result of a different number of hidden layers and the relevant neurons within hidden layers, one hidden layer with 18 neurons is selected. The absolute minimum error and the absolute maximum error between the predicted results and measured ones are 0.2% and 9.3%, respectively. This shows that the predictions of the BP network agree the measured data well and the BP network can be used to predict effectively the displacement of soil nailed structure during excavating.

REFERENCE

Basheer, I. A. & Najjar, Y. M. 1996. Predicting dynamic response of absorption columns with neural networks. *J. Comp. in Civ. Engrg.* ASCE, 10(1): 31-39.

Bruce, D. A. & Jewell, R. A. 1986. Soil nailing: application and practice-part 1. *Ground Engineering.* 19: 10-15.

Chang, W. T., Chow, Y. K. & Liu, L. F. 1995. Neural network: an alternative to pile driving formulas. *Comp. and Geotechnics*, 17: 135-156.

Chen, Z. Y. & Cui, J. H. 1998. Application of Soil Nailing in Foundation Excavation. Chinese Building Industry Publishing House, Beijing. (In Chinese)

Ellis, G.W., Yao, C., Zhao, R. & Penumadu, D. 1995. Stress-strain modeling of sands using artificial neural networks. *J. Geotech. Engrg.* ASCE, 121(5): 429-435.

Flood, I. & Kartam, N. 1994. Neural networks in civil engineering. I: Principles and understanding. *J. Comp. in Civ. Engrg.*, ASCE, 8(2): 131-148.

Gassler, G. 1990. In-situ techniques of reinforced soil. *Proc. Int. Reinforced Soil Conference*: 185-196.

Ghaboussi, J. 1992. Potential application of neural-biological computational models in geotechnical engineering. *Proc. 4th Int. Symp. on Numerical Models in Geomech.*, Swansea, U. K.: 138-149.

Ghaboussi, J., Garrett, J. H. Jr. & Wu, X. 1991. Knowledge-based modeling of material behavior with neural networks. J. Engrg. Mech., ASCE, 117(1): 132-153.

Ghaboussi, J & Sidarta, D. E. 1997. New method of material modeling using neural networks. *Numerical models in Geomechanics*, Pietruszczak & Pande Eds.: 393-400.

Goh, A.T.C. 1994. Seismic liquefaction potential assessed by neural networks. *J. Geotech. Engrg.* ASCE, 120(9): 1467-1480.

Goh, A.T.C. 1996. Neural-networks modeling of CPT seismic liquefaction data. *J. Geotech. Engrg.* ASCE, 122(1): 70-73.

Goh, A.T.C., Wong, K.S. & Broms, B.B. 1995. Estimation of lateral wall movements in braced excavations using neural networks. *Can. Geotech. J.*, 32: 1059-1064.

Karayiannis, N. B. & Venetsanopoulos, A. N. 1993. Artificial neural networks: learning algorithms, performance evaluation, and applications. Kluwer academic publishers, Boston.

Kiefa, M.A.A 1998. General regression neural networks for driven piles in cohesionless soils. *J. Geotech. Geoenviron. Engrg.* ASCE, 124(12): 1177-1185.

Lee, I. M. & Lee, J. H. 1996. Prediction of pile bearing capacity using artificial neural networks. *Comp. and Geotechnics*, 18(3): 189-200.

Mathew, A., Kumar, B., Sinha, B. P. & Pedreschi, R. F. 1999. Analysis of masonary panel under biaxial bending using ANNs and CBR. *J. Comp. in Civ. Engrg.* ASCE, 13(3): 170-177.

Rumelhart,,D.E., Hinton, G.E. & Williams, R.J. 1986. Learning representation by back-propagation errors. *Nature*, 323(9): 533-536.

Schlosser, F. 1982. Behavior and design of soil nailing. *Proc. Symp. Recent Development in Ground Improvement Techniques*, AIT, Bangkok: 389-397.

Shen, C. K., Bang, S. & Herrman, L. R.1981. Ground movement analysis of earth support system. *J. Geotech. Engrg., Div.*, ASCE, 107(12): 1609–1624.

Shi, J.S., Ortigao J.A.R. & Bai J. 1998. Modular neural networks for predicting settlements during tunneling. *J. Geotech. Geoenviron. Engrg.* ASCE, 124(5): 389-395.

Sun, H. T. & Wu, X. 1998. Study on neural networks method of deformation prediction of foundation pit based on artificial. *Rock and Soil Mechanics*, 19(4): 63-68. (In Chinese)

Tabrizi, K. S. 1996. 3-D element modeling of soil-nailed walls. PhD thesis, The State University of New Jersey, New Brunswick, New Jersey

Teh, C.I., Wong, K. S., Goh, A.T.C. & Jaritngam. 1997. Prediction of pile capacity using neural networks. *J. Comp. in Civ. Engrg.* ASCE, 11(2): 129-138.

Wong, F., Tuang, A. & Dong, W. 1992. Seismic hazard prediction using neural nets. *Proc. 10th World Conf. On Earthquake Engrg.*, 339-343.

Zhu, J.H., Zaman, M.M. & Anderson, S.A. 1998. Modeling of soil behavior with a recurrent neural network. *Can. Geotech. J.*, 35: 858-872.

Computer Methods and Advances in Geomechanics, Desai et al. (eds) © 2001 Balkema, Rotterdam, ISBN 90 5809 183 X

Prediction of diaphragm wall deflection in deep excavations using artificial neural networks

J.C.Jan, S.Y.Chi & J.C.Chern
Geotechnical Engineering Research Center, Sinotech Engineering Consultants Incorporated, Taipei, Taiwan

ABSTRACT: An Adaptive L-BFGS supervised neural network was employed to predict the diaphragm wall deflection in deep excavation. Eighteen case histories with 4 to 7 excavation stages were selected for training and verification. Meanwhile, each diaphragm wall is discretized into 18 uniform subintervals with 19 nodal points. The results show that the artificial neural network can predict the magnitude as well as the location of maximum deflection of diaphragm wall reasonably well. The average error in maximum wall deflection predicted is around 9mm. More than 87% of the predicted location of maximum deflection point is less than one nodal point from the observed location.

1 INTRODUCTION

In urban area, braced diaphragm walls are generally used for deep excavations in soft soil. A satisfactory braced diaphragm wall can not only provide a safe excavation but also minimizing deformations in the surrounding ground, which is of vital importance in order to avoid doing costly damages to adjacent buildings. Observational methods are often used in deep excavation projects to ensure the safety of construction. Peck (1969a) first compiled the observational data in deep excavations and tunneling in soft grounds. He attempted to summarize the knowledge of ability to judge whether the excavation is feasible under various methods of construction.

Ou and Tang (1994) employed optimization techniques to determine soil parameters for finite element analysis in deep excavation. Chi et al. (1999) developed an information construction approach for deep excavation. In their report, they used optimization process to back-analyze soil parameters, and the back-analyzed soil parameters were then used to predict the wall deflection of next excavation stage. Although back analysis combined with finite element analysis provides an effective numerical approach for engineers to estimate wall deflections and surface settlements in excavations, the back-analysis technique is limited by the number of soil parameters and the soil model used.

Based on the review mentioned above, it may be noted that experiences always play an important role in geotechnical problems which can not be easily solved by analysis due to the difficult in determining the material properties. However, empirical rules are difficult to establish due to the large variation in engineering and material conditions. Back analysis of monitored data, which reflect the field conditions, appears to be the most rational approach. But it is still limited by the number of soil parameters and the soil model used.

Artificial neural networks are intelligent numerical approaches that can automatically acquire knowledge from experience. The approaches have been demonstrated to be able of modeling complex problems using a certain amount of training instances. One advantage of the artificial neural network is that it can model complex problems in which the governing equations are difficult to define. Another advantage is that the artificial neural network is capable of processing uncertain information or noise data. Flood and Kartam (1994a, b) provided a discourse on the understanding, usage, and potential for application of artificial neural networks in civil engineering. They reported that artificial neural networks are capable of implementing in mapping problems, model dynamic processes, transitory problems and optimization problems. In this paper, the prediction of diaphragm wall deflection in deep excavation by using neural network is presented.

2 ARTIFICIAL NEURAL NETWORKS

Back-propagation neural network (BPN), one of supervised neural network, has widely been utilized in civil engineering. Vanluchene and Sun (1990) discussed the back-propagation neural network in structural engineering. They confirmed the feasibility of applications of BPN in structural engineering. Goh (1995) applied back-propagation neural networks to assist solving complex engineering problems. The neural network approaches, as indicated from his research, are proposed to replace empirical and semi-empirical equations in geotechnics. In principle, the performance of neural networks in engineering applications can be significantly improved by selecting a suitable representational framework to present the training input/output pattern pairs. Gunaratnam and Gero (1994) discussed the effect of representation of input/output pairs for training instances on the learning performance of the BPN learning algorithm in the problems of structural design. The dimensionless representation is reported to result in a simpler mapping function and makes it possible to train network on a small data set and still have the capability for reasonable accurate predictions.

Back-propagation neural network involves a feedforward network with optimization computing. A feedforward network is displayed in Figure 1. The feedforward network consists of one layer input nodes, one or more layers hidden nodes, and one layer output nodes. The nodes between two sequential layers are fully linked. Each node denotes an artificial neuron, and each link represents a weight. The artificial neuron consists of a computing process and a set of connected weights. In the model, the inputs, x_i, are multiplied by the connected weights, w_i, in the computing process. The weighted sum of the inputs, Z, is calculated as follows:

$$Z = \sum_{i=1}^{n} w_i x_i \tag{1}$$

Then the weighted sum of the inputs is processed by an activation function, $F(Z)$, to obtain an output

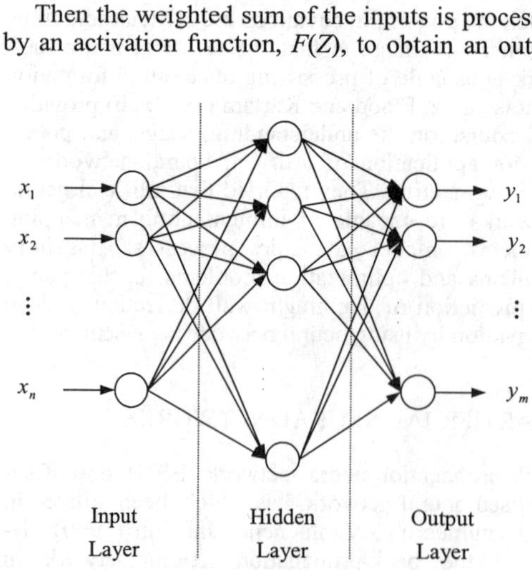

Figure 1 A feedforward network with one hidden layer

signal, y. A sigmoid function in the following form is most commonly used as an activation function.

$$y = F(Z) = \frac{1}{1+e^{-(Z+\theta)}} \tag{2}$$

where θ is a constant. The advantage of adopting the sigmoid function is that both the function and its derivative are differentiable everywhere.

Feedforward and error-feedback are two primary computing processes in the back-propagation neural network. The feedforward computing starts from the input layer through the hidden layers to the output layer so as to calculate the outputs. The error feedback computing modifies weights from output layer to input layer in order to minimize the errors between calculated and desired outputs. Notably, the mathematical optimization techniques are generally used in the error feedback computing as the learning algorithms. In the several learning algorithms, the gradient decent algorithm is the simplest method and listed as follows:

$$\Delta w^k = -\eta \frac{\partial E}{\partial w} \tag{3}$$

$$w^{k+1} = w^k + \Delta w^k + \alpha \Delta w^{k-1} \tag{4}$$

where E is the error function, η represents the learning step and α denotes the moment ratio. In order to upgrade the learning performance of back-propagation neural network, the Broydon-Fletcher-Goldfarb-Shanno (BFGS) algorithm (Nocedal, 1980), a second order optimization technique, is utilized to replace the gradient decent method. Meanwhile, a quadratic line search scheme is introduced into the BFGS algorithm, termed as an adaptive L-BFGS (Hung and Lin 1994), to predict the proper learning step.

3 KNOWLEDGE REPRESENTATION OF DIAPHRAGM WALL DEFLECTION

A diaphragm wall structure system is depicted in Figure 2. In the figure, the wall length is assumed to be 1.8 times the final excavation depth, H. The term R, called as index of observation point, is the normalized depth of measuring point. The term W denotes the thickness of diaphragm wall, and D the depth of current excavation. The monitored data is transformed into the instance (input/output), which is represented as follows:

Inputs:

Wall thickness W

Equivalent SPT-N value between the depth of D-$0.25H$ and D+$0.25H$ $\overline{N}$

Depth of excavation surface D

Index of observation point R

Wall deflection of observation point in i-3th excavation stage Δ_R^{i-3}

Wall deflection of observation point in i-2th excavation stage Δ_R^{i-2}

Wall deflection of observation point in i-1th excavation stage Δ_R^{i-1}

Output:

Wall deflection of observation point in ith excavation stage Δ_R^{i}

In the instances, it should be noted that, if i equals 1, Δ_R^{i-1}, Δ_R^{i-2} and Δ_R^{i-3} are set as zero; if i equals 2, Δ_R^{i-2} and Δ_R^{i-3} are set as zero; and if i equals 3, Δ_R^{i-3} is set as zero.

In this study, eighteen case histories of deep excavation with four to seven excavation stages each, resulting a total number of 93 sets of wall deflection, were used to establish the instance base. Each diaphragm wall was discreteized into 18 uniform subinterval with 19 nodal points. Therefore, 1767 (93×19) instances were generated for training and verification.

4 COMPUTING RESULTS

A one hidden layer feedforward network with seven input nodes, fifteen hidden nodes, and one output node was employed to solve the diaphragm wall deflection problem. The flow chart of verification is shown in Figure 3 and depicted as follows:

Step 1: Select a case history for verification.

Step 2: Train the neural network using other case histories.

Step 3:Predict the wall deflection of next excavation stage.

Step 4:After the excavation stage is completed, then append the measured wall deflection of the excavation stage to the training instances.

Step 5: Retrain the neural network.

Step 6: Repeat the Step 3 to Step 5 until the prediction is completed.

The 18 case histories of deep excavation in Taipei basin are used to verify the performance of neural network prediction. Herein, we only concern the 3-7 excavation stages, i.e., 57 sets of wall deflection were used, because the engineering failures seldom occur in the first and second excavation stages. Hence, a total number of 1083 (57×19) instances are adopted for verification. The predicted maximum deflections of the 57 sets of wall deflection are shown in Figure 4. The correlation coefficient be-

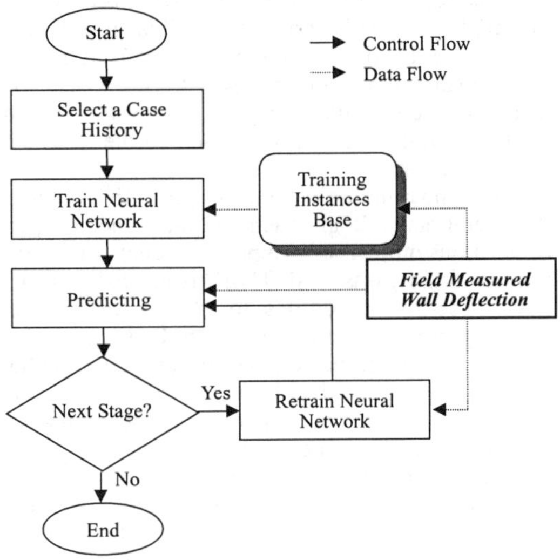

Figure 3 Flow chart of verification.

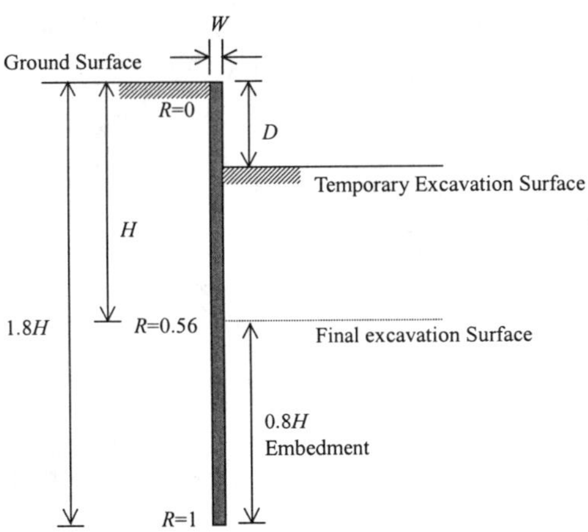

Figure 2 Illustration of diaphragm wall structure.

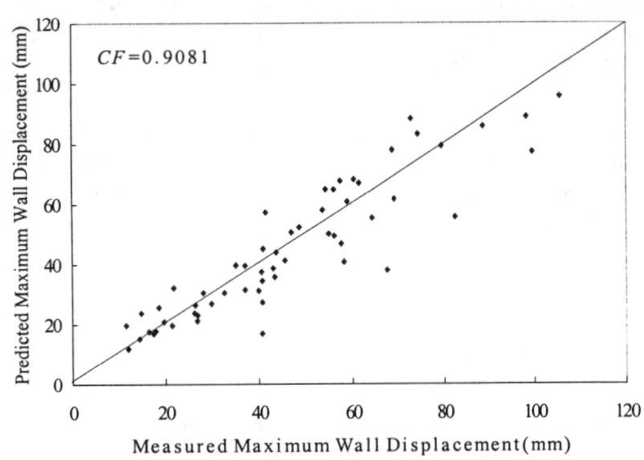

Figure 4 Results of predicted maximum wall deflection.

tween predicted and measured maximum wall deflections is 0.9081. The number of cases with relative error of predicted maximum wall deflection in the range of [0, 10%], [10%, 20%], and exceeding 20% are 28, 16, and 13, respectively. If we define that the prediction is failed when the relative percentage error exceeds 20, then 77% of cases are considered to have acceptable predictions.

Furthermore, the average error of all verification instances is 5.3mm. The average error of the seven largest deflection points of each excavation stage of case histories is 6.4mm. The average error of wall deflection of embedded points is 4.8mm. As a result, the largest prediction error occurs near the maximum wall deflection nodal point. Additionally, the 38.6% of maximum wall deflection of computed case histories occur at the same nodal point, as compared with measured case histories. The 49.1% of maximum wall deflection of computed case histories are missed by one nodal point only. This means more than 87% of the predicted maximum deflection point is less than one nodal point away from the observed location. These results imply that the wall shape of predicted wall deflection resembles to the measured wall deflection.

For comparing the accuracy of prediction between the neural network approach and other methods, the excavation project of Taipei National Enterprise Center (TNEC) was used. TNEC is located in Taipei basin. The soil conditions for the upper 33m are primarily soft clay with N value in the range of 2 to 5. The excavation was carried by top-down method in 7 stages to a maximum depth of 19.7m. It used 0.9m thick and 35m deep diaphragm wall as the soil retaining structure. The predicted results by neural network approach and the monitored values are shown in Fig. 5. The predictions by using the back-analyzed soil parameters from optimization process and soil parameters based on local empirical formulas are also presented in the figure. The neural network gives acceptable prediction of wall deflections except for excavation stages 2 and 3. The results of finite element analysis based on local empirical formulas are markedly different from measured wall deflection in all excavation stages. In addition, the results of finite element analysis using back-analyzed soil parameters provide acceptable prediction in the maximum wall deflection except for excavation stages 2 and 3. But, the predicted wall deflections of embedded points are inexact except for excavation stage 2 and 7.

5 SENSITIVITY ANALYSIS

In order to understand the effects of input parameters on the neural network prediction, the hierarchical analysis developed by Yang and Zhang (1997) was used as an approach to identify the degree of importance of the input parameters. Hereinafter, the hierarchical analysis is briefly reviewed as follows:

1. A BP-based neural network has been trained successfully.
2. Calculate the variance of output with the change of each input parameter. The variance can be replaced by the following differentiation of the

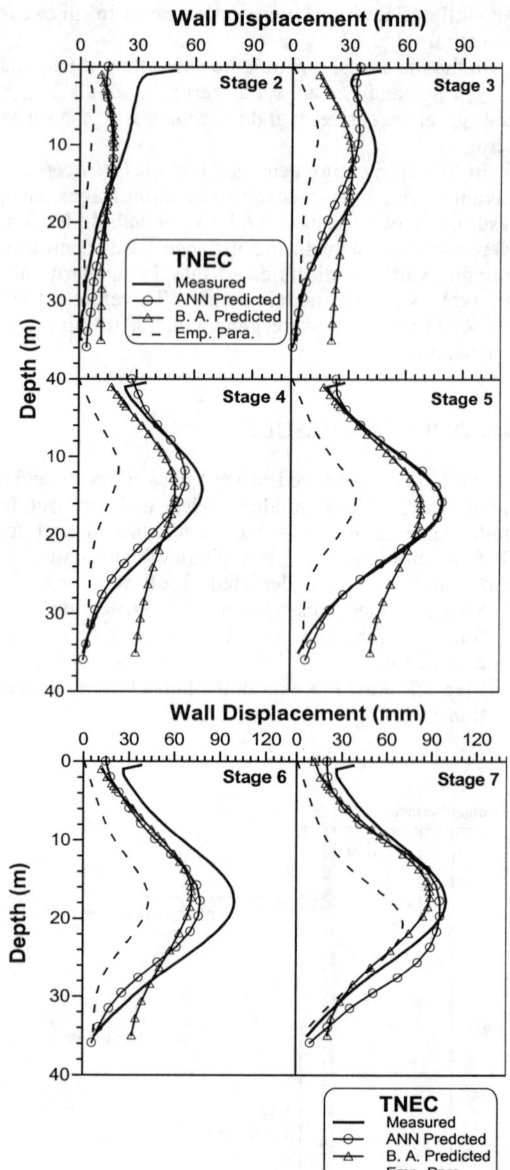

Figure 5 Comparison of Back-analysis and ANN Predictions of Case TNEC

equation

$$\frac{\partial y_k}{\partial x_i}=\sum_{j_n}\sum_{j_{n-1}}\cdots\sum_{j_1}\left(w_{j_nk}F'(Z_k)w_{j_{n-1}j_n}F'(Z_{j_n})\cdots w_{ij_1}F'(Z_{j_1})\right) \tag{5}$$

where j_n, j_{n-1},..., and j_1 denote the hidden nodes in the nth, (n-1)th, ..., and the first hidden layer, respectively. The term w_{j_nk} denotes the weight between the kth output node and the hidden node j_n. The term $w_{j_{n-1}j_n}$ denotes the weight between the hidden nodes j_{n-1} and j_n. The term w_{ij_1} denotes the weight between the ith input node and the hidden node j_1. The terms Z_k, Z_{jn} and Z_{j1} denote the weighted sums of kth output node, the hidden node j_n and j_1, respectively. In addition, the term F' denotes the differential function of F.

3. Sum the variance of training instances as a temporary variable T_{ik} of x_i and y_k.

$$T_{ik}=\sum_p\left|\left(\frac{\partial y_k}{\partial x_i}\right)_p\right| \tag{6}$$

where p denotes the pth training instances.

4. Assume I_{ik} as the importance index of x_i to y_k. I_{ik} can be calculated in the following:

$$I_{ik}=\frac{T_{ik}}{T_{\max}} \tag{7}$$

where T_{max} is the maximum sum of variance.

The results of sensitivity analysis are shown in Figure 6. The importance indexes of W, D, $\overline{N}$, R, Δ_R^{i-3}, Δ_R^{i-2} and Δ_R^{i-1} are 0.3204, 0.121, 0.1313, 0.173, 0.271, 0.3759 and 1, respectively. As indicated, the input parameters D, $\overline{N}$ and R have small influence in the prediction of diaphragm wall deflection. On the other hand, the wall deflection in last excavation stage is the most important factor for the prediction.

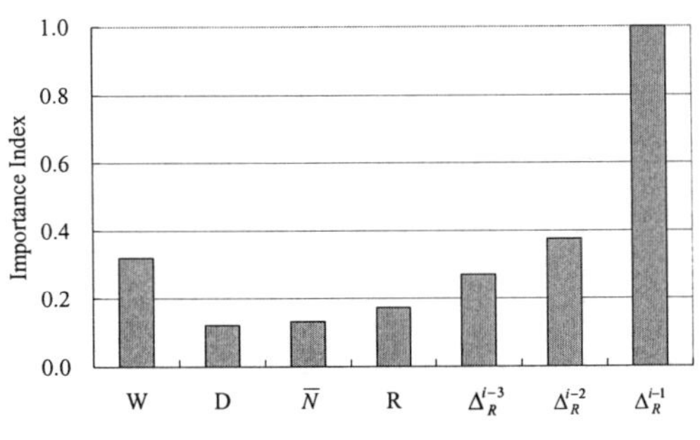

Figure 6 Results of sensitivity analysis.

Restated, the incorporation of the previous excavation stages to the knowledge representation, which reflect both the soil and construction factors for the case to be predicted, has the more important effect on the neural network prediction.

6 CONCLUSION

In this paper, the adaptive L-BFGS supervised neural network is used to predict the diaphragm wall deflection. Eighteen case histories of deep excavations with four to seven excavation stages each are employed for training and verification. The conclusions are summarized as follows:

1. This study proposes a neural network prediction for deep excavation to assess the safety of retaining system during construction stage. The advantage of using neural network prediction is that it does not need a rigorous understanding of the cause and effect relationship, which determines the behavior of the problems being modeled. Moreover, the predictions of wall deflections in deep excavations are independent on soil models.
2. The artificial neural network can reasonably predict the magnitude as well as the location of maximum deflection of diaphragm wall in deep excavation. In addition, the reliable prediction about wall deflections in the embedded position also is achieved. These results imply that the predicted wall shape matches the measured wall shape. As a result, neural network prediction allows the engineers to estimate the wall structure system prior to the next excavation stage.
3. Using wall deflections of previous three excavation stages as inputs to the network makes a more accurate prediction of the following excavation stage. Also, the importance of soil parameters is markedly reduced, as compared with other conventional approaches. The results point out that the monitoring data involving real information, which reflects the uncertainty and construction factors of field situation. Therefore, this approach can provide a reliable prediction on the possible performance of the deep excavation.

ACKNOWLEDGEMENTS

The results presented in the paper are parts of research project of Sinotech Engineering Consultants, Inc. under grant no. SEC/R-GT-99-01.

REFERENCE

Chi, S. Y., Chern, J. C., and Wang, C. C. 1999. Information construction approach for deep excavation. *Proceeding of International Symposium Geotechnical Aspects of Underground Construction in Soft Ground*: 471-476. Tokyo.

Flood, I. and Kartam, N. 1994. Neural networks in civil engineering I:principles and understanding. *Journal of computer in civil engineering*, 8 (2):131-148.
Flood, I. and Kartam, N. 1994. Neural networks in civil engineering II:principles and understanding. *Journal of computer in civil engineering*, 8 (2), 149-162.
Goh, A. T. C. 1995. Back-propagation neural networks for modeling complex systems. *Artificial Intelligence in Engineering*, 9 (1), 143-151.
Gunaratnam, D. J. and Gero, J. S. 1994. Effect of representation on the performance of neural networks in structural engineering applications. *Microcomputers in civil engineering*, 9 (2): 97-108.
Hung, S. L. and Lin, Y. L.1994. Appilcation of an L-BFGS neural network learning alforithm in engineering analysis and design. *Proceedings of the Second National Conference on Structural Engineering*, Taiwan, R. O. C. (in Chinese)
Nocedal, J. 1980. Updating quasi-newton matrix with limited storage. *Math. Compuation*, 35: 20-33.
Ou, C. Y. and Tang, Y. G. 1994. Soil parameter determination for deep excavation analysis by optimization. *Journal of the Chinese Institute of Engineers*, 17 (5): 671-688.
Peck, R. B. 1969a. Deep excavations and tunneling in soft ground. Proceeding of Seventh international conference on soil mechanics and foundation engineering. 225-290.
Vanluchene, R. D. and Sun, R. 1990. Neural networks in structural engineering. *Microcomputers in Civil Engineering*, 5 (3): 207-205.
Yang, Y. and Zhang, Q. 1997. A hierarchical analysis for rock-engineering using artificial neural networks. Rock Mechanics and Rock Engineering, 30 (4): 207-222.

Computer Methods and Advances in Geomechanics, Desai et al. (eds) © 2001 Balkema, Rotterdam, ISBN 90 5809 183 X

Estimation of air losses from tunnels driven under compressed air using neural networks

A.A.Javadi
Department of Civil and Environmental Engineering, University of Bradford, UK

ABSTRACT: This paper explores the capabilities of neural networks to predict the air losses in compressed air tunnelling. A back-propagation neural network has been trained and used to predict the air losses from the Feldmoching tunnel in Munich. In this project, compressed air was used to retain the groundwater and stabilize the tunnel face. Shotcrete was used as the temporary lining while the final permanent lining was constructed in free air. The tunnel passed through variable ground conditions ranging from coarse gravel to sand and clay. Grouting, an additional layer of shotcrete and a layer of mortar were occasionally used to control the air losses. The results of the prediction of the air losses from the tunnel using the neural network have been compared with the field measurements. Comparison of the results show that artificial neural network can be an efficient engineering tool in prediction of air losses from tunnels in compressed air tunnelling.

1 INTRODUCTION

Compressed air as a means of dewatering and stabilizing the ground has been utilized for over a hundred years and has been successfully applied to a wide variety of geological and construction circumstances.

To apply this method of tunnelling, compressed air losses must be estimated to determine the plant capacity to supply a sufficient volume of air to maintain a safe pressure whilst allowing for leakage from the tunnel. A realistic estimation of air losses can optimize the cost of air supply requirements and preparation for possible changes in air losses with tunnel advance, physical conditions and time.

Adoption of a tunnel pressure and the required volume of air to maintain that pressure is currently based on judgement or empirical formulae. There are several empirical methods for calculating tunnel pressure, P_t, and the quantity of air required, Q (Snee & Javadi, 1996 and Javadi and Snee, 1997). The accuracy of these predictive methods was discussed by Hoad and Gittoes (1992) who concluded that the methods could be appropriate as a rapid first estimate of the air losses. However, the methods do not account for real influences on air consumption such as multi-layered soils, speed of tunnelling or surface conditions.

In this paper, a neural network approach is presented for prediction of air losses from tunnel face and perimeter walls. Field data from the Feldmoching tunnel in Munich has been used to train and test the network. The results of the neural network prediction are compared to the measured data. It is shown that neural network can be used as an efficient tool in prediction of air losses from tunnels in compressed air tunnelling.

1.1 *Factors Affecting Air Losses*

The total amount of air lost from a tunnel, Q_{total}, is the sum of three main sources:

- Air loss from the tunnel face, Q_{face},
- Air loss from tunnel perimeter walls, Q_{wall},
- Other losses (including air locks, drainage, and ventilation), Q_{other}

$$Q_{Total} = Q_{face} + Q_{wall} + Q_{other} \tag{1}$$

The third item can be approximated from the size of the air lock, the pressure and frequency of use. However, it is not significant compared to face losses occurring in open ground (Hoad & Gittoes, 1992) or losses from the tunnel perimeter walls.

The main factors influencing the air losses from the tunnel face and perimeter walls are the permeability of shotcrete and the ground (which is in turn a function of degree of saturation of the soil). The permeability of shotcrete varies as it cures, so this time-dependency should be taken into account when analyzing the air losses in a tunnel where shotcrete is used as a primary or permanent support.

1.2 *Current practice in design for compressed air tunnelling*

An appropriate air pressure must be applied in the tunnel to balance the groundwater pressure and, due to leakage of air from the tunnel, a constant recharge of air must be supplied to maintain that pressure. If air leakage increases, a greater volume of air must be supplied or measures taken to reduce the air losses. However, in its most simplistic form, compressed air as a means of groundwater control, is justified on the premise that if the air pressure in the tunnel is equal to the pressure of the water in the ground, a state of equilibrium should be created.

As well as changing the effective stress conditions, there can be a secondary positive effect of reducing ground settlement. The air pressure reacts against the radial stresses at the tunnel wall which creates uplift of the periphery and supports the face. Furthermore, there is an overall reduction in the total load on the tunnel due to a lowering of the density of the ground (Strobl, 1986).

Figure 1 shows the relative pressures for an air pressure set at the tunnel invert. The resulting over pressure at the tunnel crown, P_r, is the difference between the two pressures. As long as there is no significant flow of water towards the tunnel face, the ground should achieve a satisfactory effective strength for stability at air pressures less than the full head of water.

However, it is common for layers of soil with different geological origins and properties to be encountered in driving tunnels. This inhomogenity is compounded by the flow of air through the ground which may cause some changes in the structure, state of stress, properties and the strength of the soil. Also a region of the soil is dewatered by the air flow, so the properties of the soil in this region will be different to those in the saturated soil. These conditions result in a very complex geotechnical environment.

2 NEURAL NETWORK

A neural network is an information processing technique based on the way biological nervous systems, such as brain, process information. Neural networks are composed of a large number of highly interconnected processing elements, or neurones, arranged generally in two or more layers. These layers include an input layer, a number of hidden layers and an output layer. Neurones in hidden layers are used to find associations within the input data and extract patterns than can provide meaningful outputs.

A neural network system uses the human-like technique of learning by examples to resolve problems. Just as in biological systems, learning involves adjustments to the synaptic connections that exist between the neurones.

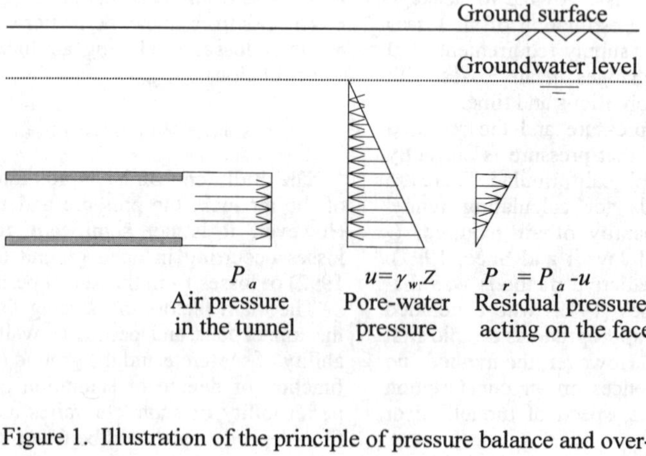

Figure 1. Illustration of the principle of pressure balance and over-pressure

The output of each neurone, responding to a particular combination of inputs, has an influence (or weight) on the overall output. Weighting is controlled by the level of activation of each neurone, and the strength of connection between individual neurones. Patterns of activation and interconnection are adjusted to achieve the desired output from the training data. Corrections are based on the difference between actual and desired output, which is computed for each training cycle. If average error is within a prescribed tolerance the training is stopped, the weights are locked in and the network is ready to use.

Back-propagation is the most widely used neural network paradigm and the one used in this study.

Figure 2 shows the structure of a back-propagation neural network. Each layer is composed of several neurones and these are fully connected to neurones of the succeeding layer.

A three-layer propagation back- network was used in this study. The input layer had seven neurones namely the tunnel length, L, equivalent permeability of the soil layers over depth of the tunnel, K / D, the product of the area of each soil layer at the tunnel face by its permeability, $K_i.A_i$, applied air pressure in the tunnel, P_{air}, and the perimeter area per unit length of the tunnel, P. The output layer had a single neurone representing the air loss, Q.

3 CASE STUDY

The Feldmoching tunnel, U8 N-8 in Munich, Germany, was used as a case study. The construction of the 635 meter long Feldmoching tunnel, U8 N-8, was started in July 92 and finished in February 94. The tunnel was constructed using a top heading and bench method. Compressed air was used to control the groundwater and shotcrete was used as temporary support. The final concrete lining was installed in free air after completion of the tunnel.

The excavation of the first few meters of the tunnel was started with open trench and sheet piling which provided access to the underground work. The first 50m of the tunnel length was constructed using a twin sidewall drift method. From 50m to 320m, the tunnel was constructed by a single sidewall drift method and from 400m to 635m, the tunnel was converted to twin tunnels. Detailed information on the geometry of the tunnel, tempo-

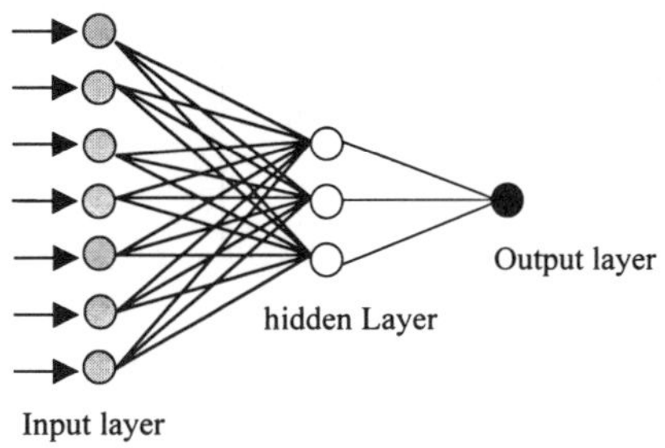

Figure 2. Architecture of a neural network

rary and permanent lining and construction methods can be found in Javadi (1997) and Javadi et al. (1999).

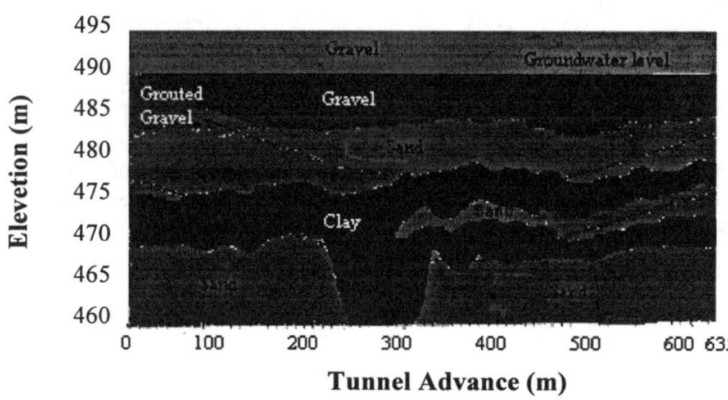

Figure 3. Geology of the tunnel route

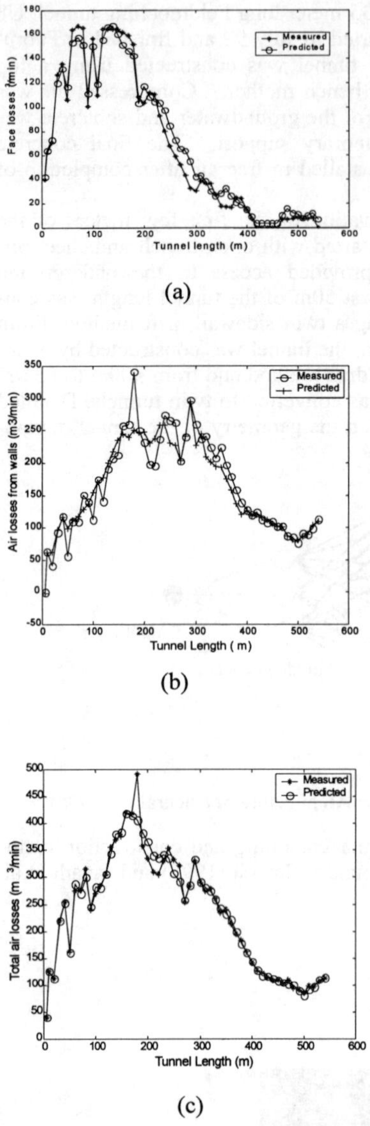

Figure 4. Typical simulation of training data (a) air losses from face (b) air losses from walls, (c) Total air losses

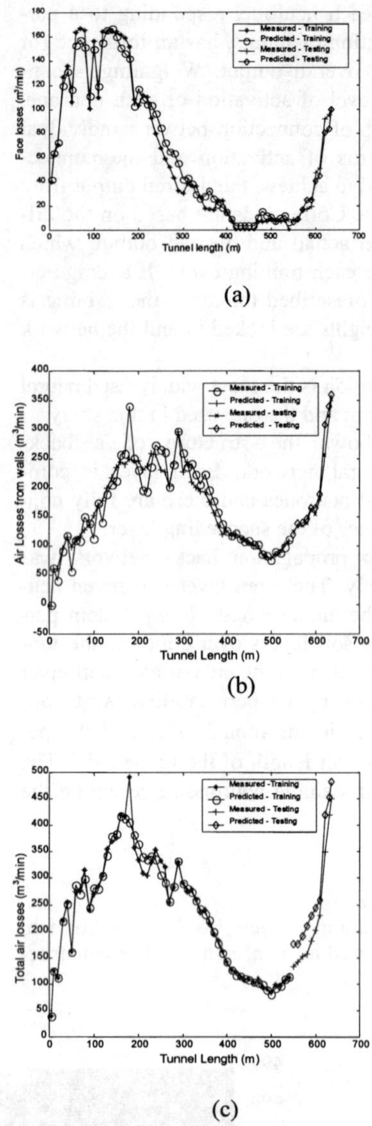

Figure 5. Simulation of training and testing data (a) air losses from face (b) air losses from walls, (c) total air losses

Figure 3 is a geological section of the tunnel route. The soil layers comprise a coarse gravel layer near to the ground surface underlain by sand. Beneath the sand is a layer of low permeability silty clay underneath which lies another layer of sand. Also, there is a lens of sand in the silty clay layer between chainage 300m and 635m. The groundwater level is approximately 5 meters below the ground surface. At the beginning of excavation work, the tunnel face was entirely in sand and gravel layers, so high air losses were expected and grouting was carried out to reduce the permeability. The tunnel face advanced into the lower permeability clay layer to approximately 300m where the whole face was in clay and the air losses were reduced. From about 470m, the tunnel face gradually moved out of the clay layer and, consequently, the air losses increased.

3.1 *Training and testing*

Field data from the Feldmoching tunnel was used for training and testing of the network. The data include the input parameters affecting the air losses and the corresponding measured air losses from the tunnel.

The network was trained using the back-propagation algorithm. The data for the tunnel lengths of 0 to 540 meters was used for training the network. The data for the tunnel lengths between 550m to 635m was used for testing.

4 RESULTS

The neural network was trained and tested for three separate cases for prediction of the air losses from the tunnel face, from the tunnel perimeter walls and the total air losses. At the end of the training phase, the measured air losses were compared to those predicted by the neural network. Comparison of the measured and predicted air losses with the tunnel advance for the three cases, is shown in figure 4 for the data used in the training phase. The figure shows that the neural network is capable of capturing and predicting the main features of the variation of the air losses with change in the ground conditions for all the three cases with a high accuracy.

The next step was to examine the robustness of the trained network in generating correct predictions for a new set of data, not already used during the training phase. The trained network was used to simulate the air losses from the tunnel for the tunnel lengths between 550m to 635m. The data from this section was not introduced to the network during its training. Furthermore, this section was notably different from the section used for training in terms of the tunnel geometry and ground conditions. Figure 5 shows the comparison between the measured air losses and those predicted by the neural network for the training and testing data for the three cases. The comparison shows that the trained neural network is also capable of prediction the air losses for the geological conditions not experienced during the training of the network.

5 CONCLUSION

A back-propagation neural network was trained and used to simulate the air in compressed air tunnelling. The network was applied to a case study and the results of the neural network prediction have been compared with the field measurements. It was shown that neural network is capable of simulating, with high accuracy, the air losses from tunnels driven under compressed air in complex geological conditions. It can be concluded that neural network can be an efficient engineering tool in prediction of air losses from tunnels in compressed air tunnelling.

6 ACKNOWLEDGEMENT

The authors would like to acknowledge their gratitude to Professor H. J. Bösch of the Technical University of Munich for his valuable co-operation in providing the field data.

7 REFERENCES

Hoad, R. S. and Gittoes, G. P. 1992. Estimation of compressed air losses. *Proceedings Conf. Engineering and Health in Compressed Air Work, CIRIA*, 157-175.

Javadi, A. A. and Snee C. P. M. 1997. New method for prediction of air losses in compressed air tunnelling. *Int. Congr. World Tunnelling'97, London, 57-69.*

Javadi, A. A. 1997. Compressed air tunnelling; geotechnical aspects and numerical modelling, Ph.D thesis, University of Bradford, UK. Javadi, A. A. 1997. Compressed air tunnelling; geotechnical aspects and numerical modelling, Ph.D thesis, University of Bradford, UK.

Javadi, A. A., Farmani, R., Toropov, V. V. and Snee, C. P. M. 1999. Identification of parameters for air permeability of shotcrete tunnel lining using a genetic algorithm. *Computers and Geotechnics,* vol. 25, No. 1, pp. 1-24.

Snee C. P. M. and Javadi, A. A. 1996. Prediction of compressed air leakage from tunnels. *Tunnelling and Underground Space Technology*, vol. 11, No. 2, 189-195.

Strobl, B. 1986. NATM under compressed air - geotechnical aspects and numerical modelling. *Proc. Int..Conf. Large Underground Openings, Firenze, Italy,* 686-692.

Computer Methods and Advances in Geomechanics, Desai et al. (eds) © 2001 Balkema, Rotterdam, ISBN 90 5809 183 X

Site characterization using generalized regression neural networks

C.H. Juang & T. Jiang
Clemson University, S.C., USA

ABSTRACT: Site characterization is an important and challenging task in geotechnical engineering practice. The ultimate goal in site characterization is to be able to estimate in-situ soil properties at any space point at a site based on limited tests. This paper presents a new approach, based on artificial neural network, for site characterization. Emphasis is placed on use of generalized regression neural networks in one- and two-dimensional site characterization.

1 INTRODUCTION

The ultimate goal in site characterization is to be able to estimate the in-situ soil properties at any space point at a site based on limited number of tests. In the conventional approach, engineers characterize the site based on limited test results and interpret them in terms of working soil profiles that are a very simplified model for the in-situ soil properties. The task of establishing a working soil profile or profiles consists of two main steps: (1) generalization of results from limited tests, and (2) estimation of design soil parameters based on generalized test results. Depending on the size of the project, knowledge of local geology and experience, engineers may incorporate different degrees of conservatism in these two steps. With this approach, however, the uncertainty in the estimated design soil parameters is not quantified.

Mapping and predicting in-situ soil properties is a very challenging task (Tabba&Yong, 1981a, b). Uncertainty about the properties of in-situ soils may come from various sources including spatial variability, measurement noise, measurement and model bias, and statistical error due to limited measurements (Baecher, 1986). To better characterize in-situ soil properties and to provide a basis for subsequent probabilistic analysis, probability theory and geostatistics have been attempted. The classical work on random field for probabilistic site characterization has been documented by VanMarcke (1983). Characterization spatial variability of soils using geostatistics has been reported (Soulie et al. 1990; Chiasson et al. 1995; DeGroot, 1996). Random field theory and geostatistics in site characterization using the National Geotechnical Experimentation Sites (NGES) data have been reported (Fenton, 1998; Kulatilake, 1998; Wu, 1998).

The approach taken by exiting probabilistic and geostatistics methods for site characterization may be viewed as a calibration approach. Limited soil test data are used to calibrate a theory or model, i.e., to estimate some undermined coefficients in the model or to derive some statistical parameters that characterize the site. In the present study, a different approach to site characterization based on artificial neural network (ANN) is taken.

2 GENERALIZED REGRESSION NEURAL NETWORK

Generalized regression neural network (GRNN) is a kind of artificial neural network, which has three layers: the input layer, the hidden layer and the output layer. In a GRNN, the hidden layer is a nonlinear local mapping. This layer contains radial basis neurons that use a Gaussian transfer function. This function is centered over various receptive fields in the input space. Receptive fields are zones in the input space, which activate the local neurons. If an input vector lies in a receptive field, corresponding to which the hidden neurons will be activated. If an input vector lies outside a receptive field, the corresponding hidden neuron will not be activated.

In a GRNN design, each training vector has a corresponding neuron in the hidden layer. The weight matrix W_H in the hidden layer is simply a collection of all training vectors presented to it. When a new input vector is presented to the network, the similarity in terms of distance between this new input vector with each of those stored in W_H is calculated:

$$dist = |p - W_H^j|, \quad \text{for } j = 1, Q \tag{1}$$

where Q = number of neurons in the hidden layer; p = input vector; $dist$ = distance between p and W_H.

The calculated distance is then adjusted by the bias b_i, which is essentially a spread parameter in the transfer function of the hidden neuron. This is done with the following equations:

$$b_i = 0.8326/s \tag{2}$$

$$n_1 = dist \times b_i \tag{3}$$

where n_1 = the adjusted distance, and s = the spread, which is a parameter in the Gaussian transfer function. Note that the spread in the GRNN is equivalent to the autocorrelation distance in a spatial analysis.

The output of the hidden neuron, a_1, is:

$$a_1 = e^{[-(n_1)^2]} \tag{4}$$

The output of the output layer is:

$$n_2 = W_O \times a_1 \tag{5}$$

where W_O = weight matrix in the output layer.

In designing GRNN model, the most important task is to select a proper spread (s). The procedure for selecting the spread is outlined in this paper for one- and two-dimensional site characterization. As previously mentioned, the spread is equivalent to the autocovariance distance in a spatial analysis or a sill distance in geostatistics.

In a spatial analysis, the square exponential function is commonly used as the autocovariance function:

$$C_x(r) = V(x)e^{(-r/r_0)^2} \tag{6}$$

where $C_x(r)$ = autocovariance at a distance r; $V(x)$ = variance of the data set; r_0 = autocovariance distance at which $C_x(r)$ decays to $C_x(r)/e$.

In geostatistics, the spherical model is commonly used:

$$C_x(r) = V(x)[1 - (3/2)(r/a) + (1/2)(r/a)^3] \tag{7}$$

where a is the sill distance beyond which the measured data are no longer correlated.

The relationships between the spread used in GRNN and the autocovariance distance, and between the spread and the sill distance, may be described below:

$$s \approx 0.2743\, a \tag{8}$$

$$s = 0.8326\, r_0 \tag{9}$$

3. SITE CHARACTERIZATION

3.1 *Problem Description*

Figure 1 shows the layout of 23 CPTs at the "sand" site of the Texas A&M University's National Geotechnical Experimentation Site (NGES). At this site, the soil from the ground surface to the depth of about 12m is predominately sands. Below the sand layer is a very hard plastic clay extended to a depth of at least 33m. For each CPT sounding, cone tip resistance (q_c) and sleeve friction (f_s) are continuously measured. In this study, only q_c is considered in the site characterization.

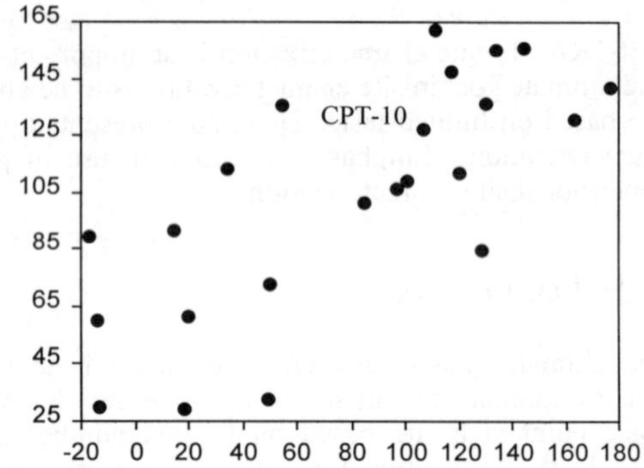

Figure 1. Layout of CPT soundings

3.2 *One-Dimension Analysis*

One-dimensional site characterization here refers to characterization of the site in vertical direction only. For a given CPT sounding, the x- and y-coordinates of the "hole" location are fixed, thus, $q_c = q_c(z)$, where z is the depth of a measured q_c data point. In the present study, GRNN is used to approximate the measured q_c-z relationship.

The spread, s, is determined using the procedure described below:

1) Divide the data of a CPT sounding into two subsets in a ratio of, roughly, 2:1 for training and testing, respectively.
2) Begin the training of the intended GRNN by assuming a small initial spread value, typically, 0.1m.
3) Evaluate the trained GRNN using the testing data set. This is done simply by comparing the GRNN-predicted q_c value with the known q_c value for each case in the testing data set. A root-mean-error-square (RMES) of all cases is calculated:

$$RMES = \sqrt{\sum_{i=1}^{n} \varepsilon^2 \Big/ n} \tag{10}$$

where ε = the difference between the known q_c value and predicted q_c value, and n = the number of cases in the testing data set.

4) Repeat Steps 2 and 3 a number of times using a gradually increased spread value. With each adopted spread, a GRNN is trained and used to predict q_c values for each case in the testing data set, and a RMES is then calculated.
5) Plot the RMES versus the corresponding spread value. The spread that yields a minimum RMES is considered as the optimum spread.

As an example, a CPT sounding is analyzed herein. Table 1 shows the training and testing data sets. These data are taken from a CPT sounding record. Here, the CPT sounding records are divided into segments of three data entries, and the first two entries are placed in the training data set, while the

Table 1. Data of sounding CPT-10

	No.	Depth m	q_c MPa	Predicted q_c MPa
Training	1	2.06	8.15	8.81
	2	2.32	9.73	8.91
	4	2.95	7.95	8.58
	5	3.23	9.01	8.71
	7	3.90	9.75	9.43
	8	4.14	9.22	9.43
	10	4.77	9.06	9.62
	11	4.99	10.21	9.84
	13	5.62	11.89	9.63
	14	5.93	6.37	8.57
	16	6.89	2.74	3.07
	17	7.18	2.60	4.16
	19	7.81	11.30	11.39
	20	8.01	14.67	12.16
	22	8.70	8.18	8.67
	23	8.96	7.92	7.95
	25	9.62	6.20	6.36
	26	9.91	6.12	6.00
	28	10.61	3.42	5.41
	29	10.86	7.69	5.86
	31	11.78	8.02	7.53
	32	12.07	6.89	7.38
Testing	3	2.63	7.30	8.76
	6	3.49	9.50	9.06
	9	4.39	11.24	9.42
	12	5.27	11.60	10.10
	15	6.17	8.96	7.31
	18	7.47	10.48	7.51
	21	8.38	9.49	10.73
	24	9.26	6.76	7.20
	27	10.18	2.05	5.54
	30	11.54	8.05	7.54

third entry is placed in the testing data set. Following the above procedure, a plot of the calculated RMES versus the spread is obtained, as shown in Figure 2. For this example, a spread of 0.44 m is determined. This optimum spread is then used to train the intended GRNN. Figure 3 shows the predicted q_c values compared to the measured q_c values in the vertical direction. The good agreement shown in Figure 3 indicates that the trained GRNN is capable of approximating the measured q_c values in a CPT sounding.

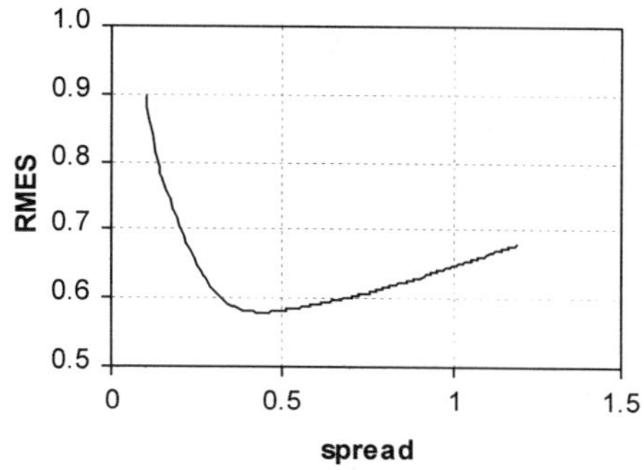

Figure 2. Searching process for spread (CPT-10)

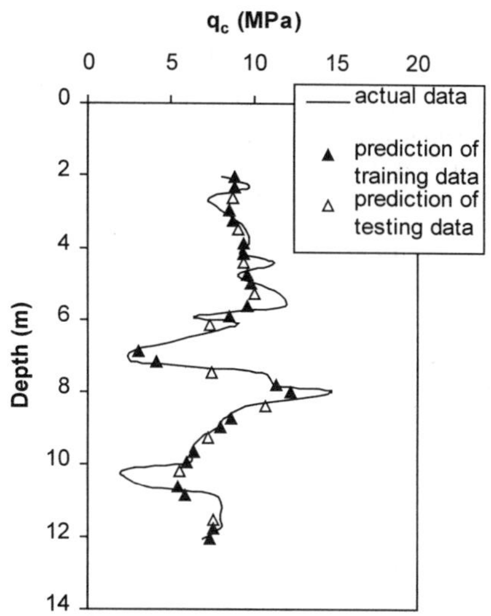

Figure 3. Comparison of actual data with predicted data

3.3 *Two-dimensional analysis*

In the two-dimensional site characterization, the z-coordinate is fixed, thus, $q_c = q_c(x,y)$. GRNN is employed to generalize this relationship from the measured data.

As an example to describe the procedure that is developed for two-dimensional site characterization, consider a set of measured q_c values at the depth of 4 m from the 23 CPTs. At all other depths, the site can be characterized in the same way. The data taken from the depth of 4 m are shown in Table 2.

Table 2. Data at depth = 4 m

No.	x m	y m	q_c Mpa
CPT1	84.63	101.10	9.33
CPT5	128.75	84.63	5.49
CPT6	130.07	136.32	11.20
CPT7	49.42	32.94	7.20
CPT8	144.27	155.75	8.44
CPT9	111.33	162.45	7.60
CPT10	106.78	127.23	7.51
CPT11	120.42	111.90	12.14
CPT12	49.98	72.70	4.82
CPT14	20.11	61.34	5.50
CPT17	100.57	109.03	8.97
CPT18	96.80	106.00	3.22
CPT21	163.13	130.72	11.72
CPT22	177.26	142.20	7.02
CPT23	18.22	29.44	8.12
CPT24	-12.89	30.12	3.74
CPT25	-14.21	60.14	9.64
CPT26	-16.87	89.34	3.25
CPT27	14.78	91.29	10.62
CPT28	34.23	113.53	6.68
CPT29	54.88	135.30	11.17
CPT31	117.66	147.41	8.89
CPT41	133.91	155.21	9.18

As in the one-dimensional analysis, the first step in establishing a GRNN model to approximate the function $q_c = q_c(x,y)$ is to determine a spread. The procedure for this task is outlined below:

1) Set up a grid according to the layout of CPTs at the site to be characterized. The spacing of the grid points depends on the distance between CPTs, and the appropriate spacing may be obtained by trail-and-error process. For the layout of the 23 CPTs used in this study, the spacing of the grid is selected to be 40 ft. The grid contains 5 rows and 6 columns and the total number of grid points is 30.
2) Train a GRNN using the 23 q_c values shown in Table 2. Initially, a small value is selected for spread for training GRNN.
3) The trained GRNN, in the form of $q_c = q_c(x,y)$, is then used to "predict" q_c at each grid point in the grid selected in Step 1.
4) Using the obtained q_c at each grid point and the same spread as the one used in Step 2 to train a GRNN, and then using the newly trained GRNN to predict q_c at each CPT location. The difference (or error) between the predicted and measured q_c at each CPT (at the depth of 4 m) is determined, and finally a RMES is calculated based on the errors at the 23 CPTs.
5) Repeat Steps 2, 3, and 4 a number of times using a gradually increased spread value. With each adopted spread, a RMES is calculated, and thus, a plot of the RMES versus the corresponding spread value can be made. The spread that yields a minimum RMES is considered as the optimum spread.

For the example data set analyzed, the optimum spread is 13 m and the corresponding RMES is 0.38 MPa (Figure 4). The obtained RMES is relatively small, compared to the range of the q_c values (ranging from 3.22 to 12.14 MPa) and the standard variation of the data set (2.66 MPa).

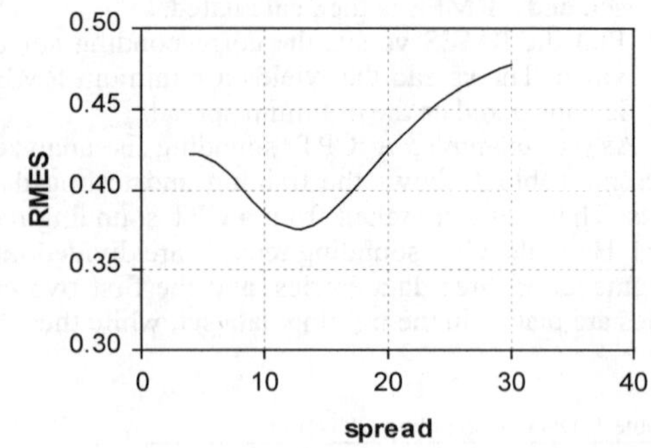

Figure 4. Searching process for spread at z = 4 m

Note that the validity of the above procedure is based on the premise that if the spread is properly selected, the q_c values at the grid points can be predicted based on the GRNN trained with the q_c values at the actual CPT locations, and vice versa.

Recall that the spread in a GRNN is related to the autocovariance distance. Based on the optimum spread determined above, its corresponding autocovariance distance is about 16 m, which is in the range of 15 m to 30 m generally accepted in geostatistics (DeGroot, 1996).

Figure 5 shows a contour of the predicted q_c values on the z = 4 m plane prepared based on the data

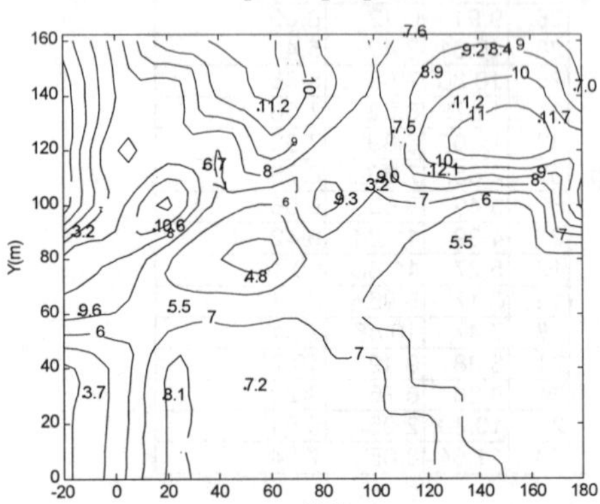

Figure 5. Contour of the predicted q_c on the z = 4m plane with 23 measured q_c values

generated from the trained GRNN. The q_c values at the 23 CPTs at the depth of 4 m are also shown.

4. CONCLUDING REMARKS

This paper provides a brief introduction of generalized regression neural network and how it works in generalizing soil properties based on limited in-situ tests. Using a suitable spread, GRNN can be trained to approximate one-dimensional relation, $q_c = q_c(z)$, as well as two-dimensional relation, $q_c = q_c(x,y)$. The procedures for determining an optimum spread for one- and two-dimensional site characterization are outlined in the paper. These results provide a basis for three-dimensional site characterization using GRNN. One possible way for training GRNN to approximate the relation, $q_c = q_c(x,y,z)$, in a three-dimensional site characterization is to adopt the spread determined from the two-dimensional analysis, coupling with a transformation of the z-direction distance based on the spread ratio. This is the subject of an ongoing research project at Clemson University.

5. ACKNOWLEDGEMENTS

This study is supported by the National Science Foundation through Grant No. CMS-9610103. This financial support is greatly appreciated. Mr. P.C. Lu participated in the early stage of this study and his contribution is appreciated.

6. REFERENCES

Baecher, G.B. (1986), Geotechnical error analysis. *Trans. Research Record* 1105, pp. 23-31

Briaud, J.L. (1997), *The National Geotechnical Experimentation Sites at Texas A&M University: Clay and Sand*, Report No. NGES-TAMU-007, September 1997, Department of Civil Engineering, Texas A&M University, College Station, Texas.

Chen, S., Cowan, F.N., and Grant, P.M. (1991), Orthogonal least squares learning algorithm for radial basis function networks, *IEEE Transaction on Neural Networks*, Vol. 2, No. 2, March 1991, pp. 302-309.

Chiasson, P., Lafleur J., Soulie M., and Law, K.T. (1995), Characterizing spatial variability of a clay by geostatistics, *Canadian Geotechnical Journal*, Vol. 32, pp. 1-10.

DeGroot, D.J. (1996), Analyzing spatial variability of in situ soil properties, *Uncertainty in the Geologic Environment: from Theory to Practice, Proceedings of Uncertainty '96, ASCE Geotechnical Special Publication* No. 58, pp. 210-238.

Demuth, H.B. and Beale, M. (1997), *Neural Network Toolbox, User's Guide*, The MathWorks, Inc., Natick, MA.

Fenton, G.A. (1998), Random field characterization of NGES data, paper presented at the *Workshop on Probabilistic Site Characterization at NGES*, Seattle, Washington, August 2, 1998.

Kulatilake, P.H.S.W. and Um, J. (1998), Spatial variation analysis using cone penetrometer data for the clay site at Texas A&M Riverside Campus, paper presented at the *Workshop on Probabilistic Site Characterization at NGES*, Seattle, Washington, August 2, 1998.

Phoon, K.K. and Kulhawy, F.H. (1996), On quantifying inherent soil variability, *Uncertainty in the Geologic Environment: from Theory to Practice, Proceedings of Uncertainty '96, ASCE Geotechnical Special Publication* No. 58, pp. 326-340.

Rumelhart, D.E., Hinton, G.E., and Williams, R.J. (1986), Learning internal represntations by error propagation, in D.E. Rumelhart and McClelland (eds), *Parallel Distributed Processing*, Vol. I, Chapt. 8, MIT Press, Cambridge, MA, pp. 318-362.

Soulie, M., Montes, P., and Silvestri, V. (1990), Modeling spatial variability of soil parameters, *Canadian Geotechnical Journal*, Vol. 27, pp. 617-630.

Specht, D.F. (1991), A general regression neural network, *IEEE Transactions on Neural Networks*, Vol. 2, No. 6, pp. 568-576.

Tabba, M.M. and Yong, R.N. (1981a), Mapping and predicting soil properties: theory, *Journal of Engineering Mechanics, ASCE*, Vol. 107, No. EM5, pp. 773-794.

Tabba, M.M. and Yong, R.N. (1981b), Mapping and predicting soil properties: applications, *Journal of Engineering Mechanics, ASCE*, Vol. 107, No. EM5, pp. 795-811.

Tsoukalas, L.H. and Uhrig, R.E. (1997), *Fuzzy and Neural Approaches in Engineering*, John Wiley & Sons, Inc., New York, NY, pp587.

VanMarcke, E.H. (1983), *Random Fields: Analysis and Synthesis*, MIT Press, Cambridge, MA.

Wasserman, P.D. (1993), *Advanced Methods in Neural Computing*, Van Nostrand Reinhold, New York, NY.

Wu, T.H. (1998), Variations in clay deposits of Chicago, paper presented at the *Workshop on Probabilistic Site Characterization at NGES*, Seattle, Washington, August 2, 1998.

Computer Methods and Advances in Geomechanics, Desai et al. (eds) © 2001 Balkema, Rotterdam, ISBN 90 5809 183 X

Overconsolidation ratio of Norwegian clays predicted by neural networks

P.U. Kurup & N.K. Dudani
University of Massachusetts, Lowell, Mass., USA

ABSTRACT: This paper examines the feasibility of using neural networks to estimate the overconsolidation ratio (OCR) of clays from piezocone penetration test (PCPT) data. A feed-forward neural network model, trained by back-propagation, was developed using the following five input variables: cone resistance, total vertical overburden pressure, pore water pressure at the cone tip, pore water pressure measured just above the cone base (reference location), and the hydrostatic pore water pressure. The model was trained using partial set of PCPT data from four Norwegian geotechnical test sites (Bakklandet, Troll east, Glava and Valoya clays). The remaining data was used to test and validate the model. The neural network model was found to give very reliable estimates of OCR when compared to the OCR predictions made by some of the existing methods.

1 INTRODUCTION

The piezocone penetrometer has become widely popular as an in situ investigation tool of choice for subsurface profiling, and determining engineering soil properties. The overconsolidation ratio (OCR) defined as the ratio of the maximum effective preconsolidation stress and the existing effective overburden stress, is an important factor governing the strength, stress-strain behavior, and the compressibility characteristics of fine-grained soils. Knowledge of OCR is hence essential in selecting relevant soil parameters for the design of geotechnical, geomechanical and pavement systems. In recent years, the piezocone penetration test (PCPT) has become very popular in estimating the OCR of cohesive soil deposits. It is faster, economical and gives continuous OCR profiles compared to the laboratory oedometer test.

Currently a number of models exist to predict OCR from PCPT data. Some are empirical or semi-empirical in nature, and many others have a theoretical background based on cavity expansion solutions or bearing capacity theories. This paper examines the feasibility of using neural networks to estimate the overconsolidation ratio of clays from PCPT data. The neural network was trained and tested using PCPT data from four well-referenced (Sandven 1990) Norwegian geotechnical test sites (Bakklandet, Troll east, Glava and Valoya clays).

2 NEURAL NETWORK MODELING

2.1 *Overview of neural networks*

Artificial neural networks (ANN) are data processing paradigms that work similar to the brain in processing information (Khanna 1989, Dowla and Rogers 1995, Hagan et al. 1999). Just like the neurons in the human brain, neural networks are made of a number of interconnected processing units (neurons) organized in layers. Neural networks have been found to be very useful in learning complex relationships between multidimensional data. A particular strength of ANN is its relative tolerance to noisy and fuzzy data that makes it more robust and flexible than mathematical models. Many types of neural network exist. These neural networks differ in the topography or architecture and the rules of learning and self-organization. Back-propagation neural networks are one of the most widely used types of ANN because of its simplicity and robustness. A feed-forward network, trained by back-propagation, was used in this research.

A typical back-propagation neural network consists of processing units (neurons) organized in layers. The connection between the neurons in the different layers may be as shown in Figure 1, where the output from one neuron is one of the inputs to all the neurons in the next layer and the inputs are the outputs from all the neurons in the previous layer. With each connection is associated

Table 1. Training Data for the Neural Network Model

Site	Depth (meters)	σ_v (kPa)	q_t (kPa)	u_1 (kPa)	U_2 (kPa)	u_0 (kPa)	Reference OCR	ANN Prediction
Glava	7	126	960	602	448	63.77	5.8	5.8
Glava	8	144	96	653	493	73.58	5.2	5.2
Glava	9	162	1019	712	549	83.39	5.4	4.4
Glava	10	180	1054	737	590	93.20	4.2	4.1
Glava	11	198	1078	763	610	103.01	4.1	4.0
Glava	12	216	1086	782	643	112.82	3.7	3.8
Glava	14	252	1197	850	704	132.44	5.9	5.9
Glava	15	270	1193	895	732	142.25	3.0	3.0
Glava	16	288	1215	895	750	152.06	3.6	3.6
Glava	17	306	1284	940	773	161.87	2.6	2.6
Troll East Area	4	72	157	178	127	39.24	2.1	2.0
Troll East Area	5	90	200	213	160	49.05	1.9	2.0
Troll East Area	6	108	231	255	187	58.86	1.8	1.9
Troll East Area	7	126	281	295	209	68.67	1.7	1.7
Troll East Area	8	144	343	321	239	78.48	1.6	1.6
Troll East Area	9	162	356	344	274	88.29	1.5	1.5
Troll East Area	10	180	398	388	288	98.10	1.5	1.5
Troll East Area	11	198	457	444	322	107.91	1.5	1.4
Troll East Area	12	216	505	491	354	117.72	1.4	1.4
Troll East Area	13	234	553	540	387	127.53	1.4	1.4
Troll East Area	14	252	600	580	418	137.34	1.4	1.4
Troll East Area	15	270	636	642	444	147.15	1.4	1.4
Troll East Area	16	288	654	690	458	156.96	1.4	1.4
Valoya	5	90	1539	455	364	19.62	5.7	5.7
Valoya	6	108	1596	827	524	29.43	9.9	9.9
Valoya	7	126	1387	884	588	39.24	11.4	11.0
Valoya	8	144	1682	932	615	49.05	5.6	5.6
Valoya	9	162	1683	997	666	58.86	10.2	10.2
Valoya	10	180	1756	1019	623	68.67	8.0	8.0
Valoya	11	198	1671	672	615	78.48	8.5	8.4
Valoya	12	216	1660	982	710	88.29	7.1	7.4
Valoya	13	234	1650	1053	710	98.10	5.9	5.9
Valoya	14	252	1757	1107	732	107.91	6.3	6.3
Valoya	15	270	1992	1180	788	117.72	7.1	7.1
Bakklandet	7	126	878	578	486	29.43	3.7	3.8
Bakklandet	8	144	942	640	516	39.24	3.0	3.8
Bakklandet	10	180	956	617	521	58.86	3.5	3.4
Bakklandet	11	198	977	617	537	68.67	2.8	2.8
Bakklandet	12	216	1012	725	598	78.48	2.2	2.3
Bakklandet	14	252	1040	713	608	98.10	2.5	2.5
Bakklandet	15	270	1251	826	699	107.91	1.9	1.7
Bakklandet	16	288	1132	764	679	117.72	2.3	2.3
Bakklandet	17	306	1425	809	710	127.53	1.9	1.9

a modifiable weight (models the synapse in the brain). Each neuron transforms the weighted sum of the inputs into a single outgoing activity that it transmits to all other neurons in the next layer. A transfer function (sigmoid function) is used to transform the input values to the hidden layer neurons.

ANNs, like people, learn by examples. Training of a neural network is conducted by presenting a series of example pattern of associated input and output values. Initially when a network is created the connection weights are set to random values (random numbers). As the training sets of inputs and outputs are presented, the weights are automatically modified by the adopted learning rule (back-propagation) until the ANN gives the desired output. Once the ANN is trained, the prediction mode simply consists of propagating the data through the network, giving immediate predictions.

2.2 *A Neural Network model for estimating OCR*

The following five input variables were used for

Table 2. Testing Data for the Neural Network Model

Site	Depth (meters)	σ_v (kPa)	q_t (kPa)	u_1 (kPa)	U_2 (kPa)	u_0 (kPa)	Reference OCR	ANN Prediction
Glava	3	54	642	286	229	25	7.4	7.4
Glava	4	72	747	433	366	34	7.2	7.2
Glava	5	90	833	534	427	44	5.4	5.7
Glava	6	108	875	489	386	54	6.3	6.3
Troll East Area	2	36	115	124	86	20	3.4	2.0
Troll East Area	3	54	142	129	101	29	2.4	2.0
Troll East Area	17	306	700	754	521	167	1.4	1.4
Troll East Area	18	324	763	565	465	177	1.4	1.4
Valoya	16	288	1807	1185	829	128	6.5	3.9
Valoya	17	306	2119	1450	1020	137	5.7	5.1
Valoya	18	324	1974	1355	949	147	4.5	3.3
Valoya	19	342	2373	1377	1010	157	2.6	1.4
Bakklandet	5	90	756	567	430	10	3.3	3.8
Bakklandet	6	108	915	584	460	20	4.7	3.8
Bakklandet	18	324	1459	967	872	137	2.1	1.6
Bakklandet	19	342	1476	967	872	147	1.8	2.0

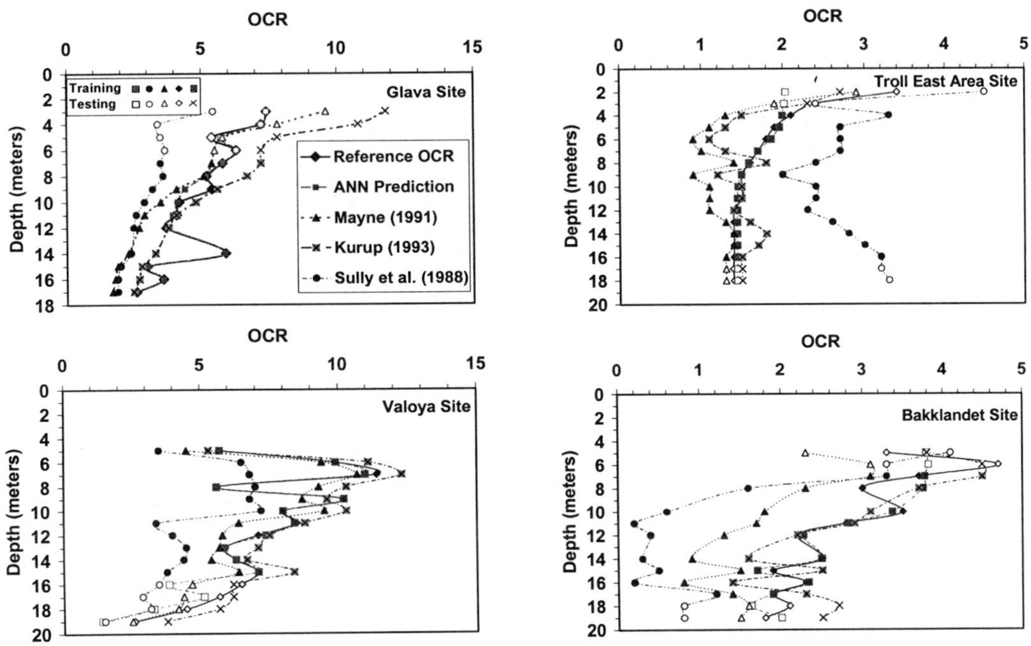

Figure 2. Predicted profiles of OCR in Glava, Troll East Area, Valoya and Bakklandet sites.

the model: corrected cone resistance (q_t), total vertical overburden pressure (σ_v), pore water pressure at the cone tip (u_1), pore water pressure measured just above the cone base (u_2), and the hydrostatic pore water pressure (u_0). Hence the input layer has five neurons. The only output is the OCR, and therefore there is only one output neuron (in the output layer). The hidden layer enables non-linear modeling of the sensor data. The number of neurons in the hidden layer is determined by a trial and error method, i.e. by training the network with different number of hidden neurons and comparing the results with the desired output. A hidden layer with six neurons gave good results for estimating the OCR of the sites. Thus a 5x6x1 network architecture was trained and tested for predicting OCR from PCPT data. The architecture of the ANN model used in this study is illustrated in Figure 1.

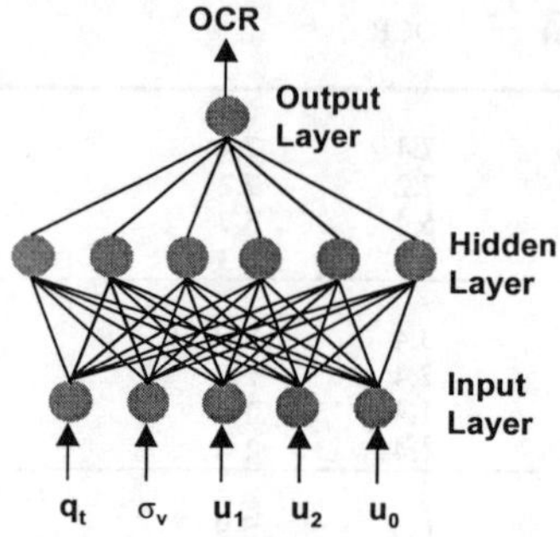

Figure 1. Architecture of the neural network model.

2.3 *Training and testing the model*

The neural network was trained using partial set (75%) of PCPT data (Table 1) from four well-referenced Norwegian geotechnical test sites (Bakklandet, Troll east, Glava and Valoya clays). The weight and bias matrices obtained after the training phase are given below.

Weights between the input and the hidden layer neurons (in the form H x I):

$$\begin{bmatrix} -8.45 & 27.42 & 3.52 & -3.21 & -4.11 \\ -13.62 & -25.06 & -11.53 & 65.47 & -24.15 \\ 11.50 & -9.17 & -1.87 & 2.40 & -7.20 \\ -15.91 & 7.10 & -3.48 & 6.43 & 4.79 \\ -7.12 & 52.91 & 373.01 & -136.10 & -136.10 \end{bmatrix}$$

Bias for the hidden layer neurons (H x 1):

$$\begin{bmatrix} -21.95 \\ -9.65 \\ 4.55 \\ 3.53 \\ 48.36 \end{bmatrix}$$

Weights between the hidden and the output layer neurons (in the form O x H):

$$\begin{bmatrix} -10.15 & 1.61 & -14.24 & 1.41 & -0.83 \end{bmatrix}$$

Bias for the output neuron (in the form O x 1): $[11.471]$

After training, the neural network was once again presented with the data that was used for training. The ANN predictions are compared with the reference OCR values obtained from laboratory oedometer tests in Figure 2. It can be seen that the ANN model was able to accurately predict the OCRs of the training data. The model was further tested using the remaining data (Table 2) that was not used in the training phase. The OCRs predicted by the ANN model, and those predicted by the methods proposed by Sully et al. (1988), Mayne (1991), Kurup (1993), and Tumay et al. (1995) are compared with the reference OCR values in Figure 2. It can be seen that the ANN model gave very reliable estimates of OCR when compared to other methods.

3 CONCLUSIONS

A feed-forward, back-propagation neural network model was developed to predict overconsolidation ratio of soil deposits from PCPT data. The ANN model was found to give very reliable OCR estimates when compared to some of the existing interpretation methods.

4 ACKNOWLEDGEMENTS

The authors acknowledge the support of the National Science Foundation, Career Award 9875037 under the direction of Dr. Fragaszy, Program Director for Geomechanics & Geotechnical Systems.

5 REFERENCES

Dowla, F.U. and L.L. Rogers. 1995. *Solving problems in environmental engineering and geoscience with artificial neural networks*. The MIT Press.

Hagan, M.T., H.B. Demuth and M. Beale. 1999. *Neural network design*. PWS Publishing Company.

Khanna, T. 1989. *Foundations of neural networks*. Addison-Wesley Publishing Company.

Kurup, P.U. 1993. *Calibration chamber studies of miniature piezocone penetration tests in cohesive soil specimens; Ph.D. Dissertation*. Louisiana State University, Baton Rouge, LA.

Mayne, P.W. 1991. Determination of OCR in clays by PCPT using cavity expansion and critical state concepts. *Soils and Foundations* 31(2): 65-76.

Sandven, R. 1990. *Strength and deformation properties of fine grained soils obtained from piezocone tests; Ph.D. Dissertation*. Norwegian Institute of Technology, Trondheim, Norway.

Sully, J.P., Campanella, R.G. & Robertson, P.K. 1988. Overconsolidation ratio of clays from penetration pore pressures. *ASCE Journal of Geotechnical Engineering Division* 114(2): 209-216.

Tumay, M.T., Kurup, P.U. & Voyiadjis, G.Z. 1995. Profiling OCR and Ko from piezocone penetration tests. *Proc. intern. symp. On Cone Penetration Testing, Linkoping, Sweden, 4-5 October 1995*. Swedish Geotechnical Society: SGF Report 3:95(2):337-342.

Computer Methods and Advances in Geomechanics, Desai et al. (eds) © 2001 Balkema, Rotterdam, ISBN 90 5809 183 X

A new wavelet neural network predicting model for AE time series of rock burst

Yunliang Tan, Hui Zhou & Xianjun Han
Engineering Department, Shandong University of Science and Technology, People's Republic of China

ABSTRACT: Based on the features of acoustic emission time series of rock burst, a new wavelet neural network model in which the transferring function such as Sigmoid function is replaced by wavelet radix which determined by flex factor and parallel factor. The numbers of hidden layer and the other parameters are given automatically by the network. itself. It is shown by the prediction test of acoustic emission time series in Mentougou Mine in China that the average predicting error of BP network is 6.7%, the average predicting error of wavelet neural network is only 1.1%, also, the training time of wavelet network is much more shorter than BP network.

1 INTRODUCTION

Rock bursts are experienced in underground mining at various locations in the world, causing death and injury to underground miners and damage to mine structures. As an efficient monitoring technology, acoustic emission is widely used in deep mines in China (Tang Chunan,1993), because acoustic emission events are useful to predict when and where the rock burst would occur Applications of Grey Model (Deng Julong,1992)and artificial neural network for the series of acoustic emission are developed rapidly in the recent years. However, many research achievements have shown that Grey Model is only suitable for the exponential increasing tendency time series prediction, BP neural network is unstable and the balance position of the network easily gets into a local infintesimal level due to the structure parameters and training parameters have to be given artificially (Lu Xieqiang,1997). In this paper, a stable and reliable new wavelet neural network for the prediction of the acoustic emission time series of rock burst is built.

2 TRANSFERRING FUNCTION OF WAVELET NEURAL NETWORK

A wavelet neural network is constructed by the combination of wavelet transformation and neural network. In the network, the traditional Sigmoid function is replaced by the wavelet radix function which determined by a flexing factor and a paralleling factor, and it can approach a definite time series excellent by selecting the parameters for the network. A square integrabel function $\varphi(t) \in L^2(R)$ which satisfies equation (1) is called mother wave.

$$\int_{-\infty}^{+\infty} \left|\hat{\varphi}(\xi)\right|^2 \left|\xi\right|^{-1} d\xi < +\infty \tag{1}$$

Where $\varphi\hat{}(\xi)$ is the Fourier transform of $\varphi(t)$. The following equation (2) is constructed as

$$\varphi_{ab}(t) = \frac{1}{\sqrt{|a|}} \varphi\left(\frac{t-b}{a}\right) \tag{2}$$

Where a and b is real number, $a \neq 0$, $\varphi_{ab}(t)$ is called continuous sub-wave which depending on a and b ,and generated by mother wave.

For a non-linear time series function $f(t) \in L^2(R)$, its wavelet transformation can be expressed as:

$$k_f(a,b) = \frac{1}{\sqrt{|a|}} \int_{-\infty}^{+\infty} f(t)\varphi\left(\frac{t-b}{a}\right)dt \tag{3}$$

Where b is the translating factor and a is the extension factor for the function $\varphi_{ab}(t)$. Furthermore, the power of discretion for the local structure of $f(t)$ can be adjusted by a and b. Similar to Fourier Transformation analysis, wavelet analysis is the method approaching a signal function by decomposing it to normal orthogonal bases. Because wavelet bases have nicer local characters, they can approach the signal function $f(t)$ optimally.

To choice such wavelet function $\varphi_i(t),(i=1, 2, \cdots\cdots, N)$ controlled by different translating factor b_i and different extension factor a_i, instead of transferring sigmoid function, we can realize the optimal approach for any function $f(t)$.

$$\varphi_i(t) = \frac{1}{\sqrt{|a_i|}}\varphi\left(\frac{t-b_i}{a_i}\right) \tag{4}$$

3 WAVELET NEURAL NETWORK PREDICATING MODEL

we make the linear congruence sum to approach the AE time series by finite wavelet base series:

$$\hat{u}(t) = \sum_{i=1}^{N} w_i \varphi_i\left(\frac{t-b_i}{a_i}\right) \tag{5}$$

Where, $\hat{u}(t)$ is the approaching value for AE time series, $u(t)$ is the actual value of AE, w_i is the weight, N is the number of wavelet base.

The wavelet network frame of one input and one output is shown in Fig.1. The structural parameters such as w_i, a_i, b_i and N can be optimized by least square energy function listed in equation (6). The arithmetic to determine limited N is to test it step by step, i.e., under the condition of allowable error ε, calculation of E_i is done form N=1 first, if $E_i < \varepsilon$, then N=i; else let N=i, and continue to calculate until $E_{N^*} < \varepsilon$ when $N = N^*$.

$$E_N = \frac{1}{2}\sum_{t=1}^{N}\left[y(t) - \hat{y}(t)\right]^2 \tag{6}$$

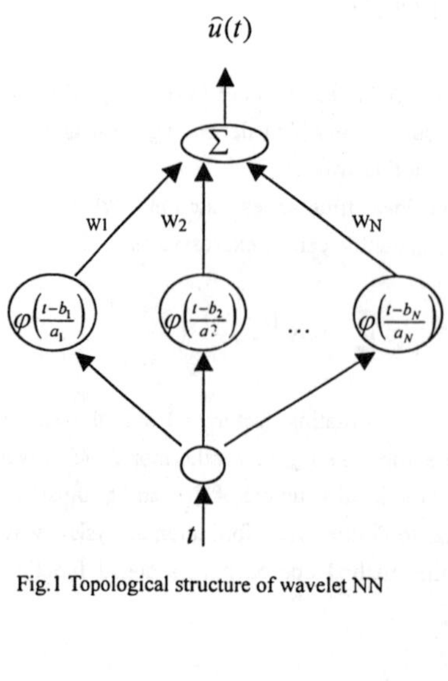

Fig.1 Topological structure of wavelet NN

For every given N, the process of determining parameters w_i, a_i, b_i can be taken as following steps.

First, to choice Morlet mother wavelet:

$$\varphi(t) = \cos(1.75t)e^{-\frac{t^2}{2}} \tag{7}$$

Let $\eta=(t-b_i)/a_i$, $\tilde{u}=-\Sigma[U(t)-\hat{U}(t)]$, the gradient of w_i, a_i, and b_i is :

$$g(w_i) = \frac{\partial E_N}{\partial w_i} = \tilde{u}\cos(1.75\eta)e^{-\frac{\eta^2}{2}} \tag{8}$$

$$g(a_i) = \frac{\partial E_N}{\partial a_i} = \frac{\tilde{u}\eta w_i}{a_i}[1.75\sin(1.75\eta) + \cos(1.75\eta)]e^{\frac{\eta^2}{2}} \tag{9}$$

$$g(b_i) = \frac{\partial E_N}{\partial b_i} = \frac{\tilde{u}w_i}{a_i}[1.75\sin(1.75\eta) + \eta\cos(1.75\eta)]e^{\frac{\eta^2}{2}} \tag{10}$$

Then, the conjugate gradient method used to optimize w_i, weight vector is stated as $w=(w_1,w_2,\ldots\ldots,w_N)$. If the searching direction of step k is assigned as $s(w)^k$, we can express it in the form:

$$s(w)^k = \begin{cases} -g(w)^k & ,k=1 \\ -g(w)^k + \dfrac{g(w)^{kN}\cdot g(w)^k}{g(w)^{(k-1)N}\cdot g(w)^{k-1}}\cdot s(w)^{k-1} & ,k\neq 1 \end{cases} \tag{11}$$

The weight vector can be adjusted in the form:

$$w^{k+1} = w^k + \alpha_w s(w)^k \tag{12}$$

Where, α_w is the searching step. The searching process stops till w converges at a definite value. The optimizing process for a_i and b_i are the same as w.

4 LIVING EXAMPLE ANALYSIS

A typical rock burst monitoring test was done in Mentougou Mine of Beijing Mining Bureau in 1986. There are three AE exploring sensors Y1 ,Y2 and Y3 being put into the wall of the gate ahead of the West No.2 working face. From June 1 to June 23, the energy rate number was counted.. First, both BP neural network and wavelet network WNN is trained with the above monitored energy rate data; then , predicting test is done. Trained time of BP network is 56 Sec., trained time of WNN network is only 26 Sec., the predicting results are shown in Fig. 2, the predicting errors are listed in Tab.1.

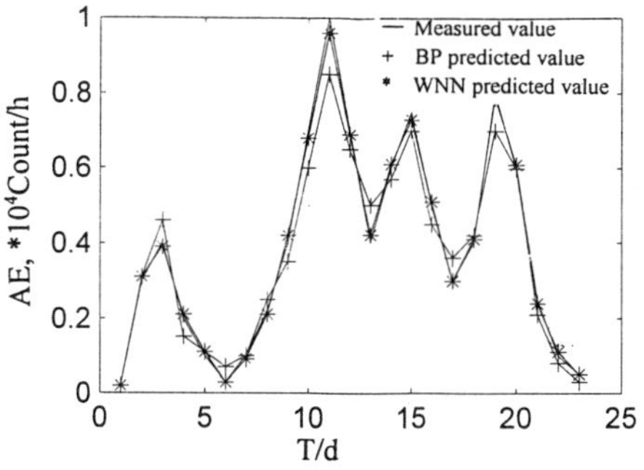

(a) Y1 sensor

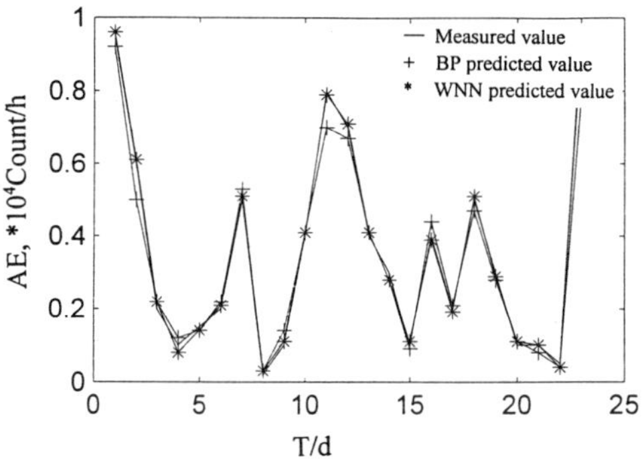

(b) Y2 sensor

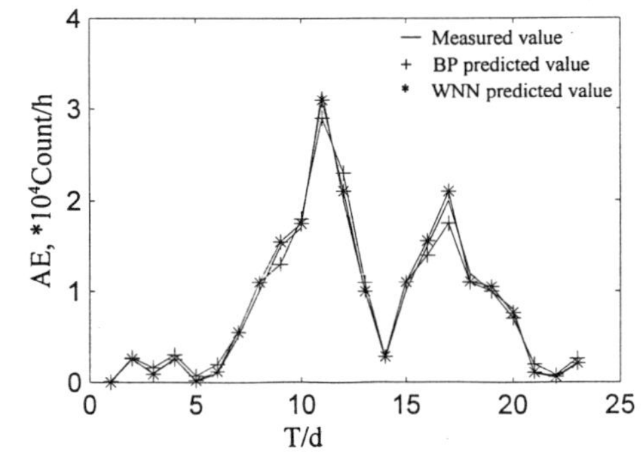

(c) Y3 sensor

Fig. 2 Predicting results of AE in Mentougou Mine

Tab. 1 Comparison of predicting errors

Sensor	Y1	Y2	Y3	average
BP network	7.8%	5.5%	6.8%	6.7%
WNN network	1.4%	0.7%	1.2%	1.1%

As shown in Tab.1, the predicting error of WNN is only 1.1%, less than that of BP network.. In other words , WNN neural network is much more accurate for the prediction of rock burst AE time series.

5 CONCLUSIONS

A new wavelet neural network model WNN is built, In which the transferring function such as Sigmoid function is replaced by wavelet radix which determined by flex factor and parallel factor, the numbers of hidden layer and the other parameters are given automatically by the network. itself. Therefore, the WNN network can avoid the disadvantage of network parameters must be chosen artificially in such neural network as BP. It is shown by the test of AE time series prediction in Mentougou Mine in China , that the average predicting error of BP network is 6.7%, the average predicting error of wavelet neural network is only 1.1%, also, the training time of wavelet network is much more shorter than BP network..

ACKNOWLEDGEMENT

The research is being supported by China National Natural Science Foundation and China Education Ministry Foundation.

REFERENCES

Tan Yunliang,Wang Xueshui 1983. AE Omen in The process of Hard Roof Collapse. *Chines J. Rock Mech. and Engineering.* 11:275-283

Tang Chunan 1993. *Catastrophe of Rock Failure*:70-73. Bejing: China Coal Publish House

Lu Xieqiang 1997. Neural Network and Its Applications in Nonlinear Time Series Prediction. *J. System Theory and Practice* .12:97-99

He Zhenya 1997. *Intelligence of Neural Network*:45-55. Hunan:Hunan Science Publish House.

Deng Julong 1992. *Grey Prediction and Decision*: 35-48 Wuhan: Huazhong University of Science and Engineering

Computer Methods and Advances in Geomechanics, Desai et al. (eds) © 2001 Balkema, Rotterdam, ISBN 90 5809 183 X

A fast radial basis function neural network model for approaching the stress-strain curve of rock

Yunliang Tan
Engineering Department, Shandong University of Science and Technology, People's Republic of China
Zhou Hui
Mining Department, Northeastern University, Shenyang, People's Republic of China

ABSTRACT: In this paper, a fast approaching model for the stress-strain curve of rock by using radial basis function neural network has been built. In the model, the transferring function such as Sigmoid function is replaced by radial basis function, a criteria for improving the generalization capability of neural network is given. It is shown that the model can approach the stress-strain curve accurately and rapidly tested by the stress-strain curve of marble experimented by Rummel F. and Fairhurst in 1970. Therefore, the model is valuable for developing the numerical method of computer to the level of high-speed in rock mechanics.

1 INTRODUCTION

The constitutive equation of stress-strain is the basis of numerical method of rock mechanics. In recent years, many researchers have made a lot achievements. For examples, Duncan-Chang (1996) put up a nonlinear model in which elastic modulus E and Possion's ratio v are taken as variables of hyperbolic function. Yin Youquan (1986)gave an elastic-plastic singualr yeild surface theory, Kawamor,T. (1988) described the stress-strain relation of rock by damage mechanics. But the above constittive models can't express the stress-strain equation of rock successfully because of the comlexity of rock fabric. And it often leads great errors between prcatical engneering and numerical calculation. Also, interative calculation of plastic constitutive equation in the numerical simulation costs too much time.Thus , we put up a new fast radial basis function neural network model RBF for approaching the stress-strain curve of rock.

2 FAST APPROACHING RBF MODEL

2.1 *Advantages of RBF neural network*

There are two methods to approach the complicated nonlinear function by neural network: overall approaching network and local approaching network. For any input , if the changing of any node weight leads to change for all output, it is called overall approaching method. Else, for any input , if there are only seldom node weights affecting the output , it is called the partial approaching method. BP neural network is a typical overall approaching model, it approaches very slowly, but partial approaching method is very fast, RBF neural network is a typical one which first was put up by Broomhead and Lowe (1988) . Another advantage of RBF network is all-right generaliztion ability. It can forcast better than BP neural network because only partial weights are adjusted in a network.

2.2 *Structure and arithmetic of RBF*

A typical RBF network is shown in Fig.1, there are 2 layers : the first layer is hidden layer whose transferring function is Gauss distribution function; the second layer is output layer whose transferring function linear function. The position and width of RBF function are defined by every neutron weight and its threshold in hidden layer. A lot of researched work showed that a neural network with certain number of neurons in RBF hidden layer and proper weights can approach arbitrary complicated no-linear function, and approaching error is minimum. It is because the mapping from input layer to hidden is non-linear and from hidden layer to output layer is linear, that the weights vector of network can be solved directly from linear equations. Just so, the RBF network can approach fast and can avoid the learning of the network get into partly minimum. The arithmetic of RBF neural network is described in the next paragraph.

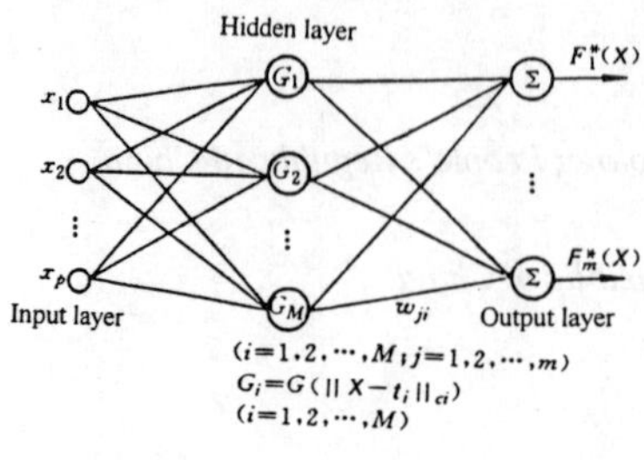

Fig 1 Frame of RBF neural network

For input $X=(x_1,x_2,\ldots\ldots,x_p)$, radial basis function of RBF network is expressed as Gaussian function:

$$\alpha_j(x) = \exp\left(-\frac{\|x - c_j\|^2}{\sigma_j^2}\right) \tag{1}$$

Where c_j is the center of jth basis function which can be determined by NN cluster analysis (Tan Yunliang, 1995), σ_j is a constant which determines the center width of RBF function $\alpha_j(x)$. $\| x-c_j \|$ is the norm of vector $x-c_j$. With the increase of $\| x-c_j \|$, $\alpha_j(x)$ decreased to zero rapidly. For given input $X\in R^n$, only few elements of the center near X are activated, so partial approaching goal arrives.

Linear mapping form hidden layer to output layer is

$$F_i^* = \sum_{j=1}^{m} w_{ij}.\alpha_j(x) \qquad i=1,2,\ldots,M \tag{2}$$

Weights are adjusted as

$$w_{ij}(L+1) = w_{ij}(L) + \beta(F_i^{*d} - F_i^*(L))\alpha_j(x)/[\alpha^T(x)\alpha(x)] \tag{3}$$

F_i^{*d} is the ith output mean value, $F_i^{*d}(L)$ is the output i at Lth calculation, β is the learning ratio, $\alpha(x)$ is the mapping vector for hidden layer.

2.3 *Improvement of generalization ability*

Generalization ability of neural network is very important. Many researches show that all weights in BP network can be adjusted for every learning process. If too much special samples are learned and minimum error is chased for the learned samples, the generalization ability of neural network will decrease rapidly, as shown in Fig.2. This is called overfitting phenomenon.

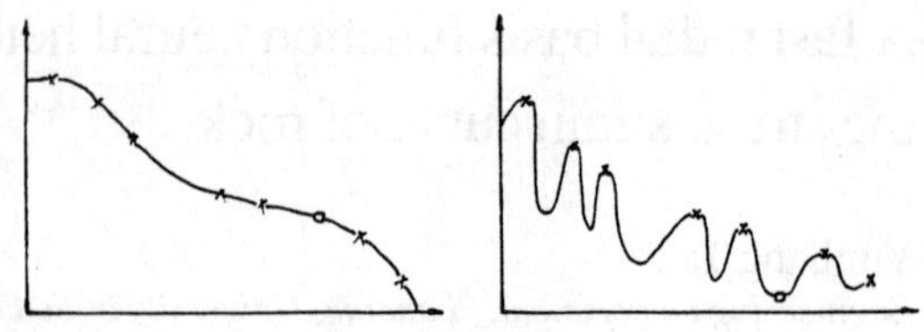

(a) All-right curve (b) Poor curve

Fig 2 Generalization capability of NN

×—Training sample, O—Generalization sample

RBF network can improve the generalization ability of neural network. In order to improve the generalization further more, judgement (4) can be added to the RBF network.

$$\Delta F \geq \frac{n\lambda}{\Delta w.2\log_2(1+\frac{F}{E})} \tag{4}$$

Where Δw is arithmetical difference of the weight for the two training time one after another. Δf is the squared root error for the predicating value of a new sample, n is the number of input nodes, λ is a constant, it is tested that $\lambda = 1 \times 10^{-4}$, F is the average output value after normalized for the trained samples, E is the squared root error output value after normalized for the trained samples.

3 TYPICAL STRESS-STRIAN APPROACHING ANALYSIS OF ROCK

3.1 *Typical stress-strain of rock*

A typical stress-strain curve of rock (Rummel and Fairst ,1970) is shown in Fig.3. When lateral stress σ_3 changes from 0Mpa to 35Mpa, the curve appears strain-soften feature; when lateral stress σ_3 increases to 50Mpa, the curve appears strain-harden feature. Therefore, by learning the curve under the conditions of different σ_3 as shown in Fig.3, the constitutive equation of marble can be simulated just well.

3.2 *Training for marble stress-strain curve by RBF network*

The parameters of RBF network for approaching stress-strain curve of marble are determined below. The training goal error is 0.002, the maximum hidden layer neuronss are 100, RBF distributing constant is σ_j=0.1.The lateral stress σ_3=0、7、28、55Mpa. The learning test shows that, there are only 11 neurons in hidden layer, training time are 11 seconds For the same training goal, BP network training time are 144 seconds, are as about 130 times much as RBF network.

The approaching results for lateral stress σ_3=0、3.5、21、35、55Mpa are shown in Fig.4, the corresponding neurons in the hidden layer are 8、11、17、26、29, the approaching error is less than 2×10^{-3}. Obviously, the approaching results are satisfied.

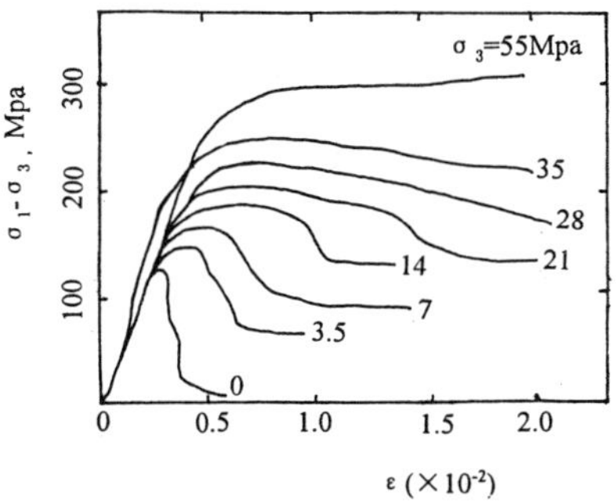

Fig 3 Stress-strain curve of marble

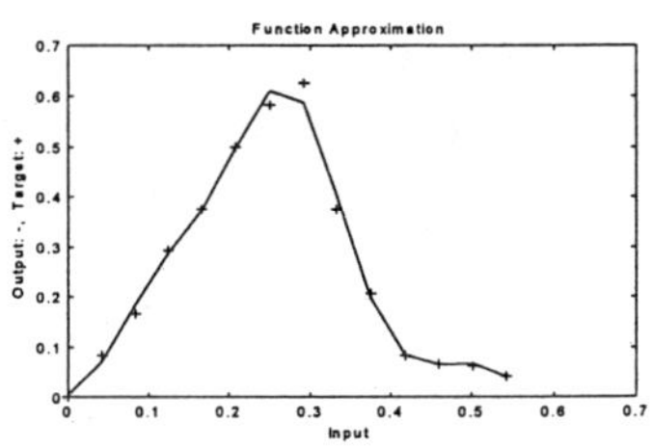

a. σ_3=0Mpa

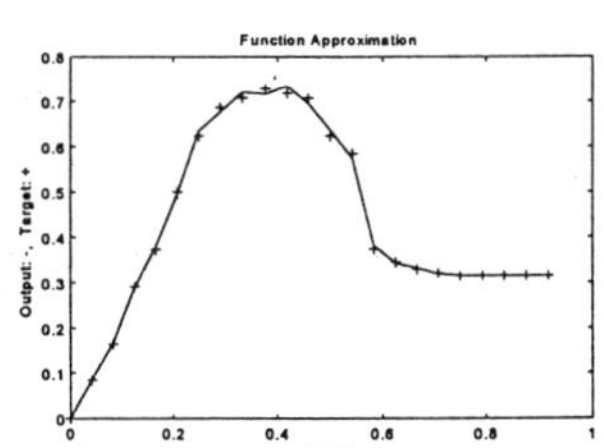

b. σ_3=3.5Mpa

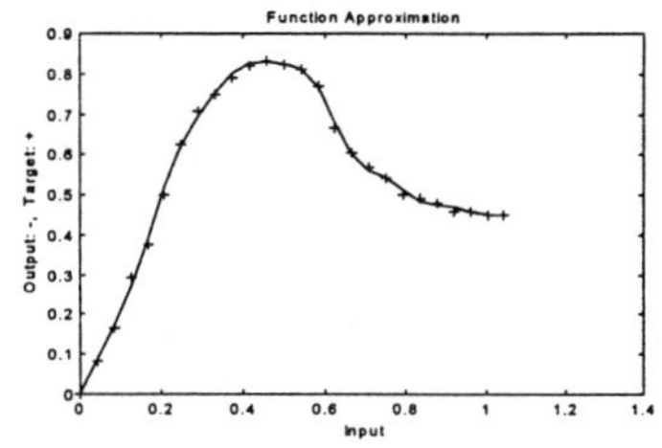

c. σ_3=21Mpa

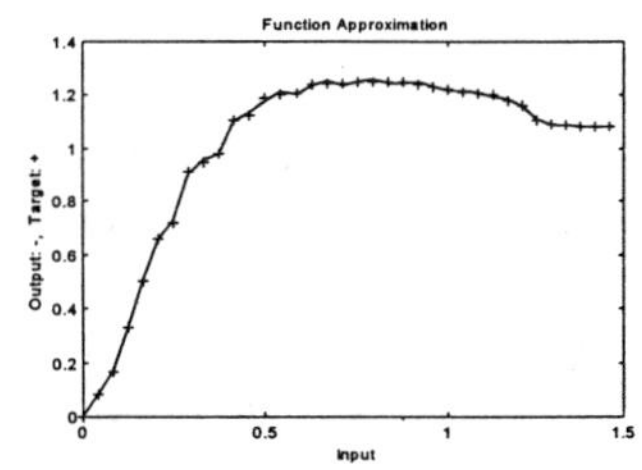

d. σ_3=35Mpa

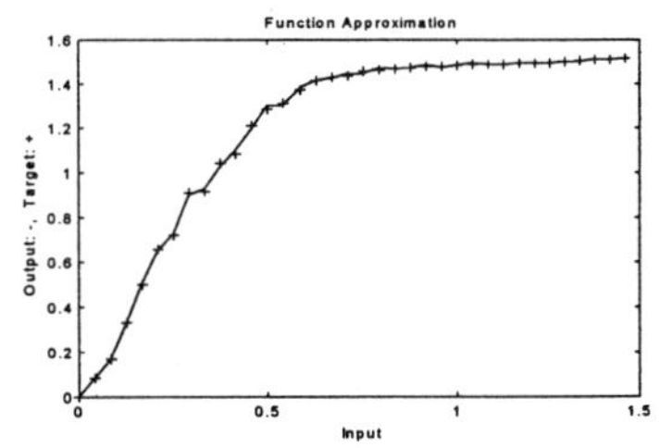

e. σ_3=55Mpa

Fig.4 Approaching result of stress-strain curve of marble

+: Experiment value, —: Approaching value.

4 CONCLUSIONS

To approach constitutive equation of rock is a long term task in rock mechanics and engineering. The model of RBF network for approaching stress-strain curve is a new way to simulate the constitutive equation of rock. It is shown by the test of laboratory data that, RBF network can approach constitutive equation of rock primely and fast, it is valuable for the rock mechanics computation.

ACKNOWLEDGMENT

The research is being supported by China National Natural Science Foundation and China Education Ministry Foundation.

REFERENCES

Duncan J. M., Chang C.Y.1996. Nonlinear analysis of and strain in soils, *J. Soil Mech. Found. Div. ASCE.*, SM5:1629~1653

Yin Youquan 1986. Expression of stress space and strain Space by a elastic-plastic singualr yeild surface. *ACTA Mechanics* 19:31-38

Kawamoto. T. 1988 Deformation and fracturing behavior of discontinuous rock mass and damage mechanics theory. *Int. J. Num. & Anal. Methods Geomech.*, 12:1~30

Tan Yunliang, Wang Yongjia 1995.Neural network model of cluster analysis for the classification indexes of gate. *Chiness journal of rock mechanics and engineering*, 14:139~144

Powell M J D. 1985 Radial basis functions for multivariable interpolation. *IMG Conference on Algorithms for the Approximation of Function and Data*:143-167. UK:shrivenham,

Broomhead DS and Lowe D. 1988. Multivariable functional interpolation and adaptive networks. *Complex System,* 16:321-355

Computer Methods and Advances in Geomechanics, Desai et al. (eds) © 2001 Balkema, Rotterdam, ISBN 90 5809 183 X

Extension to generalized method of slices for reliability analysis of reinforced slopes

V.Tandjiria, C.I.Teh & B.K.Low
School of Civil and Structural Engineering, Nanyang Technological University, Singapore

ABSTRACT: This paper discusses the reliability analysis of reinforced slopes based on the generalized method of slices. A new extension to the generalized method of slices is proposed for reinforced slope stability analysis by incorporating the reinforcement effect. The method is then applied to reliability analysis of reinforced slopes. The results obtained from the proposed method compared well with those analyzed using another limit equilibrium method. For the reliability analysis, soil strength and reinforcement strength properties are considered to be the random variables. The minimum factor of safety and the reliability index are computed efficiently using the constrained optimization approach in a spreadsheet environment. The versatility of the method is demonstrated through a stability analysis of a reinforced embankment.

1 INTRODUCTION

Slopes or embankments constructed over weak foundation soils are often strengthened by reinforcements. By placing a reinforcement across the base of a slope, the overall stiffness of the slope system increases. Some portions of stresses are transferred to the reinforcement. This reduces the mobilized shear stresses and strains in the foundation. Hence, the safety factor of the slope is also improved.

The limit equilibrium methods are often used by soil engineers due to their simplicity and good agreement in back analyses of actual slope stability problems. Rowe and Soderman (1985) analyzed reinforced embankments using the extension of the simplified Bishop slip circle analysis. Additional assumptions were needed to incorporate geotextile reinforcements. Hird (1986) adopted circular slip mechanisms using a standard equilibrium method with slight modifications to take account of the reinforcement. Low et al. (1990) applied the semi-analytical limit equilibrium method to reinforced embankments built on soft ground. Sabhahit et al. (1994) analyzed the stability of reinforced embankments using the generalized method modified from the Janbu generalized procedure of slices This method has capability to search for the critical arbitrary slip surface.

A limit equilibrium method produces the safety factor that is used to measure the performance of slopes. However, the safety factor cannot explicitly reflect the uncertainties of design parameters involved in any slope problems. A better alternative is to use a reliability index which depends on mean values of each parameter and their intrinsic variations. The reliability index formulated by Hasofer-Lind (1974) is one of the popular techniques to calculate the reliability index of a system. The Hasofer-Lind reliability index is expressed in matrix form (Ditlevsen 1981) as:

$$\beta_{HL} = \min_{x \in F} \sqrt{\left\{\frac{\mathbf{x}-\mathbf{m}}{\sigma}\right\}^T [R]^{-1} \left\{\frac{\mathbf{x}-\mathbf{m}}{\sigma}\right\}} \quad (1)$$

where $\boldsymbol{x}$ is a vector representing the set of normally distributed random variables, $\boldsymbol{m}$ their mean values, σ their standard deviations, $\boldsymbol{R}$ the correlation matrix and $\boldsymbol{F}$ the failure region. It is noted that β_{HL} is basically the shortest distance in units of directional standard deviation of the failure surface to the point where the mean values of the random variables are located.

In this paper, an extension to the generalized method of slices that incorporates reinforcement effect is proposed and applied to the reliability of reinforced slopes.

2 GENERALIZED METHOD OF SLICES FOR REINFORCED SLOPES

The method of slices developed by Chen and Morgenstern (1983) is one of the rigorous limit equilibrium methods. It can analyze general slip surfaces, satisfies both the force equilibrium and the moment equilibrium and allows varying side force orientation. Chen and Li (1998) extended the method by considering earth pressures applied to various types of supports and water pressures due to a tension crack. An extension to the method by taking into account the effect of reinforcements will be discussed in the following section.

Figures 1 and 2 show a trial failure surface and the forces working at a typical slice, respectively. The final force and moment equilibrium equations modified from Chen and Morgenstern (1983) and Chen and Li (1998) by adding the influence of reinforcement forces for a given slip surface are expressed in eqs. (2) and (3).

$$\int_a^b p(x)s(x)dx = G_m \tag{2}$$

$$\int_a^b p(x)s(x)t(x)dx = M_m \tag{3}$$

where

$$p(x) = \left(\frac{dW}{dx} + q\right)\sin\left(\varphi_e' - \alpha\right) - r_u\frac{dW}{dx}\sec\alpha.\sin\varphi_e'$$
$$c_e'\sec\alpha.\cos\varphi_e' - \frac{dT}{dx}\cos\left(\varphi_e' - \alpha\right) \tag{4}$$

$$s(x) = \sec\psi\exp\left(-\int_a^x \tan\psi\frac{d\beta}{d\xi}d\xi\right) \tag{5}$$

$$t(x) = \int_a^x (\sin\beta - \cos\beta\tan\alpha)\exp\left(\int_a^\xi \tan\psi\frac{d\beta}{d\zeta}d\zeta\right)d\xi \tag{6}$$

$$G_m = P_w - P\frac{s(b)}{\sec\psi} \tag{7}$$

$$M_m = P_w h_w - P\left[h\cos\delta + \frac{s(b)t(b)}{\sec\psi}\right] + \int_a^b \frac{dT}{dx}h_t dx \tag{8}$$

$$\psi = \varphi_e' - \alpha + \beta \tag{9}$$

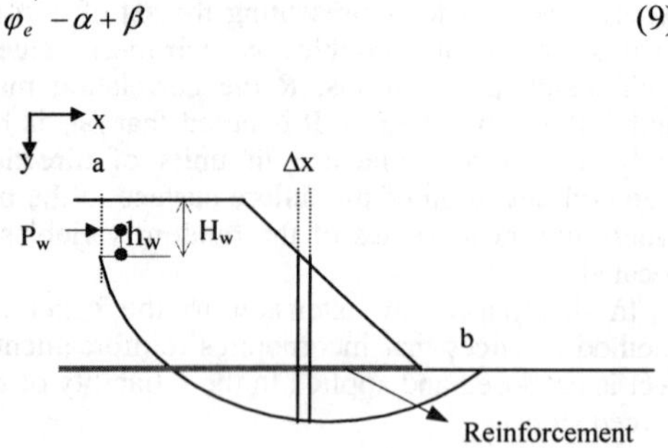

Figure 1. Boundary Conditions of Slope

The notation of the above formulations are explained as follows: α and β are the inclinations of the base and the interslice forces, respectively; dW/dx is the weight of the soil mass per unit width; q is the external vertical surface load per unit width; G and T are the interslice force working at a distance y_t to the reference level and the mobilized reinforcement force working at a distance h_t to the mid-point of the base of the slice, respectively; c_e' and φ_e' are the effective mobilized cohesion and friction angle, respectively and r_u is the pore pressure ratio. P_w is the water pressure at the tension crack. h_w is $H_w/3$ where H_w is the height of the tension crack. P is the earth pressure which is considered in analyzing the stability of retaining structures. h is the distance of P to the base of the slice at the position of the retaining structure. δ is the inclination of the earth pressure with respect to the x axis. The side force inclination β was assumed in Chen and Morgenstern (1983) as follows:

$$\tan\beta = \lambda f(x) + f_0(x) \tag{10}$$

where λ is a coefficient to be determined; $f(x)$ is an assumed function; and $f_o(x)$ is another function used to satisfy β at the end boundaries of the slipping bodies, a and b, as shown in figure 1. For this study, $f(x)$ is assumed to be a sine function and $f_o(x) = 0$.

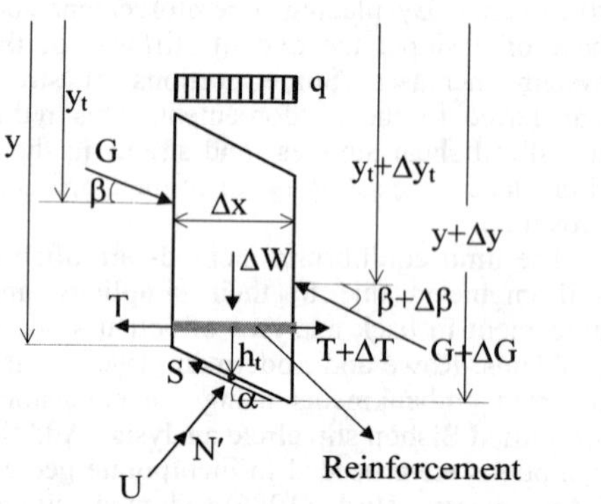

Figure 2. Force Equilibrium of a Slice

The factor of safety (FS) is considered in the effective mobilized cohesion and friction angle along the slip surface. The former is c'/FS and the latter is obtained from $\tan^{-1}(\tan\phi'/FS)$. c' and ϕ' are the actual cohesion and friction angle of the soil along the slip surface. In addition, a similar factor of safety is also considered for the mobilized tensile forces in the reinforcement ($T = T_{max}/FS$). T_{max} is the ultimate tensile strength of the reinforcement. Besides the boundary criteria, the physical admissibility criteria ($G(x)$ and $\sigma'(x) > 0$) should also be satisfied.

3 CONSTRAINT OPTIMIZATION APPROACH FOR REINFORCED SLOPES

A critical circular slip surface can be found by searching the minimum factor of safety of several combinations of slip surfaces subject to relevant boundary conditions. Similarly, the reliability of a reinforced slope can be achieved by minimizing the reliability index, which is a function of the random variables. Both the problems of finding the minimum safety factor and the reliability index become constrained optimization problems.

For deterministic analysis, the objective function is to minimize the factor of safety while the main constraints are eqs. (2) and (3) and the physical admissibility criteria. The design variables are the coordinates of slip surface, FS and λ.

For reliability analysis, the objective function is to minimize the Hasofer-Lind reliability index subject to constraints on the performance function, the physical admissibility criteria, eqs. (2) and (3). The performance function for this problem is $(FS-1) = 0$. The design variables are any random variables considered in the reinforced slope system, the coordinates of slip surface, FS and λ.

The numerical procedures for the deterministic and reliability analyses of the reinforced slope have been developed in a spreadsheet environment using both the constrained optimization tool Solver and the Visual Basic for Applications programming language in Microsoft Excel®. Eqs (2) and (3) are solved using Legendre Gauss quadrature technique. The numerical procedures are described in Low et al. (1998).

4 DETERMINISTIC ANALYSIS

The versatility of the proposed method is demonstrated through a back calculation of an embankment on Muar clay deposit in Malaysia (Low and Tang 1997). The net embankment height after settlement is 5.25 m while the gradient of the slope is 2:1. The embankment has an effective cohesion (c') of 19 kPa, a friction angle(ϕ') of 26° and a unit weight of 20.5 kN/m^3. The undrained cohesion (c_u) profile is modeled multilinear as shown in figure 3. The tension crack (h_c) is 2.0 m and the ultimate strength of reinforcement (T_{max}) is 80 kN/m.

The circular slip surface is considered in this study. To investigate the effect of reinforcements on the embankment stability, an unreinforced embankment and two reinforced embankments are considered. The force distributions along the reinforcement of the two reinforced embankments are assumed linear and cubic. The maximum force of each distribution which is set to be the maximum mobilized forces occurs at the intersection point of the slip surface and the reinforcement. In addition, the force at the intersection point of the reinforcement and the embankment toe is set equal to zero. The results of the proposed method are then compared to those obtained using the semi-analytical limit equilibrium method for reinforced embankments which is based on moment equilibrium (Low and Tang 1997).

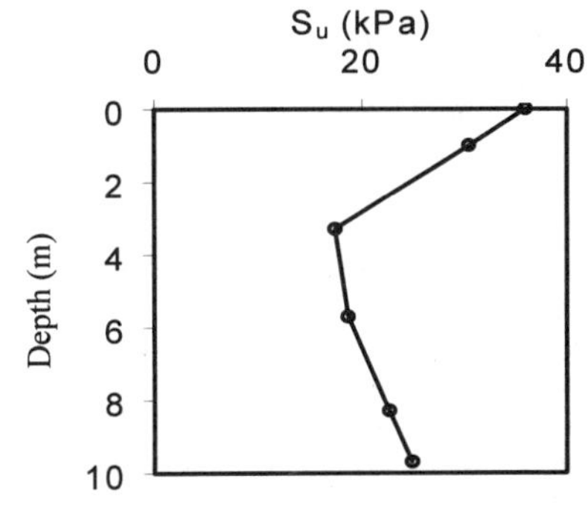

Figure 3. Undrained Shear Strength

For the unreinforced case, a safety factor of 1.144 is obtained while for the reinforced cases the factors of safety for both linear and cubic force distribution are identical (FS = 1.209). The factor of safety for the reinforced case corresponds to a mobilized tensile force of 66.19 kN/m and λ of 0.034. The assumed force distribution along the reinforcement does not influence the computed embankment's factor of safety.

For a comparison, the safety factors obtained from the semi-analytical method (with $H_w = h_c = 2$ m) are 1.138 and 1.203 for the unreinforced case and the reinforced one, respectively. This indicates that the factor of safety obtained from the proposed method compared well with that obtained from the semi-analytical method. Figure 4 shows the critical slip surface for all reinforced cases.

5 RELIABILITY ANALYSIS

The reliability index formulated by Hasofer-Lind (1974) is adopted in this section to investigate the reliability of the Muar embankment.

Ten (instead of 12 in Low & Tang (1997)) normally distributed random variables are considered, namely the embankment cohesion (c_m), the embankment friction angle (ϕ_m), the ultimate strength of the reinforcement (T_{max}), the height of tension crack

($H_w = h_c$) and the six values of the undrained shear strength of the foundation soil at six different depths. The mean values of the random variables are similar to the deterministic parameters in the previous section. The coefficients of variation of the first four parameters are assumed 0.15, 0.1, 0.1 and 0.3, respectively while the coefficient of variation of each undrained shear strength of the foundation soil is 0.15. It is also assumed that the correlation between c_m and ϕ_m is –0.3 and the correlations among the undrained shear strength obey the negative exponential model of spatial continuity, as was assumed in Low and Tang (1997).

Once the reliability index is obtained, the probability of failure (P_f) can be estimated using

$$P_f = 1 - \Phi(\beta) \tag{11}$$

Where $\Phi(.)$ is the cumulative distribution function of the standard normal variate.

The reliability indexes of the Muar embankment for the linear reinforcement force distribution and the cubic one are identical ($\beta = 1.87$). This indicates that the assumed force distribution along the reinforcement does not influence the computed embankment reliability. The above reliability index corresponds to a failure probability of 3.08 %. The calculated mobilized tensile forces are 79.12 kN/m and 79.18 kN/m for linear and cubic force distribution, respectively. The corresponding λ's are 0.019 and 0.017 for linear and cubic force distribution, respectively.

For comparison, the reliability analysis of the Muar embankment obtained using the semi-analytical method is 1.93 ($P_f = 2.68$ %). The calculated mobilized tensile force is 79.08 kN/m.

6 CONCLUSIONS

An extension to the generalized method of slices is presented for slope stability analysis by incorporating the reinforcement effect. The method is able to handle reinforced embankment problems which account for varying soil properties in the foundation soil, tension cracks, interslice forces and pore pressures. In addition, non-uniform force distribution along the reinforcement can be modeled.

The proposed method has been applied to the deterministic and reliability analyses of the Muar embankment in Malaysia via spreadsheet procedures. For such an embankment case, the assumed force distribution along the reinforcement does not influence the computed safety factor and reliability index.

The optimization solver in the spreadsheet environment is very effective and very efficient to perform numerical calculations for the deterministic and probabilistic analyses of reinforced slopes. The reinforcing force distribution model can easily be implemented via the spreadsheet procedure.

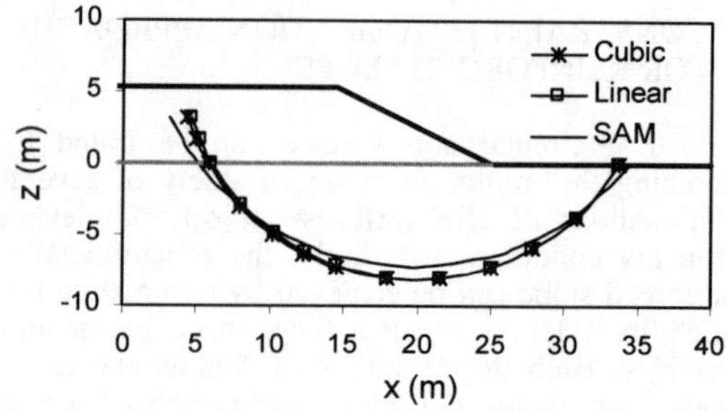

Figure 4. Critical Slip Surfaces

REFERENCES

Chen, Z., and Morgenstern, N.R. (1983). Extensions to the Generalized Method of Slices for Stability Analysis, Canadian Geotechnical Journal, Vol. 20(1), 104-119.

Chen, Z. and Li, S. (1998). Evaluation of Active Earth Pressure by the Generalized Method of Slices. Canadian Geotechnical Journal, Vol. 35, 591-599.

Ditlevsen, O. (1981). Uncertainty Modeling: with Applications to Multi-Dimensional Civil Engineering Systems. McGraw-Hill Book, New York.

Hasofer, A.M., and Lind, NC. (1974). Exact and Invariant Second-Moment Code Format, Journal of Engineering Mechanics, ASCE, 100 (1): 111-121.

Hird, C.C. (1986). Stability Charts for Reinforced Embankments on Soft Ground. Geotextiles and Geomembranes, Vol. 4, 107-127.

Low, B.K., Gilbert, R.B. and Wright, S.G. (1998) Slope Reliability Analysis Using Generalized Method of Slices., Journal of Geotechnical and Geoenvironmental Engineering ASCE, 350-362.

Low, B.K. and Tang, W.H. (1997). Reliability Analysis of Reinforced Embankments on Soft Ground. Canadian Geotechnical Journal, Vol. 34, 672-685.

Low, B.K., Wong, K.S., Lim, C. and Broms, B.B. (1990). Slip Circle Analysis of Reinforced Embankments on Soft Ground. Geotextiles and Geomembranes, Vol. 9, 165-181.

Rowe, R.K. and Soderman, K.L. (1985). An Approximate Method for Estimating the Stability of Geotextile Reinforced Embankments. Canadian Geotechnical Journal, Vol. 22, 392-398.

Sabhahit, N., Basudhar, P.K., Madhav, M.R. and Miura, N. (1994). Generalized Stability Analysis of Reinforced Embankments on Soft Clay. Geotextiles and Geomembranes, Vol. 13, 765-780.

Computer Methods and Advances in Geomechanics, Desai et al. (eds) © 2001 Balkema, Rotterdam, ISBN 90 5809 183 X

A practical approach for risk assessment employing reliability analysis and finite elements

R.Thurner & H.F.Schweiger
Institute for Soil Mechanics and Foundation Engineering, Graz University of Technology, Austria

R.Pöttler & B.Mattle
ILF-Consulting Engineers, Innsbruck, Austria

ABSTRACT: In this paper a short overview on probabilistic calculations in geotechnics and some remarks on the stochastic nature of soils are given. The numerical procedures adopted (a Taylor Series and a point estimate method (PEM)) are briefly described. For a shallow tunnel excavation finite element analyses are performed using material parameters based on stochastic considerations. From these analyses input parameters for limit state functions are obtained. With the Taylor Series the most important parameters are identified which are consequently used in the more sophisticated PEM-analysis. The obtained probabilistic results are compared with typical design value calculations and some preliminary results for the evaluation of limit state functions are shown.

1 INTRODUCTION

In the past two decades the importance of reliability and the stochastic nature of soil properties have been recognized in practical geotechnical engineering and the advantages of probabilistic tools are now generally accepted. However in practice purely deterministic methods are still dominantly used because of the additional computational efforts involved introducing stochastic methods in combination with advanced computational models.

Recent achievements in numerical analysis and the fast development of hardware have significantly reduced run times as well as pre- and post processing efforts so that a probabilistic approach based on deterministic finite element calculations seems to be a feasible way for the future and thus worthwhile to be investigated.

Figure 1 shows a typical flowchart for a deterministic calculation in geotechnics together with the additional steps (highlighted in Figure 1) required when introducing probabilistic concepts.

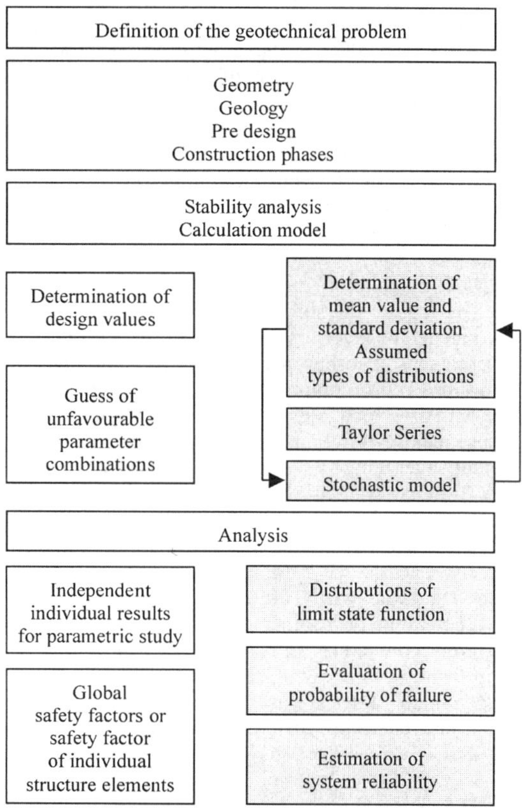

Figure 1. Scheme of typical deterministic analysis with probabilistic extensions.

2 STOCHASTIC MODELLING AND DESIGN VALUES FOR GEOMATERIALS

In general three major sources of uncertainties in soil or rock profile modelling can be distinguished (Vanmarcke 1977). The first one is the natural heterogeneity or the in-situ variability of the soil. The limited availability of information about subsurface conditions is the second one, which can be improved by additional soil testing and application of Bayes′ theorems. The third one is measurement errors, test imperfections and the human factor. The latter aspect is not considered in this paper.

In practice most geotechnical problems are still solved using a deterministic approach employing soil parameters, which are "on the safe side". In order to assess the behaviour of a geotechnical structure or the change of the deterministic factor of safety with respect to variable parameters, such as e.g. shear strength, sensitivity analyses are usually carried out.

The known uncertainty of the real value of each soil property at a certain point can be defined as a random variable described by a mean value, coefficient of variation (cov) and a probability distribution function. Common probability distribution functions are the unity, normal and the lognormal, whereas the latter one has to be preferred for variables, which do not become negative. More sophisticated distributions like Gumbel or Beta give better results for special cases, but they need much more computational effort and therefore practical applications are rare.

The coefficient of variation can be obtained from experimental results, other projects in similar soil conditions or from the literature (e.g. Lacasse & Nadim 1997, Rackwitz 2000, Orr 2000). The "3σ-rule" (Duncan 2000) is also a helpful tool for assessing the standard deviation of uncertain parameters using engineering judgment.

An important point for the practical application is the choice of the method, which is used for combining different informations. A simple tool to take different sources of information into account is the program CombInfo (1999) which is based on Bayes′ theorem.

Most calculation methods in geotechnical engineering require the soil profile to be modelled using homogeneous layers with constant soil properties. For this reason soil properties have to be defined not only for a certain point in space, but also for the entire domain which is used in the calculation process. Consequently it was found that due to the fact of spatial averaging of soil properties (Vanmarcke 1977) the coefficient of variation is reduced significantly. Typical assumptions for the determination of this autocorrelation reduction factor can be found for e.g. in Lacasse & Nadim (1997), Mostyn & Li (1993) and Rackwitz (2000).

Whereas in probability analysis these uncertainties are accounted for in a proper way, for purely deterministic design calculations a decision has to be made which values are used in the geomechanic calculations. There are different possibilities to derive the design values from given test results or distributions (Orr 2000). In some codes design values are defined as 5%-fractile. But due to the limited number of tests for most practical problems the calculation of this fractile yields to too conservative and uneconomic results. Thus, as mentioned above, some other information should be taken into account. Schneider (1999) suggests as practical approach for estimating characteristic values a reduction of the mean value by half of the standard deviation if no additional information is available.

3 SAFETY FACTOR OR RELIABILITY ANALYSIS OF GEOTECHNICAL STRUCTURES

The definition and discussion concerning safety and/or reliability of geotechnical structures is a widely discussed issue. There are a lot of different definitions concerning safety. Common to all is that safety is a relative measure and there exists no "absolute safety".

3.1 *Deterministic analysis*

In geotechnical calculations the factor of safety F_s is usually determined by using deterministic methods and can be defined as the safety margin:

$$F_s = R - S \tag{2}$$

where R is the "resistance" and S is the "action".

As shown by Li et al. (1993) and many others, in general F_s is a variant index because it depends on the interpretation of resisting and driving forces. In combination with the uncertainties of the soil properties and the fact that geotechnical engineers tend to use conservative design values, this leads to reliability levels without a sound theoretical basis.

3.2 *Probabilistic analysis*

As discussed in section 2 input parameters for geotechnical calculations are associated with uncertainties. Definition of the correct combination of the variables for the calculation process proves to be very difficult due to the lack of closed form solutions.

We define the limit state function $G(X)$:

$$G(\boldsymbol{X}) = R(\boldsymbol{X}) - S(\boldsymbol{X}) \tag{3}$$

where $R(X)$ is the "resistance", $S(X)$ is the "action", and X is the collection of random input parameters. For $G(X)<0$ failure is implied, while $G(X)>0$ means stable behaviour. The boundary defined by $G(X)=0$ separating the stable and unstable state is called the limit state boundary (Mostyn & Li 1993).

Due to the fact that with the same factor of safety one may have large differences in the probability of failure, it is desirable to work with an invariant measure for reliability. The well known reliability index β_{HL} (Hashofer & Lind 1974) is a convenient and consistent measure of the reliability level of a system or performance function.

Even if mean and standard deviation are known the probability of failure depends strongly on the probability density function which defines $G(\boldsymbol{X})$.

The probability of failure p_f is defined as:

$$p_f = P[G(\boldsymbol{X}) \leq 0] = \int_{G(\boldsymbol{X})\leq 0} f(\boldsymbol{X})dx \qquad (4)$$

where $f(\boldsymbol{X})$ is the common probability density function of the vector formed by the variables $\boldsymbol{X}$, which is usually unknown.

To solve this integral many different possibilities have been proposed in the literature. An overview is given by Petschacher (1993). In this paper we will restrict ourselves to numerical integration to determine the kth moment of G ($E[G^k(\boldsymbol{X})]$), because simulation methods like FORM (First Order Reliability Method) or SORM (Second Order Reliability Method) require the partial derivation of the performance function. These are not available from standard finite element codes.

Another key issue is the difference between the probability of failure of a given or assumed failure mechanism and the system reliability (Petschacher 1993 and Rackwitz 1998).

Furthermore an important point is the fact that all determined probabilities of failure are related to a certain period. Usually the calculated probability is applied to one year. To determine the probability of failure for the period of construction, some approaches are available, for example assuming an event occurrence of Poissonian nature. This leads to simple asymptotic formulas for the extreme value distribution for a given time period.

3.3 *Point estimate methods (PEM)*

A number of different approaches have been suggested in the literature to integrate Equation 4 (for e.g. Li 1992 and Zhou & Nowak 1988). In this work two approaches are used for assessing the statistical parameters of a function $G(\boldsymbol{X})$ of multiple random variables.

The Taylor Series (U.S. Army Corps of Engineers 1999) is a simple method to estimate the influence of each variable on the limit state function and gives a first coarse overview on the variability of the obtained results. For n uncorrelated and normal distributed variables 2n+1 calculations are required to estimate the first two statistical moments (mean μ and standard deviation σ) of $G(\boldsymbol{X})$.

In the approach proposed by Zhou & Nowak (1988) selected points in the standard normal space are used. These points must be transformed in the typically correlated and non standard normal distributed space. The integration of Equation 4 can be assumed using the non-product formula:

$$E\left[G^k(\boldsymbol{X})\right] \cong \sum_{j=1}^{m} w_j G^k(x_j) \qquad (5)$$

where m is the number of points considered, w_j are the weights and x_j are the obtained points in the non-normal space.

4 EXAMPLE

4.1 *Geometry and subsoil conditions*

The geometry of the shallow tunnel analysed in the following is given in Figure 2 including the assumed ground conditions. The tunnel is situated in a limestone formation ("rock"), with a soil layer of an average thickness of approximately 5 m on top ("soil").

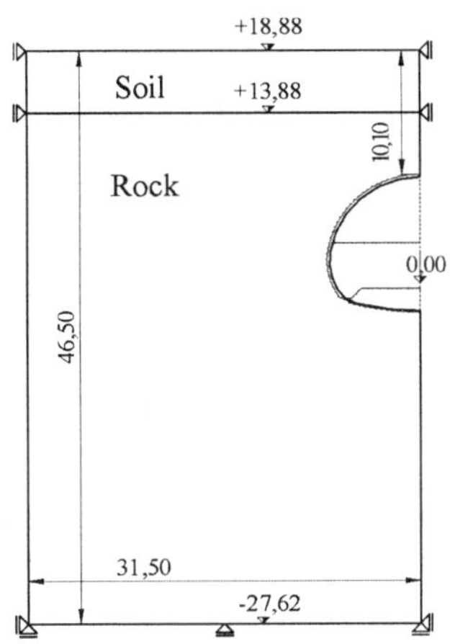

Figure 2. Geometry used for calculation.

The calculations are performed using the so-called Hardening Soil Model implemented into the finite element code PLAXIS (Brinkgreve & Vermeer 1998). The tunnel is excavated according to the principles of the New Austrian Tunneling Method (NATM) thus a shotcrete lining with rock bolts provides the support during construction stages whereas a cast concrete shell serves as final lining. Excavation modelling is made in sections and prerelaxation factors of around 25 % (for each section) according to the load reduction method has been assumed to account for 3D effects in the 2D plane strain analysis. The shotcrete lining is modelled with elastic beam elements. It is distinguished between the young and old shotcrete, to incorporate the aging effect for the stiffness in an approximate way. The goal of the analysis is a comparison between results obtained by using design values and from probability analysis. In addition the probability of failure during construction is determined.

For the given geology a first parameter evaluation resulted in 12 significant parameters for this problem, which are treated as basic variables as shown in Table 1.

Table 1. Basic variables for design value calculations and Taylor Series analysis.

	Parameter	Distribution	Mean	Stand. dev.	Unit
Soil	φ	lognormal	21	2	[°]
	c	lognormal	24	8	[kN/m²]
	E	lognormal	24000	8000	[kN/m²]
	Ko	normal	0,65	0,1	[-]
Rock	φ	lognormal	37	2,5	[°]
	c	lognormal	100	18,75	[kN/m²]
	E	lognormal	200000	31250	[kN/m²]
	Ko	normal	0,7	0,1	[-]
ΣMStage	ΣMstage1	normal	0,25	0,03	[-]
	ΣMstage2	normal	0,25	0,02	[-]
Shotcrete	G_{young}	lognormal	2000000	500000	[kN/m²]
	G_{old}	lognormal	6000000	1500000	[kN/m²]

For soil and rock the angle of internal friction, the cohesion, the coefficient of lateral earth pressure and the elastic modulus at a reference stress level are variables. ΣMStage1 indicates the prerelaxation factor for the top heading and ΣMstage2 is used for the bench. Uncertainties in thickness and stiffness of the shotcrete lining are included in the calculation of the shear moduli (G_{young} and G_{old}), which are treated as variables.

4.2 *Taylor Series and Matrix of Influence*

Using a Taylor Series analysis the influence of each variable on a certain limit state function can be quantified. For the 12 variables shown in Table 1 25 calculations are required to get an estimation of mean and standard deviation for each limit state function. In this case the settlement of the crown (Uy) and the relation between allowable and calculated normal force (N_{all}) in the shotcrete lining at the end of construction is evaluated. For the allowable normal force a definition after Schikora & Ostermeier (1988) is used. To include the utilization of the rock mass strength the maximum shear strain (gam_s) after the excavation of the top heading is considered.

Figure 3 shows the maximum normal forces after the last construction stages for all calculations. For e.g. calculation 22 and 23 shows the effects in the normal force for a variation of the stiffness of the young shotcrete.

To quantify this effect for each variable separately a relation between the obtained standard deviation of the normal forces based on all calculations

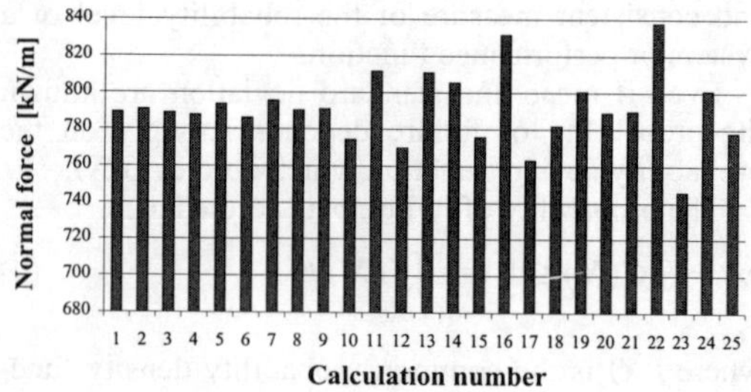

Figure 3. Maximum normal forces for 25 calculations – Taylor Series analysis.

and the standard deviation for the normal force caused by a single variable is used. Doing this with all variables a first estimation on the influence of each variable on the respective result is obtained and shown in a matrix of influence (Table 2). It can be seen that roughly half of the variables have a significant influence.

Table 2. Matrix of influence.

	N_{all}	Uy	gam_s	Total sum
φ-Soil	0	0	1	1
c-Soil	0	0	1	2
E-Soil	1	2	2	5
Ko-Soil	0	0	1	1
φ-Rock	3	4	9	16
c-Rock	4	3	6	13
E-Rock	1	5	4	10
Ko-Rock	4	9	6	19
ΣMStage 1	2	2	4	7
ΣMStage 2	0	0	1	1
G_{young}	12	9	2	22
G_{old}	1	0	0	2

At this end a decision can be made, which variables should be used in further calculations and which one can be treated as deterministic values as their influence on the result is not significant. The defined threshold value depends on engineering judgment and should be used carefully (Einstein 2000). Figure 4 shows the total relative influence value and the defined threshold value has been chosen here as approximately 4 %.

4.3 *Comparison for results obtained with probabilistic analysis and design values*

Based on the results of the previous section only 7 parameters are treated as variables for further investigations with the PEM after Zhou & Nowak (1988), i.e. 99 calculations are required ($2n^2 + 1$). The results of these calculations will be compared with deterministic analyses making different assumptions for design values.

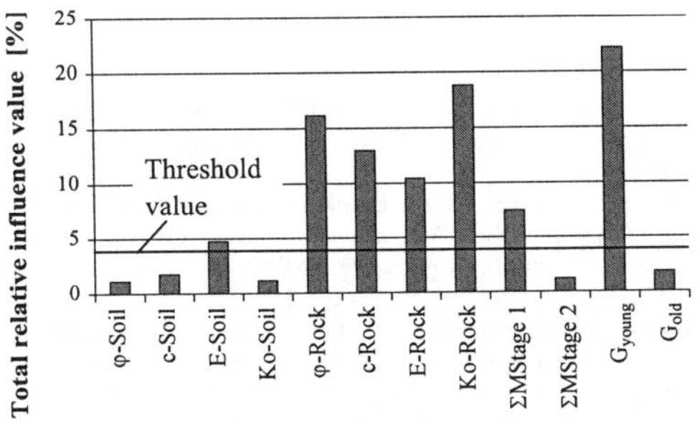

Figure 4. Total relative influence value.

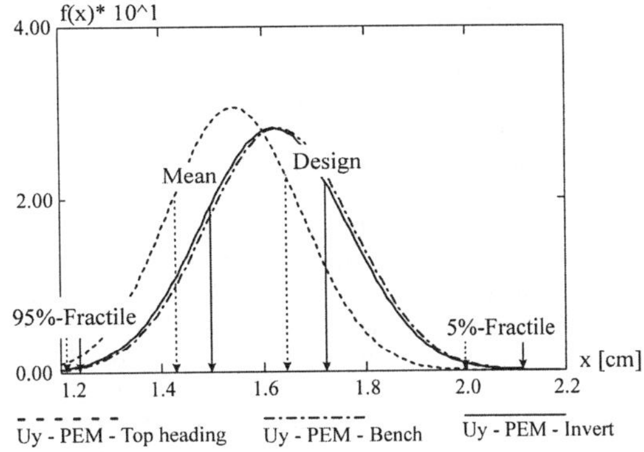

Figure 5. Density distribution for the settlement of the crown during the three construction stages.

The choice of design parameters is often based on personal experience rather than on a sound theoretical basis. In the following some possible assumptions for deterministic analysis will be investigated here, namely

- mean values: mean values are used
- characteristic design values: mean value reduced by half of the standard deviation are used (approx. 35%-fractile)
- 5%-fractile: for the rock and soil parameters 5%-fractile value are used; for the young shotcrete stiffness the 95%-fractile is used
- 95%-fractile: for the rock and soil parameters 95%-fractile value are used; for the young shotcrete stiffness the 5%-fractile is used

From the 99 calculations mean value and standard deviation for a particular result, e.g. the crown settlement, can be determined. By assuming a normal distribution (due to a small cov) the probability of obtaining a result above or below the corresponding deterministic analysis can be evaluated.

A graphical view of this comparison for the settlement of the crown is given in Figure 5 for all construction stages. The determined fractiles of exceeding the results obtained with "calculation values" are shown for the normal force and the settlement of the crown in Table 3. For e.g. using the 5%-fractiles for input parameters result in a 0.02%-fractile (for exceeding) of the settlement of the crown after the excavation of the top heading. It can be concluded that a certain input fractile results in a different fractile for the output, a result which is of course not surprising from a theoretical point of view but often not appreciated in practice.

4.4 *Evaluation of limit state functions*

With the obtained mean values and standard deviations based on the PEM limit state functions can be evaluated. These limit state functions should be related to failure modes including the shotcrete lining and the surrounding rock.

Table 3. Fractiles of exceeding the obtained results according to the various calculation values.

Fractiles [%]	Settlement of crown Uy		Normal force	
Calculation value	Top heading	Bench	Top heading	Bench
5%-Fractile	0.02	0.001	0.001	0.0001
Design value	22	18.6	23.4	23.6
Mean	81.2	87.1	29.6	27.6
95%-Fractile	99.42	99.79	99.999	99.999

For the shotcrete lining the allowable normal force will be compared with the assumed lognormal distribution for the obtained PEM results. According to the results obtained from the design value calculation 15 cm shotcrete thickness is required. The deterministic factor of safety for this case would be FOS = 1.6. The corresponding probability of failure is $p_f = 1.5E\text{-}6$. If a thickness of 20 cm is chosen it results in $p_f = 1.3E\text{-}7$ and a FOS = 2.1. For a thickness of 10 cm the FOS is below 1.0 and the obtained $p_f = 2E\text{-}4$.

Supplemental to the limit state function for the shotcrete lining some other limit state functions were investigated for the settlement of the crown and the maximum shear strain. In both cases there are some difficulties to define the threshold values, thus further investigations on this topic are under progress. For e.g. to evaluate the utilization of the surrounding rock not only the maximum value of shear strain should be evaluated but also the extent of plastic zones should be considered.

5 CONCLUSIONS

A short overview of probabilistic calculations in geotechnics and some remarks on the stochastic nature of soils were given.

For a shallow tunnel excavation the Taylor Series method was used to determine the influence of each parameter on the results of a finite element calculation. Based on this information an objective decision could be made whether a parameter was treated as variable or as a deterministic value in further calculations.

With a more sophisticated point estimate method the normal force and the settlement of the crown were evaluated based on 99 calculations. For comparison calculations with some typical "design values" based on certain fractile values were performed. It could be shown that for a certain input fractile different, not predictable fractiles of the results are obtained.

The failure of the shotcrete lining was used as example for evaluating a limit state function. The obtained probabilities of failures correspond well with the deterministic factors of safety. To include the utilization of the surrounding rock further investigations are under progress. First steps with the evaluation of the maximum shear strains and the degree of utilization of the rock strength in certain areas seem to be promising.

The results presented here indicate the potential of the proposed approach and further work in improving this concept is presently in progress.

REFERENCES

Brinkgreve, R.B.J. & Vermeer, P.A. 1998. PLAXIS Finite Element Code for Soil and Rock Analysis. Version 7.1. Rotterdam: Balkema

Duncan, J.M. 2000. Factors of safety and reliability in geotechnical engineering, J. of geotechnical and Geoenvironmental engineering, Vol. 126, No. 4: 307-316.

Einstein, H.H. 1996. Risk and Risk Analysis in Rock Engineering. Tunneling and Underground Space Technology. Vol. 1. No. 2. 141-155.

Einstein, H. H. 2000: Risikoanalyse in der Geotechnik, Seminarunterlagen für den Short Course, 17.6.2000, Geotechnical Group Graz, Graz University of Technology, Austria,

Fredlund, D.G. 1984. Analytical Methods for Slope Stability Analysis. Landslides - Proc. 4th Int. Symposium, Toronto 16–21 September 1984: 229-250. Toronto: CGS

Hashofer, A.M. & Lind, N.C. 1974. Exact and invariant second-moment code format. J. for Engineering Mechanics Division, ASCE, 100: 111-121.

Lacasse, S. & Nadim, F. 1997. Uncertainties in characterizing soil properties. NGI-Publication 201.

Li, K.S. 1992. Point estimate method for calculating statistical moments. J. for Eng. Mechanics 118(7): 1506-1511.

Li, K.S., Lee, I.K. & Lo, S-C.R. 1993.Limit state design in geotechnics. Proc. of the Conf. on probabilistic methods in geotechnical engineering. Canberra, 10-12 February 1993: 29-42. Rotterdam: Balkema

Mostyn, G.R. & Li, K.S. 1993. Probabilistic slope analysis – State-of-play. Proc. of the Conf. on probabilistic methods in geotechnical engineering. Canberra, 10-12 February 1993: 89-109. Rotterdam: Balkema.

Orr, T.L.L. 2000. Selection of characteristic values and partial factors in geotechnical designs to Eurocode 7. Computers and Geotechnics 26: 263-279.

Petschacher, M. 1993. Zuverlässigkeit technischer Systeme. Bericht Nr. 199, Institut für Baustatik und Konstruktion. Zürich: Birkhäuser.

Program CombInfo 1.0. 1999. Institute of Structural Engineering, Swiss Federal Institute of Technology ETH. Switzerland.

Rackwitz, R. 1998. Zuverlässigkeit und Lasten im konstruktiven Ingenieurbau. Vorlesungsumdruck. TU-München

Rackwitz, R. 2000. Reviewing probabilistic soils modeling. Computers and Geotechnics 26: 199-223.

Schikora, K., Ostermeier, B. 1988: Temporäre Sicherung von Tunneln mit Spritzbeton, Bauingenieur 63, 399-403.

Schneider, H.R. 1999. Determination of characteristic soil properties. Proc. XII International Conference on Soil Mechanics and Geotechnical Engineering. Vol 4. Hamburg, 2271-2274

Thurner, R., Schweiger, H.F. & Petschacher, M. 2000. Reliability analysis for geotechnical problems via finite elements – Some preliminary results from a comprehensive study. Proc. of the ICASP 8 conference. Sydney. 12-15 December 1999: 469-476 Rotterdam. Balkema.

U.S. Army Corps of Engineers: ETL 1110-2-556, „Risk-based Analysis in geotechnical engineering for support of planning studies", 1999

Vanmarcke, E.H. 1977. Probabilistic Modeling of Soil Profiles. J. of the Geotechnical Engineering Division, ASCE, 103: 1227-1246.

Vilaescusa, E. 1993. Statistical modelling of rock jointing. Proc. of the Conf. on probabilistic methods in geotechnical engineering. Canberra. 10-12 February 1993: 221-231. Rotterdam. Balkema.

Zhou, J. & Nowak, A.S. 1988. Integration formulas to evaluate functions of random variables. Structural safety 5: 267-284.

Computer Methods and Advances in Geomechanics, Desai et al. (eds) © 2001 Balkema, Rotterdam, ISBN 90 5809 183 X

Modeling the stress-strain relation for granite using finite element-neural network hybrid algorithms

H.-C.Wu, X.-B.Zhang & T.Bao
Guizhou University of Technology, Guiyang, People's Republic of China

S.H.Al-Jibouri
University of Twente, Enschede, Netherlands

ABSTRACT: In this paper, it is shown how stress-strain behavior over a wide range of confining stress can be encapsulated within a single multilayer feedforward neural network. Data used to train the neural network is directly captured from the results of biaxial compression tests for granite at the ship locks in the Three Gorges on the Yangtze River. The simulation of the stress-strain relation for the granite is conducted by means of finite element-neural network(FE-NN) hybrid algorithms. In the algorithms, force boundary conditions are applied and displacement boundary conditions are enforced through training of the NN material model for the granite. The applicability and predictive capability of the FE-NN hybrid model for the granite is checked by analyzing stability of the high granite slope at the ship locks in the Three Gorges on the Yangtze River, and some significant conclusions are drawn out.

1 INTRODUCTION

An important recent interest in neural networks (NNs) within geomechanics is the question of its coupling with other numerical methods. In the majority of reported cases, the application of NNs leads to a stand alone system .In some cases it is difficult to establish exactly how these new tools fit together with the existing tools (numerical methods). The key to maximizing the benefits of this new NN technology is in its integration into our existing tools, for example FE, thereby endowing the later with increased capability. This paper targets this issue specifically, evaluating stability for the high granite slope at the ship locks in the Three Gorges on the Yangtze River, employing FE-NN hybrid model.

2 LEARNING VERSUS MODELLING OF THE STRESS-STRAIN RELATION

Modeling is a fundamental method in research of engineering problems. Modeling of the observed phenomena, such as the stress-strain behavior of material, enables the understanding of phenomena and makes most engineering analysis possible. NNs (Fig.1) offer a new and fundamentally different method of approaching the task of the behavior modeling. First, the material behavior, as represented in the experimental data, is examined in order to identify the major features of material behavior. Once the material behavior is sufficiently well understood, and its main features identified, then a mathematical model is developed to simulate this behavior. The artificial neurons in the input layer may represent the state of stresses, the state of strains and the stress increments. The units in the output layer may represent the strain increments .In

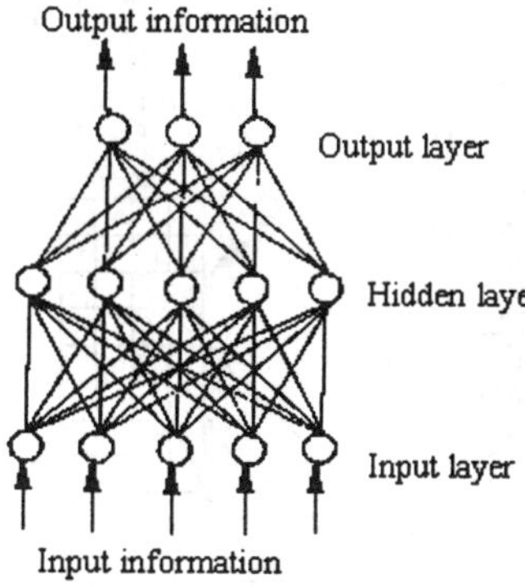

Fig.1 A three layer NN material model for granite

the new behavior modeling method, the stress-strain behavior is directly captured from the results of biaxial compression tests for ganite at the ship locks in the Three Gorges on the Yangtze River by means of the learning capabilities of NNs and is encoded within the connection strengths of a self-organizing feed-forward NN. During the training, the activation values of both the input and the output layers are specified and the learning rules are applied to modify the strengths of the connections. After a NN has been trained with a sufficient amount of test data, along different stress paths, it will be capable of giving the correct strain for any set of stress increments, at any state of stress.

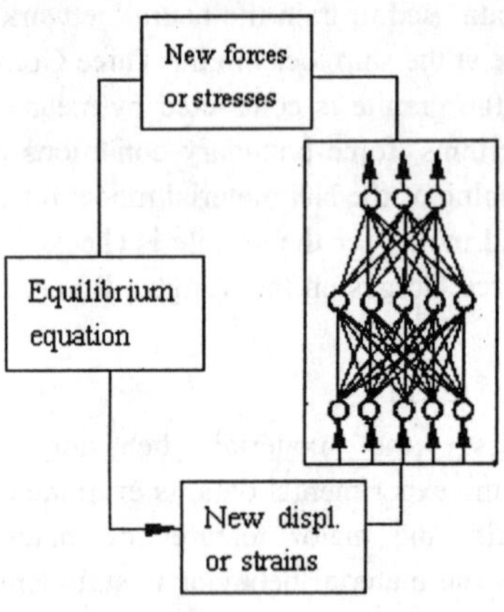

Fig.2 The concept of the FE-NN hybrid algorithms

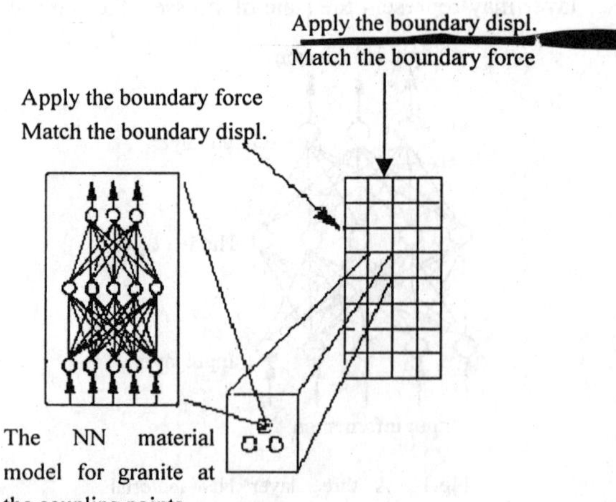

Fig.3 Training of the NN granite model within the FE

The new behavior modeling model provides the mapping from stress increments to strain increments (stress controlled model). However, in computational mechanics and finite element analysis, the inverse mapping from the strain increments to the stress increments (strain controlled model) is needed. One of the important features of the NN computational models is that the inverse mapping can be developed as easily as the direct mappings.

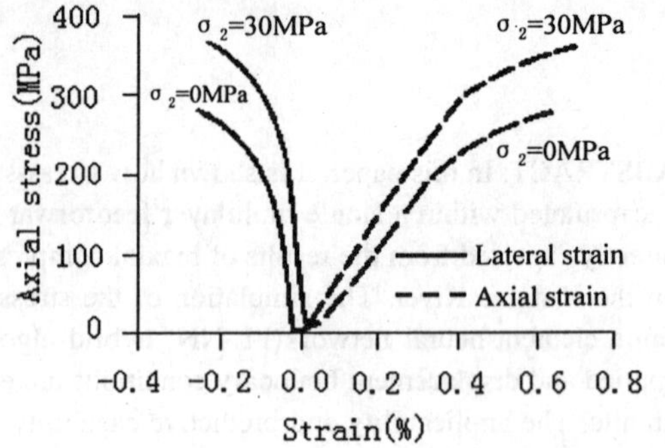

Fig.4 The axial stress-lateral(or axial) strain curves of the FE-NN hybrid model for granite under the confining stress(σ_2)

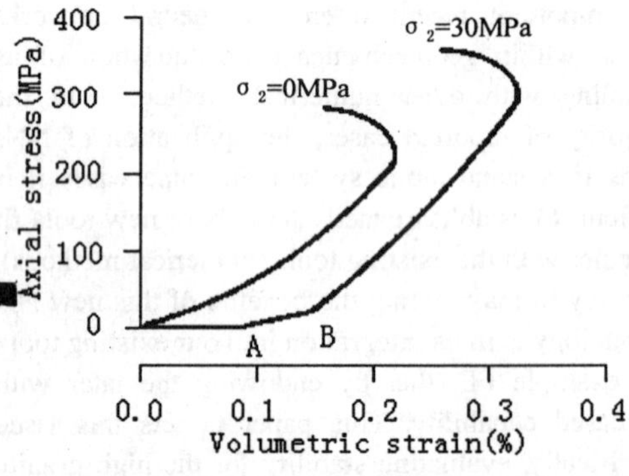

Fig.5 The axial stress-volumetric strain curves of the FE-NN hybrid model for granite under the confining stress(σ_2)

The training data for the new NN computation containing the knowledge of the nonlinear mapping to be modeled must include a sufficient number of stress paths, acquired by the network. A data set which includes the relevant knowledge is called a comprehensive data set. A NN trained with a comprehensive data set on the stress-strain behavior will be able to use the generalization capability of the NN and respond appropriately when queried about the stress paths not included in the training data set.

3 FINITE ELEMET-NEURAL NETWORK HYBRID ALGORITHMS

Fig.2 shows the concept of the FE-NN hybrid algorithms. The purpose of a NN material model employed within the FE is to use the measured boundary forces and displacements to train a NN granite model ,where the NN exactly replaces the role of the more conventional constitutive laws in the FE. The FE-NN hybrid approach allows the training of the NN granite model in a FE analysis of the biaxial conpression test (Fig.3).

material model so that the FE results match the second set of the measured boundary conditions.

The algorithm steps in which force boundary conditions are applied and displacement boundary conditions are enforced through training of the NN material model are:

Apply the measured load increment $\Delta\overline{P}_n$ and analyze the structure:

$$P_n = P_{n-1} + \Delta\overline{P}_n \tag{1}$$

$$K_t^{j-k}\delta U_n^j = p_n - I_{n-1} - \delta I_{n-1}^{j-1} \tag{2}$$

n=load increment number
j=iteration number

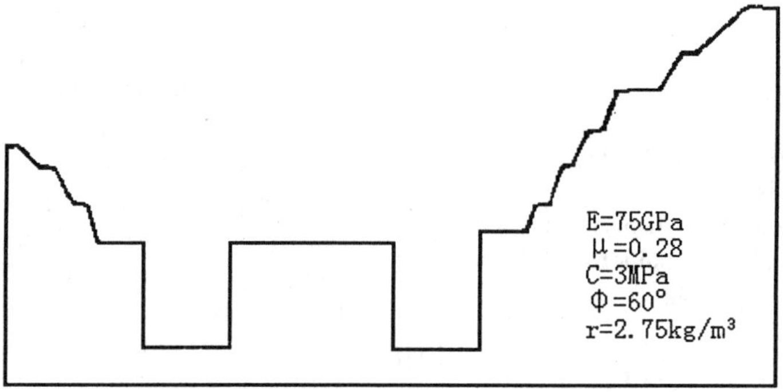

Fig.6 The analytical model of the 115m high granite slope

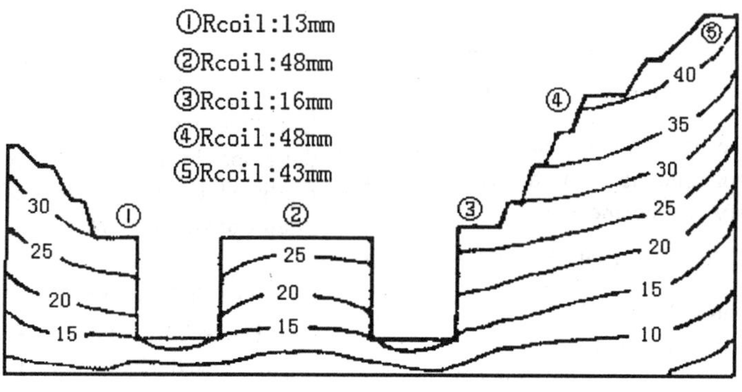

Fig.7 The horizontal displacement(mm) distribution of the high granite slope

The FE-NN hybrid algorithm is based on applying one set of the measured boundary conditions in the FE analysis and iteratively modifying the connection weights of the NN

$$\delta\varepsilon_n^j = B\delta U_n^j \tag{3}$$

$$\Delta U_n^j = \Delta U_n^{j-1} + \delta U_n^j \tag{4}$$

$$\Delta\varepsilon_n^j = \Delta\varepsilon_n^{j-1} + \delta\varepsilon_n^j \tag{5}$$

$$\Delta\sigma_n^j = \Delta\sigma_n^{j-1} + \delta\sigma_n^j \tag{6}$$

Compute the boundary displacement errors $\delta\Delta\bar{U}_n^j$, as the difference between the measured boundary displacements and the computed boundary displacements:

$$\delta\Delta\bar{U}_n^j = \Delta\bar{U}_n^m - \Delta\bar{U}_n^j \tag{7}$$

If the boundary displacement errors are less than a prescribed tolerance, then update the displacements, strains and stresses:

$$U_n = U_{n-1} + \Delta U_n^j \tag{8}$$

$$\varepsilon_n = \varepsilon_{n-1} + \Delta\varepsilon_n^j \tag{9}$$

$$\sigma_n = \sigma_{n-1} + \Delta\sigma_n^j \tag{10}$$

and go to the first step and apply the next load increment. Otherwise, continue.

Apply the boundary displacement errors as displacement boundary conditions, with no other loads acting on the FE mesh:

$$\begin{bmatrix} K_{iaa}^{j-k} & K_{iba}^{j-k} \\ K_{iba}^{j-k} & K_{ibb}^{j-k} \end{bmatrix} \begin{Bmatrix} \delta\Delta U_n^j \\ \delta\Delta\bar{U}_n^j \end{Bmatrix} = \begin{Bmatrix} P_{n-1} \\ \bar{P}_{n-1} + \delta\bar{P}_n^j \end{Bmatrix} - \begin{Bmatrix} I_{n-1} \\ \bar{I}_{n-1} \end{Bmatrix} \tag{11}$$

The strain increments, $\delta\Delta\varepsilon_n^j$, at the coupling points are computed and they are propagated through the NN granite model to determine the corresponding stress increments, $\delta\Delta\sigma_n^j$. These stress increments are the errors at the output of the NN granite model ,corresponding to the errors at the measured boundary displacements.

The error at the output of the NNs, $\delta\Delta\sigma_n^j$, is backpropagated through the NN granite model and the connection weighs are modified.

These steps are repeated until the NN granite model learns the stress-strain behavior. With the trained NN, the FE analysis will be able to produce the correct measured boundary displacements, when subjected to the measured boundary forces.

4 THE FE-NN HYBRID MODEL FOR GRANITE

The studied results show that:

1. The axial stress-lateral(or axial) strain relations of the FE-NN hybrid model for the granite under the confining stress(σ_2) are nonlinear curves(Fig.4).
2. The axial stress-volumetric strain curves of the FE-NN hybrid model for the granite under the confining stress(σ_2) have two obvious turning points which correspond respectively to the stress level of crack closing and to the level of confining stress(σ_2) (Fig.5). The deformation between the above two points is the elastic response of the rock matrix under hydrostatic pressure.

5 STABILITY SIMULATION FOR THE HIGH GRANITE SLOPE AT THE SHIP LOCKS IN THE THREE GORGES USING THE FE-NN HYBRID ALGORITHMS

Fig.6 shows the analytical model of the 115m high granite slope at the ship locks in the Three Gorges. Movement on the vertical boundaries is constrained horizontally only, while movement on the bottom boundary is constrained in both horizontal and vertical directions. A mesh consisting of 1151 nodes and 1084 rectangular elements is utilized in the analysis. Fig.7 represents the horizontal displacement(mm) distribution for the high granite slope at the Three Gorges simulated by the FE-NN model for the granite. The maximum horizontal displacement at the shoulder of the slope is 42mm, and there occur recoil deformations in the roofs of the slope. In a sense, it may be concluded that the predicted results of the FE-NN hybrid model for granite are quite good.

6 CONCLUSIONS

1. The axial stress-lateral(or axial) strain relations of the FE-NN hybrid model for granite under the confining stress(σ_2) are nonlinear curves.
2. The axial stress-volumetric strain curves of the FE-NN hybrid model for the granite under the confining stress(σ_2) have two obvious turning points which correspond respectively to the stress level of crack closing and to the level of confining stress(σ_2). The deformation between the above two points is the elastic response of the rock matrix under hydrostatic pressure.

3. The maximum horizontal displacements at the shoulder of the slope and the roof of the ship locks are respectively less than 45mm and 30mm, which will not impair the normal work of the ship locks in the future.

4. The simulated results for the high granite slope at the ship locks in the Three Gorges by using the FE-NN hybrid model for granite are in a closing agreement with the site data.

5. The FE-NN hybrid model for granite is able to use the generalization capability of the NN and is capable of giving the new results of biaxial compression tests, and update the hybrid model for granite.

REFERENCES

Ghabousi, Jamshid 1994. Neural network based modeling in geomechanics. *Computer Methods and Advances in Geomechanics*, 153-164. Rotterdam: Balkema.

Hertz, Jhon 1991. *Introduction in the theory of neural computation*. 1-228. Redwood City: California Addison-Wesley Publishing Company.

Wu, Hong-Ci 1998. Stability analysis of the high slope at the permanent ship locks in the Three Gorges on the Yangtze River by FLAC. *J. of Guizhou Univ. of Tech.*, 27(1):32-38

Wu, Hong-Ci 1996. Roof controlling in longwall face using neural networks. *Ground Pressure and Strata Control.* 3:6-9.

Wu, Hong-Ci 1996. Floor classification and roof weighting prediction in longwall face based on neural networks. *J. of Guizhou Univ. of Tech.*, 25(4):32-36.

Wu, Hong-Ci 1999. Constitutive model simulation for soil by using finite element-neural network hybrid algorithms. *J. of Guizhou Univ. of Tech.*, 28(2):95-99.

Computer Methods and Advances in Geomechanics, Desai et al. (eds) © 2001 Balkema, Rotterdam, ISBN 90 5809 183 X

Prediction of groundwater elevation of Huanglashi landslide with ANN method

Luo Xianqi
University of Three Gorge, Yichang & Institute of Rock and Soil Mechanics, Chinese Academy of Sciences, Wuhan, People's Republic of China

Jiang Qinghui & Ge Xiurun
Institute of Rock and Soil Mechanics, Chinese Academy of Sciences, Wuhan, People's Republic of China

Liu Sifeng
University of Three Gorge, Yichang, People's Republic of China

ABSTRACT: This paper will establish the nonlinear relation between landslide groundwater elevation and rainfall process with BP model of ANN (Artificial Neural Net) in order to predict the groundwater elevation of Huanglashi landslide on the condition of typical rain process which induced landslide. The result of saturated/unsaturated seepage theory with FEM is listed to compare with the predicting values. It offers fundamental hydrographic-geological data for estimating the stability and disaster of Huanglashi landslide.

1 INTRODUCTION

The Huanglashi landslide, locating in north bank of Yangzi River, in 66Km of upriver of Three Gorges Dam Site, is one of the biggest landslides along Yangzi River. It includes Dashiban, Shiliushubao, Hengping as well as many contemporary shallow landslides. Data indicated that reviving of the Huanglashi landslide resulted from rainfall.

During rainfall, the groundwater elevation is intimately related to the vegetable, the vaporizing degree of surface water, the degree of maintaining water of slope and permeability coefficient of soil etc. Their relation is highly nonlinear. At present, main studying method is FEM, which generalized the landslide into simple model, and set up their mathematics model. Due to complexity, especially the nonlinear relation between the groundwater elevation and its effecting factors, the calculation results are far disaccord with practical conditions, so the new methods have been studied.

The monitor systems that include the globe displacement measure system, the crack monitor system, the shallow surface water monitor system and the ground water level in deep hole monitor system etc., were set up on the landslide in 1994. It is significantly important in theories and practical engineering to set up interrelation among characteristic physical parameters using these monitor data to predict developing tendency of landslide. The variable process of groundwater elevation caused by rainfall process is forecasted in this paper.

BP model of ANN has broadly application perspective in nonlinear relation aspect because of fast convergence, highly error-tolerated capability and nonlinear characteristic. In this paper, the nonlinear relation between landslide groundwater level and rainfall process is established with BP model of ANN in order to offer fundamental hydrographic-geological data for estimating the stability of Huanglashi landslide.

2 BACK PROPAGATION MODEL OF ANN

2.1 *Learning algorithm of BP model*

It is assumed that BP net model has L layers. When P pieces of samples are input, the expected output is:

$$T_d=[T_{d1}, T_{d2}, \ldots, T_{dp}] \quad (1)$$

When the pth sample is input, the input and output of the jth node of the lth (l=1, 2, …, L) layer are respectively:

$$\text{Input:} \quad net_{jp}^{(l)} = \sum_{i=1}^{n_{l-1}} W_{ji}^{(l)} O_{ip}^{(l-1)} \quad (2)$$

$$\text{Output:} \quad O_{jp}^{(l)} = f(net_{jp}^{(l)}) \quad (3)$$

The purpose of the learning is to minimize each sample's error Ep, so that the total error E of the net is minimized.

$$E_p = \frac{1}{2}\sum_{j=1}^{m}(T_{jdp} - \hat{T}_{jp})^2 \quad (p=1, 2, \ldots, P) \quad (4)$$

$$E = \frac{1}{2p}\sum_{p=1}^{P} E_p \quad (5)$$

Where $T_{jdp}, \hat{T}_{jp}$ is respectively the expected output and the real output of the jth node of the output layer.

Table 1. Samples describing relationship between rainfall and water elevation without drain well

Item/NO.		1	2	3	4	5	6	7	8	9	10
Rainfall		104.5	60.7	67.7	73.3	142.5	101.1	102.9	23.6	66.5	86.5
Rain intensity		5.5	3.79	4.51	4.89	10.96	9.19	11.43	2.95	9.5	12.36
Rain type		3	2	1	1	3	3	1	1	1	1
Deviation		0.03	0.84	0.43	0.63	0.73	0.59	0.61	0.81	0.06	0.21
Status before raining		3.5	2.1	8.7	5.6	3.1	17.3	0.2	3.6	3.3	6.7
Water elevation before raining	Bk1	14.49	15.55	13.27	12.89	12.39	13.87	14.84	15.48	13.57	13.57
	Bk2	22.46	22.75	18.84	21.29	20.22	21.66	22.24	22.29	22.09	22.18
	Bk3	9.55	13.12	7.21	6.07	5.63	5.49	11.04	11.74	7.22	7.08
	Bk4	24.34	24.61	22.75	22.99	22.65	22.97	23.28	24.62	23.28	23.30
Water elevation after raining	Bk1	14.61	15.55	12.38	13.26	6.91	10.36	13.33	15.48	13.69	13.22
	Bk2	22.47	22.73	19.58	21.78	8.98	18.85	17.76	22.29	22.09	20.79
	Bk3	9.44	12.84	4.82	6.56	2.78	4.21	9.72	11.91	6.58	5.18
	Bk4	24.29	22.78	22.67	23.01	20.32	22.66	22.68	24.40	23.29	22.97

Table 2. Samples describing relationship between rainfall and water elevation with drain well

Item/NO.		1	2	3	4	5	6	7	8	9
Rainfall		191.1	150.7	88.0	181.5	80.0	74.60	300.7	394.3	77.5
Rain intensity		14.7	37.68	14.67	22.69	10.0	18.65	42.96	17.14	38.75
Rain type		3	1	1	2	2	1	2	3	1
Deviation		0.13	0.38	0.25	0.31	0.31	0.13	0.50	0.24	0.25
Status before raining		27.9	1.0	0.7	5.3	1.8	5.1	3.0	2.7	6.9
Water elevation before rainfall	Bk1	8.29	12.34	10.73	13.57	14.50	11.71	11.90	9.84	11.17
	Bk2	8.48	21.96	22.98	22.48	22.64	21.81	22.48	21.57	22.14
	Bk3	2.85	4.92	5.84	7.31	11.52	8.05	6.99	5.08	5.98
	Bk4	20.99	23.92	24.49	24.38	23.92	23.47	23.58	24.63	24.96
Water elevation after rainfall	Bk1	6.45	7.78	8.15	8.37	11.6	10.47	7.78	7.13	10.38
	Bk2	8.84	8.53	15.59	13.44	20.81	20.10	6.98	6.85	18.48
	Bk3	-0.08	-0.05	-0.15	2.35	7.01	5.41	-0.22	0.04	2.88
	Bk4	14.33	19.64	21.31	22.59	23.03	22.68	18.95	22.54	24.46

Table 3. Output of the learning samples without drain well

Item/NO.		Output of the learning samples									
		1	2	3	4	5	6	7	8	9	10
Water elevation after rainfall	Bk1	14.57	15.54	12.37	13.27	6.93	10.35	13.34	15.48	13.67	13.23
	Bk2	22.39	22.72	19.59	21.77	9.02	18.84	17.81	22.29	22.10	20.78
	Bk3	9.39	12.83	4.83	6.54	2.78	4.21	9.71	11.91	6.60	5.20
	Bk4	24.26	22.77	22.67	23.00	20.32	22.64	22.70	24.40	23.28	22.98

Table 4. Output of the learning samples with drain well

Item		Output of the learning sample								
		1	2	3	4	5	6	7	8	9
Water elevation after rainfall	Bk1	6.44	7.78	8.15	8.36	11.60	10.47	7.77	7.14	10.37
	Bk2	8.84	8.53	15.59	13.42	20.81	20.10	6.98	6.92	18.47
	Bk3	0.07	-0.04	-0.14	2.33	7.00	5.40	-0.21	0.10	2.87
	Bk4	14.33	19.63	21.31	22.59	23.03	22.68	18.95	22.51	24.45

The weight value and penalty value of the net is modified with gradient method. The iteration equation of the weight coefficient of the lth layer is:

$$W(k+1) = W(k) + \eta \Delta E_p \qquad W = \{w_{ij}\} \tag{6}$$

Where, η is learning step, k is iteration times.

$$\Delta E_p = \frac{\partial E_p}{\partial w_{ij}^{(l)}} = \frac{\partial E_p}{\partial net_{jp}^{(l)}} \frac{\partial net_{jp}^{(l)}}{\partial w_{ij}^{(l)}} = \frac{\partial E}{\partial net_{jp}^{(l)}} O_{ip}^{(l-1)} \tag{7}$$

Where, $\delta_{pj}^{(l)} = \frac{\partial E_p}{\partial net_{jp}^{(l)}}$ is the node error.

The bigger η is, the more violent the change of weight is. It causes the learning process to vacillate. To get the enough big η, but not to make vacillation, the energy item is usually added into the modified formula of the weight value.

$$w_{ij}(k+1) = (1-mc)\eta \delta_{jp} o_{jp}^{(l-1)} + mcw_{ij}(k) \tag{8}$$

2.2 Training process of BP model

① To initiate net

A set of random number is input to the initial weight value, and the learning step η, the limit error ε, as well as the net structure including the net layer number L and the node number n_l of each layer are set up in advance.

② To offer one set of samples to net

③ To circulate the net to each sample p

a. To calculate forward the input and output of each node by layer.

b. To calculate the output error E_p of the pth sample and the total error E of the net.

④ When E is less than the limit error ε, or the iteration times reach the criterion, the learning process terminated. Otherwise, the error is backward transmitted.

a. To calculate backward the error $\delta_{jp}^{(l)}$ of each node by layer.

Output layer: $\delta_{jp}^{(l)} = O_{jp}^{(l)}(1-O_{jp}^{(l)})(y_{jdp} - O_{jp}^{(l)})$ (9)

Hidden layer: $\delta_{jp}^{(l)} = O_{jp}^{(l)}(1-O_{jp}^{(l)})\sum \delta_{jp}^{(l)} w_{kj}^{(l+1)}$ (10)

b. To modify the net weight value according to (8).

3 PREDICTION OF THE GROUND WATER ELEVATION OF HUANGLASHI LANDSLIDE WITH BP MODEL

After the monitor system of Huanglashi landslide has been set up since 1994, the plentiful data is offered. Firstly, the nonlinear relationship between rainfall and the groundwater elevation is established with BP model of ANN trained by survey data. Secondly, The groundwater elevation caused by typical rainfall process is predicted with trained model to provide the fundamental data for evaluating the stability and disaster of landslide. The typical rainfall process in this paper is from June to July in 1983, which ever derive the Huanglashi landslide.

The BP net consists of input layer, output layer and hidden layer, which include only one layer in this paper. The input layer has nine nodes, which include rainfall, rain intensity, rain type, deviation, status before rain, the initial groundwater elevation of survey hole BK1, BK2, BK3 and BK4. The output layer consists of four nodes, which are respectively the groundwater elevation of survey hole BK1, BK2, BK3 and BK4 after raining. Ten nodes are determinated in the hidden layer based on the training.

Some characteristic items to describe the rain process need to be digitized. The rain type, 1 is single peak, 2 is double peak, 3 is multi-peak. Deviation is the rate of the time of peak rainfall to the total time of a rain process. Status before raining is the rate of the former rainfall to the interval time during two raining processes.

When the error between the real output and the expected output of the learning sample is less than 0.05, the training net could be used to predict the project.

On the condition of the typical rain from June to July in 1983, the groundwater elevation of the landslide is predicted by trained net. The input sample, rainfall is 448.8mm, rain intensity is 12.5mm, the rain type is 3, the deviation is 0.5, the status before raining is 3.5.

Table.1 and table.2 are respectively the learning samples with and without drain well. Table.3 and table.4 are respectively the output of the learning sample with and without drain well. Table.5 is the result of the groundwater elevation of landslide with and without drain well from June to July in 1983.

Table 5. Prediction results of the landslide groundwater elevation caused by rainfall, from June to July in 1983. (Unit: m)

Item	Without drain well	With drain well
BK1	5.8486	6.9753
BK2	4.1694	7.3629
BK3	0.4106	-0.0627
BK4	18.0192	23.1897

4 THE RESULTS OF SATURATED AND UNSATURATED SEEPAGE WITH FEM

The equation of Saturated/unsaturated seepage theory is as follow:

$$\frac{\partial}{\partial x}\left(k_x\frac{\partial h}{\partial x}\right)+\frac{\partial}{\partial y}\left(k_y\frac{\partial h}{\partial y}\right)=m_w\rho_w g\frac{\partial h}{\partial t} \qquad (11)$$

Where, k_x and k_y are respectively horizontal and vertical permeability coefficient.

$$k=\begin{cases} k_s & saturate \\ k(S_w) & unsaturate \end{cases} \qquad (12)$$

k_s is saturate seepage coefficient. S_w is degree of saturate. ρ_w is density of groundwater. g is gravity acceleration. In equation, the groundwater k_s is constant in saturate area, however it is function of degree of saturate in unsaturated area.

Where, $m_w=-\frac{\partial\theta_w}{\partial(-u_w)}$ (13)

There is no negative pore pressure in saturate area, so m_w is zero. The permeability coefficient change with $-u_w$ because it is not constant in unsaturated area. Figure 1. is the maximum groundwater elevation with/without drain well on the condition of rainfall of one hundred return years in EL175 of Yangzi River.

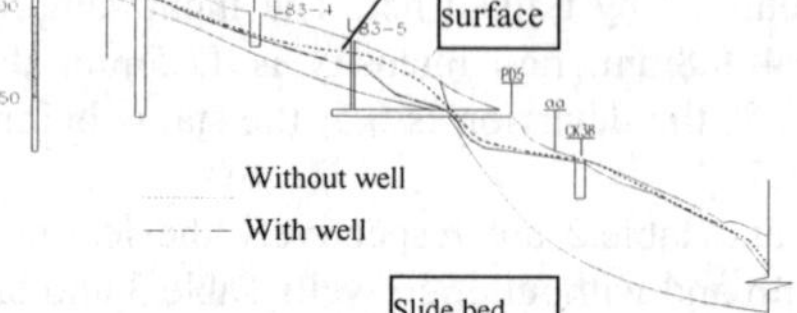

Figure 1. Groundwater curve with and without drain well

5 CONCLUSION AND SUGGESTION

The groundwater elevation with and without drain well is predicted with BP model of ANN, and the regularity of results is good. The drain effect with well is obvious.

The method of prediction of the groundwater elevation with BP model of ANN has advantages as follow: the faster convergence; the high error-tolerated capability, the highly nonlinear and the credible results etc. However, the problem is that the input samples aren't enough systematic. It is suggested that the design and the survey of the groundwater of the Huanlashi landslide should be intensive, and the integrity of the survey items should be considered.

REFERENCES

1．Luo Xianqi, Jiang Qinghui, Ge Xiurun, Liu Sifeng, Analysis of stability of Huanlashi landslide, Chinese journal of Rock Mechanics and Engineering, 2000，19（5）1-5。

2．Luo Xianqi, Zhang Xuanbing, Ge Xiurun, Back analysis of landslide groundwater based on ANN, Proceedings of the the 6th national Conference on Rock Mechanics and Engineering, in Oct. Of 2000, Wuhan, PRC.

Computer Methods and Advances in Geomechanics, Desai et al. (eds) © *2001 Balkema, Rotterdam, ISBN 90 5809 183 X*

Neural network model for predicting subsurface contamination

Seung Hi Yi
Camp Dresser and McKee Incorporated, Tampa, Fla., USA

Robert Gilbert & Dongyuan Wang
Department of Civil Engineering, University of Texas, Austin, Tex., USA

ABSTRACT: Subsurface contamination with Dense Non-Aqueous Phase Liquids such as TCE is a difficult problem to characterize due to numerous uncertainties associated with the subsurface and flow conditions. However, quantifying uncertainty in the spatial distribution (location and extent) of DNAPLs in the subsurface is necessary to design effective investigation and remediation programs. In this paper, we illustrate a non-dimensional approach for estimating the location and extent of DNAPLs using a back-propagation neural network. Distributions of DNAPL in the subsurface for a two-dimensional domain were simulated with a numerical model of multiple-phase flow and transport. The resulting "data" set was then used to calibrate a four-layered back-propagation neural network with a total of 22 nodes and 30 connection and bias weights. The advantages of the neural network over the numerical simulator are ease of use and speed. A noteworthy feature of the network calibration is that a probabilistic approach was adopted to account both for (1) inherent variability between the network predictions and the simulator results and (2) uncertainty in the parameters that describe the neural network. The result is a reasonably general methodology that can be used to predict the location and extent of DNAPLs in the subsurface with any given set of site-specific input parameters.

1 INTRODUCTION

1.1 *Motivation*

Subsurface contamination with Dense Non-Aqueous Phase Liquids (DNAPLs) such as TCE is known to be present at many sites. Many have tried to characterize DNAPL contamination in the subsurface because DNAPLs can be very toxic in aqueous concentrations well below their solubility limits (Pankow and Cherry 1996). However, success in doing so has been very limited (Mercer et. al. 1996) since uncertainty in the characteristics of the aquifer and the contamination leads to uncertainty in modeling the fate and transport of DNAPL contaminants (McGrath et. al. 1997). Yet without any methodical way of characterizing the sites contaminated with DNAPL, investigation and ultimately remediation costs become unmanageable as seen in many Superfund sites (McGrath et. al. 1997). If a general methodology that can be used to predict the location and extent of DNAPLs in the subsurface with any given set of site-specific input parameters can be developed, it may greatly help solve the screening problems present at DNAPL contaminated sites.

1.2 *Background*

A hypothetical study domain, shown in Figure 1 was developed in order to assess the variability in the spatial distribution of the Dense Non-Aqueous Phase Liquids (McGrath 1998). The study domain contains a uniform background medium of sandy aquifer, with the base of the aquifer, corresponding to the upper surface of aquitard, sloped at an angle, α. It can also contain a depression on the base where DNAPL may pool. A layer with uniform low permeability was embedded in the background medium to represent the typically layered subsurface. Along this low permeability layer, a gap may also exist. The background medium properties were assumed to be present in locations where the gap exists.

This study domain was designed to be versatile and general. The positions and sizes of the low permeability layer, the depression in the aquitard, and the gap in the low permeability layer are allowed to vary to simulate as many different cases as possible.

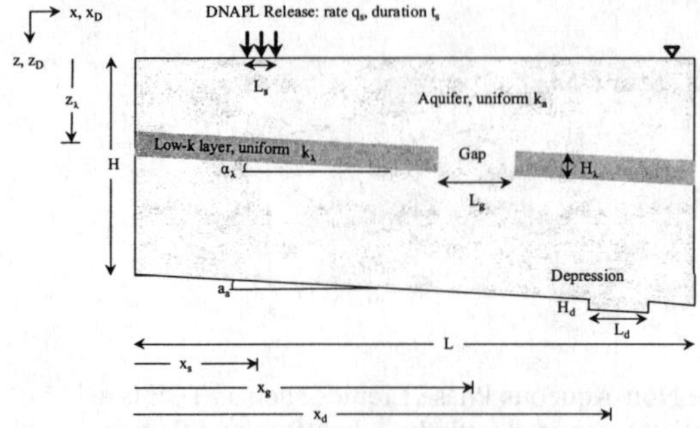

Figure 1. Hypothetical Study Domain

This hypothetical study domain was non-dimensionalized to increase its generality. An advantage of a scaled representation is that results obtained at one scale can be used to extrapolate to another when values of scaled parameters are matched for both scales. Hence, the scaled representation can increase the generality of the results obtained (McGrath 1998).

A total of 38 petrophysical and geometrical parameters, such as porosity and location of the center of the gap in the low permeability layer, described the study domain. Petrophysical parameters and properties of chlorinated solvents were limited to physically realistic values based on data from case-study sites and published values from various sources.

A two-dimensional, two-phase flow model was used to evaluate the location and distribution of DNAPL as a function of the 38 input parameters. The numerical simulator used in this work, UTCHEM, has been validated for a number of laboratory and field cases (Delshad et. al. 1996). UTCHEM solves for the pressures (P_w, P_{nw}), the concentrations (C_w, C_{nw}), and the saturations (S_w, S_{nw}) of the phases in a separate calculation at each time step in a simulation. Release of DNAPL was modeled as a set of vertical wells in UTCHEM.

With inspectional analysis and extensive sensitivity analyses, sixteen dimensionless parameters were finally retained and these sixteen dimensionless parameters became the input nodes in the neural network model. These parameters are presented in Table 1.

From the simulated saturation profile, several dimensionless output measures were selected in order to summarize important features of the distribution

Table 1. Dimensionless Input Parameters

Dimensionless Parameters (range of the parameter)		Descriptions
$R_L = \frac{L}{H}\sqrt{\frac{k_{z_a}}{k_{x_a}}}$	(4,40)	effective aspect ratio (Shook et. al. 1992)
$N_\alpha = \frac{L}{H}\tan\alpha_a$	(0.07,0.63)	dip-angle group (Shook et. al. 1992)
$M^o = \frac{k^o_{r_w}}{k^o_{r_{nw}}}\frac{\mu_{nw}}{\mu_w} = \frac{\lambda^o_{r_w}}{\lambda^o_{r_{nw}}}$	(0.1,0.7)	endpoint mobility ratio (Shook et. al. 1992)
$N_B = \frac{(\rho_{nw}-\rho_w)gH}{\sigma}\sqrt{\frac{k_{z_a}}{n_a}}$	(4,75)	bond number (Shook et. al. 1992)
$N_C = \frac{k^o_{rw}}{\mu_w}\frac{\sigma}{Hq_s}\sqrt{n_a k_{z_a}}$	(0.5,5.5)	global capillary number (Shook et. al. 1992)
$R_\alpha = \frac{\tan\alpha_\lambda}{\tan\alpha_a}$	(0.1,0.5)	slope ratio
$R_{k_z} = \frac{k_{z_a}}{k_{z\lambda}}$	(10,100)	vertical-permeability contrast ratio
t_D	(1,30)	ratio of DNAPL volume released to total DNAPL volume
$P_{D_{D\lambda}} = \frac{P_{D\lambda}}{\sigma}\sqrt{\frac{k_{z\lambda}}{n_\lambda}}$	(0.12, 0.35)	dimensionless displacement pressure for low-permeability layer
n_{nw}	(2.25,3.25)	relative-permeability exponents for nonwetting phase
$z_{\lambda_D} = \frac{z_\lambda}{H}$	(0.15,0.7)	depth of center of layer
$x_{d_D} = \frac{x_d}{L}$	(0.12,0.88)	location of center of depression
$L_{s_D} = \frac{L_s/2}{L}$	(0.01,0.05)	width of DNAPL release
$x_{gD} = \frac{x_g}{L}$	(0.25,0.75)	location of center of gap in low-permeability layer
$L_{gD} = \frac{L_g/2}{L}$	(0.04,0.25)	length of gap
$S_{nw_{r_a}}$	(0.05,0.25)	residual nonwetting-phase saturation for the aquifer

to help quantify variability in the DNAPL distribution. A total number of eight measures were selected. Five measures captured first and second moment behavior of the saturation distribution and three corresponded to the limits of DNAPL migration. The saturation of the DNAPL distribution was used to calculate the mass of DNAPL as a function of location in the study domain (Figure 1). In this paper, the focus is on a single output measure, the horizontal centroid of DNAPL saturation:

$$\overline{x}_D = \frac{\sum_{\text{all cells } i} S_{nw_i} \overline{x}_{iD}}{\sum_{\text{all cells } i} S_{nw_i}} \quad (1.1)$$

where $\overline{x}_{iD}$ is the dimensionless center of the ith grid cell in the numerical simulator.

2 METHODOLOGY

2.1 *Design of the Neural Network*

A four-layered backpropagation neural network with a total of 22 nodes and 24 connection weights and 6 bias weights was designed to predict the dimensionless horizontal coordinate of the center of mass of DNAPL. Figure 2 shows a schematic design of the network developed.

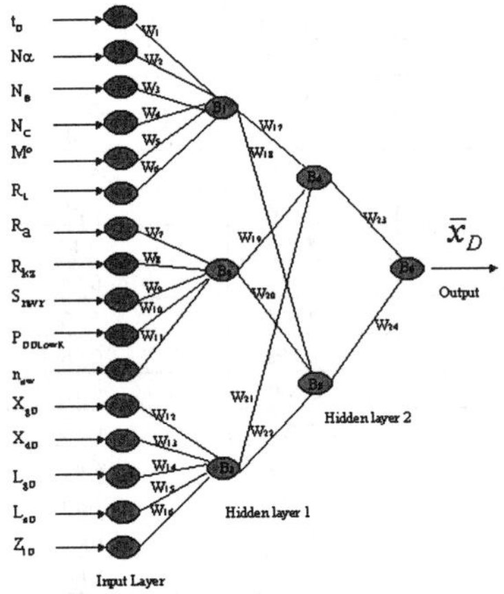

Figure 2. Schematic of the neural network

There are sixteen nodes in the input layer representing each dimensionless parameter. These sixteen dimensionless parameters were then grouped into three categories (Figure 2): characteristics of the release, characteristics of the chemical released, and parameters describing the characteristics of geometry of the hypothetical domain. This pre-classification was intended to reduce the number of hidden nodes in each hidden layer.

2.2 *Model for Inherent Variations*

Because the neural network is a simplification of the numerical simulator, there will be variations between the neural network results and the results from the numerical simulator inherent in the neural network model. Inherent variations for a single point were modeled as normally distributed about the network prediction with the following standard deviation:

$$\sigma_Y = a + b*\mu_Y \quad (2.1)$$

where σ_Y is the standard deviation for the horizontal centroid (denoted Y for simplicity), μ_Y is the neural network prediction (the mean value for Y), and a and b are model coefficients,. Therefore, the standard deviation is the inherent variations is not assumed to be constant and may increase or decrease depending on the magnitude of the neural network prediction. Correlations between inherent variations at two different points were modeled as follows:

$$\rho_{(i,j)} = \exp\{-[\sum_{\text{all p}} |p_i - p_j|/\theta_p))] + 1/\theta_n\}, \ i \neq j \quad (2.2)$$

$$\rho_{(i,j)} = 1, \ i = j \quad (2.3)$$

where i and j denote the data points, p denotes input parameters to the neural network (there are 16 input parameters with the neural network as shown on Figure 2), θ_p is the scale of fluctuation (Vanmarcke 1989) for input parameter p, and θ_n is a nugget factor.

Positive correlations between nearby data points indicate that if the neural network is underpredicting one data point is it more likely to also under-predict the nearby data point. Correlations between data points decrease as the relative distance between the data points increases. The relative distance for each parameter is computed by dividing the absolute distance between two parameters by the scale of fluctuation found for that parameter. This model is shown schematically in Figure 3 for an example with two parameters. The point of interest is closest to the data points #1 and #2, but farthest away from #3 in scales of a relative distance. Our point will therefore be highly correlated with #1 and #2, but essentially independent of point #3.

The nugget factor, θ_n, accounts for having two data points that are not perfectly correlated even with the same 16 input parameters to the neural network. This lack of correlation occurs because there

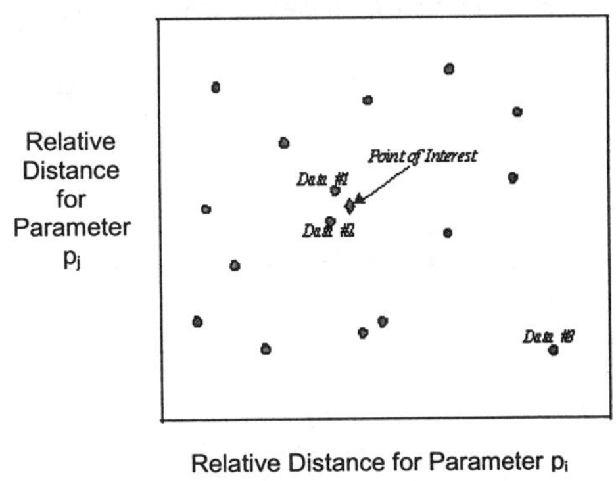

Figure 3. Illustration of the Correlation Model

are actually more than 16 parameters required to describe fully the study domain; the 16 parameters just represent the most sensitive or important parameters. As θ_n increases, the nugget correlation increases (with a maximum of 1.0 when θ_n is equal to infinity).

2.3 *Calibration of Model*

The neural network model was calibrated probabilistically using the First-Order, Second-Moment Bayesian Method, or FSBM (Gilbert 1999). This approach involves determining the set of model parameters that maximize the likelihood of for obtaining the data set. A data set with 250 numerical simulator results was used to calibrate the model. The model parameters to be calibrated include the 30 connection and bias weights in the neural network (Figure 2), the two standard deviation parameters a and b in Equation (2.1), and the 17 scales of fluctuation in Equation (2.2). The FSBM method gives updated information about the expected values for the calibrated model parameters, the standard deviations in the calibrated model parameters, and the correlation coefficients between calibrated model parameters. The standard deviations indicate how much information about these parameters can be determined from data set and the correlation coefficients indicate where there is confounding information about pairs of parameters.

It is of interest to compare this probabilistic approach for network calibration with the conventional approach. First, the conventional approach to training a neural network is to calculate the root mean square (RMS) error, and then find the set of network weights that minimizes this error (e.g., Dowla and Rogers 1995). This approach is the same as maximizing the likelihood if the standard deviation in inherent variations is a constant (if the term b is zero in Equation 2.1) and if the correlations between data points are all zero (if the scales of fluctuation in Equation 2.2 are zero). However, as will be discussed when the results are presented, neither one of these assumptions is reasonable with this data set

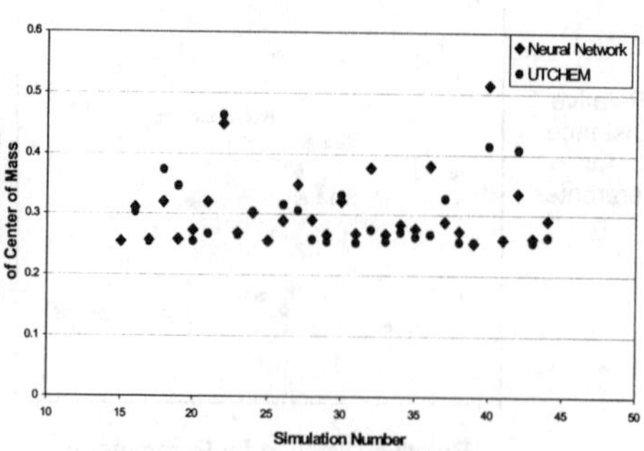

Figure 4. Example Neural Network Predictions

Table 2. Updated Information for the Neural Network Model Parameters

Model Parameter	Updated Median	Updated Coefficient of Uncertainty
W_1	-2.398	0.29
W_2	-0.684	0.41
W_3	-0.954	0.41
W_4	-1.327	0.34
W_5	0.925	0.36
W_6	2.180	0.32
W_7	-7.823	0.31
W_8	-6.784	0.35
W_9	9.464	0.27
W_{10}	-0.327	4.55
W_{11}	0.965	1.64
W_{12}	1.859	0.53
W_{13}	0.327	1.13
W_{14}	-0.262	1.33
W_{15}	-1.488	0.53
W_{16}	-0.683	0.72
W_{17}	14.620	1.66
W_{18}	2.292	0.42
W_{19}	-3.003	0.75
W_{20}	-0.211	10.36
W_{21}	0.386	4.83
W_{22}	0.262	9.23
W_{23}	-5.170	0.48
W_{24}	1.508	0.54
B_1	-1.633	1.19
B_2	-1.729	0.95
B_3	0.966	1.17
B_4	2.885	0.76
B_5	0.479	4.83
B_6	2.517	1.00
a	0.000	1000
b	0.108	0.06
θ_n	31.126	0.61
θ_{t_D}	0.058	1000
θ_{N_a}	8.184	1.33
θ_{N_B}	0.047	1000
θ_{N_C}	0.052	1000
θ_{M°	27.157	28.29
θ_{R_L}	0.041	1000
θ_{R_a}	1898.330	1000
$\theta_{R_{kx}}$	0.017	1000
$\theta_{S_{nwr_a}}$	0.861	0.57
$\theta_{P_{DD_\lambda}}$	1.963	0.52
$\theta_{n_{nw}}$	11.755	5.36
$\theta_{X_{sD}}$	0.004	1000
$\theta_{X_{dD}}$	0.038	1000
$\theta_{L_{sD}}$	1.633	0.57
$\theta_{L_{sD}}$	2.422	1.39
$\theta_{Z_{1D}}$	0.005	1000

and this neural network. Second, the conventional approach does not give information about how confidently the model parameters have been determined, which depends on the quality and quantity of data. The probabilistic approach used herein provides this information in terms of the second moments (standard deviations and correlation coefficients) for the calibrated model parameters.

3 RESULTS

The basic pattern of the numerical simulator results was fairly well matched with the neural network (Figure 4). The percentage error ranged from - 66 percent to + 27 percent.

Table 2 shows the updated median and variance values for the 49 model parameters that compose the neural network model and the inherent variability in the network predictions. The coefficient of uncertainty is defined here as the standard deviation divided by the mean value. Also, the scales of fluctuation have all been normalized so that the minimum value for an input parameter is 0.15 and the maximum value is 0.85 (see Table 1 for the minimum and maximum values). Therefore, the scales of fluctuation are all relative to a maximum range of 0.7 for each input parameter.

The results for the median values indicate the following. First, the standard deviation for the inherent variations between the network predictions and the simulator results depends on the magnitude of the network prediction (that is the median value for b is non-zero). Second, the inherent variations are correlated (that is there are numerous scales of fluctuation that are large).

The results for the coefficient of uncertainty indicate that there is significant uncertainty in numerous model parameters. This result is not surprising given that 49 model parameters are being calibrated with a set of only 250 data points. However, it is noteworthy because this model uncertainty is typically neglected. Also, while not presented due to space considerations, there are numerous correlations between the updated model parameters.

4 DISCUSSION

The calibrated neural network model can now be used to predict the horizontal center of mass for the DNAPL for any set of input parameters. It can also indicate the magnitude of uncertainty in this prediction.

Table 3 lists three example cases. The first case is when a point of interest is very far away from any of

Table 3. Updated Mean and Standard Deviation

Example	Expected Value	Inherent Standard Deviation	Total Standard Deviation
Case I	0.32266	0.035	0.040
Case II	0.33326	0.026	0.384
Case III	0.24632	0.027	0.042

the data points. In this case, the neural network output only depends on the network design parameters. The second case is when a point of interest is correlated with a few of the data points in the UTCHEM simulated data set. For this particular example, there were 63 data points that were correlated with the point of interest. The final case is when a point of interest has the exact same set of input dimensionless parameters as one of the points in the UTCHEM simulated data set. In this case, the neural network prediction equals to the UTCHEM simulated value.

The inherent standard deviation in Table 3 indicates the magnitude of uncertainty in the network prediction assuming that all of the model parameters are known with certainty. The total standard deviation includes uncertainty due to both the inherent variations and that due to uncertainty in the calibrated model parameters. Note that the uncertainty in the model parameters, which is typically neglected, is a significant source of uncertainty here.

5 CONCLUSIONS

In this paper, we illustrate a non-dimensional approach for estimating the location and extent of DNAPLs using a back-propagation neural network. Inherent variations between the neural network predictions and the data set used to calibrate the network was incorporated by implementing a probabilistic approach. The following conclusions were drawn in this application. First, the magnitude of these inherent variations depended on the magnitude of the network prediction and was not a constant between data points. Second, the inherent variations were correlated between nearby data points. Third, the contribution of uncertainty arising from uncertainty in the calibrated model parameters due to a limited quantity of data was significant. These conclusions indicate that the conventional approach for training neural networks that minimizes the square error and that neglects model uncertainty is limited.

The neural network model developed herein can be used to predict the location and extent of DNAPLs in the subsurface with any given set of site-specific input parameters and to indicate the degree of confidence in that prediction.

6 ACKNOWLEDGMENTS

The authors wish to acknowledge the Gulf Coast Hazardous Substance Research Center, the Texas Advanced Research Program and the National Science Foundation for supporting this research.

7 REFERENCES

Ang, A. H-S. and Tang, W. H. (1975), Probability Concepts in Engineering Planning and Design, Vol. 1 - Basic Principles, John Wiley & Sons, New York.

Delshad, M., Pope, G. A., and Sepehrnoori, K. (1996), "A Compositional Simulator for Modeling Surfactant Enhanced Aquifer Remediation. 1: Formulation," *Journal of Contaminant Hydrology*, Vol. 23, pp. 303-327, August, 1996.

Dowla, F. U. and Rogers, L. L. (1995), Solving Problems in Environmental Engineering and Geosciences with Artificial Neural Networks, The MIT Press, Cambridge, Massachusetts.

Gilbert, R. B. (1999), "First-Order Second Moment Bayesian Method for Data Analysis in Decision Making" Geotechnical Engineering Center, Department of Civil Engineering, The University of Texas, Austin, 50 pp.

McGrath, T. C. (1998), Value and Reliability of DNAPL Investigation Programs. Doctoral dissertation, The University of Texas, Austin, TX, 478 pp.

McGrath, T. C., Yi, S. H., Gilbert, R. B., and McKinney, D. C. (1997), "Risk-Based Decision Approach for Aquifers Contaminated with Dense, Non-Aqueous Phase Liquids (DNAPLs)" Poster Summary, The Environmental Solutions Program. The University of Texas, Austin, Texas.

Mercer, J. W., Adeel, Z., and Faust, C. R. (1996), "A Review of Modeling Approaches for Remediation," *Proceedings*, Dense, Non-Aqueous Phase Liquids (DNAPLs) in the Subsurface Environment: Assessment and Remediation, an ASCE Specialty Conference, Washington, D. C., pp. 46-65.

Mehrotra, K., Mohan, C., and Ranka, S. (1997), Elements of Artificial Neural Networks, The MIT Press, Cambridge, Massachusetts.

Pankow, J. F., and Cherry, J. A. (1996), Dense Chlorinated Solvents and Other DNAPLs in Groundwater: History, Behavior, and Remediation, Waterloo Press, University of Waterloo, Ontario.

Shook, M., Li, D., and Lake, L. W. (1992), "Scaling Immiscible Flow Through Permeable Media by Inspectional Analysis," *In Situ*, vol. 16, No. 4, pp. 311-349.

Vanmarcke, E. (1983), Random Fields: Analysis and Synthesis, The MIT Press, Cambridge, Massachusetts.

Computer Methods and Advances in Geomechanics, Desai et al. (eds) © 2001 Balkema, Rotterdam, ISBN 90 5809 183 X

Reliability of drilled shafts

L. M. Zhang
Department of Civil Engineering, Hong Kong University of Science and Technology, People's Republic of China

ABSTRACT: One advantage of drilled shafts is their ability to support one column by one shaft. The experience with the drilled shaft foundations is generally satisfactory. However, the calculated risk considering the uncertainties understood so far is relatively low. This paper attempts to look into additional factors that lead to the discrepancy between the observed and calculated risks. Two factors are identified: redundancies in failure criteria and in system effects. The impacts of the two factors on reliability of drilled shafts are evaluated using a first order reliability method. Based on the present study, the reliability of shafts may be improved by including the two factors. If no system effects can be identified, it may be necessary to increase the design factor of safety to ensure the foundation reliability.

1 INTRODUCTION

One advantage of drilled shafts is their ability to support one column by one shaft. The reliability of single drilled shafts has been studied by several investigators (Barker et al. 1991; ASSHTO 1994, Ochiai et al 1994; Kulhawy 1996; Kulhawy and Phoon 1996; Withiam et al. 1997; Long and Wysockey 1999). Through a calibration technique against current allowable stress design (ASD) practice, the reliability index β of drilled shafts is found to be in the range of 2.5 to 3.0 (Barker et al. 1991). The back calculated reliability level is somewhat low, compared with that of superstructures which is designed using a target β in the neighborhood of 3.5.

The observed performance of drilled shaft foundations designed with prevailing ASD methods is generally satisfactory. This is inconsistent with the calculated risk. This paper attempts to look into possible factors that lead to the discrepancy between the observed and calculated risks. The redundancies in failure loads and system effects are identified to be such factors. The impacts of the factors on the reliability of drilled shafts are evaluated along the line of Load and Resistance Factor Design (LRFD).

2 METHOD OF RELIABILITY EVALUATION

The first order reliability method is employed to conduct reliability analyses. The performance function follows,

$$\textit{Resistance} \geq \textit{Effects of loads} \tag{1}$$

If both resistance and load effects are log-normal variates and the only load effects to be considered are dead and live loads, the value of β in the framework of LRFD can be expressed as (Barker et al. 1991; Withiam et al. 1997)

$$\beta = \frac{\ln\left(\dfrac{\lambda_R FS\left(\dfrac{Q_D}{Q_L}+1\right)}{\lambda_{QD}\dfrac{Q_D}{Q_L}+\lambda_{QL}} \sqrt{\dfrac{1+COV_{QD}^2+COV_{QL}^2}{1+COV_R^2}} \right)}{\sqrt{\ln[(1+COV_R^2)(1+COV_{QD}^2+COV_{QL}^2)]}} \tag{2}$$

in which FS = factor of safety in the traditional allowable stress design; Q_D and Q_L = nominal values of dead and live loads, respectively; λ_R, λ_{QD}, and λ_{QL} = bias factors for resistance, dead load, and live load, respectively; COV_R, COV_{QD}, and COV_{QL} = coefficients of variation for resistance, dead load, and live load, respectively. The bias factors λ_R, λ_{QD}, and λ_{QL} are defined as the ratios of their measured values to their corresponding nominal values.

The following load statistics are adopted according to the LRFD Highway Bridges Design Specifications (AASHTO 1994): λ_{QD} = 1.08, λ_{QL} = 1.15, COV_{QD}= 0.13, and COV_{QL} = 0.18. The dead to live load ratio Q_D/Q_L is a function of a bridge's span length. Investigations by Barker et al. (1991) and McVay et al. (2000) show that the reliability index of piles is not sensitive to span length. Therefore, only a value of $Q_D/Q_L = 1.58$, which corresponds to a span length of 27 m as used by McVay et al. (2000), is adopted for analysis in this paper.

3 MODEL ERROR

The FHWA design method suggested by Reese and O'Neill (1988) is considered. This method uses an alpha method to calculate the skin friction in clay and a beta method to calculate the shaft friction in sand. The toe bearing is estimated using correlations with the undrained strength S_u in clay and with the standard penetration blow count N in sand.

Several databases of static load tests of drilled shafts are available for evaluating the model errors associated with this design method (Reese and O'Neill 1988; O'Neill et al. 1996; Long and Wysockey 1999). Figure 1 shows the measured and predicted capacities of 42 shafts in the database of O'Neill et al. (1996). The measured capacity refers to the load at a head displacement of 5% shaft diameter (D). The bias factor of the model, λ_M, defined as the ratio of the measured to predicted capacities R_m/R_n, and its coefficient of variation COV_M are 1.26 and 0.42, respectively. The database of Long and Wysockey (1999) has 122 cases. The average ratio of predicted to measured capacities (the reciprocal of λ_M for individual cases) and its COV are 0.82 and 0.49, respectively. The statistics of the two databases appear to be consistent. It should be warned, however, that several construction methods might have been involved in constructing the drilled shafts in the databases. Data entries with different construction methods should be compiled separately in future research, since construction details have potential effects on the performance of drilled shafts.

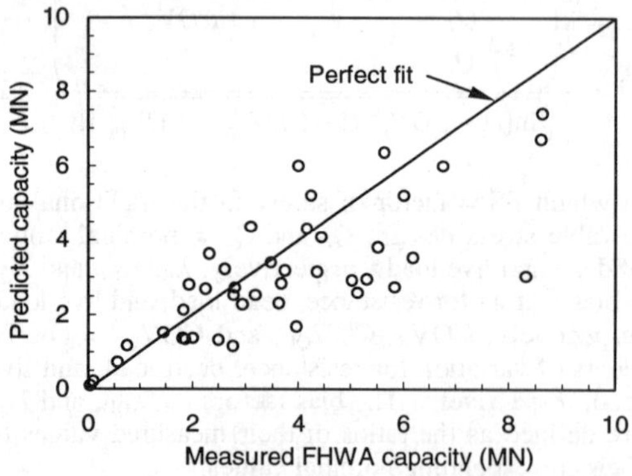

Figure 1. Comparison of measured capacity and predicted capacity with the FHWA method

4 REDUNDANCY IN FAILURE CRITERIA

The nominal capacity of piles and drilled shafts is always determined based on certain failure criteria. There are several existing criteria for determining the ultimate capacity of drilled shafts, such as the Fuller Hoy's criterion (1970), the Davisson criterion (1972), and the FHWA criterion (Reese and O'Neill 1988). The failure loads predicted by these criteria can be significantly different. In addition, a "failure" defined by any of the criteria may be different from what a real failure on site is believed to be. Generally, there is a capacity reserve between the load defined by a failure criterion and the load at punching failure of the drilled shaft. Thus, the failure load itself may be a major source of uncertainty.

Figure 2 shows the relationship between the FHWA failure load and the normalized Fuller-Hoy failure load, as well as the relationship between the FHWA failure load and the normalized failure load at 76 mm settlement. The FHWA failure load is the load at a head displacement of 5% shaft diameter. The Fuller-Hoy failure load corresponds to the load at which the slope of the load-displacement curve is 0.14 mm/kN (0.05"/ton). Both the Fuller-Hoy capacity and the failure load at 76-mm displacement are closer to the punching failure load of a drilled shaft than the FHWA failure load. The load ratios in Figure 2 can be conservatively considered as the bias factors which describe the redundancy introduced by the two failure criteria. Figure 2 also shows that the normalized failure loads and the FHWA failure load can be considered to be statistically independent.

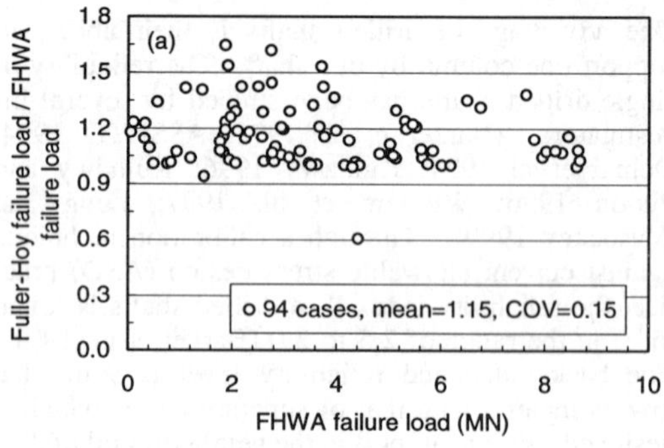

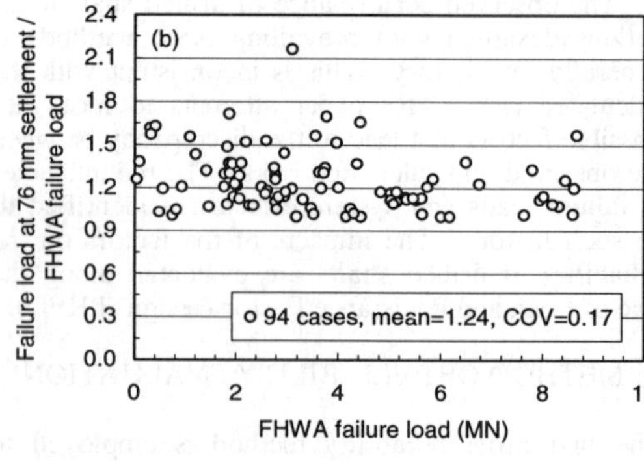

Figure 2. Relationship between the FHWA failure load and (a) the normalized Fuller-Hoy failure load, (b) the normalized failure load at 76 mm settlement

This is because the normalized failure loads do not appear to vary with the FHWA failure load. The statistics of the redundancy due to the failure criteria are listed in Table 1 for the use of reliability assessment later.

Table 1. Statistics of bias due to failure criteria with respect to the FHWA 5%D failure load

Failure criterion	Number of cases	Bias factor λ_F	COV_F
Fuller-Hoy	94	1.15	0.15
76 mm settlement	94	1.24	0.17

5 REDUNDANCY IN SYSTEM EFFECTS

Although one drilled shaft can be used to support one column, a number of drilled shafts may be connected through columns, beams, and other superstructure elements in a real scenario. System effects arise when there is any interaction between the shafts and the superstructure and there is any path to transfer loads from one shaft to another. Tang and Gilbert (1993), Aggarwal et al. (1996) and Bea et al. (1999) investigated the reliability of pile systems for offshore platforms. It was found that there was a bias introduced by system effects. This bias separates the condition where the most heavily loaded pile has been brought to its ultimate computed capacity and when all of the piles have been brought to the same state. The bias, called system factor χ, can be used to describe system effects. This factor depends on, among other factors, type of superstructure and layout of foundations.

Reported data on system factor of general pile-supported structures are rare and site specific. Bea et al. (1999) and Aggarwal et al. (1996) found a χ value of 1.2-2.0, and a COV_S value of 0.1-0.2. For a more meaningful evaluation, several χ values (i.e., 1.0, 1.5, and 2.0) are adopted for parametric study in this paper. An average COV_S of 0.17 obtained from the offshore platform studies mentioned above is used. In addition, the system factor χ is assumed to be a log-normal variate, as is the single shaft capacity. Since the foundation-superstructure interaction is usually not considered in design, i.e., a default value of $\chi = 1.0$ is assumed, the bias factor for system effects is simply $\lambda_S = \chi$.

6 RELIABILITY EVALUATION INCLUDING THE REDUNDANCIES

Given the bias factors and COVs for the model error and the redundancies in failure criteria and in system effects, the combined bias λ_R and COV_R of the ultimate capacity of a drilled shaft can be expressed as

$$\lambda_R = \lambda_M \lambda_F \lambda_S \quad (3)$$

$$COV_R = \sqrt{COV_M{}^2 + COV_F{}^2 + COV_S{}^2} \quad (4)$$

where λ_M, λ_F, and λ_S are the bias factors for model error, failure criterion, and system effects, respectively; COV_M, COV_F, and COV_S are their coefficients of variation.

In case of groups of drilled shafts, group effects may also lead to improved safety. However, group effects will not be considered in this paper, because single drilled shafts are frequently used and information on group effects of drilled shafts is limited.

Figure 3 shows the calculated relationship between the factor of safety (FS) and the β associated with the FHWA design method, with or without considering the uncertainty in failure criteria. The calculated β increases with FS. If no redundancy due to failure criteria is accounted for, the β value at a design FS of 2.5 is only 2.15. If the redundancy is considered, the β value will increase to 2.35 with respect to the Fuller-Hoy failure load, and to 2.46 with respect to the failure load at 76-mm displacement. The two β values are relatively low. This means that, if a shaft works independently and does not interact with other structure members, its reliability tends to be insufficient even if the redundancy in its failure load is considered. From this figure, a FS of 3.2 to 4.2 may be needed in order to achieve a β value of 3.0-3.5.

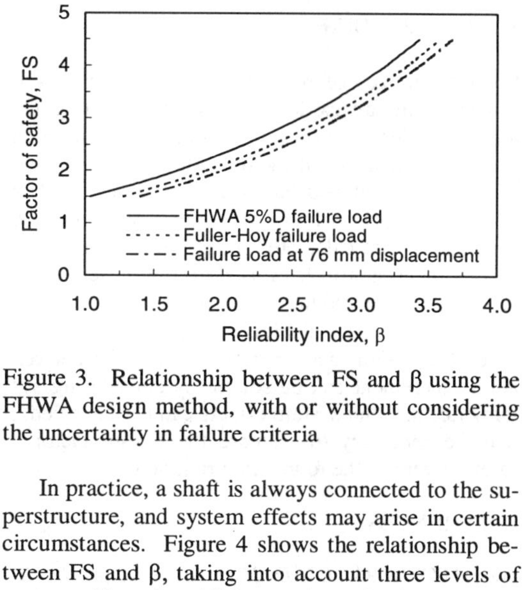

Figure 3. Relationship between FS and β using the FHWA design method, with or without considering the uncertainty in failure criteria

In practice, a shaft is always connected to the superstructure, and system effects may arise in certain circumstances. Figure 4 shows the relationship between FS and β, taking into account three levels of system effects in addition to the redundancy in the design capacity with respect to the failure load at 76 mm displacement. Significant influence of system effects on β is observed in this figure. The β values at FS = 2.5 are 2.46, 3.12, and 3.68, respectively, when system bias factors of λ_S = 1.0 (no system effects), λ_S = 1.5, and λ_S = 2.0 are considered. It should be noted that system effects depend on type of superstructures. For instance, in the study of Aggarwal et al. (1996), the eight-leg jackets showed considerable system effects, whereas the 4-leg jacket indicated no load redistribution. For a viaduct or a bridge bent supported by two shafts that are far apart, no system effects can be expected. In evaluating the reliability of foundations, system effects cannot be considered unless they are supported by adequate proofs, either by site monitoring or by numerical simulations.

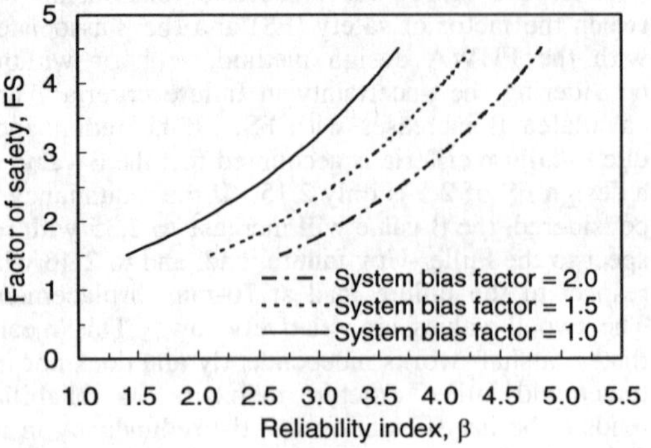

Figure 4. Influence of system bias factor on the reliability of single drilled shafts

7 CONCLUSIONS

Possible sources of redundancy in the capacity of drilled shafts are explored. Two contributing factors, i.e. failure criteria and system effects, are identified. The statistics of the redundancy due to failure criteria are developed based on a loading test database. The impacts of the two factors on reliability of single shafts are calculated using the first order second moment approach. Based on the present study, considering the redundancy due to failure criteria may increase the reliability index β from 2.15 to 2.46 at a design factor safety of 2.5. Further reliability increment can be sought by considering system effects. If no system effects can be identified, it may be necessary to increase the design factor of safety to ensure the foundation reliability.

REFERENCES

AASHTO. 1994. *LRFD Highway Bridges Design Specifications, SI Units*, first edition. American Association of State Highway and Transportation Officials, Washington, D.C.

Aggarwal, R.K., Litton, R.W., Cornell, C.A., Tang, W.H., Chen, J.H., and Murff, J.D. 1996. Development of pile foundation bias factors using observed behavior of platforms during Hurricane Andrew. *Proc., 28th Annual Offshore Technology Conference*, Vol.2, TX, 445-455, OTC 8078.

Barker, R.M, Duncan, J.M., Rojiani, K.B., Ooi, P.S.K., Tan, C.K., and Kim, S.G. 1991. *Load Factor Design Criteria for Highway Structure Foundations*. Final report, NCHRP project 24-4, Virginia Polytechnic Institute and State University, Blacksburg, VA.

Bea, R.G., Jin, Z., Valle, C., and Ramos, R. 1999. Evaluation of reliability of platform pile foundations. *Journal of Geotechnical and Environmental Engineering*, ASCE, 125(8): 696-704.

Davisson, M.T. 1972. High capacity piles. *Proc. Lecture Series Innovations in Foundation Construction*, ASCE, Illinois Section, Chicago.

Fuller, F.M., and Hoy, H.E. 1970. Pile load tests including quick-load test method conventional methods and interpretations. *HRB* 333: 78-86.

Kulhawy, F.H. 1996. From Casagrande's "calculated risk" to reliability-based design in foundation engineering. *Journal of the Boston Society of Civil Engineers*, 11(2): 43-56.

Kulhawy, F.H. and Phoon, K.K. 1996. Engineering judgement in the evolution from deterministic to reliability-based foundation design. *ASCE Geotechnical Special Publication* No. 58: 29-48. New York.

Long J.H., and Wysochey, M.H. 1999. Accuracy of methods for predicting axial capacity of deep foundations. *ASCE Geotechnical Special Publication* No. 88: 80-195. Reston, VA.

McVay, M.C., Birgisson, B., Zhang, L.M., Perez, A., and Putcha, S. 2000. Load and resistance factor design (LRFD) for driven piles using dynamic methods-A Florida perspective. *Geotechnical Testing Journal*, ASTM, 23(1): 55-66.

Ochiai, H., Li, X.-X., Otani, J., and Matsui, K. 1994. Reliability of vertical bearing resistance of bored friction piles. *Memoirs of the Faculty of Engineering*, Kyushu University, 54(1), 1-24.

O'Neill, M.W., Townsend, F.C., Hassan, K.M., Buller, A., and Chan, P.S. 1996. *Load Transfer for Drilled Shafts in Intermediate Geomaterials*. FHWA-RD-95-172, Federal Highway Administration, McLean, VA, 194p.

Reese, L.C., and O'Neill, M.W. 1988. *Drilled Shafts: Construction Procedures and Design Methods*. Publication No. FHWA-HI-88-042, Federal Highway Administration, McLean, VA, 564p.

Tang, W.H. and Gilbert, R.B. 1993. Case study of offshore pile system reliability. *Proc., 25th Offshore Technology Conference*, TX, 677-686, OTC 7196.

Withiam, J.L., Voytko, E.P., Barker, R.M., Duncan, J.M., Kelly, B.C., Musser, S.C., and Elias, V. 1997. *Load and Resistance Factor Design (LRFD) for Highway Bridge Substructures*. FHWA Report DTFH61-94-C-00098, Federal Highway Administration, McLean, VA.

3 Constitutive models: Soils and rocks

Computer Methods and Advances in Geomechanics, Desai et al. (eds) © 2001 Balkema, Rotterdam, ISBN 90 5809 183 X

An elasto-plastic description of the difference between sand and clay

A. Asaoka, M. Nakano & T. Noda
Department of Civil Engineering, Nagoya University, Japan

ABSTRACT: Naturally deposited clay/sand are mostly found in structured states. In addition those soils are usually at overconsolidated states. In order to describe mechanical behavior of structured and overconsolidated soils super-subloading surfaces are newly introduced to the original Cam-clay model. In the new soil model hardening with plastic volume expansion and softening with plastic volume compression are both found possible other than the usual hardening with compression and the usual softening with expansion.

Both the decay of structure and the loss of overconsolidation proceeds with ongoing plastic deformation, but the former proceeds at a much slower pace than the latter in clay. To the contrary, the decay of structure proceeds much faster than the loss of overconsolidation in sand. The expectation made above is positively accepted through examining numerical simulations of (1)secondary consolidation of clay, (2)undrained shear behavior of clay and sand and (3) rapid collapse of soil structure of loose sand.

1 INTRODUCTION

Fig. 1 shows a typical 1-D compression behavior of naturally deposited structured clay. The straight line in the figure denotes a normal consolidation line of the fully remolded clay. The compression test of this natural clay was conducted under a finite vertical strain rate with "permeable top and impermeable bottom" boundary conditions. Then the test results of the figure merely shows an overall compression behavior of the specimen during its consolidation procedure. The "element-wise" compression behavior of the clay under fully drained condition should differ, to a certain extent, from this figure, e.g. *see* Fig. 10. However, Fig.1 still suggests that structured clay is possible to take its state in the "impossible" region of the fully remolded destructured clay, that

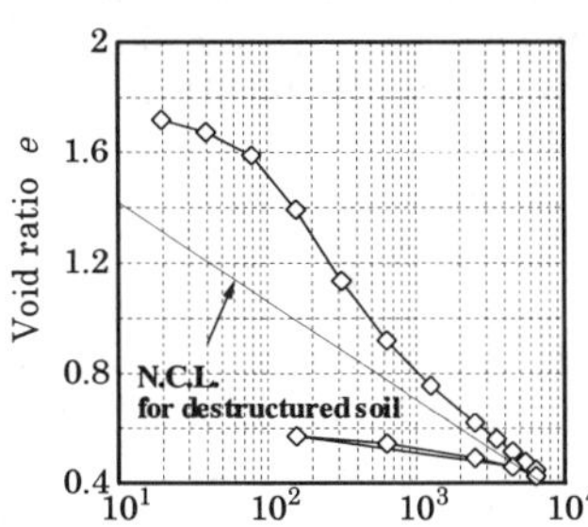

Figure 1. Typical 1-D compression behavior of natural clay

is, outside the "Roscoe surface", which identifies the demand for a "superloading yield surface" concept for structured soil.

Asaoka, Nakano and Noda (1998, 2000a, 2000b) introduced a superloading yield surface, together with Hashiguchi's subloading yield surface (Hashiguchi, 1978, 1989), to the original Cam-clay in order to describe mechanical behavior of structured and overconsolidated soils. Fundamental conceptions of the new soil model are explained here in six steps. (1) Fully remolded/destructured and normally consolidated soil is assumed to follow the original Cam-clay model. This is just for the sake of simplicity. (2) Existence of a superloading yield surface is assumed, which is geometrically similar to the Cam-clay yield surface, lying outside the Roscoe surface, denoting the impossible state for remolded soil. The similarity ratio of the Cam-clay yield surface to the superloading yield surface is denoted by R^* ($0<R^*<1$). (3) When the state of stress is on the superloading yield surface, the soil is said to be at normally consolidated state. For the plastic deformation of the normally consolidated soil, the associated flow rule is applied on the superloading yield surface. (4) In order to determine the subsequent superloading yield surface, the material time derivative of R^* is expressed in terms of ongoing plastic deformation, which is called evolution law for R^*. The evolution law describes the fact that soil structure gradually decays as plastic deformation proceeds, which is equivalent to say that R^* is gradually increasing up to 1 when loading and the superloading yield surface finally coincides with the Cam-clay yield surface after complete remolding. (5) Structured soil, initially on the superloading yield surface, becomes overconsolidated soil when unloading. Soil in such overconsolidated state, when reloading occurs, obeys the subloading yield surface through the associated flow rule. The subloading yield surface is again geometrically similar to the superloading yield surface. The similarity ratio of the subloading yield surface to the superloading yield surface is denoted by R, which is also $0<R<1$. Note here that the reciprocal of R gives OCR of the soil. (6) To determine the subsequent subloading yield surface, the evolution law for R is again needed. The evolution law describes the well-known fact that with ongoing plastic deformation overconsolidated soil gradually reaches normally consolidated state.

In the present study, the difference between sand and clay is discussed from the viewpoint of the difference between the rate of the decay of structure and the rate of the loss of overconsolidation. Essential points of the discussion are as follows: In clay, the loss of overconsolidation proceeds at a much faster pace than the decay of structure, which causes softening with plastic volume compression under a considerably low stress ratio (i.e., small q/p'). The occurrence of softening with volume compression triggers off delayed compression/secondary consolidation. To the contrary, the decay of structure proceeds at a much faster pace in sand than the loss of overconsolidation. When loose sand is the case, initial rapid decay of structure easily causes "compaction" or liquefaction. When medium dense or dense sand is the case, delayed loss of overconsolidation causes hardening with plastic volume expansion even after the occurrence of softening with plastic volume compression. It may be needless to say that hardening with plastic volume expansion and softening with plastic volume compression are both out of the scope of the classical Cam-clay model.

The limitation of the discussion given in this paper comes from the total lack of the discussion on anisotropy. If this limitation inspires open discussion in the conference, authors appreciate it very much.

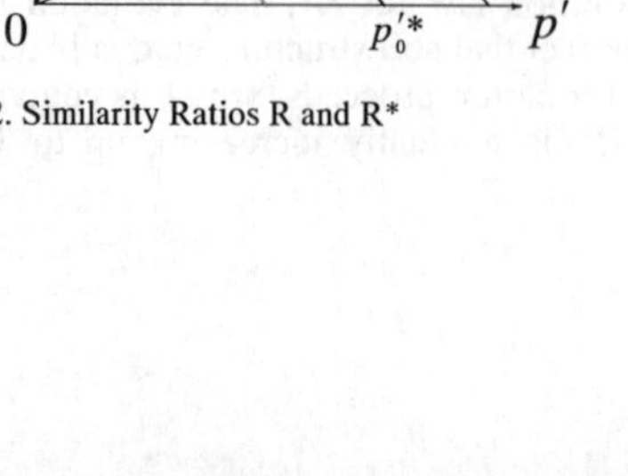

Figure 2. Similarity Ratios R and R*

2 A SKETCH OF THE MODEL EXAMINED

Key formulations are given next. For details, *see* Asaoka, Nakano and Noda (2000a). The definitions of R^* and R are simply given in Fig.2, from which

$$f(p',q)+\int_0^t J\mathrm{tr}\boldsymbol{D}^p d\tau+\mathrm{MD}\ln R^*-\mathrm{MD}\ln R$$

$$=\mathrm{MD}\ln\frac{p'}{p_0'^*}+\mathrm{D}\frac{q}{p'}+\mathrm{MD}\ln\frac{R^*}{R}+\int_0^t J\mathrm{tr}\boldsymbol{D}^p d\tau=0 \quad (1)$$

gives the subloading yield surface, where M and D are the critical state parameter of the Cam-clay and the dilatancy parameter, respectively, while D^p denotes plastic stretching. J in the equation is the Jacobian determinant of the deformation gradient tensor $\boldsymbol{F}$, and is expressed in terms of specific volume as follows:

$$J=\det\boldsymbol{F}=\frac{1+\mathrm{e}}{1+\mathrm{e}_0} \quad (2)$$

in which e and e_0 are the void ratio at current (time t) and reference (time t=0) state, respectively. The $p_0'^*$ in Eq.(1) is the mean effective stress on the Cam-clay yield surface that corresponds to the initial mean effective stress p_0' of the soil in the reference state, *see* Fig.2 again. The important thing here is that soil generally takes its stress state on the subloading yield surface.

The associated flow rule

$$\boldsymbol{D}^p=\lambda\frac{\partial f}{\partial\boldsymbol{T}'}\quad(\lambda>0) \quad (3)$$

governs the plastic response of soils, in which λ is a plastic multiplier. Prager's consistency condition

$$\frac{\partial f}{\partial\boldsymbol{T}'}\cdot\dot{\boldsymbol{T}}'+J\mathrm{tr}\boldsymbol{D}^p+\mathrm{MD}\frac{\dot{R}^*}{R^*}-\mathrm{MD}\frac{\dot{R}}{R}=0 \quad (4)$$

determines the size of subsequent yield surface.

$\dot{R}^*$ and $\dot{R}$ in Eq.(4) should be positive with ongoing plastic deformation so that R^* and R are both increasing to one gradually when loading. As a measure of ongoing plastic deformation the norm of the shear component of plastic stretching $\|\boldsymbol{D}_s^p\|$ is employed, in which

$$\boldsymbol{D}_s^p=\boldsymbol{D}^p-\frac{1}{3}\mathrm{tr}\boldsymbol{D}^p\boldsymbol{I}=\lambda\frac{3\boldsymbol{S}}{2q}\frac{\partial f}{\partial q}. \quad (5)$$

The use of $\|\boldsymbol{D}_s^p\|$ for the evolution laws of R^* and R is again just for the sake of simplicity, through which a simple closed form expression for $\mathrm{M_s}$ is obtained, *see* Eqs.(11) and (19). Thus, the evolution laws for R^* and R are expressed here

$$\dot{R}^* = JU^* \sqrt{\frac{2}{3}} \| D_s^p \| \tag{6}$$

and

$$\dot{R} = JU \sqrt{\frac{2}{3}} \| D_s^p \|, \tag{7}$$

respectively. U^* and U in these equations are both positive scalar function of R^* and R, respectively, which satisfies following constraints:

$$U^*(R^*=1)=0 \tag{8}$$

and

$$U(R=1)=0 \tag{9}$$

Substituting evolution laws into consistency condition, and applying the associated flow rule, the plastic multiplier λ is obtained in terms of stresses as follows:

$$\lambda = \frac{\frac{\partial f}{\partial \boldsymbol{T}'} \cdot \dot{\boldsymbol{T}}'}{J \frac{\mathrm{D}}{p'^2}(\mathrm{M}_s p' - q)} \tag{10}$$

in which

$$\mathrm{M}_s = \mathrm{M}(1 - \frac{\mathrm{D}U^*}{R^*} + \frac{\mathrm{D}U}{R}) \tag{11}$$

is now a new critical state parameter. Note here that, for λ being positive, there should be three different cases for loading:

$$\frac{\partial f}{\partial \boldsymbol{T}'} \cdot \dot{\boldsymbol{T}}' > 0 \text{ (Hardening) when } q<\mathrm{M}_s p' \tag{12}$$

$$\frac{\partial f}{\partial \boldsymbol{T}'} \cdot \dot{\boldsymbol{T}}' = 0 \text{ (Perfectly plastic) on } q=\mathrm{M}_s p' \tag{13}$$

$$\frac{\partial f}{\partial \boldsymbol{T}'} \cdot \dot{\boldsymbol{T}}' < 0 \text{ (Softening) when } q>\mathrm{M}_s p' \tag{14}$$

The new critical state line $q=\mathrm{M}_s p'$ is thus found to be the watershed between hardening and softening. Note also here that the critical state parameter M in a classical Cam-clay model is a material constant, while the new critical state parameter M_s is a variable. When loading, the M_s is increasing/decreasing with ongoing plastic deformation.

For the M_s being positive, Eq.(11) gives the following additional constraint condition on U^*:

$$U^*(R^*=0)=0 \tag{15}$$

The simplest form that satisfies Eqs.(8) and (15) may be

$$U^* = \frac{1}{\mathrm{D}} R^* (1 - R^{*m^*}). \tag{16}$$

As for the function U of R, following Hashiguchi (1989),

$$U = -\frac{m}{\mathrm{D}} \ln R \tag{17}$$

is employed for simplicity. The condition

$$U(R=0)=+\infty \tag{18}$$

indicates that if R is very close to zero the soil exhibits almost elastic response when reloading.

In Eqs.(16) and (17) m^* and m are degradation parameters of structured states and overconsolidated state, respectively. The greater those variables, the faster the decay of those state. Substituting Eqs.(16) and (17) into Eq.(11), one gets

$$\mathrm{M}_s = \mathrm{M}(R^{*m^*} - \frac{m \ln R}{R}) \tag{19}$$

which again clearly indicates that the parameter M_s is a variable with R^* and R.

Constitutive equation of the new soil model is expressed as follows:

$$\overset{\circ}{\boldsymbol{T}}' = \boldsymbol{E}\boldsymbol{D}^e = \boldsymbol{E}\boldsymbol{D} - \Lambda \boldsymbol{E} \frac{\partial f}{\partial \boldsymbol{T}'} \tag{20}$$

in which $\boldsymbol{E}$ and $\boldsymbol{D}^e$ denote the elastic modulus tensor and elastic stretching, respectively, while $\overset{\circ}{\boldsymbol{T}}'$ is the objective co-rotational rate tensor (Dienes, 1979). The Λ in the equation is the plastic multiplier in terms of stretching and is expressed as follows:

$$\Lambda(=\lambda) = \frac{\frac{\partial f}{\partial \boldsymbol{T}'} \cdot \boldsymbol{E}\boldsymbol{D}}{\frac{\partial f}{\partial \boldsymbol{T}'} \cdot \boldsymbol{E} \frac{\partial f}{\partial \boldsymbol{T}'} + J \frac{\mathrm{D}}{p'^2}(\mathrm{M}_s p' - q)} \tag{21}$$

Since the denominator is shown to be positive for almost all stress state, the loading criterion for the constitutive equation is now found simply as

$$\frac{\partial f}{\partial \boldsymbol{T}'} \cdot \boldsymbol{E}\boldsymbol{D} > 0. \tag{22}$$

3 HARDENING, SOFTENING AND PLASTIC VOLUME CHANGE

Suppose that $\dot{R} >> \dot{R}^*$, then only R is increasing to 1 while R^* remains almost constant. In this case, Eq.(11) and (19) shows that M_s is decreasing. This is called, in this study, the effect of the loss of overconsolidation on M_s.

To the contrary, if $\dot{R}^* >> \dot{R}$, the M_s is increasing. This is then the effect of the decay of structure on M_s.

Since q=Mp' line is the watershed between plastic volume compression and plastic volume expansion, then softening becomes possible even with plastic volume compression when M_s < M, (Fig.3) Conversely, when M_s > M, hardening may occur even with plastic volume expansion (Fig.4).

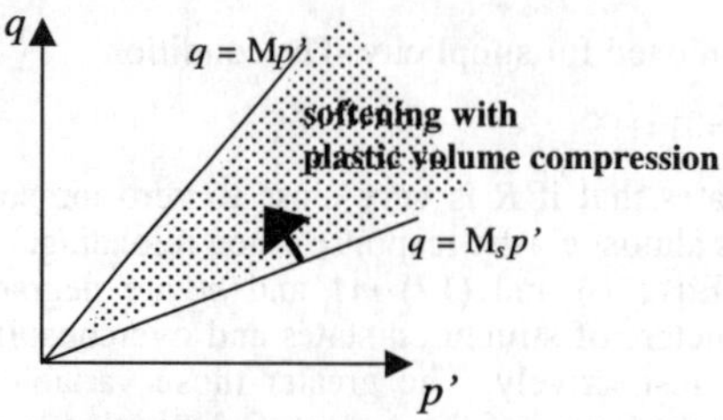

Figure 3. Softening with compression

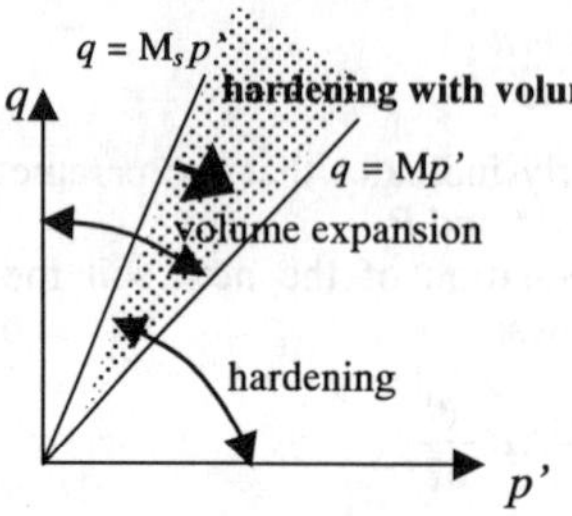

Figure 4. Hardening with expansion

4 DIFFERENCE BETWEEN CLAY AND SAND

In principle, there are two processes from structured and overconsolidated state to destructured and normally consolidated state. Clay loses its overconsolidation first and therefore M_s comes down first. After that the M_s is increasing gradually to M with the decay of structure, which may need considerably large shear deformation. This is schematically illustrated in Fig.5. Structure of clay remains long.

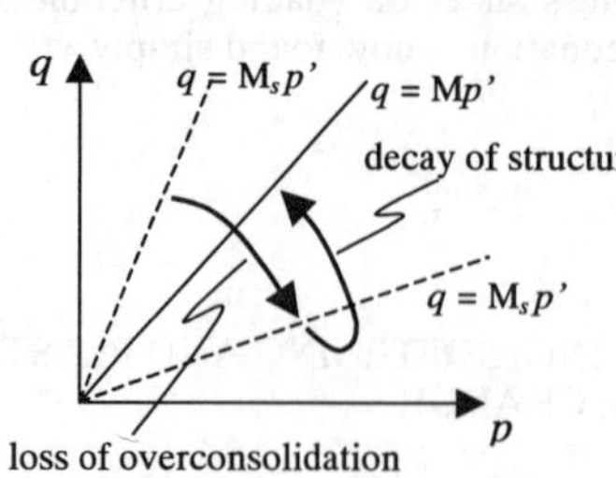

Figure 5. Typical clay behavior

In sand, to the contrary, the decay of structure proceeds very easily/rapidly even with a relatively small plastic deformation. However, overconsolidation remains long. When the sand is initially at a loose state with structure, M_s is increasing up first with a sudden collapse of structure, and then the M_s comes down very slowly to M with a gradual loss of overconsolidation, *see* Fig. 6.

Discussion made above is examined through numerical examples with some experiments.

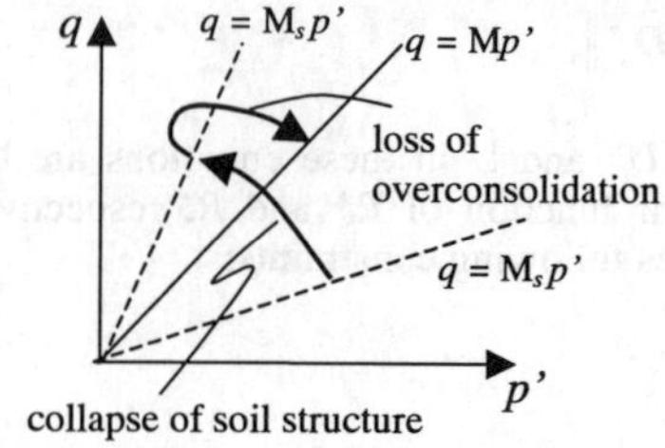

Figure 6. Typical sand behavior

5 SECONDARY CONSOLIDATION OF NATURAL CLAY

Even when the clay is initially at overconsolidated state, the M_s comes down very rapidly after some amount of plastic deformation, *see* Fig.5. Therefore, even in 1-D consolidation with relatively low stress ratio, softening becomes possible with plastic volume compression. In case of softening, "coefficient of consolidation" becomes negative and then not dissipation but generation of excess pore pressure should be observed during consolidation procedure. As the decay of structure proceeds, M_s is increasing to M gradually, that is, with the passage of time the clay now exhibits hardening. However, before hardening begins, the clay can not dissipate excess pore pressure, which causes delayed compression/secondary consolidation.

A typical oedometer test in laboratory is numerically simulated in this section. Material constants and necessary initial values of clay are tabulated in Table 1. The overall compression behavior of the clay under vertical load application with constant rate is illustrated in Fig.7. At the loading stages A, B and C in the figure, the vertical load was kept constant so that the clay specimen underwent 1-D consolidation under constant load application. Three settlement behaviors are given with time in Fig.8.

Table 1: Material constants and initial conditions

Compression index $\tilde{\lambda}$	0.131
Swelling index $\tilde{\kappa}$	0.075
Critical state constant M	1.53
Poisson's ratio ν	0.3
Degradation parameter of structure m^*	0.5
Degradation parameter of overconsolidated state m	10.0
Initial specific volume v_0	2.19
Initial value of $1/R^*(1/R_0^*)$	20.0
Initial overconsolidation ratio $1/R_0$	100.0
Soil permeability k(cm/sec)	7.8×10^{-9}

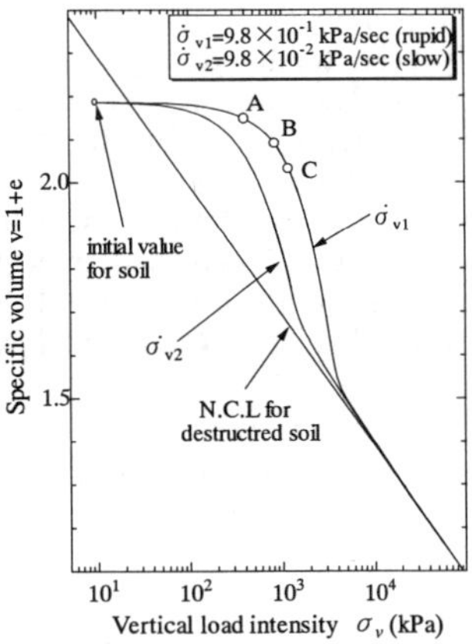

Figure 7. Three loading stages A, B and C

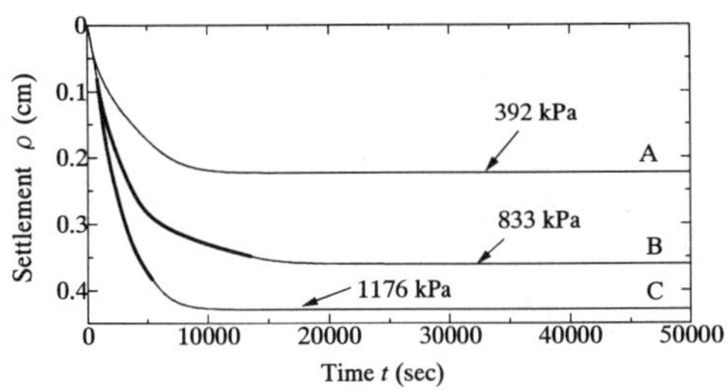

Figure 8. Secondary consolidation

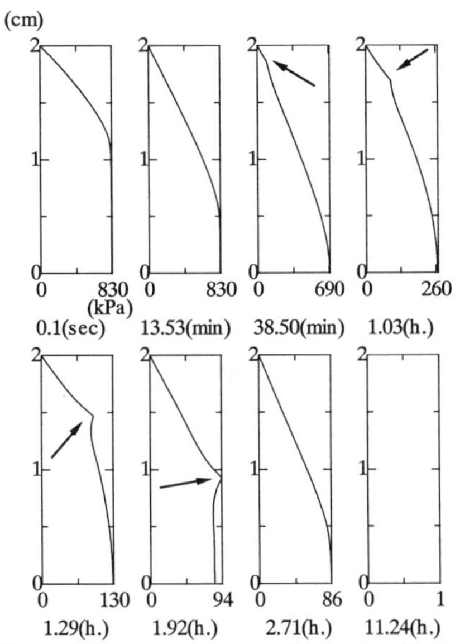

Figure 9. Excess pore pressure rise due to softening

Delayed compression can be clearly observed at loading stage B, pore pressure isochrones for which are given in Fig.9. Arrows in the figure indicate the points where softening occurs with excess pore pressure rise. An element-wise 1-D compression behavior directly obtained from the constitutive equation under fully drained condition is illustrated in Fig.10, which is superimposed upon Fig.7. As can be seen in the figure, loading magnitude of B is slightly exceeds the softening point. For detailed discussion, *see* Asaoka Nakano and Noda, 2000b.

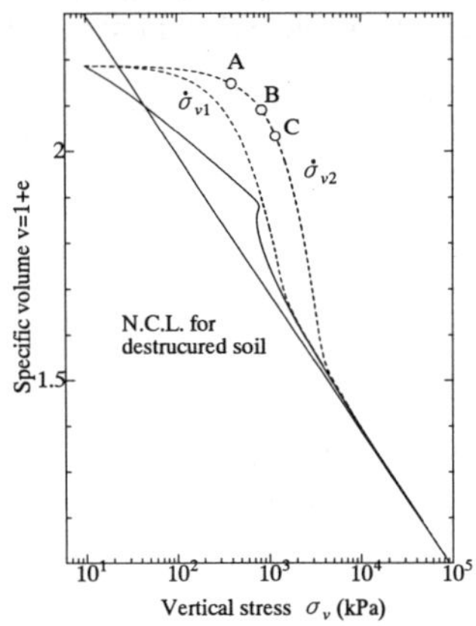

Figure 10. "Element-wise" 1-D compression behavior

6 UNDRAINED SHEAR BEHAVIOR OF CLAY AND SAND

To make examples simple, plane strain undrained shear behavior is examined. Uniform shear deformation field with constant volume is illustrated in Fig.11. The deformation field yields stretching and rotation, by which the stress response of the soil is directly computed through the constitutive equation.

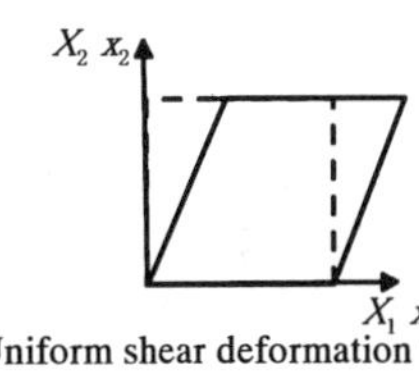

Figure 11. Uniform shear deformation with constant volume

Table 2: Material constants and initial conditions of clay and sand

	clay	sand
Compression index $\tilde{\lambda}$	0.25	0.0419
Swelling index $\tilde{\kappa}$	0.045	0.016
Critical state constant M	1.43	1.24
Poisson's ratio ν	0.3	0.3
Degradation parameter of structure m^*	0.1	1.8
Degradation parameter of overconsolidated state m	10.0	0.04
Initial specific volume v_0	2.33	1.92(1.77)[2.07]
Initial value of $1/R^*(1/R_0^*)$	8.0	0.22(1.0)[16.2]
Initial overconsolidation ratio $1/R_0$	5.7	12.7(16.2)[1.0]

Tabulated in Table 2 are the elasto-plastic soil parameters together with m^* and m. Necessary initial conditions of the soil are also given in the table. Clear distinction between clay and sand should be observed in the clear differences in m^* and m. The loss of overconsolidation proceeds at much faster pace in clay, while the decay of structure, at much faster pace in sand. Concerning the initial value of R^*, it should be noted here that, in dense sand, initial structure that had ever existed when the sand was loose has already been totally decayed due to the densification process. Therefore R^* =1 is assumed from the beginning in dense sand. This topic will be discussed in the next section.

Computed results of the undrained shear behaviors of one overconsolidated clay and three sands with different densities are given in Figs. 12, 13, 14 and 15, respectively. Typical experimental results

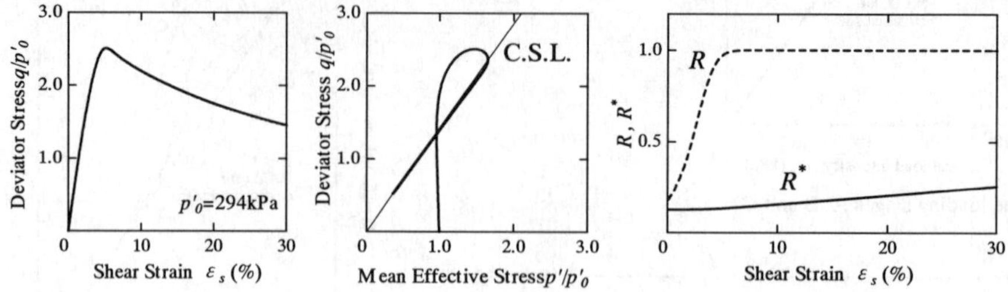

Figure 12. Undrained shear behavior of overconsolidated clay, a typical "kink"

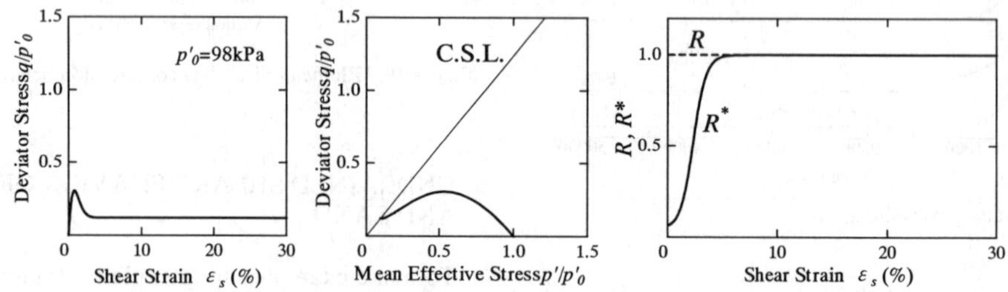

Figure 13. Undrained shear behavior of loose sand

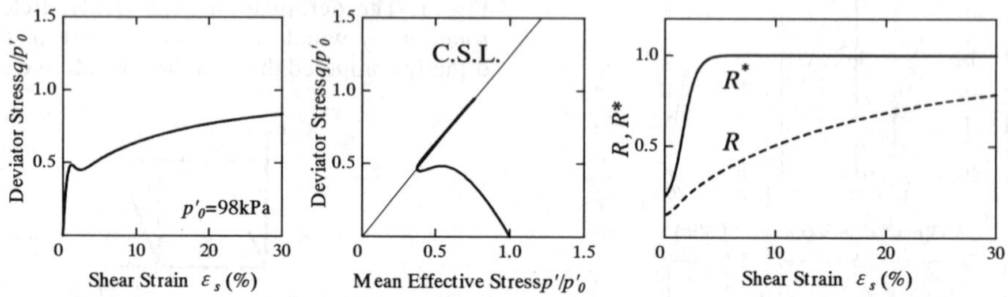

Figure 14. Undrained behavior of medium dense sand, hardening after softening

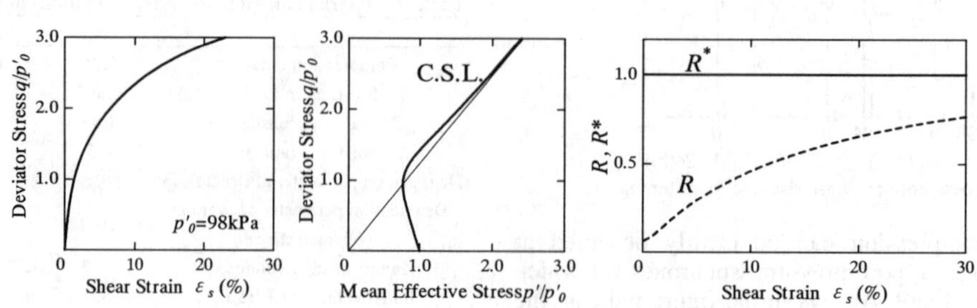

Figure 15. Undrained shear behavior of dense sand

observed in triaxial tests in laboratory with constant cell pressure are given in Figs.16 and 17. Even in a triaxial test the behavior of the soil specimen should be solved as an initial-boundary value problem. The discrepancy between computed results of Figs. 12-15 and experimental results of Figs. 16-17 should be discussed, at least, from this viewpoint, examples for that are given in Figs. 18 and 19.

7 SUDDEN COLLAPSE OF STRUCTURE IN LOOSE SAND

Two different behaviors of loose sand under repeated shear stress application are examined next. One is under fully drained condition and the other, under perfectly undrained condition. One may imagine sand compaction and sand liquefaction, respectively. Material constants with m and m^* are the same as those in Table 2. Initial overconsolidation ratio of the loose sand is assumed to be one.

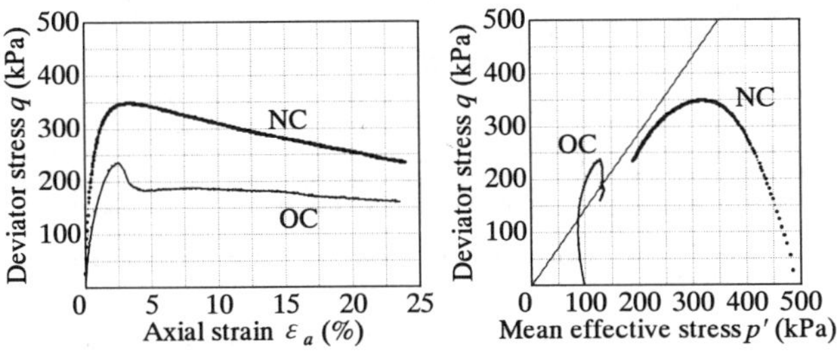

Figure 16. Undrained triaxial compression test of natural clay (Experiment)

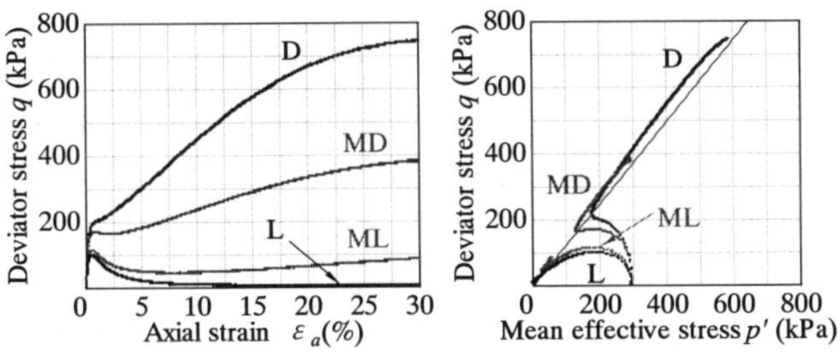

Figure 17. Undrained triaxial compression test of sand (Experiment)

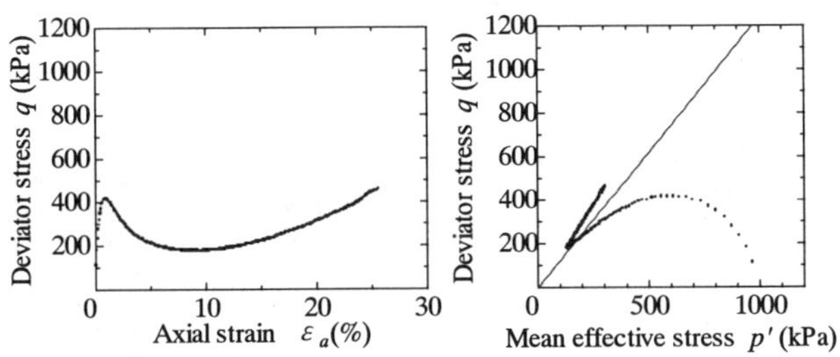

Figure 18. Undrained triaxial compression test of medium loose sand (Experiment)

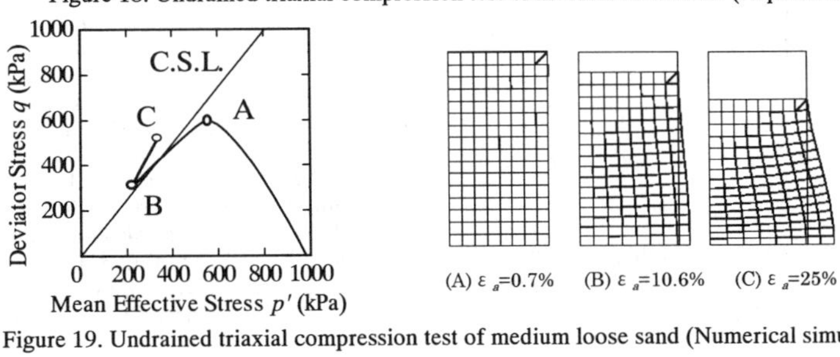

Figure 19. Undrained triaxial compression test of medium loose sand (Numerical simulation)

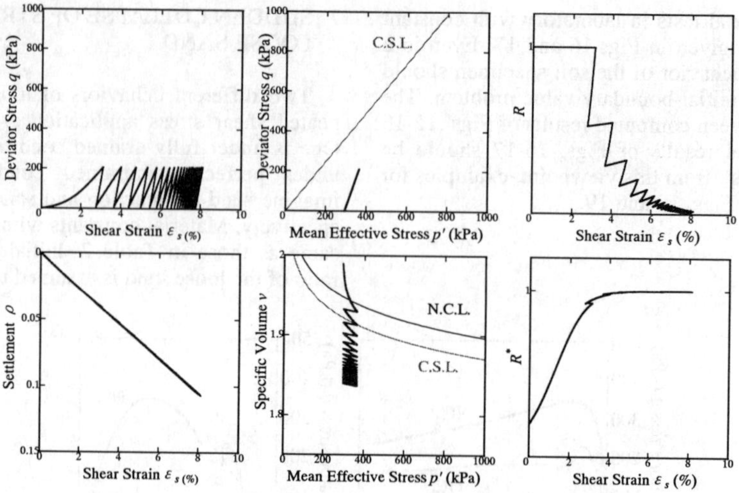

Figure 20. Repeated shear of loose sand in fully drained condition

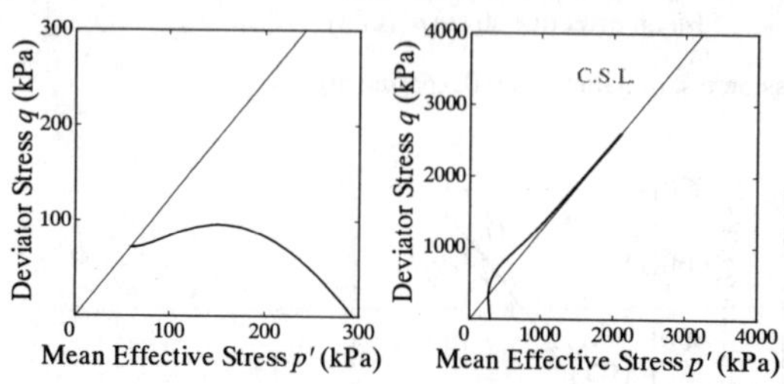

Figure 21. Undrained shear behavior before and after "compaction"

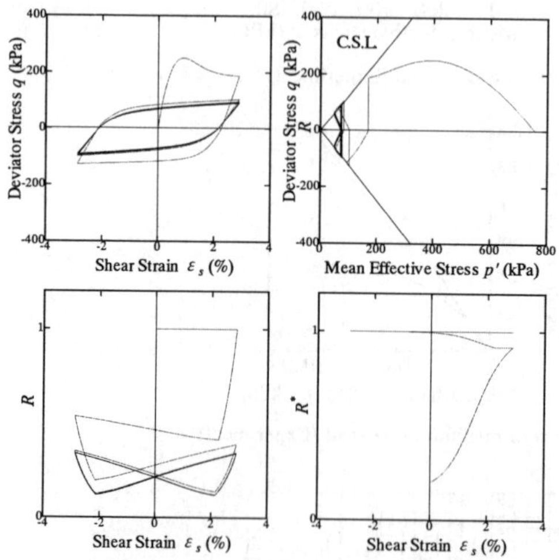

Figure 22. Repeated shear of loose sand in perfectly undrained condition

Fig. 20 shows a typical compaction procedure of loose sand. The shear stress application is repeated until the OCR exceeds 100. Necessary number of repetition was less than 30. The comparison in the undrained shear behavior of the sand before and after compaction is given in Fig. 21.

Undrained behavior of the same loose sand under repeated shear stress application is given in Fig.22. After several repetitions the plastic volume compression reaches Equilibrium State with plastic volume expansion and the sand arrives at a steady state.

8 CONCLUSION

Through numerical simulations, together with fundamental experiments, the expectation is positively accepted that in clay loss of overconsolidation proceeds much faster than the decay of soil structure and in sand, vice versa.

REFERENCES

Asaoka, A., Nakano, M. & Noda, T. 1998. *Superloading yield surface concept for the saturated structured soils*, Proc. of 4th European Conference on Numerical Methods in Geotechnical Engineering, Udine,233-242

Asaoka, A., Nakano, M. & Noda, T. 2000a. *Superloading Yield Surface Concept for Highly Structured Soil Behavior*, Soils and Foundations, 40-2:99-110.

Asaoka, A., Nakano, M., Noda, T. & Kaneda, K.2000b. *Delayed Compression/consolidation of Natural Clay due to degradation of soil structure*, Soils and Foundations, 40-3: 75-85.

Dienes, J.K. 1979. *On the Analysis of Rotation and Stress Rate in Deforming Bodies*, Acta Mechanica, 32: 217-232.

Hashiguchi, K. 1978. *Plastic Constitutive Equations of Granular Materials*, Proc. of US-Japan Seminar on Continuum Mechanics and Statistical Approaches in the Mechanics of Granular Materials (Cowin, S.C. and Satake, M. eds.), Sendai, JSSMFE: 321-329.

Hashiguchi, K. 1989. *Subloading Surface Model in Unconventional Plasticity*, International Journal of Solids and Structures, 25: 917-945.

Computer Methods and Advances in Geomechanics, Desai et al. (eds) © 2001 Balkema, Rotterdam, ISBN 90 5809 183 X

Dynamic behavior of mortar: Experimental data and modeling

N.D.Cristescu & O.Cazacu
Department of Aerospace Engineering, Mechanics and Engineering Science, University of Florida, Gainesville, Fla., USA

C.A.Ross & M.J.Schmidt
Graduate Engineering and Research Center, University of Florida, Shalimar, Fla., USA

ABSTRACT: To characterize the influence of confining pressure on mortar both quasistatic and dynamic tests have been performed. Dynamic loading tests were conducted at rates on the order of one hundred million times faster then in the quasistatic tests. The confined dynamic tests showed similar stress-strain response as the quasistatic tests. Dilatancy has been observed at high levels of the principal stress difference for both dynamic and static conditions. The dynamic compressive strengths are approximately double that of the quasistatic compressive strengths, while the shear strengths at the higher strain rates were found to be as high as five times that of the quasistatic shear strengths. A new rate dependent constitutive model for mortar has been developed. This model captures the main features of the material response such as compressibility and dilatancy, strain hardening, influence of the loading rates, and creep and relaxation.

1 INTRODUCTION

Strain rate effects on various unconfined strength properties of cementitious materials have been reported by many researchers, such as compressive data by Malvern & Ross (1986), Malvern et al.(1985), and Bischoff & Perry (1985) , tensile strength data by Cowell (1996), Ross et al.(1989; 1995; 1996), Rossi & Boulay (1989) and Brara (1999), and shear strength data by Schmidt & Ross (1999). The general trend in the unconfined tests of all the above referenced cementitious material data is a very slight increase in strength from a low strain rate of 1.0 x 10^{-7}/s up to a strain rate of approximately 1.0/s for tensile strength properties, 10/s for shear strength properties and 100/s for compressive strength properties. Above these critical strain rates a rather abrupt increase in strength occurs with a slope of ($\log_{10}$ strength/$\log_{10}$ strain rate) of approximately one-third. Representative strain rate effects on unconfined strength properties reported by Schmidt & Ross (1999) are given in Figure 1.

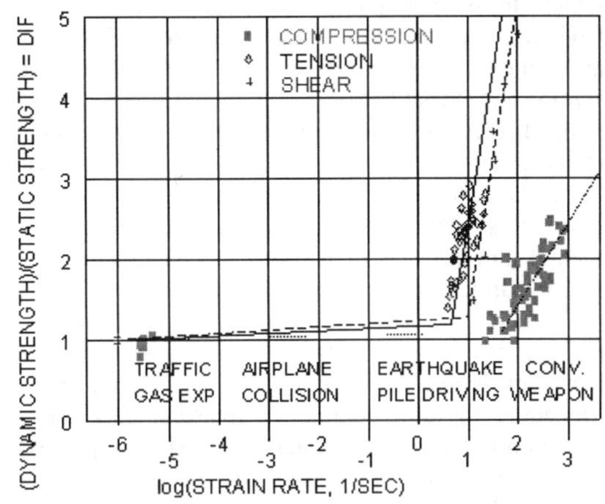

Figure 1. General strain rate effects on mortar strength.

Effects of confining pressure on compressive strength of cementitious materials and geomaterials are presented in some detail by Cristescu (1989) and Cristescu & Hunsche (1998). The general trend of increasing confining pressure on cementitious materials and geomaterials (usually tested in triaxial cells) is to produce increased deviatoric stresses relative to the axial, circumferential, radial and resultant volumetric strains. In a conventional triaxial compression test, the axial strain increases monotonically with the applied axial load. The axial compression continues until swelling occurs, known as dilatancy, at which time tensile columnar cracking begins and specimen failure follows. The volumetric strain is essential in the data analysis so it is necessary that both longitudinal axial strain and either radial or circumferential strain to be measured in the triaxial test. Since the radial and circumferential strains are equal in a cylindrical specimen only one of these strains need to be measured.

The combined effects of dynamic axial loading and hydrostatic confining pressure on concrete was studied experimentally by Malvern & Jenkins (1990) using a 7.62 cm diameter split Hopkinson pressure bar (SHPB), located at the University of Florida. It is reported that under dynamic conditions both axial strain at fracture and strain hardening are increased. In these tests the specimens were not instrumented and consequently the volumetric strain was not measured; the axial strain was determined from the strain gages of the SHPB.

Combined effects of tensile loading and lateral compression were studied by Weerheijm (1996) using a 7.62 cm diameter gravity driven SHPB located at Delft Technical University, The Netherlands. A strain gage instrumented prismatic specimen was used in a biaxial setup where the dynamic axial tension was applied by the SHPB and a brush-bearing platen was used to compression load the specimen in one of the lateral directions. For these tests the residual tensile impact strength decreased with increased lateral compression. This tensile strength decrease is consistent with columnar cracking, which forms parallel to the highest compression load, as discussed in Cristescu & Hunsche (1998) and studied experimentally by Ross (1998) and Jerome and Ross (2000).

In this paper, an experimental and theoretical study aimed at characterizing the dynamic response of mortar is presented. The material studied contains masonry sand but no large aggregates. The experimental part consists of a series of quasi-static tests and dynamic tests at several strain rates and under both unconfined and confined conditions. Based on the unconfined data an elastic/viscoplastic model has been developed. The structure of this model and a procedure for determining the material parameters based on a minimal set of data is given. Comparison between the model predictions for unconfined quasi-static and dynamic conditions and test results are good, with a degree of accuracy within the natural scatter of the data.

2 EXPERIMENTS

Mortar specimens of 7.62 cm diameter, D, and length, L, (L/D = 1.0) were subjected to quasistatic, confined and unconfined loadings using an MTS testing machine at a strain rate of approximately 1.2 10^{-6}/s. Same size mortar specimens were tested unconfined and confined in the University of Florida 7.62 cm diameter split Hopkinson pressure bar (UFSHPB) at strain rates of approximately 60/s and 160/s. The same confining cell and SHPB as used by Malvern & Jenkins (1990), was used in this study. Mix proportions and basic properties of the mortar are given in Table 1. Modulus test results for mortar conducted at a strain rate of 1.2 · 10^{-6}/s is shown in Figure 2.

Table 1: Mix proportions and basic mechanical properties.

Proportions & properties	Value
Portland Cement	450 g
Sand, sieve #4	1620 g
WRDA-19 0.5%	3.7 g
F" fly ash	297 g
Water, w/c = 0.55	411 g
Specific weight	2.08 kg/m^3
Compressive strength	42.2 MPa
Compressive modulus	26.4 GPa

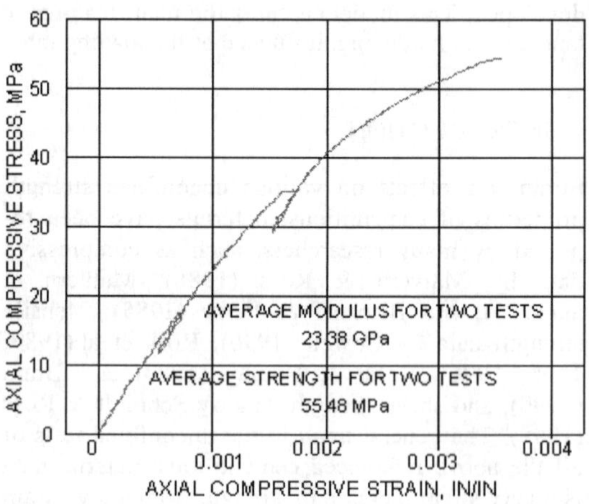

Figure 2. Quasistatic test for mortar; strain rate 1.25E-6/s.

The loading path in this unconfined test consisted of monotonic loading up to 17.5 MPa, at which level the axial stress was held constant for a time interval of 10 minutes. Short-term creep was recorded even at this low stress level. At the end of the creep stage, partial unloading followed by fast reloading showed irreversible strain. Although, some slight hysteresis was observed, the Young's modulus can be determined with good accuracy from the unloading slope. At 35 MPa another 10 minute creep-unloading-reloading cycle was performed. This test was repeated a second time utilizing a specimen instrumented with two sets of electrical resistance strain gages, each set diametrically opposed and at the middle of the specimen. Each set contained one axial and one circumferential strain gage. The average E obtained from both unconfined tests was 23.38 GPa. This setup allowed for measurement of the axial and transverse strains, which are then used in the calculation of the volumetric strain.

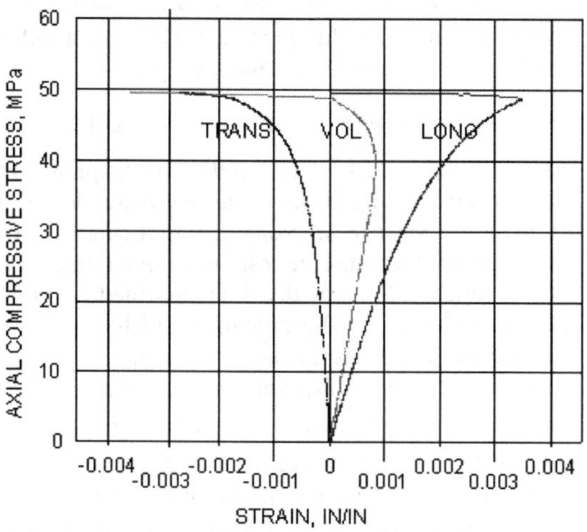

Figure 3. Longitudinal, transverse and volumetric strains for unconfined quasistatic test on mortar; strain rate1.18E-6/s.

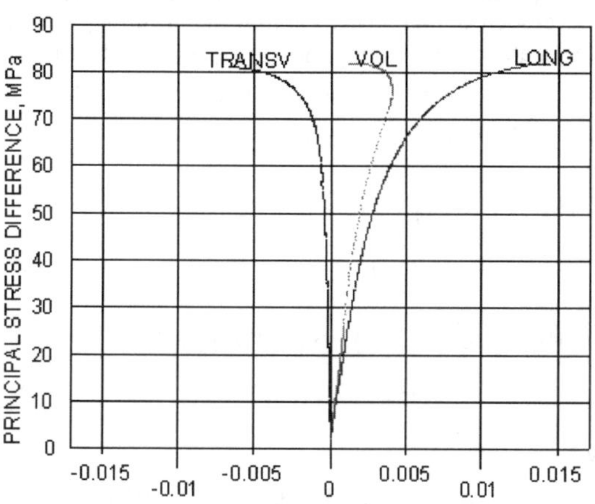

Figure 4. Longitudinal, transverse and volumetric strains for confined quasistatic test for mortar; confining pressure of 6.9 MPa, strain rate 1.25E-6/s.

The axial stress vs. volumetric strain curve obtained in this second test (Figure 3) shows compressibility up to 35 MPa and strong volume expansion (dilatancy) up to failure. The value of the bulk modulus estimated from the tangent to the quasi-linear middle

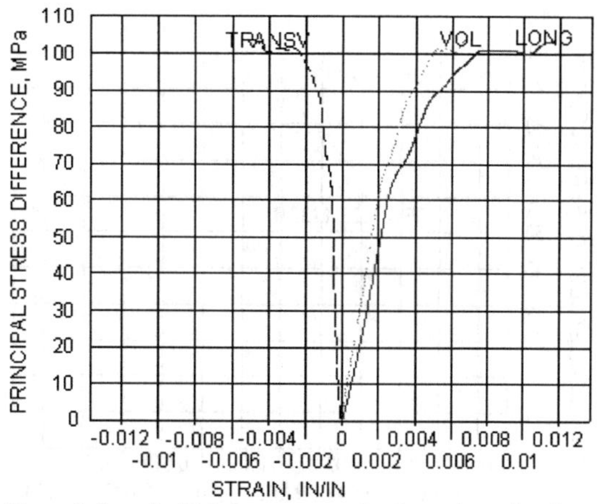

Figure 5. Longitudinal, transverse and volumetric strains for a dynamic confined test on mortar; confining pressure of 6.9 MPa, strain rate 52/s.

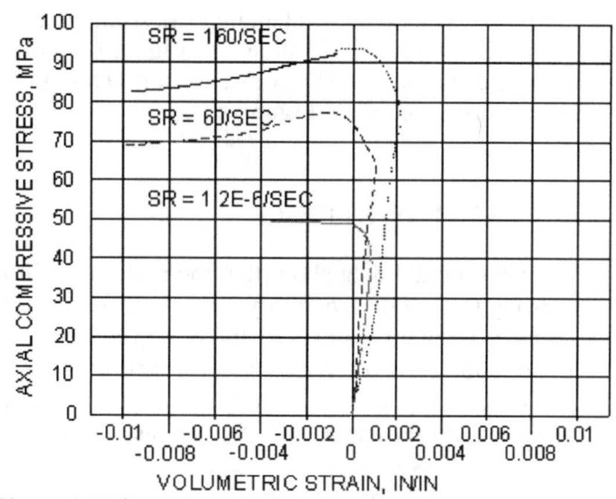

Figure 6. Volumetric strain for three different strain rates vs. axial compression for unconfined mortar tests.

portion of the curve is 13.3 GPa. A quasistatic test at 6.9 MPa confining pressure was also performed. In this test, the loading was increased monotonically (Figure 4). The deviatoric stress vs. axial strain curve shows an initial non-linear portion (generally, attributed to closing of existing microcracks) followed by a quasi-linear portion, and a final non-linear portion with opposed concavity.

The deviatoric stress vs. volumetric strain curve (Figure 4) shows that with increasing deviatoric stress the material becomes strongly dilatant. As expected the dilatancy threshold is higher than under unconfined conditions. The results of a dynamic compression test at a strain rate of approximately 50/s and the same confining pressure of 6.9 MPa are shown in Figure 5. This points out the influence of the strain rate and loading pulse length in tests using

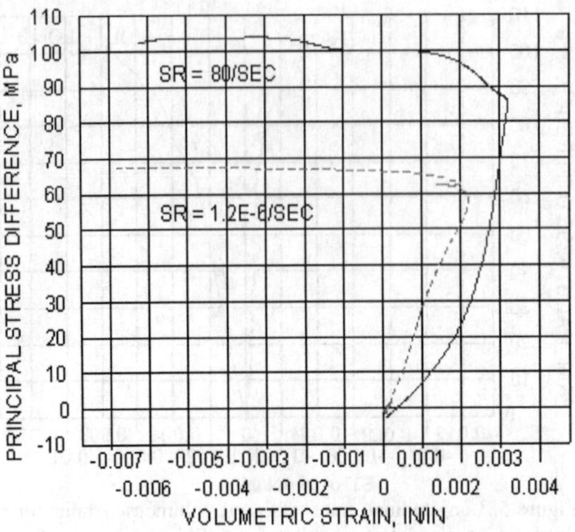

Figure 7. Strain rate effect on volumetric behavior for confining pressure of 3.45 MPa

the SHPB. The effect of changing the strain rate for unconfined mortar test is shown in Figure 6 and the effect of changing the strain rate for a given confining pressure is shown in Figure 7.

3 CONSTITUTIVE MODEL

The experimental data show that the material exhibits strong strain-rate sensitivity. A natural framework in which the main features of the observed behavior can be modeled is offered by the viscoplasticity theory. To develop a 3-D constitutive equation, confined triaxial quasistatic data for at least 3 different confining pressures are necessary. Based on the available data for mortar, a simplified elastic/viscoplastic constitutive equation that applies for axisymmetric triaxial conditions has been developed. It is shown that although simple in concept and in expression the model captures the main features of the material response such as compressibility and dilatancy, strain-hardening, influence of the loading rate, and creep and relaxation. The model can be easily extended to 3-D conditions.

We consider that the material response under axisymmetric triaxial conditions can be described as

$$\dot{\varepsilon}_1 = \left(\frac{1}{3G}+\frac{1}{9K}\right)\dot{\sigma}_1 + \left(-\frac{1}{3G}+\frac{1}{9K}\right)\dot{\sigma}_3 + h_1\langle f_1(\sigma_1,\sigma_3)-\varepsilon_1\rangle, \quad (1)$$

$$\dot{\varepsilon}_3 = \left(-\frac{1}{6G}+\frac{1}{9K}\right)\dot{\sigma}_1 + \left(\frac{1}{6G}+\frac{2}{9K}\right)\dot{\sigma}_3 + h_3\langle f_2(\sigma_1,\sigma_3)-\varepsilon_3\rangle \quad (2)$$

where, $\dot{\varepsilon}_1$ and $\dot{\varepsilon}_3$ are the axial strain rate and transverse strain rate, respectively; σ_1 and σ_3 are the axial and radial stresses, G is the shear modulus, and K is the bulk modulus. In (1) to (2)$\langle\ \rangle$is the Macauley bracket, which defines the positive part of any expression, i.e. $\langle A\rangle=\frac{1}{2}(A+|A|)$ while h_1 and h_3 are viscosity parameters. It is worthwhile to point out that by introducing a *shearing viscosity* different from the *volumetric viscosity* the particularities of the irreversible volumetric response can be described very accurately. Specifically, it is possible to capture the rate influence on the dilatancy threshold and on the extent of the dilatant zone in the material.

Although, the specific expressions of the stabilization curves $f_1(\sigma_1,\sigma_3)$ and $f_3(\sigma_1,\sigma_3)$ are most accurately determined from creep tests, data obtained in quasistatic compression at strain rate on the order of $10^{-6}/s$ can provide a good approximation of these functions. The viscosity parameters can be obtained from compression test data in a SHPB at two different strain rates. Based on the unconfined quasistatic data presented in the previous section, we have fully developed a 1-D stress version of the model. Thus, the stabilization boundaries are considered to be of the form: $\varepsilon_1 = F_1(\sigma_1)$ and $\varepsilon_3 = F_3(\sigma_1)$. A fourth order polynomial fit the data well

$$F_1(a_1,a_2,a_3,a_4,\sigma_1) = a_4\sigma_1^4 + a_3\sigma_1^3 + a_2\sigma_1^2 + a_1\sigma_1, \quad (3)$$

$$F_3(b_1,b_2,b_3,b_4,\sigma_1) = b_4\sigma_1^4 + b_3\sigma_1^3 + b_2\sigma_1^2 + b_1\sigma_1, \quad (4)$$

where,
$a_1=3.522\cdot10^{-5}$, $a_2=1.131\cdot10^{-6}$, $a_3=-6.026\cdot10^{-8}$ $a_4=1.06\cdot10^{-9}$, $b_1=-2.08\cdot10^{-6}$, $b_2=-2.08\cdot10^{-6}$, $b_3=10^{-7}$, and $b_4=-1.5\cdot10^{-9}$. The dynamic Young modulus E is 28.73 GPa, while the Poisson's ratio ν = 0.2. The viscosity coefficient h_1 can be determined from data obtained in unconfined dynamic tests performed at strain rates $\dot{\varepsilon}^{(1)}$ and $\dot{\varepsilon}^{(2)}$, respectively ($\dot{\varepsilon}^{(1)} \neq \dot{\varepsilon}^{(2)}$). Indeed, if $\sigma_1^{(1)}$ and $\sigma_1^{(2)}$ are stresses corresponding to the same value of the axial

strain ε_1, while $\dot{\sigma}_1^{(1)}$ and $\dot{\sigma}_1^{(2)}$ are the corresponding loading rates, from (1) it follows that:

$$h_1 = \frac{\left(\varepsilon_1^{(1)} - \varepsilon_1^{(2)}\right) - \left(\dfrac{\dot{\sigma}_1^{(1)} - \dot{\sigma}_1^{(2)}}{E}\right)}{F_1\left(\sigma_1^{(1)}\right) - F_1\left(\sigma_1^{(2)}\right)} \tag{5}$$

From the unconfined tests performed at the strain rates $\dot{\varepsilon}^{(1)} = 60$/s and $\dot{\varepsilon}^{(2)} = 160$/s ($\dot{\sigma}_1^{(1)} = 1.346 \cdot 10^4$ MPa/s and $\dot{\sigma}_1^{(2)} = 2.07 \cdot 10^6$ MPa/s respectively), we get $h_1 = 6 \cdot 10^3$ s^{-1}. A formula similar to (5) can be used to derive $h_3 = 20 \cdot 10^3$ s^{-1}.

Consider an unconfined compression test at a constant load rate, *i.e.* at moment $t_0 = 0$: $\varepsilon(t_0) = \mathbf{0}$, $\sigma(t_0) = \mathbf{0}$, and for $t \geq t_0$: $\sigma_1(t) = \dot{\sigma}_1^0 t$, $\sigma_3(t) = 0$. Integrating (1)-(2) we obtain

$$\varepsilon_1(\sigma_1, \dot{\sigma}_1) = \frac{1}{E}\left(1 - \exp\left[-h_1\left(\frac{\sigma_1}{\dot{\sigma}_1}\right)\right]\right)\frac{\dot{\sigma}_1}{h_1} + \sum_{i=0}^{4}(-1)^i \times \left(\frac{\dot{\sigma}_1}{h_1}\right)^i\left(F_1^{(i)}(\sigma_1) - F_1^{(i)}(0)\cdot\exp\left[-h_1\left(\frac{\sigma_1}{\dot{\sigma}_1}\right)\right]\right), \tag{6}$$

$$\varepsilon_3(\sigma_1, \dot{\sigma}_1) = \frac{-\nu}{E}\left(1 - \exp\left[-h_3\left(\frac{\sigma_1}{\dot{\sigma}_1}\right)\right]\right)\frac{\dot{\sigma}_1}{h_3} + \sum_{i=0}^{4}(-1)^i \times \left(\frac{\dot{\sigma}_1}{h_3}\right)^i\left(F_3^{(i)}(\sigma_1) - F_3^{(i)}(0)\cdot\exp\left[-h_3\left(\frac{\sigma_1}{\dot{\sigma}_1}\right)\right]\right), \tag{7}$$

In (6) and (7), $F^{(i)}$ stand for the ith derivative of F with respect to σ_1. If in (6) and (7) $\dot{\sigma}_1 \to 0$, we obtain the stabilization boundaries $\varepsilon_1 = F_1(\sigma_1)$ and $\varepsilon_3 = F_3(\sigma_1)$, respectively while for $\dot{\sigma}_1 \to \infty$ we get the corresponding elastic curves.

4 COMPARISON WITH THE DATA

In the following, we present a comparison between the model predictions and unconfined direct compression data in the SHPB. The dynamic tests were done at constant strain rates of 60/s and 160/s. The average loading rate was $\dot{\sigma}_1^{(1)} = 1.346 \cdot 10^4$ MPa/s, and $\dot{\sigma}_1^{(2)} = 2.07 \cdot 10^6$ MPa/s, respectively. Figures 8 and 9 show the theoretical $\sigma_1 - \varepsilon_1$ and $\sigma_1 - \varepsilon_3$, corresponding to the constant loading rates $\dot{\sigma}_1^{(1)}$ and $\dot{\sigma}_1^{(2)}$ against the experimental curves, and the elastic response.

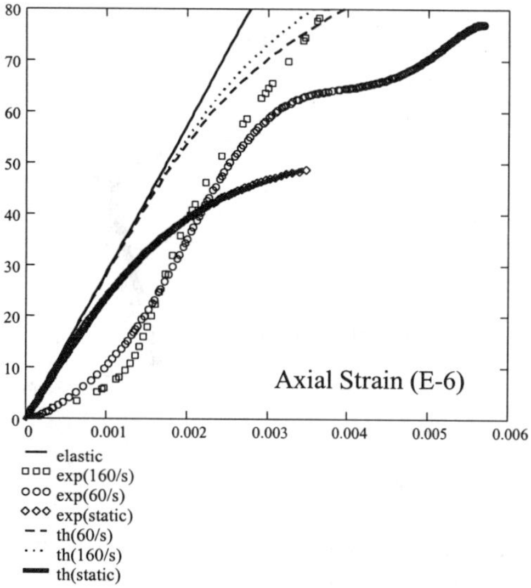

Figure 8. Comparison between the theoretical $\sigma_1 - \varepsilon_1$ curves, unconfined dynamic data and quasi-static test.

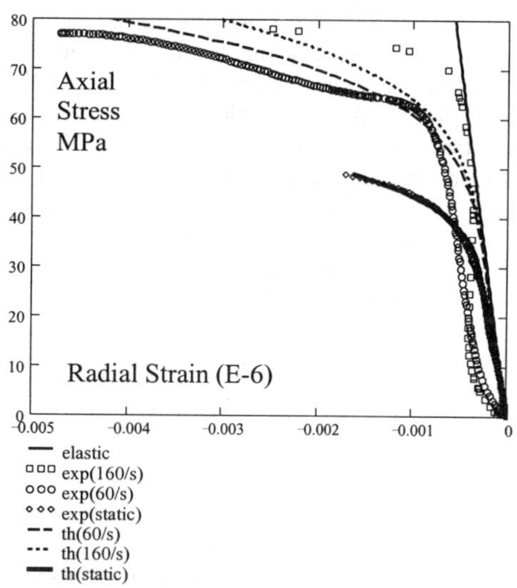

Figure 9. Comparison between the theoretical $\sigma_1 - \varepsilon_3$ curves, unconfined dynamic data, and quasi-static data.

The comparison of the model predictions and data for volumetric strains are given in Figure 10. As expected the static stress-strain curves fall below the dynamic ones. Also, the increase in the strain rate results in a "raising" of all the stress-strain curves. The lower parts of the experimental dynamic stress-axial strain curves present upward concavities, which are not due to rate effects but result from other phenomena such as the crushing of asperities at the ends of the specimen. The static stress-strain

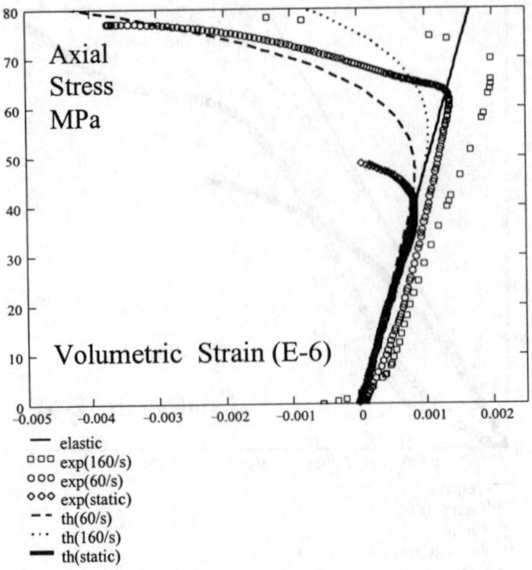

Figure 10. Comparison between the theoretical $\sigma_1 - \varepsilon_V$ curves, dynamic data, and quasi-static test.

curves are well reproduced by the model. The theoretical dynamic curves are obtained from the static ones assuming material viscosity and using the concept of overstress. The instantaneous response, i.e. the material response when the strain rate is increased to infinity is modeled as elastic. It can be noted that the rate influence is correctly described. The theoretical curves correspond to a constant loading rate (the average loading rate in the test) while in the actual tests, the loading rate was not constant. Therefore, we cannot expect a perfect agreement between the model and data. However, the general trends of the data are reproduced and for higher values of the stress the comparison is within the natural scatter of the data. The model is shown capable of reproducing the main behavior trends in the material.

ACKNOWLEDGEMENT

This material is based on research sponsored by the Munitions Directorate, Air Force Research Laboratory, under agreement number F08635-98-D-0016. The U.S. Government is authorized to reproduce and distribute reprints for governmental purposes notwithstanding any copyright notation thereon.

The view and conclusions contained herein are those of the authors and should not be interpreted as necessarily representing the official policies or endorsements, either expressed or implied, of the Munitions Directorate, Air Force Research Laboratory or the U.S. Government.

REFERENCES

Malvern, L. E. & C.A. Ross 1986. Dynamic Response of Concrete and Concrete Structures, *Two Annual Reports and a Final Report AFOSR Grant F49620-83-K007, AF Office of Scientific Research*, 1982-1986, Washington, DC.

Malvern, L.E., D.A. Jenkins, T.Tang, & C.A. Ross 1985. Dynamic Compression of Concrete, *The 2nd Symp on the Interaction of Non-Nuclear Munitions with Structures, Panama City, FL, April 1985:* 194-199.

Bischoff, P.H. & S.H. Perry 1985. Compressive Strain Rates of Concrete, *Proc. Materials Res. Soc., Fall 1985, Boston, Dec. 1985.*

Cowell, W. 1996. Dynamic Properties of Plain Portland Cement Concrete, *US Naval Civil Engineering Laboratory*, Port Hueneme, CA.

Ross, C.A., P.Y. Thompson & J.W. Tedesco 1989. Split-Hopkinson Pressure Bar Tests on Concrete and Mortar in Tension and Compression, *ACI Material Journal*,86, (5): 475-481.

Ross, C.A., J.W. Tedesco & S.T. Kuennen 1995. Effects of Strain Rates on Concrete Strength, *ACI Materials Journal*, 92(1): 37-47.

Ross, C.A., D.M. Jerome, J.W. Tedesco & M.L. Hughes, 1996. Moisture and Strain Rate Effects on Concrete Strength. *ACI Materials Jouranl.* 93(3):293-300.

Rossi, P. & C. Boulay 1989. Etude Experimentale du Couplage Fissuration-Viscosite au Sein du Beton, Project No. 9, Rapport GRECO, *Rheologies des Geomateriaux.*

Brara, A. 1999. *Etude Experimentale de Tractions Dynamique du Beton par Ecaillage*, Dissertation for Ph.D., Universite de Metz, Metz, France.

Schmidt, M. J.& C.A. Ross 1999. Shear Strength of Concrete Under Dynamic Loads, *ASME Pressure Vessel and Piping Conf., Boston, MS, August 1999.*

Cristescu, N. 1989. *Rock Rheology*. Dordrecht: Kluwer Academic Publishers.

Cristescu, N. & U. Hunsche 1998. *Time Effects in Rock Mechanics*, Chichester: John Wiley & Sons.

Malvern, L.E. & D.A. Jenkins 1990. Dynamic Testing of Laterally Confined Concrete, *Tech. Rept. ESL-TR-89-47 Air Force Engineering and Services Lab, Tyndall AFB, FL.*

Weerheijm, J. 1996. Concrete Under Impact Tensile Loading and Lateral Compression, Ph.D. Dissertation, *Delft Technology University, Delft, Netherlands, November 1996.*

Ross, C.A. 1998. The Effects of Dynamic Pre-Compression on the Dynamic Tensile Strength of Concrete and Mortar, *ASME Conf. PVP-Vol. 361, Structures Under Extreme Loading Conditions-1998, Eds. Zukas, Jerome and Shin, San Diego, CA, pp 31-36, 1998.*

Jerome, D.M. & C.A. Ross 2000. Experiments on Columnar Splitting of Concrete in Axial Compression, *2000 ASME Press. Vessel and Piping Conference.*

Computer Methods and Advances in Geomechanics, Desai et al. (eds) © 2001 Balkema, Rotterdam, ISBN 90 5809 183 X

Modeling creep behavior of granular materials

Poul V. Lade
Aalborg University, Denmark

Chi-Tseng Liu
LAW / Crandall Incorporated, Los Angeles, Calif., USA

ABSTRACT: A systematic series of creep tests on sand under triaxial stress states has been performed. The experiments showed that the nature of inelastic creep strains is similar to that of plastic strains. The time-dependent inelastic work may be normalized on values of reference plastic work and is independent of confining pressure and stress level. It varies linearly with normalized time when plotted on log-log scales. To capture the time-dependent inelastic strains, the existing single hardening model is extended by multiplying the plastic work in the hardening function by a power function of time, which requires determination of one parameter. Thus, the predicted time-dependent behavior is based on the modeled time-independent behavior. The time-dependent behavior is predicted and compared with results of creep experiments.

1 INTRODUCTION

In geotechnical engineering projects, soil creep and relaxation often cause long-term deformation and pressures on buildings, bridge abutments, earth retaining structures, and earth slopes. Excessive deformation with time may cause structures to fail. Another mode of soil collapse is instability and liquefaction: Granular materials may become unstable when exposed to certain stress paths inside the failure surface. In nature, especially fine silty sands are susceptible to liquefaction under static and cyclic loading conditions, but these fine silty sands also exhibit considerable creep behavior, and they are therefore affected by time effects. Drained creep at a given stress state causes the plastic yield surface to move out and away from the current stress state, and the sand consequently becomes stiffer in the region immediately outside the stress point. Increasing amounts of drained creep therefore require increasingly high stress differences or decreasingly low effective confining pressures to render the sand unstable. Thus, time effects play significant roles in the behavior of soils.

Prediction of creep, relaxation and loading rate effects in granular materials requires a constitutive model with a capability to account for time-dependent behavior. In this presentation the time-dependent behavior is incorporated in an existing single hardening plasticity model, whose capabilities have been demonstrated in the literature. The advantage of adding this capability to the existing model is that prediction of creep under three-dimensional conditions is immediately incorporated in the model. Because the time effects modeled here correspond to rates lower than those employed in conventional triaxial testing, emphasis has been placed on time effects at working stress conditions away from failure. Although failure may be affected by rate effects, especially at higher rates, this effect is relatively small at the rates and times considered here (see e.g. Yamamuro and Lade 1993), and only small modifications, within the scatter in the data, would be required from effects on failure. This effect has consequently not been taken into account in the study presented below.

2 PREVIOUS OBSERVATIONS

Considerable insight into the time-dependent behavior of clays has been gained from a number of studies in which drained tests have been performed, and excellent reviews have been presented in the literature (see, e.g., Cristescu 1989; Feda 1992; Mitchell 1993; Leroueil and Marques 1996; Sheahan and Kaliakin 1996).

Sands generally creep less than clays at the same stress states (Murayama et al. 1984; Lade et al. 1997), but the amount of creep increases with confining pressure, particularly after crushing becomes important at high stresses (Yamamuro and Lade 1993). More crushing occurs in loose than in dense sand (Bopp and Lade 1997). However, fine silty sands, whose structure is metastable and may be susceptible to instability and liquefaction

(Yamamuro and Lade 1997, 1998; Lade and Yamamuro 1997), also show considerable time-effects. In this particle structure the silt particles may occupy locations at or near the contact points of the larger sand particles. This particle structure tends to form in the fine range of particle sizes where the attractive and repulsive forces begin to compete with the gravitational forces (Lade et al. 1998). Thus, the finer particles push the larger particles slightly apart, and upon application of normal and shear stresses, the fine particles will tend to slide into the void spaces, allowing the larger particles to move together. Such particle structures formed in loose silty sands are precarious and unstable, and slippage occurs easily in such structures. This generates an inherently contractive volumetric tendency, which under undrained conditions may lead to instability and static liquefaction (Bjerrum 1973; Lade et al. 1997). Thus, time-dependent behavior in granular materials may be caused by crushing of sand grains or by collapse of metastable particle structures.

A study of drained creep and relaxation in Antelope Valley sand was performed by Lade and Liu (1998) to (1) investigate the nature of time effects, (2) study whether time-dependent behavior may be modeled by adding time effects within the framework provided by work-hardening plasticity theory, and (3) provide data for development of a constitutive model for soil stress-strain-time behavior. The structure of the dry deposited fine sand exhibited some collapse upon introduction of water in the pores. The fine sand had sufficient permeability to immediately dissipate pore pressures, thus allowing observation of creep deformations directly after changes in stress. The experiments showed that the nature of creep strains is similar to that of plastic strains in the sense that they may be predicted from the framework provided by hardening plasticity theory. In particular, the potential surface determined from and used for prediction of plastic strains may also be used for prediction of time-dependent creep strains. It also appears that the yield surface and the plastic potential surface move out together during creep, and the point at which to evaluate inelastic strain increment directions is at the current location of the yield surface and the accompanying plastic potential surface. Prediction of the movement and new location of the yield surface was attempted on the basis of the change in inelastic work occurring during creep (Lade 1994).

3 SINGLE HARDENING MODEL

The single hardening constitutive model, to which the time-dependent behavior will be added, was developed on the basis of a thorough review and evaluation of data from experiments on frictional materials such as sand, clay, concrete and rock (Kim and Lade 1988; Lade and Kim 1988 a, b). In this model the total strain increments observed in a material when loaded are divided into elastic and plastic components so that

$$\dot{\varepsilon}^t_{ij} = \dot{\varepsilon}^e_{ij} + \dot{\varepsilon}^p_{ij} \tag{1}$$

Both the elastic and the plastic strain increments are expressed in terms of effective stresses, and all functions for the elastic behavior, failure criterion, flow rule, yield criterion, and work-hardening/softening law are expressed in terms of stress invariants I_1, I_2, and I_3 and the stress deviator invariant J_2. The operative equations in this model were briefly reviewed by Lade (1990) and by Lade and Kim (1995). Only those relating to the yield criterion and the work-hardening relation will be used here, and they are reviewed below.

3.1 Yield Criterion and Work-Hardening Law

The model employs a single isotropic yield surface that is expressed as a contour of constant plastic work as measured from the stress origin:

$$f_p = f'_p(\sigma) - f''_p(W_p) \tag{2}$$

in which the expression for the yield surface in terms of stresses is:

$$f'_p(\sigma) = \left[\psi_1 \frac{I_1^3}{I_3} - \frac{I_1^2}{I_2}\right]\left[\frac{I_1}{p_a}\right]^h \cdot e^q \tag{3}$$

where the parameter ψ_1 (= constant for a given material at given density) acts as a weighting factor between the triangular shape (from the I_3 term) and the circular shape (from the I_2 term), p_a is atmospheric pressure in the same units as the first stress invariant I_1, h is a constant for a given soil, and q varies from zero at the hydrostatic axis to unity at the failure surface.

In Eq. 2 the expression for the yield criterion in terms of plastic work is:

$$f''_p(W_p) = \left[\frac{1}{D}\right]^{1/\rho}\left[\frac{W_p}{p_a}\right]^{1/\rho} \tag{4}$$

where ρ and D are constants for a given material, and W_p is the plastic work calculated from

$$W^p = \Sigma\, dW_p = \Sigma\, \sigma_{ij} \cdot d\varepsilon^p_{ij} \tag{5}$$

Thus, f_p" varies with plastic work only. The values of ρ and D are given by:

$$\rho = p/h \quad (6)$$

and

$$D = \frac{C}{(27\psi_1+3)^{\rho}} \quad (7)$$

in which

$$W_p = C \cdot p_a \cdot \left(\frac{I_1}{p_a}\right)^p \quad (8)$$

The values of the material parameters C and p are determined from isotropic compression tests, as discussed below.

The resulting yield surface is shaped as an asymmetric teardrop, with smoothly rounded triangular cross-sections in octahedral planes. As the plastic work increases, the isotropic yield surface inflates until the current stress point reaches the failure surface. The relation between the yield criterion and the plastic work is described by a monotonically increasing function whose slope decreases with increasing plastic work, as shown in Figure 1. Beyond failure the yield surface contracts isotropically with increasing amount of plastic work.

4 MODEL FOR CREEP EFFECTS IN SAND

The development consists of modifying the plastic hardening relation to include effects of time-dependent inelastic (or visco-plastic) work. The stress-strain-time relation therefore takes the following form:

$$d\varepsilon_{ij}^{vp} = d\lambda(W_p, t) \cdot \frac{\partial g}{\partial \sigma_{ij}} \quad (9)$$

in which $d\lambda(W_p,t)$ is determined from the hardening relation that includes plastic and time-dependent inelastic work. Once the formulation of the multiplier $d\lambda(W_p,t)$ has been determined, the stress-strain-time relation is given by the equation above. All other portions of the constitutive model have been developed and shown to work well.

The scalar factor of proportionality $d\lambda$ in Eq. 9 may be expressed as:

$$d\lambda = \frac{dW_p}{\mu g_p} \quad (10)$$

in which μ is a material parameter, g_p is the plastic potential function, and the increment in plastic work can be determined by differentiation of the modified hardening equation.

Examples of experimental results from this study were presented by Lade and Liu (1998). To compare time-dependent behavior at different stress levels, the time is normalized as follows:

$$T = \frac{t}{t_0} \quad (11)$$

in which t is the elapsed time since the change in stress, and t_0 is a reference time. Consistent patterns of soil behavior were observed in all experiments after 2 minutes of creep, but most often before. Nevertheless, because of this consistency, the reference time was taken as t_0 = 2 minutes in this study. Thus, T = 1 at 2 minutes of creep.

To express the variation of inelastic work in creep experiments, the inelastic work produced during creep (also referred to as W_p) is calculated and normalized on the inelastic work at T =1, as follows:

$$\bar{W}_p(T) = \frac{W_p(T)}{W_p(T=1)} \quad (12)$$

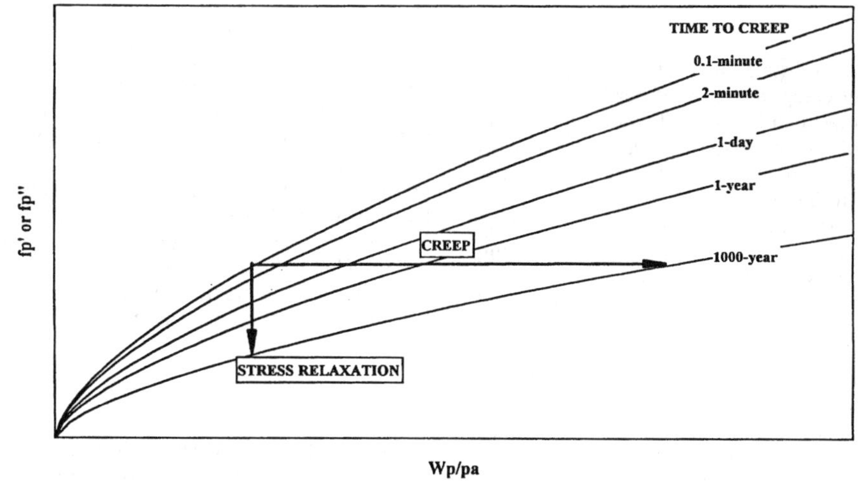

Figure 1. Schematic illustration of creep effects on the change of hardening curves with time.

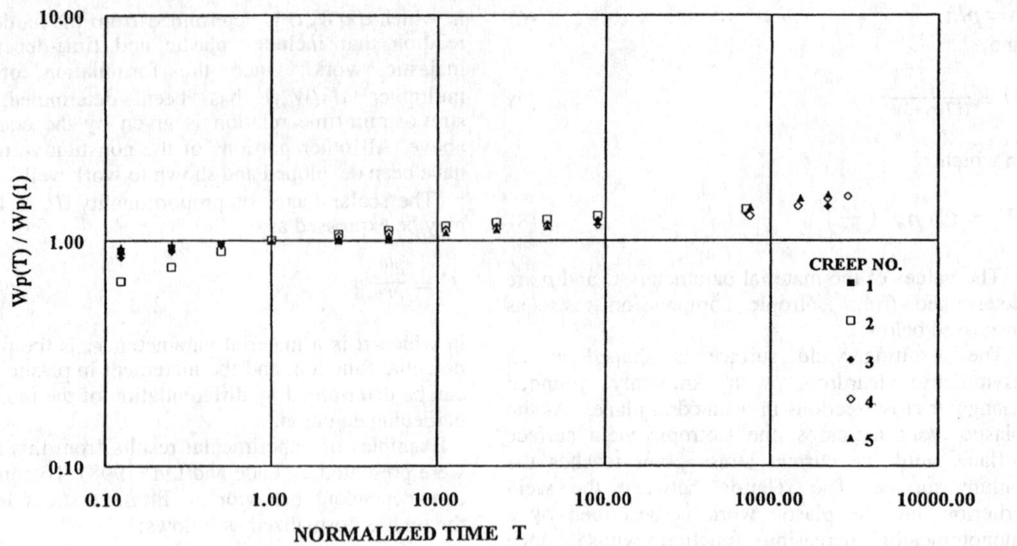

Figure 2. Normalized inelastic work versus normalized time from isotropic compression creep tests.

Figure 2 shows the normalized, accumulated inelastic work plotted versus normalized time on a log-log diagram at five different values of isotropic stress. The results of the five creep tests describe the same straight line within the scatter of the data. The creep at constant isotropic state of stress causes compression and the inelastic work therefore increases with time. Higher stresses produce larger creep strains and more inelastic work. The normalization indicated in Figure 2 captures these effects, and the straight line is expressed as:

$$\frac{W_p(T)}{W_p(1)} = T^{\kappa} \tag{13}$$

Since the state of stress and the yield function is constant during creep, a family of hardening curves for different times can be derived from Eq. 13. The consequence of the variation in plastic work with time expressed in Eq. 13 for creep and relaxation is also indicated in Figure 1.

To determine the dimensionless hardening parameters C and p in Eq. 7 the plastic work during isotropic compression, W_p/p_a, is plotted against the first stress invariant, I_1/p_a, in a log-log diagram and fitting a straight line, as shown in Figure 3. This may be done at different times T, as also shown in Figure 3. The magnitude of C is the value of W_p/p_a at I_1/p_a = 1, and the value of p is the slope of the fitted line. Apart from the initial deviations, the straight lines are essentially parallel and vary in a systematic manner with time for creep. Thus, the slope p is and the parameter C is determined at the reference time of 2 minutes (T = 1). The variation in intercept with $I_1/p_a = 1$ is then expressed by the relation in Eq. 13, and therefore according to Eq. 7:

$$C(T) = C(1) \cdot T^{\kappa} \tag{14}$$

Creep experiments were performed on Antelope Valley sand in isotropic compression and in triaxial compression tests at various stress levels (proximity to failure) following loading along conventional, constant confining pressure and along proportional loading stress paths. In order to test whether the expression in Eq. 13 is generally correct for creep at all stress states and for all stress paths leading to these stress states, the inelastic work was calculated for various creep tests, normalized on the inelastic work at t_0 = 2 minutes, and plotted versus T. Experiments were performed with creep initiated at up to five different stress states and lasting up to 10 days for each specimen. The different experiments produced the same straight line within the scatter of the results, as shown in Figure 4. The best fitting value of the exponent κ in Eq. 13 is found to be 0.0386. This is the only new parameter added to the model in order to predict time effects under general three-dimensional conditions.

5 PREDICTION OF CREEP STRAINS

All parameters for the single hardening model are independent of time effects except the value of C, whose variation with time is expressed through Eq. 14. Small time effects may also be present on failure,

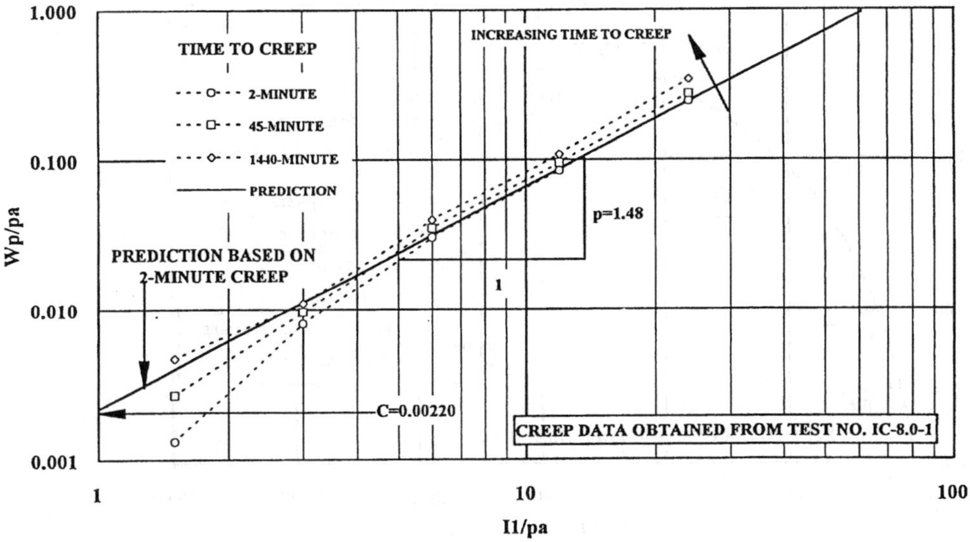

Figure 3. Variation of hardening parameters from isotropic creep tests.

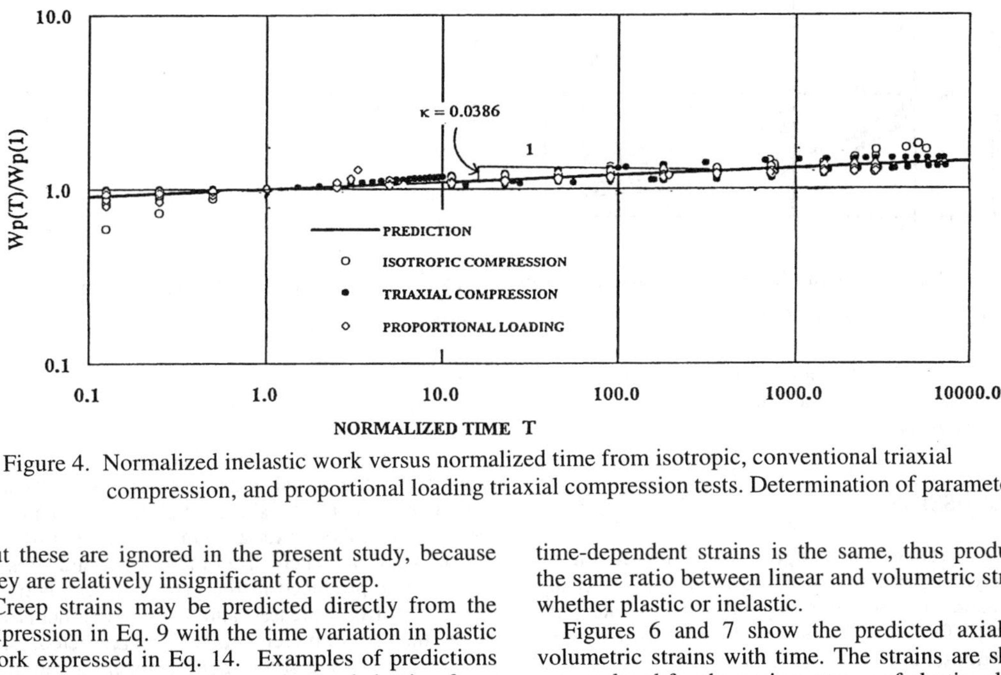

Figure 4. Normalized inelastic work versus normalized time from isotropic, conventional triaxial compression, and proportional loading triaxial compression tests. Determination of parameter κ.

but these are ignored in the present study, because they are relatively insignificant for creep.

Creep strains may be predicted directly from the expression in Eq. 9 with the time variation in plastic work expressed in Eq. 14. Examples of predictions of stress-strain and volume change behavior for a triaxial compression test on Antelope Valley sand are shown in Figure 5, with the solid curves corresponding to the reference behavior, taken to correspond to 2 minutes of creep at each point. The predictions of creep at four different stress ratios are shown on the upper diagram, while the volume change behavior in the lower diagram follows the primary loading curve. This is because the potential for plastic strains and the potential for inelastic, time-dependent strains is the same, thus producing the same ratio between linear and volumetric strains, whether plastic or inelastic.

Figures 6 and 7 show the predicted axial and volumetric strains with time. The strains are shown accumulated for the various stages of elastic, plastic, and creep behavior as predicted by the model. Thus, creep strains are not initialized at each new stage of creep. Deviations between predicted and observed behavior are therefore accumulated as the test progresses. The majority of these deviations are caused by experimental scatter and by failure of the single hardening model to predict the time-independent, plastic strains with high accuracy. If the predicted and the observed creep behavior

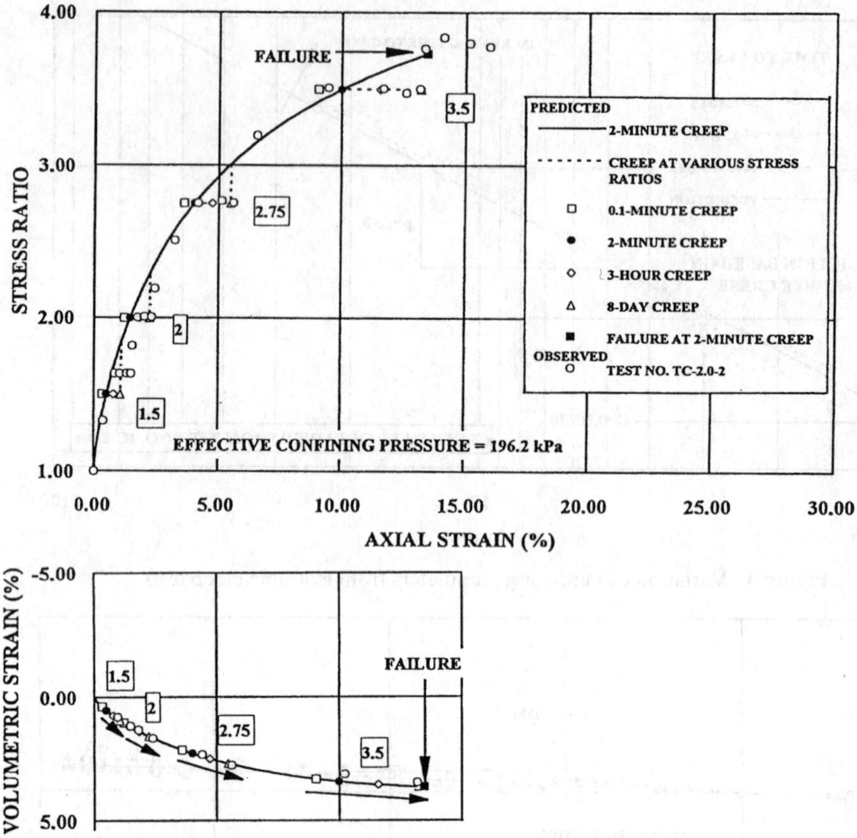

Figure 5. Comparison of observed and predicted stress-strain and volume change relations in triaxial compression test with creep stages at four different stress ratios.

were initialized at the same points at the beginning of each new creep stage, the comparisons would look quite attractive.

6 DISCUSSION

Several advantages are present when incorporating a single yield surface in a constitutive model: Using the plastic work contours as yield surfaces takes into account effects of shear as well as volumetric strains. Furthermore, the direct association between yielding and plastic work results in a unique hardening curve, and it produces a model that is mathematical consistent. Comparison with experimental evidence shows that this formulation of the single yield surface fits very well with observed behavior of frictional materials, i.e. it captures the conditions for yielding correctly for various stress paths. In addition, employment of a single yield surface makes the model more efficient for applications involving large computations.

Employment of plastic work as the hardening parameter (rather than e.g. effective strain or volumetric strain) makes the model particularly suitable for addition of time effects such as loading rate effects, creep, and relaxation. All types of strain increments (normal, shear, volumetric) and all stresses are taken into account in calculating the plastic work, and such strain increments and stresses may also be produced in the process of predictions. Thus, if current stresses and stress increments are given, then strain increments may be predicted. Or if current stresses and strain increments are given, then stress increments may be predicted from the model. In connection with time effects, this capability comes in conveniently. Pure creep, which occurs under constant stresses, produces inelastic strains and therefore inelastic work, which is very similar to plastic work. Thus, the plastic work increases with time, while the stresses stay constant, as indicated in Figure 1. Conversely, once the model has been modified to include time effects, then it is possible to keep the stresses constant, let the time pass, thereby

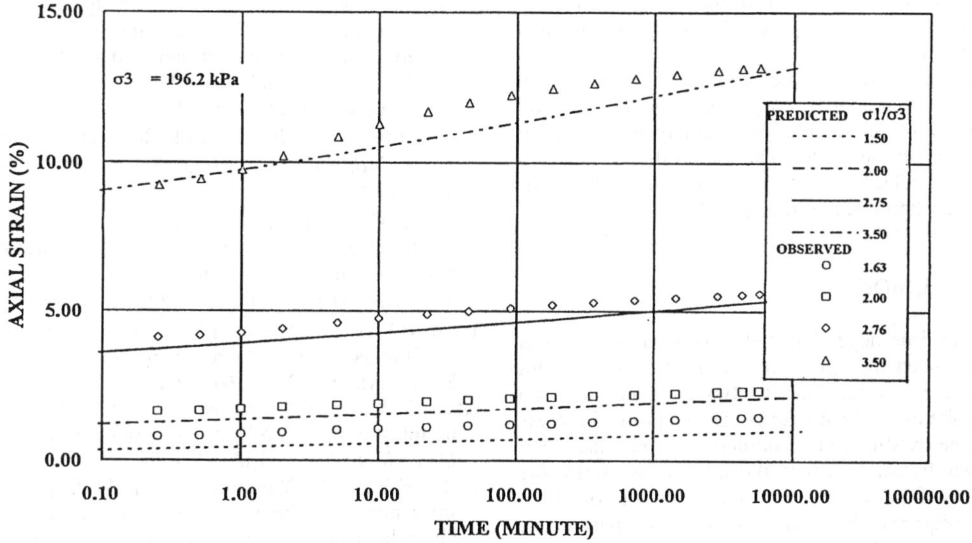

Figure 6. Comparison of observed and predicted axial strain versus time relation in triaxial compression test with creep stages at four different stress ratios.

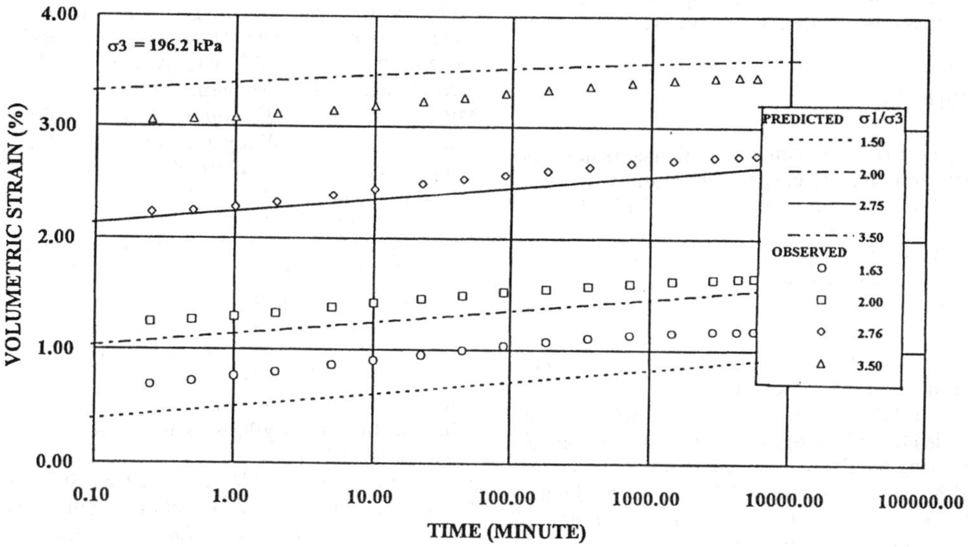

Figure 7. Comparison of observed and predicted volumetric strain versus time relation in triaxial compression test with creep stages at four different stress ratios.

producing inelastic (i.e. plastic) work, and use the model to predict the corresponding inelastic (i.e. creep) strains.

If, on the other hand, strains are prevented from taking place, then stress relaxation will occur. Without strains, inelastic or plastic work will not occur, but since the time passes, then the current stress point moves on to a lower work-hardening relationship, as indicated in Figure 1. Conversely, this feature of the model allows calculation of the relaxed stresses when the strains are fixed. Any combination of creep and relaxation may also be predicted by changing both time and stress state or work input.

The same formulation also allows the calculation of loading rate effects. As both the load and the time changes, the current point moves to another work-hardening curve, and effects of lower as well as higher strain rates may be handled by this formulation.

Thus, using the plastic (inelastic) work as the hardening parameter in the model allows incorporation and calculation of all types of known time effects in the model. The model then involves a hardening relation that is time dependent, as indicated schematically in Figure 1.

7 CONCLUSION

Modeling time-dependent behavior such as creep in sand has been incorporated into a time-independent constitutive model for frictional materials. To capture the creep occurring at various points in stress space, the existing single hardening model has been extended by multiplying the plastic work in the hardening function by a power function of time, which requires determination of one parameter. Thus, the predicted time-dependent behavior is based on the modeled time-independent behavior.

The modified model was shown to predict the observed creep behavior of Antelope Valley sand at various stress ratios with good accuracy.

REFERENCES

Bjerrum, L. 1973. Problems of soil mechanics and construction on soft clays and structurally unstable soils. *Proc. 8th Int. Conf. Soil Mech. Found. Engrg.*, 3: 111-159.

Bopp, P.A. & P.V. Lade 1997. Effects of initial density on soil instability at high pressures. *J. Geotech. and Geoenv. Engrg.,* ASCE, 123(7): 671-677.

Cristescu, N. 1989. *Rock Rheology*, Kluwer Academic Publishers, Boston.

Feda, J. 1992. *Creep of Soils and Related Phenomena*. Development in Geotechnical Engineering 68, Elsevier, Amsterdam.

Kim, M.K. & P.V. Lade 1988. Single hardening constitutive model for frictional materials, I. Plastic potential function. *Computers and Geotechnics*, 5(4): 307-324.

Lade, P.V. 1990. Single Hardening Model with Application to NC Clay. *J. Geotech. Engrg.*, ASCE, 116(3): 394-414.

Lade, P.V. 1994. Creep effects on static and cyclic instability of granular soils. *J. Geotech. Engrg.*, ASCE, 120(2): 404-419.

Lade, P.V. & M.K. Kim 1988 a. Single hardening constitutive model for frictional materials, II. Yield criterion and plastic work contours. *Computers and Geotechnics*, 6(1): 13-29.

Lade, P.V. & M.K. Kim 1988 b. Single hardening constitutive model for frictional materials, III. Comparisons with experimental data. *Computers and Geotechnics*, 6(1): 31-47.

Lade, P.V. & M.K. Kim 1995. Single Hardening Constitutive Model for Soil, Rock and Concrete. *Int. J. Solids Struct.*, Elsevier Science Ltd., 32(14): 1963-1978.

Lade, P.V., C.D. Liggio, Jr. & J.A. Yamamuro 1998. Effects of Non-Plastic Fines on Minimum and Maximum Void Ratios of Sand. *Geotech. Test. J.*, ASTM, 21(4), 336-347.

Lade, P.V. & C.-T. Liu, 1998. Experimental Study of Drained Creep Behavior of Sand. *J. Engrg. Mech.*, ASCE, 124(8): 912-920.

Lade, P.V. & J.A. Yamamuro 1997. Effects of nonplastic fines on static liquefaction of sands. *Can. Geotech. J.*, 34(6): 918-928.

Lade, P.V., J.A. Yamamuro & P.A. Bopp 1997. Influence of Time Effects on Instability of Granular Materials. *Computers and Geotechnics*, 20(3/4): 179-193.

Leroueil, S. & M.E.S. Marques 1996. State of the Art: Importance of strain rate and temperature effects in geotechnical engineering. in *Measuring and Modeling Time Dependent Soil Behavior*, Geotech. Spec. Publ. 61, ASCE, ed. by T.C. Sheahan and V.N. Kaliakin, 1-60.

Mitchell, J.K. 1993. *Fundamentals of Soil Behavior*. 2nd edition, John Wiley and Sons, New York.

Murayama, S., K. Michihiro & T. Sakagami 1984. Creep characteristics of sands. *Soils and Foundations*, 24(2): 1-15.

Sheahan, T.C. & V.N. Kaliakin 1996. *Measuring and Modeling Time Dependent Soil Behavior*, ASCE Geotech. Spec. Publ. 61, New York.

Yamamuro, J.A. & P.V. Lade 1993. Effects of strain rate on instability of granular materials. *Geotech. Test. J.*, ASTM, 16(3): 304-313.

Yamamuro, J.A. & P.V. Lade 1997. Static liquefaction of very loose sands, *Can. Geotech. J.* 34(6): 905-917.

Yamamuro, J.A. & P.V. Lade 1998. Steady-State Concepts and Static Liquefaction of Silty Sands, *J. of Geotech. and Geoenv. Engrg.*, ASCE, 124(9): 868-877.

Computer Methods and Advances in Geomechanics, Desai et al. (eds) © 2001 Balkema, Rotterdam, ISBN 90 5809 183 X

Constitutive modelling and determination of material parameters for rock mass and rockfill materials

A. Varadarajan & K.G. Sharma
Department of Civil Engineering, IIT Delhi, New Delhi, India

ABSTRACT: Triaxial tests are conducted on quartz mica schist rock samples collected from a project site. Strain controlled loading is applied using a servocontrolled loading machine to capture strain softening behaviour of the rock. Triaxial tests are also conducted on modelled rockfill material obtained from a project site. Large size triaxial cells are used for testing. Constitutive model based on disturbed state concept is used to characterise the behaviour of the two materials. Material parameters are obtained from the tests and the predictions by the model are verified. Procedures are proposed to predict the material parameters of the rock mass (with joints, discontinuities etc.) and the prototype (large size) rockfill materials to facilitate analysis and design of engineering structures.

1 INTRODUCTION

Major engineering structures such as dams and underground structures are constructed in rock mass. Rockfill materials are commonly used in the construction of many rockfill dams. Testing and modelling of the behaviour of the rock mass with such complex factors as joints and discontinuities and the rockfill with large sized particles have been a challenging task to the geotechnical engineers. This paper deals with the testing and modelling of the behaviour of intact rock and modelled (small sized) rockfill materials. Procedures for determination of material parameters for rock mass (with joints, discontinuities etc.) and prototype (large size) rock fill materials from the parameters obtained from the laboratory tests are presented.

2 LABORATORY TESTS

2.1 *Intact rock*

A schistose rock, quartz mica schist collected from Nathpa Jhakri hydropower project, a run-of-the-river project constructed in the middle reaches of the river Satluj in the Himachal Pradesh state, India has been used for the tests. The rock has a specific gravity of 2.74, dry density of 26.8 kN/m^3 and tensile strength of 8 MPa. Samples of 54.75 mm diameter with length/diameter ratio equal to 2 were used in the tests. Triaxial tests were conducted using, a high pressure triaxial cell, a MTS servo-controlled loading system specially designed to capture strain softening behaviour and a hydraulic pressure unit. Electrical resistance strain gauges were used to measure axial and lateral strains. Strain controlled triaxial compression tests with various confining pressures were conducted using a computer controlled data acquisition system. The complete details of the testing is given in Hashemi (1999).

2.2 *Modelled rockfill material*

Rockfill material obtained from Ranjit Sagar rockfill dam project, a hydropower project constructed across river Ravi in Punjab state, India has been used in the study. The material consists of rounded/subrounded particles upto 320 mm size. This contains pieces of conglomerate, sandstone, quartzite, shale, claystone and other materials of older rocks and recent alluvium. Three modelled gradation curves as derived using parallel gradation modelling technique (Lowe, 1964) having a maximum particle size of 80, 50 and 25 mm respectively were used. Drained triaxial tests at confining pressure, 350-1400 kPa were conducted on samples of 381 mm diameter and 813 mm height. The samples were compacted to a relative density of 87%. Large size triaxial cell, a hydraulic axial loading unit and a confining pressure unit at Central Soil and Materials Research Station, New Delhi were used for the tests. The axial deformation and the volume change in the sample were measured during testing. The breakage of the particles as expressed by breakage factor, Bg (Marsal, 1965) was determined after each test. The complete details of testing are given in Gupta (2000).

3 CONSTITUTIVE MODELS

The behaviour of geological materials is characterised using various models such as elastic, elastoplastic and elastoviscoplastic models. Herein, the plasticity based constitutive model using disturbed state concept, DSC, developed by Desai (1994) has been used due to its generalised framework for incorporating various features of geological materials. The constitutive model as adopted for the rock and rockfill material is briefly presented in the following. In the DSC, the material response is expressed in terms of the responses in the continuous (relatively intact, RI) and discontinuous (fully adjusted, FA) parts as

$$\sigma_{ij}^{a} = (1 - D)\,\sigma_{ij}^{i} + D\sigma_{ij}^{c} \tag{1}$$

where a, i and c superscripts denote observed, RI and FA states respectively. For the rock, the coupling (disturbance) function between the two states can be written in the form as

$$D = D_u\left(1 - \exp(-A\xi_D^Z)\right) \tag{2}$$

where, D_u is the ultimate disturbance, A and Z, disturbance parameters and ξ_D is the deviatoric part $\left(dE_{ij}^{p}\right)$ of plastic strain trajectory given by

$$\xi_D = \int \left(dE_{ij}^{p}\, dE_{ij}^{p}\right)^{1/2} \tag{3}$$

In the case of rockfill material the disturbance function is expressed as

$$D = \frac{\xi_D}{C_1 + C_2\,\xi_D} \tag{4}$$

where, C_1 and C_2 are disturbance constants.

The relatively intact state behaviour is depicted with associated elastoplastic hardening response for the rock and is characterised by Hierarchical Single Surface (HISS) model (δ_0) as

$$F = \left(\frac{J_{2D}}{P_a^2}\right) - \left[-\alpha\left(\frac{\bar{J}_1}{P_a}\right)^{n} + \gamma\left(\frac{\bar{J}_1}{P_a}\right)^{2}\right](1 - \beta S_r)^{m} \tag{5}$$

and

$$\bar{J}_1 = J_1 + 3R$$

where, J_{2D} is the second invariant of deviatoric stress tensor; J_1 is the first invariant of the stress tensor, P_a is the atmospheric pressure, γ, β and m = -0.5 are material response functions associated with the ultimate behaviour; α is the hardening function; n is the phase change parameter; R is the bonding stress and S_r is the stress ratio given by

$$S_r = \frac{\sqrt{27}}{2}\,\frac{J_{3D}}{J_{2D}^{1.5}} \tag{6}$$

in which J_{3D} is the third invariant of deviatoric stress tensor.

As a simplification, it is assumed that strains in RI and FA states are equal and the material in FA state can carry no shear stress but can carry hydrostatic stress which is same in the RI and FA states, i.e., $J_1^c = J_1^i$ (Desai, 1994).

For rockfill material, non-associative model has been used with the yield function given by Eq.(5) and the plastic potential function Q is defined as

$$Q = \left(\frac{J_{2D}}{P_a^2}\right) - \left[-\alpha_Q\left(\frac{\bar{J}_1}{P_a}\right)^{n} + \gamma\left(\frac{\bar{J}_1}{P_a}\right)^{2}\right](1 - \beta S_r)^{m} \tag{7}$$

where,

$$\alpha_Q = \alpha + \kappa\,(\alpha_0 - \alpha)(1 - r_v) \tag{8}$$

in which κ is non associative parameter, α_0 is α at the beginning of shear loading and

$$r_v = \frac{\xi_V}{\xi} \tag{9}$$

$$\xi_V = \int \frac{\left|d\,\varepsilon_v^{\,p}\right|}{\sqrt{3}} \tag{10}$$

The hardening functin α is expressed as a function of ξ as

$$\alpha = \frac{a_1}{\xi^{\eta_1}} \tag{11}$$

where, a_1 and η_1 are the material parameters and ξ is the trajectory of plastic strain,

$$\xi = \int \left(d\varepsilon_{ij}^{p}\, d\varepsilon_{ij}^{p}\right)^{1/2} \tag{12}$$

4 DETERMINATION OF MATERIAL PARAMETERS

The procedure for the determination of material parameters has been described in detail in various references (Desai, 1994,1995) and is not repeated here. The determination of only the disturbance parameters which are different for the two materials are presented herein. Rearranging Eq.(1) and the fact that σ_{ij}^{a} and $\sigma_{ij}^{(i)}$ lie on the same deviatoric plane, perpendicular to the hydrostatic axis, as $J_1^c = J_1^i$, yields

$$\sqrt{J_{2D}^{a}} = \sqrt{J_{2D}^{i}}\,(1 - D) \tag{13}$$

Therefore, the value of D_u the disturbance at the residual stage can be obtained as

$$D_u = 1 - \frac{\left(\sqrt{J_{2D}}\right)_r}{\left(\sqrt{J_{2D}}\right)_u} \tag{14}$$

where r and u denote the residual and ultimate stages corresponding to the observed and RI states respectively. The ultimate stage $\left(\sqrt{J_{2D}}\right)_u$ can be found from the RI state response (1.1 – 1.15 times the peak value). Therefore, for each test, a particular value of D_u could be calculated. Based on Eq.(13), the value of D can be determined at each observed point of stress-strain curve for every test. Conducting a least square analysis on the Eq.(2), the disturbance parameters, A and Z are determined for each experimental test.

In the case of rockfill material, the Eq.(4) for disturbance can be rearranged as

$$\frac{\xi_D}{D} = C_1 + C_2\,\xi_D \tag{15}$$

The value of ξ_D can be calculated from the stress-strain curve of a test and the value of D can be found from Eq.(13). Using a regression analysis of Eq.(15), the values of C_1 and C_2 can be determined.

The RI state for the rockfill material is determined using the breakage factor, Bg as follows. The plot between Bg and $(\sigma_1/\sigma_3)_f$ (at failure) from the observed values of experimental results can be expressed as

$$\left(\frac{\sigma_1}{\sigma_3}\right)_f = \frac{1 + B_g}{A_d B_g + B_d} \tag{16}$$

where, A_d and B_d are constants and can be determined from the observed results. At RI state, the breakage factor is assumed to be zero and Eq.(16) reduces to

$$\left(\frac{\sigma_1}{\sigma_3}\right)_f = \frac{1}{B_d} \tag{17}$$

for RI state. This stress is used to determine the stress-strain relationship of RI state. The details of the determination of the material parameters are given in Gupta (2000). All the material parameters determined for the rock, quartz mica schist and rockfill material from Ranjit Sagar Dam are presented in Table 1.

The yield surfaces and typical stress-strain curves as predicted and observed for the two materials are shown in Figs. 1 and 2. The predictions are satisfactory.

5 MATERIAL PARAMETERS FOR ROCK MASS AND PROTOTYPE ROCKFILL MATERIAL

5.1 *Rock mass*

The material parameters for the rock mass have been determined using the procedure proposed by Ramamurthy (1993) as follows:

The unconfined compressive strength and Young's modulus ratios of jointed rock mass and intact rock are expressed as

$$\sigma_{cr} = \frac{\sigma_{cj}}{\sigma_{ci}} \exp(-0.008 J_f) \tag{18}$$

$$E_r = \frac{E_j}{E_i} \exp(-0.0115 J_f) \tag{19}$$

where, σ_{cj}, E_j and σ_{ci}, E_i are unconfined compressive strength and Young's modulus for the jointed rock mass and intact rock respectively. The joint factor J_f is defined as

$$J_f = \frac{J_n}{\bar{n}\bar{r}} \tag{20}$$

where, J_n is the joint frequency, $\bar{n}$ is the inclination parameter that depends on the joint orientation (with respect to major principal stress direction) and $\bar{r}$ is the joint strength parameter which is a function of the joint condition. For the quartz mica schist at the site, $J_n = 5$, $\bar{n} = 0.38$, $\bar{r} = 0.6$, $\sigma_{ci} = 15.71$ MPa, $E_i = 8591$ MPa and $J_f = 21.9$.

For jointed rock mass, the failure criterion is

$$(\sigma_1 - \sigma_3)/\sigma_3 = B_j\,(\sigma_{cj}/\sigma_3)^{\alpha_j} \tag{21}$$

where, B_j and α_j are the material parameters. The values of B_j and α_j are found as

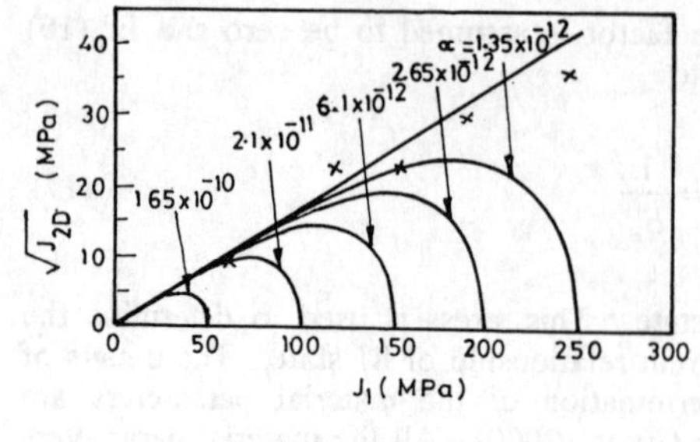

(a) Yield Surface

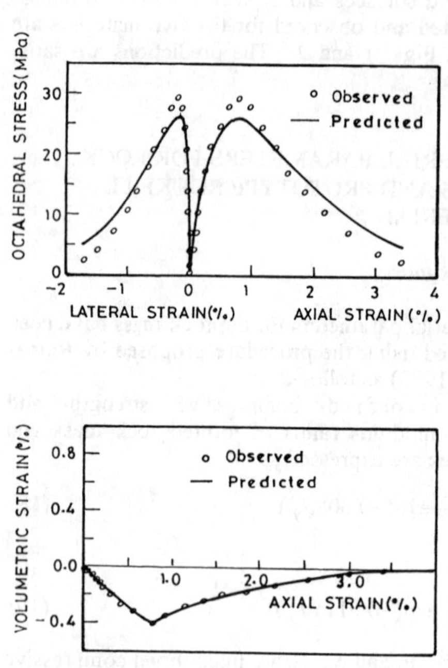

(b) Stress-Strain-Volume Change Response (σ_3=45 MPa)

Figure 1. Quartz Mica Schist

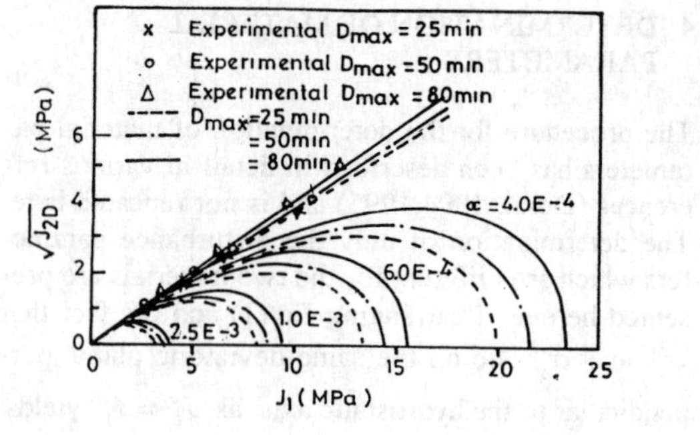

(a) Yield Surface

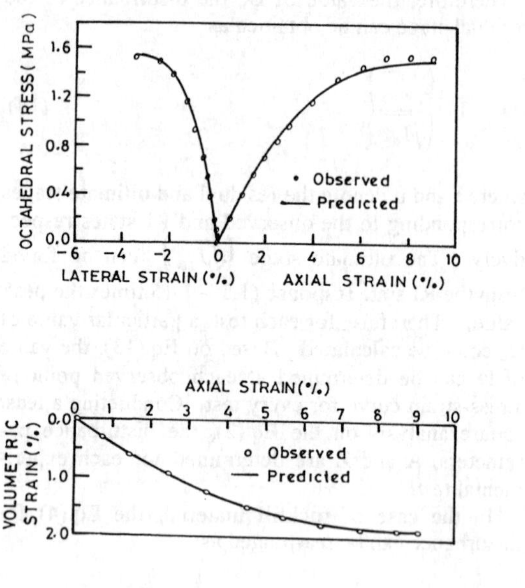

(b) Stress-Strain-Volume Change Response (σ_3=0.7 MPa) D_{max} = 50mm

Figure 2. Modelled Rockfill Material

$$B_i/B_j = 0.13 \exp\,[2.037\,\alpha_j/\alpha_i] \tag{22}$$

and

$$\alpha_j/\alpha_i = (\sigma_{cj} / \sigma_{ci})^{0.5} \tag{23}$$

where, σ_{cj}/σ_{ci} is found from Eq. (18). From Eq. (21) the values of σ_1 and σ_3 at failure for various confining pressures are determined for the rock mass. These stresses are used for the calculation of ultimate parameters of the constitutive model. The other parameters are assumed to be the same as for intact rock. The material parameters for the rock mass as determined are shown in Table 1.

5.2 *Rockfill material*

In the case of rockfill materials, the parameters for the prototype rockfill material are obtained from those of modelled rockfill materials using the equation of the form

$$P = A_1\, D_r^{A_2} \tag{24}$$

where P is the parameter for the modelled rockfill material with maximum particle size, D_{max}; D_r = Ratio of D_{max} / D_{pmax}; D_{pmax} = maximum particle size of

Table 1. Material Parameters for Quartz Mica Schist and Ranjit Sagar Dam Material

Material constants	Quartz Mica Schist		Ranjit Sagar Dam Material			
	Intact rock	Rock mass	Modelled material, D_{max} (mm)			Prototype
			25	50	80	
Elasticity						
k	29269	22746	193.69	220.34	253.63	343.35
n′	0.2645	0.2645	0.6386	0.6683	0.7146	0.8141
ν	0.2	0.2	0.31	0.30	0.29	0.30
Ultimate						
γ	0.0202	0.01352	0.0780	0.084	0.0912	0.1083
β	0.4678	0.3889	0.730	0.732	0.743	0.73
Bonding Stress, 3R (MPa)	46.99	41.90	-	-	-	-
Phase Change, n	5.0	5.0	3.0	3.0	3.0	3.0
Hardening						
a_1	0.13E-13	0.13E-13	0.65E-4	0.30E-4	0.985E-5	1.0E-6
η_1	0.6	0.6	0.4675	0.5600	0.7618	1.2865
Non-associative, κ	-	-	0.23	0.23	0.22	0.23
Disturbance						
A / C_1	220.7	220.7	0.016	0.032	0.05	0.1959
Z / C_2	1.339	1.339	6.0	7.34	8.84	13.803
D_u	0.97	0.97	-	-	-	-

prototype rockfill material; A_1 and A_2 are constants. The constants A_1 and A_2 are determined from the known values of P and D_r for modelled rockfill materials. Then by setting the value of D_r=1, the value of P for the prototype rockfill material is found. The parameters as found for the prototype rockfill material are shown in Table 1.

6 CONCLUSIONS

Triaxial tests have been conducted on quartz mica schist rock and modelled rockfill material using specialised testing equipment. Constitutive model based on disturbed state concept has been shown to satisfactorily characterise the behaviour of the two materials. Procedures have been presented to determine the parameters of the rock mass in the field and the prototype (large sized) rockfill material to facilitate analysis of engineering structures.

7 REFERENCES

Desai, C.S. 1994. Hierarchical single surface and disturbed state constitutive models with emphasis on geotechnical applications. Chapter 5 in *Geotechnical Engineering,* K.R. Saxena (Editor), New Delhi, India: Oxford & IBH Pub. Co. Ltd.

Desai, C.S. 1995. Constitutive Modelling using the disturbed state as microstructure self-adjustment concept. Chapter 8 in *Continuum models for materials with microstructure,* H.B. Muhlhaus (Editor), Chickester, U.K: John Wiley.

Gupta, A.K. 2000. *Constitutive modelling of rockfill materials.* Ph.D. Thesis, I.I.T. Delhi, India.

Hashemi, M. 1999. *Constitutive modelling of a schistose rock in the Himalaya.* Ph.D. Thesis, I.I.T. Delhi, India.

Lowe, J. 1964. Shear strength of coarse embankment dam materials. *Proc. 8th Int. Congress on Large Dams.* 3: 745-761.

Marsal, R.J. 1965. Discussion. *Proc. 6th Int. Conf. on Soil Mech. & Found. Engg.* 3: 310-316.

Ramamurthy, T. 1993. Strength and modulus reponses of anisotropic rocks. Chapter 13 in *Comprehensive Rock Engineering,* Vol. I. Oxford, U.K.: Pergamon Press.

Computer Methods and Advances in Geomechanics, Desai et al. (eds) © 2001 Balkema, Rotterdam, ISBN 90 5809 183 X

Role of stress dilatancy on sand behaviour: Fabric, cyclic and strain-localization related issues

R.G.Wan
Department of Civil Engineering, University of Calgary, Alb., Canada

P.J.Guo
Department of Civil Engineering, University of Calgary, Alb., Canada
(Presently: C-CORE, Memorial University of Newfoundland, Canada)

ABSTRACT: The paper discusses the relevance of dilatancy and associated fabric dependencies on the behaviour of sand. The focal point is the development of a stress dilatancy model with embedded fabric information as derived from micromechanical analysis and energy principles. Dilatancy is extended to cyclic loading regime using double sliding plane concept in which dilatancy is manifested along an unloading plane as well. Discussion of model simulations with reference to fabric, strain localization and cyclic loading are presented so as to illustrate the prominent role of dilatancy.

1 INTRODUCTION

Dilatancy plays a crucial role in determining the mechanical behaviour of sand due to its discrete particulate nature. In fact, dilatancy depends on fabric (geometrical grain packing) and thus can be viewed as a kinematical constraint imposed by the grains as they override each other under confinement during shear deformation. Newland and Alley (1957) followed by Rowe (1962) derived so-called stress dilatancy relationships on the basis of simple assemblies undergoing shear induced volume change and minimized dissipated energy. However, important limitations of these stress-dilatancy theories are that they are unable to capture stress, density and fabric dependencies which are key issues for modelling localization and cyclic responses of sand. There have been attempts to develop stress-dilatancy models with reference to internal fabric, see for example Tokue (1979), Matsuoka (1974), Nemat Nasser (1980) and Ueng and Lee (1990). Moreover, various experimental studies carried out by Nakai (1997), and Vaid & Sasitharam (1992) clearly demonstrate the importance of dilatancy and its dependencies on stress and void ratio levels. Models proposed by Gudehus (1996), Wan & Guo (1999a) and Gajo & Wood (1999) have been developed to address these issues.

In this paper, a stress dilatancy model with embedded constitutive features that can describe density, pressure, fabric and loading path dependencies of sand behaviour is presented. Information on fabric and its evolution during loading history are incorporated into the model based on micromechanical analysis. In the cyclic loading regime involving stress reversals, different slip planes generally form and govern the mobilization of both friction angle and dilatancy. Thus, dilatancy can be extended to cyclic loading and a constitutive model formulated in hypoplasticity. As such, the resulting model can describe drained or undrained response of sand in both monotonic and cyclic loading regimes using a single set of constitutive material parameters. Simulations of drained and undrained cyclic responses of sand with inclusion of fabric are presented. Sand behaviour along imposed strain increment ratio paths are also discussed in the light of strain localization predictions.

2 BASIC CONCEPTS AND MAIN ISSUES

In looking at Rowe's original stress dilatancy equation, it is found that it implies a 'unique' linear relationship between principal stress ratio (σ_1/σ_3) and the resulting rate of dilatancy or contractancy (D), i.e. $\sigma_1/\sigma_3 = K_\mu D$. The proportionality constant K_μ is related to an intrinsic material parameter such as the interparticle friction angle φ_u. However, it is known that the dilatancy response of sand does not follow Rowe's energy line and deviations occur based on density, stress and fabric levels. Fig. 1 shows dilatancy plot as obtained from drained triaxial compression tests on clean oil sand at two different void ratios, see Guo (2000). The relatively loose sand gives a fully contractive response for all applied stress ratios, while the denser one displays volume contractancy followed by dilatancy at higher stress ratios. Hence, the volumetric response of sand cannot be determined from a 'unique' energy line. Rather, it is thought that there is a family of energy lines that sand behaviour will follow based on its current state such as stress, density and fabric.

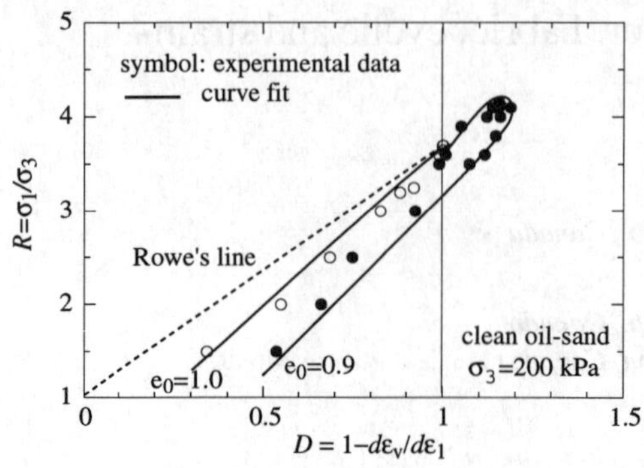

Figure 1. Dilatancy response of sand

2.1 *Microstructure based stress-dilatancy model*

The idea is to formulate stress-dilatancy based on micromechanical approach with the consideration of grain kinematics and statics. The transition from micro- to macro- scale is made possible by using volume averaging and equating rate of work dissipation due to grain slippage to rate of work input. The stress-dilatancy equation simplified into conventional triaxial stress conditions was derived in Guo (2000), i.e.

$$\sin\psi_m = \frac{4}{3}\frac{\sin\varphi_m - \sin\varphi_f}{(1-\sin\varphi_f\sin\varphi_m)} \quad (1)$$

$$\sin\varphi_f = \frac{X(F_{33}/F_{11})+\gamma^*}{a+\gamma^*}\left(\frac{e}{e_{cr}}\right)^\alpha \sin\varphi_{cv} \quad (2)$$

in which F_{11} and F_{33} are fabric tensor components in axial and radial directions respectively, γ^* is the *true* plastic shear strain that is reflective of fabric, and finally, X, a and α are constants.

The sign convention is such that a negative ψ_m refers to contraction and a positive ψ_m means dilation. It is evident from Eqs. (1) and (2) that the sign of the mobilized dilatancy angle ψ_m is determined by the relative magnitudes of the mobilized friction angle φ_m and the characteristic friction angle φ_f. The latter depends on a number of parameters such as current void ratio e, stress level through critical void ratio e_{cr}, friction angle at critical state, fabric F_{ij} and accumulated true plastic strains γ^*.

2.2 *Stress-dilatancy in cyclic loading*

Experimental observations on cyclic loading of sand suggest that different sliding planes are mobilized during loading and unloading. Since different sliding planes involve different particles undergoing local slip and overriding, it is natural to introduce the notion of mobilized friction angles φ^L and φ^U in both loading and unloading respectively. Consequently, dilatancy is manifested in both loading and unloading regimes as confirmed experimentally by Pradhan et. al. (1989) in cyclic triaxial tests on Toyoura sand.

In view of describing stress dilatancy in cyclic loading regime, Eq. (1) can be extended using the above concept. Thus, the same form of dilatancy law can be applied to both loading and unloading with mobilized friction and dilatancy angles solely dependent upon the current deformation state, i.e. plastic strain, void ratio and fabric. Therefore,

$$\sin\psi_m^{L,U} = -\frac{\dot{\varepsilon}_v^{L,U}}{\dot{\gamma}^{L,U}} = \frac{4}{3}\frac{(\sin\varphi_m^{L,U} - \sin\varphi_f^{L,U})}{(1-\sin\varphi_m^{L,U}\sin\varphi_f^{L,U})} \quad (3)$$

with superscripts L and U representing loading and unloading respectively.

2.3 *Cyclic hypoplasticity model*

In Wan & Guo (1998) a simplified version of the stress dilatancy model without fabric was implemented into a classical elasto-plasticity framework. However, in the cyclic loading regime, sands exhibit stiffness degradation with increased plastic deformations during cycles of loading, unloading and reloading. For modelling such a behaviour, it is convenient to follow a hypoplasticity framework that involves loading and memory surfaces with kinematic hardening. Also, hysteric stress-strain behaviour can be modelled by using a small elastic core which moves with the current stress attached to a loading surface f. The model is developed in terms of conventional triaxial stress and strain variables, deviatoric stress q, and mean effective stress p.

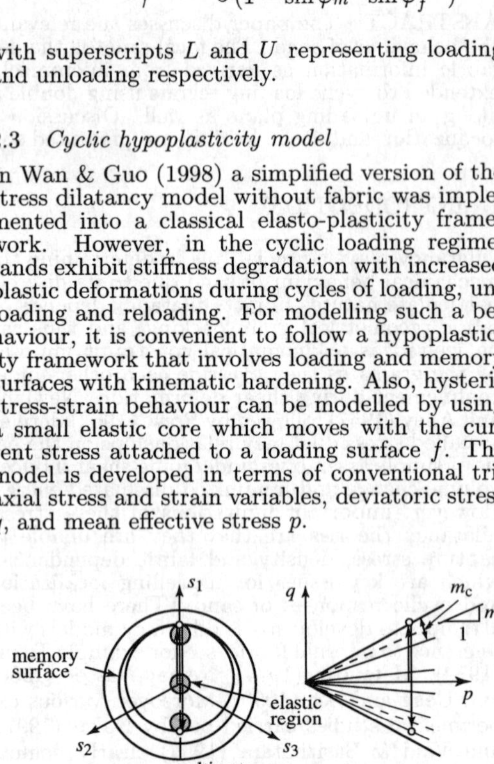

Figure 2. Evolution of loading surface and elastic core

2.3.1 *Loading surface and kinematic hardening*

In $p-q$ space, the loading surface f takes the same form as Mohr-Coulomb criterion with the mobilized friction angle φ_m controlling a parameter m and prescribing the shift of the elastic core, i.e.

$$f = (q - mp) \mp m_c p = 0 \quad (4)$$

where m_c describes the expansion of the loading surface. Fig. 2 illustrates the mobilization of friction, loading and unloading with kinematic hardening. The isotropic hardening parameter m_c describes densification and hence is controlled by changes in void ratio mainly attributed to dilatancy by way of plastic volumetric changes $\dot{\varepsilon}_v^p$. On the other hand, the translation of the loading surface is controlled by mobilization of friction described by

$$\sin\varphi_m^{L,U} - \sin\varphi^* = \frac{\bar{\gamma}^p}{a+\bar{\gamma}^p}\left(\frac{e}{e_{cr}}\right)^{-\beta} (C\sin\varphi_{cv} - \sin\varphi^*) \qquad (5)$$

in which $\bar{\gamma}^p = C(\gamma^p - \gamma_m^p)$, with γ_m^p being the plastic shear strain at the point of unloading where mobilized friction is φ^*. The parameter C is a loading index with $C = 1$ at loading/reloading and $C = -1$ at unloading. Fig. 3 illustrates the mobilization of friction angle during unloading as described by Eq. (5).

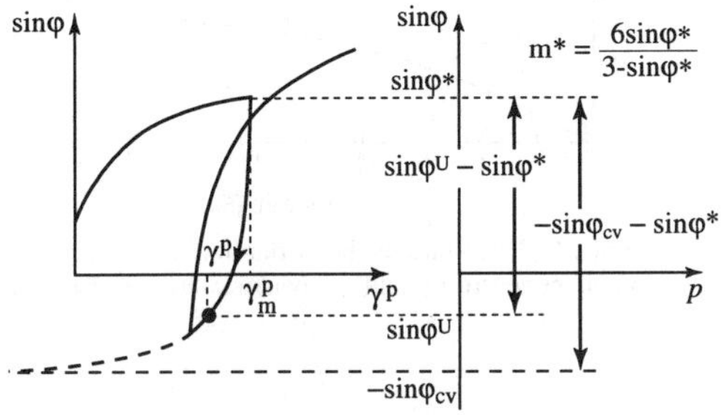

Figure 3. Mobilization of friction at loading and unloading

2.3.2 *Plastic potential and hardening modulus*

In hypoplasticity, the inelastic (plastic) strain increment $\dot{\boldsymbol{\varepsilon}}^p$ is governed by a plastic modulus H, a plastic flow direction $\mathbf{m}$, a loading direction $\dot{\boldsymbol{\sigma}}$, and the outer normal $\mathbf{n}$ to the loading surface f. Thus,

$$\dot{\boldsymbol{\varepsilon}}^p = \frac{1}{H}\langle \mathbf{n} : \dot{\boldsymbol{\sigma}} \rangle\, \mathbf{m} \qquad (6)$$

where $\langle . \rangle$ refers to MacCauley brackets. The plastic flow direction is calculated from some plastic potential function $g = q - p\sin\psi_m$, with the mobilized dilatancy angle being determined from the stress dilatancy equation, i.e. Eq. (3). The plastic modulus H at any instant during deformation history is obtained from the consistency condition $\dot{f} = 0$, with f being the loading surface. The complete derivation of H can be found in Guo (2000). An important result is that H depends on the distance $\delta_{\sigma e}$ between current state and the corresponding conjugate state on a so-called bounding surface in $p - q - e$ space. Here the bounding surface is defined as the critical state surface modified with the term $(e/e_{cr})^{-\beta}$. As such, the bounding surface moves according to the evolution of void ratio and stress level.

3 CYCLIC LOADING SIMULATION

3.1 *Drained*

It is of interest to study the slow drained cyclic behaviour of sand due to its relevance to soil-structure interaction problems such as the integral bridge, see England et. al. (1997). The numerical simulation involves an element test on Ottawa sand with model parameters given in Guo (2000). Cyclic shear strains with a given amplitude are applied to the element so that the axial stress varies cyclically while radial stress is kept constant with the possibility of principal stress direction reversals.

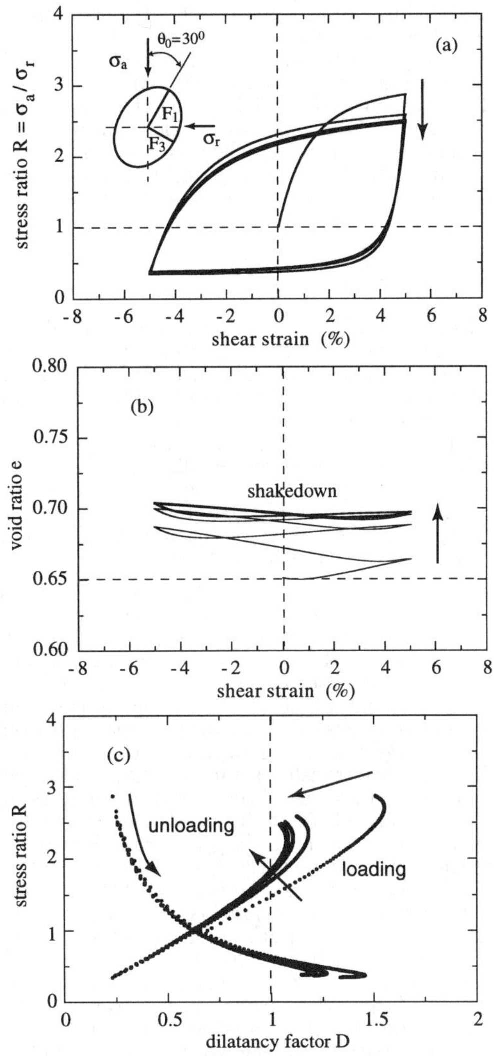

Figure 4. Drained cyclic behaviour of Ottawa sand: (a) stress ratio-shear strain, (b) void ratio-shear strain, and (c) dilatancy plot

Figs. 4a, b show stress-ratio versus shear strain and void ratio versus shear strain responses respectively considering an initial void ratio $e_0 = 0.65$ and an initial principal fabric ratio $\Omega = F_1/F_3 = 1.33$. Also, the principal fabric direction makes an angle of 30^0 with respect to the direction of principal stress. As a consequence of the dilatancy characteristics of the sand, there is net dilation over the cycles with a decrease in maximum stress ratio. A shakedown condition is finally obtained when the net volume change with a loop is zero. Fig. 4c shows the evolution of dilatancy with the maximum dilatancy rate being exceeded in each cycle. In every new cycle, the maximum dilatancy rate decreases with a shift of energy lines from right to left. If a 'unique' energy line, as implied by Rowe's model, was used, it would not have been possible to capture shakedown conditions for volumetric changes, see Fig. 4b. The current model allows the energy lines within each half cycle to shift position during deformation history until the volume changes as integrated along the loading and unloading energy lines nullify each other at shakedown conditions.

3.2 *Undrained*

The role of dilatancy changes is next highlighted in an undrained cyclic test which involves liquefaction. Excess pore pressure build up is a response to subtle variations in contractancy and dilatancy. For illustrative purposes, an initial void ratio $e_0 = 0.735$ was selected with initial fabric ratio $\Omega = 0.86$ and confining pressure of 40 kPa. Axial stresses are varied so as to produce a deviatoric stress amplitude of ±20 kPa. Figs. 5a, b show the effective stress path and stress-strain response calculated from the model.

In the first quarter cycle up to point 2 on Fig. 5a, there is tendency for compaction, but in order to maintain zero overall volume change, high excess pore pressures build up. During unloading in the next quarter cycle (2-3), particles will slide on a different plane than the one during loading in such a way that further compaction would like to take place. Again due to undrained conditions, this translates into further excess pore pressures build up. When stress reversal occurs with continued unloading, particles continue to slide (3-4) until, according to stress dilatancy, a characteristic state is reached where the material would like to dilate. Hence, the rate of excess pore pressure generation becomes negative and the effective stress path changes direction, see 4-5. Finally in the next reloading cycle, a new sliding plane is formed with compaction potentially taking place. More excess pore pressures build up moving the effective stress path close to the zero effective stress state where liquefaction takes place. Fig. 5b. shows the stress loop in deviatoric stress against shear strain plot showing collapse whenever zero effective stress is reached.

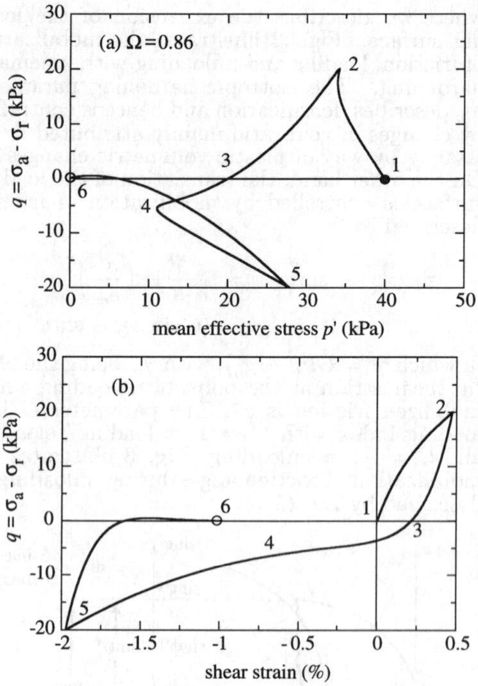

Figure 5. Undrained cyclic response of Toyoura sand: (a) effective stress path (b) deviatoric stress-shear strain

3.3 *Strain localization*

Strain localization in the form of shear band formation was studied in Wan & Guo (1997) using a version of the proposed stress-dilatancy model without fabric. Classical bifurcation analysis considering both flow and deformation theories was performed in order to predict both shear band inclinations and strains at localization. Model predictions obtained were in good agreement with published experimental data.

In this paper, the shear band analysis has been extended to the case of imposed strain increment ratio paths, see Wan & Guo (1999b). In these strain paths, the material is forced to deform at a constant strain increment ratio defined as $\vartheta = d\varepsilon_v/d\gamma$. As such, a positive value of ϑ refers to forcing the material to undergo contractive volumetric changes, while a negative ϑ corresponds to dilative volumetric changes. Chu et. al. (1993) performed these strain path experiments with the argument that localization could be avoided if the material was taken to a strain path where instability does not exist.

More recently, Vaid and Sivathayalan (1998) and Eliadorani (2000) performed similar tests in which strain increment ratio paths were maintained by injecting fluid in the specimen. The increment of volume change was basically controlled by injecting water into the sample to produce the desired ratio ϑ for a given axial strain increment.

Figs. 6a, b show model simulations along different ϑ paths with $\vartheta = 0$ coinciding with the undrained case. It is found that a variety of effective stress responses can be obtained depending on the value ϑ. Interestingly, a small inflow of water as in the case $\vartheta = -0.35$ produces a large reduction in strength when compared with the undrained case, i.e. $\vartheta = 0$, see Figs. 6a,b. This was demonstrated experimentally by Eliadorani (2000). The localization points are also shown in Figs. 6a, b. It is found that when the sample is compressed at a constant strain increment ratio, the occurrence of localization will be delayed. The higher the strain ratio increment ϑ, the larger the critical strain at localization. Fig. 7 shows the shear band inclination as a function of ϑ. A steeper shear band inclination is expected when the specimen is forced to undergo compressive volumetric changes.

4 CONCLUSIONS

In this paper we have illustrated the importance of dilatancy and fabric in the behaviour of sand by referring to cyclic loading in both drained and undrained conditions. Strain localization issues are also discussed with particular focus to the response of sand along constant strain increment ratio paths along which the material is constrained to deform a specified amount. The model presented hinges on a stress-dilatancy equation that embeds stress, density and fabric levels for both loading and unloading. The use of a kinematic hardening rule within hypoplasticity allows the description of friction mobilization during cyclic loading. Since the model systematically includes interdependencies of stress, fabric and density changes, only one set of constitutive parameters is needed to model sand over a wide range of conditions. Also, the stress dilatancy model can capture dramatic changes in constitutive responses along various constant strain ratio increment paths mimicked by inflow tests. These paths can occur in the field when fluid migrates into sand layers due to outflow from adjacent materials as result of variable permeability conditions. Any numerical modelling of liquefaction must be able to capture the correct response of sand along these paths.

5 ACKNOWLEDGEMENTS

The authors are grateful to funding received from the Natural Science and Engineering Research Council of Canada and the National Energy Board of Canada.

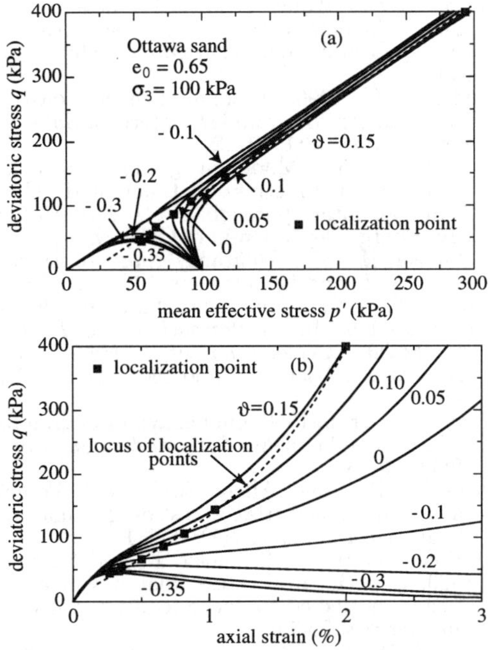

Figure 6. Localization points for various ϑ path responses: (a) effective stress path, (b) deviatoric stress-shear strain

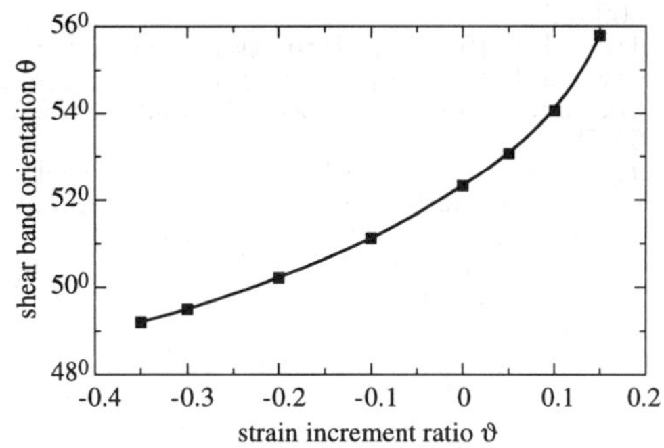

Figure 7. Shear band inclination versus ϑ

REFERENCES

Chu, J., Lo, S. C. R. and Lee, I. K. 1993. Response of granular soil in strain-path testing. *Modern Approach to Plasticity* (Kolymbas, D. ed.). 247-258.

Eliadorani, A. A. 2000. The response of sands under partially drained states with emphasis on liquefaction. *Ph.D Thesis*, University of British Columbia, Vancouver, Canada.

England, G. L, Tsang C. M, Dunstan, T and Wan, R. 1997. Drained granular material under cyclic loading with temperature-induced soil-structure interaction. *Applied Mechanics Review.* 50(10): 553-579.

Gajo, A. and Wood, D.M. 1999. Severn-Trent sand: a kinematic-hardening constitutive law; the q-p formulation. *Geotechnique.* 49(5): 595-614.

Gudehus, G. 1996. A comprehensive constitutive equation for granular materials. *Soils and Foundations.* 36(1): 1-12.

Guo, P. J. 2000. Modelling granular materials with respect to stress-dilatancy and fabric: A fundamental approach. *Ph.D. Dissertation*, Civ. Engrg. Dept., Univ. of Calgary.

Matsuoka, H. 1974. Dilatancy characteristics of soils. *Soils and Foundations.* 14(3): 55-66.

Nakai, T. 1997. Dilatancy characteristics of geomaterials. In: *Deformation and Progressive Failure in Geomechanics, IS-NAGOYA'97* (Asaoka, A., Adachi, T. and Oka, F. eds.). 899-906.

Nemat-Nasser, S. 1980. On behaviour of granular soils in simple shear. *Soils and Foundations.* 20(1): 59-73.

Newland, P.L. and Allely, B.H. 1957. Volume change in drained triaxial test on granular materials. *Geotechnique.* 7(1): 17-34.

Pradhan, T. B. S., Tatsuoka, F. and Sato, Y. 1989. Experimental stress dilatancy relations of sand subjected to cyclic loading. *Soils and Foundations.* 29(1): 45-64.

Rowe, P. W. 1962. The stress-dilatancy relation for static equilibrium of an assembly of particles in contact. *Proc. of Royal. Soc. A*, 269: 500-527.

Tokue, T. (1979). Deformation behaviours of dry sand under cyclic loading and a stress-dilatancy model. *Soils and Foundations.*19(2): 63-78.

Ueng, T. and Lee, C. 1990. Deformation behaviour of sand under shera -particulate approach. *J. Geotech. Eng., ASCE*, 6(11):1625-1640.

Vaid, Y. P. and Sasitharan, S. 1992. The strength and dilatancy of sand. *Can. Geotech. J.* 29: 522-526.

Vaid, Y. P. and Sivathayalan S. 1998. Fundamental factors affecting liquefaction susceptibility of sands. *Keynote lecture at the International Workshop on the Physics and Mechanics of Liquefaction,* Sept. 10-11, 1998.

Wan, R. G. and Guo, P. J. 1998. A simple constitutive model for granular soils: Modified stress-dilatancy approach. *Computers and Geotechnics.* 22(2): 109-123.

Wan, R. G. and Guo, P. J. 1999a. A pressure and density dependent dilatancy model for granular materials. *Soils and Foundations.* 39(6): 1-11.

Wan, R. G. and Guo, P. J. 1999b. Modelling stress dilatancy response of sand along imposed constant strain-increment-ratio paths. *4th International Conference on Constitutive Laws for Engineering Materials*, Rensselaer Polytechnic Institute, Troy, NY, USA, July 27-July 30, 1999.

Wan, R. G. and Guo, P. J. 1997. Coupled dilatancy-compaction model for homogeneous and localized deformation of granular materials. *Proceedings of Numerical Models in Geomechanics NUMOG VI,* Pietruszczak & Pande (eds), Balkema, Rotterdam. 3-8

Computer Methods and Advances in Geomechanics, Desai et al. (eds) © 2001 Balkema, Rotterdam, ISBN 90 5809 183 X

Modelling the destructuration of soft natural clays

B.A. Baudet & S.E. Stallebrass
Geotechnical Engineering Research Centre, City University, London, UK

ABSTRACT: In recent years there have been a number of detailed experimental studies of the behaviour of soft natural clays, which have investigated the effect of the natural structure of the soil on its strength and stiffness. It is clear that the behaviour of these soils cannot be predicted by classical models developed for reconstituted clay. A simple law of destructuration is used to extend an existing model developed for reconstituted clay to predict the behaviour of soft natural clays. This law uses parameters which are soil properties and can be rigorously derived from a single set of test data. Only two parameters, in addition to those used to model the reconstituted clay, are required to describe the effects of structure. The paper shows that the combination of a reliable model for reconstituted clay with a simple law of destructuration allows the main features of the behaviour of soft natural clays to be predicted.

1 INTRODUCTION

Deposits of very compressible soils are found all over the world. They are of very recent geological origin, having been formed during the past 20,000 years. It is these normally consolidated or lightly overconsolidated "soft " clays that are the subject of this paper.

The behaviour of soft clays *in situ* often differs greatly from their behaviour when reconstituted in the laboratory due to natural "structure" formed during the deposition of the clay. The difference between the structure of a natural clay *in situ* and its structure when reconstituted can be attributed to the arrangement of the particles, fabric and/or bonds between particles. In the following, the term "natural" will refer to the clay *in situ* and "intrinsic" to the clay after losing its natural structure by remoulding or reconstituting.

A common measure of structure is sensitivity, defined as the ratio of the undrained shear strength of the natural clay to its intrinsic undrained shear strength at the same water content (Terzaghi 1944). Much work has been carried out more recently to extend this method of measuring the effect of structure, in particular by Cotecchia (1996) and the "sensitivity" framework which has been derived from this work will be outlined later. However, the key issue that will be addressed in this paper is that for many soft clays the natural structure is not stable.

Coop et al. (1995) established that the natural structure of a clay may be stable or metastable, or more commonly combine these two mechanisms, depending on the cause of the natural structure, bonding or fabric. When the natural structure is metastable, the clay tends to a state on loading which in both compression and shearing is characteristic of the intrinsic rather than the natural clay. This process is termed destructuration and is most common in bonded clays.

This paper is concerned with the development of an appropriate relationship to describe destructuration that allows for both metastable and stable structure and uses parameters that can be determined directly from data from standard laboratory tests. It will be shown that this simple approach still enables the main features of the behaviour to be predicted.

2 EXPERIMENTAL OBSERVATIONS OF DESTRUCTURATION

The effects of structure on the behaviour of natural clays are often highlighted by normalising data with respect to composition. For this purpose, two normalising variables have been recently defined:

- void index I_v (Burland 1990),

  where $I_v = \dfrac{e - e_{100}^*}{C_c^*}$

- normalised volume v_n, (Coop & Cotecchia 1995),

  where $\ln(v_n) = \dfrac{N_0^l - \ln(v)}{\lambda^l}$

The variables e_{100}^*, C_c^*, N_0^l and λ^l are intrinsic properties of a soil, which depend on the liquid limit. Normalising void ratio, e, or specific volume, v, in this way leads to a unique compression curve for reconstituted clay, in each case. This is called the intrinsic compression curve (ICC).

Figure 1 illustrates the change in structure during compression using normalised one-dimensional compression curves for three soft natural clays plotted in the *lnv*$_n$*-lnp'* plane. The *in situ* states of the three clays lie outside the ICC, which is representing the position of the intrinsic state boundary surface, implying that the natural state boundary surface lies outside the intrinsic surface. After yield, the compression curve for Sibari clay follows a line parallel to the intrinsic compression curve: it has a stable structure. The compression curves of Pisa and Bothkennar clays show an abrupt yield and converge towards the intrinsic compression curve: they have a metastable structure.

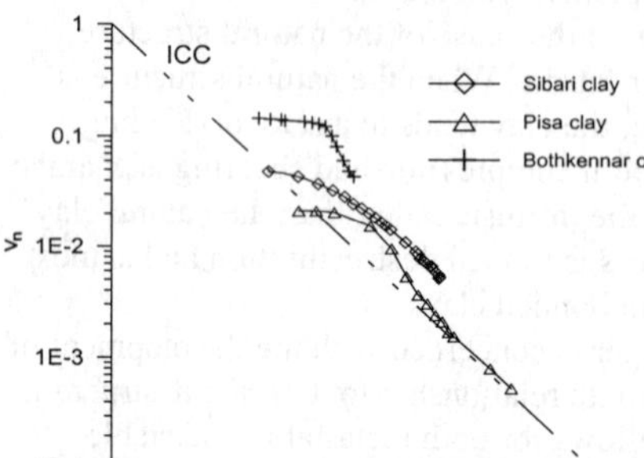

Figure 1. Normalised compression data for 3 soft natural clays

Following work by Skempton (1970), Cotecchia (1996) found that clays of different structures and origins but the same sensitivity yield at the same distance from the ICC in a normalised plot of void index against vertical effective stress. This defined a new framework for the behaviour of natural clays: the "sensitivity" framework. The new expression for sensitivity was the ratio of vertical effective stress at yield (σ'_y) to the equivalent pressure on the ICC (σ'_{vo}) (Fig. 2). This is equivalent to the sensitivity previously defined in terms of strength, by Terzaghi (1944).

The effect of metastable structure on the behaviour of natural clays loaded in shear compression is illustrated in Figure 3. In this figure data from undrained load tests on natural and reconstituted Bothkennar clay have been normalised by the equivalent pressure, p^*_{ie}. The equivalent pressure, p^*_{ie}, is defined as the mean effective stress at the intersection of the current elastic wall with the isotropic ICC. For a reconstituted soil with a state boundary surface defined by the ICC, p'/p^*_{ie} should not exceed unity. The intrinsic state boundary surface (ISBS) is shown in the figure. The comparison shows that structure leads to a higher undrained shear strength, corresponding to the peak value of q/p^*_{ie}, and that there is a dramatic reduction in strength due to subsequent destructuration. This is represented by the decrease in q/p^*_{ie}. It is this destructuration which, together with the state of the soil, controls the behaviour of the clay. Therefore it is most important that a constitutive model for soft clays predicts destructuration correctly.

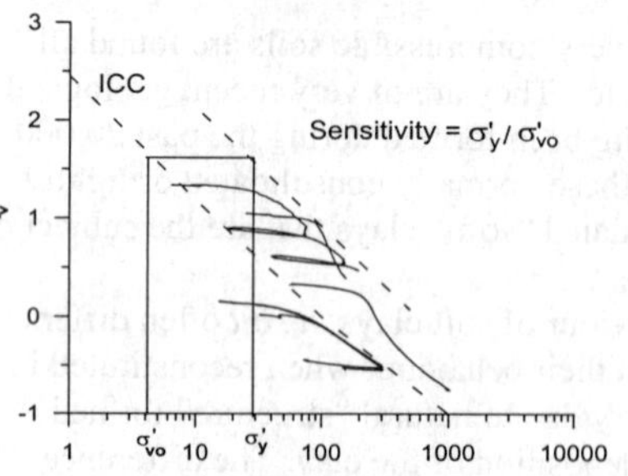

Figure 2. Normalised compression data for Pisa clay defining sensitivity

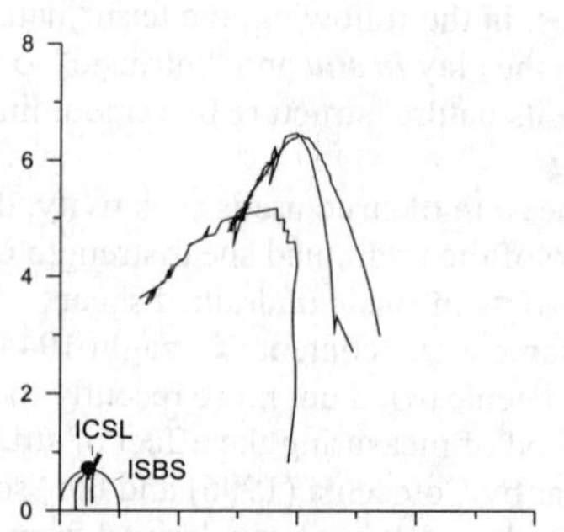

Figure 3. Normalised shearing data for Bothkennar clay

The validation of such a model relies on tests carried out in the laboratory. It is known that sampling and reconsolidation procedures cause disturbance to the soil. As a consequence the soil tested in the laboratory has a structure which has already been altered. This adds to the complexity of modelling soil behaviour in the field.

3 NUMERICAL MODELLING OF DESTRUCTURATION

A number of constitutive models for structured soils have been recently developed (Kavvadas & Amorosi 2000, Rouainia & Muir Wood 2000, Tamagnini & d'Elia 1999). These models, developed within frameworks of bounding surface and kinematic plasticity, all include a bounding surface to model the effects of natural structure. Degradation of structure is described by damage-type mechanisms causing a reduction in size of this surface. The damage function is a general expression of plastic volumetric and shear strain, involving several parameters which cannot be determined except by fitting experimental curves or by parametric studies. All the models allow elasto-plastic deformations inside the bounding surface therefore destructuration starts before the soil state reaches the surface. The empirical expressions defining destructuration in these models make their use complex in finite element analyses and require a wide range of test data for parametric studies.

An important part of the destructuration law described here is that the expression for the damage function uses parameters representing soil properties, which can be derived rigorously from a single set of test data. Firstly, in the new law the change in structure is measured by the change in sensitivity, which has the rigorous definition given earlier. Secondly the exponential damage law relating this change in sensitivity to plastic strain is formulated so that the rate of destructuration is defined by a single parameter that can be shown to be a function of measurable quantities. An exponential expression similar to those used in other models but conforming to these conditions is given below:

$$s = 1 + (s_0 - 1)\exp(-k.f(\varepsilon^p)) \tag{1}$$

where s is the current sensitivity, s_0 is the original sensitivity, k is the rate of destructuration and $f(\varepsilon^p)$ is the plastic strain.

Experimental observations of isotropic compression tests and undrained shear tests on natural clay samples suggest that plastic volumetric and shear strains are of similar importance to destructuration. Consequently, the plastic strain was assumed to be simply equal to the magnitude of the plastic strain increment vector $|\varepsilon^p|$, a definition which requires no additional constants to be introduced.

To compute k it is necessary to differentiate equation (1), which gives:

$$\frac{ds}{1-s} = k \cdot d\left|\underline{\varepsilon^p}\right| \tag{2}$$

There is thus a direct relationship between change in sensitivity and change in plastic strain. In an isotropic normal compression test, once the soil reaches a normally consolidated state, the change in $|\varepsilon^p|$ can be assumed to be the change in magnitude of the total volumetric strain. Sensitivity can be obtained directly from a graph of the logarithm of specific volume against mean effective stress.

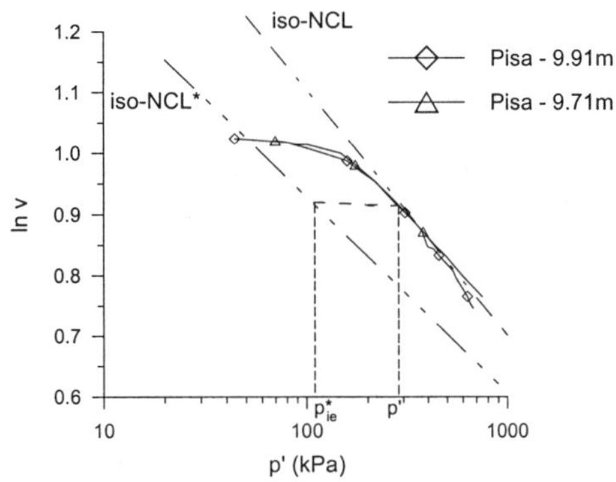

Figure 4. Definition of sensitivity

This is illustrated in Figure 4 which shows isotropic compression data from a metastable clay, Pisa clay. Current sensitivity is measured by the ratio p'/p^*_{ie} for data representing a normally consolidated state, following a definition similar to that of the Sensitivity framework. For a stable clay like Sibari clay, the rate of destructuration k would be found to be zero.

This destructuration rule is used to extend the Three-surface kinematic hardening (3-SKH) model (Stallebrass & Taylor 1997), developed for reconstituted clays, to predict the behaviour of soft clays. As in the 3-SKH model, the extended model, renamed the Structure 3-SKH (S3-SKH) model, consists of three surfaces: a yield surface, a history surface and a bounding surface, which has been renamed the structure surface (Fig. 5). Structure is included in the model following Rouainia & Muir

Wood (2000), by controlling the size of the structure surface by both change in volume and destructuration. With this configuration, only two additional parameters need to be known: the initial sensitivity which defines the initial size of the structure surface (s_0) and the rate at which sensitivity reduces with plastic straining (k).

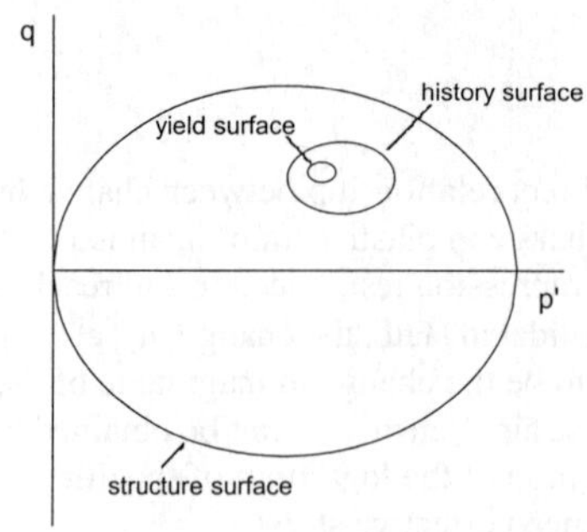

Figure 5. S3-SKH model defined in stress space.

4 EVALUATION OF MODEL

In this section, the model is evaluated against experimental data from two soft natural clays with a metastable structure.

Pisa clay is a lightly overconsolidated clay and was retrieved from the clayey saline water deposit lying below the Tower of Pisa. The data shown are from soil from the Upper Clay layer which is called Pancone clay. Bothkennar clay is also a lightly overconsolidated clay and was deposited at the Forth of Firth, in Scotland. It is characterised by three main facies: bedded, laminated and mottled, so that the behaviour observed in the laboratory may be different according to the provenance of the samples. Both clays have high plasticity indices, of the order of 40-50%, resulting from their organic content. Liquid limits vary throughout the samples, with an average of 70-80%.

The rates of destructuration for Pisa and Bothkennar clay are derived following the procedure described in section 3. In practice, as shown in Figure 6, the points obtained from the experimental data can be scattered due to the small increments involved in using equation (2). A simple solution is to fit a straight line through the test data corresponding to normally consolidated states (iso-NCL in Figure 3) and derive the value of k in this way from the equations of the isotropic natural and intrinsic normal compression curves. The straight lines of gradient k derived numerically are also shown in Figure 6. They plot in good agreement with the scattered data, confirming that this is a reasonable approach.

The initial size of the structure surface is determined from the maximum sensitivity measured in isotropic compression, with an allowance for disturbance.

All other parameters needed for the S3-SKH model are derived from tests on reconstituted clay. Thus, they can be evaluated reliably using experimental data from well controlled repeatable tests. This choice is also consistent with the basic methodology of using parameters which are soil properties.

Predictions from the S3-SKH model are evaluated here against data from shear tests which is a rigorous check because the two new parameters were determined using pure compression data.

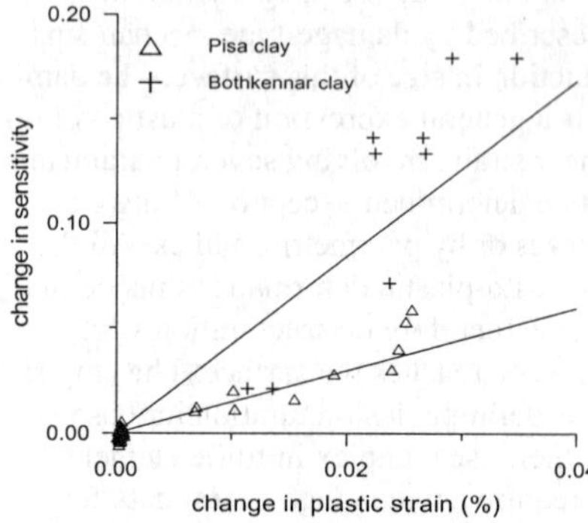

Figure 6. Change in sensitivity against change in plastic strain

The model has been used to simulate a series of drained stress probes performed on Pisa clay. The stress paths carried out during the test were reproduced in the analysis including simulating the reconsolidation history and the effect of sampling stresses. Experimental and predicted data are presented in Figure 7 normalised with respect to the equivalent pressure, p^*_{ie}, on the isotropic Intrinsic Compression Curve. With the inclusion of the destructuration law the S3-SKH model can predict the reverse in normalised stress path direction observed in the experimental data. In addition, a model without structure would predict an expanding outer surface with plastic strain, but the stress paths predicted by the S3-SKH model define a definite outer limit surface almost coincident with that determined experimentally.

Prediction of behaviour during undrained shearing allows the destructuration law to be evaluated when the plastic strains are predominantly plastic shear strains. An undrained stress path test on Bothkennar clay has been simulated again including the sampling and reconsolidation history. The analysis of the undrained test was carried out to a final strain equal to that of the experimental data. The comparison between predicted and experimental data in Figure 8 shows that although the peak strength is predicted correctly the model overpredicts the amount of destructuration. The rate of destructuration k was derived by making a general assumption that there would be negligible effect of fabric on the clay. However, the sample tested was retrieved from the laminated facies of the deposit, so it is likely that some effect of fabric may have occurred leading to the soil structure becoming stable before reaching the intrinsic behaviour. The different facies of the clay have not been distinguished in the analyses presented here. Nevertheless the comparison is reasonable, supporting the choice and derivation of the two parameters representing structure.

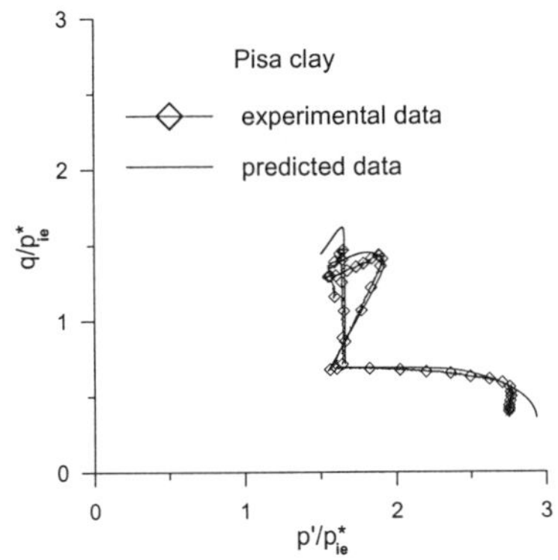

Figure 7. Data from drained stress paths tests on Pisa clay (data from Callisto, 1996)

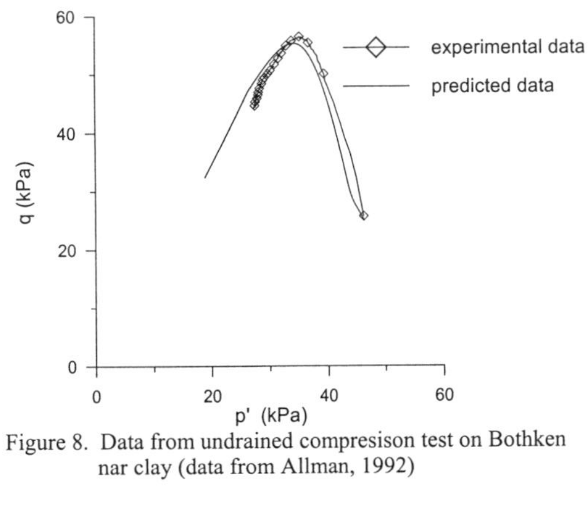

Figure 8. Data from undrained compresison test on Bothken nar clay (data from Allman, 1992)

5 SUMMARY AND CONCLUSIONS

A new model has been proposed to predict destructuration in soft natural clays. In the new destructuration law, change in structure is assumed to be equivalent to change of sensitivity, which has a rigorous definition. Sensitivity is allowed to decrease exponentially with increasing plastic strain, which is defined to be simply equal to the magnitude of the plastic strain increment vector. This law is used to extend an existing model developed for reconstituted clays to predict the behaviour of natural clays. One of the advantages of the model is that, in addition to the usual parameters for the reconstituted clay, it requires only two parameters to represent structure, which can be derived from a single set of test data. Predictions using such a simple law are consistent with the behaviour of natural soft clay, and support the assumptions made in developing the model.

ACKNOWLEDGEMENTS

The Authors would like to thank Dr Luigi Callisto for providing data from Pisa clay and Dr Matthew Coop for data from Sibari clay.

REFERENCES

Allman, M. 1992. Internal report GE/92/09, City University

Callisto, L. 1996. *Studio sperimentale su un'argilla naturale: il comportamento meccanico dell'argilla di Pisa*. PhD thesis, University La Sapienza, Rome, Italy

Coop, M.R., J.H. Atkinson & R.N. Taylor 1995. Strength and stiffness of structured and unstructured soils. *Proc. 11th ECSMFE*. 1:55-62

Coop, M.R. & F. Cotecchia 1995. The compression of sediments at the archeological site of Sibari. *Proc. 11th ECSMFE*. 8:19-26

Cotecchia, F. 1996. *The effects of structure on an Italian pleistocene clay*. PhD thesis, University of London

Kavvadas, M. & A. Amorosi 2000. A constitutive model for structured soils. *Géotechnique*. 50:3:263-274

Rouainia, M. & D. Muir Wood 2000. A kinematic hardening constitutive model for natural clays with loss of structure. *Géotechnique*. 50:2:153-164

Skempton, A.W. 1970. The consolidation of clays by gravitational compaction. *Q.J.Geologocal Soc.* 125: 373-411

Stallebrass, S.E. & R.N. Taylor 1997. The development and evaluation of a constitutive model for the prediction of ground movements in overconsolidated clays. *Géotechnique*. 47:2:235-253

Tamagnini, C. & M. d'Elia 1999. A simple bounding surface model for bonded clays. *Pre-failure deformation characteristics of geomaterials*. 565-572

Terzaghi, K. 1944. Ends and means in Soil Mechanics. *Engineering Journal*. 27:608

Computer Methods and Advances in Geomechanics, Desai et al. (eds) © 2001 Balkema, Rotterdam, ISBN 90 5809 183 X

Reconsolidation deformation of saturated soft clay and its calculation

Bai Bing & Zhou Jian
Department of Geotechnical Engineering, Tongji University, Shanghai, People's Republic of China

ABSTRACT: The reconsolidation characteristics of saturated soft clay under impact loading are studied by laboratory tests, and particularly, the influences of drainage conditions on consolidation and reconsolidation of soft clay are analyzed in detail. The uniqueness of the relationship between pore water pressure and reconsolidation volumetric strain is pointed out, and the methods for determining coefficient of reconsolidation volume compressibility and reconsolidation index are given. It is demonstrated that the shear strengths of reconsolidation of soft clay after several impact loading can be increased greatly. Besides, a model for predicting the deformation of soft clay foundation after impact loading is suggested. The calculation results by the proposed model are in good accordance with the test data.

1 INTRODUCTION

As a common method, dynamic consolidation (dynamic compaction) has been widely used in the ground treatment. The previous studies showed that this method is suitable for crushed stone, sandy soil, miscellaneous fill and collapsible loess, etc., and however, not for saturated soft clay. In recent years, the method of reinforcing saturated soft clay by both improving drainage conditions and combining preloading method and dynamic compaction method has been used successfully in china. For this reason, the basic concepts of the drainage consolidation method of combining static and dynamic loading was proposed by the authors (Bai Bing, 1998a), and furthermore, the problems concerned in design and construction have been discussed.

Based on a large number of laboratory test results, the pore water pressure generation and dissipation and reconsolidation deformation characteristics of saturated soft clay under impact loading are studied systematically in this paper, and their mechanism is also investigated. The quantitative change of residual deformation and residual pore pressure of soft clay with several factors such as confining pressure, impact energy and the numbers of impacting, etc. is also discussed. The reconsolidation deformation law and its influencing factors of soft clay after impact loading are studied, and particularly, the influence of drainage condition on consolidation and reconsolidation of soft clay are emphasized. Besides, the effect of reconsolidation on the shear strength of saturated soft clay is investigated. Based on the test results, a method for predicting the deformation of saturated soft clay foundation after impact loading is proposed.

2 SAMPLES AND TESTING PROGRAM

The clay sample used in experiments is disturbed Wuchang clay, of which physical properties are: liquid limit w_L, 36.9%, plastic limit w_P, 16.4%, plasticity index Ip, 20.5, specific gravity G, 2.71. The size of prepared sample is 3.91cm in diameter and 8cm in height, and the average water content of specimens is 37.4%. The dry unit weight γ_d=13.3kN/m^3, and the degree of saturation S_r>90%.

Specimens were isotropically consolidated under several effective confining pressures (for example σ'_{3c}=30kPa, 50kPa, 100kPa, 150kPa), and then were subjected to axial impact loading under a constant confining pressure, in which several impact loading were used. Excess pore pressure developed during this loading. After impact loading was stopped, a residual excess pore pressure remained in the specimen, which was isotropically confined by the same total pressure as before impact loading. The specimen was then allowed to drain and the residual excess pore pressure was allowed to dissipate. After that, the next series of axial impact loading are applied.

3 TYPICAL TEST RESULTS AND ANALYSYS

3.1 *The reconsolidation characteristics of saturated soft clay*

The saturated soft clay deposited naturally often has low permeability (for example, coefficient of permeability $k<1\times10^{-8}$ m/s). It has been proved by practice that the key problem of the proposed reinforcement method is to improve the drainage condition of saturated soft clay. As a laboratory test method, the various drainage conditions are simulated by installing filter paper, drainage wells (a kind of paper string with high permeability) in the middle of the samples, or their combination. Figure1 shows the processes of consolidation and reconsolidation for some typical testing. In Figure1, curve 1 shows the process of consolidation for sample *B*-10 without filter paper and drainage well (σ'_{3c}=50kPa). Curve2 shows the processes of consolidation and reconsolidation for sample *B*-9 with filter paper and drainage wells (σ'_{3c}=50kPa) after several stages impact loading and reconsolidation. Curve3 shows the process of consolidation for sample *B*-19 with filter paper and drainage wells under greater confining pressure (σ'_{3c}=150kPa) than that of sample *B*-9. Figure1 shows that, under the drainage condition being improved, the rate of consolidation is accelerated greatly, and furthermore, the rate of reconsolidation is also accelerated greatly than initial consolidation and the following reconsolidation processes also have a fast tendency than before. After three-stage reconsolidation, the reconsolidation volumetric deformation can be increased by 20% over initial consolidation. It is also showed that the samples after several stage impact loading and reconsolidation under a lower confining pressure have a surcharge precompression effect.

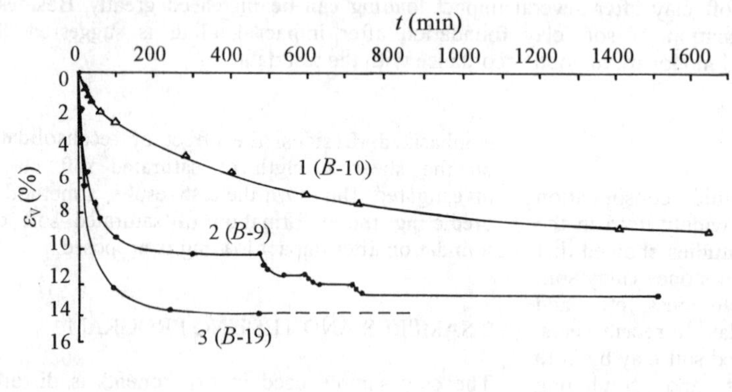

Figure1. Processes of consolidation and reconsolidation of saturated soft clay

Figure2 gives the processes of reconsolidation of saturated soft clay samples after the first stage of impacting under various drainage condition, which show that, with the drainage condition being improved gradually (from sample *B*-9, *B*-8, *B*-24 to *B*-10, σ'_{3c}=50kPa for all samples), both the quantity of reconsolidation volumetric deformation and the rate of reconsolidation are increased greatly.

It can be seen from Figure2 and Figure3 that the processes of reconsolidation after impact loading also can be classified two stages, namely, primary consolidation process and secondary consolidation process. The laboratory test results (sample *B*-8, σ'_{3c}=50kPa, with filter paper and drainage wells) in Figure3 show that, with the drainage condition being improved, the secondary consolidation volumetric deformation is also decreased greatly after three stages of impacting (Bai Bing et al, 1999). This phenomena have the same law as that of the saturated soft clay foundation soils undergoing surcharge precompression (Johnson, 1970).

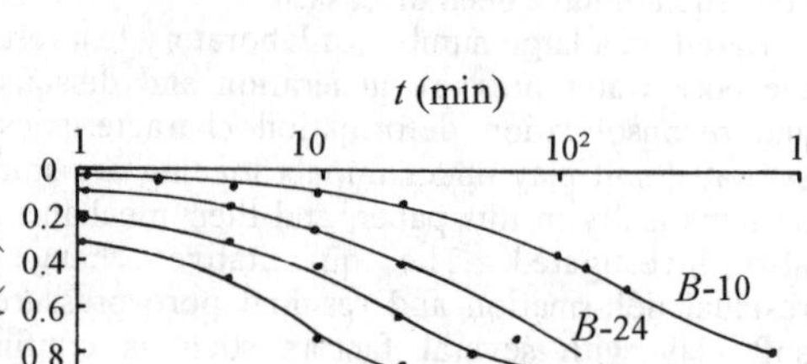

Figure2. Processes of reconsolidation after the first stage of impacting

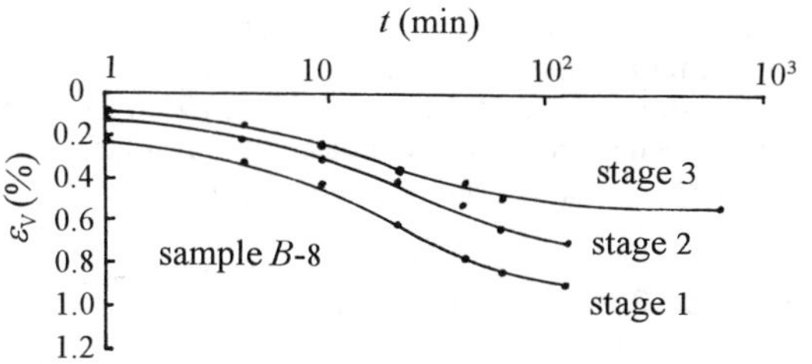

Figure3. Processes of reconsolidation under various stages of impacting

3.2 *The unique relationship between pore pressure and reconsolidation volumetric strain*

Based on a large number of laboratory test results, the relationship between pore water pressure, u, and reconsolidation volumetric strain, ε_V, is analyzed, and furthermore, the normalized curves $\varepsilon_V \sim u/\sigma'_{3c}$ are given in Figure4 and Figure5, respectively.

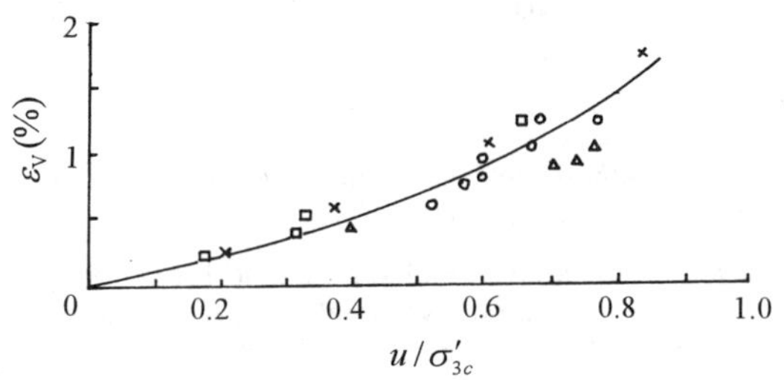

Figure4. Relation between ε_V and u/σ'_{3c} after the first stage of impacting

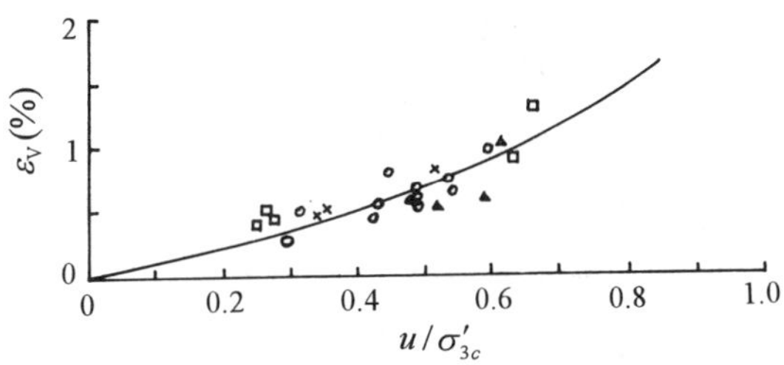

Figure5. Relation between ε_V and u/σ'_{3c} after the second and third stages of impacting

The laboratory test results show that the reconsolidation volumetric strain, ε_V, have an increase tendency with increasing the pore water pressure, u, and however , the effective confining pressure, σ'_{3c}, have a great effect on reconsolidation volumetric strain ε_V. In the other words, when pore pressure u is constant, the ε_V decrease with increasing the σ'_{3c}.

It can be seen from Figure4 and Figure5 that there is a unique relationship between pore water pressure and reconsolidation volumetric strain under any confining pressure, any impact energy and any numbers of impacting. Further studies (Bai Bing, 1998b) show that, for overconsolidated samples (OCR=1, 2, 4, 8), the relationship between ε_V and u/σ'_{3c} can also be described by the same functions. However, it should be pointed out that with the over consolidation ratio of samples increasing, the residual pore pressure generated is decreased under constant impact energy, and consequently the reconsolidation volumetric deformation will also be decreased. The above relationship can be expressed as follows:

$$\varepsilon_V = 1.34u/\sigma'_{3c} \quad (u/\sigma'_{3c} \leqslant 0.3) \tag{1}$$

$$\varepsilon_V = 1.72(u/\sigma'_{3c})^2 + 0.31(u/\sigma'_{3c}) + 0.15 \quad (u/\sigma'_{3c} > 0.3) \tag{2}$$

where ε_V =reconsolidation volumetric strain, u =residual pore pressure, σ'_{3c} =effective confining pressure. The above two equations provide a method for the additional settlement calculation of saturated soft clay foundation under several stages of impact loading and reconsolidation, and have a great practical meaning of application.

3.3 *The coefficient of reconsolidation volume compressibility*

According to the test results, an important parameter m_r is defined, or $m_r = d\varepsilon_V/du$, which is called the coefficient of reconsolidation volume compressibility and often been used for the reconsolidation deformation calculation under the conjunction of static and dynamic loads. For the purpose of practical use, according to equation (1) and equation (2), the method for determining coefficient of reconsolidation volume compressibility is given by:

$$m_{ro} = 1.34 \times 10^{-2}/\sigma'_{3c} \quad kP_a^{-1} \quad (u/\sigma'_{3c} \leqslant 0.3) \tag{3}$$

$$m_r / m_{ro} = 2.57\,(u/\sigma'_{3c}) + 0.23 \quad (u/\sigma'_{3c} > 0.3) \tag{4}$$

where m_{ro} =initial coefficient of reconsolidation volume compressibility, which is an important index that reflect the reconsolidation characteristics of saturated soft clay.

Further studies show that, for clayey soils, plasticity index Ip and initial water content have a great effect on m_{ro}, and furthermore, there is a good linear relation between m_{ro} and Ip (Bai Bing et al. 1999), which can be seen in Figure6 and expressed as:

$$m_{ro} = (0.23+0.055\,Ip) \times 10^{-2}/p'_c \quad kP_a^{-1} \tag{5}$$

where p'_c=the effective confining pressure, considering the consolidation stress state.

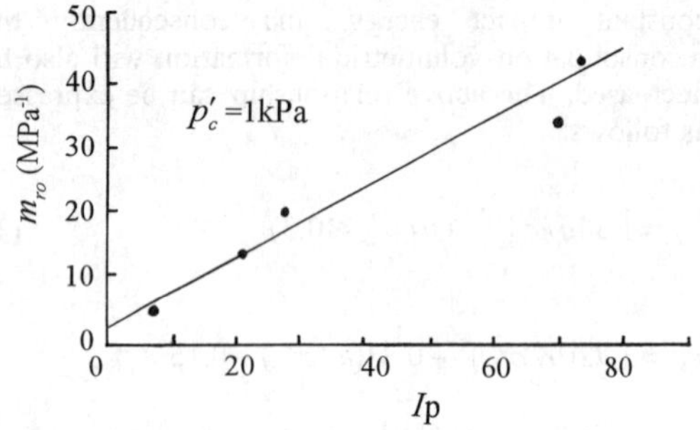

Figure6. Linear relation between m_{ro} and Ip

3.4 *The improvement of shear strength*

The effects of impact loading history on the deformation and strength of soft clay are analyzed, and it is demonstrated according to test results that the samples undergoing several stages impact loading and reconsolidation show obvious quasi-overconsolidation state, and have much greater shear strength than before. For the samples tested in this paper, we have c=8kPa, φ=18° (before impacting) and c=20kPa , φ=22° (after impacting), the details of which can be seen in the concerned reference (Bai Bing,1998a).

4 THE DEFORMATION CALCULATION

4.1 *The reconsolidation compression index*

Now, an another new parameter is defined, or the reconsolidation compression index, which shows the change of void ratio due to the change of effective stress, and can be expressed as:

$$C_{rc} = \frac{-\Delta e}{\log\sigma'_{3c} - \log(\sigma'_{3c} - u)} = \frac{\Delta e}{\log(1 - \dfrac{u}{\sigma'_{3c}})} \tag{6}$$

where C_{rc}= reconsolidation compression index, u= the residual pore pressure under impacting, σ'_{3c}=the effective confining pressure, Δe=the change of void ratio after the dissipation of pore pressure.

The previous study (Bai Bing et al, 1999) showed that C_{rc} is decreased with increasing the residual pore pressure, u/σ'_{3c}, and can be expressed as:

$$C_{rc} = 6 - 2\frac{u}{\sigma'_{3c}} \qquad (\times 10^{-2}) \tag{7}$$

Using equation (7), the reconsolidation volumetric deformation can be calculated by:

$$\varepsilon_V = -\frac{\Delta e}{(1+e_0)} \tag{8}$$

where ε_V= the reconsolidation volumetric strain, e_0 =the initial void ratio.

Considering equation (6) and equation (8), the following calculation formulae can be deduced:

$$\varepsilon_V = \frac{-C_{rc}}{(1+e_0)}\log(1 - \frac{u}{\sigma'_{3c}}) \tag{9}$$

4.2 *A method for calculating the* reconsolidation *deformation of saturated soft clay*

Based on the above results, a method for calculating the reconsolidation volumetric deformation of saturated soft clay under several impact loading and partially drained condition is suggested. As shown in Figure7, when the dissipation rate of pore water pressure is α_j (j=1,2,3…) after each stage impacting, and α_j is defined as the ratio of pore water pressure dissipated after this stage impacting to the total pore water pressure before dissipation, the next stage impacting is loaded

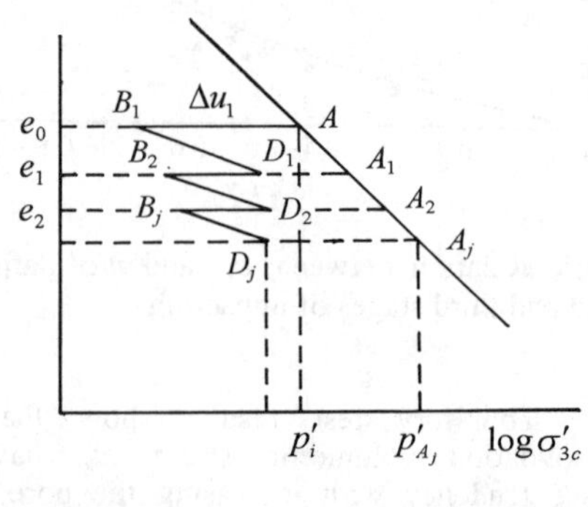

Figure7. The calculation of settlement under several stages impact loading and partially drained condition

After the first stage impact loading, when the dissipation rate of pore pressure is α_1, the change of void ratio is:

$$\Delta e_{B_1D_1} = C_{rc1}\{\log[p_i' - (1-\alpha_1)\Delta u_1] - \log(p_i' - \Delta u_1)\}$$

$$= C_{rc1}\log\left[\frac{1-(1-\alpha_1)\frac{\Delta u_1}{p_i'}}{1-\frac{\Delta u_1}{p_i'}}\right] \tag{10}$$

According to equation (9), the change of volumetric strain can be expressed as:

$$\varepsilon_{V1} = -\frac{C_{rc1}}{1+e_0}\log\left[\frac{1-(1-\alpha_1)\frac{\Delta u_1}{p_i'}}{1-\frac{\Delta u_1}{p_i'}}\right] \tag{11}$$

After the stage j impact loading, when the dissipation rate of pore pressure is α_j, the change of void ratio is:

$$\Delta e_{B_jD_j} = C_{rcj}\log\left[\frac{1-(1-\alpha_j)\frac{\Delta u_j}{p_i'}}{1-\frac{\Delta u_j}{p_i'}}\right] \tag{12}$$

where C_{rcj}= the reconsolidation compression index after the stage j impact loading, and can be determined by equation (7), Δu_j= the total pore pressure after the stage j impact loading. Consequently, the change of volumetric strain can be expressed as:

$$\varepsilon_{Vj} = -\frac{C_{rcj}}{1+e_0}\log\left[\frac{1-(1-\alpha_j)\frac{\Delta u_j}{p_i'}}{1-\frac{\Delta u_j}{p_i'}}\right] \tag{13}$$

Then, the total volumetric strain after the stage j impacting and reconsolidation can be expressed as:

$$\varepsilon_V = \sum_{k=1}^{j}\varepsilon_{Vk} = -\frac{1}{1+e_0}\sum_{k=1}^{j}C_{rck}\log\left[\frac{1-(1-\alpha_k)\frac{\Delta u_k}{p_i'}}{1-\frac{\Delta u_k}{p_i'}}\right] \tag{14}$$

Obviously, when α_k (k=1,2,···,j)=1, the total volumetric strain of reconsolidation of saturated soft clay by several stages of impact loading and reconsolidation processes from equation (14) is:

$$\varepsilon_V = \frac{1}{1+e_0}\sum_{k=1}^{j}C_{rck}\log(1-\frac{\Delta u_k}{p_i'}) \tag{15}$$

where C_{rc} can be calculated by C_{rc}=6-2(u/σ_{3c}') according to test results(Bai Bing et al. 1999).

Table 1 gives the comparison of predicted reconsolidation volumetric strain with tested one under several stages impact loading and drained condition. It can be seen from Table 1 that the calculation results by the proposed model are in good accordance with the test data.

5 CONCLUSIONS

The following conclusions are obtained from the above study:

1. With the drainage condition being improved, the rate of consolidation is accelerated greatly, and furthermore, the rate of reconsolidation is also accelerated greatly than initial consolidation and the following reconsolidation processes also have a fast tendency than before.

2. The samples of saturated soft clay after several stage impact loading and reconsolidation under a lower confining pressure have a surcharge precompression effect, and in the meantime, the

Table 1. Comparison of predicted reconsolidation volumetric strain with tested one

Sample number	B-7	B-8	B-9	B-11	B-22	B-21	B-3	B-20	B-4	B-19
Confining pressure σ_{3c}' (kPa)	50	50	50	50	50	30	100	100	150	150
Tested ε_V (%)	2.8	2.1	2.2	2.6	2.6	2.1	1.6	3.6	1.4	2.5
Predicted ε_V (%)	3.4	2.2	2.4	2.6	2.6	2.5	1.5	3.3	1.2	2.2
Relative error (%)	21.4	4.8	9.1	0	7.7	19.0	-6.3	-8.3	-14.3	-12.0

secondary consolidation deformation is also decreased greatly.

3. The test results show that there is a good uniqueness of the relationship between pore water pressure and reconsolidation volumetric strain, which has no any relation to confining pressure, impact energy, numbers of impacting and overconsolidation state.

4. Both the coefficient of reconsolidation volume compressibility and the reconsoliadation compression index are proposed, and furthermore, there is a good linear relation between m_{ro} and Ip.

5. A model for predicting the deformation of saturated soft clay foundation after impact loading is suggested in this paper. The calculation results by the proposed model are in good accordance with the test data.

These discussions provide some reliable foundations for the treatment of saturated soft clay foundation by the drainage consolidation method of combining static and dynamic loading.

ACKNOWLEDGMENTS

The research described in this paper was supported by the National Science Foundation of China under grant number 59738160. The authors are deeply grateful for advice given by professor Liu Zude of Wuhan University of Hydraulic and Electric Engineering.

REFERENCES

Bai Bing 1998a.The behavior of saturated soft clay under impact loading and its application. *Ph. D. Thesis, Wuhan University of Hydraulic and Electric Engineering*, Wuhan, China.

Bai Bing 1998b.The deformation characteristics of overconsolidated clay under impact loading. *Journal of Wuhan Transportation University,* 22(1): 82-85

Bai Bing & Liu Zude 1998. Growth and dissipation of pore water pressure in saturated soft clay under impact loading. *Rock and Soil Mechanics,* 19(2): 33-38

Bai Bing & Liu Zude 1999. The study on reconsolidation characteristics of saturated soft clay. *Chinese Journal of Geotechnical Engineering,* 21 (2): 189-195

Johnson, S.J. 1970.Precompression for improving foundation soils, *Journal of Soil Mechanics and Foundation Engineering,* 96(1): 111-144

Computer Methods and Advances in Geomechanics, Desai et al. (eds) © 2001 Balkema, Rotterdam, ISBN 90 5809 183 X

A simplified model describing the cyclic behavior of lightly overconsolidated clays in simple shear

G. Biscontin & J. M. Pestana
University of California, Berkeley, Calif., USA

ABSTRACT: The response of soft cohesive soils to a combination of static and irregular cyclic loading conditions is of paramount importance to the overall performance of many offshore structures. Of particular interest is the modeling of the generation of shear induced pore pressure and the accumulation of plastic shear strains during cyclic loading for lightly overconsolidated clays. This paper briefly describes the key elements of a new simplified effective-stress-based model used to predict the cyclic behavior of lightly overconsolidated cohesive soils in simple shear. The new model incorporates governing laws describing the effect of previous consolidation stress history on the stress-strain-strength characteristics of these soils. Model capabilities are illustrated through parametric studies on the effect of cyclic shear stress amplitude as well as non-symmetrical cyclic loading.

1 INTRODUCTION

The use of offshore structures for exploration and storage of mineral resources is gaining attention as the economic interest in offshore drilling increases (e.g., Andersen & al. 1978). More recently, the problem of stability of gently inclined slopes on the continental shelf for seismic (e.g., Puzrin & al. 1997, Biscontin & al. 2000) or wave loading conditions (e.g., Schapery & Dunlap 1978) has received worldwide attention. The typical geological profile of these sites is characterized by nearly parallel layers of soft cohesive sediments, usually normally consolidated to lightly overconsolidated (i.e., OCR < 2) clays with depths ranging from a few meters to hundreds of meters. In all of these cases, the prediction of overall performance requires the accurate representation of the stress-strain-strength characteristics of these soils subjected to irregular cyclic loading.

Numerous constitutive laws have been presented for soft cohesive materials, such as those based on the work hardening elasto-plasticity theory (e.g., Prévost 1978, Pestana 1994). However, these models often require a fully six-dimensional stress space implementation and thus are highly computationally intensive for the solution of relatively simple problems such as the one-dimensional shear wave propagation or the response of soils in simple shear conditions. A more typical approach in those cases is the use of empirical relationships derived from experimental data. For these models, the first loading stress-strain curve (i.e. backbone curve) is closely replicated by a continuous (Ramberg & Osgood, 1943) or piece-wise linear (Iwan 1967) expression. In most cases, these simplified models cannot predict the generation of excess pore pressure during cyclic loading or the effect of consolidation shear stresses on the behavior of the soil. A new simplified model was proposed to describe the response of normally to lightly overconsolidated clays in cyclic simple shear conditions by Pestana & al. (2000) and the key components and capabilities are illustrated here.

2 SIMPLE DSS MODEL

The key element for a realistic prediction of cyclic clay behavior is the implementation of an accurate stress-strain-strength relationship, which incorporates the effect of previous consolidation stress history, soil nonlinearity and the generation of excess pore pressures and accumulation of strains during irregular cyclic shearing. Pestana & al. (2000) proposed a simplified model, hereinafter referred to as SIMPLE DSS, to describe the response of normally consolidated clays in simple shear which has been used to perform site response analyses of submerged slopes (Biscontin & al. 2000, 2001).

The model uses a "plastic state surface" defined in the normal effective stress-shear stress (σ_n, τ) space to describe the undrained effective stress paths from normally consolidated (NC) states following

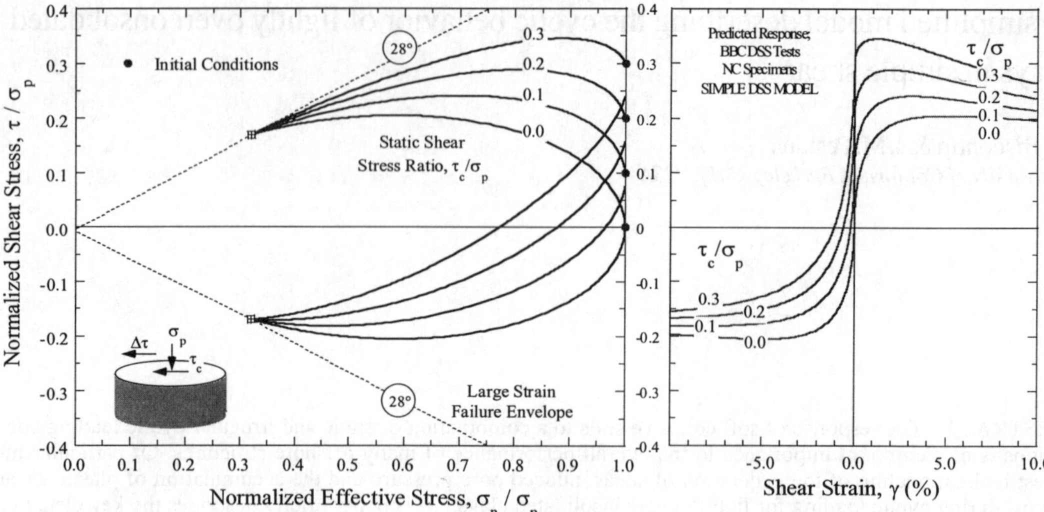

Figure 1: Effect of anisotropic consolidation history on the undrained behavior of NC clays in simple shear

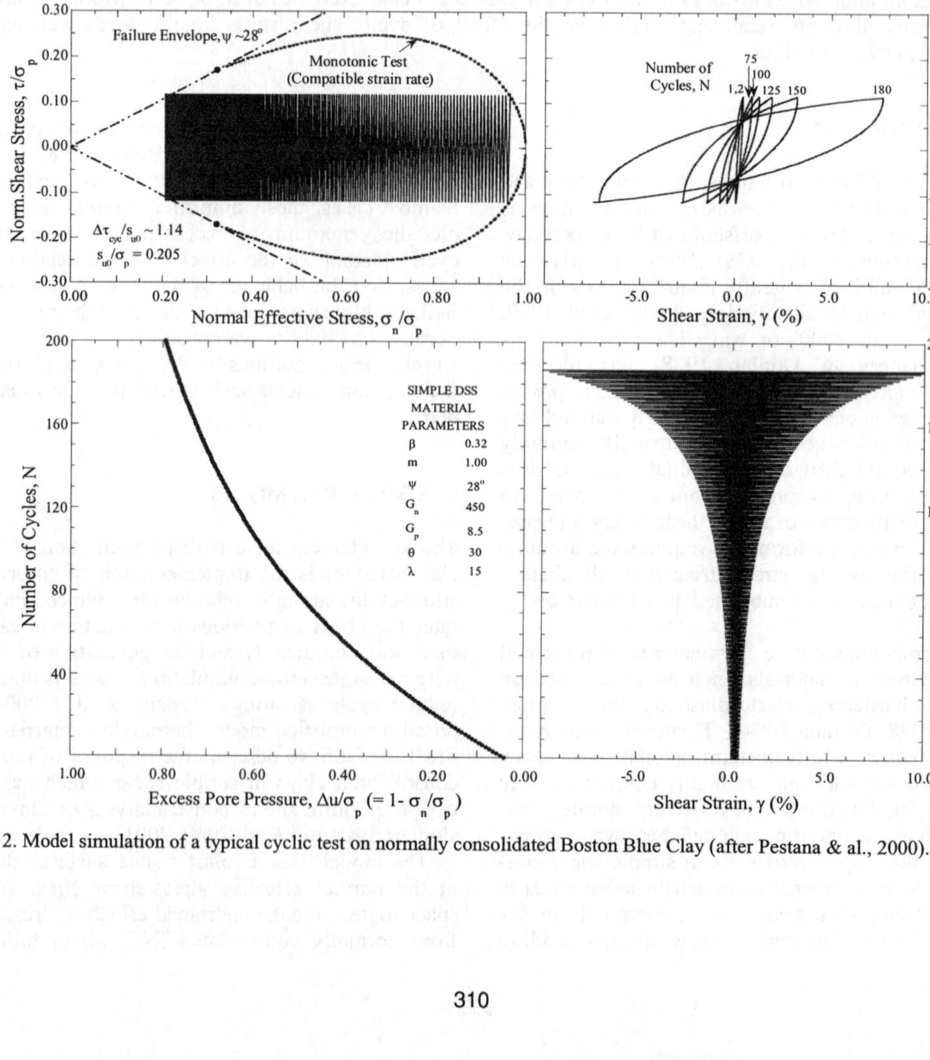

Figure 2. Model simulation of a typical cyclic test on normally consolidated Boston Blue Clay (after Pestana & al., 2000).

the development of the BEAR-CLAY model for clays (Luccioni 1999). The effective stress path is given by:

$$\tau^2 = \sigma_n{}^2 \tan^2\psi - \sigma_n{}^2 \zeta_t^2 \cdot \left(\frac{(\sigma_n/\sigma_p)^m - \beta^m}{1-\beta^m} \right) \quad (1a)$$

$$\zeta_t^2 = \left[\tan^2\psi + 0.8.\eta_{tc}{}^2 - 1.8.\eta_t \eta_{tc} \right] \quad (1b)$$

where τ and σ_n are shear and normal stress, respectively, η_{tc} represents the consolidation shear stress ratio (= τ_c/σ_p, where τ_c is the consolidation shear and σ_p is the normal effective stress for NC clays) and η_t is the current shear stress ratio (= τ/σ_n). This surface is defined by three parameters: ψ controls the maximum shear stress ratio and controls the large strain failure envelope, β controls the maximum pore pressure generation at large strains and m mostly influences the peak undrained shear strength. They can be estimated using the stress path of a standard monotonic simple shear test.

The proposed formulation describes the change in shear stress ratio using the traditional decomposition into elastic and plastic components as follows:

$$\frac{\partial \eta_t}{\partial \gamma} = \left(\frac{1}{[\partial \eta_t/\partial \gamma]^e} + \frac{1}{[\partial \eta_t/\partial \gamma]^p} \right)^{-1} \quad (2)$$

where the elastic and plastic components are given by:

$$\left[\frac{\partial \eta_t}{\partial \gamma} \right]^e = G_n \left(\frac{\sigma_p}{\sigma_n} \right)^{1-b} ; \quad b = 0.5 \quad (3a)$$

$$\left[\frac{\partial \eta_t}{\partial \gamma} \right]^p = G_p F.(\tan\psi - \mathrm{sign}(\eta_t.\dot{\gamma})\, \eta_t) \quad (3b)$$

$$F = \frac{(\tan\psi - \mathrm{sign}(\eta_{tc}.\dot{\gamma})\ \eta_{tc})}{(\eta_t - \eta_{tc})} \quad (3c)$$

where G_n (= $G_{max\ NC}/\sigma_p$) is a material parameter describing the small strain stiffness, G_{max}, for NC specimens and G_p is a material parameter characterizing first loading or backbone curve. Figure 1 shows predictions of stress paths and stress strain curves for monotonic simple shear tests at different values of the normalized initial shear stress, τ_c/σ_p = 0, 0.05, 0.075, 0.100 and 0.125. Pestana & al. (2000) show good agreement with experimental data. As the consolidation shear stress increases the undrained shear strength also increases, but the behavior becomes progressively more brittle with more softening. In addition, the peak takes place at lower strains and the difference between the initial and the maximum shear stress decreases. The consequence is that a higher initial shear stress requires less stress increase and less mobilized strain to reach unstable post-peak conditions. The shear strain (and stress) is arbitrarily defined as positive if it acts in the same direction of the consolidation shear stress (τ_c).

A "load state surface" controls the effective rate dependent stress path and thus the development of excess pore pressure during cyclic loading. Although this is a typical element of bounding surface plasticity models, in SIMPLE DSS the stress path formulation is decoupled from the stress-strain response to allow for nearly independent determination of material parameters. For the sake of mathematical simplicity, the shape of the surface is determined only by the conditions at the last reversal and a strain rate dependent parameter, θ. The plastic component of the stress-strain curve inside the plastic state surface is given by a separate expression controlled by parameter λ. Parameters θ and λ are selected by matching the curves for excess pore pressure and accumulation of plastic strains vs. number of cycles, respectively, for a cyclic simple shear test at the strain rate of interest. SIMPLE DSS gives results similar to those obtained with generalized models based on bounding surface plasticity (e.g., Pestana & Whittle 1999), but it is computationally more efficient because it is formulated in a reduced stress space.

2.1 *Parameters for Boston Blue Clay*

Parameters for Boston Blue Clay were evaluated by Pestana & al. (2000) and they are reported in Table 1. Boston Blue Clay is a low plasticity, moderately sensitive, primarily illitic clay. Engineering properties of both natural and resedimented Boston Blue Clay have been extensively studied and, as a result, there is a large database of high quality tests.

Monotonic tests were carried out at a rate of 5% strain per hour, while cyclic tests were stress controlled and performed at a rate of 0.1 Hz. The dependency of clay behavior on strain rate affects some of the parameters. Since θ and λ are determined directly from cyclic tests, their values are mostly appropriate. Parameters m and G_p are derived from monotonic tests performed at a considerably lower rate and they must be adjusted to describe the response at the strain rate of interest (Biscontin & Pestana, 2000). A value of m equal to 1.0 was used for cyclic tests, while the standard monotonic DSS tests are described by m=-0.25.

Table 1. SIMPLE DSS model parameters for Boston Blue Clay

Parameter	β	m	ψ	G_p	G_n	θ	λ
Value	0.32	1.00	28	10	450	30	15

3 PREDICTION OF BEHAVIOR

Figure 2 shows simulation of a typical cyclic simple shear test with no initial consolidation shear stress (i.e.,τ_c = 0) and a single amplitude normalized cyclic

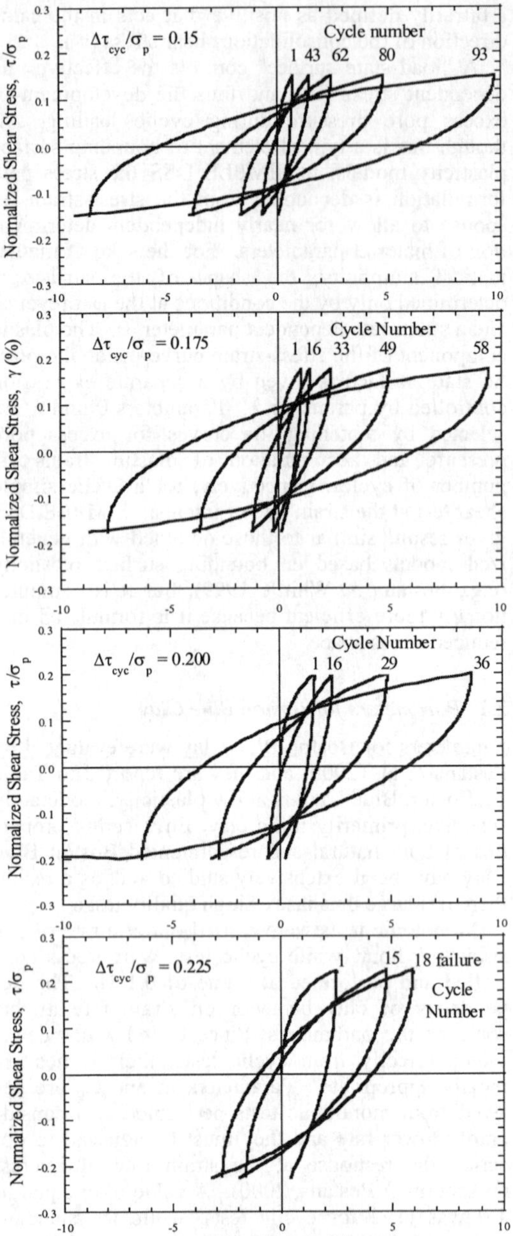

Figure 3. Effect of normalized cyclic shear stress on the cyclic behavior of NC Boston Blue Clay.

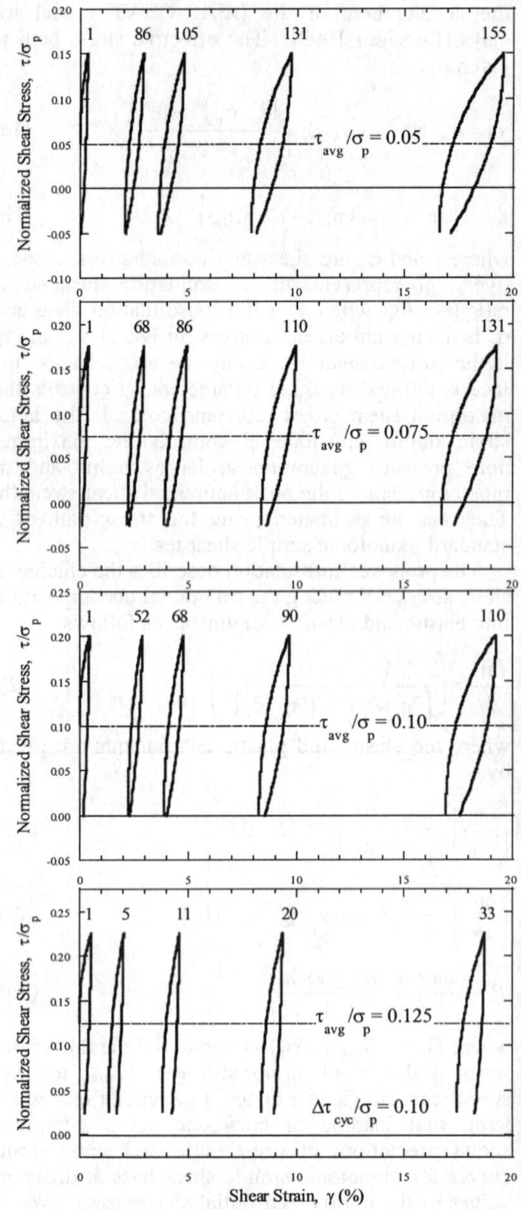

Figure 4. Effect of average shear stress ratio on the cyclic behavior of NC Boston Blue Clay- K_α effects.

shear stress ($\Delta\tau_{cyc}/\sigma_p$, where σ_p is the past maximum normal pressure) of 0.117. This test was used to calibrate the model parameters for Boston Blue Clay and is discussed in detail by Pestana & al (2000). It also illustrates the ability of the model to capture shear modulus degradation, pore pressure generation and accumulation of plastic strains during cyclic loading.

In order to assess the effect of consolidation shear stress on the cyclic behavior, a 'baseline' is established by using the SIMPLE DSS model to predict the response of Boston Blue Clay subjected to stress controlled cyclic DSS tests with increasing normalized cyclic shear stress ($\Delta\tau_{cyc}/\sigma_p$ = 0.15, 0.175, 0.20 and 0.225). Figure 3 compares the predictions of the stress-strain curves at selected cycles: the first and then at the approximate strain levels of 1.5, 2.5, 5

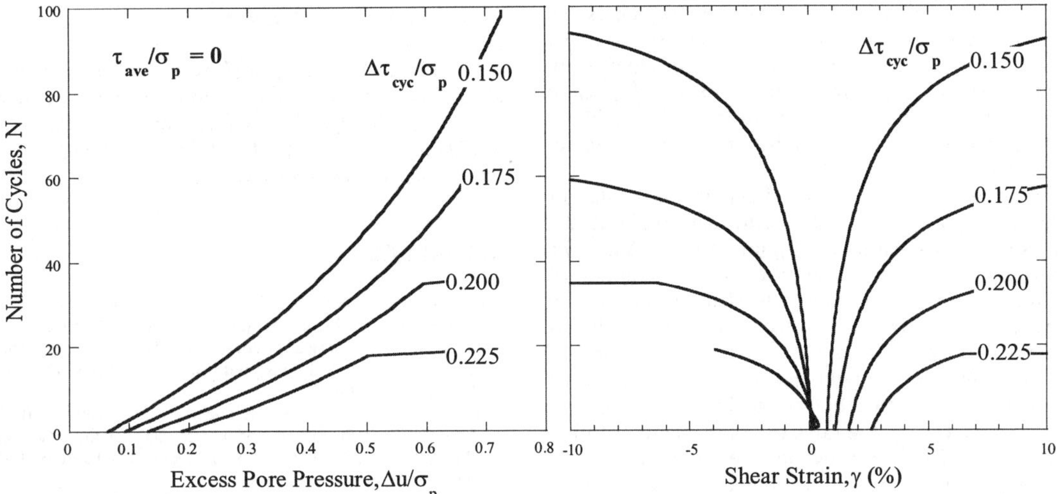

Figure 5. Effect of cyclic shear stress ratio on the behavior of symmetric cyclic DSS tests on NC clay.

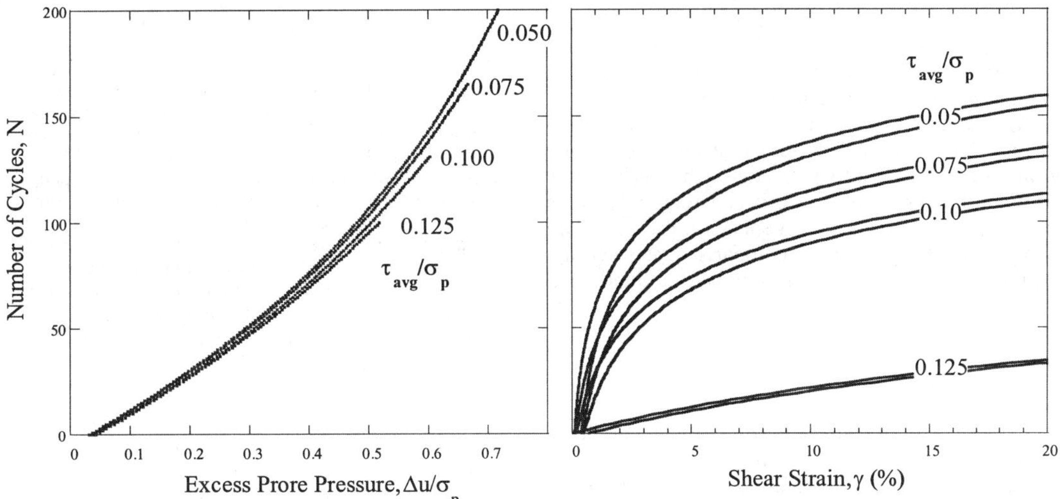

Figure 6. Effect of average shear stress on the response of NC clays in cyclic DSS.

and 10% at the positive peak of the cycle. The plastic strains are accumulating during every cycle, leading to flatter and more elongated loops centered around the origin of the shear stress-shear strain space. Although the plastic strains are increasing, the average strain is very small for symmetric cyclic loading. The rate of accumulation is increasing with the cyclic shear stress amplitude, as shown by the decrease in the cycle number at which the selected target strain levels are reached. The small asymmetry of the loops can be explained by the fact that the accumulation of plastic strains is higher as the shear stress ratio gets closer to the failure envelope. Most of the "drift" in the average shear strain results from the plastic strain developed during the first cycle of loading.

A series of cyclic stress controlled tests with non-zero initial shear stress was also simulated. The cyclic shear stress amplitude, $\Delta\tau_{cyc}/\sigma_p$, was equal to 0.1 for all tests. To simulate the response of an element on a slope, the average shear stress was selected to correspond to the initial consolidation shear stress, τ_c/σ_p was 0.050, 0.075, 0.100 and 0.125, reaching the same maximum levels as in the previous tests. Figure 4 shows the stress-strain curves for selected cycles. The loops are not centered around the zero strain axis, instead as the plastic strains accumulate, they progressively shift to higher levels of strain (i.e., movement of the element downslope).

The shape of the loops does not change as much as it did in the $\Delta\tau_c = 0$ case, although increasing hysteresis is present. Even if the amplitude is the same, the number of cycles required to reach the "threshold" strains decreases with increasing consolidation shear stress. This is expected because the maximum shear stress increases with τ_c and the accumulation of plastic strains is faster for higher shear stress ratios.

Figure 5 and 6 are a summary of the results for the two groups of tests, with and without the consolidation shear stress. Figures 5 and 6 show the development of pore pressure vs. the number of cycles. In the first group of tests the samples with higher cyclic shear stress reach the failure in a low number of cycles. In the second group the limit of 20% strain –that would constitute the end of an actual test –is reached before the failure envelope.

The increase in the rates of accumulation of plastic strains for higher cyclic stresses is easily observed in Figure 5. The asymmetric behavior mentioned previously is also apparent in this figure. In Figure 6 the average strain during a cycle of loading shifts away from zero with increasing rate for higher consolidation shear stresses. Failure or end of test conditions are reached at a lower number of cycles in the tests of Figure 5 than in those of Figure 6.

Although the maximum shear stress is the same for couples of tests from the first and the second group, the normalized amplitude of the cyclic shear stress for the tests with $\tau_c = 0$ is different and always larger than 0.1 (constant value used in the tests with $\tau_c \neq 0$). For this reason the comparison between the results of the two groups can only be of a qualitative nature.

4 CONCLUSIONS

The presence of an initial consolidation shear stress has a considerable influence on the behavior of soft, cohesive normally consolidated soils, both in monotonic and cyclic simple shear tests. A new simplified effective-stress-based model, called SIMPLE DSS, was developed to describe the response of this type of soils for the use in seismic site response analysis. With the use of this model, the effect of the consolidation shear stress was investigated in this paper. During monotonic loading higher τ_c causes the response to become more brittle, the peak to shift to lower strain levels and the additional stress to the maximum to decrease. This implies that a lower amount of stress and strain needs to be mobilized for the specimen to reach potentially unstable consitions. During cyclic loading, the presence of an initial shear stress forces the average shear strain to increase, shifting the stress-strain loops away from zero. The shape of the loops remains similar, even with increasing number of cycles, with only minor increases in hysteresis compared to tests with no τ_c.

5 ACKNOWLEDGMENTS

This work was sponsored by the Research Council of Norway and by the National Science Foundation, NSF, through the Faculty Early Development, CAREER award CMS-9623979. This support is most gratefully appreciated.

REFERENCES

Andersen, K.H., Hansteen, O.E., Hoeg, K. and Prevost, J.H. (1978) "Soil deformations due to cyclic loads on offshore structures," Numerical Methods in Offshore Engineering, eds. Zienkiewicz, Lewis and Stagg, J. Wiley & Sons, Chichester, 413-452.

Andersen, K.H., Dyvik, R., Kikuchi, Y. and Skomedal, E. (1993) "Clay behavior under irregular cyclic loading," Publication 189, Norwegian Geotechnical Institute, Norway, 1-15.

Biscontin, G., Pestana, J.M., Nadim, F. and Andersen, K. (2000) "Seismic triggering of submarine slides in soft cohesive soil deposits," Prediction of Underwater Slide and Slump Hazards, Watts, Synolakis and Bardet, eds. Balkema (in press).

Biscontin, G., Pestana, J.M., Nadim, F. and Andersen, K. (2001) "Seismic response of normally consolidated cohesive soils in gently inclined submerged slopes," 4th International Conference on Recent Advances in Geotechnical Earthquake Engineering and Soil Dynamics.

Biscontin, G. and Pestana, J.M. (1999) "Influence of peripheral velocity on undrained shear strength and deformability characteristics of a bentonite-kaolinite mixture", Geotechnical Engng Report No. UCB/GT/99-19, Dep. of Civil & Environ. Engng., University of California, Berkeley.

Frydman, S., M. Talesnick, G. Almagor & G. Wiseman 1988. Simple Shear Testing for the study of the earthquake response of clay from the Israeli continental slope. *Marine Geotechnology* 7(3): 143-171.

Iwan, W.D. (1967) "On a class of models for the yielding behavior of continuous and composite systems,' J. Appl. Mech., ASME, 34 (3), 612-617.

Luccioni, L.X. (1999) "Numerical development and implementation of a constitutive model for clays with application to deformations around a deep excavation," Ph.D. Thesis, University of California, Berkeley, CA.

Pestana, J.M., Biscontin, G., Nadim, F. and Andersen, K. (2000) " Modeling Cyclic Behavior of Lightly Overconsolidated Clays in Simple Shear," Soil Dynamics and Earthquake Engineering, (in press).

Pestana, J.M. and Whittle, A.J. "Formulation of a unified constitutive model for clay and sand," Int. J. Num. Anal. Meth. in Geomech., 1999, 23 (4), 1215-1243.

Prévost, J.H. "Plasticity theory for soil stress-strain behavior," J. Engng. Mech. Div., ASCE, 1978, 104 (EM5), 1177-1194.

Puzrin, A., Frydman, S. &Talesnick, M. (1997). Effect of degradation on seismic response of Israeli continental slope. *J. Geotch. And Geoenv. Eng., ASCE* 121(12):85-93.

Ramberg, W. and Osgood, W.T. (1943) "Description of stress-strain curves by three parameters," Technical note 902, National Advisory Committee for Aeronautics.

Schapery, R.A. and Dunlap, W.A. (1978) "Prediction of Storm-Induced Sea Bottom Movement and Platform Forces," 10th Offshore Tech. Conf., Houston, OTC No. 3259.

Computer Methods and Advances in Geomechanics, Desai et al. (eds) © 2001 Balkema, Rotterdam, ISBN 90 5809 183 X

On the thermomechanical interpretation of plastic flow rules

K. Hashiguchi
Department of Agricultural Engineering, Kyushu University, Fukuoka, Japan

ABSTRACT: The pertinence of the non-associated flow rule for frictional materials has been insisted by the thermomechanical studies. It would be of importance to examine the pertinence and/or the validity of the thermomechanical approach to the formulation of plastic constitutive equations, whilst various works on this approach have been reported up to the present. In this article it is shown that both the associated and the non-associated flow rules are derived independent of frictional property of materials from the dissipation functions which are more general than the ones postulated in the past studies. Thus, it is concluded that which flow rule of them is pertinent cannot be judged by the thermomechanical approach so that this approach would be powerless to the formulation of plastic constitutive equations in the isothermal process.

1 INTRODUCTION

The *non-associated flow rule* causes the following defects.

a. Constitutive equations become to be complicated since not only the yield surface but also the plastic potential surface are incorporated.
b. Constitutive equations exhibit the peculiar mechanical response (Hashiguchi, 1991, 1997).
c. Stiffness tensor loses the symmetry.
d. Analysis of boundary value problems becomes to be difficult by these faults of constitutive equation.

The typical model with the non-associated flow rule is the *Drucker-Prager model* (Drucker & Prager, 1952) for over-consolidated soils, while the associated flow rule is adopted in the *von Mises yield condition* for metals and the *original* and the *modified Cam-clay models* for normal-consolidated soils. Which is pertinent the associated or the non-associated flow rule for plastic constitutive equations of geomaterials has been discussed but is the unsolved problem up to the present.

The non-associated flow rule would be adopted by misunderstanding that

1. the plastic volumetric strain is predicted to be unrealistically expansive,
2. the critical load in plastic instability problems is predicted to be unrealistically high

if the associated flow rule is adopted for over-consolidated soils.

In addition, the *thermomechanical approach* have been adopted in order to verify the pertinence of the non-associated flow rule for frictional materials (Houlsby, 1979, 1981, 1982, 1993, Modaressi *et al.*, 1994, Collins & Houlsby, 1997), and they concluded that *the concepts of friction and of non-associated flow rule are seen to be intimately linked* and that *by comparison with metals, geomaterials differ in that they exhibit frictional behavior, they undergo plastic changes of volume and need to be modeled by plasticity theories with 'non-associated' flow* which is quoted from the article of Collins & Houlsby (1997). It would be of importance to examine the pertinence and/or the validity of the thermomechanical approach to the formulation of plastic constitutive equations, while various works have been reported on this approach up to the present.

In this article it is shown that both the associated and the non-associated flow rules are derived from the dissipation functions which are more general than the ones postulated in the past studies (Houlsby, 1981, 1982, Modaressi *et al.*, 1994, Collins & Houlsby, 1997) for the von Mises yield condition for metals and the Drucker-Prager model (Drucker & Prager, 1952), the original Cam-clay model (Schofield & Wroth, 1968) and the modified Cam-clay model (Roscoe & Burland, 1968) for geomaterials. Thus, it will be concluded that which flow rule of them is pertinent cannot be judged from the aspect of the fulfillment of thermomechanical requirement. Besides, it will be indicated that the adoption of the non-associated flow rule would be caused by the deficiency of constitutive equations falling within the framework of the conventional plasticity (Drucker, 1988).

The signs of a stress (rate) and a stretching (a symmetric part of velocity gradient) components are taken to be positive for tension, and the stress for soils is meant to be the *effective stress*, i. e. the stress excluded a pore pressure from the total stress throughout this article.

2 THERMOMECHANICAL APPROACH

The *dissipation function* D in the isothermal process of rate-independent materials is to be the function of the plastic stretching (plastic strain rate) $\boldsymbol{D}^p$, the stress $\boldsymbol{\sigma}$ and the scalar- or tensor-valued internal variable $\boldsymbol{H}$ and has to be homogeneous of degree one in $\boldsymbol{D}^p$ for rate-independent materials so that it follows by the Euler's theorem that

$$D(\boldsymbol{D}^p, \boldsymbol{H}, \boldsymbol{\sigma}) = \mathrm{tr}\left(\frac{\partial D}{\partial \boldsymbol{D}^p}\boldsymbol{D}^p\right). \tag{1}$$

On the other hand, the dissipation function can be written as

$$D = \mathrm{tr}(\boldsymbol{\chi}\boldsymbol{D}^p), \tag{2}$$

where $\boldsymbol{\chi}$ is the *dissipation* (or *generalized*) *stress*. From Eqs. (1) and (2) it follows that

$$\mathrm{tr}\left\{\left(\boldsymbol{\chi} - \frac{\partial D}{\partial \boldsymbol{D}^p}\right)\boldsymbol{D}^p\right\} = 0 \tag{3}$$

which means that the difference between $\boldsymbol{\chi}$ and $\partial D/\partial \boldsymbol{D}^p$ must be orthogonal to $\boldsymbol{D}^p$. Ziegler (1981, 1983; cf. also Ziegler & Wehrli, 1987) argued that this difference is in fact zero for a very wide class of constitutive equations, so that

$$\boldsymbol{\chi} = \frac{\partial D}{\partial \boldsymbol{D}^p} \tag{4}$$

which is usually referred to as '*orthogonality principle*'.

The function $\bar{f}$ fulfilling

$$D(\boldsymbol{D}^p, \boldsymbol{H}, \boldsymbol{\sigma}) + \bar{f}(\boldsymbol{\chi}, \boldsymbol{H}, \boldsymbol{\sigma}) = \mathrm{tr}(\boldsymbol{\chi}\boldsymbol{D}^p) \tag{5}$$

is the *Legendre dual function* of the dissipation function and has the properties:

$$\bar{f}(\boldsymbol{\chi}, \boldsymbol{H}, \boldsymbol{\sigma}) = 0, \ \boldsymbol{D}^p = \lambda\frac{\partial \bar{f}(\boldsymbol{\chi}, \boldsymbol{H}, \boldsymbol{\sigma})}{\partial \boldsymbol{\chi}} \tag{6}$$

by the singular Legendre transformation due to the homogeneous degree one in $\boldsymbol{D}^p$ of the dissipation function D. λ is the positive proportionality factor. Eq. (6) means that the function $\bar{f}$ is to be the yield function in the *dissipation stress space* and the associated flow rule holds in this space.

Hereinafter, we limit our discussion to the decoupled material, i. e. the material the instantaneous elastic compliance of which is independent of a plastic deformation and thus it holds that

$$\boldsymbol{\chi} = \boldsymbol{\sigma} - \boldsymbol{\rho}, \tag{7}$$

where $\boldsymbol{\rho}$ is the *shift stress* (cf. Collins & Houlsby, 1997).

The yield function f in the *true stress space* is given by

$$f(\boldsymbol{\sigma}, \boldsymbol{H}) = \bar{f}(\boldsymbol{\chi}(\boldsymbol{\sigma}), \boldsymbol{H}, \boldsymbol{\sigma}). \tag{8}$$

It holds that

$$\frac{\partial f(\boldsymbol{\sigma}, \boldsymbol{H})}{\partial \boldsymbol{\sigma}} = \frac{\partial \bar{f}(\boldsymbol{\chi}, \boldsymbol{H}, \boldsymbol{\sigma})}{\partial \boldsymbol{\sigma}} + \frac{\partial \bar{f}(\boldsymbol{\chi}, \boldsymbol{H}, \boldsymbol{\sigma})}{\partial \boldsymbol{\chi}} \tag{9}$$

because of $\partial\boldsymbol{\chi}/\partial\boldsymbol{\sigma} = \boldsymbol{I}$ (identity tensor) for Eq. (7). Hence, substituting Eq. (9) into Eq. $(6)_2$, the plastic stretching $\boldsymbol{D}^p$ can be rewritten as

$$\boldsymbol{D}^p = \lambda\left\{\frac{\partial f(\boldsymbol{\sigma}, \boldsymbol{H})}{\partial \boldsymbol{\sigma}} - \frac{\partial \bar{f}(\boldsymbol{\chi}, \boldsymbol{H}, \boldsymbol{\sigma})}{\partial \boldsymbol{\sigma}}\right\} \tag{10}$$

which means that the associated flow rule does not hold in general in the true stress space.

In what follows one considers materials without the shift stress, i. e. $\boldsymbol{\rho} = \boldsymbol{0}$, whilst the shift stress would not be needed for geomaterials which instead exhibits the rotational hardening (cf. Hashiguchi & Chen, 1998). Therefore, the true stress formulation can be derived from the dissipation stress space formulation by simply replacing $\boldsymbol{\chi}$ with $\boldsymbol{\sigma}$.

3 PLASTIC FLOW RULES

Collins & Houlsby (1997) assumed the following dissipation function D and derived the Coulmb-Mohr failure criterion for cohesionless materials.

$$D = \mu p\left\|\boldsymbol{D}^{p*}\right\| + \Lambda(D_v^p - \beta\left\|\boldsymbol{D}^{p*}\right\|), \tag{11}$$

where

$$p \equiv -\frac{1}{3}\mathrm{tr}\boldsymbol{\sigma}, \tag{12}$$

$$D_v^p \equiv \mathrm{tr}\boldsymbol{D}^p. \tag{13}$$

$(\ )^*$ stands for the deviatoric component and $\|\ \|$ the magnitude. Λ is the *Lagrange multiplier* in order to incorporate the dilatancy property $D_v^p/\left\|\boldsymbol{D}^{p*}\right\| = \beta$, where β is the *dilatancy coefficient*. μ is the material constant which is the difference between the friction and dilatancy coefficients as it will be known from the fact that $\beta+\mu$ is later derived to be the coefficient of internal friction.

Let the dissipation function of Eq. (11) be extended as follows:

$$D = (c + \mu p)\left\|\boldsymbol{D}^{p*}\right\| + \Lambda(D_v^p - \beta\left\|\boldsymbol{D}^{p*}\right\|), \qquad (14)$$

where $c(H)$ is the material function of the hardening variable H (the equivalent strain for metals and the plastic volumetric strain for clays), which corresponds to the cohesion as will be known later.

The dissipation stress $\boldsymbol{\chi}$ for Eq. (14) is given by Eq. (4) by

$$\boldsymbol{\chi} = (c + \mu p - \beta\Lambda)\frac{\boldsymbol{D}^{p*}}{\left\|\boldsymbol{D}^{p*}\right\|} + \Lambda\boldsymbol{I}, \qquad (15)$$

from which one has

$$\chi_p = -\Lambda, \quad \left\|\boldsymbol{\chi}^*\right\| = (c + \mu p - \beta\Lambda), \qquad (16)$$

where

$$\chi_p \equiv -\frac{1}{3}\mathrm{tr}\boldsymbol{\chi}. \qquad (17)$$

The yield condition in the dissipation stress space is derived by eliminating Λ from Eq. (16) as

$$\bar{f}(\boldsymbol{\chi}, \boldsymbol{\sigma}, H) = \left\|\boldsymbol{\chi}^*\right\| - (c + \mu p) - \beta\chi_p = 0 \qquad (18)$$

for which the associated flow rule holds regarding $\boldsymbol{\sigma}$ to be independent of $\boldsymbol{\chi}$ as described in Eq. $(6)_2$ and thus it holds that

$$D_v^p / \left\|\boldsymbol{D}^{p*}\right\| = -\mathrm{tr}\frac{\partial\bar{f}}{\partial\chi_p} \Big/ \frac{\partial\bar{f}}{\partial\left\|\boldsymbol{\chi}^*\right\|} = \beta. \qquad (19)$$

The yield condition in the true stress space is given from Eq. (18) by replacing $\boldsymbol{\chi}$ with $\boldsymbol{\sigma}$ as

$$f(\boldsymbol{\sigma}, H) = \left\|\boldsymbol{\sigma}^*\right\| - c - (\mu + \beta)p = 0 \qquad (20)$$

which is the yield surface of the *Drucker-Prager model* (Drucker & Prager, 1952). It holds for Eq. (20) that

$$-\mathrm{tr}\frac{\partial f}{\partial p} \Big/ \frac{\partial f}{\partial\left\|\boldsymbol{\sigma}^*\right\|} = \beta + \mu. \qquad (21)$$

Eqs. (19) and (21) means that the associated flow rule does not hold in general but it holds for μ=0, whilst the case of c=0 has been studied by Houlsby (1981, 1993), Chandler (1985) and Collins & Houlsby (1997) deriving the Coulomb-Mohr failure criterion without a hardening for cohesionless (purely-frictional) materials. The yield surface and the flow rule given by Eqs. (18)-(21) are illustrated in Figure 1. The von Mises yield condition is obtained in the case of $\mu+\beta$=0 for which the associated flow rule does not hold except the case of $\mu=\beta$=0. Thus, the non-associated flow rule is accepted thermomechanically for the non-frictional (purely-cohesive) material, undergoing the plastic volume change. It cannot be insisted from the thermomechanical aspect that the non-associated flow rule is linked with the frictional property.

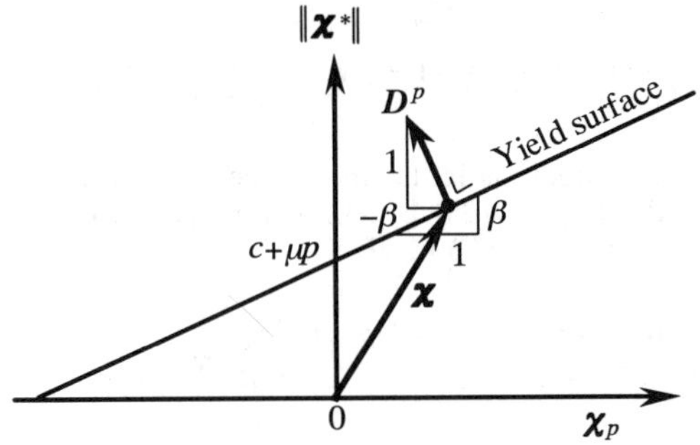

(a) Dissipation stress space

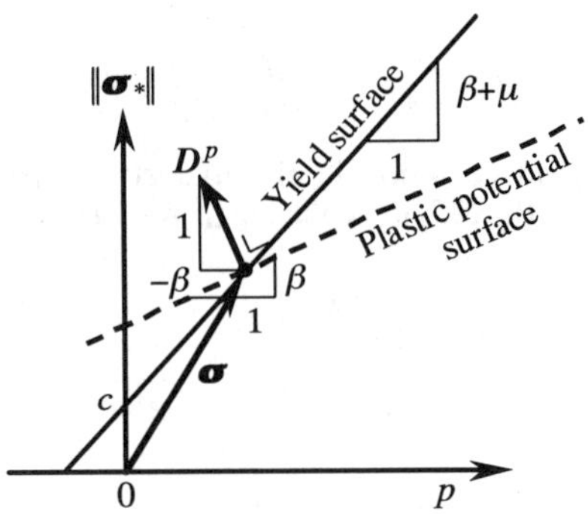

(b) True stress space

Figure 1. The yield surface and the plastic stretching in the Drucker-Prager model.

Further, assume the dissipation function

$$D = mF\exp\left(-\frac{\kappa}{m}\frac{D_v^p}{\left\|\boldsymbol{D}^{p*}\right\|} - 1\right)\left\|\boldsymbol{D}^{p*}\right\| + (\kappa - 1)pD_v^p, \qquad (22)$$

where $F(H)$ is the isotropic hardening function and κ is the material constant. m is the inclination of the critical state line in the $(p, \left\|\boldsymbol{\sigma}^*\right\|)$ space, which depends on the Lode angle in general but let it be regarded to be constant limiting our discussion only to the stress state with a constant Lode angle in this article. The dissipation stress $\boldsymbol{\chi}$ for Eq. (22) is given by Eq. (4) as follows:

$$\boldsymbol{\chi} = -\left\{\kappa F\exp\left(-\frac{\kappa}{m}\frac{D_v^p}{\left\|\boldsymbol{D}^{p*}\right\|} - 1\right) - (\kappa - 1)p\right\}\boldsymbol{I}$$

$$+mF\Big(\frac{\kappa}{m}\frac{D_v^p}{\|\boldsymbol{D}^{p*}\|}+1\Big)\exp\Big(-\frac{\kappa}{m}\frac{D_v^p}{\|\boldsymbol{D}^{p*}\|}-1\Big)\frac{\boldsymbol{D}^{p*}}{\|\boldsymbol{D}^{p*}\|} \tag{23}$$

from which one has

$$\chi_p=\kappa F\exp\Big(-\frac{\kappa}{m}\frac{D_v^p}{\|\boldsymbol{D}^{p*}\|}-1\Big)-(\kappa-1)p\,,$$

$$\|\boldsymbol{\chi}^*\|=mF\left|\frac{\kappa}{m}\frac{D_v^p}{\|\boldsymbol{D}^{p*}\|}+1\right|\exp\Big(-\frac{\kappa}{m}\frac{D_v^p}{\|\boldsymbol{D}^{p*}\|}-1\Big). \tag{24}$$

The yield condition in the dissipation stress space is derived from Eq. (24) by eliminating $\boldsymbol{D}^p$, i. e. the term $(\kappa/m)D_v^p/\|\boldsymbol{D}^{p*}\|+1$ as

$$\bar{f}(\boldsymbol{\chi},\boldsymbol{\sigma},H)=\{\chi_p+(\kappa-1)p\}\exp\left[\frac{\|\boldsymbol{\chi}^*\|}{\frac{m}{\kappa}\{\chi_p+(\kappa-1)p\}}\right]-\kappa F=0\,. \tag{25}$$

It holds form Eq. $(6)_2$ with Eq. (25) or from Eq. (24) that

$$D_v^p/\|\boldsymbol{D}^{p*}\|=\frac{\|\boldsymbol{\chi}^*\|}{\chi_p+(\kappa-1)p}-\frac{m}{\kappa}. \tag{26}$$

The yield condition in the true stress space is given from Eq. (25) by replacing $\boldsymbol{\chi}$ with $\boldsymbol{\sigma}$ as

$$f(\boldsymbol{\sigma},H)=p\exp\left(\frac{\|\boldsymbol{\sigma}^*\|}{mp}\right)-F=0 \tag{27}$$

which is the yield surface of the *original Cam-clay model*. It holds from Eqs. (26) and (27) that

$$D_v^p/\|\boldsymbol{D}^{p*}\|=\frac{1}{\kappa}\Big(\frac{\|\boldsymbol{\sigma}^*\|}{p}-m\Big),$$

$$-\mathrm{tr}\frac{\partial f}{\partial p}\Big/\frac{\partial f}{\partial\|\boldsymbol{\sigma}^*\|}=\frac{\|\boldsymbol{\sigma}^*\|}{p}-m \tag{28}$$

which means that the associated flow rule does not hold in general but it holds only for the special case of κ=1 which has been studied by Houlsby (1981), Modaressi *et al.* (1994) and Collins & Houlsby (1997). The inequality $\kappa\geq 1$ is required in order that the plastic volumetric change is suppressed from the case of the associated flow rule. The yield surface and the flow rule given by Eqs. (22)-(28) are illustrated in Figure 2.

Furthermore, assuming the dissipation function

$$D=\kappa\frac{F}{2}\Big(\sqrt{\big(\tfrac{m}{\kappa}\big)^2\|\boldsymbol{D}^{p*}\|^2+D_v^{p^2}}-D_v^p\Big)+(\kappa-1)pD_v^p, \tag{29}$$

the yield surfaces in the dissipation and true stress spaces for the *modified Cam-clay model* are derived in the same manner as follows:

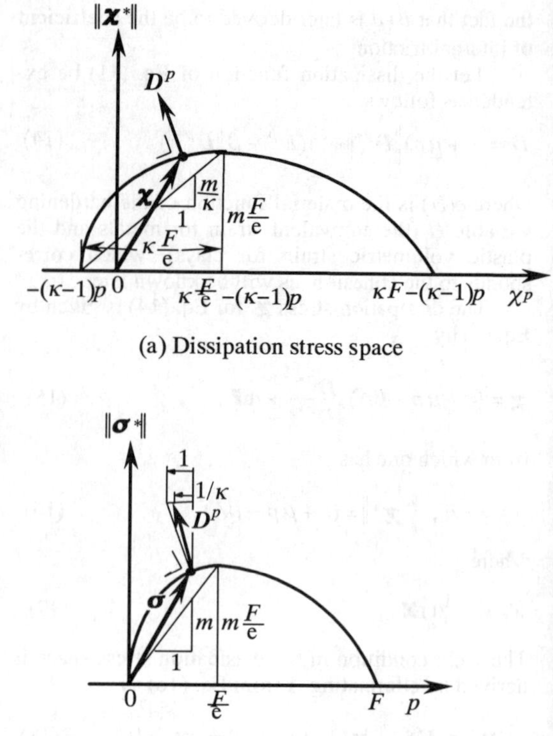

(a) Dissipation stress space

(b) True stress space

Figure 2. The yield surface and the plastic stretching in the original Cam-clay model.

$$\bar{f}(\boldsymbol{\chi},\boldsymbol{\sigma},H)=\left[\frac{\chi_p-\{\kappa\frac{F}{2}-(\kappa-1)p\}}{\kappa\frac{F}{2}}\right]^2+\left(\frac{\|\boldsymbol{\chi}^*\|}{m\frac{F}{2}}\right)^2-1=0\,. \tag{30}$$

$$f(\boldsymbol{\sigma},H)=\left(\frac{p-\frac{F}{2}}{\frac{F}{2}}\right)^2+\left(\frac{\|\boldsymbol{\sigma}^*\|}{m\frac{F}{2}}\right)^2-1=0\,. \tag{31}$$

for which it holds that

$$D_v^p/\|\boldsymbol{D}^{p*}\|=\frac{\frac{F}{2}-p}{\frac{\kappa}{m^2}\|\boldsymbol{\sigma}^*\|},$$

$$-\mathrm{tr}\frac{\partial f}{\partial p}\Big/\frac{\partial f}{\partial\|\boldsymbol{\sigma}^*\|}=\frac{\frac{F}{2}-p}{\frac{1}{m^2}\|\boldsymbol{\sigma}^*\|} \tag{32}$$

which means that the associated flow rule does not hold in general but it holds for κ=1 which has been studied by Houlsby (1981), Modaressi *et al.* (1994) and Collins & Houlsby (1997). The yield surface and the flow rule given by Eqs. (30)-(32) are illustrated in Figure 3.

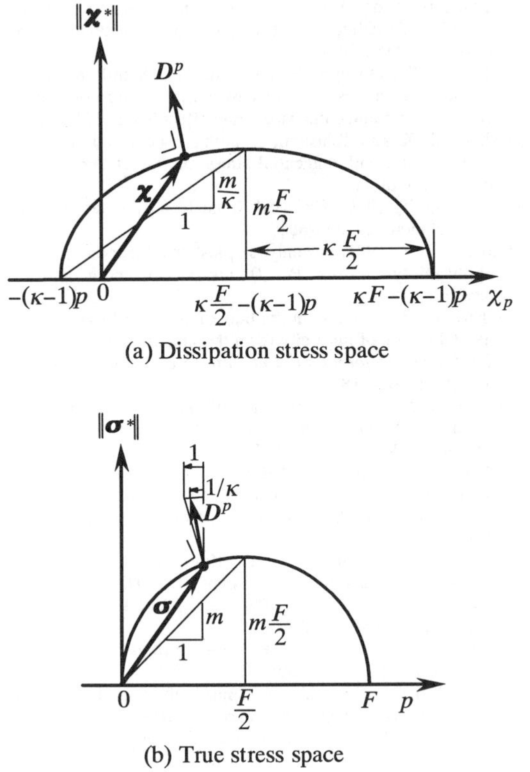

Figure 3. the yield surface and the plastic stretching in the modified Cam-clay model.

4 DEFECTS OF THE CONVENTIONAL PLASTICITY

Drucker (1988) has reviewed comprehensively the development of plasticity theory and classified plastic constitutive equations into the *conventional plasticity* and the *unconventional plasticity*, where the interior of the yield surface is assumed to be the elastic domain in the former and the elastoplastic domain in the latter. While the traditional plasticity represented by the Prandtl-Reuss equation for metals and the Drucker-Prager and the Cam-clay models for geomaterials is to be the conventional plasticity, he evaluated the *subloading surface model* as the typical unconventional model which is extended so as to describe the plastic deformation induced by the rate of stress within the yield surface. The Drucker-Prager model falls within the framework of the conventional plasticity. However, the conventional plasticity has the following defects.

a. The *smoothness condition* (Hashiguchi, 1993b, 1997, 2000) is violated at the moment when the stress reaches the yield surface and thus the unrealistic stress-strain curve with a sharp bending is predicted even in the monotonic-proportional loading process. On the other hand, the subloading surface model fulfilling the smoothness condition describes the smooth *elastic-plastic transition* (Hashiguchi, 1980).
b. The judgement on the fulfillment of the yield condition in addition to the judgement on the direction of stretching or stress rate is required in the *loading criterion*. On the other hand, the former judgement is not required in the subloading surface model falling within the framework of the unconventional plasticity (Hashiguchi, 1993a, 1993b, 1994, 1997, 2000).
c. Needless to say, the *cyclic loading behavior* can never be described by the conventional plasticity which predicts only a repetition of elastic deformation without a strain accumulation in the cyclic loading.

The deformation behavior of over-consolidated soils exhibiting the softening behavior can never be described pertinently by the conventional plasticity and thus the plastic volumetric expansion and the critical load in plastic instability problems are predicted to be unrealistically high by the associated flow rule. However, one should notice the following facts.

1. The plastic volumetric strain is predicted realistically even for over-consolidated soils by the subloading surface model with the associated flow rule (cf. Hashiguchi & Ueno, 1977; Hashiguchi, 1980; Hashiguchi & Chen, 1998; Hashiguchi *et al.*, 2000a).
2. The critical load in plastic instability problems would be predicted realistically by the tangential plasticity (Hashiguchi, 1998; Hashiguchi, *et al.*, 2000b, Hashiguchi & Tsutsumi, 2000), i. e. the subloading surface model extended so as to describe the inelastic stretching induced by the stress rate component tangential to the subloading surface.

Thus, the adoption of the non-associated flow rule would be caused by the deficiency of constitutive equations in the framework of the conventional plasticity.

5 CONCLUDING REMARKS

There would not exist the thermomechanical reason why the non-associated flow rule has to be adopted for frictional materials, as it is verified above that both the associated and the non-associated flow rules can be derived for the yield surface of the von Mises yield surface and the yield surfaces of the Drucker-Prager and the Cam-clay models by the thermomechanical approach. Eventually, it could be concluded as follows:

1. Which is pertinent the associated flow rule or the non-associated one cannot be judged from the aspect of the fulfillment of thermomechanical requirement, and thus both the associated and the non-associated flow rules could be accepted from the thermomechanical aspect.
2. The fact that the non-associated flow rule is adopted for over-consolidated soils would be caused by the deficiency of constitutive equations for over-consolidated soils.

The thermomechanics in the present level would be helpless to formulate plastic constitutive equations in the isothermal process. While the thermodynamics is to be the most general and fundamental principle of the nature, one should use the principle in the pertinent way suitable to it.

ACKNOWLEDGEMENT

The author wishes to thank Prof. Ian F. Collins, University of Auckland, New Zealand for his detailed explanation of thermomechanics for solid materials during his staying form Dec. 1, 1999 to April 1, 2000 in my laboratory.

REFERENCES

Chandler, H. W. 1985. A plasticity theory without Drucker's postulate, suitable for granular materials. *J. Mech. Phys. Solids* 33: 215-226.

Collins, I. F. & Houlsby, G. T. 1997. Application of the thermomechanical principles to the modelling of geotechnical materials. *Proc. Royal Soc. London A* 453, 1975-2001.

Drucker, D. C. 1988. Conventional and unconventional plastic response and representation. *Appl. Mech. Rev.* (ASME) 41: 151-167.

Drucker, D. C. & Prager, W. 1952. Soil mechanics and plastic analysis or limit design. *Quart. Appl. Math.* 10: 157-164.

Hashiguchi, K. & Ueno, M. 1977. Elastoplastic constitutive laws of granular materials. *Constitutive Equations of Soils. Proc. 9th ICFSME, Spec. Session 9*, Tokyo, 73-82. Tokyo, JSSMFE.

Hashiguchi, K. 1980. Constitutive equations of elastoplastic materials with elastic-plastic transition. *J. Appl. Mech.* (ASME) 47: 266-272.

Hashiguchi, K. 1991, Inexpedience of the non-associated flow rule. *Int. J. Numer. Anal. Meth. Geomech.*, 15: 753-756.

Hashiguchi, K. 1993a. Fundamental requirements and formulation of elastoplastic constitutive equations with tangential plasticity. *Int. J. Plasticity* 9: 525-549.

Hashiguchi, K. 1993b. Mechanical requirements and structures of cyclic plasticity models. *Int. J. Plasticity* 9: 721-748.

Hashiguchi, K. 1994. On the loading criterion. *Int. J. Plasticity* 10: 871-878.

Hashiguchi, K. 1997. The extended flow rule in plasticity. *Int. J. Plasticity* 13: 37-58.

Hashiguchi, K. 1998. The tangential plasticity. *Metals and Materials* 4: 652-656.

Hashiguchi, K. 2000. Fundamentals in constitutive equation: continuity and smoothness conditions and loading criterion. *Soils and Foundations* 40(4), in ress.

Hashiguchi, K. & Chen, Z.-P. 1998. Elastoplastic constitutive equations of soils with the subloading surface and the rotational hardening. *Int. J. Numer. Anal. Meth. Geomech.* 22: 197-227.

Hashiguchi, K., Saitoh, K., Okayasu, T. & Tsutsumi, S. 2000a. Comparison of Drucker-Prager's and subloading surface models for prediction of soil deformation behavior. *Proc. Plasticity 2000*: 297-299.

Hashiguchi, K., Tsutsumi, S., Okayasu, T. & Saitoh, K. 2000b. Shear band analysis in soils by elastoplastuc constitutive equation with tangential stress rate effect. idem: 300-302.

Hashiguchi, K. and Tsutsumi, S. 2000. Elastoplastic constitutive equation with tangential stress rate effect. *Int. J. Plasticity* 16, in press.

Houlsby, G. T. 1979. The work input to a granular material. Geotechnique 29: 354-358.

Houlsby, G. T. 1981. A study of plasticity theories and their applicability to soils. *Ph. D. thesis, University of Cambridge*.

Houlsby, G. T. 1982. A derivation of the small-strain incremental theory of plasticity form thermodynamics. *Proc. IUTAM Conf. Deformation and Failure of Granular Materials*, Delft, 109-118.

Houlsby, G. T. 1993. Interpretation of dilatation as a kinematic constraint. In D. Kolymbas (ed.), *Modern Approaches to Plasticity* 19-38. New York: Elsevier.

Modaressi, H., Laloui, L. & Aubry, D. 1994. Thermodynamical approach for Camclay-family models with Roscoe-type dilatancy rules. *Int. J. Numer. Anal. Meth. Geomech.* 18: 133-138.

Roscoe, K. H. & Burland, J. B. 1968. On the generalized stress-strain behaviour of 'wet' clay". In J. Heyman & F. A. Leckie (eds), *Engineering Plasticity*: 535-608. Cambridge: Cambridge University Press.

Schofield, A. N. & Wroth, C. P. 1968. *Critical State Soil Mechanics*, London : McGraw-Hill.

Ziegler, H. 1981. Discussion of some objections to the thermomechanical orthogonality. *Ingenier-Arch*. 50: 149-164.

Ziegler, H. 1983. *An Introduction to Thermomechanics*, 2nd edition, Amsterdam: North-Holland.

Ziegler, H. & Wehrli, C. 1987. The derivation of constitutive relations from the free energy and the dissipation function. *Advances in Appl. Mech.* 25: 183-257.

Computer Methods and Advances in Geomechanics, Desai et al. (eds) © 2001 Balkema, Rotterdam, ISBN 90 5809 183 X

Application of discontinuum analysis to cyclic plastic deformation of coarse granular materials

T. Ishikawa
East Japan Railway Company, Tokyo, Japan

Y. Ohnishi
Kyoto University, Japan

ABSTRACT: This paper shows an analytical study to elucidate cyclic plastic deformation of coarse granular materials. Some simulations of triaxial compression tests were performed regarding its particle as an element of discontinuum analysis. The applicability of discontinuum analysis to cyclic plastic deformation of coarse granular materials and the deformation characteristics such as stress-induced anisotropy and repeated effect were discussed. As the results, it was shown that the fabric transformation of particle alignment accompanied by sliding between grain particles caused cyclic plastic deformation of coarse granular materials. Therefore, discontinuum analysis is effective in analyzing the complex behavior of the materials during cyclic loading.

1 INTRODUCTION

"Track deterioration", a phenomenon frequently observed at ballasted track, is one of the principal assignments in railway engineering because it has serious consequences on the safety of train operations. Track deterioration is the phenomenon of track irregularity at rail level gradually worsening beneath the train loads. Its dominant factor is supposed to be the cyclic plastic deformation of "railroad ballast"; a track component of ballasted track composed of crushed stone (ballast). Therefore, the research on the cyclic plastic deformation characteristics of ballast is significant to elucidate the mechanism of track deterioration.

In general, the mechanics of granular materials is usually used in analyzing the complex behavior of coarse granular materials such as ballast during cyclic loading. From the viewpoint of the mechanics of granular materials, the cyclic plastic deformation of ballast is caused by the movement of ballast particles. Therefore, a method for analyzing the complex behavior of ballast needs to evaluate its granular properties such as the shape of grain particles or the grain size distribution. In this case, discontinuum analysis seems to be effective in simulating the behavior of ballast because it regards a ballast particle as an element of discontinuum analysis.

2 OBJECTIVES OF RESEARCH

The objectives of this paper are:

- to examine the applicability of discontinuum analysis to cyclic plastic deformation of coarse granular materials.
- to examine the cyclic deformation characteristics of coarse granular materials in terms of the mechanics of granular materials.

In this paper, the simulations of monotonic loading and cyclic loading triaxial compression tests are performed with discontinuum analysis where a ballast particle is regarded as a block (Here, the term block is used in the same way as the term element.) in order to discuss the above assignments.

3 OUTLINE OF ANALYSIS

3.1 *Analytical methods*

Discontinuous Deformation Analysis (DDA, proposed by Shi & Goodman, 1985) is used for simulating triaxial tests. DDA is a kind of discontinuum analysis. Discontinuum analysis, which can treat the problem of discontinuous boundaries, has recently been used extensively in rock mechanics and the mechanics of granular materials. Especially in the case of granular materials, the movement of grain particles might be traced by using discontinuum analysis because each particle is treated as an element of discotinuum analysis. DDA is based on the principle of minimum potential energy like FEM (Finite Element Method), but DDA is different from FEM in the respect that DDA is based on discontinuum mechanics like DEM (Distinct Element Method).

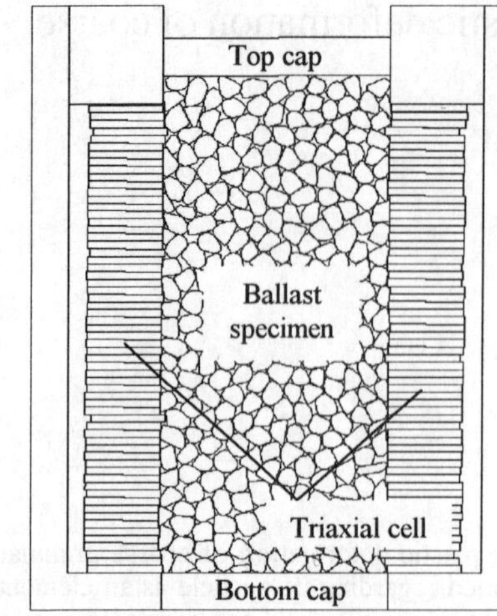

Figure 1. Element mesh of DDA model

Table 1. Features of DDA Model

Property	Analysis	Test
Number of Ballast Blocks	489	—
Uniformity Uc	1.23	1.70
Mean Grain Size D_{50}	42mm	41mm
Porosity n	15.9%	38.2%

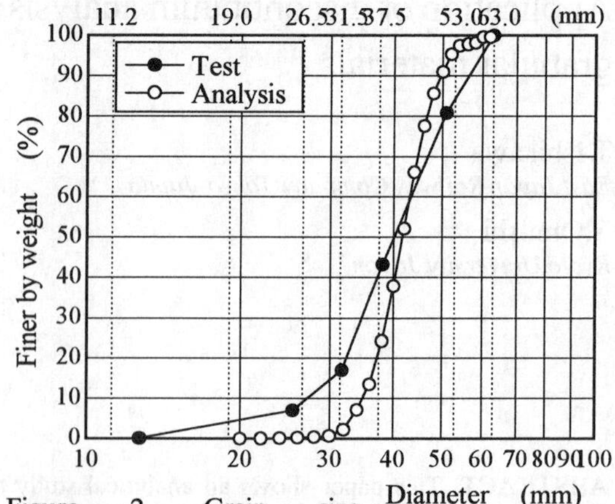

Figure 2. Grain size distribution

Table 2. Material properties of DDA model

Property	Ballast	Cap	Cap
Unit Mass γ	2.77tm^{-3}	1.00tm^{-3}	1.00tm^{-3}
Young's Modulus E	20GPa	1000GPa	0.1GPa
Poisson's Ratio ν	0.10	0.10	0.10
Cohesion Cu	0	0	0
Friction Angle ϕu	55°	0	0

3.2 *Modeling*

The simulations of triaxial tests were performed with a two-dimensional DDA model using linear elastic blocks with plane strain. In DDA, each block is separated by its boundaries and moves individually. Figure 1 shows an element mesh of DDA models before loading. Each polygon block expressing a ballast particle was divided by the method of Voronoi tessellation in view of both the diversity of ballast particles and the angularity of ballast particles. Furthermore, analytical specimens of ballast blocks were made by falling ballast blocks one by one into a frame of triaxial cells to express the random particle alignment in real test specimens. Subsequently stability analysis was done by gravity force. Figure 1 shows the state of DDA model after stability analysis.

The gradation curve for analytical specimens of ballast blocks are shown in Figure 2 and Table 1 shows the feature of DDA models in comparison with test conditions. Seeing Figure 2 and Table 1, the average block size of analytical specimens is nearly as large as that of test specimens. However, the porosity of DDA models is much smaller than that of test specimens as shown in Table 1. This phenomenon seems to be caused by the reason that our DDA models are two-dimensional models though the laminated state of test specimens is three-dimensional. In this simulation, analytical specimens are assumed to be the densest state compacted by vibrator as well as test conditions. Therefore, it is considered that the difference in porosity between analytical specimens and test specimens originates in the difference in dimension between both.

3.3 *Analytical conditions*

Table 2 shows the material properties of blocks and the interface properties of block edges. The interface properties of block edges are characterized by block friction angle (ϕu) and cohesion of surface (Cu) in DDA. Here, the reason why ϕu between ballast blocks was set equal to 55° is that the internal friction angle derived from the analytical results in case of ϕu=55° indicated the best fit to test results in simulating triaxial tests with DDA (Ishikawa et al, 1997). As for ϕu between a ballast block and other kind of blocks, the value was set equal to zero. Besides, as for Cu between all materials, the value was set equal to zero considering that ballast is coarse granular material. On the other hand, the material properties of DDA blocks were characterized by unit mass (γ), Young's modulus (E) and Poisson's ratio (ν). As for the material properties, the parameters were set referring to the original material properties of andesite ballast.

The simulations of triaxial tests were performed

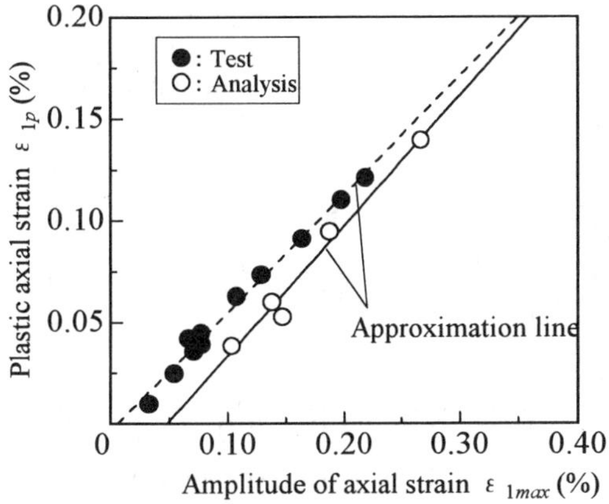

Figure 3. Property of plastic deformation

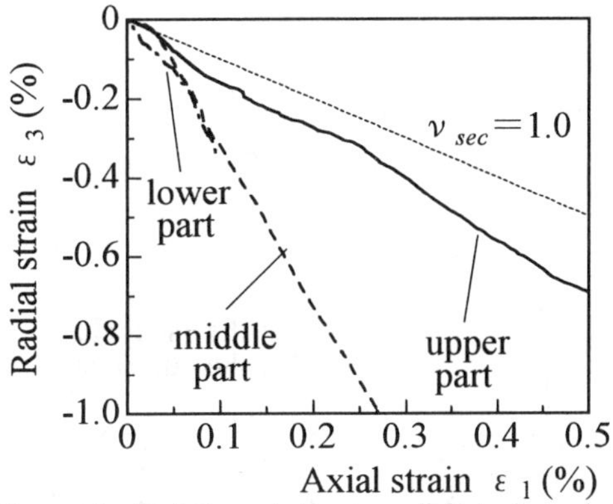

Figure 4a. Relations between radial strain and axial strain in test results

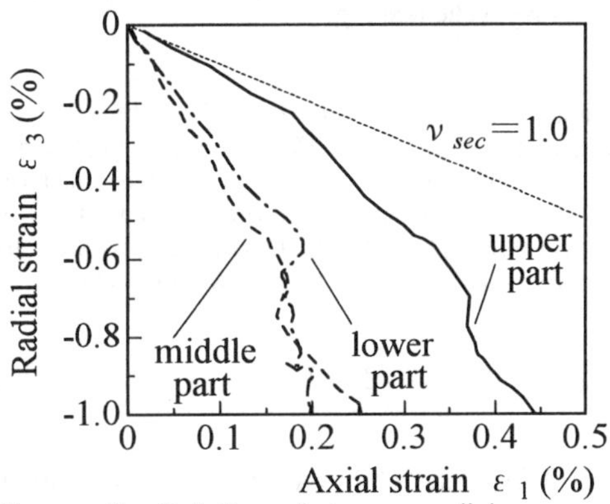

Figure 4b. Relations between radial strain and axial strain in analytical results

as follows; first, analytical specimens after stability analysis were isotropically consolidated under the confining pressure (σ_3) of 19.6, 39.6 and 58.9kPa. Subsequently monotonic loading (ML) tests were performed at a constant axial stress rate of (σ_1 / σ_3 =1)/sec at each σ_3. On the other hand, cyclic loading (CL) tests were also performed after isotropic consolidation. Axial deviatoric loads ($q = \sigma_1 - \sigma_3$) were applied five times to the specimen at the load frequency of 0.1Hz. The cyclic compression load q of triangular wave was cycled in the range of principal stress ratio (σ_1 / σ_3) from 1 to 6. Here, the gravity force was applied to analytical specimens through the simulation.

4 ANALYTICAL RESEALTS

4.1 *Applicability of discontinuum analysis*

First, the mechanism of plastic deformation in discontinuum models is discussed. Figure 3 shows the relations between plastic axial strain (ε_{1p}) and amplitude of axial strain (ε_{1a}) at the 1st loading comparing analytical results with test results. Seeing Figure 3, ε_{1p} of analytical results increases with the increment of ε_{1a} as well as test results. Furthermore, comparing the approximation line derived from analytical results with that of test results, the slope of the analytical line is nearly as large as that of the experimental line. From these results, it is considered that discontinuum analysis is capable of simulating plastic deformation of ballast even without setting any constitutive relationship to represent plasticity. These results indicate that the plastic deformation of coarse granular materials is caused by the movement of grain particles.

Next, the tendencies for ballast particles to move are discussed. Figure 4 shows the relations between axial strain (ε_1) and radial strain (ε_3) at a ML test comparing analytical results and test results. Here, the upper strain was locally measured at a third upper part of specimens, the middle strain was done at a third middle part, and the lower strain was done at a third lower part. Seeing Figure 4, the relation of each part in analytical results differs one from another as well as in test results. Furthermore, each relation of analytical results is similar to that of test results. A relation between ε_1 and ε_3 seems to express the average tendency for ballast particles to move at the part of specimens. In other words, the particle movement varies from part to part inside ballast specimens and as a result the local deformation occurs in ballast specimens. Therefore, it is considered that discontinuum analysis is capable of simulating the particle movement of coarse granular materials under cyclic loading to some extent.

4.2 *Mechanism of cyclic deformation*

First, the stress-induced anisotropy of coarse granular materials is discussed in terms of the mechanics

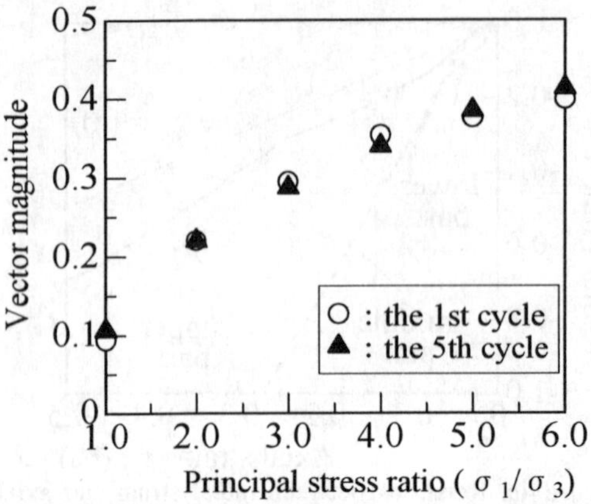

Figure 5. Relations between VM and principal stress ratio during cyclic loading

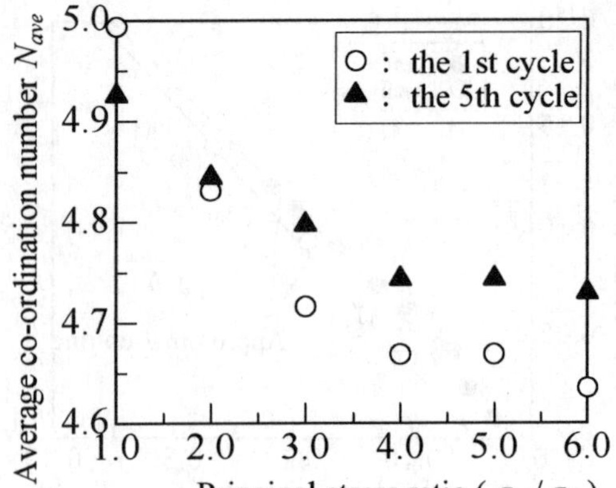

Figure 6. Relations between *Nave* and principal stress ratio during cyclic loading

of granular materials. Figure 5 shows the relations between vector magnitude (VM) and principal stress ratio (σ_1/σ_3) during cyclic loading in analytical results. Here, VM expresses the concentration of contact force between grain particles toward the principal axis as shown in the following equation:

$$VM=\frac{1}{M}\left\{\left(\sum_{k=1}^{M}\cos 2\theta_k\right)^2+\left(\sum_{k=1}^{M}\sin 2\theta_k\right)^2\right\}^{0.5} \qquad (1)$$

where M is a total of contact points between grain particles. Seeing Figure 5, VM gradually increases with the increment of σ_1/σ_3 and the increasing tendency at the 5th cycle is similar to the one at the 1st cycle. According to our triaxial test results using ballast specimens, it is revealed that the stress-induced anisotropy becomes stronger with the increment of σ_1/σ_3. Therefore, these results seem to indicate that the stress-induced anisotropy has a relationship to vector magnitude. This is because the particle alignment of coarse granular materials varies so as to support external force efficiently and as a result the contact force between ballast particles concentrates in the direction of principal axis.

Next, the repeated effect is also discussed. Figure 6 shows the relations between average co-ordination number (N_{ave}) and σ_1/σ_3 during cyclic loading in analytical results. Here, N_{ave} expresses the average total of contact points per particle. Seeing Figure 6, N_{ave} gradually decreases with the increment of σ_1/σ_3. However, the decreasing tendency at the 5th cycle is different from the one at the 1st cycle. Furthermore, according to Field (1963), N_{ave} may be expressed by the following equation:

$$N_{ave} = 12/(1+e) \qquad (2)$$

where e is the void ratio. In short, N_{ave} expresses the void ratio indirectly. Therefore, if the average co-ordination number converges in some constant value during cyclic loading, the deformation characteristics of ballast specimens also become constant because ballast particles do not move much due to the increased density. These results indicate that the repeated effect such as cyclic plasticity has a relationship to average co-ordination number.

5 CONCLUSIONS

The following conclusions can be obtained;

1. Discontinuum analysis is capable of simulating the cyclic plastic deformation of coarse granular materials even without setting any particular constitutive relation except the frictional law.
2. Cyclic plastic deformation of coarse granular materials originates in the particle movement accompanied by the fabric transformation of particle alignment during cyclic loading.
3. Stress-induced anisotropy is concerned with the concentration of contact force between grain particles, and repeated effect is concerned with the variation of average co-ordination number.

REFERENCES

Field, W.G., (1963), Towards the statistical definition of a granular mass, *Proc. 4th A. and N. Z. cont. on Soil Mech.*: 143-148.

Ishikawa, T., Ohnishi, Y. & Namura, A., (1997), DDA applied to deformation analysis of coarse granular materials (ballast), *Proc. of ICADD-2*, Kyoto, 253-262.

Shi, G.H. & Goodman, R.E., (1985), Two dimensional discontinuous analysis, *Int. J. Num. Anal. Methods. Geomech*, 9: 541-556.

Computer Methods and Advances in Geomechanics, Desai et al. (eds) © 2001 Balkema, Rotterdam, ISBN 90 5809 183 X

Simulations of soft clay behaviour with S-CLAY1 and MCC

Minna Karstunen & Simon Wheeler
University of Glasgow, UK

Anu Näätänen
Helsinki University of Technology, Finland

ABSTRACT: The paper presents simulations of normally consolidated clay samples subjected to various stress paths under axisymmetric conditions. Two constitutive models have been used. S-CLAY1 is a recently developed elasto-plastic model, which incorporates a rotational component of hardening to represent development and erasure of soil anisotropy. Corresponding simulations have also been performed with Modified Cam Clay (MCC). Loading paths simulated include: isotropic loading followed by an anisotropic loading; axial and lateral compression; and loading at constant deviatoric stress q. It is shown that the greatest differences are expected for stress paths with constant q, for which S-CLAY1 is superior to MCC.

1 INTRODUCTION

During the last decade considerable developments have been made in modelling the anisotropy of soft soils (e.g. Banerjee & Yousif 1986, Dafalias 1987, Davies & Newson 1993). Many of these models, however, have not been systematically verified against triaxial tests data on natural soft soils and some of them incorporate hardening laws which make sense only within a limited region in stress space. On the other hand, models such as those proposed by Whittle & Kavvadas (1994) and Pestana & Whittle (1999) attempt to represent features such as small strain non-linearity and hysteretic behaviour, as well as anisotropy, in a single modelling framework. Consequently, these models are very complex.

Accurate description of the yielding of soft clays is essential for safe and economic design of structures on soft deposits. This requires use of models that can describe realistically features such as anisotropy. The models need to be relatively simple so that they are easy to understand and it is possible to determine the model input parameters from standard laboratory tests. Only then can the practising geotechnical engineers be confident about using the models in geotechnical design.

The recently proposed model S-CLAY1 (Näätänen et al. 1999, Wheeler et al. in press) was developed by keeping simplicity and reality in mind. The model has two additional model constants over Modified Cam Clay (MCC), and the initial value of one additional parameter, describing the anisotropy, is also required. Two of these values can estimated very simply based on knowledge of the normally consolidated value of K_0 (or friction angle), while for the third one a default value has been suggested (Wheeler et al. 1999). Comparisons of S-CLAY1 simulations with triaxial test data from several Finnish clays (Wheeler et al. 1999, Näätänen & Lojander 2000, Wheeler et al. in press) have shown that S-CLAY1 gives superior predictions over MCC.

The first part of this paper gives a short description of the model. The second part shows some single element simulations from S-CLAY1 and MCC of normally consolidated soft clay samples subjected to various stress paths.

2 MODEL FORMULATION

The main equations describing the model formulation are given below (for more detailed descriptions see Wheeler et al. 1999). S-CLAY1 describes the behaviour of normally consolidated or lightly overconsolidated soft soils. It takes into account the initial anisotropy of the soil as well as the changes of the anisotropy due to plastic straining. Although the model has been generalised for 3D (Wheeler et al. in press) it is presented here in triaxial stress space. Elastic behaviour is assumed to be isotropic, similar to the MCC model, while the yield function is a sheared ellipse:

$$f = (q - \alpha p')^2 - (M^2 - \alpha^2)(p'_m - p')p' = 0 \quad (1)$$

where p' and q are the mean effective stress and deviatoric stress, M is the critical state value of stress ratio η (where $\eta = q/p'$) and the parameters p'_m and α define the size and the inclination of the yield curve (Fig. 1). For the case of isotropy, the yield curve reduces to the same as that of MCC. An associated flow rule is assumed and evidence supporting this was presented by Näätänen & Lojander (2000).

The model incorporates two hardening laws. The first one describes changes in size of the yield curve and it is identical to that of MCC (Eq. 2). The second hardening rule describes the changes of orientation of the yield curve produced by plastic straining (Eq. 3). This form was developed based on a large testing programme with Otaniemi clay (Näätänen et al. 1999).

$$dp'_m = \frac{vp'_m \, d\varepsilon_v^p}{\lambda - \kappa} \quad (2)$$

$$d\alpha = \mu\left[\left(\frac{3\eta}{4} - \alpha\right) < d\varepsilon_v^p > + \beta\left(\frac{\eta}{3} - \alpha\right) | d\varepsilon_d^p |\right] \quad (3)$$

According to Eq. 3 plastic volumetric strains attempt to drag α towards a target value of 3η/4 and plastic shear strains attempt to drag α towards a target value of η/3. μ and β are new soil constants. The constant μ controls the absolute rate at which α heads towards its current target value. The constant β controls the relative effectiveness of plastic shear strains and plastic volumetric strains in yield curve rotation.

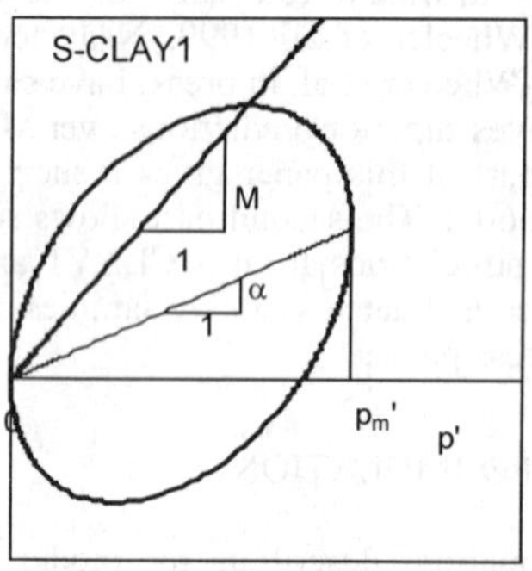

Figure 1. Anisotropic yield curve of S-CLAY1.

3 MODEL SIMULATIONS

Model simulations offer a convenient way to explore how the model behaves with respect to laboratory tests and other constitutive models. Therefore, a single element simulation program for stress-controlled drained loading in axisymmetric conditions was developed (Wheeler et al. 1999). In this paper simulations have been performed to demonstrate typical predictions of S-CLAY1 and to investigate in which stress paths would give the most radical differences between the predictions of S-CLAY1 and MCC.

3.1 *Isotropic loading with anisotropic reloading*

The first simulation is for a naturally anisotropic soil sample subjected first to isotropic loading (η=0.0), and then unloading followed by subsequent reloading with η= 0.67 (Fig. 2). The initial yield curves of S-CLAY1 and MCC have been fitted to coincide at the K_0-axis, as this would be common in design practice. The initial S-CLAY1 yield curve was defined with p_m'=20 kPa and α=0.42. The following material parameters and model constants, representative of Otaniemi clay, have been used: λ=0.44, κ=0.04, M=1.1, ν=0.2, e_0=2.2, β=0.67 and μ=20 (Näätänen et al. 1999). The simulation started with p'=10 kPa at the isotropic axis.

The stress paths and the initial and subsequent yield curve locations predicted by S-CLAY1 are presented in Fig. 2a. After the onset of yielding, the yield curve is expanded and rotated towards an increasingly isotropic orientation (Fig. 2a & b). With the chosen parameters, loading to p'= 60 kPa has not quite rotated the yield curve to the target value (α=0), as shown in Fig. 2b. During re-loading at η=0.67, the yield curve continues to expand and rotates back to an increasingly anisotropic orientation.

As expected, MCC predicts a lower yield stress than S-CLAY1 during the first loading stage, and hence predicts lower values of volumetric strains (Fig. 2c). During the first loading stage the S-CLAY1 yield curve has rotated into an almost isotropic orientation, and therefore the values of yield stress predicted in the second loading stage by the two models are very similar. S-CLAY1 predicts negative shear strains during the first loading stage, whereas according to MCC shear strains are zero (Fig. 2d). After the second yield, the strain directions initially coincide, but then increasingly deviate as the yield curve predicted by S-CLAY1 rotates. Hence, S-CLAY1 predicts much smaller shear strains than MCC.

3.2 *Other loading paths*

The next simulations represent axial compression ($d\sigma_3$=0) and lateral compression ($d\sigma_1$=0), which are examples of idealised stress paths relating to construction and passive earth pressure problems. All simulations have been started at normally consolidated K_0 conditions, where the yield curves of S-CLAY1 and MCC have been fitted to coincide.

For axial compression, the differences between the two model predictions are notable within the region of working loads (Fig. 3a). The MCC model approaches failure more slowly than S-CLAY1. The same conclusions apply to lateral compression (Fig. 3b), where the differences are even more significant. Both simulations were continued to the critical state, although the results presented have been limited to 60 % shear strains.

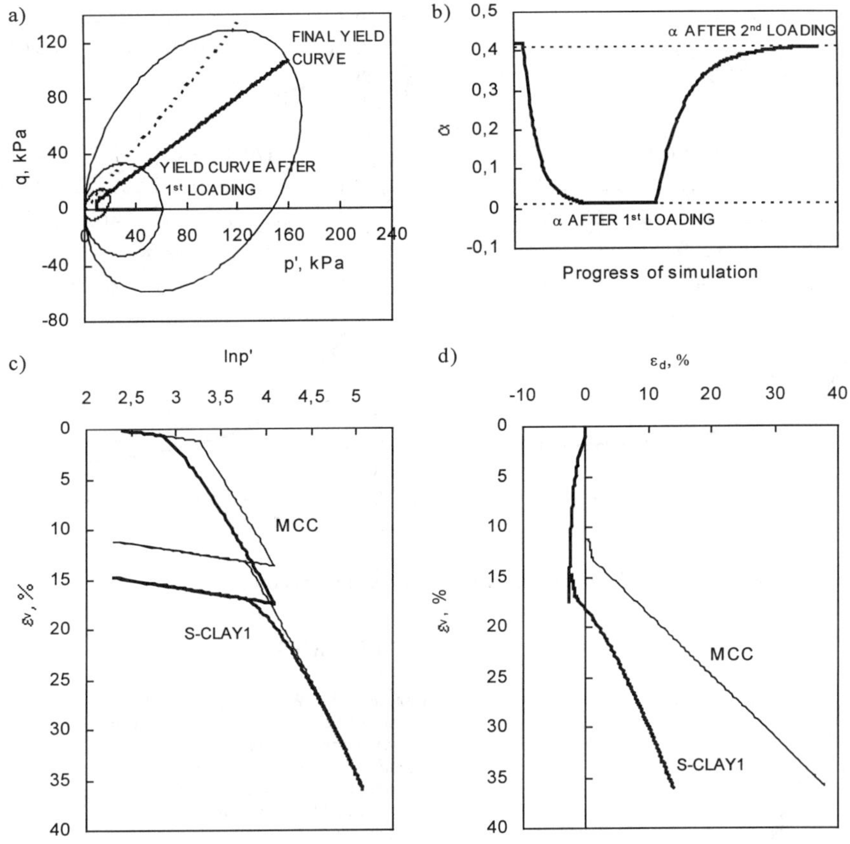

Figure 2. Model simulation. Isotropic loading with anisotropic reloading.

Loading at constant q, with p' increasing, is a stress path where the differences between the two models are very great. A simulation was performed of a constant q loading test on Paimio clay CAD737 (Karstunen 1991).

The initial yield curve was defined with p_m'=48.5 kPa and α=0.45 (Näätänen & Lojander 2000). The material parameters and model constants used were: λ=0.9, κ=0.05, M=1.2, ν=0.1, e_0=2.54, β=0.76 and μ=20. The simulation was started at an initial stress state of p'=41.6 kPa and q=37.5 kPa. The initial stress state is hence slightly above the K_0 line.

The pattern of straining is presented in Figure 4. The agreement between S-CLAY1 and the experimental results is very good, while the MCC model overestimates the shear strains significantly.

4 CONCLUSIONS

Model simulations with S-CLAY1 and MCC have been presented for various stress paths. Previous work (Näätänen & Lojander 2000, Wheeler et al. in press) has shown that natural clays have significantly anisotropic yield behaviour. The predicted initial yield points with MCC and S-CLAY1 will only correspond if the initial yield curves are fitted to coincide for the corresponding stress path. In practice, when different elements of soil are following different stress paths under foundation loading, this will not be possible. If, as usually would be done, the yield curves are fitted at the K_0 -axis, predictions of the first yield will be in error for all other stress paths if MCC is used.

For many stress paths, such as axial and lateral compression (Fig. 3), the differences between the two models are notable. In tests where the sample is sheared to failure, MCC approaches the critical state more slowly than S-CLAY1.

The biggest differences were found in simulations with constant q loading. Comparison with experiments showed that S-CLAY1 is superior to MCC in these kinds of load situations. In practical problems such as embankments and foundations on soft clays the stress paths followed often fall between a constant q loading and axial compression. The

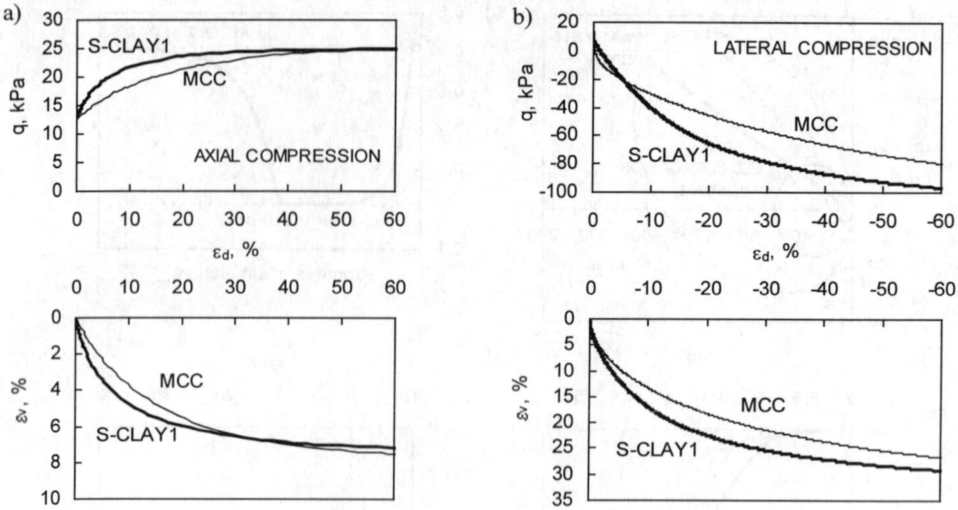

Figure 3. Model simulations. a) Axial compression. b) Lateral compression.

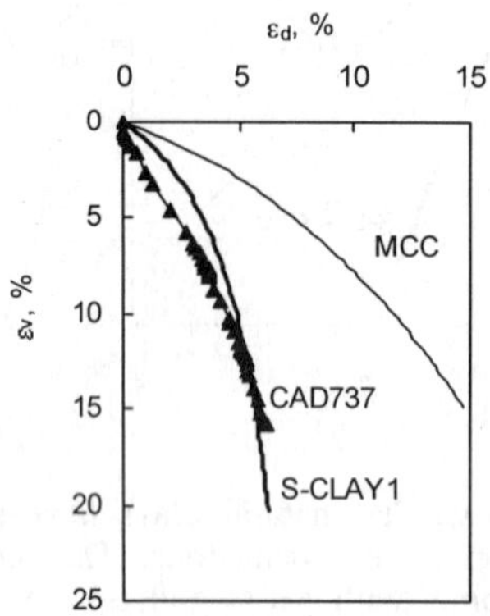

Figure 4. Constant q loading. Model simulation and CAD737.

simulations presented in this paper suggest that S-CLAY1 could potentially give improved predictions for these sorts of loading conditions without much added complexity.

ACKNOWLEDGEMENTS

This research was supported by the Finnish Academy. The collaboration has been carried out as a part of a Research Training Network on "Soft Clay Modelling for Engineering Practice" supported by the European Community through the specific research and technological development programme "Improving the Human Research Potential and the Socio-Economic Knowledge Base".

REFERENCES

Banerjee, P.K. & Yousif, N.B. 1986. A plasticity model for the mechanical behaviour of anisotropically consolidated clay. *Int. J. for Num. and Anal. Meth. in Geomech.* 10, 521-541.

Dafalias, Y.F. 1987. An anisotropic critical state clay plasticity model. *Proc. 2nd Int. Conf. on Constitutive Laws for Engineering Materials*, Tucson, Arizona 1987. Vol. 1: 513-521. Elsevier.

Davies, M.C.R. & Newson, T.A. 1993. A critical state constitutive model for anisotropic soil. Predictive Soils Mechanics, 219-229. London: Thomas Telford.

Karstunen, M. 1991. The visco-plastic model of Paimio clay. MSc thesis. Helsinki University of Technology.

Karstunen, M., Näätänen, A., Lojander, M. 1995. The effect of natural anisotropy of clays. *Proc. 11th ECSMFE*, Copenhagen. Vol. 6: 6.83-6.88. Danish Geotechnical Society, dgf-Bulletin 11.

Näätänen, A. & Lojander, M. (2000). Modelling anisotropy of Finnish clays. *Proc. 7th Finnish Symposium on Mechanics*, Tampere, May 25-26, 2000: 589-598.

Näätänen, A., Wheeler, S., Karstunen, M., Lojander, M. 1999.Experimental investigation of an anisotropic hardening model for soft clays. *Proc. 2nd Int. Symp. on Pre-failure Deformation Characteristics of Geomaterials*, Torino: 541-548. Rotterdam: A.A.Balkema.

Pestana, J.M. & Whittle, A.J. (1999). Formulation of a unified constitutive model for clays and sands. Int. J. Num. Anal. Meth.Geomech. 23:1215-1243.

Wheeler, S. 1997. A rotational hardening elasto-plastic model for clays. *Proc. of the 14th ICSMFE*, Hamburg, Vol. 1: 431-434. Rotterdam: A. A. Balkema

Wheeler, S., Näätänen, A., Karstunen, M. 1999. Anisotropic hardening model for normally consolidated soft clays. *Proc. NUMOG VII*, Graz: 33-40. Rotterdam: A. A. Balkema.

Wheeler, S., Näätänen, A., Karstunen, M., Lojander, M. (in press) Anisotropic elasto-plastic model for natural soft clays. Submitted to Géotechnique.

Whittle, A.J. Kavvadas, M.J., 1994. Formulation of MIT-E3 constitutive model for overconsolidated clays. *J. of Geotech. Eng.* 120 (1): 199-224.

Computer Methods and Advances in Geomechanics, Desai et al. (eds) © 2001 Balkema, Rotterdam, ISBN 90 5809 183 X

Evaluation and modeling of response of clay to loading rate

D.R. Katti, J.Tang & S.Yazdani
Department of Civil Engineering, North Dakota State University, Fargo, N.Dak., USA

ABSTRACT: The accurate prediction of ground response to varying loading rate is important for safe, reliable and economical design of structures In this work, a new approach to predict the stress-deformation response of clays to varying strain rate is described. We present the results of our study wherein we have conducted an in depth quantitative evaluation of observed experimental response of clays to varying loading rate. Our study indicates that the loading rate response of clay could be modeled as an overconsolidated clay response. We have quantitatively estimated the change in overconsolidation ratio (OCR) due to change in loading rate. Our results show that both shear strength and excess pore water pressure response provide quantitatively similar magnitudes of change in OCR with increasing loading rate. This suggests that the relationship between loading rate and change in OCR can be considered as an inherent material characteristic. A relationship between the change in OCR and change in loading rate is described.

1 INTRODUCTION

The undrained stress-deformation behavior of clays is significantly affected by the loading rate. The soil response to loading rate has been experimentally studied by a number of researchers (Richardson and Whitman, 1963; Lefebvre and Lebueuf, 1987; Sheahan et al., 1996; Lefebvre and Pfendler, 1996; David and Campanella, 1997; Benenetto and Tatsuoka, 1997). These experiments confirm that the strength of the soils increase with increase in the rate of loading.

The clay samples used in the above tests were generally normally consolidated. Recently, Sheahan et al., (1996) conducted tests under varying strain rate (0.05, 0.5, 5 and 50 %/hour) and different OCR values (1,2,4 and 8). The development of constitutive models requires a clear understanding of the material response and mechanisms responsible for the response. To develop a clear understanding of the material response a data set was selected from reported experimental work from literature (Sheahan, 1991; Sheahan et al., 1996). The experimental study consisted of a number of undrained triaxial compression tests on resedimented Boston clay. The strain rates used in this study were 0.005%/h, 0.05%/h, 0.5%/h, 5%/h, and 50%/h. Details of the experimental program and the results of the experiments can be found in references cited above. For consistency, we have used the data from tests conducted with the same end conditions, which consisted of lubricated ends.

The undrained response of clay to varying loading rate is influenced by such microscopic characteristics of the clay such as clay fabric, adsorbed water around the particles and hydraulic conductivity. At very slow strain rates, the deformation of the clay structure (redistribution of the particles) and the subsequent migration of the pore water within the clay matrix and the generation of shear induced excess pore water pressure is expected to occur in synchronization with the external loading rate. However, as the strain rate increases, the clay skeleton will become less compliant, and the magnitude of shear generated excess pore water pressure will be smaller.

2. RESPONSE OF CLAYS TO VARYING STRAIN RATE

2.1 *OCR - shear strength relationship under varying strain rates*

In this paper, the shear strength of the clay is defined as the peak deviator stress. We have represented the deviator stress in terms of the square root of the second invariant of the deviatoric stress tensor, and is represented by $\sqrt{J_{2D}}$. In terms of the three principal stresses, σ_1, σ_2, σ_3, $\sqrt{J_{2D}}$ is mathematically expressed as

$$\sqrt{J_{2D}} = \frac{1}{6}\sqrt{(\sigma_1-\sigma_2)^2+(\sigma_2-\sigma_3)^2+(\sigma_1-\sigma_3)^2} \quad (1)$$

We have normalized the strength by dividing the strength with the first invariant of the effective stress J_{1c}' at the end of consolidation phase of the test. J_{1c}' is mathematically defined as,

$$J_{1c}' = \sigma_{1c}' + \sigma_{2c}' + \sigma_{3c}' \quad (2)$$

where σ_{1c}', σ_{2c}' and σ_{3c}' are the effective principal stresses at the end of consolidation phase.

The peak undrained shear strength of a saturated clay increases with increase in overconsolidation ratio. To understand the influence of strain rate on the peak strength of the clay, the peak undrained shear strength at various strain rates have been plotted with respect to the OCR values in Figure 1. The plot shows that the peak strength increases with increasing strain rate and for all magnitudes of OCR's in the data set.

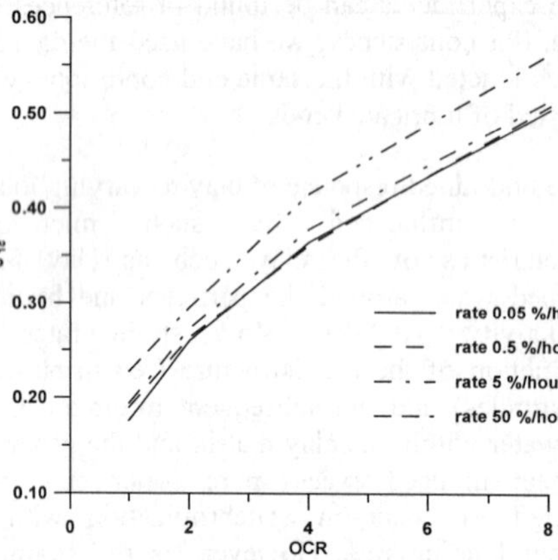

Figure 1 Plot of peak undrained shear stress at various strain rates versus OCR

2.2 *OCR-excess pore water pressure at peak strength under varying strain rates*

In an undrained shear condition, the pore water pressure may increase or decrease depending upon the tendency of the clay skeleton to compress or dilate. The normalized shear induced excess pore water pressure, ΔP during shear is defined as,

$$\Delta P = \left(\Delta U - \frac{\Delta\sigma_1}{3}\right) \Big/ J_{1c}' \quad (3)$$

with

$$\Delta U = U_{peak} - U_c \quad (4)$$

where U_{peak} is the pore water pressure at the peak strength, U_c is the pore water pressure at the end of the consolidation phase, and $\Delta\sigma_1/3$ is the change in octahedral normal stress. The shear induced pore water pressure has been shown to be a reliable and consistent measure of excess pore water pressure for tests with varying strain rate (Sheahan ,1991).

The excess pore water pressure versus OCR at different strain rates is plotted in Figure 2.

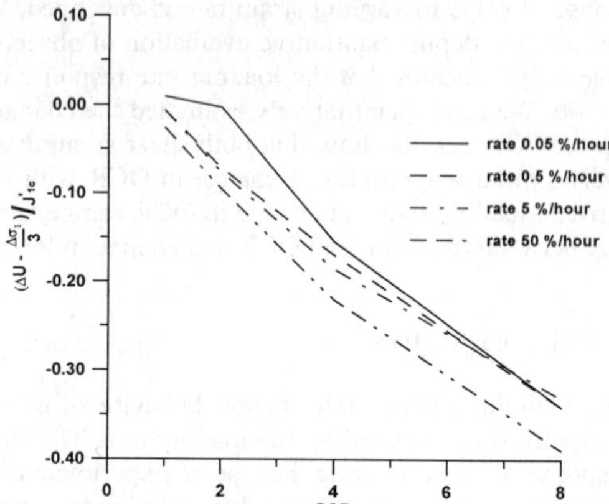

Figure 2. Plot of shear induced excess pore water pressure at peak stress versus OCR at various strain rates

The excess pore water pressure decreases with increasing strain rate. As mentioned earlier, the magnitude of excess pore water pressure response during shear depends upon the deformation characteristics of the clay skeleton. If the clay skeleton has a tendency to compress significantly (more compliant) in an undrained condition, the excess pore water pressure will be high. On the other hand, if the soil skeleton has less tendency to compress (less compliant), the change in excess pore water pressure will be small. For very stiff skeleton, the excess pore water pressure become negative. This observation along with the observation for peak strength with strain rate described in the previous section indicates that the clay skeleton becomes less compliant with increasing strain rate for all OCRs.

2.3 Effective stress path

In our analysis, we have represented the effective normal stress as J_1'/J_{1c}' where J_1' is the first invariant of effective stress tensor, J_{1c}' is the first invariant of effective stress tensor at the end of consolidation. The deviatoric stress axis is represented by $\sqrt{J_{2D}}/J_{1c}'$ In the effective normal stress versus deviatoric stress space, for a normally consolidated clay, since the pore water pressures during the test are positive, the entire stress path is located to the left of the total stress path curve which is a straight line for a monotonic cylindrical triaxial compression test. For highly overconsolidated clay, due to the development of negative pore water pressure during shear, the major portion of the effective stress path lies to the right of the same total stress path. For OCR in between, the shape of the effective stress path lies between the effective path of normally consolidated clay and that of the highly overconsolidated clay. The shape of the effective stress path also indicates whether the clay is normally consolidated or overconsolidated.

To understand the effect of strain rate on the effective stress path of clay samples sheared in a triaxial shear device under consolidated undrained conditions, we have analyzed the effective path of the laboratory tests described previously. The effective stress path for normally consolidated clay (OCR=1) is shown in Figure 3.

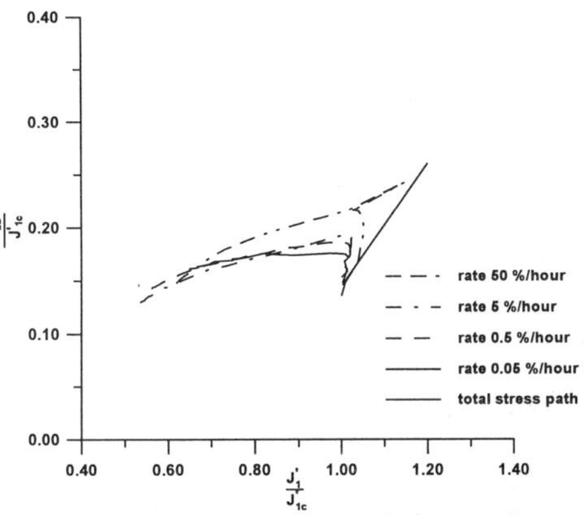

Figure 3. Effective stress path for normally consolidated clay in cylindrical triaxial tests at various strain rates. The solid line is the total stress path.

It can be seen that, with increasing strain rate, the effective stress path moves towards the right indicating generation of lower excess pore water pressure compared to the slow test. These stress paths at higher strain rates are similar to stress path expected for clays with higher OCR values. Similar observations can be seen for higher OCRs as the stress path approaches the peak stress. However, due to possible inaccuracies in the measurement of small changes in excess pore water pressure for OC samples, the early stages of the effective stress paths for OC samples do not show a clear distinction in magnitudes with strain rate. However, as the stress path approaches peak stress, the apparent overconsolidation effect is clearly observed. These effective stress paths indicate to us that the clay at a given OCR will respond similar to clay at a higher OCR with increasing strain rate.

Since similar behavior is observed for the peak shear strength with strain rate, excess pore water pressure at peak strength with strain rate as described in the previous sections, there is a good indication that the strain rate response in clays can be modeled as an apparent overconsolidated clay response.

3. CHANGE IN OCR VERSUS CHANGE IN STRAIN RATE RELATIONSHIPS

In this section, we have quantitatively estimating the change in OCR due to change in strain rate. In addition, we will propose a relationship between the change in OCR and change in strain rate. For this, we are using a strain rate of 0.05 %/hour, the lowest strain rate in the data set, as a reference strain rate. OCR values for high strain rates will be evaluated from reference peak stress versus OCR relationship as well as independently from shear induced excess pore water pressure at peak stress versus OCR.

The change in OCR values obtained independently from the two methods should be similar if the relationship between change in strain rate and resulting change in apparent OCR is an inherent material characteristic.

3.1 OCR-Strain Rate Relationship Based On Reference Peak Shear Strength versus OCR Response

A plot of peak shear strength versus normalized strain rate represented by $\dot{\varepsilon}/\dot{\varepsilon}_0$, where $\dot{\varepsilon}_0$ is the reference strain rate, and $\dot{\varepsilon}$ is the strain rate is shown in Figure 4.

The experimental data for individual OCRs show a logarithmic relationship between shear strength and normalized strain rate. Based on this range of OCR values, this logarithmic relationship appears quite reasonable. However, it is possible that at very high OCRs, the threshold strain rate may be higher than the reference strain rate, which we have assumed in this analysis. At very high OCRs, the

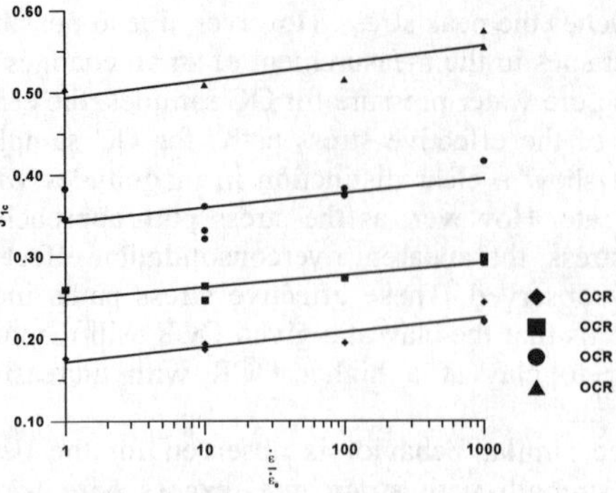

Figure 4. Plot of peak shear stress versus normalized strain rate at various OCR.

Table 1. Material parameters for Boston clay

OCR	A1	A2	B1	B2
1	0.0077	0.1717		
2	0.0063	0.2478		
4	0.0075	0.3424		
8	0.009	0.4935	0.1555	0.1612

threshold strain rate will become the reference rate.

The relationships between the peak shear strength and normalized strain rate is proposed below:

$$\sqrt{J_{2D}} \Big/ J_{1c}' = A_1 \ln(\frac{\dot{\varepsilon}}{\dot{\varepsilon}_0}) + A_2 \tag{6}$$

where, A_1 and A_2 are material parameters, and are functions of OCR. The values of A_1 and A_2 for this particular clay are given in Table I. The parameter A_2 is the normalized peak stress at reference strain rate and the parameter A_1 reflects the rate of change of normalized peak stress with change in strain rate. Similar forms of the equation have been used before to relate undrained shear strength with strain rate, for example Soga and Mitchell (1996).

The relationship between the peak shear stress versus OCR at the reference rate is shown in Figure 5 and can be expressed as,

$$OCR = e^{\left(\frac{\sqrt{J_{2D}}}{J_{1c}'} + B_1\right) \Big/ B_2} \tag{7}$$

where, B_1 and B_2 are material parameters. The parameter B1 is the normalized peak stress for normally consolidated clay at a reference strain rate. Parameter B2 reflects the change in normalized peak stress as a result of change in OCR at the reference strain rate. For Boston clay, the values for B1 and B2 are shown in Table I.

We have used this relationship at the reference strain rate between the peak shear strength and OCR. Further, in order to estimate the apparent increase in OCR as a result of change in strain rate, the following method is used. For a given strain rate, the peak stress is known from the experimental data set. From the OCR versus peak stress relationship at the reference strain rate (Figure 5), the OCR corresponding to the peak stress at the higher strain rate is obtained. Mathematically, the same procedure is accomplished by combining equations 6 and 7, and obtaining the

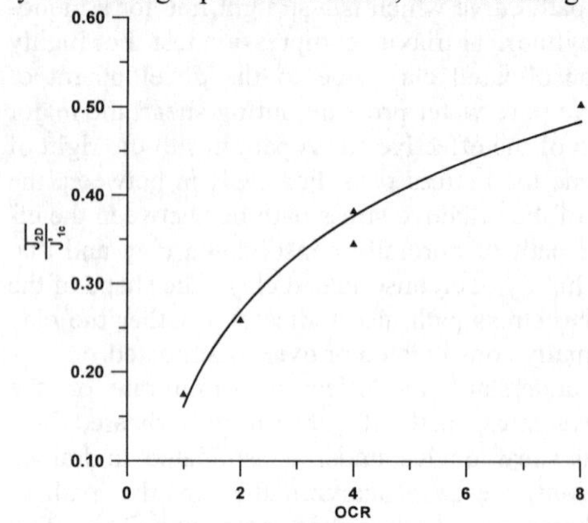

Figure 5. Plot of peak shear stress versus OCR at reference strain rate

following relationship between OCR and strain rate,

$$OCR = e^{\left[A_1 \ln(\frac{\dot{\varepsilon}}{\dot{\varepsilon}_0}) + A_2 + B_1)\right] \Big/ B_2} \tag{8}$$

3.2 OCR-Strain Rate Relationship Based On Reference Excess Pore Water Pressure Versus OCR Response

A plot of excess pore water pressure versus normalized strain rate represented by $\dot{\varepsilon}_0 / \dot{\varepsilon}$ is shown in Figure 6. The relationships between the excess pore water pressure and normalized strain rate is given below:

$$(\Delta U - \frac{\Delta\sigma_1}{3}) \Big/ J_{1c}' = C_1 \ln(\frac{\dot{\varepsilon}}{\dot{\varepsilon}_0}) + C_2 \tag{9}$$

where, C_1 and C_2 are material parameters, and are functions of OCR. The parameter C_2 is the normalized shear induced excess pore water pressure at peak stress at the reference strain rate and C_1 reflects

Table 2. Material parameters for Boston clay

OCR	D1	D2	C1	C1
1	-0.0059	0.0162		
2	-0.0093	-0.0333		
4	-0.0103	-0.1343		
8	-0.0084	0.4935	-0.1669	0.0517

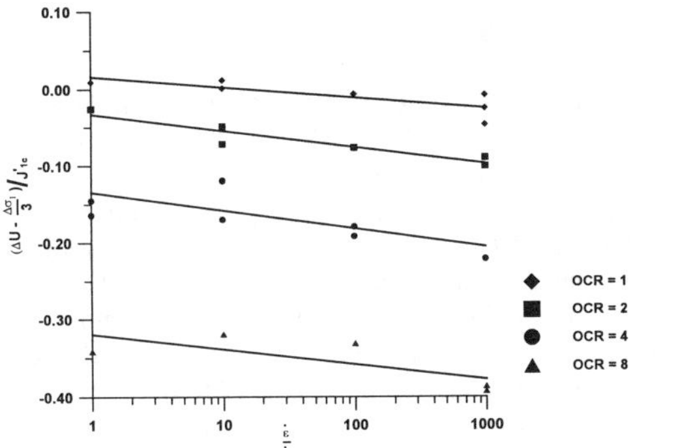

Figure 6 Plot of shear induced excess pore water pressure at peak stress versus normalized strain rate at various OCR.

the change in the shear induced excess pore water pressure as a result of change in strain rate. The values of C_1 and C_2 for the Boston clay are given in Table II.

The relationship between the excess pore water pressure versus OCR at the reference rate is below.

$$OCR = e^{\left(\frac{\Delta U - \frac{\Delta\sigma_1}{3}}{J_{1c}'} + D_1\right) / D_2} \tag{10}$$

where D_1 and D_2 are material parameters. The parameter D_1 is the normalized shear induced excess pore water pressure at the peak stress for the normally consolidated clay sheared at the reference strain rate. The parameter D_2 reflects the change in the shear induced excess pore water pressure as a result of change in OCR at the reference strain rate. For Boston clay, the values of D_1 and D_2 are shown in Table II.

We have used the relationship in equation 10 as the reference relationship between the excess pore water pressure and OCR. To evaluate the apparent increase in OCR as a result of change in strain rate, the following method is used. For a given strain rate, the shear induced excess pore water pressure at peak stress is known from the experimental data. From the OCR versus shear induced excess pore water pressure at peak stress relationship at the reference strain rate (Figure 9), the OCR corresponding to the excess pore water pressure at the higher strain rate is

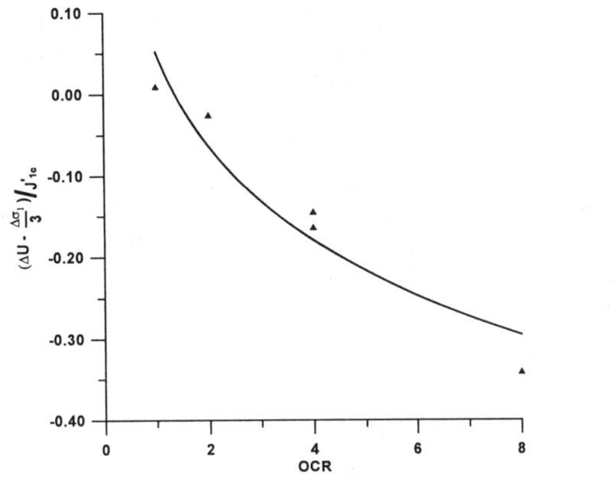

Figure 7. Plot of shear induced excess pore water pressure at peak stress versus OCR at the reference rate.

obtained. Mathematically, the same procedure is accomplished by combining equations 9 and 10, and obtaining the following relationship between OCR and strain rate,

$$OCR = e^{[C_1 \ln(\frac{\dot{\varepsilon}}{\dot{\varepsilon}_0}) + C_2 + D_1)] / D_2} \tag{11}$$

3.3 Comparison of OCR values calculated from OCR-Peak stress and OCR-excess pore water pressure relationships

In the previous sections, the change in OCR as a result of change in strain rate is calculated independently based on two approaches, (1) peak stress, and (2) shear induced excess pore water pressure at the peak stress. The corresponding relationships developed for the two cases are shown in Equations 8 and 11. These relationships are based on experimental results and have a similar mathematical form. We compared the magnitudes of the new apparent OCR values as a result of increase in strain rate from the two approaches. The calculated OCR values are presented in Table III.

Equation 8 represents the OCR values computed from the peak shear strength - OCR relationship and Equation 11 represents the values computed from the excess pore water pressure – OCR relationship. The results in table III show that the results are quit similar in magnitude. Since both the shear strength response and the excess pore water pressure response provided similar magnitudes of OCR, this relationship between rate and OCR could be considered as an inherent material characteristic and is quantifiable.

Table III. Comparison of computed OCR from peak strength - OCR relationship and excess pore water-pressure – OCR relationship

OCR	Strain Rate	OCR calculated using equation 8 (peak stress - OCR relationship)	OCR calculated using equation 11 (excess pore water pressure – OCR relationship)
1	0.05%/hour	1	1
	0.5%/hour	1.20	1.34
	5%/hour	1.34	1.46
	50%/hour	1.50	1.58
2	0.05%/hour	2	2
	0.5%/hour	1.92*	1.89*
	5%/hour	2.10	2.15
	50%/hour	2.30	2.45
4	0.05%/hour	4	4
	0.5%/hour	3.58	3.51
	5%/hour	4.00	4.05
	50%/hour	4.48	4.69
8	0.05%/hour	8	8
	0.5%/hour	9.68	10.38
	5%/hour	11.06	11.66
	50%/hour	12.64	13.09

* inconsistency.

4. CONCLUSIONS

The undrained response of clay with increasing strain rate shows an apparent overconsolidation behavior. As strain rate increases, the apparent overconsolidation ratio increases. We have proposed a relationship between strain rate and apparent overconsolidation ratio using lower strain rate (0.05 % /hour) tests conducted at various OCRs as a reference strain rate response. The magnitudes of the apparent overconsolidation ratio calculated using reference strain rate peak stress versus OCR response or using excess pore water pressure at peak stress versus OCR response are similar. In clays, with increase in overconsolidation ratio, the clay skeleton becomes less compliant due to reduction in the void ratio and consequently the magnitude of the shear generated excess pore water pressure is smaller. In a macroscopic sense, the clay can be overconsolidated just enough to mimic the shear strength response or excess pore water pressure response of the clay to a given strain rate. This novel approach would significantly simplify the development of constitutive models and laboratory testing for evaluation of parameters for clays subject to varying strain rate.

5. ACKNOWLEDGEMENTS

This work was completed with partial support from ND EPSCoR (grant # 8301). Experimental data used in this work is from publications and doctoral dissertation of Dr. Thomas C. Sheahan.

REFERENCES

Benenetto H., Tatsuoka F. 1997. Small strain behav ior of geomaterials: modeling of strain rate effects. Soils and Foundations, 37(2), 127-138.

David . P. and Campanella, R. G. 1997. Time-dependent behavior of undisturbed clay." Journal of Geotechnical Engineering. 103(GT7):693- 709.

Katti D.R., and Desai C.S. 1995. Modeling and Testing of Cohesive Soil Using Disturbed State Concept, J. Eng. Mech.. ASCE 121(5):648-658.

Lefebvre G. and LeBoeuf D. 1987. Rate Effects and Cyclic Loading of Sensitive Clays, J. Geo tech. Engr. ASCE 113(5):476-489.

Lefebvre G. and Pfendler P. 1996. Strain Rate and Preshear Effects in Cyclic Resistance of Soft Clay, J. of Geotech. Eng. ASCE 122(1):21-26.

Richardson M. Jr. and Whitman, R.V. 1963. Ef fect of Strain-rate Upon Undrained Shear Resistance of a Saturated Remoulded fat Clay. Geotechnique 13:310-324.

Sheahan T. C. 1991. An Experimental Study of the Time-Dependent Undrained Shear Behavior of Resedimented Clay Using Automated Stress Path Triaxial Equipment, Ph.D. Dissertation, Massachusetts Institute of Technology.

Sheahan T.C., Ladd C.C., and Germaine J.T. 1996. Rate-Dependent Undrained Shear Behavior of Saturated Clay. J. of Geotech. Eng. ASCE 122(2):99-108.

Soga K. and Mitchell J.K. 1996. Rate Dependent Deformation of Structured Natural Clays. Proc. ASCE Geotech. Special Pub. No. 61, pp. 243-257.

Computer Methods and Advances in Geomechanics, Desai et al. (eds) © 2001 Balkema, Rotterdam, ISBN 90 5809 183 X

An anisotropic elastoplastic constitutive model for natural soils

M. Kavvadas & G. Belokas
National Technical University of Athens, Greece

ABSTRACT: The paper presents the mathematical formulation of an inviscid elasto-plastic constitutive model for natural sedimentary soils, i.e., soils which possess structure due to post-depositional bonding and/or stress-history. Emphasis is given in the description of anisotropy and its evolution due to gradual destructuration caused by recent stressing. Initial (fabric-induced) and subsequent (stress-induced) anisotropy are described using a rotated distorted hyper-ellipsoid (to describe the structure envelope / yield surface) and an appropriate kinematic hardening rule. The destructuration process is controlled by the isotropic hardening rule and depends on both the volumetric and deviatoric plastic strains. At the fully destructured state, the isotropic hardening rule reduces to a Cam-Clay type hardening, with an improvement in the definition of the Intrinsic Compression Curves under anisotropic compression.

1 INTRODUCTION

Structure has been recognized as an agent differentiating the mechanical behavior of natural (i.e., in-situ) and recently reconstituted soils since the late-seventies (e.g. Leroueil et al 1979). However, general-purpose frameworks for the qualitative description of soil structure have only been developed in the last decade (e.g. Vaughan 1988, Leroueil & Vaughan 1990, Burland 1990), mainly as a result of the accumulated knowledge from experimental work on undisturbed and reconstituted soils. Since then, significant contributions have been made in the development of quantitative behavioral frameworks leading in the introduction of a State Boundary Surface (SBS) for structured soils in the context of mathematical plasticity theory (e.g. Smith et al 1992, Jardine 1992, Leroueil 1997, Kavvadas 1998). Following these developments, constitutive models incorporating the structure-SBS as a yield surface have been proposed in an attempt to model features of structured soil behavior not possessed by reconstituted soils (such as the their existence in meta-stable states not attainable by reconstituted soils). Most of these models are based on the Critical State concept (e.g. Muir-Wood 1995, Lagioia & Nova 1993, Kavvadas and Amorosi 2000).

Models for natural soils should be able to describe the following:

- Increased stiffness (see Leroueil & Vaughan 1990) and strength (e.g. Smith et al 1992, Allman & Atkinson 1992; Fig. 1) compared to those of the reconstituted (structureless) soils.
- Gradual destructuring with the accumulation of plastic strains, resulting in a reduction of both stiffness (Fig. 2) and normalized strength (Fig. 1), or equivalently in reduction of the "bond strength" (Vaughan et al 1988).
- Stress-induced anisotropy upon loading and structure-induced anisotropy due to preferable orientation of soil particles, which result in different compressive and extensive strength (see Leroueil 1997).

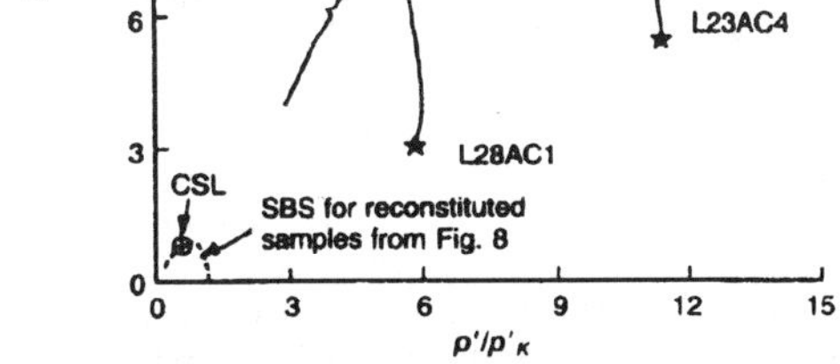

Figure 1. Intrinsic SBS and natural soils' undrained stress path (from Allman & Atkinson 1992).

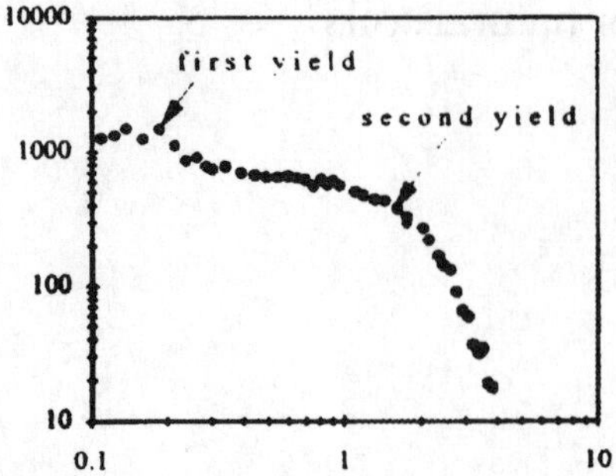

Figure 2. Tangent stiffness vs strain of an artificially bonded soil (from Malandraki & Toll 1996).

The proposed model manages not only to encapsulate the above key features of natural soil behavior, but it also uncouples the peak shear strength from the critical state strength. Moreover, the model includes an empirical relationship to uniquely define the state of a structureless normally consolidated soil (following the state parameter concept). The paper focuses on these concepts, which actually differentiate the proposed model from other constitutive models. Finally, the model is formulated in general stress and strain tensors which describe the isotropic (σ and ε) and deviatoric hyper-plane (**s** and **ε**) components (where: $\boldsymbol{\sigma}=\mathbf{s}+\sigma\mathbf{I}$ and $\boldsymbol{\varepsilon}=\mathbf{e}+\varepsilon\mathbf{I}$).

2 LIMIT STATE ENVELOPES

To account for structure and its effects it is necessary to reconsider the Limit State Envelopes. The model uses four separate envelopes. The first is the State Boundary Envelope (SBE, Fig. 3) defining the region of the permissible states of a structureless soil (as defined by Gens & Potts 1982). This surface can be determined from drained tests under constant stress ratios s/σ (e.g. isotropic and anisotropic compression tests in triaxial apparatus). The second is the Intrinsic Structure Strength Envelope (ISSE, Fig. 3) which is a bounding surface and accounts for the anisotropy of the structureless material. In conventional Critical State Soil Mechanics, the ISSE and the SBE envelopes coincide since this theory was developed for isotropic materials. The introduction of distinct ISSE and SBE envelopes permits undrained stress paths which do not lie on the SBE (e.g. Lewin & Burland 1970, Gens and Potts 1982, and Graham et al 1988) and is in agreement with the observation that undrained stress paths of isotropically and anisotropically consolidated soils differ (Fig. 4). The third is the classical "yield surface" (called Plastic Yield Envelope - PYE, Fig. 3) where appreciable plastic volumetric and/or shear strains start to develop.

States outside the SBE (called "metastable states") are permissible only in structured soils. To account for this, the Structure Strength Envelope (SSE), a surface larger than the ISSE, is introduced (Fig. 3) and models the additional strength and stiffness due to soil structure. Upon loading, plastic strains can develop before the SSE is reached, especially when the material state reaches the metastable domain. In structured soils, appreciable plastic strains are associated with the process of destructuring and a state can exist (at large strains) where all structure is lost. This means that in a normalized q/p'_e-p'/p'_e plot, the SSE shrinks, finally coinciding with the ISSE.

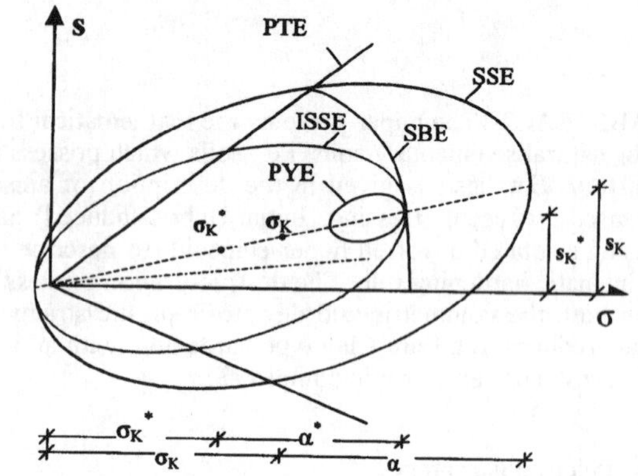

Figure 3. The proposed scheme of state envelopes

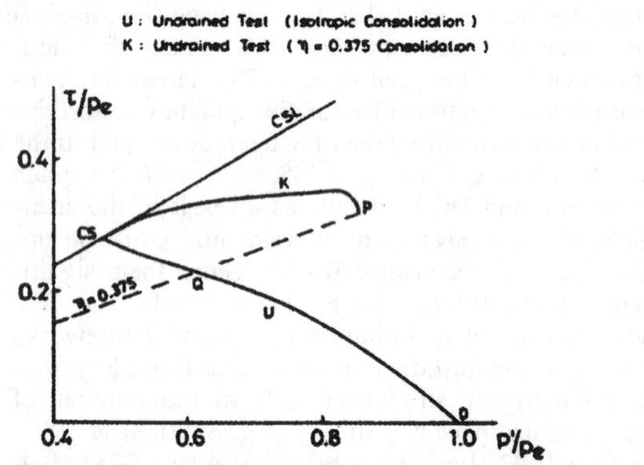

Figure 4. Undrained stress paths for isotropically (U) and anisotropically (K) consolidated samples (from Gens & Potts 1982)

3 INTRINSIC COMPRESSION CURVES OF STRUCTURELESS SOILS

According to Rendulic's postulate (1937) there is a unique SBE, which is a function of the current stress and the specific volume. In classical plasticity, the SBE is assumed to be the yield surface where plastic strains start to develop. On the contrary, in structured soils this is not always true, and the SBE represents only the possible stress states of normally loaded structureless soils. For such states, it is assumed that for a radial stress path (i.e. constant stress ratio s/σ) there is a unique Intrinsic Compression Curve (ICC) on the v-p′ plane (Fig. 5). Consequently, there exist infinite Intrinsic Compression Curves, the position of them being a function of a parameter N_n (the value of specific volume for p′=1kPa) according to the equation:

$$v = N_n - \lambda^* \ln p \tag{1}$$

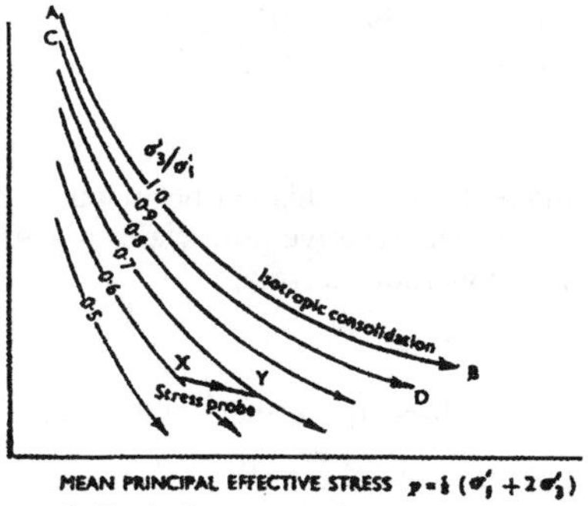

Figure 5. Intrinsic compression curves with respent to stress ratio σ_3'/σ_1' (from Lewin & Burland 1970).

Literature review (Fig. 6) shows that the parameter N_n is given by a relationship of the form:

$$N_n = N_{iso} - (N_{iso} - \Gamma)\cdot\left[\frac{(q/p')}{M}\right]^n \tag{2a}$$

where $N_n=N_{iso}$ for the Intrinsic Isotropic Compression Curve (IICC), $N_n=\Gamma$ for the Critical State Curve and M is the ratio of q_f/p_f' at critical state. This relationship is equivalent to the following one which makes use of the normalised state parameter (takes values from 0 to 1):

$$\frac{N_n - \Gamma}{N_{iso} - \Gamma} = 1 - \left[\frac{(q/p')}{M}\right]^n \tag{2b}$$

or, more generally (subscript f denotes critical state failure):

$$N_n = N_{iso} - (N_{iso} - \Gamma)\cdot\left[\frac{(\mathbf{s}:\mathbf{s}/(\sigma')^2)}{(\mathbf{s}:\mathbf{s}/(\sigma')^2)_f}\right]^n \tag{3}$$

A fairly good estimate of the exponent (n), irrespectively of soil type, is 2 to 3.

The above relationship holds in cases where the ICC is uniquely defined by the current specific volume and the ratio of deviatoric over mean effective stress (i.e., clayey soils). The formulation needs to be expanded for soils for which N_n depends also on the initial density before yield (e.g. sandy), while more research is required for clayey soils too. It should be emphasized that equation 3 has been produced without any assumption with respect to the shape of ISSE and SBS but using only the concept of state parameter; thus, it is generic and can apply to any constitutive model.

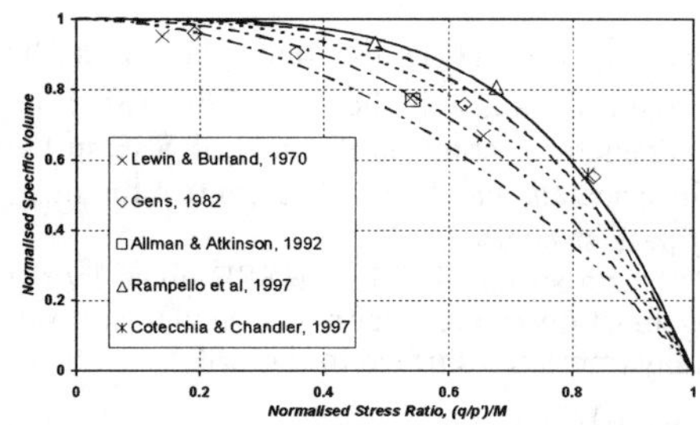

Figure 6. Normalized state parameter versus normalized stress ratio from anisotropic compression tests on reconstituted soils.

4 THE PROPOSED MODEL

The presented model is a two-surface elastoplastic constitutive model which incorporates the Critical State concepts and the behavioural framework for Structured Soils in an incremental plasticity formulation.

4.1 *Characteristic Surfaces*

The model adopts the twin yield surface scheme (Fig. 3), as proposed by Vaughan et al (1988) and Malandraki & Toll (1994): the Structure Strength Envelope (SSE) and the Plastic Yield Envelope (PYE) and one reference surface, the Intrinsic Structure Strength Envelope (ISSE). The SBE is not defined independently if equations 1 and 3 are used (as in paragraph 4.3.1). The SSE represents the margin beyond which significant breakdown (i.e. destructuration) of interparticle bonds appears, while the PYE is an inner surface which encloses the region of purely elastic behavior (i.e. serves the purpose of the classical "yield surface"). Stress states exist either inside (elastic response) or on (elasto-plastic response) the PYE.

The SSE is a rotated and distorted hyper-ellipsoid, in order to describe stress- and structure- induced anisotropy. Due to the rotation of the SSE, the model can predict different compressive and extensive peak strengths. Initial anisotropy is described by selecting the initial inclination (with respect to isotropic axis σ) of the SSE, while following stress-induced anisotropy is controlled by the kinematic hardening rule described below. The size of the SSE surface depends on the magnitude of soil structure and is described by the isotropic hardening rule. The SSE is described by the following function (Kavvadas 1998) where the symbol ":" indicates a summation of the products:

$$F(\boldsymbol{\sigma};\boldsymbol{\sigma}_K,\alpha)\equiv \frac{1}{c^2}(\mathbf{s}-\frac{\sigma}{\sigma_K}\mathbf{s}_K):(\mathbf{s}-\frac{\sigma}{\sigma_K}\mathbf{s}_K)+(\sigma-\sigma_K)^2-\alpha^2=0 \quad (4)$$

where $\boldsymbol{\sigma}_K=\mathbf{s}_K+\sigma_K\mathbf{I}$ ($\mathbf{I}$ is the unit tensor) is the center of the SSE, and α is the size of the SSE (Fig. 3). The parameter c controls the size of SSE in the deviatoric space and can be used to control the magnitude of peak strength.

The PYE is a surface inside the SSE given by the following function (which defines a similar and homothetically translated surface to the SSE):

$$f(\boldsymbol{\sigma};\boldsymbol{\sigma}_K,\boldsymbol{\sigma}_L,\alpha)\equiv \frac{1}{c^2}(\mathbf{s}-\frac{\sigma-\sigma_L}{\sigma_K}\mathbf{s}_K-\mathbf{s}_L):(\mathbf{s}-\frac{\sigma-\sigma_L}{\sigma_K}\mathbf{s}_K-\mathbf{s}_L) \quad (5)$$
$$+(\sigma-\sigma_L)^2-(\xi\alpha)^2=0$$

where $\boldsymbol{\sigma}_L=\mathbf{s}_L+\sigma_L\boldsymbol{I}$ the center of the PYE and ξ the similarity ratio between PYE and SSE.

Finally, the reference surface ISSE is described by the function (the asterisk denotes intrinsic values or properties of the structureless soil):

$$F^*\left(\boldsymbol{\sigma}^*;\boldsymbol{\sigma}^*{}_K,\alpha^*\right)\equiv\frac{1}{c^2}(\mathbf{s}^*-\frac{\sigma^*}{\sigma^*{}_K}\mathbf{s}^*{}_K):(\mathbf{s}^*-\frac{\sigma^*}{\sigma^*{}_K}\mathbf{s}^*{}_K) \quad (6)$$
$$+(\sigma^*-\sigma^*{}_K)^2-\alpha^{*2}=0$$

4.2 *Flow rule*

An incrementally linear non-associated flow rule is used for the plastic strains:

$$\dot{\boldsymbol{\varepsilon}}^p=\dot{\Lambda}_f\cdot\mathbf{P}_f \quad (7a)$$

where

$$\dot{\Lambda}_f=\frac{1}{H_f}\cdot(\mathbf{Q}_f:\dot{\boldsymbol{\sigma}})=\frac{1}{H_f}\cdot(Q_f\cdot\dot{\sigma}+\mathbf{Q}'_f:\dot{\mathbf{s}}) \quad (7b)$$

is a scalar measure of the magnitude of the plastic strains, while $\mathbf{P}_f$ is the plastic potential tensor giving also the direction of plastic strains. H_f is the plastic hardening modulus, while $\mathbf{Q}_f$ is the gradient of the yield surface given by ($\mathbf{Q}_f=\mathbf{Q}'_f+Q_f:\mathbf{I}$).

$$Q_f=2(\sigma-\sigma_L)-\frac{2}{c^2}\frac{1}{\sigma_K}\left(\mathbf{s}-\frac{\sigma-\sigma_L}{\sigma_K}\mathbf{s}_K-\mathbf{s}_L\right):\mathbf{s}_K \quad (8a)$$

$$\mathbf{Q}'_f=\frac{2}{c^2}\left(\mathbf{s}-\frac{\sigma-\sigma_L}{\sigma_K}\mathbf{s}_K-\mathbf{s}_L\right) \quad (8b)$$

The deviatoric ($\mathbf{P}_f'$) and isotropic (P_f) components of the plastic potential tensor ($\mathbf{P}_f=\mathbf{P}_f'+P_f:\mathbf{I}$) are:

$$\mathbf{P}'_f=\frac{1}{c^2}\mathbf{s} \text{ and } P_f\equiv-\frac{\overline{\lambda}_1}{\alpha}h(\boldsymbol{\sigma}) \quad (9a)$$

where

$$h(\boldsymbol{\sigma})=\frac{1}{k^2}(\mathbf{s}-\sigma\hat{\xi}):(\mathbf{s}-\sigma\hat{\xi})-\sigma^2=0 \quad (9b)$$

is a rotated and distorted hyper-cone called the Phase Transformation Envelope (PTE). The PTE delimits three regions in the stress space (Fig. 3):

a) Inside the PTE, where contractant behaviour under drained conditions and positive pore water pressures under undrained conditions develop:

$$P_f>0\Rightarrow\dot{\varepsilon}_v>0$$

b) Outside the PTE, where dilatant behaviour under drained conditions and negative pore water pressures under undrained conditions develop:

$$P_f<0\Rightarrow\dot{\varepsilon}_v<0$$

c) On the PTE, where the critical state condition can be achieved : $P_f=0\Rightarrow\dot{\varepsilon}_v=0$

In the expression for $h(\boldsymbol{\sigma})$, $\hat{\xi}$ is a deviatoric tensor representing the "rotation" of the cone-axis giving critical state strength anisotropy, while k represents the offset of any scanline of the cone relatively to cone-axis. With the introduction of the non-associated flow rule, critical state strength (controlled by the PTE) and peak strength (controlled by SSE) have been uncoupled and are independently controlled.

4.3 *Hardening rules*

Hardening rules describe the evolution of the hardening variables which control the size and position of the characteristic surfaces (i.e. α; $\boldsymbol{\sigma}_K$, $\boldsymbol{\sigma}_L$). Following standard plasticity it is assumed that the material does not harden during elastic deformation. The proposed model employs two hardening rules for isotropic and kinematic hardening.

4.3.1 *Isotropic hardening*

The isotropic hardening rule controls the size of the SSE (i.e., it describes the evolution of material structure) according to the equation:

$$\dot{\alpha}=\alpha^*\left(\frac{1+e}{\lambda^*-\kappa^*}\right)\dot{\varepsilon}^p_v-(\alpha-\alpha^*)(A_v\,\dot{\varepsilon}^p_v+A_q\,\dot{\varepsilon}^p_q) \quad (10a)$$

where, e the void ratio (v=1+e), λ^*, κ^* are the intrinsic compressibility parameters, $\varepsilon^p{}_v$ and $\varepsilon^p{}_q$ are the accumulated plastic volumetric and deviatoric strains,

$A_v=\zeta_v\cdot exp(-\eta_v\cdot\varepsilon^p_v)$,

$A_q=\theta_q+\zeta_q\cdot exp(-\eta_q\cdot\varepsilon^p_q)$,

$\dot{\varepsilon}_v{}^p \equiv \dot{\boldsymbol{\varepsilon}}^p\!:\!\mathbf{I}$ is the plastic volumetric strain increment,

$\dot{\varepsilon}_q{}^p \equiv \sqrt{\tfrac{2}{3}\dot{\mathbf{e}}^p\!:\!\dot{\mathbf{e}}^p}$ is the modulus of plastic deviatoric strain increment ($\dot{\mathbf{e}} \equiv \boldsymbol{\varepsilon}^p - \varepsilon^p{}_v\,\mathbf{I}$),

ζ_v, η_v the volumetric structure degradation components and ζ_q, η_q, θ_q are the deviatoric structure degradation parameters.

According to this scheme, destructuring occurs with plastic strains (i.e., for stress states on the PYE). At full destructuration $\alpha=\alpha^*$ and the rule reduces to the hardening rule of the Cam-Clay model. The isotropic hardening rule (Eq. 10a) has two components :

- the first component describes the hardening of the structureless material by controlling the size of the ISSE with a Cam-Clay type volumetric hardening:

$$\alpha^*\left(\frac{1+e}{\lambda^* - \kappa^*}\right)\dot{\varepsilon}_v^p \tag{10b}$$

- the second component describes the destructuration process by controlling the difference in the size of the Intrinsic and Structure Strength Envelopes, using both volumetric and shear hardening:

$$(\alpha - \alpha^*)(\mathrm{A}_v\,\dot{\varepsilon}_v^p + \mathrm{A}_q\,\dot{\varepsilon}_q^p) \tag{10c}$$

The size of ISSE is independently controlled by using the following expression:

$$\alpha^* = \frac{1}{2}\left[\sigma^* + \alpha\cdot(1 - \sigma_K/\alpha)\right] \tag{11}$$

where, $\sigma^* = exp\left(\dfrac{\mathrm{N_n} - v}{\lambda^*}\right)$

$\mathrm{N_n}$ is given by equation 3, in which it is assumed that:

$$\frac{\left(\mathbf{s}\!:\!\mathbf{s}/(\sigma')^2\right)}{\left(\mathbf{s}\!:\!\mathbf{s}/(\sigma')^2\right)_f} = \frac{\left(\mathbf{s}_K\!:\!\mathbf{s}_K/(\sigma'_K)^2\right)}{\left(\mathbf{s}\!:\!\mathbf{s}/(\sigma'_K)^2\right)_f},\ \mathbf{s}_K/\sigma_K = \mathbf{s}^*{}_K/\sigma^*{}_K$$

and thus the ISSE rotates with the SSE.

With an appropriate selection of the parameters of the above hardening law, various types of destructuring can be portrayed, including collapse-type behavior. Moreover, since only volumetric hardening exists in structureless states, λ_1 can be used to predict the value of the $\mathrm{K_o}$ coefficient (or the corresponding $\mathrm{N_n}$) and consequently the position of the ICC's for any stress ratio.

4.3.2 *Kinematic hardening*

The kinematic hardening rule governs the motion of the characteristic surfaces in the stress space (by controlling the motion of their centers K and L) whenever there is plastic stain, thus describing stress induced anisotropy.

4.3.2.1 Kinematic hardening of the SSE

The following equations control the evolution of structure anisotropy:

a) For material states inside the SSE:

$$\dot{\boldsymbol{\sigma}}_K = \frac{\dot{\alpha}}{\alpha}\boldsymbol{\sigma}_K \tag{12}$$

(i.e. the center K of SSE moves along a radial path passing through the origin O and center K). In this case, anisotropy does not change, as *primary anisotropy tensor* $\mathbf{b}_K = \mathbf{s}_K/\sigma_K$ remains constant.

b) For material states on the SSE:

$$\dot{\boldsymbol{\sigma}}_K = \frac{\dot{\alpha}}{\alpha}\left[\boldsymbol{\sigma}_K + \psi\left(\mathbf{s} - \chi\frac{\sigma}{\sigma_K}\mathbf{s}_K\right)\right] \tag{13}$$

(i.e. the center K of SSE moves along a radial path and shifts in the deviatoric stress space according to the $\psi(\mathbf{s} - \chi(\sigma/\sigma_K)\mathbf{s}_K)$ component). In this case, anisotropy evolves and adjusts to a preferred direction (i.e. both *primary anisotropy tensor* $\mathbf{b}_K = \mathbf{s}_K/\sigma_K$ and radial stress path $\mathbf{s}/\sigma$ stabilize) when $\mathbf{s}/\sigma = \chi(\mathbf{s}_K/\sigma_K)$ (i.e. $\mathbf{b}_K = (\mathbf{s}/\chi\sigma)$).

4.3.2.2 Kinematic hardening of the PYE

The center of PYE moves according to the following rules:

a) For material states on the SSE, the two characteristic surfaces are in contact at the point of the current stress state ($\boldsymbol{\sigma}$):

$$(\boldsymbol{\sigma} - \boldsymbol{\sigma}_L)/\xi\alpha = (\boldsymbol{\sigma} - \boldsymbol{\sigma}_K)/\alpha \Rightarrow \boldsymbol{\sigma}_L = (1-\xi)\boldsymbol{\sigma} - \xi\,\boldsymbol{\sigma}_K \tag{14}$$

b) For material states inside SSE, the centre $\boldsymbol{\sigma}_L$ moves in such a way so that current stress state (point M) moves towards conjugate point M'. The geometric similarity of the two surfaces permits the definition of conjugate points (M and M') on PYE and SSE respectively, such that the normal vectors at these points are parallel. The stress at the conjugate point M' is $\boldsymbol{\sigma}(M') = \boldsymbol{\sigma}_K + (\boldsymbol{\sigma} - \boldsymbol{\sigma}_L)/\xi$. The translation rule of center L is :

$$\dot{\boldsymbol{\sigma}}_L = \frac{\dot{\alpha}}{\alpha}\boldsymbol{\sigma}_L + \dot{\mu}\cdot\boldsymbol{\beta} \tag{15}$$

where:

$\boldsymbol{\beta} \equiv MM' = \boldsymbol{\sigma}(M') - \boldsymbol{\sigma}(M) = (\boldsymbol{\sigma} - \boldsymbol{\sigma}_L)/\xi - (\boldsymbol{\sigma} - \boldsymbol{\sigma}_K)$

is the direction vector and $\dot{\mu}$ a factor evaluated from consistency condition (i.e. the requirement that upon plastic strains, stress state is always on PYE) :

$$\dot{\mu} = \frac{\mathbf{Q}_f\!:\!\dot{\boldsymbol{\sigma}} - 2\dfrac{\dot{\alpha}}{\alpha}\left(\dfrac{1}{c^2}\left(\mathbf{s} - \dfrac{\sigma - \sigma_L}{\sigma_K}\mathbf{s}_K - \mathbf{s}_L\right)\!:\!\left(\mathbf{s} - \dfrac{\sigma}{\sigma_K}\mathbf{s}_K\right) + (\sigma - \sigma_L)\sigma\right)}{\mathbf{Q}_f\!:\!\boldsymbol{\beta}} \tag{16}$$

4.4 *Elasticity*

The elastic component describes the behavior inside PYE. Poro-elasticity and hyper-elasticity can be incorporated as outlined by Kavvadas & Amorosi (2000).

4.5 *Plastic hardening modulus H*

For material states on the SSE, the H-modulus is given from the expression:

$$H_o \equiv H = 2 \cdot \xi \cdot R \cdot T \tag{17}$$

where the consistency condition (i.e., the requirement that the material state remains on the SSE) :

$\dot{F} = 0$ and the relationship (manipulating Eq. 10)

$\dot{\alpha} = \alpha \Lambda R$

are being used to determine T and R resulting in:

$$T = (\sigma - \sigma_K)\sigma + \frac{1}{c^2}\left(\mathbf{s} - \frac{\sigma}{\sigma_K}\mathbf{s}_K\right) : \left[\mathbf{s} - \frac{\sigma}{\sigma_K}\mathbf{s}_K + \frac{\sigma}{\sigma_K}\psi\left(\mathbf{s} - \chi\frac{\sigma}{\sigma_K}\mathbf{s}_K\right)\right] \tag{18}$$

$$R = \frac{\alpha^*\left(\frac{1+e}{\lambda^* - \kappa^*}\right)P - (\alpha - \alpha^*)(A_v P + A_q(\operatorname{sign}\dot{\Lambda})\sqrt{\tfrac{2}{3}(\mathbf{P}':\mathbf{P}')})}{\alpha} \tag{19}$$

The interpolation mapping rule proposed by Kavvadas & Amorosi (2000) can be used for the H-modulus calculation for material states on PYE but inside SSE so that there is a smooth transition of H.

5 CONCLUSIONS

The proposed model can describe many aspects of the behavior of structured soils and especially the process of destructuring and the evolution of material anisotropy. The versatility of the model lies on its formulation in the generalized stress-strain space. Its importance lies on the fact that concepts of behavior of structured and structureless soils have been combined to produce a rational framework.

Possible steps to improve the above model formulation, which is now under research, are: the following: a) to change c in equations 4 to 6 from constant to variable so that it depends on the previous compression stress ratio as well, b) to improve the failure criterion by introducing a Matsuoka & Nakai (1982) type of failure criterion, c) to improve the deviatoric part of the flow rule so that it is does not depend only on the current stress but also on the anisotropy of the material, d) include a relationship of the form $\ln v = N_n - \lambda \ln \sigma'$ in view of the advantages highlighted by Hashiguchi (1995).

REFERENCES

Allman M.A. & Atkinson J.H. 1992. Mechanical properties of reconstituted Bothkennar soil. *Géotechnique*. 42(2):289-301.

Burland J.B. 1990. On the compressibility and shear strength of natural clays. *Géotechnique*. 40(3):329-378.

Cotecchia F. & Chandler R.J. 1997. The influence of structure on the pre-failure bahaviour of a natural clay. *Géotechnique*. 47(3):523-544.

Gens A. & Potts D.M. 1982. A theoretical model for describing the behaviour of soils not obeying the Rendulic's principle. *Int. Symp. on Numerical Methods in Geomechanics*. Zyrich: 24-32. Rotterdam: Balkemma.

Graham J., Crooks, J.H.A & Lau, S.L.K. 1988. Yield envelopes: identification and geometric properties. *Géotechnique*. 38(1):125-134.

Hashiguchi K. 1995. On the linear relations of V-ln$p$ and ln$v$-ln$p$ for isotropic consolidation of soils (Short communication). *Int. Journal for Numerical and Analytical Methods in Geomechanics*. 19:367-376.

Jardine R.J. 1992. Some observations on the kinematic nature of soil stiffness. *Soils and Foundations*. 32(2):111-124.

Kavvadas M. 1998. General Report: Modelling the soil behaviour - Selection of soil parameters. *The Geotechnics of Hard Soils - Soft Rocks*. Naples. Rotterdam: Balkema.

Kavvadas M. & Amorosi A. 2000. A Constitutive Model for Structured Soils. *Géotechnique*. 50(3): 263-274.

Lagioia R. & Nova R. 1993. A constitutive model for soft rocks. *Proc. Int. Symp. on the Geotechnical Engineering of Hard Soils – Soft Rocks*. in Athens. Vol.1:625-632. Rotterdam: Balkema.

Leroueil S. 1997. Critical state soil mechanics and the behaviour of real soils. Keynote lecture. *Symp. on Recent Developments in Soil and Pavement Mechanics*. Rio de Janeiro.

Leroueil S. & Vaughan P.R. 1990. The general and congruent effects of structure in natural soils and weak rocks. *Géotechnique*. 40(3):467-488.

Leroueil S., Tavenas F., Brucy F., la Rochelle P. & Roy M. 1979. Behavior of Destructured Natural Clays. *Journal Geotechnical Engrg Div., ASCE*. 105(GT6):759-779.

Lewin P.I. & Burland J.B. 1970. Stress-probe experiments on saturated normally consolidated clay. *Géotechnique*. 20(1):38-56.

Malandraki V. & Toll D.G. 1994. Yielding of a weakly bonded artificial soil. *Proc. Pre-failure Deformation of Geomechanics*:315-320. Rotterdam: Balkema.

Matsuoka H. & Nakai T. 1982. A new failure criterion for soils in three-dimensional stresses. *Proc. IUTAM Conf. on Deformation and Failure of Granular Materials*. Delft:253-263.

Muir Wood D. 1995. Kinematic hardening model for structured soil. *Proc. Int. Symp. on Numerical Models in Geomechanics (NUMOG V)*. Davos:83-88. Rotterdam: Balkema.

Rampello S., Viggiani G. & Amorosi A. 1997. Small strain stiffness of reconstituted clay along constant triaxial effective stress ratio paths. *Géotechnique*. 47(3):475-489.

Rendulic L. 1937. Ein Grundgesetz der Ton-mechanik und sein experimentaller Beweis. *Bauingenieur*. 18:459-467.

Smith P.R., Jardine R.J. & Hight D.W. 1992. The yielding of Bothkennar clay. *Géotechnique*. 42(2):257-274.

Vaughan P.R. 1988. Characterising the mechanical properties of in-situ residual soil. *Proc. 2nd Int. Conf. Geomechanics in Tropical Soils*. Singapore. Vol.2:469-487.

Vaughan P.R, Maccarini M. and Mokhtar S.M. 1988. Indexing the engineering properties of residual soils. *Quart. J. Engng Geology*. 21(1):69-84.

Computer Methods and Advances in Geomechanics, Desai et al. (eds) © 2001 Balkema, Rotterdam, ISBN 90 5809 183 X

A constitutive model for thermo-hydro-mechanical analysis in unsaturated porous media

N. Khalili
School of Civil and Environmental Engineering, The University of New South Wales, Sydney, N.S.W., Australia

B. Loret
Institut de Mécanique de Grenoble, Laboratoire Sols, Solides, Structures, France

ABSTRACT: A rigorous and unified treatment of the theory of non-isothermal flow and deformation in unsaturated porous media is presented. The governing equations based on the equations of equilibrium, the effective stress concept, Darcy's law, Fourier's law and the conservation equations of mass and energy are derived using a systematic macroscopic approach.

1 INTRODUCTION

Constitutive models for heat and mass transport in porous media are often derived based on restrictive assumptions such as spontaneous thermal equilibrium between the phases, diffusion dominated transport processes, and irreversibility of thermal coupling effects. They are invariably aimed at a specific application involving a heat generating canister located in a low permeability formation (e.g. see Thomas & He 1995, Gens et al 1998, Zhou et al 1998).

The main objective in this paper is to present a more complete and rigorous treatment of the theory of heat and mass transport in deformable unsaturated porous media. The governing equations are derived based on a systematic macroscopic approach, satisfying the equations of balance of mass, momentum, energy, and equilibrium. The physical processes considered include: mechanical deformation, thermal expansion, fluid advection, vapour diffusion, heat conduction, liquid flux due to temperature gradient, heat flux due to pressure gradient, phase exchange (vaporisation, condensation), heat advection (by moving liquid and moving gas), heat exchange between the phases, heat of wetting, heat due to phase compression and heat transfer due to phase exchange (latent heat). Thermal non-equilibrium between the phases is assumed throughout.

2 DEFORMATION MODEL

To describe deformation in an unsaturated thermoelastic medium, consider a representative elementary volume of the medium, subject to external (total) stresses, σ_{ij}. The equations of equilibrium for this elemental volume can be written as,

$$\frac{\partial \sigma_{ij}}{\partial x_j} + F_i = 0 \qquad i = 1 \text{ to } N \tag{1}$$

where F_i is the body force per unit volume, x_i, $i = 1$ to N are the orthogonal spatial coordinates, and N is the space dimension. Extending Terzaghi's effective stress to unsaturated soils, and considering the effects of pore gas and pore liquid pressures separately, the effective stress, σ'_{ij}, in an incremental form is expressed as,

$$d\sigma'_{ij} = d\sigma_{ij} + \alpha_l dp_l \delta_{ij} + \alpha_g dp_g \delta_{ij} \tag{2}$$

where p_l is the pore liquid pressure, p_g is the pore gas pressure (positive for compressive pressure and negative for tensile or suction), δ_{ij} is the Kronecker's delta , and α_l and α_g are the incremental effective stress parameters, defined as (Khalili & Khabbaz 1995 and Loret & Khalili 2000a),

$$\alpha_l = \frac{c_m}{c} - \frac{c_s}{c} \qquad \text{and} \qquad \alpha_g = 1 - \frac{c_m}{c} \tag{3}$$

where c_s is the compressibility of solid grains, c is the drained compressibility of the soil structure, and c_m is the compressibility of the soil structure with respect to a change in matric suction $(p_g - p_l)$.

For a thermoelastic medium the stress-temperature-strain relationship is written as,

$$d\sigma'_{ij} = D_{ijkl} d\varepsilon_{kl} - \gamma dT_s \delta_{ij} \qquad (4)$$

where D_{ijkl} is the drained stiffness of the soil skeleton, ε_{ij} is the strain, T_s is the temperature of the solid constituent, and $\gamma = -1/3(\partial\sigma'_{mm}/\partial T_s)_{\varepsilon_{kk}}$. Notice that implicit in Eqn. (4) is the assertion that the thermal expansion of the solid skeleton is controlled only by the temperature of the solid constituent (Khalili & Loret, 2000).

Now, using the relationship between the strain and displacement we have,

$$d\varepsilon_{ij} = \frac{1}{2}\left(\frac{\partial du_i}{\partial x_j} + \frac{\partial du_j}{\partial x_i}\right) \qquad (5)$$

where u_i denotes the displacement components of the solid skeleton. Combining Eqns. (1) to (5) yields,

$$\frac{\partial}{\partial x_j}[D_{ijkl}\frac{\partial du_k}{\partial x_l} - (\alpha_l dp_l + \alpha_g dp_g + \gamma dT_s)\delta_{ij}] + dF_i = 0 \qquad (6)$$

Eqn. (6) is the differential equation governing the deformation in unsaturated thermoelastic media.

3 FLOW MODEL

Non-isothermal flow through unsaturated porous media is modelled through a system of two interacting continua: one representing the liquid flow and the other the gas flow. The interaction between the two flow regions is established through the phase exchange term, describing transfer of water from liquid phase to gas phase in a vapour form, and vice versa, and through the soil water characteristic curve describing the dependency of the volume fraction occupied by each phase (or the degree of saturation) to suction at a given temperature. The coupling between the flow and deformation fields is established through the effective stress parameters, α_l and α_g.

3.1 *Liquid Phase*

Flow of liquid through unsaturated porous media can be described by combining Darcy's law and the conservation equation of mass. Darcy's law, including the coupling thermal effects, can be written as (Khalili & Loret, 2000),

$$v_{li}^r = -\frac{k_{lij}}{\mu_l}\left(\frac{\partial p_l}{\partial x_j} + \rho_l g_j\right) - k_{lTij}\frac{\partial T_l}{\partial x_j} \quad i=1 \text{ to } N \qquad (7)$$

where k_{lij} is the intrinsic permeability of the water phase and k_{lTij} is the coefficient of thermal coupling for liquid flux. μ_l and ρ_l are the dynamic viscosity and the density of liquid respectively, g_i is the ith component of gravitational acceleration, and v_{li}^r is the relative velocity of liquid defined as,

$$v_{li}^r = n_l(v_{li} - v_{si}) \qquad (8)$$

where n_l is the liquid volume fraction, v_{li} is the absolute liquid velocity and v_{si} is the solid skeleton velocity,

$$v_{li} = \frac{\partial u_{li}}{\partial t} \quad \text{and} \quad v_{si} = \frac{\partial u_i}{\partial t} \qquad (9)$$

where u_{li} and u_i are the components of displacements of fluid and solid, respectively.

Satisfying the conservation of fluid mass yields,

$$-\frac{\partial}{\partial x_i}(\rho_l n_l v_{li}) = \frac{\partial}{\partial t}(n_l \rho_l) + \rho_l j_{lg} \qquad (10)$$

where j_{lg} denotes the rate of moisture transfer between the liquid phase and the gas phase (negative for condensation and positive for vaporisation), which can be described using Dalton's equation as,

$$j_{lg} = \eta(p_{sv} - p_v) \qquad (11)$$

in which, $\eta > 0$ is the liquid phase transfer coefficient, and p_{sv} is the saturated vapour pressure which can be obtained using the psychrometric law (Kelvin's equation),

$$p_{sv} = p_{svo} \exp\left[\frac{-(p_g - p_l)M_l}{\rho_l R T_g}\right] \qquad (12)$$

where M_l is the molecular mass of liquid, R is the universal gas constant, p_g is the gas pressure, and p_{svo} is the saturated vapour pressure in contact with the liquid over a planar surface. p_{svo} is only a function of temperature, and can be obtained using the vapour pressure data published in the literature, or calculated based on Clausius-Clapeyron equation,

$$p_{svo} = \varsigma \exp(\frac{-LM_l}{RT_g}) \quad (13)$$

in which L is the latent heat of vaporisation, and $\varsigma = p^*_{svo} \exp(LM_l / RT^*)$, with T^* and p^*_{svo} being the temperature and the saturated vapour pressure in contact with water over a planar surface, respectively, at a reference configuration.

Now, substituting Eqn. (8) into (10),

$$-\frac{\partial}{\partial x_i}(\rho_l v^r_{li}) - \frac{\partial}{\partial x_i}(\rho_l n_l v_{si}) = \frac{\partial}{\partial t}(\rho_l n_l) + \rho_l j_{lg} \quad (14)$$

and introducing the Lagrangian total derivative concept with respect to moving solids, and moving liquid, Eqn. (14) can be rearranged to,

$$-\rho_l \frac{\partial}{\partial x_i}(v^r_{li}) = n_l \frac{d_l \rho_l}{dt} + \rho_l \frac{d_s n_l}{dt} + \rho_l n_l \frac{\partial v_{si}}{\partial x_i} + \rho_l j_{lg} \quad (15)$$

Considering a barotropic liquid $\rho_l = \rho_l(p_l, T_l)$,

$$\frac{d_l \rho_l}{dt} = \rho_l c_l \frac{d_l p_l}{dt} - \rho_l c_{lT} \frac{d_l T_l}{dt} \quad (16)$$

in which $c_l = \frac{1}{\rho_l}\frac{\partial \rho_l}{\partial p_l}\Big|_{T_l}$ and $c_{lT} = -\frac{1}{\rho_l}\frac{\partial \rho_l}{\partial T_l}\Big|_{p_l}$ are the coefficients of compressibility and thermal expansion for liquid, respectively. Using the definition of volumetric liquid content, $n_l = V_l / V$,

$$\frac{d_s n_l}{dt} = \frac{1}{V}\left(\frac{d_s V_l}{dt} - n_l \frac{d_s V}{dt}\right) \quad (17)$$

Eqns. (7), (15), (16) and (17) can be combined to yield,

$$\frac{\partial}{\partial x_i}\left[\frac{k_{lij}}{\mu_l}\left(\frac{\partial p_l}{\partial x_j} + \rho_l g_j\right) + k_{lTij}\frac{\partial T_l}{\partial x_j}\right] = n_l c_l \frac{d_l p_l}{dt} + \frac{1}{V}\frac{d_s V_l}{dt} - n_l c_{lT}\frac{d_l T_l}{dt} + j_{lg} \quad (18)$$

In Eqn. (18), $d_s V_l / V$ represents the change in the pore liquid space over the current volume of the porous medium, which for non-isothermal conditions can be calculated from (Loret and Khalili, 2000b),

$$\frac{d_s V_l}{V} = (c_m - c_s) d_s(\frac{\sigma_{ii}}{3} + p_g) - c'_m d_s(p_g - p_l) - n_l c_s d_s p_g + \xi_{lg\,T} d_s T + n_l c_T d_s T_s \quad (19)$$

in which c'_m represents the change in the volumetric pore liquid content with suction, $\xi_{lg\,T}$ denotes the change in the pore liquid space due to a change in the curvature of the liquid-gas interface, arising from the temperature dependency of the interfacial surface energy. T is the average temperature acting on the liquid-gas interface, approximated as $T = n_s T_s + n_g T_g + n_l T_l$, in which n_s is the volumetric solid content, and n_g is the volumetric gas (dry air and vapour) content. T_s is the temperature of the solid constituent, T_g is the gas temperature, T_l is the liquid temperature.

Using Eqns. (2) to (4), Eqn. (19) can be rearranged to,

$$\frac{d_s V_l}{V} = \alpha_l d_s \varepsilon_{ii} + (c'_m - \alpha_l^2 c)\ d_s p_l - (c'_m - \alpha_l^2 c - \alpha_l c_s + n_l c_s) d_s p_g - (\alpha_l - n_l) c_T d_s T_s + \xi_{lg\,T} d_s T \quad (20)$$

Substituting (20) into (18), noting $\frac{1}{V}\frac{d_s V}{dt} = \frac{\partial v_{si}}{\partial x_i}$, and introducing the approximation $d_s(.)/dt \approx \partial(.)/\partial t$ yield the differential equation governing non-isothermal liquid flow through an unsaturated porous medium,

$$\frac{\partial}{\partial x_i}\left[\frac{k_{lij}}{\mu_l}\left(\frac{\partial p_l}{\partial x_j} + \rho_l g_j\right) + k_{lTij}\frac{\partial T_l}{\partial x_j}\right] = (n_l c_l + c'_m - \alpha_l^2 c)\frac{\partial p_l}{\partial t} - (c'_m - \alpha_l^2 c - \alpha_l c_s + n_l c_s)\frac{\partial p_g}{\partial t} + \alpha_l \frac{\partial^2 u_i}{\partial t \partial x_i} - n_l c_{lT}\frac{\partial T_l}{\partial t} - n_l c_{lT} v_{li}\frac{\partial T_l}{\partial x_i} + n_l c_l v_{li}\frac{\partial p_l}{\partial x_i} - (\alpha_l - n_l) c_T \frac{\partial T_s}{\partial t} + \xi_{lg\,T}\frac{\partial T}{\partial t} + j_{lg} \quad (21)$$

3.2 *Gas Phase*

Treating air and water vapour mixture as an ideal gas, the mass averaged advective velocity of air-vapour mixture with respect to the moving solids, due to both pressure and temperature gradients, is written as,

$$v_{gi}^{r} = -\frac{k_{gij}}{\mu_g}\frac{\partial p_g}{\partial x_j} - k_{gTij}\frac{\partial T_g}{\partial x_j} \tag{22a}$$

where k_{gij} and μ_g are the permeability and the dynamic viscosity of air-vapour mixture, respectively, and k_{gTij} is the coefficient of thermal coupling for gas flux. In (22a), v_{gi}^{r} is defined as,

$$v_{gi}^{r} = n_g(v_{gi} - v_{si}) \tag{22b}$$

with v_{gi}, the mass averaged velocity of the gas,

$$v_{gi} = \frac{1}{\rho_g}(\rho_{da}v_{dai} + \rho_v v_{vi}) \tag{22c}$$

where v_{dai} is the absolute velocity of the dry air, v_{vi} is the absolute velocity of the vapour, and

$$\rho_g = \rho_{da} + \rho_v \tag{22d}$$

ρ_g is the density of the gas, ρ_{da} is the density of the dry air, and ρ_v is the density of the vapour. n_g is volumetric gas content defined as,

$$n_g = n - n_l = \frac{\rho_l}{\rho_l - \rho_v}[(1 - S_r)n - HnS_r] \tag{23}$$

where H is the coefficient of solubility of gas in liquid defined by Henry's law, S_r is the degree of saturation defined as the volume of water (i.e. volume of liquid plus condensed vapour) over the volume of voids $= V_w/V_v$, and n is the porosity$= V_v/V$.

The diffusive velocities of the dry air and the vapour with respect to the mass averaged velocity of gas are defined as,

$$\hat{v}_{dai} = v_{dai} - v_{gi} = \frac{1}{n_g}(v_{dai}^{r} - v_{gi}^{r}) \tag{24a}$$

$$\hat{v}_{vi} = v_{vi} - v_{gi} = \frac{1}{n_g}(v_{vi}^{r} - v_{gi}^{r}) \tag{24b}$$

with,

$$\rho_{da}\hat{v}_{dai} + \rho_v\hat{v}_{vi} = 0 \tag{24c}$$

in which v_{vi}^{r} and v_{dai}^{r} are the relative vapour and dry air velocities with respect to the moving solid.
Satisfying the conservation of mass for the dry air and vapour, we have,

Dry air

$$-\frac{\partial}{\partial x_i}(\rho_{da}n_g v_{dai}) = \frac{\partial}{\partial t}(n_g\rho_{da}) \tag{25}$$

Vapour

$$-\frac{\partial}{\partial x_i}(\rho_v n_g v_{vi}) = \frac{\partial}{\partial t}(n_g\rho_v) - \rho_l j_{lg} \tag{26}$$

The conservation equation of mass for the gas phase is simply obtained by summing the balance Eqns. (25) and (26),

$$-\frac{\partial}{\partial x_i}(\rho_g n_g v_{gi}) = \frac{\partial}{\partial t}(n_g\rho_g) - \rho_l j_{lg} \tag{27}$$

Now, following the same procedure as for the liquid phase, Eqn. (27) can be expanded and re-arranged to,

$$\begin{aligned} -\rho_g\frac{\partial}{\partial x_i}(v_{gi}^{r}) = n_g\frac{d_g\rho_g}{dt} + \\ \rho_g\frac{d_s n_g}{dt} + \rho_g n_g\frac{\partial v_{si}}{\partial x_i} - \rho_l j_{lg} \end{aligned} \tag{28}$$

Considering the dry air-vapour mixture as an ideal gas, we have,

$$\rho_g = \frac{P_g M_g}{RT_g} \tag{29}$$

where M_g is the molecular mass of air-vapour mixture, with $\frac{\rho_g}{M_g} = \frac{\rho_v}{M_l} + \frac{\rho_a}{M_a}$, $P_g = p_g + p_{atm}$ is the absolute gas pressure. The derivative of ρ_g with respect to the moving gas in Eqn. (28) can be written as,

$$\frac{d_g\rho_g}{dt} = \rho_g c_g\frac{d_g p_g}{dt} - \rho_g c_{gT}\frac{d_g T_g}{dt} \tag{30}$$

in which $c_g = 1/P_g$, and $c_{gT} = 1/T_g$. Substituting Eqns. (22a) and (30) into Eqn. (28) gives,

$$\frac{\partial}{\partial x_i}\left(\frac{k_{gij}}{\mu_g}\frac{\partial p_g}{\partial x_j}+k_{gTij}\frac{\partial T_g}{\partial x_j}\right)=$$
$$n_g c_g \frac{\partial p_g}{\partial t}+\frac{1}{V}\frac{d_s V_g}{dt}-n_g c_{gT}\frac{\partial T_g}{\partial t}-$$
$$n_g c_{gT} v_{gi}\frac{\partial T_g}{\partial x_i}+n_g c_g v_{gi}\frac{\partial p_g}{\partial x_i}-\frac{\rho_l}{\rho_g}j_{lg} \tag{31}$$

Introducing the constitutive relationship,

$$\frac{d_s V_g}{V}=\alpha_g \varepsilon_{ii}+[c'_m-\alpha_l^2 c-(\alpha_l-$$
$$\alpha_g+n_g-n_l)c_s]\ d_s p_g-[c'_m-$$
$$\alpha_l^2 c-(\alpha_l-n_l)c_s]d_s p_l-(\alpha_g-$$
$$n_g)c_T d_s T_s-\xi_{lgT}d_s T \tag{32}$$

and assuming $v_{si}\frac{\partial(.)}{\partial x_i}<<\frac{\partial(.)}{\partial t}$, Eqn. (31) can be written as,

$$\frac{\partial}{\partial x_i}\left(\frac{k_{gij}}{\mu_g}\frac{\partial p_g}{\partial x_j}+k_{gTij}\frac{\partial T_g}{\partial x_j}\right)=[n_g c_g+$$
$$c'_m-\alpha_l^2 c-(\alpha_l-\alpha_g+n_g-n_l)c_s]\frac{\partial p_g}{\partial t}-$$
$$[c'_m-\alpha_l^2 c-(\alpha_l-n_l)c_s]\frac{\partial p_l}{\partial t}+\alpha_g\frac{\partial^2 u_i}{\partial t\partial x_i}-$$
$$n_g c_{gT}\frac{\partial T_g}{\partial t}-n_g c_{gT}v_{gi}\frac{\partial T_g}{\partial x_i}+n_g c_g v_{gi}\frac{\partial p_g}{\partial x_i}-$$
$$(\alpha_g-n_g)c_T\frac{\partial T_s}{\partial t}-\xi_{lgT}\frac{\partial T}{\partial t}-\frac{\rho_l}{\rho_g}j_{lg} \tag{33}$$

4 VAPOUR TRANSPORT

The vapour transport model is derived based on two processes: diffusion due to the vapour density gradient, and advection due to the bulk flow of the gas; i.e. the vapour is treated as a constituent within the gas phase. The link between the vapour transport model and the liquid flow model is established through the mass exchange term and the psychrometric law. The coupling with the deformation model is achieved through the gas phase and the effective stress parameter α_g. Following the Fickian concept of mass transport by molecular diffusion, the flux due to the vapour gradient is written as,

$$q_{vi}^d=\rho_v n_g \hat{v}_{vi}=\rho_v(v_{vi}^r-v_{gi}^r)=-n_g D_{vij}\frac{\partial\rho_v}{\partial x_j} \tag{34}$$

in which q_{vi}^d is the diffusive vapour mass flux, and D_{vij} is the molecular diffusivity of vapour in the pore gas. The advective vapour mass flux with respect to the moving solid is,

$$q_{vi}^a=\rho_v v_{gi}^r \tag{35}$$

Rewriting the conservation of vapour mass, we obtain,

$$-\frac{\partial}{\partial x_i}(q_{vi}^d+q_{vi}^a)-\frac{\partial}{\partial x_i}(n_g\rho_v v_{si})=$$
$$\frac{\partial}{\partial t}(n_g\rho_v)-\rho_l j_{lg} \tag{36}$$

Notice that the second term on the left-hand side of Eqn. (36) describes the conservation of vapour mass equation with respect to a fixed reference system rather than the moving solid. In view of using the constitutive equations, Eqn. (36) can be rearranged to,

$$-\rho_v\frac{\partial}{\partial x_i}[\frac{1}{\rho_v}(q_{vi}^d+q_{vi}^a)]=n_g\frac{d_v\rho_v}{dt}+$$
$$\rho_v\frac{d_s n_g}{dt}+\rho_v n_g\frac{\partial v_{si}}{\partial x_i}-\rho_l j_{lg} \tag{37}$$

Approximating, $\frac{\partial V_g}{\partial t}\approx\frac{d_s V_g}{dt}$, considering vapour as a perfect gas $\rho_v=P_v M_l/RT_g$, and using Eqns. (32), (34) and (35), Eqn. (37) can be further expanded in terms of the primary variables,

$$\frac{\partial}{\partial x_i}(\frac{1}{\rho_v}n_g c_v D_{vij}\frac{\partial p_v}{\partial x_j})+\frac{\partial}{\partial x_i}[(k_{gTij}-$$
$$\frac{1}{\rho_v}n_g c_{vT}D_{vij})\frac{\partial T_g}{\partial x_j}]+\frac{\partial}{\partial x_i}(\frac{k_{gij}}{\mu_g}\frac{\partial p_g}{\partial x_j})=$$
$$n_g c_v\frac{\partial p_v}{\partial t}+[c'_m-\alpha_l^2 c-(\alpha_l-\alpha_g+n_g-$$
$$n_l)c_s]\frac{\partial p_g}{\partial t}-[c'_m-\alpha_l^2 c-(\alpha_l-n_l)c_s]\frac{\partial p_l}{\partial t}$$
$$+\alpha_g\frac{\partial^2 u_i}{\partial t\partial x_i}-n_g c_{vT}\frac{\partial T_g}{\partial t}-n_g c_{vT}v_{vi}\frac{\partial T_g}{\partial x_i}+$$
$$n_g c_v v_{vi}\frac{\partial p_v}{\partial x_i}-(\alpha_g-n_g)c_T\frac{\partial T_s}{\partial t}-\xi_{lgT}\frac{\partial T}{\partial t}-\frac{\rho_l}{\rho_v}j_{lg} \tag{38}$$

in which, $P_v = p_v + p_{atm}$ is the absolute vapour pressure, $c_v = 1/P_v$, and $c_{vT} = 1/T_g$.

5 HEAT TRANSPORT

The heat transfer model is based on the assumption that energy is transferred through all three phases and that the phases are not at thermal equilibrium. Vapour and dry air are assumed to be at the same temperature.

The energy balance equation for the solid phase, assuming small changes in temperature, is written as (Bowen & Chen 1975),

$$\frac{\partial}{\partial x_i}\left(n_s\lambda_{sij}\frac{\partial T_s}{\partial x_j}\right) = n_s\rho_s C_s\left(\frac{\partial T_s}{\partial t} + v_{si}\frac{\partial T_s}{\partial x_i}\right) - T_s[(\alpha_l - n_l)c_T - n_s\xi_{lgT}]\frac{\partial p_l}{\partial t} - T_s[(\alpha_g - n_g)c_T + n_s\xi_{lgT}]\frac{\partial p_g}{\partial t} + T_s\gamma\frac{\partial^2 u_i}{\partial t\partial x_i} + \kappa_{sl}(T_s - T_l) + \kappa_{sg}(T_s - T_g) \tag{39}$$

in which λ_{sij} is the thermal conductivity of solid phase, C_s is the heat capacity of the solid, κ_{sl} is the coefficient of heat exchange between the solid and liquid phase, and κ_{sg} is the coefficient of heat exchange between the solid and gas phase. In Eqn. (39), the first term on the right hand side represents the change in the heat capacity due to a change in the temperature, the second term denotes the heat transfer due to convection, the third and the fourth terms represent the change in the energy due to a change in the matric suction (heat of wetting), the fifth term represents the change in the internal energy of the solid phase due to volume change of solid skeleton, and the last two terms represent the exchange of energy between the solid phase and the liquid phase and between the solid phase and the gas phase, respectively. The energy balance equation for the liquid phase is in turn written as (Khalili & Loret 2000),

$$\frac{\partial}{\partial x_i}\left(n_l\lambda_{lij}\frac{\partial T_l}{\partial x_i}\right) = n_l C_l\rho_l\frac{\partial T_l}{\partial t} - n_l(c_{lT} - \xi_{lgT})T_l\frac{\partial p_l}{\partial t} - n_l\xi_{lgT}T_l\frac{\partial p_g}{\partial t} + L\rho_l < j_{lg} > - C_l\rho_l j_{lg}T_l + \kappa_{sg}(T_l - T_s) + \kappa_{lg}(T_l - T_g) + n_l C_l\rho_l v_{li}\frac{\partial T_l}{\partial x_i} \tag{40}$$

in which λ_{lij} and C_l are the thermal conductivity and the heat capacity of the liquid phase, respectively. κ_{lg} is the coefficient of heat exchange between the liquid phase and the gas phase, and L is the latent heat of vaporisation. Notice that the fourth term on the right hand side, representing the energy loss due to phase exchange, which will only be present during vaporisation.

Similarly, the energy balance equation for the gas phase can be written as,

$$\frac{\partial}{\partial x_i}\left(n_g\lambda_{gij}\frac{\partial T_g}{\partial x_i}\right) = n_g C_g\rho_g\frac{\partial T_g}{\partial t} + n_g\xi_{lgT}T_g\frac{\partial p_l}{\partial t} - n_g(c_{gT} + \xi_{lgT})T_g\frac{\partial p_g}{\partial t} - L\rho_l < -j_{lg} > + C_g\rho_l j_{lg}T_g + \kappa_{sg}(T_g - T_s) + \kappa_{lg}(T_g - T_l) + n_g C_g\rho_g v_{gi}\frac{\partial T_g}{\partial x_i} \tag{41}$$

in which λ_{gij} and C_g are the thermal conductivity and the heat capacity of the gas phase, respectively. Notice that the terms on the right hand side of Eqns. (39) to (41) are positive when they correspond to loss of energy for the corresponding phases.

REFERENCES

Bowen, R. M. & Chen, P. J. 1975, "Planar progressive waves in a binary mixture of linear elastic material", *J. Appl. Mech.*, 45, 493-499.

Gens, A., Garcia-Molina, A.J., Olivella, S., Alonso, E.E. & Huertas, F. 1998 , "Analysis of full scale in situ test simulating repository conditions", *Int. J. Numer. Anal. Meth. Geomech.*, 22, 515-548.

Khalili, N., & Khabbaz, M.H., 1995 , "On the theory of three-dimensional consolidation in unsaturated soils", *Proceedings of 1st Int. Conf. on Unsaturated Soils*, Alonso & Delage (Eds.), Paris, France, 745-750.

Khalili, N. & Loret, B. 2000, "Heat and mass flow in deformable unsaturated porous media: theory", Submitted.

Loret, B. & Khalili, N. 2000a , "A three phase model for unsaturated soils". In press. *Int. J. Num. Analy. Methods in Geomechanics.*

Loret, B. & Khalili N. 2000b, "Thermo-Mechanical potentials for unsaturated soils", Course CISM, *Advanced Numerical Applications and Plasticity in Geomechanics*, Udine June 26-30, 2000, to be published by Springer Verlag, eds. D.V. Griffiths and G. Gioda.

Thomas H.R., & He Y., 1995 , "An analysis of coupled heat, moisture and air transfer in a deformable unsaturated soil", *Géotechnique*, 45(4), 677-689.

Zhou, Y. Rajapakse, R.K.N.D. & Graham, J. 1998, "Coupled heat-moisture-air transfer in deformable unsaturated media", *J. Eng. Mech., ASCE*, 124(10), 1090-1099.

Computer Methods and Advances in Geomechanics, Desai et al. (eds) © 2001 Balkema, Rotterdam, ISBN 90 5809 183 X

A conceptual framework for modelling the mechanical behaviour of structured soils

M. D. Liu & J. P. Carter
Department of Civil Engineering, The University of Sydney, N.S.W., Australia

ABSTRACT: A conceptual framework for describing the mechanical behaviour of structured soils has been proposed in a four-dimensional space, consisting of the current stress state, stress history, voids ratio and a measure of soil structure. Within this framework, the stress-strain behaviour of soil is divided into four regions in stress space, *i.e.*, an elastic region, a stable deformation region, a sub-yielding region and a virgin yielding region. The general modelling of soil behaviour in each of the four regions and the influence of soil structure on this behaviour are investigated. The concept of the critical state of deformation is extended to include the behaviour of soil with potentially enough platy particles to form a residual state of deformation after sustained shearing.

1 INTRODUCTION

Numerical analyses have been increasingly applied to the solution of geotechnical problems. However, the success of these numerical applications depends to a large degree on understanding the key mechanisms of the problem, which are usually determined by the mechanical and hydraulic properties of the geo-materials involved (Vaughan, 1994; Carter, 1996). It is well known that many geo-materials possess a natural structure, which means that they may have properties different from the same material in a reconstituted state. Hence, understanding the behaviour of structured soils should be a key component of any attempt to develop accurate simulation of the performance of geotechnical structures involving naturally structured soils. Although during the past five decades a large amount of fruitful research work has been carried out on the stress-strain behaviour of natural soils (*e.g.*, Burland 1990; Leroueil and Vaughan 1990; Gens and Nova 1993; Whittle, 1993; Wheeler, 1997; Tatsuoka *et al*, 1997; Rouainia and Muir Wood, 2000), the models proposed to date may be regarded as being at the pioneering stage.

In this paper, the current structure of soil is introduced into the description of the mechanical behaviour of soil and a conceptual framework for structured soils is developed in four dimensions, *viz.*, the current stress state, stress history, voids ratio and a measure of soil structure. The stress-strain behaviour of soil is then divided into four regions in stress space. The general modelling of soil behaviour in each of the four regions and the influence of soil structure on this behaviour are investigated.

2 STRUCTURED SOIL BEHAVIOUR

The formation and development of soil structure normally produce soil anisotropy. Destructuring usually leads to a reduction of anisotropy. In order to concentrate on introducing the physical concepts of the framework and to avoid the complexity of mathematical details, only the isotropic effects of soil structure are considered here. The stress and strain quantities are defined in the usual way, *e.g.*, Liu and Carter (2000a). Only a brief introduction of the proposed framework is presented here; a more detailed description can be found in a report by Liu and Carter (2000b).

2.1 Four regions of soil behaviour

In the proposed model, the stress-strain behaviour of soil is divided into four regions in stress space, *i.e.*, an elastic region, a stable deformation region, a sub-yielding region and a virgin yielding region. The last two identified regions have been described elsewhere (Liu & Carter, 2000a). The remained are described below.

Based on experimental observations involving a large body of test data of cyclic loading (*e.g.*, El-Sohby, 1964; Andersen *et al*, 1980; Wood, 1982), it is often found that the response of soil to repeated loading along well defined and fixed stress paths (not involving failure) is basically stable after about the fifth cycle of loading. For the first five cycles the magnitude of the plastic deformation reduces steadily. After about the fifth cycle, the deformation is essentially stable, *i.e.*, the strain of the soil is independent of the number of cycles. The behaviour of soil in the stable region of stress space is obviously not purely elastic because hysteresis may occur. Consequently, a new region of deformation, the stable deformation region, is introduced to describe this particular type of behaviour. The stable deformation region generally includes the elastic region.

2.2 Surfaces for structured soil in p′ – q space

Five different surfaces in stress space are introduced to describe the deformation of structured soil. These are illustrated in Figure 1 and descriptions of each follow. It should be noted that the size of each surface is defined by its intersection with the isotropic stress (p') axis.

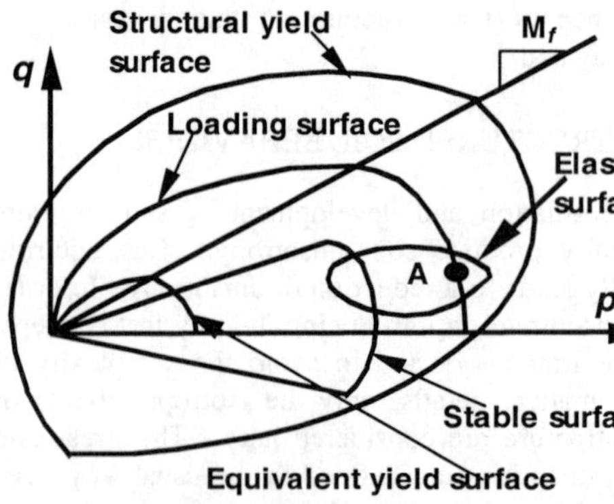

Figure 1 Five surfaces for structured soil

Structural Yield Surface: The yield surface for a structured soil is referred to as the structural yield surface. The size of a structural yield surface is denoted by $p'_s(S)$. The value of p'_s is determined by the current soil structure.

Equivalent Yield Surface: The equivalent yield surface for a structured soil is the yield surface for the corresponding reconstituted soil with the same voids ratio and the same stress state as the structured soil. The size of the equivalent yield surface is denoted by $p'_o(e)$. For soil of a given mineralogy p'_o is determined by the current voids ratio and the stress state.

Loading Surface: The surface on which the current stress state always stays is described as the loading surface. It is assumed that plastic deformation is associated with the change of the loading surface. The size of the loading surface is denoted by $p'_c(\sigma)$. The value of p'_c is determined entirely by the current stress state.

Elastic Surface: The pure elastic deformation region of a soil is defined by the elastic surface. The size of an elastic surface is denoted by p'_e, which is mainly determined by the mineralogy of the soil and may also be dependent on the current soil structure and the stress state.

Stable Surface: The stable surface for a soil defines the region where the deformation of the soil after about five cycles of repeated loading is stable. The size of a stable surface is denoted by $p'_n(M^*_{cs})$. p'_n is mainly determined by intrinsic soil properties and may also depend on soil structure.

Four types of soil behaviour are identified according to the location of the current stress state, as indicated in Table 1. A subscript to indicate the stress point with which a surface is associated is also suggested. For example, $p'_{s,I}$ represents the size of the structural yield at a soil state *I*.

Table 1. Types of behaviour

Behaviour	Condition
Elastic	Stress state inside the elastic surface, p'_e
Virgin yielding	$p'_c \geq p'_s$ and $dp'_c > 0$
Subyielding	Stress state inside p'_s and out side p'_e and p'_n
Subyielding during the first few cycles	Stress state inside p'_n and out side p'_e
Stable (after about 5 cycles)	Stress state inside p'_n and out side p'_e

Based on the types of soil behaviour listed in Table 1, a classification of the types of loading is proposed, as indicated in Table 2.

Table 2. Loading types

Loading Type	Condition
First loading	$p'_c = p'_c\|_{max}$ and $dp'_c > 0$
Reloading	$p'_c < p'_c\|_{max}$ and $dp'_c > 0$
Unloading	$p'_c < p'_c\|_{max}$ and $dp'_c < 0$

In Table 2 $p'_c|_{max}$ is the maximum loading surface the soil has ever experienced.

2.3 Hardening and destructuring

Liu and Carter (2000a) presented a detailed study of the plastic volumetric hardening of reconstituted soils. The following conclusions were drawn for a reconstituted soil.

1. The plastic volumetric strain increment is dependent on the change in size of the yield surface, but independent of the stress path.
2. There is a one-to-one relationship between the yield surface and the plastic volumetric strain.
3. Because the elastic volumetric strain depends only on the stress increment and the current stress state, the size of the yield surface for a given soil can be linked uniquely to the current voids ratio, and the current stress state.

Similarly, an assumption that both the hardening and the destructuring of a naturally structured soil are dependent on the plastic volumetric deformation is adopted. This assumption leads to the following consequences.

1. The magnitude of plastic volumetric deformation of the natural soil is dependent on the change in the size of the yield surface, irrespective of the stress path.
2. The structural yield surface is dependent on the current soil structure, the current voids ratio, and the current stress state.

2.4 Progressive destructuring and the development of soil structure

There are many factors that may alter the structure of a soil, such as ageing, leaching, change of the chemical components of the pore fluid, loading and weathering. These factors may be divided into two categories. The first category encompasses stress variations, which result only in destructuring. The process of destructuring is monotonic and irrecoverable. The second category includes all other factors, which may cause either destructuring or the development of structure.

2.4.1 Modelling destructuring

A sketch of the proposed destructuring model is illustrated in Figure 2. Suppose the initial state for a soil in situ is at point A, and its initial structural yield surface is indicated by the mean stress $p'_{y,i}$. For loading inside the structural yield surface, either elastic or elastoplastic deformation occurs. The structural yield surface may vary as the stress state varies, because both the hardening and the destructuring of the yield surface are normally induced by plastic deformation. As the loading continues along stress path AB, the loading surface expands. The loading surface and the structural yield surface coincide at point B, that is, $p'_{c,B} = p'_{s,B}$. Virgin yielding starts then at point B and continues for loading along the stress path BC. Generally speaking, $p'_{c,B} \neq p'_{y,i}$.

Suppose unloading occurs at point C along stress path CD, *i.e.*, $dp'_c < 0$. The current stress state turns inside the maximum loading surface. Either elastic or elastoplastic deformation then occurs, according to the location of the current stress state and the type of loading. Suppose the stress path changes direction at point D, with $dp'_c > 0$. In this case reloading takes place and continues along stress path DE. The structural yield surface may vary during both unloading and reloading. Suppose at point E the loading surface coincides with the structural yield surface, *i.e.*, $p'_{c,E} = p'_{s,E}$. In this case virgin yielding will recommence at point E and will continue for loading along stress path EF. Generally speaking $p'_{s,E} \neq p'_{s,C}$, because of hardening and destructuring.

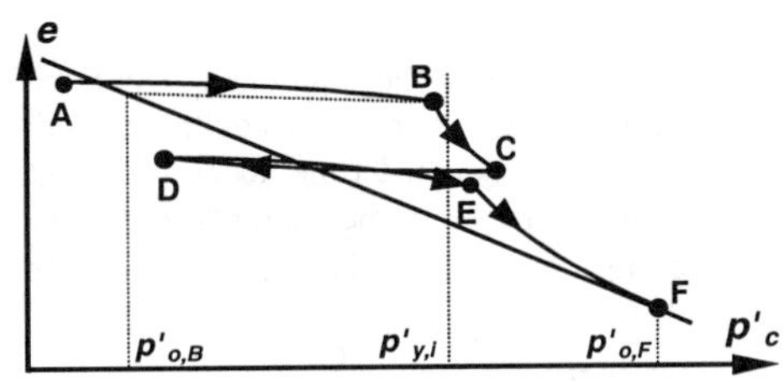

(a) Behaviour of structured soil in e-p'_c space

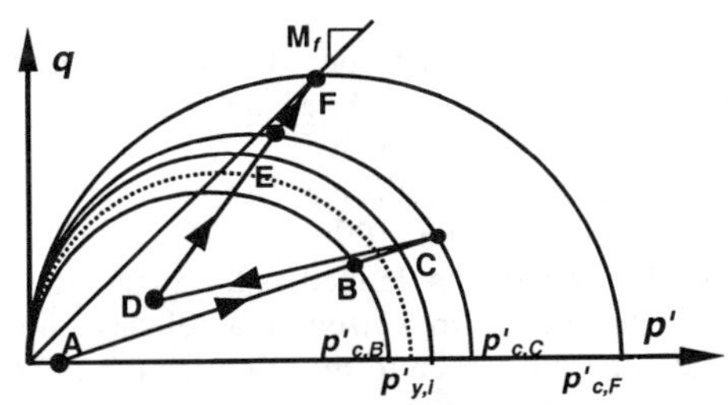

(b) Variation of surfaces during loading

Figure 2 A destructuring model

The variation of the structural yield surface is determined by two factors: the hardening mechanism and the destructuring mechanism. Both mechanisms are dependent on plastic volumetric deformation. It is assumed that the effects of the two mechanisms can be modelled independently and the change in the current yield surface can be expressed in terms of a hardening effect dH and a destructuring effect dR, as given in the following equation:

$$dp'_s = dH - dR \quad (1)$$

Various hardening and destructuring functions can be formulated within the proposed framework for the sub-yielding and virgin yielding behaviour of soil.

2.4.2 Modelling the development of soil structure

As only the isotropic effects of soil structure are considered in this paper, the influence of soil structure development on the yield surface is described by the change in size of the structural yield surface. For example, the development of soil structure due to ageing has been widely observed (*e.g.*, Leroueil and Vaughan, 1990). Based on the work by Mesri and Shahien (1997), an approximate equation for describing the ageing effect on the structural yield surface may be proposed as follows:

$$p'_{y,i} = p'_o \left(\frac{t}{t_p} \right)^{\beta} \quad (2)$$

where p'_o is the size of the equivalent yield surface corresponding to the current voids ratio and the current stress state, t_p is the time needed for primary compression, t is the time measured from the end of primary compression, and β is a material constant.

2.5 *Soil strength*

The final state of a soil usually corresponds to the critical state of deformation or the residual state of deformation, depending on the platy particles content of the soil. A brief description of the residual strength is given here. A more detailed report on the two strengths can also be found in the paper by Liu and Carter (1999).

Residual strength is defined as the shear strength of a soil that can be mobilised on a polished sliding surface, after it has been formed through the soil due to the alignment of its platy particles. For any given soil it is the minimum strength attainable. There are four major aspects of the residual strength: (1) the soil must have enough plate-like particles so that a smooth slickensided surface can be formed; (2) the sliding surface of well-aligned soil particles must exist for the residual strength to be mobilized; (3) the sliding surface of well-aligned platy particles can facilitate residual failure only along that surface; (4) the residual sliding surface once formed is usually not modified by subsequent deformations of relatively small magnitude.

For most practical problems, the dominant factor influencing the residual strength will be the clay fraction, defined as the weight of the clay particles less than 0.002 mm in size over the total weight of the soil sample. Skempton (1985) observed that for a soil with a clay fraction less than 20% the final strength is the critical state strength. Otherwise, the final strength of the soil is its residual strength. The final strength of a soil, expressed as a friction angle, may vary from 32° to as low as 8° with the variation of the clay fraction.

The change of soil strength after the critical state is uniquely dependent on the orientation of soil particles along a potential sliding surface. A new soil parameter ω_o, the degree of particle orientation, is introduced. If there is no preferred orientation of soil particles then $\omega_o = 0$. However, if the soil reaches the residual strength state then $\omega_o = 1$.

It is convenient here to express soil strength in terms of the ratio of the generalised deviator stress to mean effective stress (q/p'). If the soil strength is assumed to vary linearly with particle orientation ω_o, then the mobilised stress ratio may be written as

$$M_f = M^*_{cs} - (M^*_{cs} - M^*_r)\omega_o \quad (3)$$

where M_f is the maximum shearing resistance in the sliding plane after the soil has reached the critical state strength, M^*_{cs} is the critical state strength, and M^*_r is the residual strength.

For convenience in engineering applications, it is necessary to link soil parameters with conventionally measurable mechanical quantities. It is proposed therefore that ω_o is linked with the plastic work associated with shear deformation along a potential sliding surface, W^p_s, by the following equation

$$\omega_o = \begin{cases} \dfrac{\ln W^p_s}{\ln W^p} & \text{for } W^p_s \le W^p \\ 1 \text{ for} & W^p_s > W^p \end{cases} \quad (4)$$

where W^p is the magnitude of plastic work associated with shear deformation required to form the sliding surface for residual state. It is possible to measure W^p experimentally.

3 CONCEPTUAL FRAMEWORK

The deformation of soil can be classified as elastic deformation, stable deformation, sub-yielding and virgin yielding, according to the conditions defined in Table 1. In this section, the proposed critical state conceptual framework will be described in terms of the following the following seven components:

1. Equivalent yield surface, structural yield surface and the loading surface;
2. Elastic surface;
3. Stable surface;
4. Flow rule for virgin yielding;
5. Hardening and destructuring of the yield surface;
6. Stable deformation; and
7. Transition to the residual state of deformation.

The mathematical form of the elastic deformation is well established, and the sub-yielding behaviour of soil under a very large number of cycles of loading has been modelled by Liu and Carter (2000a). That work can be used as base for describing the elastic and sub-yielding behaviour of structured soils.

3.1 Equivalent yield surface, structural yield surface and loading surface

Because only isotropic effects of soil structure are considered here, the influences of soil structure on the yield surface is described fully by the change in size of the structural yield surface. Hence, the equivalent yield surface and the structural yield surface are of the same shape, *i.e.*, the mathematical form of these two equations is the same. The equivalent yield surface is also the yield surface for a reconstituted soil. According to Schofield (1980), the yield surface of reconstituted soils is actually made up of three parts: the plastic yielding surface, the Hvorslev failure surface and the fracture failure surface. Burland *et al.* (1996) demonstrated that the yield surface for naturally structured clays is also made up of the same three parts (Fig. 3a). Consequently, the following mathematical form is proposed for the equivalent surface (Fig. 3b):

$$f(p'_o)=\begin{cases}\dfrac{q^2}{4}+\dfrac{M_f^2}{3}\left(p_o'^2-4p'p'_o+3p'^2\right)=0 & \text{for } p'\ge 2p'_o/3\\ q^2-M_f^2p'\left(4p'_o/3-p'\right)=0 & \text{for } p'<2p'_o/3\end{cases} \quad (5)$$

The equivalent yield surface is made of two quarters of two different ellipses. The two quarter-ellipses meet at the apex C ($2p'_o/3$, $2M_fp'_o/3$) in Figure 3b. It may be seen from the equation (5) that dq/dp' is continuous at the intersection point.

The structural yield surface is assumed to possess the same shape as the yield surface for the corresponding reconstituted soil, and is given by

$$f(p'_s)=\begin{cases}\dfrac{q^2}{4}+\dfrac{M_f^2}{3}\left(p_s'^2-4p'p'_s+3p'^2\right)=0 & \text{for } p'\ge 2p'_s/3\\ q^2-M_f^2p'\left(4p'_s/3-p'\right)=0 & \text{for } p'<2p'_s/3\end{cases} \quad (6)$$

It is assumed that the loading surface is of the same shape as the equivalent yield surface and is given by

$$f(p'_c)=\begin{cases}\dfrac{q^2}{4}+\dfrac{M_f^2}{3}\left(p_c'^2-4p'p'_c+3p'^2\right)=0 & \text{for } p'\ge 2p'_c/3\\ q^2-M_f^2p'\left(4p'_c/3-p'\right)=0 & \text{for } p'<2p'_c/3\end{cases} \quad (7)$$

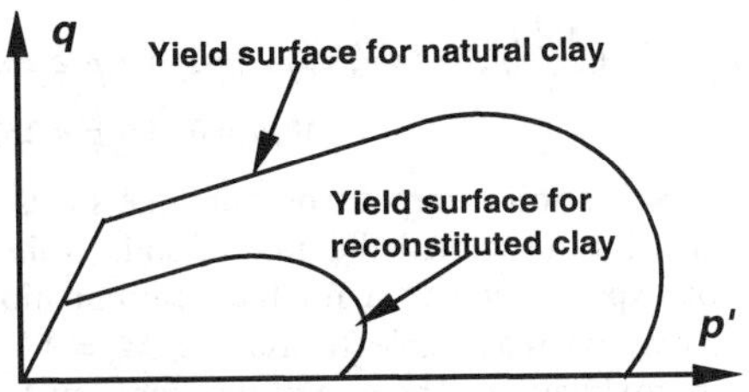

(a) A sketch of the yield surface for natural clay and reconstituted clay (after Burland, 1996)

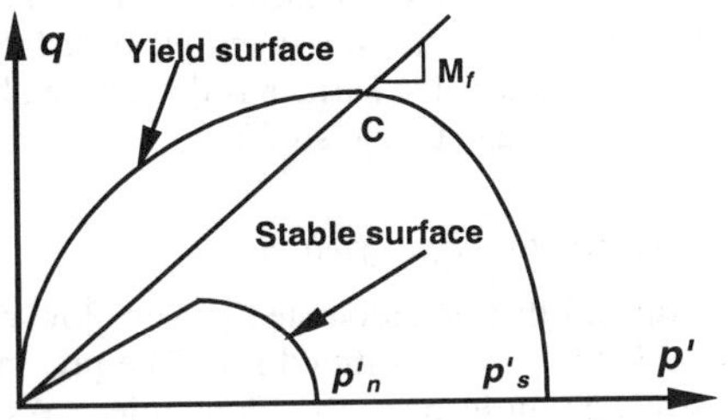

(b) Proposed yield surface and stable surface

Figure 3 Yield surface and stable surface for structure soil and reconstituted soil

3.2 Elastic surface

The elastic surface and the elastic deformation parameters are generally influenced by the stress state, the strain state, the voids ratio, and the soil structure (Burland *et al.*, 1996; Tatsuoka *et al.*, 1997). The elastic surface may be assumed to be an ellipse, however its position and size are strongly dependent on the stress history as well as the soil structure. For a reconstituted soil the position of the elastic surface is dependent on the stress history only. For a structured soil the surface is dependent on the soil structure and the stress history. One of the simpler ways to determine the position of the surface is to assume that the translation of the elastic surface follows the rules proposed by Mroz (1969). For simplicity, the aspect ratio of the elastic surface may be assumed as equal to M_f. For more accurate modelling, a description of the variation of the aspect ratio of the elliptical elastic surface with bonding of the soil may be required, as may be seen from the report by Carter & Airey (1994).

3.3 Stable surface

The region of stress space in which the soil remains stable under repeated loading depends mainly on two factors: the stress ratio and the size of the structural yield surface. A sketch of the stable surface is shown in Figure 3b and a simple equation for it is proposed as

$$f(p'_n)=\begin{cases}\dfrac{q^2}{4}+\dfrac{M_n^2}{3}\left(p_n'^2-4p'p'_n+3p'^2\right)=0 & \text{for } p'\geq 2p'_n/3\\ q-M_n p'=0 & \text{for } p'<2p'_n/3\end{cases} \quad (8)$$

where M_n is a soil parameter defining the stress ratio boundary for the stable behaviour of soil. If there is lack of experimental data for the determination of M_n it may be reasonable to assume $M_n = M_f$ as a first approximation. The size of the stable surface is assumed to depend on the size of the current structural yield surface. If destructuring occurs rapidly the structural yield surface shrinks and may become identical to the equivalent yield surface. In this situation, the stable surface is determined by the size of the equivalent yield surface.

3.4 Flow rule for virgin yielding

It is assumed that an associated plastic flow rule is applicable to the reconstituted soil. The presence of soil structure influences the flow rule. As only isotropic effects of soil structure are considered, the variation of the yield surface size may be used as a parameter for quantifying the degree of soil structure. It is proposed that the flow rule for a structured soil can be expressed as that for the corresponding reconstituted soil multiplied by a modification function ψ, where ψ is assumed to be a function of p'_s/p'_o. Therefore the following flow rule can be obtained by differentiating equation (6):

$$\frac{d\varepsilon_d^p}{d\varepsilon_v^p}=\begin{cases}\dfrac{3q\psi\left(\dfrac{p'_s}{p'_o}\right)}{4M_f\left[2\sqrt{p'^2-\dfrac{3}{4}\left(\dfrac{q}{M_f}\right)^2}-p'\right]} & \text{for } p'\geq 2p'_c\\[2ex] \dfrac{2qp'}{q^2-M_f^{\,2}p'^2}\psi\left(\dfrac{p'_s}{p'_o}\right) & \text{for } p'<2p'_c\end{cases} . \quad (9)$$

The multiplier $\psi(p'_s/p'_o)$ should be determined from experimental data.

3.5 Hardening and destructuring

A basic assumption of the proposed theoretical framework is that both the hardening and the destructuring of a naturally structured soil are dependent on the plastic volumetric deformation. Both dH and dR may be expressed in terms of $d\varepsilon_v^p$. It is likely that the hardening function, H, will be dependent on the current stress state and the voids ratio. It is assumed that the destructuring function, R, will be dependent on the current stress state, the voids and the size of the structural yield surface. Therefore

$$dH = H(\sigma', e)d\varepsilon_v^p \quad (11)$$

and

$$dR = R(\sigma', e, p'_s)d\varepsilon_v^p . \quad (12)$$

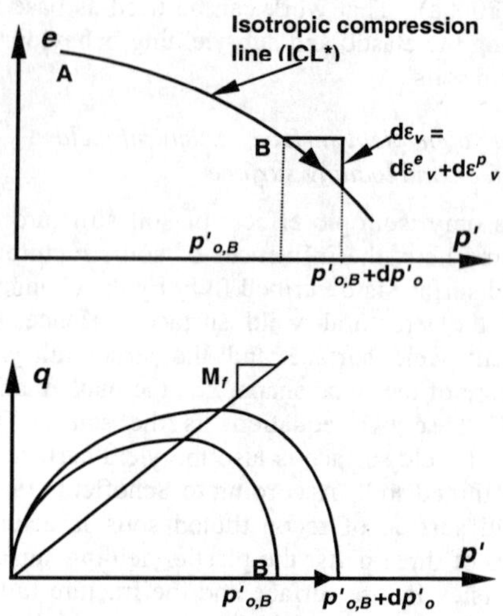

Figure 4 Hardening of soil

The hardening and destructuring functions are to be determined from experimental data. For a reconstituted soil, structure has been removed completely. Consequently, there is no destructuring effect during the plastic deformation of the soil, *i.e.*, $dR \equiv 0$. In this case the variation of the yield surface during virgin yielding is completely determined by the hardening function.

A useful way to obtain the hardening function for a reconstituted soil is to examine its isotropic compression behaviour (Fig. 4). For a reconstituted soil during virgin yielding, the equivalent yield surface, the structural yield surface and the loading surface are all identical. The size of the equivalent yield surface, expressed by equation (5), is also numerically equal to the mean effective stress on the surface, *i.e.*, point B in Figure 4. Therefore, during isotropic compression $p'_o = p'$. Suppose that soil behaves as a virgin yielding material at point B, the total volumetric strain increment may be expressed as

$$d\varepsilon_v = \left(\frac{l}{1+e}\right)dp' = \left(\frac{l}{1+e}\right)dp'_o \tag{13}$$

where l is the gradient of soil deformation at point B in the e - p' space. The elastic volumetric strain increment can be calculated based on Hooke's law with the Young's modulus E and Poisson's ratio ν. Consequently the following equation is obtained

$$d\varepsilon_v^p = \left[\frac{l}{1+e} - \frac{3(1-2\nu)}{E}\right]dp'_o . \tag{14}$$

Equation (14) is also the hardening function for a reconstituted soil. The following points should be noted.

1. Parameters l, E and ν are usually not constant but vary with the current stress state, voids ratio and soil structure.
2. The hardening function for a structured soil may not be the same as that for the corresponding reconstituted soil. Modification of equation (14) may be needed for a structured soil.

The destructuring function can be determined experimentally, because the variation of the structural yield surface dp'_s can be measured in experiments. Knowledge of the hardening and destructuring functions allows the plastic volumetric strain increment to be calculated from a virgin loading stress increment. Based on the flow rule given by equation (9), the plastic strain increment for a structured soil during virgin yielding can be computed.

Clearly, all soil properties are dependent on soil structure and are influenced by destructuring as well as the development of soil structure. Properties that are usually most sensitive to soil structure are E, ν, the size and position of the elastic surface, the size of the stable surface, and the flow rule for virgin yielding and sub-yielding. Ideally, the influence of destructuring on all these properties should be studied individually, perhaps in a manner similar to that suggested to asses the variation of the structural yield surface.

3.6 Stable deformation

The behaviour of soil in the stable region involves hysteresis and the mathematical description of the hysteresis loop has been well established. One straightforward and accurate method is the Mroz theory of a field of working-hardening moduli (Mroz, 1969). For the purpose of predicting the response of geotechnical structures during a loading with a large number of cycles, simplified methods may be employed since the total residual deformation of a soil element, which traverses the stable region, is zero.

3.7 Transition to residual state

The orientation of platy particles occurs only with large shear deformation. It is suggested that the effect of the orientation of platy particles on soil deformation needs to be considered only after the soil has reached the strength defined by the parameter M_f and with the distortional strain reaching 10%. Before this point of deformation, there is no particular orientation of the soil particles if the soil has never previously experienced any large shear deformation. On the contrary, if the soil has been previously sheared with a large accumulated shear deformation, there is a preferred orientation of the soil platy particles and this orientation remains unchanged.

By differentiating equations (3) and (4), and assuming that there is no volumetric change during the transition from the strength M_f to the residual state of deformation, the following stress and strain relationship can be obtained:

$$\begin{cases} d\varepsilon_d^p = -\dfrac{W_s^p \ln W^p}{(M^*_{cs} - M^*_r)q} dM_f \\ d\varepsilon_v^p = 0 \end{cases} . \tag{15}$$

For soil with less than 20% of platy particles, the final state of the soil is the critical state of deformation and there is no such transition.

4 CONCLUSION

A conceptual framework for describing the mechanical behaviour of structured soils has been proposed. The stress-strain behaviour of soil is divided into four regions in stress space, *i.e.*, an elastic region, a stable deformation region, a sub-yielding region and a virgin yielding region. The influence of soil structure on the mechanical behaviour of structured soil has been discussed.

The concepts proposed in this paper have not yet been subjected to rigorous testing, involving their application to the available experimental data for structured soils. However, validation of the proposed modelling framework will be the subject of future research work.

5 ACKNOWLEDGEMENT

The work described in this paper is supported by a Large Grant awarded by the Australian Research Council. Partial funding was also provided by the Special Research Centre for Offshore Foundation Systems, established and supported under the Australian Research Council's Research Centres Program.

6 REFERENCES

Andersen K. H., Pool J. H., Brown S. E. & Rosenbrand W. F. (1980), "Cyclic and static laboratory tests on Drammen clay", *J. the Geotechnical Engineering Div.*, ASCE, Vol. 106(5), pp.499-529.

Burland J. B. (1990), "On the compressibility and shear strength of natural soils", *Géotechnique*, Vol. 40(3), pp.329-378.

Burland J. B., Rampello S., Georgiannou V. N. & Calabresi G. (1996), "A laboratory study of the strength of four stiff clays", *Géotechnique*, Vol. 46(3), pp.491-514.

Carter J. P. & Airey D. W. (1994), "The engineering behaviour of cemented marine carbonate soils", *Geotechnical Engineering: Emerging Trends in Designs and Practice*, Saxena (*ed*), pp.65-101.

El-Sohby M. A. (1964), The Behaviour of Particulate Materials Under Stress, Ph.D. thesis, Manchester University.

Gens A. & Nova R. (1993), "Conceptual bases for a constitutive model for bonded soils and weak rocks", *Geotechnical Engineering of Hard Soils – Soft Rocks*, Anagnostopoulos *et al* (*ed*), Vol. 1, pp.485-494.

Leroueil S. and Vaughan P. R. (1990), "The general and congruent effects of structure in natural soils and weak rocks", *Géotechnique*, Vol. 40(3), pp.467-488.

Liu M. D. & Carter J. P. (1999), "A failure criterion for intact and fissured clay", *Proc.* 8th Australia-New Zealand Geomechanics Conference, Hobart, Vol. 2, pp.861-867.

Liu M. D. & Carter J. P. (2000a), "On the volumetric deformation of reconstituted soils", *Int. J. for Numerical and Analytical Methods in Geomechanics*, Vol. 24(2), pp.101-133.

Liu M. D. & Carter J. P. (2000b), "A conceptual framework for the behaviour of structured soils", Research Report, University of Sydney.

Mesri G. & Shahien M. (1997), "Geotechnical characteristics and compression of Pisa clay", *Proc.* 14th Int. Conference Soil Mechanics and Foundation Engineering, Vol. 1, pp.373-379.

Mroz Z. (1969), "An attempt to describe the behaviour of metal under cyclic loads using a more general workhardening model", *Acta Mechanica*, Vol. 2, pp.199-212.

Rouainia M. and Muir Wood D. (2000), "A kinematic hardening model for natural clays with loss of structure", *Géotechnique*, Vol. 50(2), pp.153-164.

Schofield A. N. & Wroth C. P. (1968), *Critical State Soil Mechanics*, MacGraw-Hill, London.

Schofield A. N. (1980), "Cambridge centrifuge operations", *Géotechnique*, Vol. 30(3), pp.227-268.

Skempton A. W. (1985), "Residual strength of clays in landslides, folded strata and the laboratory", *Géotechnique*, Vol. 35(1), pp.1-18.

Tatsuoka F., Jardine R. J., Lo Presti D., Di Benedetto H., and Kodaka T. (1997), "Theme lecture: charaterising the pre-failure deformation properties of geomaterial", *Proc.* 14th Int. Conference Soil Mechanics and Foundation Engineering, Vol. 4, pp.2129-2164.

Vaughan P. R. (1994), "Assumption, prediction and reality in geotechnical engineering", *Géotechnique*, Vol. 44(4), pp.573-609.

Wheeler S. J. (1997), "A rotational hardening elasto-plastic model for clays", *Proc.* 14th Int. Conference Soil Mechanics and Foundation Engineering, Vol. 1, pp.431-434.

Whittle A. J. (1993), "Evaluation of a constitutive model for overconsolidated clays", *Géotechnique*, Vol. 43(2), pp.289-314.

Wood D. M. (1982), "Laboratory investigations of the behaviour of soils under cyclic loading: a review", *Soil Mechanics-Transient and Cyclic Loading*, Pande *et al* (*ed*), pp.513-582.

Computer Methods and Advances in Geomechanics, Desai et al. (eds) © 2001 Balkema, Rotterdam, ISBN 90 5809 183 X

Discontinuous-nonlinear deformation analysis method and applications in geotechnics

Maotian Luan & Yong Li

Department of Civil Engineering and State Key Laboratory of Coastal and Offshore Engineering, Dalian University of Technology, People's Republic of China

ABSTRACT: In this paper, the numerical method for discontinuous deformation analysis of multi-body interaction systems developed by the authors is extended to elasto-plastic problems. Nonlinear characteristics arising from contacts of interfaces and constituent media are taken into consideration. The fundamental formulations for discontinuous-inelastic coupling deformation analysis is established based on the extended divisional parametric variational principle. The resulting equations are mathematically reduced to a standard linear-complementary problem to be solved numerically with use of Lemke's algorithm. The problem of earth pressure on rigid retaining wall is analyzed to illustrate applicability of the proposed method. The effects of soil strength parameters on earth pressures in the static state, active and passive limit states are studied through numerical computations. It is shown that the proposed method is a new and potentially powerful numerical method for discontinuous and elasto-plastic coupling problems in geotechnical engineering.

1 INTRODUCTION

Both structural or tectonic discontinuity and material nonlinearity are key features of earth structures with natural joints, cuts or artificial interfaces. In essential, the important aspect in discontinuous deformation analysis is treatment of contacts. The analysis methods and numerical models should be capable to capture these essential characteristics. With the rapid development of computer technology and numerical techniques, discontinuous deformation simulation has received vast investigations. However usually the nonlinear behavior of constituent materials is overlooked while the discontinuous feature is incorporated. There are two main classes of methodology, i.e., Lagrangian methods and penalty function methods, for dealing with the contact or discontinuous deformation problems. Lagrangian methods include Lagrange's multiplier methods (Carpenter et al. 1991) and augmented Lagrangian methods (Heege et al. 1996) of which Lagrange's multiplier represents contact force or contact stress. The methods of penalty functions contain interface-element methods (e.g., Ngo & Cordelis 1967, Goodman et al. 1968), distinct element methods (Cundall et al. 1989), discontinuous deformation analysis (Shi 1988) and numerical manifold method (Shi 1996). In later methods, the contact forces or stresses cannot be defined directly. Based on the generalized finite elements and contact-force elements, a new numerical method for discontinuous deformation analysis of multi-body system has been developed by the authors (Luan et al. 2000, Li & Luan 2000). As independent variables additional to unknown displacements, the contact forces are directly incorporated in the governing equations and can be simultaneously determined. In this paper, an extension of the proposed method to the problems involving nonlinear or inelastic behavior of materials is made. The fundamental formulations are established based on extended divisional parametric variational principle. The resulting equations of the system are reduced to a linear-complementary problem to be numerically solved by virtue of Lemke's algorithm. In order to illustrate applicability of the proposed method, the soil-wall interaction problem is analyzed. The static state, active and passive states of backfill behind rigid retaining wall are investigated. Through numerical computations, it is shown that the proposed method is a potentially powerful numerical method for discontinuous and elasto-plastic coupling problems.

2 COMPUTATIONAL MODEL

Firstly, the basic governing equations and boundary conditions together with the constraint conditions of boundary-value problems with discontinuity are formulated. For multi-body interaction systems, the discontinuous deformation computational mechanics model (DDCMM) has been developed (Luan et al. 2000; Li & Luan 2000). In this methodology, the generalized finite element model is used for modeling various possible deformation and movement patterns of comprising bodies with different characteristics. The contact-force element model is utilized to simulate nonlinear behavior of the interfaces, joints, faults, or other discontinuous cuts. The model is extended to elastoplastic framework as following.

2.1 *Governing equations of discontinuous-inelastic deformation systems*

For two-dimensional problems, a multibody system is composed of m bodies with different characteristics. At time t, its current configuration is $\Omega=\sum_{i=1}^{m}\Omega_i$ with the boundary $S(\Omega)=\bigcup_{i=1}^{m}S(\Omega_i)$, with each body occupied by Ω_i with its boundary $S(\Omega_i)=S_i^u\cup S_i^q\cup S_i^c$, where S_i^u and S_i^q respectively represents known displacement and traction boundaries, S_i^c is contact boundary. The boundary-value problem can be sated as that under the current configuration at time t, incremental displacements $\boldsymbol{u}$, incremental stresses $\boldsymbol{\sigma}$, contact stress increments $\boldsymbol{p}^c$ are to be defined. Those unknown variables must fulfill equilibrium conditions, kinematic equations and constitutive relation, i.e.,

$$\widetilde{\boldsymbol{B}}^{\mathrm{T}}\left(\tfrac{\partial}{\partial x},\tfrac{\partial}{\partial y}\right)\boldsymbol{\sigma}_i=\boldsymbol{b}_i-\rho_i\boldsymbol{a}_i-\eta_i\boldsymbol{v}_i \tag{1}$$

$$\boldsymbol{u}_i=\boldsymbol{u}_i^{\mathrm{r}}+\boldsymbol{u}_i^{\mathrm{d}} \tag{2a}$$

or

$$\boldsymbol{\varepsilon}_i=\boldsymbol{\varepsilon}_i^{\mathrm{r}}+\boldsymbol{\varepsilon}_i^{\mathrm{d}}=\boldsymbol{\varepsilon}_i^{\mathrm{d}}=-\widetilde{\boldsymbol{B}}\left(\tfrac{\partial}{\partial x},\tfrac{\partial}{\partial y}\right)\boldsymbol{u}_i^{\mathrm{d}}=-\widetilde{\boldsymbol{B}}\left(\tfrac{\partial}{\partial x},\tfrac{\partial}{\partial y}\right)\boldsymbol{u}_i \tag{2b}$$

$$\boldsymbol{\sigma}_i=\boldsymbol{D}_i\left(\boldsymbol{\varepsilon}_i-\boldsymbol{\varepsilon}_i^{\mathrm{p}}\right) \tag{3a}$$

$$f_i^{\mathrm{p}}\left(\boldsymbol{\sigma}_{ti}+\boldsymbol{\sigma}_i,\boldsymbol{\varepsilon}_{ti}^{\mathrm{p}}+\boldsymbol{\varepsilon}_i^{\mathrm{p}},\boldsymbol{k}_i^{\mathrm{p}}\right)\le 0 \tag{3b}$$

$$\boldsymbol{\varepsilon}_i^{\mathrm{p}}=\left(\frac{\partial g_i^{\mathrm{p}}}{\partial\boldsymbol{\sigma}_i}\right)^{\mathrm{T}}\left(\delta_i^{\mathrm{p}}-\delta_{ti}^{\mathrm{p}}\right) \tag{3c}$$

$$\delta_i^{\mathrm{p}}\begin{cases}\ge 0 & (f_i^{\mathrm{p}}=0)\\ =0 & (f_i^{\mathrm{p}}<0)\end{cases} \tag{3d}$$

where $\boldsymbol{b}_i$, $\boldsymbol{v}_i=\dot{\boldsymbol{u}}_i$ and $\boldsymbol{a}_i=\dot{\boldsymbol{v}}_i=\ddot{\boldsymbol{u}}_i$ are respectively body force, velocity and acceleration of the i-th body; $\boldsymbol{u}_i^{\mathrm{r}}$ and $\boldsymbol{u}_i^{\mathrm{d}}$ respectively represent rigid movement and deformation; $\boldsymbol{D}_i$ is stiffness coefficient matrix; f_i^{p} and g_i^{p} respectively denote plastic yielding function and potential function; $\boldsymbol{\varepsilon}_i^{\mathrm{p}}$ is plastic strain increment; δ_{ti}^{p} and δ_i^{p} are respectively plastic parameters at time t and $t+\Delta t$, $\boldsymbol{k}_i^{\mathrm{p}}$ is hardening parameter to memory deformation history.

At the same time, along the interface between bodies, the following constraints for controlling the contact constitutive behavior should be satisfied.

$$\boldsymbol{f}_{gij}^{\mathrm{c}}\left(\boldsymbol{g}_{ij}^{\mathrm{c}}+\boldsymbol{g}_{tij}^{\mathrm{c}}\right)\cdot\boldsymbol{f}_{pij}^{\mathrm{c}}\left(\boldsymbol{p}_{tij}^{\mathrm{c}}+\boldsymbol{p}_{ij}^{\mathrm{c}}\right)=0 \tag{4a}$$

in which

$$\boldsymbol{f}_{gij}^{\mathrm{c}}\left(\boldsymbol{g}_{ij}^{\mathrm{c}}+\boldsymbol{g}_{tij}^{\mathrm{c}}\right)\ge 0 \tag{4b}$$

$$\boldsymbol{f}_{pij}^{\mathrm{c}}\left(\boldsymbol{p}_{tij}^{\mathrm{c}}+\boldsymbol{p}_{ij}^{\mathrm{c}}\right)\le 0 \tag{4c}$$

$$\boldsymbol{g}_{ij}^{\mathrm{c}}=\left[g_{ijn}^{\mathrm{c}}\ \ g_{ij\tau}^{\mathrm{c}}\right]^{\mathrm{T}}=\left(-\boldsymbol{L}_i\boldsymbol{u}_i\ \ \boldsymbol{L}_i\boldsymbol{u}_j\right) \tag{4d}$$

where $\boldsymbol{p}_{tij}^{\mathrm{c}}$ and $\boldsymbol{p}_{ij}^{\mathrm{c}}$ respectively represents contact force at time t and its increment along the interface S_{ij}^{c} between body i and body j. $\boldsymbol{g}_{tij}^{\mathrm{c}}$ and $\boldsymbol{g}_{ij}^{\mathrm{c}}$ are relative gap of contact points at time t and relative gap increment respectively; $\boldsymbol{f}_{gij}^{\mathrm{c}}(\ )$ and $\boldsymbol{f}_{pij}^{\mathrm{c}}(\ )$ are gap function for describing relative flow of contact and contact yielding function, $\boldsymbol{L}_i$ is unit outward normal vector of body i on contact interface. Equation (4) physically implies that when contact stress of two bodies along interface cannot reach yielding condition, their flowing gap is zero and consequently they lock each other, otherwise, while their flow gap is greater or equal to zero, they will detach or debound, slide.

In addition, various boundary conditions must be imposed, e.g.,

$$\boldsymbol{u}_i=\overline{\boldsymbol{u}}_i \quad \text{on } S_i^u \tag{5a}$$

$$\boldsymbol{p}_i=\widetilde{\boldsymbol{B}}^{\mathrm{T}}\left(n_{ix},n_{iy}\right)\boldsymbol{\sigma}_i=-\overline{\boldsymbol{p}}_i \quad \text{on } S_i^q \tag{5b}$$

where $\boldsymbol{p}_i=\left[p_{ix}\ \ p_{iy}\right]^{\mathrm{T}}$, $\overline{\boldsymbol{p}}_i=\left[\overline{p}_{ix}\ \ \overline{p}_{iy}\right]^{\mathrm{T}}$ is known face force, n_{ix} and n_{iy} represent direction cosine of outward normal vector of the boundary of body i respectively with respect to x and y.

2.2 *Elastoplastic generalized finite element*

In general, the displacement interpolation function of a generalized finite element can be written as

$$\boldsymbol{u}_i=\boldsymbol{N}_i\boldsymbol{U}_i \tag{6}$$

where $\boldsymbol{N}_i$ is interpolation shape function, $\boldsymbol{U}_i$ is the generalized degrees of freedom. Then stresses in element i can be given by Equation (3). In order to assess the deformation state of element, average stresses of the element can be formulated as below

$$\overline{\boldsymbol{\sigma}}_i=\boldsymbol{D}_i\left(\overline{\boldsymbol{\varepsilon}}_i-\overline{\boldsymbol{\varepsilon}}_i^{\mathrm{p}}\right) \tag{7}$$

where

$$\overline{\boldsymbol{\varepsilon}}_i=\left(\iint_{\Omega_i}-\widetilde{\boldsymbol{B}}\left(\tfrac{\partial}{\partial x},\tfrac{\partial}{\partial y}\right)\boldsymbol{N}_i\,\mathrm{d}x\,\mathrm{d}y\Big/\iint_{\Omega_i}\mathrm{d}x\,\mathrm{d}y\right)\boldsymbol{U}_i$$

$$\overline{\boldsymbol{\varepsilon}}_i^{\mathrm{p}}=\left(\frac{\partial g_i^{\mathrm{p}}}{\partial\overline{\boldsymbol{\sigma}}_i}\right)^{\mathrm{T}}\left(\delta_i^{\mathrm{p}}-\delta_{ti}^{\mathrm{p}}\right)$$

Analog to Equation (3b), the plastic yielding function of element can be expressed in terms of its average stresses and strains

$$\bar{f}_i^{\mathrm{P}}\left(\bar{\sigma}_{ti}+\bar{\sigma}_i,\bar{\varepsilon}_{ti}^{\mathrm{P}}+\bar{\varepsilon}_i^{\mathrm{P}},\boldsymbol{k}_i^{\mathrm{P}}\right)\le 0 \tag{8}$$

For Mohr-Coulomb criterion, Equation (8) can be rewritten

$$\bar{f}_i^{\mathrm{P}}=\sqrt{\left(\tilde{\sigma}_{ix}-\tilde{\sigma}_{iy}\right)^2+4\tilde{\tau}_{ixy}^2}-\left(\tilde{\sigma}_{ix}+\tilde{\sigma}_{iy}\right)\sin\phi_i-2c_i\cos\phi_i\le 0 \tag{9}$$

where $\tilde{\sigma}_i=\bar{\sigma}_{ti}+\bar{\sigma}_i$, c_i and ϕ_i are cohesion and friction angle of material in element i.

Plastic potential function in accordance with yielding function Equation (9) can be expressed as

$$\bar{g}_i^{\mathrm{P}}=\sqrt{\left(\tilde{\sigma}_{ix}-\tilde{\sigma}_{iy}\right)^2+4\tilde{\tau}_{ixy}^2}-\psi\left(\tilde{\sigma}_{ix}+\tilde{\sigma}_{iy}\right)\sin\phi_i+c_0 \tag{10}$$

where c_0 is a constant, ψ is swelling factor. $\psi=1$ indicates that the plastic flow rule is associated with Mohr-Coulomb criterion. Otherwise it is non-associated. Specially, the plastic flow for $\psi=0$ is non-associated without plastic swelling and it is non-associated with plastic shrinkage for $\psi<0$.

It is obvious that Mohr-Coulomb criterion is nonlinear function of average stresses. The linearization is made as following

$$\bar{\boldsymbol{f}}_i^{\mathrm{P}}=\boldsymbol{M}_i^{\mathrm{P}}\left(\bar{\sigma}_{ti}+\bar{\sigma}_i\right)-\boldsymbol{K}_i^{\mathrm{fp}}+\boldsymbol{\lambda}_i^{\mathrm{P}}=0 \tag{11a}$$

$$\bar{\boldsymbol{g}}_i^{\mathrm{P}}=\boldsymbol{G}_i^{\mathrm{PT}}\left(\bar{\sigma}_{ti}+\bar{\sigma}_i\right)+\boldsymbol{c}_0 \tag{11b}$$

$$\bar{\varepsilon}_i^{\mathrm{P}}=\boldsymbol{G}_i^{\mathrm{P}}\left(\delta_i^{\mathrm{P}}-\delta_{ti}^{\mathrm{P}}\right) \tag{11c}$$

$$\lambda_i^{\mathrm{P}}\cdot\delta_i^{\mathrm{P}}=0\quad \lambda_i^{\mathrm{P}}\ge 0\quad \delta_i^{\mathrm{P}}\ge 0 \tag{11d}$$

in which

$$\boldsymbol{M}_i^{\mathrm{P}}=\begin{bmatrix}1-e_1 & -e_1 & -1-e_1 & -1-e_1 & -e_1 & -1-e_1\\ -1-e_1 & -e_1 & 1-e_1 & 1-e_1 & -e_1 & -1-e_1\\ 2e_0 & 4e_0 & 2e_0 & -2e_0 & -4e_0 & -2e_0\end{bmatrix}\frac{1}{\cos\phi_i}$$

$$\boldsymbol{G}_i^{\mathrm{P}}=\begin{bmatrix}1-e_2 & -e_2 & -1-e_2 & -1-e_2 & -e_2 & -1-e_2\\ -1-e_2 & -e_2 & 1-e_2 & 1-e_2 & -e_2 & -1-e_2\\ 2e_0 & 4e_0 & 2e_0 & -2e_0 & -4e_0 & -2e_0\end{bmatrix}\frac{1}{\cos\phi_i},$$

$$\boldsymbol{K}_i^{\mathrm{fp}}=\left[2c_i\ \ 2c_i\ \ 2c_i\ \ 2c_i\ \ 2c_i\ \ 2c_i\right]^{\mathrm{T}},$$

$$\delta_i^{\mathrm{P}}=\left[\delta_{i1}^{\mathrm{P}}\ \ \delta_{i2}^{\mathrm{P}}\ \ \delta_{i3}^{\mathrm{P}}\ \ \delta_{i4}^{\mathrm{P}}\ \ \delta_{i5}^{\mathrm{P}}\ \ \delta_{i6}^{\mathrm{P}}\right]^{\mathrm{T}}$$

$$\lambda_i^{\mathrm{P}}=\left[\lambda_{i1}^{\mathrm{P}}\ \ \lambda_{i2}^{\mathrm{P}}\ \ \lambda_{i3}^{\mathrm{P}}\ \ \lambda_{i4}^{\mathrm{P}}\ \ \lambda_{i5}^{\mathrm{P}}\ \ \lambda_{i6}^{\mathrm{P}}\right]^{\mathrm{T}}$$

where λ_i^{P} is relaxation variable of stress strength which corresponds to residual strength, $e_1=\sin\phi_i$, $e_2=\psi\sin\phi_i$, $e_0=1/\sqrt{3}$.

Substituting Equation (11c) into Equation (7) leads to

$$\bar{\sigma}_i=\boldsymbol{D}_i\left(\boldsymbol{E}_i\boldsymbol{U}_i-\boldsymbol{G}_i^{\mathrm{P}}\left(\delta_i^{\mathrm{P}}-\delta_{ti}^{\mathrm{P}}\right)\right) \tag{12}$$

where

$$\boldsymbol{E}_i=\iint_{\Omega_i}-\widetilde{\boldsymbol{B}}\left(\tfrac{\partial}{\partial x},\tfrac{\partial}{\partial y}\right)\boldsymbol{N}_i\,\mathrm{d}x\,\mathrm{d}y\Big/\iint_{\Omega_i}\mathrm{d}x\,\mathrm{d}y$$

Equation (11a) may finally take the form of

$$\boldsymbol{M}_i^{\mathrm{P}}\boldsymbol{D}_i\boldsymbol{E}_i\boldsymbol{U}_i-\boldsymbol{M}_i^{\mathrm{P}}\boldsymbol{D}_i\boldsymbol{G}_i^{\mathrm{P}}\left(\delta_i^{\mathrm{P}}-\delta_{ti}^{\mathrm{P}}\right)+\lambda_i^{\mathrm{P}}=\boldsymbol{K}_i^{\mathrm{fp}}-\boldsymbol{M}_i^{\mathrm{P}}\bar{\sigma}_{ti} \tag{13}$$

2.3 *Contact force element*

Contact force element is composed of contact pairs, in which the variation of contact stresses along interface are stipulated by a given pattern and it is subjected to the constraints imposed by both yielding criterion and flow rule of contact. Contact stresses of interface between generalized finite elements i and j can be generally represented

$$\boldsymbol{p}_{ij}^{\mathrm{c}}=\boldsymbol{T}_{ij}\boldsymbol{R}_{ij}=\sum_{k=1}^{n_{cij}}\boldsymbol{T}_{ijk}\boldsymbol{R}_{ijk} \tag{14}$$

where n_{cij} is the number of contact pairs, $\boldsymbol{T}_{ij}$ and $\boldsymbol{R}_{ij}$ are respectively the interpolation shape function and generalized degrees of freedom of contact stresses along interface ij; $\boldsymbol{T}_{ijk}$ and $\boldsymbol{R}_{ijk}$ are interpolation shape function and contact stress of its k-th contact pair. Linear pattern of distribution of contact stress can be used for simplification (Luan et al. 2000). Linearization of Mohr-Coulomb yielding criterion in Equation (4) may be expressed as

$$\boldsymbol{f}_{\mathrm{p}}^{\mathrm{c}}\left(\boldsymbol{R}_{tij}^{\mathrm{c}}+\boldsymbol{R}_{ij}^{\mathrm{c}}\right)=\sum_{k=1}^{n_{cij}}\left(\boldsymbol{M}_{ijk}^{\mathrm{c}}\left(\boldsymbol{R}_{tijk}+\boldsymbol{R}_{ijk}\right)-\boldsymbol{K}_{ijk}^{\mathrm{fc}}+\boldsymbol{\lambda}_{ijk}^{\mathrm{c}}\right)=\boldsymbol{0} \tag{15}$$

in which

$$\boldsymbol{M}_{ijk}^{\mathrm{c}}=\begin{bmatrix}-1 & -\tan\phi_{ij} & -\tan\phi_{ij}\\ 0 & 1 & -1\end{bmatrix}^{\mathrm{T}},$$

$$\boldsymbol{K}_{ijk}^{\mathrm{fc}}=\left[-\sigma_{tij}\ \ c_{ij}\ \ c_{ij}\right]^{\mathrm{T}},\quad \boldsymbol{\lambda}_{ijk}^{\mathrm{c}}=\left[\lambda_{ijk}^{\mathrm{n}}\ \ \lambda_{ijk}^{s+}\ \ \lambda_{ijk}^{s-}\right]^{\mathrm{T}}$$

where σ_{tij}, ϕ_{ij} and c_{ij} are respectively tensile strength, friction angle and cohesion of contact interface ij, $\lambda_{ijk}^{\mathrm{n}}$, λ_{ijk}^{s+} and λ_{ijk}^{s-} are normal component, positive and negative tangential components of residual strength of the k-th contact pair of contact interface ij respectively.

Flow rule associated with Mohr-Coulomb criterion of friction contact may be written as

$$\boldsymbol{f}_{\mathrm{g}}^{\mathrm{c}}\left(\boldsymbol{g}_{ij}^{\mathrm{c}}+\boldsymbol{g}_{0ij}^{\mathrm{c}}\right)=\sum_{k=1}^{n_{cij}}\left(\boldsymbol{H}_{ijk}\boldsymbol{\delta}_{ijk}^{\mathrm{c}}-\left(\boldsymbol{G}_{ijk}^{\mathrm{c}}\boldsymbol{U}_{ij}+\boldsymbol{G}_{tijk}^{\mathrm{c}}\right)\right)=\boldsymbol{0} \tag{16}$$

in which

$$\boldsymbol{H}_{ijk}=\begin{bmatrix}1 & 0 & 0\\ 0 & -1 & 1\end{bmatrix}^{\mathrm{T}},$$

$$\boldsymbol{G}_{ijk}^{\mathrm{c}}=\left[-\boldsymbol{L}_i\boldsymbol{N}_{ik}\ \ \boldsymbol{L}_i\boldsymbol{N}_{jk}\right],\qquad \boldsymbol{G}_{tijk}^{\mathrm{c}}=\boldsymbol{G}_{ijk}^{\mathrm{c}}\boldsymbol{g}_{tijk}^{\mathrm{c}},$$

$$\boldsymbol{U}_{ij}=\left[\boldsymbol{U}_i\ \ \boldsymbol{U}_j\right]^{\mathrm{T}},\qquad \boldsymbol{\delta}_{ijk}^{\mathrm{c}}=\left[\Delta_{ijk}^{\mathrm{n}}\ \ \Delta_{ijk}^{s+}\ \ \Delta_{ijk}^{s-}\right]^{\mathrm{T}}$$

where $\Delta_{ijk}^{\mathrm{n}}$, Δ_{ijk}^{s+} and Δ_{ijk}^{s-} are normal, positive and negative tangential components of flow gap of the k-th contact pair of contact interface ij. Interrelation between yielding criterion and flow rule can be given by the following complementary conditions

$$\boldsymbol{\lambda}_{ijk}^{\mathrm{c}}\cdot\boldsymbol{\delta}_{ijk}^{\mathrm{c}}=\boldsymbol{0},\quad \boldsymbol{\lambda}_{ijk}^{\mathrm{c}}\ge\boldsymbol{0},\quad \boldsymbol{\delta}_{ijk}^{\mathrm{c}}\ge\boldsymbol{0} \tag{17}$$

2.4 *Extended divisional parametric variational principle and mathematical formulations*

The divisional parametric variational principle for discontinuous deformation systems (Luan & Li 2000) can be extended to elasto-plastic condition. In this principle, the plastic flow parameters and contact forces are taken as parametric variables without participating variational process and hence the governing equations will be subjected to constraints for these variables. Therefore, resulting governing equations can be finally formulated

$$KU - C_{\mathrm{p}}\left(\delta^{\mathrm{p}} - \delta_t^{\mathrm{p}}\right) - C_R R = F \tag{18a}$$

$$H\delta^{\mathrm{c}} - \left(G^{\mathrm{c}}U + G_t^{\mathrm{c}}\right) = 0 \tag{18b}$$

$$M^{\mathrm{c}}\left(R_t + R\right) - K^{\mathrm{fc}} + \lambda^{\mathrm{c}} = 0 \tag{18c}$$

$$\lambda^{\mathrm{c}} \cdot \delta^{\mathrm{c}} = 0,\ \lambda^{\mathrm{c}} \geq 0,\ \delta^{\mathrm{c}} \geq 0 \tag{18d}$$

$$M^{\mathrm{p}}DEU - M^{\mathrm{p}}DG^{\mathrm{p}}\left(\delta^{\mathrm{p}} - \delta_t^{\mathrm{p}}\right) + \lambda^{\mathrm{p}} = K^{\mathrm{fp}} - M^{\mathrm{p}}\overline{\sigma}_t \tag{18e}$$

$$\lambda^{\mathrm{p}} \cdot \delta^{\mathrm{p}} = 0,\ \lambda^{\mathrm{p}} \geq 0,\ \delta^{\mathrm{p}} \geq 0 \tag{18f}$$

in which

$$K = \sum_{i=1}^{n} K_i,\quad C_{\mathrm{p}} = \sum_{i=1}^{n} C_{\mathrm{p}i},\quad C_R = \sum_{i=1}^{n} C_{Ri},\quad F = \sum_{i=1}^{n} F_i,$$

$$R = \sum_{i=1}^{n-1}\sum_{j=1}^{n_{\mathrm{e}i}}\sum_{k=i}^{n}\sum_{l=1,i=k,j\neq l}^{n_{\mathrm{e}k}}\sum_{m=1}^{n_{\mathrm{c}ijkl}} R_{ijklm},\ H = \sum_{i=1}^{n-1}\sum_{j=1}^{n_{\mathrm{e}i}}\sum_{k=i}^{n}\sum_{l=1,i=k,j\neq l}^{n_{\mathrm{e}k}}\sum_{m=1}^{n_{\mathrm{c}ijkl}} H_{ijklm},$$

$$M^{\mathrm{c}} = \sum_{i=1}^{n-1}\sum_{j=1}^{n_{\mathrm{e}i}}\sum_{k=i}^{n}\sum_{l=1,i=k,j\neq l}^{n_{\mathrm{e}k}}\sum_{m=1}^{n_{\mathrm{c}ijkl}} M_{ijklm}^{\mathrm{c}},\ R_t = \sum_{i=1}^{n-1}\sum_{j=1}^{n_{\mathrm{e}i}}\sum_{k=i}^{n}\sum_{l=1,i=k,j\neq l}^{n_{\mathrm{e}k}}\sum_{m=1}^{n_{\mathrm{c}ijkl}} R_{tijklm},$$

$$K^{\mathrm{fc}} = \sum_{i=1}^{n-1}\sum_{j=1}^{n_{\mathrm{e}i}}\sum_{k=i}^{n}\sum_{l=1,i=k,j\neq l}^{n_{\mathrm{e}k}}\sum_{m=1}^{n_{\mathrm{c}ijkl}} K_{ijklm}^{\mathrm{fc}},\ \lambda^{\mathrm{c}} = \sum_{i=1}^{n-1}\sum_{j=1}^{n_{\mathrm{e}i}}\sum_{k=i}^{n}\sum_{l=1,i=k,j\neq l}^{n_{\mathrm{e}k}}\sum_{m=1}^{n_{\mathrm{c}ijkl}} \lambda_{ijklm}^{\mathrm{c}},$$

$$\delta^{\mathrm{c}} = \sum_{i=1}^{n-1}\sum_{j=1}^{n_{\mathrm{e}i}}\sum_{k=i}^{n}\sum_{l=1,i=k,j\neq l}^{n_{\mathrm{e}k}}\sum_{m=1}^{n_{\mathrm{c}ijkl}} \delta_{ijklm}^{\mathrm{c}},\quad G^{\mathrm{c}} = \sum_{i=1}^{n-1}\sum_{j=1}^{n_{\mathrm{e}i}}\sum_{k=i}^{n}\sum_{l=1,i=k,j\neq l}^{n_{\mathrm{e}k}}\sum_{m=1}^{n_{\mathrm{c}ijkl}} G_{ijklm}^{\mathrm{c}},$$

$$G_t^{\mathrm{c}} = \sum_{i=1}^{n-1}\sum_{j=1}^{n_{\mathrm{e}i}}\sum_{k=i}^{n}\sum_{l=1,i=k,j\neq l}^{n_{\mathrm{e}k}}\sum_{m=1}^{n_{\mathrm{c}ijkl}} G_{tijklm}^{\mathrm{c}},\qquad M^{\mathrm{p}} = \sum_{i=1}^{n}\sum_{j=1}^{n_{\mathrm{e}i}^{\mathrm{p}}} M_{ij}^{\mathrm{p}},$$

$$G^{\mathrm{p}} = \sum_{i=1}^{n}\sum_{j=1}^{n_{\mathrm{e}i}^{\mathrm{p}}} G_{ij}^{\mathrm{p}},\qquad D = \sum_{i=1}^{n}\sum_{j=1}^{n_{\mathrm{e}i}^{\mathrm{p}}} D_{ij},\qquad E = \sum_{i=1}^{n}\sum_{j=1}^{n_{\mathrm{e}i}^{\mathrm{p}}} E_{ij},$$

$$\overline{\sigma}_t = \sum_{i=1}^{n}\sum_{j=1}^{n_{\mathrm{e}i}^{\mathrm{p}}} \overline{\sigma}_{tij},\qquad \lambda^{\mathrm{p}} = \sum_{i=1}^{n}\sum_{j=1}^{n_{\mathrm{e}i}^{\mathrm{p}}} \lambda_{ij}^{\mathrm{p}},\qquad \delta^{\mathrm{p}} = \sum_{i=1}^{n}\sum_{j=1}^{n_{\mathrm{e}i}^{\mathrm{p}}} \delta_{ij}^{\mathrm{p}},$$

$$\delta_t^{\mathrm{p}} = \sum_{i=1}^{n}\sum_{j=1}^{n_{\mathrm{e}i}^{\mathrm{p}}} \delta_{tij}^{\mathrm{p}},\qquad K_i = \sum_{j=1}^{n_{\mathrm{e}i}} K_{ij},\qquad C_i^{\mathrm{p}} = \sum_{j=1}^{n_{\mathrm{e}i}^{\mathrm{p}}} C_{ij}^{\mathrm{p}},$$

$$F_i = F_{bi} + F_{a_t i} + F_{v_t i} + F_{qi} + F_{pi} + F_{\sigma_t i} + F_{\varepsilon_t i},\quad R_i = \sum_{j=1}^{n_{\mathrm{e}i}}\sum_{k=1}^{m}\sum_{l=1,k=i,l\neq j}^{n_{\mathrm{e}k}} R_{ijkl},$$

$$K_{ij} = K_{ij}^{\mathrm{d}} + \overline{\beta}K_{ij}^{\mathrm{c}}/\Delta t + \overline{\alpha}K_{ij}^{\mathrm{m}}/(\Delta t)^2,\qquad C_{\mathrm{R}i} = \sum_{j=1}^{n_{\mathrm{e}i}}\sum_{k=1}^{m}\sum_{l=1,k=i,l\neq j}^{n_{\mathrm{e}k}} C_{\mathrm{R}ijkl},$$

$$K_{ij}^{\mathrm{d}} = \iint_{A_{ij}} \widetilde{B}^{\mathrm{T}}\left(\tfrac{\partial}{\partial x},\tfrac{\partial}{\partial y}\right)N_{ij}^{\mathrm{T}}D_{ij}\widetilde{B}\left(\tfrac{\partial}{\partial x},\tfrac{\partial}{\partial y}\right)N_{ij}\,\mathrm{d}x\mathrm{d}y,\quad K_{ij}^{\mathrm{m}} = \iint_{A_{ij}} \rho_i N_{ij}^{\mathrm{T}}N_{ij}\,\mathrm{d}x\mathrm{d}y,$$

$$K_{ij}^{\mathrm{c}} = \iint_{A_{ij}} \eta_i N_{ij}^{\mathrm{T}}N_{ij}\,\mathrm{d}x\mathrm{d}y,\qquad F_{bi} = \sum_{j=1}^{n_{\mathrm{e}i}}\iint_{A_{ij}} N_{ij}^{\mathrm{T}}b_i\,\mathrm{d}x\mathrm{d}y,$$

$$F_{qi} = \sum_{j=1}^{n_{\mathrm{e}i}}\int_{S_{ij}^{q}} N_{ij}^{\mathrm{T}}\overline{p}_{ij}\,\mathrm{d}s,\qquad F_{pi} = \sum_{j=1}^{n_{\mathrm{e}i}}\int_{S_{ij}^{p}} N_{ij}^{\mathrm{T}}f_{ij}\,\mathrm{d}s,$$

$$F_{a_t i} = -\sum_{j=1}^{n_{\mathrm{e}i}}\iint_{A_{ij}} \rho_i\overline{\alpha}N_i^{\mathrm{T}}\left(v_{tij}/\Delta t + \tfrac{1}{2}a_{tij}\right)\mathrm{d}x\mathrm{d}y,$$

$$F_{v_t} = -\sum_{j=1}^{n_{\mathrm{e}i}}\iint_{A_{ij}} \eta_i N_{ij}^{\mathrm{T}}N_{ij}^{\mathrm{T}}\left(\overline{\beta}v_{tij} - \left(1 - 0.5\overline{\beta}\right)a_{tij}\Delta t\right)\mathrm{d}x\mathrm{d}y,$$

$$F_{\varepsilon_t i} = -\sum_{j=1}^{n_{\mathrm{e}i}}\iint_{A_{ij}} \widetilde{B}^{\mathrm{T}}\left(\tfrac{\partial}{\partial x},\tfrac{\partial}{\partial y}\right)D_i\varepsilon_{tij}\,\mathrm{d}x\mathrm{d}y,\ F_{\sigma_t i} = -\sum_{j=1}^{n_{\mathrm{e}i}}\iint_{A_{ij}} \widetilde{B}^{\mathrm{T}}\left(\tfrac{\partial}{\partial x},\tfrac{\partial}{\partial y}\right)\sigma_{tij}\,\mathrm{d}x\mathrm{d}y,$$

$$C_{ij}^{\mathrm{p}} = \iint_{A_{ij}} -\widetilde{B}^{\mathrm{T}}\left(\tfrac{\partial}{\partial x},\tfrac{\partial}{\partial y}\right)N_{ij}D_{ij}G_{ij}^{\mathrm{p}}\,\mathrm{d}x\mathrm{d}y,\ C_{Rijkl} = -\int_{S_{ijkl}^{c}} L_{ij}^{\mathrm{T}}N_{ij}^{\mathrm{T}}T_{ijkl}\,\mathrm{d}s$$

where $\overline{\alpha} = 1/\alpha$, $\beta = \beta/\alpha$; α and β are the parameters of Newmark's time integration scheme. It is obvious that Equation (18) is a linear complimentary problem (LCP) with free variables and equality constraints to be solved by the following algorithm.

2.5 *Computer algorithm*

The system equation (18) can be solved by virtue of a series of transformations. From Equation (18a), the displacements can be solved

$$U = K^{-1}\left(C_{\mathrm{p}}\left(\delta^{\mathrm{p}} - \delta_t^{\mathrm{p}}\right) + C_R R + F\right) \tag{19a}$$

Substituting it into Equation (18b) yields

$$R = K_{\mathrm{R}}\left(H\delta^{\mathrm{c}} - G^{\mathrm{c}}K^{-1}C_{\mathrm{p}}\left(\delta^{\mathrm{p}} - \delta_t^{\mathrm{p}}\right) - G^{\mathrm{c}}K^{-1}F - G_t^{\mathrm{c}}\right) \tag{19b}$$

in which

$$K_{\mathrm{R}} = \left(G^{\mathrm{c}}K^{-1}C_{\mathrm{R}}\right)^{-1}$$

Substitution of Equation (19b) into Equation (18c) gives

$$\lambda^{\mathrm{c}} + X_{\mathrm{cc}}\delta^{\mathrm{c}} + X_{\mathrm{cp}}\delta^{\mathrm{p}} = Y^{\mathrm{c}} \tag{19c}$$

where

$$X_{\mathrm{cc}} = M^{\mathrm{c}}K_{\mathrm{R}}H,\quad X_{\mathrm{cp}} = -M^{\mathrm{c}}K_{\mathrm{R}}G^{\mathrm{c}}K^{-1}C_{\mathrm{p}},$$

$$Y^{\mathrm{c}} = K^{\mathrm{fc}} - M^{\mathrm{c}}R_t + M^{\mathrm{c}}K_{\mathrm{R}}\left(G^{\mathrm{c}}K^{-1}F + G_t^{\mathrm{c}}\right) + X_{\mathrm{cp}}\delta_t^{\mathrm{p}}$$

Then by substituting Equations (19a) and (19b) into Equation (18e), following equation can be obtained

$$\lambda^{\mathrm{p}} + X_{\mathrm{pc}}\delta^{\mathrm{c}} + X_{\mathrm{pp}}\delta^{\mathrm{p}} = Y^{\mathrm{p}} \tag{19d}$$

where

$$X_{\mathrm{pc}} = M^{\mathrm{p}}DEK^{-1}C_{\mathrm{R}}K_{\mathrm{R}}H,$$

$$X_{\mathrm{pp}} = M^{\mathrm{p}}DEK^{-1}C_{\mathrm{p}} - M^{\mathrm{p}}DEK^{-1}C_{\mathrm{R}}K_{\mathrm{R}}G^{\mathrm{c}}K^{-1}C_{\mathrm{p}} - M^{\mathrm{p}}DG^{\mathrm{p}},$$

$$Y^{\mathrm{p}} = K^{\mathrm{fp}} - M^{\mathrm{p}}\overline{\sigma}_t - M^{\mathrm{p}}DEK^{-1}F + M^{\mathrm{p}}DEK^{-1}C_{\mathrm{R}}K_{\mathrm{R}}\left(G^{\mathrm{c}}K^{-1}F + G_t^{\mathrm{c}}\right) + X_{\mathrm{pp}}\delta_t^{\mathrm{p}}$$

Combining Equations (19c) and (19d) along with Equations (18d) and (18f), the following standard complementary equations can be gained

$$\lambda + X\delta = Y \tag{20a}$$

$$\lambda \cdot \delta = 0,\ \lambda \geq 0,\ \delta \geq 0 \tag{20b}$$

where

$$\lambda = \begin{Bmatrix} \lambda^{c} \\ \lambda^{p} \end{Bmatrix}, \quad \delta = \begin{Bmatrix} \delta^{c} \\ \delta^{p} \end{Bmatrix}, \quad X = \begin{bmatrix} X_{cc} & X_{cp} \\ X_{pc} & X_{pp} \end{bmatrix}, \quad Y = \begin{Bmatrix} Y^{c} \\ Y^{p} \end{Bmatrix}$$

For this standard LCP problem, i.e., Equation (20), Lemke's complementary rotating algorithm is preferred to employ because of its higher efficiency and numerical stability.

3 NUMERICAL EXAMPLE

To illustrate the proposed method, the interaction system of retaining wall and backfill with a smooth wall-back as shown in Figure 1(a) is studied. The retaining wall is taken as a rigid body and the computational model of the system is given in Figure 1(b). In order to investigate the effects of soil parameters on earth pressures, different cases are considered by respectively varying cohesion ((e.g., c=0, 6.25, 12.5, 25.0, 37.5kPa for $\phi = 20°$) and friction angle (e.g., $\phi = 0°, 10°, 20°, 30°$ for c=25kPa). Three states, i.e., static state, active and passive limit states, of the system are respectively analyzed.

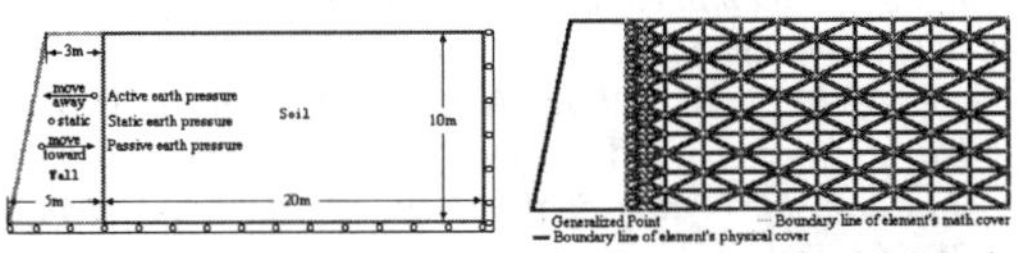

(a) Wall-soil system (b) Computational Model

Figure 1. Retaining wall and backfill soil system.

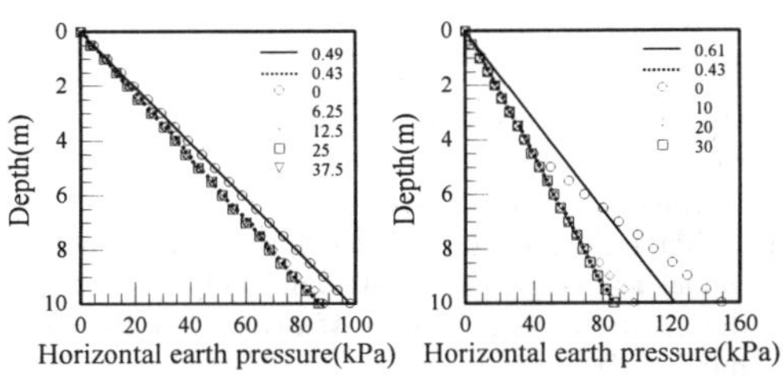

(a) $\phi = 20°$ (b) $c = 25$ kPa

Figure 2. Distribution of static earth pressures.

3.1 *Static earth pressures*

Shown in Figure 2(a) is distribution of the computed static earth pressures for different values of cohesion under $\phi = 20°$. The variation of static earth pressure along depth is basically linear. For a given value of soil friction angle, static pressure decreases with increase of soil cohesion, and however it will keep a constant while cohesion increases to a certain value. The effect of soil cohesion on static earth pressure becomes considerable especially when cohesion is very small. It seems that the static earth pressures locate between two lines, of which earth pressure coefficients K_0 at rest are respectively equal to 0.49 and 0.43. As shown is Figure 2(b), for pure cohesive backfill (e.g., $c = 25$ kPa), static earth pressures vary nonlinearly along depth since soils under a certain depth enter into plastic state. Compared with empirical curve with $K_0 = 0.61$, there is a certain degree of deviation, however their total static earth pressures are almost identical. The degree of non-linearity decreases with increase of friction angle of soil. When soil friction angle increases to a certain degree, static earth pressures keep constant and agree well with empirical line with $K_0 = 0.43$.

3.2 *Active earth pressures*

Distribution of active earth pressures is shown in Figure 3. It is observed that active earth pressure is approximately linear function of depth and cohesion. With increase of soil cohesion, tension-crack depth increases, however earth pressure decreases. Distribution of active earth pressure has some deviation at the neighborhood of tension-crack depth compared with Rankine's theory. However with increase of depth, it is very close to Rankine's solution. This is because that the proposed method can naturally guarantee active earth pressure to be zero up to tension-crack depth, however in Rankine's theory it is artificially regarded as zero in fact that it is less than zero. It is obvious from Table 1 and Table 2 that the computed total active earth pressures are slightly lower than Rankine's solutions. For a given cohesion of backfill (e.g., $c = 25$ kPa), active earth pressure is nonlinearly related to friction angle of soil. With increase of friction angle, tension-crack depth increases, and active earth pressure decreases.

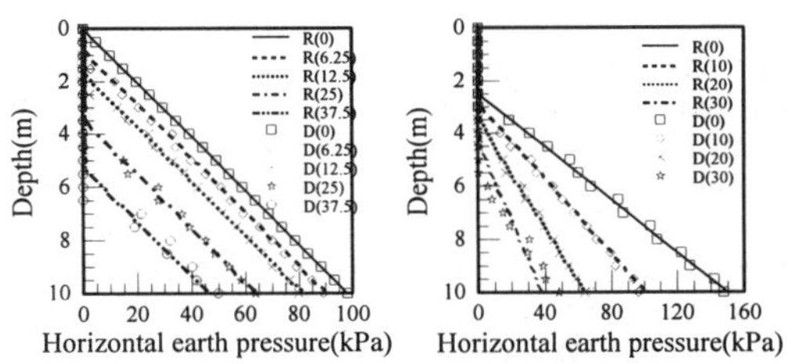

(a) $\phi = 20°$ (b) $c = 25$ kPa

Figure 3. Distribution of active earth pressures.

Table 1 Comparison of earth pressures ($\phi = 20°$)

c	P_a (kPa)		P_p (kPa)	
(kPa)	Rankine	DDCMM	Rankine	DDCMM
0	490.29	490.04	2039.61	2042.07
6.25	406.67	405.01	2218.13	2220.16
12.5	330.86	330.01	2396.65	2399.81
25	202.69	200.01	2753.68	2756.02
37.5	105.77	105.01	3110.72	3113.79

Table 2 Comparison of earth pressures ($c = 25$ kPa)

ϕ	P_a (kPa)		P_p (kPa)	
(°)	Rankine	DDCMM	Rankine	DDCMM
0	562.5	560.4	1500.00	1501.48
10	347.04	345.02	2016.15	2017.07
20	202.69	200.01	2753.68	2756.02
30	107.16	105.01	3866.03	3849.6

3.3 *Passive earth pressures*

Distribution of the computed passive earth pressures is shown in Figure 4. Obviously, the passive earth pressure varies linearly along depth. It increases proportionally with cohesion and agrees well Rankine's solution. For a given cohesion of backfill, e.g., $c = 25$ kPa, it is indicated that passive earth pressure is nonlinearly associated with friction angle of soil. The passive earth pressure becomes larger for higher friction angle. From Table 1 and Table 2, it can be seen that total passive earth pressure is exactly the same as Rankine's theory.

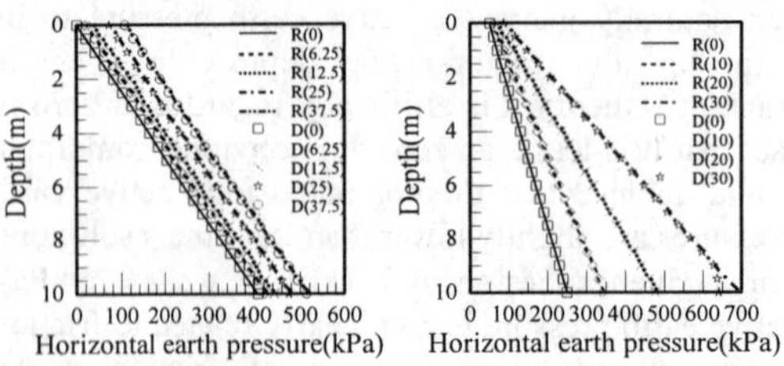

(a) $\phi = 20°$ (b) $c = 25$ kPa

Figure 4. Distribution of passive earth pressures.

4 CONCLUSIONS

In this paper, the discontinuous deformation computational mechanics model (DDCMM) for multi-body interaction system with contacts is extended to consider elasto-plastic behavior of geo-materials. The fundamental theory and mathematical formulations are given. Then the method is numerically implemented and the interaction system of retaining wall and backfill with a smooth wall-face is studied. Through numerical computations and analyses, it is shown that the proposed method is potentially capable to simulate complex mechanical behavior of multi-body systems with structural discontinuity and material nonlinearity.

ACKNOWLEDGMENTS

The financial support from the Trans-Century Training Programme Foundation for the Talents offered by the Ministry of Education of China is gratefully acknowledged.

REFERENCES

Carpenter, N.J., R.L. Taylor & M.G. Katona 1991. Lagrange constraints for transient finite element surface contact. *International Journal for Numerical Methods in Engineering*, 32:103-128.

Cundall, P.A & R.D. Hart 1989. Numerical modeling of discontinuation. In G.G.W. Mutoe, M. Henriksen & H-P Huttelmaier (eds), *Proceedings of the First U.S. Conference on Discrete Element Methods, USA, 1989*:1-19, CSM Press.

Goodman, R.E., R.L. Taylor & T.L. Brekke 1968. A model for the mechanics of jointed rock. *Journal of the Soil Mechanics and Foundations Division, ASCE*, 94(SM3): 637-659.

Heege, A. & P. Alart 1996. A frictional contact element for strongly curved contact problems. *International Journal for Numerical Methods in Engineering*, 39:165-184.

Li, Yong & Maotian Luan 2000. Fundamental and formulations of discontinuous deformation mechanics model and its linear programming algorithm. *Journal of Dalian University of Technology*, 40(3):351-357.

Luan, Maotian, Yong Li & Qing Yang 2000. The discontinuous deformation computational mechanics model and its application in stability analysis of rock slope. *Chinese Journal of Rock Mechanics and Engineering*, 19(3):289-294.

Luan, Maotian & Yong Li. The computational mechanics model and applications of multi-body discontinuous deformation system. *Acta Mechanica Sinica* (to be published).

Ngo, D. & A.C. Cordelis 1967. Finite element analysis of reinforced beam. *ACI Journal*, 64(14):152-163.

Shi, Genhua 1988. Discontinuous deformation analysis: a new numerical model for the statics and dynamics of block system. *Ph.D. Thesis*, University of California, Berkeley, California.

Shi, Genhua 1996. Manifold method. In M.Reza Salami & Don Banks (eds), *Proceedings of the First International Forum on Discontinuous Deformation Analysis (DDA) and Simulations of Discontinuous Media, Berkeley, California, USA, June 1995*: 52-204, New Mexico:TSI Press.

Computer Methods and Advances in Geomechanics, Desai et al. (eds) © 2001 Balkema, Rotterdam, ISBN 90 5809 183 X

Grain-shape effect on peak strength of granular materials

T. Matsushima
Institute of Engineering Mechanics and Systems, University of Tsukuba, Japan

K. Konagai
Institute of Industrial Science, University of Tokyo, Japan

ABSTRACT: The effect of irregular grain shape on the peak strength of granular materials is studied by means of the Discrete Element Method (DEM). A series of simple shear tests is simulated with elliptic elements (EL), regular polygonal elements (RP), and circular elements (CR). The results suggest that overall shape, which can be represented by EL, does not greatly affect the strength of granular assembly if compared with CR with the same void ratio. On the other hand, the effect of surface roughness, represented by RP, is dominant, because the grain column formed in the assembly is stable due to a large moment transmission at the contact points. In order to verify these effects on real sands, grain shape is analyzed for several sands. The results of this analysis, combining with the existing experimental results, appear to be consistent with the numerical results.

1 INTRODUCTION

A granular material is an assembly of many discrete grains, and its mechanical characteristics are necessarily affected by microscopic parameters such as grain size, grain shape, and crushability. Among them, it was reported that the shear strength of poorly graded sand under the usual confining pressure differs in different sands mainly due to differences in grain shape (Oda 1971, Yoshida 1994, Yoshimura and Ogawa 1994). This paper focuses on this grain-shape effect.

The natural grain of sands and gravels has an irregular, complicated shape, and it is therefore difficult to define a shape parameter that properly represents its mechanical contribution. Various parameters have been proposed by many researchers so far (Wadell 1932, Lees 1964, Zingg 1935, Yoshida 1994, Yoshimura 1994). According to Oda (Oda 1978), the existing shape parameters are roughly classified as follows:

(1) Overall shape (average shape): sphericity (Krumbein 1941), elongation ratio and flatness ratio (Lees 1964) are included in this category.
(2) Surface roughness: roundness (Wadell 1932) and angularity (Lees 1964) are included in this category.
(3) Very small surface unevenness due to the molecule arrangement, which is related to the mineral contents of the material.

The mathematical interpretation of this classification can be possible by means of Fourier expansion (Zahn and Roskies 1972, Yoshida 1994); its lower-order terms correspond to the overall shape, the relatively higher-order terms to the surface roughness, and the very high-order terms to the very small surface unevenness.

Based on the above classification, some experimental studies have been made. Oda et al. (Oda et al. 1971) measured maximum and minimum void ratios (void-ratio ranges) and shear strengths by direct shear tests for various sands and suggested that surface roughness affects both the shear strength and void-ratio range to a considerable extent, while the other two shape parameters are not influential. Yoshimura and Ogawa (Yoshimura and Ogawa 1994) obtained the similar results by a series of Triaxial compression tests. On the other hand, Miura et al. (Miura et al. 1998) conducted similar triaxial tests with various sands, and reported that the effect of crushability is more significant than that of angurality.

In spite of these extensive studies, the mechanical role of grain shape is still not clear, and the quantitative estimation of the grain-shape effect is not possible, mainly due to the complexity of the microscopic parameters of sands. These parameters are mutually dependent and it is difficult to study the effect of a specific parameter by a physical experiment.

On the contrary, a numerical experiment based on the Discrete Element Method (DEM) is a powerful tool to study such effects because it allows changing a specific microscopic parameter while keeping another parameter unchanged. Moreover, this method provides micro-mechanical information including contact forces and grain motions. Therefore, many numerical studies have been made in recent decades to discuss the grain-shape effect on the macro behavior of granular materials (Mahboubi et al. 1997, Mirghasemi et al. 1997). Among them, Rothenberg and Bathurst (Rothenberg and Bathurst 1992) conducted a series of bi-axial test with elliptic elements of different aspect ratios, and showed that the maximum peak strength was exhibited with an aspect ratio of 0.7 to 0.8. However, DEM can treat only idealized shapes such as ellipses and polygons, which are far from the real grain shape of sands, and there is no comprehensive research including physical and numerical observations.

This paper deals with three types of elements, namely circular, elliptic, and regular polygonal elements, which are denoted by CR, EL, and RP, respectively. The effect of overall grain shape can be studied by comparing the behavior of EL with that of CR, while surface-roughness effect is related with the RP behavior. The effect of the very small surface unevenness is not discussed in this paper though it can be studied in DEM by changing the friction coefficient between elements.

A series of simple shear simulations is conducted with various void ratios. Based on both macroscopic and microscopic results, the overall-shape effect and the surface-roughness effect are discussed. In particular, the importance of moment transmission at the grain contacts is discussed.

The numerical results are, then, examined by the behavior of real sands. Digital images of grains are collected for several sands, and their shape parameters based on the Fourier expression are calculated. Together with the existing PSC test results, the grain-shape effect of sands is discussed.

2 A 2-D DEM WITH ELLIPTIC AND REGULAR-POLYGONAL GRAINS

In order to study the effect of two grain-shape paramers, namely overall shape and surface roughness, two programs, respectively denoted as EL-DEM and RP-DEM, were developed. The former deals with EL and CR grains, and the latter treats RP grains.

Both programs are based on the explicit-time marching scheme, as is usual in the Discrete Element Method (Cundall 1971). To verify the accuracy of the simulation, all kinds of mechanical energies in the whole system are calculated, from which the numerical error, err, is checked at every time step according to the following equations:

$$err = E_{err} / E_{sh} \quad (1),$$

$$E_{err} = E_{int} - E_{ext} = (E_v + E_k + E_c + E_f + E_p + E_{wp} + E_{wc}) - E_{sh} \quad (2),$$

where E_{err} is the error of the total energy balance, E_v is the total kinetic energy of the grains, E_k is the total strain energy at the contact points, E_c is the energy dissipation by damping at contact points, E_f is the energy dissipation by friction at contact points, E_p is the total potential energy of grains by gravity force (=0 in this study), E_{wp} is the potential energy with respect to the pressure-controlled wall, E_{wc} is the energy dissipation by the damping of the pressure-controlled wall, and E_{sh} is the applied energy by shearing.

The contact-detection algorithm of EL-DEM is made analytically by Euler's method for a 4-th order equation (Matsushima 1992). In the program, however, some additional branches of the algorithms are necessary for the case in which the coefficient of the 4-th order term is close to zero. Consequently the calculation time is three to four times as long as that of the simplest DEM with CR assembly.

In an RP-DEM algorithm (Matsushima 1997), the following two advantages make the computation efficient in comparison with the usual DEM with polygonal elements:

(1) Each element has only one additional parameter, the number of angles, in contrast to circular elements.

(2) A preliminary contact judgment between the circumscribed circles of an RP element works quite effectively.

As a consequence, its computation time is only three to four times as long as that in the case of CR assembly.

3 DEM SIMPLE SHEAR SIMULATION

3.1 DEM parameters

Table 1 shows the DEM parameters used in this study. Grain-size distribution is shown in Figure 1(a). It is based on a normal distribution, which has a mean diameter, D_{50}, of 2.0mm. Grain-shape distributions for RP and EL are shown in Figure 1(b) and (c), respectively. In Figure 1(b), the average element-shape is a regular octagon, and this assembly is denoted as RP(8) in this study. In the same fashion, the assembly shown in Figure 1(c) is denoted as EL(0.8), since the average aspect ratio, b/a, is 0.8.

Table 1 DEM parameters used in this study

Parameter	Value
Density of grain	2.5 (g/cm^2)
Spring constant (normal)	9.6e7 (N/cm)
Spring constant (shear)	2.4e7 (N/cm)
Damping coefficient (normal)	2.0 (kg/s)
Damping coefficient (shear)	1.0 (kg/s)
Friction coefficient ($\tan\phi_\mu$)	0.51 (ϕ_μ=27 deg.)
Time increment	2.0e-6 (s)

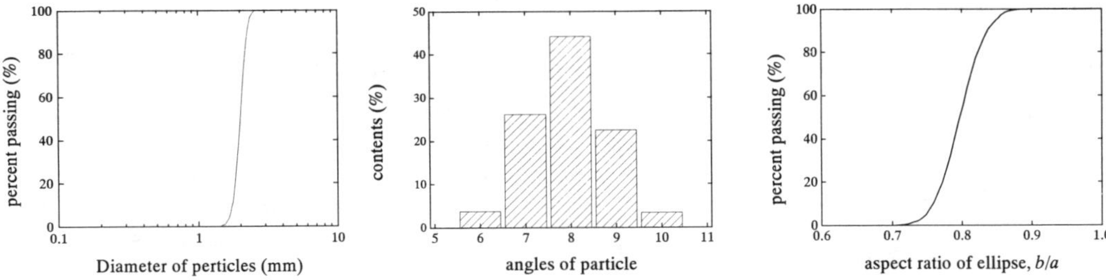

(a) Grain-size distribution (b) Grain-shape distribution (RP) (c) Grain-shape distribution (EL)

Figure 1 Size and shape distributions used in this study

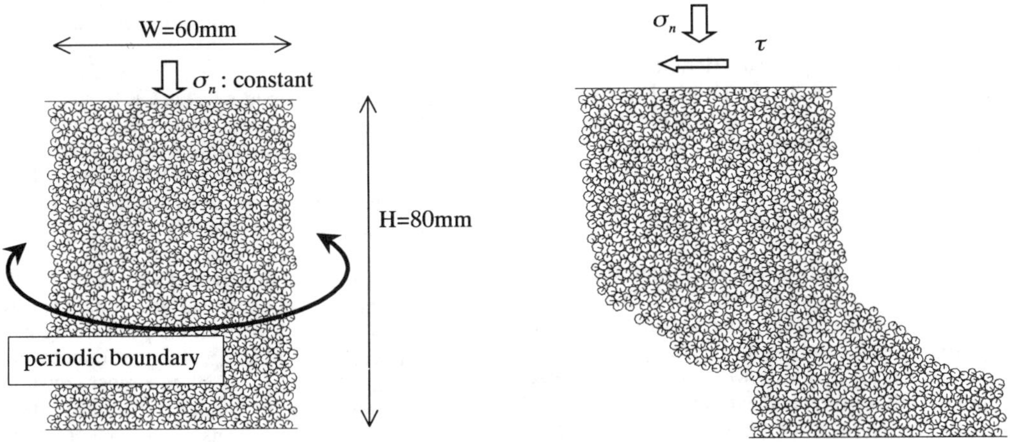

(a) The initial state (b) A specimen after the shear

Figure 2 A specimen before and after the shear deformation

3.2 *An Outline of simple shear simulation*

In order to avoid the effect of inherent anisotropy, a specimen is packed in a circular domain by a gravity force, which is directed to the center of the domain. Then the specimen for the simple shear test is cut out of it in a rectangular shape. After that, introducing periodic boundaries as well as wall elements, a simple shear specimen that contains about 1100 elements is prepared as shown in Figure 2(a). Keeping the confining pressure, σ_n, constant ($\sigma_n \approx 0.28$ N/cm), the bottom wall element is moved in a horizontal direction. Accordingly, the specimen is sheared and a shear band is built up as shown in Figure 2(b).

Figure 3 shows an example of a macroscopic result, namely the variation of stress ratio, τ / σ_n, to the global shear strain, $\gamma_{ave} = d_s / h$, and the variation of the global normal strain, $\varepsilon_{ave} = d_n / h$, to γ_{ave}, where τ is the shear stress, d_n and d_s are normal and shear displacements of the wall element, respectively, and h is the height of the specimen. From the peak value of the stress ratio, the internal friction angle for simple shear, $\phi_{peak} = \tan^{-1}(\tau / \sigma_n)$, can be calculated.

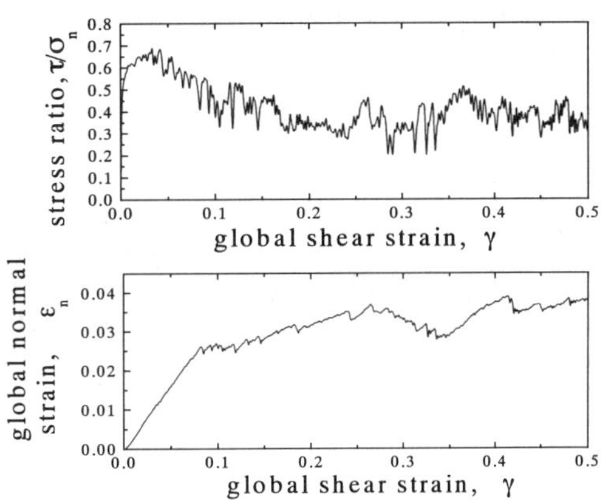

Figure 3 An example of a stress-strain relation (top) and a dilation curve (bottom)

Figure 4 shows an example of the energy balance during a shear test. The energy given by the external shear force is consumed mainly by the dilation of the

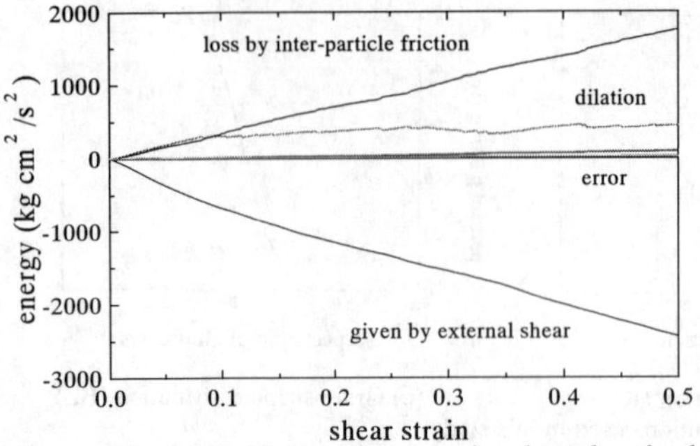

Figure 4 An example of energy conservation through a simple shear test

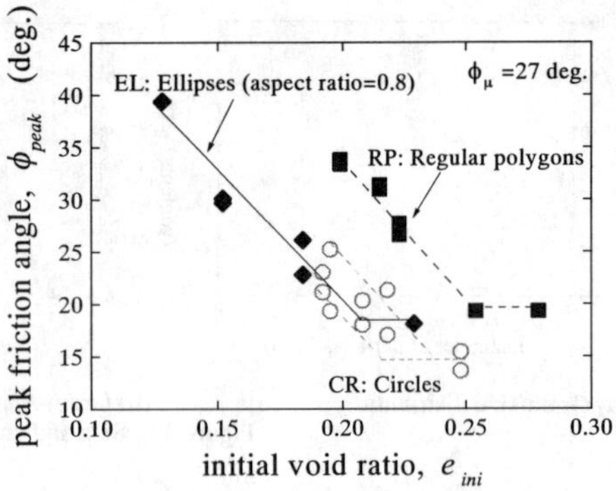

Figure 5 Variation of a peak friction angle to an initial void ratio

material and the inter-particle friction. This implies that the damping at the contact points does not play a significant role. The error of the calculation is small enough (around 2% of the energy given by shear).

3.3 Numerical results and discussion

Several specimens with different initial void ratios were prepared for the simulation. The densest specimen for each element type was made keeping the friction coefficient zero during the packing process. In most cases, two tests were simulated for each specimen with different shear directions. Figure 5 shows the variation of the peak friction angle, ϕ_{peak}, (peak strength) with the initial void ratio, e_{ini}, and the grain type. This figure indicates the following two important points:

(1) A dense EL exhibits a very high ϕ_{peak} due to its relatively high density in comparison with that with CR. However, if ϕ_{peak} at the same e_{ini} is compared, EL has the almost same ϕ_{peak} as CR.

(2) RP gives a higher ϕ_{peak} than CR at the same e_{ini}. However RP cannot be compacted as densely as EL, and the peak strength of dense RP is lower than that of dense EL. As for the loose specimen, very loose specimen can be made with RP.

These results imply that the overall shape represented by EL affects the peak strength indirectly via the void ratio, while the surface roughness represented by RP directly acts on the peak strength.

If the variation of ϕ_{peak} to e_{ini} is concerned, the decreasing rate of ϕ_{peak} to e_{ini} is almost constant and does not depend on the element type except the cases of the very loose specimens. This tendency seems to be consistent with the experimental results and the theories (Caquot and Kerisel 1956, Mogami 1965). This effect may be connected with the coordination number (the number of contact points per a grain), as indicated by Rothenberg and Bathurst (Rothenberg and Bathurst 1992).

The micro-mechanism of the surface-roughness effect can be explained with Figure 6, which shows grains transmitting the normal contact forces higher than the average force in the simulations on a dense RP and a dense CR. These figures also include information about the directions of grain rotations. These clockwise and anti-clockwise rotations tend to take place

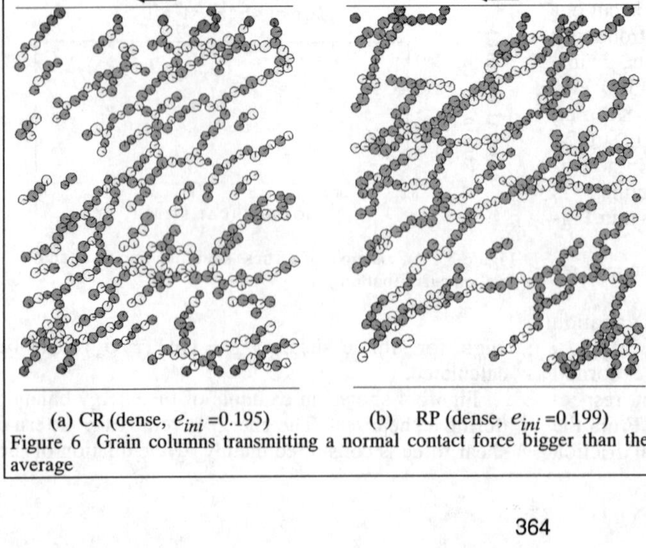

(a) CR (dense, e_{ini} =0.195) (b) RP (dense, e_{ini} =0.199)

Figure 6 Grain columns transmitting a normal contact force bigger than the average

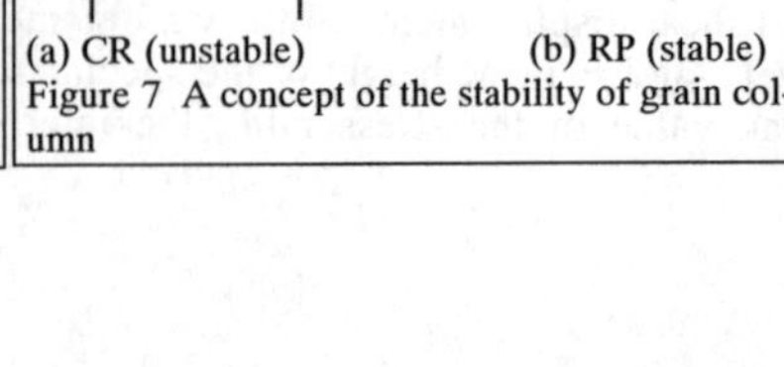
(a) CR (unstable) (b) RP (stable)

Figure 7 A concept of the stability of grain column

by turns like gear motion in CR, while a couple of grains contacting to each other tend to rotate in the same direction like a rigid column (or a rigid mass) in RP.

This difference in the microstructure should be mainly caused by the existence of moment transmission at the contact point, which avoids a relative rotation between the contacting grains. Moreover the average magnitude of the contact moment should be related to the surface roughness of the grains. An angular grain tends to exhibit a relatively higher contact-moment than a round grain, resulting in a stable grain column (Figure 7).

4 A MESUREMENT OF GRAIN-SHAPE OF SANDS AND ITS EFFECT

4.1 An Image analysis of grain-shapes of real sands (Yoshida 1993)

In order to study the grain-shape effect of real sands, an image analysis of sand grains was conducted for four sands including S.L.B. sand, Ottawa sand, Toyoura sand, Wakasa sand (Figure 8). The Images of 50 grains for each sand were taken with a digital microscope. The contour of each grain was then analyzed by an edge-detection algorithm, and its radius function $r(\theta)$ (Figure 9) was obtained. Moreover, $r(\theta)$ was expanded in a Fourier series as follows:

$$r(\theta) = \frac{b_0}{2} + \sum_{k=1}^{\infty}\{a_k \sin(k\theta) + b_k \cos(k\theta)\} \qquad (3),$$

where a_k and b_k are coefficients of each frequency, k, and they can be expressed as follows:

$$a_k = \frac{1}{\pi}\sum_{i=1}^{ndot}\{r(\theta_i)\sin(k\theta_i)(\theta_i - \theta_{i-1})\} \qquad (4),$$

$$b_k = \frac{1}{\pi}\sum_{i=1}^{ndot}\{r(\theta_i)\cos(k\theta_i)(\theta_i - \theta_{i-1})\} \qquad (5),$$

and $ndot$ is the number of pixels of the contours. Finally Fourier spectrum for each k is:

$$c_k = \sqrt{a_k^{\,2} + b_k^{\,2}} \qquad (6)$$

The spectrum c_0 represents the average radius, and c_1 should be equal to zero if the radius function is defined based on the center of the grain. The spectrum c_2 represents the overall grain shape as the aspect ratio, and the spectra for higher k correspond to the surface roughness of a grain.

4.2 Results and discussions

Figure 10 plots PSC-test results for the four sands collected from the literature (Tatsuoka et al. 1987, Yoshida 1994, Park 1993). In order to remove the effect of the void ratio, a peak friction angle at the

Figure 8 Photographs of the sand grains taken with a digital microscope

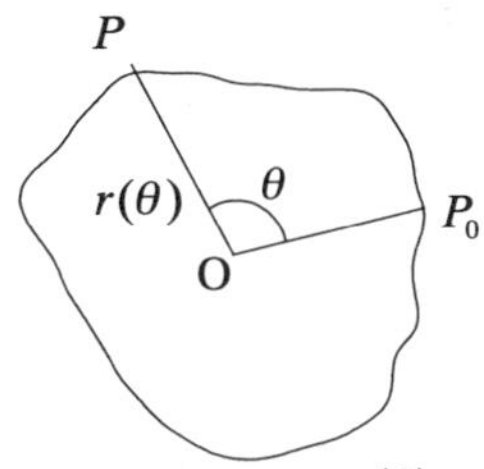

Figure 9 Radius function, $r(\theta)$

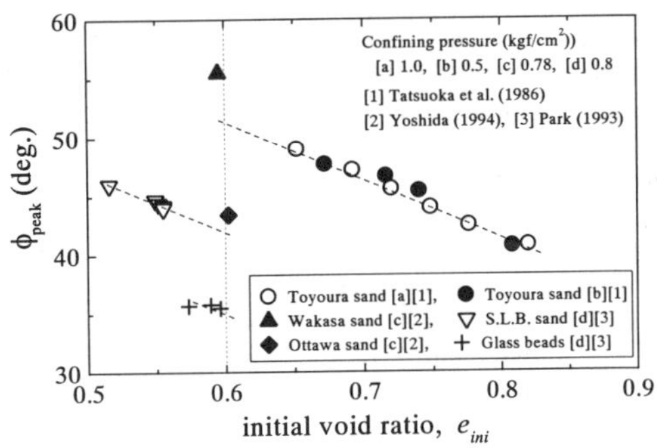

Figure 10 Variation of ϕ_{peak} to e_{ini} in sands

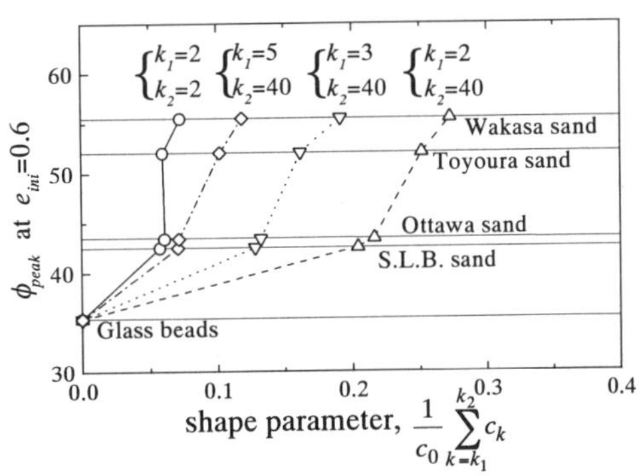

Figure 11 Variation of ϕ_{peak} at $e_{ini} = 0.6$ to a shape parameter based on Fourier expression

void ratio of 0.6 is determined assuming a linear relation between ϕ_{peak} and e_{ini}.

Comparing Figure 10 with Figure 5, the following considerations on the grain-shape effect are possible:
(1) Surface roughness is significant in Toyoura sand and Wakasa sand.
(2) S.L.B. sand should be more elliptic than Toyoura sand.
These tendencies may also be recognized from Figure 8.

Moreover, Figure 11 shows the relationship between ϕ_{peak} at the same e_{ini} and the shape parameters, which are defined by the summations of Fourier spectra normalized by the grain radius, c_0. The parameter with k_1=2 and k_2=2 corresponds to the overall-shape parameter, c_2/c_0, and there is little variation with different sands. On the other hand, the surface-roughness parameter by the summation of spectra higher than c_5, clearly increases with increasing the peak strength, which is consistent with the numerical results.

5 CONCLUSIONS

The grain-shape effect on the peak strength of granular materials was studied both numerically and experimentally. DEM simple shear simulations with regular polygonal elements, elliptic elements and circular elements were conducted with several specimens of different initial void ratios. The findings are summarized as follows:
(1) Overall shape affects the void ratio range but is not significant for the peak strength in a direct manner.
(2) Surface roughness makes the grain-column structure stable because of the contact-moment transmission, which results in higher peak strength.

The above results were verified with experimental results with real sands.

ACKNOWLEDGMENT

Authors gratefully acknowledge Prof. Tatsuoka of University of Tokyo and Dr. Yoshida of Kajima Corporation for invaluable discussions throughout the research, Mr. Sato of the University of Tokyo for providing the sand materials, Prof. Tateishi of Nagoya University for the rental of a digital microscope, and Mr. Kaneko, a former student of the University of Tsukuba, for the image analysis. This work was supported in part by a grant-in-aid of the Ministry of Education, Science, Sports and Culture, Japan.

REFERENCES

Caquot, A. and Kerisel, J. 1956. Traite de mechanique de sols. 3e edition, Gauthier Villars.

Cundall, P. A. 1971 A computer model for simulating progressive, large-scale movements in blocky rock systems. Symp. ISRM, Nancy, France. Proc., 2: 129-136.

Krumbein, W. C. 1941 Measurement and geologic significance of shape and roundness of sedimentary particles, J. of Sedimentary Petrology 11: 64-72.

Lees, G. 1964. A new method for determing the angularity of particles, Sedimentology 3: 2-21.

Mahboubi, A., Cambou, B. and Fry, J-J. 1997 Numerical modeling of the mechanical behavior of non spherical, crushable particles. Powders & Grains 97, Behringer & Jenkins (eds): 139-142.

Matsushima, T. 1992. Study on dynamic failure process of particle assembly by Discrete Element Method, Master thesis, University of Tokyo (in Japanese).

Matsushima, T. 1997. Grain-to-grain interaction affecting dynamic deformation of granular slope. Doctoral dissertation, University of Tokyo (in Japanese).

Mirghasemi, A. A., Rothenburg, L. and Matyas, E. L. 1997. Numerical simulations of assemblies of two-dimensional polygon-shaped particles and effects of confining pressure on shear strength. Soils and Foundations 37(3): 43-52.

Mogami, T.1965. A statistical approach to the mechanics of granular materials. Soils and Foundations 5(2): 26-36.

Oda, M., Enomoto, F. and Suzuki, T. 1971. A Study on Effects of grain-shape and mineralogy on geotechnical characteristics of sands. Tsuchi-to-Kiso, JGS, 12(4): 17-24 (in Japanese).

Oda, M. 1978. Fundamental characteristics of granular materials and their geotechnical significance. Tsuchi-to-Kiso, JGS, 26(8): 63-70 (in Japanese).

Park, C-S. 1993. Deformation and strength characteristics of a variety of sands by plane strain compression tests. Doctoral dissertation, University of Tokyo.

Rothenberg, L. and Bathurst, R. J. 1992. Micromechanical features of granular assemblies with planer elliptical particles. Geotechnique 42(1): 79-95.

Tatsuoka, F., Sakamoto, M., Kawamura, T. and Fukushima, S. 1936. Strength and deformation characteristics of sand in plane strain compression at extremely low pressures. Soils and Foundations 26(1): 109-118.

Wadell, H. 1932. Volume, shape and roundness of rock particles, J. Geology 40: 443-451.

Yoshida, T. 1994. Strain localization and shear banding during failure of sands. Doctoral dissertation, University of Tokyo.

Yoshimura, Y. and Ogawa, S. 1994. The influence of grain shape on isotropic consolidation and shear characteristics of sands. Proc. JSCE 487(III-26): 187-196 (in Japanese).

Zahn, C. T. and Roskies, R. Z. 1972. Fourier descriptions for plane closed curves, IEEE Trans. Computers. c-21: 269-281.

Zingg, T. 1935. Beitrag zur Schotteranalyse, Petrographische. Mitteilungen 15: 39-140.

Computer Methods and Advances in Geomechanics, Desai et al. (eds) © 2001 Balkema, Rotterdam, ISBN 90 5809 183 X

Shear behavior of sand under monotonic and cyclic loadings and its elastoplastic modeling

T.Nakai, M.Hinokio, T.Hoshikawa & H.Yoshida
Nagoya Institute of Technology, Japan

E.Q.Chowdhury
University of Calgary, Alb., Canada

ABSTRACT: A kinematic hardening elastoplastic model for sand, which has been named kinematic t_{ij}-sand model, is extended to simulate the behavior of sand under different density and under cyclic loading in general stress systems. Inclusion of the parameter '*d*' in the proposed model can express the influence of density and confining pressure. Incorporating the hardening limit surface, stabilization of strains and increase of stiffness can be simulated. The proposed model predicts well the observed behavior of sand in monotonic and cyclic triaxial and true triaxial tests. Stress-strain behavior, particularly anisotropy and dilatancy of sand, under monotonic and cyclic loadings is then discussed on the basis of experimental and calculated results.

1 INTRODUCTION

A kinematic hardening elastoplastic model for sand named t_{ij}-sand model (Nakai, Fujii and Taki; 1989) has been proposed, which can describe not only the influence of the intermediate principal stress on the deformation and strength and the influence of stress path on the direction of plastic flow but also the development of the anisotropy of sand due to stress path history. However, some of the material parameters still should vary with the initial density and/or confining pressure. It is also known as an experimental fact that the magnitude of the strains decreases and finally stabilizes with the increasing number of cyclic loading under drained condition. To improve these points, the previous kinematic hardening model is extended. The validity of the extended model is then checked by cyclic triaxial and true triaxial tests.

2 OUTLINE OF THE PREVIOUS MODEL

The previous kinematic hardening model (Nakai, Fujii and Taki, 1989) has been developed by introducing the kinematic hardening rule in stress ratio space into the isotropic hardening model based on t_{ij}-concept (Nakai, 1989). Since the detailed formulation of the model has been described in the paper mentioned above, we will show here only outline of the model.

To consider the influence of stress path, the plastic strain increment is divided into two components, namely, the component satisfying associated flow rule $d\varepsilon_{ij}^{p(AF)}$ and the component of isotropic compression $d\varepsilon_{ij}^{p(IC)}$.

$$d\varepsilon_{ij}^{p} = d\varepsilon_{ij}^{p(AF)} + d\varepsilon_{ij}^{p(IC)} \tag{1}$$

These two components are calculated as follows:

$$d\varepsilon_{ij}^{p(AF)} = \Lambda \frac{\partial f}{\partial t_{ij}} \tag{2}$$

$$d\varepsilon_{ij}^{p(IC)} = K\langle dt_N \rangle \frac{\delta_{ij}}{3} \tag{3}$$

Here, t_{ij} is the modified stress introduced instead of ordinary stress σ_{ij} (Nakai and Mihara; 1984.) to con-

Table 1 Comparison between ordinary concept and tij-concept

ordinary concept	t_{ij} - concept
σ_{ij}	$t_{ij}=\sigma_{ik}a_{kj}$
δ_{ij} (*unit tensor*)	a_{ij}
$p=\sigma_{ij}\delta_{ij}/3$	$t_N=t_{ij}a_{ij}$
$s_{ij}=\sigma_{ij}-p\delta_{ij}$	$t'_{ij}=t_{ij}-t_N a_{ij}$
$q=\sqrt{(3/2 s_{ij}s_{ij})}$	$t_S=\sqrt{(t'_{ij}t'_{ij})}$
$\eta_{ij}=s_{ij}/p$	$x_{ij}=t'_{ij}/t_N$
$\eta=q/p=\sqrt{(3/2\,\eta_{ij}\eta_{ij})}$	$X=t_S/t_N=\sqrt{(x_{ij}x_{ij})}$
$\eta_{ij}^{*}=(s_{ij}-s_{ij0})/p$	$x_{ij}^{*}=x_{ij}-n_{ij}$
$\eta^{*}=\sqrt{(3/2\,\eta_{ij}^{*}\eta_{ij}^{*})}$	$X^{*}=\sqrt{(x_{ij}^{*}x_{ij}^{*})}$
$d\varepsilon_v=d\varepsilon_{ij}\delta_{ij}$	$d\varepsilon_{SMP}^{*}=d\varepsilon_{ij}a_{ij}$
$de_{ij}=d\varepsilon_{ij}-d\varepsilon_v\delta_{ij}/3$	$d\varepsilon'_{ij}=d\varepsilon_{ij}-d\varepsilon_{SMP}^{*}a_{ij}$
$d\varepsilon_d=\sqrt{(2/3\,de_{ij}de_{ij})}$	$d\gamma_{SMP}^{*}=\sqrt{(d\varepsilon'_{ij}d\varepsilon'_{ij})}$

sider the influence of intermediate principal stress on the deformation and the strength. The yield surface f is formulated as a function of mean stress t_N and stress ratio Z based on the t_{ij}-concept. The comparison of the stresses and the strain increment parameters between the ordinary concept such as Cam-clay model and the t_{ij}-concept is shown in Table 1.

$$f = \ln t_N + \zeta(Z) - \ln t_{N1} = 0 \quad (Z = X^* + n) \tag{4}$$

and $\zeta(Z)$ is expressed by the following form:

$$\zeta(Z) = \frac{-\alpha}{1-\alpha}\ln\left(1-(1-\alpha)\frac{Z}{\mathrm{M}^*}\right) \tag{5}$$

Here, n in Equation (4) is the norm of n_{ij}, ($n=\sqrt{(n_{ij}\, n_{ij})}$), and n_{ij} represents the center of the yield surface in the stress ratio x_{ij} space. We assume the tensor n_{ij} is movable with keeping constant distance $X^*=\xi$ between the present stress ratio tensor x_{ij} and the back stress ratio tensor n_{ij}. In this model, plastic work $W^{*p}=\int t_{ij} d\varepsilon_{ij}^p$ is used as the plastic hardening parameter to describe positive and negative soil dilatancy.

3 EXTENSION OF MODEL

The yield function f represented by Equations (4) and (5) is obtained from the linear relation between stress ratio t_S/t_N and plastic strain increment ratio $d\varepsilon_{SMP}^{*p(AF)}/d\gamma_{SMP}^{*p(AF)}$ in Figure 1. The shape of yield surface in $t_N - t_S$ plane is shown by the dotted curve in Figure 2. However, this yield surface has singular points on the t_N axis and at the tip. To avoid this problem, the following stress ratio – plastic strain increment ratio relation (solid line in Fig.1) is assumed (Chowdhury and Nakai, 1998):

$$-\frac{d\varepsilon_{SMP}^{*p}}{d\gamma_{SMP}^{*p}} = \frac{Z^{\beta} - \mathrm{M}^{*\beta}}{Z^{\beta-1}} \tag{6}$$

This relation leads to the following yield function, which is represented by the solid line in Figure .2.

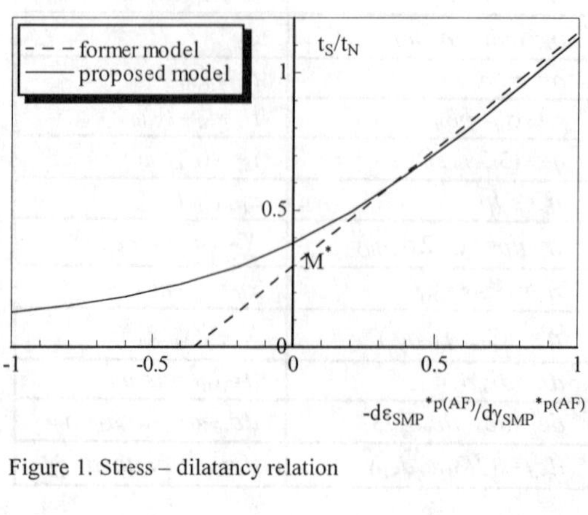

Figure 1. Stress – dilatancy relation

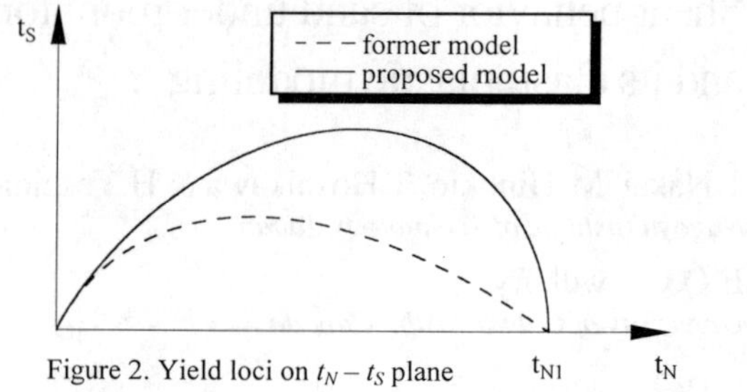

Figure 2. Yield loci on $t_N - t_S$ plane

$$f = \ln t_N + \frac{1}{\beta}\left(\frac{Z}{\mathrm{M}^*}\right) - \ln t_{N1} = 0 \tag{7}$$

In order to describe the stress-strain behavior of sand uniquely using the same material parameters regardless of the density, we introduce a new value d, which represents the difference between the present void ratio and the void ratio at normally consolidated state in the same stresses, as illustrated in Figure 3. In this figure, the solid line is the subloading surface (SLS), which always expands or contracts through the present stress state. The broken line is the normal yield surface (NYS), which corresponds to the usual yield surface of normally consolidated soil in conventional plasticity. Using these two surfaces, we can take into account the influence of density (Hashiguchi, 1980). Here, t_{N1} and t_{N1e} are the values of these surfaces at t_N axis. The difference d can be expressed as

$$d = (\lambda - \kappa)\ln\left(\frac{t_{N1e}}{t_{N1}}\right) \tag{8}$$

where d=0 denotes the normally consolidated state (the loosest state), and d>0 the denser state. Further, we use plastic volumetric strain ε_v^p as the strain hardening parameter instead of W^{*p} in the former

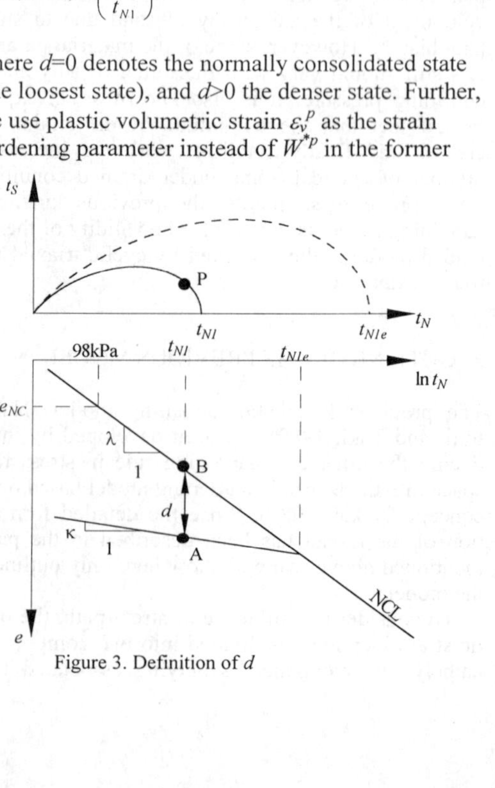

Figure 3. Definition of d

model. We give the plastic multiplier Λ in Equation (2) by the following form, introducing the valuable d and referring to subloading surface concept (Hashiguchi, 1980).

$$\Lambda = \frac{\dfrac{\partial f}{\partial \sigma_{lm}} d\sigma_{lm}}{\dfrac{1}{C_p}\left(\dfrac{\partial f}{\partial t_{kk}} + a \cdot \dfrac{d^2}{t_N}\right)} \tag{9}$$

The above formulation makes it possible to describe the following points: (1)positive and negative soil dilatancy regardless that plastic volumetric strain ε_v^p is used as the strain hardening parameter, (2)density dependency of the strength and stiffness and (3)strain softening .

The above subloading type model obeying kinematic hardening can describe gradual stabilization of strains, but the rate of stabilization is slower than that of experimental observation. To rectify this problem, a new surface named hardening limit surface is introduced, and the plastic coefficient Λ of Equation (9) is modified by the following manner. We introduce two new surfaces in addition, i.e., hardening limit surface (HLS) and maximum loading surface (MLS). Here, the hardening limit surface (HLS) is similar in shape with the normal yield surface and is equal to or larger than the normal yield surface in size. This surface controls the limit of plastic strain hardening – when the normal yield surface coincides with the hardening limit surface, plastic modulus becomes infinite and the sand behaves almost elastically. Next, the maximum loading surface (MLS), which keeps the track of the subloading surface has a similar shape with the subloading surface. This surface coincides with the subloading surface and expands with the subloading surface, when the subloading surface tries to expand beyond the preceding maximum loading surface. The values of HLS and MLS at t_N axis are denoted t_{NIL} and t_{NIU} as shown in Figure 4. To describe the stabilization of the stress-strain curve under cyclic loading, the plas-

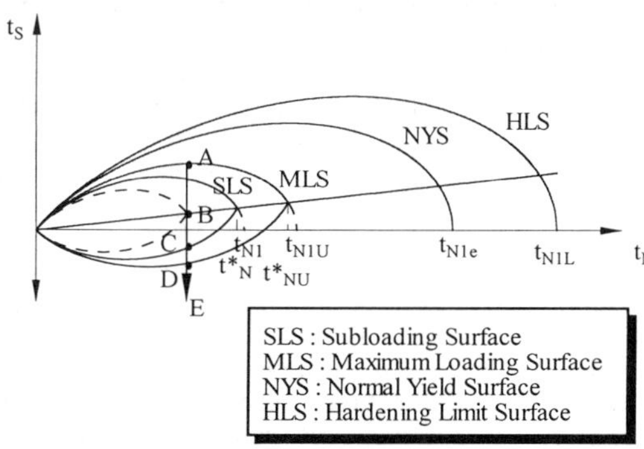

Figure 4.Explanation of four surfaces

tic multiplier Λ in Equation (9) is modified as follows:

$$\Lambda = \frac{\dfrac{\partial f}{\partial \sigma_{lm}} d\sigma_{lm}}{\dfrac{1}{F}\left\{\dfrac{1}{C_p}\left(\dfrac{\partial f}{\partial t_{kk}} + a\dfrac{d^2}{t_N}\right)\right\}} \tag{10}$$

As it is clear from this equation, the same plastic strain with that in Equation (9) occurs when F=1, no plastic strain occurs when F=0 and smaller strain occurs when 0<F<1. We then formulate the function F so as to satisfy the condition that F=0 when $t_{NIe}=t_{NIL}$, F=1 when $t_{NIe}<t_{NIL}$ & $t_{NI}=t_{NIU}$ and F<1 when $t_{NIe}<t_{NIL}$ & $t_{NI}<t_{NIU}$ and fit the observed trends.

$$F = \left(\frac{t^*_{N1}}{t^*_{N1U}}\right)^2 \left(\frac{t_{N1L} - t_{N1e}}{t_{N1L} - t_{N1e}\left(\dfrac{t_{N1}}{t_{N1U}}\right)^{0.2}}\right) \tag{11}$$

$$t_{N1L} = t_{N1e0}\left(b1 \cdot \left(\frac{t_{N1U} - t_{N1U0}}{t_{N1U0}}\right)^2 + b2\right) \tag{12}$$

4 EXPERIMENTAL VERIFICATION

The validity of the extended model is checked by triaxial tests and true triaxial tests under monotonic and cyclic loadings. Every test was carried out under constant mean principal stress (p=196kPa). Toyoura sand was used in the experiments. Table 2 shows the values of material parameters for Toyoura sand. These parameters are independent of the initial density and the initial confining pressure.

Table 2 Material parameters for Toyoura sand

λ	0.035
κ	0.0045
e_{NC}	1.024
$R_{CS}=(\sigma_1/\sigma_3)_{CS(comp.)}$	2.8
β	2
a	50
ξ	0.4
$b1$	0.03
$b2$	1.1

4.1 *Monotonic loading tests*

Since plastic multiplier of Equation (10) is the same as that of Equation (9) under monotonic loading (F=1), only the effect of d is discussed here. Figure

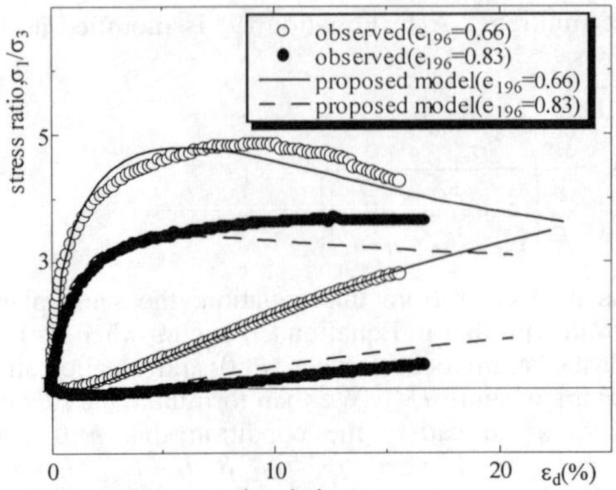

Figure 5. Stress- strain relations

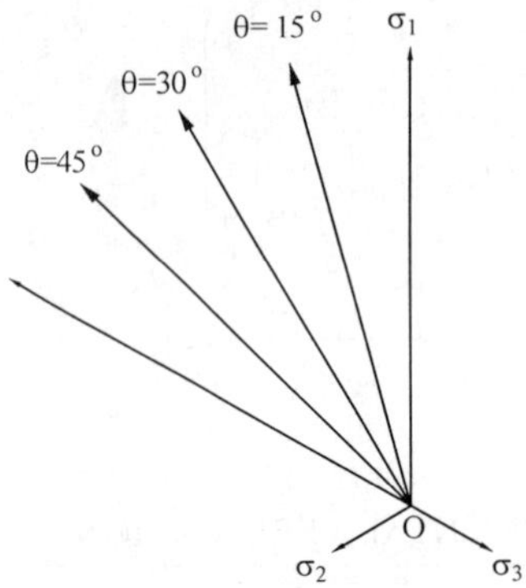

Figure 6(a). Stress paths on octahedral plane

5 shows the principal stress ratio – deviatoric strain – volumetric relation in triaxial compression tests on sands with different initial void ratio. We can see that the curves calculated by proposed model predict well the differences of the stiffness, strength and dilatancy between loose and dense sands.

Figure 6 shows the stress paths on the octahedral plane in true triaxial tests and the observed results (dots) and calculated curves of the tests. We can see that the model describes properly the stress- strain relation under three-different principal stress conditions as well.

4.2 *Cyclic loading tests*

Cyclic triaxial compression and extension tests were carried out along the stress path in *p-q* planes as illustrated in diagram (a) in Figures 7-9. The variations of volumetric strain and deviatoric strain with the change of principal stress ratio are shown in diagrams (b) and (c), respectively. In these figures, dots denote the observed results; solid curves the result calculated by the proposed model and the broken curves the results by the model based on Equation (9) (without the effect of *F*). We can see from Figure 7 that the proposed model (solid curves) predicts

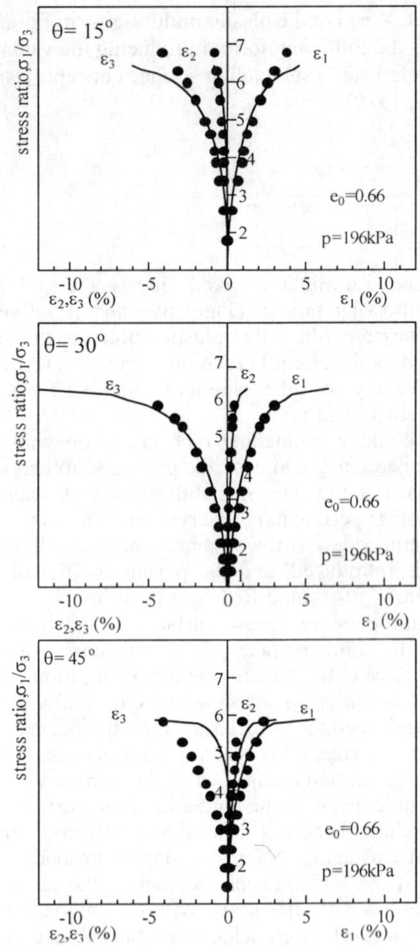

Figure 6(b). Stress- strain relations

well the stabilization of strains after several cycles. On the other hand, the model without the effect of *F* (broken curves) can not describe the observed stabilization of strains. In the test with smaller amplitude of stress ratio (σ_a/σ_r and σ_r/σ_a=3), the observed strains stabilize drastically in a few cycles and their accumulative strain is much smaller. Such tendency also can be predicted by the proposed model. Figure 9 shows the results of cyclic test with three kinds of constant amplitudes (σ_a/σ_r and σ_r/σ_a=2, 3 and 4 in order). Five cycles loadings are applied in each amplitude. The proposed model shows similar trend with the observed results as a whole.

As shown in Figure 10(a), three cycles triaxial compression test using true triaxial apparatus is performed. First, sand is sheared to σ_a direction and unloaded. Next, it is sheared and unloaded to σ_r direction and σ_l direction in order. Diagrams (b) and (c) are the observed and calculated directions of strain increment vectors on the octahedral plane in loading process along every direction. We can see

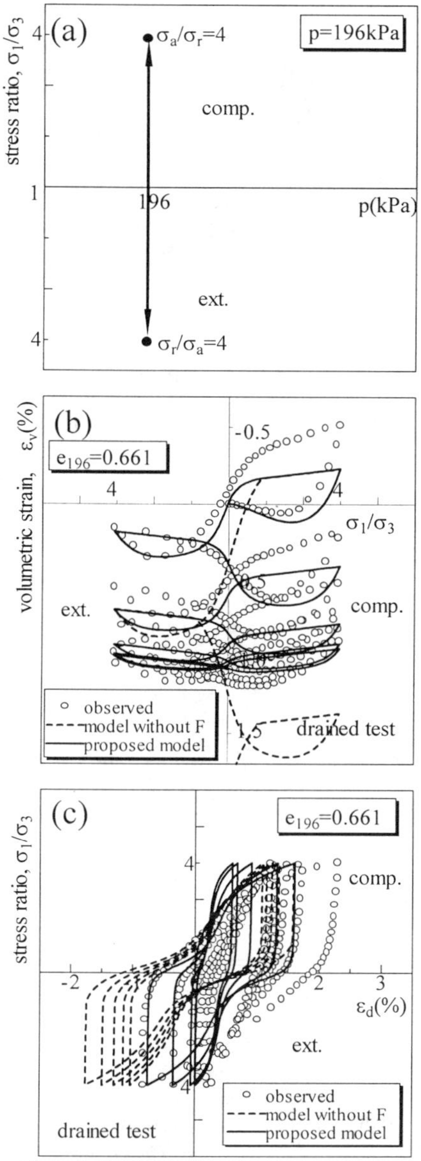

Figure 7. Cyclic triaxial test

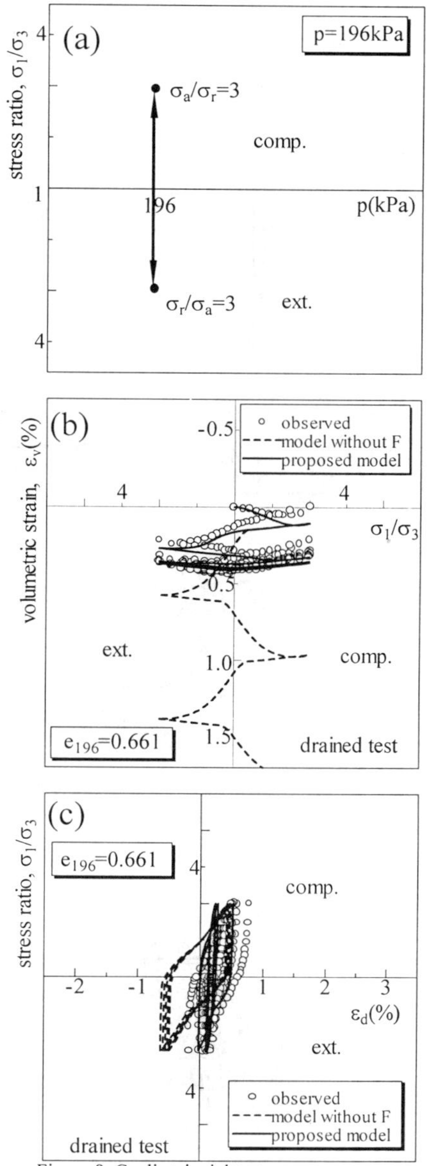

Figure 8. Cyclic triaxial test

from diagram (b) that though the observed strains are influenced by the preceding stress path at the beginning of the successive stress path, their influence is faded with successive loading. Such tendency is predicted by the model.

5 CONCLUSIONS

The main results of the present study are summarized as follows:

(1)A kinematic hardening model for sand proposed before is extended to one that can take into account the influence of density and the stabilization of strains under cyclic loading. The influence of density is considered by introducing a new density value *d* and considering subloading concept. Stabilizations of the strains are expressed by introducing the hardening limit surface and the maximum loading surface in the modeling.
(2)The validity of the extended model is checked by the triaxial and true triaxial tests. The proposed model using unified material parameters predicts

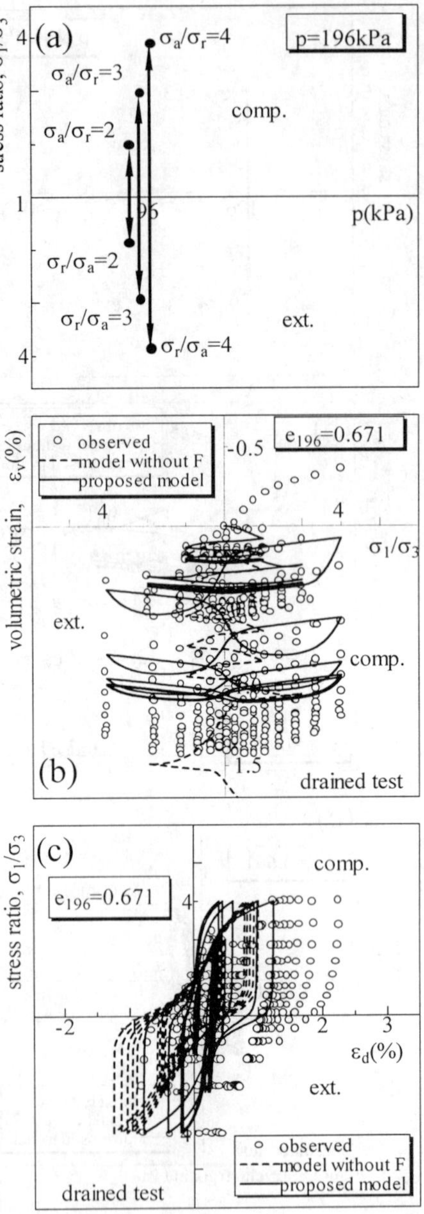

Figure 9. Cyclic triaxial test

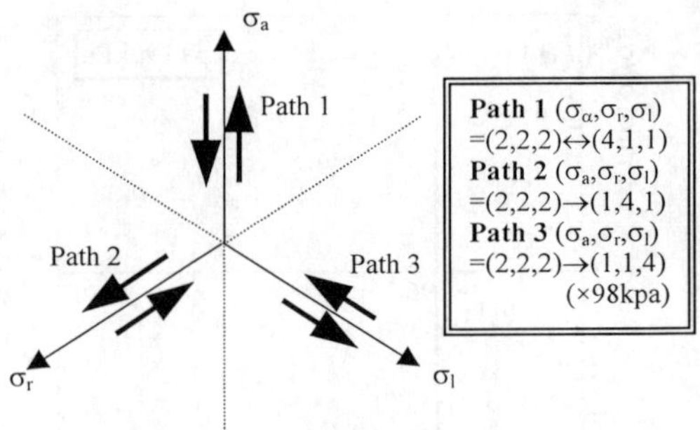

Figure 10(a). Stress paths on octahedral plane

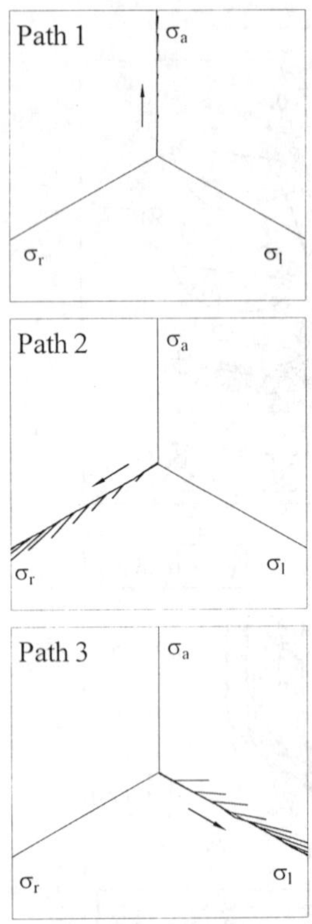

Figure 10(b). observed strain increment vectors

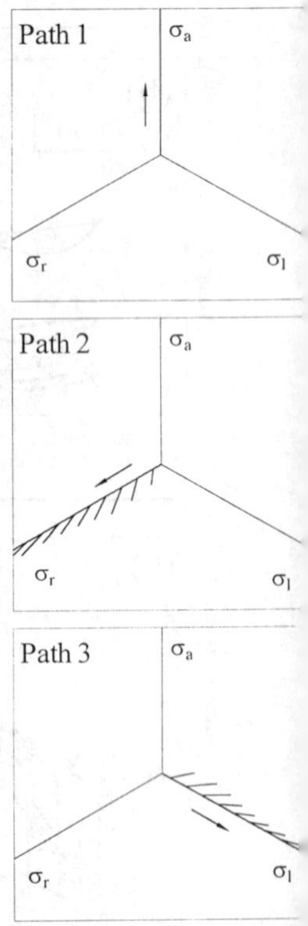

Figure 10(c). calculated strain increment vectors

well not only the deformation and strength characteristics under monotonic loading but also the development of anisotropy and the stabilization of strains under cyclic loading.

REFERENCES

Nakai, T. 1989. An isotropic hardening elastoplastic hardening model sand considering the stress path dependency in three-dimensional stresses, *Soils and Foundations*, 29(1), 119-137.

Nakai, T., Fujii, J &Taki, H. 1989. Kinematic extension of an isotropic hardening model for sand, *Proc.of NUMOG III*, 1, 36-45.

Nakai, T. & Mihara, Y. 1984. A new mechanical quantity for soils and its application to elastoplastic constitutive models, *Soils and Foundations*, 24(2), 82-94.

Chowdhury, E.Q. & Nakai, T. 1998. Consequence of the tij-concept a new modeling approach, *Computers and Geotechnics*, 23, 131-164.

Hashiguchi, K. 1980. Constitutive equation of elastoplastic materials with elasto-plastic transition, *Jour. of Appl. Mech., ASME*, 102(2), 266-272.

Computer Methods and Advances in Geomechanics, Desai et al. (eds) © 2001 Balkema, Rotterdam, ISBN 90 5809 183 X

An evaluation of soft soil models based on trial embankments

H. P. Neher & M. Wehnert
Institute of Geotechnical Engineering, University of Stuttgart, Germany

P. G. Bonnier
PLAXIS B.V., Delft, Netherlands

ABSTRACT: Two constitutive models are introduced briefly and used for back-analysis of two test embankments in order to assess their performance. The Soft-Soil model based on the modified Cam-Clay model, is employed as a reference model. The Soft-Soil-Creep model is an extension which includes time and strain rate effects. It simply needs one additional input parameter compared to the Soft-Soil model. Therefore the performance of these two models can be compared relatively easily. The first embankment considered is the well-known Boston trial embankment. For this embankment, however, measurements of settlements and horizontal displacements have not been continued long enough to include secondary compression. Consequently, for the second 2D FE-analysis an embankment with a long period of secondary settlements has been considered. This is an embankment from the Skå Edeby test site in Sweden.

1 INTRODUCTION

In recent years more and more constructions like buildings and embankments have been built on soft soils. Under soft soils the authors consider almost normally consolidated clays, clayey silts and peats. The special feature of these materials is their high compressibility. Considering the tangential oedometer stiffness moduli at a reference vertical pressure of 100 kPa, Janbu (1985) reports for normally consolidated clays $E_{oed} = 1$ to 3 MPa. For normally consolidated sands he determined values in the range of 10 to 50 MPa. Hence, in oedometer tests normally consolidated clays behave ten times softer then normally consolidated sands.

Another feature of soft soils is the linear stress-dependency of their oedometer stiffness. When plotting a stress-stiffness curve one gets a line of the form $E_{oed} = \sigma/\lambda^*$. That behaviour leads to the well-known logarithmic compression law

$$\varepsilon - \varepsilon_0 = \lambda^* \cdot \ln\left(\frac{\sigma}{\sigma_0}\right) \quad (1)$$

where compression is regarded positive both for stresses as well as strains. These considerations are valid for normal consolidated stress states and do not include secondary compression.

But all soils exhibit some creep. Hence primary compression is always followed by a certain amount of secondary compression. Assuming that the secondary compression is a small percentage of the primary compression, it is clear that creep plays an important part in problems involving large primary settlements, i.e. soft soils. This is for instance the case for the Boston and Skå Edeby trial embankment on soft soils considered here.

The construction of those embankments and their subsequent behaviour were simulated involving the deformation and pore-pressure. Two constitutive laws are used to show the effects of taking creep into account or not. One is the Soft-Soil model (SS-model), which is based on the modified Cam-Clay model, using the Mohr-Coulomb (MC) criterion to describe failure. The other, the Soft-Soil-Creep model (SSC-model), is an extension to account for creep.

2 MODEL BASICS

The above mentioned two constitutive models are briefly introduced in this section. They are both implemented in the FE-Code PLAXIS.

2.1 Soft-Soil model

The SS-model is based on the modified Cam-Clay model. A logarithmic relationship between the volumetric strain ε_v and the mean effective stress p' is assumed. Because of the usage of the volumetric strain ε_v instead of the void ratio, Eq. (2) involves the modified compression index λ^* instead of the index λ as used by Burland (1965). For virgin isotropic compression it yields

$$\varepsilon_v - \varepsilon_{v0} = \lambda^* \cdot \ln\left(\frac{p'}{p'_0}\right). \tag{2}$$

For isotropic unloading/reloading situations the elastic volume strain is formulated as

$$\varepsilon_v^e - \varepsilon_{v0}^e = \kappa^* \cdot \ln\left(\frac{p'}{p'_0}\right). \tag{3}$$

The parameter κ^* is the modified swelling index which determines soil behaviour during unloading and reloading. This behaviour is assumed to be elastic and is described by Hooke's law of elasticity. Eq. (3) implies the following linear stress dependency of the tangent stiffness modulus.

$$E_{ur} = 3\,(1 - 2\nu_{ur}) \cdot \frac{p'}{\kappa^*} \tag{4}$$

The subscript ur is used to specify that the parameters are related to unloading and reloading. In the SS-model κ^* and ν_{ur} are used as input parameters to compute the elastic strains.

For triaxial stress states the yield function of the SS-model is defined as

$$f = p^{eq} - p_p^{eq} \tag{5}$$

where p^{eq} is related to the actual stress state and p_p^{eq} is the equivalent pre-consolidation stress, see Fig. 1.

$$p^{eq} = \frac{q^2}{M^2\,(p' + c' \cdot \cot\varphi')} + (p' + c' \cdot cot\varphi') \tag{6}$$

This stress p_p^{eq} is a function of the plastic strain.

$$p_p^{eq} = p_{p0}^{eq} \exp\left(\frac{\Delta\varepsilon_v^p}{\lambda^* - \kappa^*}\right) \tag{7}$$

The yield function f (Eq. (5)) can be described as ellipses in the p'-q-plane. The tops of the ellipses are located on a line with the inclination M. In the modified Cam-Clay model (Burland 1965, 1967) the M-line represents the critical state line, which describes the stress states at post peak failure. It should be noted that in the SS-model the MC-criterion with the strength parameters φ' and c' is used to describe the failure. Both the MC-line and the M-line are given the same shift of $c' \cdot cot\varphi'$ away from the origin. This is illustrated in Fig. 1 and is taken into account in (Eq. (6)).

The total yield contour shown in Fig. 1 by the bold lines is the boundary of the elastic area. The MC-failure line is fixed, but the cap (ellipse with p_p^{eq}) may increase due to primary compression.

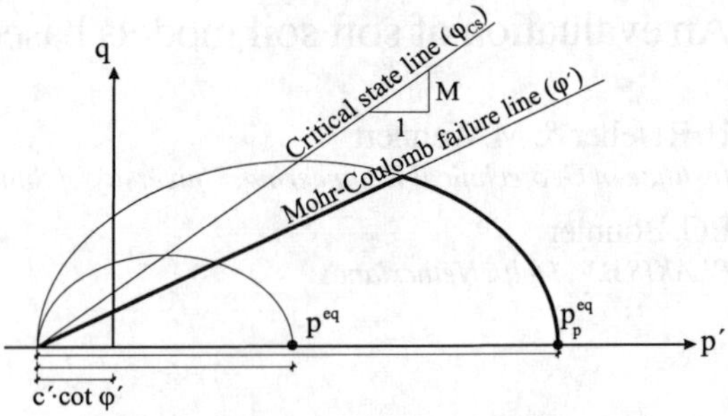

Figure 1. Yield surfaces of the SS-model in p'-q-plane

2.2 Soft-Soil-Creep model

Buisman (1936) was probably the first to propose a constitutive law for creep after observing that soft soil settlements cannot be fully explained by classic consolidation theory. In the framework of 1D-secondary compression other researchers like Bjerrum (1967), and Garlanger (1972) should be mentioned. More mathematical lines of research on 3D-creep were followed by, for example, Sekiguchi (1977), Adachi and Oka (1982) and Borja and Kavaznjian (1985).

Butterfield (1979) proposes a creep equation of the form

$$\varepsilon^H = \varepsilon_c^H + \mu^* \cdot \ln\left(\frac{\tau_c + t'}{\tau_c}\right) \tag{8}$$

where the deformation during consolidation is given by the expression ε_c^H. The modified creep index μ^* describes the secondary compression per logarithmic time increment. In this equation ε^H is the logarithmic strain. The superscript $'H'$ is applied because the logarithmic strain measurement was originally used by Hencky. In case of large strains both Butterfield (1979) and Den Haan (1994) showed that logarithmic strain supersedes the traditional engineering strain.

Please note, that the time τ_c is not the consolidation time t_c. The time τ_c is not a material parameter either as it not only depends on the consolidation but also on the geometry of the tested sample. Janbu (1969) developed a construction for evaluating the parameter μ^* and the time τ_c from experimental data. This and the difference between τ_c and t_c is shown in Fig. 2.

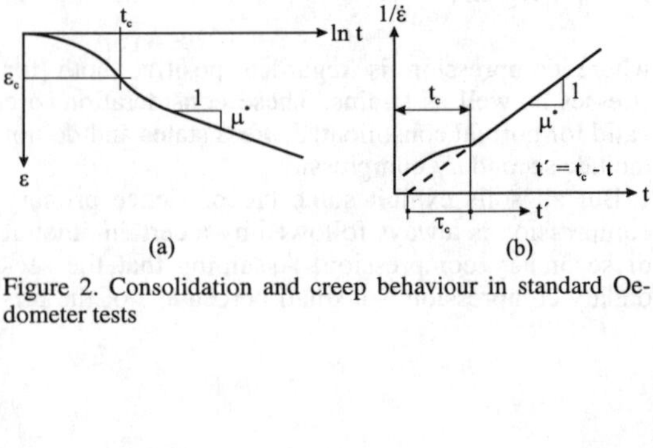

Figure 2. Consolidation and creep behaviour in standard Oedometer tests

The strain rate as derived from Eq. (8) reads

$$\dot{\varepsilon} = \frac{\mu^*}{\tau_c + t'} \quad \text{or inversely} \quad \frac{1}{\dot{\varepsilon}} = \frac{\tau_c + t'}{\mu^*}. \tag{9}$$

For isotropic stress states one finds the same modified creep index μ^*. Therefore one can include the secondary compression part and combine Eqs. (2) and (3) to get the total volumetric strain as

$$\begin{aligned} \varepsilon_v &= \varepsilon_v^e + \varepsilon_v^{cr} = \varepsilon_{vc}^e + \varepsilon_{vc}^{cr} + \varepsilon_{vac}^{cr} \\ &= \kappa^* \ln\left(\frac{p'}{p'_0}\right) + (\lambda^* - \kappa^*) \ln\left(\frac{p'_{pc}}{p'_{p0}}\right) \\ &\quad + \mu^* \ln\left(\frac{\tau_c + t'}{\tau_c}\right) \end{aligned} \tag{10}$$

where ε_v is the total volumetric strain due to an increase in mean effective stress from p'_0 to p' in a time period of $t_c + t'$. The total volumetric strain is divided into an elastic and a visco-plastic creep part, denoted by *e* and *cr* respectively. The visco-plastic part can be separated into a part during consolidation and a part after consolidation. This is given by the subscripts *c* and *ac*. The subscript *c* has also been added to the elastic strain to indicate that this is also related to consolidation. Fig. 3 clarifies these relations.

One should note that the isotropic consolidation line (IC-line) is not reached after the end of consolidation, but after some creep has occurred. In fact the IC-line is the line obtained while applying one load step per day. This is based on the assumption that consolidation occurs in less than one day.

Eq. (10) is valid for a constant mean effective stress, but for transient or continuous loading problems it is necessary to formulate a constitutive law in differential form. Moreover there is still the question of how to express the time τ_c analytically. Therefore the basic idea is adopted that all inelastic strains are time dependent. Additionally following Bjerrum (1967) it is supposed that the pre-consolidation stress

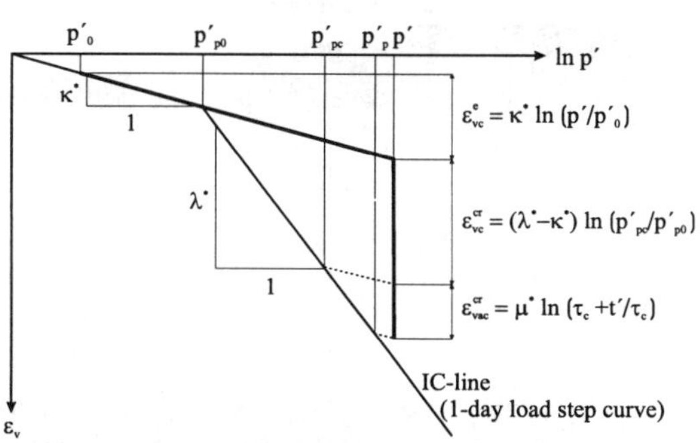

Figure 3. Logarithmic relationship between volumetric strain and mean stress including creep

depends entirely on the amount of the creep strain being accumulated by time. Eq. (10) can then be written as

$$\begin{aligned} \varepsilon_v &= \varepsilon_v^e + \varepsilon_v^{cr} \\ &= \kappa^* \ln\left(\frac{p'}{p'_0}\right) + (\lambda^* - \kappa^*) \ln\left(\frac{p'_p}{p'_{p0}}\right) \end{aligned} \tag{11}$$

where

$$p'_p = p'_{p0} \exp\left(\frac{\Delta \varepsilon_v^{cr}}{\lambda^* - \kappa^*}\right). \tag{12}$$

Combining Eqs. (10) with (11) one obtains

$$\begin{aligned} \varepsilon_{vac}^{cr} &= \varepsilon_v^{cr} - \varepsilon_{vc}^{cr} = \\ &\quad (\lambda^* - \kappa^*) \ln\left(\frac{p'_p}{p'_{pc}}\right) = \mu^* \ln\left(\frac{\tau_c + t'}{\tau_c}\right). \end{aligned} \tag{13}$$

Assuming that in conventional testing the load is increased stepwise and each load step is maintained for a constant period of $t_c + t' = \tau$, where τ is precisely one day, the so-called IC-line with $p'_p = p'$ is then obtained. Eq. (13) leads for $OCR = \frac{p'_p}{p'} = 1$ to

$$(\lambda^* - \kappa^*) \ln\left(\frac{p'}{p'_{pc}}\right) = \mu^* \ln\left(\frac{\tau_c + \tau - t_c}{\tau_c}\right). \tag{14}$$

With respect to τ the difference $\tau_c - t_c$ is very small. So Eq. (14) can be simplified as

$$\frac{\tau}{\tau_c} = \left(\frac{p'}{p'_{pc}}\right)^{\frac{\lambda^* - \kappa^*}{\mu^*}} \quad \text{or} \quad \tau_c = \tau \cdot \left(\frac{p'_{pc}}{p'}\right)^{\frac{\lambda^* - \kappa^*}{\mu^*}}. \tag{15}$$

Having derived a expression for τ_c it is now possible to formulate the differential creep equation.

$$\dot{\varepsilon}_v = \dot{\varepsilon}_v^e + \dot{\varepsilon}_v^{cr} = \kappa^* \cdot \frac{\dot{p}'}{p'} + \frac{\mu^*}{\tau_c + t'} \tag{16}$$

where $\tau_c + t'$ can be eliminated by means of Eq. (13) to obtain

$$\dot{\varepsilon}_v = \dot{\varepsilon}_v^e + \dot{\varepsilon}_v^{cr} = \kappa^* \cdot \frac{\dot{p}'}{p'} + \frac{\mu^*}{\tau_c} \left(\frac{p'_{pc}}{p'_p}\right)^{\frac{\lambda^* - \kappa^*}{\mu^*}} \tag{17}$$

where p'_p is defined as in Eq. (12). Entering Eq. (15) into Eq. (17) one gets

$$\dot{\varepsilon}_v = \dot{\varepsilon}_v^e + \dot{\varepsilon}_v^{cr} = \kappa^* \cdot \frac{\dot{p}'}{p'} + \frac{\mu^*}{\tau} \left(\frac{p'}{p'_p}\right)^{\frac{\lambda^* - \kappa^*}{\mu^*}}. \tag{18}$$

Assuming the same yield function as in the SS-model (Eq. (5)) the strain rate can then be defined as

$$\underline{\dot{\varepsilon}} = \underline{\underline{D}}^{-1} \underline{\dot{\sigma}}' + \frac{1}{\alpha} \frac{\mu^*}{\tau} \left(\frac{p^{eq}}{p_p^{eq}}\right)^{\frac{\lambda^* - \kappa^*}{\mu^*}} \frac{\partial p^{eq}}{\partial \underline{\sigma}'} \tag{19}$$

where

$$p_p^{eq} = p_{p0}^{eq} \exp\left(\frac{\Delta\varepsilon_v^{cr}}{\lambda^* - \kappa^*}\right) \tag{20}$$

and α is defined as $\alpha = \frac{\partial p^{eq}}{\partial p'}$. Hence the SSC-model is an extension of the SS-model that takes creep into account.

3 ANALYSIS AND RESULTS

In this paragraph, the results of a back-analysis of two embankments are presented. Note that for both applied models the input parameters are always the same. The only difference is that the modified creep index μ^* is not used in the SS-model. Therefore creep is not taken into account in this model.

3.1 Boston Trial Embankment

An extension of the Interstate highway I-95 north of Boston was built in 1965. A portion of the embankment was instrumented with piezometers, settlement rods and inclinometers. The embankment has a height of 11 m with a crest width of 28 m and a footing width of 84 m. It is founded on a 3 m thick sand-layer, which has an underlying 41 m thick deposit of Boston Blue Clay (BBC). The BBC layer is divided into 12 sub-layers with different compressibility parameters and over-consolidation ratios (OCR). Under the embankment the top peat layer is replaced by fill. The parameters used for the calculation are given in Tab. 1 for sand, fill and peat respectively. The constitutive law applied to the sand and fill is the Hardening Soil model (Brinkgreve and Vermeer 1998). The peat was computed with the MC-model. The following parameters are the same for all 12 sub-layers of BBC.

$$M = 1.86 \qquad c = 1\,kPa$$
$$\nu_{ur} = 0 \qquad \varphi' = 33.4°$$
$$\psi = 0°$$

Table 1. Soil parameters for sand, fill and peat

	sand	fill	peat
$\gamma_r\ [kN/m^3]$	18	20	11,8
$E_{50}^{ref}\ [MPa]$	40	40	-
$E_{ur}^{ref}\ [MPa]$	120	120	-
E_{oed}^{ref} resp. $E_{oed}\ [MPa]$	18.5	19	0,208
$m\ [-]$	0.5	0.5	-
ν_{ur} resp. $\nu\ [-]$	0.2	0.2	0.3
$c\ [kPa]$	1	5	5
$\varphi'\ [o]$	37	35	25
$\psi\ [o]$	0	0	0
$OCR\ [-]$	1	1	1
$K_0\ [-]$	0.426	0.398	0.577
$K_0^{NC}\ [-]$	0.45	0.398	0.577

In Tab. 2 the other parameters for the 12 sub-layers of BBC are plotted. All parameters were determined according to the detailed subsoil description of Ladd, Whittle and Legaspi (1994). The chosen ratio between the modified soil parameters is $\lambda^*/\kappa^* = 4$ and $\lambda^*/\mu^* = 35$.

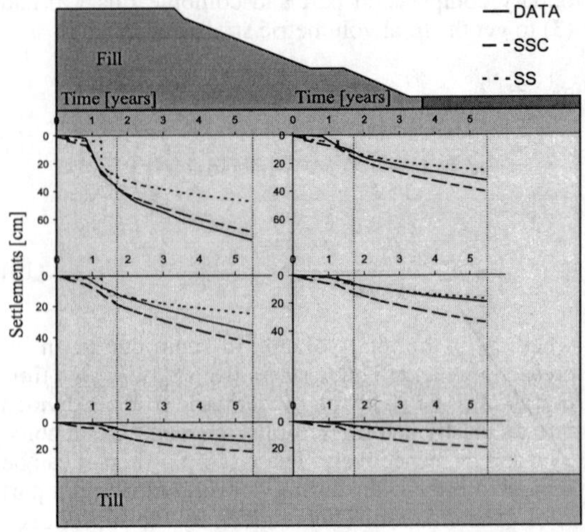

Figure 4. Vertical displacements at different depths under the embankment

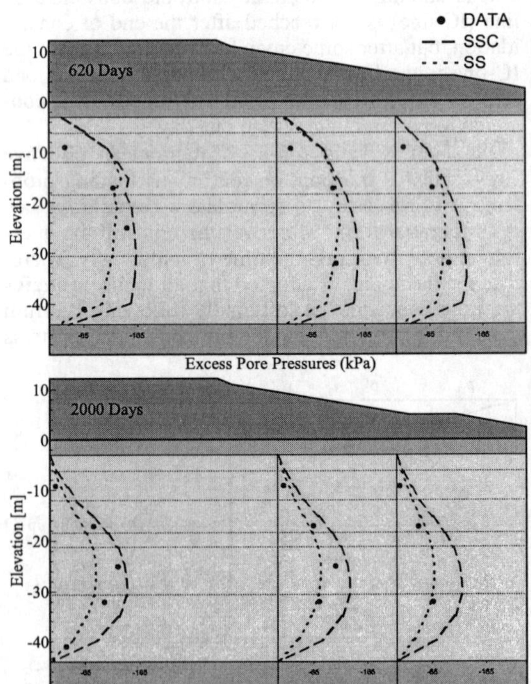

Figure 5. Pore-pressure distribution in different sections

Table 2. Soil parameters for Boston Blue Clay

layer	A	B1	B2	C1	C2	D	E1	E2	E3	F1	F2	F3
γ_r $[kN/m^3]$	19			17.7								
k_h/k_v $[-]$	2			1.5								
k_v $[10^{-10}m/s]$	15.8	10.4		6.3		7.1	7.8			5.9		
λ^* $[10^{-1}]$	0.522	0.652		0.870		1.304	1.957			1.522		
OCR $[-]$	8.34	4.6	3.07	2.25	1.77	1.44	1.25	1.21	1.18	1.16	1.14	1.11
K_0 $[-]$	1.35	1.08	0.85	0.80	0.65	0.60	0.57	0.57	0.56	0.56	0.56	0.55

The inclination M of the critical state line is significantly steeper than the MC-failure line to prevent the overprediction of horizontal deformation. One gets this inclination by choosing a M-value, which yields to $K_0^{NC} = 1 - \sin\varphi'$. The overprediction of horizontal displacements is well known in the framework of the classical critical state mechanics. Moreover the value for $\nu_{ur} = 0$ is chosen to uncouple the elastic strain. This is done to reduce the magnitude of the horizontal displacements as well.

The high OCR-values (Tab. 2) indicate that creep will not be so important because about half of the BBC layer has an OCR-value of at least two, decreasing at the bottom of the layer to a value near one. For that type of soil the amount of creep will be small in comparison to the rest of the settlements.

Figs. 4 to 6 show the calculation results and the measurements. The solid lines indicate the measurements, the dashed lines are the results from the SSC-model and the dotted lines are the analysis with the SS-model. In Fig. 4 one can recognize that consolidation is not over after the measurement period of 2000 days, because the settlement curves still have an inclination. That fact is supported by Fig. 5, one can see that in the center of the BBC there is still a pore-water-pressure of around 120 kPa after 2000 days. The SSC-model "slightly" overestimates in the most cases the vertical displacements. Moreover the calculated pore-water-pressures are slightly higher then those measured. The horizontal displacements are overestimated as well. In most cases the SS-model matches the measurements better, especially for the deep settlements. So it seems that if there are high OCR values in the soft soils the SSC-model is not so useful.

3.2 Skå Edeby test embankment

The test embankment was built on the test site of Skå Edeby about 25 km west of Stockholm in the year 1961. The embankment has a height of 1.5 m with a crest width of 4 m and slopes 1 : 1.5. The fill was instrumented with piezometers and settlement markers at different locations under and outside the fill. The soil under the embankment consists of soft clay with a thickness of 15 m on top of till or rock. The soft clay layer is divided into 9 sub-layers with different compressibility parameters and OCRs. The following parameters are for all 9 sub-layers the same.

$$M = 1.61 \qquad c = 1\,kPa$$
$$\nu_{ur} = 0.15 \qquad \varphi' = 30\,°$$
$$\psi = 0\,°$$

The other parameters used for the calculations are given in Tab. 3. All parameters were taken from the literature (Larsson 1997). The ratio between the modified soil parameters is $\lambda^*/\kappa^* = 6$ and $\lambda^*/\mu^* = 15$. The inclination of the critical state line M is here chosen to match the K_0^{NC}-value correctly, as well.

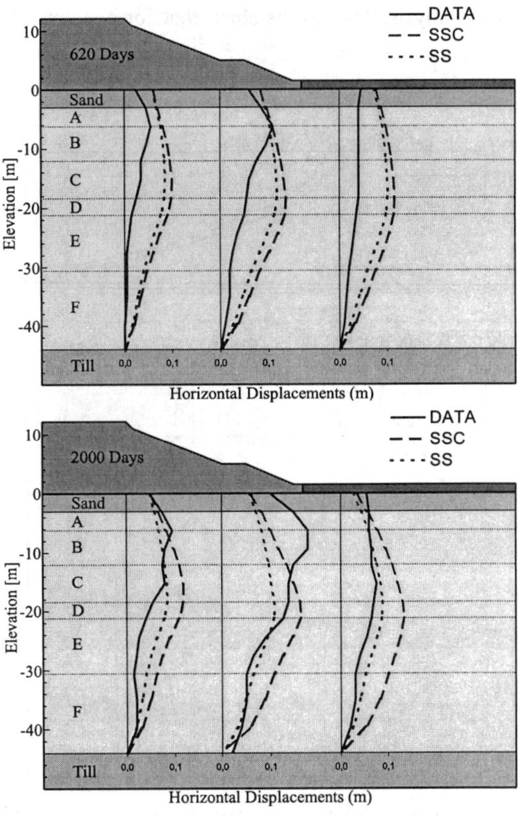

Figure 6. Horizontal displacements in different sections

Table 3. Soil parameters for Skå Edeby

layer	γ_r $[kN/m^3]$	k $[m/d]$	λ^* [−]	OCR [−]
A	14.2	$8.64 \cdot 10^{-5}$	0.106	14.1
B	14.4		0.091	2
C		$7.0 \cdot 10^{-5}$	0.083	1.2
D	16.1	$6.22 \cdot 10^{-5}$	0.076	1.0
E		$5.44 \cdot 10^{-5}$		
F		$5.01 \cdot 10^{-5}$		
G		$4.75 \cdot 10^{-5}$		
H			0.069	

As shown in Tab. 3 the soft clay is almost normally consolidated ($OCR = 1$). Only the upper 2 m have a high OCR value caused by a crust. Hence, a large amount of creep is to be expected. Furthermore it demonstrates that the settlements as well as the pore-pressures and the horizontal displacements were strongly underestimated by the SS-model whereas the results from the SSC-model and the measured data agree fairly well.

4 CONCLUSIONS

After a brief introduction of the SS-model and the SSC-model two test embankments were used for back-analysis. The results show that for overconsolidated soils the SSC-model has no advantages compared to the SS-model, because in such case creep doesn't play an important role. However in normally consolidated soft clays like in Skå Edeby creep becomes significant and the SSC-model exceeds the SS-model. This should be kept in mind when simulating the construction of any building on soft soils.

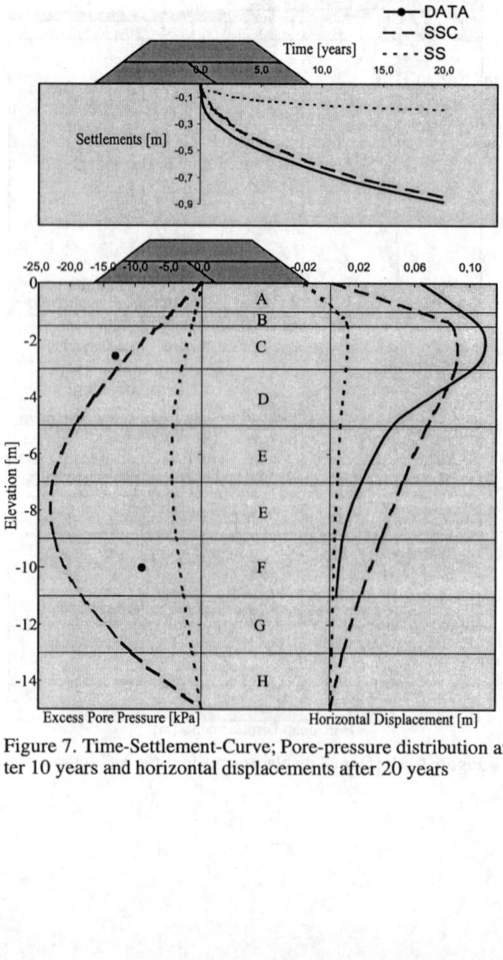

Figure 7. Time-Settlement-Curve; Pore-pressure distribution after 10 years and horizontal displacements after 20 years

REFERENCES

Adachi, T. and F. Oka (1982). Constitutive equation for normally consolidated clays based on elasto-viscoplasticity. *Soils and Foundations 22*, 57–70.

Bjerrum, L. (1967). 7th rankine lecture: Engineering geology of norwegian normally-consolidated marine clays as related to settlements of buildings. *Géotechnique 17*(1), 81–118.

Borja, R. I. and E. Kavaznjian (1985). A constitutive model for the stress-strain-time behaviour of 'wet' clays. *Géotechnique 35*(3), 283–298.

Brinkgreve, R. B. J. and P. A. Vermeer (Eds.) (1998). *PLAXIS Finite Element Code for Soil and Rock Analyses*. Rotterdam: A.A.Balkema.

Buismann, K. (1936). Results of long duration settlement tests. In *Proceedings 1st International Conference on Soil Mechanics and Foundation Engineering*, Volume 1, Cambridge, Massachusetts, pp. 103–107.

Burland, J. B. (1965). The yielding and dilation of clay (correspondence). *Géotechnique 15*(2), 211–214.

Burland, J. B. (1967). *Deformation of Soft Clay*. Ph. D. thesis, Cambridge University.

Butterfield, R. (1979). A natural compression law for soils (an advance on e-log p')(technical note). *Géotechnique 29*(4), 469–480.

Den Haan, E. J. (1994). *Vertical Compression of Soils*. Ph. D. thesis, Delft University.

Garlanger, J. E. (1972). The consolidation of soils exhibiting creep under constant effective stress. *Géotechnique 22*(1), 71–78.

Janbu, N. (1985). 25th rankine lecture: Soil models in offshore engineering. *Géotechnique 35*(3), 239–281.

Ladd, C. C., A. J. Whittle, and D. E. Legaspi (1994). Stress-deformation behaviour of an embankement on boston blue clay. In *Vertical and Horizontal Deformations of Foundations and Embankments Proceedings of Settlement '94*, College Station Texas, pp. 1730–1759.

Larsson, R. (1997). Consolidation of soft soils. Report 29, Swedish Geotechnical Institute.

Sekiguchi, H. (1977). Rheological characteristics of clays. In *Proceedings 9th International Conference on Soil Mechanics and Foundation Engineering*, Volume 1, Tokyo, pp. 289–292.

Stolle, D. F. E., P. A. Vermeer, and P. G. Bonnier (1997). A soft soil model and experiences with two integration schemes. *Numerical Models in Geomechanics*, 123–128.

Vermeer, P. A. and H. P. Neher (1999). A soft soil model that accounts for creep. In R. B. J. Brinkgreve (Ed.), *Proceedings of the International Symposium "Beyond 2000 in Computational Geotechnics"*, Amsterdam, pp. 249–261. Balkema.

Vermeer, P. A., D. F. E. Stolle, and P. G. Bonnier (1997). From classical theory of secondary compression to modern creep. In J.-X.Yuan (Ed.), *Computer Methods and Advances in Geomechanics*, Volume 4, Wuhan, pp. 2469–2478. Balkema.

Computer Methods and Advances in Geomechanics, Desai et al. (eds) © 2001 Balkema, Rotterdam, ISBN 90 5809 183 X

Application of Monot to the modelling of spatially varying soils

C.Onisiphorou & M.A.Hicks
Manchester School of Engineering, University of Manchester, UK

ABSTRACT: The Monot double-hardening soil model has been calibrated against a database of 74 triaxial tests on Erksak sand, for a wide range of relative densities. By expressing important material parameters as functions of state parameter, the calibration is suitable for modelling material heterogeneity. This has been illustrated by analysing a simple drained deformation problem: i.e. random fields of state parameter have been converted to sets of Monot parameter values; then used in finite element computations of biaxial compression.

1 INTRODUCTION

Traditionally, geotechnical analyses have followed a deterministic approach, in which so-called uniform deposits are represented by constant (e.g. mean) material properties; in the case of stability, this results in a single factor of safety. An alternative approach is to take account of the spatial variations in material properties (i.e. heterogeneity), which, in reality, exist in all geomaterials. In this instance, each material property is represented by its mean and standard deviation, and by some relationship defining the degree of spatial correlation; for stability problems, this stochastic-type of approach leads to reliability (i.e. the probability that failure will not occur).

For many problems, the simpler deterministic approach may be appropriate; however, this paper represents the first stage in analysing problems for which a stochastic approach may be necessary. In particular, the aim is to investigate the influence of heterogeneity in problems of static liquefaction (e.g. Hicks & Boughrarou 1998, Hicks & Onisiphorou 2000).

For such problems, a major consideration is the method for modelling the variability of very complex material behaviour: e.g. Popescu et al. (1997) generated cross-correlated, bi-variate, property fields, and Fenton & Vanmarcke (1998), multivariate fields (some cross-correlated), for the simulation of liquefaction due to cyclic loading. This paper proposes a strategy based on generating univariate fields of state parameter (Been & Jefferies 1985), from which constitutive model parameters may be easily derived. The paper briefly reviews the constitutive model Monot; describes its calibration against an extensive laboratory database; then, finally, its application for a simple problem involving material variability.

2 MONOT SOIL MODEL

Figure 1 shows a cross-section through the Monot soil model (Molenkamp 1981) in the s-t plane (in which s and t are effective isotropic and deviatoric stress invariants, respectively). This shows that two isotropically hardening yield surfaces subdivide stress space (within the failure surface) into four distinct regions. They are: an associated compressive yield surface, mainly for modelling plastic strains due to increasing pressure; and a non-associated deviatoric yield surface, for modelling plastic strains due to shearing. Therefore, including a non-linear elastic component, there are three model components, which are assumed to be additive.

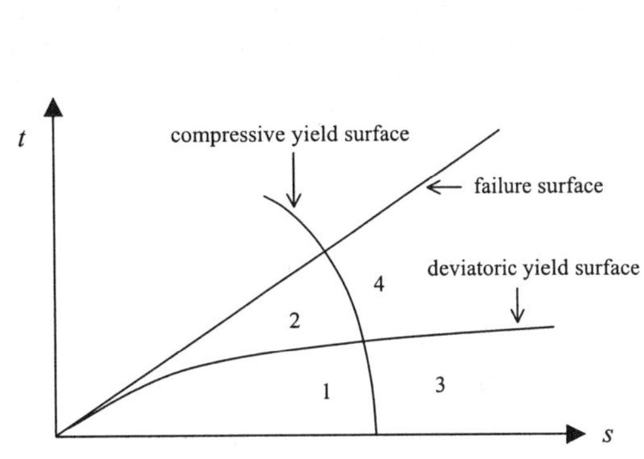

Figure 1. Principal features of Monot.

3 MONOT CALIBRATION

3.1 *Material parameters*

Although Molenkamp (1981) proposed 21 material parameters, in practice, most of these are never, or hardly ever, calibrated, as reported by Hicks (2000) who reviewed a large number of previous calibrations to material data, many of these being used in practical engineering applications (e.g. Hicks & Boughrarou 1998).

Indeed, only 8 parameters have regularly been calibrated (the remainder being given standard default values). These eight are listed in Table 1, along with possible ranges of values. *A* and *AP* are stiffness scalar and curvature parameters for the elastic component, while *B* and *BP*, and *E* and *EP*, relate to the plastic compressive and plastic deviatoric model components, respectively. *C* is a scalar defining failure surface size, and, in the triaxial compression plane, is related to the Mohr-Coulomb peak friction angle ϕ_{peak} by

$$C = \frac{2\sqrt{2}\sin\phi_{peak}}{3 - \sin\phi_{peak}} \tag{1}$$

FICV is the constant volume friction angle ϕ_{cv}.

Table 1. Summary of Monot parameters to be calibrated.

Model component	Parameter	Type	Range
Elastic	*A*	scalar	>0.0
	AP	curvature	0.0-1.0
Plastic compressive	*B*	scalar	>0.0
	BP	curvature	0.0-1.0
Plastic deviatoric	*C*	scalar	>0.0
	FICV	ϕ_{cv}	$\geq\phi_{\mu}$ *
	E	scalar	>0.0
	EP	curvature	≥1.0

* ϕ_{μ} is the interparticle friction angle.

Hicks (2000) further showed that calibrated values for *AP*, *BP* and *EP* showed little variation, and suggested these so-called 'luxury' parameters be given the default values listed in Table 2, the remaining five parameters being adequate for most practical purposes. However, for this investigation, all eight parameters in Table 1 have been calibrated, with the derived values for the three curvatures then being compared to the suggested values in Table 2.

Table 2. Relative importance of Monot parameters.*

Luxury parameters	Default value	Essential parameters
AP	0.4	*A*
BP	0.6	*B*
EP	3.0	*C*
		FICV
		E

* Hicks (2000).

3.2 *Overview of calibration*

Table 3 summarises the calibration method (Hicks 1990, 2000). It involves calibrating one or two parameters at a time, using specific stress paths to isolate different model components. This is most easily done using drained data, as indicated. However, for accurate simulation of pore pressures, undrained data is also used; hence, the current calibration has been checked against undrained data after 'Stage 5'. (Note that, for practical applications such as liquefaction, the modelling of undrained data takes precedence over drained data.)

Table 3. Summary of calibration sequence.

Stage	Parameters calibrated	Test data
1	*A, AP*	Isotropic unload/reload
2	*B, BP*	Isotropic loading
3	*C*	Drained triaxial compression
4	*FICV*	Drained triaxial compression
5	*E, EP*	Drained triaxial compression

For this investigation, Monot has been calibrated against a laboratory database of 74 tests on Erksak sand: i.e. 3 isotropic consolidation tests, and 71 triaxial compression tests (31 drained and 40 undrained). The relationship between parameter values and soil relative density has been established by subdividing the tests, for calibration purposes, into five groups, according to state parameter value; i.e. from dense (Group 1) to very loose (Group 5).

3.3 *Calibration results*

Table 4 lists the calibrated parameter values. Those for the three curvatures agree closely with the values suggested in Table 2. Of the other five, *A* and *E* are expressed as functions of ψ, but not *B* (note that the plastic compressive component has little or no influence on undrained response); *C* is highly dependent on ψ, and is derived by combining Equation 1 with

$$\phi_{peak} = -49.81\psi + 30.8 \tag{2}$$

FICV showed minimal variation with ψ and has therefore been assigned a constant value.

Table 4. Summary of calibrated Monot parameters.

Monot parameter	Value
A	$0.0063\psi+0.002$
AP	0.37
B	0.00054
BP	0.58
C	$f(\psi)$*
FICV	28°
E	$0.1\exp(12.7\psi)$
EP	2.6

* from Equations 1 and 2.

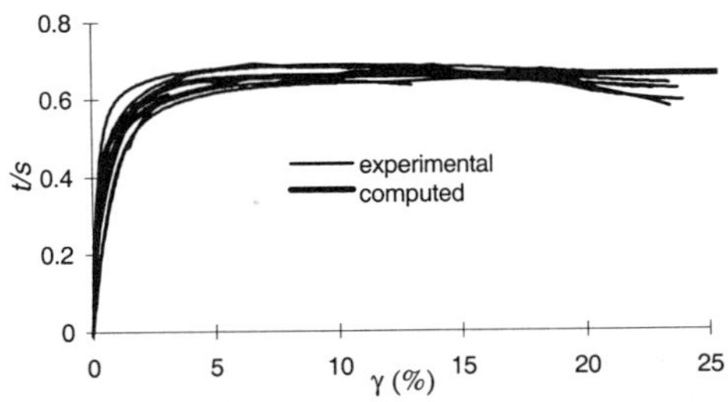

(a) shear stress ratio vs. shear strain invariant

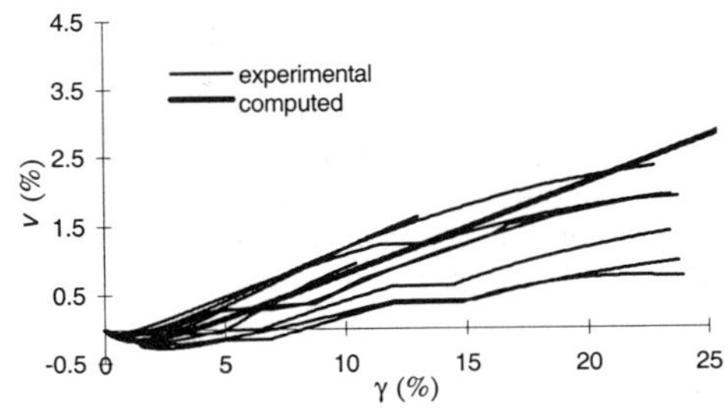

(b) volumetric strain invariant vs. shear strain invariant

Figure 2. Calibrated fit to drained triaxial compression data for Erksak sand (Group 3, ψ=-0.077).

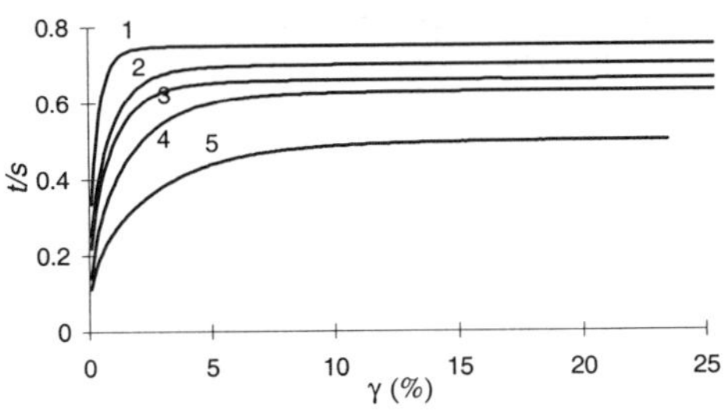

(a) shear stress ratio vs. shear strain invariant

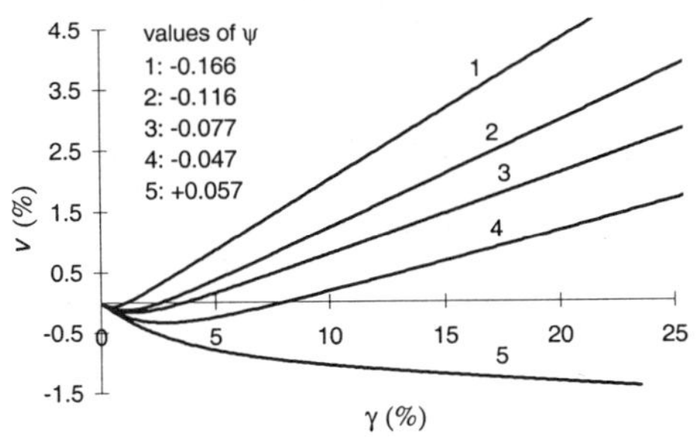

(b) volumetric strain invariant vs. shear strain invariant

Figure 3. Drained triaxial compression test responses computed for various state parameters.

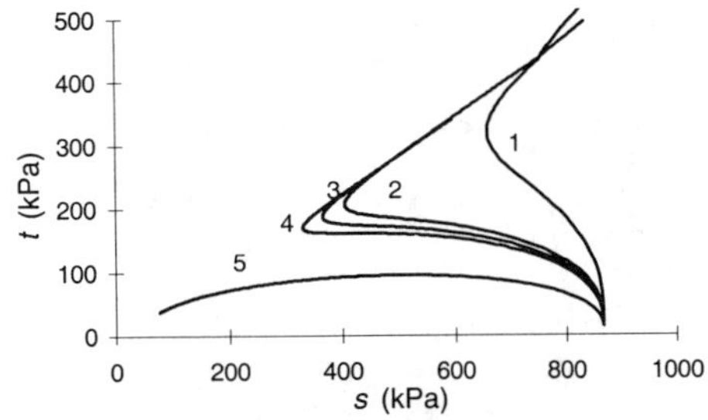

Figure 4. Undrained triaxial compression test responses computed for various state parameters.

Figure 2 shows a typical fit to experimental data, while Figures 3 and 4 summarise the calibration by showing computed drained and undrained triaxial compression responses, for all five groups.

4 APPLICATION TO BIAXIAL COMPRESSION

Monot's application to spatial variability is simply illustrated in Figure 5, for the case of drained biaxial compression. A square specimen, side length D, is subjected to loading in the vertical direction via perfectly smooth and rigid end-platens. Figure 5a shows the spatial variation of state parameter, in which light and dark zones indicate looser and denser materials, respectively. This isotropic random field has been generated using 'Local Average Subdivision' (Fenton & Vanmarcke 1990), and a 'cell size' of $D/64$: i.e. the domain has been subdivided into 64×64 square cells, each with a different value of ψ. These values are a function of the spatial scale of fluctuation θ, equal to $D/10$ in this example, and the state parameter point statistics: i.e. mean μ and standard deviation σ, assuming a normal distribution (Hicks & Onisiphorou 2000). For the spatial variation in Figure 5a, two cases have been considered: μ=-0.1, σ=0.01; and μ=-0.1, σ=0.1.

The problem has been analysed by finite elements using the code developed by Hicks (1995): i.e. the domain has been discretised using a mesh of 32×32, 8-node, square elements, with each element using 2×2 Gaussian integration. Hence, each element occupies the same space as four random field cells, and each element integration point is assigned a different value of ψ, as determined from the random field. This is obviously more efficient than having a 64×64 element mesh and assigning constant properties over an element, and is possible due to the accuracy of the reduced integration points for 8-node elements. Once the values of ψ have been mapped onto the mesh, the distribution of Monot parameters is determined from Table 4.

Figures 5b, c illustrate algorithm performance by showing incremental shear strain invariant contours

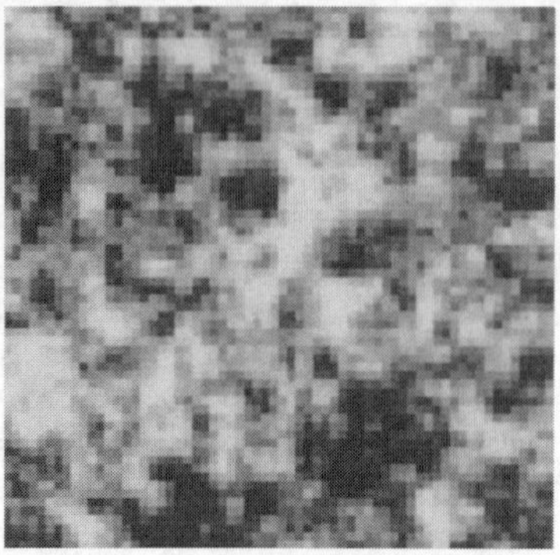

(a) state parameter random field (θ/D= 0.1)

(b) incremental shear strain invariant (μ=-0.1, σ=0.01)

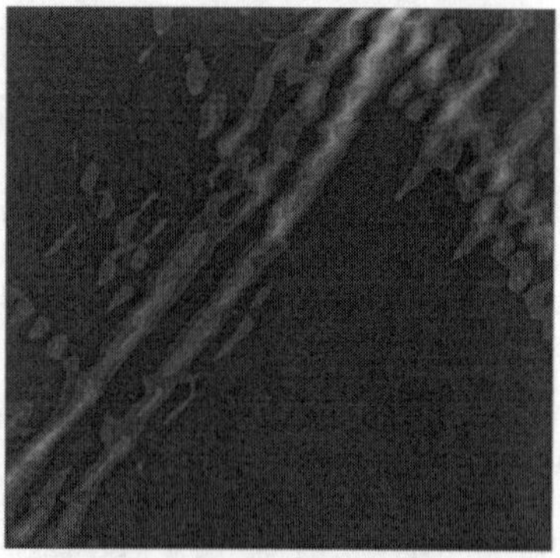

(c) incremental shear strain invariant (μ=-0.1, σ=0.1)

Figure 5. Simulation of drained biaxial compression.

at the end of loading (i.e. peak load), for low and high σ, respectively. Despite the frictionless boundary conditions, non-uniform strain fields develop due to material heterogeneity. For low σ, multiple shear planes can sometimes be seen (Fig. 5b); higher values of σ usually result in fewer, more dominant, planes (Fig. 5c).

5 CONCLUSIONS

A convenient procedure for describing the spatial variability of material properties using a complicated soil model has been proposed. This is based on defining the model calibration in terms of state parameter (Been & Jefferies 1985). Material heterogeneity can then be modelled by generating uni-variate random fields of state parameter, from which model parameter values can be rapidly back-calculated for use in finite element analysis.

The procedure has been demonstrated for the Monot soil model (Molenkamp 1981), by calibrating against a database of 74 triaxial tests and analysing the problem of drained biaxial compression. The proposed method can be readily used in practical stochastic investigations (e.g. Hicks & Onisiphorou 2000); these involve multiple realisations of material variability as part of a Monte Carlo simulation process.

ACKNOWLEDGEMENT

This work has been funded by the UK Engineering and Physical Sciences Research Council (EPSRC Grant GR/L34662).

REFERENCES

Been, K. & Jefferies, M.G. 1985. A state parameter for sands. *Géotechnique* 35: 99-112.

Fenton, G.A. & Vanmarcke, E.H. 1990. Simulation of random fields via Local Average Subdivision. *ASCE J. Eng. Mech.* 116(8): 1733-1749.

Fenton, G.A. & Vanmarcke, E.H. 1998. Spatial variation in liquefaction risk. *Géotechnique* 48: 819-831.

Hicks, M.A. 1990. Numerically modelling the stress-strain behaviour of soils. *PhD thesis*. University of Manchester, UK.

Hicks, M.A. 1995. MONICA—A computer algorithm for solving boundary value problems using the double-hardening constitutive law Monot: I. Algorithm development. *Int. J. Num. Anal. Meth. Geomech.* 19:1-27.

Hicks, M.A. 2000. Experience in calibrating the double-hardening soil model Monot. (submitted for publication).

Hicks, M.A. & Boughrarou, R. 1998. Finite element analysis of the Nerlerk underwater berm failures. *Géotechnique* 48: 169-185.

Hicks, M.A. & Onisiphorou, C. 2000. Stochastic analysis of saturated soils using finite elements. *Final Report to EPSRC, Grant GR/LR34662*.

Molenkamp, F. 1981. Elasto-plastic double hardening model Monot. *LGM Report CO-218595*. Delft Geotechnics.

Popescu, R., Prevost, J.H., & Deodatis, G. 1997. Effects of spatial variability on soil liquefaction: some design recommendations. *Géotechnique* 47: 1019-1036.

Computer Methods and Advances in Geomechanics, Desai et al. (eds) © 2001 Balkema, Rotterdam, ISBN 90 5809 183 X

A new constitutive model for granular materials

F. Peng, F. Tatsuoka & M.S.A. Siddiquee
Department of Civil Engineering, University of Tokyo, Japan

S.J.M. Yasin
Bangladesh University of Engineering and Technology, Bangladesh

ABSTRACT: A new elasto-plastic constitutive model is proposed, represented by one unique stress history-independent work-hardening rule. A unique relationship between a strain energy quantity and a stress parameter, which is independent of stress history, is presented. The model is based on results from a series of drained plane strain compression tests on saturated dense Toyoura sand with precise stress and strain measurements along various stress paths. The proposed model is capable of simulating the effects on the deformation characteristics of stress history and stress path. The effects of pressure-level, anisotropy and work-softening associated with strain localization into a shear band are also taken into account.

1 INTRODUCTION

In the elasto-plastic models for soils, the hardening function controls the development of yield locus by yielding of soil. The hardening parameter involved in the hardening function is usually a state parameter, which is independent of previous stress history and current stress path direction. For soils, certain plastic strain component, or plastic strain energy, is often employed as the hardening parameter.

Tatsuoka et al. (1986) proposed to use plastic shear strain as the stress history-independent strain hardening parameter based on some test observations. Later, Tatsuoka et al. (1993), Siddiquee et al. (1994) and Yoshida et al. (1994) employed this assumption in modeling the elasto-plastic behavior of Toyoura sand. They performed a series of FEM numerical simulation of the results from plane strain bearing capacity model tests of strip footing. The constitutive model for sand that was used in the analysis was obtained based on results from a series of plane strain compression (PSC) tests performed at constant σ_c, while considering anisotropy in the strength and deformation characteristics (Tatsuoka et al., 1991; 1993). Although the peak footing load could be rather accurately simulated in the FEM analysis, the pre-peak footing settlement was always underestimated (Siddiquee, 1999; Kotake, 1998). Yasin and Tatsuoka (2000) considered that this discrepancy can be attributed, at least partly, to the use of plastic shear strain as the hardening parameter, and for that reason, shear strains for stress paths traced in the ground were underestimated.

The FEM analysis also showed that the local stress paths in the failing zones in reinforced soil in PSC tests and the boundary value loading tests, such as bearing capacity tests, are similar to those of anisotropic compression at stress ratios close to that at failure (Kotake, 1998; Peng et al., 2000). It may not be relevant to assume in the FEM analysis of such cases described above that: a) plastic shear strain is the hardening parameter that is independent of stress history; and b) the shear strain potential can be determined by ordinary PSC tests at constant confining pressure.

The objective of this study is, therefore, to propose an elasto-plastic constitutive model with a non-associated flow rule and isotropic work hardening properties with a unique stress history-independent hardening rule for granular geomaterials. The model is capable of simulating the strength and deformation characteristics for a wide variety of stress histories and stress paths. The model also takes into account cross-anisotropic elasticity, effects on the pre-peak as well as post-peak stress-strain behavior of pressure level, anisotropy and density and strain localization into shear band.

2 GENERAL DESCRIPTION OF THE PROPOSED MODEL

The proposed material model is coupled with an isotropically work-hardening and -softening, non-associated, elasto-plastic material description. The yield function F (Eq. 1) and the plastic potential function G (Eq. 4) are of, respectively, Mohr-Coulomb and Drucker-Prager type.

$$F = -\eta I_1 + \frac{1}{g(\theta)}\sqrt{J_2} - k_1 = 0 \quad (1)$$

where I_1 is the first invariant of stress (i.e., hydrostatic stress component, positive in compression), J_2 is the second stress invariant (i.e., deviatoric stress) and $g(\theta)$ is the Lode angle function, which is defined as:

$$g(\theta) = \frac{3-\sin\phi_{mob}}{2\sqrt{3}\cos\theta - 2\sin\theta\sin\phi_{mob}} \quad (2)$$

where η is the deviatoric stress at $\theta = 30^o$ (on the π-plane) (i.e., the triaxial compression with $\sigma_1 > \sigma_2 = \sigma_3$), which is related to the mobilized angle of internal friction ϕ_{mob} as:

$$\eta = \frac{2\sin\phi_{mob}}{\sqrt{3}(3-\sin\phi_{mob})} \quad (3)$$

k_1 in Eq. 1 (and k_2 in Eq. 4) is the cohesion intersect, which is zero in the present analysis.

The potential function has the similar form as the yield function Eq. 1 with $g(\theta) = 1.0$ in Eq. 2, as:

$$G = -\alpha' I_1 + \sqrt{J_2} - k_2 = 0 \quad (4)$$

Eq. 4 has an advantage of having differentiability at all the stress states. The factor α' is defined depending on the type of analysis. For plane strain conditions, α' is defined as:

$$\alpha' = \frac{\tan\psi}{\sqrt{9+12\tan^2\psi}} \quad (5)$$

where ψ is the mobilized angle of dilatancy, which is given as;

$$\psi = \arcsin[-d\varepsilon^p_{vol} / d\gamma^p] \quad (6)$$

where $d\gamma^p$ and $d\varepsilon_{vol}{}^p$ are the plastic shear and volumetric strain increments (positive in compression). In the present study, the value of ψ was determined from the following equation:

$$q / p' = m\left(-d\varepsilon_{vol}{}^p / d\gamma^p\right) + c \quad (7)$$

where m and c are the material constants.

The work-hardening parameter, W^{p*}, was defined as:

$$W^{p*} = \int [\sigma_{ij} d\varepsilon_{ij}{}^p / (p'/p'_a)^n] \quad (8)$$

where p' is the effective mean stress, p'_a is the atmospheric pressure and n is a material constant. The hardening function for the pre-peak regime is assumed as follows (see Fig. 1):

$$X\left(W^{p*}\right) = a \times \ln\left(W^{p*}/b + h\right) \quad (9)$$

where X is the function of stress level, given as:

$$X = q / p' + r \cdot \ln\left(p'/p'_a\right) \quad (10)$$

where *a, b, h* and *r* are the material constants.

The above work-hardening function for pre-peak regime (Eq. 9) has not zero tangent stiffness at the peak state. So, Eq. 10 is needed to modified to alleviate this drawback. The following equation (Eq. 11) was used to be fitted the data for an arbitrarily chosen small range before the peak:

$$X\left(W^{p*}\right) = X_{max} \times \left(\frac{2.0\sqrt{W^{p*}W_f}}{W^{p*} + W_f}\right)^K \quad (11)$$

where X_{max} is the peak strength parameter (i.e., X at the peak state) and W_f is the energy parameter W^{p*} at the peak state, which depends on the pressure level, the angle of bedding plane orientation and void ratio. The power K is a parameter introduced for smooth transition. By using Eq. 11, smooth transition near the peak from the pre-peak toward post-peak regime is attained by defining the power K so that Eq. 11 passes through a given pre-peak point. In the present study, this point was selected at $X = 0.8 \cdot X_{max}$, at which most of the measured work-hardening relation start deviating from their fitted relations. Then, using a three-point fitting method, the parameter K is obtained as follow:

$$K = \log\left(\frac{X_1}{X_{\max}}\right) / \log\left(\frac{2\sqrt{W_1 \cdot W_f}}{W_1 + W_f}\right) \tag{12}$$

where W_1 is W^{p^*} obtained by substituting $X_1 = 0.8 \cdot X_{\max}$ into the measured work-hardening equation (see Fig. 1).

Therefore, the plastic hardening modulus can be expressed as follow:

$$H' = \frac{\partial F}{\partial W^{p^*}} \sigma_{ij} \frac{\partial G}{\partial \sigma_{ij}} / (p / p_a)^n \tag{13}$$

In the present study, it was assumed that the deformation of a given sand element under uniform boundary stress conditions be homogeneous in the pre-peak regime, and that strain localization into a shear band starts suddenly at the peak stress state. The shear banding was introduced by using a strain localisation parameter S in the additive decomposition of total strain increment as follows (Tanaka & Sakai, 1993);

$$d\varepsilon_{ij} = d\varepsilon_{ij}{}^{e} + S \cdot d\varepsilon_{ij}{}^{p} \tag{14}$$

where $S = F_b / F_e$, in which F_b is the area of shear band in each finite element and F_e is the area of the finite element. To model the post-peak stress-strain properties inside a shear band, a work-softening function is introduced as follow (see Fig. 1):

$$X\left(W^{p^*}\right) = X_r + (X_{\max} - X_r)\exp\left[-\left(\frac{W^{p^*} - W_f}{W_r}\right)^2\right] \tag{15}$$

where W^{p^*} is the work softening parameter, X_r is the stress state parameter at the residual state and W_r is the work-softening parameter, which determines the rate of decrease in the stress from the peak strength towards the residual strength. Hence, the elasto-plastic stress-strain relationship incorporating shear banding can be expressed as;

$$d\sigma_{ij} = \left[D_{ijkl}{}^{e} - \frac{S \cdot D_{ijkl}{}^{e} b_{kl} a_{ij}{}^{T} D_{ijkl}{}^{e}}{H' + S a_{ij} D_{ijkl}{}^{e} b_{kl}}\right] d\varepsilon_{ij} \tag{16}$$

where $D^{e}{}_{ijkl}$ is the elastic matrix; $a_{ij} = \partial F / \partial \sigma_{ij}$, $b_{ij} = \partial G / \partial \sigma_{ij}$.

Most granular materials deposited naturally or artificially, or compacted vertically, exhibit cross-anisotropic deformation properties. In the present study, therefore, cross-anisotropic elasticity is considered, which is based on some physical tests with respect to anisotropic elastic deformation properties of sand (Hoque, 1996; Hoque & Tatsuoka, 1996, 1998).

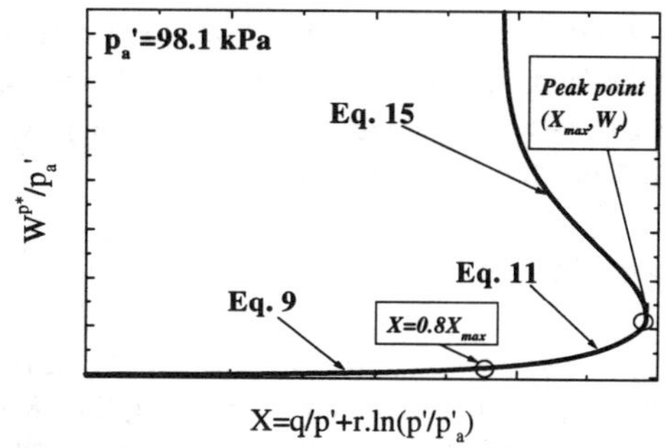

Fig. 1. Proposed work hardening-softening

3 WORK-HARDENING CHARACTERISTICS

Yasin and Tatsuoka (1999, 2000) performed a series of drained stress path PSC tests on saturated dense specimens of Toyoura sand with precise stress and strain measurements (see Fig. 2). It has been shown that all the strain increments (i.e., axial, lateral, shear and volumetric) that occurred by loading between two stress states were noticeably dependent on the intermediate stress paths.

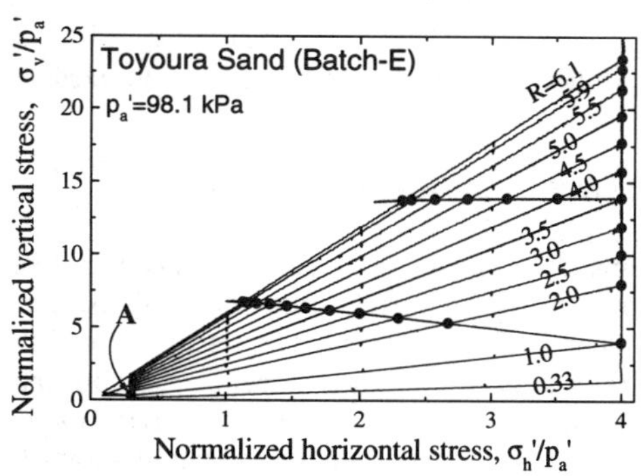

Fig. 2. Measured stress paths in PSC tests

Based on the above physical test results, it has been found that one form of energy function (Eq. 8) could be stress history-independent. For the plane strain case, the plastic work increment dW^p was normalized by being divided by the stress variable, $s[=(\sigma_v+\sigma_h)/2]$, with some power of, n, as follows:

$$W^{p*} = \int\left[\left(t\cdot d\gamma^p + s\cdot d\varepsilon^p_{vol}\right)/\left(s/p_a\right)^n\right] \quad (17)$$

where $t=(\sigma_v'-\sigma_h')/2$, $s=(\sigma_v'+\sigma_h')/2$, $p_a'=98\,kPa$ and W^{p*} is equal to 0 at stress state A of $\sigma_v'/p_a'=\sigma_h'/p_a'=0.3$. The measured relationship between $Y=W^{p*}/p_a'$ and $X=t/s+r\cdot\ln(s/p_a)$ was fitted by the following equation:

$$X = a\times\ln(Y/b+h) \quad (18)$$

where, *a=0.1288, b=3.4505×10⁻⁵, h=0.4310, n=0.9* and *r=0.09*, which were obtained by fitting Eq. 18 to the data from the PSC tests of Toyoura sand shown in Fig. 3.

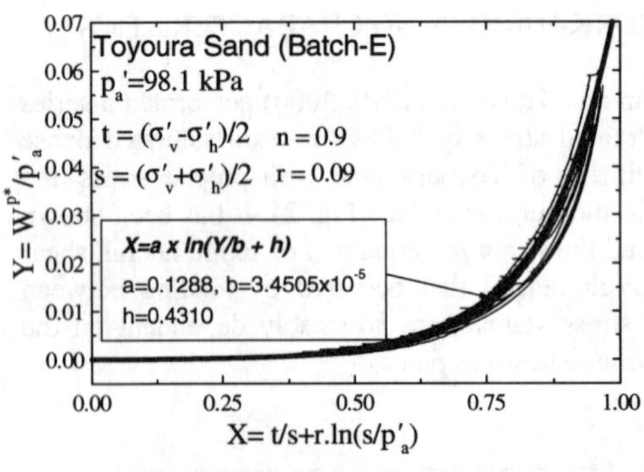

Fig.3 Fitted work-hardening function at pre-peak

4 STRESS-DILATANCY BEHAVIOR

Yasin and Tatsuoka (2000) also examined the effects of stress path on the stress-dilltancy relationship under plane strain conditions. The result showed that the effects of stress path direction is noticeably smaller in the plots of plastic strain increments than in those of total strain increments. Fig. 4 shows the stress-dillatancy ($t/s \sim d\varepsilon^p_{vol}/d\gamma^p$) relations from the PSC tests. The average relationship between $t/s \sim d\varepsilon^p_{vol}/d\gamma^p$ was fitted as follows:

$$t/s = m\cdot\left(-d\varepsilon^p_{vol}/d\gamma^p\right)+c \quad (19)$$

where m and c are the material constants which are, respectively, *0.596* and *0.639* for Toyoura sand in this present study.

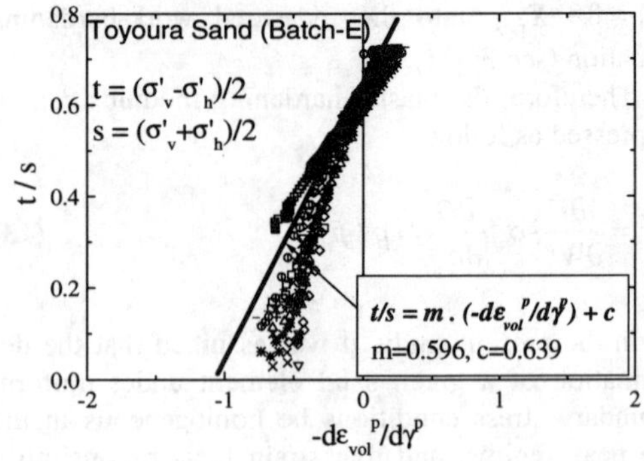

Fig. 4. Fitted average stress-dilatancy relationship

5 SOME RESULTS OF FEM SIMULATION

A FEM code incorporating the proposed constitutive model for sand was developed. A set of non-linear equations was solved by the Dynamic Relaxation technique (Tanaka & Kawamoto, 1988), which has a reputation in solving highly non-linear equations, especially for high friction angle materials as the one analyzed in the present study. The integration of the elasto-plastic equation was done by the Return Mapping scheme (Ortiz & Simo, 1986), which is a first-order approximated Euler backward integration. A single four-node quadrilateral element (*2 cm × 2 cm*) was selected and the load control increment was adopted for FEM simulation. We will show below some examples of FEM simulation based on the proposed model for Toyoura sand.

Fig. 5 compares the relationships between the stress ratio, $R=\sigma_v'/\sigma_h'$, and the shear strain, $\gamma=\varepsilon_v-\varepsilon_h$, obtained from the PSC tests for Toyoura sand and the FE analysis of one element. It is seen that the proposed model, incorporated in the FE code, can well simulate the strength and deformation characteristics for a wide variety of stress paths. Note, however, that viscous effects, which are not negligible in the physical tests, are not modeled in the present study.

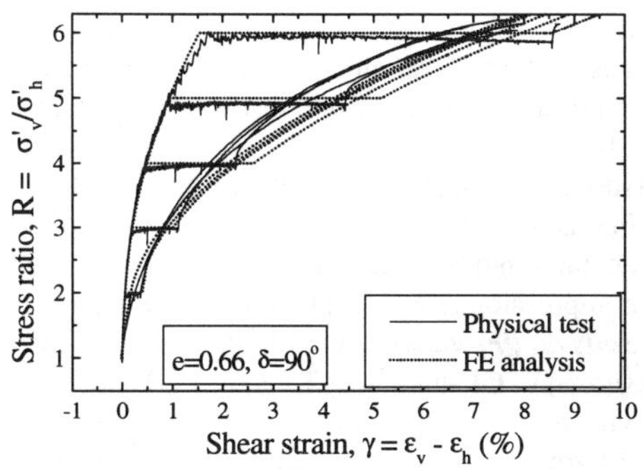

Fig. 5(a). Comparison of stress ratio vs. shear strain relations between FEM and physical test

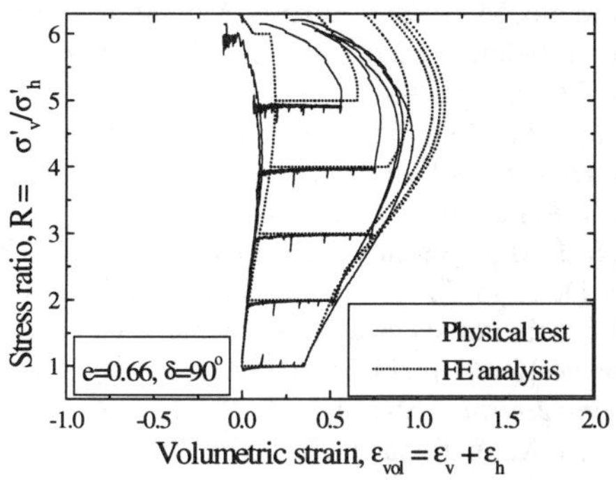

Fig. 5(b). Comparison of stress ratio vs. volumetric strain relations between FEM and physical test

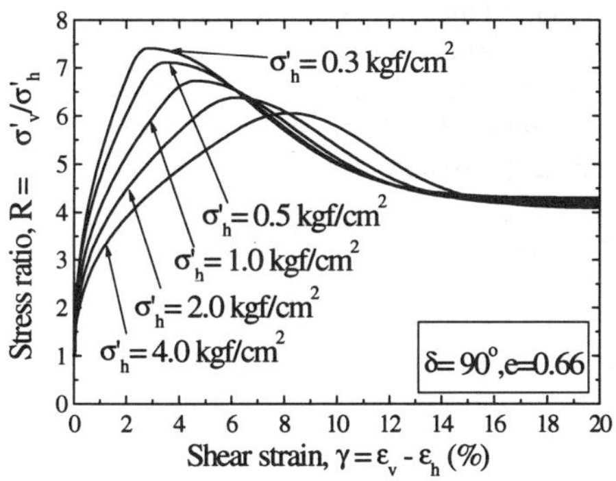

Fig. 6. Dependency on pressure level (confining pressure) by the proposed model (FEM)

Figs. 6 through 8 show the simulation of the effects of pressure level (confining pressure, σ_c), the angle

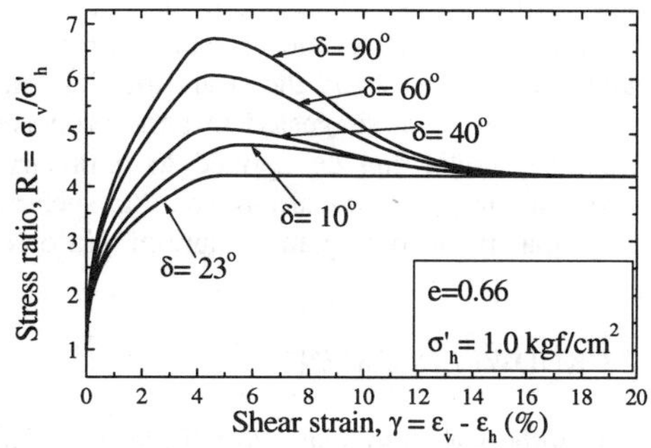

Fig. 7. Dependency on strength anisotropy by the proposed model (FEM)

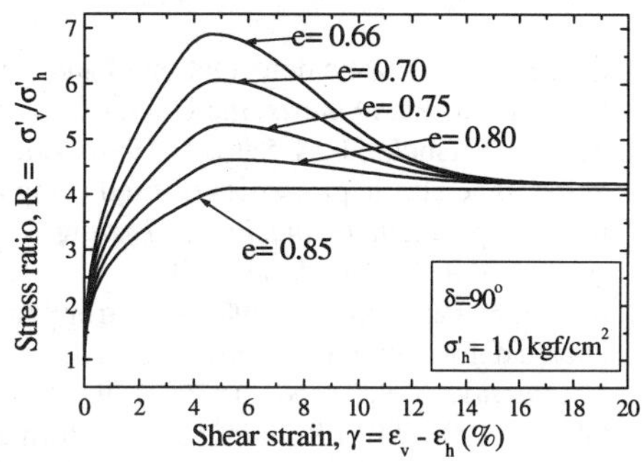

Fig. 8. Dependency on initial void ratio by the proposed model (FEM)

of bedding plane orientation (δ) and void ratio (e). It is seen that these factors can be reproduced satisfactorily by the proposed model for Toyoura sand.

6 CONCLUSIONS

The quantity $W^{p*} = \int\left[\left(t \cdot d\gamma^p + s \cdot d\varepsilon_{vol}^p\right)/(s/p_a')^n\right]$ $(p_a' = 98kPa)$ is essentially stress history-independent and a rather unique function of the stress parameter $X = t/s + 0.09 \cdot \ln(s/p_a')$ for different stress histories and stress paths in PSC tests on Toyoura sand. By using this quantity as the hardening parameter, a new elasto-plastic constitutive model with a non-associated flow rule and isotropic work-hardening and -softening properties for granular geomaterials has been proposed. The proposed model is capable of simulating the strength and deformation characteristics for various stress histories and stress paths, while taking into account cross-anisotropic elasticity, the effects of pressure level, anisotropy, density and the strain localization into a shear band. It is suggested that the proposed constitutive model can be applied to boundary value problems in geotechnical engineering, such as bearing capacity of footing and failure of reinforced soil.

ACKNOWLEDGEMENTS

The authors wish to thank Prof. Tadatsugu Tanaka at the University of Tokyo for his advice to the non-linear solution method in the present study.

REFERENCES

Hoque, E. 1996. Elastic deformation of sands in triaxial test. *Ph. D thesis*, the Univ. of Tokyo.

Hoque, E., Tatsuoka, F. & Sato, T. 1996. Measuring anisotropic elastic properties of sand using a large triaxial specimen. *Geotechnical Testing Journal*, ASTM, Vol. 19, No. 4: 411-420.

Hoque, E. & Tatsuoka, F. 1998. Anisotropy in the elastic deformation of materials. *Soils and Foundation*, Vol. 38, No. 1: 163-179

Kotake, N. 1998. FEM Simulations of deformation and failure of reinforced soil. *Doctoral thesis*, the Univ. of Tokyo

Ortiz, M. & Simo, J. C. 1986. An analysis of a new class of integration algorithms for elastoplastic constitutive relations. *Int. J. Numer. Methpds Eng.*,Vol. 23: 353-366.

Peng, F. L., Kotake, N., Tatsuoka, F., Hirakawa, D. & Tanaka, T. 2000. Plane strain compression behaviour of geogrid-reinforced saand and its numerical analysis. *Soils and Foundations*, Vol. 40, No. 3: 55-74.

Siddiquee, M. S. A. 1994. FEM simulation of deformation and failure of stiff geomaterials based on element test results. *Ph. D. thesis*, the Univ. of Tokyo.

Siddiquee, M. S. A., Tanaka, T., Tatsuoka, F., Tani, K. & Morimoto, T. 1999. FEM simulation of scale effect in bearing capacity of strip footing on sand. *Soils and Foundations*, Vol. 39, No. 4: 91-109

Tanaka, T. & Kawamoto, O. 1988. Three dimensional finite element collapse analysis for foundations and slopes using dynamic relaxation. *Proceedings of Numerical Methods in Geomechanics*: 1213-1218. Insbruch.

Tanaka, T. & Sakai, T. 1993. Progressive failure and scale effect of trap-door problems with granular materials. *Soils and Foundations*, Vol. 33, No.1: 11-22.

Tatsuoka, F., Sakamoto, M., Kawamura, T. & Fukushima, S. 1986. Strength and deformation characteristics of sand in plane strain compression at extremely low pressures, *Soils and Foundations*. Vol. 26, No. 1: 65-86.

Tatsuoka, F. Okahara, M., Tanaka, T., Tani, K., Morimoto, T. & Siddiquee, M. S. A. 1991. Progress sive failure and particle size effect in bearing capacity of a footing on sand. *Geotechnical Engineering Congress (Mclean et al. eds.)*, Vol. 2: 788-802.

Tatsuoka, F., Siddiquee, M. S. A., Park, C. S., Sakamoto, M. & Abe, F. 1993. Modelling stress-strain relation of sand, *Soils and Foundations*. Vol. 33, No. 2: 60-81.

Yasin, S. J. M. 1998. Strength and deformation characteristics of sands in plane strain shear tests. *Ph. D thesis*, the Univ. of Tokyo.

Yasin, S. J. M., Umetsu, ., Tatsuoka, F., Arthur, J. R. F. & Dunstan, T. 1999. Plane strain strength and deformation of sands affected by batch variations in two different types of apparatus. *Geotechnical Testing Journal*, ASTM, Vol. 21: 80-100.

Yasin, S. J. M. & Tatsuoka, F. 2000. Stress history-dependent deformation characteristics of dense sand in plane strain. *Soils and Foundations*, Vol. 40, No.2: 77-98.

Yoshida, T., Tatsuoka, F., Siddiquee, M. S. A., Kamegai, Y. & Park, C. S. 1994. Shear banding in sand in plane strain compression. *Proceedings of Third International Workshop on Localisation and Bifurcation Theory for Soils and Rock (Chambon et al., eds.)*: 165-170. France: Grenoble (Aussois)

Computer Methods and Advances in Geomechanics, Desai et al. (eds) © 2001 Balkema, Rotterdam, ISBN 90 5809 183 X

Coupled modeling using DSC and TFA theoretical process

P.P. Procházka
Civil Engineering, Department of Structural Mechanics, CTU Prague, Czech Republic

J.Trčková
Institute of Structure and Mechanics of Rocks, Czech Academy of Sciences, Czech Republic

ABSTRACT: Eigenstresses and eigenstrains act out a very important role in many branches of applied mechanics. The eigenparameters may represent plastic strains or relaxation stresses (or also prestresses, change of temperature, etc.), and may serve as free parameters for improving numerical models to get the computed quantities that should be as close as possible to the real state. In order to get a reasonable agreement between these quantities when comparing both computed and measured values, a special variational formulation can be stated, dealing with the minimum variance of differences between measured and computed values. Using a very useful treatment, transformation field analysis (TFA), having recently been proposed by Dvorak & Procházka, the problem leads to a linear system of algebraic equations. In this case, we start with distinct state concept (DSC) having been proposed by Desai and describing nonlinear material behavior of the rock, and the nonlinear material properties are "tuned" by the eigenparameters.

1 INTRODUCTION

The most important problem in underground structure assessment there is a description of mechanical behavior of surrounding rock. The coupled modeling may serve a bridge for overcoming this problem. It consists of mutual influence of real situation, scale models, and mathematical models. The material behavior in the mathematical model is estimated by results from measurements in the other models.

It is well known that the changes in material properties during the loading process may be involved in eigenparameters, which stand for plastic strains, relaxation stresses, etc. The idea of transformation analysis (Dvorak, G. J., & Procházka, P. 1996, Dvorak, G. J., Procházka, P. & Srinivas, S. 1999) consist of *a priori* relating internal parameters, included in stress, strain, or displacement field, and eigenparameters on the other hand. The computation may then start with elastic states and from comparison with the experiment, or in site measurement, the proper values of eigenparameters are obtainable.

A similar approach, the disturbed state concept (DSC) was proposed by Desai, see (Desai, C. S. 1974, Desai, C. S. 1994). He uses certain artificial *disturbance function* and simulates various mechanical states in a body (damage, elastoplasticity, viscoplasticity, creep, cracking, etc.). Our goal is to generalize the TFA and to start with an estimation of physically nonlinear behavior of the body in the sense of the DSC. Then the algorithm may be applied to the overall behavior of the body as well as to a local observation of the material properties. The disturbance function may either be estimated or determined more precisely from large amount of experiments. Much easier is to state the eigenparameters, which are also more reliable. The Unified Function Concept (DSC & TFA) seems to be prospective.

In this paper, we concentrate our attention on the stability of tunnel face. Nonlinear mechanical behavior of this system will be involved in a numerical procedure, results from experiments will be used, and the DSC & TFA concept will be applied. The approximation of the correct stress-strain relation is obtained from comparative computations with the measurement "in situ" or with the results from physical modeling. Since the TFA concept is not appropriate enough when dealing with stress-strain relations, which are described by multivalued functions (microcracking in concrete, localization of stresses, some kinds of damage, etc.), the new concept will enable one to overcome this restrictions.

The above-said concept DSC & TFA will be applied to description of behavior of rock mass, the underground structures (such as tunnel lining, foundations, underground power stations, etc.) will be considered elastic. Obviously, the generalization is possible.

2. FORMULATION OF THE TFA

In this section, our aim is to formulate the general procedure for the TFA. This may be done in terms of many modern numerical methods. It seems that the BEM is the most appropriate numerical method in this case, but the FEM is also admissible.

First, let us consider that the body (part of a structure, element, system of more elements, composite) behaves linearly; i. e. Hooke's linear law is valid in entire body (this assumption admits among others an application of the BEM). When the problem is correctly posed, the displacement vector, strain tensor and stress tensor can be obtained from the Navier equations, kinematical equations, and the linear Hooke's law.

In the second step we select points, where the measured values are at our disposal, either from experiments in laboratory, or from in situ measurements. We also select points or regions (subdomains) from the body under study, and apply there successively unit eigenparameter impulses (either eigenstresses or eigenstrains) to get an influence tensor (matrix). In order to precise this statement, denote A_i, $i=1,...,n$, either the points or regions where the eigenparameters will be applied. Let, moreover, the set of points where the measured values are known be B_j, $j=1,...,m$. Then the real stress at B_j is a linear hull of stress σ^{ext} at B_j due to external loading and eigenstrains μ and ε^{pl}, or eigenstress λ and relaxation stress σ^{rel} at A_i:

$$\sigma = \sigma^{ext} + P\,\mu + Q\,\varepsilon^{pl}, \text{ or } \sigma = \sigma^{ext} + R\,\lambda + T\,\sigma^{rel} \quad (1)$$

or in differential (incremental) form:

$$d\sigma = d\sigma^{ext} + P\,d\mu + Q\,d\varepsilon^{pl},$$
$$\text{or}$$
$$d\sigma = d\sigma^{ext} + R\,d\lambda + T\,d\sigma^{rel} \quad (2)$$

where the influence matrices P, Q, and R and T may be identical, as any eigenparameter may stand for the plastic or relaxation parameter (say, eigenstrain may stand for plastic strain, which is obvious from (1)). The dimensions of σ, σ^{ext}, μ, ε^{pl}, λ, and σ^{rel} $m \times 6$ (because of symmetric stress and strain tensors) and the dimensions of P and Q are $m \times 6 \times n$. In the classical TFA the values of μ, or ε^{pl} are calculated from minimization of variance of computed and measured stresses, cf. Fig. 1, where eigenstress λ characterizes the nonlinear behavior of the material. Note that it holds: $\lambda = -L\,\mu$, where L is the elastic material stiffness matrix.

Note that the first relations in (1) and (2) describe the initial strain method while the second relations in those equations formulate the initial stress method The eigenparameters may generally stand for plenty of phenomena like change of temperature, swelling, watering, etc. This is why we could split the eigenparameters in (1) into two parts: eigenparameters themselves and the quantities connected with physically nonlinear behavior of the material.

3. DISTURBED STATE CONCEPT (DSC)

The idea for this theory was originally proposed by Desai and the theory characterized behavior of overconsolidated clays. Since then, Desai and coworkers have developed and successfully applied this concept to other materials.

The DSC is a unified modeling theory for mechanical behavior of material and interfaces. It allows incorporation of the internal changes on interfacial boundaries of phases (both micro- and macrolevel) and the resulting mechanism in a deforming material into the constitutive description. Initially, the material under external loading is in relative intact state (RI). Using such theories as elasticity, plasticity and viscoplasticity may theoretically treat the intact state, i. e. no cracking is considered in this state. After increasing the external loading, the material transforms from the RI state to the fully adjusted state (FA) or critical state, which is an asymptotic state, the material at that may no longer carry certain or all stresses. For example, microcracking and subsequent softening are such disturbances.

Desai uses a scalar disturbance function D, having different expressions depending of mechanical properties in the model under consideration. The equilibrium equation for a material element in terms of stresses is derived as

$$D_u\,S_{ij} = (D_u - D)\,\sigma^{IS}_{ij} + D\,\sigma^{FA}_{ij} \quad (3)$$

where S_{ij} stands for average (observed) response, D_u is max D and is in most cases equal to one. Using the incremental method, the differentiation of the last relation yields

$$D_u\,dS_{ij} = (D_u - D)\,d\sigma^{IS}_{ij} + D\,d\sigma^{FA}_{ij} + dD\,(\sigma^{FA}_{ij} - \sigma^{IS}_{ij}) \,. \quad (4)$$

The first term of the right hand side of the last equation expresses continuum constitutive law for elastoplastic (visco-plastic) behavior, the second term obeys the classical Kachanov damage formulation (Kachanov 1986), and the third term in (4) indicates different stresses in the two parts.

The incremental constitutive equations for the RI part and the FA part are expressed as:

$$d\sigma^{IS}_{ij} = C^{IS}_{ijkl}\,\varepsilon^{IS}_{kl}\,, \quad d\sigma^{FA}_{ij} = C^{FA}_{ijkl}\,\varepsilon^{FA}_{kl} \qquad (5)$$

where C^{IS}_{ijkl} are in our case the components of ideally plastic constitutive tensor with hardening for IS part, furthermore C^{IS}_{ijkl} are the components of elastoplastic constitutive tensor for IS part, superscripts IS and FA indicate the phases. For more details concerning the DSC see Desai.

In order to make more transferable the TFA and the improved concept using Desai idea, a uniaxial stress-strain distribution is depicted in Fig. 1 and in Fig. 2. In Fig. 1 one can observe that we start with computation of elastic medium, where even not too precise values of material properties are required. From Fig. 2 it is seen that nonlinear material behavior is estimated and the relaxation stresses improve the constitutive law in accordance with measurement.

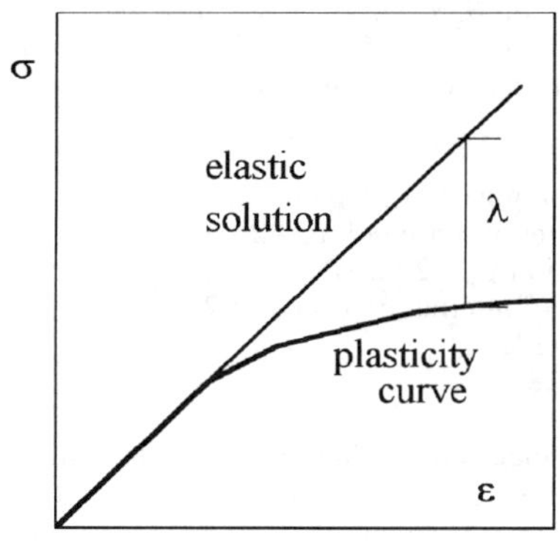

Fig. 1: TFA method

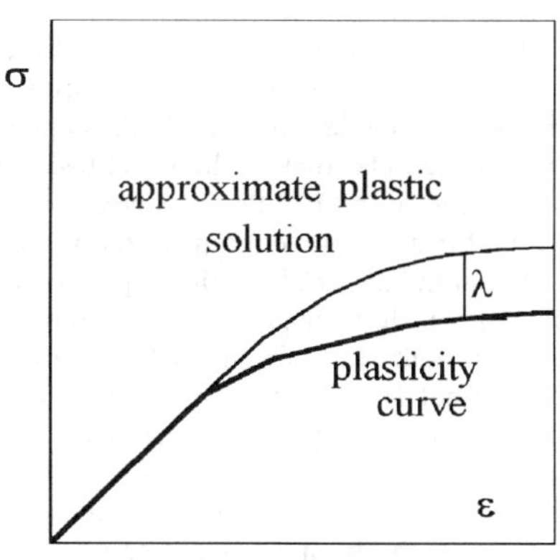

Fig. 2: DSC & TFA

4. UNIFIED FUNCTION CONCEPT

Recall first the TFA involving the DSC. The transformation field analysis consists in expressing the stress σ at an arbitrary point ξ of the domain by virtue of superposition of stress $\sigma^{ext}(\xi)$ at ξ due to external loading which is applied to an plastic, viscoplastic, or others matrerials, and a linear hull of, say, the eigenstresses λ nd plastic strains ε^{pl} at other points x. Since we assume that at each point six values of stress, plastic stress, and eigenstress tensors are prescribed, the relation stresses σ^k at the points B_k, k = 1,...,m, and the eigenstresses and plastic stresses λ^l, l=1,...,n and σ^{pl}, l = 1,..., m at A_l becomes (to simplify the expressions the vector notation for tensors is used), cf. (1):

$$(\sigma_i)^k = (\sigma^{ext}_i)^k + \sum_{j=1}^{6}\sum_{l=1}^{m}(R_{ij})^{kl}(\lambda_j)^l +$$

$$\sum_{j=1}^{6}\sum_{l=1}^{m}(T_{ij})^{kl},\ (\sigma^{rel}_j)^l, \quad i = 1,...,6,\ k = 1,...,m. \quad \{6\}$$

The stresses $(\sigma^{ext}_j)^l$ have now the meaning of a current initial stress (say elastic, tangential, secant, etc.).

The first and the third terms of the right hand side of (6) have to be equal to S_{ij} in (3), which is the current stress. Using the notation $(S_i)^k$ instead of $(S_{ij})^k$, the relation (6) can be written as:

$$(\sigma_i)^k = (S_i)^k + \sum_{j=1}^{6}\sum_{l=1}^{n}(R_{ij})^{kl}(\lambda_j)^l, \quad i = 1,...,6,$$
$$k = 1,...,m. \quad \{7\}$$

The differentials of current stresses become:

$$d(\sigma_i)^k = d(S_i)^k + \sum_{j=1}^{6}\sum_{l=1}^{n}(R_{ij})^{kl}d(\lambda_j)^l, \quad i = 1,...,6,$$
$$k = 1,...,m. \quad \{8\}$$

where $d(S_i)^k$ is controlled by (4).

A natural requirement is to assure that the difference between measured $(\sigma^{meas}_i)^k$ and computed stresses be as small as possible. This means that the variance

$$I[(\lambda_j)^l] = \sum_{i=1}^{6}\sum_{k=1}^{m}[(\sigma_i)^k - (\sigma_i^{meas})^k]^2 \rightarrow \text{minimum} \qquad (9)$$

Differentiating I with respect to (λ_α) leads to a system of linear algebraic equations for unknown $(\lambda_j)^l$:

$$\sum_{j=1}^{6} \sum_{l=1}^{n} (A_{\alpha j})^{\beta l} (\lambda_j)^l = Y_\alpha{}^\beta, \ \alpha = 1,...,6, \ \beta = 1,...,m, \tag{10}$$

where

$$(A_{\alpha j})^{\beta l} = \sum_{i=1}^{6} \sum_{k=1}^{m} (R_{ij})^{kl} (R_{i\alpha})^{k\beta}$$

$$Y_\alpha{}^\beta = -\sum_{i=1}^{6} \sum_{k=1}^{m} (S_i)^k - (\sigma^{meas}{}_i)^k + $$

$$+ \sum_{j=1}^{6} \sum_{l=1}^{m} (R_{ij})^{kl} (\lambda_j)^l (R_{i\alpha})^{k\beta}$$

In order to get $d(\sigma_i)^k$ in (8), we need to calculate $d(\lambda_j)^l$. This is the difference between $(\lambda_j)^l$ from (10) and the same quantities from previous step of incremental method.

The procedure deserves a closer attention. In the first step the influence matrices are created, as described in the above explanation (Section 2). The distribution of the disturb function D is determined mostly from laboratory tests, or in our case, estimated from scale experiments in stands. The incremental method is mostly used when applying the DSC.

Let us start with some load of the trial body.

At the beginning the FA state will be most probably not reached. The intact state follows the von Mises-Huber-Hencky law. When increasing the load, the IS is attained and the DSC has to involve both IS and FA states into the computation. Then, the stresses are split into IS and FA parts. Increments of both these parts can be done from the DSC and the total stress for both parts is given by adding the increments to the previous steps. The same is valid for the total current stress; see (7).

Since the relation (10) is linear, we substitute $(S_i)^k$ and $(\lambda_i)^k$ there. Then applying the minimum condition for the additional eigenstresses, the improvement of current stresses or error for the DSC is obtained.

5. EXAMPLE

The procedure was applied to investigation of stability of tunnel face. In order to compare the values from experiments (substituting in this case the "in situ" measurements) with mathematical formulation and results, a model from physically equivalent materials was prepared in glass stands. Steel rods to keep the sidewalls in vertical position reinforced the sides of the stand. Since the stresses can be measured in a very complicated way, the principal quantity to us was horizontal displacements at the tunnel face.

The model stand was prepared in such a way that the scale with respect to reality was 1:100. In the stand of dimensions 500 x 500 x 570 mm^3 (width x length x height) a tunnel of cylindrical shape was modelled. The tunnel had inner diameter 76 mm and length of 173 mm. The horizontal axis of the tunnel was positioned at 191.5 mm above the stand bottom symmetrically with respect to the sidewalls of the stand.

The equivalent of the tunnel concrete lining was modeled by a tube from hardened paper (outer diameter 83 mm, thickness of the lining 3.5 mm). After positioning the tube into the model, the entire stand was filled by the model material that substituted the rock mass. The equivalent material was a mixture of sand with grain 0.2 mm, bentonite and fat A00 in ration 65% + 29% + 6%. The material properties were determined by standard laboratory methods used in soil mechanics with the following results:

Unit weight $\gamma = 1.55$ g/cm^3
Compressive strength $\sigma_t = 26.9$ kPa
Cohesion $C_p = 20.2$ kPa
Angle of internal friction $\phi_p = 27.5^0$
$E = 4$ GPa
$\nu = 0.28$

The model was loaded by external loading on the terrain. The metal weights in layer were used for this purpose. After changing the external load by adding successively one weight layer, on vertical tunnel face at selected points, displacements were measured.

For this purpose, an equipment was prepared (Fig. 3), which enabled one to find out the values of displacements with accuracy of 0.05 mm. Six lighten duralumin needles went freely through holes in two circular Plexiglas disks, which stabilized their position. One end of the needles was equipped by a small circular heads to prevent a penetration of needles into the model material. At the opposite end of the needles, the scale measurement was glued. Before the experiment tests, the needles were leaned against the tunnel face. In order to ensure free deformation of the tube, very soft rubbers were placed along the outer boundary of the disks. Using thedolite, changes of the position of the needles – their movement out of the tunnel – are determined with respect to the initial state of the needles. The function of this equipment was tested in a model stand.

The pressure at the horizontal tunnel axis was calculated from the initial stress state (no opening considered) in dependence of terrain loading. The largest difference is between last and last but one stage, then the failure arrives. In the sense of scale modelling, the admissible load and face deformation can be calculated for real structure. In Fig. 4 we can see the overview at the stand with the special measuring equipment. In front of the side with needles, theodolite was positioned for measuring the displacements.

The physical model is described in details (including the values measured during the experiment) in (Procházka, P. & Trčková, J., 2000).

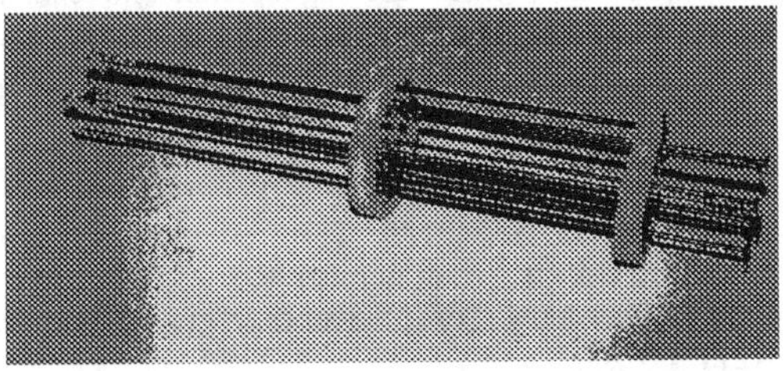

Fig. 3 Equipment for measuring displacements

Fig. 4 View at the stand with instrumentation

Numerical experiments started with 3D isoparametric finite elements with 24 DOF. Mixed formulation for von Mises-Huber-Hencky elastoplastic material behavior of rock with hardening was applied to numerical computation of intact phase of Desai's model. A regular mesh of elements was created and the vertical cut in the tunnel axis is depicted in Fig. 5. In this picture the subdomains where eigenparameters were considered are also shown. Since the number of components of eigenparameter tensor may not exceed the total number of measured values, certain components have to be aggregated. In our case we introduced:

$$\lambda_x = \lambda_y = \frac{\nu}{1-\nu}\lambda_z$$

The more accurate computation should be carried out, the more points of measurement have to be considered and, consequently, more components of eigenparameter tensor could be taken into consideration.

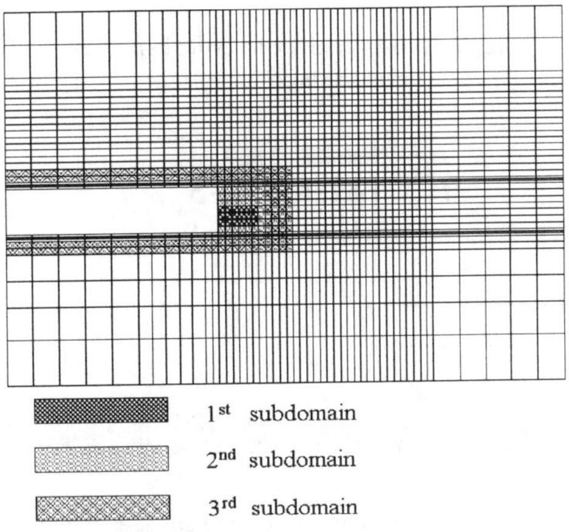

Fig. 5 Vertical cut of FEM mesh with subdomains for introducing eigenparameters

Fig. 6 shows the distribution of vertical stresses in elastic state of the rock mass in vertical cut crossing the tunnel axis. There is small concentration of stresses at "vertices" of the rock and the maximum compressive stress attains 0.8 MPa.

In Fig. 7 the vertical stresses are depicted in vertical cut. In this case, the M-H-H material behavior of rock is considered in intact state of the DSC, and no improvement by the TFA is introduced. Comparing these results with Fig. 8, where the DSC & TFA model is applied, the results are quite different. This is not an impact of wrong conception of the DSC, but rather wrong guess of hardening rule in M-H-H hypotheses, and of a wrong estimate of the material parameters.

On the other hand, the peaks in vertical stresses are nearly same, and are concentrated near upper and lower part of the tunnel lining. It follows from this that the decisive results do not require TFA improvement. If observation of the stress distribution in the rock is required, the TFA improvement delivers more accurate results than the DSC itself. Generally, the linked model is more accurate providing reliable measurement results are at disposal.

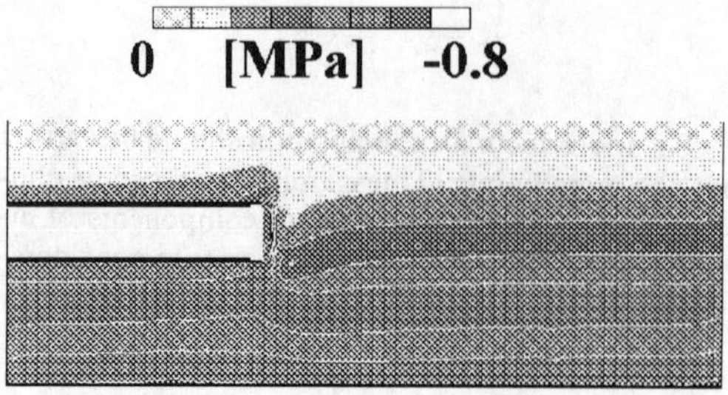

Fig. 6 Hypsography of vertical stresses at elastic stage

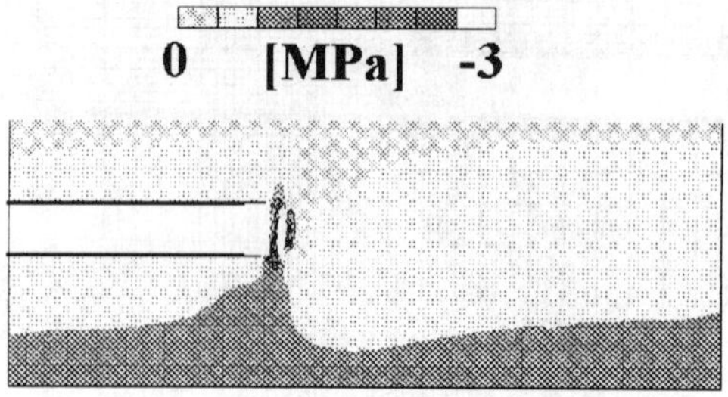

Fig. 7 Hypsography of vertical stresses when introducing the DSC stages

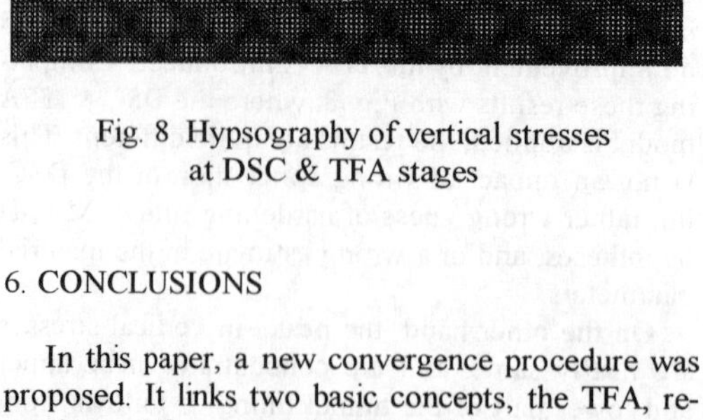

Fig. 8 Hypsography of vertical stresses at DSC & TFA stages

6. CONCLUSIONS

In this paper, a new convergence procedure was proposed. It links two basic concepts, the TFA, recently used for composite and soil continua, and the DSC proposed by Desai, who prevailingly applied it to soils. On a coupled mathematical-experimental example, it was proved that this procedure might bring about certain new views to the theoretical behavior of soils, and has a good application in solving practical problems.

The eigenparameters (eigenstrains or eigenparameters) play role of additional quantities that improve estimates of constitutive laws. They improve the values of plastic strains (plastic stresses), but also give the error of originally computed plastic strains (relaxation stresses). They may prompt that the selected plastic rule is wrong and show the way to recover it. In addition, they may warn the researchers that the experiments provide wrong values from measurement or that not appropriated apparatus was used.

In any case, when the eigenparameters exhibit large values, the experiments should be carried out more accurately or the selected constitutive laws could be suspicious.

ACKNOWLEDGEMENT: This research was supported by GA ČR, grant number 103/99/1035 and by Czech Academy of Sciences, grant number A2119001/00.

REFERENCES

Desai, C. S. 1974. A Consistent Finite Element Technique for Work-Softening Behavior. In: J. T. Oden et al (eds.), *Int. Conf. on Comp. Meth. In Nonlinear Mech.* Univ. of Texas at Austin.

Desai, C. S. 1994. *Constitutive Modeling Using the Disturbed State Concept*, Chapter 8, *Continuum Models for Materials with Microstructure*, ed. H. Mulhaus, John Wiley & Sons, UK.

Dvorak, G. J., & Procházka, P. 1996. Thick-walled Composite Cylinders with Optimal Fiber Prestress, *Composites*, Part B, 27B, 643-649.

Dvorak, G. J., Procházka, P. & Srinivas, S. 1999. Design and Fabrication of Submerged Cylindrical Laminates, Part I and Part II, *Int. J. Solids & Structures,* 1248-1295.

Kachanov, L. M. 1986. *Introduction to Continuum Damage Mechanics.* Martinus Nijhoff Publishers, Dordrecht, Netherlands.

Procházka, P. & Trčková, J. 2000. Coupled modeling of concrete tunnel lining, to appear in *Our World Concrete and Structures,* Singapore.

Computer Methods and Advances in Geomechanics, Desai et al. (eds) © 2001 Balkema, Rotterdam, ISBN 90 5809 183 X

Constitutive modeling of unsaturated soils: Stress variables and strain increment parameters

R.S.Sharma
Department of Civil and Environmental Engineering, University of Bradford, UK

ABSTRACT: Selection of appropriate stress variables plays a vital role in success of a constitutive model for unsaturated soils. Although net stress and suction have been used as two stress variables by many researchers, various other combinations of stress variables have been suggested. Equally important input for a constitutive model is selection of strain increment parameters. This latter aspect did not receive much attention in the development of constitutive models for unsaturated soils. In this paper, issue of appropriate stress variables along with strain increment parameters is addressed. Experimental results are used to illustrate the use of stress variables and conjugate strain increment parameters.

1 UNSATURATED SOILS

Unsaturated soils are three phase materials, containing soil particles, water and air. In saturated soils the void spaces contain a single liquid (typically water), whereas in unsaturated soils the voids are filled with a mixture of a liquid and a gas (typically water and air). The presence of air along with water in the soil voids gives rise to two types of pore pressures: pore air pressure u_a and pore water pressure u_w. The pore air pressure is higher than the pore water pressure due to surface tension effects (see, for example, Sharma, 1998). The difference in the values of the pore air pressure and the pore water pressure is known as matric suction.

There are many important fields of application in which an understanding of the engineering behaviour of unsaturated soils is required. For example, climatic conditions mean that in over half of the world the water table is at a considerable depth and therefore construction activity occurs within unsaturated soils. In such situations the vast majority of geotechnical problems arise from ground movements caused by wetting or drying of the unsaturated soil under load. In addition, any construction activity involving compacted fills require an understanding of unsaturated soil behaviour.

The difficulty involved in understanding and modeling the complexities of unsaturated soil behaviour have, in the past, often acted to discourage widespread research in this area. Advances of recent years have, however, suggested possible ways forward (see Sharma, 2000). In the last decade, a number of elasto-plastic constitutive models for unsaturated soils have been proposed.

Selection of appropriate stress variables plays a vital role in success of a constitutive model for unsaturated soils. Although net stress $\sigma - u_a$ and matric suction $u_a - u_w$ have been used as two stress variables by many researchers, various other combinations of stress variables have been suggested. Equally important input for a constitutive model is selection of strain increment parameters. This latter aspect did not receive much attention in the development of constitutive models for unsaturated soils. In this paper, issue of appropriate stress variables along with strain increment parameters is addressed. Experimental results are used to illustrate the use of stress variables and strain increment parameters.

2 SELECTION OF STRESS STATE PARAMETERS FOR UNSATURATED SOILS

It was recognized in the early 1960s that the mechanical behaviour, especially the volumetric response, of an unsaturated soil cannot be described in terms of a single stress state parameter, but an additional stress parameter is required. Several researchers including Bishop and Blight (1963) and Burland (1964) suggested the use of two independent stress variables. Matyas and

Radhakrishna (1968) adopted the "net stress" $\sigma - u_a$ and the matric suction $u_a - u_w$ as the independent stress variables to describe the mechanical behaviour of unsaturated soils.

Fredlund and Morgenstern (1977) argued, on the basis of equilibrium analysis and experimental results, that any two of the three stress parameters $\sigma - u_a$, $\sigma - u_w$, and $u_a - u_w$ would be sufficient to describe fully the stress state of an unsaturated soil. The first two of these stress parameters $\sigma - u_a$ and $\sigma - u_w$ are tensor quantities, whilst the third parameter $u_a - u_w$ is a scalar quantity. It is therefore sensible to select either $\sigma - u_a$ and $u_a - u_w$ or $\sigma - u_w$ and $u_a - u_w$ as the appropriate stress variables, because use of the two tensor variables $\sigma - u_a$ and $\sigma - u_w$ would lead to unnecessary complexity and redundant information. The combination of "net stress" $\sigma - u_a$ and matric suction $u_a - u_w$ is usually selected, because u_a is commonly zero (relative to atmospheric pressure) in the field, and the net stress and the matric suction then simplify to the total stress and the negative pore water pressure respectively. In addition, the pore water pressure in an unsaturated soil is generally negative, which is often difficult to measure accurately. This means an element of uncertainty will be involved in only one stress variable if $\sigma - u_a$ and $u_a - u_w$ are selected, whereas the uncertainty will affect both the variables if $\sigma - u_w$ and $u_a - u_w$ are adopted.

Considerable research has been done using the net stress $\sigma - u_a$ and the suction $u_a - u_w$ as the two variables, but other alternative sets of stress variables have been proposed by a number of researchers, including Kohgo *et al* (1993 *a* and *b*), Modaressi and Abou-Bekr (1994), Jommi and Di Prisco (1994) and Bolzon *et al* (1996). The search for alternative stress variables seems to be prompted by an objective to adopt stress variables with stronger physical significance than the set of net stress $\sigma - u_a$ and suction $u_a - u_w$. To achieve it, a stress variable is quite commonly selected in such way that it defines uniquely the variation of shear strength of an unsaturated soil. For a full description of the remaining aspects of mechanical behaviour, such as, volumetric response and yielding, suction or some function of suction is adopted as another stress variable. The general form of the proposed variables is: $\sigma - u_a + f_1(s)$ and $f_2(s)$, with $f_1(s)$ and $f_2(s)$ being functions of suction. A comprehensive summary of stress variables used by various researchers is given by Gens (1996) and critical evaluation is presented by Wheeler and Karube (1996). It is important to note that the use of alternative stress variables should be justified by either:

- describing additional important aspects of unsaturated soil behaviour that cannot be represented by using the net stress $\sigma - u_a$ and the suction $u_a - u_w$ as the two stress variables; or
- resulting in simpler constitutive relationships between stresses and strains than can be achieved with models described in terms of net stress $\sigma - u_a$ and suction $u_a - u_w$.

The potential advantages in using the alternative stress variables should be substantial rather than merely resulting in another way of expressing the existing constitutive models. This is because the alternative stress variables are generally more complex than the net stress $\sigma - u_a$ and the suction $u_a - u_w$ and working in terms of the alternative variables can pose new problems. One example of such potential problems is representation of stress paths. Wetting and drying paths under constant applied loading or isotropic loading and unloading paths at constant suction are simple stress paths in terms of net stress $\sigma - u_a$ and suction $u_a - u_w$ (with one or the other variable being held constant), whereas these stress paths can become complicated and difficult to define in terms of the alternative stress variables. This in turn can lead to two potential problems. Firstly, devising simple experimental tests to measure soil parameters can be difficult. Secondly, it can be difficult for a user of the model to "think" in terms of the alternative stress variables. It is, however, important to note that these arguments are cautionary only, and the issue of selecting stress variables should not be considered as settled, because there are important aspects of unsaturated soil behaviour that it has, so far, not been possible to describe by the models formulated in terms of the net stress $\sigma - u_a$ and the suction $u_a - u_w$ (see, Sharma and Wheeler, 2000).

If net stress $\sigma - u_a$ (a tensor variable) and suction $u_a - u_w$ (a scalar variable) are selected as appropriate stress variables, the appropriate stress variables to use for the special conditions of the triaxial test $(\sigma_2 = \sigma_3)$ are given by Matyas and Radhakrishna (1968) as the mean value of net stress $p - u_a$, the deviator value of net stress q and matrix suction s, and can be defined as follows:

$$p'' = p - u_a = \frac{1}{3}(\sigma_1 + 2\sigma_3) - u_a \quad (1)$$

$$q = (\sigma_1 - u_a) - (\sigma_3 - u_a) = \sigma_1 - \sigma_3 \quad (2)$$

$$s = (u_a - u_w) \quad (3)$$

where σ_1 and σ_3 are the axial total stress and radial total stress respectively.

3 WORK CONJUGATE STRAIN PARAMETERS

For saturated soils, the state of deformation can be fully defined, within the special case of a triaxial test ($\varepsilon_2 = \varepsilon_3$), by the volumetric strain ε_p and shear strain ε_q, defined by:

$$\varepsilon_p = \varepsilon_1 + 2\varepsilon_3 \tag{4}$$

$$\varepsilon_q = \frac{2}{3}(\varepsilon_1 - \varepsilon_3) \tag{5}$$

where ε_1 and ε_3 are the axial and radial strains respectively.

For a saturated soil the volumetric strain ε_p and shear strain ε_q are work conjugate with the mean effective stress p' and deviator stress q in that the increment of work input per unit volume of soil dW is given by:

$$dW = p'd\varepsilon_p + qd\varepsilon_q \tag{6}$$

An increment of volumetric strain $d\varepsilon_p$ can be related to the change of specific volume v:

$$d\varepsilon_p = -\frac{dv}{v} \tag{7}$$

where the specific volume v is the volume of soil (soil particles and voids) containing unit volume of soil particles, related to the void ratio e by:

$$v = 1 + e \tag{8}$$

For saturated soils, the volumetric state is uniquely defined by the value of v.

For an unsaturated soil an additional volumetric parameter is required, in addition to v, in order to specify fully the relative volumes of soil particles, pore air and pore water. This additional volumetric parameter could be selected as the degree of saturation S_r, the water content w or specific water volume v_w, introduced by Wheeler (1991) and defined as the volume of solid particles and water within a body of soil containing unit volume of soil particles:

$$v_w = 1 + eS_r = 1 + wG_s \tag{9}$$

From Equation 9, it is clear that the specific water volume v_w tends to specific volume v as the degree of saturation S_r approaches unity.

Houlsby (1997) presented a comprehensive theoretical analysis of the work input to an unsaturated granular material. This analysis shows that, as stated by Wheeler and Sivakumar (1995), if $p - u_a$, q, and $u_a - u_w$ are selected as the stress variables, the work input equation can be written as:

$$dW = (p - u_a)d\varepsilon_p + qd\varepsilon_q + (u_a - u_w)d\varepsilon_w \tag{10}$$

where the strain variable $d\varepsilon_w$ is defined by:

$$d\varepsilon_w = -\frac{dv_w}{v} \tag{11}$$

Alternatively, if $p - u_w$, q, and $u_a - u_w$ are selected as stress state variables, Equation 10 can be re-arranged to:

$$dW = (p - u_w)d\varepsilon_p + qd\varepsilon_q + (u_a - u_w)d\varepsilon_a \tag{12}$$

where the strain variable $d\varepsilon_a$ is defined by:

$$d\varepsilon_a = \frac{d(v - v_w)}{v} \tag{13}$$

Houlsby (1997) also shows that Equation 10 can be re-arranged to:

$$dW = (p - S_r u_w - (1 - S_r)u_a)d\varepsilon_p + qd\varepsilon_q - n(u_a - u_w) \tag{14}$$

where n is the porosity.

Equation 14 indicates that an alternative choice of stress state variables would be $(p - S_r u_w - (1 - S_r)u_a)$, q, and $n(u_a - u_w)$. For convenience, these new stress variables are denoted as:

$$p^* = (p - S_r u_w - (1 - S_r)u_a) \tag{15}$$

$$s^* = n(u_a - u_w) \tag{16}$$

The conjugate strain parameters corresponding to the stress variables p^*, q *and* s^* are then volumetric strain ε_p, shear strain ε_q, and the decrease of degree of saturation. Other forms of re-arrangement of Equation 10 are also possible. It is clear that in terms of a valid work input equation, various possible combinations of stress state variables are acceptable, and each combination of stress state variables has a different set of conjugate strain parameters.

4 AN EXAPMLE OF EXPERIMENTAL RESULTS IN TERMS OF WORK CONJUGATE PARAMETERS

To illustrate the use of new stress variables p^* and s^* along with conjugate strain parameters, results from a test on a sample prepared from a mix of kaolin and bentonite are presented in Figure 1. The test was carried out using the axis-translation technique. The test was performed under isotropic stress conditions. The sample was subjected to wetting/drying cycles at a constant value of mean net stress p'' of 10 kPa. This means that holding p''constant, suction was varied. It should, however, be noted that p^* was not held constant.

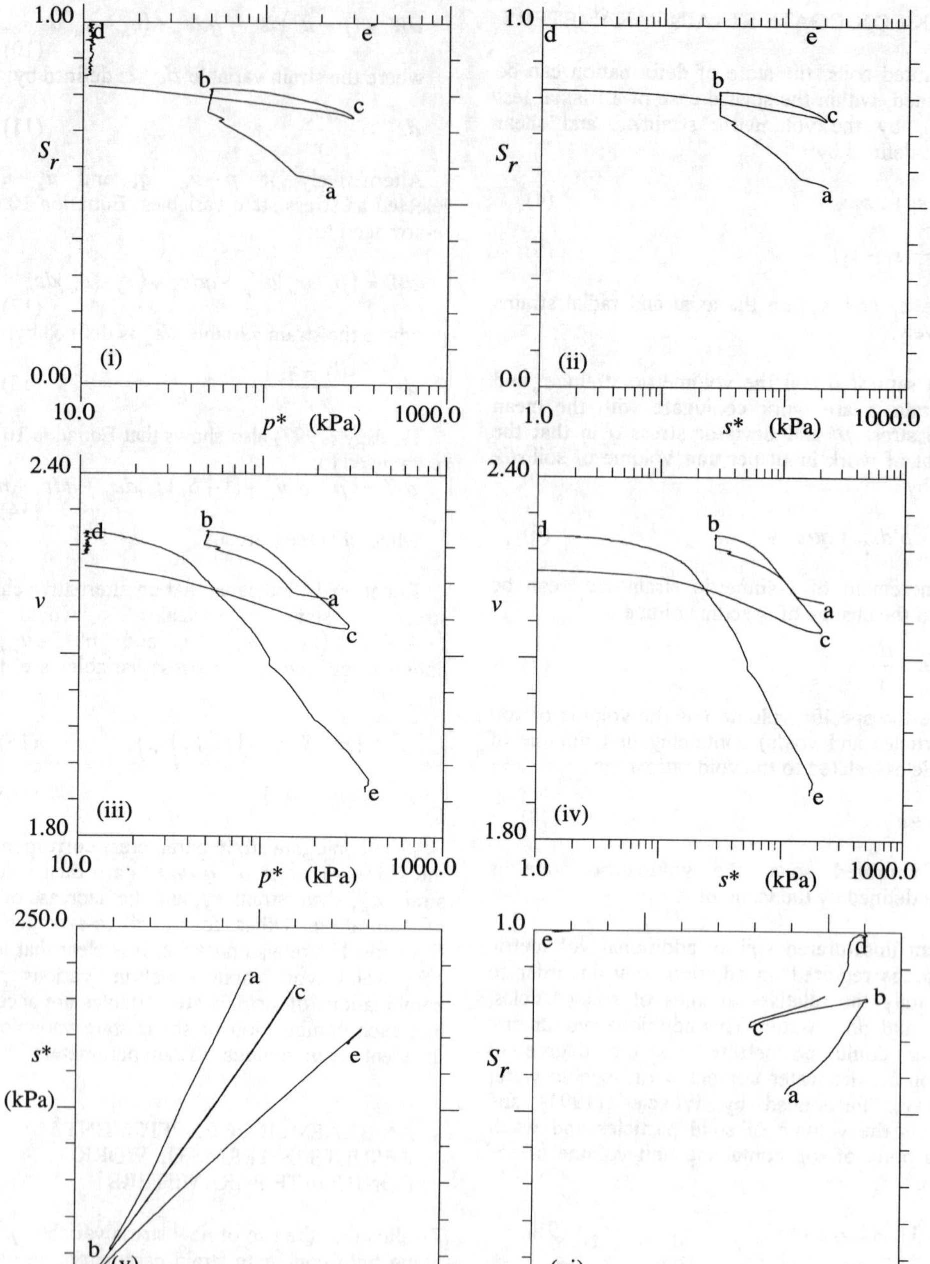

Figure 1: Experimental results for an unsaturated soil in terms of work conjugate parameters

It is useful to note the format in which experimental results are presented here. Six plots are used (see Figure 1) . The first four plots (i, ii, iii, iv) are laid out in such a way that the two axes of a plot are common to other plots. This layout is useful in gaining an overall picture by being able to project from one plot to another. Plot (i) shows the test path in the $S_r : p^*$ plane. A logarithmic scale is used for p^* in this plot. Plot (ii) gives the soil-water characteristic curve in the $S_r : s^*$ plane. The variation of specific volume v is plotted against p^* and s^* in plots (iii) and (iv) respectively. Finally the stress path in the $s^* : p^*$ plane and the strain path in $S_r : v$ plane are given in plots (v) and (vi) respectively.

The results are not discussed here in detail. It is nevertheless important to point out that the results presented in Figure 1 could account for the features of soil behaviour those could not be explained using the conventional stress variables p'' and s (see Sharma, 1998). Furthermore the stress variables and the conjugate strain increment parameters used here give better insight into the soil behaviour, especially note the format used for presentation of results.

An important point to note is that the issue of stress variables should not be considered in isolation of the conjugate strain increment parameters. It is, however, not the intention here to suggest that other alternative stress variables cannot be used.

5 CONCLUSIONS

In this paper selection of stress variables for unsaturated soils has been linked with the appropriate conjugate strain increment parameters. Use of stress variables along with conjugate strain increment parameters has been illustrated using experimental results from controlled suction tests on an unsaturated soil. The additional and important features of unsaturated soil behaviour that can be discerned by linking stress variables and conjugate strain increment parameters suggest that the selection of stress variables should not be considered in isolation of the appropriate strain parameters.

Most of the existing elasto-plastic constitutive models for unsaturated soils are presented in terms of stress variables without any consideration of the conjugate strain parameters. This aspect obviously needs to be integrated into the constitutive models. Work currently in progress is attempting to use stress variables and conjugate strain increment parameters in the development of an improved constitutive model for unsaturated soils.

ACKNOWLEDGEMENTS

The experimental results presented in this paper formed part of a research project funded by the U.K. Engineering and Physical Sciences Research Council. The author would like to thank Professor S J Wheeler, University of Glasgow and Professor G T Houlsby, University of Oxford for their help in developing a framework of stress variables and conjugate strain parameters.

REFERENCES

Bishop, A.W. and Blight, G.E. (1963). Some aspects of effective stress in saturated and partly saturated soils. Géotechnique, Vol 13(3), 177-197.

Bolzon, G., Schrefler, B.A. and Zienkiewicz, O.C. (1996). Elastoplastic soil constitutive laws generalized to partially saturated states. Géotechnique, Vol.46(2), 279-289.

Burland, J.B. (1964). Effective stress in partly saturated soils. Géotechnique, Vol.14, 64-68.

Fredlund, D.G. and Morgenstern, N.R. (1977). Stress state variables for unsaturated soils. J. Geotech. Eng., Div., ASCE, Vol.103(GT5), 447-466.

Gens, A. (1996). Keynote lecture: Constitutive modeling: Application to compacted soils. Proc. 1st Int. Conf. Unsaturated Soils, Paris, Vol.3, 1179-1200.

Houlsby, G.T. (1997). The work input to an unsaturated granular material. Géotechnique, Vol.47(1), 193-196.

Jommi, C. and di Prisco, C. (1994). Un semplice approcio teorico per la modellazione del comportamento meccanico di terreni granulari parcialmente saturi. Conf, Il ruolo dei fluidi nei problemi di ingegneria geotecnica, Mondovi, 167-188.

Kohgo, Y., Nakano, M. and Miyazaki, T. (1993 *a*). Theoretical aspects of constitutive modeling for unsaturated soils. Soils and Foundations, Vol 33(4), 49-63.

Kohgo, Y., Nakano, M. and Miyazaki, T. (1993 *b*). Verification of generalized elasto-plastic model for unsaturated soils. Soils and Foundations, Vol 33(4), 64-73.

Matyas, E.L. and Radhakrishna, H.S. (1968). Volume change characteristics of partially saturated soils. Géotechnique, Vol.18(4), 432-448.

Modaressi, A. and Abou-Bekr, N. (1994). Constitutive model for unsaturated soils: validation on a silty material. 3rd Eur. Conf. Numerical Methods in Geotech. Eng., Manchester, 91-96.

Sharma, R.S. (1998). Mechanical behaviour of unsaturated highly expansive clays. DPhil thesis, University of Oxford, U.K.

Sharma R.S. (2000). Recent advances in constitutive modeling of unsaturated soils. Keynote paper: Proceedings of International Conference on Geotechnical, Geoenvironmental Engineering, and Management in Arid Lands, Al Ain, UAE (in press).

Sharma R.S. and Wheeler S.J (2000). Behavior of an unsaturated highly expansive clay during wetting/drying cycles. Proc. Int. Conf. on Unsaturated soils (UNSAT-ASIA 2000), Singapore, 721-726 .

Wheeler, S.J. (1991). An alternative framework for unsaturated soil behaviour. Géotechnique, Vol.41(2), 257-261.

Wheeler, S.J and Karube, D. (1996). State of the art report: Constitutive modeling. Proc. 1st Int. Conf. Unsaturated Soils, Paris, Vol.3, 1323-1356.

Wheeler, S.J. and Sivakumar, V. (1995). An elasto-plastic critical state framework for unsaturated soil. Géotechnique, Vol.45(1), 35-53.

Computer Methods and Advances in Geomechanics, Desai et al. (eds) © 2001 Balkema, Rotterdam, ISBN 90 5809 183 X

Modeling time-dependent stress-strain behavior of stiff geo-materials and its applications

M.S.A.Siddiquee & F.Tatsuoka
Department of Civil Engineering, University of Tokyo, Japan

ABSTRACT: A modeling of time-dependent (viscous) shear stress and shear strain behavior of clean sands in drained plane strain compression tests is presented. The general three-component model is used as the framework for constitutive modeling. A viscous model is developed to incorporate decaying properties of the viscous response of sand taking place by a change in the strain rate, called the TESRA (**T**emporary **E**ffects of **S**train **R**ate and **A**cceleration) model.

1 INTRODUCTION

In order to assess the long-term performance of earth structures and ground, it is necessary to predict the time-dependent ground deformation, including creep deformation and stress relaxation. A number of constitutive models describing the intrinsic viscous stress-strain characteristics of geomaterials has been proposed (Tatsuoka et al., 2000a). Most of them are too simplistic models to capture the rate effects with real soil.

The paper describes a general elasto-viscoplastic model, called the TESRA model, formulated within the framework of classical three-component model and explains its implementation within the scope of generalized elasto-plastic isotropic strain-hardening Finite Element code. The model is applicable only to shear deformation in its present formulation. The TESRA model states the scaling of the shear stress of soil according to the instantaneous rate of irrecoverable strain and strain rate and decay of the viscous effects. The model can be easily integrated analytically or numerically in one-dimensional cases. But the integration scheme in FEM analysis needs some specific considerations. Choice of suitable invariant parameters is the key to the implementation of the model into a FEM code. The model has been successfully integrated with an existing FEM code. Some example problems have been solved with FEM. The FEM code can predict the time-dependent behavior of a wide variety of geomaterials quite accurately without spending any extra computational time or storage.

2 FUNDAMENTAL THEORY

2.1 Three-component viscous model

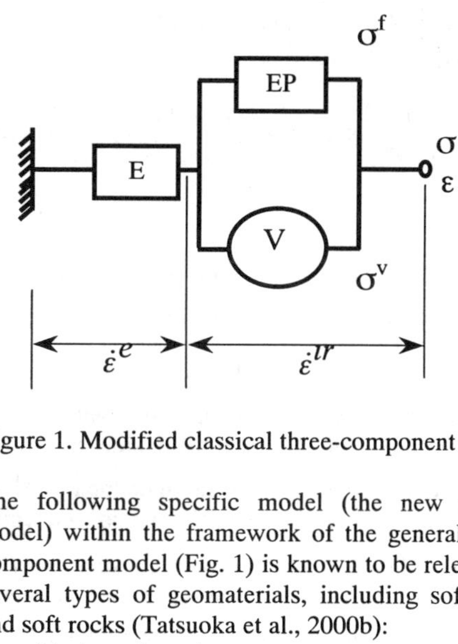

Figure 1. Modified classical three-component model.

The following specific model (the new isotach model) within the framework of the general three-component model (Fig. 1) is known to be relevant to several types of geomaterials, including soft clays and soft rocks (Tatsuoka et al., 2000b):

$$\sigma = \sigma^f\left(\varepsilon^{ir}\right) + \sigma^v\left(\varepsilon^{ir}, \dot{\varepsilon}^{ir}\right) \quad (1a)$$

$$\sigma^v\left(\varepsilon^{ir}, \dot{\varepsilon}^{ir}\right) = \sigma^f\left(\varepsilon^{ir}\right) \cdot g_v(\dot{\varepsilon}^{ir}) \quad (1b)$$

$$\sigma = \sigma^f\left(\varepsilon^{ir}\right) \cdot \{1 + g_v(\dot{\varepsilon}^{ir})\} \quad (1c)$$

where $g_v(\dot{\varepsilon}^{ir})$ is the viscosity function, which is always zero or positive and is given as follows for any strain (ε^{ir}) or stress (σ^f) path (with or without cyclic loading):

$$g_v(\dot{\varepsilon}^{ir}) = \alpha\cdot[1-\exp\{1-(\frac{|\dot{\varepsilon}^{ir}|}{\dot{\varepsilon}_r^{ir}}+1)^m\}] \quad (\geq 0) \tag{2}$$

2.2 Formulation of TESRA model

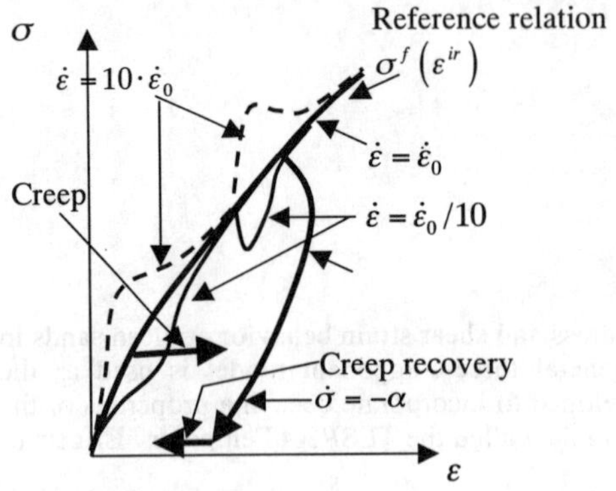

Figure 2. Behavior of the TESRA model.

Eq. 1b can be rewritten as follows:

$$\sigma^v = \int_{\tau=\varepsilon_1^{ir}}^{\varepsilon^{ir}} \left[d\sigma^v\right]_{(\tau)} = \int_{\tau=\varepsilon_1^{ir}}^{\varepsilon^{ir}} \left[d\{\sigma^f\cdot g_v(\dot{\varepsilon}^{ir})\}\right]_{(\tau)} \tag{3a}$$

$$= \int_{\tau=\varepsilon_1^{ir}}^{\varepsilon^{ir}} \left[\left(\frac{\partial\sigma^f}{\partial\varepsilon^{ir}}\right)\cdot g_v(\dot{\varepsilon}^{ir}) + \sigma^f\cdot\left(\frac{\partial g_v(\dot{\varepsilon}^{ir})}{\partial\dot{\varepsilon}^{ir}}\right)\cdot\frac{\ddot{\varepsilon}^{ir}}{\dot{\varepsilon}^{ir}}\right]_{(\tau)} \cdot d\tau \tag{3b}$$

$(\sigma^v \geq 0)$

where $\sigma^v(\geq 0)$ is the current viscous stress component, where $\varepsilon^{ir} = \varepsilon^{ir}$; $[d\sigma^v]_{(\tau)}$ is the viscous effect on the stress that took place when $\varepsilon^{ir} = \tau$; ε_1^{ir} is the strain at the start of integration, where $\sigma^v = 0$; the first term of the right-hand side of Eq. 3b represents the component of the increment $d\sigma^v$ as the effect of the irreversible strain increment $d\varepsilon^{ir}$ when loading continues at a constant $\dot{\varepsilon}^{ir}$ (i.e., the effect of irreversible strain rate); and the second term represents the component of the increment $d\sigma^v$ as the effect of the increment of irreversible strain rate $d\dot{\varepsilon}^{ir}$ applied at a certain constant ε^{ir} (i.e., the effect of irreversible strain acceleration).

2.2.1 Introducing the decay function

Granular material always shows a tendency with straining of decay of the stress jump that takes place by a strain-jump (Fig. 2). To model this phenomenon, the decay function $g_{decay}(\varepsilon^{ir}-\tau) = r_1^{(\varepsilon^{ir}-\tau)}$ (Fig. 3) is incorporated in to Eq. (3) to obtain:

$$\sigma^v = \int_{\tau=\varepsilon_1^{ir}}^{\varepsilon^{ir}} \left[\left(\frac{\partial\sigma^f}{\partial\varepsilon^{ir}}\right)\cdot g_v(\dot{\varepsilon}^{ir}) + \sigma^f\cdot\left(\frac{\partial g_v(\dot{\varepsilon}^{ir})}{\partial\dot{\varepsilon}^{ir}}\right)\cdot\frac{\ddot{\varepsilon}^{ir}}{\dot{\varepsilon}^{ir}}\right]_{(\tau)} \cdot r_1^{(\varepsilon^{ir}-\tau)}\cdot d\tau \tag{4}$$

where r_1 is a positive constant less than 1.0. In case r_1= 1.0, Eq. 4 becomes Eq. 3, therefore Eq. 1.

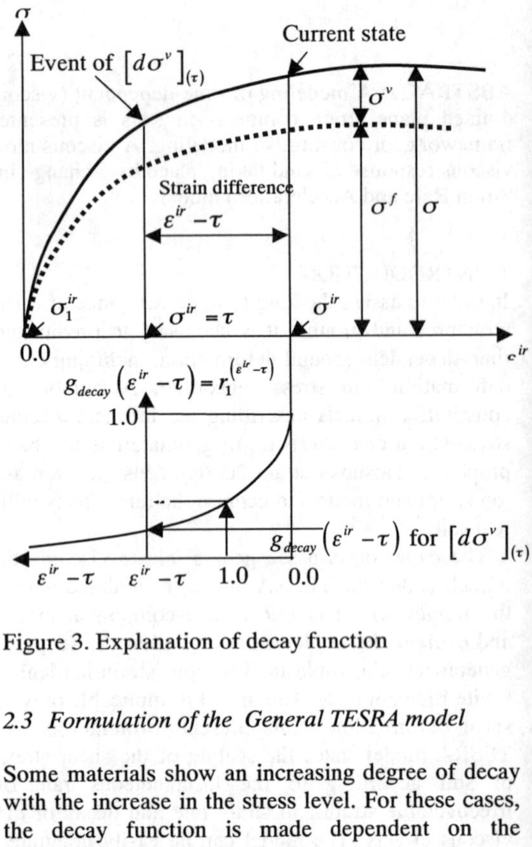

Figure 3. Explanation of decay function

2.3 Formulation of the General TESRA model

Some materials show an increasing degree of decay with the increase in the stress level. For these cases, the decay function is made dependent on the irreversible strain as:

$$\sigma^v = \int_{\tau=\varepsilon_1^{ir}}^{\varepsilon^{ir}} \left[\left(\frac{\partial\sigma^f}{\partial\varepsilon^{ir}}\right)\cdot g_v(\dot{\varepsilon}^{ir}) + \sigma^f\cdot\left(\frac{\partial g_v(\dot{\varepsilon}^{ir})}{\partial\dot{\varepsilon}^{ir}}\right)\cdot\frac{\ddot{\varepsilon}^{ir}}{\dot{\varepsilon}^{ir}}\right]_{(\tau)} \cdot r(\varepsilon^{ir})^{(\varepsilon^{ir}-\tau)}\cdot d\tau \tag{5}$$

where $r(\varepsilon^{ir})$ means that the decay parameter r decreases with ε^{ir}.

The following function was chosen for $r(\varepsilon^{ir})$ to simulate the primary loading stress-strain behaviour:

$$r(\varepsilon^{ir}) = r_i \qquad \text{at } \varepsilon^{ir} = 0 \tag{6a}$$

$$r(\varepsilon^{ir}) = \frac{r_i + r_f}{2} + \frac{r_i - r_f}{2} \cdot \cos\left\{\pi \cdot \left(\frac{\varepsilon^{ir}}{c}\right)^n\right\} \quad \text{for} \quad 0 \le \varepsilon^{ir} \le c \qquad (6b)$$

$$r(\varepsilon^{ir}) = r_f \qquad \text{for } \varepsilon^{ir} \ge c \qquad (6c)$$

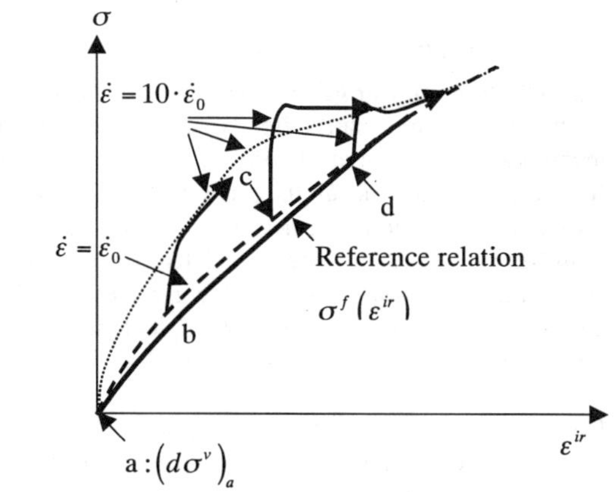

Figure 4. Stress-strain behavior of the general TESRA model with variable decay function.

where r_i and r_f are the initial and ultimate values of r, where $0 \le r_f \le r_i \le 1.0$; and c and n are positive constants (see Fig. 5).

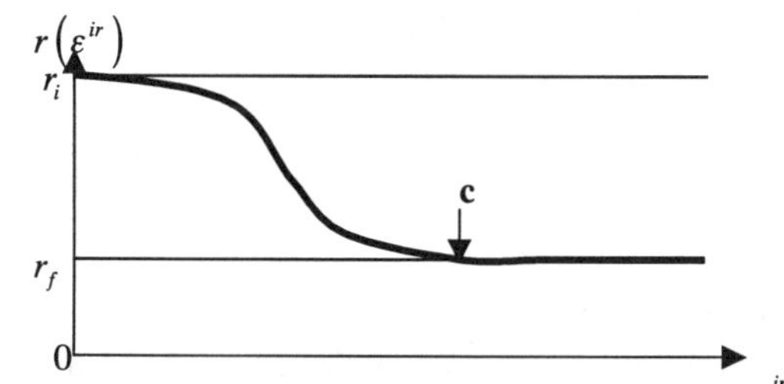

Figure 5. Variable decay function.

For primary loading cases, all of the new isotach model, the TESRA model and the general TESRA model can be represented by a set of common equations (Eqs. 4, 5 and 6) using the following different values of r_i and c:

$c = \infty$ and $r_i = 1.0$ (always $r = 1.0$): no decay (the new isotach model);

$c = \infty$ and $0 < r_i < 1.0$ (always $r = r_i < 1.0$) (the TESRA mode); and

$0 < c < \infty$ and $0 < r_f < r_i < 1.0$ (the general TESRA model).

3 VALIDATION OF THE MODEL

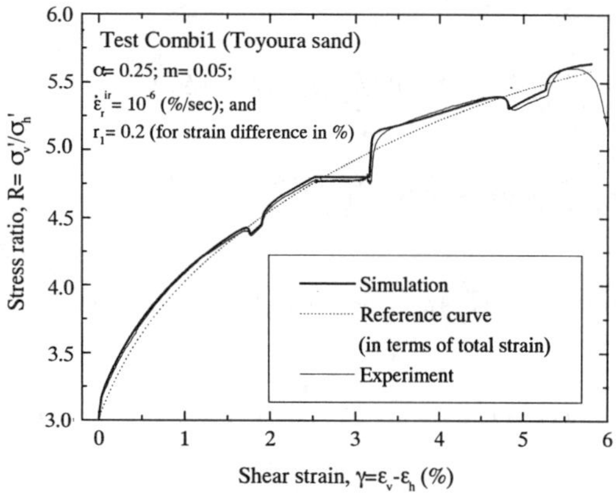

Fig. 6 Measured stress-strain relations for a drained PSC test (**e_0**= 0.730) on Toyoura and its simulation by the TESRA model.

Fig. 6 shows the relationship between the stress ratio $R = \sigma'_v/\sigma'_h$ and the shear strain $\gamma = \varepsilon_v - \varepsilon_h$ from a PSC test on Toyoura sand and its simulation by the TESRA model. Toyoura sand, from Japan, is a quartz-rich sub-angular poor-graded medium-sized sand with D_{50}= 0.18 mm, U_c= 1.64, G_s= 2.65, e_{max}= 0.99 and e_{min}= 0.62. The PSC specimen was rectangular-prismatic of 20 cm high, 16 cm long and 8 cm wide (in the σ_3 direction) with lubricated top and bottom ends. The specimen was prepared by pluviating air-dried sand particles through air and were then made saturated. The specimen was anisotropically consolidated at a stress ratio R= 3.0 and at a constant rate of axial strain, $\dot{\varepsilon}_v$ = 0.0125 %/min., from the initial stress state with $\sigma'_h = 29$ kPa towards the final consolidation stress state with $\sigma'_h = 392$ kPa. The axial strain rates during PSC loading was basically either 0.0005%/sec or 0.006%/sec. At several intermediate stages, the axial strain rate was increased and decreased at a constant rate, while one creep test was performed for about 24 hours. It can be seen that the test result is simulated very well by the TESRA model. The model parameters are shown in Fig. 6. The validation of the TESRA model is fully reported by Di Benedetto et al. (2000) and Tatsuoka et al. (2000b).

4 IMPLEMENTATION DETAILS

4.1 *Numerical integration scheme*

The basic TESRA equation (Eq. 4) can be generalized to include unloading-reloading cases as:

$$
\begin{aligned}
\sigma^{v} &\cong \int_{\tau=\varepsilon_1^{ir}}^{\varepsilon^{ir}-\Delta\varepsilon^{ir}} \left[d\sigma^{v}\right]_{(\tau,\varepsilon^{ir},h)} + \Delta\sigma^{v} \\
&= \int_{\tau=\varepsilon_1^{ir}}^{\varepsilon^{ir}-\Delta\varepsilon^{ir}} \left[d\{(\sigma^{f}-\sigma_r)\cdot g_v(\dot{\varepsilon}^{ir})\}\right]_{(\tau)} \cdot g_{decay}(D)\cdot d\tau + \Delta\sigma^{v} \\
&= \left[\int_{\tau=\varepsilon_1^{ir}}^{\varepsilon^{ir}-\Delta\varepsilon^{ir}} \left[d\{(\sigma^{f}-\sigma_r)\cdot g_v(\dot{\varepsilon}^{ir})\}\right]_{(\tau,h)} \cdot r_1^{(D-|\Delta\varepsilon^{ir}|)}\cdot d\tau\right]\cdot r_1^{|\Delta\varepsilon^{ir}|} + \Delta\sigma^{v} \\
&= \left[\sigma^{v}\right]_{(\varepsilon^{ir}-\Delta\varepsilon^{ir})}\cdot r_1^{|\Delta\varepsilon^{ir}|} + \Delta\{(\sigma^{f}-\sigma_r)\cdot g_v(\dot{\varepsilon}^{ir})\}\cdot r_1^{\frac{|\Delta\varepsilon^{ir}|}{2}}
\end{aligned}
\tag{7}
$$

$$
g_{decay}(D) = r_1^{D}; \qquad D = \int_{\tau}^{\varepsilon^{ir}} \left|d\varepsilon^{ir}\right| \tag{8}
$$

where;

1. $|d\varepsilon^{ir}|$ is the absolute value of irreversible strain increment $d\varepsilon^{ir}$;
2. σ_r is the value of σ^{f} at the latest stage where the sign of $d\varepsilon^{ir}$ was changed;
3. $\left[\sigma^{v}\right]_{(\varepsilon^{ir}-\Delta\varepsilon^{ir})}$ is the viscous stress at one step before where the irreversible strain is equal to $\varepsilon^{ir}-\Delta\varepsilon^{ir}$;
4. $\Delta\varepsilon^{ir}$ is the difference in the irreversible strain between the current step and the step immediately before $\Delta\varepsilon^{ir}$; and
5. $\Delta\{(\sigma^{f}-\sigma_r)\cdot g_v(\dot{\varepsilon}^{ir})\}$ is the difference in the value of $(\sigma^{f}-\sigma_r)\cdot g_v(\dot{\varepsilon}^{ir})$ between the current step and the step immediately before, given as

$$
\Delta\{(\sigma^{f}-\sigma_r)\cdot g_v(\dot{\varepsilon}^{ir})\} = \Delta(\sigma^{f}-\sigma_r)\cdot g_v(\dot{\varepsilon}^{ir}) + (\sigma^{f}-\sigma_r)\cdot \Delta g_v(\dot{\varepsilon}^{ir}) \tag{9}
$$

The term $\Delta\{(\sigma^{f}-\sigma_r)\cdot g_v(\dot{\varepsilon}^{ir})\}\cdot r_1^{\frac{|\Delta\varepsilon^{ir}|}{2}}$ in Eq. 7 is the viscous effects that took place for this range of $\Delta\varepsilon^{ir}$ that has decayed by an amount of $r_1^{\frac{|\Delta\varepsilon^{ir}|}{2}}$.

So, for known values of $\left[\sigma^{v}\right]_{(\varepsilon^{ir}-\Delta\varepsilon^{ir})}$, $\Delta\varepsilon^{ir}$, Δt (i.e., $\dot{\varepsilon}^{ir} = \dfrac{\Delta\varepsilon^{ir}}{\Delta t}$) and ε^{ir}, the value of σ^{v} at the current state, where $\varepsilon^{ir} = \varepsilon^{ir}$, can be obtained by Eq. 7, without doing an integration.

Repeating this procedure from the start of loading, the whole stress-strain-time relationship can be obtained for any given history of strain or stress with or without cyclic loading.

4.2 *Implementation into FEM Code*

An existing nonlinear FEM code will be used, which is based on the matrix-free dynamic relaxation technique (Siddiquee et al., 1999), highly optimized for very fast and accurate computation of highly nonlinear equations emerging from material nonlinearity. Due to the specialty of the DR (Dynamic Relaxation), the following steps to be considered carefully for relevant implementation of the TESRA model into the FEM code.

Computation of irrecoverable strain rate

In case of modeling granular materials, obviously for the rate of effective plasticity parameter, *k*, the primary candidate is the rate of the second irrecoverable strain invariant. This is a scalar quantity and it is already being calculated in the FEM code. But there is one problem with this; i.e., as it is always positive, it cannot be used in unloading situations. So the following variable is used in the present formulation:

$$
k = \varepsilon_1^{ir} - \varepsilon_3^{ir} \qquad \dot{k} = \frac{k}{dt} \tag{10}
$$

Selection of current stress variable

Stress ratio, $R = \sigma_1/\sigma_3$ is selected to represent the current shear stress level.

4.3 *Pseudo-algorithm*

The following steps are followed to calculate the viscous stress:

1. Calculation of trial strain
2. Stress increment calculation by the hypo-elastic constitutive law (Hoque, 1998):

   $$E_h = E_0\sigma_h^{n}, \quad E_v = E_0\sigma_v^{n}$$

3. Elasto-plastic solution (without viscous effects) is calculated. Plastic consistency conditions and a flow rule are applied and each trial stress is integrated back to the inviscid yield surface, f^{f}.

$$df^f(\sigma_{ij},k)=0,\quad d\varepsilon_{ij}^p=\lambda\frac{\partial\varphi}{\sigma_{ij}}$$

(a) The current strain rate is calculated from the plastic strain rate generated in the previous step.
(b) The iterating inviscid stress is further corrected by the TESRA Model Eq. (7).

4. Nodal forces are calculated and compared with the applied load and the iteration continues, starting from 1 again, until the equilibrium is reached.

5 NUMERICAL SIMULATION

5.1 *Problem definition*

In order to perform numerical simulations, the following one-element configuration has been devised. The material is Toyoura sand with a void ratio, e=0.66 and a confining pressure, σ_3=98 kPa.

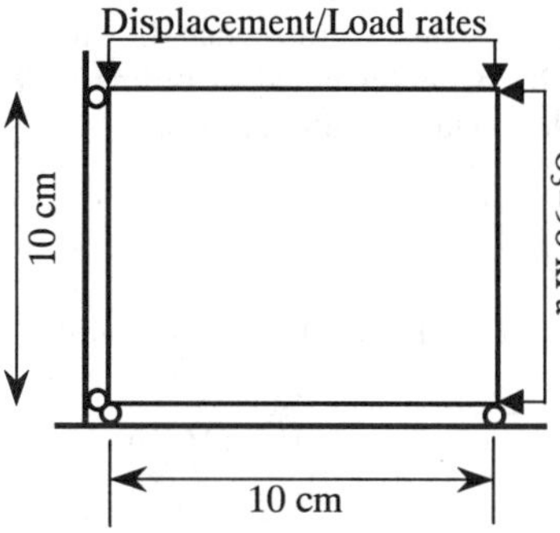

Figure 7. Definition of the problem.

5.2 *Material model*

An isotropically strain-hardening-softening elasto-visco-plastic model is used here (Siddiquee 1999). The yield function is described by the Mohr-Coulomb failure criterion and plastic potential by the the Drucker-Prager criterion. Essentially this is a non-associated flow analysis and the relevant dilatancy is modeled by adapting the Rowe's stress-dilatancy relationship. The growth function represented by strain hardening is modeled by the General Hyperbolic Equation (GHE) (Siddiquee 1999). The viscous model parameters those were employed: $\alpha = 0.23$, m=0.045, $\dot{\varepsilon}_0^{ir} = 0.0001$ and r_1=0.3 (the TESRA model).

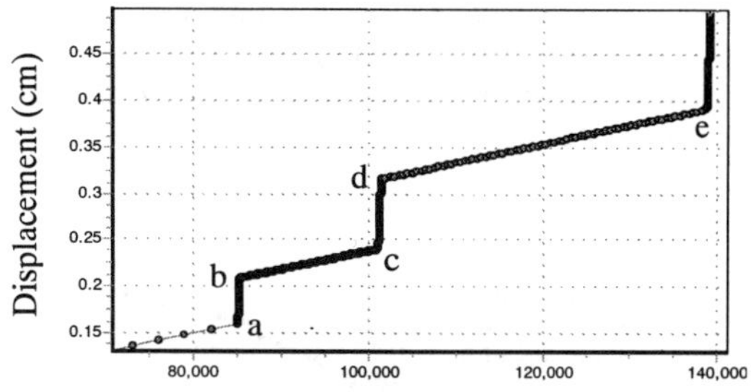

Figure 8. Displacement vs. time relationship for displacement control loading

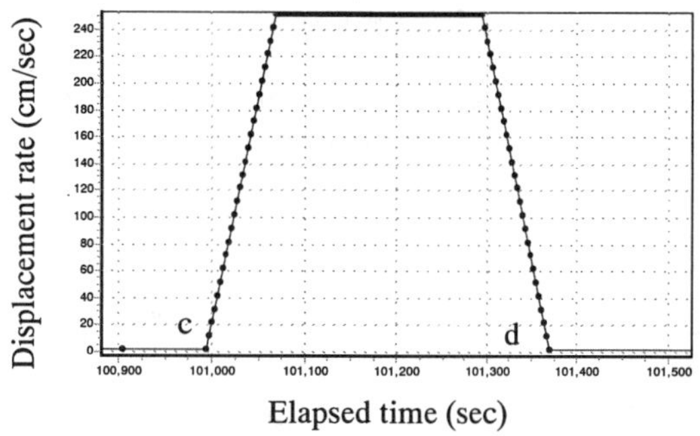

Figure 9. Applied rate of displacement

5.3 *FEM Analysis*

The following two cases were simulated :

1) a displacement control test with varying displacement rates (Figs. 8 & 9); and
2) a load control test with varying load rates (Figs. 11 & 12).

It may be seen from the results from the analysis (Figs. 10 and 13), while referring to Fig. 6, that the simulation is very realistic, in particular with the following phenomena;

1) temporary over- and under-shooting in the stress by a sudden increase and decrease in the displacement rate (i.e., strain rate) or the loading rate (i.e., stress rate); i.e., a sudden change in the stiffness immediately after a sudden change in the strain or stress rate, followed by a sudden change in the tangent stiffness and rejoining to the original stress-strain behaviour with subsequent loading;
2) creep behaviour, followed by very stiff behaviour when loading is resumed at a constant stress rate (Fig. 13); and continuation of the development of shear strain for a some stress range after the shear

stress has started decreasing (from point d in Fig. 13), followed by a switching from an increase to a decrease in the shear strain in the course of stress decreasing.

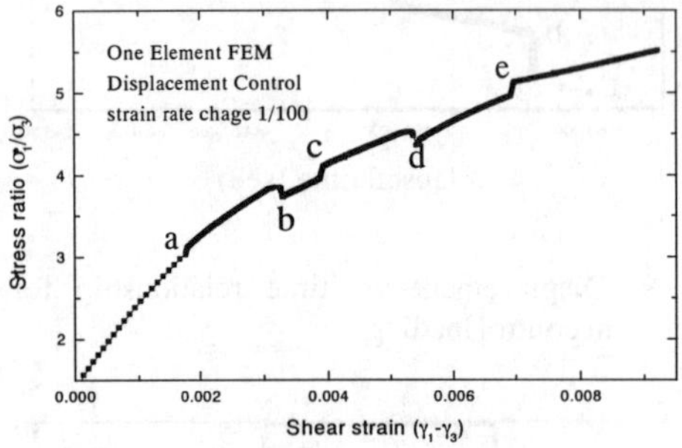

Figure 10. Stress-strain result obtained from FEM computation (displacement-control).

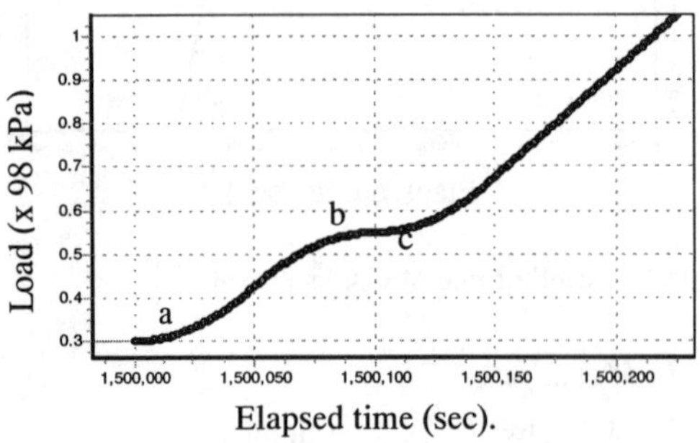

Figure 11 Load time relationship for load control analysis

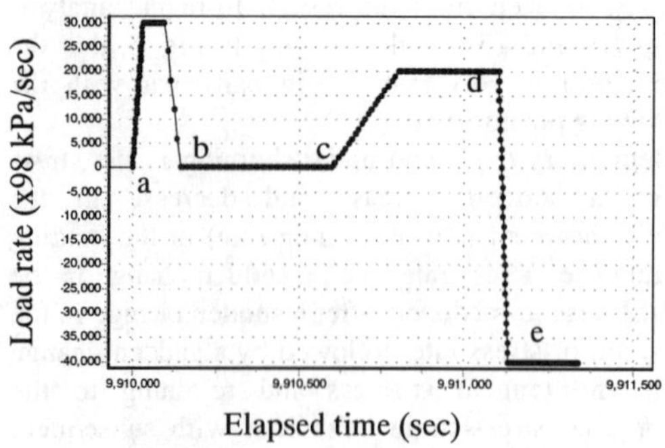

Figure 12. Applied rate of loading

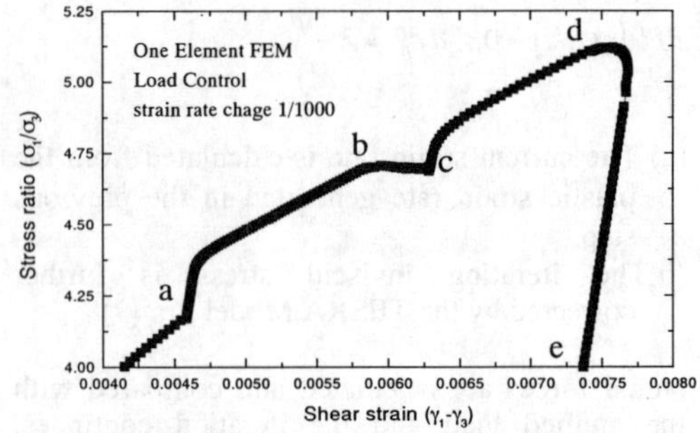

Figure 13. Stress-strain result obtained from FEM computation (load control)

6 CONCLUSIONS

The following conclusions can be drawn:

1. Time-dependent stress-strain behavior of granular materials is a function of previous and instantaneous strain-rate and acceleration.
2. With granular materials, the viscous effects on the stress value decay with the increase in the irreversible strain.
3. The above viscous features can be modeled and have been implemented in ordinary FEM codes without major modifications.

7 REFERENCES

Di Benedetto,H. and Tatsuoka,F. 1997. Small strain behaviour of geomaterials: modelling of strain effects, *Soils and Foundations,* Vol.37, No.2, pp.127-138

Di Benedetto,H., Tatsuoka,F. and Ishihara,M. 2000b.: Time-dependent deformation characteristics of sand and their constitutive modelling, *Soils and Foundations* (submitted).

Hoque,E. and Tatsuoka,F. 1998. Anisotropy in the elastic deformation of materials", Soils and Foundations", *Soils and Foundations*, Vol.38, No.1, pp.163-179.

Siddiquee,M.S.A., Tanaka,T., Tatsuoka,F., Tani,K. and Morimoto,T. 1999. FEM simulation of scale effect in bearing capacity of strip footing on sand", *Soils and Foundations,* Vol.39, No.4, pp.91-109.

Tatsuoka,F., Uchimura,T., Hayano,K., Di Benedetto,H., Koseki,J. and Siddiquee,M.S.A. 2000a. Time-dependent deformation characteristics of stiff geomaterials in engineering practice", the Theme Lecture, *Proc. of the Second International Conference on Pre-failure Deformation Characteristics of Geomaterials, Torino, 1999* (Jamiolkowski et al., eds.) , Balkema, Vol. 2 (to appear).

Tatsuoka,F., Ishihara,M., Di Benedetto,H. and Kuwano,R. 2000b. Time-dependent deformation characteristics of geomaterials and their simulation, *Soils and Foundations* (submitted).

Yasin,S.J.M. and Tatsuoka,F. 2000. Stress history-dependent deformation characteristics of dense sand in plane strain", *Soils and Foundations*, Vo.40, No.2, pp.55-74.

Computer Methods and Advances in Geomechanics, Desai et al. (eds) © 2001 Balkema, Rotterdam, ISBN 90 5809 183 X

An anisotropic hardening elastoplastic model for clay and sand and its application to numerical analysis

D.A. Sun & H. Matsuoka
Department of Civil Engineering, Nagoya Institute of Technology, Japan

Y.P. Yao
Department of Civil Engineering, Beijing University of Aeronautics, People's Republic of China

H. Ishii
Central Japan Railway Company, Nagoya, Japan

ABSTRACT: The Sekiguchi-Ohta model is extended to be a unified elastoplastic model for clay and sand by introducing a transformed stress tensor $\tilde{\sigma}_{ij}$ and a new hardening parameter H. The model can describe the negative and positive dilatancy characteristics of soils with the initially stress-induced anisotropy in 3D stresses. In order to apply the model to practical engineering by using FEM, the elastoplastic constitutive tensor is derived. An example is shown to demonstrate usefulness of the model in the finite element analysis for predicting deformation of anisotropically consolidated clay and sand layers under embankment loading.

1 INTRODUCTION

Sekiguchi-Ohta's model (1977), which can be considered the extension of the original Cam-clay model to the normally K_0-consolidated clay, is widely incorporated into the finite element computation to predict the stress-deformation of earth structure in engineering practice in Japan. Being the same as the Cam-clay model, the Sekiguchi-Ohta model also uses the Extended Mises criterion as shear yield and failure criteria, which is not appropriated for soil behavior. The SMP criterion is one of the best criteria for describing shear yield and failure behavior of soils. So, the inviscid Sekiguchi-Ohta model has been revised by the SMP criterion (Sun et al. 1998) through a transformed stress tensor $\tilde{\sigma}_{ij}$ (Matsuoka et al. 1999), which is deduced from what makes the SMP criterion become a cone with an axis at the space diagonal line in the transformed principal stress space. That is, in the revised Sekiguchi-Ohta model, the SMP criterion is employed as the criterion of shear yield or failure, while the model parameter is the same as the original model. In this paper, after reviewing the stress-dilatancy relation of the Sekiguchi-Ohta model, the revised model is extended to be a unified elastoplastic model for clay and sand by introducing a new hardening parameter H, which can describe the negative and positive dilatancy characteristics of soils with the initially stress-induced anisotropy. Model complexity is controlled through only adding one material parameter to those of the Sekiguchi-Ohta model. The comparisons with experimental data from drained shear tests in different stress paths show that the proposed model gives excellent predictions for normally K_0-consolidated clay and sand.

In order to apply the model to practical engineering by using the finite element method, its elastoplastic constitutive tensor is derived. An example is shown to demonstrate usefulness of the model in the finite element analysis for predicting deformation of anisotropically consolidated clay and sand layers under embankment loading.

2 STRESS-DILATANCY RELATION OF THE SEKIGUCHI-OHTA MODEL

Before describing the proposed elastoplastic model, the stress-dilatancy relation of the Sekiguchi-Ohta model is given for good understanding of our model. If we define η_1 as Eq. (1), the stress-dilatancy relation of the Sekiguchi-Ohta model can be obtained as Eq. (2).

$$\eta_1 = \frac{3}{2\eta^*}(\eta_{ij} - \eta_{ij0})\eta_{ij} \quad (1)$$

$$\eta_1 = M - \frac{d\varepsilon_v^p}{d\varepsilon_d^p} \quad (2)$$

where ε_v^p and ε_d^p are the plastic volumetric and deviatoric strains. η_{ij}, η_{ij0} and η^* are the same as de-

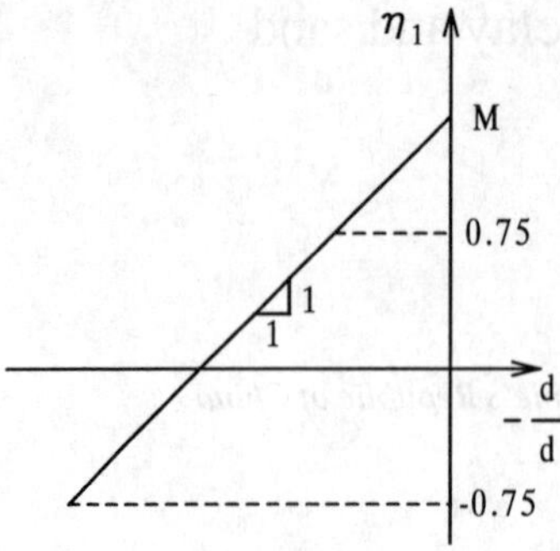

Figure 1. Stress-dilatancy relation of Sekiguchi-Ohta's model

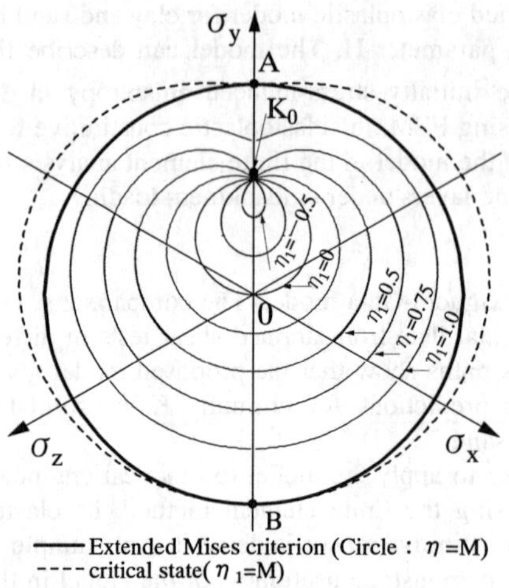

Figure 2 Contour lines of η_1 in π-plane

scribed in the original paper (Sekiguchi and Ohta 1977). Eq. (2) can be drawn as shown in Figure 1 when K_0=0.5. Figure 2 shows the contour lines of η_1 in the π-plane when K_0=0.5. From K_0 to the triaxial compression side point A, η_1 changes from 0.75 to M, and from K_0 to the triaxial extension side point B, η_1 changes from -0.75 to M. Here M is the value of q/p at the critical state in triaxial compression stress. In the other case, η_1 changes from some value in the range of -0.75~0.75 to M.

3 AN ANISOTROPIC HARDENING ELASTOPLASTIC MODEL FOR CLAY AND SAND

As explained before, the Extended Mises criterion with the initial stress-anisotropy (η^*=constant) is

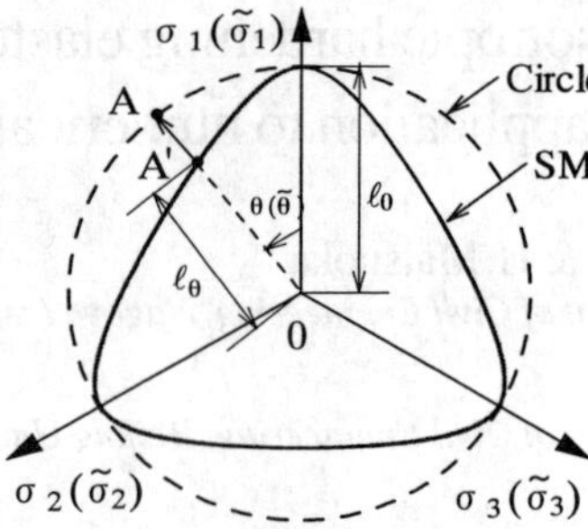

Figure 3. The SMP criterion in π-plane (solid line) and $\tilde{\pi}$-plane (dotted line)

adopted as shear yield criterion in the Sekiguchi-Ohta model. As is well known, the Extended Mises criterion is not appropriated for describing soil behavior and the SMP criterion is one of the best criteria for describing shear yield and failure behavior of soils. In order to introduce the SMP criterion into the Cam-clay model concisely, a transformed stress $\tilde{\sigma}_{ij}$ has been proposed (Matsuoka et al 1999). As shown in Figure 3, the transformed stress is deduced from what makes the SMP curve in the π-plane become a circle with the center being the origin in the $\tilde{\pi}$-plane. The transformed stress can be calculated from the conventional stress as follows:

$$\tilde{\sigma}_{ij} = p\delta_{ij} + \frac{\ell_0}{\sqrt{s_{kl}s_{kl}}} s_{ij} \tag{3}$$

where s_{ij} is the deviatoric stress, and δ_{ij} Knonecker's delta. ℓ_0 is written by (Figure 3)

$$\ell_0 = 2\sqrt{\frac{2}{3}}\frac{I_1}{3\sqrt{(I_1I_2 - I_3)/(I_1I_2 - 9I_3)} - 1} \tag{4}$$

where I_1, I_2 and I_3 are the first, second and third stress invariants.

If we take the place of the stress tensor σ_{ij} with the transformed stress tensor $\tilde{\sigma}_{ij}$, and introduce the new hardening parameter H to the Sekiguchi-Ohta model, the following yield function can be obtained.

$$f = \frac{\lambda - \kappa}{1 + e_0}\left(\ln\frac{\tilde{p}}{\tilde{p}_0} + \frac{\tilde{\eta}^*}{M}\right) - H = 0 \tag{5}$$

where, $\tilde{p}_0$ is the value of $\tilde{p}\,(=\tilde{\sigma}_{ii}/3)$ at the initial shear state with the void ratio e_0, λ and κ are slopes

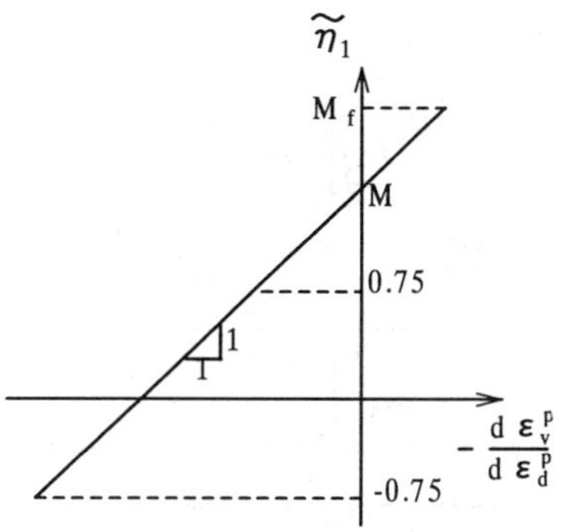

Figure 4. Proposed Stress-dilatancy relation

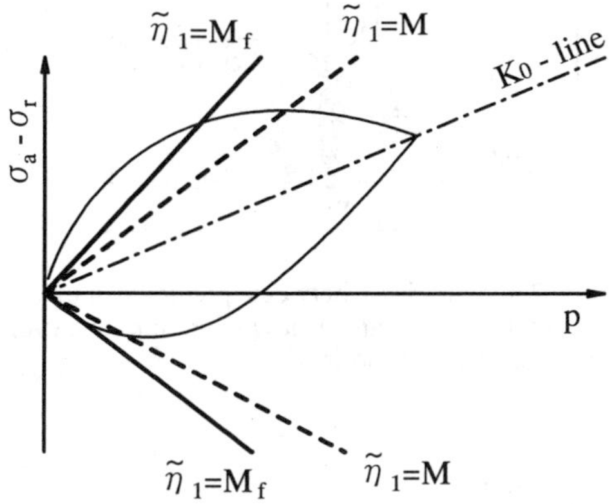

Figure 5 Proposed yield curve (fine solid line), failure state lines (bold solid line) and phase transformation lines (dotted line) in triaxial stress

of the normal compression and swelling lines in e~ ln p plane, and $\tilde{\eta}^*$ is defined as

$$\tilde{\eta}^* = \sqrt{\frac{3}{2}(\tilde{\eta}_{ij} - \tilde{\eta}_{ij0})(\tilde{\eta}_{ij} - \tilde{\eta}_{ij0})} \tag{6}$$

in which

$$\tilde{\eta}_{ij} = \frac{\left(\tilde{\sigma}_{ij} - \tilde{p}\delta_{ij}\right)}{\tilde{p}} \tag{7}$$

$\tilde{\eta}_{ij0}$ is the value of $\tilde{\eta}_{ij}$ at the initial shear state. The increment of the hardening parameter H is defined as

$$dH = \frac{M^4}{M_f^4}\frac{M_f^4 - \tilde{\eta}_1^{\ 4}}{M^4 - \tilde{\eta}_1^{\ 4}} d\varepsilon_v^p \tag{8}$$

where M and M_f are the values of q/p at the phase transformation ($d\varepsilon_v^p = 0$) and the failure state

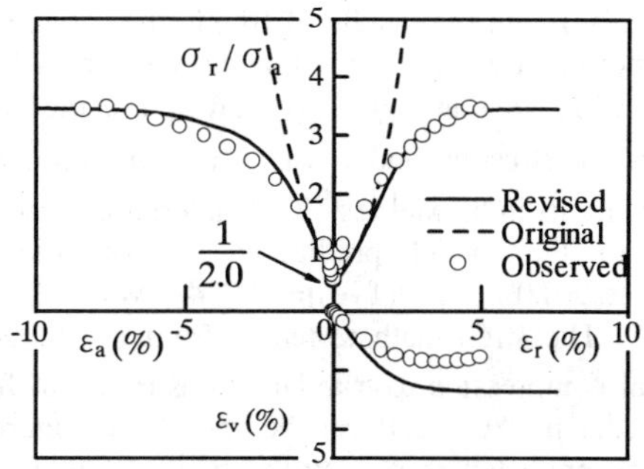

Figure 6 Comparison between predicted and experimental stress-strain behavior of clay shearing from anisotropic consolidation

($\varepsilon_d^p \to \infty$) in triaxial compression stress. $\tilde{\eta}_1$ is defined as the same in the form as Eq. (1).

$$\tilde{\eta}_1 = \frac{3}{2\tilde{\eta}^*}(\tilde{\eta}_{ij} - \tilde{\eta}_{ij0})\tilde{\eta}_{ij} \tag{9}$$

The proposed stress-dilatancy relation is

$$\tilde{\eta}_1 = M - \frac{d\varepsilon_v^p}{d\varepsilon_d^p} \tag{10}$$

Figure 4 shows the shape of Eq. (10) when K_0=0.5.

Figure 5 shows the proposed yield curve (Eq. (5), fine solid curve), failure state lines ($\tilde{\eta}_1 = M_f$, bold solid line) and phase transformation lines ($\tilde{\eta}_1$=M, dotted line) in triaxial compression and extension stresses. From Eq. (8), we can know that when M_f=M the failure state line becomes the same as the phase transformation line, and the hardening parameter H becomes the plastic volumetric strain ε_v^p, which is widely used as the hardening parameter in the elastoplastic model such as the Cam-clay model and the Sekiguchi-Ohta model for normally consolidated clay.

4 COMPARISON OF MODEL PREDICTION WITH EXPERIMENTAL RESULTS

When M_f=M, the proposed model becomes the Sekiguchi-Ohta model revised by the SMP criterion (briefly named the revised Sekiguchi-Ohta model; Sun et al. 1998). The model parameters of the revised Sekiguchi-Ohta model are the same as those of the original Sekiguchi-Ohta model. Figure 6 shows comparisons of the predicted results by the Sekiguchi-Ohta model (dotted line) and its revised model (solid line) with the results of triaxial tests (sign ○) on a non-dilative soil (experimental data from Ichihara 1997). In that figure, σ_a and σ_r are the axial and radial stresses, and ε_a and ε_r are the axial and radial strains in triaxial test. The material parameters used in the model prediction are as follows: $\lambda/(1+e_0)=0.021$, $\kappa/(1+e_0)=0.0034$, M=1.32 and $\nu=0.3$. The stress path tested is from $\sigma_a/\sigma_r=2$ in triaxial compression to triaxial extension up to failure under p=980kPa. It can be seen from Figure 6 that the predicted results by the revised Sekiguchi-Ohta model is better than by the Sekiguchi-Ohta model, being compared with the test results.

Figure 7 shows the experimental and predicted results of triaxial compression and extension tests from K_0 (=0.5)-consolidation state on Toyoura sand. The stress paths tested are from $\sigma_a/\sigma_r=2$ in triaxial compression to triaxial compression and triaxial extension up to failure under p=196kPa. The material parameters are $\lambda/(1+e_0)=0.0508$, $\kappa/(1+e_0)=0.0112$, M=0.95, M_f=1.66 and $\nu=0.3$. It can be seen from Figures 6 and 7 that the proposed model can predict well the stress-strain relation of not only clay but also sand under initially anisotropic stress.

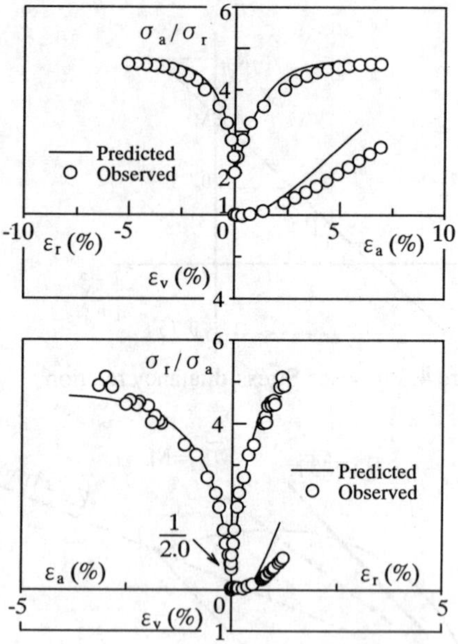

Figure 7 Comparison between predicted and experimental stress-strain behavior of sand shearing from anisotropic consolidation

5 ELASTOPLASTIC CONSTITUTIVE TENSOR

The elastoplastic constitutive tensor of the proposed model used in solving elastoplastic problem by means of the finite element method is derived here. The elastic part of the stress-strain relation can be written in the incremental form as

$$d\sigma_{ij} = D^e_{ijkl} d\varepsilon^e_{kl} = D^e_{ijkl}\left(d\varepsilon_{kl} - d\varepsilon^p_{kl}\right) \tag{11}$$

Where D^e_{ijkl} is the elastic constitutive tensor. ε_{kl} and ε^p_{kl} are the total strain tensor and its plastic component respectively. Because the associated flow rule is assumed in the $\tilde{\sigma}_{ij}$-space.

$$d\varepsilon^p_{ij} = \Lambda \frac{\partial f}{\partial \tilde{\sigma}_{ij}} \tag{12}$$

where the yield function f is the function of the stress tensor and the hardening parameter. From the consistency condition,

$$df = \frac{\partial f}{\partial \sigma_{ij}} d\sigma_{ij} + \frac{\partial f}{\partial H} dH = 0 \tag{13}$$

Substituting Eqs. (8), (11) and (12) into Eq. (13) gives

$$\Lambda = \frac{\dfrac{\partial f}{\partial \sigma_{ij}} D^e_{ijkl} d\varepsilon_{kl}}{X} \tag{14}$$

where

$$X = \frac{M^4}{M_f^4}\frac{M_f^4 - \tilde{\eta}_1{}^4}{M^4 - \tilde{\eta}_1{}^4}\frac{\partial f}{\partial \tilde{\sigma}_{ij}}\delta_{ij} + \frac{\partial f}{\partial \sigma_{ij}} D^e_{ijkl}\frac{\partial f}{\partial \tilde{\sigma}_{kl}} \tag{15}$$

Substituting Eqs. (12) and (14) into Eq. (11), we get a general form of the proposed model.

$$d\sigma_{ij} = D_{ijkl} d\varepsilon_{kl} \tag{16}$$

where the elastoplastic constitutive tensor is

$$D_{ijkl} = D^e_{ijkl} - D^e_{ijmn}\frac{\partial f}{\partial \tilde{\sigma}_{mn}}\frac{\partial f}{\partial \sigma_{st}} D^e_{stkl} / X \tag{17}$$

Introducing the elastic constitutive tensor of Hooke's law into Eq. (15) and (17), we can obtain

$$X = \frac{\lambda - \kappa}{1 + e_0}\frac{M^3}{M_f^4}\frac{M_f^4 - \tilde{\eta}_1{}^4}{(M^2 + \tilde{\eta}_1{}^2)(M + \tilde{\eta}_1)\tilde{p}} + L\frac{\partial f}{\partial \sigma_{ii}}\frac{\partial f}{\partial \tilde{\sigma}_{jj}} + 2G\frac{\partial f}{\partial \sigma_{ij}}\frac{\partial f}{\partial \tilde{\sigma}_{ij}} \tag{18}$$

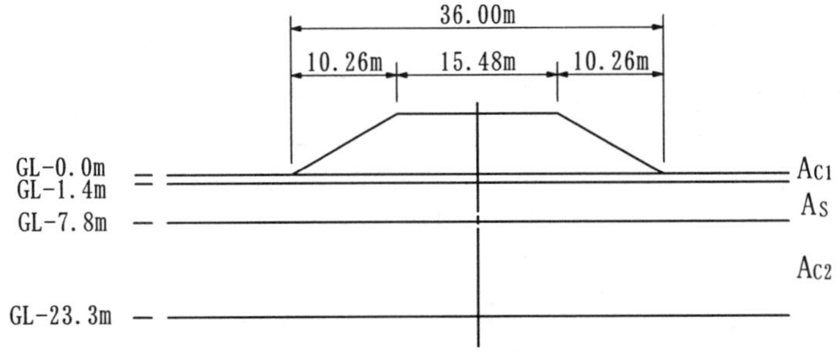

Figure 8. Profile of embankment and ground

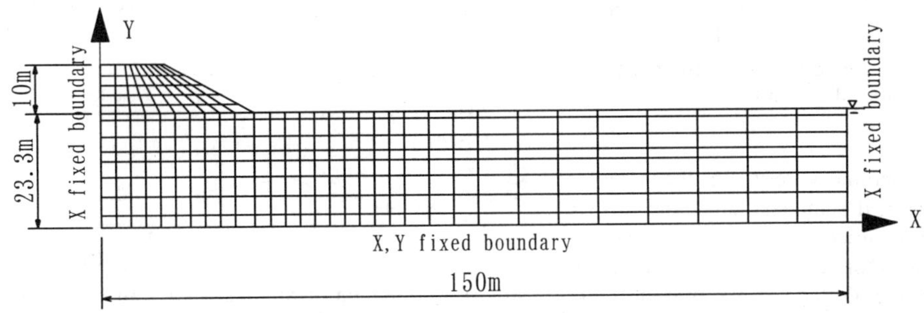

Figure 9. Finite element meshes and boundary conditions

Table 1 Material parameters used in FE computation

	Ac2(Clay)	As(Sand)
$\lambda/(1+e_0)$	1.30×10^{-1}	9.44×10^{-3}
$\kappa/(1+e_0)$	1.30×10^{-2}	4.94×10^{-3}
M	1.40	1.45
M_f		1.79
ν	0.30	0.30
k(m/day)	2.0×10^{-4}	1.0×10^{-2}

$$D_{ijkl} = L\delta_{ij}\delta_{kl} + G(\delta_{ik}\delta_{jl} + \delta_{il}\delta_{jk}) - \left(L\frac{\partial f}{\partial\tilde{\sigma}_{mm}}\delta_{ij} + 2G\frac{\partial f}{\partial\tilde{\sigma}_{ij}}\right)\left(L\frac{\partial f}{\partial\sigma_{nn}}\delta_{kl} + 2G\frac{\partial f}{\partial\sigma_{kl}}\right)/X \quad (19)$$

where L and G are Lame's constant, which can be calculated by

$$G = \frac{E}{2(1+\nu)} = \frac{3(1-2\nu)(1+e_0)}{2(1+\nu)\kappa}p \quad (20)$$

$$L = \frac{E}{3(1-2\nu)} - \frac{2}{3}G = \frac{(1+e_0)}{\kappa}p - \frac{2}{3}G \quad (21)$$

where E and ν are the elastic modulus and Poisson's ratio respectively.

6 FINITE ELEMENT COMPUTATION

The proposed model is introduced into a popular FEM program DASCAR developed by Iizuka and Ohta (1987). The governing equations describing the behavior of the soil skeleton coupled with the pore water flow are formulated by Biot (1941). The technique employed in discretizing the pore water flow was originally proposed by Akai and Tamura (1978). As an example of the application of the proposed model to finite element computation, as shown in Figure 8, a natural ground under embankment loading (Iizhuka and Ohta 1987) is analyzed. The ground consists of Ac1 layer (surface clay layer: GL0~-1.4m), As layer (sand layer: GL-1.4~-7.8m) and Ac2 layer (uniform marine clay layer, GL-7.8~-23.3m). The embanked material is the dredged sand with unit weight being 1.73 tf/m^3. Figure 9 shows the finite element meshes and boundary conditions. Only half section is computed for symmetry. The upper and lower boundaries are assumed to drain. Table 1 shows the values of the material parameters used in finite element computation. These values are

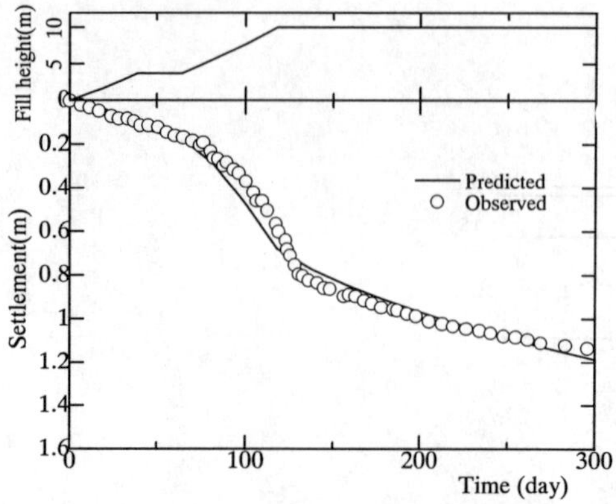

Figure 10. Loading process and settlement of ground surface at center of embankment

determined form laboratory tests on the samples taken from the site of the embankment. The surface clay layer (Ac1) is assumed as an elastic material.

The observed and computed settlement at center of the ground surface are plotted against time in Figure 10. The plot ○ is the observed results and the solid line is the results computed by the proposed model. The excess pore water pressures induced by embankment loading at different depths of central line are plotted against time in Figure 11. From Figures 10 and 11, it can be seen that the settlement and the excess water pressure computed by the proposed model fairly agree with the observed results. It can be seen from Figure 11 that the computed excess pore water pressure at depth 21.5m is less than the observed value. This is due to adopting the lower boundary as the drained boundary in finite element computation.

7 CONCLUSIONS

1. A unified elastoplastic model for clay and sand has been proposed, which is capable of describing deformation and strength characteristics of clay and sand with initially anisotropic consolidation state in 3D stress.

2. For non-dilative soil, the model parameters are the same as those of the Cam-clay model or the Sekiguchi-Ohta model. For dilative soil, one model parameter is only necessary to be added to the model parameters of the Cam-clay model or the Sekiguchi-Ohta model.

3. By using the given elastoplastic constitutive tensor, the model can be easily applied to finite element computation.

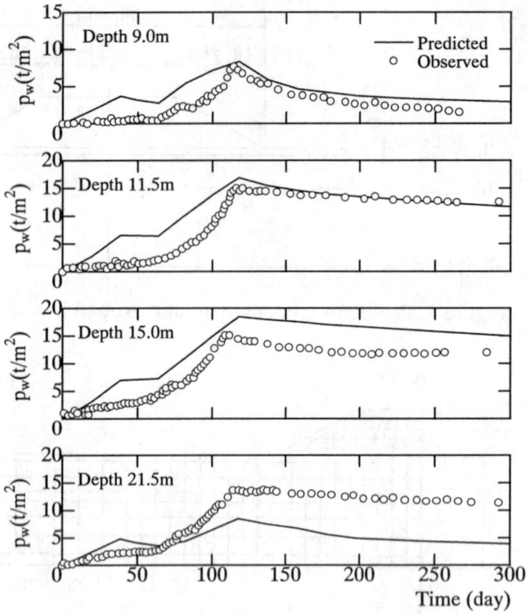

Figure 11. Generation and dissipation of excess pore water pressure at different depths of central line

REFERENCES

Akai, K. & Tamura, T. 1978. Numerical analysis of multi-dimensional consolidation accompanied with elasto-plastic constitutive equation. Proc. JSCE 269: 95-104 (in Japanese).

Biot, M. A. 1941. General theory of three-dimensional consolidation. Journal of Applied Physics 12:155-164.

Icihara, W. 1997. Deformation and strength characteristics of intermediate soil under triaxial stress condition. Bachelor's thesis at Nagoya Institute of Technology (in Japanese).

Iizuka, A. & Ohta, H. 1987. A determination procedure of input parameters in elasto-viscoplastic finite element analysis. Soils and Foundations 27(3): 71-87.

Matsuoka, H. Yao, Y. P & D. A. Sun 1999. The Cam-clay models revised by the SMP criterion. Soils and Foundations 39(1): 81-95.

Sekiguchi, H. & Ohta, H. 1977. Induced anisotropy and time dependency in clays. Proc. 9th ICSMFE, Speciality Session 9: 229-238.

Sun, D. A., Matsuoka, H. Yao, Y. P. & H. Ishii 1998. Sekiguchi-Ohta's model revised by SMP criterion, Proc. 33th Japan National Conf. on Geotechnical Engineering: 603-604 (in Japanese).

Computer Methods and Advances in Geomechanics, Desai et al. (eds) © 2001 Balkema, Rotterdam, ISBN 90 5809 183 X

Inverse problems in modeling the constitutive relations of rock and soil

Jing Tao Wang & Mei Ying Ding
School of Civil and Architectural Engineering, Huazhong University of Science and Technology, Wuhan, People's Republic of China

ABSTRACT: setting up the constitutive model of rock and soil essentially is an inverse problem, which belongs to the model identification in inverse problem theory. Since the artificial neural network (ANN) has powerful nonlinear mapping and self-learning abilities, it provides a new approach for solving this inverse problem, The authors et al. have set up four nonlinear ANN models of constitutive relations for sand and clay. This research have shown that many difficulties in setting up the constitutive models for rock and soil could be overcome by using ANN, especially, finding the mathematical expressions of models could be avoided and the process of determining the model parameters is greatly simplified. In addition, The ANN model would easily reflect the influence of stress path, volumetric expansion and shrinkage of soil caused by shear stress and strain softening. The ANN model also has strong fault-tolerance function.
Key Words: Constitutive model of rock and Soil inverse problem neural network

1 INTRODUCTION

Now many investigators proposed more than one hundred constitutive models for rock and soil. However, there are only a few models that are widely applied in engineering. There are two chief reasons for that:

1. Rock and soil are geological materials and situated in natural environment, which have suffered various actions. Therefore, it is difficult to describe the complex deformation of rock and soil by using a simple mathematical model that has only a few parameters;
2. There also exists a basic paradox in traditional modeling method. One respect requires the models are able to describe the constitutive relation of rock and soil accurately; The other requires the model parameters are as less as possible and also can be easily determined in tests.

The inverse problem theory (Tarantola, 1987) is a new field that studies the inverse processes of various physical phenomena. Back inferring the constitutive law of rock and soil from their mechanical response measured in tests in fact is an inverse problem. Based on the inverse problem theory, the first author has investigated various possible pathes for solving this inverse problem and proposed an ANN approach.

ANN (Rumelhart et al. 1986) is an information processing system with a capability of massive parallel processing, which is set up by imitating the neural architecture of the human brain. The unique capability of ANN is to acquire the knowledges from real world, i.e. self-learning ability.

The authors et al. have set up four nonlinear ANN models of constitutive relations for sand and clay. The study results have shown that ANN is good at dealing with the inverse problems in modeling the constitutive law of rock and soil.

2 INVERSE PROBLEM THEORY

In the process of knowing the world, the people discover the general relations of cause and effect. Inferring the effect from cause is called the forward problem. However, back inferring the cause from effect is an inverse problem. The main task of the inverse problem theory is to set up a mapping from the observable parameters to model parameters. In fact, modeling the constitutive law of rock and soil is back inferring their inherent stress-strain relation from their mechanical behaviors. The process of setting up the constitutive model for rock and soil can be divided into the following four steps:

1. Determine of basic model frame: first determining the basic frame of model. According to the features of constitutive relations of rock and soil and the requirements of engineering, the model of different types can be selected. For example, elastic model, nonlinear model, elastoplastic model and so on;

2. Parametrization of the model: based on the model frame, finding a minimal set of model parameters whose values completely characterize the constitutive relations of rock and soil. These parameters are called the model parameters;

3. Inverse modeling: use of the actual results of observable parameters measured in some designed tests, to infer the actual values of the model parameters;

4. Forward modeling: use of the set up model to predict the results of measurements on some observable parameters. That is the verification for the model.

Here it should be pointed out strong feed-backs exist between these steps.

3 NEURAL NETWORK APPROACH

Hecht-Nielsen (Hecht-Nielsen 1990) and Hornik et al. (Hornik et al. 1989) have shown that B-P networks are capable of approximating measurable function from one finite dimensional space to another to any desired degree of accuracy. Therefore, it is possible to characterize the highly nonlinear constitutive relation of rock and soil by using ANN. Here a B-P network is used and its architecture is shown in Figure 1. It have an input layer, an output layer and at least one hidden layer. Each layer is composed of several processing units, which are fully connected to processing units of the succeeding layer. The network is trained by samples measured in tests. Afterwards, the properly trained network is inserted in the FEM program instead of the mathematical model. The research results carried out by the first author et al. are introduced as follows.

Based on the model of Nanjing Hydraulic Research Institute (Huang 1983), the stress-strain relation is expressed by using two nonlinear functions.

$$\varepsilon_v = f_1(p,q) \quad (1)$$

$$\bar{\varepsilon} = f_2(p,q) \quad (2)$$

where ε_v is the volumetric strain, $\bar{\varepsilon}$ is the shear strain of octahedron, p and q are normal stress and shear stress of octahedron respectively . The effect of q on ε_v and that of p on $\bar{\varepsilon}$ are reflected by functions $f_1(p,q)$ and $f_2(p,q)$ whose types are not designed. Here the key problem is how to determine the two functions through tests. The increment expressions of equations (1) and (2) are written as follows,

$$d\varepsilon_v = \frac{\partial f_1}{\partial p}dp + \frac{\partial f_1}{\partial q}dq \quad (3)$$

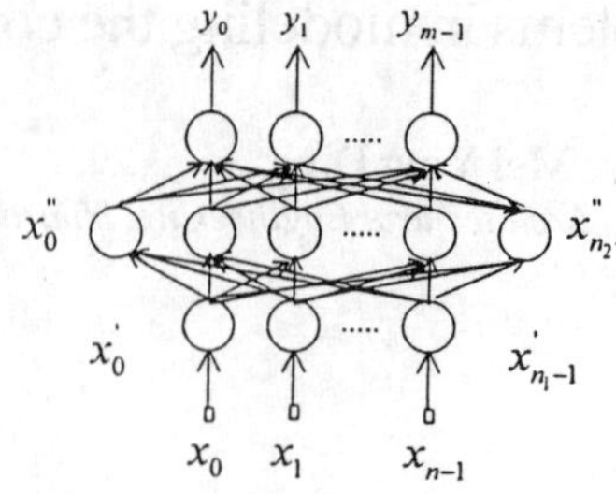

Figure1. B-P network.

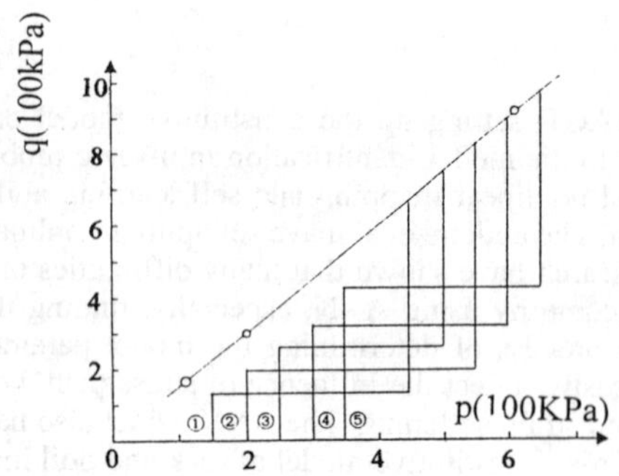

Figure 2. Stress pathes in tests.

$$d\bar{\varepsilon} = \frac{\partial f_2}{\partial p}dp + \frac{\partial f_2}{\partial q}dq \quad (4)$$

The four coefficients in Equations (3) and (4) can be determined through a series of triaxial tests, in which the stress path can be chosen. The training samples for the B-P network are directly obtained from the test curves of triaxial tests, $(\varepsilon_v, \bar{\varepsilon}) \sim p$ and $(\varepsilon_v, \bar{\varepsilon}) \sim q$. A series of drained compression tests (Wu & Wang 2001) (Li & Wang 2000) for fine sand and medium sand were carried out in laboratory, in which the chosen stress pathes are shown in Figure 2. The curves ε_v~q for p= constant and the curves $\bar{\varepsilon} \sim p$ for q= constant for medium sand in different stress pathes are described in Figure 3. It can be seen from Figure 3 there exist the effects of q on ε_v and those of p on $\bar{\varepsilon}$ in the deformation process of medium sand. Also there exist such effects of q on ε_v and those of p on $\bar{\varepsilon}$ for fine sand as shown in Figure 4. The test results show that the volumetric variation for medium sand is monotonic shrinkage but that for fine sand is from shrinkage to expansion. From a series of test curves for fine sand and medium sand , 126 and 70 test points are chosen as training samples for network respectively.

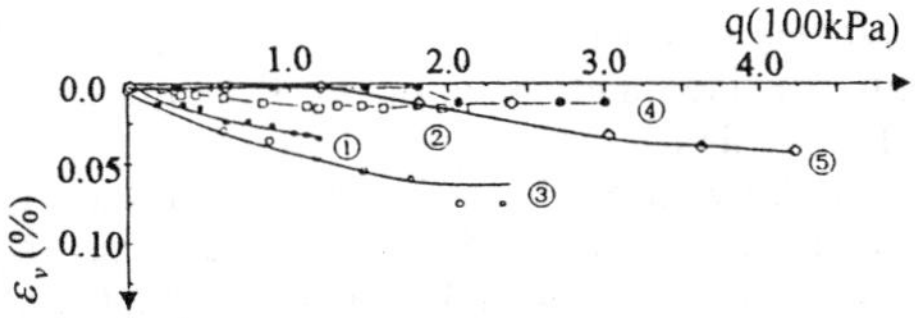

(a) P=constant

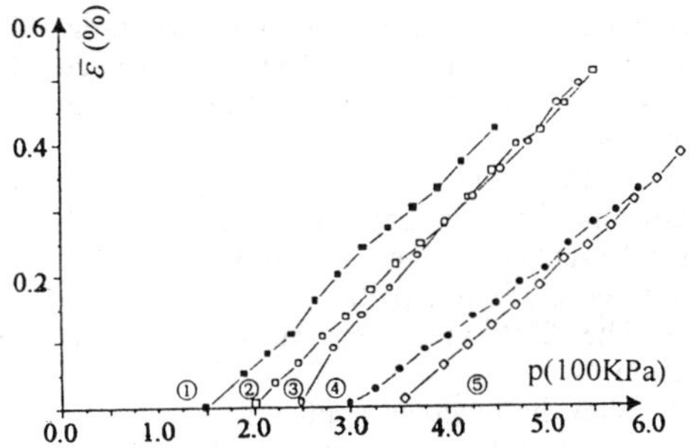

(b) q= constant

Figure3. Test curves $\varepsilon_v \sim q$ (a) and $\bar{\varepsilon} \sim p$ (b) for medium sand.

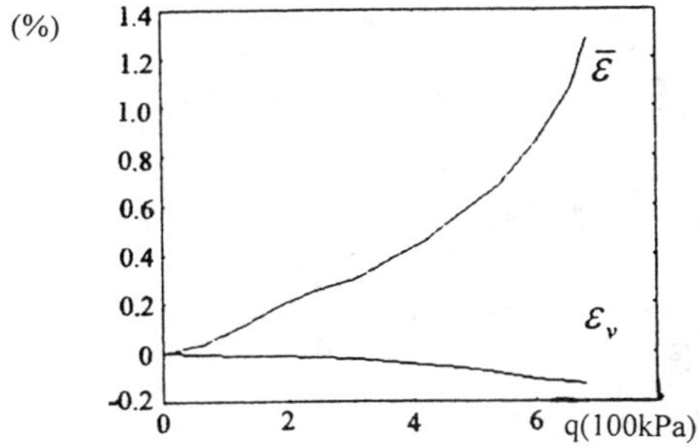

(a) p=500KPa

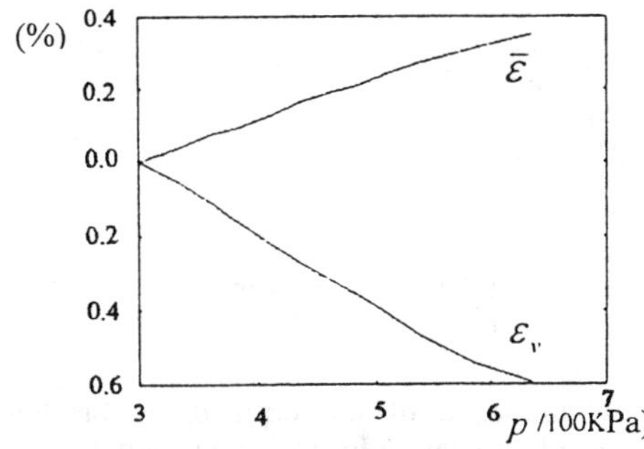

(b) q=400KPa

Figure 4. Test curves $(\bar{\varepsilon}, \varepsilon_v) \sim q$ (a) and $(\bar{\varepsilon}, \varepsilon_v) \sim p$ (b) for fine sand.

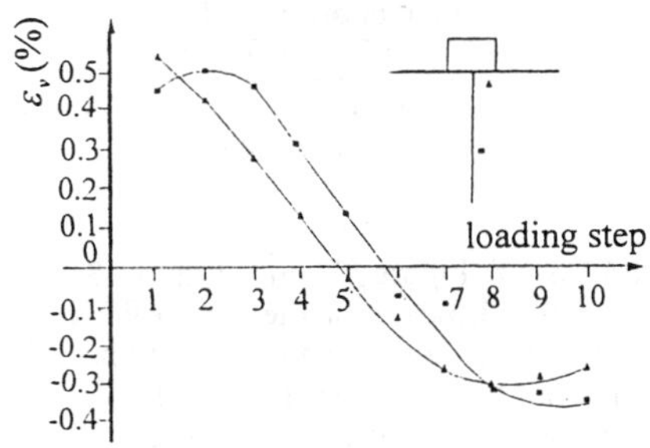

Figure 5. The volumetric strains of two points in fine sand beneath the foundation.

The trained networks have been implanted into a FEM program to calculate a circular foundation, whose diameter is 2.438m, The load applied on the foundation is 270KPa and divided into 10 loading steps. The fine sand and medium sand used in tests are respectively taken as the soil underneath the foundation. The volumetric strains of two points in the fine sand beneath the foundation are calculated and shown in Figure 5. Those in the medium sand are given in Figure 6. The calculated results have shown that the variation of volumetric strains for medium sand are monotonic shrinkage with the increase of load but those for fine sand are from shrinkage into expansion, which are consistent with the above results of triaxial tests.

Therefore, it has been verified that the ANN models for fine sand and medium sand could well reflect the volumetric variation features for sand .

In addition, the triaxial tests for medium sand were carried out(Zeng & Wang 2000), in which the stress path is according to σ_3=constant that is used in the Duncan's model (Duncan & Chang 1970). Based on these test results, a Duncan's model and an ANN model for medium sand were set up. In order to examine the fault-tolerance function of ANN model, i.e. the error of an individual datum can not influence on the final model accuracy severely, a large eeror of initial modulus E_i is introduced into the above test data. A Duncan's model and an ANN model were set up according to the modified data. Afterward, the settlements of the circular foundation were respectively calculated by using the above four models, The settlements of the foundation bottom calculated by using the Duncan's models are illustrated in Figure 7 and those by using the two ANN models are given in

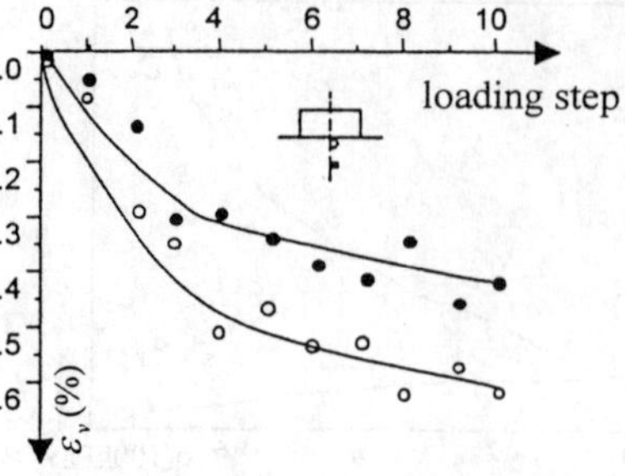

Figure 6. The volumetric strains of two points in medium sand beneath the foundation.

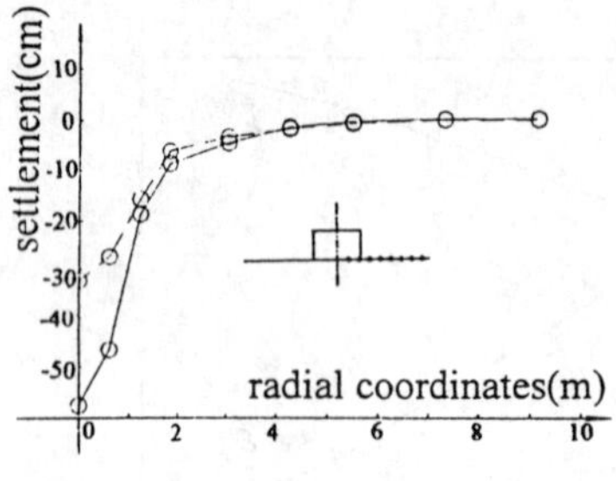

Figure 7. Influence of the error of E_i on the settlements of foundation for Duncan's model.

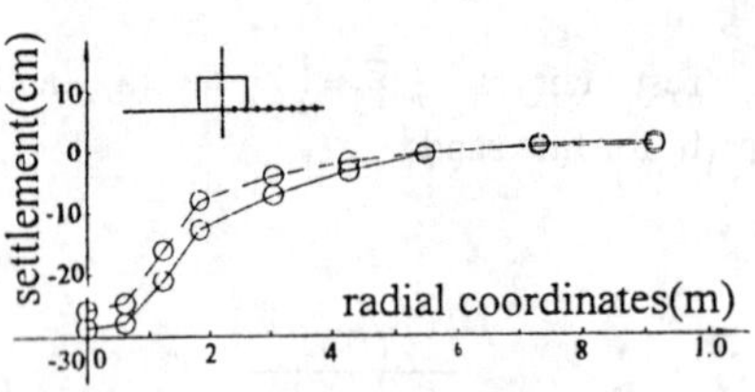

Figure 8. Influence of the error of E_i on the settlements of foundation for ANN model.

Figure 8.

Comparison of Figure 7 and Figure 8, it is obvious that the influence of the error of E_i on the Duncan's model is larger than that on the ANN model. This is because the Duncan's model is determined only by three parameters, E_i, C (cohesion) and φ (friction angle) but the ANN model has so more than 40 parameters that the error of an individual datum could only cause local effect.

4 CONCLUSION

From the process of setting up the four nonlinear ANN models for sand and clay, it is concluded that the ANN model is capable of characterizing the soil constitutive law and superior to the traditional modeling method.

1. Making use of ANN model instead of general mathematical formulae to describe the constitutive law of soil and rock could overcome the difficulties to find the mathematical expressions;

2. The parameters of ANN model can be automatically generated through training the network by directly using test data. Hence, the process of determining model parameters is greatly simplified;

3. In the process of determining ANN model parameters, the stress path can be arbitrarily chosen, which provides a effective approach for studying the influence of stress path;

4. The ANN model has strong fault-tolerance function;

5. With the practical applications of model and the increase of learning samples, the ANN model can be continually improved and perfected.

REFERENCES

Duncan, J. M. & C.Y. Chang 1970. Nonlinear analysis of stress and strain in soil. J. of the soil mechanics and foundations division, proceeding of the ASCE, 96(SM5).

Hecht-Nielsen, R. 1990. Neuro computing. Addison-Wesly Publishing Company, Reading.

Hornik, K., M. Stinchcomebe & H. White 1989. Multilayer Feedforward networks are universal approximaters. Neural Network. Vol.2: 356-366.

Huang, W.X. 1983 The engineering properties for soil. Water conservancy & hydropower Publishing , China.

Li, Q. & J.T. Wang 2000. A neural network model for the soil constitutive relationship considering the coupling relations between p, q and ε_v, $\bar{\varepsilon}$. Chinese J. of soil engineering and foundation. 14(2).

Rumelhart, D.E., G.E. Hinton & R.J. Williams 1986. Parallel distributed process. Vol.1: Foundations, Cambridge. MA.USA: MIT Press.

Tarantola, A. 1987. Inverse problem theory. Amsterdam: Elsevies Science Publishing.

Wu, X.F. & J.T. Wang 2001. Neural network model of constitutive law of soil with reflecting strain softening. Chinese J. of rock mechanics and Engineering 20(1). (in press).

Zeng, J & J.T.Wang 2000. Neural network model for the nonlinear constitutive relationship for soil. Third Int. Conf. on Advances of computer Methods in Geotechnical and Geoenvironmental Eng. Moscow.

Computer Methods and Advances in Geomechanics, Desai et al. (eds) © 2001 Balkema, Rotterdam, ISBN 90 5809 183 X

A new implementation of elastic-viscoplastic block theory for rock masses

W.M.Wang & S.H.Chen
Wuhan University of Hydraulic and Electric Engineering, People's Republic of China

ABSTRACT: A new block theory for rock masses is formulated, in which the elastic-viscoplastic characteristics of both rock blocks and discontinuities are taken into account. Using the shape functions of p-version hierarchical finite element method, the displacement fields of rock blocks can be expressed as functions of so called general degree of freedoms. The equilibrium equations of the rock block system are deduced on the basis of the virtual work principle. A numerical example is studied which indicates that the proposed new block theory is prospective because of its simple pre-processing and high accuracy.

1 INTRODUCTION

Rock masses delimited by faults, joints and cracks etc, are typical discontinuous media. FEM is undoubtedly the most sophisticated and widely used numerical method. However, it is a time-consuming and arduous work to generate meshes for complicated rock masses. While in DEM (Cundall 1988), DDA (Shi 1992), and the block theory (Chen 1994) etc., rock blocks are studied as discrete elements and no artificial mesh is needed, so the pre-processing is much easier than that of FEM. In this point of view, these numerical methods based on the discontinuous medium mechanics theory are of great potentiality.

But at the current stage of their developments, there are still many problems deserved further studies, such as the improvement of the numerical stability, the yielding and flow of rock blocks at the stress or strain localization zone. In this paper, the shape functions of p-version hierarchical FEM (Cheng et al. 1999) are introduced into rock blocks, and a new implementation of block theory considering the elastic-viscoplastic characteristics of both rock blocks and discontinuities is put forward. A numerical example is given to show the accuracy of the proposed method.

2 FORMULATION OF THE NEW ELASTIC-VISCOPLASTIC BLOCK THEORY

Three types of coordinate systems are needed for the formulation: global coordinate system, local coordinate system of every rock block, as well as local coordinate system of every discontinuity.

The X-axis of the global coordinate system is horizontal and eastward; the Y-axis is also horizontal but northward; the Z-axis is upright.

The local coordinate system of the ith rock block is defined as follows: its origin is located at the centroid of the block's minimum encircling cuboid; its ξ_i, η_i, and ζ_i-axis are paralleled to global X, Y, and Z-axis respectively; the transformation between the global and local coordinate systems of the ith block is:

$$\{\xi_i \;\; \eta_i \;\; \zeta_i\}^T = \{k_{ix}(x - x_{i0}) \;\; k_{iy}(y - y_{i0}) \;\; k_{iz}(z - z_{i0})\}^T \tag{1}$$

where:

$k_{it} = (t_{max} - t_{min})/2$, $t_{i0} = (t_{max} + t_{min})/2$, and $t \in (x, y, z)$

The local coordinate system of the ith discontinuity is defined as follows: its Z_i-axis is perpendicular to the plane and upward; its X_i and Y_i-axis are projections of the strike and dip directions to the plane respectively. The transformation between the global and local coordinate systems of the ith discontinuity can be noted as follows:

$$\{x_i \;\; y_i \;\; z_i\}^T = [L]_i \{x \;\; y \;\; z\}^T \tag{2}$$

in which the transformation matrix is function of the dip direction and dip angle of the ith discontinuity.

Choose the b1th block as representative block in the deduction, and denote its neighbor block as b2th block, which connects with the b1th block through the ith discontinuity. The deformation and stress increments of the ith discontinuity are denoted as:

$$\{\Delta\delta\}_{di} = \{\Delta\delta_x \ \Delta\delta_y \ \Delta\delta_z\}_{di}^{T} \tag{3}$$

$$\{\Delta\sigma\}_{di} = \{\Delta\tau_{zx} \ \Delta\tau_{zy} \ \Delta\sigma_z\}_{di} \tag{4}$$

The displacement, strain, and stress increments of the b1[th] block are expressed as:

$$\{\Delta u\}_{b1} = \{\Delta u \ \Delta v \ \Delta w\}_{b1}^{T} \tag{5}$$

$$\{\Delta\varepsilon\}_{b1} = \{\Delta\varepsilon_x \ \Delta\varepsilon_y \ \Delta\varepsilon_z \ \Delta\gamma_{xy} \ \Delta\gamma_{yz} \ \Delta\gamma_{xz}\}_{b1}^{T} \tag{6}$$

$$\{\Delta\sigma\}_{b1} = \{\Delta\sigma_x \ \Delta\sigma_y \ \Delta\sigma_z \ \Delta\tau_{xy} \ \Delta\tau_{yz} \ \Delta\tau_{xz}\}_{b1}^{T} \tag{7}$$

2.1 *The displacement shape functions of block*

The minimum encircling cuboid of each rock block is defined as its mother element. Since the mother element is a regular cuboid, all the probable shape functions of hexahedron in high order isoparametric and p-version FEM can be adopted here. In this paper, the functions of Serendipity family based on standard Legendre polynomials are chosen as the shape functions of blocks' mother elements, and their displacement fields are expressed as functions of the so-called general degree of freedoms. The hierarchic characteristics and better numerical stability of the Serendipity functions are very important for the further research of p-version adaptivity.

The shape functions of Serendipity family (Cheng et al. 1999) consist of point, edge, face and body shape functions.

The point shape functions are represented as:

$$N_i = \frac{1}{8}(1+\xi_i\xi)(1+\eta_i\eta)(1+\zeta_i\zeta) \quad (i=1,2,\ldots,8) \tag{8}$$

in which ξ_i, η_i and ζ_i are local coordinates of the ith node of the mother element.

The edge shape functions are represented as:

$$E_i^p = \frac{1}{4}\Phi_p(\xi)(1+\eta_i\eta)(1+\zeta_i\zeta) \quad (i=1,2,3,4)$$
$$E_i^p = \frac{1}{4}(1+\xi_i\xi)\Phi_p(\eta)(1+\zeta_i\zeta) \quad (i=5,6,7,8)$$
$$E_i^p = \frac{1}{4}(1+\xi_i\xi)(1+\eta_i\eta)\Phi_p(\zeta) \quad (i=9,10,11,12) \tag{9}$$

in which:

$$\Phi_p(t) = \int_{-1}^{t} L_{p-1}(\tau)d\tau \quad (p \ge 2)$$

L_p is p-order standard Legendre polynomial, ξ_i, η_i and ζ_i are local coordinates of the ith edge of the mother element.

The face shape functions are represented as:

$$F_i^{p(l,m)} = \frac{1}{2}\Phi_l(\xi)\Phi_m(\eta)(1+\zeta_i\zeta) \quad (i=1,2)$$
$$F_i^{p(l,m)} = \frac{1}{2}(1+\xi_i\xi)\Phi_l(\eta)\Phi_m(\zeta) \quad (i=3,4)$$
$$F_i^{p(l,m)} = \frac{1}{2}\Phi_l(\xi)(1+\eta_i\eta)\Phi_m(\zeta) \quad (i=5,6) \tag{10}$$

in which $l,m \ge 2$, $l+m=p$ ($p \ge 4$), ξ_i, η_i and ζ_i are local coordinates of the ith face of the mother element.

The body shape functions are represented as:

$$B^{p(l,m,n)} = \Phi_l(\xi)\Phi_m(\eta)\Phi_n(\zeta) \tag{11}$$

in which $l,m,n \ge 2$, $l+m+n=p$ ($p \ge 6$).

The form of the matrix of the shape functions is determined by p, the order of the block's displacement functions. If p equals 2, the matrix of the shape functions is made up of $N_i(i=1,2,\ldots,8)$ and $E_i^2(i=1,2,\ldots,12)$. If p equals 3, $E_i^3(i=1,2,\ldots,12)$ must be added to the matrix. If p equals 4, $E_i^4(i=1,2,\ldots,12)$ and $F_i^{4(2,2)}(i=1,2,\ldots,6)$ must also be added. Another case is that the orders of the block's displacement functions along different directions are not all equal to each other. For example, if the orders along x, y and z directions are equal to 3, 2 and 1 respectively, then the matrix is made up of $N_i(i=1,2,\ldots,8)$, $E_i^2(i=1,2,\ldots,8)$ and $E_i^3(i=1,2,3,4)$. The rest may be deduced by analogy. The matrix of lower order is a submatrix of that of higher order, which is called the hierarchical feature.

The displacement increments in the b1[th] block can be represented in terms of the general degree of freedoms of its mother element as below:

$$\{\Delta u\}_{b1} = [N]_{b1}\{\Delta U\}_{b1} \tag{12}$$

in which:

$$[N]_{b1} = [N_1 \ N_2 \ \ldots \ N_n]_{b1}$$

$$[N_i]_{b1} = \varphi_i[I] \quad (i=1,2,\ldots,n)$$

$$\{\Delta U\}_{b1} = \{\Delta U_1 \ \Delta V_1 \ \Delta W_1 \ldots \Delta U_n \ \Delta V_n \ \Delta W_n\}_{b1}^{T}$$

$[N]_{b1}$ is the matrix of the shape functions, $\{\Delta U\}_{b1}$ is the general displacement increments vector corresponding to the general degree of freedoms, n is the number of the general degree of freedoms, φ_i is a point, edge, face, or body shape function as the case may be, and $[I]$ is the 3-dimensional unit matrix.

2.2 *The deformation Compatibility Equation*

The strain increments of the b1th block are:

$$\{\Delta\varepsilon\}_{b1} = [B]_{b1}\{\Delta U\}_{b1} \tag{13}$$

in which $[B]_{b1}$ is the strain matrix.

$$[B]_{b1} = [B_1 \ B_2 \ \dots \ B_n]_{b1}$$

$$[B_i]_{b1} = \begin{bmatrix} \varphi_{i,x} & 0 & 0 \\ 0 & \varphi_{i,y} & 0 \\ 0 & 0 & \varphi_{i,z} \\ \varphi_{i,y} & \varphi_{i,x} & 0 \\ 0 & \varphi_{i,z} & \varphi_{i,y} \\ \varphi_{i,z} & 0 & \varphi_{i,x} \end{bmatrix}_{b1} \quad (i = 1,2,\dots,n)$$

The relative displacement increments on the ith discontinuity plane is:

$$\{\Delta\delta\}_{di} = [L]_{di}([N]_{b1}\{\Delta U\}_{b1} - [N]_{b2}\{\Delta U\}_{b2}) \tag{14}$$

2.3 *The constitutive equation*

If an explicit time stepping scheme (Owen 1980) is adopted, the stress and strain increments in the b1th block can be expressed as:

$$\{\Delta\sigma\}_{b1} = [D]_{b1}\{\Delta\varepsilon\}_{b1} + \{\Delta\sigma^{vp}\}_{b1} \tag{15}$$

in which the elastic matrix is the function of the Young's modules and Poisson's ratio; the viscoplastic stress increments are given according to the flow law in which the Drucker-Prager model is adopted for the yielding of the rock block.

At the same time, the deformation and stress increments at any point on the ith discontinuity plane will have the following relationship:

$$\{\Delta\sigma\}_{di} = [D]_{di}\{\Delta\delta\}_{di} + \{\Delta\sigma^{vp}\}_{di} \tag{16}$$

in which the elastic matrix is the function of the normal and tangential stiffness factors; the visco-plastic stress increments are also given according to the flow law, while the Mohr-Coulomb yielding criterion is adopted for the discontinuity.

2.4 *The equilibrium equations*

The virtual work principle for block system can be expressed as:

$$\sum_{bi=1}^{bn} W_{bi} + \sum_{di=1}^{dn} W_{di} = W_F \tag{17}$$

where W_{bi} is the virtual work of the ith block, W_{di} is the virtual work of the ith discontinuity plane, W_F is the virtual work of external forces, bn and dn are the total numbers of blocks and discontinuities in rock masses.

If the b1th and b2th block have virtual displacement $\{\Delta u^*\}_{b1}$ and $\{\Delta u^*\}_{b2}$ respectively, and the corresponding virtual general displacement vectors are $\{\Delta U^*\}_{b1}$ and $\{\Delta U^*\}_{b2}$, the virtual strain vectors are $\{\Delta\varepsilon^*\}_{b1}$ and $\{\Delta\varepsilon^*\}_{b2}$, the virtual deformation on the ith discontinuity, is $\{\Delta\delta^*\}_{di}$, then they must have the following relationships:

$$\{\Delta u^*\}_{b1} = [N]_{b1}\{\Delta U^*\}_{b1} \tag{18}$$

$$\{\Delta u^*\}_{b2} = [N]_{b2}\{\Delta U^*\}_{b2} \tag{19}$$

$$\{\Delta\varepsilon^*\}_{b1} = [B]_{b1}\{\Delta U^*\}_{b1} \tag{20}$$

$$\{\Delta\varepsilon^*\}_{b2} = [B]_{b2}\{\Delta U^*\}_{b2} \tag{21}$$

$$\{\Delta\delta^*\}_{di} = [L]_{di}([N]_{b1}\{\Delta U^*\}_{b1} - [N]_{b2}\{\Delta U^*\}_{b2}) \tag{22}$$

The contribution of the b1th block to the total virtual work is:

$$W_{b1} = \iiint_{b1} \{\Delta\varepsilon^*\}_{b1}^T \{\Delta\sigma\}_{b1} d\Omega \tag{23}$$

The contribution of the ith discontinuity is:

$$W_{di} = \iint_{di} \{\Delta\delta^*\}_{di}^T \{\Delta\sigma\}_{di} dS \tag{24}$$

The virtual work of external forces is:

$$W_F = \sum_{bi=1}^{bn} \{\Delta U^*\}_{bi}^T \{\Delta f\}_{bi} \tag{25}$$

in which $\{\Delta f\}_{bi}$ is the general load vector of the ith block. The form of $\{\Delta f\}_{bi}$ will be discussed in the load transferring later in this paper.

Substitute Equations 12-16, 18-25 into Equation 17, arrange the both sides according to the virtual general displacement vector $\{\Delta U^*\}_{b1}$, and note that

$\left\{\Delta U^*\right\}_{b1}$ is a virtual vector, then a set of equations will be obtained:

$$\{\Delta f\}_{b1} = \iiint_{b1}[B]_{b1}^{\mathrm{T}}[D]_{b1}[B]_{b1}\{\Delta U\}_{b1}d\Omega + \iiint_{b1}[B]_{b1}^{\mathrm{T}} \left\{\Delta\sigma^{vp}\right\}_{b1}d\Omega + \sum_{di,b1}\iint_{di}\left([N]_{b1}^{\mathrm{T}}[L]_{di}^{\mathrm{T}}[D]_{di}[L]_{di} [N]_{b1}\{\Delta U\}_{b1} - [N]_{b1}^{\mathrm{T}}[L]_{di}^{\mathrm{T}}[D]_{di}[L]_{di}[N]_{b2} \{\Delta U\}_{b2} + [N]_{b1}^{\mathrm{T}}[L]_{di}^{\mathrm{T}}\left\{\Delta\sigma^{vp}\right\}_{di}\right)dS \tag{26}$$

in which the subscript di runs over the discontinuities around the b1[th] block, and the subscript b2 runs over the neighbor blocks around the b1[th] block corresponding to the ith discontinuity.

Combining all such equations of every block, the equilibrium equations of the block system will take the following form:

$$[K]\{\Delta U\} = \{\Delta F\} \tag{27}$$

in which $[K]$ is the global stiffness matrix whose elements are given by:

$$[K]_{bi,bi} = \iiint_{bi}[B]_{bi}^{\mathrm{T}}[D]_{bi}[B]_{bi}d\Omega + \sum_{di,bi}\iint_{di}[N]_{bi}^{\mathrm{T}}[L]_{di}^{\mathrm{T}}[D]_{di}[L]_{di}[N]_{bi}dS \tag{28}$$

$$[K]_{bi,bj} = -\iint_{di}[N]_{bi}^{\mathrm{T}}[L]_{di}^{\mathrm{T}}[D]_{di}[L]_{di}[N]_{bj}dS \tag{29}$$

$\{\Delta U\}$ is the global general displacement increments vector:

$$\{\Delta U\} = \left\{\{\Delta U\}_{b1}^{\mathrm{T}}\ \{\Delta U\}_{b2}^{\mathrm{T}}\ \ldots\ \{\Delta U\}_{bn}^{\mathrm{T}}\right\}^{\mathrm{T}} \tag{30}$$

and $\{\Delta F\}$ is the global general load vector:

$$\{\Delta F\} = \left\{\{\Delta F\}_{b1}^{\mathrm{T}}\ \{\Delta F\}_{b2}^{\mathrm{T}}\ \ldots\ \{\Delta F\}_{bn}^{\mathrm{T}}\right\}^{\mathrm{T}} \tag{31}$$

$$\{\Delta F\}_{bi} = \{\Delta f\}_{bi} + \left\{\Delta f^{vp}\right\}_{bi} \tag{32}$$

$$\left\{\Delta f^{vp}\right\}_{bi} = -\iiint_{bi}[B]_{bi}^{\mathrm{T}}\left\{\Delta\sigma^{vp}\right\}_{bi}d\Omega - \sum_{di,bi}\iint_{di}[N]_{bi}^{\mathrm{T}}[L]_{di}^{\mathrm{T}}\left\{\Delta\sigma^{vp}\right\}_{di}dS \tag{33}$$

2.5 *The load transferring*

There are two types of load increments: one is the equivalent initial load from the viscoplastic deformation of both blocks and discontinuities $\left\{\Delta f^{vp}\right\}_{bi}$, which has been given above; another is the general external load $\{\Delta f\}_{bi}$, which can be obtained by load transferring based on the virtual work principle (Chen 1994).

If a point force $\{\Delta q\} = \left\{\Delta q_x\ \ \Delta q_y\ \ \Delta q_z\right\}^{\mathrm{T}}$ actint on the b1[th] block, and the coordinate of acting point is $\left(x_q\ \ y_q\ \ z_q\right)$, then its virtual work is:

$$\left\{\Delta u^*\right\}_{b1}^{\mathrm{T}}\{\Delta q\} = \left\{\Delta U^*\right\}_{b1}^{\mathrm{T}}[N]_{b1}^{\mathrm{T}}\Big|_{\left(x_q\ y_q\ z_q\right)}\{\Delta q\} \tag{34}$$

It must be equal to the virtual work of the equivalent general load:

$$\left\{\Delta U^*\right\}_{b1}^{\mathrm{T}}\{\Delta f\}_{b1} = \left\{\Delta U^*\right\}_{b1}^{\mathrm{T}}[N]_{b1}^{\mathrm{T}}\Big|_{\left(x_q\ y_q\ z_q\right)}\{\Delta q\} \tag{35}$$

Therefore:

$$\{\Delta f\}_{b1} = [N]_{b1}^{\mathrm{T}}\Big|_{\left(x_q\ y_q\ z_q\right)}\{\Delta q\} \tag{36}$$

The surface load $\{\Delta p\} = \left\{\Delta p_x\ \ \Delta p_y\ \ \Delta p_z\right\}^{\mathrm{T}}$ acting on the ith discontinuity of the b1[th] block can be transferred into equivalent general load in the same way, and its result is:

$$\{\Delta f\}_{b1} = \iint_{di}[N]_{b1}^{\mathrm{T}}\{\Delta p\}dS \tag{37}$$

The volume force $\{\Delta v\} = \left\{\Delta v_x\ \ \Delta v_y\ \ \Delta v_z\right\}^{\mathrm{T}}$ of the b1[th] block can also be transferred as follows:

$$\{\Delta f\}_{b1} = \iiint_{b1}[N]_{b1}^{\mathrm{T}}\{\Delta v\}d\Omega \tag{38}$$

3 NUMERICAL EXAMPLE

A gravity dam, whose profile is shown in Figure 1, has been studied. The parameters of the basement rock and dam concrete are listed in Table 1. The parameters of the discontinuities as well as the basal face are listed in Table 2.

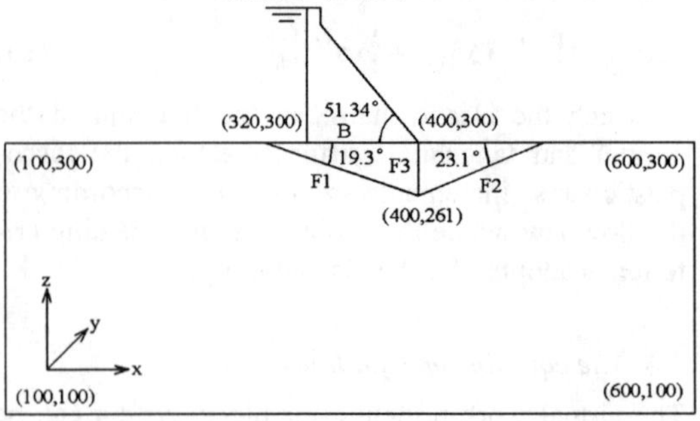

Figure 1. The profile of the gravity dam

Table 1. Parameters of the basement rock and dam concrete.

Parameters	Rock	Concrete
Young's modules (GPa)	15	20
Poisson's ratio	0.20	0.15
Friction angle ($^{\circ}$)	45.0	45.0
Coherence (MPa)	1.2	1.2
Strength of tension (MPa)	0.2	0.2

Table 2. Parameters of the discontinuities.

Parameters	F1	F2	F3	B
Normal stiffness (GN/m^3)	0.6	1.6	1.8	2.1
Tangential stiffness (GN/m^3)	0.3	0.7	0.8	0.9
Friction angle ($^{\circ}$)	26.6	31.0	38.7	45.0
Coherence (MPa)	0	0.02	0.02	0.2
Strength of tension (MPa)	0	0.01	0.01	0.02

The numerical example is analyzed with two different methods, elastic-viscoplastic block theory proposed in this paper and elastic-viscoplastic FEM. The meshes of FEM are shown in Figure 2.

In both methods, the initial stresses caused by the gravity of dam and basement are taken into account, but the displacements are neglected. In the computation of the block theory, 5-order shape functions are adopted along the x and z-direction, but only 1-order shape functions along the y-direction. While in FEM, 904 elements are generated totally. The displacement distributions obtained from the block theory and FEM are shown in Figure 3 and Figure 4, and the principal stresses are in Figure 5 and Figure 6 respectively.

Both the displacement and the stress distributions obtained from the block theory are anastomotic with that of FEM. It proves that our new elastic-viscoplastic block theory is correct and practical. It also can be inferred that the new block theory is efficient and convenient in the deformation and stability analysis of rock masses.

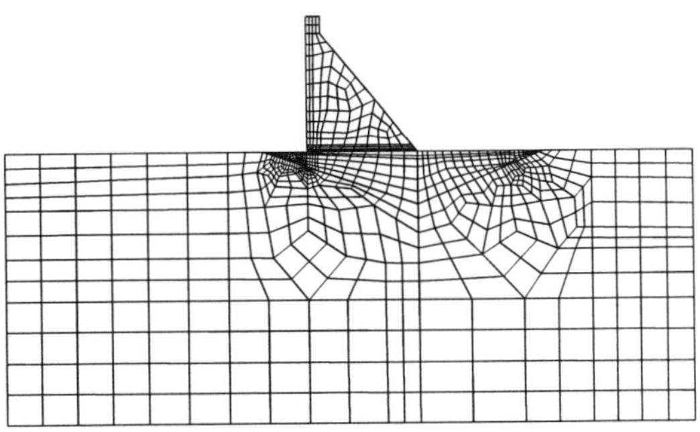

Figure 2. The meshes of FEM

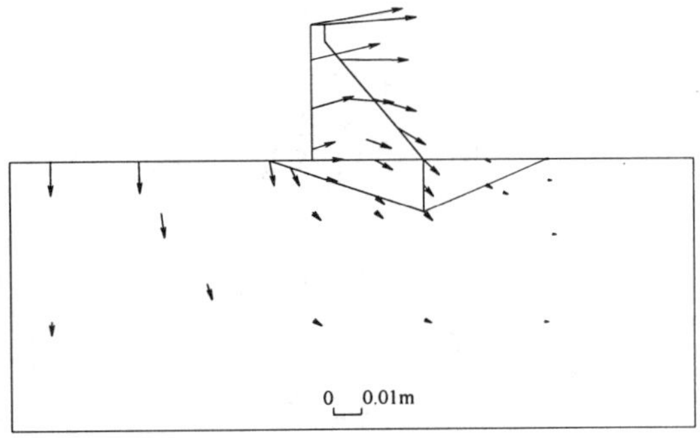

Figure 3. The displacements calculated by block theory

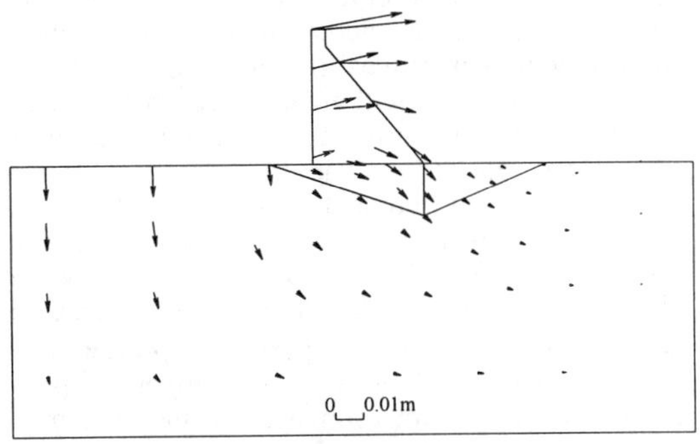

Figure 4. The displacements calculated by FEM

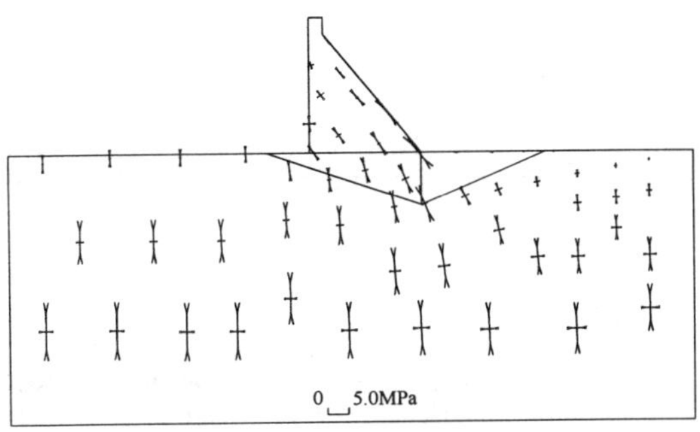

Figure 5. The principal stresses calculated by block theory

4 CONCLUSION

A new implementation of elastic-viscoplastic block theory considering the elastic-viscoplastic characteristics of both rock blocks and discontinuities has been proposed in this paper, which can be used to

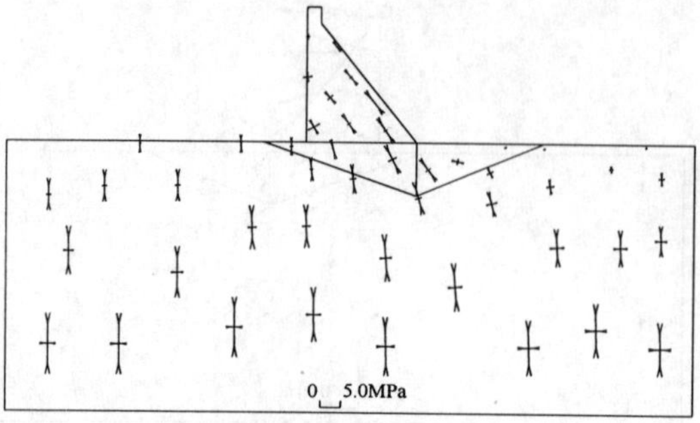

Figure 6. The principal stresses calculated by FEM

analyze the deformation and stability of discontinuous rock masses. The adoption of the hierarchical shape functions improves the numerical stability of the computation, and will greatly simplify the course of p-adaptivity. Furthermore, it is very convenient to adopt shape functions of various orders for different blocks or along different directions in order to ensure accuracy with small computational consumption. Additionally, the consideration of the viscoplastic flow of rock blocks increases the accuracy of analysis and will further expand the application of the block theory. However, some important problems such as the error estimator and the implementation of p-adaptivity are to be further studied before the methodology can be completely introduced into the engineering practice.

ACKNOWLEDGEMENT

The support of the National Science Foundation of China (NSFC) under contract No. 59979021 is gratefully appreciated.

REFERENCES

Chen S.H. 1994. A new development in elastic-viscoplastic block theory of rock masses. *Proc. 8th Int. Conf. Computer Meth. & Adv. in Geomech.* :2235-2240.

Cheng Z. & S.H. Chen 1999. 3-D hierarchical FEM analysis for hydraulic structure. *J. Hydraulic Engineering (in Chinese).* (12): 53-58.

Cundall P.A. 1988. Formulation of a three dimensional distinct element model. *Int. J. Rock Mech. Min. Sci. & Geomech.* 25:107-125.

Owen D.R.J. & E. Hinton 1980. *Finite Elements in Plasticity: theory and practice.* Pineridge Press Ltd. (UK).

Shi G.H. 1992. Discontinuous deformation analysis: a new numerical model for the statics and dynamics of deformable block structures. *Eng. Comput.* 9:157-168.

Computer Methods and Advances in Geomechanics, Desai et al. (eds) © 2001 Balkema, Rotterdam, ISBN 90 5809 183 X

Dynamic behavior of saturated sands based on cyclic torsional shear test

Gancheng Xu & Hongcai Bai
Design and Research Bureau of Air Force Engineering, Beijing, People's Republic of China

ABSTRACT: This paper performs a number of cyclic torsional shear tests on samples of saturated sands by using both stress-controlled and strain-controlled loading under undrained conditions. According to the experimental results obtained by the two loading approaches, the paper establishes the degradation models for shear stress-strain backbone curve which occurs as pore water pressure builds up. A comparison of both degradation models obtained under stress-controlled and strain-controlled conditions shows that their instantaneous patterns for any value of pore water pressure present a reasonable uniqueness. In practical boundary value problems, the calculated accuracy of test results for either the stress-controlled or strain-controlled experiments can always meet the standard within practical engineering requirements.

1 INTRODUCTION

In making a dynamic response analysis of soil deposits subject to earthquake excitation, we know that seismic loading acting on soil deposits resulting from the upward propagation of shear waves is of a stress controlled cyclic loading manner, while in some other cases possibly encountered in practical engineering, the actual dynamic loading exerted on soil deposits is of a strain controlled cyclic loading manner. For the two loading manners that have been in existence in practical engineering problems in the last 30 years, many researchers, both at home and abroad, have conducted both stress-controlled and strain-controlled cyclic loading experiments to study the dynamic behavior of sand material, such as the increase in pore water pressures and stress-strain behavior during cyclic loading. Therefore, for engineering application the experimental results are obtained either by stress-controlled cyclic loading test or by strain-controlled cyclic loading test, including cyclic triaxial test, cyclic simple shear test and cyclic torsional shear test. Because of the limited conditions of test equipment and labor power, usually only one type of the tests is performed, either stress-controlled or strain-controlled, and then the obtained experimental results are directly applied to practical engineering problems. But there remains a problem: of the two types of experimental results, which one can better describe the real dynamic behavior of sand? It is therefore necessary to study the different effects on dynamic behavior of sands by the two different cyclic loading approaches, for instance, the inherent relations in pore water pressure increase and stress-strain backbone curve. No one has so far conducted any systematic test study on this subject, nor has any conclusion been obtained for practical application.

In regard to the pore water pressure build-up regularity in saturated sands under cyclic loading the author (1992) once proposed two models, one is a strain pore water pressure model based on strain controlled cyclic torsional test results, which considers the length of effective shear strain path as an endochronic parameter and the other is a stress pore water pressure model based on stress-controlled cyclic torsional test results, which is expressed by effective shear stress path. A comparison of the two models shows that there is a reasonable inherent relation between them.

By still using the stress-controlled and strain-controlled cyclic torsional loading tests, this paper presents a systematic test study on stress-strain properties of saturated sands, and proposes their degradation models of stress-strain backbone curves which occurs as pore water pressure builds up, and investigates into the inherent relations between the instantaneous patterns of both degradation models for the backbone stress-strain curves.

2 BACKBONE CURVE DEGRADATION MODEL IN STRAIN-CONTROLLED TEST

A lot of strain-controlled cyclic torsional tests are performed on saturated sands in order to establish a stress strain backbone curve model of degradation that results from pore water pressure increases. Those tests were conducted by using DTC-199 torsional servo hydraulic closed loop testing apparatus built in Japan. A hollow cylindrical specimen with 10cm in height, 6cm in inner diameter and 10cm in outer diameter was enclosed in a triaxial cell, and cyclic torsional loads were applied on top of the specimen. In any test in this investigation, a saturated densely packed specimen of fresh Fuji river sand

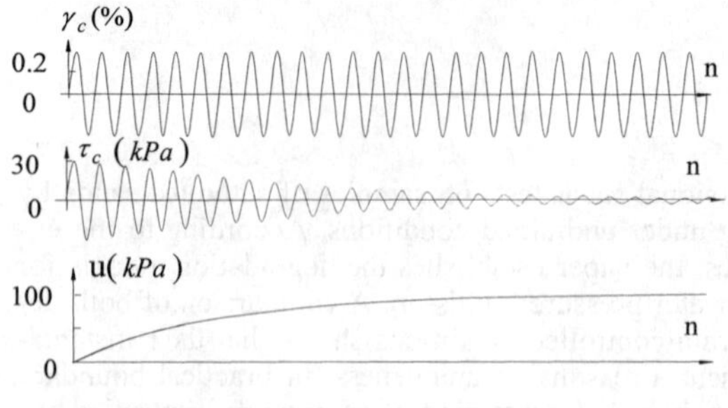

Figure1. Typical records during strain-controlled cyclic torsional tests.

was sheared under undrained condition. Physical properties of the sand are: $\gamma_{dmax} = 17.26\ kN/m^3$, $\gamma_{dmin} = 14.02\ kN/m^3$, $d_{60} = 0.42$, $d_{50} = 0.4$, $d_{10} = 0.27$, $C_u = 1.56$. Specimens were prepared by pouring fresh deaired Fuji river sand into a mold filled with deaired water with a spoon. Densification was achieved by tapping the mold with a wooden hammer. In this test, the confining pressures inside and outside the hollow specimen, $\sigma_{30}{}'$, are the same. The tests were conducted with D_r= 48%, f=1Hz, $\sigma_{30}{}' = 98.1kPa$. The applied constant shear strain amplitudes in the test are (%): 0.0275, 0.05, 0.1, 0.12, 0.202, 0.285 and 0.35. While the cyclic torsional tests are conducted for each constant shear strain γ_c, both pore water pressure u and shear stress τ_c generated in saturated sands during cyclic loading are measured, the corresponding number of load cycles n are recorded. A typical result of cyclic torsional shear test for γ_c= 0.202% is shown in Figure 1. It indicates that as a result of the action of cyclic loading γ_c, shear stress τ_c generated in sand samples decreases gradually, and when pore water pressure u reaches the initial confining pressure $\sigma_{30}{}'$, i.e. the occurrence of liquefaction, τ_c approaches the limit zero with the loss of shear strength of sands. Shown in Figure 2 are the relationship between the shear stresses generated in sands during cyclic torsional shear test for various constant strain amplitudes ranging from γ_c = 0.0275% to 0.35% and the number of load cycles. The corresponding pore water pressures are given in Figure 3 plotted against the number of load cycles.

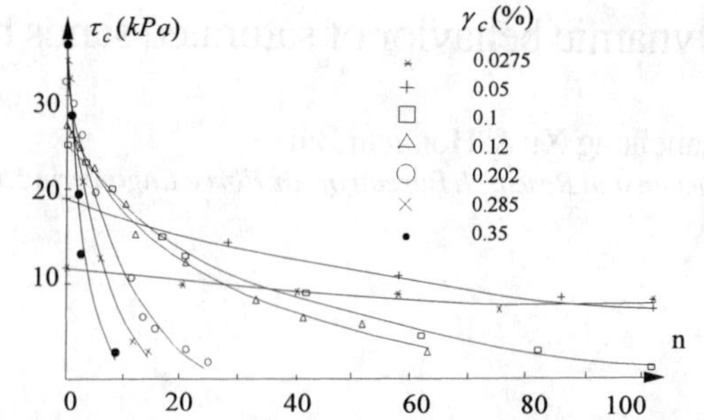

Figure 2. Measured shear stresses under cyclic torsional tests.

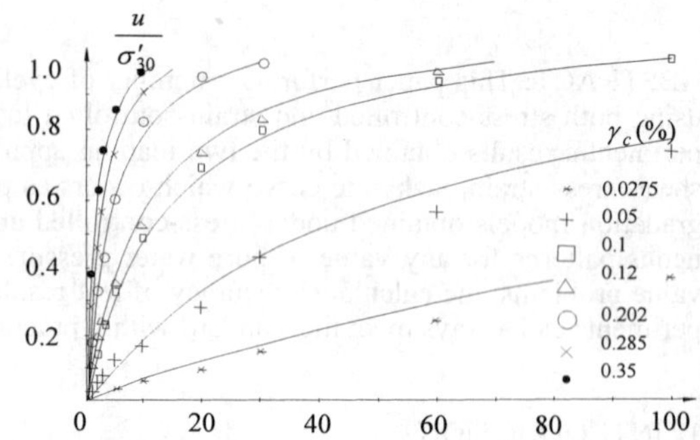

Figure 3. Experimental data on pore water pressures.

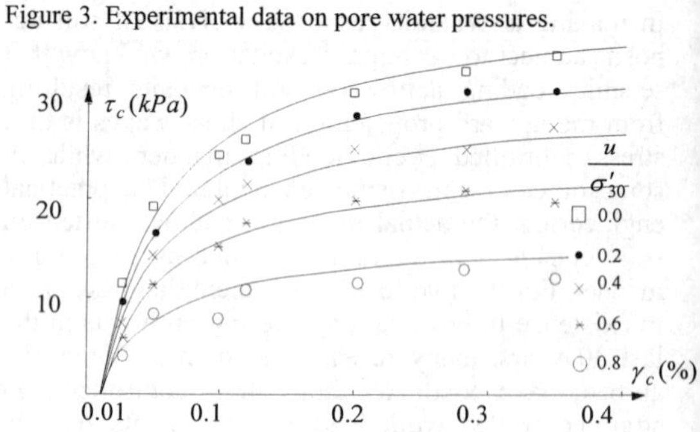

Figure 4. Shear stresses versus shear strains.

In Figure 3, horizontal straight lines are drawn with $u/\sigma_{30}{}'$ = 0.0, 0.2, 0.4, 0.6, 0.8 respectively. The number of cycles n corresponding to the intersection points between horizontal straight lines and the curves of γ_c = constants are found, and the vertical lines of the obtained n = constants are drawn in Figure2. By determining the intersection points between the lines of n = constants and the curves of γ_c = constants, together with reading off the values of the ordinates corresponding to the intersection points, the shear stresses τ_c are determined.

For various pore water pressure values equal to 0,0.2,0.4,0.6,0.8, the measured shear stresses can be plotted against the given shear strains as shown in Figure 4. From Figure 4, it seems clear that for vari-

Table 1.

γ'_c(%) / u/σ'_{30}	0.0175	0.04	0.09	0.11	0.192	0.275	0.34
0.0	0.15217	0.21053	0.36923	0.42308	0.62276	0.82707	0.97143
0.2	0.19178	0.2406	0.40134	0.460733	0.66609	0.88	1.088
0.4	0.2373	0.2832	0.45	0.51462	0.75098	1.1	1.23636
0.6	0.29787	0.3678	0.6	0.62857	0.96962	1.3333	1.7325
0.8	0.4516	0.4923	1.3	1.04762	1.74091	2.15686	2.9247

Table 2.

u/σ'_{30}	a	b	τ_{max}	G_{max}
0.0	2.540338	0.124764	0.393648	8.015133
0.2	2.742765	0.144319	0.364596	6.929095
0.4	3.220032	0.16376	0.310556	6.106497
0.6	4.340417	0.187355	0.230393	5.337461
0.8	7.359068	0.311991	0.135887	3.405221

* unit: b(10^{-2}), τ_{max}(10^2kPa), G_{max}(10^4kPa).

ous pore water pressures, the relations between shear stresses and shear strains can be expressed by hyperbolic loading curves:

$$\tau_c = \frac{\gamma'_c}{b + a\gamma'_c} \tag{1}$$

in which $\gamma_c' = \gamma_c - \gamma_t$, γ_t is threshold shear strain, for the test done on Fuji river sand γ_t =0.01%, and a, b are constants for a given sand at a given density. Equation (1) can be rewritten as follows:

$$\frac{\gamma'_c}{\tau_c} = a\gamma'_c + b \tag{2}$$

From the experimental data shown in Figure 4, the ratio γ_c'/τ_c corresponding to various pore water pressures and γ_c' can be calculated (Table 1). From the calculated data listed in Table 1 the ratios, γ_c'/τ_c, are plotted against γ_c' for various pore water pressures (Figure 5).

Equation (1) can be written into equation (3):

$$\tau_c = \frac{G_{max}\gamma'_c}{1 + \frac{G_{max}}{\tau_{max}}\gamma'_c} \tag{3}$$

where, $\tau_{max} = 1/a$, $G_{max} = 1/b$, which are respectively referred to as maximum shear stress and maximum shear modulus of sand material under consideration. According to the straight lines of u/σ_{30}' =constants shown in Figure 5, the constants a, b in equation (2) and corresponding τ_{max}, G_{max} in Table 2 are plotted respectively against (1- u/σ_{30}') in the double logarithmic coordinates as shown in Figure 6(a) and (b).

We obtain the expressions for G_{max} and τ_{max} by using a least-squares curve fitting method as follows:

$$G_{max} = g_0(1 - \frac{u}{\sigma'_{30}})^{m_1} \tag{4}$$

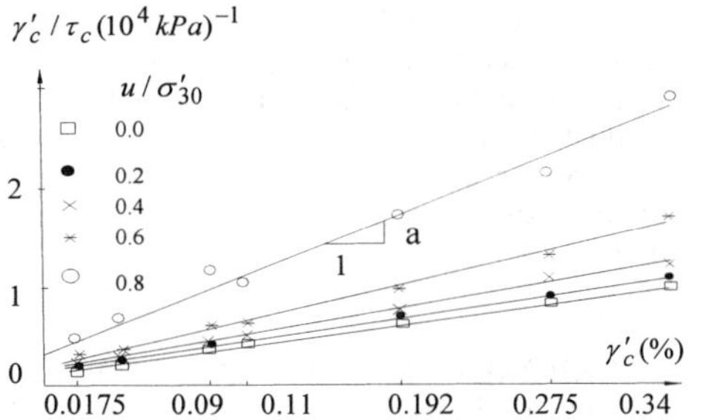

Figure 5. Computed relationship between γ_c'/τ_c and γ_c'.

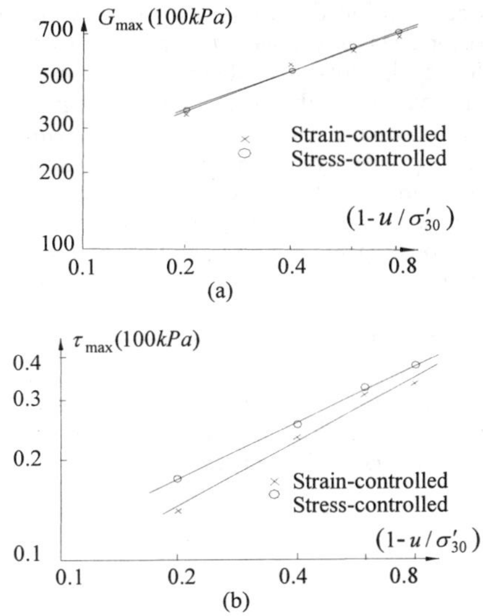

Figure 6. Computed relationship between G_{max}, τ_{max} and (1- u/σ_{30}').

$$\tau_{max} = \tau_0(1 - \frac{u}{\sigma'_{30}})^{m_2} \tag{5}$$

where $g_0 = 79.2289MPa$, τ_0 = 40. $6kPa$,$m_1 = 0.501$, $m_2 = 0.584$.

With the substitution of G_{max} and τ_{max} in equation(3), we have the following degradation model of stress-strain backbone curve along with pore pressure increase:

$$\tau_c = \frac{g_0(1-\frac{u}{\sigma'_{30}})^{m_1}(\gamma_c-\gamma_t)}{1+\frac{g_0(1-\frac{u}{\sigma'_{30}})^{m_1}}{\tau_0(1-\frac{u}{\sigma'_{30}})^{m_2}(\gamma_c-\gamma_t)}} \tag{6}$$

where γ_c is expressed in percentage. A comparison of the calculated results obtained by equation (6) and the experimental data is given in Figure 4. It appears that equation (6) is capable of making reasonable predictions for shear stress-strain behavior in saturated sands during strain-controlled cyclic torsional loading.

3 BACKBONE CURVE DEGRADATION MODEL IN STRESS-CONTROLLED TEST

To explore the effect of both stress-controlled and strain-controlled cyclic loading on shear stress-strain behavior in saturated sands, the paper, in addition to the above mentioned strain-controlled cyclic torsional shear tests, also conducts a number of stress-controlled cyclic torsional shear tests on saturated sands and establishes the corresponding degradation model of backbone curve based on experimental data. The applied physical properties and the initial stress conditions of the sands in the tests are the same as those in the strain-controlled cyclic torsional tests. The constant shear stress amplitudes applied in the tests are (*kPa*): 30.8, 24.5, 19.3, 16.1, 11.6, 9.1. During cyclic torsional loading for each shear stress amplitude τ_c, the pore water pressure u/σ_{30}' and shear strain γ_c generated in saturated sands are measured and the corresponding number of load cycles *n* are recorded. Figure 7 shows the typical results of cyclic torsional shear test for

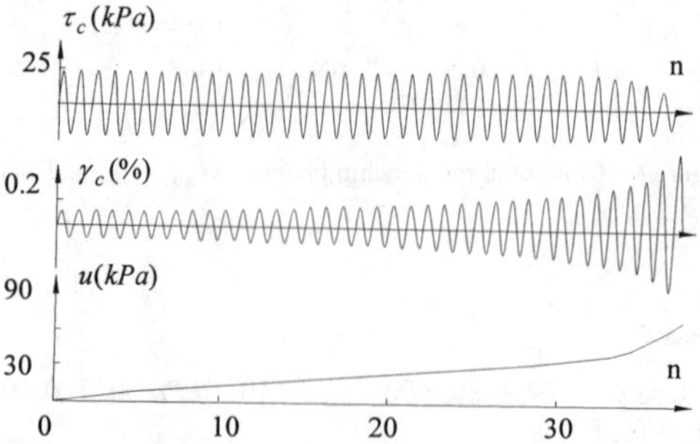

Figure 7. Typical records during stress-controlled cyclic torsional tests.

τ_c=24.5*kPa*, in which the cyclic shear strain γ_c and pore water pressure u/σ_{30}' are plotted respectively, against the number of cycles. It is seen that shear strain γ_c increases with the number of cycles, and pore water pressure gradually builds up, but the curve pattern of the increasing pore water pressure is quite different from that as obtained under strain-controlled cyclic loading. In Figure 8 and 9, shear strain γ_c and pore water pressure u/σ_{30}' generated in saturated sands during cyclic loading at various constant stress amplitudes ranging from τ_c = 9.1 to 30.8*kPa* are shown respectively versus the number of cycles.

In a way similar to the strain-controlled loading situation, by drawing the horizontal straight lines of u/σ_{30}'=0.0, 0.2, 0.4,0.6,0.8, respectively in Figure 9, we can determine the intersection points between the lines of u/σ_{30}' = constants and the curves of τ_c =constants. By reading off the values of the abscissas corresponding to the intersection points, the number of load cycles *n* are determined. By drawing the vertical lines of the obtained number of cycles *n* = constants and determining the intersection points between the lines of *n* = constants and the curves of τ_c= constants in Figure 8, together with reading off the values of the ordinate corresponding to the intersection points, we can then obtain the shear strains generated in stress-controlled cyclic torsional shear on saturated sands. For pore water pressures ranging from u/σ_{30}' = 0.0 to 0.8, the given shear stresses are shown in Figure 10 versus the corresponding measured shear strains. From the experimental data shown in Figure 10 it can be seen that the response of the saturated sand during stress-controlled cyclic

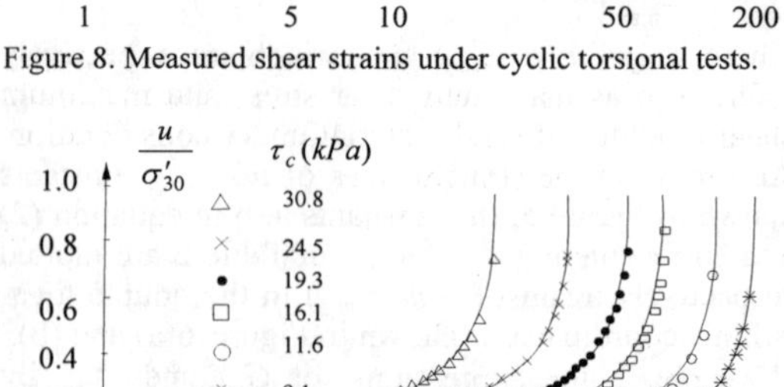

Figure 8. Measured shear strains under cyclic torsional tests.

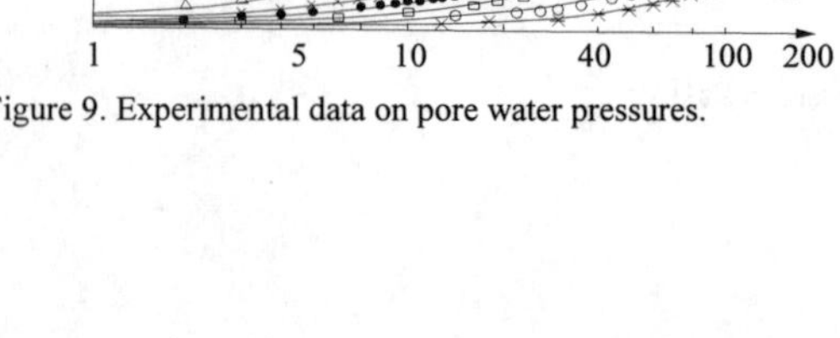

Figure 9. Experimental data on pore water pressures.

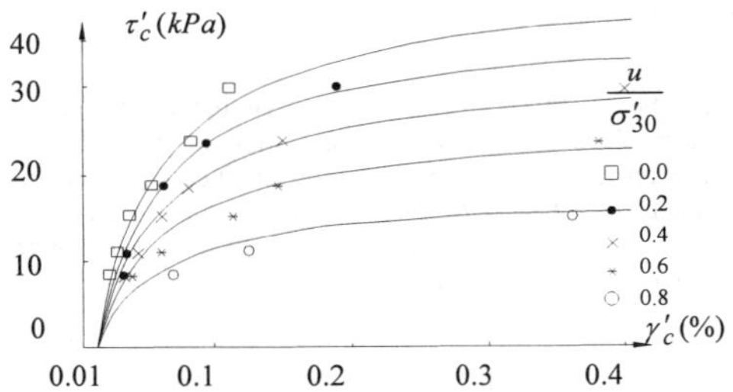

Figure 10. Shear stresses versus shear strains.

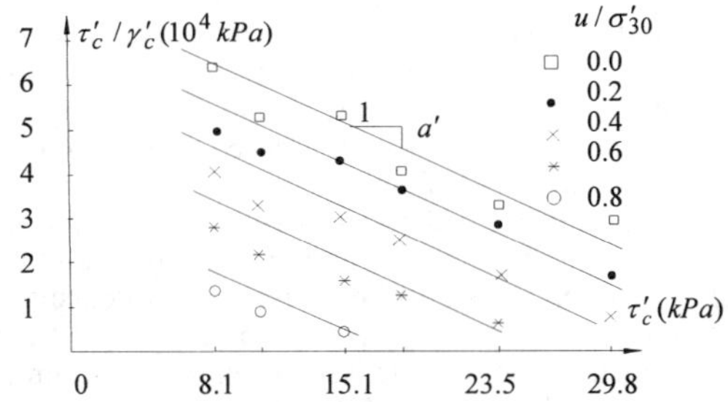

Figure 11. Relationship between τ_c'/γ_c' and τ_c'.

Table 3.

u/σ_{30}' \ τ_c'	0.081	0.106	0.151	0.183	0.235	0.298
0.0	6.48	5.25	5.3	4.067	3.279	2.98
0.2	4.84	4.386	4.25	3.66	2.776	1.656
0.4	4.05	3.3	3.02	2.52	1.678	0.764
0.6	2.8	2.17	1.589	1.262	0.635	
0.8	1.35	0.902	0.431			

torsional loading also follows the hyperbolic stress-strain relation:

$$\tau_c' = \frac{\gamma_c'}{b + a\gamma_c'} \tag{7}$$

where $\tau_c' = \tau_c - \tau_{tp}$, $\gamma_c' = \gamma_c - \gamma_t$. Similar to the threshold shear strain γ_t, τ_{tp}, which depends on the maximum elastic shear modulus and γ_t of sand material, is referred to as a threshold shear stress in this paper. As for Fuji river sand under consideration, $\tau_{tp}=1kPa$.

Equation (7) can be rewritten as

$$\frac{\tau_c'}{\gamma_c'} = b' + a'\tau_c' \tag{8}$$

where $a' = -a/b$, $b' = 1/b$. From the experimental data shown in Figure 10, the ratios τ_c'/γ_c' corresponding to various pore water pressures and γ_c' can be calculated(Table 3). From the calculated data listed in Table 3 the ratios, τ_c'/γ_c', are plotted against τ_c' for various pore water pressures(Figure 11) . According to the obtained straight lines in Figure 11, a', b' in equation (8) and their corresponding a, b can determined with τ_{max} and G_{max} calculated by $\tau_{max}=1/a$ and $G_{max}=1/b$, as listed in Table 4. G_{max}, and τ_{max} are plotted respectively against $(1-u/\sigma_{30}')$ in the double logarithmic coordinates(also see Figure 6(a)and (b)). The expressions for G_{max} and τ_{max} are the same as equation (4) and (5) except for $g_0 = 79.945MPa$, $\tau_0 = 42.3kPa$, $m_1 = 0.511$, $m_2 = 0.551$. Then the final degradation model of shear stress-strain backbone curve in stress-controlled cyclic torsional shear on saturated sands is as follows:

Table 4.

u/σ_{30}'	a	b (10^{-2})	τ_{max} $(100\,kPa)$	G_{max} $(10^4\,kPa)$
0.0	2.3614	0.1251	0.423470	7.994499
0.2	2.6701	0.1402	0.374518	7.132199
0.4	3.1283	0.1624	0.319664	6.156283
0.6	3.9106	0.1999	0.255713	5.003221
0.8	5.7275	0.2849	0.174596	3.509748

$$\tau_c' = \frac{g_0(1-\frac{u}{\sigma_{30}'})^{m_1}(\gamma_c-\gamma_t)}{1+\dfrac{g_0(1-\frac{u}{\sigma_{30}'})^{m_1}}{\tau_0(1-\frac{u}{\sigma_{30}'})^{m_2}(\gamma_c-\gamma_t)}} \tag{9}$$

The calculated results obtained by equation(9) are also shown in Figure 10, we can obtain the ratios τ_c'/γ_c', which are listed in Table 3. The corresponding ratios τ_c'/γ_c' are plotted against τ_c' for the pore water pressures equal to 0.0,0.2, 0.4,0.6,0.8 so that we can obtain a set of the results calculated by equation (9) show encouraging agreement with the results of laboratory tests on saturated sands.

4 CONCLUSIONS

This paper examines the stress strain behavior of saturated sands during both strain controlled and stress-controlled cyclic torsional shear loading. Two degradation models for shear stress-strain backbone curve as a function of pore water pressure have been established respectively as shown in equations (6) and (9). Predictions resulted from both degradation models are shown in Figure 4 and 10 respectively. By comparing Figure 4 with Figure 10, we can reach the following conclusions:

During cyclic loading, no matter what value the pore water pressure in saturated sands is, the initial shear modulus obtained by stress-controlled cyclic loading is completely equal to that by strain-controlled cyclic loading, i.e., when shear strain γ_c is very small ($\gamma_c \leqslant 0.07\%$) for a given pore water pressure, both stress-strain backbone curves under the above mentioned two types of cyclic loading conditions are completely the same. This is because the mechanical behavior of saturated sands is still in an elastic state when γ_c is very small. For elastic saturated sand material, the stress-controlled cyclic shear on it is completely identical with the strain-controlled cyclic shear on it.

When $\gamma_c > 0.07\%$, τ_c generated in sands under stress-controlled cyclic loading for a given pore water pressure is slightly greater than the τ_c under strain -controlled cyclic loading, a little discrepancy appears between two backbone degradation models with a tolerance of less than 5%. The author believes that in conducting cyclic torsional shear tests on saturated sands, many factors, including testing apparatus factor and some artificial factors, may have some influence on experimental data. Therefore, the above mentioned slight discrepancy may well have been caused by a series of accumulated experimental errors, rather than different loading controlled approaches.

From the above mentioned comments, it can be seen that for any value of pore water pressure, the instantaneous patterns between two stress-strain backbone curves obtained under stress-controlled and strain-controlled cyclic loading conditions possess a reasonable uniqueness, in other words, the stress-strain behavior during cyclic loading is an intrinsic property of sand material, which is independent of whether the stress-controlled cyclic loading or the strain-controlled cyclic loading is used on sand material.

In both theoretical research into cyclic properties on saturated sands and practical engineering application, experimental results are obtained either by strain-controlled cyclic loading or by stress-controlled cyclic loading, including cyclic triaxial test, cyclic simple shear test and cyclic torsional shear test. Because of the uniqueness for backbone degradation curves as concluded in this paper, the experimental results obtained by either strain-controlled or stress-controlled cyclic loading can always meet the standard within practical engineering requirements.

REFERENCES

Xu, Gancheng 1992. Dynamic stress-strain behavior of saturated sand and study on elastoplastic modelling. *Ph. D. the sis*, Xian University of Science and Engineering. Xian.

Xu, Gancheng 1993. *Elastoplastic modelling study on cyclic stress-strain behvior of saturated sands. Proc. intern. symp. on plasticity.* Beijing.

Computer Methods and Advances in Geomechanics, Desai et al. (eds) © 2001 Balkema, Rotterdam, ISBN 90 5809 183 X

Elastoplastic modelling of saturated sands based on pore water pressure

Gancheng Xu & Hongcai Bai
Design and Research Bureau of Air Force Engineering, Beijing, People's Republic of China

Yuan Chen
Air Force Logistical Institute, People's Republic of China

ABSTRACT: Taking pore water pressure as a hardening parameter, this paper suggests a dynamic elastoplastic constitutive model and presents experimental verification of the proposed model based on the concept of anisotropic kinematic hardening modulus field theory for the description of fundamental dynamic properties of saturated sands under cyclic loading. The model predictions as verified through torsional experiment for saturated sands indicate that the model proposed in the paper is correct and feasible.

1 INTRODUCTION

In practical engineering, almost all geotechnical engineering problems are treated as solving boundary value problems on certain conditions. Whether the solutions of this kind of problems fit engineering practice depends to a large extent on the choice of an appropriate constitutive law which is capable of proper simulation of several fundamental aspects of soil response to cyclic loading. With sophisticated modern electronic computers and increasingly matured numerical methods, a research into constitutive equations of soil material appears to become more and more important. In this area of study, constitutive models based on elastoplastic theory have enjoyed broad attention, and researches of this kind have gradually moved from static problems to dynamic ones. A lot of laboratory tests and in situ observations indicate that the behavior of soil material under load conditions is basically of elastoplasticity. Therefore, for dynamic boundary value problems of geotechnical engineering, the constitutive models of soil material based on elastoplastic theory both fit the mechanical behavior of soil material and meet the need for solving practical dynamic boundary value problems.

Although the traditional elastoplastic model based on plasticity mathematical theory can successfully describe a series of soil properties during static loading, it cannot sufficiently describe residual pore water pressure and permanent deformation caused by unloading and reloading. The traditional theory holds that soil responds elastically during unloading and reloading with no plastic deformations produced .For soil material, in fact, both elastic and plastic deformations occur during unloading, reloading and reversal loading, accompanied by anisotropic phenomenon caused by plastic stains. Over the past two decades, many researchers in foreign countries have been searching for and developing constitutive models which can meet the above mentioned demands, and have proposed dozens of dynamic constitutive models such as the multiyield-surface model suggested by Prevost(1977), Phillips(1972), Mroz, Noris and Zienkiewicz (1981) and Mroz and Pietruszczak(1983). But in these models there are some weaknesses such as the difficulties in deciding model parameters, the absence of physical meaning, and the complexity of model theory formulation. That is why they have not found wide application in engineering practice. Some researchers attempted to apply only the yield surface and bounding surface to form a two-yield-surface model, with the yield surface field in-between being replaced by interpolation function. Such models are those proposed by Ghaboussi and Momen(1979), Poorooshasb and Pietruszczak(1986). But the two-yield-surface model needs in advance to select a hardening rule of translation and expansion or contraction of loading surface, therefore, this type of models have their inherent limitations.

This paper applies a combination of isotropic and kinematic hardening rule, taking dynamic residual pore water pressure as a hardening parameter, with cyclic behavior of saturated sands being formulated respectively in the stress space of octahedral plane and hydrostatic axis. Thus, a set of systematic descriptive approaches are established, which describe how nested yield surfaces expand, contract and translate in the stress space, and how the elastoplastic moduli corresponding to the nested yield surfaces vary with the defined hardening parameter.

Finally the paper carries out the numerical simulation computation for dynamic torsional test results to verify the correctness and feasibility of the presented model.

2 ELASTOPLASTIC INCREMENTAL STRESS-STRAIN EQUATIONS

Assume that the elastic behavior of soil material is linear and isotropic, and nonlinearity and anisotropy are dealt with in the plasticity of soil. The elastic behavior is described by the general Hooke's law:

$$d\sigma_{ij} = E_{ijkl}(d\varepsilon_{kl} - d\varepsilon_{kl}^{p}) \tag{1}$$

$$E_{ijkl} = (B-2G/3)\delta_{ij}\delta_{kl} + G(\delta_{ik}\delta_{jl} + \delta_{il}\delta_{jk}) \tag{2}$$

in which B = elastic volume modulus, and G =elastic shear modulus.

In the course of plastic deformation, the consistency condition requring a stress point to remain on the yield surface is as follows:

$$df = \frac{\partial f}{\partial \sigma_{ij}} d\sigma_{ij} + \frac{\partial f}{\partial \alpha_{ij}} d\alpha_{ij} + \frac{\partial f}{\partial \kappa} d\kappa = 0 \tag{3}$$

where : $f(\sigma_{ij}, \ \alpha_{ij}, \ \kappa) = 0$ is a yield surface function. The flow rule is given by the following equation:

$$d\varepsilon_{ij}^{p} = d\lambda P_{ij} \tag{4}$$

where: $d\lambda$ = the magnitude of plastic strain increment, P_{ij} = the unit vectors normal to the potential surface g. The vector P_{ij} can be resolved into deviatoric and dilational components so that :

$$P_{ij} = P_{ij}^{S} + P_{kk}^{V}\delta_{ij} \tag{5}$$

The unit vectors normal to the yield surface *f*, Q_{ij} can also be written as that similar to equation (5).

In order to make the hardening rule suitable for complicated loading path, the paper also uses a combination hardening of plastic volume strain and length of deviatoric plastic strain path. Here the linear combination is used, i.e.

$$d\alpha_{ij} = \left\{ ad\varepsilon_{kl}^{p}\delta_{kl} + b\left[\frac{2}{3} de_{kl}^{p} de_{kl}^{p}\right]^{1/2} \right\}\mu_{ij} \tag{6}$$

$$d\kappa = cd\varepsilon_{kl}^{p}\delta_{kl} + d\left\{\frac{2}{3} de_{kl}^{p} de_{kl}^{p}\right\}^{1/2} \tag{7}$$

With the substitution of equations (6) and (7) in equation (3) we have:

$$d\lambda = \frac{Q_{ij}E_{ijkl}d\varepsilon_{kl}}{H_0 - \frac{1}{N}\left[\beta_1 \frac{\partial f}{\partial \alpha_{ij}}\mu_{ij} + \beta_2 \frac{\partial f}{\partial \kappa}\right]} \tag{8}$$

where :

$$H_0 = BQ_{ij}\delta_{ij}P_{kl}\delta_{kl} + 2GP_{kl}^{S}Q_{kl}^{S}$$

$$\beta_1 = 3aP_{kk}^{V} + b\left\{\frac{2}{3}P_{kl}^{S}P_{kl}^{S}\right\}^{1/2}$$

$$\beta_2 = 3cP_{kl}^{V} + d\left\{\frac{2}{3}P_{kl}^{S}P_{kl}^{S}\right\}^{1/2}, \quad N = \left(\frac{\partial g}{\partial \sigma_{ij}}\frac{\partial g}{\partial \sigma_{ij}}\right)^{1/2}$$

Now let :

$$H' = -\frac{1}{N}\left[\beta_1 \frac{\partial F}{\partial \alpha_{ij}}\mu_{ij} + \beta_2 \frac{\partial f}{\partial \kappa}\right] \tag{9}$$

then, the equation (8) can be written as:

$$d\lambda = \frac{Q_{ij}E_{ijkl}d\varepsilon_{kl}}{H' + H_0} \tag{10}$$

where: H' is referred to as plastic modulus.

According to equations (1), (4) and (10) we have:

$$d\sigma_{ij} = \left(E_{ijkl} - \frac{E_{ijkl}P_{kl}Q_{ij}E_{ijkl}}{H' + H_0}\right)d\varepsilon_{kl} \tag{11}$$

Equation (11) is a general expression for elastoplastic stress-strain relation. If specific yield function and hardening rule are given, the stresses and strains during cyclic loading can be obtained from equation (11).

So far as the undrained saturated sand discussed in the presend paper is concerned, anisotropic kinematic hardening modulus field theory is applied, and material failure surface equation takes the following from:

$$F = \left\{\frac{3}{2}s_{ij}s_{ij}\right\}^{1/2} - I_1 k_2 = 0 \tag{12}$$

where:

$$I_1 = \sigma_{kk}, \ s_{ij} = \sigma_{ij} - \tfrac{1}{3}\delta_{ij}I_1, \ k_2 = g(\theta)k_1$$

$$g(\theta) = \frac{3 - \sin\varphi}{2\sqrt{3}(\cos\theta - 1/\sqrt{3}\sin\theta\sin\varphi)}$$

where: k_1 = a material failure parameter, θ = the Lode angle in the octahedral plane.

According to the equality principle concerning area surrounded by both the Mohr-Coulomb yield curve and the equivalent circular yield curve, θ, can be defined by the following equation:

$$\theta = \sin^{-1}\left[\frac{-\frac{2A}{3}\sin\varphi\sqrt{\frac{4A^2}{9}\sin^2\varphi - 4\left(\frac{\sin^2\varphi}{3}+1\right)\left(\frac{A^2}{3}-1\right)}}{2\left(\frac{\sin^3\varphi}{3}+1\right)}\right] \tag{13}$$

in which $A = \sqrt{\frac{\pi(9-\sin^2\varphi)}{6\sqrt{3}}}$.

Assuming that the shape of yield surface is the same as that of failure surface, and the nested yield surfaces are introduced , the yield function is:

$$f = \left\{\frac{3}{2}(s_{ij}-\alpha_{ij}^{(m)})(s_{ij}-\alpha_{ij}^{(m)})\right\}^{1/2} - I_1 k_2^{(m)} = 0 \tag{14}$$

where: $\alpha_{ij}^{(m)}$, $k_{ij}^{(m)}$ are the kinematic and isotropic hardening parameters of the yield surface f_m. The geometric shape of the nested yield surfaces expressed by equation (14) is shown in Figure1. Since this paper applies a method by which the movement of yield surfaces is described in the stress space of octahedral plane and in the space of hydrostatic stress axis respectively it can be assumed that within a certain time step the mean effective stress tensor varies very little, and that within this time step, yield surfaces translate along with a stress point only within the octahedral plane corresponding to the above mentioned mean effective stress tensor.

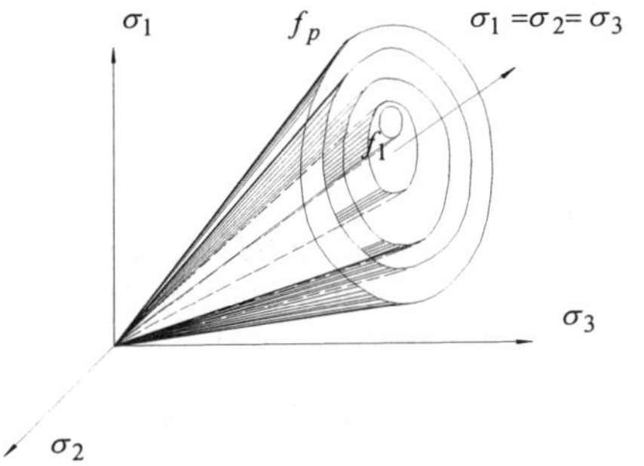

Figure 1. Distributions of the nested yield surfaces in the stress space.

Therefore, the following related formulas will be discussed in the above mentioned octahedral plane, and the associated flow rule is applied. Now, $\partial f/\partial\sigma_{ij}$, $\partial f/\partial\alpha_{ij}$ and $\partial f/\partial(I_1k_2)$ are obtained by differentiating equation (14), then the substitution of them in equation (9) results in:

$$H'^{(m)} = \beta_1 Q_{ij}^{(m)}\mu_{ij}^{(m)} + \sqrt{\frac{2}{3}}\beta_2 \tag{15}$$

According to the consistency condition (3), $d\lambda$ becomes:

$$d\lambda = \sqrt{\frac{2}{3}}\frac{H_0}{H'^{(m)}+H_0}\frac{(s_{ij}-\alpha_{ij}^{(m)})d\varepsilon_{ij}}{I_1k_2^{(m)}} \tag{16}$$

in which $H_0 = 2G$. The incremental stress-strain relation (11) can be written as:

$$d\sigma_{ij} = \left(B-\frac{2G}{3}\right)d\varepsilon_{kk}\delta_{ij} + 2Gd\varepsilon_{ij}$$

$$-(2G-H^{(m)})\frac{3}{2}\frac{s_{ij}-\alpha_{ij}^{(m)}}{\left[I_1k_2^{(m)}\right]^2}(s_{kl}-\alpha_{kl}^{(m)})d\varepsilon_{kl} \tag{17}$$

or

$$de_{ij} = \frac{ds_{ij}}{2G} + \frac{3}{2H'^{(m)}}\frac{s_{ij}-\alpha_{ij}^{(m)}}{\left[I_1k_2^{(m)}\right]^2}(s_{ij}-\alpha_{kl}^{(m)})ds_{kl} \tag{18}$$

in which $H^{(m)}$ is called the elastoplastic shear modulus, and

$$H^{(m)} = \frac{1}{1/2G + 1/H'^{(m)}} \tag{19}$$

3 MOVEMENT OF YIELD SURFACE IN STRESS SPACE, HARDENING PARAMETER AND CONVENTION CYCLIC TEST

First, in the space of hydrostatic stress axis, when residual pore water pressure caused by cyclic loading in soil increases, effective stresses in soil gradually decrease, and the first invariant of effective stress tensor I_1 also decreases correspondingly, and the yield surfaces translate towards the origin of the stress space until liquefaction occurs in saturatd sands. Therefore, the expansion and contraction of nested yield surfases in hydrostatic stress axis are controlled by the above mentioned pore water pressure.

Second, in the octahedral stress space, nested yield surfaces translate along with changes in deviatoric stresses, and the translation of nested yield surfaces is assumed to satisfy the following condition:

$$\frac{s_{ij}-\alpha_{ij}^{(m)}}{[I_1k_2^{(m)}]} = \frac{s_{ij}-\alpha_{ij}^{(m-1)}}{[I_1k_2^{(m-1)}]} = \cdots = \frac{s_{ij}-\alpha_{ij}^{(1)}}{[I_1k_2^{(1)}]} \tag{20}$$

As shown in Figure2, the instantaneous translation of yield surface f_m is:

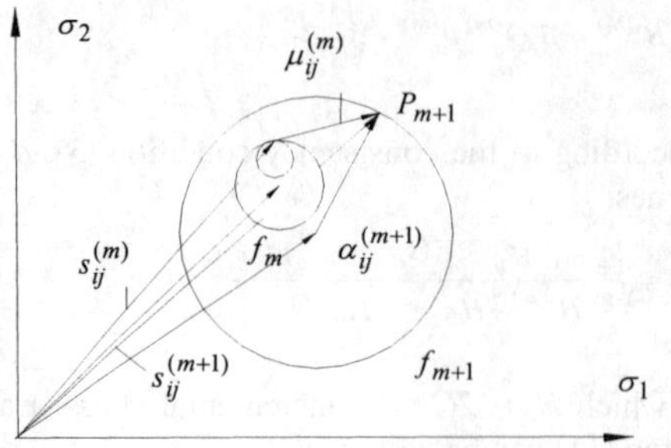

Figure 2. Translation rule of yield surfaces in π plane

$$d\alpha_{ij}^{(m)} = d\xi\mu_{ij}^{(m)} \tag{21}$$

where $d\xi$ and $\mu_{ij}^{(m)}$ are the magnitude of translation distance of yield surface f_m and its translation direction tensor.

$$\mu_{ij}^{(m)} = s_{ij}^{(m+1)} - s_{ij}^{(m)} \tag{22}$$

The substitution of $s_{ij}^{(m+1)}$ and $s_{ij}^{(m)}$ obtained from equation (20) in equation (21) results in:

$$\mu_{ij}^{(m)} = \frac{1}{I_1 k_2^{(m)}}\left(I_1 k_2^{(m+1)} - I_1 k_2^{(m)}\right)s_{ij} - \frac{1}{I_1 k_2^{(m)}}\left(\alpha_{ij}^{(m)} I_1 k_2^{(m+1)} - \alpha_{ij}^{(m+1)} I_1 k_2^{(m)}\right) \tag{23}$$

$d\xi$ in equation (21) is obtained from equation (3):

$$d\xi = \frac{(3H^{(m)}/2)(s_{ij} - \alpha_{ij}^{(m)})de_{ij} - I_1 k_2^{(m)} d[I_1 k_2^{(m)}]}{[I_1 k_2^{(m+1)}][I_1 k_2^{(m)}] - \frac{2}{3}(s_{ij} - \alpha_{ij}^{(m+1)})(s_{ij} - \alpha_{ij}^{(m)})} \tag{24}$$

On the other hand, if a stress point retreats from the yield surface, f_m, because of unloading , the stress point first moves within f_l , then again reaches f_l in the opposite direction, and then f_l translates towards $f_2, f_3, \cdots f_m$ with the stress point. The corresponding plastic modulus become $H'^{(1)}$, $H'^{(2)}$, $\cdots H'^{(m)}$ successively. At this time, equation (17) is still used to compute the stress increment tensor during unloading and reversal loading, or equation (18) is used to compute the strain increment, but corresponding plastic moduli during unloading and reversal loading need to be used.

The expansion, contraction and translation of yield surfaces can be represented by a hardening parameter that reflects stress-strain history as experienced by the soil in the course of cyclic loading. This hardening parameter should be a monotonically increasing variant. This paper neglects the part of elastic strain in total strain, and therefore takes $\bar{\varepsilon} = \int\left\{\frac{2}{3}de_{ij}de_{ij}\right\}^{1/2}$ as the hardening parameter. As we know, residual pore water pressure induced in soil during cyclic loading is also a monotonically increasing variant, and it seems that it can also be used for the hardening parameter. As a matter of fact, $\bar{\varepsilon}$ can be fully represented by residual pore water pressure.

If cyclic strain-controlled tests are performed on saturated sands, $\bar{\varepsilon}$ can be represented by cyclic strain amplitude. For cyclic triaxial tests, we have:

$$d\bar{\varepsilon} = |d\varepsilon_c| = \frac{2}{3}d\gamma_c \tag{25a}$$

$$\bar{\varepsilon} = \frac{8}{3}n\gamma_c \tag{25b}$$

where: n = number of cycles, $\gamma_c = 1.5\,\varepsilon_c$, ε_c = an axial strain amplitude.

For cyclic direct simple shear or torsional shear tests, we have:

$$d\bar{\varepsilon} = \frac{1}{\sqrt{3}}d\gamma_c \tag{26a}$$

$$\bar{\varepsilon} = \frac{4}{\sqrt{3}}n\gamma_c \tag{26b}$$

in which γ_c = a shear strain amplitude.

For cyclic stress-controlled tests, the equation similar to equation (25) and (26) can be obtained with the accumulated increment method. As the model of pore water pressure proposed by many researchers can also be expressed by the number of cycles n and shear strain γ_c, $\bar{\varepsilon}$ can be related with residual pore water pressure, and finally, the residual pore water pressure can be taken as the hardening parameter.

In the conventional cyclic tests such as triaxial, direct simple shear and tortional shear tests, equation (14) is simplified to:

$$\left\{\left[(\sigma_y - \sigma_x) - (\alpha_y^{(m)} - \alpha_x^{(m)})\right]^2 + (\sqrt{3}\tau_{xy} - \sqrt{3}\alpha_{xy}^{(m)})^2\right\}^{1/2} - I_1 k_2^{(m)} = 0 \tag{27}$$

Similarly, equation (18) is simplified to equations (28) and (29):

$$d\bar{\gamma} = \frac{2}{3H^{(m)}}dq \tag{28}$$

$$d\gamma_{xy} = \frac{2}{H^{(m)}}d\tau_{xy} \tag{29}$$

where : $\bar{\gamma} = 2/3(\varepsilon_y - \varepsilon_x)$, $q = \sigma_y - \sigma_x$.

With the help of the suggested dynamic elastoplastic model, the paper now discusses its numerical analysis about cyclic tosional shear results. In calculation, a loading cycle is divided into 160 time steps. At the same time, the initial backbone curve corresponding to the zero pore water pressure is divided into 100 segments, that is, 100 nested yield surfaces are distributed over the stress space surrounded by the failure surface.

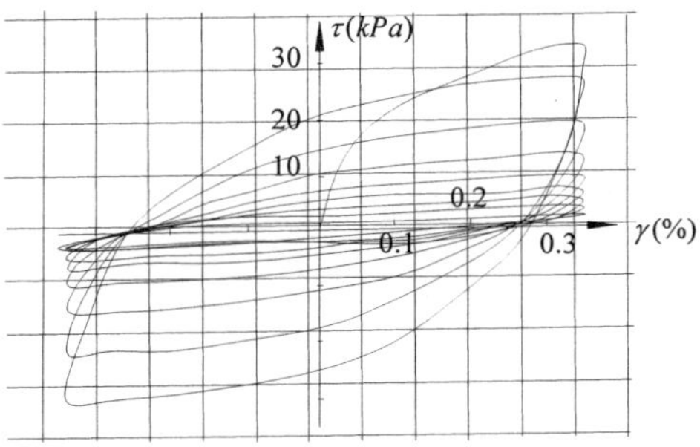

Figure 3. Stress-strain loops generated during torsional shear tests(γ_c=0.35%)

The torsional tests are carried out under strain-controlled and undrained conditions. The tests done for this study were performed by using DTC-199 torsional servo-hydraulic closed-loop testing apparatus built in Japan, hollow specimen with 10*cm* in outside diameter, 6*cm* inside diameter, and 10*cm* in height, and Fuji river sand is applied. Its related indexes are as follows: $\gamma_{dmax}=17.26kN/m^3$, $\gamma_{dmin}=14.02kN/m^3$, $d_{60}=0.42$, $d_{50}=0.4$, $d_{10}=0.27$, $C_u=1.56$, $D_r=48\%$, $f=1Hz$, $\sigma'_{30}=98.1kPa$.

Figure3 shows the measured results of the stress-strain hysteretic curves obtained from cyclic torsional shear tests when $\gamma_c=0.35\%$.

The corresponding model predictions are shown in Figure4 in which the number of nested yield sur-

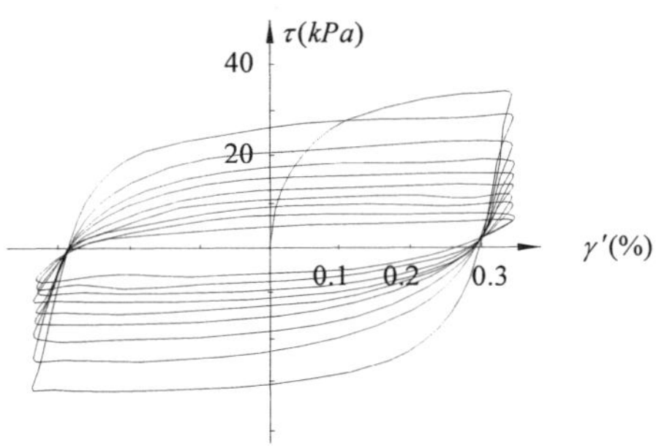

Figure 4. Stress-strain loops generated with the model.

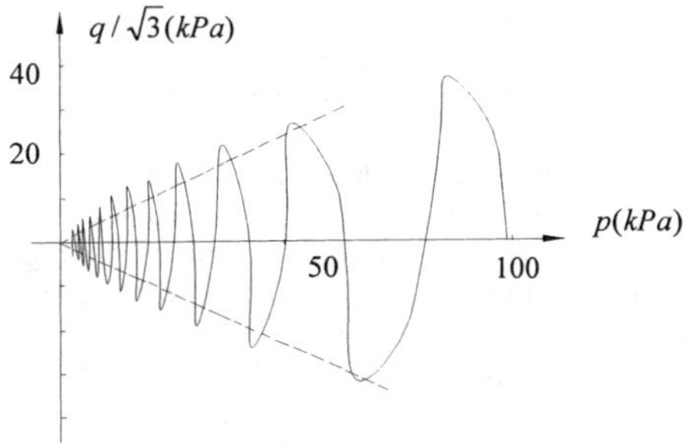

Figure 5. Stress path computed by the model.

faces = 100, $\phi'=0$, $c'=0$, $\sigma'_{30}=98.1kPa$, $\Delta t=1/60s$, $\gamma_c=0.35\%$, $g_0=792.3$, $\tau_0=0.406$, $m_1=0.501$, $m_2=0.584$. Comparisons of model predictions with experimental results show good agreement both iquality and quantity, indicating that the model concepts are sound and feasible.

The stress path computed by the model is shown in Figure 5.

4 CONCLUSIONS

The dynamic elastoplastic model as presented in this paper on the basis of anisotropic kinematic hardening modulus field theory is capable of appropriate simulations of several fundamental aspects of soil response to cyclic loading history. Among these are nonlinear hysteretic stress-strain behavior, plastic deformation induced by unloading, reloading and reversal loading, anisotropy, path-dependent stress-strain properties and liquefaction.

Combining isotropic and kinematic hardening rules, taking pore water pressure as a hardening parameter and formulating cyclic properties of saturated sands respectively in the stress space of octahedral plane and hydrostatic axis are all favorable for establishing a set of systematic approaches which describe how nested yield surfaces expand, contract and translate in stress space, and how elstoplastic moduli corresponding to yield surfaces vary with the defined hardening parameter. The model predictions show sound agreement with experimental results as verified for Fuji loose sand.

REFERENCES

Prevost, J.H. 1977. Mathematical modelling of monotonic and cyclic undrained clay behavior. *Int. J. Numer. Anal. Meth. Geomech.* 195-216.

Phillips. A. 1972. On rate independent continuum theories of graphite and their experimental verification. *Nuc. Eng. and Design*. 143-156.

Mroz, Z. , Norris, V. A. & Zienkiewicz, O. C. 1981. An anisotropic critical state model for soils subject to cyclic loading. *Geotechnique*. 451-469.

Mroz, Z. & Pietruszczak, S. A. 1983 A constitutive model for sand with anisotropic hardening rule. *Int. J. Numer. Anal. Meth. Geomech.* 305-320.

Ghaboussi, J. & Momen, H. 1979. Plasticity model for cyclic behavior of sands. *Proc. 3rd Intern. Conf. Numer. Anal. Meth. Geomech.*

Poorooshasb, H. B. & Pietruszczak, S. A. 1986. A generalized flow theory for sand. *Soil and Foundation.* 1-15.

Computer Methods and Advances in Geomechanics, Desai et al. (eds) © 2001 Balkema, Rotterdam, ISBN 90 5809 183 X

A new 3-D elastic visco-plastic model for time-dependent stress-strain behaviour of clayey soils and its applications

Jian-Hua Yin
Department of Civil and Structural Engineering, Hong Kong Polytechnic University, People's Republic of China

ABSTRACT: This paper presents a newly developed 3-D elastic visco-plastic (EVP) model for the time-dependent stress-strain behaviour of clayey soils and applications in consolidation analysis of geotechnical structures on soft ground. The paper gives a brief introduction to a specific new 3-D EVP model. The paper presents applications of the 3-D EVP model in a case study and a parametric study. The increase of excess porewater pressure under constant loading is an interesting phenomenon. For one-dimensional straining condition, there are no Mandel-Cryer effects. The creep mechanism is the sole factor causing the porewater pressure increasing. This paper extends the explanation for 1-D straining case to interpret the porewater increase in the clay in a general stress condition. In general, no mater how complicated the stress states are, for rising excess porewater pressure to occur, the volume strain change (compression) caused by excess porewater dissipation must be smaller than that caused by creep compression.

1 INTRODUCTION

In many cases, the geotechnical structures are constructed on and/or installed in soft clayey soil ground. The performance of the geotechnical structures (deformation, porewater pressure and stability) is a big concern of the engineers and the owners. The key to predicting the performance is a representative constitutive model for the time dependent strain-strain behaviour of the clayey soils. This paper presents a newly developed 3-D elastic visco-plastic (EVP) model for the time-dependent behaviour of clayey soils and applications in consolidation analyses of test embankments and an artificial island on soft ground. The paper will give a brief introduction to fundamental concepts used in the EVP modelling and its framework proposed by Yin and Graham (1989, 1994, 1999). One specific new 3-D EVP model is introduced. The paper presents applications of the 3-D EVP model in a case study and a parametric study.

2 A NEW ELASTIC VISCO-PLASTIC MODEL

Yin and Graham's (1999) 3-D Elastic Visco-Plastic (3-D EVP) constitutive relationship can be expressed as

$$\dot{\varepsilon}_{ij} = \dot{\varepsilon}_{ij}^{e} + \dot{\varepsilon}_{ij}^{vp} = \left(\frac{1}{2G}\dot{s}_{ij} + \frac{\kappa}{3V}\frac{\dot{p}'}{p'}\delta_{ij}\right) +$$

$$\frac{\psi}{Vt_0}\exp\left[\left(\varepsilon_{vm0}^{r} + \frac{\lambda}{V}\ln\frac{p'_m}{p'_{mo}} - \varepsilon_{vm}\right)\frac{V}{\psi}\right]\frac{1}{2p' - p'_m}\frac{\partial F}{\partial \sigma'_{ij}} \tag{1}$$

where $\dot{\varepsilon}_{ij}$ is the total strain rate (i=1,2,3; j=1,2,3); $\dot{\varepsilon}_{ij}^{e}$ and $\dot{\varepsilon}_{ij}^{vp}$ are the elastic and visco-plastic strain rate; σ'_{ij} is effective stress; the mean effective stress p' is defined as $p' = \sigma'_{kk}/3$; $\dot{s}_{ij}$ is the deviator stress rate; the deviator stress s_{ij} is defined as $s_{ij} = \sigma'_{ij} - \delta_{ij}\sigma'_{kk}/3$, where $\delta_{ij} = 0$ if i≠j, $\delta_{ij} = 1$ if i=j; G is elastic shear modulus; κ/V (V is specific volume), ψ/V and t_o, λ/V, p'_{mo} and ε_{vm0}^{r} are model parameters.

The F in Eqn.(1) is the function of visco-plastic flow surface (Yin and Graham 1999). According to Yin and Graham (1999), the surface is expressed as

$$F = p'^2 + \frac{q^2}{M^2} - p'p'_m = 0 \tag{2}$$

where M is the slope of strength envelop in q-p' plane; q is the generalised deviator stress and $q = \sqrt{\frac{3}{2}s_{ij}s_{ij}}$. In Eqn.(1) and Eqn.(2), p'_m is the mean effective stress value at which the flow surface in

Eqn.(2) intercepts the mean effective p'–axis in q-p' plane. The sub-index "m" stands for the mean stress or volume strain under isotropic stressing condition or $p' = p'_m$ with q=0. For example, ε_{vm} in Eqn.(1) is the total volumetric strain under isotropic stressing. According to Yin and Graham (1999), the rate of ε_{vm} can be expressed as

$$\dot{\varepsilon}_{vm} = \frac{\kappa}{Vp'_m}\dot{p}'_m + \frac{\psi}{Vt_0}\exp\left[\left(\varepsilon^r_{vm0} + \frac{\lambda}{V}\ln\frac{p'_m}{p'_{mo}} - \varepsilon_{vm}\right)\frac{V}{\psi}\right] \tag{3}$$

The total strain ε_{vm} in Eqn.(1) can be obtained by solving Eqn.(3) numerically. Eqns.(1) and (3) are differential equations of the 3-D EVP model for describing the time-dependent stress-strain behaviour of clays. Readers can refer to Yin & Graham (1999) for more details of the 3-D EVP model, the method for determination of mode parameters and verification of the model.

3 CONSOLIDATION MODELLING OF CLAYS UNDERNEATH TARSIUT ISLAND

Figure 1 shows a cross-section of Tarsiut island and foundation soils (Becker et al. 1985). The Tarsiut island was a man-built artificial island, located approximately 60 kilometres offshore with sea water depth approximately 21m. This artificial island consisted of four independent concrete box caissons set on an underwater sand berm. As shown in Figure 1, Zone 1 is soft to medium stiff marine clay and Zone 2 is stiff clayey silt. Below Zone 2 was Zone 3 of frozen soils. The sand berm construction started on July 5, 1981 and was completed on September 9, 1981. The island construction (caissons and sand core fill) was completed on October 21, 1981, following the sand berm construction.

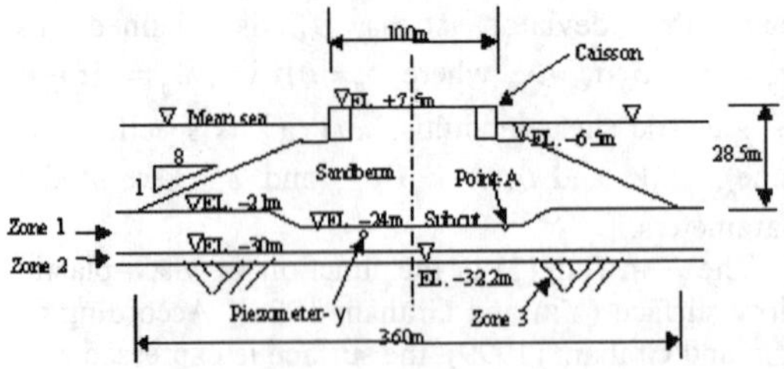

Figure1. Geometry of a caisson retained island – Tarsiut island (after Yin and Zhu 1999)

A piezometer was installed in the clay (at the point of piezometer-Y, see Figure 1, in Zone 1), 2m below the base (EL.-24m) of the sand fill and approximately 28m from the central symmetric line, at the time of two months after completion of the island construction. Measured data of excess porewater pressure at the point of piezometer-Y were reported by Becker et al. (1985), as shown in Figure 2. The porewater pressure increased for about 4 months (120days) following the piezometer installation, and reached a peak of 190kPa. It is noted that after the island construction was completed, the vertical load on the foundation soils was constant. This increase in porewater pressure under constant vertical loading could not be explained by traditional analysis. Is this caused by the Mandel-Cryer effects? Conlin et al. (1985) established a finite element (FE) model for fully coupled consolidation simulation of the porewater pressure and deformation behaviour of the foundation clay at Tarsiut island using the Modified Cam-Clay constitutive relationship. However, the FE consolidation model failed to simulate the phenomenon of porewater pressure increase. This implies that the Mandel-Cryer effects were not the cause of the porewater pressure increase in the clay underneath the Tarsiut island. Conlin et al. (1985) believed that creep might be the cause. However, their FE consolidation model could not consider the creep nature of soils.

The 3-D EVP model (Yin and Graham 1999) in Eqns.(1), (2) and (3) was incorporated into a finite element consolidation program based on fully coupled Biot's consolidation theory. Using the FE program, an axisymmetric finite element model, with1367 four-node elements and 1462 nodes is established for modelling the consolidation behaviour of the foundation clays underneath Tarsiut island.

In the FE model both the right and left boundaries are undrained hydraulically and constrained in horizontal direction but free to move in vertical direction. The bottom boundary is fixed in both vertical and horizontal directions, and is assumed to be undrained (below the boundary were frozen soils). The top boundary, i.e., the surface of Zone 1, is freely drained (overlying Zone 1 was sand fill).

The buoyant unit weight of foundation soils and sand berm was 8 kN/m^3 and 8.1 kN/m^3, respectively. The average unit weight of concrete caissons with sand core (above sea level) was 18.2 kN/m^3. The behaviour of sand fill and clayey silt in Zone 2 are assumed to be linear elastic. For the sand (loose), Young's modulus and Poisson's ratio are E=2000kPa and ν=0.3. The sand is assumed to be a completely free drainage material. For the clayey silt in Zone 2, E=20000 kPa, ν=0.32 and permeability k=1.7×10^{-9}m/s. These values are the same as those used by Conlin et al. (1985).

The behaviour of the clay in Zone 1 is considered to be elastic visco-plastic and described by Yin and Graham's (1999) EVP model. The EVP model parameters for the clay in Zone 1 (the clays in Zone 1 were divided into six sub-layers) are listed in Table 1. Approaches for determining the model parameters can be found in Yin and Graham (1999). The values of parameters e_o, K_o, ϕ', ν for the six sub-layers are taken from Conlin et al. (1985). The strength line slope M is calculated by $M=6\sin\phi'/(3-\sin\phi')$. The two parameters κ/V and λ/V are calculated using the same values of κ, λ and $V=1+e_o$ from Conlin et al. (1985). The elastic bulk modulus is equal to $K=p'/(\kappa/V)$. The shear modulus G is calculated as $G=1.5K(1-2\nu)/(1+\nu)$ using K and ν in Table 1.

Table 1: EVP model parameters for six sub-layers of clay in Zone 1

Soil layers Elevation in metre	e_o	K_o	κ/V	λ/V	p'_{mo} kPa	ε^r_{vm0}	ψ/V	t_o day	φ'	M	ν
EL. –21 (or EL–24 in subcut) to EL. –25	1.01	0.7	.0065	0.060	1	0	.006	1	29	1.16	0.45
EL. –25–EL. –26	0.95	0.7	.0056	0.058	1	0	.006	1	29	1.16	0.45
EL. –26–EL. –27	0.90	0.7	.0047	0.052	1	0	.005	1	29	1.16	0.45
EL. –27–EL. –28	0.85	0.7	.0038	0.047	1	0	.005	1	29	1.16	0.45
EL. –28–EL. –29	0.79	0.7	.0028	0.044	1	0	.005	1	29	1.16	0.45
EL. –29–EL. –30	0.74	0.7	.0017	0.037	1	0	.005	1	29	1.16	0.45

The hydraulic permeability k of clays in Zone 1 is a key parameter for consolidation modelling. Becker et al. (1985) reported that the k value in vertical direction for clays in Zone 1 was $k_v=1\times10^{-10}$m/s. In the FE analysis of the Tarsiut island, Conlin et al. (1985) used a k value of 3×10^{-10}m/s for top two sub-layers and 5×10^{-10}m/s for the following four sub-layers. Conlin et al. (1985) noted that the permeability k might not be a constant.

For the clay in Zone 1, no test data are available about possible dependence of permeability on void ratio or effective stress. The authors tried a range of constant k values from 2.0×10^{-9} m/s to 0.5×10^{-10} m/s in the consolidation analysis. The FE results show, in general, that when the k value is smaller than 7.0×10^{-10} m/s, significant increase in porewater pressure at piezometer-Y is found. However, the computed curve of porewater pressure vs. time is not in good match with the measured results. The authors tried to vary gradually the k value from 2.0×10^{-9} m/s to 0.5×10^{-10} m/s. The computed curve is in good agreement with the measured data as shown in Figure 2. The k value range of 2.0×10^{-9} m/s to 0.5×10^{-10} m/s obtained in this way (by matching to the measured results) is considered to be the best-estimated values from this simple back-analysis. The geometric average value of the k values is $\sqrt{2.0\times10^{-9}\times0.5\times10^{-10}}$ = 3.16×10^{-10}m/s which is closed to the k value used by Conlin et al. (1985).

It is seen in Figure 2 that the computed porewater pressure values are in good agreement with the measured results. The most significant finding, however, is not the good agreement but the possibility and capability of simulating the porewater pressure rising from the proposed FE consolidation analysis using Yin and Graham's EVP model (1999).

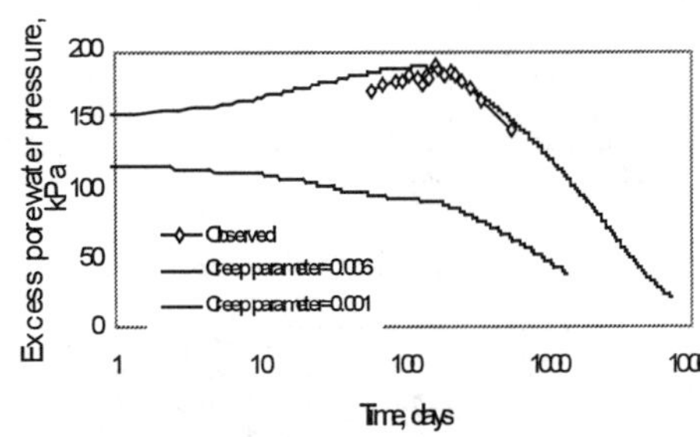

Figure 2. Comparison of observed and simulated porewater pressures at piezometer-Y (after Yin and Zhu 1999)

For comparison, the creep parameter ψ/V value is reduced from 0.006 to 0.001 with all other parameter values unchanged. The computed curve of porewater pressure vs. time using ψ/V=0.001 has no peak. This implies that the creep behaviour of the soil has significant effects on the porewater pressure increasing.

4 PARAMETRIC STUDY OF A STRIP FOOTINH ON CLAY

Typical parameters used in the FE modeling are h=8 m, L=40 m, a=8 m, k=6.15×10^{-10} m/s (see Fig.3). The EVP model parameters are listed in Table 1. In the following modeling studies, the same values of these parameters are used except where different

Table 1. EVP model parameters of Hong Kong (HKMD) and the typical value range for

	κ/V	λ/V	t_0 (day	ψ/V	p'_{mo} (kPa	ε^r_{vmo}	M	ν
HKM:	0.01	0.079	1	.002	15.2	0	1.26	0.4
Relation:	$\frac{C_r}{\ln(10)V}$	$\frac{C_c}{\ln(10)V}$	24 hrs	$\frac{C_{\alpha e}}{\ln(10)V}$	$p'_{c,24}$	$\varepsilon_{c,24}$	$\frac{6\sin\varphi'}{3-\sin\phi'}$	0.2 to 0.45

values are indicated.

Due to the page limitation, only presented in this paper are typical deformation and porewater pressure at three points, that is, points A (x=0, z=h/2), B (x=0, z=0), and C (x=a, z=0) as shown in Fig.3. The results of the FE analysis are shown in Fig.3 for different values of the creep parameter.

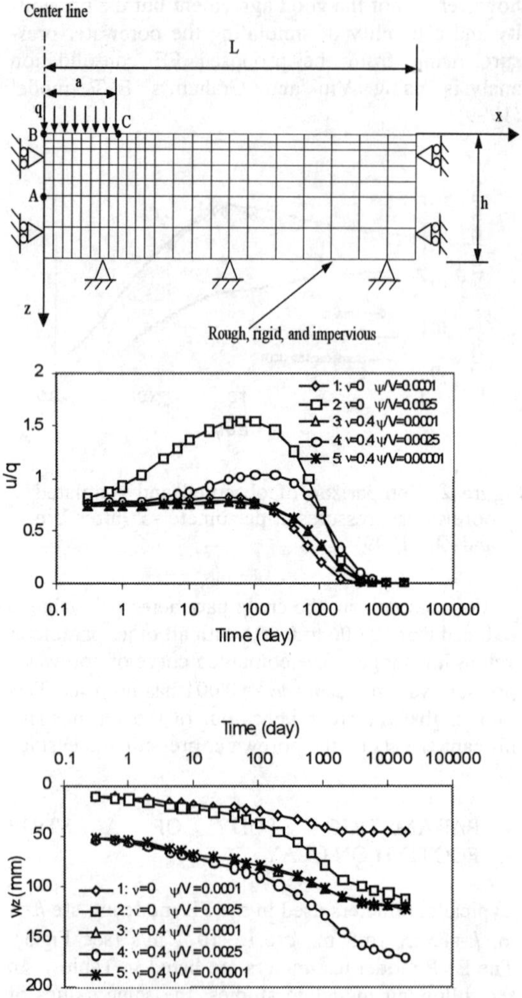

Figure 3. FE mesh (top), porewater pressure (middle) and settlement (bottom) for different creep parameter ψ/V

5 CONCLUSIONS

From the study in this paper, the following conclusions may be drawn.

(a) The proposed finite element consolidation model with Yin and Graham's (1999) EVP constitutive relationship can simulate reasonably well the consolidation behaviour of the clay underneath Tarsiut island including the rising of excess porewater pressure.

(b) The porewater pressure increase at Tarsiut island is not due to the Mandel-Cryer effects, but mainly due to the creep compression nature of the clay.

(c) To model the creep induced porewater pressure, a constitutive relationship describing well the time-dependent stress-strain (or creep) behaviour of the soil must be used.

(d) It is found in this case study that the increase in excess pore water pressure in the middle of the clay layer underneath embankment is larger than the total stress increment at the initial stage of loading. This phenomenon is due to the creep nature of the clay and is simulated in this paper for a layered soil profile.

ACKNOWLEDGEMENT

Financial supports from RGC grant (PolyU 5065/97E) of the University Grants Council of Hong Kong SAR Government are acknowledged

REFERENCES

Becker, D.E., Jefferies, M.G., Shinder, S.B. and Crooks, J.H.A. (1985). "Porewater pressure in clays below caisson islands". ASCE Arctic 85 Conf., San Francisco, March 1985, pp. 75-83.

Conlin, B.H., Jefferies, M.G. and Maddock, W.P. (1985). "An assessment of the behaviour of foundation clay at Tarsiut N-44 caisson retained island". Proceedings of 17th Offshore Technology Conference, Houston, Texas, USA, May 6-9, 1985, pp. 379-387.

Yin, J.-H. and Graham, J. (1989). "Visco-elastic-plastic modelling of one-dimensional time-dependent behaviour of clays". *Canadian Geotechnical Journal*, Vol. 26, pp. 199-209.

Yin, J.-H. and Graham, J. (1994). "Equivalent times and one-dimensional elastic visco-plastic modelling of time-dependent stress-strain behaviour of clays". *Canadian Geotechnical Journal*, Vol. 31,pp. 42-52.

Yin, J.-H. and Graham, J. (1999). "Elastic visco-plastic modelling of the time-dependent stress-strain behaviour of soils". *Canadian Geotechnical Journal*, Vol.36, No.4, pp.736-745.

Yin, J.-H. and Zhu, J.G. (1999). Elastic visco-plastic consolidation modelling and interpretation of porewater pressure responses in clay underneath Tarsiut island. Canadian Geotechnical Journal, Vol.36, No.4, pp.708-717.

Computer Methods and Advances in Geomechanics, Desai et al. (eds) © 2001 Balkema, Rotterdam, ISBN 90 5809 183 X

An elasto-plastic model for soft clay under cyclic loading

Jian Zhou, Maosong Huang, Shiming Wu & Yong Chi
Department of Geotechnical Engineering, Tongji University, Shanghai, People's Republic of China

ABSTRACT: A generalized elastoplastic model for Shanghai soft clay is presented, where an existing elasto-plastic bounding surface model proposed by Dafalias & Herrmann (1982) is modified in order to capture the elasto-plastic behavior of the soil mass during unloading. The model is numerically implemented and simulations of undrained unloading response for isotropically consolidated soils are provided. The model predictions compared favorably with experimental results.

1 INTRODUCTION

For monotonic loading, many currently used models perform satisfactorily for soft clays. In particular, the critical state model based on classical isotropic strain hardening plasticity theory is capable of dealing with drained, undrained, and normally consolidated conditions. The most obvious limitation of the critical state model is that it is not applicable to cyclic loading conditions. However, in practical engineering, repeated loading and unloading occur in the problems concerning earth excavation, earthquake response or in structures subjected to wave action. It may be argued that plasticity models based on isotropic plastic hardening rules is not adequate for situations in which repeated loading or unloading occurs. Here standard isotropic plastic hardening forms fail to give any permanent densification or hysteretic effects as observed in the laboratory tests. In order to take account of those behaviours, more elaborate plastic models based on the combination of isotropic and kinematic plastic hardening rules have been proposed (Provest 1977, Pietruszczak & Mroz 1983). Those models can achieve good results but often at a price of complexity. A relatively simple approach to achieving the desired effect is through the use of bounding surface models. The first bounding surface model was introduced by Dafalias & Popov (1975), and Dafalias & Herrmann (1982). Here classical plasticity is used to describe the bounding surface and for all other points inside the bounding surface the loading and unloading direction as well as moduli are defined directly by a simple interpolation rule from values at the bounding surface. It is noted that no plastic deformation on unloading was considered in Dafalias and Herrmann's model. However, experimental studies have shown that plastic deformation occurs during unloading. In this paper, we will introduce a relatively simple modification of Dafalias and Herrmann's model which accounts reasonably well for the phenomena observed under unloading tests for Shanghai soft clay.

2 BOUNDARY SURFACE EQUATION AND ITS EVOLUTION

As shown in Figure 1, an elliptical boundary surface is selected in the $p'-q$ stress space, intersecting the positive p'-axis at point A(p_c,0) and the projection of the critical state line in the p-q space at point B. The projection of point B on the p'-axis is located at C(α_c,0). Introducing a parameter R defining the eccentricity of ellipse, i.e. $R=\alpha_c/p_c$, we can write for the bounding surface

$$F=\left(\frac{\bar{p}'}{p_c}\right)^2+(R-1)^2\cdot\left(\frac{\bar{q}}{M\cdot p_c}\right)^2-\frac{2}{R}\cdot\frac{\bar{p}'}{p_c}+\frac{2-R}{R}=0 \quad (1)$$

where M is the slope of critical state line in the $p'-q$ space with $M=M_c$ for $q>0$ and $M=M_e$ for $q<0$.

Assuming that only plastic volumetric strain ε_v^p controls hardening, dependence of the boundary surface $F = 0$ upon ε_v^p is realized through the relationship between p_c and where p_c reflects the

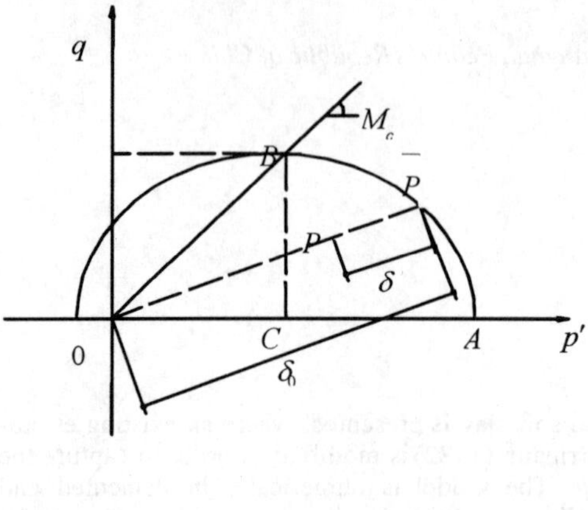

Figure 1. Boundary surface and the radial mapping rule

pre-consolidation history of the material and is used to define the hardening behavior. The hardening function in rate form is given by

$$\dot{p}_c = \frac{1+e_0}{\lambda - k} p_c \dot{\varepsilon}_v^p \qquad (2)$$

where e_0 is the initial void ration of the soil mass. The strain hardening law uses the consolidation parameter λ and k obtained by isotropic loading and unloading tests of normally consolidated soil. The parameter λ is the slope of *e-ln p′* during loading, while k is the initial value of the slope during rebound.

Boundary surface encloses all stress states experienced by the soil. For a given stress point $P(p',q)$ inside the bounding surface, a corresponding point $\overline{P}(\overline{p}',\overline{q})$ on the bounding surface can be found through a simple radial mapping rule from the origin. Therefore, the position of the point P at $p'-q$ stress space can be expressed as

$$\overline{p}' = \gamma p_c \qquad \overline{q} = \gamma\eta p_c \qquad (3)$$

where $\eta = \overline{q}/\overline{p}' = q/p'$ and

$$\gamma = \frac{1+(R-1)\sqrt{1+R(R-2)(\eta/M)^2}}{R[1+(\eta/M)^2+R(R-2)(\eta/M)^2]} \qquad (4)$$

Representing the unit normal vector at the image point $\overline{P}(\overline{p}',\overline{q})$ on the boundary surface by $\boldsymbol{n}$, the components of $\boldsymbol{n}$ in the directions of p' and q are

$$n_p = \frac{1}{g}(\gamma - \frac{1}{R}) \qquad n_q = \frac{1}{g}\cdot\eta\cdot\gamma\cdot\left(\frac{R-1}{M}\right)^2 \qquad (5)$$

where $g = \sqrt{\left(\gamma - \frac{1}{R}\right)^2 + \eta^2\gamma^2\left(\frac{R-1}{M}\right)^4}$.

3 FORMULATIONS OF ELASTO-PLASTIC DEFORMATION ON UNLOADING AND RELOADING

Consider a rate-independent elastoplastic medium. The constitutive relation can be specified at $p'-q$ stress space as

$$\dot{p}' = K(\dot{\varepsilon}_v - \dot{\varepsilon}_v^p) \qquad \dot{q} = 3G(\dot{\varepsilon}_s - \dot{\varepsilon}_s^p) \qquad (6)$$

where K is the elastic bulk modulus and G is the elastic shear modulus. Determination of loading and unloading, which is differentiated by the sign of the loading function L, is same as that in Dafalias and Herrmann's model, and L is defined as

$$L = \frac{1}{H}(\dot{p}'n_p + \dot{q}n_q) = \frac{1}{\overline{H}}(\dot{\overline{p}}'n_p + \dot{\overline{q}}n_q) \qquad (7)$$

In the above equation, H is the loading plasticity modulus and $\overline{H}$ is the "boundary" plasticity modulus. Representing the plasticity moduli for loading and unloading by H_L and H_U respectively, the plastic strains of the soil mass can be expressed by

$$\dot{\varepsilon}_{vL/U}^p = \frac{1}{H_{L/U}}(\dot{p}'n_p + \dot{q}n_q)n_{pL/U} \qquad (8a)$$

$$\dot{\varepsilon}_{sL/U}^p = \frac{1}{H_{L/U}}(\dot{p}'n_p + \dot{q}n_q)n_{qL/U} \qquad (8b)$$

in which ε_{vL}^p and ε_{sL}^p are the plastic volumetric and deviatoric strains for loading, ε_{vU}^p and ε_{sU}^p are the plastic volumetric and plastic strains for unloading, and (n_{pL}, n_{qL}) and (n_{pU}, n_{qU}) are the unit vectors in the direction of plastic flow for loading and unloading respectively. When L=0, load changing is neutral and purely elastic deformations occur.

Employing the associative flow rule, the direction of plastic flow during virgin loading and reloading is therefore identical to the outward normal direction of the boundary surface, that is

$$n_{pL} = n_p \qquad n_{qL} = n_q \tag{9}$$

Experimental results from the undrained cyclic loading tests for the normal consolidated clay show that the pore water pressure as well as the effective stress vary continuously during loading and unloading. This indicates that plastic volumetric strain is still developing on unloading, behaving as a volumetric contraction or expansion under drained conditions. Figure 2 shows the effective stress path of soft clay under undrained cyclic loading. It can be founded that there exists a "phase-transformation line" with the slope *m* during unloading. Unloading which occurs over this line leads to volumetric expansion instead of contraction.

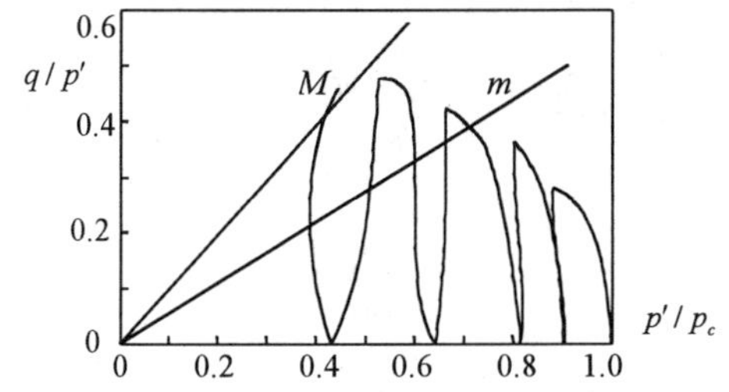

Figure 2. Effective stress path for undrained cyclic deviatoric loading (after Dafalias & Herrmann 1982)

Conventional triaxial CU tests for Shanghai soft clay carried out by the authors clearly indicate that the volume of the soil specimen was contracting during loading while the subsequent unloading still behaves volumetric contraction. It must be pointed out here that the phenomenon was observed from the tests which were carried out slowly enough to eliminate the effect of time lag. Volumetric contraction of a soil specimen in a CD test reflects increase in the pore water pressure in a CU test as shown in Figure 3.

Based on the above experimental results, it is reasonable to define the direction of plastic flow as

$$\boldsymbol{n}_U = (n_{pU}, n_{qU}) \tag{10}$$

where $n_{pU} = \begin{cases} -n_p & \left|\dfrac{\eta_u}{m}\right| < 1 \\ |n_p| & \left|\dfrac{\eta_u}{m}\right| \geq 1 \end{cases}$ and

$n_{qU} = -n_q$.

Now all that remains is to define the plastic moduli for reloading and unloading. From the consistency condition $\dot{F} = 0$ and equations (1), (2), (5), (7) and (8), the "boundary" plastic modulus $\overline{H}$ can be written in the following

$$\overline{H} = \frac{1+e_0}{\lambda - k} \cdot \frac{p_c}{R} \cdot \frac{1}{g^2}\left(\gamma - \frac{1}{R}\right)(\gamma + R - 2) \tag{11}$$

The uniqueness of the bounding surface plasticity theory lies on that the plastic modulus depends not only on the evolution of the boundary surface, but also is a function of the distance δ between the current stress point and its "image" point on the bounding surface. Many variants regarding the relationship between the plastic modulus for reloading H_L and the "boundary" plastic modulus $\overline{H}$ exist in the literature. The simplest interpolation is of the following form

$$H_L = \overline{H} + p_a h\left[1 + |M/\eta|^{0.2}\right] \frac{\delta}{\delta_0 - \delta} \tag{12}$$

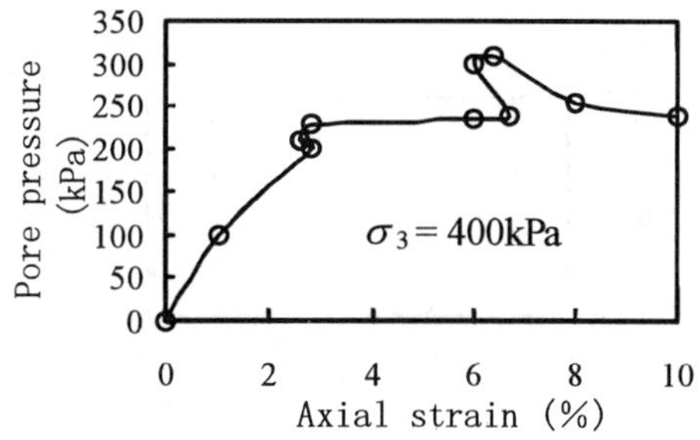

Figure 3. Pore water pressure curve of CU test for Shanghai soft clay

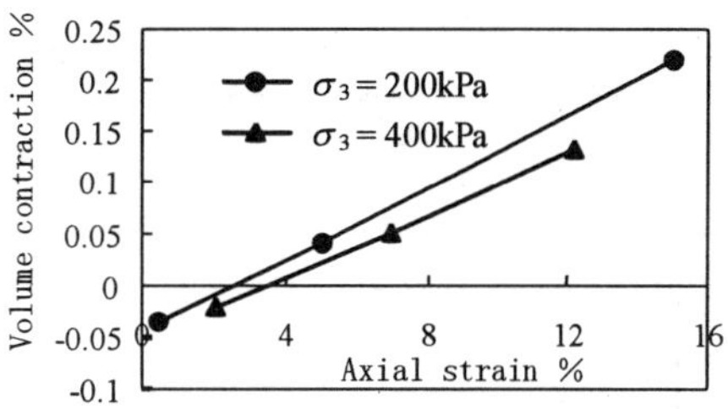

Figure 4. Volumetric contraction of clay during unloading (after Jiao & Chen 1994)

According to experimental studies reported in Jiao & Chen (1994), there are four features concerning the volumetric contraction of the clay during unloading:

(1) The volumetric contraction is directly proportional to the magnitude of axial strain at which the unloading take places.
(2) The volumetric contraction increases along with the decrease of the confining pressure as shown in Figure 4.
(3) The volumetric contraction increases along with the increase of the effective strength index.
(4) The volumetric contraction increases along with the increase of σ_1/σ_3. If σ_1/σ_3 equals to 1, then the volume expands during unloading.

All the four features are related to the magnitude of σ_1/σ_3.

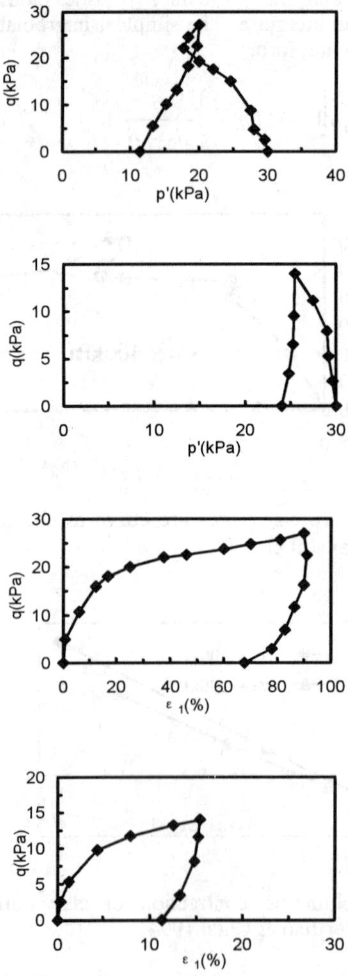

Figure 5. Deformation characteristics of soft clay during loading-unloading

Summarizing the above factors, one can find that, along with the increase of the stress ratio at which unloading takes place, the transition from dilatancy to compaction occurs and the volumetric compaction increases beyond the transition. Figure 5 shows two sets of experimental results from undrained unloading tests for Shanghai soft clay.

It can be seen that unloading from a higher stress ratio results in higher plastic volumetric strain. Indeed unloading can be considered as an active loading process and thus the unloading plastic modulus should also be a function of the distance between the current stress point and the "image" point on the bounding surface. Assuming a point yield surface for unloading (Zienkiewicz, Leung & Pastor 1985), the unloading plastic modulus can be expressed simply by

$$H_U = H_{u0} \cdot \frac{\delta_0 - \delta}{\delta} \cdot \left(\frac{|\eta_u - M|}{M} \right)^{-r_u} \qquad \left| \frac{\eta_u}{M} \right| < 1 \qquad (13)$$

$$H_U = H_{u0} \qquad \left| \frac{\eta_u}{M} \right| \geq 1 \qquad (14)$$

where η_u is the stress ratio from which unloading takes place and is often referred to as the unloading stress ratio

$$\eta_u = \left(\frac{q}{p'} \right)_u \qquad (15)$$

and H_{u0} and γ_u being appropriate constants. There are three additional parameters introduced for describing the deformation characteristic of soft clay during unloading and they can be determined from unloading tests under different stress levels. It must be pointed out that the bounding surface is no more kept unchanged during unloading due to the plastic deformation.

4 VERIFICATION TESTS

In order to verify the performance of the proposed model, several undrained triaxial tests were carried out on Shanghai soft clay. All the tests were conducted on HX-100 servo controlled tri-axial apparatus under undrained conditions. The specimen were initially consolidated isotropically and loading-unloading-reloading was simulated by increasing and decreasing the pressure in the axial direction without altering the later confining pressure. The tests can be divided into two groups:

(1) Unloading tests. – The specimen were loaded in compression to an axial strain range of 1-4% and unloaded by reducing the axial pressure until complete collapse occurred.

(2) Reloading tests. – The specimen were compressed to an axial strain range of 1–2% and unloaded to the initial isotropic state; then reloaded by raising the pressure in the axial direction until the specimen ruptured.

The model parameters involved in the numerical simulations are calibrated and listed in Table 1. The numerical predictions shown in Figures 6 and 7 along with the measured data exhibit a reasonably good agreement with the experimental results. The capability of the proposed model in reproducing well the hysteresis associated with cycling is also demonstrated. In addition, a simulation of the test data shown in Figure 5 is presented in Figure 8. It can be seen that the numerical prediction agrees favorably with the experimental results regarding the influence of the unloading stress ratio. In fact, if the unloading occurred at a lower stress ratio, the soil mass would have behaved almost elastic, which reflects a very minor build-up of the pore water pressure.

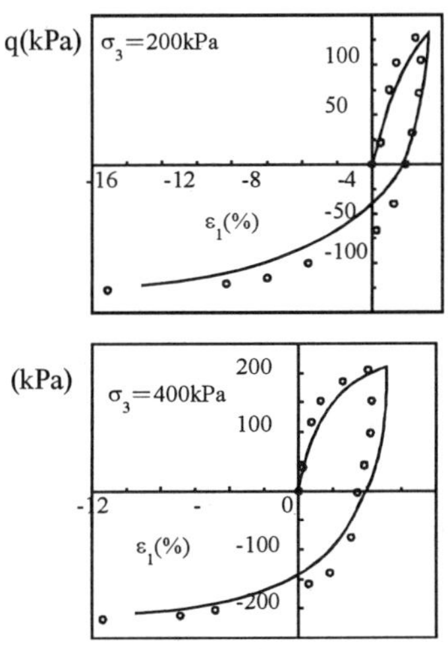

Figure 6. Deformation characteristics of Shanghai soft clay during unloading

Table 1. Model parameters used

R	h_e	h_c	k	λ	M_c	M_e	m	H_{u0}	γ_u
2.75	14	18	0.06	0.14	0.95	0.80	0.74	1250	4

5 CONCLUSIONS

In this paper, an elasto-plastic bounding surface model for Shanghai soft clay has been formulated by introducing plastic effect on unloading into the model initially developed by Dafalias and Herrmann (1982). The loading direction vectors, direction vectors of plastic flow and plastic moduli corresponding to loading, unloading and reloading can be summarized as follows:

(1) For a loading path: $(n_{pL}, n_{qL}) = (n_p, n_q)$, $H = \overline{H}$

(2) For an unloading path:

$$n_{pU} = \begin{cases} -n_p & \left|\dfrac{\eta_u}{m}\right| < 1 \\ |n_p| & \left|\dfrac{\eta_u}{m}\right| \geq 1 \end{cases}, \quad n_{qU} = -n_q, \quad H = H_U.$$

(3) For a reloading path: $(n_{pL}, n_{qL}) = (n_p, n_q)$,

$$H = \overline{H} + \hat{H}\frac{\delta}{\delta_0 - \delta}$$

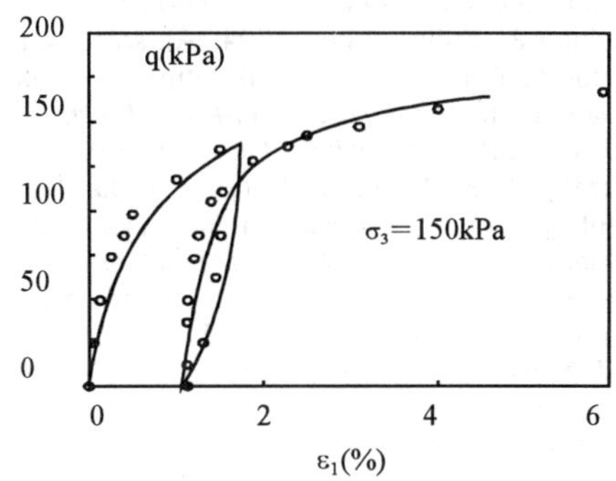

Figure 7. Deformation characteristics of Shanghai soft clay during unloading-reloading

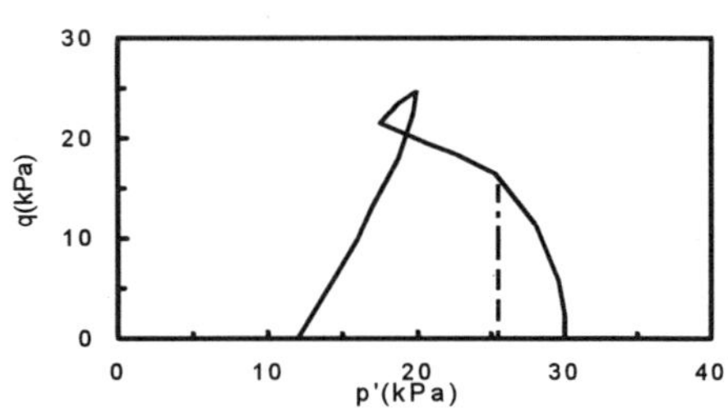

Figure 8. Effective stress path for Shanghai soft clay during loading-unloading

This is important acknowledging the fact that only three more parameters are required for elasto-plastic formulation on unloading. This is certainly a small number and they can be easily determined unloading tests under different stress levels. The model is numerically implemented and its validity has been examined by simulating the undrained unloading tests for Shanghai soft clay.

Despite the simplicity of the presented constitutive model, it can reproduce behavior of normally consolidated and lightly overconsolidated clays under certain loading conditions. Further studies concerning heavily overconsolidated clays and memory of past stress history are required.

REFERENCES

Dafalias, Y. F. & Herrmann, L. R. 1982. Bounding surface formulation of soil plasticity. In G N Pande and O C Zienkiewicz (eds), *Soil Mechanics -Transient and Cyclic Loads*, Wiley , New York , 253-282.

Dafalias, Y. F. & Popov, E. P. 1975. A model of non-linearly hardening materials for complex loadings. *Acta Mechanica*, **21**, 173-192

Dequan Jiao & Yujing Chen 1994. Anisotropy and volume-contraction of soil due to axial unloading in CD test. *Chinese Journal of Geotechnical Engineering* , **16**(4), 9—16.

Pietrusczak, S. & Mroz, Z. 1983. On hardening anisotropy of K_0 –consolidated clays. *Int J. Numer. and Analyt. Methods in Geomech.*, **7**, 19.

Prevost, J. H. 1977. Mathematical modeling of monotonic and cyclic undrained clay behavior. *Int J. Numer. and Analyt. Methods in Geomech.*, **1**, 195-216.

Zienkiewicz, O. C., Leung, K. H. & Pastor, M. 1985. Simple model for transient soil loading in earthquake analysis I. Basic model and its application. *Int J. Numer. and Analyt. Methods in Geomech.*, **9**, 453-476.

4 Constitutive models: Interfaces and joints

Computer Methods and Advances in Geomechanics, Desai et al. (eds) © 2001 Balkema, Rotterdam, ISBN 90 5809 183 X

Interface elements in slope stability analysis

J.M.Dłużewski
Warsaw University of Technology, Poland

K.Ciuhak
Institute of Meteorology and Water Management, Poland

ABSTRACT: In the paper the slope stability problem is approached by means of finite element method. In the case where the location of a sliding line is known the interface elements can be very helpful in the numerical analysis. Such situation takes place where the sliding line is initiated or in the post failure analysis. The idea came from the back analysis of the accident of the slope failure in Dychów, Poland. The slope was a part of the dam close to the water inlet. The moving mass of the soil broke the wall of the power plant building and damaged turbogenerators. The accident happened without any earlier sign of danger. It has been the largest geotechnical accident during latest years in Poland.

1 INTRODUCTION

1.1 *Interface elements*

In the finite element modeling the interface elements are widely used to describe the soil behavior in the thin zones which are created between stiff steel or concrete structure and soil. To model interaction between the structure and the soil various types of the interface elements are used. In the paper the zero thickness interface elements are applied (Goodman & Dubois, 1972). The elasto visco-plastic constitutive law is used at the discontinuity line.

1.2 *Slope stability analysis*

The finite element method is very helpful in the analysis of the slope stability problems. Most often the c-φ reduction procedure is used to calculate the safety factor and the first damaged mechanism. The safety factor is equal to

$$F = \frac{c}{c^F} = \frac{tg\varphi}{tg\varphi^F} \tag{1}$$

where c and φ are the cohesion and friction angle respectively and c^F and φ^F are the cohesion and friction angle when the failure occurs.

The c-φ reduction procedure can be also efficiently used in back analysis of the accidents of the slope stability failures. In such cases the location of the sliding line are known. The numerical calculations can be done applying the finite element mesh with the interface elements inserted along the initiated sliding line. The zero thickness interface elements can be used. Next the slope stability analysis is performed by means of the c-φ reduction procedure at the interfaces line. The solution is compared with the result obtained for the c-φ reduction procedure applied for the whole analyzed area.

2 BASIC EQUATIONS

2.1 *Virtual work principle*

The virtual work principle for the continuum with the displacement discontinuity surface can be written in the following form

$$\int_V \delta\boldsymbol{\varepsilon}^T \boldsymbol{\sigma} dV + \int_{S_C} \delta(\Delta\mathbf{u}_{s-n})^T \mathbf{t}_c dS_c = \int_V \mathbf{f}^T \delta\mathbf{u} dV + \int_S \mathbf{s}^T \delta\mathbf{u}\, dS \tag{2}$$

where σ is the stress tensor, $\delta\mathbf{u}$, $\delta\varepsilon$ are the virtual displacements and strains, $\mathbf{t}_c$ is the traction transmitted at the interface, **f**, **s**, are the body forces and the surface traction, $\delta(\Delta\mathbf{u}_{s-n})$ is the virtual displacement discontinuity. Large deformation effects are not considered here.

$$\mathbf{t}_c = \begin{bmatrix} \tau \\ \sigma_n \end{bmatrix}, \quad \Delta\mathbf{u}_{s-n} = \begin{bmatrix} \Delta\mathbf{u}_s \\ \Delta\mathbf{u}_n \end{bmatrix} = \begin{bmatrix} u_s^l - u_s^r \\ u_n^l - u_n^r \end{bmatrix} \tag{3}$$

2.2 *Constitutive equations*

The Coulomb-Mohr yield criterion is accepted for the interface element

$$f(\mathbf{t}_c)=|\tau|+\sigma_n \tan\varphi - c, \tag{4}$$

where φ and c are the friction angle and cohesion, respectively. The rate of plastic slip inside the interface depends on the plastic potential

$$g(\mathbf{t}_n)=|\tau|+\sigma_n \tan\psi \tag{5}$$

where ψ is the dilatation angle. The non-associated flow rule is used $\psi = 0$.

The constitutive equations for the interface can defined by the viscoplastic flow rule. The relative rate of the viscoplastic displacements is defined

$$\Delta\dot{\mathbf{u}}^{vp}_{s-n}=\gamma < \phi(f) > \frac{\partial g}{\partial \mathbf{t}_c} \tag{6}$$

where g is the potential, f is the yield function, γ is a viscosity coefficient and $< \phi(f) >$ is a special function. The brackets < > stand for an on/off switch. If $f > 0$, then the switch is on, if $f < 0$, the switch is off. We assume that the function $\phi(f)$ takes one of the following forms

$$\phi(f)=f^n, \quad n \geq 1, \tag{7}$$

The choice of the function $\phi(f)$ has an influence on the rate of the viscoplastic displacements. The stresses at the discontinuity line are calculated as follows

$$\mathbf{t}_c=\mathbf{D}\ (\Delta\dot{\mathbf{u}}_{s-n} - \Delta\dot{\mathbf{u}}^{vp}_{s-n}) \tag{8}$$

where $\mathbf{D}$ is the elastic material array, $\Delta\mathbf{u}^{vp}_{s-n}$ is the increment of viscoplastic relative displacements. The array $\mathbf{D}$ for the elastic behaviour is taken without coupling terms between shearing and normal stresses.

$$\mathbf{D}=\begin{bmatrix} d_s & 0 \\ 0 & d_n \end{bmatrix}, \tag{9}$$

where in the elastic case d_s and d_n are the shear and bulk moduli, respectively.

The details of the interface formulation can be found in the paper (Dłużewski, 1997). The examples are solved by means of HYDRO-GEO program.

3 EXAMPLES

3.1 *Embankment*

The slope stability problem is analyzed, Fig.1. The material parameters are shown in the table 1. First the safety factor and the damage mechanism is obtained by means of the various strip methods. The results are collected, table 2. Next the finite element method and the classical c-φ reduction procedure is used to calculate the safety factor and to obtain the damage mechanism, Fig.2-4. To solve the elasto-plastic problem the pseudo-visco plastic iterative

Table.1 Material parameters

No	Material	E [MPa]	ν	γ [kN/m³]	φ [°]	c [kPa]
1	clay	20	0.3	19.0	25	15
2	sand	40	0.3	18.5	34	5
3	clay	80	0.3	19.5	25	40

γ - volumetric weight, ν - Poisson's ratio, E - Young modulus, φ - friction angle, c - cohesion

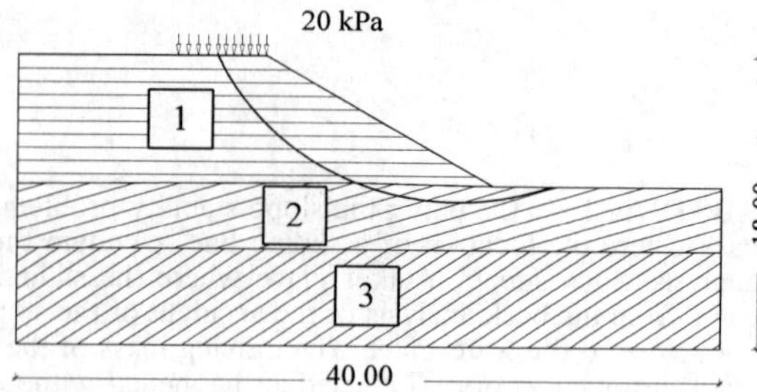

Figure 1. Slope stability problem

Table.2 Safety factors

No	Method	F	F FEM c-φ reduction
1	Fellenius - standard	1.513	
2	Fellenius -modified	1.513	
3	Bishop - standard	1.605	1.602
4	Bishop - modified	1.606	
5	Morgenstern-Price	1.597	

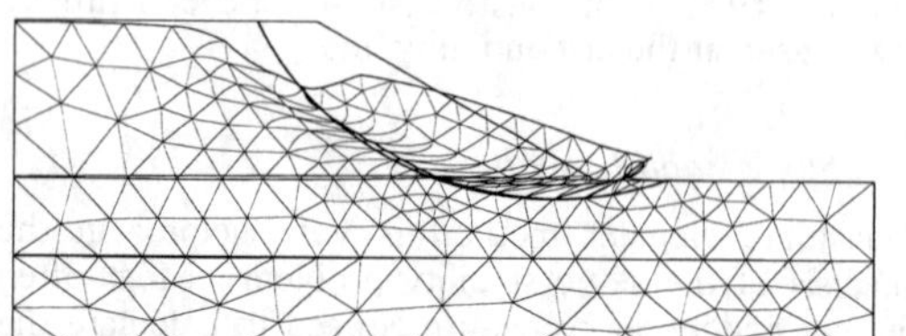

Figure 2. Damage mechanism

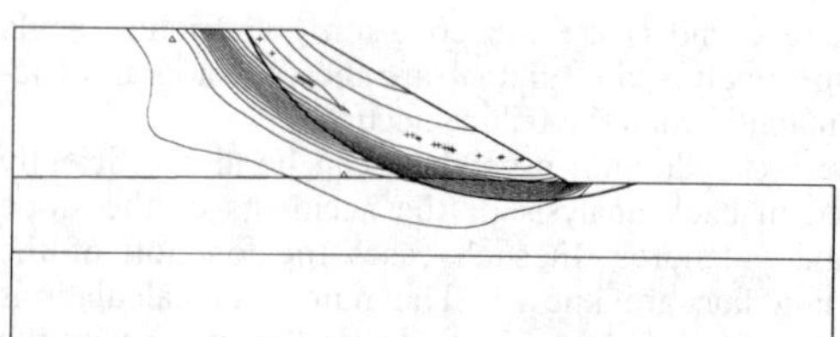

Figure 3. Izolines of the displacements

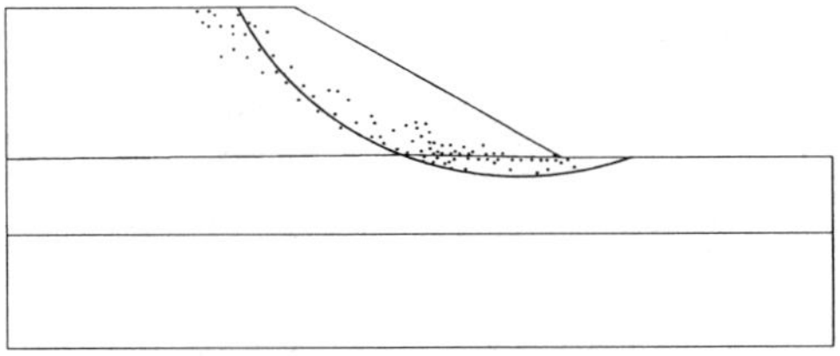

Figure 4. Plastic points

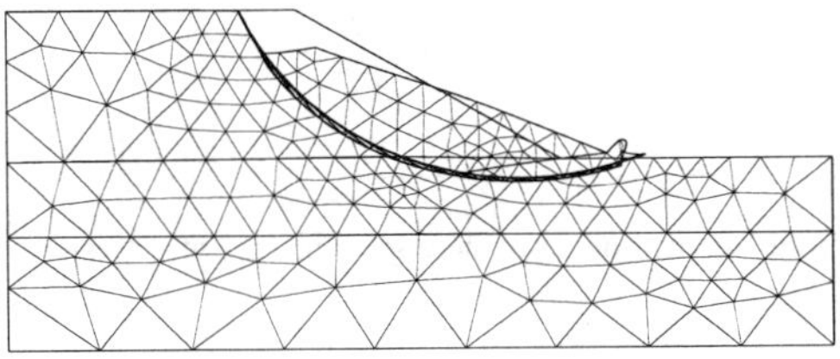

Figure 5. Damage mechanism for the c-φ reduction at the interfaces line

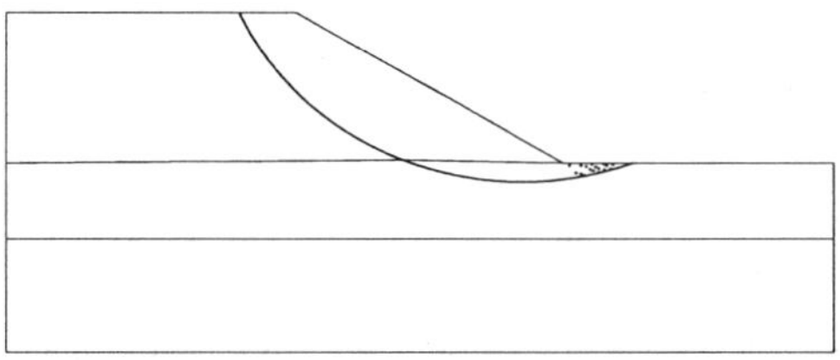

Figure 6. Plastic points for the c-φ reduction at the interfaces line

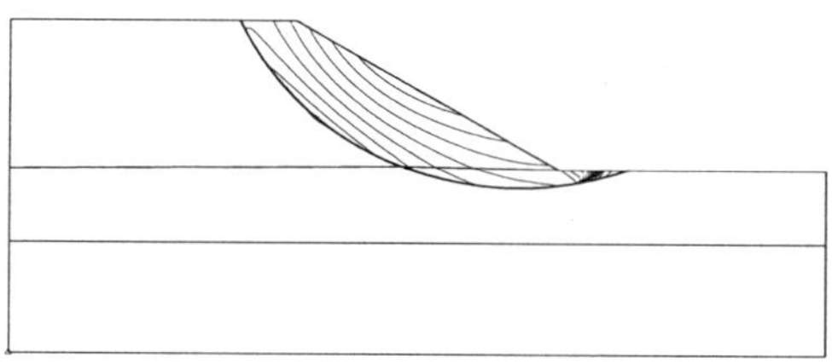

Figure 7. Izolines of the displacements located at the sliding line

approach is used. The c-φ reduction procedure is activated for all soils involved into analysis. The obtained damaged is different comparing the sliding line obtained form the strip methods. Next the interface elements are activated along the sliding line obtained form the Fellenius standard method. The c-φ reduction procedure is activated for the interface elements only. The obtained damaged mechanism is a bit different comparing the solution form the Fellenius method, Fig.5-7. Especially at the ground level close to the toe point.

3.2. *Slope failure in Dychów*

The slope failure at Dychów dam in Poland (Dłużewski et al. 1998). is studied herein. The accident happened in 1997. The slope failure caused the soil movement towards the nearby power plant. The road embankment and the downstream part of the dam lost their stability. The soil damaged the wall of the power plant making the great hole. The power plant with accompanying infrastructure (the water inlet and pipelines) were located in unfavorable geological conditions. The power plant is located at the valley of the river Bóbr. The valley is composed of two terraces. The upper terrace is composed of Quaternary formation (Holocene). Starting from the top, there are fine and medium sands, partly with inclusions of silt and gravel. Below, Tertiary formations are found: clay and silt as well as thin layers of brown coal. The excavated soil, during construction have been mixed and reused in construction of the embankment. The embankment has been risen from loose, cohesionless soils with interlayers of silt and construction's rubbish. The material paramerers are given in the table 3.

Table.3 Material parameters

No	Material	γ [kN/m^3]	ν	E [MPa]	φ [°]	c [kPa]
1	Concrete	24.0	0.167	32 400	-	-
2	Sheet pile walls	25.0	0.30	100 000	-	-
3	Embank-ments	19.0	0.30	60	32	0
4	Sand + gravel+stone	19.0	0.30	100	40	0
5	Grey mud	14.0	0.30	10	15	15
6	Clay + silt	19.0	0.30	40	30	17

γ - volumetric weight, ν - Poisson's ratio, E - Young modulus, φ - friction angle, c - cohesion

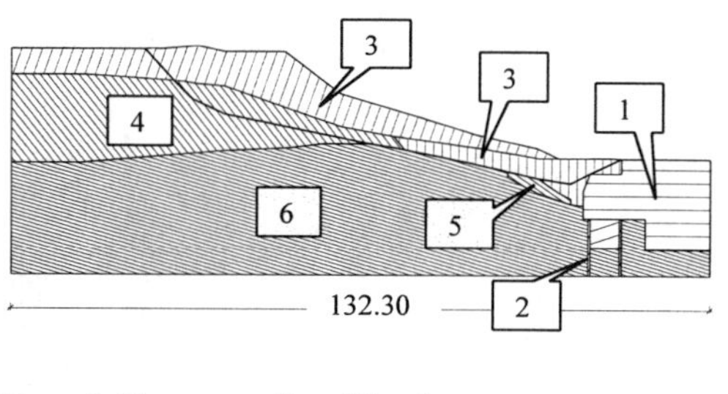

Figure 8. The cross section of the slope

To study the reasons of the accident several stability calculations are done. First the classical c-φ reduction procedure for all elements is calculated. The damaged mechanism in the form of the deformed mesh is shown in the Fig. 9.

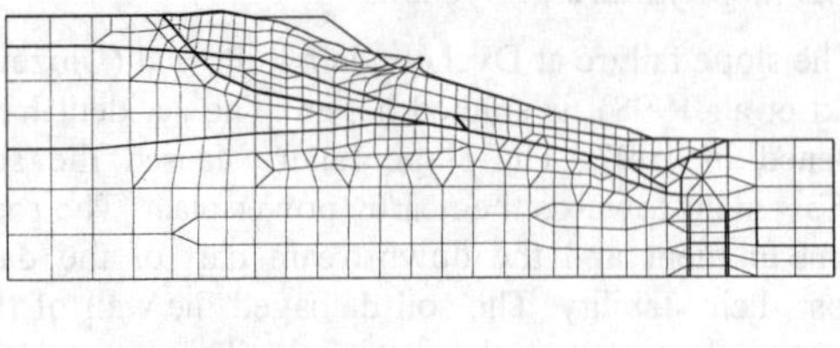

Figure 9. The damaged mechanism, F=1.305

The obtained damaged mechanism is different comparing the sliding line in situ. Next the c-φ reduction procedure is applied to the chosen zones where the suffosion (internal erosion) is supposed. The obtained damaged mechanism is much alike to the one observed in situ, Fig.10. The isolines of the displacements are shown in the Fig. 11 and the plastic points in the Fig.12. At the end the c-φ reduction procedure is applied to the interface line only. The obtained damaged mechanism is exactly the same as in site, Fig.13, Fig.14, Fig.15 but the safety factor is much higher comparing to the first one.

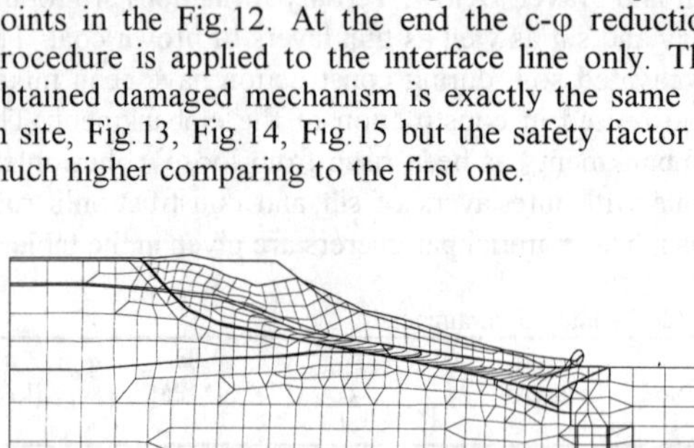

Figure 10. The damaged mechanism (internal erosion), F=1.728

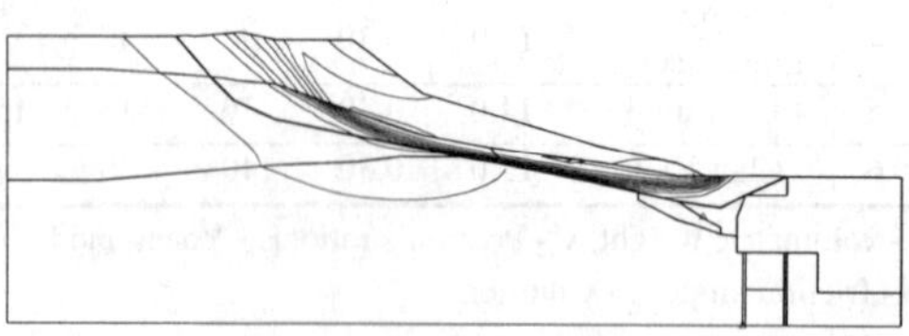

Figure 11. The isolines of the displacements (internal erosion)

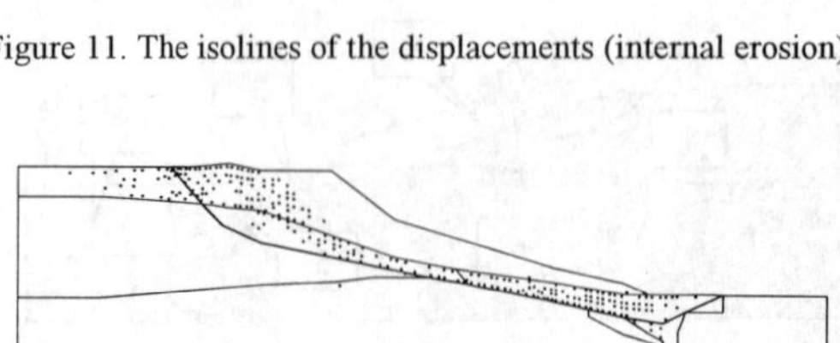

Figure 12. The plastic points (internal erosion)

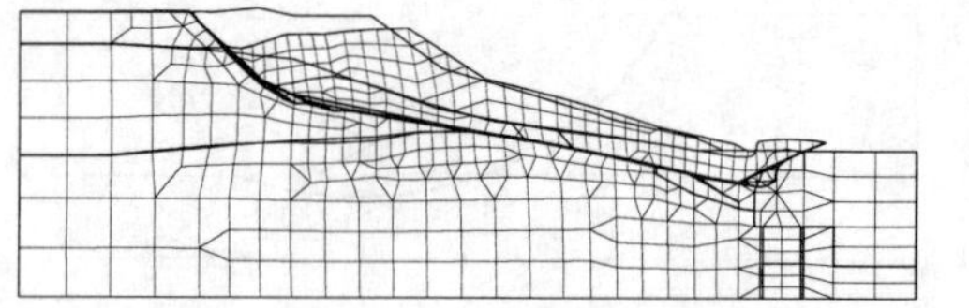

Figure 13. The damaged mechanism (interfaces), F=2.066

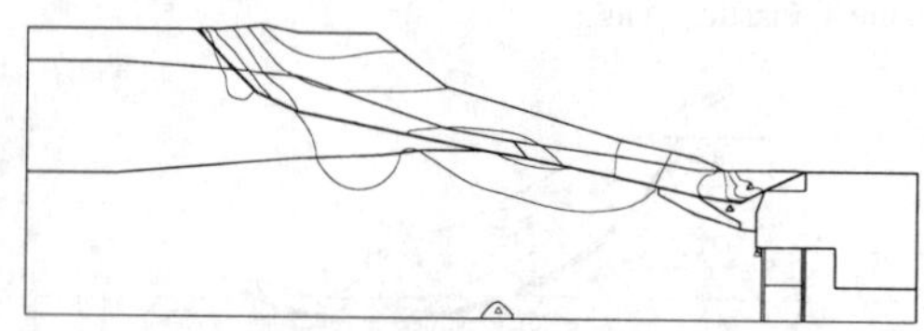

Figure 14. The isolines of the displacements (interfaces)

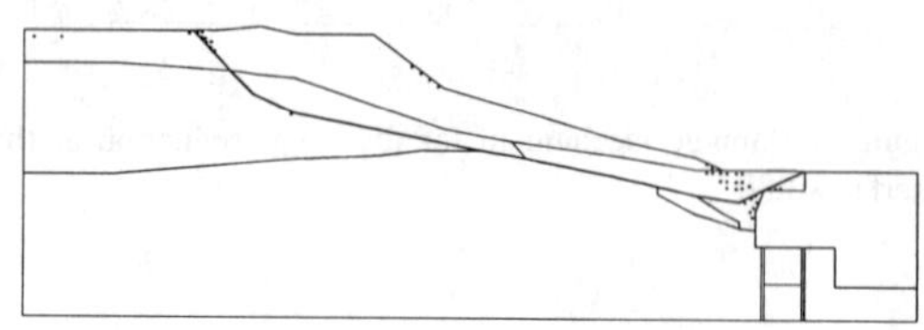

Figure 15. The plastic points (interfaces)

4. CONCLUSIONS

The interface element can be efficiently applied in the post failure – back analysis of the slope stability problems. Such formulation makes a very efficient tool for the engineers in searching for reasons of the accidents of soil structures failures. It also gives a possibility of reconstruction of an accident in the numerical model and study different elements which influence the slope failure.

REFERENCES

Dłużewski J. M. 1997 *HYDRO-GEO – finite element code for geotechnics, hydrotechnics and environmental engineering*, (in Polish), Warsaw University of Technology, Warsaw, Poland

Goodman R. E. & Dubois J. 1972 Duplication of dilatancy in analysis of jointed rocks. *Journal of the Soil Mechanics and Foundation Division, ASCE,98,SM4*

Dłużewski J. M. & Hrabowski W. 1998 Slope failure at Dychów dam - analytical and numerical study of the accident, *Proceedings of the International Symposium on New Trends and Guildelines on Dam Safety*, Balkema Barcelona Spain

Computer Methods and Advances in Geomechanics, Desai et al. (eds) © 2001 Balkema, Rotterdam, ISBN 90 5809 183 X

Some considerations on failure mechanism of rock joint

Masanobu Doi
Department of Civil Engineering, Takamatsu National College of Technology, Japan

Satoru Ohtsuka
Department of Civil and Environmental Engineering, Nagaoka University of Technology, Japan

ABSTRACT: Rock structure often includes various kinds of joints. The mechanical property of rock joint affects the stability of rock structure. In this study, authors employed the numerical procedure based on the lower bound theorem by introducing the contact condition along rock joint to throw new light upon the failure mechanism of rock joint. By simplifying the complex real geometry of rock joint, the effect of the two types triangular asperity shapes on joint strength was investigated. One has only the first order asperity, and the other has both the first order and the second order asperity. The results by the analytical procedure showed that the difference between the two types of asperity produced the mobilization of shear strength along rock joint.

1 INTRODUCTION

The existence of rock joint is well known to affect the stability of rock structure. In general, rock mass includes various kinds of joints, which widely range from micro to huge ones. The effects of rock joints on the stability of rock structure depend on their size, geometrical configuration, shear strength characteristics and so on. The design of rock stability should be taken these factors into consideration, especially the shear strength of rock joint has been investigated by many researchers(e.g. Patton(1966); Barton(1972); Ladanyi & Archambault(1970); Rengers(1970)). Shear strength of rock joints has been clarified experimentally to depend on the contact friction angle and the joint roughness.

Natural rock joint has very complicated shape. The shape of rock joint can be regarded as assemblage of various order asperities. It is commonly observed that the shear strength of rock joint is dependent on the normal stress along joint and is non-linear against that. The purpose of this study is to investigate analytically the shear strength properties of rock joint by considering the first order and the second order asperities and make clear the mechanism of the shear strength mobilization of rock joint.

The analytical procedure, which has been developed and applied to rock structure stability problem(e.g. Ohtsuka et. al., (1997); Ohtsuka & Doi, (1998)), is employed in this study. The method of this analysis is originally based on the lower bound theorem of the shakedown analysis(Koiter,1960). With the use of finite element discretization, the analytical method is cast into a simple linear programming problem as Maier(1969) proposed. By introducing the contact condition along the joint is well considered in the analysis of stability problem for the generation of plastic deformation in rock structure and/or both sliding and detachment along rock joint.

Focusing on the effects of asperity along rock joint, the resultant shear strength of joint is investigated for the two types of asperity shape. One has only the first order asperity, and the other has both the first order and the second order asperity. In this study, the plane strain compression test of rock specimen including a joint is simulated. The numerical simulation can consider the ideal condition of rock joint different from experiments.

2 OUTLINE OF ANALYTICAL PROCEDURE

2.1 *Lower Bound Theorem in Shakedown Analysis with Linear Programming Problem*

The analytical procedure employed in this study is based on the lower bound theorem in the shakedown analysis(Koiter,1960). The main feature of the analysis is to deal with the residual stress corresponding to the plastic deformation, which is defined as the difference between the true stress σ and the elastic stress σ^e such as

$$\sigma^r = \sigma - \sigma^e \tag{1}$$

σ and σ^e satisfy the equilibrium equation, respectively so that σ^r is self-equilibrate. The rock mass, which is an intact rock, is modeled as an elastic perfectly plastic material.

The lower bound theorem in plasticity assures that a rock mass is stable against the external $F(t)$ if any time independent residual stress σ^r can be found everywhere in the rock mass satisfying

$$f(\sigma^e(t)+\bar{\sigma}^r)<0 \tag{2}$$

where f is a yield function of rock mass. If a rock mass is stable for the applied load, the behavior of it is proven to shakedown to be elastic against any repeat of prescribed load.

The lower bound analysis is formulated as a maximization problem. When the external force is a monotonically increasing force, the shakedown analysis coincides with the limit analysis. With the use of linear yield function as

$$N^T\left(\rho\sigma^e+\bar{\sigma}^r+\sigma_0\right)\le K \tag{3}$$

the shakedown analysis can be formulated as a linear programming problem(Maier,1969). In the above equation, the finite element discretization of stress is conducted. The matrix N is assemble of the outward normal unit vectors n_i and K is the assemblage vector of threshold values k_i as $n_i\sigma\le k_i$. The relationship between n_i and k_i is illustrated in Fig.1. When the yield function is non-linear, it is linearized piece-wisely, as shown in this figure. σ_0 denotes the initial stress in the rock mass.

The analysis against the external force F can be formulated with a load factor ρ for F as follows:

$$s=\max\left\{\rho\ \middle|\ \begin{array}{c} N^T\left(\rho\sigma^e+\bar{\sigma}^r+\sigma_0\right)\le K \\ B^T\sigma^e=F \\ B^T\bar{\sigma}^r=0. \end{array}\right\} \tag{4}$$

where s indicates the ultimate load intensity. B is a matrix correlating the stress vector with the external vector. The second and third equations express the equilibrium equations on elastic and residual stresses, respectively. It should be noted that the redistribution of stress is considered with the residual stress, which is determined by solving the boundary value problem.

2.2 *Condition of Contact Along Rock Joint*

The traction q along joint has two components of the normal stress q_n and the shear stress q_s. As the normal stress, the extension in stress is usually not permitted as $q_n\le 0$ when the extension stress is defined as positive. Even if there is certain cohesion c_n along joint, the condition of contact on q_n is followed by $q_n\le c_n$. On the contrary, the condition of contact on the shear stress along joint can be defined by introducing shear models. In the case of the cohesion model, it is defined as $-c_s\le q_s\le c_s$ where c_s is the cohesion along joint. In the same way, the condition in the friction and cohesion model is described by $q_n\tan\phi_\mu-c_s\le q_s\le -q_n\tan\phi_\mu+c_s$, where ϕ_μ is the angle of friction along joint.

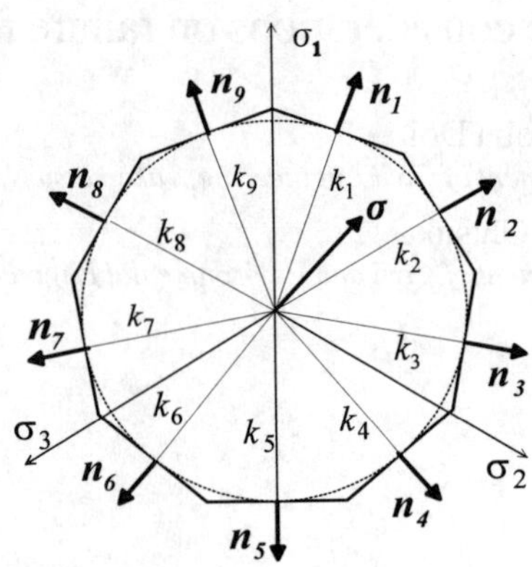

Figure 1. Piecewise linear yield fuction.

These conditions of contact are expressed in the following equation:

$$\begin{bmatrix} 1 & \tan\phi_\mu \\ -1 & \tan\phi_\mu \\ 0 & 1 \end{bmatrix}\begin{Bmatrix} q_s \\ q_n \end{Bmatrix}-\begin{Bmatrix} c_s \\ c_s \\ c_n \end{Bmatrix}\le 0 \tag{5}$$

The conditions of contact along joint indicates the constraint conditions on possible stress field in rock mass so that Eq.(3) is replaced by

$$\bar{N}^T\left(\rho\begin{Bmatrix} \sigma^e \\ q^e \end{Bmatrix}+\begin{Bmatrix} \bar{\sigma}^r \\ \bar{q}^r \end{Bmatrix}+\begin{Bmatrix} \sigma_0 \\ q_0 \end{Bmatrix}\right)\le \bar{K} \tag{6}$$

with the contact condition of Eq.(5) along joint.

2.3 *Analysis Taking Account of Contact Condition Along Rock Joint*

The traction q is a stress vector along joint, which is introduced into the equilibrium equation by employing the joint element developed by Goodman *et. al.*(1970). The lower bound analysis for rock mass including joint is formulated by using the constraint conditions Eq.(6) and the equilibrium as follows:

$$s = \max\left\{\rho \left| \begin{array}{c} \overline{N}^T\left(\rho\begin{Bmatrix}\sigma^e\\ q^e\end{Bmatrix}+\begin{Bmatrix}\overline{\sigma}^r\\ \overline{q}^r\end{Bmatrix}+\begin{Bmatrix}\sigma_0\\ q_0\end{Bmatrix}\right)\leq \overline{K} \\ \overline{B}^T\begin{Bmatrix}\sigma^e\\ q^e\end{Bmatrix}=F \\ \overline{B}^T\begin{Bmatrix}\overline{\sigma}^r\\ \overline{q}^r\end{Bmatrix}=0. \end{array}\right.\right\} \quad (7)$$

The conditions of contact along joint are considered as the yield function for the joint elements. The sliding and detachment along joints are taken as the plastic deformation of joint elements. The redistribution of stress and traction is taken into account with the residual stress σ^r and the residual traction $\overline{q}^r$ along joint which are determined by solving the boundary value problem.

The joint element method introduces two springs to rock joint as the normal stiffness k_n and the shear stiffness k_s. There exists a problem of unclearness on physical meanings of stiffness for these springs. However, the effect of joint deformation on elastic stress field is not so much by taking a large stiffness on k_n and k_s. The equilibrium equation has a great advantage to be visibly expressed with stress σ and traction q. By taking a large value for the joint stiffness, the rational result is obtained from the viewpoint of rigid plastic assumption on rock joint behavior.

3 ANALLYTICAL INVESTIGATION ON JOINT SHEAR STRENGTH AND DISCUSSION

3.1 *Conditions of Calculation*

In this study, the plane strain compression tests are simulated to reveal the failure mechanism of rock joint. Figure 2 illustrates schematically the jointed rock model under the confining sress σ_c and the deviator sress σ_d. The employed analytical procedure gives the maximum value of σ_d at the limit state. The mean plane of joint crosses the horizontal plane at angle of θ. The inclination of asperity is defined as the angle i between the asperity and mean plane of joint. This figure illustrates the case of the first order asperity.

The material constants employed for intact rock and rock joint in this study are shown in Table 1. The yield function of Drucker-Prager is employed for intact rock. The corresponding strength parameters of cohesion c and angle of shear resistance ϕ under the plane strain condition are exhibited in the table. The stiffness parameters for the joint element are assumed very large as explained before.

θ and ϕ are fixed at angle of 45° and 0°, respectively, through this study. Therefore, the direction of the mean plane of joint basically coincides with the direction of failure plane for intact rock since θ equals to $45^\circ + \phi/2$. The shear strength of joint is estimated on the mean joint plane by considering the normal stress σ_n and shear strength τ_f along the prescribed plane. These stresses are determined by the principal stresses of $\sigma_1 = \sigma_c + \sigma_d$, $\sigma_3 = \sigma_c$ and the direction of mean joint plane θ.

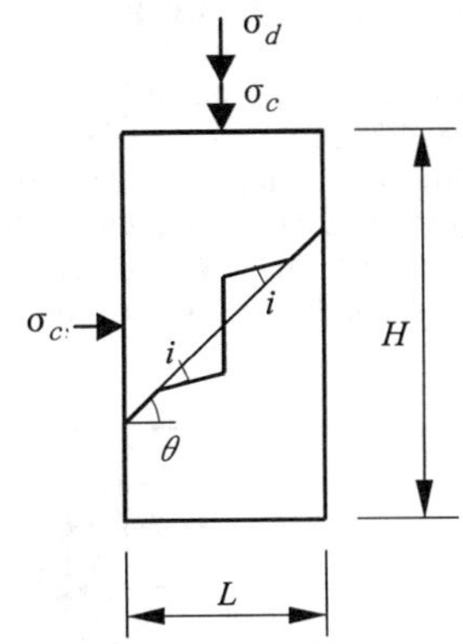

Figure 2. Schematic view of jointed rock model under plane strain condition.

Table 1. Material constants for intact rock and joint.

rock	*Young's modulus*	E	*5000.0* MPa
	Poisson's ratio	ν	*0.2*
	Cohesion	c	*5.0* MPa
	Ang. of int. friction	ϕ	*0°*
joint	*Spring constants*	k_n *and* k_s	10^{14} kPa/m
	Cohesion	c_n *and* c_s	*1.0* kPa
	Ang. of sliding friction	ϕ_μ	*0°*

3.2 *Effect of First Order Asperity on Joint Shear Strength*

Patton(1966) carried out the direct shear box tests on many artificial specimens. They have some regular triangle asperities. He derived the following conclusions from his experimental results. The failure envelopes for the shear strength of joint possessing triangular asperities could be well approximated to bilinear relationship. Namely, the inclination of primary portion in failure envelopes is equal to $i+\phi_\mu$. ϕ_μ is the angle of frictional sliding resistance for joint surface. The inclination of secondary portions is close to ϕ_r which is defined as the angle of residual strength. The changes in failure mode are found related to the physical properties of asperities along joint. However, the actual shear strength of rock joint apparently shows non-linearity against the normal stress acting along joint. This study will explain this relationship from the viewpoint of failure mode of joint microstructure.

The finite element mesh used in this analysis is illustrated in Fig.3. The size of jointed rock model is equal to 6.0*cm* long and 12.0*cm* height. The inclination of asperity is kept constant as $i=20^{\circ}$ and the angle of frictional sliding resistance along joint is also kept constant as $\phi\mu=0^{\circ}$ to evaluate only the effect of asperity shape on the shear strength of joint. Figure 4 shows the results of computation on the case of the rock joint including the first order asperity. On the whole, it is readily seen that the failure envelope is close to bi-linear relationship. The inclination of primary portion of this failure envelope is exactly identical to the angle of $i+\phi\mu=20^{\circ}+0^{\circ}=20^{\circ}$. The inclination of secondary portion becomes much smaller.

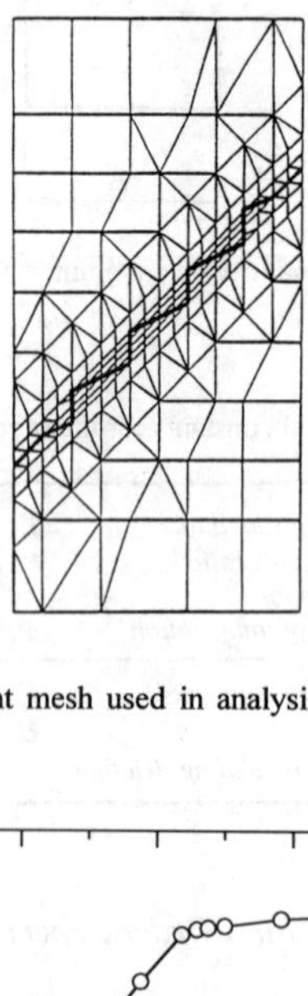

Figure 3. Finite element mesh used in analysis (only the first order asperity).

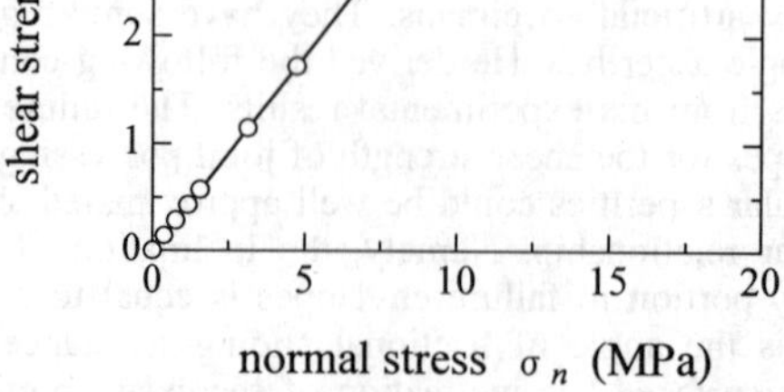

Figure 4. Effect of the first order asperity on joint shear strength.

This result indicates the possible change in failure mode with increasing the normal stress along joint. That is to say, the sliding along joint takes place at low normal stress and asperity is fully sheared at high normal stress. The combined failure mode of the partly sliding along joint and shearing asperity almost never can be observed. Although there are some differences on the basic conditions between Patton's experiments and this analysis, the both results are found to be almost same.

3.3 *Effect of Second Order Asperity on Joint Shear Strength*

In the case of rock joint including the first order asperity, only the bi-linear shear strength behavior is observed as mentioned above. It is commonly observed that the shear strength properties of rock joint show non-linearity against the normal stress along joint. In writers' opinion, the non-linearity of the shear strength of joint reflects the complex real geometry of asperity. The shape of rock joint can be regarded as assemblage of various order asperities.

To estimate the effect of the asperity shape on the shear strength property of rock joint, the analytical procedure proposed in this study is applied to jointed rock which have both the first order and the second order asperity as illustrated in Fig.5. In this figure, the angle i_1 and i_2 are the inclinations of the first order asperity and the second order asperity, respectively. These angles are kept constant as angle $i_1=i_2=20^{\circ}$ in here. Figure 6 shows the finite element mesh used in this analysis. The size of jointed rock model is the same one as shown in Fig.3.

Figure 7 indicates the effect of the second order asperity on the shear strength of joint. The same analytical result as shown in Fig.4 is also illustrated. In the case of the first order asperity without the second one, the shear strength of joint could be a bi-linear relationship against the normal stress along joint. However, in the case of the joint with the second order asperity, the shear strength is obtained as a tri-linear relationship for the normal stress.

The inclination of primary portion in failure envelopes is exactly equal to $i_1+i_2=40^{\circ}$. This result shows that the sliding failure mode along the second order asperity is appeared when the normal stress is low. The inclination of secondary portion is close to 20°. The magnitude of this angle is the same inclination as the first order asperity inclination. It can be thought that this behavior is related with the sliding failure mode along the first order asperity. Needless to say, the base of the second order asperity is fully sheared off. The inclination of the last portion is becomes much smaller and the magnitude of the shear strength is almost same as the case of only the first order asperity. From this result, it is clear that the failure mode transfers from the joint failure mode into the shear failure of the intact rock.

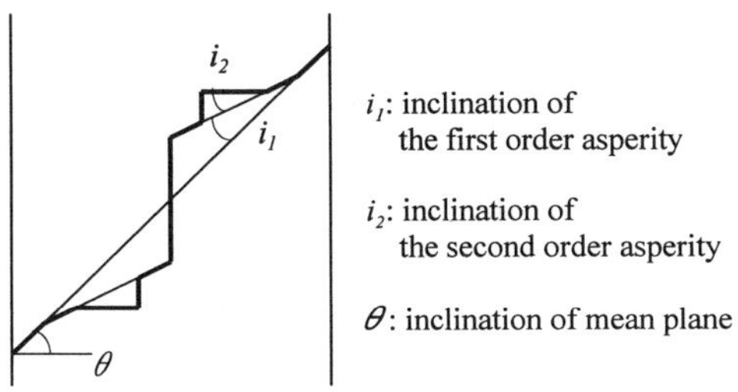

Figure 5. Schematic view of rock joint including the first order and the second asperity.

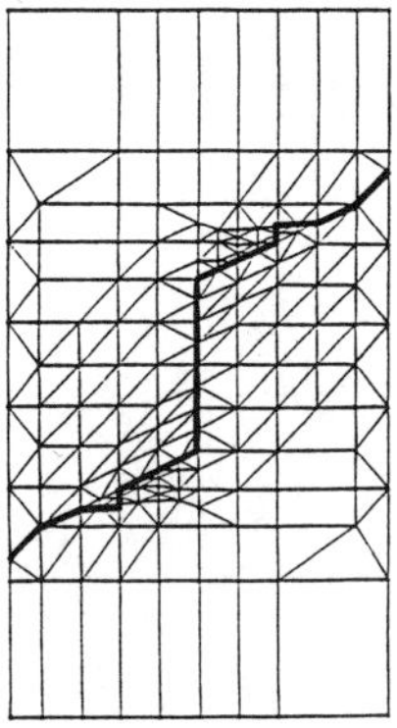

Figure 6. Finite element mesh used in analysis (joint including the first order and the second asperity).

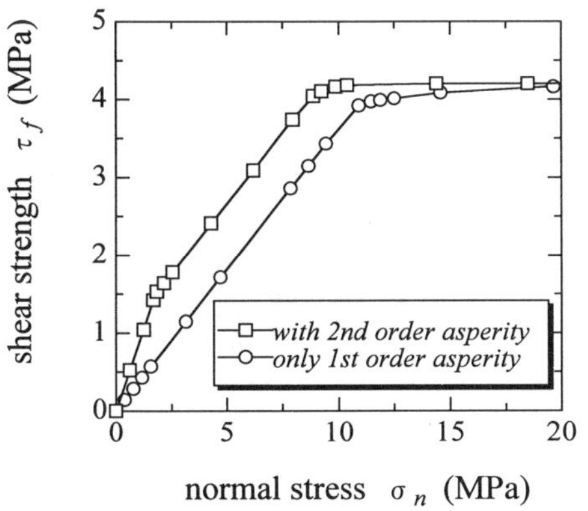

Figure 7. Effect of the first order and the second order asperity on joint shear strength.

As mentioned above, the obtained results suggest that the non-linearity in rock joint shear strength against the normal stress can be caused by the change in failure mode of rock joint. With the increase in the normal stress, the angle of frictional sliding resistance of rock joint seems to decrease and the magnitude of cohesion of rock joint seems to increase. The non-linearity in real rock joint reflects the change in failure process of various order of asperity depending on the normal stress along joint.

4 CONCLUSIONS

This paper employed the analytical procedure based on the lower bound theorem in plasticity by introducing the contact condition along joint to clarify the mechanical property of joint shear strength. By simplifying the complicated real geometry of rock joint, the effect of triangular asperity on joint shear strength was investigated. The followings are concluded in this study.

1. The shear strength of joint with the first order asperity was investigated. It was clear that the failure envelope was close to bi-linear relationship against the normal stress along joint. The obtained results was almost same with the experimental results by Patton.
2. To investigate the effect of the asperity shape on the shear strength property of rock joint, the proposed analytical procedure is applied to jointed rock which have both the first order and the second order asperity. In this case, the shear strength was obtained as a tri-linear relationship against the normal stress.
3. Through some numerical analyses, the non-linearity of joint shear strength against the normal stress was shown to be properly expressed by the change in failure process of various order of asperity depending on the normal stress along joint.

ACKNOWLEDGEMENTS

The authors are grateful to Mr. R. Motegi, S. Matsuoka, N. Mifune and H. Yoshida of Nagaoka University of Technology for thier suggestions and help in computation to conduct this study.

REFFERENCES

Barton,N.R. 1972. A model study of rock joint deformation, *Proc. of ASCE*, 94, SM3, pp.637-659.

Goodman,R.E., Taylor,R.L. & Brekke,T. 1968. Amodel for the mechanics of jointed rock, *Proc. of ASCE*, 94, SM3, pp.637-659.

Koiter,W.T. 1960. General theorems for elastic plastic solids, *Geotechnique*, Rankine Lecture, Vol.21.

Ladanyi,B. & Archambault,G. 1970. Simulation of shear behaviour of a jointed rock mass, *Proc. of 11th Symp. Rock Mech., AIME*, pp.105-125.

Maier,G. 1969. Shakedown theory in perfectelastoplasticity with associated and nonassociated flow-laws: a finite element linear programming, *Meccanica* Vol.4, No.3, pp.1-11.

Ohtsuka,S., Yamada,E. & Matsuo,M.1997. Bearing capacity analysis of rock structures including cracks, *Proc. of 9th Int. Conf. of Int. Assoc. for Comp. Mech. Adv. on Geomech.*,Vol.1, pp.739-744.
Ohtsuka,S. & Doi,M. 1998. Stability analysis of jointed rock slope, *Proc. of* 3rd *Int. Conf. on Mech. of Jointed Rock* and *Faulted Rock*, pp.523-528.
Patton,F.D. 1966. Multiple modes of shear failure inrock, *Proc. 1st Cong. ISRM.*, Lisbon, pp.509－513.
Rengers,N. 1970. The influence of surface roughness on friction properties of rock planes, *Proc. 2nd Cong. ISRM.*, Belgrade, Vol.1, pp.1－31.

Computer Methods and Advances in Geomechanics, Desai et al. (eds) © 2001 Balkema, Rotterdam, ISBN 90 5809 183 X

Non persistent discontinuity modelling of rock masses using Numerical Limit Analysis (NLA)

A.F. Durand
Laboratory of Civil Engineering, Universidade Estadual do Norte Fluminense (UENF), Campos, Brazil

E.A. Vargas Jr & L.E. Vaz
Department of Civil Engineering, Pontíficia Universidade Católica do Rio de Janeiro (PUC-Rio), Brazil

ABSTRACT: It is very often found when analyzing the stability of rock masses, that joints occur in a non-persistent fashion. The present paper presents an application of Numerical Limit Analysis (NLA) techniques for the determination of collapse load factors when such situations arise. Additionally, failure mechanisms through velocity fields as well as stress distributions can be determined. The paper presents fundamentals of the NLA formulation used and its specialization to the analysis of non-persistent joints in a fractured rock mass. Illustrative and validation examples are presented and discussed.

1. INTRODUCTION

In many geotechnical problems such as the ones found when determining the bearing capacity of foundations, retaining structures, slope stability and underground excavations, a primary objective consists in the determination of a collapse load, a maximum load the geotechnical system is able to support before it collapses. These loads are generally determined by limit equilibrium methods or by elastoplastic finite element analyses. In the present work, use is made of numerical limit analysis (NLA), an alternative, often very advantageous technique in relation to the ones described but in general seldom used in practice. Slope stability analysis of slopes in fractured rocks is a common feature in Geotechnical Engineering. When doing so, it is very often that one considers joints to be persistent. The reasons for doing so are the difficulties in establishing the exact locations of the joints and the difficulties found in the analysis itself. There are no specific analytical methods to deal with this situation. The present paper presents an application of Numerical Limit Analysis (NLA) for the solution of such problems. Initially, fundamentals of the NLA as used in the context of the present work are presented. Subsequently, illustrative and validation examples are presented and discussed, emphasizing advantages and disadvantages of the method.

2. LIMIT ANALYSIS AS A MATHEMATICAL PROGRAMMING PROBLEM

Limit Analysis is based in the so-called limit theorems (lower and upper bound) of Plasticity Theory. In the lower bound formulation, more practical from the engineering point of view, the mechanical behaviour of rigid plastic solids can be represented by equations that describe equilibrium, and strength (yield). This system when loaded by a proportional set of loads (in relation to the initial loads) reaches a limit state. The loads corresponding to this limit state are the collapse loads. The problem can therefore be described as one of finding the maximum load that the system can sustain, at the same time obeying the equations of equilibrium, and nowhere violating the strength (yield) criterion.

These statements can be translated into the following equations :

$$\text{maximize } \lambda \text{ (load factor)} \quad (1)$$

subjected to :

$$\mathbf{C}\,\sigma = \mathbf{q}_p + \mathbf{q}_a\,\lambda \quad \text{(equilibrium)} \quad (2)$$

$$\mathbf{D}\,\sigma \le c_j \quad \text{(discontinuity restrictions)} \quad (3)$$

$$f(\sigma) \le 0 \quad \text{(strength restrictions)} \quad (4)$$

where :

σ : Stress field in the system domain .

C : Matrix containing coefficients of the equilibrium and compatibility equations.

D: Matrix containing coefficients of the discontinuities.

$\mathbf{q}_p$:Vector containing permanent loads.

$\mathbf{q}_a$: Vector containing variable loads.
$f(\sigma)$: Yield function for the material.

Equations 1, 2, 3 and 4 establish a mathematical programming (MP) problem, one which contains equalities and inequalities. This problem has linear equality constraints and non-linear inequality constraints. The solution of the non linear MP problem can be obtained through the use of available efficient computer routines/algorithms (Durand, 2000 and Araújo, 1997) and provide as its solution, the load factor λ, the stress field $\boldsymbol{\sigma}$ and through the dual MP problem (Durand, 2000 and Araújo,1997), the velocity field $\dot{\mathbf{u}}$.

3. NUMERICAL FORMULATION

The static equilibrium equations in 2D space is:

$$\begin{bmatrix} \frac{\partial}{\partial x_1} & 0 & \frac{\partial}{\partial x_2} \\ 0 & \frac{\partial}{\partial x_2} & \frac{\partial}{\partial x_1} \end{bmatrix} \begin{Bmatrix} \sigma_{11} \\ \sigma_{22} \\ \sigma_{12} \end{Bmatrix} + \begin{Bmatrix} q_1 \\ q_2 \end{Bmatrix} = \begin{Bmatrix} 0 \\ 0 \end{Bmatrix} \quad (5)$$

Considering

$$\nabla = \begin{bmatrix} \frac{\partial}{\partial x_1} & 0 & \frac{\partial}{\partial x_2} \\ 0 & \frac{\partial}{\partial x_2} & \frac{\partial}{\partial x_1} \end{bmatrix} \quad (6)$$

(5) can be rewritten as :

$$\nabla \sigma + \mathbf{q} = 0 \quad (7)$$

The deformation rates are related to the velocity field (Chou & Pagano, 1992) by

$$\dot{\varepsilon} = \nabla^T \dot{\mathbf{u}} \quad (8)$$

where:

$$\dot{\varepsilon} = \begin{Bmatrix} \dot{\varepsilon}_{11} \\ \dot{\varepsilon}_{22} \\ \dot{\varepsilon}_{12} \end{Bmatrix} \quad ; \quad \dot{\mathbf{u}} = \begin{Bmatrix} u_1 \\ u_2 \end{Bmatrix} \quad (9)$$

where :
σ : Stress field vector
$\mathbf{q}$: Vector of forces
∇ : Differential operator matrix
$\dot{\varepsilon}$: Deformation rates vector
$\dot{\mathbf{u}}$: Velocities vector

Using the principle of virtual displacements (velocities in this case) one can state that:

$$\int_v \delta\dot{\varepsilon}_{ij}^T \sigma_{ij} dv = \lambda \left[\int_v \delta u_i q_i^B dv + \int_s \delta u_i q_i^S ds \right] \quad (10)$$

Considering the discretization

$$\sigma = \begin{Bmatrix} \sigma_{11} \\ \sigma_{22} \\ \sigma_{12} \end{Bmatrix}; \qquad \sigma = \mathbf{H}_\sigma \hat{\sigma} \quad (11)$$

$$\mathbf{u} = \mathbf{H}_u \hat{\mathbf{u}} \;\Rightarrow\; \dot{\varepsilon} = \nabla^T \mathbf{H}_u \hat{\mathbf{u}} = \mathbf{B} \hat{\mathbf{u}} \quad (12)$$

one gets (Durand, 2000) by substituting (11) and (12) in (10) and operating:

$$\int_v \mathbf{B}^T \mathbf{H}_\sigma dv \; \hat{\sigma} = \lambda \left[\int_v \mathbf{H}_u^T \mathbf{q}^B dv + \int_s \mathbf{H}_u^T \mathbf{q}^S ds \right] \quad (13)$$

or:

$$\mathbf{C}\,\hat{\sigma} = \lambda\,\mathbf{p}\big|_{\text{Active}} + \mathbf{p}\big|_{\text{Permanent}} \quad (14)$$

where:

$$\mathbf{C} = \int_v \mathbf{B}^T \mathbf{H}_\sigma dv \;\; ; \; \mathbf{B}^T = \mathbf{H}_u^T \nabla \quad (15)$$

$$\mathbf{p} = \left[\int_v \mathbf{H}_u^T \mathbf{q}^B dv + \int_s \mathbf{H}_u^T \mathbf{q}^S ds \right] \quad (16)$$

$\hat{\sigma}$: Nodal stresses
$\hat{\mathbf{u}}$: Nodal velocities
$\mathbf{H}_\sigma$: Interpolation matrix for stresses
$\mathbf{H}_u$: Interpolation matrix for velocities
λ : Load factor

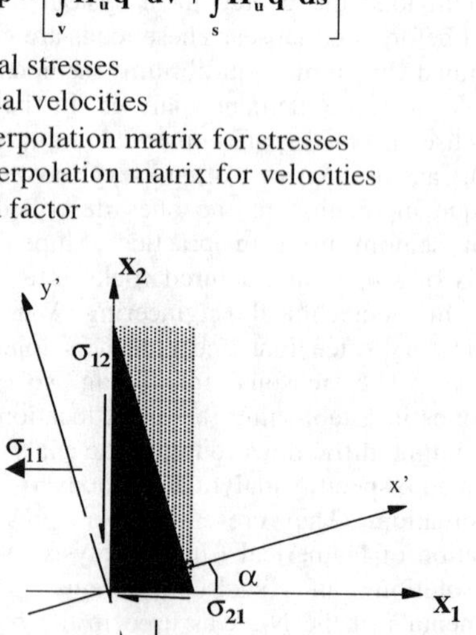

Figure 1: Two-dimensional rotation of state of stress

$$\begin{Bmatrix} \sigma_n \\ \tau \end{Bmatrix} = \begin{bmatrix} \sin^2\alpha & \cos^2\alpha & -2\sin\alpha\cos\alpha \\ -\sin\alpha\cos\alpha & \sin\alpha\cos\alpha & (\cos^2\alpha - \sin^2\alpha) \end{bmatrix} \begin{Bmatrix} \sigma_{11} \\ \sigma_{22} \\ \sigma_{12} \end{Bmatrix} \qquad (17)$$

$$|\tau| + \sigma_n \tan(\phi_j) \leq c_j \qquad (18)$$

where:
σ_n : Normal stress to α plane (see Fig. 1)
τ : Shearing stress in α plane (see Fig. 1)
ϕ_j : Joint friction angle
c_j : Joint cohesion

$$\mathbf{D} = \begin{bmatrix} -\sin\alpha\cos\alpha + \sin^2\alpha\tan\phi_j & \sin\alpha\cos\alpha + \cos^2\alpha\tan\phi_j & \cos^2\alpha - \sin^2\alpha - 2\sin\alpha\cos\alpha\tan\phi_j \\ \sin\alpha\cos\alpha + \sin^2\alpha\tan\phi_j & -\sin\alpha\cos\alpha + \cos^2\alpha\tan\phi_j & -\cos^2\alpha + \sin^2\alpha - 2\sin\alpha\cos\alpha\tan\phi_j \end{bmatrix} \qquad (19)$$

$$\mathbf{D}\,\boldsymbol{\sigma} \leq c_j \qquad (20)$$

In order to complete the equations for the general problem as established in equations 1, 2, 3 and 4, an appropriate criterion to represent the strength of intact rocks is needed. In the context of the present work, the Mohr-Coulomb criterion was used. This criterion can be written as:

$f(\sigma)$:

$$\sqrt{\left(\frac{\sigma_{11}-\sigma_{22}}{2}\right)^2 + \sigma_{12}^2} + \left(\frac{\sigma_{11}+\sigma_{22}}{2}\right)\sin(\phi) - \mathbf{c}\cos(\phi) \leq 0 \qquad (21)$$

where:
ϕ : Friction angle of intact material
c : Cohesion of intact material

4. VALIDATION AND ILLUSTRATIVE EXAMPLES

Two examples are presented for illustrative and validation purposes. The first concerns the compression loading of a specimen containing two inclined flaws and the second the stability analysis of a slope containing non-persistent joints.

4.1 Compression test of prismatic specimens with pre-existing flaws

The results shown in this section are an attempt to reproduce numerically the experimental results obtained by Reyes (1991). Reyes performed compression tests on specimens made of an artificially made material, gypsum based, in an attempt to study the effect of flaws on failure mechanisms (fracture propagation) and on the strength of rocks. The experimental work comprised a considerable number of tests performed with various geometries of flaw orientation and distribution in the specimen. For the purposes of comparison with the numerical results using NLA, only one of such test geometries will be considered. It concerns a specimen containing two flaws inclined at 45 degrees to the horizontal and spaced as shown in Figure 2. Although not explicitly stated in Reyes (1991), it was possible to estimate the strength properties of the intact material, φ(friction angle) = 55 degrees and c (cohesion) = 3.64 MPa. The numerical exercise consisted in the determination of the collapse load of the specimen. It was considered that both friction angle and cohesion are equal to zero along the flaw surface. Two ways of imposing this condition were tried. In the first, the finite element mesh explicitly contained the flaws as openings. Equal normal and shear stresses were imposed on the elements adjacent to the flaws. In the NLA formulation used, these were considered as additional equality constraints. The results corresponding to this situation are shown in Fig 2. The load factor obtained with NLA for the specimen corresponds to a strength of 18,3 MPa. This value fits into the range of strength values obtained experimentally although it can not be directly compared to the experimental results as the strength properties could only be estimated. Figure 2 also shows stresses and velocities. The stress plots show the effect of the flaws in the distribution of tensile stresses. The stress vectors show that in the region between the flaws, concentration of tensile stresses occur, leading to the conclusion that there should be a coalescence of them during failure.

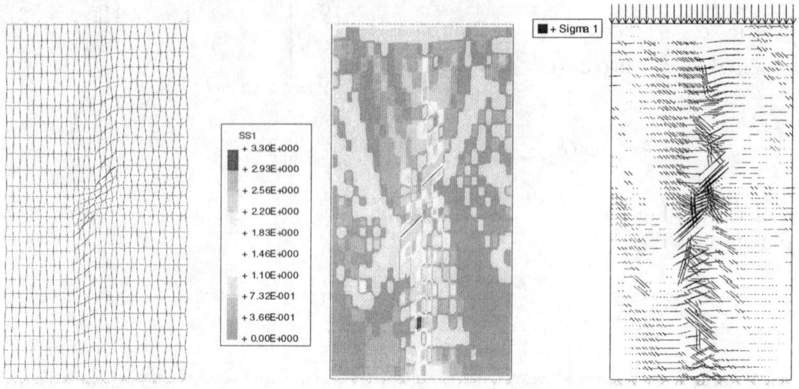

Figure 2: Mesh with explicit flaws (ϕ =54,9°; c = 3,64 MPa; ϕj = 0° ; cj = 0,0 MPa)

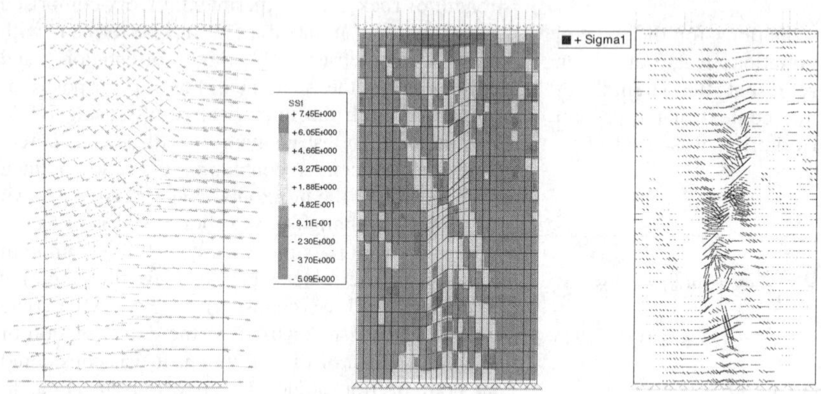

Figure 3: Mesh with non explicit flaws (ϕ =54,9°; c = 3,64 MPa; ϕj = 0° ; cj = 0,0 MPa)

The velocity plot points to a splitting mechanism of the specimen. Figure 3 shows the analysis of the same case shown in Figure 2 but now using an alternative means of representing the flaws in NLA. In this case, no explicit representation of the flaws in the finite element mesh was done although a line of elements parallel to the flaws were created. As before, restrictions were imposed in the normal and shear stresses on the adjacent to the flaws elements The results obtained in terms of strength (load factor), stress distribution, velocities are very similar to the ones obtained with the previous representation of flaws. This implicit representation of flaws has the advantage of simplifying mesh preparation.

4.2 Slope Stability in a Fractured Rock Mass

This example concerns the analysis, using NLA, of a rock slope containing joints in either persistent or non-persistent fashion. Two cases were considered in analysis. The first case considered the existence of a sole persistent joint as shown in the left-hand side FE mesh of Figure 6. The second case considered the existence of two non-persistent joints as shown in the right hand side FE mesh of the same Figure. It is clear in this latter case that a rock bridge develops which should increase the strength of the slope as compared to the persistent joint case. NLA analysis considered two forms of accounting for the presence of joints.

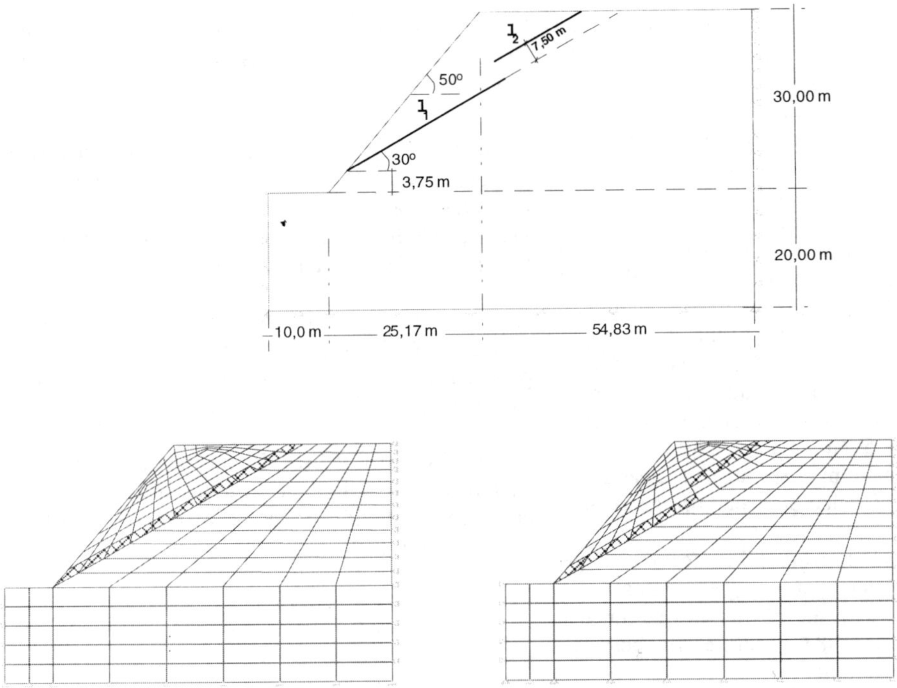

Figure 4: Geometry of the analysed slope and FE meshes used in NLA

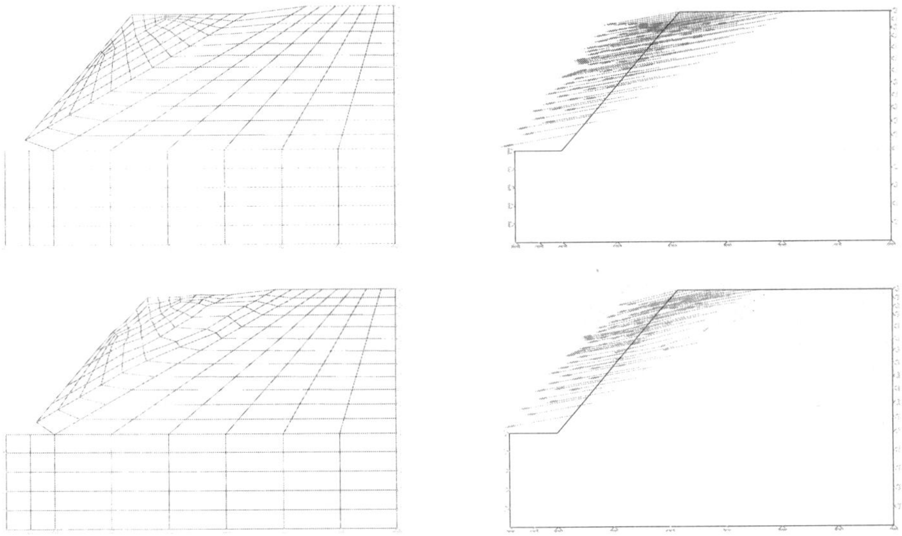

Figure 5: NLA analysis of fractured slope: deformed meshes and velocity vectors.

In the first, it was assumed that the joints crossed a line of elements and therefore the restrictions on strength were imposed on them. In the second case, the strength restrictions were imposed on an average of stresses between lines of elements above and below the joints.

As for material properties it was considered that the joint strength parameters were c(cohesion) = 0,10 MPa and friction angle = 20 degrees whereas the intact rock parameters were c = 1,0 MPa and friction angle = 35 degrees. Figure 5 summarizes the results. Regarding the load factor (Table 1), a simple stability calculation provides a load factor of 3,40 for the persistent joint case, value which agrees reasonably well with both cases of analysis as shown in the table of Figure 5. No analytical solution exists for the non-persistent joint case but it is clear form the results that the existence of a rock bridge greatly strengthens the slope.

The obtained results using NLA for the slope with non persistent joints were in agreement, at least in qualitative terms, with the results and comments made by Einstein et al (1983) regarding the possible coalescence of the two joints.

Table 1: Load factor for the analyzed cases.

	Restrictions	
	$\|\tau\| \leq -\sigma \tan\phi_j + c_j$	$\sigma = (\sigma^a + \sigma^b)/2$ $\tau = (\tau^a + \tau^b)/2$ $\|\tau\| \leq -\sigma \tan\phi_j + c_j$
Persistent Discontinuity	**λ = 3,18**	**λ = 3,60**
Non Persistent Discontinuity	**λ = 6,65**	**λ = 7,24**

5. CONCLUSIONS

The results presented in the paper are encouraging towards the use of NLA for the solution of stability problems of rock masses containing non-persistent joints. Further analyses are necessary to verify certain specific aspects as for example better ways to represent joints in a discrete or smeared way.

6. REFERENCES

Araújo, L.G., 1997. Numerical study of stability problems in Rock Mass using Limit Analysis (in Portuguese). PhD thesis, Pontifícia Universidade Católica do Rio de Janeiro, Rio de Janeiro, Brazil.

Chou. P.C. & Pagano, N.J., 1992. *Elasticity Tensor, Dyadic, and Engineering Approaches*. New York: Dover Publications. 290 p.

Durand, A.F., 2000. Applications of limit analysis to geotechnical problems modelled as conventional and Cosserat continua.(in Portuguese). PhD. thesis. Dept. of Civil Engineering, Pontifícia Universidade Católica do Rio de Janeiro, Rio de Janeiro, Brazil.

Einstein, H. H., Veneziano, D., Baecher, G. B. & O'Reilly, K. J., 1983. The effect of discontinuity persistence on rock slope stability. Int. J. Rock Mech. Min. Sci. Vol 20, n° 3, pp 227 – 236.

Reyes, O. M., 1991. Experimental study and analytical modelling of compressive fracture in brittle materials. PhD thesis, Massachussets Institute of Technology, USA.

Computer Methods and Advances in Geomechanics, Desai et al. (eds) © 2001 Balkema, Rotterdam, ISBN 90 5809 183 X

Disturbed state modelling of sand-steel interfaces under dynamic loading

Inn-Joon Park & Soo-Il Kim
Department of Civil Engineering, Yonsei University, Seoul, Korea

Chandra S. Desai
Department of Civil Engineering and Engineering Mechanics, University of Arizona, Tucson, Ariz., USA

Ji-Hyeung Yoo
Department of Civil Engineering, Kyungil University, Kyungsan, Korea

ABSTRACT: In this research, the behavior of saturated interfaces between sand and steel has been studied. A constitutive model based on the disturbed state concept (DSC) is utilized to represent this behavior. In particular, the focus is the cyclic axial loading of piles in saturated sands. The model parameters for saturated sand and a sand-interface are evaluated using data from cyclic shear test (truly tri-axial tests and CYMDOF tests). Finally, a typical boundary value problem is solved using nonlinear dynamic finite element program in which the DSC model with the foregoing parameters is implemented.

1. INTRODUCTION

Major problems associated predicting soil-structure interaction behavior are based on the fact that soil is neither homogeneous nor elastic and soil properties are spatially variable. There also exist significant coupling problems between the behavior of the structure and geologic media. When designing foundation or other geotechnical structures, the load transfer between the soil and structure takes place at the contact surface. This surface is called the *interface*. The behaviors of interface play an important role in the load-deformation response of soil-structure interaction systems.

The purpose of this study is to develop improved model for simulation of saturated sand-structure interface behavior based on a general concept, called the *disturbed state concept* (DSC), which can characterize the behavior of saturated soils. The initial idea for this theory was introduced by Desai (1974) to characterize the softening response of an over-consolidated soil by expressing the observed response in terms of its response in its normally consolidated state as the reference state. DSC model has been successfully verified with respect to other materials such as dry sands (Armaleh and Desai, 1990) and liquefaction of saturated sand (Desai *et al.*, 1998; Park, 1997). In this study, DSC model is calibrated with respect to laboratory tests such as cyclic multi-degree-of-freedom shear test (CYMDOF) for the saturated interface. The model can be implemented in a dynamic nonlinear finite element procedure for coupled soil-fluid response toward solution of a typical boundary value problem such as simulation of an axially loaded pile.

2. DISTURBED STATE CONCEPT MODEL

The DSC is based on the idea that any stage during deformation of a material element is composed of a mixture of materials in two stages; relative intact (RI) stage and fully adjusted (FA) stage. The disturbances due to external load can involve particle motions and self-adjustment of the microstructure leading to micro cracking and softening, and stiffening. As a consequence, the material in the RI state is transforming into the FA state at randomly disturbed locations (see Figure 1). The coalescence of the FA clusters can lead to critical extent or to shear bands, causing liquefaction and failure.

The RI response can be characterized by using continuum theories such as elasticity and

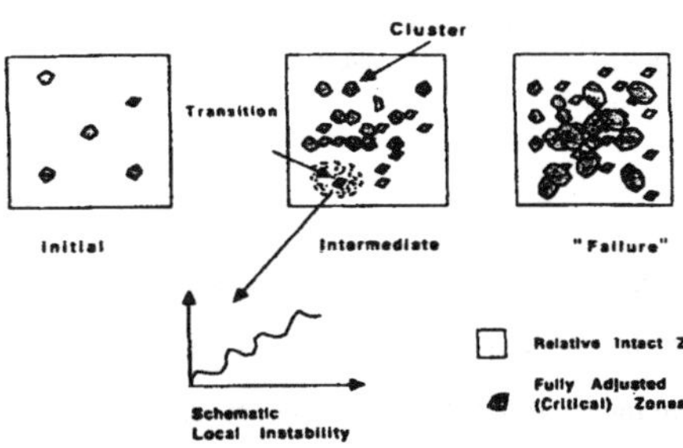

Figure 1. Relative intact and fully adjusted state in DSC (Desai, *et al.*, 1998)

elasto-plasticity. The FA state can be defined as critical state concept that can carry shear reached up to that state for given initial hydrostatic pressure and deform in shear under constant volume, like a constrained liquid-solid (Roscoe, et al., 1958).

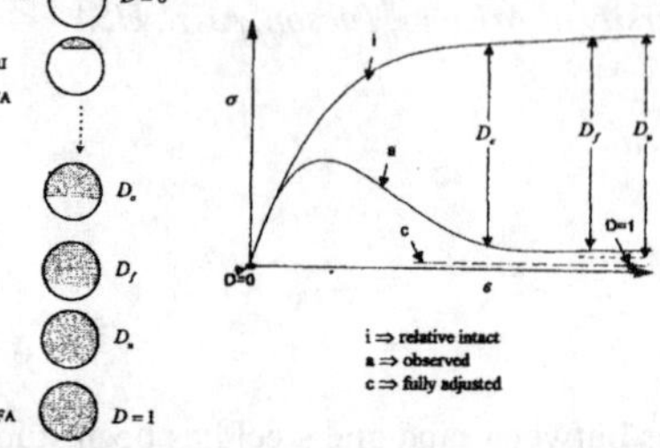

Figure 2. Representation of DSC (Desai, *et al.*, 1998)

2.1 *Relative intact state*

In order to evaluate the disturbance, it is necessary to define the behavior of the RI material, which can be characterized by using a continuum theory such as linear or non-linear elasticity and elasto-plasticity. When the elasto-plastic theory is used, the incremental stress-strain equation is expressed as,

$$d\sigma_{ij}^{i} = C_{ijkl}^{ep(i)} d\varepsilon_{kl}^{i} \tag{1}$$

Where $d\sigma_{ij}^{i}$ and $d\varepsilon_{ij}^{i}$ are the RI incremental stress and strain tensor, and $C_{ijkl}^{ep(i)}$ is the elasto-plastic constitutive tensor, which will depend on the model used, e.g. von Mises, Mohr-Coulomb, or hierarchical single surface (HISS) plasticity (Desai, 1980; Desai *et al.*, 1886). Here, the HISS model is used in which the yield function, F (see Figure 3.) is given by

$$F = J_{2D} - (-\alpha J_1^n + \gamma J_1^2) = 0 \tag{2}$$

Where J_{2D} is the second invariants of the deviatoric stress tensor, S_{ij}, J_1 is the first invariant of the stress tensor, γ is the parameter related to the ultimate yield behavior and shape of yield surfaces, respectively, n is phase change parameter, and α is the hardening function expressed as

$$\alpha = \frac{h_1}{\xi^{h_2}} \tag{3}$$

Where h_1 and h_2 are the hardening parameters, ξ is the trajectory of plastic strains $= \int \sqrt{d\varepsilon_{ij}^{p} d\varepsilon_{ij}^{p}}$, and $d\sigma_{ij}^{j}$ is the increment of plastic strain tensor.

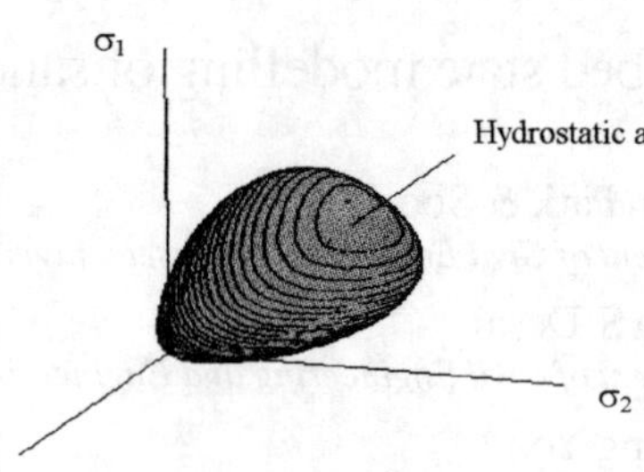

Figure 3. (a) The yield surface of HISS model

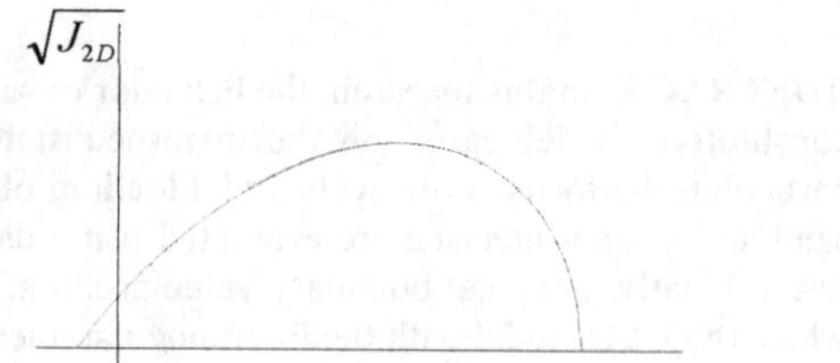

Figure 3. (b) The yield surface of HISS model in $\sqrt{J_{2D}} - J_1$ space

2.2 *Fully adjusted state*

The FA state is modeled using critical state concept. At the critical state condition, shear deformation can continue without further changes in volume. Such critical state often lies on a straight line called *critical state line* (CSL) with the slope $\overline{m}$, given by

$$\sqrt{J_{2D}^{c}} = \overline{m} J_1^{c} \tag{4}$$

During shearing, the materials change in volume and initial void ratio changes. Finally, at the critical state the void ratio, e^c, is given by

$$J_1^{c} = 3 p_a \times e^{\left(\frac{e_0^c - e^c}{\lambda}\right)} \tag{5}$$

Where e_0^c is the value void ratio at critical state corresponding to $J_1^c = 3p_a$ and λ is a slope of the critical state line. In this study, the behavior of the material at the FA or critical state is defined using Eq. (4) and (5).

2.3 *Disturbance function*

The disturbance function can be defined as

$$D = \frac{M_s^c}{M_s} \tag{6}$$

where M_s^c is the mass of solids in the FA state and M_s is the total mass of solids in the materials. Initially with no disturbance the material is assumed to be entirely in the RI state, therefore D equals to zero. With full disturbance the material can be assumed to be fully in the FA state, and at the ultimate state, $D = D_u \leq 1$.

The disturbance function, D, used in the current study is given by Armaleh and Desai (1990):

$$D = D_u\left[1 - e^{\left(-A\xi_D^Z\right)}\right] \tag{7}$$

where A, Z, and D_u are material parameters. Where D_u is often assumed to be unity.

3 PARAMETERS USED FOR SIMULATION OF AXIALLY LOADED PILE

The foundation soil is assumed as medium dense (D_r=60%) Ottawa Sand. Truly triaxial test data reported by Gyi (1996) is used to evaluate DSC model parameters (see Table 1). CYMDOF-P test results and parameters for interface between saturated sand and steel reported by Alanazy (1996) are used to define the behavior of interfaces. Two parameters--A and Z-- are also redefined to capture the behavior of interface in this study (see Table 1).

Table 1. Material parameters for numerical simulation

	Parameter	Saturated Sand Parameters	Interface Parameters
Elastic Moduli	E	193000 (kPa)	20531 (kPa)
	ν	0.380	0.384
Plasticity Parameters	n	2.450	3.400
	γ	0.123	0.274
Hardening Parameters	h_1	0.8450	0.8250
	h_2	0.0215	0.3180
Critical State Parameters	$\overline{m}$	0.150	0.35
	λ	0.020	0.0134
	e_o^c	0.601	0.580
DSC Function Parameters	A	**4.22**	**1.239**
	Z	**0.43**	**0.64**
	D_u	**0.99**	**1.00**

The parameters presented in Table 1 are employed for simulation of the axially loaded pile.

4 BOUNDARY VALUE PROBLEM

The performance of the proposed DSC model in boundary value problems is demonstrated by applying it to dynamic analysis of axially loaded pile with dimensions and loading similar to that in a field model pile test. The model pile test on the pile segment load test was conducted by The Earth Technology Corporation (1986) under a joint research project with The University of Arizona (Wathugala and Desai, 1990). The simulation of axially loaded pile is modified in order to characterize the behavior of interface between steel pile segment and saturated sands by using the finite element program based on DSC model. It is assumed that the foundation soil is made of fully saturated Ottawa sand with $D_r = 60\%$. Analysis is carried out herein for the pile segment of 0.076m and 1.895m length. Notice that, except the assumption of foundation soil, the dimensions of pile segment and soil around pile used for the finite element analysis are same as the original test program.

4.1 *Finite Element Mesh and Boundary Conditions*

The finite element mesh consists of total of 192 elements and 225 nodes. Four node isoparametric elements are used for soil and interface elements. Eight elements are used for interface zone. Notice that the interface elements are located within soil zone having a thickness equal to 0.012m and interface parameters. Detailed meshes for FEM analysis around interfaces are shown in Figure 4.

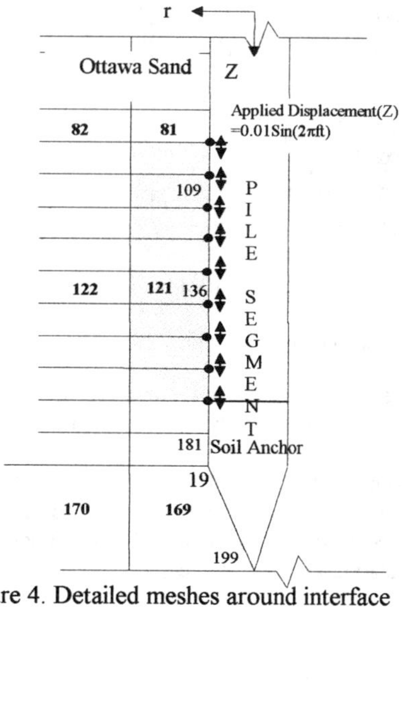

Figure 4. Detailed meshes around interface

For the boundary conditions, the left and right vertical sides of the boundary of the mesh are restrained in the r-direction and are free in the z-direction. The bottom boundary is restrained in the z-direction, but free in the r-direction, and the top boundary is free both in directions, r and z.

4.2 *Results of Simulation of Axially Loaded Pile*

The solutions of this boundary problem are evaluated by *DSC_DYN2D* program. The following a boundary problem is studied in order to demonstrate the effect of interface behavior under cyclic loading with interface element.

A purpose of this simulation is to define the characteristic of liquefaction around interface. Therefore, pore pressures and disturbances around interface elements are found and represented from the results of finite element analysis.

The results of finite element analysis in this case are shown from Figure 5 to Figure 7. And Figure 8 shows the disturbance contours at typical times around the loaded elements to define the behavior of saturated sand included interfaces under cyclic shearing.

Figure 5 shows the displacements in z direction at nodes 136 and 137. The difference of displacements between two nodes is about 0.006m. Because the element 121 is considered as interface element, relative displacement between interface and soil can occur under cyclic shearing.

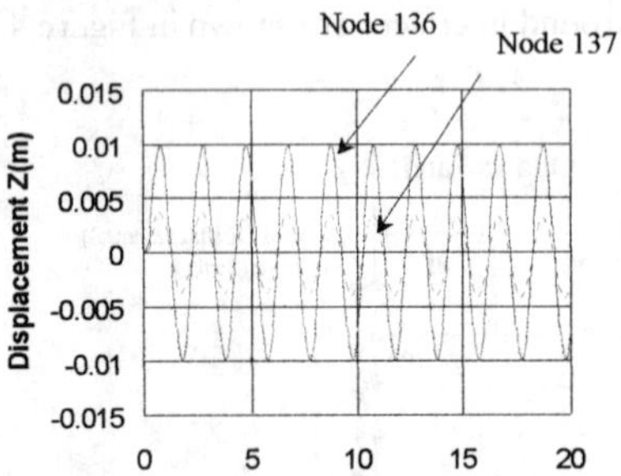

Figure 5. Displacements in typical nodes

Figure 6 shows the pore pressure from finite element analysis. From this figure, initial liquefaction occurs at about 2sec (1 cycle) based on the conventional definition—initial liquefaction occurs when initial confining stress of 164.5kPa is close to excess pore pressure of 163.8kPa. Therefore, initial liquefaction occurs at 2sec.

Figure 7 shows the disturbances in elements 121 and 122. When D in saturated sand having initial confining stress of 164.5kPa becomes to 0.84, initial liquefaction can occurs at 2sec. Based on this result, D becomes 0.84 about 2sec in element 121 and around 10sec in element 122 (Figure 7). From these results, analysis of pile segment can define the liquefaction phenomenon and behavior of interface.

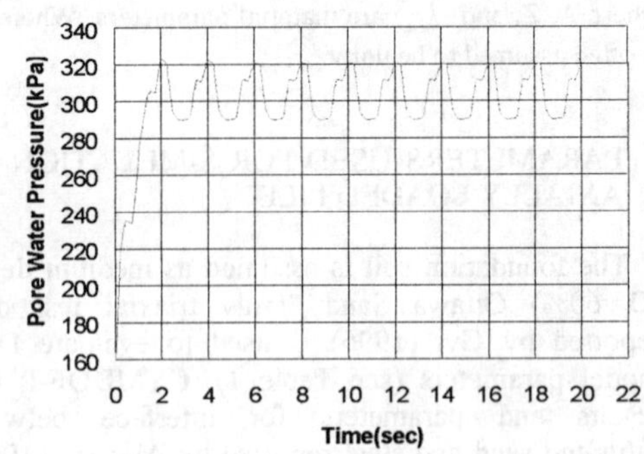

Figure 6. Pore water pressure in element 121

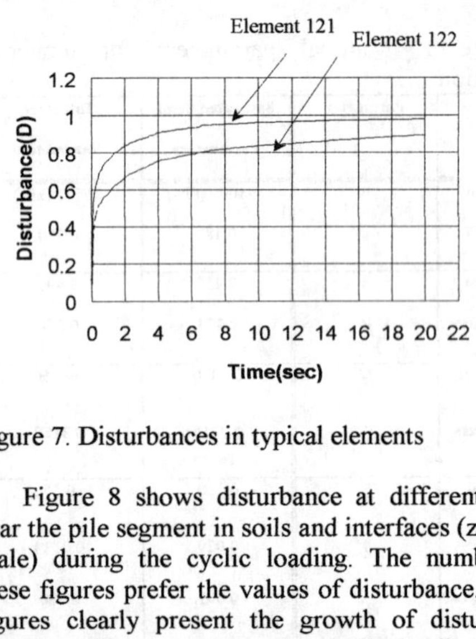

Figure 7. Disturbances in typical elements

Figure 8 shows disturbance at different times near the pile segment in soils and interfaces (zoom in scale) during the cyclic loading. The numbers in these figures prefer the values of disturbance. These figures clearly present the growth of disturbance from interfaces to soils under cyclic loading. Also, theses figures show that the rate of growth of disturbance is faster around top and bottom of pile segment than that in other areas. Because the stress concentration occurs around theses two zones, sands in these zones can be disturbed faster. Especially, liquefaction in interfaces occurs at 2sec when the disturbance, D, is greater than 0.82.

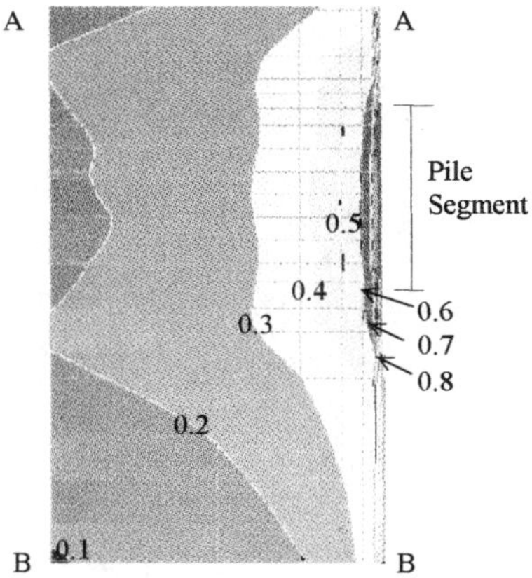

Figure 8. (a) Disturbance Contours (Zoom In Around Interface) at Time=6sec

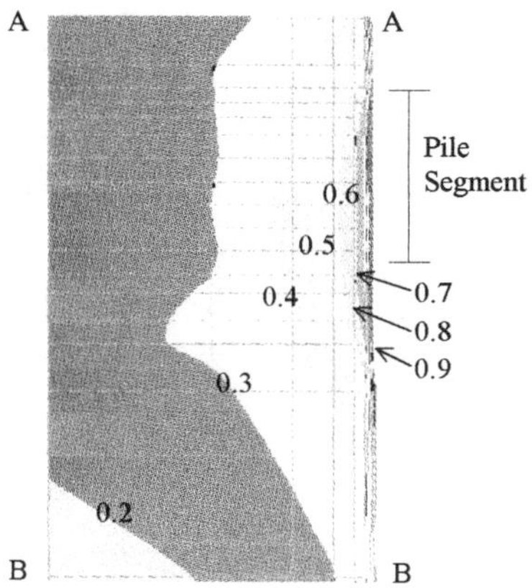

Figure 8. (b) Disturbance Contours (Zoom In Around Interface) at Time=10sec

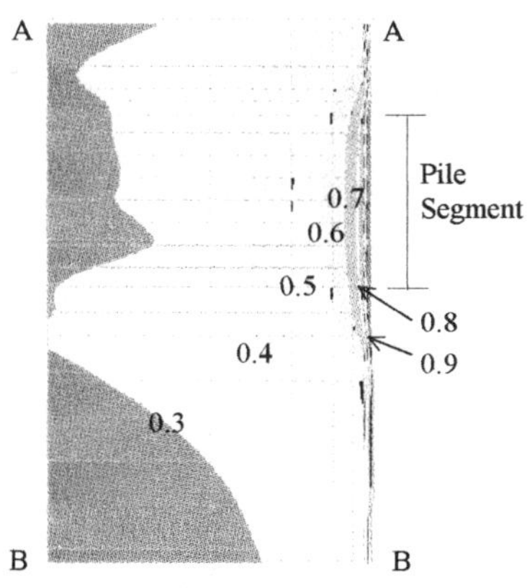

Figure 8. (c) Disturbance Contours (Zoom In Around Interface) at Time=20sec

5 CONCLUSIONS

The Disturbed State Concept model can provide a powerful procedure to characterize plastic hardening, strain softening, excess pore water pressure, and liquefaction response of saturated sands and interfaces under static and dynamic loading.

The Disturbed State Concept model, with HISS model as the intact state and critical state as the fully adjusted state, were modified for interface and implemented in two phase dynamic finite element program, DSC-DYN2D. The simulation of axially loaded pile has demonstrated the applicability of this model.

It is shown that the finite element program, DSC-DYN2D, is capable for the saturated interfaces and dynamic analysis in geotechnical engineering problems. Therefore, the DSC allows for models for soils and interfaces within the same mathematical framework, and permits the analysis of important soil-structure interaction effects.

REFERENCES

Alanazy, A. S. 1996. Testing and Modeling of Sand-Steel Interfaces Under Static and Cyclic Loading. *Ph.D. Dissertation*, Dept. of Civil Engrg. And Engrg. Mech., Univ. of Arizona, Tucson, Arizona.

Armaleh, S. H. & Desai, C. S. 1990. Modeling Include Testing of Cohesionless Soils Under Disturbed State Concept. *Report to the NSF*, Dept. of Civil Engrg. And Engrg. Mech., Univ. of Arizona, Tucson, Arizona.

Desai, C. S. 1974. A Consistent Finite Element Technique for Work-Sofening Behavior. *Proc. Int. Conference on Computational Methods in Nonlinear Mechanics*, Univ. of Texas at Austin, Austin, Texas.

Desai, C. S. 1980. A General Basis for Yield, Failure, and Potential Functions in Plasticity. *Int. J. Num. Analyt. Meth. In Geomech.*, Vol. 4, pp. 361-375.

Desai, C. S., Somasundaram, S., & Frantziskonis, G. 1986. A Hierarchical Approach for Constitutive Modeling of Geologic Materials. *Int. J. Num. Analyt. Meth. In Geomech.*, Vol. 10, No. 3, pp. 225-257.

Desai, C. S., Park, I. J. & Shao, C. 1998. Fundamental yet simplified model for liquefaction instability. *Int. J. Num. Analyt. Meth. In Geomech.*

Gyi, M. M. 1996. Multiaxial Cyclic Testing of Saturated Ottawa Sand *M. S. Thesis*, Dept. of Civil Engrg. And Engrg. Mech., Univ. of Arizona, Tucson, Arizona.

Roscoe, K. H., Schofield, A. & Wroth, C. P. 1957. On the Yielding of Soils. *Geotechnique*, Vol. 8, pp. 22-53.

Wathugala G. W. & Desai C. S. 1993. Constitutive Model for Cyclic Behavior of Clay, I: Theory. *J. of Geotech. Engrg.* ASCE Vol. 119, No. 4, pp.715-729.

Computer Methods and Advances in Geomechanics, Desai et al. (eds) © 2001 Balkema, Rotterdam, ISBN 90 5809 183 X

Strain-hardening approach of jointed rock using Hoek-Brown criteria

G. Swoboda & A. Elhashimy
University of Innsbruck, Austria

ABSTRACT: This paper presents the application of the isotropic strain-hardening rule to jointed rock using Hoek-Brown failure criteria. These yield criteria have been converted into convenient form for use in elasto-plastic finite element analysis. Also, implementation of the criteria in the finite element analysis program FINAL has been worked out and tested for plain strain problems. A comparison between the results of the strain-hardening behavior of a plane strain problem of a strip footing resting on a rock mass and the results of the same footing when using perfectly plastic behavior has been proposed.

1 INTRODUCTION

Most of jointed rock exhibits hardening behavior. Which means an increasing of the stress state beyond the initial yield surface and into the hardening range results in both elastic and plastic deformations. With the development of plastic deformation, a new yield surface called *subsequent yield* or *loading surface* evolves. A yield surface changes its current configuration during the loading process, so that the stress point always lies on it. However, there are an infinite number of evolutions of the yield surface that meet this condition. The rule governing the changes of the subsequent yield surfaces is called the *hardening rule*. Several such rules have been proposed in the past for use in plastic analysis. The response of materials after initial yielding can differ considerably, depending on the particular hardening rule employed. When an initial yield surface is known, the rule of hardening defines its modification during the process of plastic flow.

2 HARDENING RULES

A number of hardening rules have been proposed. The most widely used rules are those of *isotropic hardening*, *Kinematic hardening* and a combination of both, the so-called *mixed hardening*. The general form of the loading function is shown in Eq.(1).

$$f(\sigma_{ij}, \epsilon^p_{ij}, K) = F(\sigma_{ij}, \epsilon^p_{ij}) - K^2(\epsilon^p) = 0 \quad (1)$$

in which the hardening parameter K^2 represents the size of the yield surface, while the function $F(\sigma_{ij}, \epsilon^p_{ij})$ defines the shape of that surface. The parameter K^2 is expressed as a function of ϵ^p, called the *effective strain*, which is an integrated increasing function of the plastic strain increments but not the plastic strain itself. The value of ϵ^p depends on the loading history or the plastic strain path.

2.1 *Isotropic hardening rule*

This rule is based on the assumption that the yield surface expands uniformly without distortion or translation during a loading process, as schematically shown in Figure 1.

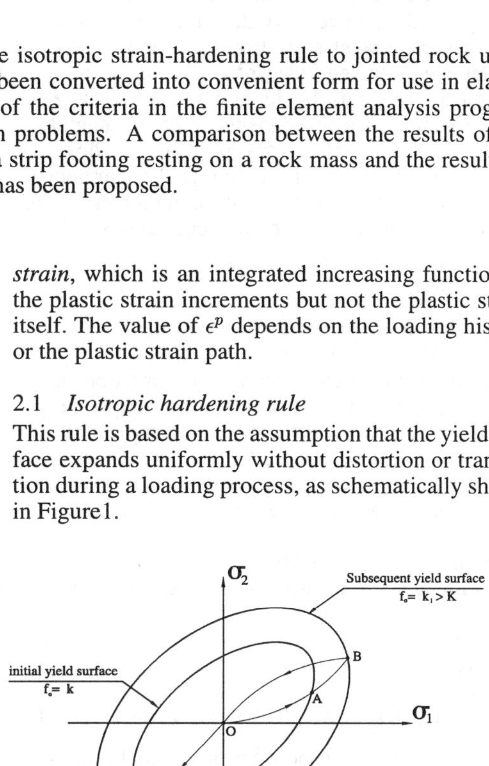

Figure 1: Subsequent yield surface for isotropic-hardening material.

The mathematical expression for the yield surface can therefore be written as follows:

$$f(\sigma_{ij}, K) = f_o(\sigma_{ij}) - K(\epsilon_p) = 0 \quad (2)$$

Where $f_o(\sigma_{ij})$, a function defined by the yield surface stress level, $K(\epsilon_p)$, a hardening or growth function, defines the size of the surface. ϵ_p is a hardening parameter whose value represents the plastic loading history of the material , which is known as the effective plastic strain. The size of the yield surface is now governed by the value K which can be considered as the effective stress σ_e. Since plastic deformation is an anisotropic process, it cannot be expected that the theory of isotropic hardening will produce realistic results when complex loading paths with stress reversal are considered.

3 FLOW RULE

Plastic strains are produced when the current stress state σ with the stress increment σ+dσ is outside the elastic region. $d\epsilon_{ij}^p$ is the outward vector at the boundary of the elastic region. Plastic strains are produced during a loading process. To describe the stress-strain relationship for an elastic-plastic deformation, the direction and the magnitude of the plastic strain increment vector $d\epsilon_{ij}^p$ must be defined. It can be defined as the ratio among the components and their magnitude against the stress increments $d\sigma_{ij}$. Thus, the flow rule is the rule defining the ratio among the components of $d\epsilon_{ij}^p$, or the direction of the strain increment vector and it can be defined as shown in equation(3)

$$d\epsilon_{ij}^p = d\lambda \frac{\partial f}{\partial \sigma_{ij}} \quad (3)$$

where $d\lambda$ is a non-negative scalar function

The gradient $\partial f / \partial \sigma_{ij}$ defines the direction of the plastic strain vector $d\epsilon_{ij}^p$, which is normal to the potential surface f = 0 at the current stress point, and thus the flow rule is also called the *normality condition*. On the other hand, the length or the magnitude of the plastic strain increment vector is determined by $d\lambda$.

Since the plastic strain develops along the normal to the current loading surface, the normality condition of equation(3) will assure a unique solution for a given boundary-value problem using any stress-strain relations developed on this basis.

4 EFFECTIVE PLASTIC STRESS AND STRAIN

Since the hardening response is observed during plastic deformation and the degree of hardening depends on the plastic loading history, the plastic loading history must first be recorded in order to describe the hardening behavior. Second, a description of the dependence of the hardening behavior on the plastic loading history must be introduced. For the former, a hardening or growth function K was introduced, while for the latter, a monotonically increasing scalar called the *hardening parameter k* was introduced. The hardening function K is a function of the hardening parameter k, and its functional form is material-dependent. The most common material test is performed under the uniaxial loading condition. As always, the parameters necessary to describe hardening behavior in general loading are identified inthis type of experiment. Therefore, for the sake of convenience, we define the *effective stress* σ_e and the effective plastic strain ϵ_p, which are reduced to the stress and plastic strain of the uniaxial stress test, respectively. The hardening function K is related to the effective stress σ_e, and the effective plastic strain ϵ_p may be taken to be the hardening parameter k itself. σ_e is a function of ϵ_p, and the specific form of this function is determined from the uniaxial test data, as shown in Figure 2.

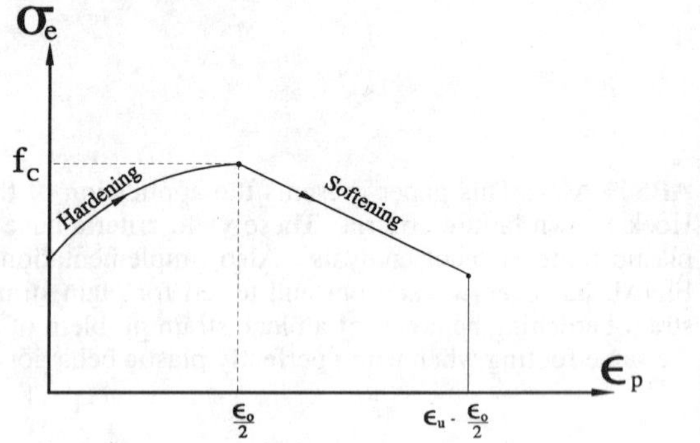

Figure 2: Effective Stress - Effective Plastic Strain Relationship.

In the present work the relationship between the effective stress and the effective plastic strain is extrapolated from the uniaxial stress-strain relationship using the conventional *madrid parabola*.

$$\sigma = E_o \epsilon - \frac{1}{2} \frac{E_o}{\epsilon_o} \epsilon^2 \quad (4)$$

where

E_o	initial elasticity modulus,
ϵ	total strain,
ϵ_o	total strain at peak stress.

Substituting the elastic strain component $\epsilon_e = \sigma / E_o$ in Eq.(4)

$$\sigma = -E_o \epsilon_p + \sqrt{2 E_o^2 \epsilon_o \epsilon_p} \quad (5)$$

where ϵ_p is the plastic strain component, and ϵ_o can be taken to be equal to $2 f_c / E_o$ for intact rock materials. Using the current effective plastic strain in Eq.(5) gives the effective stress level defining the current loading surface position.

4.1 *Effective stress*

Since the loading function $f_o(\sigma_{ij}, \epsilon_{ij}^p, K) = 0$ by definition, determines whether additional plastic flow takes place or not and is also a positively increasing function, it can be used as a truly significant stress variable to define the effective stress. Consider the case of isotropic hardening, in which the loading function takes the form of Eq.(2)or the function $f_o(\sigma_{ij})$ is used to define the effective stress. Since the effective stress should decrease to the stress σ_1 in the uniaxial test, it follows that the loading function $f_o(\sigma_{ij})$ must be some constant A times the effective stress σ_e to some power$n$.

$$f_o(\sigma_{ij}) = A\sigma_e^n \tag{6}$$

where A and n are determined from the condition that σ_e reduces to the stress σ_1 in the uniaxial test.

4.2 *Effective plastic strain*

In this study the plastic deformation is usually represented by the so-called effective plastic strain ϵ_p. The materials complying with this hypothesis are often termed strain-hardening materials. The parameter ϵ_p may be called the hardening parameter that can be denoted generally by k.
The effective plastic strain, ϵ_p, is defined using some simple combination of plastic strain increments. The value of ϵ_p is always positive and increasing. The simplest combination of this type with the correct dimension is shown in Eq.(7)

$$d\epsilon_p = C\sqrt{d\epsilon_{ij}^p d\epsilon_{ij}^p} \tag{7}$$

where the positive constant C is determined from the condition that ϵ_p is reduced to the absolute value of ϵ_{11}^p for the uniaxial loading test. Using the flow rule Eq.(3) we find.

$$d\epsilon_{11}^p = \frac{\partial f}{\partial \sigma_{11}} d\lambda \ , \qquad d\epsilon_p = C\sqrt{\frac{\partial f}{\partial \sigma_{ij}}\frac{\partial f}{\partial \sigma_{ij}}} d\lambda \tag{8}$$

Under the uniaxial loading condition, dϵ_p is equal to dϵ_{11}^p by definition. Thus, from Eq.(8), we obtain.

$$C = \frac{\frac{\partial f}{\partial \sigma_{11}}}{\sqrt{\frac{\partial f}{\partial \sigma_{ij}}\frac{\partial f}{\partial \sigma_{ij}}}} \tag{9}$$

where all terms should evaluated under the condition of uniaxial loading. The flow vector $\frac{\partial f}{\partial \sigma_{ij}}$ can be expressed in a form suitable for numerical computation.

$$a^T = \frac{\partial f}{\partial \sigma_{ij}} = \frac{\partial f}{\partial I_1}\frac{\partial I_1}{\partial \sigma_{ij}} + \frac{\partial f}{\partial \sqrt{J_2}}\frac{\partial \sqrt{J_2}}{\partial \sigma_{ij}} + \frac{\partial f}{\partial \theta}\frac{\partial \theta}{\partial \sigma_{ij}} \tag{10}$$

where

$$\sigma_{ij}^T = \{\ \sigma_{11} \quad \sigma_{22} \quad \sigma_{33} \quad \tau_{23} \quad \tau_{31} \quad \tau_{21}\ \}$$

$$sin3\theta = \frac{-3\sqrt{3}}{2}\frac{J_3}{J_2\sqrt{J_2}} \tag{11}$$

differentiating θ by σ_{ij} we obtain.

$$\frac{\partial \theta}{\partial \sigma_{ij}} = \frac{-\sqrt{3}}{2cos3\theta}\left[\frac{1}{J_2\sqrt{J_2}}\frac{\partial \sqrt{J_3}}{\partial \sigma_{ij}} - \frac{3J_3}{J_2^2}\frac{\partial \sqrt{J_2}}{\partial \sigma_{ij}}\right] \tag{12}$$

Substitute Eq.(12) in Eq.(10), then write.

$$a = C_1 a_1 + C_2 a_2 + C_3 a_3 \tag{13}$$

where

$$a_1^T = \frac{\partial I_1}{\partial \sigma_{ij}} \ , \quad C_1 = \frac{\partial f}{\partial I_1}$$

$$a_2^T = \frac{\partial \sqrt{J_2}}{\partial \sigma_{ij}} \ , \quad C_2 = \frac{tan3\theta}{\partial \sqrt{J_2}}\frac{\partial f}{\partial \theta}$$

$$a_3^T = \frac{\partial \sqrt{J_3}}{\partial \sigma_{ij}} \ , \quad C_3 = \frac{-\sqrt{3}}{2cos3\theta}\frac{1}{J_2\sqrt{J_2}}\frac{\partial f}{\partial \theta}$$

Only the constants C_1, C_2 and C_3 are then necessary to define the yield surface. Thus, we can achieve a simplicity of programming as only these three constants have to be varied between one yield surface and another.

5 HOEK-BROWN CRITERION

The Hoek Brown criterion is mainly for rock materials, where the yield function of this type of material depends mainly on experimental data. It can be shown as

$$f = 4J_2cos^2\theta + \frac{m\sigma_c\sqrt{J_2}}{\sqrt{3}}(sin\theta + \sqrt{3}cos\theta) - \frac{m\sigma_c I_1}{3} \tag{14}$$

where
σ_c uniaxial compressive strength of the intact rock,
m,s material constant.
For the uniaxial loading test in X_1 direction, where σ_{11} is the only nonzero component, we obtain.

$$\frac{\partial f}{\partial \sigma_{11}} = 2\sigma_{11} + \frac{2m\sigma c}{3} \tag{15}$$

and

$$a = \begin{Bmatrix} (-2(\frac{m\sigma c}{3} + \sigma_{11}) \\ (\sigma_{11} - \frac{m\sigma_c}{6}) \\ 0 \\ (\sigma_{11} - \frac{m\sigma_c}{6}) \end{Bmatrix} \tag{16}$$

substituting Eq.(15) and Eq.(16) in Eq.(9), we obtain the positive constant C for Hoek- Brown material in the form.

$$C = \frac{2\sigma_{11} + \frac{2m\sigma_c}{3}}{\sqrt{6\sigma_{11}^2 + 2m\sigma_c\sigma_{11} + \frac{1}{2}m^2\sigma_c^2}} \tag{17}$$

The complete elasto-plastic incremental stress-strain relationship can be expressed as follows:

$$d\sigma = [D_{ep}]d\epsilon \tag{18}$$

and the elasto-plastic D_{ep} matrix can be expressed as

$$[D_{ep}]=[D]-\frac{[D]\{a\}\{a\}^T[D]^T}{H'+\{a\}^T[D]^T} \tag{19}$$

where H' is the hardening modulus and can be given by.

$$H'=-C\frac{\partial f}{\partial k}\frac{dk}{d\epsilon_p}\sqrt{\frac{\partial f}{\partial \sigma_{ij}}\frac{\partial f}{\partial \sigma_{ij}}} \tag{20}$$

6 NUMERICAL EXAMPLE

The example considered here is a plane strain problem of a strip footing resting on an interlocked, partially disturbed rock soil mass with multifaceted angular blocks formed by four or more discontinuity sets and subjected to a prescribed load intensity of 100 MPa.

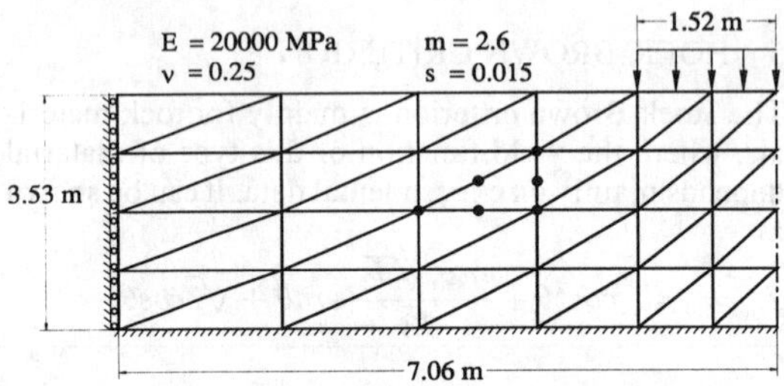

Figure 3: Strip footing soil system under plane strain condition.

This footing was analyzed several time under the same load intensity but with different uniaxial compressive strength of intact pieces of rock σ_c using both hardening plastic and perfectly plastic behavior. The finite element discretization, the geometrical and material data are presented in Figure 3.

A comparison of the central settlement of the surface point for various increments in pressure intensity with different σ_c is presented in Figure 4 for both hardening and perfect plastic behavior.

7 CONCLUSIONS

1. An attempt is made to discuss the Hoek-Brown failure criteria with its recent development of strain-hardening behavior for intact rock masses.
2. Using the proposed approach the yield criteria were converted into convenient forms for use in the finite element analysis.
3. An attempt was also made to generalize computation of the plastic flow vector through the definition of constants C_i
4. At any uniaxial compressive strength for the intact rock both the hardening and perfectly plastic behavior have the same trend.
5. The hardening plasticity behavior of the rock

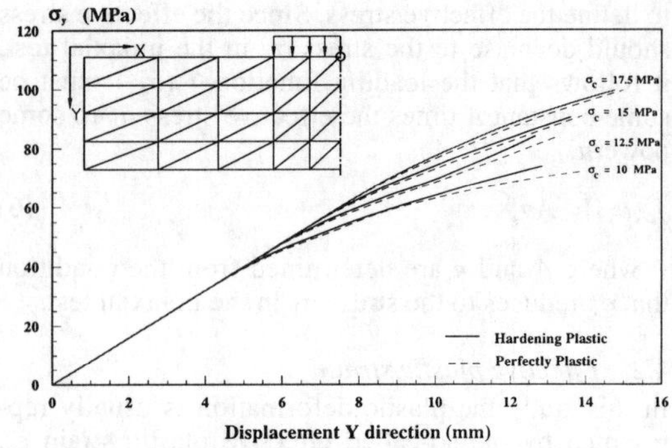

Figure 4: Stress-Displacement characteristics of strip footing with different σ_c.

materials always has a higher load-carrying capacity than does perfectly plastic behavior at the same load intensity.

REFERENCES

Viladkar, N.M. & J. Godbole 1993. Convenient forms of yield criteria in elasto-plastic analysis of geological materials. it Computer and Structure. 55:115-130.

Chen, W. 1994. *Constitutive equations for engineering materials(Part B)*. Elsevier, studies in applied mechanics 37.

Baz, M. 1997. Deep tunnel analysis: practical and theoretical considerations. Dissertation. University of Innsbruck

Computer Methods and Advances in Geomechanics, Desai et al. (eds) © 2001 Balkema, Rotterdam, ISBN 90 5809 183 X

An elastic-plastic solution for collinear joints in rock mass stressed

Cheng Wang
Department of Civil Engineering, Chongqing Jiaotong University, People's Republic of China

Anfu Deng
School of Civil Engineering, Chongqing University, People's Republic of China

ABSTRACT: The strength and stability of rock mass are determined by the properties of the joints in it. In this paper the mechanical problem of collinear joints in rock mass is simplified as the one of single joint in finite dimension rock mass stressed. The joint damage is analyzed by the improved near crack line analysis method in which the assumption of the small scale yielding at the crack tip has been abandoned. The influence of frictional and cohesive strength on the joint surfaces to the stress fields is considered in rock mass stressed. The equation between length of the plastic region on the joint line and the stress fields is given. In addition, the failure criterion of the joint rock mass is obtained. The data of a model test that is consisted of plaster stone and silicon soil has been used to verify the analytical results in this paper.

1 INTRODUCTION

In the earlier years, the fracture mechanics of metal material was applied to the rock fracture mechanics. Up to now the acquired results are not satisfactory because of the complicated properties of rock material. Nevertheless, the importance of the compressive shear fracture in rock mass slope; earthquake and mining engineering etc. have been noticed extensively. So far researchers have unfolded tentative studies about the compressive shear fracture of the rock mass, and the aspect on the criterion of compressive shear fracture has been acquired progresses. However, the study about mechanism of compressive shear fracture in rock mass is rarely seen to report. Now it is generally considered that the reason of compressive shear fracture is to link up each other small crack paralleled to the direction of maximal principal stress, but the process linked up the small cracks or joints cannot be explained reasonably. Therefore in this paper through the fundamental equations with Mohr-coulomb equation, the problem of compressive shear fracture in rock is solved in terms of the crack line field method. The elastic-plastic stress fields and the plastic zone size on joint line are found out. Then the damaged mechanism and the failure rule are explained better.

2 EQUILIBRIUM EQUATIONS AND YIELD CRITERION

There is a set of collinear joints in rock mass, shown as Fig.1. In order to obtain their analytical solutions of stress fields on the joint line, the collinear multi-joint problem can be simplified as the single

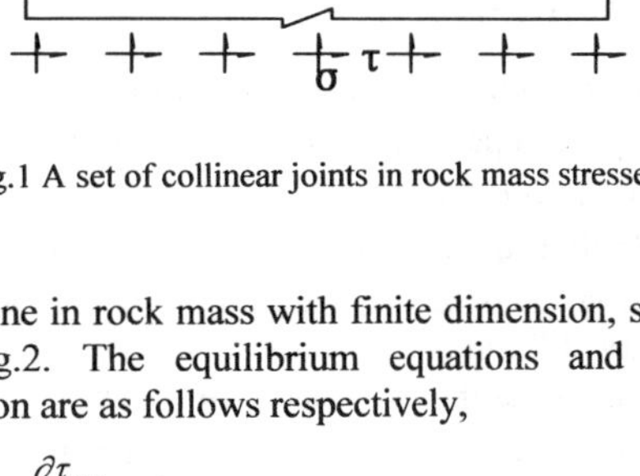

Fig.1 A set of collinear joints in rock mass stressed

joint one in rock mass with finite dimension, shown as Fig.2. The equilibrium equations and yield criterion are as follows respectively,

$$\frac{\partial \sigma_x}{\partial x} + \frac{\partial \tau_{xy}}{\partial y} = 0 \tag{1}$$

$$\frac{\partial \tau_{xy}}{\partial x}+\frac{\partial \sigma_y}{\partial y}=0 \quad (2)$$

$$(\sigma_x-\sigma_y)^2+4\tau_{xy}^2=[2c\cos\varphi+(\sigma_x+\sigma_y)\sin\varphi]^2 \quad (3)$$

where the c= cohesion of rock material, φ=frictional angle of rock material。

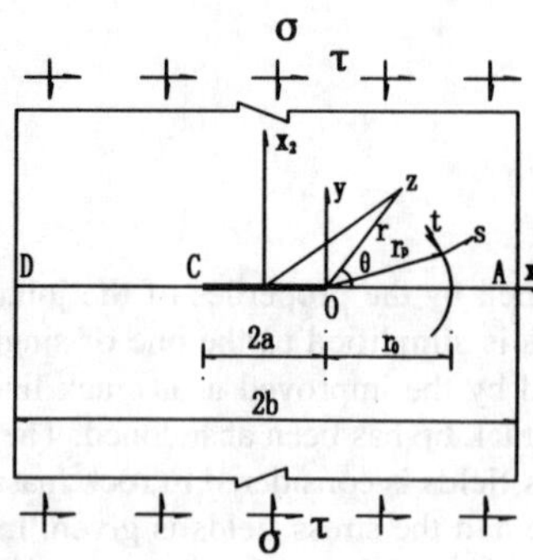

Fig.2 A joint in finite rock mass stressed

3 PLASTIC STRESS FIELDS NEAR JOINT LINE

As usual, the solutions of the plastic stresses in full ranges are required to find the set of partial differential equations consisted of equilibrium equations and yield condition, and the analytical solutions can be not obtained. However, near the joint line the difficult problem can be transformed into the set of ordinary differential equations to be solved. In the plastic region the stresses near the joint line can be expressed in the form:

$$\sigma_x^p=p_0(x)+p_1(x)y \quad (4)$$

$$\sigma_y^p=q_0(x)+q_1(x)y \quad (5)$$

$$\tau_{xy}^p=s_0(x)+s_2(x)y^2 \quad (6)$$

Where we have omitted the higher-order terms of y and τ_{xy}^p is considered to be symmetric with respect to y=0. Substitution of Eqs (4), (5), (6) into Eqs (1), (2), (3) after $q_0(x)$ is assumed to be a constant and collecting terms of the same order y yields a set of ordinary differential equations. Solving the set of equations gives the expressions of $p_0(x)$、$p_1(x)$、q_0、$q_1=0$、$s_0(x)$ and $s_2(x)$. Then the plastic stresses can be obtained by substituting them into equations of (4), (5), (6)

$$\sigma_x^p=2c\,tg\varphi+(\sec^2\varphi+tg^2\varphi)q_0-\frac{4s_0}{\cos^2\varphi}\frac{y}{x+A} \quad (7)$$

$$\sigma_y^p=q_0 \quad (8)$$

$$\tau_{xy}^p=s_0-\frac{2s_0}{\cos^2\varphi}\frac{y^2}{(x+A)^2} \quad (9)$$

where $s_0=c+q_0tg\varphi$, A is a integral constant , A and q_0 will be determined by the matching conditions on elastic-plastic boundary. From equations (7) the plastic stresses on tip of joint are not infinite just because of the constant A.

4 ELASTIC STRESS FIELDS NEAR JOINT LINE

For a discrete joint loaded by the compressive shear in an infinite body, the elastic stresses have been found by using Westergaard's complex stress function. However, the elastic stresses on the joint line in a finite body, shown as Fig.2, are very difficult to be found. As we were done (WANG Cheng, LI Xiaohong &DENG Anfu, 1998), the elastic stresses can be assumed in the form to be similar to the one in the infinite body, as follows:

$$\sigma_x^e=\sigma-\frac{1}{\sqrt{r(2a+r)}}[(a+r)M+\frac{3a^2}{2a+r}N\theta] \quad (10)$$

$$\sigma_y^e=-\frac{1}{\sqrt{r(2a+r)}}[(a+r)M-\frac{a^2}{2a+r}N\theta] \quad (11)$$

$$\tau_{xy}^e=\frac{1}{\sqrt{r(2a+r)}}[(a+r)N-\frac{a^2}{2a+r}M\theta] \quad (12)$$

Where M and N= the parameters related to the dimension 2b, the length of the joint 2a, stresses etc. They will be determined by reasonable boundary conditions on the joint surface. The conditions can be established by considering the force equilibrium of a part after cutting the body in two along the joint lines AO and CD in Fig.2, included the frictional and cohesive strength on the joint surface. The following equations are obtained:

$$2a\sigma+2q_0r_0-2\int_{r_0}^{b-a}(\sigma_y^e)_{\theta=0}dr=2b\sigma \quad (13)$$

$$2a(c_j+\sigma tg\varphi_j)+2s_0r_0+2\int_{r_0}^{b-a}(\tau_{xy}^e)_{\theta=0}dr=2b\tau \quad (14)$$

where c_j, φ_j =cohesion and frictional angle on the joint surface. The M and N can be found out by substituting Eq. (11) and Eq. (12) into Eq. (13) and

Eq. (14) as follows:

$$M = \frac{q_0 r_0 - (b-a)\sigma}{\sqrt{b^2 - a^2} - \sqrt{r_0(2a + r_0}} \quad (15)$$

$$N = \frac{b\tau - s_0 r_0 - a(c_j + \sigma tg\varphi_j)}{\sqrt{b^2 - a^2} - \sqrt{r_0(2a + r_0)}} \quad (16)$$

By substituting Eq. (15) and Eq. (16) into (10)~(12), the elastic stresses on the joint lines of satisfying all boundary conditions are obtained.

5 MATCHING RESULTS OF ELASTIC-PLASTIC FIELDS ON THEIR BOUNDARY

At the elastic-plastic boundary, the normal and tangent traction components are the following forms:

$$\sigma_n = \sigma_x n_x^2 + \sigma_y n_y^2 + 2\tau_{xy} n_x n_y \quad (17)$$

$$\sigma_t = (n_x^2 - n_y^2)\tau_{xy} + (\sigma_y - \sigma_x) n_x n_y \quad (18)$$

From the conditions that σ_n and σ_t are continuous across the elastic-plastic boundary (WANG Cheng &DENG Anfu 1997), by using Eq. (10)~(12) and (7)~(9) and collecting terms of same order of θ^0 and θ^1 four equations are obtained about $q_0, A, r_0, r_2/r_0$. Solving the four equations and letting dimensionless variables $\eta_q = 1 + \frac{q_0}{c} tg\varphi$, $\eta_\tau = \frac{\tau}{c}$, $\eta_\sigma = \frac{\sigma}{c}$, $\xi = \frac{r_0}{a}$, $\lambda = \frac{b}{a}$, $k = \frac{c_j}{c}$ yield

$$\eta_\tau = \frac{k + \eta_\sigma tg\varphi_j}{\lambda} + \frac{1 + m\xi}{m\lambda} \frac{[m(\lambda-1)+1]\eta_\sigma tg\varphi + m\xi + 1}{m\xi + 1 + 2tg^2\varphi} \quad (19)$$

$$\frac{A}{a} = [\frac{4}{3\cos^2\varphi}(1+\xi)(2+\xi) - 1]\xi \quad (20)$$

$$\frac{q_0}{c} = \frac{[m(\lambda-1)+1]\eta_\sigma - 2tg\varphi}{m\xi + 1 + 2tg^2\varphi} \quad (21)$$

$$2\frac{r_2}{r_0} = 1 + \frac{(1-\eta_q)m\xi + m(\lambda-1)\eta_\sigma tg\varphi}{(1+\xi)(2+\xi)(2\eta_q tg^2\varphi - \eta_\sigma tg\varphi)} \quad (22)$$

$$m = \frac{1+\xi}{\sqrt{\xi(2+\xi)(\lambda^2-1)} - \xi(2+\xi)} \quad (23)$$

The formula (19) denotes that the length of plastic region on joint line is varied with σ, τ. Fig. 3 shows four sets of curves about $\xi \sim \eta_\tau$ at $\varphi = 35°$ when η_σ= 0.2, 0.5, 1.0, 1.5 respectively. At the same reason, the other $\xi \sim \eta_\tau$ curves with respect to different φ、η_σ can be obtained.

Eq. (23) requires $\xi \le \lambda - 1$. When $\xi = \lambda - 1$, the collinear joints are linked up to the critical state of rock mass failure. Then Eq. (19) can be simplified as follows

$$(\eta_\tau)_{cr} = (k + \eta_\sigma tg\varphi_j)\frac{1}{\lambda} + (1 + \eta_\sigma tg\varphi)(1 - \frac{1}{\lambda}) \quad (24)$$

where $\frac{1}{\lambda}$=*a/b* denotes the ratio linked up, Eq. (27) means that the shear strength of rock mass is consisted of the joint surface strength and rock bridge strength.

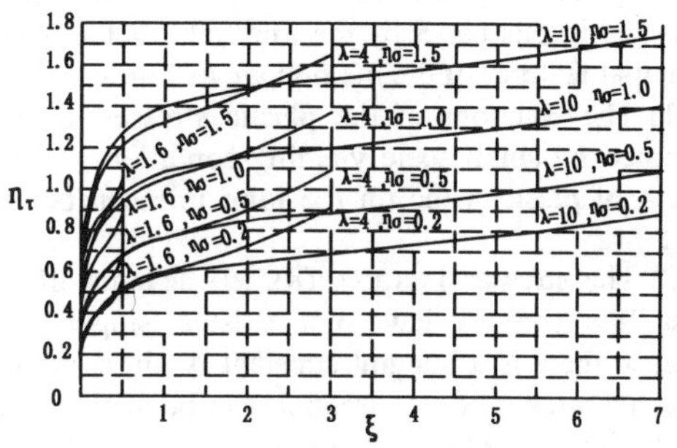

Fig.3 Curves between ξ and η_τ

6 COMPARISONS WITH MODEL TEST

The mechanical parameters of materials in the model (Xu Jinnan, 1998) are as follows: axial compressive strength $\sigma_c = 2.05 Mpa$; axial tensional strength $\sigma_t = 0.27 Mpa$; cohesion $c = 0.53 Mpa$; the fractional angle $\varphi = 35°$; the cohesion and fractional angle on the joint surface are $c_j = 0.1 Mpa$ and $\varphi_j = 10°$. When vertical normal stress on the joint surface was kept constant, the applied horizontal load was gradually increased to failure. The comparisons with model test are listed in Table 1. It is known from the Tab.1 that the calculated values are satisfactory agreement with the tested ones.

Table 1. Comparisons between calculated values and tested ones.

a/b	Stress (0.1Mpa)	Shear strength (0.1Mpa)		a/b	Stress (0.1Mpa)	Shear strength (0.1Mpa)	
		Tested	Calculated			Tested	Calculated
0%	3.0	7.34	7.40	43%	3.5	5.43	5.11
0%	4.0	8.56	8.10	44%	3.0	5.36	4.82
0%	5.0	8.99	8.80	45%	3.5	5.03	4.99
9%	0.8	5.76	5.44	46%	4.0	5.63	5.16
19%	4.0	7.86	6.88	46%	6.0	6.10	6.08
13%	3.0	6.43	6.64	50%	5.0	5.35	5.34
13%	4.0	6.80	7.27	60%	0.8	3.52	3.03
13%	6.0	8.06	8.53	60%	2.0	3.65	3.49
20%	5.0	7.49	7.42	62%	3.0	4.34	3.76
42%	5.0	5.91	5.89	62%	4.0	4.14	4.14
43%	0.8	4.39	3.83	62%	6.0	4.95	4.89

7 CONCLUSIONS

The elastic-plastic stresses near joint line in rock mass loaded by compressive shear have been analyzed in this paper by the line field method. The principal features of the method are as follows:

(1) The solutions of the plastic stresses are not confined by small scale yielding conditions, so the length of plastic zone on the joint line can be any size;

(2) The stresses Eqs (7), (8), (9) near tip of the joint line ($r \rightarrow 0$) have not existed singularities because there is a integral constant A, therefore the maximal stress and its direction can be analyzed by the stresses at the tip of the joint;

(3) The damage size on the joint line must satisfy the relation $0 \le \xi \le \lambda - 1$. When $\xi = \lambda - 1$, the collinear joints are linked up. Then failure criterion of the rock mass can be defined by the formula (24);

(4) If the other yielding criterion of rock is used in the fundamental equations, the corresponding results can be obtained by the method of this paper.

REFERENCE

YI Zhijian etc. Exact Solutions of Near Crack Line Fields for Mode I Crack under Plane Stress in an Elastic-Perfectly Plastic Solid. *Applied Mathematics and Mechanics,* 1996,17(4): 335- 342.

WANG Cheng & YI Zhijian 1998. Elastic-plastic analysis of discrete joint with mode II load. *China Journal of rock mechanics and engineering.* 17 (2): 153-157.

WANG Cheng & DENG Anfu 1997. elastic-plastic analysis of discrete joint loaded by compressive shear. *Journal of Chongqing Jianzhu University.* 19 (2): 31-34.

WANG Cheng, DENG Anfu, Damage Mechanics Analysis of Fissures in Rock mass Loaded by Shear. *Proc. Ninth Int. Conf. Of IACMAG*, A.A. Balkema Publishers, Netherlands, Dec. 1997. 725-728

WANG Cheng, LI Xiaohong & Deng Anfu 1998. Study Of Shear Strength On Collinear Multi- cracked Plate, *Proceedings Of International Symposium On Strength Theories---Applications And Developments & Prospects for 21st Century 1998: 841-846*

WANG Cheng & ZHENG Lukun 1998. Near Crack Line Elas- tic-Plastic Analysis For A Finite Plate Loaded By Two Pairs Of Anti-plane Point Forces, *Applied Mathematics And Mechanics* 98.6: 547- 554

XU Jingnan & ZHU Weishen 1995. The Strength of collinear cracks failure under shearing- compressive stress. Chinese Journal of rock mechanics and engineering. 14 (4): 306~311

Computer Methods and Advances in Geomechanics, Desai et al. (eds) © 2001 Balkema, Rotterdam, ISBN 90 5809 183 X

A static relaxation solution for rock joints and interfaces

Chen Wensheng
Institute of Rock and Soil Mechanics, Chinese Academy of Sciences, Wuhan, People's Republic of China

Ge Xiurun
Institute of Rock and Soil Mechanics, Chinese Academy of Sciences, Wuhan, and Institute of Geotechnical Engineering, Shanghai Jiao Tong University, People's Republic of China

Helmut F. Schweiger
Institute for Soil Mechanics and Foundation Engineering, Technical University of Graz, Austria

ABSTRACT: Based on Static Relaxation Lagrangian Analysis(SRLA) method proposed by authors (1997), a method to solve rock joints and interfaces is proposed in this paper. The method can simulate complex deformation caused by rock joints and interfaces. The relevant formulae of the presented method are deduced. Some typical examples are given in this paper.

1 INTRODUCTION

Rockmass is a king of material with various discontinuous interfaces, including rock stratum, faults, joints and other similar textures, different properties, damaged zone caused by excavation steps, etc,. The aim of a numerical method for rockmass is mainly to simulate the important mechanics characteristics mentioned in above, because nonlinear deformation and latent torsion deformation are usually caused by mentioned characteristics.

The finite element method and boundary element method are often adopted to do numerical analysis of rockmass. Some computer codes of these methods developed joint elements or slide lines (Zienkiewicz O. C., Valliappan S. and King I. P. 1968, Ghaboussi J., Wilson E. L. and Isenberg J. 1973, Heuze F. E. and Barbour T. G. 1982). Lately, the conception of discrete element was introduced (Cundall P. A. 1971), the simulation of the complex deformation or movement along joints and interfaces becomes much easier than before. Generally, the numerical methods to simulate discontinuities fall into two classes, the first class is based on continuum mechanics while the second is on discrete model. Joint element in finite element method obviously belongs to the first class. Because the joint element has some hypothesis and is sensitive to global equations of the system, it is difficult to model large scale displacement and complex deformation of jointed rockmass. It can only model the deformation of discontinuities at the same grade as continuous rock blocks. The second class can simulate complex deformation behaviors of discontinuities. However, the discrete rockmass like discrete model seldom exist in practical problem and it is more difficult to discrete the zone into blocks than finite element method. This is the reason why the results obtained from this discrete element class methods are quite different from measured *in-situ* and the second class has not as widely applied as the first class such as finite element method so far.

In this paper, a new method to solve joints and interfaces in rockmass problem is proposed based on the static relaxation Lagrangian analysis method (SRLA) in reference (Chen. etc., 997). It takes the advantages of both the first class and the second one. The method here can simulate deformations of rockmass especially the complex deformations along joints and interfaces more approximately to what rockmass really behaves.

2 DISCONTINUOUS INTERFACE AND NODES ON THE INTERFACE

The deformation of rockmass consists of the deformation of continuous rock blocks, which are separated by joints or interfaces, and the deformation along discontinuities such as joints and interfaces. It is a combination of continuous deformation (of rock block) and discontinuous deformation (along joints and interfaces). Hibion S in Japan pointed out that the opening deformation of rockmass is about 63 percent of the total deformation after observed the damaged zone in large scale underground chamber excavation (Hibion S. and Motejma M. 1995). Obviously the opening deformation is the discontinuous part of deformation. However, the finite element method based on continuum mechanics principles can not present this part of deformation satisfyingly.

In order to get better results closing to what it should be when mechanics analysis of geotechnical project is carried out, these two parts deformation of rockmass and their respective characteristic in mechanical behaving should be taken into account. In this paper, some presuppositions are made for the discontinuities:

(1)Rock blocks separated by discontinuities are continuous medium and can be calculated by the static relaxation method proposed by authors.

(2)The stress between interface satisfied no tension principle and Mohr-Coulomb principle described as below:

$$P_n \le 0, \qquad |P_\tau| \le C - F \cdot P_n \qquad (1)$$

where C is the cohesion and $F = tg\varphi$, φ is the friction angle of the joint surface.

(3)No overlap and zero thickness between joint surface. The joint surfaces are straight line or segments of straight lines.

The nodes on joint surface usually have two cases as shown in Figure 1.

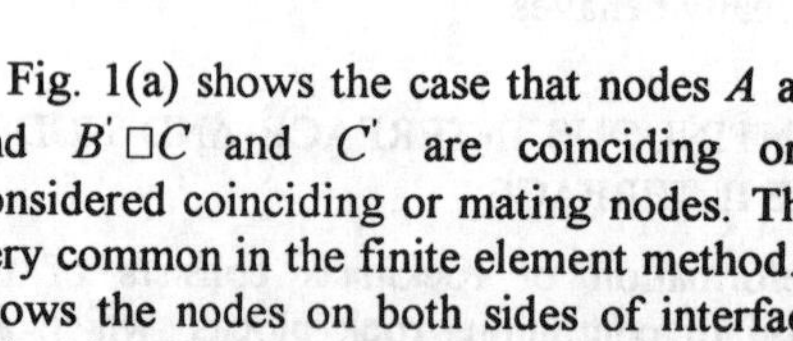

(a) (b)

Fig.1 Nodes on joint surface

Fig. 1(a) shows the case that nodes A and A' □B and B' □C and C' are coinciding or can be considered coinciding or mating nodes. This case is very common in the finite element method. Fig. 1(b) shows the nodes on both sides of interface do not coincide or are not mating nodes. It is the most common situation when a zone is made to mesh element. Fig. 1(a) can also be taken as a special form of Fig. 1(b). The method proposed by authors (1997) is a node by node relaxation method, the mechanics analysis is thus simplified to perform analysis of the nodes on the interface.

If one node is on one side of the interface, its relative nodes and element are only in the same side as itself. For example, in Fig.1, the relative nodes and elements of node B are nodes A,C, etc,. and elements e_1 □ e_2. Then, the nodes on the interface can be calculated by the relaxation method introduced by authors(1997).

In Fig. 1(b), from one node such as B to draw a line vertical to the other side, the intersection point B'' can be obtained. If B'' is on the line segment $A'B'$, segment $A'B'$ is taken as relative segment of node B. Nodes A' and B' are called interface relative nodes of node B. In Fig. 2, if B'' is close to (or coincides with) node A' (or B'), the relative segment can be judged by the direction in which point B'' closes to node A'. Segment $B''B$ is the distance of node B to its interface relative segment. Taking the normal direction of $A'B'$ as positive, when:

$$B''B > \delta \quad \text{No contact on node B}$$
$$B''B \le \delta \quad \text{Contact established} \qquad (2)$$

where δ is a permitted positive value.

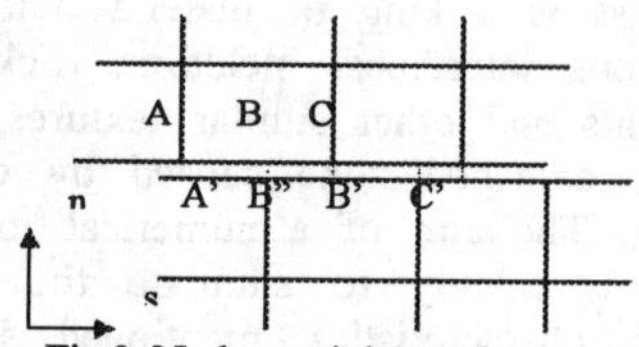

Fig.2 Nodes on joint surface

Fig.3 shows some special cases of interface. These cases can be solved as follow:

Case a

Nodes A□A' are on the same position (node A' does not exist during coding). Node A is at the crack tip. Obviously the relative elements of node A are e_1, e_2, e_3, e_4 shown in Fig. 3(a). During node A is relaxed, no interface effect will occur. But node A is interface relative node of node B or B'.

Case b

This case will often take place when interface is through to the boundary of rockmass. In Fig. 3(b), node A' has no interface relative segment but it is interface relative node of node A. Then node A can be regarded as relative node of node A'. Relative nodes on interface are reciprocal.

Case c

Interface forks and nodes A and B are on both interfaces 1 and 2. Then nodes A and B have their relative nodes and segment of interface 1 and 2, respectively.

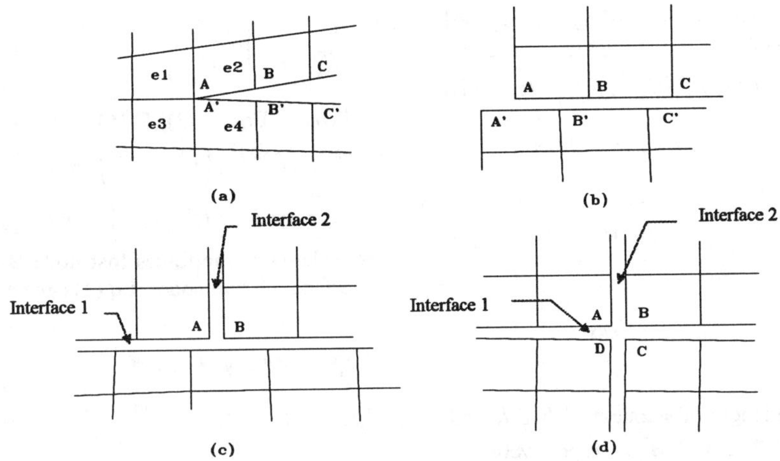

Fig.3 Some special cases of interface.

Case d

Joint 1 and 2 intersect each other. Nodes A, B, C and D are at the intersecting point and locate both on joints 1 and 2. They have their relative nodes and segment on joint 1 and 2 respectively.

Fig.2 and Fig.3 show all cases of node on joint interfaces.

3 STATIC RELAXATION SOLUTION OF NODES ON JOINT INTERFACE

Fig.4 shows Fig.2 in details. During node B is to be relaxed, its relative nodes are A, B, C,D, E and F, relative elements are e_1, e_2 and its interface relative segment is $A'B'$.

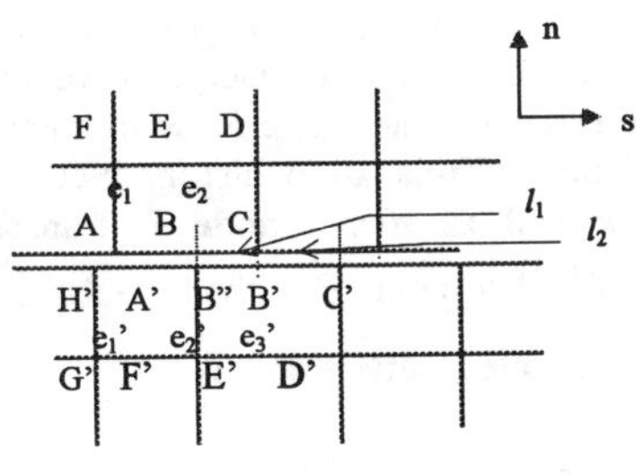

Fig.4 Node B on interface and its relative nodes and elements, relative segment

The distance from node B to line $A'B'$ is $B''B$. The contact can be checked according to Eq. (2). If node B contacts with segment $A'B'$, B will act on $A'B'$. Otherwise, node B will calculated by the method proposed by the authors(1997). Supposing a unbalance force $F_B\left(f_x^B, f_y^B\right)$ is acting on node B□force effect coefficients of B are $\left(I_{u_B}^x, I_{v_B}^x\right)$, $\left(I_{u_B}^y, I_{v_B}^y\right)$ and of A' and B' are $\left(I_{u_{A'}}^x, I_{v_{A'}}^x\right)$, $\left(I_{u_A}^y, I_{v_A}^y\right)$, $\left(I_{u_B}^x, I_{v_B}^x\right)$, $\left(I_{u_{B'}}^y, I_{v_{B'}}^y\right)$ respectively. When node B contacts with $A'B'$, it cannot be calculated by the method put out by Chen(1998) because a part of unbalance force of B will be transferred to segment $A'B'$. Nodes A' and B' will get equivalent forces from B.

Supposing the force transferred to B'' from B is $F_B^{B''}\left(f_x^{BB''}, f_y^{BB''}\right)$ and the unbalance force left on B is $F_B^O\left(f_x^{BO}, f_y^{BO}\right)$

$$\begin{cases} f_x^{BO} = f_x^B - f_x^{BB''} \\ f_y^{BO} = f_y^B - f_y^{BB''} \end{cases} \tag{3}$$

where F_B^O will cause node B to have a displacement increment $(\Delta x_B, \Delta y_B)$ during current relaxation step. According to the force distribution method, the following formulae can be obtained:

$$\begin{Bmatrix} \Delta x_B \\ \Delta y_B \end{Bmatrix} = \begin{bmatrix} I_{u_B}^x & I_{u_B}^y \\ I_{v_B}^x & I_{v_B}^y \end{bmatrix} \cdot \begin{Bmatrix} f_x^{OB} \\ f_y^{OB} \end{Bmatrix} \tag{4}$$

The force transferred to B'', $F_B^{B''}$, will be transferred to nodes A' and B' according to equivalent principle. Let the forces transferred to A' and B' are $F_B^{A'}\left(f_x^{BA'}, f_y^{BA'}\right)$ and $F_B^{B'}\left(f_x^{BB'}, f_y^{BB'}\right)$ respectively, define $m = \frac{l_1}{l}$, $n = \frac{l_2}{l}$, $(n+m=1)$ □then:

$$\begin{cases} f_x^{BA'} = nf_x^{BB''} \\ f_y^{BA'} = nf_y^{BB''} \end{cases} \quad (5)$$

$$\begin{cases} f_x^{BB'} = mf_x^{BB''} \\ f_y^{BB'} = mf_y^{BB''} \end{cases} \quad (6)$$

where l is the length of segment $A'B'$ □l_1 and l_2 are the length of $B''A'$ and $B''B'$ respectively.

The displacement increment of node A' can be obtained from Eqs. (5) and (6)

$$\begin{Bmatrix} \Delta x_{A'} \\ \Delta y_{A'} \end{Bmatrix} = \begin{bmatrix} I_{u_{A'}}^x & I_{u_{A'}}^y \\ I_{v_{A'}}^x & I_{v_{A'}}^y \end{bmatrix} \cdot \begin{Bmatrix} f_x^{BA'} \\ f_y^{BA'} \end{Bmatrix} \quad (7)$$

Similarly, the displacement increment of B' is

$$\begin{Bmatrix} \Delta x_{B'} \\ \Delta y_{B'} \end{Bmatrix} = \begin{bmatrix} I_{u_{B'}}^x & I_{u_{B'}}^y \\ I_{v_{B'}}^x & I_{v_{B'}}^y \end{bmatrix} \cdot \begin{Bmatrix} f_x^{BB'} \\ f_y^{BB'} \end{Bmatrix} \quad (8)$$

From Eqs.(7) and (8) we can obtain the displacement increment of point B'' using linear interpolation

$$\begin{Bmatrix} \Delta x_{B''} \\ \Delta y_{B''} \end{Bmatrix} = n \cdot \begin{Bmatrix} \Delta x_{A'} \\ \Delta y_{A'} \end{Bmatrix} + m \cdot \begin{Bmatrix} \Delta x_{B'} \\ \Delta y_{B'} \end{Bmatrix} \quad (9)$$

Various solutions are given for different cases as follows:

(1) Eq. (1) (no tension and Mohr-Coulomb principle) is satisfied on node B. Relative displacement in shear direction and opening displacement in normal direction will not occur on B. Then we have

$$\begin{Bmatrix} \Delta x_{B''} \\ \Delta y_{B''} \end{Bmatrix} = \begin{Bmatrix} \Delta x_B \\ \Delta y_B \end{Bmatrix} \quad (10)$$

Substituting Eqs. (4)~(9) into (10) we get

$$n \cdot \begin{bmatrix} I_{u_A}^x & I_{u_A}^y \\ I_{v_A}^x & I_{v_A}^y \end{bmatrix} \cdot \begin{Bmatrix} f_x^{BA} \\ f_y^{BA} \end{Bmatrix} + m \begin{bmatrix} I_{u_B}^x & I_{u_B}^y \\ I_{v_B}^x & I_{v_B}^y \end{bmatrix} \cdot \begin{Bmatrix} f_x^{BB} \\ f_y^{BB} \end{Bmatrix}$$

$$= \begin{bmatrix} I_{u_B}^x & I_{u_B}^y \\ I_{v_B}^x & I_{v_B}^y \end{bmatrix} \cdot \begin{Bmatrix} f_x^{BO} \\ f_y^{BO} \end{Bmatrix} \quad (11)$$

From Eqs. (3),(5),(6),(11) and $n+m=1$, $\left\{f_x^{BO},\ f_y^{BO}\right\}^T$, $\left\{f_x^{BA'}, f_y^{BA'}\right\}^T$ and $\left\{f_x^{BB'}, f_y^{BB'}\right\}^T$ can be obtained. $\left\{f_x^{BA'}, f_y^{BA'}\right\}^T$ and $\left\{f_x^{BB'}, f_y^{BB'}\right\}^T$ are considered as the forces that node B transfers to its interface relative nodes. Eq.(3) can be further written as

$$\begin{cases} f_x^{BO} = f_x^B - f_x^{BA'} - f_x^{BB'} \\ f_y^{BO} = f_y^B - f_y^{BA'} - f_y^{BB'} \end{cases} \quad (12)$$

Eq.(12) shows the transferring force process more clearly.

(2) The force transferred from B to its relative segment, we obtained in case (1), can not satisfy the shear stress condition in Eq. (1). Node B will occur relative shear displacement along segment $A'B'$. Then two conditions must be satisfied: The first is no overlap at point B or in other words, node B has the same normal displacement increment as node B'' □The second is that the shear force at node B satisfies the equal-sign in Eq. (1). From the first condition the following equation can be obtained:

$$-\sin\varphi \cdot \Delta x_{B''} + \cos\varphi \cdot \Delta y_{B''} = -\sin\varphi \cdot \Delta x_B + \cos\varphi \cdot \Delta y_B \quad (13)$$

here φ is the angle of segment $A'B'$ crossing with the horizontal axis(counterclockwise is positive). One thing must be reminded is that the force acting on the interface at node B is just the force which node B transfers to its interface relative segment or interface relative nodes. The force in current step is accumulation of relaxation history. Set the current step is n and at step i node B transfers force $\left(f_{x,i}^{BB''},\ f_{y,i}^{BB''}\right)$ to point B'' □then the force acting on interface at node B after step n is

$$\left.\begin{aligned} \sum_{i=1}^{n-1} f_{x,i}^{BB''} + f_{x,n}^{BB''} = \sum f_x^{BB''} \\ \sum_{i=1}^{n-1} f_{y,i}^{BB''} + f_{y,n}^{BB''} = \sum f_y^{BB''} \end{aligned}\right\} \quad (14)$$

the first item at the left of equal-sign is the accumulation force that node B transfer to its interface relative segment, the increment

$\left(f_{x,n}^{BB''},\ f_{y,n}^{BB''}\right)$ is just $\left(f_x^{BB''},\ f_y^{BB''}\right)$ in Eq. (3). $\left(\sum f_x^{BB''},\ \sum f_y^{BB''}\right)$ is the force after current step. As the shear force satisfies the equation of Mohr-Coulomb principle, set shear force is p_τ, it can be expressed as

$$p_\tau = \cos\varphi \cdot \sum f_x^{BB''} + \sin\varphi \cdot \sum f_y^{BB''} \qquad (15)$$

then we get

$$|p_\tau| - c + tg\varphi \cdot \left(-\sin\varphi \cdot \sum f_x^{BB''} + \cos\varphi \cdot \sum f_y^{BB''}\right) = 0 \qquad (16)$$

here C is the cohesion of segment $A'B'$.

From Eqs. (16) and (13), the force $F_B^{B''}\left(f_x^{BB''},\ f_y^{BB''}\right)$ transferred from B to B'' is obtained and the incremental displacements of relative nodes can be obtained successively.

(3) The force transferred from B to its relative segment, we obtained in case (1), can not satisfy the no tension condition in Eq. (1). We have

$$-\sin\varphi \cdot \sum f_x^{BB''} + \cos\varphi \cdot \sum f_y^{BB''} \geq 0 \qquad (17)$$

then the contact is cancelled at node B, the force transferred is

$$f_x^{BB''} = f_y^{BB''} = 0 \qquad (18)$$

Meanwhile, the accumulation force in relaxation history $\left(\sum_{i=1}^{n-1} f_{x,i}^{BB''},\ \sum_{i=1}^{n-1} f_{y,i}^{BB''}\right)$ should be released and added to the unbalance force of node B. The force on interface relative nodes A' and B' must cut the corresponding parts.

(4) If both case (2) and (3) exist□we can do as case (3).

From above, we can mind that normal and shear contact stiffnesses K_n and K_s are not needed in the computational process□the nodes on interface can slide along the surface and be separated from the surface. This is the main feature of the proposed method in this paper.

4 TEST EXAMPLES

(1) Fig.5 (a-1) and (a-2) show a same cantilever. There is a perfectly smooth interface between the two layers of elements. The element size is 1×1m^2. The concentrating force P is 20×10^{3}KN shown in Fig.5. Poisson's ratio ν is 0.3 and mass is omitted. Fig.5 (a-2), (a-3) and (a-4) show the opening process of interface. Fig.5 (b-2), (b-) and (b-5) show the sliding process.

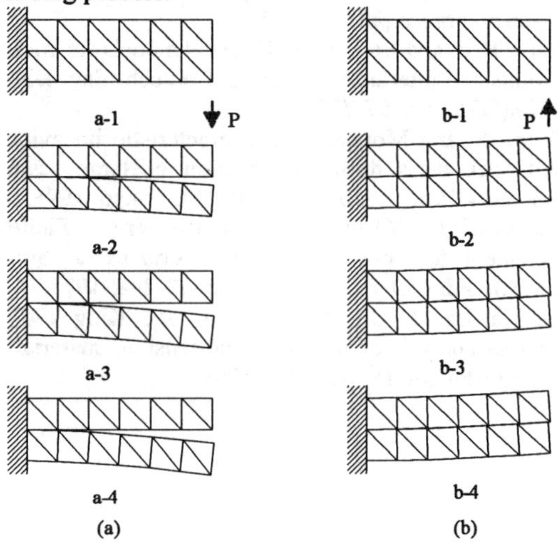

Fig.5 Test example (1)

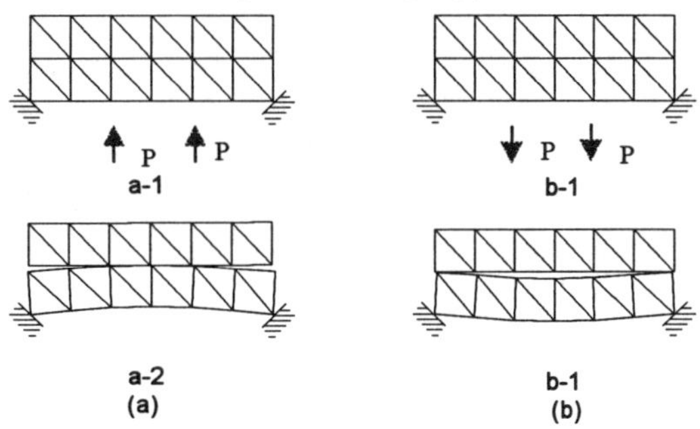

Fig.6 Test example (2)

(2) Fig 6 shows the same beam whereas the two ends are fixed. P is 80×10^{3}KN. Location and direction of P is shown in Fig.6. Fig. 6(a-2) and (b-2) show more complex deformation behaviors of the interface.

REFERENCES

Chen Wensheng. Ge Xiurun & H.F Schweiger. A new computational scheme in geotechnical engineering analysis. 9th Int. Comp. Meth. & Adv. Geomech. pp391-396.1997.11 Wuhan China.

Chen Wensheng. Research on relaxation numerical methods in rock and soil mechanics and its application. Ph.D thesis. Institute of Rock and Soil Mechanics, Chinese Academy of Sciences. 1998 (in Chinese)

Coetzee M.J., et al. FLAC Basic. Itasca Consulting

Group Inc. 1993.

Cundall P. A. A computer model for simulating progressive large scale movement in blocky rock systems. Symp. Int. Society of rock mechanics. Nancy. France.ppII-8. 1971.

Heuze F. E. and Barbour T. G., New models for rock joints and interfaces, ASCE. J. Geotech. Eng. Div. 108(GT5). pp757-776. 1982.

Hibion S. and Motejma M. Characteristic behavior of rock mass during excavation of large scale caverns. Proc. 8th, Int. Conf. ISRM. Tokyo. 1995.

Ghaboussi J., Wilson E. L. and Isenberg J. Finite element for rock joints and interfaces. J. Soil Mech. Fdns. Div., ASCE. 99. pp833-848. 1973

Zienkiewicz O. C., Valliappan S. and King I. P. Stress analysis of rock as a 'no-tension' material. Geotechnique 18. pp56-66. 1968.

Computer Methods and Advances in Geomechanics, Desai et al. (eds) © 2001 Balkema, Rotterdam, ISBN 90 5809 183 X

Time-dependent behavior of rock joints in high slope of shiplock, TGP

Ding Xiuli & Zhou Huoming
Yangtze River Scientific Research Institute, Wuhan, People's Republic of China
Yang Chunhe & Bai Shiwei
Institute of Rock and Soil Mechanics, Chinese Academy of Sciences, Wuhan, People's Republic of China

ABSTRACT: The results of a large-scale shear creep test on rock joints without filling are described. The investigation was made in order to determine the creep properties and behavior of the discontinuities in the high slope of ship lock at Three Gorges Project (TGP), China. The shear stress-displacement-time responses of joints were obtained for different normal compressive stresses. The effects of loading history on shear displacement of joint surfaces were analyzed. Based on the experimental results, the creep constitutive equations of joints in shear were proposed. The relation between parameters in equations and stress was also discussed.

1 INTRODUCTION

The assessment of rock slope stability often relies on rock short-term behavior. The time plays a significant role in the majority of processes occurring in the rock mass of a rock slope. The knowledge about the time-dependent behavior of rock mass is essential in predicting the long-term stability and deformation of rock slope. Since the time-dependent deformation and strength of rock mass is controlled in large part by the creep behavior of rock joints, it is important to understand the creep properties of joints.

The creep behavior of natural joints is rather complicated. It depends not only on the characteristics of joint surfaces itself, such as the degree of surface roughness, the joint wall compressive strength, the degree of weathering, the presence or absence of filler, but also on the external conditions like stress regime, moisture content, temperature etc. Previous investigations were largely performed on artificial joints under laboratory conditions. Many valuable results were obtained [1][2][3]. In situ creep tests can reflect some true creep features of actual rock joints.

The high slope of ship lock at TGP is formed by deep excavation in rock mass, with a maximum depth of 170m and aperture width of 350m. The slope consists of mainly hard and competent granite. In view of the large area disturbed by deep excavation, the stress regime is strongly altered. The stress redistribution would occur throughout the rock mass. The process of stress equilibration includes both instantaneous and time-dependent components. The results Monitoring on the rock mass behavior during excavation of the slope showed show that the strains of rock mass have exhibited a pronounced time-dependency. This necessitates us to gain a better understanding of rock mass creep movements in the high slope, especially the creep properties of rock joints.

The rock joints in the ship lock area at TGP are classified into two groups of joints without soft and loose filling and with weak filler material. The largest parts of discontinuities are joints without filling. The joints without filling and with slight asperity surfaces exist widely in the rock mass of slope, they are main objective tested and studied in this investigation.

2 IN SITU CREEP TEST

2.1 Test situation and description of joint samples

The shear creep tests on rock joints in site were carried out in an exploratory at 100m below the ground surface, which crosses the north slope of shiplift and the south slope of permanent shiplock. The experimental field was located at about 360-375m from the adit entrance, where six of rock samples containing similar joint surfaces were formed in floor by cutting artificially. The rock mass tested was fresh and hard amphibole-plagioclase granite. According to the post-test description on joint surfaces morphology, the surfaces were generally planar, with slight asperity microscopically, and the amplitude of asperities was 1-5mm, without filling. All tested joint surface sizes were 500×500mm. The test adit kept a almost constant temperature(around 19℃) and humidity(nearly 100%).

2.2 Experimental setup

Figure 1 shows the experimental setup. A groove was formed on the top of the rock joint sample by means of chiseling artificially. The hydraulic pillow was mounted to apply a normal load. Then the rock mass on two sides of the sample was cut off, and the shear loading mechanism was assembled. Four measuring marks were embedded near two ends of the sample, and 50mm away from the shear surface, where eight dial gauges were fixed with an accuracy of 0.001mm. They were used to monitor the relative motions of opposing joint surfaces, four for normal relative displacement and four for shear relative displacement, assuming the rock mass beneath the joint surface remained stationary. In order to guarantee the maintenance of constant loads throughout the experiment, the loading devices capable of holding a constant pressure were employed. The normal loading system consists of nitrogen bomb, pressure reducing valve, oil/nitrogen accumulator, and hydraulic pillow. The shear loading apparatus consists of hydraulic pump, pressure maintaining tank, filling valve, and jacks etc.

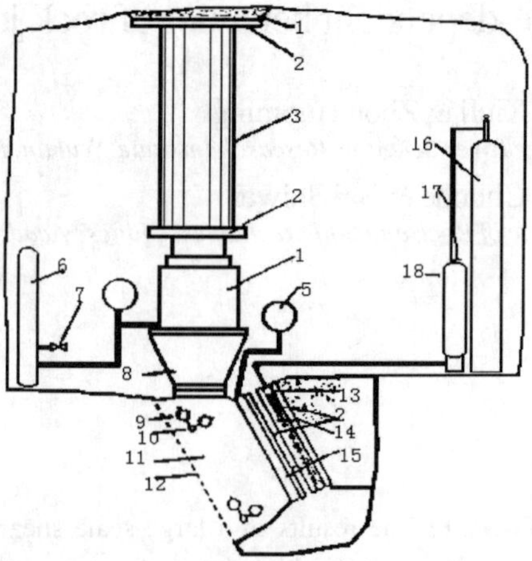

1-mortar; 2-steel plate; 3-loading pillar; 4-jacks; 5-oil pressure gauge; 6- pressure maintaining tank; 7-incoming oil valve; 8-transferring pressure tank; 9-dial gauge; 10- measuring marks; 11-sample; 12- joint surfaces; 13- pressure pillow; 14-spherical guide; 15- hydraulic pillow; 16- nitrogen bomb; 17-pressure reducing valve; 18- pressure maintaining tank

2.3 Test procedure

The inclined push method was used in this creep experiment. The shear direction was parallel to joint surface of samples. Differential normal compressive stresses were rapidly applied to six specimens, which ranged from 0.3 to 1.8 MPa. Throughout the creep tests, normal stresses remained constant. When the normal deformation of samples reached an approximate invariant value, the initial shear loads were applied (within about 20 seconds), and the values of shear stress were 20% of the estimated peak shear strength on joint surfaces. Afterwards, the shear loads were increased step by step, and at each load step the shear stresses were maintained constant. The pressures exerted on the samples were given by:

$$q = \tau / \cos\theta \tag{1}$$

$$p = (\sigma - q)\sin\theta \tag{2}$$

where q is the pressure in the vertical direction; p is the normal pressure; θ is the angle of the joint surface with the vertical direction; τ is the shear stress and σ is the normal stress.

After each step shear stress was applied, the instantaneous displacements were recorded immediately. The creep displacements were recorded in different time intervals: 1/12 hours, 1/6 hours, 1/4 hours, 1/2 hours, 1 hours, 2 hours, 4 hours, 8 hours, 12 hours, 16 hours, and 24 hours. 24 hours later. The displacements with respect to time were recorded every for 8 hours. If the magnitude of displacements every day (24 hours) was less than 0.001mm, the shear displacements of joints can be regarded as relatively constant, and the shear loads were increased to the next level. For each of shear load level, the duration of the test was not less than 120 hours. The tests were terminated until the sample failed.

During the creep experiments, the changes of the normal pressures and shear pressures were adjusted at any time so that they were maintained constant. The variation in stress was not more than 2 of the prescribed stress on the sample.

3 TEST RESULTS AND ANALYSIS

The shear stress-displacement-time responses of joints at various normal compressive stresses were obtained, The experimental results for three samples are shown in Figures 2. At the moment when each level of shear stress was applied, the instantaneous shear displacements of the joint surfaces were observed. Afterwards, the time-dependent shear displacements (creep displacements) increased with time under constant loads, and the creep rate decreased with time. The test results show that the displacements along joint surfaces tend towards invariant with time under most levels of shear stress. This process typically need 5-10 days. Under high levels of shear stress, quasi-linear creep could be observed for some samples. When shear stress increased up to a certain value, the joint surface occurred slip and the shear displacements increased rapidly.

Because incremental loading method was adopted in the creep tests, irrecoverable plastic deformation in tangential direction caused by the former step shear loading will have an effect on shear deformation corresponding to next step load. The shear creep deformation may decrease with increasing shear stress on the contrary for the same duration, and the experimental curves obtained from creep tests breach of the mechanical behavior of discontinuities in creep process. For the purpose of describing true creep properties of joint surfaces at various shear stress levels, the effect of loading history must be considered in the process of dealing with the experimental datum. Fig.3 shows the creep curves corrected by accounting for the influences of loading history. The processed curves can reflect much the intrinsic creep mechanical behavior of joints.

Based on the experimental results, an empirical equation describing the relationship between shear displacement and time for joints in shear was proposed as:

$$u(t) = A + B \ln t + ct \qquad (3)$$

where u is the shear displacement, t is time, and A, B, and c are constants for given load. The second term and the third term in equation (3) on the right-hand side represent primary transient creep and secondary steady-state creep respectively. The tertiary or accelerating creep almost couldn't be observed in tests. The stage creep isn't considered. A regressive analysis is finished by means of the least square method. The creep constitutive equations for the various levels of shear stress and all parameters in equation (3) are determined.

In order to give the relationship between coefficient A, B, c and applied stress, we introduce a new term τ_p, which is named the shear creep strength of joints. In a physical sense,

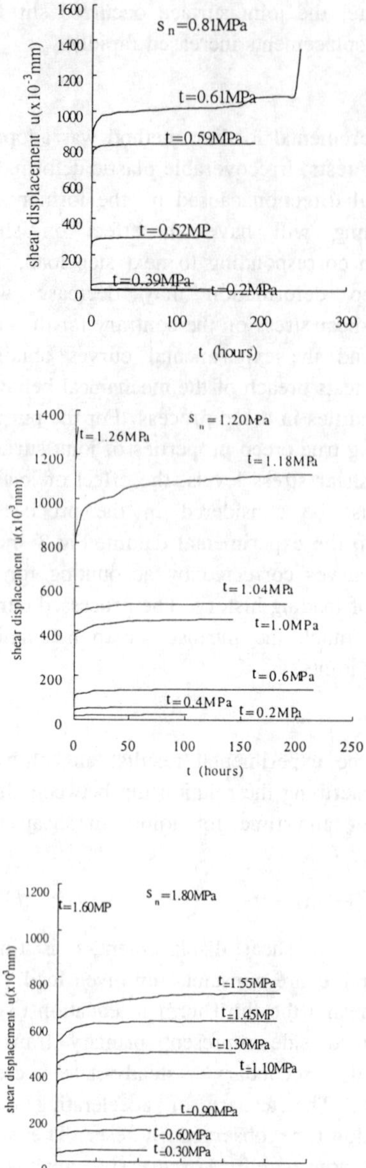

Fig. 2 Shear stress-displacement-time responses of joints

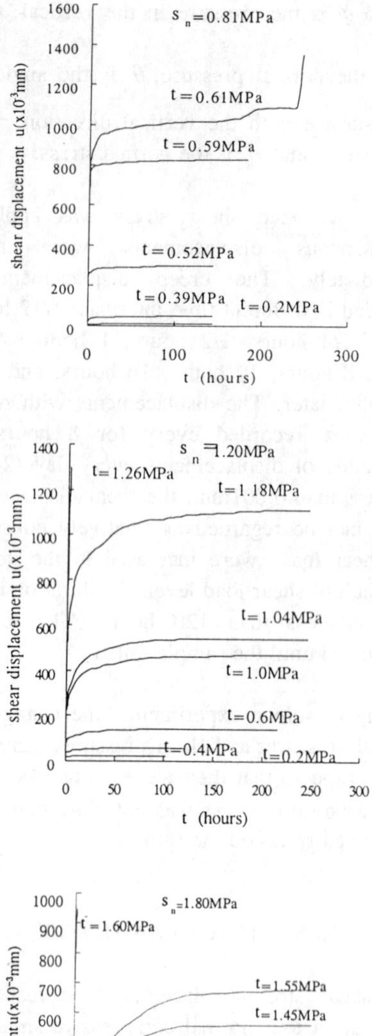

Fig. 3 Shear stress-displacement-time responses of joints considering the effects of loading history

τ_p is shear strength related to time. For specified joint surfaces, the magnitude of τ_p depends on normal stress σ_n largely. The shear creep displacement u is a function of τ/τ_p, where τ

is shear stress applied to sample. The parameters A, B, and c are related to τ/τ_p. The last step shear stress in creep tests (i.e. the shear stress when sample occurs failure) is approximately regarded as shear creep strength τ_p. The expression is given as follows:

$$H = ke^{m(\tau/\tau_p)} \tag{4}$$

where H represents A, B, or c; k and m are experimental constants, which are governed by the property and asperity of joint surfaces. Fitting curves on A, B, or c and τ/τ_p are shown in figure 4. These curves exhibit that coefficient A, B, and c continuously increase with increasing ratio of τ/τ_p, and when the ratio of τ/τ_p approaches to 1, the value of A, B, and c increases largely, shear creep displacement u increases rapidly as well.

4 CONCLUSION

The experimental results showed the shear behavior of joints without filling during creep exhibited a pronounced transient primary creep, under most of stress levels, the rate of shear displacement decreased with time and the displacement would tend to a stable value at last. At high stresses, quasi-linear secondary creep could be observed. Tertiary creep stage almost couldn't be observed in tests. An empirical equation describing both primary and secondary creep of joints in shear was presented, which reflects the relationship between shear displacement and time, the value of each coefficient in the equation is related to the applied stress level.

ACKNOWLEDGMENT

This research was supported by the National Science Foundation of China under grant No. 59493600-III-4.

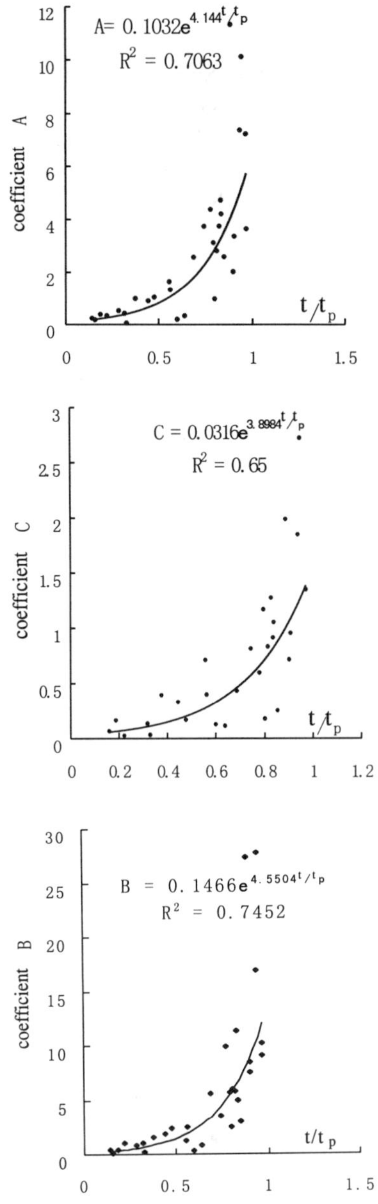

Fig. 4 Fitting curve on coefficient A, B, or C and τ/τ_p

REFERENCES

Bowden, R.K.(1984) "Time-dependent Behavior of Joints in shale", M.A.Sc.Thesis, University of Toronto,126 pp.

Curran, J.H. and A.M. Crawford (1980) "A

Comparative Study of Creep in Rock and Its Discontinuities", Proceedings of the 21st U.S. National Rock Mechanics Symposium, Rolla, Missouri, May 28-30.pp.596-603
Ding Xiuli & Liu Xiongzhen (1999) "A Study on Creep properties of Rock Joints", Journal of Underground Space, Vol.19, No.5

5 Microcracking, localization, fracture, Nano-micro-macro correlation

Computer Methods and Advances in Geomechanics, Desai et al. (eds) © 2001 Balkema, Rotterdam, ISBN 90 5809 183 X

Propagation of strong discontinuity in elastoplastic solids

Ronaldo I. Borja
Department of Civil and Environmental Engineering, Stanford University, Calif., USA

ABSTRACT: The mechanical response of solid media changes drastically as the deformation evolves from a diffuse state to a highly localized state. This metamorphosed behavior is known as strain localization or localized deformation, which has received much research attention lately. This paper focuses on the propagation of localized zone of deformation in the form of strong discontinuity, or displacement jump, in elastoplastic frictional solids. Specifically, the analysis investigates the initiation and evolution of strong discontinuity in elastoplastic solids under active and passive loading.

1 INTRODUCTION

Material instability is generally regarded as a precursor to failure in geomaterials. It manifests itself in the form of shear band, a narrow zone of intense straining to which most softening phenomena are attributed. Strain localization often leads to abrupt failure and total collapse of the structure. It is therefore of importance to be able to predict when a shear band forms and how this zone of intense deformation is oriented within the material.

Much research effort has focused lately on developing mathematical and numerical models for capturing experimentally observed shear banding phenomena where the zone of intense shearing has a finite and measurable characteristic thickness. Such zone of deformation is termed "weak discontinuity." In contrast, the thickness of the shear band may be too small in some cases that the continuum essentially experiences a displacement jump, or "strong discontinuity." This paper focuses on the formulation, implementation, and application of a strong discontinuity approach to analyzing strain localization phenomena within the finite element framework.

A classic manifestation of strain localization in geotechnical application occurs behind a retaining wall in the form of slip surfaces. For horizontal loose sandy backfill supported by a smooth vertical wall, Rankine's (1857) theory predicts that all points in the backfill will yield at the same time when either horizontally stretched (active loading) or horizontally compressed (passive loading). In this paper we re-analyze the retaining wall problem using the strong discontinuity approach and show that the localization condition is first satisfied at the ground surface. Then, we show that the discontinuity propagates downward at an angle that depends on the state of stress, active or passive, at the onset of localization.

2 LOCALIZATION THEORY

2.1 *Summary of previous works*

A complete development of the strong discontinuity model is presented in a number of recently published references (Regueiro & Borja 1999; Borja 2000; Borja & Regueiro 2000; Borja et al. 2000; Regueiro & Borja 2000). The formulation proposed in the above references has similarities to those found in other previous works (Simo et al. 1993; Simo & Oliver 1994; Armero & Garikipati 1995; Larsson et al. 1996; Oliver et al. 1999), except for a few but significant differences. Among the major differences include the fact that the new formulation clearly distinguishes between pre-localization and post-localization responses, where the latter response is handled through the notion of a damage yield model. Furthermore, a specific finite element formulation based on the standard Galerkin approximation has been proposed by Borja (2000) which does not require a static condensation procedure to be performed on the element level.

2.2 *Localization condition*

We consider an elastoplastic continuum shown in Fig. 1. Let us assume that the constitutive response prior to localization is described by a yield function

F and a plastic potential function Q, with stress gradients $\boldsymbol{f}$ and $\boldsymbol{q}$, respectively. Further, let $\boldsymbol{c}^e$ denote the elastic tangential moduli tensor having the property of major and minor symmetries, as well as positive-definiteness. The condition for the onset of displacement jump, or strong discontinuity, along a potential slip surface S with unit normal vector $\boldsymbol{n}$ is governed by the vanishing of the determinant of the elastic-perfectly plastic acoustic tensor $\boldsymbol{A}$, and takes the form

$$\det(\boldsymbol{A}) = 0, \quad \boldsymbol{A} = \boldsymbol{n} \cdot \boldsymbol{c}^{ep} \cdot \boldsymbol{n} \tag{1}$$

where

$$\boldsymbol{c}^{ep} = \boldsymbol{c}^e - \boldsymbol{c}^e : \boldsymbol{q} \otimes \boldsymbol{f} : \boldsymbol{c}^e / (\boldsymbol{f} : \boldsymbol{c}^e : \boldsymbol{q}) \tag{2}$$

Here, $\boldsymbol{A}$ is obtained from the continuum elastic-perfectly plastic tangential moduli tensor, $\boldsymbol{c}^{ep}$, which is devoid of the continuum plastic modulus term H.

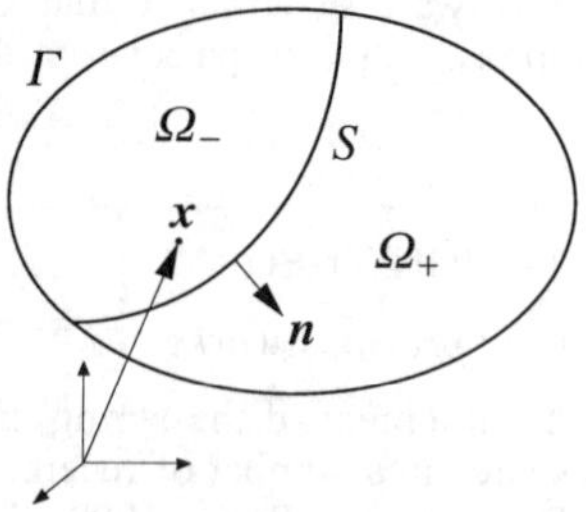

Figure 1. Elasto-plastic solid cut by a shear band.

The idea behind Equations 1&2 is that a discontinuous displacement field, or a singular strain field, is permissible provided that the traction vector inside the band is bounded. This constraint is imposed most conveniently by putting the formulation within the framework of theory of distribution, see Stakgold (1998). Note that although the plastic modulus H does not appear explicitly in Equation 1 the determinant expression is still a function of the stress state, and is therefore influenced by the value of H used to predict that particular stress state.

Bifurcation analysis seeks a configuration where stresses are favorable to initiate a displacement jump based on the satisfaction of Equation 1. A by-product of the eigenvalue analysis is the unit vector $\boldsymbol{n}$ characterizing the orientation of the normal to the shear band at the point where the determinant condition is first satisfied. In addition, a unit eigenvector $\boldsymbol{m}$ may be derived from the eigenvalue problem defining the instantaneous direction of the velocity jump vector at the onset of localization. The unit vectors $\boldsymbol{n}$ and $\boldsymbol{m}$ are shown pictorially in Fig. 2(a). At the onset of localization the vector $\boldsymbol{m}$ makes an angle ψ, called the mobilized angle of dilatancy, relative to the shear band. The angle ψ characterizes the constitutive behavior of the shear band at post-localization and should not be confused with the continuum angle of dilation used to describe the plastic volume change behavior of an intact continuum.

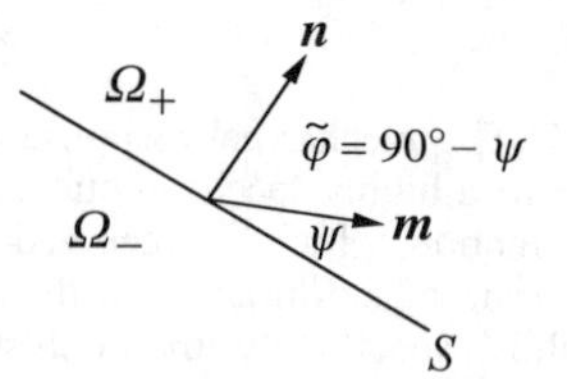

Figure 2. Significance of unit vectors $\boldsymbol{m}$ and $\boldsymbol{n}$.

2.3 *Specialization of localization condition to nonassociated Drucker-Prager model*

Consider an elastoplastic frictional solid yielding homogeneously according to a nonassociated Drucker-Prager model with no cohesion. In this case, the yield function F takes the form

$$F = (3/2)^{1/2}\|\boldsymbol{s}\| + (3)^{1/2}\beta p = 0 \tag{3}$$

where $\boldsymbol{s}$ is the deviatoric component of stress, p is the mean normal stress, and β is a model parameter. Furthermore, assume a plastic potential function of the form

$$Q = (3/2)^{1/2}\|\boldsymbol{s}\| + (3)^{1/2} b p \tag{4}$$

where b is the dilatancy parameter. Clearly, the model is nonassociated if $b \neq \beta$.

For the above Drucker-Prager model, the localization condition in plane strain takes a much simpler form (Runesson et al. 1991). Here, the determinant condition (1) is equivalent to the following scalar expression

$$h = (1-\nu)(a+\beta+b)^2 - 2(\beta-b)^2 = 0 \tag{5}$$

where ν is the elastic Poisson's ratio, and

$$a = 3(2)^{1/2} s_3 / \|\boldsymbol{s}\| \tag{6}$$

in which s_3 is the off-plane normal component of the deviatoric stress tensor $\boldsymbol{s}$. The localization condition can be detected by checking to see if the sign of h reverses during a given loading increment. If it does, then h must have crossed the value zero within this loading increment, and thus the element deformation may be enhanced to accommodate the displacement jump. Throughout this paper we shall refer to h as the localization function.

2.4 *Damage model*

We complete the development by introducing a damage model describing the yield behavior of the material at post-localization. Here, we assume that yielding is concentrated to the band and that elastic unloading takes places outside the band. In this case, we can consider a damage function characterizing the yield condition on the band of the form

$$G = \mathrm{sym}(\boldsymbol{t} \otimes \boldsymbol{n}):\boldsymbol{T} - [c - (\boldsymbol{n} \otimes \boldsymbol{n}):\boldsymbol{T} \tan \phi] = 0 \quad (7)$$

where $\boldsymbol{t}$ is the unit tangent vector to the surface of discontinuity, $\boldsymbol{T}$ is the Cauchy stress tensor, c is the mobilized cohesion, and ϕ is the mobilized friction angle on the surface of discontinuity. Note that Equation 7 is nothing else but a simple cohesive-frictional interface law on the surface of discontinuity.

The model for bifurcated response represented in Equation 7 takes effect immediately at the onset of localization, where the condition $F = 0$ is also satisfied (since the point is yielding instantaneously as a continuum). Assuming ϕ is constant, we can compute the instantaneous value of the cohesion c necessary to satisfy the condition $G = 0$. Let $\boldsymbol{T}^*$ denote the value of the stress tensor at the onset of strong discontinuity; then $G = 0$ requires that

$$c = \mathrm{sym}(\boldsymbol{t} \otimes \boldsymbol{n}):\boldsymbol{T}^* + (\boldsymbol{n} \otimes \boldsymbol{n}):\boldsymbol{T}^* \tan \phi \quad (8)$$

A softening response can now be captured by allowing c to decrease with plastic slip on the band according to a softening modulus H_S ["cohesion softening" according to the terminology of Vermeer & de Borst (1984)]. This is possible provided that c never becomes negative. Once the cohesion c has reached the value zero, yielding becomes purely frictional in nature. Thus, ϕ has the physical significance of being the friction angle at residual state.

3 NUMERICAL SIMULATIONS

The finite element mesh for the problem is shown in Fig. 3. The mesh is composed of 100 constant strain triangle (CST) elements, which is unstructured in the sense that no attempt was made to align element sides to expected orientation of the shear band. We assume the following material parameters: before localization, $E = 26{,}000$ kPa, $\nu = 0.3$, $\beta = 0.4$, $b = 0.2$ (non-associated) and $H = 0$ (perfect plasticity); after localization, $\phi = 20°$, $H_S = -2{,}000$kPa (cohesion softening modulus on the band), and $c = 0$ at residual state (purely frictional). Perfect plasticity for the continuum model results in the localization criterion for weak discontinuity being identical to that for strong discontinuity (Borja et al. 2000). The vertical faces of the mesh are supported on horizontal rollers (smooth wall).

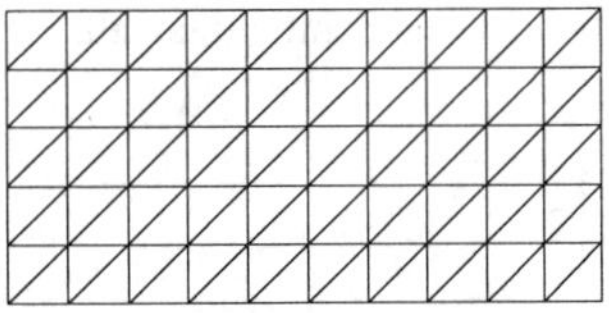

Figure 3. Plane strain FE mesh for smooth retaining wall problem: backfill has height = 5 m, width = 10 m.

To establish the initial condition, a gravity load of 20 kN/m^3 is applied to the mesh of Fig. 3, assuming the lateral supports to be unyielding. With the material parameters stated earlier, this load is enough to cause all points in the mesh to yield plastically, i.e., all stress points initially lie on the yield surface prior to any imposed wall movement. This guarantees that the localization function h have values at all Gauss points, determined according to Equation 5. The variation of initial h with depth is shown in Figs. 4 and 5, and corresponds to the curve $\Delta x = 0$ m. The open circles have heights of 0.5, 1.5, 2.5, 3.5, and 4.5 m, while the h-values of the circles are the averages of the two localization function values derived from the corresponding CST elements at those heights. Note that the dimensionless h-values are nearly uniform with an average of around $h = 3.6$ at the end of the gravity load imposition phase, implying that it is not possible for localization to take place initially.

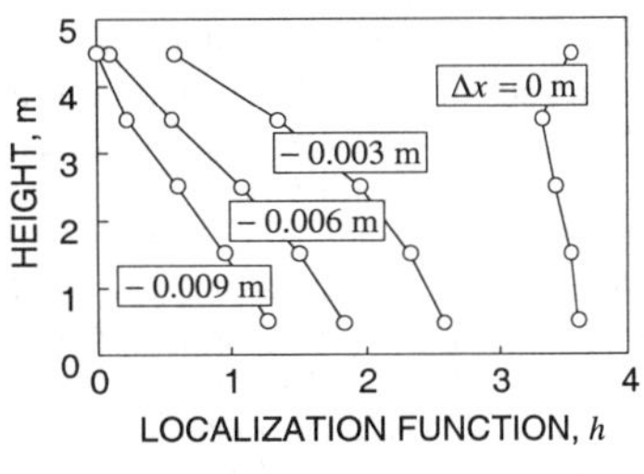

Figure 4. Variation of localization function with wall movement: active loading.

The mesh is then stretched horizontally by letting the wall move away from the backfill (active loading). Figure 4 shows that all the Gauss points in the continuum continue to yield plastically as the mesh is stretched horizontally, allowing the solution to provide values of the localization function h. However, h decreases faster near the ground surface. The Gauss points near the ground surface satisfy the localization condition $h = 0$ at a wall movement of about –0.009 m. This means that at this wall movement a displacement discontinuity can possibly occur at the top, and then propagate downward as the wall continues to move away from the backfill. This is in contrast to Rankine's theory for a loose granular backfill which predicts simultaneous yielding (and failure) of the material behind the wall.

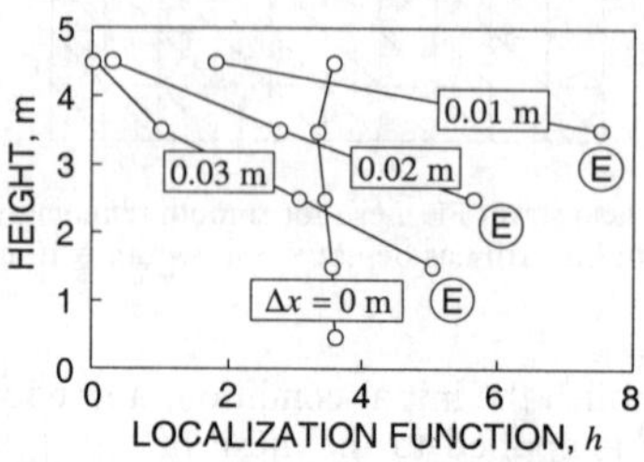

Figure 5. Variation of localization function with wall movement: passive loading.

From the same initial stress state the mesh is now compressed horizontally by forcing the wall to move toward the backfill (passive loading). Figure 5 shows that this causes the backfill material near the bottom of the wall to unload elastically, as evidenced by the fact that the solution could not provide a localization function value below the points marked "E" (for "elastic") in Fig. 5. Where the backfill material is yielding plastically, the localization function increases with depth from the ground surface. With continued wall compression the localization function value approaches zero at a wall movement of about 0.03 m. Again, this implies that at this wall movement a displacement jump is possible anywhere near the ground surface. For this problem the wall movement necessary to initiate a displacement jump at the ground surface is larger for passive loading (0.03 m) than for active loading (0.009 m).

The above bifurcation analyses also produce inclination angles for the slip surfaces. For active loading the orientation of the emerging shear band is calculated to be about 54° from the horizontal, while for passive loading the emerging shear band orientation is calculated to be about 38°, also with respect to the horizontal. These angles are influenced by the Drucker-Prager parameters β and b, by the Poisson's ratio ν, and by the values of stresses that the continuum point is subjected to at the time of localization (Borja and Regueiro 2000). Partially propagated shear bands are shown in Figs. 6 and 7 for active and passive loading conditions, respectively, wherein the shaded regions are represented by elements traced by the shear bands.

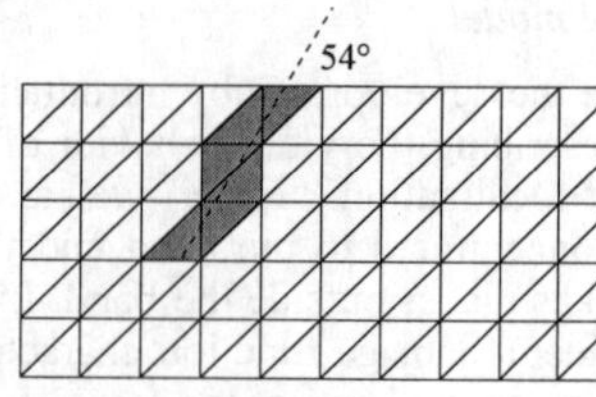

Figure 6. Traced elements: active loading.

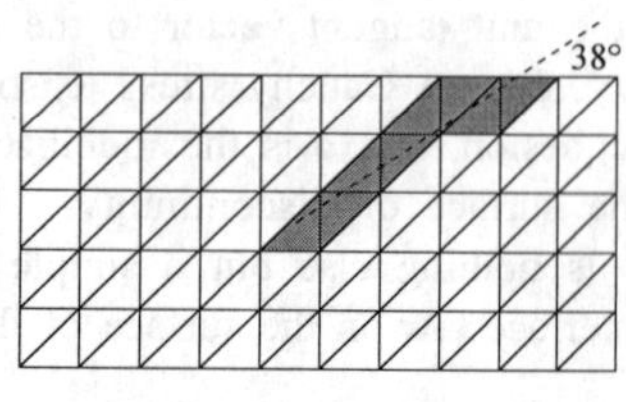

Figure 7. Traced elements: passive loading.

4 SUMMARY

A mathematical model for localized deformation analysis in the form of strong discontinuity, or displacement jump, has been presented. The model is illustrated with examples addressing the propagation of strong discontinuity in elastoplastic frictional backfill material behind a smooth retaining wall in both active and passive loading.

ACKNOWLEDGMENTS

Financial support for this research was provided by the G3S Division of the U.S. National Science Foundation under Contract No. CMS-9700426, through the program of Dr. Priscilla P. Nelson.

REFERENCES

Armero, F & Garikipati, K. (1995). Recent advances in the analysis and numerical simulation of strain localization in inelastic solids. In D.R.J. Owen, E. Oñate & E. Hinton (eds.), *Proceedings of computational plasticity IV*, International Center for Numerical Methods in Engineering (CIMNE), Barcelona, Spain: 547-561.

Borja, R.I. 2000. A finite element model for strain localization analysis of strongly discontinuous fields based on standard Galerkin approximation. *Computer Methods in Applied Mechanics and Engineering*, in press.
Borja, R.I. & Regueiro, R.A. 2000. Strain localization of frictional materials exhibiting displacement jumps. *Computer Methods in Applied Mechanics and Engineering*, in press.
Borja, R.I., Regueiro, R.A. & Lai, T.Y. 2000. FE modeling of strain localization in soft rock. *Journal of Geotechnical and Geoenvironmental Engineering*, ASCE 126(4): 324-343.
Larsson, R., Runesson, K. & Sture, S. 1996. Embedded localization band in undrained soil based on regularized strong discontinuity theory and FE analysis. *International Journal of Solids and Structures* 33: 3081-3101.
Oliver, J., Cervera, M. & Manzoli, O. 1999. Strong discontinuities and continuum plasticity models: The strong discontinuity approach, *International Journal of Plasticity* 15: 319-351.
Rankine, W.J.M. 1857. On the stability of loose earth. *Philos. Trans. Roy. Soc. London* 147(1): 9-27.
Regueiro, R.A. & Borja, R.I. 1999. A finite element model of localized deformation in frictional materials taking a strong discontinuity approach. *Finite Elements in Analysis and Design* 33(4): 283-315.
Regueiro, R.A. & Borja, R.I. 2000. Plane strain finite element analysis of pressure sensitive plasticity with strong discontinuity. *International Journal of Solids and Structures*, in press.
Runesson, K., Peric, D. & Sture, S. 1991. Discontinuous bifurcations of elastic-plastic solutions at plane stress and plane strain, *International Journal of Plasticity* 7: 99-121.
Simo, J.C., Oliver, J. & Armero, F. 1993. An analysis of strong discontinuities induced by strain-softening in rate-independent inelastic solids. *Computational Mechanics*, 12: 277-296.
Simo, J.C. & Oliver, J. 1994. A new approach to the analysis and simulation of strain softening in solids. In: Z.P. Bazant, Z. Bittnar, M. Jirázek & J. Mazars (eds.), *Fracture and damage in quasi-brittle structures*, E&FN Spon, London: 25-39.
Stakgold, I. 1998. *Green's function and boundary value problems*. 2nd Ed., Wiley, New York.
Vermeer, P.A. & de Borst, R. 1984. Nonassociated plasticity for soils, concrete and rock. *Heron*, 29(3): 1-64.

Computer Methods and Advances in Geomechanics, Desai et al. (eds) © 2001 Balkema, Rotterdam, ISBN 90 5809 183 X

Modeling Ithaca's natural hydraulic fractures

B.J. Carter & A.R. Ingraffea
Cornell University, Ithaca, N.Y., USA

T. Engelder
Penn State University, University Park, Pa., USA

ABSTRACT: Joints within the Catskill Delta complex are believed to be natural hydraulic fractures and abundant indirect evidence supports this claim. The indirect evidence is based on many observations of joint surface morphology and joint spacing among other features. The Alleghanian orogeny left a unique imprint on the joint surfaces also, indicating a rotating horizontal compressive stress field. Numerical simulations illustrate that the formation and propagation of the joints can be described by hydraulic fracturing mechanisms. The numerical simulations include hydraulic fracture propagation from sedimentary structures at layer interfaces and fracture twisting in response to the reorientation of the remote stress field after initial joint propagation.

1 INTRODUCTION

Hydraulic fractures occur naturally in the Earth's crust. These range in size from mm as small pressure partings to km as magmatic dike intrusions. Ithaca, New York, which lies in the heart of the Finger Lakes region, is a great place to see intermediate-scale natural hydraulic fractures. Glaciation and erosion have led to large outcrops of middle and upper Devonian clastic rocks. These rocks have experienced two major orogenic events in addition to uplift and erosion all of which have left a pattern of brittle fracture in the rocks as seen today. A brief geological history of the area is necessary to set the stage for the discussion of natural hydraulic fractures in these rocks.

Outcrops within the Finger Lakes are part of the Devonian Catskill Delta Complex, a clastic wedge that is more than 2 km thick in the Finger Lakes region. The complex was deposited on a continental shelf with an open sea to the south and consists of layers of black shales, gray shales, silstones and sandstones, and an occasional limestone. Uplift accompanying the Acadian orogeny to the east provided a large source of sediment for the clastic wedge that was prograding into the Appalachian Basin. Deposition rates reached 170m per Ma. At the close of the Paleozoic, during the Pennsylvanian and Permian, the Alleghanian orogeny caused widespread folding, faulting and jointing in eastern North America. This was the last major orogeny to affect this part of the continent.

Engelder and others (see Engelder et al. 2000) have studied the joints and other geologic features within the Catskill Delta Complex for many years. There is an abundance of evidence that indicates that many of these joints developed as natural hydraulic fractures caused by abnormally high fluid pressures. The Alleghanian orogeny also left a unique imprint as many joints show a clockwise sequence in response to the reorientation of stress as the African continent collided with North America.

There is a significant body of literature that describes the joints and tries to interpret their origin and formation (e.g. Pollard & Aydin 1988). However, there have been few numerical simulations to illustrate the theorized processes. The aim of this paper is to examine various components of these joints using numerical simulations to determine whether current formation theories are reasonable or valid.

2 GEOLOGIC EVIDENCE OF NATURAL HYDRAULIC FRACTURES AND STRESS ROTATIONS

The current theories of joint formation in the Catskill Delta Complex depend upon visual inspection of many joints and joint sets (Engelder et al. 2000). The first bit of evidence comes from the surface morphology of many of the joint surfaces in the siltstone.

Cracks tend to start at points of higher stress concentration (McConaughy & Engelder 2000). In the case of sedimentary beds, the largest stress concentrations are in the form of concretions, fossils, ripples, groove casts, and similar features. Figure 1 shows a joint surface in the Ithaca Formation siltstone that starts at a groove cast, a location where the siltstone penetrates the shale layer, formed during erosion of the muddy substrate and deposition of the silt. The figure also shows the plumose surface morphology of the joint surface. The plumose structures and the related arrest lines are well described by Bahat & Engelder (1984). Barbs, the darkened lines in Figure 1, radiate from the plume axis and indicate the direction of fracture growth. Arrest lines mark the termination of crack propagation increments. Natural hydraulic fractures do not have a steady supply of fluid (Lacazette & Engelder 1992). As the crack propagates, the pressure decays and the crack arrests until the pressure builds up again.

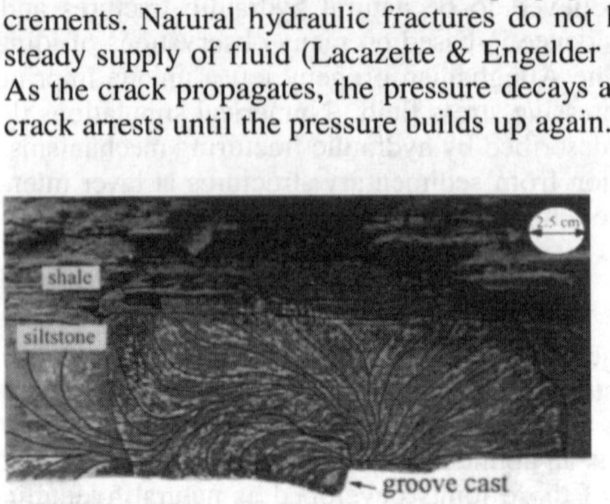

Figure 1. Plumose joint surface starting from a groove cast.

It can also be seen in Figure 1 that the fracture initially stays within the siltstone layer. It has been found that in sedimentary basins where the horizontal stress is less than the vertical, the shale layers have a higher horizontal stress than the siltstones (Evans et al. 1989, Warpinski 1989). Assuming that the pore pressure is the same in both the shale and siltstone layers, then the fractures would form in the siltstones first. This will be more evident in the next figure.

The difference in horizontal stress leads to jointing in the siltstones followed by jointing in the shale. Joints in the shale beds usually start at the edges of existing joints in the siltstone; the joints act as conduits for the pore fluid. If there is no rotation of the principal horizontal stresses, joints will develop in the shale in plane with those in the sandstone. However, if the horizontal stress does rotate, then en-echelon or fringe cracks will develop (Pollard et al. 1982). Figure 2 shows a set of fringe cracks that start at the interface between the siltstone and shale. These fractures propagate downward. The numerous small cracks at the immediate interface reduce to a few large cracks as propagation continues.

Figure 2. Multiple en-echelon cracks propagate down from the siltstone-shale interface and are rotated from the parent joint.

There are many more features of the joints that offer indirect evidence of natural hydraulic fractures, including joint spacing, slip on open joints, crosscutting and abutting relationships, and evidence of a pressure seal that could have led to overpressurization of the formation. However, it is beyond the scope of this paper to discuss all of this evidence. Instead, numerical simulations are used to illustrate that natural hydraulic fracturing processes can explain some of the features described herein.

3 HYDRAULIC FRACTURING SOFTWARE

The 3D numerical simulations that are described in the following sections are performed using HYFRANC3D and accompanying software (Carter et al. 2000a,b). In addition, ANSYS (ANSYS 2000) and FRANC2D (CFG 2000) are used for some 2D analyses.

The 3D software consists of OSM, FRANC3D, HYFRANC3D, and BES. OSM is used to create the initial geometric models. FRANC3D is a fracture analysis code with pre- and post-processing capabilities. HYFRANC3D builds upon FRANC3D by adding a fluid flow module that couples the elastic structural response with fluid flowing in the fracture. BES is a linear elastic boundary element program.

BES provides an elastic influence matrix that relates the set of unit pressures (p) at the nodes on the crack surface to the elastic displacements (w). The fluid flow formulation is based on the elasticity, lubrication and continuity equations. For a Newtonian fluid, the stress ahead of the crack front has a 1/3 singularity, as does the fluid pressure inside the crack (Carter et al. 2000b). Physically, the fluid pressure cannot be singular, thus, a fluid lag must develop at the crack front (Figure 3). The finite element formulation in HYFRANC3D combines an analytical solution for the pressure and crack opening displacement at the crack front with a standard finite element formulation of the lubrication equation for the bulk of the fracture. The analytical solution at the crack front provides an elegant way to overcome the singular behavior and limits the required number of elements at the crack front. Hydraulic fracture simulation proceeds as a series of crack advance increments, requiring an elastic solution and coupled flow analysis at each step.

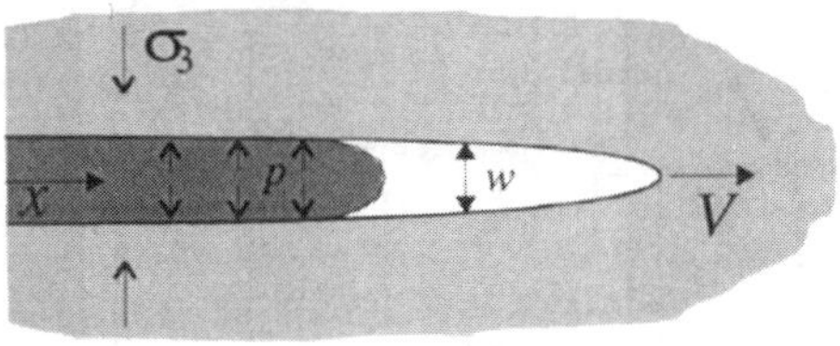

Figure 3. Crack front region with fluid lag; p is pressure, w is crack opening displacement, V is crack front velocity, σ_3 is closure stress and x is the local coordinate axis along the crack.

4 SIMULATION OF HYDRAULIC FRACTURE PROPAGATION FROM A GROOVE CAST

Cracks tend to start at pre-existing flaws and points of high stress concentration. The groove cast in Figure 1 is modeled to evaluate the stress intensification and to determine the hydraulic fracture propagation behavior from this location. The intent of these simulations is to show that hydraulic fracture is a plausible mechanism; thus, realistic boundary conditions are necessary, but accurate values are not essential to provide qualitative results.

The numerical model consists of the siltstone layer only. McConaughy & Engelder (2000) describe the difficulties in determining and modeling the correct boundary conditions. They conclude that the interface between the shale and siltstone must allow slip. To simplify the 3D model, the upper and lower shale layers are not modeled; a vertical constraint is applied to the upper surface. A far-field compressive stress is applied in the horizontal direction only; the vertical stress is neglected. The compressive stress perpendicular to the crack surface is 50 MPa and the stress parallel to the crack plane is 100 MPa. The elastic modulus is set to 14 GPa and the Poisson's ratio is set to 0.1.

The initial plane strain ANSYS™ stress analysis clearly shows that the groove cast alters the horizontal stress (Fig. 4). A tensile stress is induced at the groove cast. An initial crack is nucleated in the 3D numerical model at this location and hydraulic fracture propagation is simulated.

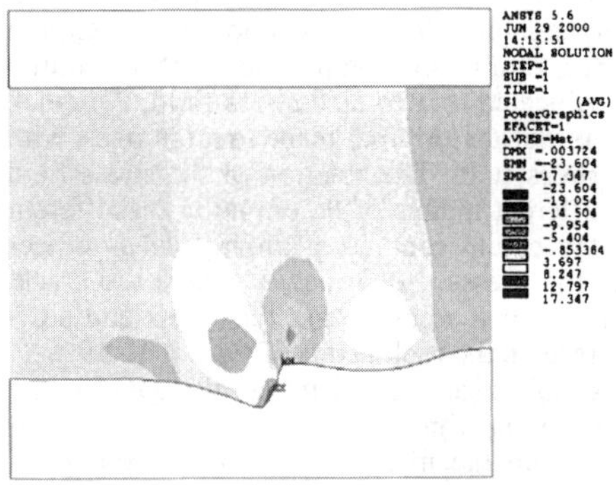

Figure 4. ANSYS™ contours of maximum principal stress around a groove cast in a siltstone layer; tension is positive and values are in MPa.

Figure 5 shows the shape of the hydraulic fracture after 17 steps of propagation. The fluid boundary conditions consist of a crack mouth flow rate equal to 1.0e-6 m^3/s and zero flow rate at the crack front. These boundary conditions do not match the presumed conditions of periodic fluid pressure build-up, but it is assumed that this will only affect the fracturing time and the surface features. The shape of the fracture to this stage matches that seen in the field reasonably well, although the barbs and arrest lines cannot be reproduced. The simulation can be continued such that the fracture breaks through the upper surface of the silt layer, but this is not complete yet.

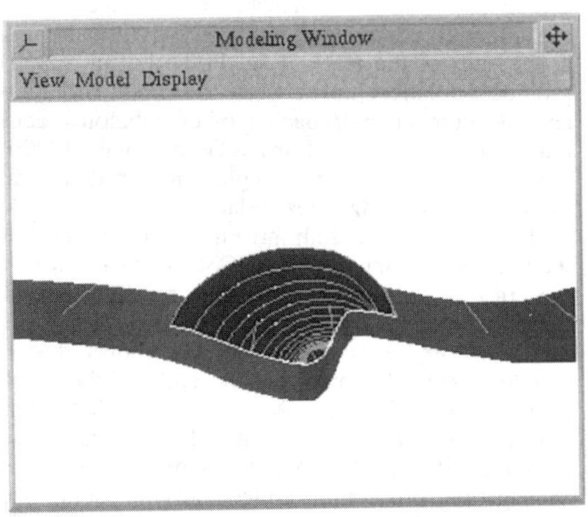

Figure 5. Shape of a hydraulic fracture from a groove cast after 17 steps of propagation.

5 SIMULATION OF HYDRAULIC FRACTURE IN A ROTATING STRESS FIELD

Hydraulic fractures in these brittle rocks tend to grow perpendicular to the direction of maximum tension (or minimum compression). If an initial crack is subjected to a rotated stress field, the crack will kink out of the original plane to align itself with the new stress field. The rotation of the stress field induces different modes of deformation on different parts of the crack front. The kinking behavior depends on the pressure distribution inside the crack, the geometry of the crack, and the stress and stiffness contrasts in the rock layers (Carter et al. 1999).

As a simple example, an initial elongated radial crack (emanating from a horizontal borehole) is oriented 70° from the maximum horizontal compressive stress (Fig. 6). The crack is nucleated in a single layer model with an elastic modulus of 27.6 GPa and a Poisson's ratio of 0.25. The vertical stress is 69 MPa, the maximum horizontal stress is 62 MPa, and the minimum horizontal stress is 48 MPa. The crack mouth flow rate is set to 0.0795 m^3/s.

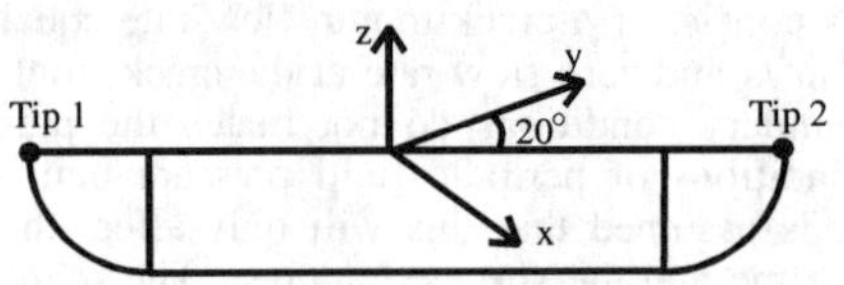

Figure 6. The initial crack shape is 9 m long and 0.3 m deep; σ_{Hmax} is aligned with the x-axis (not to scale).

Hydraulic fracture is simulated and stress intensity factors (SIF's) are computed. Figure 7 shows the distribution of SIF's along the initial crack front. The mode I values are highest, but there is a significant mode III component; mode II is significant only at the ends of the crack front. Pollard et al. (1982) provide a relationship between the mode III/I ratio and the number and spacing of en-echelon cracks that form along the crack front. Germanovich (2000) is conducting experiments on glass and brittle plastic to further characterize these relationships.

The combination of all three modes is studied less well. However, the HYFRANC3D software allows the crack to propagate while ignoring the mode III SIF. Under these circumstances, the crack front does not break up, although, it quickly shows that it should. Figure 8 shows that the ends of the crack turn to align with σ_{Hmax}. Most of the crack front is not aligned with σ_{Hmax}. Instead, "large" ripples can be seen. The ripples are a result of the numerical propagation procedure, which only accounts for modes I and II. As the fracture propagates, the mode II and mode III SIFs increase rather than decrease. Due to the elongated shape, more effort is required to turn the entire crack front as compared to a more circular shape crack (Carter et al, 1999). As the crack front does not break, the combination of modes I and II produces initially minor oscillations in the crack front. These escalate as the crack continues to grow, leading to the ripples.

The mode III SIF value in the middle of the crack front reaches 10 MPa·√m at the final stage (15th step of propagation) while the mode I value remains relatively constant for each step, ranging from 20 to 25 MPa·√m along the crack front. Rotation and break up of the crack front is seen in many locations in the Ithaca area; Figure 9 shows a typical example.

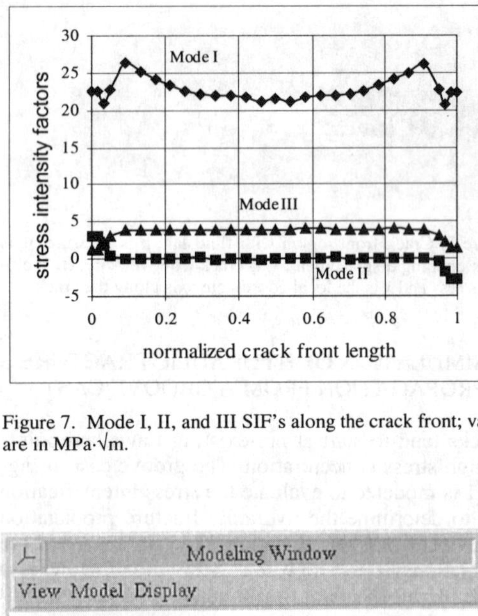

Figure 7. Mode I, II, and III SIF's along the crack front; values are in MPa·√m.

Figure 8. Ignoring mode III SIF during propagation leads to "large" ripples in the crack front indicating it should break up. The ends of the crack turn to align with the stress field.

Figure 9. Joint in a siltstone indicating a rotated stress field as the crack surface shows obvious breaks; this might be seen as a set of en-echelon cracks when looking from above.

6 SIMULATION OF MULTIPLE HYDRAULIC FRACTURES

Multiple parallel cracks tend to reduce in number and increase in spacing as they propagate. It is often assumed that the length to spacing ratio should be about 1.0 for a stable system of parallel cracks (Hwang, 1999). Figure 2 shows an example of a large number of closely spaced fractures at the shale-siltstone interface. As the cracks propagate down into the shale, most of the cracks arrest very quickly. Only a few, relatively widely spaced cracks remain at the base of the shale layer. Based on this figure, the length to spacing ratio is significantly greater than 1.0. Hwang (1999) has studied systems of parallel cracks, although not for 3D hydraulic fractures. Both 3D hydraulic fracture simulations and 2D simulations will be performed here.

A set of three hydraulic fractures is used to gain insight into the behavior of a system of many parallel hydraulic fractures. The initial model is shown in Figure 10. The model is symmetrically constrained. The initial cracks are 1 mm deep and 6.35 mm apart. The crack mouth flow rate is 1.0e-6 m^3/s and the material properties are the same as in section 4. In reality, more cracks should be modeled, but the simulations become quite time consuming as more cracks are added; three cracks should provide adequate insight.

HYFRAN3D computes pressure distributions and crack opening displacements in the 3D fractures along with SIF distributions along the crack front. The usual procedure for growing multiples cracks is to scale the crack growth for all cracks based on the relative values of the stress intensity factors. Using this technique, the 3D hydraulic fractures are propagated for 10 steps. Although the mode I SIF for the center crack is slightly less than that for the two outer cracks (14.5 versus 15.2 MPa·√m for the initial crack), it continues to propagate. At step 10, the crack length is 10.5 mm. The length to spacing ratio is 1.7 and the center crack shows no signs of arresting based on the relative mode I SIF values, 3.4 versus 3.5 MPa·√m for the center and outer cracks respectively. Based on Figure 2, one might expect that the length to spacing ratio could be much higher than this. To verify that the crack growth is still stable, 2D simulations are performed.

FRANC2D is able to compute SIF's, energy release rates, and rates of energy release rates. The latter allows the program to decide whether crack growth is stable and to decide which of the cracks are arrested. Although FRANC2D does not simulate fluid flow, a static pressure distribution of arbitrary shape can be applied to the crack surfaces (edges). It was shown previously (Carter et al, 1999) that if the correct pressure distribution, based on the 3D fluid flow analysis in HYFRANC3D, is applied to the crack surfaces of the 2D model, the correct hydraulic fracture behavior is captured. Thus, the pressure profile through the middle of each 3D crack is applied to the 2D cracks to evaluate the rates of energy release rates.

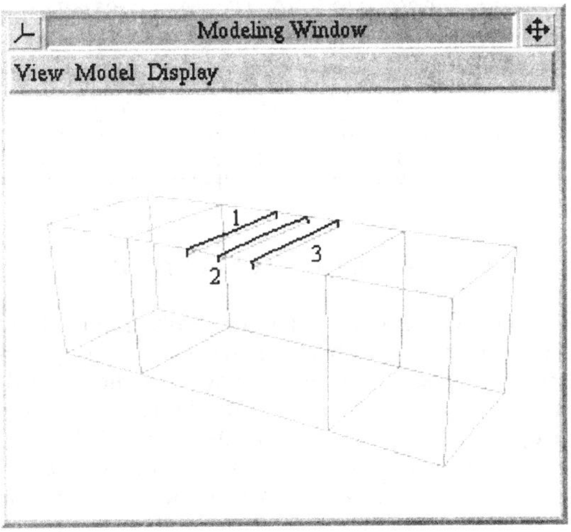

Figure 10. Initial 3D model of multiple parallel cracks.

The initial FRANC2D analysis predicts that the center crack has a slightly lower rate of energy release rate than the two outer cracks. The values are presented in matrix form in Table 1. The off-diagonal terms define the interaction between neighboring cracks. The negative values indicate that crack growth is stable for all cracks. The influence of neighboring cracks is relatively small, indicating a weak interaction.

Table 1. Matrix of rates of energy release rate for 1 mm cracks.

	Crack 1	Crack 2	Crack 3
Crack 1	-0.6537E-01	-0.8141E-02	-0.2088E-02
Crack 2	-0.8141E-02	-0.6184E-01	-0.8141E-02
Crack 3	-0.2088E-02	-0.8141E-02	-0.6537E-01

FRANC2D simulations corresponding to steps 5 and 10 of the 3D model are performed to examine the change in the rates as the cracks grow. The pressure profiles from the corresponding 3D HF simulations are used in the 2D simulations. The rates of energy release rate are provided in Tables 2-3. It is evident that the rate for the center crack is much less than the rate for the outer cracks at step 10, almost an order of magnitude. Also, the interaction between neighboring cracks is much stronger than for the initial configuration. However, the rate is still non-zero and crack growth is still stable. The simulations are being continued to determine the final length to spacing ratio at which the center crack arrests.

Table 2. Matrix of rates of energy release rate at 3.7 mm.

	Crack 1	Crack 2	Crack 3
Crack 1	-0.2531E-01	-0.5444E-02	-0.2027E-02
Crack 2	-0.5444E-02	-0.1514E-01	-0.5443E-02
Crack 3	-0.2027E-02	-0.5443E-02	-0.2534E-01

Table 3. Matrix of rates of energy release rate at 10.7 mm.

	Crack 1	Crack 2	Crack 3
Crack 1	-0.1474E-01	-0.7807E-02	-0.4911E-02
Crack 2	-0.7807E-02	-0.2951E-02	-0.7807E-02
Crack 3	-0.4911E-02	-0.7807E-02	-0.1473E-01

7 SIMULATION OF EN-ECHELON CRACKS

En-echelon cracks are often the result of crack front break up due to high mode III stress intensity factors. The model in section 5 shows that the mode III stress intensity factor should not be ignored during propagation. Crack front break up is difficult to model, however, and is still not well understood.

A simplified approach is taken here to examine crack front break up under a mode III loading. The model is shown in Figure 11. A single edge crack in a unit cube is subjected to a shear force while the crack faces are prevented from overlapping by the inclusion of a static fluid pressure. The elastic modulus is 10 GPa and Poisson's ratio is set to zero. The SIFs are shown in Figure 12, which indicates the relatively high mode II and mode III SIF.

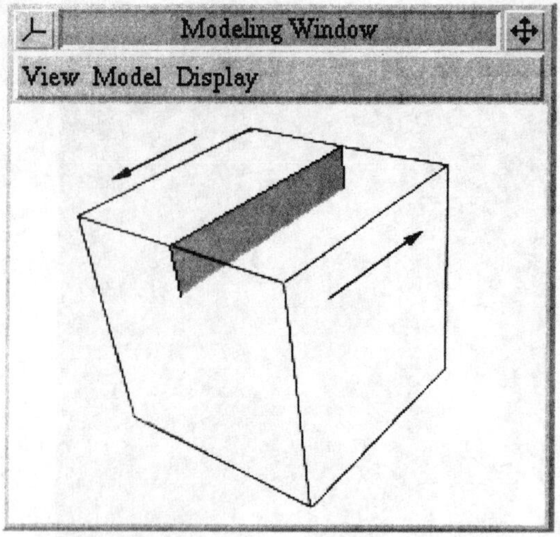

Figure 11. Single edge crack subjected to mode III shear.

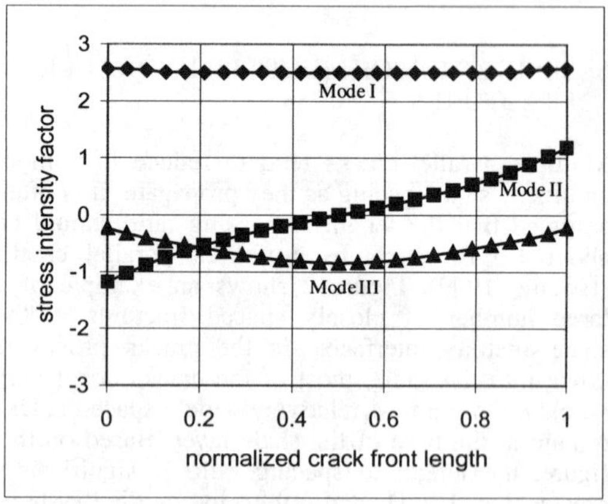

Figure 12. SIF's for single edge crack under torsion.

Pollard et al. (1982) provide the following relationship between the mode III/I ratio and the en-echelon crack orientation:

$$\beta = 1/2 \tan^{-1}\left[\frac{K_{III}}{K_I(1/2-\nu)}\right] \quad (1)$$

$$K_I > 0$$

The inequality ensures an open crack. The equation defines the orientation of a plane in front of the parent crack that has zero shear stress and the maximum tensile stress. If the mode III SIF is zero, the crack grows in its own plane. Based on the values in Figure 12, β is –21° in the middle portion of the crack front. The initial spacing of the en-echelon cracks is less well defined and Germanovich (2000) continues to work on this.

The initial crack front break up is modeled by manually creating a set of three small en-echelon "petal" cracks at the crack front as described by Germanovich (2000). Figures 13 and 14 show the side and plan views of the crack configurations. They are spaced 0.05 units apart with a radius of 0.02 units.

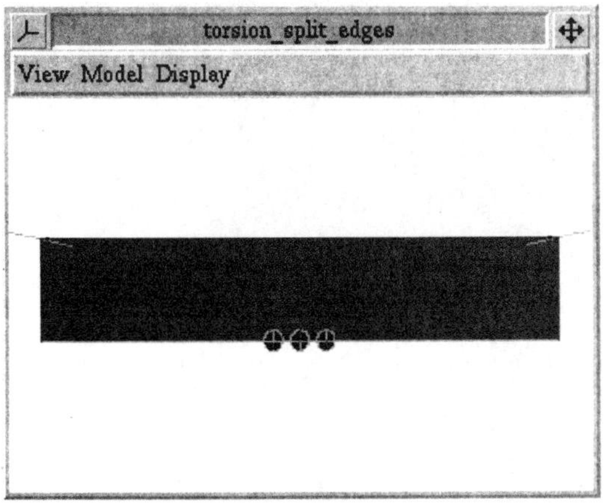

Figure 13. Side view of 3 en-echelon cracks simulating the crack front break up.

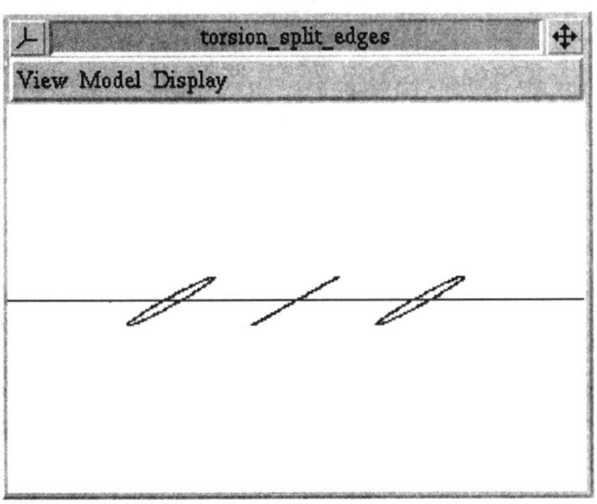

Figure 14. Plan view of 3 en-echelon cracks simulating the crack front break up.

The initial stress analysis was performed without computing the elastic influence matrix. The maximum principal stress is concentrated at the intersection of the main and "petal" cracks (Figure 15). Efforts to define the crack growth criterion and propagation behavior for this crack configuration are continuing and further results will be presented elsewhere.

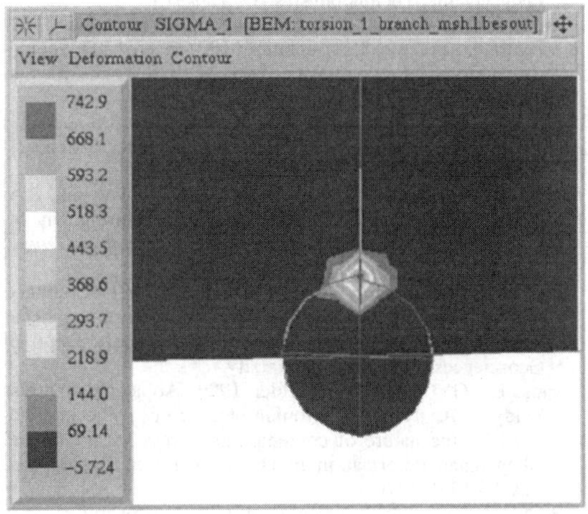

Figure 15. Maximum principal stress is concentrated at the intersection of the main and "petal" cracks.

8 CONCLUSIONS

The indirect evidence for natural hydraulic fractures is abundant in the Ithaca, NY region. The proof might depend upon the interpretation, however. Numerical hydraulic fracture simulations lend additional insight to the natural processes. Hydraulic fracture simulation from a sedimentary interface is modeled to examine the influence of natural stress concentrations, the propagation of multiple fractures, and the re-orientation of fractures in a rotating stress field.

The simulations are not complete yet, but they do indicate that hydraulic fracturing is a plausible mechanism for describing the joints seen in the rocks in the Ithaca, NY region. The simulations are continuing and the final results will be presented later.

ACKNOWLEDGEMENTS

The authors would like to acknowledge the past and present financial support of the NSF (EIA-9972853; EIA-9726388; CMS-9625406). This research was conducted using the resources of the Cornell Theory Center, which receives funding from Cornell University, New York State, federal agencies, and corporate partners. Penn State's Seal Evaluation Consortium (SEC) supported the fieldwork.

REFERENCES

ANSYS™ 2000. http://www.ansys.com/.

Bahat, D. & T. Engelder 1984. Surface morphology on cross-fold joints of the Appalachian Plateau, New York and Pennsylvania, Tectonophysics 104:299-313.

Carter, B.J., X. Weng, J. Desroches & A.R. Ingraffea, 1999. Hydraulic Fracture Reorientation: Influence of 3D Geometry, Hydraulic Fracture Workshop, 37th US Rock Mechanics Symposium, Vail, CO.

Carter, B.J., P.A Wawrzynek & A.R. Ingraffea 2000a. Automated 3D Crack Growth Simulation, Gallagher Special Issue Int. J. Num. Meth. Engng. 47:229-253.

Carter, B.J., J. Desroches, A.R. Ingraffea & Wawrzynek, P.A. 2000b. Simulating fully 3d hydraulic fracturing. In M. Zaman, G. Gioda & Booker, J. (eds), *Modeling in Geomechanics*, 525-557, Wiley Publishers.

Engelder, T. & students 2000. The Catskill Delta Complex: Analog for modern continental shelf and delta sequences containing overpressured sections. SEC Report, Dept. of Geosciences, Penn State University.

Evans, K., G. Oertel, T. Engelder 1989. Appalachian Stress Study 2: Analysis of Devonian shale core: some implications for the nature of contemporary stress variations and Alleghanian deformation in Devonian rocks. J. Geophys. Res. 94:1755-1770.

FRANC2D 2000. http://www.cfg.cornell.edu/software/.

Germanovich, L. 2000. Pers. Comm.

Hwang, C. 1999. Virtual crack extension method for calculating rates of energy release rate and numerical simulation of crack growth in two and three dimensions. PhD Thesis, Cornell University, Ithaca, NY.

Lacazette, A., & Engelder, T. 1992. Fluid-driven cyclic propagation of a joint in the Ithaca siltstone, Appalachian Basin, New York, in B. Evans & T-F. Wong, (eds.), *Fault mechanics and transport properties of rocks*. 297-324, London: Academic Press Ltd.

McConaughy, D.T. & Engelder, T. 2000. The nature of stress concentrations associated with joint initiation in bedded clastic rocks. Journal of Structural Geology (in press).

Pollard, D.D. & Aydin, A. 1988. Progress in understanding jointing over the past century. Geological Society of America Bulletin, 100:1181-1204.

Pollard, D.D., P. Segall & P.T. Delaney 1982. Formation and interpretation of dilatant echelon cracks. Geol. Soc. Bull. 93:1291-1303.

Warpinski, N.R. 1989. Determining the minimum in situ stress from hydraulic fracturing through perforations. Int. J. Rock Mech & Min. Sci. 26:523-532.

Computer Methods and Advances in Geomechanics, Desai et al. (eds) © 2001 Balkema, Rotterdam, ISBN 90 5809 183 X

A three-dimensional validation of crack curvature in muscovite mica

J.C. Hill
Carnegie Mellon University, Pittsburg, Pa., USA

J.W. Foulk III, P.A. Klein & E.P. Chen
Sandia National Labs, Livermore, Calif., USA

ABSTRACT: Experimental and computational efforts focused on characterizing crack tip curvature in muscovite mica. Wedge-driven cracks were propagated under monochromatic light. Micrographs verified the subtle curvature of the crack front near the free surface. A cohesive approach was employed to model mixed-mode fracture in a three-dimensional framework. Finite element calculations captured the crack curvature observed in experiment.

1 INTRODUCTION

The mechanical structure of rock can be characterized on multiple length scales. At each length scale, fracture along known interfaces can dominate the global response of the geotechnical structure (Jaeger, 1979). Therefore, the development and validation of numerical methods capable of predicting the fracture of major joints as well as crack growth along grain boundaries are clearly needed. Moreover, these tools must be developed for the inherent, three-dimensional nature of rock.

The fracture of rock can be addressed within the context of a cohesive zone. Cohesive zone approaches provide a general framework for modeling the fracture process (Barenblatt, 1959, Dugdale, 1960, Willis, 1967, Needleman, 1987, Needleman, 1990a,b, Gonzalo, et al., 2000). Of particular interest is the variance in crack tip curvature with confining pressure.

To simplify matters, the authors have chosen to investigate, muscovite mica, a mineral that cleaves along known, crystallographic planes. The goal of the effort was to not only document the crack tip curvature experimentally, but also predict curvature within the context of the finite element method.

2 EXPERIMENTAL EFFORT

2.1 *Material selection*

Muscovite mica, an alumina silicate, is widely used in brittle fracture and interface studies because of its optical and cleavage properties (Obreimoff, 1930, Bryant, et al., 1963, Bailey, 1967, Wan, et al., 1990). In 1930, Obreimoff validated the Griffith fracture criterion for brittle materials in a constant displacement double-cantilever beam (DCB) experiment. Most recently, Wan (Wan, et al., 1992) studied environmental effects such as the relative humidity of the atmosphere on the crack propagation and fracture energy of mica.

2.2 *Specimen geometry*

In this study, crack lengths, and thus energy release rates, were measured for the DCB geometry. Displacement boundary conditions were applied via a narrow wedge. Unlike a constant load or constant moment DCB test, the wedge enforces a constant displacement and yields a constant, stable crack length. The energy release rate, G, is only a function of the wedge height, beam geometry, modulus, and crack length.

2.3 *Specimen preparation*

Mica is composed of silica tetrahedra that are strongly bonded in layers to aluminum cations. A layer of potassium ions weakly bonds silica-aluminum-silica layers together. Because of its layered crystal structure, mica cleaves along the weak plane of potassium bonds illustrated in Figure 1. Its cleavage properties and optical transparency make mica an ideal brittle material for fracture studies.

To prepare the specimens for the fracture testing, 50 mm by 50 mm by 0.15 mm single crystal muscovite mica sheets were waxed together and cut using a precision saw into 10 mm by 50 mm specimens. The wax was removed in a series of fifteen-minute sonic baths of trichloroethylene, acetone, and isopropanol. Because of the layered structure of mica, edge damage was a significant problem and would negatively affect the testing. Therefore, specimens with significant edge damage were discarded.

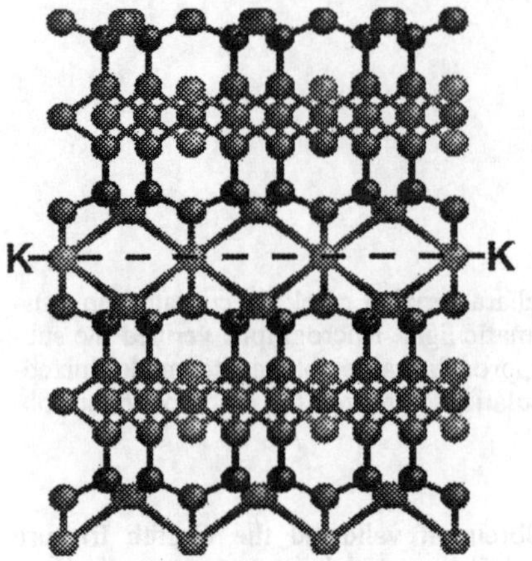

Figure1. Layered crystal structure of muscovite mica.

2.4 *Test procedure*

A crack was initiated in the specimen using a 267 μm blade. The specimens were significantly thinner than the blade; thus, cleaving the mica into two, symmetric beams was difficult. Samples with severely asymmetric beams were not used. Figure 2 represents a schematic of the fracture chamber.

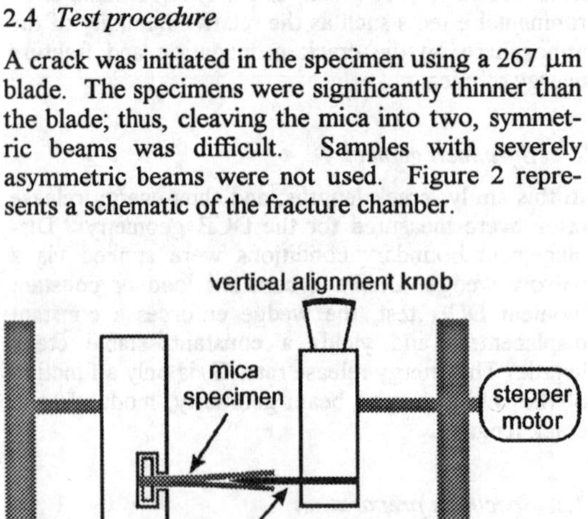

Figure 2. Schematic of the fracture chamber.

The cleaved mica was placed into a specimen holder and the wedge was adjusted vertically to produce a symmetric crack. A stepper motor pushed the mica specimen onto the blade in increments of one micron at user-specified velocities. The crack propagation was viewed with an optical microscope using green light with a wavelength of 550 nm. The crack lengths, measured from the tip of the blade, were measured using a micrometer attached to the microscope stage.

2.5 *Experimental results*

In Figure 3, the edge effect on the crack tip curvature was readily apparent. In this result, the beam heights were 81 μm and 133 μm. The Young's modulus for muscovite mica was taken to be 169 GPa (McNeil, 1993).

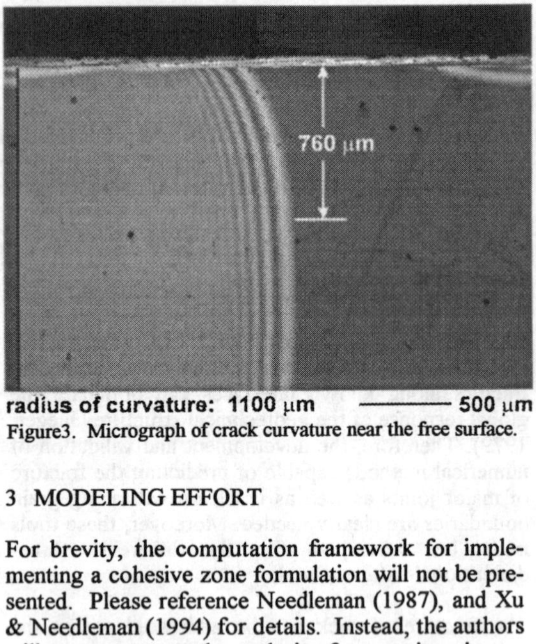

Figure3. Micrograph of crack curvature near the free surface.

3 MODELING EFFORT

For brevity, the computation framework for implementing a cohesive zone formulation will not be presented. Please reference Needleman (1987), and Xu & Needleman (1994) for details. Instead, the authors will concentrate on the analysis of muscovite mica.

3.1 *Bulk and interface models*

Because experimental evidence indicated a preferential cleavage plane, one could easily separate the bulk and interface models. The bulk model was chosen to be linear elastic. As noted before, the elastic modulus of muscovite mica is 169 GPa (McNeil, 1993). The poisson's ratio was taken to be 0.25.

The interface model employed in this work was proposed by Xu & Needleman (1994). The cohesive tractions are linear-exponential with increasing separation. The local gap is resolved into normal and tangential components, Δ_n and Δ_t. The characteristic length scales associated with normal and tangential de-adhesion are δ_n and δ_t, respectively. For brittle fracture, the critical energy release rate, G_{Ic} can be equated to the normal work of separation, ϕ_n. The two-dimensional expressions are detailed in Equation 1 and Equation 2. Mixed-mode separation is

$$T_n = -\frac{\phi_n}{\delta_n}\exp\left(-\frac{\Delta_n}{\delta_n}\right)$$
$$\left\{\frac{\Delta_n}{\delta_n}\exp\left(-\frac{\Delta_t^2}{\delta_t^2}\right)+\frac{1-q}{r-1}\left[1-\exp\left(-\frac{\Delta_t^2}{\delta_t^2}\right)\right]\left[r-\frac{\Delta_n}{\delta_n}\right]\right\} \quad (1)$$

$$T_t = -\frac{\phi_n}{\delta_n}\left(2\frac{\delta_n}{\delta_t}\right)\frac{\Delta_t}{\delta_t}\left\{q+\left(\frac{r-q}{r-1}\right)\frac{\Delta_n}{\delta_n}\right\}$$
$$\exp\left(-\frac{\Delta_n}{\delta_n}\right)\exp\left(-\frac{\Delta_t^2}{\delta_t^2}\right) \quad (2)$$

addressed through the model parameters q and r,

$$q=\frac{\phi_t}{\phi_n},\ r=\frac{\Delta_n^*}{\delta_n}$$
(3)

where q relates the normal work of adhesion to the shear work of adhesion and r is a measure of the normal opening associated with complete shear separation. While G_{1c} was an input parameter taken from experiment, 0.82 J/m^2, the characteristic length scales were chosen to be 100 nm for both normal and tangential separation.

Because the beams were different heights, the neutral axis did not coincide with the fracture plane. Thus, the crack propagated under mixed-mode conditions. Since detailed measurements of the cohesive behavior have not been preformed, we assumed q = 1 and r = 0. Future studies should consider if G_{1c} is truly equal to G_{2c}.

3.2 *Spatial discretization*

As noted in the experimental section, the mica was pushed onto the wedge at a prescribed rate. Because mica is perfectly brittle, the crack length ahead of the wedge remains constant. This condition produces a "steady-state" crack length of roughly 7 mm. In order to reflect the experiment, a beam length of 10 mm was chosen to span the "steady-state" crack length.

Element spacing along the length was chosen based on the size of the cohesive zone. Defining this zone to span from $0.1\delta_n$ to $4.0\delta_n$, simulations yielded a cohesive zone of 200 microns. An element size of 25 microns guaranteed 8 elements in the zone for integration purposes. Because eight-noded hexagonal elements were employed in bending, ten elements span the thickness. Finally, spacing along the thickness was driven by the curvature shown in Figure 3. Notice the crack curves over approximately 760 microns. In order to place at least ten elements within the area of curvature, elements were chosen to be 75 microns wide. Mesh depth also corresponded to Figure 3, roughly 2.1 mm. These constraints resulted in a mesh with 255,838 nodes and 235,200 elements. Figure 4 illustrates the decomposed mesh for parallel processing. A deformed view of the mesh is displayed in Figure 5.

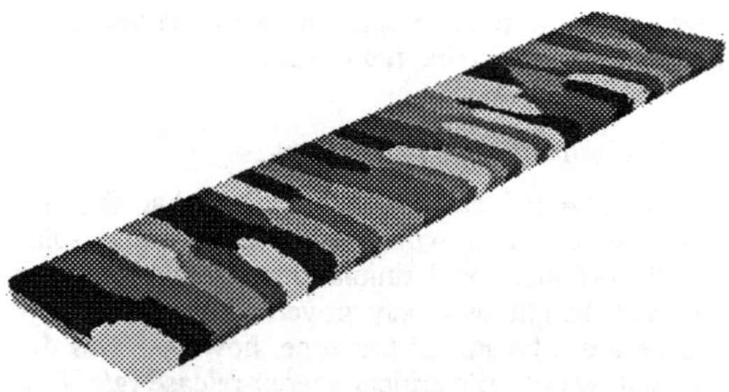

Figure 4. DCB mesh decomposed for 64 processors.

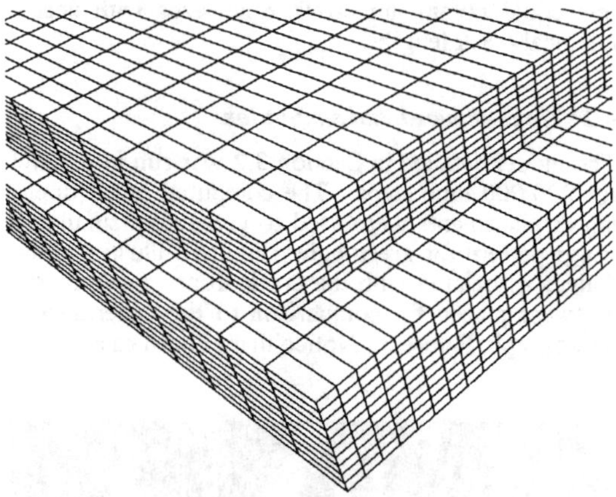

Figure 5. Required mesh spacing for DCB mesh.

3.3 *Boundary conditions*

While the boundary conditions reflect the experiment, no attempt was made to enforce the wedge geometry. Rather, an end displacement was applied to the central surface of each beam. The far end of the beam was pinned at the lower corner. Symmetry was applied along the thickness. Although the detailed shape of the blade edge would be important during initiation, our effort focused on crack curvature during "steady state" propagation.

4 RESULTS

Because future work will involve dynamic processes, all simulations presented were performed using explicit dynamics time integration with loading rates selected to minimize inertial effects.

4.1 *Two-dimensional simulations*

Two-dimensional simulations verified that the imposed velocity approximated the quasi-static solution. Two-dimensional simulations also verified that the crack length was only governed by the energy release rate. The size of the zone, however, was determined by both the critical energy release rate, G_{Ic}, and the length scales associated with de-adhesion, δ_n and δ_t. Keeping G_{Ic} constant and varying δ_n and δ_t, one could obtain numerous zone sizes with nearly identical crack lengths.

4.2 *Three-dimensional simulations*

The model detailed in Section 3.2 was run in parallel for 200,000 time steps. The computed crack opening displacements were used to generate a simulated fringe pattern for green light. Figure 6 illustrates the fringes associated with the crack front. Scale markers were placed for comparison to the experimental micrograph, Figure 3. Notice that in both cases, the

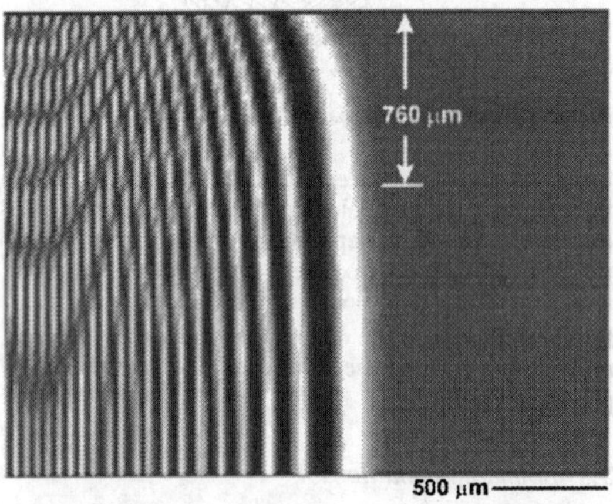

Figure 6. Simulated crack front curvature in muscovite mica.

crack curved over approximately 760 μm. This curvature can be attributed to a corresponding variation in the local energy release rate.

The first few fringes in both the simulation and experiment were indicative of the local crack opening displacement profile. Because the experimentally observed crack front was "smeared" over a simulated cohesive zone size of 200 μm, fringes observed in experimental were narrower than those generated by the prediction. Smaller zone sizes could be obtained through a reduction in δ_n and δ_t. A smaller zone size would produce narrower fringes and clarify crack front curvature. However, as noted in Section 3.2, the mesh size would be changed accordingly. More realistic zone sizes would require models on the order of several million degrees of freedom.

5 CONCLUSION

A cohesive approach to fracture was validated through comparison to crack curvature in the muscovite mica mineral system. Experimental results clearly illustrated the nature and length scale of crack front curvature. Because mica cleaved along predetermined planes, mixed-mode fracture was successfully modeled through a layer of cohesive zone elements. Three-dimensional finite element simulations yielded a curved crack front that compared favorably with experimental findings.

More work is needed. Future simulations should explore a reduction in the cohesive zone size and its effect on the local crack opening displacement profile and crack curvature.

REFERENCES

Barenblatt, G.I. 1959. The formation of equilibrium cracks during brittle fracture: general ideas and hypotheses, axially symmetric cracks. *Applied Mathematics and Mechanics* 23: 622-636.

Bailey, A.I. & Kay, S.M. 1967. A direct measurement of the influence of vapour, of liquid, and of oriented monolayers on the interfacial energy of mica. *Proceedings of The Royal Society of London* A301: 1464.

Bryant, P.J., Taylor, L.H., & Guthshall, P.L. 1963. Cleavage studies of lamellar solids in various gas environments. *Transactions Of the 10th National Vacuum Symposium* New York: Macmillan

Dugdale, D.S. 1960. Yielding of steel sheets containing slits, *Journal of the Mechanics and Physics of Solids* 8: 100-104.

Gonzalo, R., Ortiz, M., Pandolfi, A. 2000. Three-dimensional finite element simulation of the dynamic Brazillian tests on concrete cylinders. *International Journal for Numerical Methods in Engineering* 48(7): 963-994.

McNeil, L.E. 1993. Elastic moduli of muscovite mica. *Jounral of Physics: Condensed Matter* 5(11): 1681-1690.

Obreimoff, J.W. 1930. The splitting strength of mica. *Proceedings of The Royal Society of London* A217: 905.

Needleman, A. 1987. A continuum model for void nucleation by inclusion debonding. *Journal of Applied Mechanics* 54:525-531.

Needleman, A. 1990a. An analysis of decohesion along an imperfect interface. *International Journal of Fracture* 42: 21-40.

Needleman, A. 1990b. An analysis of tensile decohesion along an interface. *Journal of the Mechanics and Physics of Solids* 38: 289-324.

Wan, K.T., Lathabai, S., & Lawn, B.R. 1990. Crack velocity functions and thresholds in brittle solids. *Journal of the European Ceramic Society* 6(4): 259-268

Wan, K.T., Lawn, B.R., & Horn, R.G. 1992. Repulsive interaction between coplanar cracks in the double-cantilever geometry. *Journal of Materials Research* 7(6): 1584-1588.

Willis, J.R. 1967. A comparison of the fracture criteria of Griffith and Barenblatt. *Jounral of the Mechanics and Physics of Solids* 15: 151-162.

Xu, X.-P. and Needleman, A. 1994. Numerical simulations of fast crack growth in brittle solids. *Journal of the Mechanics and Physics of Solids* 42: 1397-1434.

Computer Methods and Advances in Geomechanics, Desai et al. (eds) © 2001 Balkema, Rotterdam, ISBN 90 5809 183 X

Particle in cell and discrete element models for granular materials

H.-B. Mühlhaus, H. Sakaguchi & L. Moresi
CSIRO Exploration and Mining, Nedlands, W.A., Australia

N. Graham
SORD Technologies, Perth, W.A., Australia

ABSTRACT: The modelling of geotechnical problems frequently presents the need to track material composition, pressure, temperature and stress history through extreme levels of deformation including thermal convection. Suitable algorithms for this kind of modeling are usually complicated and slow. We present two different approaches: a particle-in-cell finite element method which combines many of the advantages of traditional finite elements with the geometrical flexibility of pure particle methods and a 3D discrete element formulation for solids and particulates involving arbitrary particle shapes and complex particle-structure interactions.

1 INTRODUCTION

One of the most successful ways to understand and control complex deformation, fracture and flow processes is forward modelling from the proposed initial state fol-lowing material interfaces and the temperature/pres-sure/stress history of material elements. Simulation algorithms must therefore be able to deal with very large strains associated with predominantly creep controlled or fluid dynamical processes while faithfully tracking material history and interfaces. Versatility and robustness are usually associated with the various formulations of the Finite Element Method (FEM). The need to track material history strongly suggests a Lagrangian formulation which provides a reference frame locked with the material itself. Unfortunately, large deformations and evolving fine structure and fractures are quite difficult to handle within the FEM because mesh distortion and remeshing are required to maintain optimal element configurations. The Particle-In-Cell (PIC) scheme is a hybrid numerical method which falls somewhere between the Finite Element Method (FEM) and a purely Lagrangian particle method such as the Discrete Element Method (DEM).

The PIC scheme attempts to combine the versatility of the continuum FEM with the geometrical flexibility of DEM. In PIC we use Lagrangian particles together with an Eulerian mesh. The mesh is used to solve nodal point unknowns using almost exactly the same formulation as the standard FEM, however, the mesh is not required to track material deformation avoiding the problem of distortion. Our formulation, and implementation, is derived from granular and fluid flow applications in which fluid and granular velocities are solved on the mesh, and material strains are recorded by the Lagrangian particles. The limit of applicability of any continuum method, even when combined with the most advanced computational schemes, is reached if the shape of individual particles strongly affect the flow properties (see Figure 5 and the related discussion) and or if the model involves granulation, cutting and comminution - type processes. In such cases one has to resort to a suitable discrete element (DEM) formulation. In DEM, solids and particulates are described as assemblies of bonded and unbonded spherical particles (Figure 2).

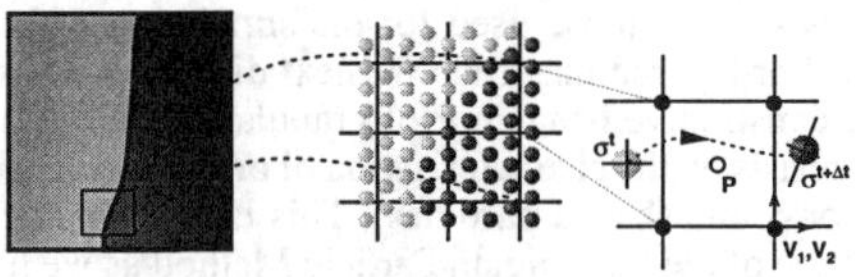

Figure 1: Schematic of the LPM

The procedure for the establishment of the domains of each boundary value problem, and the procedure for the solution of the equations of motion are similar to the ones employed in the discrete element method originally developed by Cundall & Strack (1979). However, in the contact search algorithm, we take advantage of the fact that most of the contacts within the macro particles are permanent (see Sakaguchi &

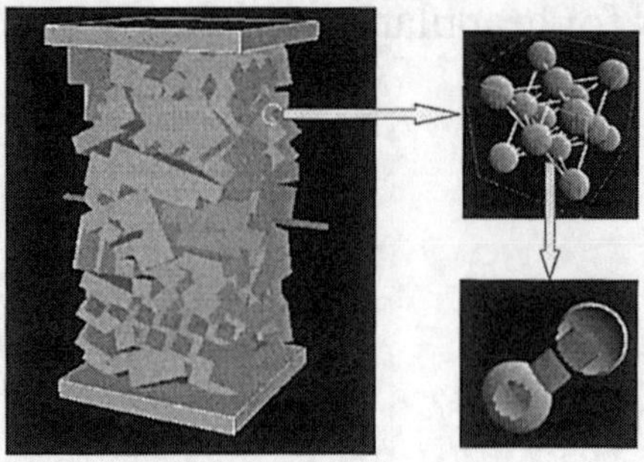

Figure 2: Schematic of the DEM

Mühlhaus 1997 for details). The paper is structured as follows: We first give brief outlines of the governing equations for the continuous and the discrete models. In the continuum description we assume the framework of a Cosserat Continuum and point out the relationship to Director Continua (e.g. Chandrasekar 1992). The use of a director as the basic variable rather than the rotation tensor as in the Cosserat theory may be advantageous in the case of oriented particles such as those used for the simulation of laminated or layered material. We next discuss aspects of the constitutive behaviour of granular media with an emphasis on the characterisation of oriented (e.g. pencil or penny shaped particles). This is followed by an outline of the Lagrangian Particle Method as we have implemented it. Finally the potential of the methods is illustrated by means of example solutions related to block caving , hopper flows, a proto-type of a discrete element simulations of a 3D DEM model for a remotely controll underground mining vehicule.

2 FORMULATION

2.1 *Continuum Model*

In the following we give a brief outline of the salient balance equations of the Cosserat continuum. We use rectangular, Cartesian coodinates x_i, $(i = 1, 2, 3)$, and v_i and ω_i^c are the translational and rotational velocities of a material point at position x_i and time t. In general ω_i^c is not equal to the spin

$$\omega_i = -\epsilon_{ikl} W_{kl} \tag{1}$$

where

$$W_{kl} = \frac{1}{2}(v_{k,l} - v_{l,k})$$

of an infinitesimal element of the continuum. In Equation (1), ϵ_{ikl} designates the permutation symbol and $(\cdot)_{,k} = \partial(\cdot)/\partial x_k$ are partial derivatives. The mass balance equation

$$\frac{\partial \rho}{\partial t}(\rho v_i)_{,i} = 0 \tag{2}$$

and the balance of linear momentum

$$\sigma_{ij,j}\rho(b_i - \dot{v}_i) = 0 \tag{3}$$

of the standard continuum and the Cosserat continuum are identical. (ρ is the density.) In the Cosseart theory the angular momentum balance of the standard continuum

$$\frac{d}{dt}\int_V \rho\epsilon_{ikl}x_k v_l dV = \int_S \epsilon_{ikl}x_k t_l dS \int_V \rho\epsilon_{ikl}x_k b_l dV \tag{4}$$

is extended in two repects. Firstly, a volume moment ρm_i^V and a surface momoent m_i^S are introduced in addition to the moments of the volume force ρb_i and the traction t_i. That is, the right hand side of Equation (4) is supplemented by

$$\int_V \rho m_i^V dV \int_S m_i^S dS \tag{5}$$

Secondly, we have to include the angular momentum ρd_i of the particles in the integral on the left hand of Equation (4). With these assumptions, the angular momentum balance is obtained as

$$\frac{d}{dt}\int_V \rho(\epsilon_{ikl}x_k v_l d_i)dV =$$

$$\int_S (\epsilon_{ikl}x_k t_l m_i^S)dS \int_V \rho(\epsilon_{ikl}x_k b_l m_i^V)dV \tag{6}$$

For the derivation of the local form of Equation (6), we use the relationships

$$t_i = \sigma_{ij}n_j \text{ and } m_i^S = m_{ij}n_j \tag{7}$$

where σ_{ij} and m_{ij} are stress and couple stress tensors, respectively. Application of Gauss' theorem and consideration of Equations (2) and (3) yields

$$m_{ij,j} - \epsilon_{ikl}\sigma_{kl}\rho(m_i^V - \dot{d}_i) = 0 \tag{8}$$

We conclude that the Cauchy stress tensor σ_{kl} is nonsymmetric in general, though symmetry could be retained if m_i^V, $\dot{d}_i$ and $m_{ij,j}$ were to form an equlibrated system by themselves. Thus the non-symmetry might not be of importance in applications where the exchange of momentum between translational and rotational freedoms is weak.

For a more complete outline of the Cosserat theory as applied to granular material we refer to Mühlhaus & Hornby (1997) and Mühlhaus & Hornby (1999). The papers addresses, among other things, statistical aspects, granular heat and the energetics of granular flow.

For finite element computations we require the weak form of the momentum and angular momentum balance equations. The weak form is obtained in the ususal way multiplying (3) and (8) with the test functions w_i and φ_i respectively. Integration over the volume and application of Gauss' theorem yields:

$$\int_V \sigma_{ij}(w_{i,j}\epsilon_{ijk}\varphi_k)m_{ij}\varphi_{i,j}dV =$$

$$\int_V \rho(b_i - \dot{v}_i)w_i(m_i^V - \dot{d}_i)\varphi_i dV \int_A t_i w_i m_i^S \varphi_i dA \tag{9}$$

Equation(9) must hold for all kinematically admissible w_i, φ_i vanishing on those parts of the boundary where displacements and/or rotations are prescribed.

If the granulate is composed of oriented particles (rod or disc shaped) it is sometimes advantage to use the director $\boldsymbol{a}$, with $|\boldsymbol{a}| = 1$ instead of the rotation tensor $\boldsymbol{R}^c$, where $\boldsymbol{W}^c = \dot{\boldsymbol{R}}^c\boldsymbol{R}^T$ to describe the microkinematics of the granulate. In case of rods or discs, $\boldsymbol{a}$ signifies the averaged local rod axis on the average unit normal of the disc surfaces.

A theory based on $\boldsymbol{v}$, grad$\boldsymbol{v}$, $\boldsymbol{a}$, grad$\boldsymbol{a}$ was proposed by Chandrasekar 1992 (see also de Gennes & Prost 1995) in the context of a continuum theory for liquid crystals. It should be noted that in directors theories the torsional moment about the director axis usually vanishes. For 2D problems, however, Cosserat and director theories are fully equivalent. In section 3 we propose a continuum model for 2D flows of rod-shaped particles within the framework of a director theory.

2.2 *Discrete Element Model*

We consider two granules i and j having translational velocities $\boldsymbol{v}^i$ and $\boldsymbol{v}^j$ and angular velocities $\boldsymbol{s}^i$ and $\boldsymbol{s}^j$. Continuum versions of the following formulations have been derived by Mühlhaus & Oka 1996 and Mühlhaus & Hornby 1997. The relative velocity at the contact is

$$\boldsymbol{g}^{ji} = \boldsymbol{v}^{ji} + d^{ji}\boldsymbol{n}^{ji} \times \boldsymbol{s}^{ji}, \tag{10}$$

where $\boldsymbol{v}^{ji} = \boldsymbol{v}^j - \boldsymbol{v}^i$, $\boldsymbol{s}^{ji} = \boldsymbol{s}^j + \boldsymbol{s}^i$, $\boldsymbol{n}^{ji}$ is the unit vector along the connecting line between the mass centres of the granules and d^{ji} is the distance between the mass centres.

In the traditional manner, we calculate the normal component of the contact force vector, $\boldsymbol{F}_n^{ji}$, at each instant in time, from the overlap, $d^{ji} - D^{ji}$, where d^{ji} is the current distance between particles j and i and D^{ji} is the sum of their radii. Note that non-spherical particles are represented as assemblies of spherical "micro-particles" using the macro-particle generator of Sakaguchi and Mühlhaus 1997. Accordingly, $\boldsymbol{F}_n^{ji}$ is always parallel to $\boldsymbol{n}^{ji}$

The tangent force, $\boldsymbol{F}_t^{ji}$ and the contact moments are much less significant than $\boldsymbol{F}_n^{ji}$ for the macroscopic properties of the granulate, but unfortunately much more complicated to calculate. We write the corrotational rates of $\boldsymbol{F}_t^{ji}$ and $\boldsymbol{M}^{ji}$ as

$$\overset{\nabla}{\boldsymbol{F}_t^{ji}} = \dot{\boldsymbol{F}}_t^{ji} - \boldsymbol{W}^{ji}\boldsymbol{F}_t^{ji}, \tag{11}$$

$$\overset{\nabla}{\boldsymbol{M}^{ji}} = \dot{\boldsymbol{M}}^{ji} - \boldsymbol{W}^{ji}\boldsymbol{M}^{ji}, \tag{12}$$

$$W_{\alpha\beta}^{ji} = \frac{1}{d^{ji}}\left(v_\alpha^{ji}n_\beta^{ji} - v_\beta^{ji}n_\alpha^{ji}\right) \tag{13}$$

Where $\dot{\boldsymbol{n}}^{ji} = \boldsymbol{W}^{ji}\boldsymbol{n}^{ji}$ and $\dot{\boldsymbol{t}}^{ji} = \boldsymbol{W}^{ji}\boldsymbol{t}^{ji}$; $\boldsymbol{t}^{ji}$, with $\boldsymbol{n}^{ji} \cdot \boldsymbol{t}^{ji} = 0$, is the unit vector in the direction of $\boldsymbol{F}_t^{ji}$.

The contact forces and moments are related to the relative velocities (10) and relative rotations though suitable elasto-visco-plastic constitutive relationships. For a detailed outline we refer to Sakaguchi et al. (2000).

The configurations of the particle system are determined as usual by integration of the equations of motion

$$\sum_{\text{contacts}} \boldsymbol{F}^{ji}\boldsymbol{\gamma}^i - \boldsymbol{m}^i\dot{\boldsymbol{v}}^i = 0 \tag{14}$$

and

$$\sum_{\text{contacts}} \left(\frac{D^i}{2}\boldsymbol{n}^{ji} \times \boldsymbol{F}^{ji}\boldsymbol{M}^{ji}\right) + \boldsymbol{\mu}^i - \frac{1}{8}m^i {D^i}^2\dot{\boldsymbol{s}}^i = 0, \tag{15}$$

where m^i and D^i are the mass and diameter of the i-th particle; $\boldsymbol{\gamma}^i$ and $\boldsymbol{\mu}^i$ are external particle forces and particle moments respectively. Equations (14) and (15) are the discrete particle versions of (3) and (8) respectively.

On the scale of grains and micro cracks the structure of rock appears highly disordered and complex in general. A major advantage of the present model is that such complexities can be easily dealt with. A simple algorithm for the generation of complex

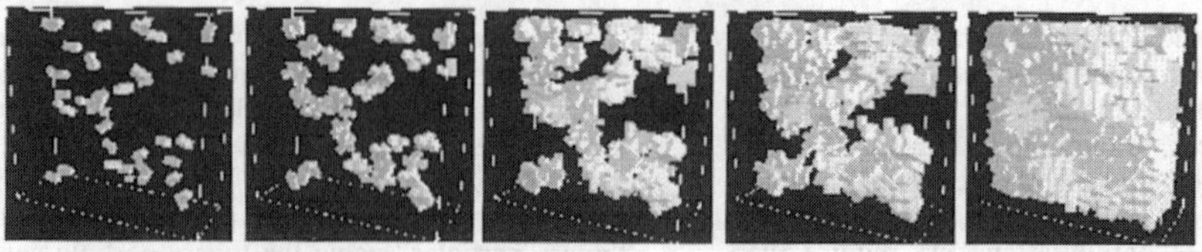

Figure 3: Growth of three dimensional macro-particles. *see* Sakaguchi & Mühlhaus (1997)

shaped macro-particles is described in Sakaguchi & Mühlhaus (1997).

Initially, seed particles are generated in a given domain by a uniform random process in which coordinate pairs are continuously determined until the total number of the seed particles reaches a preset value. Once all the seed particles are positioned, macro-particles are grown according to the random process from seed particles to fill the space. The resulting particles shapes are usually irregular (Figure 3) with a significant proportion of non-convex particle shapes. However, the proportion of convex particle shapes can be also controlled by using choosing growing direction (see Sakaguchi and Mühlhaus 1997).

Assemblies of bonded macro-particles may represent, for instance, particular rock or ceramics fabrics, while assemblies of loose macro-particles are considered in fragmentation, comminution and flow problems involving angular granules.

3 LAGRANGIAN PARTICLE METHOD

Initially, particles are evenly distributed within each element so that the element mass can be portioned out equally to the individual particles: $m_p = \rho_0 V_e / N_{ep}$ where m_p is the mass of particle p, ρ_0 is the fluid density, V_e is the volume of element e, and N_{ep} is the number of particles in element e. The discrete density distribution is then represented by

$$\rho(x,t) = \sum_{p=1}^{N_{ep}} m_p \delta(x - x_p) \tag{16}$$

The shape functions for the underlying mesh are used to interpolate the nodal velocity and rotation fields to any location

$$v_i = \sum_{\alpha=1}^{N_{en}} N_\alpha v_i(\boldsymbol{x}_\alpha), \qquad \omega_i^c = \sum_{\alpha=1}^{N_{en}} N_\alpha \omega_i^c(\boldsymbol{x}_\alpha) \tag{17}$$

and also

$$w_i = \sum_{\alpha=1}^{N_{en}} N_\alpha w_i(\boldsymbol{x}_\alpha), \qquad \phi_i = \sum_{\alpha=1}^{N_{en}} N_\alpha \phi_i(\boldsymbol{x}_\alpha) \tag{18}$$

where N_{en} is the number of element nodes, N_α is the shape function associated with node α, and $\boldsymbol{x}_\alpha$ is the location of node α.

Specific stresses and moment stresses are defined as

$$s_{ij} = \frac{1}{\rho}\sigma_{ij}, \qquad \mu_{ij} = \frac{1}{\rho}m_{ij} \tag{19}$$

Substituting (17),(19) into (9), and dropping acceleration terms as required by the infinite Prandtl number approximation,

$$\sum_{\text{elements}} \left(\int_{V_e} \rho \sum_{\alpha=1}^{N_{en}} (\mathbf{B}_\alpha \mathbf{W}_\alpha)^T \mathbf{s} dV - \int_{V_e} \rho \sum_{\alpha=1}^{N_{en}} (N_\alpha \mathbf{W}_\alpha)^T \boldsymbol{g} dV \right) - \sum_{\text{elements}} \int_{\partial V_e} \sum_{\alpha=1}^{N_{en}} (N_\alpha \mathbf{W}_\alpha)^T \boldsymbol{s}_s dA = 0 \tag{20}$$

The vectors and matrices appearing in (20) are defined as follows:

$$\mathbf{W}_\alpha{}^T = \{w_i, \phi_i\}_{\boldsymbol{x}=\boldsymbol{x}_\alpha} \qquad \mathbf{s}^T = \{s_{ij}, \mu_{ij}\} \tag{21}$$

$$\mathbf{g}^T = \{b_i, m_i^v\} \qquad \mathbf{s}_s{}^T = \{t_i, m_i^s\} \tag{22}$$

In 2D we have

$$\mathbf{B}_\alpha = \begin{bmatrix} N_{\alpha,1} & 0 & 0 \\ 0 & N_{\alpha,2} & 0 \\ N_{\alpha,2} & 0 & N_\alpha \\ 0 & N_{\alpha,1} & -N_\alpha \\ 0 & 0 & N_{\alpha,1} \\ 0 & 0 & N_{\alpha,2} \end{bmatrix} \tag{23}$$

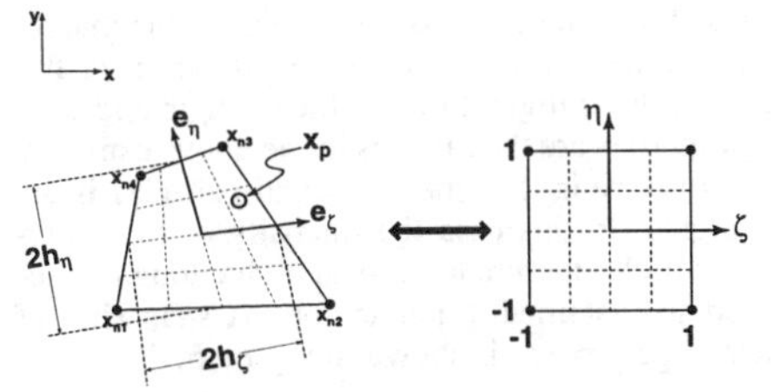

Figure 4: Coordinate systems in global mesh, distorted element, and master element reference frames

and the W and s pseudo-vectors are defined as

$$\mathbf{W}_\alpha{}^T = \{w_1, w_2, \omega_3^c\}, \tag{24}$$

$$\mathbf{s}^T = \{s_{11}, s_{22}, s_{12}, s_{21}, \mu_{31}, \mu_{32}\} \tag{25}$$

The definitions in (23) and (25) are restricted to 2D (x_1, x_2 plane); (20) – (22) are general.

This derivation is, so far, identical to that for a standard, grid-based finite element discretization. However, when (16) is inserted into (20), we obtain terms of the form

$$\mathbf{f}_\alpha^e = \int_{V_e} \left(\sum_{p=1}^{N_{pe}} m_p \delta(x - x_p) \mathbf{B}_\alpha^T \right) dV = \sum_{p=1}^{N_{pe}} m_p \mathbf{B}_\alpha^T \Big|_{x=x_p} \tag{26}$$

where $\mathbf{f}_\alpha^e$ is the element force vector referred to the grid but now computed using the particle locations and properties.

3.1 *Element matrices and particle properties*

In a standard finite element formulation, the stiffness matrix for an element, $\boldsymbol{k}^e$ is built up in a segregated form, often written as:

$$k_{ab}^e = \int_{\Omega^e} \boldsymbol{B}_a^T \boldsymbol{D} \boldsymbol{B}_b d\Omega^e \tag{27}$$

where $\boldsymbol{B}_a$ is a matrix comprising shape function derivatives (associated with node a) obtained from the constitutive relationship, and $\boldsymbol{D}$ is a matrix of material properties. For our formulation, the material property matrix is considered to be an attribute of an individual particle together with its current state and history and not a property of the mesh. Therefore, in the context of the numerical integration scheme based on the locations of individual particles,

$$k_{ab}^e = \sum_{p=1}^{n_{ep}} w_p \boldsymbol{B}_a^T(\boldsymbol{x}_p) \boldsymbol{D}_p \boldsymbol{B}_b(\boldsymbol{x}_p) \tag{28}$$

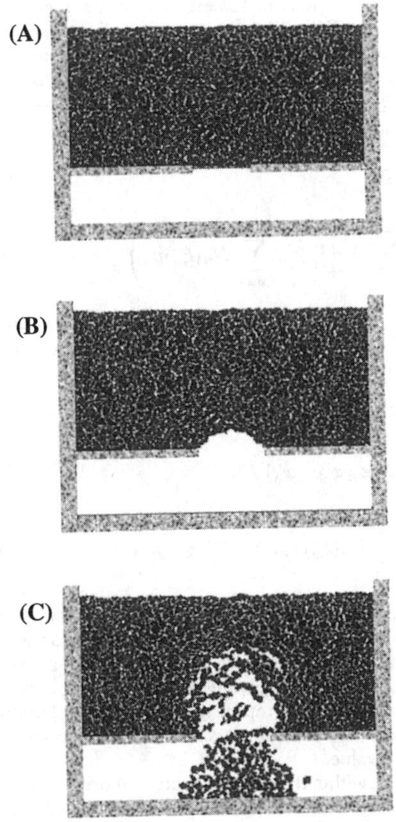

Figure 5: Block caving simulation. (A) Initial state, (B) Stable equilibrium configuration, (C) Flow pattern for fracturing particles.

3.2 *Element inverse mapping*

It is usual to change variables in the element integrals such as (27) to a regular master element. This greatly simplifies the computation, but in our case there is an additional step required before we can map to a master element. The particle positions are known in the global coordinate system and must first be mapped into the element local coordinate system. Zhao *et al.* (1999) found an algebraic mapping for bilinear quadrilateral elements, but a more general approach is required for higher order elements and for 3D.

The notation is given in Figure 4 for the 2D case — the extension to three dimensions is straightforward. e_ξ and e_η are unit vectors in the 'natural directions' of the distorted element which map to the ξ and η axes respectively in the master element. h_ξ and h_η are characteristic dimensions of the element in the appropriate directions. We wish to map the coordinates, $\boldsymbol{x}_p$ of the particle p to the coordinates in the master element, $\boldsymbol{\xi}_p$. The procedure is iterative: we first guess the initial value of $\boldsymbol{\xi}_p$ and use this to predict the global coordinates $\boldsymbol{x}_p^0$ as follows

$$\boldsymbol{\xi}_p = (0,0) \tag{29}$$

$$\boldsymbol{x}_p^0 = \left(\sum_{n=1}^{n_{en}} N_n(\boldsymbol{\xi}_p) x_n, \sum_{n=1}^{n_{en}} N_n(\boldsymbol{\xi}_p) y_n \right) \tag{30}$$

where n_{en} is the number of nodes in the element, and $\boldsymbol{x}_n$ are their coordinates. We compute $\boldsymbol{\xi}_p$ through a number of corrector steps:

$$\xi_p \leftarrow \xi_p \beta \left(e_{\xi x} x_p^i e_{\xi y} x_p^i \right) / h_\xi \tag{31}$$

$$\eta_p \leftarrow \eta_p \beta \left(e_{\eta x} x_p^i e_{\eta y} x_p^i \right) / h_\eta \tag{32}$$

$$\boldsymbol{x}_p^i = \left(\sum_{n=1}^{n_{en}} N_n(\boldsymbol{\xi}_p) x_n, \sum_{n=1}^{n_{en}} N_n(\boldsymbol{\xi}_p) y_n \right) \tag{33}$$

i is the iteration index, β is a relaxation parameter which we chose to be 1.0 for the first iteration and 0.9 for subsequent iterations. When $\boldsymbol{e}_\xi$ and $\boldsymbol{e}_\eta$ are orthogonal, the iteration completes in one step, otherwise it is necessary to repeat the correction step until the predicted $\boldsymbol{x}_p^i$ is within a satisfactory tolerance of the known value.

We use $\boldsymbol{\xi}$ within the element search algorithm since the master element coordinates give an immediate determination of whether the particle lies inside or outside the element even for highly distorted elements (if the element is too distorted for this procedure to work, then it is also too distorted to use to create the element matrices)

For detailed benchmarks of the algorithm, see Moresi *et al* (2000).

4 EXAMPLES

4.1 *DEM model of granular flow*

Let us consider a rectangular domain, which has a trap door located at the centre of the bottom side, filled with randomly shaped, cohesionless particles (Figure 5(A)) generated by the macro-particles generator which briefly described above. This example could be a simplified model for mining of an ore body in block caving. Upon opening of a trap door centered at the lower side of the domain, granular flow set in and stops after a stable arch has formed above the trapdoor. Figure 5(B) shows the final equilibrium configuration. In the following example we demonstrate that the stability of the arch is determined by the ratio of the strength of the asperities of the macro-particles to the weight of the elemental particles. The initial particle configuration and the strength and elasticity properties are the same as in the above example. However the weight of the elemental particles is ten times higher. In this case the intensity of the asperity and particle fracture process is high enough to accommodate sustained granular flow. A snap shot of the discharge process is shown in Figure 5(C).

4.2 *3D DEM model for a remotely controlled underground mining vehicle robot*

The simulation is based entirely on a discrete element realisation (see Figures 6(a) and (b)) of the robot head.

The discrete element method is usually applied for simulations of granular flows on all scales ranging collisional momentum transfers in the rings of Saturn to the simulation of phase transitions in nano-crystals. For the simulation of structural elements, we have developed novel, panel elements. Kinematically, the panel elements represent geometrical constraints for the motion of the surrounding soil and the ground water. Each panel element senses the resulting force and moment from the local pore pressures and the particles currently in contact with the panel. The effect of wall friction is also considered. The current mining vehcle head realisation consists of more than 8000 panel elements, whereby each tine is made up of 6 panels. Textures of any desired kind of the drive belts can also be considered. Each drive belt consists of approximately 600 panels in the present configuration. Each kinematic component has it own mass centre whereby the mass centres of the tines and the drives are continuously moving relative to the resulting mass centre of the robot head. Forces and moments from all panel locations can be recalled during all times during the running of a simulation. The motion and trajectory of the head is controlled by individually controllable drive belts and the dive planes (4 dive planes, two in front and two in the back). In the model, each dive plane consists of about 50 panels. An important task of the simulation will be to test the functionality of the various kinematic controls. At present a fully operational, virtual mining robot has been developed along the lines described above.

4.3 *LPM model of discharging silo*

We model a silo using a logically rectangular mesh of bilinear-velocity, constant-pressure quadrilateral elements. The silo shape is obtained by distorting the lower portion of the mesh. We assume symmetry about the mid-plane to save computational effort, and apply a reflecting boundary condition there. The sidewalls are given a no-slip boundary condition, and the bottom and top are unconstrained. Particles are ini-

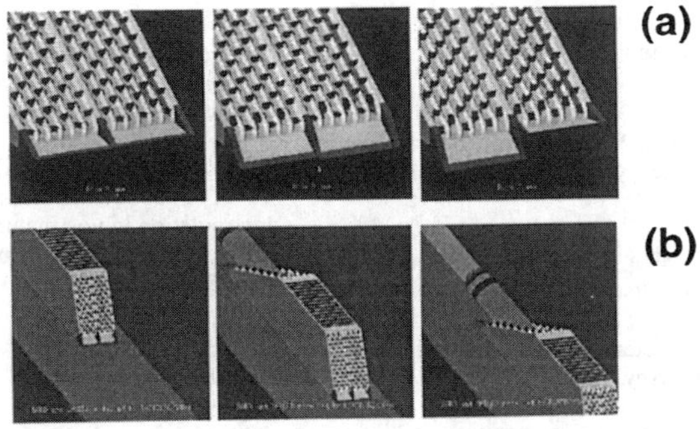

Figure 6: Remotely controlled underground mining vehicle

tially distributed evenly within elements such that the silo is three-quarters full of material. A rectangular grid of dyed particles is used to measure the deformation during the run. The material is assumed to be rigid-plastic, obeying a Drucker-Prager yield criterion (Mühlhaus *et al.* 1999). Figure 7 shows a sample run in which the material had a friction coefficient of 0.7. The initial filling level is shown by a horizontal line, the light grey grid lines are the strain-markers, and the intensity of the shading indicates the strain rate. Interacting shear bands develop after very small deformation has occurred. As the deformation increases, the surface begins to slump, in particular above the opening. The longer-term behaviour is shown in the inset taken from a lower resolution simulation.

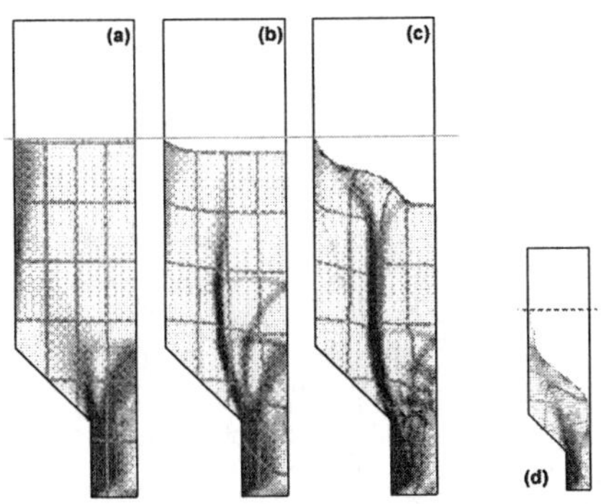

Figure 7: Silo discharge model using the LPM (A) Initial state with grid of deformation markers, (B,C) discharge showing bands of high strain rate evolving with the free surface. (D) Near complete discharge.

5 CONCLUDING REMARKS

Classical field theories often describe the relations and interactions between macroscopic observables that arise from some sort of averaging over microscopic complexions of the system. It is also the case that many systems which are quite different at the microscopic scale possess essentially the same description at the macroscopic level leading to a relatively small number of field theories that model a wide variety of materials. By contrast, generalized continuum theories and discrete element methods of the type we have described provide a richer physical framework to describe and distinguish between different material microstructures. Starting from the virtual power expression for dense assemblies of spherical granules, expressions for the average stress tensor, momentum and angular momentum balances and corresponding variational boundary conditions can be derived. Turning to computational aspects, we derived a symbolic solution of a general, mixed boundary value problem for Cosserat materials within the framework of a finite element based particle-in-cell method. The potential of the theory and the particle-in-cell model is demonstrated by means of two dimensional silo flow and trapdoor problems.

ACKNOWLEDGEMENT

The financial support by SORD Technologies Ltd, and by the Centre for Offshore Foundation System (COFS) is greatfully acknowledged.

REFERENCES

Barde, J.P. & J. Proubet, J. 1991. A numerical investigation into the structure of persistent shear bands in granular media. *Géotechnique* 41 (4)L 5: 99–613.

Chandrasekar, S. 1992. Liquid Crystals, 2nd Edition, (Cambridge University Press, 1992)

Chapman, S. & Cowling, T.G. 1995. The mathematical theory of non-uniform gases: Cambridge University Press, Cambridge.

Cundall, P.A. & Strack, O.D.L. 1979. A discrete numerical model for granular assemblies. *Géotechnique.* 29: 47–65.

de Gennes, P.G. & Prost, J. 1995. The physics of liquid crystals, 2nd Edition: Clarendon Press, Oxford.

Hughes, T.J.R. 1987. The finite element method: Prentice Hall, Englewood-Cliffs, New Jersey.

Moresi, L., Mühlhaus, H.-B. & Dufour, F. 2000. Particle - in -cell solutions for creeping viscous flows with internal interfaces, *Proc. 5th Int. Workshop on Bifurcation and Localisation in Geomechanics*, Perth, Australia: Balkema, in press.

Mühlhaus, H.-B. & Oka, F. 1996. Dispersion and wave propogation in discrete and continuous models for granular materials. *Int. J. Solids Structures*, 33: 2841–2858.

Mühlhaus, H.-B. & Hornby, P.1997. On the reality of antisymmetric stresses in fast granular flows. *Proc. IUTAM Conf. on Granular and Porous Media*, eds. Fleck and Cocks: Kluwer Academic Publishers, 299–311.

Mühlhaus, H.-B. & Hornby, P. 1999. Polar continua and the micromechanics of granular materials. Chapter 2.2 *Mechanics of granular materials, an introduction*, eds. M. Oda and K. Iwashita: A. A. Balkema, Rotterdam, 86–106.

Mühlhaus,H.-B., Moresi,L. & Freij-Ayoub, R. 1999. Lagrangian particle modelling of granular flows, 2nd Australian Congress on Applied Mechanics, Canberra, 10-12: 482-487.

Sakaguchi, H. & Mühlhaus, H.-B. 1997. Mesh free modelling of failure and localization in brittle materials. In: *Deformation and progressive failure in geomechanics, Proc. IS-Nagoya '97* eds. Asaoka, A., Adachi, T. and Oka, F.: Pergamon, 15–21.

Sakaguchi, H. & Mühlhaus, H.-B. 2000. Hybrid Modelling of Coupled Pore Fluid-Solid Deformation Problems, *Pure appl. geophys.* 157: (in press).

Savage. S.B. 1998. Analysis of slow, high concentration flows of granular materials. *J. Fluid Mech.* 377: 1–26.

Sulsky, D.,Chen,Z. & Schreyer. H.L.1994. A particle method for history dependent materials, *Comput. Methods Appl. Mech. Engrg.*, 118: 179–196.

Thornton, C. 2000. Microscopic approach contributions to constitutive modelling. In: *Constitutive modelling of granular materials* ed. D. Kolymbas: Springer-Verlag, 193–208.

van Mier,J.G.M., Schlangen, E & Vervuurt, A. 1995. Lattice type fracture models for concrete. In : *Continuum Models for Materials with Microstructure* ed. H.B.Mühlhaus: Wiley, 341–377.

C. Zhao, C., Hobbs,B.E.,Mühlhaus, H.-B. & Ord, A. 1999. A consistent point-searching algorithm for solution interpolation in unstructured meshes consisting of 4-node bilinear quadrilateral elements, *Int J. Numer. Math. Engng* 45: 1509–1526.

Computer Methods and Advances in Geomechanics, Desai et al. (eds) © 2001 Balkema, Rotterdam, ISBN 90 5809 183 X

Gradient dependent viscoplastic constitutive models and strain localization analysis of water saturated cohesive soil

Fusao Oka, Mingjing Jiang & Yosuke Higo
Department of Civil Engineering, Kyoto University, Japan

ABSTRACT: Deformation behavior of clay specimen modeled as a viscoplastic model with second order strain gradient during shear is numerically analyzed by a soil-water coupled finite element method. In addition, effect of strain gradient term on growth rate of fluctuation were obtained by the linear instability analysis. It is found that the strain localization pattern and stress-strain curve greatly depend on the initial non-homogeneous property, strain gradient and the permeability.

1 INTRODUCTION

It is well known that the treatment of heterogeneity of geomaterials is an important problem in geomechanics. The present paper studies the effects of initial heterogeneity and permeability on the deformation pattern including strain localization of cohesive soil modeled by second order strain gradient dependent poro-viscoplastic constitutive equations. As for the material heterogeneity, we have considered random distribution of material parameters such as strength and viscoplastic constants. For the effects of permeability, we have numerically simulated clay behavior with different permeability coefficients under the globally undrained conditions.

2 GRADIENT DEPENDENT VISCOPLASTIC MODEL FOR CLAY

Adachi et al.[1987] have shown that the overstress type viscoplastic model with strain hardening as proposed by Adachi and Oka[1982] , cannot describe the acceleration creep behavior of clay. To overcome this problem of the conventional overstress type constitutive model, Adachi et al.[1987] and Oka et al.[1994] introduced a second material function into the model. The viscoplastic flow rule is given by

$$D_{ij}^{vp} = \eta \langle \Phi_1(F) \rangle \Phi_2(\xi) \frac{\partial f}{\partial \sigma'_{ij}} \quad , \quad F = \frac{f - \kappa_s}{\kappa_s} \quad (1)$$

$$\sigma'_{ij} = \sigma_{ij} - u_w \delta_{ij}, \quad (2)$$

where D_{ij}^{vp} is the viscoplastic strain rate, η is the viscosity parameter,σ_{ij} is the total stress tensor, σ'_{ij} is the Terzaghi's effective stress tensor, u_w is the pore water pressure, f is the dynamic yield function, δ_{ij} is the Kronecker's delta,

Φ_1 is a material function accounting for strain rate effects, Φ_2 is a second material function, <> is the Macauley brackets, $F = 0$ denotes the static yield function, and κ_s is the hardening parameter. The second material function is introduced to describe the vanishing of rate dependency at the failure state. In the present study, the following form for the second material function is adopted.

$$\Phi_2 = 1 + \xi. \quad (3)$$

The internal variable ξ express the deterioration of the material and obeys the following evolutional equation.

$$\dot{\xi} = \frac{M^{*2}}{G_2^*(M^* - \eta^*)^2}\dot{\eta}^*, \quad (4)$$

where $M^*$is the value of the stress ratio η^* at failure , G_2^* is a material parameter and η^* is the stress invariant ratio defined by

$$\eta^* = \frac{\sqrt{2J_2}}{\sigma'_m}, \quad (5)$$

where J_2 is the second invariant of the deviatoric stress tensor $S_{ij}(S_{ij}$ =deviatoric part of $\sigma_{ij})$ and σ'_m is the mean effective stress.

It has been experimentally found that the shear strength and deformation characteristics of clay depend on the volumetric strain. The volumetric plastic strain is used as a hardening parameter in the well known Cam-clay model [1963]. The volumetric inelastic strain due to both consolidation and dilatancy is a measure of the deterioration of granular materials. On the other hand, Oka 1995, Mühlhaus and Oka, 1995, Oka et al, 1998 demonstrated that the higher order gradients may be attributed to the fact that the soil is discrete. Frantziskonis[1993] has also shown that the material inhomogeneity can be described by the constitutive model with higher order strain gradient. Thus, in the present paper, we have introduced the second order gradient of volumetric viscoplastic strain into the constitutive model to describe more accurately and sufficiently the deformation of clay by considering the non-local and viscoplastic effects of the material. In particular, static yield function includes the Laplacian of the viscoplastic volumetric strain is proposed as follows;

$$f_s - \kappa_s = \frac{\sqrt{2J_2}}{M^* \sigma'_m} + \ln \frac{\sigma'_m}{\sigma'_{my}} - a_3 \nabla^2 \varepsilon_v^{vp} = 0, \quad (6)$$

where κ_s is the hardening parameter, ε_v^{vp} is the viscoplastic volumetric strain ($=\int D_{kk}^{vp} dt$), $a_3 \nabla^2 \varepsilon_v^{vp}$ is the gradient term with a_3 defined as a material constant, J_2 is the second invariant of the deviatoric stress tensor S_{ij} (S_{ij}=deviatoric part ofσ_{ij}) and σ'_m is the mean effective stress. The quantity σ'_{my} is a hardening parameter.

Following a Perzyna's theory[1963], we assume that the dynamic yield function is the same as the static yield function in the following way.

$\eta \Phi_1(F)$ in Eq.(1) is given by

$$\eta \Phi_1(F) = c \cdot M^* \sigma'_m \exp\{m' (\frac{\sqrt{2J_2}}{M^* \sigma'_m} + \ln \frac{\sigma'_m}{\sigma'_{me}} - \frac{1+e}{\lambda - \kappa} \varepsilon^{vp} - a_3 \nabla^2 \varepsilon_v^{vp})\}, \quad (7)$$

$$c = c_0 \exp\{m' \ln \frac{\sigma'_{me}}{\sigma'_{my0}}\}, \quad (8)$$

where M^*, m' and c are material constants and the gradient coefficient a_3 is is also assuming to be constant. In Eq.(7), σ'_{me} is the initial value of σ'_m, σ'_{my0} is the initial value of hardening parameter, λ is the consolidation index, κ is the swelling index, and e is the void ratio.

The elastic strain rate is given by the isotropic Hooke's law i.e.

$$D_{ij}^e = \frac{1}{2G} \dot{S}_{ij} + \frac{\kappa}{3(1+e)\sigma'_m} \delta_{ij} \dot{\sigma}'_m, \quad (9)$$

where $D_{ij}^e (\dot{\epsilon}_{ij}^e)$ is the elastic strain rate tensor, G is the elastic shear modulus, $\dot{S}_{ij}$ is the deviatoric stress rate tensor and δ_{ij} is Kronecker's delta.

The total strain rate $D_{ij} (= \dot{\varepsilon}_{ij})$ is decomposed into elastic and viscoplastic parts.

3 TWO-DIMENSIONAL LINEAR STABILITY ANALYSIS

In this section, we study the instability of a two dimensional motion under plane strain undrained conditions and determine the preferred orientation of the localized zone, where the small strain fluctuations will grow.

For simplicity, we assume that clay is rigid plastic for deviatoric behavior ,i.e. $D_{ij}^e = 0$. Then,

$$D_{ij} = D_{kk} \delta_{ij} + D'_{ij} \quad (10)$$

where D'_{ij} is the deviatoric part of D_{ij}.

For plane strain undrained(isochoric) conditions, we then have

$$D_{33} = 0, \quad D_{11} + D_{22} = 0 \quad (11)$$

We only consider the case for which the effective stress components are given by the expression

$$[\sigma'_{ij}] = \begin{bmatrix} \sigma'_1 & 0 & 0 \\ 0 & \sigma'_2 & 0 \\ 0 & 0 & \sigma'_3 \end{bmatrix} \quad (12)$$

From the undrained or isochoric condition , Eqs.(9) and (10), and the initial condition ($\varepsilon^{vp} = 0$), after integration we have

$$\varepsilon^{vp} = -\frac{\kappa}{1+e} \ln \frac{\sigma'_m}{\sigma'_{me}}, \quad (13)$$

where ε^{vp} is the viscoplastic volumetric strain and σ'_{me} is the initial value of σ'_m.

Next, we consider the equilibrium equations, undrained conditions and constitutive equations in the perturbed configuration. The equilibrium equations can be written as follows

$$\sigma_{ij,j} = S_{ij,j} + \sigma'_{m,j} \delta_{ij} + u_{w,j} \delta_{ij} = 0 \quad (14)$$

where, as usual, comma denotes differentiation with respect to spatial coordinates and and u_w is the pore water pressure.

The perturbations of the pore water pressure u_w, velocities v_i, and the volumetric viscoplastic strain $\varepsilon^{vp} = \int D_{kk}^{vp} dt$ are assumed to be of the periodic form as

$$\begin{bmatrix} \tilde{u}_w \\ \tilde{v}_i \\ \tilde{\varepsilon}^{vp} \end{bmatrix} = \begin{bmatrix} u_w^* \\ v_i^* \\ \varepsilon^{vp*} \end{bmatrix} \exp[iq(n_i x_i) + \omega t], \quad (15)$$

where q is the wave number($= 2\pi/l$, l: wave length), ω is the speed of the fluctuation growth and n_i is the component of the unit vector $\boldsymbol{n}$ which is normal to shear band (see Fig.1).

$$\mathrm{n} = (-\sin\theta, \ \cos\theta) = (n_1, \ n_2) \quad (16)$$

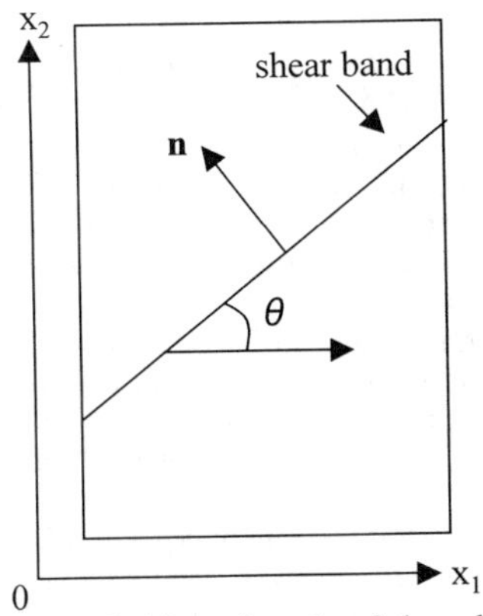

Fig.1 Definition of angle of shear band θ.

Under the undrained (isochoric) condition given by Eq.(12) , we have

$$v_1^* n_1 + v_2^* n_2 = 0. \tag{17}$$

From the constitutive equations, we obtain

$$\dot{\varepsilon}^{vp} = (M^* - \eta^*) I_2, \tag{18}$$

where $I_2 = \sqrt{D_{ij} D_{ij}} = \sqrt{D_{ij}^{vp'} D_{ij}^{vp'}}$ and $D_{ij}^{vp'}$ is the deviatoric part of D_{ij}^{vp}.

As a result, we have

$$\dot{\tilde{\varepsilon^{vp}}} = \omega \varepsilon^{\tilde{v}p}, \tag{19}$$

$$\varepsilon^{\tilde{v}p} = \frac{1}{\omega}(M^* - \eta^*)\tilde{I^2} \tag{20}$$

The perturbed equilibrium equations are

$$\tilde{\sigma}_{ij,j} = \tilde{S}_{ij,j} + \tilde{\sigma}'_{m,j}\delta_{ij} + \tilde{u}_{w,j}\delta_{ij} = 0, \tag{21}$$

and by substituting in it the constitutive equations, we obtain

$$B_1 D_{ij} n_j D_{kl} n_l v_k^* + B_2 v_i^* + B_3 D_{kl} n_l n_j v_k^* \delta_{ij}$$
$$+ iqn_j u_w^* \delta_{ij} = 0, \tag{22}$$

where D_{ij} denotes as before the strain rate tensor.

$$S = (M^* - H_0)(M^* - \eta^*)\sigma'_m \frac{(1+e)}{\kappa}$$
$$- \frac{M^{*3}\eta^*}{m'(G_2^*(M^* - \eta^*) + M^*\eta^*)} \frac{(1+e)\sigma'_m}{\kappa}$$
$$+ \frac{(1+e)}{\lambda - \kappa}(M^* - \eta^*)M^*\sigma'_m + a_3 M^* \sigma'_m q^2 (M^* - \eta^*)$$

$$= \Big[(M^* - \eta^* - M^* ln(\sigma'_m/\sigma'_{me})\frac{\lambda}{\lambda - \kappa}\Big)\Big(M^* - \eta^*)\Big)$$
$$- \frac{M^{*3}\eta^*}{m'(G_2^*(M^* - \eta^*) + M^*\eta^*)} + \frac{\kappa}{\lambda - \kappa} M^*(M^* - \eta^*)$$
$$+ \frac{\kappa}{1+e} a_3 M^* q^2 (M^* - \eta^*)\Big] \frac{(1+e)\sigma'_m}{\kappa}, \tag{23}$$

$$H_1 = \frac{\sigma'_m M^*}{m' I_2}. \tag{24}$$

We can rewrite Eqs.(17) and (22) in a matrix form as:

$$[A]\{y\}^T = \{0\} \tag{25}$$

in which

$$[A] = \begin{bmatrix} A_{11} & A_{12} & iqn_1 \\ A_{21} & A_{22} & iqn_2 \\ n_1 & n_2 & 0 \end{bmatrix} \tag{26}$$

$$\{y\} = \{v_1^*, \quad v_2^*, \quad u_w^*\} \tag{27}$$

From Eq.(25), the eigenvalue (dispersion equation) is obtained from the condition $det[A] = 0$

$$\omega = \frac{-S}{\frac{\sqrt{2J_2}}{2I_2} Y_1 \cot^2 2\theta + H_1}, \quad Y_1 > 0 \tag{28}$$

Since the denominator of Eq.(41) is positive, the sign of ω depends on the sign of S.

When the stress ratio η^* is small, S is positive. At this stage, ω is negative and the viscoplastic system is stable. On the other hand, When $M^* - \eta^*$ becomes close to zero, S can be negative and ω becomes positive. Consequently, the system becomes unstable. If G_2^* is infinite, the second term of S becomes zero and the sign of S is positive. The sign of ω, therefore, becomes negative and the system is always stable. In addition, it is worth noting that the model with the second gradient of viscoplastic volumetric strain becomes stable even if G_2^* is finite since the sign of S can be positive when the magnitude of positive α_3 is large enough and the wave length is very small(q is large). The viscoplastic system may become unstable when the wavelength of fluctuation $2\pi/q$ is large enough even if α_3 is positive. If the sign of α_3 is negative, the viscoplastic system may easily be unstable. From Eq.(28), it is evident that the preferred orientation of localization is $\theta = 45°$.

Table 1: Material parameters for strain localization analysis with different permeability coefficients

Parameter	Value
Compression index λ	0.172
Swelling index κ	0.054
Initial void ratio e_0	1.28
Initial mean effective stress σ'_{me}	200 (kPa)
Coefficient of earth pressure at rest K_0	1.0
Viscoplastic parameter m'	21.5
Viscoplastic parameter C	4.5×10^{-8}(1/s)
Stress ratio at failure M^*_f	1.05
Elastic shear modulus G	5500 (kPa)
Softening parameter G^*_2	10000 (kPa)
Gradient parameter a_3	0.0/0.5 (cm^2)
Coefficient of permeability $k_x=k_y=k$	1.54×10^{-4}, $\times10^{-6}$, $\times10^{-10}$(cm/s)

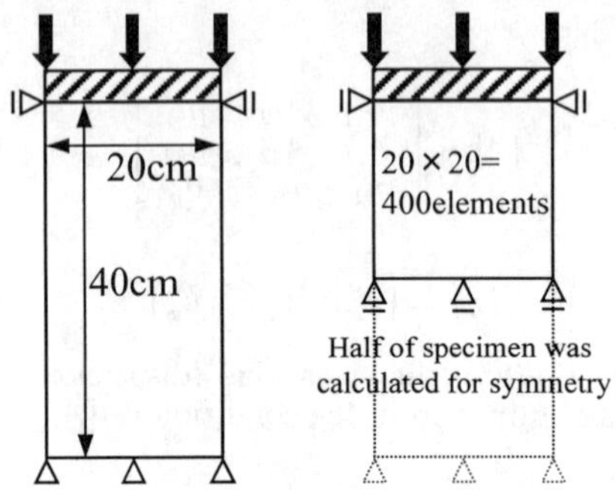

Fig.2 Boundary conditions and size of specimen

4 FINITE ELEMENT ANALYSIS OF STRAIN LOCALIZATION BY A GRADIENT DEPENDENT ELASTOVISCOPLASTIC MODEL

In the finite element analysis, the updated Lagrangian method with the objective Jaumann rate of Cauchy stress is used. An eight-noded quadlilateral element with a reduced Gaussian (2×2) integration is used to eliminate the shear locking and to reduce the tendency of appearance of a spurious hourglass mode. The viscoplastic volumetric strain is also defined at eight nodes using a quadratic interpolation function. On the other hand, the pore water pressure is defined at four corner nodes. The weak form of the continuity equation is integrated with (2×2) full integration. By using this combination of the spatial integration scheme, the effective stresses, the pore water pressure and the strains are calculated at the same integration points in each element.

In the numerical study, we consider the weak form of equilibrium equations, the second gradient-dependent dynamic yield function and the weak form of continuity equation for pore fluid. As for the boundary conditions, we assumed first gradient of viscoplastic strain is equal to zero [see Aifantis et al.,1999],because of no particular force at the boundary. In Fig.2 is shown the boundary conditions for plane strain problem used for numerical analysis, while in Table 1 are shown the parameters of Adachi-Oka model used in the computation. Gradient parameter, in principle, can be determined by the width of shear band. The strain rate of compression is 1.0%/min.

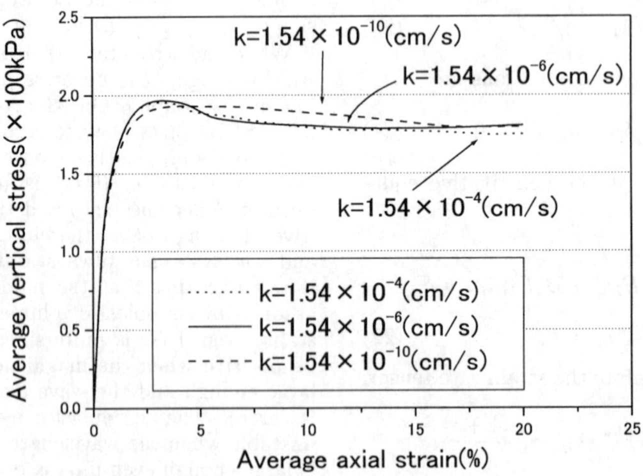

Fig.3 Average vertical stress strain relationships with different coefficient of permeability

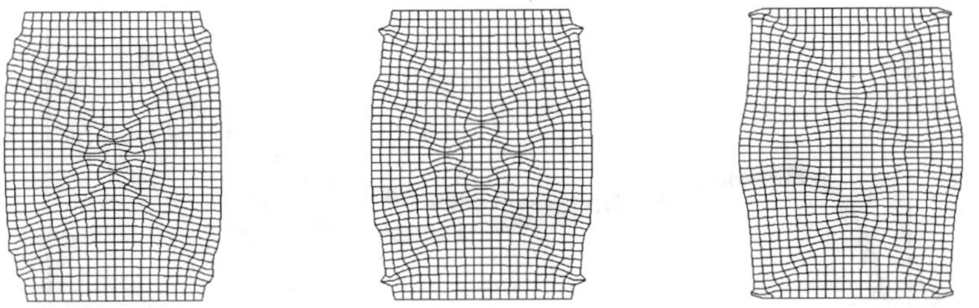

$k = 1.54\times10^{-4}$cm/s $k = 1.54\times10^{-6}$cm/s $k = 1.54\times10^{-10}$cm/s

Fig.4 Deformed mesh at an average axial strain of 15%

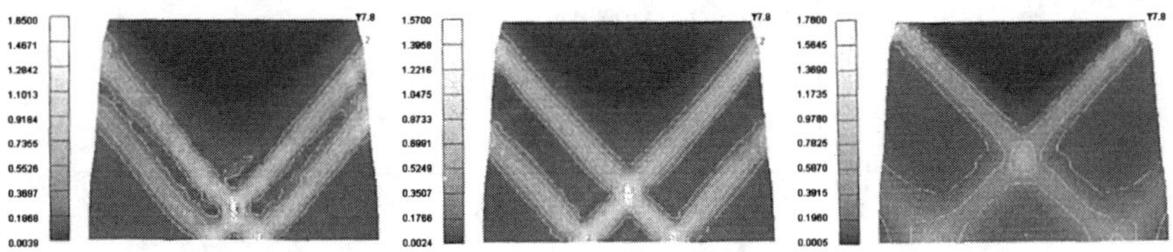

$k = 1.54\times10^{-4}$cm/s $k = 1.54\times10^{-6}$cm/s $k = 1.54\times10^{-10}$cm/s

Fig.5 Distribution of accumulated viscoplastic shear strain at an average axial strain of 15%

The average vertical stress vs. strain relationships with different coefficients of permeability are shown in Fig.3. In the early stage of loading, i.e. in the hardening range, it is seen that the material system with low permeability is less stable because of low average stress. On the other hand, in the strain softening range, the material with low permeability is less unstable because the average stress is larger than those of other cases. These behaviors are consistent with the theoretical consideration mentioned in section 3.

Fig.4 shows the deformed meshes at an average axial strain of 15% with different coefficients of permeability k. It is shown that pore fluid has an apparent influence on the formation of shear band. It is found that four shear bands are symmetrically formed in all cases. The space between two shear bands with same direction becomes larger with decrease of permeability k.

Fig.5 shows the distributions of accumulated viscoplastic shear strain $\gamma^p (= \int d\gamma^p, d\gamma^p = (de^p_{ij} de^p_{ij})^{1/2})$at an average axial strain of 15% with different coefficients of permeability k. It can be found out from the figures that the thickness, inclination of shear band, distance between shear bands, and the extent of localization of strain depend on permeability k. The accumulated strain is more extensively localized when permeability k is rather small.

Fig.6 shows the boundary conditions and distribution of stress ratio at failure, M^*_f (range of perturbation R=5%), using pseudo-random number. In Fig.7, the stress strain curves obtained by the different range of perturbation of M^*_f. Average axial stress becomes larger with an increase of R. Fig.8 shows typical stress paths in elements. In the case of homogeneous material, deviator stress always decreases along the failure line. On the other hand, in the case of random distribution of M^*_f(stress ratio at failure), stress path of some elements deviates from failure line with an increase of q. This is the reason for the increase of average axial stress at large strains.

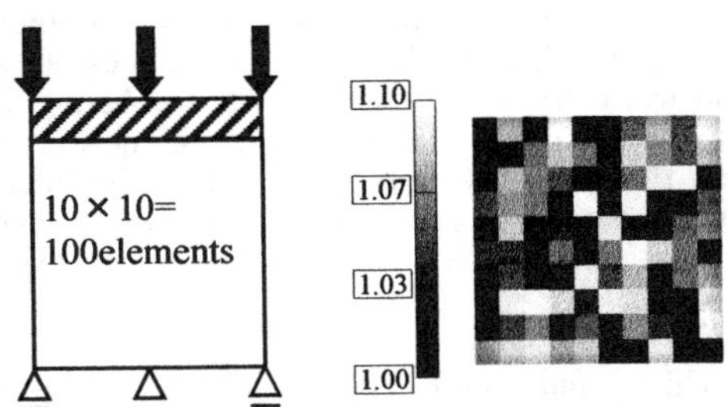

Fig.6 Boundary conditions and distribution of M*

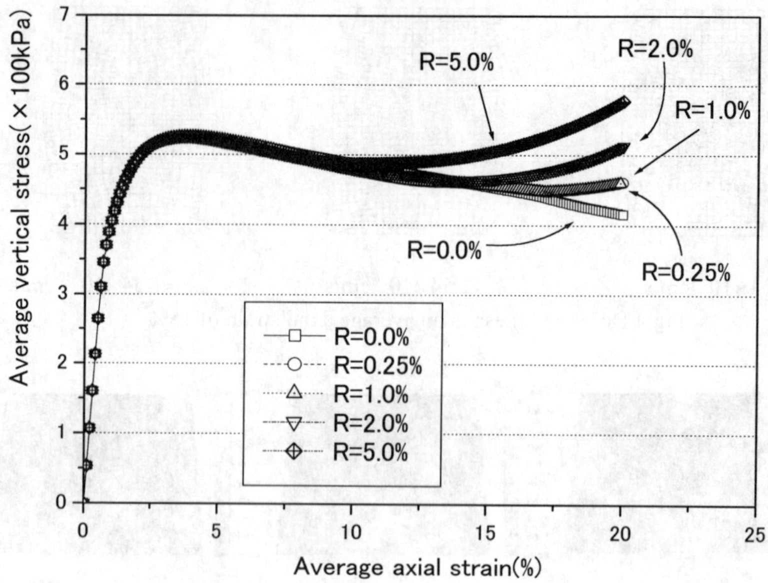

Fig.7 Stress-strain relationships obtained by different range of perturbation of M*

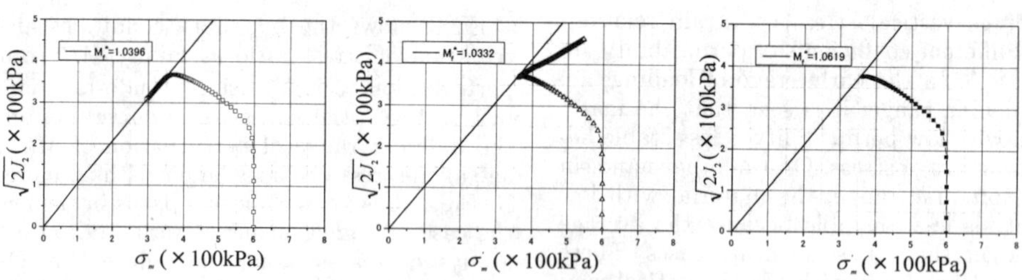

Fig.8 Three types of stress path

5 CONCLUSIONS

From the present study, it is found that the accumulated strain is more localized when gradient term a_3 is rather small, while the thickness shear band will increase with gradient term. As for the deformation pattern, permeability, namely, pore water transport has a great influence on the formation of shear bands. Material inhomogeneity makes material more stable than the homogeneous one.

REFERENCES

Adachi, T. and Oka, F. (1982), Constitutive equations for normally consolidated clay based on elasto-viscoplasticity, Soils and Foundations, 22, 4,57-70.

Aifantis, E.C., F.Oka, A.Yashima and T.Adachi. (1999), Instability of gradient dependent elasto-viscoplasticity for clay, Int. J. Num. Anal. Meth. in Geomechanics, 23, 10, 973-994.

Frantziskonis, G.(1993), Crack pattern related universal constants, Probabilities and Materials, Tests, Models and Applications, ed. by D.Breysse, Kulwer Academic Pub., 361-376.

Oka, F., H.B.Mühlhaus, A.Yashima, K.Sawada (1998), Quasi-static and dynamic characteristics of strain gradient dependent non-local constitutive models, in Material Instabilities in Solids, De Borst and Van Der Giessen eds., John Wiely & Sons.

Computer Methods and Advances in Geomechanics, Desai et al. (eds) © 2001 Balkema, Rotterdam, ISBN 90 5809 183 X

Numerical investigation of strain localization in a hypoplastic Cosserat material under shearing

Erich Bauer & Wenxiong Huang
Institute of General Mechanics, Graz University of Technology, Austria

ABSTRACT: This paper presents numerical investigations of strain localization in a plane granular layer between two parallel rough plates under monotonic shearing. A micro-polar approach is formulated within the framework of a hypoplastic Cosserat continuum to describe the essential properties of a dry and cohesionless granular material like sand. The constitutive model is based on incremental non-linear tensor-valued functions and captures the influence of the pressure, the void ratio, the mean grain diameter and the rotation resistance of the grains on the evolution of the stresses and couple stresses. The finite element calculations show that location and thickness of the localized zone is mainly affected by the initial state quantities, the boundary conditions and the rotation resistance of particles at the interface between the granular layer and the surfaces of the plates.

1 INTRODUCTION

It is experimentally evident that the phenomenon of strain localization in narrow zones called shear bands is essentially related to the micro-structure of the material. For large shearing dilatancy, strain softening and polar effects, i.e. pronounced grain rotations, can be observed within the localized zone (e.g. Bogdanova-Bontcheva & Lippmann 1975, Oda 1993). The results of shear tests with sand specimens show that the thickness of a localized deformation zone is not a material constant. The thickness mainly depends upon the grain size distribution, grain shape, pressure level, initial density, dilatancy resistance of the boundaries and the roughness of the surfaces adjoining the boundaries of the granular body (e.g. Mühlhaus & Vardoulakis 1987, Uesugi et al. 1988, Vardoulakis & Sulem 1995, Tejchman 1997, Oda & Kazama 1998, Gudehus 2000). For investigations of strain localization using the finite element method an internal length must be included in the constitutive model. Otherwise the predicted thickness of the localized zone is scaled by the element size and the boundary value problem becomes mathematical ill-posed in the post-bifurcation regime (e.g. de Borst et al. 1992).

The focuse of this paper is on studying the influence of the boundary conditions on the evolution of strain localization within a granular layer between two parallel plates under plane shearing (Figure 1) . To this end the mechanical behavior of the granular material is described with a hypoplastic Cosserat model developed by Tejchman 1997 and Bauer & Huang 2000. The model is called hypoplastic since it describes the behavior of plastic materials without yield surface and plastic potential (Kolymbas & Wu 1993). The model takes into account Cosserat rotations, couple stresses, the mean grain diameter as the internal length and the current void ratio which is related to the pressure-dependent maximum, minimum and critical void ratios. The Cosserat boundary conditions are suitable to model the effect of the grain size and the surface roughness of the plates on the interface behavior of the granular material at the boundary of the plates.

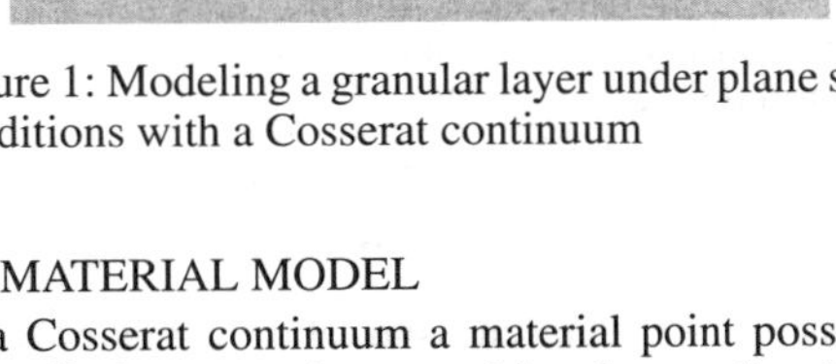

Figure 1: Modeling a granular layer under plane strain conditions with a Cosserat continuum

2 MATERIAL MODEL

In a Cosserat continuum a material point possesses three displacement degrees of freedom u_i $(i = 1, 2, 3)$

and three rotational degrees of freedom which are called Cosserat rotations ω_i^c $(i = 1,2,3)$. The gradient of the Cosserat rotations $\partial\omega_i^c/\partial x_j$ corresponds to the curvatures κ_{ij} which are associated with the couple stresses μ_{ij}. The rate of deformation and the rate of curvatures are defined as $\dot{\varepsilon}_{ij} = \partial\dot{u}_i/\partial x_j + \epsilon_{kij}\dot{\omega}_k^c$ and $\dot{\kappa}_{ij} = \partial\dot{\omega}_i^c/\partial x_j$ respectively, where $\partial\dot{u}_i/\partial x_j$ denotes the velocity gradient and ϵ_{ijk} denotes the permutation tensor. The proposed hypoplastic Cosserat model includes three state variables, i.e. the non-symmetric stress tensor σ, the couple stress tensor μ and the void ratio e. The evolution of the state variables are described by the following rate type equations (Tejchman & Bauer 1996, Tejchman 1997, Bauer & Huang 2000):

$$\mathring{\sigma}_{ij} = f_s\big[\hat{a}^2\,\dot{\varepsilon}_{ij} + (\hat{\sigma}_{kl}\,\dot{\varepsilon}_{kl} + \hat{\mu}_{kl}\,\dot{\kappa}_{kl})\,\hat{\sigma}_{ij} + f_d\,\hat{a}\,(\hat{\sigma}_{ij} + \hat{\sigma}_{ij}^*)\sqrt{\dot{\varepsilon}_{kl}\,\dot{\varepsilon}_{kl} + \dot{\bar{\kappa}}_{kl}\,\dot{\bar{\kappa}}_{kl}}\,\big], \qquad (1)$$

$$\mathring{\mu}_{ij} = f_s\,d_{50}\big[a_c^2\,\dot{\bar{\kappa}}_{ij} + \hat{\mu}_{ij}\,(\hat{\sigma}_{kl}\,\dot{\varepsilon}_{kl} + \hat{\mu}_{kl}\,\dot{\kappa}_{kl} + 2\,f_d\,a_c\sqrt{\dot{\varepsilon}_{kl}\,\dot{\varepsilon}_{kl} + \dot{\bar{\kappa}}_{kl}\,\dot{\bar{\kappa}}_{kl}})\big], \qquad (2)$$

$$\dot{e} = (1+e)\,\dot{\varepsilon}_{kk}\,, \qquad (3)$$

with the abbreviations:
$\hat{\sigma}_{ij} = \sigma_{ij}/\sigma_{kk}$, $\hat{\sigma}_{ij}^* = \hat{\sigma}_{ij} - \delta_{ij}/3$, $\bar{\kappa}_{ij} = d_{50}\,\kappa_{ij}$, $\hat{\mu}_{ij} = \mu_{ij}/(d_{50}\,\sigma_{kk})$. Herein δ_{ij} denotes the Kronecker delta and d_{50} denotes the mean grain diameter, which is used as an internal length. The influence of the mean pressure and the current void ratio on the incremental stiffness, the dilatancy behaviour and the peak stress ratio is taken into account with the stiffness factor f_s, i.e.

$$f_s = \left(\frac{e_i}{e}\right)^{\beta} \frac{h_s}{n\,h_i\,(\hat{T}_{kl}\,\hat{T}_{kl})} \frac{(1+e_i)}{e_i} \left(-\frac{T_{kk}}{h_s}\right)^{1-n}, \qquad (4)$$

and the density factor f_d (Gudehus 1996, Bauer 1996), i.e.

$$f_d = \left(\frac{e - e_d}{e_c - e_d}\right)^{\alpha}. \qquad (5)$$

Herein α, β, n an h_s are constitutive constants. The scalar h_i in (4) can be obtained from a consistency condition (Gudehus 1996):

$$h_i = \frac{8\sin^2\varphi_c}{(3-\sin\varphi_c)^2} + 1 - \frac{2\sqrt{2}\sin\varphi_c}{3-\sin\varphi_c}\left(\frac{e_{io} - e_{do}}{e_{co} - e_{do}}\right)^{\alpha}.$$

In (4) and (5) the current void ratio e is related to the maximum void ratio e_i, the minimum void ratio e_d and the critical void ratio e_c. This limit void ratios decrease with an increase of the mean pressure T_{kk}, i.e. $e_i/e_{i0} = e_d/e_{d0} = e_c/e_{c0} = \exp[-(-T_{kk}/h_s)^n]$ where e_{i0}, e_{d0}, e_{c0} are the corresponding values for $T_{kk} = 0$ (Figure 2). Factors $\hat{a}$ and a_c in (1) and (2) are related

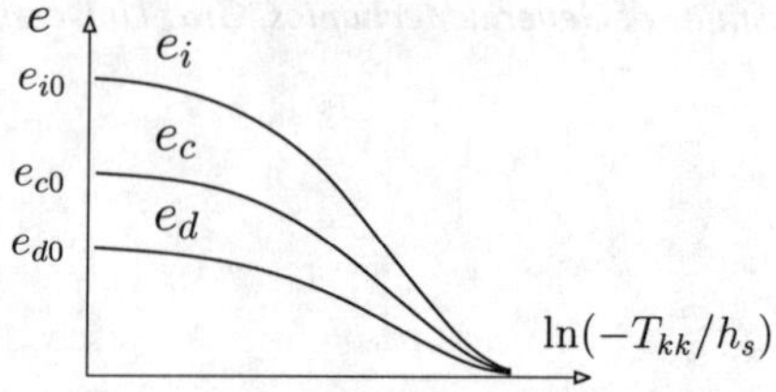

Figure 2: Decrease of e_i, e_c and e_d with T_{kk}

to stationary states, i.e. a_c is assumed to be a constant and $\hat{a}$ depends on the so-called angle of internal friction φ_c and on $\hat{T}_{kl}^{*s}$ which denotes the symmetric part of the normalized stress deviator $\hat{T}_{kl}^*$. The proposed representation for $\hat{a}$ reads:

$$\hat{a} = \frac{\sin\varphi_c}{3-\sin\varphi_c}\left[\sqrt{b} - \sqrt{\hat{T}_{kl}^{*s}\hat{T}_{kl}^{*s}}\right], \qquad (6)$$

with

$$b = \frac{8/3 - 3\,(\hat{T}_{kl}^{*s}\,\hat{T}_{kl}^{*s} + \hat{T}_{kl}^{*s}\,\hat{T}_{lm}^{*s}\,\hat{T}_{mk}^{*s})}{1 - 3\,\hat{T}_{kl}^{*s}\,\hat{T}_{lm}^{*s}\,\hat{T}_{mk}^{*s}/(\hat{T}_{kl}^{*s}\,\hat{T}_{kl}^{*s})}.$$

The constitutive model includes ten constants which can be estimated from the grain size distribution, the grain hardness, the grain shape and roughness (Herle & Gudehus 1999). For the present numerical investigations the following material constants for a medium sand are used:
$\varphi_c = 30°$, $e_{i0} = 1.2$, $e_{d0} = 0.51$, $e_{c0} = 0.82$, $n = 0.4$, $h_s = 190$ MPa, $\alpha = 0.11$, $\beta = 1.05$, $d_{50} = 0.5$ mm, $a_c = 1.0$.

3 INTERFACE BEHAVIOR

The slide and rotation resistance of particles at the surface of the plates is mainly determined by the surface roughness of the plates and the size, shape and roughness of the grains. Very rough plates can capture the adjoining grains so that neither sliding nor rotating may occur. Then the relative displacement and the Cosserat rotation at the boundary of the granular body is zero. For medium rough surfaces and quasi-static processes an empirical relation between the Cosserat rotation ω^c and the relative displacement u_r was proposed by Tejchman (1997), i.e.

$$\omega^c = f_\omega\,u_r/d_{50} \qquad (7)$$

where the dimensionless factor f_ω reflects the rotation resistance depending on the roughness of the surface and the size and shape of the grains.

4 NUMERICAL INVESTIGATIONS

The present hypoplastic Cosserat model was implemented in the finite element code ABAQUS and for plane strain conditions a four-node element with bilinear shape functions was used to describe the displacements and Cosserat rotations within the element. For the case of plane strain only two displacement degrees of freedom and one rotational degree of freedom remain for each node (Figure 1). In order to alleviate volumetric locking a so-called selective reduced integration technique was adopted (Nagtegaal et al. 1974). A Newton iteration procedure was used to solve the non-linear equation system and an implicit time integration technique was applied to compute the stress and couple stress increments.

The focus of the present numerical investigations is on studying the influence of the boundary conditions on the evolution of strain localization and void ratio within a granular layer under plane shearing. All calculations are carried for a shear layer height of 4 cm starting from same homogeneous and isotropic initial state with $e_o = 0.6$, $\sigma_o = -100$ kPa and $\mu_o = 0$. At the bottom nodes the displacement degrees of freedom and the rotation degree of freedom are locked, i.e. $u_1 = u_2 = 0$ and $\omega^c = 0$. At the top boundary the vertical stress $\sigma_{22} = -100$ kPa is kept constant, the Cosserat rotation is locked and a shear displacement u_{1T} is initiated by prescribed horizontal node displacements while the vertical displacement is obtained as a result of the dilatancy behaviour within the whole specimen.

In order to model the lateral boundary conditions of the classical plane simple shear test a linear displacement field must be prescribed on both sides of the specimen, i.e. with respect to the height h of the layer $u_1(x_2) = u_{1T} x_2 / h$. The numerical investigation shows (Figure 3.a) that in contrast to a classical continuum the distribution of the state quantities within the specimen obtained for a polar continuum becomes strongly heterogeneous from the beginning of shearing. The brighter bands indicate a higher void ratio and consequently the location where failure starts. The zones are eccentrically located and slightly inclined, which also reflects the experimental observation made with the so-called simple shear apparatus of the Cambridge type on sand specimens. For shearing under a constant lateral pressure the deformation field is also rather heterogeneous as shown in Figure (3.b). It is obvious that the lateral boundary conditions assumed in Figures (3.a, 3.b) have a significant influence on the evolution of the state quantity, i.e. the distribution of the void ratio changes with the horizontal co-ordinate and it is also influenced by the size of the specimen. The results are independent of the co-ordinate in the direction of shearing if the symmetry conditions for an infinite shear layer are introduced (Bauer & Huang 1999). The symmetry condition can be modeled by applying constraint conditions to the side nodes of the finite element mesh, i.e. each node on the left boundary is controlled to have the same displacements and Cosserat rotation as the corresponding node with the same vertical co-ordinate on the right boundary.

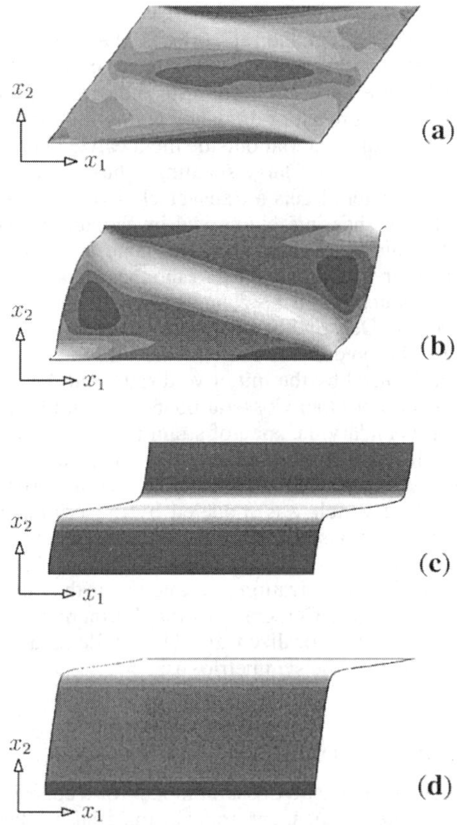

Figure 3: Plane shearing between two parallel plates under a constant vertical stress $\sigma_{22} = -100$ kPa. Influence of the boundary conditions on the evolution of the void ratio for: (a) lateral linear displacement fields; (b) lateral constant pressures; (c) part of an infinite plane layer with locked Cosserat rotations at the bottom and top surfaces; (d) part of an infinite plane layer with locked Cosserat rotations at the bottom and a certain coupling between the Cosserat rotation and the horizontal displacement at the top surface, i.e. $\omega_T^c = -0.025\, u_{1T}/d_{50}$

In the following the influence of the Cosserat boundary conditions at the top surface of an infinite shear layer is discussed. For symmetric Cosserat boundary conditions, e.g. for locked Cosserat rotations at the bottom and top surfaces, the deformation is localized within a narrow zone in the middle of the layer (Figure 3.c). The light strip indicates a higher void ratio as a result of dilatancy within the localized zone while the material outside the localized behaves like a rigid body for large shearing. Thus, an initially isotropic material gets a transversely isotropic structure during shearing. Numerical investigations show that the thickness of the localizes zone increases almost linear with an increase of the mean grain diameter d_{50} which is in accordance with experimental observations (Oda & Kazama 1998) (Bauer & Huang 2000). The predictions indicate that the thickness is also influenced by the initial void ratio and the pressure level. For other Cosserat boundary conditions at the top boundary the zone of strain localization is no longer located in the middle of the shear layer. The results in Figure (3.d) were obtained for a smoother top surface which was modeled by prescribing the Cosserat rotations at the top boundary using relation (7) with $\omega_T^c = -0.025\, u_{1T}/d_{50}$. It is obvious that the zone of strain localization is located near the boundary with a higher Cosserat rotation. Furthermore, the thickness of the localized zone is smaller than the value found for the symmetric case.

5 CONCLUSIONS

A special hypoplastic Cosserat model for a cohesionless and rate-independent granular material has been considered. The model captures the influence of the pressure, the void ratio, the mean grain diameter and the rotation resistance of the grains on the evolution of the stresses and couple stresses. The mean grain diameter serves as the internal length, thus the results of finite element calculations are mesh independent provided the element size is small enough. Finite element calculations were performed to simulate the behavior of an plane strip of granular material under monotonic shearing for different boundary conditions. For a finite width of the shear layer the results are strongly influenced by the assumed lateral boundary conditions and the size of the layer. To simulate the symmetry conditions of an infinite shear layer proper constraints were prescribed at the lateral boundaries of the mesh. Then the results are independent of the co-ordinate in the direction of shearing. From the numerical investigations it follows that polar effects are noticeable from the beginning of shearing. For large shearing the deformation is localized within a narrow zone. For symmetric Cosserat boundary conditions at the top and at the bottom of the shear layer the localized zone occurs in the middle of the shear layer, otherwise it is located near to the smoother boundary.

REFERENCES

Bauer, E. (1996). Calibration of a comprehensive hypoplastic model for granular materials. *Soils and Foundations 36*(1), 13–26.

Bauer, E. & Huang, W. (1999). Numerical study of polar effects in shear zones. In P. Pande & Schweiger (Eds.), *Proc. of the 7th Int. Symp. on Num. Models in Geomechanics*, pp. 133–138. Balkema.

Bauer, E. & Huang, W. (2000). A numerical study of polar effects in shear zones. In *Int. Conf. on Localization and Bifurcation, Australien, Perth.* in press.

Bogdanova-Bontcheva, N. & Lippmann, H. (1975). Rotationssymmetrisches, ebenes Fließen eines granularen Modellmaterials. *Acta Mechanica 21*, 93–113.

de Borst, R., Mühlhaus, H. B., Pamin, J., & Sluys, L. (1992). Computational modelling of localisation of deformation. In *Proc. 3th Int. Conf. Comp. Plast*, pp. 483–508. Pineridge Press.

Gudehus, G. (1996). A comprehensive constitutive equation for granular materials. *Soils and Foundations 36*(1), 1–12.

Gudehus, G. (2000). Forced and spontaneous polarization in shear zones. In Mülhaus (Ed.), *Int. Conf. on Bifurcation and Localization, Perth.* Balkema.

Herle, I. & Gudehus, G. (1999). Determination of parameters of a hypoplastic constitutive model from grain properties. *Mech. Cohes.-Frict. Mater. 4*(5), 461–486.

Kolymbas, D. & Wu, W. (1993). Introduction to hypoplasticity. In Kolymbas (Ed.), *Modern Approaches to Plasticity*, pp. 231–232. Elsevier.

Mühlhaus, H.-B. & Vardoulakis, I. (1987). The thickness of shear bands in granular materials. *Géotechnique 37*(3), 271–283.

Nagtegaal, J. C., Parks, D. M., & Rice, J. R. (1974). On numerically accurate finite element solutions in the fully plastic range. *Comp. Meth. Appl. Mech. Engng.* (4), 153–177.

Oda, M. (1993). Micro-fabric and couple stress in shear bands of granular materials. *Powders and Grains* (3), 161–167.

Oda, M. & Kazama, H. (1998). Microstructure of shear bands and its relation to the mechanisms of dilatancy and failure of dense granular soils. *Géotechnique 48*(4), 465–481.

Tejchman, J. (1997). *Modelling of shear localization and autogeneous dynamic effects in granular bodies.* Number 140. Veröff. Inst. Boden- u. Felsmechanik Univ. Karlsruhe.

Tejchman, J. & Bauer, E. (1996). Numerical simulation of shear band formation with a polar hypoplastic constitutive model. *Computers and Geotechnics 19*(3), 221–244.

Uesugi, M., Kishida, H., & Tsubakihara, Y. (1988). Formation of shear bands in sand bodies as a bifurcation problem. *Soils and Foundations 28*(1), 107–118.

Vardoulakis, I. & Sulem, J. (1995). *Bifurcation Analysis in Geomechanics.* Blackie Academic & Professional, London.

Computer Methods and Advances in Geomechanics, Desai et al. (eds) © *2001 Balkema, Rotterdam, ISBN 90 5809 183 X*

Multiscale material characterization and application to artificially created microstructures

C. Haynie & G. Frantziskonis
Civil Engineering and Engineering Mechanics, University of Arizona, Tucson, Ariz., USA

ABSTRACT: Wavelet analysis is being used to rationalize information at various scales in several branches of science. However, this powerful technique has not been applied to characterizing microstructures of materials, even though many critical questions could be addressed. The paper addresses the quantitative characterization of microstructural constituents at various scales of observation, and utilizes artificially created simple microstructures for illustrating the multiscale characterization process in a lucid fashion. Dominant length scales are identified as those regions in the scale-space where the energy of the wavelet transform and/or peaks of local concentration dominate, thereby providing a measure of when stress concentrations associated with microstructural features initiate microcracks.

1 INTRODUCTION

Material modeling under service loading conditions aims at understanding the mechanisms which control the material response, and thus predicting and ultimately controlling their behavior outcome. Several models based at the macroscopic level have been developed for this purpose. Such models are typically based on the observed behavior, and do not explain or give insight on how to control the response of the material. This lack of knowledge has pointed out the importance of studying the behavior of materials at the microscopic level and this has proven beneficial in understanding material behavior at the macroscopic level. There remains however an efficient means of linking the information on material structure as it appears at various scales. Investigators have in the past taken the properties determined at the micro level and have attempted to (often empirically) correlate them to those at the macro level based on the observed behavior. For example, atomistic simulations can be used to feed the constitutive behavior of finite elements, which are then used to simulate large-scale problems [e.g., Tadmor et al, 1996]. Through this process, models are typically tailored to predict the behavior of a material under certain loading conditions. However, such models are rather material specific. In other words, there is no general process available that can link information on material behavior at several scales concurrently. If such a process was available then we could know the role of each structural feature and thus understand and, where applicable, possibly alter material behavior.

Since microstructure is multiscale for (most) materials, length scale becomes very important. If only a certain scale is investigated, one may overlook material features at other scales that might actually be more important for the behavior. This scaling problem leads to the need to research many different scales concurrently. The central theme in this paper is an examination of the wavelet technique as a tool or means to effective multiscaling and in particular to that of identifying a dominant scale, meaning the scale at which the most useful (dominant) information is contained. In order to illustrate the process, artificial material microstructures are studied herein. Numerous microstructures in 1-D and 2-D were simulated. The simulated microstructures where then scaled to various different resolutions, from 1x to 512x and the wavelet transformation were performed at each resolution. The energy of the wavelet coefficients where calculated and plotted versus magnification. By analyzing the plots a dominant scale can be determined, as is illustrated.

Wavelet-based multiscaling has been recently introduced and examined. For a general description of the process we refer to [Frantziskonis & Deymier, 2000] and [Frantziskonis, 2000]. For simulated microstructures, see [Haynie, 2000] where the complete details on the results presented herein can also be found. For other relevant applications, see [Frantziskonis et al, 2000; Frantziskonis and Hansen, 2000].

2 WAVELET TRANSFORM OF SIMULATED MICROSTRUCTURES

Wavelet theory has become an important tool in many engineering and science disciplines in the past few years. It has been used in data compression, image analysis, the study of ocean currents, magnetic resonance, electrocardiography, the study of large scale distribution of matter in the universe, the analysis of the shape of neural cells to mention a few, and the list goes on. However, the application to mechanics of materials and in particular the problem of scale in microstructure analysis has been limited and is only recently being investigated.

A simple definition of a wavelet is a small wavy function constructed so that it contains certain mathematical properties. A set of wavelets is constructed from what is referred to as a mother wavelet. This set of functions then can be used to represent any number of other functions in time and space. The time-frequency localization is one of the most important features of wavelets. This important property leads to the notion of multi-resolution. It is said that wavelets is a mathematical microscope; signals, pictures, data or functions can be analyzed at varying levels of focus. For example, imagine yourself miles from the surface of the earth and focusing in on a region that contains an object of interest. At this level an individual may be able to make out the geological features of the area. Now if your where to move in closer you would be able to pick the object, say a building, out from other buildings in the area. Now as you move closer you can see that this building is red in color. Now if you moved even closer you would be able to ascertain that the building is made of masonry units. As you move in closer you will be able to ascertain that these masonry units where placed in a running bond pattern. As you move in very close you will be able to see the small grains that make up the mortar and as we move even closer the molecular structure of those sand particles are visible. The basic framework of all these views is essentially the same using wavelets. The general trend can be recognized as well as the intricate details.

A wavelet $\psi(x)$ transforms a function $f(x)$ according to equation 1 below:

$$W_f(a,b) = \int_R f(x)\psi_{a,b}(x)dx \tag{1}$$

The two-parameter family of functions, Equation 2,

$$\psi_{a,b}(x) = \frac{1}{a^{D/2}}\psi\left(\frac{x-b}{a}\right) \tag{2}$$

is obtained from a single one, ψ, called the mother wavelet, through dilations by the factor a^{-1} and translations by the factor b. Here, D represents the dimensionality of the problem and $a \in \mathbb{R}_+$, $a \neq 0$ and $b, x \in \mathbb{R}^D$. The factor $1/a^{D/2}$ is included for normalization purposes and, with it all the wavelets have the same energy. Other types of normalization, however, are possible. In evaluating the simulated microstructures presented in the following the Daubechies wavelets [Daubechies, 1992] and the corresponding scaling functions with five vanishing moments were used. It is noted here that other wavelets were compared with the Daubechies wavelets, and no significant changes in the results were observed. Only five vanishing moments [Daubechies, 1992] were used because of the simplicity of the simulated microstructures. Once the wavelet transforms were performed on each scale of the simulated microstructure, the energy of the details to move from a lower scale to the current scale was calculated. In order to accomplish this the dot product of each set of coefficients with its self was performed. The square root of this result was then determined, thus giving an energy representation of the details required to move from one scale to the next. These results are then combined in a plot of energy vs. scale, and a dominant scale can be determined.

3 SIMULATED MICROSTRUCTURES

The two-dimensional simulated microstructures were computer-generated, and were produced to be 512 points x 512 points. The pattern imposed was square and repeated itself throughout the entire 512-point square. One can imagine a checkerboard or brick layout. The lines that make up the squares have their own internal structure (e.g., mortar in the brick structure). To create these microstructures the "mortar" pattern was first developed at a resolution of 512x (scale 10) the original. In order to reach scale 1 a way to average the points was developed. The averaging scheme can best be described if we look at Figure 1 which represents the following 8 points in the top left hand corner of the 512 x 512 microstructure representation:

	X1	X2	X3	X4
Y1	1	2	5	6
Y2	3	4	7	8

Figure 1. Simulated microstructure value and position

In order to get the values for the next coarser scale the four values in x1, y1; x2, y1; x1, y2; x2, y2 are averaged to get 2.5. This value is then assigned to position x1, y1 in the next coarsest scale. Then the values in x3, y1; x4, y1; x3, y2; x4, y2 are averaged to get 6.5, this value is then placed in the x2, y1 position. This scheme proceeds throughout the micro

Original Matrix												
	x1	x2	x3	x4	x5	x6	x7	x8	x9	x10	x11	x12
y1	1.00	1.00	1.00	1.00	1.00	1.00	1.00	1.00	1.00	1.00	1.00	1.00
y2	1.00	0.00	0.00	0.00	0.00	0.00	0.00	0.00	1.00	0.00	0.00	0.00
y3	1.00	0.00	0.00	0.00	0.00	0.00	0.00	0.00	1.00	0.00	0.00	0.00
y4	1.00	0.00	0.00	0.00	0.00	0.00	0.00	0.00	1.00	0.00	0.00	0.00
y5	1.00	0.00	0.00	0.00	0.00	0.00	0.00	0.00	1.00	0.00	0.00	0.00
y6	1.00	0.00	0.00	0.00	0.00	0.00	0.00	0.00	1.00	0.00	0.00	0.00
y7	1.00	0.00	0.00	0.00	0.00	0.00	0.00	0.00	1.00	0.00	0.00	0.00
y8	1.00	0.00	0.00	0.00	0.00	0.00	0.00	0.00	1.00	0.00	0.00	0.00
y9	1.00	1.00	1.00	1.00	1.00	1.00	1.00	1.00	1.00	1.00	1.00	1.00
y10	1.00	0.00	0.00	0.00	0.00	0.00	0.00	0.00	1.00	0.00	0.00	0.00
y11	1.00	0.00	0.00	0.00	0.00	0.00	0.00	0.00	1.00	0.00	0.00	0.00
y12	1.00	0.00	0.00	0.00	0.00	0.00	0.00	0.00	1.00	0.00	0.00	0.00
Next Coarser Matrix												
	x1	x2	x3	x4	x5	x6						
y1	0.75	0.50	0.50	0.50	0.75	0.50						
y2	0.50	0.00	0.00	0.00	0.50	0.00						
y3	0.50	0.00	0.00	0.00	0.50	0.00						
y4	0.50	0.00	0.00	0.00	0.50	0.00						
y5	0.75	0.50	0.50	0.50	0.75	0.50						
y6	0.50	0.00	0.00	0.00	0.50	0.00						

Figure 2. Simulated microstructure and resulting courser scale

structure. Figure 2 demonstrates this averaging scheme on a larger matrix. As can be seen the matrix at the next coarser scale is ¼ of the size of the original. In order to complete the matrix either the pattern is repeated in each of the other three quadrants or the base value is applied to the other three quadrants or some sort of combination of the two. i.e. the pattern may only be applied to the edges of two of the quadrants, thus giving the microstructure the pattern on two sides of the 512 x 512 matrix. The remaining area would then be filled in with the base value. This allows for a high number of different patterns within another pattern. This scheme will carry on until the scale 1 is created. Several simulated microstructures where created for 1-D and 2-D cases. We only present some 2-D results here; further ones can be found in [Haynie, 2000].

4 CASE STUDIES OF 2-D MICROSTRUCTURES

Two of the many examples [Haynie, 2000] are discussed here. The first one contains a square brick pattern within "mortar joints" and then is repeated starting at scale six, and then repeated each successive scale. Figures 3 and 4 show scale 10 and scale 6 respectively. The wavelet transforms and the energy of the respective scales was determined. These results were then plotted and are presented in Figure 5. In evaluation of the plot the energy decreases as the resolution increases until scale 5. This is a result of the overall pattern becoming less important, or we can say it is no longer in the field of view. At scale 5 the energy begins to increase. This is due to the finer pattern starting to come in to view. Thus, the highest energy value, or dominant scale occurs at scale 10.

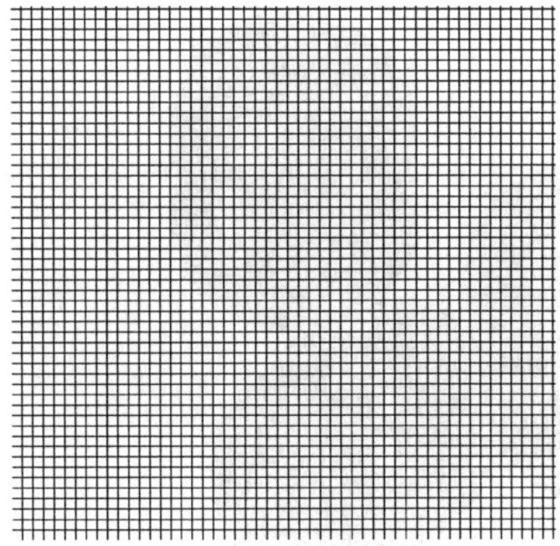

Figure 3. Scale 10 – Brick Pattern, initial microstructure

Figure 4. Scale 6 – Brick pattern, introduction of the overall pattern

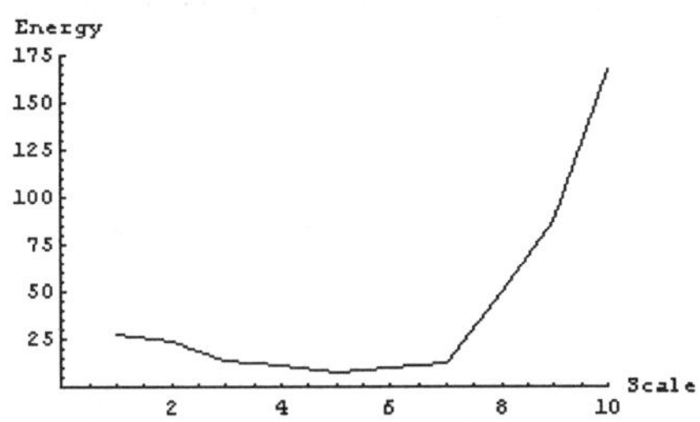

Figure 5. Energy vs. Scale – Brick pattern

The second microstructure presented is one in which a porous type microstructure was created. The finest scale was computer-generated and using the above-mentioned averaging scheme the coarser scales were produced. In this case a total of thirteen scales were analyzed. Figures 6 & 7 show representations of the microstructure at scale 13 and scale 10 respectively. An overall pattern was introduced at the latter scale.

The results of the energy vs. scale are presented in Figure 8. In examination of the plot one can see that the highest energy occurs at scale 5, thus the dominant scale. However, one cannot ignore scale 4 or

Figure 6. Scale 13 – Porous type microstructure, finest scale

Figure 7. Scale 10 – Porous type microstructure, introduction of overall pattern

Figure 8. Energy vs. Scale – Porous type microstructure

scale 13, both of which may contain valuable insight in how the material might behave.

5 CONCLUSIONS

It has been shown that wavelet theory can be a very important tool in mechanics of materials. As research strives to completely understand material behavior, scale becomes very important. To date there has been no rigorous means to determine which scales should be investigated. Even though the subject area is in its infancy, this investigation shows how wavelet theory can determine which scale contains the most information or the scale at which characteristics of the material can be determined. Models can then be developed to predict the behavior of materials by combining information at different scales proposed recently [e.g., Frantziskonis & Deymier, 2000]. Even though this investigation looked at simulated microstructures the technique can be applied to real microstructures. One way to accomplish this is through the use of microscopy techniques at numerous scales. The wavelet transform can then be used to predict the dominant scale or range of scales.

ACKNOWLEDGEMENT - support from NSF grant CMS 9812834 is gratefully acknowledged

6 REFERENCES

Daubechies I., *Ten Lectures on Wavelets,* SIAM, Philadelphia, Pennsylvania, 1992.

Frantziskonis G., L.B. Simon, J. Woo & T.E. Matikas, "Multiscale Characterization of Pitting Corrosion and Application to an Aluminum Alloy," *Eur. J. Mechanics, A/Solids,* 19, 309-318, 2000.

Frantziskonis G., P.A. Deymier, "Wavelet Methods for Analyzing and Bridging Simulations at Complementary Scales – the Compound Wavelet Matrix and Application to Microstructure Evolution," *Modeling Simul. Mater. Sci. Eng.*, to appear.

Frantziskonis G., "Wavelet-based Multiscaling - Application to Material Porosity and Identification of Dominant Scales," under review (2000).

Frantziskonis G. & A. Hansen, "Wavelet-based Multiscaling in Self-affine Random Media," *Fractals,* under review (2000).

Haynie C., "Multiscale Analysis in Mechanics of Materials and Application to Simulated Microstructures," MSc Thesis, University of Arizona, May 2000.

Tadmor E.B, R. Phillips and M. Ortiz, "Mixed Atomistic and Continuum Models of Deformation in Solids," *Langmuir,* 12, 4529 (1996).

Computer Methods and Advances in Geomechanics, Desai et al. (eds) © 2001 Balkema, Rotterdam, ISBN 90 5809 183 X

Evolution of microstructure during swelling in expansive clays

D.R. Katti & V. Shanamugasundaram
Civil Engineering, North Dakota State University, Fargo, N.Dak., USA

ABSTRACT: The present work elucidates the results of our study on the evolution of microstructure of saturated swelling clay bentonite with swelling. A new controlled uniaxial swelling (CUS) cell was designed and fabricated to allow expansive clay samples to be swollen to predetermined levels and undisturbed removal for microanalytical testing using scanning electron microscopy (SEM) techniques. The void and particle gray scales were identified in the SEM images using energy dispersive x-ray analysis and digital image analysis of the images using custom developed macros and Scion Image™ software. An increase in swelling and reduction of swelling pressure due to swelling is accompanied by reduction in particle size possibly due to breakdown of clay agglomerates into smaller sized particles. A significant effect of swelling on microstructure of swelling clays may have large impact on stress deformation as well as hydraulic properties of the clay.

1 INTRODUCTION

Accurate modeling of response of swelling soils to external loading conditions I the environment is critical for the safe and economical design of structures. Swelling soils have severe economic ramifications on civil engineering projects in expansive soil areas. (Rama Rao et al. 1988, Jones and Holtz 1973). Current research reflects significant progress in the development of continuum-based models for soils. These models range from the original limit state approach of deformation-based models using theory of elasticity to advanced elasto-plastic and visco-plastic models. (Katti and Desai 1995, Desai et al. 1993, Mroz, et al. 1979, Banerjee and Yousif 1986) These continuum-based models are based on bulk response have for the most part overlooked the effect of clay mineralogy and microstructure, thus making predictions from such models vulnerable to type of clay minerals, structure, type of ions in the pore fluid, stabilization techniques etc. Further, these models are not linked to the intrinsic material properties, they are not general and consequently not adequately robust. Concurrently, research has taken place in the fundamental understanding of mechanics at particulate level. (Martin, et al. 1997, Zaidman and Castaneda 1996, Chen and Anandrajah 1996, Bedford and Windle 1993; Axelrad and Muschik, 1988, Katti 1987, Katti and Katti 1996, Katti et al. 2000, Yong et al. 1984) The stress deformation response of swelling soils are significantly affected by the type of clay minerals, amount of clay mineral, soil fabric, presence of various types of ions in the pore fluid etc. Researchers have also investigated correlation between crystallographic parameters for clays with swelling pressure (Handy and Turgut 1987, Low 1973). Modern analytical electron microscopy techniques offers the scanning electron microscope (SEM) as a useful tool for the study of microstructural changes in clays with high spatial resolution. The usefulness of this technique lies in simultaneous microstructural characterization at high spatial resolution concurrent with elemental analysis using energy dispersive x-ray spectroscopy (EDS). Earlier work by Collins (Collins 1984) on the study of microfabric forms in natural expansive soils using SEM indicated that accurate characterization of microfabric of the clay is the key to properly assess the swelling behavior of expansive soils. The main focus of this paper is to quantitatively elucidate the correlation between swelling and swelling pressure and the clay microstructure. A new controlled uniaxial swelling (CUS) device is used in combination with scanning electron microscopy techniques.

2 EXPERIMENTAL

A new controlled uniaxial swelling device was designed to allow a predetermined amount of swelling, measure the corresponding swelling pressure and allow for easy removal of the sample for consecutive microanalysis in the SEM. Details of the construction of the CUS cell can be seen in previous work (Katti and Shanamugasundaram, 2000).

2.1 *Materials and methods*

Bulk bentonite clay (Natural Gel Wyoming Bentonite) was obtained from Wyo-Ben Inc. (Billings, MT) The physical properties of the bentonite samples found in our laboratory are shown in Table I.

Table I. Physical properties of bentonite samples.

Physical Property	Value
Specific Gravity	2.67
Initial Void ratio	2.145

The chemical composition of the clay mineral was confirmed to be sodium montmorillonite using x-ray diffraction analysis at the NDSU Materials characterization laboratory. Phases were identified by comparison with in-house reference diffractograms, as well as by computer search-match procedures which employ the International center for diffraction data (ICDD) powder diffraction file. Deaired and deionized water was used during the swelling and saturation phase of the experiments.

2.2 *Mechanical testing*

Bentonite samples were dry compacted in the cells using static compaction to a height of 14-mm and a diameter of 70mm. An initial void ratio of 2.145 (dry unit weight of 8.33 kN/m^3) was used for all samples. The piston of the CUS cell was locked in position to maintain a zero volume change during saturation. The cell, immersed in a water tank, was then placed in a SATEC 22 EMF universal testing machine. The piston was maintained at zero volume change position by maintaining the position of the loading ram of the load frame through closed loop electronics. The swelling pressure at no volume change condition was recorded.

The loading ram was then moved up to a predetermined position of 7 mm for 50 percent swelling and 10.5 mm for 75 percent swelling. The percent swelling is defined as the change in the volume of the sample relative to the initial dry volume. Swelling pressure with time for each of the swelling conditions is recorded.

In our attempt to ascertain the time required for complete saturation, a dry sample identical to that used for microstructural analysis was placed in an oedometer which was then positioned in the load frame and allowed to saturate at zero volume change. The variation of swelling pressure was recorded over a period of 21 days. An asymptotic approach of the swelling pressure at 10 to 11 days was observed. This fact was further confirmed by the calculation of degree of saturation on removal of the sample from the oedometer after 21 days.

2.3 *Microstructural characterization*

The 0 percent, 50 percent and 75 percent swollen samples were removed from the cells and 5 mm x 5mm x 1 mm thick samples were axially sliced from the central region using a thin ceramic knife. The sample slices were prepared for SEM analysis using cyrofixation in liquid nitrogen. Samples were then placed in cold stage of JEOL 6300 (JEOL, Peabody, MA) analytical SEM operated at 5kV maintained at –165 ºC. Elemental analysis was performed using EDS detector (NORAN Instruments, Middleton, WI) on the SEM. The high concentration of silicon in the solid phase as compared to the void allowed for the delineation between solids and voids. Micrographs were digitally recorded at 50X, 650X, 2,000X and 10,000X magnifications. The micrographs obtained were then analyzed using SCION IMAGE™ (Scion Corp., Fredrick, MD; windows version of NIH Image™, National Institutes of Health).

Custom modules (Katti and Shanamugasundaram 2000) were developed to digitally delineate the void and solid regions in the electron micrographs and calculate the particle size dimensions.

3 RESULTS AND DISCUSSION

The variation of swelling pressure of dry bentonite clay sample during saturation was recorded over a period of 21 days. The swelling pressure rapidly increases to about 120 kPa within one day. This is followed by a gradual increase to about 160 kPa within 9-10 days. After a period of 10 to 11 days, the swelling pressure attains an almost constant magnitude of 163 kPa. On removal of the sample from the oedometer after 21 days on completion of the test, the degree of saturation was found to be about 100 %. A small drop in swelling pressure observed after about one day of wetting may be attributed to the reorientation of particles within the sample during saturation of the dry sample.

The magnitude of swelling pressure of bentonite clay found in our laboratory experiments fall within the range of swelling pressure values for bentonite clay samples reported in the literature. The swelling pressure of these clays is a function of the initial void ratio and is shown to increase exponentially with an increasing dry density (decreasing void ratio) (Chen, 1988; Fredlund and Rahardjo 1993).

Based on the above experiment, we used a minimum saturation time of 30 days for all samples in the CUS cell. Further, a fully saturated sample in the CUS cell was placed in the load frame and swelling pressure was measured at 0, 1, 2, 4, 10, 17.5, 25, 50, 75 and 100 percent swelling. Relative swelling pressure versus swelling percentage is shown below.

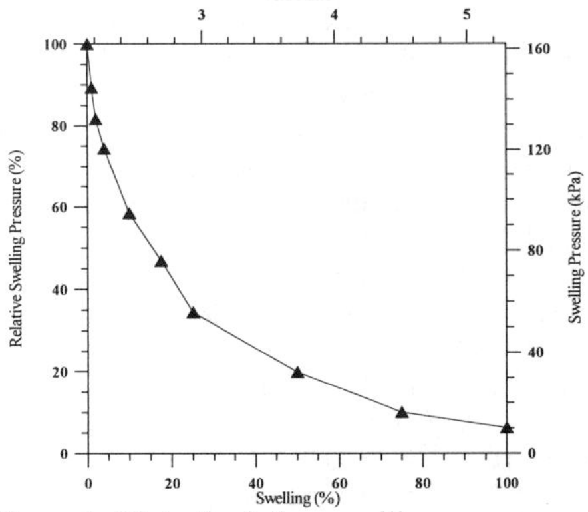

Figure 1. Plot of relative swelling pressure vs swelling.

The swelling pressure is expressed relative to the maximum observed swelling pressure of 162 kPa. Clearly, as shown in Figure 3, the maximum swelling pressure is observed at 0 % swelling which decreases to about 6 % of the maximum at 100 % swelling. Further, the swelling pressure reduces to 50 % of the maximum at 15 to 20 % swelling which decreases to 10 % at 75 % swelling.

In order to understand the effect of swelling and swelling pressure on microstructural changes in the bentonite samples, three identical samples of bentonite clay were each swollen to 0 %, 50 % and 75 % of their original volume uniaxially. These samples were removed from the cell and sectioned/frozen for microstructural analysis as described earlier. These samples were mounted in a cryogenic stage of the SEM. Figures 2 a, b and c show the obtained images.

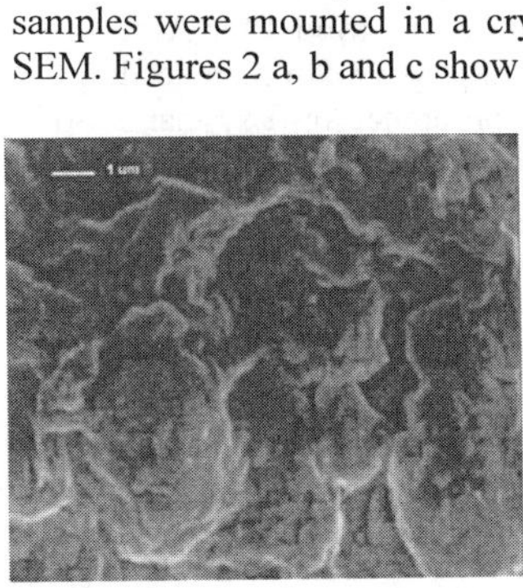

a)
(0%swelling)

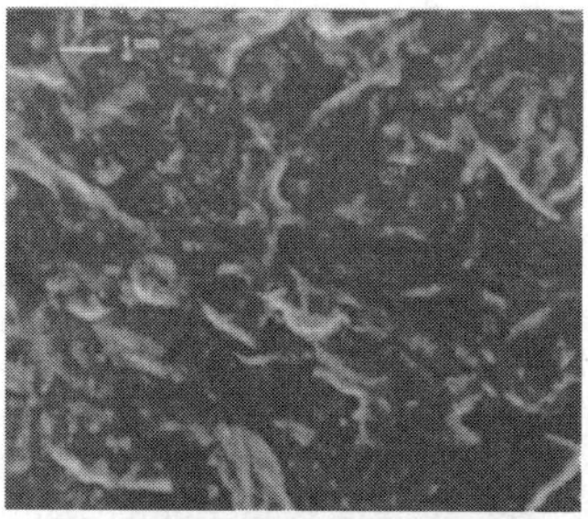

b)
(50% swelling)

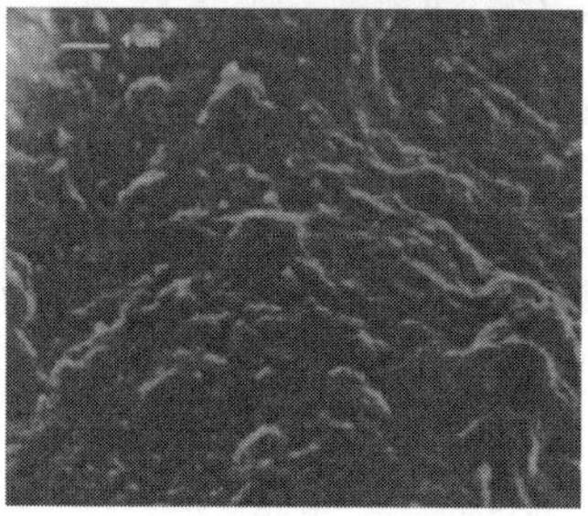

c)
(75% swelling)

Figure 2. Scanning electron micrographs of bentonite samples at a)0% b)50% and c)75% swelling.

As shown, large particle sizes are observed for the 0 % swelling sample and the particle sizes appear to diminish in the 75 % swelling sample. In our attempt to quantitatively determine the changes in particle sizes, we used the Scion Image™ software in combination with custom modules. The first use the digitized images to obtain boundaries of particles in the images followed by counting the particles and measurement of number of pixels along the narrow axis for each particle. A calibration factor for pixel size enables the measurement of mean particle sizes. Plots illustrating the variation of particle size with swelling and relative swelling pressure are shown as,

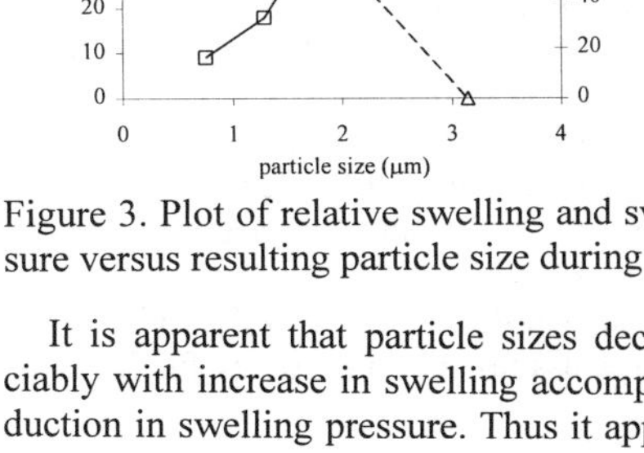

Figure 3. Plot of relative swelling and swelling pressure versus resulting particle size during swelling.

It is apparent that particle sizes decrease appreciably with increase in swelling accompanied by reduction in swelling pressure. Thus it appears that on swelling, as water enters the interlayers in montmorillonite, large agglomerates break into smaller particle sizes.

5 CONCLUSIONS

The new CUS device allowed controlled swelling of clay samples and measurement of swelling pressure at controlled and predetermined extent of swelling of the expansive clay samples when inserted in a closed loop mechanical testing load frame. Samples at predetermined swelling were removed for microanalytical testing in a SEM. The microstructural analysis of the swelling clays followed by digital image analysis showed that the particle size of clay samples reduced appreciably with increase in swelling and swelling pressure. These studies shed light on the correlation between microstructure and two macroscopic properties, swelling and swelling pressure. This represents a step towards better understanding of fundamental mechanisms responsible for swelling and swelling pressure effects in expansive clays. Our subsequent work focuses on incorporation of microstructural parameters into mathematical models to eventually yield more robust constitutive models of expansive clays.

REFERENCES

Axelrad, D.R., Muschik,W. 1988. Constitutive laws and microstructure. Seminar at Wissenschaftskolleg-institute for advanced study, february 23-24, berlin, pp 1-204.Springer-verlag, Berlin,Germany.

Banerjee, P.K., and Yousif, N.B. 1986. A Plasticity Model for the Mechanical Behavior of Anisotropically Consolidated Clays. International Journal of Numerical Analytical Methods, Geomechanics, 10:521-541.

Bedford, S.E. and Windle, A.H., 1993. Modeling of Microstructure in Mesophases. Liquid Crystals, 15(1):31-63.

Chen, F.H. 1988. Foundations on Expansive Soils, 2nd ed., Elsevier, Amsterdam, 463-469.

Chen, J. And Anandarajah, A. 1996. Van der Waals Attraction Between Spherical Particles.Journal of Colloid and Interface Science, 180(2):519-523.

Collins, K. 1984. Characterization of Expansive Soil Microfabric. *In* Proceedings of 5th International Conference on Expansive Soils, Adelaide, South Australia, May, 37-41.

Desai, C.S., Wathugala, G.W., and Matlock, H.1993. Constitutive Model for Cyclic Behavior of Clays. Journal of Geotechnical Engineering, ASCE, 119(4):730-748

Fredlund, D.G., and Rahardjo, H. 1993. Soil Mechanics for Unsaturated Soils. Wiley Publications, New York.

Handy, R.L., and Turgut, D. 1987. Swelling Pres sure versus D-Spacing of Montmorillonite. *In* Proceedings of the 6th International Conference on Expansive Soils, New Delhi, India, pp161-166.

Jones, D.E., and Holtz, W.G. 1973. Expansive Soils The Hidden Disaster. Civil Engineering, ASCE, 43(CE8): 49.

Katti D.R. 1987. Role of CNS on Passive Resis tance of Saturated Expansive Soil, 6th International Conf. on Expansive Soil-87, New Delhi.

Katti D.R., and Desai C.S. 1995. Modeling and Testing of Cohesive Soil Using Disturbed State Concept. Journal of Engineering Mechanics, ASCE, 121(5): 648-658.

Katti D.R., and Shanamugasundaram V. 2000. Influ ence of Swelling on Microstructure of Expansive Clays. accepted for publication in Canadian Geotechnical Journal.

Katti R. K., Katti A.R. 1996. Behavior of Saturated Expansive Soils and Control Methods. IBH Oxford, New Delhi.

Katti R.K., Katti A.R., Katti D.R. 2000. Influence of gravity on granular soil mechanics, In press IBH Oxford Ltd. New Delhi, India.

Low, P.F. 1973. Fundamental Mechanisms Involved in the Expansion of Clays, as Particularly Related to Clay Mineralogy. Proceedings of Workshop on Expansive Clays and Shales in Highway Design and Construction, Federal HighwayAdministration, Denver, CO, 1: 70-91.

Martin, U., Muhle, U., And Oettel, H., 1997. Micro structure and Modelling of the Deformation Behaviour of superalloys at High Temperatures. Computational Materials Science, 9:92-98.

Mroz, Z., Norris, V.A., and Zienkiewicz., O.C. 1979. Application of an Anisotropic Hardening Model in the Analysis of Elasto-Plastic Deformation of Soils. Geotechnique, 29: 1-34.

Rama Rao, R., Rahardjo, H, and Fredlund, D.G. 1988. Closed Form Heave Solutions for Expansive Soils. Journal of Geotechnical Engineering, ASCE, 114(5): 573-588.

Yong, R.N., Sadana, M.L., and Gohl, W.B. 1984. A Particle Interaction Energy Model for Assessment of Swelling of an Expansive Soil, Proceedings of 5th International Conference on Expansive Soils, Adelaide, Australia, 4-12.

Zaidman, M. and Castaneda P. 1996. The Finite Deformation of Nonlinear Composite materials-II. Evolution of Microstructure. International Journal of Solids and Structures, 33(9):1287-1303.

Computer Methods and Advances in Geomechanics, Desai et al. (eds) © 2001 Balkema, Rotterdam, ISBN 90 5809 183 X

Nano-Meso-Macro scale response simulation of biomimetic nanocomposites

K.S. Katti & D.R. Katti
Civil Engineering, North Dakota State University, Fargo, N.Dak., USA

J.M. Sopp, W.M. Mercer & M. Sarikaya
Materials Science and Engineering, University of Washington, Seattle, Wash., USA

ABSTRACT: Nacre (mother-of-pearl), the inner layer of seashells is a ceramic laminated biocomposite with exceptional mechanical properties of fracture toughness and strength. Nacre microarchitecture is described as 'brick and mortar' with $CaCO_3$ bricks and organic mortar. In this work, a multiscale modeling approach for nacre using three dimensional finite element (FE) modeling was developed. Nanoscale parameters (elastic modulus and hardness) were obtained from nanoindentation experiments using atomic force microscopy. Experimental macro and nanoscale parameters were incorporated into FE models. The yield stress of the organic was maintained at 40 $\times10^{-6}$ N/μm^2, 50 $\times10^{-6}$ N/μm^2, 60 $\times10^{-6}$ N/μm^2, 80 $\times10^{-6}$ N/μm^2, 120 $\times10^{-6}$ N/μm^2, 240 $\times10^{-6}$ N/μm^2, 320 $\times10^{-6}$ N/μm^2 and 400 $\times10^{-6}$ N/μm^2 (40 to 400 MPa). Tensile tests were simulated for each of these values of yield stress of organic phase. The resulting yield stress of nacre was compared to experimentally obtained value from literature. Our results indicate that a very high yield stress of organic is necessary to obtain the experimentally obtained yield stress of nacre.

1 INTRODUCTION

Currently, it is becoming increasingly known that "nanoscience and nanoengineering will lead to unprecedented understanding and control over the fundamental building blocks of all physical things" (Report of Interagency Working Group on Nanoscience Engineering and Technology). Biological structures represent the very essence of nanoscience wherein complex architectures are engineered at the molecular scale. Living organisms produce materials with physical properties that surpass those of analogous synthetic technological materials with similar phase compositions. This is mainly due to the fact that biological materials are highly organized from the molecular, to the nanometer, and to the macro scale, often, in a hierarchical manner providing an ordered continuity at all dimensional scales of the structure (Sarikaya and Aksay 1995, Simkiss and Wilbur 1989, Winwright 1982).

A truly multiscale simulations regime, implies simulations over all length scale from atomistic macro. The inter-relationship between mechanics at the nanoscale (nanomechanics) and materials design is an important issue for a realistic simulation of different scales from molecular to macroscopic. (Chong 1999) Nanomechanics and scale effects (scaling-up from nm to mm, cm or more) is addressed in recent studies on metallic and semiconductor material systems (Bastawros and Kim 2000, Kim 2000, Cho and Joannopoulos 1996). Due to the complexity of interfaces in biomolecules and biological nanocomposites (such as nacre) a truly multiscale approach for such 'natural' materials with the current state of knowledge is a distant reality. Our efforts focus on applying advanced nanoscale characterization techniques to continuum based numerical models to lead a biomimetic materials design with simulations. The purpose of this study is to develop a modeling procedure that incorporates nanoscale responses at surfaces and interfaces in nacre into numerical methods.

2 NACRE AS A BIOMIMETIC MODEL

Molecular control of mineralization by organic phases in nanocomposite material systems has shown great promise for the design of new high performance smart hybrid nanocomposites (Winwright 1982, Sarikaya and Aksay1995, Simkiss and Wilbur 1989, Belcher et al. 1996, Addadi and Weiner 1992, Mann 1989, Mann 1993, Sarikaya and Aksay 1992, Sarikaya 1999). The main reason for interest in this field lies in the ability to provide simple routes to design high performance composites. The fundamental understanding of the molecular mechanisms and interactions during the process of biomineralization has been a major research thrust recently. Biomimetic efforts have been an important research focus in the scientific community, particularly for hierarchical materials such as seashells.

Nacre is a ceramic laminated composite consisting of highly organized polygonal shaped aragonitic platelet layers of thickness of about 0.5 μm separated by thin 30 nm layers of organic composed mainly of proteins and polysaccharides. This particular microarchitecture is similar to 'brick and mortar' (Gunnison et al. 1990). A transmission electron micrograph and schematic of brick-mortar structure is shown in Figure1.

Figure 1. TEM image of 'brick and mortar' structure of nacre with inset showing a schematic of the same.

The aragonite platelets are 0.2 to 0.8 μm thick and 5 to 10 μm long, depending on the mollusc species. The platelets are mostly pseudohexagonal, but they can also be trigonal, square, and pentagonal. The organic matrix is generally uniform (10-15 nm thick) throughout the thickness, made of mostly ordered proteins and polysaccharides within an amorphous matrix, and, therefore, forms a nanocomposite (Heur et al. 1992, Sarikaya and Aksay 1995).

The aragonitic $CaCO_3$ structure also occurs in deposits of geologic origin and exhibits identical crystallographic structure but small differences are detected in the electronic structure possibly from the influence of organic (Katti et al. 1999). The high fracture toughness of nacre cannot be explained entirely by the small volume fraction of the organic component. The organic/inorganic interface in nacre is a critical issue in determining the overall mechanical properties in nacre. This interface is biochemically, structurally, and mechanically compatible. The organic phase typically represents 1 to 5 % by weight of nacre. The specific role of interfacial and interphasial mechanical relationships at the organic-inorganic interfaces on bulk mechanical behavior of nacre are lacking in literature (Schaffer et al. 1997)

Nacre structure has a combination of high strength both in compression and bending (sarikaya et al. 1990). The average four-point bend fracture toughness K_{IC}, (10 $MPa/m^{1/2}$) and specific flexural strength, σ_F (180 MPa) values of nacre of red abalone and pearl oyster are some 20 to 50 times those of synthetically produced monolithic $CaCO_3$ polycrystals and are also better than most monolythic ceramics (such as Si_3N_4, SiC, and ZrO_2). Several toughening mechanisms operate depending on the loading conditions, and simultaneously contribute to the toughening of the nacre. These include crack blunting/branching, microcrack formation, plate pull-out, crack bridging, and sliding of platelets all adding a significant energy absorption during crack propagation leading to high toughness (Gunnison et al. 1992) Theoretical calculations based on the Griffith criterion suggest a maximum strength of 50 MPa assuming that aragonite platelet thickness (~0.5 μm) as the maximum flaw size in the structure. Thus, strength of the bonding at organic/inorganic interface appears to play an important role for the overall strengthening of the composite.

An accurate prediction of mechanical behavior of advanced nanocomposites is of great interest for design of such high performance composites. Yet, completely accurate material models are lacking for such nanocomposites. One approach, as followed by several researchers, is to model complex biologically based materials with existing constitutive models for similar materials such as synthetic polymers and natural rubbers, which include elastic, non-linear elastic or visco-elastic approaches. Such approach is followed for modeling biocomposites such as bone, teeth, and soft tissue (Borgersen and Sakaguchi 1994, Brodland and Clausi 1994)

The existing material models are not adequate for an accurate prediction of mechanical response under complex loading conditions. This is because of the distribution of the organic phase, its intricate local and interface structures are not well understood.

Finite element analysis is an important tool to predict material response based on discretization of the material into finite 'elements' that simulates the real continuous surroundings (Desai and Abel 1972). Often, complex tests and simulations can be conducted using FEM, which may be impossible or difficult to conduct in the laboratory. This is particularly true for complex nanocomposite biological systems. Numerical models to predict mechanical response of nanocomposite systems are lacking in literature. Another major problem is that these models also require quantitative information about the local structures and properties of the individual components of a nanocomposite, data that are often lacking mainly because of the technical difficulties of obtaining such information at small dimensions. We overcame these difficulties and determined nanomechanical properties of the individual components of nacre using scanning probe techniques (Mercer et al. 2000). In this work we incorporate nanoscale experimentally obtained parameters into numerical models to predict response in nacre.

3 METHODOLOGY

3.1 *Construction of 3D FE model of nacre*

As a first step, the individual 3D hexagonal aragonitic platelets were constructed based on microstructural parameters shown in Table I. Each aragonitic brick was constructed from 4 eight noded isoparametric elements. Further, the organic layer was attached to all the faces of the aragonitic brick. The thickness of the organic layer was taken to be 0.015 μm which is half of the reported thickness of organic layer between aragonitic bricks. This constitutes the basic unit in the finite element analysis. A 10X10X10 3D model was constructed. Details of the process are shown elsewhere (Katti et al. 2000, Katti and Katti 1999). Mechanical tests are simulated on such a structure. FEM software, MARC™ (MSC MARC Corp.) was used along with a pre and post processor, MENTAT™ (MSC MARC Corp.)

Table 1. Elastic Properties and Microstructural Parameters of Constituents of Nacre

Property	Organic Phase	Inorganic Phase
Volume %	~5	~95
Composition	Polysaccarides and proteins	Aragonitic $CaCO_3$
Micro-structure	~20 nm layers around aragonitic platelets	Hexagonal platelets of 0.5 μm height and 2.8 μm sides
Elastic Modulus	20 MPa	99.5 GPa

Tensile and compression tests were simulated by setting nodal displacements of the first or lowest face of the 3D model to zero and applying face pressure in the z direction to the top face of the 3D model. These boundary conditions were applied directly to aragonite surfaces. All the nodal displacements on the top aragonite face were equalized by linking all surface nodes to central node. This procedure effectively minimizes the 'edge effects' commonly observed in FEM analyses.

3.2 *Determination of Nano Mechanical Properties of Biocomposite Nacre*

The use of numerical models requires the precise knowledge of the properties of the individual components of the composite in addition to their micro- and nano-structural details. The mechanical properties of the aragonite platelets as well as the organic matrix were obtained using ultrafine indentations provided by the atomic force microscopy techniques (Mercer et al. 2000). In this case, we used vertical indentation using a Hysitron Picoindentation™ system that is attached to an AFM (Park CP Scanning probe System). This system allows vertical loads up to 9000 μNewton with displacement as much as 1.0 μm with a precision of less than 1.0 nm. Two different indenters were used because of the differences in properties of the hard and the soft components. For the biogenic aragonite, a sample of nacre was cleaved and the surface was polished to a roughness <1 nm. Using a sharp diamond-shaped Berkowich tip (diamond, in <111> axis), force (F) - depth (d) curves were obtained. From the slope of the unloading curve and the area of impression the elastic modulus (E) and the hardness (H) values, respectively, were determined. The methodology for determining nanoscale E and H values using the setup are well documented (Oliver and Pharr 1992, Doerner and Nix 1986)

For the organic matrix, the nacre was cleaved, and the native surface was directly indented face-on. The organic matrix is thin (only a few 10s of nm) and considerably softer than biogenic aragonite. Berkovich tips are sharp and, therefore, they concentrate too much stress in the indentation zone, penetrating the organic film even at the lowest loads that can be applied. Therefore, we engineered broad-tipped metallic (W) indenters with semispherical tips to optimize load resolution per unit area and, at the same time, increase the total contact area. This tip, then, provided a better sampling of the organic film while minimizing the applied stress. Schematic of the test geometry is shown in Figure 2.

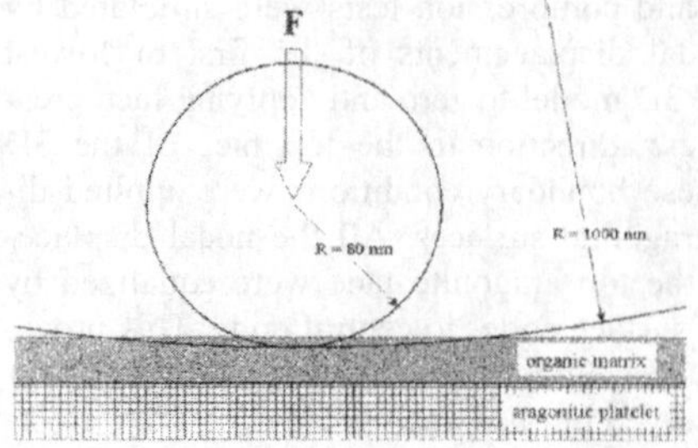

Figure 2. Schematic of test geometry for organic phase indentation experiments.

3.3 Modeling Experiments

Linear: Previous work showing tensile test simulations at low load (well within the elastic regime) show that both higher modulus of organic phase and platelet-platelet contact can result in a net higher modulus of nacre (Katti and Katti 2000, Katti and Katti 1999). Since the reported modulus of nacre is 25-50 GPa (Jackson et al. 1998, Currey 1980, Currey 1974), our preliminary results indicate that a combination of higher composite modulus of the organic phase and direct transfer of load between a portion of aragonitic platelets can possibly result in the observed mechanical response of nacre under small stresses.

Non-Linear Behavior: Initially, a smaller 3D model composed of 4x4x4 elements was used for this analysis. The inorganic layers are modeled by elastic models but organic phase is assumed to exhibit elasto-plastic response. Yield stress (σ_y) of organic was varied from 40 x 10^{-6} N/μm^2 (40 MPa) to 400 x 10^{-6} N/μm^2 (400 MPa). Bulk experiments in literature indicate the onset of nonlinearity in nacre at approximately 40 x 10^{-6} N/μm^2 (40 MPa). [31,32] We used Von Mises yield criterion and an isotropic work hardening slope(Π). This work hardening slope is defined as the slope of stress increment over corresponding plastic strain increment. Although Π for organic may change with stress, we have assumed it to be constant in this analysis. Thus, to find Π, we have used results of bulk stress-strain behavior with an assumption that only organic component has yielded. Total strain is given by sum of plastic and elastic components as shown as:

$$\varepsilon = \varepsilon^p + \varepsilon^e \qquad (1)$$

where, ε is the total strain, and ε^p and ε^e represents plastic and elastic strains. Newton-Rhapson integration scheme was used for numerical integration. Further, large deformation finite strain and updated lagrangian procedure was used in the analysis to maintain a generality for subsequent research. A tensile stress path of 0 to 60 x 10^{-6} N/μm^2 (60 MPa) is applied to the model in increments of 2.5 x 10^{-6} N/μm^2 for a varying yield stress of organic phase. The yield stress of the organic phase was maintained at 40 x10^{-6} N/μm^2, 50 x10^{-6} N/μm^2, 60 x10^{-6} N/μm^2, 80 x10^{-6} N/μm^2, 120 x10^{-6} N/μm^2, 240 x10^{-6} N/μm^2, 320 x10^{-6} N/μm^2 and 400 x10^{-6} N/μm^2 (40 to 400 MPa). The choice of initial value of yield stress of organic phase is the onset of nonlinearity in nacre response at that value. For each value of organic phase yield stress stress-strain response of nacre is plotted. The resulting yield stress of nacre was compared to experimentally obtained value.

4 RESULTS AND DISCUSSION

Tensile tests were conducted on the 10x10x10 model with single load increment of 2 x 10^{-6} N/μm^2 (2MPa). This load was estimated to occur well within the elastic range based on the experimental data from literature (Currey 1980). The reported mechanical behavior corresponds to yield strength of 56 x 10^{-6} N/μm^2 (or 56 MPa). For a 10x10x10 element, the net displacement was observed to be 4.473X10^{-4} μm. Figure 3 shows contour band plots of displacements for this test. This displacement corresponds to an elastic modulus of 25 GPa. Similar low load tensile simulations were conducted on 4x4x4 model. A scale-down over two orders of magnitude results in a modest lowering of elastic modulus from 25 GPa to 20 GPa.

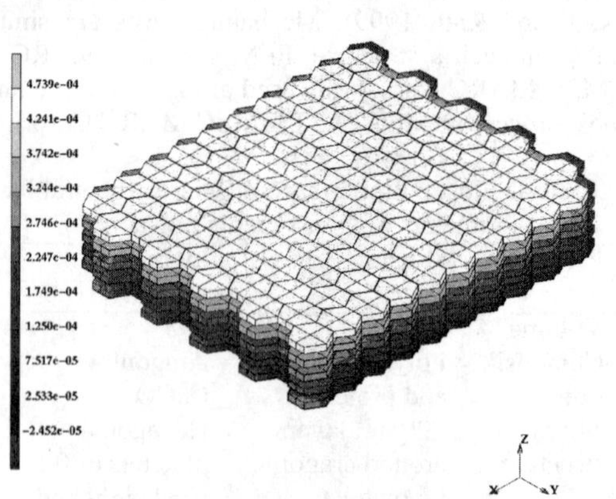

Figure 3. Contour band plots of displacements under simulated tensile test at single load increment of 2MPa. Grayscale shows displacement in z direction.

All nonlinear simulations were conducted on 4x4x4 models. The work hardening slope, Π is found to be 2.857 x 10^{-4} N/μm^2 (285.7 MPa). The slope for organic phase yield stress of 40 x 10^{-6} N/μm^2 is shown in figure 4. The same work hardening slope was used for all non-linear simulations.

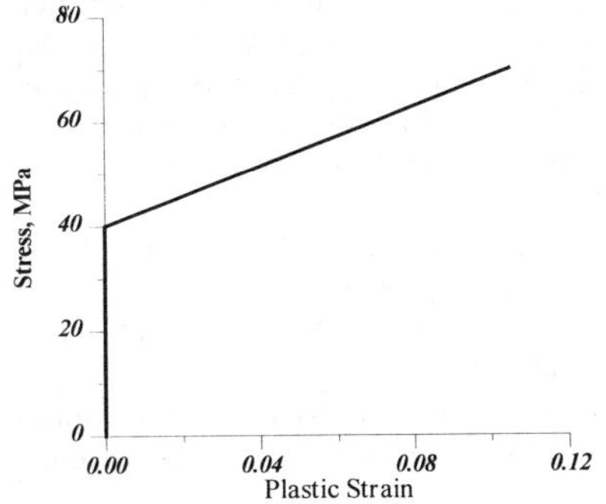

Figure 4. Work hardening slope.

Tensile tests were simulated under identical loading conditions as described in previous section. Simulated stress-strain behavior of nacre for a varying organic phase modulus is shown in Figure 5. Clearly, as seen in the figure, yield stress of nacre increases with yield stress of organic phase. The yield stress of nacre varies from approximately 4.4 x 10^{-6} N/μm^2 to 39 x 10^{-6} N/μm^2 as the yield stress of organic phase varies from 40 x 10^{-6} N/μm^2 to 400 x 10^{-6} N/μm^2. The above simulation is for a 4x4x4 model. Similar simulation are conducted on scaling-up the model to 10x10x10. Figure 6 shows the simulated stress-strain behavior of nacre for both 4x4x4 and 10x10x10 models as compared to that obtained from macroscopic experiments in literature for the organic phase yield stress of 400 x 10^{-6} N/μm^2.

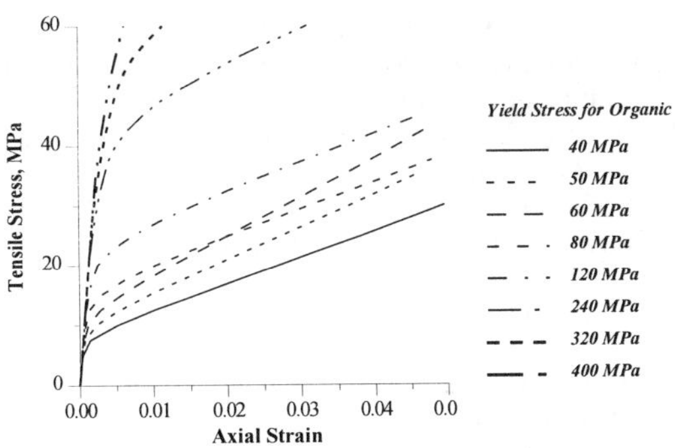

Figure 5. Simulated stress-strain behavior of nacre at varying yield stress of organic phase. Loading condition is identical to all tests. (0 to 60 x10^{-6} N/μm^2 in increments of 2.5 x10^{-6} N/μm^2)

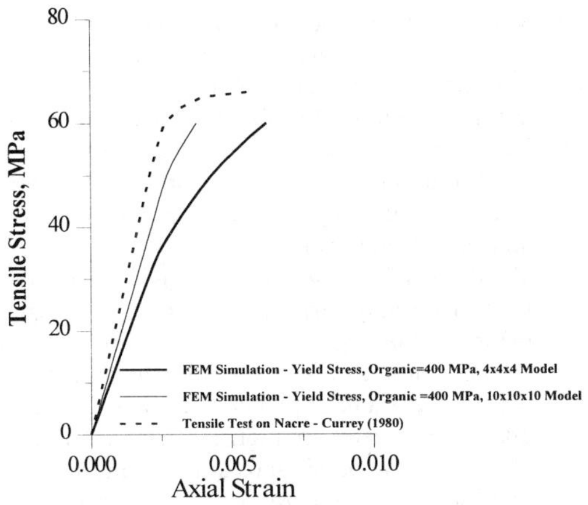

Figure 6. Comparison of simulated stress-strain response of nacre at organic phase yield stress = 400 x 10^{-6} N/μm^2 with experimental for 4x4x4 and 10x10x10 models.

This indicates that a much higher yield stress of organic phase is necessary to obtain the experimentally obtained yield stress of nacre. Furthermore scaling-up to a 10X10X10 model suggests a better match with experiment.

5 CONCLUDING REMARKS

This work represents a multiscale approach towards modeling mechanical responses in a nanocomposite biological system, nacre. 3D finite element model has been constructed for nacre using the material parameters obtained using nanoindentation techniques and also microstructural parameters. Simulated tensile tests were conducted on this model at low and high loads. Nanoindentation experiments yield elastic moduli of inorganic and organic components as 99.5 GPa and 20 GPa respectively. Using these experimentally obtained elastic moduli, tensile tests at low load yield a modulus of nacre as 25 to 20 GPa. The modest variation in modulus results from scale-effects on using 10x10x10 and 4x4x4 models.

Nonlinear responses at higher loads are simulated under the assumption that nonlinearity in nacre results entirely from elasto-plastic response of organic. The inorganic component is modeled as an elastic solid. Simulated tensile tests under identical loading conditions indicate that a very high (400 MPa) yield stress of organic phase is necessary to obtain the experimentally obtained stress-strain behavior and yield stress of nacre. The reason for the necessary high yield stress of nacre results from higher effective local stresses in organic phase than applied. In this work the organic phase is modeled as an isotropic homogeneous material. A high yield stress of organic phase suggests the following possibilities.

1. The inorganic phase is modeled as an elastic solid implying and infinitely large yield stress. In reality the inorganic phase may also exhibit an elastoplastic response.
2. Yield stress of organic phase is indeed very high. This may also result from a composite nature of the organic phase with a high yield stress component. Also, possible hard contacts between aragonite platelets through organic phase may result in higher stress in organic phase than applied and thus result in high yield stress of the 'composite' organic phase.

Further, as more information on the molecular level processes at the organic-inorganic interface and molecular structure and properties of the organic layer are revealed, these may be incorporated into modeling the behavior of individual phases and interfaces. In addition, complex tests can be simulated using the FEM model of nacre which may not be feasible in the laboratory.

REFERENCES

Addadi L, and Weiner S. 1992. Control and Design Principles in Biological Mineralization . Agnew Chem. Int. Ed Engl. 31:153-160.

Currey J D, The Mechanical Properties of Biological Materials edited by J. F. V. Vincent and J. D. Currey Cambridge University Press, London. 75.

Belcher A M, Wu X H, Christensen R J, Hansama P K, Stucky G D., Morse D. E. 1996. Control of Crystal Phase Switching and Orientation by Soluble Mollusc-Shell Proteins. Nature381:56-58.

Currey J. D. 1974. The Mechanical Behavior of Some Molluscan Hard Tissues J. Zool. 173:1974-80

Desai C S, Abel J F. 1974. Introduction to the Finite Element Method. New York: Van Nostrand Reinhold.

Doerner M. F., and Nix W. D.1986. A method for interpreting the data from depth-sensing indentation instruments. J. Mater. Res. 1(4):601-611.

Gunnison K. E., Sarikaya M., and Aksay I. A.1992. Structure-Mechanical Property Relationships in a Biological ceramic-Polymer Composite: Nacre. Mat. Res. Soc. Proc. 1992;255: 625.

Heuer A. H., Fink D. H., Larajai D. J., Arias J. L. Calvert P., Kendall K., Messing G. L., Blackwell J., Rieke P. C., Thompson D. H., Wheeler A. P., Veis A., and Kaplan A. I..1992. *Science* 255 :1098-1100.

Jackson A P, Vincent J F V, Turner R M, Proc. R. Soc. Lond. 1998;B234:415.

Katti D. R., Katti K. S., 2000. 3D finite element modeling of nacre micro architecture: relationship to mechanical properties Proc. Mater. Res. Soc. 599.

Katti D. R., and Katti K. S. 1999. Modeling Microarchitecture and Mechanical Behavior of Nacre Using 3D Finite Element Techniques, Part I: Elastic Properties. submitted to J. Mater. Sci.

Katti K. S., Qian M., Frech D. W., Sarikaya M.1999. Low-Loss Electron Energy-loss Spectroscopy and Dielectric Function of Biological and Geological Polymorphs of $CaCO_3$ Microscopy and Microanalysis.5:358-364.

Mann, S. Biomineralization: Chemical and Biological Perspectives edited by S. Mann, J. Webb and R. J. P. Williams (VCH Verlagsgesselschaft, Weiheim, 1989) 35.

Mann, S. 1993. Molecular Tectonics in Biomineralization and Biomimetic Materials Industry. Nature 365:499

Mann S. 1988. Molecular Recognition in Biomineralization. Nature 33:119-120.

Mercer W. N., Sopp J. M., Fong H., Katti K.S., Katti D.R. and & Sarikaya M 2000. Nanomechanical Properties of a Biocomposite, Mollusk shell Nacre. *Proc. Microscopy and Microanalysis*, MSA 58th annual meeting.

Shaping the World Atom by Atom. 2000. Report of Interagency Working Group on Nanoscience, Engineering and Technology (IWGN)). Washington DC.

Oliver W C, Pharr G M. 1992. An improved technique for determining hardness and elastic modulus using load and displacement sensing indentation experiments J. Mater. Res. 7(6): 1564.

Sarikaya M. 1999. Biomimetics: Fabrication of Materials through Biology. *PNAS* 96 (25) :14183-14185.

Sarikaya, M., Gunnison, K. E., Yasrebi, M., Milius, D. L., and Aksay, I. A. Mat. Res. Soc. Proc. 1990;174:109.

Sarikaya M, Aksay I. A. 1995. Biomimetics: Design and Processing of Materials. New York:American Institute of Physics. 34-90

Schaffer T. E., Ionescu-Zanetti C., Proksch R., Fritz M., Walter D. A., Almqvist N., Zaremba C. M., Belcher A. M., Smith B. L Stucky G. D., Morse D. E., Hansma P. K. 1997. Does Abalone Nacre Form by Hetroepitaxial Nucleation or by Growth through Mineral Bridges?. Chem. Mater. 9:1731-1141.

Simkiss K, Wilbur K M, Biomineralization. 1989.New York:Academic Press.

Winwright S A (eds.) 1982. Mechanical Design in Organisms, N.J.:Princeton Univ. Press.

Computer Methods and Advances in Geomechanics, Desai et al. (eds) © 2001 Balkema, Rotterdam, ISBN 90 5809 183 X

Micro-mechanical analysis of excavation disturbed zones around tunnels

H. Konietzky & L. te Kamp
Itasca Consultants GmbH, Gelsenkirchen, Germany

P. Blümling
NAGRA, Wettingen, Switzerland

J.C. Mayor
ENRESA, Madrid, Spain

ABSTRACT: The **P**article **F**low **C**ode (PFC) was used to investigate the hydro-mechanically coupled development of the **E**xcavation **D**isturbed **Z**one (EDZ) around tunnels. The modeling has given detailed insight into the complete stress and deformation fields, the damage process of the EDZ in space and time, the fracture pattern inside the EDZ, the development of the pore water pressure and the dewatering process. The results can be summarized as follows. The EDZ extends to a distance of app. 2.5 m from the tunnel wall. Tensile cracking is dominant with only minor contribution of shear cracking. The damage decreases non-linear with distance from the borehole wall. 50% to 70% of the damage occurs immediately after the tunnel creation, the rest within app. half a year. The maximum pore water pressure reaches values of up to 2.5 MPa (initial value 2.0 MPa) at a distance of 2.5 m from the tunnel wall shortly after the tunnel creation. Locally, the micro-mechanical behavior in terms of pore water pressures or displacements can significantly deviate from the macroscopic behavior inside the EDZ. Locally and short-termed, water pressure increases by a factor of 10 were observed during dewatering, as well as movements opposite to the general trend of borehole convergence. Dewatering in combination with micro-cracking lead to a complete disappearance of the pore water pressure peaks inside the EDZ within half a year.

1 INTRODUCTION

Numerical modeling was carried out to get an understanding of the hydro-mechanically coupled behavior of the EDZ at the ED-B section of the New Gallery of the Mt. Terri project. The micro-mechanical damage (micro-cracks) in the immediate vicinity of the tunnel contour and the pore pressure development in the damaged area were of special interest.

The modeling was performed in two steps, (1) the back-analysis of laboratory tests for the calibration of the micromechanical parameters, and (2) the modeling of the development of the EDZ.

The modeling was performed using the 2-dimensional **P**article **F**low **C**ode PFC^{2D} (Itasca 1999), which is based on the Distinct Element Method using an explicit time-marching calculation scheme. PFC models the movement and interaction of circular (2D) or spherical (3D) particles and wall elements. Contacts are detected automatically, including abolished or newly created contacts. Material strength is implemented via a bond contact model. PFC has some unique capabilities which allow to investigate the macro- and the micro-mechanical behavior as well as the damaging process simultaneously. A short description of the physical-mathematical background is given by te Kamp and Konietzky (1999).

2 THE TUNNEL MODEL

The numerical model represents a cross section perpendicular to the tunnel axis of 20 m x 20 m, with a tunnel radius of 1.8 m. Figure 1 shows a quarter of the numerical model. To minimize model size and calculation times, the model consisted of concentric areas with particle radii between 3 cm and 14.4 cm. Within each concentric area, the particles are placed at random with uniformly distributed radii, using a ratio of maximum to minimum of 1.2 for all areas. In total, app. 25.405 particles were used. The porosity was set to 15%.

Scaling laws were used for the micro-mechanical parameters, so that the response of the numerical model is not depending on the size of individual particles. The macroscopic behavior of the undisturbed model was nearly homogeneous and isotropic and corresponded to the observed stress and deformation characteristics (te Kamp and Konietzky 1999).

3 HYDRAULIC COUPLING AND NUMERICAL CALIBRATION

The calibration of the mechanical parameters has been performed in the same way as already described by te Kamp and Konietzky (1999), with the latest laboratory and field test results. The microscopic parameters for the numerical model have been set following a uniform distribution with mean value and a range. For the normal bond strength (tensile failure), a range of 14% of the mean value was used. For the shear bond strength (shear failure), the range was set to 4.5 % of the mean value.

The PFC code has been extended by several functions to incorporate a full hydro-mechanical coupling into the calculation scheme. A domain or pore is defined as a space surrounded by a continuos chain of particles (Figure 2). The pore pressure is calculated at the center of the domains. Two different mechanisms are involved, mechanical pore pressure changes due to pore space changes, and flow due to pore pressure gradients.

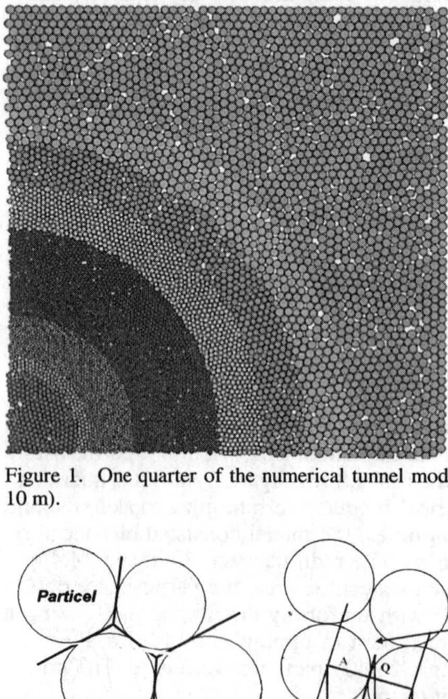

Figure 1. One quarter of the numerical tunnel model (10 m x 10 m).

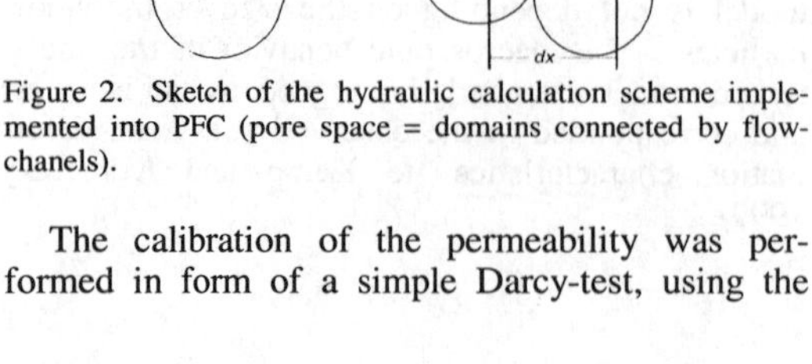

Figure 2. Sketch of the hydraulic calculation scheme implemented into PFC (pore space = domains connected by flowchanels).

The calibration of the permeability was performed in form of a simple Darcy-test, using the specimen which has already been used for the calibration of the mechanical parameters (te Kamp and Konietzky 1999).

The implementation of the hydraulic coupling into the Particle Flow Code leads to different permeabilities for intact and broken material. A separate calibration of the permeability for the intact and broken material has not been possible due to the lack of data and the unknown influence of the crack closure pressure on the permeabilities. Therefore, the calibration of the permeability has then been performed under the assumption that the experimental results described the permeability of the intact material.

4 MODELING OF THE EDZ

The tunneling process has been modeled under dry conditions as well as with hydro-mechanical coupling. The mechanical parameters in all modeling cases were identical, but different random distributions of the strength parameters were used.

The tunnel has been created by the deletion of the particles inside the tunnel area. A procedure similar to the 'load reduction method' was applied to take into account the 3D-effect during tunnel advance. The initial stress field is characterized by 6 MPa vertical principal stress and 4 MPa horizontal principal stress.

Three different runs were made with hydro-mechanical coupling, using the calibrated permeability and permeabilities reduced by a factor of 10 and 100, respectively (Table 1).

Table 1: Modeling cases under wet conditions

Wet_1	Calibrated permeability ($1{,}3 \cdot 10^{-12}$ m/s)
Wet_2	Reduced by factor 10
Wet_3	Reduced by factor 100

The pore pressure at the tunnel wall was fixed at 0 MPa. The pore pressure outside the model area was fixed to the initial pore water pressure of 2 MPa.

5 RESULTS OF HYDRO-MECHANICAL COUPLED TUNNEL MODELING

The three runs with hydro-méchanical coupling have qualitatively revealed the same general behavior:

- the tunnel convergence is anisotropic according to the stress field and reaches maximum values in the order of 30 mm.
- the EDZ reaches a distance of app. 2.5 m from the tunnel wall.
- slightly inhomogeneous fracture pattern, with dominant damage parallel to the minimum principal virgin stress (horizontal).
- most of the damage (50 to 70%) occurs immediately after the tunnel opening, the rest

develops due to the hydro-mechanical coupling and fracture development with time.

- dominant tensile fractures inside the EDZ.
- decreasing fracture density with increasing distance from the tunnel wall.
- the maximum damage in terms of broken to total bonds reaches maximum values at the tunnel wall of 1.
- the damage state is a highly non-linear function of the distance from the tunnel contour.

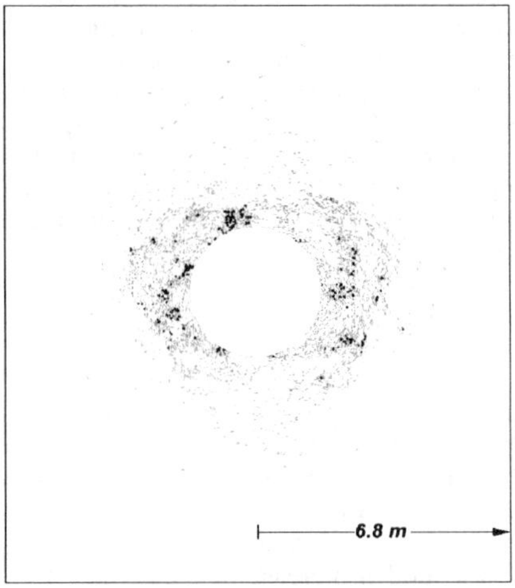

Figure 3. Crack pattern around the tunnel after 6 months. Light: tensile cracks. Dark: shear cracks.

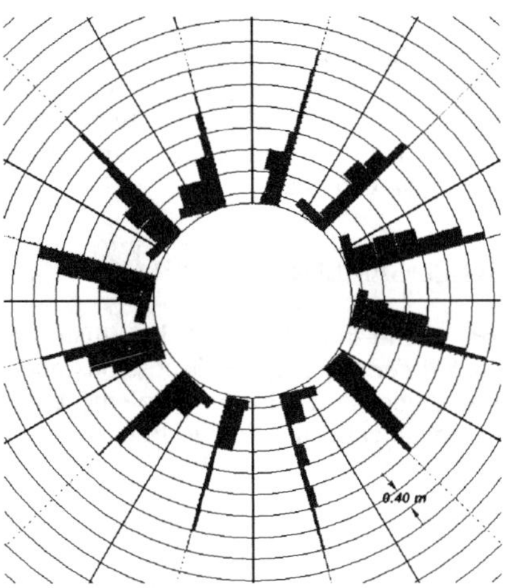

Figure 4. Damage state after 6 months. Number of cracks divided by the total number of particle contacts in each observation cell for wet_1.

Compared to numerical runs without pore pressure, where the dominant fracture type is also tensile,

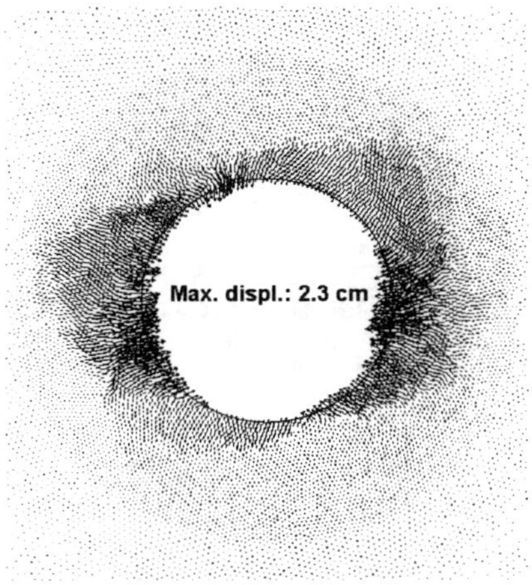

Figure 5. Displacement field for case wet_1 after 6 months.

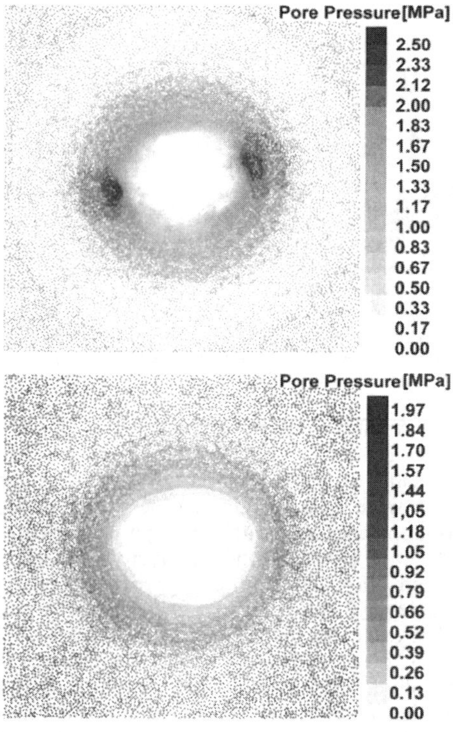

Figure 6. Pore pressure distribution (in MPa) for wet_1 after 1 day (top) and 6 months (bottom).

the ratio of the number of tensile to shear cracks is smaller, being in the order of 10 in the case of the coupled modeling. Figure 3 shows the final fracture pattern for the case wet_1 after 6 months. Figure 4 illustrates the damage state for the same case in terms of the ratio of broken to total contacts.

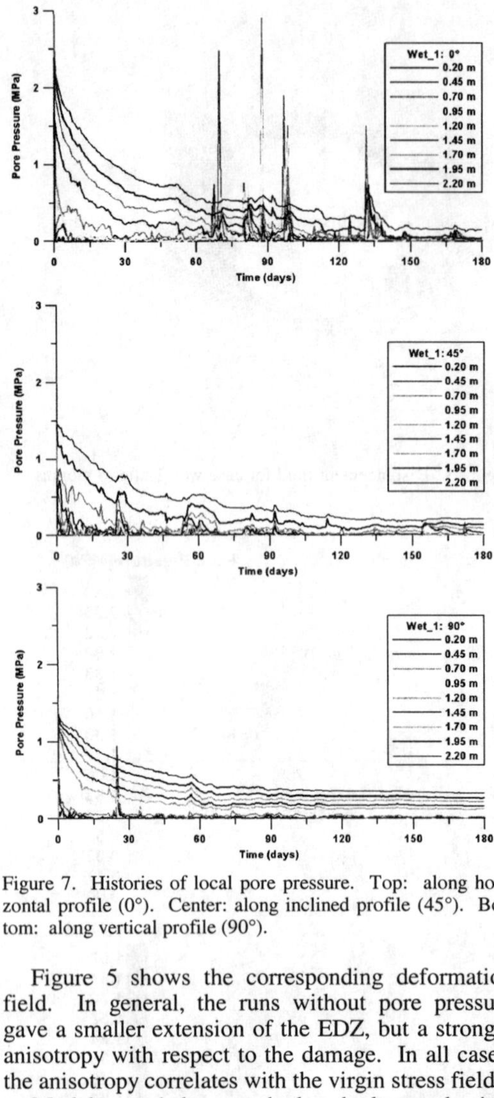

Figure 7. Histories of local pore pressure. Top: along horizontal profile (0°). Center: along inclined profile (45°). Bottom: along vertical profile (90°).

Figure 5 shows the corresponding deformation field. In general, the runs without pore pressure gave a smaller extension of the EDZ, but a stronger anisotropy with respect to the damage. In all cases, the anisotropy correlates with the virgin stress field.

Model wet_1 has reached a hydro-mechanical equilibrium after app. 6 months, whereas models wet_2 and wet_3 show ongoing micro-cracking and water flow due to the much lower permeabilities.

Parallel to the maximum principal stress (vertical), the pore water pressure decreases continuously towards the tunnel wall. Parallel to the minimum principal stress (horizontal), the pore water pressure reaches a maximum value above the initial pore pressure value in app. 2.5 m distance from the tunnel wall. This location coincides with the area of reduced pore space and increased secondary stresses.

Figure 6 shows the pore water pressure distribution for case wet_1 after 1 day and 6 months respectively. Local pore pressure histories were taken along different profiles (Figure 7).

Figure 6 and Figure 7 indicate the general nonlinear trend of pore water pressure reduction due to dewatering. Contrary to the general pore pressure reduction, strong local and short-termed pore water pressure peaks were observed. Pore space collapse leads to a sudden local increase of the pore water pressure (up to a factor of 10 of the actual average value). The spikes declined within a short period of time because of the high pressure gradients.

The spikes move to neighboring areas and lead to a coalescence of micro-cracks to a more macroscopic crack. Similar observations were made regarding the porosity and strain profiles. Both show dominant trends (in agreement with continuum theory), but locally and short-termed a reverse response has been observed (dominant dilatant behavior, but locally compressive). Some of the high-precision measurements at Mt. Terri strongly support these numerical observations.

6 DISCUSSION OF RESULTS

The general trend of the numerical results corresponds well with the field and lab tests. The hydro-mechanical coupled modeling has shown that a strong local variation of pore water pressures can occur. This is caused by the collapse of pore spaces, which lead to a sudden increase of pore water pressures in these spore spaces and decrease in the pore pressure in neighboring pore spaces. This corresponds well to the earlier lab tests (te Kamp and Konietzky 1999) and latest lab tests on clay material (Jung 1999), where a back-analysis indicated, that local pore water pressures can temporarily reach values up to 20 times the average value.

A comparison with the in situ results and observations (stress and deformation field, fracture pattern and extension, pore water pressure distribution) shows a satisfying agreement.

7 ACKNOWLEDGEMENT

The authors would like to extend their thanks to Dr. Peter Cundall for the helpful discussions about the different strategies and the final implementation of the hydro-mechanical coupling into PFC.

8 LITERATURE

Itasca Consulting Group: PFC^{2D} (Vers. 2.0), Minneapolis, Minnesota, USA, 1999.

Jung, D.: personnel communication.

te Kamp, L., Konietzky, H. et.al.: Micomechanical back-analysis of laboratory tests on rock,Proc. NUMOG VII, Numerical models in Geomechanics, A.A.Balkema, Rotterdam, 1999, p. 411 – 416.

Computer Methods and Advances in Geomechanics, Desai et al. (eds) © 2001 Balkema, Rotterdam, ISBN 90 5809 183 X

The numerical simulation of discrete crack propagation in masonry dam structures

C.M.König
GKW Ingenieurgesellschaft mbH, Bochum, Germany

C.Könke
RWTH, Aachen University of Technology, Germany

V.Bettzieche
Ruhrverband Essen, Germany

ABSTRACT: The numerical simulation of stress and pressure fields in fractured porous media presents a highly non-linear coupled problem of structural and fluid mechanics. The stress field depends on the fluid pressure field and the fluid pressure field is influenced by the changing boundary conditions along propagating crack trajectories. This "hydrofracturing" problem, which has been already intensively studied in the area of oil exploitation simulations, can not be appropriately represented by a smeared crack model. The physical boundary condition along the free crack surfaces can only be approximated with a discrete crack model.
An incremental two-step method has been developed to realise a coupled calculation of pressure field and crack propagation. The method will be demonstrated by a hydrofracture-simulation of a masonry dam.

1. INTRODUCTION

Many reservoirs in Germany are dammed up by masonry dams, being established almost 100 years ago. These dams possess a triangular shaped cross-section, which was bricked up by hand using quarry stones sized 0,50 x 0,50 m and larger. The stones were placed particularly careful on the exterior surfaces of the dams and fixed in mortar as waterproof as possible. Within the dam structure itself the stones were partly tilted, but still any existing gaps were filled up and carefully fixed in mortar together with gravel. The dams were founded on an existing bedrock, released from the top gravel revealed down to the stable rock. The proof of stability for these kind of dams can be executed by numerical methods. For a close-to-reality projection of the coupled hydro-mechanical processes, the methods used must be able to illustrate temperature stresses and hydraulic pressures as well as the crack propagation.

The methods used for the calculation of the crack propagation can basically be separated into smeared and discrete crack models. The approach presented here has been based on a discrete crack propagation model. The physical effects are thereby substantially better illustrated. The model supplies a picture of the macroscopically large defects of a numerically stable method without 'stresslocking effects', often obeserved with smeared crack models.

The performance and efficiency of the models is to be demonstrated on the basis of the discrete crack simulation, coupled to a hydraulic pressure calculation for a dam.

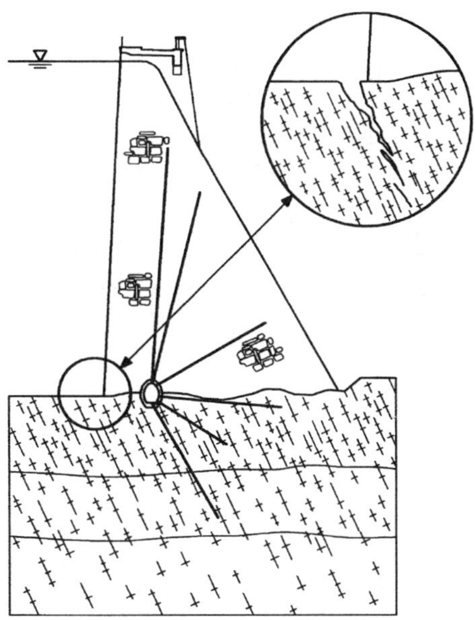

Figure 1. Crack in dam foundation

2. SIMULATION OF DISCRETE CRACK INITIATION

Due to fissures and stratification in the subsoil of a dam as a result of sedimentation and loading history, the rock shows an unisotropic material behaviour. These fissures however only have a small tensile strength, resulting from mineralogization within the fissure surfaces and from cohesion. If this tensile strength is exceeded, then the fissure tears apart and a crack develops.

The dam foundation, which is situated well below reservoir level, usually is the area where structural failures occur (see fig. 1). The inner forces of a crack, resulting from water penetrating into the crack, have to taken into consideration when simulating the propagating crack. These forces depend on the crack aperture. With an open crack the hydrostatic water pressure causes direct pressure loads. The decreasing hydrostatic pressure in the crack propagation zone results in a decrease of the external pressure loads along the crack faces and thus in an increasing stress within the material (see fig. 2). Failure of material occurs, as soon as a critical fracture energy is exceeded.

If the crack propagation has to be calculated for a rock with fissures, then an unisotropic material law (e.g. Mohr-Colomb), defining the failure stress and residual strength in dependence of the main fissure direction, must be taken as a basis.

3. MACRO CRACK PROPAGATION CALCULATION

The modelling of discrete crack propagation problems by means of the finite-element method (FEM) has been presented for the first time at the beginning of the 80's by Saouma and Ingraffea irrespective of an initial discretization. These simualtion algorithms have furtehron developed by Swenson in 1985/87 [3] and Ingraffea/Wawrzynek and coworkers [3]. Today discrete crack propagation simulations have been applied to three-dimensional problems of linear-elastic fracture mechanics [5]. In this method the crack is modelled by discrete crack increments in the structure. However in order to be able to use this procedure for linear-elastic problems, an "initial crack increment" must be given in the model. In many cases the location of the initial crack can be predicted when considering the stress distribution in the undamaged structure.

The initial crack is introduced into the FE model as discrete crack, i.e. the two crack surfaces are considered as free crack surfaces. The crack tip

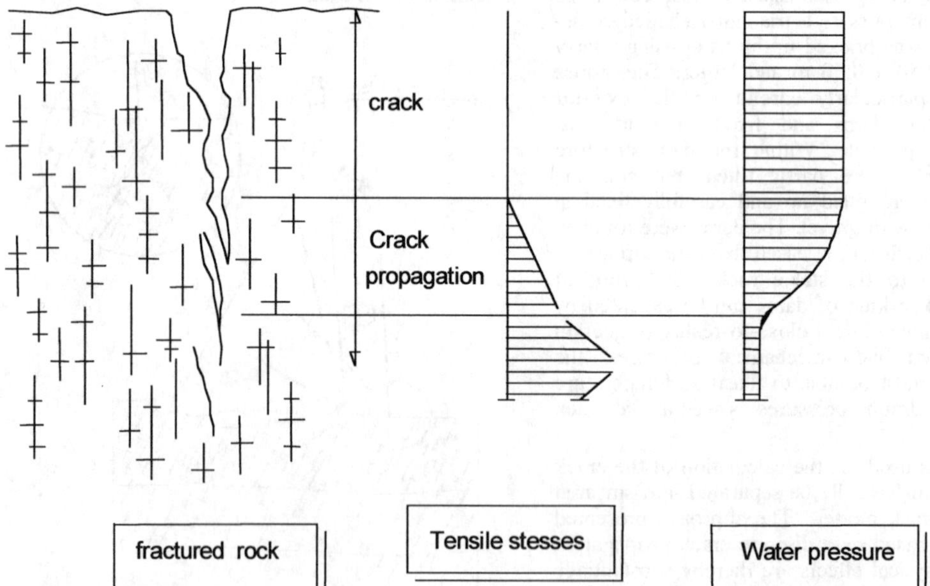

Figure 2: Process of the crack propagation

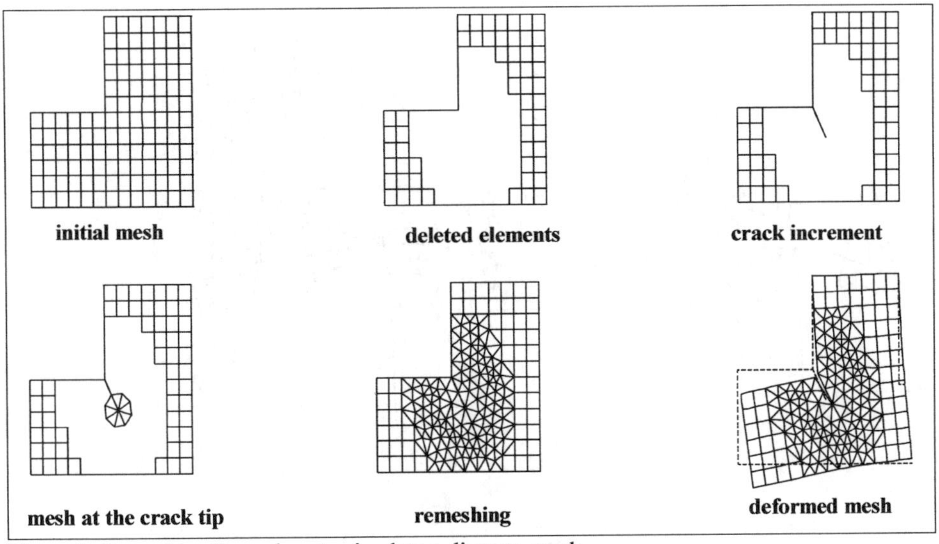

Figure 3: Remeshing Procedure to simulate a discrete crack

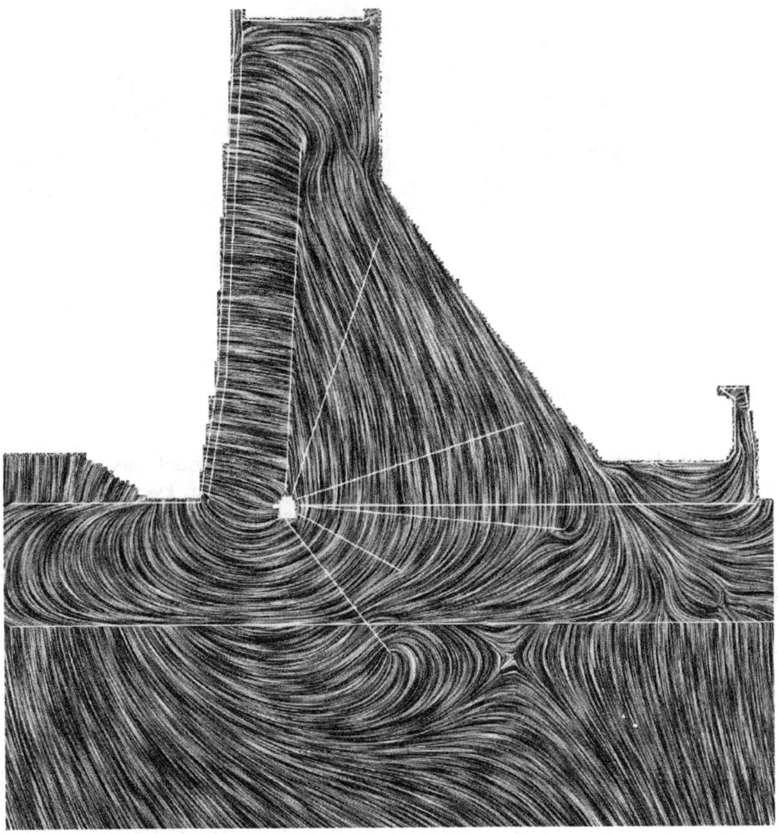

Figure 4. Seepage within the dam

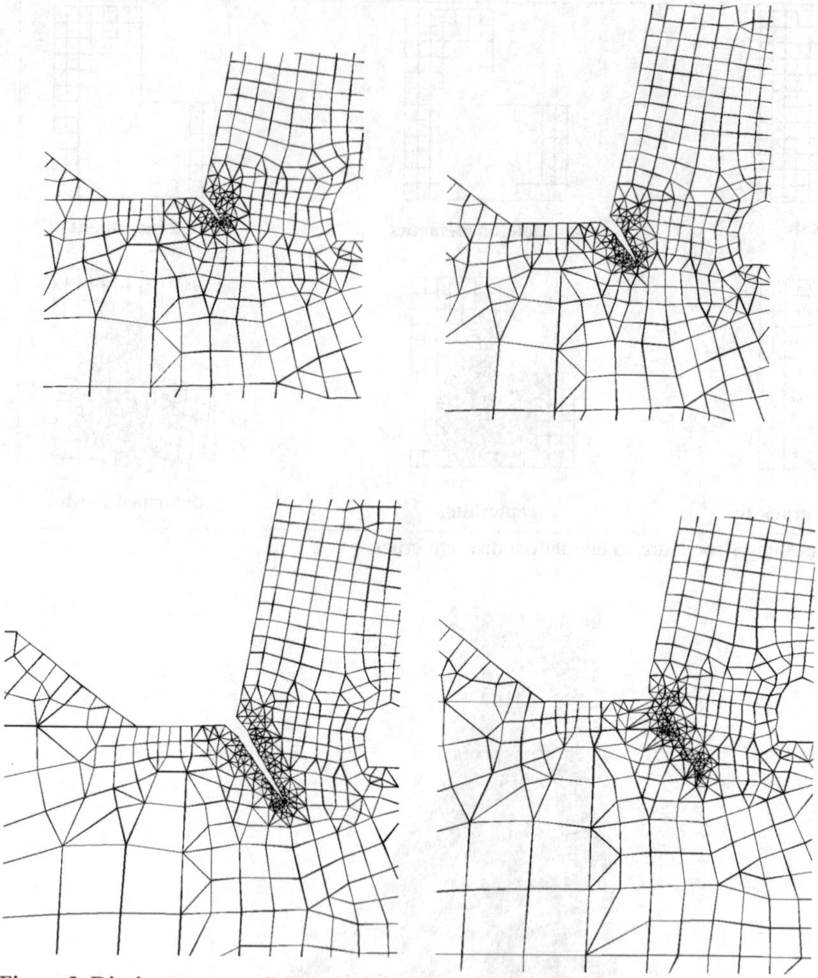

Figure 5. Displacement conditions during crack propagation

respectively the crack front can be modelled with special crack tip features, such as stress singularities in case of linear-elastic fracture mechanics. Therefore special degenerated quarter-node crack tip elements have been developed. Therewith the stress-singularity of linear fracture mechanics can be modelled. For an existing crack, a prediction of the direction and the size of the next crack increment can be made with the help of fracture-mechanic characteristics, such as the stress intensity factors or the J-integral calculus.

For problems of durability the appropriate phenomenological crack propagation laws - Paris or Foreman laws – can be analysed [6].

The basic procedure of the discrete crack propagation simulation is prototypically represented in fig. 3. It shows that the FE mesh, as well as the geometrical model must be adapted to the new crack geometry after every crack increment. The crack propagation due to quasistatic loading is incrementally pursued. The direction of the crack propagation is thereby determined as perpendicular to the main tensile stress.

4. COUPLING OF CRACK AND FLOW CALCULATION

In case of "hydrofracturing" problems the forces, which are responsible for crack expansion and which result from water penetrating into the crack, must be considered. A new flow computation with altered boundary conditions must be done for every crack increment, which supplies the new compressive forces effecting the crack for the further crack

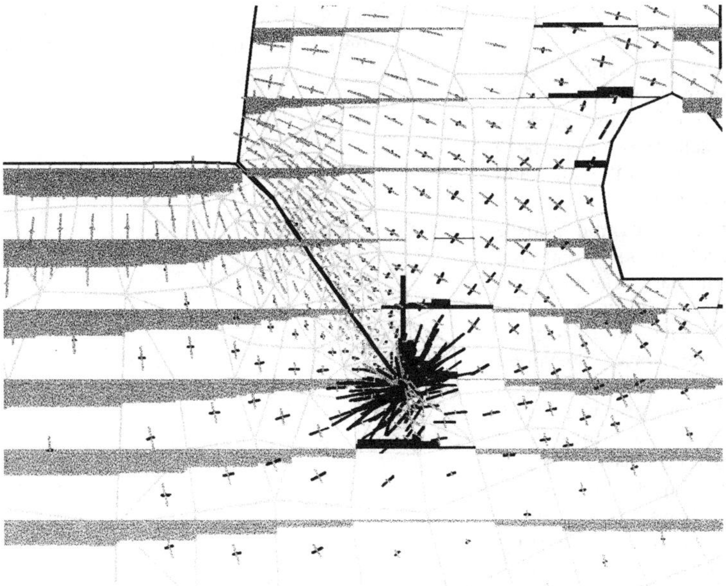

Figure 6: Stress trajectories for final crack state

propagation calculation. This two-step procedure has been automatized for the FEM calculation within the software SPRING [8].

At first there is a flow computation and a further calculation based on the undamaged state of tension. An initial crack is then determined according to this calculation. This initial crack is placed in the area of the largest main tensile stress, perpendicularly to it. As described in the previous chapter, the area of the initial crack is again supplied with a new mesh. Subsequently a flow computation is executed with newly defined boundary conditions along the crack trajectory. It supplies the compressive force for the following step of stress calculation. The new crack increment is the result of this stress calculation. After remeshing the area of the crack increment another flow computation with changed boundary conditions is performed. This iteration is continued until the crack no longer expands, i.e. the critical fracture energy will be no longer exceeded.

5. APPLICATIOIN

The efficiency and stability of the algorithms presented above are examined in an exemplary application. The dam represented in fig. 4 possesses a drainage gallery and drainage boreholes. The complex flow area in the dam and the effect of cracks, which can not be excluded for extreme loading conditions, may be illustrated close-to-reality only with the methodology of a coupled flow and crack calculation.

The flow problem has been considered in the undamaged status at the beginning of the calculation. A boundary condition for seeping on the downstream face of the dam and the drainage installations inside the dam were set, as well as a fixed potential boundary condition for the upstream face of the dam. The resulting flow area and its resulting side issuer is illustrated in fig. 4.

The result of the mechanical problem shows large horizontal tensile stresses at the upstream base of the dam. This result has been confirmed by results from literature. The position of the initial crack is selected at the transition between dam base and fissured subsoil (fig. 5). The following iteration of the flow and crack propagation calculation results in a crack, which propagates diagonally under the base of the dam into the bedrock. Fig. 6 illustrates the principal stress trajectories and the vertical stress component along horizontal lines at the end of the calculation. Green colour is showing negative vertical stresses and red colour is showing areas of tension stresses.

The crack with a final length of several meters led to a stress redistribution within the dam base area. The tensile stresses at the crack tip have been reduced and the fracture energy criterion now detects an arresting crack. Even for this final situation the dam does not fail.

6. SUMMARY

A coupled flow-crack-model with a discrete crack propagation simulation has been presented for the proof the load carrying capacity of a masonry dam. The crack propagation calculation has been influenced by the complex migrating water load and changed local flow pressure conditions.

As known from literature, horizontal tensile stresses result in vertical relief cracks within the rock at the upstream base of the dam. The direction and expansion of these cracks are defined by the material properties of the rock. The history of rock origin as well as disintegration have developed separation layers, such as stratification and fissures, within the rock. The material is thereby weakened, since no tensile stresses can be transferred perpendicularly to these separation layers and a possible shearing force is reduced in direction of the fissure. The crack trajectory follows the separation layers beginning at the base of the dam, propagating into the subsoil, until the crack until the stress redistribution leads to an arresting crack.

7. REFERENCES

[1] Rashid, Y.R.: Ultimate Strength Analysis of prestressed concrete pressure vessels, Nuclear Engineering and Design 7 (1968), 334-344.

[2] Rots, J.G.: Computational Modeling of Concrete Fracture, PH.D. Thesis, University Delft, 1988.

[3] Wawrzynek, P.A., Ingraffea A. R.: Discrete Modeling of Crack Propagation. Theoretical Aspects and Implementation Issues in Two and Three Dimensions, Department of Structural Engineering, Report Number 91-5, Cornell University, Ithaca, New York, 1991.

[4] Cervenka, J.: Discrete Crack Modeling in Concrete Structures, Ph.D. Thesis, Department ofCivil, Environmental and Architectural Engineering, University of Colorado, Boulder, 1994.

[5] Potyondy, D.O.: A Software Framework for Simulating Curvilinear Crack Growth in Pressurized Thin Shells, School of Civil and Environmental Engineering,Report Number 93-5, Cornell University, Ithaca, New York, 1993

[6] Broek, D.: Elementary engineering fracture mechanics, Noordhoff International Publishing, 1974.

[7] Bettzieche, V.; Temperaturmessungen und -berechnungen am Beispiel einer alten Gewichtsstaumauer, Wasserwirtschaft 87 (1997); www.ruhrverband.de.

[8] SPRING; SPRING User's Manual Version 1.1 (1999) GKW Ingenieurgesellschaft, Bochum.

Computer Methods and Advances in Geomechanics, Desai et al. (eds) © 2001 Balkema, Rotterdam, ISBN 90 5809 183 X

Micromechanical modelling of rock strength under dynamic compression

H.B.Li & T.J.Li
Institute of Rock and Soil Mechanics, Chinese Academy of Sciences, Wuhan, People's Republic of China

C.H.Dowding
Department of Civil Engineering, Northwest University, Evanston, Ill., USA

ABSTRACT: A micormechanical model is employed to simulate rock strength under dynamic compression. The model is based on interacting sliding cracks uniformly distributed in rock. Stress intensity factor of the sliding crack array considered is estimated by pseudo-traction method. A dynamic crack growth criterion is used in the modelling. It is shown the modelled rock strength increase with increasing strain rate and confining pressure. The rising rates of the modelled rock strength with increasing strain rate decrease with the increasing confining pressure. The rising rate of the modelled strength with increment of confining pressure are unaffected by strain rate. The modelled compressive strength agrees with the experimental results.

1 INTRODUCTION

Dynamic compressive strength of rock is one of the important parameters in assessing stability of rock cavern and study wave propagation under dynamic loads. Researches on this topic are mainly through experiments (Donath and Fruth, 1971; Masuda et al, 1987). Those experimental studies indicate that the compressive strength of rock generally increases with increasing strain rate and confining pressure.

Efforts have also been paid to theoretically modelling the rate dependent properties of rock under dynamic compression. These works are mostly based on visco-elastic or visco-plastic model for rock (Chong et al, 1980; Blaton, 1981). After 1980s, with the utilisation of scanning electron microscope equipment, useful works have been conducted to microscopically investigate failure process of brittle rock under compression. These works indicated that the dominant failure mechanism of brittle rock is the growth and nucleation of the pre-existing microcracks under compression. Thereafter, researchers started to pay more attention to analyse the behaviour of microcrack pre-existing in rock in an effort to model the macroscopic properties of rock.

In the present paper, the emphasis is to model the rock strength under dynamic compression by micromechanical model. The modelled results are further compared to the experimental results.

2 MICROMECHNICAL MODEL

Various micromechanical models, such as cylindrical pore model, dislocation pile-up model and sliding crack model, have been introduced in an effort to simulate the failure process and the macroscopic mechanical properties of rock under compression. Among these models, sliding crack model is widely accepted.

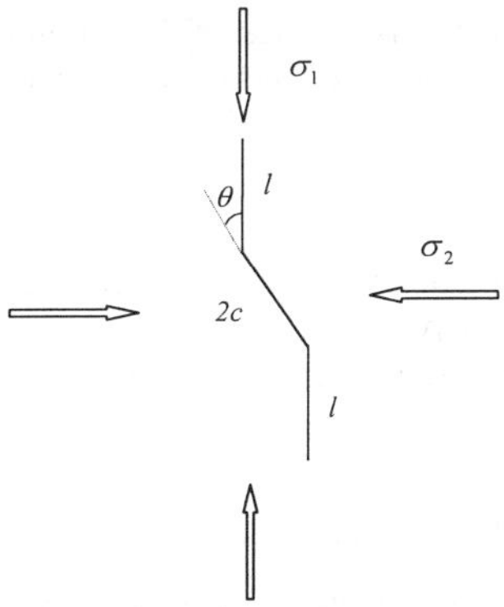

Figure 1. Sliding crack model

The sliding crack model, as shown in Figure 1, was first proposed by Brace and Bombolakis (1963), as a mechanism to study non-elastic dilation of rock under compression (axial compressive stress σ_1 and lateral compressive stress σ_2). It contains an initial crack, at length of *2c*, oriented at an angle θ with respect to σ_1 direction, and a pair of tensile cracks. The tensile cracks are caused by the sliding of initial crack under compressive loads.

In the present paper, an array of sliding cracks, as shown in Figure 2 is presented in order to model the shear failure of rock under compression. In Figure 2, the axis of the sliding crack array lies an angle α to the direction of axial stress. The spacing between two cracks is *2w*.

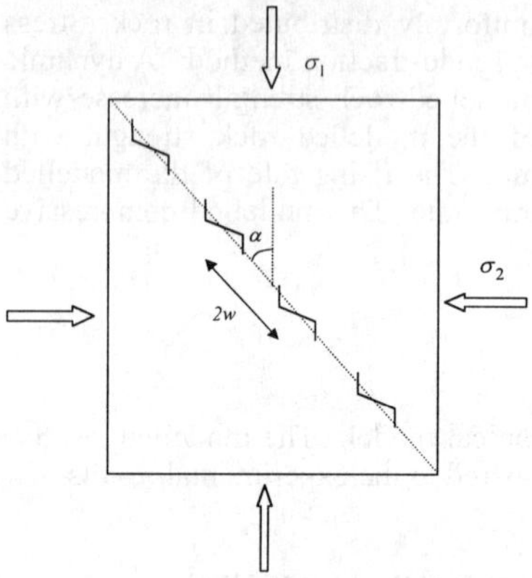

Figure 2. Sliding crack array under compression

Similar to the works by Deng and Nemat-Nasser (1994), Figure 2 can be simplified and decomposed as crack configuration in Figure 3(a) and Figure 3(b). In Figure 3 (a), an array of tensile cracks, at length of $2l$, and crack spacing of *2w*, respectively, are loaded at its centre to a pair of splitting forces F. The forces F represent the effect on the tensile crack of the initial crack's sliding. When cohesive stress is neglected, F can be estimated by:

$$\tau^* = \tfrac{1}{2}(\sigma_1 - \sigma_2)\sin 2\theta - \tfrac{1}{2}\mu[\sigma_1 + \sigma_2 - (\sigma_1 - \sigma_2)\cos 2\theta]$$

$$F = 2c\tau^* \tag{1}$$

where μ is the frictional coefficient of crack faces.

In Figure 3(b), an array of tensile cracks are subjected to axial stress (σ_1) and confining pressure (σ_2).

3 STRESS INTENSITY FACTOR

Pseudo-traction method is used to determine the stress intensity factor of crack array in Figure 3(a) and 3(b). Detail of the method can be referenced to Deng and Nemat-Nasser,1994; Horri and Nemat-Nasser,1985;Li 1999.

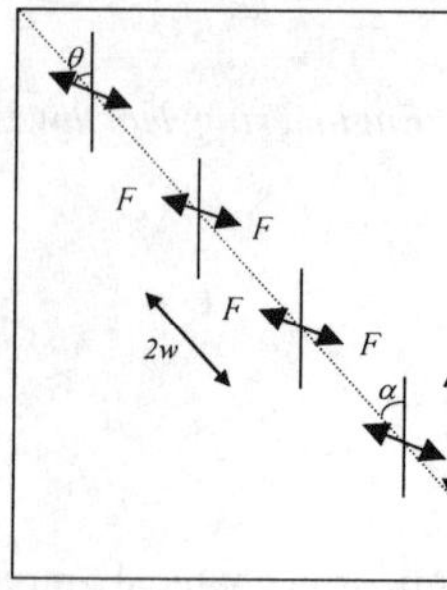

(a)

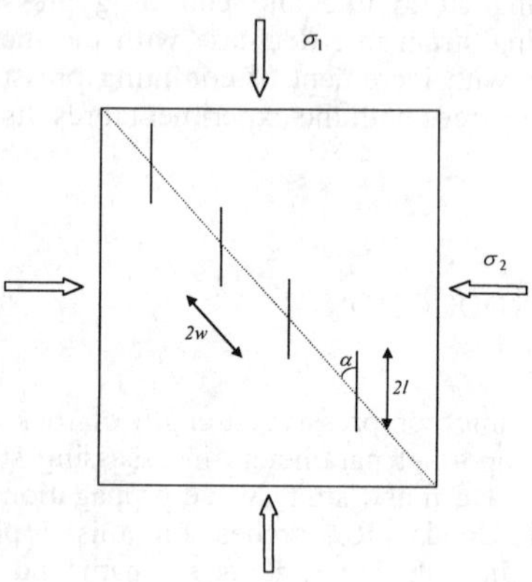

(b)

Figure 3. Decomposition of Figure 2

Consider an elastic solid containing two cracks, as shown in Figure 4(a). Each crack is loaded at its centre to a pair of normal forces P^1 and P^2, as well as shear forces Q^1 and Q^2. By the pseudo-traction method, the original problem in Figure 4(a) can be decomposed into two subproblems, as shown in Figure 4(b). In each subproblem, an elastic body contains only one crack. The crack is loaded by P^i and Q^i $(i = 1,2)$, as well as unknown pseudo tractions σ^{pi} and τ^{pi} $(i = 1,2)$, which are induced from the near crack.

The pseudo tractions can be expanded into Taylor series as (Deng and Nemat-Nasser,1994; Horri and Nemat-Nasser,1985;Li 1999):

$$\sigma^{pj} - i\tau^{pj} = \sum_{n=0}^{\infty}(\sigma_n^{pj} - i\tau_n^{pj})(x^j / l^j)^n \qquad (j = 1,2) \tag{2}$$

where n is the term of Taylor series, σ_n^{pj} and τ_n^{pj} are n term components of pseudo tractions.

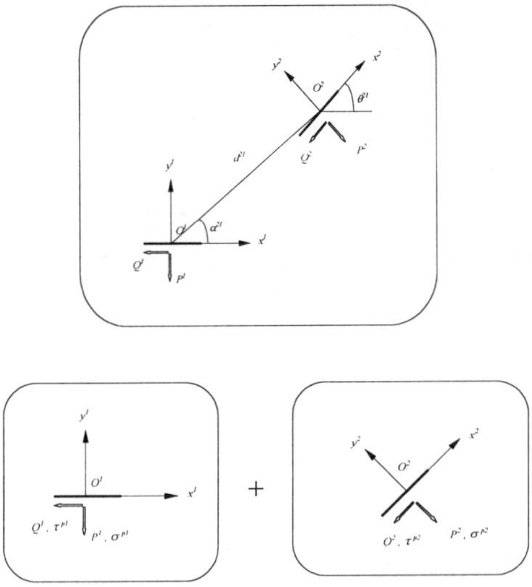

Figure 4. (a): A elastic body containing two cracks; (b): Two subproblems of Figure 4(a)

The stress intensity factor of the crack array in Figure 4(a) can then be calculated as:

$$K_I^j = \sqrt{\pi l^j}\left(\frac{P^j}{\pi l^j} + \sum_{k=0}^{\infty} 2g_k\sigma_{2k}^{pj} \pm \sum_{k=1}^{\infty} 2g_k\sigma_{2k-1}^{pj}\right) (j = 1,2) \tag{3}$$

Considering crack array in Figure 3(a), as pseudo tractions on each crack are symmetric, it is easy to obtain:

$$\sigma_{2n-1}^{pk} = \tau_{2n-1}^{pk} = 0, \; (n = 1,2,......N; \; k = 0,\pm1,\pm2,......\pm\infty) \tag{4}$$

In addition, as a result of periodicity, the pseudo tractions in each crack are identical. Therefore, if we consider $2M+1$ cracks in Figure 3(a), $2N$ term approximation of pseudo tractions are given by (Deng and Nemat-Nasser,1994; Horri and Nemat-Nasser,1985):

$$\sigma_{2n}^{p} = \sum_{m=0}^{N}\left\{\left(2\sum_{k=1}^{M} A_{2n2m}\right)\sigma_{2m}^{p} + \left(2\sum_{k=1}^{M} B_{2n2m}\right)\tau_{2m}^{p}\right\} + \left(2\sum_{k=1}^{M} E_{2n}\right)P + \left(2\sum_{j=1}^{M} F_{2n}\right)Q$$

$$\tau_{2n}^{p} = \sum_{m=0}^{N}\left\{\left(2\sum_{k=1}^{M} B_{2n2m}\right)\sigma_{2m}^{p} + \left(2\sum_{k=1}^{M} D_{2n2m}\right)\tau_{2m}^{p}\right\} + \left(2\sum_{k=1}^{M} F_{2n}\right)P + \left(2\sum_{j=1}^{M} H_{2n}\right)Q$$

(5)

where, $P = F\sin\theta \quad Q = F\cos\theta$. The definition of A_{2n2m}, B_{2n2m}, etc., can be referenced to Deng and Nemat-Nasser,1994; Horri and Nemat-Nasser,1985.

According to Equation (3), the stress intensity factor of crack array in Figure 3(a) is:

$$K_I^{array} = \frac{1}{\sqrt{\pi l}}\left(P + \sum_{k=0}^{\infty} 2g_k\sigma_{2k}^{p}\right) \tag{6}$$

Similarly, for the crack array in Figure 3(b), the mode I stress intensity factor is then given by:

$$K_I^{array} = \sqrt{\pi l}\left(\sigma_2 + \sum_{k=1}^{\infty} 2g_k\sigma_{2k}^{p}\right) \tag{7}$$

The Pseudo tractions are given by (Deng and Nemat-Nasser,1994):

$$\sigma_{2n}^{p} = \sum_{m=0}^{N}\left\{\left(2\sum_{k=1}^{M} A_{2n2m}\right)\sigma_{2m}^{p} + \left(2\sum_{k=1}^{M} B_{2n2m}\right)\tau_{2m}^{p}\right\} + \left(2\sum_{k=1}^{M} A_{2n0}\right)\sigma_2$$

$$\tau_{2n}^{p} = \sum_{m=0}^{N}\left\{\left(2\sum_{k=1}^{M} B_{2n2m}\right)\sigma_{2m}^{p} + \left(2\sum_{k=1}^{M} D_{2n2m}\right)\tau_{2m}^{p}\right\} + \left(2\sum_{k=1}^{M} B_{2n0}\right)\sigma_2$$

(8)

In the present paper, we only consider 2 ($N = 1$) term of pseudo traction, and neglect pseudo traction components containing terms of orders higher than $(l/w)^2$ for the simplicity, the model I stress intensity factor of crack array in Figure 2 can then be expressed as:

$$K_I^{array} = \frac{1}{\sqrt{\pi(l+l^*)}}\left[\frac{1 - 2\sum_{k=1}^{M} A_{00} + 2\sum_{k=1}^{M} E_0 + 2\sum_{k=1}^{M} F_0 ctg\theta}{1 - 2\sum_{k=1}^{M} A_{00}}\right]P - \frac{1}{1 - -2\sum_{k=1}^{M} A_{00}}\sigma_2\sqrt{\pi l} \tag{9}$$

where $l^* = 0.27c$, which ensure the validity of Equation (9) when the tensile crack length is very small (Horii and Nemat-Nasser, 1986).

4 CRACK GROWTH CRTITERION

The commonly used crack growth criterion under static loads is:

$$K_I = K_{IC} \tag{10}$$

where K_I is stress intensity factor (SIF), and K_{IC} is the fracture toughness of rock. The criterion states that when stress intensity factor reaches the fracture toughness of rock, cracks start to grow.

Under dynamic loads, crack growth velocities have great influence on dynamic SIF. In most case, dynamic SIF can be written as (Nemat-Nasser and Deng, 1984):

$$K_{ID} = k(v)K_I \tag{11}$$

where K_{ID} is dynamic SIF, K_I is static SIF, and $k(v)$ is the function of crack growth velocity (v). The function $k(v)=1.0$, when $v=0$, and $k(v)=0$, when v approaches to the limit velocity of crack growth, which is normally regarded as the velocity of Rayleigh wave in that material.

The crack growth criterion under dynamic loads is then (Nemat-Nasser and Deng, 1994):

$$K_{ID} = k(v)K_I = K_{IC}^d \quad (12)$$

where K_{IC}^d is the dynamic toughness of rock.

As it is very difficult to precisely determine the function $k(v)$, several approximations of the function are presented. The most commonly used estimation of the function $k(v)$ is presented by Freund (1990) as:

$$k(v) = \frac{v_r - v}{v_r - 0.75v} \quad \text{when tensile crack is subjected to its center to a pair of splitting forces}$$

$$k(v) = \frac{v_r - v}{v_r - 0.5v} \quad \text{when tensile crack is subjected to far field compressive loads} \quad (13)$$

where v_r is the velocity of Rayleigh wave.

Crack growth criterion of crack array in Figure 2 under dynamic compression can then be described as:

$$\frac{v_r - v}{v_r - 0.75v}\frac{1}{\sqrt{\pi(l+l^*)}}\left[\frac{1-2\sum_{k=1}^{M}A_{00}+2\sum_{k=1}^{M}E_0+2\sum_{k=1}^{M}F_0 ctg\theta}{1-2\sum_{k=1}^{M}A_{00}}P\right] - \frac{v_r - v}{v_r - 0.5v}\frac{1}{1--2\sum_{k=1}^{M}A_{00}}\sigma_2\sqrt{\pi l} = K_{IC}^d \quad (14)$$

where K_{IC}^d is the dynamic toughness for rock, which is determined by Li et al (2000).

Further work by Li (1999) indicates that under the strain rate from 10^{-4} to 10^0 s^{-1}, the crack growth velocity have very small effect on the crack growth and the rock failure. Equation (14) can be simplified as:

$$\frac{1}{\sqrt{\pi(l+l^*)}}\left[\frac{1-2\sum_{k=1}^{M}A_{00}+2\sum_{k=1}^{M}E_0+2\sum_{k=1}^{M}F_0 ctg\theta}{1-2\sum_{k=1}^{M}A_{00}}P\right] - \frac{1}{1--2\sum_{k=1}^{M}A_{00}}\sigma_2\sqrt{\pi l} = K_{IC}^d \quad (15)$$

5 MODELLING RESULTS

According to Equation(12), the modelled rock strength can be easily obtained. Table 1 lists the microscopic coefficients in the present modelling.

Table 1 Microscopic parameters in the modelling

α	20°
θ	45°
2c	1.0mm
2w	3.0mm
μ	0.3

Figure 5 plots the change of modelled strength of a granite at strain rate from 10^{-4} to 10^0 s^{-1}, under confining pressures of 20,50 and 80 MPa. It is shown that the modelled rock strengths generally increase with increasing strain rate and the rising rate decrease with increasing confining pressure. The modelled results are consistent with the experimental results reported by Li et al (1999).

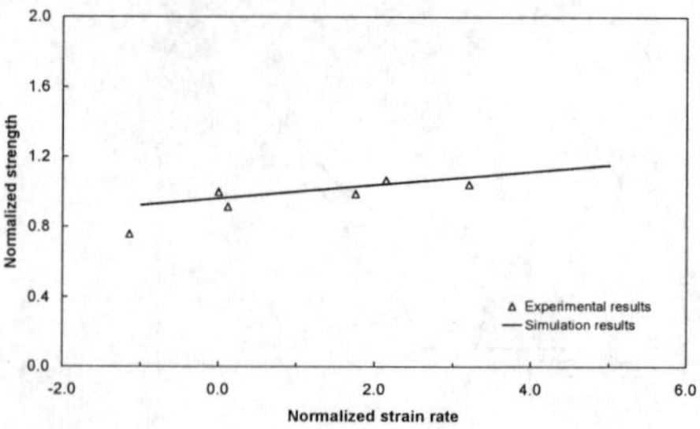

(a) Confining pressure: 20 MPa

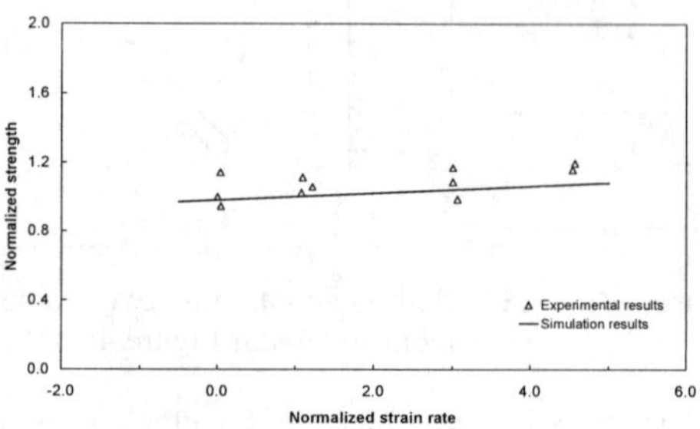

(b) Confining pressure: 50 MPa

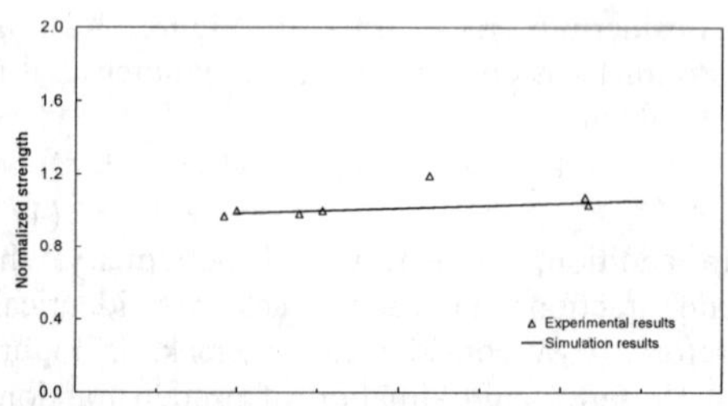

(c) Confining pressure :80 MPa

Figure 5. Change of strength with strain rate under different confining pressures (strain rate is normalised by the minimum strain rate and strength is normalised by the strength at the minimum strain rate)

The modelled strengths of the granite increase with the increasing confining pressures, as shown in Figure 6. The changing rates of modelled strengths with confining pressure are almost identical at different strain rates and agree with the experimental results.

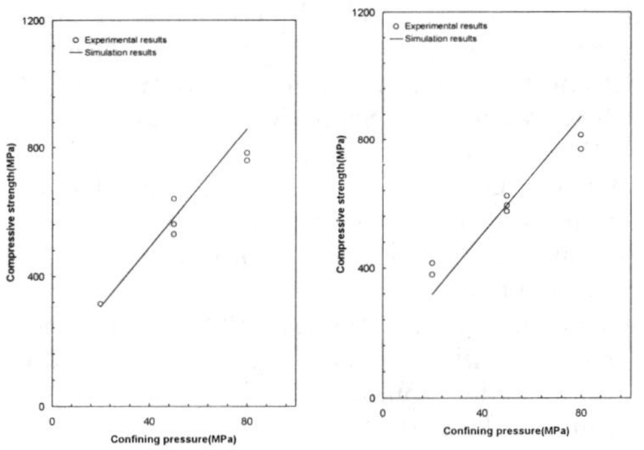

(a): strain rate: $10^{-4}s^{-1}$ (b): strain rate: $10^{-3}s^{-1}$

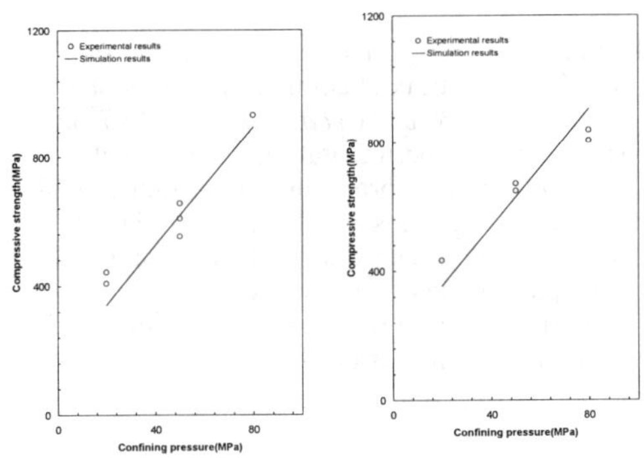

(c): strain rate: $10^{-1}s^{-1}$ (d): strain rate: $10^{0}s^{-1}$

Figure 6. Change of strength with confining pressures under different strain rates

6 DISCUSSION AND CONCLUSION

Sliding crack model is employed to model the rock strength under dynamic compression in the present paper. The modelled results agree with the experimental results. Sliding crack model is initially a hypothetical model to analyse mechanical properties of brittle material under compression. With the development of SEM, it is now revealed that sliding cracks are actually exist in brittle material under compression, and the growth results in failure of brittle material (Horii and Nemat-Nasser, 1985). For this case, sliding crack model is widely used to study mechanical properties of brittle material under static compressive loads. This paper presents a preliminary study on rock strength under dynamic compression using the sliding crack model. By using the model, the paper basically presents a good explanation of the rate and confining pressure dependent properties of rock in microscopic scale.

It can be concluded from Equation (14),that the crack growth velocity and the rate dependence of fracture toughness result in the rock strength's increment with increasing strain rate. In addition, the impediment of confining pressure on the growth of sliding crack causes the increment of compressive strength with increasing confining pressure.

According to the micromechancial model, the influence factors on the rate and confining pressure dependent properties of rock can be also revealed. As an example, Figure 7, 8, 9 and 10 illustrate the change of modelled rock strength with micromechanical parameters. It can be seen that the modelled rock compressive strength clearly increase with decreasing initial crack length and increasing crack spacing. The rising rates of rock strength with increment of strain rate have a trend to increase with decreasing initial crack length and increasing crack spacing.

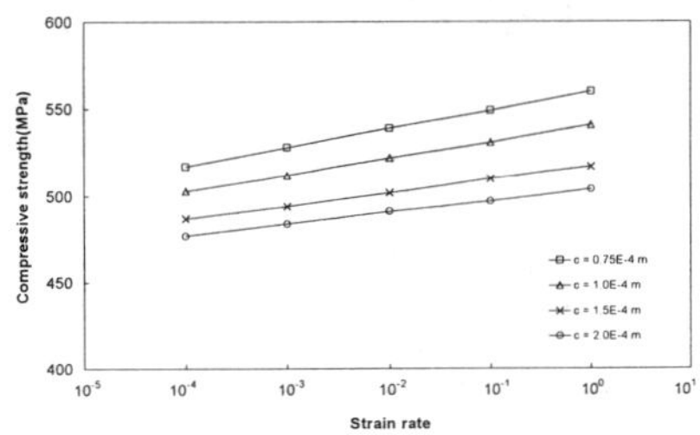

Figure 7. Change of modelled rock strength with strain rate under different initial crack length (confining pressure: 50 MPa , $w/c = 3$, $\theta = 45^{\circ}$, $\mu = 0.3$)

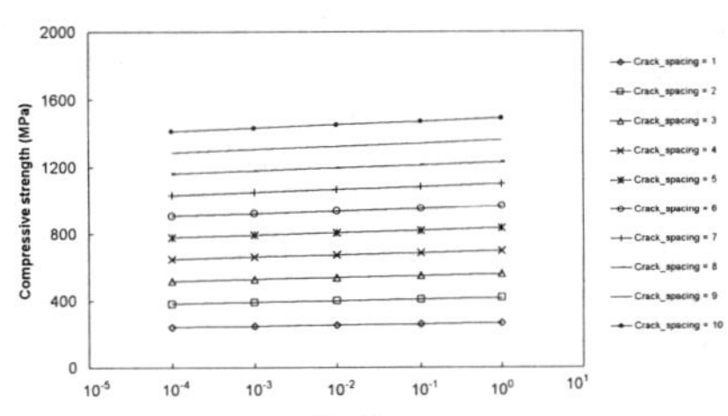

Figure 8. Change of modelled rock strength with strain rate under different crack spacing (confining pressure: 50 MPa, $2c = 1.0mm$, $\theta = 45^{\circ}$, $\mu = 0.3$, crack spacing is normalised by initial crack length)

The changing rate of strength with confining pressures seems unaffected by the initial crack length, but clearly increase with the increasing crack spacing, as shown in Figure 9 and Figure 10.

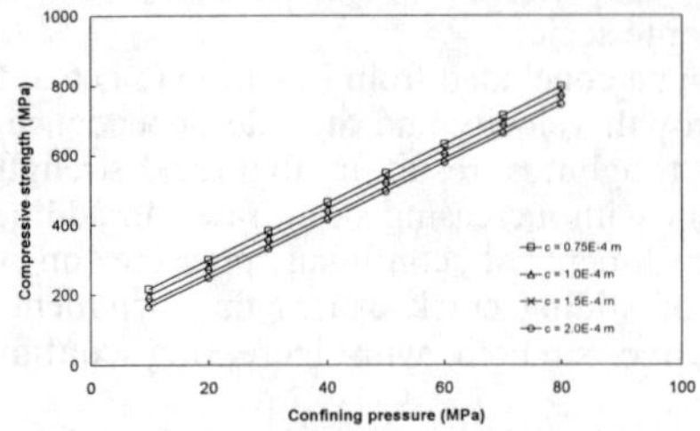

Figure 9. Change of modelled rock strength with confining pressure under different initial crack length (strain rate:$10^{-1}s^{-1}$, $w/c = 3$, $\theta = 45^\circ$, $\mu = 0.3$)

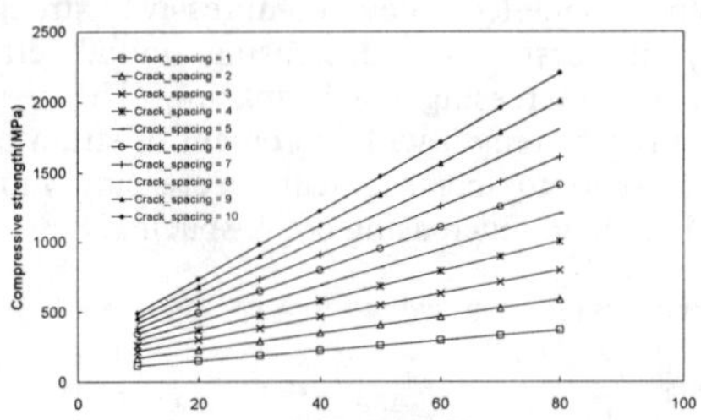

Figure 10. Change of modelled rock strength with confining pressure under different crack spacing (strain rate:$10^{-1}s^{-1}$, $2c = 1.0mm$, $\theta = 45^\circ$, $\mu = 0.3$, crack spacing is normalised by initial crack length)

REFERENCES

Donath,F.A and Fruth,L.S. 1971. Dependence of strain-rate effects on deformation mechanism and rock type. *J. Geol. Soc.* 79:347-371.

Masuda,K., Mizutani,H and Yamada,H. 1987. Experimental study of strain-rate dependence and pressure dependence of failure properties of granite. *J.Phys.Earth*.35:37-66.

Brace,W.F and Bombolakis,E.G. 1963. A note on brittle crack growth in compression. *J. Geophys. Res*.68:3709-3713.

Horii,H and Nemat-Nasser,S.1985. Elastic fields of interaction inhomogeneties. *Int. J. Solids Structures*. 21(7):731-745.

Deng,H. and Nemat-Nasser,S.1994. Microcrack interaction and shear fault failure. *Int. J. Damage Mech*. 3:3-37.

Horri,H and Nemat-Nasser,S.1986. Brittle failure in compression: splitting, faulting, and brittle-ductile transition. *Phil. Trans. Royal. Soc. London.* 319:337-374

Li,H.B., Zhao,J. and Li,T.J.1999. Triaxial compression tests of a granite at different strain rates and confining pressures. Int. J. Rock. Mech. Min. Sci.36(8):1057-1063.

Li,H.B., Zhao,J. and Li,T.J. 2000. Micromechanical modelling of mechanical properties of a granite under dynamic uniaxial compressive loads. *Int. J. Rock Mech. Min. Sci.*(in press).

Horri, H. and Nemat-Nasser, S.1985. Compression induced micro crack growth in brittle solids :axial splitting and shear failure. *J. Geophys. Res*. 90:3105-3125.

Chong, K.P., Hoyt, P.M., Smith, J.W. and Paulsen, B.Y.1980. Effects of strain rate on oil shale fracturing. *Int. J. Rock. Mech. Min. Sci.* 17:35-43.

Blanton, T.L.1981. Effect of strain rate from 10^{-2} to 10 sec^{-1} in triaxial compression tests on three rocks. *Int. J. Rock. Mech. Min. Sc.* 18:47-62.

Li.H.B. 1999. Experimental and theoretical study on mechanical properties of granite under dynamic compressive loads. Ph.D thesis. Institute of rock and Soil Mechanics, The Chinese Academy of Sciences.

Freund, L.B.1990. Dynamic fracture mechanics, Cambridge University Press.

Computer Methods and Advances in Geomechanics, Desai et al. (eds) © 2001 Balkema, Rotterdam, ISBN 90 5809 183 X

Finite element investigation of localization in porous media

X. Liu, A. Scarpas & J. Blaauwendraad
Section of Structural Mechanics, Faculty of Civil Engineering and Geosciences, Delft University of Technology, Netherlands

ABSTRACT: Finite element dynamic strain localization analyses of a saturated soil specimen are presented. Numerical modelling of the interaction between the mechanical behavior of the solid phase and the flow characteristics of the liquid phase has been formulated on the basis of Biot's theory. The theory of dynamic plasticity has been chosen as the most suitable material constitutive framework for modelling the solid material response in the nonlinear range. With this model, the influence of soil-water interaction on the localization band formation and propagation has been investigated.

1 INTRODUCTION

Localized deformations in the form of narrow shear bands are often observed to develop after large inelastic deformation in materials. Within the localization band, the material behaviour is inelastic. The analysis of strain localization is of importance in engineering practice because localization is a precursor to sudden failure. Typical examples can be found in frictional materials such as concrete, rocks and soils.

Strain localization has been extensively studied in recent years, in particular in connection with single-phase solids. Among others Needleman (1988), Loret et. al. (1990), Sluys (1992) etc., have demonstrated that the use of classical rate-independent continuum models for material localization simulation can result in a loss of hyperbolicity of the equations of the dynamic initial value problem. Then, wave speeds within the body of the solid become imaginary and wave propagation becomes impossible. The numerical manifestation of this pathological behaviour, is the localization of deformations in the finite element mesh in zones whose thickness tends to zero as the element mesh is refined. Within these zones no energy consumption takes place.

Rate or gradient dependency of the constitutive model has been shown to naturally introduce a internal length-scale parameter into the solution of the initial value problem that can prevent loss of hyperbolicity. Mesh independence can then be ensured.

Loss of hyperbolicity can also occur in the simulation of strain localization in multi-phase materials, Loret et. al. (1991), Schrefler et al. (1995). Because of the presence of the velocity gradient term in the fluid continuity equation, an internal length scale is introduced naturally. This internal length scale depends on several material parameters and in particular on the permeability. For the case of one-dimensional fully or partially saturated porous medium under dynamic loading, an internal length scale has been derived by Zhang et al. (1999).

In this **contribution** the findings of Zhang et al (1999) are combined with a dynamic plasticity formulation of the flow surface proposed by Desai (1990) and applied to the numerical investigation of the dynamic response of a multi-phase medium. Interaction between the phases is accounted for by means of Biot's theory.

In the following, some aspects of the governing equations of the porous medium and of the material model are first presented. Then, the influence of soil-water interaction on the numerical response of a saturated soil specimen subject to impact type loading conditions is examined. In accordance with the results of Zhang et al. (1999) for the one-dimensional case, it will be shown that in three-dimensional simulations also, the presence of an internal length scale ensures mesh independence.

2 GOVERNING EQUATION OF POROUS MEDIUM

The multi-phase medium is treated as a two-phase continuum consisting of a solid component with voids that are filled with fluid. Biot's theory is utilized to set up the governing equations of the dynamic response of the material. The air phase in partially saturated zones is assumed to stay continuous (no bubbles existing) and is always at atmospheric pressure. The deformation of the porous skeleton is controlled by the effective stress $\boldsymbol{\sigma}'$ that is defined as:

$$\boldsymbol{\sigma}' = \boldsymbol{\sigma} + \mathbf{I} \cdot \tilde{p} \tag{1}$$

in which $\boldsymbol{\sigma}$ is the total stress vector, vector $\mathbf{I}$ corresponds to the Kronecker delta, $\tilde{p}$ is the averaged water pressure defined as $\tilde{p}=S_w \cdot p_w$; S_w is the water saturation and p_w is the excess water pressure. Tensile stresses and strains are assumed positive.

With these assumptions, the virtual work principle can be expressed, Zienkiewicz et al. (1984) and Schrefler et al. (1995) as

$$\begin{aligned}
&\int_V \delta\mathbf{u}^T \cdot \rho \cdot {}^t\ddot{\mathbf{u}} \cdot dV + \int_V \delta\mathbf{u}^T \cdot c \cdot {}^t\dot{\mathbf{u}} \cdot dV \\
&+ \int_V \delta\boldsymbol{\varepsilon}^T \int_{t-\Delta t}^{t} \dot{\boldsymbol{\sigma}}' \cdot dt \cdot dV - \int_V \delta\boldsymbol{\varepsilon}^T \int_{t-\Delta t}^{t} \mathbf{I} \cdot \dot{\tilde{p}} \cdot dt \cdot dV \\
&= \int_V \delta\mathbf{u}^T \cdot \mathbf{b} \cdot dV + \int_A \delta\mathbf{u}^T \cdot \mathbf{t} \cdot dV \\
&- \int_V \varphi \cdot S_w \cdot \rho_w \cdot \mathbf{a}_{ws} \cdot dV - \int_V \delta\boldsymbol{\varepsilon}^T \cdot {}^{t-\Delta t}\boldsymbol{\sigma} \cdot dV
\end{aligned} \tag{2}$$

in which: $\mathbf{b}$ is the vector of applied body forces; $\mathbf{t}$ is the vector of applied surface tractions; ρ is the material density of the multiphase medium defined as: $(1-\varphi)\rho_s + \varphi S_w \rho_w$; ρ_s is the mass density of the solid phase; ρ_w is the mass density of water phase; φ is porosity of the medium; c is the damping parameter, a_{ws} is the relative acceleration of water.

Also, the continuity condition of the liquid phase can be expressed as:

$$\begin{aligned}
&-\nabla^T \left\{ \mathbf{k} \frac{k_{rw}(S_w)}{\mu_w} \left(\nabla p_w + \rho_w \cdot (\ddot{\mathbf{u}} + \mathbf{a}_{ws}) + \rho_w g h \right) \right\} \\
&+ \varphi \frac{\partial S_w}{\partial t} + \frac{\varphi S_w}{K_w} \frac{\partial p_w}{\partial t} + S_w \cdot \mathbf{I}^T \cdot \frac{\partial \boldsymbol{\varepsilon}}{\partial t} = 0
\end{aligned} \tag{3}$$

in which g is the gravity, h is the head above some arbitrary datum, $\mathbf{k}$ is the absolute permeability matrix of the medium, μ_w is the dynamic viscosity of the fluid, $k_{rw}(S_w)$ is the relative permeability which is a function of water saturation and K_w is the bulk modulus of water.

Eq. (2) and Eq. (3) represent the general forms of the governing equations which constitute the basis of the finite element dynamic simulation of a fully/partially saturated medium. They have been implemented in INSAP-PM, Liu et al. (2000). The solution of these governing equations is obtained numerically by means of a Newton-Raphson iterative procedure.

3 CONSTITUTIVE MODEL

The theory of dynamic plasticity has been chosen as the constitutive framework. By retaining the fundamental to classical plasticity notions of flow surface, decomposition of strains, hardening and/or softening, the theory has emerged as an attempt to provide a realistic, unified, phenomenological approach for materials independent of whether they exhibit or not strain rate dependent inelastic deformations.

3.1 *Constitutive framework*

In similarity to inviscid plasticity, the plastic rate is defined as:

$$\dot{\boldsymbol{\varepsilon}}_p = \dot{\lambda} \cdot \frac{\partial f}{\partial \boldsymbol{\sigma}} \tag{4}$$

with λ a constant of proportionality and f a response surface associated with a locus of states of stress corresponding to a certain magnitude of inelastic response. Evolution of plastic flow is determined by the consistency condition:

$$\dot{f}(\boldsymbol{\sigma}, \dot{\varepsilon}_{eff}, \kappa) = 0 \tag{5}$$

in which $\dot{\varepsilon}_{eff}$ is the effective strain rate and κ is some measure of hardening/softening, Scarpas et al. (1998). The standard Kuhn-Tucker conditions apply.

3.2 *Hardening response*

At this phase of the investigation utilization of a single flow surface for hardening response simulation was deemed sufficient.

3.2.1 *Flow surface characteristics*

In order to focus on the influence of the length scale introduced by the solid-fluid coupling, a strain rate independent form of the flow surface proposed by Desai (1990), has been chosen. In stress invariant space it is defined as:

$$f_D = \frac{J_2}{p_a^2} - F_b \cdot F_c = 0 \tag{6}$$

in which:

$$F_b = \left[-\alpha \cdot \left(\frac{I_1 + R}{p_a} \right)^n + \gamma \cdot \left(\frac{I_1 + R}{p_a} \right)^2 \right] \tag{6a}$$

$$F_c = 1 \tag{6b}$$

I_1, J_2 are the stress invariants and p_a is the atmospheric pressure.

Eq. (12) represents a closed surface in the $\left(I_1, \sqrt{J_2}, \theta\right)$ space, Fig. 1, alleviating thus the need for definition of additional cap surfaces along the hydrostatic axis.

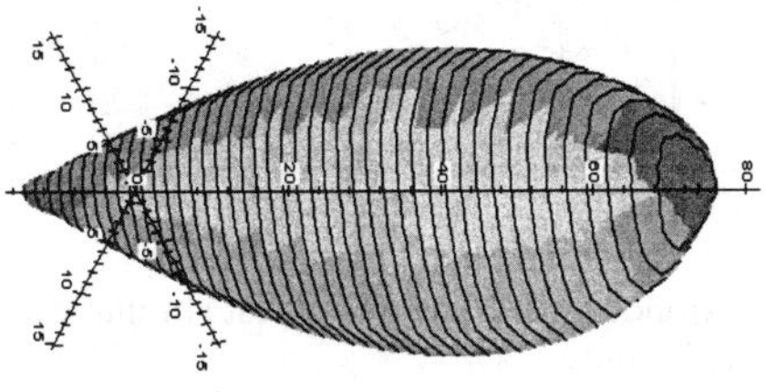

Fig. 1 Schematic of Desai response surface

The hardening response of the material is controlled by parameter α. As α decreases, the size of the flow surface increases. By defining α as a decreasing function of plastic strain or plastic work, the characteristics of the hardening response of the material can be simulated.

On a physical basis α can be associated with some measure of plastic deformation, plastic work W_p, equivalent plastic strain ξ etc. In this contribution, it is defined as:

$$\alpha = \alpha(\xi) \tag{7}$$

in which the equivalent plastic strain ξ is defined by:

$$\xi = \int d\varepsilon_p = \int \sqrt{d\varepsilon_{ij}^p \cdot d\varepsilon_{ij}^p} \tag{8}$$

In this investigation, the following relationship was utilized for α, Scarpas et al. (1998) :

$$\alpha = \frac{\alpha_0 (\xi_{lim} - \xi)}{1 + c_0 \cdot \xi} \tag{9}$$

in which ξ_{lim} is the experimentally determined effective plastic strain at maximum response, α_0 defines the limit of elastic response and c_0 is an experimentally determined constant.

The slope of the ultimate response surface in the $I_1 - \sqrt{J_2}$ plane is controlled by parameter γ. It is attained for $\alpha = 0$. In this investigation, the value of γ as determined by Frantziskonis et al. (1986) was utilized. Parameter n determines the apex of the sface on the $I_1 - \sqrt{J_2}$ plane. It defines the state of stress beyond which the material begins to dilate.

3.3 *Response degradation*

In the model, for deformation levels beyond those corresponding to $\alpha = 0$, an isotropic measure of response degradation has been introduced by means of specifying γ in the post fracture regime, γ_{pf}, as a decaying function of the monotonically increasing equivalent post fracture plastic strain ξ_{pf}:

$$\gamma_{pf} = \eta \cdot \gamma + (1 - \eta) \cdot \gamma_r \tag{10}$$

$$\eta = \left(1 + \kappa_1 \cdot \xi_{pf}\right) \cdot e^{-\kappa_2 \cdot \xi_{pf}} ; \tag{10a}$$

$$\gamma_r = \kappa_3 \cdot \gamma \tag{10b}$$

and κ_1, κ_2 and κ_3 are material parameters. For the purposes of this investigation $\kappa_1 = \kappa_2 = \kappa_3$ was specified.

4 NUMERICAL INVESTIGATION

By means of INSAP-PM, the influence of soil permeability and mesh refinement on the formation and propagation of localization bands in a fully saturated soil material have been investigated. Two permeability values and two different density meshes were examined.

4.1 *Specimen Geometry and Material Characteristics*

A soil specimen 50 mm wide, 100 mm high and 1 mm thick was selected for the numerical simulations. Free drainage was assumed at the horizontal boundaries of the mesh, Fig. 2. All other boundaries were assumed impermeable. The bottom surface of specimen was fixed. The displacements of the front and back planes of the specimen were also restrained in the thickness direction. The finite element mesh consisted of 20-nodded brick elements. In order to verify the mesh insensitivity of the model, two types of mesh were examined, one consisted of 300 elements and another of 580 elements.

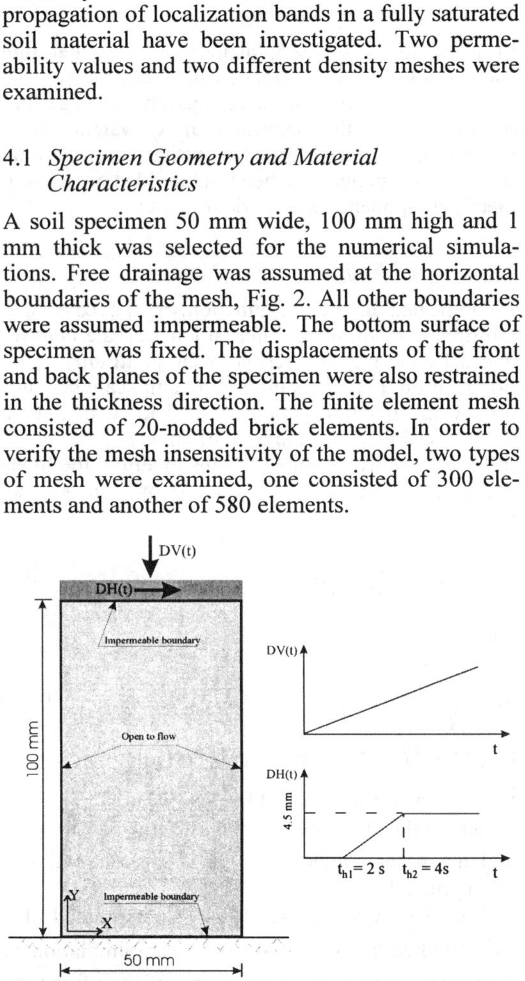

Fig. 2 Schematic of specimen geometry and loading functions

In order to initiate localization, the top of the specimen was subjected to both, a small amount of horizontal displacement and monotonically increasing vertical displacements at a rate of 7.5 mm/sec. Details of the loading functions are shown in Fig. 2.

In this study, the material properties of 'Munich' sand, Frantziskonis et al. (1986), in combination with the dynamic plasticity model described in the previous section were chosen for the simulation. The main material parameters are shown in Table 1.

Table. 1 Main material parameters

E (Mpa)	63.4	n	3.2
ν	0.21	α_1	0.1231
ρ_s (Kg/m³)	2.0e-03	c_0	4055
ρ_w (Kg/m³)	1.0e-03	ξ_{lim}	0.065
kx=ky=kz (m/s)	5.0e-03~5.0e-06	κ_1	10.8
φ	0.2	γ	0.105

4.2 *The definition of damage*

In plasticity and viscoplasticity related literature various definitions of damage can be encountered. In the framework of this investigation, damage has been defined as the magnitude of equivalent plastic strain in the material, Eq. (8). Furthermore, the total amount of damage has been subdivided in a volumetric component and a deviatoric one.

4.3 *Internal length scale prediction*

As indicated in the introduction, in the case of a multiphase medium, an internal length scale is introduced by the gradient terms in the continuity equation (see Eq. (3)). This internal length scale depends on several material parameters and in particular on the permeability. For the case of an one-dimensional fully or partially saturated porous medium, the internal length scale has been derived by Zhang et al. (1999) as:

$$l \cong \frac{2c_m\eta}{k \cdot K^2 \cdot \bar{\alpha}^2 \cdot S_w \cdot Q^*} \tag{11}$$

where

$$c_m = \sqrt{\frac{E_2}{\rho}}, \quad \eta = \frac{E_2}{Q^*}, \quad E_2 = \bar{\alpha}^2 S_w Q^* + \bar{H} \tag{11a}$$

in which $\bar{\alpha}$ is Biot's Constant, K is the wave number, $k = k_w/\rho_w g$, k_w is permeability, Q^* is a parameter related to the compressibility of the solid and the water components. $\bar{H}$ is the material softening modulus.

For the 'Munich' sand it was estimated that Q^*=1000 MPa. As shown in Fig. 3, simulation of the monotonic response by means of the aforementioned solid material constitutive model, indicated an average softening modulus $\bar{H} = -0.71\text{MPa}$.

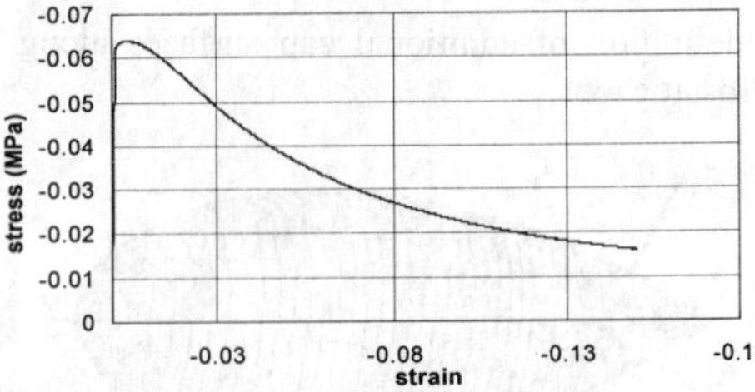

Fig. 3 Monotonic compression test simulation for "Munich" sand

On the basis of these values and for permeability values of 5.0e-03 m/s and 5.0e-06 m/s, the length scale from Eq. (11) is computed as 0.0028 mm and 2.8mm, respectively, indicating the strong influence of permeability on regularization.

4.4 *In time response of the specimen*

The specimen with permeability of 5.0e-03m/s was investigated first. During the simulation, a shear band initiated from the right corner of the bottom surface and gradually propagated along the diagonal direction of specimen, Fig. 4. It can be seen that the elements within the localization band dilate.

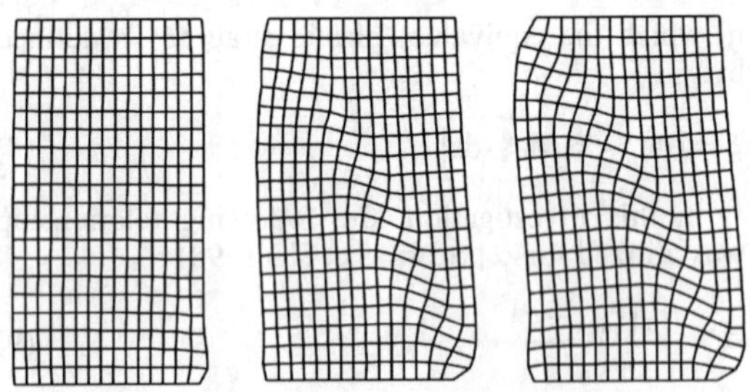

Fig. 4 Localization band propagation in specimen (k=5.0e-03 m/s)

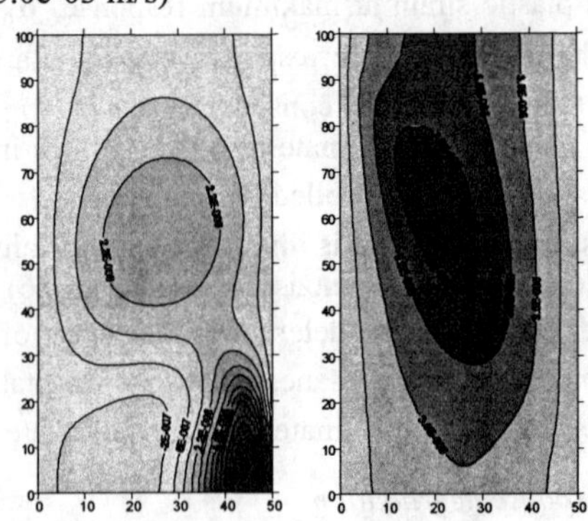

(a) (b)

Fig. 5 Pore pressures distribution within the specimen (k=5.0e-03 m/s)

The distribution of excess pore pressures within the specimen at initiation of the localization band, Fig. 5(a) and, after full propagation of the band, Fig. 5(b), indicate the development of a suction field along the band. (positive pore pressure implies suction). At the vertical edges of the specimen the pore pressure drops to zero.

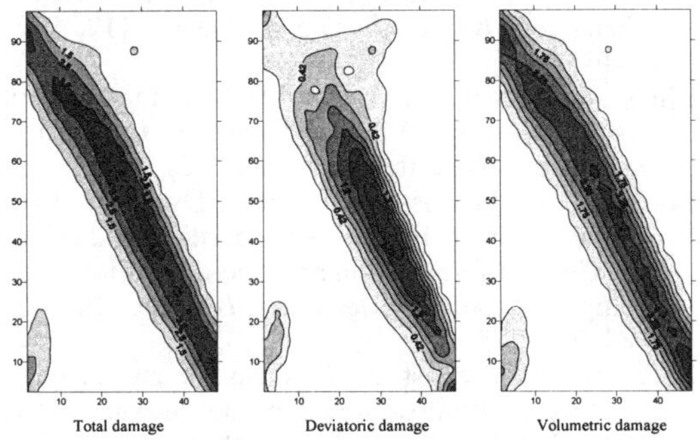

Fig. 6 Damage patterns within the specimen (k=5.0e-03 m/s)

As expected, within the specimen, damage accumulated primarily among the localization band, Fig. 6. Because of the large material dilatancy, the magnitude of the volumetric damage component is higher than the deviatoric one.

In order to investigate the influence of soil permeability on the conditions leading to initiation of the localization phenomenon, a specimen with lower permeability (k=5.0e-06 m/s) was examined.

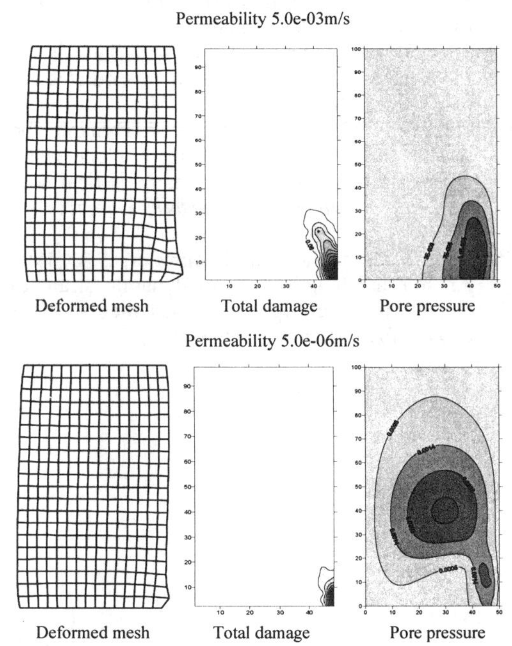

Fig. 7 Influence of soil permeability on localization band formation

The response of both specimens, at the same level of vertical displacement, is compared in Fig. 7. For the lower permeability specimen, higher water suction increases the strength of the soil material and results to reduced levels of damage at band initiation. Also, after band formation, at the same level of vertical displacement as in Fig. 6, lower values of damage can be seen for the specimen with lower permeability in Fig. 8. Similar effects have also been observed by by Loret et al. (1991) and Schrefler et al. (1995).

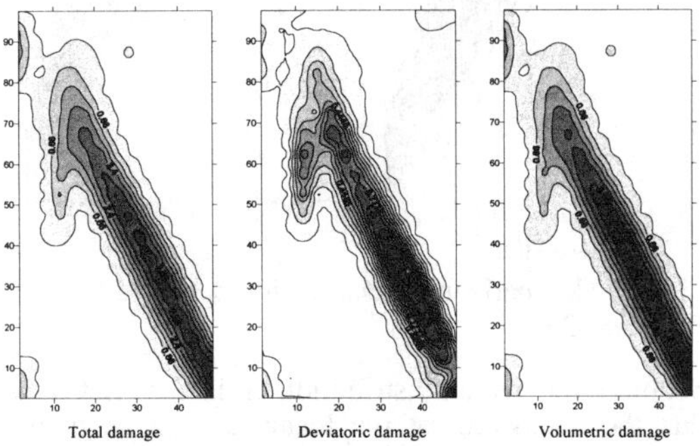

Fig. 8 Damage patterns within the specimen (k=5.0e-06 m/s)

In order to verify mesh insensitivity of the results, another mesh with k=5.0e-03 m/s and consisting of 580 elements was also examined. At same vertical displacement level, the deformed meshes for both cases are shown in Fig. 9.

It can be observed that the localization-bands pattern and their widths span the same geometric region of the specimen. Nevertheless, the finer mesh is shown to be more capable of simulating the field of large displacements associated with the free boundaries of the localization band.

Damage patterns corresponding to the two finite element meshes are shown in Fig. 10. Even though the maximum values of damage at the center of the localization band are almost identical, careful exami-

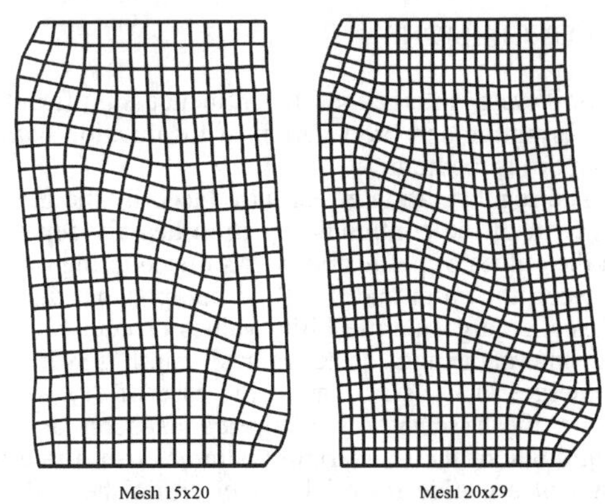

Fig. 9 Comparison of mesh deformations for (k=5.0e-03 m/s)

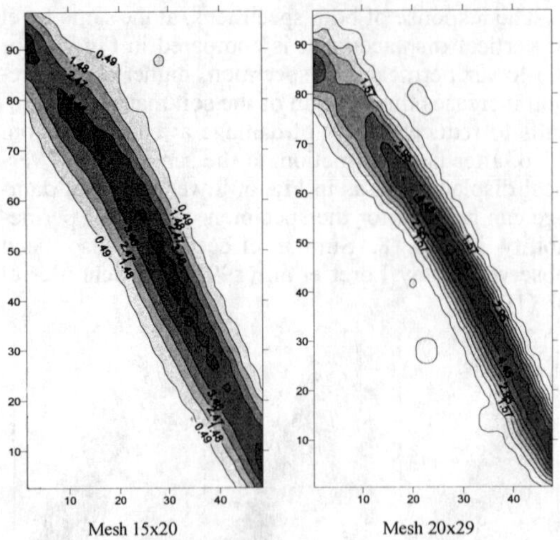

Fig. 10 Comparison of localization bandwidths (k=5.0e-03 m/s)

nation of the overall distribution of damage indicates that the bandwidth of the localization zone in the finer mesh is somewhat smaller signifying thus a possible tendency for mesh dependency.

As mentioned earlier, for the purposes of this series of numerical tests, the strain rate sensitivity capability of the solid material model was not activated. This option, in combination with the low value of the internal length scale for the investigated multiphase medium (k=5.0e-03 m/s), have been known to constitute the necessary conditions for possible mesh sensitivity.

Results by Zhang et al. (1999) have also indicated a tendency for mesh sensitivity for the same range of permeability values and internal length scales.

In a following publication, the influence of larger internal length scale values with and without the beneficial effect of a rate sensitive model will be presented.

5 CONCLUSIONS

Soil-water interaction can influence significantly the formation and propagation of localization bands in a porous medium.

The regularization effect introduced by the gradient term of the continuity equation heavily depends on the values of permeability. As the coupling effect between the solid and the fluid components diminishes (i.e. larger permeability values or equivalently low internal length scale values), mesh sensitivity reemerges. This is similar to the case of dry solids where mesh sensitivity reappears as soon as the magnitude of the regularizing parameter of a rate or gradient sensitive material model diminishes, Sluys (1992).

REFERENCES

Desai, C.S., 1990. Modeling and testing: implementation of numerical models and their application in practice. *CISM Courses and Lectures*, 311. New York: Springer Verlag.

Frantziskonis, G., Desai, C. S., and Somasundaram, S., 1986. Constitutive model for nonassociative behavior, ASCE, *J. Engng. Mech.*, 112, 932-946.

Liu X., Scarpas A., 2000. INSAP-PM Finite Element System – User's Manual. Faculty of Civil Engineering, TU-Delft, The Netherlands.

Loret, B. and Prevost, J. H., 1990. Dynamic strain localisation in elasto(visco-)plastic solids, Part 1. General formulation and one-dimensional examples, *Comput. Mech. Appl. Mech. Engng.*, 83, 247-273.

Loret, B. and Prevost, J. H., Dynamic strain localisation in fluid-saturated porous media. 1991. ASCE, *J. Engng. Mech.*, 11, 907-922.

Needleman, A. 1988. Material rate dependence and mesh sensitivity in localization problems. *Comput. Methods Appl. Mech. Engrg.*, 67, 69-85.

Scarpas, A. and Blauuwendraad, J., 1998. Expermental calibration of a constitutive model for asphaltic concrete. Proceedings of the Euro-C Conference on the Computational Modelling of Concrete Structures, Badgastein, Austria, March 31- April 3, Rotterdam: Balkema, 193-202.

Sluys, L.J., 1992. Wave propagation, localisation and dispersion in softening solids. *Ph.D. Thesis*, Faculty of Civil Engineering, Delft University of Technology, Netherlands.

Schrefler, B.A., Majorana, C. E. and Sanavia, L., 1995. Shear band localization in saturated porous media. *Arch. Mech.*, 47, 577-599.

Zienkiewicz, O.C., Shiomi, T., 1984. Dynamic behaviour of saturated porous media; The generalized Biot formulation and its numerical solution. *Int. J. Numer. Anal. Meth. Geomech.*, 8, 71-96.

Zhang, H.W., Sanavia, L. and Schrefler, B.A., 1999. An internal length scale in dynamic strain localization of multiphase porous media. *Mech. Cohes.-Frict. Mater.*, 4, 443-460.

Computer Methods and Advances in Geomechanics, Desai et al. (eds) © 2001 Balkema, Rotterdam, ISBN 90 5809 183 X

Continuous finite element analysis of strain localization and failure in geostructures

A. Modaressi
Ecole Centrale Paris, MSSMat Laboratory, CNRS UMR 8579, France

K.C. Vossoughi & D. Gouvenot
Ecole Centrale Paris, MSSMat Laboratory, and Soletanche-Bachy, Nanterre, France

ABSTRACT: This paper presents a continuous finite element analysis of strain localization and failure in the domain of retaining structures notably the diaphragm walls during excavation. Several crireria which could be admitted as precursors of failure such as Rice's condition of strain localization, contours of incremental distortion or variation of incremental displacement vector, the non positiveness of the second order work and finally the non-convergence of the computation are examined to evaluate the most conservative or realistic results. In order to pursue the numerical algorithm until failure, the constitutive behavior is progressively adapted and the material is assumed to behave in an anisotropic manner following the potential shear band directions.

1 INTRODUCTION

The use of retaining structures has become a common and unavoidable solution in urban sites. Though the wall deflection and the ground surface settlement are the key factors which cause damage, the classical design methods such as analytical limit equilibrium or the pressure diagram do not provide any information concerning these factors. The finite element method provides attractive solutions for such problems, but the standard finite element solutions fail to converge when the load limit is reached. Classical design methods are based on the notion of failure but the definition of such a concept is not at all evident when using numerical methods of analysis such as finite element. Furthermore, it is well known that the mesh refinement and the alignement of the elements change the results obtained by the standard FE method.

In practical geotechnical engineering, it is considered that if in a numerical computation the non-convergency appears the structure is in failure, and consequently studying the post-limit behavior of geotechnical structure is useless. Because, it is assumed that once the limit load reached, the soil is already failed and there is no need to analyze its behavior any more. But it has been proved that, in soils as well as in many solids, the presence of the modes of intense deformation restricted to narrow regions are observed to develop as a precursor to failure when the limit load is reached. These deformation patterns are considered to be the cause of ultimate failure and can be strong discontinuities for the materials exhibiting softening behavior (Simo et al. 1993) and (Garikipati 1996). But in reality, the geotechnical structure could contain a slip surface and still be in a pre-limit state (Regueiro and Borja 1997). Hence, the study of the post-limit behavior of the geostructurs seems to have great interests.

On the other hand, in an industrial point of view, one can see the philosophy of ultimate limit state (ULS) design in Eurocode 7 (part of the Eurocodes that deals with geotechnical design). This philosophy takes for each geotechnical problem a minimum number of limit states to be analyzed. Each limit state is a failure mode of the geostructure which depends on the kinematics of the failure and should be modeled. Because of the complexity of failure modeling and the other verifications demanded by the code, only a numerical method such as the finite element can be used. Hence, the correct modeling of failure with the numerical methods is becoming more and more important and more performant numerical tools are needed.

This paper presents a continuous finite element analysis of strain localization and failure in the domain of retaining structures notably the diaphragm walls during excavation. Sevaral criteria which could be admitted as precursors of failure such as Rice's condition of strain localization, concentrated contours of incremental distorsion, or the variation of incremental displacement directions, the non positiveness of the second order work and finally the non-convergence of the numerical algorithm are examined to evaluate the most conservative or realistic results.

In addition, the phenomena of failure in granular materials can be viewed as a continuous precedure which is involved with two particular phenomena. First, in the homogeneous strain and stress field, the condition of strain localization is fulfilled resulting in non-homogeneous deformation accompanied by shear banding. Once this condition reached, even for an initially isotropic soil, the behaviour can no more be considered as isotropic, as the appearance of such shear bands induces a certain anisotropy to the behaviour.

In practice, the bifurcation condition is not reached simultainuously everywhere and even so, they are rarely continuously aligned following a surface. To circumvent these shortcomings and to obtain a continuous strategy which can result in a realistic response, the progressive achivement of bifurcation condition is accompanied by the introduction of anisotropy to the material behaviour.

So, in order to pursue the computation until failure, the constitutive behavior is progressively adapted and the material is assumed to behave in an anisotropic manner following the potential shear band directions. Then the ability of the progressive evolution of the constitutive model to reach failure is illustrated by a geometrically and mechanically simple case. Several numerical applications of excavation problems with diaphragm walls are presented and the above conditions and the chronology of their achivement for different failure mechanisms or limit states are illustrated.

In the first section the theory of the second order work, bifurcation analysis and strain localization are recalled. Next, the constitutive model used in the computation and its evolution is presented. At the end, the finite element applications in the domain of retaining structures are demonstrated.

2 STABILITY ANALYSIS AND THE SECOND ORDER WORK

One of the methods used for the stability analysis is the theory of second order work. Based on the works of Hill and Drucker the stability condition could be translated as the following. Considering a stress- strain state for a material that has submitted a given strain history, a sufficient condition for the stability is the positiveness of the second order work for any couple of stress and strain increments (Darve and Laouafa 1999):

$$\forall d\underline{\underline{\sigma}}, d\underline{\underline{\varepsilon}} \qquad d^2w = d\underline{\underline{\sigma}} : d\underline{\underline{\varepsilon}} > 0 \tag{1}$$

so, the loss of this condition is equivalent to instability and could be used as a failure precursor.

3 BIFURCATION ANALYSIS AND DISCONTINUITY

Basical concepts of the bifurcation analysis and the condition for discontinuity as introduced by (Rice 1976) are largely discussed in literature (Aubry 1988) and (Ortiz et al. 1991). The phenomen of strain localization in an elastoplastic system represents surfaces along which the normal derivation of the deformation field becomes discontinuous. The conditions in which the discontinuity surfaces appear could be found using the theory of singular surfaces regarding to associated generalized derivatives and the distribution theory. We limit our study briefly to the weak form of discontinuity which is the jump in deformation field.

3.1 *Jump Condition and Direction of Discontinuity*

It has been supposed that before the onset of discontinuity both deformations and stresses are continuous in the domain and besides the discontinuity surface the domain is in plastic loading condition. So the elastoplastic tensor $\mathbf{C}^{ep}$ and the acoustic elastoplastic tensor $\mathbf{Q}^{ep}$ are continuous across the surface of discontinuity. This leads to write the compatibility condition regarding to equilibrium as the following. This equation leads to find the eigenvalue of the discontinuity mode $[d_n \underline{v}]$:

$$\mathbf{Q}^{ep}(\underline{n}).[d_n\underline{v}] = 0 \tag{2}$$

where:

$$\mathbf{Q}^{ep}(\underline{n}) = \underline{n} \cdot \mathbf{C}^{ep}(\underline{n}) \cdot \underline{n} \quad \text{or} \quad Q^{ep}_{ij} = C^{ep}_{ijkl} \underline{n}_i \underline{n}_l \tag{3}$$

with $\underline{n}$ the normal vector to the discontinuity surface. The necessary condition for the equation (2) to have solutions is defined as below :

$$\mathbf{det}(\mathbf{Q}^{ep}(\underline{n})) = 0 \tag{4}$$

This condition is the well known Rice's condition for localization. It verifies the begining of the loss of ellipcity in the elastoplastic system of equations.

4 THE EVOLUTIVE CONSTITUTIVE MODEL

An isotropic elastoplastic model with linear elasticity and Mohr-Coulomb yield function is chosen for the unlocalized material. The following set of constitutive equations describes the material behaviour before strain localization.

$$\dot{\underline{\underline{\sigma}}} = \mathbf{C}_i(\dot{\underline{\underline{\varepsilon}}} - \underline{\underline{\varepsilon}}^p_i) \tag{5}$$

$$\underline{\underline{\dot{\varepsilon}}}_i^p = \dot{\lambda} \ \frac{\partial g}{\partial \underline{\underline{\sigma}}} \quad (6)$$

where $\underline{\underline{\sigma}}$ is the stress tensor, $\underline{\underline{\varepsilon}}$ is the strain tensor, $\mathbf{C}_i$ the elastic moduli which is supposed to be isotropic, $\dot{\lambda}$ is the plastic multiplier found using the consistency equation:

$$\dot{f}(\underline{\underline{\sigma}}) = 0 \quad (7)$$

where $f(\underline{\underline{\sigma}})$ is the Mohr-Coulomb yield function given by :

$$f(\underline{\underline{\sigma}}) = \mid \sigma_1 - \sigma_3 \mid + (\sigma_1 + \sigma_3)\sin\varphi - 2c\cos\varphi \quad (8)$$

φ and c are friction angle and cohesion of the material, and σ_1 and σ_3 are major and minor principal stresses. $g(\underline{\underline{\sigma}})$ is the flow potential given by :

$$g(\underline{\underline{\sigma}}) = \mid \sigma_1 - \sigma_3 \mid + (\sigma_1 + \sigma_3)\sin\psi \quad (9)$$

which can result in non-associated plasticity if $\varphi \neq \psi$. Once the bifurcation condition reached, it is supposed that the bifurcation directions present priviledged directions for shear. So, the material is considered to present anisotropic elasticity and plastic strain with respect to the shear band. The last condition implies the introduction of additional plastic mechanisms governed by yield functions taking into account the stress vector applied on the shear band. As two potential bifurcation directions exist, two plastic mecanisms are to be added. Let identify each shear surface with the vector normal to it ($\underline{n}_1$ or $\underline{n}_2$). The stress vector on each surface could be defined as:

$$\underline{t}_k = \underline{\underline{\sigma}} \cdot \underline{n}_k \quad \text{with} \quad k = 1, 2 \quad (10)$$

resulting in the normal stress ($\sigma_k^{'}$) and shear stress vector ($\underline{\tau}_k$) with:

$$\sigma_k^{'} = \underline{n}_k \ \underline{\underline{\sigma}} \ \underline{n}_k \quad (11)$$

$$\underline{\tau}_k = \underline{t}_k - \sigma_k^{'} \ \underline{n}_k \quad (12)$$

Now, the total plastic strain rate will be the sum of the plastic strain rates mobilized by each yield function:

$$\underline{\underline{\dot{\varepsilon}}}^p = \sum_k (\underline{\underline{\dot{\varepsilon}}}_k^p) + \underline{\underline{\dot{\varepsilon}}}_i^p \quad (13)$$

where $\underline{\underline{\dot{\varepsilon}}}_k^p$ is given according to the Coulomb friction law which results in a tangential strain on the shear band and,

$$\underline{\underline{\dot{\sigma}}} = \mathbf{C}_a(\underline{\underline{\dot{\varepsilon}}} - \underline{\underline{\dot{\varepsilon}}}^p) \quad (14)$$

where $\mathbf{C}_a$ is the elasticity moduli tensor which takes into account the anisotropy direction introduced by the shear bands. The yield function corresponds to the Coulomb's criterion and is :

$$f_k = \mid \underline{\tau}_k \mid + \sigma_k^{'} \tan\varphi_k - c_k \quad (15)$$

where $\mid . \mid$ denotes the norm of the vector and φ_k and c_k are effective friction angle and cohesion which can decrease during deformation and reach their residual values φ_r and c_r. So that:

$$\tan\varphi_k = r_k^{\varphi} \ \tan\varphi \quad (16)$$

$$c_k = r_k^c \ c \quad (17)$$

where r_k^{φ} and r_k^c are intermediate hardening variables varying between unity and their residual values ($\tan\varphi_e = r_{lim}^{\varphi} \tan\varphi$) depending on the plastic strain on k plane given by:

$$r_k^c = \frac{1}{a^c \mid \underline{\gamma}_k^p \mid + 1} + r_{lim}^c \quad (18)$$

$$r_k^{\varphi} = \frac{1}{a^{\varphi} \mid \underline{\gamma}_k^p \mid + 1} + r_{lim}^{\varphi} \quad (19)$$

and a^c and a^{φ} are two parameters which control the rate of softening. An associated flow rule is considered for the plastic shear strain so:

$$\underline{\dot{\gamma}}_k^p = \dot{\lambda}_k \frac{\underline{\tau}_k}{\mid \underline{\tau}_k \mid} \quad (20)$$

The role played by the dilatancy on the normal strain is not considered at present. Now, a multipotential plasticity problem is to be solved for which the number of active mechanisms is variable.

5 NUMERICAL APPLICATIONS

To illustrate the performance of the methodology proposed a geometrically simple model incorporating an active soil failure due to the movement of a rigid wall is modeled. Then some excavation problems with variable failure mechanisms are studied.

5.1 *Application of the enhanced model*

This numerical test consists of a rectangular soil mass subjected to prescribed uniform horizontal displacements at its upper left boundary. The base and the right hand boundaries are fixed. 4 node elements with reduced integration are used. The material properties are $E = 100\ MPa$, $\nu = 0.3$, $\phi = 30^0$, $\psi = 5^0$, $\gamma = 18.9 kN/m^3$ and $c = 10\ kPa$. No property degradation is considered (constant C and φ). Rice's condition of localization predicts discontinuity surfaces making between 47^0 to 52^0 with the horizontal direction. In Figure 1, the contours of incremental distorsions are given. As it can be seen, the enhanced model results in sharper strain localization.

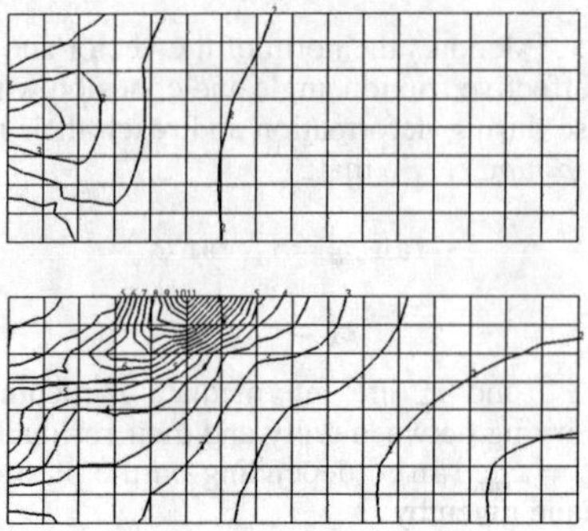

Figure 1: Contours of incremental distorsion computed using classical finit element (upper) and enhanced model (lower). Each contour level represents a variation of 0.6%.

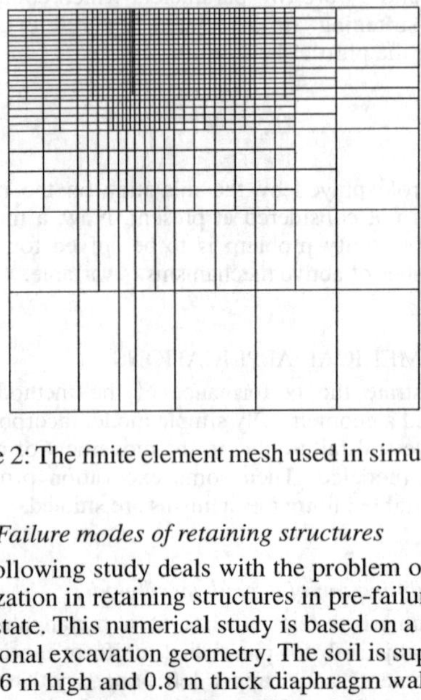

Figure 2: The finite element mesh used in simulations.

5.2 *Failure modes of retaining structures*

The following study deals with the problem of strain localization in retaining structures in pre-failure or in limit state. This numerical study is based on a two dimensional excavation geometry. The soil is supported by a 16 m high and 0.8 m thick diaphragm wall.

The two dimensional quadrilateral four and five node elements are used to mesh the soil. The two sides of the wall are meshed very finely using only the four node elements to have a good precision in these areas. At the bottom and at the rigth side of the mesh, for decreasing the number of nodes and elements a combination of four and five node elements is used.

The diaphragm wall is considered to be constructed in a massive of soil with 100 m high and 100 m length. The retaining structure is located 28 m far from the left side of the massive considered.

The experience of the authors has shown that the role of the model dimensions and the mesh type and mesh size are not negligible. In such problems, in order to exclude the non-realistic shear stresses in the soil, a mesh with a minimum height of 5h (h is the retaining structure's height) must be considered (Modaressi and Vossoughi 2000). The nodes at the lateral and lower boundaries of the mesh are fixed. The finite element mesh used in the analyses is shown in Figure 2.

Two configurations of the retaining structure are considered. The first one is a cantilever retaining structure. The second one is an externally braced retaining structure. The soil is assumed to behave in a fully drained manner with zero pore fluid pressures everywhere. Plane strain analyses are performed.

In order to simplify the model, only one type of material is considered for the massive of soil. A sandy granular material with a very small value of cohesion is assumed. The retaining structure is considered to be a diaphragm wall made by reinforced concrete. The material properties are the same as used in the simple example while the concrete properties are $E_c = 21000\ MPa$, $\nu_c = 0.2$.

The analyses assume a real chronology of excavation consedering the brace (or anchor) executation. The construction sequence is simplified to fit the numerical model. The simulation consists of the following steps: first the stresses are initialized, then the soil is excavated to 3 m depth. After that, the wall is braced. In this step, no excavation is done simultainsely with the bracing execution. Once the bracing in place, the excavation continues. The excavation is made by steps of 2 m. At each step one row of elements is excavated.

The initial stresses in both the soil and the beam have been estimated by the means of a primary computataion in which the only applied forces are their weights ($k_0 = 0.5$).

Full convergency is supposed to be achieved when the norm of unbalanced forces are less than 1% of the applied ones and the norm of incremental displacments is less than 0.1% of the norm of the displacement.

6 SIMULATION RESULTS

For both cases two types of results are given. Figures 3 and 5 present the directions of strain localization verified via Rice's condition. On the same figures, we have illustrated the sign of the second order work. Black-colored integration points are those in which the second order work is negative while the gray ones show that this value is still possitive. In each figure two excavation heights are presented. The upper case corresponds to the last step at which the above mentionned convergency criteria is verified. However, the computation has been continued and the lower cases illustrate results obtained for less sever convergency criteria on unbalanced forces (about 4%). Displacement convergency has always been reached. For the cantilever diaphragm wall the last step with full convergency corresponds to $10m$ of excavation while $12m$ of excavation shows the major instability of the soil in front and behind the wall. For the diaphragm wall braced externally at $3m$ far from the walls head, $14m$ and $16m$ of excavation heights are obtained. As it can be noted, at the last converged step the zone in which the condition of strain localization is reached does not present a continuous zone. This zone becomes continuous for the step with more unbalanced forces.

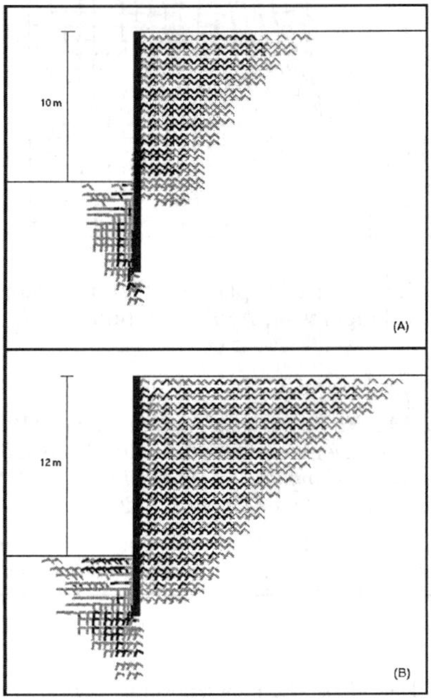

Figure 3: Strain localization directions and the second order work for the cantilever diaphragm wall; A) $H_{exc} = 10m$, B) $H_{exc} = 12m$.

Figures 4 and 6 show the increments of displacements for the same cases. As it can be noted, the last converged step does not present any sharp variation of the vectors which could correspond to a discontinuity creation. On the other hand, in the less converged cases the apparition of the failure is much more apparent.

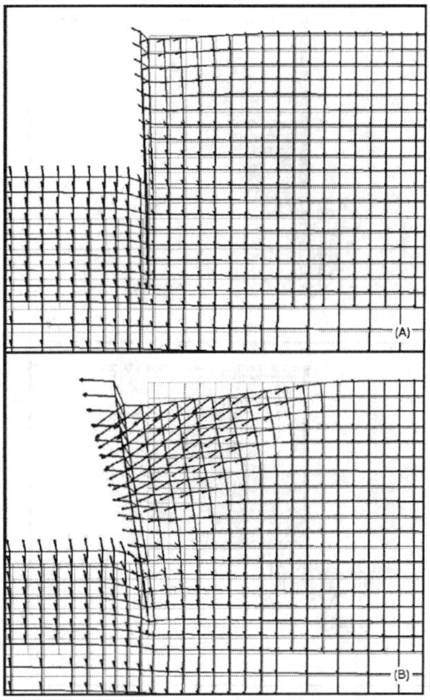

Figure 4: Increment of displacement for the cantilever diaphragm wall; A) $H_{exc} = 10m$, B) $H_{exc} = 12m$.

7 CONCLUSIONS

In this study the capacity of the finite element method to model the limit state of the structures using an enhanced method is presented. The bifurcation analysis and the onset of the localization via Rice's condition could be a precursor of a limit state of a structure. The concept of security and the different limit states for a structure are to be studied more deeply. The second order work too can be another precursor of instability of materials.

Other simulations done on different configurations of problem, the anchored retaining wall for example, satisfy the procedures developped in this study.

It can be noted that the strain localization condition is always reached before the instability condition. In addition this last condition covers a more limited area.

8 ACKNOWLEDGEMENT

The authors wish to thank the Federation Nationale des Travaux Publics (FNTP) for its financial aid.

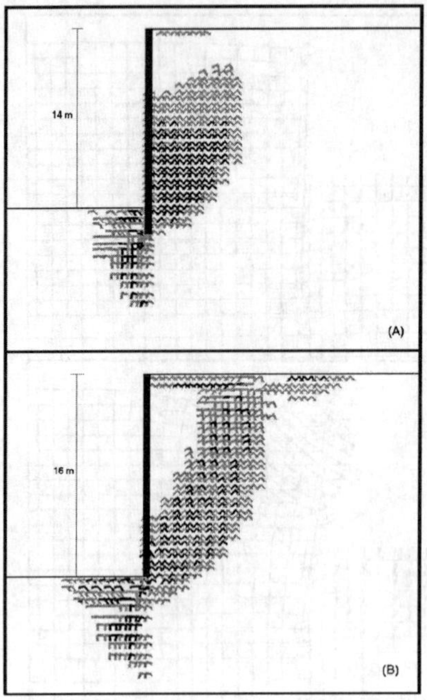

Figure 5: Strain localization directions and the second order work for the externally braced diaphragm wall; A) $H_{exc} = 14m$, B) $H_{exc} = 16m$.

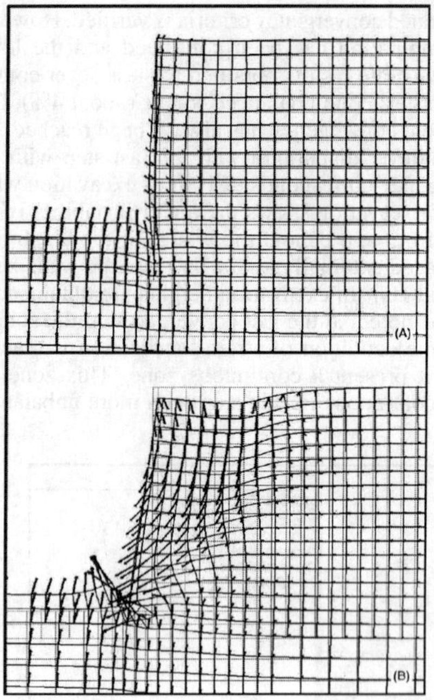

Figure 6: Increment of displacement for the externally braced diaphragm wall; A) $H_{exc} = 14m$, B) $H_{exc} = 16m$.

REFERENCES

Aubry, D. (1988). Transition from localized to discontinuous behaviour in porous dilatant materials. In J. Mazars and Z. Bazant (Eds.), *Proceedings of the France-US Workshop on Strain Localization and Size Effect due to Cracking and Damage (Cracking and Damage)*, Cachan, France, pp. 259–268. Elsivier Applied Science.

Darve, F. and F. Laouafa (1999). Plane strain instabilities in soil. appliction to slope stability. In *Proceedings of the Int. Symp. on Numerical Models in Geomechanics*, Graz, Austria.

Garikipati, K. (1996). *On strong discontinuities in inelastic solids and their numerical simulation*. Ph. D. thesis, Stanford University, Department of Aeronautics and Astronautics, USA.

Modaressi, A. and K. Vossoughi (2000). The use of nuimerical modeling in retaining wall design. In *Proceedings of the Third International Conference on Advances of Computer Methods in Geotechnical and Geoenvironmental Engineering (Geoecology and Computers)*, Moscow Russia, pp. 327–332. A.A. Balkema.

Ortiz, M., Y. Leroy, and A. Needleman (1991). A finite element method for localized failure analysis. *Computer Methods in Applied Mechanics and Engineering 7*, 1–27.

Regueiro, R. and R. Borja (1997). Continuum finite element analysis of strain localization in slopes. In Pietruszczak and Pande (Eds.), *Numerical Models in geomechanics*, pp. 213–218. Balkema, Rotterdam.

Rice, J. (1976). The localization of plastic deformation. *Theoretical and Applied Mechanics*, 207–220.

Simo, J., J. Olivier, and F. Armero (1993). An analysis of strong discontinuities induced by strain-softening in rate-independent inelastic solids. *Computational Mechanics 12*, 277–296.

Computer Methods and Advances in Geomechanics, Desai et al. (eds) © 2001 Balkema, Rotterdam, ISBN 90 5809 183 X

A new approach of modeling generation and progress of cracks in discontinuous rock masses

M. Nakagawa
CRC Research Institute Incorporated, Osaka, Japan

Y. Jiang
Faculty of Engineering, Nagasaki University, Japan

ABSTRACT: An approach is proposed for expressing the generation and progress mechanism of the new cracks due to shear and tension failure in the discontinuous rock masses, together with behavior of the existing discontinuities, by using distinct element method(DEM). First, the principles and procedure for modeling the generation and progress of new cracks, and the rational method to define the potential crack planes are presented. Next, excavation simulations of underground opening in the discontinuous rock masses are shown as applications to field problem. Based on above, validity of the proposed analytical approach for estimating the stability of rock structures is also discussed.

1 INTRODUCTION

Most of deformation and failure of rock slopes, foundations, underground openings in rock masses is governed by the characteristics of geometrical distribution of the existing discontinuities and the mechanical behavior in a single discontinuity. Furthermore, it is also thought that stability of above rock structures extremely depend on new cracks, which are generated and progress due to loading and/or excavation. A perfect analytical approach has been proposed for expressing the generation and progress mechanism of the new cracks due to shear and tension failure in the matrix, together with behavior of the existing discontinuities, by using discontinuous analysis(Nakagawa and Jiang 1999). Distinct element method(DEM) is used in the study. The validity of the formulation has also been shown in comparison with some closed form solusion. By the way, validity of the results by this approach are dependent on the way of setting of potential crack planes, which need to be set in the DEM model in advance. In this paper, validity of application of the proposed analytical approach to estimate stability of rock structures is discussed, by the excavation simulation of underground opening in the discontinuous rock masses, based on the potential crack planes defined with the rational way shown later.

2 MODELING GENERATION AND PROGRESS OF NEW CRACKS BY USING DEM

In DEM, rock joints are represented by boundaries of rock blocks, thus, each rock block is represented as intact rock surrounding with rock joints(Cundall 1971). The intact rock within the rock block is assumed elastic body in the proposed approach. Generation location and progress direction, distribution frequency of new cracks, which will be thought to generate potentially, must be defined in advance in this approach, so as to make use of boundaries of blocks as potential crack planes, as shown in Figure 1. The potential crack planes are hypothetical and show the continuous behavior with the surrounding intact rock by means of defining equivalent stiffness and strength to surrounding intact rock before satisfying the fail-

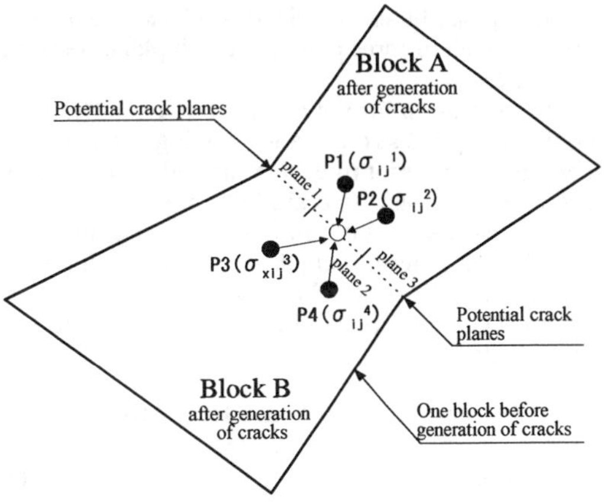

Figure 1. Definition of potential crack planes and estimation of principal stresses for failure criteria.

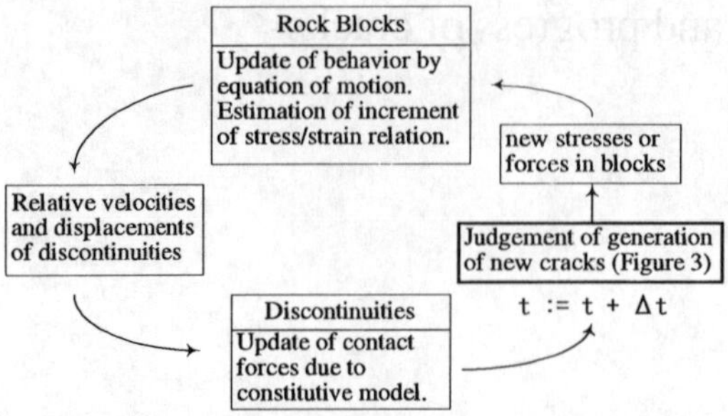

Figure 2. Calculation cycle for the distinct element method. (Δ t denotes infinitesimal increment of time for one cycle)

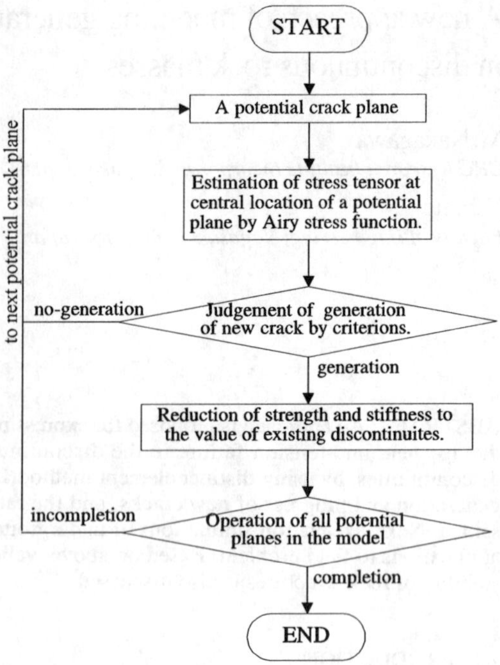

Figure 3. Process for judgement of generation of new cracks.

ure criteria. Physical new cracks are generated when the stresses acting on a potential crack plane reach at the limitation of shear strength or tension strength. Its properties are reduced automatically to the values of the existing discontinuities at the same time. For example in Figure 1, a lump of rock block which is consist of block A and B(bold line) is kept to be continuous before generation of cracks. After generation of cracks, discritized potential crack planes(three in this figure) become to discontinuous planes successively. A lump of rock block is divided by block A and block B just after all of three potential crack planes become new cracks.

Failure criteria for not a plane but a solid(matrix) should be applied as criteria for judging generation of new cracks. In order to define the failure modes for each potential crack, two kind of criteria are applied based on calculations of principal stresses at the location of each potential crack, as shown in Figure 1. Criterion functions for shear failure(f_s) and tension failure(f_t) can be expressed as equations (1) and (2), respectively.

$$f_s = (1.0 - \sin\phi)\sigma_1 - (1.0 + \sin\phi)\sigma_3 - 2.0\,\mathrm{c}\cos\phi \quad (1)$$

$$f_t = \sigma_t - \sigma_3 \quad (2)$$

where c = cohesion, ϕ = internal friction angle. σ_t = tension strength of intact rock. Principal stresses on a potential crack planes can be evaluated by using the Airy stress function (3), and components of stress are given by equation (4). The application of this function is adequate becuase each rock block is treated as elastic body. Coefficients a, b, c and d in equation (3) are defined according to the continuous condition of stresses around the potential crack planes.

$$\Phi = \frac{a}{6}X^3 + \frac{b}{2}X^2Y + \frac{c}{2}XY^2 + \frac{d}{6}Y^3 \quad (3)$$

$$\sigma_x = \frac{\partial^2\Phi}{\partial y^2} \qquad \tau_{xy} = -\frac{\partial^2\Phi}{\partial x\partial y} \qquad \sigma_y = \frac{\partial^2\Phi}{\partial y^2} \quad (4)$$

Maximum and minimum principal stresses can be calculated from the three stress components in equation (4).

Judgement of generation and progress of new cracks in DEM model are carried out based on two failure criteria for each calculation cycle as shown in Figure 2. The object of the judgement is only for the prior defined potential crack planes and the process of the judgement is finished when all of the potential cracks have been evaluated. This process is shown in Figure 3. The mechanical properties of the potential crack planes satisfied one of both failure conditions are decreased as ones of the existing joints.

New cracks are permited to behave as rock discontinuities (i.e. slip, separation). If failure criteria are satisfied successively in the nearby region to cracks generated previously by redistribution and concentration of stresses, those are thought to present progress of new cracks.

3 APPLICATION TO SIMULATION OF UNDERGROUND OPENING EXCAVATION

3.1 *Summary of simulation*

The excavation of a deep underground opening is used as an example of simulation by using the distinct element method with the proposed approach. The model is consist of exiting joints and intact rock shown in Figure 4. The top of the opening is 400m depths from the ground surface, height is 55m and width is 25m. The rock masses around this opening are composed of sand stone which is fresh and less weathering. Mechanical properties of intact rock are shown in Table 1. Coulomb slip model is adopted as mechanical model of discontinuities based on the result of laboratory shear test(Jiang and Nakagawa 1999) and the properties of them are shown in Table 2.

After the initial in-situ stress distribution is reproduced based on the coefficient of lateral pressure Ko=0.6, bench excavation of opening from arch is carried out without any supports.

Figure 5 presents the distribution of only the opened joints around the opening at the time when excavated the 5th bench. The intact rock is modeled as the elasto-plastic body in this case. The opening is thought to be judged stable because it can be found that no key blocks appear around the opening.

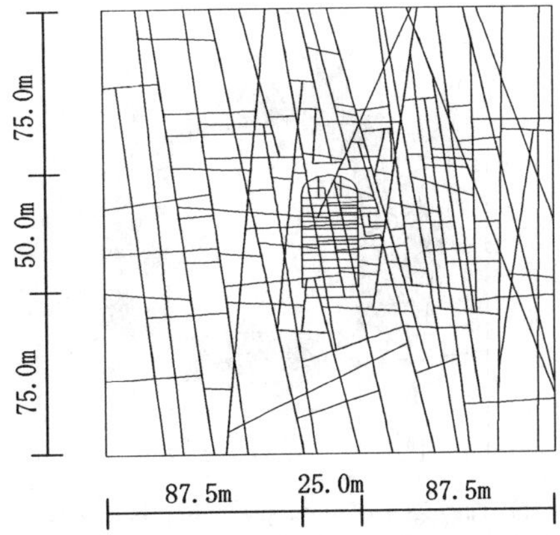

Figure 4. DEM model of the underground opening. (exiting joints and intact rock)

Table 1: Properties of intact rock around the opening.

Young's modulus(E)	30.0(GPa)
Poisson's ratio(ν)	0.17
Density(ρ)	2.40(g/cm^3)
Cohesion(c)	1.00(MPa)
Friction angle(ϕ)	40.0(°)
Tension strength(σ_t)	3.5(MPa)

Table 2: Properties of existing joints around the opening.

Normal stiffness (Kn)	1000.0(GPa/m)
Shear stiffness (Ks)	0.8(GPa/m)
Joint cohesion(c^j)	0.0(MPa)
Joint friction angle(ϕ^j)	30.0(°)
joint tension strength(σ_t^j)	0.0(MPa)

3.2 *Definition of potential crack planes*

The potential crack planes are defined by using both the analytical method and the rational judgments. First, the region of the setting of potential crack planes are intended to enclose the plastic region from the preparatory elasto-plastic calculations shown in Figure 6(a). Next, as for defining direction of potential crack planes, the five regions around the opening, that the direction of principal stress vectors shown in Figure 6 (b) may be regarded as almost the same, are assumed as shown in Figure 6(c).

Lateral walls of openings are thought that slip failure is potential in general, so the application of "slip angle"(45° + ϕ/2) to the direction of generation of new cracks is thought to be appropriate, the potential cracks are defined as 65° because of ϕ =40° . On the other hand, collapse due to tension failure around the crown of the opening is in general thought to be potential. Though the extent of development of tension stress is known to tend to depend on coefficient of lateral pressure, the direction of new cracks are judged to coincide with direction of principal axis shown in Figure 6(c) against tension failure. Thus the potential crack planes are set at this angle around the crown of the opening.

Based on above, the DEM model which is able to represent generation and progress of new cracks against the initial model (Figure 4) is created as shown in Figure 6(d).

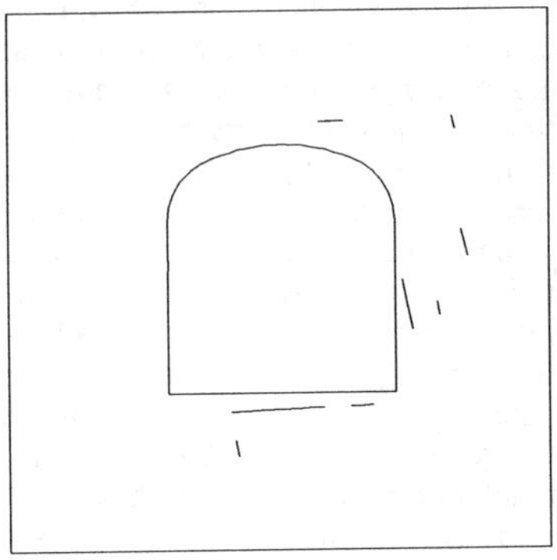

Figure 5. Opened joints around the opening. (intact rock are modeled as elast-plastic body)

3.3 *Result and discussion*

Excavation simulations are carried out step by step from the part of the arch downword to the bottom of the opening. The parameters of c , ϕ , σ_t shown in Table 1 are used as the parameters for the criteria of the equa-

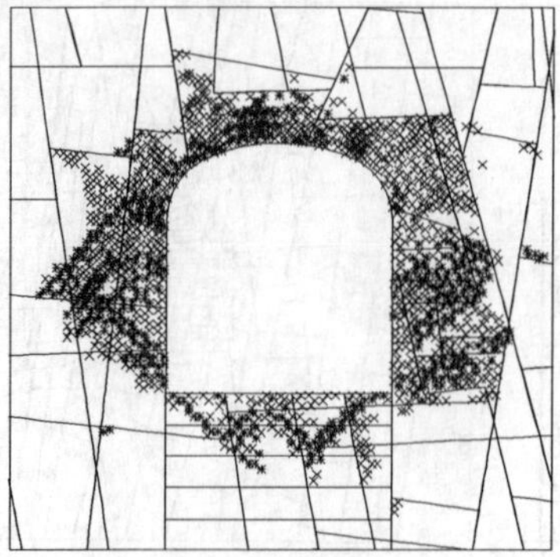

(a) The region of development of plastic failure.

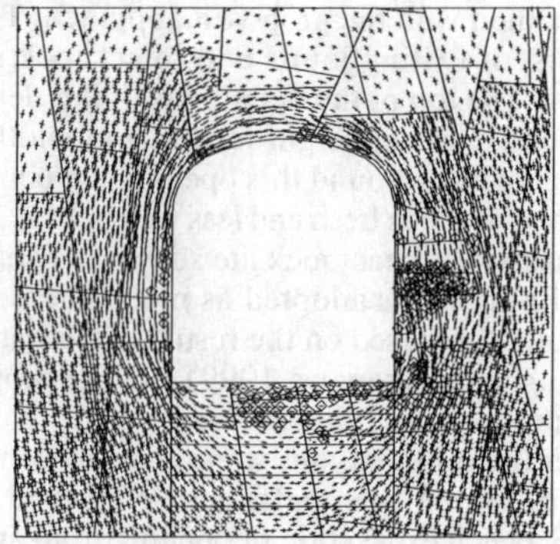

(b) The distribution of principal stress vectors. (◇ denots tention)

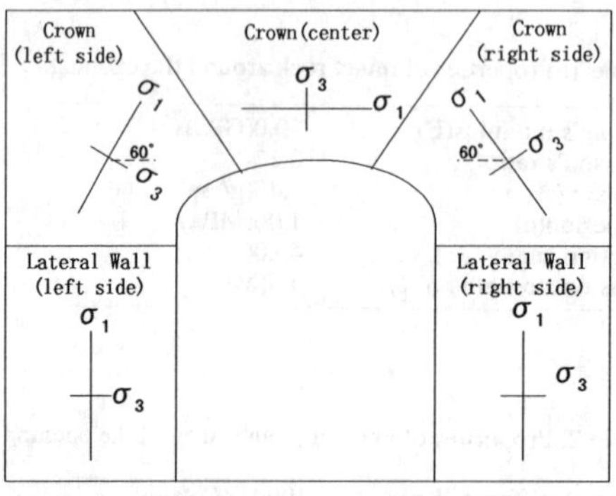

(c) The decision of the direction of potential crack planes based on principal axes.

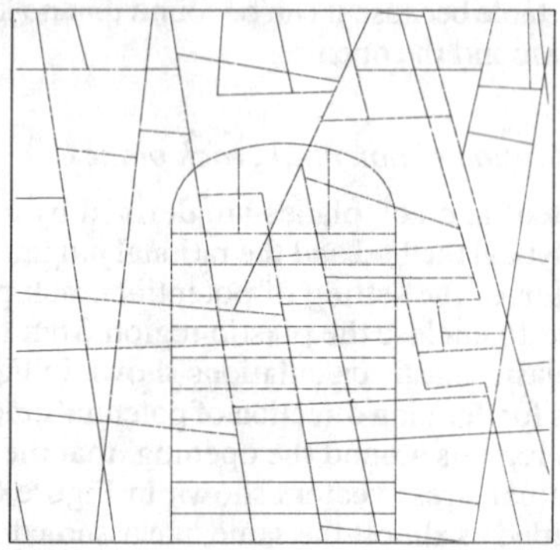

(d) The difinition of potential crack planes.

Figure 6. The process of definition of potential crack planes.

tions (1) and (2) to evaluate new cracks.

Figure 7(a) represents the process of generation and progress of new cracks at the each excavation step of the arch. Figure 7(b) represents all of the opened joints for both existing joints and generated new cracks complexly at the same excavation step as Figure 7(a). It can be seen that the new cracks due to shear failure develop and make progress toward the outside of the excavated part as the excavation steps are going on. Moreover, it can be seen that a bit of behavior as rock joints is necessary for new cracks to be opened by making the comparison of Figure 7(a) and Figure 7(b) . Because generation of key blocks are found at the right and left corners above the arch at the 3rd excavation step, it can be said that existence of dangerous regions are predicted.

Figure 8 shows the state of generation and development of new cracks when excavated the 5th bench. It is found that some rock masses formed key blocks appear and are falling down though the development of key blocks has not been predicted before excavation. It is thought to indicate the possibility for slip failure at the further excavation step due to the expansion of new cracks around the lateral wall. By the way, the comparison about displacements at the excavation periphery at the end of the 5th excavation step based on the difference of failure modes in the intact rock is shown in Table 3. One of the failure modes is the generation and progress of new cracks. The other is plastic failure based on the elast-plastic constitutive model. It is found that 3-5 times larger displacements are obtained in case of the former failure mode than those in case of the latter failure mode. It is thought that larger displacements in the former failure mode are caused by the generation of many new cracks around the periphery.

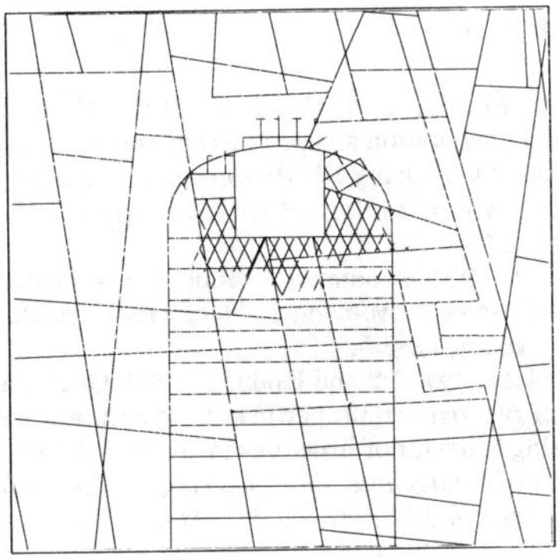

(Excavation step 1)

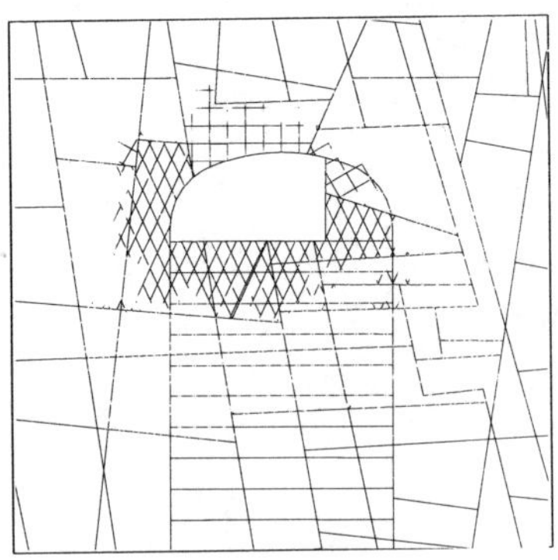

(Excavation step 2)

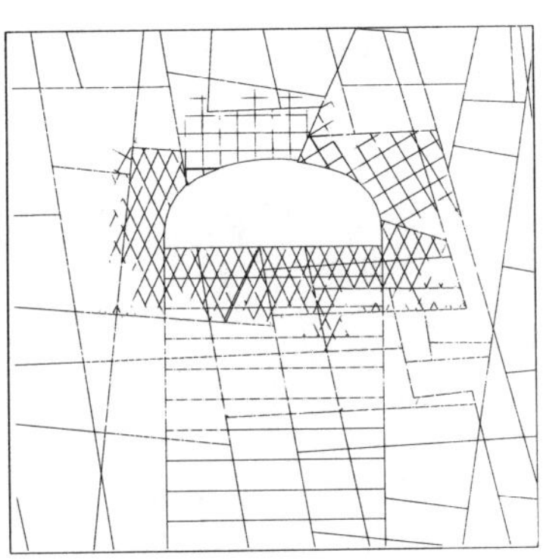

(Excavation step 3)

(a) Process of generation and progress of new cracks

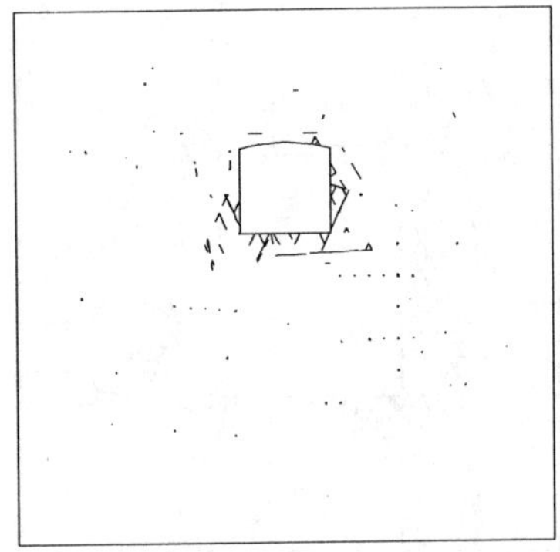

(Excavation step 1)

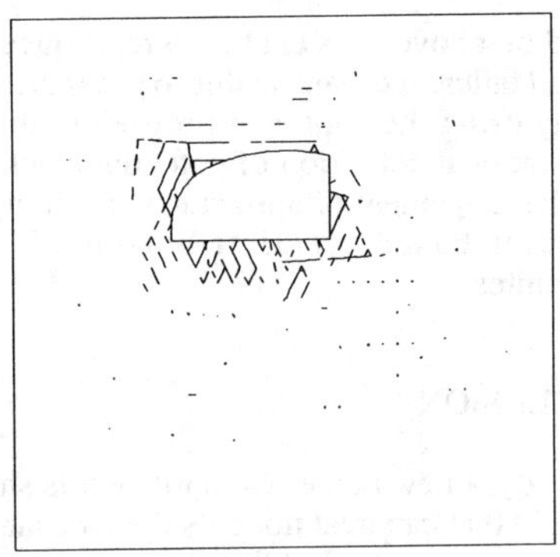

(Excavation step 2)

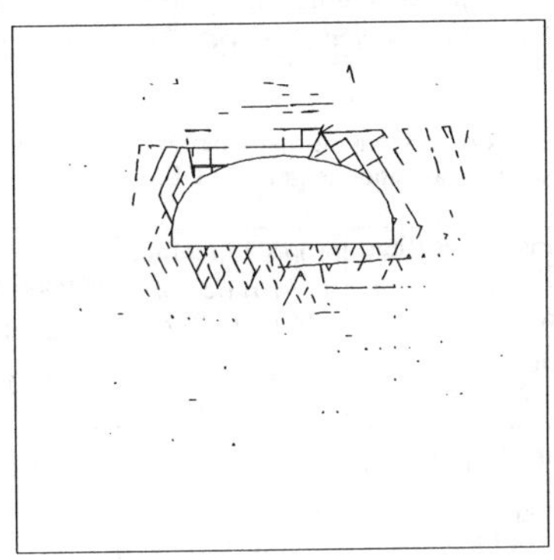

(Excavation step 3)

(b)Generation of opened joints.

Figure 7. The progress of failure of the rock masses due to arch excavation.

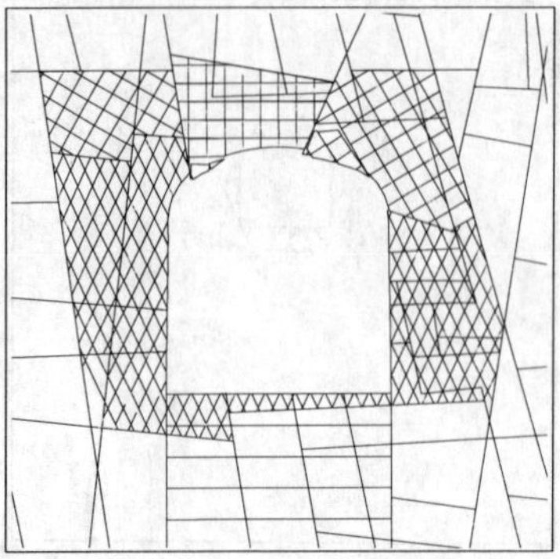

Figure 8. The state of generation and grogress of new cracks at the 5th excavation step.

Based on above, it is enable to reproduce the behavior and failure mechanism due to generation of new cracks by using the approach propsed in this paper with the rational definition of potential crack planes, even if the judgement of apparance of failure blocks is difficult based on distribution of natural discontinuites.

4 CONCLUSION

In this study, a new numerical approach is shown by using DEM that can treat not only the mechanical behavior of existing joints but that due to generation and progress of new cracks at the part of intact rock. Furthermore, this approach is applied to the excavation simulation of a underground opening in the discontinuous rock masses with the potential crack planes defined by the proposed method. As a result, it can be

Table 3 The comparisons of displacements obtained by using the proposed method and the general DEM.

analysis method / locations of measurement		proposed method (generation of new cracks)	general DEM
settlement of center of crown		-158.6	-33.1
horizontal displacements at the center of lateral wall	right side	-86.1	-28.1
	left side	86.1	17.3

unit(mm)

seen that behavior and failure mechanism caused by generation and progress of new cracks can be reproduced exactly. The presented numerical approach has the ability to represent the actual phenomenon more directly and is useful and applicable to rock engineering design.

REFERENCE

Nakagawa, M., Jiang, Y. Esaki, T. 1999. New approach of modeling generation and progress of cracks in discontinuous rock masses by using DEM: *Journal of Geotechnical Engineering, Japan Society of Civil Engineers, No.631/3-48, pp.397-410.*

Cundall, P.A. 1971. A computer model for simulating progressive large scale movements in blocky rock systems: *Proc.Int. Symp., Nancy, pp.2-8.*

Jiang, Y., Nakagawa, M. and Esaki, T. 1999. Quantitative evaluation of mechanical properties of the natural rock joints for analyzing behavior of structures in discontinuous rock masses: *Journal of Geotechnical Engineering, Japan Society of Civil Engineers, No.624/3-47, pp.231-243.*

Computer Methods and Advances in Geomechanics, Desai et al. (eds) © 2001 Balkema, Rotterdam, ISBN 90 5809 183 X

3D block analysis for fractured rock

Y.Ohnishi & Q.Yu
Geofront Laboratory, Department of Civil Engineering, Kyoto University, Japan

ABSTRACT: This study presents a method to identify 3D blocks enclosed by rock fractures and excavations. The fractures may be stochastic (simulated) or deterministic. The fractures are finite-sized, and are assumed to be disk-shaped. When simulating stochastic fractures an inverse method is used. The excavation faces may be convex or concave polygonal faces with or without interior holes. Volume element (voxel) propagation method is used in block identification. An actual example is presented to illustrate the application of this study.

1 INTRODUCTION

Generally, for most structures in hard rocks, their host rocks are of high-level stability from global perspective; potential instability of local-scale blocks formed by different scales of fractures and excavation faces are the major concerns. In order to counteract the threat from block instability, a great amount of rock blocks will have to be reinforced. When a project needs a wide scope of rock excavation, the researches on the number, location, shape, and dimensions of rock blocks become very important.

In most of the earliest works on rock block identification, fractures were assumed to be infinite, e.g., Goodman & Shi (1982), and Hoek et al. (1995). Some preliminary discussions on block identification that incorporate fracture geometries were presented by Lin et al. (1988), and Ikegawa & Hudson (1992), and Yu et al (1999). Works that predict 3D blocks for actual excavations have yet rarely been reported.

For this study, the block identification has just the same methodology for both stochastic and deterministic fractures. The only difference is that the geometric parameters (location, orientation and size) of deterministic fractures are input data; and the parameters of stochastic fractures are generated random numbers by Monte Carlo simulation.

2 SIMULATING STOCHASTIC FRACTURES

2.1 *Spatial location and population of fracture*

The most often used model to define the spatial location of stochastic fractures is Poisson model. In a Poisson model, each fracture is generated independently; spatial coordinates of its center are generated through a uniform distribution disregarding the locations of formerly generated fractures. How many fractures would be generated is controlled by only one parameter, a density parameter that specifies the average number of fracture centers within unit volume of rock mass.

USA Committee for Rock Mechanics (1996) suggested a simple relationship to determine the number of the fractures that would be generated within a specified volume of rock: $N_L = N_v A\cos\theta$, where N_L is fracture frequency along sample line (or borehole); A is mean fracture area; θ is the angle between the fracture poles and sample line. According to this relationship, when fractures are disc-shaped, it may be demonstrated,

$$d_3 = \frac{4d_1}{\pi E(D^2)} \tag{1}$$

where d_1 = one-dimensional fracture density, i.e. the number of fractures encountered by unit length of borehole that is drilled along the mean direction of fracture unit normal; d_3 = 3D fracture density, i.e. the number of fracture centers within unit volume of rock; $E(D^2)$ is the mean value of the squared fracture diameter.

If the volume of rock mass of modeled region is V, the 3D density of fracture set i is d_3^i, the number of generated fractures within the modeled region would be $N = V\sum d_3^i$. After the generation of simulated fracture network with Equation (1), one may check the fracture density of the simulated network once again. It should be not unusual that the simulated fracture network leads to a density d_l that differs from the measured one (i.e. the one which has been used in determining the number of generated fractures when simulating the fracture network). This may be ascribed to the fact that: (a) Equation (1) assumes that fractures are disc-shaped, but actual fractures are not the case; and d_l is an in situ parameter, and (b), in the equation, only the mean of areas and orientations are respected; 3D density is not only controlled by the mean of area and orientation, but also by the distribution patterns of them.

To overcome this difficulty, we add a correcting coefficient C_i, to Equation (1). That is,

$$N = \sum \frac{4VC_i d_l^i}{\pi E(D_i^2)} \tag{2}$$

Coefficient C_i is one of the calibrated parameters. It will be determined through inverse method.

To determine fracture population, an outcrop or an excavation may provide more accurate information than a borehole or a scan line. When the fractures of a rock mass is mapped on some excavation surfaces (or outcrop) and some boreholes which are not sub-parallel to the excavation, we recommend the following equation,

$$N = \sum C_i \cdot d_2 \cdot V \cdot f_l \tag{3}$$

where C_i is a correcting parameter, d_2 is the number of fractures centers within unit area of excavation, and f_l is the fracture frequency along the boreholes.

Naturally, either Equation (2) or (3) only provides an initial estimation of fracture numbers. The accurate number of fractures in a given volume of rock mass will be determined through the inverse approach.

2.2 *Fracture shape and size*

To build a 3D model, geometric shape and spatial extent of fractures must be addressed. This paper assumes that a fracture is disk-shaped.

Under in situ condition, the size 3D distribution of fractures cannot be observed directly. Based on the assumption that fractures are disc-shaped, the probability distribution of fracture diameter may be inferred from the distribution of trace length, which can be measured through excavation surfaces or natural outcrops. But the direct estimation of trace length are often affected by varied biases, and the distribution inferred from trace length involves an integral that can only be evaluated numerically, and this also becomes a difficulty when building a statistical model with Monte Carlo simulation. To avoid these objections, this study first assumes that fracture diameter and trace length have the same distribution forms (e.g., lognormal or exponential distribution), then determines the mean and standard deviation of fracture diameters through model calibration.

2.3 *Simulating process of fracture network*

The process to build a statistical model of fracture network includes the following major steps:

1 Construct initial variable generator for each geometric parameter.

2 Generate fracture network with Monte Carlo method.

3 Sample the simulated fractures in the same way as that the in situ actual fractures are sampled; investigate the trace length and density of sampled fractures.

4 Compare the simulated fractures with the measured with regarding to trace length, and 1D & 2D density.

5 If the simulated and measured fractures do not make a satisfying match, improve the mean and the standard deviation of fracture diameter.

6 Repeat Step 2 through Step 5 until the simulated and measured fractures make a satisfying match

Through above steps, the final variable generator for each of the geometric parameters is determined which will be used to generate stochastic fractures for block analysis.

3 3D-BLOCK IDENTIFICATION

When the size and shape of individual fractures are incorporated, it is impossible to construct a general analytical method for block identification. In the following sections, first unconnected fractures will be removed from the fracture network, and then each connected fracture will be divided into independent patches enclosed by its intersections with other fractures. In the process of block identification a fracture patch will be taken as an elemental unit.

3.1 *Elimination of unconnected fractures*

In a fracture network, not all fractures (which may be deterministic or generated by stochastic simulation) are connected; some fractures do not intersect other fractures, and some fractures intersect too few fractures. Before block identification, unconnected fractures should be identified and be eliminated.

A fracture is referred to as connected only when it plays a part in block formation (i.e., it forms a surface of some block). A connected fracture must feature the 2 following characteristics: (1) it is connected to at least 3 fractures (including excavation faces), (2) within the planar disk of the considered fracture, the intersections between the considered fracture and other connected fractures (or excavation faces) must form at least one connected loop.

By checking every fracture against the two characteristics mentioned above, all unconnected fractures can be found out and eliminated. It should be noted that this is an iteration procedure. Some fractures which look like connected may be found out to be unconnected after some unconnected fractures are eliminated.

3.2 *Identification of fracture patches*

A connected fracture is divided into patches by the intersections on it. It is a fracture patch and not a whole fracture that forms a surface of block. Therefore, after the elimination of unconnected fractures and the calculation of intersections among fractures, the patches of each fracture will be identified. For instance, the fracture shown in Figure 1a includes 3 patches A1, A2, and A3, as it is shown in Figure 1b.

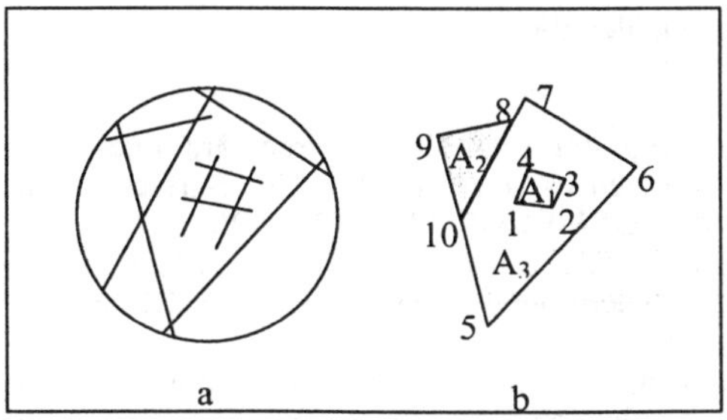

Figure 1. Patches of a connected fracture

One individual patch may be represented by its boundary nodes arranged in clockwise direction or anti-clockwise direction. For example, patch A1 may be represent by a node series 1-> 2-> 3-> 4.

A big patch may include some little patches; the included patches are called the son-patches of the big patch. The patch including son-patches is represented by its boundary nodes and the boundary nodes of its son-patches. For instance, patch A3 includes one son-patches (A1), may be represented with three node series: 5-> 6-> 7-> 8-> 10, and 4-> 3-> 2->1. Consequently, a fracture is represented by a number of patches; a patch is represented by the nodes on its boundaries and the nodes on the boundaries of its son-patches; and a node is described by three-dimensional coordinates.

3.3 *Block identification*

The algorithm of block identification may be called "voxel (volume element) propagation". It is similar to the method of spatial-occupancy enumeration which is often used when dealing with the problem of computer graphic, and in which a solid composed of known faces is decomposed into cells arranged in a fixed, regular grid.

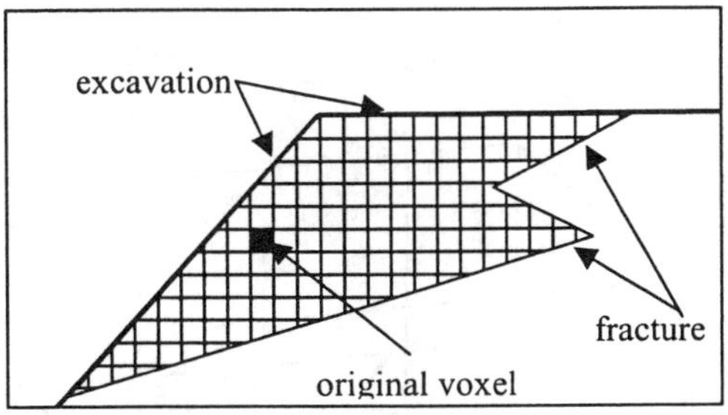

Figure 2. Voxel propagation process

The basic procedures of voxel propagation method are shown in Figure 2. First an original voxel is positioned at some location, and then let the voxel propagate. A voxel is assumed to be cube-shaped. In the propagation process, each voxel will continue to propagate until all the 6 sides of it is bounded with neighboring voxels or fracture patches. All the fracture patches encountered during the propagation procedure compose a block.

By judiciously positioning a number of original voxels, all the blocks composing rock mass of modeled domain can be identified. But, for the problems such as rock block stability in rock excavation, it is commonly the case that only the blocks exceeding a certain dimension and near excavation surface are of high engineering importance hence are interested.

To make sure that a propagation process identifies an interested block, the original voxel is positioned with the prompt of interactive graphics. The position of an original voxel is first determined in two-dimensional space by picking some point with mouse on the 2D graphic of a given excavation face. Then, the picked point is transformed from two-dimensional space into three-dimensional space, and is made a little distance of shift along the direction perpendicular to the excavation face to make sure that the point be within studied rock mass.

3.4 *Example of application*

An artificial slope within the dam site area of the

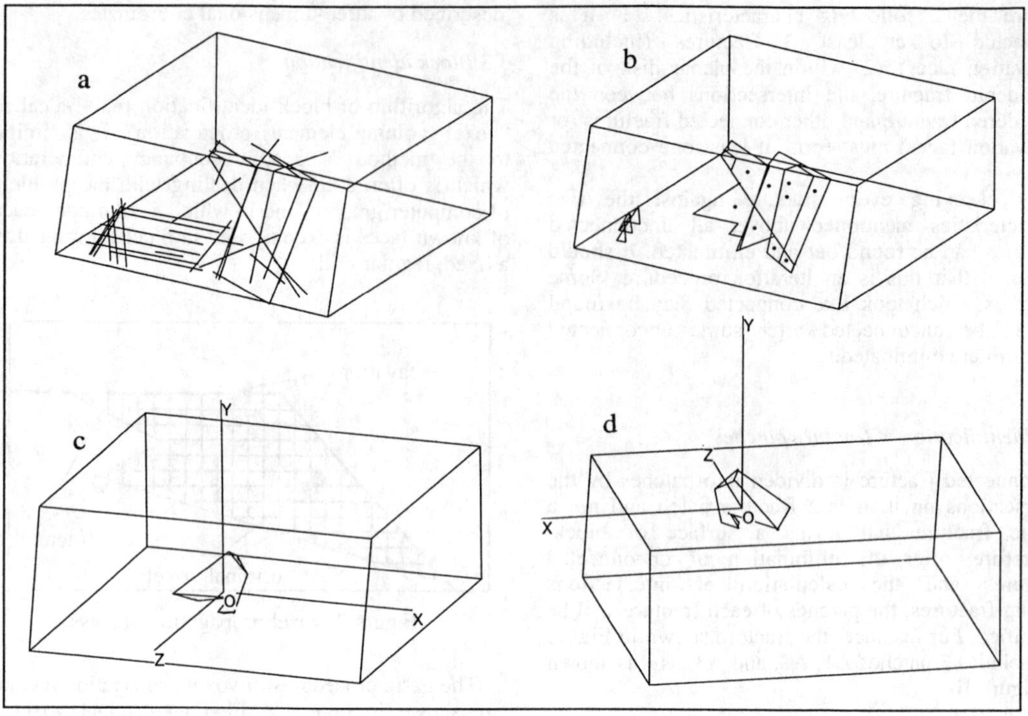

Figure 3. An example of block identification

Three Gorges Project was mapped in detail; Part of the mapped fractures is shown in Figure 3a. The slope is about 25m high, with a dip direction of 255^0, and a dip angle of 75^0. In the following block identification, the slope is supposed to be cut to 14m high. Some results during block identification are shown in Figure 4b-4d. Figure 4b shows the 3D intersections among the block-faces and excavations. Figure 4c & Figure 4d show one of the identified blocks when observing from different directions.

4 CONCLUSION

The 3D bloc method presented in this study can identify the rock block formed by deterministic or stochastic fractures; the geometric parameters of fractures, including fracture locations, orientations and sizes may be taken into account. This is confirmed through an actual example.

REFERENCES

Goodman, R.E. & G.H. Shi 1982. Geology and rock slope stability – application of the key block concept for rock slopes, Proc. 3rd International Conferences on Stability in Surface Mining (AIME/SME) 347-373. NEW YORK.

Hoek, E., P.K. Kaiser & W.F. Bawden 1995. Support of Underground Excavation in Hard Rock. A.A. Balkema Publishers.

Ikegawa, Y. & J.A. Hudson 1992. A novel automatic identification system for three-dimensional multi-block systems. Engineering Computing. 9:169-179.

Lin, D., C. Fairhurst & A.M. Starfield 1988. Geometrical identification of three-dimensional rock block system using topological techniques. Int. J. Rock Mech. Sci. Min & Geomech. Abstr. 25:139-147.

National Research Council 1996. Rock Fracture and Fluid Flow: Contemporary Understanding and Applications, National Academy Press, Washington, D. C.

Yu, Q., Y. Ohnishi, G. Xue & D. Chen 1999. An approach to the identification of three-dimensional rock block. Proceedings of International Symposium on Imaging Applications in Geology:261-264. Liege, Belgium.

Computer Methods and Advances in Geomechanics, Desai et al. (eds) © 2001 Balkema, Rotterdam, ISBN 90 5809 183 X

Modeling particle damage in discrete element simulations

M.E. Plesha, T.B. Edil & P.J. Bosscher
University of Wisconsin, Madison, Wis., USA

R.P. Jensen
Sandia National Laboratory, Albuquerque, N.Mex., USA

N. Ben Kahla
Ecole Nationale D'Ingénieurs De Gabès, Tunisia

ABSTRACT: Using the discrete element method (DEM) with clustering, a novel means of numerically modeling damage of particles is presented. Damage is treated by allowing clusters to break apart according to a failure criterion based upon sliding work. Simulations show that a very distinct shear zone is usually evident when particle damage was considered and the degree of damage was shown to be related to the angularity of the clusters.

1 INTRODUCTION

Soils are highly discontinuous media and are well-suited to being modeled by the discrete element method (DEM). Cundall and Strack (1979) pioneered the application of DEM to soils. A long recognized problem however has been the limited ability of simple-shape DEM elements, such as disks in two-dimensions and spheres in three-dimensions, to adequately represent the roughness of natural particles. This paper describes an enhancement to the two-dimensional discrete element method called *clustering*, including the ability for clusters to break apart, thus allowing for modeling of particle damage such as grain crushing and abrasion. More detailed results are given in Jensen, Bosscher, et al. (1999), Jensen, Edil, et al. (2000) and Jensen, Plesha, et al. (2000).

2 CLUSTERING

The basic idea of clustering is very simple. Several simple shape DEM elements are combined to form one particle of complex shape, called a *cluster*. This allows the geometry of a natural particle to be more closely represented, while simultaneously retaining the numerical efficiency associated with the simple shape elements. Shown in Figure 1 are several new two-dimensional cluster shapes. Clustering can be easily implemented in a DEM code using the existing contact detection and contact force computation algorithms that are present, except that intra-cluster contacts are forced to be linear elastic. Hence, a cluster is a semi-rigid collection of elements which, for practical purposes translates and rotates as a rigid body. In creating these new cluster types, we require each DEM element in a cluster to have at least two contacts so that the intracluster element contacts would not require bending stiffnesses in order to be semi-rigid.

3 PARTICLE DAMAGE

To implement damage within DEM, we choose one or more criteria that specify when one or more individual DEM elements of a cluster disassociate, and we call this process *declustering*. When one of these criteria is met for a DEM element in a cluster, the cluster is broken (partially or fully) by changing the links in the computer program that identify the DEM element as being a member of a particular cluster. Since all of the DEM elements are recognized by

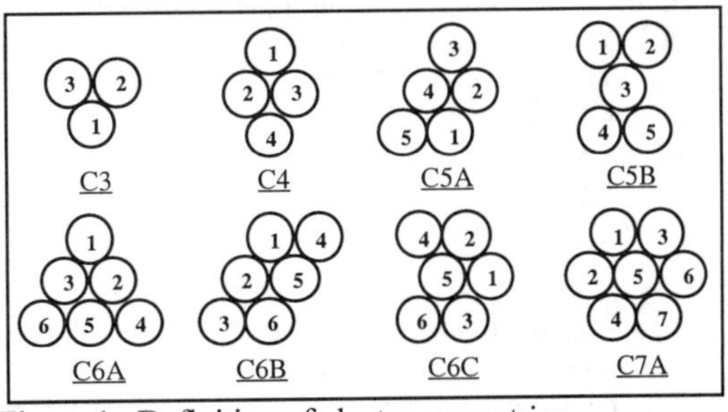

Figure 1. Definition of cluster geometries.

the program as separate entities having independent degrees of freedom, the DEM elements that at one time comprise a cluster at some subsequent time may model the fragments of a cluster.

There appears to be two fundamental mechanisms by which particles can experience damage. The first is abrasion, or wear, such as occurs when one particle frictionally slides past another and is gradually abraded in the process. The second is damage due to overstressing in which a crack propagates through a particle, breaking it into two or more smaller particles. Because shear zone behavior is always accompanied by considerable localized frictional sliding, in this paper we emphasize abrasion damage. Alternatively, damage due to fracture may be more appropriate for modeling a comminution process or a problem involving large compression loading and small relative displacements such as occur in wave propagation problems.

A useful measure of the energy dissipated during frictional sliding of one particle against another is the sliding work (or plastic work), W. Because frictional forces always oppose the direction of relative sliding, the sliding work is nonnegative and is a nondecreasing function of time. We use an energy density criterion to determine if enough work has been accumulated by a particular DEM element of a cluster to warrant its separation (breakage) from the cluster. This requires a user-specified critical energy density, W_0. For an element i of a cluster, the product of W_0 with the DEM element's volume V_i gives the sliding energy that the element can absorb, W_i^{max}, before breakage occurs

$$W_i^{max} = W_0 V_i \tag{1}$$

Such a criterion allows particles, or clusters, of different sizes to be used in a model of a medium. The greater the value of W_0, the more resistant particles are to damage due to abrasion. Furthermore, a larger DEM element in a cluster requires more work, or energy, for breakage to occur, than for a smaller DEM element.

During the course of computation, the increment of sliding work done on a particular DEM element i is easily computed at each time step by summing the work increments for all *extra-cluster* contacts for the element

$$dW_i = \sum_{\substack{\text{number} \\ \text{extra-cluster} \\ \text{contacts}}} f_t \, d\delta_t^s \tag{2}$$

where f_t is the tangential force at the contact, $d\delta_t^s$ is the increment of relative tangential plastic (sliding) displacement between the DEM element and a contacting neighbor, and the summation excludes all intra-cluster contacts. The total work is obtained by integrating (summing) the work increments

$$W_i = \int dW_i \tag{3}$$

As long as $W_i < W_i^{max}$ the cluster remains intact. If $W_i \geq W_i^{max}$ then the DEM element separates from the cluster. The exact manner in which this separation occurs, and potential impact on other DEM elements within the cluster, depends on the geometry of the cluster, position of the separating DEM particle within the cluster, and previous particle separations for the cluster.

To develop scenarios for DEM element separations from a cluster, we begin by requiring every element within a cluster to have at least two intra-cluster contacts. Hence, the simplest cluster in our scheme consists of three DEM elements, as shown by cluster C3 in Figure 1. The merit of this requirement is that a semi-rigid arrangement of DEM elements can always be constructed using only normal and tangential contact springs.

As clusters degrade, we require each DEM element remaining in the cluster to also maintain at least two intra-cluster contacts. For example, in the three particle cluster C3 shown in Figure 1, when the first DEM element separates, the remaining two elements would each have only one intra-cluster contact and hence would not be semi-rigid. Under these circumstances we allow these DEM elements to also immediately separate in which case the cluster is fully destroyed leaving three DEM elements, each with independent motion. For clusters comprised of more than three DEM elements, there are several different scenarios in which individual elements can break off and each of these scenarios must be anticipated and accounted for. Figure 1 shows several different cluster types that we have implemented, and a local particle numbering system for each cluster. Table 1 lists the schemes and order in which the DEM particles could break off of a cluster. The table is to be used in conjunction with Figure 1 and is read as follows. The first column identifies the cluster type. The second column identifies the local number of the DEM particle that is breaking off the cluster. The next set of columns, entitled "Declustering Scheme", details the possible cluster configurations that remain after a particular particle has broken off. For example, if particle number 1 of a C6A cluster were to break off, the result would be one C5A cluster and one single particle. The C5A cluster that was left over now obeys the declustering scenarios listed in that column of Table 1 where we note that its constituent DEM particles will likely have accumulated some sliding energy from earlier deformations.

Table 1. Declustering schemes for modeling particle damage with clusters.

Cluster Type	Particle Breaking Off	Declustering Scheme					
		Single	C3	C4	C5A	C5B	C6C
C3	1 or 2 or 3	3					
C4	1 or 4	1	1				
	2 or 3	4					
C5A	1 or 2	2	1				
	3 or 5	1		1			
	4	5					
C5B	1 or 2 or 4 or 5	2	1				
	3	5					
C6A	1 or 4 or 6	1			1		
	2 or 3 or 5	3	1				
C6B	1 or 6	2		1			
	3 or 4	1			1		
	2 or 5	3	1				
C6C	4 or 6	1			1		
	2 or 3	2		1			
	1	1				1	
	5	6					
C7A	1 or 2 or 3 or 4 or 6 or 7	1					1
	5	7					

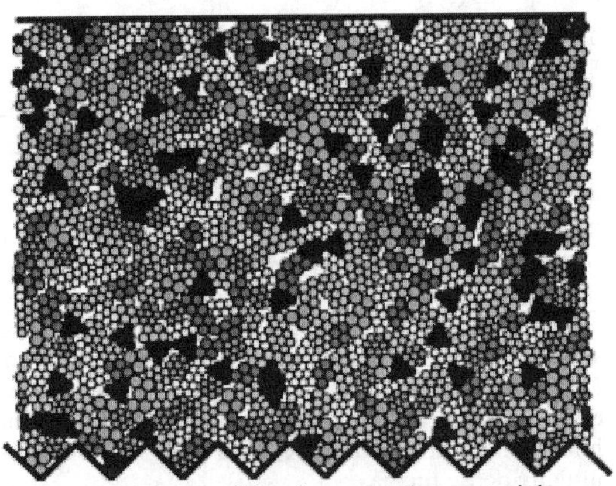

Figure 2. DEM model of a granular material employing 500 clusters of various size and shape. Vertical boundaries at left and right are periodic boundaries.

4 SIMULATIONS

4.1 *Description of model*

Numerical experiments were conducted to demonstrate the utility of declustering for modeling particle damage, and to study the effects of particle damage on structure-media interfaces. These simulations were performed in two-dimensions using the discretization shown in Figure 2. The domain consists of a rough horizontal structure surface at the bottom, a horizontal boundary at the top, and two vertical periodic boundaries. The medium consists of 500 clusters as follows: 50 three particle clusters, and 75 clusters each of the C4 four particle clusters, C5A five particle clusters , the C6A six particle cluster, the C6B six particle cluster, the C6C six particle cluster, and the C7A seven particle cluster. The clusters are generated with initially random position, and are then consolidated by applying a compressive force to the upper horizontal boundary. The width of the domain is 15.0 mm and the approximate height after consolidation is also 15 mm. Once the consolidation is complete, the upper boundary is subjected to either a prescribed compressive load (for a constant normal stress test), or a prescribed constant displacement (for a constant volume test). For the constant normal stress tests, the upper horizontal boundary was subjected to a normal force of 33.3 N which corresponds to a 35 kPa normal stress. As described in Jensen et al. (1999), the periodic boundaries allow DEM particles that pass out of one boundary to be automatically reintroduced at the opposite boundary, thus effectively modeling a domain with infinite horizontal extent. Because of the periodic boundaries, the domain can be viewed as representing a two-dimensional vertical section of a ring shear device.

The coefficient of friction for contact between two DEM elements or a DEM element and a surface is 0.4, similar to the coefficient of friction for quartz particle-quartz particle contact. The normal and shear stiffnesses for particle-particle and particle-structure contacts are 1.2 x 10^6 N/m. These stiffnesses were computed using Hertz theory for contact between two spherical bodies with the modulus of elasticity for quartz. For the simulations that include damage, the critical energy density is taken as $W_0=10^7$ J/m^3. As described in Jensen, Plesha, et al. (2000), this value was determined by comparing numerically computed breakage factors (a measure of particle damage) with experimentally determined values reported by Hoteit (1990) and Zeghal (1993).

The shearing surface (i.e., structure surface) has sawtooth-shaped roughness with ±45°angles of inclination with respect to the horizontal. To represent far-field effects, the DEM particles in contact with the upper horizontal surface are "glued" to that surface by restricting them to have no rotation, no horizontal displacement, and the same vertical displacement as the boundary. Such treatment is used to help insure that if a shear zone forms, it will not be immediately along this boundary. For all analysis cases, the rough structure surface was given a uniform slow shear velocity up to a total shear displacement of 6.3 mm.

The simulations include an initially "loose" medium and an initially "dense" medium. The "loose" medium models were created using the consolidation procedure described above. By taking the loose medium models, and temporarily setting the interparticle coefficient of friction to zero, additional consolidation is achieved leading to the "dense" medium models. After the dense medium models were fully compacted, the interparticle coefficient of friction was then set back to 0.4 for the subsequent simulation.

4.2 *Results*

Shown in Figure 3 are montages of graphs for both initially loose and initially dense media, with no damage and with damage. Presented are shear force versus tangential displacement, normal displacement (dilation) versus tangential displacement, average displacement of DEM elements within a horizontal layer versus the vertical position of the horizontal layer, and the average rotation of the DEM elements within a horizontal layer versus the vertical position of the layer. Each domain was divided into ten horizontal layers of equal depth. The average DEM element displacement within a horizontal layer is taken to be the mean displacement of all elements within the layer. Similarly, the average rotation of the DEM elements in a horizontal layer is taken as the mean rotation of the elements within the layer.

The shear stress versus shear displacement results for the simulations with no damage show greater post-peak strength than the simulations with damage. Furthermore, once peak strength is achieved, both the initially loose and initially dense media show similar response.

With no damage, both the initially loose and initially dense media dilate, with the initially dense media showing greater dilation, as expected. When damage is included, the initially dense medium shows early dilation that reaches a steady state value, while the initially loose medium shows early dilation, followed by compaction. These results are in good qualitative agreement with response of natural materials.

The remaining plots show average particle displacement and average particle rotation through the depth of the sample, and are useful for identifying the formation and location of shear zones. The

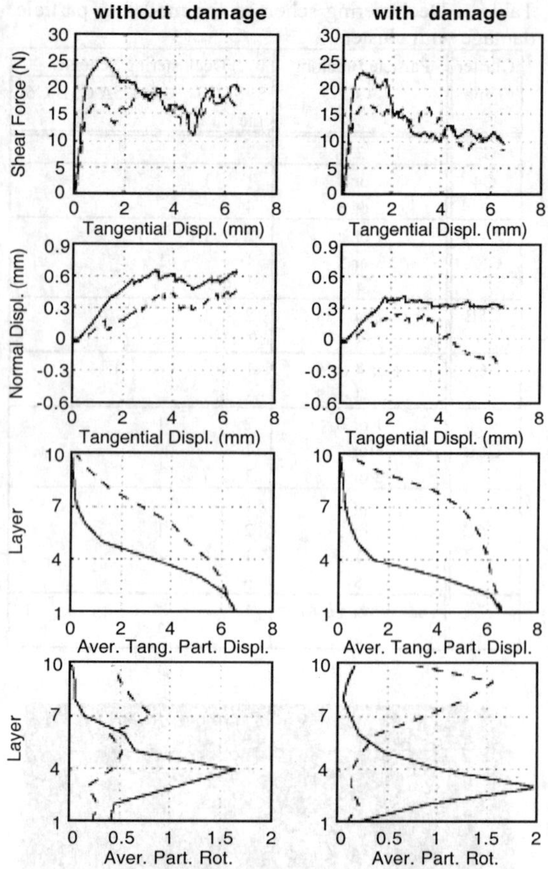

Figure 3. Constant compressive stress DEM simulations. Graphs of shear force at shearing surface vs. tangential displacement of shearing surface, dilation of top surface vs. tangential displacement of shearing surface, horizontal layer vs. average displacement of DEM particles in layer, and horizontal layer vs. average rotation (radians). Dashed lines represent the initially loose media and the solid lines represent the initially dense media.

initially dense media simulations, with damage and without, both show a very distinct shear zone. These are evident from both the particle displacement versus depth plots and the particle rotation versus depth plots. The initially loose simulation with no damage does not show a distinct shear zone, although there is a mild localization of particle rotations that is perhaps evidence of a process zone forming. For the initially loose simulation with damage, a shear zone is much more distinct, and forms near the upper surface of the domain, as opposed to the lower structural surface. The studies reported in Jensen, Plesha, et al. (2000) show that the most significant particle damage (declustering) occurs in regions of shear localization. Thus, as shear deformation progresses, particles tend to loose roughness and the shear zone evolves into a region that has lower shear resistance than the remainder of the medium, which may explain why the shear stress versus shear displacement plots for the simulations with damage show a significant decrease of strength as the shear displacement grows.

5 CONCLUSIONS

An enhancement to the discrete element method that effectively models particle damage in particulate media has been presented. The enhancement is based on *clustering* whereby particles of complex shape are modeled by combining several DEM particles of simple shape into a semi-rigid assemblage. The individual DEM particles that constitute a cluster are allowed to *decluster*, or break off according to a sliding energy criteria and thereby model grain crushing. Numerical "experiments" simulating ring shear tests were performed to demonstrate the importance of damage and the effectiveness of declustering for modeling grain crushing.

ACKNOWLEDGEMENTS

This research was performed with financial support from National Science Foundation Grant CMS-9302281, U.S. Air Force Office of Scientific Research Grant MIPR-93-0031, and AASERT Grant F49620-94-1-0325.

REFERENCES

Cundall, P.A., and O.D.L. Strack 1979. A discrete numerical model for granular assemblies', *Geotechnique*, **29**, 47-65.

Hoteit, N. 1990. *Contribution a l'Etude du Comportment d'Interface Sable Inclusion at Application au Frottement Apparent*, Doctorate Thesis, Institut National Polytechnique de Grenoble, France.

Jensen, R.P., P.J. Bosscher, M.E. Plesha, and T.B. Edil 2000. DEM simulation of granular media - structure interface: effects of surface roughness and particle shape," *Int. j. numer. anal. methods geomech.*, **23**, 531-547.

Jensen, R.P., T.B. Edil, P.J. Bosscher, M.E. Plesha, and N. Ben Kahla 2000. Effects of particle shape on interface behavior of DEM-simulated granular materials," submitted for publication.

Jensen, R.P., M.E. Plesha, T.B. Edil, P.J. Bosscher, , and N. Ben Kahla 2000. DEM Simulation of Particle Damage in Granular Media - Structure Interfaces "Effects of particle shape on interface behavior of DEM-simulated granular materials, submitted for publication.

Zeghal, M. 1993. *Modeling of Sand-Structure Interfaces*, Ph.D. Thesis, University of Wisconsin-Madison, Department of Civil and Environmental Engineering.

Computer Methods and Advances in Geomechanics, Desai et al. (eds) © 2001 Balkema, Rotterdam, ISBN 90 5809 183 X

Fracturing and failure in rocksalt

S. Saeb, T. H. Taylor & C.T. Francke
RockSol Consulting Group Incorporated, Boulder, Colo., USA

R.C. Carrasco
Westinghouse Government Environmental Services, Carlsbad, N.Mex., USA

ABSTRACT: The creep and viscoplastic behavior of rocksalt has been studied in the literature. However, the fracturing and failure of rocksalt have not received sufficient attention. Rocksalt subjected to low strain rates demonstrates viscoplastic behavior. Under higher strain rates rocksalt shows brittle behavior. In the latter case, fracturing develops at the grain scale level and propagates as visible fractures on the surface of the excavations. The results of many years of observation at the underground excavations at WIPP will be used to study the fracturing and failure of rocksalt. Numerical models will be used to study the post-peak behavior of rocksalt. It will be shown that the rocksalt behavior at the material and structural level is strongly related to the frictional characteristics between the grains.

1 INTRODUCTION

All geomaterials including concrete and rock undergo fracturing and, ultimately, failure under increasing load and displacement. Although it must be considered in the design of any load-bearing structure, fracturing and failure is a very complicated process to define. To develop a fracturing or failure criteria, the behavior of the rock has to be observed, conceptual models of failure need to be developed, and numerical models simulating the conceptual model need to be verified. Viscoplastic deformation of rocksalt has been studied in depth and various models are used to successfully simulate and predict this behavior. Power laws are the most popular form of the constitutive models to simulate the continuous viscoplastic deformation of salt (Krieg, 1984). However the fracturing and failure of rocksalt has not received sufficient attention. Because salt is a top candidate for geologic nuclear waste isolation around the world, the development of fracture criteria is of great importance.

The inelastic behavior of rocksalt is inherently complex and is a function of its mechanical history as well as the current state of stress and strain and the loading rate. Therefore, the empirical models based on curve fitting expressions of experimental results obtained under simple loading conditions are of limited use.

The work presented here was completed for the United States Department of Energy (DOE), which was authorized by Public Law 96-164 to provide a research and development facility for demonstrating the safe permanent disposal of transuranic (TRU) wastes from national defense activities and programs of the United States exempted from regulations by the U.S. Nuclear Regulatory Commission (NRC). The Waste Isolation Pilot Plant (WIPP), located in southeastern New Mexico near Carlsbad, was constructed to determine the efficacy of an underground repository for disposal of TRU wastes

2 FRACTURING IN ROCKSALT

Carter (1992) tested potash-salt rock from the Saskatoon area of Saskachewan. The uniaxial compressive strength of this rock was 24 megapascals (Mpa) with an average elastic modulus of 17 gigapascals (Gpa) and Poisson's ratio of 0.3, and the average tensile strength between one and two MPa. Carter showed that the crack initiation, volumetric strain reversal and strength of rocksalt is loading rate dependent (Figure 1). At high loading rates, potash behaves in a brittle manner. At the strain rate of 25 microstrain per second ($\mu\varepsilon / s$), the material shows dilatancy after reaching the yield point indicating a brittle behavior. However at the strain rate of 0.035 $\mu\varepsilon / s$ the uniaxial stress did not rise above 16 MPa. After three months, the specimen was 20 percent shorter and had numerous axial cracks. There is no point of volumetric strain reversal for this time period indicating the material behaves fully ductile.

Sluys (1992) studied the softening and localization in geomaterials including rock. These materials exhibit softening behavior as they undergo increasing

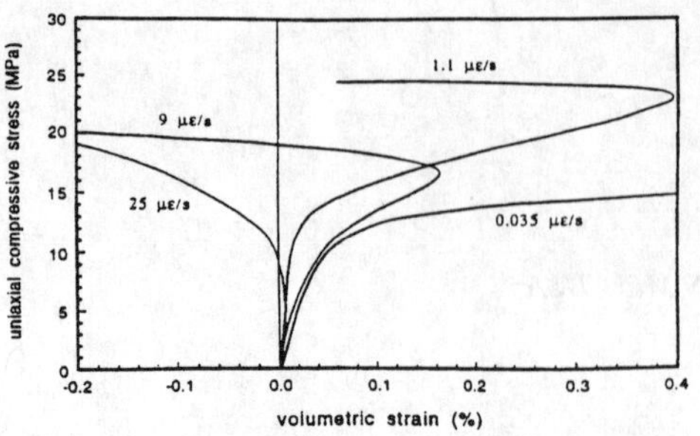

Figure 1. Carter results showing load rate dependency of rocksalt.

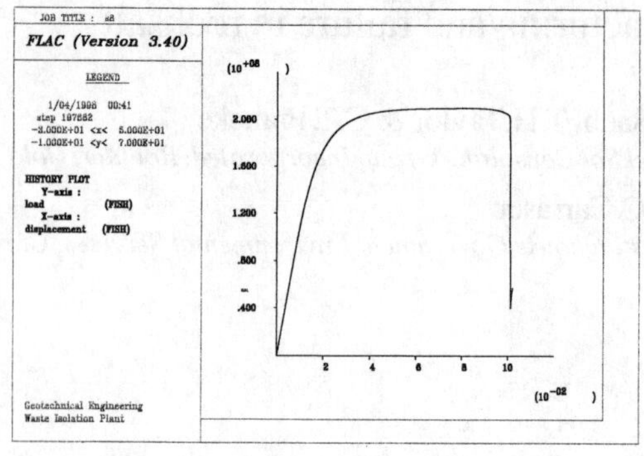

Figure 4. Results for viscoplastic sample S8.

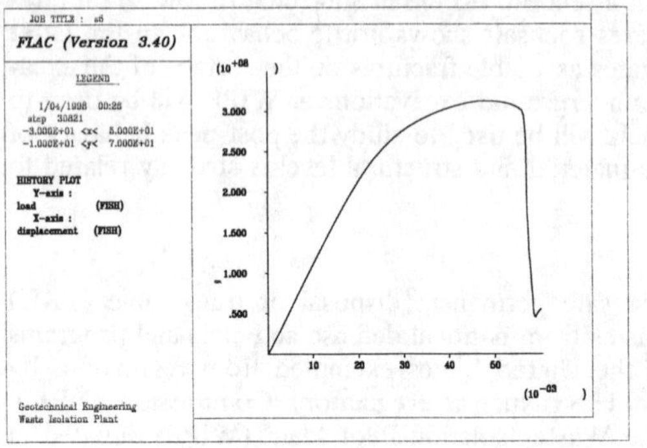

Figure 2. Results for viscoplastic sample S5.

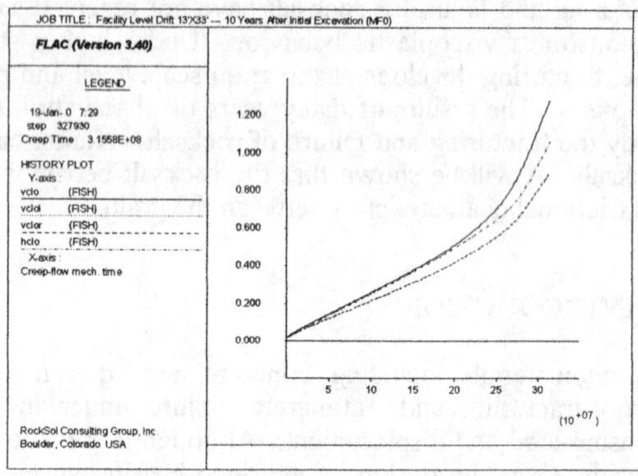

Figure 5. FLAC results showing accelerated displacement caused by softening.

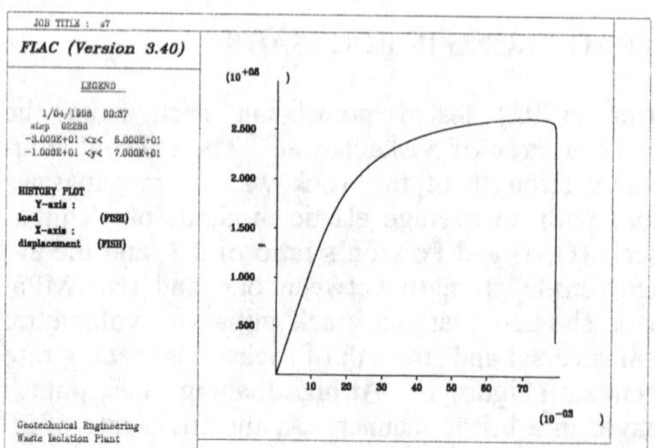

Figure 3. Results for viscoplastic sample S7.

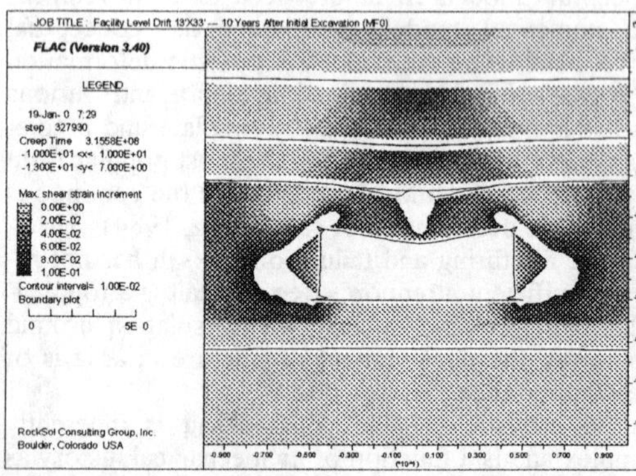

Figure 6. FLAC results showing strain concentration around the excavation. White = greater than 10%.

deformation. This behavior can be best described as the reduction of the load-carrying capacity accompanied by increasing localized deformation after reaching the limit load. In a continuum description, the load displacement characteristics exhibit a descending branch in the form of a negative slope on the stress-strain diagram.

The softening phenomenon can be observed under deformation controlled loading. First the load increases approximately linearly with the displacement up to about 60% of the peak. The cracks develop between the interface of the aggregates and mortar and the displacement starts to increase more than proportionally with respect to the load. A steep drop in the load occurs with increasing displacement and a descending branch in the diagram appears. Softening is usually a precursor to failure and all deformation localizes in small fracture zones.

Sluys observed that localization regions at a macro-scale have a finite size. The width of the localization zone w was determined experimentally for concrete as $w = 2.7d_a$ in which d_a is the maximum aggregate size, and for sand $w = 10d_g - 20d_g$ in which d_g is the mean grain size diameter. The scale of the localization may differ by some order of magnitude. In metals, the localization bands can be as small as one millimeter and in rock outcrops can be as large as several meters. The consumption of energy in the softening zone is finite also. Furthermore, in wave propagation problems a partial reflection and a partial transmission of waves on softening or localization zones can be observed.

Under tension, the softening can be demonstrated by experimental results. Under compressive loading, softening may occur at the end of the plastic range and can be considered as a precursor to failure. Whether softening is a true material property that can be used in a stress-strain relation is still under debate. However, in the numerical modeling, the descending branch in the stress-deformation curve is attributed to the material and has been incorporated in the constitutive equation.

3 VISCOPLASTIC MODELING

The phenomenological strain-softening models include crack models, based on a total or a decomposed strain concept, plasticity models with yield limit degradation, and continuous damage models, based on the degradation of the stiffness. FLAC is used to model the behavior of rocksalt at WIPP. FLAC uses a version of the finite difference method to solve the equilibrium equation. From those mentioned above, the only model that could be implemented in FLAC was a viscoplastic model in the form of combination of a creep and a plasticity model. No special treatment was used to resolve the strain-softening deficiencies such as mesh dependency, in FLAC. FLAC uses the WIPP reference creep law for the creep model which is a power law (Krieg 1984).

Itasca (1996) implemented the strain-softening plasticity model in FLAC. The Drucker-Prager model was most compatible with the WIPP reference creep law because both models are formulated in terms of the second invariant of the deviatoric stress tensor. Viewed in the octahedral plane, both models exhibit responses that depend only on radial distance from the isotropic-stress locus. In this model it is assumed that the associated strain rates act in series.

At the first trial, a linear softening model was used. When the loading rate is fast enough strain localizes in two series of symmetrical patterns at the top and the bottom of the sample. When the sample was subjected to a lateral strain controlled loading, the localization pattern is very similar to that observed in the roof beam of the underground excavations. This showed that localization is a good tool to predict the fracturing around underground openings.

One deficiency of this model was that it used a linear model for the softening branch. Experimental results of the geomaterials show that the softening branch is a concave curve. Furthermore the results of the further studies showed that at slower loading rates although the peak more stretched, the peak strength is not reduced with increasing rate. This is not compatible with the results reported by Carter, among others. Experimental results indicate that the peak strength is reduced with the reduction of the strain rate. Therefore one expect that the softening should be a function of the plastic strain as well as the viscoplastic strain. For these reasons a softening coefficient in the following form was selected to reduce the strength of salt:

$$e^{-(\beta.\varepsilon_p + \gamma \cdot \varepsilon_{vp})/\alpha}$$

where α, β, and γ are material parameters to be calibrated to the experimental results. At this point experiments showing the post peak behavior of the rocksalt were not available. Therefore an initial attempt was made to select these values to demonstrate the model properties.

Cases S5, S7, and S8 simulate constant strain unixial compressive tests. The tests use material parameters $\beta = 0.5$, $\gamma = 2.0$. and $\alpha = 2.0\text{e-}3$. β and γ can also be considered as weighting functions for the plastic and viscoplastic strains. Cases S5, S7 and S8 are the same except that the upper and lower platen displacement rates are as follows: 1.0e-7, 1.0e-8, and 3.0e-9 m/s respectively. The results of these test cases are shown in Figure 2 through 4.

As can be seen by slowing down the rate of loading the samples demonstrate more ductile behavior as was expected. Also by reducing the loading rate, the value of the peak force is reduced. This is due to the fact that viscoplastic effect is accounted in defining softening. At the slower loading rates the sample has

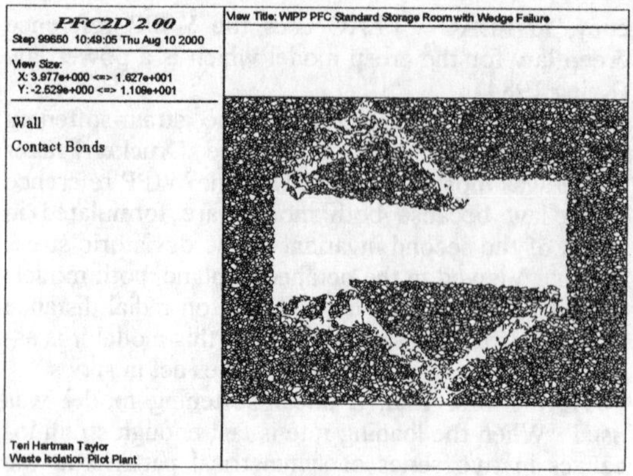

Figure 7. PFC results showing failure bands around the excavation.

enough time to deform viscoplasitcally. The viscoplastic deformation is more spread out through the sample. In case S8, the loading rate is very low therefore the sample undergoes a long period of continued deformation under almost a constant stress. The sample reaches failure after going through a long period of secondary creep and the failure represent the tertiary creep.

These softening parameters were applied to a FLAC model of a WIPP waste disposal room mined in rocksalt formations nearly 650 meters beneath the ground surface. The model was allowed to creep for ten years after the room was excavated. Figure 5 shows the vertical (vclo, vclol, vclor) and horizontal (hclo) convergence data from the model. This data shows accelerating convergence when softening occurs. Figure 6 shows shear strain contours around the room. Localized bands of high shear strain develop near the corners of the room and on a diagonal line above the room. The shear localization above the room creates a wedge shaped slab which is often seen in the field at WIPP. This demonstrates that a strain-based failure criterion for rocksalt could explain the fracture patterns seen in the field.

4 PFC MODELING

Particle Flow Code (PFC) uses a series of cylindrical elements to model the rock mass. These particles are rigid themselves and are connected through contacts which are assigned shear and normal bond and stiffness between the cylinders. The contact deformation as well as sliding and opening causes the mass to deform. Since the structure modeled by PFC is comprised of discontinuous elements, fracturing is relatively easier to model. Fracturing can be interpreted as the broken bond between the particles. The broken bonds could be normal or shear bonds or both.

PFC is a valuable tool for modeling and predicting the fracturing and failure of rock mass. PFC is more suited to modeling in grain scale. Therefore it becomes computationally prohibitive for larger models. The viscoplastic and creep behavior also has to be modeled to accurately predict mechanical behavior of rocksalt.

PFC was used to model a standard WIPP disposal room mined in rocksalt formations nearly 650 meters beneath the ground surface. Due to limitations in computer speed and memory size, only the first few meters around the four meter tall by ten meter wide rooms were included in the model. Boundary conditions that mimic the displacement rates seen in the field were applied to the model. Figure 7 shows the fracturing pattern around the excavation predicted by PFC. These results closely match fracture patterns seen in the field. Further laboratory and numerical experiments are required to refine the PFC rocksalt fracture model.

5 SUMMARY AND CONCLUSIONS

Fracturing and failure in rocksalt was studied through reviewing the experimental results reported in literature and applying this to numerical modeling. The general trend and behavior of fractures in rock-salt-potash samples was studied. Available models and methods to model the fracturing were reviewed. A viscoplastic softening model was proposed and tested and the results were examined qualitatively. Due to the lack of suitable experimental results, the material parameters were chosen based on experience to demonstrate the general performance of the model. Different loading rates and the effect of loading rates on the strength of the salt was considered. The new constitutive law can model the reduction of strength by reducing the loading rate and also the tertiary creep. Also the capabilities of the PFC code in modeling fracturing was examined and the results showed the capability of the model in simulating the fracturing of rocksalt.

6 REFERENCES

Carter, B.J. 1992. *Physical and Numerical Modeling of Fracture in Rock.* Ph.D. Dissertation. Winnipeg, Manitoba, Canada: University of Manitoba.

Itasca Consulting Group, Inc. 1996. *The Addition of New Material Models to FLAC, FLAC3D, PFC2D, and PFC3D, Combining Creep, Elasticity and Plasticity.* Minneapolis, Minnesota: Itasca Consulting Group, Inc.

Krieg, R.D. 1984. *Reference-Stratigraphy and Rock Properties for the Waste Isolation Pilot Plant Project.* SAND83-1908. Albuquerque, New Mexico: Sandia National Laboratories.

Sluys, L. J. 1992. *Wave Propagation, Localization and Dispersion in Softening Solids.* Ph.D. Dissertation. Delft, South Holland, Netherlands: Delft Technical University.

Computer Methods and Advances in Geomechanics, Desai et al. (eds) © 2001 Balkema, Rotterdam, ISBN 90 5809 183 X

Size effect with consideration of stress gradients

Chengxue She
Wuhan University of Hydraulic and Electric Engineering, People's Republic of China

W.E. Banford
University of Melbourne, Vic., Australia

J.R. Enever
CSIRO Petroleum, Australia

ABSTRACT: The size effect with the consideration of stress gradients on the fracture strength of the rock around a borehole is studied. On the basis of Bazant's size effect theory, the fracture strength of the rock when crack occurs on the borehole wall is derived first with the influence of the stress gradients being considered. Then the fracture strength predicted by the proposed formula and that by the Bazant's are compared with a set of model test data showing that the proposed formula is more coincident with the model test results. And at last some conclusions are drawn.

1. INTRODUCTION

In petroleum industry, the stability of wellbores is very important. Model tests with small scaled boreholes such as hollow cylinders are used to analyze the stability of the wellbore hole under various actions. However, when results of these tests are analyzed, the size effect is apparently exhibited for specimens of similar geometry but different size.

Many researchers have worked on the size effect of rock e.g., Bazant et al (1984,1990), Kooijman,et al(1991), Sulum et al(1987). For the geometry considered here the results of tests by Ringstad(1993) show that under the water pressure on the outer wall of Red wildmoor hollow cylinder sample, a moderate but significant size effect was observed. The size effect model proposed by Bazant in 1984 lay the foundation for the analysis of size effect. It has successfully explained the size effect in strength of quasi-brittle materials. Gettu et al (1998) reviewed the size effect model and examined its range of applicability through comparisons with data from tests on notched and unnotched beams of normal and high strength concrete. These works have led to a framework for simulating the failure of concrete, rock and other quasi-brittle materials.

But in some cases, the stress gradients have an important role in the size effect for hollow cylinder tests(Ringstad et al,1993). The study using the Cosserat Continuum Model by Sulem et al(1987) shows the effect of stress gradients on the fracture strength.

In this paper, by using the Banzant's size effect model, the fracture strength formula suitable for the hollow cylinder model tests are proposed with the stress gradients being considered. The fracture strength predicted by the proposed formula and that by the Bazant's are compared with a set of model test data showing that the proposed formula is more coincident with the model test results.

2 FRACTURE STRENGTH WITH SIZE EFFECT

Fig.1 is the schematic cross-section of the hollow cylinder sample. R1 and R2 are the radius of the hole and the outer wall respectively. In the hole, the hole

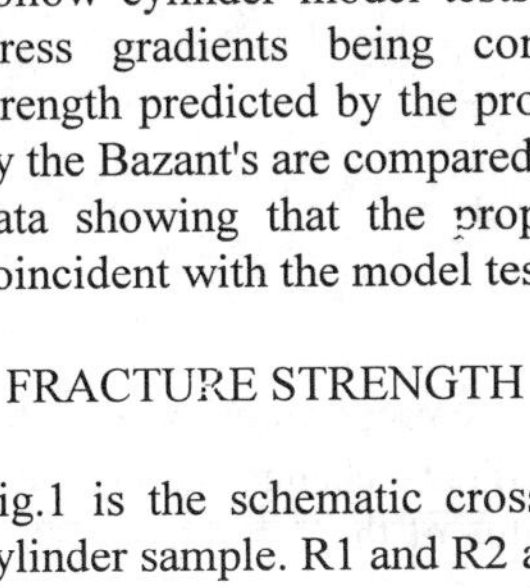

Fig.1 Schematic cross-section of hollow cylinder sample

wall is subjected to water pressure p. By increasing p to a certain value, the hollow cylinder cracks to failure. The p corresponding to the initiation of crack on hole wall is the peak value of water pressure. By calculating the circumferential stress at pressure p, the fracture strength of rock is obtained. Keeping the ratio of R1/R2 a constant, which means the samples are geometrically similar, changing R_1 and R_2 and doing the same test, different fracture strength is then got.

If the rock is assumed to be a kind of linear elastic material, the tangential and radial stresses in the hollow cylinder can be expressed as

$$\begin{cases} \sigma\theta = \dfrac{p\ e^2}{1-e^2}\ (1+\dfrac{R_2^2}{r^2}) \\ \sigma_r = \dfrac{p\ e^2}{1-e^2}\ (1-\dfrac{R_2^2}{r^2}) \end{cases} \tag{1}$$

where σ_θ is tangential stress and σ_r radial stress. $e=R_1/R_2$, p is the water pressure on the hole wall and r is the distance from hole centre to a certain point in hole wall(see Fig.1).

By derivation of σ_θ and σ_r with respect to r, the first order of derivatives of σ_θ and σ_r are

$$\begin{cases} \dfrac{\partial \sigma\theta}{\partial r} = \dfrac{2pe^2R_2^2}{1-e^2}\ \dfrac{1}{r^3} \\ \dfrac{\partial \sigma_r}{\partial r} = \dfrac{2pe^2R_2^2}{1-e^2}\ \dfrac{1}{r^3} \end{cases} \tag{2}$$

in which the derivatives are proportional to r^{-3}, this means that the stress gradients change sharply with r. On the hole wall the value of the derivatives can be written as

$$\begin{cases} \left.\dfrac{\partial \sigma\theta}{\partial r}\right|_{r=R_1} = -\dfrac{c_{11}}{R_1}p \\ \left.\dfrac{\partial \sigma_r}{\partial r}\right|_{r=R_1} = \dfrac{c_{11}}{R_1}p \end{cases} \tag{3}$$

in which $c_{11}=2/(1-e^2)$. For fixed R_1/R_2, the smaller the R_1, i.e. the smaller the sample size, the bigger the stress gradients.

According to Bazant(1984), in rock and other quasi-brittle materials, fracture is preceded by a gradual dispersed micro cracking that occurs within a relatively large fracture process zone ahead the tip of a continuous crack, the width ω_c of the crack band front can be treated as a material property.

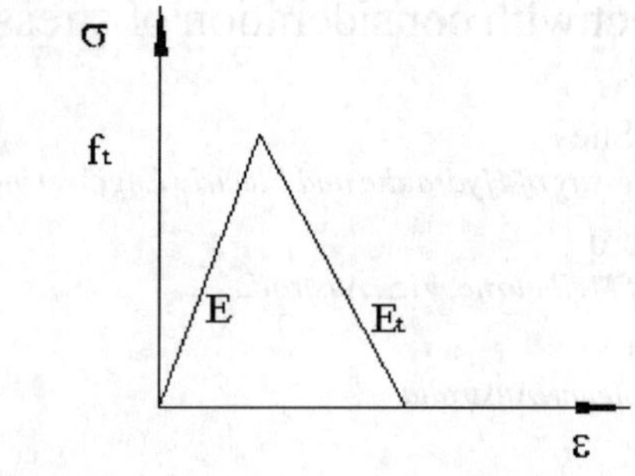

Fig. 2 Relationship between tensile stress σ and strain ε

$\omega_c = nd_g$, where d_g is the grain size in rock and n=5 for rock material. For uniaxial situation, the stress strain relationship is considered to be a bilinear stress--strain diagram shown in Fig.2, characterized by elastic modulus E, tensile strength f_t, and strain softening modulus E_t. The energy consumed per unit advance of the crack band, called the fracture energy can be expressed as

$$G_f = nd_g\ (1+\frac{E}{|E_t|})\ \frac{f_t^2}{2E} \tag{4}$$

The strain along the direction of σ_θ is

$$\varepsilon\theta = \frac{1}{E}\ (\sigma\theta - \gamma\sigma_r) \tag{5}$$

in which γ is the poisson's ratio. If the analysis is confined in the condition of a$\prec\prec$ d, in which a is the crack length and $d = R_2 - R_1$, then within the crack scope the stresses can be seen to be uniform, σ_θ and σ_r can be written as

$$\begin{cases} \sigma\theta = c_1\ p \\ \sigma_r = -\ p \end{cases} \tag{6}$$

in which $c_1 = \dfrac{1+e^2}{1-e^2}$, By substituting (6) into (5), ε_θ can be rewritten as

$$\varepsilon\theta = \frac{c_1+\gamma}{E}p \tag{7}$$

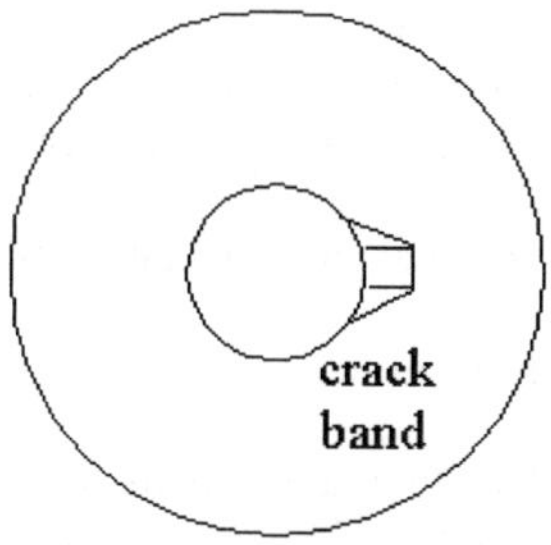

(a)

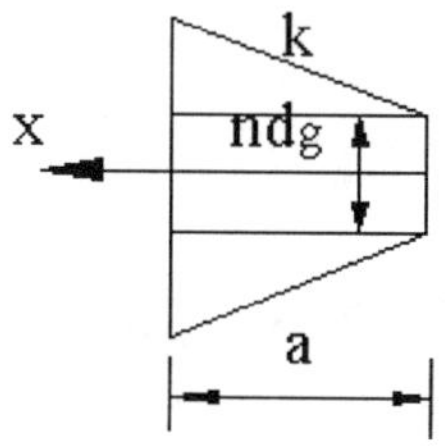

(b)

Fig.3 Schematic diagram of crack band

Fig.3 (a) shows the crack band in the vicinity of the hole wall when crack propagation occurs. If the curvature around the crack band is neglected, the crack band can approximately be simplified as Fig.3(b).The value of stress when crack propagates can be determined by the requirement of energy balance.

In the following the fracture strength with consideration of the stress gradient will be derived

Fig.3 (b) is the schematic crack band diagram for calculation, at x from the origin o, stress σ_θ can be expressed as

$$\begin{cases} \sigma_{\theta x} = \sigma_\theta - \dfrac{c_{11}}{R_1} p(a-x) \\ \sigma_{rx} = \sigma_r + \dfrac{c_{11}}{R_1} p(a-x) \end{cases} \tag{8}$$

where σ_θ and σ_r are the same as those in (6).

The circumferential strain is

$$\varepsilon_{\theta x} = \frac{1}{E}\left[(c_1+\gamma)-(1+\gamma)\frac{c_{11}}{R_1}(a-x)\right]p \tag{9}$$

By integration, the total energy release when the crack propagates can be written as

$$W=\int_0^a\left(k_1 x\,\sigma_{\theta x}\varepsilon_{\theta x}+\frac{1}{2}n d_g \sigma_{\theta x}\varepsilon_{\theta x}\right)dx \tag{10}$$

Using the requirement of energy balance:

$$\frac{\partial W}{\partial a} = G_f \tag{11}$$

the fracture strength of rock with consideration of the influence of stress gradients is derived, it is

$$\sigma_N = B \frac{1}{\sqrt{c + \dfrac{\lambda}{\lambda_0}}} f_t \tag{12}$$

in which B, λ_0 , λ and d are

$$\begin{cases} B = \sqrt{1+\dfrac{E}{|E_t|}} \\ \lambda_0 = \dfrac{n}{2k_1}\dfrac{d}{a} \approx 2.5\dfrac{d}{a} \\ \lambda = \dfrac{d}{d_g} \\ d = R_1 - R_2 = \dfrac{1-e}{e}R_1 \end{cases} \tag{13}$$

and

$$c = 1 - \frac{4(1+\gamma+e^2)}{\left[(1+\gamma)+e^2(1-\gamma)\right](1+e^2)}\,\frac{1-e}{e}\,\frac{a}{d} \tag{14}$$

According to Gettu(1998), the parameter λ_0 in expression (12) can be written as

$$\lambda_0 = \frac{EG_f}{B^2 d_g g(\alpha_0)} \tag{15}$$

in which E, G_f, B and d_g are the same as stated above. $g(\alpha_0)$ is a constant value of the non-dimensional energy release rate at the relative initial crack length α_0. It depends only on the specimen geometry, not on its size. For many typical specimen geometries, the values of $g(\alpha_0)$ are listed in handbooks(H.Tada,1985, Y.Murakami,1987), or it can be obtained by linear elastic finite element analysis. Let

$$d_0 = \frac{E G_f}{B^2 g(\alpha_0)} \tag{16}$$

then expression (12) can be transferred to be

$$\sigma_N = B\frac{1}{\sqrt{c+\dfrac{d}{d_0}}} f_t \tag{17}$$

and the parameter c in (17) can also be changed to be

$$c = 1 - \frac{10(1+\gamma+e^2)}{(1+\gamma+e^2(1-\gamma))(1+e^2)} \frac{(1-e)}{e} \frac{d_g}{d_0} \tag{18}$$

For comparison, the Bazant's size effect formula (1984) is also listed here

$$\sigma_N = B\frac{1}{\sqrt{1+\dfrac{d}{d_0}}} f_t \tag{19}$$

The difference between expression (17) and (19) is in (17) the parameter c subsitutes 1 in (19). The value c reflects the influence of the stress gradients on the fracture strength of rock. Because $c \prec 1$, for a certain rock material and geometric similar samples, the fracture strength predicted by (17) is bigger than that by (19). This fact means the stress gradient will enhance the fracture strength.

3 COMPARISON WITH TEST RESULTS

To check the influence of the stress gradients on the fracture strength of rock, it is necessary to do a series of model tests to measure the fracture strengths of rock for different size but geometry similar samples. Then use the expression (17) and (19) to predict the fracture strengths of these samples. By comparing these values, the influence of stress gradients on the fracture strength of rock and the difference between the calculated values and the measured values of model test can be obtained.

A series of model tests of hollow cylinders under the action of water pressure on hole wall were conducted (by J.R..Enver). The specimens are synthetic sandstone with uniform grain size . The test samples have the same geometry but different size. By increasing p to a certain value, the hollow cylinder cracks to failure. The value of p can be transferred to the circumferential stress. This stress and the geometric sizes were then recorded, see the upper most line in Fig. 4.

The values of the parameters for calculation are:

dg=0.1 mm B=1.0
d0=50.2 mm e=0.2
ft=17.2 Mpa

The results by calculation are shown also in Fig.4. The mid and the lower lines are the results calculated by expression (17) and (19) respectively. It can be seen that the results predicated by expression (17), which takes consideration of the influence of stress gradient, is fit better with the test

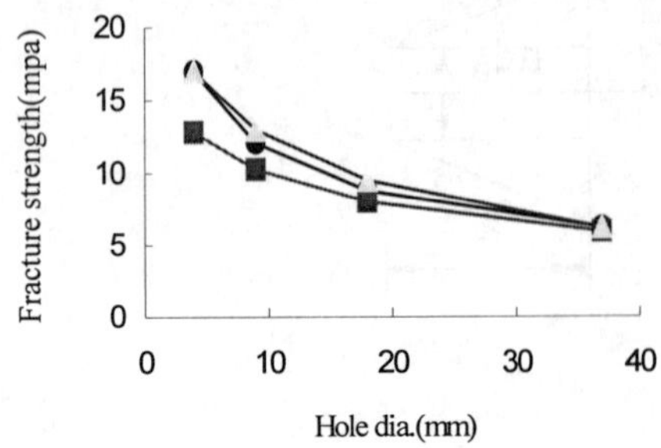

Fig.4 Fracture strength vs hole diameter

results, while the values predicated by expression (19) has a relatively bigger difference with the test results. The smaller the sample the bigger the difference. The biggest difference is up to 32.5%. The difference between the results of model test and the predication by expression (17) is very small and can be omitted. This confirms the importance of taking into consideration the effect of stress gradients in certain condition such as small sample size.

With the increasing of the size of samples, the difference between the results predicated by expression (17) and (19) decreases and both are coincident with the model test results.

4 CONCLUSION

On the basis of the theory of Bazant's size effect, for borehole samples the expression (17) predicting the fracture strength of rock relating to the sample size is derived, in which the effect of stress gradients on the fracture strength is taken into consideration.

Because $c \prec 1$, for a certain rock material and geometric similar samples, the fracture strength of rock predicted by (17) is bigger than that by (19). This fact means the stress gradients will enhance the fracture strength.

By comparison with the test results, it is confirmed that the result predicated by the expression with consideration of the stress gradients fits the test results better. This means that stress gradients really have effect on the fracture strength, and in certain condition it can't be omitted. Taking consideration of the stress gradients will be important to correctly predict the fracture strength.

In this paper, the influence of triaxial stress state is not considered. This problem is complicated and will be discussed in the future.

REFERENCE

Bazant Z. P. Et al 1994. Fracture size effect: review of evidence for concrete structure. J. Struct. Engng. 120: p2377.

Bazant Z. P. 1984. Size effect in blunt fracture: concrete, rock, metal. Journal of Engineering Mechanics.110: p518

Bazant Z. P. 1990. Determination of fracture energy, process zone length and brittleness number from size effect, with application to rock and concrete. International Journal of Fracture. 44:111-131.

Kooijman A.P. et al 1991. Hollow cylinder collapse: measurement of deformation and failure in an X-ray CT scanner, observation of size effect. Proc. 32nd US Rock Mech. Symp., Rock Mechanics as a Multidisciplinary Science. p657 Rotterdam:Balkema.

Sulem J. et al 1989. Simplified bifurcation analysis of deep boreholes in rocks with microstructure. Proc. Int. Symp.. p845. Rotterdam:Balkema.

C.Ringstad et al 1993. Scale effects in hollow cylinder tests. Scale Effects in Rock Masses 93 (Pinto da Cunha, ed.). p75. Rotterdam,:Balkema.

Ravindra Gettu, Hector Saldivar 1998. Implications of the size effect method for analyzing the fracture of concrete. Int. J. Solids Structures. 35: p4121

H. Tada et al 1985.The stress analysis of cracks handbook(2nd ed.). Paris Productions Inc., St. Louis, Mo.

Y.Murakami 1987. Stress intensity factors handbook. Pergamon Press.

Computer Methods and Advances in Geomechanics, Desai et al. (eds) © 2001 Balkema, Rotterdam, ISBN 90 5809 183 X

Plastic spin and gradient theory for modeling large strain behavior of soils

C.R. Song & G.Z. Voyiadjis
Department of Civil and Environmental Engineering, Louisiana State University, Baton Rouge, La., USA

ABSTRACT: This study incorporates the micro behavior of soils into the macro behavior by utilizing the plastic spin and gradient theory. In order to consider the microstructural grain rotations, the plastic spin is used. For considering the microstructural grain interaction, the gradients theory is used. This approach allows for the simulation of large strain problems in soils such as shear band problems and cone penetration test.

1. INTRODUCTION

The behavior of soils is governed by many factors such as stiffness, strain rate and stress history, etc. The effect of micro-structural changes is especially important for soils that are subjected to large strains and rotations. This study incorporates the effects of micro-structural changes by utilizing the plastic spin and gradient theory. The plastic spin is used to incorporate the non-coaxiality of micro-structural rotation in reference to the macro-structural rotation. The gradient theory is used for the micro-structural interaction between the micro scale material particles. For this study, an anisotropic modified Cam Clay model is used. The gradient theory is formulated in two length scales: the meso-scale and macro-scale. Meso-scale gradient is formulated in terms of micro level back stress gradient and volumetric strain gradient. Macro-scale gradient is formulated in terms of the macro level back stress gradient and volumetric strain gradient.

From this study, it is expected that the effects of micro-structural changes (plastic spin) come into play when shear strains are large. The effects of micro-structural interaction (gradient) are also significant in the case of large strains. The effects of the gradient are also important in localization problems. The gradient results in well-posedness of the numerical solution and smoothening of the strain distribution in the shear band.

2. GRADIENT

Voyiadjis & Deliktas (2000) derived a multi scale gradient theory. In this paper, it is applied to geomaterials using an anisotropic modified Cam Clay model. The thermoelastic Helmholtz free energy may be expressed in terms of internal state variables such that (Voyiadjis & Deliktas, 2000, Voyiadjis et al. 2000):

$$\Psi = \Psi(\varepsilon^e, \mathbf{T}, \mathbf{A}^{(p)}_{(k)}, \nabla\mathbf{A}^{(p)}_{(k)}, \nabla^2\mathbf{A}^{(p)}_{(k)}, \nabla\hat{\mathbf{A}}^{(p)}_{(k)}, \overline{\nabla^2\hat{\mathbf{A}}^{(p)}_{(k)}}) \quad (1)$$

In equation (1), Ψ is Helmholtz free energy, ε^e is the elastic strain, $\mathbf{T}$ is the temperature, superscript p represents plasticity, $\mathbf{A}^{(p)}_{(k)}$ is the macroscale internal state variables such as isotropic hardening factor and kinematic hardening factor. ∇ represents the first order gradient, ∇^2 represents the second order gradient, over-bar represents the average and $\hat{\mathbf{A}}^{(p)}_{(k)}$ represents the microscale internal state variables. When one assumes symmetrical distribution of the gradients, the odd order gradient terms vanish, and equation (1) becomes:

$$\Psi = \Psi(\varepsilon^e, \mathbf{T}, \mathbf{A}^{(p)}_{(k)}, \nabla^2\mathbf{A}^{(p)}_{(k)}, \overline{\nabla^2\hat{\mathbf{A}}^{(p)}_{(k)}}) \quad (2)$$

Voyiadjis & Deliktas (2000) formulation couples plasticity with damage. However, the effects of damage are excluded in this study. The first order and second order gradient of the macro and micro internal variables are defined as follows (Voyiadjis & Deliktas, 2000):

$$\nabla^2\mathbf{A}^{(p)}_{(k)} = \left\{\frac{\partial^2\mathbf{A}_{ij}}{\partial\mathbf{x}_k\partial\mathbf{x}_k}\right\} \quad (3)$$

$$\overline{\nabla^2 \hat{\mathbf{A}}^{(p)}_{(k)}} = \frac{1}{\mathbf{V}_{\mathbf{RVE}}} \int_{V_{RVE}} \frac{\partial^2 \hat{\mathbf{A}}^{(p)}_{(k)}}{\partial \hat{\mathbf{x}}_k \hat{\mathbf{x}}_k} \mathbf{dV}_{\mathbf{RVE}} \tag{4}$$

The value of the thermodynamic forces can be obtained through the evolution relations of the internal state variables. However, it should be pointed out that there are two types of evolution equations that need to be developed, one at the macroscale and the other at the mesoscale level. The former can be obtained by assuming the physical existence of the dissipation potential at the macroscale. The later one can be obtained by a micromechanical or phenomenological approach.

One may also consider that the evolution equations of the internal state variables $\dot{\mathbf{A}}^{(p)}_{(k)}$, can be obtained by integrating the evolution equations of the local internal state variables at the mesoscale, that is $\hat{\dot{\mathbf{A}}}^{(p)}_{(k)}$ over the domain of the RVE. However, integration of the $\hat{\mathbf{A}}^{(p)}_{(k)}$ is cumbersome task since at the mesoscale, $\hat{\mathbf{A}}^{(p)}_{(k)}$ is a function of many different aspects of material inhomogeneities such as interaction of defects, size of defects, spacing between them, and distribution of defects within the sub RVE. Therefore, in this work evolution equations of the macroscale internal state variables are obtained through the use of the generalized normality rule of thermodynamics. In this regard the macroscale dissipation potential is defined only in terms of the macroscale flux variables disregarding temperature effects as follows

$$\Theta = \Theta(\dot{\varepsilon}^{\mathbf{p}}, \dot{\mathbf{A}}^{(p)}_{(k)}, \nabla^2 \dot{\mathbf{A}}^{(p)}_{(k)}, \overline{\nabla^2 \hat{\dot{\mathbf{A}}}^{(p)}_{(k)}}) \tag{5}$$

By using Legendre-Fenchel transformation of the dissipation potential (Θ), one can obtain complementary laws in the form of the evolution laws of flux variables as function of the dual variables:

$$\Theta^* = \Theta^*(\dot{\sigma}, \dot{\mathbf{B}}^{(p)}_{(k)}, \nabla^2 \dot{\mathbf{B}}^{(p)}_{(k)}, \overline{\nabla^2 \hat{\dot{\mathbf{B}}}^{(p)}_{(k)}}) \tag{6}$$

Equation (6) is also expressed as follow for anisotropic modified Cam Clay model:

$$\Theta^* = \mathbf{F}(\dot{\sigma}, \sum_{i=1}^{N} \dot{p}_{o,i}, \sum_{i=1}^{N} \dot{\bar{\alpha}}_i) \tag{7}$$

In equation (7) $p_{o,i}$ *and* $\bar{\alpha}_i$ are the isotropic and kinematic hardening forces respectively. The evolution equation of $p_{o,i}$ *and* $\bar{\alpha}_i$ are obtained from the normality rule as follows:

$$\mathbf{B}^{(p)}_{(k)} = C\dot{\mathbf{A}}^{(p)}_{(k)} = C\frac{\partial \Psi}{\partial \mathbf{B}^{(p)}_{(k)}} \tag{8}$$

In equation (8), C is a constant. Using equation (8) and flow rule, one has following evolution equations:

$$\dot{\varepsilon}^p_v = -\dot{\Lambda}\, tr\left|\frac{\partial F}{\partial \sigma_{ij}}\right| = -\dot{\Lambda}\, tr\left|\frac{\partial f}{\partial \sigma_{ij}}\right| \tag{9.a}$$

$$\dot{\bar{\alpha}}_{ij} = -C_1 \dot{\Lambda} \frac{\partial F}{\partial \bar{\alpha}_{ij}} = -C_1 \dot{\Lambda} \frac{\partial f}{\partial \bar{\alpha}_{ij}} \tag{9.b}$$

$$\overline{\nabla^2 \dot{\hat{\alpha}}_{ij}} = -C_2 \dot{\Lambda} \frac{\partial F}{\partial \overline{\nabla^2 \hat{\alpha}_{ij}}} = -C_2 \dot{\Lambda} \frac{\partial f}{\partial \overline{\nabla^2 \hat{\alpha}_{ij}}} \tag{9.c}$$

$$\overline{\nabla^2 \dot{\hat{p}}_o} = -C_3 \dot{\Lambda} \frac{\partial F}{\partial \overline{\nabla^2 \hat{p}_o}} = -C_3 \dot{\Lambda} \frac{\partial f}{\partial \overline{\nabla^2 \hat{p}_o}} \tag{9.d}$$

In equation (9) $\bar{\alpha}_{ij}$ represents the back stress, and $\dot{\Lambda}$ is a plastic multiplier. Using the anisotropic modified Cam Clay model expressed by Voyiadjis & Song (2000), the following relation is obtained:

$$\begin{aligned} f = p^2 - pp_o + \frac{3}{2M^2}\{&(s_{ij} - p\alpha_{ij})(s_{ij} - p\alpha_{ij}) \\ &+ (p_o - p)p\alpha_{ij}\alpha_{ij}\} = 0 \end{aligned} \tag{10}$$

In equation (11), p is the mean principal stress, p_o is the initial mean principal stress, M is the slope of critical state line (CSL) in p vs. q space, s_{ij} is the deviatoric stress, and α_{ij} is the rotation of yield surface with respect to p axis (dimensionless back stress factor). When one applies the gradient to the isotropic hardening factor p_o and anisotropic hardening factor α_{ij} the following expressions are obtained

$$\dot{\bar{p}}_o = \dot{p}_o - a_1 \nabla^2 \dot{p}_o - b_1 \overline{\nabla^2 \dot{\hat{p}}_o} \tag{11}$$

$$\dot{\bar{\alpha}}_{ij} = \dot{\alpha}_{ij} - a_2 \nabla^2 \dot{\alpha}_{ij} - b_2 \overline{\nabla^2 \dot{\hat{\alpha}}_{ij}} \tag{12}$$

In equations (12.a) and (12.b) $\dot{\bar{p}}_o$, $\dot{p}_o$ are global and local mean principal stresses) for isotropic hardening and $\dot{\bar{\alpha}}_{ij}$, $\dot{\alpha}_{ij}$ are global and local dimensionless back stresses for kinematic hardening, respectively. In equations (12.a) and (12.b), the parameters a_1, b_1, a_2, b_2, also take into account the dimension discrepancies between a quantity and the second order gradient of that quantity.

One obtains the evolution equation for $\dot{\alpha}_{ij}$ from equation (10.b) (note that $\dot{\alpha}_{ij} = \dot{\bar{\alpha}}_{ij} / p$). One also obtains the evolution equation for $\overline{\nabla^2 \dot{\hat{\alpha}}_{ij}}$ from equation (10.c). Here one needs to note that the evolution equation of $\overline{\nabla^2 \dot{\hat{\alpha}}_{ij}}$ is used rather than that of $\dot{\hat{\alpha}}_{ij}$. The partial derivatives of f with respect to α_{ij} and $\overline{\nabla^2 \dot{\hat{\alpha}}_{ij}}$ are given below:

$$\frac{\partial f}{\partial \alpha_{ij}} = \frac{3}{M^2} p\{p_o \alpha_{ij} - s_{ij}\} \tag{13.a}$$

$$\frac{\partial f}{\zeta \nabla^2 \partial \hat{\alpha}_{ij}} = \frac{3}{M^2} p\{p_o \alpha_{ij} - s_{ij}\} \tag{13.b}$$

In equation (13.b), ζ accounts for the dimensional discrepancy. b_2 in equation (12.b) accounts for dimensional discrepancy problem and the material characteristics. One also obtains the evolution equation for $\dot{\bar{p}}_o$ from equation (10.a) as follows:

$$\dot{\bar{p}}_o = -\frac{1+e_o}{\lambda - \kappa} \frac{p_o(p_o + \Delta p)}{\Delta p} \dot{\varepsilon}_v^p = K \dot{\varepsilon}_v^p \tag{14}$$

Using relation (13.b), the following equation is obtained.

$$\frac{\partial f}{\xi \nabla^2 \partial \hat{p}_o} = -p + \frac{3}{2M^2} p_o \alpha_{ij} \alpha_{ij} \tag{15}$$

In equation (15), ξ is a similar parameter to ζ in equation (13.b).

The plastic multiplier is obtained as follows. From relation (11) one obtains the following differential expression:

$$df = \frac{\partial f}{\partial p} dp + \frac{\partial f}{\partial s_{ij}} s_{ij} + \frac{\partial f}{\partial \varepsilon_v^p} d\varepsilon_v^p + \frac{\partial f}{\partial \alpha_{ij}} d\alpha_{ij} = 0 \tag{16}$$

Equation (16) is the typical expression for continuity with respect to local stresses. When one considers the gradient, one should include the gradient terms due to both the isotropic hardening and kinematic hardening terms. $d\varepsilon_v^p$ has isotropic hardening characteristics and $d\alpha_{ij}$ has kinematic hardening characteristics. Therefore one should incorporate the gradient in $d\varepsilon_v^p$ and $d\alpha_{ij}$ as shown in equations (10.a) and (12.b), respectively. One may define B_{ij} as follows:

$$B_{ij} = \frac{\partial f}{\partial \sigma_{ij}} = \frac{\partial f}{\partial p} \frac{\partial p}{\partial \sigma_{ij}} + \frac{\partial f}{\partial s_{kl}} \frac{\partial s_{kl}}{\partial \sigma_{ij}} \tag{17}$$

Using the additive decomposition of the strain rate, one obtains the following expression:

$$\dot{\sigma}_{ij} = C_{ijkl} \dot{\varepsilon}_{kl} - C_{ijkl} \dot{\Lambda} B_{kl} \tag{18}$$

In equation (18), $\dot{\Lambda}$ is the plastic multiplier. Using equations (16) and (18) one obtains:

$$\begin{aligned} & B_{ij}(C_{ijkl}\dot{\varepsilon}_{kl} - C_{ijkl}\dot{\Lambda}B_{kl}) \\ & + \frac{\partial f}{\partial \alpha_{ij}}(\dot{\alpha}_{ij} - a_1 \nabla^2 \dot{\alpha}_{ij} - b_1 \nabla^2 \dot{\hat{\alpha}}_{ij}) \\ & + \frac{\partial f}{\partial \varepsilon_v^p}(\dot{\varepsilon}_v^p - a_2 \nabla^2 \dot{\varepsilon}_v^p - b_2 \nabla^2 \dot{\hat{\varepsilon}}_v^p) = 0 \end{aligned} \tag{19}$$

Using equation (19), the following expression for the plastic multiplier $\dot{\Lambda}$ is obtained.

$$\dot{\Lambda} = \frac{B_{ij} C_{ijkl} \dot{\varepsilon}_{kl}}{\Gamma + \Pi + T} \tag{20}$$

The terms in equation (20) are defined as follows:

$$\Gamma = B_{ab} C_{abcd} B_{cd} \tag{20.a}$$

$$\Pi = \frac{\partial f}{\partial \varepsilon_v^p}(B_{mm} - a_1 \nabla^2 B_{mm} - b_1 \nabla^2 \hat{B}_{mm}) \tag{20.b}$$

$$T = \frac{\partial f}{\partial \bar{\alpha}_{ab}}(A_{ab} - a_2 \nabla^2 A_{ab} - b_2 \nabla^2 \hat{A}_{ab}) \tag{20.c}$$

$$A_{ab} = \varepsilon_v^p \frac{c}{p}(s_{ab} - x\, p\, \alpha_{ab}) \tag{20.d}$$

$$\nabla^2 \hat{A}_{ab} = -C_2 \frac{\partial f}{\partial \nabla^2 \hat{\bar{\alpha}}_{ij}} p \tag{20.e}$$

$$\nabla^2 \hat{B}_{mm} = (-p + \frac{3}{2M^2} p_o \alpha_{ij} \alpha_{ij}) / K \tag{20.f}$$

Substituting equation (20) into equation (18), one obtains the following stress-strain relation.

$$d\sigma_{ij} = C_{ijkl}(1 - \frac{B_{mn} C_{mnop} B_{op}^T}{B_{ab} C_{abcd} B_{cd} + H_p^v + H_p^a}) d\varepsilon_{kl} \tag{22}$$

In equation (21), H_p^v and H_p^a are defined as follows:

$$H_p^v = \frac{\partial f}{\partial \varepsilon_v^p}(B_{mm} - a_1 \nabla^2 B_{mm} - b_1 \nabla^2 \hat{B}_{mm}) \tag{23.a}$$

$$H_p^a = \frac{\partial f}{\partial \bar{\alpha}_{ij}}(A_{ab} - a_2\nabla^2 A_{ab} - b_2\nabla^2 \hat{A}_{ab}) \qquad (23.b)$$

Equations (20) and (22) show the direct effects of the gradients on the stiffness tensor.

3. PLASTIC SPIN

To understand the physical concept of the plastic spin, one needs to go back to the stress co-rotation or a similar concept. The stress transformation in a different coordinate system can be easily expressed as follows:

$$\mathbf{T^* = QTQ^T} \qquad (24)$$

In equation (24), T* is the stress tensor transformed to a new coordinate system, T is the stress tensor in the original coordinate, and Q is the transformation matrix. However, in the case of the stress rate, the coordinate transformation is not that easy as that shown in equation (24). The rate is expressed as shown in equation (25):

$$\dot{\mathbf{T}}^* = \dot{\mathbf{Q}}\mathbf{TQ^T} + \mathbf{Q}\dot{\mathbf{T}}\mathbf{Q^T} + \mathbf{QT}\dot{\mathbf{Q}}^\mathbf{T} \qquad (25)$$

In equation (25), the upper dot represents the time rate. In expression (25), there is no guarantee that the first and third terms are zero. Therefore one introduces the spin tensor W to compensate for those non-zero terms as shown in equation (26):

$$\dot{\mathbf{T}}^* = \dot{\mathbf{T}} - \mathbf{WT} + \mathbf{TW} \qquad (26)$$

In equation (26), W is termed the 'spin tensor'. The spin tensor W corrects the constitutive equation for the rigid body rotation. However, even with the spin tensor W, the stress-strain response of the material at large strains may become unstable. The reason for this is the lack of the proper incorporation of the micro-mechanical behavior of the material. This micro-mechanical activity introduced additional stress change and consequently on needs to modify the spin tensor. The modified spin tensor is expressed as **W*** or Ω such that

$$\dot{\mathbf{T}}^* = \dot{\mathbf{T}} - \mathbf{W^*T} + \mathbf{TW^*} = \dot{\mathbf{T}} - \mathit{\Omega}\mathbf{T} + \mathbf{T}\mathit{\Omega} \qquad (27)$$

In equation (27), ***W'*** is defines as follows:

$$\mathbf{W^*} = \mathit{\Omega} = \mathbf{W} - \mathbf{W^P} \qquad (28)$$

In equation (28), $\mathbf{W}^p$ is the plastic spin. Thus one can use the plastic spin in order to incorporate the non-coaxiality of micro rotation and macro rotation. The plastic spin $\mathbf{W}^p$ is expressed in the general form as suggested by Voyiadjis & Kattan (1991) as follows:

$$\mathbf{W}_i^{s''} = \xi \mathbf{N}_{i,abcd} \mathbf{d}_{cd}^s \qquad (29)$$

where,

$$\mathbf{N}_{i,abcd} = \xi(\alpha_{i,am}\mathbf{M}_{mbcd} - \mathbf{M}_{ancd}\alpha_{i,nb})\mathbf{d}_{cd}^s \qquad (30)$$

$$M_{mnij} = c\, E'^{ss}_{klmn} n_{kl} n_{ij} / H \qquad (31)$$

$$c = \frac{H}{H + E'^{s}_{abcd} n_{ab} n_{cd}} \qquad (32)$$

H = Hardening Modulus

α_i = back-stress

n_{ij} = normal to the yield surface

4. CONCLUSIONS

From this study, the formulations for the two kinds of micro-mechanical phenomena grain rotation and grain interaction are presented. The well posedness of the numerical solution is also expected. The effects of these micro-mechanical behaviors are especially important for large strain problems, such as in the case of shear bands, the gradients will act as a homogenizing factor.

5. REFERENCES

Voyiadjis, G.Z. and Deliktas, B. 2000. Multi-scale Analysis of Multiple Damage Mechaisms Coupled with Inelastic Behavior of Composite Materials. *Mechanics Research Communications Journal*, Vol. 27, No.3, pp.295-300

Voyiadjis, G.Z., Deliktas, B. and Aifantis, E.C. 2000. Multi-scale Analysis of Multiple Damage Mechaisms Coupled with Inelastic Behavior of Composite Materials. *Computational Methods in Applied Mechanics and Engineering.* submitted for publication

Voyiadjis, G.Z. and Kattan, P.I. 1991. Phenomenological evolution equations for the backstress and spin tensors. *Acta Mechanica*, Vol.88, pp.91-111

Voyiadjis, G.Z. and Song, C.R. 2000. Anisotropic Modified Cam Clay Model with Plastic Spin for Finite Strains. *J. of EM Div. ASCE.* (in press)

Computer Methods and Advances in Geomechanics, Desai et al. (eds) © 2001 Balkema, Rotterdam, ISBN 90 5809 183 X

Rate dependent gradient theory for shear band analysis

G.Z.Voyiadjis & C.R.Song
Department of Civil and Environmental Engineering, Louisiana State University, Baton Rouge, La., USA

ABSTRACT: The gradient theory is commonly used for the homogenization of the localized high strain problems. In this work, the rate dependency of the material is added as another homogenizing mechanism. In practice, these two mechanisms occur simultaneously. The separate formulation and coupling of these two mechanisms is presented here for a more rational approach to localized high strain problems as in the case of shear bands.

1. INTRODUCTION

Shear bands undergo localized concentrated high shear strains. These high shear strains cause instability in the numerical solution, typically known as Neumann's boundary value problem or Dirichlet's boundary value problem. In physical sense, the highly concentrated shear strains in the shear band cause the volume change of the material, and results in the stress transfer to the surrounding area. For numerical solutions that do not account for this volume change and do not consider numerical instability, the obtained results for the shear bands are incorrect. Typically the results over-predict the strain level.

One approach to obtain a more accurate solution for this localized behavior is to use the gradient theory. The gradient theory considers the stress(strain) transfer between the shear band and the surrounding area, and essentially homogenizes (regularizes) the highly localized shear strains of the shear band.

Another factor affecting the homogenizing mechanism is the rate dependency of the material. In the shear band, the shear strains are much higher than the surrounding area. These high shear strains occur in the identical elapsed time as the shear strains that occur in the surrounding area. The shear strain rates in the shear band are much higher than those of the surrounding area. At higher strain rates, the viscous material exhibits higher shear resistance at the same shear strain level. The constitutive relations that have rate dependency will recognize the higher shear resistance in the shear band, and will naturally homogenize(regularize) the localized high shear strain behavior to some degree.

The desirable constitutive relations, therefore, should have both gradient theory and rate dependency for more reliable modeling of the shear band. In this study both the gradient theory and the rate dependency are incorporated in an anisotropic modified Cam Clay model.

2. GRADIENT THEORY

A classical example of this micromechanical interaction in geo-materials is the 'energy correction' concept (Taylor, 1948) for the direct shear of dense dry sands. As the shear strain increases, the sand particles role over each other, causing an increase in the normal stress. The energy used for the normal stress increase is reported as much as 16% of the total energy used (Taylor, 1948). This phenomenon is mainly due to the grain re-alignment or re-orientation.

The gradient theory allows one to incorporate the inter-granular stress into the continuum mechanics. When materials with a simple cubic array are subjected to deformation as shown in Figure 1-a, grains will fall into the void space and will reach the compacted condition as shown in Figure 1-b. This process will reduce the volume and nullify or reduce the contact stress for the adjacent layer of grains on the top.

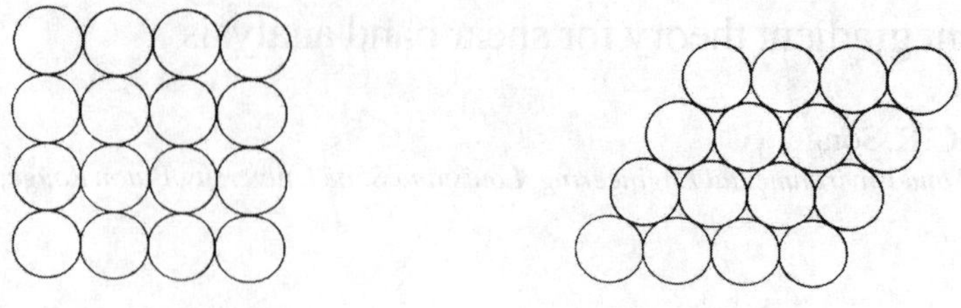

(a) Simple cubic array (b) Cubic Tetrahedral Array

Figure 1 Grain arrangement during shear

When materials with a cubic tetrahedral array are subjected to deformation as shown in Figure 1-b, grains will climb over the adjacent grains and reach the loose packing condition as shown in Figure 1-a.. This process increases the volume. If the volume expansion is restricted, this process increases the contact stresses for the adjacent layer of grains at the top.

This behavior indicates that whenever one includes the stress (strain) gradient, it develops stress (strain) transfer to the neighboring grains. This kind of stress (strain) transfer is important when one has a locally concentrated high strain region, such as a shear band. Along the shear band, micro-structural stress redistribution is needed for modeling the accurate behavior of the material. This is the fundamental concept of the gradient theory. The gradient theory is therefore related to the inter-granular quantities that may be related to the back stresses, inter-granular normal forces, tangential forces, etc.

The formulation for the gradient theory is presented by Voyiadjis & Deliktas (2000) and Voyiadjis et al. (2000). The components of the macroscale gradient terms and the norm of the averaged mesoscale gradient terms of the macroscale internal state variables of plasticity may be used as additional higher order internal variables. Through these internal variables one can retain information regarding the mesoscale configuration, non-local material behavior, and localization effects. In such an approach, the thermoelastic Helmholtz free energy may be expressed in terms of these sets of internal state variable such that:

$$\Psi = \Psi(\boldsymbol{\varepsilon}^e, T, \mathbf{A}^{(r)}_{(k)}, \nabla^2 \mathbf{A}^{(r)}_{(k)}, \tilde{\nabla} \mathbf{A}^{(r)}_{(k)}, \overline{\hat{\nabla}^2 \hat{\mathbf{A}}^{(r)}_{(k)}}, \overline{\tilde{\nabla} \hat{\mathbf{A}}^{(r)}_{(k)}}) \tag{1}$$

In equation (1) $\mathbf{A}^{(r)}_{(k)}$ represents the macroscale internal state variables such that:

$$A^{(r)}_{(k)} = p, \alpha \tag{2}$$

In equation (2) p, α variables characterize the isotropic and kinematic hardening in plasticity, respectively. The isotropic hardening variable of plasticity, p, is a scalar quantity and is expressed in terms of the second order tensor, ε^p, describing the plastic strain rate. The higher order internal state variables are introduced by averaged mesoscale gradient terms, which are a measure of the mesostructural variability within the RVE (Representative Volume Element). The bracket subscripts or superscripts do not indicate the tensorial nature of variables. In the above equations the bracketed superscript "p" implies plasticity internal state variables. The bracketed subscript or subscript "k" indicates the number of different state variables in plasticity and damage. With the higher order representation of the material inhomogeneity, it is possible to evaluate the inelastic flow, and time dependent deformation. Such higher order representation of material state is defined by constitutive equation (1). The corresponding gradient terms are defined as follows:

$$A^{(r)}_{(k)} = \frac{1}{\mathbf{V}_{\mathbf{RVE}}} \int_{\mathbf{V}_{\mathbf{RVE}}} \hat{A}^{(r)}_{(k)} \, \mathbf{dV}_{\mathbf{RVE}} \tag{3a}$$

$$\nabla^2 \mathbf{A}^{(r)}_{(k)} = \left\{ \frac{\partial^2 A^{(r)}_{(k)ij}}{\partial x_j x_j} \right\} \tag{3b}$$

$$\tilde{\nabla} \mathbf{A}^{(r)}_{(k)} = \nabla \mathbf{A}^{(r)}_{(k)} . \nabla \mathbf{A}^{(r)}_{(k)} = \left\{ \frac{\partial \mathbf{A}^{(r)}_{(k)}}{\partial \hat{x}_j} . \frac{\partial \mathbf{A}^{(r)}_{(k)}}{\partial \hat{x}_j} \right\} \tag{3c}$$

$$\overline{\hat{\nabla}^2 \hat{\mathbf{A}}^{(r)}_{(k)}} = \frac{1}{V_{RVE}} \int_{V_{RVE}} \frac{\partial^2 \hat{\mathbf{A}}}{\partial x_j x_j} dV_{RVE} \tag{3d}$$

$$\overline{\tilde{\hat{\nabla}}} \hat{\mathbf{A}}^{(r)}_{(k)} = \frac{1}{V_{RVE}} \int_{V_{RVE}} \hat{\nabla} \hat{\mathbf{A}}^{(r)}_{(k)} : \hat{\nabla} \hat{\mathbf{A}}^{(r)}_{(k)} dV_{RVE} \tag{3e}$$

where, V_{RVE} represents the volume of RVE. In the case of gradient terms with reference to the back-stress, this characterizes the internal embedded stress variations introduced by grain interaction, etc. One can express the time derivative of equation (3) in terms of its higher order state variables as follows:

$$\begin{aligned}\dot{\Psi} &= \frac{\partial \Psi}{\partial \varepsilon^e} : \dot{\varepsilon}^e + \frac{\partial \Psi}{\partial T}\dot{T} + \frac{\partial \Psi}{\partial A^{(r)}_{(k)}} : \dot{A}^{(r)}_{(k)} \\ &+ \frac{\partial \Psi}{\partial \nabla^2 A^{(r)}_{(k)}} : \nabla^2 \dot{A}^{(r)}_{(k)} + \frac{\partial \Psi}{\partial \tilde{\nabla} A^{(r)}_{(k)}} : \tilde{\nabla} \dot{A}^{(r)}_{(k)} \\ &+ \frac{\partial \Psi}{\partial \overline{\hat{\nabla}^2 \hat{A}^{(r)}_{(k)}}} : \overline{\hat{\nabla}^2 \dot{\hat{A}}^{(r)}_{(k)}} + \frac{\partial \Psi}{\partial \overline{\tilde{\hat{\nabla}} \hat{A}^{(r)}_{(k)}}} : \overline{\tilde{\hat{\nabla}} \dot{\hat{A}}^{(r)}_{(k)}}\end{aligned} \tag{4}$$

By substituting equations (4) into the Clausus-Duhem inequality one obtains:

$$\begin{aligned}&\left(\sigma - \rho\frac{\partial \Psi}{\partial \varepsilon^e}\right) : \dot{\varepsilon}^e + \sigma : \dot{\varepsilon}^p - \left(s - \rho\frac{\partial \Psi}{\partial T}\right)\dot{T} - \rho\frac{\partial \Psi}{\partial A^{(r)}_{(k)}} : \dot{A}^{(r)}_{(k)} \\ &- \rho\frac{\partial \Psi}{\partial \nabla^2 A^{(r)}_{(k)}} : \nabla^2 \dot{A}^{(r)}_{(k)} - \rho\frac{\partial \Psi}{\partial \tilde{\nabla} A^{(r)}_{(k)}} : \tilde{\nabla} \dot{A}^{(r)}_{(k)} \\ &- \rho\frac{\partial \Psi}{\partial \overline{\hat{\nabla}^2 \hat{A}^{(r)}_{(k)}}} : \overline{\hat{\nabla}^2 \dot{\hat{A}}^{(r)}_{(k)}} - \rho\frac{\partial \Psi}{\partial \overline{\tilde{\hat{\nabla}} \hat{A}^{(r)}_{(k)}}} : \overline{\tilde{\hat{\nabla}} \dot{\hat{A}}^{(r)}_{(k)}} \\ &- \frac{\vec{q}}{T} \cdot \overrightarrow{\mathbf{grad}}\, T \geq 0\end{aligned} \tag{5}$$

In equation (5) the higher order mesoscale gradient terms, $\overline{\hat{\nabla}^2 \hat{A}^{(r)}_{(k)}}$, $\overline{\tilde{\hat{\nabla}} \hat{A}^{(r)}_{(k)}}$, are dependent on the macroscale variables, $A^{(r)}_{(k)}$. However, the random periodic boundary condition ensures that there is no net flux of mesoscale gradients across the RVE boundary. Such a constraint would effectively prevent coupling between macroscale gradient and mesoscale gradient terms.

The first two terms in equation (5) represent the thermoelastic laws of thermodynamics and the remaining terms represent the total dissipation process due to plasticity. This can be expressed as the plastic dissipation as follows:

$$\begin{aligned}\Pi^p &= \sigma : \dot{\varepsilon}^p - Q^{(p)}_{(k)} : \dot{A}^{(p)}_{(k)} - U^{(p)}_{(k)} : \nabla^2 \dot{A}^{(p)}_{(k)} \\ &- V^{(p)}_{(k)} : \tilde{\nabla} \dot{A}^{(p)}_{(k)} - W^{(p)}_{(k)} : \overline{\hat{\nabla}^2 \dot{\hat{A}}^{(p)}_{(k)}} - Z^{(p)}_{(k)} : \overline{\tilde{\hat{\nabla}} \dot{\hat{A}}^{(p)}_{(k)}}\end{aligned} \tag{6}$$

In equations (6) $\mathbf{Q}^{(r)}_{(k)}$, $\mathbf{V}^{(r)}_{(k)}$, $\mathbf{U}^{(r)}_{(k)}$, $\mathbf{W}^{(r)}_{(k)}$, and $\mathbf{Z}^{(r)}_{(k)}$ are defined as the thermodynamic forces and are expressed below:

$$\mathbf{Q}^{(r)}_{(k)} = \rho\frac{\partial \Psi}{\partial \mathbf{A}^{(r)}_{(k)}} \tag{7a}$$

$$\mathbf{U}^{(r)}_{(k)} = \rho\frac{\partial \Psi}{\partial \nabla^2 \mathbf{A}^{(r)}_{(k)}} \tag{7b}$$

$$\mathbf{V}^{(r)}_{(k)} = \rho\frac{\partial \Psi}{\partial \tilde{\nabla} \mathbf{A}^{(r)}_{(k)}} \tag{7c}$$

$$\mathbf{W}^{(r)}_{(k)} = \rho\frac{\partial \Psi}{\partial \overline{\hat{\nabla}^2 \hat{\mathbf{A}}^{(r)}_{(k)}}} \tag{7d}$$

$$\mathbf{Z}^{(r)}_{(k)} = \rho\frac{\partial \Psi}{\partial \overline{\tilde{\hat{\nabla}} \hat{\mathbf{A}}^{(r)}_{(k)}}} \tag{7e}$$

One can now expresses the analytical form of the Helmholtz free energy as the quadratic form of its internal state variables given below:

$$\begin{aligned}\rho\Psi &= \frac{1}{2}\varepsilon^e : E(\phi) : \varepsilon^e + \frac{1}{2}\mathbf{a}^{(r)}_{(k)}\mathbf{A}^{(r)}_{(k)} : \mathbf{A}^{(r)}_{(k)} \\ &+ \frac{1}{2}\mathbf{b}^{(r)}_{(k)}\nabla\mathbf{A}^{(r)}_{(k)} : \nabla\mathbf{A}^{(r)}_{(k)} + \frac{1}{2}\mathbf{c}^{(r)}_{(k)}\nabla\nabla\mathbf{A}^{(r)}_{(k)} : \nabla\nabla\mathbf{A}^{(r)}_{(k)} \\ &+ \frac{1}{2}\mathbf{d}^{(r)}_{(k)}\overline{\left\|\hat{\nabla}\hat{\mathbf{A}}^{(r)}_{(k)}\right\|\left\|\hat{\nabla}\hat{\mathbf{A}}^{(r)}_{(k)}\right\|}\end{aligned} \tag{8}$$

Using equations (7) and (8) the following definitions can be obtained for the thermodynamic forces.

$$V^{(r)}_{(k)} = a^{(r)}_{(k)} A^{(r)}_{(k)} \tag{9a}$$

$$U^{(r)}_{(k)} = b^{(r)}_{(k)} \nabla A^{(r)}_{(k)} \tag{9b}$$

$$W^{(r)}_{(k)} = c^{(r)}_{(k)} \nabla\nabla A^{(r)}_{(k)} \tag{9c}$$

$$Z^{(r)}_{(k)} = d^{(r)}_{(k)} \overline{\left\|\hat{\nabla}\hat{A}^{(r)}_{(k)}\right\|} \tag{9d}$$

The value of the thermodynamic forces can be obtained through the evolution relations of the internal state variables. However, it should be pointed out that there are two types of evolution equations that need to be developed such as one at the macro and the other at the mesoscale level. The former can be obtained by assuming the physical existence of the dissipation potential at the macroscale. The later one can be obtained by a micromechanical or phenomenological approach.

One may consider that the evolution equations of the internal state variables, $A^{(r)}_{(k)}$, can be obtained by integrating the evolution equations of the local internal state variables at the mesoscale, that is $\hat{A}^{(r)}_{(k)}$ over the domain of the RVE. However, integration of the $\hat{A}^{(r)}_{(k)}$ is more cumbersome task mathematically since at the mesoscale $\hat{A}^{(r)}_{(k)}$ is a function of many different aspects of material inhomogeneities. Therefore, in this work evolution equations of the macroscale internal state variables are obtained through the use of the generalized normality rule of thermodynamics. In this regard the macroscale dissipation potential is defined only in terms of the macroscale flux variables as follows:

$$\Theta = \Theta(\dot{\varepsilon}^p, \dot{A}^{(r)}_{(k)}, \nabla\dot{A}^{(r)}_{(k)}, \nabla\nabla\dot{A}^{(r)}_{(k)}, \overline{\left\|\hat{\nabla}\hat{A}^{(r)}_{(k)}\right\|}) \tag{12}$$

By using the Legendre-Fenchel transformation of the dissipation potential (Θ), one can obtain complementary laws in the form of the evolution laws of flux variables as function of the dual variables.

$$\Theta^* = \Theta^*(\sigma, V^{(r)}_{(k)}, \nabla V^{(r)}_{(k)}, \nabla\nabla V^{(r)}_{(k)}, \overline{\left\|\hat{\nabla}\hat{V}^{(r)}_{(k)}\right\|}) \tag{13}$$

It is assumed here that distribution of the gradient function is symmetrical, and the modified plastic multiplier is given below:

$$\dot{\Lambda} = \frac{B_{ij} C_{ijkl} \dot{\varepsilon}_{kl}}{\Gamma + \Pi + T} \tag{14}$$

where,

$$B_{ij} = \frac{\partial f}{\partial \sigma_{ij}} \tag{15.a}$$

$$\Gamma = B_{ab} C_{abcd} B_{cd} \tag{15.b}$$

$$\Pi = \frac{\partial f}{\partial \varepsilon_v^p}(B_{mm} - a_1\nabla^2 B_{mm} - b_1\nabla^2 \hat{B}_{mm}) \tag{15.c}$$

$$T = \frac{\partial f}{\partial \overline{\alpha}_{ab}}(A_{ab} - a_2\nabla^2 A_{ab} - b_2\nabla^2 \hat{A}_{ab}) \tag{15.d}$$

$$A_{ab} = \varepsilon_v^p \frac{c}{p}(s_{ab} - x\, p\, \alpha_{ab}) \tag{15.e}$$

$$\nabla^2 \hat{A}_{ab} = -C_2 \frac{\partial f}{\partial \nabla^2 \hat{\overline{\alpha}}_{ij}} p \tag{15.f}$$

$$\nabla^2 \hat{B}_{mm} = (-p + \frac{3}{2M^2} p_o \alpha_{ij} \alpha_{ij}) / K \tag{15.g}$$

Using equation (14) and additive decomposition of the strain rate, the following stress-strain relationship is obtained.

$$d\sigma_{ij} = C_{ijkl}(1 - \frac{B_{mn} C_{mnop} B^T_{op}}{B_{ab} C_{abcd} B_{cd} + H^v_p + H^a_p}) d\varepsilon_{kl} \tag{16}$$

In equation (16), H^v_p and H^a_p are defined as follows:

$$H^v_p = \frac{\partial f}{\partial \varepsilon^p_v}(B_{mm} - a_1\nabla^2 B_{mm} - b_1\nabla^2 \hat{B}_{mm}) \tag{17.a}$$

$$H^a_p = \frac{\partial f}{\partial \overline{\alpha}_{ij}}(A_{ab} - a_2\nabla^2 A_{ab} - b_2\nabla^2 \hat{A}_{ab}) \tag{17.b}$$

Equations (14) and (16) show the direct effects of the gradients on the constitutive relations.

3. RATE DEPENDENCY DUE TO VISCOSITY

Rate dependency of the material is due to the viscous property of the material. For localized large strain problems, such as shear band problems, the strain rate at the shear band is much larger than the strain rate outside the shear band. For example, the strain rate at the shear band may be 100 times that outside the shear band. When the overall strain level of the material is not large, the shear band typically does not develop, therefore, severe strain rate differences in the material do not develop. However, in situations of localized large strain problems such as the cone penetration problem, slope stability problem, etc. the incorporation of the rate dependency in the constitutive relations is required.

Typically a material shows higher strength at higher strain rates. This results in the mobilization of relatively higher shear strength of the material at the shear band than outside the shear band. Therefore the incorporation of rate dependency reduces the strain level at the shear band. This strain reduction is called regularization or homogenization.

When one incorporates the gradients in dense material, the gradients usually reduce the stress and strain level at the shear band, since the stresses and strains are transferred to the neighboring particles. Both the rate dependency of the material and the stress transfer by gradients act as homogenization mechanisms. When one has only one homogenization mechanism in the model, one may obtain –reasonably good results. However, the model calibration may not be correct for this case, and the extension of the model shall be very limited.

The homogenization effects by rate dependency is obtained by applying Perzyna's (1966) over stress approach. In this work, the creep is not addressed. The additive decomposition of total strain rate into elastic and visco-plastic parts gives the following relation.

$$\dot{\varepsilon}_{ij} = \dot{\varepsilon}^{e}_{ij} + \dot{\varepsilon}^{vp}_{ij} \tag{18}$$

In equation (18), $\dot{\varepsilon}_{ij}$ is the total strain rate, $\dot{\varepsilon}^{e}_{ij}$ is the elastic strain rate, and $\dot{\varepsilon}^{vp}_{ij}$ is the visco-plastic strain rate. When one applies the anisotropic modified Cam Clay model, and Perzyna's (1966) elasto-visco-plastic theory, one obtains the following expression for the visco-plastic strain.

$$\dot{\varepsilon}^{vp}_{ij} = <\Phi(F)> \frac{\partial f_d}{\partial \sigma_{ij}} \tag{19}$$

In equation (19), <> are the McCauly Brackets, f_d is a dynamic yield function incorporated with rate dependency, and $\Phi(F)$ is a visco-plastic multiplier, that is a function of F which is defined as follows:

$$F = \frac{f_d - \kappa_s}{\kappa_s} \tag{20}$$

where, κ_s is a hardening factor in static loading. Equation (20) is rewritten as follows:

$$f_d = (1+F)\kappa_s \tag{21}$$

When one uses an anisotropic modified Cam Clay model (Dafalias, 1987, Voyiadjis & Song, 2000), the yield criterion for static condition appears as follow:

$$f_s = p^2 - pp_o + \frac{3}{2M^2}\{(s_{ij} - p\alpha_{ij})(s_{ij} - p\alpha_{ij}) + (p_o - p)p\alpha_{ij}\alpha_{ij}\} = 0 \tag{22}$$

In equation (22), p is the mean principal stress, p_o is the initial mean principal stress, M is the slope of the critical state line (CSL) in p vs. q space, s_{ij} is the deviatoric stress, and α_{ij} is the rotation of yield surface with respect to p axis (dimensionless back stress factor). When one separates the isotropic hardening factor p_o in equation (22), one obtains:

$$\frac{p^2 + \frac{3}{2M^2}\{(s_{ij} - p\alpha_{ij})(s_{ij} - p\alpha_{ij}) - p^2\alpha_{ij}\alpha_{ij}\}}{p(1 - \frac{3}{2M^2}\alpha_{ij}\alpha_{ij})} = p^s_o \tag{23}$$

From equations (21) and (23), one obtains:

$$f_d = p^s_o(\frac{A}{p^d_o}) \tag{24}$$

where, A is expressed as the left term of equation (23) with respect to the rate dependent loading. The rate dependent yield criterion similar to equation (22) is given as follows:

$$\frac{\tilde{p}^2 + \frac{3}{2\tilde{M}^2}\{(\tilde{s}_{ij} - \tilde{p}\tilde{\alpha}_{ij})(\tilde{s}_{ij} - \tilde{p}\tilde{\alpha}_{ij}) - \tilde{p}^2\tilde{\alpha}_{ij}\tilde{\alpha}_{ij}\}}{\tilde{p}(1 - \frac{3}{2\tilde{M}^2}\tilde{\alpha}_{ij}\tilde{\alpha}_{ij})} - p^d_o = 0 \tag{25}$$

where, symbol ~ and superscript d indicate rate dependent loading. The functional form of Φ is assumed similar to Oka (1981) as follows:

$$\Phi(F) = Cexp(m\frac{p^d_o}{p^s_o}) \tag{26}$$

where, C and m are the material constants. In equations (25) and (26), the gradients are included in α_{ij}, p^s_o and p^d_o. Equations (18) to (26) are for the rate dependent gradient theory. Through this approach, the incorporation of both homogenization mechanisms is achieved.

4. RATE DEPENDENCY DUE TO FLOW IN POROUS MEDIA

The rate dependency of porous media depends not only on the viscous material properties but also on the pore fluid flow characteristics of the material during deformation. For time dependent loading for the saturated soils, the pore water in the soil is pressurized and tends to escape from the pressurized zone to the unpressurized zone. This is a natural procedure due to entropy. Due to the limited hydraulic conductivity of the soil, this pore water escape procedure (so called "dissipation") takes time. Therefore another time dependent behavior need to be considered here. The coupled theory of mixtures deals with this kind of flow problems in the porous media. Here, Prevost's (1980) coupling formulation for saturated granular material is adopted as follows:

$$-div\ [(n^w/\rho^w)\ \mathbf{K}^{ws}(grad\ P_w - \rho_w\mathbf{b})] + div\ \mathbf{v}^s = 0 \tag{27}$$

where, ρ^s is the mass density of the soil, ρ^w is the mass density of the water, $\mathbf{K}^{ws}$ is the permeability tensor, ρ_w is the density of water, $\mathbf{v}^s$ is the solid velocity, P_w is the pore water pressure, and $\mathbf{b}$ is the body force vector. For the application of rate dependent gradient theory to saturated geomaterials, equation (26) needs to be incorporated in the equilibrium equation and coupled with equation (27). This coupling effect is expected to be more prominent in the large strain problems. The incremental equilibrium equation from the principle of virtual force is given as follows:

$$
\begin{aligned}
{}^{n+1}R = & \int_{{}^nV} ({}^n\sigma'^{s}_{AB} + {}^nP_w\delta_{AB})\delta^n e_{AB} d^nV \\
& + \int_{{}^nV} ({}^n\sigma'^{s}_{AB} + {}^nP_w\delta_{AB})\delta^n \eta_{AB} d^nV \\
& + \int_{{}^nV} D^{***}_{ABCD} (\Delta_n e_{AB} + \Delta^n \eta_{AB})\delta^n e_{AB} d^nV \\
& + \int_{{}^nV} D^{***}_{ABCD} (\Delta_n e_{AB})\delta^n \eta_{AB} d^nV \\
& + \int_{{}^nV} J^s X^s_{A,a} X^s_{B,b} w_{ab} \Delta P_w ({}^n e_{AB} + {}^n \eta_{AB}) d^nV
\end{aligned}
\quad (28)
$$

where, D^{***}_{ABCD} is the elasto-plastic stiffness tensor, σ'^{s}_{AB} is the effective stress tensor, P_w is the pore water pressure, δ represents the increment, e_{AB} is the linear strain tensor, and η_{AB} is the non-linear strain tensor. The incremental form of equation (27) is expressed as follows:

$$
\begin{aligned}
& J^s C^s_{ij} \dot{\varepsilon}_{ij} - J C^{s-1}_{ij} C^{s-1}_{ij} X^s_{D,a} \frac{\partial}{\partial X_p} \\
& \times [\frac{n_w}{\rho_w} K^{ws}_{AB} X^s_{a,A} (\frac{\partial P_w}{\partial X_B} - \rho_w B_B)] = 0
\end{aligned}
\quad (29)
$$

where, $C^s_{ij} = X^s_{K,I} X^s_{K,I}$, $\dot{\varepsilon}_{ij}$ is the strain rate tensor, J is the Jacobian, and $X^s_{a,A} = \partial^{n+1} X_a / \partial^n X_A$, $B_b = b_b / X^s_{b,B}$.

Combining equations (28) and (29), one can express these relations in a matrix form such that:

$$
\begin{bmatrix} {}_nK & -{}_n\Omega \\ -{}_n\Omega^t & {}_n\Psi\delta t \end{bmatrix} \begin{bmatrix} \Delta U \\ \Delta W \end{bmatrix} = \begin{bmatrix} {}_n\Phi \\ {}_n\Pi \end{bmatrix} \quad (30)
$$

In equations (30), $_nK$ is the stiffness matrix, $_n\Omega$ is the coupling matrix, $_n\Psi$ is the flow matrix (which has hydraulic conductivity tensor in it), ΔU is the incremental nodal displacement, ΔW is the incremental pore water pressure, and δt is the incremental time.

5. CONCLUSIONS

The coupled equations of gradient-viscosity, and pore water flow are incorporated in this paper. The the more realistic analysis of the saturated geomaterials is therefore anticipated from this formulation.

6. REFERENCES

Dafalias, Y.F. 1987. An Anisotropic Critical State Clay Plasticity Model. *Constitutive Laws for Engineering Materials: Theory and Applications*, Eds. C.S. Desai et al. pp.513-521

Oka, F. 1981. Prediction of Time-Dependent Behavior of Clay, *10th Proceedings of International Conference on Soil Mechanics and Foundation Engineering*. pp.215-218

Perzyna, P. 1966. Fundamental Problems in Viscoplasticity. *J. of GE, ASCE*, Vol. 104, No. GT8, pp. 1075-1090

Prevost, J.H. 1980. Mechanics of continuous porous media. *International Journal of Engineering Science*. Vol. 18, pp.787-800

Taylor, D.W. 1948. *Fundamentals of Soil Mechanics*, John Wiley & Sons, pp.333-351

Voyiadjis, G.Z., Deliktas, B. 2000. Multi-scale Analysis of Multiple Damage Mechanisms Coupled with Inelastic Behavior of Composite Materials. *Mechanics Research Communications Journal*, Vol. 27, No.3, pp.295-300

Voyiadjis, G.Z., Deliktas, B. and Aifantis, E.C. 2000. Multi-scale Analysis of Multiple Damage Mechaisms Coupled with Inelastic Behavior of Composite Materials. *Computational Methods in Applied Mechanics and Engineering*. submitted for publication

Voyiadjis, G.Z. and Song, C.R. 2000. Anisotropic Modified Cam Clay Model with Plastic Spin for Finite Strains. *J. of EM Div. ASCE*. (in press)

Computer Methods and Advances in Geomechanics, Desai et al. (eds) © 2001 Balkema, Rotterdam, ISBN 90 5809 183 X

An investigation of element-size-dependency in strain localization

T.Y.Yap & M.A.Hicks
Manchester School of Engineering, University of Manchester, UK

ABSTRACT: Element-size-dependency, for strain-softening geomaterials, has usually been demonstrated for problems involving rapidly-developing failure mechanisms. This paper considers a practical geotechnical problem in which failure develops more gradually: i.e. passive earth pressure failure behind a 5m retaining wall. Results are presented for a wide range of uniform mesh densities, showing that a converged load-displacement response is possible when using a standard continuum. Although fine mesh discretisations are needed, as would be the case for a non-standard continuum, convergence can occur for shear band thicknesses much larger than observed in practice. Indeed, if only peak load convergence is required, much cruder meshes may be adequate.

1 INTRODUCTION

Many investigators have emphasised the problems encountered when modelling localization in non-associated strain-hardening, or strain-softening, geomaterials (e.g. Needleman 1988): i.e. for a standard continuum, there is a loss of elipticity in the constitutive equations, leading to element-size-dependency in finite element analysis (as seen in non-converged shear band widths and load-displacement responses). This has prompted the development of non-standard continuum theories for restoring the well-posedness of constitutive relations (Adachi et al. 1998).

So far, element-size-dependency has mainly been shown for problems in which failure mechanism development is rapid: e.g. as in brittle materials, or for simple loading tests such as biaxial compression. In these situations, it is difficult, if not impossible, to obtain a converged solution using standard continuum theory and the case for non-standard continua is, therefore, clearly demonstrated.

This paper investigates element-size-dependency for a practical problem in which failure develops more gradually; specifically, passive earth pressure failure, behind both smooth and rough retaining walls. Finite element results for different mesh densities have been compared, to see if a converged response is possible when using only a standard continuum.

2 SOIL MODEL

The non-associated Mohr-Coulomb soil model proposed by Hicks et al. (2000) has been used. This is based on linear elasticity, linear cohesion hardening/softening, and perfect plasticity. Figure 1 shows that the yield surface is a function of the accumulated plastic shear strain invariant $\bar{\varepsilon}^p$, where

$$\bar{\varepsilon}^p = \frac{\sqrt{2}}{3}\left((\varepsilon_1^p - \varepsilon_2^p)^2 + (\varepsilon_2^p - \varepsilon_3^p)^2 + (\varepsilon_3^p - \varepsilon_1^p)^2\right)^{1/2} \quad (1)$$

It is also a function of c_i, c_p and c_r: the initial, peak and residual cohesions, respectively. The hardening/softening moduli are given by $h = \mathrm{d}c/\mathrm{d}\bar{\varepsilon}^p$, in which $h = h_h$ for hardening and $h = h_s$ for softening. Plastic volumetric strains are defined by a constant dilation angle ψ.

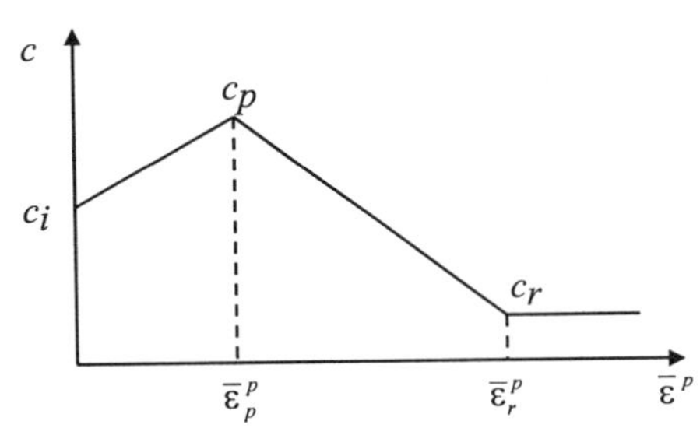

Figure 1. Cohesion as a function of accumulated plastic shear strain invariant.

3 FINITE ELEMENT ANALYSIS

Figure 2 shows the problem geometry: i.e. a 5m wall is translated horizontally into a drained soil mass, of dimensions 15m × 7.5m. The mesh base is fixed and the right-hand boundary is on rollers allowing only vertical movement, while the top boundary, and left-hand boundary below the wall, are free. Perfectly smooth and perfectly rough walls have both been considered.

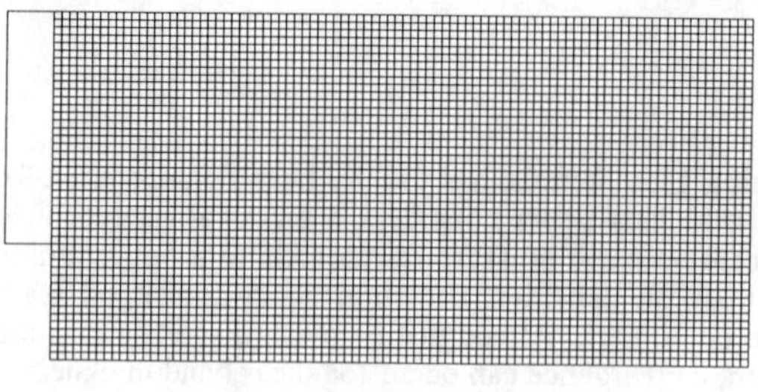

Figure 2. Passive earth pressure problem (mesh M8).

Table 1 lists the material parameter values and, Table 2, the nine uniform meshes which have been considered, these covering a wide range of element sizes. Each mesh uses 8-node square elements and 2 × 2 Gaussian integration. One of the finer meshes (M8) is shown in Figure 2.

Table 1. Summary of material parameters.

Parameter	Symbol	Value
Young's modulus	E	10^5kPa
Poisson ratio	ν	0.3
Friction angle	ϕ	30°
Dilation angle	ψ	0, 10*, 20, 30°
Initial cohesion	c_i	5kPa
Peak cohesion	c_p	25kPa
Residual cohesion	c_r	5kPa
Hardening modulus	h_h	8700kPa
Softening modulus	h_s	-580kPa

* Main dilation angle considered.

Table 2. Uniform meshes used in analysis.

Code	No. of elements	Elements along wall
M1	72	4
M2	162	6
M3	450	10
M4	882	14
M5	1458	18
M6	2178	22
M7	3042	26
M8	4050	30
M9	5832	36

For each analysis, the starting stresses have been generated using a unit weight of 20kN/m^3 and a coefficient of earth pressure at rest, K_0=1.0. The wall has been loaded by applying horizontal, equal-sized, displacement increments. (Parametric studies have been carried out to ensure that increment sizes, for different mesh densities, have been small enough (Hicks et al. 2000).)

3.1 *Smooth retaining wall*

For the smooth wall, the nodes attached to the wall are free to move in the vertical direction. Figure 3 shows the horizontal load-displacement responses for all meshes listed in Table 2, and compares peak and residual loads with Rankine limit loads based on c_p and c_r. The figure shows load-displacement convergence at higher mesh densities.

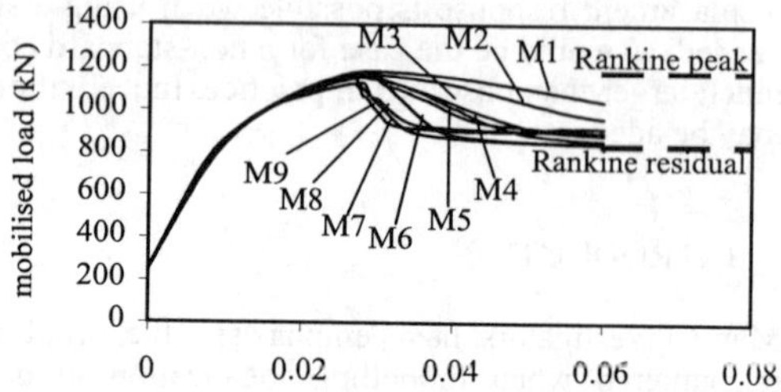

Figure 3. Influence of element size on smooth wall load-displacement response (ψ=10°).

3.2 *Rough retaining wall*

For the rough wall, the nodes attached to the wall are fully restrained in the vertical direction. Figure 4 compares the nine load-displacement responses, with peak and residual loads now being compared with the limiting solutions of Caquot and Kerisel (once again, based on c_p and c_r). As for the smooth wall, the solution converges for finer meshes.

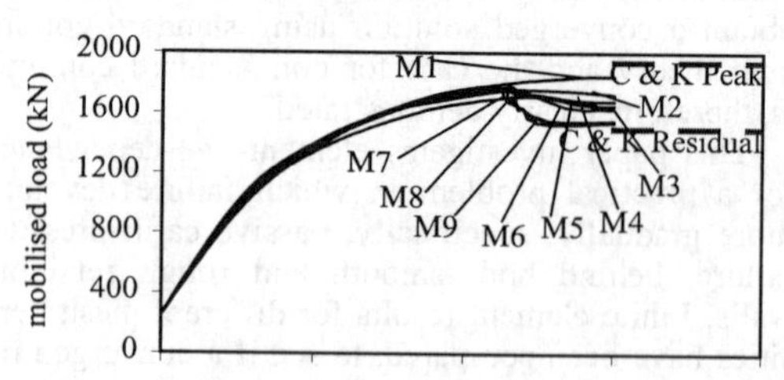

Figure 4. Influence of element size on rough wall load-displacement response (ψ=10°).

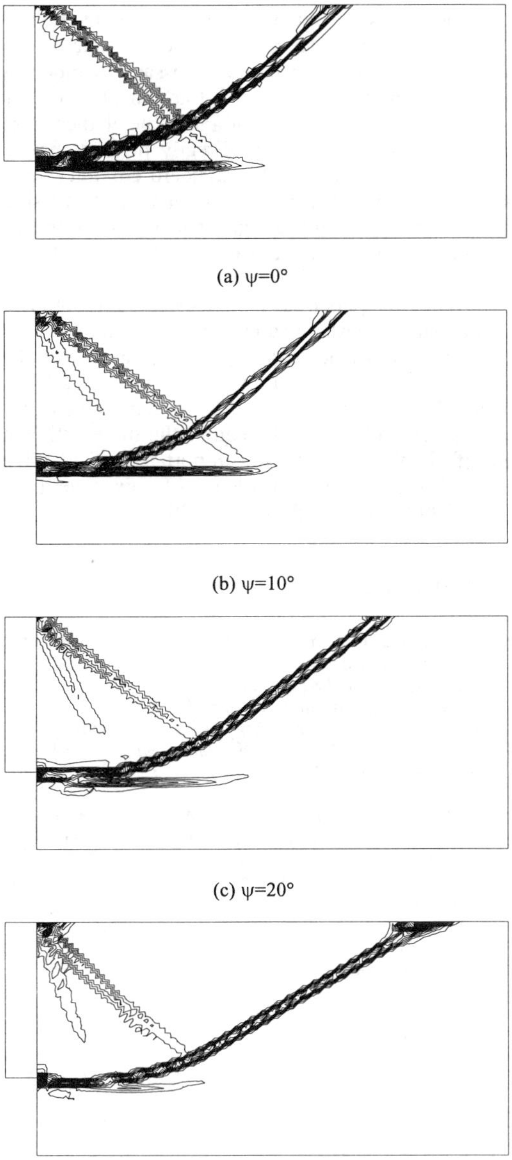

(a) ψ=0°

(b) ψ=10°

(c) ψ=20°

(d) ψ=30°

Figure 5. Influence of dilation angle on accumulated shear strain invariant contours at end of rough wall analysis (mesh M8).

Mechanism development for the rough wall has also been studied, as illustrated in Figure 5, which shows accumulated shear strain invariant contours at the ends of four analyses, using mesh M8 and various dilation angles. These plots highlight the complex nature of mechanism development: initially, two planes develop, one horizontally from the wall base, the other at an oblique angle from the wall top; then, at around peak load, a primary plane takes over, this originating from the wall base and curving upwards before exiting the ground surface at an angle approximately equal to the Roscoe solution, $\theta=45°-\psi/2$ (refer to Table 3). The primary plane becomes the dominant plane, as demonstrated in Figure 6, which shows incremental shear strain invariant contours at the end of the analysis for $\psi=10°$ (cf. Figure 5(b)).

Table 3. Comparison of theoretical and computed failure plane orientations for rough wall (exit angle).

Dilation angle	Roscoe solution	Computed result
0°	45°	43°
10°	40°	40°
20°	35°	36°
30°	30°	30°

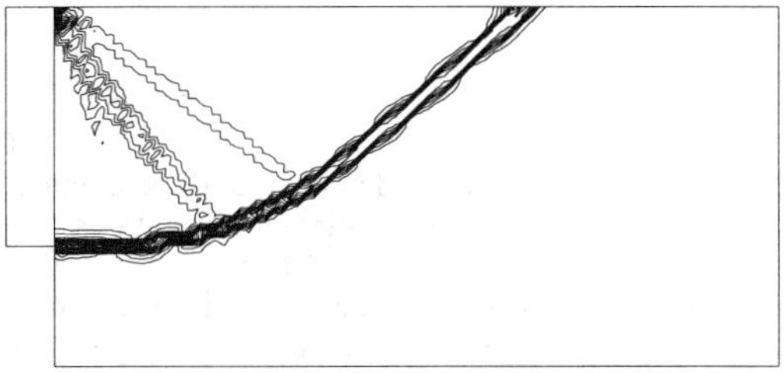

Figure 6. Contours of incremental shear strain invariant at end of rough wall analysis for ψ=10° (mesh M8).

4 SUMMARY OF RESULTS

Figure 7 summarises the results for smooth and rough walls, by plotting peak load, and load at the end of the analysis, against mesh density (as indicated by the number of elements directly behind the wall). For the rough wall, peak load convergence is very rapid: i.e. the difference in peak load between meshes M2 and M9 is less than 1%. For the smooth wall, the rate of convergence is slower, but still reasonable.

For both smooth and rough walls, post-peak convergence is also attainable, but this requires much finer meshes than for peak load convergence. Note that Figure 7 is slightly misleading, because it implies that the residual load converges more rapidly for the smooth wall case; however, this is due to the smooth wall analyses being taken further beyond peak load than the rough wall analyses (compare Figures 3 and 4). Indeed, complete post-peak convergence (i.e. residual load and post-peak slope) occurs more rapidly for the rough wall.

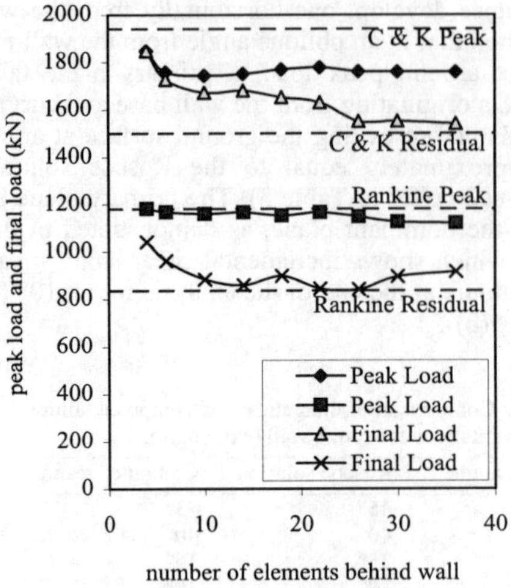

Figure 7. Influence of number of elements along wall on peak and final loads (for smooth and rough walls).

5 CONCLUSIONS

Previous researchers have demonstrated the element-size-dependency of finite element solutions, for strain-softening geomaterials, when using standard continua; this has mainly been for brittle materials, or for small problems, such as biaxial compression, in which failure mechanism development is rapid (due to the near-uniform stress conditions prior to failure).

This paper has demonstrated that element-size-dependency also exists for larger scale problems, in which failure mechanisms develop more gradually (i.e. progressive failure). However, unlike analyses of biaxial compression, using a standard continuum, converged load-displacement responses have been obtained when using fine mesh discretisations. For the 5m wall in this investigation, 26-30 8-node elements have been needed along the wall, corresponding to an element size of around 0.2m (i.e. a nodal spacing of 0.1m). This gives a shear band thickness of approximately 0.3m (i.e. 1.5 × element size); hence, although small elements are needed, convergence of the load-displacement response may be possible for shear band thicknesses much greater than observed in practice. Indeed, it has been shown that peak load convergence may be obtained with far fewer elements; i.e. for the rough wall case, only 6 elements have been needed behind the wall. This is relevant for geotechnical problems in which failure is load-induced (e.g. due to gravity).

For standard continua, there is no convergence of shear band thickness with decreasing element size. However, in engineering practice, shear band thickness is of secondary importance, once it has ceased to influence global behaviour (i.e. once convergence of the load-displacement response has been achieved). Should it become necessary to model a specific shear band thickness, this may be crudely done through specification of a suitable element size (as shear band thickness is directly proportional to element size). However, for standard continua there is no control over the number of elements traversing a band; hence, the advantage of using non-standard continua for detailed analyses of shear bands themselves.

Finally, this paper has emphasised the importance of adequate mesh discretisation for localization problems involving strain-softening geomaterials; this is equally important for both non-standard and standard continua. Fulfilling this requirement for large scale problems will require the more efficient use of elements; hence, the importance of adaptive mesh refinement for problems of this type (e.g. Hicks 1997, 2000, Hicks et al. 2000).

REFERENCES

Adachi, T., Oka, F. & Yashima, A. (eds.) 1998. Localization and bifurcation theory for soils and rocks. *Proc. 4th Int. Workshop on Localization and Bifurcation Theory for Soils and Rocks*. Gifu, Japan. Rotterdam: Balkema.

Hicks, M.A. 1997. Implementation of adaptive mesh refinement in MONICA. *Proc. 6th Int. Symp. Num. Models Geomech*: 409-414. Montreal, Canada. Rotterdam: Balkema.

Hicks, M.A. 2000. Coupled computations for an elastic-perfectly plastic soil using adaptive mesh refinement. *Int. J. Num. Anal. Meth. Geomech.* 24: 453-476.

Hicks, M.A., Yap, T.Y. & Abu Bakar, A. 2000. Adaptive and fixed mesh study of localization in a strain-softening soil. (submitted for publication) *Proc. 5th Int. Workshop on Bifurcation and Localisation in Geomechanics*. Perth, Australia. Rotterdam: Balkema.

Needleman, A. 1988. Material rate dependence and mesh sensitivity in localization problems. *Computer Meth. Appl. Mechanics and Eng.* 67: 69-85.

6 Applications, validation, construction processes

Computer Methods and Advances in Geomechanics, Desai et al. (eds) © 2001 Balkema, Rotterdam, ISBN 90 5809 183 X

Performance of deep excavation in overconsolidated clay: Measurements and numerical analysis

Y. El-Mossallamy
Ain Shams University, Cairo, Egypt

ABSTRACT: Deep excavation pits up to 20 m and deeper are required to build basements and underground car parks in the centers of large towns such as Frankfurt, Germany. This is due to the lack of available building sites and the increased need for more office space during the course of the city's development into a European center of finance. The main geotechnical challenge during the excavation works is to keep the deformation of the pit excavation support systems within acceptable limits to minimize damage on adjacent structures. The paper describes the performance of the approx. 16.0 m deep pit excavation required for the construction of the high-rise building "Japan Center" in Frankfurt. The unique feature of this project was the use of relief wells in conjunction with staggered pile lengths to minimize the water pressure on the retaining structure in order to achieve an economical design that fulfills the serviceability and the environmental requirements. A finite element analysis was conducted to model the behavior of the pit excavation coupling both the hydraulic and the deformation calculations. The results of the measurements and of the numerical analysis will be presented and compared in a short overview illustrating the benefits that can be achieved.

1 INTRODUCTION

During the course of its development into a European center of finance, the Rhein-Main metropolis Frankfurt has developed a skyline unique in Europe with regard to its prominent high-rise buildings.

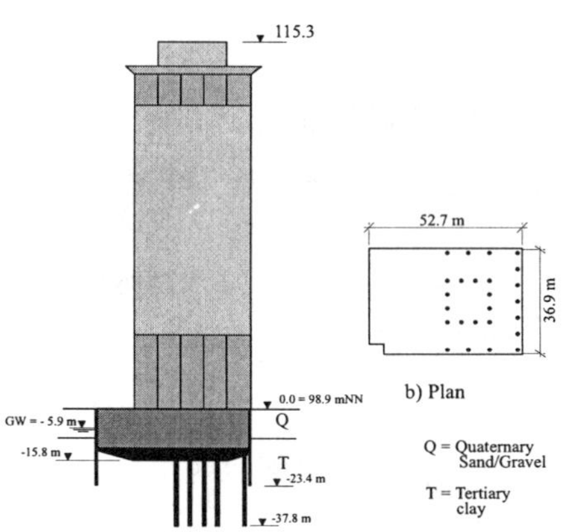

Figure 1: General layout of the high rise building Japan-Center and its foundation

A deep excavation pit of about 16 m was required to build the basements and underground car parks of the Japan building. The main tower of the building complex has an area of about 37 x 37 m and it is planed eccentric to the basement area (Fig. 1). The foundation of both the main tower and the side basements could be constructed without settlement joints applying a piled raft foundation. The piles were placed eccentrically below the tower to minimize the building tilting to acceptable limits.

2 GEOTECHNICAL CONDITION

In the town center the subground consists of quaternary sand and gravel down to about 5 to 7 m followed mainly by the so-called Frankfurter Clay which developed 2 to 10 million years ago as a result of the sedimentation in the tertiary sea in the Mainz basin. This clay includes limestone banks, lignite coal lenses and layers of calcareous sand. The groundwater level is mostly just above the clay surface and circulates in the fissured limestone banks and sand lenses resulting in different confined aquifer pressures. The clay is geologically overconsolidated through older, already eroded sediments and volcanic rock from the Vogelsberg volcanoes. As a result of this development, the subsoil is highly horizontally stressed. This influences the deformation and the failure behavior of this overconsolidated clay. The main geotechnical parameters of different soil layers are summarized in Table 1.

3 PIT EXCAVATION SUPPORT SYSTEM

Due to the above mentioned subground conditions and due the presence of underground contamination in some areas, lowering of the groundwater table must be prevented as far as possible to minimize the risk of excessive settlements of neighboring buildings and to avoid environmental impacts. Under the subsoil conditions of Frankfurt this is achievable effectively by applying impermeable walls round the pit excavations and connecting the lower end of them to the clay layer beneath the water leading quaternary layer.

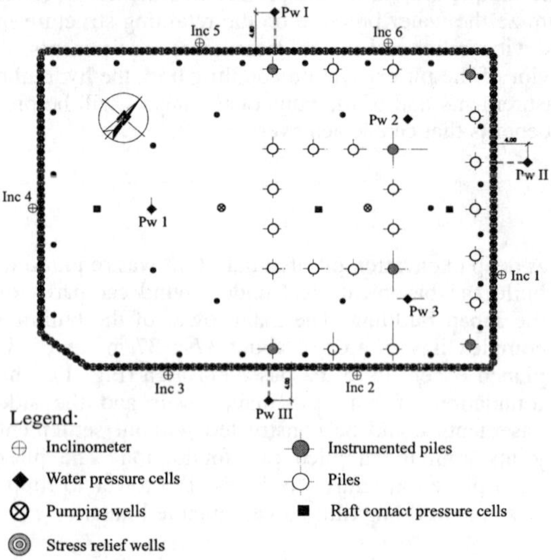

Figure 2: Plan with measuring devices

The unique feature of the retaining structure used to support the pit excavation of the Japan Center project was the use of relief wells in conjunction with staggered pile lengths (Figs. 2,3) to minimize the water pressure on the retaining structure in order to achieve an economical design that fulfills the serviceability and the environmental requirements. The relief wells must be distributed over the pit area to prevent bottom heave due to upward directed hydraulic gradients. The retaining walls consisted of secant bored piles with a diameter of 0.9 m. Plain concrete piles were first constructed 1.5 m apart with the pile tips just beneath the final excavation depth. Reinforced concrete piles were then constructed between the plain piles with 0.15 m overlapping to guarantee the water tightness. The reinforced piles had an embedded length of about 8 m below the final excavation depth (Fig. 2).

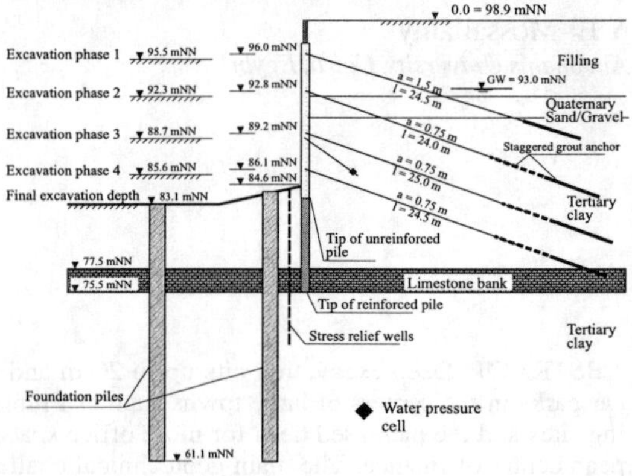

Figure 3: Layout and dimensions of the pit excavation

4 NUMERICAL ANALYSIS

A finite element analysis was conducted to model the behavior of the pit excavation coupling both the hydraulic and the deformation calculations. The soil was idealized using an elasto-plastic constitutive law (the hardening soil model, HSM Plaxis software). The main advantage of this model is its ability to distinguish between un-/reloading and primary loading conditions. The soil stiffness depends in this model on the stress path and the applied stress level. In the presented work the drained soil parameters were applied to study the long term behavior of the pit excavation. For the design purposes both short and long term conditions should be checked.

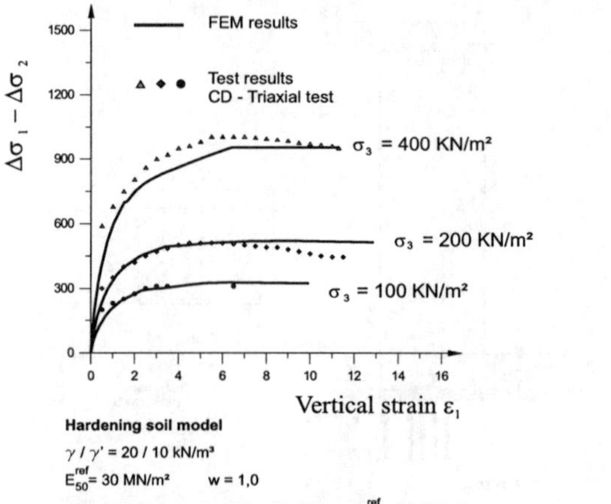

Figure 4: Back calculation of the conducted triaxial tests on undisturbed samples of Frankfurt clay

A traditional soil exploration as well as an extensive in-situ investigation program were conducted to obtain a reliable estimation of the behavior of the pit excavation support system. Series of triaxial tests were carried out on undisturbed samples of Frankfurt clay. Figure 4 shows the results of the triaxial tests and the numerical results using the hardening soil model. This comparison verifies the applied constitutive law. El-Mossallamy & Stahlmann (1999) have shown that the hardening soil model is a very realistic model to deal with problems of excavation due to its ability to consider the plastic strain due to shear stresses below the Mohr-Coulomb failure surface (shear hardening).

Table 1: Geotechnical parameters

a- Hardening soil mode

Soil parameter	Filling	Quaternary Sand/Gravel	Overconsolidated clay
E_{50}^{ref} [MN/m^2]	20	30	35
E_{ur}^{ref} [MN/m^2]	50	75	105
ν_{ur} [-]	0.2	0.2	0.2
m [-]	0.5	0.5	1.0
R_f [-]	0.9	0.9	0.9
γ / γ' [kN/m^3]	18/8	19/11	20/10
k_x [m/sec]	10^{-3}	10^{-3}	2.5×10^{-5}
k_y [m/sec]	10^{-3}	10^{-3}	0.01 k_x
c´ [kN/m^2]	-	-	20
φ´ [°]	30	35	20
K_o [-]	0.5	0.43	0.8

where:

E_{50}^{ref}	Primary loading stiffness
E_{ur}^{ref}	Unloading/reloading stiffness
ν_{ur}	Unloading/reloading Poisson's ratio
m	Power in stiffness laws
R_f	Failure ratio
γ/γ'	Total / Effective unit weight of soil
k_x, k_y	Permeability coefficient in the horizontal and vertical direction
c'	Cohesion
φ'	Angle of internal friction
K_o	Coefficient of earth pressure at rest

b- Mohr-Coulomb model

Soil parameter	Limestone
E [MN/m^2]	750
ν [-]	0.3
γ / γ' [kN/m^3]	20/10
k_x [m/sec]	10^{-3}
k_y [m/sec]	10^{-3}
c´ [kN/m^2]	200
φ´ [°]	35

c- Structure elements

Piles: $E_{concrete}$ = 30000 MN/m² and ν = 0.2
Anchor tendons: E_{steel} = 195000 MN/m²

The required soil parameters of different applied constitutive laws were determined based on the conducted laboratory and in-situ tests as well as on experience gained in similar projects. These parameters are given in Table 1.

A plane strain finite element analysis was conducted using isoparametric triangular elements with 15 nodes. Figure 5 shows the used finite element mesh. The bored pile wall was modeled as linear elastic material. The pressed tieback was modeled using a special element that can carry only tension forces and has no bending stiffness. The free length of the anchor was modeled with a node to node truss element that carries only axial forces.

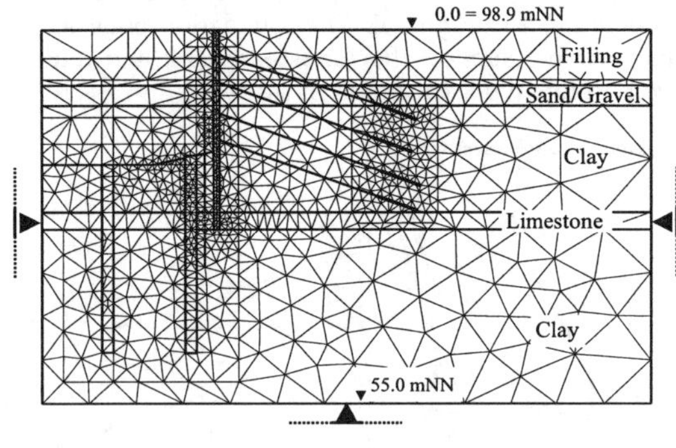

Figure 5: Finite element mesh

Since the stress history has a significant effect on the soil behavior, the different construction stages were modeled as follows:

Stage 1: Initial stresses
Stage 2: Construction of the bored pile wall
Stage 3: First excavation phase
Stage 4: Insertion and prestressing of the first anchor
Stage 5: Second excavation phase with seepage analysis considering the hydraulic conditions of the soil, the wall and stress relief wells
Stage 6: Insertion and prestressing of the second anchor
Stage 7: Third excavation phase with seepage analysis
Stage 8: Insertion and prestressing of the third anchor
Stage 9: Fourth excavation phase with seepage analysis
Stage 10: Insertion and prestressing of the fourth anchor
Stage 11: Fifth excavation phase to reach the final excavation depth with seepage analysis

Due to the limited space, only the results of the final stage of construction will be presented and discussed.

Figure 6 demonstrates a comparison between the calculated and the measured horizontal movements of the bored pile wall. To investigate the effect of different parameters, the following cases were calculated:

Case 1: Considering neither the effect of the limestone bank nor the effect of foundation piles

Case 2: Considering the effect of the limestone bank and neglecting the foundation piles

Case 3: Considering the effect of both the limestone bank and the foundation piles

The good agreement between measured and calculated movements (Case 3) can be recognized. The relatively stiff limestone bank reduces the horizontal displacement of the wall significantly as can be seen from curve 2 in Figure 6. The foundation piles act as anchors that support the limestone bank and reduce the horizontal displacement of the wall.

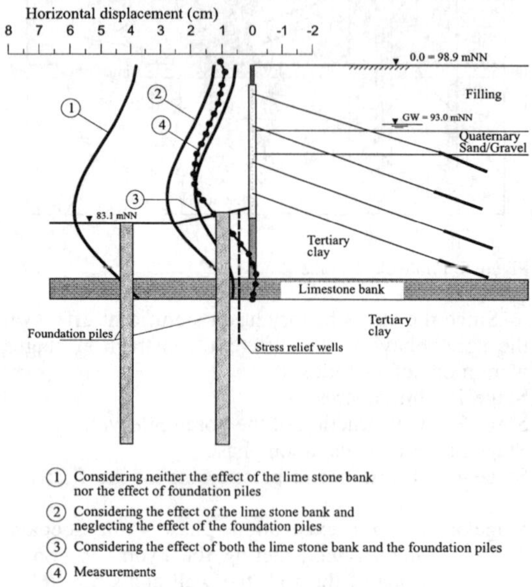

Figure 6: Comparison between observed and calculated horizontal displacements of the excavation support wall

Figure 7 compares the calculated settlement beside the pit excavation with the measured values at different buildings in the close vicinity of Japan Center. In general there is a good agreement between calculated and observed values. It can be observed that the pit excavation effects the displacement of neighboring buildings in a relatively large area in its vicinity. This is mainly due to the increase of the soil effective stresses due to seepage forces. There is also an observed heave specially in the domain of the anchor grout in contrast to the calculated values. This discrepancy is due to the relatively high pressure applied by the grouting of the anchor body during its installation and the secondary injection one day after the installation. This grouting pressure was not considered in the numerical modeling.

The water pressure acting on such a water tight retaining structure has a pronounced effect on the structural design and on the deformation behavior. In traditional analyses this water pressure is in most cases considered as hydrostatic pressure. Regarding the seepage, the actual water pressure acting on the wall is smaller than the hydrostatic pressure (Fig. 8).

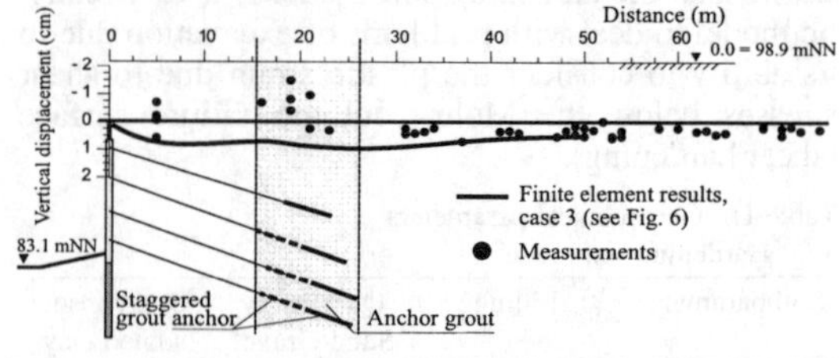

Figure 7: Comparison between observed and calculated vertical displacements beside the pit excavation

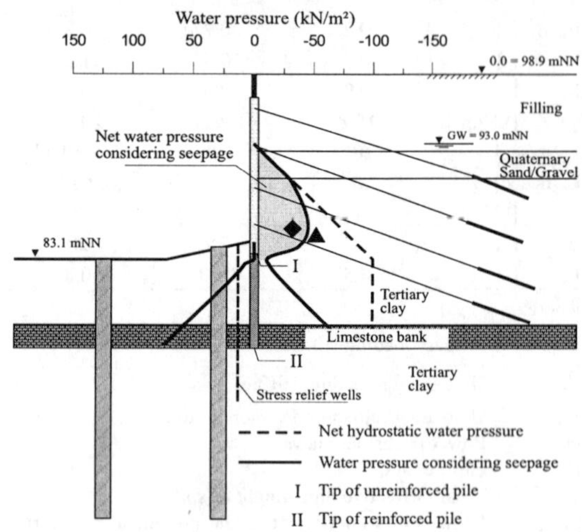

a- Water pressure distribution

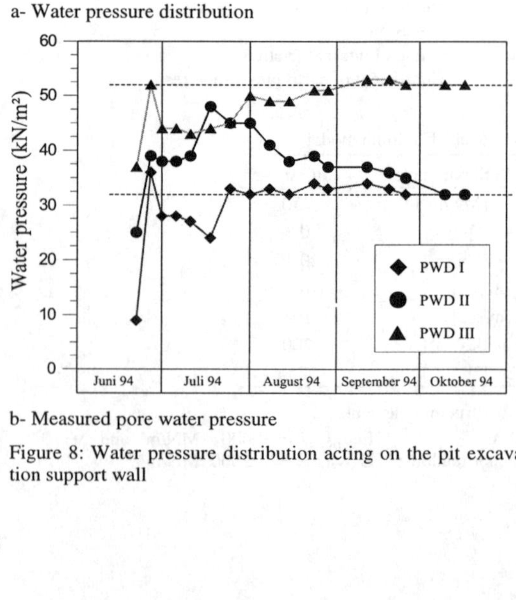

b- Measured pore water pressure

Figure 8: Water pressure distribution acting on the pit excavation support wall

The finite element results show that taking the seepage analysis into consideration in conjunction with convenient hydraulic conditions (staggering the pile lengths and installing stress relief wells) reduces the water pressure acting on the wall significantly. This leads to a more economic design regarding the required embedded length of the wall, the internal stresses of the wall and the anchor forces. Figure 8b shows the measured water pressure behind the wall. The measured values ranged between 32 kN/m² and 54 kN/m² which agree very well with the calculated values. This supports the use of staggering piles to decrease the water pressure on the wall and to enhance the deformation behavior of the system.

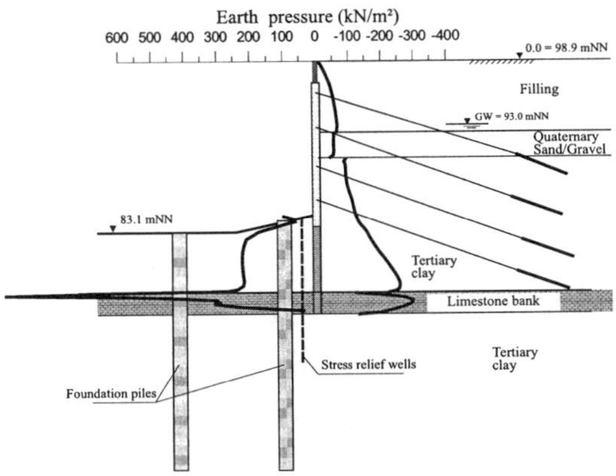

a- Earth pressure distribution acting on the wall (FEM)

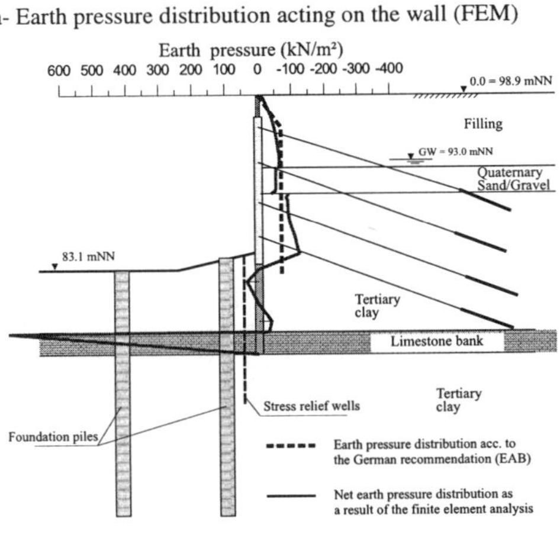

b- Comparison between the results of FEM and the German recommendation EAB

Figure 9: Earth pressure distribution acting on the pit excavation support wall

Figure 9a shows the earth pressure distribution acting on the wall as a result of the finite element analysis (FEM). The net earth pressure acting on the wall is demonstrated in Figure 9b. The earth distribution acting on the wall till the zero loaded point according to the German recommendation for pit excavation (EAB) is added for comparison. This comparison shows a good agreement between the results of the finite element analysis where the earth pressure is an output of the analysis and the earth pressure distribution according to the classical design where it is given as input applied load. The more important feature by the classical design is the choice of a reliable earth pressure distribution to obtain a more reliable design of the wall. The suggested values given by the German recommendation EAB seem to be reasonable.

5 CONCLUSION

Understanding of the effects of the interaction between the pit excavation support system and the subsoil based on appropriate theoretical knowledge and on experienced application of measurement techniques and numerical modeling together with tested and proven implementation design methods are the necessary qualification for a safe and economic design.

The field monitoring and feed-back of this information into the design procedure of similar projects by a suitable analytical model is a challenging geotechnical engineering task. Both the existing knowledge and the design method can be improved by such an interactive process between observation and numerical modeling.

The conducted finite element analysis supports the classical methods that are based more or less on the experience gathered over a long time. Nevertheless, this example shows the advantages of the finite element analysis to obtain a more realistic estimation of the deformation behavior coupling both the seepage and the deformation analysis in a unique model taking more complex effects such as the presence of stiff layers and the foundation piles into consideration. This type of coupling analysis leads to a more economic design which fulfils all stability and serviceability requirements.

6 REFERENCES

PLAXIS 1999. *Finite element code for soil and rock analysis.* Version 7.11, Manual, Plaxis B.V, Balkema.

Vermeer, p. & Meier, Ch. 2000. Deformation analysis for deep excavations. *Proceeding of the fourth international geotechnical engineering conference:* 153-171. Cairo, Egypt.

Sills, G.C, Burland, J.B., und Czechowski, M.K. 1977. Behavior of an anchored diaphragm wall in stiff clay. *The tenth international conference on soil mechanics and foundation engineering volume* 2: 147-154. Tokyo.

Breth, H. & Romberg. W. 1973. Messungen an einer verankerten Baugrubenwand *Forschungsbericht 10/1971, TU Darmstadt, Germany.*

El-Mossallamy, Y. & Stahlmann, J 1999. Monitoring and numerical analysis of tunnels in complex geological conditions *Proceeding of the International Symposium Beyond 2000 in Computational Geotechnics:* 143-151. Balkema, Rotterdam.

Computer Methods and Advances in Geomechanics, Desai et al. (eds) © 2001 Balkema, Rotterdam, ISBN 90 5809 183 X

Stability analysis of high slope rock mass in the Three Gorges permanent shiplock during engineering unloading

Wu Gang
Department of Architecture and Civil Engineering, Shanghai Jiaotong University, People's Republic of China

ABSTRACT: This paper summarizes the properties and the failure characteristics of the high slope rock mass in the Three Gorges shiplock under engineering excavation. Based on the mechanical mechanism of rock mass failure caused by unloading in the excavating process and use of a designed finite element program, a numerical simulating research on stability of the shiplock high slope in Three Gorges Project has been performed in terms of the present design scheme. Results of the study show that the present design scheme is safe and reliable.

1 GENERALIZATION OF THE PROJECT

Three Gorges Project on Yangtze River represents the largest hydroelectric project in the world at present, which can bring enormous benefits of comprehensive utilization in preventing flood, generating electricity, shipping, breeding, irrigation. The permanent shiplock in Three Gorges Project, which is a double-route five-grade shiplock, can assure that the Yangtze River golden waterway will be easily accessible, so that permits cargo ships of ten-thousand tons level to arrive in Wuhan and Chongqing directly, which is key to shipping through Three Gorges Project.

The Three Gorges permanent shiplock system, the total length of which is 6442m, lies on the left side of Yangtze River. The length of the shiplock is 1617m. The effective dimension of a shiplock room is 280×34×5m. A rock mass pier that is 60m wide lies between shiplock rooms. The whole shiplock rooms are built in the granite hill by excavating deeply. The maximum deepness reaches up to 174.5m. After excavating, the maximum height of the north slope is 137.8m, while 157.8m to the south slope. The slopes on both sides of the pier are about 50m high, the maximum height of which is 70m. As Three Gorges Project is the largest hydroelectric structure in the world, the shiplock high slope in Three Gorges Project, which involves factors such as length, height, gradient and importance, is a rare rock slope fit for excavating deeply in domestic and abroad.

The stability problem is much more difficult and important for the shiplock high slope in Three Gorges Project than that for common projects, for it is built by a large-scale excavating in rock mass subjected to certain stresses in the ground stress field. Stability of the high slope largely depends on the construction style, as well as geological conditions, configurations and environmental factors. Hence, it is of great importance to carry out researches on stability of the shiplock high slope in Three Gorges Project during engineering unloading. It is helpful that to optimize the design, establish efficient and rational excavating procedures for construction, set up an economical and reliable remedy scheme, found a security-supervise system which can monitor the behavior of rock mass comprehensively during slope construction and performance. Many Chinese scholars have made vast studies on the stability problem of the shiplock high slope in Three Gorges Project, and certain results have been obtained (Dong, 1994; Ha & Li, 1994; Xu et al. 1994).

Based on the mechanical mechanism of rock mass unloading failure and use of the designed finite element program, author studied on the constructional excavating project of the representative geological section 20-20' of the shiplock high slope in Three Gorges Project have been preformed according to the present design.

2 NUMERICAL ANALYSIS ON STABILITY OF THE HIGH SLOPE ROCK MASS IN THE THREE GORGES SHIPLOCK UNDER EXCAVATION UNLOADING

To study changes in deformations and stresses of rock mass of the high slope under the effect of excavating unloading. A numerical simulation analysis, which is for the highest representative geological section 20-20' of the shiplock high slope, has been carried out by using the designed finite element program.

2.1 *Partition of the section and its calculated meshes*

2.1.1 *General conditions of the section*

The section 20-20' that takes on a hypsography of north-lowness and south-tallness lies in No.3 gate-room, it is the section that has the largest slope height during excavating. It distances to the beginning of the gate-room is about 79m. After excavating, the height of the slope is about 166m, the hatch on the ground surface is 366m in width.

In this section, there is an entirely strong-weathering stratum which is about 18m under the land surface, followed with a feeble-weathering stratum about 20m in depth, below which is tiny-weathering and fresh rock mass respectively. The stratums contain β_μ γ dikes and schist xenoliths.

2.1.2 *Partition of the calculated region and meshes*

Considering the specific conditions of the original section, together with simplicity and convenience in calculation, it is appropriate to expand the calculated region as large as possible if the calculation conditions permit, so that the effect of boundary on the result can be reduced. The calculated area is described as follows: It is 360m from the left and 420m from the right to the center line of the middle shiplock pier which is treated as the axis. The total length of the horizontal and the perpendicular is 780m and 320m respectively (namely the altitude to Wusong is 260m), and the borderline is defined according to the present design scheme.

In terms of the characteristics of material and the procedures of the excavating steps, rock mass can be divided into 12 different element groups by using the equal-parameter element of two- dimension quadrangle. Total elements and nodes are 779 and 820 respectively. Meshes are be divided by the Super-SAP program as shown in Figure 1.

The main boundary nodes on the excavating surface of the section 20-20' are shown in Figure 2.

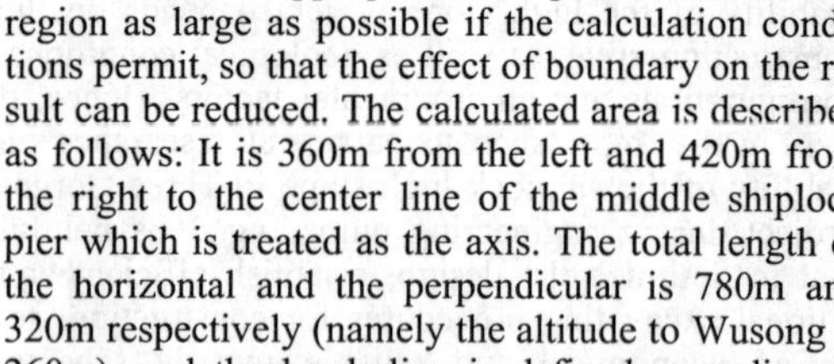

Figure 1. Schematic of calculated meshes of Section 20-20'.

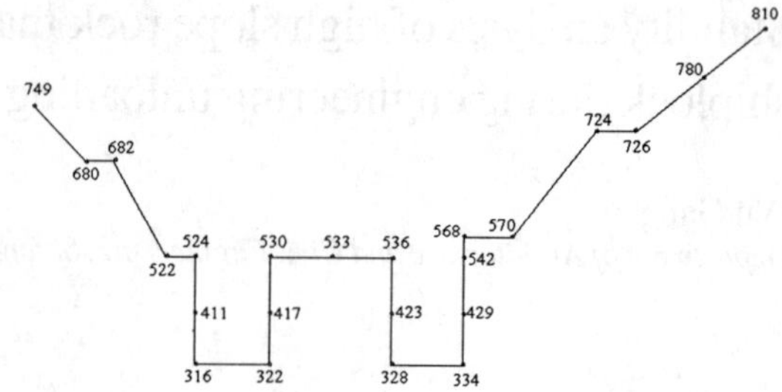

Figure 2. Schematic of boundary nodes of Section 20-20'.

2.2 *Selection of calculation method and model parameters for material*

2.2.1 *Calculating program*

The designed finite element program is chosen as the calculating program. The program allows for effects of the gravitational stress and the tectonic stress on rock mass. It can be used in the excavating calculation of slope and simulate the construction process of the slope.

2.2.2 *Calculation scheme*

Rock mass is subjected to the gravitational stress and the horizontal tectonic stress, while osmotic pressure is not included. Displacements and stresses of rock mass can be evaluated according to the excavating steps.

The process of the excavation is accomplished in five steps. The first step, excavating to the top of the middle pier (the altitude to Wusong is 140m). The second step, excavating down 30m in the left gate-room; thirdly, excavating down 30m in the right gate-room to make its deepness equal to that of the left; fourthly, excavating down 20m in the left gate-room (the altitude to Wusong is 90m); fifthly, excavating down 20m in the right too. The project of the excavation is as shown in Figure 3.

2.2.3 *Selection of model and its parameters for material*

Rock mass can be treated as elastic material. When

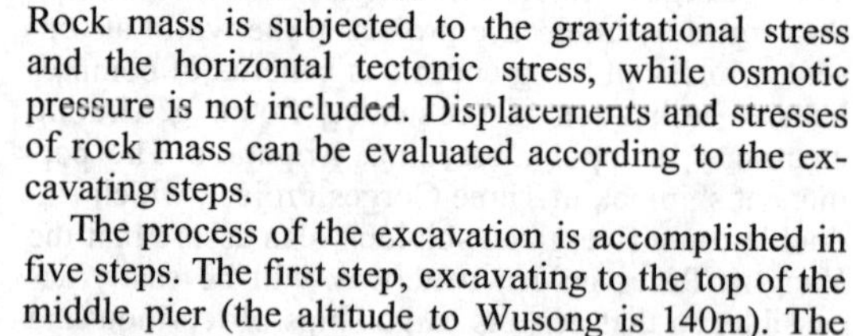

Figure 3. Schematic of the excavation steps of Section 20-20'.

the extension principle stress appears in rock mass, and if it exceeds tensile strength of material, tension damage will occur in the principle direction (unidirection or multi-direction).

According to the report on mechanical parameters for rock mass of the shiplock region in Three Gorges Project made by Yangtze River Scientific Research Institute (1992), calculating parameters for rock mass which this paper adopts are given in Table 1.

2.3 *Ground stress field and boundary conditions*

With respect to survey data of geological conditions of the shiplock region in Three Gorges Project, the loads that rock mass is subjected to are the gravitational stress and the tectonic stress, namely. (Yangtze River Scientific Research Institute, 1992)

Table 1. Calculating parameters for rock mass.

classification of Rock mass	Entirely strong-weathered rock mass	Feeble-weathered rock mass	Fresh rock mass
Thickness (m)	18	20	
Dry unit weight (KN/m^3)	26	26.8	27
Elastic modulus(GPa)	0.98	15	40
Poisson's Ratio	0.30	0.24	0.22
Tensile strength(MPa)	0	1.0	1.5
Compressive strength(MPa)	14	50	100

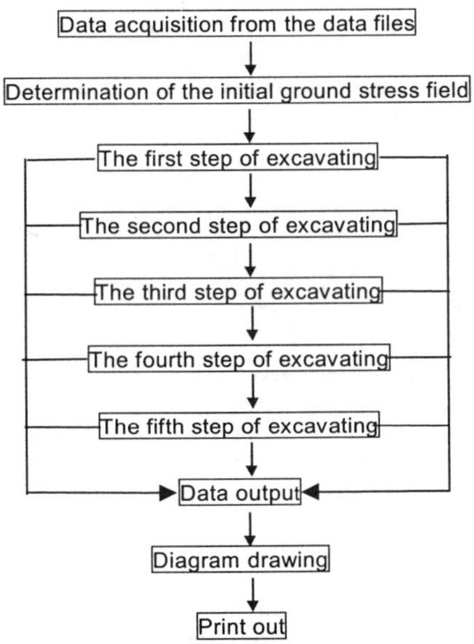

Figure 4. The calculating program schematic of slope rock mass during excavation unloading.

Table 2. Correlation between displacements of the part boundary nodes on Section 20-20' and the excavating steps.

Node number	Displacement	The First step	The second step	The third step	The fourth step	The fifth step
749	u_x	55.219	63.317	63.565	70.359	70.43
	u_y	18.476	17.476	17.754	16.334	16.517
	u	58.228	65.684	65.998	72.230	72.253
682	u_x	30.591	38.075	38.860	45.291	45.503
	u_y	-1.725	-2.231	-2.856	-3.765	-4.593
	u	30.640	38.140	38.956	45.447	45.734
524	u_x	3.422	17.369	18.229	24.117	24.641
	u_y	3.063	-4.476	-3.284	-6.599	-5.779
	u	4.593	17.936	18.522	25.004	25.310
530	u_x	0.843	-11.717	-3.689	-11.395	-3.382
	u_y	5.481	2.083	5.221	3.796	6.145
	u	5.545	11.901	6.393	12.011	7.014
536	u_x	-1.852	-7.002	3.689	-4.333	3.354
	u_y	5.665	6.472	5.746	7.279	6.669
	u	5.960	9.535	6.828	8.471	7.465
568	u_x	-7.806	-10.187	-17.444	-18.412	-21.611
	u_y	3.111	4.193	-0.997	0.103	-2.012
	u	8.403	11.016	17.472	18.412	21.704
724	u_x	-17.877	-18.871	-21.938	-22.429	-24.835
	u_y	0.470	1.057	-0.259	0.212	-1.120
	u	17.883	18.901	21.940	22.430	24.860
810	u_x	-68.343	-69.015	-71.649	-71.943	-74.131
	u_y	-0.696	-7.152	-1.398	-1.436	-2.163
	u	68.347	69.385	71.663	71.957	74.163

Note: It is assumed that the horizontal displacement is positive in the right direction, and the vertical displacement is positive in the upward direction; the unit of displacement is mm.

The gravitational stress $\sigma_y = \gamma H$ 1

The tectonic stress σ_x 3.901+0.002H 2

Where γ is the specific gravity of rock mass, H is the deepness from the earth's surface to the calculating point. The boundary conditions are described as follows: here, it is assumed that two endpoints on the underside of the foundation are fixed points (the altitude to Wusong is -60m), while other points which are restricted in the vertical direction can move freely in the horizontal direction. Points on the two side boundaries are restricted to move in the horizontal direction, but free in the vertical direction. The earth's surface and the manual boundaries formed by excavating are free boundaries on which no external loads have effect.

2.4 *Calculating process*

The calculating process in this paper can be illustrated by the following schematic of program.

2.5 *Calculated results*

2.5.1 *Deformations of slope rock mass during unloading due to excavating*

Displacements of boundary nodes on the section 20-20' of the slope at each excavating step are showed in Table 2 with respect to the calculated results.

Displacements of Section 20-20'of the slope at each excavating step are shown in Figures 5-9(the dimension of displacements is 500 times larger than that of graphics).

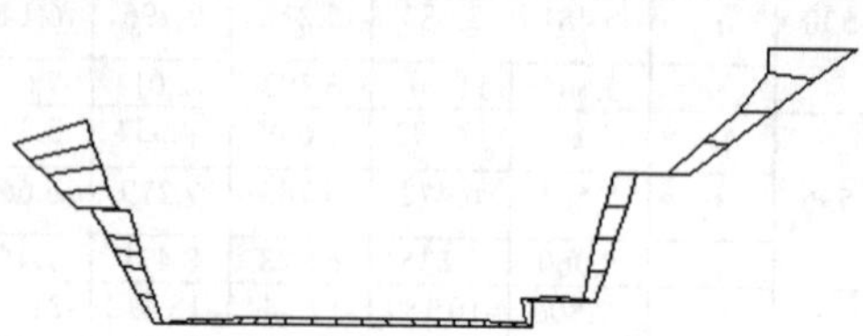

Figure 5. Deformation schematic of Section 20-20' at the first step of excavating.

In the figure, the maximum horizontal displacement is 55.219mm,the minimum horizontal displacement is -68.343mm,the maximum vertical displacement is 18.476mm, the minimum vertical displacement is -1.225 mm, and the maximum displacement is 68.347 mm.

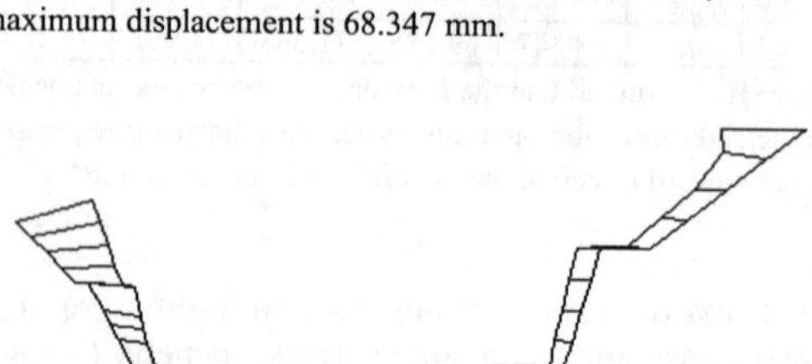

Figure 6. Deformation schematic of Section 20-20' at the second step of excavating.

In the figure, the maximum horizontal displacement is 63.317mm, the minimum horizontal displacement is -69.015mm, the maximum vertical displacement is 17.4766mm, the minimum vertical displacement is -7.152mm, and the maximum displacement is 69.385mm.

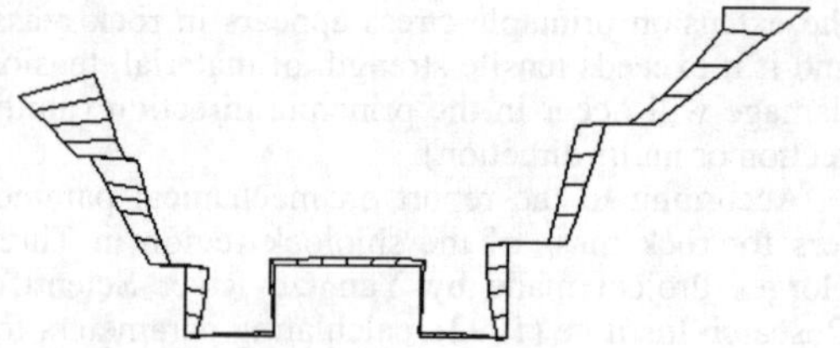

Figure 7. Deformation schematic of Section 20-20' at the third step of excavating.

In the figure, the maximum horizontal displacement is 63.565mm, the minimum horizontal displacement is -71.649mm, the maximum vertical displacement is 17.754mm, the minimum vertical displacement is -3.284mm, and the maximum displacement is 71.663mm.

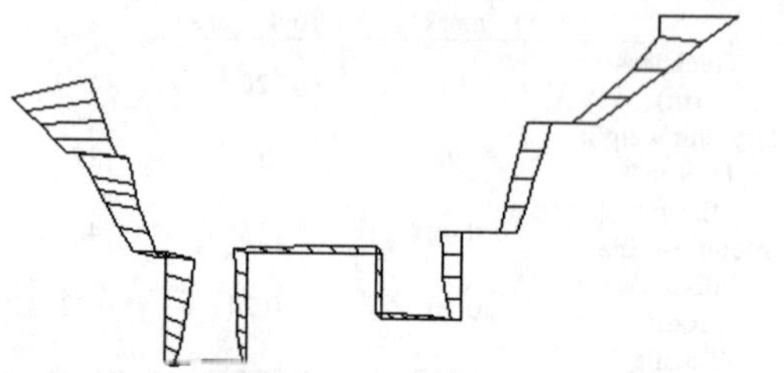

Figure 8. Deformation schematic of Section 20-20' at the fourth step of excavating.

In the figure, the maximum horizontal displacement is 70.359mm, the minimum horizontal displacement is -71.943mm, the maximum vertical displacement is 16.334mm, the minimum vertical displacement is -6.599mm, and the maximum displacement is 72.230mm.

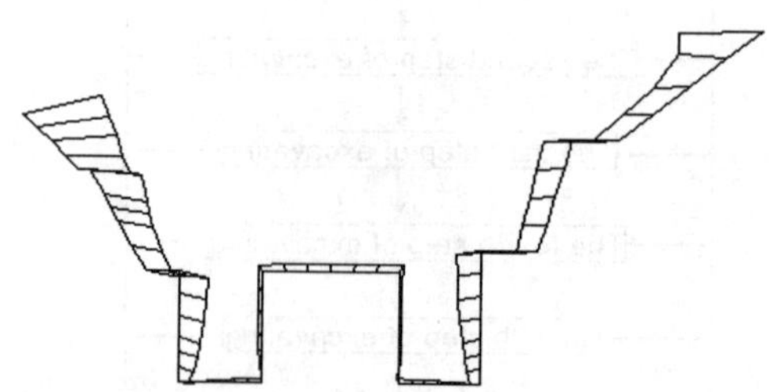

Figure 9. Deformation schematic of Section 20-20' at the fifth step of excavating.

In the figure, the maximum horizontal displacement is 70.430mm,the minimum horizontal displacement is -74.131mm, the maximum vertical displacement is 16.517mm, the minimum vertical displacement is -5.779mm, and the maximum displacement is 74.163mm.

Table 3. Correlation between the stresses of the part boundary nodes on Section 20-20' and the excavating steps.

Node number	Stress	Step1	Step2	Step3	Step4	Step5
749	σ_1	0.262	0.268	0.827	0.272	0.272
	σ_2	-2.553	-2.520	-2.396	-2.498	-2.499
	α	-27.51	-27.91	-22.83	-28.18	-28.16
680	σ_1	-0.183	0.027	-0.179	-0.186	-0.194
	σ_2	-6.835	-6.870	-6.674	-6.581	-6.587
	α	43.54	43.77	43.68	43.87	43.87
522	σ_1	-2.880	-0.364	-0.367	-0.305	-0.305
	σ_2	-11.70	-6.681	-6.703	-5.640	-5.680
	α	33.71	33.06	33.08	29.26	29.37
524	σ_1	-0.499	-0.479	-0.471	-0.392	-0.390
	σ_2	-9.324	-0.855	-0.889	-0.739	-0.755
	α	-8.021	39.93	39.61	43.85	43.43
316	σ_1	-3.201	-5.090	-4.427	-5.801	-5.574
	σ_2	-6.161	-8.478	-8.090	-18.41	-17.76
	α	-19.59	-22.88	-30.70	-39.08	-40.67
322	σ_1	-2.283	-2.932	-1.618	-1.942	-1.213
	σ_2	-6.585	-8.662	-7.656	-9.767	-7.008
	α	-7.719	-3.712	-9.780	18.29	10.51
530	σ_1	-0.164	0.000	0.000	-0.083	-0.072
	σ_2	-6.249	-0.184	-0.154	-0.150	-0.207
	α	-1.160	36.67	37.13	-27.53	-36.46
533	σ_1	-0.150	-0.014	-0.001	-0.007	-0.146
	σ_2	-5.962	-1.352	-0.293	-0.326	-0.305
	α	0.186	20.71	4.912	9.095	7.145
536	σ_1	0.174	-0.091	-0.000	-0.015	-0.007
	σ_2	-6.289	-3.265	-0.129	-0.141	-0.243
	α	1.346	7.653	15.98	8.363	31.05
328	σ_1	-5.662	-2.378	-1.556	0.328	-0.902
	σ_2	-9.377	-6.297	-7.681	-2.084	-6.520
	α	-27.60	9.022	12.39	-21.65	14.66
334	σ_1	-3.150	-2.404	-4.146	-3.372	-4.537
	σ_2	-6.219	-6.085	-7.931	-7.452	-15.68
	α	18.93	26.14	28.03	35.16	38.98
570	σ_1	-0.898	-0.843	-0.470	-0.465	-0.377
	σ_2	-8.384	-7.899	-5.256	-5.233	-4.277
	α	-40.62	-39.58	-31.82	-31.65	-28.38
724	σ_1	-0.105	-0.098	-0.077	-0.075	-0.064
	σ_2	-0.864	-0.873	-0.816	-0.826	-0.785
	α	-36.47	-36.85	-36.58	-36.80	-36.39
810	σ_1	-0.527	0.420	-0.302	-0.305	-0.304
	σ_2	-3.064	-3.163	-3.279	-3.274	-3.269
	α	20.79	20.01	25.31	25.21	25.25

Note: it is assumed in list 3 and following figures that the tensile stress is positive, the compressive stress is negative. α is the angle between the direction of σ_1 and the horizontal direction. The stress unit is MPa.

Here, results in Table 2 and Figures 5- 9 indicate that: after the first step of excavating, there is a rebounce in rock mass, the south slope moves in the northerly direction (left), the north slope displaces southwards (right), the top of the middle pier dis-

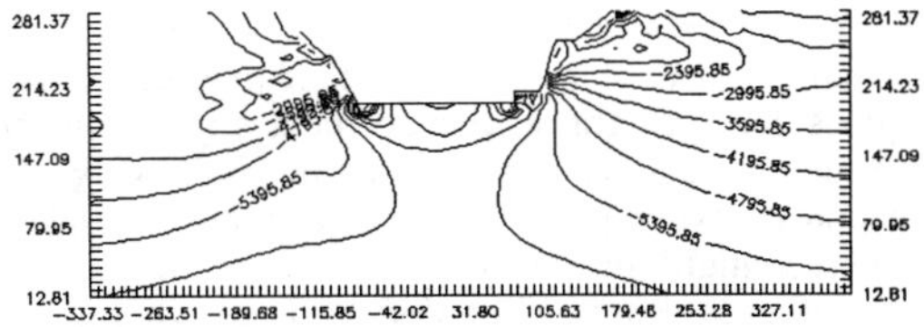

Figure 10. Stress isolines of Section 20-20'at the first excavating step.

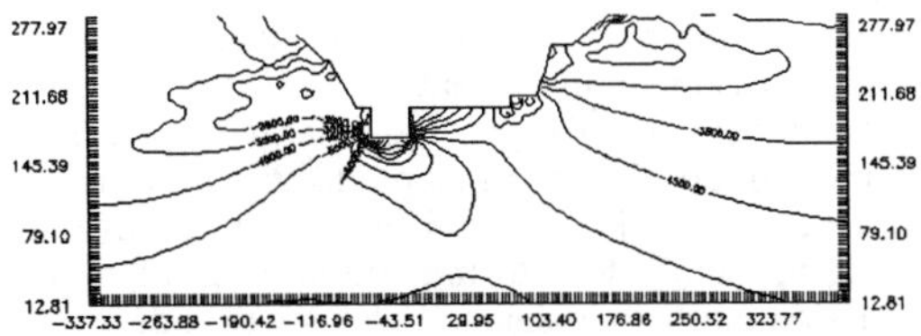

Figure 11. Stress isolines of Section 20-20' at the second excavating step.

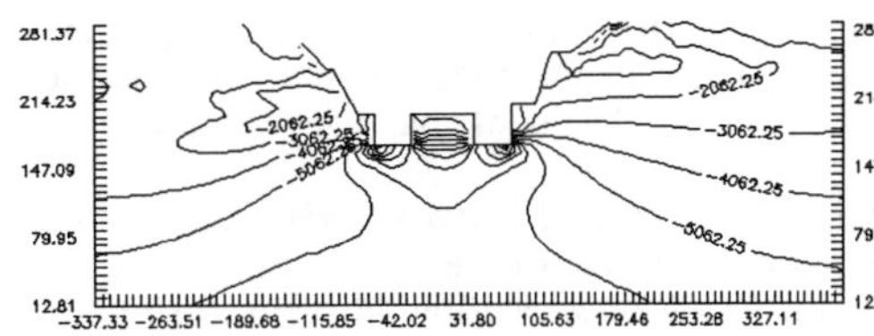

Figure 12. Stress isolines of Section 20-20' at the third excavating step.

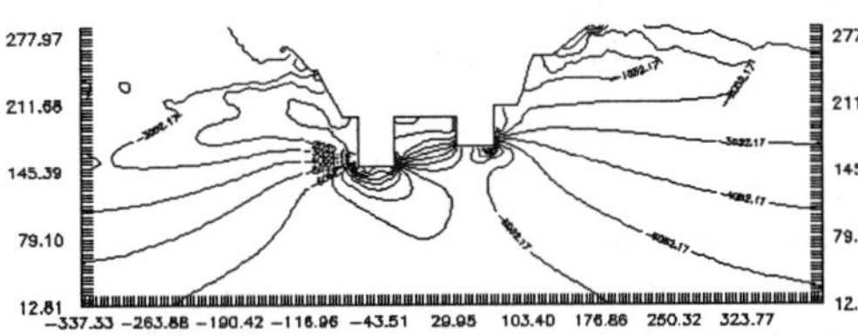

Figure 13. Stress isolines of Section 20-20'at the fourth excavating step.

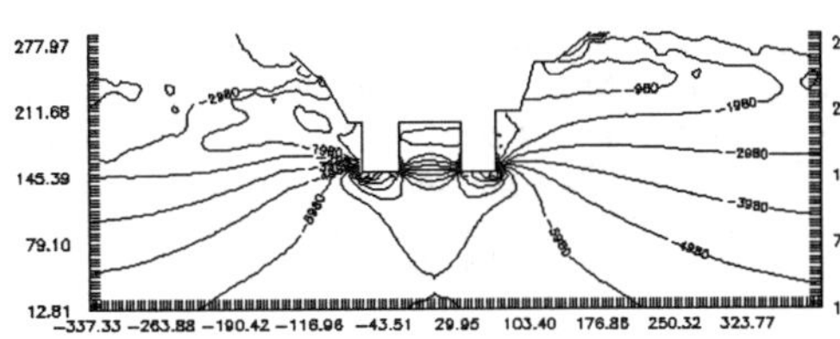

Figure 14. Stress isolines of Section 20-20' at the fifth excavating step.

places upwards. As for the whole slope, the displacement of the left side is smaller than that of the right side. The largest displacement appears in the entirely strong-weathering stratum. Following the second excavating, deformations of major boundary nodes continue to develop on the base of the first step of excavating. After the third excavating, the displacement increment of the north slope reduces, while the displacement increment of the south slope increases, the pier displaces to the top-south. Following the fourth step of excavating, the displacement increment of the north slope increases, while the displacement increment of the south slope is relatively smaller, the pier has a large displacement to the top-left. The north slope displaces to the bottom-right caused by the fifth step of excavating, while the south slope to the bottom-left, the middle pier continues to upheave. The south side of the whole slope has a total displacement approximately equal to that of the north side. The displacement in the entirely strong-weathering stratum of the slope is 50~75 mm, the displacement in the feeble-weathering stratum of the slope is 26~50 mm, the displacement in the fresh stratum of the slope is 15~26 mm, in the middle pier is 6~12 mm.

2.5.2 *Stresses in rock mass of the slope during unloading due to excavating*

By the means of converting the element stresses into the node stresses which are subsequently converted into the principle stresses, correlation between the stresses of the part boundary nodes on Section 20-20' and the excavating steps are given in Table 3.

Stress isolines of Section 20-20' under different excavating conditions are shown in Figures 10-14 (the stress unit in each figure is KPa).

It can be shown from Figures 10-14 that the ground stress almost remains the same in the areas far from the excavating region along with the continuous increase of excavating depth. However, stresses in rock mass change constantly due to release of the ground stress in the excavating surface and its adjoining areas, the compression stress decreases while the tension stress increases, the area under tension also gradually expands. After excavating, the upside part of the erective wall and the top part of the middle pier in the shiplock would occur a large-scale range tension area, the maximum tension stress is 0.827Mpa. A considerable accumulation of the compression stress would occur at the soleplate of the left and the right gate-room and the corner of the erective wall, where the maximum compression stress reaches up to 18.41Mpa.

3 EVALUATION OF STABILITY OF THE HIGHSLOPE ROCK MASS UNDER EXCAVATION UNLOADING CONDITION

This paper carries out a numerical simulation analysis through a five-step-excavating process on the highest representative section of the shiplock high slope by using the finite element program according to the present design scheme. The results show that:

(1) As a whole, geological conditions of the shiplock region in Three Gorges Project is good. Hence, the problem of large-scale destruction doesn't exist in the shiplock region.

(2) A multistep excavating numerical modeling analysis on Section 20-20' which has the maximum height by using the reprogrammed finite element program has been carried out, the results of researches show that the main damage form of slope rock mass, caused by unloading is tension or tension-shear damage.

(3) Changes in deformations and stresses of slope rock mass during unloading along with the process of excavating have been illustrated through figures, it can be seen that satisfactory results have been achieved in this way. The results are similar to available solutions.

(4) As for Section 20-20' of the shiplock in the Three Gorges project that is a high slope, sliding damage may occur in the entirely strong-weathering stratum. Generally speaking, the other part of the slope is stable except that tension or tension-shear damage perhaps emerges in some areas (for example the top platform of the gate, the erective wall and the upside of the pier). Stability of the slope can be assured as long as appropriate reinforcing measures have been taken. Hence, the present design scheme is safe and reliable.

REFERENCES

Dong, X.S. 1994. Research on high rock slope of TGP locks. *Journal of Yangtze River Scientific Research Institute* 11(3):20~25. (In Chinese)

Ha, Q.L. & Li, J.L. 1994. A study on finite element analysis of shiplock high slope and its structure optimization in Three Gorges Project. *Chinese Journal of Rock Mechanics and Engineering* 13(2):95~108.

Xu, B.Y. et al. 1994. Rock slope engineering and solid mechanics. *Advances in Mechanics* 24(4): 518-524. (In Chinese)

Yangtze River Scientific Research Institute (eds). 1992, Studying theses anthologies of the Three Gorges Project (first part, second part). Wuhan, China.

Computer Methods and Advances in Geomechanics, Desai et al. (eds) © 2001 Balkema, Rotterdam, ISBN 90 5809 183 X

Effects of a deep excavation on an adjacent shield-driven tunnel

X.W.Ge & K.M.Lee
Department of Civil Engineering, Hong Kong University of Science and Technology, People's Republic of China

ABSTRACT: A detailed field instrumentation program was conducted in a deep excavation site in Shanghai, China. The ground movements and pore water pressure changes outside the excavation pit and the deflection of an adjoining shield-driven tunnel are interpreted and analyzed. The deformation mechanism of the tunnel in response to the deep excavation is examined based on the field measurement and analytical results.

1 INTRODUCTION

In most urban downtown areas, deep excavation sites are often inevitably located in the close proximity of existing structures and underground facilities. As most of the subway tunnels have very low tolerance for ground movements, therefore, it is of great importance to assess the effects of deep excavation on adjacent tunnel facilities and identifying sections where special remedial construction measures are required. Despite of the increasingly frequent occurrence of this kind of construction problems, little research work has been conducted to examine the fundamental behavior of this kind of soil-tunnel interaction problems.

Through an instrumented case history recorded in the Wanxiang International Square, Shanghai, the effects of a deep excavation on an adjacent shield-driven subway tunnel is examined.

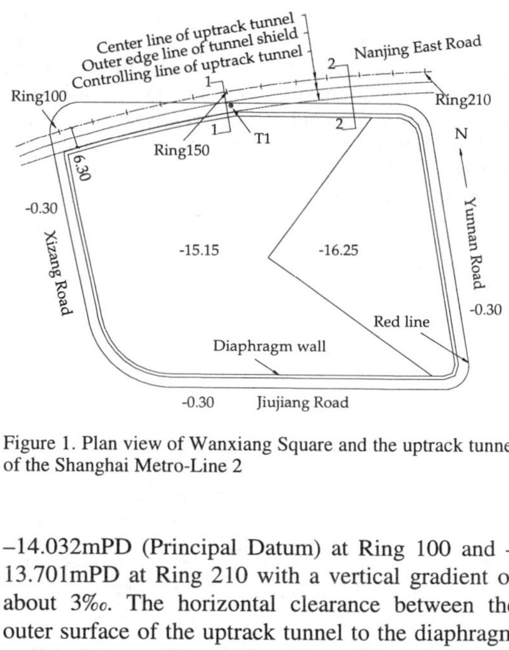

Figure 1. Plan view of Wanxiang Square and the uptrack tunnel of the Shanghai Metro-Line 2

2 SITE GEOMETRY

The Wanxiang International Square (about 100m ×70m) is located in a busy commercial area in the vicinity of the Shanghai Metro-Line 2. Figure 1 is the plan view of the deep excavation and the uptrack tunnel of the Shanghai Metro-Line 2. Rings 100 to 210 are located approximately parallel to the pit. The outer diameter of tunnel rings is 6.2m, while each tunnel ring is composed of 6 pieces of lining block with 0.35m in thickness, and 1.0m in length. Every two rings are connected in non-staggering format by 17 longitudinal steel bolts, while every two lining blocks in circumferential direction is connected by 2 bolts. The centerline of uptrack tunnel is located at –14.032mPD (Principal Datum) at Ring 100 and –13.701mPD at Ring 210 with a vertical gradient of about 3‰. The horizontal clearance between the outer surface of the uptrack tunnel to the diaphragm wall is 3.2m at Ring 100 to Ring 150, and 9.0m at Ring 200.

The excavation depth is 16.05m at the main building area and 14.85m at the podium area as shown in Figure 1. Reinforced concrete (RC) diaphragm wall with temporary RC struts was adopted as earth retaining structures in the site. The construction reference elevation ± 0.00 m is equivalent to +3.30mPD. The thickness of the panels along north side where the tunnel situates underneath, is 1.2m and the toe level at the main building area and podium area are respectively -34.30m and –31.80m. The thickness and the toe depth of the panels along other three sides are thinner and shallower, respectively. The profile of Cross-section 1-1 which is in the middle of north side of the excavation pit is shown in Figure 2. Three sets of strutting systems made of cast-in-place RC were applied to support the wall. These three sets of strutting systems were supported by vertical props that situated on bearing piles. Jet grouting and cement-soil mixing was conducted preceding the excavation inside the pit along north side, while only cement-soil mixing was conducted along other 3 sides of the pit. The raft-pile foundation was employed and the superstructure was supported by a total of 824 bore piles with diameters of 850mm or 650mm cast *in situ* with toe level at about -43.20m.

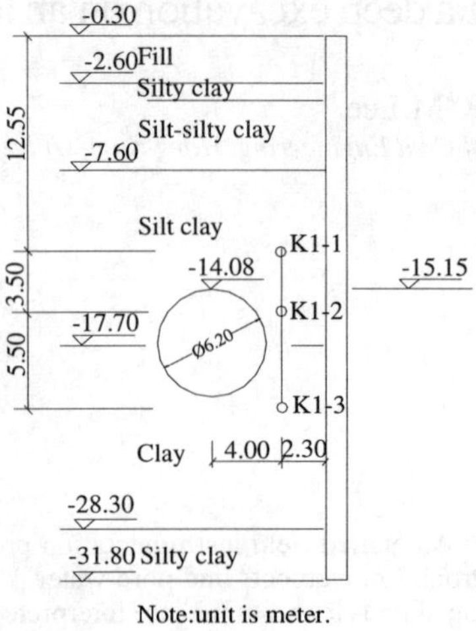

Figure 3. Layout of piezometers along depth in Section1-1

3 GEOLOGICAL CONDITION IN THE SITE

Shanghai ground mainly consists of alluvial and marine sediments. The soils in the upper 30 to 40 m are very soft to soft silty clays underlain fine sand and silty clay strata. Shanghai clay can be differentiated into a number of uniformly layered and well-defined strata based on physical properties and soil behaviors as shown in Table 1. The elevation of soil layers can be found in Figure 3. The water table in ground is normally at -1.05m~-1.37m.

4 CONSTRUCTION SEQUENCE

The diaphragm wall, bore piles and soil improvement were carried out between Feb. and May 1998, and the uptrack tunnel was excavated adjacent to the site during Jun. 1998. The construction sequence from dewatering to compensation grouting is shown at Table 2. The water table inside the pit was kept at 0.5 to 1.0m below the excavation surface, while the water table fluctuation outside the pit was controlled around ±1.0m in order to reduce the effect on surrounding facilities. The total excavation was divided into 4 major steps according to the locations of strutting systems. Compensation grouting was employed to prevent further tunnel deflection after finishing the construction of bottom slab.

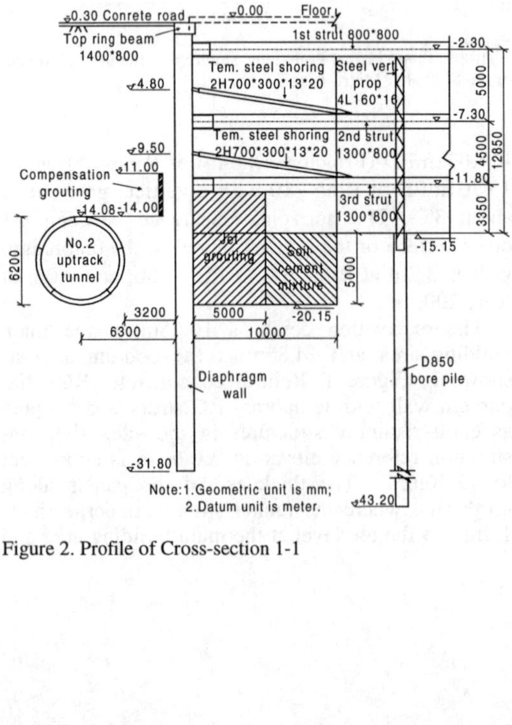

Figure 2. Profile of Cross-section 1-1

Table 1. General soil parameters obtained from *in situ*

Layer No	w(%)	γ (kN/m^3)	c_{cu}(kPa)	φ_{cu} (°)	K_o	c_u(kPa)
2	34.8	18.7	13	14.8	0.46	76
3	40.5	18.0	7	18.7	0.52	36
4	51.1	17.0	10	9.8	0.57	32
5-1	34.8	18.3	10	15.9	0.50	36
5-2	33.0	18.3	10	19.1	0.41	119
6	23.8	19.9	30	12.8		142
7-1	22.5	20.1	6	26.3		

Table 2. Construction sequence

Steps	Duration (day)	Time period
Dewater	126	Aug. 8~Dec. 11
1st excavation & 1st strut	19	Aug. 18~Sep. 5
2nd excavation & 2nd strut	40	Sep. 6~Oct. 15
3rd excavation & 3rd strut	30	Oct. 16~Nov. 14
4th excavation & bottom slab	27	Nov. 15~Dec. 11
Compensation grouting	32	Dec. 12~Jan. 12, 99

5 FIELD INSTRUMENT ARRANGEMENT

A number of monitoring points were set up around the pit area and along the uptrack tunnel which include: inclination of diaphragm wall; ground movement; displacement of vertical props; axial force in struts; water table fluctuation; pore water pressure change (the layout of three piezometers along depth in Section 1-1 is shown in Figure 3); ground surface settlement outside the excavation pit at Section 1-1; settlement and horizontal displacement markers in the tunnel; and the convergence of the tunnel. Details about the field instrumentation arrangement are given in Ge's Ph.D. thesis (Ge 2000).

6 INTERPRETATION OF FIELD INSTRUMENTATION DATA

6.1 *Ground movement beside the diaphragm wall*

The horizontal ground movement obtained by inclinometers installed at T1 (see Figure 1) during excavation is shown in Figure 4. It can be observed that the bottom slab played an important role in restraining the development of horizontal ground movements. At the end of excavation steps, the maximum horizontal displacement of soil occurred below the bottom of excavation and corresponding to approximately the mid-height of the diaphragm wall.

The total displacement vectors incorporating inclinometer and extensometer readings along depth at T1 at the end of excavation are shown in Figure 5. The upper layers of soil tended to settle resulting from the large inward displacement of diaphragm wall, while the lower layers of soil was observed to heave upward which is corresponding to the inward and upward deformation occurred at the lower portion of the diaphragm wall. It is also observed that the middle layers, which were almost in the same level as pit bottom, only had slightly settlement.

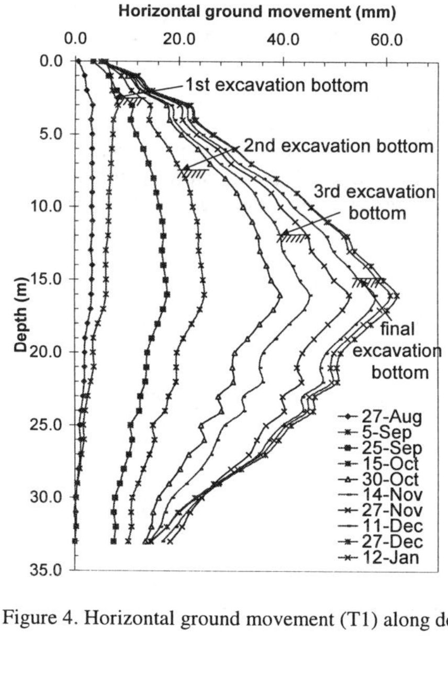

Figure 4. Horizontal ground movement (T1) along depth

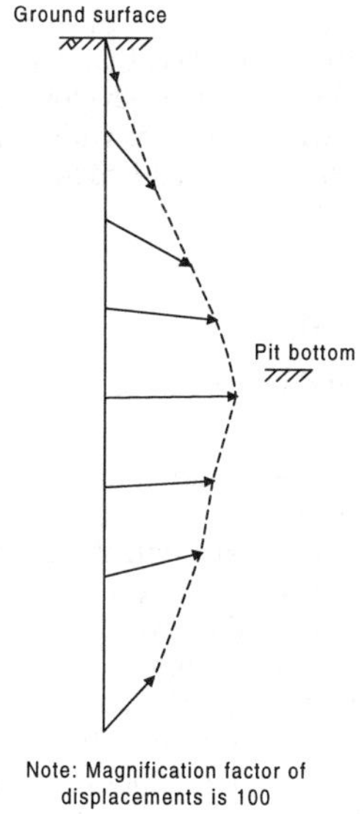

Figure 5. Displacement vectors in the end of excavation

6.2 *Pore water pressure changes outside the diaphragm wall*

The decrease of water table outside the excavation pit was observed to be less than 1.2m during excavation. The total pore water pressure (TPWP) together with hydrostatic pressure with time at K1 are shown in Figure 6. During the whole excavation period, the TPWP decreased at a rate of 0.20 to 0.60kPa/day at the depth of 12m to 22m. The rate of pore water pressure reduction is higher at deeper

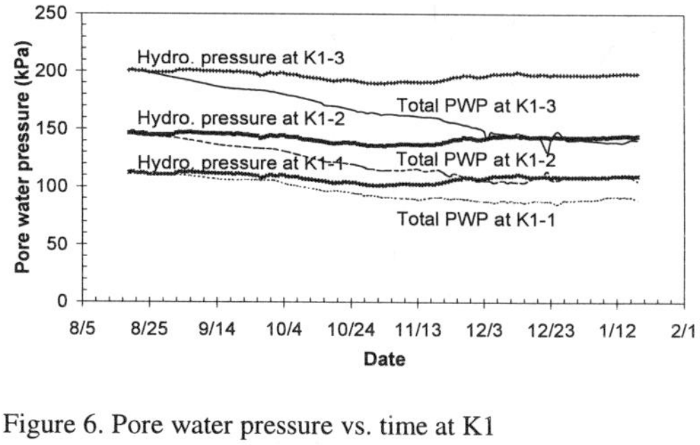

Figure 6. Pore water pressure vs. time at K1

levels than at shallow levels. At the end of excavation, the TPWP is about 20 to 26% less than the hydrostatic pressure, thus resulted in the development of negative excessive pore water pressure with depth (about –20kPa at the depth of 12m and reduced to about –54kPa at the depth of 22m).

7 MECHANISM ANALYSIS

7.1 *Pore water pressure and stress changes around the tunnel lining*

In order to study the effect of excavation on the pore water pressure (PWP) and the earth pressure developed around the tunnel lining, analytical seepage analyses around the excavation pit were conducted. The changes in PWP are indicated by two factors: 1) seepage induced by the dewatering inside the excavation pit at various steps of excavation, and 2) the decrease of horizontal earth pressure resulting from excavation process. It should be noted that the water table outside the pit did not change too much during the excavation process, it can be assumed as constant during each excavation step.

With the use of Mandel's mapping function (Azizi 1999), the relationship of depth y (with the coordinate system as shown in Figure 7) and the velocity potential Φ_y can be expressed by the following equation

$$y = \beta \frac{\Phi_y}{\lambda} - \alpha \sin \frac{\Phi_y}{\lambda} \quad (1)$$

where α, β and λ are three positive real numbers. The pore water pressure at any depth y can be obtained by

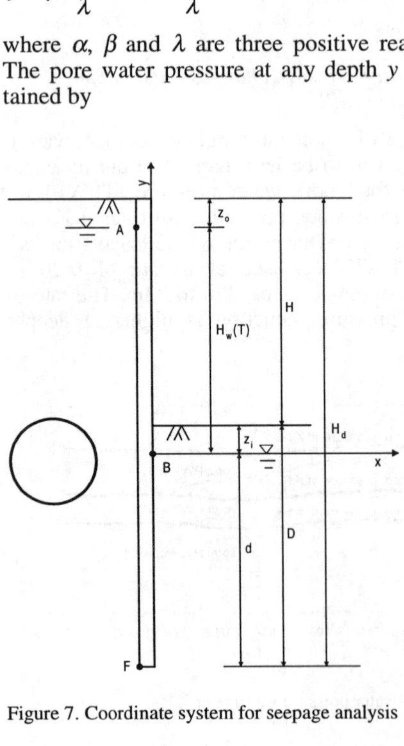

Figure 7. Coordinate system for seepage analysis

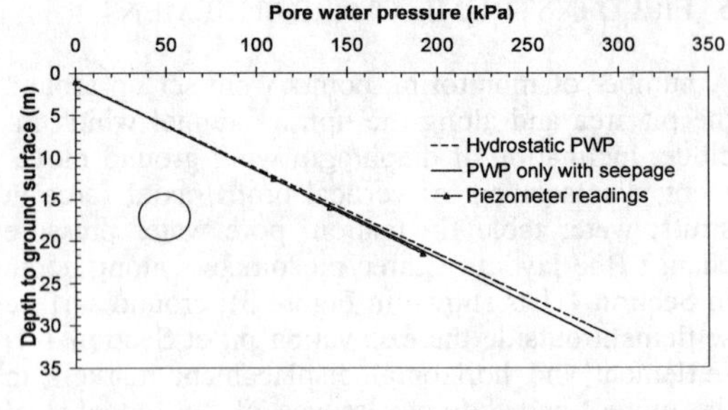

Figure 8. Pore water pressure in the end of 1st excavation step

$$u = \gamma_w (\Phi_y - y) \quad (2)$$

The hydrostatic PWP, the analytically predicted PWP induced by the seepage effect (without consideration of the effect of stress release caused by excavation) and the piezometer readings obtained by K1-1 to K1-3 at the end of the first and the fourth excavation steps are plotted in Figures 8 and 9. It can be seen that the PWP decreases from its hydrostatic value during dewatering and excavation. The changes of PWP can be divided into two parts: one is due to seepage, Δu_s, and the other is due to the decrease of lateral earth pressure, Δu_e, although they always occurred accompanying with each other. Seepage played an important role in reducing the PWP throughout the entire excavation processes, while the decrease of lateral pressure only accounted for a small effect at the beginning of excavation. The effect of the later part seems to increase with the excavation in progress. As shown in Figure 9, after the fourth excavation step, Δu_s is about 17 to 19% of the hydrostatic pressure from the depth of 12m to 22m, while Δu_e is 2 to 7% of the hydrostatic pressure. It should be noted that the total change of PWP is about 19 to 26% of the hydrostatic pressure.

The surrounding pressure of the tunnel can be determined based on the geological condition and piezometer readings. With dewatering and excavation in progress, the decrease of horizontal

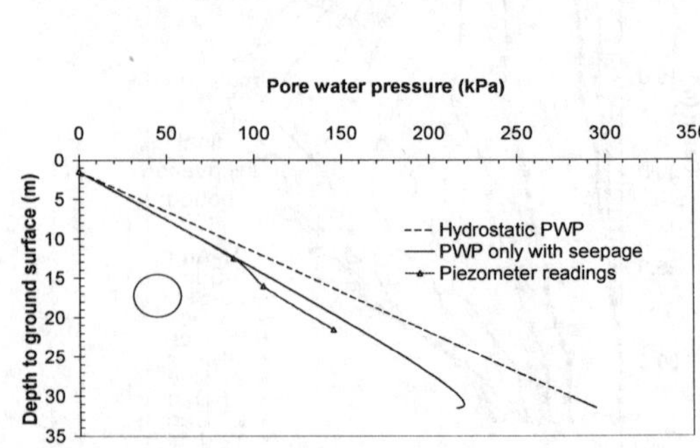

Figure 9. Pore water pressure in the end of 4th excavation step

lateral stress, $\Delta\sigma_h$, can be derived by the following equation

$$\Delta\sigma_h = \frac{\Delta u_e}{1-A} \tag{3}$$

where A is the PWP parameter. A_f (A at failure) value is relative stable (assumed as 0.8) and can be adopted to approximately determine the change of horizontal total stress at final step.

7.2 *The bending moment and convergence in tunnel lining*

During excavation, the lateral earth pressure on the diaphragm wall gradually decreased from the earth pressure at rest and finally approached active pressure. Furthermore, these earth pressure changes resulted in unbalanced stress redistribution and ground movement in the vicinity of diaphragm wall. The deformation of the tunnel ring was a combination of several kinds of movements, such as translation, rotation and squash deformation (convergence). In this paper, only the tunnel convergence and the induced bending moment are discussed, as this effect is considered to be the most significant factor affecting the stress state within the tunnel lining.

An analytical approach which takes the joint effect into consideration was proposed to determine the bending moment distribution and diameter change of segmental tunnel lining by Lee et al. (in press). The joint flexural stiffness K_θ (kN-m/rad) is defined as the bending moment per unit length required to develop a unit rotation angle along a joint of the assembled segments. Based on the field observation of earth pressure distributions acting around segmental tunnels in soft clay (Lee & Ge, submitted), the earth pressures can be expressed as illustrated in Figure 10, where p_1 is the vertical overburden soil pressure, p_2 is the reaction pressure at the bottom of the lining, p_3 is the total lateral earth pressure developed at the crown level of the tunnel lining, p_4 is the additional lateral earth pressure developed at the tunnel invert level, p_5 is the self-weight of the tunnel lining and p_6 is the soil resistance pressure. Here p_6 is assumed to be distributed over the range of 45° to 135° with respect to the vertical direction around the tunnel and to act perpendicular to the tunnel with a parabolic pattern as defined by the following equation:

$$p_6 = p_h(1-2\cos^2\varphi) \tag{4}$$

Here, p_h is the soil resistance developed at the spring-line of the tunnel and φ is the angle measured form the vertical direction around the tunnel. By adopting Winkler's type of soil reaction, p_h, can be estimated by

$$p_h = K_s\Delta_h \tag{5}$$

where K_s is the soil resistance coefficient and Δ_h is the horizontal displacement at the spring-line of the tunnel.

The lateral earth pressures can be obtained from the previous section, and the bending moment and tunnel diameter change during excavation can be calculated by the analytical approach. A program was employed to carry out the calculation (Ge 2000). The bending moment and horizontal diameter change of tunnel lining at Section 1-1 during the entire excavation process can be plotted in Figure 11. For parametric study, a number of cases for the tunnels with different joint stiffness are studied and the results are also shown in Figure 11. From this figure, it can be noted that the less flexible the tunnel is, the higher the maximum bending moment occurs with same horizontal diameter change during excavation. In order to describe the sensitivity of bending moment to tunnel convergence, a parameter ξ (kN-m/mm), namely bending moment sensitivity ratio is proposed as following

$$\xi = \frac{dM}{d\Phi} \tag{6}$$

where M is the maximum bending moment in tunnel lining, and Φ is the diameter of the tunnel. A

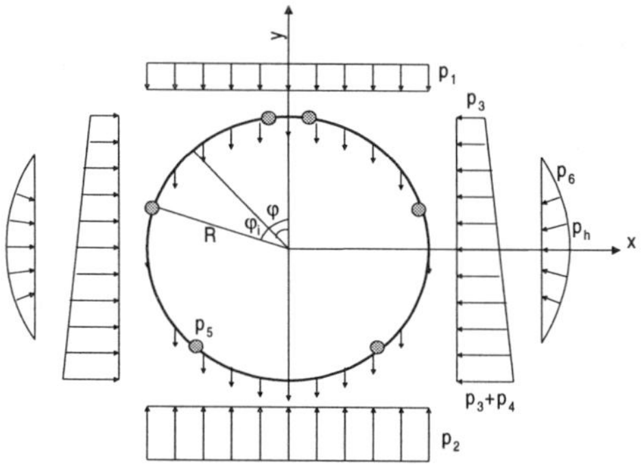

Figure 10. Model diagram of a jointed tunnel lining

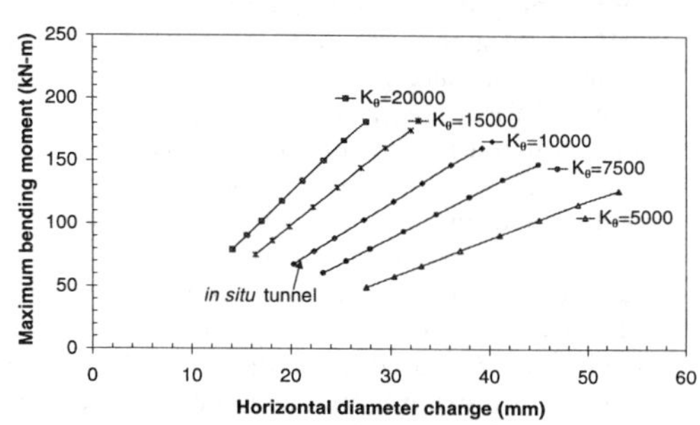

Figure 11. Maximum bending moment and horizontal diameter change during excavation

regression equation based on Figure 11 can be obtained as following:

$$\xi = 8\log\frac{K_\theta l}{EI} + 14 \tag{7}$$

where l is the calculation length usually taken as 1m to represent a typical unit length of lining segment, EI is the bending rigidity of the tunnel lining. This equation is proved to be less affected by embedded depth of the tunnel.

The horizontal diameter change obtained by the analytical approach at the end of excavation is 39.2mm as shown in Figure 11 which is very close to the value back-calculated from field observation, 33.1mm. The maximum bending moment, 160kN-m, is much larger than the critical value 78.0kN-m which is corresponding to the state when the concrete at outside or inside surface reach its tension strength. It is concluded that further measures such as compensation grouting between the diaphragm wall and the tunnel are necessary.

8 CONCLUSIONS

As the soil inside the excavation pit tended to heave due to vertical stress relief during excavation, and the diaphragm wall together with the strutting system were slightly displaceable, therefore, ground movement took place and caused the translation, rotation and shape change of the adjacent tunnel. Field instrumentation and relevant analyses provided the opportunity to understand the effects of deep excavation on the adjacent tunnel. The findings or conclusions are summarized as following:

(1) Bottom slab played an important role in restraining the development of ground movement. The maximum horizontal displacement of soil occurred approximately corresponding to the mid-height of the diaphragm wall.

(2) During the entire excavation period, the total pore water pressure decreased at certain rate, while this rate at deeper level is larger than that at shallow level.

(3) Seepage played an important role in decreasing the PWP throughout the entire excavation process, while the decrease of lateral pressure only accounted for a small effect at the beginning of excavation and this effect seems to increase with the excavation in progress. The earth pressure around the tunnel can be determined based the calculation of pore water pressure.

(4) The tunnel diameter change and induced bending moment can be determined by an analytical approach. The less flexible the tunnel is, the higher the maximum bending moment occurs with same horizontal diameter change during excavation. The bending moment sensitivity ratio ξ can be employed to describe the sensitivity of bending moment to tunnel convergence during excavation.

(5) The methodology presented herein can be adopted in solving similar problems in future projects.

REFERENCES:

Azizi, F. 1999. *Applied Analyses in Geotechnics*. New York: E & FN Spon.

Ge, X. W. 2000. Effects of construction activities on existing underground tunnels. Ph. D. thesis. The Hong Kong University of Science and Technology.

Lee, K. M., Hou, X. Y., Ge, X. W. & Tang, Y. 2000. An analytical solution for a jointed shield-driven tunnel lining. *International Journal of Analytical and Numerical Methods in Geomechanics* (in press).

Lee, K. M. & Ge, X. W. 2000. The equivalence of a jointed shield-driven tunnel lining into a continuous ring structure. *Canadian geotechnical journal* (Submitted).

Computer Methods and Advances in Geomechanics, Desai et al. (eds) © 2001 Balkema, Rotterdam, ISBN 90 5809 183 X

Estimates of construction-induced ground deformation at a deep excavation in San Francisco, California

P.Golesorkhi, C.J. Divis & J.S. Saunders
Treadwell and Rollo Incorporated, San Francisco, Calif., USA

ABSTRACT: The Four Seasons Hotel and Tower project, in San Francisco, California, required a 21-m deep excavation for five levels of underground parking. A 36-story tower is currently being constructed over the underground parking. Adjacent structures include a 30-story Marriott Hotel, an 8-story masonry office building, a tall one-story historic former brick power substation, and a promenade for local major public transit tunnels (BART and MUNI). The tunnels underlie a major automobile and surface trolley public transit artery and run parallel to the excavation about 10 m away along the northern boundary. The subsurface conditions to the excavated depth consisted of predominantly medium dense to very dense sand. The wall system for the subsurface structure included an approximately 33 to 37 m deep permanent concrete diaphragm wall (slurry wall construction technique). The wall was used as the temporary shoring, groundwater cutoff, and the permanent wall. This paper compared the results of a finite element method (FEM) analysis of the wall system with the measured deformation of the wall. In general the FEM provided reasonable estimates of the maximum deformations of the wall, however, the deformations in the top 10 m of the wall were overestimated by a factor of about 1.5 to 2.

1 INTRODUCTION

The Four Seasons Hotel and Tower is currently under construction on a 5260 m^2 parcel in downtown San Francisco (see Figure 1). Existing buildings, each with their own basements, flank the site on the west, south, and east. Market Street, underlain by the BART and MUNI metro tunnels, is to the north.

Construction at the site includes a 36-story high-rise tower on the northern half of the site, a 4-story low-rise structure on the southern half of the site, and five levels of underground parking. The tower will be a multi-use structure with the hotel occupying 11 floors and the remaining floors slated for condominiums and retail space. The five levels of underground parking required a 21-m deep excavation, one of the deepest in San Francisco history.

The massive below-grade construction required a dewatering system capable of depressing the water table at least 10 m below the bottom of the excavation with minimal impact on adjacent structures and facilities. Because of the site conditions, depth of excavation, and proximity to critical public transit structures, a concrete diaphragm cutoff wall was constructed to minimize movements. The wall was constructed using the slurry construction technique and was designed to act as temporary shoring, cutoff wall, and permanent basement wall. The wall was constructed in 4.5 to 6 m long, 1-m wide tongue and groove concrete panels. The depth of each panel varied depending on the depth to the top of the Old Bay Clay, a stiff relatively impermeable marine clay. The total wall depth ranged from about 33 to 37 m. The wall was keyed 3 m into the Old Bay Clay to provide the necessary cutoff for dewatering.

2 SUBSURFACE CONDITIONS

The generalized subsurface conditions are shown on Figure 2. They consist of about 1.5 to 4.5 m of sandy fill over 3 to 4 m of medium dense sand, locally referred to as Dune Sand. A thin, relatively compressible, layer of marsh deposit, between 0.6 to 1.2 m thick, is below the Dune Sand. Dense Colma sand, which is underlain by Old Bay Clay, is present to depths of about 30 to 34 m below the ground surface. Groundwater was encountered at a depth of about 9 to 10 m below the ground surface.

3 CONSTRUCTION SEQUENCE

The construction of the below-grade parking garage included activities associated with the various elements of excavation, shoring, dewatering, and

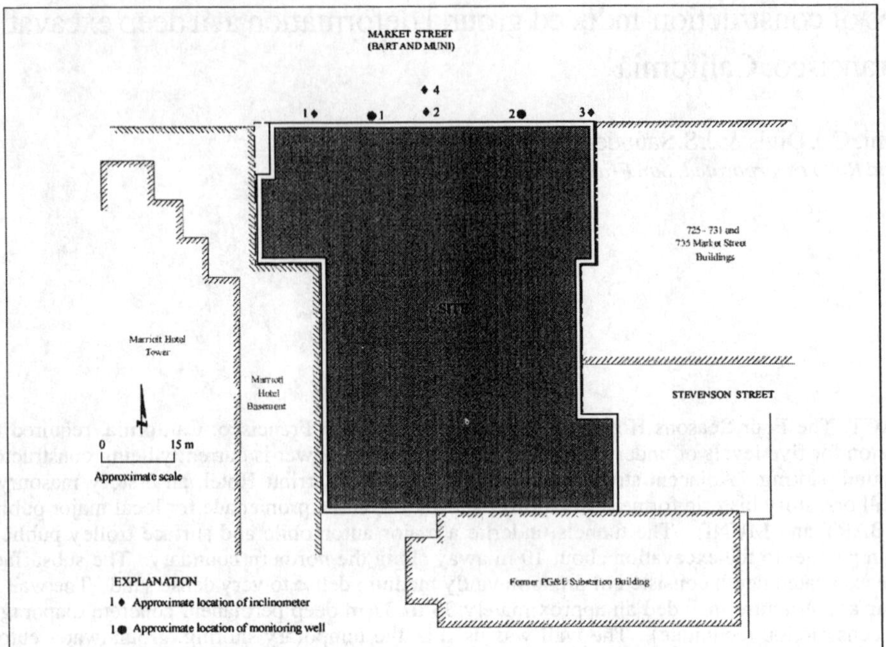

Figure 1. Site Plan

foundation construction. The concrete diaphragm wall had to be constructed before any dewatering or excavation could be performed.

Part of the construction of the wall included demolition and removal of obstructions along the perimeter of the site. Obstructions included: old foundations from previous structures at the site and the temporary shoring (soldier beams and tiebacks) for the existing building to the west.

After the perimeter of the site was cleared of obstructions, a guidewall was installed to construct the slurry wall. Once the guidewall was completed the construction of the diaphragm wall began.

The next sequence of operations, following completion of the diaphragm wall, included soil excavation and construction of the additional shoring elements. The shoring system included tiebacks and internal bracing along the northern side of the site

Table 1. Soil Parameters

	Symbol	Units	Modeled Soil Layers					
			Sandy Fill (SP/SM)	Dune Sand (SP)	Marsh Deposit (CL)	Clayey to Silty Sand (SC/SM)	Old Bay Clay (CL)	Silty Sand (SM)
Material Model	-	-	M-C	M-C	M-C	M-C	M-C	M-C
Dry Unit Weight	γ_d	kN/m^3	100	105	115	110	70	110
Total Unit Weight	γ_t	kN/m^3	115	120	135	130	105	130
Permeability	k_x,k_y	m/s	1.0E-07	1.0E-06	1.0E-09	1.0E-08	1.0E-09	1.0E-07
Modulus of Elasticity	E_{70}	kN/m^2	20,700	50,700	8,300	101,400	20,700	101,400
Poisson's Ratio	υ	-	0.20	0.30	0.45	0.30	0.45	0.35
Internal Friction	ϕ	Degs.	32	38	0	40	0	40
Cohesion	c	kN/m^2	0	0	57.5	24.0	96.0	48.0
Coefficient for Lateral Earth Pressure	K_o		0.5	0.5	0.8	0.5	0.6	0.5

due to the presence of the BART and MUNI tunnels. Four levels of tiebacks were installed along the east and south walls. One level of tiebacks was installed below the wall of the adjacent building along the

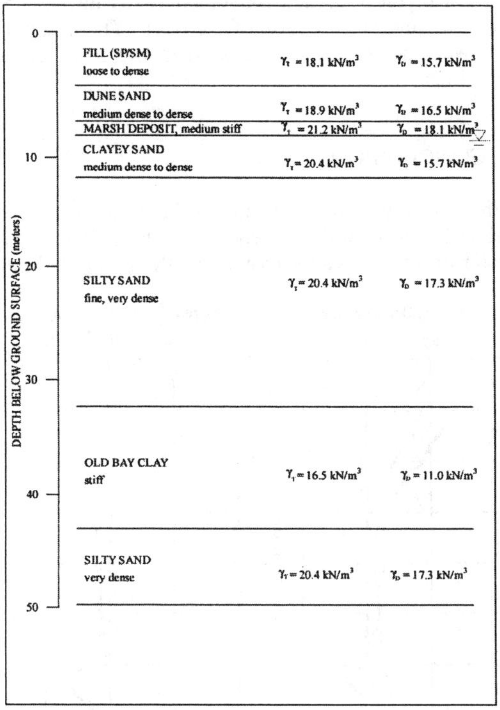

Figure 2. Idealized Subsurface Profile

western side, and two levels of diagonal struts were installed along the northern side. The dewatering system, which included six interior perimeter wells, was installed after the site was excavated to the first level of bracing (about 4.5 m below adjacent grade). Dewatering began immediately after the installation of the wells. The final level of bracing included one level of rakers, using the central portion of the mat as the kicker block. An intermediate shoring step between the installation of the rakers and construction of the mat included the use of a 3-m wide soil bench about 0.6 m below the lowest level of shoring and a 2H:1V soil berm to the bottom of the excavation. Following the construction of the central portion of the mat, the rakers were installed and stressed. Subsequently, the perimeter soil bench and berm were excavated and the foundation mat completed.

4 GEOTECHNICAL MONITORING PROGRAM

As part of this project, a geotechnical monitoring program was implemented. It consisted of four inclinometers and two clusters of piezometers, which were constructed along the Market Street side of the excavation. Figure 1 shows the locations of these instruments. Inclinometers I-1 through I-3 were installed at a distance of about 2 m away from the wall, while I-4 was installed at a distance of about 7 m away from the wall. The purpose of installing I-4 was to measure the attenuation of lateral ground deformation with increasing distance from the wall. Piezometers P-1 and P-2 were installed to measure the fluctuations in the groundwater level during the construction of the wall and the subsequent dewatering of the site. Additionally, ten surface settlement markers were installed along Market Street and monitored.

5 FINITE ELEMENT MODEL AND RESULTS

The purpose of our analyses was to determine if a relatively simple finite element model could effectively predict lateral wall movements for an excavation this deep. A two-dimensional finite element program, PLAXIS, was used to perform the analyses. The soil was modeled using Mohr-Coulomb and elastic parameters. The diaphragm wall and struts and rakers, were modeled as elastic beams and fixed-end anchors, respectively.

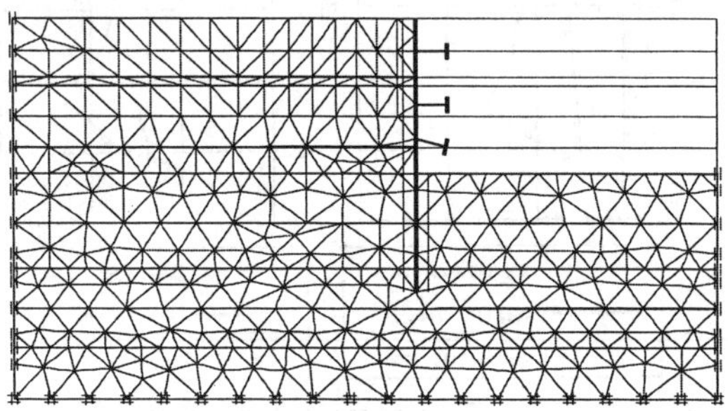

Figure 3. FEM Mesh at Stage 4

The Mohr-Coulomb soil model used, requires parameters which are readily obtained from a standard geotechnical investigation and laboratory testing program. As such, this approach provides a relatively simple practical way of modeling soil behavior.

The soil parameters used in our model are shown in Table 1. The unit weight, friction angle, and cohesion values were selected based on local experience and laboratory tests. Existing data for the site included SPT penetration resistance, unconfined compressive strength, unconsolidated/undrained triaxial compression, sieve analysis, consolidation, and direct shear tests. The modulus of elasticity and Poisson's ratio were estimated based on published values. Several iterations were necessary to appropriately define the parameters for the model.

The diaphragm wall was modeled as a beam element with an effective thickness of 1 m, an axial stiffness (EA) of 3.9×10^7 kN/m and flexural rigidity (EI) of 3.6×10^6 kN-m^2/m. The struts were modeled as fixed-end anchors with an elastic modulus of 2×10^8 kN/m^2.

As with previous studies by others (Plaxis, 1999), we found the simple Mohr-Coulomb approach to be sensitive to the elastic parameters chosen for the soil. Average elastic moduli, based on published data (EPRI, 1990), were assigned to each soil layer. When compared to observed movements of the wall, the model over- and under-predicted movements calculated at the top and bottom, respectively.

Figure 4 shows the comparison of the results of

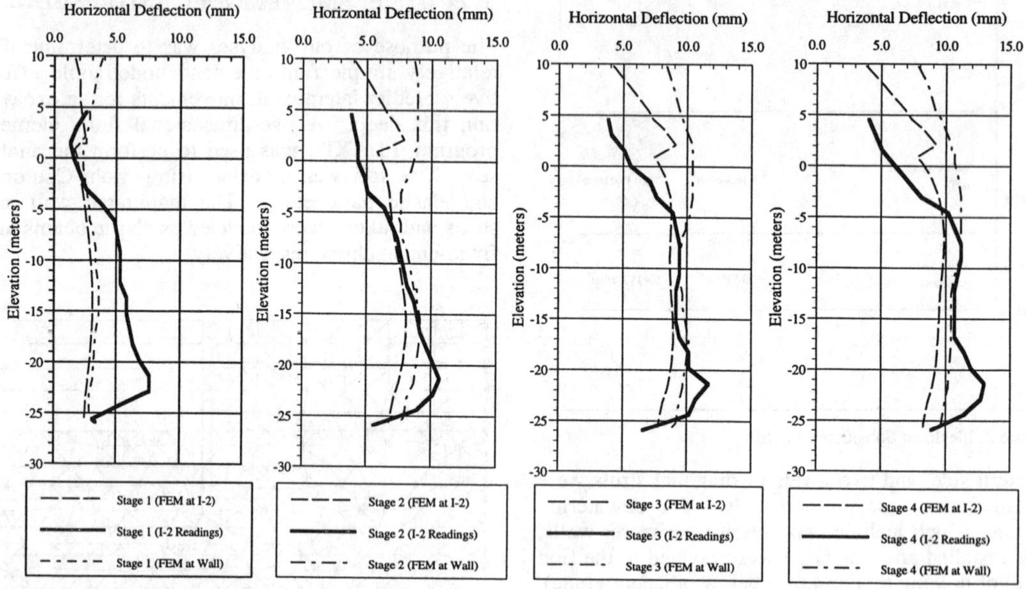

Figure 4: Results of FEM compared to Inclinometer I-2 data.

the FEM analyses at the wall, at a distance equivalent to the location of Inclinometer I-2, and the actual readings from I-2. The stages noted on the figure relate to the following construction activities:

- Stage 1 – Following excavation of 4 m of soil.
- Stage 2 – After installation of level 1 bracing and excavation to a depth of 11 m.
- Stage 3 – After installation of level 2 bracing and excavation to a depth of 16 m.
- Stage 4 – After installation of rakers and excavation to a depth of 21 m (maximum excavation depth).

The maximum computed deformation at I-2 was about 10 mm (stage 4); the measured movement at the same depth was over 11 mm. The maximum measured movement at I-2 occurred near the bottom of the wall and was about 13 mm (stage 4) while the computed deformations were about 8 mm. The measured maximum deformation occurred at a depth of about 30 m below the ground surface, not 16 m as calculated by the model. It should be noted that a portion of the recorded movements was due to dewatering and construction of the wall. These activities were not explicitly modeled in the FEM computations.

6 CONCLUSIONS

While it may difficult to model the construction of the wall (slurry wall sequence) and the effect of dewatering on the wall movements, the computed deformation profile for the first stage provided a reasonable estimate of the measured deformations. For stages 2 through 4, the FEM over-estimated the deformations of the wall within the top 15 m. This overestimation was by a factor of about 1.5 to 2 in the top 10 m. The movements computed at the mid-height of the wall agreed well with the measured deformations. However, near the bottom of the wall where the influence of the construction of the wall and the dewatering of the site on the measured movements were greatest, the FEM underpredicted wall movements.

The finite element method using the Mohr-Coulomb soil model appears to provide a reasonable estimate of the maximum movements of the wall. However, the results presented in this paper are based on an iterative process where the initial estimates of the input parameters, based on empirical data, resulted in excessive computed wall deformations. Subsequently, the parameters had to be refined to obtain more reasonable results. Although in our case the actual movements of the wall were known before the FEM analysis, it would have been possible to develop a similar model solution based on engineering judgement.

REFERENCES

AGS 1988. Geotechnical Investigation Report, Yerba Buena Gardens Center, Office Tower, San Francisco, California, November.

PLAXIS Finite Element Code for Soil and Rock Analyes 1998, Version 7, User's Manual, Edited by Brinkgreve, R.B.J., and Vermeer, P.A., A.A. Balkema/Rotterdam/Brookfield.

PLAXIS 1999, Course Notes in Computational Geotechnics, Boulder, Colorado, August.

Treadwell & Rollo, Inc. 1997, Final Report Geotechnical Investigation CB-1 Tower, San Francisco, California, December.

Electric Power Research Institute (EPRI) 1990, Manual on Estimating Soil Properties for Foundation Design, August.

Computer Methods and Advances in Geomechanics, Desai et al. (eds) © 2001 Balkema, Rotterdam, ISBN 90 5809 183 X

Numerical simulation of the construction process on deep shaft by three-dimensional elasto-plastic finite element analysis

T. Hagiwara & T. Fujii
Nishimatsu Construction Company Limited, Tokyo, Japan

F. Cai & K. Ugai
Department of Civil Engineering, Gunma University, Kiryu, Japan

ABSTRACT: In order to investigate the earth pressures acting on shaft lining and deformation behaviour of the nearby ground, a series of finite element analyses was carried out. The program used was based on an axisymmetric elasto-plastic model combined with shaft excavation process. This paper describes numerical simulations of the previous centrifuge model tests of the shaft excavation. Particular attention is given to the effects of shaft rigidity on the shaft lining stresses during excavation. The results of strains and earth pressures acting on the shaft from the numerical analysis were favorably consistent with those of the centrifuge model tests.

1 INTRODUCTION

In recent years, as the scale and depth of underground structures have increasingly grown, there has been a growing engineering interest in deep shafts. In particular, the increasing number of deep shafts has been made near existing structures and at newly placed landfill areas in Japan. The construction at these areas makes it more and more difficult to meet the criteria such as the allowable deformation because of their hardness. Therefore, accurate and practical design method for the deep shaft must be established as soon as possible in more safe and economical ways. However, there are very few case histories of detailed earth pressure and deformation measurements of deep shafts so far.

In this study, in order to investigate the earth pressures acting on deep shaft lining and deformation behaviour of the nearby ground, a series of finite element analyses was carried out. Centrifuge model tests to access the effects of the shaft rigidity on earth pressures acting on circular shafts in a dry sand, by taking into account the shaft excavation process, have been conducted by Ueno et al.(1996).

This paper deals with accurate numerical simulations of their centrifuge model tests, which were performed using finite element analysis. Particular attentions are given to the change of shaft lining stresses due to shaft excavation and the effects of shaft rigidity on the shaft lining stresses. The general trend of the theoretical results showed a good agreement with the values from the centrifuge model tests.

2 FINITE ELEMENT SIMULATION OF CENTRIFUGE MODEL TESTS

A series of centrifuge model tests has been carried out in an attempt to investigate the effects of shaft stiffness on the earth pressure acting on circular shafts in a dry sand by Ueno et al.(1996). Figure 1 shows a schematic diagram of centrifuge model. In their centrifuge model, complex three-dimensional soil-structure interaction problems, including excavation processes of the shaft, were successfully modeled. All tests were performed in a 100-G acceleration field using model cylindrical shafts made of copper and/or acrylic.

Numerical analysis was performed to simulate the centrifuge model tests as far as possible. The geometory of the numerical model with the finite element mesh is given in Figure 2. Shaft excavation process was modeled as an axisymmetric problem. Finite elements used here are isoparametric rectangular elements consisting of 8 nodal points. Four Gauss points are used for numerical integration. The Mohr-Coulomb failure criterion was applied to the soil and Drucker-Prager failure criterion was used as the plastic potential for the soil. The calculation was carried out under the non associated law with $\psi=9.5^{\circ}$ (ψ:dilation angle)(Bolton,1987). The soil and shaft parameters for the analysis are depicted in Table 1.

Numerical simulation was carried out to predict the results of centrifuge tests for three cases of apparent Young's modulus Et/R (E:Young's modulus of the shaft lining, t: the thickness of the shaft, t: the

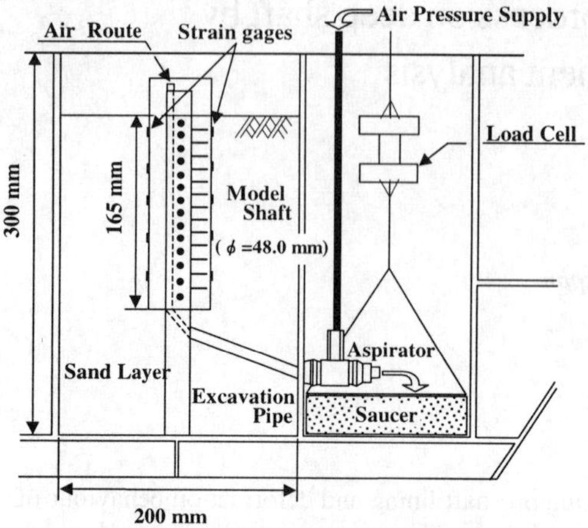

Fig. 1 Schematic diagram of centrifuge model for shaft excavation (Ueno, et al.,1996)

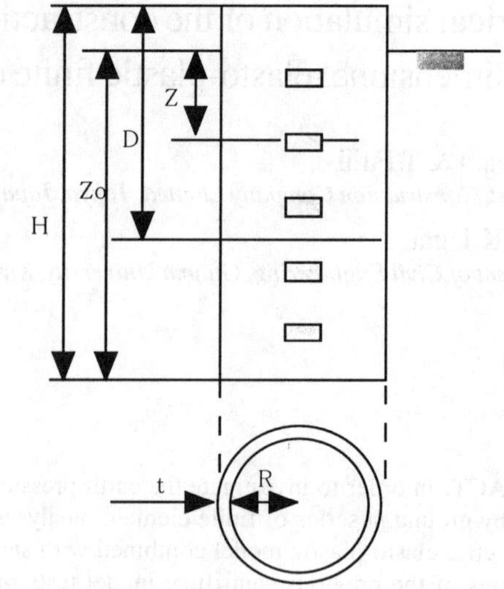

Fig. 2 Model shaft

Table 1 Soil and shaft parameters for the analysis

	Toyoura sand	Shaft
Modulus of elasticity, E	a	129 GPa[b]
Poisson ratio, ν	0.26	0.35
Unit weight, γ	15.68 kN/m^3	
Cohesion, c	0.0 kPa	
Internal friction angle, ϕ	40.5°	
Dilatancy angle, ψ	9.5°	

[a]Toyoura sand:

$$G = 800\frac{(2.17-e)^2}{1+e}Pa\left(\frac{\sigma_m}{Pa}\right)^{0.5}$$

[b] OE30AC: E=3.12 GPa

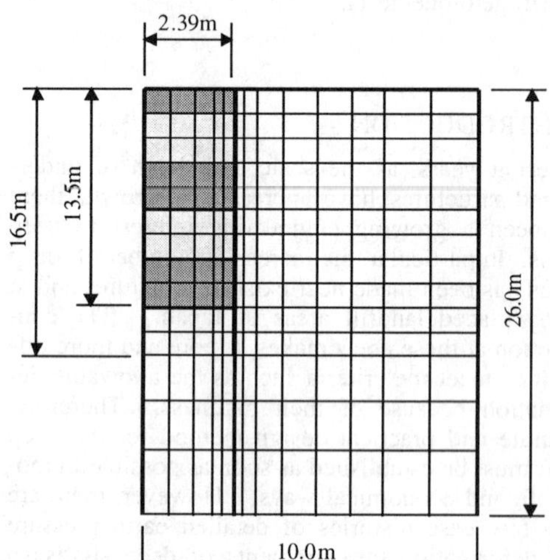

Fig. 3 Finite element mesh used for FE analysis

Table 2 Model shaft for centrifuge tests

Test code	Mat.	R_m (mm)	t_m (mm)	z_{om} (mm)	Et/R (Gpa)
OE2550	Copper	23.9	0.25	16.5	1.40
OE5050	Copper	23.0	0.50	16.5	2.58
OE30AC	Acrylic	23.9	3.00	16.5	0.39

thickness of the shaft, R: the inner radius of the shaft) which is defined hereby to represent the stiffness of the shaft, as shown in Figure 2. The values are listed in Table 2, together with Test code. These values including aspect ratio cover the range of apparent Young's modulus of actual cast-in-place reinforced concrete shafts.

3 RESULTS AND DISCUSSIONS

Figure 4 shows changes in distributions of the radial strain due to shaft excavation for the test OE2550 in Table 2. From numerical results it is found that as the excavation process proceeds, the point which shows the highest strain moves to the deeper one. This is because that the inner wall of the shaft was supported by the internal sand and radial strains at the points deeper than excavation depth were small.

From Figure 4 it is made clear that the numerical results reasonably and accurately predict the centrifuge results, particularly concerning the change of radial strains due to excavation process.

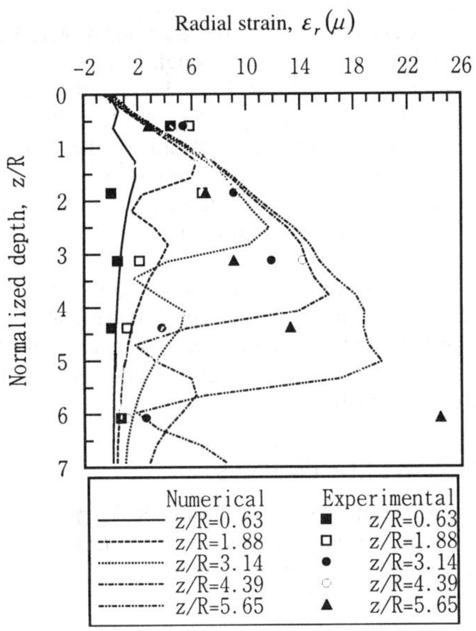

Fig. 4 Change of radial strains during shaft excavation

Figure 5 shows the comparison of the normalized radial earth pressure distribution obtained from three model shafts with different apparent Young's modulus Et/R. The estimated values of the active earth pressure predicted by means of the Berezantzev (1958), Rankine method based on two dimensional posutilate and the Jaky's at-rest earth pressure are also presented in this figure. The values of the earth pressures tend to become greater as Young's modulus increases. This tendency is clearly observed over the shallower half. The values obtained from OE30AC model shaft are relatively small and close to the Berezantzev's value. This trend is quite similar to that of the test

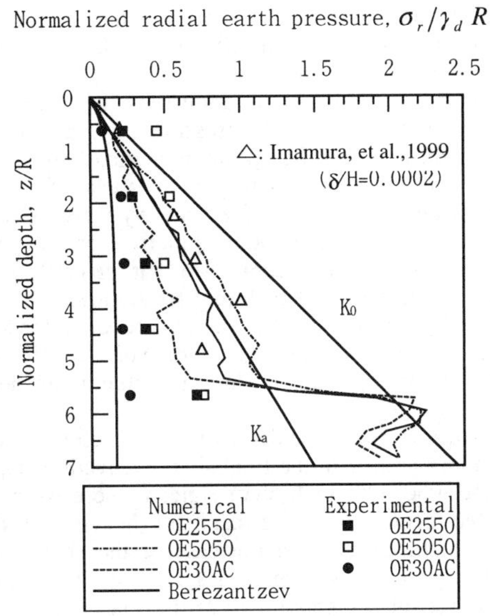

Fig. 5 The effects of shaft stiffness on horizontal earth pressures

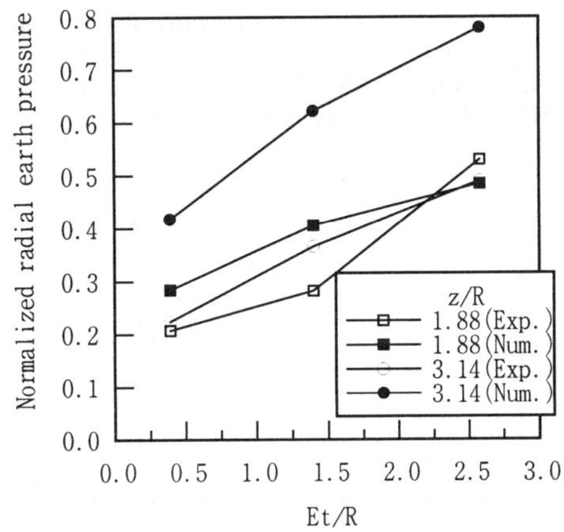

Fig. 6 Shaft rigidity vs earth pressure

results.

Furthermore, the numerical results of high rigid shaft (OE5050) are nearly consistent with measured results from the centrifuge tests, which have been carried out under the condition of rigid shaft model (Imamura, et al.,1999).

These results imply that the stiffness of the shaft dominates how much radial pressure acts on the shaft. It appears from Figure 5 that the earth pressures both calculated and measured are much smaller than Rankine's result which overestimates the active earth pressures on an axisymmetric shaft at the deeper points. One of the reasons for this is considered to be that the redistribution of stress at deeper points is made by means of the arching effects (three-dimensional effects), which develop with the shaft excavation.

Figure 6 indicates relationships between the normalized radial earth pressure (α) and Et/R. Although α values based on numerical results at z/R= 3.14 are relatively larger than those of centrifuge results, the values of α linearly increase with increasing of Et/R in both numerical and experimental results. It is confirmed that the earth pressures acting on the shaft lining are significantly affected by the shaft rigidity. Therefore, the effects of shaft stiffness should be effectively considered when building rational design methods of deep shafts.

Figure 7 shows the displacements vectors of surrounding ground in the final shaft excavation stage. In OE5050 with high shaft rigidity, there were no visible displacements, except for heaving at the bottom. In contrast, in the case of the shaft with high rigidity as shown in Figure 7(a), horizontal inward displacements mode of the shaft wall has a characteristic of showing could be seen near the shaft. The displacement depth of the shaft height, swelling toward excavation side. Since the stiffness of the shaft for practical use is fairly large enough to restrict radial deformations of the shafts, the soil ground surrounding the shafts may not fully yielded, as shown in Fig. 7(b). This analogy is also confirmed from the result of the earth pressure

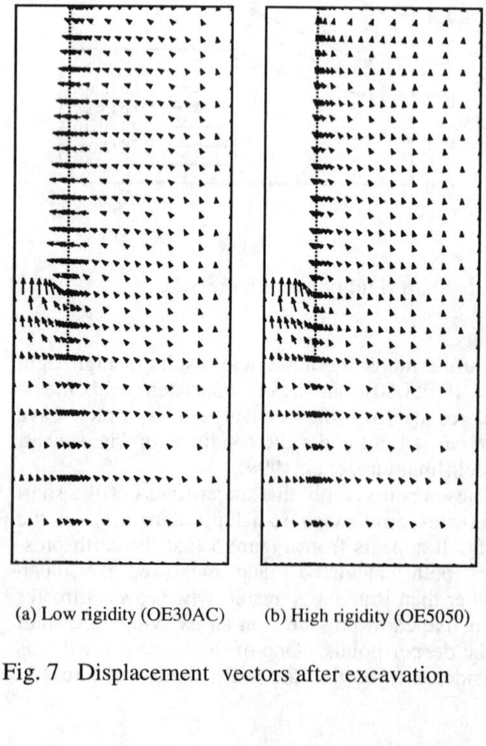

(a) Low rigidity (OE30AC) (b) High rigidity (OE5050)

Fig. 7 Displacement vectors after excavation

distribution of OE5050, which exists at around Rankine's active earth pressure distribution, as shown in Figure 5.

4 CONCLUSIONS

Numerical analysis based on the axisymmetric FEM was performed and compared with the previous centrifugal model results of the cylindrical deep shaft excavation. The knowledge gained by this study is summarized as follows:

1. Numerical results well predicted the radial strains and sectional stresses acting on the shaft wall during shaft excavation obtained from the centrifuge model tests and reliability of the proposed numerical approach in this study was validated.
2. The three-dimensional effects, mainly acted as the arching action, gave smaller lateral earth pressures than those based on the two-dimensional model, especially at deeper depth of the shaft. More rational design method for deep shafts would be made possible by taking into account the three-dimensional effects of shafts structures, including the effect of stress level.
3. As the shaft rigidity decreases, the earth pressures acting on the shaft and its surrounding ground movements increase. The crucial effects of the shaft rigidity on the shaft behaviour were confirmed.

5 ACKNOWLEDGEMENTS

The authors would like to thank Dr. K. Ueno of Tokushima University for his encouragement and helpful suggestions in the preparation of this paper.

REFERENCES

Berezantzev, V.G. 1958. Earth pressure on the cylindrical retaining walls, *Brussels Conf. on Earth Pressure Problems*:21-27

Botlon, M.D. 1987. The strength and dilatancy of sands, *Geotechnique, Vol. 36, No.1*: 65-78.

Imamura, S. et al. 1999. Earth pressures acting on a deep circular shaft and the movements of adjacent ground in sandy ground, *Proc of Geotechnical Aspect of Underground Construction in Soft Ground (edited by Kusakabe, Fujita and Miyazaki), IS-Tokyo 99*:100-106.

Ueno, K., Yokoyama, Y., Ohi, A. and Fujii, T. 1996. Earth pressures acting on flexible circular shafts in sand, *Proc. of Geotechnical Aspect of Underground Construction in Soft Ground (edited by Mair and Taylor)*:237-242.

Computer Methods and Advances in Geomechanics, Desai et al. (eds) © 2001 Balkema, Rotterdam, ISBN 90 5809 183 X

A non-linear finite element procedure for evaluation of settlement induced by tunnel excavation

S. Hernández & L. E. Romera
School of Civil Engineering, University of La Coruña, Spain

ABSTRACT: A numerical model to simulate the excavation procedure with a full face tunnelling boring machine and evaluate the settlements in soft ground together with forces developed in the steel shield support and the lining is presented in this paper. The problem is solved by means of the software FESTE (Finite Element Software for Tunnelling Excavations), written by the authors, at University of La Coruña, considering several nonlinearities such as the contact problem due to interaction soil-structure, geometric discontinuities and elastoplastic materials. The constructive process is reproduced with two-dimensional and three-dimensional models, applying these methodology to several sections of the new lines of the Madrid underground railways excavated during years 1996-1999, the numerical results are in agreement with the available instrumental measurements of soil displacements.

1 INTRODUCTION

The construction of bored tunnels in soft ground is a complex operation that inevitably causes ground movements. The prediction of settlements in the urban environment is an important aspect of the problem to preclude damage to buildings or other underground structures.

From a qualitative point of view, the main causes of the soil movements and their relationship with the tunnel construction stages are well known. Figure 1 shows the usual profiles of vertical and horizontal movements in surface (Fig. 1a), the variation of vertical and horizontal movements with depth (Fig. 1b), and the horizontal movements in the tunnel axis direction (Fig. 1c).

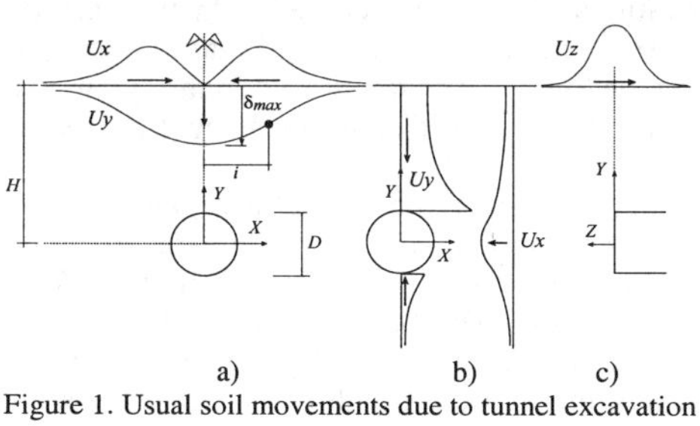

Figure 1. Usual soil movements due to tunnel excavation

For the case of a single tunnel in green field conditions, the surface settlement developed above advancing tunnel heading has been studied by many authors (Attewell et al., 1986, and Uriel & Sagaseta, 1989). The transverse settlement is well-described by a Gaussian distribution curve (Peck, 1969), and the longitudinal profile of settlement along tunnel axis form their distribution function. The problem is 3-D in the vicinity of the tunnel face, whereas far from there, on achieve a plane strain state.

The prediction methods of soil movements can be classified in empirical, experimental, analytical and numerical. Empirical methods are based on field observations of settlements. Among the experimental methods, it can be emphasized the test of scaled models in geotechnical centrifuges (Nomoto et al., 1999).

The analytical methods are based in different hypothesis about the deformations produced in the vicinity of a circular hole in a semi-infinite domain. It can be mentioned the work of Sagaseta (1987), Verruijt (1997), and Loganathan & Poulos (1998).

Traditionally, the numerical methods are based in two-dimensional models, but in the last years, three-dimensional models have been developed. Among others, it can be emphasized the work of Ghaboussi et al. (1983), Rowe et al. (1983), Lee et al. (1991), and Benmebarek et al. (1999).

2 EXCAVATION PROCEDURE

In the case of underground railways works studied in the city of Madrid, the excavation was carried out by a full face tunnelling boring machine (T.B.M.) equipped with a rotary excavator (Fig. 2).

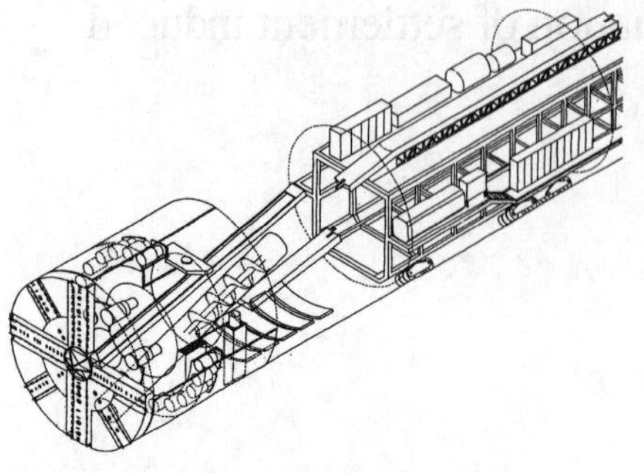

Figure 2. Tunnelling machine and rear convoy

The whole system consists of a steel shield containing the extraction machinery and the hydraulic jacks, the stabilization of the excavation front is obtained by an earth pressure balance system (E.P.B.). The shield is centered with respect to excavated hole by steel plates located on the lower part of the shield.

Segment lining elements of pre-cast concrete segments compose the final tunnel structure. They are placed at the tailpiece of the machine and grouting mortar is used to fill up the void between the concrete elements and the ground. Once the lining is finished the machine is advanced by thrusting on the tunnel structure. A detailed scheme with dimensions is presented in Figure 3.

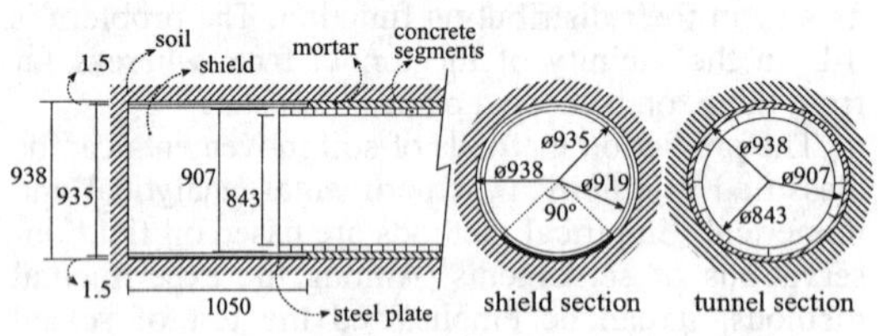

Figure 3. System diagram and general dimensions (cm)

3 ORIGIN OF SETTLEMENTS

The soil movements are maximum near to the tunnel excavation and decrease far off tunnel axis.

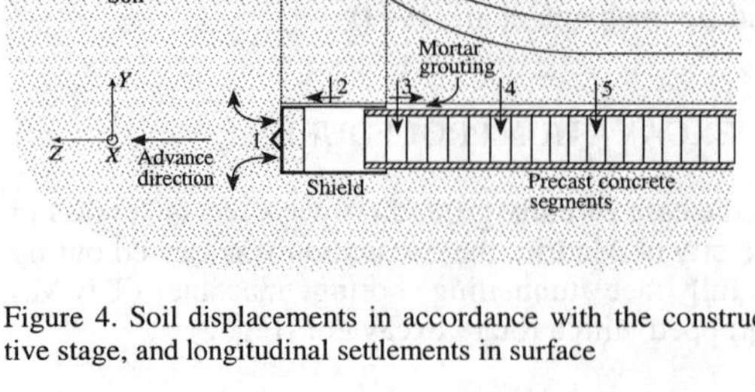

Figure 4. Soil displacements in accordance with the constructive stage, and longitudinal settlements in surface

The principal ground displacement sources associated with shield tunnelling stages (Fig. 4) are the follows (Romo, 1997):

Stress changes in the tunnel face surroundings (detail 1 in Fig. 4). If the pressure on tunnel face is lower than the free field stresses, then the soil will move toward the shield. On the other hand, in the opposite case the soil ahead the tunnel face will move away from it.

The over-cutting in the passage of the shield creates an annular gap between the tunnel wall and the shield, that will lead to ground movements (detail 2 in Fig. 4) which are mainly perpendicular to tunnel axis. Another annular gap develops between the soil and the concrete segments which form the lining (detail 3 in Fig. 4), producing new inwards ground movements. Although, if the injection pressure of mortar is greater than the overburden then outward soil displacements are developed (Bernat et al., 1997). When the mortar solidified, additional movements appear due to lining deflection as ground loading develops (detail 4, in Fig. 4). Finally, movements due to consolidation and creep appear (detail 5, Fig. 4).

The ground movements ahead the tunnel face are reduced using the E.P.B. system, even they can be null. In this case the three-dimensional effects of the construction process are small. On the other hand, the Madrid soil have in general a high stiffness, and the creep effects are very small. So the most important effects in soil movements are due to the clearance generated in the excavation process.

4 FINITE ELEMENT MODELLING

4.1 *Program description*

The type of analysis is static, employing a displacements formulation. Among the principal features of the code FESTE are the consideration of contact nonlinearities to resolve soil-structure interaction, and the employment of mapped quadratic infinite elements (Marques & Owen, 1984) to model boundary conditions in unbounded media. In the case of circular excavations the pole of infinite elements is placed in the axis of excavation, and in the symmetry plane. The value of the displacement field in the infinite is considered null.

The analysis is incremental iterative, and the solution in each step is obtained with the Newton-Raphson method or Newton-Raphson modified. Three types of nonlinearities are considered.

Geometric nonlinearity due to construction process is taken into account dividing the analysis into a series of stages. At each stage there is a definition of elements activated or constructed, and disactivated or excavated. Only the contributions due to active elements are assembled in the stiffness matrix.

Contact nonlinearity to resolve the interaction between soil and constructive elements. The theory used allows node-segment contact and consists at a flexibility matrix linked to the contact area. This approach, which works in an iterative way and can be also combined with materials with nonlinear constitutive curves. The contact algorithm includes Coulomb friction, and in each step of the analysis calculates the contact zone that develops and the contact forces produced.

Material nonlinearity is included considering elastoplastic materials with the Drucker-Prager plastification criteria, and using different elasticity modulus in zones under compression or extension. The code incorporates the error estimator Z^2 (Zienkiewicz & Zhu, 1987). If the presence of water is to be taken into account, the limit cases in saturated soil with null drainage at zero time and full drainage at an infinite time can be considering through the use of the Naylor method (Naylor et al., 1981).

4.2 *Two-dimensional approach*

4.2.1 *Model description*

In this case the problem is modeling in plane strain, under the hypothesis that 3-D effects are very much attenuated due to earth pressure balance system. Finite element meshes as presented in figure 5 were used in this case. The model includes:

– Tunneling shield: after an extensive analysis an equivalent value of ten times the elasticity module of the steel was defined to represent the real stiffness of the tunnel shield.

– Pre-cast concrete segments: linear elastic values were defined for the mechanical properties of this material.

– Mortar grouting: this material becomes stiffer few hours after being injected into the tail void.

– Ground materials: with properties related to horizontal layers. The upper and lower height for each layer is defined, and is possible to define a linear variation of the elasticity modulus *E'* and the coefficient of earth pressure at rest K_0 with depth.

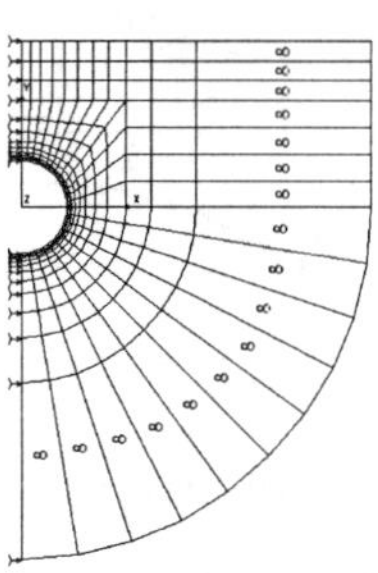

Figure 5. Finite element mesh and boundary conditions

The tunnelling procedure was separated into three different phases (figure 6). In the first stage the full section excavation and setting up the shield is taken into account. The model contains the shield, the lower steel plates of it and a void between the soil and shield was considered with a value of 3 cm based on the average working condition (fig. 6a). Interaction between soil and shield was included.

In the second stage the shield is taken out of the model and the concrete elements set is included as shown in figure 6b. Also an injection pressure introduced by the mortar and producing the distribution indicated in figure 6d is considered on the soil and with opposite direction on the concrete segments. Interaction between soil and concrete segments was included.

Finally, in the third stage, solid mortar grouting elements are added to the structural model and a distribution field of pressures opposite to the pressures mentioned in the second stage, is modeled between the soil and the lining.

Additionally, at each stage, gravitational loads corresponding to excavation devices are considered.

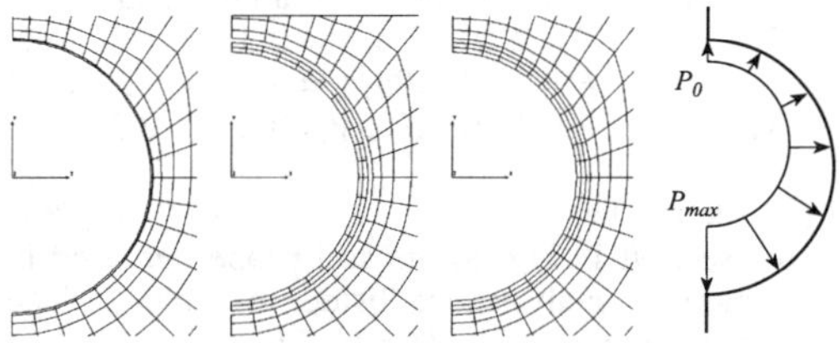

a) b) c) d)

Figure 6. Set of excavation stages (a, b, c), and injection pressure field (d)

To select the mesh used in the analysis, on comparing the results in stress error and maximum movements in several meshes as shown in Figure 7, varying the mesh density and mesh dimensions. The results obtained with these meshes, applied in the same excavation case, appears in table 2. The mesh 2 was choused for the subsequent analysis.

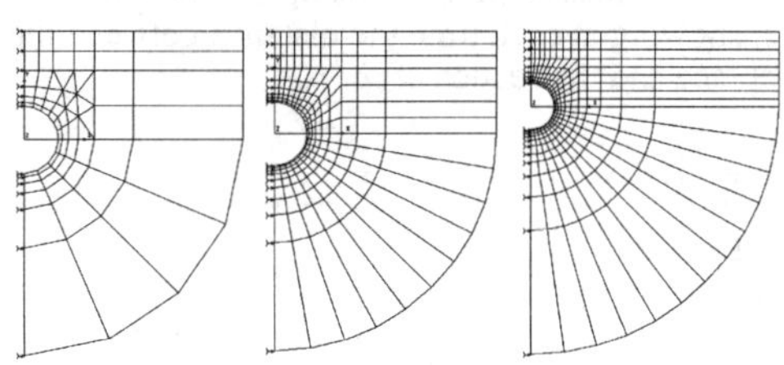

Mesh 1 Mesh 2 Mesh 3

Figure 7. Set of meshes compared

Table 1. Numerical results

	D.O.F*	A.P.E*	M.E.E*	U_x(mm)	U_y(mm)
Mesh 1	166	8.99	24.21	–4.98	–19.7
Mesh 2	484	5.04	15.43	–5.25	–20.3
Mesh 3	652	4.01	11.78	–5.30	–20.2

* D.O.F: Degrees of freedom
A.P.E: Average percentage error
M.E.E: Maximun elemental error

4.2.2 *Numerical results*

Three different examples of settlement measured in the extension of underground railways system of Madrid were compared with numerical results obtained from the methodology presented.

The soil layers for each one of the three sections studied appear in Figure 8. The presence of water is taken into account only in the first section.

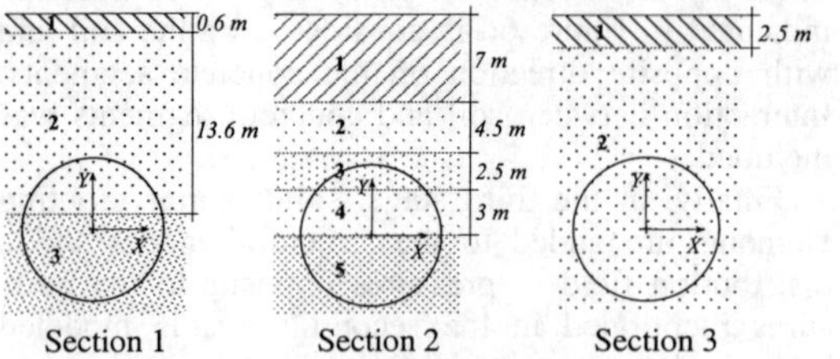

Figure 8. Soil layers in each section

The properties of steel shield, concrete segments and solidified mortar considered in all cases are:

- Shield: $E = 2{\cdot}10^9$ kPa, $\nu = 0.3$, $\gamma = 7.85$ t/m^3
- Concrete: $E = 3{\cdot}10^7$ kPa, $\nu = 0.25$, $\gamma = 2.5$ t/m^3
- Mortar: $E = 1{\cdot}10^5$ kPa, $\nu = 0.2$, $\gamma = 2.2$ t/m^3

4.2.2.1 Section 1

For the section 1 (Fig. 8) the soil properties used in the numerical analysis are listed in table 2. The depth of the tunnel axis is $H = 15.5$ m.

Table 2. Soil properties

Layer	E	ΔE	ν	K_0	c	ϕ	γ
	kPa	kPa/m			kPa	°	t/m^3
1	$1{\cdot}10^4$	0	0.3	0.4-0.4	10	28	1.8
2	$1.35{\cdot}10^5$	0	0.3	0.4-0.6	100	33.5	2.0
3	$1.75{\cdot}10^5$	3000	0.3	0.6-0.4	150	32.5	2.1

In this section, the injection pressure measured at the axis level was $P_{i,eje} = 215$ kPa, and the volume of injected mortar was $V_m = 5.76$ m^3/m. Figure 9 presents the measured settlement at six different times, showing at each time the distance between the tunnel face and the section.

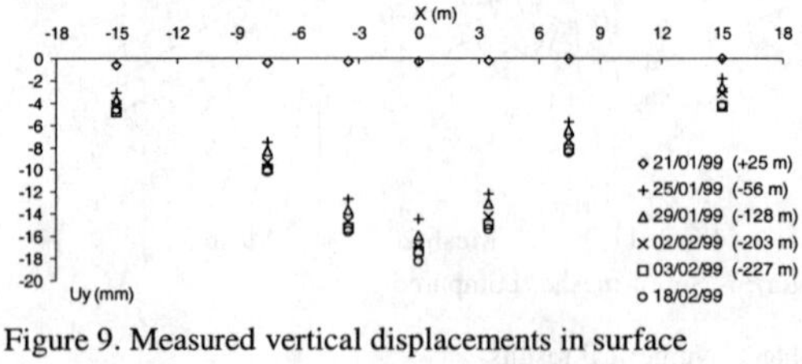

Figure 9. Measured vertical displacements in surface

The approximate instantaneous settlement was 14.5 mm, and the maximum settlement measured was 18.25 mm.

The results obtained with the numerical model in each stage appear in Figure 10, together with the instantaneous measured settlement (shown as •). In this case only geometrical and contact nonlinearities are considered, with elastic ground without water, and $P_{i,eje} = 150$ kPa.

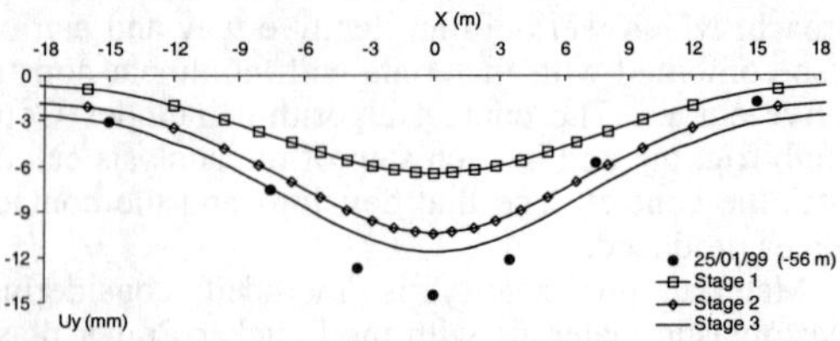

Figure 10. Vertical displacements in surface at each stage

One of the most important parameters is the injection pressure. In figure 11 the final results obtained using the previous pressure and the measured pressure of 215 kPa are contrasted. In the studied sections a reduction about 30% in injection pressures improve the results obtained, this fact can be justified due to pressure drop-off in the pass of fluid mortar through concrete segments.

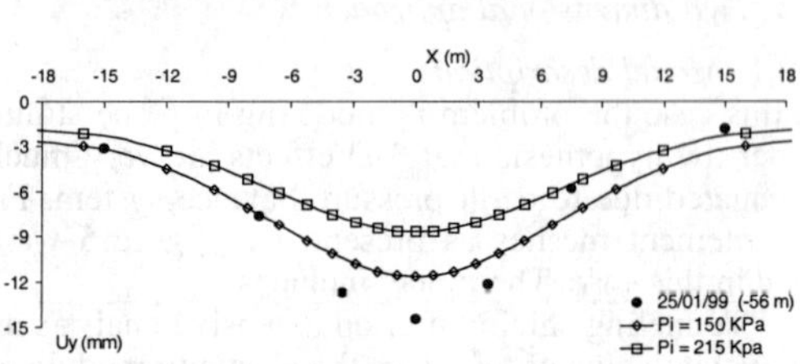

Figure 11. Final displacements with two different injection pressures

The effects of material nonlinearity are reduced. In figure 12 the numerical solution considering elastic soil (Case 1) is contrasted with the solution obtained including a reduced value from the elasticity modulus in zones under extension (Case 2), with a reduction factor of 2.5, and considering elasto-plastic soil model (Case 3). The material nonlinearity improve the numerical solution, but a close resemblance between the three sets of results can be observed.

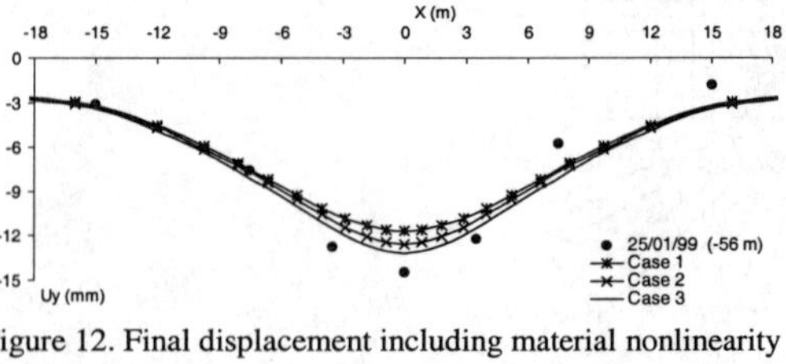

Figure 12. Final displacement including material nonlinearity

In the case 3, the volume of injected mortar calculated is 4.32 m^3/m, considering the water level at 0.6 m of depth, this volume is 3.85 m^3/m in agreement with the measured data. In Figure 13 the final settlements considering geometric, contact, material nonlinearities, with null drainage and full drainage are contrasted with the instantaneous and final settlements measured.

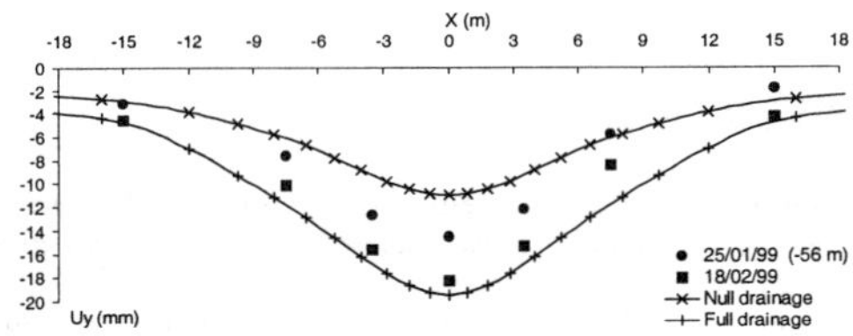

Figure 13. Vertical displacements in surface including the effect of water

4.2.2.2 Section 2

For this section (Fig. 8) the soil properties used in the numerical analysis are listed in table 3. The depth of the tunnel axis is $H = 17$ m.

Table 3. Soil properties

Layer	E	ΔE	ν	K_0	c	ϕ°	γ
	kPa	kPa/m			kPa	°	t/m^3
1	$1{\cdot}10^4$	0	0.3	0.4-0.4	10	28	1.8
2	$1.25{\cdot}10^5$	0	0.28	0.4-0.4	100	33	2
3	$7.5{\cdot}10^4$	0	0.3	0.4-0.5	75	35	2
4	$1.25{\cdot}10^5$	0	0.28	0.5-0.6	100	33	2
5	$1.75{\cdot}10^5$	5000	0.3	0.6-0.4	150	32.5	2.1

In Figure 14 the final results for vertical displacements in surface and measured data are presented. In Figure 15 appears the variation with depth of vertical displacements along the vertical axis of symmetry. In both cases, a close resemblance between the numerical and the measured data can be observed.

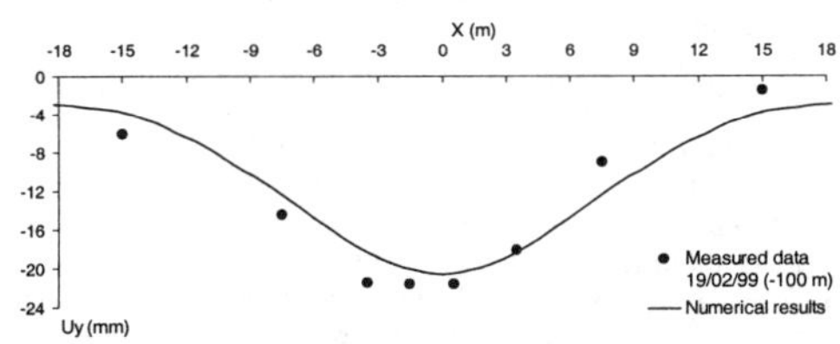

Figure 14. Vertical displacements in surface

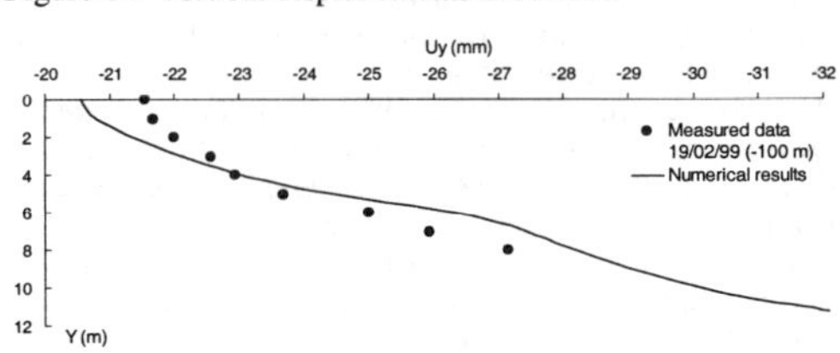

Figure 15. Vertical displacements at $x = 0$

4.2.2.3 Section 3

For this section (Fig. 8), the soil properties used in the numerical analysis are listed in table 4. The depth of the tunnel axis is $H = 15.2$ m, and the measured volume of injected mortar was $V_m = 4.4$ m^3/m.

Table 4. Soil properties

Layer	E	ΔE	ν	K_0	c	ϕ°	γ
	kPa	kPa/m			kPa	°	t/m^3
1	$1{\cdot}10^4$	0	0.3	0.4-0.4	10	28	1.8
2	$1.35{\cdot}10^5$	0	0.3	0.4-1.5	137	33.5	2
3	$1.35{\cdot}10^5$	5000	0.3	1.5-1.0	150	33.5	2

In this case the vertical displacements in surface are small elevations, together with horizontal displacements in depth towards the tunnel axis. The second and third layers of soil are overconsolidated clays with $K_{0max} = 1.5$, and the injection pressure at the axis level considered is $P_{i,eje} = 335$ kPa. The results obtained with the model, and the measured data appear in Figures 16 and 17. The volume of injected mortar calculated is $V_m = 4.47$ m^3/m.

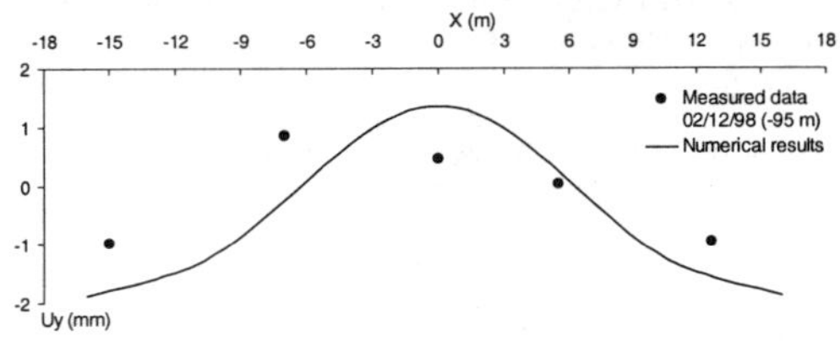

Figure 16. Vertical displacements in surface

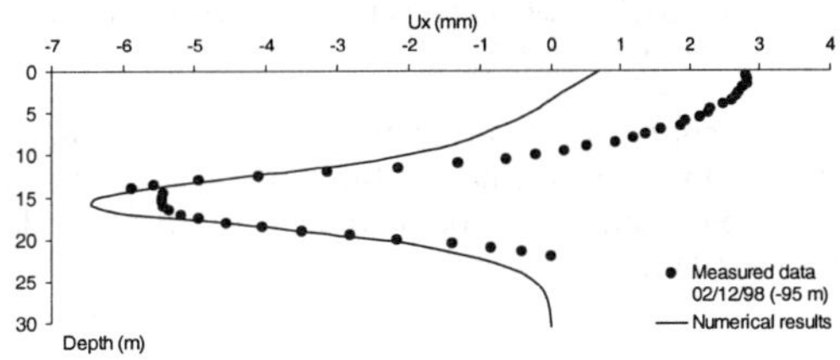

Figure 17. Horizontal displacements at $x = 6$ m

4.3 *Three-dimesional approach*

4.3.1 *Model description*

The 3-D model that has been developed appears in Figure 18. In it, infinite elements are incorporated. Moreover, geometrical and contact nonlinearities are considered so that twelve excavations advances have been carried out in direction z, with a varying advance length.

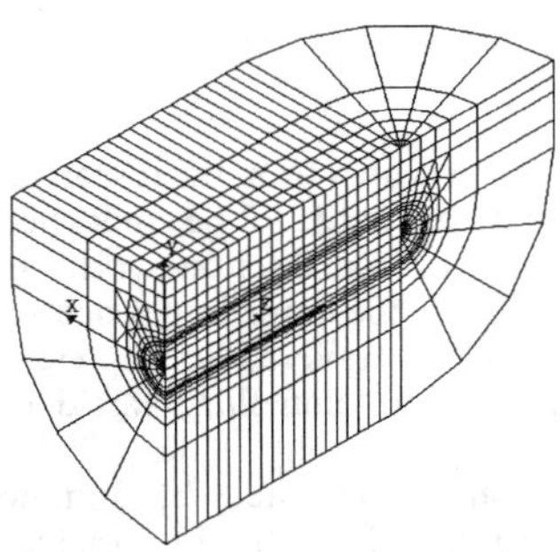

Figure 18. 3-D model

4.3.2 *Numerical results*

The model is applied to section 1, with properties listed in Table 2. Considering elastic soil and without water. Some of the results obtained appear in Figures 19 and 20.

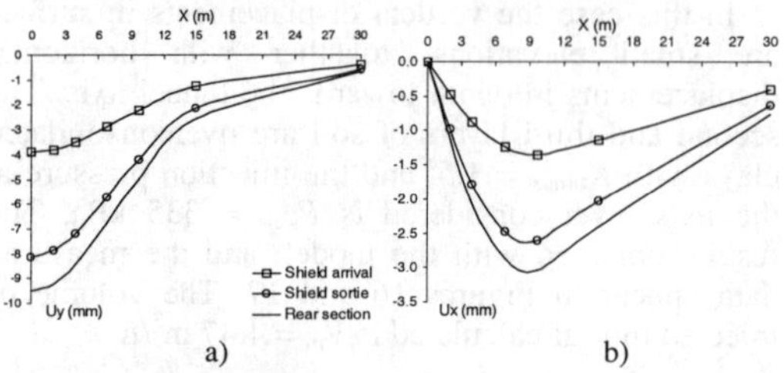

Figure 19. Transversal sections of vertical (a) and horizontal (b) displacements in surface

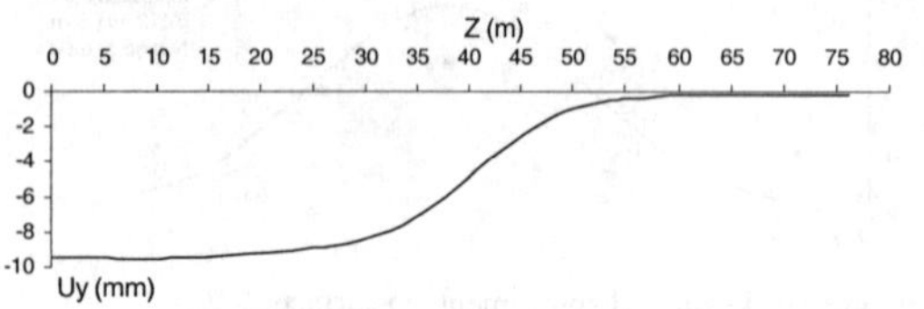

Figure 20. Longitudinal profile of vertical displacements in surface above the tunnel axis

In Figure 21, a comparison has been made between the final results of the vertical settlements in the rear section *xy* obtained with the 3-D model and the 2-D model.

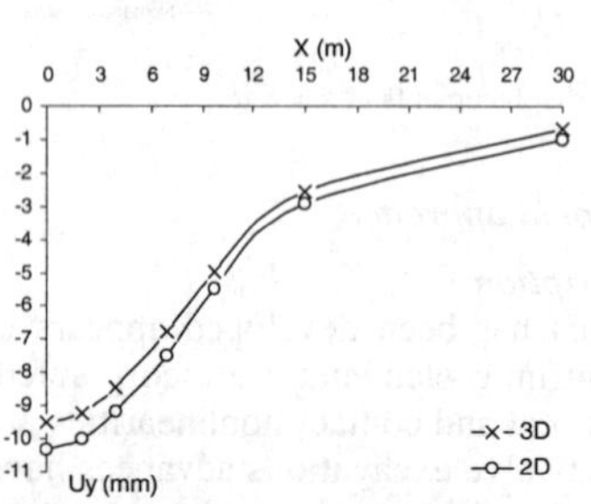

Figure 21. Comparison of vertical displacements in surface

5 CONCLUSIONS

Several conclusions can be drawn from the research:

– Excavation procedures need to be modelled including several phases of construction works in order to reproduce accurately the tunnelling process.

– Non-linear analysis is requiered to represent contact between the soil and tunnelling shield or tunnel structure.

– Boundary conditions represented by infinite elements can lead to a reduced number of variables in the problem.

– The methodology presented can reproduce adequately the tunnel construction process, with solutions in good agreement with settlement measurements of actual tunnelling works.

6 REFERENCES

Attewell, P.B. Yeates, J. & Selby, A.R. 1986. *Soil Movements Induced by Tunnelling and their effects on Pipelines and Structures*. Glasgow: Blackie.

Benmebarek, S., Kastner, R., Ollier, C. & Charif, A. 1998. Auscultation et modélisation numérique du processus de creusement à l'aide d'un tunnelier. *Géotechnique* 48(6): 801-818.

Bernat, S., Cambou, B., Kastner, R., Ollier, C., Dubois, P. & Guibert, E.G. 1997. Shallow tunnel – In-situ monitoring and modelling, *Proc. 14th Int. Conf. on Soil Mech. and Found. Eng.*, Hamburgo, Vol. 3: 1493-1496.

Ghaboussi, J., Hansmire, W.H., Parker, H.W. & Kim, K.J. 1983. Finite element simulation of tunneling over subways. *J. Geotech. Eng.* 109 (3): 318-327.

Inokuma, A. & Fujimoto, A. 1996. *An improvement of F.E.M. analysis in ground settlement prediction in shield tunneling.* Geotechnical Aspects of Underground Construction in Soft Ground. Mair & Taylor (eds.). Rotterdam: Balkema.

Lee, K.M. & Rowe, R.K. 1991. An analysis of three-dimensional ground movements: the Thunder Bay Tunnel. *Canadian Geotech. J.* 28: 25-41.

Loganathan, N. & Poulos, H.G. 1998. Analytical Prediction for Tunneling-induced ground movements in Clays. *J. Geotech. and Geoenv. Eng.* 124(9): 846-856.

Marques, J.M.M.C. & Owen, D.R.J. 1984. Infinite Elements in Quasi-Static Materially Nonlinear Problems, Computers and Structures 18(4): 739-751.Nomoto, T., Imamura, S., Hagiwara, T., Kusakabe, O., & Fujii, N. 1999. Shield Tunnel Construction in Centrifuge. *J. of Geotech. and Geoenv. Eng.* 125(4): 289-300.

Naylor, D.J., Pande, G.N., Simpson, B. & Tabb, R. 1981. *Finite Elements in Geotechnical Engineering*. Swansea, U.K: Pineridge Press.

Peck, R.B. 1969. Deep excavations and tunnelling in soft ground. State of the Art Report. *Proc. 7th Int. Conf. on Soil Mech. and Found. Eng.*, México: 225-290.

Romo, M.P. 1997. Soil movements induced by slurry shield tunneling, *Proc. 14th Int. Conf. on Soil Mech. and Found. Eng.*, Hamburgo, Vol. 3: 1473-1481.

Rowe, R.K., Lo, K.Y. & Kack, G.J. 1983. A method of estimating surface settlement above tunnels constructed in soft ground. *Canadian Geotechnical Journal* 20: 11-22.

Sagaseta C. 1987. Analysis of undrained soil deformation due to ground loss. *Géotechnique* 37(3): 301-320.

Uriel, A.O. & Sagaseta, C. 1989. Selection of design parameters for underground construction. *Proc. 12th Int. Conf. on Soil Mech. and Found. Eng.*, Rio de Janeiro, Brasil. Gen. Report Session 9.9.4: 2521-2551.

Verruijt, A. 1997. A complex variable solution for a deforming circular tunnel in an elastic half-plane. *Int. J. Num. and Anal. Methods in Geomech.* 21(2): 77-89.

Zienkiewicz, O.C. & Zhu, J.Z. 1987. A Simple Error Estimator and Adaptive Procedure for Practical Engineering Analysis. *Int. J. Num. Methods in Eng.* 24: 337-357.

Computer Methods and Advances in Geomechanics, Desai et al. (eds) © 2001 Balkema, Rotterdam, ISBN 90 5809 183 X

Analysis of hydro-mechanical interaction of rock salt in underground waste repositories

Z. Hou & K.-H. Lux
TU Clausthal, Clausthal-Zellerfeld, Germany

ABSTRACT: This article will as a part of geotechnical assessment introduce a new concept of hydro-mechanical coupling with the main parameters of "pore pressure, seepage force and effective stress" based on the *Hou/Lux* material model and using additively the relation $k = k(\varepsilon_{Vol}, \sigma_3)$ according to *Stormont (1992)* to numerically simulate the hydro-mechanical interaction in case of a flooded repository and to evaluate the efficiency of sealing constructions. As example, a dam construction taking into account the interactions between hydraulic and mechanical processes is examined.

1 INTRODUCTION

The objective of underground waste repositories is to isolate e.g. the radioactive and other toxic particles from the biosphere. Therefore among others, the behaviour of the disturbed rock zone (DRZ) in conjunction with geotechnical barriers such as drift or shaft seals and interaction between hydraulics and mechanics as well as the efficiency of the geological barrier are important for geomechanical safety analysis of repositories.

In regard to geotechnical safety analysis and assessment as well as in regard to interactions between mechanical and hydraulic processes in case of a flooded repository, material models are required, which not only describe transient and stationary creep, but also are capable of adequately describing structural damage, dilatancy, creep rupture and tertiary creep. A corresponding *Hou/Lux* material model, based on the *Lubby2* material model according to *Lux (1984)* and based on the continuum damage mechanics, e.g. according to *Kachanov (1986)* or *Lemaitre (1992)*, *Hou (1997)* and *Hou & Lux (1998, 1999)*, is applied for the analysis of the behaviour of the DRZ and the interaction between hydraulics and mechanics.

2 MECHANICAL MODEL *HOU/LUX* WITH CREEP RUPTURE CRITERION AND TERTIARY CREEP

The *Hou/Lux* material model is a material model, which includes in general the effects of the deformation mechanisms of diffusion and dislocation, strain hardening and recovery as well as damage and (to some extend) damage healing, and which is thus capable of fulfilling, at least in principle, the most important demands made on a material model. The above mentioned mechanisms directly contribute to the formation of inelastic strain rates and thus to the current state of deformation and state of stress. Below, the main components of the *Hou/Lux* material model are briefly introduced:

(1) *Hou/Lux* material model for the total strain rate with submodels *Hou/Lux*-ODS and *Hou/Lux*-MDS ->

- total strain rate:

$$\dot{\varepsilon}_{ij} = \dot{\varepsilon}^{e}_{ij} + \dot{\varepsilon}^{i}_{ij} \tag{1}$$

- elastic strain rate:

$$\dot{\varepsilon}^{e}_{ij} = \frac{1}{2G}\cdot\frac{\dot{s}_{ij}}{1-D} + \left(\frac{1}{9K} - \frac{1}{6G}\right)\cdot\frac{I_1}{1-D}\cdot\delta_{ij} \tag{2}$$

- inelastic strain rate:

$$\dot{\varepsilon}^{i}_{ij} = \dot{\varepsilon}^{vp}_{ij} + \dot{\varepsilon}^{d}_{ij} + \dot{\varepsilon}^{h}_{ij} \tag{3}$$

(2) Submodel *Hou/Lux*-ODS (without direct damage) for the viscoplastic strain rate induced by diffusion, dislocation, strain hardening and recovery –>

$$\dot{\varepsilon}^{vp}_{ij} = \frac{3}{2}\left[\frac{1}{\overline{\eta}^{*}_{K}\cdot\exp\left(k_2\cdot\frac{\sigma_v}{1-D}\right)}\cdot\left(1-\frac{\varepsilon_{tr}}{\max\varepsilon_{tr}}\right) + \frac{1}{\overline{\eta}^{*}_{m}\cdot\exp\left(m\cdot\frac{\sigma_v}{1-D}\right)\cdot\exp(1\cdot T)}\right]\cdot\frac{s_{ij}}{1-D} \tag{4}$$

$$\max\varepsilon_{tr} = \frac{1}{G_K}\cdot\frac{\sigma_v}{1-D} \tag{5}$$

$$G_K = \begin{cases} \overline{G}^*_K \cdot \exp\left(k_1 \cdot \frac{\sigma_v}{1-D}\right) \cdot \exp(l_1 \cdot T) & \varepsilon_{tr} < \max\varepsilon_{tr} \\ \overline{G}^*_{KE} \cdot \exp\left(k_{IE} \cdot \frac{\sigma_v}{1-D}\right) \cdot \exp(l_{IE} \cdot T) & \varepsilon_{tr} < \max\varepsilon_{tr} \end{cases} \quad (6)$$

$$\overline{G}^*_{KE} \cdot \exp\left(k_{IE} \cdot \frac{\sigma_v}{1-D}\right) \cdot \exp(l_{IE} \cdot T) \leq \overline{G}^*_K \cdot \exp\left(k_1 \cdot \frac{\sigma_v}{1-D}\right) \cdot \exp(l_1 \cdot T) \quad (7)$$

(3) Damage material model *Hou/Lux*-MDS (with direct damage) for the strain rate induced by damage and healing ->

$$\dot{\varepsilon}^d_{ij} = \dot{\varepsilon}^{ds}_{ij} + \dot{\varepsilon}^{dz}_{ij}$$

$$= a3 \cdot \frac{\left\langle \frac{F^{ds}}{F^*} \right\rangle^{a1}}{(1-D)^{a2}} \cdot \frac{\partial Q^{ds}}{\partial \sigma_{ij}} + a3 \cdot \frac{\left\langle \frac{F^{dz}}{F^*} \right\rangle^{a1}}{(1-D)^{a2}} \cdot \frac{\partial Q^{dz}}{\partial \sigma_{ij}} \quad (8)$$

$$\dot{\varepsilon}^h_{ij} = \dot{\varepsilon}^h_{ij}(D, h, \sigma_v, \sigma_3, \theta, T) \quad (9)$$

(4) Flow functions and potential functions for the identification of structural damages (pore and crack formation) caused by shear stress and tensile stress ->

$$F^{ds} = \sqrt{3J_2} - \beta_{Dil}(\sigma_3, \theta) = \sqrt{3J_2} - \eta_{Dil}(\sigma_3) \cdot \beta(\sigma_3, \theta) \quad (10)$$

$$F^{dz} = a11 \cdot \left[\langle -\sigma_3 \rangle + \langle -\sigma_2 \rangle + \langle -\sigma_1 \rangle - a12\right]^{a13} \quad (11)$$

$$Q^{ds} = \sqrt{3J_2} - a0 \cdot \beta_{Dil}(\sigma_3, \theta) \quad (12)$$

$$Q^{dz} = a14 \cdot \left[\langle -\sigma_3 \rangle + \langle -\sigma_2 \rangle + \langle -\sigma_1 \rangle\right] \quad (13)$$

$$\eta_{Dil}(\sigma_3) = 1 - a4\exp(-a5 \cdot \sigma_3) \quad (14)$$

$$\beta(\sigma_3, \theta) = \beta^{TC}(\sigma_3, \theta) \cdot k(\sigma_3, \theta)$$
$$\beta^{TC}(\sigma_3, \theta) = a6 - a7\exp(-a8 \cdot \sigma_3) \quad (15)$$

$$k(\sigma_3, \theta) = \left[\frac{1}{\cos\left(\theta + \frac{\pi}{6}\right) + a9 \cdot \sin\left(\theta + \frac{\pi}{6}\right)}\right]^{\exp(-a10 \cdot \sigma_3)} \quad (16)$$

(5) Kinetic equation for the damage development (pore and crack formation) ->

$$\dot{D} = a15 \cdot \frac{\left[\left\langle \frac{F^{ds}}{F^*} \right\rangle + \left\langle \frac{F^{dz}}{F^*} \right\rangle\right]^{a16}}{(1-D)^{a17}} \quad (17)$$

where

$\dot{\varepsilon}^{vp}_{ij}$ viscoplastic strain rate tensor induced by the deformation mechanisms of diffusion, dislocation, strain hardening and recovery, in 1/d,

$\dot{\varepsilon}^d_{ij}, \dot{\varepsilon}^{ds}_{ij}, \dot{\varepsilon}^{dz}_{ij}$ viscoplastic strain rate tensor induced by damages resulting from shear stress and tensile stress, in 1/d,

$\dot{\varepsilon}^h_{ij}$ viscoplastic strain rate tensor induced by damage healing, in 1/d,

$\max\varepsilon_{tr}$ maximum strain of transient creep, in - ,

F^{ds}, F^{dz} flow function due to shear stress and tensile stress, in MPa,

Q^{ds}, Q^{dz} potential function due to shear stress and tensile stress, in MPa,

$\beta, \beta^{TC}, \beta_{Dil}$ strength function, TC-strength and dilatancy strength function, in MPa,

$k(\sigma_3, \theta)$ correction function for the consideration of the stress geometry θ, in -,

σ_3 minimum principle stress, in MPa,

η_{Dil} relation between dilatancy strength and short-term strength, in -,

D, h damage parameter, damage healing parameter, $0 \leq D \leq D_f < 1$, in -,

D_f damage parameter at failure, in -,

F^* = 1 MPa.

Further parameters belong to the *Lubby2* material model and can be found as well as values for the parameters in *Hou (2000)*.

The *Hou/Lux* material model has the following general mechanical characteristics:

- If the damage stresses F^{ds} and F^{dz} remain negative, there are and will be no damages in the course of time, because the damaging stress is smaller than the strength of the structure (= dilatancy strength).
- If the damage stresses F^{ds} and F^{dz} are positive, but still relatively small, damage will occur and further develop very slowly. The time until creep rupture in this case can reach extremely high values.
- If the damage stresses F^{ds} and F^{dz} are positive and also relatively high, the damage increases rapidly. As a consequence, tertiary creep and creep rupture will also occur relatively fast.
- The minimum stresses only influence the creep rate significantly, if the dilatancy strength is exceeded.
- Furthermore, the material model is capable of realistically describing and determining short-term and long-term strength, the beginning and the development of damaging, the beginning and the development of dilatancy, creep deformation induced directly by damages or indirectly by a reduction of the load-bearing cross-sectional area, real stress resulting from a reduction of the load-bearing cross-sectional area, transient, stationary and tertiary creep, creep rupture, the relation between time-to-rupture and minimum stress as well as the coefficient of utilisation. Details can be found in *Hou (1997)*.

Applications of the *Hou/Lux* material model can be found in literature references, e.g. *Hou (1997)* and *Lux & Hou (2000)*.

3 INTERACTION BETWEEN HYDRAULIC AND MECHANICAL SIMULATION

3.1 *A New Concept of Hydro-mechanical Coupling*

Drilling of underground excavations and induced stress redistribution disturb the mechanical integrity of rock salt in the near field and lead to a permeability increase. This phenomenon associated with damage may induce significant hydro-mechanical coupled effects, which may make the safety assessment difficult. In order to analyse the hydro-mechanical interaction induced by damage, a new concept of hydro-mechanical coupling with the main parameters of "pore pressure, seepage force and effective stress" is introduced, based on the *Hou/Lux* material model and the relation $k = k(\varepsilon_{Vol}, \sigma_3)$ according to *Stormont (1992)*:

$$k = A \cdot \left(\frac{\sigma_3}{F^*}\right)^{\lambda} \cdot (\varepsilon_{Vol})^{s} \qquad (18)$$

where k = permeability in m^2, A, s, λ = parameters in m^2, - and -, ε_{Vol} = dilatancy in -, $F^* = 1$ MPa, σ_3 = minimum principle stress in MPa.

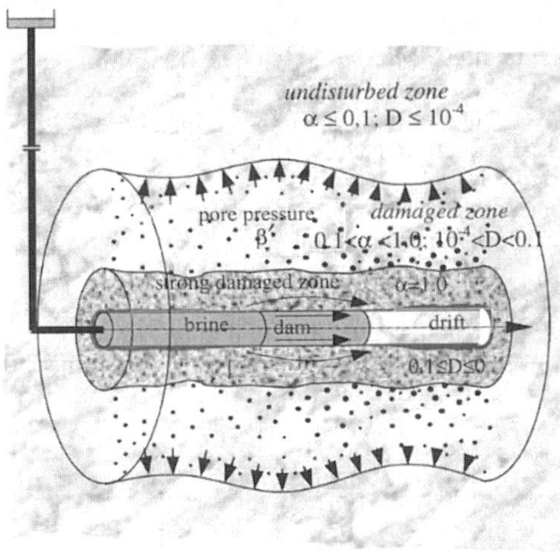

Figure 1. Simplified application of the new concept around a geotechnical barrier

According to this concept with a certain degree of abstraction there are 4 different zones to be taken into consideration around a geotechnical barrier (Figure 1):

- a fracture zone with spalling
- a strongly damaged zone with *Biot's* coefficient $\alpha = 1$, damage parameter $0.4 \geq D \geq 0.1$ and seepage force

  $$F_{xi} = \partial p/\partial x_i \qquad (19a)$$

- a damaged zone with *Biot's* coefficient $1 > \alpha > 0.1$, damage parameter $0.1 > D > 10^{-4}$ and seepage force

  $$F_{xi} = \alpha \partial p/\partial x_i + p \partial \alpha/\partial x_i \qquad (19b)$$

- as well as an undisturbed zone with *Biot's* coefficient $0 \leq \alpha \leq 0{,}1$, damage parameter $10^{-4} \geq D \geq 0$, pore pressure p = 0 and seepage force

  $$F_{xi} = 0 \qquad (19c)$$

where p = pore pressure in MPa, α = Biot's coefficient in -, F_{xi} = seepage force in MN/m^3, $x_i = x_1, x_2, x_3$ coordinate in m.

The permeability changes as a result of the redistribution of the minimum principal stresses and the volume changes, determined by the *Hou/Lux* material model. Via the change of the hydraulic gradient this change in permeability leads to a change in the seepage force as per equations 19a – 19c, which in turn causes a new state of stress.

With the implementation of the *Hou/Lux* material model into the *MISES3* FEM program, *Hou (2000)*, and together with the relation according to *Stormont (1992)*, the extended *MISES3* FEM program can now also numerically investigate the interaction between hydraulics and mechanics in a first rough approximation. The complete concept will be realised numerically in the near future.

The following paragraphs describe some individual parameters of this concept.

3.2 *Damage Parameter D*

For a representative rock element the damage parameter D describes the relation of the damaged area to the total area and takes values between D = 0 (fully intact material) and D = 1 (fully ruptured material). The value D = 1 can never be reached, because a fracture or creep rupture must occur before this value is reached. Some experience from numerical calculations together with lab tests for example declare a value of D = 0.4 as a limit value for the occurrence of creep rupture. A value of D = 0.1 characterizes the beginning of the tertiary creep phase while D = 0.0001 describes the beginning of the change in the permeability. Accordingly, the rock formation surrounding the geotechnical barrier in Figure 1 is divided into four zones, a fracture zone, a severely damaged zone, a damaged zone and an intact zone. Spalling occurs at a damage parameter larger than 0.4 in the first zone.

3.3 *Biot's Coefficient α*

The amount of pore pressure that has to be subtracted from the total compressive stresses σ for the calculation of the effective stresses σ' is determined by *Biot*'s coefficient α:

$$\sigma' = \sigma - \alpha p \qquad (20)$$

The *Biot*'s coefficient α can be determined by the matrix bulk modulus K_M and by the material bulk modulus K_B including pore space:

$$\alpha = 1 - K_B/K_M \qquad (21)$$

In case of no pore space $K_B = K_M$ it follows $\alpha = 0$. Within the classical hypotheses used in soil mechanics, the matrix is assumed to be incompressible and $\alpha = 1$. The effective elastic stress in this case is equal to that of *Terzaghi*:

$$\sigma' = \sigma - p \quad (22)$$

From the above description it can be derived that the higher the amount of pores and cracks (= damage), the smaller the bulk modulus of the material K_B and thus the *Biot*'s coefficient α runs from $\alpha = 0$ to $\alpha = 1$. Thus the *Biot*'s coefficient α could be a function of the damage parameter D, which in turn can be calculated with the *Hou/Lux* material model. This function $\alpha = f(D)$ has yet to be verified or substantiated in the laboratory by means of corresponding tests.

3.4 *Strength in "Wet" Condition β´*

In order to investigate the influences of the pore water pressure on the strength, strength tests with and without rubber jackets were carried out e.g. in *Thorel & Ghoreychi (1996)*. The main results are summarized below.

Whatever the confining pressure (σ_3 = 20, 40 and 60 MPa) and the type of the confining fluid (oil and saturated brine) are, the strength remains in the range of 27 to 42 MPa. The mean value of 33,5 MPa is very close to the uniaxial strength of MDPA rock salt (about 30 MPa).

In "wet" condition (triaxial tests without jacket), the confining pressure does not improve the strength as it is observed in dry condition: the failure remains more or less brittle.

For rock mass areas near failure, the concept of effective stress can be applied using a *Biot's* coefficient value α close to 1:

$$\sigma_3' = \sigma_3 - p, \quad \sigma_1' = \sigma_1 - p \quad (23)$$

It can thus be concluded that the fracture strength β' in "wet" condition can also be calculated using equations 15 & 16, if the effective minimum principle stress σ_3' is used instead of the total stress σ_3:

$$\beta'(\sigma_3',\theta) = \beta'^{TC}(\sigma_3',\theta)\cdot k(\sigma_3',\theta)$$
$$\beta'^{TC}(\sigma_3',\theta) = a6 - a7\exp(-a8\cdot\sigma_3') \quad (24)$$
$$\sigma_3' = \sigma_3 - \alpha\cdot p$$

$$k(\sigma_3',\theta) = \left[\frac{1}{\cos\left(\theta+\frac{\pi}{6}\right) + a9\cdot\sin\left(\theta+\frac{\pi}{6}\right)}\right]^{\exp(-a10\cdot\sigma_3')} \quad (25)$$

3.5 *Hydraulic pressure with supporting effect*

The hydraulic pressure that supports the rock mass is located at the border between dry rock with no or slight damage, where the primary permeability is still effective so that no brine migration with pore pressure builds up, and the saturated rock formation. The numerical value of this supporting pressure at a certain point on this borderline corresponds to the pore pressure at that point.

3.6 *Seepage force F*

In literature, the seepage force F_{xi} is mostly calculated in the direction x_i as per equation (19a). However, this equation assumes that the *Biot*'s coefficient α in the rock formation is constant and that $\alpha = 1$, which applies for soil mechanics. In the severely damaged zone equation (19a) can be used, because the *Biot*'s coefficient α will be close to one ($\alpha = 1$) as a result of severe damage or high porosity in this zone.

For the general case of $0 < \alpha \neq \text{const} < 1$ within the damaged zone, the seepage force is to be determined as per equation (19b). It is obvious that the equations (19a and 19c) can be derived from equation (19b), if the *Biot*'s coefficient $\alpha \equiv 1$ and $\alpha \equiv 0$ is correspondingly inserted in to equation (19b).

In case of a constant *Biot*'s coefficient α, which is, however, smaller than one ($\alpha < 1$), the seepage force F_{xi} as per equation (19b) is then:

$$F_{xi} = \alpha\partial p/\partial x_I \quad (26)$$

That means that only a part of the seepage force determined as per equation (19a) is effective within the rock mass in case of $\alpha = \text{const} < 1$.

Correspondingly, the seepage force is calculated individually in the three different zones as per equations (19a – 19c) and inserted for the determination of the new states of stress.

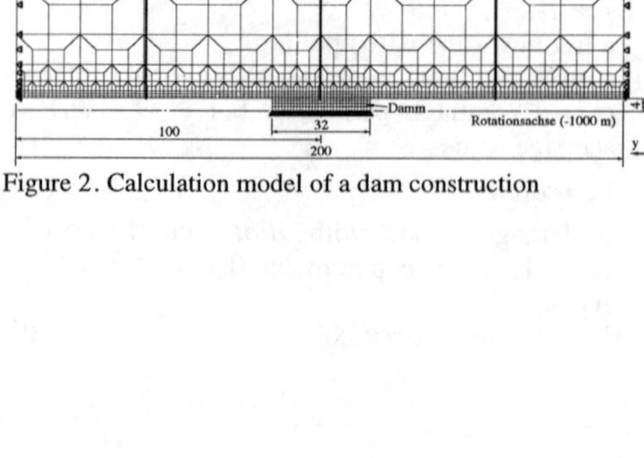

Figure 2. Calculation model of a dam construction

4 DAM CONSTRUCTION INVESTIGATIONS TAKING INTO ACCOUNT THE INTERACTIONS BETWEEN MECHANICAL AND HYDRAULIC PROCESSES

With the subsequent implementation of the *Hou/Lux* material model into the *MISES3* FEM program, *Hou (2000),* and together with the relation $k_f = k_f$ (ε_{Vol}, σ_3, σ_v), with the help of the extended *MISES3* FEM program the interactions between hydraulics and mechanics can now also numerically be investigated in a first rough approximation. The new concept of hydro-mechanical coupling with the main parameters of "pore pressure, seepage force and effective stress" will be numerical realized in the next time.

As a first practice-oriented example a dam construction located 1000 m below the surface with a length of l = 32 m and a diameter of d = 8 m is investigated using the *MISES3* FEM program (primary stress: $\sigma_x = \sigma_y = \sigma_z = 22$ MPa). Figure 2 shows a rotational model with a length of 200 m and an outer diameter of 130 m created for this purpose. 50 years after the excavation, the dam is built in. 10 years after, the brine entry is planned with a pressure of P_{Fl} = 10 MPa on the left-hand side of the drift. The initial permeability k of the saliniferous rock formation and the dam construction material is $k = 10^{-15}$ m/s and $k = 10^{-8}$ m/s. The *Biot's* coefficient α of the rock salt and the dam construction material must be assumed constant $\alpha \equiv 1$ because of the limitation of the *MISES3* FEM program .

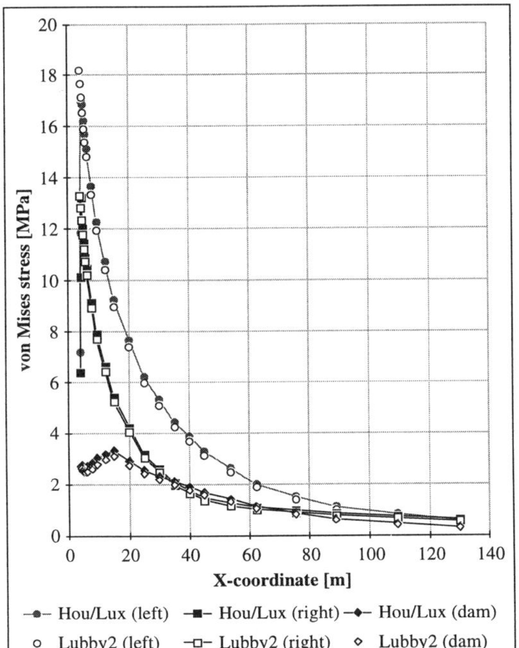

Figure 3. Comparison of the von Mises stress shown in sections right, left and dam according to the *Lubby2* and *Hou/Lux* material models taking into account mechanical and hydraulic interactions (t = 110 years)

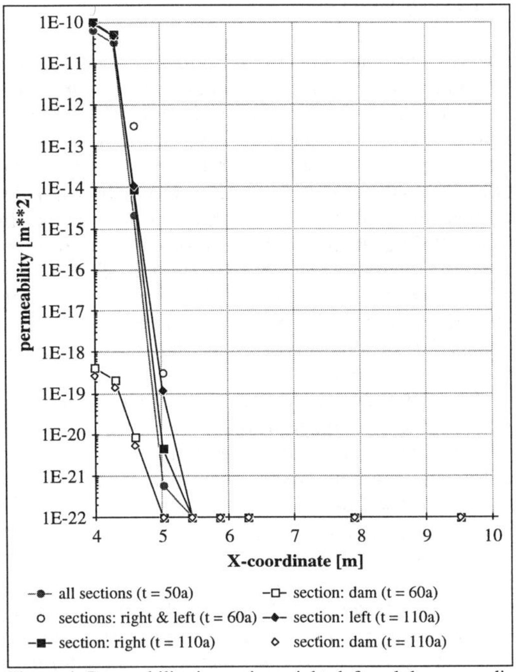

Figure 4. Permeability in sections right, left and dam according to the *Hou/Lux* material model taking into account mechanical and hydraulic interactions (t = 50, 60 and 110 years)

Figure 3 compares the von Mises stresses in the sections on the air (right) and brine (left) face of the drift as well as over the dam in time t = 110a.

Clear differences in the rock formation stresses are visible, which are detected with the *Lubby2* and *Hou/Lux* material models and under consideration of hydraulic and mechanical processes. A destrengthened damage zone that reaches approx. 1 m into the rock formation in the sections right and left can only be calculated using the *Hou/Lux* material model. It is furthermore evident that this destrengthening process is taking into account the interaction between hydraulic and mechanical processes increases. The von Mises stresses in the dam section have almost the same small values of 0.5 – 3.5 MPa with the *Lubby2* and *Hou/Lux* material models because of the supporting effect of the dam construction and the rock formation convergence. This tendency can be also seen in Figure 4. The permeability in the dam section reduces from t = 50 years to t = 60 and 110 years. In the sections right (air) and left (brine), however, the permeability increases due to the increasing damage of the rock formation (→ dilatancy).

5 CONCLUSIONS

The *Hou/Lux* material model considers most of the factors that are regarded as necessary to be included in a material model for ductile salt rocks, thus eliminating, at least in principle, some of the deficits still remaining in the field of material model formulations for rock salts. In particular, the also newly developed non-linear failure criterion introduced in *Hou (1997)* as well as the creep rupture criterion introduced in *Hou (1997)* and *Hou & Lux (2000)* have been integrated into the *Hou/Lux* material model. The *Hou/Lux* material model thus offers new possibilities for the determination of the mechanical behaviour of salt rocks:

The spatial and temporal development of the softening zones can be determined.

Intensities of the damage within a softening zone and resulting spalling over time can be calculated.

Further consequences of the damage to the mechanical structure such as e.g. increased porosity and permeability, can generally be represented in dependence of the dilatancy.

A new concept of hydro-mechanical coupling with the main parameters of "pore pressure, seepage force and effective stress" based on the *Hou/Lux* material model and using additively the relation $k = k(\varepsilon_{Vol}, \sigma_3)$ according to *Stormont (1992)* is developed to numerically simulate the hydro-mechanical interaction in case of a flooded repository and to evaluate the efficiency of sealing constructions. With the subsequent implementation of the *Hou/Lux* material model into the *MISES3* FEM program, *Hou (2000),* and together with the relation $k_f = k_f\ (\varepsilon_{Vol}, \sigma_3, \sigma_v)$, with the help of the extended *MISES3* FEM program the interactions between mechanical and hydraulic processes can now also numerically be investigated in a first rough approximation.

A practice-oriented calculation example shows the necessity of considering rock damages, material softening and dilatancy as well as the interaction between mechanical and hydraulic processes.

6 REFERENCES

Hou, Z. 1997. Untersuchungen zum Nachweis der Standsicherheit für Untertagedeponien im Salzgebirge. *Dissertation an der TU Clausthal.*

Hou, Z. 2000. Numerical Implementation of the Material Model *Hou/Lux* for Rock Salt Based on Continuum Damage Mechanics and it's Numericals Simulations. *Proceedings of EUROCK 2000 Symposium*, Aachen, March 2000: 745 – 750. Essen: Glueckauf – Verlag.

Hou, Z. & Lux, K.-H. 1998. Ein neues Stoffmodell für duktile Salzgesteine mit Einbeziehung von Gefügeschädigung und tertiärem Kriechen auf der Grundlage der Continuum-Damage-Mechanik. *Geotechnik 21 (1998) Nr. 3*: 259 - 263.

Hou, Z. & Lux, K.-H. 1999. A material model for rock salt including structural damages as well as practice-oriented applications. *Proceedings of 5th Conference on Mechanical Behavior of Salt (Preprints)*, Bucharest, August 1999: 55-59.

Hou, Z. & Lux, K.-H. 2000. Ein Schädigungsmodell mit Kriechbruchkriterium für duktile Salzgesteine auf der Grundlage der Continuum-Damage-Mechanik. *Bauingenieur Band 75 (2000)*: 300-307.

Kachanov, L. M. 1986. *Introduction to continuum damage mechanics.* Dordrecht: Martinus Nijhoff Publishers.

Lemaitre, J. 1992. *A course on damage mechanics.* Berlin: Springer - Verlag.

Lux, K.-H. 1984. *Gebirgsmechanischer Entwurf und Felderfahrungen im Salzkavernenbau.* Stuttgart: Ferdinand Enke Verlag.

Lux, K.-H., Hou, Z., Düsterloh, U. & Xie, Z. 2000. Approaches for validation and application of a new material model for rock salt including structural damages. *Proceedings of 8th World Salt Symposium*, Hague, May 2000: 271 – 277. Amsterdam: Elsevier.

Stormont, J. C., Daemen, J. J. K. & Desai, C. S. 1992. Prediction of dilation and permeability changes in rock salt. *International Journal for Numerical and Analytical Methods in Geomechanics. Vol. 16*: 545 - 569.

Thorel, L. & Ghoreychi, M. 1996. Rock salt damage – Experiment results and interpretation. *Proceedings of 3th Conference on Mechanical Behavior of Salt, Paris, 1993*: 75 – 90. Clausthal-Zellerfeld: Trans. Tech. Publications, 1996

Computer Methods and Advances in Geomechanics, Desai et al. (eds) © 2001 Balkema, Rotterdam, ISBN 90 5809 183 X

Stability analysis of braced excavation with base improvement

M. Katagiri, T. Ishii, K. Ohishi & K. Saitoh
Nikken Sekkei Nakase Geotechnical Institute, Kawasaki, Japan

K. Azuma
Electric Power Development, Tokyo, Japan

ABSTRACT: The Finite Difference Method (FDM) combined with Shear Strength Reduction Method (SSRM) was applied to the braced excavation, in order to evaluate its global safety factor and the effectiveness of base improvement. The effectiveness of base improvement has already been confirmed by the centrifuge model test. This paper presents that the SSR-FDM can reasonably calculate the global safety factor to the centrifuge model tests, and that the deformation behavior of the ground at the failure may change due to the base improvement.

1 INTRODUCTION

It is serious to evaluate the stability of the deep braced excavations in soft clay grounds. The advantage of the base improvement in such an excavation work is the increase of safety and reduction of cost for the retaining structure. Now, some simple methods have been proposed. The methods, however, may not be applied to the ground including a cement-improved part because the failure mechanism such a ground may be complicated. For example, Tanaka (1993) reported that a large heave was observed at the cross section with base improvement that was assessed to be stable by the circular failure method

In this paper, the Finite Difference Method (FDM) was applied to the centrifuge model tests on the braced excavation with the base improvement. Using the FDM combined with the Shear Strength Method (SSRM), the global safety factor against the basal heave was calculated and the effectiveness of the ground improvement was evaluated.

2 OUTLINE OF CENTRIFUGE MODEL TESTS ON BRACED EXCAVATION

Ohishi et al. (1999) performed the centrifuge excavation tests on the simplified models composed of clay and heavy liquid, and confirmed the effectiveness of ground improvement in the excavation bottom for the stability of the deep braced excavation. The modeling points were as follows;

i) Three types of base conditions, untreated model (case-1) and improvement grounds with q_u = 110 (case-2) and 368 kPa (case-3) were set in the model grounds.

ii) A two dimensional braced excavation with 16 m deep and 25 m wide was modeled with a scale of 1 to 100. The model was gained 100 g of acceleration.

iii) Number of strut in the model was changed to two, simplified the prototype model with five struts. The position of the lowest strut, however, did not change.

iv) A model ground was combined with the soil layer below the excavation and heavy liquid parts above the base of the excavation. The heavy liquid was set at 1.6 g/cm^3 as same as total unit mass of the clay ground.

3 NUMERICAL MODEL AND CALCULATION PROCEDURE

Numerical mesh modeled a half of the centrifuge model and material properties of each component are shown in Figure 1. As the excavation region and the parts behind the retaining wall were liquid, the stresses acting on the clay ground, improved part and retaining wall were determined by the level of heavy liquid in the excavation part.

The stress-strain relations of soils including improved part were assumed elastic-perfectly plastic models. The failure criteria of the soils were defined

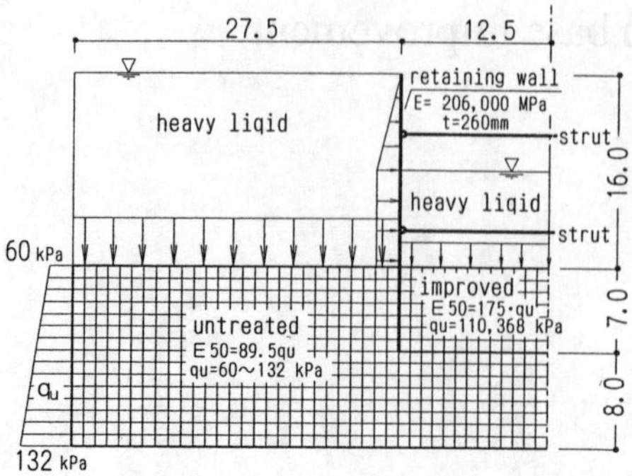

Figure 1 Numerical model (unit: m)

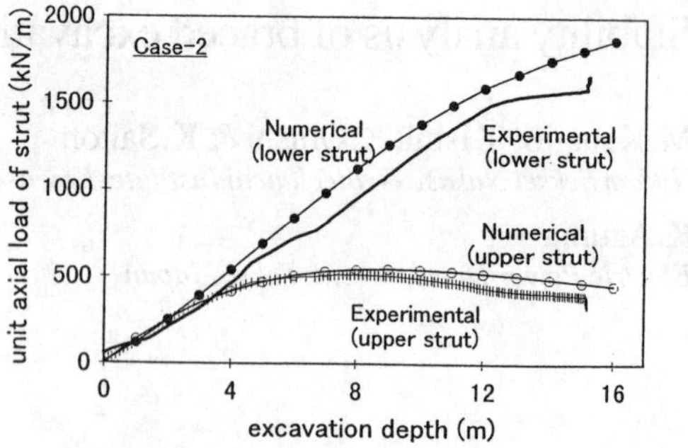

Figure 2 Axial load response of struts during excavation

by the Mohr-Coulomb's equations. The coefficients of deformation for the clay ground and improved part were respectively set at 89.5•q_u and 175•q_u from the results of unconfined compression tests. The stress-strain relations for the retaining wall and strut were assumed linear elastic models. Those coefficients for their components were used the value for steel, 206,000 MPa.

In the numerical analysis, the FDM proposed by Cundall and Board (1988) was combined with the SSRM by Ugai (1990). In the FDM, stress strain relations of soil and structures were calculated by the explicit finite difference and dynamic relaxation methods, respectively.

In the SSRM, the safety factor was defined as the ratio of the shear strength parameters of soils at the failure to those at the actual condition. The shear strength parameters were reduced by a shear strength reduction factor *Fr* until the failure occurred. The initial value of *Fr* was assumed to be unity because the failure did not occur. Then the value of *Fr* was increased step by step. At a given value of *Fr*, the reduced shear strength became too small to sustain the retaining wall, and the displacement in the ground was not convergent. This condition was defined as the failure of the ground. The shear strength reduction factor at immediately preceding failure was considered as the minimum global safety factor.

In this analysis, an increment of *Fr* was fixed at 0.01, and an iteration number was 2,000 in a normal condition. If the calculated total internal stress is not consist with the acting external force, the condition is judged not to be convergent. In such a case, the failure of the ground was judged by the followings;

i) The distributions of yield condition and deformation rate vector in the ground were drawn.

ii) The failure was determined by the continuity of the yield region and by the same magnitude of the displacement rate along with a slip line.

iii) When the failure was not occurred, *Fr* was increased.

4. EVALUATION OF NUMERICAL RESULTS

4.1 Axial load response of strut during excavation

Figure 2 shows the reaction of axial loads of two struts during excavation in the case-2. The numerical results agree with the centrifuge test results. In the case-1, the numerical result on the upper strut is slightly larger than the test one, but the numerical response of the lower strut is consistent with the test result. In the case-3, the numerical relations in both the struts show about half magnitudes obtained from the test, but the trends of response of axial load in both the numerical and experimental results are similar. It is therefore worth discussing the deformation behavior in the ground obtained by the FDM.

4.2 Deformation behavior and failure condition

Figure 3 shows the yield region and deformation rate vectors before and at the failure in the case-2 under the condition of complete excavation obtained by the SSR-FDM. In the case where *Fr* was 1.27, the yield region developed around the toe of the retaining wall, but was not continuously reached to the surface. In *Fr* = 1.28, the yield region was formed from the excavation part to the backfill side, the deformation rate vectors along with the yield region had significant values and the rate were almost the same. These mean the sliding in the yield region occurs. The safety factor, therefore, is judged as 1.27. The sliding zone formed by connecting the yield region had 3 m in width, and passed about 3 to 6 m below the toe of retaining wall.

Figure 4 shows the displacement vectors at the end of excavation in the case-2 by the centrifuge. The maximum displacement at the basal heave

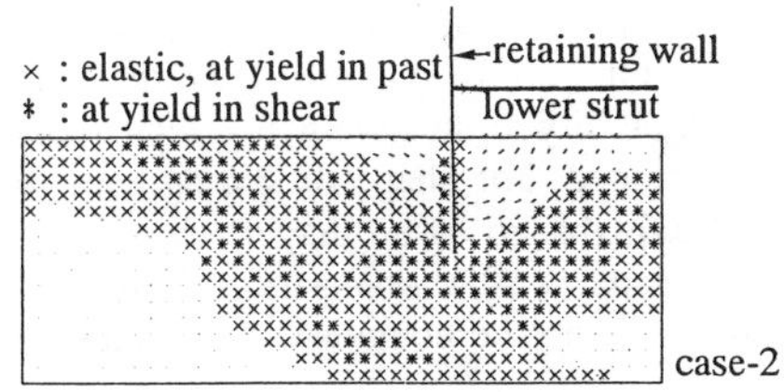

(a) Fr = 1.27

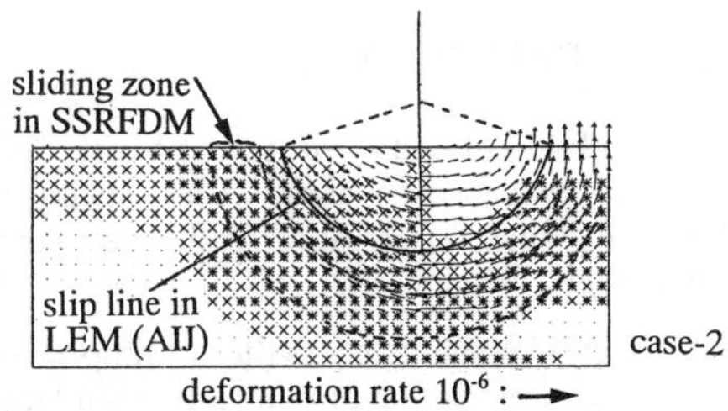

(b) Fr = 1.28

Figure 3 Yield region and deformation rate vectors

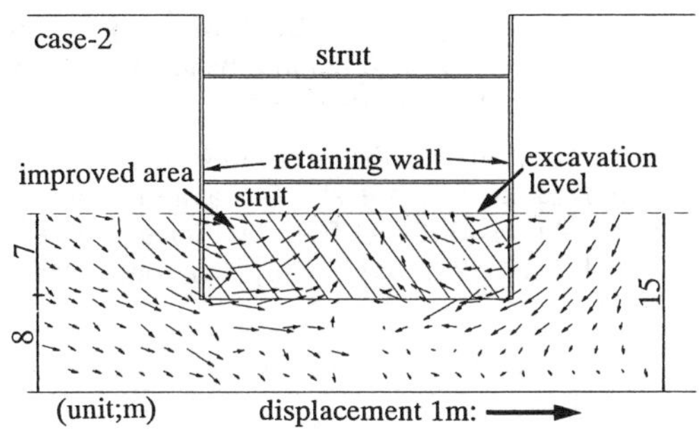

Figure 4 Deformation vectors of case-2 at final excavation

reached 20 cm in prototype scale and relatively large deformations in the ground occurred. This condition, however, may not reach the failure because the deformation of the ground is smaller than that in the case-1 as mentioned next.

Figure 5 shows the yield region and displacement rate vectors at the failure by the SSRFDM and displacement vectors by centrifuge test in the case-1 at 16 m of excavation. The *F*r by the SSR-FDM was calculated as 1.06. The safety factor in this case becomes therefore 1.05. The maximum deformation of the base of excavation at this excavation level reached 50 cm or more in prototype scale, and was larger than that in the case-2. This case was considered as the failure. The sliding zone by the SSR-FDM in Figure 5(a) is regarded a half ring with 2m in thickness and passes below the toe of the retaining wall. The sliding zone imaged from the centrifuge tests in Figure 5(b) is also formed a circular band passed around the toe of retaining wall. Both the sliding zones are almost the same.

Figure 6 shows the same relations as Figure 5 in the case-3. About 7 cm in prototype scale of deformation in the base of the excavation was measured at the end of excavation, and was judged as the stable condition. In the SSR-FDM at *F*r = 1.42, the yield region in the ground was generated and continue. The safety factor is determined as 1.41.

4.3 Comparison of Safety factor

The safety factors by the SSR-FDM are compared with those by the conventional method (LEM (AIJ))

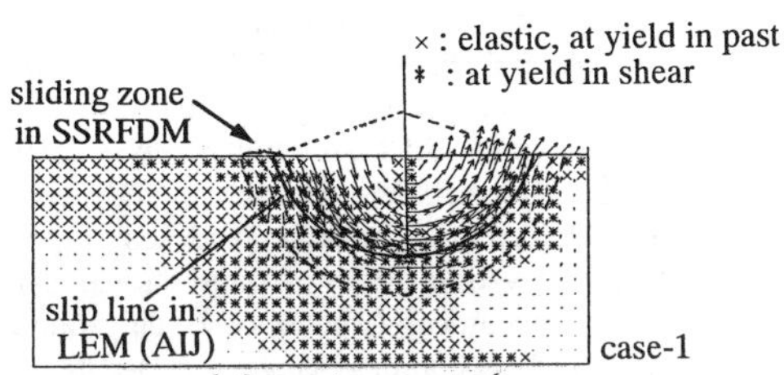

(a) Yield region and deformation rate vector with Fr = 1.06

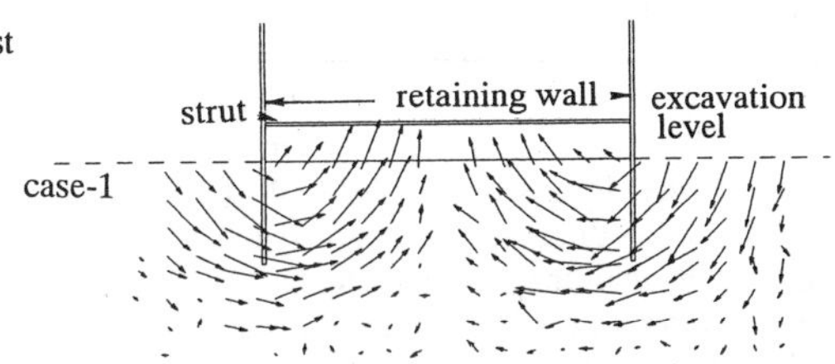

(b) Deformation vectors from centrifuge test

Figure 5 Comparison of deformation behavior in case-1

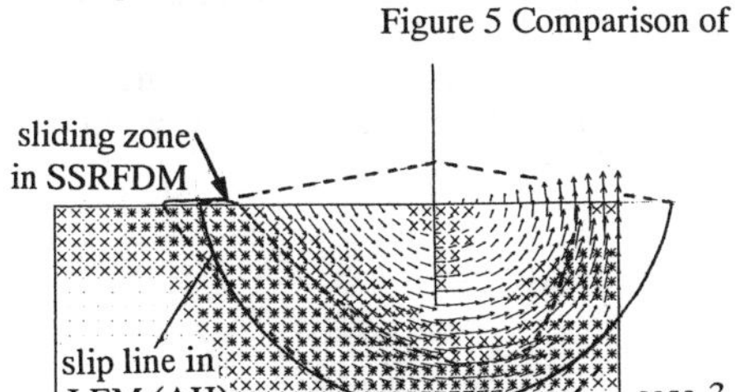

(a) Yield region and deformation rate vector with Fr = 1.42

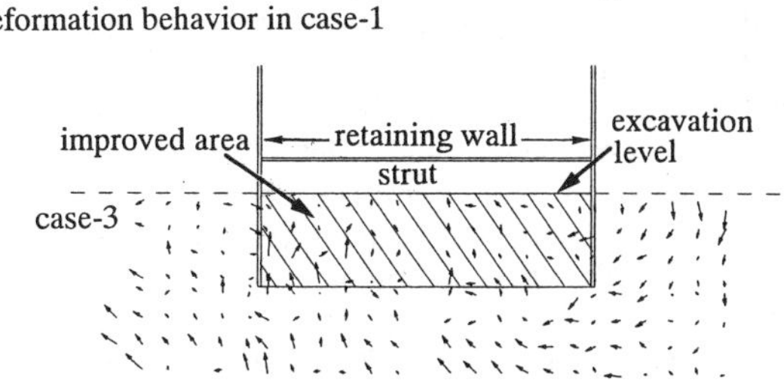

(b) Deformation vectors from centrifuge test

Figure 6 Comparison of deformation behavior in case-3

Table 1 Comparison of safety factor

Case	Excavation Depth (m)	Safety Factor	
		SSR-FDM	LEM(AIJ)*
1 : untreated	12	1.37	1.20
1 : untreated	14	1.14	1.00
1 : untreated	16	1.05	0.90
2 : q_u=110 kPa	16	1.27	1.05
3 : q_u=368 kPa	16	1.41	1.74

*: After Ohishi et al. (1999)

based on the limit equilibrium method (Architectural Institute of Japan, 1988). In this method, the circle with its center at the lowest strut is assumed to pass below the retaining wall and the rigidity of retaining wall was ignored. The safety factor is driven from the equilibrium of resisting and sliding moments. This method recommends more than 1.2 of safety factor for the stability analysis of heaving.

Table 1 shows the safety factors of some situations in the test cases by the SSR-FDM and LEM (AIJ). The results by the LEM (AIJ) were referred to Ohishi et al. (1999). In the case-1, the values for several excavation levels are indicated. At the excavation level of 14 m, about 20 cm of heave in the excavation part was observed. This level was judged not to fail although the large deformation occurred. At 16 m, more than 50 cm was observed, and it was judged that the failure occurred. In the case-2, about 20 cm of heave was also observed at the end of excavation, and was judged not to reach the failure. These behaviors were represented by the safety factors obtained from the SSR-FDM.

In the case-1 and 2, the safety factors by the SSR-FDM are larger than those by the LEM (AIJ), respectively. The reason is thought that the rigidity of the retaining wall is ignored in the LEM (AIJ), and that the size of sliding block assumed in the LEM (AIJ) is smaller than that imaged by the yield region and deformation rate vectors. In the case-1 mentioned before, the sliding zone assumed by the deformation vectors from the centrifuge test was close to that by the SSR-FDM.

In the case-3, however, the safety factor by the SSR-FDM is smaller than that by the LEM (AIJ). The difference of safety factor is thought to be largely dependent on the size and shape of the sliding block. The position of sliding block imaged by the SSR-FEM is located inside the slip line assumed by the LEM (AIJ). The shape of sliding zone imaged by the yield region and deformation rate vectors in Figure 6(a) is not considered a circle arc. In particular, an unclear part in the center of the model is recognized. This region may be influenced by the behavior of the opposite side. As the failure condition in the case-3 could not be obtained from the centrifuge test, the suitability of the assumed sliding zone could not be discussed. The shape of sliding zone imaged by the SSR-FDM, however, may be more reasonable than the shape of slip line assumed by the LEM (AIJ), considering the boundary condition of centrifuge test.

5. CONCLUDING REMARKS

Using the FDM combined with the SSRM, the stability of the braced excavation with base improvement was analyzed. From the analysis, the following can be drawn;

(1) The FDM constructed with a simple stress strain-relation can approximately simulate the axial loading response and the deformation behavior of ground.

(2) The safety factor by the SSR-FDM can be corresponded to the deformation behavior of the ground from the centrifuge test.

(3) In the case where the excavation bottom was improved strongly, the slip line assumed by the LEM (AIJ) is different from that imaged by the SSR-FDM. The safety factor, therefore, differs largely.

REFERENCES

Architectural Institute of Japan (1988), Recommendation for design and construction practice of earth retaining for excavation, (in Japanese)

Cundall, P.A. and Board, M. (1988), A microcomputer program for modeling large-strain plasticity problems, Proc. of 6th IACMAG, Innsbruck ,3:2101-2108.

Ohishi, K., Katagiri, M., Saitoh K. and Azuma K. (1999), Deformation behavior and heaving analysis of deep excavation, Proc. of International Symposium on Geotechnical Aspects of Underground Construction in Soft Ground, Tokyo, 1:693-698.

Tanaka, H. (1993), Behavior of braced excavations stabilized by deep mixing method, Soils and Foundations, 33-2:105-115.

Ugai, K. (1990), Availability of shear strength reduction method in stability analysis, Tsuchi-to-kiso, 38-1:67-72, JSSMFE (in Japanese).

Computer Methods and Advances in Geomechanics, Desai et al. (eds) © 2001 Balkema, Rotterdam, ISBN 90 5809 183 X

Some truths and half-truths on use of geotechnical computational models

Fred H. Kulhawy
Cornell University, Ithaca, N.Y., USA

ABSTRACT: In geotechnical engineering, there have been countless "applied" studies since the 1960s that have stressed the use of various computational models (material, analytical, and/or numerical) to evaluate specific field problems. Commonly in these studies, selective comparisons are presented with field observations to illustrate the "validity" of the model simulation. Not surprisingly, these comparisons are almost always very good. A poor comparison is a true rarity. But are all of these models really equally good? Clearly the answer is no, in the absolute sense. Calibration has been done to ensure that the forcing function (load, gradient, etc.), imposed on the selected and coupled calculation and material property models, yields a system response that is acceptable, as outlined in Figure 1.

Figure 1. Components of Geotechnical Prediction (Kulhawy 1992).

In this paper, some of the realities of modeling are examined candidly to focus attention on more consistent and informative representation of geotechnical models. The examples used are drawn from foundation engineering, but the underlying themes are applicable to all of geotechnical engineering. The fundamental underlying premise is that modeling of geotechnical engineering problems is no trivial task. Many combinations of models and properties can give "acceptable" results, but only if limited aspects of system response are considered. At present, all models have some inherent limitations. However, when used appropriately, many available models can be powerful analysis and design tools. The very simple models have many limitations and can be subject to much uncertainty, while more sophisticated models may have less restrictions. However, regardless of the modeling sophistication, much attention must he paid to addressing the boundary conditions and the actual soil response that will occur. Finally, pertinent to all geotechnical modeling is the need to define very precisely the assumptions of the model, the material properties, the boundary conditions, the solution methodology, and the details of the calibrations done or employed to establish the proper range of validity of the model. Additional details are given elsewhere (Kulhawy 1994, 2000).

REFERENCES

Kulhawy, FH 1992. On the Evaluation of Static Soil Properties. In R.B. Seed & R.W. Boulanger (eds), *Stability & Performance of Slopes & Embankments-II (GSP 31)*: 95-115. New York: ASCE.

Kulhawy, FH 1994. Some Observations on Modeling in Foundation Engineering. In HJ Siriwardane & MM Zaman (eds), *Proc. 8th Intl. Conf. Computer Methods & Advances in Geomech.(1)*: 209-214. Balkema.

Kulhawy, FH 2000. Some Modeling Issues in Geotechnical Engineering. In *Proc. John Booker Memorial Symp.*, Sydney (Balkema - in press).

Computer Methods and Advances in Geomechanics, Desai et al. (eds) © 2001 Balkema, Rotterdam, ISBN 90 5809 183 X

Finite element analysis of a deep excavation employing jet grouting techniques: Results from a case study

C. Kummerer & H. F. Schweiger
Institute for Soil Mechanics and Foundation Engineering, Graz University of Technology, Austria

E. Falk
Keller Fondazioni S.r.l., Verona, Italy

ABSTRACT: This paper presents a novel approach of constructing a 30 m deep excavation utilizing an advanced jet grouting technique. The design of the wall is based on both limit equilibrium methods and comprehensive finite element analyses. By means of these investigations the client could be convinced of the feasibility of the proposed construction concept. Besides the investigations for working load conditions the finite element method was applied for assessing the stability of the construction by the reduction of strength parameters. These results are compared with results from limit equilibrium analyses. A comparison of the finite element analyses with geodetic survey and anchor force measurement is provided.

1 INTRODUCTION

Economical restraints and problems with well established techniques often lead to innovations in construction techniques. Besides the economic aspect the safety of the construction is of main importance. With increasing depths of building pits new approaches concerning both the construction technique and the calculation method have to be developed. The most common techniques for supporting deep excavations as the one discussed here are diaphragm walls, large diameter piles and sheet pile walls supported by struts or anchors. Due to the specific conditions encountered in this project these methods proved to be not applicable.

Therefore a jet grouting technique has been suggested as an alternative method for the construction of the 30 m deep excavation discussed in this paper. For the design of the retaining wall the finite element method was utilized. Compared to classical limit equilibrium methods numerical methods have the advantage that a prediction of deformation in the ground, wall deformations, settlements behind the wall and the stresses in the supporting wall can be obtained. A basic requirement for a reliable prediction is that appropriate constitutive laws are employed. The finite element code PLAXIS utilized for the analyses provides a material law which allows a reasonable modelling of the behaviour of soils without excessive computational effort and a straightforward parameter identification. Thus the model is well suited for solving practical problems in geotechnical engineering.

2 PROBLEM DESCRIPTION AND SUBSOIL CONDITIONS

Based on a design of Swiss architect Mario Botta a new casino is built at the eastern shore of lake Lugano in Switzerland. A deep excavation of 30 m depth in a slope adjacent to existing buildings is required for the construction of the casino.

2.1 *Subsoil and groundwater conditions*

The subsoil consists of gravely, silty sand down to 1.3 m below the ground surface. Underneath a dominant layer of silty, sandy gravel containing boulders with more than 1 m in diameter is situated. Locally small areas of cohesive material were found. At a depth of about 42 m below the top of the planned Soilcrete wall rock is located under the debris mass. The surface of the bed rock dips with an angle of about 30° to the lake.

The ground water level was found at an elevation of 269 m, i.e. about 1 m below the final excavation depth and therefore had not to be taken into account in the design of the support system.

2.2 *Construction technique*

Various techniques were considered for the wall construction. Due to expected problems with conventional construction techniques such as diaphragm walls and bored pile walls caused e.g. by the inclined construction area, drilling deviation and presence of boulders a jet grouting technique (double system by Keller Fondazioni S.r.l.) was utilized. This technique can be easily adjusted on site to account for last minute changes e.g. in the geometry. This was actually the case in this project when the client required a change of the depth of excavation from 26.5 to 29.3 m, the jet grouting being in progress already. This could be achieved by increasing the thickness of the wall and providing 11 instead of 9 rows of permanent prestressed anchors.

In the central area the wall (see cross section B-B in Figure 1) consists of two parts, the upper part 1.5 m in diameter and the lower part with a thickness increasing from 2.0 m at its top to 2.5 m at the base of the wall. The lower part of the wall is put in place from the fifth excavation level and is composed of three single jet grouting columns. In front of the wall a Soilcrete body of 14 m width and an average depth of about 10 m is put in place in order to stabilize the wall and provide a footing for the future building.

For the topmost three anchor rows prestressed anchors with an ultimate load of 1099 kN were employed. A prestress force of 580 kN was applied. From the fourth anchor row down the wall is supported by grouted anchors with an ultimate load of 1318 kN which were prestressed to 725 kN. The horizontal spacing between the anchors is 2.8 m. To ensure a sufficient load distribution horizontal concrete walings of 80 cm height and 40 cm width are cast in place.

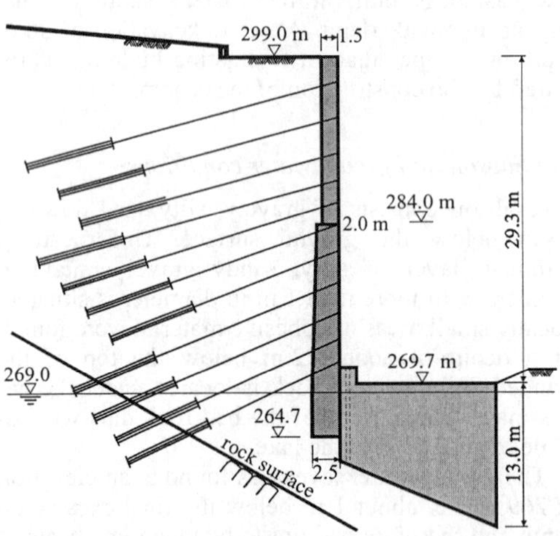

Figure 1. Cross section B-B.

Adjacent to this central part 25 m deep sections A-A and C-C are excavated (see Fig. 2). The Soilcrete wall is composed of an 15 m long upper part with 1.5 m diameter and a lower part of 16 m length and 1.8 m diameter and anchored with nine rows of permanent anchors.

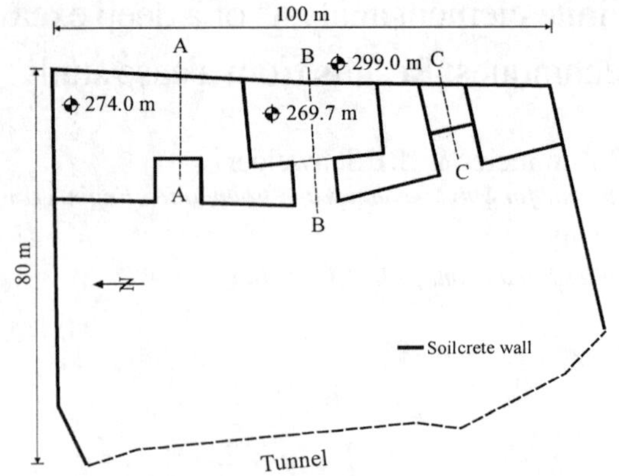

Figure 2. Ground plan of excavation pit.

Figure 3 depicts the construction works for the tenth excavation stage of northern area of the building pit. It can be seen that the surface between the horizontal walings is sealed with shotcrete including a wired mesh in order to support local inhomogeneities.

Figure 3. Construction work for tenth excavation level.

3 CALCULATIONS

3.1 Design concept

For the design of the wall traditional limit equilibrium methods as well as finite elements analyses were employed.

Investigations with limit equilibrium methods were performed in order to provide a factor of safety engineers are familiar with. Critical inner and outer slip circles and combined slip lines were investigated and these calculations yielded the required number of anchor rows, anchor forces and length of anchors. For this type of calculations the program STAND (Otta 1997) was applied.

Comprehensive finite element analyses have been undertaken utilizing the program PLAXIS (Brinkgreve & Vermeer 1998) to check the anchor design resulting from the limit equilibrium calculations. Two aspects were of main importance: The finite element calculation had to prove that the expected deformations can be minimized to acceptable values. Furthermore, the layout of the prestressed anchors, especially the prestress forces and the vertical spacing between the anchors, was optimized with respect to stresses in the Soilcrete wall. Zones of critical stress concentrations, in particular tensile stresses, had to be limited because the Soilcrete wall has no reinforcement.

For the mathematical description of the behaviour of the soil the so-called "Hardening Soil Model" was employed. The basic features of this elasto-plastic model are:

- stress dependent stiffness according to a power law
- hyperbolic stress-strain relationship based on the model presented by Duncan & Chang (1970), modified however to be suitable for implementation within the framework of plasticity
- distinction between primary deviatoric loading and unloading/reloading.
- failure according to Mohr-Coulomb criterion

In addition to shear hardening this model also accounts for compression hardening and can therefore be applied to both soft soils and stiff soils. For a more detailed description of this constitutive law see Brinkgreve & Vermeer (1998).

The stiffness and strength parameters for the gravel layer are given in Tables 1 and 2. E_{50}^{ref} and E_{oed}^{ref} represent the stiffness for primary loading and E_{ur}^{ref} for unloading at a reference stress p^{ref} respectively. The power m defines the dependence of the stiffness with stress level. The input parameters for the "Hardening Soil Model" were chosen according to previously performed finite element calculations under similar conditions and values given in the literature (e.g. von Soos 1995) because no specific data from site investigation has been available.

Table 1. Stiffness parameters for gravel.

Soil layer	E_{50}^{ref}	E_{oed}^{ref}	E_{ur}^{ref}	ν_{ur}	p^{ref}	m
	kN/m²	kN/m²	kN/m²	-	kN/m²	-
Gravel	50000	50000	150000	0.2	100	0.5

The rock layer was modelled using the elastic perfectly plastic Mohr-Coulomb model.

Table 2. Strength parameters for gravel.

Soil layer	c^{ref}	φ	ψ	R_f
	kN/m²	°	°	-
Gravel	0.1	37	7	0.9

The Soilcrete body was assumed to behave elastic with a Young's modulus E = 9.5E6 kN/m² and a Poisson ratio $\nu = 0.2$. These input parameters have been determined from experimental investigations.

Ground anchors are modelled using bar elements for the free anchor length and special membrane elements, termed geotextile elements in PLAXIS, for the grouted part. This provides a reasonably realistic modelling of the behaviour of the prestressed anchors.

The finite element mesh consists of about 4000 6-noded triangular elements with quadric shape function (Fig. 4).

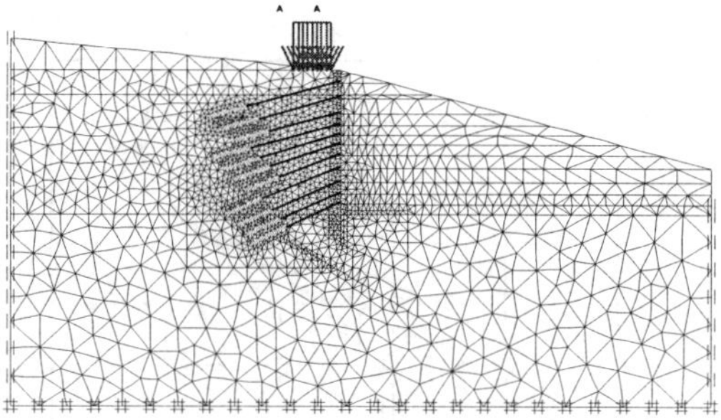

Figure 4. Finite element mesh.

All finite element calculations were performed under plane strain and drained conditions. The influence of the construction procedure of the Soilcrete wall was not considered. The load cases have been calculated as follows:

- primary stress state according to geometric conditions ($K_0 = 1 - \sin\varphi$)
- boundary load representing a static load of 15 kN/m² on the road behind the wall
- activation of upper part of Soilcrete wall
- excavation steps 1 – 4, activation of anchor located 0.5 m above corresponding excavation level and application of prestress forces
- activation of lower part of Soilcrete wall
- excavation steps 5 – 12, activation of anchor located 0.5 m above corresponding excavation level and application of prestress forces

The activation of the Soilcrete body in front of the wall is modelled at excavation level 287.5 m and 275.6 m respectively.

3.2 Results from calculations

As mentioned earlier a comprehensive site investigation programme has not been performed for this project. Thus the exact position of the rock surface was not explored in advance. In order to assess the influence of the rock layer on the expected deformations a numerical investigation was undertaken. The results presented below refer to calculations without taking into account the rock layer.

Due to the stiffness of the Soilcrete body in front of the wall and the high anchor prestress forces the calculated horizontal displacements of the Soilcrete wall are small. At the final excavation step the displacement of the top of the wall is less than 1 cm, the maximum displacement in the middle of the wall slightly more than 1 cm. Introducing the stiff rock in the analysis the maximum displacement is less than 1 cm. The lateral displacement of selected points for the excavation steps are presented in Figure 5.

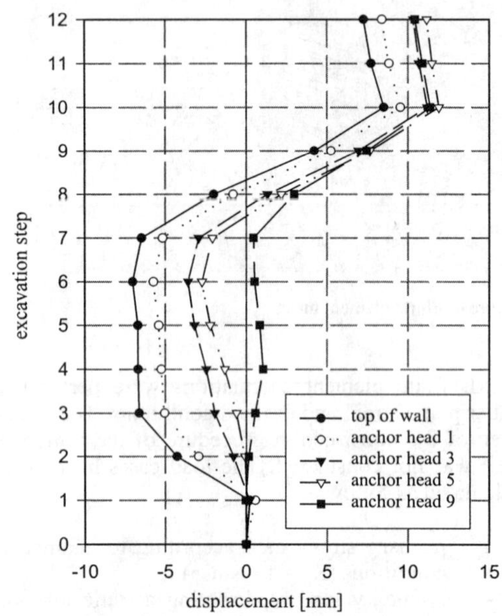

Figure 5. Horizontal displacements of selected points of the Soilcrete wall.

Finite element analyses were also performed for cross sections A-A and C-C. The maximum calculated wall displacements were again in the range of about 1 cm.

As mentioned above one important aspect of the success of the concept was that the stresses in the Soilcrete wall are within a tolerable range considering that there is no reinforcement available. A relative simple approach is adopted here namely to transfer the state of stress obtained from the finite element calculations for critical cross sections of the wall into a Mohr's circle diagram. Compared to the strength parameters resulting from laboratory investigations it was proved that the safety margin is sufficient (Fig. 6). Of course this procedure is rather crude and further investigations on suitable constitutive models for Soilcrete material are presently under progress.

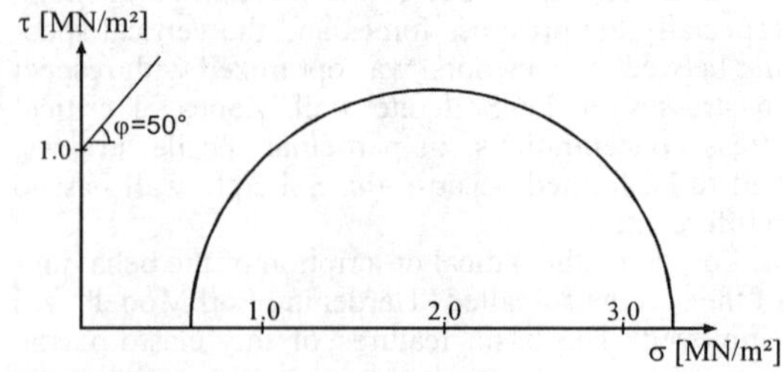

Figure 6. Critical stress state in the wall.

Figure 7 shows the force distribution in geotextile elements representing the grout body of prestressed anchors. It is assumed that there is a perfect bond between geotextile and surrounding soil. Thus a continuous load transfer is achieved and this concept has been successfully applied to model prestressed ground anchors with sufficient accuracy at least for working load conditions.

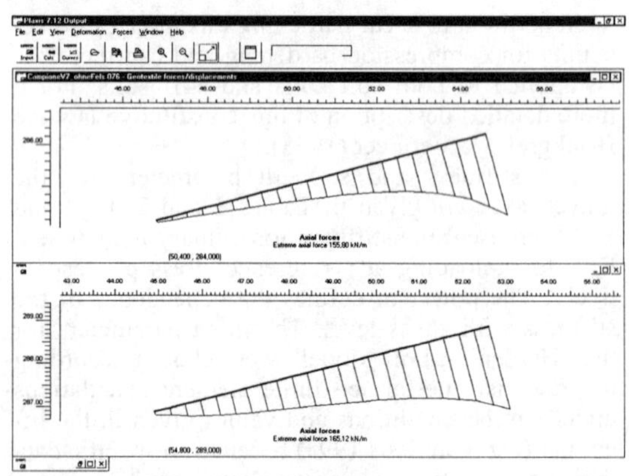

Figure 7. Axial forces of geotextile elements.

For assessing the stability of the construction the finite element code PLAXIS offers the so-called phi – c reduction procedure. In this procedure the strength parameters tan φ and c are reduced simultaneously up to the point where the system fails numerically. Structural elements are not concerned by the strength reduction. The factor of safety ΣMsf obtained from the strength reduction is defined as:

$$\sum \text{Msf} = \frac{\tan \varphi_{\text{input}}}{\tan \varphi_{\text{reduced}}} = \frac{c_{\text{input}}}{c_{\text{reduced}}} \tag{1}$$

Considering relatively simple structures (e.g. homogenous slope) this approach is similar to conventional methods and therefore yields the failure mechanism with the minimum factor of safety. For complex structures as shown here the calculated factor of safety is not automatically the minimum one. Nevertheless this procedure is a valuable tool in getting a better understanding what may happen at the edge to the limit state of the structure.

Performing the phi - c reduction procedure it becomes apparent that the anchor forces increase from bottom to top due to the decrease of strength. Once the anchor force reaches the ultimate value this force remains constant because an elastic-perfectly plastic constitutive law for the node-to-node anchors is applied. A failure of the anchors at that ultimate load cannot be modelled at present. Therefore a failure line cannot cut through the free tendon length but has to move behind the bonded length of the anchors. The factor of safety obtained from this type of calculation is higher than the one resulting from limit equilibrium method, even if the grouted length of anchors is introduced in the latter method.

Figure 8 presents the ΣMsf-factor, Figures 9 - 11 show contour lines of total shear strains at different stages of the phi – c reduction indicating the development of a failure mode. For comparison the critical failure line from the limit equilibrium analysis applying Janbu´s method is depicted in Figure 12.

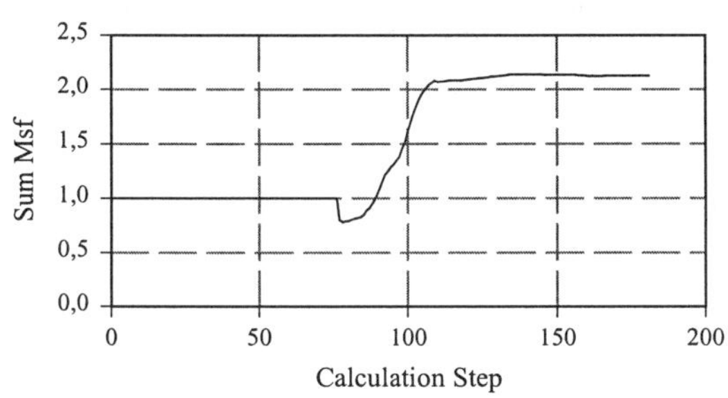

Figure 8. Factor of safety obtained from phi – c reduction.

4 MEASUREMENTS

The geotechnical measurement comprises geodetic survey, inclinometer and anchor force measurements.

The geodetic survey system is installed at the old casino to monitor the horizontal displacements of the wall. As these measurements were not started at the very beginning of the construction phase and the

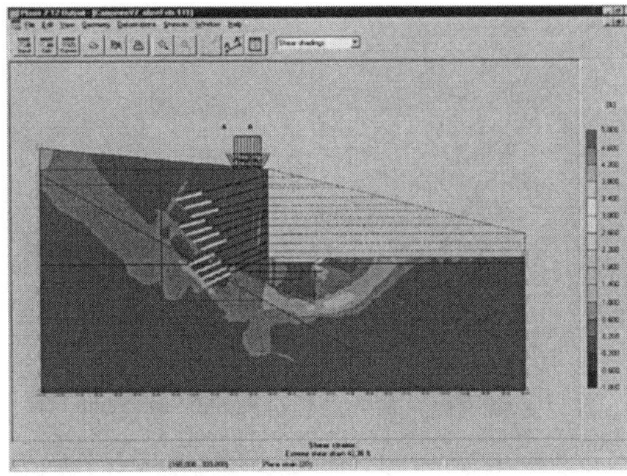

Figure 9. Contour lines of shear strains resulting from phi-c reduction at calculation step 111.

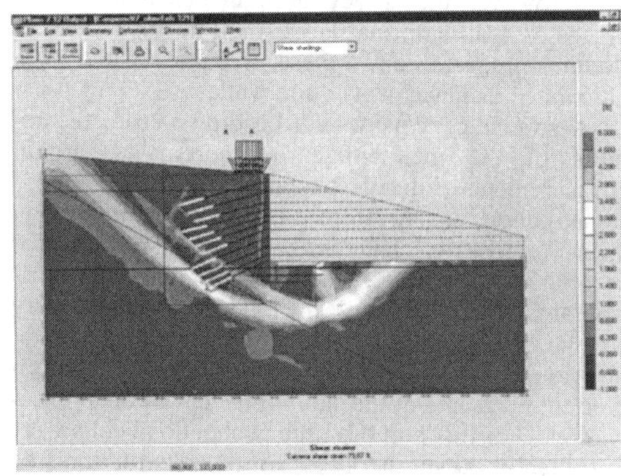

Figure 10. Contour lines of shear strains resulting from phi-c reduction at calculation step 121.

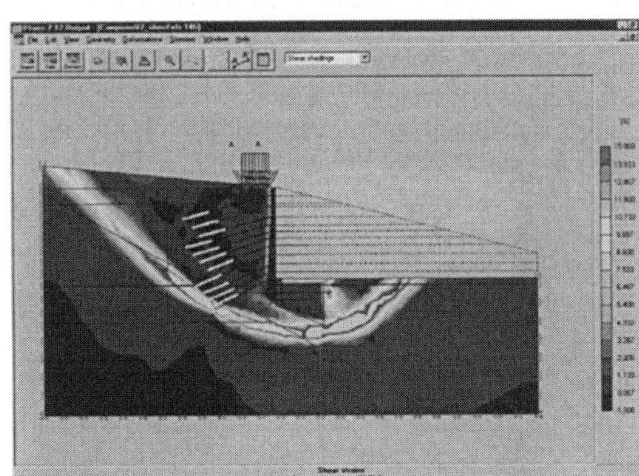

Figure 11. Contour lines of shear strains resulting from phi-c reduction at calculation step 146.

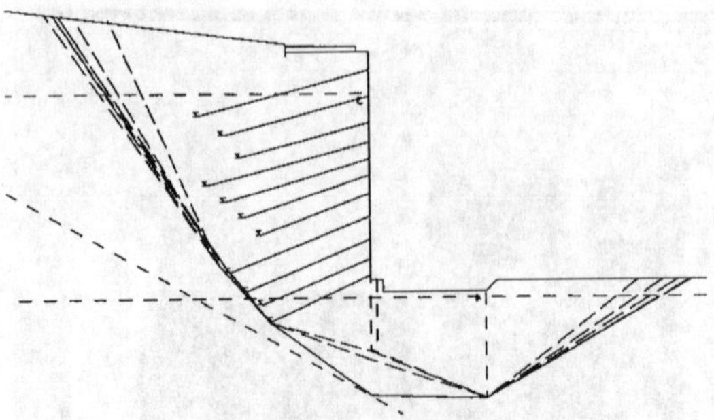

Figure 12. Critical slip line from limit equilibrium method calculation.

displacements are in the range of the accuracy of the measuring devices a detailed comparison between calculated and observed displacements is not possible. However the surveying indicates that the increase of wall deformation is small during the excavation progress and these results match the calculated displacements quite well.

A similar conclusion can be drawn from the results of the inclinometer measurements. Again measurements did not start from the beginning of construction and therefore a one to one comparison measurements – calculations is not available. But it is obvious also from these measurements that horizontal displacements of the wall are indeed very small.

Monitoring of the anchor forces is one of the important measures in the safety concept of deep excavations. For this purpose nine anchor-load cells were installed in the northern part of the construction site. In the southern area the anchors in one cross section will be observed by eleven anchor-load cells. These measurements show that the anchor forces change only to a small extent compared with prestress forces applied. The results of the numerical prediction match the measured anchor forces well and indicate the accuracy of the utilized model. A comparison between calculated and measured forces for anchor rows 3 and 4 is given in Table 3.

Table 3. Comparison between calculated and measured anchor forces.

Anchor row	calculated	measured
	kN/anchor	kN/anchor
3	567	559-570
4	719	702-706

5 CONCLUSIONS

A jet grouting technique was utilized to construct a 30 m deep excavation. For the design of the retaining wall comprehensive finite element analyses were carried out. These calculations proved the feasibility of the proposed construction concept with respect to deformations and stresses of the Soilcrete wall. A comparison between finite element results and in situ measurements is presented to the extent available.

In addition to limit equilibrium methods the finite element method was also applied to estimate the stability of the support system by means of successively reducing the strength parameters of the soil. The latter proved to be a valuable tool for assessing the behaviour of the structure at the edge to failure. However the factor of safety obtained from finite element analyses resulted in higher factors of safety compared to limit equilibrium analyses. This is not surprising however because the assumptions made inherently in the simple limit equilibrium analyses are not reflected in the finite element analyses. It is believed that the finite element model is superior to conventional methods in the respect that it represents more closely the actual behaviour in the field. However more investigations in this type of problems are required before definite conclusions can be drawn as far as reliable factors of safety are concerned.

6 REFERENCES

Brinkgreve, R.B.J. & Vermeer, P.A. 1998. Finite Element Code for Soil and Rock, Version 7. Rotterdam: Balkema.

Duncan, J.M. & Chang, C.Y. 1970. Nonlinear analysis of stress and strain in soils. J. of the Soil Mechanics and Foundations Division, Vol. 56, 1629-1653.

Otta, L.J. 1997. Softwarepaket STAND Version 5.1, Benutzerhand.

Soos, P. von 1995. Eigenschaften von Boden und Fels; ihre Ermittlung im Labor. Grundbautaschenbuch, Band 1, 87-153. Berlin: Ernst & Sohn (in German).

Computer Methods and Advances in Geomechanics, Desai et al. (eds) © 2001 Balkema, Rotterdam, ISBN 90 5809 183 X

Material Information Management System

J.N. Meegoda
Department of Civil and Environmental Engineering, New Jersey Institute of Technology, Newark, N.J., USA

C.Tang
Institute of Transportation, New Jersey Institute of Technology, Newark, N.J., USA

ABSTRACT: A computerized Material Information Management System (MIMS) is developed. This intranet-based information management system is a state-of-the-art research project where material engineers working closely with computer and operations research specialists to develop a comprehensive and fully functional Material Information Management System for the New Jersey Departments of Transportation (NJDOT). It reduces the paper-workloads significantly and provides the capability to organize relevant data rapidly. The implementation of MIMS also shortens the time between project completion and final closeout mandated by the federal highway administration for federally funded projects. The project has a great potential to be a showcase for the nation and to be applied by other state Departments of Transportation.

1 INTRODUCTION

The New Jersey Department of Transportation (NJDOT) has taken a leadership role in developing computerized database information management systems. It has been recognized that needs exist to: a) reduce paperwork, b) expedite the material approval process, c) shorten the time between project completion and final closeout, and d) analyze the functional relationships between material testing data and material performances in order to identify significant trends in time.

Currently, most material test results are applied on a micro scale and paper-filed. To integrate the operational functionality of the material data filing, processing, and transfer, a computerized local/remote Intranet network management system should be developed. This system will standardize the data entry procedures, define performance evaluation measures, certify material test data, analyze the relationships between testing data and actual material performance, streamline the project closeout process, generate summary reports, and be able to communicate with NJDOT existing information systems, such as PMIS. Fully developed, it is envisioned that the system will connect 20 field offices, 3 regional offices, and the central office in Trenton to transfer information, trail material samples, and periodically replicate the material database. During an evaluation period, both the current paper-based filing system and the electronic intranet-based MIMS will work side-by-side as co-existing systems. Eventually, MIMS will replace the current system.

2 SYSTEM CONFIGURATION

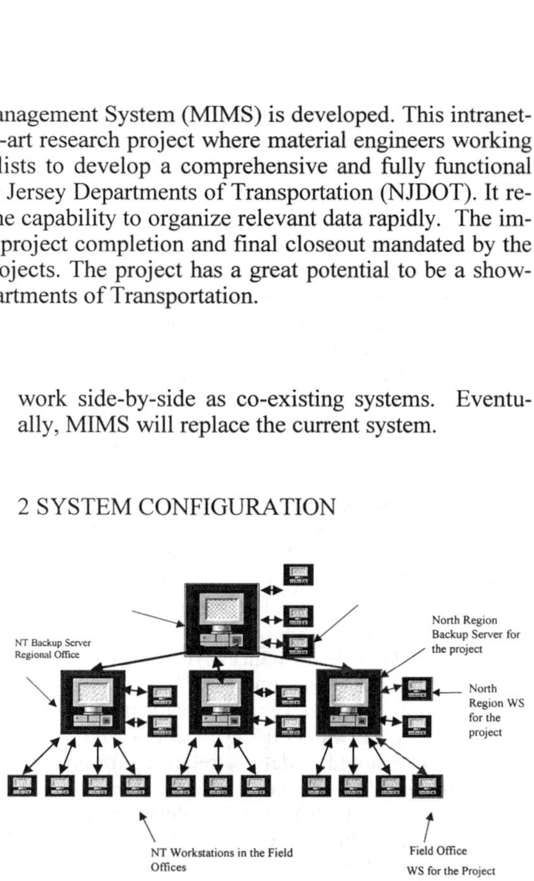

Figure 1 Final Network Configuration for MIMS

NJIT research team is using a server based on Intel chip (low cost) with Windows NT operating system (security) and SQL database (graphical interface using Visual Basic). During the first phase of this project, NJIT research team would install a primary domain controller (PDC) running Windows NT with SQL server in the Trenton office or the head office. NJIT research team will finally implement the project on a dedicated directory in the NJDOT NT server. During the second phase of this project, another server will be utilized in the North region as a backup domain controller (BDC). A computer running Windows 98 in one field office selected by NJDOT bureau of material will be connected to the PDC via a 56K.V90 modem pool or digital subscriber line (DSL). NJIT research team will provide instructions for adding two more BDC for the remaining regional offices. Figure 1 shows the computer configuration when the system is fully implemented. The PCD and BDC should have the capabilities of disk duplexing and tape backup support to maximize the reliability and fault tolerance of the network. In order to manage the vast amount of data, NJIT research team plans to develop a way to archive the data older than a specific time, say five years.

The Microsoft SQL server version 7.0 has reduced administrative overhead from its current version 6.5 and hence SQL version 7.0 is be used in this project. Many server configuration options have been simplified in SQL version 7.0. For example, a database increases allocated resources when necessary without ever committing them and decreases the resources used when they are no longer needed. The SQL 7.0 databases now reside on operating system files instead of on SQL Server logical devices. Database files can expand automatically, eliminating the need for administrators to make manual adjustment of these settings.

SQL Server 7.0 works with Windows NT Workstation and Microsoft Windows® 95/98 as well as Windows NT server. The SQL version 7.0 accessed using Windows 95/98 is fully featured except for a few limitations imposed by the operating system. For example, SMP, named pipes, full-text search, asynchronous I/O, and integrated security are not supported on Windows 95/98. However, this will not hinder the MIMS performance if the system design avoids using these limitations.

In addition, SQL Server provides a host of new features for data warehousing. Summarized information can be stored in data warehouses for analytical applications to use conveniently. Furthermore, using Microsoft English Query, a development tool gives users the ability to query a SQL Server database in natural-language (English) questions, users can obtain information about particular tables, fields, and data of the SQL database without having to know how to write query statements.

3. RESEARCH SCOPE

Currently, there are more than 110 forms, reports, memos, and labels and cards in use. Many of these forms have the same usage with different layouts. Each one of them has to be filed manually. It is a very time consuming and labor intensive process. Also, the transfer of these documents among related units of NJDOT causes delays in decision making. The Material Information Management System (MIMS) will streamline and simplify the entire filing procedure of the Bureau of Materials, thus relevant information will be integrated into an electronic data processing and management system automatically.

The design and the development of the MIMS system require considerable research efforts. The major study topics include:

(1) How to identify and synthesize the operational logic and business rules in the material testing processes so as to search for the best ways to organize and unify the forms, reports, approvals and all relevant documents;
(2) How to normalize the MIMS relational database (SQL) and apply Intranet techniques to secure data integrity and manage data flows within NJDOT existing networks;
(3) How to convert and integrate the existing OIT electronic data for concrete cylinders;
(4) How to handle data approvals and notification of approvals to appropriate users;
(5) How to protect and archive the data;
(6) How to accommodate both Metric and English units;
(7) How to improve the system flexibility such that new materials and new information can be inserted;
(8) How to minimize the data entry errors through cross checking;
(9) How to minimize the routine workload and improve the efficiency of the data processing with customized client application programs; and
(10) How to automate the filing processes and speed-up the closeout procedures by scheduled replication tasks and alert message service of the SQL server. In other words, the entire closeout procedure will be monitored by the system and the responsible person will be notified of the current project status.

The proposed SQL based MIMS system will support the following features when applicable:

- Registration integrity of material samples, products, customer and vendor specifications
- Multilevel limit checking for specification limits, warning limits, outer limits, and user defined limits
- Automatic calculations of compliance, penalties, etc.
- Automatic approval/validation of testing results
- Conditional event triggers and automatic electronic mail for testing results
- Incorporate penalty/bonus adjustments to project expenses

- Generation of summary reports
- Metric/English units conversion
- Bar-code reading support and label printing
- Direct data downloading from nuclear gage and concrete compression tester
- Instrument Maintenance/Calibration scheduling & recording

4 OPERATION PROCEDURE

The MIMS data entry is central to all samples no matter whether of raw material, in process, intermediate, or finished product. By MIMS design, the data entry can be done manually (basic information initialization), automatically (calculated and transferred results), or be triggered (data downloading) by other device.

Each sampling process has its own logical life from beginning to the end. All data entry forms in MIMS will be created according to their sampling logic and unified by their virtual usage. It is the integrated design of the sampling cycles that makes MIMS such a flexible and reliable information management system.

First, after the approval a sampling template form is filled by inspectors with the basic information such as project ID, job number, material type, testing date, location, producer, customer, inspectors and operators. This basic information needs to be entered into MIMS only once and can be automatically pulled up as part of any sampling forms in the future.

Second, laboratory supervisors outline the tests to be completed and field engineers and sampling workers will correspondingly generate specific worksheets from MIMS and will fill the necessary information after samples were taken for testing. Each sample is uniquely coded for identification using a user defined sample ID number. Using this information, all relevant data about the sample can be retrieved immediately at any stage of its entire sampling process form any location in the system provided the user has the required authority. The sample-taking process is accomplished when the worksheets are returned to the local laboratory and the identification information has been entered into MIMS database.

Third, samples are tested at a local laboratory or sent to the central laboratory in Trenton with labels. Since all the sampling ID information has been input into the system, the labs can pop-up the required testing result form onto a computer screen based on the label received and fill the form with the tested results. MIMS will automatically check the specific limits and calculate the results so as to generate Pass/Fail reports immediately, and will calculate the pay factors. If all results pass specification, MIMS will recommend a sample completion status of Complies. Comments are entered in the event that a Conditional Approval status is assigned to a sample. Once the reports are generated, the local and central labs as well as other units of NJDOT will be able to use them for further analysis.

Meanwhile, other related operational reports, business memos, and closeout documents are also ready to be created by people in charge whenever the required information has been available in MIMS. MIMS will provide the capabilities of customized data querying, standardized report generation as well as comprehensive graphics utility to display result trends.

Finally, all the sampling data are secured in the relational SQL database with integrity for reliable and flexible retrieval and application. The schematic diagram for concrete cylinder tests shown below illustrates the operational procedure discussed previously.

The MIMS will accommodate test results for all 53 of the materials tested by NJDOT's Bureau of Materials. Following, however, is a list of the highest priority materials.

1. Portland Cement
2. Portland Cement Concrete (Cast-In-Place)
3. Asphalt Cement (Asphalt Binder)
4. Bituminous Concrete (Hot Mix Asphalt)
5. Aggregate Sources (Coarse, Fine, & Mineral Filler)
6. Soil Aggregate
7. Dense Graded Aggregate Base Course (Virgin & Recycled Materials)
8. Structural Concrete (Prestressed Beams, Piles, and Noisewall and Precast Culverts)
9. Structural Steel Bridge Members (Box Beams, I-Beams, Diaphragms, etc.)
10. Asphaltic Emulsions
11. Asphaltic Oils (Cutbacks)
12. Concrete Deck Overlay Protective Systems
13. Concrete Pipe
14. GO & GA Signs (Overhead & Adjacent Guide Signs)
15. Guiderail Materials (Rail, Posts)
16. Overhead and Cantilevered Sign Structures
17. Paint – Structural
18. Paint – Traffic (Epoxy)
19. Precast Concrete (Retaining Wall, Noisewall, Inlets, etc.)
20. Steel Reinforcement (Epoxy Coated Rebar and Spiral)
21. Steel Reinforcement (Uncoated Rebar and Spiral)
22. Welded Wire Mesh

5 IMPLEMENTATION

In order for proper implementation a mission statement was developed, based on which the MIMS is formulated. It is given below.

To provide a system for creating and gathering material related business and testing information about NJDOT material bureau personnel and facilities; contracts and awards; contractors, suppliers, producers and plants; product mix design; production quality control; field sampling operation as well as Lab. test results via Intranet of NJDOT Materials Bureau. Also this system should be able to generate analysis reports based on their collected information and maintain all relevant data documents in a SQL server database. Ideally, this system should facilitate the management of material testing and project close out process via secured, user-friendly WebPages and VB 6.0 user graphical interfaces.

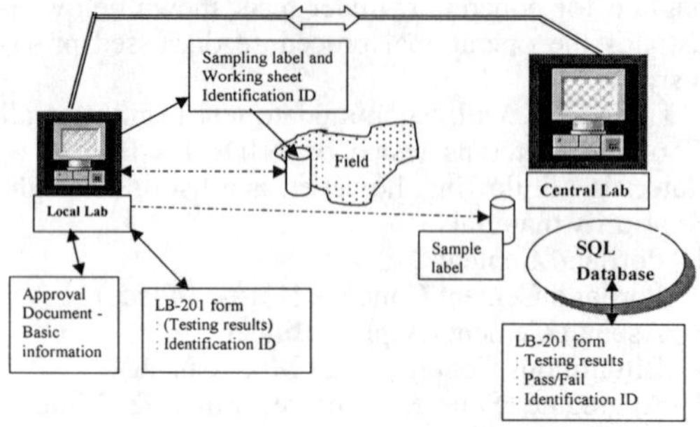

Figure 2-The schematic diagram showing the implementation for concrete cylinder tests.

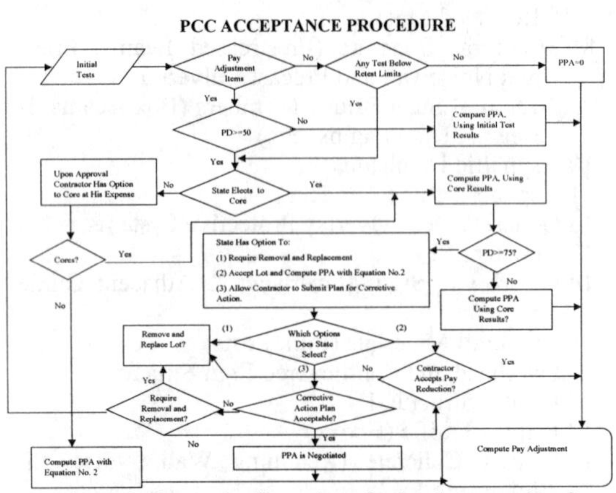

Figure 3 shows the flow chart for implementation for concrete cylinder tests.

The research project is at the initial stages where the database is built to implement the testing of concrete cylinders, which certifies the fresh concrete poured. Based on tonnage used, the fresh concrete is the highest consumed material for NJDOT. Figure 3 is a schematic diagram showing the implementation for concrete cylinder tests. Figure 3 shows the flow chart for implementation with the pay factors for not achieving the specified concrete strength values. The other research issues incorporated in the database include the security of the system. Once an NJDOT employee is logon into the system based on the username the system assign the read, write, edit and delete privileges based on the position and the location of the user. For instance a technician entering raw data in a field office can only modify the data entered in that office. Once his or her supervisor approves the data no one is allowed to edit, modify or delete. For the concrete mix designs the existing database based on FoxPro is already imported to SQL database and hence once the mix design ID is entered all the fields in the LB 201 form used for the concrete cylinder test.

6 CONCULSIONS

Currently, most state department of transportation material test results are applied on a micro scale and paper-filed. Each lot is assessed separately for compliance, penalties, etc. All the test result reports are manually generated, distributed, and filed away individually. The filing processes of approving projects, preparing material test result reports, and searching relevant closeout paper-documents are very labor intensive and time-consuming. The need exists to replace the current paper-based filing and transportation system by an intranet-based electronic filing system. The needs also exist to streamline the process of approval and compiling federal highway agency closeout packages as well as to analyze the functional relationships between the testing data and the materials' performance. NJIT is currently developing a SQL based material information management system to computerize all material data. The project is at the initial stage where the database is ready to implement concrete cylinder tests, which is for the largest consumed material for NJDOT.

7 ACKNOWLEDGEMENTS

New Jersey Department of Transportation and National Center for Transportation and Industrial Productivity jointly sponsored this research. The program manager at NJDOT is Mr. Robert Sasor and for NCTIP is Dr. Lazar Spasovic. Although NJDOT and NCTIP funded the research described in this paper, it has not been subjected to both Agencies' review process for the technical content, quality assurance/quality control, or administrative review. Therefore, it does not necessarily reflect the views of both agencies. Authors would also like to acknowledge the NJDOT task force headed by Ms. Eileen Connolly and Mr. John Zim.

Computer Methods and Advances in Geomechanics, Desai et al. (eds) © 2001 Balkema, Rotterdam, ISBN 90 5809 183 X

Ground collapse with buckling of arch type underground structures

Y. Nakamura
Ministry of Agriculture, Forestry and Fisheries of Japan, Japan

T. Tanaka
Graduate School, University of Tokyo, Japan

M. Kinouchi
Graduate School, Meiji University, Japan

ABSTRACT: The buckling mechanism of arch type underground structures was studied by model experiment and finite element analysis. Infinitesimal deformation analysis considering only material non-linearity could not predict the limit load, so finite deformation analysis was indispensable. In the analysis, using the internal friction angle of the soil layer at the interface element, collapse loads tend to be evaluated excessively. Therefore, the interfacial friction angle between the soil layer and arch structure has to be taken into account in the analysis. The maximum shear strain obtained by finite deformation analysis showed that shear band occurred from the side to the top of the arch as was also seen in the model experiment.

1 INTRODUCTION

The behavior of arch type underground structures is closely related to the characteristics of the surrounding ground. As the displacements increase, the effect of geometrical non-linearity by finite deformation cannot be ignored. The material non-linearity of the surrounding soil becomes conspicuous. Regarding the problem of the interaction of structures and the ground, it is very important to find the buckling load when evaluating the safety of large underground structures. There has been little research related to the numerical analysis of buckling of underground structures.

The aim of this study is to grasp quantitatively the mechanism of buckling destruction of arch type underground structures. A model experiment and elastoplastic finite element analysis were compared.

Infinitesimal deformation analysis, considering only material non-linearity, and finite deformation analysis, were carried out.

2 MODEL EXPERIMENT

2.1 Test apparatus and buckling load experiment

The test apparatus for the buckling model experiment is shown in Fig.1. The surroundings of the arch are shown in Fig.2. Plane strain conditions were maintained in this experiment.

An arch structure of span 12.0cm and center angle 180 degrees was made of hard chlorinated vinyl sheet (Young ratio E=2,943,000kN/m^2).

The sand used in this test was Toyoura sand. The dry density was 1.600-1.619g/cm^3 (relative density

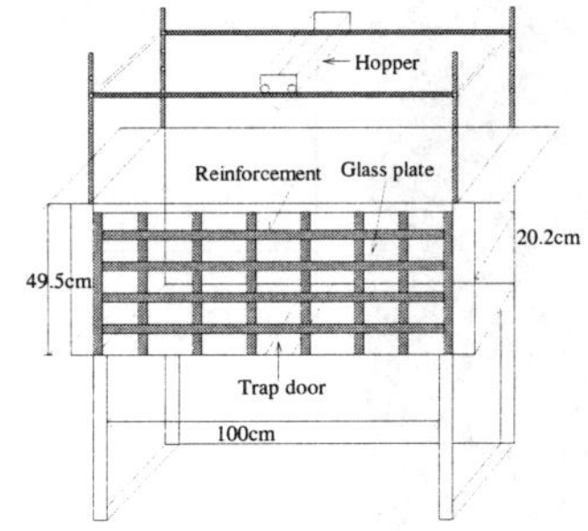

Fig.1. Test apparatus.

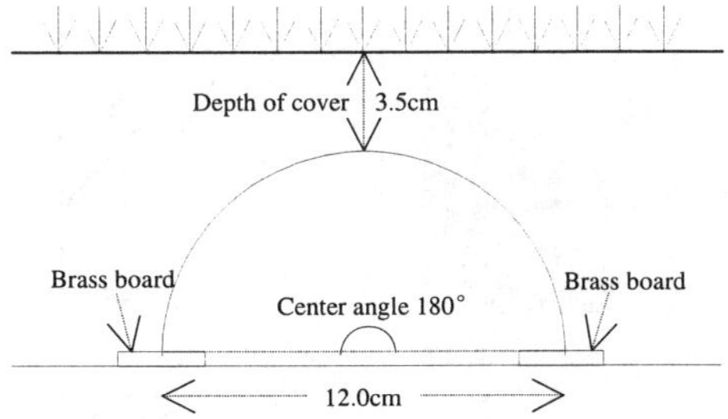

Fig.2. Surroundings of arch.

D_r=88-93%). The sand mass was prepared by pouring air-dried sand into the testing box through the hopper. For lubrication to reduce the sidewall friction, a latex rubber membrane and a thin layer of silicone grease were used. When the distributed load on the sand layer was increased, the arch structure collapsed, suddenly and catastrophically, along with the foundations.

The buckling load was 52.97-72.59kN/m^2 (62.24kN/m^2 of 19 tests average).

2.2 Image processing

In order to observe the model experiment in detail, ground collapse was observed by a high-speed camera as image processing.

The high-speed camera used for this research can take 240 frames/sec at a shutter speed of 1/10,000 second. With the high-speed camera, the buckling conditions were captured in 10 to 12 frames. It was confirmed that the collapse occurred in 0.05 seconds.

Fig.3 shows the process of the ground failure. These pictures were taken at intervals of 1/80 second, starting just before buckling. The collapse started by the arch crown, followed by one of the arch springs protruding inwards.

Fig.4 shows the process of the occurrence of shear band. It was confirmed that shear band propagated from the side to the top part of the arch.

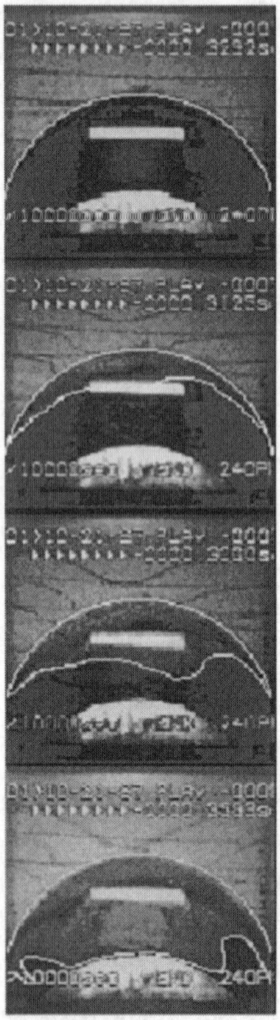

Fig.3. The process of the ground transformation.

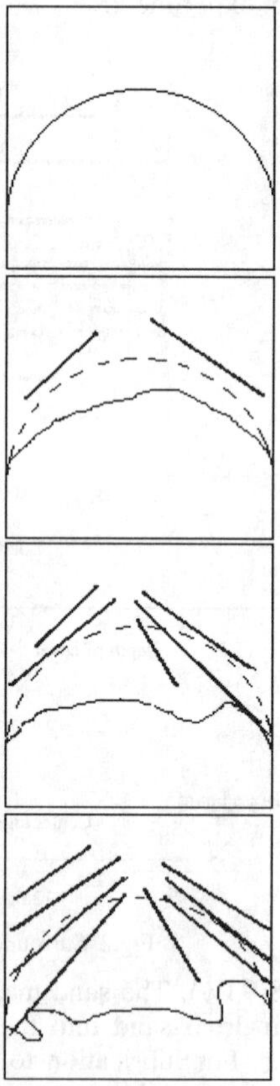

Fig.4. The process of the occurrence of shear band.

3 FINITE ELEMENT ANALYSIS

3.1 Dynamic relaxation method

In this study, formulation that combined the return mapping method and the dynamic relaxation method was applied (Tanaka et al., 1987).

3.2 Formulation of a mixed type composition model

The yield function (f) and the plastic potential function (Φ) are given by;

$$f = \alpha I_1 + \frac{\bar{\sigma}}{g(\theta_L)} = 0 \tag{1}$$

$$\Phi = \alpha' I_1 + \bar{\sigma} = 0 \tag{2}$$

where

$$\alpha = \frac{2\sin\phi}{\sqrt{3}(3-\sin\phi)} \tag{3}$$

$$\alpha' = \frac{2\sin\psi}{\sqrt{3}(3-\sin\psi)} \tag{4}$$

where I_1 is the primary invariant of deviatoric stresses and $\bar{\sigma}$ is the second invariant of deviatoric stress. In the case of the Mohr-Coulomb model, $g(\theta_L)$ takes the following form;

$$g(\theta_L) = \frac{3-\sin\phi}{2\sqrt{3}\cos\theta_L - 2\sin\theta_L \sin\phi} \tag{5}$$

ϕ is the mobilized friction angle and θ_L is the Lode angle. The frictional hardening-softening functions are expressed as;

$$\alpha(\kappa) = \left(\frac{2\sqrt{\kappa\varepsilon_f}}{\kappa+\varepsilon_f}\right)^m \alpha_p$$

$$(\kappa \le \varepsilon_f): \text{hardening-regime} \tag{6}$$

$$\alpha(\kappa) = \alpha_r + (\alpha_p - \alpha_r)\exp\left\{-\left(\frac{\kappa-\varepsilon_f}{\varepsilon_r}\right)^2\right\}$$

$$(\kappa \ge \varepsilon_f): \text{softening-regime} \tag{7}$$

where m, ε_f and ε_r are material constants.

The unit weight (γ), residual friction angle (ϕ_r), Poisson's ratio (ν) and initial shear modulus (G_o) are chosen based on the data from the test of air-dried dense Toyoura sand. γ=16.1kN/m^3, G_0=78,400kPa, ϕ_r=33(degrees), ν=0.3. The elastic moduli are estimated using the following equations.

$$G = G_o \frac{(2.17-e)^2}{1+e}\sqrt{\sigma_m} \tag{8}$$

$$K = \frac{2(1+\nu)}{3(1-2\nu)}G \tag{9}$$

The peak friction angle (ϕ_p) is estimated from the empirical relations proposed by Bolton (1987), based on the plane strain compression test on dense Toyoura sand by Tatsuoka et al. (1986).

$$I_r = D_r\left[5 - ln\left(\frac{\sigma_m}{150}\right)\right] - 1 \quad (\sigma_m \geq 150\text{kN/m}^2) \tag{10}$$

$$I_r = 5D_r - 1 \quad (\sigma_m \leq 150\text{kN/m}^2) \tag{11}$$

$$\phi_p = \phi_r + 5I_r \tag{12}$$

The dilatancy angle (ψ) is estimated from Rowe's stress-dilatancy relation.

The material constants are fitted by a trial and error method based on the data set of triaxial compression tests. Material constants m, ε_f and ε_r are assumed to have the following values.

$m = 0.3, \varepsilon_f = 0.1, \varepsilon_r = 0.6$

3.3 Finite Deformation Analysis

Formulations of finite deformation analysis based on total-Lagrangian method (Hibbitt et al. (1970)) and updated-Lagrangian method (Nagtegaal & De Jong (1981)) were applied. In both of solution methods, dynamic relaxation was applied.

4 FINITE ELEMENT ANALYSIS RESULTS

Regarding arch structure in soil mass, elastoplastic finite element analysis with geometrical non-linearity was performed. The results were compared with the image processing results of the model experiment.

In finite element analysis I, comparisons of infinitesimal deformation analysis and finite deformation analysis methods were made, and in finite element analysis II, the influences of interfacial friction between arch and soil layer were investigated.

The material constants of Toyoura sand used for calculation were as follows.
D_r=90%, G_o=78,480kN/m^2, ν=0.3, ϕ_r=34(deg), ε_f=0.1, m=0.3, ε_r=0.6, ε_d=0.2, S.B.(shear band thickness)=0.3cm

G_o was determined by Iwasaki et al. (1977, 1978), ϕ_r, ε_f, m, ε_r and ε_d were determined by a plane strain compression test by Tatsuoka et al. (1993), the shear band thickness was determined by a model experiment (anchor model test by Sakai et al. (1995)).

4.1 Finite element analysis I

Finite element analysis was performed using a finite element mesh, as shown in Fig.5 (number of elements 443, number of nodes 485). In the analysis, distributed load increment 1.962kN/m^2 was applied.

Infinitesimal deformation analysis, finite deformation analysis (total-Lagrangian) and finite deformation analysis (updated-Lagrangian) were applied. The material constants of the arch were E=2,943,000kN/m^2 and ν=0.3. The thickness of the arch was t=0.02cm.

Fig.6 shows the relation between the load and arch crown displacement. The buckling load could not be obtained from infinitesimal deformation analysis. With finite deformation analysis, however, in both total-Lagrangian and updated-Lagrangian, the buckling load was obtained. From these results, it was clarified that infinitesimal deformation analysis was not suitable for the problem of underground structures and that analysis taking into account geometric non-linearity was required.

In finite deformation analysis, results were largely the same for both total-Lagrangian and updated-Lagrangian. Based on this, the simpler total-Lagrangian was used for further finite deformation analysis.

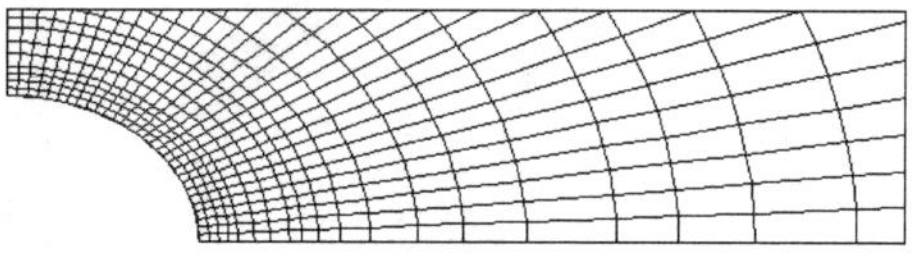

Fig.5. Finite element mesh.

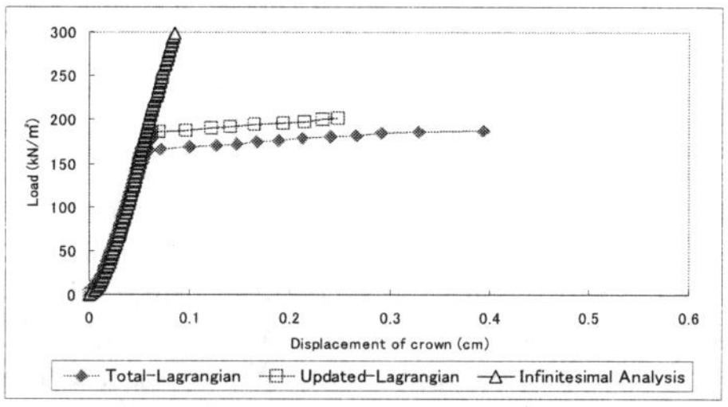

Fig.6. Relation between load and displacement by finite element analysis I.

4.2 Finite element analysis II

In finite element analysis II, the influence of the interfacial friction between the arch structure and soil layer was examined.

In analyses, without the internal friction angle, collapse loads tend to be evaluated excessively. Therefore, the interfacial friction angle between the soil layer and arch structure has to be taken into account.

The interfacial friction angle was changed: 5, 10, 15, 20, 25, and 34 degrees. The finite element mesh shown in Fig.6(number of elements 443, number of nodes 485) was used for the analysis.

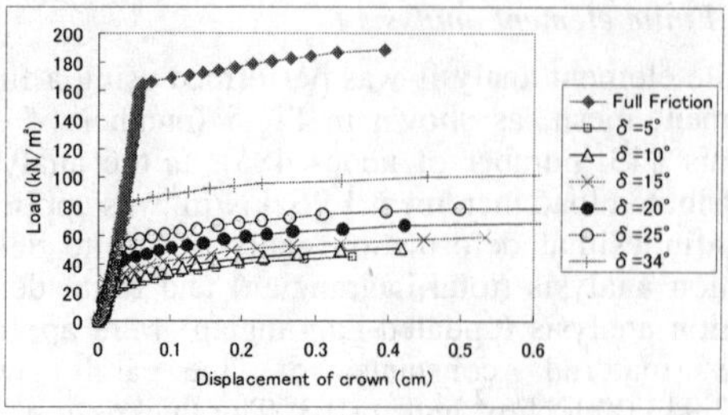

Fig.7. Relation between load and displacement by finite element Analysis II.

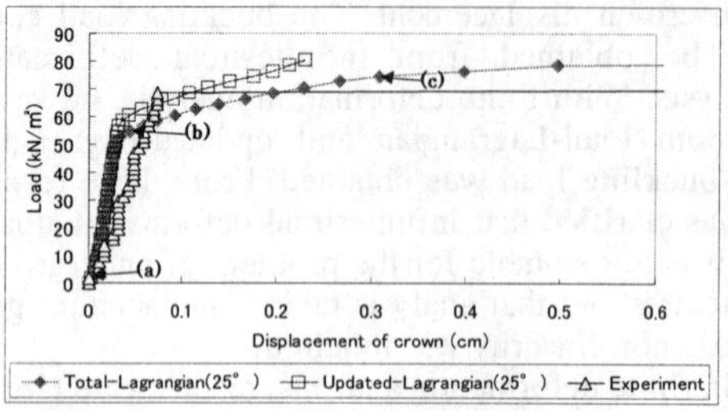

Fig.8. Relation between load and displacement by finite element Analysis II. (Interfacial friction 25 degrees)

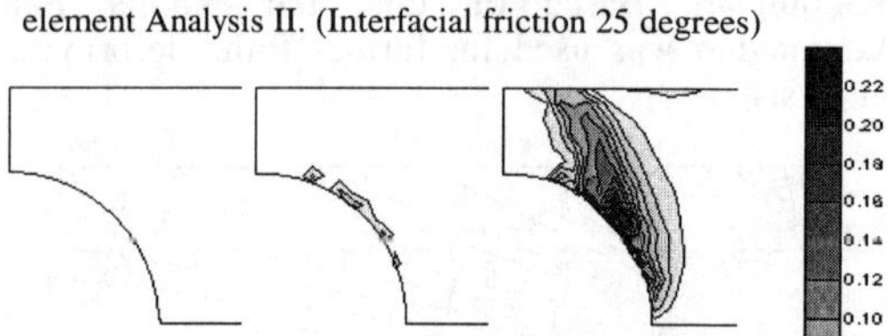

(a) Load=5.886kN/m^2 (b) Load=54.936kN/m^2 (c) Load=74.556kN/m^2

Fig.9. Contour maps of maximum shear strain.

Fig.7 shows the relation between the load and arch crown displacement obtained from finite element analysis. The collapse load became higher as the interfacial friction angle increased. In the analysis, which applied the internal friction angle of the soil layer as the interfacial friction angle, the collapse load was about 170 kN; 3 times higher than that of the model experiment. When an interfacial friction angle of 25 degrees was applied, the collapse load was close to that of the experiment.

Analyses applying an interfacial friction angle of 25 degrees were performed both by total-Lagrangian and updated-Lagrangian. Fig.8 shows the relation between the load and arch crown displacement. A collapse load close to that of the experiment was obtained taking interfacial friction into consideration with the updated-Lagrangian as well as the total-Lagrangian.

The interfacial friction angle between the soil layer and arch structure has to be taken into account in the analysis.

Contour maps of the maximum shear strain at (a), (b), (c) in Fig.8 are shown in Fig.9. At the initial stage of loading, a minute strain spread in the soil layer. However, at the time of buckling, it was confirmed that shear band occurred from the side to the top of the arch, as was also observed in the image processing of the model experiment.

5 CONCLUSIONS

In this study, the buckling mechanism of arch type underground structures was studied by image processing of a model experiment and finite element analysis. The main conclusions drawn from this study are as follows.

(1) Regarding the buckling problem of arch type underground structures, infinitesimal deformation analysis considering only material non-linearity could not predict the limit load. Finite deformation analysis taking into account geometrical non-linearity was indispensable.

(2) The interfacial friction angle between the soil layer and arch structure has to be taken into account in the analysis.

(3) The maximum shearing strain evaluated by finite deformation analysis showed that shear band occurred from the side to the top of the arch, as was also observed in the image processing of the model experiment.

REFERENCES

Bolton, M.D. 1987. The Strength and Dilatancy of Sands, Geotechnique 36(1), p.65-78., 1986. Discussion on the Strength and Dilatancy of Sands by Tastuoka, Geotechnique 37, p.219-226.

Hibbitt, H.D., Marcal, P.V. and Rice, J.R. 1970. A finite element deformation for problems of large strain and displacement , Int.J.Solids and Struct. 6, p.1069-1086.

Iwasaki, T. and Tatsuoka, F. 1977. Effects of Grain Size and Grading on Dynamic Shear Moduli of Sands, Soils and Foundations 17(3), p.19-35.

Iwasaki, T., Tatsuoka, F. and Takagi, Y. 1978. Shear Moduli of Sands under Cyclic Torsional Shear Loading, Soils and Foundations 18(1), p.39-56.

Nagtegaal, J.C. & De Jong, J.E. 1981. Some Computational Aspects of Elastic-Plastic Large Strain Analysis, Int. J. Nume. Meth. Engng. Vol.17, p.15-41.

Sakai, T., Miyauchi, T. and Hotta, T. 1995. An Effect of Grain Size on Anchor Problem, Trans. of JSIDRE No.175, p.73-80. (in Japanese)

Tanaka, T. and Kawamoto, O. 1987. Three-dimensional Finite Element Analysis for Collapse Load of Foundations and Slopes Using Dynamic Relaxation, Trans. of JSIDRE No.131, p.79-86. (in Japanese)

Tatsuoka, F. Sakamoto, M. Kawamura, T. and Fukushima, S. 1986. Strength and Deformation Characteristics of Sand in Plane Strain Compression at Extremely Low Pressures, Soils and Foundations 26(1), p.65-84.

Tatsuoka, F., Siddiquee, M.S.A., Park, C.S., Sakamoto, M. and Abe, F. 1993. Modeling Stress-Strain Relations of Sand, Soils and Foundations 33(2), p.60-81.

Computer Methods and Advances in Geomechanics, Desai et al. (eds) © 2001 Balkema, Rotterdam, ISBN 90 5809 183 X

An efficient finite element methodology for nonlinear problems in geomechanics

L. E. Romera & S. Hernández
School of Civil Engineering, University of La Coruña, Spain

ABSTRACT: In this paper the software FESTE (Finite Element Software for Tunnelling Excavations), written by the authors at the University of La Coruña, will be presented. This code can be used for solving a great variety of geomechanic structural problems, having been applied in the simulation of tunnel construction by means of shield tunneling with earth pressure balance. Among the principal features of the code are the employment of infinite elements to model boundary conditions in unbounded media, the consideration of contact nonlinearities to resolve soil-structure interaction, geometric nonlinearities permitting the simulation of every type of constructive process, and materials nonlinearities. The first part of the paper describes the features of the code. Later the paper will present a benchmark of cases to validate FESTE and, finally, one example of the application of tunneling excavation by means of a full face tunneling boring machine (TBM), with a two-dimensional and three-dimensional simulation of the process.

1 INTRODUCTION

Finite element models in geomechanics need to include several important features in order to be appropriate for foundation analysis or tunnel excavation. Initial stress fields, nonlinear behavior for interaction problems, nonlinear material models of soil, capability of modeling construction and excavation, consideration of water effects, and modeling unbounded media are a short list among the requirements for efficient software.

A great number of commercial packages useful for engineering applications in mechanical or structural analysis are useless in geomechanics because they lack the aforementioned properties. In the computer program that is presented, the basic features of a code applicable to geotechnical problems have been incorporated. However, special attention has been paid to the modelisation efficient of contact non-linearity to resolve soil-structure interaction problems, as well as the modelisation problem of unbounded media. The code includes a two-dimensional analysis module for resolving plane strain and plane stress problems, in addition to a three-dimensional analysis module.

As for the material's non-linearity elastoplastic models have been considered in small displacements and strains; one of the recent fields of research involves the consideration of more realistic laws for soil behaviour.

2 SOFTWARE DESCRIPTION

The code that has been developed concentrates on the calculation phase and solution of a problem (Fig. 1). To generate the model and visualize the results, the Geostar graphic processor is used (Cosmos/m, 1995).

The type of analysis chosen is static, employing a displacements formulation of the problem. In Figure 2, the program's general structure is presented.

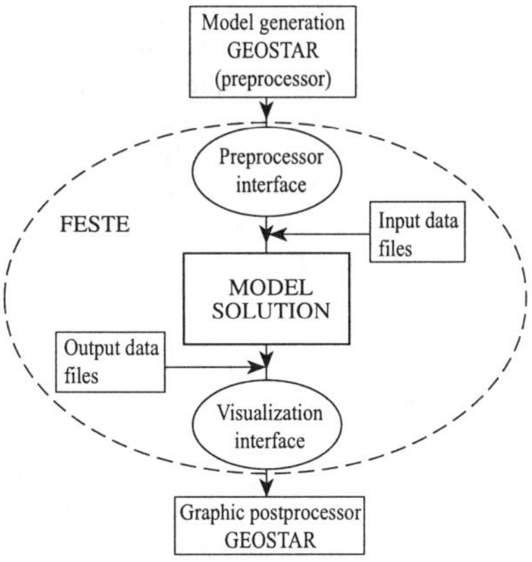

Figure 1. Graphic interfaces between the FESTE code and Geostar.

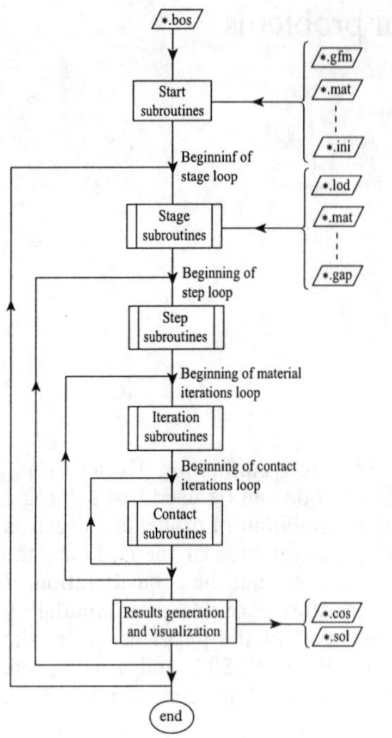

Figure 2. Program layout

In the subroutine block of the set-up phase, the mesh, initial conditions, materials, and the characteristics of each phase are defined.

To permit the representation of construction process, the simulation is divided into a series of phases consecutive in time called "stages". At each stage, there is a definition of the elements that are activated or constructed and disactivated or excavated with respect to the initial reference domain, external loads, possible contact zones, new materials and boundary conditions. Only the contributions due to active elements are assembled in the stiffness matrix (Ghaboussi et al. 1983).

So that the convergence of the analysis is facilited, each stage is divided into steps. In these, the stage loads are applied incrementally according to the time variation curve associated with each load. The solution for each step is obtained in an iterative fashion with the Newton-Raphson method or Newton-Raphson modified (Bathe & Cimiento, 1980), resolving the non-linearity of the material.

Within each material iteration, for the displacements solution obtained, the contact subroutines check if the contact conditions have been met. If these have not been met, the contact subroutines calculate in an iterative way a new solution in movements and the interaction forces generated, as represented in Figure 3, for step t, iteration i, and being $\mathbf{a}$ = displacements vector; $\mathbf{K}_T$ = tangent stiffness matrix; $\mathbf{F}$ = external forces vector; $\mathbf{F}_c$ = contact forces vector; and $\mathbf{R}$ = internal forces vector.

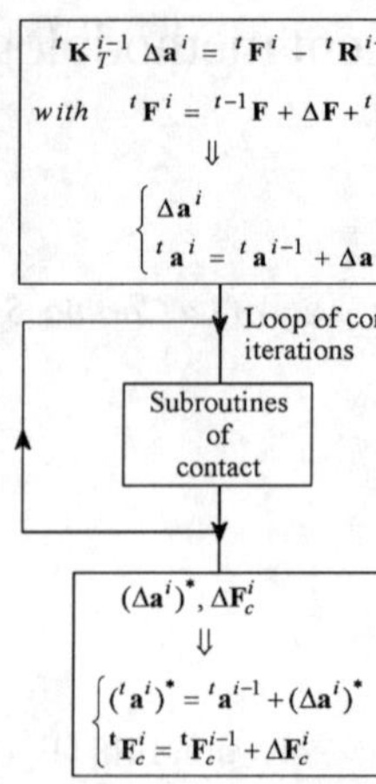

Figure 3. Analysis process with material and contact nonlinearities.

Defining the tangent stiffness matrix (1) and the internal forces vector (2) as:

$$ {}^t\mathbf{K}_T^{i-1} = \int_\Omega \mathbf{B}^T\ {}^t\mathbf{D}_T^{i-1}\ \mathbf{B}\, d\Omega \quad and \quad \mathbf{D}_T = \left(\frac{\partial\sigma}{\partial\varepsilon}\right) \tag{1} $$

$$ {}^t\mathbf{R}^{i-1} = \int_\Omega \mathbf{B}^T\ {}^t\boldsymbol{\sigma}^{i-1}\, d\Omega \tag{2} $$

where $\mathbf{B}$ = deformation matrix; $\mathbf{D}_T$ = tangent constitutive matrix; $\boldsymbol{\sigma}$ = stress vector; and $\boldsymbol{\varepsilon}$ = strain vector.

The finite elements incorporated are compatible and linear: quadrangular and triangular types in 2-D, and hexahedron or tetrahedron in 3-D. In order to improve their performance in incompressible conditions and avoid the effects of shear locking (Cook et al. 1989), selective numerical integration and the incompatible versions of the quadrangular and hexahedron elements have been incorporated (Taylor et al. 1976).

The code incorporates the error estimator in stresses Z^2, developed by Zienkiewicz and Zhu (1987), measuring global and elemental errors.

Defining the materials associated with the ground is done through property groups, normally related to horizontal layers. The upper and lower heigth for each layer is defined, and it is possible to define a linear variation of the elasticity modulus *E*' and the coefficient of earth pressure at rest K_0 with the depth. If the presence of water is to be taken into account, the program allows the user to considerer – through the use of the Naylor method (Naylor et al. 1981)- the limit cases in saturated soil with null drainage at zero time and full drainage at an infinite time.

The loads under consideration are: consistent gravitational loads, concentrated forces, pressure loads that are constant or with linear variations, and initial stress states calculated from K_0.

2.1 *Infinite elements*

With the aim of improving the code's efficiency in modeling the unbounded media commonly found in geotechnical problems, mapped quadratic infinite elements (Zienkiewicz et al. 1983, Marques & Owen, 1984) have been added. In the case of 2-D, defining the geometric transformation as:

$$\mathbf{x} = \sum_{i=1}^{4} M_i(\xi,\eta)\mathbf{x}_i \tag{3}$$

$$with \begin{cases} M_1 = \frac{1}{2}(1-\xi)\frac{-2\eta}{1-\eta} \\ M_2 = \frac{1}{2}(1+\xi)\frac{-2\eta}{1-\eta} \end{cases} \quad \begin{cases} M_3 = \frac{1}{2}(1-\xi)\frac{(1+\eta)}{1-\eta} \\ M_4 = \frac{1}{2}(1+\xi)\frac{(1+\eta)}{1-\eta} \end{cases} \tag{4}$$

while maintaining the approximation of the habitual displacement field.

The real geometry of the element becomes infinite, placing two of its nodes in the infinite in the case of the 2-D element with six nodes (Fig. 4), or with four nodes in the infinite in the case of the infinite hexahedron element with twelve nodes.

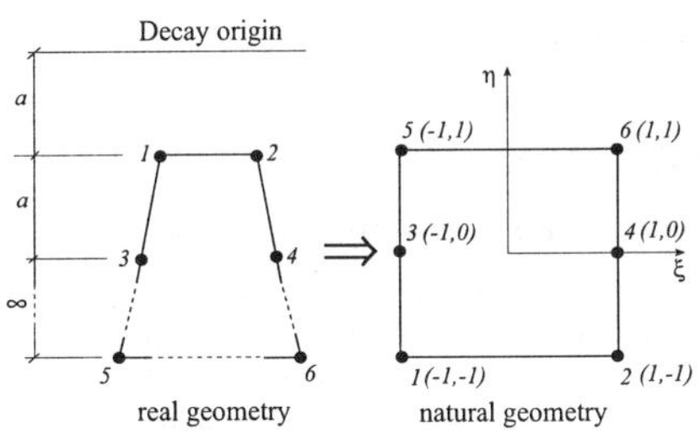

Figure 4. Two-dimensional infinite element

In this kind of element, the position of its pole marks the origin of its decay. The value of the displacement field in the infinite is considered null.

2.2 *Modeling the non-linearity of the material*

The Drucker-Prager plastification criteria (5) are used, considering isotropic elastoplastic materials, and adjusting the constants α and K according to the Mohr-Coulomb criteria (Chen, 1994).

$$f(I_1, J_2) = \alpha I_1 + \sqrt{J_2} - K = 0 \tag{5}$$

being I_1 = first invariant of stress tensor; J_2 = second invariant of deviatoric stress tensor; and α, K the material's constant.

Also included is the possibility that the material will behave differently under compression and extension, defining a reduced value from the elasticity modulus in the latter case and calculating at each step of the analysis the zones under extension through an elastic prediction of the stresses.

2.3 *Modeling contact nonlinearity*

The mechanical problem of contact between solids is resolved by applying the F.E.M through a discrete approach, defining possible contacting or hitting nodes and their target segments or zones of possible contact. In Figure 5, one will find the hitting node and target segment defined for the 2-D and 3-D examples, in the case of the linear finite elements.

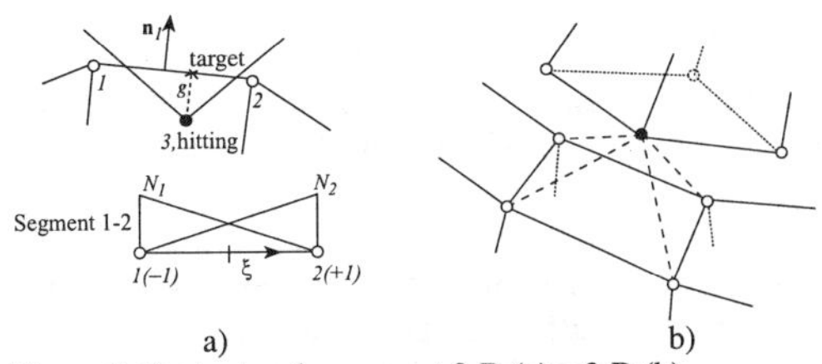

Figure 5. Contact node-segment 2-D (a) y 3-D (b)

In each step of the analysis the contact algorithm calculates the contact zone that develops and the contact forces produced, in an iterative fashion, using trial and error techniques.

It is possible to impose the contact conditions by defining the scalar gap function in each contact segment, such as the signed distance between the contact node and the segment on which it rests, through the normal segment vector and the deformed geometry. To do so, the following restrictions apply: $g(\mathbf{x}) \geq 0$ (null penetration), and $f_{cn}(\mathbf{x}) \geq 0$ (normal contact forces always of compression), or in a combined form $g(\mathbf{x})f_{cn}(\mathbf{x}) = 0$. If the Coulomb friction is also considered $f_{ct}(\mathbf{x}) \leq \mu f_{cn}(\mathbf{x})$, with μ being the friction coefficient, and f_{cn} and f_{ct} the normal and tangential contact forces.

The application of the F.E.M. and a matricial formulation of the previous equations lead to these expressions in step t of the calculation (Zhong, 1993):

$${}^{t}\mathbf{K}_T \Delta\mathbf{a} = {}^{t}\mathbf{F} - {}^{t-\Delta t}\mathbf{R} \quad with \quad \Delta\mathbf{G} = \mathbf{N}\Delta\mathbf{a} + {}^{t-\Delta t}\mathbf{G} = \mathbf{0} \tag{6}$$

bringing together in the second term the set of restrictions due to the segments that have closed. **G** becomes the global clearance vector and **N**, a matrix dependent on the geometry of the contact zone that is produced.

The most common way to solve the previous equations is to impose the restrictions through Lagrange multipliers (Parisch, 1989), penalty methods (Rizzo, 1991) or a modification of these, such as perturbed Lagrangian or augmented Lagrangian (Simo & Laursen, 1992). In the program presented here, an algorithm of contact with Coulomb friction has been developed. It is based on a flexibility method devised for the contact between two nodes by Sachdeva and Ramakrishnan (1981), but has been generalised to include the most general case of contact between nodes and segments.

In each step of the analysis, an initial calculation is made without taking contact restrictions into account. The penetrations vector $\mathbf{a}_c$ is defining by referring to the penetrations produced in the segments. With the contact points obtained in each segment, the flexibility matrix $\mathbf{X}_c$ is calculated by imposing unitary loads (Fig. 6a) and obtaining the relative movements of each contact pair:

$$\begin{cases} X_{cii} = \left(\mathbf{a}_{i3} - \mathbf{a}_{p_i}\right)\mathbf{n}_i \\ X_{cji} = \left(\mathbf{a}_{j3} - \mathbf{a}_{p_j}\right)\mathbf{n}_j \end{cases} \tag{7}$$

Supposing that m are the segments with penetrations, the iterative resolution of the equation:

$$\mathbf{X}_c^m \, \mathbf{f}_c^m = \mathbf{a}_c^m \tag{8}$$

provides the contact forces $\mathbf{f}_c$, in r segments ($r \le m$), that remove the calculated penetrations.

To avoid this calculation in each step of the flexibility matrix, due to the variations of contact points positions, on advance the calculation of a generalized flexibility matrix $\mathbf{Xg}_c$:

$$\mathbf{Xg}_c = \begin{bmatrix} \mathbf{Xg}_{11} & \cdots & \mathbf{Xg}_{1m} \\ \vdots & & \vdots \\ \mathbf{Xg}_{m1} & \cdots & \mathbf{Xg}_{mm} \end{bmatrix} \text{ with } \mathbf{Xg}_{ij} = \begin{bmatrix} Xg_{ij}^{11} & Xg_{ij}^{12} \\ Xg_{ij}^{21} & Xg_{ij}^{22} \end{bmatrix} \tag{9}$$

Then the unitary loads states defined in Figure 7b are used.

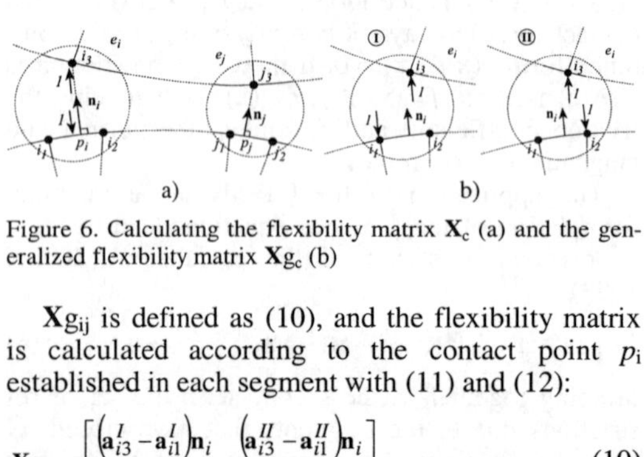

Figure 6. Calculating the flexibility matrix $\mathbf{X}_c$ (a) and the generalized flexibility matrix $\mathbf{Xg}_c$ (b)

$\mathbf{Xg}_{ij}$ is defined as (10), and the flexibility matrix is calculated according to the contact point p_i established in each segment with (11) and (12):

$$\mathbf{Xg}_{ij} = \begin{bmatrix} \left(\mathbf{a}_{i3}^{I} - \mathbf{a}_{i1}^{I}\right)\mathbf{n}_i & \left(\mathbf{a}_{i3}^{II} - \mathbf{a}_{i1}^{II}\right)\mathbf{n}_i \\ \left(\mathbf{a}_{i3}^{I} - \mathbf{a}_{i2}^{I}\right)\mathbf{n}_i & \left(\mathbf{a}_{i3}^{II} - \mathbf{a}_{i2}^{II}\right)\mathbf{n}_i \end{bmatrix} \tag{10}$$

$$X_{cij} = \mathbf{N}_i^T \, \mathbf{Xg}_{ij} \, \mathbf{N}_j \quad i, j = 1, \ldots, m \tag{11}$$

$$\mathbf{N}_i^T = \left\lfloor N_1^i\left(\xi_{p_i}\right), N_2^i\left(\xi_{p_i}\right)\right\rfloor \tag{12}$$

where N_1 y N_2 are the shape functions of the segment, which, in the 2-D example dealt with earlier, are unidimensional and linear.

Figure 7 provides a scheme of the contact algorithm developed.

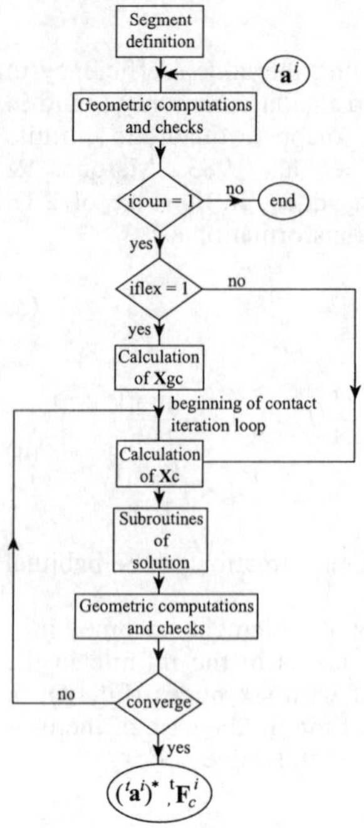

Figure 7. Scheme of the contact algorithm

3 VALIDATION EXAMPLES

3.1 *Strip footing in a semi-infinite domain*

A linear variation of the transversal elasticity module G with depth, and an incompressible domain is supposed (Fig. 8). Figures 9 and 10 present the results obtained with two meshes and compatible as well as incompatible elements, along with the problem's theoretical solution (Poulos & Davis, 1991) to the variation of vertical displacement.

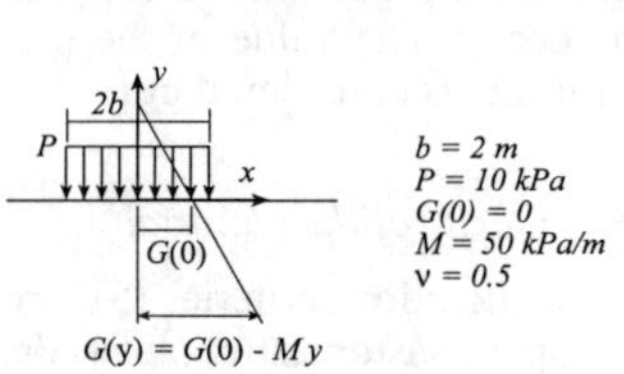

Figure 8. Problem statement

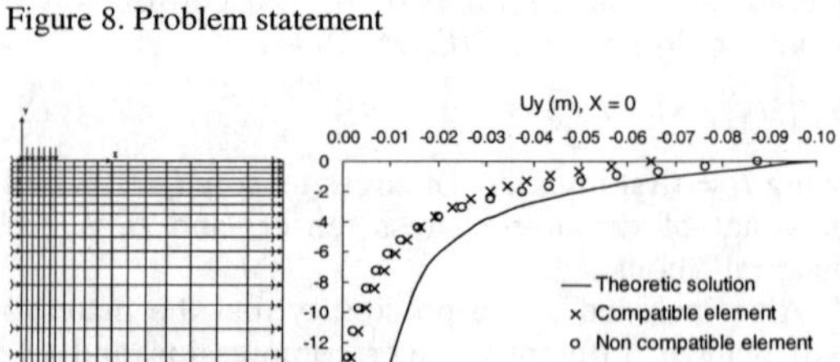

Figure 9. Vertical displacement solution in x = 0 with mesh 1: 722 degrees of freedom, without infinite elements

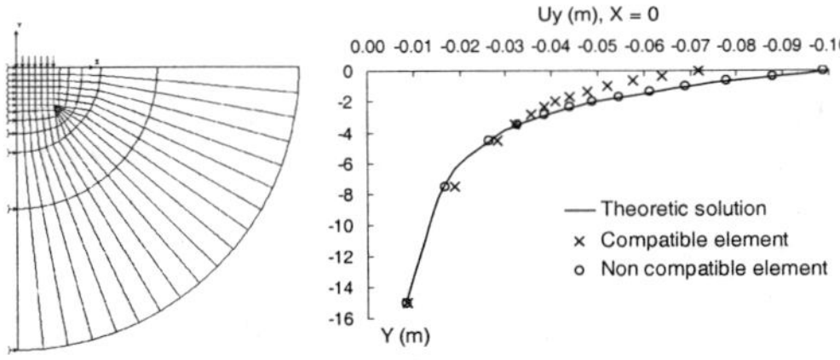

Figure 10. Vertical displacement solution in x = 0 with mesh 2: 326 degrees of freedom, with infinite elements

An observable trend is how the infinite elements improve the solution obtained; the same can be said of incompatible elements.

3.2 *Footing collapse analysis*

This involves a strip load on an elastoplastic soil, with the properties and dimensions indicated in Figure 11a. The solution for the critical load, 1090 kPa (Chen, 1994), coincides with the one obtained in the analysis (Fig. 11b).

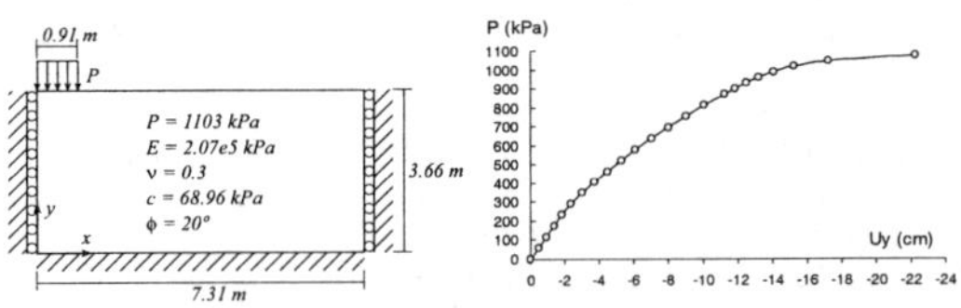

a) b)

Figure 11. Problem stament and numerical solution

3.3 *Spherical excavation in an infinite domain*

Figure 12 presents the F.E. model, symmetrical along the three coordinate plains, and with the initial stress state. The stress results are compared alongside the theoretical solutions (Poulos & Davis, 1991) in Figure 13.

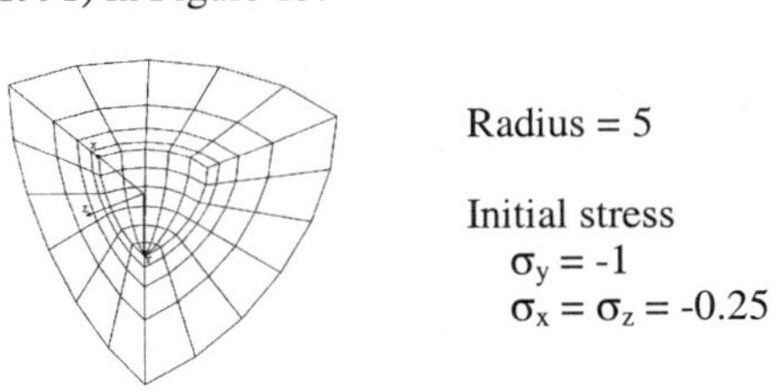

Figure 12. Three-dimensional mesh

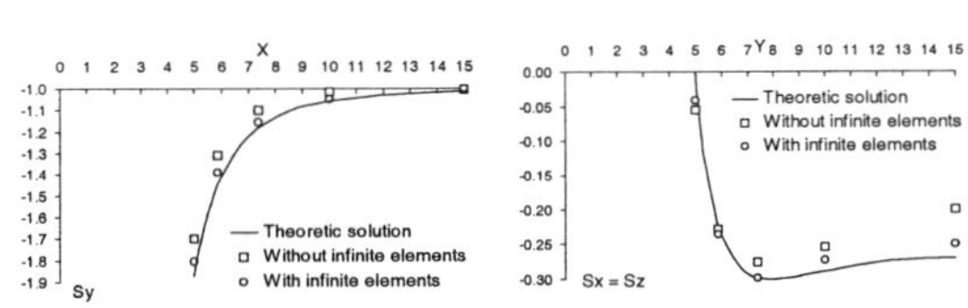

a) b)

Figure 13. Stress solution for σ_y (a) and $\sigma_x = \sigma_z$ (b), with and without infinite elements

3.4 *Contact between a cylinder and an infinite plane*

This problem, dealing with a frictionless hertzian contact, is solved with the mesh represented in Figure 15a. Infinite elements are used to model the plain. In figure 15b the numerical solution obtained for the contact stresses is contrasted with the theoretical one (Pilkey, 1994). A close resemblance between the two sets of results can be observed.

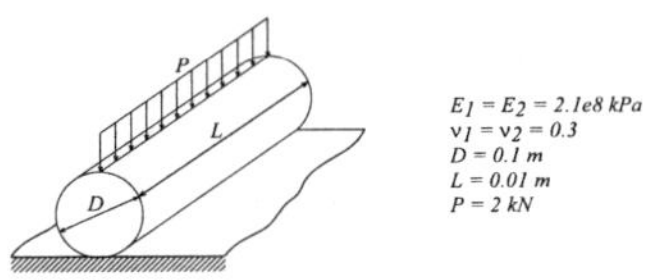

Figure 14. Problem statement

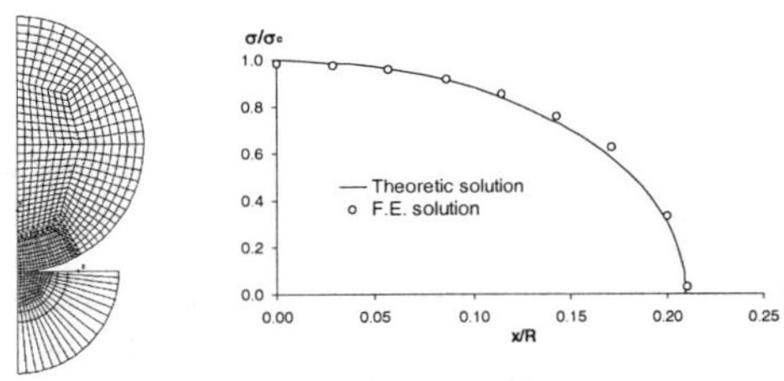

a) b)

Figure 15. Finite element mesh a), and contact stress results b)

4 APPLICATION EXAMPLES

Serving as an application example, some results obtained for the modelisation of the construction process of a tunnel by means of a full-face TBM with earth pressure balance system are presented. These models were applied to instrumented tunnel sections carried out during the enlargement of the underground system of the Madrid metro (1995-1999).

A 2-D model appears in Figure 16a, the soil is modeled with finite and infinite elements, and three construction stages corresponding with: full section excavation and setting up the shield tunneling, subsequently installing the concrete segments with mortar injection and removing the shield, and the final stage, allowing the mortar to solidified. Due to clearance, produced by the excavation process between the soil and the shield tunneling or lining, the problem of contact is considered in the first two stages.

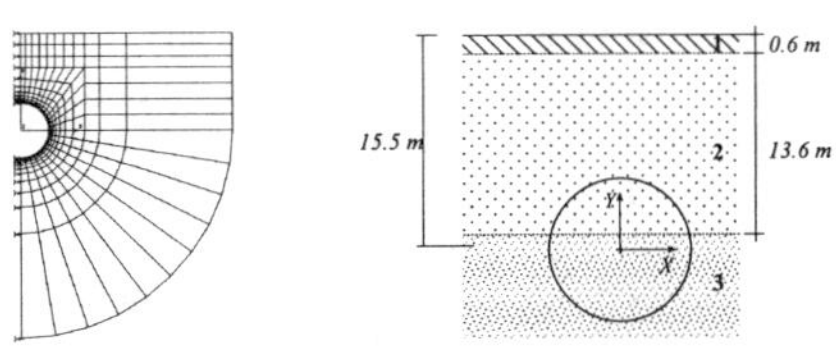

a) b)

Figure 16. F.E. model a), and section layers b)

For the section represented in Figure 16b, using the soil properties listed in Table 1, the results obtained with the model appear in Figure 17.

Table 1. Soil properties

Layer	E	ΔE	ν	K_0	c	φ	γ
	kPa	kPa/m			kPa	°	t/m^3
1	$1 \cdot 10^4$	0	0.3	0.4-0.4	10	28	1.8
2	$1.35 \cdot 10^5$	0	0.3	0.4-0.6	100	33.5	2.0
3	$1.75 \cdot 10^5$	3000	0.3	0.6-0.4	150	32.5	2.1

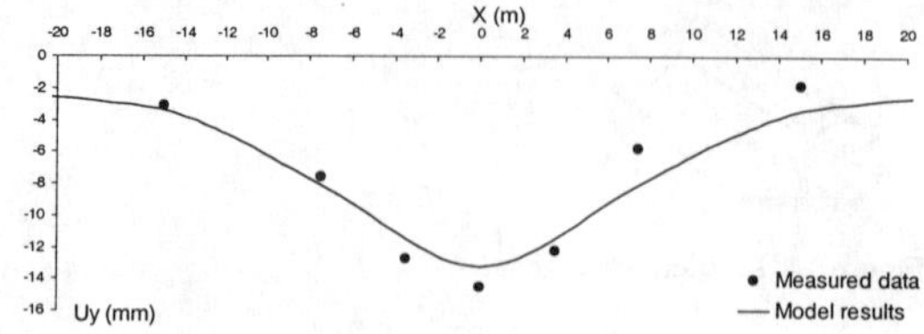

Figure 17. Vertical settlements in surface

The 3-D model that has been developed appears in Figure 18a. In it, infinite elements are incorporated. Moreover, geometrical and contact nonlinearities are considered so that twelve excavations advances have been carried out in direction *z*, with a varying advance length. In Figure 18b, a comparison has been made between the final results of the vertical settlements in the rear section *xy* obtained with the 2-D model and the 3-D model.

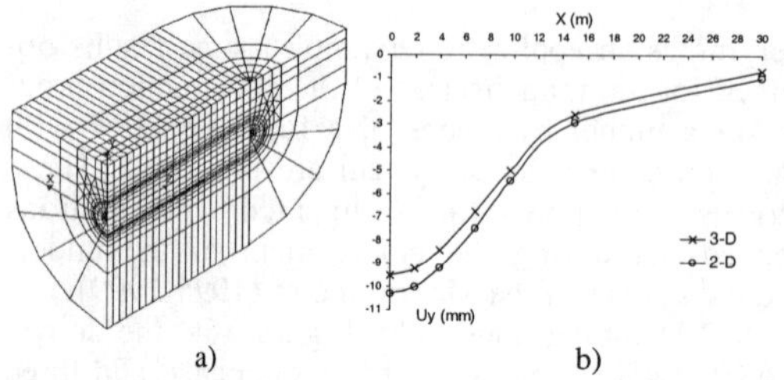

Figure 18. Tridimensional model a), and vertical settlements b)

5 CONCLUSIONS

The program that has been presented allows the user to combine geometric and material nonlinearities with a realistic modelisation of the contact problem based on a flexibility method.

The contact algorithm developed offers an advantage alongside the more commonly used methods of Lagrange multipliers or the penalty method, because the contact iterations are made with a reduced system of equations in which only the nodes associated with the contact zone are involved. Furthermore, this method maintains a precision on par with that of Lagrange multipliers.

The incorporation of infinite elements allows for the efficient modelisation of the geotechnical problems, noticeably reducing the dimensions and degrees of liberty of the meshes that are employed, as well as increasing the precission of the results.

The consideration of gaps between soil and tunnel shield and tunnel structure, in the modelisation of the excavation process with TBM, allows a suitable representation of this kind of process on a two- as well as three-dimensional level.

6 REFERENCES

Bathe, K.J. & Cimiento, A.P. 1980. Some practical procedures for the solution of nonlinear finite element equations. *Computer Methods in Applied Mechanics and Engineering* 85: 22-59.

Chen, W.F. 1994. *Constitutive Equations for Engineering Materials. Vol. 2: Plasticity and Modeling.* Elsevier.

Cook, R.D., Malkus, D.S. & Plesha, M.E. 1989. *Concepts and Application of Finite Element Analysis,* 3ª ed., John Wiley and Sons, New York.

Cosmos/m, v.1.75 A, 1995. Finite Element Analysis System. Structural Research & Analysis Corporation. (4 vol.).

Ghaboussi, J., Hansmire, W.H., Parker, H.W. & Kim, K.J. 1983. Finite element simulation of tunneling over subways. *Journal of Geotechnical Engineering* 109 (3): 318-327.

Marques, J.M.M.C. & Owen, D.R.J. 1984. Infinite Elements in Quasi-Static Materially Nonlinear Problems, Computers and Structures 18(4): 739-751.

Naylor, D.J., Pande, G.N., Simpson, B. & Tabb, R. 1981. *Finite Elements in Geotechnical Engineering.* Swansea, U.K: Pineridge Press.

Parisch, H. 1989. A Consistent Tangent Stiffness Matrix for Three-Dimensional Non-linear Contact Analysis. *International Journal for Numerical Methods in Engineering* 28: 1803-1812.

Pilkey, W.D. 1994. *Formulas for stress-strain and structural matrices,* Chapter 9, Contact stress,: 403-438. Jhon Wiley & Sons, New York.

Poulos, H.G. & Davis, E.H. 1991. *Elastic solutions for soil and rock mechanics.* Centre for Geotechnical Research, University of Sidney: 193-194.

Rizzo, A.R. 1991. FEA Gap Elements: Choosing the Right Stiffness. *Mechanical Engineering*, June: 57-59.

Sachdeva, T.D. & Ramakrishnan, C.V. 1981. A Finite Element Solution for the Two-Dimensional Elastic Contact Problems with Friction. *International Journal for Numerical Methods in Engineering* 17: 1257-1271.

Simo, J.C. & Laursen, T.A. 1992. An Augmented Lagrangian Treatment of Contact Problems Involving Friction. *Computers & Structures* 42(1): 97-116.

Taylor, R.L., Beresford, P.J. & Wilson E.L. 1976. A non-Conforming Element for Stress Analysis, *International Journal for Numerical Methods in Engineering* 10: 1211-1219.

Zhong, Z.H. 1993. *Finite Element Procedures for Contact-Impact Problems.* New York: Oxford University Press.

Zienkiewicz, O.C. & Zhu, J.Z. 1987. A Simple Error Estimator and Adaptive Procedure for Practical Engineering Analysis. *International Journal for Numerical Methods in Engineering* 24: 337-357.

Zienkiewicz, O.C., Emson, C. & Bettess, P. 1983. A Novel Boundary Infinite Element. *International Journal for Numerical Methods in Engineering* 19: 393-404.

Computer Methods and Advances in Geomechanics, Desai et al. (eds) © 2001 Balkema, Rotterdam, ISBN 90 5809 183 X

Numerical modeling in forensic engineering

W. H. Roth, E. M. Dawson & S. Nesarajah
URS Corporation, Los Angeles, Calif., USA

ABSTRACT: The role of numerical modeling in geotechnical failure analysis is discussed in general and, specifically, in the form of three case histories. It is argued that conventional analysis methods, by ignoring soil-structure interaction, tend to err on the conservative side. Hence, they are ill suited for failure analysis where the objective is to find the most realistic behavior with little room for error on either side. Presented case histories include (1) failure of an EPB-TBM mucking system, (2) failure of a temporary tunnel liner, and (3) failure of excavation-support struts. It is concluded that modeling-based visual presentations can be very effective in clarifying complex failure mechanisms to non-engineers. This is particularly important in court settings where such presentations are more likely to be accepted as evidence than judgement-based artist renderings which tend to be dismissed as speculative manipulations.

INTRODUCTION

Numerical modeling has long been a vital tool for failure analysis in aeronautics and related mechanical-engineering fields. However, its potential for investigating the probable cause of soil-related failures, has yet to be fully recognized. Many such investigations are performed in the context of design or construction disputes, the resolution of which tends to be entrusted to experienced geotechnical practitioners used to relying on engineering judgement, time-tested empirical rules, and conventional simplified methods of analysis. After all, in geotechnical engineering, perhaps more so than in any other engineering field, experience-based professional judgement is widely acknowledged to be more important than "fancy modeling techniques." Certainly, the latter should never be relied on without the benefit of the former, and Terzaghi's words of wisdom are as valid today as they were over half a century ago: *"in soil mechanics the accuracy of computed results never exceeds that of a crude estimate"* (Goodman 1999).

Notwithstanding the above, when it comes to failure investigations, empirical rules and conventional methods of geotechnical analysis face severe limitations. Perhaps the most serious shortcoming is the inability to take into account soil-structure interaction effects. For design purposes, most engineers still ignore this important aspect by separately analyzing soil and structure. The justification frequently given for this approach is that it errs on the conservative side. While this may be so for design, failure analysis has little room for error on either side. Its objective is to find out what actually happened, as opposed to merely ensuring that a structure is safe (i.e. far removed from the state of failure). Any analysis which involves the de-coupling of a structure from the soil it is embedded in, or founded on, is likely to produce questionable conclusions in the context of a failure investigation.

The three case histories discussed below were all analyzed with the explicit finite difference code FLAC (Itasca 1998). Soils were modeled using relatively simple constitutive models with Mohr-Coulomb yield criteria enhanced with volume-yield caps where appropriate; and structures were represented by linear-elastic beam and cable elements formulated to cutoff bending moments and axial forces, respectively, at user-specified yield limits. FLAC and its predecessor codes have been extensively verified and utilized in the last 20 years for analyzing a wide spectrum of static and dynamic problems in geotechnical engineering (Bureau et al. 1996 & 1985; Inel et al. 1993; Roth 2000; and Roth et al. 1997, 1996, & 1993).

CASE 1: FAILURE OF EPB-TBM MUCKING SYSTEM

Difficult subsurface conditions of mixed-face tunneling through siltstone and coarse alluvium below the groundwater table, prompted the utilization of an Earth Pressure Balance TBM for the construction of the North Outfall Replacement Sewer (NORS) tunnel in Los Angeles. With an overburden of about 25 m, the muck pressure required to maintain a stable tunnel face was estimated to be about 250 kPa. Because cobbles encountered were too large to pass through a Screw Conveyer (Figure 1, right), the contractor chose a Pressure Relief Gate to remove the muck from the face chamber (Figure 1, left). In contrast to a Screw Conveyor where muck is *extracted* from the chamber, a Pressure Relieve Gate requires the muck to be *extruded* through a trap-door opening.

When excessive ground losses at the tunnel face caused significant surface settlements, a changed-site-conditions claim was submitted. Suspecting that fundamental shortcomings of the Pressure Relieve Gate, rather than unexpected ground conditions, were responsible for the experienced difficulties, the owner rejected the contractor's claim. As part of the claim defense, numerical modeling was performed simulating muck removal by Screw Conveyor versus Pressure Relief Gate (Roth et al. 1996). The muck was modeled as elastic-plastic Mohr Coulomb material, and TBM advance was simulated with a velocity boundary pushing into the muck chamber. Muck extraction through the Screw Conveyor was accomplished with a velocity boundary (Figure 2), while a pressure boundary was applied at the Pressure Relieve Gate opening (Figure 3) simulating muck extrusion against the hydraulic-jack-loaded trap door.

The analysis results plotted in Figures 2 and 3 show that with the Screw Conveyor it was possible to maintain the face pressure of 250 kPa without having to apply excess jacking forces. In contrast, with the Pressure Relief Gate the muck began "locking up" in the chamber - even when the trap door was left open (i.e. zero boundary pressure at the trap-door opening). The computed force necessary to advance the TBM quickly exceeded the capacity of the hydraulic jacks. This finding collaborated inspectors' observations of muck being removed by simply letting it drop through the open trap door. Therefore, it was concluded that the ground losses experienced during tunneling were due to using a faulty mucking system which was incapable of maintaining the required pressure in the muck chamber.

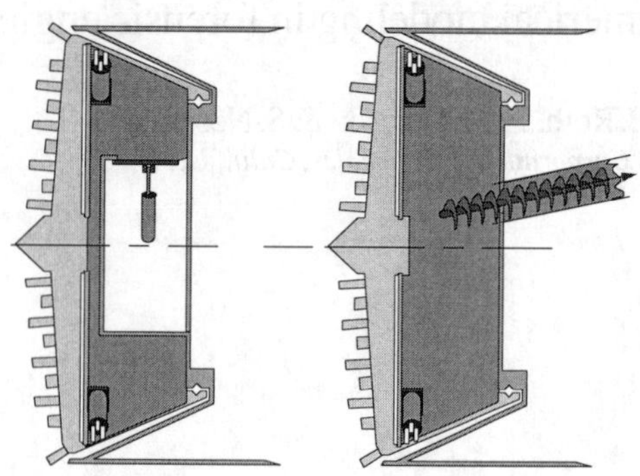

Figure 1. Earth Pressure Balance (EPB) TBM with Pressure Relieve Gate (left), and Screw Conveyor.

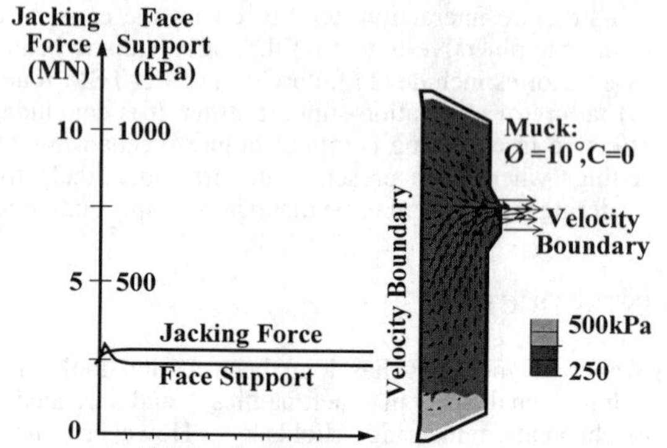

Figure 2. Simulation of muck removal from EPB face chamber, by Screw Conveyor.

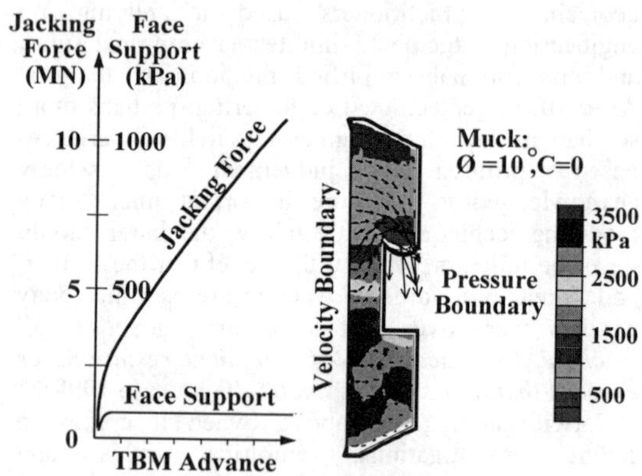

Figure 3. Simulation of muck removal from EPB face chamber, by Pressure Relieve Gate.

CASE 2: INVESTIGATION OF TUNNEL-LINER FAILURE

A day after the leading digger shield excavating LA Metro Redline's twin tunnels along Hollywood Boulevard had passed Hudson Avenue, water was observed seeping through the expanded pre-cast tunnel liner. Ground settlements, which initially had reached about 5 cm at this location, stabilized after reaching 10 cm over the next 3 weeks. Because there was no immediate danger to structures, the decision was then made to allow the trailing shield to pass. However, when the second shield reached this location, the temporary support of the first tunnel nearly collapsed as the hardwood wedges in the 10-o'clock-position expansion gap were crushed in several consecutive rings and crown segments dropped up to 18 cm. They nearly cleared the edges of the underlying springline segments narrowly avoiding a total collapse of the tunnel crown. As street settlements reached up to 28 cm, Hollywood Boulevard was closed, and the crown segments were emergency supported by vertical steel posts (Figure 4).

Figure 4. Emergency support after crushing of hardwood wedges in the liner expansion gap at Hollywood & Hudson.

Because the same hardwood wedges had been used previously for many miles of tunneling without incident, the question was "why did they fail at this particular location?" To solve this puzzle, a numerical simulation of the construction sequence (Figure 5) was performed with the model mesh shown in Figure 6. Extensive monitoring data from settlement markers and extensometers allowed fine-tuning of the model at every stage, up to the point of failure. The sandy, silty alluvium was modeled as elastic-plastic Mohr Coulomb material with a volume-yield cap. The latter was calibrated to contract upon wetting, based on consolidation test data indicating a slight tendency for hydro-compaction. The liner segments were modeled with elastic beams hinged at 8- and 4-o'clock positions, and non-linear springs were used to simulate the hardwood wedges in the 10- and 2-o'clock expansion joints. The force-deformation characteristics of these springs where derived from the results of laboratory compression tests on the hardwood wedges stored on site for future use.

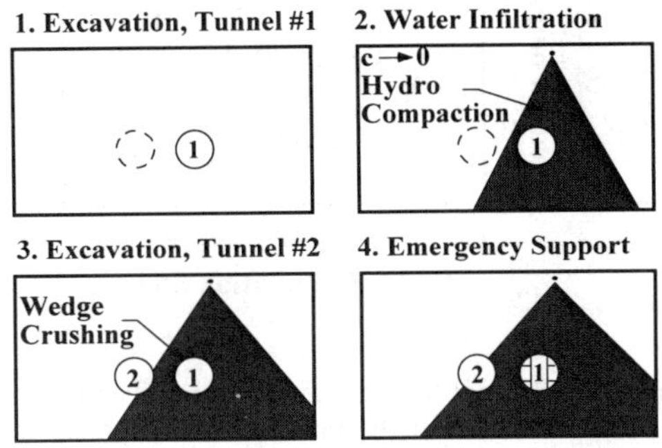

Figure 5. Construction sequence modeled with FLAC.

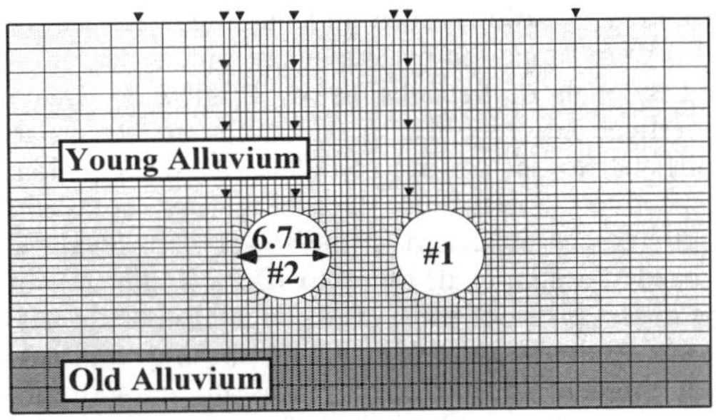

Figure 6. Model mesh with locations of settlement plates and multi-anchor extensometers (shown as triangles).

The analysis results confirmed that under "dry" conditions, the tunnel support remained stable for all construction stages. However, when the soil was wetted after excavating the first tunnel, the hardwood wedges on the pillar side of the first tunnel collapsed upon excavating the second tunnel (Figure 7). Additional "what-if" analyses indicated that the tunnel liner would have withstood the hydro-compaction induced loading, had it not been for the fact that the actual compressive strength of the hardwood wedges turned out to be less than half of the specified strength.

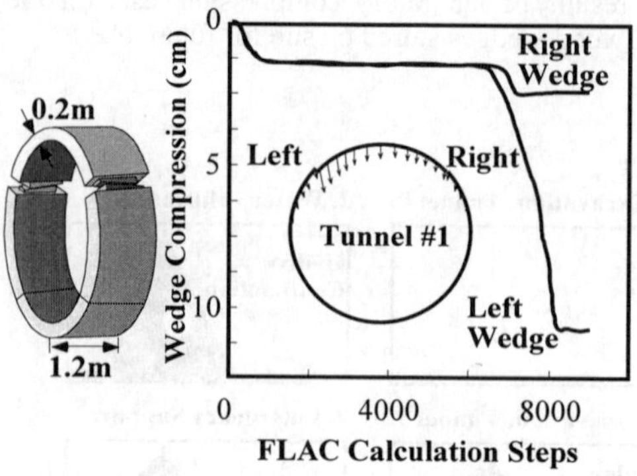

Figure 7. Computed vertical compression of hardwood wedges in tunnel #1 in response to excavating tunnel #2.

CASE 3: FAILURE OF EXCAVATION SUPPORT STRUTS

Temporary excavation support for LA Metro Redline's Civic Center station in siltstone was provided by soldier piles and lagging. The soldier piles were supported by 3 levels of pipe struts on top and 3 levels of tiebacks below (Figure 8). Design specifications required pre-loading of struts and tiebacks to 50% and 100% of design load, respectively. Even before reaching final excavation depth, forces measured in the lower strut level "C" exceeded design loads by up to 100% (Figure 9), and the gusset plates of some struts began to buckle. As a result of the strut problems, the shoring contractor was directed to immediately install an additional strut level between "B" and "C."

Arguing that strut overloading most likely had been caused by the contractor's failure to properly pre-tension the tiebacks, the owner withheld payment for the installation of the additional struts. This position was based on incidental reports from the field about some tiebacks having been locked off 10 to 20 percent short of the specified design load. Bearing the nearly $1M cost for this emergency measure was a considerable financial burden for the contractor, who subsequently submitted a claim for changed site conditions. As it happened, the results of the failure investigation discussed below turned out to support the contractor's claim of changed site conditions.

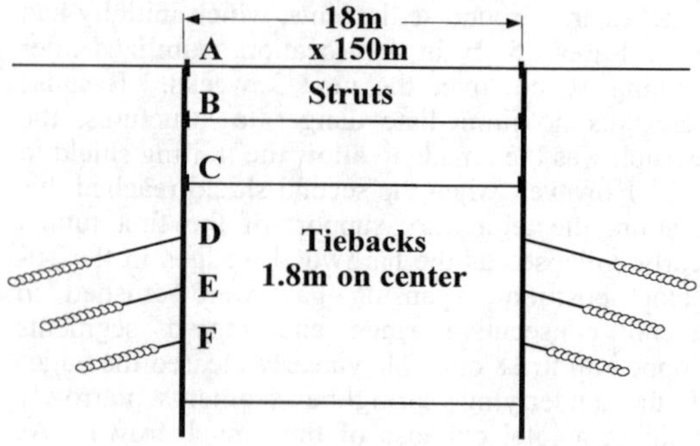

Figure 8. Cross section of Civic Center station excavation.

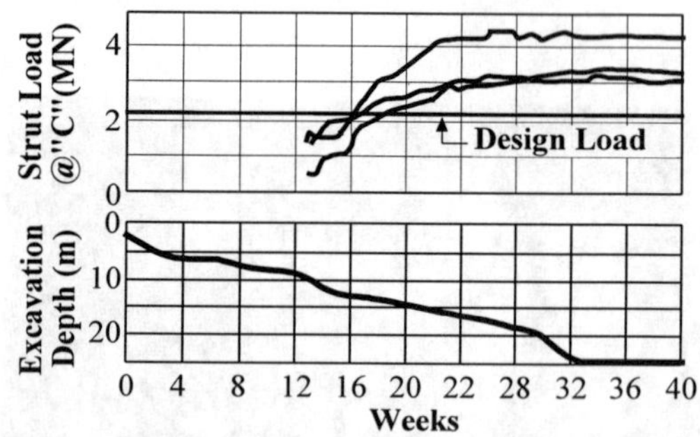

Figure 9. Measured strut loads at support level "C."

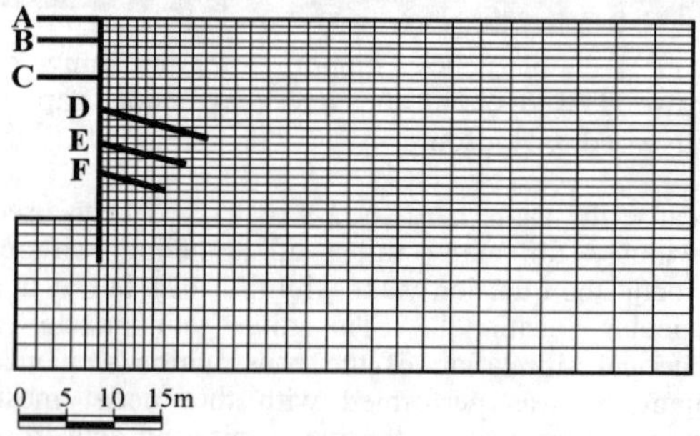

Figure 10. Model mesh for Civic Center station excavation.

The failure analysis was performed by numerical modeling utilizing the mesh shown in Figure 10 (Roth et al. 1993). The excavation sequence was simulated in stages with intermittent installation and pre-loading of struts and tiebacks. Geologic information suggested that the site was located within a compressional tectonic regime. This was supported by the fact that the field-measured 100% overloading of the "C" struts could only be duplicated in the model when horizontal and vertical pre-excavation stresses were assumed to be equal (i.e. Ko=1.0).

To clarify whether insufficient tensioning of the tiebacks could have possibly caused the strut failure, comparative analyses were performed with 100% and 70% pre-loaded tiebacks. As shown in Figure 11, both of these cases resulted in the same amount of strut overloading (Figure 11); they also produced computed wall deflections which agreed quite well with measured data. Based on these analysis results, insufficient tieback tensioning was eliminated as a possible cause for the strut failure.

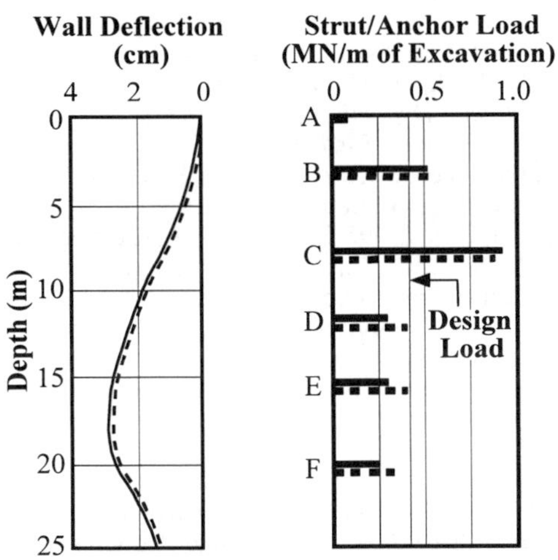

Figure 11. Analysis results of as-built, mixed-support case, with tiebacks tensioned to 70% (solid lines) and 100% of design load.

Next, a comparison was made between supporting the excavation by struts only versus tiebacks only, whereby the struts were installed without pre-loading. Analysis results are presented in Figure 12, where the solid and dashed lines represent the tiebacks-only and struts-only case, respectively. Even though the computed strut forces did exceed design values in the lower portion of the excavation, the analysis results suggested that neither support system would have failed. Hence, it was concluded that mixing the rigid strut support with the inherently more flexible tieback system most likely was to blame for the experienced strut failure.

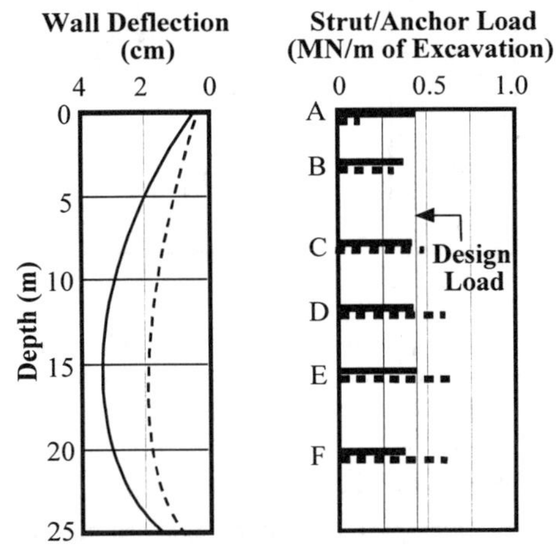

Figure 12. Comparison of excavation supported by tiebacks only (solid lines) vs. struts only.

A third case was analyzed in order to investigate if the problems with the as-built, mixed support system could have been corrected more economically than by installing the costly additional struts mentioned above. This question was of interest, because one alternate remedy suggested at the time, was to simply relieve the strut loads by allowing the excavation walls to give. Because of the great strength and stiffness of the siltstone, it was estimated that relieving the strut loads to acceptable levels would result in no more than a few millimeters of additional wall deflection. This argument was further supported by the fact that no increase in wall deflection was detected after some of the "C" struts' gusset plates had buckled.

Relieving the strut load in the numerical model was simulated with a special support element programmed to yield upon reaching the design load. The analysis results, shown as solid lines in Figure 13, are compared with the as-built baseline case. Relieving the strut load resulted in only 5 mm of additional deflection in the upper portion of the wall, while the maximum wall deflection remained unchanged altogether.

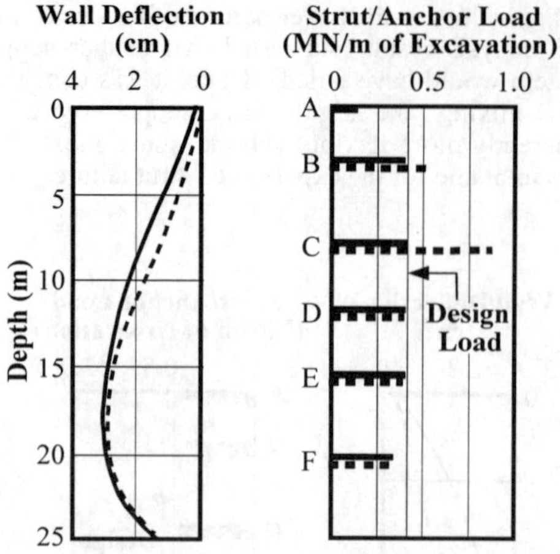

Figure 13. Comparison of mixed excavation support where struts are allowed to "give" (solid lines) vs. as-built "rigid" case.

DISPUTE RESOLUTION

The three examples of failure analysis discussed above succeeded in producing convincing results and conclusions about the most probable failure mechanism and cause. However, such technical issues are only part of the story in a legal dispute about failure liability. The analysis of the EPB mucking system (Case 1), for example, managed to convince the judge that malfunctioning of the mucking system, rather than changed site conditions, were to blame for the tunneling problems experienced by the contractor. The presented analysis results were so effective, in fact, that the judge put part of the blame on the owner for allowing the contractor to use a TBM with "such obvious shortcomings."

The findings pertaining to the crushed hardwood wedges of the tunnel liner (Case 2) were recognized by most parties involved as the most plausible explanation for the liner failure. However, since the results of this investigation pointed a finger at just about all the parties involved, the case was settled without contesting the analysis findings.

Finally, the outcome of the strut-failure analysis (Case 3) turned out to support the contractor's claim that earth pressures provided in the contract specifications were not applicable for the given site conditions. They not only failed to consider the compressional tectonic environment, but also overlooked soil-structure interaction effects leading to force concentrations in the mixed support system of this particular excavation.

CLOSURE

Numerical models, combined with today's advanced graphics software, are particularly well suited for producing highly effective presentations understood by engineers and non-engineers alike. Numerical-analysis-based cartoons or animations can be very persuasive and are likely to be accepted as evidence in court. In contrast, artist renderings of hypothetical failure scenarios based on engineering judgement or empirical rules are likely to be dismissed as speculative manipulations.

REFERENCES

Bureau, G., Inel, S., Davis, C. A., & Roth, W.H. 1996. *Seismic Response of Los Angeles Dam, California, During the 1994 Northridge Earthquake.* USCOLD 1996 Annual Meeting and Lecture, Los Angeles: California.

Bureau, G., Volpe, R.L., Roth, W.H., & Udaka, T. 1985. *Seismic Analysis of Concrete Face Rockfill Dams.* ASCE International Symposium on Concrete Face Rockfill Dams, Detroit: Michigan.

Inel, S., Roth, W.H., & de Rubertis, C. 1993. *Nonlinear Dynamic Effective-Stress Analysis of Two Case Histories*. 3rd International Conference on Case Histories in Geotechnical Engineering, St. Louis: Missouri.

Goodman, R.E. 1999. Karl Terzaghi: The Engineer as Artist, ASCE Press, Reston: Virginia.

Itasca Consulting Group, Inc. 1998. FLAC Version 3.4: Fast Lagrangian Analysis of Continua-User's Manual

Roth, W.H. 2000. *Numerical Modeling Pays Off for LA's Alameda Corridor.* AUA News, Vol. 15, No. 12.

Roth, W.H., Inel, S., Radwanski, R., & Stirbys, T. 1997. *Analysis of Tunnel Response to Adjacent Shaft Excavation.* World Tunnel Congress '97, Vienna: Austria.

Roth, W.H., Babendererde, S., & Inel, S. 1996. *An Evaluation of EPB Muck Removal Systems.* North American Tunneling '96 Conference, Washington: D.C.

Roth, W.H., Stirbys, A., de Rubertis, C., & Ellis, R. 1993. *Performance of a Braced Excavation in Silt Stone*. 3rd International Conference on Case Histories in Geotechnical Engineering, St. Louis: Missouri.

Computer Methods and Advances in Geomechanics, Desai et al. (eds) © 2001 Balkema, Rotterdam, ISBN 90 5809 183 X

Lateral bearing capacity design of deep foundation on slope by FEM

S. Sakajo & T. Nishioka
Kiso-jiban Consultants Company Limited, Tokyo, Japan

ABSTRACT: This paper describes a case study result of the lateral bearing capacity design of deep foundation. Appearance of peak strength always can not be expected on the soft ground and the design codes could not predict well the actual value. Finite element analysis is now a good tool to re-produce the stress and strain relation of soil realistically. From the comparisons between the observation and numerical analysis, it is conclude that only finite element method could be expected to predict the bearing capacity of deep piles very well.

1 INTRODUCTION

This research is on lateral bearing capacity of deep foundation on slope, which is now a hot geo-technical issue on the electric-power cable tower design. Because its cost reduction is key as same with the other construction designs. A proper enough geo-technical information may be useful to establish the best design. Therefore, Tokyo Electric Power Co., Ltd. carried out lateral load tests of deep foundation on slope of Nagaoka site Niigata Japan, with enough soil investigations (Tokyo Electric Power Co., Ltd., 1982).

However, many conventional designs in the world can be used on little information of soils, where many design methods on the limit equilibrium method have been theoretically proposed. Even in the field test for establishing these design codes, few soil investigations have been conducted. In such designs, slope angle of the slip line is key, which can not be fitted with observations actually. Furthermore on the lateral-bearing capacity on slope, the larger displacement may be yielded than on the horizontal ground.

Therefore, the maximum bearing capacity is difficult to be obtained from the in-situ load test. Thus, the only advanced numerical analysis like FEM on enough soil investigation could design the lateral bearing capacity of deep foundation on sloped ground. The cost reduction of construction of deep foundation on slope can achieved by the numerical analysis.

2 SITE CONDITIONS AT NAGAOKA AND TOHOKU

Figure1(a) shows the model of Nagaoka site. The length of the deep foundation is 10 m and the diameter is 3.0 m, where the ground consists of A, B and C-layers. A-layer and C-layer are sand. B-layer is silt. On the other hand, Figure1(b) shows the model of Tohoku site. The length of the deep foundation is 23.5 m and the diameter is 2.5 m, where ground is volcanic clay.

Site conditions are summarized and material numbers are summarized in Table 1 and Table 2. The deep foundation at Tohoku site is roughly twice longer than that at Nagaoka site. The slopes at both sites are same with 30 degrees. It can be understood that the ground at Tohoku site is much weaker than that at Nagaoka site. The stress and strain relations of soil element were obtained from a series of the tri-axial compression tests at both sites. The soil parameters were obtained from the several test results. Only at Ngaoka site, the Young's modulus was specially confirmed by plate loading tests. These values are in the elastic region of a cyclic loading and they seem very reliable to represent the in-situ. Before using the soil parameters in the simulation, they were assessed by comparing the soil elemental tests. It is interesting to investigate the difference of lateral bearing capacity between two different sites.

Table 1. Site conditions

Site Name	Slope	Soil material	Deep Foundation	Design
Nagaoka	30°	Sand and silt	Diameter:3m Embedment depth:10m	Field test on Express Highway Research Foundation and Tokyo Electric Power Co., Ltd.
Tohoku	30°	Volcanic clay	Diametre:2.5m Embedment depth:23.5m	

Table 2. Material numbers

Site Name		Frictional Angle	Dilatancy Angle	Young's Modulus	Poisson's Ratio	Cohesion	Unit weight
		Degrees	Degrees	kN/m²		kN/m²	kN/m³
Nagaoka	A-Layer	34	5.0	36,260	0.4	39.2	17.64
	B-Layer	37	5.0	24,500	0.4	49.0	18.62
	C-Layer	34	5.0	36,260	0.4	17.64	17.64
	Foundation	-	-	2.45×10^7	0.167	-	-
Tohoku	Volcanic-clay	23.5	0.0	19,110	0.4	27.44	15.03
	Foundation	-	-	1.95×10^7	0.167	-	-

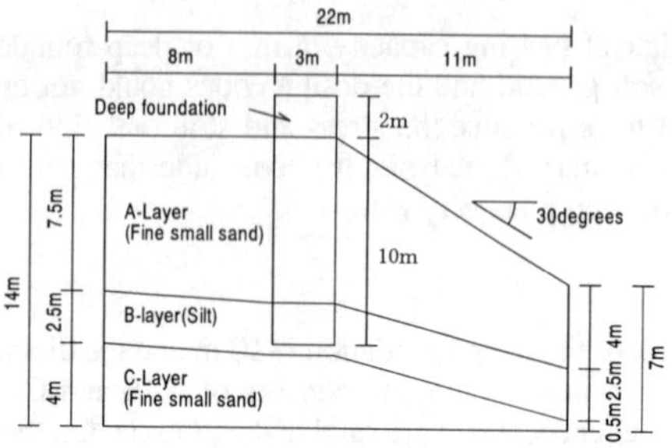

(a) Nagaoka site

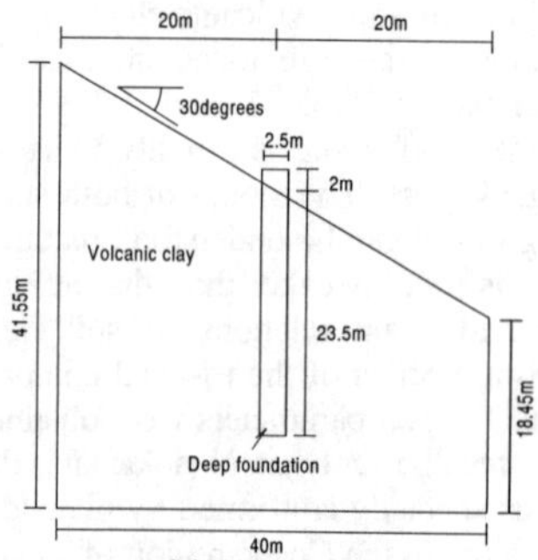

(b)Tohoku site

Figure 1. Model sections

3 DESIGHN METHOD

3.1 *Current Method and comparing method*

Figure2 shows the schematic figure of lateral loading model in the current design by Express Highway Research Foundation Method and the comparing method by Tokyo Electric Power Co., Ltd. As shown in this figure, the lateral loading resistance can be calculated by the basis of passive earth pressure zone. Therefore, the slip line angle α is very important in these designs. Figure3 shows the slip surface of each method on the horizontal plane. Figure3(a) shows the design of the Express Highway Research Foundation and Figure3(b) shows the design of Tokyo Electric Power Co., Ltd. From these figures, it can be found that there is a large difference between the two designs on the concept to determine the angle α. Eventually, the spreading of the slip surface on the horizon plane β becomes very different between them.

Table 3. Comparisons of calculation methods

	Current method Express Highway Research Foundation	Comparing method Tokyo Electric Power Co., Ltd.
Limit value	$Rq = a_0 \cdot \frac{W^{*2}(\cos\alpha + \sin\alpha \cdot \tan\phi) + C \cdot A}{\sin\alpha - \cos\alpha \cdot \tan\phi}$	
	a_0=1.0	a_0=0.6
Shape of slip zone	α : an angle that the lateral bearing capacity is minimum value β : $\phi/3$	α : 45° $+\phi/2+\theta$ β : 45°
Limit bearing capacity after being plastic*1	$C=0, \phi=\frac{2}{3}\phi$ (reduction of ϕ)	

*1: Optional case to require a special techniques
*2: Weight of the sliding mass

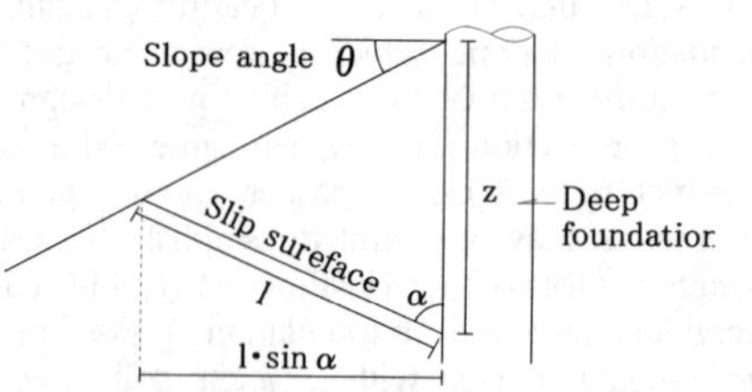

Figure 2. Conventional lateral loading conventional design (on the cross-section)

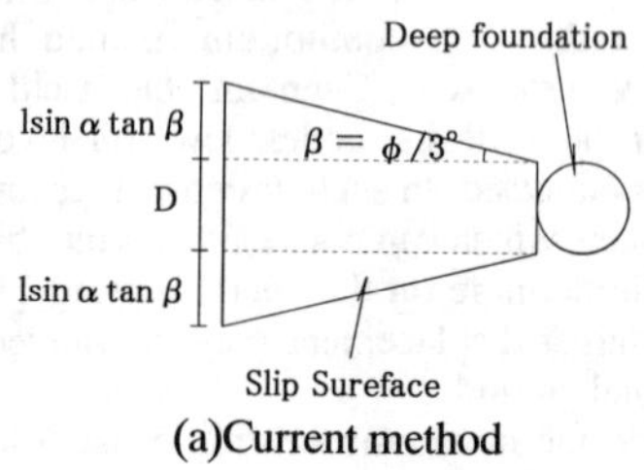

(a)Current method

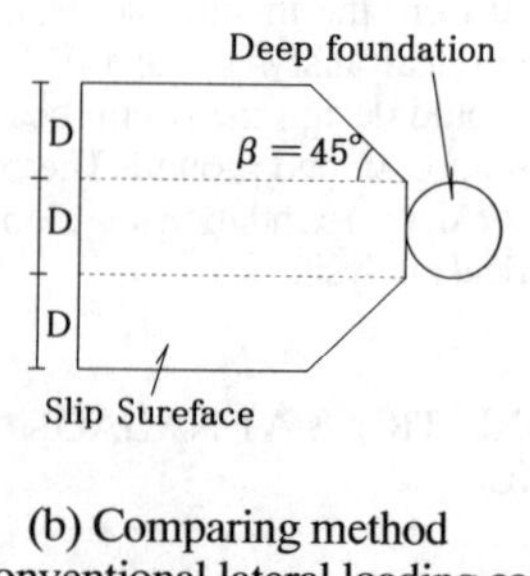

(b) Comparing method

Figure 3. Conventional lateral loading conventional design (on the horizon plane)

3.2 *FEM analysis method*

The authors adopted the same procedure on finite element method proposed by Prof. Ugai at Gunma University. A simple elasto-plastic model is used where Drucker-Prager type constitutive equation combined with Mohr-Coulomb failure criteria as shown in Figures 4(a) and (b). The finite element model is three-dimensional as shown in Figure4(c).

The initial stress of the sloped ground was obtained by static gravity analysis. By giving the lateral load at the deep foundation top, the lateral displacement was obtained. The non-linearity was calculated by the iterative analysis. The maximum displacements in calculation were set to be 20 cm at Nagaoka site and 30 cm at Tohoku site respectively.

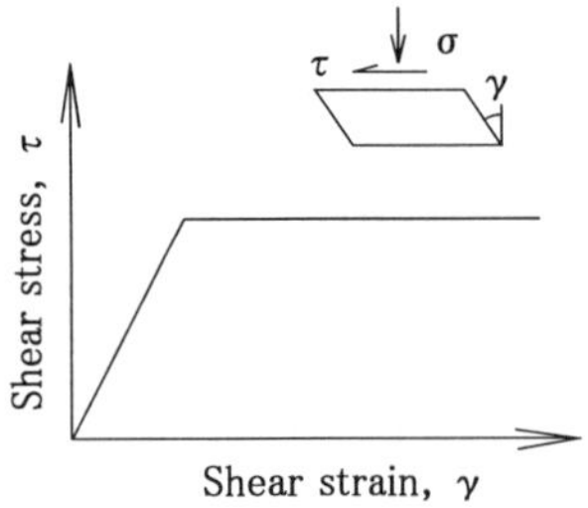

(a) Stress-strain relations

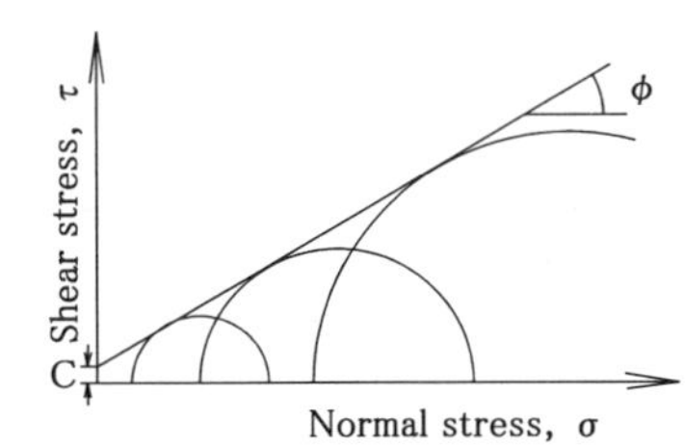

(b) Mohr-Coulomb failure criteria

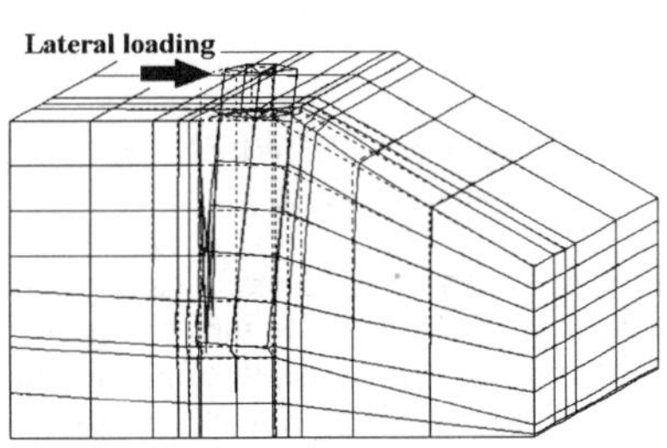

(c) Model of lateral loading

Figure 4. Schematic figure of this numerical procedure

4 NUMERICAL RESULTS

Figure 5 shows the computed load and displacement curves for Nagaoka site and Tohoku site. Only Nagaoka site has the experimental result to be compared. At Nagaoka, the initial tangency of the experimental curve was under-estimated by the computation. However, the experimental curve could be simulated well by the computation on average. At the 20 cm displacement, the computed resistance agrees with the experiment very well. The computation for Tohoku site shows a weaker tangency at the initial part of the curve and it has more bearing capacity than that Nagaoka site. It is because the length of deep foundation at Nagaoka site is longer than that at Tohoku site. Furthermore, it is relating that the soil Young's modulus at Nagaoka site is weaker than that at Tohoku site as explained in Section 2. The experimental and computed results do not show clear peaks on load resistance.

Figure 6 shows the shear failure zone on the cross section at Nagaoka site and Tohoku site. Figure 6(a) and (b) show the observed shear failure zones at Nagaoka site respectively. The computed shear failure zone is represented by the safety factor contours, which is at 20 cm displacement. The contour line with the safety number of 1.05 coincided well with the observed shear failure zone. The computed contour comes in the area of existing slip surface. Figure 6(c) shows computed shear failure zone at 30 cm displacement at Tohuku site. It is also represented by the safety contour. The angle α is about 60° at both sites.

Figure 7 shows the failure surfaces on the horizontal plane at Nagaoka site and Tohoku site. Figures 7(a) and (b) show the observed and computed failure surfaces at Nagaoka site respectively. Figure 7(c) shows computed failure surfaces at Tohoku. In these figures, the contour of safety number of 1.05 also may represent the failure surface. From these figures, it can be seen that the length and width of the computed failure surface is similar to the observed one at Nagaoka site. The length and width are 7m and 4.5m respectively. However, it can be also noticed that the computed failure surface is widened a little larger than the observed one, if the zone with safety number of up to 2.0 is taken into account. Furthermore, the computed depth of the failure zone start reaches to the 8m, which is a little deeper than the observed one. This depth is that the compressive in front of the deep pile starts. The design codes can take assumption that this depth denoted by the Z is the 2/3 of the full embedment depth of the deep foundation. Eventually, the computed depth could coincide with the observation well although the computation overestimates the observations a little.

From the computation at Tohoku site comparing with that at Nagaoka site, the shear failure zone is occurred relatively near the ground surface because the embedment depth of the deep foundation at Tohoku site is much longer than at Nagaoka site. However, the failure surface is similar to each other. The deformation mode of the deep foundation at Tohoku site seems more inclined than that at Nagaoka site. It implies that the stiffness of the deep foundation must be assessed with the ground stiffness, in order to yield a more lateral bearing capacity.

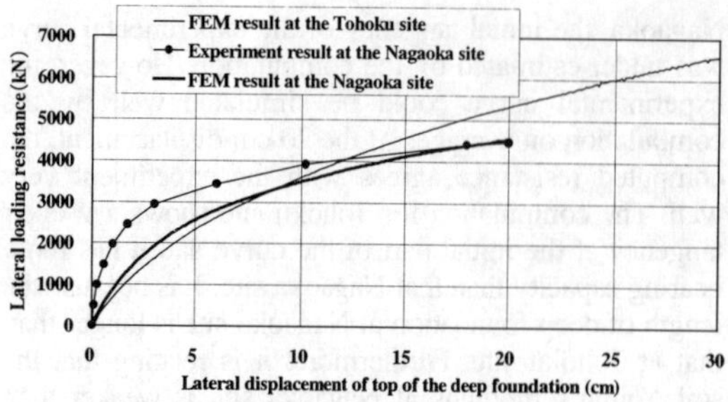

Figure 5. Numerical results compared with the experiments at Nagaoka and Tohoku.

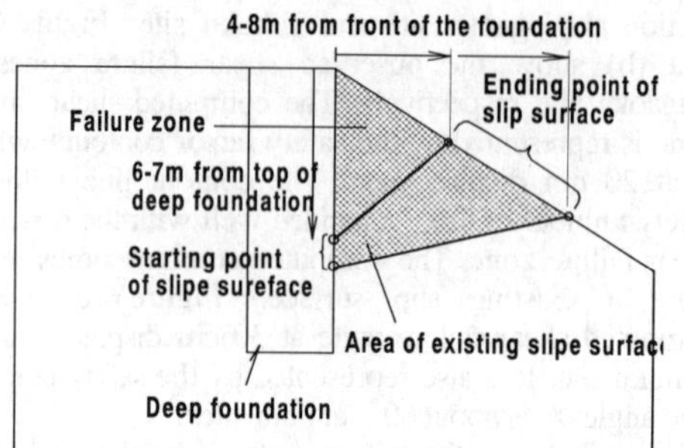

(a) Nagaoka experiment

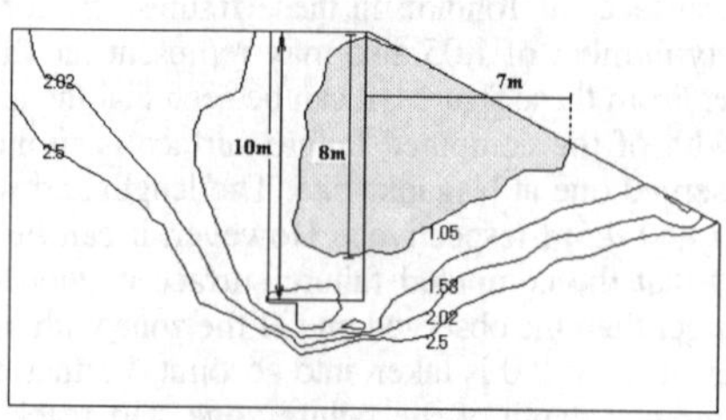

(b) Nagaoka computation

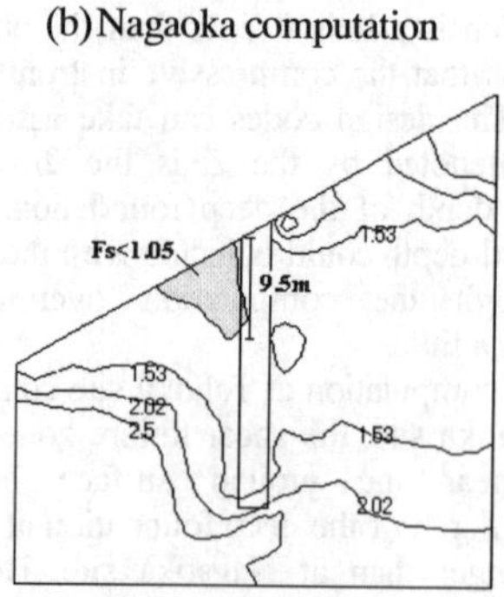

(c) Tohoku computation

Figure 6. Failure zone at Nagaoka and Tohoku (cross-section)

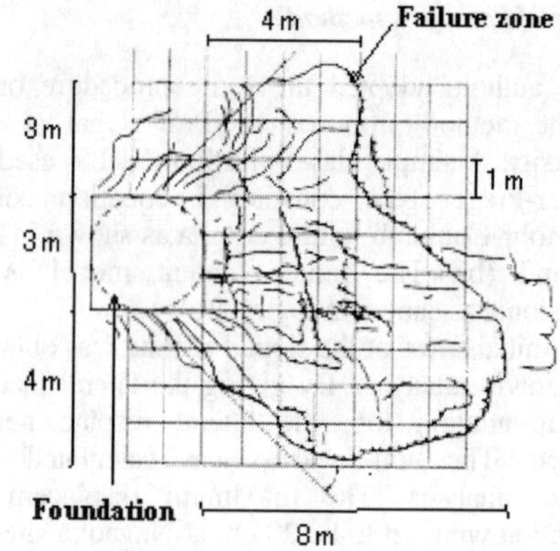

(a) Nagaoka experiment

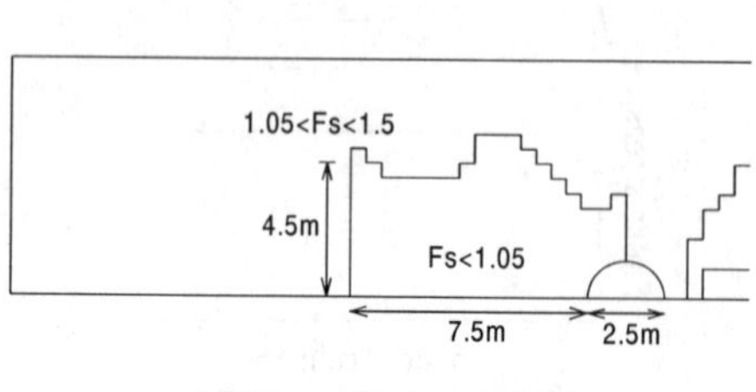

(b) Nagaoka computation

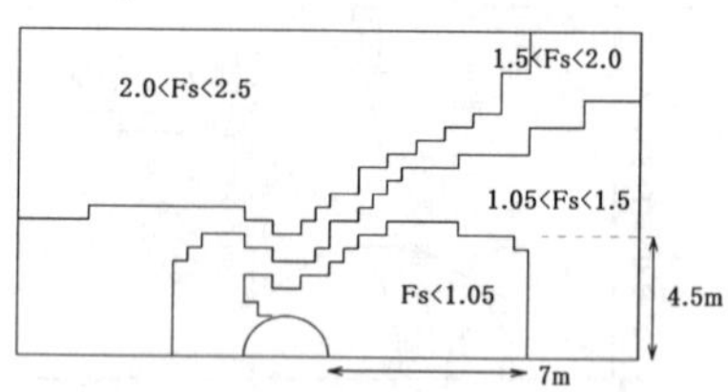

(c) Tohoku computation

Figure 7. Failure zone at Nagaoka and Tohoku (on horizontal plane)

5 COMPARISONS BETWEEN EXPERIMENTS, FEM AND DESIGN CODES

Table4 summarizes all lateral-bearing capacities. In this table, each design code defines the depth Z to be the 2/3 of the full embedment depth of pile foundation. The Z is set to be 6.5 m at Nagaoka site and the Z is is set to be 15m at Tohoku site.

From this table, FEM analysis at Nagaoka site gives a close answer to fit with the observation. On the other hand, it is not realistic to calculate the bearing capacity by the design codes, Express Highway Research Foundation and Tokyo Electric Power Co., Ltd. The design by Express Highway Research Foundation gives 1,421 kN, which is much lower value than that by Tokyo Electric Power Co., Ltd.

The main reason can be found that the failure zones of

the design codes are smaller than the finite element analysis, which coincided the observation well. Another reason is that the method by Tokyo Electric Power Co., Ltd. tends to give a larger value of α than that by Express Highway Research Foundation as shown in Table 3.

Figure 8 shows the computed relation between lateral bearing capacity and the α. From this figure, the both codes give the minimum value at the α of 75° (The α of Tokyo Electric Co., Ltd. should be constant actually). Figure 9 shows difference of α among the two codes. In this figure, (a) and (b) corresponds to Express Highway Research Foundation and Tokyo Electric Co., Ltd. And (c) denotes the finite element analysis or the observation. From this figure, it can be seen that there is a large discrepancy in the failure modes among the codes and observation.

Table4. Lateral bearing capacity compared to the observed results

Method	Nagoka	Tohokou
Experiments	4,479	-
FEM	4,551	5,331
Express Highway Research Foundation	1,421	4,232
Tokyo Electric Power Co., Ltd.	3,943	8,130

(unit:kN)

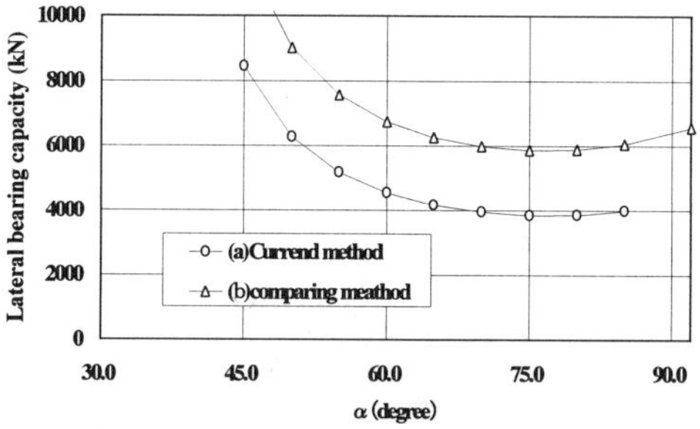

Figure 8. Relations between the slope angle α and bearing capacity

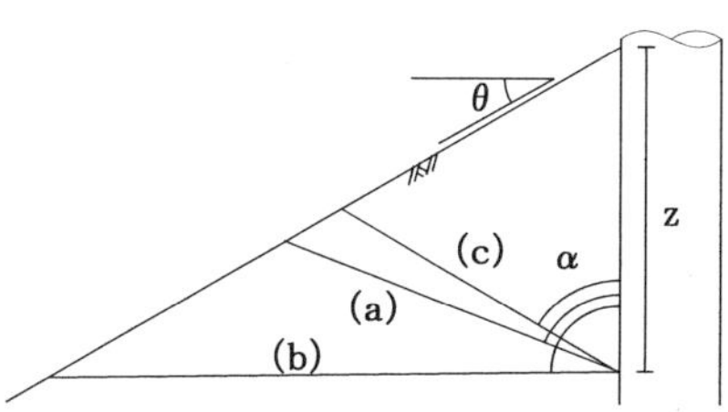

Figure 9. Definition of α of the conventional codes

6 CONCULSIONS

Lateral bearing capacity of deep foundation can be calculated from many codes. However, the calculated values are changed with a large variety on the specific sites because the codes were derived on the basis of the different assumptions on the different soil conditions from the sites. Through a case study at Nagaoka site and others, it suggests that only finite element analysis could be the only method to evaluate the lateral bearing capacity of deep foundation on slope. From the above things, the following main conclusions ware developed:

1) The code by Express Highway Research Foundation defines more conservative value than the code by Tokyo Electric Power Co., Ltd.

2) FEM analysis with a proper constitutive equation and could give a good answer to fit with the observation, even if the bearing capacity can not appear clearly.

3) The actual lateral bearing capacity may be larger that the designed values and the design by FEM analysis may reduce the construction cost.

4) Young's modulus obtained from plate loading test in the elastic region by the cyclic loading may underestimate the initial tangency of the loading and displacement curve.

7 ACKNOWLEDGEMENT

The authors would like to show sincere appreciation to Tohoku Electric Co., Ltd. Japan and Prof. K. Ugai at Gunma University Japan in cooperation with this project.

REFERENCES

Yoshii, S., Ogawa, T. and Takeuchi, T.,1984. Development of Rational Design Method for Deep Foundation of Steel Towers for Passing Electric Power —Examination of Lateral bearing Capacity—, Journal of Civil Engineering of Electric Power(189):43-53

Reports of Research and study for Foundation Construction,1952, Express Highway Research Foundation52(3)

Ugai, K., D. Leshchinsky, 1995. Three-dimensional limit equilibrium and finite element analysis: a comparison of results. Soils Found. 35(4):1-7

Wakai, A., Ugai, K. and Goes ,S. ,1995. The 3-D FE analysis of model group piles embedded in sand, Proc. of International symposium on Numerical Models in Geomechanics, Davos, Switzerland :613-618

Computer Methods and Advances in Geomechanics, Desai et al. (eds) © 2001 Balkema, Rotterdam, ISBN 90 5809 183 X

Numerical model in slope instability: A case history

L. Scesi
Department of I.I.A.R., Polytechnic of Milan, Italy

ABSTRACT: The study is aimed at a better comprehension of the causes originating two landslides in the Airolo commune (Switzerland) and of the mechanisms ruling their progression. It was carried out using both traditional techniques and numerical models, among which the UDEC calculation code, also in order to check its potentiality. This code has been seldom used in the study of rocky slopes stability, in particular in presence of much water. According to what emerged from the study, it can be ascertained to what extent numerical models represent a sound support to traditional techniques, as they allow the identification of moving surfaces and of blocks of the slide, a better evaluation of the geologic risk that is associated to particular zones of the territory and the choice of the best protective measures.

1 FORWORD

The town of Airolo in Switzerland presents two areas where there have been very important landslides. These areas, Sasso Rosso (Red Stone) and I Dragoni (The Dragons), represent both a direct and a potential danger to the town and its infrastructures. The Sasso Rosso area was interested by an important landslide in 1898 and, more recently by smaller, local landslides. In the I Dragoni area, some mass wastings occurred accompanied by modest landslides and joint openings. This study is aimed at better understanding the causes which gave origin to the two accidents and mechanisms ruling their progression. Traditional techniques were used as well as numerical models. In particular, the UDEC calculation code (Cundal et al. 1985), seldom used in the study of rock slope stability especially where much water is present, was implemented to check on its potentiality .

2 GEOLOGICAL CONTEXT

The area under examination is located in the northern part of the Canton Ticino in the town of Airolo, Switzerland, between Val Canaria (Canaria Valley) on the east and Val Bedretto (Bedretto Valley) on the west, to the orographic left of the Ticino River. Attention was focused on the Sasso Rosso and I Dragoni areas at an altitude that extends from 1600 meters to 2000 meters because of their potential danger to Airolo and the valley floor infrastructures such as the streets, highway and Gothard (Gottardo) railway (Fig. 1). Two important structural units characterize the area: the Gothard massif to the north and the Lucomagno unit (overthrusted on the previous) to the south. The Gothard massif can be divided into three zones: the southern zone characterized by a prevalence of Mesozoic sedimentary rocks sinking toward the north, the central zone with a prevalence of gneiss and mica schist with subvertical foliation, and the northern zone characterized by a prevalence of paragneiss and schists (Ercinic) sinking toward the south. The Airolo area is rich in dolomite, calc-schists, cellular dolomites and anhydrites (Novena zone) which belong to the Mesosoic overburden and in gneiss and mica schist which are represented by the formations of the Tremola Series (Serie della Tremola) (Valenti 1992).

3 SLOPE INSTABILITY FACTORS

There are essentially three determinant elements in slope stability:

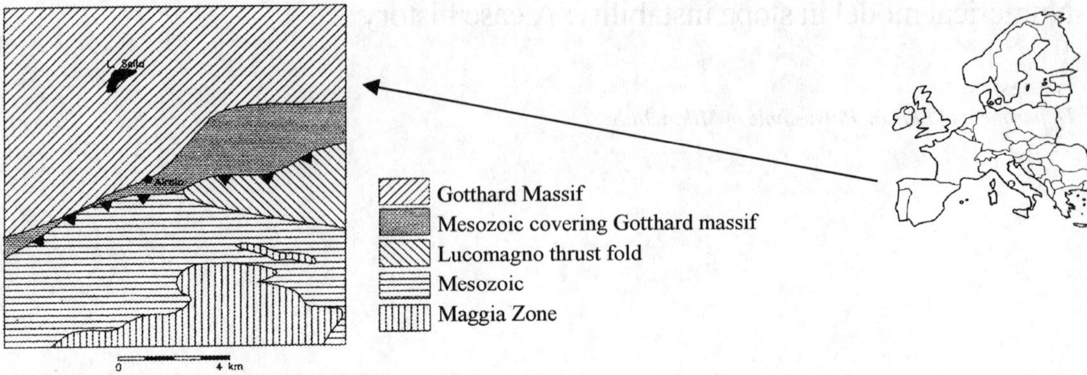

Figure 1. Location of studied area and geological map of Airolo zone.

a) Hackenwurf, which is a phenomenon of gravitational origin and is linked to the presence of Triassic dolomites and cellular dolomites which outcrop at the slope base. The weight of metamorphic units standing above determine the overturning of the mica schists toward the valley. The direct consequence is the change in the position of foliation layers whose inclination, on the surface, can often be approximated on the horizontal (Fig. 2).

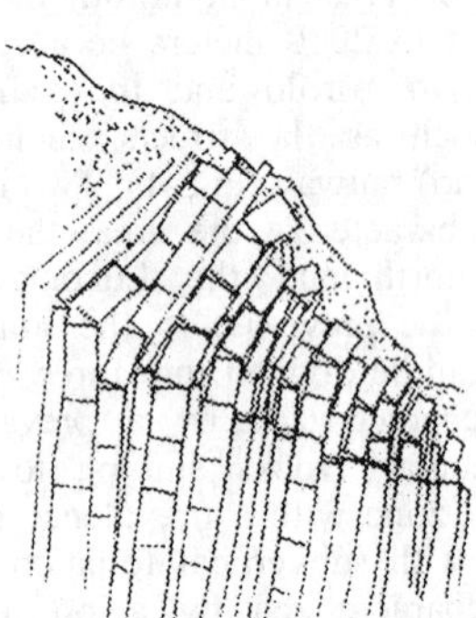

Figure 2. "Hackenwurf" phenomenon (E. Hoek & J.W. Bray, 1981).

b) Discontinuity systems which control the structural evolution of the slope. Two systems, System K1 and System K 2, oriented to one another almost orthogonally were detected:

1. system K1 generally sinks toward SSE at 160° with an inclination between 40° and 70°. It represents the basal breaking of unstable rocky masses.

2. system K2 sinks mainly toward WSW usually with an important inclination between 75° and 80°. They represent the most dangerous kinematic side breakings in case of accidents.

c) The presence of "weak rocks" (mylonitic layers and shear zones) distributed on the slope subparallely oriented with respect to the foliation areas. The mechanical characteristics of the rocks decrease greatly in these zones due to important tectonic stresses that the area underwent. The movement is mainly focused in these parts of the mass.

4 GEOMECHANIC CHARACTERIZATION OF THE ROCK MASS

Tests performed over the years such as geodesic measurements, bore holes, refraction and reflection shoting, lab tests on samples, and the surface geologic and geologic-structural surveys made possible a geomechanical characterization of the rock masses.

In particular, the Beniawki classification and the Romana method made evident the potential slope instability (average RMR ≥ 40 – average SMR = 25 ÷ 40 - poor rocks and unstable slope-).

Moreover, rock samples were taken to perform Point Load and triaxial tests. Triaxial tests imposed axial deformations that were directed both perpendicular to and parallel to the foliation layers, in an effort to emphasize the different rock behavior. The sample breaking under different confinement pressures was then assessed (0, 100, 200 MPa) The setting of the cohesion value and the rock friction angle followed. The elastic module value E (E = 6.35GPa) was obtained from the test performed with no confinement pressure.

Geomechanical surveys and tests performed on samples extracted at different quotas pointed out that the most endangered strip is located between 1700 and 2000 m. asl for both Sasso Rosso and I Dragoni. Such a confirmation came from the monitoring network data that detected superficial shifts of 1.5 – 2 cm/ year. In particular, a tension crack is located below the main scarp of the main break of the famous Sasso Rosso slide at 1850 – 1880m. It is positioned along a discontinuity of System K1 and it isolates a block of rock of at least 300.000m3. Two levels of effective danger were detected in the area in that situation. The first can be connected to the moving of superficial blocks free from the discontinuity systems K1 and K2. The scale is small to average involving volumes ranging from 102 and 103 m3. The second situation involves bigger blocks and seems to be linked to the general decay of the slope.

5 NUMERICAL MODELLING

The studies undertaken did not allow for the definition of the entity of the potentially unstable masses and the landslide kinematisms that could start in the slopes with good approximation. It was therefore decided to implement a mathematical simulation which could give indications on the slope behavior - especially concerning the critical strip previously identified -, on the real instability and size of the moving masses in relation to different hydraulic conditions (Lee et al. 1993).

The following hypotheses were set for the modeling: an elastic-linear mechanic behavior for intact blocks; an elastic behavior for discontinuities until the available shear strength is reached, a plastic behavior for bigger movements. The Coulomb criteria was used for the soil shear strength of discontinuities, with residual friction angle equal to about 20°, as direct soil shear strength tests were performed on the joints (Barton et al. 1997).

UDEC solves the problem of water circulation by considering the water movement within the discontinuities between impermeable blocks. According to the properties of the fluid and the joint opening value, UDEC draws the permeability (k) and, as a consequence, the specific flow (Louis 1969).

5.1 Parameters used

The following parameters were used for the modelling:

a) rock:

density = 2600 kg/m^3
E *(Young's Modulus)* = 6.35 GPa
K *(volume modulus)* = 6.00 GPa
G *(shear modulus)* = 2.40 GPa
cohesion = $3.7 \cdot 10^7$ Pa
friction angle = 21°

b) discontinuity:

normal stiffness K_n = $1 \cdot 10^{10}$ Pa
tangential stiffness K_t = $1 \cdot 10^9$ Pa
cohesion = $4 \cdot 10^5$ Pa
friction angle = 20°
discontinuity opening on the surface (first 90 m) = 0.002 m
discontinuity opening in depth (over 90 m) = 0.00001 m

Data concerning the technical characteristics of rocks and discontinuities were obtained from in situ and lab tests, whereas stiffness parameters (both normal and tangential) describing the elastic behavior of joints were calculated using the following formulas:

$$K_n = \frac{E_m E_r}{s(E_r - E_m)} \quad (1)$$

$$K_s = \frac{G_m G_r}{s(G_r - G_m)} \quad (2)$$

where E_m = Young's modulus of the rock mass; E_r = Young's modulus of the intact rock; G_m = shear modulus of the rock mass; G_r = shear modulus of the intact rock; s = discontinuity spacing.

Different simulations were made for each model, according to the hydraulic conditions set for the surrounding area – supply domain, water pressure in the joints, hydraulic gradient. Three different cases were considered concerning the supply domain – 1) no recharge (dry slope), 2) lateral recharge (constant recharge from the top), 3) upper recharge (supply from heavy rainfalls and melting snow). The hydraulic gradient modeling used the following data to simulate how the change in gradient can influence the slope behavior: 1) the horizontal hydraulic gradient is equal to 0.001 to simulate flow from the top; 2) the horizontal hydraulic gradient is equal to 0.01; 3) the vertical hydraulic gradient is equal to 0.01. Afterwards, in the presence of water, the following hypotheses were made – the maximum hydrostatic pressure is equal to 3.01 MPa in the case of lateral recharge and 0.26 MPa in the case of upper recharge.

5.2 "I Dragoni" Modeling

Studies carried out on the area show a substantial stability on the "I Dragoni" slope. No trace of breaking mechanism was recorded that could affect a wide area of that section due to the peculiarity of the slope asset. It has discontinuity layers pointing downslope in a dip that is higher than the slope. All simulations showed that the most important movements are focused close to the topographic surface in two different zones. One zone is located around 2000 meters and shows an abrupt change in dip, whereas the other is at 1800 meters.

The movement modula of the check points measure about 2 cm and are located in one of the above described areas. The presence of water in the slope discontinuities causes changes in the movements of the check points only in the intermediate calculation phase without affecting the final modulus. There are two types of movement mechanisms: translational and complex (Fig. 3). It can be deducted that the slope behavior is strongly influenced by the structural asset. At the end of all simulations, a condition of static equilibrium was reached, indicating the slope stability. Locally, tension cracks could originate right above 2000 m. They could set small and middle blocks free causing isolated landslides, therefore, the only phenomenon characterizing the evolution of this area in time is the retreat of the crown.

5.3 *"Sasso Rosso" modelling*

Similarly to what was done for "I Dragoni", stability conditions of the "Sasso Rosso" slope were evaluated with no water as well as with different types of water supply. Some interesting results were drawn due to the results of the modelling (Fig. 4).

1) The most important movements are recorded in the peak zone of the slope where breaking causes the retreat of the main scarp. This is due to the sliding of blocks that belong to the crown of the ancient slide. The blocks originate from the combination of foliation layers and the discontinuities perpendicular to them. The phenomena is limited to the first few meters in depth, but it could be accentuated in spring and autumn due to the melting of snow or to heavy rainfalls.
2) Other important movements are recorded between 1750 and 1900 m., where translational phenomena take place along discontinuity layers close to the topographic surface. The model also showed how the presence of water accelerates the movements and increases the shifting modulus reaching values close to the decimeter.
3) In time, the evolution of those movements could interest big masses (25-30,000 m^3), putting at great risk the town of Airolo.

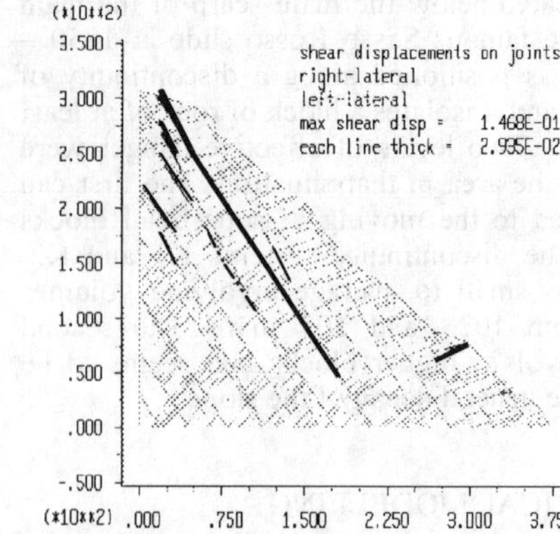

Figure 3. Shear displacements on joints at the end of simulation with lateral water recharge. In the figure the joints reaching the peak shear strength are in evidence.

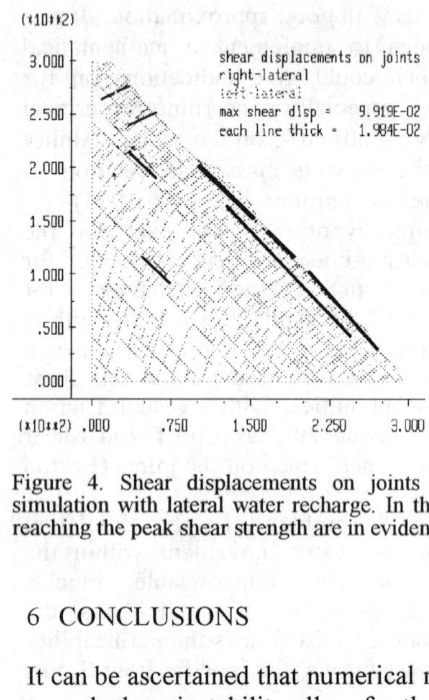

Figure 4. Shear displacements on joints at the end of simulation with lateral water recharge. In the fig. the joints reaching the peak shear strength are in evidence.

6 CONCLUSIONS

It can be ascertained that numerical models applied to rock slope instability allow for the identification of the origin and the main causes of movements with a certain degree of precision. In the upper area of "I Dragoni", tension cracks were correlated to shear faults due to local stress conditions and did not indicate the possibility of important phenomena development, whereas the "Sasso Rosso" instability was identified as translational movements along discontinuity layers close to the topographic surface. Those results, however, cannot be considered definitive due to the relative (machine time), rather than the absolute value time has in modelling.

The model could be calibrated by correlating the real movements to the simulated ones in order to foresee the future evolution of the slope. That would be possible if the monitoring data of representative section spots were to have a high sampling frequency over a long period of time.

It can be affirmed how the use of a numerical model is an important support to traditional study techniques, despite the fact that the introduction of simplifications is required in numerical modelling. Actually, it allows for the identification of moving surfaces and of blocks of the slide with a certain degree of precision, a better evaluation of the geological risk associated to particular zones of the territory and allowing for the best protective measures to be chosen.

ACKNOWLEDGEMENT

We want to thank the Istituto Geologico Cantonale of Cadenazzo (Canton Ticino, Switzerland). A special thanks goes to G. Beatrizotti and G. Valenti and the Studio di Consulenza Geologica ed Ambientale for the data and the cooperation provide.

REFERENCES

Barton N. & Choubey V. 1977. The shear strength of joints in theory and practice. *Journal of Rock Mechanics*; 12/1977, Vol. 10, No. 1,2.

Cundall P.A. & Hart R.D. 1985. *Development of Generalized 2-D and 3-D Distinct Element Programs for Modelling Jointed Rock.* Itasca Counsulting Group, Inc..

Lee Cheng-Haw & Farmer Ian 1993. *Fluid Flow in Discontinuous Rocks*. Chapman & Hall.

Louis C. (1969*). A study of groundwater flow in jointed and its influence on the stability of rock masses*. Imperial College Rock Mechanics Research Report, 10,.

Valenti G. 1992. *Comune di Airolo - Zone di Pericolo*. Istituto Geologico Cantonale.

Computer Methods and Advances in Geomechanics, Desai et al. (eds) © 2001 Balkema, Rotterdam, ISBN 90 5809 183 X

Comparison of finite element results obtained for a geotechnical benchmark problem

H.F.Schweiger
Institute for Soil Mechanics and Foundation Engineering, Graz University of Technology, Austria

ABSTRACT: One of the aims of the working group 1.6 "Numerical Methods in Geotechnics" of the German Society for Geotechnics is to provide recommendations for numerical analyses in geotechnical engineering. In addition benchmark examples are specified and the results obtained by various users employing different software are compared. In this paper example No. 3, a tied back diaphragm wall in Berlin sand, will be discussed. Whereas in previous benchmarks material properties have been given and the type of constitutive law has been predefined, here results from laboratory tests (triaxial and one-dimensional compression tests) have been given and the choice of the constitutive model has been left to the user. Thus a wide range of material models have been used ranging from simple elasto-perfectly plastic Mohr-Coulomb models to hypoplastic formulations. As expected the scatter of results is significant. In this paper the results are compared and critically assessed.

1 INTRODUCTION

So far relatively little attention has been paid in the literature on validation and reliability of numerical models in general and on specific software in particular although some attempts have been made (e.g. Schweiger 1991). However, as will be shown in this paper there is a strong need for defining guidelines and procedures to arrive at reliable numerical models in practical geotechnical engineering.

Another aspect, also very often ignored in the literature but equally important, is the identification of input parameters, which represent the properties of the ground in situ, from available experiments.

Benchmarking is, at least to the author's opinion, of significant importance in geotechnical engineering, probably more so than in other disciplines such as structural engineering. The reason for that may be summarized as follows

- the domain to be analyzed is generally not clearly defined by the structure
- it is not always clear whether continuum or discontinuum models are more appropriate for the problem at hand
- a wide variety of constitutive models exist in the literature but there is no "approved" model for each type of soil
- in most cases construction details cannot be modelled very acurately in time and space (excavation sequence, prestressing of anchors etc.) within the financial and time constraints given in practice
- soil/structure interaction is often important and may lead to numerical problems (e.g. certain types of interface elements)
- implementation details and solution procedures may have a significant influence on the results of certain problems but may not be important for others
- there are no approved implementation and solution procedures for commercial codes (implicite vs explicite strategies, return algorithms etc.)

All of these aspects leave ample room for personal preferences both on developers and users side respectively. From a practical point of view this situation is by no means desirable because it proves to be very difficult to obtain objective results because of the numerous modelling assumptions involved. Modelling techniques vary depending on the personal experience of the user and, to a certain extent, also on the code utilized. So far no clear guidelines exist and thus results for a particular problem may vary significantly if analysed by different users, even for working load conditions.

These problems have been addressed amongst others by the working group 1.6 "Numerical Methods in Geotechnics" of the German Society for Geotechnics and it is the aim of this group to provide recommendations for numerical analyses in geotechnical engineering. The group has published general recommendations (Meissner 1991), recommendations for numerical simulations in tunnelling (Meissner 1996) and recommendations for deep excavations will be published shortly. In addition benchmark examples are specified and the results obtained by various users employing different software are compared.

So far three examples have been specified by the working group and discussed in two workshops. The first two of them, a tunnel excavation and a deep excavation problem, have been rather idealized problems with very tight specification so that little room for interpration was left to the analysts. Despite the simplicity of the examples and the rather strict specifications significant differences in the results are obtained even in cases where the same software has been utilized by different users (Schweiger 1998).

This contribution will be dedicated to the example No. 3 which represents an actual application (a tied back diaphragm wall in Berlin sand) which has been only slightly modified in order to reduce the computational effort. Whereas in the first two examples constitutive model and parameters have been prespecified in the latter one the choice of constitutive model has been left to the user and parameters had to be taken from literature or determined from laboratory tests which had been included in the specifications.

2 SPECIFICATION OF PROBLEM

The example is closely related to an actual project in Berlin. Slight modifications have been introduced in modelling the construction sequence such as the groundwater lowering which has been performed in various steps in situ but is modelled here in one step prior to excavation. The constitutive model to be used has not been prescribed but was left to the analysts. Some basic material parameters have been taken from the literature and additionally results from one-dimensional compression tests on loose and dense samples have been given together with triaxial tests on dense samples. Thus the exercise represents closely the situation one faces in practice.

As it is the aim of this section to demonstrate the necessity of performing these kind of exercises only the most relevant part of the specification will be given here (Fig. 1), a detailed description can be obtained from the author. For the same reason only a limited number of results will be presented.

2.1 *General assumptions postulated*

- plane strain
- influence of diaphragm wall construction may be neglected, i.e. initial stresses without wall, then wall "wished-in-place" (weight of wall e.g. as difference to soil weight)
- diaphragm wall modelling: beam or continuum elements
- interface elements between wall and soil (wall friction is specified as $\varphi/2$)
- domain to be analysed (suggested: see Figure 1)
- horizontal hydraulic cut off at –30.00 m not to be considered as structural support
- given anchor forces are design loads

2.2 *Material parameters*

Some reference values for stiffness and strength parameters from the literature frequently used in the design of excavations in Berlin sand are given below (z = depth below surface):

$E_s \approx 20\,000\sqrt{z}$ kN/m^2 for $0 < z \leq 20$ m
$E_s \approx 60\,000\sqrt{z}$ kN/m^2 for $z > 20$ m
$\varphi = 35°$ (medium dense)
$\gamma = 19$ kN/m^3
$\gamma' = 10$ kN/m^3
$K_o = 1 - \sin\varphi$

The given stiffness moduli E_s may be used to arrive at approximate values for the Young's modulus for calculations with linear elastic - perfectly plastic material models (Mohr-Coulomb) assuming an appropriate Poisson's ratio.

Properties for diaphragm wall (linear elastic):

$E = 30\,000 \times 10^3$ kN/m^2
$\nu = 0.15$
$\gamma = 24$ kN/m^3

In addition to these values results from oedometer tests (on loose and dense samples) and triaxial tests (confining pressure σ_3 = 100, 200 and 300 kN/m^2) have been provided. It was not possible to include a significantly large number of test results and thus the question arose whether the stiffness values obtained from the oedometer test have been representative. If, for example, the constitutive model requires a tangential oedometric stiffness at a reference pressure of 100 kN/m^2 as an input parameter, a value of only $E_s \approx 12\,000$ kN/m^2 was found based on these experiments. If a secant modulus for a pressure range beyond 200 kN/m^2 is determined a value of about 40 000 kN/m^2 is obtained. This was considered as to low by many authors and indeed other test results from Berlin sand in the literature indicate higher values. For example from Ohde (1951) values of about 35 000 to 45 000 kN/m^2 could be estimated as reference loading modulus of a medium dense sand at a reference pressure of 100 kN/m^2. This aspect will be addressed again later when discussing the results.

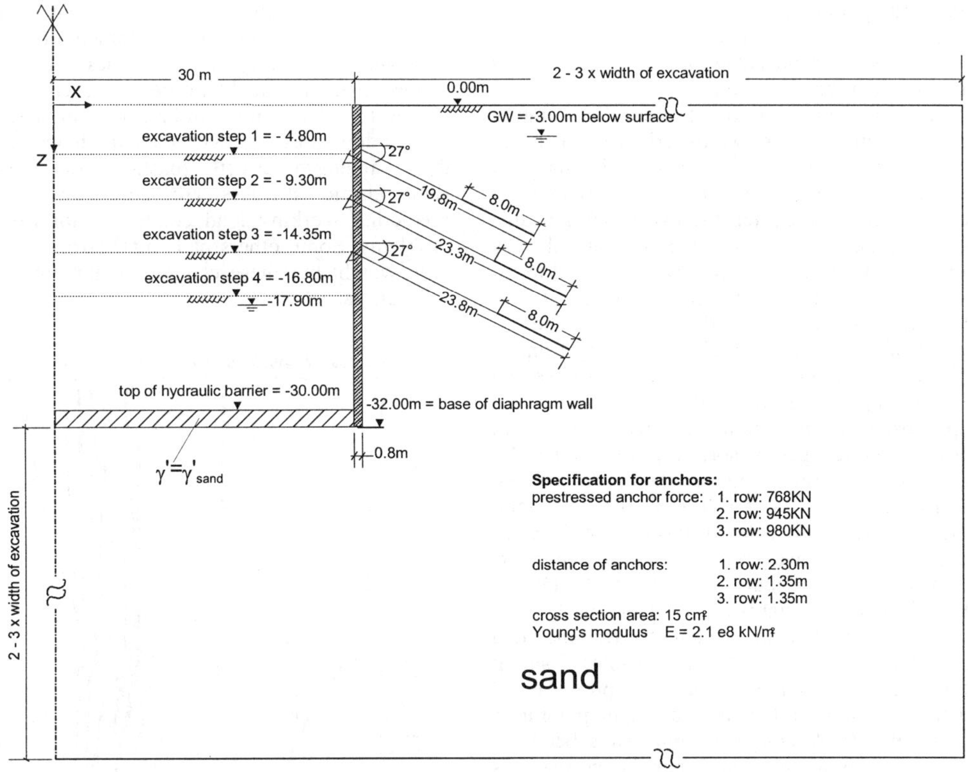

Figure 1. Geometry of benchmark example No.3 of the working group 1.6 of the German Society for Geotechnics.

The following computational stages had to be performed in order to simulate construction:

stage 0: Initial stress state is given by $\sigma_v = \gamma H$, $\sigma_h = K_o \gamma H$ (wall "wished-in-place", deformations = 0)

stage 1: GW-lowering to –17.90 m

stage 2: excavation step 1 (to level -4.80 m)

stage 3: activation of anchor 1 at level –4.30 m and prestressing

stage 4: excavation step 2 (to level -9.30 m)

stage 5: activation of anchor 2 at level –8.80 m and prestressing

stage 6: excavation step 3 (to level -14.35 m)

stage 7: activation of anchor 3 at level –13.85 m and prestressing

stage 8: excavation step 4 (to level –16.80 m)

Distance and prestressing loads for anchors follow from Figure 1.

3 COMMENTS ON SUBMITTED SOLUTIONS

A wide variety of programmes and constitutive models has been employed to solve this problem. Simple elastic-perfectly plastic material models such as the Mohr-Coulomb or Drucker-Prager failure functions, still widely used in practice have been chosen by a number of authors. Several entries utilized the computer code PLAXIS (Brinkgreve & Vermeer 1998) with the so-called Hardening Soil Model. This constitutive model is based on the well known formulation by Duncan & Chang (1970), but formulated within the theory of plasticity. It incorporates shear hardening and volumetric hardening, a stress dependent stiffness for primary loading and unloading/reloading and the stress dilatancy theory by Rowe (1962). Other similar plasticity models have been used, one submisssion used a simplified small strain stiffness for the elastic range. Three entries employed a hypoplastic formulation (one without and two with intergranular strains) and it should be realized that conventional parameters such as angle of internal friction or Young's modulus are no longer input parameters when using this concept. For reasons mentioned earlier only a limited number of analysts used the provided laboratory test results to calibrate their material model.

Most of the analysts used data from the literature from Berlin sand or their own experience to arrive at input parameters for their analysis. Only marginal differences exist in the assumptions of strength parameters (everybody trusted the experiments in this respect), the angle of internal friction φ was taken as 36° or 37° and a small cohesion was assumed to increase numerical stability by some authors. A significant variation is observed however in the assumption of the dilatancy angle Ψ, ranging from 0° to 15°. An even more significant scatter is observed in the assumption of the stiffness parameters although most analysts assume an increase with depth either by introducing some sort of power law similar to the formulation presented by Ohde (1951) which in turn corresponds to the formulation by Janbu (1963), or by defining different layers with different Young's moduli. However the choice of the reference moduli for primary loading and unloading/reloading varied significantly. One of the reasons for this was that the oedometer test included suggested a rather low value for primary loading modulus at a reference vertical pressure at 100 kN/m^2, a value often used as input parameter (see section 2.2). Although it would have been desirable to have more test data it is believed that this represents the situation in practice where usually only a very limited number of experimental results is available and engineering judgement is essential in order to arrive at the required input parameters which describe the behaviour of the in situ ground adequately. However as will be shown later it turned out that this affected only a small number of solutions submitted and could not be identified as the main source of discrepancies in results.

Additional variation is introduced by different formulations for the interface elements, element types, domains analysed and modelling of prestressed anchors. A comprehensive summary of all features of the submitted solutions is beyond the scope of this paper and can be found in Schweiger (2000).

4 SELECTED RESULTS

Figure 2 shows the deflection curve of the diaphragm wall for all solutions submitted and it follows that the results scatter in a range which is by no means acceptable. The horizontal displacement of the top of the wall varies between -229 mm and +33 mm (-ve means displacement towards the excavation). Looking into more detail of Figure 2 it can be observed that entries B2, B3, B9a and B7 are extremely off the "mainstream" of results. B2, B3, B9a are the ones which derived there input parameters mainly from the provided oedometer tests which however showed very low stiffnesses as compared to values given in the literature. B7 used a constant Young's modulus and a Mohr-Coulomb failure criterion. It compares well with the experience of the author that elasto-perfectly plastic models with constant elastic stiffness are not well suited for predicting working load conditions for excavation problems. Some others had small errors in the specific weight but these cannot account for the large differences.

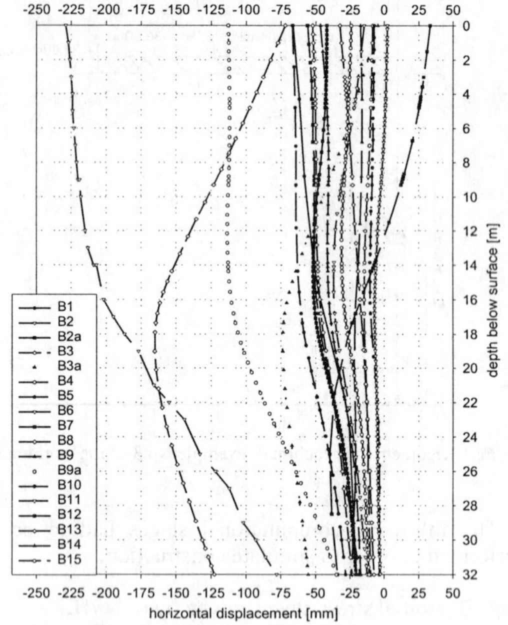

Figure 2. Wall deflection at final construction stage for all solutions submitted.

In situ measurements are available for this project and although the example here has been slightly modified in order to facilitate the calculations the order of magnitude of displacements can be assumed to be known. Figure 3 shows measured wall deflections for the final construction stage together with calculated results whereas only calculations which are "near" the measured values are included. The scatter is still significant. It should be mentioned that measurements have been taken by inclinometer readings but unfortunately no geodetic survey of the wall head is available. It is very likely that the base of wall is not fixed as assumed by the inclinometer measurement and that a parallel shift of the measurement of about 5 to 10 mm would reflect the in situ behaviour more correctly. This is confirmed by other measurements under similar conditions in Berlin. The calculated maximum horizontal wall displacement still varies between 7 and 57 mm, the shape of the deflection curve is also quite different. Comparing the results of the calculations with measurements it has to be pointed out though that the simplification introduced in modelling the groundwater lowering (one step lowering instead of stepwise lowering according to excavation progress) leads to higher horizontal displacements. Further studies revealed that the difference in calculated displacement due to the difference in modelling the groundwater lowering is strongly dependent on the constitutive law employed and ranges in the order of 5 to 15 mm. This fact explains also some of the observed differences in the results.

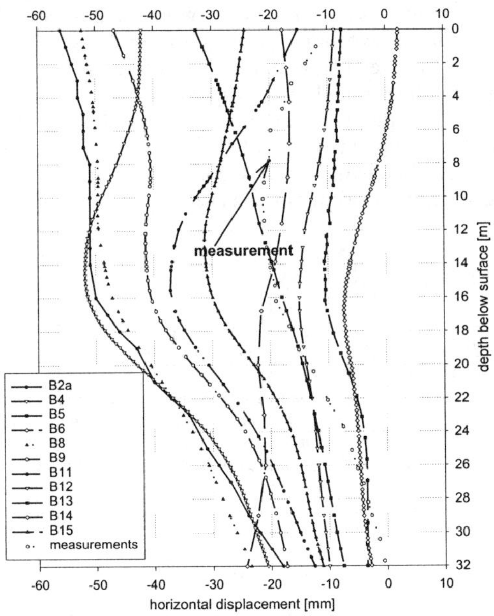

Figure 3. Wall deflection at final construction stage for selected solutions and inclinometer measurements.

Figure 4 depicts the calculated surface settlements, again only for solutions presented in Figure 3. Settlements of 50 mm have to be compared with a heave of about 15 mm. Considering the fact that calculation of surface settlements is one of the main goals of such an analysis the need for recommendations and guidelines in order to minimize unrealistic modelling assumptions is obvious.

Figure 5 shows the development of anchor forces for the upper row of anchors. Maximum anchor forces for the final excavation stage range from 106 to 634 kN/m. A careful inspection of Figure 5 reveals that some of the analyses did not model the prestressing of the anchors correctly because they do not show the specified prestressing force in the respective construction step. Bending moments, important from a design aspect, also differ significantly from 500 to 1350 kNm/m.

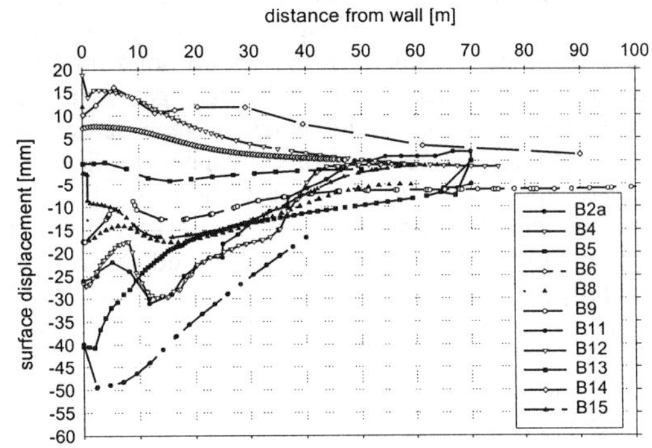

Figure 4. Surface displacements at final construction stage for selected solutions.

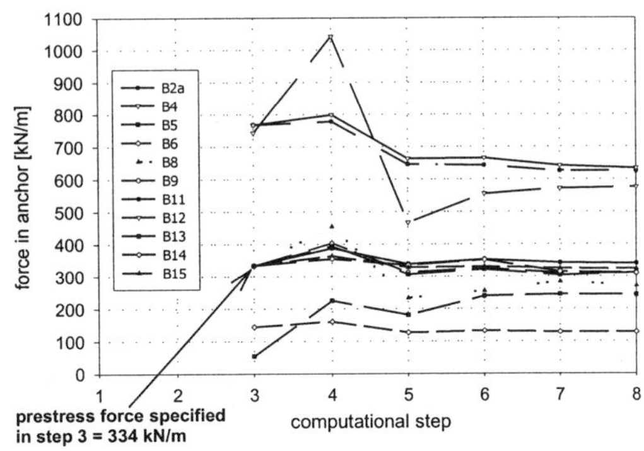

Figure 5. Forces in first row of anchors for selected solutions.

It follows from Figure 3 that no conclusions are possible with respect to the constitutive model or assumptions concerning element types and so on. B4 is a Drucker-Prager model, B5, B6, B9 and B12 are Mohr-Coulomb models, B13 is a hypoplastic formulation including intergranular strains and B14 a shear hardening plasticity model including a small strain stiffness formulation in the elastic range. Comparing the three entries employing a hypoplastic formulation it became evident that hypoplasticity without considering intergranular strains (Niemunis & Herle 1997) does not give reliable results, at least not for these types of problems.

Even with the same code (PLAXIS) and the same constitutive model (Hardening Soil Model) significant differences in the results are observed entirely depending on the personal interpretation of the stiffness parameters from the information available. B2 and B9a relied on the oedometer stiffness included in the specification, B2a, B8, B11 and B15 choose the parameters based on the literature and personal experience (Fig. 6). Although the latter four analyses are within a reasonable range if one consideres the uncertainties inevitably apparent in geotechnical analysis the differences in maximum horizontal wall deformation is still about 200%. A slightly different conclusion can be drawn when surface settlements are looked at. Here the more advanced models tend to give more realistic results than the elasto-perfectly plastic models. Further studies and a more detailed examination of all solutions submitted is presently under progress.

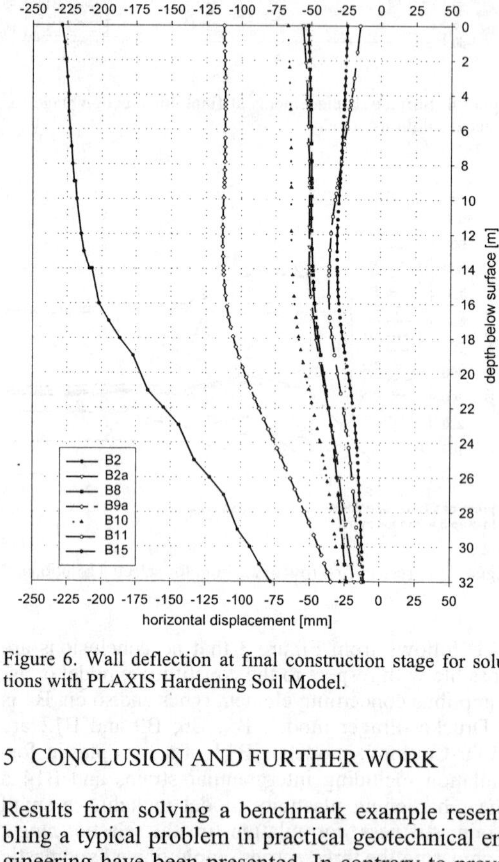

Figure 6. Wall deflection at final construction stage for solutions with PLAXIS Hardening Soil Model.

5 CONCLUSION AND FURTHER WORK

Results from solving a benchmark example resembling a typical problem in practical geotechnical engineering have been presented. In contrary to previously defined examples the choice of constitutive law and parameter identification was left to the analyst. As expected this lead to unacceptable differenecs in results the reasons for that could not be fully identified. The experimental results from oedometer tests included in the specification showed relatively low stiffness values at a reference pressure of 100 kN/m^2, a value many people use as reference parameter for their analysis. This was one source of discrepancies in results because many other entries relied on past experience or values from the literature for determining their input parameters. Even amongst these submissions a wide scatter in wall deflection, surface displacements and bending moments could be observed.

However it is believed that this exercise closely represents the situation in practice and reveals once again that numerical modelling in geotechnics is a difficult task. A lot of experience is required in order to make sensible assumptions for material and model parameters. The exercise presented here clearly indicates the need for guidelines and training for numerical analysis in geotechnical engineering in order to achieve reliable solutions for practical problems.

The working group 1.6 of the German Society for Geotechnics will continue to work in this area. In addition the working group A of the COST-Action of the European Community also makes some effort in providing guidelines for applications of numerical models in geotechnical engineering. Finally ERTC7 and in particular TC12 of the International Society of Soil Mechanics and Geotechnical Engineering have started working in this area.

REFERENCES

Brinkgreve, R.B.J. & P.A. Vermeer 1998. PLAXIS*: Finite Element Code for Soil and Rock Analyses*, Version 7, Balkema.

Duncan, J.M. & C.Y. Chang 1970. Nonlinear analysis of stress and strain in soils. *Journal of the Soil Mechanics and Foundations Division* 56: 1629-1653.

Janbu, N. 1963. Soil compressibility as determined by oedometer and triaxial tests. Proceedings *European Conf. On Soil Mechanics and Foundation Engineering*, Wiesbaden, Germany. 19-25.

Meissner, H. 1991. Empfehlungen des Arbeitskreises 1.6 "Numerik in der Geotechnik", Abschnitt 1, "Allgemeine Empfehlungen", *Geotechnik*, 14: 1-10. (in German)

Meissner, H. 1996. Tunnelbau unter Tage – Empfehlungen des Arbeitskreises 1.6 "Numerik in der Geotechnik", Abschnitt 2. *Geotechnik*, 19: 99-108. (in German)

Niemunis, A. & Herle, I. (1997). Hypoplastic model for cohesionless soils with elastic strain range. *Mechanics of cohesive-frictional materials*, 2: 279-299.

Ohde, J. 1951. Grundbaumechanik (in German), *Huette*, BD. III, 27. Auflage.

Rowe P.W. 1962. The stress-dilatancy relation for static equilibrium of an assembly of particles in contact. *Proc. Roy. Soc.* A269. 500-527.

Schweiger, H.F. 1991. Benchmark Problem No. 1: Results. *Computers and Geotechnics*, 11: 331-341.

Schweiger, H. F. 1998. "Results from two geotechnical benchmark problems", *Proc. 4th European Conf. Numerical Methods in Geotechnical Engineering*, Cividini, A. (ed.). 645-654. Springer.

Schweiger, H. F. 2000. Ergebnisse des Berechnungsbeispieles Nr. 3 "3-fach verankerte Baugrube". *Tagungsband Workshop "Verformungsprognose für tiefe Baugruben"*, Stuttgart, 16. und 17.3.2000. 7-67. (in German)

Computer Methods and Advances in Geomechanics, Desai et al. (eds) © 2001 Balkema, Rotterdam, ISBN 90 5809 183 X

Simulation of initial stresses for tunnel models

G. Swoboda & M.A. Mahmoud
University of Innsbruck, Austria

I. Hladik
Dolexpert-Geotechnics, Praha, Czech Republic

ABSTRACT: Numerical modelling has been an important design tool for tunnels for many years. The quality of the results obtained with modelling depends on the knowledge of the physical data of the rock and the initial stress field. Here a new technique is presented for analyzing the initial stress field for situations, where the topography of steep mountains exerts a considerable influence on the stress field. The dimension of tunnel is very small compared with the dimension of the mountain. Modelling the initial stress field of the entire mountain region would require a huge numerical model. The new technique divides the problem into two structural systems. System one, also called a micro-model, deals with the tunnel, and system two, called a macro-model, with the entire mountain. The micro-model is a standard two- or three-dimensional finite element model. The macro-model is divided into coarse mesh elements. The micro-model lies inside the macro-model, and the stresses are mapped from a small number of macro-elements to the micro-model.

1 INTRODUCTION

The most common means of estimating the displacements and the forces of tunnel excavation is the numerical model, where the impact of the dead load above the tunnel and the tunnel excavation are solved at the same time. In any underground structure the load and the resulting displacement due to excavation of the rock is a function of the initial stress field. Here a method is presented where this stress field is generated in an additional model, as shown in Figure 1.

The initial stresses are the stresses produced by

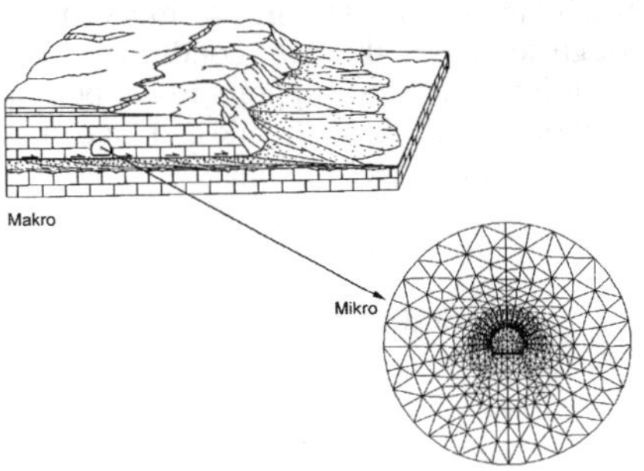

Figure 1: Relation between micro- and macro model

the own weight of overburden soil and rock and by geological movements of the region. Measurement of initial stresses for deep tunnels is generally very demanding and the scattering of measurement data makes it is very difficult to be used directly as inputs for the numerical model. There are two methods to simulate the stress field in a numerical model. One is based on the dead weight of the overburden material and the second on the input of a known initial stress field.

1.1 *Dead weight method*

In this method the dead weight of all elements of the overburden region must be activated as shown in Figure 2. The effect of irregular and inclination of the surface can be taken into consideration. The nodal forces $\{F\}$ can be defined as

$$\{F\} = \int [N]^T \{P\} d(vol) \tag{1}$$

where [N] is the shape functions

$\{P\} = \left\{ \begin{array}{c} 0 \\ \gamma \end{array} \right\}$ and γ is the density.

The resulting forces for each element are activated and must be applied on a model with vertical free boundary conditions to allow the assumed vertical displacement during this loading case. However, these boundary conditions are true only for a geological situation with a horizontal surface. The resulting displacements of this loading must be eliminated for consideration of the total settlement due to the underground excavation. For excavation simulation in the

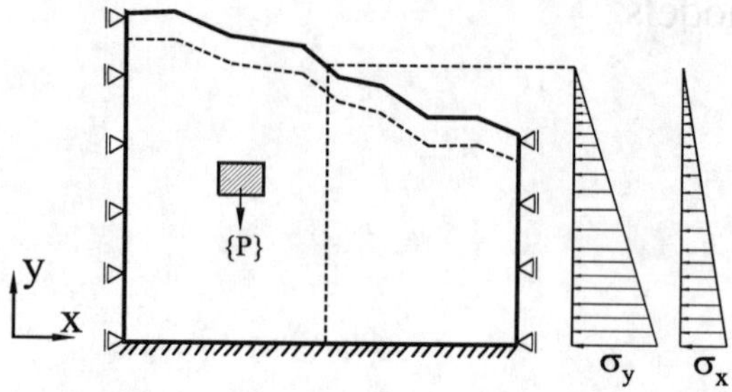

Figure 2: Stress distribution in case of dead load

numerical model the support condition must be modified, because at the farfield the nodes have no movement and should be considered in the model as fully supported nodes. The loading "dead load" will give vertical stresses σ_y and horizontal stresses σ_x. The stresses σ_x are a function of Poisson's ratio ν. This means that ν is considered an artificial constant to model the horizontal stress field.

1.2 Initial stress method

The second method is based on the assumption that the density of the ground is known. If the surface is horizontal the vertical stress field σ_y is known. It is a linear distribution of the form $\sigma_y = \gamma.h$, where h is the height of overburden. With the help of k, the earth pressure at rest, which depends on the soil parameters and geological conditions, the horizontal stresses are defined with the relation:
σ_x = k $.\sigma_y$ as shown in Figure 3.

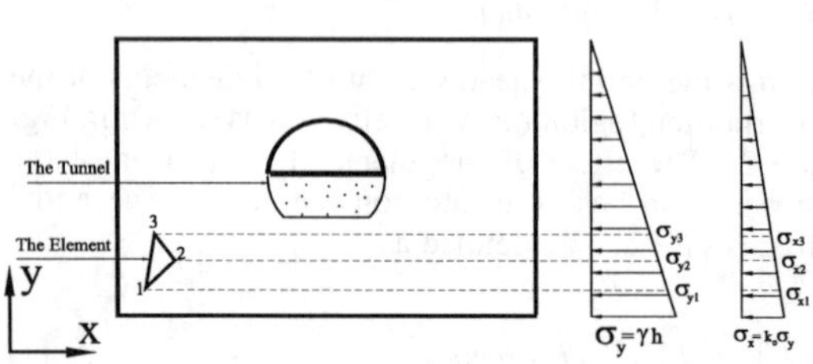

Figure 3: Stress distribution in case of initial stress

2 STRESS-MAPPING METHOD

The stress-mapping method separates the determination of the initial stress field and the excavation of an underground structure. In the so-called macro-model, as shown in Figure 1, the stress situation under a mountain, including geological movements, is calculated. This model must describe a large region around the tunnel and represent the stress field of the entire region in which the tunnel will be excavated. It also must describe the geological situation up to the surface. Within the macro-model a suitable micro-model should be identified. The purpose of this model is only to describe the change of the stress field around the tunnel. Usually the size of such a model must consider a region of three times the diameter of the excavated structure. In the stress-mapping method the stress field of the macro-model is mapped with suitable shape functions to the elements of the micro-model. The macro-model can be a three-dimensional (3D) model using 20-node brick elements (IPQS elements) or a two-dimensional (2D) model using LST elements. The micro-model can be a three-dimensional model or a two dimensional model. The following situations must be considered:

1. macro-model is 3D and micro-model is 3D
2. macro-model is 3D and micro-model is 2D
3. macro-model is 2D and micro-model is 2D

This paper takes case 2 into consideration. This case has the advantage that the position of micro-model can be changed inside the macro-model according to the value of z coordinate. Thus, different positions of the 2D model inside the macro-model can be considered. The calculation is based on two steps, namely step 1 in which the macro-model is solved by applying the dead load, after which the stresses are mapped to the micro-model using the shape functions, and step 2 in which the micro-model load is the initial stress field and the well known procedure of excavation.

2.1 Macro-model (mountain)

Figure 4 shows the typical shape of the macro-model and the applied coordinate system. The model is based on a coarse three-dimensional mesh of brick elements (IPQS-elements). The macro-model must be large enough to satisfy the field conditions. The boundary conditions are fixed supports in the bottom and roller supports in all vertical sides. Plane A-A shows the position of the micro-model.

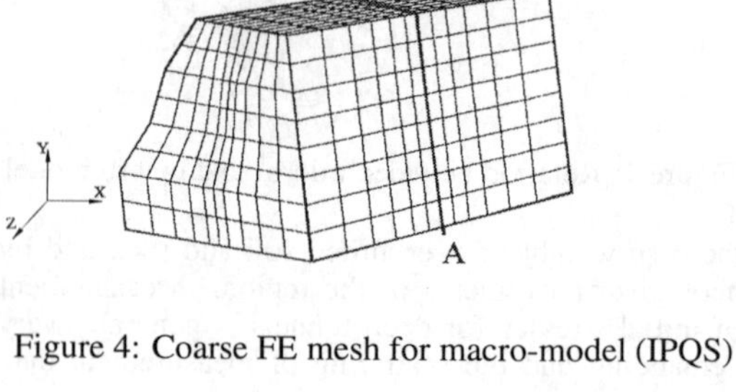

Figure 4: Coarse FE mesh for macro-model (IPQS)

2.2 *Micro-model (tunnel)*

Figure 5 shows the 2D tunnel model and the chosen surrounding soil before tunnel excavation. The soil is simulated using a fine 2D (LST) elements. The body of the tunnel was divided into BEAM 6 elements. The exact position of the micro-model with respect to the vertical cross-section of the macro-model is shown in Figure 6.

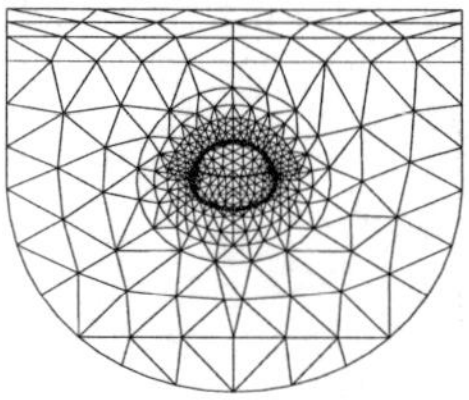

Figure 5: Fine FE mesh for micro-model (LST)

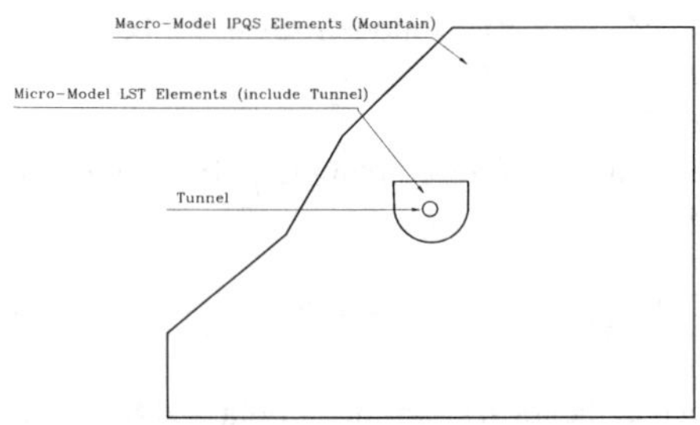

Figure 6: Position of micro-model in plane A-A

3 SHAPE FUNCTIONS

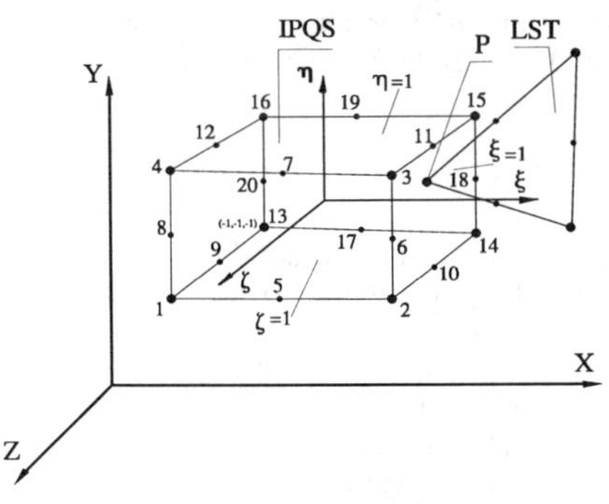

Figure 7: Relation macro-element E (IPQS) and micro-element (LST)

In Figure 7 the relation of one point P of the micro-model LST elements inside the macro-model IPQS element is shown. With the help of shape functions the initial stresses of the macro-model are mapped from known nodal stresses to the stresses of the one point of the LST element. It should be noted that transformation from the local to the global coordinate system which employs the finite element shape functions is used in all finite element formulations and can be found elsewhere [Zienkiewicz (1977)]. On the other hand, no treatment of inverse problem for the search algorithm between local and global coordinate system is followed in the present formulation.

3.1 *Problem description*

Given point $P = \{x^P, y^P, z^P\}^T$ (arising from the micro-model) in the global Cartesian coordinate system, find the stress vector $\{\sigma^P\}$ at point P using known macro-model nodal stresses $\{\sigma\}$. for the sake of simplicity let us here consider a 20 nodes isoparametric brick element. Our solution procedure of the problem proceeds in the following two steps:

1. Find an element E of the macro-model and the corresponding local coordinates $\{l_1^P, l_2^P, l_3^P\}$ of point P such that P is inside element E.
2. Let $\{X_i\}^T = \{x_i, y_i, z_i\}$ for $i = 1, \ldots, 20$ are nodes of element E. Given stress vectors $\{\sigma^i\}$ at nodes X_i, approximate stress vector $\{\sigma^P\}$ at point P.

Requirements of item 1) can be expressed in matrix form as follows:

$$\{P\} = [N(l_1^P, l_2^P, l_3^P)]\{X\} \quad \text{and} \quad -1 \leq l_1^P, l_2^P, l_3^P \leq 1 \tag{2}$$

where the corresponding entries of matrix $[N] \in \Re^{3\times 60}$ are equal to N_i or 0 with $N_i(l_1, l_2, l_3)$ $(i = 1, \ldots, 20)$ being the FE shape functions of element E and vector $\{X\}^T = \{\{X_1\}^T, \ldots, \{X_{20}\}^T\}$. Shape functions N_i are naturally defined in the local coordinate system.

Mapping $\mathcal{T} : \Re^3 \longrightarrow \Re^3$ from the local to the global coordinate system is defined in the standard FE formulation as follows:

$$\mathcal{T}(l_1, l_2, l_3) = [N(l_1, l_2, l_3)]\{X\}. \tag{3}$$

If there is a one-to-one correspondence in the mapping $\mathcal{T}$ from the local to the global coordinate system, then the inverse mapping $\mathcal{U} = \mathcal{T}^{-1}$ exists.
We shall outline here how to construct approximation $\bar{\mathcal{U}}$ to mapping $\mathcal{U}$. It is assumed that point P lies in the region where $\mathcal{U}$ exists and is sufficiently smooth.

Eq. (3) defining mapping $\mathcal{T}$ as a polynomial in local coordinates (l_1, l_2, l_3) such that for $i = 1, \ldots, 20$ we have

$$\mathcal{T}(l_1^i, l_2^i, l_3^i) = \{X_i\} \tag{4}$$

where $\{\mathcal{L}^i\}^T = \{l_1^i, l_2^i, l_3^i\}$ are local coordinates of nodes X_i. Thus, it is naturally to define mapping $\bar{\mathcal{U}} : \Re^3 \longrightarrow \Re^3$, where $\bar{\mathcal{U}} = \{\bar{\mathcal{U}}_1, \bar{\mathcal{U}}_2, \bar{\mathcal{U}}_3\}^T$ is an approximation to $\mathcal{U}$, as a polynomial in the global coordinates:

$$\bar{\mathcal{U}}_j(x,y,z) = \{A^j\}^T\{\varphi(x,y,z)\}, \qquad j = 1,2,3 \quad (5)$$

where

$$\{A^j\}^T = \{a_1^j, \ldots, a_{20}^j\} \quad (6)$$

and

$$\{\varphi_j(x,y,z)\}^T = \{1, x, y, z, xy, yz, xz, xyz, x^2, y^2, z^2, x^2y, x^2z, y^2x, y^2z, z^2x, z^2y, x^2yz, y^2xz, z^2xy\}. \quad (7)$$

In order to uniquely determine coefficient vectors $\{A^j\}$ and to relate the values of $\bar{\mathcal{U}}$ to $\mathcal{U}$, the following condition is imposed Eq. (4)):

$$\bar{\mathcal{U}}_j(\{X_i\}) = \mathcal{U}_j(\{X_i\}) = l_j^i \qquad i = 1, \ldots, 20 \;\; j = 1,2,3. \quad (8)$$

For given local coordinate direction j, condition Eq. (8) is a linear system of equations with unknown coefficient vector $\{A^j\}$. Note that the left-hand side of Eq. (8) is independent of j. Thus, having solved Eq. (8) for $\{A^j\}$, mapping $\bar{\mathcal{U}}$ is defined by Eqs. (5)-(8). Local coordinates corresponding to point P are now obtained as follows:

$$l_j^P = \bar{\mathcal{U}}_j(\{P\}) = \{\varphi(\{P\})\}^T\{A^j\} \qquad j = 1,2,3. \quad (9)$$

Point P is defined as to *be inside* element E if $-1 \leq l_j^P \leq 1$ for each $j = 1,2,3$. Note that for local coordinates computed from Eq.(9) the standard finite element local to global transformation Eq.(2) is satisfied with some approximation. Nevertheless, error in Eq. (2) related to the size of element E was of order 1.0E-4 in the test examples, which is quite satisfactory for practical purposes.

Given macro-model stresses $\{\sigma^i\}$ at nodes X_i of element E, , the micro-model stress vector $\{\sigma^P\}$ at point P can be approximated using the FE shape functions Eq. (2):

$$\sigma_k^P = \sum_{i=1}^{20} N_i(l_1^P, l_2^P, l_3^P)\sigma_k^i \qquad k = 1, \ldots, 6. \quad (10)$$

Another alternative for computing $\{\sigma^P\}$ is to construct element shape functions $\bar{N}_i(x,y,z)$ in the global coordinates of the form (Eqs. (5)-(7)):

$$\bar{N}_i(x,y,z) = \{A^i\}^T\{\varphi(x,y,z)\}, \quad i = 1, \ldots, 20 \quad (11)$$

and computing the coefficient vectors $\{A^i\}$ from the condition

$$\bar{N}_i(\{X_k\}) = \delta_{ik} \qquad i = 1, \ldots, 20 \;\; k = 1, \ldots, 20. \quad (12)$$

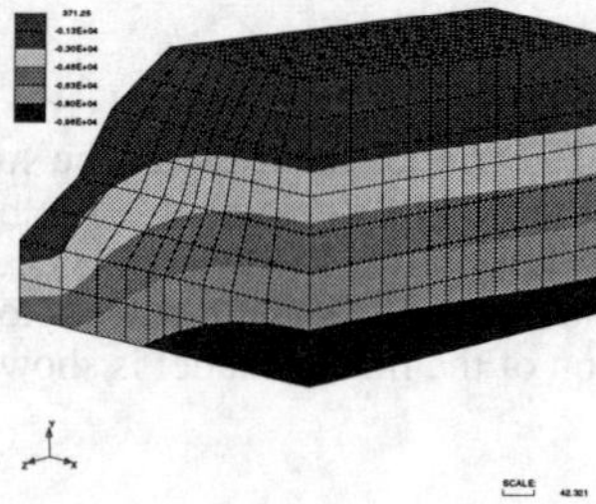

Figure 8: Stress distribution of σ_y for macro-model

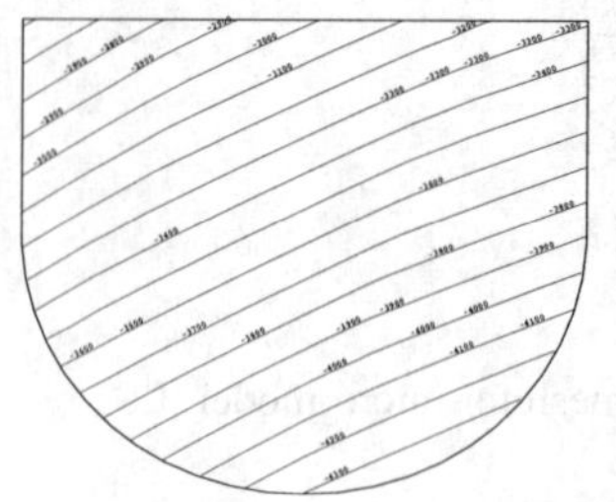

Figure 9: Mapped micro-modell stress σ_y, before excavation

Local coordinates corresponding to point P are computed as

$$l_j^P = \sum_{i=1}^{20} \bar{N}_i(\{P\})l_j^i \qquad j = 1,2,3. \quad (13)$$

It is seen that the local coordinates obtained from Eqs. (9) and (13) are the same.
Using the "global" shape functions $\bar{N}_i$, the micro-model stress vector $\{\sigma^P\}$ can be approximated as:

$$\sigma_k^P = \sum_{i=1}^{20} \bar{N}_i(\{P\})\sigma_k^i \qquad k = 1, \ldots, 6. \quad (14)$$

As far as our test examples are concerned, the relative difference between $\{\sigma^P\}$ obtained from Eqs.(10) and (14) was found negligible for practical purposes.

4 APPLICATION

The macro-model was solved under the effect of dead load and the resulting distribution of vertical stress σ_y is shown in Figure 8.
The mapped stress of the micro-model is shown in Figure 9. Excavation of the tunnel was simulated in three steps: excavation of the crown, first bench excavation and final bench excavation as shown in Figure 10. The stress in the micro-model following excavation is as shown in Figure 11 and the normal force in tunnel BEAM 6 elements is as shown in Figure 12. The results obtained with the stress-mapping method were compared with the results obtained by solving

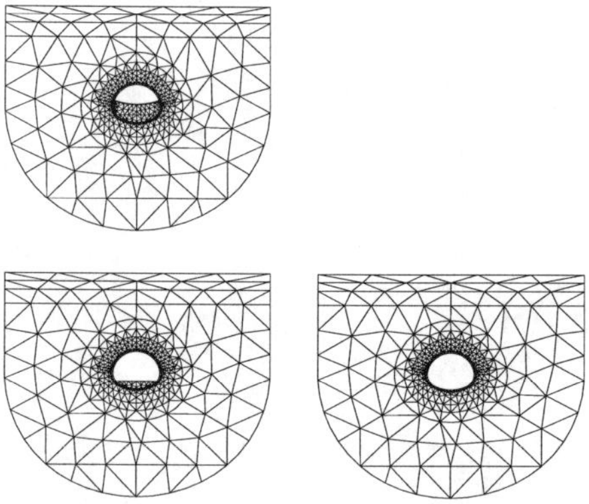

Figure 10: Steps of excavation simulation for micro-model

the micro-model solely by applying initial stress. In the second case only a stress field with vertically oriented principal stresses can be considered. The results for normal force after excavation is shown in Figure 13.

From these figures it was noticed that the results obtained with the new technique agree approximately with the results obtained by solving the micro-model only. The inclination of isoline stresses in Figure 9 is due to the inclination of ground surface of macro-model.

5 CONCLUSIONS

1. Using the stress-mapping method two coordinate systems for two- and three-dimensional models can be used at the same time.

2. The stress field of the micro-model can be generated at any cross section with the help of the Z-coordinate. Thus, the critical section for the design can be found.

3. The method improves the application of two-dimensional models, because it is based on a realistic initial stress field.

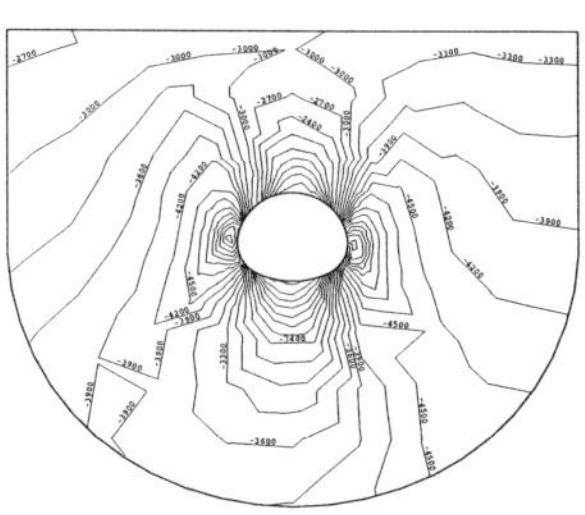

Figure 11: Micro-macro mapping, σ_y, after excavation

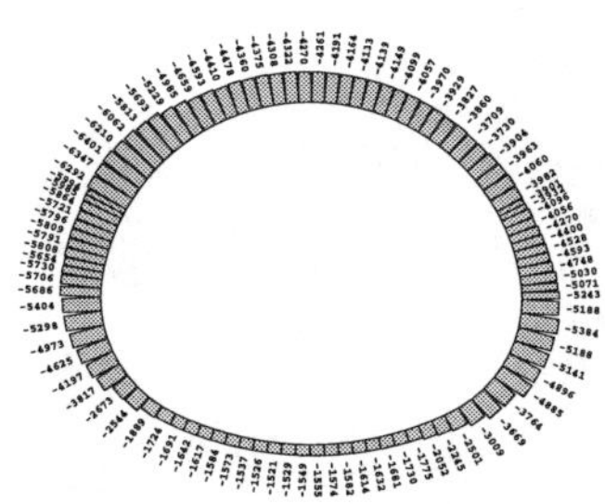

Figure 12: Micro-macro mapping, N, after excavation

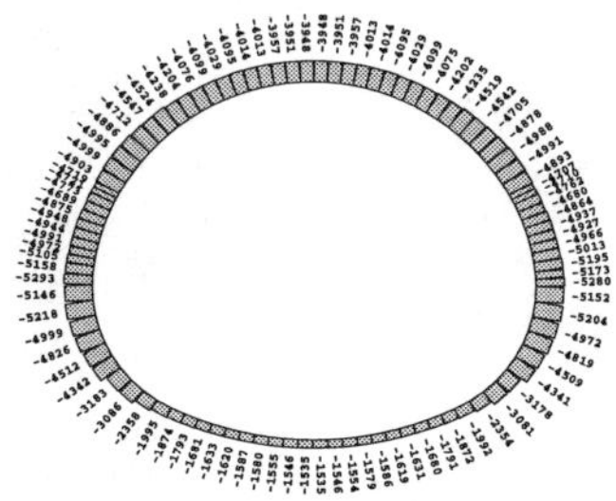

Figure 13: Initial stress method, N, after excavation

ACKNOWLEDGMENTS

The work reported here was performed under contract No. 800507 of the Austrian Industrial Research Fund (Forschungsfrderungsfonds fr die gewerbliche wirtschaft).

REFERENCES

Zienkiewicz, O. C. 1977. The Finite Element Method, Third Edition, Mc Graw-Hill, New York, 802pp.

Computer Methods and Advances in Geomechanics, Desai et al. (eds) © 2001 Balkema, Rotterdam, ISBN 90 5809 183 X

Database management of a large, fast-track field exploration program

D.C.Ward, P.M.Godlewski & R.A.Robinson
Shannon and Wilson Incorporated, Seattle, Wash., USA

G.R.Fischer
Shannon and Wilson Incorporated, Denver, Colo., USA

ABSTRACT: Large, fast-track field exploration programs present unique challenges to geotechnical engineers. Successfully accomplishing these programs requires managing the large amount of data associated with the explorations so that it is useable to multiple sources on a continuous, almost immediate basis. A custom database application called Project Track was developed to track all aspects of the exploration program for the Central Puget Sound Light Rail Transit Project, from initially selecting boring locations to tabulating readings from instrumentation installed in the completed borings. The information stored in Project Track was used for multiple purposes: to generate daily, weekly, and monthly progress reports to task managers and other members of the design team; to assist in reconciling invoices by tracking drilling costs; to track field and laboratory results; to track our Quality Control procedures; to rapidly develop updated profiles and cross sections; and to assist in preparation of reports.

1 INTRODUCTION

The proposed Central Puget Sound Light Rail Transit System (Link) includes alignments both north and south of downtown Seattle, Washington as well as an alignment in Tacoma, Washington, totaling 32.8 km of new construction. Shannon & Wilson, Inc., completed limited explorations for the conceptual and preliminary engineering phases for the design team in 1998 and 1999. In March 1999, Shannon & Wilson, Inc., was awarded the contract for performing more comprehensive geotechnical engineering services for about 7.2 km of tunnel from the existing Westlake Station to an underground termination just north of the University of Washington (referred to as the North Corridor Tunnel) and about 1.6 km of tunnel through Beacon Hill (referred to as the Beacon Hill Tunnel).

The North Corridor Tunnel will include three deep-mined stations, two elevator shafts at each of the three deep-mined stations, two mined cross-overs, one cut-and-cover station, one cut-and-cover cross-over, cross passages between the twin-tubes of the running tunnel at a spacing of about 240 m, and one midline ventilation shaft between the Capitol Hill Station and the Pacific Street Station. The Beacon Hill Tunnel will include two portal structures, one elevator shaft, and a shell for a future deep-mined station. The depths of the mined stations will range from about 50 to 80 m, which, when constructed, will make them among the deepest subway stations in soil in the world. The engineering required to construct these stations will require a significant understanding of the geotechnical properties and behavior of the soils and groundwater.

As a result of the preliminary engineering phase, it was estimated that to quantify the properties and behavior of the soils and meet the schedule milestones, over 7,300 m of drilling and sampling (97 borings), approximately 30 borehole pressuremeter tests, 10 downhole seismic tests, 5 aquifer pumping tests, approximately 50 slug tests, and six major reports (one Geotechnical Data Report, one Geotechnical Interpretive Report and one Geotechnical Baseline Report for each of two separate contracts) would be required in 6 months. At the peak of the project over 100 people would be required to work on the project. In addition, to meet the fast-track schedule, up to 8 drill rigs working simultaneously would be required.

In preparation for conducting a large, fast-track field exploration program, a custom database application called Project Track was developed to manage the large amount of data associated with the explorations. Project Track was used to track all aspects of the exploration program, from initially selecting boring locations, to tabulating readings from instrumentation installed in the completed borings and developing finished profiles, cross sections and data plots.

2 STORING DATA

Spreadsheets are often used to track the progress and present the results of a geotechnical exploration program. As many as a dozen spreadsheets may be required to track such pertinent data and activities as boring locations, utility checks, drilling rates, drilling costs, instrumentation, lab test results, scheduling, and budget. In dealing with over 2,000 samples and specimens, with at least one laboratory test for each sample, maintaining the proper Quality Control (QC) would be an almost impossible task without a database system. Some of the other problems associated with maintaining separate data files and spreadsheets are data redundancy, inaccessible data, lack of standards, and lost productivity. Each of these is described below.

If multiple spreadsheets are used to track data, redundant data will likely be stored in the spreadsheets, which can lead to apparent discrepancies in the data. For example, the interpreted geologic unit initially assigned to a given sample by a field representative changed as the understanding of the geology of Seattle evolved (the boring logs went through a 7 step QC process), and additional laboratory test results became available. Given the number of people working on the project, it would have been an arduous task to update every spreadsheet to ensure each user was aware of the change in the geologic unit assigned to a sample.

Further, data stored in spreadsheets on a desktop computer is considered inaccessible.

Even if data is entered on time and stored in spreadsheets on a computer network, it may be inaccessible if it is entered in a non-standard manner. Multiple spreadsheets created by multiple people often lack a "standard" feel. The storage of data, the manipulation of data, and the calculations in spreadsheets created by another person are often not easy to follow and can impede the free exchange of data and ideas.

Lost productivity, in this model, can result from periodic checking for data redundancy and discrepancies, seeking the correct up-to-date spreadsheet, or trying to discern the way data is stored in another user's spreadsheet before a task can begin.

The solution to the data management problem associated with using multiple data files and spreadsheets to store data was to create a centralized database which would integrate the data and data presentation and manipulation tools available in ArcView – a Geographic Information System (GIS) program, AutoCAD, Geotechnical Integrator (gINT), Microsoft EXCEL, Microsoft PROJECT, and a proprietary in-house laboratory test program written in Visual Basic. The goal was to link the data files from these programs together to avoid the problems of dealing with out-of-date information and conflicting data.

3 PROJECT TRACK

3.1 *Development*

Project Track was developed at Shannon & Wilson, Inc., using Microsoft ACCESS and was linked to gINT – a popular geotechnical database and log creation program. The data tables in gINT were linked into Project Track so that by opening Project Track, all of the data in gINT and Project Track were accessible in one location.

The version of Microsoft ACCESS used to create Project Track had a different database engine than the one used in gINT. Therefore, although the data from gINT could be linked to Project Track, the gINT data could not be edited in Project Track. This meant that the data had to be stored in two separate data files. Consequently, care was taken in the development phase to avoid redundant data fields in the two separate data files. For example, the depth of a boring is a required field in gINT. So rather than duplicate the field in Project Track, when the depth was need in a report, a query was used to select the depth from a linked gINT data table.

3.2 *Data Sources*

Project Track was used to store information on costs, drilling rates, schedule, permitting, groundwater instrumentation installations, in-situ tests, laboratory strength tests, and methane tests. Information on boring depths, locations, sampling methods, index laboratory tests, geology, samples, and specimens was stored in gINT. Data was entered into Project Track daily from drilling summary sheets completed by our field representatives or by task managers. Data also was entered into gINT from information on the field boring logs and laboratory index property test results. Queries were used to retrieve the information stored in the two data files. Functions were used to evaluate the data.

3.3 *Queries and Functions*

Queries were used to create datasheets which were used for analyzing and reporting data. The queries often linked the data stored in the two data files. For example, the data from in situ tests (e.g. pressuremeter tests and downhole seismic tests) were stored in Project Track while the interpreted geologic data specific to the borings in which the testing was performed was stored in gINT. When analyzing in situ tests results, queries were used to assign the most up-to-date interpreted geologic unit to the test results.

Several functions were written in Project Track to evaluate the data. For example, one function calculated the percent clay, based on the 0.002-mm grain size criteria, from the hydrometer data. Another

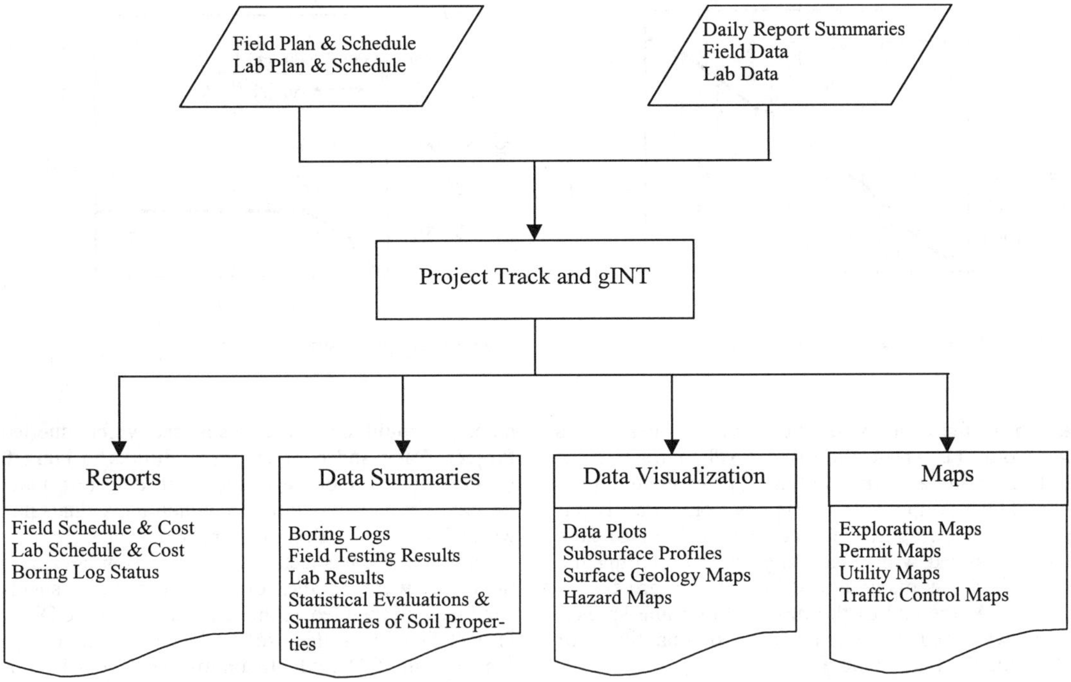

Figure 1. Relationship between various data sources, Project Track and gINT, and the output.

function searched the database to find the nearest laboratory test result to given depth in boring. This function was helpful for looking for correlations between laboratory test results and in-situ test results (e.g. plasticity index and limit pressure from pressuremeter testing).

3.4 *Output*

The output from Project Track was designed to help multiple users identify the status of various aspects of the project, potential problems with the field exploration program, track costs, budgets, and schedules, look for oddities in the data, look for correlations among various different data sets, and generate reports while always using the most up-to-date and reliable data. The output from Project Track can be divided into four general categories: reports, data summaries, data visualization, and maps. A schematic illustrating the relationship of the various data sources and the output is shown on Figure 1.

One output report from Project Track was the Inspector's Sheet. An inspector's sheet was produced for each boring and included such pertinent information as the approximate location and elevation of the boring, the planned depth, the zone in the boring with the closer 0.75 m sampling interval, any in-situ testing scheduled for the boring, the initial plan for installing a groundwater monitoring device, site contacts (whether a private right-of-entry [ROE] or government agency), a record of the utilities contacted by the Utility Locate Service, and a place for additional notes. This form provided the inspector with the most commonly requested data and enabled them to readily identify potential problems such as unmarked utilities.

Project Track was also used to check for potential conflicts associated with the planned installation of groundwater level measuring devices and in-situ testing (downhole seismic testing and tomography). For example, because tomography testing required that the borehole be full of water during the test, an observation well could not be installed in a boring selected to be used for tomography. Queries generated in Project Track would notify the users if there was a potential conflict.

A record of the drilling costs was generated using Project Track. During each shift, field representatives filled out a "Daily Drilling Summary" form which contained a detailed list of pay items, the footage drilled for the day, and a summary of the depths and thickness of the interpreted geologic units encountered. At the end of the shift, the driller and field representative would agree on the pay items and the driller would sign the form. The representative would turn the form in that evening and the next morning a data entry clerk would enter the data into Project Track. Every Monday, a spreadsheet developed in Microsoft EXCEL, containing a pivot table, would be used to query Project Track

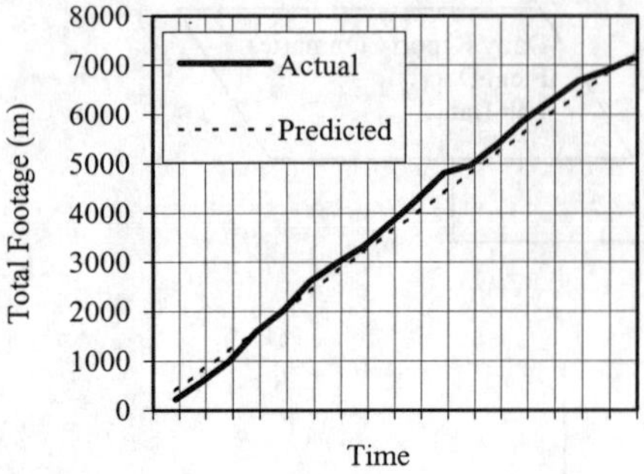

Figure 2. Actual versus predicted drilling rates.

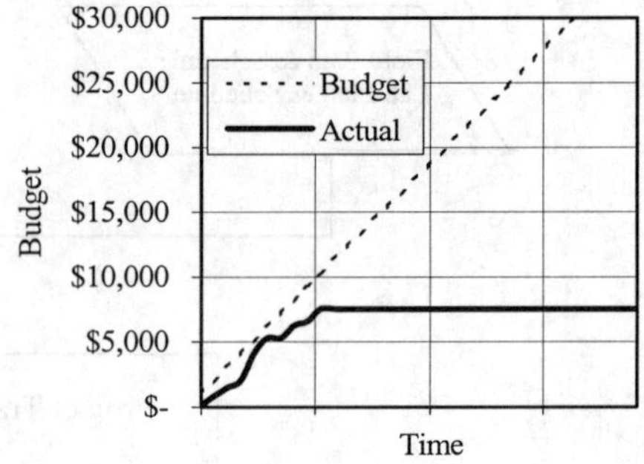

Figure 3. Budget versus actual costs.

and print the charges for the previous week. This list would be sent to the drilling subcontractors for billing approval. In addition, the estimated total would be used to track drilling costs on a weekly basis. By reviewing the charges on a weekly basis, the labor involved with generating and reconciling the invoices for as many as 8 drill rigs was greatly reduced. At the end of the month, this same spreadsheet was used to generate a list of quantities for checking the final invoice.

The data entered from the "Daily Drilling Summary" forms was also used to generate a daily progress report containing the footage drilled and the geologic units encountered. This report was e-mailed to the design team who was proceeding with preliminary engineering of the tunnel and stations. It was also used to generate a weekly report listing the borings started, completed, and those estimated to be started and completed the following week. In addition, the data was used to create a monthly progress report summarizing the field exploration program. A graph showing predicted drilling rate versus the actual drilling rate is presented on Figure 2.

The same data used to generate this graph was used to update weekly schedule projections by altering the estimated duration of the borings not yet completed to match the trends observed in the data. For scheduling purposes, an initial conservative estimated of 11 m/day for drilling and sampling of the borings (including mobilization to the site, well installation, and site clean-up) was used. When queries of the data indicated that a given driller was regularly achieving closer to 15 m/day, the schedule was adjusted accordingly.

Another output was a series of graphs used to track the budgets of individual tasks. Significant tasks included field, administration, in-situ testing, groundwater, laboratory, subsurface characterization, and reports. Because the data from our time sheets is stored in a Microsoft ACCESS compatible form, the labor costs for the previous week were generally available every Monday. An example task tracking graph is presented on Figure 3. A task manager could open a spreadsheet which queried Project Track and compare expenditures to budget. In addition, by looking at the same data in a table format, the manager could see which individuals had worked on the task during the previous week.

Individual datasheets for every laboratory and field test were another output. These data sheets were used in the Geotechnical Data Reports (GDRs). Project Track was also used to develop the table presented in the GDR which summarized every laboratory test completed. The table for the North Corridor Tunnel contained a matrix over 1,700 samples long by about 20 tests wide. It also included information about each sample such as boring designation, depth, sample number, interpreted geologic unit, and a soil description. By linking tables and using the database structure, data discrepancies and duplications were avoided.

Another output was a set of queries used to look for correlations in laboratory results. For example, queries were written to examine if there was a correlation between liquid limit and shear strength for each of the 3 major fine-grained geologic units. The queries could also be used to double-check the geologic interpretation for samples. For example, if the plastic limit for a given sample fell outside of a certain range, the sample was re-examined to check if it was identified correctly.

Project track also allowed for the development of stick logs for use in AutoCAD to generate subsurface profiles. The stick logs contained blow counts, sampler information, data on gas monitoring, in-situ testing, groundwater elevation, USCS classification, and interpreted geologic unit. These stick logs could be quickly generated allowing rapid updating of the subsurface conditions. The rapid updating of the subsurface profiles allowed the tunnel designers to evaluate the proposed alignment using the most up-to-date data available. Consequently, revised profiles and cross sections could be developed frequently with greatly reduced effort and always using the most current available data.

One of the map outputs made possible by linking Project Track and ArcView were site vicinity maps which graphically illustrated the progress of the field exploration program. The colors used on the Exploration Status Maps allowed for an easy means to distinguish among borings which were completed, in-progress, ready, and not-ready. Because Project Track was linked to ArcView, the user could "identify" a boring in ArcView and be shown which steps needed to be accomplished before the status of boring could be upgraded from "not-ready" to "ready".

4 RESULTS

By storing the data in a relational database structure, the problems normally associated with multiple data files and spreadsheets were eliminated. Data redundancy and data discrepancies were eliminated by using queries to construct datasheets. For example, the laboratory test results from a Consolidated-Undrained (CU) Triaxial Compression Test were stored based on boring number and depth. When the datasheet for a CU test was opened, the boring number and depth were used to query the data table containing the interpreted geologic unit in order to assign the most up-to-date value.

One drawback to storing data in a centralized database is that the ability for one person to enter data can be dependent on another person having already entered other data for which they are responsible. For example, the data related to the well construction had to be entered before water level data could be entered. This meant that if a round of groundwater measurements was taken soon after a well was installed, the data pertaining to that well might not have been able to be entered immediately.

On the other hand, the database was stored on an office-wide network so the data was always accessible to all project personnel. The linking of tables allowed task managers to create queries which were used to check and see if data was being entered in a timely manner. For example, a query was used to check the elapsed time between completion of a boring and each of the seven steps in the QC process for the boring logs. If the elapsed time exceeded a pre-determined threshold, the boring was flagged, and the cause of the delay investigated.

The database structure also enforced standards. By assigning a strict formatting scheme to each column in the data table, the consistency of boring designations, the number of decimal places, the units (e.g. cm or mm), and geologic unit terminology could be maintained.

Moreover, loss in productivity was minimized by limiting data redundancy and data discrepancy, checking for accessibility of data, and maintaining standards.

The centralized data structure also reduced the amount of time it took to make changes to reports. For example, when two previously separate contract sections were combined into one, only one data field per a boring had to be changed in order for all of the data associated with that boring to be reported in queries for the new contract section.

Statistical evaluation and data visualizations made possible using Project Track, along with the results of laboratory testing on over 17,000 samples and the results of in situ testing, enabled the development of a database of soil properties which has facilitated a previously unparalleled understanding of the nature and behavior of the soils along the proposed tunnel alignment.

5 CONCLUSION

The use of database management tools allowed for a large, fast-track field exploration program to be successfully completed ahead of schedule and on budget. It provided technical, scheduling, and cost information as well as project tracking tools that were at the disposal of the various task leaders.

The productivity achieved using a centralized database is dependent on each person promptly entering the data for which they are responsible. However, unlike spreadsheets, it is easy to tell if the data is being entered promptly and in the proper format.

Computer Methods and Advances in Geomechanics, Desai et al. (eds) © 2001 Balkema, Rotterdam, ISBN 90 5809 183 X

The solution of conceptual problems linked with the construction of a tunnel

Kamila Weiglova
Institute of Geotechnics, Faculty of Civil Engineering, Brno University of Technology, Czech Republic

ABSTRACT: In view of the fact that for building a railway tunnel on the track Žatec - Chomutov a test section was determined (an experimental tunnel in the quarry Libouš II) in which extensive measurements of the strain and deformations in situ were carried out and in our laboratory a physical model was built in the scale 1 : 10, it is possible to compare the model methods with the actually existing magnitudes and to use the for the determination of the technological process of building the tunnel.

1. INTRODUCTION

For the selection of the technological procedure of tunnel building, when we are interested in the relation between the dimensions of the stope and its lining on the one hand and the deformation, strain and stability of the rock on the other hand, it is possible to start not only from laboratory experiments and the in situ measurements, but, if possible, also from the physical and/or mathematical model.

This comparison was possible in the railway tunnel on the track Žatec - Chomutov. The tunnel at the length of 1 640 m is designed as a single-track one with the cross-section area of 41.1 m^2 and uniform ratio of inclination of 9 $^{0}/_{00}$. Its vertical line is 10 - 34 m below the present surface of the field.

Due to the fact that for the implementation of this project an experimental section (a pilot gallery) was determined in which extensive measurements of strain and in situ deformations were carried out and in our laboratory a physical model was implemented at the scale of 1 : 10, it was possible to compare the results of the modelling methods with the actual magnitudes which were used for the determination of the technological procedure of the tunnel building.

2. CONTROL GALLERY DEEP AND IN SITU MEASUREMENTS

The building of the tunnel near Březno on the shifting of tracks on the track Žatec - Chomutov will be implemented under complex engineering-technological conditions represented by tunnelling in fine grained soils. The soils can be characterised as saturated, overconsolidated, torn clays which have specific properties, such as high plasticity, considerable fracturing, volume instability mostly dilatant behaviour, considerable anisotropy and sensitivity to progressive deformation.

For the prepared tunnel building it is necessary to know not only the properties of the individual soils and rocks, but also the behaviour of the massif as a whole within the reach of the stope carried out, when the size of deformations affects, besides others, the way of development of the tunnel.

That is why, as part of a detailed engineering-geological investigation, a control gallery deep was made that permitted extensive geotechnical measurements.

The length of the control gallery deep was selected to be 37.6 m, the circular section of the gallery of the diameter of 2.5 m. The height of the overlying beds is graded, to about 5 m of overlying beds in the first part of the gallery, in the middle part from 5 to 10 m with respect to the gradient of the slope, ad in the remaining part 10 m of overlying beds.

3. SUBSTANTIATION OF THE IMPLEMENTATION OF THE PHYSICAL MODEL

For the choice of the technological procedure the determining factor is stability of the rocks (in the overlying beds) both from the point of view of the safety of work, from the point of view of mechanisation of working methods as well as in view of the maximum efficiency of building.

It is easy to understand the effort of opening the whole tunnel cross-section at once or divide the stope as little as possible and maximally utilise the carrying capacity of the rock.

An apposite forecast of the stability is, however, very difficult, because it is necessary to take into consideration a whole number of factors influencing each other. They are above all the dimensions of the untimbered tunnel, the strength of the rock, its original strain, the force action carried into the rock by the break, the rigidity of the provisional lining of the tunnel, the direction and inclination of the planes of discontinuity in the rock to the axis of the tunnel and, in the case of clay, also the time factor is much significant. From those factors we usually know only the dimensions of the front face of the tunnel and the rigidity of the outfit. The knowledge of the other factors depends on the extent of the geotechnical investigation of the given territory. For these and a number of further reasons the stability of the rock is usually determined by estimate and on the basis of experiments from building other tunnels.

An exact finding of the rock stability is possible on physical models. An indisputable advantage of physical models is the fact that, with correctly chosen input data, they create the random conditions themselves. That is why at present the results of physical modelling can be considered relatively the most reliable.

4. MODEL GEO-BRNO

After a detailed analysis and statistical evaluation of laboratory experiments the method of modelling was elaborated which respected the following basic points:

- the relation between the transversal dimension of the stope and the rock stability
- the state of strain in the rock environment
- rock deformation around the tunnel breaks
- the state of strain in the provisional outfit
- the evaluation of the jetcrete lining

These assumptions were applied by means of the methods of similarity and dimensional analysis on the physical model Geo-Brno 97 which was determined above all for measuring the changes in the strain in the surroundings of the circular cross-section in the course of the technological procedure of tunnelling.

The model was built in the laboratory of the physical modelling of the Institute of Geotechnics in the scale 1 : 10 using the suitably selected equivalent material.

For the model a stand was constructed, dimensions 130 by 100 by 39.2 cm. The main requirement at the end of the stand is a perfect rigidity, which is necessary with respect to the method of measuring the magnitudes of strain and deformations on the model. It was possible to load the model by hydraulic cushions from all sides (with the exception of the front part - glass - and the upper part, where the load is simulated mechanically). By changing the pressure in the individual cushions in the course of the experiment it was possible to imitate almost every situation of loading. In essence it was a triaxial apparatus (see Fig. 1).

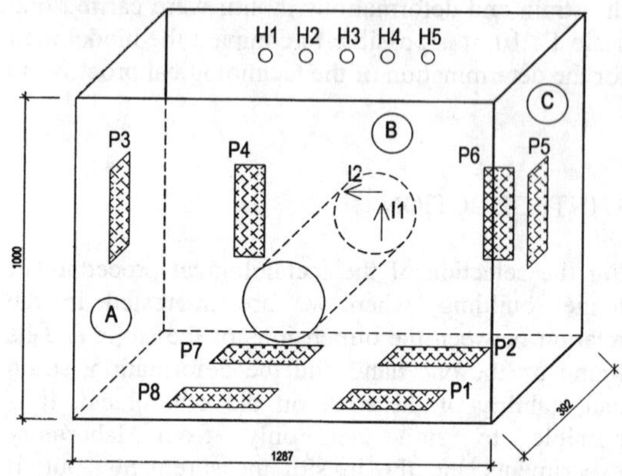

scale 1 : 10

I_1 - I_2 - the measuring gauge of convergence

H_1 - H_5 - the measuring gauge of vertical deformation

P_1 - P_8 - the hydraulic measuring gauge of the state of strain

Fig. 1 Stand Geo Brno - location of hydraulic measuring gauges of the state of strain and measuring gauges of deformation

The model was intended for measuring the changes in strain during the technological procedure of tunnelling in the surroundings of the circular cross-section of the diameter of 25 cm (i.e. 2.5 m) with he height h = 50 cm (5 m), i.e. for low overlying beds in the pilot gallery. The width of the model was 39.2 cm, so that the tunnelling was carried out in 4 advances of 10 cm each and then the model was loaded until the failure.

In the technological procedure of tunnelling the methodology of lining was simulated in the model of shifting the coat outfit. This method reminds of the mechanical shield of tunnelling, but it has also common steps with the method of outfit by means of jetcrete. The outfit met the conditions of the similarity in structure rigidity (jetcrete) and thus the possibility of advance tunnelling was made possible to the length of 10 cm (1 m). Unlike the method of shield tunnelling, under our model conditions there was a perfect co-action of the structure with the neighbourhood, i.e. there was no free space left between the outfit and the rock.

5. TO THE ANALYSIS OF RESIDUAL LATERAL SOIL PRESSURES EVOKED BY THE COMPACTION OF EQUIVALENT MATERIALS

The model tests proper were started only after a thorough analysis of the effect of loading, relieving and reloading cycles on the model. This analysis is essential for determining the size of the residual lateral earth pressures due to the compaction of the equivalent material.

The compaction (the effect of loading, relieving and loading) of equivalent materials causes the increase in lateral pressures. The increment of strain due to compaction can have far reaching consequences in the models Geo-Brno 97, because the mechanical behaviour of equivalent materials depends on the force action of equivalent materials and the repeated loading can substantially affect these effects.

The increment strains due to compaction evoke deformations in the yielding structure and therefore a perfectly rigid shield was used.

For a hysteresis model it was necessary to determine: the coefficient of relieving $0 < a < 1$, the coefficient of re-loading $0 < b < 1$ and the coefficient of the lateral earth pressure at rest $0 < K_r < 1$.

The experiment has shown that the process of compaction in the model is a certain form of reconsolidation, where the strain due to the effect of temporary loading to a certain extent remains preserved in equivalent material even after removing the load.

In our model Geo-Brno the lateral residual strain was 28.3 % of the total horizontal strain.

6. THE MODEL TESTS PROPER

For the solution of the model Geo-Brno 97 we started from the following main assumptions:

- rocks in the massif are homogeneous (clay to claystone)
- during the technological procedure of tunnelling three regions can be formed, elastic,. Elastically plastic and the region of failure (this assumption is confirmed by the results of in situ measurement)
- the transformation properties in different directions are not changed (not completely fulfilled)
- it is possible to solve the concurrence of the underground structure and its neighbourhood on models (this assumption is confirmed by the existence of invariants of similarity and the possibility of expressing physical equations by means of them)

The objective of technological tunnelling in the model was the determination of the size of pressures inside the tunnel necessary for maintaining the stability of the working face for different advance lengths. The fact whether the working face will cave in or not depends on the working face geometry and the size of the developed negative pressures.

For solving the stability of the model Geo-Brno we started from the theory of limit states. The solution was by means of dimensionless parameters that can be derived from dimensional analysis.

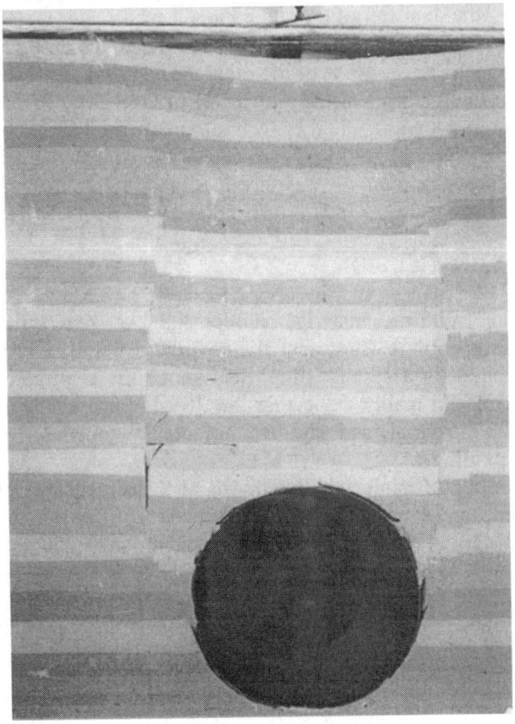

Fig. 2. The state after the failure - model Geo-Brno 97 - detail

For the formulation of the solution of the stress necessary for maintaining the stability of the tunnel with low overlying beds it is necessary to explain the following dimesionless parameters:

1. P_v - the ratio of the height of the overlying beds to the diameter of stope; $P_v = 2$
2. P_c - the value of the share of the volume mass of the material, multiplied by the diameter of the stope and cohesion; $P_c = 0.679$, $P > 0$
3. P_{tm} - the ratio of the maximum total stress in the axis of the tunnel before tunnelling to cohesion; $P_{tm} = 3.1$
4. P_{hl} - the ratio of the length of the unlined stope (l = 10 cm, l_2 = 20 cm, l_3 = 30 cm, l_4 = 40 cm) to the diameter of the stope (d = 25 cm); P_{hl} = 0.4; 0.8; 1.2; 1.6

Since the complete solution of the problems of the stability of the model Geo-Brno 97 by means of dimensionless magnitudes is difficult, the experiment was divided into four parts (l_1 - l_4), each of them being important for the stability. The results of the model Geo-Brno 97 can be summarised into the following points:

- the ratio of the unlined stope to the diameter of the stope P_{hl} - 1
- the ratio of the decline of the top of the stope to the diameter of the stope P_{vd} = 1.6 %
- the ratio of the critical stress to the cohesion P_{kn} = 0.5 (i.e. 16.08 % of the maximum stress)

On the basis of the model test of Geo-Brno 97 the harmonogram of the technological procedure of building the railway tunnel on the railway track Žatec-Chomutov was determined

7. CONCLUSION

In building material models it is practically impossible to represent the reality in detail. Every model is a certain simplification of the reality, of course a reality respecting the decisive properties of the material and the technological procedure of building.

In our opinion the intentions of the dimensions of the break up (i..e. for the diameter of 250 cm the length of the untimbered stope is as much as 250 cm) in using the method of lining by jetcrete. Jetcrete as lining material does not take only the mechanical functions, but in a physical action it reinforces the superficial layer of the rock and makes it take an active part in the stabilisation of the stope.

The performed tests on the models Geo-Brno 97 have demonstrated that modelling by means of equivalent materials is a suitable method for studying the processing taking place by the failure of the stability of the rock massif.

REFERENCES

1. Weiglová, K.: Concluding report to the Grant of the GE CR 103/95/1265 „Stability problems of the tunnel working face“
2. Weiglová, K., Nazari, F.> The geotechnical parameters of clay for the choice of the technological procedure in building the railway tunnel Žatec - Chomutov. 2nd Geotechnical Conference STU Bratislava 1966
3. Weiglová, K.: Material relations considering the dilatant behaviour of soils, XI International Scientific Conference, Technical University, Faculty of Civil Engineering in Brno, 1999

The contribution was elaborated as part of the grant project GA 103/99/1135 „Technological methods of influencing the stability of the tunnel working face“ and the research intention of the Faculty CEZ: J 22/98: 26 11 00 007

Computer Methods and Advances in Geomechanics, Desai et al. (eds) © 2001 Balkema, Rotterdam, ISBN 90 5809 183 X

An improved method for designing an anchored sheet pile wall

A.Yao & F.Tang
Faculty of Engineering, China University of Geosciences, Wuhan, People's Republic of China

I.M.Smith
Division of Civil Engineering, University of Manchester, UK

ABSTRACT: The type of sheet pile wall is widely been selected to support deep excavation in China due to its convenience of construction and cost saving. The mostly used method for designing the support is a conventional method such as free earth support or fixed earth support. However, the design results by either of free earth support or fixed earth support method are far consistent with measured data from practical field. In this paper, authors propose a simpler method or called the improvement of fixed earth support method. The method is the same as the fixed earth support except using force balance condition rather than moment balance condition to find the length from the section where active earth pressure equals to passive earth pressure to the bottom of the pile. The comparison of the displacement by FEM indicates that by this method a safe and economical design can easily be obtained.

1. INTRODUCTION

With the rapid development of high-rises in recent years in China, the projects of deep excavations are more and more. To support the excavation, the type of sheet pile wall is widely been selected due to its convenience of construction and cost saving. However, many serious failures occurred and more than 40 percent of them resulted from the design according to professor Tang's and Dr Zeng's statistics (Y. Q. Tang 1999, X. M. Zeng 1999).

The methods of designing the anchored sheet pile wall can be divided into three categories: conventional method, column beam method and finite element method (Briaud, J. L. et al, 1998). The most currently used methods in China are conventional methods such as free earth support and fixed earth support method. It is because that national and several local codes stipulated to do so (e.g. Construction Bureau of Hubei Province, 1998).

However, the design results by either free earth method or fixed earth support method are far consistent with measured data from practical field. The bending moment computed by the free earth support method is so great that designer has to reduce the value by some way(Cernica, J. N.). The length of the pile computed by the fixed earth support method is too long to save cost. In this paper, we propose a simpler method or called the improvement of fixed earth support method and attempt to achieve the goal of cost saving and safety for designing anchored sheet pile wall.

2. THE ANALYSIS OF FREE AND FIEXD EARTH SUPPORT METHODS

2.1 *Free earth support method*

The free earth support method takes the support structure of excavation as a free body. The active pressure and passive pressure as well as anchor forces are assumed in equilibrium. From force balance $F=0$ and moment balance $M_c=0$ (The C section is the bottom end of piles as shown in Fig. 1.), the depth of pile and anchor force can be calculated. The moment at C is zero, which is equivalent to a hinge at the point. From a point of theoretical mechanics, the free earth support method sounds reasonable. From a point of material mechanics, however, the resultant of

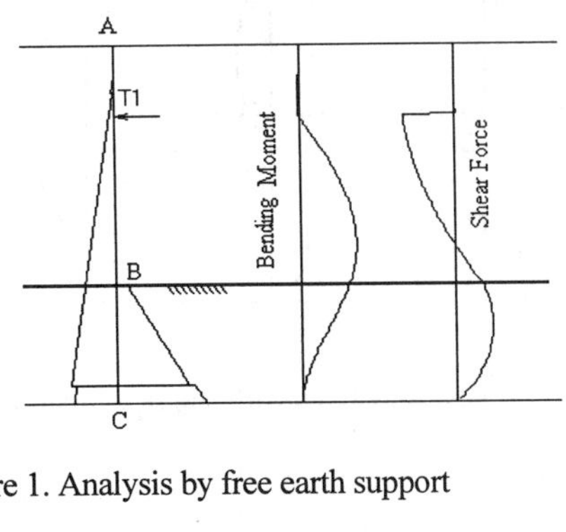

Figure 1. Analysis by free earth support

shear forces at the bottom end of the structure equals zero and correspondingly the bending moment at this point approaches to the maximum value (zero). The amplitude of negative bending moment somewhere on the pile must be too great. In other words, the distribution of bending moment is much uneven. The maximum value of the negative bending moment resulted from the big anchor force. Designing the section of the support pile on this value is not acceptable.

Many researchers have studied the failures of the support structure for excavation (e.g. Hashash, Y. M. A., et al, 1996; Yao, A. et al, 2000; Zeng, X. M. et al, 1999). The failure forms are different from case to case. Some slipped in curved surface. Some failed in horizontal movement. Some overturned around some point. For most cases, the failure form seemed to be a combination among above three forms. However, the failure form as overturning around the bottom of the piles as assumed for free earth support method may hardly been found. The results computed by elasto-plastic finite element method indicate that the support designing by free earth support method tends to fail in a form of "leg stretching". The top of the support moved backward and the bottom moved forward to excavation. That the anchor force is too great can be figured out from this failure form. And the conclusion can be drawn that taking the bottom of the piles as the moment balance point is unreasonable.

2.2 *Fixed earth support method*

Fixed earth support method is also called equivalent beam method. It simply assumed that the moment at the point D (as shown in Fig. 2) where the active earth pressure equals passive earth pressure is zero. From

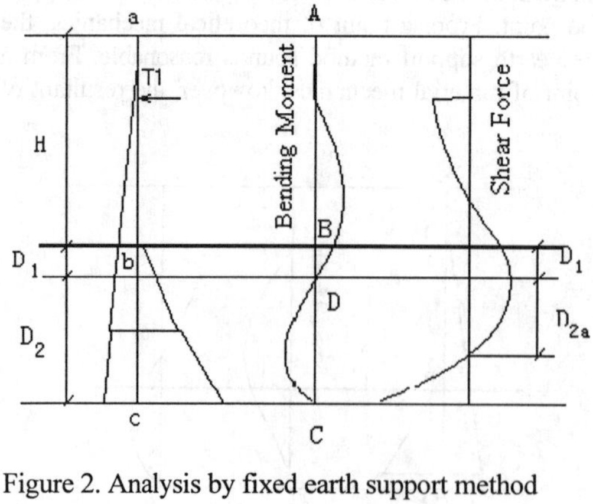

Figure 2. Analysis by fixed earth support method

the resultant of earth pressure being zero, D_1 can be determined. Then consider AD and DC as two free bodies. From force balance and moment balance for AD part, the anchor force and the force acting at the point D can be found. From balance condition $M_c=0$ for DC part, D_2 can be calculated. Changing the value and position of the anchor force, the absolute maximum value of bending moment on AD can be adjusted to approach to that on DC parts. Designing the pile section on this value can take full advantages of the support.

In mechanics, there are two problems in the fixed earth support method. One is assuming the moment at the point where active pressure equals passive earth pressure to be zero. The other is computing the D_2 by the balance condition of moment. Although first problem is not strictly consistent with the mechanical theory, but in lots of cases, the position is not far away from observation. For the second problem, however, is not logical in terms of mechanics and results in that computed the length is too long to save cost.

3. THE IMPROVED METHOD

On the theory of theoretical mechanics, a free body in equilibrium in plane needs meeting force and moment balance conditions. If taking the pile AC as a free

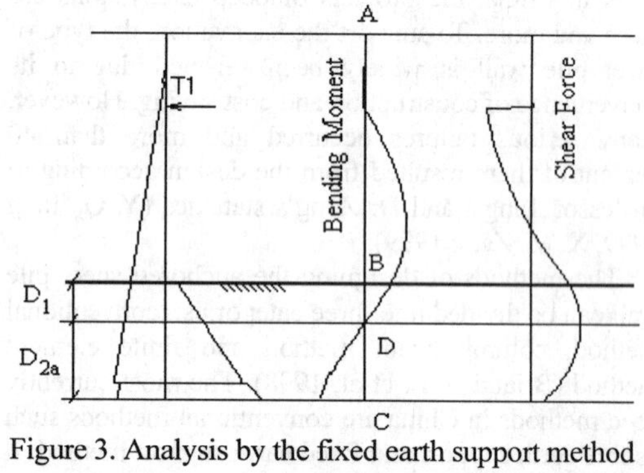

Figure 3. Analysis by the fixed earth support method

body, considering F=0 and $M_d=0$, the two unknowns can be determined. If the assumption that at the hinge point D, active earth pressure equals passive earth pressure is adopted, three unknowns including D_1, D_{2a} as shown in Fig. 3 and anchor force T_1 can be calculated. The results by this method are the same as the fixed earth support method except D_{2a}.

The steps of the method is described as following:

(1). Finding the point D at which the active pressure is equal to passive pressure.

(2). Calculating the distance from the point to the bottom of the excavation as D_1.

(3). Calculating the anchor force by moment balance around the point.
(4). Calculating the D_{2a} under force balance condition.
(5). Summing (D_1 D_{2a}) and multiplying a safety factor as bedded depth of the pile.

To demonstrate the computing procedure, an example of excavation in sand soil is discussed as following.

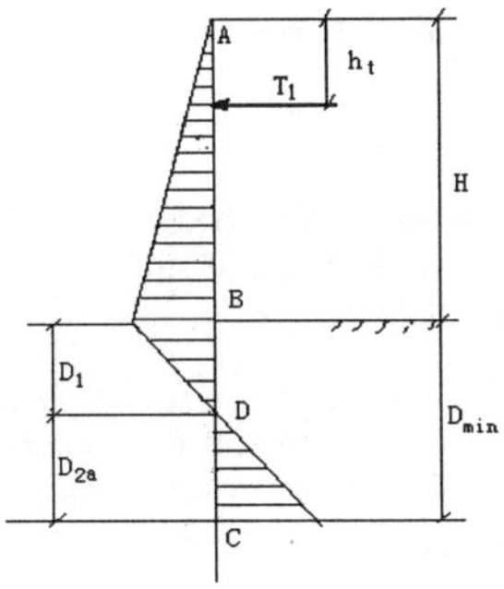

Figure 4. The example of analysis in sand soil by the improved method.

(a) Finding D_1;

$$D_1 = \frac{Hk_a}{k_p - k_a} \tag{1}$$

where H=depth of excavation; k_a=the factor of active pressure; k_p= the factor of passive pressure.
(b) Computing T_1 by

$$\sum m_d = 0$$

$$T_1 = \frac{\gamma(k_a(D_1+H)^3 - k_p D_1^{\ 3})}{6(H - h_t + D_1)} \tag{2}$$

where h_t=distance from the top of pile to anchored position; γ=gravity of soil.
(c) Computing D_{2a} by

$$\sum X = 0$$

$$D_{2a} = \sqrt{\frac{k_a(H+D_1)^2 - k_p D_1^{\ 2} - T_1}{(k_p - k_a)}} \tag{3}$$

(d) Find bedded depth of piles

$$D_{\min} = f(D_1 + D_{2a}) \tag{4}$$

The step (c) is a key step for this method. The D_{2a} by this method is shorter than D_2 by the fixed earth support method. In sand soil as does in this case, the relation between D_2 and D_{2a} is

and the minimum bedded depth of piles computed by this method approaches to that by free earth support

$$D_2 = \sqrt{3} D_{2a} \tag{5}$$

method. The computing procedure by this method is simpler than the fixed earth support method because to calculate D_{2a} from force balance condition only needs solving a quadratic equation rather than solving a cubic equation as in fixed earth support method.

4 COMPARISONS AMONG THREE METHODS

To compare this method with free and fixed earth support methods, we have computed several practical projects (e.g. Yao, A. et al, 2000a). At first we calculated the anchor force, bedded depth of piles, and than analyzed the displacements of the support structure by elasto-plastic finite element method (EPFEM) for free earth support, fixed earth support and the improved method respectively. The displacement comparison indicated that the design by this method seemed to be a shorter pile with less displacement.

Here is another example. The excavation is 8 m, and an anchor is set at 3 m from ground surface. The properties of soils are shown in Table 1. Calculated results display in Table 2. The displacements are computed by the EPFEM as shown in Fig. 5. We only plot out the displacements of the pile wall so as to compare easily.

Compared the improved method with free earth support method, we can see from Table 2 and Fig. 5 that the length of pile computed by the former is approached to that by the latter. However, the anchor force and absolute bending moment by the former are much smaller than the latter, that means a longer and larger anchor, and a larger section of piles are needed for the latter. The displacements from the former are less than the latter and the safety factor of former is higher than the latter. In other words, design by the improved method is economical and safe in comparison with the free earth support method.

The calculated the anchor force and absolute maximum bending moment are the same between fixed and improved fixed earth support method, but the length calculated by the latter is shorter than the former. On experience, the safety factor as 1.3-1.5 can meets the requirement of safety for support structure. The safety factor 1.6 is enough, but 2.0 from the fixed earth support method is too safe to save cost.

Table 1: Properties of soils

Layer No.	Dept h (m)	γ (kN/m³)	φ (°)	c (kPa)	E (MPa)	μ (-)
1	2	18.00	28.6	7.26	4.27	0.3
2	14	18.00	10.00	10.	2.7	0.3
3	30	20.00	20.00	0	8.5	0.3

Table 2: Calculated results

Method	Anchor force (kN) at 3.00 m	Maximum bending moment (kNm)	Minimum bending moment (kNm)	Length of pile (m)	Factor of safety (-)
Free earth support	202.85	4.88	-626.49	15.63	1.3
Fixed earth support	166.04	429	-441	18.76	2.0
Improvement	166.04	429	-441	15.36	1.6

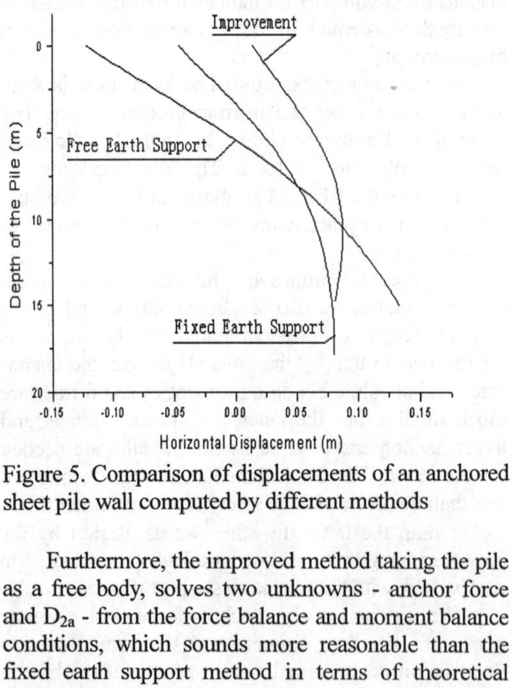

Figure 5. Comparison of displacements of an anchored sheet pile wall computed by different methods

Furthermore, the improved method taking the pile as a free body, solves two unknowns - anchor force and D_{2a} - from the force balance and moment balance conditions, which sounds more reasonable than the fixed earth support method in terms of theoretical mechanics.

From the analysis above, a conclusion can be drawn that the improvement of fixed earth support can easily achieve the goal of safety and cost saving, and that is why we call the method an "improved" method.

5 CONCLUSIONS

After analyzing the drawbacks of free and fixed earth support, an improvement of the fixed earth support method is proposed on the basis of force balance and the moment balance around the point at which the active earth pressure equals passive earth pressure. Calculated results show that the design by this method tends to be safe and economical. Manually computing by this method is not so difficult because only a quadratic equation needs solving.

REFERENCES

Briaud, J. L. et al, 1998, Beam-column method for tieback walls: *J. of Geotech. & Geoenvir. Engrg.*, USA, (67)

Cernica J. N., 1995, *Foundation Design*: John Wiley&Sons Inc,

Construction bureau of Hubei province, 1998, *Code of Hubei province for deep excavation*, DB42/159-1998, (in Chinese)

Hashash, Y. M. A., et al, 1996, Ground movement prediction for deep excavations in soft clay, *J. Geotech. and Geoenvir. Engrg.* ASCE, 122(6)

Tang, Y. Q. et al 1999, *Analyses and treatment of problems in excavation projects*: China Construction Industry Publishing House, (in Chinese)

Yao, A. et al, 2000a A Comprehensive Method for Designing Support Structure of Excavation: *GeoEng 2000*, Melbourne, Australia

Yao, A. et al, 2000b, Analyzing the factors affecting the displacements of supports for excavation by the simulation of FEM, *Coal geology & exploration* (in press, in Chinese)

Zeng, X. M. et al, 1999, *Warning and reminding records of problems in excavation and slopes*: China Construction Industry Publishing House, (in Chinese)

Computer Methods and Advances in Geomechanics, Desai et al. (eds) © 2001 Balkema, Rotterdam, ISBN 90 5809 183 X

Computer optimization of monitoring programs in geomechanics

S.A.Yufin, M.G.Zertsalov, O.K.Postolskaya, A.V.Manko & D.V.Ustinov
Moscow State University of Civil Engineering, Russia

ABSTRACT: Authors propose an integrated approach for working out the design of underground structures and to further monitor the behavior of the object in a unified way, combining capabilities of the advanced FEM computer code with GIS and DBMS. It may be especially effective and useful for such projects like underground facilities for radioactive waste storage or underground structures of urban environment, however there are no limitations for practical implementations.

1 INTRODUCTION

With increasing computer power information technologies and numerical models are able to handle problems of extreme complexity. More and more information has to be analyzed and prepared as input data for such models, numerous results has to be interpreted, compared and understood. Furthermore, the results have to be explained comprehensively to politicians and the public.

Efficiency of implementation of a numerical procedure in practical engineering apparently depends on at least four basic factors:

- Mechanical - i.e. on our understanding of the processes involved in construction and further operation of the project and ability to properly describe these processes in the form of constitutive laws;
- Numerical, which includes formalization of the constitutive laws along with many supplementing algorithms, starting with data preparation and not exactly ending with presentation of the results of analyses;
- Human factor, which probably is the most important in the results of engineering activities, and
- Legal factor, which may be decisive in the whole project, but is beyond the scope of the proposed paper.

It is a natural and logical fact, that there is no such thing in the world as a completed software product, i.e. that it would be impossible to further improve it. However, let us assume that we deal with only such type of software, which performs perfectly within the limitations specified in the user's manual - the fact, which makes the code being worth to be commercially distributed. In this case several questions may be posed in relation to the human factor in computer-aided support of engineering decisions:

- Is the selection of the required software well based?
- Is the set of input data on the level with software capabilities both in quality and quantity?
- Is the possible sensitivity of a selected model to different types of input parameters taken care of?
- Is the engineering task force qualified and experienced to the level of responsibility for solutions they may propose?
- Is there a guarantee that the contractor will proceed with construction exactly as it was considered in the computer model?

The list of questions may be prolonged, but getting a strict "YES" for all of them is practically unreachable. A working compromise lies somewhere between excellent computer work and practical experience of designer and contractor.

The importance of the listed problems become especially clear in the following two types of projects, which draw the utmost attention of both officials and rock/soil mechanics professionals:

- underground structures for radioactive waste disposal, and
- underground structures of urban environment.

In both cases uncertainty of spatial distribution of principal geological data and many aspects of the future behavior of the structure tend to implementation of the comprehensive monitoring systems, starting from the very first stages of the excavation work on the site or even earlier.

Numerical models coupled with Geographic Information Systems (GIS) and powerful Database Management Systems (DBMS) offer the possibility to integrate heterogenous, spatial data, to connect these data and visualize and analyze them.

An integrated approach for working out the design of underground structures and to further monitor the behavior of the object in a unified way, combining capabilities of the advanced FEM computer code with GIS and DBMS is needed for many real projects. Let us consider some practical cases.

2 PRESSURE TUNNEL

The hydraulic pressure tunnel of the Ingouri hydropower station in Georgia, former USSR Republic, was constructed in early seventies and put into operation in 1978 (Yufin et al, 1999). The rock mass along the tunnel is represented by thinly layered jointed hard siltstones, limestones and to a lesser degree – sandstones. The tunnel has an inner diameter of 9.0m with water head ranging up to 120m, the tunnel has a reinforced concrete lining of 0.7 m thickness.

After safe operation of the tunnel for slightly more than 10 years, water leakage was noticed on the steep banks of the valley of the small river, crossing the axis of the tunnel at an angle of approximately 90°. The depth of the tunnel in this place is only 32m from the river bottom. A steel lining was designed for this valley-crossing region, but was not installed for the reasons now untraceable.

Quite logically the missing steel lining was blamed for the leakage in the valley and the problem of proving the opposite became very important for contractor. Such proof, for obvious reasons, could have been based on the monitoring data, but no systematic observations were made in the region.

Numerical modeling of the situation looked like a logical solution. However, rock bedding approximately along the tunnel axis but with dip of 60° and different angles of incline of the V-shaped valley sides, did not permit use of even one plane of symmetry and called for full 3D modeling.

For the analysis the rock mass was represented as a transversly-isotropic medium with the plane of isotropy directed along the tunnel axis and inclined by 60° to the horizontal surface. Both rock and concrete lining were considered elasto-plastic with Mohr-Coulomb type laws. The reinforcement of the lining was ignored as an additional factor of safety.

The analyses showed, that the influence of the river valley sides provided additional vertical stress in the most shallow part of the tunnel/valley crossing. This feature, which could not have been accounted for in 2D modeling, provides purely elastic deformation of the concrete lining even in the case where the groundwater pressure outside the tunnel was not regarded as active (Barton et al, 1994b). Such results were confirmed by the later inspection of the tunnel lining – the event everybody tried to avoid because of the high cost of the lowering water level of the reservoir and interruption of the power production. The source of the leakage appeared to be the reservoir, created by the high arch Ingouri dam approximately 1 km away from the region in question. The situation led to costly grouting of rock mass near the reservoir, after which the leakage in the valley stopped.

The numerical portion of the problem has been recently revisited for verification purposes as a portion of a Ph.D. thesis and on a request from the station owner. FEM modeling has being done with the use of the Z_SOIL® PC code (Z_SOIL, 1999). Different constitutive models for the rock mass, different soft- and hardware, accounting for the groundwater in the slopes and base of the river valley and a more detailed FEM mesh resulted in principally the same observed phenomena and conclusions.

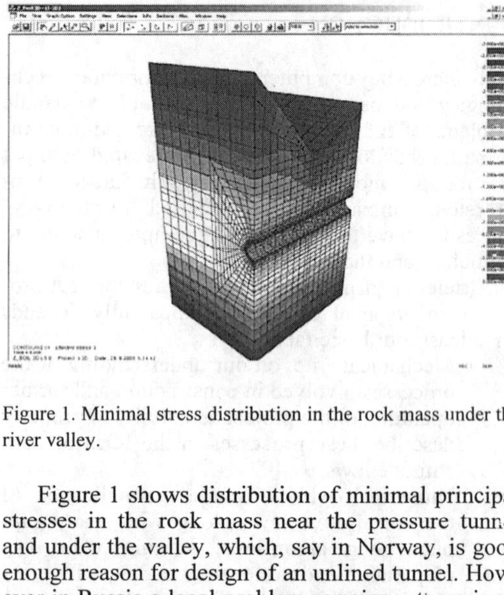

Figure 1. Minimal stress distribution in the rock mass under the river valley.

Figure 1 shows distribution of minimal principal stresses in the rock mass near the pressure tunnel and under the valley, which, say in Norway, is good enough reason for design of an unlined tunnel. However in Russia a legal problem remains – there is no working code of practice on the territory of the former USSR, which would permit design of a pressure tunnel lining without steel in the described conditions – overburden less than half of the water head. Statistics from several monitoring programs in similar situation would have been of great help for establishing new codes of practice, which would account more closely for such specific cases like river valley crossing by pressure tunnels, not a rare situation in hydraulic engineering.

3 SET OF CAVERNS

Another analysis cases also start to have a historic coloring. The Tehri hydropower scheme in Uttar Pradesh in the Himalayas in India should incorporate a set of underground caverns for the hydropower station and at a later stage analogous amount of openings for installation of machinery of the pumped storage station. Tasks of numerical analyses of these caverns sporadically come to hands of authors for the period of over 10 years.

The geology of the region is described by Singh & Goel (1999a). The rock mass in the vicinity of all caverns is composed of phyllites of different degree of jointing and fracturing. The set of caverns is situated at the depth of approximately 300 m.

At the design stage some 10 years ago 2D static and dynamic FEM analyses were performed for two most important parallel caverns – those of machine hall (MH) and transformer hall (TH) (Fink et al, 1991). For obvious reasons the geological model taken for the analysis at that time could not be so detailed as at the recent period of excavation.

The results showed high sensitivity of the set of two openings to the adopted sequence of excavation/anchoring of caverns. For the case of dynamic loading the primarily importance was given to the absence of the rigid vault structures in both caverns. Accounting for different design of the future powerhouse cavern of the pumped storage station in a 3D case (Barton et al., 1994a) until now has no practical consequences.

Recent analyses has been performed for three principal cases: 3D for real complex geological structure of the site, 3D for average uniform geological structure and 2D for one of the representative cross-sections of the real 3D case. Two last cases were analyzed as the reference basis both for the main case and previous analyses of these and independent authors.

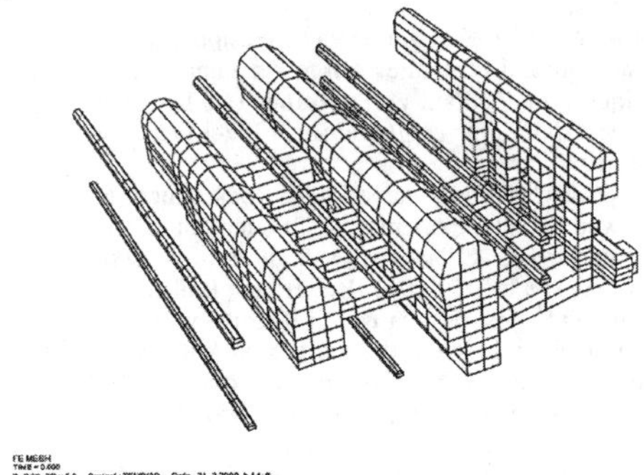

Figure 2 Set of openings taken for the analyses

The set of openings taken for analyses is shown on Figure 2. It included both MH and TH caverns, two openings of the surge chambers, busducts between MH and TH and drainage galleries. A later case accounts also for pressure turbine inlets and outlets, positioned at an angle to main caverns.

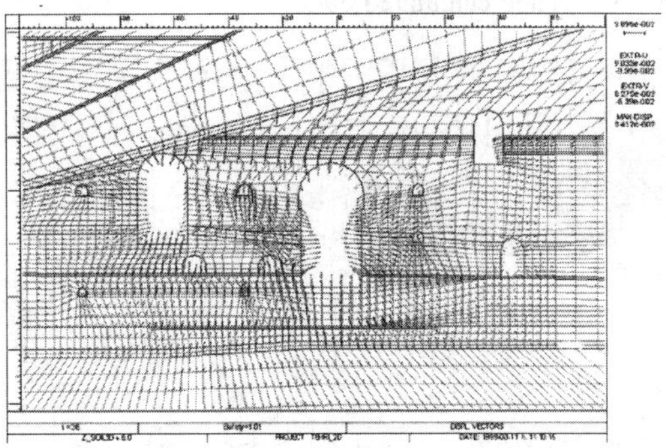

Figure 3 Displacement vectors for the last stage of excavation in the 2D analyses case.

The 2D case produces 2-3 times larger displacements than 3D cases (Yufin & Postolskaya, 2000) in good accordance with less sophisticated ten-years-old calculations (Fink et al., 1991) and results of other authors (Singh & Goel, 1999b).

The in-situ measurements of the displacements of the walls of two principal caverns by means of extensometers give displacement values, which are in good accordance with respective calculated values in the principal 3D case. This shows that 3D modeling and realistic geomechanical model of the rock mass together with reliable software provide reliable insight to the rock mass behavior for the period of excavation of caverns. Regretfully the monitoring program on the site was introduced with great delay as compared to the time of the very first excavation work in the caverns, thus diminishing the value of recorded data. However, interpretation of the results of measurements is easier with available prognosis data and this experience as a whole may be considered as one of the first steps in practical implementation of the proposed system.

4 FUTURE APPLICATION

Full-scale application of the system is expected at the site of the underground radioactive waste disposal facilities to be constructed in the North-West Region of Russia (Melnikov et al., 1999).

The construction of the facilities should start with excavation of two parallel Underground Rock Laboratory caverns, where a planned set of geomechanical and related tests will be conducted prior to the construction of the storage caverns and connecting tunnels. The monitoring system is being designed on the base of the prognosis made by numerical modeling. Positioning of extensometers and other continuous measuring devices is being selected according to the results of modeling and introduced into GIS for further account and comparison of the measured and calculated data.

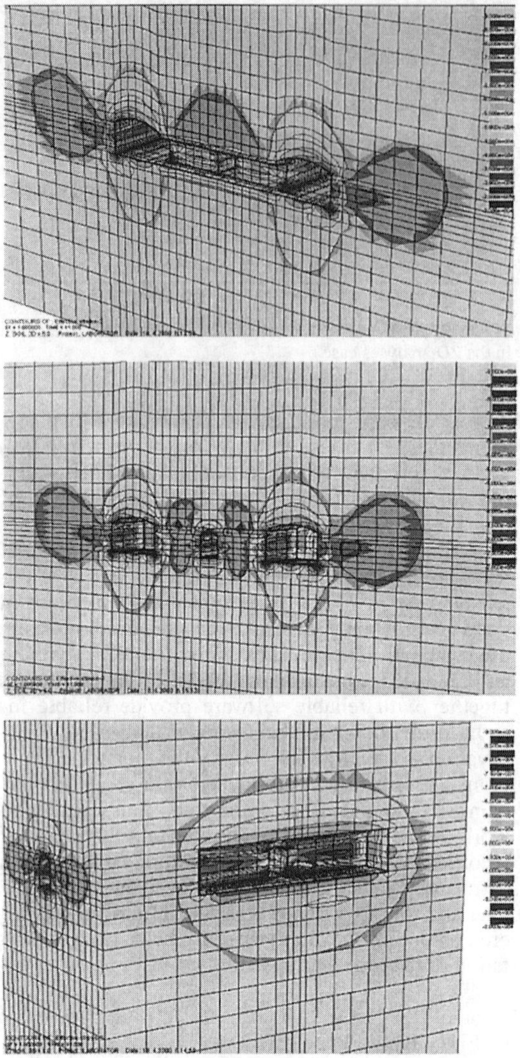

Figure 4. Minimal principal stress distribution in different cross sections of two laboratory caverns and connecting tunnels.

Figures above show stress distribution in the crystalline rock mass surrounding two laboratory caverns and connecting tunnels. Several different sequences of cavern excavation were studied to select number, types, length and position of extensometers for the mine-by and other in-situ experiments preceding excavation of operation waste storage caverns of the designed facility. The GIS and set of databases is being prepared for industrial application. Integration of the system on the data file level permits the joint use of industrial and home-made software.

5 CONCLUSIONS

Practical application of the proposed computer system, which incorporates industrial well-recommended FEM package, specialized analysis codes, GIS and DBMS will permit at the earliest stages to work out a strategy for the construction and long-term operational periods of the storage facility.

REFERENCES

Barton, N., S.A.Yufin, P.Chryssanthakis, T.L.By & M.L.Kagan 1994a. Direct and back analysis numerical techniques in rock engineering. In Lin Yunmei (ed), *New development in rock mechanics and engineering:* 23-28. Shenyang: North-eastern University Press.

Barton, N., S.A.Yufin & G.Swoboda 1994b. Engineering decisions in rock mechanics based on numerical modeling. In H.J.Siriwardane & M.M.Zaman (eds). *Computer methods and advances in geomechanics:* 2221-2228. Rotterdam: Balkema.

Fink, A.K., L.P.Mikhailov, A.B.Yelenich & S.A.Yufin 1991. Static and dynamic analyses of rock caverns. In W.Wittke (ed), *Proceedings 7th ISRM Congress*: 1905-1908. Rotterdam: Balkema.

Melnikov, N.N., V.P Konukhin. & S.A.Yufin 1999. Disposal of radioactive waste in the geological formations of the North-West region of Russia. In S.Saeb & Ch.Francke (eds.), *Proc. of the international workshop on the Rock Mechanics of Nuclear Waste Repositories*: 143-158. ARMA.

Singh, B. & R.K.Goel 1999. *Rock mass classification. A practical approach in civil engineering*: 190-199(a), 133(b). Amsterdam: Elsevier.

Yufin, S.A. 1979. Numerical analysis of rock structures considering material nonlinearities. *Proc. 20th US Rock Mechanics Symposium:* 265-272. Austin, Tx.

Yufin, S.A. 1993/1995. General report to Eurock'93. In L.Ribeiro e Sousa & N.F.Grossmann (eds). *Safety and environmental issues in rock engineering*: 1029-1041. Rotterdam: Balkema.

Yufin, S.A. & O.Yu.Kalugin 1995. Decision support system for underground structures based on numerical modeling techniques. In P.J.Pahl & H.Werner (eds), *Computing in civil and building engineering*: 1327-1332. Rotterdam: Balkema.

Yufin, S.A., & O.K.Postolskaya 2000. Rock structure stability as viewed from experience of 3D modeling. In Girard, Liebman, Breeds & Doe (eds.). *Pacific Rocks 2000. "Rock around the rim"*: 1059-1063. Rotterdam: Balkema

Yufin, S.A., O.K.Postolskaya & V.I.Rechitsky 1993. Stability of rock caverns as viewed from the back analysis data. In L.Ribeiro e Sousa & N.F.Grossmann (eds). *Safety and environmental issues in rock engineering*: 751-758. Rotterdam: Balkema.

Yufin, S.A., V.I.Rechitsky & V.K.Gorbushina 1999. Tracing the fortunes of Russian hydro caverns. Aftermath of the USSR. *International Water Power & Dam Construction.* v.51, #11: 36-38. Dartford: Wilmington Publishing Ltd.

Z_SOIL 1999 User manual. Zace Services Ltd Report 1985-1999. Lausanne: Elmepress International.

7 Geoenvironmental engineering and coupled processes

Computer Methods and Advances in Geomechanics, Desai et al. (eds) © 2001 Balkema, Rotterdam, ISBN 90 5809 183 X

New developments in the application of the discrete element method to coupled process problems

L.S.Costin, D.C.Preece & R.P.Jensen
Sandia National Laboratories, Albuquerque, N.Mex., USA

B.K.Cook
Massachusetts Institute of Technology, Cambridge, Mass., USA

ABSTRACT: The discrete element method (DEM) has been in development for a number of years. As the method matures, it is being extended to address a wider range of problems. Rapid increases in computational capacity have also led to the ability to treat more complex phenomena. One area where significant progress has been made in the past several years is in the simulation of coupled fluid flow-particle motion phenomena. Applications for coupled fluid flow and particle motion codes range from surface mine blasting design to the simulation of flow in porous media with applications to oil and gas production, environmental restoration, and ceramics production. This paper presents a summary of recent developments in coupled process modeling with DEM, provides a number of examples of current applications, addresses some of the problems with current approaches, and discusses future developments that are needed to address even more complex problems.

1 INTRODUCTION

Continuum-based methods for solving mechanics problems that involve unconsolidated, disaggregated materials or consolidated materials composed of many distinct particles or blocks, have always been problematic. The development of discrete element methods (DEM) was largely motivated by the need to simulate the behavior of these materials, and address problems with large material motion, disaggregation, and multi-particle kinematics (Cundall and Strack, 1979). As the method has matured, it is being extended to address a wider range of problems, and the rapid increases in computational capacity have also led to the ability to treat more complex phenomena. One area where significant progress has been made in the past several years is in the simulation of coupled fluid flow-particle motion phenomena. Applications for coupled fluid flow and particle motion codes range from surface mine blasting design (macroscale) to the simulation of flow in porous media (microscale), with applications to oil and gas production, environmental restoration, and ceramics production. This paper reviews some recent developments in the application of DEM to the simulation of coupled fluid-solid interaction phenomena. Other papers in this session will provide more detail on the specific examples cited here.

2 DISCRETE-ELEMENT METHOD

DEM is a suite of numerical algorithms that allow for the efficient modeling of materials as a collection of rigid (or in some cases, deformable) bodies that interact through surface collisions and inter-particle bonds. Each particle in a DEM simulation is modeled as a geometrically simplified object, using either a functional (e.g. sphere) or discrete representation. Particles are usually assumed to be rigid; however, some codes include the option to model the particles as deformable bodies by overlaying them with a finite element (FE) mesh and, using standard FE solution methods, with a selection of constitutive models, to solve for the particle deformation. Bonding between particles is modeled with numerical links that are effectively springs with a user-specified stiffness and tensile strength. During each model iteration, contacts between objects are found using spatial sorting algorithms and the contact forces resolved. Particle position is then updated according to Newton's second law, the conservation of linear and angular momentum, with particle forces equal to the sum of the contact forces and, in the case of coupled processes, other relevant forces such as the dynamic fluid force, and the buoyancy (body) force.

3 COUPLED EXPLOSIVE GAS FLOW AND ROCK MOTION MODELING

One major application of DEM methods has been the simulation of rock blasting. Blasting is often used both to break up rock masses and to move large quantities of rock so that underlying ore or coal seams can be exposed. This application requires the coupling of blast forces and gas pressure generated by the explosive to the solid mass, which is usually modeled as a stacked set of consolidated disks (2-D) or spheres (3-D). An example of this is the simulation of blast-induced muck movement using a 2-D discrete element code called DMC. Early versions of DMC have been used to model rock motion associated with blasting in a cratering environment (Preece and Taylor, 1990), and in confined volume blasting associated with in-situ oil shale retorting (Preece, 1990 a & b). In these applications, DMC treated the explosive loading as force versus time functions on specific spheres which were adjusted to obtain observed face velocities. However, it was recognized that a significant deficiency in this approach was the inability to simulate the explosive gas flow and the lack of coupling between gas flow and the rock motion. This problem has now been addressed by executing a finite difference code that computes gas flow through a porous media (Baer, 1989) in conjunction with DMC. The marriage of these two capabilities in a revised code are now called DMC_BLAST (Preece and Knudsen, 1991; Preece, 1993).

A key element in coupling gas flow and rock motion is the calculation of momentum transfer from the flowing gas to the spherical elements. The fact that the computational techniques for gas flow (finite difference) and rock motion (DEM) are different, dictates a unique approach to calculating the gas loading on the spheres. Pressure gradients from the gas simulation are used to calculate forces on the spheres. The normal discrete element time integration scheme is then applied where particle forces are used to obtain accelerations, which are integrated to obtain velocities, which are then integrated again to give particle locations. As the spheres begin to move under the loading, the porosity/permeability of the gas flow grid is modified to reflect the open space created. The typical sphere/gas computational grid arrangement is shown in Figure 1.

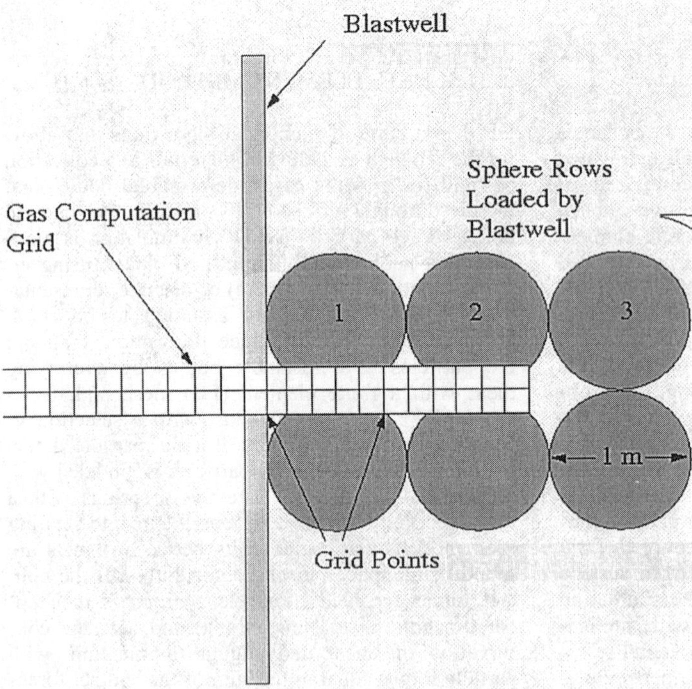

Figure 1. Typical sphere/gas computation grid arrangement illustrating the local coupling of the discrete elements with the gas computation grid. For simplicity, only the discrete elements that are loaded by the gas expansion are shown.

One challenge in this procedure is the difference in size between the spheres and the finite difference cells. The size of the finite difference cells is controlled by the diameter of the blastwell (i.e. borehole). Another major challenge in coupling schemes is the difference in time step dictated by the physics associated with the two different phenomena being coupled. In these calculations the time-step for the gas flow calculation is typically several orders of magnitude less than the time-step required for sphere motion (Preece and Knudsen, 1992).

Examples of this coupled gas-flow/rock-motion capability for simulations of a coal mine bench blast are given in another paper presented by Preece et al. (2001) in this theme session titled: *Modeling Coal Seam Damage in Cast Blasting.*

4 FLUID FLOW IN POROUS MEDIA

DEM methods are ideally suited for the simulation of multiphase systems at the grain scale. This approach allows full resolution of the interaction of individual solid particles with other solid particles and the surrounding fluid. Two implementations of coupled fluid flow - particle motion formulations have been developed in recent years. The first assumes that fluid flow through the pore space in a collection of solid particles can be modeled using Darcy's equation for flow through porous media, whereas, the second uses the Lattice-Boltzmann method to simulate a Newtonian fluid. In addition to the fluid coupling, these simulations require more complexly shaped particles to realistically simulate the effect that asperities, particle angularities, and particle roughness have on particle-particle and fluid-particle interactions.

In the first method, fluid flow simulation is accomplished by using finite element techniques to solve the 2-D Laplace formulation for Darcy-Flow given in Equation (1),

$$k_x \frac{\partial^2 \phi}{\partial x^2} + k_y \frac{\partial^2 \phi}{\partial y^2} = 0 \qquad (1)$$

where k_x and k_y are porous media permeabilities in the x and y directions and ϕ is the pressure (potential) function, which is the only variable (or degree-of-freedom) in the problem (Klosek, 1995). Current implementations usually assume that k_x and k_y are equal, requiring only one input, k. Two methods for treating the permeability are being explored. The first assumes that the permeability remains constant throughout the simulation in which case the fluid-flow conditions are solved for only once. The second method allows the discrete element distribution to modify the permeability.

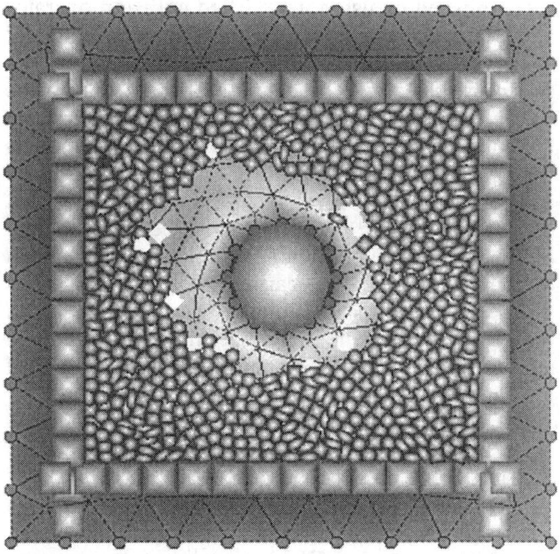

Figure 2. Simulation of sand production into a perforation channel with steady state fluid flow from the boundary to the central channel. Note that particles have been torn loose from the matrix and dragged into the channel by the fluid.

A detailed development of the coupled method with examples of applications to the problem of sand production (Figure 2) are given in O'Connor et al. (1997) and in the paper *Modeling sand production with Darcy-flow coupled with discrete elements*, by Jensen et al. (2001) presented in this theme session. While these early results are encouraging, some deficiencies remain. Because the boundary could influence the solution, a large number of particles need to be used in the simulations. It is also likely that Darcy fluid flow is not the best description of the physics where particles are being transported by the fluid. Further, two-dimensional analyses are always affected because of the constraints against out-of-plane motion. Particle behavior can be quite different when the motions are not constrained to movement within a plane.

To address some of these difficulties, an alternative approach is being pursued. This involves the application of the Lattice-Boltzmann method (LBM) to simulate the fluid flow. LBM is a numerical method for simulating a Newtonian fluid. The method has been employed to analyze the detailed behavior of gases and liquids for a number of years. Although this method is founded on the kinetic theory of dilute gases, it has been shown to accurately reproduce the Navier-Stokes equations (Chen & Doolen 1998) for fluid flow. The fluid is modeled as packets of particles that move about a regular lattice, or grid, defined with the appropriate boundary conditions. Collision and redistribution (or streaming) of fluid packets occur at the lattice sites according to specific relationships that conserve mass and momentum. Coupling with discrete elements is accomplished by using a modified collision operator at the lattice sites covered by the elements (Noble & Torczynski 1998). Momentum is transferred between the fluid and the discrete element during the collision-tracking step at all lattice sites covered by the solid particle. This coupling accurately models fluid flow in the close vicinity of the solid particles and reproduces the local fluid-induced stresses that drive solid particle motion. A detailed development of the coupled method with examples of applications to coupled fluid flow – particle motion problems is given by Cook et al. (2000). Three-dimensional, massively parallel implementations of this method are currently being developed to take advantage of the natural parallelism of the LBM formulation.

5 CURRENT DIFFICULTIES AND FUTURE ADVANCEMENTS

The state of coupled DEM modeling has advanced considerably in the last few years, but numerous difficulties remain. The essential part of DEM that has not been well addressed thus far is the particle-to-particle bonding. The nature of the inter-particle bonding is the key to the accurate simulation of porous material deformation, disaggregation processes, and localization phenomena that commonly occur in porous materials (shear and compaction banding). While the current descriptions of the bonding are sufficient to elicit good qualitative behavior, accurate quantitative results are more difficult. Few, if any data exist on grain-to-grain bonding in materials like sandstone or soils, making it very difficult to resolve this issue.

The second issue that is perpetually troubling is the adequacy of boundary conditions. In coupled problems, specifying realistic boundary conditions is nearly impossible. The tact of making the model large enough so that the boundaries do not affect the process of interest is also riddled with difficulties, primarily because it is computationally very expensive. Some progress has been made regarding the application of mechanical boundary conditions by embedding the DEM problem within a finite element mesh so that the effect of distant boundaries can be simulated by the mechanical behavior of the finite element structure. However, much remains to be done to make this type of coupling computationally attractive.

Finally, to achieve even true qualitative results requires three-dimensional simulation capability. This presents two fundamental problems. First, is the computational problem of dealing with huge numbers of particles and less efficient contact search methods. The growth of massively parallel computing technology will, in time, overcome some of the difficulty. A massively parallel 3-D implementation of coupled models will be able to address a broad range of important but currently intractable geomechanical and industrial challenges. Fundamental insights are expected into the disaggregation, transport, and clustering processes underlying such outstanding problems as sand production, proppant design, sedimenting and fluidized suspensions, and lubricated transport of viscous materials. Second, is the problem of visualization of the model and the results in 3-D. This problem is not unique to DEM, but is perhaps exaggerated because, for DEM, the essence of the results is a visualization of the motion of particles in space.

In conclusion, it is apparent that coupled DEM methods have become a powerful tool in addressing a number of problems that are virtually intractable by continuum modeling methods. Continued development of the method and more widespread extension to fully three-dimensional formulations will greatly enhance the impact of DEM on analysis of geotechnical problems.

ACKNOWLEDGEMENTS

The work summarized in this paper was performed at Sandia National Laboratories and funded by the U.S. DOE under Contract No. DE-AC04-AL85000. Sandia is a multi-program laboratory operated by Sandia Corporation, a Lockheed Martin Company, for the U.S. DOE.

REFERENCES

Baer, M. R., and Gross, R. J. 1989. A Two Dimensional Flux-Corrected Transport Solver for Convectively Dominated Flows, Sandia National Laboratories, *SAND85-0613*.

Chen, S. & G. Doolen 1998. Lattice Boltzmann method for fluid-flows. *Annual Review of Fluid Mechanics* 30: 329-364.

Cook, B. et al. 2000. Direct Simulation of Particle-laden Fluids. *Proceedings of the 4th North American Rock Mechanics Symposium*. Balkema, Rotterdam.

Cundall, P.A. & Strack, O.D.L. 1979. A Discrete Numerical Model for Granular Assemblies. *Geotechnique* 29(1): 47-65.

Jensen, R. P. et al, 2001. Modeling sand production with Darcy-Flow coupled with discrete elements. *Proceedings of the 10th International conference on Computer Methods and Advances in Geomechanics, Tucson, AZ.*

Klosek, J.T. 1995. *The integration of fluid dynamics with a discrete-element modeling system: Algorithms, Implementation, and Applications*. Masters Thesis, MIT Dept. of Civil and Env. Eng. Cambridge, MA.

Noble, D. & Torczynski, J 1998. A Lattice Boltzmann method for partially saturated computational cells. *International Journal of Modern Physics C 9(8): 189-201.*

O'Connor, R.M., Torczynski, J.R., Preece, D.S., Klosek, J.T., and Williams, J.R. 1997. Discrete element modeling of sand production. *International Journal of Rock Mechanics and Mining Sciences. 34(3-4): 373.*

Preece, D. S. 1990a. Rock Motion Simulation and Prediction of Porosity Distribution for a Two-Void Level Retort, *Proceedings of the 23rd U. S. Oil Shale Symposium*, Colorado School of Mines, Golden, Colorado.

Preece, D. S.1990b Rock Motion Simulation of Confined Volume Blasting, *Proceedings of the 31st U. S. Symposium on Rock Mechanics*, Colorado School of Mines, Golden, Colorado.

Preece, D. S. 1993. Momentum Transfer From Flowing Explosive Gases to Spherical Particles During Computer Simulation of Blasting-Induced Rock Motion, *Proceedings of the Ninth Annual Symposium on Explosives and Blasting Research*, International Society of Explosives Engineers, January 31-February 4, 1993, San Diego, CA.

Preece, D. S. and Chung, S. H. 2001. Modeling coal seam damage in cast blasting. *Proceedings of the 10th International conference on Computer Methods and Advances in Geomechanics, Tucson, AZ.*

Preece, D. S., and Knudsen, S. D. 1991. Blasting Induced Rock Motion Modeling Including Gas Pressure Effects, *Proceedings of the 24th U. S. Oil Shale Symposium*, Colorado School of Mines, Golden, Colorado.

Preece, D. S., and Knudsen, S. D. 1992 Computer Modeling of Gas Flow and Gas Loading of Rock in a Bench Blasting Environment, *Proceedings of the 33rd U. S. Symposium on Rock Mechanics,* Santa Fe, New Mexico.

Preece, D. S. and Taylor, L. M. 1990 Spherical Element Bulking Mechanisms for Modeling Blasting Induced Rock Motion, *Third International Symposium on Rock Fragmentation by Blasting, Brisbane, Queensland, Australia.*

Computer Methods and Advances in Geomechanics, Desai et al. (eds) © 2001 Balkema, Rotterdam, ISBN 90 5809 183 X

Analysis of gas phase transport phenomena in compacted clay barriers

A.Gens, S.Olivella & B.Valleján
Geotechnical Engineering Department, Technical University of Catalunya, Barcelona, Spain

ABSTRACT: A proper understanding of the behaviour of compacted clay barriers requires not only the study of liquid flow but also a careful examination of the transport processes that occur in the gas phase. This need is enhanced in the case of engineered barriers for high-level radioactive waste that are subjected to the simultaneous effects of thermal loading, hydration from the host rock and gas generation from corrosion and other sources. The paper describes a formulation that takes into account these transport processes. The flow of each gas phase component is the sum of advection, diffusions and dispersion fluxes, each term having its corresponding constitutive relation. Special attention is given to the changes of pore structure that occur during hydration of compacted clays. They result, even in absence of total porosity changes, in variations of intrinsic permeability that must be taken into account in the analyses. A simulation of the behaviour of an engineered barrier and immediately adjacent rock in a high level waste repository is described. It demonstrates the capability of coupled thermo-hydro-mechanical analyses to model the processes that occur in the barrier and near field. The effects of taking into account the variation of intrinsic permeability are highlighted.

1 INTRODUCTION

Unsaturated porous media can be considered as an ensemble of three phases: solid, liquid and gas. These phases are not homogeneous but may contain more than one species. For instance the liquid phase may have air (and other gases) in dissolution whereas the gas phase can be a mixture of various gases, including water vapour. Transport phenomena occurring in the gaseous phase is of major concern in a number of environmental engineering problems. A typical example refers to the design of engineered barriers in underground repositories for radioactive high-level waste disposal. In most designs, the canisters containing the waste are placed in either vertical boreholes or horizontal drifts surrounded by an engineered barrier generally composed of compacted swelling clay. Therefore, the material is, at least initially, in an unsaturated state. A good understanding of the behaviour of the engineered barrier and immediately adjacent rock is of paramount importance to ensure an adequate degree of long-term safety.

High-level radioactive waste has a large heating power potential and the engineered barrier is subjected to a very significant temperature increases especially the material close to the canisters. Temperatures well above 100°C are often reached. These high temperatures cause considerable evaporation/condensation and vapour migration phenomena. Vapour migration controls the distribution of humidity in the barrier and, via thermo-mechanical coupling, it also influences the distribution of temperatures and the development of swelling stresses.

Another important phenomenon that must be analyzed in this situation is the migration of gas through the barrier and surrounding rock. Many processes may give rise to gas generation ranging from radiolysis to microbial action. In high-level radioactive waste, however, the main source for gas production is usually corrosion, especially if carbon-steel canisters are adopted. The main product is hydrogen.

Gas may be transported across the engineered barrier by diffusion in the liquid phase, if the quantities produced are small. However, the carrying capacity of this mechanism is very limited and, typically, transport in the gaseous phase must take place. On occasions, gas pressure may reach values high enough to fracture hydraulically barrier. This phenomenon is not considered in this paper.

An engineered barrier subjected to thermal loading from the waste, to hydration from the surrounding rock and to gas generation effects gives rise to a complex situation with a great number of phenomena interacting with each other. In addition, the barrier also interacts with the surrounding rock. A proper modelling of this problem requires the performance of coupled thermo-hydro-mechanical analyses in which he gaseous phase is explicitly considered. In this paper, the main features of such a formulation as related to transport phenomena in the gas phase are described. Afterwards the significance of the relationship between the changes in clay microstructure and gas permeability is discussed and a model proposed. To demonstrate its applicability, a modelling example is finally presented.

2 GAS TRANSPORT AND VAPOUR MIGRATION IN AN UNSATURATED SOIL

2.1 *Mechanisms and formulation*

The treatment of the gas phase transport phenomena presented in this Section are part of a general coupled thermo-hydro-mechanical formulation described elsewhere (Olivella et al. 1994, Gens et al. 1998b). In general the movement of any component of the gas phase (including vapour) is computed as the sum of advection, diffusion and dispersion fluxes. For simplicity only two components will be considered here, water vapour and a mixture of the rest of the gases considered as a single species.

Advection is described by a generalized Darcy's law (Bear 1972):

$$\mathbf{q}_g = \frac{-k_{rg}\mathbf{k}}{\mu_g}\left(\nabla P_g + \rho_g g \nabla z\right) \tag{1}$$

where ρ_g is gas density, $\mathbf{k}$ is the intrinsic permeability tensor, k_{rg} is gas relative permeability, μ_g is gas viscosity, g is gravity and P_g is gas pressure. The viscosity of the gas phase is a function of its composition and temperature. The advection of a component (e.g. water vapour) will be given by:

$$\left(\mathbf{i}_g^w\right)_{advection} = \rho_g^w \mathbf{q}_g = \omega_g^w \rho_g \mathbf{q}_g \tag{2}$$

where ω_g^w is mass fraction of water in gas. Equation (2) is generally valid both for a gas phase composed by pure vapour ($\omega_g^w=1$) and for a mixture of gas and vapour ($\omega_g^w<1$).

The equation for advective flux above has been written under the assumption of laminar flow. When the pores are small and the pressure is also low, the mean free path of gas molecules may be comparable to the pore size. When this occurs, slip takes place and the assumption of purely laminar flow is no longer valid. A simple modification of Equation (1) to account for this Knudsen diffusion effect involves the modification of the apparent intrinsic permeability according to (Wa'il Abu-El-Sha'r & Abriola 1997):

$$\mathbf{k}_a = \mathbf{k} + \mathbf{I}\frac{D_k \mu_g}{P_g} \tag{3}$$

where D_k is the Knudsen diffusion coefficient and $\mathbf{I}$ is the identity matrix.

The diffusive flux component is calculated as:

$$\left(\mathbf{i}_g^w\right)_{diffusion} = -\left(\mathbf{D}_g^w\right)_{diffusion} \nabla \omega_g^w \tag{4}$$

$$\left(\mathbf{D}_g^w\right)_{diffusion} = \tau \phi D \mathbf{I} S_g \rho_g \tag{5}$$

where τ is tortuosity, ϕ is porosity, D is a molecular diffusion coefficient for vapour and S_g is degree of saturation of gas phase. Equation (4) corresponds to a generalised form of Fick's law for molecular diffusion in porous media.

Finally dispersive flux is calculated as:

$$\left(\mathbf{i}_g^w\right)_{dispersion} = -\left(\mathbf{D}_g^w\right)_{dispersion} \nabla \omega_g^w \tag{6}$$

$$\left(\mathbf{D}_g^w\right)_{dispersion} = \rho_g\left(d_t\left|\mathbf{q}_g\right|\mathbf{I} + (d_l - d_t)\frac{\mathbf{q}_g\mathbf{q}_g^t}{\left|\mathbf{q}_g\right|}\right) \tag{7}$$

where d_l and d_t are dispersivities. In Eq. (6) Fick's law is used but with the tensor corresponding to the dispersion of a component of a phase flowing in a porous medium (Bear 1972). When the gas phase becomes a single component phase, both diffusion and dispersion vanish.

Finally the total flux of vapour can be written as:

$$\mathbf{j}_g^w = \left(\mathbf{i}_g^w\right)_{advection} + \left(\mathbf{i}_g^w\right)_{diffusion} + \left(\mathbf{i}_g^w\right)_{dispersion} = \omega_g^w \rho_g \mathbf{q}_g - \mathbf{D}_g^w \nabla \omega_g^w \tag{8}$$

where the diffusive and the dispersive terms have been lumped together. Considering the other gas component (i.e. the rest of the gas phase excluding water vapour) a similar expression holds:

$$\mathbf{j}_g^a = \left(\mathbf{i}_g^a\right)_{advection} + \left(\mathbf{i}_g^a\right)_{diffusion} + \left(\mathbf{i}_g^a\right)_{dispersion} = \omega_g^a \rho_g \mathbf{q}_g - \mathbf{D}_g^a \nabla \omega_g^a \tag{9}$$

The gas phase is then considered as a binary mixture of two species in which one component exerts an influence on the motion of the other component via diffusion phenomena. According to Bird et al. (1960), in a binary system the mass fluxes of species with respect to the mass averaged velocity of the phase are zero. If it can be assumed that the mass averaged velocity coincides here with $\mathbf{q}_g$ then it should follow that $\mathbf{D}_g^w=\mathbf{D}_g^a$. If, on the other hand, Darcy's flux ($\mathbf{q}_g$) gives the volume (or molar) averaged velocity (Bear & Bachmat 1986) then mass fractions can be substituted by molar fractions and the same result is obtained (i.e. $\mathbf{D}_g^w=\mathbf{D}_g^a$). The specific problems associated with phase changes and vapour migration at high temperatures have been discussed in detail in Olivella & Gens (2000).

2.2 *Pressure gradients induced by vapour migration in a mass closed system*

To observe the influence of vapour migration on liquid and gas pressures, it is useful to consider a simplified one-dimensional case of a mass closed system under steady state conditions. Combining equations (8) and (9), the expression

$$\mathbf{j}_g^w = \left(\frac{\omega_g^w}{1-\omega_g^w}\right)\mathbf{j}_g^a - \left(\frac{1}{1-\omega_g^w}\right)\mathbf{D}_g^w \nabla \omega_g^w \tag{10}$$

is obtained. Neglecting dispersion, the assumption of steady state conditions yields:

$$\mathbf{j}_g^a = \omega_g^a \rho_g \mathbf{q}_g - \mathbf{D}_g^a \nabla \omega_g^a = 0 \tag{11}$$

$$\rho_g \mathbf{q}_g = \frac{1}{\omega_g^a}\mathbf{D}_g^a \nabla \omega_g^a = -\frac{1}{\left(1-\omega_g^w\right)}\mathbf{D}_g^w \nabla \omega_g^w \tag{12}$$

$$\nabla P_g = \frac{\tau}{k}\frac{\mu_g \phi S_g D}{(1-\omega_g^w) k_{rg}} \nabla \omega_g^w = \frac{\tau}{k} A \nabla \omega_g^w \tag{13}$$

Equation (13) indicates that the ratio between the material parameters tortuosity (τ) and intrinsic permeability (k) controls the gas pressure gradient.

In a similar way, liquid pressure gradient induced by vapour migration can be obtained as a function of vapour mass fraction. From the condition of mass of water conservation at steady state regime (again dispersion is neglected), the following equations are obtained:

$$\mathbf{j}_l^w + \mathbf{j}_g^w = 0 = \rho_l \mathbf{q}_l - \left(\frac{1}{1-\omega_g^w}\right)\mathbf{D}_g^w \nabla \omega_g^w \tag{14}$$

$$\rho_l \mathbf{q}_l = \left(\frac{1}{1-\omega_g^w}\right)\mathbf{D}_g^w \nabla \omega_g^w \tag{15}$$

$$\nabla P_l = -\frac{\tau}{k}\frac{\mu_l \phi \rho_g S_g D}{\rho_l k_{rl}\left(1-\omega_g^w\right)} \nabla \omega_g^w = -\frac{\tau}{k} B \nabla \omega_g^w \tag{16}$$

where the vapour flux ($\mathbf{j}_g{}^w$) has been calculated from Equation. (10). It can be seen that liquid pressure gradient induced by vapour migration is again proportional to the ratio between tortuosity (τ) and intrinsic permeability (k).

3 VARIATION OF INTRINSIC PERMEABILITY IN COMPACTED CLAYS

3.1 *Experimental observations*

In the previous Section, the dominant influence of the value of intrinsic permeability was demonstrated. There is a tendency to consider intrinsic permeability as a constant for a material at a given value of porosity. However, this is not the case for compacted swelling clays subject to hydration. Even if overall porosity remains constant, pore size distribution changes drastically during hydration and, therefore, intrinsic permeability will also vary significantly. When this parameter is measured under water saturated conditions, a very low intrinsic permeability is found. In contrast, much higher values are observed when, the permeability to gas is measured before saturation. Figures 1 and 2 show examples of measurements in compacted Boom clay and in compacted FEBEX bentonite for dry densities ranging from 1.65 to 1.7 g/cm^3. It is clear that a single intrinsic permeability can not be used for these clays. For instance, if the value of intrinsic permeability for water was used, the gas permeability that would be obtained is several orders of magnitude lower than the actual measured values.

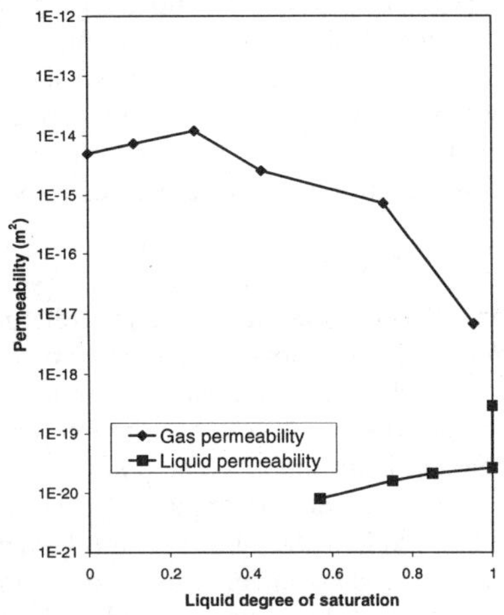

Figure 1. Intrinsic permeability of compacted Boom clay to gas and water flow (Volckaert et al. 1994).

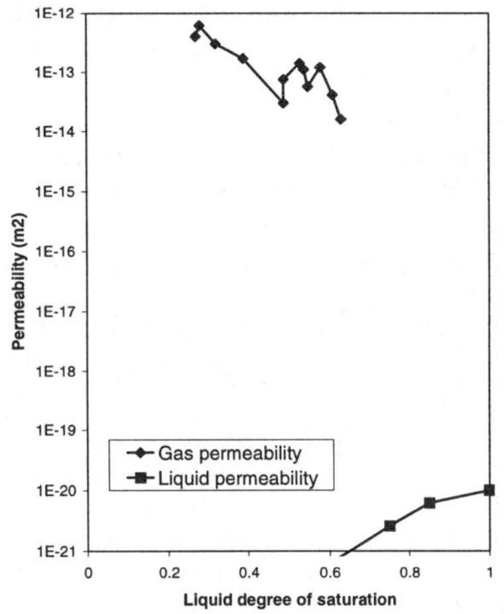

Figure 2. Intrinsic permeability of compacted Febex clay to gas and water flow (Villar 1998)

3.2 *A model for the variation of intrinsic permeability in compacted clays*

A model that attributes very different mobility to gas and water is necessary to account for the variation of intrinsic permeability in compacted clays. A simple way to achieve this objective is to assume a double porosity structure. Intrinsic permeability can be considered a function only of the macroporosity that changes as hydration progresses.

If M stands for macro and m for micro, the following relationships are proposed:

$$\phi = \phi_M + \phi_m \qquad \phi_M = \phi \exp(-\beta S_l) \qquad (17)$$

where an exponential function has been introduced as a first approximation for the dependence of macroporosity on degree of saturation. It is assumed that when the medium is nearly dry, practically all porosity is macroporosity while when the medium tends to saturation, all porosity becomes microporosity. Using this definition, intrinsic permeability is calculated as:

$$\mathbf{k} = \mathbf{k}(\phi_M) = \mathbf{k}(\phi_o \exp(-\beta S_l)) \qquad (18)$$

In equation (18) any classical function relating intrinsic permeability to porosity can be adopted. Since permeability is a function of macroporosity only, as a first approximation, water in the microporosity is assumed to be immobile. The double structure concept used here is conceptually consistent with that used to explain the observed mechanical behaviour of swelling clays under stress and suction changes (Alonso et al. 1999, Gens & Alonso 1992).

4 EXAMPLE OF APPLICATION

To demonstrate the applicability of the formulation presented above, the analysis of a scheme for vitrified radioactive waste disposal is presented. It corresponds to the design envisaged for the Est site in France where the host rock is mudstone. The concept for storage in this site is based on the construction of parallel drifts with a diameter of 6 m at 70 m intervals. These drifts will be excavated at a depth of 490 m. Every 18 m, two horizontal boreholes will be drilled on opposite sides of the drift with a length of about 19 m. Eight canisters containing vitrified waste will be placed in each borehole. These canisters are located inside a steel liner 1 cm thick. The space between the liner and the borehole wall will be backfilled with an engineered barrier 0.80 m thick made up of compacted bentonite. The diameter of the canister is 0.54 m and its effective length is 1.71 m. Therefore, eight canisters have a total length of 13.68 m, the rest of the borehole will be completely backfilled with compacted bentonite. Figure 3 shows the scheme for the emplacement of the canisters.

The aim of the coupled THM analyses presented is the simulation of the behaviour of the engineered barrier and immediately surrounding rock when subjected to the simultaneous effects of heating from the radioactive waste, hydration from the host rock and gas generation. Because of the limited extent of the area of interest and the significant length of the drifts, a one-dimensional geometry with radial symmetry has been adopted for the analysis.

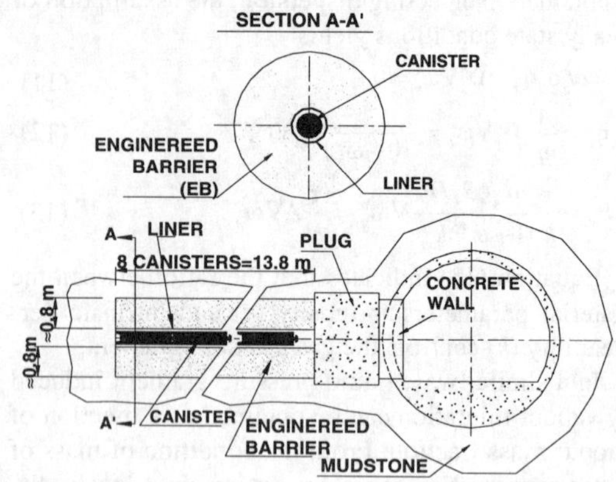

Figure 3. Scheme of the underground repository for high level radioactive waste

The following operational phases are considered:

- Isothermal phase without canisters (0-5 years). The boreholes have been drilled and the liner and bentonite have been installed. The bentonite in the barrier starts to hydrate under isothermal conditions.
- Ventilation phase (5-35 years). The canisters have been emplaced but the access drifts have not been backfilled yet. The drift walls are assumed to remain at a constant temperature due to ventilation. This enhances heat dissipation.
- Backfilled phase (35-10000 years). The drifts have been backfilled and they are no longer ventilated.

A constant gas generation rate of 5 mols of hydrogen per canister and year has been assumed. It corresponds to a typical corrosion rate for carbon steel canisters. The analyses have been performed using the computer code CODE_BRIGHT (Olivella et al. 1996). Detailed information on parameters, initial and boundary conditions are given in Gens et al. (1998c)

The variations of intrinsic permeability and gas and liquid permeabilities with degree of saturation adopted using the double porosity model are shown in Figure 4. It can be noted that they are close to the experimental values obtained from two different compacted clays. A parallel analysis using a constant intrinsic permeability has also been performed to highlight the effect of the double porosity model. In this alternative analysis, the gas intrinsic permeability has been equated to the saturated liquid permeability.

The computed evolution of the degree of saturation at four selected points is shown in Figure 5a. Two of the points are located in the engineered barrier, one adjacent to the canister and the other close to the host rock. The two other points are located in the host rock at two different distances from the drift axis. It can be observed that a strong drying occurs near the canister whereas, closer to the rock, a constant increase in degree of saturation is obtained. Some desaturation is also observed in the rock. This pattern of behaviour corresponds well to that observed in large scale in situ tests carried out simulating repository conditions (Gens et al., 1998a).

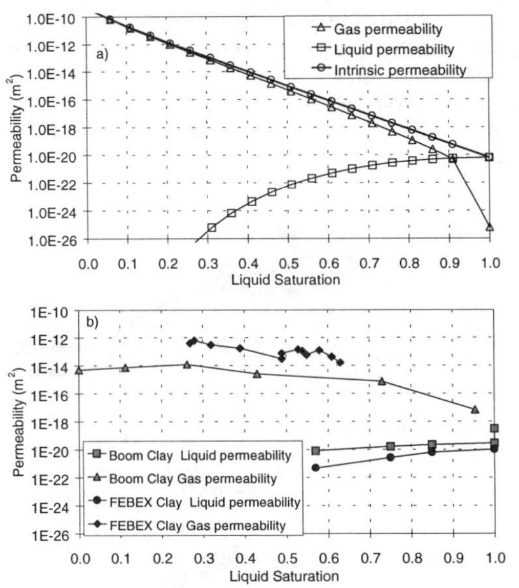

Figure 4. a) Permeability variations with degree of saturation according to the double porosity model. b) Observed variation of gas and liquid permeability for two compacted clays (Volckaert et al. 1994, Villar 1998).

In contrast, the analysis with constant intrinsic permeability (Figure 5b) shows a quite different hydraulic behaviour of the barrier with a very limited drying close to the canister, contrary to observations. The low gas permeability limits the operation of binary diffusion reducing therefore the magnitude of vapour migration. In consequence, the intensity of drying close to the canister is smaller. Because of the lack of water transfer by vapour, hydration of the barrier close to the rock is somewhat slower.

The evolution of temperatures at the same four selected points is shown in Figure 6. The sudden change of temperature at the end of the isothermal phase (5 years) is clearly visible. The end of the ventilation phase (35 years) is also noticeable but its effect is not large. It can be noted that temperature close to the canister reach very high values, above 200°C. Temperatures naturally reduce with increasing distance to the canister but they are still quite high, especially in the engineered barrier. These results demonstrate the capability of the formulation to deal with high temperatures.

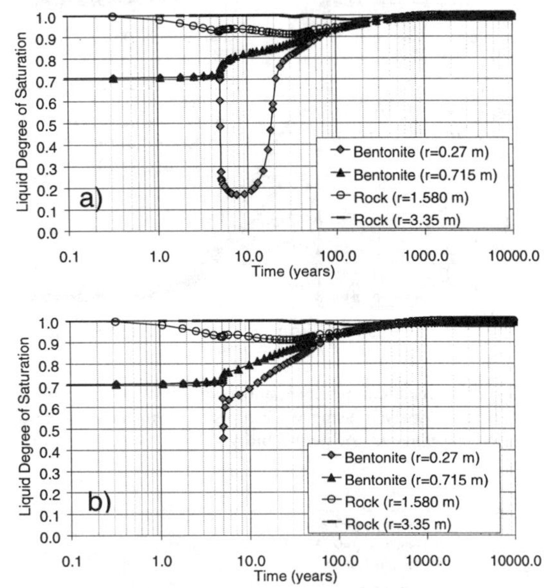

Figure 5. Evolution of the degree of saturation with time for four selected points. a) Analysis using double porosity model b) Analysis using constant intrinsic permeability.

The analysis with the double porosity model results in a somewhat higher maximum temperature than the constant intrinsic permeability one. This is a consequence of the strong drying of the material close to the canister. Drying causes a reduction in thermal conductivity that leads to a higher temperature. It is an interesting instance of thermo-hydraulic coupling.

The evolution of gas pressure for the two analysis is presented in Figure 7. It can be observed that there are two episodes of gas pressure rise. The first one occurs at relatively early times (up to 30 years) and it is related to the increase of vapour pressure due to temperature. As temperature reduces and vapour migrates the gas pressure diminishes. It can be noticed that the gas pressure in the analysis with constant intrinsic permeability reaches a higher value at this early stage. This a natural consequence from the fact that this calculation does not take into account the change in pore structure of the material and, therefore, it assumes a much lower gas permeability when the material is far from saturation.

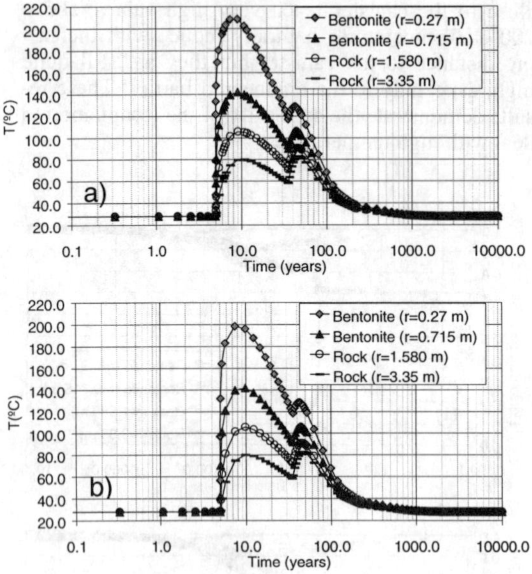

Figure 6. Evolution of temperatures with time for four selected points. a) Analysis using double porosity model b) Analysis using constant intrinsic permeability.

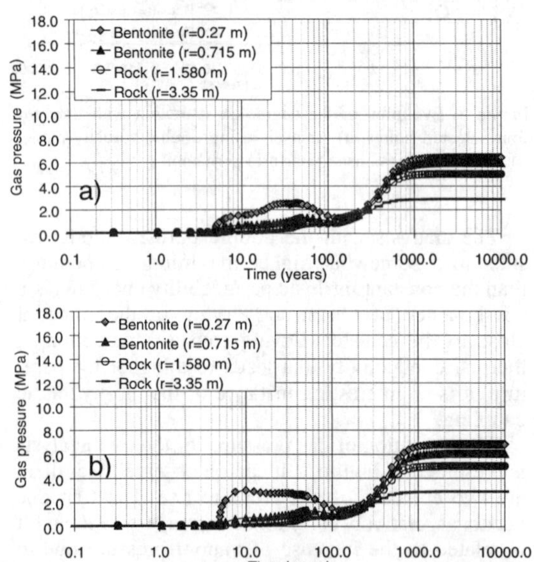

Figure 7. Variation of gas pressures with time for four selected points. a) Analysis using double porosity model b) Analysis using constant intrinsic permeability.

After the gas pressure reduction associated with the drop in temperature, a second episode of gas pressure increase is observed. This increase is due to the build up of gas production due to corrosion, finally reaching a steady value that allows a gas migration flow equal to the gas generation term. In reality, the gas pressure would finally drop because gas production would reduce, as the corroding material becomes exhausted after about 10000 years. In this case, the gas pressures in the two analyses are similar. The barrier material is very close to saturation and the effective gas permeabilities have comparable values for the two analyses.

The variation of gas saturation with time is shown in Figure 8. Note that, for clarity, the gas saturation scale is logarithmic. Initially the material is unsaturated and, in the double porosity analysis, drying drives the degree of saturation to very low values close to the canister. Eventually, however, hydration overcomes the transient temperature effects and a steady reduction of gas saturation with time is computed. Finally it reaches a very low value that is sufficient, however, to allow gas passage. This agrees with many observations that gas migration requires only a small desaturation of the barrier material to occur.

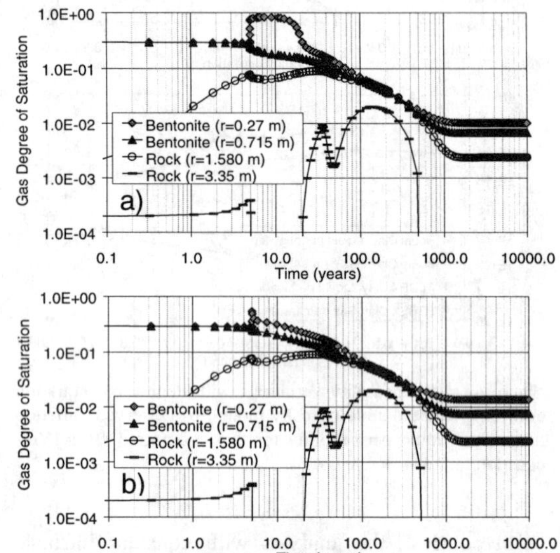

Figure 8. Variation of gas saturation with time for four selected points. a) Analysis using double porosity model b) Analysis using constant intrinsic permeability.

Finally Figure 9 presents the development of the net radial stresses (i.e. the difference between total stresses and gas pressures) in the barrier. They are the consequence of the swelling potential of the clay responding to the development of the hydration in the material.

It can be observed that the bentonite close to the rock develops swelling pressure earlier because it hydrates before the inner part. In fact, the differences between the two analyses reflect the stronger drying

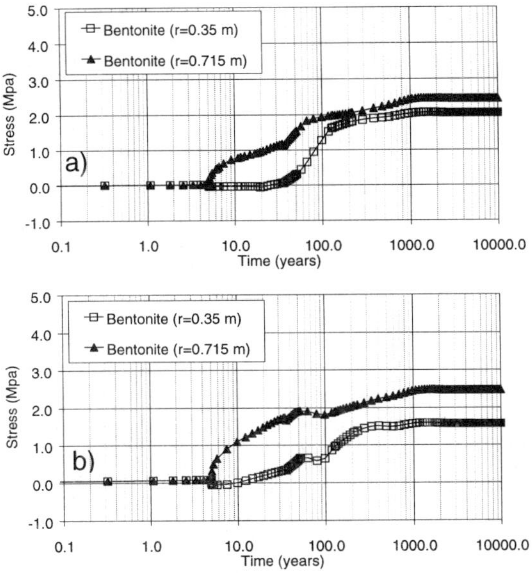

Figure 9. Variation of net radial stresses in the bentonite with time. a) Analysis using double porosity model b) Analysis using constant intrinsic permeability.

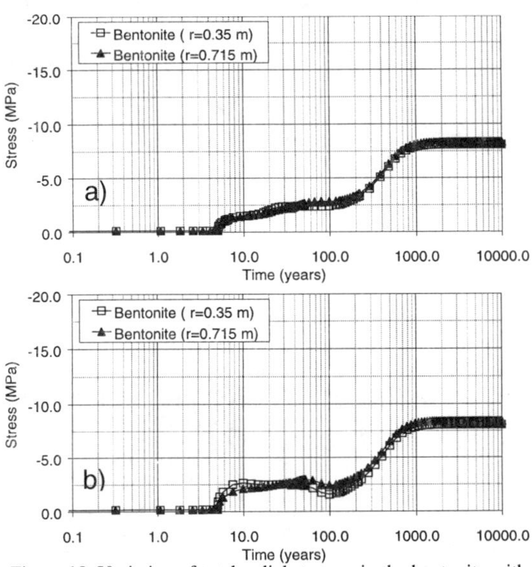

Figure 10. Variation of total radial stresses in the bentonite with time. a) Analysis using double porosity model b) Analysis using constant intrinsic permeability.

of the double porosity case that delays hydration of the inner barrier and, therefore, retards also the development of swelling pressure. The net stress reached at the end of the analysis is lower than the swelling pressure, which is about 4.5 MPa. There are two reasons. In the first place, the host rock (mudstone) is relatively deformable and, hence, it does not provide perfect confinement. In addition, gas migration prevents the full saturation of the barrier, hindering the clay from developing its full swelling potential. Figure 10 shows the evolution of total radial stresses in the barrier where the effect of the gas pressure increase can be readily observed. Now, the differences between the two analyses appear smaller because of the dominant effect of the variation of gas pressure, especially at long times.

5 CONCLUSIONS

The successful design of confining clay barriers requires a good understanding of their behaviour with regard not only to liquid flow but also to the various processes occurring in the gas phase. This is especially so in the case of engineered barriers for high-level radioactive waste where the effects of thermal loading, hydration and gas generation and transport may occur simultaneously. Coupled thermo-hydro-mechanical analyses that take into account the gas phase are required to simulate the problem in an adequate manner. Gas movement must account for, not only advective fluxes, but diffusive/dispersive flow as well. This is especially important in the treatment of water vapour movement that is often dominant at high temperatures.

The importance of taking into account the changes in pore structure that inevitably occur during hydration of compacted swelling clays has been highlighted. They result in extreme changes in intrinsic permeability that may span several orders of magnitude.

The formulation developed has been applied to a conceptual design of an underground repository in mudstone. This example of application demonstrates the capability of the formulation and computer code to deal with gas phase transport phenomena. The use of a double porosity model to take into account intrinsic permeability changes leads to correct prediction with respect to the hydration patterns of the barrier, including drying close to the canister. The differences in hydration pattern cause, via the corresponding couplings, differences in maximum temperatures and swelling stress development. The differences between analyses reduce in the long term when the barrier is much closer to saturation.

ACKNOWLEDGEMENTS

The authors acknowledge the financial support for this work through a research contract with ANDRA (Agence Nationale de Dechets Radioactifs). The authors also thank Frederic Plas, Gauthier Vercoutere and Sylvie Berger of ANDRA for very fruitful discussions.

REFERENCES

Alonso, E.E., Vaunat, J. & Gens A. 1999, Modelling the mechanical behaviour of expansive clays. *Engineering Geology*, 54: 173-183.

Bear, J. 1972. *Dynamics of fluids in porous media*, Elsevier.

Bear, J. & Bachmat, Y. 1986. Macroscopic Modelling of Transport Phenomena in Porous Media. 2. Applications to Mass, Momentum and Energy Transport, *Transport in Porous Media*, 1: 241-269.

Bird, R. B., Stewart W.E. &. Lightfoot. E.N. 1960. *Transport Phenomena*, New York: John Wiley.

Gens A.& Alonso E.E. 1992. A framework for the behaviour of unsaturated expansive soils. *Canadian Geotechnical Jnl.*, 29:1013-1032.

Gens A., Alonso, E.E., Huertas, F., Santiago, J.L., Garcia-Molina, A.& Guimares, L. Do N. 1998a. Febex large scale in situ test. Modelling and interpretation. *High-Level Radioactive Waste management Conf.*: 119-121. La Grange Park, Illinois: American Nuclear Society.

Gens, A., Garcia-Molina, A.J., Olivella, S., Alonso, E.E. & Huertas, F. 1998b. Analysis of Full Scale In-situ Heating Test Simulating Repository Conditions. *Int. Journal for Numerical and Analytical Methods in Geomechanics*, 22: 515-548.

Gens, A, Olivella S., & Buxo, P. (1998c). *Modelling of the THM behaviour of an Engineered Barrier. C-waste canisters.* C RP 0UPC 98-001. Paris: ANDRA.

Olivella, S., Carrera, J., Gens A. & Alonso, E.E. 1994. Non-isothermal Multiphase Flow of Brine and Gas through Saline media. *Transport in Porous Media*, 15: 271-293.

Olivella S. & Gens A. 2000. Vapour transport in low permeability unsaturated soils with capillary effects. *Transport in Porous Media*, 40,: 219-241.

Olivella, S., Gens, A., Carrera, J. & Alonso, E.E. 1996. Numerical Formulation for a Simulator (CODE_BRIGHT) for the Coupled Analysis of Saline Media. *Engineering Computations*, 13: 87-112.

Villar, M. V. 1998. *Ensayos para el proyecto FEBEX*, Report 70-IMA-L-5-51, Madrid: CIEMAT.

Volckaert, G, L. Ortiz, P. De Cannière, M. Put, S.T. Horseman, J.F. Harrington, V. Fioravante & M. Impey, 1994. Modelling and Experiments on Gas Migration in Repository Host Rocks, *MEGAS Project, Final Report, Phase 1*, Brussels: European Commission.

Wa'il Abu-El-Sha'r &. Abriola, L.M. 1997. Experimental assessment of gas transport mechanisms in natural porous media. Parameter estimation. *Water Resources Research*, 33: 505-516.

Computer Methods and Advances in Geomechanics, Desai et al. (eds) © 2001 Balkema, Rotterdam, ISBN 90 5809 183 X

Chemical softening and hardening of geomaterials in situ

T. Hueckel & F. Tao
Duke University, Durham, N.C., USA

G. Cassiani & A. Pellegrino
ENI-AGIP, Milan, Italy

ABSTRACT: Dependence of mechanical properties of geomaterials on the variable chemical environment or advancement of chemical or mineralogical reactions is known to be a crucial factor in many cases of mechanical stability of geo-masses. Processes of mineral dissolution usually results in material weakening while precipitation of minerals such as of silica in aging or calcite in clay fractures may lead to material strengthening. Experimental background of the process of aging of deep sediments is presented. It is postulated that aging consist of two processes of dissolution and precipitation affecting two different structural components. Their respective chemical softening and chemical hardening within plasticity theory is described. The joint response of the two components to further loading is then modeled.

1 INTRODUCTION

In many long-term engineering endeavors involving geomaterials, and whenever natural geological scale processes affect engineering works, engineers tend to resort to study creep, to take into account "time effect". However, the very term "time effect" may appear in various circumstances to be an excessive shortcut. Instead, one may use, and in many cases there is a sufficient information for it, an intrinsic time, or a process, or reaction advancement measure, to follow the evolution of the mechanical behavior over long-term histories.

It what follows we will focus on one specific phenomenon related to the "time effect", namely aging of petroleum/gas bearing sediments. This problem is socially important because of the serious consequences in terms of subsidence of the sediments that the exploitation of the resources may have on environment and on on-shore human habitat, in particular. Aging of soils is known as a set of processes of evolution of the mechanical properties of soils subjected to a long – term duration constant load (Schmertmann, 1991). Because of an impracticality of reconstructing the geological history of the sediments, on one hand, and of the need to quantify the progress of aging with the testing of the current state, it is proposed to try to link specific chemical processes resulting in the observed mechanical property changes.

2 EXPERIMENTAL EVIDENCE

Materials used for testing were: a predominantly clayey sand: DALIA 1, and a clay (50% smectitic), TEA 1, coming respectively from the depth of 1210 m. and 3270m of the prospective natural gas recovery fields in Northern Adriatic Sea. The most significant test result is derived from the comparison of a monotonic, incremental loading oedometric test and what will be called "aging test", which includes a monotonic loading to the stress equal to that *in situ*, followed by a pause at a constant loading of duration of 14 days, followed then by resumption of the monotonic loading. While it is realized that the "aging test" does not reproduce the loading history of the actual sediments exactly, in summarizes its essential aspects. The *in situ* effective stress was determined from stratigraphic data as 12.6 MPa for DALIA 1, and 35.04 MPa for TEA 1. Figure 1 shows two tests for comparison for DALIA 1. Surprising as it may be, TEA 1 clay behavior presents very similar characteristics. The 14 days the aging episode resulted in a modest secondary compression strain. The stress-strain curves below the aging episode are nearly identical, but they diverge significantly after it, showing a visible stiffening of the material. It appears that the amount of the stiffening in sand and clay is comparable, despite the difference in terms of depth, amount of the *in situ* stress, void ratio, and the total amount of strain reached. From the engineering point of view, it is clearly the range of stress above the *in situ* stress that is of direct interest. Indeed, in the case of oil/gas extraction the decrease in pore pressure of the extracted fluid will induce a corresponding effective stress increase. For that reason the stress – strain curves portion above the *in situ* stress, or the post-aging behavior is focused on in what follows.

On the basis of the above results four principal characteristic aspects of the post-aging behavior of undisturbed material were identified, which are crucial for the subsidence prediction (as shown for TEA 1 in Figure 2). These are: (i) decrease in incremental compressibility moduli in the post-aging stress excess range; (ii) relative reduction in strain in the post-aging stress excess range; (iii) a stress excess range affected by (i), (iv) strain gain during the aging period. The first three characteristics affect directly the numerical prediction of subsidence, the fourth one may be a factor in the experimental determination of the three former ones. There is another important characteristics of the post-aging behavior, which is (v) the total post-aging stress excess or total strain reduction range, defined as the total range in which the stress is above the comparable stress, and the strain is below that in reference monotonic tests. However, the values were not measured in the performed tests, because they were much higher than expected, and resulted to be outside the test range.

The above characteristics (i) and (ii) of the post-aging behavior, both expressing the decrease in deformability have a different connotation. The decrease in the incremental moduli expresses a local effect, which is stress dependent. The strain reduction is an integrated effect and gives a global measure of the aging impact, over certain stress range above the aging stress level. The standardized stress excess value corresponds to the value at which the incremental moduli decrease falls to zero. The stress range is defined as that in which the incremental compressibility moduli are lower than in the reference tests, as opposed to the total stress range (v) where the stress value is above that in monotonic stress.

To convey the sense of the effects of aging numerically, relative (normalized) values calculated for the three of the characteristics discussed above are summarized in Table 1, in terms of the average values below.

Table 1

material	relative* reduction in rate uniaxial compressibility modulus, c_M	relative** strain reduction due to aging	relative*** stress excess range affected by aging
DALIA	49.4%	38.0%	12.0%
TEA	69.9%	44.0%	45.0%

All relative changes are devided by the respective variables at the moment of the onset of aging.

Several other aging history effects were addressed in the experiments on remolded material First, it appears that sediments have the memory of their previous aging episodes. Tests on remolded DALIA-1 B8 and B10 samples show that when aging is repeated at the same stress level after a cycle of a partial unloading, the amount of strain per unit time during the "second aging" is more than one order of magnitude lower than during the "first aging". Secondly, the rate of straining changes significantly after 24 hours ('switching' from a pore pressure driven consolidation to secondary compression), but it is also clear that the rate of secondary compression does not decrease significantly after 14 days. However, stiffening of the material seems to occur much faster. Two accidental, short duration, aging episodes were investigated. An episode of a 2½ hour duration (Figure 3) in test B4 on DALIA-1 sand shows a significant decrease in deformability modulus (75%). However, the stress range affected by the short-term aging is relatively small, 6% of the aging stress level, against 28%, for a longer term aging. Finally, it was found that for longer (7 months) duration tests, the mean strain developed during aging was 0.759%, against 0.187% in 14-day aging (Figure 4). However, the decrease in compressibility modulus over 7- month is from $4.61\text{x}10^{-4}\text{cm}^2/\text{kg}$ to $1.935\text{x}10^{-4}$ cm^2/kg, that is 57.9%, against analogous decrease for 14- day

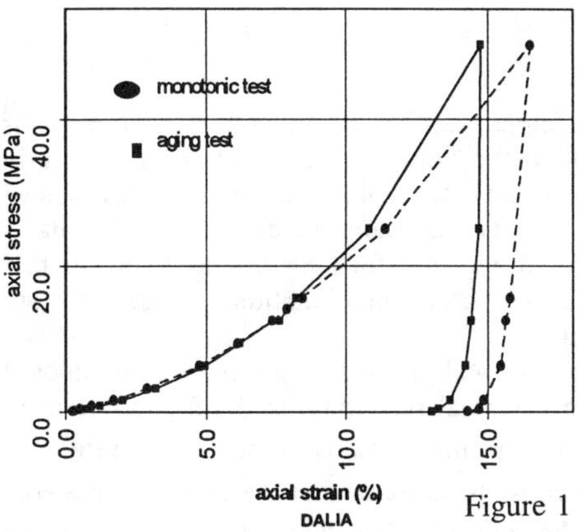

Figure 1

DALIA 1 - TEST B4

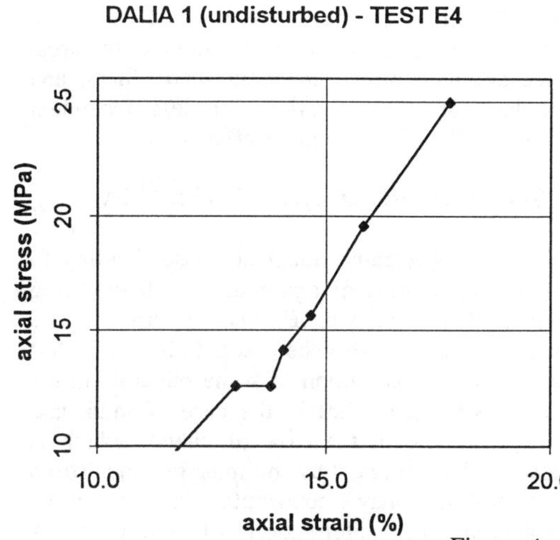

Figure 4

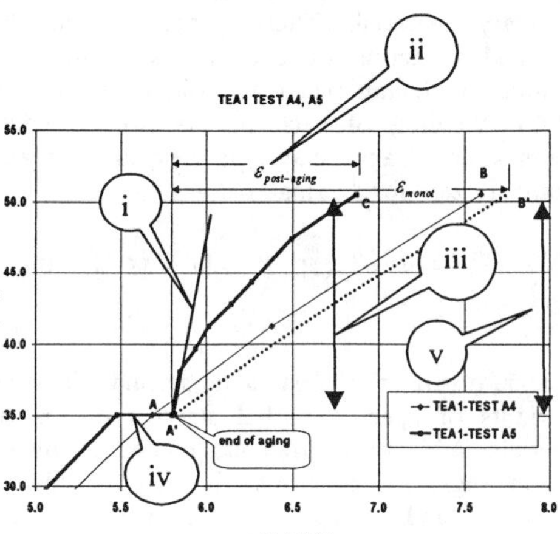

Figure 2

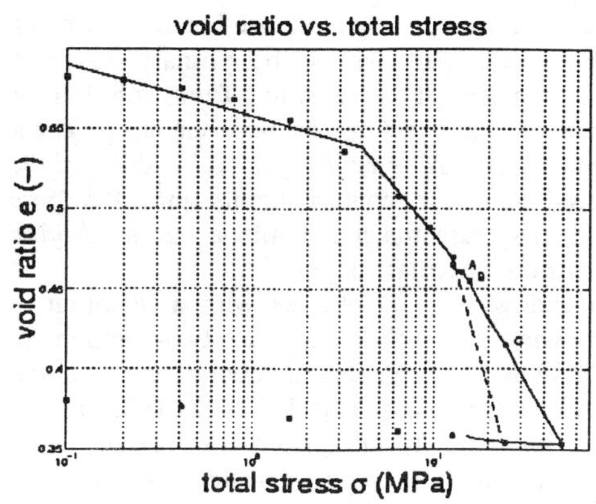

Figure 5a

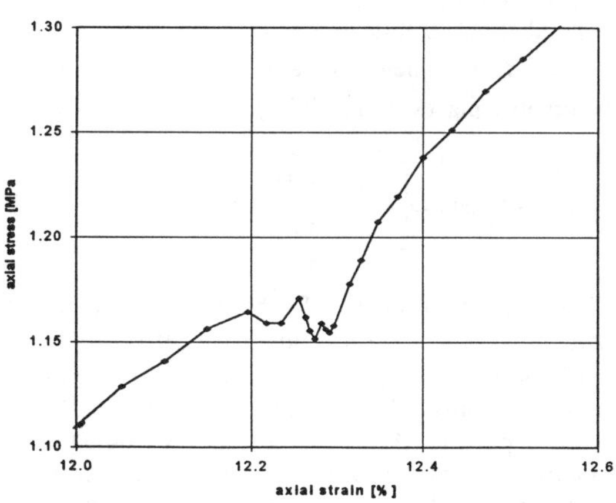

Figure 3

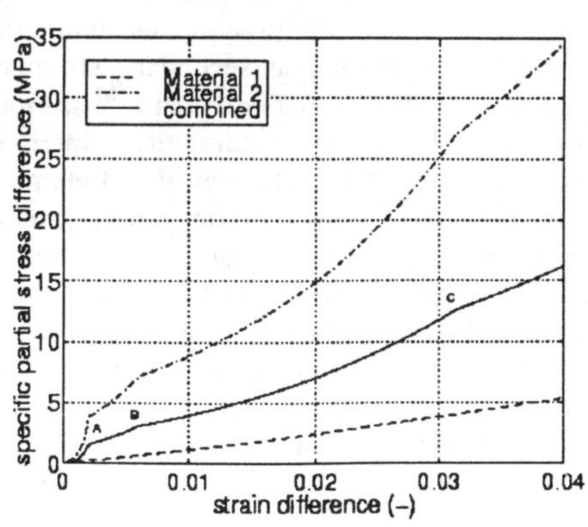

Figure 5b

tests of 52%, in average. This showed that some effects of aging such as strain amount, or stress range are more time dependent than others, and that decrease in compressibility is less dependent on the test duration than other effects.

3 SIMULATIONS OF AGING OF DALIA 1

The actual physico-chemical processes leading to the overall process of aging are far from being even sketchily known. Earlier hypotheses and some experiments (Mitchell and Solymer, 1984) suggested that dissolution of some minerals in the primary structure (silica in the case of sand, and montmorillonite in the case of clays), which is known to be facilitated by the inter-granular stress concentration, may contribute to an initial weakening and deformation of sediment. A subsequent change in electrolytic properties of the pore liquid (gel formation) are seen as contributing to building a secondary structure bonding by silica acid gel film particles of the primary structure through precipitation of silica from solution or suspension and acting as a cementing agent between them. Alternatively, the dissolution of the stressed solids at contact points into the pore liquid, may turn the liquid into a stress carrying suspension, flocculate or gel.

To endow such hypotheses with a quantitative framework, we shall use a two-component, evolving soil microstructure model for the aging conditions outlined by Hueckel et al., 1999. In this model an initial (primary) structure exists in an unloaded soil consisting of a system of grains. At the constant *in situ* stress linked to the overburden, dissolution of the stressed mineral at grain contacts of the primary structure may lead to its irreversible strain. Further during aging the dissolved silica (or other mineral dissolved) precipitates onto this primary structure forming a secondary structure. The partial stress components $\sigma^{(I)}$ and $\sigma^{(II)}$ are total stresses carried by the individual fractions, calculated per total area of the considered element, whereas $\sigma^{(1)}$ and $\sigma^{(2)}$ are specific partial stresses occurring in the material of each fraction, calculated per area of that fraction. Any additional compressive stress is distributed between the two components, because the two materials now deform jointly. Thus, a strain rate is a (weighed) sum of the rates of the partial stresses,

$$\dot{\sigma}_{ij} = \dot{\sigma}_{ij}^{(I)} + \dot{\sigma}_{ij}^{(II)} = \dot{\sigma}_{ij}^{(1)}\alpha + (1-\alpha)\dot{\sigma}_{ij}^{(2)} \qquad (1)$$

In a simplest option coefficient α, $0 < \alpha < 1$ is related to the volume fraction of the material in the (constant) total volume of solids. It is assumed for simplicity that the secondary material remains unstressed and undeforming during aging, or that dissolution and precipitation processes are sequential.

The chemical process as we see it is composed of two chemical reactions: of dissolution of mass from the primary material with the rate $\dot{\xi}_d$, followed by precipitation of the dissolved material with the rate $\dot{\xi}_p$. The dissolution is assumed to mechanically weaken the primary material. The precipitation is assumed to strengthen the secondary material. These processes will be described in terms of chemical softening and chemical hardening plasticity, following Hueckel (1992). Yielding of each of the two materials occurs when a Cam – clay type yield function and its rate are zero for that material,

$$f^{(i)} = (p^{(i)})^2 - p^{(i)}p^{(i)}{}_c(\varepsilon_v^{pl},\xi^{\alpha}) + \left(q^{(i)}/M^{(i)}\right)^2 = 0;$$
$$df^{(i)} = 0 \qquad (2)$$

p' and q are the first and second deviatoric invariants of specific partial stresses of the two materials. $M^{(i)}$ are coefficients related to internal friction angle for each material. To describe the weakening and strengthening of the considered components, the apparent maximum past isotropic stress describing hardening $p_c^{(i)}$, is a function of the respective degrees of advancement of reactions defined through time integral of the reaction rates $\dot{\xi}_\alpha, \alpha = d, p$. They are also functions of usual volumetric plastic strain (joint).

$$p^{\alpha} = p^{\alpha}{}_{c0}\, exp\left[\frac{(1+e_0)\varepsilon_v^{pl}}{(\lambda-\kappa)} + r^{\alpha}\varepsilon_v^{a}\xi_{\alpha}\right] \qquad (3)$$

r^{α} is the coefficient of chemical softening or hardening, $p_{c0}{}^{\alpha}$ is the maximum past effective mean partial stress, and $\varepsilon_v{}^{a}$ is the strain value at the onset of aging. Using relative values of stress and strains with respect to those at the onset of aging is a modification of the formula used by Hueckel et al. (1999) to enable us to use the data from remolded specimens to simulate undisturbed material, based on previous findings. λ and κ are bulk moduli of Cam-clay model.

It is assumed that the material element is a closed system, i.e. that no mass is exchanged with the external environment during the reactions. Therefore, the reactions are governed by the kinetic rate equations of the dissolution or precipitation of the respective minerals. We furthermore propose that

$$\dot{\xi}_p = -\dot{\xi}_d \quad (4)$$

This means that the mass of primary material is dissolved at the same rate as it precipitates to form the secondary material.

Following Hueckel et al. (1999) we will exploit a postulate by Leroueil et al., (1996) and Peret (1995), that during aging there is an unusual growth (about 3 times higher) of shear wave velocity across the material and that this indicates structuration of the material. We shall thus use the change in initial shear modulus as a measure to quantify the precipitation reaction rate and indirectly to measure the development of the secondary material structure.

It is recognized that the processes in the primary and the secondary material are completely different also from the plasticity point of view. The primary material undergoes chemical softening compensated completely by the strain hardening needed to fulfill two conditions: that of yielding and of constant effective specific partial stress in the primary material, $df = 0$ and $d\sigma'=0$. This leads to the possibility of measuring the rate of dissolution in terms of plastic strain rate,

$$\frac{d\xi_d}{d\varepsilon_v^{pl}} = -\left(\frac{\partial f}{\partial \xi_d} \bigg/ \frac{\partial f}{\partial \varepsilon_v^{pl}} \right) \quad (5)$$

Knowing both the advancement of dissolution reaction and the plastic strain rate one may identify the chemical softening constant, r^d

$$dp^1 = dp_{c0}^1 = p^1 \left[\frac{1+e_0}{\lambda-\kappa} d\varepsilon_v^{pl} + r^d \varepsilon_v^a d\xi_d \right] = 0 \quad (6)$$

However, while dissolution rate is not known, an alternative is to use the hypothesis that $\dot{\xi}_p = -\dot{\xi}_d$ Furthermore, we shall assume a linear form of the Leroueil – Peret postulate change in the elastic shear modulus, G_0, (initial modulus of unloading, or "dynamic modulus"), with respect to that at the onset of the secondary compression $G^a{}_0$

$$\dot{\xi} = \xi_0 \dot{G}_0 / G^a{}_0 \quad (7)$$

Thus the coefficient of chemical softening can be found from (6) and (7) as

$$r_\xi{}^d = r^d \xi_0 = -\frac{1+e_0}{\lambda-\kappa} \frac{d\bar{\varepsilon}_v^{pl}}{d\bar{G}_0} \quad (8)$$

where $\bar{\varepsilon}_v^{pl} = \varepsilon_v^{pl} / \varepsilon_v^a$ and $\bar{G}_0 = G_0 / G_0^a$

if the rate of change of shear modulus and of the volumetric plastic strain is measured. It should be noted that since the primary partial stress is constant, so is the elastic strain, and that also the secondary partial stress is by definition equal to zero. The latter statement however does not imply that the secondary material is inert during aging and it undergoes a strain-less hardening due to the precipitation of minerals with the rate $\dot{\xi}$ at an elastic state. Thus,

$$dp_{c0}^2 / p_{c0}^a = r^p \varepsilon_v^a d\xi_p > 0 \quad (9)$$

Using again the Leroueil-Peret postulat, knowing from the experiment the aging induced total stress excess, and from the properties of the primary material being able to calculate the secondary partial stress difference (proportional to the yield surface size growth in the secondary material), we can deduce the precipitation hardening constant, r^p.

In the tests on DALIA 1 material, the changes in the velocity of the shear waves were measured for remolded specimens and at a (fictitious) aging stress of 0.808 MPa. Table 2 shows corresponding void ratio and calculated shear modulus variations.

Table 2

B5		B6	
void ratio	shear modulus G (MPa)	void ratio	shear modulus G (MPa)
0.6678	185.4	0.6797	205.2
0.6673	219.0	0.6794	218.7

The first row data refer to the aging onset. With the data contained in Table 2, the mean value of the two cases for dissolution softening constant $r_\xi{}^d$ from eq.(8) is obtained as equal to 0.678 with $\lambda = 0.0538$ and $\kappa = 0.0082$ for the primary material. For the secondary material, with the stress excess of 0.023 MPa', and the elastic material constants back-calculated as $\kappa = 0.0102$, with $\alpha = 0.629$ the precipitation hardening coefficient is obtained as $r\xi^p = r^p \xi_0 = 5.81$. These coefficients, obtained from tests on remolded material will now be used for the simulations of the behavior of undisturbed specimens, under the premise that when properly normalized the behavior of both materials exhibits the same characteristic, at least in some processes.

The objective of this exercise is to evaluate the roles of the primary and secondary structure in the post-aging phase. The aging strain is taken directly from the experiments. The material parameters for the primary material were identified on the basis of the pre-aging phase behavior. The constants for the secondary material are back-calibrated. The chemical softening/hardening effect is based on the portion immediately preceding and two portions immediately after the aging episode.

The main result concerns the stress partition. Fig. 5a shows the results in terms of void ratio and logarithmic stress. Fig. 5b presents the re-calculated specific partial stress difference-strain difference relationship above the aging stress in the linear stress scale. The partial stress difference in the secondary material is clearly higher than the corresponding partial stress difference, above the aging stress in the primary material. However, the total partial stress in material *(1)* is the stress difference seen, plus its "initial" aging stress, that is 12.59 MPa, whereas in material *(2)* the initial stress is zero. Thus, the absolute partial stress in material *(2)* is lower than that in material *(1)* until the strain difference of 0.05 above the end of aging. The specific partial stress in primary material is larger than that in the secondary material up to the strain of about 0.03. Finally, the primary material is in the plastic state in the whole range above the aging stress. The secondary material reaches yielding at the value of 4.2 MPa of the specific partial stress, corresponding to the partial stress of 2.18 MPa and to the total stress of 14.77 MPa. Thus, even if in terms of stress differences material *(2)* carries more load than material *(1)*, in terms of the absolute stress values, this is not the case. Actually, in terms of pre-consolidation stress, material *(2)*, at the moment of its yielding, carries less than ¼ of the specific partial stress that is carried by the primary material.

4 CONCLUSIONS

From the engineering point of view, it is the load above the *in situ* stress that is important. In carrying this load the role of the secondary material is more important than of the primary one. Consequently, also its yielding point is extremely important, because it defines what was earlier called as "stress excess" due to aging. Thus, understanding of its strength is crucial for the prediction of the stress range in which the decreased compressibility occurs.

5 ACKNOWLEDGEMENT

The first two authors acknowledge the support granted to Duke University by Eni, Divisione Agip, Milan, Italy.

6 REFERENCES

Hueckel T. 1992, *Can. Geotech. J.*, 29, 1071-1086

Hueckel T., Fan Tao, G. Cassiani and A. Pellegrino 1999, in: *Constitutive Laws for Eng. Mat., 4th Int. Conf*, Troy, NY

Leroueil, S., Perret, D., and Locat, J. in: *Measuring and Modeling Time Dependent Soil Behavior*, GSP No. 61, 137-150.

Mitchell, J.K. and Z.V. Solymer, 1984, *J. Geotech. Engineering*, 110, 11, 1559-1576

Perret, D, 1995, Diagenese mecanique precoce des sediments fins du Fjord du Saguenay, *Ph.D. th., Universite Laval*

Schmertmann, J.H., 1991 *J. Geot. Eng*, 117, 9, 1288 - 1312

Computer Methods and Advances in Geomechanics, Desai et al. (eds) © 2001 Balkema, Rotterdam, ISBN 90 5809 183 X

Modelling of biologically induced clogging of granular media

R.K. Rowe & A.J. Cooke
Department of Civil and Environmental Engineering, University of Western Ontario, Canada

ABSTRACT: A problem recently gaining attention in the design and operation of solid waste landfills has been the buildup of organic and inorganic material in the leachate collection system due to the passage of leachate. This paper summarizes a model designed to predict the results of experimental research and ultimately predict the clogging of leachate collection systems. The model represents the leachate flow path as a series of separate fixed film reactors. Acetate, propionate, butyrate, calcium and suspended microorganisms are tracked through the system. The methanogenesis of acetate and acetogenesis of propionate and butyrate is simulated, with the products being biomass fixed to the porous media, acetate, and, using a relationship between acid degradation and calcium carbonate precipitation, fixed inorganic material. Attachment and detachment is also simulated. An example is shown which demonstrates the inputs required by the model and the results which can be generated.

1 INTRODUCTION

In order to minimize the contamination of surface and ground water, modern municipal solid waste landfills are often constructed with a barrier system containing a leachate collection system (or systems), and either a natural or engineered liner (or liners). The liner provides resistance to the outward flow of leachate. To be effective, however, the driving forces responsible for migration of contaminants through the liner must be controlled. The leachate collection system, therefore, is intended to perform two main functions: control the leachate head on the liner, thus reducing advective velocities of out-flowing leachate, and collect most, if not all, of the leachate produced by the landfill, thus reducing the contaminant levels of the landfill.

Typical leachate collection systems consist of a series of perforated pipes in a granular layer of crushed stone, gravel or sand. This layer is virtually impossible to clean, and, due to the passage of leachate, will undergo some degree of clogging due to the build-up of organic and inorganic material with time. The accumulation of clog material decreases the pore space available for flow and causes decreases in the hydraulic conductivity of the granular material. If conditions are good for clogging, the pore space can decrease to such an extent that the hydraulic conductivity of the granular layer is no longer sufficient to transport the leachate to the pipe without an increase in the driving force (head), and hence a leachate mound begins to develop.

If clogging continues it will result in further decreases in hydraulic conductivity and further increases in the height of the leachate mound. If the mound exceeds design levels, the collection system may be considered to be clogged since it is no longer performing its primary function of controlling the leachate level. However, although clogged, the leachate collection system at this point is not impermeable, and can still transmit leachate through the liner system as well as some leachate to the pipes.

Predicting the rate of clogging is essential to the modelling of the potential impact of the landfill on the environment since it is associated with the removal of contaminant mass from the landfill and the advective velocity of contaminant transport through the liner. The ability to predict the rate of clogging for a given system can also be used to allow the design of systems for a required service life. If the leachate collection system will operate without failure for long enough, the concentration of contaminants in the leachate will reduce to levels that will have a negligible impact if they migrate out of the landfill after eventual failure of the collection system (or a decision to terminate its operation) (see Rowe, 1991, 1995; Rowe et al., 1995). If there is premature failure of the leachate collection system, there will be a decrease in removal of contaminants by the system, and increased advective transport of contaminants through the liner, resulting in the transport of contaminants at concentrations that may cause unacceptable impact on the environment.

An examination of samples of clog material from landfill leachate collection systems, by Brune et al. (1991) and Fleming et al (1999) found that while it was made up of both inorganic and organic matter, calcium and carbonate were the chief components, and comprised over 50% of the dry mass. Due to the presence of bacterial slime with all precipitate accumulations, Brune et al. (1991) postulated that the formation of this clog material was largely due to microbial activity occurring in the collection system. Chemical data from laboratory reactors containing coarse gravel and permeated with landfill leachate (Rittmann et al., 1996) showed that the anaerobic removal of organic acids was directly related to the precipitation of calcium carbonate.

This paper outlines a model that has been developed to predict the rate of clogging of gravel material in laboratory column tests, mesocosm tests (cells operated with simultaneous unsaturated downward flow and horizontal saturated flow through actual field material), and in the field. The application of the model will then be illustrated with respect to a column test on a uniform 25mm gravel material to demonstrate the types of predictions that can be made with the model.

2 MODEL DESCRIPTION

2.1 *Model Requirements*

Based on the findings of Brune et al (1991), Rittmann et al (1996) and Fleming et al (1999), the model was developed based on clogging being caused by the growth of biomass and precipitation of minerals on the porous media. Leachate chemistry (volatile fatty acid concentrations, mineral concentrations, etc) and drainage properties (gravel size, permeability, flow rate, etc.) both control, and are controlled by the accumulation of biomass and precipitates. Since calcium carbonate was found to dominate in clog material it was assumed as a first approximation that it could be used as a surrogate for the mixture of mineral precipitates detected in the inorganic clog matter.

The model was developed such that for any location or time, the following interdependent properties can be calculated:

- quantity of fixed biomass
- organic acid concentrations (particularly acetate, propionate, butyrate)
- calcium concentration
- quantity of precipitate
- porosity and/or hydraulic conductivity

In order to meet these requirements, the model combines concepts associated with transient anaerobic, fixed film biological processes (such as those considered in wastewater treatment modelling) with concepts of geotechnical engineering. The model uses a time marching algorithm to model the evolution of the influent and effluent organic concentration, active biofilm, inert biofilm, mineral precipitate and porosity at any position or time. Since acetate, propionate, and butyrate are the substrates primarily responsible for biological processes causing clogging, the model considers the microbial activity corresponding to any mixtures of these three acids. The following sections provide a brief description of the key components of the BioClog model.

2.2 *Flow Distribution*

The model was designed for predicting field cases and column studies. In either case, the drainage system is divided into segments, each sufficiently small that within each segment each property may be assumed homogeneous without significant error (see Figure 1). Each segment is assumed to act as a separate, fixed film reactor through which there may be either unsaturated (percolation) or saturated flow. The proper combination of segment types is predetermined by the model as discussed below.

To model a one-dimensional column experiment, the segments are linked into a series in which the influent concentrations to the column are applied to the first segment in line, the effluent concentrations of each subsequent segment are the influent concentrations to the next, and the effluent concentrations of the last segment in line are the effluent concentrations of the column as shown in Figure 1.

When modelling the field case, the leachate collection system is taken to consist of a continuous, uniformly thick granular blanket and perforated leachate collection pipes. It is assumed that the granular layer is lain on an impervious clay liner in a sawtooth pattern with uniform slopes. The model was designed to simulate the leachate treatment, clog accumulation, and mounding that occurs along the leachate flow path from the high point down to the pipe, as shown in Figure 1. Infiltrating leachate from the waste above is distributed over the length of the flow path. While the lower portion of the drainage layer transmits a saturated flow of leachate

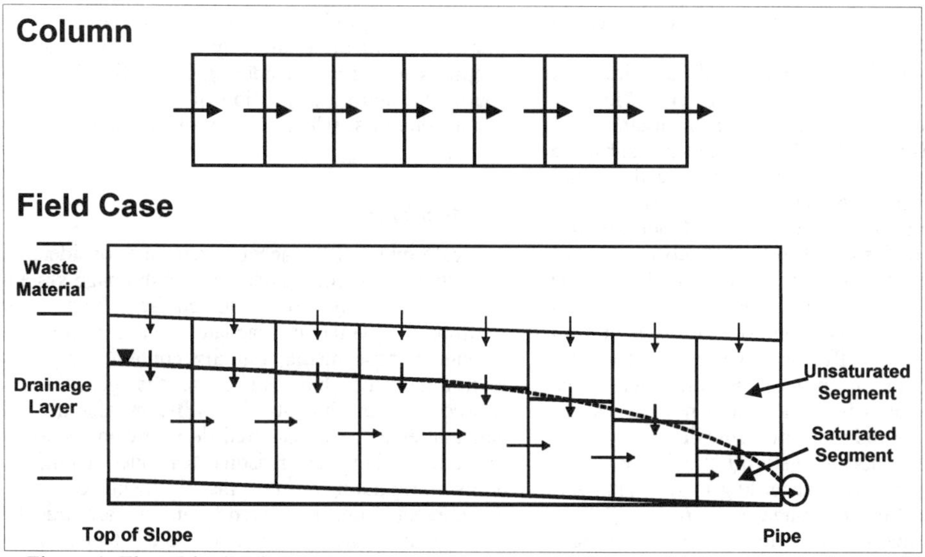

Figure 1: Flow Distribution

downslope to the pipe, the upper portion may be unsaturated, with leachate percolating downward. In this case, there is an unsaturated segment above a saturated segment at each location, with multiple sets of segments making up the drainage path. Special consideration is given to modelling the downward flow in the unsaturated segments. This effluent is then mixed with effluent of upstream segments before being applied to the horizontally flowing saturated segment. Each time step, the saturated height is recalculated based on the infiltration rate, slope, and hydraulic conductivity, the latter being linked to the porosity in each segment. The height of the boundary between the unsaturated and saturated segments at any location varies according to changes in infiltration rate or hydraulic conductivity of the collection layer. Thus, as clogging is predicted, the development of the mound of leachate may also be predicted.

2.3 *Representation Of The Porous Media*

The porous media in each segment is represented using ideal spheres of equal diameter. This facilitates the geometric calculation of porosity and surface area as a function of the film thickness on the porous medium. For the preliminary evaluation of the model (currently ongoing), this simple geometrical representation is effective since the column experiments conducted for this purpose utilize equal-sized spherical glass beads. Applicability of this assumption to irregularly shaped natural drainage materials has also been examined experimentally.

Acting as a reactor, each segment reduces the concentration of substrate in the fluid, and the chemical energy is partially converted into bacterial cell mass. This growth of biomass forms an active film layer on the porous media, with a separate active biofilm formed for each substrate. As explained later, this bacterial activity controls the formation of a second film, an inactive film, which consists of inert biomass and mineral precipitate. It is assumed that the inactive films grow on the porous media surface, beneath the active films. A stagnant liquid diffusion layer is assumed to exist between the active films and the bulk liquid.

The specific surface (the media surface area per unit volume) and porosity in each segment varies with thickness of the clog films. These properties are determined using a modified version of a set of geometrical equations for porous media assuming uniform film growth on equal-sized, ideal spheres by Taylor et al (1990). To accommodate for thick films (relative to particle size) such as those encountered in field and experimental study of drainage system clogging, the original equations were modified by Cooke and Rowe (1998).

2.4 *Biofilm Growth Model*

Acetogenesis of propionate and butyrate and the methanogenesis of acetate, and the consequent growth of biomass are represented in the model using a nonsteady-state biofilm model. The model was adapted from the work of Rittmann and McCarty (1981) and Rittmann and Brunner (1984).

The model is summarized here, and described in greater detail by Cooke et al (1999).

Firstly, a modified version of the iterative procedure developed by Rittmann and McCarty (1981) is used to calculate the flux of organic acid into the biofilm. In this procedure substrate undergoes mass transport resistance by a the liquid diffusion layer surrounding the biofilm, and then is utilized by the biofilm according to Monod kinetics and molecular diffusion. The flux of substrate out of the bulk fluid (the leachate) into the biofilm is dependent on the diffusion layer thickness, biofilm thickness, bulk substrate concentration, the biofilm density, substrate diffusion coefficients for free water and biofilm, and bio-kinetic parameters. The substrate does not diffuse into, and is not utilized by, the inactive films and attachment media.

The growth of new active biofilm is based on the calculated substrate flux according to Rittmann and McCarty (1980) and Rittmann and Brunner (1984). The loss of biomass is determined using a first order biofilm loss rate coefficient. In the model discussed here, the loss coefficient includes decay losses and detachment losses, where the detachment losses are based on shearing effects and/or growth effects.

2.5 *Mineral Precipitation Model*

From the preliminary data of laboratory experiments in which leachate collection was simulated using landfill materials infiltrated with actual landfill leachate, Rowe et al. (1995) suggested a linear relationship between COD removal and $CaCO_3$ precipitation. Rittmann et al. (1996) qualitatively discussed the chemical relationship between the microbial oxidation of COD to inorganic carbon to the mass of calcium carbonate precipitated out of the landfill leachate in these experiments. Fleming et al (1999) reported a similar linear relationship observed in raw landfill leachate effluent samples.

By utilizing the biofilm model to predict the removal of COD from the leachate, and the available concentration of calcium, the model uses a calcium carbonate "yield coefficient" to derive the mass of calcium removed from the leachate and mass of calcium carbonate precipitated within the drainage media. The calcium carbonate yield coefficient, Y_c, is a parameter unique to this model which is derived from the slope of a plot of COD removed versus $CaCO_3$ precipitated for the same or similar leachate. The work of Cooke et al (1999) contains a presentation of column study data, the method of determining Y_c , equations used to calculate the inactive film thickness from Y_c and an estimated Y_c value of 0.21 mgCaCO$_3$/mgCOD for the columns studied.

It is assumed that the inactive portion of the clog material contains a relatively small accumulation of inert biomass produced by decay of the active film. The amount of these non-degradable organic solids is approximated by considering the biofilm loss rate coefficient and an estimation of the fraction of the active biomass which is non-degradable due to decay.

2.6 *Attachment*

Typical raw landfill leachates transport considerable quantities of microorganisms to the drainage system where they may attach to surfaces and form biofilms. Synthetic leachate, when permeated through porous media, initially contains very few microorganisms, but can cause the growth and subsequent detachment of significant amounts of cell matter to be transported downstream, possibly to reattach. For these reasons suspended degraders, either originating from the influent or from detachment from the fixed biomass, are tracked through the system. To model the attachment of suspended degraders a first-order rate coefficient, K_{att}, was used. K_{att} is calculated from the porous media particle diameter, collision efficiency, porosity, and Darcy velocity (Clement et al, 1997). Equations by Tien et al (1997), which include the effects of a representative cell's density and size, and the temperature, density and viscosity of the leachate, are used to calculate the collector efficiency parameter.

2.7 *Mass Balance Equations*

Mass balances are used to calculate the concentration of chemical constituents in the effluent from a particular segment based on losses and additions according to the processes previously described. There is a mass balance for each of the three organic acids, each of the three suspended degraders, and one for calcium. The mass balances account for the degradation of propionate and butyrate to acetate by both the fixed and the suspended degraders, suspended degraders undergo growth and loss, and calcium precipitation corresponds to the activity of both fixed and suspended degraders.

3 MODEL DEMONSTRATION

3.1 *Model Parameters*

A hypothetical column flow rate study will be used to demonstrate some of the features of the model. In the study, three identical columns will be permeated with leachate and the effect different flow rates will be predicted by the model. It is assumed that the columns are 2500 mm long, 250 mm in diameter, and filled with 25 mm diameter (on average) gravel. An initial, uniform gravel porosity of 0.4 was assumed. The properties of the influent leachate are: 7,500 mgCOD/L acetic acid, 7,500 mgCOD/L propionic acid, 2,000 mgCOD/L butyric acid, 1,000 mg/L calcium, and 200 mgVSS/L suspended microorganisms. As presented in Cooke et al (1999), a calcium carbonate yield coefficient of 0.21 $mgCaCO_3$ / mgCOD was applied. Biofilm growth and loss coefficients for the VFAs were obtained from literature. The flow rates used in the study were 24.5 L/d, 49.1 L/d, and 98.2 L/d.

3.2 *Model Results*

Since the model results are for demonstration purposes, they are not intended to thoroughly analyze the effects of flow rate on drainage material clogging, but may be viewed as general trends which may be expected under the assumed conditions.

The change in normalized total effluent COD concentration with time is shown in Figure 2. It can be seen that in all three cases there is a steep initial decline in effluent COD, followed by a plateau, and then a further decrease. The figure shows that at higher flow rates, the effluent COD concentration is higher, thus indicating less treatment of the leachate.

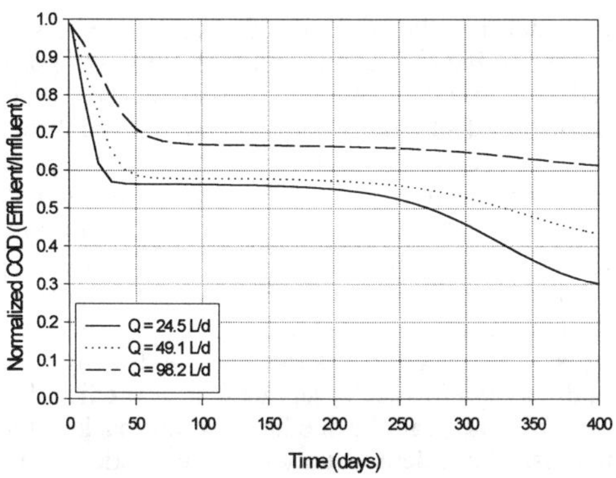

Figure 2: Calculated normalized effluent COD concentration for column tests fed raw leachate with flow rates of 24.5 L/d, 49.1 L/d, and 98.2 L/d.

At higher flow rates, leachate contact time with the fixed-film (and therefore treatment time) is decreased, and the rate at which fixed-film is detached from the porous media (most of which is removed from the system) increases, resulting in less treatment.

The reason for the two separate decreases in effluent COD in Figure 2 may be deduced from Figure 3, which illustrates the predicted concentrations of the three VFAs in the effluent. By comparing Figures 2 and 3 it can be assumed that the majority of the early loss in effluent total COD is due to bacterial utilization of acetate (an organic acid which is rapidly degraded), while the later, further decline in COD is the result of the additional degradation of propionate (an organic acid with slower degradation kinetics). The butyrate undergoes relatively little degradation within the period of study. The small increase in effluent acetate which occurs after approximately 250 days is likely due to the production of acetate from propionate degradation which also begins in this period. It may also be in some part due to the detachment and subsequent removal from the system, of biofilm which at this period of time would be mostly degraders of acetate.

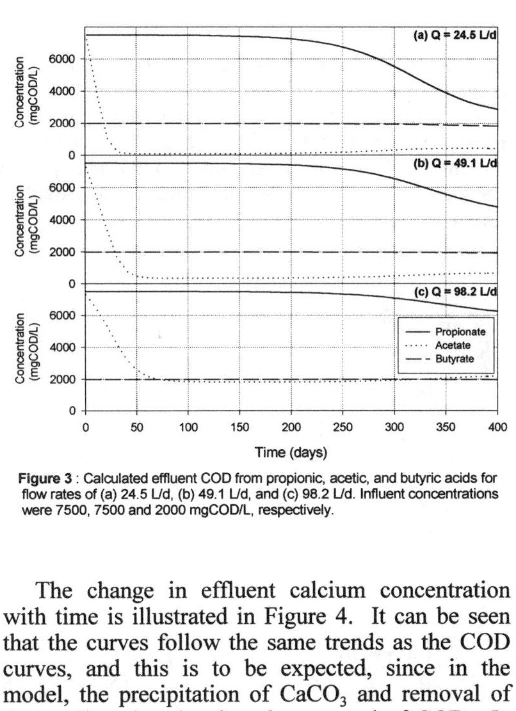

Figure 3 : Calculated effluent COD from propionic, acetic, and butyric acids for flow rates of (a) 24.5 L/d, (b) 49.1 L/d, and (c) 98.2 L/d. Influent concentrations were 7500, 7500 and 2000 mgCOD/L, respectively.

The change in effluent calcium concentration with time is illustrated in Figure 4. It can be seen that the curves follow the same trends as the COD curves, and this is to be expected, since in the model, the precipitation of $CaCO_3$ and removal of Ca are linearly related to the removal of COD. In the column with the lowest flow rate (Q = 24.5 L/d) the calcium is predicted to be completely removed from the influent after about 325 days.

The removal of COD and calcium correspond to growth of biological and mineral films, both of which decease the porosity of the column. The porosity profiles along the length of the columns are shown for 100, 200, 300 and 400 days in Figure 5. It can be seen that the column with the greater flow rate experienced a more rapid and uniform growth of clog material, while the column with lower flow underwent rapid clog accumulation only at the inlet end of the column. Interestingly, considering only

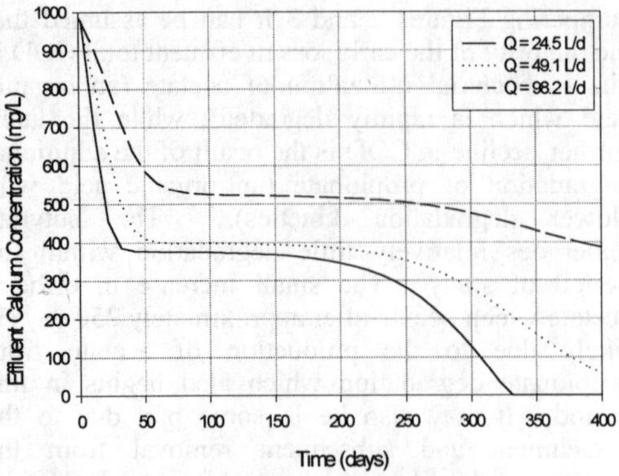

Figure 4 : Calculated effluent calcium concentration for column tests fed raw leachate with flow rates of 24.5 L/d, 49.1 L/d, and 98.2 L/d.

this porosity very near to the inlet shows that it has actually decreased a little more rapidly than the higher flow rates, since treatment and growth is more efficient at lower flow rates. The acids and calcium in this slower column, though, are quickly depleted, leaving little for clog accumulation downstream.

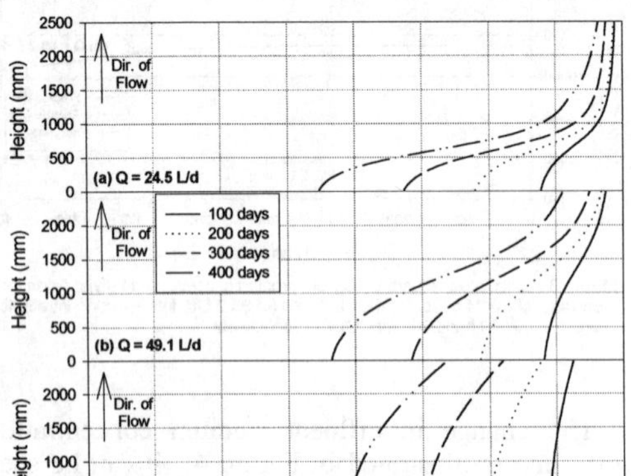

Figure 5 : Porosity profiles at 100, 200, 300 and 400 days for flow rates of (a) 24.5 L/d, (b) 49.1 L/d, and (c) 98.2 L/d (Note: initial porosity = 0.40).

Figure 6 illustrates the predicted idealized profiles of the inactive film thickness and total film thickness after 400 days. The difference, or thickness between the two films is the active film thickness. It is important to note that the majority of the film is inactive (mineral) film.

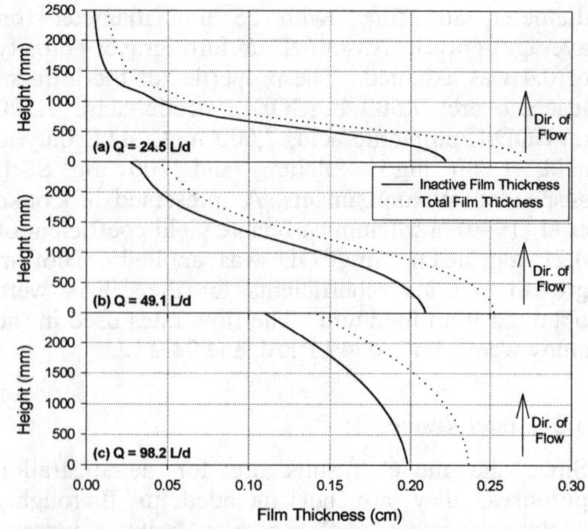

Figure 6 : Film thickness profiles at 400 days for flow rates of (a) 24.5 L/d, (b) 49.1 L/d, and (c) 98.2 L/d.

4 ACKNOWLEDGMENTS

Funding for the research was provided by the Natural Sciences and Engineering Research Council of Canada.

5 CONCLUSIONS

A model has been designed which predicts the growth of biological and mineral films on porous media. Acetic, propionic, and butyric acids, calcium, and suspended biomass are tracked through a series of segments, each of which act as a separate, fixed-film reactor. The film growth is predicted using a biofilm growth and loss model, and a linear relationship between VFA degradation and calcium carbonate precipitation. Detachment, transport, and attachment of microorganisms is also included. Porosity, specific surface, and hydraulic conductivity of each segment are predicted by assuming the granular media acts like packed ideal spheres. As a demonstration of the model, a short example is presented which shows the effects of increasing flow rate on leachate chemistry and clogging of a hypothetical column experiment.

6 REFERENCES

Brune, M., Ramke, H.G., Collins, H. J., and Hanert, H. H. (1991). Incrustation Processes in Drainage Systems of Sanitary Landfills, *Proceedings Sardinia 91, Third International Landfill Symposium*. CISA, Environmental Sanitary Engineering Centre, Cagliari, Italy, pp. 999-1035.

Clement, T.P., Peyton, B.M., Skeen, R.S., Jennings, D.A., J.N., Petersen (1997). Microbial growth and transport in porous media under denitrification conditions: experiment and simulations, *J. Cont. Hydrol.*, Vol 24, pp. 269 - 285.

Cooke, A. J. and Rowe, R. K. (1998). Extension of porosity and surface area models for uniform porous media, *J. Envir. Engrg., ASCE*, Vol 125, No. 2, pp. 126-136.

Cooke, A. J., Rowe, R. K., Rittmann, B. E. and Fleming, I. R.(1999). Modelling biochemically driven mineral precipitation in anaerobic biofilms, *Water, Science and Technology*. Vol. 39, No. 7, pp. 57-64.

Fleming, I. R., Rowe, R. K., and Cullimore, D. R. (1999). Field observations of clogging in a landfill leachate collection system, *Can. Geotech. J.* Vol. 36, No. 4, pp 685-707.

Rittmann, B. E., and Brunner, C. W. (1984). The non-steady state biofilm process for advanced organics removal, *Jour. WPCF*, Vol. 56, No. 7, pp. 874-880.

Rittmann, B. E. and McCarty, P. L. (1980). A model of steady-state biofilm kinetics, *Biotechnology and Bioengineering*, Vol 22, pg. 2343.

Rittmann, B. E. and McCarty, P. L. (1981). Substrate flux into biofilms of any thickness, *J. Environ. Eng. Div., ASCE*, Vol. 107 (EE4), pp. 831-849.

Rittmann, B. E., Fleming, I. R., and Rowe, R. K. (1996). Leachate Chemistry: Its Implications for Clogging, *North American Water and Environment Congress '96*, Anaheim, CA, June, paper 4 (CD-ROM) 6p, Session GW-1, Biological Processes in Groundwater Quality.

Rowe, R. K. (1991). Contaminant impact assessment and the contaminant lifespan of landfills. *Can. J. Civil Engrg.*, Vol. 18, No. 2, pp. 244-253.

Rowe, R. K. (1995). Leachate characterization for MSW landfills. *Proc. Fifth Int. Landfill Symp.*, Sardinia, Italy, Vol. 2, pp. 327-344.

Rowe, R. K., Fleming, I. R., Cullimore, D.R., Kosaric, N., and Quigley, R. M. (1995). *A research study of clogging and encrustation in leachate collection systems at municipal solid waste landfills.* Prepared for Interim Waste Authority Ltd. Available from the Geotechnical Research Centre, Department of Civil and Environmental Engineering, The University of Western Ontario.

Taylor, S. W., Milly, P. C. D., and Jaffe, P.R. (1990). Biofilm growth and the related changes in the physical properties of a porous medium, 2, Permeability, *Water Resour. Res.*, Vol. 26, No. 9, pg. 2161 - 2169.

Tien, C. (1989). *Granular Filtration of Aerosols and Hydrosols*. Butterworth Publishers, MA, USA.

Computer Methods and Advances in Geomechanics, Desai et al. (eds) © 2001 Balkema, Rotterdam, ISBN 90 5809 183 X

Thermal/Hydraulic/Chemical/Mechanical (THCM) behaviour of partly saturated soil

H.R.Thomas, P.J.Cleall & A.A.Hashm
Geoenvironmental Research Centre, Cardiff School of Engineering, Cardiff University, UK

ABSTRACT: The work presented in this paper describes progress in the development of a model for the coupled thermal, hydraulic, chemical and mechanical (THCM) behaviour of partly saturated soil. In particular the results of recent research work on the inclusion of geochemical effects in thermal hydraulic mechanical (THM) behaviour is presented. The effects of changes in the chemical solute concentration on the mechanical response are investigated, with an example demonstrating the capability of the approach proposed. Also, research on the validation of the hydraulic-chemical component is presented with the work applied to simulate a series of experimental tests.

1 INTRODUCTION

The engineering performance of high level nuclear waste disposal repositories is a problem of international major importance. Currently deep geological repositories, in which waste will be placed within a multibarrier system employed to control the migration of the radionuclides, are under consideration by a number of organisations (Swedish Nuclear Fuel Supply Co, 1983, DETR, 1999, EU 1999).

The proposed concept involves the placement of a partly saturated engineered clay barrier to form one layer of the multibarrier system. The performance of this barrier is affected by a number of phenomena including i) elevated temperatures due to heat energy emitted by the waste, ii) ground water recovery after closure tending to resaturate the engineered barrier, iii) a build up of gas pressure as the barrier resaturates iv) geochemical behaviour of both the barrier and the host environment. To carry out a suitable assessment of the barrier system each of these phenomena needs to be taken into account.

The behaviour of partly saturated soil in relation to i) the flow of moisture, heat, gases and chemicals and ii) the stress-strain response of the medium is therefore of great importance. An analysis of the bi-directional non-linear coupling between each of the flow fields and the stress strain field is required. For example the effects of the chemical composition of the system on the hydraulic and mechanical response need to be addressed.

The authors and co-workers have been actively involved in the development of such models for a number of years. Initially the prediction of the coupled movement of heat and moisture was addressed (Thomas and King, 1991). The effects of a varying air pressure were then included (Thomas and Sansom, 1995). A coupled flow and deformation model, utilising a non-linear elastic constitutive relationship was then developed (Thomas and He, 1995). Later an elasto-plastic constitutive relationship was included in the model (Thomas and He, 1998) allowing collapse and swelling of the soil under wetting to be simulated.

The next step was to attempt to include the effect of chemicals on the soil behaviour, to extend the THM models to THCM. This development is necessary because of possible effects of the geochemistry on the flow and deformation behaviour of bentonitic soils (Swedish Nuclear Fuel and Waste Management Co, 1997, Yong, 1999). It should be noted that the work presented in this paper is a first attempt at representing this very complex problem.

In the approach proposed this stage requires three main areas of development: i) flow of chemical solutes in pore fluid, ii) effect of the presence of solutes on the stress–strain behaviour of the soil and iii) geochemical interaction of solute with other species and soil solid.

The first component has been attempted herein via a coupled advection-diffusion mass balance equation (Bear and Verruijt, 1987, Yong et al, 1992).

An initial attempt at including the second component based on an osmotic potential term, has been presented previously (Thomas and Cleall, 1999).

The final component, geochemical interaction of the solute with other species and the soil solid, is attempted herein via the integration of the existing flow model with a geochemical model. Moisture and chemical solute flow are linked in a sequential manner to a suitable geochemical model. As part of the validation of this step this model has then been applied to the modelling of the flow and fate of heavy metals.

This paper presents the theoretical approach developed, the numerical methods employed and some examples demonstrating the capabilities of the model.

2 THEORY

Partly saturated soil can be considered a three-phase porous medium consisting of solid, liquid and gas. In this paper, the liquid phase is considered to be pore water and the gas phase as pore air. Solutes are defined as any chemical constituent dissolved in the fluid phase, whether as ions or electrically neutral complexes. A set of coupled governing differential equations can now be developed to describe the flow and deformation behaviour of the soil. The primary variables of the model are pore water pressure, u_l, pore air pressure, u_a, temperature, T, chemical solute concentration, c_s and displacement $\mathbf{u}$. The governing equations for heat and dry air transfer have been dealt with in detail elsewhere (Thomas & Sansom 1995). The development of the remaining equations is dealt with below.

2.1 *Moisture transfer*

The governing equation for moisture transfer in an partly saturated soil can be expressed as

$$\frac{\partial(\rho_l n S_l)}{\partial t}+\frac{\partial(\rho_v n(S_l-1))}{\partial t}=-\rho_l\nabla.\mathbf{v}_l-\rho_l\nabla.\mathbf{v}_v -\nabla.\rho_v\mathbf{v}_a \qquad (1)$$

where S_l is the degree of saturation. The velocities of pore liquid, $\mathbf{v}_l$, and pore air, $\mathbf{v}_a$, are based on a generalised Darcy's law. The pore liquid velocity can also include an osmotic term. The definition of vapour velocity follows the approach presented by Thomas and King (1991).

2.2 *Chemical solute migration*

Two main components are addressed in the representation of chemical solute migration, namely flow behaviour and geochemical interactions.

2.2.1 *Solute transfer*

The governing equation for solute transport equation may be written in a multi-chemical species form with the sink term expanded, to account for mass accumulation from reactions due to the sorption process as (Bear and Verruijt, 1987)

$$\frac{\partial(nS_l c_s^i)}{\partial t}+\frac{\partial(nS_l S^i)}{\partial t}+\nabla\cdot[c_s^i\mathbf{v}_l]-\nabla\cdot[nS_l D_h\nabla(c_s^i)]-\nabla\cdot[c_s^i D_h\nabla(nS_l)]=0 \qquad (2)$$

where S is the sorbed chemical concentration and the superscript i represents the species. D_h is the coefficient of hydrodynamic dispersion, which can be expressed as:

$$D_h = D_d + D \qquad (3)$$

where D is the coefficient of mechanical dispersion and D_d is the molecular diffusion coefficient..

2.2.2 *Geochemical modelling*

In the approach adopted by the authors a geochemical model has been employed to represent the chemical interactions between the components of the different phases of the soil system. The considered chemical reactions are ion complexation in the aqueous solution, sorption on solid surfaces and precipitation/dissolution of solids. In particular MINTEQA2 (Westall et al., 1976) has been implemented to calculate the second term of equation (2). MINTEQA2 is based on an equilibrium approach and offers both electrostatic and non-electrostatic models to represent sorption. The mass action expression for a system of n independent components can be given for m species as

$$K_i = \{S_i\}\ \Pi\, X_j^{-a_{ij}} \qquad (4)$$

where K_i is the equilibrium constant for the formation of species i, $\{S_i\}$ the activity of species i, X_j the activity of component j, a_{ij} the stoichiometric coefficient of component j in species i and $\prod$ indicates the product over all components in species i.

The set of n independent components is governed by n mass balance equations of the form

$$Y_j = \Sigma\, a_{ij}\, C_i - T_j \qquad (5)$$

where T_j is the total concentration of component j and Y_j is the difference between the calculated total concentration of component j and the known analytical total concentration of component j.

For the electrostatic models, charge balance equations are defined as

$$Y_\sigma = \Sigma\, a_{i\sigma}C_i - T_\sigma \qquad (6)$$

where $a_{i\sigma}$ is the stoichiometry of the electrostatic component pertaining to σ in species i where σ is the surface charge.

2.3 *Stress-strain relationship*

The total strain, ε, is assumed to consist of components due to suction, temperature, chemical and stress changes. This can be given in an incremental form, without loss of generality, as

$$d\varepsilon = d\varepsilon_\sigma + d\varepsilon_c + d\varepsilon_s + d\varepsilon_T \tag{7}$$

where the subscripts σ, c, T and s refer to net stress, chemical, temperature and suction contributions.

The stress strain relationship can therefore be expressed as

$$d\sigma'' = \mathbf{D}\left(d\underline{\varepsilon} - d\underline{\varepsilon}_c - d\underline{\varepsilon}_s - d\underline{\varepsilon}_T\right) \tag{8}$$

where σ'' is the net stress and $\mathbf{D}$ is the elastic matrix.

A number of constitutive relationships have been implemented to describe the contributions shown in equation (8). In particular for the net stress, temperature and suction contributions both elastic and elastoplastic formulations have been employed (Thomas & He, 1995, Thomas & He, 1998). To describe the contribution of the chemical solute on the stress-strain behaviour of the soil, as a first approximation, an elastic state surface concept was proposed (Thomas & Cleall, 1999) which described the contribution of the chemical solute via a elastic relationship based on osmotic potential.

2.4 *Coupled equations*

The series of governing differential equations described above can be expanded and expressed in terms of the primary variables of the model as;

$$C_{Tl}\frac{\partial u_l}{\partial t} + C_{TT}\frac{\partial T}{\partial t} + C_{Ta}\frac{\partial u_a}{\partial t} + C_{Tu}\frac{\partial \mathbf{u}}{\partial t} = \nabla[K_{Tl}\nabla u_l] + \nabla[K_{TT}\nabla T] + \nabla[K_{Ta}\nabla u_a] + V_{Tl}\nabla u_l + V_{TT}\nabla T + V_{Ta}\nabla u_a + V_{Tc_s}\nabla c_s + J_T$$

$$C_{ll}\frac{\partial u_l}{\partial t} + C_{lT}\frac{\partial T}{\partial t} + C_{la}\frac{\partial u_a}{\partial t} + C_{lu}\frac{\partial \mathbf{u}}{\partial t} = \nabla[K_{ll}\nabla u_l] + \nabla[K_{lT}\nabla T] + \nabla[K_{la}\nabla u_a] + \nabla[K_{lc_s}\nabla c_s] + J_l$$

$$C_{al}\frac{\partial u_l}{\partial t} + C_{aT}\frac{\partial T}{\partial t} + C_{aa}\frac{\partial u_a}{\partial t} + C_{au}\frac{\partial \mathbf{u}}{\partial t} = \nabla[K_{al}\nabla u_l] + \nabla[K_{aa}\nabla u_a] + \nabla[K_{ac_s}\nabla c_s] + J_a$$

$$C_{c_sl}\frac{\partial u_l}{\partial t} + C_{c_sa}\frac{\partial u_a}{\partial t} + C_{c_sc_s}\frac{\partial c_s}{\partial t} + C_{c_su}\frac{\partial \mathbf{u}}{\partial t} = \nabla[K_{c_sl}\nabla u_l] + \nabla[K_{c_sT}\nabla T] + \nabla[K_{c_sa}\nabla u_a] + \nabla[K_{c_sc_s}\nabla c_s] + J_{c_s}$$

$$C_{ul}du_l + C_{uT}dT + C_{ua}du_a + C_{uc_s}dc_s + C_{uu}d\mathbf{u} + d\mathbf{b} = 0 \tag{9}$$

where K_{ij} and C_{ij} represent the corresponding terms of the governing equations (i,j=l,T,a,c_s,u).

3 NUMERICAL SOLUTION

3.1 *Numerical formulation*

The numerical solution of the governing equation is achieved by a combination of the finite element method for the spatial discretisation and a finite difference time stepping scheme for temporal discretisation.

The Galerkin weighted residual method is employed to formulate the finite element discretisation. For the governing equation described in section 2, this method yields, in a concise notation,

$$\mathbf{A}\phi + \mathbf{B}\frac{\partial \phi}{\partial t} + \mathbf{C} = \{0\} \tag{10}$$

where ϕ represents the variable vector $\{u_l, T, u_a\ c_s, u\}^t$ and $\mathbf{A}$, $\mathbf{B}$ and $\mathbf{C}$ are matrices and vectors representing the discretised components of the governing equation.

To solve equation (10) a numerical technique is required to achieve temporal discretisation. In this case an implicit mid-interval backward difference algorithm is implemented, since it has been found to provide a stable solution for highly non-linear problems (Thomas & King, 1991). This approach yields

$$\phi^{n+1} = \left[\mathbf{A}^{n+1/2} + \frac{\mathbf{B}^{n+1/2}}{\Delta t}\right]^{-1}\left[\frac{\mathbf{B}^{n+1/2}\phi^n}{\Delta t} - \mathbf{C}^{n+1/2}\right] \tag{11}$$

where the superscript represents the time level. A solution can be found to equation (11) via a predictor-corrector iteration scheme (Thomas & King, 1991).

For the elasto-plastic based stress equilibrium equations a stress return algorithm is required. A backward Euler procedure is employed (Crisfield 1991) with the tangent cutting method (Ortiz & Simon, 1986,) used to improve the accuracy of this algorithm.

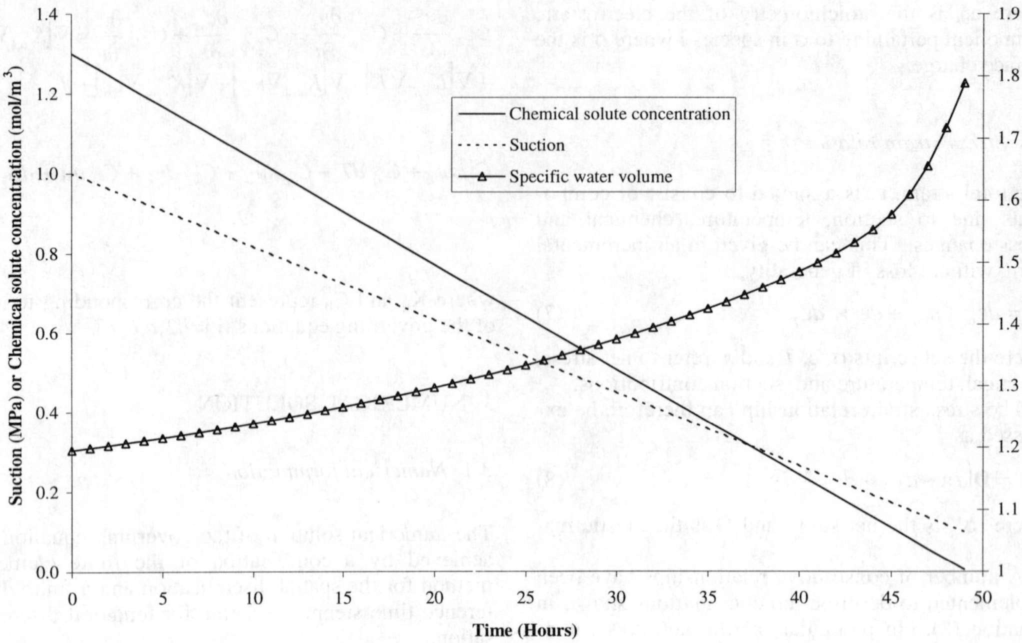

Figure 1 Load path and specific water volume during test.

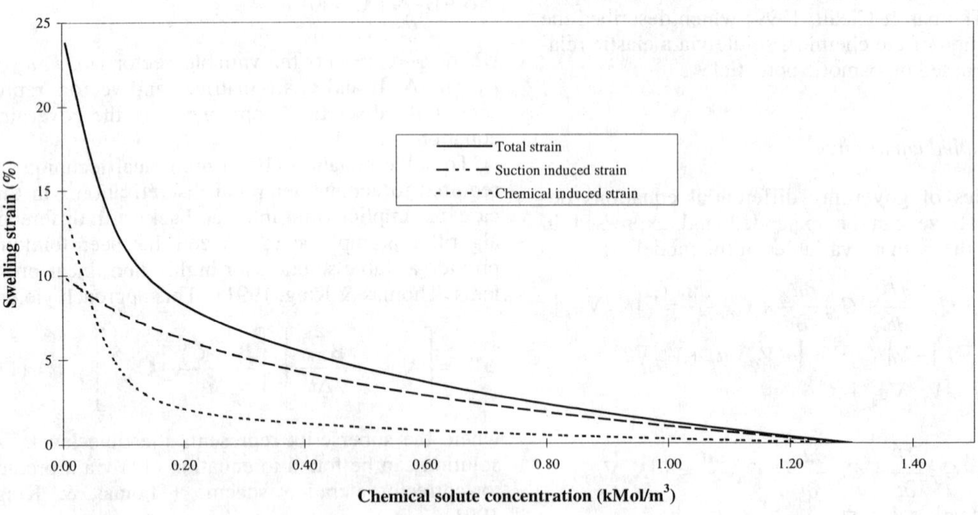

Figure 2 Chemical solute concentration versus swelling strain

A solution of the chemical interaction equations is achieved via a Newton-Raphson approximation method (Westall et al, 1976). Coupling of the two sections of the overall solution process, the transport equations and the chemical interaction equations, is accomplished via a sequential iteration approach. This method sequentially solves the flow and interaction equations during each iteration of the predictor correct scheme mentioned above.

With appropriate initial and boundary conditions the set of non-linear coupled governing differential equations can be solved via the numerical solution presented above.

3.2 *Computer code*

A computer code known as COMPASS (COde for Modelling PArtly Saturated Soil) has been developed to implement the numerical approach detailed above. This includes an 'engine' written in FORTRAN 90 and a graphic user interface written in VISUAL BASIC. The engine includes an extensive library of finite elements and allows both steady state and transient 1,2 and 3D problems to be analysed. In addition to this a number of methods to solve the set of simultaneous equations are available including frontal, LU, LDLT, Cholesky and a number of conjugate gradient and preconditioner solvers (Hinton and Owen, 1977; Bathe, 1996).

During the course of the development of COMPASS, an extensive verification and validation program has been undertaken. It is recognised, however, that validation of the code is a process rather than a series of events and as such, continued work on this aspect is envisaged.

4 APPLICATIONS

4.1 *Hydraulic-Chemical-Mechanical*

In proposed high level nuclear waste repositories it is possible that the buffer materials will be exposed to saline ground water. This will affect the performance of the clay barrier with an expected reduction in the swelling pressures produced during water uptake. In this example a non-linear elastic model of expansive soils has been used to investigate the effect of chemical solute concentration on deformation (Cleall, 1998, Thomas & Cleall 1999). Experimental results presented by Richards et al, (1984), show the development of strain in an partly saturated sample of Pleistocene clay when subject to variations of pore fluid solute concentration, in this case NaCl. Although the reported results are not comprehensive enough to allow the complete development of a state surface for void ratio there are enough data to allow the description of deformation modulii at a vertical stress of ≈70kPa.

As a test example, a load path has been specified, shown in Figure 1 and COMPASS has been used to predict the deformation response of the soil. The predicted deformation response of the soil is shown in Figure 2. It can be observed that the total volumetric strain of the sample at the conclusion of the load path was 23.8%. The total strain can be separated into a strain of 13.9% due to chemical solute concentration changes and a strain of 9.9% due to the suction change. Figure 1 also shows the increase in the volume of water in the sample in terms of specific water volume. The flow of water can be observed to increase as the swelling of the sample increase as the test progresses.

This example illustrates that experimental results can be used to predict the stiffness of the soil due to changes in chemical solute concentration. Also shown is the importance of including the effect of a variation of chemical solute concentration in the soil.

The THM model has been previously applied to simulate laboratory scale experiments where a confined sample of soil has been subjected to a heating source (Cleall and Thomas, 1999). The model utilised an elasto-plastic constitutive relationship for the stress-strain behaviour of the soil. The transfer processes represented included the flow of heat, moisture and air. An encouraging correlation between the observed behaviour and the simulated results was found. The results demonstrated the ability of the model to represent the main mechanisms of flow in such soils and the dominant features of the stress-strain behaviour.

4.2 *Hydraulic-Chemical*

In this section the hydraulic-chemical component of the model is applied to simulate a series of experimental tests. In terms of THCM modelling this exercise aims to validate the capability of the model in predicting the soil's coupled hydraulic-chemical behaviour and as such this work forms part of the validation programme mentioned in section 3.2. In particular the model's ability to represent the influence of the change in the chemical composition of the soil and pore water on both the hydraulic and solute transport performance is examined.

In the example considered the reactivity of the solute is addressed via the incorporation of the geochemical model MINTEQA2. The geochemical model considers the different chemical mechanisms of interaction between the solid and the liquid phases and consequently the partitioning of the different chemical species between these phases at equilibrium.

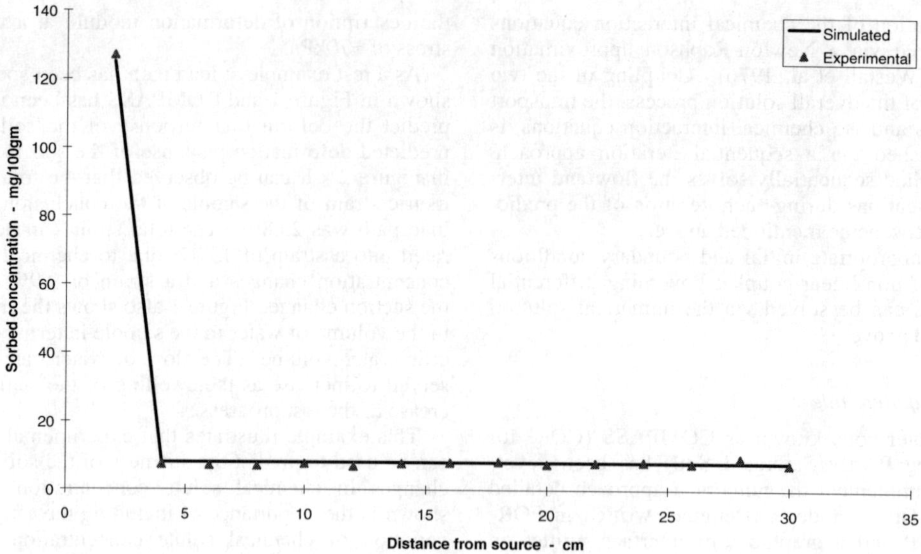

Figure 3. Simulated and experimental profiles of Lead sorbed concentration Sr=0.2 Cc=500ppm pH=7.6

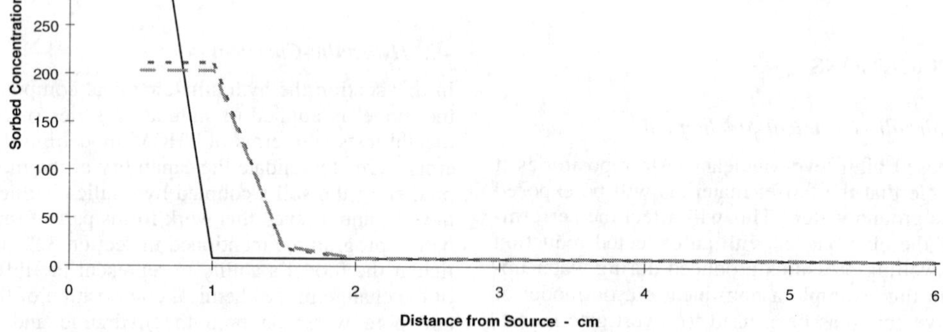

Figure 4. Simulated sorbed concentrations for different pH

The experiment investigated the effect of soil pH on the retention of lead in a suction driven flow horizontal leaching cell. Batch Equilibrium tests were conducted to obtain the sorption isotherms and retention data necessary for the simulation. The Langmuir model, which is one of seven sorption models available in MINTEQA2, was found to offer good representation of the batch experimental data. The moisture retention capacity and hydraulic conductivity of the soil were also determined experimentally.

The experiment shows deterioration in the retention capacity of the soil with a reduction in its pH. The model's ability to simulate this behaviour was examined by comparing experimental results with the numerical. Figure 3 shows good agreement between the experimental and numerical retention profile. Figure 4 shows the model simulation of the change in the retention capacity of the soil under different soil pH. These results were found to be in good agreement with the experimental results (Hashm, 1999).

5 CONCLUSIONS

The work presented in this paper describes recent research performed in the development of a model for the coupled flow of heat, moisture, air and chemical solute in a deformable partly saturated soil. In particular the effect of chemical solute concentration on the mechanical behaviour of the soil has been explored via a non-linear elastic constitutive relationship. Also, work on the validation of the hydraulic-chemical component of the model is presented with the model applied to simulate a series of experimental tests. In terms of the deformation behaviour and the coupled hydraulic chemical behaviour the simulations and examples show some encouraging results.

6 ACKNOWLEDGEMENTS

The work carried out has been part funded by an EPSRC research studentship. This support is gratefully acknowledged.

REFERENCES

Bear, J. & A. Verruijt 1987. *Modeling groundwater flow and pollution.* Kluwer Academic Publishers Group.

Bathe, K. J. 1996. *Finite element procedures.* Prentice Hall, New Jersey.

Cleall, P.J., 1998. An investigation of the thermo / hydraulic / mechanical behaviour of unsaturated soils, including expansive soils. Ph.D. Thesis, University of Wales, Cardiff, U.K.

Cleall, P.J. & H.R. Thomas 1999. Application of a coupled thermal-hydraulic-mechanical model for unsaturated soil. *Proceedings of the 7th International Conference on Numerical Methods in Geomechanics*: 237-242, Balkema.

Crisfield M.A. 1991. *Non-linear finite element analysis of solids and structures*, John Wiley & Sons.

DETR 1999. An R&D strategy for the disposal of high level radioactive waste and spent nuclear Fuel. *Department Environment Transport and Regions* U.K.

EU 1999. Calculation and testing of behaviour of unsaturated clay. *Commission of the European Union, Contract Report, EC Contract FI4WCT950003*

Hashm, A A 1999. A study of the transport of a selection of heavy metals in unsaturated soil. PhD Thesis, School of Engineering, Geoenvironmental Research Centre, University of Wales, Cardiff.

Hinton, E & D.R.J. Owen 1977. Finite element programming, *Academic Press, London.*

Ortiz M. & J.C. Simon 1986. An analysis of a new class of integration algorithm of elasto-plastic constitutive relations, *Int. J. of Num. Meth. in Eng.*, 23

Richards, B.G., P. Peter & R. Martin 1984. The determination of volume change properties in expansive soils. *Proc. 5th. Int. Conf. Exp. Soils. Adelaide*: 179-186.

Swedish Nuclear Fuel and Waste Management Co 1997. Bentonite swelling pressure in strong NaCl solutions: Correlation between model calculations and experimentally determined data.. *Swedish Nuclear Fuel and Waste Management Co Technical Report* 97-31.

Swedish Nuclear Fuel Supply Co/ Division KBS 1983. Final storage of spent nuclear fuel - KBS-3, III Barriers. *Swedish Nuclear Fuel Supply Co/Division, KBS Report*: 9:1-16:12.

Thomas, H.R. & S.D. King 1991. Coupled temperature/capillary potential variations in unsaturated soil. *ASCE, Journal of Eng. Mech.* 117(11): 2475-2491.

Thomas, H.R & Y. He 1995. Analysis of coupled heat, moisture and air transfer in a deformable unsaturated soil. *Geotechnique*. 45(4): 677-689.

Thomas, H.R. & M.R Sansom 1995. Fully coupled analysis of heat, moisture and air transfer in unsaturated soil. *ASCE, Journal of Eng. Mech.* 121(3): 392-405.

Thomas, H.R & Y. He 1998. Modelling the behaviour of unsaturated soil using an elasto plastic constitutive relationship. *Geotechnique.* 48(5): 589-603.

Thomas, H.R. & P.J. Cleall 1999. Inclusion of expansive clay behaviour in coupled thermo hydraulic mechanical models. *Int. Jour. of Engineering Geology* 54: 93-108.

Westall, J C, Zachary, J C. & F M Morel 1976. MINQEQL: A computer programme for the calculation of chemical equilibrium composition of aqueous systems. Cambridge, Massachuesetts, Department of Civil Engineering, Massachusetts Institute of Technology. Tech. Note 18.

Yong, R.N., A.M.O. Mohamed & B.P. Warkentin 1992. *Principles of contaminant transport in soils.* Elsevier Science B.V, Amsterdam.

Yong, R.N. 1999. Soil suction and soil-water potentials in swelling clays in engineered clay barriers. *Int. Jour. of Engineering Geology* 54: 3-13.

Computer Methods and Advances in Geomechanics, Desai et al. (eds) © 2001 Balkema, Rotterdam, ISBN 90 5809 183 X

Unsaturated soil mechanics: Constitutive modelling and coupling effects

L.Vulliet & L.Laloui
Swiss Federal Institute of Technology, Lausanne, Switzerland

ABSTRACT: This theme paper addresses various issues concerning the modelling of unsaturated soils. Starting from general considerations based on thermodynamics permits the derivation and discussion of the general form of field equations. Experimental evidences are then presented and constitutive modelling is treated in the context of elasto-plasticity based on the HISS family of models; it includes post-peak behaviour using the disturbed state concept (DSC). Finally, coupling effects are considered by presenting the finite element implementation and some applications, including the behaviour of unsaturated slopes.

1 INTRODUCTION

Unsaturated Soil Mechanics is becoming a major issue in modern soil mechanics. Several of the leading research teams around the world have now implemented research programmes in this area. A first reference text book was published by Fredlund & Rahardjo already in 1993. International conferences are now specifically dedicated to this topic (e.g. Int. Conf. On Unsaturated Soils, Paris 1995 and Beijing 1998).

The driving forces behind this movement are diverse:

- New needs for advanced modelling in geoenvironmental engineering (waste disposal, site remediation, energy production).
- Better equipped soil mechanics laboratories and the development of new testing and measuring devices.
- Improvements of constitutive models in elasto-plasticity.
- Increased computational power and programming comfort.

The challenges are great in all of the above-mentioned categories and many questions remain unanswered. Beginning with the basic physics of the phenomenon, the geometry of porous media is rather complex and presents multi-scale porosity affecting surface tension effects between air and water phases, complex distribution of the phases in space, and possible discontinuity of the phases (trapped air bubbles or discontinuous water drops). Therefore, extreme simplifications are needed to solve such problems in practice. As an example, phases are usually assumed to be continuous. Field equations can then be rigorously derived in the context of thermodynamics (Hutter et al., 1999). However, their form will depend on further assumptions and the validity of well-known approaches (such as the Biot theory (Biot, 1955) or Bishop equations (Bishop, 1959), etc.) may be questioned.

Also, laboratory experiments are more complex for unsaturated soils than they are for saturated or dry samples (as in conventional Soil Mechanics). One additional field variable must be measured: the air pressure or suction, depending on the testing equipment. Volume change measurements require special techniques (Geiser et al., 2000). The time required for consolidation may be rather large (several weeks for a silt using the air-pressure technique) and ad-hoc techniques, such as the osmotic method, have been developed for fine-grained (clayey) soils (Cui & Delage, 1996).

Experimental evidence reveals the complex mechanical response of unsaturated soils. Desaturating a sample, under deviatoric conditions, increases the strength and the stiffness of the soil, up to a certain degree of saturation. For low levels of degree of saturation (almost dry samples), the behaviour again tends towards that of a fully saturated sample. The volumetric behaviour under isotropic stress conditions is non-trivial. Contraction, expansion and structural collapse are observed depending on the stress/humidity history and loading conditions.

Elasto-plastic constitutive models have been developed for unsaturated soils. Numerous questions are still open, however, such as the type and number of stress state variables and internal variables to use, the formulation in terms of effective vs. net mean stress, post-peak behaviour, etc.

Finally, the introduction of such models in a boundary-value problem requires specific numerical techniques to treat the fully-coupled, non-linear and transient nature of the system of equations.

This theme paper addresses some of these issues, however, it obviously does not pretend to solve, or even touch on, all of them. Starting from general considerations based on thermodynamics permits the derivation and discussion of the general form of field equations. Then, constitutive modelling is treated in the context of elasto-plasticity based on the HISS family of models; it includes post-peak behaviour using the disturbed state concept (DSC). Finally, coupling effects are considered by presenting the finite element implementation and some applications, including the behaviour of unsaturated slopes and the compaction of agricultural soils caused by heavy machinery.

2 THERMODYNAMICS

For many years, the different theoretical concepts used in soil mechanics were primarily based on experimental observations and phenomenological approaches (effective stress, coupling effects...). The theoretical foundation of these concepts as obtained from basic thermodynamics was not clear, since the constitutive theory of continuum mechanics, as well as the porous media theory, were not as systematically developed as they are now.

Broadly speaking, a soil is a mixture of a granular solid matrix with voids filled with a fluid and/or a gas, and possibly other constituents. From now on, we use the term « soil » to denote the mixture and not merely its solid component.

The theories describing the deformation of soils under external forces are continuous mixture models of these constituents, with various degrees of complexity. In this paper, it will be postulated that the volume fractions of the various constituents add up to unity (massless constituents can also be treated as empty space; in such a case, the sum of the volume fractions would be less than unity and the system of equations would be modified, see Hutter et al., 1999).

Based on a structured thermodynamic theory of mixtures developed by Svendsen & Hutter (1995) in which the constituent volume densities play the role of internal variables, a continuum approach was developed by Hutter, Laloui & Vulliet (1999) to obtain field equations for compressible unsaturated soils. This formulation is based on the fact that: (i) mass interactions among the constituents may occur, (ii) some or all of the constituents are "incompressible" in the sense that they possess constant intrinsic mass densities, (iii) the constituents are continuous and, (iv) constituent energy interactions are ignored.

Let us consider the case of an isothermal three-phase mixture of grains (s), water (w) and air (g). Any field quantity ϕ can be expressed either as intrinsic (per unit volume of constituent α) and written as ϕ_α, or partial (per unit mixture volume) and written with a bar $\bar{\phi}_\alpha$. Partial constituent fields are related to intrinsic constituent fields via the constituent volume density of infinitesimal volume fraction n_α:

$$\bar{\phi}_\alpha = n_\alpha \phi_\alpha , \tag{1}$$

As a gas by its very definition fills all the space available to it, we have $n_s+n_f+n_g=1$. In addition, we assume that the grains and water are incompressible and the gas is compressible.

Mass Conservation

The mass balance laws for the constituents soil, water and gas and the balance relation for the gas-volume-density evolution take, in this case, the forms:

$$\frac{\partial n_\alpha}{\partial t} + \mathrm{div}\left(n_\alpha \mathbf{v}_\alpha\right) = 0, \ (\alpha = s, w) \tag{2}$$

$$\frac{\partial \left(n_g \rho_g\right)}{\partial t} + \mathrm{div}\left(n_g \rho_g \mathbf{v}_g\right) = 0, \tag{3}$$

$$\frac{\partial n_g}{\partial t} + \mathrm{div}\left(n_g \mathbf{v}_g\right) = \nu_g . \tag{4}$$

The volume production rate density ν_g of the compressible constituent gas must be prescribed as a constitutive relationship.

$\bar{\rho}_\alpha$ is the partial mass density of the constituent α and $\mathbf{v}_\alpha$ is their (differentiable) constituent spatial velocity.

The volume fractions can be identified, in this case, using the definition $n=n_f+n_g$:

$$n_f = \frac{V_f}{V} = nS_r , n_g = \frac{V_g}{V} = n\left(1 - S_r\right), n_s = \frac{V_s}{V} = 1 - n. \tag{5}$$

where V_α is the volume occupied by the phase α (α = s,f,g) and S_r is the degree of water saturation.

Momentum Equilibrium

The Cauchy stress $\bar{\mathbf{T}}_\alpha$ (normal stresses are positive in tension) is expressed as:

$$\bar{\mathbf{T}}_\alpha = -\bar{\varpi}_\alpha \mathbf{I} + \Sigma_\alpha \qquad (6)$$

where Σ_α is the dynamic contribution of $\bar{\mathbf{T}}_\alpha$, $\bar{\varpi}_\alpha$ the thermodynamic equilibrium stress, and $\mathbf{I}$ the identity matrix. For the special case of a ternary mixture, Eqn. 6 reduces to (Laloui et al., 1998):

$$\begin{aligned} \mathbf{T}_s &= -(1-n)\left(\beta_s + \varpi\right)\mathbf{I} + \Sigma_s, \\ \mathbf{T}_f &= -nS_r\left(\beta_f + \varpi\right)\mathbf{I}, \\ \mathbf{T}_g &= -n(1-S_r)\varpi\mathbf{I}. \end{aligned} \qquad (7)$$

of which the third expression identifies ϖ as the intrinsic pressure in the gas phase, p^g. The water and gas are represented by perfect fluids, $\Sigma_f = \Sigma_g = \mathbf{0}$.

The configuration pressure β_α is function of the Helmholtz free energy ψ_I given by:

$$\beta_\alpha = \rho \frac{\partial \psi_I}{\partial n_\alpha} =: \beta, \quad \forall \alpha = 1,\dots,n. \qquad (8)$$

ψ_I is, among other things, a function of n_s and n_f: $\psi_I = \tilde{\psi}_I\left(n_s, n_f, \cdot\right)$. Adopting n and S_r as independent variables, $\psi_I = \hat{\psi}_I\left(n, S_r, \cdot\right)$, yields:

$$\beta_f = \frac{\rho}{n} P(S_r, \cdot) = p^w - p^g, \quad \beta_s = S_r\left(p^w - p^g\right) \qquad (9)$$

where

$$P(S_r, \cdot) = \frac{\partial \hat{\psi}_I(S_r, \cdot)}{\partial S_r} \qquad (10)$$

with p^w being the pore water pressure. Thus, we have now identified all pressure-like variables arising in the general theory for this case. ϖ may be identified by the intrinsic pressure and p^g in the gas phase; β_f is what is termed matric suction in soil science (the classical notation is $\beta_f = s$) and $\beta_s = S_r \beta_f$.

The partial stress expressions are then:

$$\begin{aligned} \mathbf{T}_s &= -(1-n)\left(S_r s + p^g\right)\mathbf{I} + \Sigma_s, \\ \mathbf{T}_f &= -nS_r p^w \mathbf{I}, \\ \mathbf{T}_g &= -n(1-S_r)p^g \mathbf{I} \end{aligned} \qquad (10)$$

and

$$\mathbf{T} = -S_r p^w - p^g(1-S_r) + \Sigma_s \qquad (11)$$

for the total stress.

Eqn. 11 is equivalent to the Terzaghi effective stress concept relation (valid for a saturated binary mixture) generalised to the domain of a ternary mixture by Bishop (1959), if S_r is replaced by a weighting parameter χ. This Bishop effective stress relation has also been derived with a hybrid mixture theory (averaging theory) by Gray & Hassanizadeh (1991) and by Schrefler (1995).

It is proven here that Bishop's generalisation of the effective stress concept is meaningful provided that the weighting parameter χ equals the degree of saturation S_r. However, this result is obtained only if (i) the mixture is comprised of density-preserving constituents, (ii) both fluids are perfect and, (iii) both fluids (water and gas) are continua and the gas is compressible.

Eqn. 11 also shows the necessity of having two « effective » stresses, as proposed by Coleman (1962) and Fredlund *et al.* (1978), to describe the behaviour of unsaturated soils properly. The first stress is related to the total stress $\mathbf{T}$ in excess of the pore-air pressure p^g (also called *net mean pressure* p^*, $p^*=p+p^g$), with p being the mean total stress ($p = \frac{1}{3}\mathrm{tr}\,\mathbf{T}$). The second stress is the matric suction ($s=p^w-p^g$). This theoretical result corresponds to experimental observations (Fredlund & Morgenstern, 1977 ; Laloui et al., 1995, 1997).

3 CONSTITUTIVE MODELLING

3.1 *Experimental evidence*

The main features of unsaturated soils can be summarised as follows (Laloui et al., 1997, Geiser, 1999);

The *hydric behaviour* can be investigated by subjecting samples to drying and wetting cycles. Five main patterns may be observed: (i) on a drying path, the samples remain saturated until a certain suction s_e (termed air entry suction) is reached; (ii) an initial increase in suction induces irreversible deformation until a suction close to s_e is reached. Beyond this point, the behaviour becomes quasi-reversible; (iii) a hysteretic behaviour is observed on drying-wetting cycles; (iv) for some soils on a wetting path, a collapse phenomenon is observed (e.g. Jennings and Burland 1962, Matyas and Radhakrishna 1968); (v) the mechanical stress level influences the hydric behaviour.

The *mechanical behaviour* under isotropic loading conditions presents the following features (see also Matyas & Radhakrishna 1968, Alonso et al. 1990, Maâtouk et al. 1995, Sivakumar 1993): (vi) the preconsolidation pressure increases as the suction increases; (vii) suction-induced stiffening effects can be seen. Deviatoric paths reveal: (viii) peak strength increases with suction. This may be interpreted in the sense of a cohesion increase, while maintaining a constant friction angle; (ix) at a constant initial net mean pressure, volumetric deformation on deviatoric paths of partially saturated soils decrease slightly as suction is increased (Cui & Delage 1996). However, some authors have observed variable trends (Maâtouk et al. 1995, Sivakumar 1993); (x) suction contributes to a small increase in the elastic rigidity of the soils for both stress combinations; (xi) for high suction or small net mean pressure, an important loss of strength (brittle failure) is observed; (xii) a unique critical state line can be observed in the stress plane for any value of suction.

3.2 *Constitutive models*

Several elasto-plastic constitutive models have been developed for unsaturated soils, incorporating the above-mentioned features to various extents (see Section 3.1). Alonso et al. (1990) first proposed a critical state framework (based on Cam-Clay) involving two independent sets of stress variables, namely the net stress and the suction. This approach was further developed by Wheeler & Sivakumar (1995) and Wheeler (1996). Simultaneously, other models based on a single effective stress were also developed. Their main advantage, when compared to models involving the net stress as a state variable, is a straightforward transition from the saturated to the unsaturated state (Kogho et al., 1993, Abou-Bekr, 1995, Bolzon et al., 1996).

Following performance testing of the Alonso et al. model (Geiser et al., 1997a), a new model, $\delta_{1\text{-unsat}}$, was proposed by Geiser, Laloui & Vulliet (1999), based on a different combination of stress variables, namely, the saturated effective stress σ' (=σ-p^w) and the suction s.

The mathematical formulation of this model is based on a decomposition of the strain increment into an elastic and a plastic part:

$$\dot{\varepsilon}_{ij} = \dot{\varepsilon}_{ij}^{e} + \dot{\varepsilon}_{ij}^{p} \tag{12}$$

The elastic increment $\dot{\varepsilon}_{ij}^{e}$ is composed of a mechanical and a hydric strain increment:

$$\dot{\varepsilon}_{ij}^{e} = \dot{\varepsilon}_{ij}^{em} + \frac{1}{3}\dot{\varepsilon}_{v}^{eh}\delta_{ij} = \mathbf{D}_{ijkl}^{e-1}\dot{\sigma}_{kl}' + \frac{1}{3}A^{-1}\dot{s}\delta_{ij} \tag{13}$$

where $\dot{\varepsilon}_{ij}^{em}$ is the elastic mechanical strain increment induced by the variation of the effective stress σ', $\frac{1}{3}\dot{\varepsilon}_{v}^{eh}$ is the reversible hydric strain increment, $\mathbf{D}^e$ is the classical elastic tensor and A is a proportionality coefficient which describes the hydric behaviour.

The plastic strain increment is also deduced from mechanical and hydric considerations.

The model has two yield surfaces: F_1 represents the plastic limit in the effective stress plane and F_2 gives the yield limit in the suction against mean effective stress plane.

The mechanical yield surface F_1 is expressed as:

$$F_1 \equiv \frac{J_{2D}}{p_a^2} - \left[-\alpha\left(\frac{J_1' + R}{p_a}\right)^n + \gamma\left(\frac{J_1' + R}{p_a}\right)^2 \right] F_s \tag{14}$$

where $F_s = \left[1 - \beta.\left(\sqrt{27}/2.J_{3D} \cdot J_{2D}^{-3/2}\right)\right]^{-0.5}$

J_1' is the first invariant of the effective stress tensor; J_{2D} is the second invariant of the deviatoric stress tensor; J_{3D} is the third invariant of the deviatoric stress tensor; p_a is the atmospheric pressure; R is a bonding stress parameter related to the cohesion; γ and β are ultimate parameters related to the friction angle and the critical pressure, and n is the phase change parameter. α is the hardening parameter related to the volumetric plastic deformation and suction (Geiser et al., 1999).

The hydric yield surface F_2 is expressed as:

$$F_2 \equiv -\left[-\alpha_h\left(\frac{3s}{p_a}\right)^n + \gamma\left(\frac{3s}{p_a}\right)^2 \right] F_s \tag{15}$$

where α_h is the hardening parameter related to the hydric plastic deformation.

The model is associated for the hydric path (F_2) and non-associated for the mechanical path (F_1).

3.3 *Examples of numerical predictions*

The $\delta_{1\text{-unsat}}$ model requires 15 parameters, five of which relate to unsaturated aspects: three for the hydric behaviour, one for the isotropic mechanical behaviour and one enabling the evolution of the peak strength. Details of the determination of material parameters are presented in Geiser et al. (1997a),

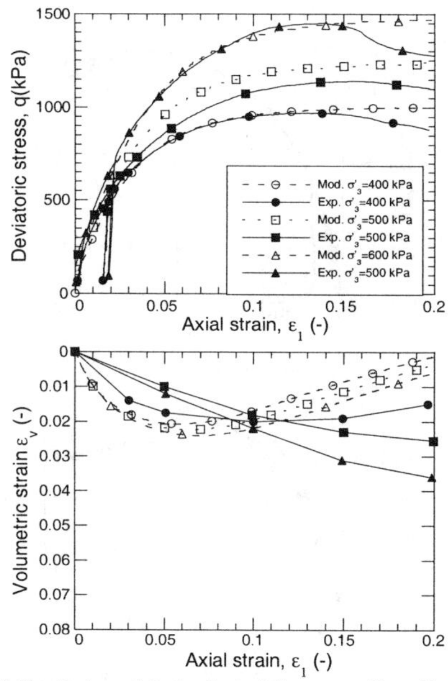

Figure 1. Prediction of drained triaxial tests on Sion silt - Initial suction of 100 kPa and varying σ'_3 (400, 500 & 600 kPa).

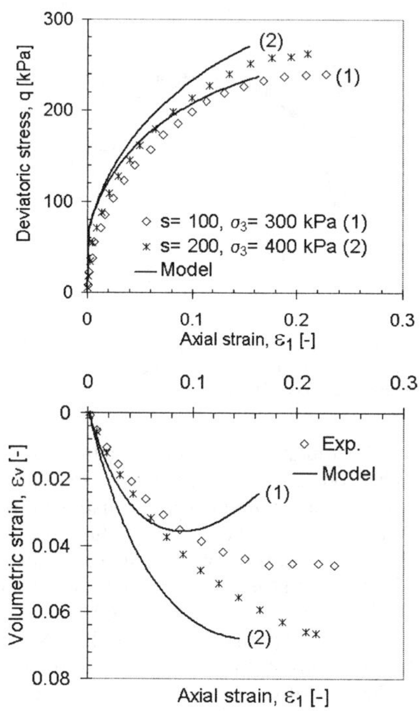

Figure 2. Prediction of triaxial tests on kaolin - Constant mean effective stress (experimental data from Maâtouk, 1993).

Geiser (1999), Rifa'i et al. (2000).

Figure 1 illustrates a prediction of triaxial tests at three deviatoric stress levels and constant suction on a Sion silt. Figure 2 shows a prediction at two suction levels with two confining pressures on a Kaolin. The proposed $\delta_{1\text{-unsat}}$ model is able to reproduce most of the features of unsaturated soils on mechanical and hydric paths. The predictions are quantitatively good, except on deviatoric paths, where the predicted volume dilatancy is slightly excessive.

3.4 *Extension of the model with DSC*

The post-peak behaviour may be modelled in the general framework of the disturbed state concept (DSC, e.g. Desai, 1995). The DSC is based on the idea that a deforming material element can be treated as a mixture of two constituent parts in the relative intact (RI) and fully-adjusted (FA) states, referred to as reference states. During external loading, the material experiences internal changes in its microstructure due to a self-adjustment process and, as a consequence, the initial RI state transforms continuously to the FA state. The observed stress tensor σ_{ij} is defined as:

$$\sigma_{ij} = (1 - D)\sigma^{i}_{ij} + D\,\sigma^{c}_{ij} \tag{16}$$

where σ^{i}_{ij} is the RI stress, σ^{c}_{ij} the FA stress and D the disturbance function. D varies between 0 (intact state) and 1 (adjusted state).

The DSC approach allows different choices for the description of the FA and RI states. The RI state can be represented by using the theory of elasticity, plasticity, viscoplasticity, etc. (Desai et al., 1996, Geiser et al., 1997b). Numerical tests have shown that assuming an elastic RI state leads to problems in the formulation, as the complete non-linear behaviour has to be included in the disturbance function. Consequently, the following approach was chosen (Fig. 3):

- the intact state is represented by the previously described $\delta_{1\text{-unsat}}$ model
- the fully-adjusted state corresponds to a saturated state with a modified initial mean effective stress $p_o'_{mod.}$:

$$p_o'_{mod.} = (p_0' - p^g) + S_r(p^g - p^w) \tag{17}$$

The observations of the experimental tests with brittle failure on the Sion silt (Geiser, 1999, Cui, 1993) led to this modified stress. It was shown that,

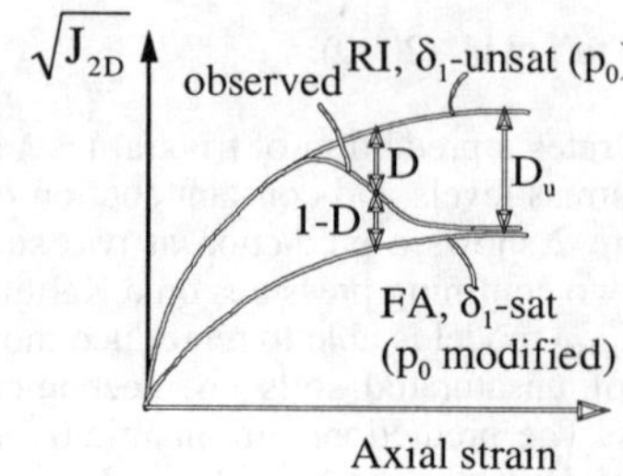

Figure 3. Determination of D as a function of $\sqrt{J_{2D}}$.

after failure, the soil tends to a smaller ultimate state than for an intact soil.

The disturbance function D is expressed here as a scalar function of the deviatoric stress invariant only, since brittle failure mainly appears in the deviatoric plane (ε_1-q). The disturbance function is then:

$$D = \frac{\sqrt{J_{2D}}^{i} - \sqrt{J_{2D}}}{\sqrt{J_{2D}}^{i} - \sqrt{J_{2D}}^{c}} \tag{18}$$

with the exponent "i" corresponding to the RI state and "c" corresponding to the FA state (see Fig. 3).
The disturbance is assumed to be a function of the trajectory of the deviatoric plastic strains ξ_D and suction s (Geiser et al., 1997b). It is expressed as:

$$D = D_u(1 - e^{-B\xi_D^Z}) \tag{19}$$

where D_u is the ultimate value of D (a material parameter), which may be assumed to be independent of suction; B and Z are material parameters, functions of suction.
Finally, the extension of the model to DSC involves the determination of three new parameters B, Z and D_u.

Performances of the DSC extension of the model are shown for three triaxial tests carried out at an initial effective confining pressure of 600 kPa. Figure 4 shows the simulations conducted at three different suction levels.

The test conducted at a low degree of saturation (Fig. 4c, s=280 kPa) is peculiar, as the sample never reached the peak value obtained for other tests at the same effective confining pressure. For this high level of suction, the brittle failure in the sample starts almost at the beginning of the test. Despite this, the simulation is very good.

4 COUPLING EFFECTS

Based on the momentum and mass balance equations presented in Section 2, a coupled hydro-

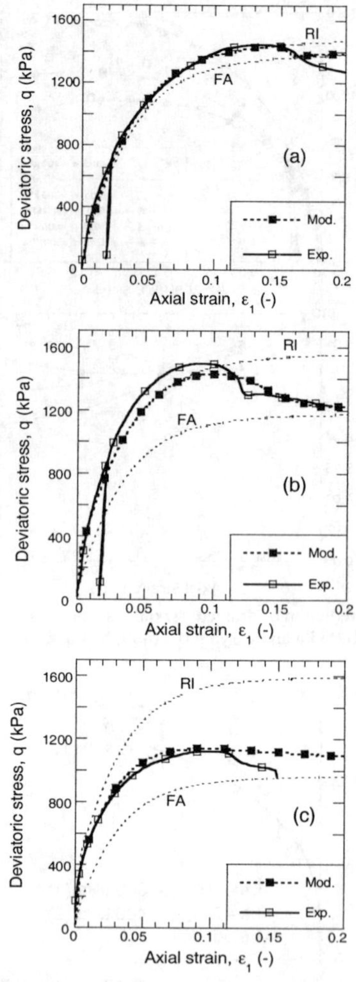

Figure 4. Simulation of triaxial shearing tests at a confinement pressure of σ_3'=600 kPa incorporating the DSC - (a) s=100 kPa, (b) s=200 kPa and (c) s=280 kPa.

mechanical formulation is derived for the simulation of the behaviour of unsaturated soils. Similar formulations can be obtained also by using works of Hassanizadeh & Gray (1979) or Schrefler (1995). For details of the derivation of our equations from those in Section 2 to the following ones, the reader is referred to Klubertanz et al. (1997 a & b).

Following the hypothesis presented in Section2, the gas phase is supposed to be governed by the ideal gas law. The two fluid phases flow freely and a non-linear pore pressure - saturation relation is used. The resulting system of equations is discretised in space using the finite element technique and in time via an explicit Euler scheme.

4.1 *Mathematical formulation*

From Section 2, the momentum balance of the three phase mixture reads:

$$\nabla\bullet[\sigma' - S^w p^w \mathbf{I} - (1- S^w) p^g \mathbf{I}] + \rho\mathbf{g} = 0 , \quad (20)$$

where $\mathbf{g}$ is the gravity vector and ρ the total density of the mixture ($\rho = (1\text{-}n) \rho^s + n S^w \rho^w + n S^g \rho^g$). S^w and S^g are the degrees of water and air saturation defined as the volume ratios of water and air, respectively, divided by the total pore volume of the control volume, and $\nabla\bullet$ is the divergence operator.

Similarly, the combined mass balance equations for water/solid or air/solid read, with some assumptions and using Darcy's law for both fluid phases:

$$\left(\frac{n}{S^w}\frac{\partial S^w}{\partial s}+\frac{n}{\rho^w}\frac{\partial \rho^w}{\partial p^w}\right)\frac{\partial p^w}{\partial t}-\frac{n}{S^w}\frac{\partial S^w}{\partial s}\frac{\partial p^g}{\partial t}+ \nabla\bullet v^s+\nabla\bullet\left(\frac{k_{rw}K}{\mu^w}\left(\nabla p^w-\rho^w\mathbf{g}\right)\right)=0 \quad (21)$$

for the water/solid mass balance and

$$\left(\frac{n}{1-S^w}\frac{\partial S^w}{\partial s}+\frac{n}{\rho^g}\frac{\partial \rho^g}{\partial p^g}\right)\frac{\partial p^g}{\partial t}-\frac{n}{1-S^w}\frac{\partial S^w}{\partial s}\frac{\partial p^w}{\partial t}+ \nabla\bullet v^s+\nabla\bullet\left(\frac{k_{ra}K}{\mu^g}\left(\nabla p^g-\rho^g\mathbf{g}\right)\right)=0 \quad (22)$$

for the air/solid mass balance.

The relative permeability k_{rw} of water is assumed to be a function of the degree of water saturation and porosity, the relative permeability of air k_{ra} being a function of porosity only (Seker, 1983). K is the intrinsic permeability and μ the dynamic viscosity. v^s is the (Lagrangian) velocity of the solid skeleton. A non-linear capillary pressure - saturation relation obtained experimentally on different materials was applied (Seker, 1983).

Equations (20), (21) and (22) represent a set of three equations for the three unknowns: solid displacement u_i, air pressure p^g and water pressure p^w (u_i being, of course, linked to σ'_{ij} in (1) via a constitutive equation and to v^s via derivation in time). Other choices for the set of unknowns are, of course, possible.

4.2 *Discretised system*

Using a standard Galerkin procedure and an explicit Euler method for discretisation, we obtain the following system of equations:

$$\begin{aligned}
&\mathbf{L}\, \mathbf{du}_{n+1} + \mathbf{L}^w\, dp^w_{n+1} + \mathbf{L}^g\, dp^g_{n+1} = \mathbf{F}_{n+1}\text{-}\nabla\bullet\mathbf{T}_n\text{-} \mathbf{L}^w p^w_n - \mathbf{L}^g p^g_n \\
&\mathbf{L}^{wT}\, \mathbf{du}_{n+1} + (\mathbf{M}^w + \Delta t\ \mathbf{H}^w)\, dp^w_{n+1} + \mathbf{N}^{wg}\, dp^g_{n+1} = - \Delta t\, \mathbf{H}^w p^w_n + \mathbf{F}^w_{n+1} \\
&\mathbf{L}^{gT}\, \mathbf{du}_{n+1} + (\mathbf{M}^g + \Delta t\ \mathbf{H}^g)\, dp^g_{n+1} + \mathbf{N}^{gw}\, dp^w_{n+1} = - \Delta t\, \mathbf{H}^g p^g_n + \mathbf{F}^g_{n+1}
\end{aligned} \quad (23\ a\text{-}c)$$

In (23 a-c) $\mathbf{L}$ is the rigidity matrix, $\mathbf{L}^w$ and $\mathbf{L}^g$ the liquid and gas flow matrices, respectively, and $\mathbf{F}^\alpha$ the loading vectors. $\mathbf{T}$ is the Cauchy stress tensor.

The system (23) is evaluated in a two-step approach. Based on the results at time step n, the system of equations is solved and the matrices are updated using the new values for stresses and pressures. The solution at time step n+1 is found using the updated matrices. This approach proved to be stable and accurate for reasonably large values of the time step Δt. Convergence behaviour was generally good (Klubertanz et al., 1999).

5 APPLICATIONS

5.1 *Debris flow initiation*

In the context of a Swiss debris flow research program (BSFIT, 1999), numerical parametric studies using the previous formulation and constitutive model were carried out to improve the understanding of debris flow initiation mechanisms and to permit the prediction of the failure location and initial debris volume.

As an example, the effect of the timing of the water supply on pore water pressure generation is shown in Figure 6 (Klubertanz et al., 2000). Three calculations are presented using the same maximum value of infiltration rate q_w but different infiltration durations t_{max} (Table 1). For all of these cases, the slope angle was set to α=25° (see FE mesh Fig. 5).

Figure 6 shows that case 5 results in the most critical reaction of the slope, despite the fact that in case 6 more water was supplied. Case 5 yields higher pore pressures and larger displacements at the surface than all of the other cases. If we accept the relation between precipitation and water supply as in the given example, the observations of intensity - duration thresholds are partially confirmed: water supply over a short time range for the same intensity. In the case of the landslide-type initiation of debris flows this numerical approach permits the consideration of the strong coupling among the three phases of a soil. Severe structural degradation negatively affects, or even impedes, plant growth.

It is especially suited to reproduce any effect that an unsaturated zone - often present in debris flow initiation zones - may have. Compared to conventional

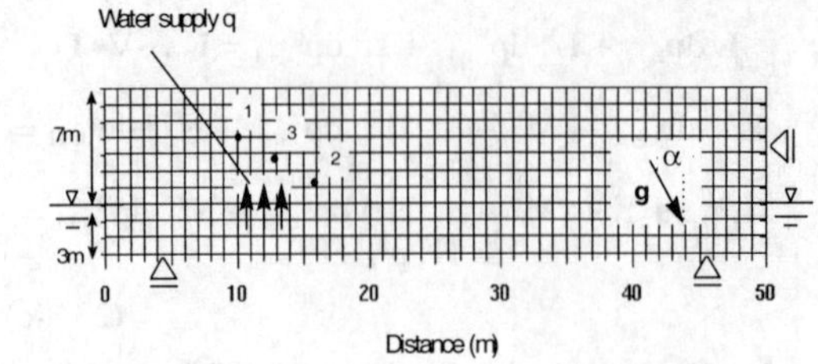

Figure 5. Example of a slope stability problem using unsaturated soil mechanics. Geometry, mesh and boundary conditions. Note the inclination of the gravity vector **g**.

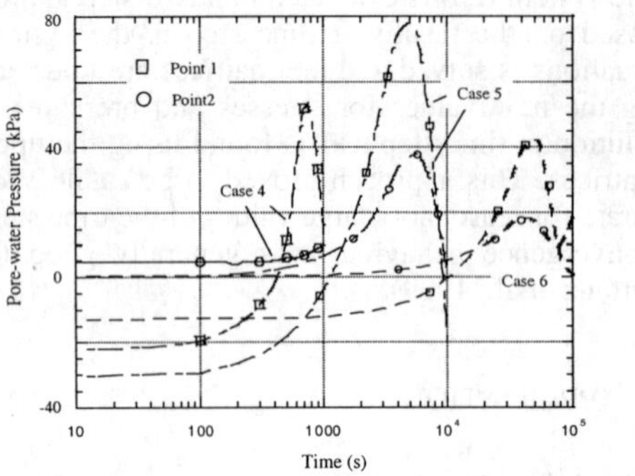

Figure 6. Pore-water pressure, vs. time (logarithmic scale) at points 1 and 2.

Table 1. Parameters varied

Case no	α [°]	t_{max} [s]	q_w [$m^3/(m^2\,s)$]
4	25	1000	2 10^{-4}
5	25	10,000	2 10^{-4}
6	25	100,000	2 10^{-4}

two-phase analyses, the ability to model stability behaviour in saturated and unsaturated conditions is improved due to constitutive laws which take into account the influence of the degree of saturation and the air-pressure.

5.2 *Soil compaction*

Heavy agricultural machinery causes structural degradation in agricultural subsoils.

The effects of the impact of a single pass with a sugar beet harvester on the soil properties of an unploughed eutric cambisol were investigated by means of field measurements and laboratory testing. In addition, 2D calculations of strain, stress and subsequent compaction were carried out using our three-phase model for unsaturated soil (Klubertanz et al., 1999, Gysi et al.).

Due to the pass of the machinery, the soil was compacted down to a depth of between 0.15 and 0.25 m. This compaction was evidenced by an increase in bulk density and pre-consolidation pressure

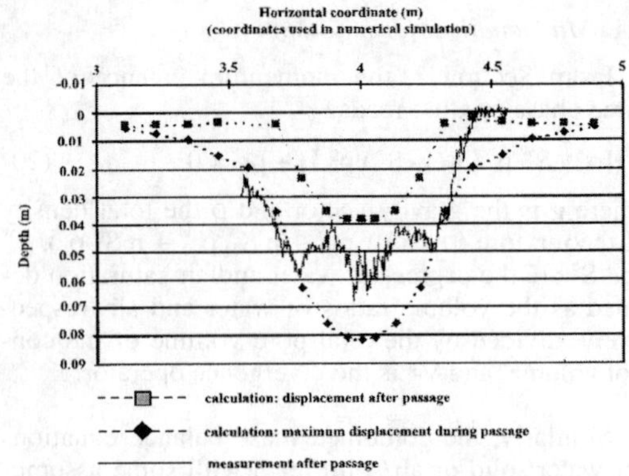

Figure 7. Example of a compaction problem using unsaturated soil mechanics - Measured (after passage) and calculated (during and after passage) vertical surface displacements.

as well as by a decrease in total porosity and macroporosity. The surface displacement measured in the field is consistent with the calculated model data (Fig. 7).

Here again, an unsaturated soil mechanics approach provides additional information and includes improved physics in the modelling.

6 CONCLUSIONS

The modelling of unsaturated soil has been improved considerably over the last ten years on different fronts: thermodynamics, laboratory experiments, constitutive modelling, numerical implementation and case studies. Some general comments were discussed in this paper; in particular,

- Some of the well-known coupled theoretical formulations describing the deformation of soils were thermodynamically analysed. For example, we established a thermodynamic basis for effective stress in three phase mixtures and delineated the applicability of Terzaghi's effective stress concept in unsaturated soils.
- A constitutive model, δ_1-unsat, was described by means of two independent stress variables, σ-p_w and s. Particularly, it was shown that the hydro-mechanical behaviour was qualitatively and quantitatively well predicted in the studied range of suctions and stresses. This also applies to water undrained conditions (not shown here).
- The incorporation of the DSC allowed the prediction of brittle failures for high levels of suction under deviatoric conditions.
- Since a saturated effective stress σ-u_w was adopted as a state variable in this constitutive model, the transition from a saturated to an unsaturated state is straightforward.

- Based on the momentum and mass balance equations obtained by thermodynamics, a coupled hydro-mechanical formulation was derived for the simulation of the behaviour of unsaturated soils. The resulting coupled system of equations with three field unknowns was discretised in space using the finite element technique and in time via an explicit Euler scheme.
- Results of two boundary value problems showed that the proposed approach provides more information and includes improved physics in the numerical modelling.

The major effects of unsaturation can now be accounted for – at least in principle - in practical applications and case studies. However, this is only the beginning of a new field. To reach a level comparable to conventional – saturated – soil mechanics, large additional efforts are required in several (partly contradictory) directions, e.g. a better understanding of the basic physics, improved constitutive modelling, simpler testing procedures, and user-friendly and robust computer programs. Above all, applications are needed and thus close contact with the civil engineering industry is needed. Technology transfer programs and continuing education should now begin to incorporate some aspects of unsaturated soil mechanics.

7 ACKNOWLEDGEMENTS

This research was funded in part by the Swiss National Science Foundation, the Swiss Federal Institute of Technology Lausanne (EPFL), the Board of Directors of the Swiss Federal Institutes of Technology and the Swiss Federal Research Station for Agricultural Economics and Engineering (FAT). It was partly carried out in the context of ALERT-Geomaterials, a network of European laboratories, with the Swiss participation being financed by the Federal Office for Education and Science (OFES).

Several colleagues and co-workers are to be thanked for their invaluable contributions over the past years: Dr. F. Geiser, Dr. G. Klubertanz, Prof. K. Hutter, Prof. C.S. Desai, Mr. G. Gruaz, Mr. M. Gysi, Mr. A. Rifa'i and Mr. Ph. Gachet.

REFERENCES

Abou-Bekr, N. (1995). Modélisation du comportement mécanique et hydraulique des sols partiellement saturés. Doctoral thesis, Ecole Centrale Paris.

Alonso, E.E., A. Gens & A. Josa (1990). A constitutive model for partially saturated soils. Géotechnique 40(3): 405-430.

Biot, M. A. (1955). Theory of elasticity and consolidation for a porous solid. J. Appl. Phys 26(2): 182-185.

Bishop, A.W (1959). The principle of effective stress. Tecknisk Ukeblad 106(39), pp 859-863.

BSFIT, Board of the Swiss Federal Institutes of Technology (1999), Synthesis report of the project debris flow, 36 p.

Bolzon, G., Schrefler B. A. & Zienkiewicz O. C. (1996). Elastoplastic soil constitutive laws generalized to partially saturated states. Géotechnique, 46 (2), pp. 279-289.

Coleman, J.D. (1962). Stress strain relations for partly saturated soil, Geotechnique 12(4), pp 348-350.

Cui, Y. J. (1993). Etude du comportement d'un limon compacté non saturé et de sa modélisation dans un cadre élastoplastique. Doctoral thesis ENPC Paris.

Cui, Y.J. and P. Delage (1996). Yielding and plastic behaviour of an unsaturated compacted silt. Géotechnique, 46(2), 291-311.

Desai, C.S. (1995). Constitutive modelling using the disturbed state as microstructure self adjustment concept. Continuum models for materials with microstructures. H.B. Mülhaus Ed., John Wiley & Sons.

Desai, C.S., L. Vulliet, L. Laloui & F. Geiser (1996). Disturb state concept for constitutive modeling of partially saturated porous materials Internal report EPFL- Switzerland.

Fredlund, D.G. & H. Rahardjo (1993). Soil mechanics for unsaturated soils. New-York: John Wiley & Sons.

Fredlund, D.G, Morgenstern, N.R, Widger, A. (1978). Shear strength of unsaturated soisl, Canadian Geotechnical Journal 15(13), pp. 313-321.

Fredlund, D.G. & N.R. Morgenstern (1977). Stress state variables for unsaturated soils. ASCE, J. Geotech. Eng. Div. 103(GT5): 447-466.

Geiser F., Laloui L., Vulliet L. (2000). On the volume measurement in unsaturated triaxial test. 1st Asian conf. on Unsaturated Soils, Singapore, pp. 669-674.

Geiser F. (1999). Comportement mécanique d'un limon non saturé. Etude expérimentale et modélisation constitutive. Thèse de doctorat EPFL no 1942, 244 p.

Geiser F, Laloui L. & Vulliet L. (1999). Unsaturated soil modelling with special emphasis on undrained conditions. Proc. 7th Int. Symp. on Numerical Models in Geomechanics, NUMOG, pp. 9-14

Geiser F., Laloui L. & Vulliet L. (1997a). Constitutive modelling of unsaturated sandy silt. Proc. 9th Int. Conf. on Comp. Meth. And Advances in Geomech., IACMAG, Vol. 2, pp. 899-904

Geiser F., Laloui L., Vulliet L., Desai C.S. (1997b). Disturbed state concept for partially saturated soils. Proc. 6th Int. Conf. on Num. Meth. in Geomech., NUMOG, pp. 129-133

Gray W. G. & Hassanizadeh S.M. (1991). Unsaturated flow theory including interfacial phenomena. Water Resources Research, **27**(8), 1855-1863.

Gysi M., Klubertanz G., Vulliet L. (to be published). Compaction of an Eutric Cambisol under heavy wheel traffic in Switzerland - field data and modelling. J. of Soil and Tillage Research

Hassanizadeh, M. and W.G. Gray (1990). Mechanics and thermodynamics of multiphase flow in porous media including interphase boundaries', Adv. in Water Resources, **13**, 169-186.

Hassanizadeh, M. and W.G. Gray (1979). General conservation equations for multi-phase systems: 3. Constiutive theorie for porous media flow, Adv. in water resources, 2, pp. 25 - 40

Hutter K., Laloui L., Vulliet L. (1999). Thermodynamically based mixture models of saturated and unsaturated soils. Journal of Mech. of cohesive -frictional Materials, vol. 4. No 4, pp. 295-338.

Jennings, J.E.B, Burland, J.B. (1962). Limitations to the use of effective stresses in partly saturated soils, Geotechnique 12(2), pp 125-144.

Klubertanz G., Laloui L., Vulliet L. (2000). On the use of multiphase modelling to simulate the initiation of slope movements.8th Int. Symp. on Landslide ISL, pp.831-836.

Klubertanz G., Laloui L., Vulliet L. (1999). On the Implementation and Performance of a Coupled Multiphase Model for Porous Media : Numerical Studies and Applications. Proc. of the first European Conference on Computational Mechanics (ECCM'99), 19p.

Klubertanz G., Vulliet L., Gysi M., Ott A. (1999). A study on soil compaction due to agricultural traffic: computational results and field data. Proc. 7th Int. Symp. on Num. Models in Geomech., NUMOG, pp. 621-626.

Klubertanz G., Laloui L., Vulliet L. (1997a). Numerical modelling of the hydro-mechanical behaviour of unsaturated porous media. Proc. NAFEMS world congress, Vol. 2, pp. 1302-1313.

Klubertanz G., Laloui L., Vulliet L. (1997b). Numerical modeling of unsaturated porous media as a two and three phase medium: a comparison. Proc. 9th Int. Conf. On Comp. Meth. and Advances in Geomech., IACMAG, Vol. 2, pp. 1159-1164.

Kogho, Y., M. Nakano & T. Myazaci (1993). Theoretical aspects of constitutive modelling for unsaturated soils. Soils and Foundations 33(4): 49-63.

Laloui L., Hutter K., Vulliet L. (1998). Thermodynamics of Saturated and Unsaturated Soils. Biot Conf. on Poromechanics, , Thimus et al. (eds), Balkema/Rotterdam, pp. 93-97

Laloui L., Geiser F., Vulliet L., Li X.L., Bolle A., Charlier R. (1997). Characterization of the Mechanical Behavior of an Unsaturated Sandy Silt. Proc. XIVth Int. Conf. on Soil Mech. & Found. Engrg, Vol. 1, pp. 347-350

Laloui L., Vulliet L., Gruaz G. (1995). Influence of Suction on the Mechanical Behaviour of Sandy Silt. Proc. 1st Int. Conf. on Unsaturated Soils, Balkema and Presses ENPC, Vol. 1, pp. 133-138

Maâtouk, A. (1993). Application des concepts d'état limite et d'état critique à un sol partiellement saturé effondrable.Doctoral thesis, University of Laval.

Maâtouk, A., S. Leroueil & P. La Rochelle (1995). Yielding and critical state of collapsible unsaturated silty soil. Géotechnique 45(3): 465-477.

Matyas, E.L. and H.S. Radhakrishna (1968). Volume change characteristics of partially saturated soils. Géotechnique, 18(4), 432-448.

Rifa'i A., Laloui L., Vulliet L. & Geiser F. (2000). Performances of HiSS $\delta_{1\text{-unsat}}$ model under saturated and unsaturated conditions. 1st Asian conf. on Unsaturated Soils, Singapore, pp. 255-260.

Schrefler, B.A. (1995). F.E. in environmental engineering: coupled thermo-hydro-mechanical processes in porous media including pollutant transport. Archives of Computational Methods in Engineering, Vol. 2,3, 1-54.

Seker, E. (1983). Etude de la déformation d'un massif de sol non saturé, Doctoral thesis EPFL n° 492, Lausanne.

Sivakumar, V. 1993. A critical state framework for unsaturated soils. PhD thesis, University of Sheffield.

Svendsen B. & Hutter K. (1995).. On the thermodynamics of a mixture of isotropic materials with constraints. *Int. J. Eng. Sci.* , 33, 2021-2054

Wheeler, S. J. & V. Sivakumar 1995. An elasto-plastic critical state framework for unsaturated soil. Géotechnique 45(1): 5-53.

Wheeler, S. J. 1996. Inclusion of specific water volume within an elasto-plastic model for unsaturated soil. Canadian Geotechnical Journal 33: 42-57.

Computer Methods and Advances in Geomechanics, Desai et al. (eds) © 2001 Balkema, Rotterdam, ISBN 90 5809 183 X

Analysis of fractionation of sediments caused by an acoustic field

N. Aboobaker & J. N. Meegoda
Department of Civil and Environmental Engineering, New Jersey Institute of Technology, Newark, N.J., USA
D. Blackmore
Department of Mathematical Sciences, New Jersey Institute of Technology, Newark, N.J., USA

ABSTRACT: Fractionation and segregation of fine particles on the basis of the contrast factor between the particles and the suspending fluid has many industrial applications. A new method for fractionation of sediments in the micron size by using acoustic standing waves is presented. An experimental set up was built to verify the applicability of the technology. The acoustic forces acting on a particles in the experimental system mathematically modeled. The resulting second order ordinary differential equation could not be numerically solved and do not have a closed form solution either. An approximate solution is proposed that would calculate the particle trajectories and concentration at collection planes in an acoustic field taking into account all the forces acting on the particles. The proposed approximate solution is both quantitatively and qualitatively closer to the observed results.

1 INTRODUCTION

The fractionation of particles by size or other physical properties such as density has many applications in a variety of technologies including material processing, polymer recycling, food processing, and the fuel industry. Fractionation of particles can be used in engineering fields such as decontamination of dredged sediments.

Dredging in the US requires long-term alternatives as more than 230 million cubic meters of dredged material per year need to be disposed of. Recently, the Ocean Dumping Act prevented the ocean dumping and inland disposal of contaminated dredged residues from ports such as New York, New Jersey and Boston. This is due to high levels of contaminants such as metals, Dioxins, PAHs, and PCB in dredged sediments. Small size particles, neutral buoyancy, or uniform electromagnetic charged surface pose difficulties with fractionation. Based on the above discussion there is a need for cost effective technology for fine soil particles. Decontamination and separation of fine sediments is useful in treating sediments, leaving contaminated water to be treated separately, and cleaned sediments to be reused as construction material returned to the sea.

In this new technology, particles are discriminated, based on density and compressibility differences rather than the particle size. The technology used in this research is based on an acoustic force field.

2 ACOUSTIC FORCE FIELD

King, 1934, first proposed the force acting on spherical particle suspended in a standing acoustic wave field. This analysis was restricted to a rigid particle with a radius much smaller than the wavelength of sound, and for a standing wave field created by oppositely travelling, single frequency, sinusoidal waves. Figure1 shows forces acting on a particle in an acoustic field. These forces are:

1. Primary Axial Acoustic Radiation Force (F_{PARF})
2. Primary Transverse Acoustic Radiation Force (F_{PTRF}), and
3. Secondary Acoustic Radiation Force (F_S)

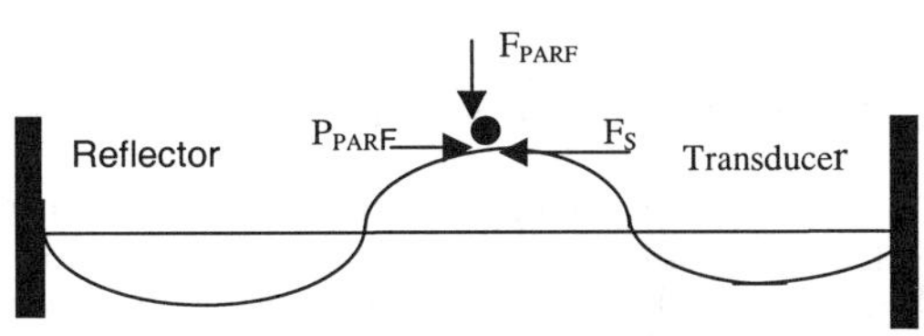

Figure 1. Acoustic forces on the particle

2.1 *Primary axial acoustic force*

Almost immediately following the application of an acoustic field, particles experience a time-averaged primary axial acoustic force, F_{ac} (F_{PARF}), which is generated by the interaction between particles and the primary wave field. The primary axial radiation force drives dispersed particles toward the velocity antinodes of the resonance field. The magnitude of the force depends on the difference in compressibility and density between the particle and the medium. The primary acoustic force was derived by Yosioka and Kawasima and given below.

$$F_{ac} = V_0 E_{ac} k G \sin(2kx)\underline{\mathbf{k}} \qquad (1)$$

The acoustic energy density, E_{ac}, is a measure of the energy residing in a wave field. V_o is the volume of one particle, $k= 2\pi/\lambda$ is the wave number of the acoustic radiation, λ is the sound wavelength, z is the axial distance from a pressure node, and $\underline{\mathbf{k}}$ is the unit vector in the axial direction. The response of any solid suspended in a fluid, to a resonant acoustic field depends on the acoustic contrast factor, G. For any solid particle suspended in fluid with size $r<<\lambda$, the acoustic contrast factor is given by:

$$G=\frac{\left(\beta_f - \beta_P\right)}{\beta_f}+\frac{3\left(\rho_P - \rho_f\right)}{\left(\rho_f+2\rho_P\right)} \qquad (2)$$

where, β_f and β_p are the compressibility of fluid and particle respectively, and ρ_f and ρ_p are the densities of fluid and particle respectively.

2.2 *Primary and secondary acoustic force*

The Primary transverse force acts normal to the wave propagation. Of the three forces, the primary axial radiation force generally has the greatest magnitude.

3 APPLICATION OF NEWTON'S SECOND LAW

The proposed technology can be implemented in a rectangular narrow channel of upward fluid flow. Two walls of the channel can be made with a piezoelectric transducer and rigid reflecting surface, as shown in Figure 2. When the transducer is energized at the proper frequency to maintain a resonant acoustic field, there will be a pressure node located on the mid-plane of the chamber, and pressure antinodes located on the chamber walls. The acoustic force on a suspended particle results from the particle-fluid interaction that arises when the particle and suspending fluid have different acoustic properties. This force is maximum when the particle is at the pressure node at quarter wavelength, mid-plane of the chamber width. For a dilute suspension, the secondary radiation forces, the body forces and the hydrodynamic interactions are neglected. Change of particle momentum is equal to:

$$(\rho_p V_0 + 0.5\rho_f V_0)\ dv/dt = F_{PARF} + F_{PTRF} + V_0(\rho_p - \rho_f)\, g + F_D \qquad (3)$$

$$m_v a = F_{PARF} + F_{PTRF} + V_0(\rho_p - \rho_f)\, g + F_D \qquad (4)$$

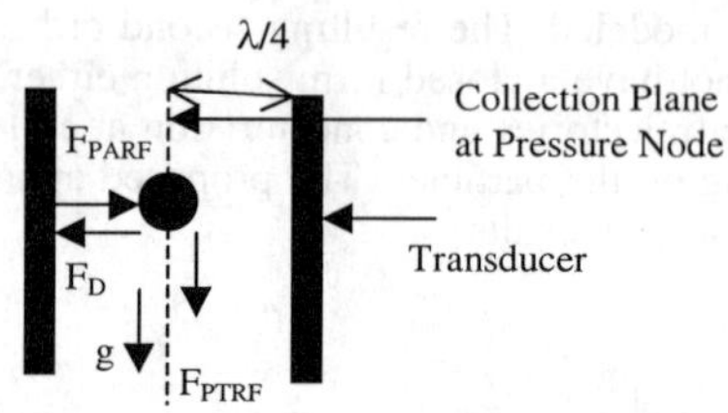

Figure 2. Particle in an acoustic chamber

Where v is the particle velocity and g is the gravitational acceleration. The mass in the momentum term is the "virtual mass" of the particle, m. Since the particle's velocity is v, the drag force is given by Stokes' Law ($F_D=-6\pi\mu r v$). Where μ is the viscosity of the fluid and r is the radius of the particle in suspension. The summation of the forces in the direction of the acoustic wave propagation gives:

$$F_{ac} = F_{PARF} = m_v a + 6\pi\mu r v = V_0 E_{ac} k G \sin(2kx) \qquad (5)$$

Thus the acoustic force, F_{ac}, on a particle in an acoustic field is due to the primary axial radiation force and can be used to calculate the particle trajectories.

4 MATHEMATICAL ANALYSIS

By expanding and rearranging equation (5), the migration of any particle in response to a resonant acoustic field can be determined.

$$-6\pi\mu R\frac{dx}{dt} + 4\pi R^3 k E_{ac} F \sin(2kx) = (\rho_p V_0 + \rho_f V_0)\frac{dv}{dt}$$

The above equation can be rearranged and written in terms of constants:

$$x'' + cx' - k\sin\beta x = 0 \tag{6}$$

For a particle of micron size suspended in a liquid fluid and a frequency of 300 kHz, the parameters *c, k* and β are all positive constants having the following orders of magnitude:

$$c=O(10^8),\ k=O(10^9),\ \beta=O(10^{-3}) \tag{7}$$

Note that (6) has the form of a damped nonlinear spring (or pendulum) equation. As $1<<c$ and $1<<k$, one can view (6) also as a singular perturbation ODE of the form:

$$\varepsilon x'' + c^{\sim} x' - k^{\sim}\sin\beta x = 0 \tag{8}$$

where $\varepsilon<<1$; in fact, $\varepsilon = O(10^{-8})$, $c^{\sim} = 1 = O(1)$ and $k^{\sim} = k/c = O(10)$. In the literature, researchers assume instantaneous viscous relaxation (which often leads to significant errors in singular perturbation problems) and simply use the solutions of the form given below to approximate the solution of (6).

$$c x' - k\sin\beta x = 0 \tag{9}$$

We shall show that although this approximation produces results that are qualitatively correct, quantitative errors are incurred that can be significant for some applications, in particular for the use of acoustic fields for fine particle segregation treatment.

4.1 *Preliminary analysis*

We have not found a closed form solution of (6), so we have had to rely on qualitative and approximate analysis techniques that we shall discuss in this section. Useful information on the nature of the solutions (6) can be obtained by recasting it as a 2-dimensional system and performing a careful phase plane analysis. To do this, we first set $v = x'$ and then rewrite (6) as the following system in the *x, v*-plane:

$$x' = v,\ v' = k\sin\beta x - cv <=> (x', v') = F(x, v) \tag{10}$$

The fixed (critical or stationary) points of (10) are those points in the *x, v*-plane satisfying $F(x, v) = 0 \Leftrightarrow v = 0,\ k\sin\beta - cv = 0$. Hence these points are:

$$\left\{\left(m\frac{\pi}{\beta}, 0\right) : m \in Interger\right\} \tag{11}$$

To find the local behavior of the solutions (10) near the fixed points, we compute F' (for even and odd integer respectively) at these fixed points:

$$F'\left(\frac{m\pi}{\beta}, 0\right) = \begin{bmatrix} 0 & 1 \\ \pm\beta k & -c \end{bmatrix} \tag{12}$$

The eigenvalues of (12) are the roots of the equation:

$$\mu^2 + c\mu - \beta k = 0 \tag{13}$$

$$\mu = \frac{-c \pm \sqrt{c^2 + 4\beta k}}{2} \tag{14}$$

Hence there is one negative eigenvalue μ_-^e and one positive μ_+^e eigenvalue. Where the approximate formulas follow from (7). The eigenspaces associated to these eigenvalues are calculated to be:

$$E(\mu_+^0) : \begin{bmatrix} \mu_+^0 & -1 \\ \beta k & \mu_+^0 + c \end{bmatrix} \begin{bmatrix} \xi \\ \eta \end{bmatrix} = \begin{bmatrix} 0 \\ 0 \end{bmatrix} \Rightarrow \begin{bmatrix} \xi \\ \eta \end{bmatrix} = const \begin{bmatrix} 1 \\ \mu_+^0 \end{bmatrix} \tag{15}$$

$$E(\mu_-^0) : \begin{bmatrix} \mu_-^0 & -1 \\ \beta k & \mu_-^0 + c \end{bmatrix} \begin{bmatrix} \xi \\ \eta \end{bmatrix} = \begin{bmatrix} 0 \\ 0 \end{bmatrix} \Rightarrow \begin{bmatrix} \xi \\ \eta \end{bmatrix} = const \begin{bmatrix} 1 \\ \mu_-^0 \end{bmatrix} \tag{16}$$

We conclude that the fixed points of the form $(2m\pi/\beta, 0)$ are all saddle points having local phase plane behavior that looks like:

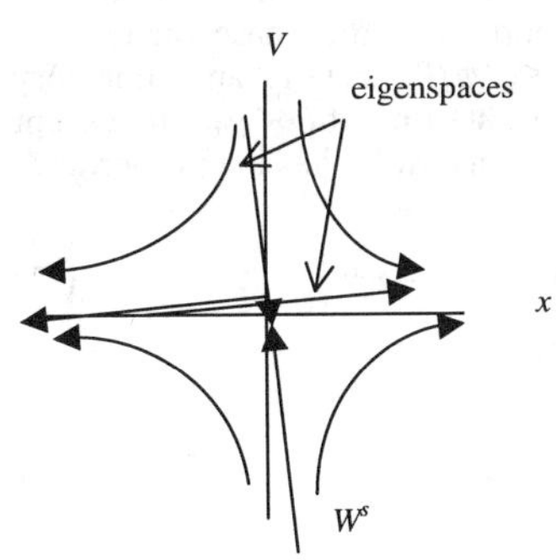

Figure 3. Phase Plane near $(2\,m\pi/\beta, 0)$

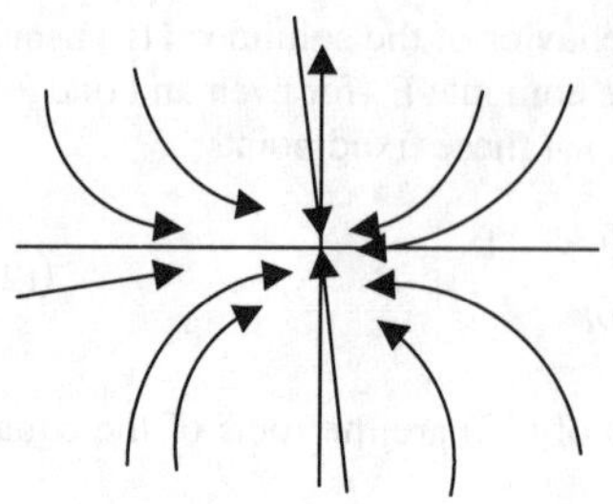

Figure 4. Phase Plane near $((2m+1)\pi/\beta, 0)$

On the other hand, the fixed points of the form $((2m+1)\pi/\beta, 0)$ are all sinks with local phase plane behavior shown in figure 4. Collecting all the results above, it is not hard to show that the entire phase plane of (10) looks like:

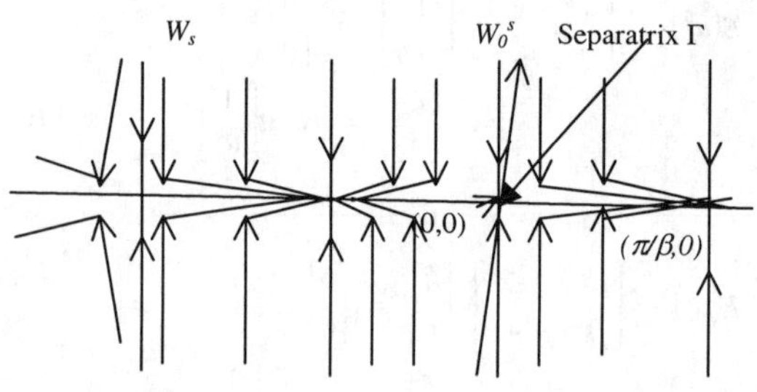

Figure 5. Phase Portrait of (eq. 10)

Quite a bit of information on the solutions of (6) (and equivalently (10)) can be extracted from the above analysis, which is illustrated in Figure 5. The physically relevant portion of the phase plane is of course $R=\{(x, v): 0\leq x\leq \pi/\beta\}$. Clearly any trajectory $\theta_t(x_o, v_o) = (x(t), v(t))$ starting at $(x_0, v_0)\in R$, except for points on the stable manifold, has the properties:

$$\big(x(t), v(t)\big)\rightarrow \left(\frac{\pi}{\beta},0\right) \quad as\ t\rightarrow\infty \tag{17}$$

$$\frac{dv}{dx}=\frac{x''}{x'}\rightarrow -\frac{\beta k}{c} \quad as \quad t\rightarrow\infty \quad (or\ x\rightarrow\frac{\pi}{\beta}) \tag{18}$$

In fact, all of these trajectories converge very rapidly to the separatrix Γ (which is just the portion of the unstable manifold for the origin, W^u_0, that is contained in R). This separatix appears to be quite close to the graph of $v= k/c \sin\beta x$, and so we can infer the following result that shall be proved in the sequel:

$$\big(x(t),v(t)\big)\rightarrow\left(x,\frac{k}{c}\sin\beta x\right) \tag{19}$$

$$as\ t\rightarrow\infty\left(or\ x\rightarrow\frac{\pi}{\beta}\right)$$

We also see from our above analysis that the often used approximation (9) is not really adequate to represent the solutions of (6) (⇔ (10)) since it is just a single curve in the phase plane region R beginning at (0,0) and terminating at $(\pi/\beta, 0)$; albeit a curve to which (almost) all of the solutions (trajectories, orbits) of (6) converge to rather rapidly as $t\rightarrow\infty$ $(x\rightarrow\pi/\beta)$.

Since (6) (⇔(10)) apparently has no closed form solution, it is natural to try to compute numerical approximate solutions using a standard ODE solver such as the Runge-Kutta method. However, an attempt to apply the Runge-Kutta method to (6) directly leads to rather surprising results that are completely unsatisfactory. To explain what happens, we first note that the system form of (6), namely (10), is such that there is an enormous discrepancy between the magnitude and the magnitude of the derivative of the first component $F_1 := v$ and the corresponding magnitudes of the second component $F_2 := \sin\beta x - cv$; in fact, the ratios of the second set of magnitudes to the first is at least $O(10^8)$. In the presence of such enormous magnitudes, one might expect to encounter some problem with conventional ODE solvers. More specifically, in order to obtain good accuracy using the Runge-Kutta method one should take a time step size of no more than about h= 0.1 in order to benefit from the global approximation error of the Runge-Kutta method of $O(h^4)$. If one uses this increment, it is found that there is little movement in the trajectory after several hundred steps, so if one starts in R near $x = 0$, very little progress is made in moving across R to the point $(\pi/\beta, 0)$ even after hundreds of steps. Furthermore, if one continues this for thousands of steps, round off errors accumulate to the point where the approximate numerical solution starts behaving in ways that are not consistent with the properties of the solution that we deduced from our above analysis.

In fact, the conclusions we have already reached regarding the solutions of (6) (⇔ (10) actually reveal the source of the problem encountered using numerical methods. Suppose the trajectory starts at a point (x_0, v_0) in R near 0 = (0,0). We have shown that this trajectory converges to the separatrix Γ very rapidly, in fact, the distance between $\theta_t(x_0, v_0)$ and Γ decreases in time like e^{-ct}. Once θ_t is very close to Γ (which is $R\cap W^u_0 \cup \{(\pi/\beta, 0)\}$), it moves across R towards $(\pi/\beta, 0)$ like $e^{\beta kt/c}$. Hence, the time required to get very close to $(\pi/\beta, 0)$ is roughly $e^{\beta kt/c} = \pi/\beta \Rightarrow$

$t=c/\beta k \log(\pi/\beta)$, thus it follows from (7) that the time required is roughly $O(10^3)$. So with a time step of about h=0.1, the number of steps required to come close to seeing the asymptotic nature of the solution as $t\to\alpha$ is $O(10^4)$. Using this many steps is not only computationally expensive; it also makes the contamination of the approximate numerical data by round off errors almost inevitable. Moreover, we see from the intrinsic nature of the solutions described above that the basic problems with numerical solutions cannot be ameliorated through the use of standard rescaling techniques. We infer from all of this that we shall need to employ other methods to determine useful approximate trajectories of (6) that stretch all the way from an initial point (x_0, v_0) to a very small neighborhood of the sink $(\pi/\beta, 0)$ and that these methods must be of an asymptotic nature for $t\to\infty$.

4.2 *Simpler approximate formulas*

The approximate solution we have obtained so far are still rather unwieldy due to the very large and very small magnitude of some parts of the formulas. From equation (6), we first define $y:=\pi/\beta - x$ and then substitute this into (6) to obtain the following equivalent equation in the new dependent variable:

$$y'' + c\,y' + k\sin\beta y = 0 \tag{20}$$

The idea is to do some scaling to facilitate an asymptotic expansion in terms of some small parameter ε. Using the Taylor series expansion for sinx, and rescaling the dependent and then independent variables:

$$u'' + 10^4 u' + \lambda u = \varepsilon\, u^3\left[\frac{1}{3!} - \frac{1}{5!}u^2 + \frac{1}{7!}u^4 - ....\right] \tag{21}$$

Where:

$$u = \beta\, y, \quad \lambda = k\beta\left(\frac{10^4}{c}\right)^2, \quad \varepsilon = k\beta\left(\frac{10^4}{c}\right)^2$$

Then the solution of (21) is expressed in the form:

$$u = u_0 + \varepsilon u_1 + \varepsilon^2 u_2 + ... \tag{22}$$

Substituting this in (21) and equating ε^0 terms using the initial condition we obtain:

$$u_0 = c_1 e^{r_1 t} + c_2 e^{r_2 t} \tag{23}$$

Similarly one can obtain the ε^1 and ε^2 terms using the method of undetermined coefficients, which yields solutions of the form:

$$u_1 = d_1 e^{r_1 t} + d_2 e^{r_2 t} + U_1(\tau) \tag{24}$$

$$u_2 = p_1 e^{r_1 t} + p_2 e^{r_2 t} + U_2(\tau) \tag{25}$$

where $U_1(\tau)$ and $U_2(\tau)$ are a linear combinations of exponential functions. The same procedure can be used to obtain approximations to $x'(t)=v(t)$, that are well suited for obtaining approximation solutions to the concentration equation. Defining the velocity as:

$$v = w + v_s = w + \frac{k}{c}\sin\beta x \tag{26}$$

and substituting into (20), one obtains:

$$\left(w + \frac{k}{c}\sin\beta x\right)\frac{dw}{dx} + c\left(1 + \frac{\beta k}{c^2}\sin\beta x\right)w + \frac{1}{2}\beta\left(\frac{k}{c}\right)^2\sin 2\beta x = 0 \tag{27}$$

with initial condition:

$$w(x_0) = v(x_0) - \frac{k}{c}\sin\beta x_0 = v^0 - \frac{k}{c}\sin\beta x_0 \tag{28}$$

Now the independent variable is solved by defining $\xi=\beta x$, where $0<\varepsilon<<1$ is a small parameter ($\varepsilon=\beta$ in the actual approximate solutions). For the analysis, the solution is assumed to have the form:

$$w = w_0 + \varepsilon w_1 + \varepsilon^2 w_2 + ... \tag{29}$$

Solving the equations with the initial conditions and then equating terms of orders ε^0, ε^1 and ε^2 gives rise to an approximate solution. For example:

$$w_0(\xi) = w(\xi_0)e^{\xi_0 - \xi}\left[\left(\frac{1-\cos\xi_0}{1-\cos\xi}\right)\times\left(\frac{1+\cos\xi}{1+\cos\xi_0}\right)\right]^{\frac{c^2}{2\beta k}} \tag{30}$$

It is found that w_1 and w_2 are insignificant compared to w_0. Thus a very good approximation to $v(x)$ is:

$$v(x) \approx \frac{k}{c}\sin\beta x + (v^0 - \frac{k}{c}\sin\beta x_0)\times e^{\beta(x_0-x)}\left[\frac{\tan\left(\frac{\beta x_0}{2}\right)}{\tan\left(\frac{\beta x}{2}\right)}\right]^{\frac{c^2}{\beta k}} \tag{31}$$

$$v(x) \approx v_*(x) := \frac{k}{c}\sin\beta x + (v^0 - \frac{k}{c}\sin\beta x_0)\left(\frac{\pi-\beta x}{\pi-\beta x_0}\right)^{\frac{c^2}{\beta k}} \tag{32}$$

Using v_* to approximate v in the concentration equation, we obtain:

$$\frac{\partial n}{\partial t} + v_*\frac{\partial n}{\partial x} = -\left(\frac{dv_*}{dx}\right)n \tag{33}$$

where $n=n(x, t)$ denotes the concentration and $n_0=n(x,0)$ is the constant initial concentration. Using the method of characteristics on this Cauchy problem yields the following approximate formula for the concentration:

$$\frac{n}{n_0} \cong \frac{v_*\left(\frac{2}{\beta}\tan^{-1}\left(e^{\frac{-\beta kt}{c}}\tan\frac{\beta x}{c^2}\right)\right)}{v_*(x)} \tag{34}$$

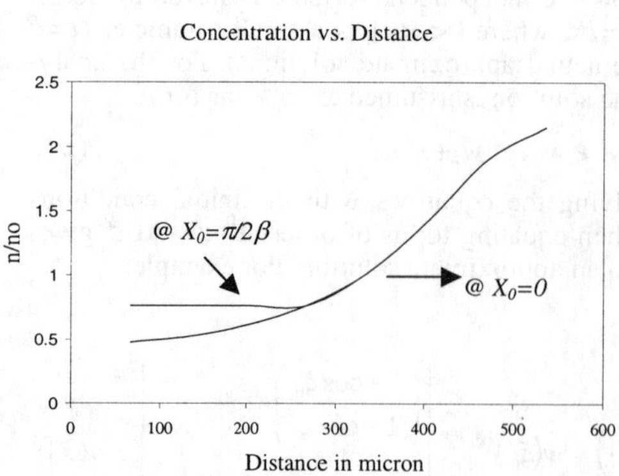

Figure 6. Predicted concentration profiles at t=2 seconds

Note that this actually depends on x_0 as parameter in addition to the variables x and t. for example, when $x_0=0$ we get exactly the result reported in the literature. On the other hand, if say $x_0= \pi/2\beta$ we obtain the approximate concentration as shown in the Figure 6, which has the same basic shape as for x_0=0. In order to eliminate x_0, we must return to the fundamental concentration and reformulate by an additional averaging operation over x_0.

5 CONCULSIONS

The new technology using an acoustic standing wave field in the flow field is proposed to fractionate and segregate suspended particles. This technology can be used to remediate contaminated fine dredged sediments. The fractionation process of suspended particles due to the effect of acoustic standing wave and laminar flow fields has been mathematically modeled through the calculation of particle trajectories and displacement. The particle trajectories move towards the pressure node planes, lying at half wavelength intervals, so that the average distances between the particles are decreased considerably.

Calculating the trajectories and concentration of the micron size particles suspended in a fluid in an acoustic standing waves taking into account all the forces acting on a particle, produces in correct qualitative and quantitative results. Because no exact solution can be found and there are extremely large parameters in the governing equations, it is necessary to supplement standard numerical procedures with approximate asymptotic methods. This approach yields approximate solutions that are in closer qualitative and quantitative agreement with observed results than other methods used in the literature.

6 REFERENCES

Aboobaker, N., Meegoda, J., N., Blackmore, D., "Fractionation and Segregation of Suspended Particles using Acoustic and Flow Fields," Proceedings Hazardous and Industrial Waste_32nd Mid_Atlantic Industrial and Hazardous Waste Conference, Technomic Publishing Company, INC. Lancaster, PA, pp. 659-669, 2000.

Dwayne, A. Johnson, Feke, L. Donald, "Methodology for Fractionation Suspended Particles using Ultrasonic Standing Wave and Divided Flow Fields," Separation Technology, Vol. 5, pp. 251-258, 1995.

Mandralis, I. Zenon, Feke, L. Donald, "Fractionation of Suspensions Using Synchronized Ultrasonic and Flow Fields," AIChE Journal, Vol. 39, No. 2, pp. 197-205, 1993.

Whitworth, Glenn and Coakley, W. T., "Particle Column Formation in a Stationary Ultrasonic Field," Acoustic Society of America, Vol. 91, No. 1, pp. 79-85, 1991.

Woodside, M. Steven, and Piret, M. James, "Acoustic Force Distribution in Resonators for Ultrasonic Particle Separation," AIChE Journal, Vol. 44, No. 9, pp. 1976-1984, 1998.

Computer Methods and Advances in Geomechanics, Desai et al. (eds) © 2001 Balkema, Rotterdam, ISBN 90 5809 183 X

A finite element method for fully coupled thermo-hydro-diffusion problems and its applications to geo-science and geo-engineering

Ö.Aydan
Tokai University, Department of Marine Civil Engineering, Shimizu, Japan

ABSTRACT: Coupled problems are commonly encountered in geo-science and geo-engineering such as radioactive nuclear waste disposal, petroleum and geothermal exploitations. Many experimental, theoretical and numerical studies have been undertaken in literature. Particularly the studies concerned with radioactive waste disposal have been concerned with thermo-hydro-mechanical aspects of the phenomenon and diffusion phenomenon of the radioactive substances is mostly omitted. Since the diffusion phenomenon should be quite important in long term, a mechanical model based on the mixture theory is described to couple thermal, hydrological and diffusion fields. In this model, Duffour and Soret effects, which are mostly neglected in previous studies, are also considered to couple the thermal and diffusion fields. Then a finite element formulation of the derived theoretical model is given and a series of numerical analyses was carried out for the simulation of laboratory tests is presented. Furthermore, some parametric studies are performed to investigate the coupling effects of Duffour and Soret effects on thermal and diffusion fields.

1 INTRODUCTION

The coupled problems are commonly encountered in geo-science and geo-engineering. The safe disposal of nuclear wastes, petroleum and geothermal exploitations are some of examples for such problems in geo-engineering. Similar problems are also encountered in earthquake prediction studies in geo-science. The radioactive nuclear waste disposal particularly receives great attention and one may find many studies in literature. Most of these studies on radioactive nuclear waste disposal are based on thermo-hydro-mechanical concepts and it is very rare to see any study, which includes diffusion phenomena of radioactive substances. Therefore, a mechanical model, which includes 4 different aspects, should be the most appropriate approach for radioactive waste disposal. Although the variation of stress field around the disposal site may be of great concern for cavity stability in very near field, its effect is expected to be quite limited after back-filling the excavated space in long term. In this article, the mechanical effect is omitted in this coupling model presented herein and a theoretical formulation, based on the mixture theory is described for the thermo-hydro-diffusion phenomena. In the theoretical formulation, Duffour and Soret effects are considered for coupling the thermal and diffusion fields with each other. Then, a finite element formulation of the coupled model is presented and it is used for numerical analysis of some laboratory tests and compared with experimental results. Furthermore, a series of parametric numerical analyses are performed to investigate

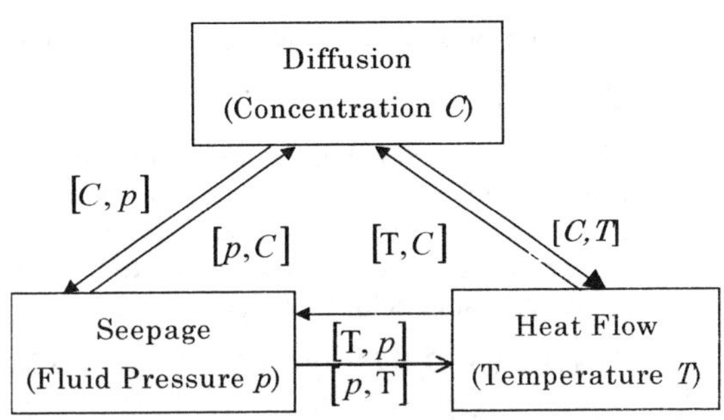

$[C,p]$:Effect of Diffusion on Seepage (density variation)
$[p,C]$:Effect of Seepage on Diffusion (Advection of Concentration due to fluid flow)
$[T,p]$:Effect of Heat Flow on Seepage (Density variation due to Temperature variation)
$[p,T]$:Effect of Seepage on Heat flow(Advection of Temperature due to fluid flow)
$[T,C]$:Effect of Heat Flow on Concentration(Duffour effect))
$[C,T]$:Effect of Diffusion on thermal field(Soret effect))

Figure 1 Coupling model

a) Simulation of solute transport in rock under laboratory conditions,
b) Temperature field of geo-thermal fields under forced seepage
c) Non-isothermal advective moisture transport through buffer materials, and
d) Parametric studies on the consideration of Dufour and Soret laws in purely coupled hydro-thermo-diffusion problems.

2 MECHANICAL MODELLING

The mechanical modelling of thermo-hydro-diffusion phenomena is based on the mixture theory (Trusdell and Toupin 1960, Eringen and Ingram 1965). Figure 1 illustrates how three fields are coupled. Following the principles of the mixture theory, and some appropriate constitutive relations, one can easily derive that the fundamental equations for each field as follows (Aydan 2001):

Seepage field

If the density variation of solid phase and the porosity variation of skeleton are negligible, the fundamental equation for seepage field may be re-written together with D'Arcy's law as:

$$n\left(\frac{1}{K_f}\frac{\partial p}{\partial t}-\beta_{fT}\frac{\partial T}{\partial t}-\beta_{fC}\frac{\partial C}{\partial t}\right)=-\nabla\cdot\left(\frac{k}{\eta}\nabla p\right) \quad (1)$$

where k is permeability and η is viscosity coefficient of fluid, K_f is fluid compressibility, β_{fT} is thermal expansion coefficient and β_{fC} is diffusive expansion coefficient.

Diffusion field

Fick's law is often employed as a constitutive law for diffusion problems. Extending this law by associating concentration flux of solid phase with gradients of temperature and concentration, the following equation for diffusion field is obtained:

$$\frac{\partial C}{\partial t}=\nabla\cdot(D_C\nabla C+D_T\nabla T)-\mathbf{v}_r\cdot\nabla C \quad (2)$$

where D_C and D_T are defined as $D_T=(1-n)D_{sT}+nD_{fT}$ and $D_C=(1-n)D_{sC}+nD_{fC}$. D_{sC} and D_{fC} are the diffusion coefficient of solid phase and fluid phase, D_{sT} is thermal diffusivity of solid phase and D_{sC} is Dufour's coefficient (Bear 1988), respectively. The second term on the right-hand side is the advective term.

Thermal field

Relating internal energies of solid and fluid phases to temperature field with the use of specific heat coefficients c_s and c_f, and the appropriate constitutive laws, one easily gets the fundamental equation for thermal field as

$$\rho c\frac{\partial T}{\partial t}=\nabla\cdot(\lambda_T\nabla T+\lambda_C\nabla C)-\rho_f c_f\mathbf{v}_r\cdot\nabla T \quad (3)$$

where $\lambda_T=(1-n)\lambda_{sT}+n\lambda_{fT}$, $\lambda_C=(1-n)\lambda_{sC}+n\lambda_{fC}$, and $\rho c=(1-n)\rho_s c_s+n\rho_f c_f$. λ_{sT} is thermal conductivity coefficient, λ_{sC} is Soret's coefficient of solid, λ_{fT} is thermal conductivity coefficient, λ_{fC} and is Soret's coefficient of fluid, respectively.

3 FINITE ELEMENT FORMULATION

3.1 *Weak Forms of Fundamental Equations*

Seepage field

The governing equation of seepage field is assumed to be subjected to the following boundary conditions

Pressure boundary condition

$$p=p_0 \text{ on } S_p \quad (4)$$

Fluid flux boundary

$$-(n\rho_f\frac{k}{\eta}\nabla p)\cdot\mathbf{n}=\hat{q} \text{ on } S_q \quad (5)$$

Taking a variation on pressure field δp and integrating by parts, one gets the weak form of Eq. (1) as

$$n\left(\int_V\delta p\frac{n}{K_f}\frac{\partial p}{\partial t}dV-\int_V\delta p\beta_{fT}\frac{\partial T}{\partial t}dV-\int_V\delta p\beta_{fC}\frac{\partial C}{\partial t}dV\right)+\int_V\nabla(\delta p)\cdot\frac{k}{\eta}\nabla p dV=\int_{S_q}\delta p\hat{\mathbf{q}}dS \quad (6)$$

Diffusion field

The diffusion equation is assumed to be subjected to the following boundary conditions

Concentration boundary

$$C=C_0 \text{ on } S_C \quad (7)$$

Concentration flux boundary

$$-(D_C\nabla C+D_{CT}\nabla T)\cdot\mathbf{n}=\hat{f} \text{ on } S_f \quad (8)$$

Taking a variation δC and applying the integration by parts to Eq. (2), we have the weak form of Eq.(2) as

$$\int_V\delta C\frac{\partial C}{\partial t}dV+\int_V\nabla\delta C\cdot D_C\nabla CdV+\int_V\nabla\delta C\cdot D_C\nabla TdV+\int_V\delta C\mathbf{v}_r\nabla CdV=\int_{S_q}\delta C\hat{\mathbf{f}}dS \quad (9)$$

Thermal field

Eq. (3) is assumed to be subjected to the following boundary conditions

Temperature boundary

$$T=T_0 \text{ on } S_T \quad (10)$$

Heat flux boundary

$$-(\lambda_T\nabla T+\lambda_{TC}\nabla C)\cdot\mathbf{n}=\hat{h} \text{ on } S_h \quad (11)$$

Taking a variation δT on temperature field and applying the integration by parts to the first term of Eq. (3), one easily gets its weak form as:

$$\int_V \delta T \rho c \frac{\partial T}{\partial t} dV + \int_V \nabla \delta T \cdot \lambda_T \nabla T dV + \int_V \nabla \delta T \cdot \lambda_{TC} \nabla C dV + \int_V \delta T \rho_f c_f \mathbf{v}_r \cdot \nabla T dV = \int_{S_q} \delta T \hat{h} dS \quad (12)$$

3.2 *Discretization of Weak Forms*

3.2.1 *Discretization in Physical Space*

Pressure, concentration and temperature variables are interpolated in a typical finite element with the use of shape functions as given below:

$$p = [N]\{P\},\ C = [N]\{\phi\},\ T = [N]\{\chi\} \quad (13)$$

Inserting these relations into each respective weak form and after some manipulations, one easily gets the following equation system for each field as

Seepage field

$$[M]_{PP}\{\dot{P}\} + [M]_{PC}\{\dot{\phi}\} + [M]_{PT}\{\dot{\chi}\} + [K]_{PP}\{P\} = \{Q\}_P \quad (14)$$

where

$$[M]_{PP} = \int_V \frac{n}{K_f}[N]^T[N]dV;$$

$$[M]_{PC} = -\int_V n\beta_{fC}[N]^T[N]dV;$$

$$[M]_{PT} = -\int_V n\beta_{fT}[N]^T[N]dV;$$

$$[K]_{PP} = -\int_V \frac{k}{\eta}[B]^T[B]dV;\ \{Q\}_P = \int_{S_q} N^T \hat{q} dS$$

Diffusion field

$$[M]_{CC}\{\dot{\phi}\} + [K]_{CC}\{\phi\} + [K]_{CT}\{\chi\} = \{Q\}_C \quad (15)$$

where

$$[M]_{CC} = \int_V [N]^T[N]dV;$$

$$[K]_{CC} = \int_V D_C[B]^T[B]dV - \int_V \frac{k}{\eta}[N]^T([B]\{P\})^T\{B\}dV$$

$$[K]_{CT} = \int_V D_{CT}[B]^T[B]dV;\ \{Q\}_C = \int_{S_f} N^T \hat{f} dS$$

Thermal field

$$[M]_{TT}\{\dot{\chi}\} + [K]_{TT}\{\chi\} + [K]_{TC}\{\phi\} = \{Q\}_T \quad (16)$$

where

$$[M]_{TT} = \int_V \rho c[N]^T[N]dV$$

$$[K]_{TT} = \int_V \lambda_T[B]^T[B]dV - \int_V \rho_f c_f \frac{k}{\eta}[N]^T([B]\{P\})^T[B]dV$$

$$[K]_{TC} = \int_V \lambda_{TC}[B]^T[B]dV;\ \{Q\}_T = \int_{S_h} N^T \hat{h} dS$$

Finally the following simultaneous equation system is obtained for whole domain as

$$[M]\{\dot{X}\} + [K]\{X\} = \{Y\} \quad (17)$$

where

$$[M] = \begin{bmatrix} [M]_{PP} & [M]_{PC} & [M]_{PT} \\ [0] & [M]_{CC} & [0] \\ [0] & [0] & [M]_{TT} \end{bmatrix};$$

$$[K] = \begin{bmatrix} [K]_{PP} & [0] & [0] \\ [0] & [K]_{CC} & [K]_{CT} \\ [0] & [K]_{TC} & [K]_{TT} \end{bmatrix}$$

$$\{\dot{X}\} = \begin{Bmatrix} \dot{P} \\ \dot{\phi} \\ \dot{\chi} \end{Bmatrix};\ \{X\} = \begin{Bmatrix} P \\ \phi \\ \chi \end{Bmatrix};\ \{Y\} = \begin{Bmatrix} \{Q\}_P \\ \{Q\}_C \\ \{Q\}_T \end{Bmatrix}$$

3.2.2 Discretization in Time Domain

Although there are several techniques for discretization in time domain, θ -method is chosen in this study. Since Eq. (17) holds at any time, one may easily write the following relation for a time step ($m + \theta$) as

$$[M]\{\dot{X}\}_{(m+\theta)} + [K]\{X\}_{(m+\theta)} = \{Y\}_{(m+\theta)} \quad (18)$$

With the use of Taylor expansions of variables and after some manipulations, Eq. (18) takes the following form:

$$[C^*]\{X\}_{m+1} = \{Y^*\}_{m+1} \quad (19)$$

where

$$[C^*] = \left[\frac{1}{\Delta t}[C] + \theta[K]\right]$$

$$\{Y^*\}_{m+1} = \left[\frac{1}{\Delta t}[C] - (1-\theta)[K]\right]\{X\}_m + \theta\{Y\}_{m+1} + (1-\theta)\{Y\}_m$$

It should be noted that matrices $[K]_{CC}$ and $[K]_{TT}$ contains unknown vector variable $\{P\}$. Therefore, the resulting equation system is non-linear. However, if time step is sufficiently small, it can be linearized with the use of variable $\{P\}$ of the previous time step.

4 ANALYSES AND DISCUSSIONS

The first example is concerned with the simulation of diffusion-seepage tests on a sandstone sample carried out by Igarashi & Tanaka (1998) under isothermal condition. Bear, J. (1988): *Dynamics of Fluids in Porous Media.* Elsevier, New York. The pressure gradient in the sample was set to 260MPa/m. Figure 2 shows pressure distribution through the sample at different time steps. Since the pressure distribution attains the steady state in a short period of time, this implies that the diffusion takes place under constant fluid velocity field through the sample. Material properties used in the analyses are given in Table 1. The computations were carried out for two different situations, specifically, 1) Advection+Conduction and 2) Conduction only. Figure 3 compares the breakdown curves for computed two situations with the experimental one. In the analyses, the effect of diffusion on seepage field is neglected. As seen from the figure, the best fit to the experimental response was obtained from the computation for the advection+conduction situation while much longer period of time is required for conductive diffusion. This fact simply implies that if the diffusion coefficient, which is obtained from a conductive diffusion model under different pressure gradients, will differ from each other.

Table 1 Material properties used in analyses

Property	Value
Fick's diffusion coefficient of solid (m^2/day)	$5.1x10^{-5}$
Fick's diffusion coefficient of fluid (m^2/day)	$2.1x10^{-4}$
Porosity (%)	16
Permeability (m^2)	$3.1x10^{-12}$
Sample length (mm)	50
Sample diameter (mm)	50

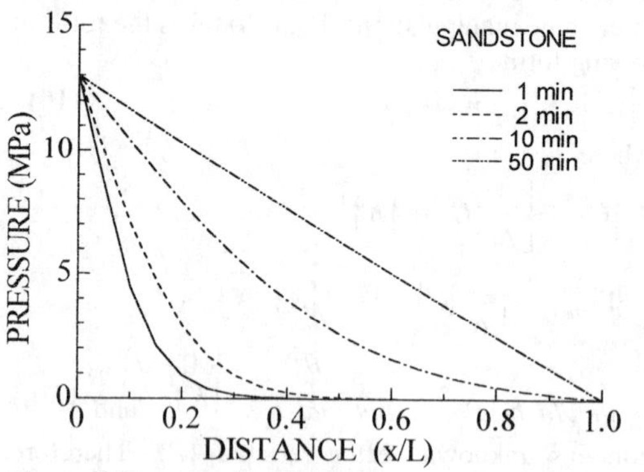

Figure 2 Pressure distribution in a sandstone sample

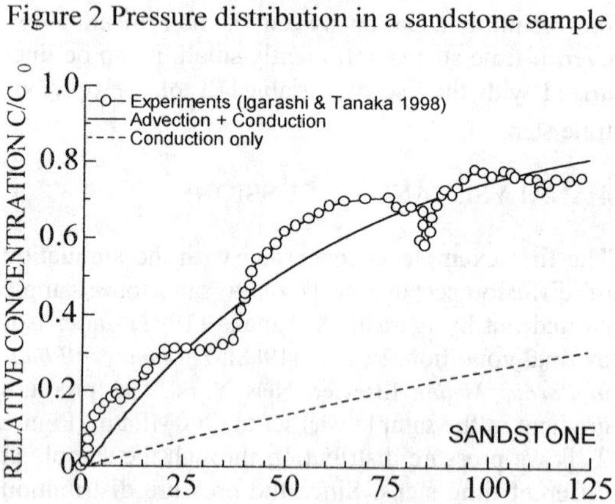

Figure 3 Comparison of computed breakdown curves with the experimental curve

The method presented in the previous sections is applied to analyse the temperature distribution measured at Armutlu hot springs in the NW Turkey. The measurements were carried out by Pfister et al. (1997). They used some models based on the steady state equation of heat transport with advective upward flow to interpret the temperature measurements. For the interpretation of the temperature distribution, an upward velocity of 0.5m/year was assumed at a depth of 150m. However, The general temperature profiles of rock mass and nearby thermal spring channels may be different since the upward fluid velocity could vary due to the permeability rock masses. Three different situations are considered, specifically:

CASE 1: no advection

CASE 2:advection with an upward fluid velocity of 30mm/year, and

CASE 3:advection with an upward fluid velocity of 100mm/year.

CASE 1 corresponds to purely conductive heat transport while CASE 2 and CASE 3 do the advective heat transport. In the analyses, an air element next to the ground surface is introduced since the temperature of the surface of rock may not be the same as that of surrounding air (Aydan et al. 1985). The parameters used in numerical analyses are given in Table 2. The thermal properties of micaschists and calcerous schist are assumed to be the same while the thermal conductivity of diabase was taken 0.2 times that of schists. This value is less than the value reported by Pfister et al. (1997). Since the initial temperature distribution was unknown, the whole domain was initially set to 76^{0}C and the computations were carried out for 25000 years. Figure 4 shows the time history of the temperatures of some selected points. As noted from the figure, the temperature becomes almost steady after 7000 years. To compare temperature distributions for three cases, the temperature profiles at 25000 years are selected as the steady state temperature of the vicinity of the well at Armutlu. The temperature profile for CASE 1 closely predicts the the lower bound values of measured temperature profile. On the other hand, the temperature profile for CASE 2 predicts the upper bound values of the measured profile. The actual situation in the vicinity of the Armutlu hot spring may be just in between two predictions. The temperature profile for CASE 3 is very close to the temperature profile of hot water in the well, although the profile near the ground surface is a bit different from the actual measurements. Nevertheless, such a situation may be simulated if the rock material is replaced with steel and water in order to simulate the effect of the metal casing of the drill-hole. Furthermore, it should be noted that the upward fluid velocity should be much less than that reported by Pfister et al. (1997).

The next example is concerned with moisture transport through a buffer material for radioactive waste disposal under different ambient temperatures, reported by Kanno et al. (1999). Recently Basha and Selvadurai (1998) theoretically solved this problem as a spherically symmetric problem.

Table 2 Properties of materials used in numerical analyses

Parameter	Schists	Diabase	Air
Thermal conductivity of solid ($Wm^{-1}K^{-1}$)	2.7	0.54	-
Thermal conductivity of fluid ($Wm^{-1}K^{-1}$)	0.6	0.6	0.026
Density of solid (kgm^{-3})	2600	2600	-
Density of fluid (kgm^{-3})	1000	1000	1.2
Specific heat of solid ($kJkg^{-1}K^{-1}$)	0.89	0.89	-
Specific heat of fluid ($kJkg^{-1}K^{-1}$)	4.187	4.187	1.0
Porosity (%)	10	10	100
Heat production rate ($W m^{-3}$)	$1.6x10^{-6}$	$1.6x10^{-6}$	-

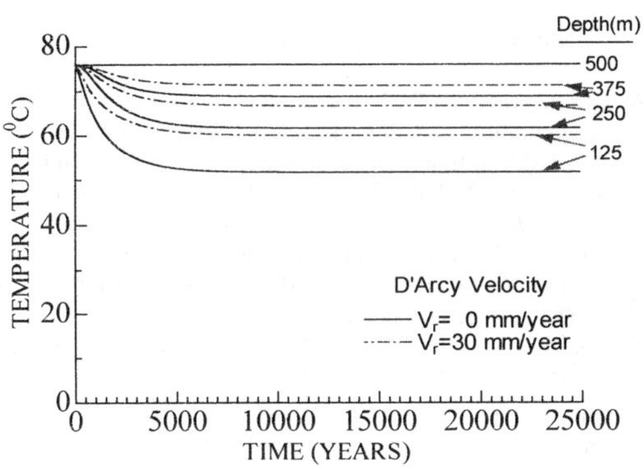

Figure 5 Variation of temperatures of some selected points with time

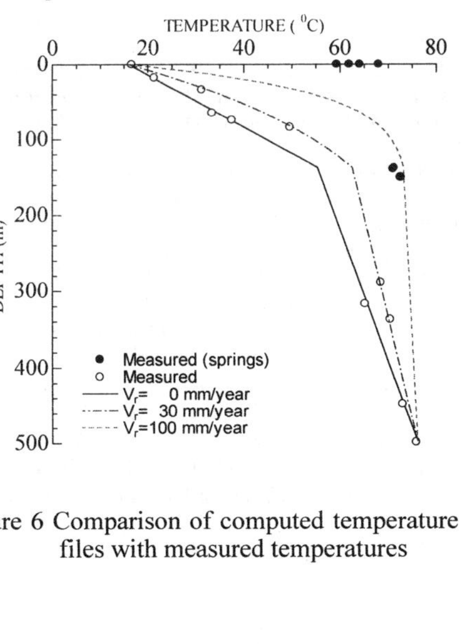

Figure 6 Comparison of computed temperature profiles with measured temperatures

Kanno et al. (1999) used bentonite as a buffer material in their tests and the ambient temperature was varied from 25^0 to 60^0C. They determined Fick's diffusion coefficient of bentonite for moisture transport using a conductive diffusion model. Figure 7 shows a simulation of moisture concentration distributions through the sample at different time steps, carried out at an ambient temperature of 25^0C. They found that the diffusion coefficients obtained for different ambient temperatures differ from each other.

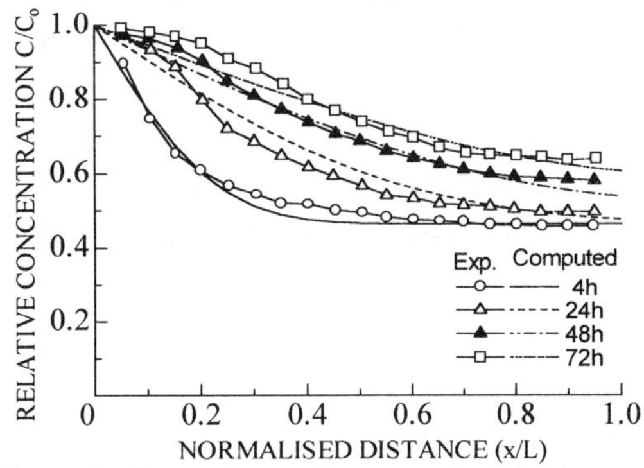

Figure 7 Comparison of computed moisture concentration distributions through the sample with experimental distributions at different time steps

With the experimental finding of Kanno et al. (1999) in mind, a series of parametric studies were carried out. Specifically the following four cases were considered in parametric studies:

CASE 1:Dufour coefficient and Soret coefficient are nil (thermal and diffusion fields are uncoupled)
CASE 2:Dufour coefficient is not nil, but Soret coefficient is nil
CASE 3:Dufour coefficient is nil, but Soret coefficient is not nil
CASE 4:Dufour coefficient and Soret coefficient are not nil (fully coupled)

Material properties used in analyses are given in Table 3 and computed results are shown in Figures 8-9 for CASE 1 and CASE 4 only. For all cases, temperature distributions through the sample become uniform and the effect of diffusion field on temperature distributions are not observed. CASE 4 is probably the most close situation to actual situations. Particularly, the effects of Duffour coefficient and Soret coefficient are quite remarkable. However, there are almost no experimental results reported on these coefficients to confirm the findings from computations. Therefore, it may be stated that re-assessment of existing experiments with the con-

Table 3 Material properties used in numerical analyses

Parameter	CASE 1	CASE 4
Solid's thermal conductivity ($Wm^{-1}K^{-1}$)	0.277	0.277
Fluid's thermal conductivity ($Wm^{-1}K^{-1}$)	0.600	0.600
Dufour's coefficient of solid ($Wcm^{-1}kg^{-1}$)	0	72
Dufour's coefficient of fluid ($Wcm^{-1}kg^{-1}$)	0	72
Soret's coefficient of solid ($cm^2kg\ s^{-1}K^{-1}$)	0	$3.6x10^{-5}$
Soret's coefficient of solid ($cm^2kg\ s^{-1}K^{-1}$)	0	$3.6x10^{-5}$
Solid's density ($kg\ m^{-3}$)	1600	1600
Fluid's density ($kg\ m^{-3}$)	1000	1000
Solid's specific heat ($kJ\ kg^{-1}K^{-1}$)	5.0	5.0
Fluid's specific heat ($kJ\ kg^{-1}K^{-1}$)	1.0	1.0
Fick's diffusion coefficient of solid (cm^2/s)	$6.1x10^{-3}$	$6.1x10^{-3}$
Porosity (%)	46.3	46.3

sideration of Duffour coefficient and Soret coefficient is urgently necessary.

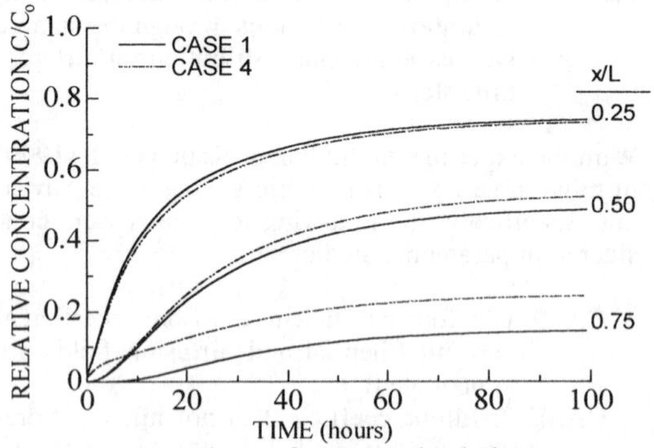

Figure 8 Comparison of computed relative concentration versus time curves for CASE 1 and CASE 4

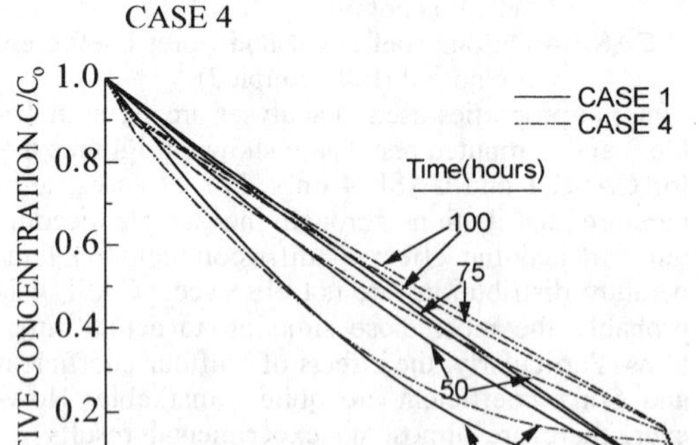

Figure 9 Comparison of computed relative concentration distributions at different time steps for CASE 1 and CASE 4

5 CONCLUSIONS

In this article, a new mechanical model for fully coupled thermo-hydro-diffusion phenomena in geoscience and geo-engineering on the basis of mixture theory and its finite element representation are presented. The validity of this model is checked through some experiments, which correspond to its special forms. From these comparisons, some shortcomings of determination of Fickian diffusion coefficients from conductive diffusion models are pointed out. Parametric studies on the effect of Duffour and Soret coefficients which are generally neglected in diffusion models indicated that these coefficients must be taken into account in numerical analyses. Such considerations can explain the dependency of Fickian diffusion coefficient on ambient temperature and fluid velocity. Furthermore, there is an urgent experimental necessity to obtain the actual values of Duffour and Soret coefficients for a meaningful assessment of fully coupled thermo-hydro-diffusion phenomena.

REFERENCES

Aydan Ö., A. Ersen, Y. Ichikawa, and T. Kawamoto (1985). Temperature and thermal stress distributions in mass concrete shaft and tunnel linings during the hydration of concrete (in Turkish). *The 9th Mining Science and Technology Congress of Turkey*, Ankara, 355-368.

Aydan Ö., (2001). Modelling and analysis of fully coupled thermo-hydro-diffusion phenomena. *IS-SHIZUOKA*, Shizuoka.

Basha, H.A. and Selvadurai, A.P.S. (1998): Heat induced moisture transport in the vicinity of a spherical heat source. *International Journal for Numerical and Analytical Methods in Geomechanics,* 22, 969-981.

Bear, J. (1988): *Dynamics of Fluids in Porous Media.* Elsevier, New York.

Eringen, A.C. and Ingram, J.D. 1965. A continuum theory of chemically reacting media-I. *Int. J. Eng. Sci.*, 3, 197-212.

Igarashi, S. and Tanaka, Y. (1998): The effect of a discontinuity on solute transport through sandstone. *10th National Rock Mechanics Symposium,* Tokyo, JRMS, 359-364.

Kanno, T., Fujita, T., Takeuchi, S., Ishikawa, H., Hara, K. and Nakano, M. (1999): Coupled thermo-hydro-mechanical modelling of bentonite buffer material. *International Journal for Numerical and Analytical Methods in Geomechanics,* 23, 1281-1307.

Trusdell, C.A. and Toupin, R.A. 1960. The classical field theories. In S. Flügge (Ed.) *Handbuch der Physik III/1,* Berlin, Springer Verlag, 226-793.

Pfister, M., Rybach, L. and Şimşek, Ş. (1997): Geothermal reconnaissance of the Marmara Sea Region (NW Turkey). *Active Tectonics of Northwestern Anatolia-The Marmara Poly-Project*, vdf Hochschulverlag AG an der ETH Zürich, 503-535.

Computer Methods and Advances in Geomechanics, Desai et al. (eds) © 2001 Balkema, Rotterdam, ISBN 90 5809 183 X

Artificial calculated boundary conditions and numerical simulation of temperature and moisture variation in subsoil of building

Wen-Hua Chen
Department of Civil Engineering, Northern Jiaotong University, Beijing, People's Republic of China

ABSTRACT: In order to improve the accuracy of numerical simulation of temperature and moisture transfer in subsoil of building, the interactions of temperature and moisture has been untied, some key parameters affected variation of heat and moisture from solid phase, liquid phase and gas phase, are totally taken into consideration. Especially, some artificial boundary conditions are studied and compared to compute the affection of upper structure, a discrete artifact boundary conditions is introduced to simulate the interaction of the extra dominion medium to inter dominion medium, the method can improve the precision and the stability of calculation.

1 INTRODUCTION

There are many differences between level ground and subsoil of building on transfer energy and water, soil is very complex, like inhomogeneous, discontinuity, added the affection of upper structure, it is difficult to describe the movement status of temperature and moisture. By today there is no analytic resolution of this question, the classic and typed method is suit to free field. To find the number resolution of multi-dimension of temperature and moisture Of unfree field ,In this paper, some questions of simulation are analyzed. According to the theory of transfer moisture and heat in porous medium, the simplified analysis model are built, selecting some key parameters, like λ and ζ (λ is the special heat, ζ is the coefficient of transfer heat)must think of the interactions of temperature and moisture, some affected factors, like porosity and water et al., have been analyzed. FEM(Finite Element Method)is used to simulate heat and moisture variation in subsoil of building , because the size of calculate model is limited, so we must select a limited dominion to replace the unlimited dominion to take into consideration of the affection from upper structure, in order to describe correctly the case of exchange of energy and moister in boundary, a artifact boundary conditions must be used to simulate interaction of the external dominion medium to inter dominion medium. Three sorts of boundary conditions have been compared, Finally some interested conclusions can be found.

2 THE SIMPLIFIED MODEL OF MOISTURE FIELD AND TEMPERATURE FIELD

The temperature of earth's surface change with seasonal exchange, in different depth of layers, there are different period temperature change strip, just like day-changed layer, year-changed layer and years-changed layer, day-changed layer depth of soil is 1~2m, which is a stable temperature field changing with irradiation of the sunlight, year-change temperature about 15~30m under ground .

2.1 The compound functions of temperature and moisture of unsaturated soil

The compound functions of temperature and moisture of unsaturated soil, in plane questions, which taking into consideration of interaction and effect of temperature and moisture(x along the surface ,z: along depth direction):

$$\frac{\partial T}{\partial t} = \frac{\lambda}{\varsigma}\left(\frac{\partial^2 T}{\partial x^2} + \frac{\partial^2 T}{\partial z^2}\right) - c_w \rho_w \left(v_x \frac{\partial T}{\partial x} + v_z \frac{\partial T}{\partial z}\right)$$

$$v_x = -k_x \frac{\partial C}{\partial x} \qquad v_z = -k_z \frac{\partial C}{\partial z} \tag{1.1}$$

$$K(C)\left(\frac{\partial^2 C}{\partial x^2} + \frac{\partial^2 C}{\partial z^2}\right) \pm \frac{\partial K(C)}{\partial t} - \frac{\partial C}{\partial t} = 0 \tag{1.2}$$

K(C) and C are parameters of seepage and pressure of water thinking of suction, C is the moisture ,T is temperature.

Those equations are very complex and have many parameters, a simplified equation must be suggested to think about the affection of porosity and other phased medium to physical character of heat and moisture in soil.

2.2 The equation of heat transfer and simplified model

The base relation of two dimension temperature field is listed as equation(2) :

$$\frac{\partial T}{\partial t} = \frac{\lambda}{\varsigma\rho}\left(\frac{\partial^2 T}{\partial x^2} + \frac{\partial^2 T}{\partial z^2}\right) \tag{2}$$

ζ :special heat of soil systems((10^2J/kg℃), ρ :density of soil, λ : :parameters of transfer hear of earthy (W/m/℃) .

$$\zeta = \zeta_s\theta_s + \zeta_w\theta_w + \zeta_a\theta_a \tag{3}$$

$$\lambda = \frac{f_s\theta_s\lambda_s + f_w\theta_w\lambda_w + f_a\theta_a\lambda_a}{f_s\theta_s + f_w\theta_w + f_a\theta_a} \tag{4}$$

$\varsigma_s,\varsigma_w,\varsigma_a$ are the special heat of solid phased ,liquid phased and gas phased, $\theta_s,\theta_w,\theta_a$:the volume percent of each phased, $\lambda_s,\lambda_w,\lambda_a$ parameters of transfer heat of each phased medium, f_s,f_w,f_a :right gene of calculation.

The character of transfer heat of soil is related to the component of mine, the size of particle and the quantity of water in soil; also porosity and density et al, the bigger porosity dry soil, λ is small ,while bigger porosity saturated soil ,transfer heat very easily, λ in z orientation is bigger than that of in x orientation ,temperature change on a day, and seasonal change of weather will change the quantity of water of soil, then induced the change of .temperature-grads improve the movement of water and vapor, water and gas transfer from a high temperature to a low place, the seasonal change temperature of temperature will bring water and vapor to move in a large region, at last it will lead the change of volume of soil ,just like the swell and craze of the highway ,road of airport, or some shallow foundations, ς is decided by the three phased medium of solid liquid and gas, water has a great affection on ς.

2.3The analysis model of moisture

Water in soil has two status of liquid water and vapor ,water in soil underground distributed in unsaturated strip(filled with vapor) and saturated strip, the strip filled with vapor is combined of earth-water strip , seepage strip and seepage strip, and capillary strip, the suction of soil and grads of temperature will lead water to move ;suction of soil change with sort of soil and moisture, suction of dry soil is bigger than that of saturated soil, water move from a high moisture to a low place, if the water-level of underground is shallow in subsoil, the capacity of the bear the weight cut down ,the bottom of building become moisture ,it will affect the safety of structure and hurt the health of people. According the theory of transfer moisture in porous medium(simulation of the water and gas moving in fiber),the analysis model of moisture is suggested [3](the simplified equation of water transfer in soil is in two dimension):

$$D_A\left(\frac{\partial^2 C}{\partial x^2}+\frac{\partial^2 C}{\partial z^2}\right)\pm\frac{\partial K(C)}{\partial t}-\frac{\partial C}{\partial t}=0 \tag{5}$$

$D_A = 2.302\,\frac{0.98}{P}\left(\frac{T}{256}\right)^{1.81}(m^2/\mathrm{sec})$ P is the pressure of air , T is the temperature of soil , K(C) is relate to suction.

The above equation is complete , because the K(c) can be get by test , in the same time ,if you do not think of the the change of volume of soil or foundations, the K(c) maybe be regarded as K(c) =0.0 , the equation (5) is :

$$D_A\left(\frac{\partial^2 C}{\partial x^2}+\frac{\partial^2 C}{\partial z^2}\right)-\frac{\partial C}{\partial t}=0$$

the above equation had been researched by motion from 1996 to 1998 in East-northern in China , it is a right equation.

2.4 The discontinuity physical characters of subsoil

In general, a type non-uniformity foundation is layered, parameters change along z direction, one layer is regarded as a kind if porous medium, in this paper, we think of inhomogeneous in landscape orientation, the variable of density and porosity can be different in x direction in the same layer, In order to research the temperature and moister. Porosity-change is bring randomly in each layer, the datum of porosity is limited in a range, for example ,the Ith layer, the porosity is submitted in range of $\theta_I \sim \theta_I'$,in FEM, each element the θi can be born by random function, the random datum can be easily born by computer, to geotechnical engineering ,the average porosity and the range of porosity of each soil layer can be tested, in fact, the computer only make a random data of 0~1.0,like the follow equation:

$$\theta_{I,e}=\theta_{I,av}+\theta_{I,range}\times Random(0\sim 1) \tag{6}$$

$\theta_{i,e}$:the porosity of i element, $\theta_{I,av}$:the average porosity of the Ith layer, $\theta_{I,range}$:the change range of the Ith layer, Combined with the equation (3)and (4), some discontinuity parameters will be computed.

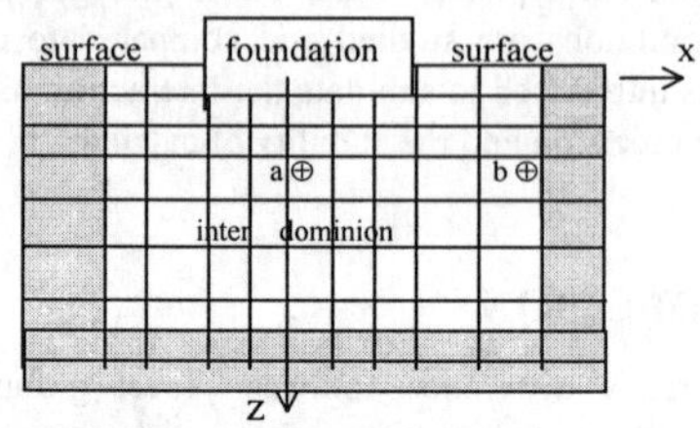

□ Artificial calculate boundary

Figure1. Discrete calculated model of subsoil

3 ARTIFICIAL CALCULATED BOUNDARY CONDITIONS,SIMULATION OF MOISTURE AND TEMPERATURE IN SUBSOIL

3.1 Selection of calculate model and calculate method

It is difficult to describe the movement status of moisture and temperature in subsoil, it is a multi-dimensions question of transfer heat and moisture in unfreedom field.

The numerical resolution of multi-dimension of moisture and temperature of freedom field are founded by numerical simulation, those methods including the method of separate variable, the method of integral transform, the method of finite difference and the method of flab variable, In this paper, the explicit finite difference method is used, this method change differential coefficient equations of temperature or moisture into difference equations, including the equations of boundary. and the girding of difference method is showed in Figure.1,the size of model can be decided by the need of analysis of questions, the model is separated into inter-dominion and boundary dominion, to a difference equations of inter-dominion and boundary dominion are listed as equations(7)~(10):

(A) the difference equations of inter-dominion of the temperature transfer equation of (2) and the moisture variation equation(5)are those:

$$\frac{T_{i,j}^{m+1}-T_{i,j}^{m}}{\tau}=\frac{\lambda}{\varsigma\rho\Delta h^2}\left(T_{i+1,j}^{m}-4T_{i,j}^{m}+T_{i-1,j}^{m}+T_{i,j+1}^{m}+T_{i,j-1}^{m}\right) \tag{7}$$

$$\frac{C_{i,j}^{m+1}-C_{i,j}^{m}}{\tau}=\frac{D_A^2}{\Delta h^2}\left(C_{i+1,j}^{m}-4C_{i,j}^{m}+C_{i-1,j}^{m}+C_{i,j+1}^{m}+C_{i,j-1}^{m}\right) \quad (8)$$

(B) the difference equations of boundary dominion of a question of temperature and moisture and the start condition are listed as equations(9)~(10)

$$\begin{cases} T(x,z,0)=T_0-\frac{T_0-T_3}{H}z & 0<z<H \\ T(x,0,t)=T_2 & foundation \\ T(x,0,t)=T_0+T_1\sin(\omega t), free-surface, t>0 \end{cases} \quad (9)$$

$$\begin{cases} C(x,z,0)=0.0 & 0<z<H \\ C(x,0,t)=C_2 & foundation, t>0 \\ C(x,0,t)=C_0+C_1\sin(\omega t), free-surface, t>0 \end{cases} \quad (10)$$

3.2 Artificial calculated boundary conditions of moisture and temperature of subsoil of building

1, A describe of boundary conditions based on the questions of heat transfer:

The boundary conditions of temperature field fitting for level freedom field are those in analytic: (1) the datum of temperature have been given in any time of boundary. this is to say that the distribution of temperature of the surface with time is known: $T_s=f(x,\tau)$,(2)the density of heat transfer upright direction, the surface function of time in each point are given $q_s(\tau)$ is the density of heat transfer of upright direction $q_s(\tau)=\lambda(\frac{\partial T}{\partial n})$,(3)any palace of boundary, according to the law of heat transfer and Newton cooled law, the density of energy transfer form the surface to the around medium each time is $q_s(\tau)=h(T_s(\tau)-T_A(\tau))$, (4) the equations of heat exchange.

2.Artificial discrete boundary conditions of temperature of a discrete model of subsoil of building

There is period-rule in heat moving in soil, There are three distributed strips where temperature change period, day-change-temperature layer, year-changed-temperature layer, heat move like wave in these strips ,so, when the surface is acted by a period wave temperature field ,the case of temperature transfer in subsoil is like the status of wave propagation with damp. To the question of calculating the unlimited dominion ,in general ,an plane questions of half uniformity infinite dominion is replaced by the calculated result of limited dominion ,to analysis a complex question of the temperature field and moister field in subsoil of building ,the range of numerical analysis and size of calculate model are limited, added the affection of upper structure ,the question of unlimited dominion must be simplified into limited dominion's ,importing a artifact boundary conditions to simulate interaction of the extra dominion medium to inter dominion medium .so the status of energy and water moving through the artifact boundary in limited dominion like in unlimited dominion , let the water and energy to diffuse or transfer to forever.

The method of different calculation of temperature field and moisture field are to discrete the variable of temperature and moisture in time dominion and geometry place dominion step by step with time ,there are some artificial boundary conditions of local region to the questions of wave propagation in time dominion: Clayton-Enquist boundary conditions, Highdon boundary conditions, Higdon –Keys boundary conditions, Transmission boundary conditions and absorb boundary conditions[5,6](a stickiness boundary conditions), there are no research harvest and experiments on dealing with the temperature field and moisture field ,in this paper a discrete artificial calculated boundary conditions is introduced and compared with other boundary conditions. These boundary conditions are briefly described as following: 1) using one-dimension analytic solution of free-field to modeling the boundary conditions, 2) Author suggested a discrete artificial calculated boundary conditions(a sort of boundary conditions including one rank and two rank precision boundary conditions),three methods are introduced and analyzed as follow.3) immobility boundary conditions.

(1) As the temperature changes only along the depth and have nothing to do with the level orientation, so the temperature field of forever place can be replaced by the analytic result of one dimension question, if the size of model is big enough to deal with the boundary as one dimension heat transfer question: :

: $$\frac{\partial T}{\partial t}=\frac{\lambda}{\varsigma\rho}\frac{\partial^2 T}{\partial z^2}$$

the number of temperature of boundary can be written as the simplified expression of analytic equation[4]:

$$T(z,t)=T_{om}(t)\exp\left(-\sqrt{\frac{n\varsigma\rho\pi}{\lambda T_0}}z\right)\cos\left(\frac{2n\pi}{T_0}-\sqrt{\frac{n\varsigma\rho\pi}{\lambda T_0}}z\right) \quad (11)$$

$T_{om}(t)$: and $T_0$are the top and period of temperature on the surface ,n=1or2,other parameters are same in equation (1).

(2) The discrete artificial calculated boundary conditions are briefly those[6]:

T(x,z,t) is a point of artificial calculated boundary conditions of temperature field, x is the level direction ,z is the upright orientation, then:

$$T(x,z,t)=\sum_{j=1}^{N}(-1)^{j+1}c_j^N u(x-jW_A\Delta t,z,t-(j-1)\Delta t) \quad (12.1)$$

(the equation of moisture field is:

$$C(x,z,t)=\sum_{j=1}^{N}(-1)^{j+1}c_j^N c(x-jQ_A\Delta t,z,t-(j-1)\Delta t) \quad (12.2)$$

W_A and Q_A are the artificial confirm parameters(like the artificial wave velocity of wave propagation),N is positive Integer, N>1, c_j^N is a binomial coefficient,

One rank precision boundary conditions for temperature field is: $T(x,z,t+\Delta t)=T(x-W_A\Delta t,z,t)$ (13.1)

One rank precision boundary conditions for moisture field is:

$$C(x,z,t+\Delta t)=C(x-Q_A\Delta t,z,t) \quad (13.2)$$

two rank precision boundary conditions for temperature field is: $T(x,z,t+\Delta t)=2T(x-W_A\Delta t,z,t)-T(x-2W_A\Delta t,z,t-\Delta t)$ (14.1)

two rank precision boundary conditions for moisture field is:

$$C(x,z,t+\Delta t)=2C(x-Q_A\Delta t,z,t)-C(x-2Q_A\Delta t,z,t-\Delta t) \quad (14.2)$$

(3) Immobility boundary conditions

If the capability of computer is very large and the feature of explicit calculated method, the calculated size can be very big, the temperature and moisture in boundary in this status will be regarded to being held, suppose a point(x, z, t)of boundary ,so the boundary conditions are:

T(x,z,t)=constant(C(x,z,t)=constant) (15)

3.3 Stability of calculation of interior joints and boundary

It is a very complex case that the artificial boundary conditions will break the stability of calculation, in order to reduced the affections form artificial boundary conditions to interior joints and improve the precision of calculation, avoid the losing-stability, Stickiness damp are added in the boundary in discrete model, in the same time a physical smooth technology is used in boundary. The conditions of stability of the interior joints difference format of equation(2) is: $\lambda\Delta\tau/\varsigma\rho\Delta h^2 \le 1/4$, Δt is the time step distance Δh is the step distance of place, ρ is the density of soil.

4 CALCULATED EXAMPLE AND ANALYSIS OF THE RESULT

4.1 Some base parameters and start conditions

The subsoil is sand ,the width of foundation is 20 m, there is a source of heat on earth's surface: $T_0 + T_1\sin(\omega t)$,the temperature of the room keep 20°C, the level of water underground is 5m ,the average porosity of soil under foundation 5m is 0.55,the contain weight of soil is 1600kg/m^3, the average porosity of soil under foundation 5-15m is 0.45,the contain weight of soil is 1800kg/m^3(seen in figure 2), of different layers can be calculated by the

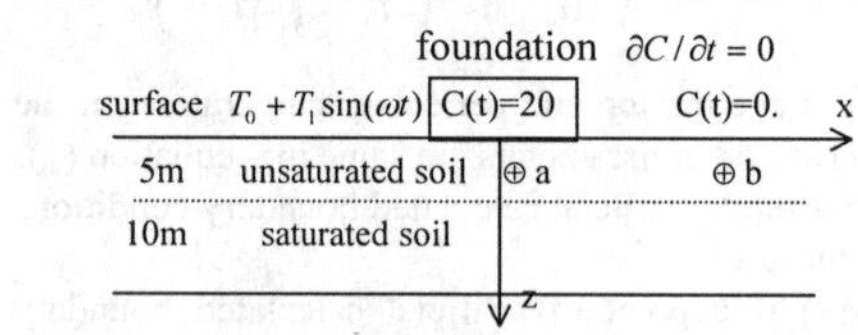

Figure 2. The diagram of subsoil

equation (3)and (4),the foundation has been designed to prevent water transfer into, the moisture of earth-surface is C(t)=0.0,the stripe is C(t)=1.0,the length of model is 200m,the depth of model is 100m, Δτ=0.05hour, Δh=0.2m, $\omega = \pi/12$. the calculated result are showed in Figure 3 and Figure 4.Discrete boundary conditions is used.

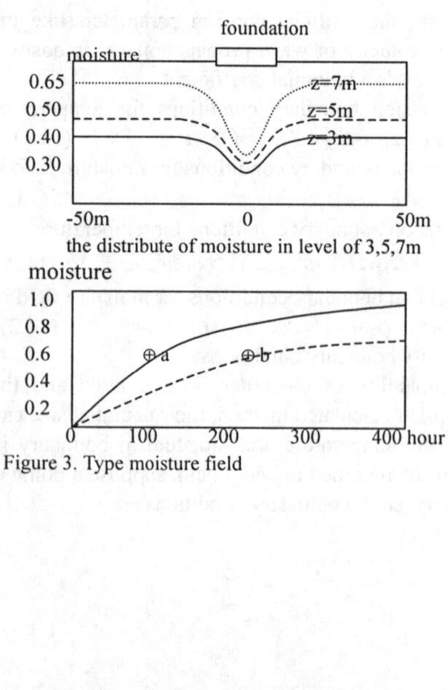

Figure 3. Type moisture field

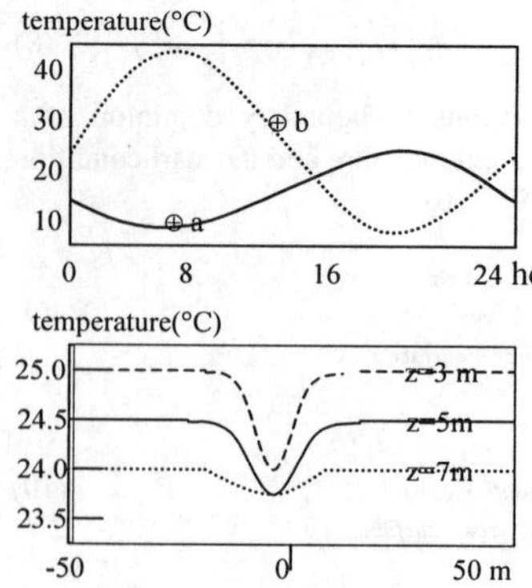

Figure.4 Type temperature field

4.2 Calculate example and a analysis of result of three boundary conditions

The comparisons among three sorts of boundary conditions, are put into the calculation of temperature. In order to simplify analysis method. the structure is symmetrical ,the coordinate space position of some points, which have been compared, are unchanged in spite of the size of model ,one half model is being studied ,we select a large sized of model (800m length,100m depth),the compute result of three boundary conditions in very large sized model are almost like, so the results are selected as a gist of comparison. form the result of calculation we can find some question:(1) if the size of model is big ,the compute result of three methods are relatively consistent in interior joint.(2)Decrease the size of model, the result of discrete artificial boundary conditions is consistent to the reference, there is a little small, it flatly explain that the energy can be transfer red completely in the boundary. the result of under foundation are relatively consistent by the an analytical result of one dimension or by immobility boundary conditions, but beyond the foundation, the result are some difference or mistake, the calculated mistake become big when the size of model decreased, the results of boundary are instability and illogicality. the results become very big quickly in the boundary using an analytical solution of one dimension question or using immobility boundary conditions, and the results exhale, maybe it is caused by a great accumulate of energy in the boundary. seen in figure.6. (3) the discrete boundary conditions is reasonable compared with front two kind of boundary conditions, the results in boundary jump clearly than that of inter-dominion using other two kinds of boundary conditions, there is no clearly jump of conclusion in boundary using discrete artificial boundary conditions, the curve is very smoothly, as shown in figure.5 .

5 CONCLUSION

Through computation and analysis above, some useful results maybe be suggested as:

(1) There is a good applicant effect of discrete artificial boundary conditions in finite dominion to simulate the

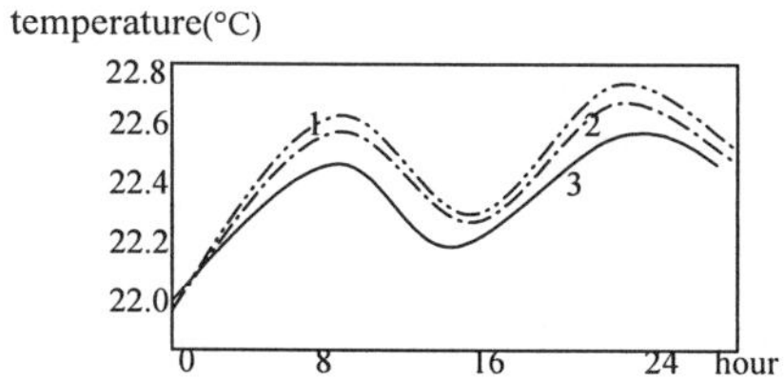

Chart of Temperature time in the across point of 4 m level under ground under foundation

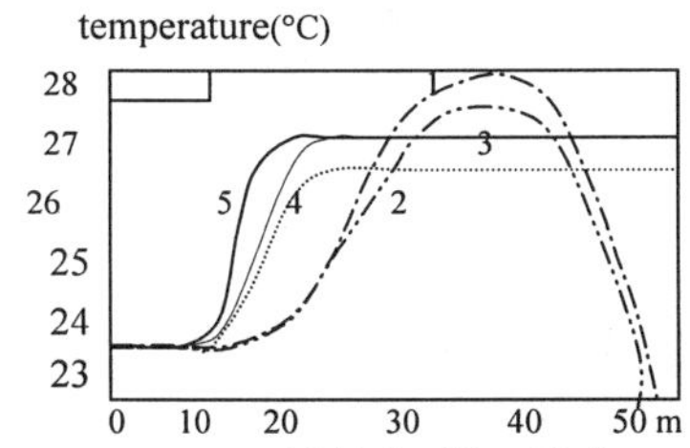

model A(size:100m×30m)

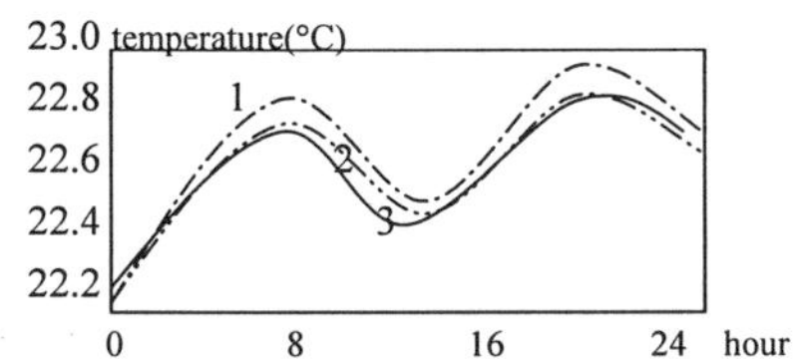

chart of temperature-time at the across point of 4m level under ground under foundation

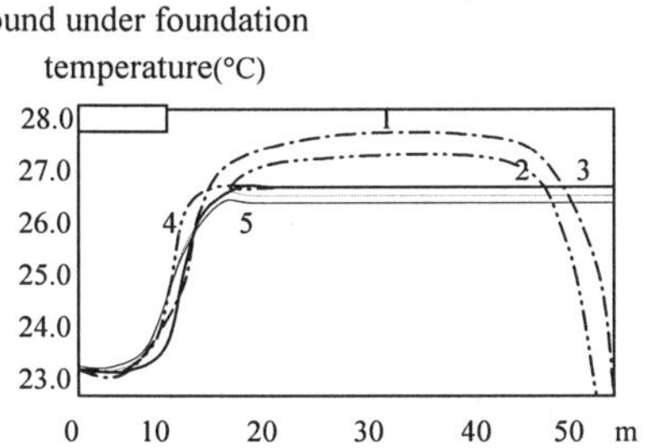

model B (size:200m×40m)

(0 stand for the symmetry line of foundation)

1: immobility boundary conditions, 2:an analytical solution of one dimension 3: reference, 4: one rank discrete boundary 5: two rank discrete boundary

Figure.5 Compared analysis chart of three sort of boundary conditions in two sized model,

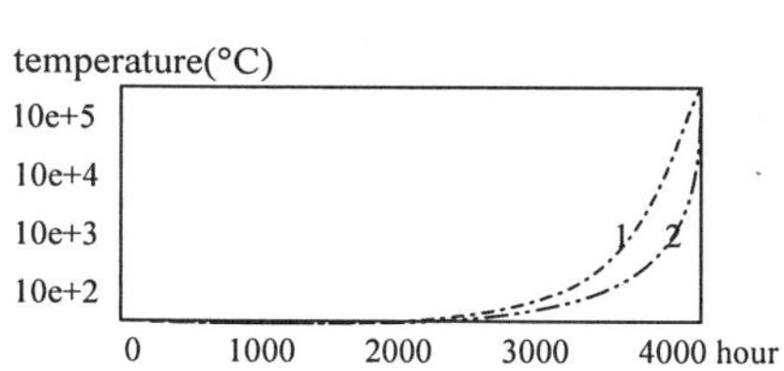

1: immobility boundary conditions,2:an analytical solution of one dimension

Figure 6. Result exhale in boundary using first and third method

temperature field and moisture field in half infinite dominion foundation under the action of period change temperature field, the method can improve the calculated precision of points of interior dominion, this conditions is superiority than other sort of boundary conditions, and there are a good stability of calculation using the discrete boundary conditions.(2) The affections and interactions of three phased medium of soil (of water, porosity and gas) have been taken into consideration dealing with heat transfer, the across-relation of temperature and moisture is untied when the temperature field being calculated, the question is simplified. (3) The affection of temperature to moisture has been taken into consideration, it is need to be researched.. (4) The upper buildings have some affections on temperature and moisture transfer, it is different from freedom field, there are some questions to be done. we must pay more attention to. (5) The volume of subsoil has been researched leaded by the change of temperature and moisture. (6) Some computed methods need to be founded to simulate the across questions of moisture field and temperature field in three dominions. just like equation (1).

ACKNOWLEDGEMENT:

This research have been supported by the fund of science of china

REFERENCES

Fredlund , D.G.et al. 1997. *Mechanics of unsaturated soil* , (translated Chinese).*Beijing.*

Mu-zeng,Lin, 1991.*The heat physical characters of rock an -d its application,* Chongqin , 8~20

Fu-lu,Men,Wen-hua,Chen et al,1998.Transfer and disaster of water of subsoil, *J. Natural disaster(in Chinese)*, 17:117-121

Aketer. ER.G.1983. *Analysis of transfer medium and heat* (translated Chinese). :Beijing.

Xiu-ming,Shao, 1995. The absorb boundary conditions of equations of wave in anisotropism medium, *Geophy - sics*,(in Chinese) 56~78

Wen-hua,Chen. 2000. Question of boundary to calculate variation of heat and water in subsoil (in Chinese) J.Geo.Eng., 22

Computer Methods and Advances in Geomechanics, Desai et al. (eds) © 2001 Balkema, Rotterdam, ISBN 90 5809 183 X

Massively parallel direct simulation of multiphase flow

B.K.Cook
Department of Civil and Environmental Engineering, Massachusetts Institute of Technology, Cambridge, Mass. & Geomechanics Department, Sandia National Laboratories, Albuquerque, N.Mex., USA

D.S.Preece
Geomechanics Department, Sandia National Laboratories, Albuquerque, N.Mex., USA

J.R.Williams
Department of Civil and Environmental Engineering, Massachusetts Institute of Technology, Cambridge, Mass., USA

ABSTRACT: Our understanding of multiphase physics and the associated predictive capability for multiphase systems are severely limited by current continuum modeling methods and experimental approaches. This research will deliver an unprecedented modeling capability to directly simulate three-dimensional multiphase systems at the particle-scale. The model solves the fully coupled equations of motion governing the fluid phase and the individual particles comprising the solid phase using a newly discovered, highly efficient coupled numerical method based on the discrete-element method and the Lattice-Boltzmann method. A massively parallel implementation will enable the solution of large, physically realistic systems.

1 INTRODUCTION

Our understanding of multiphase physics and the associated predictive capability for multiphase systems are severely limited by current continuum modeling methods and experimental approaches. For example, continuum models overlook the microscale solid-fluid interactions from which macroscopic system properties emerge, while experimental inquiries have been thwarted by the fine scale and complexity of many-body systems.

An alternative modeling approach is to simulate multiphase systems at the grain-scale, fully resolving the interaction of individual solid particles with other solid particles and the surrounding fluid. Until recently, the direct simulation of these systems has proven computationally intractable.

In this research, the fully coupled equations of motion governing the fluid phase and the individual particles comprising the solid phase are solved using a newly discovered, highly efficient coupled numerical method based on the discrete-element method and the Lattice-Boltzmann method. Forcing mechanisms represented (to at least the first order) in the model include fluid-induced stresses on particle surfaces, intergranular bonding between particles, intergranular forces from particle collisions and static formation stresses, and gravity and buoyancy forces. Laminar flow through a particle assemblage is fully resolved, including such detailed structure as vortex formation in the particle wake. Because the coupled method accurately models fluid flow in the close vicinity of the solid particles, it accurately reproduces the local fluid-induced stresses that drive solid particle motion.

2 DISCRETE-ELEMENT METHOD

The discrete-element method is a suite of numerical algorithms that allows for the efficient modeling of granular materials as a collection of arbitrarily shaped, rigid bodies that interact through surface collisions and bonds. Each particle in a granular system is modeled as a geometrically simplified object using either a functional (e.g. sphere) or discrete representation. Particles are assumed to be rigid; however, a soft contact model based on the assumption of local deformation during contact is used with normal and shear contact forces modeled by a spring-dashpot system, and shear force magnitude limited by friction (Rege 1996). Cement bonding between particles is modeled with numerical links that are effectively springs with a user-specified stiffness and tensile strength. During each model iteration, potential contacts between objects are found using efficient spatial sorting algorithms (Perkins 1999). Objects in contact are identified and the contact forces resolved. Particle position is then updated according to Newton's second law, the conservation of linear and angular momentum, with particle forcing equal to the sum of the contact forces, the dynamic fluid force, and the buoyancy (body) force. More detailed discussions of the development and application of discrete-element models can be found in Jensen et al. (1999) and Preece et al. (1999).

3 LATTICE-BOLTZMAN METHOD

Lattice-Boltzmann is a naturally parallel numerical method for simulating a Newtonian fluid. The method has been employed to analyze the detailed behavior of gases and liquids for a number of years. Although this method is founded on the kinetic theory of dilute gases, it has been shown to accurately reproduce the Navier-Stokes equations (Chen & Doolen 1998), which govern fluid flow in the geomechanics problems of interest in this study. The fluid is modeled as packets of particles that move about a regular lattice, or grid, defined with the appropriate boundary conditions. Collision and redistribution (or streaming) of fluid packets occur at the lattice sites according to specific relationships that conserve mass and momentum. Most of the computational cost for each time-step update is associated with the collision phase, which is local in nature (in that it requires no additional information from neighboring nodes). The only data that is exchanged between lattice nodes occurs during the streaming of the distributions. Herein lies one of the primary advantages of the Lattice-Boltzmann method over other traditional methods in computational fluid dynamics, such as the finite-element method.

Coupling with discrete elements is accomplished by using a modified collision operator at the lattice sites covered by the elements (Noble & Torczynski 1998). With this scheme, the lattice-Boltzmann collision operator is modified to include a dependency on lattice cell porosity, allowing for the sub-grid resolution of solid particle boundaries. Momentum is transferred between the fluid and the discrete element during the collision-tracking step at all lattice sites covered by the solid particle. The coupled method accurately models fluid flow in the close vicinity of the solid particles and reproduces the local fluid-induced stresses that drive solid particle motion. A detailed development and validation of the coupled method can be found in Cook et al. (2000). The two-dimensional applications reported by Cook reproduced such complex phenomena as drafting-kissing-tumbling (Joseph et al. 1987) in multi-particle sedimentation simulations.

4 DISCUSSION

An innovative coupled numerical model using the Lattice-Boltzmann method for fluid flow and the discrete-element method for particle motion has been developed and validated in our earlier work (Cook et al. 2000). Because the geomechanical problems of interest are indisputably three-dimensional in nature, current work is focused on the extension of the model to three dimensions. Despite the relatively high numerical efficiency of the Lattice-Boltzmann and discrete-element methods, the numerical analysis of these large, three-dimensional multiphase systems will be extremely expensive computationally. To make these problems tractable, the natural parallelism of the Lattice-Boltzmann method will be exploited to develop a massively parallel implementation of the model. Extremely efficient massively parallel implementations of the Lattice-Boltzman method have been achieved because the method is primarily local in space and explicit in time (O'Connor & Fredrich 1999).

A massively parallel implementation will address a broad range of important but currently intractable geomechanical and industrial challenges. Fundamental insights are expected into the disaggregation, transport, and clustering processes underlying such outstanding problems as sand production, proppant design, sedimenting and fluidized suspensions, and lubricated transport of viscous materials.

REFERENCES

Chen, S. & G. Doolen 1998. Lattice Boltzmann method for fluid-flows. *Annual Review of Fluid Mechanics* 30: 329-364.

Cook, B. et al. 2000. Direct Simulation of Particle-laden Fluids. *Proceedings of the Fourth North American Rock Mechanics Symposium*. Balkema, Rotterdam.

Jensen, R. et al. 1999. Coupled fluid-flow/particle-motion computer modeling of sand production in oil wells, SPE No. 56630. Society of Petroleum Engineers Annual Conference, Houston, TX.

Joseph, D. et al. 1987. Nonlinear mechanics of fluidization of beds of spheres, cylinders, and discs in water. In G. Papanicolau (ed.), *Advances in Multiphase Flow and Related Problems*, SIAM.

Noble, D. & J. Torczynski 1998. A Lattice Boltzmann method for partially saturated computational cells. *International Journal of Modern Physics C* 9(8): 189-201.

O'Connor, R. & J. Fredrich 1999. Microscale flow modeling in geologic materials. *Phys Chem. Earth (A)* 24: 611-616.

Perkins, E. 1999. *Spatial reasoning for general n-body physics – discrete element algorithms* (Masters thesis). Massachusetts Institute of Technology: Cambridge, MA.

Preece, D. et al. 1999. Sand production modeling using superquadric discrete elements and coupling of fluid flow and particle motion. 37th US Rock Mechanics Symposium (Vail Rocks), Vail, CO.

Rege, N. 1996. *Computational Modeling of Granular Materials* (Ph.D. thesis). Massachusetts Institute of Technology: Cambridge, MA.

Computer Methods and Advances in Geomechanics, Desai et al. (eds) © 2001 Balkema, Rotterdam, ISBN 90 5809 183 X

A three-dimensional finite element program capable of calculating reinforcement response in soil nailed excavations

A.L.de Ambrosis & J.C.Small
Department of Engineering, University of Sydney, N.S.W., Australia

ABSTRACT: The main purpose of shotcrete facing in soil nailed excavations is generally held to be the control of local failure in the soil between the nails. This relatively non-essential role has meant that facing design has remained reasonably rudimentary. In this paper shotcrete facings in soil nailed excavations have been analysed using a purpose built three-dimensional finite element program. A full-scale experimental soil nailed wall, constructed as part of the French National Research Project Clouterre, has been modelled using the program, and comparisons between calculated and observed behaviour are presented and discussed. The parameters used in the simulations are justified using published data. Generally good agreement is found between calculated and published displacements, nail forces, stresses and failure mechanisms.

1 INTRODUCTION

Ease of construction, flexibility of design and economic competitiveness have meant that soil nailed excavations have become increasingly accepted as a feasible means of excavation support. However, despite there being a large pool of published data describing actual walls constructed and at least three widely accepted design codes, (Japanese design code (JHPC 1987), recommendations of Clouterre (FHWA 1993), British standards (BS 8006:1995)), there still remains much to be learned regarding working stresses within the reinforcement system. In particular, the earth pressures experienced by the shotcrete facing are largely unquantified.

Facing design for soil nailed excavations is presently based upon a number of simplifications of soil behaviour, or empirical rules of thumb. The existing design methods utilise three main techniques:

- Assuming the earth pressures at the face are equal to a percentage of the Coulomb's active earth pressure.
- Assuming that the effect of overburden is negated by the action of the soil nails and designing the facing to retain an active soil wedge forming between the nails.
- Assuming the maximum nail tension forces are equal to the expected total load at the face.

Little is understood about which parameters have the greatest effect on facing loadings and moments.

This paper presents a three-dimensional (3D) finite element technique capable of directly calculating the reinforcement response in a soil nailed excavation. Results for the moments per unit length that are generated in the shotcrete facing are presented, and the progam's ability to accurately model deflections, nail forces and failure modes experienced by a soil nailed excavation will be verified through comparisons of observed and calculated responses.

2 PROGRAM DESCRIPTION

For the analysis, a 3D finite element program, capable of simulating construction of soil nailed excavations was developed. A 3D simulation was considered necessary, so that the nails could be shown as bars, (a 2D simulation represents the reinforcements as continuous flat sheets). When considering the moments induced in the facing, it was important that the 'point' restraint provided by the soil nails be reflected in the simulation.

Three types of element are used by the program; a twenty noded elasto-plastic 'soil' element, an eight noded isoparametric Mindlin Shell 'shotcrete' element and a two noded beam 'nail' element. The reinforcement elements share nodes with the main soil elements. As such, slip between the soil and reinforcing elements is not modelled. Inclusion of separate soil nail and shotcrete elements has meant that induced moments and forces in the support elements can be directly calculated.

Soil behaviour is characterised by assuming an elasto-plastic (elastic - perfectly plastic) response. The Sloan and Booker (1986), modified Mohr-Coulomb yield surface is used to describe the failure

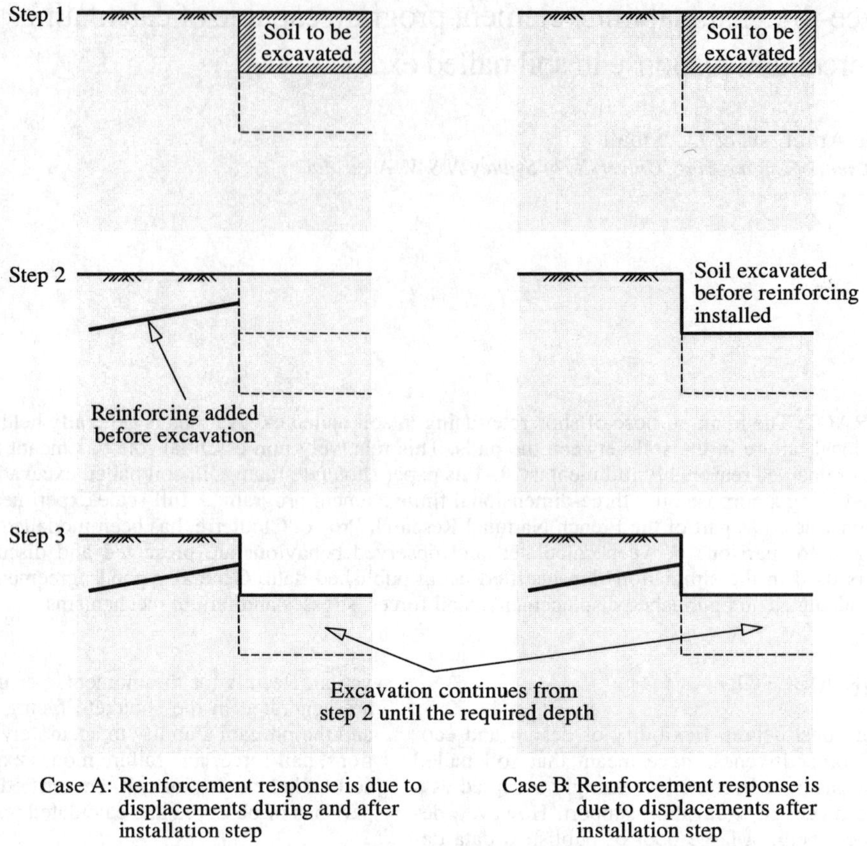

Figure 1. Excavation simulation

criterion, and a Von Mises' plastic potential surface is used to define strain orientation at failure. Use of Von Mises' plastic potential means that soil dilation is not considered.

Excavation is simulated in steps, using Brown and Booker's (1985) virtual work solution to calculate the induced forces. A skyline solver has been incorporated, in order to reduce the memory required to run the program.

3 EXCAVATION SIMULATION

For simulation of excavation in elasto-plastic soils, it is essential that excavation be carried out in steps that reflect the actual excavation process. There are two reasons why this is necessary in soil nailed systems.

- During a stepped excavation the soil will experience stress states, which are different to its final stress state. As such, the simulated excavation process needs to properly reflect the actual excavation process, so that any plastic deformation and resulting permanent strain experienced in intermediate construction stages is incorporated in the total excavation response.
- Without a stepped simulation process it would not be possible to properly model the influence of the incremental construction process. The layered nature of soil nailed wall construction, means that each layer of reinforcement has a unique zero stress position. This phenomenon needs to be properly mirrored in any simulation for meaningful results.

The incremental nature of a soil nailed wall's construction, has some bearing on the applicability of any simulation of a soil nailed system. In reality, the shotcrete and nail grouting are placed wet onto the deforming excavation. All stresses within the reinforcement are induced by displacements after the reinforcements have hardened. In the simulation, the time dependent nature of the displacements is not

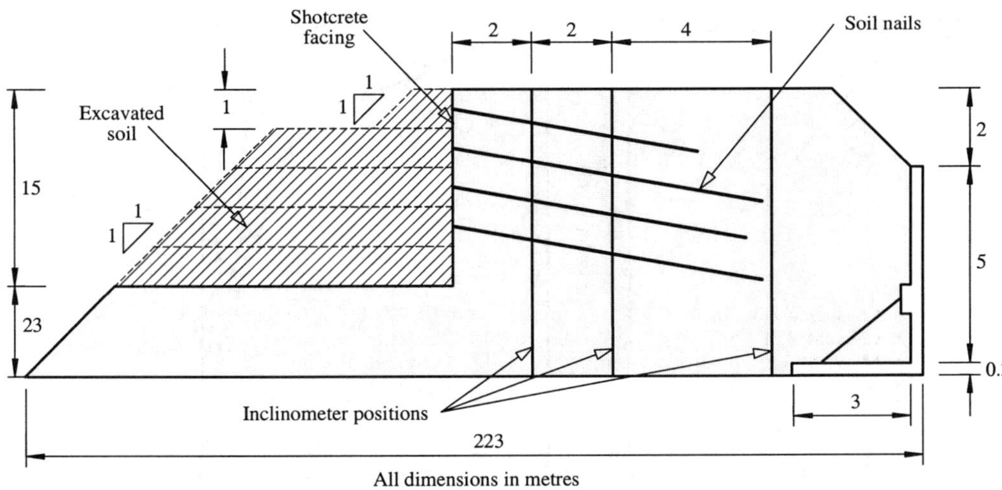

Figure 2. Embankment dimensions.

modelled. Either, the reinforcements are placed before the current excavation step is completed or alternatively, they are placed after all the displacements associated with the current step have occurred.

Figure 1 illustrates the two simulations for one excavation step. These two cases represent the limits of the actual behaviour. For case A, the stresses in the reinforcements are induced by displacements associated with the current step as well as displacements induced by following steps. For case B, the displacements due to the current step are said to have occurred before the reinforcing is placed. As such, the stresses in the reinforcement are only due to displacements induced by following steps. The actual behaviour of an excavation most likely sits between these two limits.

For this simulation, case B is taken to represent the excavation's behaviour, because it is believed that this case more closely represents reality.

4 FULL SCALE EXPERIMENTAL WALL

Plumelle et al. (1990), Schlosser et al. (1992) and Unterreiner et al. (1997) have presented the results of a full scale experimental soil nailed wall constructed as part of the French National Research Project CLOUTERRE. Figure 2 shows a cross section of an embankment built for the project. The soil nail wall was constructed by cutting into the embankment material.

Table 1. Soil parameters used

	Reported	Simulation	Method of determination
φ	36°-40°	38°	Triaxial tests
c	3-4 (kPa)	3 (kPa)	Triaxial tests, water content
K_o	0.38	0.38	$K_o = 1\text{-}\sin\varphi$
γ	16.1 (kN/m^3)	16.1 (kN/m^3)	Avg. unit weight

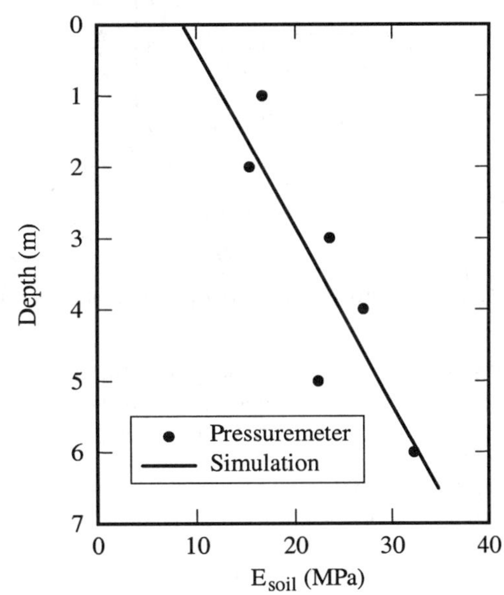

Figure 3. Measured and simulated soil modulus

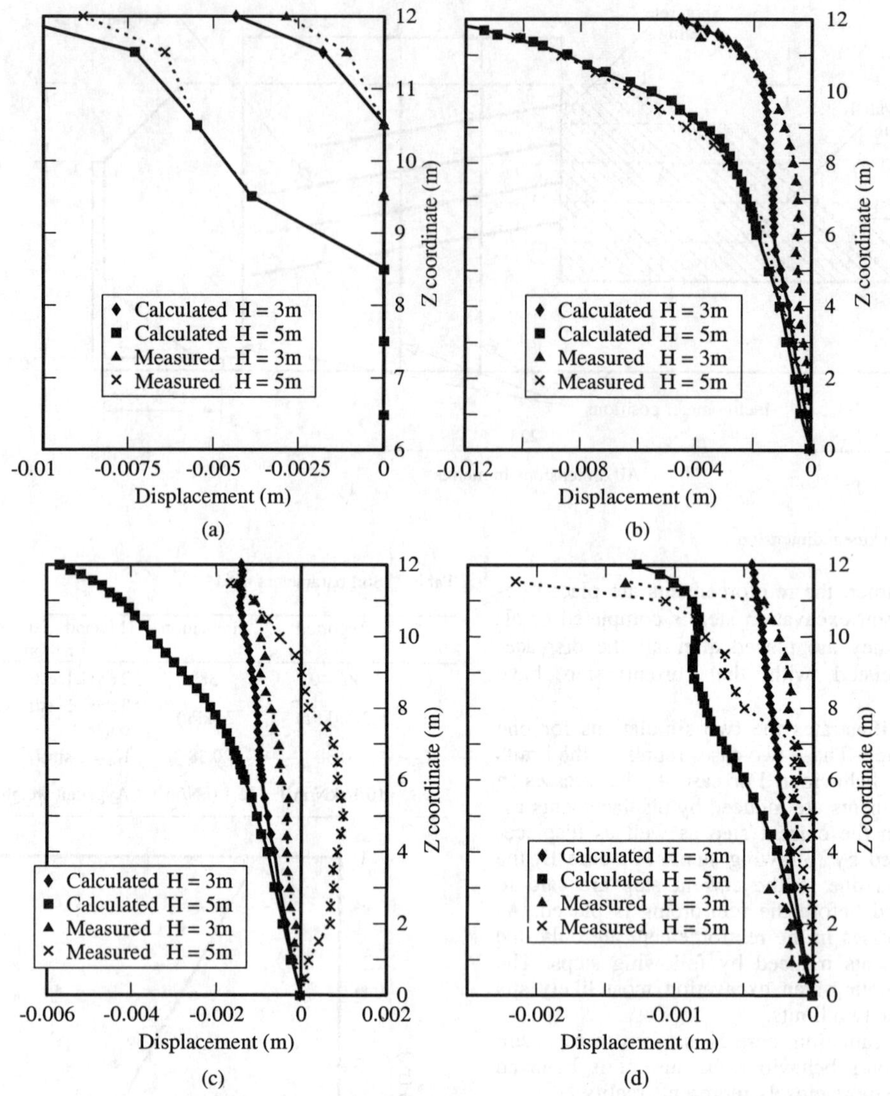

Figure 4. Lateral movement of inclinometers (a) at facing (b) 2m behind facing (c) 4m behind facing (d) 8m behind facing.

Construction was incremental, with 1m high layers of the wall being built at a time. Nails were inclined at 10° to the horizontal, with nail lengths ranging from 6 to 8m. The facing was constructed of 100mm thick mesh reinforced shotcrete. The homogeneity and density of the backfill used to construct the embankment was tightly controlled. Excavation induced displacements were measured using inclinometers, positioned as shown in Figure 2.

Extensive testing was used to determine soil strength parameters. A Menard pressuremeter was used to calculate soil modulus, and triaxial testing was used to determine the angle of shearing resistance ϕ of the soil.

4.1 *Simulation of experimental wall*

The large amount of testing conducted as part of the Clouterre project, and the controlled nature of the embankment fill in which the soil nailed wall was built, meant that a high degree of confidence could be placed in the soil parameters reported by the Clouterre project. As such these parameters are used wherever possible in the simulation. Table 2

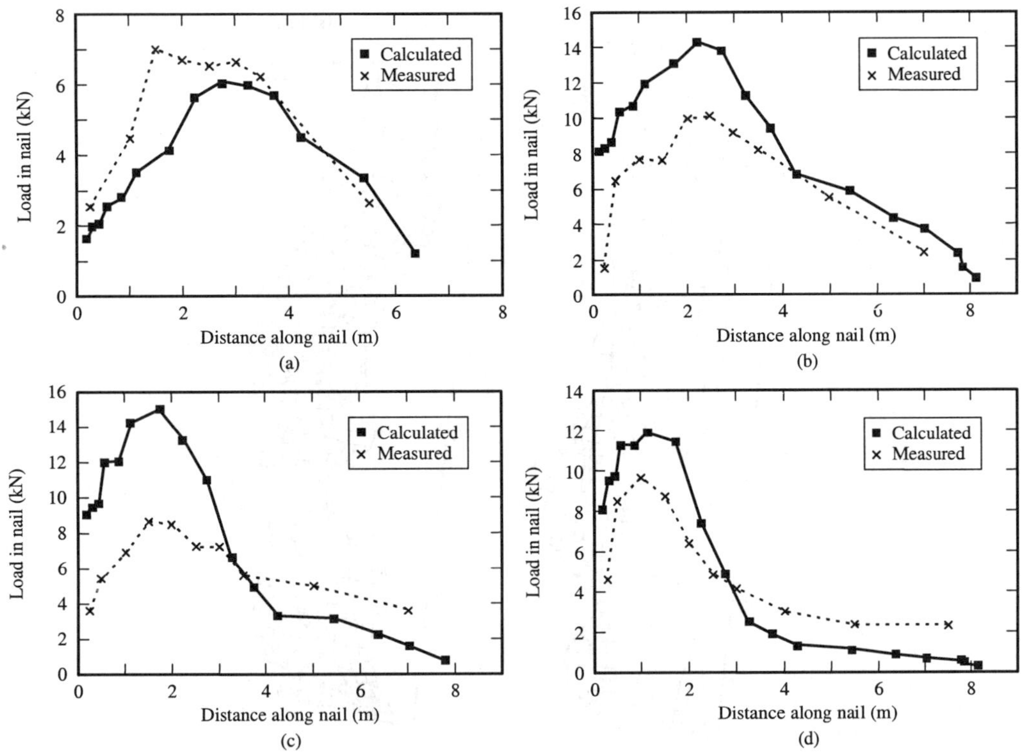

Figure 5. Forces in soil nails (a) top nail (b) second from top (c) third from top (d) bottom nail.

shows comparisons of reported and adopted soil parameters. Figure 3 shows the measured and selected Young's Modulus variation with depth. Initial stresses were considered to be geostatic, with K_o calculated using the relationship shown in Table 1.

4.2 *Comparison of results*

Figures 4(a)-(d) show a comparison of calculated and observed horizontal displacements. Results for two excavation stages (wall heights, 3m and 5m) are shown in each figure. Figures 4(a) and 4(b) (displacements for the facing and 2m behind the facing) show excellent agreement. In figures 4(c) and 4(d), the agreement of calculated and observed readings is not as good, particularly the observed displacements for the 5m deep excavation shown in Fig 4(c) where the measured values show a backward movement of the inclinometer.

Figures 5(a)-(b) show comparisons of observed and calculated nail forces. Only fair agreement is found for the size and distribution of nail loads in the second and third nails from the top, but for the other nails, the agreement is better.

There is also good agreement between the positions of observed and calculated nail force maximums, showing that the program is properly simulating the observed failure mechanism.

4.3 *Moments in facing*

Moments may also be computed in the shotcrete facing that is represented by 8 noded isoparametric shell elements. The computed moments per unit length are presented in Figure 6 which shows moments per unit length in the horizontal direction M_{xx} and in the vertical direction M_{yy}.

It can be seen from the plots that the moments peak at the locations of the bolts that are located at either edge or at the centre of the strip of facing material analysed.

The moments in the vertical direction M_{yy} can be seen to be greatest at the third row of bolts down from the top of the excavation. The smaller diameter bolts at the top of the excavation do not induce much moment into the facing because they are less stiff and because the soil pressures on the facing are smaller at shallow depth.

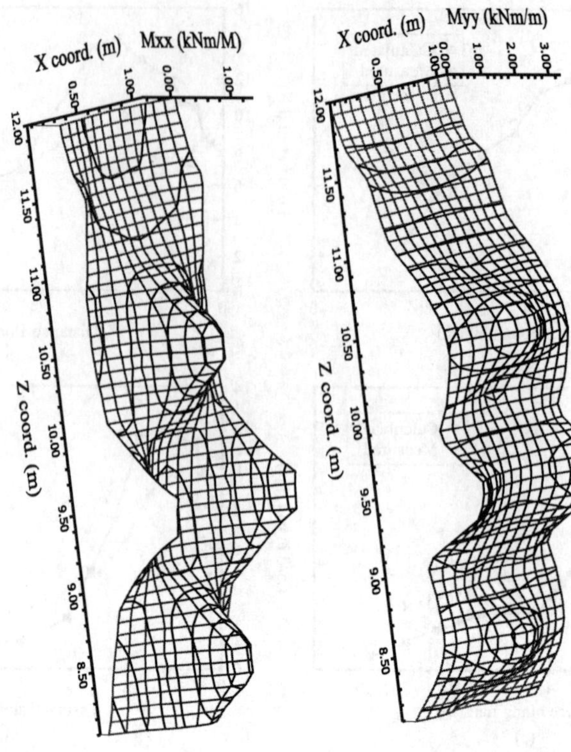

Figure 6. Moments in shotcrete facing.

5 CONCLUSIONS

Comparisons between observed and calculated displacements, nail loads and failure mechanisms show that soil nailed excavations can be simulated using the 3D finite element program that has been presented. This leads to confidence in predicting the moments that are likely to occur in the shotcrete facing of a soil nailed excavation.

REFERENCES

British Standards Institution 1995. *Code of practice for strengthened/reinforced soils and other fills*. HMSO, London, BS 8006, 156.

Brown P.T. & Booker J.R. 1985. Finite Element Analysis of excavation. *Computers and Goetechnics*, 1: 207-220.

Federal Highways Administration (FHWA) 1993.*Recommendations Clouterre 1991* (English translation). Report FHWA-SU-026, Washigton DC, 302.

Japan Highway Public Corporation 1987. *Guide for design and construction on reinforced slope with steel bars*. Tokyo.

Plumelle, C., Schlosser F. Delage P. Knochenmus G. 1990, French National Research Project on soil nailing: Clouterre. *Design and Performance of earth retaining structures, Proc. ASCE Conf.*, 660-675, Ithaca, New York.

Schlosser, F., Plumelle, C., Unterreiner P. & Benoit, J. 1992. Failure of a full scale experimental soil nailed wall by reducing the nail lengths (French research project CLOUTERRE). *Earth reinforcement practice*, Ochiai, Balkema, 531-535.

Sloan S.W. & Booker J.R. 1986 Removal of singularities in Tresca and Mohr-Coulomb yield functions, *Communications in applied numerical methods*, 2: 173-179.

Unterreiner P., Benhamida, B. & Schlosser, F. 1997. Finite element modelling of the construction of a full scale experimental soil-nailed wall. French National Research Project, CLOUTERRE. *Ground Improvement*, 1: 1-8.

Computer Methods and Advances in Geomechanics, Desai et al. (eds) © 2001 Balkema, Rotterdam, ISBN 90 5809 183 X

A three phase model for contaminated groundwater disinfection

A. El Nimr & A. Dif
Mansoura University, Egypt

S. Mansour
National Water Research Center, Cairo, Egypt

ABSTRACT: A three phase optimization mathematical model is developed to study the controlled injection of disinfecting compounds in the soil. The model treats the groundwater flow, the transport of both the contaminant and the disinfecting compound through the soil as well as the resulting chemical and biological changes of the groundwater quality. The two dimensional model applies to the full and partial saturated conditions. The developed model is applied on an area west of Cairo. The area is bounded from one side by a long contamination source. The groundwater in the region is heavily contaminated both biologically and chemically. The developed mathematical model is proved to be very effective in finding the solution to improve the quality. The contamination content could be reduced to the required low levels.

1 INTRODUCTION

Water pollution is a serious hazard that threatens the existing water resources. Together with environmental necessity, remediation as well as rehabilitation of contaminated sites are considered essential components of water management policy.

The popular techniques for containing and treating contaminated groundwater involve physical contaminant control techniques as well as chemical, and biological treatment processes. In situ treatment of contaminated groundwater can be advantageous over both techniques. Injecting the proper chemical or biological disinfecting compounds into substrata cleans the water body totally rather than partly and reduces considerably the environmental impacts. In this work, a procedure is presented to calculate the optimum location and rates of decontaminant inje - tion using a combined three phase finite element-optimization model.

2 MODELING PROCEDURE

2.1 *Mathematical Fundamentals*

The mathematical formulation of the problem consists of three parts. In the first part, the fluid flow and mass transport through porous media equations are presented. The second part is concerned with the correlation between the contaminant and the deco - taminating compound. The third part is devoted for the optimization process correlation.

The two dimensional fluid flow through porous media can represented by the following equation

$$\nabla[\rho K(\theta)\nabla\Phi]=\partial(\rho\theta)/\partial t \quad (1)$$

where ρ = fluid density; K = hydraulic conductivity; θ = water content; Φ is the total energy; and t is the time dimension

The transport of contaminant and decontaminating compound through saturated porous media are represented as follows:

$$\partial C_i/\partial t = \mathrm{div}[C_i v - \rho\, D_d\, \mathrm{grad}(C_i/\rho)] \quad (2)$$

where C_i = mass concentration; i=1 for contaminant; i=2 for the decontaminant; and v = fluid velocity.

The decontaminating compound C_2 reacts with contaminant C_1 as they meet to produce a new product that is not considered within the class of contaminating compounds. The form of the function F that correlates C_1 with C_2 is, dependent on the type of contaminant and the used disinfecting compound. It is, therefore, given the general form below, which has to be specified for each case independently. Hence:

$$C_2 = F(C_1) \quad (3)$$

Through the optimization process O, it is intended to approach the ideal locations L(x, y) and rate of i - jection at that location $R_L(x, y, t)$ of the decontaminant. This is dependent on many factors such as price of land at different locations P_L, price of decontaminant P_c, and cost of environmental impacts P_e. Thus, a specific formula has to be established for a variable G that is a function of the previously mentioned parameters. The optimization process is conducted to approach the final solution G_{min} which is the minimum value of the function G. Or:

$$G_{min} = \min G(x, y, P_L, P_c, P_e, \ldots) \quad (4)$$

Usually the cost condition in the optimization fo - mula will force the decontaminant injection locations to be limited to a small number of nodes while the other nodes will get a permanent zero rate of injection

2.2 *Method of Analysis*

The method of finite element is used to solve the fluid and mass transport problems. To achieve this, the study area is divided into subregions according to soil and fluid properties, availability for constru - tion of injection wells, cost of land, etc. as illustrated in Figure 1.Each subregion is divided into a convenient number of elements. The Galerkin method is used to conduct the numerical solution.

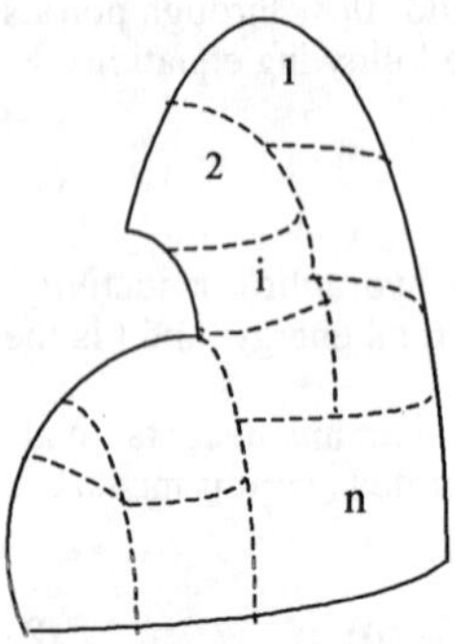

Figure 1 Division of area of study into subregions

The reaction result between the contaminant and the disinfecting compound is calculated using Equation 3. The optimization procedure is conducted by solving the problem for all possible conditions and using an iteration process to reach the optimum solution of equation 4.

2.3 *Program Flowchart*

The input data of the program includes the soil and fluid properties, the initial and boundary conditions of the study region, the correlation between the contaminant and decontaminating compound as well as their maximum and minimum concentration values that are allowed in the region. . It includes also the optimization formula and its parameters.

The solution is conducted such that the deco - taminant can be of any rate between zero and a given maximum value. It can be applied at a single node or simultaneously at several nodes. The iteration process is conducted systematically starting at the lowest node in the far left.

The rate of decontaminant concentration is increased by a given infinitesimal value more than the previous iteration process until the maximum allowable concentration is reached at one or more nodes within the specified ranges. The iteration process moves from the position point of view from left to right and upwards.

The developed program is written in basic language and its flow chart is shown in Figure 2.

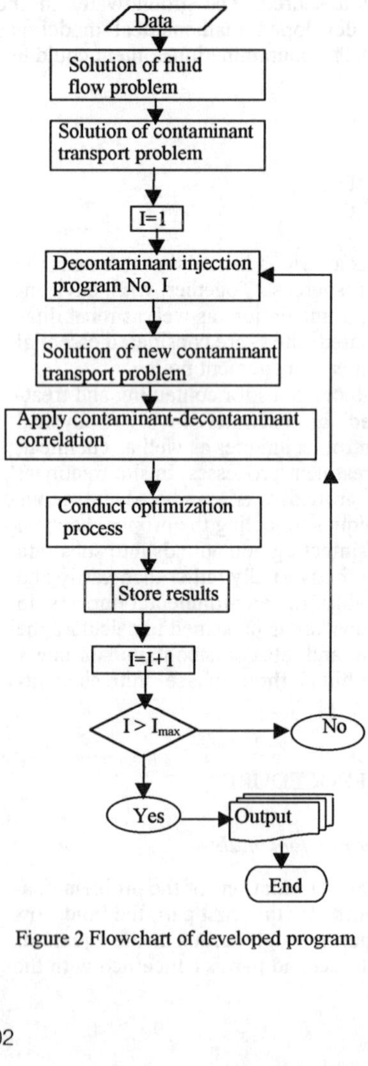

Figure 2 Flowchart of developed program

The program is calibrated and tested. Sensitivity analyses regarding the influence of different parameters on the solution are made.

3 CASE STUDY

The model is applied in a highly contaminated area north–west of Cairo, Egypt. The names of water bodies that bound the simulated region from all sides are written in italic font as shown in Figure 3. The study region has a total area of about 18 kilometer square. The main source of contamination is a drain that bounds the region from the southern side. The drain is used for disposal of domestic liquid wastes of nearby villages.

The subsurface geology, groundwater levels and quality and the hydraulic properties are determined through the execution of deep boreholes and the adoption of a field and laboratory testing program. It is found that the top 50-m of the soil consists of gravely sand that includes few thin lenses of clay at levels about 30 m below the ground level. The groundwater level in the area lies between 1.5 m and 3 m below the ground level and it has biological Oxygen Demand (BOD) values 96-218 mg/l. The measured higher values are near the contamination source. The vertical hydraulic conductivity as calculated from field and laboratory tests is found to be between 0.005m /d and .017 m/d while that in the horizontal direction is 0.035 - 1.5 m/d while the average water content is found to be 0.48.

The area is separated into 13 subregions according to the hydraulic properties. Each subregion is divided into 8 node type square elements. The side length of each element is 125 m. Figure 4 includes. the subregions as well as the locations of the two sections L-L and T-T.

In the analysis, the nodes that are located in highly populated districts are not allowed to be used as injection locations. The maximum allowable BOD value is taken to be 30 mg/l for all nodes that are at distances greater than 125 m from contaminant source. In the calculation it is assumed that every 1 mg of BOD needs 0.5 mg of chlorine gas to be cleaned. The price of land is assumed to be co - stant within the study area.

It is found that the optimum solution can be reached by injecting chlorine at the 8 locations that are indicated by the letter X in Figure 5. The total daily injection is 213 kg/d. The calculated steady state groundwater level contours are illustrated in the same figure.

The variation with time of contaminant concentration at nodes A, B, C, D& E (as indicated in Figure 5) is shown in Figure 6. The BOD values, before starting the application of the decontaminant (t=0) and after 200 days along the two sections L-L and T-T are shown in Figure 7 and Figure 8, respectively.

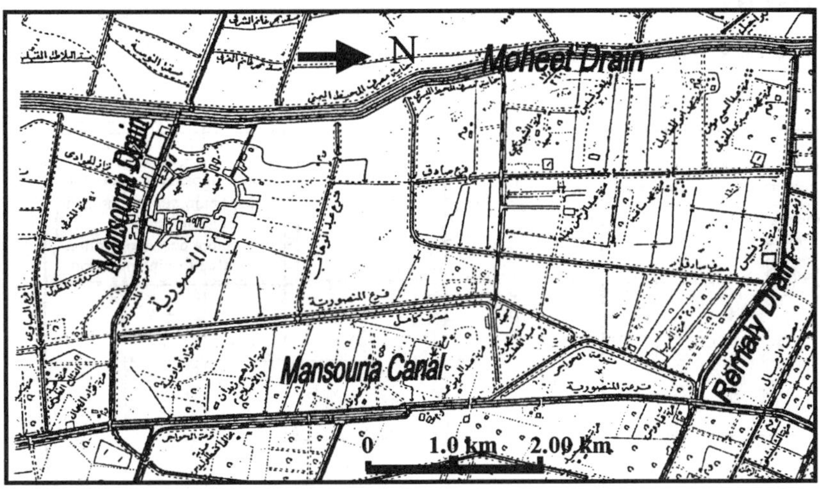

Figure 3. Plan of study area

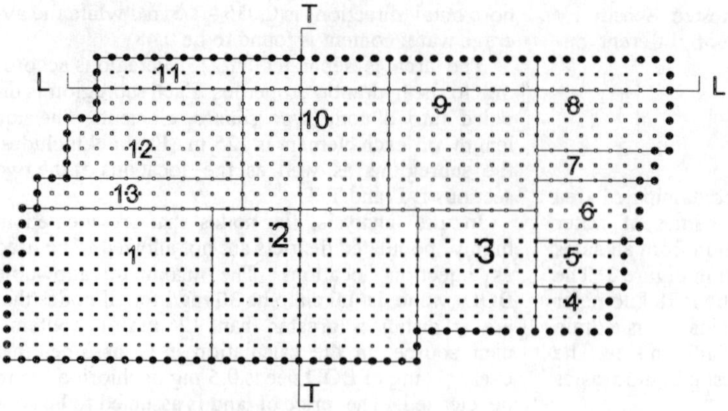

Figure 4 Subregions of study area

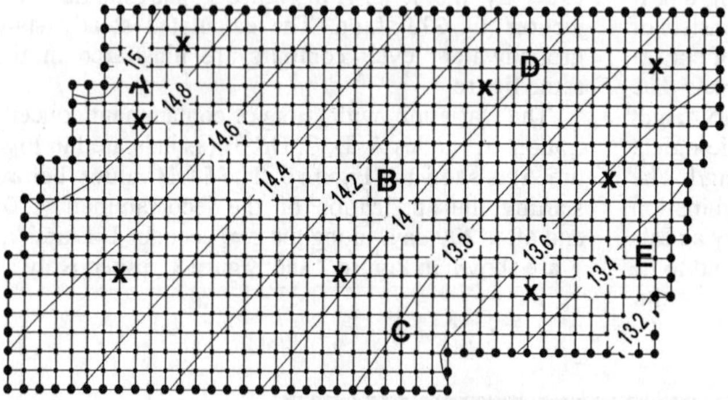

Figure 5 Location of injection nodes and groundwater contours

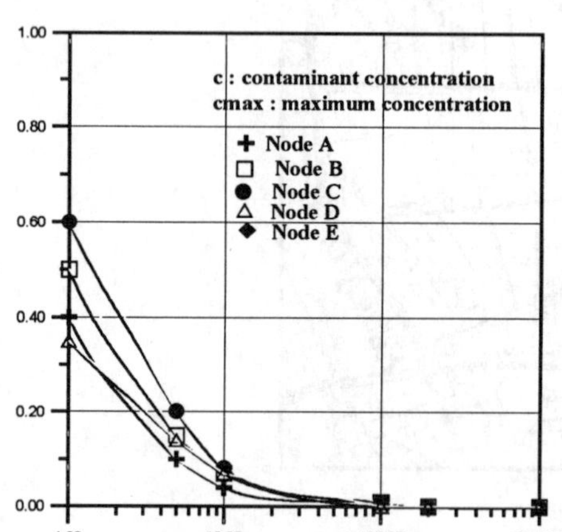

Figure 6 Variation of contaminant concentration with time

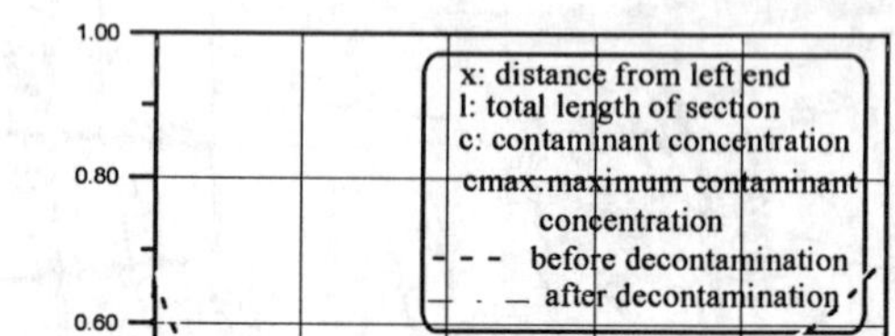

Figure 7 Contaminant concentration before and after application of decontaminant along section L-L

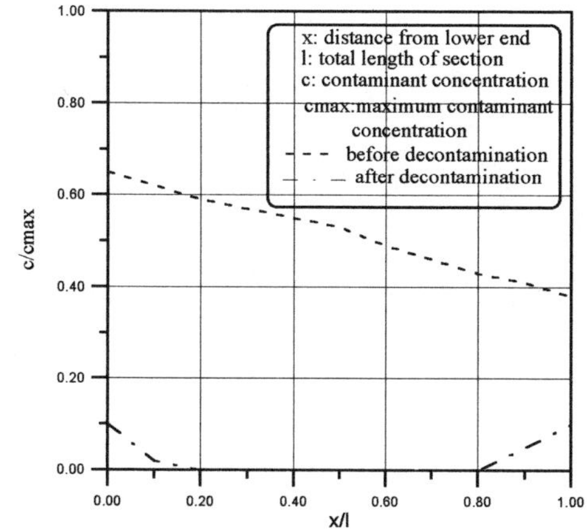

Figure 8 Contaminant concentration before and after applic - tion of decontaminant along section T-T

4 CONCLUSIONS

Conducting the process of disinfection of groundwater in situ is advantageous especially from the point of environmental protection. The developed model can help in finding the most feasible rates and locations of decontaminant injection according to the nature of contaminant and soil properties.

The application of the program in the case study indicated that it can be easily applied for practical conditions and that the results agree considerably with actually measured values. The program includes also the possibility of existence of adsorption and decay. They are not applied in the presented case study due to lack of field data.

REFERENCES

Bear, J. 1987. *Modeling of groundwater flow and pollution*. Dordrecht, Boston, USA; D. Reidel.

Brent S.E.& Jonathan S.F. 1993. Compositional simulation of groundwater Contamination by organic compounds, 1, model development and verification, *Water Resource. Res.*, 29 (6), 1697-1708.

Mansour, S..M. 2000. *Study of in situ treatment of contam - nated groundwater*. Ph.D. thesis. Mansoura, Egypt.

Metcalf& Eddy Inco.1991. *Wastewater engineering treatment. Disposal, reuse*. Snide Ed. New York, USA; McGraw-Hill.

Sandras, J., and Fulmer, M.,1996. *New Technology to decrease Cost, increase effectiveness of groundwater remediation,* University of Georgia Savannah River Ecology Laboratory News Release

Sun, Ne-Zenung . 1995. *Mathematical Modeling of Groundwater Pollution.* Translated from Chinese by Pengfei& Dehong. New York, USA; Springer, United States Code.1994. Title 42, *The Public Health and Welfare*, Chapter 103, 42 USC Sec 9621, Cleanup Standards

Yen G.T& Tripathi V. S. 1991 A Model for simulating tran - port of reactive multipecies components: model development and Demonstration, *Water Resource. Res.*, 27 (12), 3075-3094

Computer Methods and Advances in Geomechanics, Desai et al. (eds) © 2001 Balkema, Rotterdam, ISBN 90 5809 183 X

Numerical modelling of pressuremeter tests in unsaturated soil

D.Gallipoli
Universitat Politecnica de Catalunya, Barcelona, Spain

M.Karstunen & S.J.Wheeler
University of Glasgow, UK

ABSTRACT: The paper presents finite element simulations of pressuremeter tests in a normally consolidated unsaturated clay. The mechanical behaviour is represented by the Alonso et al. (1990) elasto-plastic model while for the degree of saturation variation, the relationship proposed by Gallipoli (2000) is adopted. The aim is to explore the coupling of liquid and gas flows with deformations at different loading rates and the rate effect on the computed cavity pressure-cavity strain relationship. The range of simulated loading rates covers the whole spectrum of conditions from undrained to drained. For the usual loading rates adopted in practice, the soil response is predicted as close to undrained with respect to the liquid and drained with respect to the gas. The predicted cavity stress-cavity strain relationships, however, change very little over the entire range of loading rates simulated. Therefore, for the conditions assumed in this work, the cavity stress-cavity strain relationship of a undrained pressuremeter test can be reasonably approximated by that of a fully drained (constant suction) test.

1 INTRODUCTION

Pressuremeter testing is widely used as an in-situ method for determining soil parameters in saturated soils, but the application of the technique to unsaturated soils is still highly tentative because of difficulty in the interpretation of test results.

A number of numerical studies of pressuremeter tests in saturated soils have been presented in the literature, including simulations of fully drained tests as well as cases where coupling between flow and deformation took place. Often the objective of these studies was to fit the numerical simulations to pressuremeter test data obtained in-situ in order to determine the values of the mechanical parameters of the soil.

Questions relating to the correct numerical modelling of pressuremeter tests in unsaturated soils are still to be answered. The few studies published to date tend to assume that the expansion of the pressuremeter device is a fully drained process (Consoli et al. 1997, Schnaid & da Silva 1997) so that the value of suction within the soil remains constant during expansion (suction is the difference between pore air pressure and pore water pressure). This hypothesis simplifies the problem, because it avoids the need for fully coupled flow-deformation analysis. Moreover, it allows the same constitutive models employed for drained analyses in saturated soils to be used for unsaturated soils (based on the assumption that unsaturated soils at constant suction behave qualitatively in the same way as saturated soils) as long as a dependency of the model parameters on the level of suction is accounted for.

Suction monitored pressuremeter tests in unsaturated granite residual soils of southern Brazil by Schnaid (2000) seem to validate the assumption of constant values of suction in the zone of soil around the cavity during pressuremeter expansion. Constant values of suction were measured by tensiometers at radial distances from the borehole axis of 30 cm and 60 cm. The hypothesis of fully drained conditions may, however, be acceptable for residual soil with fissures, as these increase the overall permeability, but it is debatable if this is valid for all types of soil. Further investigation is needed, therefore, to explore under which circumstances the expansion process could be considered as fully drained.

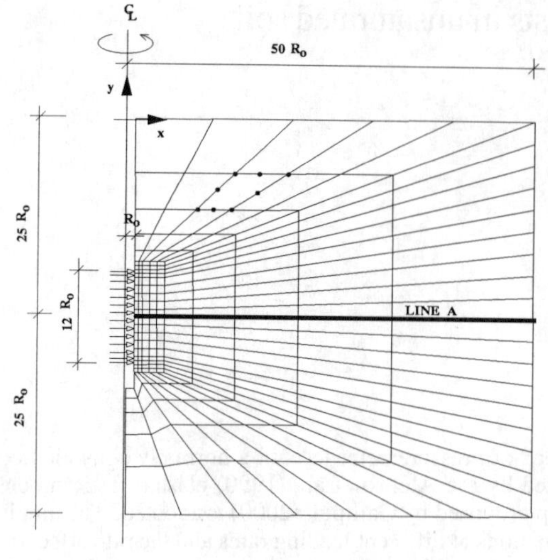

Figure 1: Mesh and pressuremeter loads

This paper presents numerical simulations of pressuremeter tests in a normally consolidated unsaturated clay soil at different rate of expansion. The aim is to explore the effect of the loading rate on the coupling of flow and deformation and on the cavity stress-cavity strain relationship.

2 MODEL DESCRIPTION

2.1 *Geometry, mesh and boundary conditions*

The type of test modelled is a self-boring pressuremeter test with geometry of the Cambridge self-boring pressuremeter (Wroth & Hughes 1972). This device has a ratio between the length of the expanding section and the radius of approximately 12. Pressuremeter expansion was modelled as a bi-dimensional axi-symmetric problem with the finite element code "Compass" (Thomas & He 1998) as improved by Gallipoli (2000). The mesh consisted of 240 elements and a total of 781 nodes (Fig. 1). A finer mesh than the one presented was also tested to ensure the adequacy of the refinement in the zones where highest gradients of displacement and suction occur (i.e. borehole wall and extremities of pressuremeter membrane). Quadrilateral eight-noded isoparametric elements were used, with the same shape functions for all the nodal unknowns (displacement, pore air pressure and pore water pressure). Figure 1 also shows the top, bottom and outer boundaries, which have been chosen to be sufficiently distant from the pressuremeter not to influence the numerical predictions.

Modelling of the boundary conditions at the borehole wall must take into account the geometry of the pressuremeter device. The pressuremeter has above and below the membrane two cylindrical steel sections of the same diameter as the deflated testing section. These two steel sections prevent the soil from moving radially towards the borehole but they do not impose any restraint on the outward radial displacements. Therefore, ideally a varying boundary condition should be imposed at the borehole wall. While the radial net stress (net stresses are the difference between total stresses and pore air pressure) at this interface is greater than zero, the boundary condition should be displacement-controlled (zero value of radial displacement) and, for the liquid and gas flows the boundary should be impermeable. When the radial net stress becomes zero any further reduction is not possible, as tension cannot occur at the interface between the soil and steel. Separation of the soil from the probe may, therefore, occur and the boundary condition changes to stress-controlled (zero value of radial net stress). For the boundary conditions on the liquid and gas flows, additional hypotheses on the presence of liquid and gas in the space between the probe and soil after separation need to be introduced. The problem becomes even more complex if the friction between the probe and the soil is considered. Due to the complexities outlined above, in simulations of pressuremeter tests the simplified assumptions of either a stress-controlled boundary condition (Fioravante et al. 1994), or a displacement-controlled one (Houlsby & Carter 1993) have been used. Both types of boundary conditions at the borehole wall were tested in this work and different net stress states were predicted in a limited zone of soil located just below the pressuremeter depending on the boundary condition. The overall response of the test, however, was practically the same in both cases. In all the analyses presented in this paper a displacement-controlled impermeable boundary was assumed because this seemed closer to the real condition during a pressuremeter test, and also, because it reduced the number of nodal unknowns. The radial displacements of the first two elements above and below the loading section were not restrained and a stress-controlled boundary condition (with radial net stress equal to the in-situ value) was assumed instead.

2.2 *Constitutive laws and in-situ stress state*

The mechanical behaviour of the soil was represented by the elasto-plastic model of Alonso et al. (1990). The values for the relevant parameters, reported in Table 1, were selected on the basis of laboratory test results for compacted unsaturated Speswhite Kaolin (Sivakumar 1993, Zakaria 1995) as described in Gallipoli (2000). In particular, Gallipoli (2000) discusses the choice of a value greater than 1 for the parameter r (to be used in combination with a value of the reference pressure p^c sensibly higher than the range of interest for the hardening parameter $\overline{p}_o^*$), whereas

Table 1: Parameter values for stress-strain model.

Parameter	Value
ν, Poisson's ratio	0.3
κ_s, swelling index for changes of s	0.020
κ, swelling index for changes of $\overline{p}$	0.025
M, slope of CSL at constant s	0.9
k, strength increase with s	0.5
p^c, reference pressure	2 MPa
$\lambda(0)$, slope of NCL at s=0	0.13
$N(0)$, v on NCL at s=0 for $\overline{p}$=p^c	1.66
r, ratio of NCL slopes at $s \to \infty$ and s=0	1.5
β, param. for NCL slope change with s	10 MPa^{-1}

Table 2: Parameters for degree of saturation model.

Parameter	Value
ϕ	0.02691 kPa^{-1}
ψ	8.433
m	0.03586
n	3.746

Table 3: Parameters for permeability relationships.

Parameter	Value
S_{res}, residual degree of saturation	0.4
k_s, coeff. water permeability at S_r=1	10^{-9} m s^{-1}
k_d, coeff. air permeability at S_r=S_{res}	10^{-8} m s^{-1}

Alonso et al. (1990) suggested, in the development of the model, values smaller than 1 for r (in combination with relatively small values of p^c).

Two modifications were introduced to the modelling of the elastic behaviour. Firstly, a constant value of Poisson's ratio ($\nu = 0.3$), was assumed instead of a constant value of shear modulus (G). With a constant shear modulus (G), negative values of Poisson's ratio were predicted in a localized zone of soil at the extremities of the pressuremeter membrane due to a significant decrease of mean net stress, which resulted in very low values of bulk modulus (K). Clearly, the choice of a constant value of Poisson's ratio has the drawback of implying a non-conservative elastic relationship, because the shear modulus depends on the mean net stress (Zytynski et al. 1978). This option seemed, however, to provide a more representative mode of deformation in the elastic domain.

Secondly, a lower limit of bulk modulus ($K = 200$ kPa) was enforced in the code in order to avoid an ill-conditioned system of equations when values of mean net stress close to zero were predicted in very circumscribed regions of soil located at the two extremities of the pressuremeter membrane. The ill-conditioning was due to values of bulk modulus and specific volume tending respectively to zero and infinity when the value of mean net stress tended to zero, as predicted by the Alonso et al. (1990) model.

The variation of the degree of saturation followed the relationship proposed by Gallipoli (2000), which introduces a dependency on the specific volume (v) into the van Genuchten (1980) expression:

$$S_r = \left(\frac{1}{1+(\phi\,(v-1)^{\psi}s)^n}\right)^m \tag{1}$$

where ϕ, ψ, m and n are model parameters. This relationship, if combined with an elasto-plastic stress-strain model, such as the Alonso et al. (1990) model, predicting the variation of specific volume, is able to reproduce elasto-plastic variations of degree of saturation in a fashion consistent with the stress-strain behaviour. The parameter values for the degree of saturation relationship of Equation 1 are reported in Table 2. They were selected on the basis of the tests on compacted unsaturated Speswhite Kaolin performed by Sivakumar (1993) and Zakaria (1995) (see Gallipoli (2000) for details on the selection process).

The coefficients of water permeability (k_w) and air permeability (k_a) were assumed to vary with degree of saturation according to the reduced Brooks & Corey (1964) relationship (where the pore size distribution index is sufficiently great to have no influence on the permeability relationship):

$$k_w = k_s {S_e}^3 \tag{2}$$

$$k_a = k_d(1-S_e)^2(1-S_e) \tag{3}$$

where S_e is the effective degree of saturation defined as:

$$S_e = \frac{S_r - S_{res}}{1 - S_{res}}. \tag{4}$$

The values chosen for the relevant parameters (Gallipoli 2000) are reported in Table 3.

The initial distributions of net stress and suction were assumed constant throughout the soil domain. The vertical net stress was 100 kPa and the horizontal net stresses (radial and circumferential net stress) were calculated as 60 kPa by using a coefficient of earth pressure at rest, K_{nc} of 0.6. The coefficient K_{nc} was computed according to the Jâky's simplified formula ($K_{nc} = 1 - \sin\phi'$) for normally consolidated saturated soils, by assuming an angle of friction, ϕ' of 23 degrees. This value is consistent with the value of M reported in Table 1 for triaxial compression. The chosen in-situ net stress state corresponds to a soil depth between 5 and 6 metres.

An initial suction of 200 kPa was assumed in all analyses by imposing a pore air pressure of zero (relative to atmospheric pressure) and a pore water pressure of -200 kPa (relative to atmospheric pressure). Atmospheric pressure was assumed to be 100 kPa and it only influenced the compressibility of the gas phase.

In practice net stresses increase with depth and suction is also likely to vary with depth. The assumption of constant initial values, however, allowed a clearer interpretation of the effect of pressuremeter expansion, as spatial variations of net stresses and suction were entirely attributable to the expansion process.

As the soil is in a normally consolidated state, the stress state defined above must belong to the yield locus. Therefore the initial value of the hardening parameter, $\overline{p}_o^*$ in the Alonso et al. (1990) model was calculated as 15.6 kPa.

3 NUMERICAL RESULTS

Finite element analyses were performed by applying the same increase of cavity pressure (130 kPa) over six different periods of time: 90 s (1.5 minutes), 900 s (15 minutes), 9000 s (2.5 hours), 90000 s ($\approx$1 day), 900000 s ($\approx$10 days), 9000000 s ($\approx$104 days). In each analysis the loading rate was constant throughout the expansion. The loading rate corresponding to the time of 900 s can be regarded as similar to a pressuremeter test carried out according to the Ménard specifications, as explained in Gallipoli (2000).

Figure 2 shows the variations of pore water pressure, pore air pressure and suction at the borehole wall at the mid-height of the pressuremeter membrane for the different loading times. Inspection of Figure 2a suggests that a pressuremeter test, carried out at a loading rate similar to that of a Ménard test, can be regarded as almost undrained with respect to the liquid phase. A much longer time of loading of the order of 100 days, is needed in order to avoid substantial excess pore water pressure.

For the pressuremeter tests in unsaturated residual soils performed by Schnaid (2000) at Universidade Federal do Rio Grande do Sul, Brazil the ultimate cavity strain of 41% was attained in a time between 1 and 2 hours. The loading rate was of the same order of magnitude as in the numerical analysis simulating a test duration of 9000 s. Such a loading rate, in combination with the soil parameters values adopted in this work, would produce a response closer to undrained with respect to the liquid rather than drained. There may be cases, as these numerical results indicate, where the execution of drained pressuremeter tests in unsaturated soils is not a realistic option.

For the two tests simulations employing faster rates of loading (i.e. 90 s and 900 s) a clear, although very small, increase of pore air pressure can be observed in Figure 2b indicating a moderate degree of coupling between gas flow and deformation (for all slower rates of loading the test is effectively drained with respect to the gas). This increase of pore air pressure implies that the difference between the variations of suction computed for the two faster loading rates (Fig. 2c) is greater than the difference between the corresponding variations of pore water pressure (Fig. 2a).

Figure 3 shows the ultimate excess pore water pres-

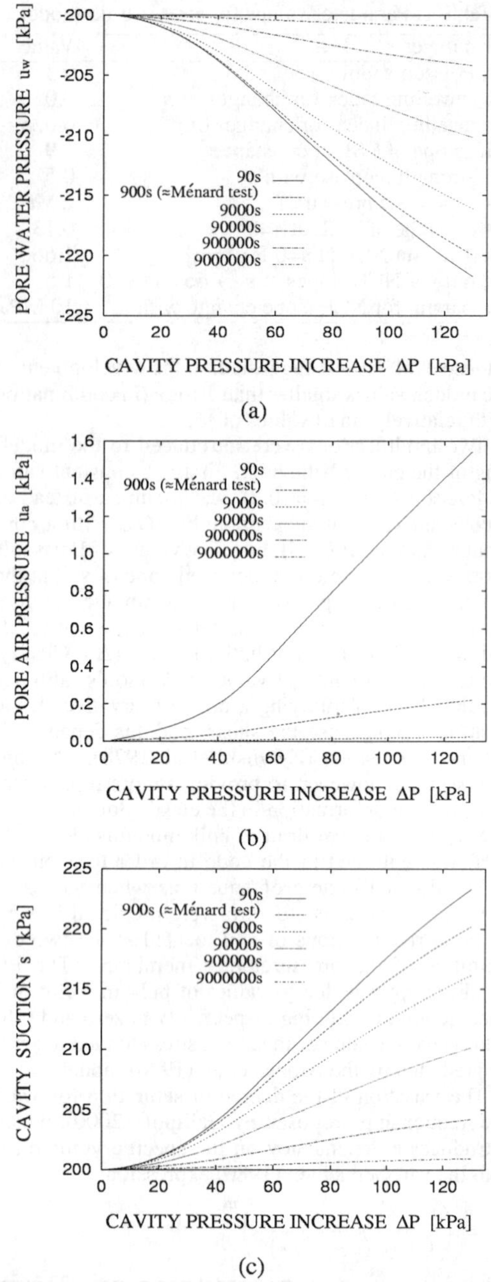

Figure 2: Influence of time of loading on the predicted: (a) cavity pressure-water pressure curve (b) cavity pressure-air pressure curve (c) cavity pressure-suction curve

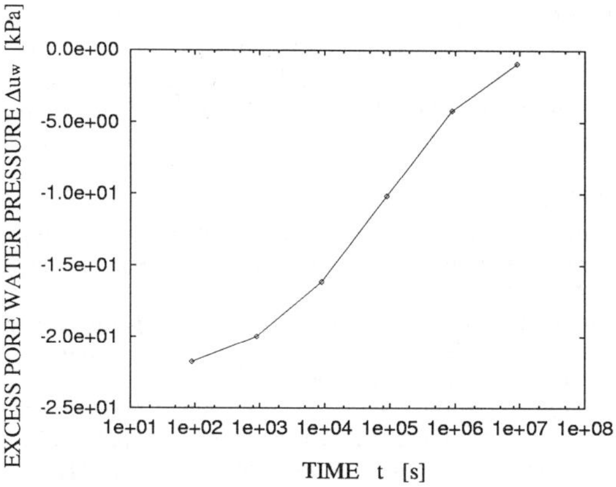

Figure 3: Influence of time of loading on the predicted ultimate excess pore water pressure

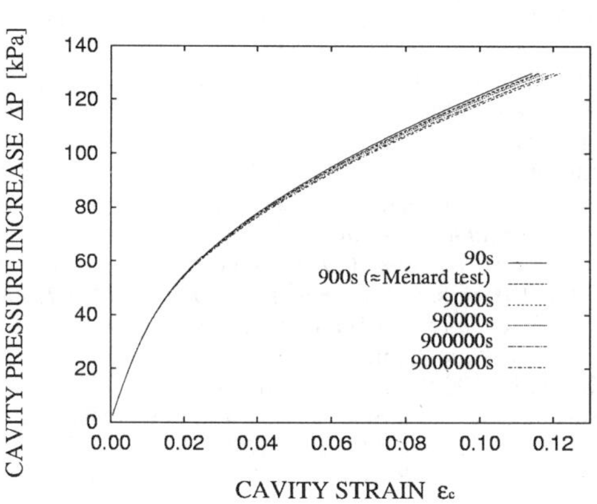

Figure 4: Influence of time of loading on the predicted cavity pressure-cavity strain curve

sure at the borehole wall at the mid-height of the pressuremeter membrane, for the six different times of loading. The S-shaped curve in Figure 3 confirms that the two analyses simulating the fastest and slowest loading rates are approaching the limit conditions of undrained and drained expansion (with respect to the liquid phase) respectively.

Figure 4 shows the cavity pressure-cavity strain relationship for different loading times. The curves in Figure 4 represent a range of loading rates which covers the whole spectrum of conditions from undrained to drained. Nevertheless, Figure 4 indicates that the loading rate does not significantly affect the overall computed cavity pressure-cavity strain response. This significant result suggests that, in some circumstances, it is reasonable to treat results from pressuremeter tests, which in reality are undrained with respect to the liquid phase, as if the tests were drained. The cavity pressure-cavity strain curve from a drained pressuremeter expansion, for which numerical and analytical methods of interpretation have been proposed based on solutions for saturated cases (Consoli et al. 1997, Schnaid & da Silva 1997), could be approximated by a similar relationship obtained from a test carried out at a significantly faster loading rate. If this result, obtained for the initial conditions and material properties of the present analyses, could be proven valid under more general circumstances, it would provide a justification to the extension of methods of interpretation of pressuremeter tests proposed for drained saturated soils to unsaturated soils.

The sensitivity of the soil response to the loading rate is expected to depend on the initial suction. In the saturated limit (zero suction) relatively greater excess pore water pressure would be predicted during the expansion process than in the case presented. Therefore, as the initial suction tends from 200 kPa to zero, greater excess pore water pressure and, consequently, greater excess suction should also be computed. Moreover, the Alonso et al. (1990) stress-strain model suggests that, for smaller values of in-situ suction than that assumed, the cavity pressure-cavity strain curves at different loading rates should exhibit greater differences, as the straining becomes more sensitive to suction changes. The insensitivity of the stress-strain response to the loading rate is also a consequence of the particular relationship for degree of saturation assumed (Eq. 1). Similar simulations were performed by changing the constitutive model for degree of saturation to the state surface expression proposed by Lloret & Alonso (1985) (see Gallipoli (2000) for further details), and significantly stronger dependency of the cavity pressure-cavity strain relationship on loading rate was obtained.

Figure 5 shows the variations of suction and principal net stresses along the radius at mid-height of the pressuremeter membrane (line A in Figure 1) at the end of the loading process for the simulation of the 900 s test. Inspection of Figure 5 reveals that variations of suction with respect to the initial in-situ value (200 kPa) are limited to a cylinder of soil having a radius from the borehole axis of less than about 20 cm. For the net stress state, the in-situ values are restored at a radial distance of about 50 cm. This numerical result agrees with experimental data from suction monitored pressuremeter tests in unsaturated granite residual soils of southern Brazil by Schnaid (2000) where constant suction was measured during expansion by tensiometers at radial distances from the borehole axis of 30 cm and 60 cm.

4 CONCLUSIONS

The paper investigated, by means of finite element analyses, the effect of loading rate on the coupling

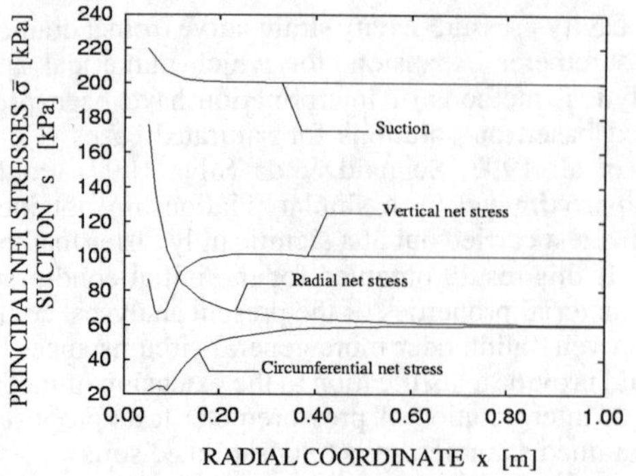

Figure 5: Radial variation of net stresses and suction

of flow and deformation during pressuremeter tests in a normally consolidated unsaturated soil. For the problem studied, a very slow loading rate ($\approx$ 1 kPa per day) is needed to achieve a drained response with respect to the liquid. For a loading rate similar to a Ménard test, the response is close to undrained and, even for a slower rate as that used by Schnaid (2000) for tests in unsaturated residual soils of southern Brazil, the behaviour is closer to undrained than drained. The cavity strain-cavity pressure relationships, for a range of loading rates from undrained to drained, were, however, all very similar. This suggests that the cavity stress-cavity strain relationship from undrained pressuremeter tests can be approximated by that from drained tests. This seems to validate the extension of methods of interpretation for drained pressuremeter tests in saturated soils to the unsaturated case.

As in-situ suction tends to zero, more significant suction changes and a stronger dependency of the stress-strain response on loading rate should be expected. A parametric study of the various factors involved is, therefore, needed to determine under which circumstances a pressuremeter test can be idealized as drained. The insensitivity of the cavity pressure-cavity strain curve to the loading rate was also a consequence of using the elasto-plastic relationship (Eq. 1) for degree of saturation proposed by Gallipoli (2000). When the state surface expression for degree of saturation of Lloret & Alonso (1985) was employed, noticeably larger differences were predicted between different rates. This confirms the critical role of improved modelling of the variation of degree of saturation for accurate predictions of suction and deformation in boundary value problems.

5 ACKNOWLEDGMENTS

The Authors thank H.R. Thomas (University of Cardiff, UK) for providing the finite element code "Compass" and F. Schnaid (Universidade Federal do Rio Grande do Sul, Brazil), A. Gens and J. Vaunat (Universitat Politecnica de Catalunya, Spain) for useful discussions. The financial support of the European Commission, in the form of a Marie Curie Fellowship awarded to the first Author, is also acknowledged.

REFERENCES

Alonso, E., A. Gens, & A. Josa (1990). A constitutive model for partially saturated soils. *Géotechnique 40*(3), 405–430.

Brooks, R. & A. Corey (1964). Hydraulic properties of porous media. *Hydrol. Paper no.3, Civil Engineering Dep., Colorado State Univ., Fort Collins.*

Consoli, N., F. Schnaid, & F. Mantaras (1997). Numerical analysis of pressuremeter tests and its application to the design of shallow foundations. In *Proceedings International Workshop on Applications of Computational Mechanics in Geotechnical Engineering, Rio de Janeiro*, pp. 25–34.

Fioravante, V., M. Jamiolkowski, & R. Lancellotta (1994). An analysis of pressuremeter holding test. *Géotechnique 44*(2), 227–238.

Gallipoli, D. (2000). *Constitutive and numerical modelling of unsaturated soils.* Ph. D. thesis, University of Glasgow, UK.

Houlsby, G. & J. Carter (1993). The effects of pressuremeter geometry on the results of tests in clay. *Géotechnique 43*(4), 567–576.

Lloret, A. & E. Alonso (1985). State surfaces for partially saturated soils. In *Proceedings International Conference Soil Mechanics and Foundation Engineering, San Francisco*, Volume 2, pp. 557–562.

Schnaid, F. (2000). Private communication.

Schnaid, F. & G. da Silva (1997). Ensaios presiométricos em um solo não saturado da região sul do Brasil. In *Anais Simpósio Brasileiro de Solos Não Saturados, Rio de Janeiro.*

Sivakumar, V. (1993). *A Critical State Framework for Unsaturated Soil.* Ph. D. thesis, University of Sheffield, UK.

Thomas, H. & Y. He (1998). Modelling the behaviour of unsaturated soil using an elasto-plastic constitutive model. *Géotechnique 48*(5), 589–603.

van Genuchten, M. (1980). A closed-form equation for predicting the hydraulic conductivity of unsaturated soil. *Soil Sci. Soc. Am. J. 44*, 892–898.

Wroth, C. & J. Hughes (1972). An instrument for the in situ measurement of the properties of soft clays. Technical Report CUED/D, Soils TR 13, University of Cambridge, UK.

Zakaria, I. (1995). *Yielding of unsaturated soil.* Ph. D. thesis, University of Sheffield, UK.

Zytynski, M., M. Randolph, R. Nova, & C. Wroth (1978). Short communication on modelling the unloading-reloading behaviour of soils. *International Journal for Numerical and Analytical Methods in Geomechanics 2*, 87–94.

Computer Methods and Advances in Geomechanics, Desai et al. (eds) © 2001 Balkema, Rotterdam, ISBN 90 5809 183 X

Prediction of heave using a general partial differential equation solver

V.Q. Hung & D.G. Fredlund
Department of Civil Engineering, University of Saskatchewan, Saskatoon, Sask., Canada

ABSTRACT: This paper presents a method for prediction of the heave based on the general volume change theory for unsaturated soils. The proposed method makes use of the general purpose partial differential equation solver, PDEase2D, along with non-linear elastic modulus functions. The elastic modulus functions are calculated from conventional oedometer test results. Typical heave problems and case histories are analyzed using the proposed method. The solutions agree well with the analytical solutions and measured results. General partial differential equation solver shows excellent promise for solving volume change problems involving highly non-linear unsaturated soil properties.

1 INTRODUCTION

The volume change of an expansive soil, due to the increase of water can distress to engineered structures. Numerous one-dimensional methods have been proposed to predict heave of expansive soils (Fredlund and Rahardjo, 1993). These methods generally make use of a linear relationship between void ratio (or vertical strain) and the log of matric suction (or total stress) in predicting heave. Fredlund et al. (1980) presented a method for the prediction of one-dimensional heave based on the volume change theory of unsaturated soils. The volume change behavior of an unsaturated soil can be discussed in terms of two independent stress state variables; namely the net normal stress $(\sigma - u_a)$ and matric suction $(u_a - u_w)$, where σ is total normal stress, u_a is pore-air pressure and u_w is pore-water pressure. It is assumed that the air phase is continuous and at atmospheric pressure.

This paper proposes an extension of the Fredlund et al. (1980) method for the prediction of heave to two-dimensions. The proposed method makes use of the finite element method, along with non-linear elastic modulus functions. The soil is assumed to be isotropic, non-linear and elastic. The elastic modulus functions (i.e., elastic modulus with respect to net normal stress, E, and elastic modulus with respect to matric suction, H) are computed from oedometer test results (i.e., volume change indices).

The theoretical formulation of the heave prediction method is based on equilibrium equations for the soil structure and the constitutive equations for the unsaturated soil. The solution of a volume change problem is obtained through the use of a general-purpose, partial differential equation solver (i.e., PDEase2D).

2 HEAVE FORMULATION

The force equilibrium of a soil element for K_0-loading condition (i.e., volume change in the y-direction only) can be expressed in the following form:

$$\frac{\partial \sigma_y}{\partial y} + \rho g = 0 \tag{1}$$

where ρ = the density of the soil, g = acceleration due to gravity.

The soil structure constitutive relation for K_0-loading condition can be written as follows (Fredlund and Rahardjo, 1993):

$$(\sigma_y - u_a) = \frac{E(1-\mu)}{(1+\mu)(1-2\mu)}\left(\varepsilon_v - \frac{(1+\mu)}{H(1-\mu)}(u_a - u_w)\right) \tag{2}$$

where E = elastic modulus for the soil structure with respect to net normal stress; H = elastic modulus for the soil structure with respect to matric suction; and μ = Poisson's ratio.

The differential expression of normal strain in terms of displacement in the y-direction (i.e., v) is as follows:

$$\varepsilon_y = \frac{\partial v}{\partial y} \tag{3}$$

where v = vertical displacement.

The governing differential equation for the K_0-loading condition can be obtained by substituting expression for net normal stress (Eq. 2) into the force equilibrium equation (i.e., Eq. 1), assuming the pore-air pressure is atmospheric.

$$\frac{\partial}{\partial y}\left(\frac{E(1-\mu)}{(1+\mu)(1-2\mu)}\left(\frac{\partial v}{\partial y}-\frac{(1+\mu)}{H(1-\mu)}(u_a-u_w)\right)\right)+\rho g=0 \tag{4}$$

The elastic moduli E and H can be calculated from swelling indices, initial void ratio and assumed Poisson's ratio (Hung, 2000; Hung and Fredlund, 2000) as follow:

$$E = n_t(\sigma - u_a)_{ave} \tag{5}$$

$$H = n_m(u_a - u_w)_{ave} \tag{6}$$

where n_t = the coefficient that relates net normal stress with elastic modulus E; n_m = coefficient that relate matric suction with elastic modulus H; $(\sigma - u_a)_{ave}$ = average of the initial and final net normal stress for an increment; and $(u_a - u_w)_{ave}$ = average of the initial and final matric suction for an increment.

$$n_t = \frac{(1+\mu)(1-2\mu)}{(1-\mu)}\,\frac{(1+e_0)}{0.4343C_t} \tag{7}$$

$$n_m = \frac{(1+\mu)}{(1-\mu)}\,\frac{(1+e_0)}{0.4343C_m} \tag{8}$$

where C_t and C_m = volume change indices with respect to net normal stress and matric suction, respectively; and e_0 = initial void ratio.

Equation 4 can be used to compute the vertical displacement due to net normal stress changes and/or matric suction changes.

3 GENERAL-PURPOSE PARTIAL DIFFERENTIAL EQUATION SOLVER

A number of mathematical computer programs have emerged over the last two decades for the solution of math related problems. PDEase2D is one of the general-purpose partial differential equation solvers, and is marketed by Macsyma Inc. The focus of the software package is on ensuring proper converging solutions where highly non-linear equations are involved.

The user of the general partial differential equation software must specify the partial differential equation to be solved. The equation contains the variables to be solved along with a series of material properties. The boundary conditions around a designated region must also be specified.

Equation 4 is non-linear since the elastic moduli are functions of the stress state variables. An incremental procedure is used within PDEase2D to solve these equations (Desai and Christian, 1977), as shown in Figure 1. The total load (or suction change) is divided into increments and one load (or suction) incremental step is applied at a time. During each incremental step, the elastic moduli are assumed to be unchanged. New moduli are selected at the beginning of each new incremental step dependent upon the stress level, and the displacements from each incremental step are accumulated to give to total displacements.

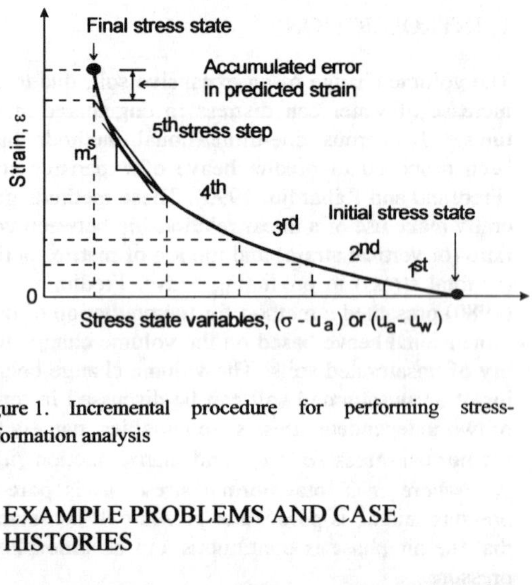

Figure 1. Incremental procedure for performing stress-deformation analysis

4 EXAMPLE PROBLEMS AND CASE HISTORIES

Several typical heave problems and case histories presented by Fredlund and Rahardjo (1993) have been analysed using the proposed method for the heave prediction (Hung, 2000). The predicted results from the proposed method are then compared to existing solutions.

The first example is associated with the total heave of an expansive soil profile below an impermeable cover. The soil below the cover undergoes an increase in pore-water pressure with time. The second example is associated with the heave of an expansive soil profile, when a part of the expansive soil profile is excavated and replaced with an inert granular material.

The first case history is associated with the monitoring of movements of a floor slab in a light industrial building in north central Regina, Saskatchewan, Canada (Yoshida et al, 1983). The second case history is associated with the heave problems in the Eston area of Saskatchewan, Canada (Ching and Fredlund, 1984). There are also measured results that are used to verify the PDEase2D solutions.

The analysis of the first example problem associated with heave due to the increase of pore-water pressure below an impermeable cover is described in Figure 2. The ground surface of an expansive soil is covered with an impermeable layer such as asphalt. With time, the negative pore-water pressure in the soil below the asphalt will increase as a result of the discontinuance of evaporation and evapotranspiration. The heave associated with changes in negative pore-water pressure (i.e., matric suction) is computed.

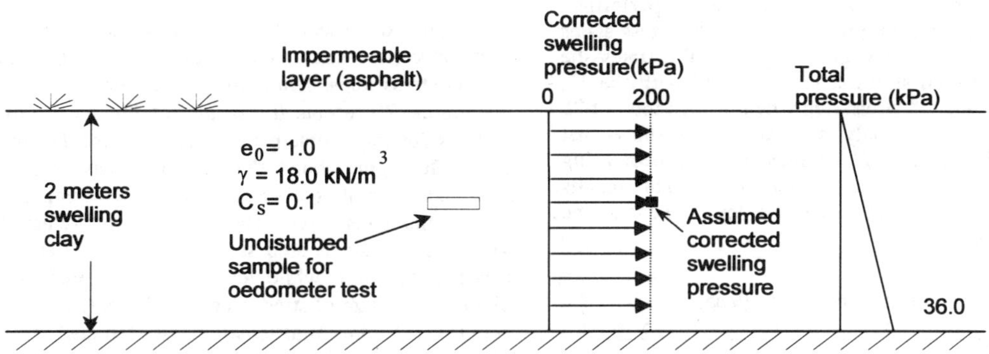

Figure 2. Problem definition for the first example

The final pore-water pressure is assumed to be zero throughout the entire depth. Soil properties and dimensions are shown on Figure 2.

In the Fredlund et al. (1980) method, the heave analysis can be performed by subdividing the active depth into a number of layers. The amount of heave in each layer is computed by considering the stress state changes at the middle of the layer. Figure 3 presents the results of the sensitivity analysis using Fredlund et al. (1980) method. It can be seen that the computed total heave does not change significantly when using more than 25 layers.

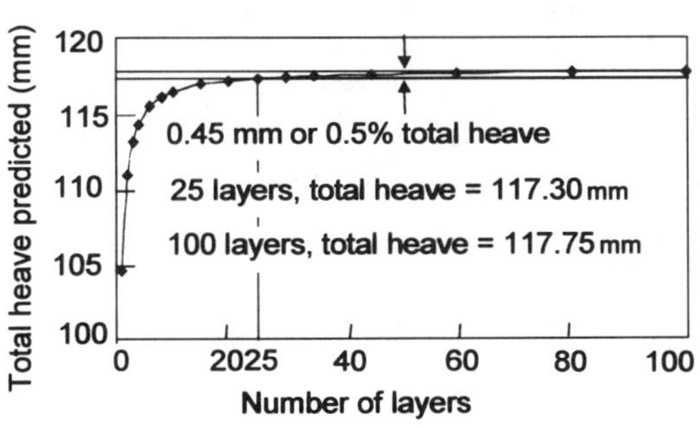

Figure 3. Total heave calculated using Fredlund et al. (1980) method versus number of layers of soils

For the prediction of heave using the finite element method, the elastic modulus function must be evaluated. The matric suction equivalent is used instead of the *in situ* suction. As a result, the measured swelling index on the net normal stress plane, C_{ts}, is used instead of the swelling index with respect to matric suction, C_{ms}. The calculation of elastic modulus was presented in Hung and Fredlund (2000).

For the calculation of the elastic modulus function, the average of the initial and final stress state for an increment must be used in place of the average of matric suction. With an initial void ratio equal to 1.0, swelling index, C_s, equal to 0.1 and assumed Poisson's ratio equal to 0.3, Equation 6 can be rewritten as follows:

$$H = 85.52(u_a - u_w)_{ave} \text{ or } H = 85.52(ST)_{ave} \qquad (9)$$

where ST = stress state accounting for overburden pressure and matric suction equivalent.

The stress state is expressed as the sum of overburden pressure and the matric suction equivalent. The elastic modulus function, H, is presented in Figure 4, showing that a constant swelling index on a semi-log plot providing a linear elastic modulus function.

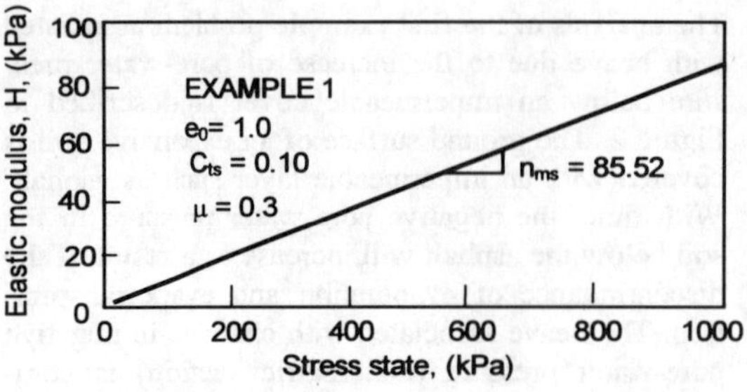

Figure 4. Elastic modulus function, H

The stress-deformation analysis is performed using an incremental procedure with the stress states subdivided into 25 steps. Figure 5 illustrates the initial and final stress state associated with the 1st, 10th, 15th and 25th stress step when using a total of 25 stress steps in the analysis. Figure 5 also shows that the elastic modulus, *H,* decreased with a decreasing stress level near to the ground surface. This results in a greater heave at low stress levels, near to the ground surface.

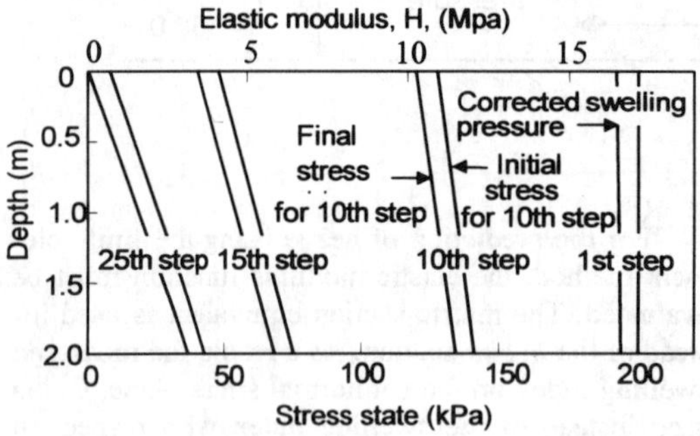

Figure 5. Stress state versus depth

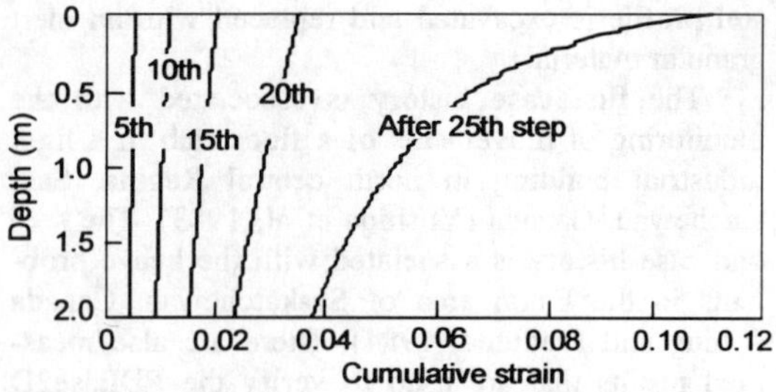

Figure 6. Cumulative strain versus depth at different stress steps

Figure 6 presents the cumulative strain versus depth at various stress steps. More strain is obtained at later stress steps, near to the ground surface.

Figure 7 presents the results of the sensitivity analysis for the problem using the PDEase2D program. The figure shows that the total heave calculated increases almost linearly with decreasing the size of the stress step. The total heave predicted when using a stress step size of 1 kPa (200 stress steps) is 117 mm while the total heave predicted when using a size of stress step of 7.2 kPa (25 stress steps) is 115 mm. With small stress step such as 1 kPa, the error is about 0.5% from the correct solution (Fig. 7). The variation in the number of stress increments used from 25 increments to 200 increments results in a change of about 2% in the calculated total heave, for this example. The smaller the size of a stress increment, the more stress steps that are needed. However, it is not practical to use extremely fine size stress steps to perform an analysis. It can be seen that, with a limited number of stress steps, the method would under-estimate the total heave (e.g., by about 3% the correct amount of heave).

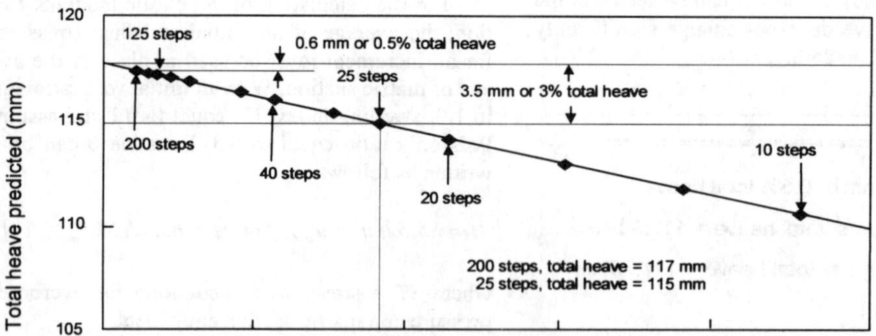

Figure 7. Total heave versus size of stress step for the first example

The values of accumulated heave obtained from the analytical method, as well as from PDEase2D program, are presented in Figure 8. Greater amounts of heave are predicted at later time steps, near to the ground surface where the elastic modulus, H, is small. The amount of heave that PDEase2D predicted when using 25 stress steps is 115 mm. The solution of PDEase2D matches well with that of the Fredlund et al. (1980) method.

The results of analysis and measured data for the case history associated with Regina clay are presented in Figure 9. The calculated results agree well with the measured data. Figure 10 presents the comparison of the total heave predicted for all heave problems considered in this study. Figure 10 shows that in all cases, the heave predicted using the finite element method agrees well with that obtained from the Fredlund et al. (1980) method.

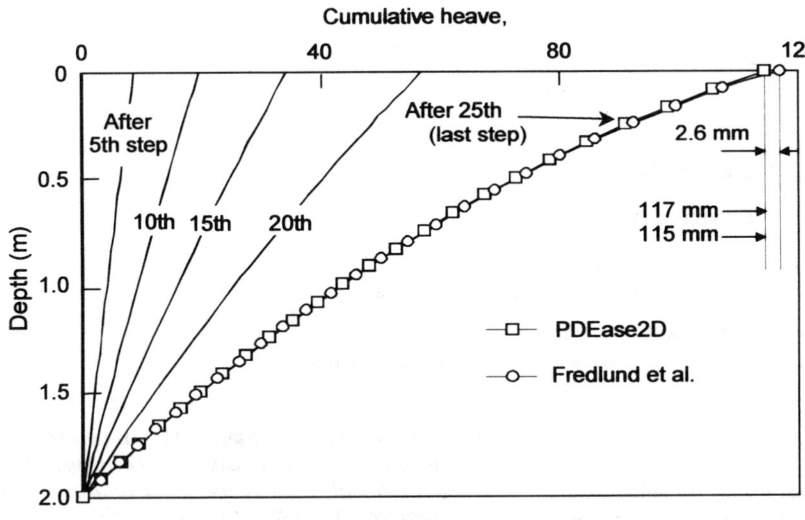

Figure 8. Comparison of the total heaves predicted for the first example

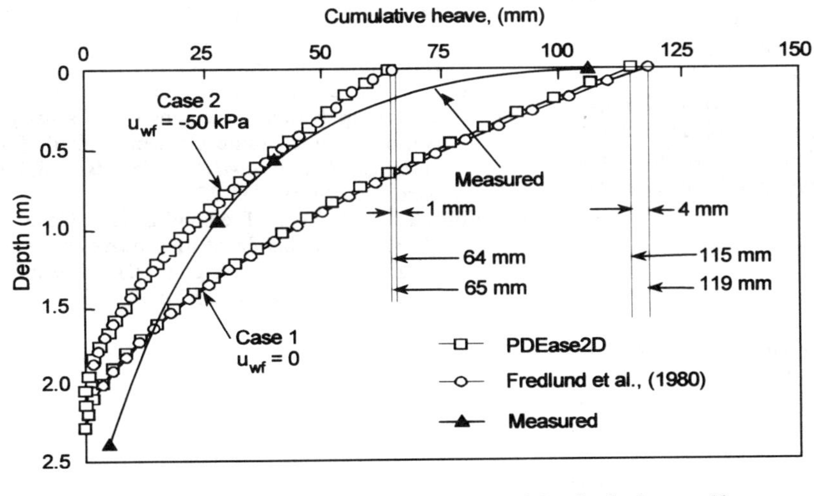

Figure 9. Comparison of the total heaves predicted and measured data for the first case history

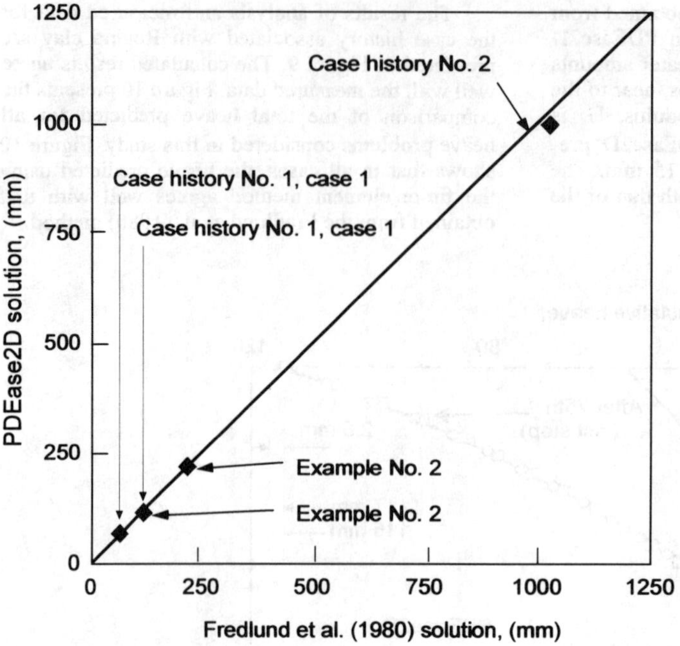

Figure 10. Comparison between the total heave predicted for all examples and case histories

5 CONCLUSION

It can be concluded that elastic modulus functions for an expansive soil can be calculated from one-dimensional oedometer test results and used to predict one-dimensional heave with the use of PDEase2D finite element solver. Comparison of the total heave predicted for all heave problems considered in this study shows that the heave predicted using the finite element method agrees well with that obtained from the Fredlund et al. (1980) method and the measured data. General-purpose partial differential equation solver such as PDEase2D shows excellent promise for solving heave problems with highly non-linear unsaturated soil properties.

REFERENCES

Ching, R.K.H. & D.G. Fredlund 1984. A Small Saskatchewan town copes with swelling clay problems. *Proceedings, 5th International Conference on Expansive Soils, Adelaide, Australia*: 306-310.

Desai, C.S., & J.T. Christian 1977. *Numerical methods in geotechnical engineering.* McGraw-Hill, New York, 783 p.

Fredlund, D.G., J.U. Hasan & H. Filson 1980. The prediction of total heave. *Proceedings, 4th International Conference on Expansive Soils, Denver, CO., June 16-18*, (1): 1-17.

Fredlund, D.G., & H. Rahardjo 1993. *Soil mechanics for unsaturated soils*. John Wiley & Sons, New York, 560 p.

Hung, V.Q. 2000. Finite element method for the prediction of volume change in expansive soils. *M.Sc. Thesis, University of Saskatchewan, Saskatoon, SK, Canada.*

Hung, V.Q. & D.G. Fredlund 2000. Volume change predictions in expansive soils using a two-dimensional finite element method. *Proceedings, Unsaturated Soil for Asia, Singapore, May 18-19:* 231-236.

Yoshida, R. T., D. G. Fredlund & J. J. Hamilton 1983. The prediction of total heave of a slab-on-grade floor on Regina clay. *Canadian Geotechnical Journal*, 20(1): 69 – 81.

Computer Methods and Advances in Geomechanics, Desai et al. (eds) © 2001 Balkema, Rotterdam, ISBN 90 5809 183 X

Modeling sand production with Darcy-Flow coupled with discrete elements

R.P.Jensen & D.S.Preece
Geomechanics Department, Sandia National Laboratories, Albuquerque, N.Mex., USA

B.K.Cook & J.R.Williams
Department of Civil and Environmental Engineering, Massachusetts Institute of Technology, Cambridge, Mass., USA

ABSTRACT: The discrete element method has been in development for several years. As the method matures, it is being extended and applied to a wider range of applications. As our computational capacity increases, the complexity of the phenomena that are being modeled also increases. The move toward more accurately modeling the physics of the problem is requiring the coupling of different numerical techniques. Coupled physical processes are commonly found in the petroleum industry, in particular, coupling between solids and fluids. One such problem is the occurrence of sand production in oil wells. As a result of the in situ stresses and fluid flow, poorly consolidated sandstone in some formations tend to disaggregate. This is a costly problem that is mitigated by very conservative approaches. This paper explores the coupling of 2-D DEM and a finite element implementation of the 2-D Laplace formulation for Darcy-Flow as a technique for assessing the potential for sand production.

1 INTRODUCTION

Sand production in oil wells is a costly and prevalent phenomenon that is not well understood. Sand production is the disaggregation of the formation due to a combination of in-situ stresses and fluid flow. It tends to be a progressive process once started and, as a result, the oil companies tend to take a conservative approach to the use of preventative measures such as gravel-pack filters or screens down-hole. Sometimes this equipment is unnecessarily installed as a preventative measure, increasing the cost of production and decreasing the productivity of the well. At the present time, the prediction of sand production is empirically based. A better understanding of the root cause of sand production is necessary to improve the prediction of the production zones and wells that are susceptible to sand production. As is often the case with micromechanical processes, it is difficult to be a neutral observer when experimentally investigating the causes of sand production. Numerical tools can provide insights and qualitative measurements that cannot be obtained experimentally.

In order to better investigate such micromechanical processes, a discrete element method (DEM) code MIMES has been developed. (Rege et al 1996) By coupling MIMES with Darcy fluid flow, this DEM code can be used as a means to study the phenomena that drive sand production as well as other coupled fluid-flow/particle motion processes.

2 DISCRETE ELEMENT METHOD (DEM)

The granular material application of DEM was originally developed about twenty years ago. (Cundall et al 1979) The basic idea of DEM, and as it is implemented in MIMES, is straightforward. Particles are modeled as independent, unique objects. These objects come into contact with neighboring particles. The interaction of these particles is modeled as shown in Figure 1 with a spring and dashpot in the normal direction and a spring, dashpot, and frictional slider in the tangential direction. Forces of each contact are computed based on the distance of particle overlap and the magnitude of the spring stiffness. When the tangential contact force reaches a predetermined limit, which can be related to the normal force, such as Coulomb friction, sliding occurs, thereby modeling interparticle friction. The interparticle contact relationship is based upon a simple linear-elastic constitutive model with option of other more sophisticated contact models, such as a nonlinear Hertzian contact model. The resultant contact force acting on the particle is computed and the ac-

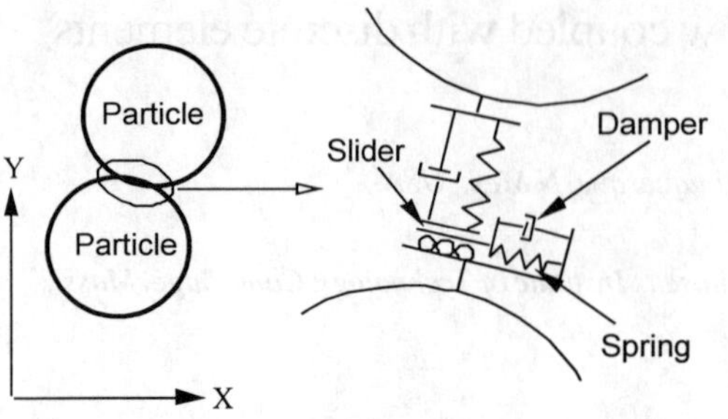

Figure 1. Particle contact model as idealized in DEM.

celeration of the particle is found using Newton's law. Using explicit time integration, the new velocity and displacement of the particle is calculated. From this new position, new contacts and contact forces are found, restarting the cycle.

In most implementations of DEM, particle-shape is a simple disc or sphere due to the fact that very robust, fast contact detection schemes require this simple geometry. However, several researchers have found that particle shape has a distinct influence on the response of the particles. (Jensen et al and Ting et al 1993) Spherical and disc shaped particles cannot capture the inherent geometry-dependent behavior that asperities, particle angularities, and particle roughness imparts. In order to capture some of these effects, the basic shape used by MIMES is an n-sided polygon. For ease of input of a wide variety of shapes, the superquadric equation is used as a template to define the particle boundaries. (Williams et al 1992)

3 FLUID-FLOW IMPLEMENTATION

3.1 *Darcy finite element formulation*

In MIMES, fluid flow simulation is accomplished by using finite element techniques to solve the 2-D Laplace formulation for Darcy-Flow given in Equation 1. (Klosek 1995)

$$k_x \frac{\partial^2 \phi}{\partial x^2} + k_y \frac{\partial^2 \phi}{\partial y^2} = 0 \qquad (1)$$

where k_x and k_y are porous media permeabilities in the *x* and *y* directions and ϕ is the pressure (potential) function, which is the only variable (or degree-of-freedom) in the problem. (Segerlind 1984) The current implementation assumes that k_x and k_y are equal, requiring only one input, *k*. Two methods for treating the permeability have been incorporated in MIMES. The first assumes that the permeability remains constant throughout the simulation in which case the fluid-flow conditions are solved for only one time. The second method allows the discrete element distribution to modify the permeability. This will be discussed in detail in a later section. In cases where the discrete elements are allowed to modify the permeability, a fluid solution can be obtained at a user-specified frequency, *n*, which is the number of discrete element time steps between fluid-flow solutions. Fluid pressure integration on each discrete element is based on pressure gradient and is described by Preece et al. (1999)

3.2 *Discrete element distribution effect on permeability*

An important capability of MIMES is Darcy flow permeability modification based on location of the discrete elements relative to the fluid flow mesh, thus completing the coupling between fluid-flow and discrete element movement. The porosity of the discrete elements can be easily calculated and can be related to the fluid-flow finite element based on the coverage of the element by discrete particles. The porosity is then related to permeability (Das 1990 and Holtz et al 1981) as given in Equation 2 below,

$$k = C\frac{e^3}{1+e} \qquad (2)$$

where k is the permeability, C is a constant, dependent on the material from which the porosity is created, and e is the porosity of the medium. Currently, the porosity of each triangular finite element is determined by intersecting the bounding circle for the discrete element with the triangle. The bounding circle approach causes some problems if the discrete elements have significant aspect ratio causing the bounding circle to affect the porosity of adjacent finite elements that it does not cover.

4 SAND PRODUCTION SIMULATION

A series of three coupled fluid-flow/particle-motion simulations of sand production from an oil well perforation channel were performed. All of the simulations had the same initial conditions except the magnitude of the tensile bonds, which represents the cohesive strength of the particle assembly. The simulation is of the cross-section of the perforation channel. Figure 2 shows the initial state of the packed particles, the boundary particles, and the fluid mesh. The perforation channel where freed sand particles are drawn away from the formation into the borehole is modeled by a sink particle. As a particle comes in contact with the sink particle, it is eliminated from the simulation, thereby modeling the

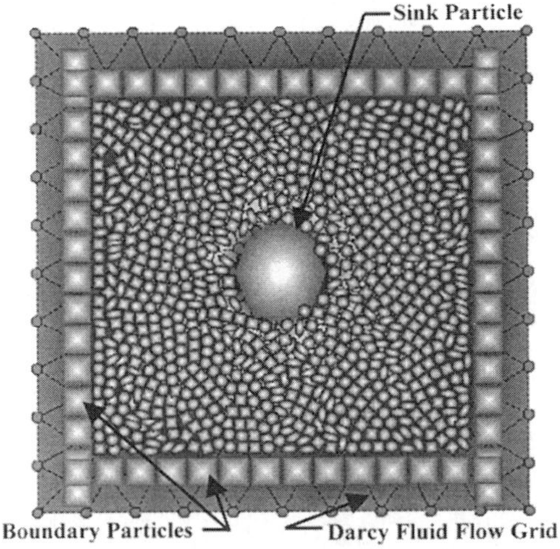

Figure 2. Geometry of perforation channel illustrating the fluid flow mesh, boundary particles, and sink particle. The sink particle represents a perforation channel. A constant force is applied to the boundary particles in order to simulate in situ stresses. Initial state of packed particles under in situ stress. The particles are bonded with tensile bonds at all contact points.

perforation. Applying a constant force to the boundary elements, which move as a rigid body, simulates an in situ stress since the simulation is 2-D. The particles are bonded with a tensile bond at all contact points. The means of bonding the particles in DEM is detailed by O'Connor et al. (1997) The fluid-flow solution was assumed to be steady state and was calculated only at the initial time step. Therefore the capability for particle-modified permeability was not utilized. This improvement will be added in future simulations. The cohesive strengths used in the different simulations are 1000 N, 1700 N, and 3000 N, for simulations A, B, and C, respectively.

Figure 3 shows simulation B in a steady state. At this point it can be seen that some of the particles have broken off from the matrix, but the formation is still stable. Simulation C, which had the highest level of particle-to-particle cohesion, was similar to B but with fewer particles broken from the matrix. In simulation A, which had the weakest level of particle-to-particle cohesion, the formation collapsed and all of the particles moved to the center of the simulation and were eliminated from the simulation, which is analogous to being drawn to the borehole through the perforation. The simulations showed that as the value of the point-to-point cohesion decreases, the number of particles breaking from the matrix increases. This follows the same trend as is found in production wells in more poorly consolidated sandstone where problems with sand production tend to exist. This strengthens the assumption that sand production occurs in more poorly consolidated sandstone due in part to weaker particle bonding.

5 CONCLUSION

This paper discusses the successful coupling of fluid-flow with a DEM system using a Darcy fluid-flow formulation implemented into the DEM code MIMES. A series of simulations were discussed which demonstrate the ability of the coupled fluid-flow/particle-motion capability of MIMES to analyze complex problems such as sand production. In these simulations, it was seen that as the strength of the cohesive bonds decreased, the number of particles breaking free from the matrix increased. These simulations seem to indicate that fluid-flow coupling with DEM can be used as a tool to increase the understanding of the mechanisms that cause sand production.

While the results were encouraging, some deficiencies remain. Because deleterious boundary effects could influence the solution, a much larger number of particles need to be used in the simulations. It could also be argued that Darcy fluid flow is not the best description of the physics where particles are being transported by the fluid. Furthermore, two-dimensional analyses in a three-dimensional world are always plagued by questions of applicability. Particle behavior can be quite different when the motions are not constrained to movement within

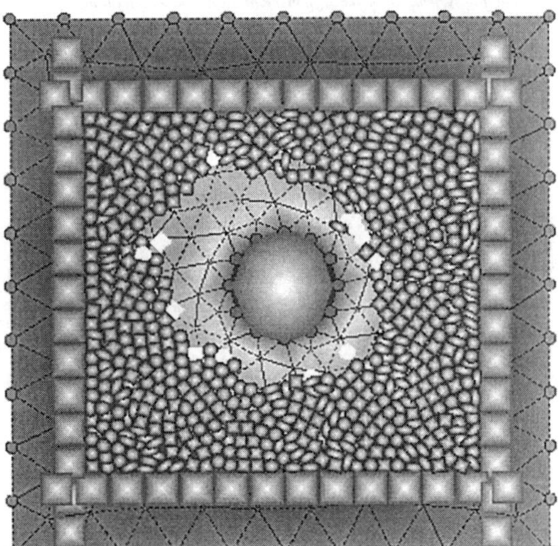

Figure 3. Simulation B. Point-to-point particle cohesion is 1700 N. The simulation is at a steady state. Note the loss of particles near the center of the packed particles.

plane. Porosity and permeability are also difficult to define in two dimensions.

In order to address some of these issues, present and future work include parallelization of the code, a Lattice-Boltzmann implementation of the fluid flow, and extension of the code into 3-D.

6 ACKNOWLEDGEMENTS

The Delaunay triangulation mesh generation technique was implemented from the code "EasyMesh v1.4 written by Bojan Niceno and is available for download at the following Website:

http://wwwdinma.univ.trieste.it/~nirftc/research/easymesh

This research was performed with support from the Advanced Computational Technology Initiative of the U.S. Department of Energy.

Cundall, P. A. & Strack, O.D.L. 1979 A Discrete Numerical Model for Granular Assemblies. *Geotechnique* 29 (1): 47-65

Das, B.M. 1990. *Principles of Geotechnical Engineering.* 2nd Ed. Boston: PWS-Kent Publishing

Holtz, R. D. and Kovacs, W.D. 1981. *An Introduction to Geotechnical Engineering.* Englewood Cliffs, NJ: Prentice Hall, Inc.

Jensen, R.P., Bosscher, P.J., Plesha, M.E. & Edil, T.B. "DEM Simulation Of Granular Media - Structure Interface: Effects of Surface Roughness and Particle Shape, "*International Journal for Numerical and Analytical Methods in Geomechanics 23".*

Klosek, J.T. 1995. *The integration of fluid dynamics with a discrete-element modeling system: Algorithms, Implementation, and Applications.* Masters Thesis, MIT Dept. of Civil and Env. Eng., Cambridge, MA.

O'Connor, R.M., Torczynski, J.R., Preece, D.S., Kosek, J.T., and Williams, J.R.,. Discrete element modeling of sand production. *International Journal of Rock Mechanics and Mining Sciences 34 (3-4): 373*

Preece, D.S., Jensen, R.P., Perkins, E.D. and Williams, J.R. 1999 Sand Production Modeling Using Superquadric Discrete Elements and Coupling of Fluid Flow and Particle Motion. *37th US Rock Mechanics Symposium (Vail Rocks).* Amsterdam, Balkema.

Rege, N. 1996 *Computer modeling of granular materials* Ph.D. Thesis, MIT Dept of Civil and Env. Eng., Cambridge, MA.

Segerlind, L.J. 1984. *Applied Finite Element Procedures.* 2nd ed. New York: John Wiley & Sons.

Ting, J.M., Khwaja, M., Meachum, L.R., & Rowell, J.D. 1993. An ellipse-based discrete element model for granular materials. *International Journal for Numerical and Analytical Methods in Geomechanics.* 17:379-398.

Williams, J.R., O'Connor, R.M. & Rege, N. 1996 Discrete element analysis and granular vortex formation. *Electronic Journal of Geotechnical Engineering,* 1, No. 1

Williams, J.R. & Pentland, A. 1992. Superquadrics and model dynamics for discrete elements in interactive design. *Engineering Computations* 9: 115-127

Computer Methods and Advances in Geomechanics, Desai et al. (eds) © 2001 Balkema, Rotterdam, ISBN 90 5809 183 X

Heat flow analyses to trace groundwater flow in Hokkaido, Japan

H. Kurikami
Graduate School of Engineering, Kyoto University, Japan

M. Ikeda
Hokkaido Engineering Consultants Company Limited

K. Miura
School of Civil Engineering, Asian Institute of Technology, Thailand

ABSTRACT: One of the most important subjects in groundwater hydrology is to clarify groundwater flow systems. Environmental tracers have recently been used to characterize the flow systems without disturbing natural groundwater flow. In this paper, groundwater temperature is analyzed as one of the environmental tracers by the finite element method, and was applied to both the Bibi River groundwater basin and the Tokachi Plain located in Hokkaido, Japan. The former is the groundwater flow system with very complicated aquifer including high permeable layers in pyroclastic plateau, while the latter treats groundwater flow in a tectonic groundwater basin. The existence of the volcanic effect on the temperature is also discussed. Good agreement between observed and calculated temperature shows the effectiveness of this method.

1 INTRODUCTION

Groundwater survey is one of the most important items in applied geology. Artificial tracer method is useful for small areas but it causes disturbance of natural groundwater gradient and velocity especially when applied to broader areas. On the other hand, environmental tracers that include physical or chemical information in the groundwater itself are useful to estimate regional groundwater flow system.

Groundwater temperature is one of the most effective environmental tracers (Domenico & Palciauskas 1973). The principle is that temperature in discharge areas is higher than average and lower than average in recharge areas at the same depth, because of heat advection. The numerical approach using the finite element method (FEM) was carried out in order to clarify the groundwater flow system in the Bibi River groundwater basin and in the Tokachi Plain located in Hokkaido, Japan.

2 METHODOLOGY

The FEM program was developed in order to simulate thermal advection-dispersion phenomenon (Ikeda et al. 1999a). The governing equation describing the steady state groundwater flow can be written as

$$\left(k_{ij}h_{,j}\right)_{,i} - q = 0 \tag{1}$$

where h = groundwater potential head; k_{ij} = the permeability tensor; and q = groundwater flux. The groundwater velocity v_i for the direction i can be determined from

$$v_i = -k_{ij}h_{,i} \tag{2}$$

And the heat transport equation describing the unsteady state temperature distribution in a saturated porous media is given by

$$\rho_{av}C_{av}\dot{T} + \rho_w C_w v_i T_{,i} - \left(K_{avij}T_{,j}\right)_{,i} - \psi = 0 \tag{3}$$

where T = temperature; ρ_{av} = density of mixture of soils and groundwater; C_{av} = heat capacity of mixture; ρ_w = density of groundwater; C_w = heat capacity of groundwater; K_{avij} = thermal conductivity tensor; and ψ = heat flux. In this method, local thermal equilibrium between the soils and the groundwater is required.

In the FEM program developed using discrete model of the equations (1) and (3), the groundwater potential distribution is firstly obtained in a steady state and then the temperature distribution is calculated. Unsteady analysis is applied only to one case of the analyses of the Tokachi Plain with pump, in which the finite difference method is applied to time domain. In this study, storage of groundwater is neglected because groundwater potential head change is conveyed much more rapidly than temperature change by saturated media. Although the Tokachi Plain was simulated in three-dimensional model, the Bibi River basin was simulated in two-dimensional model because of the difficulty of modeling and cal-

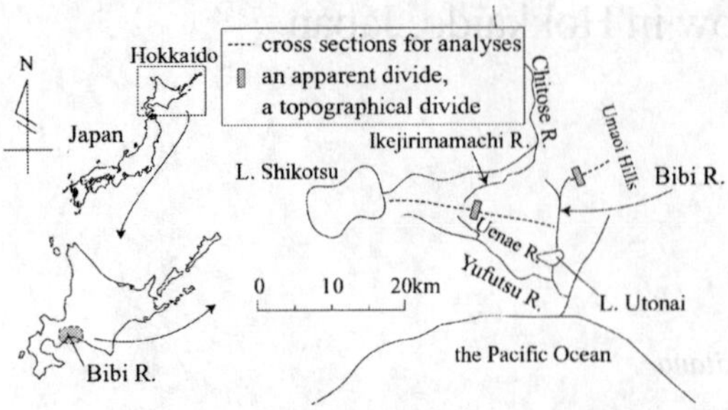

Figure 1. Location of the Bibi River and the cross sections for analyses.

culation caused by the complicated geological structure including high permeable layer.

3 THE BIBI RIVER GROUNDWATER BASIN

3.1 *Background*

The Bibi River is located in the south Hokkaido, the northernmost island of Japan (Fig. 1). The groundwater basin is mainly composed of pyroclastic flow deposits which produced Lake Shikotsu, caldera lake, ca. 40 Ka, including high permeable layer of pumice fall deposits. Therefore, the groundwater temperature survey was conducted by Ikeda et al. (1999b) in order to estimate the groundwater flow through the complicated aquifer and the extent of the groundwater basin.

The locations of the E-W cross sections for the analyses are shown in Figure 1. In the right bank of the Bibi River, there are two streams, the Ikejirimamachi River and the Uenae River, running north and south in the middle area of the section. The temperature distribution suggests that these two streams are the discharge area of the piedmont zone of Lake Shikotsu, and that topographical divide is the recharge area of the Bibi River groundwater basin. This study examines the validity of the extent of the groundwater basin estimated and the possibility that the groundwater around the Lake Sikotsu flows into the Bibi River.

On the other hand, there is the apparent divide close to the headwaters in the left bank. The investigation of water balance at the headwaters (Ikeda et al. 1999b) shows that the discharge rate measured requires the groundwater basin about 10 times as large as that equivalent to the divide. Therefore, the possibility of the extension of the basin to the Umaoi Hills is examined.

Table 1. Main physical properties used in the simulation.

Geological formations	Symbol	Horizontal permeability k_h (m/s)	Vertical permeability k_v (m/s)
Shikotsu pumice	$Spfl_2$	2.0×10^{-6}	2.0×10^{-6}
flow deposits	$Spfl_3$	1.0×10^{-6}	1.0×10^{-6}
Shikotsu pumice	$Spfa_{1,2}$	5.0×10^{-3}	5.0×10^{-3}
fall deposits	$Spfa_{7,10}$	7.0×10^{-4}	7.0×10^{-4}
Ausari formation	As-c	1.0×10^{-8}	1.0×10^{-8}
Fumonke formation	Fm-c	1.0×10^{-8}	1.0×10^{-8}

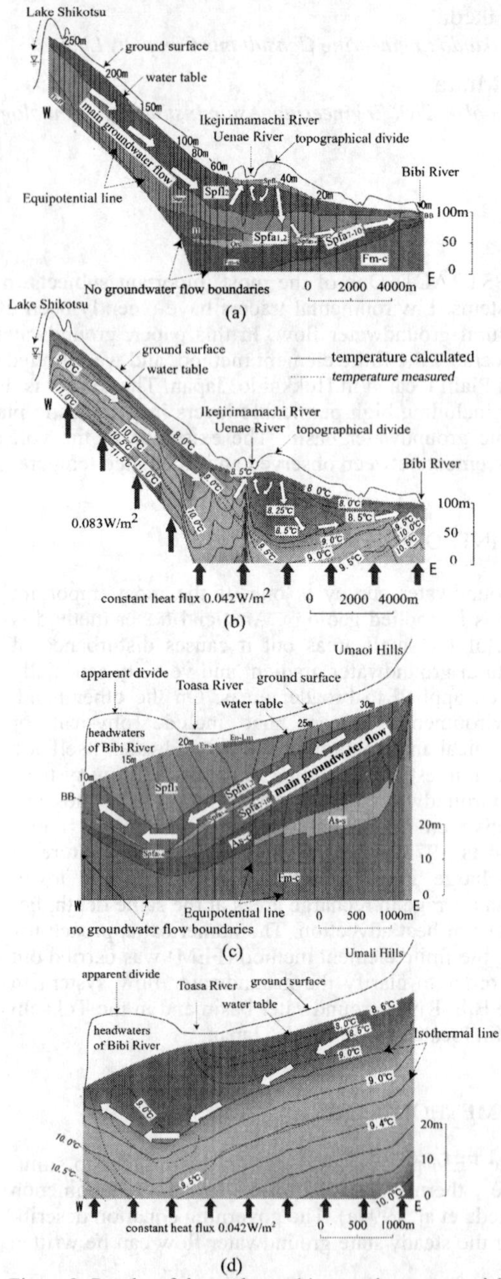

Figure 2. Results of the analyses: (a) groundwater potential head distribution in the right bank: (b) temperature distribution in the right bank: (c) groundwater potential distribution in the left bank: (d) temperature distribution in the left bank, respectively.

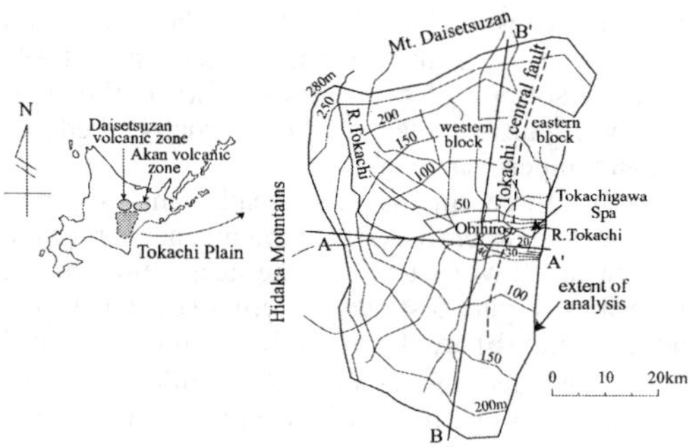

Figure 3. Location of the Tokachi Plain

3.2 *Boundary conditions and results of analyses*

The cross sections and the results of the analyses are shown in Figure 2, where (a) and (b) are of the right bank and (c) and (d) are of the left bank. For (a) and (c), showing the groundwater potential head distributions, the upper boundary is the water table and the others are no flow boundaries. On the other hand, for (b) and (d), the temperature distributions, the annual mean temperature is specified along the upper boundary. The constant heat flux 0.042 W/m^2 are provided across the lower boundary except for the western part of the right bank, where the value of heat flux 0.083 W/m^2 was specified because of the 'Green Tuff' area, higher heat flow zone in the Japanese islands occurred during the Miocene volcanic activity. The main aquifers consist of volcanic ashes and the Formation $Spfa_{1,2}$ is the most permeable layer in particular, whose coefficient of permeability is about 5.0×10^{-3}m/s. The main properties of the materials are shown in Table 1. The measured temperature distributions are overlapped in Figure 2(b) and 2(d). In spite of the existence of the high permeable layer, almost stable solution was obtained.

It can be said from the analysis of the right bank that the topographical divide is the recharge area because low temperature zone exists, and that where the two streams begin is the discharge area because high temperature zone exists. It indicates that the groundwater from Lake Shikotsu does not flow into the Bibi River except for a slight part of groundwater flow below the Formation Spfa. The temperature simulated is, however, higher than that from the measurement under the two streams. That may be because of ignorance of the three-dimensional effect. On the other hand, the analysis of left bank clarifies the expansion of the groundwater basin to the Umaoi Hills caused by the high permeable layers.

4 THE TOKACHI PLAIN

4.1 *Background*

The Tokachi Plain is a sedimentary basin located in southeast Hokkaido (Fig. 3). The plain is bordered by Mt. Daisetsuzan volcanic zone in the north and the Hidaka Mountains in the west, and is divided into eastern and western blocks by the Tokachi central fault extending from north to south. Around the central area of the plain where the groundwater discharges, temperatures are higher than average and some hot springs exist. Before our study, hot spring supply from the Daisetsuzan volcanic zone was vaguely imagined.

Recently, there has been increased interest in the groundwater flow system and the heat source of the hot springs, because the hot spring temperatures and discharge amounts have decreased. Therefore, in order to clarify the contribution of heat advection to the higher temperature zone, three-dimensional FEM analysis was applied and compared with the analysis of stable isotopic ratios reported by Ikeda et al. (2000). The results of stable isotopic ratios clarified the groundwater flow system in the Tokachi Plain as follows:

1. The stable isotopic ratio distribution of deuterium is shown in Figure 4.

2. The origin of groundwater in the western block is meteoric. In addition, no isotopic shift due to magmagnetic water was found.

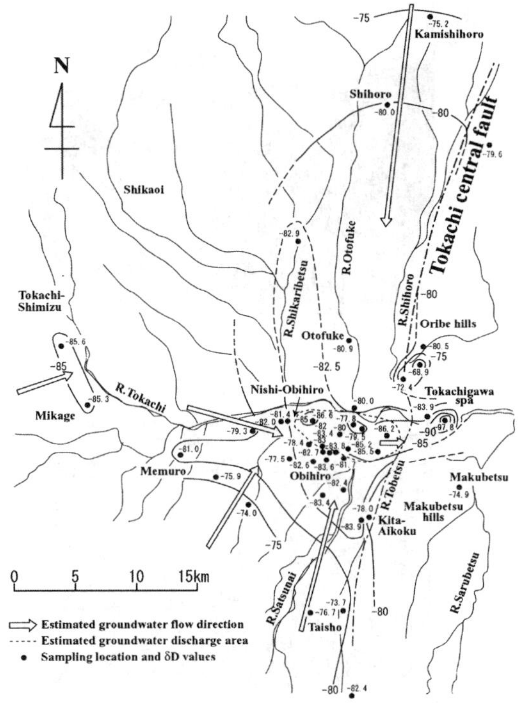

Figure 4. Stable isotopic ratio distribution of deuterium in the Tokachi Plain deep groundwater

3. The altitude effect on stable isotopic ratios shows that the groundwater in the deeper aquifer has mainly been recharged in the higher areas.

Table 2. Main physical properties used in the simulation.

Geological formations		Density $\rho_w \times 10^3$ (kg/m^3)	Horizontal permeability k_h (m/s)	Vertical permeability k_v (m/s)
Shibusan formation		1.7	2.0×10^{-5}	2.0×10^{-8}
Osarushinai formation		1.7	5.0×10^{-5}	2.0×10^{-8}
Ikeda formation	1	1.8	1.7×10^{-6}	2.0×10^{-8}
	2	1.8	1.7×10^{-5}	2.0×10^{-8}
Nukanai formation		1.8	5.0×10^{-6}	2.0×10^{-8}
Miocene, Pre-cretaceous		2.0	1.0×10^{-8}	1.0×10^{-8}

4. The groundwater recharge in piedmont areas of the plain is concentrated radially to the middle area of the plain and discharge near the Tokachi central fault.

4.2 *Boundary conditions and result of analysis*

The Tokachi Plain was modeled for the numerical analysis. The extent of the analysis is shown in Figure 3. No groundwater flow or heat flow is allowed through the lateral boundaries. The upper boundary is the water table, and is set at the annual mean temperature. In addition, the lower boundary is the bottom of the Nukanai Formation with no groundwater flow and a constant heat flux, 0.05 W/m^2. The main properties of the plain materials are shown in Table 2. The anisotropy of aquifer is caused by the sedimentary condition, that is, the alternative beds of sand and impermeable layers.

Firstly, three-dimensional steady analysis was carried out in order to evaluate the mechanism of the groundwater flow system and the heat source of the hot springs. Figure 5 shows the cross sections of the result. (a) and (b) are the groundwater potential head distributions and temperature distributions of the cross section A-A', respectively, and (c) and (d) are those of the cross section B-B'. They indicate that the groundwater recharge in the surrounding mountains is concentrated radially to the central area of the plain. This trend is in good consistency with the result of the stable isotopic analysis. And the groundwater flow system causes heat concentration in the middle part of the Tokachi central fault. The temperatures simulated were consistent with the temperatures observed at the points ① to ④ in Figure 6. Good agreement between simulated and observed temperatures suggests that the temperature distribution of the Tokachi Plain and the heat source

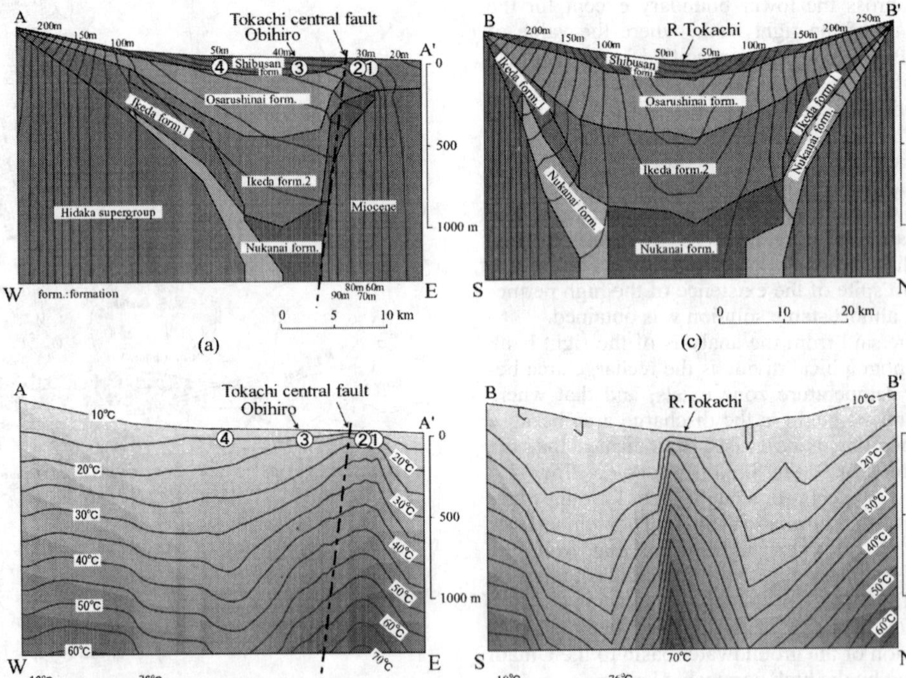

Figure 5. Result of the simulation: (a) groundwater potential head (A-A'): (b) temperature distribution (A-A'): (c) groundwater potential (B-B'): (d) temperature distribution (B-B')

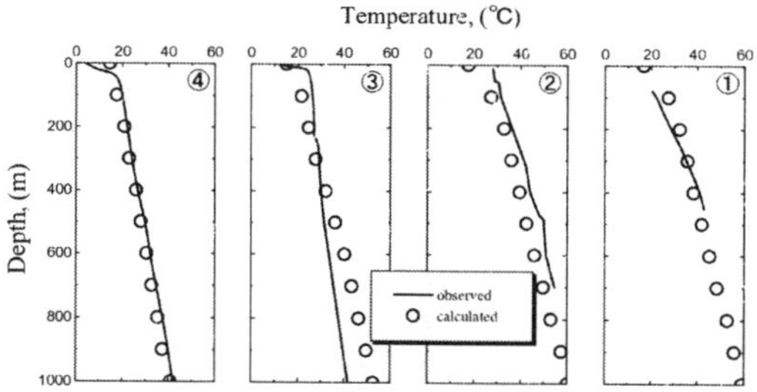

Figure 6. Comparison between observed and simulated temperatures (The location of ① to ④ is shown in Figure 4).

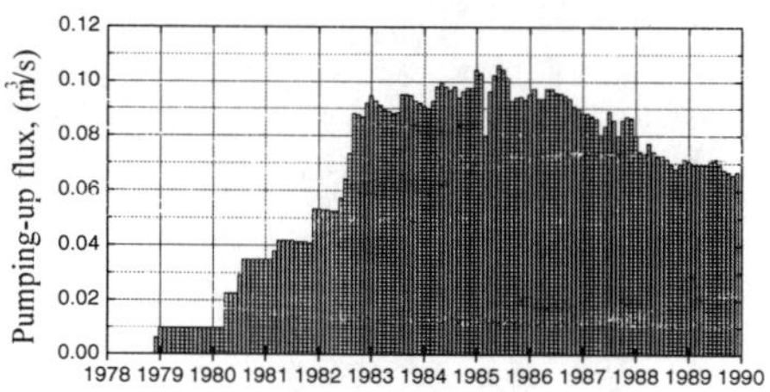

Figure 7. Actual measurement of pumping rate

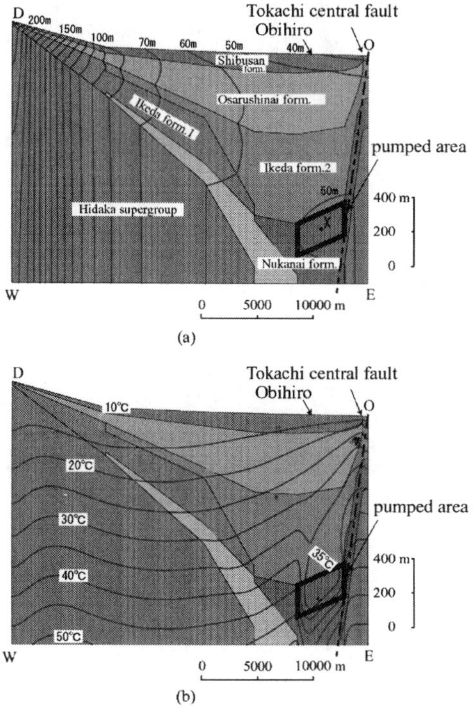

Figure 8. Result of the unsteady analysis: (a) groundwater potential distribution: (b) temperature distribution in 1985

of the hot springs are mainly explained by heat advection except for the Tokachigawa Spa.

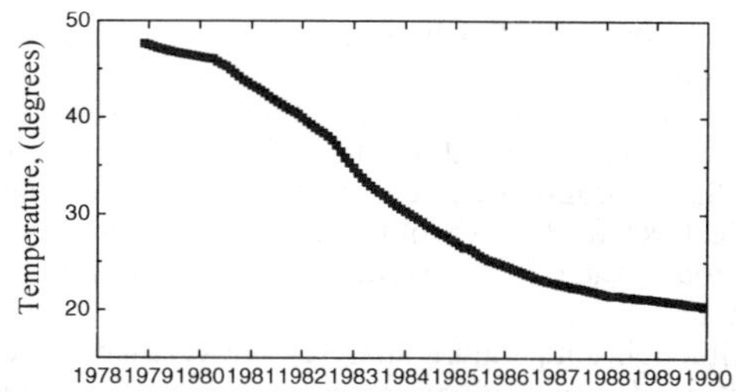

Figure 9. Time history of the groundwater temperature at the point X

Secondly, axisymmetric unsteady analysis with pumping was carried out in order to clarify the influence of pumping on the temperature decreasing. The reason of using axisymmetric model is for saving the computer storage and calculation time. The axisymmetric model in steady state has been verified by comparing three-dimensional model by Kurikami (2000). For the initial condition, the result of the analysis without pumping was applied. Figure 7 shows the pumping rate (Oka 1990) applied to the Nukanai Formation. At the same time, heat energy loss following pumping has to be considered as follows:

$$E = -\rho_w C_w T_e q_W \quad (4)$$

where E = heat energy loss due to pumping; T_e = water temperature; and q_W = pumped water volume. The time step was set for one month since 1978 to 1990. Figure 8 is the simulated temperature distribution in 1985. Groundwater flow from the Ikeda Formation to the Nukanai Formation causes the temperature decreasing. Figure 9 shows the time history of the groundwater temperature at the point X shown in Figure 8(a). With the increase of pumping rate, the temperature has decreased. The rate of temperature decreasing is, however, larger than the measured values. This may be caused by inaccuracy of the pumping rate, material properties and dropped water level.

5 CONCLUSIONS

This paper discussed the utilization of numerical thermal advection-dispersion analyses in order to estimate groundwater flow. For the case study, both the Bibi River basin and the Tokachi Plain are simulated.

The Bibi River basin has very complicated geological structure including the high permeable layers of volcanic ashes, which prevents stable temperature solution. The utility of temperature distribution as an environmental tracer was verified qualitatively by the application to the Bibi River basin. Further progress of precise modeling will make quantitative analyses possible.

On the other hand, temperature decreasing by pumping was expressed by the analyses in the Tokachi Plain. Groundwater basin management for the whole Tokachi Plain should be instituted to preserve the groundwater flow system and the hot springs in the plain.

Although temperature analyses are verified to be effective, decision of parameters is very difficult. In a groundwater survey, several ways such as potential investigation and analyses of stable isotopic ratios should be carried out.

6 ACKNOWLEDGMENT

We would like to express our thanks to Professor Y. Ohnishi of Kyoto university for his precious advice.

REFERENCES

Domenico, P. A. & Palciauskas, V. V. 1973. Theoretical analysis of forced convective heat transfer in regional groundwater flow. Geol. Soc. Am. Bull 84: 3803-3814.

Ikeda, M., Kurikami, H. & Kinya, M. 2000. Groundwater flow system in the Tokachi Plain, Hokkaido, estimated from groundwater temperature and stable isotopic ratio. journal of Japanese association of groundwater hydrology (in Japanese) 42(1): 3-26.

Ikeda, M., Kurikami, H. & Miura, K. 1999a. FEM analysis of temperature field affected by thermal advection in groundwater basin. (in Japanese) Tuchi-to-kiso Japanese Geol. Soc. 47(8): 25-28.

Ikeda, M., Miura, K. & Kurikami, H. 1999b. Modeling groundwater flow around the Bibi River in Hokkaido using groundwater temperature. (in Japanese) Jour. Japan Soc. Eng. Geol. 40(2): 70-85.

Kurikami, H. 2000. A study on numerical analysis of heat conduction and advection in groundwater flow system. (in Japanese) Master thesis of Hokkaido university.

Oka, T. 1990. Development of hot springs in the Tokachi Plain and the hot spring resource management. (in Japanese) Proc. of 1990 annual meeting of JAEG: 169-172.

Computer Methods and Advances in Geomechanics, Desai et al. (eds) © 2001 Balkema, Rotterdam, ISBN 90 5809 183 X

Thermo-mechanical modelling of the behaviour of MC clay

L. Laloui, C. Cekerevac & L. Vulliet
Swiss Federal Institute of Technology, Lausanne, Switzerland

ABSTRACT: This paper presents the capability of a cyclic thermo-elastoplastic constitutive model (LTVP) to simulate the behaviour of fine soils submitted to coupled thermo-mechanical loading. The model formulation decomposes the strain rate into non-linear thermo-elasticity and thermoplasticity based on two hardening functions (mechanical and thermal) depending on mechanical and thermal yield limits. Performances of this model are shown through numerical predictions of experimental results on different combinations of thermo-mechanical loading paths.

1 INTRODUCTION

For the range of temperatures between 5°C and 90°C, thermal effects on fine saturated soils are characterised by non-linearity and plasticity (Campanella & Mitchell 1968). Very few models treating the numerical simulation of such thermo-plastic behaviour exist in the literature. Hueckel & Borsetto (1990) presented a paper summarising their work on the modelling of the thermo-mechanical behaviour of clays and shales. The proposed model was, to our knowledge, the first constitutive thermo-mechanical law for soils explicitly including thermo-plasticity. Starting from the Cam-clay model, they introduced the thermal effect following Prager's thermo-plasticity theory. This effect is incorporated at two levels: (i) reversible dilatation is added to the elastic component of the strain tensor and (ii) a state variable is introduced in order to follow the evaluation of the yield surface with temperature. The obtained model treats normally consolidated (NC) conditions with thermo-elasticity, while thermo-plasticity is used for overconsolidated (OC) conditions.

Robinet *et al.* (1996) proposed a thermo-elasto-plastic model based on a modified Cam-clay law with mechanical and thermal plastic mechanisms. The introduction of an irreversible thermal strain remains the principal difference between this model and the Hueckel & Borsetto model.

No particular validations of these models have been performed on real, coupled thermo-mechanical paths or for cyclic solicitations.

Laloui (1993) and Modaressi & Laloui (1992, 1997) proposed a cyclic thermo-viscoplastic constitutive model (LTVP) able to represent the principal thermo-mechanical characteristics of fine soils for a wide range of loading conditions. This LTVP model handles, in the same framework, NC and OC conditions. It is based on non-linear elasticity and four kinematic plastification mechanisms: one isotropic and three deviatoric. Temperature acts on the essential rheological parameters. With this structure, the LTVP model is the first one able to take into account the thermal effect on anisotropic (initial and induced) clays and to simulate cyclic (mechanical) behaviour, even at high temperature.

Parametric studies carried out with the help of this model show that the coupling of elasto-plastic parameters with temperature produces qualitatively different results from those of a thermo-elastoplastic model without coupling (Félix *et al.* 1996).

In this paper, performances of the thermo-plastic version of the LTVP model for the prediction of behaviour are analysed. Special attention is given to the coupled thermo-mechanical paths with loading-unloading.

After a brief presentation of the main aspects of the thermo-plastic version of the LTVP model, the employed experimental results are introduced. Model parameters and back-predictions are given. The model predictions are then discussed in the final section.

2 CYCLIC THERMO-PLASTIC CONSTITUTIVE MODEL (LTVP)

The considered cyclic thermo-plastic constitutive model is the time-independent version of the cyclic thermo-viscoplastic model LTVP (Laloui 1993). It was developed a few years ago (Modaressi & Laloui 1992; Modaressi & Laloui 1997; Modaressi *et al.* 1991) based on a Hujeux model originally developed for the simulation of the hydro-mechanical behaviour of granular materials, which has undergone many experimental validations for clays as well as for sands (Aubry *et al.* 1982).
Due to the limitation of the length of this paper, the reader is invited to find details of the model in the mentioned references. In the following paragraph, only the particular points of the model will be presented.

2.1 *Thermo-elastic and Thermo-plastic formulations*

Temperature increase produces variation of void ratio (from contractant with irreversibility for normally consolidated state to dilatant with reversibility for overconslidated state). Strains are partly thermo-elastic and partly thermo-plastic. The thermo-elastic strain rate is expressed as:

$$\dot{\varepsilon}_v^{Te} = \frac{\dot{p}}{K} + \beta'_s \dot{T} \tag{1}$$

$$\dot{\varepsilon}_d^{e} = \frac{\dot{q}}{G} \tag{2}$$

where $\dot{\varepsilon}_v^{Te}$ is the volumetric strain rate and $\dot{\varepsilon}_d^{e}$ the deviatoric strain rate; $\dot{q}$ is the deviatoric effective stress and $\dot{p}$ the effective mean pressure. *K* and *G* are bulk and shear moduli, respectively, which could also depend on temperature. $\dot{T}$ is the temperature increment. The thermal expansion coefficient of the solid skeleton, β'_s, varies strongly with temperature and slightly with pressure according to:

$$\beta'_s = (\beta'_{s0} + \zeta T)\xi \tag{3}$$

with β'_{s0}, the isotropic thermal expansion coefficient at reference temperature, ζ a material parameter and ξ a function of initial density and main pressure. This formulation enables the thermo-elastic strain rate to increase with overconsolidation ratio and temperature. The thermo-plastic strain rate is given by:

$$\dot{\varepsilon}^{Tp} = \lambda(\boldsymbol{\sigma},T,\boldsymbol{\alpha}).\boldsymbol{\Psi}(\boldsymbol{\sigma},T,\boldsymbol{\alpha}) \tag{4}$$

with $\lambda(\boldsymbol{\sigma},T,\boldsymbol{\alpha})$ the plastic multiplier and $\boldsymbol{\Psi}(\boldsymbol{\sigma},T,\boldsymbol{\alpha})$ the direction of the strain increment. $\boldsymbol{\alpha}$ represents the internal variables modifying the yield surfaces $f(\boldsymbol{\sigma},T,\boldsymbol{\alpha})$. With the two state parameters (stress tensor and temperature), the thermo-plastic strain rate is given by:

$$\dot{\varepsilon}^{Tp} = \frac{1}{H}(\partial_T f.\dot{T} + \partial_{\boldsymbol{\sigma}} f : \dot{\boldsymbol{\sigma}}).\boldsymbol{\Psi} \tag{5}$$

where $H(\boldsymbol{\sigma},T,\boldsymbol{\alpha})$ is a scalar quantity representing the thermo-plastic modulus, which can be determined from the consistency equation.
In the above relationship, temperature appears in the numerator as a state parameter and in the denominator as an internal variable.
From Equation 5, the final constitutive expression is:

$$\dot{\boldsymbol{\sigma}} = \mathbf{D}^{ep}(\dot{\boldsymbol{\varepsilon}} - \frac{\beta'_s}{3}\dot{T}\boldsymbol{\delta}) - \mathbf{D}^{Tep}\dot{T} \tag{6}$$

where $\dot{\boldsymbol{\varepsilon}}$ is the total strain increment and $\boldsymbol{\delta}$ the Kroenecker symbol. $\mathbf{D}^{ep}(\boldsymbol{\sigma},T,\boldsymbol{\alpha},\boldsymbol{\varepsilon})$ is the elasto-plastic matrix in adiabatic conditions and $\mathbf{D}^{Tep}(\boldsymbol{\sigma},T,\boldsymbol{\alpha})$ the thermo-plastic matrix.

The mechanical behaviour is treated by coupling isotropic and deviatoric solicitations.

The deviatoric yield limit f_d can be represented by a Mohr-Coulomb criterion:

$$f_d = q - pFr\sin\phi \tag{7}$$

with two hardening functions *F* and *r*. The volumetric hardening/softening function $F(p, \varepsilon_v^{Tp})$ depends on the thermo-plastic volumetric strain induced either by mechanical or thermal loading:

$$F = 1 - b\left(\ln\frac{p}{p_{c0}} - \beta\varepsilon_v^{Tp}\right) \tag{8}$$

with *b* a numerical parameter, *β* the plastic compressibility modulus and p_{c0} the critical state pressure for the initial state at reference temperature. *r* induces the effect of the plastic deviatoric strain rate on the yield limit as a shear hardening.
ϕ is the critical internal friction angle depending on temperature.

The isotropic yield limit is represented as:

$$f_i = p - dp_{c0}\,exp^{\beta\varepsilon_v^{Tp}}\, r_{init} \tag{9}$$

where r_{init} corresponds to the size of the elastic domain. In "e-Ln p" plane, *d* represents the distance between the perfectly plastic line at ambient temperature and the isotropic consolidation line for the considered temperature. The temperature dependency of this parameter is expressed as:

$$d = d_0 \exp^{-\beta\Delta e} \tag{10}$$

where Δe is the void ratio variation with respect to temperature:

$$\Delta e = \frac{\Delta T}{a_t \Delta T + b_t} \tag{11}$$

a_t and b_t being material parameters.

The irreversible variation of void ratio due to thermal loading induces an increase in preconsolidation pressure as well as in shear strength. At the same time, some mechanical characteristics are affected by the thermal effect. Based on experimental observations, the friction angle depends on temperature by:

$$\phi = \phi_0 - gT \tag{12}$$

where ϕ_0 is the friction angle at the critical state and reference temperature, g, the slope of variation of friction angle with temperature.

2.2 *Undrained conditions*

Undrained thermo-mechanical tests were simulated using a continuum approach based on mass conservation of solid and fluid phases. The pore-water pressure generation $\dot{u}$ is given by (Laloui 1993):

$$\theta\beta'_f\,\dot{u} = -\dot{\varepsilon}_v + \left[\theta\beta'_f + (1-\theta)(\beta'_{s0} + \varsigma T)\xi\right]\dot{T} \tag{13}$$

where θ is the effective porosity of the solid skeleton and β'_f the compressibility of free water.

Therefore, the variation of the pore-water pressure is attributed to two factors: the variation of skeleton volume (compaction or dilation) and the thermal dilations of water.

The model has 20 parameters from which 5 are related to the thermal aspect and to the undrained conditions.

3 NUMERICAL SIMULATION OF THERMO-MECHANICAL COUPLED PATHS

Experimental results obtained by Kuntiwattanakul (1991; Kuntiwattanakul *et al.* 1995) were used in this study. Tests were performed on reconstituted samples from clay slurry made by mixing MC (Kaolin) clay and water. The aim of the research was the study of the effect of high temperature (up to 90°C) on the undrained shear characteristics of clay in normally and overconsolidated states. The properties of the MC clay used are given in Table 1.

Table 1. Properties of MC clay

w_l(%)	w_p(%)	I_p	%<2μm	%<5μm	γ_s(kN/m^3)
70	41	29	49	91	26.98

Details of sample preparations and of the loading paths can be found in Kuntiwattanakul (1991).

3.1 *Experimental paths*

For normally consolidated samples (Figure 2)

Path 1(0→1): This is the conventional undrained triaxial test at room temperature: consolidation from 20 to 196 kPa at 20°C followed by shearing under undrained conditions.

Path 2(0→1→2): Identical consolidation as for path 1, followed by a temperature increase up to 90°C under drained conditions before shearing.

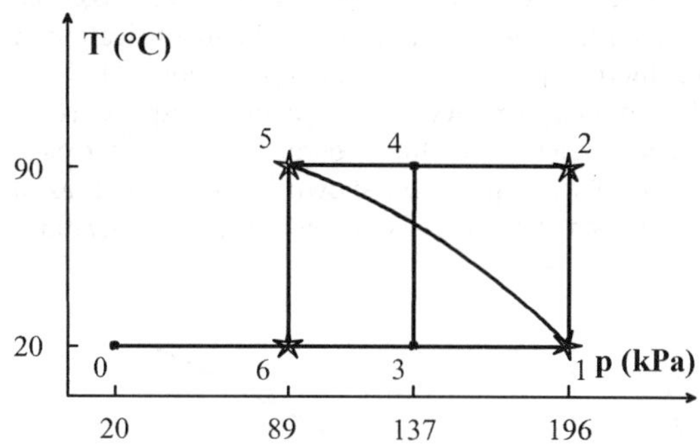

Figure 2. Experimental paths; stars mark points at which samples are sheared

Path 3(0→3→4→2): The sequence is: consolidation from 20 to 137kPa; temperature increase up to 90°C under drained conditions; consolidation pressure increasing up to 196 kPa; shearing.

Path 4 (0→1→2→1): The sequence is: consolidation up to 196kPa; heating up to 90°C; cooling under drained conditions; shearing under undrained conditions.

For overconsolidated samples (Figure 2)

Path 5 (0→1→6→5): The sequence is: consolidation up to 196 kPa; unloading (by increase of back pressure) to produce OCR=2.2; temperature increase up to 90°C under drained conditions; shearing.

Path 6 (0→1→2→5): The sequence is: consolidation up to 196 kPa; temperature increase up to 90°C under drained conditions; unloading to 89 kPa; shearing at this OCR of 2.2.

Path 7 (0→1→5): The sequence is: consolidation up to 196 kPa; temperature increase up to 90°C under undrained conditions, which induces an increase in pore pressure and yields an OCR of about 2.2; shearing.

Path 8 (0→1→6): Same as path 5 but without heating.

Path 9: thermal loading in oedometer conditions (not on the figure).

Experimental data for paths 1 to 4 and path 8 are given twice.

3.2 *Calibration of model parameters and back-predictions*

Material parameters of the MC clay for the LTVP constitutive model were determined using four tests: two triaxial shear tests in NC and OC states at room temperature (20°C, paths 1 and 8), one triaxial test at high temperature (90°C, path 2) and one thermal loading (path 9). Figure 3 shows the identification of the thermal parameters a_t and b_t (Equation 11).

The model parameters are given in Appendix 1. Back-predictions of the experimental results (made twice for each path) are shown in Figure 4 (triaxial shear tests in NC and OC states at room temperature,

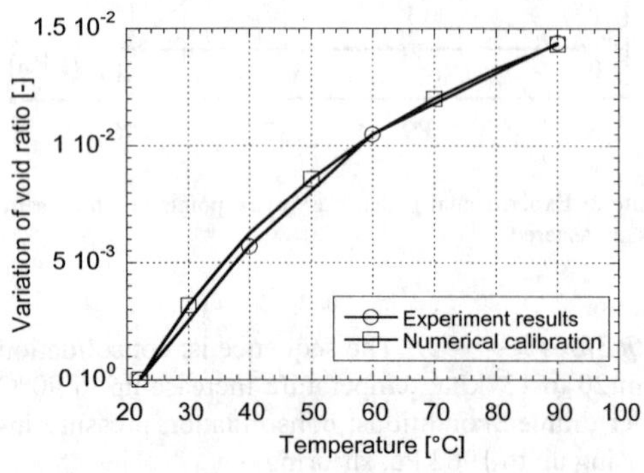

Figure 3. Variation of void ratio versus temperature (experimental results from Kuntiwattanakul, 1991)

20°C, paths 1 and 8) and in Figure 5 (triaxial shear tests in NC state at room and high temperatures, paths 1 and 2).

One can easily notice the good quality of the back-prediction except for the pore-water generation at small strains.

3.3 *Numerical prediction of thermo-mechanical loading tests*

The goal of the predicted experimental results is the analysis of the influence of the thermo-mechanical loading path on the undrained shear strength.

Numerical predictions and experimental results on paths 3 to 7 are given in Figures 6 to 8. Results are given in the plane "deviatoric stress q – axial strain ε_1"; and "*part of* excess pore pressure $\Delta u(q)$ - axial strain ε_1".

Pore pressure in triaxial conditions is generated either by mean pressure or by deviatoric stress (Henkel 1958):

$$\Delta u = \Delta u(p) + u(q) \tag{14}$$

Then,

$$\Delta u = (\Delta\sigma_1 + 2\Delta\sigma_3)/3 + \eta\sqrt{2(\Delta\sigma_1 - \Delta\sigma_3)^2} \tag{15}$$

where $\Delta\sigma_1$ and $\Delta\sigma_3$ are increments of the principal stress components in triaxial conditions and η a

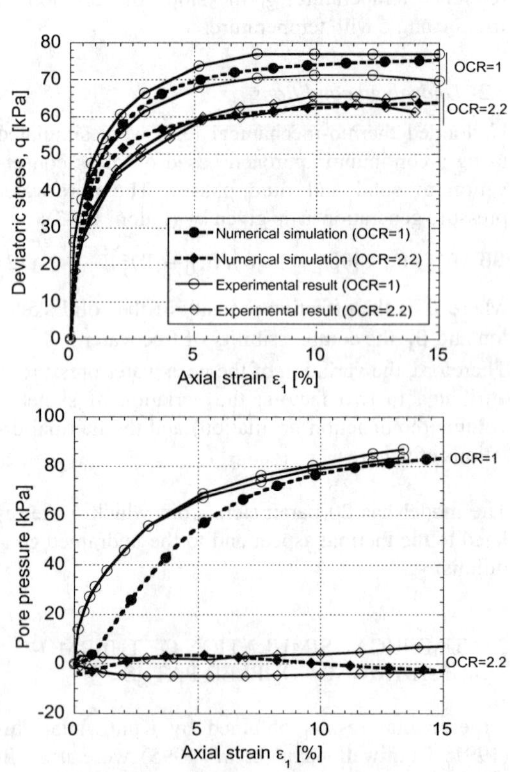

Figure 4. Comparison between experimental results (undrained shear tests at room temperature in NC and OC states; paths 1 & 8) and numerical back-prediction (experimental results from Kuntiwattanakul, 1991)

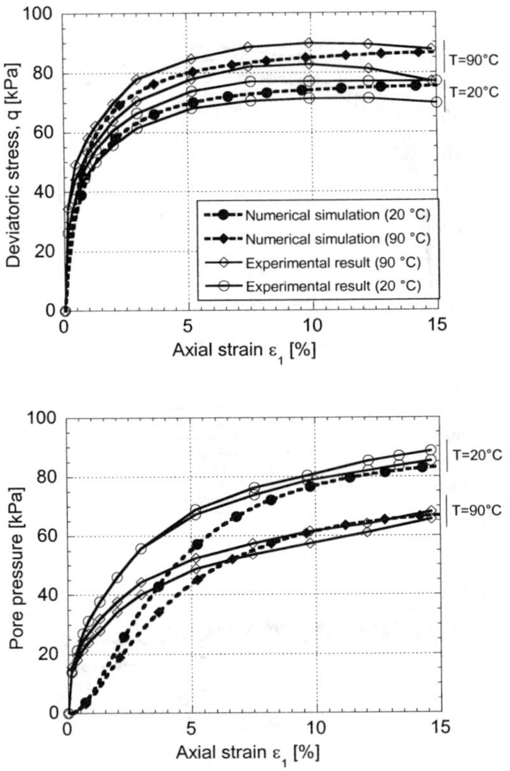

Figure 5. Comparison between experimental results (undrained shear tests at different temperatures, paths 1 & 2) and numerical back-prediction (experimental results from Kuntiwattanakul, 1991)

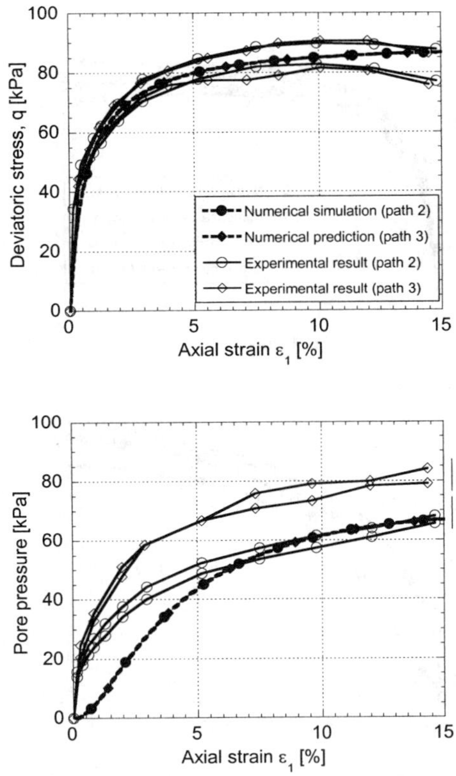

Figure 6. Comparison between experimental results (undrained shear tests, paths 2 & 3) and numerical predictions (experimental results from Kuntiwattanakul, 1991)

function which determines the increase of pore pressure due to changes in shear stress. Rearranging this equation, and taking into account that $\Delta\sigma_3 = 0$, yields:

$$\Delta u(q) = \Delta u - \Delta\sigma_1 / 3 \qquad (16)$$

Kuntiwattanakul and co-workers (Kuntiwattanakul *et al.* 1995) present their results only in terms of change of pore pressure induced by deviatoric stress, $\Delta u(q)$. In order to respect this representation, the numerical results are given in the same way.

Three cases are described in the analysis of the performance of the considered model.

Case 1 (paths 2 & 3, Figure 6): the goal here is to identify the influence of the path followed on the undrained mechanical behaviour of the clay at high temperature (90 °C). In path 2, consolidation induces a mechanical hardening, then the increase in temperature produces thermal hardening. In path 3, the mechanical hardening is induced until 137 kPa at ambient temperature then the thermal hardening is activated by heating until 90°C. This thermo-plastic hardening increases the mechanical yield surface without any change in the stress tensor. For this condition, the yield point is inside the yield surface and the next consolidation imposed is initially elastic (until the yield point reaches the yield limit) then elasto-plastic. No differences can be found in shear strength for both cases, indicating the negligible influence of the thermal state on mechanical hardening.

Case 2 (paths 1 & 4, Figure 7): the aim here is to show the influence of the heating/cooling cycle on undrained shear behaviour. The experimental results indicate that no significant thermal effect affects the mechanical behaviour in this case. Such a result is difficult to understand for this normally consolidated state, where thermal loading produces irreversible strain (Campanella & Mitchell 1968).

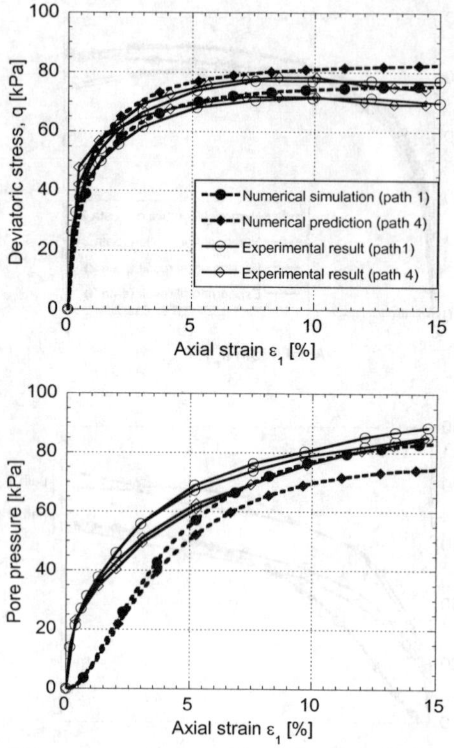

Figure 7. Comparison between experimental results (undrained shear tests at room temperature without heating, path 1, and after heating/cooling, path 4) and numerical predictions (experimental results from Kuntiwattanakul, 1991)

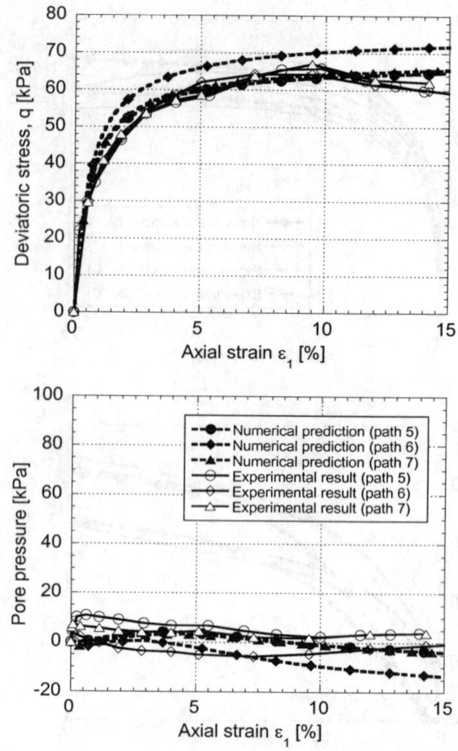

Figure 8. Comparison between experimental results (undrained shear tests at an OC state, prepared using paths 5, 6 and 7) and numerical predictions (experimental results from Kuntiwattanakul, 1991)

The model shows that thermal loading generates hardening. This produces a densification, resulting in higher shear strength (path 4, Figure 7).

Case 3 (Paths 5, 6 & 7, Figure 8): this case treats three thermo-mechanical loading paths in OC condition (OCR=2.2).

The three paths show similar experimental results for both shear strength and pore pressure. Numerical predictions compare well with experimental results.

4 CONCLUSIONS

The LTVP model is found to predict the behaviour of a clay submitted to different thermo-mechanical loading paths adequately for NC as well as for OC states.

For initial-boundary value problems, the LTVP model is available in a general 3D Finite Element code GEFDYN (Aubry *et al.* 1985) with a coupled thermo-hydro-mechanical formulation (Laloui 1993).

5 REFERENCES

Aubry, D., Hujeux, J. C., Lassoudière, F., & Memoin, Y. (1982). "A double memory model with multiple mechanisms for cyclic soil behaviour." *International Symposium on Numerical Models in Geomechanics, NUMOG*, 3-13.

Aubry, D., Chouvet, D., Modaressi, A., Modaressi, H. (1985) "GEFDYN 5 logiciel d'analyse du comportement statique et dynamique des sols par éléments finis avec prise en compte du couplage sol-eau-air." Rapport scientifique Ecole Centrale Paris.

Campanella, R. G., & Mitchell, J. K. (1968). "Influence of temperature variations on soil behaviour." *Journal of the Soil Mechanics and Foundation Division*, 94(SM3), 709-734.

Félix, B., Lebon, P., Migurez, R., & Plas, F. (1996). "A review of the ANDRA's research programmes on the thermo-hydomechanical behaviour of clay in connection with the radioactive waste disposal project in deep geological formations." *Engineering Geology*, 41, 35-50.

Henkel, D. J. (1958). "The correlation between deformation, pore water pressure and strength characteristics of saturated clays." Doctoral thesis, University of London.

Hueckel, T., & Borsetto, M. (1990). "Thermoplasticity of saturated soils and shales: Constitutive equations." *Journal of Geotechnical Engineering*, 116(2), 1765-1777.

Kuntiwattanakul, P. (1991). "Effect of High Temperature on Mechanical Behaviour of Clays." Doctoral thesis, University of Tokyo.

Kuntiwattanakul, P., Towhata, I., Ohishi, K., & Seko, I. (1995). "Temperature effects on undrained shear characteristics of clay." *Soils and Foundation*, 35(1), 147-162.

Laloui, L. (1993). "Modélisation du comportement thermo-hydro-mécanique des milieux poreux anelastiques." Doctoral Thesis, Ecole Centrale Paris.

Modaressi, H., & Laloui, L. (1992) "A cyclic thermoviscoplastic constitutive model for clays." *Numerical Models in Geomechanics - NUMOG IV*, 125-134.

Modaressi, H., & Laloui, L. (1997). "A thermo-viscoplastic model for clays." *Journal for Numerical and Analytical Methods in Geomechanics*, 21, 313-335.

Modaressi, H., Laloui, L., & Aubry, D. (1991). "Loi de comportement cyclique thermo-viscoplastique pour les géomatériaux." *Les cahiers de Rhéologie*, Vol. 9 No 1.

Robinet, J.-C., Rahbaou, A., Plas, F., & Lebon, P. (1996). "A constitutive thermomechanical model for saturated clays." *Engineering Geology*, 41, 145-169.

Appendix 1: Model parameters for MC clay

Elastic	K=80 MPa, G=37 MPa (Eq. 1 & 2) n=0.5 (non-linear elasticity exponent)*
Plastic	ϕ'=18°, β=11., p_{co}=0.1 MPa (Eq. 7 & 8)
Hardening	b=0.90, d_0=2.1 (Eq. 8 & 10) a_m=0.006,a_c=0.005,c_m=0.06,c_c=0.1(*)
Domains	r_{el}=0.001, r_{hys}=0.2, r_{mbl}=0.9 (*)
Thermal	β's=3.0E-5 1/°C, β'_f=4.0E-4 1/°C (Eq. 3 & 13) a_t=37, b_t=2222, g=35 E-5 (Eq. 11 & 12)

* Not presented in this paper.

Computer Methods and Advances in Geomechanics, Desai et al. (eds) © 2001 Balkema, Rotterdam, ISBN 90 5809 183 X

Finite-element simulation on the water migration through a nonuniform unsaturated zone

K.S. Lee
Division of Civil and Environmental Engineering, Kyonggi University, Korea

ABSTRACT: The water migration through an unsaturated zone undergoing an evaporation at the ground surface has been theoretically analyzed to develop a numerical model for the process and gain further insight into the physical process involved. The model was devised to overcome limited applicabilities of previous models to uniform media and analyze the effects of various factors on the migration. This study focuses on the mathematical description of the water migration process and the development of a numerical model. The developed model allows representing the effects of size distribution of soil particles, characteristics between soil surface and water, and initial water saturation on the migration velocities through a nonuniform unsaturated zone. This resulting model incorporates a multiphase fluid flow equation in a porous medium and correlation equations on the absolute permeability, relative permeability, and capillary pressure. Based on the nonlinear partial differential equation including terms of physical properties of pores and physicochemical characteristics of soil particles, this model is capable of predicting the effects of pore size distribution, absolute permeability, interfacial tension, contact angle, relative permeability, and capillary pressure. The water migration was simulated with a finite-element method in an unsaturated zone which was assigned realistic physical and chemical properties and boundary conditions to model one-dimensional flow. The numerical simulations were focused on the profiles of water saturation in the simulated nonuniform porous media. The simulation studies conducted in this study has shown that the model is very useful in analyzing the effects of physical and chemical characteristics in the soil environments on the water migration process.

1 INTRODUCTION

Under evaporation at the ground surface, the water in the unsaturated soil migrates toward surface. This phenomenon is caused by changes in a capillary pressure due to saturation changes. Changes in a saturation profile affect directly on the migration velocities of water in the unsaturated zone and evaporation at the surface. Quantitative evaluation on the effects of physicochemical characteristics of an unsaturated zone on the water migration, therefore, has become a matter of concern in recent years.

Several researchers have studied this problem. Dougherty *et al.* (1993) performed experimental and numerical studies on the migration of solvent through uniform media during in-situ decontamination of soils. A series of wind tunnel tests with uniform or simple nonuniform media were carried out to investigate effects of soil particles on the evaporation in soil (Lee and Chung, 1997; Lee *et al.*, 1998). Most of previous studies, however, derived conclusions based on the experimental results under limited ranges of parameters. Actual migration of water is affected by chemical properties of soil surfaces as well as physical characteristics such as porosity and saturation. Especially, effective permeability and capillary pressure are functions of saturation. To understand and predict the migration process and saturation profile more effectively, it is necessary to have mechanistic description and numerical model of water transport in soil.

This study focuses on the mathematical description of the water migration process and the development of a numerical model. The developed model allows representing the effects of size distribution of soil particles, characteristics between soil surface and water, and initial water saturation on the migration velocities through a nonuniform unsaturated zone. This resulting model incorporates a multiphase fluid flow equation in a porous medium and correlation equations on the absolute permeability, relative permeability, and capillary pressure. The nonlinear partial differential equation was solved by finite-element method. The results of theoretical and numerical studies can provide useful information on the water migration in the unsaturated zones which is characterized by various soil environments.

2 MATHEMATICAL FORMULATION

To understand the physical process involved in water flow through an unsaturated soil zone, a theoretical analysis has been undertaken. This involved deriving and solving a partial differential equation describing the water velocity and saturation in the nonuniform soil column. Writing this equation in a manageable form requires certain simplifications.

One major assumption is that the fluid behavior in the soil column can be adequately represented by including only two fluid phases in the model. The model described here includes only a liquid phase and a vapor phase. Capillary pressure and relative permeability as functions of saturation are then relatively easy to estimate using one of several correlations found in the literature. Another simplification made for the purpose of modeling was that of an isothermal system. Thus, the effects of a thermal gradient in the soil environment were not taken into account by use of an energy balance.

2.1 *Continuity equation*

To solve for the water saturation and velocity profile, a one-dimensional continuity equation for water mass in the unsaturated zone can be written as follows:

$$\varepsilon \frac{\partial S}{\partial t} = -\frac{\partial}{\partial x}(\varepsilon S v) \tag{1}$$

where ε is porosity [L^3/L^3], S is the water saturation [L^3/L^3], and v is the interstitial water velocity [Lt^{-1}, m/sec].

2.2 *Darcy's law*

Momentum balance for the movement of the water in the soil column can be written in the form of Darcy's law. For multiphase flow, the law relates the water velocity to the effective permeability and pressure gradient.

$$\varepsilon S v = -\frac{k_e}{\mu}\frac{\partial p}{\partial x} \tag{2}$$

where k_e is the effective permeability of the water [L^2, m^2] and μ is the viscosity of water [$mL^{-1}t^{-1}$, Pa-sec].

The effective permeability can be defined the product of the intrinsic permeability (k) of the porous medium and a relative permeability (k_r) to account for the effect of changing saturation level. The effect of gravity over relatively small depth of the soil column is much less than that of the capillary forces and has been assumed to be negligible in this analysis. Taking into account the fact of $p = -p_c$ in the unsaturated zone, one can derive

$$\varepsilon S v = \frac{k k_r}{\mu}\frac{\partial p_c}{\partial x} \tag{3}$$

where p_c is capillary pressure.

2.3 *Correlation equations*

Several correlations are available in the literature which provide estimates of the relative permeability. The correlation used in this study was that of the form suggested by Brooks and Corey (Dracos, 1991; Helmig, 1997).

$$k_r = \left(\frac{S - S_r}{1 - S_r}\right)^{\frac{2+3\lambda}{\lambda}} \tag{4}$$

where S_r is the irreducible saturation at which the water is in the form of isolated pendular rings. The λ parameter usually lies between 0.2 and 3.0. A very large λ describes a single grain size material, while a very small value indicates a highly nonuniform material.

To find the variation of capillary pressure with position, Eq. 3 can be rewritten in terms of the gradient in saturation.

$$\varepsilon S v = \frac{k k_r}{\mu}\frac{\partial p_c}{\partial S}\frac{\partial S}{\partial x} \tag{5}$$

The relationship of capillary pressure to saturation is assumed to be adequately represented also by Brooks and Corey such that

$$p_c(S) = p_d \left(\frac{S - S_r}{1 - S_r}\right)^{-\frac{1}{\lambda}} \tag{6}$$

The entry pressure p_d is considered as the capillary pressure required to displace the wetting phase from the largest occurring pore. It is a parameter that is related to the size of largest pore and interfacial tension between water and soil particles by the following equation.

$$p_d = \frac{2\sigma \cos\theta}{r_{\max}} \tag{7}$$

where $r_{\max}$ is the radius of largest pore [L, m], σ is an interfacial tension [mt^{-2}, kg/sec^2], θ is a contact angle.

The gradient of capillary pressure in saturation, therefore, can be expressed as

$$\frac{\partial p_c}{\partial S} = -\frac{1}{\lambda}\frac{p_d}{(1 - S_r)^{-\frac{1}{\lambda}}}(S - S_r)^{-\frac{\lambda+1}{\lambda}} \tag{8}$$

The correlation equation, that is derived from capillary pressure-saturation curves, was used to determine the absolute permeability of an unsaturated zone. Several authors including Purcell (1949), Swanson (1981), and Thomeer (1983) has proposed correlation equations approximating the absolute permeability with capillary pressure data. Swanson's correlation was adopted in this study because it utilize only the shape of capillary pressure-saturation curves without any other parameters describing geometrical characteristics of pore networks. Transformation of the Swanson's correlation, that is derived originally in a British unit system, into a SI unit system results in

$$k = 0.17813\left(\frac{1-S}{p_c}\right)_A^{2.005} \tag{9}$$

In the equation, the subscript A represents the point where the ratio $\left(\dfrac{S}{p_c}\right)$ has a maximum value.

2.4 *Flow equation*

Equations 1 and 2 were combined into one equation.

$$\frac{\partial S}{\partial t} = \frac{k}{\mu\varepsilon}\frac{\partial p_c}{\partial S} \left[\left(\frac{S-S_r}{1-S_r}\right)^{\frac{2+3\lambda}{\lambda}}\frac{\partial^2 S}{\partial x^2} + \frac{1}{(1-S_r)^{\frac{2+3\lambda}{\lambda}}}\left(\frac{S-S_r}{1-S_r}\right)^{\frac{2(\lambda+1)}{\lambda}}\left(\frac{\partial S}{\partial x}\right)^2\right] \tag{10}$$

The solution of Eq. 10 yields saturation and velocity profiles as functions of time and location.

2.5 *Initial and boundary conditions*

The above equation can be solved with appropriate initial and boundary conditions. At the bottom of column, the flux of water is set to be zero to represent a situation of an evaporation experiment.

$$\varepsilon S v = 0 \ \text{at}\ x = 0 \tag{11}$$

At the top of column, the boundary condition used was based on the amount of water lost by evaporation. The rate of evaporation R_{evap} [mL^{-1}, kg/sec] from fully saturated surface to the dry air is a function of wind speed. The liquid velocity at the surface can be found by relating the rate of evaporation to the surface area of the column.

$$\dot{M}_o = \frac{R_{\text{evap}}}{\rho A} \tag{12}$$

where ρ is the density of water [mL^{-3}, kg/m^3]. Assuming linear evaporation efficiency from an unsaturated surface, this boundary condition can be expressed

$$\varepsilon S v = \dot{M}_o S \ \text{at}\ x = l \tag{13}$$

The initial condition used for the solution of the equation was that the water saturation was a known constant value at the start of the test.

$$S = S_o \ \text{at}\ t = 0 \tag{14}$$

3 RESULTS AND ANALYSIS

3.1 *Finite-element analysis*

To accomplish the goal of this study, analyzing the effects of various factors on the water migration through an unsaturated zone, a numerical model using a finite-element method has been developed. For all calculations, the following set of parameter values was used.

depth of unsaturated zone (l) = 0.2 m
evaporation rate at surface ($\dot{M}_o$) = 1×10^{-7} m/sec

Constant water density and viscosity are assumed due to an isothermal system and relatively small pressure changes. The porosity of the soil column is taken to be 0.4. The geometrical properties of soil particles consisting an unsaturated zone are characterized by entry pressure and Brooks-Corey parameter which represent largest size of particles or pores and size distribution, respectively. Table 1 gives a summary of input parameters used in the simulation.

Table 1. Summary of input data for parameters used in the simulation study.

Parameters	Input Data
Viscosity (μ)	1.0 cp (1.0×10^{-12} Pa-sec)
Porosity (ε)	0.4
Irreducible saturation (S_r)	0.05
BC parameter (λ)	0.2, 0.5, 1.0, 2.0, 3.0
Entry pressure (p_d)	5×10^4, 1×10^5, 2×10^5, 5×10^5 Pa
Initial saturation (S_o)	0.95
$\dot{M}_o$	1×10^{-6} m/sec

Figure 1 shows calculated values of absolute permeabilities as functions of entry pressure and Brooks-Corey parameter. A combination of larger values of Brooks-Corey parameter and smaller values of entry pressure, which represents a uniform medium of large particles, results in larger absolute permeabilities. In the range of entry pressure given in Table 1, the absolute permeabilities range from 1.4×10^{-11} m^2 to 2.7×10^{-15} m^2.

Figure 2 presents the relative permeability of water as a function of saturation. The larger Brooks-Corey parameter, the larger relative permeability becomes.

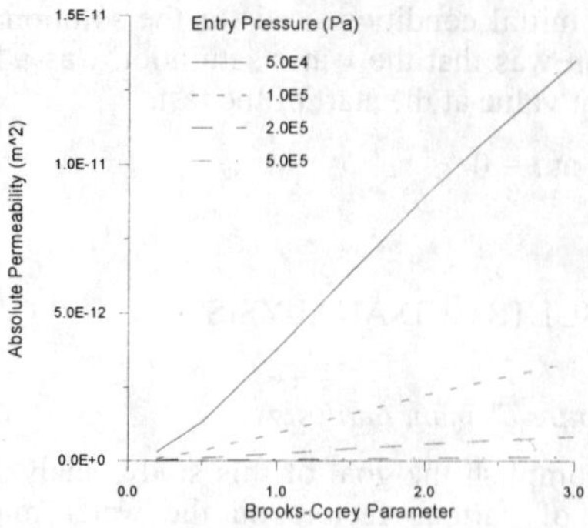

Figure 1. Absolute permeability calculated from Swanson's correlation.

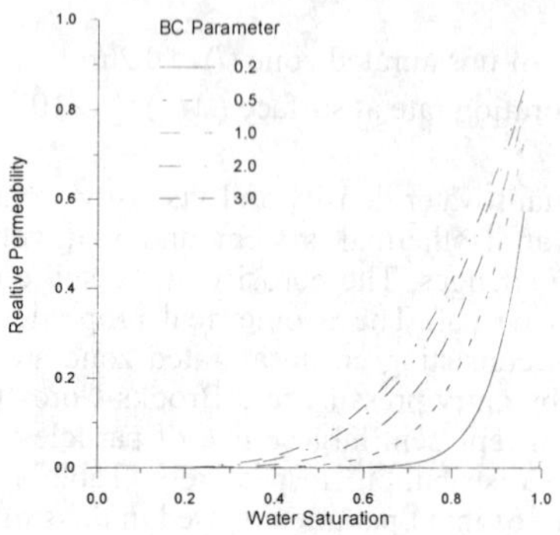

Figure 2. Relative permeability of water as a function of water saturation calculated from Brooks-Corey correlation.

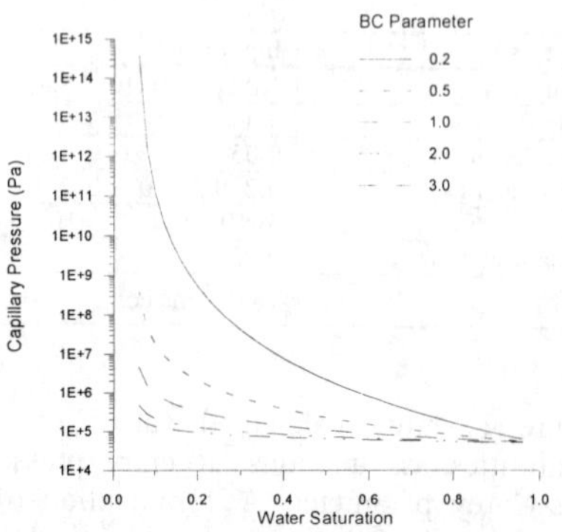

Figure 3. Capillary pressure as a function of water saturation calculated from Brooks-Corey correlation ($p_d = 5\times10^4$ Pa).

Figure 3 gives changes in capillary pressures as a function of saturation when the entry pressure is 5×10^4 Pa. Larger values of capillary pressure are obtained with smaller values of Brooks-Corey parameter. Though the shape of curves remains unchanged, capillary pressure values become larger for larger entry pressure.

3.2 *Numerical examples*

Extensive preliminary tests were conducted to verify the accuracy of the developed finite-element model. To determine how well the numerical model represented the physical situation, it was applied to a uniform soil column that had been exposed to evaporation for 100 hours. Results of these preliminary numerical calculations were compared against results of experiments and numerical analysis reported in Dougherty *et al.*'s (1993) paper. Because detailed information on the conditions of experiments was unavailable, a direct comparison was impossible. Overall shape of water saturation curve for a uniform medium after 100-day evaporation, therefore, examined. The general shape of the curve is appropriate with the highest saturation at the bottom of the column. Also, larger value of initial saturation results in larger amount of evaporation at the surface.

Calculations were performed to investigate the effects of nonuniformity of soil columns on the water migration. Figure 4 represents the water saturation after 100 hours for the cases of an initial saturation ranging from 0.5 to 0.9 for entry pressures of 5×10^4 and 5×10^5 Pa. As Brooks-Corey parameter approaches to 0.2, the saturation near the surface reduces rapidly. This trend becomes clearer as the initial saturation becomes smaller and entry pressure larger. For the case of initial saturation of 0.9, the difference in saturation distribution was less than 0.1 % even for highly nonuniform media ($\lambda = 0.2$). Reduction of initial saturation to 0.5 results in a rapid increase of saturation difference up to 100%. In summary, the effects of nonuniformity on the saturation difference increase as pore size and initial saturation become smaller.

4 CONCLUSIONS

The water migration through a nonuniform unsaturated zone has been theoretically analyzed to produce a numerical model for the process and gain further insight into the physical process involved. Through incorporating various correlation equations for relative permeability, capillary pressure, and nonuniformity into continuity equation of multiphase flow, the model is capable of predicting the effects of pore size distribution, absolute permeability, interfacial tension, contact angle, and water saturation. Numerical simulations on the flow through nonuniform media indicate that the effects of nonuniformity become larger as pore size and initial saturation become smaller. Larger initial saturation and smaller pore size result in a larger velocity near the surface. The finite-element simulation studies conducted in

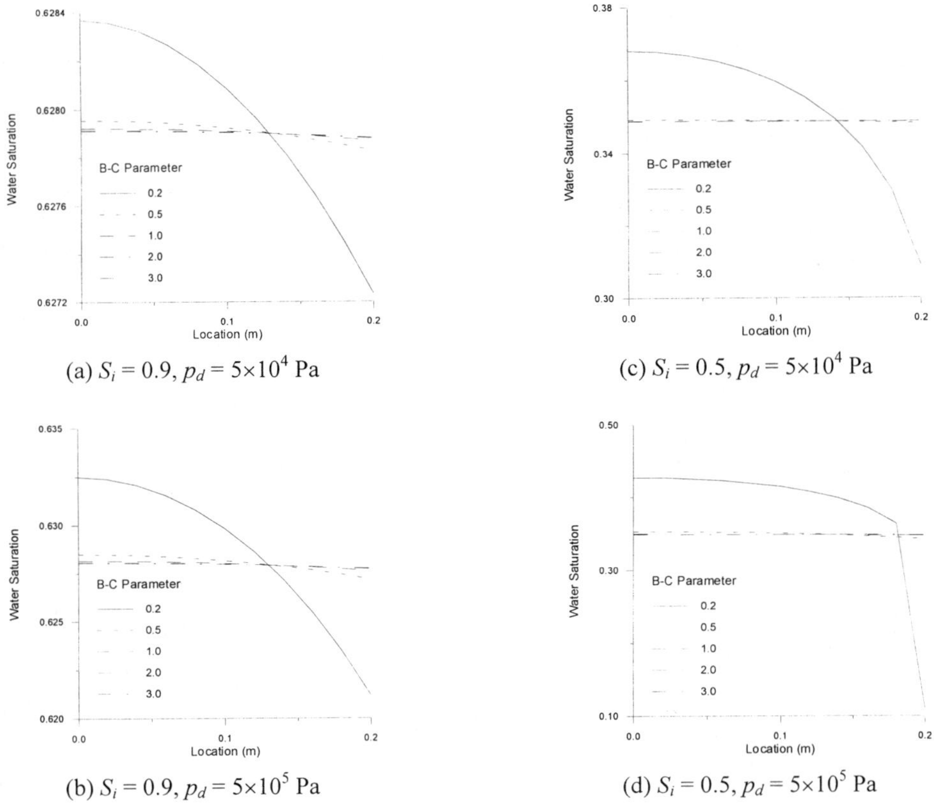

Figure 4. Saturation profiles for the cases of nonunifrom grains with different sizes after 100 hours.

this study have shown that the model is very useful in analyzing the effects of physical and chemical characteristics in the soil environments on the water migration process.

REFERENCES

Dracos, T. 1991. Multiphase Flow in Porous Media. *Modelling and Application of Transport Phenomena in Porous Media.* Dordrecht, The Netherlands: Kluwer Academic Publishers.

Dougherty, E.J., A.L. McPeters, M.R. Overcast & R.G. Carbonell 1993. Theoretical Analysis of a Method for in Situ Decontamination of Soil Containing 2,3,7,8-Tetrachlorodibenzo-p-dioxin. *Environ. Sci. Tech.* 27. 505-515.

Helmig, R. 1997. *Multiphase Flow and Transport Processes in the Subsurface: A Contribution to the Modeling of Hydrosystems.* Berlin, Germany: Springer-Verlag.

Lee, S.-D. & I.-H. Chung 1997. Study on Evaporation from the Soil Affected by the Particle Size and Soil Water Content. *J. of Kor. Soc. Env. Eng.* 19 1-8. (in Korean)

Lee, S.-D., S.-C. Park & K.-I. Lee 1998. A Study on the latent heat Exchange Model between Soil and Atmosphere. *J. of Kor. Soc. Env. Eng.* 20 1511-1522. (in Korean)

Purcell, W.R. 1949. Capillary Pressures-Their measurement Using Mercury and the Calculation of Permeability Therefrom. *Trans.* AIME 186 39-48.

Swanson, B.F. 1981. A Simple Correlation between Permeabilities and Mercury Capillary Pressures. *J. of Pet. Tech.* 2498-2504.

Thomeer, J.H. 1983. Air Permeability as a Function of Three Pore-Network Parameters. *J. of Pet. Tech.* 809-813.

Computer Methods and Advances in Geomechanics, Desai et al. (eds) © 2001 Balkema, Rotterdam, ISBN 90 5809 183 X

Succession-of-pseudosteady-states modeling on the long-time behavior of real gas flow in a closed reservoir

K.S. Lee
Division of Civil and Environmental Engineering, Kyonggi University, Korea

ABSTRACT: This paper presents a new succession-of-pseudosteady-states approach in calculating the long-time behavior of real gas flow in a closed reservoir and examined the applicability and limitations of the approach. As a first step to extend the existing pseudosteady-state approach, a linearized form of gas diffusivity equation was derived. The derivations are accomplished by introducing normalized pseudovariables. The linearized diffusivity equation allows one to apply pseudosteady-state method and estimate late-time behavior of a closed gas reservoir. It is intent of this paper to provide a guideline on the use of pseudosteady-state approach to real gas flow problems. Extensive numerical simulation study has then been performed to quantify effects of various reservoir and operational parameters which may develop nonlinearities in gas diffusivity equation. The parameters include permeability, pressure, flowrate, and gas properties. Results of simulation studies have been examined with cumulative error analyses. For most cases, a very good agreement was observed between the predictions of the model and conventional reservoir simulation. Error analyses on the simulation results demonstrate that the accuracy of a succession-of-pseudosteady-states approach mainly depends on the permeability and reservoir pressure, weakly on the flowrate, and is independent of specific gravity. Except for the cases of very low permeabilities, however, computations indicate that neglecting infinite-acting flow behavior does not introduce serious error in the analysis procedure. The validity and computational efficiency of the succession-of pseudosteady-states approach for the analysis of gas reservoirs has been confirmed. Such a model can be used for situations where large numbers of runs are desirable, for example for stochastic simulation or risk analysis.

1 INTRODUCTION

In a reservoir where there is no flow across drainage boundaries, a pseudosteady-state flow period follows the short-lived infinite-acting and transition periods under constant terminal rate conditions at the well. Once pseudosteady state has been attained, the rate of pressure change over the entire system becomes nearly constant at all points in the reservoir. For a real gas reservoir, the rate of pressure decline cannot rigorously be equal to be a constant. Therefore, it is more appropriate to use a boundary-dominated flow which refers to the time period when wellbore response is controlled by the no-flow outer reservoir boundary.

The pseudosteady-state approach has been mathematically established and shown to be an effective tool for directly calculating long-time performance of a bounded reservoir with wells producing at constant rates (Lee *et al.*, 1996). However, there appears to be a lack of information regarding the influence of various reservoir and operational parameters on the long-time depletion performance of gas reservoirs.

To apply pseudosteady-state approach to gas flow in a closed reservoir, one has to derive a linearized diffusivity equation. As is well known, the flow of gas through a permeable medium is represented by the diffusivity equation. The nonlinearity in the diffusivity equation, however, has long been recognized as one of the problems in dealing with gas flow.

In the past (Aronofsky and Jenkins, 1953), assumptions of ideal gas behavior led to the use of "pressure squared (p^2)" as the variable for analysis of gas flow rate. Later, it was shown that the use of pressure (p) instead of p^2 was appropriate under certain conditions for gas flow relationships (Matthews and Russel, 1967). Al-Hussainy *et al.* (1966) proposed a third variable, p_p, known as the pseudopressure or real gas potential defined as follows:

$$p_p = \int_0^p \frac{2p}{\mu z} dp \tag{1}$$

Pseudopressure allowed for variations in μ and z, the viscosity and compressibility factor of natural gas.

Because the diffusivity equation for gas flow in terms of the pseudopressure involves the least number of assumptions, it is considered to be the most rigorous among the three treatments (p, p^2, p_p). However, even though p_p accounts for variations in μz, the resulting differential equation still contains a nonlinear term, μc_t.

$$\nabla^2 p_p = \frac{\phi \mu c_t}{k} \frac{\partial p_p}{\partial t} \quad (2)$$

where ϕ is porosity, k is permeability, and t is time.

Because μc_t is a function of p_p, this equation is nonlinear and cannot be solved analytically. Al-Hussainy and Ramey (1966) assumed μc_t to be constant at initial conditions $\mu_i c_{ti}$ and solved the resulting linear differential equation. Unfortunately, the real gas potential approach does not appear to be as good at late times as in the initial transient period; that is, the dimensionless pseudopressure does not satisfy the standard liquid type pseudosteady-state flow equation. Al-Hussainy *et al.* (1966) showed that nonlinear effects became dominant at late times. During this period, the wellbore pressure depends on the physical properties of the produced gas and also on the flowrate. The long-time solutions obtained by Wattenbarger and Ramey (1968) also confirmed their results. Therefore, the rate of pseudopressure decline cannot rigorously be equal to a constant. These results indicate the need for a time transformation to incorporate variations in the viscosity and compressibility factor.

In this paper, a gas diffusivity equation has been linearized by introducing a pseudotime. The derived diffusivity equation allows one to apply pseudosteady-state method and estimate late-time behavior. Then, effects of all the parameters affecting long-time performance are examined with cumulative error analyses on the results from an extensive simulation study. It is intent of this paper to provide a quantitative guideline on the use of pseudosteady-state approach to real gas flow problems.

2 MATHEMATICAL FORMULATION

2.1 *Pseudovariables*

The derivation of the linearized gas diffusivity equation starts by using pseudopressure and normalized pseudotime. Pseudotime, which is an empirical function, has been proposed in various forms by many authors like Agarwal (1979), Meunier *et al.* (1987), Reynolds *et al.* (1987), Finjord (1989), and Ding *et al.* (1990).

In order to correlate the real gas pseudopressure solution with an analogous liquid solution during boundary-dominated flow, we normalized Agarwal's pseudotime through the following equation:

$$t_{pn} = \mu_i c_{ti} \int_0^t \frac{dt}{\mu c_t} \quad (3)$$

Because p_p is the variable that was most widely used, it will be used in preference to p, p^2, or any other pseudovariables.

2.2 *Linearized gas diffusivity equation*

Use of pseudopressure and normalized pseudotime produces a theoretical basis for deriving a linearized gas diffusivity equation:

$$\nabla^2 p_p = \frac{\phi \mu_i c_{ti}}{k} \frac{\partial p_p}{\partial t_{pn}} \quad (4)$$

The dimensionless pseudovariables can be defined as

$$p_{pD} = \frac{\pi k h T_{sc}}{q_{sc} p_{sc} T} \left(p_{pw} - p_p\right) \quad (5)$$

$$t_{pnD} = \frac{k t_{pn}}{(\phi \mu c_t)_i r_w^2} \quad (6)$$

where h is reservoir thickness, r_w is wellbore radius, q is flowrate, T is temperature, subscript D stands for dimensionless variable, and sc standard condition (14.7 psia, 60°F).

Use of Eqs. 5 and 6 in the gas diffusivity Eq. 4 results in an equivalent liquid equation:

$$\nabla^2 p_{pD} = \frac{\partial p_{pD}}{\partial t_{pnD}} \quad (7)$$

The close analogy between the diffusivity equation of liquid flow and the linearized real gas flow suggests that the liquid and gas solutions can be correlated at late times provided that we graph the gas solution in terms of t_{pnDA}. For constant rate production, the long-time performance of a gas reservoir can be expressed as

$$\bar{p}_{pD} = 2\pi t_{pnDA} \quad (8)$$

and

$$\bar{p}_{pD} - p_{pwD} = -\frac{1}{2} \ln \frac{4A}{e^{\gamma} C_A r_w^2} \quad (9)$$

where C_A is shape factor, γ is Euler's constant ($\approx$ 0.5772), and t_{pnDA} is a dimensionless pseudotime based on the drainage area defined as follows:

$$t_{pnDA} = \frac{k t_{pn}}{(\phi \mu c_t)_i A} \quad (10)$$

In this case, the long-time condition refers to constant decline rate of pseudopressure with respect to a normalized pseudotime. Therefore, pseudopressure is a linear function of normalized pseudotime, as shown in Eq. 8.

2.3 *Material balance calculation*

To obtain the correct equation for the late-time response, one must obtain $\bar{p}_{pD}$. Because the pseudotime approach requires information on the average reservoir pressure and fluid properties, it is impossible to obtain an expression for average pseudopressure in a closed form. For a practical purpose, the average real gas potential in our approach is conveniently evaluated from overall system material balance considerations. For a single-phase gas reservoir of initial gas-in-place G, the material balance can be expressed in terms of hydrocarbon pore volume (Dake, 1978):

$$\frac{\int_0 q(\tau)d\tau}{G} = 1 - \frac{\overline{(p/z)}}{(p/z)_i} \tag{11}$$

To investigate this approach, a table of gas properties and real gas pseudopressure as function of pressure for any given gas and reservoir conditions is generated. A pressure increment of 25 psi was used in the table since preliminary calculations indicate that reducing the pressure increment below 30 psi has a negligible effect on the results. Once $\bar{p}$ is known from Eq. 11 as function of time, p_p can be computed from the p_p versus p table. p_w can then be computed from the rearrangement of Eq. 9 and the p_p versus p table.

3 NUMERICAL SIMULATION STUDY

The pseudosteady-state equation was implemented in a finite-element program to get a pseudopressure drawdown, $\bar{p}_{pD} - p_{pwD}$. Here, the analysis procedure of the previous section is applied to pressure data obtained from gas reservoir simulator in order to illustrate a comparison of the pseudosteady-state solutions with results generated using a conventional finite-difference reservoir simulator. The simulator BOAST III, developed by US Department of Energy, was used in this work (US DOE, 1997). This numerical model has been used extensively in previous research and shown to yield highly accurate numerical solutions.

The single-phase isothermal flow of a real gas is considered. A horizontal homogeneous porous medium of constant thickness h with impermeable upper and lower boundaries, and a well of radius r_w located at the center of a square reservoir constitute the flow system.

The accuracy of pseudosteady-state approach and conventional reservoir simulation were tested for reservoirs with different parameters. Table 1 gives a reservoir description including physical properties of reservoir rock and fluid and other input parameters used in the simulation. The late-time performance of the gas field was investigated under different permeability, initial reservoir pressure, and specific gravity. A wide range of permeability and reservoir pressure values is considered in order to span variations of the parameters of practical interest.

Table 1. Input data for simulation of gas reservoirs for the comparison of conventional reservoir simulation and pseudosteady-state approach.

Reservoir Configuration	Length (a) (ft) Width (b) (ft) Thickness (h) (ft)	2,000 2,000 50
Well Data	Position (x_w, y_w) Radius (r_w) (ft)	$\frac{1}{2}a, \frac{1}{2}b$ 0.5
Reservoir Properties	Permeability (k) (md) Porosity (ϕ) Compressibility (c_r) (psi^{-1}) Temperature (T_r) (°F)	1/5/10/20/50 0.3 1×10^{-6} 195
Fluid Property	Gravity (γ_g)	0.5/0.8
Conditions	Well (q_g) (Mscf/day) Initial pressure (p_i) (psia)	5,000/10,000/15,000 1,000/2,000/3,000/5,000
Run Parameters	Maximum press. change (Δp_{max}) (psia) Maximum time step size (Δt_{max}) (day) Minimum time step size (Δt_{min}) (day)	15 1 0.001

Many preliminary runs were made to select a grid and run parameters that would give the best results. 729 cells (27×27) of nonuniform sizes were used and the maximum pressure change over a time step should be less than 15 psia to yield reliable results. As a quantitative criterion to determine the accuracy of pseudosteady-state approach, a cumulative relative error between solutions, defined as follows, were used.

$$\mathrm{CRE}(t)=\frac{\int_0^{t_c}\left|p_{wf,RS}-p_{wf,pss}\right|d\tau}{\int_0^{t_c}p_{wf,RS}\,d\tau} \qquad (12)$$

To apply this equation, one needs to set t_c, the upper limit of t during CRE computation. In this study, t_c is set to be 10 days before which a boundary-dominated flow begins for most cases. Therefore, relative error becomes negligible after t_c.

Figure 1 shows effects of flowrates and permeabilities on the CRE's in p_{wf} for the case of $\gamma_g = 0.5$ and $p_i = 5{,}000$ psia. High values of q_{sc} and k produce low value of CRE. This is due to the fact that combination of high q_{sc} and k results in less time to reach to boundary-dominated flow period. Except for the cases of extremely low permeability reservoirs, the CRE's between the two solutions are relatively small (less than 2%). Low permeability or tight reservoirs require a very large time to attain pseudosteady-state flow behavior. This fact indicates the lack of applicability of the methodology to tight gas formations. Such reservoirs may require months to years for the infinite-acting or transition to end. However, most low-permeability wells produce against low, constant wellhead pressure to maximize production.

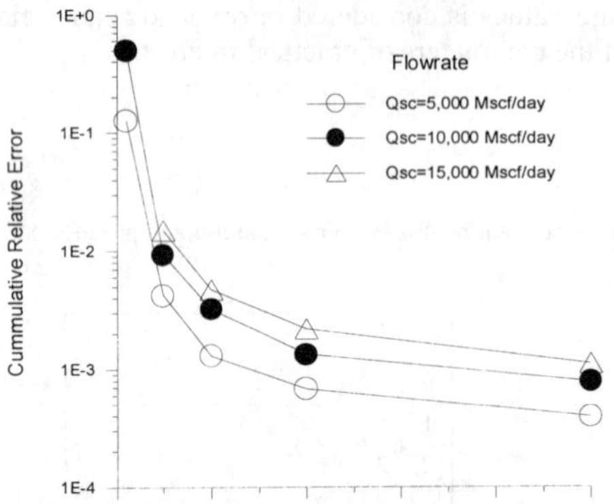

Figure 1. Cumulative relative error between solutions during calculation of flowing well pressure ($\gamma_g = 0.5$, $p_i = 5000$ psia).

Figure 2 illustrates the comparison of CRE under different flowrates and initial pressures. The CRE depends greatly on the reservoir pressure. The higher pressure makes the CRE smaller, which means the nonlinear behavior of gas flow becomes significant at low pressure. The results of Fig. 2, however, indicate that the pseudosteady-state approach is correct within 2%. Examples of low-pressure case show the applicability of the approach to both high- and low-pressure reservoirs.

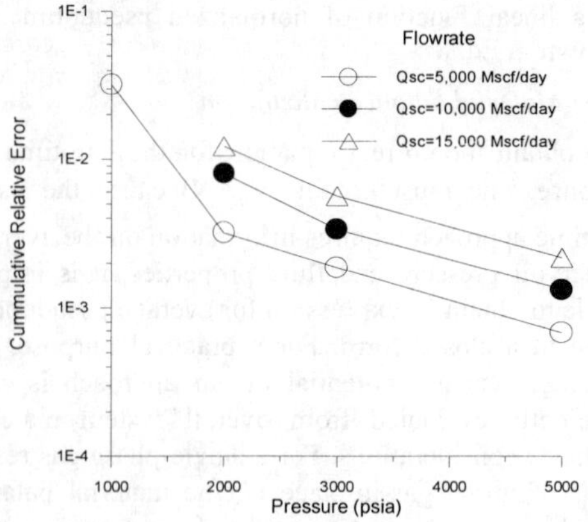

Figure 2. Cumulative relative error between solutions during calculation of flowing well pressure ($\gamma_g = 0.5$, $k = 20$ md).

The results of Fig. 3 indicate that specific gravity of gas does not affect the accuracy of the pseudosteady-state approach. As stated earlier, reservoir pressure is shown to be much more important factor.

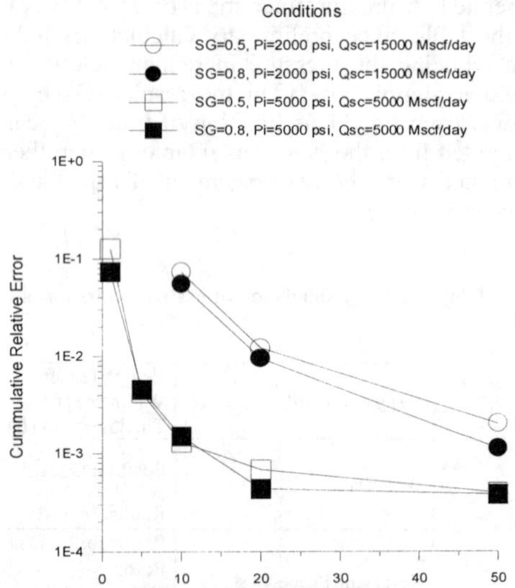

Figure 3. Cumulative relative error between solutions during calculation of flowing well pressures.

4 CONCLUSIONS

In this study, a linearized diffusivity equation in terms of pseudovariables has been derived to apply pseudosteady-state approach in calculating a long-time behavior of a closed gas reservoir. We considered the applicability and limitations of the pseudosteady-state approach to real gas depletion problems from an extensive simulation study and comparison against results of conventional reservoir simulation. The use of an overall material balance equation combined with direct pseudosteady-state calculation was shown to yield accurate results over a wide range of parameters. Error analyses on the simulation results demonstrate that the accuracy of a succession-of-pseudosteady-states approach mainly depends on the permeability and reservoir pressure, weakly on the flowrate, and is independent of specific gravity. The approach, however, has a limitation in applying to the gas flow through an extremely tight formation.

REFERENCES

Agarwal, R. G. 1979. Real Gas Pseudotime: A New Function for Pressure Buildup Analysis for MHF Wells. paper SPE 8279 presented at Annual Technical Conference and Exhibition, Las Vegas, NV, September 23-26.

Al-Hussainy, R., H.J. Ramey & P.B. Crawford 1966. The Flow of Real Gases Through Porous Media. *J. Pet. Tech.* 624-636.

Al-Hussainy, R. & H.J. Ramey 1966. Application of Real Gas Flow Theory to Well Testing and Deliverability Forecasting. *J. Pet. Tech.* 637-642.

Aronofsky, J. S. & R. Jenkins 1954. A Simplified Analysis of Unsteady Radial Gas Flow. *Trans.*, AIME 149-154.

Dake, L.P. 1978. *Fundamentals of Reservoir Engineering*. Developments in Petroleum Science, Amsterdam, The Netherlands: Elservier Scientific Publishing Co.

Ding, W., M. Onur, & A.C. Reynolds 1990. Analysis of Gas Well Late-Time Pressure and Rate Data. *J. Pet. Sci. Eng.* 293-307.

Finjord, J. 1989. An Analytical Study of Pseudotime for Pressure Drawdown in a Gas Reservoir. *SPE Form. Eval.* 303-311.

Lee, K.S., M.A. Miller & K. Sepehrnoori 1996. Succession-of-States Model for Calculating the Long-Time Performance of Depletion Reservoirs. paper SPE 37030 presented at SPE Asia Pacific Oil & Gas Conference, Adelaide, Australia, October 28-31.

Matthews, C.S. & D.G. Russel 1967. *Pressure Buildup and Flow Tests in Wells*. Richardson, TX: Soc. Pet. Eng.

Meunier, D.F., C.S. Kabir & M.J. Wittmann 1987. Gas Well Test Analysis: Use of Normalized Pseudovariables. *SPE Form. Eval.* 629-636.

Raghavan, R. 1993. *Well Test Analysis*. Englewood Cliffs, NJ: Prentice-Hall, Inc.

Reynolds, A.C., R.B. Bratvold, & W. Ding 1987. Semilog Analysis of Gas Well Drawdown and Buildup Data. *SPE Form. Eval.* 657-670.

Wattenbarger, R.A. & H.J. Ramey 1968. Gas Well Testing with Turbulence, Damage, and Wellbore Storage. *J. Pet. Tech.* 877-887.

US DOE 1997. *"BOAST 3" A Modified Version of BOAST II with Post Processors B3PLOT2 and COLORGRID*. Bartlesville, OK: US DOE.

Computer Methods and Advances in Geomechanics, Desai et al. (eds) © 2001 Balkema, Rotterdam, ISBN 90 5809 183 X

Soil suction and infiltration comparison between measured and computer prediction

Saravanan Mariappan, Faisal Haji Ali & Low Tian Huat
Civil Engineering Department, University of Malaya, Kuala Lumpur, Malaysia

ABSTRACT: Slope stability is an important aspect of man made or natural slope fringing human activities. Various expect of design need to consider before cost effective and stable slope is proposed. Tropical countries like Malaysia, which is exposed to heavy monsoon rainfall, slope failure is a common and serious event. Many researches have been carried out to study the slope failure mechanism to prevent catastrophic failures. Studies like risk and hazard mapping database analysis of slope failure with rainfall intensity and pattern, behavior of soil suctions etc. At University of Malaya, Kuala Lumpur research was carried out to study the influence of rainfall on soil matric suction. A full-scale model of rainfall simulation system was built at the field to conduct the study. The infiltration of rainwater into slope, reducing matric suction at the unsaturated zone was monitored in the research. Theoretically, the infiltrating rainwater into the slope reduces matric suction and increases pore pressure, resulting in gradual drop in shear strength to a point where stability can no longer sustained in the slope. This research was engineered to study the matric suction changes at the field and correlate the results with computer model.

1 INTRODUCTION

The research involves comparison study of field test data and computed results using engineering software with regards to rainfall and its influences on soil matric suction. In order to conduct this study a cut slope was utilized a long the link road to Kuala Lumpur International Airport with the assistance of Malaysian Public Works Department. The geological formation of the slope consists of various weathering grades, typical of Malaysian sedimentary residual soil. The 30m high cut slope exposed weathering grades ranging between II to VI. In this paper test results performed at grade V soil are presented. Figure 1.0 shows the description given by Geological Society of Engineering Group for residual soil weathering grades. Figure 2.0 indicates the geological map showing the subsoil stratification and zones of soil weathering grades of the said cut slope.

2 UNSATURATED SOILS

The mechanism of suction caused failure of slope is due to water infiltration that causes reduction of matric suction in soil under unsaturated state. This leads to decrease in effective stress of the soil strength to a point where equilibrium can no longer be sustained in the slope. The equation for unsaturated shear strength is written in terms of the stress state variables with an extension for saturated soils. (Fredlund D.G. et al , 1978).

$$\tau = c' + (\sigma - u_a) \tan \phi' + (u_a - u_w)\tan\phi^b \quad \text{----------(1)}$$

where:

c' = effective cohesion , σ = total stress ,
u_a = pore -air pressure, u_w = pore water pressure,
ϕ' = effective angle of internal friction ,
$(u_a - u_w)$ = matric suction,
ϕ^b = angle indicating the rate of increase in shear strength with respect to changes in $(u_a - u_w)$ when $(\sigma - u_a)$ is held constant.

The above equation assumes a planar failure envelope, the internal friction angle ϕ', remains constant under saturated and unsaturated condition. The angle ϕ^b , which quantifies the effect of suction is measured from the τ vs $(u_a - u_w)$ plot. The cohesion intercepts c_1, c_2 and c_3 due to the applied suction $(u_a - u_w)$ vary if the angle of internal friction ϕ' remains constant at different suction levels.

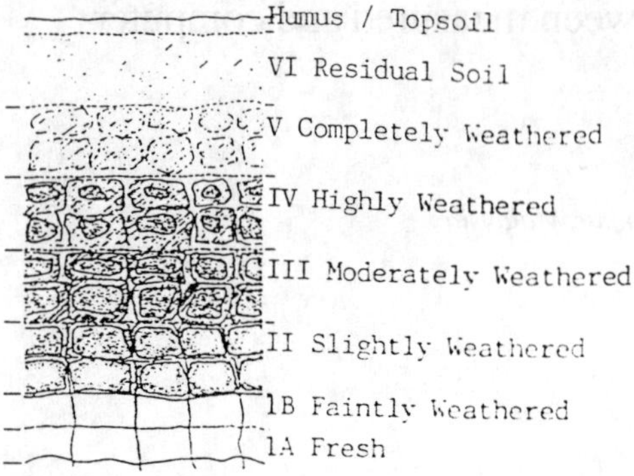

Figure 1 : Schematic representation of tropical soil weathering profiles.

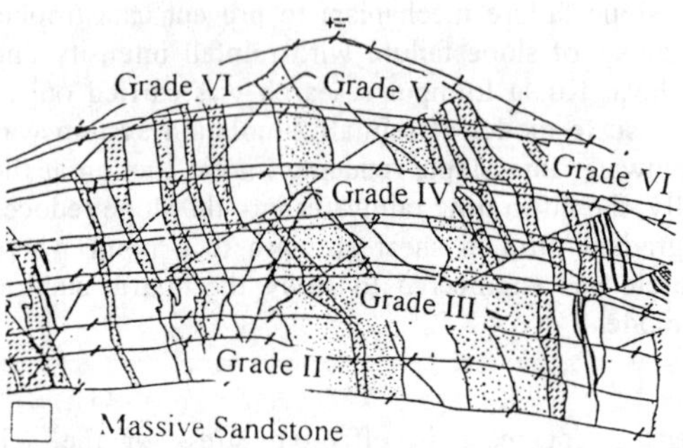

Figure 2 : Geological map of the cut slope

3 SOIL SUCTION

Soil suction commonly called `total suction' can be subdivided into two component on the basis of forces responsible for the differences in energy between the water in question and the reference state. The two components are:

i) Matric Suction, ii) Osmotic Suction

The total matric, osmotic suction can be defined as follows (Aitchison, 1965), "Matric or capillary component of the free energy''. – In suction terms, it is the equivalent suction divided from the measurement of partial pressure of water vapor in equilibrium with soil water, relative to the partial pressure of water vapor in equilibrium with a solution identical in composition with soil water."

"Osmotic (suction) component of the free energy".- In suction terms, is the equivalent suction derived from the measured partial pressure of water vapor in equilibrium with a solution identical in composition with soil water, relative to the partial pressure of water vapor in equilibrium with free pure water'.

"Total suction of free energy of soil water". In suction terms the equivalent suction derived from the measured partial pressure of water vapor in equilibrium with a solution identical in composition with soil water, relative to partial pressure of water vapor in equilibrium with free pure water''.

The above definitions clearly state that the total suction corresponds to the free energy of soil water, while the matric and osmotic suction are the components of the free energy, Thus this can be written in an equation: -

$$\psi = (U_a - U_w) + \pi$$

ψ = Total Suction, $(U_a - U_w)$ = Matric Suction
U_a = Pure Air Pressure, U_w = Pure Water Pressure,
π = Osmotic Suction

However osmotic suction is related to the salt content in the pure water which is present in the both saturated and unsaturated soils. Hence changes in osmotic suction is less significant. In this research much attention was given in measuring matric suction at the field.

4 FIELD MEASUREMENT

For field suction measurement three type of tensiometer were used, namely:

i. Jet fill tensiometer,
ii. Small tip tensiometer
iii. Quick draw tensiometer

Jet fill and small tip tensiometer were installed on to the slope to monitor the soil suction changes with rainfall. Quickdraw tensiometer was used to study the soil-water characteristic at various weathering Grades. The schematic drawing of the above tensiometer is shown in Figure 3.0

As part of the computer model requirement, insitu soil infiltration rate was investigated. Infiltrometer P-88 from Geonor was used to obtain the infiltration capacity of the soil. Figure 4.0 shows the infiltrometer used.

5 SOIL-WATER CHARACTERICTIC

The energy / soil matric suction relationship for water is an important physical properties of unsaturated soil. The soil – water characteristic curve describes the relationship between the amount of water held / moisture in soil as a function of suction to serve the basis of much more important consideration, the magnitude and engineering behavior of soils. It has long been recognized, as energy relation (suction moisture content relationship) of soil for water in multiple value functions (hysteresis effect) that depends upon the moisture history of the sample. The data for the soil – water characteristic is a combination of soil suction measurement with moisture content at both laboratory and filed. Modified Rowe Cell with suction measurement was used to study the soil moisture content at various suction values in the laboratory. Likewise quickdraw tensiometer was used to measure matric suction with various soil moisture content in the field. Figure 5.0 and 6.0 illustrates the modification Rowe Cell and a plot of soil water characteristic curve obtained at study location. The data of soil-water characteristic curve was used as an input in the computer model to simulate the change in suction with given infiltration rate, time and rainfall intensity.

6 FIELD TEST METHOLOGY

An artificial rainfall simulation sprinkler system was set up at the field using steel frame parallel to the slope. The frame was sized at 15 feet by 15 feet with four rows of PVC host fitted with six sprinkler heads on each row. The schematic diagram of the frame and sprinkle head location is shown in Figure 7.0. The required intensity of artificial rainfall was calculated based on the highest rainfall recorded in Malaysia. The inlet water supply was measured and monitored to be constant during the test using a flow meter. Whereby the surface run off water was collected using V-notch collecting drain and measured periodically.

To monitor the suction changes 7 numbers of small tip tensiometer and 4 numbers of jet fill tensiometers were installed. The small tip tensiometers were installed at 10 inches depth perpendicular to the slope surface within the sprinkling frame.Wherelse jet fill tensiometers were installed at 0.5m, 1.0m, 2.0m and 3.0m near to the slope toe and perpendicular to the slope surface. The schematic layout of the installed tensiometer positions with respect to the sprinkling frame and slope shown in Figure 8.0.

The test was performed at constant volume of water supply with artificial rainfall intensity of 2.66m^{-5} m/s. The soil suction and surface run off data were measured at 10 minutes intervals. The artificial rainfall was maintained for two and half-hours. And the subsequent time, the suction data were monitored till no further changes in reading.

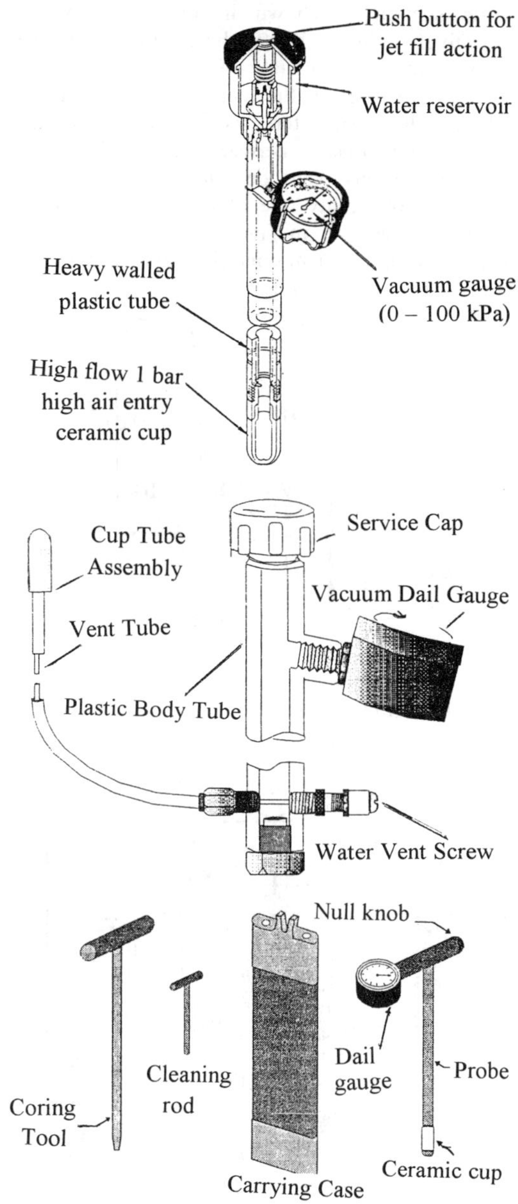

Figure 3 : Schematic drawing of Jet Fill, Small Tip and Quick draw Tensiometer.

7 COMPUTER MODEL

The computer simulation was performed using finite element, seepage analysis software, "SEEP-W" developed by Geo-Slope International Ltd. The field experiment data shown in Figure 6.0 is converted into input parameter of Soil Conductivity and Volume Water Content Function with respect to soil matric suction as shown in Figure 9 and 10 respectively. Figure 11.0 shows the developed computer model to simulate the site condition. Finer messes were given at the location where small tip and jet fill tensiometers were installed. Hence, relative comparison between field data and computed results can be analysed. The field recorded permeability rate of 2.31 X 10^{-6} m/s has also incorporated in the model. The analysis was executed using transient control mode with total run time of 2 hours 30 minutes.

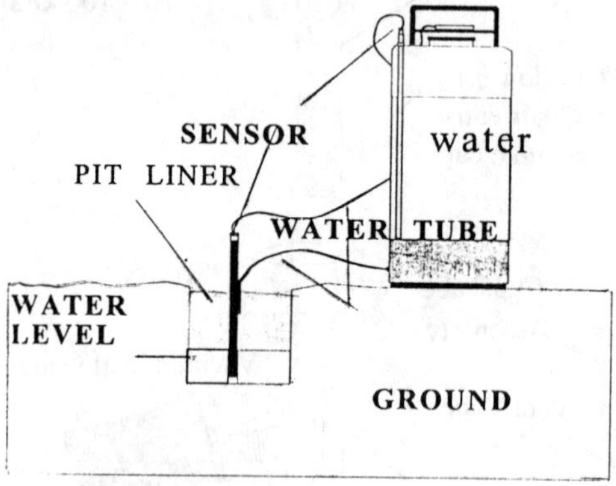

Figure 4.0 : Infiltrometer

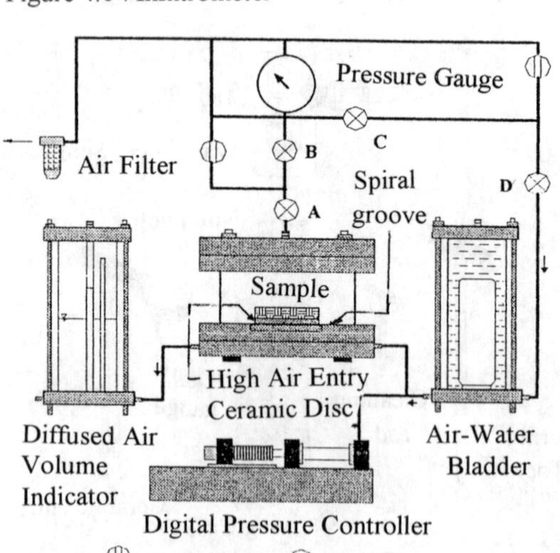

Figure 5 : Modified Rowe Cell

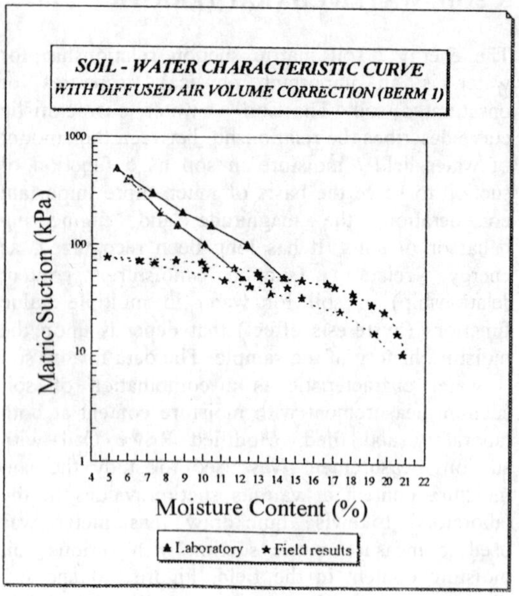

Figure 6 : Soil water characteristic curve obtained at test location

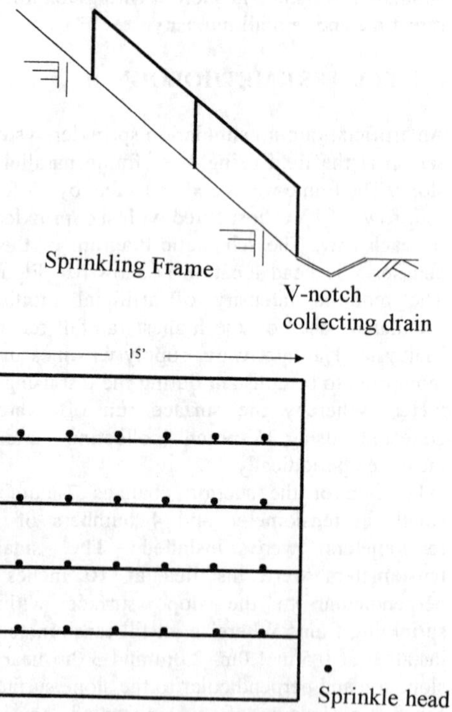

Figure 7: Diagram of the sprinkling system

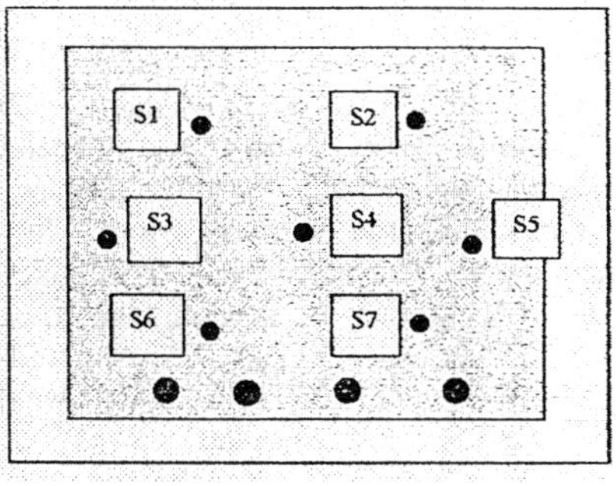

Figure 8 : Tensiometer layout

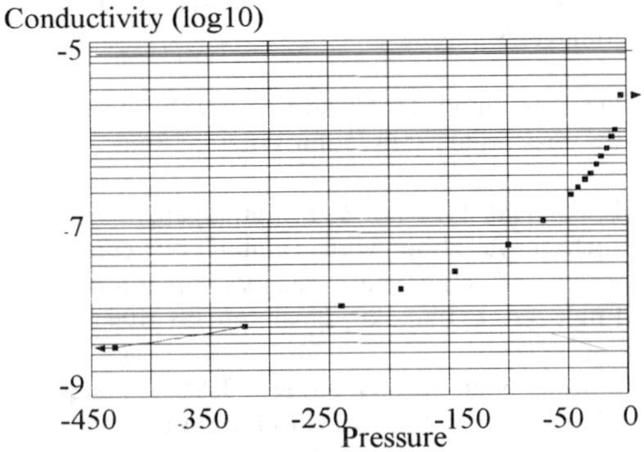

Figure 9 : Soil conductivity function

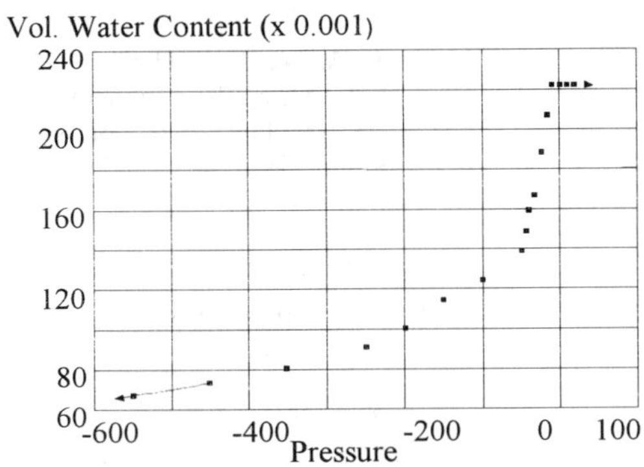

Figure 10 : Volume Water Content Function

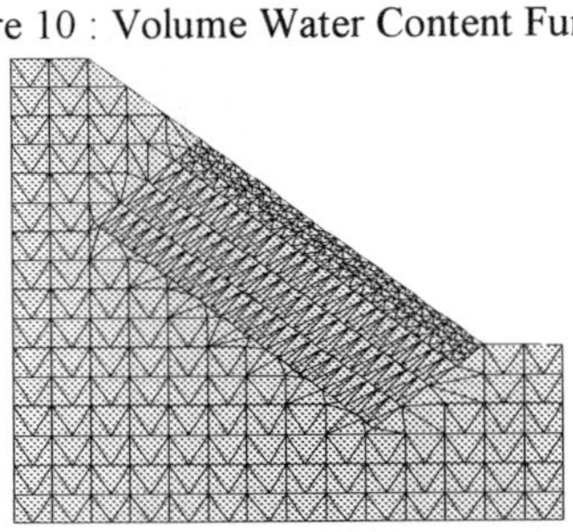

Figure 11 : Computer model of the site

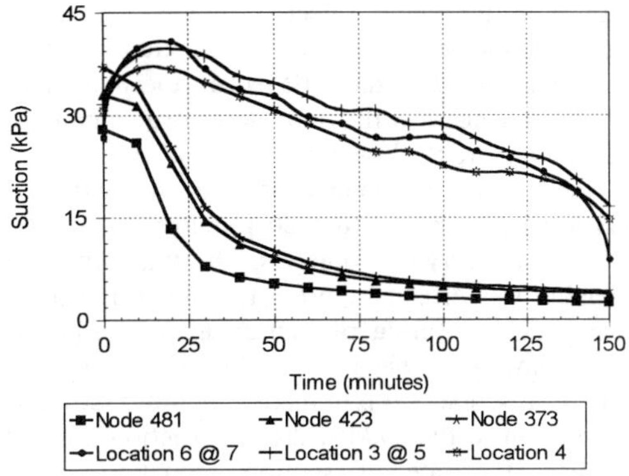

Figure 12 : Comparison plot between field and computed data

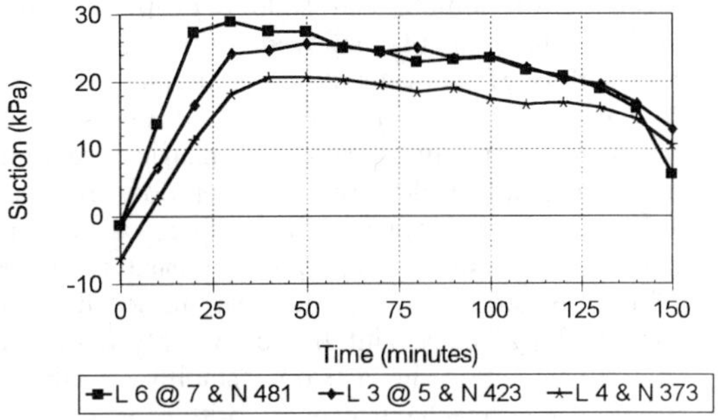

Figure 13 : Plot of net differences between field and computed data

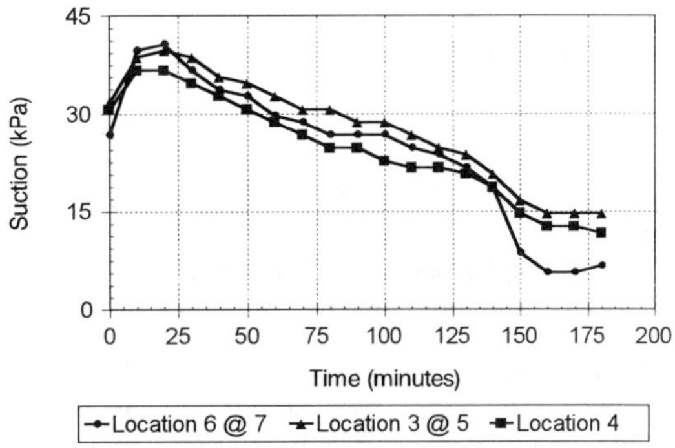

Figure 14 : Plot of continous monitoring of soil suction at the field

8 COMPARISON BETWEEN FIELD AND COMPUTED RESULTS

Relative comparison between field tensiometers data and the computed results were made by plotting the changes in matric suction with time. Figure 12.0 shows the plot of field and computed matric suctions with time for some selected points. Figure 13.0 illustrates the net differences of soil suction between field and computed data. Based on the plot, rate of suction changes with time after 30 minutes of rainfall is more the less same.

However a phenomena of sudden increase of matric suction in the first 20 minutes of rainfall is recorded in every tensiometers. The software did not compute this effect. Figure 14.0 shows the plot of continuos monitoring of suction after rainfall. The figure shows that when the rainfall is stopped at 150 minutes a sudden drop in matric suction takes place. In addition to the sudden drop in suction after 150 minutes, the rate of suction reduction is also recorded high for the next 30 minutes before stabilization take place and drying process begins. The sudden increase of matric suction in the beginning of rainfall, and sudden drop in matric suction after rainfall, believed to be caused by lamina flow of rainwater at the slope surface. The flow creates lower pressure zone at the surface, which respectively increases matric suction. However a sudden drop of rainwater flow, releases the lower pressure zone at the surface, causing an quick drop in soil suction and increase in water infiltration rate into the slope. The sudden drop in suction after a heavy rainfall is commonly not taken into consideration in a slope stability analysis. Hence, a new mechanism or algorithm need to be developed considering the above mention phenomena in the slope stability analysis methodology.

9 CONCLUSION

The test results obtained from this experiment have indicated some differences between theoretical and measured data. The different could have derived from:

1. Much higher surface runoff at site.
2. Less water infiltration due to surface cover.

However the intermediate values have shown that the rate of suction changes with time is more the less same for both field and computer prediction.

Currently continuos research are being conducted at both field and laboratory to understand the infiltration pattern of rainwater. The present infiltration algorithm used in computer software may require to be improved further by considering the influences of soil capillary and lamina flow of rainwater at the surface.

10 ACKNOWLEDGEMENT

The authors would like to acknowledge the Road Section of Public Works Department of Malaysia and Malaysian Government R & D Department (IRPA) for providing the research grants. Mr. Lee Yik Seong, Mr. K. Theva Kumar ,Mr. Eugine Lai and Mr. Teo Chong Chai in assisting the research works.

REFERENCES

Affendi A., (1996). Field and laboratory study on unsaturated residual soils in relation to slope stability analysis. Ph.D. Thesis. University of Malaya, Malaysia.

D.G. Fredlund, H. Rahardjo,(1993). Soil mechanics for unsaturated soils, John Wiley & Sons.

L. T. Huat, F.H. Ali & S.Mariappan, (1999), "Field suction variation with rainfall on cut slope in weathered sedimentary residual soil",Slope Stability Engineering, Shikoku, Japan. pp 399-404.

S. Mariappan M, F.H. Ali & L.T."Soil-Water Characteristics Relationship for Unsaturated Residual Soils", World Engineering Congress 1999, July 1999, Kuala Lumpur. –

S. Mariappan M, F.H. Ali & L.T. Huat, (1999), "Determination of shear strength parameters of unsaturated sedimentary residual soils for slope stability analyses", Slope Stability Engineering, Shikoku, Japan. pp 687-691.

Yong, R. Wakentin, B. P. (1975), Soil water behavior of soil,: Chapter 6, pp 127-150.

Computer Methods and Advances in Geomechanics, Desai et al. (eds) © 2001 Balkema, Rotterdam, ISBN 90 5809 183 X

Numerical solution for 1-D coupled sedimentation and consolidation

S. Masala & D. Chan
Department of Civil and Environmental Engineering, University of Alberta, Edmonton, Alb., Canada

ABSTRACT: A model for one-dimensional simultaneous sedimentation and consolidation has been derived as a mixture theory, using permeability as the unifying concept. The numerical solution presented is based on an explicit finite difference procedure in material coordinates, with Euler forward-marching scheme in time. The solution abandons the problem of discontinuities through a combination of convenient theoretical assumptions and the use of Lagrangian coordinates. This procedure is presently capable of solving a "closed system" with one impermeable boundary; i.e. it is confined to problems without percolation of the fluid through the system. The model is validated with a cohesive material data set. It is shown that a very good match of experimental data, both surface settlement and concentration profile development with time, can be obtained.

1 INTRODUCTION

Sedimentation is the process of settling of solid particles dispersed in a liquid. The process is driven by gravity and is caused by the inability of a fluid to sustain long-term shear stresses. During sedimentation the solid particles gradually form a sediment, a saturated soil layer, at the bottom of the suspension column. Consolidation is the time-dependent compaction of the soil skeleton in a sedimented bed as a result of a load: self- weight as well as other loads exerted on it. Both processes are of theoretical and practical interest in diverse scientific and technological fields, from chemistry and material engineering to geology and geotechnical engineering.

Historically, two major theories describing these processes have emerged independently: sedimentation has been a concern of chemists, while consolidation has traditionally remained a geotechnical subject. It is not surprising, therefore, that until recently there have been no attempts to integrate the separately developed models into one unifying theory.

Sedimentation became a geotechnical concern in recent times when modern mining industries began to produce enormous quantities of waste material in the form of a slurry. These slurries, or tailings, have a high proportion of fine particles of colloidal size and exhibit exceptionally small rates of sedimentation; therefore, the sedimentation from the suspension and the consolidation of the sedimented material occur simultaneously.

In geotechnical engineering, sedimentation has often been treated as the consolidation of very loose soils, making use of non-linear finite strain consolidation theories. However, significant differences are encountered when the results of such analyses are compared with the observed or experimental responses. Thus the need has arisen for a method that will be able to simulate and successfully predict the essential elements of the sedimentation and consolidation considered as a coupled process.

2 PHENOMENOLOGY

Extensive experimental work on this subject (Been & Sills, 1981) has clarified the basics of a continuous process of sedimentation and consolidation of solid particles from an initially suspended state.

Three modes, or regimes, of settling behaviour were identified: free settling in dilute suspensions, hindered settling in concentrated suspensions and self-weight consolidation. Among these three modes, the hindered settling phenomenon is relatively unfamiliar to an engineer. It is characterized by the sedimentation of a suspension as a whole, as if the particles were in a spatial network, but without stable direct contacts and measurable effective stresses.

During settling, zoning occurs in an initially uniform suspension. Three principal zones are distinguished: almost clear water at the top, a suspension in the middle and sediment at the bottom (Figure 1). The settling behaviour of these zones in general corresponds to the above three settling regimes.

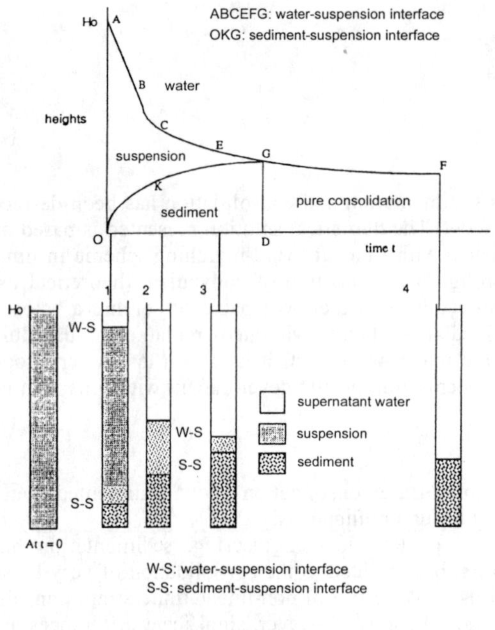

Figure 1. Settling behaviour of a suspension

The settling velocity of the particles in suspension is found to be dependent on their concentration as the main variable. The settling velocity of solids in the sediment depends on the permeability and the compressibility of the solid structure (skeleton), which are both functions of the porosity (concentration) of the solid phase.

3 MODELLING

A model for simultaneous sedimentation and consolidation in 1-D was developed by Masala (1998). Only the assumptions, a brief description and the governing equations are presented here.

The models for both phenomena were derived in Eulerian coordinates, beginning from the same theoretical basis – two laws of continuum mechanics: the law of conservation of mass and the principle of linear momentum.

3.1 *The conservation of mass (continuity) equation*

The mass balance (or continuity) equation is derived considering the suspension as a mixture composed of solid and fluid phases. Both are assumed incompressible so that the change in the total density of a mixture is completely convective—a plain result of the change in the proportion of the components within a control volume due to various fluxes over its boundaries.

The same continuity equation is valid over both suspension and sediment regions.

3.2 *Equilibrium equation*

It is assumed that the acceleration terms in the equations of motion vanish (a "quasi-static" conditions). As a consequence, the application of the method is restricted to "slowly settling" suspensions ("non-segregating" dispersions, colloidal suspensions, etc.).

The derivation of the equilibrium equation in a one-dimensional case is elementary, see for example Gibson et al. (1967). Considering the settling system as a mixture, the total axial stress is expressed as the sum of the stresses acting in the solid and the liquid phase. The total density is also expressed as the sum of partial densities of the mixture components.

Two separate equilibrium equations must be written in this case. In the region of settling suspension, the effective stresses in solids vanish and the total stress is equal to the fluid pressure: $\sigma = u$, i.e. a suspension behaves as a "heavy" liquid. In the region of consolidating sediment, the total stress is the sum of fluid pressure and a finite effective stress: $\sigma = \sigma' + u$.

3.3 *Connecting continuity and equilibrium*

The missing equation to complete the problem is a definition of the mutual interaction force between the solids and the liquid. In this derivation we have introduced the permeability as a measure of the interaction in a solid-liquid mixture, using Darcy-Gersevanov's law.

It is worth noting here that the concept of permeability has already been extended to the sedimentation stage and a suspended state of solids (Pane & Schiffmann 1997); in other words, it is assumed that the expression for the coefficient of permeability applies smoothly over a range of void ratios which encompasses both soil and suspension, regardless of whether effective stresses are present or absent. Permeability thus appears as a measure of hydrodynamic interaction between solid and liquid in a suspension. In a very general sense, it may be understood as a "constitutive law" for the solid-fluid interaction in suspensions.

3.4 *Governing equation for sedimentation*

The following governing equations were derived for sedimentation and consolidation, respectively:

$$\frac{\partial c}{\partial t}+\frac{\partial}{\partial x}\left\{c^2 k\left(G_s-1\right)\right\}=0 \tag{1}$$

$$\frac{\partial c}{\partial t}+\frac{\partial}{\partial x}\left\{c^2 k\left(G_s-1\right)-c\frac{k}{\gamma_w}\frac{\partial\sigma'}{\partial c}\frac{\partial c}{\partial x}\right\}=0 \tag{2}$$

where: c = solids concentration; G_s = specific weight of solids, γ_w = unit weight of liquid, $k\,(c)$ = permeability; $\sigma'(c)$ = compressibility, both being the functions of porosity, i.e. solid concentration c.

Both equations are expressed in terms of volumetric solids concentration c as the independent variable. The use of solids concentration is preferred over void ratio e, which is typical in soil mechanics, because of its mathematical convenience. In the sedimentation stage e becomes infinite when concentration decreases to zero, i.e. when all particles have settled.

It is easily seen that the governing equation for sedimentation may be obtained from the governing equation for consolidation by simply setting σ' = 0. This shows that the concept of permeability offers a base for the unification of theories of sedimentation and consolidation.

4 SOLUTION

The governing equation for sedimentation (1) is a hyperbolic partial differential equation (PDE) while the consolidation equation (2) is of the parabolic type. Both equations were derived in the same form – the conservation law form, which again gives evidence for the fact that both processes have the same underlying physics.

4.1 *Problems with numerical solution of the coupled system*

The “weak link” in the governing system is the sedimentation equation, as a hyperbolic PDE. The solution of such equations may have discontinuities which originate from the discontinuities in the initial and boundary data or, in the case of a non-linear PDE, equation (1), discontinuities may be developed even from smooth initial data at some later time.

In our problem discontinuities such as the water-suspension and the suspension-sediment interface normally appear in a solution, and they are an essential part of a solution. The discontinuities propagate along characteristic lines, and in any particular sedimentation problem the upper boundary of the suspension layer is a characteristic line. The sediment-suspension interface is not a discontinuity in its original meaning in PDE theory, but rather a consequence of the “redundant” boundary conditions at the bottom boundary. Nevertheless, it behaves as a discontinuity in a numerical scheme.

It was first attempted to apply the finite difference methods that were developed in the Eulerian coordinate frame, i.e. on a fixed grid. The results were discouraging. Even the procedures which had previously claimed to behave well in the presence of discontinuities, like TVD and ENO schemes, were incapable of handling the discontinuities properly. It is important to mention that two phenomena, numerical dispersion (artificial oscillations) and numerical diffusion (artificial dissipation, change of shape of the solution profile), are responsible for the deviations of a particular numerical solution from an actual one.

4.2 *Description of the solution*

Although the governing equations were derived in Eulerian coordinates, the successful numerical procedure in fact makes use of the Lagrangian finite difference grid, which follows the motion of the material. The idea was borrowed from the moving mesh method, which uses an initially uniform grid that deforms with time and the nodal points which move relative to each other adapting their spatial distribution to the requirements of an optimal description of the deformation process. When each nodal point remains connected to the same material point to which it has been connected in the initial state, then one says that the grid is a Lagrangian one.

The simplest Eulerian time-forward marching scheme was chosen for advancing the solution in time. Although it is not very sophisticated by itself, it behaved well in the examples calculated. Sporadic problems with numerical dispersion (oscillations) were removed by reducing the time step. This worked well for reasonably smooth initial conditions (approximately uniform suspension in the initial state).

In the solution presented, the water-suspension discontinuity is dealt with explicitly, while the suspension-sediment interface is actually “smeared off” between the two consecutive nodes of a grid which bound a temporary position of this discontinuity. The current computer program is able to deal with closed systems only (e.g. tank settling) without percolation of a fluid (seepage) through the system.

Discretization of the initial settling column is shown in Figure 2. The sedimenting system is divided into an array of equal cells with centered nodal points. The origin of the spatial coordinate x is placed at the surface of the suspension.

The state variable is the volumetric concentration $c(x,t)$, i.e. in the nodal points c_i^n, where the subscript i denotes the spatial position (node), while the superscript n denotes the time instant (Figure 3). It is assumed that the initial state is known, i.e. that $c(x,0) = c_0(x)$.

The solids fluxes over the cell boundaries are denoted by a fraction in the subscript, for example (Figure 3):

$$(\phi_s)_{i+\frac{1}{2}} \tag{3}$$

denotes the solids flux between the cells i and $i+1$.

The system of governing equations is expressed in the form of a conservation law for both regions of sedimentation (the suspension) and consolidation (the sediment):

$$\frac{\partial c}{\partial t}+\frac{\partial(\phi_s)_{sed}}{\partial x}=0 \;;\quad \frac{\partial c}{\partial t}+\frac{\partial(\phi_s)_{cons}}{\partial x}=0 \tag{4}$$

and:

$$(\phi_s)_{sed}=c\cdot(v_s)_{sed} \;;\quad (\phi_s)_{cons}=c\cdot(v_s)_{cons} \tag{5}$$

where: v_s = settling velocity of solid particles. Its value is calculated according to two different expressions, depending on the region:

$$(v_s)_{sed}=(G_s-1)\cdot c\cdot k \tag{6}$$

$$(v_s)_{cons}=(G_s-1)\cdot c\cdot k-\frac{k}{\gamma_w}\frac{d\sigma'}{dc}\frac{\partial c}{\partial x} \tag{7}$$

The calculation cycle consists of the following steps:

(1) calculate concentrations at cell boundaries as the averages of concentrations in adjacent nodes:

$$c^n_{i+\frac{1}{2}}=\frac{1}{2}\left(c^n_{i+1}+c^n_i\right) \;;\quad c^n_{i-\frac{1}{2}}=\frac{1}{2}\left(c^n_i+c^n_{i-1}\right) \tag{8}$$

(2) calculate boundary velocities and solids fluxes

$$(v_s)^n_{i+\frac{1}{2}}=v_s\left(c^n_{i+\frac{1}{2}}\right) \;;\quad (\phi_s)^n_{i+\frac{1}{2}}=c^n_{i+\frac{1}{2}}\cdot v_s\left(c^n_{i+\frac{1}{2}}\right) \tag{9}$$

(3) update geometry:

$$x^{n+1}_{i+\frac{1}{2}}=x^n_{i+\frac{1}{2}}+(v_s)^n_{i+\frac{1}{2}}\cdot\Delta t \;;\quad \Delta x_i^{n+1}=x^{n+1}_{i+\frac{1}{2}}-x^{n+1}_{i+\frac{1}{2}} \tag{10}$$

(4) update concentrations (the state variable):

$$c_i^{n+1}=\frac{(V_s)_i^0}{\Delta x_i^{n+1}} \tag{11}$$

with

$$(V_s)_i^0=c_i^0\cdot\Delta x_i^0 \tag{12}$$

being the total volume of solids in cell i at time $t = 0$.

The boundary conditions for the whole settling column consisted of prescribed zero solids fluxes at the outer boundaries of the first and the last cell. This was transformed into the prescribed concentrations in two imaginary cells before the first and after the last cell, and two additional imaginary nodes in these cells, denoted as nodes *(0)* and *(N+1)*:

$$c_0^n=-c_1^n \Rightarrow (\phi_s)^n_{\frac{1}{2}}\equiv 0 \tag{13}$$

$$c_N^n=-c_{N+1}^n \Rightarrow (\phi_s)^n_{N+\frac{1}{2}}\equiv 0 \tag{14}$$

thus keeping the boundary fluxes at the ends of the settling column identically equal to zero.

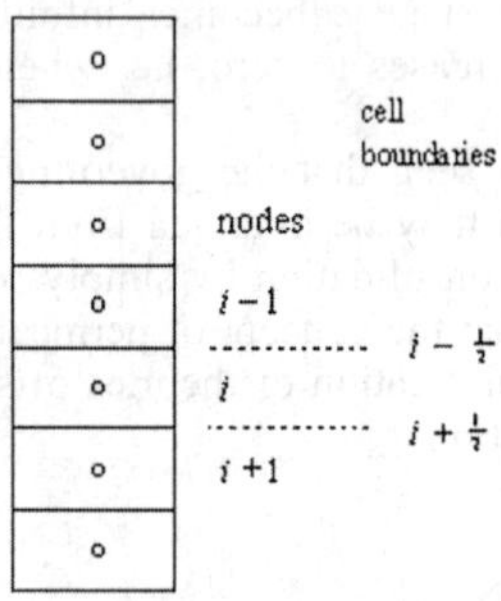

Figure 2. Numerical scheme: discretization

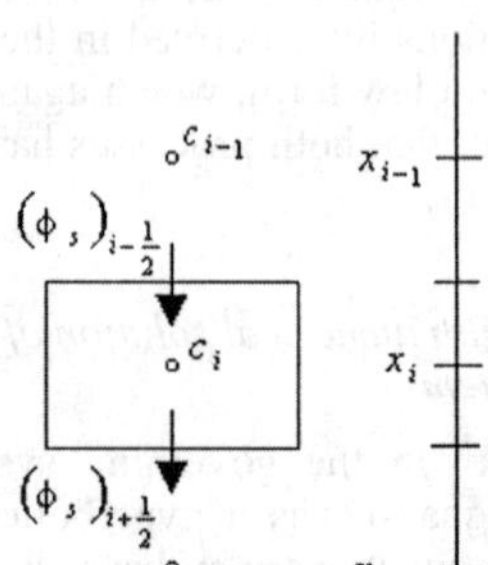

Figure 3. Numerical scheme: principal variables – concentrations and fluxes

Since the x-coordinate origin was set at the suspension surface, the settlement of the water-suspension interface was merely the displacement of the left boundary of cell *1*. The internal boundary between suspended and sedimented material in the settling column was determined in each step by comparing the nodal concentrations with the critical concentration c_{cr} .

5 EXAMPLE

The behaviour of the model can be demonstrated by the simulation of a laboratory sedimentation and self-weight consolidation test. The model behaves equally well in both non-cohesive material (rigid particles, stiff sediment) and cohesive material scenarios. The material presented here is a cohesive material suspension.

The test selected was carried out by Bowden in 1988 on Combwich mud, a natural mud from the River Parret estuary (Toorman 1999).

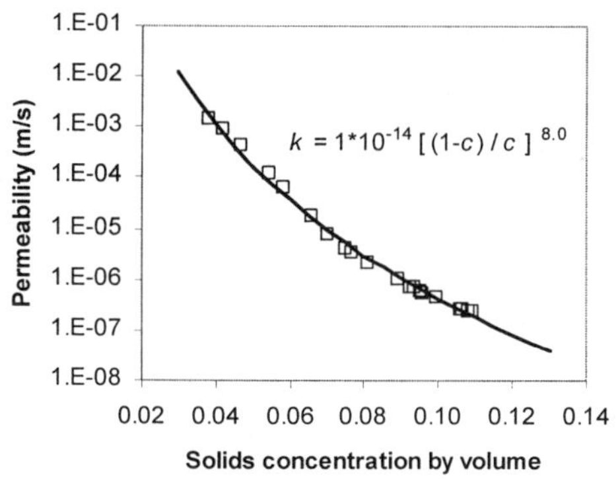

Figure 4. Permeability vs. concentration (Combwich mud)

The experimental data for the settling rate as a function of the excess density were first converted to the concentration – permeability relationship $k = k(c)$ and then fitted by a power law function for $k(c)$, Figure 4. Therefore, the main input data were obtained independently from the settling test itself.

The settlement of the suspension surface is shown in Figure 5. The testing and simulation occured over more than 300 hours, however only the first 12 hours are shown here. It can be seen that the sedimentation stage is quite short and that consolidation is the prevailing phenomenon. Although a good match with experimental data was obtained, it should be noted that much simpler models can also fit this curve.

The evolution of concentration (density) profiles is shown in Figure 6. The model both qualitatively and quantitatively predicts the typical features of the concentration distribution, and it can be shown that the maximum deviation between model and measurement is about twice the experimental error. The capacity of a model to correctly predict the density profile development with time (and the pore pressure distribution as well) is of a much greater value from an engineering point of view than to merely match measured settlement curves.

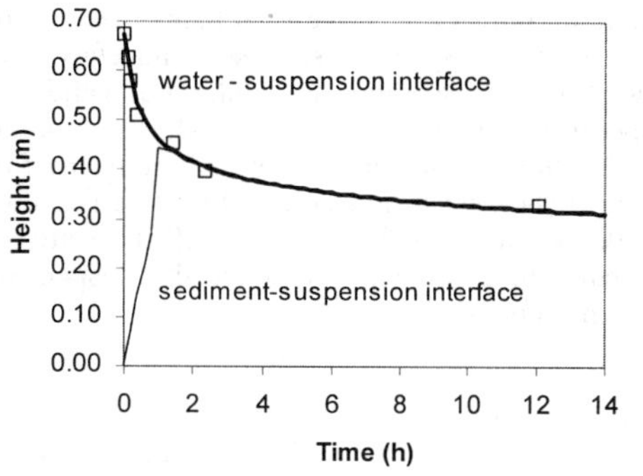

Figure 5. Suspension settlement curves

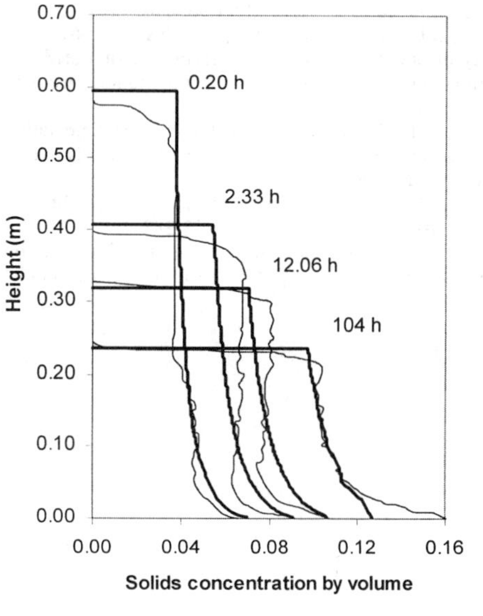

Figure 6. Concentration (density) profile evolution

6 CONCLUDING REMARKS

A model for coupled sedimentation and consolidation was derived in the form of a mixture theory, with permeability as the unifying concept. The solution method for coupled PDE governing equations was derived as a finite diference scheme with Lagrangian grid. It was capable of dealing with the discontinuities – interfaces, which appear as a regular and an essential part of the solution.

The solution abandons the problem of discontinuities through a combination of convenient theoretical assumptions and the use of Lagrangian coordinates. Namely, determination of the position of the water-suspension interface was eliminated by the use of Lagrangian coordinates which are connected to material points and move with them. Determination of the suspension-sediment interface was avoided: (1) by assuming the continuous validity of the permeability law *k(e)* in both sedimentation and consolidation regions, and (2) by adopting the fluid nature of the suspension zone with effective stress $\sigma' = 0$, and the gradual increase of effective stress in the sediment from zero value at the suspension-sediment interface.

REFERENCES

Been, K. & G.C. Sills 1981. Self-weight consolidation of soft soils: an experimental and theoretical study. *Geotechnique* 31: 519-535.

Ganzha, V.G. & E.V. Vorozhtsov 1996. *Numerical Solutions for Partial Differential Equations: Problem Solving Using Mathematica*. Boca Raton, Florida: CRC Press.

Gibson, R.E., G.L. England & M.J.L. Hussey 1967. The theory of one-dimensional consolidation of saturated clays, I. Finite non-linear consolidation of thin homogeneous layers. *Geotechnique* 17: 261-273.

Masala, S. 1998. Numerical Simulation of Sedimentation and Consolidation of Fine Tailings, M.Sc. thesis, University of Alberta, Edmonton.

Pane, V. & R.L. Schiffman 1997. The permeability of clay suspensions. *Geotechnique* 47: 273-288.

Toorman, E.A. 1981. Sedimentation and self-weight consolidation: constitutive equations and numerical modelling. *Geotechnique* 49: 709-726.

Computer Methods and Advances in Geomechanics, Desai et al. (eds) © 2001 Balkema, Rotterdam, ISBN 90 5809 183 X

Block parallel formulation for coupled geotechnical problems

F.Okulicka-Dłużewska
Warsaw University of Technology, Poland

ABSTRACT: In the paper the coupled geotechnical problems are analyzed. The finite element formulations of the coupled problems lead to the block form of the global equations sets. The algorithms of the block parallel solvers are studied in the paper. The consolidation problem is analyzed as the example of he coupled problem, which leads to the block global equations set. The proposed formulation is general and can be used for large equations systems. In such situation the time of calculations is shorter comparing the sequential version of the program.

1 INTRODUCTION

The application of the parallel algorithms allows to speed up the process of calculations. The parallel approach is worth using in the case of large systems. The finite element method, widely used to solve engineering problems, is easy to parallellize. The chosen steps of the algorithms can be calculated in parallel. The heart of the program is solver of the set of linear equations. There are many parallel solvers, which can be applied. In the paper the coupled problems are taken under consideration. The structure of such problem leads to the block solvers.

The finite element package oriented at hydro and geotechnical problems is used. The sequential version of the program is the starting point for developing the parallel version step by step. The package is composed of the three main programs: the preprocessing for mesh generation and preparing the data, processor for main mechanical calculation and graphical post processor. These programs are running under the management program. The engineering problems of coupled nonlinear consolidation are studied for a few chosen level of parallelization. The speed up tests for the non-steady stress-flow problems are carefully studied. As the engineering example the dam building, filling of the reservoir and consolidation of the subsoil, by means of finite element approach in parallel version, are analyzed. The speed up is analyzed depending on the scale of the problem i.e how many subproblems are generated during the calculations.

The results are obtained due to the computational resources owned by Computing Center Warsaw University of Technology.

2 COUPLED PROBLEMS

If we consider the thermal flow in the saturated soil the system of the differential equations describing the behavior of such system is coupled with the mechanical part.

When the finite differences method is used for the solution of the initial value problem (Lewis,R.W. & B.A.Schrefler 1998) the time stepping process became highly time consuming.

2.1 Finite Element Formulation for a Two-Phase Medium

The notation used herein follows Zienkiewicz and Taylor (1991) (considering the basic equations) and Lewis and Schrefler (1998) (considering the consolidation).

Introducing the finite element method interpolation equations can be written as follows

$$\mathbf{u}^e = \mathbf{N}\mathbf{u}, \qquad \mathbf{e} = \mathbf{B}\mathbf{u} \tag{1}$$

$$\mathbf{p}^e = \overline{\mathbf{N}}\mathbf{p} \qquad \mathbf{x}^e = \overline{\overline{\mathbf{N}}}\mathbf{x} \tag{2}$$

where **N** - shape function array for displacement approximation, **B** - geometrical array containing the linear parts of the strain tensor, $\overline{\mathbf{N}}$ - shape function array for pore-water pressure approximation, $\overline{\overline{\mathbf{N}}}$ - shape function array for coordination approximation for 'inconsistent' formulation.

The forms of $\mathbf{N}, \overline{\mathbf{N}}, \overline{\overline{\mathbf{N}}}$ depend on the type of the finite element, type of the shape function applied to the approximation of displacements, pore-water pressure and coordinates. In consolidation problems the 'consistent' and 'inconsistent' formulations are considered very often (Dłużewski 2000).

The equilibrium equation for a two phase medium takes the following form

$$\mathbf{K}_T \frac{d\mathbf{u}}{dt} + \mathbf{L}\frac{d\mathbf{p}}{dt} = \frac{d\mathbf{P}}{dt} \quad (3)$$

In the above equation the following notation is used: **P** - vector of load, **u** - vector of displacement, **p** - vector of pore-water pressure, **L** – coupling array, $\mathbf{K}_T$ – tangent array.

$$\mathbf{L} = \int_v \mathbf{B}^T \alpha\, \mathbf{m}\, \overline{\mathbf{N}}\, dv \quad (4)$$

$$\mathbf{K}_T = \int_v \mathbf{B}^T \mathbf{D}\mathbf{B}\overline{\mathbf{N}}\, dv \quad (5)$$

where **B** - geometrical array, $\overline{\mathbf{N}}$ - shape function array for pore-water pressure approximation, **m** - one-zero vector (zero for shearing stresses, one for the remaining ones), α- Biot constant.

The continuity equation serves as the source for formulation of the finite element approximation for the liquid phase. Let us introduce the trial function in the form $p_e = \overline{\mathbf{N}}\mathbf{p}$ where N_i are the elements of the array. The inconsistent formulation is applied (i.e. the same shape functions are used for the approximation of displacements and the pore-water pressure). Introducing the matrix notation and applying the Galerkin method to solve the continuity equation we get

$$\mathbf{H}\mathbf{p} + \mathbf{S}\frac{d\mathbf{p}}{dt} + \mathbf{L}^T\frac{d\mathbf{u}}{dt} = \mathbf{Q} \quad (6)$$

where the following definitions are applied: **H** - array responsible for flow properties, **S** - array responsible for water compressibility, **L** - coupling array, **Q** - nodal discharges vector, ∇ - differential operators vector, $\overline{\mathbf{N}}$ - shape functions array, **k** - permeability array, **m** - one-zero vector, **q** - discharges at the boundary.

$$\mathbf{H} = \int_V (\nabla\overline{\mathbf{N}})^T \frac{\mathbf{k}}{\mu} \nabla\overline{\mathbf{N}}\, dV \quad (7)$$

$$\mathbf{S} = \int_V \overline{\mathbf{N}}^T s \overline{\mathbf{N}}\, dV \quad (8)$$

The detail definitions of the arrays can be found in the book by Lewis and Schrefler (1998).

The nodal equilibrium equation (3) is coupled with the mass balance equation (6) via volumetric changes. Both equations can be linked in the following form

$$\begin{bmatrix} \mathbf{K}_T & \mathbf{L} \\ \mathbf{L}^T & -\mathbf{S} \end{bmatrix} \begin{bmatrix} \frac{d\mathbf{u}}{dt} \\ \frac{d\mathbf{p}}{dt} \end{bmatrix} = \begin{bmatrix} \mathbf{0} & \mathbf{0} \\ \mathbf{0} & -\mathbf{H} \end{bmatrix} \begin{bmatrix} \mathbf{u} \\ \mathbf{p} \end{bmatrix} + \begin{bmatrix} \frac{d\mathbf{P}}{dt} \\ \mathbf{Q} \end{bmatrix} \quad (9)$$

The above set of equations describes the behaviour of the two-phase medium in terms of the finite element approximation.

3 PARALLEL PROCESSES IN FINITE ELEMENT METHOD

The typical finite element method package consists of three different parts:
1. Pre-processor with mesh generation, mesh optimalization and data input
2. Processor with stiffness matrix calculation, solver and analysis of strains and stresses
3. Postprocessor for graphical and numerical representation of the computed results

The most time consuming part of the program is the processor thus the attention is focussed at this part of the method.

The finite element method package HYDRO-GEO (Dłużewski 1997) oriented at hydro and geotechnical problems is used. The program has been developed at Warsaw University of Technology in the sequential form and next extended to allow the parallel calculations. The sequential version of the program has been the starting point for the step by step development of the parallel version.

The parallel processor algorithm for single stages of construction building can be written in the following way:

Start
Data reading, initial computations

<u>**Loop for**</u> *each stage of the construction and each increment of load* <u>**do**</u>

Loop for *each subset of the set of elements* **do in parallel**

Calculate the initial stresses

Calculate the local stiffness matrix, coupling matrix, flow matrix

End do { *for each subset of the set of elements*}

Calculate the global set of fully coupled system

First part of the solver (forward substitution) **-parallel calculation**

Loop for *each plastic points* **do**

Second part of the solver (backward substitution) **-parallel calculation**

Loop for *each subset of the set of element* **do in parallel**

Calculation of strains and stresses

End do

{ *for each subset of the set of elements*}

End do{ *for each plastic points* }

End do{ *for each stage of the construction and each increment of load* }

Stop

4 BLOCK ALGORIHMS FOR SOLVING THE LINEAR EQUATIONS

Consider the system of linear equations

$$AX=B \qquad (10)$$

We divide our matrix A into m blocks, where each A_{ij} is the matrix on $n_i \times n_j$ dimensions.

$$\begin{bmatrix} A_{11} & A_{12} & \dots & A_{1m} \\ A_{21} & A_{22} & \dots & A_{2m} \\ . & & & . \\ . & & & . \\ . & & & . \\ A_{m1} & A_{m2} & \dots & A_{mm} \end{bmatrix} * \begin{bmatrix} X_1 \\ X_2 \\ . \\ . \\ . \\ X_m \end{bmatrix} = \begin{bmatrix} B_1 \\ B_2 \\ . \\ . \\ . \\ B_m \end{bmatrix} \qquad (11)$$

X and B are divided onto subvectors

$$X = \begin{bmatrix} X_1 & X_2 & \cdots & \cdots & X_m \end{bmatrix}^T$$
$$B = \begin{bmatrix} B_1 & B_2 & \cdots & \cdots & B_m \end{bmatrix}^T \qquad (12)$$

B_i and X_i , for i=1,...,m, are vectors.

Two methods of solving such system are described: direct – block Gaussian elimination and iterative – block Jacobi method. The parallel calculations are possible due to array operations. Generally most of the parallel concurrent numerical calculations are done on matrixes.

4.1 Parallelization of Gaussian Elimination

The method described by Demmel 1997 is used. The well-known Gaussian elimination is rewrote using array operations. The block Gaussian elimination on block array from the equation (11) is done. In general, after the (k-1)-th stage of elimination the system of linear equations is following

$$A^{(K-1)}X=B^{(K-1)} \qquad (13)$$

where

$$A^{(k-1)} = \begin{bmatrix} A_{11} & A_{12} & \dots & A_{1,k-1} & A_{1,k} & \dots & A_{1,m} \\ 0 & A_{22}^{(1)} & \dots & A_{2,k-1}^{(1)} & A_{2k}^{(1)} & \dots & A_{2,m}^{(1)} \\ \vdots & \ddots & \ddots & \vdots & \vdots & & \vdots \\ \vdots & & \ddots & A_{k-1,k-1}^{(k-2)} & A_{k-1,k}^{(k-2)} & \dots & A_{k-1,m}^{(k-2)} \\ \vdots & & & 0 & A_{kk}^{(k-1)} & \dots & A_{k,m}^{(k-1)} \\ \vdots & & & \vdots & \vdots & \ddots & \vdots \\ 0 & \dots & \dots & 0 & A_{mk}^{(k-1)} & \dots & A_{mm}^{(k-1)} \end{bmatrix} \qquad (14)$$

and

$$B^{(k-1)} = \begin{bmatrix} B_1 & B_2^{(1)} & \cdots & B_{k-1}^{(k-2)} & B_k^{(k-1)} & \cdots & B_m^{(k-1)} \end{bmatrix}^T \qquad (15)$$

At the k-th stage X_k is eliminated from the block equations $i=k+1,k+2,...,m$, by the subtraction of $A_{ik}^{(k-1)} (A_{kk}^{(k-1)})^{-1}$ times block equation k from each of the equations (13) in which the changed block elements are given by the following matrix calculations

$$A_{ij}^{(k)} = A_{ij}^{(k-1)} - A_{ik}^{(k-1)} (A_{kk}^{(k-1)})^{-1} A_{kj}^{(k-1)}$$
$$B_i^{(k)} = B_i^{(k-1)} - A_{ik}^{(k-1)} (A_{kk}^{(k-1)})^{-1} B_k^{(k-1)} \qquad (16)$$

for $i,j=k+1,k+2,...,m$.

After $m-1$ elimination stages

$$A^{(m-1)}X=B^{(m-1)} \qquad (17)$$

where $A^{(m-1)}$ is the block upper triangular matrix.

4.2 Backward substitution – block algorithm

To solve the block upper triangular system of equation (17), the equation can be written as follows

$$\begin{bmatrix} A_{11} & A_{12} & A_{13} & \dots & A_{1m} \\ 0 & A_{22}^{(1)} & A_{23}^{(1)} & & A_{2m}^{(1)} \\ 0 & 0 & A_{33}^{(2)} & & A_{3m}^{(2)} \\ \vdots & & \ddots & \vdots & \\ 0 & 0 & \dots & 0 & A_{mm}^{(m-1)} \end{bmatrix} * \begin{bmatrix} X_1 \\ X_2 \\ X_3 \\ \vdots \\ X_m \end{bmatrix} = \begin{bmatrix} B_1 \\ B_2^{(1)} \\ B_3^{(2)} \\ \vdots \\ B_m^{(m-1)} \end{bmatrix} \tag{18}$$

Backward substitution is calculated by matrix operations:

$$X_m = B_m^{(m-1)} (A_{mm}^{(m-1)})^{-1}$$

and

$$X_i = \left(B_i^{(i-1)} - \sum_{j=i+1}^{m} A_{ij}^{(i-1)} X_j \right) (A_{ii}^{(i-1)})^{-1} \tag{19}$$

for

$i = m-1, m-2, \dots, 1$

Example

The set of equations for consolidation problem

$$\begin{bmatrix} K & L \\ L^T & W \end{bmatrix} \cdot \begin{bmatrix} \Delta u \\ \Delta p \end{bmatrix} = \begin{bmatrix} A \\ B \end{bmatrix} \tag{20}$$

after block Gauss elimination has the following form:

$$\begin{bmatrix} K & L \\ 0 & W - L^T K^{-1} L \end{bmatrix} \cdot \begin{bmatrix} \Delta u \\ \Delta p \end{bmatrix} = \begin{bmatrix} A \\ B - L^T K^{-1} A \end{bmatrix} \tag{21}$$

4.3 Block Jacobi method

The iterative Jacobi method is analyzed. The (k+1) th step of iteration in block Jacobi method for equation (11) can be written as

$$A_{ii} X_{ii}^{k+1} = B_i - \sum_{j \neq i} A_{ij} \cdot X_{ij}^{k} \tag{22}$$

where

$$X^{k+1} = \begin{bmatrix} X_1^{k+1} & X_2^{k+1} & \dots & \dots & X_m^{k+1} \end{bmatrix}^T \tag{23}$$

is a solution calculated in this step.

Example

The Jacobi method for the set of equations for consolidation problem (21) leads to the following iterative formula:

$$\begin{aligned} K \cdot \Delta u^{k+1} &= A - L p^k \\ W \cdot \Delta p^{k+1} &= B - L^T u^k \end{aligned} \tag{24}$$

5 COMPUTATIONAL RESULTS

5.1 Supercomputer and parallel software

The parallel algorithms are tested on Sun E10000 HPC. Its architecture is characterised as follows:

- 12 processors 400MHz,
- 2GB RAM
- 72GB of the disk memory.

The three types of execution time can be compared: system time which is needed for operations like input/output and threads maintenance, real time which is equal the period observed by the user from the console and user time equal the time of processor used for calculations.

5.2 Engineering problem

As the engineering example the Mietków dam is modelled by means of finite element method calculated and analysed. The dam has been risen on the muiltilayer subsoil. Below the ground level the mud layer ranging from 0.5 to 2.5m is found at the old river beds. Next, the gravel with stones form the layer of 5m thickness. Below the gravel the clay formation is located. The finite element mesh is presented in Figure 1.

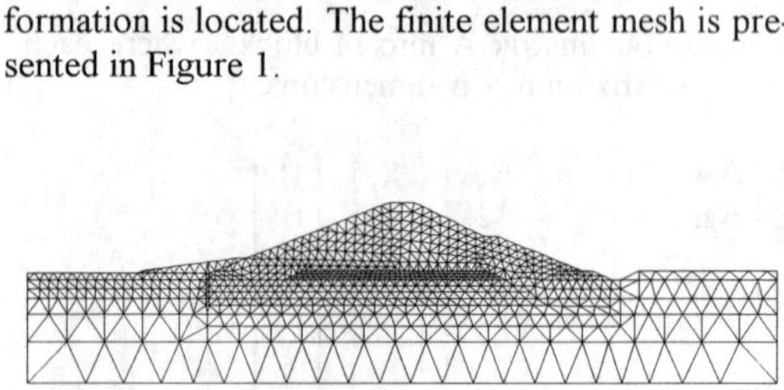

Figure 1. Finite element mesh – 6 noded izoparametric elements

The dam is built from sand-gravel mix. The concrete plates are performed from the upper water side. The clay-concrete screen of 0.8 m thickness is performed. The height of the screen ranges from 15 to 28 m. The length of the dam is more than 3 km. The dam is partly located on the old river beds where the mud layer is formed. Before rising the dam the initial stresses are introduced into the subsoil. In the second stage the soil dam with two trenches have been performed to make teeth for better interaction between the dam and subsoil.

The calculations are performed in three stages. In the first stage the initial stresses are introduced into subsoil. In the second stage dam is risen by adding the elements. In the third stage the loading causes by filling the reservoir is applied. The settlement of the dam with foundation in the form of izolines is shown in Figure 2.

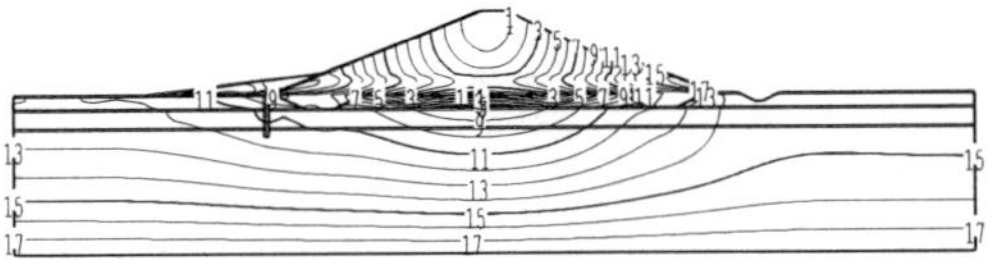

Figure 2. Izolines of the vertical displacements.

5.3 Results

The calculation are performed for different number of threads created during execution of the program on the supercomputer Sun. E10000. The parallel Gaussian elimination solver are implemented.

The speed-up of calculation times: real and user as a function of the number of parallel processes created during execution of the program is presented in Figure 3.

6 CONCLUSIONS

The speedup of execution time of the whole processor (main program) is much smaller than the speedup of the linear equation solver because quite considerable part of the system is still sequential i.e. data reading, calculation of the global set of fully coupled system.

The quite big speed up for 2 parallel processes are reached due to the optimalization of the code during compilation process. The compiler does it before splitting the code onto separate threads.

The number of the processes created concurrently during the execution of the parallel program can be bigger than the number of processors but in such

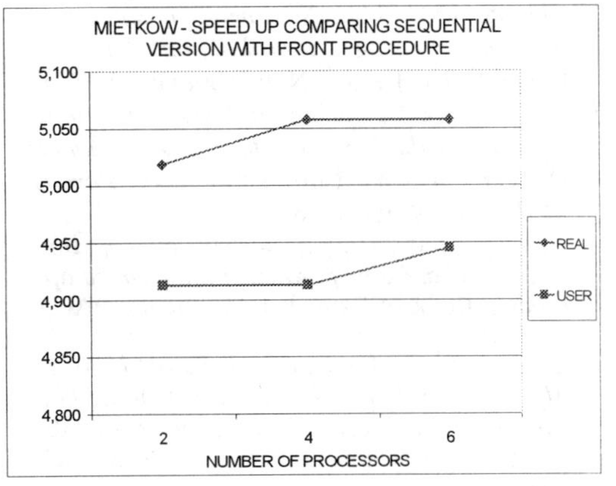

Figure 3. The speed-up of calculation times: real and user as a function of the number of parallel processes created during execution of the program.

case we can decrease performance. The Unix system creates one process more for managing purpose. SUN E10000 HPC is a multi-user system. The supercomputer is available at the same time for a group of users. The speed up depends on the number of processes created by all executed programs at the time. The result from Figure 3 is the average of 5 execution of the program.

REFERENCES

Andersen, B.S. & F..Gustavson & A.Karaivanov & J.Waśniewski & P.Y.Yalomov 1999. *LAWRA: Linear Algebra with Recursive Algorithms*, Proceedings of the Third International Conference on Parallel Processing and Applied Mathematics, Czestochowa. Poland.

Demmel, J. 1997. *Applied numerical linear algebra.* Springer-Verlag.

Dłużewski, J.M. 1997. *HYDRO-GEO – finite element code for geotechnics, hydrotechnics and environmental engineering.* (in Polish). Warsaw University of Technology. Warsaw.

Dłużewski, J.M. 1998. *Large strain consolidation for elasto-plastic soils*, Application of numerical methods to geotechnical problems, Proceedings of the fourth European Conference on Numerical Methods in Geotechnical Engineering. NUMGE'98. Udine. Italy.

Dłużewski, J.M. 1997, *Nonlinear consolidation in finite element modeling*, Proceedings of the Ninth International Conference on Computer Methods and Advances in Geomechanics, Balkema: Wuhan, China.

Dłużewski, J.M. 2000, *Nonlinear consolidation problems during consolidation process,* CISM, Udine, Lecture Notes, Springer-Verlag

Lewis, R.W. & B.A.Schrefler 1998. *The Finite Element Method in the Static and Dynamic Deformation and Consolidation of Porous Media*. John Wiley & Sons.

Okulicka, F. 2000, *High-performance computing in geomechanics by parallel finite element approach*, Book of Extended Abstracts o PARA 2000, Bergen, Norway.

Okulicka, F. 2000. *Application of Parallel Matrix Algorithms in Parallel Finite Element Method Computing in Geomechanics*, Book of Abstracts of Workshop PMAA 2000, Neuchatel, Swiss.

Topping, B.H.V. & A.I.Khan 1996, *Parallel Finite Element Computations*, Saxe-Coburg Publications, Edinburg.

Zienkiewicz, O.C. & Taylor, R. 1991, *The Finite Element Method*, vols. 1&2, McGraw-Hill, New York.

Computer Methods and Advances in Geomechanics, Desai et al. (eds) © 2001 Balkema, Rotterdam, ISBN 90 5809 183 X

Analysis on a coupled model of seepage in fractured rock mass

Zheng Shaohe, Zhu Weishen & Ping Yang
Institute of Rock and Soil Mechanics, Chinese Academy of Sciences, Wuhan, People's Republic of China

ABSTRACT: Based on the different size scales and conductivity of fractures, a coupled model of seepage including discrete model (DM) and equivalently continuous model (ECM) is presented. The coupled model can reveal the strong conductivity of large, through-going fractures (dominant fractures) and the capability to store water of network fractures. By integrating the two models (DM and ECM) into a same finite element grid in numerical analysis, some problems exited in iterative procedure can be overcome. According to its application, the results obtained from the project demonstrate the coupled model of seepage to be practical and reasonable.

1 INTRODUCTION

It is well known that seepage has an important effect on practical geotechnical engineering. So as a quantitative approach to understand groundwater flow, seepage models have been studied by many scholars (Yang D 2000, Bai.M & Elsworth.D 1993, Long.J.C.S. et al. 1982, Rouleau.A & Gale.J.E 1987) for many years. Historically, discrete model and equivalently continuous model respectively has been used to study groundwater flow in fractured rock mass. It is clear that each approach represents an extreme condition. Discrete model is assumed to be applicable when the location and orientation of fractures are known. For an equivalently continuous model, it is assumed that by averaging over a suitable representative volume of rock mass, the flow and transport properties may be represented by continuous field variables. But we should notice that when there are a number of large, through-going fractures (dominant fractures) in the area of interest, the equivalently continuous model is not accurate enough to describe groundwater flow. And if too many sets of fractures (network fractures) exist in the area of interest, it is not easy to calculate with discrete model because of heavy work. Furthermore, it is very difficult to accurately define the geological factors of each fracture in practical projects.

2 CONCEPT OF THE COUPLED MODEL OF SEEPAGE

Fractures in rock mass can be classified as dominant fractures and network fractures. The dominant fractures usually extent large and have big size scales, which are the major water transmissible conduits and have strong permeability. The main effect of them is to transmit water. Compared with dominant fractures, the size scales of network fractures are small and their permeability is weak. However, the number of network fractures is so numerous that it can provide much room to store water, which is the main storage room for groundwater. Because of these reasons, the groundwater in network fractures flows slowly and has apparently hysteretic effect. The coupled model of seepage presented in the paper can describe groundwater flow with time in dominant and network fractures respectively. The flow in dominant fractures is simulated by discrete model (DM), which can reveal the strong effect of transmitting water and overcome the problem of inaccurate precision existed in using equivalently continuous model (ECM). Flow in network fracture is simulated by equivalently continuous model, which can avoid heavy work of individually calculating fractures. The two models are coupled with constant hydraulic head and flux at the interface between the two models.

3 COUPLED MODEL AND COUPLED METHOD

The following basic assumptions are introduced to facilitate constructing the model and for completeness of the theory.

(1) The flow in dominant fractures complies with Darcy's law:

$$q = k_f \frac{\partial p}{\partial s} \qquad (1)$$

Where $k_f = -\frac{gb^2}{12\mu}$, g is gravitational acceleration, μ is kinematic viscosity, b is aperture of the fracture, p is hydraulic head, s is local coordination of fracture element.

(2) The flow in network fractures complies with linear Darcy's law on a small stress gradient.

$$q = k\frac{\partial p}{\partial x} \qquad (2)$$

Where k is coefficient of permeability .

(3) The seepage flux keeps consistent and the lost hydraulic head is zero in a circle.

3.1 *Discrete seepage model*

Considering the conservation of water in any controlled element, we can obtain (Zheng 2000):

$$div(\rho q) + \frac{\partial(n\rho)}{\partial t} + W_D + F = 0 \qquad (3)$$

$$\frac{\partial}{\partial s}\left(k_f \frac{\partial p}{\partial s}\right) + \beta\phi\frac{\partial p}{\partial t} + W_D + F = 0 \qquad (4)$$

Where ϕ is porosity rate of fracture, β is coefficient of water compressibility, s is local coordination of fracture element, W_D is the item of source and sink, F is the exchanged flux from network fractures.

3.2 *Equivalently continuous model*

If the principal directions of hydraulic conductivity tensor are consistent with the whole coordinate, the equivalently continuous model of seepage can be written as:

$$\frac{\partial}{\partial x}\left(K_{xx}\frac{\partial p}{\partial x}\right) + \frac{\partial}{\partial y}\left(K_{yy}\frac{\partial p}{\partial y}\right) = S\frac{\partial p}{\partial t} + W_E \qquad (5)$$

Where K_{ij} is the components of hydraulic conductivity tensor, S is specific storage, t is time, p is total hydraulic head, W_E is the item of source and sink of equivalently continuous model.

3.3 *Coupled model of seepage in fractured rock mass*

Considering the specific boundary condition and combining equation (4) and (5), We can obtain:

$$\left.\begin{array}{l}
\frac{\partial}{\partial s}\left(k_f\frac{\partial p}{\partial s}\right) + \beta\phi\frac{\partial p}{\partial t} + W_D + F = 0 \\
\frac{\partial}{\partial x}\left(k_{xx}\frac{\partial p}{\partial x}\right) + \frac{\partial}{\partial y}\left(k_{yy}\frac{\partial p}{\partial y}\right) = S\frac{\partial p}{\partial t} + W_E \\
p(x,y,t_0) = p_0(x,y) \quad t=t_0 \quad (x,y)\in\Omega \\
p(x,y,t) = p_1(x,y,t) \quad t\geq t_0 \quad (x,y,)\in\Gamma_1 \\
k_{xx}\cos(n,x)\frac{\partial p}{\partial x} + k_{yy}\cos(n,y)\frac{\partial p}{\partial y} = q_0 \ (x,y)\in\Gamma_2 \\
p(x,y)|_t = z|_t \quad t\geq t_0 \quad (x,y)\in\Gamma_3
\end{array}\right\} \qquad (6)$$

Where Ω is flow domain, $\Gamma_1,\Gamma_2,\Gamma_3$ is boundary condition of first, second type and free surface, p is hydraulic head, p_0 is initial water head at $t = 0$, p_1 is prescribed head of Γ_1, q_0 is prescribed flux of Γ_2.

3.4 *The numerical simulation of the coupled model of seepage*

Because of such complicated mathematical model, the only effective way is to use numerical method to simulate groundwater flow.

Using Galerkin method, we can obtain the discrete equation including discrete and equivalently continuous models. The detail analysis of such discretation can be referred to much reference (Yang 2000, Zheng 2000).

The discrete equation can be defined as:

$$[T]\{P\} + [S]\left(\frac{\partial P}{\partial t}\right) + \{Q\} = 0 \qquad (7)$$

Where [T] is global permeable matrix, [S] is storage matrix, {P} is global column matrix of hydraulic head, {Q} is column matrix of flux, t is time.

For the discrete model of dominant fracture, using simplified Wittke model, we can obtain the elemental matrix (Zheng 2000):

$$[T]_e = \begin{bmatrix} \frac{bk_f}{L} & -\frac{bk_f}{L} \\ -\frac{bk_f}{L} & \frac{bk_f}{L} \end{bmatrix}$$

$$[S]_e = -A\begin{bmatrix} \frac{bL}{3} & \frac{bL}{6} \\ \frac{bL}{6} & \frac{bL}{3} \end{bmatrix} \quad (8)$$

$$\{Q\}_e = [-\frac{bL}{2}(W_D + F) - \frac{Lq_0}{2}]\begin{Bmatrix} 1 \\ 1 \end{Bmatrix}$$

Where $A = \frac{b}{K_n}[1 + \beta K_n]$, L is the length of fracture element, b is the aperture of fracture element, k_f is the coefficient of conductivity.

For the equivalently continuous model, the above items can be defined as:

$$[T]_e = \int_{-1}^{+1}\int_{-1}^{+1}[B]^T[K][B]|J|d\xi d\eta$$

$$[S]_e = \beta n\int_{-1}^{+1}\int_{-1}^{+1}[N]^T[N]|J|d\xi d\eta \quad (9)$$

$$\{Q\}_e = W_E\int_{-1}^{+1}\int_{-1}^{+1}[N]^T|J|d\xi d\eta + q_0\int_{L_e}[N]^T d\Gamma_2$$

Where $[K]$ is permeable tensor, $|J|$ is Jacobi matrix, ξ, η is local coordinate, $[N]$ is basic function, n is the porous rate of equivalently continuous medium.

Generally, the coupled model of seepage can be solved when used iterative methods. The coupling conditions are as follows: For the equivalently continuous medium, the hydraulic head is equal at the interface between the two models; For the discrete model, the exchanged flux from equivalently continuous model is the coupling item. But it should be pointed out that the item F (exchanged flux) is very difficult to be accurately solved and the procedure of coupling analysis will cost much computation time, especially in coupling with stress.

To solve the problem of inaccurate computation and to save much time in numerical analysis, four steps presented in the paper are as follows:

(1) The fractures which is large and through-going can be considered as linear elements, and the other parts are simulated by rectangle or triangle elements (Figure 1)

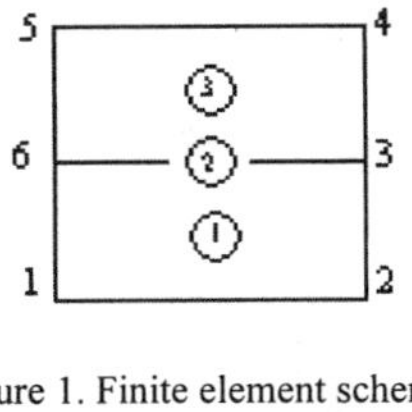

Figure 1. Finite element schematic diagram

There are three elements in Figure 1. ② is fracture element and ① and ③ are equivalently continuous elements. They are defined as:

Element	Node			
①	1	2	3	6
②	3	6	0	0
③	3	4	5	6

(2) The elemental permeable matrix is calculated respectively.

Although the linear element aperture is zero, the aperture should be given a value, which is not zero. In coupling analysis with stress, the aperture can be obtained from stress analysis.

(3) Combining the two kinds of elemental permeable matrix into the global permeable matrix [T], We can obtain the constitutive equation of seepage in every interval time Δt:

$$[T]\{P\} = \{Q\} \quad (10)$$

Where [T] is global permeable matrix, {P} is global column matrix of water head, {Q} is column matrix of flux.

Equation (10) has the following characteristics:

a. [T], {P}, {Q} have the two model's information, which can reveal the different conductivity characteristics of the two models at the same time.

b. The node flux of fracture element in {Q} has already contained the flux, which flows from equivalently continuous model to discrete model. It is F, the exchanged flux.

Take Figure 1 as an example, when calculating the hydraulic head of No.6 node of ② element, we use the flux of No.6, Q6. From Figure 1, No.6 node belongs to three elements ①, ② and ③. Q6 includes not only the flux of ② element (fracture element) but also the flux of ① and ③ elements. The latter is the exchanged flux F flowing from equivalently continuous model to discrete model.

c. Because the nodes of interface belong to the two mediums at the same time, that the hydraulic head of these nodes are equal can be guaranteed.

(4) Iterative calculated with time, the steady flow can be obtained when the precision meets the need.

4 APPLICATION

Take the injection into $3^{\#}$ minimal layer of Wang Zhuang colliery as an example, the coupled model of seepage proposed in the paper is used to simulate groundwater flow. There are 3 dominant fractures F1, F2 and F3 in flow domain. The aperture is 3mm, 3mm and 5mm respectively. Two sets of fractures exist in the equivalently continuous medium, whose average aperture is 0.2mm. A is the injection point, which is at the cross of F1 and F2. The constant hydraulic head of A is P=10.0MPa. The finite element mesh is Figure 2, Figure 3 and Figure 4 are hydraulic head contour at t=22 min and 57min respectively.

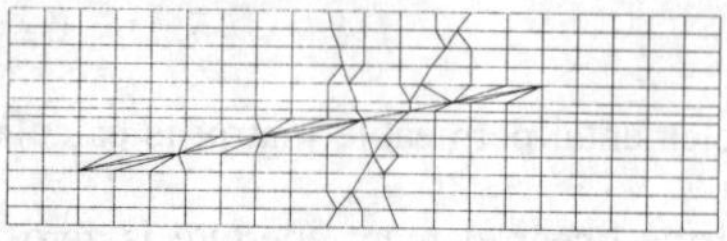

Figure 2. Finite element mesh

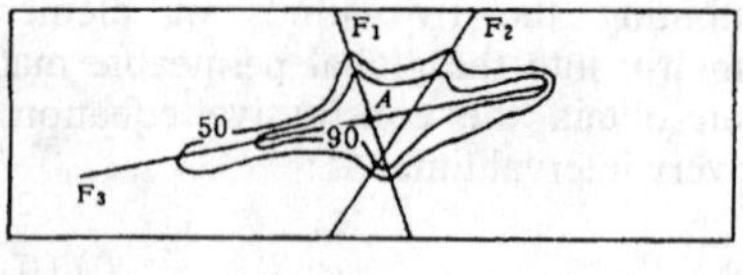

Figure 3. The hydraulic head contour at t=22min

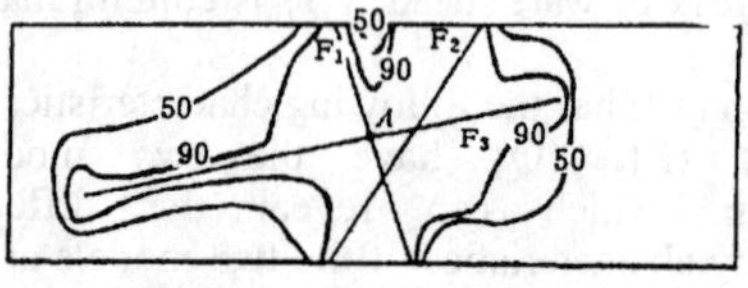

Figure 4. The hydraulic head contour at t=57min

From Figure 3 and Figure 4 , we can see that during the unsteady flow, the hydraulic head rise rapidly along the dominant fracture, which reveal the dominant fracture's strong permeability. Compared with the dominant fractures, the conductivity of equivalently continuous medium is weak and the water head rises slowly, which reveal the latter medium have apparently hyperetic effects. It is consistent with the practical situation.

5 SUMMARU AND CONCLUSION

The coupled model of seepage presented in the paper overcomes some problems exited in a single discrete or equivalently continuous model: the inaccurate precision of equivalently continuous model and the heavy work of discrete model. The model shows the strong conductivity of dominant fracture and the storage and hyperetic effect of network fractures that reveal the seepage mechanism of fractured rock mass. The method to keep the two models into a uniform discrete equation accelerates the calculation and avoids directly calculating the exchanged flux. According to the practical project, the results obtained show that the coupled model and coupled method is effective to simulate the groundwater flow in fractured rock mass.

REFERENCES

Yang Dong, Zhao Yangsheng & Zheng Shaohe 2000. The Hydraulic Model of Rock Mass with Generalized Double Porosity Media and its FEM Simulation. *Chinese Journal of Rock Mechanics and Engineering*.19(2): 182-185.

Bai.M & Elsworth.D 1993. Multiporosity and Multipermeability approach to the Simulation of Naturally Fractured Reservoirs. *Water Resource Res* 29(6): 1621-1639.

Long.J.C.S., Remer.J.S.&Wilson.C.R 1982. Porous media equivalents for networks of discontinuous fractures. *Water Resource Res*.18: 645-658.

Rouleau.A & Gale.J.E 1987. Stochastic discrete fracture simulation of groundwater flow into an underground excavation in granite. Int. J. Rock Mech. Min. Sci. & Geomech. Abstr. 24: 99-112

Zheng Shaohe 2000. Research on coupling theory between seepage and damage of fracture rock mass and its application to engineering.[Ph.D dissertation]. Institute of Rock and Soil Mechanics. Chinese Academy of Sciences. Wuhan, P.R.China.

Computer Methods and Advances in Geomechanics, Desai et al. (eds) © 2001 Balkema, Rotterdam, ISBN 90 5809 183 X

Solvent behaviour in a diffuse-double layer

David W. Smith
The University of Newcastle, N.S.W., Australia

ABSTRACT: Clay soils are known to be capable of sustaining high suctions, while cavitation theory suggests that large suctions should not be sustainable. It is therefore uncertain as to why clay soils behave as they do. By means of cavitation and diffuse-double layer theories, this paper suggests a resolution to the paradox by showing that when a diffuse double layer is present, the water next to the particle surface is actually under high positive pressure. This observation can be employed to explain how the observed behaviour of clay soils is compatible with established cavitation theory.

1. INTRODUCTION

It is well known that both saturated and unsaturated clay soils can sustain large negative pressures indefinitely. Evidence of so-called pore water 'suction' (that is, pore water tension) is provided whenever saturated clay soils are brought to the surface from deep underground and extracted from their sampling tube. Even when the total stress on the soil sample is zero, the effective stress is positive because of pore water suction. This explains why clay soils have an apparent cohesion in an unconfined compression test.

On the other hand, it is well known that 'water' cannot sustain a tension greater than about one atmosphere because of so-called cavitation phenomena. Cavitation is the main limiting factor to the satisfactory performance of many hydraulic components and so is critically important in many engineering applications. To give one practical example, this means that the inlet to a pump cannot be located more than about seven to ten metres above the water source. The question therefore arises, how do we reconcile these conflicting observations, water usually cavititating at low suctions, but sometimes not at high suctions?

To attempt to answer this question, we first need to consider cavitation and diffuse double-layer theories. It will be suggested here that diffuse-double layer theory can be employed to provide a means for reconciling the conflicting observations. To do this, we will need to investigate the behaviour of the solvent in the diffuse-double layer. It is noted here that while diffuse-double theory is well established, almost invariably attention is focussed on the ions and their behaviour in the double layer, not on the solvent and its behaviour.

2. CAVITATION

Vapourisation is the name given to a change of phase that converts a liquid to a gas. Vapourisation in a liquid may be induced by the addition of heat to provide the energy required for a phase change. This phenomena is called boiling. Alternatively, a reduction in pressure under adiabatic conditions will also lead to vapourisation and simultaneously a reduction in temperature of the fluid. Cavitation is the name given to vaporisation of a fluid upon depressurisation.

The formation of a cavity in a liquid obviously involves the rupture of the liquid (Knapp, 1958). It has been established experimentally that *pure* water can withstand tensile stresses of the order of tens or even hundreds of megapascals (Knapp 1958, Winterton 1977). To explain cavitation in water, one must assume the existence of 'weak spots' in the liquid. Because pure water is able to sustain very large tensions, the 'weak spots' are sought at the interface of the water with another material, that is, at a solid, liquid or gas phase interface of any material (other than water), with the water. It turns out there are two main types of heterogeneous nucleation sites (i) free gas bubbles suspended in the liquid, and (ii) cavities in solid surfaces which are not filled by liquid (Lecoffre, 1994). In the following, only the latter is considered.

The tensile strength of the interface between a solid surface and the liquid depends on the 'degree of wetting' of the solid by the liquid. If water molecules are strongly attached to solid surface (as they are for hydrophilic materials), there is little likelihood of cavitation. Even for relatively weak bonds between liquid and solid, the adhesive force is probably higher than the effective tensile strengths observed in normal liquids (Knapp 1958). Therefore provided solid and liquid are in contact, there is little likelihood of cavitation occurring.

This simple model may explain why the pore fluid in clay soils can sustain tensile stresses of the order of hundreds of kilopascals. However, there is no compelling reason why clay particle surfaces should be completely wetted and other surfaces are not (it is known, for example, that a siloxane surface is relatively hydrophobic, suggesting it would not be readily wetted). The foregoing discussion suggests that the reason cavitation occurs is that surfaces are not completely wetted. However, we are no closer to the mechanism explaining why 'water' cavitates at tensions approximately equal to its vapour pressure.

An answer is provided by Harvey's model (Harvey et al 1947). Harvey suggests that there are cavities in solid surfaces in contact with water, and that some of these cavities have not been completely filled by the water, that is, there remains a 'pocket' of gas and vapour in the surface cavity. The incompletely filled cavities are the nucleation sites (the 'weak spots'). Since it was proposed some 50 years ago, alternative models have been suggested, but this model remains the most plausible explanation for the long term stability of cavitation nuclei. Harvey's model is based upon well known physical principles, and has since been supported by the findings of many later researchers (eg. Knapp 1958, Winterton 1977). For these reasons, it is Harvey's model that is now described.

Consider the cavities in a solid surface. We are interested in cavities with geometric shapes that are usually described as cracks or crevices (Figure 1. shows an idealised crack, partially filled with liquid, in the surface of a solid). Harvey's model suggests that it is the cracks partially filled with liquid, that can act as the nucleation sites. Winterton (1977) explains the improbability of this explanation:

> 'With low temperatures and low dissolved gas content in the liquid, the vapour and gas pressures in the cavity can be made negligibly small, and at first sight it appears that the imposition of a small pressure should be sufficient to fill the cavity with liquid and deactivate it....In practice it is not possible to deactivate surface nucleation sites even by imposing large pressures.'

To see why it is not possible to easily deactivate a nucleation site in the surface, consider a crack partially filled with liquid as shown in Figure 1. Provided the crack is sufficiently sharp and the surface of the crack is hydrophobic, then a consideration of vertical equilibrium of forces across the meniscus as the crack leads to the equation,

$$p = p_v + p_g + \frac{2\gamma \cos\theta}{r} \quad (1)$$

where p is the pressure in the fluid, p_v is the vapour pressure in the cavity, p_g is the gas pressure in the cavity, γ is the surface tension of the fluid, r is the radius of meniscus (+ve curvature taken when the meniscus is concave towards the liquid) and θ is the contact angle (as shown in Figure 1.)

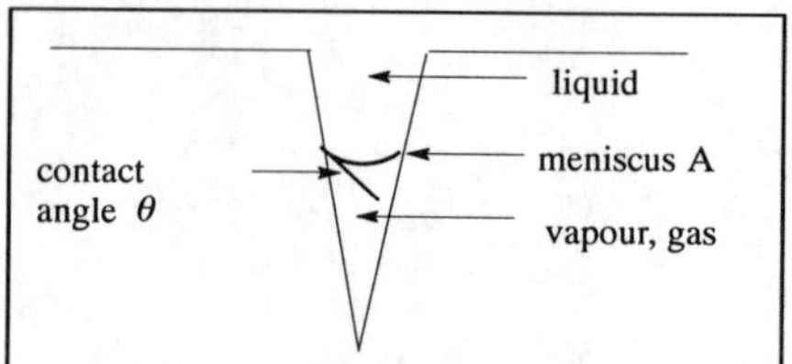

Figure 1. Harvey's model: hydrophobic crack in solid surface.

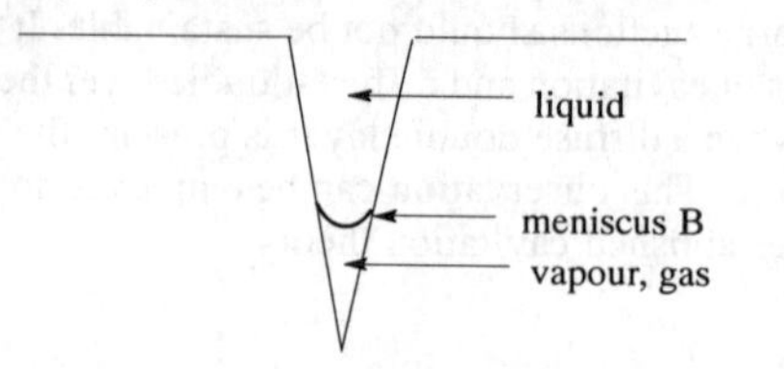

Figure 2. Harvey's model: increased pressure advances the meniscus.

As the pressure increases and the crack is filling, $\theta = \theta_a$, the 'advancing' contact angle. At the end of filling the meniscus is concave towards the liquid. In this case, the meniscus acts to sustain a lower pressure in the gas and vapour phase compared to the pressure in the liquid.

Now if p is reduced (that is, depressurisation take place) or p_v is increased (say due to an increase in temperature from an external heat source), then the meniscus will change shape, becoming first flat (the third term on the RHS of equation (1) will be zero) and then convex on the liquid side, so $\cos\theta$ becomes negative. Now, the surface tension acts to sustain a higher pressure in the gas and vapour phase compared to the pressure in the liquid. When $\theta = \theta_r$, the 'receding' contact angle, the maximum resistance to higher pressure in the gas and vapour phase relative to the liquid is reached, and cavitation is on the verge of occurring.

Equation (1) can explain for example, the sudden appearance of bubbles when a champagne bottle is opened (in this case p suddenly decreases and $p_v + p_g - p > 2\gamma \cos(\theta_r - \pi/2)/r$ leading to bubble formation on the sides of the container. A continuous stream of bubbles is sometimes seen to orginate at a nucleation site because gas dissolved in the champagne continues to diffuse into the crack cavity). Equation (1) can also be employed to explain why water at sea-level boils at 100 °C (note that in this case, as the temperature of the water increases, the water forms convection cells, and the pressure of the gas dissolved in the water decreases rapidly so as $T \rightarrow 100\ ^{\circ}C$, $p_g \rightarrow 0$ while $p_v \rightarrow 1$ atmosphere).

It has been shown by Knapp (1958) and more recently by Ridley and Burland (1999) that prior pressurisation of water, even for short periods (of the order of approximately 10 minutes) increases the tension necessary in the bulk fluid at which cavitation occurs. This observation can also be explained using Harvey's model in the following way.

As the pressure in the bulk fluid increases, the meniscus advances down the crack (eg. the meniscus at A in Figure 1. advances to the position of the meniscus at B as shown in Figure 2.), compressing the gas and vapour ahead of it. However in water, equilibrium of gas and vapour in the crack cavity is quickly re-established with the surrounding fluid. Upon sudden release of pressure, the meniscus will first become flat and change sign (because the pressure in the liquid is then lower than the sum of gas and vapour pressures in the crack cavity). However, once again equilibrium of gas and vapour in the crack cavity is quickly re-established with the surrounding fluid and the meniscus resumes its positive curvature.

Now the meniscus will only recede out of the crack when the maximum resistance to higher pressure in the gas and vapour phase relative to the liquid is reached i.e. when $\theta = \theta_r$ at the smaller radius. Assuming the temperature is constant and the dissolved gas concentration is in equilibrium, this implies an increased suction in the pore fluid is necessary before cavitation occurs. The time duration of the pressurisation is unimportant because the gas and vapour in the cavity quickly reaches equilibrium with the surrounding water. Predictions that may be made based upon this explanation are confirmed experimentally.

3. THE DIFFUSE DOUBLE LAYER

To explain why cavitation is less likely to occur if there is a diffuse-double layer present, we need to analyse the behaviour of the solvent in the double-layer. A traditional continuum model is chosen. Consider a Representative Elementary Volume (REV) in the pore fluid of a saturated porous media. Following Euler, we recognise that the rate of change of momentum equals the net force acting on the mass of fluid within the REV and so,

$$d(mv)/dt = \boldsymbol{F} \qquad (2)$$

where d/dt denotes the material or total derivative, m is the mass of fluid in the REV and $\boldsymbol{F}$ is the net force acting on the mass of fluid.

The net force $\boldsymbol{F}$ is the sum of four forces, namely, gravity ($\boldsymbol{F}_g$), pressure ($\boldsymbol{F}_{press}$), electrical ($\boldsymbol{F}_{elec}$) and viscous ($\boldsymbol{F}_{vis}$) forces. It is assumed that there is no flow in the porous medium and that gravitation forces are negligible compared to the remaining forces. We then find,

$$\boldsymbol{F}_{elec} + \boldsymbol{F}_{press} = 0 \qquad (3)$$

Now focus on a single clay particle and assume that the length and breadth of the particle are large compared to its thickness, so that a one dimensional approximation is reasonable. The x coordinate is taken normal to the particle surface with the origin at the surface. A diffuse-double layer is assumed to be present due to permanent and variable charges on the particle surface (as shown in Figure 3.). Consider an REV located near the particle surface within the diffuse-double layer. As the REV is within the diffuse—double layer, there is a net distributed positive charge (ρ^*) within the REV, attracted to the negatively charged surface. This distributed charge acts as a body force within the pore fluid of the REV. The electrical force per unit volume is given by the product of the charge density (ρ^*) times the electric field strength ($E = d\psi/dx$) and so,

$$\boldsymbol{F}_{elec} = -\rho^*(d\psi/dx) \qquad (4)$$

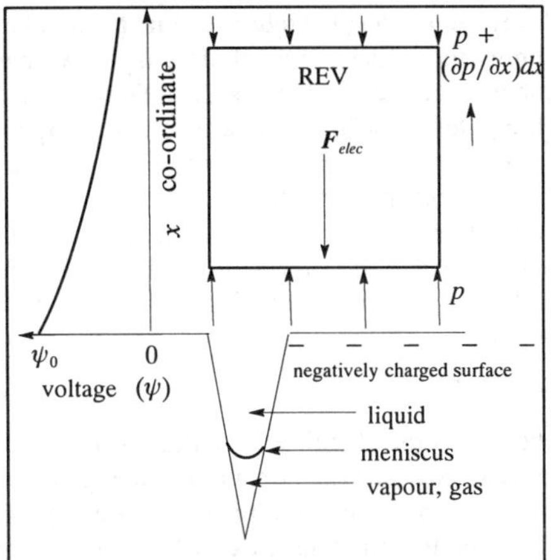

Figure 3. Clay particle showing negatively charged surface, voltage from surface in x direction, REV and crack.

where ψ is the volatge potential. This is equilibrated by the net pressure force acting on the REV,

$$\boldsymbol{F}_{press} = -dp/dx \qquad (5)$$

where p is the pore fluid pressure. Now substituting equations (4) and (5) in equation (3) leads to the equilibrium equation,

$$\rho^*(d\psi/dx) + dp/dx = 0 \qquad (6)$$

We would like to remove ρ^* from equation (6). To do this, we turn to standard electrostatic theory, and find that the variation of charge with distance from the particle surface is given by the Poisson equation and so,

$$\partial^2\psi/\partial^2 x = -\rho^*/\epsilon \quad (7)$$

where ϵ is the permittivity of the pore fluid. Substituting equation (7) in equation (6) then reveals,

$$\epsilon(d^2\psi/dx^2)(d\psi/dx) - dp/dx = 0 \quad (8)$$

Equation (8) is equivalent to,

$$(d/dx)[\epsilon(d\psi/dx)^2 - p] = 0 \quad (9)$$

Integrating this equation reveals that,

$$p = (\epsilon/2)(d\psi/dx)^2 + \text{constant} \quad (10)$$

For the case where there is no overlap with another diffuse-double layer, the constant provides the link with fluid pressure measurements made in the laboratory, as the constant represents the pressure in the bulk solution (p_b). In other words, if the pressure in the bulk solution is zero, then the constant is also zero. Where there is overlap with another diffuse-double layer, the constant represents the pressure transmitted between the particles.

Clearly, equation (10) indicates that to calculate the fluid pressure in the diffuse-double layer, we need to estimate the gradient of the electric field potential. The electric potential with distance from the charged surface is shown in Figure 3. Now we proceed to estimate this electric potential gradient.

At equilibrium the electrochemical potential of the ions must be constant everywhere; or to it put another way, the electrical and diffusional forces on the ions must balance. That is,

$$\nabla\mu_i = -z_i e\nabla\psi \quad (11)$$

where μ_i is the chemical potential of the ith ion, z_i is the valence of the this ion and e is the charge on an electron. For a flat surface, the electrostatic potential and the chemical potential are constant in planes parallel with the surface. Therefore in one spatial dimension we have,

$$\partial\mu_i/\partial x = -z_i e\partial\psi/\partial x \quad (12)$$

By definition of chemical potential (for the ith ion species) we have,

$$\mu_i = \mu_i^0 + kT\ln n_i \quad (13)$$

where μ_i^0 is the chemical potential under standard standard conditions, k is Boltzmann's constant, T is the absolute temperature and n_i is the number of the ith ions per unit volume. Differentiating equation (13) and substituting in (12) shows,

$$\frac{d\ln n_i}{dx} = -\frac{z_i e d\psi}{kTdx} \quad (14)$$

Integrating this equation from a point in the bulk solution where the voltage is zero and the ith ion concentration n_i^0 leads to the Boltzmann equation,

$$n_i = n_i^0\exp(-z_i e\psi/kT) \quad (15)$$

The charge density is related to the ion concentration by

$$\rho^* = \Sigma z_i e n_i = \Sigma z_i e n_i^0\exp(-z_i e\psi/kT) \quad (16)$$

Equations (7) and (16) may now be used to find the Poisson-Boltzmann equation:

$$\partial^2\psi/\partial x^2 = \Sigma z_i e n_i^0\exp(-z_i e\psi/kT) \quad (17)$$

Solution of the equation leads to (Heimenz and Rajagopalan 1997),

$$\Gamma = \Gamma_0\exp(-\kappa x) \quad (18)$$

where κ^{-1} is known as the diffuse double layer thickness (although it sometimes the Debye length). This is-given by,

$$\kappa^2 = (e^2/\epsilon kT)\sum_i z_i^2 n_i^0 \quad (19)$$

and Γ is defined by,

$$\Gamma = [\exp(ze\psi/2kT) - 1]/[\exp(ze\psi/2kT) + 1] \quad (20)$$

Γ_0 is evaluated with $\psi = \psi_0$. The exponentials in equation (20) may be expanded in a series. For relatively large values of x, equation (18) simplifies to,

$$ze\psi/4kT = \Gamma_0\exp(-\kappa x) \quad (21)$$

It is shown by Heimenz and Rajagopalan (1997) that the exact solution (equation (18)) is bracketed by the solution predicted by equation (21) and the Debye-Huckel equation. As the solutions predicted by equation (21) and the Debye-Huckel lie fairly close together, it follows that either equation (21) or the Debye-Huckel equations are a fairly good approximation to the exact solution. For simplicity, we chose to proceed with the Debye-Huckel equation, which is given by,

$$\psi = \psi_0\exp(-\kappa x) \quad (22)$$

It therefore follows that

$$(\partial\psi/\partial x)^2 = \kappa^2\psi_0^2\exp(-2\kappa x) \quad (23)$$

Now using equation (23) in equation (9) shows that the pressure in the pore fluid within the diffuse-double is approximately given by,

$$p = (\epsilon/2)\kappa^2\psi_0^2\exp(-2\kappa x) \quad (24)$$

and so the fluid pressure at the surface of the charged particle (p_0) is given by,

$$p_0 = (\epsilon/2)\kappa^2\psi_0^2 \quad (25)$$

Finally, including the bulk pressure (see equation (10)) leads to the equation for the fluid pressure next to the particle surface, viz,

$$p_0 = (\epsilon/2)\kappa^2\psi_0^2 + p_b \tag{26}$$

4. ESTIMATION OF FLUID PRESSURE

Let us now estimate the pressure in the solvent at the surface of a particle with a diffuse-double layer. It is assumed that the solution is a 1:1 electrolyte at a concentration of 0.01M, and that the fluid is water at 25°C. Then $\epsilon = \epsilon_R\epsilon_0 = 78.5 \times 8.85 \times 10^{-12}\ C^2J^{-1}m^{-1}$ and $\kappa = 3.29 \times 10^8\ m^{-1}$. Assuming the the surface charge is 100 mV (a reasonable value for clay particles) equation (25) indicates,

$$p_0 = (78.5 \times 8.85 \times 10^{-12}) \times (3.29 \times 10^8)^2 \times (0.1)^2$$
$$\approx 750\ kPa \tag{27}$$

while equation (26) shows that if the bulk pressure is included then,

$$p_0 \approx 750\ kPa + p_b \tag{28}$$

5. DISCUSSION

The finding that the fluid in the diffuse-double layer is compressed is reasonable from a molecular dynamics viewpoint. The diffuse-double layer results from the thermal agitation of the ions in solution. When the agitation of the ions and water molecules is of sufficient magnitude, the ions break free of the surface and form a diffuse ion swarm in the solution. The ions are only maintained at some average distance from the surface if the probability of collision between water molecule and hydrated ion is greater on the side of the ion closest to the surface. This greater probability of collision on the side closer to the surface means that the water molecules are more densely packed closer to the surface or in other words, the water molecules are under greater pressure on the side closer to the particle surface.

The continuum analysis of the diffuse double-layer made here may be criticised for a number of omissions including (i) the finite volume of the ions, (ii) the modification of the permittivity of the medium to include the effects of the double layer field and the ions present (iii) the effect of ion polarisation in the double-layer field, and (iv) the self-atmosphere effect. However, Hunter (1981) points out that:

> ...the first two corrections act in the opposite direction to the second two which explains why it is essential to attempt to take into account all the significant effects at the same time. When one single effect is singled out for examination quite erroneous conclusions may be drawn. When all the major effects are taken into consideration the overall correction to the *Poisson-Boltzmann equation* is quite small. For 1:1 electrolytes at modest concentrations (i.e. $\leq 10^{-2}$M) the correction to the simple theory is less than 2% for $\psi_0 \leq 100$ mV.

This gives some reassurance that the estimated surface pressure is at least of the order of several hundreds of kilopascals.

Equation (28) clearly indicates that a substantial negative pressure in the bulk fluid is required to reduce the fluid pressure at the surface to the order of minus 100 kPa and therefore move into the range where cavitation may be expected to occur. It is apparent that if a gas filled crack were present in surface of a particle with a double-layer, then the water in the crack may also be under substantial fluid pressure (see Figure 3. for a schematic representation).

It is therefore suggested here that this substantial positive fluid pressure due to the diffuse double layer may be an important reason why cavitation of the pore fluid is not usually observed in clay soils. If cavitation is prevented from occurring within the soil, soil will only desaturate when a gas enters pore spaces from the soil surface. The gas pressure required for the intrusion of air into the void spaces is usually referred to as the 'air-entry pressure.' In this case, the air-entry pressure will be controlled by the surface tension of the liquid and the radii of the menisci in the pore spaces at the soil surface. Clearly radii of the menisci depend on the pore size and geometry, and may expected to be influenced by the stress history of the soil prior to testing.

Of course it will be immediately obvious that in the presence of a crack, the electric field from the particle surface will require two or three spatial dimensions for its characterisation, thereby invalidating one of the primary assumptions made in the analysis presented here (i.e. that one dimensional conditions apply). However, provided the crack is not too wide, it is reasonable to presume the electric field will be one dimensional to a first approximation, and so in this case the central contention is up-held, namely, the fluid in the crack is under substantial positive pressure.

It seems reasonable to suppose that microscopic cracks exist in the surface of clay particles. These can be seen from scanning electron micrographs of clay particles. However, it is acknowledged that the process of crack formation may be important, and the existence of cracks themselves is not enough to confirm the proposed theory. Harvey's model may rely upon the cracks being present before the solid particles were wetted. Clearly some clay soils are dried and then wetted, thereby providing an obvious process leading to gas filled cracks, and so fulfilling conditions for Harvey's model to apply.

However, cracks may be formed by chemical reactions occurring on the surface of the clay particle. These cracks are presumably liquid filled as the crack forming chemical reactions generally occur in a solution. However, there may be conditions where these cracks would become gas filled. It is noted again that cracks that are completely wetted are unlikely to cavitate (Knapp 1958). It is uncertain whether cracks formed by mechanical loading of the clay particle in the presence of a liquid would be liquid or gas filled, or some combination of both. It seem probable there are conditions where both outcomes are possible.

6. CONCLUSIONS.

Clay soils have a diffuse-double layer. While attention is usually focussed on the ions themselves in the diffuse-double layer, here I have considered the behaviour of the solvent. It has been demonstrated that the double-layer is responsible for the fluid close to the particle surface being under a positive pressure of considerable magnitude, a pressure which grows larger closer to the surface. At the interface of water and many clay particles, it is probable that the fluid pressure is of the order of hundreds of kilopascals. It has been postulated that this fluid pressure may be responsible for suppressing the likelihood of cavitation in clay soils where the mechanism of cavitation is the Harvey model. As both the Harvey model of cavitation and diffuse-double layer theory are well established and can explain many experimental findings individually, it seems reasonable to combine the two theories to provide a plausible explanation for the observed behaviour of clay soils.

7. ACKNOWLEDGMENTS

This research has been carried out with financial support from the Mine Subsidence Board of New South Wales, Robert Carr and Associates and the Australian Research Council.

8. REFERENCES

Guan, Y., Fredlund, D.G. and Gan, K-M. (199), 'Behaviour of water subject to high tensile stresses',

Harvey, E.N., Elroy, M.D. and Whitely, A.H. (1947), 'On cavity formation in water', *Journal of Applied Physics*, Vol. 18. p.162-172.

Hiemenz, P.C. and Rajagopalan R. (1997). *Principles of Colloid and Surface Chemistry.* 3rd Edition. Pub. Marcel Dekker Inc.

Hunter, R.J. (1981) *Zeta Potential in Colloid Science.* Pub. Academic Press.

Knapp, R.T. (1958), 'Cavitation and nuclei', *Transactions ASME*, Vol. 80. p. 1315-1324.

Lecoffre, Y. (1994) *Cavitation: bubble trackers.* English translation by M.M. Oberai, 1997. Pub. Balkema 1999.

Ridley, A.M. and Burland, J.B. (1999), 'use of tensile strength of water for the direct measurement of high soil suction', *Canadian Geotechnical Journal*, Vol. 36, p. 178-180.

Winterton, R.H.S. (1977), 'Nucleation of boiling and cavitation', *Journal of Physics D: Applied Physics* Vol. 10 p. 2041-2056.

Computer Methods and Advances in Geomechanics, Desai et al. (eds) © 2001 Balkema, Rotterdam, ISBN 90 5809 183 X

Numerical simulation for the coupled problem of temperature and seepage fields in cold region dams

Lai Yuanming, Wu Ziwang & Zhu Yuanlin
State Key Laboratory of Frozen Soil Engineering, Cold and Arid Regions, Environmental and Engineering Research Institute, Chinese Academy of Sciences, Lanzhou, People's Republic of China

Liu Songyu
Institute of Geotechnical Engineering, Southeast University, Nanjing, People's Republic of China

ABSTRACT: In this paper, a mathematical mechanical model and the governing differential equations of the coupled problem of temperature and seepage fields, with phase change, are first derived from the theory of heat transfer and the theory of seepage. Then the finite element formulae of this problem are obtained from Galerkin's method. Finally, considering the seepage influence, an illustrative example of the temperature field in cold regions dam is provided. The example shows that the effect of seepage field on the temperature field of cold region dams is large. So the effect of this factor on the dam temperature field should be taken into account in cold regions dam engineering design.

1 INTRODUCTION

There are lots of dams in northeastern and northwestern China, Canada and Russia. These dams are located in cold regions, and they all have the different degree of damage caused by frostaction. Firstly, the shelters of dams are freezing-thawing, and get deformation because of frostaction. Then, the cores of the dams are frozen and thawn with time past. Because of these effects, some of the dams can be destroyed. Bonacina et al (1973) presented a numerical solution in order to solve heat conduction problems with phase change. Comini et al (1974) investigated nonlinear heat conduction problems with special reference to phase change by finite element methods. Li (1998) made heat stability analysis of embankment (roadbed) on the degrading permafrost district in the East of the Tibetan Plateau in China. Bathe and Khoshgoftaar (1979) proposed a finite element procedure for free surface seepage analysis without mesh iteration. Zhao and Valliappan (1993) worked out seepage problems in infinite media by transient infinite element. Zhang et al (1988) presented initial flow methods for seepage analysis with free surface. Until now, the temperature fields of dams in cold regions have neither been measured methodically, nor has the research of effect of seepage on the temperature field of dam been made. In order to solve this problem in a practical engineering manner and to provide a computational method for dam designs in cold regions, the mathematical mechanical model and the governing differential equations of the coupled problems of temperature and seepage, with phase change, are presented from the theory of heat transfer and the theory of seepage. In this paper, the finite element formulae for this problem are obtained using Galerkin's method, and the temperature field of a dam in northern Quebec has been analyzed.

2 GOVERNING DIFFERENTIAL EQUATIONS AND THEIR FINITE ELEMENT FORMULEA

This study concerns cold regions dams located at sites where the yearly average temperatures of the dams are below 0 ℃. The freezing-thawing of dam shelter occurs annually, thus, the problem discussed in this paper becomes a problem of heat transfer with phase change. Taking into account the effect of seepage on the temperature field, from the theory of heat transfer and the theory of seepage, the differential equations of the coupled problem are given by

In Ω_f,

$$C_f \frac{\partial T_f}{\partial t} = \frac{\partial}{\partial x}\left(\lambda_f \frac{\partial T_f}{\partial x}\right) + \frac{\partial}{\partial y}\left(\lambda_f \frac{\partial T_f}{\partial y}\right) - C_w \rho_w \left(V_x^f \frac{\partial T_f}{\partial x} + V_y^f \frac{\partial T_f}{\partial y}\right) \quad (1a)$$

$$S_f \frac{\partial H_f}{\partial t} = \frac{\partial}{\partial x}\left(K_x^f \frac{\partial H_f}{\partial x}\right) + \frac{\partial}{\partial y}\left(K_y^f \frac{\partial H_f}{\partial y}\right) + Q \quad (1b)$$

$$V_x^f = -K_x^f \frac{\partial H_f}{\partial x}, V_y^f = -K_y^f \frac{\partial H_f}{\partial y} \tag{1c}$$

and in Ω_u,

$$C_u \frac{\partial T_u}{\partial t} = \frac{\partial}{\partial x}\left(\lambda_u \frac{\partial T_u}{\partial x}\right) + \frac{\partial}{\partial y}\left(\lambda_u \frac{\partial T_u}{\partial y}\right) - C_w \rho_w \left(V_x^u \frac{\partial T_u}{\partial x} + V_y^u \frac{\partial T_u}{\partial y}\right) \tag{2a}$$

$$S_u \frac{\partial H_u}{\partial t} = \frac{\partial}{\partial x}\left(K_x^u \frac{\partial H_u}{\partial x}\right) + \frac{\partial}{\partial y}\left(K_y^u \frac{\partial H_u}{\partial y}\right) + Q \tag{2b}$$

$$V_x^u = -K_x^u \frac{\partial H_u}{\partial x}, V_y^u = -K_y^u \frac{\partial H_u}{\partial y} \tag{2c}$$

where superscripts f and u represent the frozen and the unfrozen states, respectively. T_f, C_f, λ_f are the temperature, volumetric heat capacity and thermal conductivity of surrounding rock in the frozen area, respectively. H_f, S_f, K_x^f and K_y^f are water head, water-supply degree, hydraulic conductivities in the x and y directions, respectively. C_w and ρ_w are the specific heat and density of water. Parameters with superscript u are the corresponding physical components in the unfrozen area.

At phase front position s(t), the continuous condition and the conservation of energy should be met, i.e.,

$$T_f(s(t),t) = T_u(s(t),t) = T_m \tag{3}$$

$$\lambda_f \frac{\partial T_f}{\partial n} - \lambda_u \frac{\partial T_u}{\partial n} = L \frac{ds(t)}{dt} \tag{4}$$

where L is the latent heat per unit volume.

The conditions at the fixed boundaries are

$$T = T_w, \ -\lambda \frac{\partial T}{\partial n} = \alpha(T - T_a) \tag{5}$$

where α is a coefficient of convective heat transfer and T_a is the equilibrium temperature for which no convection occurs.

Initial conditions are

$$T_u|_{t=0} = T_0, \ \Omega_u|_{t=0} = 0 \tag{6}$$

Water head H in seepage field should meet the following two kinds of boundary conditions:

(1) The first kind of boundary condition, i.e., the boundary of water head:

$$H = H_0 \tag{7}$$

(2) The second kind of boundary condition, i.e., the boundary of drainage quantity:

$$K_x \frac{\partial H}{\partial x}\cos(n,x) + K_y \frac{\partial H}{\partial y}\cos(n,y) = -q_0 \tag{8}$$

Eqs.(1) — (8) are the governing differential equations, and initial and boundary conditions of the coupled problem of the temperature field and seepage field. Because the specific heat and thermal conductivity of the dam versus temperature and the phase front position is to be determined, and hence, the condition of energy conversation at the phase front position is nonlinear. Moreover, the free surface of seepage field is also to be determined. Therefore, this problem is a heavily nonlinear one.

Using the method of sensible heat capacity (Bonacina, 1973), it is assumed that the phase change occurs in a range of temperatures $T_m \pm \Delta T$. When the equivalent heat capacity is constructed, the effect of the interval temperature ΔT should be included. Suppose that C_f, C_u, λ_f and λ_u do not depend on temperature T. Then in the interval: $T_m - \Delta T \le T \le T_m + \Delta T$ the following definitions may be assumed.

$$C^* = \begin{cases} C_f & T < (T_m - \Delta T) \\ \frac{L}{2\Delta T} + \frac{C_f + C_u}{2} & (T_m - \Delta T) \le T \le (T_m + \Delta T) \\ C_u & T > (T_m + \Delta T) \end{cases} \tag{9}$$

and

$$\lambda^* = \begin{cases} \lambda_f & T < (T_m - \Delta T) \\ \lambda_f + \frac{\lambda_u - \lambda_f}{2\Delta T}[T - (T_m - \Delta T)] & (T_m - \Delta T) \le T \le (T_m + \Delta T) \\ \lambda_u & T > (T_m + \Delta T) \end{cases} \tag{10}$$

For the hydraulic conductivities, the following definitions may be assumed

$$K_x^* = \begin{cases} \left(\frac{W_u}{W}\right)^\beta \cdot K_x^u & T \le T_m \\ K_x^u & T > T_m \end{cases} \tag{11}$$

and

$$K_y^* = \begin{cases} \left(\frac{W_u}{W}\right)^\beta \cdot K_y^u & T \le T_m \\ K_y^u & T > T_m \end{cases} \tag{12}$$

where W and W_u are volumetric fractions of the moisture content and the unfrozen water content. β is a constant, determined by experiments.

Using Eqs.(9) — (12), Eqs.(1) — (4) can be written as

$$C^* \frac{\partial T}{\partial t} = \frac{\partial}{\partial x}\left(\lambda^* \frac{\partial T}{\partial x}\right) + \frac{\partial}{\partial y}\left(\lambda^* \frac{\partial T}{\partial y}\right) - C_w \rho_w \left(V_x^* \frac{\partial T}{\partial x} + V_y^* \frac{\partial T}{\partial y}\right) \tag{13}$$

$$S^{*}\frac{\partial H}{\partial t}=\frac{\partial}{\partial x}\left(K_{x}^{*}\frac{\partial H}{\partial x}\right)+\frac{\partial}{\partial y}\left(K_{y}^{*}\frac{\partial H}{\partial y}\right)+Q \qquad (14)$$

where V_x^* and V_y^* are water filtration velocities. Their formulae are

$$V_x^* = -K_x^*\frac{\partial H}{\partial x}, \quad V_y^* = -K_y^*\frac{\partial H}{\partial y} \qquad (15)$$

Since this kind of problem is a heavily nonlinear one, its analytical solution can not be obtained. We obtained its numerical solution by using a numerical analytical method. Using Galerkin's method, the following finite element formulae can be obtained.

$$[M]\left\{\frac{\partial T}{\partial t}\right\}+[K]\{T\}=\{F\} \qquad (16)$$

$$[N]\left\{\frac{\partial H}{\partial t}\right\}+[G]\{H\}=\{P\} \qquad (17)$$

where

$$K_{ij}=\sum\int_{\Omega^e}\lambda^*\left(\frac{\partial N_i}{\partial x}\cdot\frac{\partial N_j}{\partial x}+\frac{\partial N_i}{\partial y}\cdot\frac{\partial N_j}{\partial y}\right)d\Omega+\sum\int_{\Gamma_2^e}\alpha N_i N_j d\Gamma$$

$$+\sum\int_{\Omega^e}C_w\rho_w\left(V_x^*\frac{\partial N_j}{\partial x}+V_y^*\frac{\partial N_j}{\partial y}\right)N_i d\Omega \qquad (18)$$

$$M_{ij}=\sum\int_{\Omega^e}C^*N_iN_jd\Omega \qquad (19)$$

$$F_i=\sum\int_{\Gamma_2^e}\alpha T_a N_i d\Gamma \qquad (20)$$

$$G_{ij}=\sum\int_{\Omega^e}\left(K_x^*\frac{\partial N_i}{\partial x}\cdot\frac{\partial N_j}{\partial x}+K_y^*\frac{\partial N_i}{\partial y}\cdot\frac{\partial N_j}{\partial y}\right)d\Omega \qquad (21)$$

$$N_{ij}=\sum\int_{\Omega^e}S^*N_iN_jd\Omega \qquad (22)$$

$$P_i=\sum\int_{\Omega^e}QN_id\Omega \qquad (23)$$

Using the method presented by Bathe (1979), the free surface of the seepage problem is determined. Then, using the Crank-Nicolson method, Eqs. (16) and (17) are solved step by step for every time interval Δt. The numerical results for the problem then can be obtained.

3 NUMERICAL SIMULATION

In order to study effect of seepage on dam temperature field, we should construct an appropriate model and need the in situ fundamental conditions as a foundation. Here we used the conditions at the scene of the dam in northern Quebec, Canada, which was constructed in September, 1994. The cross section of this dam is shown in Figure 1. AB=23.5m, AF=334m, DE=9m. The vertical distance from B to D is 76.5m. This dam is composed of seven parts. Part 1 and part 7 are shells. Parts 2 and 6 are filters. Parts 3 and 4 are cores. Part 5 is foundation rock. Their thermal physical parameters and hydraulic conductivities are given in Table 1.

Point D is determined by the water level in reservoir (shown in Figure 2). Boundary BC contacts water. The measured water temperatures vs time is given in Table 2.

Boundary CDEF is in contacts with air. The yearly-average air temperature is -3.973 ℃. The following air temperature curve can be obtained from the meteorological data around the dam site

$$T_a=-3.973+17\sin\left(\frac{2\pi}{8760}t_h+\frac{3\pi}{4}\right) \qquad (24)$$

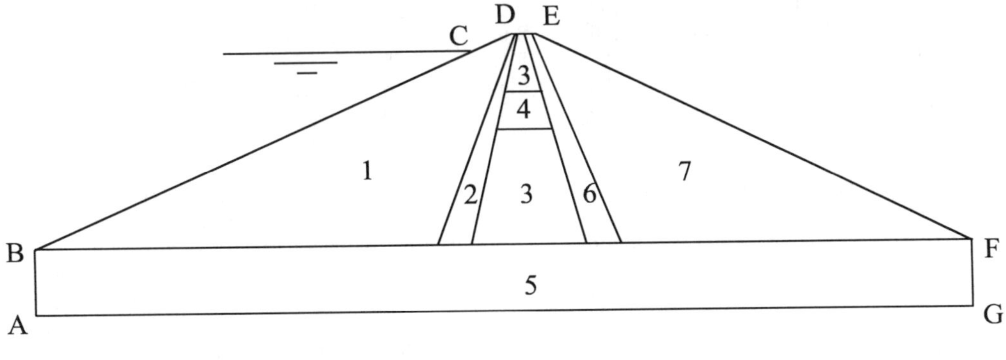

Figure 1 The cross section of dam and the computational domain.

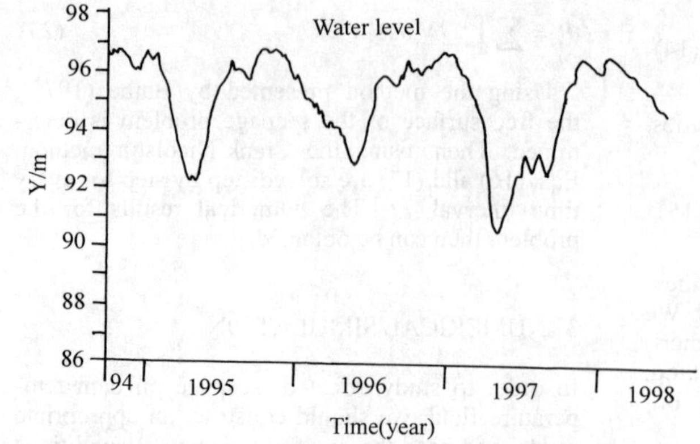

Figure 2 The water level variation in reservoir vs time.

Table 1 The thermal physical parameters and hydraulic conductivities

No. of Parts	W (%)	K_x^u and K_x^y (m/s)	λ_u (W/m℃)	C_u (10^6 J/m^3℃)	λ_f (W/m℃)	C_f (10^6 J/m^3℃)
1 and 7	13.5	5×10^{-5}	3.00	2.52	3.50	1.96
2 and 6	1.0	2×10^{-5}	1.00	1.49	1.00	1.49
3	11.1	1×10^{-7}	2.70	2.44	2.20	1.96
4	11.1	2×10^{-6}	2.70	2.44	2.20	1.96
5	1.0	1×10^{-6}	1.27	1.71	1.27	1.71

Table 2 The measured water temperatures vs time (℃)

Y(m) / Time	0-10	11-20	21-30	31-40	41-50	51-60	61-70	70-80
Jan	3.20	3.00	2.50	2.15	1.50	0.50	0.30	0.15
Feb	3.06	2.85	2.50	2.17	0.50	0.17	0.10	0.10
Mar	3.06	2.85	2.50	1.50	0.30	0.17	0.10	0.10
Apr	3.06	2.85	2.50	2.17	0.50	0.17	0.10	0.10
May	3.20	3.00	3.50	2.15	1.50	0.50	0.30	0.15
June	4.00	4.20	4.34	4.67	5.20	7.17	12.00	13.00
July	4.00	4.20	4.34	4.67	5.20	7.17	12.00	13.00
Aug	4.34	4.50	4.67	5.00	5.67	6.50	10.30	11.00
Sept	6.83	7.30	7.80	8.30	8.60	8.90	9.00	9.50
Oct	9.00	8.92	8.67	8.40	8.12	7.65	7.50	7.50
Nov	5.70	5.70	5.64	5.62	5.59	5.46	5.14	4.90
Dec	3.70	3.50	3.50	3.30	2.30	1.00	0.70	0.70

The convective heat transfer coefficient between the air and the shell of the dam is $\alpha = 15.0\,(W/m^2 \cdot ℃)$ (Zhang, 1989). This dam construction was completed on Sept.30, 1994. The initial temperature in computational domain is 3.5℃. The boundary conditions of seepage are that boundary AG is impermeable boundary, the water head at boundary ABC is equal to water level in reservoir shown in Figure 2 and the water head at boundary GF is equal to 50 m.

The computational domain Shown in Figure 1 is divided by 2684 nodes and 5160 tri-lateral elements. Substituting the above-mentioned parameters into the finite element procedure of this paper, we can obtain the calculated results by computers. When the seepage is not taken into account, the temperature fields of the dam are described in Figure 3a and b, when this dam has been used for three years and three months, three years and eight months, respec-

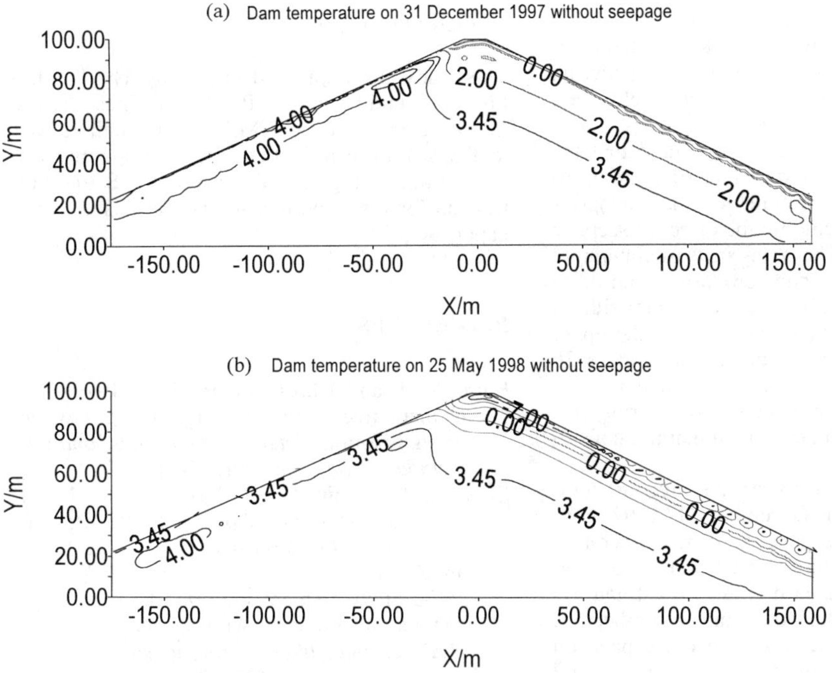

Figure 3 The temperature distribution of dam without seepage

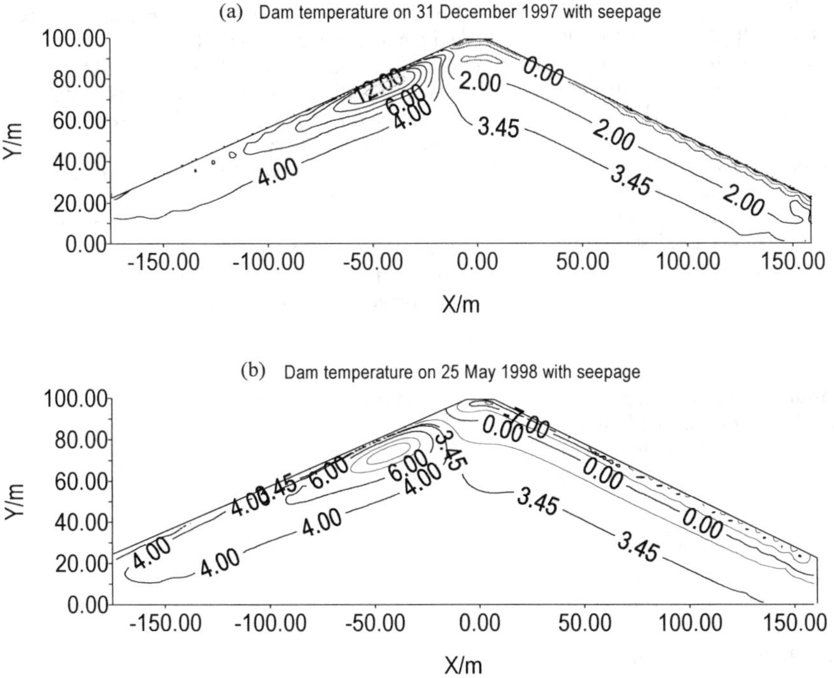

Figure 4 The temperature distribution of dam with seepage

tively, i.e., the dam temperature distributions on December 31, 1997 and May 25, 1998. The temperature fields of the dam on the same times, considering the effect of seepage on temperature fields, are shown in Figure 4a and b, respectively.

From Figures 3 and 4, it can be found that the temperature distributions of dam including the influence of seepage on temperature fields are very different from those not including the effect of seepage on temperature fields. The areas in iso-temperature lines of this dam with seepage are larger than those without seepage, especially on upper stream side of the dam. The reason is that water warms the upper stream side of dam. For example, from Figure 3b and 4b, it can be seen that area in the calculated iso-temperature line of 4℃ in this dam considering seepage factor is very much larger than that not considering seepage factor.

The calculated highest, lowest and average temperatures of this dam on December 31, 1997, considering the effect of seepage on dam temperature field, are 14.82℃, -15.62℃ and 0.751℃, respectively. While the calculated highest, lowest and average temperatures of this dam on December 31, 1997, not considering the influence of seepage on dam temperature field, are 5.38℃, -15.63℃ and 0.526℃, respectively. The calculated highest, lowest and average temperatures of this dam on May 25, 1998, including the effect of seepage on dam temperature field, are 9.58℃,-8.33℃ and 0.194℃, respectively, while the calculated highest, lowest and average temperatures of this dam on May 25, 1998, not including the influence of seepage on dam temperature field, are 4.18℃, -8.30℃ and 0.045℃, respectively.

From the mention above, it is easily concluded that the effect of seepage on temperature field in cold regions engineering is very large and the influence of this factor on the temperature field should be taken into account in the engineering design.

4 CONCLUSIONS

1. The finite element formulae, including the effect of seepage field on the temperature field in the cold regions engineering, are presented. The computational theory also is provided for the engineering design.

2. The calculated results show that the effect of seepage field on the temperature field in cold region engineering is very large. Thus, the influence of this factor on the temperature field should be taken into account in the engineering design.

5 ACKNOWLEDGEMENTS

This study was supported in part by The Foundation of "Hundred People Plan" of Chinese Academy of Sciences (to Dr.Y.M.Lai). The authors wish to thank Professor J.M.Konrad of Department of Civil Engineering, Laval University, Sainte-Foy, Canada for his providing us the data used in this paper.

REFERENCES

Bathe K. J, and Khoshgoftaar M. R. 1979. Finite element free surface seepage analysis without mesh iteration. *Inter. J. Numerical analytical methods in Geomechanics.* 3(1): 13-22.

Bonacina C.,Comini G. and Fasano A, et al. 1973. Numerical solution of phase-change problems. *Int. J. Heat Mass Transfer*. 16(6): 1852-1832.

Comini G., Guidice S.del and Lewis R.W. et al. 1974. Finite element solution of nonlinear heat conduction problems with special reference to phase change. *Inter. J. for numerical methods in engineering.* 8(6):613-624.

Li D. Q.,Wu Z.W. and Fang J.H. et al. 1993. Heat stability analysis of embankment on the degrading permafrost district in the East of the Tibetan Plateau, China. *Cold regions science and technology.* 28(3):183-188.

Zhang Y. T., chen P. And Wang L. 1988. Initial flow methods for seepage analysis with free surface. *Chinese Journal of Hydraulic Engineering.* 12(1):18-26.

Zhang Z. R. 1989. *Heat Transfer*. Beijing: High Educational Press.

Zhao C.B. and Valliappan S. 1993. Transient infinite elements for seepage problems in infinite media. *Inter. J. Numerical analytical methods in Geomechanics.* 17:323-341.

Computer Methods and Advances in Geomechanics, Desai et al. (eds) © 2001 Balkema, Rotterdam, ISBN 90 5809 183 X

Author index

COMPUTER METHODS AND ADVANCES IN GEOMECHANICS
VOLUME 2

PROCEEDINGS OF THE TENTH INTERNATIONAL CONFERENCE
ON COMPUTER METHODS AND ADVANCES IN GEOMECHANICS
TUCSON/ARIZONA/USA/7 – 12 JANUARY 2001

Computer Methods and Advances in Geomechanics

Edited by

Chandra S. Desai, Tribikram Kundu, Satya Harpalani, Dinshaw Contractor & John Kemeny

The University of Arizona, Tucson, Arizona, USA

VOLUME 2

8 *Laboratory and field testing, parameter determination;* 9 *Dynamics, earthquake analysis, liquefaction;* 10 *Soil/Rock-structure interaction;* 11 *Ground improvement and geosynthetics;* 12 *Petroleum geomechanics, hydraulic fracturing, geological simulation;* 13 *Pavement geomechanics;* 14 *Foundations, footings, piles;* 15 *Slopes, dams, walls, landslides* 16 *Tunnels, underground works, mining;* 17 *Flow, seepage, consolidation;* 18 *Education.*

A.A. BALKEMA/ROTTERDAM/BROOKFIELD/2001

The texts of the various papers in this volume were set individually by typists under the supervision of each of the authors concerned.

Published by
A.A. Balkema, P.O. Box 1675, 3000 BR Rotterdam, Netherlands
Fax: +31.10.413.5947; E-mail balkema@balkema.nl; Internet www.balkema.nl
A.A. Balkema Publishers, 2252 Ridge Road, Brookfield, VT 05036-9704, USA
Fax: 802.276.3837; E-mail: info@ashgate.com

For the complete set of two volumes, ISBN 90 5809 183 X
For Volume 1, ISBN 90 5809 184 8
For Volume 2, ISBN 90 5809 185 6

Printed in the Netherlands

Computer Methods and Advances in Geomechanics, Desai et al. (eds) © 2001 Balkema, Rotterdam, ISBN 90 5809 183 X

Table of contents

8 *Laboratory and field testing, parameter determination*

* Theme papers

12 *Petroleum geomechanics, hydraulic fracturing, geological simulation*

15 *Slopes, dams, walls, landslides*

16 *Tunnels, underground works, mining*

17 *Flow, seepage, consolidation*

8 Laboratory and field testing, parameter determination

Computer Methods and Advances in Geomechanics, Desai et al. (eds) © 2001 Balkema, Rotterdam, ISBN 90 5809 183 X

On uniqueness of identification of rockmass mechanical parameters using inverse displacement back analysis

Xia-Ting Feng & Zhiqiang Zhang
College of Resources and Civil Engineering, Northeastern University, Shenyang, People's Republic of China
Tingjie Li
Institute of Rock and Soil Mechanics, Chinese Academy of Sciences, Wuhan, People's Republic of China

ABSTRACT: The paper suggests several kinds of methods to solve uniqueness of solutions of inverse back analysis. One is to perform sensitivity analysis to determine whether a parameter can be determined using an inverse back analysis or not . Another is to use neural network to represent nonlinear relationship between displacement and parameters to be recognized. The third is to perform optimal search of parameter values at a global space using a genetic algorithm. Application to recognition of deformation modulus of various zones and coefficient of geo-stress components indicates that the methodology is effective.

1 INTRODUCTION

Inverse displacement back analysis is a popular method to determine rock or rock mass mechanical parameters from the field monitored displacement data (Kirsten 1976, Gioda 1981, Sakurai 1983, 1986, 1987). However, in order to obtain satisfying results, there are three points at least shall be addressed. One is that problem of parameter sensitivity to deformation. The parameters, which are not sensitive to deformation, can be determined through inverse back analysis of displacement. Another is that the relationship between displacement and mechanical parameters is highly nonlinear. How to obtain reasonably the nonlinear relationship? The third is that for the given relationship between parameters and displacement, search of parameters is in large space and highly multi-modal (Table 1). Different value of parameters results in the closed fitness. How to find a global optimal combination of mechanical parameters to have minimum fitness at the given geological and engineering conditions?

Table 1. Multi-modal property of parameters

Deformation modulus (GPa)				Coefficients in geo-stress field		Fitness $F(P)$
A	B	C	D	a_x	a_y	
7.52	9.68	18.95	32.10	3.96	1.59	0.630
7.52	9.84	18.95	36.30	4.80	1.59	0.431
8.45	10.72	18.95	33.13	4.80	1.44	0.313
11.29	9.68	18.95	32.12	4.79	1.58	0.312
7.52	13.85	18.95	32.10	4.79	1.46	0.295
7.52	9.68	16.56	32.08	4.79	1.59	0.294

A: Moderately weathered zone, B: Damaged zone, C: Unloading deformation zone, D: Slightly weathered or fresh zone.

In this paper, recognition of rockmass mechanical parameters of permanent shiplock at Three Gorges Project in China is used as an example to show how to solve uniqueness of parameter recognition in inverse back analysis using measured displacement data. The main strategies are applications of intelligent technology.

2 SOME CONSIDERATIONS OF PARAMETER RECOGNITION FOR PERMANENT SHILOCK ROCK MASS AT THREE GORGES PROJECT

The permanent shiplock at Three Gorges Project is a double five stages shiplock. The slope at head section of the third shiplock room is the highest at permanent shiplock area. Geological structures are the most complex and there is a fault F_{215} in this area (Figure 1). Therefore, it forms a key region to be taken an attention in permanent shiplock. Any excess deformation would result in invalidation of the navigation building. In order to estimate deformation of the region during the excavation or at the working period, mechanical parameters in various zones shall be reasonably recognized in advance.

According to weathered degree, rock masses are divided into four zones such as heavy weathered zone, moderately weathered zone, slightly weathered zone, and fresh zone. Considering effect of damage and unloading due to excavation, rock masses near surface of the slope and wall of shiplock rock are damaged. And it forms an unloading deformation zone between the damaged zone and undistributed zone. Mechanical behaviors and parameters of zones, such as deformation modulus, cohesion, friction angle, etc., shall be determined before reasonable estimation of deformation behaviors due to excavation or creep at long term.

For damaged, unloading deformation and fresh rock masses, the geo-stress equations can be written by:

$$\sigma_x = a_x + 0.01168 \times H$$

$$\sigma_y = a_y + 0.03039 \times H$$
$$\sigma_z = 4.7152 + 0.01027 \times H \quad (1)$$
$$\tau_{xy} = 0 \quad \tau_{yz} = 0 \quad \tau_{zx} = 0.0$$

The geo-stress equations for the heavy weathered rock mass can be expressed as:

$$\sigma_x = \mu\gamma H, \sigma_y = \gamma H, \sigma_z = \mu H/(1\text{-}\mu),$$
$$\tau_{xy} = 0, \tau_{yz} = 0, \text{ and } \tau_{zx} = 0.0.$$

where γ is unit weight of rock mass ($\times 10^6 \text{N/m}^3$), H is depth below surface, a_x, a_y are coefficients to be determined.

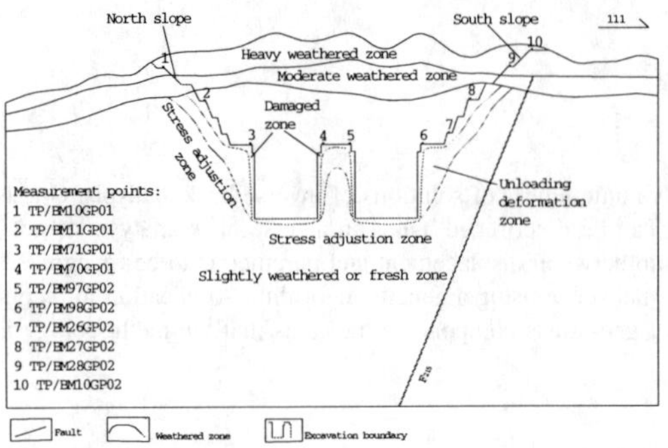

Figure 1. Engineering geology zones and arrangement of displacement monitoring points for the head section of the third shiplock room of permanent shiplock of Three Gorges Project.

3 INTELLIGENT METHOGOLOGY

3.1 *Sensitivity analysis of parameter values*

Not all parameter values can be determined by an inverse displacement back analysis. Whether a parameter is carried out a back analysis can be determined through sensitivity analysis. The sensitivity analysis is conducted by increasing 10% of value to watch its effect to deformation behaviors of rock masses. Table 2 shows modes for the parameter sensitivity analysis. The produced deformation fluctuation is shown in Table 3. The following points can be drawn from Table 3:

(1) Deformation modulus of slightly weathered zone, fresh zone, damaged zone, and unloading deformation zone, and coefficients in horizontal geo-stress components are the most sensitive to horizontal deformation of rockmass, particularly at top of the straight wall. Therefore, values of these parameters can be determined through back analysis of horizontal deformation.

(2) Friction and cohesion in various zones are not sensitive to horizontal deformation of rock mass. Therefore, values of these parameters can't be directly determined through back analysis of horizontal deformation.

(3) Deformation modulus of the faults F_{215} is sensitive to horizontal displacement of its near monitoring points such as TP/BM98GP02, TP/BM26GP02 and TP/BM27GP02. Considering complexity of fault effectiveness, deformation modulus of the fault F_{215} is determined by experimental data and engineering experience.

Table 2. Change schemes of parameter values

Parameters	Basic value	Increase 10% value of parameter with indicator *
Deformation modulus of slightly weathered zone (GPa)	10.0	*..................
Deformation modulus of damaged zone (GPa)	12.0	.*.................
Deformation modulus of unloading deformation zone (GPa)	20.0	..*................
Deformation modulus of fresh zone (GPa)	35.0	...*...............
a_x (MPa)	4.3982	....*..............
a_y (MPa)	1.6628	.....*.............
Friction angle of slightly weathered zone	1.30	......*............
Friction angle of damaged zone	1.40	.......*...........
Friction angle of unloaded zone	1.50	........*..........
Friction angle of fresh zone	1.70	.........*.........
Cohesion of slightly weathered zone (MPa)	1.0	..........*........
Cohesion of damaged zone (MPa)	1.0	...........*.......
Cohesion of unloaded zone (MPa)	1.6	............*......
Cohesion of fresh zone (MPa)	2.0	.............*.....
Deformation modulus of heavy weathered zone (GPa)	0.30	..............*....
Deformation modulus of fault F_{215} (GPa)	3.0	...............*...
Friction angle of heavy weathered zone	0.70	................*..
Friction of fault F_{215}	1.0	.................*.
Cohesion of heavy weathered zone (MPa)	0.10	..................*.
Cohesion angle of fault F_{215} (MPa)	0.50	...................*

3.2 *A neural network representation of nonlinear relationship between displacement and parameters to be back analyzed*

Considering nonlinear relationship between displacement and parameters to be determined, a multi-layer feedback neural network model is thus built as:

$$\text{NN}(n, h_1, \cdots, h_k, m) : R^n \rightarrow R^m$$
$$D = \text{NN}(n, h_1, \cdots, h_k, m)(P) \quad (1)$$

$$P = (p_1, p_2, \ldots, p_n), D = (d_1, d_2, \ldots, d_m)$$

Where p_i is the ith parameters, i=1,2,⋯, n. For example, $(P)=\{\sigma_x,\sigma_y,\tau_{xy},E,\cdots\}$, σ_x,σ_y, and τ_{xy} are components of geo-stress, E is deformation modulus, d_j is the jth component of displacement, j=1,2, ..., m.

Table 3. Change of horizontal displacement component at different monitoring points due to change of parameter values

No.	Monitoring point no.								
	TP/BM10GP01	TP/BM11GP01	TP/BM71GP01	TP/BM98GP02	TP/BM26GP02	TP/BM27GP02	TP/BM28GP02	TP/BM29GP02	Average change of displacement
1	0.07	0.08	0.06	0.05	0.03	0.05	0.08	0.08	0.063
2	0.01	0.01	1.00	0.90	0.10	0.06	0.05	0.03	0.270
3	0.11	0.12	0.29	0.28	0.13	0.05	0.07	0.07	0.140
4	1.31	1.33	1.69	1.72	1.48	1.16	1.12	1.11	1.365
5	1.59	1.57	3.29	3.60	2.24	1.54	1.41	1.43	2.084
6	0.03	0.01	0.07	0.12	0.01	0.01	0.00	0.01	0.033
7	0.02	0.02	0.01	0.00	0.01	0.01	0.01	0.00	0.010
8	0.03	0.03	0.36	0.21	0.02	0.06	0.07	0.08	0.108
9	0.01	0.01	0.16	0.11	0.01	0.01	0.00	0.01	0.040
10	0.00	0.02	0.10	0.02	0.02	0.01	0.01	0.00	0.023
11	0.03	0.01	0.02	0.02	0.00	0.00	0.01	0.00	0.011
12	0.00	0.01	0.41	0.30	0.35	0.04	0.03	0.04	0.148
13	0.02	0.00	0.21	0.31	0.04	0.00	0.00	0.01	0.074
14	0.00	0.00	0.00	0.00	0.00	0.00	0.00	0.00	0.000
15	0.02	0.02	0.05	0.03	0.01	0.00	0.01	0.00	0.018
16	0.01	0.01	0.04	0.39	0.28	0.12	0.01	0.06	0.115
17	0.02	0.01	0.01	0.01	0.01	0.00	0.01	0.00	0.009
18	0.02	0.01	0.01	0.00	0.00	0.00	0.00	0.00	0.005
19	0.01	0.01	0.01	0.01	0.01	0.00	0.00	0.00	0.006
20	0.01	0.01	0.01	0.00	0.00	0.00	0.00	0.01	0.005

In this new algorithm, the topology of the network is encoded in a binary string. The network may be fully connected. Not only nodes on neighboring layers but also nodes on different layers are all connected. Granularity bits define size of connection. The binary string will be evolved using genetic operations such as reproduction, crossover, and mutation to obtain global optimum model. This evolutionary process will not finish until the minimum error for prediction of new cases is achieved.

To check "over-training" of the network, a set of testing samples is needed. These testing sample data set can be constructed using even design method. This is considering that experimental points are even distributed at the investigated range. For an experiment on x factors with n value levels, it only needs n times of experiment. The neural network model is established when it gives good output both for learning cases and testing cases.

3.3 *Recognition of parameter value in global space using genetic algorithm*

A genetic algorithm, a global optimal search, is adopted to perform search of parameters in global large space. Fitness of the parameter set can be evaluated by

$$F(P)=\sqrt{\frac{1}{m}\sum_{i=1}^{q}\sum_{j=1}^{w}\left(\mathrm{NN}(n,h_1,\cdots,h_k,m)_{ij}(P)-u_{ij}\right)^2} \quad (1)$$

where (P) is a set of parameters to be recognized, $\mathrm{NN}(n,h_1,\cdots,h_k,m)_{ij}(P)$ is calculated displacement of the component j of the ith monitoring points, u_{ij} is measured displacement, q is number of monitoring points, w is number of displacement component, $w\times q=m$.

As progress of search using the genetic algorithm, values of parameters will be concentrated at a set having minimum fitness (Table 4). Thus, the stable solution in global optimum can be obtained.

Table 4. Stability of searched parameter values

Deformation modulus (GPa)				Coefficients in geo-stress field		Fitness $F(P)$
A	B	C	D	a_x	a_y	
7.512	9.68	18.95	32.1	3.958	1.586	0.63
7.512	9.842	18.95	36.299	4.795	1.586	0.431
7.512	9.81	18.95	36.25	4.795	1.586	0.429
7.512	9.696	16.867	32.083	4.378	1.586	0.403
7.526	9.68	18.934	34.167	4.792	1.586	0.336
7.512	9.696	18.95	32.083	4.587	1.586	0.321
8.449	10.72	18.95	33.125	4.795	1.443	0.313
11.29	9.68	18.95	32.116	4.792	1.584	0.311
11.26	9.68	18.95	32.083	4.795	1.537	0.309
7.512	9.712	18.918	32.132	4.792	1.086	0.298
7.512	9.68	14.775	32.116	4.792	1.586	0.298
7.512	9.68	16.867	32.116	4.899	1.461	0.296
7.512	13.85	18.95	32.1	4.792	1.461	0.295
7.512	9.68	16.558	32.083	4.792	1.586	0.294
7.541	9.712	16.834	32.083	4.792	1.586	0.293
7.512	9.696	16.867	32.116	4.792	1.599	0.293
7.512	13.85	18.82	32.116	4.792	1.586	0.293
7.512	13.85	18.853	32.116	4.792	1.586	0.293
7.512	13.85	18.934	32.083	4.792	1.584	0.292
7.512	9.81	17.909	32.083	4.795	1.586	0.291
7.512	9.68	18.95	32.116	4.792	1.492	0.291
7.512	11.76	18.918	32.083	4.792	1.586	0.291
7.512	9.81	18.95	32.344	4.896	1.57	0.291
7.512	9.68	18.95	32.083	4.896	1.537	0.291
7.512	10.74	18.934	32.1	4.844	1.57	0.291
7.98	9.68	18.95	32.083	4.792	1.58	0.29
7.512	9.826	18.429	32.083	4.795	1.586	0.29
7.512	9.68	18.95	32.344	4.792	1.552	0.29
7.629	9.696	18.95	32.083	4.896	1.586	0.29
7.512	10.72	18.95	32.083	4.795	1.582	0.29
7.512	10.72	18.95	32.1	4.798	1.586	0.29
7.512	9.81	18.95	32.083	4.808	1.548	0.29
7.512	10.72	18.95	32.083	4.792	1.586	0.29
7.512	9.696	18.918	32.083	4.899	1.586	0.29
7.512	9.81	18.918	32.083	4.792	1.554	0.29
7.512	9.68	18.918	32.083	4.795	1.554	0.29
7.541	9.68	18.95	32.344	4.792	1.586	0.29
7.512	9.68	18.942	32.344	4.795	1.584	0.29
7.512	9.729	18.95	32.344	4.795	1.586	0.29

7.512	9.68	18.95	32.36	4.795	1.586	0.29
7.512	9.68	18.95	32.083	4.792	1.554	0.29
7.512	9.696	18.95	32.344	4.792	1.586	0.29
7.512	9.68	18.69	32.083	4.792	1.586	0.289
7.512	9.68	18.95	32.083	4.844	1.586	0.289
7.629	9.68	18.95	32.083	4.792	1.586	0.289
7.512	9.859	18.918	32.116	4.792	1.586	0.289
7.512	9.859	18.95	32.083	4.808	1.586	0.289
7.526	9.68	18.95	32.1	4.818	1.586	0.289
7.512	9.68	18.95	32.165	4.792	1.582	0.289
7.512	9.729	18.918	32.1	4.805	1.586	0.289
7.512	9.68	18.885	32.083	4.798	1.584	0.289
7.512	9.81	18.918	32.083	4.792	1.584	0.289
7.512	9.68	18.853	32.083	4.792	1.586	0.289
7.512	9.81	18.95	32.116	4.795	1.586	0.289
7.512	9.68	18.95	32.1	4.805	1.582	0.289
7.512	9.68	18.95	32.083	4.818	1.586	0.289
7.512	9.68	18.942	32.116	4.798	1.584	0.289
7.512	9.68	18.942	32.148	4.792	1.586	0.289
7.512	9.696	18.95	32.148	4.792	1.586	0.289
7.512	9.68	18.95	32.116	4.792	1.582	0.289
7.512	9.81	18.934	32.083	4.792	1.586	0.289
7.512	9.68	18.885	32.083	4.792	1.586	0.289
7.512	9.81	18.95	32.083	4.795	1.586	0.289
7.512	9.68	18.95	32.083	4.811	1.586	0.289
7.526	9.68	18.918	32.083	4.792	1.586	0.289
7.512	9.81	18.95	32.083	4.792	1.586	0.289
7.512	9.68	18.95	32.132	4.792	1.586	0.289
7.512	9.68	18.918	32.083	4.792	1.584	0.289
7.512	9.712	18.918	32.083	4.792	1.586	0.289
7.512	9.68	18.95	32.116	4.795	1.586	0.289
7.512	9.745	18.934	32.083	4.792	1.586	0.289
7.512	9.68	18.95	32.116	4.792	1.586	0.289
7.512	9.68	18.918	32.083	4.792	1.586	0.289
7.512	9.68	18.934	32.083	4.795	1.586	0.289
7.512	9.696	18.95	32.083	4.792	1.584	0.289
7.526	9.68	18.95	32.083	4.792	1.586	0.289
7.512	9.712	18.95	32.083	4.792	1.586	0.289
7.512	9.68	18.95	32.083	4.795	1.586	0.289
7.512	9.696	18.95	32.083	4.792	1.586	0.289
7.512	9.68	18.95	32.083	4.792	1.586	0.289
7.512	9.68	18.95	32.116	4.792	1.599	0.289
7.512	9.68	18.95	32.083	4.792	1.599	0.289

Table 5. Rock mechanical parameters determined using intelligent methodology for the head region of the third shiplock room at permanent shiplock of Three Gorge Project

Deformation modulus (GPa)				Coefficients in geo-stress field	
Slightly weathered or fresh zone	Unloading deformation zone	Damaged zone	Moderately weathered zone	a_x	a_y
32.1	18.95	9.68	7.51	4.79	1.6

4 DISCUSSION AND CONCLUSIONS

Using the intelligent methodology proposed above, rock mechanical parameters for the head region of the third shiplock room at permanent shiplock of Three Gorges Project are successfully recognized as shown in Table 5 (Feng et al, 2000). Predicted displacement of lower strata using the parameters recognized using displacement data of upper strata agrees with the measurement (see Table 6). Therefore, the proposed method is effective to solve uniqueness of parameter recognition at larger space.

Table 6. Comparison of measured displacement with predictions from FLAC and neural network model with input of mechanical parameters determined by intelligent back method

	Horizontal displacement increase at monitoring points due to excavation of the shiplock room (mm)								Average absolute error
	TP/BM 10GP01	TP/BM 11GP01	TP/BM 71GP01	TP/BM 98GP02	TP/BM 26GP02	TP/BM 27GP02	TP/BM 28GP02	TP/BM 29GP02	
1	16.32	19.11	23.38	17.91	20.76	16.71	19.10	16.71	
2	17.30	17.68	26.01	23.94	23.59	16.62	14.88	15.25	2.46
3	17.36	17.79	—	—	23.64	16.85	14.91	15.29	1.83

Note: 1: Measurement, 2: Predictions from $FLAC^{2D}$ with input of recognized parameters, 3: Predictions of neural network model with input of recognized parameters.

ACKNOWLEDGEMENT

Financial supports by National Science Foundation of China, Fok Ying Tung Education Foundation under Grant no.710048, and the Teaching and Research Award Program for Outstanding Young Teachers in Higher Education Institutions of MOE, P R China are greatly acknowledged.

REFERENCES

Kirsten, HAD. 1976. Determination of rock mass elastic moduli by back analysis of deformation measurement, *Proc. Symp. On Exploration for Rock Eng.*, Johannesburg.

Gioda, G., Jurina, L. 1981. Numerical identification of soil structure interaction pressures, *Int. J. For Num. & Anal. Meth in Geomech.*, 5:33-56

Sakurai, S., Takeuchi K. 1983. Back analysis of measured displacements of tunnels, *Rock Mechanics and Rock Engineering*, 16:173-180

Sakurai, S., Dees Wasmongkol N., Shinji M. 1986. Back analysis for determining material characteristics in cut slopes, *Proc. Int. Symp. on ECRF*, Beijing, 11:770-776

Sakurai, S. 1987. Interpretation of the results of displacement measurements in cut slopes, *Proc. 2nd Int. Symp. on FMGM87*, Kobe, 2: 528-540

Feng, Xia-Ting, Zhang, Zhiqiang and Shen, Qian, Identifying rock mass mechanical parameters of Three Gorges Project permanent shiplock using intelligent displacement back analysis method, *Int. J. Rock Mech. and Min. Sci.* in press.

Computer Methods and Advances in Geomechanics, Desai et al. (eds) © 2001 Balkema, Rotterdam, ISBN 90 5809 183 X

Study on the Osterberg-cell procedure for pile testing

F.M.Abdrabbo, R.M.El-Hansy & T.M.Abdel-Aziz
Structural Engineering Department, Faculty of Engineering, Alexandria University, Egypt

ABSTRACT: The most reliable method for determining the bearing capacity of a pile is the static pile-loading test. But, there are many difficulties in using this method in case of large capacity piles, which require very large counterweights. For this reason, a new testing method using Osterberg load cell was developed. A complete design and development of O-cell, which is used for testing model piles in laboratory was carried out. Seven pairs of pile-loading tests were conducted and used to evaluate the validity of the Osterberg technique. Each test was performed twice, by the conventional and the O-cell methods. The paper outlines and discus the different methods implemented in the extrapolation of the pile base load-displacement relationship and recommend an appropriate extrapolation procedure.

1 INTRODUCTION

The new method was developed for testing both bored and driven piles and becomes easy to overcome the difficulties associated with the conventional method.

During testing, the shaft and base resistance were measured separately by pressurizing a hydraulic jack installed near the base of the tested pile, and creating an upward force on the shaft and an equal downward force to the base. A test load up to 135 MN was applied (Osterberg 1998). Survey conducted by FHWA at mid-1994 (Baker 1994), indicated that 65% of bored pile- loading tests were conducted using the O-cell method. The popularity of the method is increasing especially in U.S.A, Japan, and Western Europe. In Manila, Philippines, Fellenius et al. (1999) reported that a static loading test using O-cell test method was performed, to study the barrette capacity and deformation behavior. The maximum applied O-cell load during the tests was 11.6 MN. The authors reported that the O-cell test results represent the behavior of the barrette in a conventional head - down load test. In Japan, a drilled shaft of diameter 1200 mm was tested by O-cell up to a load of 10 MN (Karkee & kishida 1999). A suitable supplementary loading frame was designed to provide the additional reaction after the friction resistance was fully mobilized. Fujioka & Yamada (1994) reported that the concept of the system was invented in Japan in 1969 and patented, for bored piles, by Nakayama &Fujiseki in 1973 and, for driven piles, by Sumii in 1978.

A complete historical development of the method was reported elsewhere (Abdel-Aziz 1999).

2 TESTING EQUIPMENT AND MATERIAL

A rigid soil bin of diameter 750 mm and 900 mm in height was made from steel plates 10 mm in thickness to contain the soil. The bin was stiffened by steel ring beams. The pile model length (L) of 740 mm was machined from an aluminum tube with an external diameter (D) of 50 mm and 1.50 mm wall thickness, and provided with a head and O-cell model. To roughen the surface of the pile, rough sandpaper was stuck to it.

The conventional loading system used for testing the pile was designed and manufactured to facilitate applying incremental vertical load to the pile head using frictionless lever system (Figure1). The system consisted of two parallel vertical rigid frames made of steel strips, 50 mm width and 7 mm thickness, bolted to the top edge beam of the soil bin. The two frames are accommodating two fabricated steel channels No.140 which in turn supporting the lever and guiding the loading shaft. A lever made of a fabricated steel channel No.100, and provided with a frictionless ball bearing served to apply the load to the pile model through a loading shaft.

The loading system for testing the pile by the Osterberg technique is mainly consisted of oil supply unit, conduit and cell (Figure 2). The oil supply unit consisted of a reservoir, manual pump, pressure gauge and a system of control, suction and

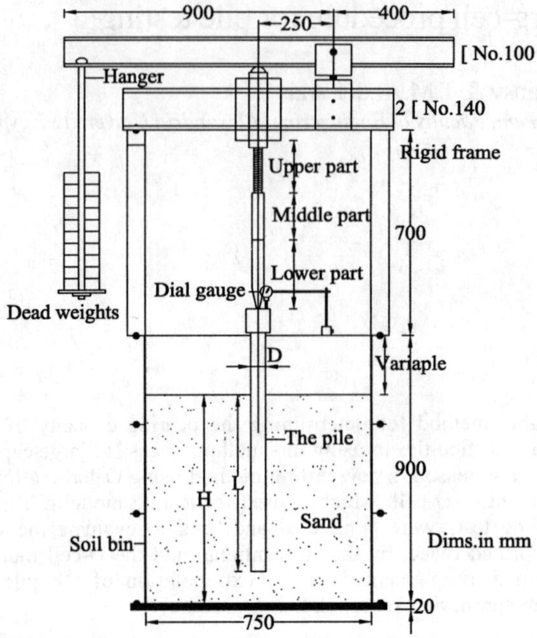

Figure 1. Complete set-up of conventional testing procedure

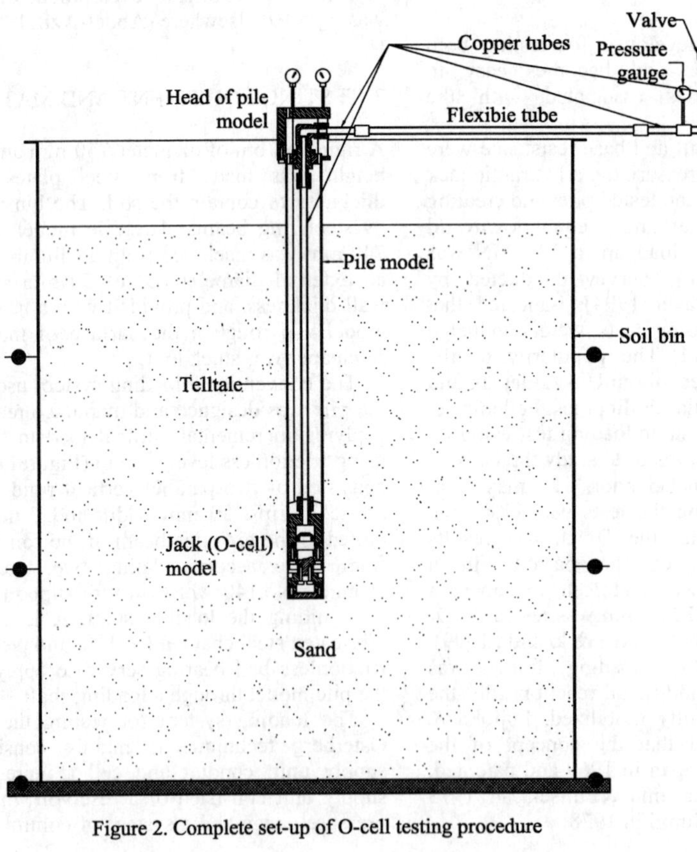

Figure 2. Complete set-up of O-cell testing procedure

delivery valves. The reservoir, filled with hydraulic oil, was machined from aluminum with a diameter of 77 mm, and a height of 144 mm. The manual pump was machined from steel and manually operated, to build up pressure inside the cell. The pressure gauge attached to the system, had a capacity of 4 N/mm^2, with an accuracy of 0.10 N/mm^2. An oil valve was connected to the system to allow releasing the pressure from the O-cell. The conduit is mainly used for transferring the oil pressure from the manual pump to the cell and consisted of tubes, nuts, tee connections, nipple and elbow. The nipples, elbow, tee-connections and conduit inside the pile were machined from copper while, the part of the conduit between the pump and the nipple near the pile head from rubber. The O-cell model is a small hydraulic jack consisted of casing incorporated a movable piston and designed to apply two equal and opposite forces. It had an external diameter of 50 mm; the same as the pile diameter, and a height of 60 mm. The cell had a stroke of 10 mm and a maximum capacity of 4300N.

When testing the pile following the conventional procedure, the downward displacement of the pile was measured by using a mechanical dial gauge of 0.01mm accuracy (Figure 1). While, when testing the pile by the O-cell method, the downward displacement of the pile base was measured via a steel telltale screwed to the lower sliding part of the O-cell, attached to a mechanical dial gauge (Figure 2). Another mechanical dial gauge was used to measure the upward displacement of the pile shaft.

The conventional loading system was calibrated to get a direct relationship between the dead weights placed at the end of the lever and the corresponding imposed load on the pile head. A load cell with capacity of 2.0 kN was used in the calibration. The own weight of the system was considered in calibration. The accuracy of the conventional loading system is 7.57 N. The O-cell was also calibrated. The accuracy of the cell was 14.25 N. The efficiency of the cell was 86.25% (efficiency = actual load exerted by the cell / theoretical load). The efficiency is a measure of the friction between the movable piston and the casing of the cell.

Dry siliceous clean sand was used to form homogeneous sand deposit of height (H). The soil was formed after placing the pile in its correct position at relative densities (Dr) of 60.80, 70.0 and 80.30 %. The corresponding angles of internal friction (ϕ) were determined using direct shear box apparatus, as 35° $30'$, 36° $30'$ and 37° $30'$ respectively. The used sand had maximum and minimum dry unit weights of 18.25 and 14.54 kN/m^3 respectively. The angle of friction between the surface of the pile (sandpaper), and the soil, was measured using direct shear box apparatus. The values of wall roughness were 34° $15'$, 35° $15'$ and 36° respectively.

3 TESTING PROCEDURE

A testing programme was set up, in which the pile was tested at seven different conditions as shown in Table 1. For each case, two tests were performed, one following the conventional procedure, and the other by the O-cell method.

When testing the pile following the conventional procedure, the maintained load test method reported in the Egyptian code of soil mechanics and foundation (1995) was, implemented. While, when testing the pile using O-cell, the ASTM quick test procedure D1143 was implemented. Each test was performed twice. In the first, the load exerted by the O-cell was increased up to failure takes place along the pile shaft. In the second, the shaft displacement was ceased just after the failure had taken place along the pile shaft using reaction rod attached to the loading frame, and the test was completed up to a base displacement equal to 10% of the pile diameter.

Table 1. Testing programme.

Case No.	H/L	L/D	Dr, %
1	1.50	12	60.80
2	1.50	12	70.00
3	1.50	12	80.30
4	1.50	10	60.80
5	1.50	8	60.80
6	1.20	12	60.80
7	1.10	12	60.80

4 TEST RESULTS

Due to limitations of space, some of the achieved results will be presented, meanwhile all results are documented elsewhere (Abdel-Aziz 1999). During testing the piles by O-cell, failure along the shaft takes place, while the load transferred to the base is far below the ultimate base value. In order to get the load-head displacement relationship of the pile, the load displacement relationship of the pile base should be extrapolated up to the ultimate value. The work presented here is confined to the methods of extrapolation.

5 METHODS OF EXTRAPOLATION

Six methods were used for extrapolation (Table 2), the first two methods are recommended by Osterberg (1998), while the writers proposed and implemented the other four methods.

The measured values of base load (Q_b) and the downward displacement (S_b) were fitted in the extrapolation relationship to calculate the mathematical constants. Then, the base load-displacement relationship is extrapolated. The load settlement relationship of the pile was obtained by superposing the shaft and the base loads corresponding to the same settlement. The ultimate pile load (Q_u) was taken as the load corresponding to a displacement equal to 10% of the pile diameter.

Table 3 shows the values of the ultimate capacities of the tested piles resulting from the different extrapolation methods, and that obtained from conventional testing procedure. The ratio between the values of the ultimate capacities of the tested piles obtained from the extrapolation methods, and that obtained from the conventional testing procedure are shown in Table 4. Also, the ratio of the ultimate shaft load (Q_{su}) to the ultimate base load (Q_{bu}) for each test is calculated and presented.

Table 2. Methods of extrapolation.

Method No.	Method
1	Hyperbolic rule, $(S_b/Q_b) = a_o + b_o (S_b)$, (Osterberg 1998).
2	Conservative, assuming that the ultimate bearing load is attained simultaneously when failure takes place along the pile shaft, (Osterberg 1998).
3	Exponential relationship, $S_b = (a)(10)^{bQ_b}$
4	Parabolic relationship, $S_b = a'(Q_b)^2 + b'(Q_b) + c'$
5	Mean values of load predicted by exponential and parabolic relationships, $Q_b = 0.50[(\log S_b - \log a)/b + (-b' + \sqrt{b'^2 - 4a'(c' - S_b)})/2a']$
6	$(S_b/Q_b) = A' + B'(Q_b)$

where,
S_b = the pile base settlement,
Q_b = the pile bearing load and
a_o, b_o, a, b, a′, b′, c′, A′ & B′ = constants to be obtained from the measured values.

Table 3. Test results.

Case No.	Q_u, (N) Conv.* method	Extrapolated method 1	2	3	4	5	6
1	970	814	817	874	1190	1032	1067
2	1460	1180	1087	1416	2020	1718	1677
3	2180	1703	1625	1962	2843	2402	2386
4	1505	814	727	1066	1995	1530	1518
5	1620	806	727	973	2205	1589	1751
6	1704	894	856	1017	1857	1437	1547
7	-	-	-	-	-	-	-

* Conv. = conventional

From Tables 3, 4, it was noticed that, for tests No. 1 to 6, where $H/L \geq 1.2$, the results, of Q_u, obtained using the O-cell were in acceptable agreement with those of conventional testing procedure, provided that an appropriate method of extrapolation is implemented. But, for test No. 7, where H/L =1.1, the equipment of the conventional test is incapable to perform the test up to failure load. Consequently, the test was terminated at a load corresponding to settlement ratio, S/D = 0.02. At the mean time, the O-cell test procedure was found unsuitable to express the true response of the pile. This may be attributed to the effect of rigid base on the response of the pile tested by the conventional procedure. While, in case of testing the pile by the O-cell, the effect of rigid base on the settlement due to shaft load was eliminated and only affects the displacement due to base load. So, the settlement of a pile tested by the O-cell is more than that the actual value.

Also, it was noticed that, the fourth proposed method (method No. (6)) is the most appropriate between the six studied methods, especially for cases with Q_{su}/Q_{bu} = 0.30 to 0.70. Extrapolation methods, numbered 1,2 in Table 2, recommended by Osterberg (1998), may be used for cases having ratio $Q_{su}/Q_{bu} > 0.70$ and $H/L \geq 1.20$.

But, the correct ratio of Q_{su}/Q_{bu} cannot be assessed before running the extrapolation process. So, it would be useful to have an approach to predict this ratio beforehand.

By observing the results obtained from tests, it was noticed that the ratio Q_{su}/Q_{bu} was directly proportional to the ratio S_{bsu}/S_{max}.,

where, Q_{su} = the ultimate shaft load, Q_{bu} = the ultimate base load, S_{bsu} = the base settlement corresponding to the shaft failure load and S_{max} = the maximum expected settlement (0.10 of the pile diameter).

Figure 3 presents the relationship between the Q_{su}/Q_{bu} and S_{bsu}/S_{max}. From this figure, it is clear that, this relationship is linear. Also, the ratio $Q_{su}/Q_{bu} > 0.70$ is corresponding to $S_{bsu}/S_{max} > 0.30$.

So, if the base settlement, S_{bsu}, corresponding to the shaft failure load >0.030 of the pile diameter, the two methods of extrapolation, numbered 1,2 in Table 2 and recommended by Osterberg (1998) may be used. If the base settlement, S_{bsu}, corresponding to shaft failure load varies between 0.007 and .030 of the pile diameter, the recommended method of extrapolation, numbered 6 in Table 2, may be used.

Table 4. Comparison between the conventional and the Osterberg method.

Case No.	Q_{su}/Q_{bu}	R* for extrapolation method No. 1	2	3	4	5	6
1	0.70	0.84	0.84	0.90	1.22	1.06	1.10
2	0.59	0.81	0.74	0.97	1.38	1.18	1.14
3	0.51	0.78	0.75	0.90	1.30	1.10	1.09
4	0.31	0.54	0.48	0.71	1.33	1.02	1.01
5	0.30	0.50	0.45	0.60	1.36	0.98	1.08
6	0.34	0.52	0.50	0.60	1.09	0.84	0.91
7	-	-	-	-	-	-	-

*R = Q_u obtained from extrapolation method /Q_u obtained from conventional test.

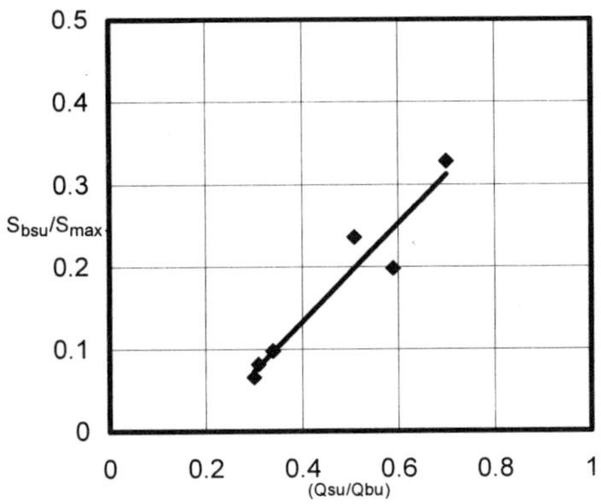

Figure 3. (Q_{su} / Q_{bu}) versus (S_{bsu} / S_{max}.).

6 CONCLUSIONS

The main conclusions drawn from the study are given below:

1.The methods of extrapolation recommended by Osterberg are appropriate for piles having $Q_{su}/Q_{bu}> 0.70$.

2.The recommended method of extrapolation by the writers is appropriate for piles having Q_{su} /Q_{bu} varies from 0.3 to 0.70.

3.To help the user of the O-cell to predict an accurate value of the ratio Q_{su} /Q_{bu} , and then choose the appropriate method of extrapolation, it is found that the relation between S_{bsu} /S_{max} and Q_{su} /Q_{bu} is linear. (where, S_{bsu} = the base settlement corresponding to shaft failure load). So, when S_{bsu} > 0.030D, the methods of extrapolation recommended by Osterberg are appropriate, and when $0.007D \leq S_{bsu} \leq 0.030$ D, the recommended method of extrapolation is appropriate.

REFERENCES

Abdel-Aziz, T.M. 1999. Static and dynamic response of pile-soil system. *A thesis submitted to Alexandria University In partial fulfillment of the requirements for the M.Sc. degree.*

Baker, C.N. 1994. Current U.S. design and construction practices for drilled piers. *1st international conference on design and construction of deep foundation.* U.W.Federal Highway Administration. 1: 305-323.

Egyptian code of soil mechanics and foundation Eng. 1995.

Fellenius, B.H. et al. 1999. O-cell testing and FE analysis of 28m deep barrette in Manila, Philippines. *JL of geotechnical and geoenvironmental engineering.* 125 (7): 566-575.

Fujioka, T. & Yamada, K. 1994. The development of a new pile load-testing system. *Proceedings international conference on design and construction of deep foundation* . 2: 670-684.

Karkee, M.B. & Kishida, H. 1999. Dynamic, intermediate and static pile loading tests and application of the test results in design practice. *4th international conference on deep foundation practice.* Singapore. 13-26.

Osterberg, J.1998. The Osterberg load test method for bored and driven piles the first ten years. *7th Internal conference and exhibition on piling and deep foundation.*

Computer Methods and Advances in Geomechanics, Desai et al. (eds) © 2001 Balkema, Rotterdam, ISBN 90 5809 183 X

A numerical model for the analysis of the Piezocone Penetration Test

M.Y. Abu-Farsakh
Louisiana Transportation Center, La., USA

M.T. Tumay
College of Engineering, Louisiana State University, La., USA

ABSTRACT: The piezocone penetrometer is widely acknowledged as the preferred in-situ testing for site characterization and evaluation of soil parameters. Therefore, a proper interpretation of the piezocone test data is required. This paper presents a new numerical model for the analysis of the piezocone penetration test (PCPT) in clayey soils. The numerical modeling is accomplished in two stages. In the first stage, the piezocone is expanded radially from zero to the piezocone radius, r_o. The second stage simulates the continuous penetration of the piezocone by applying incremental vertical displacements of the nodes representing the piezocone boundary. The numerical analyses are compared with the miniature piezocone penetration tests (MPCPT) conducted at Louisiana State University (LSU) calibration chamber. The predicted and measured values are in good agreement. The model is used to study the effect of the lateral stresses and overconsolidation ratio, OCR, on the piezocone soundings. The variation in hydraulic conductivity coefficient on the dissipation of excess pore pressure is also investigated.

1 INTRODUCTION

Recently, there is an increased emphasis toward the evaluation of geomedia parameters using in-situ testing. In-situ tests that are performed under existing in-situ stresses and boundary conditions can provide more accurate and reliable results than the laboratory tests in assessing the actual soil properties. The piezocone penetrometer is widely acknowledged as the preferred in-situ testing for site characterization. The piezocone penetrometer test (PCPT) is robust, simple, fast, economical, and provides continuous soundings with depth. The measurements of the piezocone penetrometer can be effectively utilized for detailed soil stratigraphy and the evaluation of soil properties.

Several approaches have been proposed to analyze the piezocone penetration problem. These approaches are based on the bearing capacity theories (Meyerhof, 1961), the cavity expansion theories (i.e. Torstensson, 1977), and the strain path method (i.e. Levadoux & Baligh, 1986). The finite element technique was also adapted by some researchers (i.e. Kiousis et al., 1988; Abu-Farsakh et al., 1998) to analyze the piezocone penetration problem. Teh & Houlsby (1991) proposed a combination of the strain path method with finite element technique to model the piezocone penetration.

The cavity expansion methods cannot model the strain path followed by the soil elements (Levadoux & Baligh, 1986) during cone penetration and are unable to distinguish between different soil deformation modes due to different cone tip geometries. The strain path methods assume that the soil around the cone follows a certain deformation pattern that is different from the actual displacement field resulting from penetration (Kiousis et al., 1988) with the resulting stresses not being in equilibrium. The use of finite element technique has many advantages over other methods: properly model the piezocone geometry, satisfy the equilibrium equations, and model the soil-piezocone interface friction.

Analysis of the piezocone penetration is difficult, since it involves large deformations, high nonlinear material and interface behavior. During penetration, soils adjacent to the piezocone undergo large deformations with very high strains (Acar & Tumay, 1986; Abu-Farsakh et al., 1998). To accurately analyze the piezocone penetration tests, it is necessary to have a model that takes into consideration the large deformation of the soil, nonlinear soil behavior, change of the boundary conditions with penetration, soil-piezocone interface friction and the actual in-situ stresses of the soils around the piezocone.

In this paper, a new numerical model is introduced and used in the analysis of the miniature piezocone penetration tests (MPCPT) in cohesive soils conducted at the LSU calibration chamber (Tumay & de Lima, 1992). The proposed model is used to study the effect of the lateral stresses, and overconsolidation ratio, OCR, on the cone tip resistance and the penetration generated excess pore pressure around the piezocone. The variation in the horizontal and vertical hydraulic conductivity coefficients on the dissipation of excess pore pressure is also investigated.

2 FINITE ELEMENT FORMULATION

An elasto-plastic coupled system of equations that are based on the principle of virtual work and theory of mixtures for inelastic porous media is used in this study to describe the time-dependant deformation of the saturated cohesive soils. The governing equations were developed for large deformation with finite strains in an updated Lagrangian formulation. Detailed description of the formulation of governing equations is presented elsewhere (Abu-Farsakh, 1997; Voyiadjis & Abu-Farsakh, 1997). A brief description of these equations is presented below:

For finite deformation, the elasto-plastic constitutive equation is given by

$$\overset{o}{\sigma}{}'^{s}_{ij} = D^{s}_{ijkl}\, d^{s}_{kl} \tag{1}$$

where $\mathbf{D}^s$ is the elasto-plastic stiffness matrix, and $\overset{o}{\sigma}{}'^{s}$ is the Jaumann effective stress rate tensor given as:

$$\overset{o}{\sigma}{}'^{s}_{ij} = \dot{\sigma}'^{s}_{ij} - w^{s}_{ik}\sigma'^{s}_{kj} + w^{s}_{kj}\sigma'^{s}_{ik} \tag{2}$$

where $\dot{\sigma}'^{s}$ is the effective Cauchy stress rate tensor, σ'^{s} is the effective Cauchy stress tensor, and $\mathbf{d}^s$ is the deformation rate given as follows:

$$d^{s}_{ij} = \frac{1}{2}(v^{s}_{i,j} + v^{s}_{j,i}) \tag{3}$$

and

$$w^{s}_{ij} = \frac{1}{2}(v^{s}_{i,j} - v^{s}_{j,i}) \tag{4}$$

is the spin tensor, and $\mathbf{v}^s$ is the velocity of the solid grain particles.

The incremental stress-strain relationship can be given as follows (Voyiadjis & Abu-Farsakh, 1997):

$$\delta s_{ij} = D^{*}_{ijkl}\,\delta\varepsilon_{kl} + J^{s} X^{s}_{i,I} X^{s}_{j,J}\,\delta_{IJ}\,\delta P_{w} \tag{5}$$

where δs is the incremental second-Piola Kirkchoff, $\delta\varepsilon$ is the material incremental strain tensor, $\mathbf{D}^*$ is the modified elasto-plastic stiffness matrix, δP_w is the incremental pore water pressure, δ_{IJ} is the kronecher delta, $\mathbf{J}^s$ is the Jaccobian of solid grains, and $\mathbf{X}^s$ is the deformation gradient for the solid grains.

Using the finite element discretization, the incremental strains can be expressed as follows:

$$\delta\mathring{a} = (\mathbf{B}_L + \mathbf{B}_{NL})\delta\mathbf{U}_d \tag{6}$$

where $\mathbf{B}_L$ and $\mathbf{B}_{NL}$ are the linear and nonlinear strain-displacement matrices, $\delta\mathbf{U}_d$ is the incremental displacement.

Applying the principle of virtual work in an updated Lagrangian frame with the use of equations 5 and 6, the first governing equation can be derived.

The balance of mass in terms of soil porosity n^w, solid velocity $\mathbf{v}^s$, and water velocity $\mathbf{v}^w$ can be expressed as follows (Prevost, 1980)

$$\operatorname{div}\mathbf{v}^{s} - \operatorname{div}\mathbf{v}^{w} = \frac{1}{n^{w}}[\operatorname{div}(\mathbf{v}^{s}) + (\mathbf{v}^{w} - \mathbf{v}^{s}).\nabla n^{w}] \tag{7}$$

The flow of water through porous medium is given by

$$(\mathbf{v}^{w} - \mathbf{v}^{s}) = \frac{1}{\gamma_{w}}\mathbf{K}^{ws}[\nabla(P^{w}) - \rho_{w}\mathbf{b}] \tag{8}$$

where $\mathbf{K}^{ws}$ is the permeability tensor, $\mathbf{b}$ is the body force, and γ_w is the unit weight of water.

The second governing equation in an updated Lagrangian reference frame can be obtained as follows (Voyiadjis & Abu-Farsakh, 1997)

$$J^{s} C^{s^{-1}}_{ij}\dot{\varepsilon}_{ij} - J^{s} C^{s^{-1}}_{ij} C^{s^{-1}}_{ij} X^{s}_{D,a}\frac{\partial}{\partial X_{D}}.\left[\frac{n^{w}}{\gamma_{w}} K^{ws}_{AB} X^{s}_{a,A}\left(\frac{\partial P_{w}}{\partial X_{B}} - \rho_{w} B_{B}\right)\right] = 0 \tag{9}$$

The coupled system of equations for the saturated soil can be expressed as follows:

$$\begin{bmatrix} {}_{n}K & -{}_{n}\Omega \\ -{}_{n}\Omega^{T} & -\delta t\,{}_{n}\Psi \end{bmatrix}\begin{Bmatrix}\delta U_{d} \\ \delta U_{p}\end{Bmatrix} = \begin{Bmatrix}{}_{n}\Phi \\ {}_{n}\Pi\end{Bmatrix} \tag{10}$$

where ${}_{n}\mathbf{K} = {}_{n}\mathbf{K}_{L} + {}_{n}\mathbf{K}_{NL} + {}_{n}\mathbf{K}^{T}_{NL} + {}_{n}\mathbf{K}^{G}$ is the stiffness matrix, ${}_{n}\mathbf{K}_{L}$, ${}_{n}\mathbf{K}_{NL}$ and ${}_{n}\mathbf{K}^{G}$ are the linear, nonlinear, and geometric nonlinear stiffness matrices, ${}_{n}\Omega$ is the coupling matrix, ${}_{n}\Psi$ is the flow matrix, and $\delta\mathbf{U}_p$ is the incremental nodal excess pore pressures.

3 NUMERICAL MODEL

The piezocone penetration is numerically simulated using an axi-symmetric elasto-plastic large deformation finite element analysis. The finite element simulation of the piezocone intrusion is accomplished in two stages as described in Figure 1. In the first stage, the piezocone is expanded radially from zero to the piezocone radius, r_o (piezocone volume expansion). This stage is introduced to overcome the assumption of an existing pre-bored cone hole with the surrounding in-situ stresses remaining unchanged. This stage provides the initial conditions for the second stage. The second stage simulates the continuous penetration of the piezocone by applying incremental vertical displacements at the nodes representing the piezocone boundary. The vertical displacement is applied at the rate of 2 cm/sec allowing for partial pore pressure dissipation during penetration. Once the steady state condition is reached, penetration is stopped and the excess pore pressure is allowed to dissipate with time.

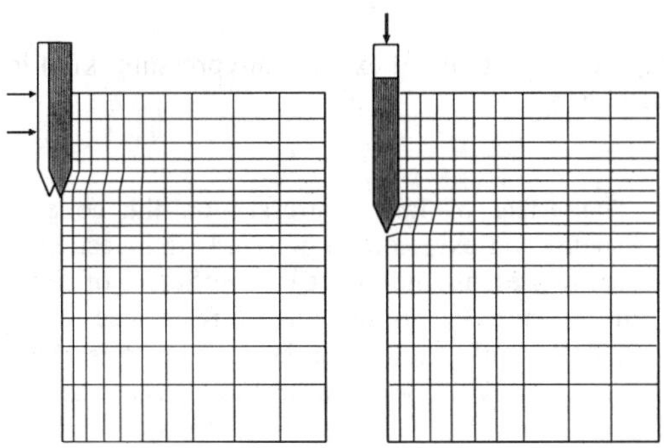

Figure 1. Numerical simulation of the piezocone penetration problem.

The piezocone penetration problem is treated as a boundary problem that includes changing of nodal boundary conditions as the piezocone penetrometer advances. Therefore, during each incremental penetration, the nodes along the piezocone boundary are continuously checked and their boundary conditions are appropriately adjusted. Based on the previous boundary condition at load increment (*k*-1) and the load-displacement criteria, a new boundary condition is assumed for load increment (*k*). The validity of the new boundry condition must be checked before one proceeds to the next load increment.

The soil-piezocone interface friction is modeled using a simple constrained approach at the nodal level. During the piezocone penetration, three interface states can be identified: fixed, slip and free states. At the beginning of penetration, all the nodes along the inclined conical surface and along the piezocone shaft are initially fixed until the sliding potential occurs. Once the sliding potential is reached for any node, the node is then allowed to slide along the surface. The Mohr-Coulomb frictional model is used to define the sliding potential for the fixed state. For more details one can refer to Abu-Farsakh et al. (1998).

4 ANALYSIS AND RESULTS

The proposed numerical model is used to analyze the penetration of 1 cm^2 and 1.27 cm^2 miniature piezocone penetrometers with an apex angle of 60°. Eight-noded, isoparametric quadrilateral finite element, Q8P4, is used in this analysis.

The results of this model are compared with the miniature piezocone penetration tests (MPCPT) in cohesive soils conducted at the LSU calibration chamber. Two different types of soil specimens were prepared. K_{50} specimens were prepared by mixing 50 % kaolinite and 50% fine sand (by weight), while the K_{33} soil specimens were prepared by mixing 33 % kaolinite and 67% fine sand. The specimens are consolidated against a backpressure of 138 kPa. Full details of specimen preparation and test procedure can be found in Kurup et al. (1994) and Lim (1999). Table 1 presents summary of the stress history of the soil specimens tested in the calibration chamber.

Table 1. Summary of stress histories of the soil specimens (Kurup et al., 1994; Lim, 1999).

Spec. No.	Soil Type	Chamber Cons.	OCR	Effective stress (kPa)	
				Vertical	Horizontal
1	K_{50}	Isotropic	1	207	207
2	K_{50}	Isotropic	5	41.5	41.5
3	K_{50}	k_o-aniso	1	207	107.6
4	K_{33}	Isotropic	1	207	207
5	K_{33}	k_o-aniso	1	207	86.2
6	K_{33}	Isotropic	1	262.2	262.2
7	K_{33}	k_o-aniso	1	262.2	104.8
8	K_{33}	k_o-aniso	10.9	24.2	40.71

The predicted cone tip resistance, q_c, profiles for the K_{33} soil specimens (4,5,6,7 and 8) are compared with measured values obtained at the steady state conditions as shown in Figure 2. The cone tip resistance values predicted at the steady state are close to the MPCPT measured values. A complete comparison between the calculated and the measured cone tip resistance, q_c, and the associated excess pore pressures, at the tip (u_1) and near the base (u_2), are presented in Table 2.

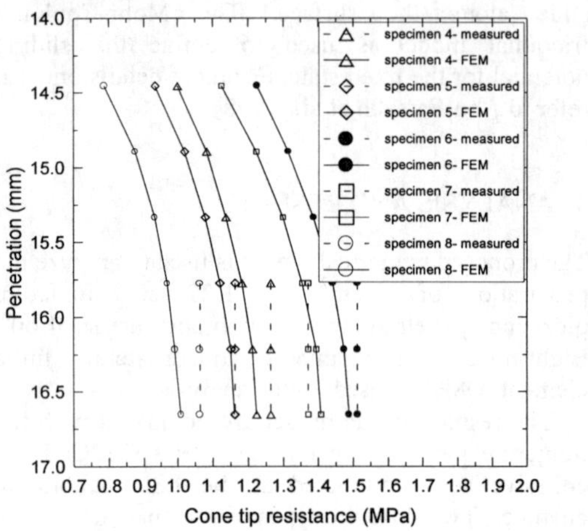

Figure 2. Measured versus predicted cone tip resistance, q_c, profile for K_{33} soil.

Table 2. Comparison between measured and predicted results at the steady state condition.

Spec. No.	q_c (MPa)		Excess pore pressure (MPa)			
			Tip		Base	
	FEM	Exp.	FEM	Exp.	FEM	Exp.
1	1.196	1.19	0.523	0.562	0.479	0.58
2	0.582	0.63	0.471	0.48	0.334	0.39
3	0.822	0.69	0.427	0.41	0.31	0.36
4	1.224	1.264	0.612	0.649	0.448	0.418
5	1.161	1.163	0.522	0.467	0.437	0.483
6	1.489	1.512	0.753	0.794	0.585	0.779
7	1.408	1.372	0.598	0.535	0.446	0.523
8	1.01	1.064	0.386	0.308	0.194	0.266

The distribution of excess pore pressures around the piezocone for specimen 3 is presented in Figure 3. The excess pore pressures developed around the piezocone during penetration is attributed to the change in the mean effective stress and the octahedral shear stress through the following relation:

$$\Delta u = \Delta\sigma_{oct} + \alpha\,\Delta\tau_{oct} \qquad (11)$$

Acceptable agreement can be seen between the predicted and measured values, and consequently, the proposed model is capable of analyzing the miniature piezocone penetration with acceptable accuracy.

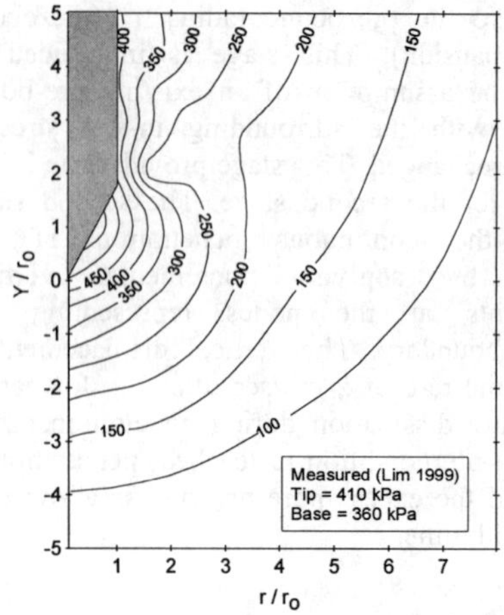

Figure 3. Contours of excess pore pressure (kPa) for specimen 3.

The effect of lateral stresses on the cone tip resistance and excess pore pressures is investigated. Figure 4 presents the effect of coefficient of lateral earth pressure, k_o, on the cone tip resistance for K_{33} soil. This clearly shows that the cone tip resistance is almost linearly proportional to the lateral stress, σ_h. This indicates the fact that the cone tip factor, N_k, should be normalized using both vertical and lateral stresses for more accurate estimation of the undrained shear strength of the soil utilizing the PCPT data.

Figure 5 depict the effect of lateral stresses on the developed excess pore pressures for K_{50} soil. The excess pore pressures around the cone tip and the base increase with the increase of σ_h.

The effect of overconsolidation ratio, OCR, on the developed excess pore pressures was also investigated for K_{33} soil specimen 5 (k_o=0.42) and presented in Figure 6. In overconsolidated clays, the coefficient of lateral earth pressure, $(k_o)_{oc}$, is interdependent on the OCR, therefore, the $(k_o)_{oc}$ has to be adjusted accordingly using the relation:

$$(k_o)_{OC} = (k_o)_{NC}\sqrt{OCR} \qquad (12)$$

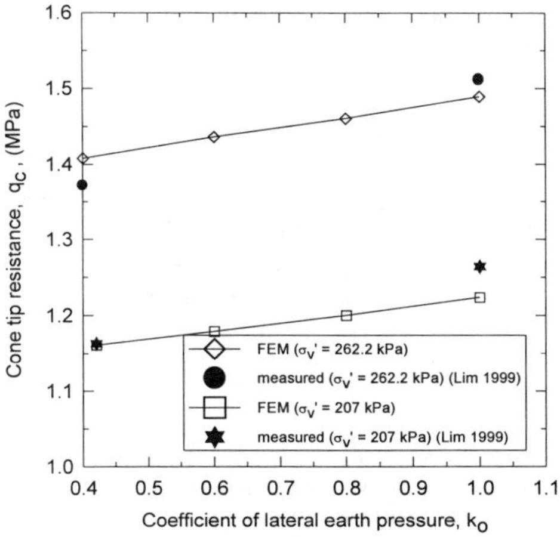

Figure 4. Effect of lateral pressure on the cone tip resistance (K_{33} soil).

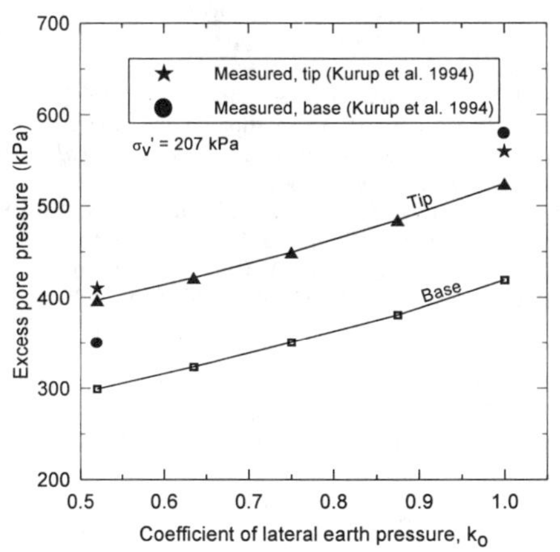

Figure 5. Effect of lateral pressure on excess pore pressure (K_{50} soil).

Sully et al. (1988) showed that the pore pressures measured at different locations (cone tip and cone base) can be related to the change in stress history. They related the OCR to a normalized pore pressure difference, PPD, defined as follows:

$$PPD = (u_1 - u_2)/u_o \tag{13}$$

The following relation between PPD and OCR was suggested:

$$OCR = 0.49 + 1.5\ PPD \tag{14}$$

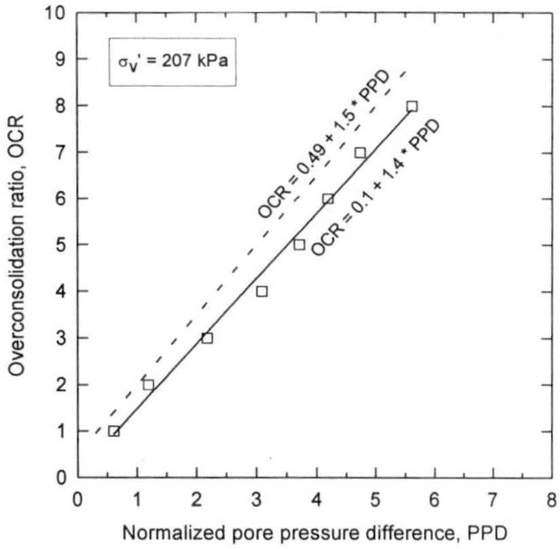

Figure 6. Variation between PPD and OCR.

The variation between the PPD and OCR is presented in Figure 6. Based on simple regression analysis, the following modified relation is obtained:

$$OCR = 0.1 + 1.4\ PPD \tag{15}$$

It should be realized that the above expression needs to be verified for different soil types with different stress histories before being adopted.

The predicted dissipation of excess pore pressure with time around the cone tip and base for specimens 1 is compared with the measured dissipation test as shown in Figure 7. The predicted and measured dissipation curves show good agreement. At early stages of dissipation, the excess pore pressure near the cone base tends to increase slightly before dissipation starts. This phenomenon is mainly due to the redistribution of pore pressures around the piezocone, immediately after penetration stops, resulting from the presence of high pore pressure gradient around the cone base.

The effect of the variation in the horizontal to vertical hydraulic conductivities (k_x, k_y) on the shape of the dissipation curves of excess pore pressures at the tip and behind the base is investigated and presented in Figure 8. The increase in k_x or k_y tends to accelerate the dissipation of excess pore pressures around the piezocone as expected. However, the effect of k_x is higher than k_y and the dissipation of excess pore pressure is more sensitive to k_x value. Consequently, the shape of the dissipation curve is mainly dominated by the value of the horizontal hydraulic conductivity. In all cases, the excess pore pressure at the tip dissipates at a faster rate than behind the base, especially at the early stages of dissipation due to the same reason mentioned earlier.

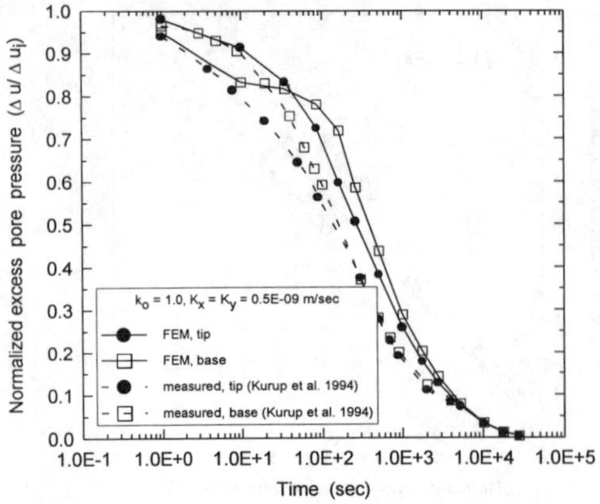

Figure 7. Dissipation of excess pore pressure.

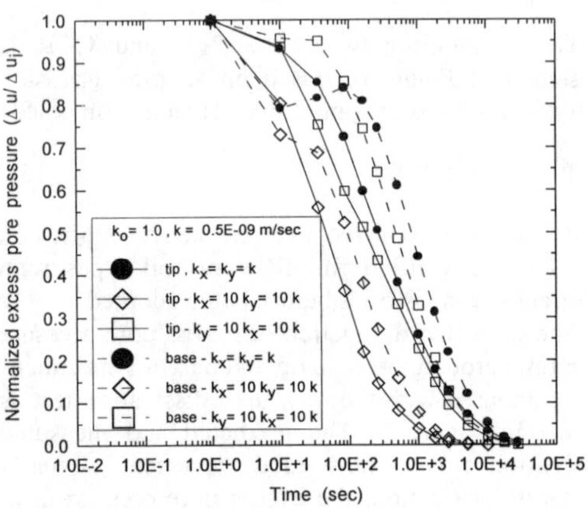

Figure 8. Effect of hydraulic conductivity on the dissipation of excess pore pressure.

5 CONCLUSIONS

1. The predicted values using the proposed model are in good agreement with the measured values.
2. The cone tip resistance and excess pore pressures are linearly proportional to the lateral stresses.
3. The excess pore pressure difference is linearly proportional to the OCR. Regression analysis was carried out, and a correlation equation is presented to predict the OCR.
4. Due to pore pressure redistribution, the excess pore pressure near the base tends to increase slightly before dissipation starts.
5. Excess pore pressure at the tip usually dissipates at a faster rate than behind the base
6. The shape of the dissipation curve is dominated by the horizontal hydraulic conductivity.

ACKNOWLEDGEMENT

The research is supported by the National Science Foundation under grant CMS-9907951. The authors also acknowledge the support and help provided by the Louisiana Transportation Research Center.

REFERENCES

Abu-Farsakh, M. Y., Voyiadjis, G. Z. & Tumay, M. T. 1998. Numerical Analysis of the Miniature Piezocone Penetration Tests (PCPT) in Cohesive Soils. *Int. J. Numer. Analyt. Meth. Geomech*., Vol. 22, pp. 791-818.

Abu-Farsakh, M. Y. 1997. Coupled Field Equations for Saturated Soils and Its Application to Piezocone Penetration and Shield Tunneling. *Dissertation*, Dept. of Civil and Envir. Engrg, Louisiana State University, LA, 200 p.

Acar, Y. B. & Tumay, M. T. 1986. Strain Field around Cones in Steady Penetration. . *J. of Geotech. Engrg*., . *J. of Geotech. Engrg*., ASCE 120, pp. 81-107. Vol. 112, No. GT2, pp. 207-213.

Kiousis, P. D., Voyiadjis, G. Z. & Tumay, M. T. 1998. A Large Strain Theory and Its Application in the Analysis of the Cone Penetration Mechanism. *Int. J. Numer. Analyt. Meth. Geomech*., Vol. 12, pp. 45-60.

Kurup, P. U., Voyiadjis, G. Z. & Tumay, M. T. 1994. Calibration Chamber Studies of Piezocone Test in Cohesive Soils. *J. of Geotech. Engrg*., ASCE 120, pp. 81-107.

Levadoux, J. N. & Baligh, M. M. 1986. Consolidation after Piezocone Penetration. *J. of Geotech. Engrg*., ASCE 112, pp. 707-726.

Lim, B. 1999. Determination of Consolidation Characteristics in Fine Soils Evaluated by Piezocone Tests. *Dissertation*, Dept. of Civil and Envir. Engrg, Louisiana State University, LA, 243 p.

Meyerhof, G. G. 1961. The Ultimate Bearing Capacity of Wedge-shaped Foundation. *Proc. 5th Int. Conf. on Soil Mechanics and Foundations*, Vol. 2, pp. 103-109.

Prevost, J. H. 1980. Mechanics of Continuous Porous Media. *Int. J. of Engrg. Science*, Vol. 18, pp. 787-800.

Sully, J. P., Campanella, R. G. & Robertson, P. K. 1988. Interpretation of Penetration Pore Pressures to Evaluate Stress History in Clays. *Proc. 1st Int. Symp. on Penetration Testing (ISOPT-1),* Vol. 2, Orlando, Florida, pp. 993-999.

Teh, C. I. & Houlsby, G. T. 1991. An Analytical Study of the Cone Penetration Test in Clay. *Geotechnique* 41, No. 1, pp. 17-34.

Torstensson, B. A. 1977. The Pore Pressure Probe. Paper 34, *Geotechnical Meeting*, Norwegian Geotechnical Society, Oslo, Norway, pp. 34.1-34.15.

Tumay, M. T. & de Lima, D.C., 1992. Calibration and Implementation of Miniature Electric Cone Penetrometer and Development, Fabrication and Verification of the LSU In-situ Testing Calibration Chamber (LSU/CALCHAS). *LTRC/FHWA* Report No. GE-92/08, 240 p.

Voyiadjis, G. Z. & Abu-Farsakh, M. Y., 1997. Coupled theory of mixtures for clayey soils. *Computer and Geotechnics*, Vol. 20, No. 3/4, pp. 195-222.

Computer Methods and Advances in Geomechanics, Desai et al. (eds) © 2001 Balkema, Rotterdam, ISBN 90 5809 183 X

Numerical optimization search techniques for the determination of creep parameters of salt rocks

A.J.Campos de Orellana
El Escorial, Madrid, Spain

ABSTRACT: Numerical optimization search techniques were applied to field data for the determination of creep parameters of salt rocks.The excellence and advantages of this method was proved when this method was applied to the determination of creep parameters of the seven saline horizontal layers existing in an underground potash operation (Iberpotash), sited nearby the town of Sallent, Barcelona, Spain (Catalonia Potash Basin), where 276 field creep curves were obtained during a 14 yr instrumentation campaign, installed in 32 underground control stations.

1 INTRODUCTION

In the early 80´s D.E.Munson & P.R. Dawson (1981) presented for the first time constitutive equations to model the flow of salt rocks within the framework of deformation mechanism maps, based on theoretical thermodynamic studies, microstructural work and laboratory creep tests.

During the following years, creep deformation mechanisms have been a matter of tremendous interest, considerable controversy and debate. Mechanisms such as cross-slip(Wawersik & Zeuch1986)(Skrotzki & Haasen 1988) , dislocation climb (Carter& Hansen 1983), and pressure solution (Spiers et al.1988),have been recently widely accepted(Spiers & Carter 1998) as those models controlling , depending on steady-state strain rate, state of stress, temperature, water content and grain size, the creep behavior of salt in mining and industrial applications (underground storage of oil and natural gas,nuclear waste repositories, oil wells, etc);that is under equivalent stresses (σ_e), temperatures and strain rates,below 25-30 MPa, 200° C and 10^{-7} sc^{-1}, respectively.

In these three creep mechanisms, the constitutive equations or creep laws, relating creep strain (ε) with stress (σ_e) and time (t), are of the power type:

$$\varepsilon = K \ \sigma_e^m \ t^n \qquad (1.1)$$

where **K** (creep constant), **m** (stress-exponent) and **n** (time-exponent) are specific for each particular saline material and functions of the stress,temperature and strain rate prevailing in the material.

In the determination of creep parameters by laboratory creep tests, equivalent stresses are easily controlled but, on the other hand, tremendous difficulties are encountered in controlling strain rates below 10^{-9}-10^{-10} sc^{-1}, which are the rates observed in underground openings and cavities. This makes in- situ creep tests preferable for the determination of creep laws , in spite of the disadvantage of not knowing, beforehand, the acting equivalent stresses.

This disadvantage of the field creep curves can be overcome by means of the application of numerical optimization search techniques and regression analysis (Beveridge & Schechter 1974) to field creep data

This work deals with the development of the analytical procedures of this technique and its validation by numerical simulations of underground openings in a potash mining operation.

2 ANALYTICAL PROCEDURE

If **i** creep curves obtained from multiple point borehole extensometers (MPBX) are available for a particular salt rock, an average value of **n** (time-exponent) and its corresponding standard deviation can be easily defined by means of regression analysis applied to log-log plots of strain-time data.

During the regression analysis, **i** values of ($K\sigma_e^m$) are also obtained.

The quotient of any two values,out of of the **i** values of ($K\sigma_e^m$)available and derived in the regression analysis of the **i** creep curves,allows the definition of stress exponent (m)values, providing that the corresponding equivalent stresses (σ_e) for each creep curve, are known or assumed.

However, any equivalent stress (σ_e) existing in a saltrock-mass ranges from a minimum of zero up to a maximum one limited by the Octahedral Shear Strength (K_o) of that material.

Octahedral Shear Strengths (K_o) intervene in the definition of the double Drucker-Prager/Von Mises Plasticity Criteria of salt rocks (Serata 1968,1992)(Horseman 1992), (Dreyer 1974), (Russell 1979), Russell et al. 1988), (Campos de Orellana 1996, 1998), (Cristescu 1998), with values rarely exceeding 15 MPa. Equivalent (σ_e) and Octahedral Shear Stresses(τ_o) are related by the expression:

$$\sigma_e = 3 / 2^{1/2}\ K_o \qquad (2.1)$$

By means of numerical optimization search techniques, **j** values of equivalent stresses (σ_e), within the established range of possible values, are chosen with any desired stress-interval ($\Delta\sigma_e$) (4, 2, 1,...MPa).

Creep curves with a correlation coefficient lower than 0.9, sholud be disregarded as nonrepresentative curves. Reading mistakes, extensometer malfunctions, defective installations and damaged heads, may lead to excessively low correlation coefficients in the regression analysis, below acceptable values.

For each set of equivalent stresses (σ_e) assigned to each of the **i** values of ($K\sigma_e^m$) of a particular saline material, C_i^2 stress-exponents (m) are calculated and therefore an average mean value for this parameter (**m**) and its standard deviation can be derived.

Equivalent stresses (σ_e) are modified, within the established range and intervals, in a repetitive computting process, until all predetermined equivalent stresses (σ_e) have been assigned to each set of ($K\sigma_e^m$) values that is, to each of the **i** creep curves of the material under study.

In this computing process **j**^{i} mean values (**m**) of the stress-exponent with their corresponding standard deviations,are obtained.The mean stress-exponent(**m**) out of the **j**^{i} computed values that shows the lowest standard deviation, is considered the stress-exponent (m) that better represents the real value of that parameter for the particular saltrock under study,in other words, the value that best fits the field creep data.

This so-defined stress-exponent (m) is bound to a set of fixed and settled values of equivalent stresses for each of the **i** creep curves or ($K\sigma_e^m$) values. This makes possible to determine the creep constant (K) and its standard deviation.

We can express these concepts mathematically by taking the standard deviation (s) of the stress-exponent (m),as the objective function to be optimized

$$s = \left(\Sigma(m_k - m_l)^2 / k \right)^{1/2} \qquad (2.2)$$

being:

$$m_k = \log A_k / \log(\sigma_{ex}/\sigma_{ey}) \qquad (2.3)$$

$$m_l = \Sigma\, m_k / k \qquad (2.4)$$

$$A_k = (K\sigma_{ex}^{mk}) / (K\sigma_{ey}^{mk}) = (\sigma_{ex}/\sigma_{ey})^{mk} \qquad (2.5)$$

$$\sigma_{ex}, \sigma_{ey} = 0, \Delta\sigma_e, 2\Delta\sigma_e, 3\Delta\sigma_e, ...3K_o/2^{1/2} \qquad (2.6)$$

k ranges from 1 up to C_2^i

l ranges from 1 up to j^i

x and y range from 1 up to i

C_2^i, denotes the number of ways in which 2 values of ($K\sigma_e^m$) can be chosen from the available **i** values of ($K\sigma_e^m$), irrespective of the order in which they are arranged.

The standard deviation function (s) is therefore, function of the equivalent stresses (σ_e) acting in the **i** creep curves, which has to be optimized (minimized) in the i-dimensional equivalent stress space.

Once this first step or level of optimization is completed, subsequent steps of optimization can be carried out starting with the mean stress-exponent (**m**) and with the stress values optimized in the previous step, until the mean value of the stress-exponent (**m**) is obtained with the desired accuracy

Obviously the maximum achievable value for this accuracy depends on the reliability of the creep curves, i.e. on the correlation coefficients derived in the regression analysis and on the number of available creep curves (**i**).

3 RESULTS AND VALIDATION

The above described analytical procedure was applied to the seven saline horizontal layers (halite, sylvinite, carnallite and rocksalt) of the Saline Formation (Catalonia Potash Basin) existing in an underground mine operated by Iberpotash and sited in the township of Sallent, Barcelona, Spain.

Details of the geology and elasto-plastic parameters of these seven fine-grained layers (1-2 mm.), are reported by A.J.Campos de Orellana (1996,1998).

Octahedral Shear Strengths (K_o) for the seven beds range from 14 MPa to 21 MPa, and the available number of creep curves (**i**) with correlation coefficients greater than 0.9, between 1 and 29 depending on the saline material.

Application of regression analysis and numerical optimization search techniques led to creep parameters for the seven salt rocks,in the range 0.36-0.90 for the time-exponents (n),1.03-1.71 for the stress-exponents (m) and 5.473426×10^{-4}- 1.024864×10^{-2} for the creep constants (K), measuring strains (ε) in mm/m,equivavalent stresses (σ_e) in MPa and time (t) in days.

These creep parameters were utilized in a numerical finite element simulation of 4 control stations, where roof sags, floor heaves and horizontal convergences were monitored for periods ranging from 102 to 250 days.

Control stations were set up in four galleries with different cross sections, cut in distinct geological configurations and located, one to each other, at horizontal distances of 1,000-1,500 meters apart.

Creep parameters were modified until differences between calculated and measured roof sags, floor heaves and horizontal convergences in the steady-state creep were of the order of 10^{-3} mm (read.unit.sensitiv. 10^{-2} mm).During the first 5 days after the gallery was opened,this difference reached an average value of 15 % and from the fifth day to the transition time (change from temporary to steady-state creep), 7%.

Modifications introduced during the numerical simulations, in the creep parameters of the seven analyzed salt rocks,with respect to the values determined by optimization techniques, were in the following ranges:

Creep Parameter (K): 0.59% / 0.90%.

Stress-Exponent (m): 4.63% / 7.45%.

Time-Exponent (n): 0.00%.

except for the Lower Salt layer (rocksalt),where these modifications reached up to 62.84%, due to the fact that only 1 curve was available for this material. A minimim of 3 curves are required for this technique. Therefore, creep parameters for the Lower Salt layer were estimated, not derived by optimization techniques, which explains the high modifications introduced in the creep parameters of this material during the numerical simulation.

The excellent approximation and good fit between initial and final creep parameter values,that is before and after the numerical match simulation,support the validity and accuracy of numerical optimization techniques in the determination of creep parameters of salt rocks and the outstandingefficiency obtained by this method in terms of reliability and precision.

The utilization of laboratory creep tests, as an alternative method to the determination of creep parameters of salt rocks, introduces serious shortcomings due to the already mentioned technical limitations of conventional testing devices, at strain rates below 10^{-10} sc^{-1} and to the nonrepresentative values of creep parameters obtained in laboratory tests compared to those of the rock-mass in-situ (effects of sample sizes, sample disturbances, etc.)

4 CONCLUSIONS

Application of regression analysis and numerical optimization search techniques to in-situ creep tests, for the determination of creep parameters of salt rocks, has been proved to be of a tremendous assistance, precision and accuracy, when field creep curves are obtained from multiple point borehole extensometers (MPBX).

A minimum of 3 field creep curves with a recommended correlation coefficient greater than 0.9, are needed for the application of this technique.

This implies reliable extensometer read- out equipments, careful readings and good installations.

This technique allows to analyze in deep and to obtain reliable creep parameters of salt rocks, without the need of troublesome, doubtful and misleading laboratory tests, making possible to forecast and to predict long-term behavior of underground salt rock openings with confidence.

REFERENCES

Beveridge, G.S.G. & Schechter, R.S. 1974. Optimization: theory and practice.New York: Mc Graw-Hill.

Campos de Orellana, A.J.1996.Pressure solution creep and non-associated plasticity in the mechanical behavior of potash mine openings. Int.Jr. Rock.Mech. & Min. Scie.; vol. 33, 347-370.

Campos de Orellana, A..J. 1998. Non-associated pressure solution creep in salt rock mines. 4th .Int. Conf. On The Mechanical Behavior of Salt, Montreal, Jun. 1996.Clausthal-Zellerfeld:Transtech. 429-444.

Carter, N.L. & Hansen, F.D.1983. Creep of rocksalt. Tectonophysics, vol. 92, 275-333.

Cristescu, J.J. & Hunsche,U.1998.A new elastic/viscoplastic model for rock salt. 4th. Int. Conf. On The Mechanical Behavior of Salt, Montreal, Jun.1996. Clausthal-Zellerfeld: Transtech. 249-262.

Dreyer, W.1974.Results of recent studies on the stability of crude oil storage in salt cavities. 4th.Salt Symp. Cleveland: The Northern Ohio Geological Society, vol.2, 65-92.

Horseman, S.T,Russell, J.E., Handin, J.& Carter, N.L. 1992. Slow experimental deformation of Avery Island salt. 7th Salt Symp.,Tokio.Amsterdam: Elsevier. 67-74.

Munson, D.E.& Dawson, P.R.1981. Salt constitutive modeling using mechanism maps.1st Int. Conf. on The Mechanical Behavior of Salt, Pennstate Univ. Nov.1981.Clausthal-Zellerfeld:Transtech.717-737

Russell, .E.1979.A creep model for salt.5th.Salt Symp. Hamburg. Cleveland: The Northern Ohio Gelogical Society. Vol. 2, 65-92.

Russell, J.E., Handin, J.& Carter, N.L.1988. Modified mechanical equations of state of salt. 2nd. Int. Conf. on The Mechanical Behavior of Salt, Hannover, Sep. 1984. Clausthal-Zellerfeld: Transtech. 409-430.

Serata, S. 1968. Application of continuum mechanics to design of deep potash mines in Canada. Int. Jr. Rock. Mech. & Min. Scie., vol.5, 293-314.

Serata, S & Fuenkajorn, K.1992.Formulation of constitutive equations for salt. 7th Salt Symp.,Tokio. Amsterdam: Elsevier. 483-488.

Skrotzki, W. & Haasen, P.1988.The role of cross-slip in the steady-state of creep salt.2nd. Int.Conf. on The Mechanical Behavior of Salt, Hannover, Sep. 1984. Clausthal-Zellerfeld: Transtech. 69-81.

Spiers, C.J., Urai, J.L .& Lister, G.S.1988. The effect of brine (inherent or added) on rheology and deformation mechanisms in salt rocks.2nd.Int.Conf. on The Mechanical Behavior of Salt, Hannover, Sep.1984.Clausthal-Zellerfeld:Transtech. 89-102.

Spiers, C.J.& Carter, N.L.1998.Microphysics of rocksalt flow in nature.4th Int.Conf.on The Mechanical Behavior of Salt, Montreal, Jun.1996 .Clausthal-Zellerfeld: Transtech. 115-128.

Wawersik, W.R., & Zeuch, D.H. 1986. Modelling and mechanistics interpretation of creep of rock salt below 200° C. Tectonophysics, vol. 121. 125-152.

Computer Methods and Advances in Geomechanics, Desai et al. (eds) © 2001 Balkema, Rotterdam, ISBN 90 5809 183 X

In-situ rock stress measurement by triaxial and hydrofracture test

J.Chaubey
Nathpa-Jhakri Power Corporation Limited, Shimla, India

ABSTRACT: Underground rock masses are subjected to compressive stress which increases, in general with depth. The rate of increase, however, varies depending upon various factors. The natural state of stress, original and undisturbed is known as " in situ stress ." In order to asses the in situ rock stress condition Triaxial rock stress measurements at proposed Power house site involving strain gauges and overcoring methods were carried out using " Hiltscher Rock Stress Measuring system".

During construction, rock stress measurement with Hydrofracturing was also carried out. The MeSy manufactured straddle assembly " Perfrac " with nylon reinforced packer element was used for fracture initiation and extension. Similarly, stress conditions around the excavated caverns were also estimated using Finite Element Calculations. The present paper deals in detail about the Triaxial Test and the Hydrofracture Test which were carried out for Uri Hydroelectric project. The paper also discusses, though in short, the Finite Element calculations.

1.0 INTRODUCTION

Underground rock masses are subjected to compressive stresses which increases in general with depth. The stresses can not be measured directly and therefore rock stress determination techniques rely on the measurement of displacement strain, deformation, that is induced by a disturbance of the rockmass.Tincelin, Obert et.al., Haimson & Fairhurst and Leeman have made significant effort in developing methods of stress measurement.

1.1 *About the project:*

The 480 MW Uri Hydroelectric project on river Jhelum in Jammu & Kashmir involved a 20m. high barrage, a 1.2 km. long Power channel, a 10.7 km. long Head Race Tunnel, two vertical pressure shafts and an underground Power house having four generating units with a 2.2 km. long tail race tunnel taking the water back to river Jhelum.

1.2 *Geology of the project:*

The area shows a rugged topography with lofty mountains of the Himalayas. The elevations vary from 1200m. to 3400m. The Barrage and the Power channel lie in Fluvio–glacial deposits consisting of boulders and cobbles in sand silt and clay. The u/s half of the Head race tunnel lie in Schist whereas the d/s end passes through Volcanic which have been completely chloritised and posses well developed schistosity. The Tail race tunnel traverses partly through Volcanic and partly through variegated Shale and Limestone.

Structurally, all the rock units in the area are intensively folded and faulted associated with the Himalayan Uplift. The area is seismically highly active and several high magnitude earthquakes have been recorded. Two major folds one anticlinal and one synclinal have been recorded in the area. Two conspicuous Himalayan lineaments namely Panjal thrust and the Murree thrust have been identified apart from many faults.

2.0 THE TRIAXIAL TEST

2.1 *General:*

This method is intended to determine the stress tensor in rock by measuring the strains that occur in the walls of a drill hole when the stresses are relieved by overcoring. The method is restricted to rocks whose behaviour is not significantly different from that of a homogeneous perfectly clastic material. The state of stress and pre-existing fracturing of the rock at the measurement locations should be such that it is possible to obtain relatively long piece of unbroken core. Because of the difficulties in obtaining satisfactory strain gauge bonding when the rock is wet, this method is usually impractical for holes in water saturated ground drilled downward.

2.2 *Equipment used :*

The in situ rock stress measurement involving strain gauges and overcoring method was carried out by using " Hiltscher Rock stress Measuring System." The assembly consists of a tube shaped probe, in the lower end of which the carrier strain gauge rosette is placed, which essentially consists of three plastic tongues 120^0 apart. The rosette carries three strain gauges. The rosette is submerged in a glue pot. A central cone activated by a fall weight in the tube of the probe, presses the carrier tongues with the gauges at a suitable moment against the wallof the bore. The vertical probe lays in a combined carrying and measuring cable. The strain gauges submerged in glue at the bottom end of the probe and when the pin point touches the bottom of the larger hole, the mechanism is released and the cone is pushed downwards. To establish the orientation of the probe, a magnetic compass is pushed with the probe. The glue pot sinks to the bottom of the small bore hole nd the strain gauges are pressed to the bore hole wall by the cone. Measurements are made from the ground surface and the switching between the strain gauge is made by the electronic unit inside

2.3 *Methodology:*

The test was carried out in the d/s end of a cross cut in an exploratory drift to the underground Power house cavern. The " Hiltscher Method " of Triaxial stress measurement involves basically drilling of a hole by using overcoring technique. In this method the stress is measured in vertical bore hole. At required test depth the core is broken by pulling only. If the angle of break with the horizontal is greater than 45^0, drilling is continued till the angle of break is less than 45^0. Acrylic resins (glue) are prepared and the strain gauge rosette is diiped into it and the probe is lowered in the hole. The strain gauges are fixed at a required depth in 36mm. hole. Initial reading before cementing is ensured in 9 directions. After a lapse of 2hrs. the reading of strain gauges is measured for about 1hr. with an interval of 15min. Then the probe carrier is hoisted and overcoring with 76mm. bit is carried out.

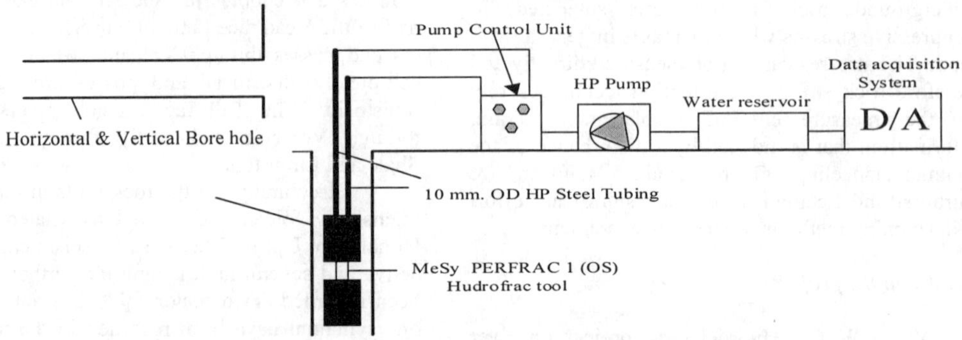

Schematic diagram showing the hydrofrac experimental set up

2.4 *Results :*

Calculated Principal Stresses (in MPa) with their direction

Point	Sigma 1	Beta	Bearing	Sigma 2	Beta	Bearing	Sigma 3	Beta	Bearing
1	12.9	15.1	246.9	5.9	29.6	345.7	5.5	56.0	133.2
1	13.8	38.0	313.4	2.0	15.4	210.9	-4.4	47.9	103.2
Mean	11.1	37.2	291.6	6.1	2.7	23.6	0.8	52.7	117.1

Beta- slope angle between the principal stress & the horizontal plane
Bearing- angle between principal stress in the horizontal plane & North.

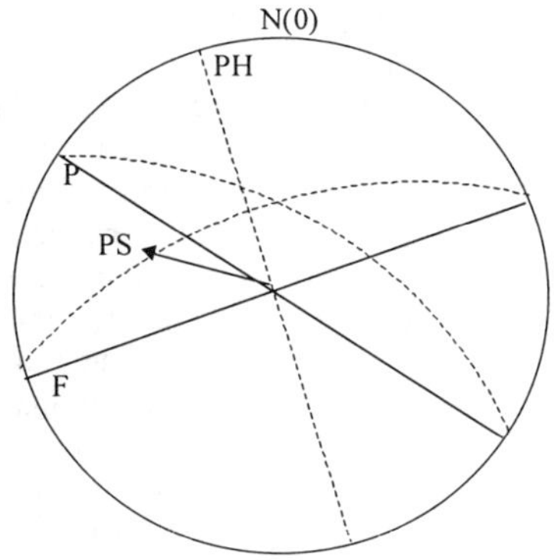

Correlation of Orincipal Stresses with attitude of foliation & trend of Panjal Thrust

The in situ rock stress measurement conducted shows that the horizontal stress components are greater than the vertical components. The greatest horizontal component along direction 103/293 is subnormal to the center line of the Power house cavern thereby indicating that the side walls of the Power house are likely to experience greater stresses than the roof . After correlating this stress field with respect to the contiguous lineaments and the topography, it was realised that the trend of the maximum horizontal stress is sub parallel to the strike of the Panjal thrust and is evidently also sub parallel to the ridge in which the underground Power house is situated.

3.0 THE HYDROFRACTURE TEST

3.1 *General :*

In situ rock stress measurement by Hydrofracture was conducted at Adit 4 at the crossing with HRT. All the test showed only reopening and shut in phases due to the presence of pre existing fracture in Panjal volcanics. Because of the presence of anisotropism, a more sophisticated interpretation method which uses the observed shut in pressure value was applied. The influence of topography and the excavation at the test site was checked with numerical modeling with 2D – UDEC code (Itasca) using measured stress K= S_h /_S_H as input data. A series of 25 tests were conducted in one vertical and three horizontal holes drilled from a cross cut at Adit 4. The bore holes had a depth of 15m. with a dia. of 48mm.

3.2 *Set up and Test Procedure :*

The MeSy manufactured straddle packer assembly PERFRAC 1(OS) with nylon reinforced packer element was used for fracture initiation and extension. Packer and internal pressure (UNIMESS pressure transducers UPA-1, 20 MPa and HBM pressure transducers P 9 AV, 100 MPa) as well as the flow rate (UNIMESS flow meter QPT 04, 0-10 lts/min) were measured uphole. After positioning the assembly at the test section the packer were inflated to a pressure of 10 MPa. The testing was started with a permeability test, for which the interval was instantaneously pressurised to a pressure level of 2 MPa and the pressure decrease in the closed system was observed. In tight sections, the injection pressure was then increased until a hydraulic fracture was

Greatest and Smallest Stresses in horizontal plane , Vertical Stresses and Error estimation

Point	Sigma A	Sigma B	Bearing Sigma A	Sigma Z	Error
1	12.4	5.8	66.5	6.1	72265.4
2	7.5	1.0	144.8	3.0	7171.9
Mean	7.3	6.0	103.4	4.5	

initiated or a pre existing fracture was opened up. Between the injection cycles the system was vented. Finally, the packer were deflated and the tool was moved for the next test.

3.3 *Stress Evaluation:*

The in situ stress measurement was conducted with following test situations :
i) pronounced topography.
ii) proximity of large excavations.
iii) presence of highly fractured Panjal volcanic.

Due to the above aspects, the classical hydrofracture hypothesis suggested by Hubert & Will (1957) could not be applied. Therefore, the data analysis was made
with the method , namely, the interpretation of measured normal stress acting across arbitrary oriented fracture planes in which the Shut in pressure Psi is used to measure the normal stress component under the assumption that the vertical is the principal stress and the vertical stress S_v is equal to the height of the overburden.
The analysis program GENSIM was used to calculate the magnitude and the direction of principal stress on the basis of the following equation:

$$S_h = (Psi - n^2 * S_v) / (m^2 + l^2 * (S_H / S_h))$$

Where l,m,n are the cosines of the direction of the induced fracture plane related to the principal stress axis.
i) Stress in vertical direction S_v is gravitational with a gradient of 0.0285 MPa/m. along the depth and the horizontal stress S_h is such that K_h (S_h / S_v) is 0.259.
ii) The out of plane horizontal stress (S_h) is such that K_H (S_H / S_v) is 1.04

3.4 *Results :*

The GENSIM program which were used for the calculation of stress tensor around the test cross cut, is based on the interpretation of measured normal stress (shut in pressure) acting across the arbitrary oriented fracture planes. The determination of shut in pressure is straight when a sharp break is observed in the pressure time curve after the fast pressure decline. Direction of max. horizontal principal stress = 130^0
Vertical stress S_v = 8.84 MPa
The stress tensor around the test tunnel with an overburden of 310m. can be given as follows :

S_h = 2.29 $\pm$ 0.0537 MPa (min. horizontal principal stress)

S_H = 9.16 + 0.2148 MPa (max. horizontal principal stress)

The vertical stress corresponds to an overburden of 310m. and density of rock being 2.91gm /cm3. The direction of max horizontal compression $S_H = 130^0$, is not in agreement with the focal mechanism solutions. The influence of topography and excavations were investigated by 2D – UDEC code (Itasca) in a plain strain mode. The result indicate that the stress at the measurement site is not influenced by either topography or excavation. The variation in the direction of maximum horizontal stress from the regional direction can be attributed to the effect of some local structure.

4.0 FINITE ELEMENT CALCULATIONS :

4.1 *Introduction :*

The scope of calculation was to estimate the stress coordination around the excavated caverns for

Comparision of Numerical results with measured results

Site	$K_{(m)}$	XX MPa	YY MPa	ZZ MPa	S_h MPa	S_v MPa	$K_{(n)}$
A	0.259	3.123	10.03	9.605	3.107	10.05	0.309
B	0.259	2.262	7.094	9.032	1.748	7.608	0.230

$K_{(m)}$ = measured ratio between S_h/S_v $K_{(n)}$ = ratio between S_h/S_v by numerical analysis

The Final stress tensor calculated by GENSIM

Stress	In-situ stress in MPa		
	Upper bound	**Middle bound**	**Lower bound**
Min. horizontal stress S_h	2.2912+0.0537	2.0040+0.1300	1.0038+0.7491
Max. Horizontal stress S_H	9.1468+0.2148	8.0660+0.5200	4.0152+2.9964
Direction of S_H	130^0	130^0	130^0

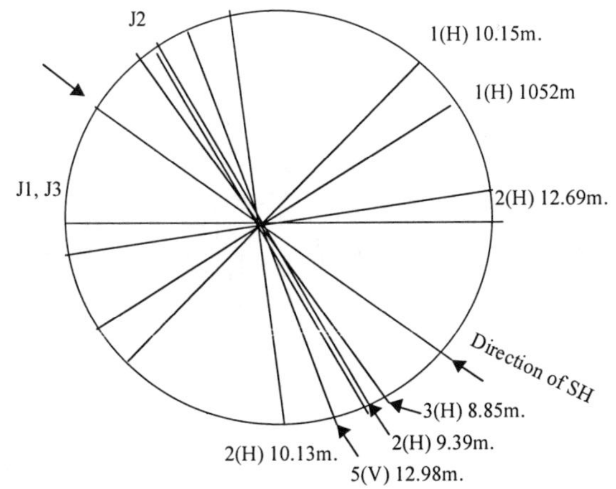

diagram showing orientations of the induced fractures vis - a -vis prefrential orientationsof all joint sets and orientation of maximum horizontal stress

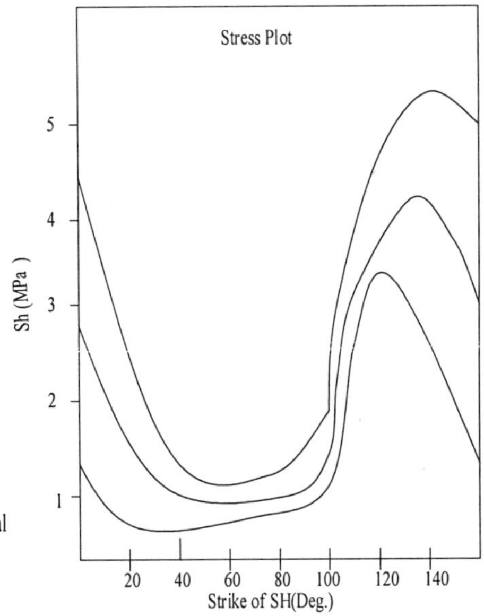

Plot showing minimum standard deviation for Sh at 130 strike direction of SH

the Power House and Transformer hall. The calculations were very preliminary and the aim was only to give a rough data of the stability of caverns with special attention being paid to the pillar stability.

Only elastic condition were assumed and the effect of the rock support was disregarded. The excavation was simulated in two steps, first, the excavation of Power House and secondly the excavation of Transformer Hall. For the calculation, a Young's modulus of 15 Gpa and a Poisson's ratio of 0.2 were assumed to be valid for the rock mass. The initial stress field was set at 15 MPa parallel to the slope and 5 MPa perpendicular. The slope inclination was assumed to be 45^0.

4.2 *Results.*

The degree of mobilisation of a certain shear strength of the rock mass for the calculated stresses was calculated in order to evaluate potential failure zones. Two different values of the rock mass strength corresponding to an unconfined compressive strength of 5 and 10 MPa was checked. The calculation showed that only minor zones of tensile stress will occur close to the caverns. The pillar is heavily loaded and a plastic zone is expected to go through the whole pillar even for the higher value of the rock mass strength. The pillar is therefore expected to require a heavy rock support with pre stressed anchors.

For normal rock conditions corresponding to a rock mass strength of 10 MPa , moderate rock support seems to be required which applies to the rest of the cavern. If poorer rock conditions are encountered for the cavern both heavy rock support of the roof and outer walls seems to be required.

REFERENCES

1. Geotechnical report on Uri- Hydroelectric Project- Unpublished report
2. In situ rock stress measurement by Hydrofracture test – Unpublished report.
3. In situ Triaxial rock stress measurement at Power House site- Unpublished report.
4. Vutukuri & Lama - Mechanical Properties of Rocks, Vol. IV

Computer Methods and Advances in Geomechanics, Desai et al. (eds) © 2001 Balkema, Rotterdam, ISBN 90 5809 183 X

Back analysis on mechanical behavior of joints, using a new method based on virtual work principle

Desheng Deng & Duc Nguyen Minh
Laboratoire de Mécanique des Solides, Ecole Polytechnique, Palaiseau, France

ABSTRACT: A nonlinear back analysis method is proposed for the characterization of the elastoplastic parameters of joints. Based on the virtual work principle, the proposed method is robust and works well for nonlinear problems. In this paper, the elastic stiffness, cohesion, and the angle of friction of joint are identified according to measured displacements. A numerical example and an application to a gravity dam foundation are illustrated.

1 INTRODUCTION

In engineering practice, joints are often encountered in underground excavation, foundation problem, and in huge concrete structures. In certain cases displacements along joints may take an important part in the deformation of the structure (Deng & Xia 1992; Constantinescu et al. 1998; Akutagawa & Sakurai 1997). However, it is difficult to estimate the deformational characteristics of joints, even if the best is made during the geological investigations. The elastoplastic parameters obtained by laboratory tests or in situ tests often deviate from the real ones. In order to overcome this shortcoming, back analysis is employed. Using displacement measurements, the uncertain deformation behavior of joints can be back calculated. In previous works (Deng et al. 1998; Deng et al. 1999; Deng & Nguyen-Minh 2000), a new method for back analysis has been proposed on the basis of the virtual work principle, for identifying load parameters in tunnel analysis and elastic parameters of layers in mining. This paper proposes a new form of back analysis designed to identify nonlinear characteristics of joints. This method is achieved by comparing the original structure model through an iterative process to a modified model in which the field measurements are prescribed. A numerical example and an application to Água Vermelha gravity dam foundation are presented to illustrate the nature of the proposed method. This back analysis method adopts the same numerical model used for stress analysis, within the framework of a simple iterative procedure. The results indicate that the elastoplastic parameters of joints can readily be identified with proper measurements.

2 MATHEMATICAL FORMULATION

2.1 *Original model*

A joint Γ_j separates the domain Ω into two subdomains Ω_1 and Ω_2 (figure 1), $\Omega = \Omega_1 \cup \Omega_2 \cup \Gamma_j$. During a certain time interval $[0,\, t_0]$, the initial state

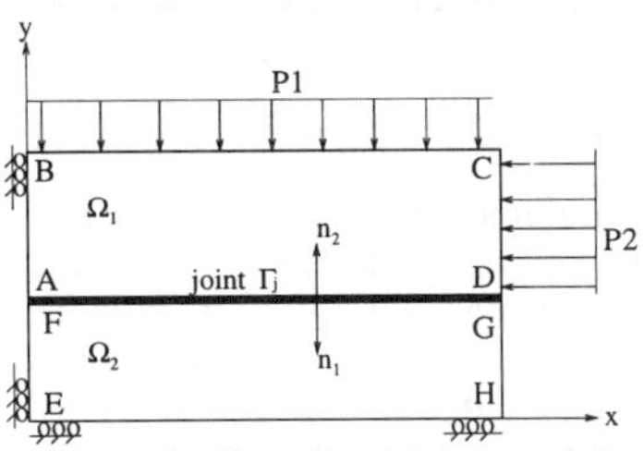

Figure 1: *Numerical joint model*

and the history of displacement limit condition (on $\Gamma_u \times [0,\, t_0]$), and the surface force $\mathbf{T}$ (on $\Gamma_T \times [0,\, t_0]$) are known. At time $t = 0$, the initial state in the domain is:

$$\begin{cases} \mathbf{u}(x,\,0) = 0 & x \in \Omega \\ \boldsymbol{\sigma}(x,\,0) = \boldsymbol{\sigma}_0(x) & x \in \Omega \\ \mathbf{T}(x,\,0) = \mathbf{T}_0(x) & x \in \Gamma_T \end{cases} \tag{1}$$

$\mathbf{u}$ and $\boldsymbol{\sigma}$ are displacement and stress respectively.

In this time interval, the displacement field has to be kinematically admissible with the limit condition of displacement on the border Γ_u, and the stress field has to be statically admissible with the limit condition of stress on the border Γ_T.

Let us consider a virtual displacement field $\hat{\mathbf{u}}(x)$ and the corresponding deformation field $\hat{\boldsymbol{\epsilon}}(x)$, which is continuous and continuously differentiable in Ω_1 and Ω_2. Supposing the stress field $\boldsymbol{\sigma}$ is continuous on Γ_j,

$$\begin{cases} \mathbf{T}_{\Omega_1}(x,t) = \boldsymbol{\sigma}(x,t)\cdot\mathbf{n}_1(x) & x\in\Gamma_j \\ \mathbf{T}_{\Omega_2}(x,t) = \boldsymbol{\sigma}(x,t)\cdot\mathbf{n}_2(x) & x\in\Gamma_j \end{cases} \quad (2)$$

and considering $\mathbf{n}_1(x) = -\mathbf{n}_2(x)$, the virtual work principle gives the following formula (Salençon 1994):

$$\int_{\Omega'} \boldsymbol{\sigma}(x,t) : \hat{\boldsymbol{\epsilon}}(x) d\Omega - \int_{\Gamma_T} \mathbf{T}(x,t)\cdot\hat{\mathbf{u}}(x) ds$$

$$+ \int_{\Gamma_j} [\![\hat{\mathbf{u}}]\!]\cdot\boldsymbol{\sigma}(x,t)\cdot\mathbf{n}(x) ds = 0 \qquad t\in[0,t_0] \quad (3)$$

Where $\Omega' = \Omega_1 \cup \Omega_2$, $\mathbf{n}(x) = \mathbf{n}_2(x)$, and $[\![\hat{\mathbf{u}}]\!]$ is the relative displacement between the joint interface,

$$[\![\hat{\mathbf{u}}]\!] = \hat{\mathbf{u}}_1(x) - \hat{\mathbf{u}}_2(x) \quad x\in\Gamma_j \quad (4)$$

with $\hat{\mathbf{u}}_1$ the displacement on Γ_j on the side of Ω_1 and $\hat{\mathbf{u}}_2$ the displacement on Γ_j on the side of Ω_2.

2.2 *Modified model and back analysis principle*

Suppose, on the actual structure, that displacement measurement $\mathbf{u}^m$ is obtained at instant t_0, on the border $\Gamma'_T \subset \Gamma_T$

$$\mathbf{u}(x,t_0) = \mathbf{u}^m(x) \quad x\in\Gamma'_T \quad (5)$$

Let then consider the modified model, in which displacement on border Γ'_T is prescribed. Solving the modified problem for given values of the unknown parameters (to be back analysed) will give for instant t_0, the solutions $\mathbf{u}(x)$, $\boldsymbol{\epsilon}(x)$, $\boldsymbol{\sigma}(x)$.

These solution fields $\mathbf{u}$, $\boldsymbol{\epsilon}$, $\boldsymbol{\sigma}$ will agree with the prescribed measurement, but not necessarily with the limit conditions of the original problem. In order to tend towards the solution of the original problem, the unknown parameters in the modified problem should be such that ***the stress and displacement fields of the modified problem should satisfy at best the original limit conditions.***

Taking $\mathbf{u}(x)$ and $\boldsymbol{\epsilon}(x)$ as the virtual displacement and virtual deformation respectively, applying the virtual work principle to the stress fields $\boldsymbol{\sigma}_0(x)$ and $\boldsymbol{\sigma}(x)$, with the force limit condition of the original model, we obtain:

$$\int_{\Omega'} (\boldsymbol{\sigma}(x) - \boldsymbol{\sigma}_0(x)) : \boldsymbol{\epsilon}(x)\, dx - \int_{\Gamma_T} (\mathbf{T}(x) - \mathbf{T}_0(x))\cdot\mathbf{u}(x)\, dx$$

$$+ \int_{\Gamma_j} [\![\mathbf{u}]\!]\cdot(\boldsymbol{\sigma}(x) - \boldsymbol{\sigma}_0(x))\cdot\mathbf{n}(x) = 0 \quad (6)$$

This is a relationship between updated parameter values and the previous estimated ones. In general case, this is a kind of fixed point problem which needs iterations supposed to converge. If N sets of measurements are provided, N parameters can be back analysed simultaneously. As only based on the virtual work principle, this method works well for nonlinear and elastoplastic problems.

In practice, absolute displacement measurements are not easy to obtain. Generally, we can have relative displacement between points, obtained by extensometer, convergencemeter, or inclinometer. Moreover, the measurement points may be situated inside the rockmass. These in situ observations appear as restraints, which can be easily considered in the modified problem (Deng et al. 1998).

2.3 *Back analysis*

In elastic phase, the constitutive equation of the Goodman elastic joint is defined as (Goodman 1976)

$$\boldsymbol{\sigma}\cdot\mathbf{n}|_{\Gamma_j} = \mathbf{K}[\![\mathbf{u}]\!] \quad (7)$$

$\mathbf{K}$ is the matrix of elastic moduli of joint, characterized by the normal and tangential stiffness (k_n and k_s) of joint, and $[\![\mathbf{u}]\!] = (u_n, u_s)^T$, with u_n and u_s the normal and tangential displacement components.

At failure, the standard Mohr-Coulomb plastic criterion

$$\sigma_f = c + \sigma_n tg\phi \quad (8)$$

is used. Where σ_f is the failure limit of the tangential stress σ_s, σ_n is the normal stress component, ϕ and c are respectively the friction angle and cohesion of the joint interface.

After each solution of the modified problem during the iterative procedure, the elastic zone Γ_j^e and plastic zone Γ_j^p of the joint are determined. Thus, from equation (6) et (7), the basic equation of this back analysis is obtained as following:

$$\int_{\Omega'} (\boldsymbol{\sigma} - \boldsymbol{\sigma}_0) : \boldsymbol{\epsilon}\, dx - \int_{\Gamma_T} (\mathbf{T} - \mathbf{T}_0)\cdot\mathbf{u}\, dx$$

$$+ \int_{\Gamma_j^e} k_n u_n^2\, dx + \int_{\Gamma_j^p} (\sigma_n - \sigma_n^0) u_n\, dx + \int_{\Gamma_j^e} k_s u_s^2\, dx$$

$$+ \int_{\Gamma_j^p} (\sigma_n tg\phi + c - \sigma_s^0) u_s\, dx = 0 \quad (9)$$

One parameter can be back analysed if the other parameters are supposed to be known. For example, from equation (9), k_s is back calculated as:

$$k_s = \frac{1}{\int_{\Gamma_j^e} u_s^2\, dx}[\int_{\Gamma_T} (\mathbf{T} - \mathbf{T}_0) \cdot \mathbf{u}\, dx - \int_{\Omega'} (\boldsymbol{\sigma} - \boldsymbol{\sigma}_0) : \boldsymbol{\epsilon}\, dx$$

$$- \int_{\Gamma_j} (\sigma_n - \sigma_n^0) u_n\, dx - \int_{\Gamma_j^p} (\sigma_n tg\phi + c - \sigma_s^0) u_s\, dx] \quad (10)$$

Other parameters (k_n, c, and ϕ) are obtained in the same way. The algorithm of back analysis consists in:

- initialize the parameter to be identified
- solve the modified problem with given parameter
- update the parameter by equation (10)
- control the iteration convergence

If we want to back analyse simultaneously two parameters, two sets of measurements are needed. Suppose the displacement data $\mathbf{u}^m$ on border $\Gamma_T^{'} \subset \Gamma_T$ at time t_0, and the relative displacement data $\triangle\mathbf{u}^m$ on border $\Gamma_T^{''} \subset \Gamma_T$ during time interval $[t_0, t_f]$ are given. Note here that the two borders $\Gamma_T^{'}$ and $\Gamma_T^{''}$ are not necessarily the same. The back analysis process is presented in (Deng & Nguyen-Minh 2000).

3 NUMERICAL TEST

A block (with dimension 1×2) with a joint in figure 1 is charged by uniform pressure P_1 on BC and uniform pressure P_2 on CD (stress is un-dimensional in this numerical test), with displacement condition limit

$$u_x|_{BE} = 0, \quad u_y|_{EH} = 0 \quad (11)$$

The initial state is

$$T_0(x): \quad P_1 = 9.0, \; P_2 = 0.0 \quad (12)$$

Displacements from a direct calculation are used as measurements in the back analysis. The mechanical parameters are chosen as follows:

elastic block $E = 1000.0, \quad \nu = 0.1,$

joint $k_n = 1 \times 10^5$, $k_s = 10.0$, $c = 2.0$, $\phi = 45^0$.

The loading is applied linearly on P_2 from 0 to 650.0, with P_1 unchanged. The instant t_0 is chosen as $P_2 = 400.0$ and the instant t_f as $P_2 = 650.0$.

For one parameter back analysis, the displacement at t_0 (plasticity occurs within 30% of the joint) on BC is given as the measurement. Figure 2 is the iteration convergence of back analysis c for different initial values. It can be seen that the non linear back analysis of one parameter converges very fast. Only $3 - 4$ iterations are needed to reach the solution with less than 5% error.

For back analysis of two parameters (ϕ and c), the displacements on BC at t_0 and t_f (plasticity occurs within 50% of the joint) are used as the given measurements. Figure 3 shows the performance of back analysis of ϕ and c with the initial values $\phi = 15^0$, $c = 10.0$. After 20 iterations, the back analysed parameters are obtained within 5% discrepancy from the real values.

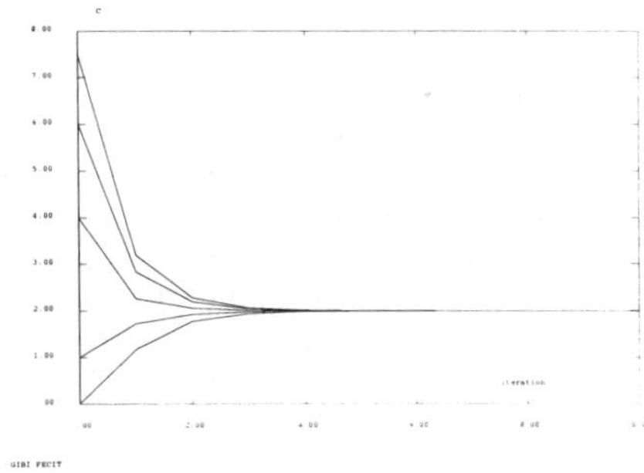

Figure 2: *Iteration convergence of c for different initial values*

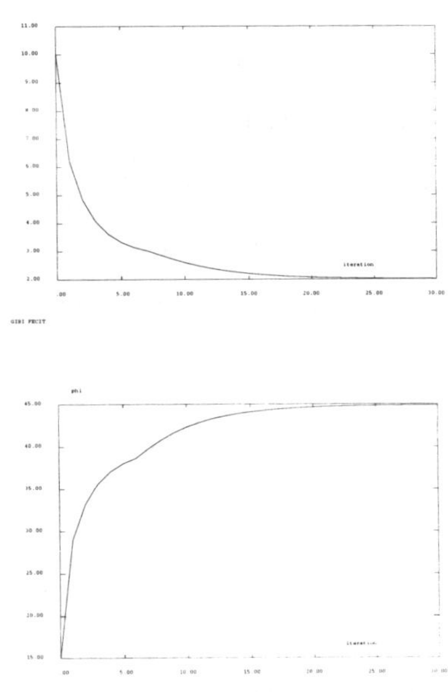

Figure 3: *Iteration convergence of ϕ and c with initial values: $\phi = 15^0$, $c = 10.0$*

4 APPLICATION TO A DAM FOUNDATION

The Água Vermelha Hydroelectric Project was built on the Rio Grande river with $1380MW$ installed capacity. The project comprises a $90m$ maximum height and $600m$ long concrete gravity dam and two lateral embankments (BCOLD 1988). The designed water level is $383.3m$. At the foundation of the concrete dam, there exist some discontinuities. A major horizontal joint (with clayey fillings) located at the elevation of $307.3m$ which was very important for the deformation and safety of the dam (Silveira et al. 1978; LNEC 1975). The dam foundation was instrumented with multiple position extensometers, direct and inverted pendulums, piezometers, and joint displacement devices, during the construction period and reservoir filling. Special care was dedicated to the monitoring of Central Wall and Spillway foundation.

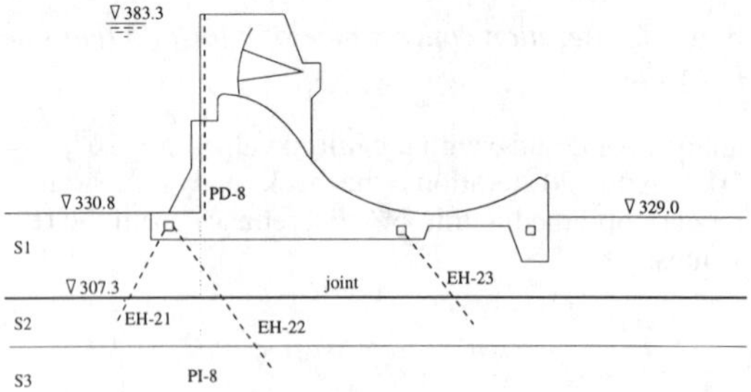

Figure 4: *Section VS-8 of Água Vermelha dam*

In this study, block section $VS-8$ (figure 4) of the spillway section is analysed with the hypothesis of plane strain problem. The concrete and rock masses are taken as elastic linear, and the joint is elastoplastic with Mohr-Coulomb plastic criterion. The displacement measurements during the reservoir filling period are listed in table 1. Mechanical parameters are listed in table 2, according to in situ and laboratory tests.

Table 1: *Displacement measurements during reservoir filling (mm), EH – extensometer, (+) extension, (-) compression; PD, PI – pendulums, (+) downstream, (-) upstream*

instru- ment	water level (m) 341.48 - 360	 360 - 370	 370 - 376
EH-21	0.0	0.05	0.20
EH-22	0.0	0.0	-0.05
EH-23	-0.20	-0.45	-0.25
PD-8	-1.0	-0.7	-0.6
PI-8	0.20	0.1	

Table 2: *Mechanical parameters*

materials	Young's modulus (MPa)	Poisson	density (T/m^3)
concrete	30000.0	0.20	2.5
layer S1	10000.0	0.20	2.9
layer S2	7000.0	0.20	2.9
layer S3	15000.0	0.20	2.9
joint	$\phi = 39^0$, $c = 0.055MPa$		

4.1 Back analysis of elastic parameters of joint

Using the above measurements, the elastic stiffness k_n and k_s of the joint are back analysed. The initial state is selected that the gravity weight and $341.48m$ upstream water level were applied. Relative measurements from water level $341.48m$ to $360m$, and from water level $360m$ to $376m$ are selected to carry out this back analysis. Figure 5 gives the convergence of the back analysis procedure. After 25 iterations, the elastic joint stiffness are obtained as $k_n = 7.6 \times 10^8 N/m^3$ and $k_s = 1.0 \times 10^8 N/m^3$.

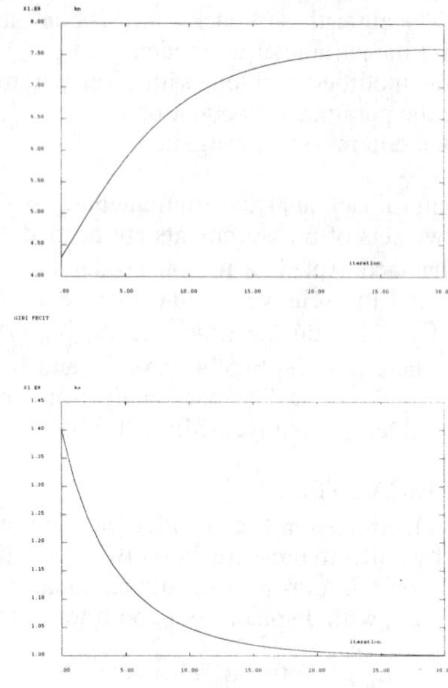

Figure 5: *Iteration convergence of k_n and k_s with initial values:* $k_n = 4.3 \times 10^8$, $k_s = 1.4 \times 10^8$

4.2 Back analysis of elastoplastic parameters of joint

For the back analysis of elastoplastic parameters of the joint, because plasticity did not happened under water level $376m$, the above measurements can not be used. Thus, a direct calculation, with mechanical parameters in table 2 and the above back analysed elastic stiffness of the joint, is carried out with upstream water level linearly augmented to $484.7m$ (plasticity occurs within 47% of the joint part at the bottom of the dam). And the displacements at half the upstream surface (top part) and the surface of the dam top at water levels $464.7m$ (plasticity occurs within 35% of the joint part at the bottom of the dam) and $484.7m$ are selected as measurements.

In figure 6, the elastoplastic parameters ϕ and c are simultaneously back analysed for initial values: $\phi = 45^0$, $c = 35000.0Pa$. Within 40 iterations, the two parameters are identified with less than 5% error.

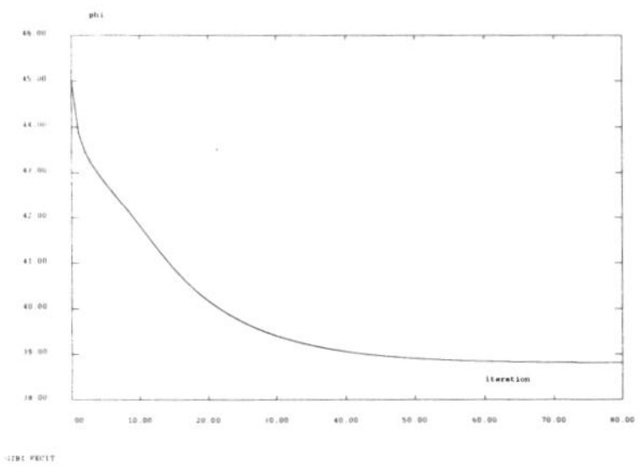

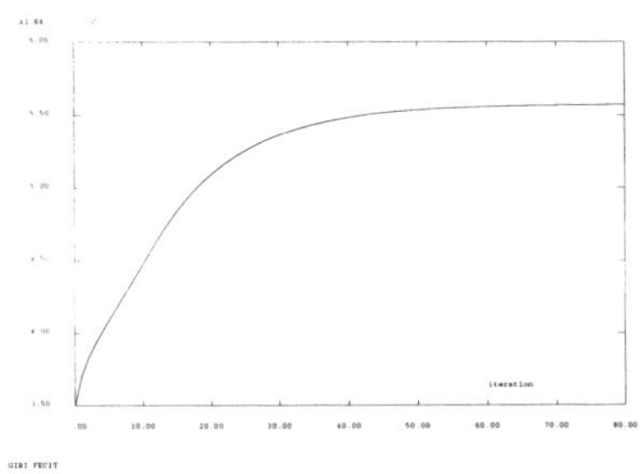

Figure 6: *Iteration convergence of ϕ and c with initial values:* $\phi = 45^0$, $c = 35000.0Pa$

5 CONCLUSIONS

The problem of parameter estimation for elastoplastic joint has been formulated in this paper. Based on the virtual work principle, this back analysis method is achieved by comparing the original calculation model to a modified model in which the field measurements are prescribed. The plastic properties of materials and joints can be properly considered in analysis, and relative measurements obtained by instruments in practical use, such as extensometer, pendulum, inclinometer, etc. are easily applied as prescribed displacements in the back analysis procedure. With sets of measurements at different loading levels, parameters can be identified simultaneously. A numerical example and application to a dam foundation illustrate that this method of non-elastic back analysis is very robust and effective, with good convergence and high stability.

6 ACKNOWLEDGEMENT

The authors would like to thank L. R. Sousa from LNEC, Portugal, and Companhia Energética de São Paulo (CESP), Brazil, who have kindly provided all the monitoring and other information on Água Vermelha Dam.

REFERENCES

Akutagawa, S. & Sakurai, S. (1997). Back analysis of linear and nonlinear deformational behaviors for multiple cross sections in discontinuous rock masses of a large underground power house cavern. In H. K. Lee, H. S. Yang, & S. K. Chung (Eds.), *Environmental and safety concerns in underground construction*, Seaul, South Korea, pp. 853 – 856. Balkema.

BCOLD (1988). Technical data about Agua Vermelha Dam. In J. P. Avila, R. I. Bicudo, & F. P. Luiz (Eds.), *Main Brazilian dams: design, construction and performance*, Brazil, pp. 377–407. BCOLD publications committee.

Constantinescu, A., Nguyen-Minh, D., & Deng, D. (1998). Joint back-analysis in mining subsidence. In *3rd workshop: Applications of computational mechanics in geotechnical engineering*, Porto, Portugal. Balkema (to be published).

Deng, D. & Nguyen-Minh, D. (2000). A back analysis method based on virtual work principle – application to some geotechnical problems. In *4th NARMS, Pacific rocks: Rock around the Rim*, Seattle, USA. Balkema (to be published).

Deng, D., Nguyen-Minh, D., & Constantinescu, A. (1998). Identification of deconfinement ratio in shallow tunnel excavation. In *3rd workshop: Applications of computational mechanics in geotechnical engineering*, Porto, Portugal. Balkema (to be published).

Deng, D., Nguyen-Minh, D., Constantinescu, A., Vieira, A., & Sousa, L. R. (1999). Back-analysis method for loading parameters in 'CV-CF' tunnel calculation – application to a case history. In *AFTES - Journées d'études internationales de Paris*, Paris, France, pp. 189–195.

Deng, D. & Xia, S. (1992). Analysis of arch dams with joints – the O-F method. In *International symposium on arch dams*, Nanjing, China, pp. 197–200. Hohai University Press.

Goodman, R. E. (1976). *Methods of geological engineering in discontinuous rocks*. West publishing company.

LNEC (1975). Estudo da barragem de Agua Vermelha pelo metodo elementos finitos. In *Serviço de barragens, Divisão de estudos especiais*, LNEC, Lisboa, Portugal.

Salençon, J. (1994). *Mécanique des milieux continus*. Paris, France: Ecole Polytechnique.

Silveira, J. F., Miya, S., & Yendo, M. (1978). Geomechanical parameters computed from instrumentation measurements at Agua Vermelha dam foundation. In M. A. Kanji & R. A. Abrahao (Eds.), *International symposium on rock mechanics related to dam foundation*, Rio de Janeiro, Brazil, pp. II.103–II.121.

Computer Methods and Advances in Geomechanics, Desai et al. (eds) © 2001 Balkema, Rotterdam, ISBN 90 5809 183 X

Study on the mechanism of Kaiser effect on the uniaxial compression test of rock

Y.X. Fan, D.L. Zhang & Q.S. Chen
China University of Geosciences, Beijing, People's Republic of China

ABSTRACT: An experimental program was undertaken to investigate the Kaiser effect (KE) of acoustic emission (AE), aiming at the relationship between KE and the damage of the specimen tested. Specimens of two different rocks were used in cyclic loading, namely: Granodiorite and Marble. It has been observed that the rock tested has the ability to memorize previous damage. Furthermore, compared with other parameters such as stress and strain, the damage within the specimen is the key nature of KE.

1 INTRODUCTION

The study on memorial content of Kaiser effect in rock under uniaxial compression conditions (Fan 1998) indicates that the KE can memorize pre-stress, pre-axial-strain and pre-transverse-strain of last loading, and the three parameters have the same advanced or hysteresis memory characters. The memorial accuracy changes with the process of loading. The study on nature of KE in rock under uniaxial compression test (Fan 2000) shows that memorial essence of KE is to memorize the physical character within the rock. It can memorize pre-volume strain, pre-Young's modulus and pre-Poisson's ratio, of which the second is memorized best and steadiest, the memorial accuracy is the same in different loading stages

The essence of KE is to memorize the previous physical character in rocks, however, how to demonstrate these characters by AE parameter and to determine the relationship between AE and rock dynamical parameters? This paper tries to study the forming mechanism of KE from the analysis of AE sources under uniaxial compression test of granodiorite and marble.

2 ANALYSIS OF AE SOURCES

In the view of statistic, rock is of isotropy. Although deformation on the scale of individual grains is decidedly anisotropic, behavior of a large group of grains may be considered isotropic. Obviously, the cracks, which have a random orientation and play an important role in rock's deformation, must also be isotropic (Li & Nardlund 1993). Therefore, we can regard rock as a composition of numerous groups in which there are various grains, cementing material and cracks. All the groups are arrayed in certain style to form a fixed special texture, just as how atoms reform metal (Yuan 1984), and acoustic emission generated in rock just as in metal because of dislocation. The differences between the rock groups and atoms exist that the former is mixture while the later is atom, and the former is connected by molecule bond while the later by metallic bond.

Rock groups are connected by molecule bond, at the same time, all grains, cementing material and cracks in the groups are also connected by molecule bond. We know the energy of molecule bond is far less than that of metallic bond among atoms. With applied force, the molecule bond can slid and be destroyed more easily than other chemical bonds.

At low stress, the inborn cracks inside the rock are open, and as stress is raised cracks are closed gradually. Therefore the whole process of cracks closing will lead the previous rocks group structure to change the molecule bond to slip and break, at last to give out acoustic emission events. Acoustic emission can be attributed mainly to cracks.

Although cracks may be closed under uniaxial compression, some sliding motion of opposing crack surface is still possible. While all cracks inside rock are closed, friction between cracks surface becomes another main reason to release acoustic emission.

By analogy with behavior under low level pressure, the acoustic emission resources of rocks under uniaxial compression can be attributed to inborn cracks, and it can be explained by the cracks' closing and the effect of friction between crack surfaces.

3 DETERMINATION OF DAMAGE FACTOR

The presence of acoustic emission of rocks under uniaxial compression can be attributed to the inborn cracks. The concept of rock groups is a viewpoint of statistics, under low pressure, especially cracks closing and elastic reform stage. Acoustic emission occurs because of the internal cracks, so we can describe the quantity of acoustic emission events through the mean length of cracks. The longer the mean length of cracks is, the more acoustic emission occurs. The average crack length has been definited (Walsh 1965), that is

$$\bar{c} = k\sqrt[3]{\frac{E}{E_{eff}} - 1} \tag{1}$$

In the equation, $\bar{c}$ is the definition of average crack length, E_{eff} is the effective Young's modulus for a rock with many oriented cracks. E is the Young's Modula of uncracked solid, and it can be got from the initial slope of the unloading curve for the rock which has been stressed to the linear portion of the stress-strain curve. k is coefficient that can be described as k_1 during open cracks period and k_2 during closed cracks period. They are

$$k_1 = \frac{1}{\sqrt[3]{\frac{4\pi}{3\bar{v}}}}$$
$$k_2 = \frac{1}{\sqrt[3]{\frac{4\pi}{15\bar{v}}\left(\frac{2+3\mu^2+2\mu^4}{(1+\mu^2)^{3/2}} - 2\mu\right)}} \tag{2}$$

$\bar{v}$ is average volume of all cracks' and arctg(1/μ) is the friction angle of cracks' surface while sliding.

In the two successive tests, both k_1 and k_2 have the same effect on the final value of previous loading test and Kaiser value, they do not affect the study of KE. So we can adopt the average length of cracks to describe damage. Therefore, the damage factor D can be defined as

$$D = \sqrt[3]{\frac{E}{E_{eff}} - 1} \tag{3}$$

Thus,

1) When material has no crack, as $E_{eff}=E$, then $D=0$. It proved that the material has no damage.

2) Before loading, rock bears cracks, as $E_{eff0}<E$, then $D_0>0$. It proved the material has native damage that has been caused in the forming period.

3) When material is pressed, cracks close gradually, rock becomes fine, then E_{eff} increases and $E_{eff0}<E_{eff}<E$, $D_0>D>0$, it proved the material's damage degree decrease gradually. The decrease can be regarded as supplement to native damage by stress doing work.

4) When applied force increases, there will be some new cracks and old cracks will extend, the average of crack increases, E_{eff} decreases, D increases, and it proved there appear new damages.

5) When material is in limiting destroy, the E_{eff} is thought as infinitely small and D is infinity and it proved the material develops from microscopic damage to macroscopic destroy.

4 RESEARCH ON DAMAGE MEMORY OF KE

Twenty-four granodiorite and marble specimens were cyclic loaded in this study. Load level range from 8 to 80MPa, which was less than 52% of failure. For cracks-bearing rock, the loading range evolves from crack opening to crack closing.

In this study, Zhang's regression analysis method was employed to determine Kaiser point (Zhang 1984) and Hardy's concept of Felicity ratio (FR) (Hardy 1993) was also used to describe accuracy of cyclic reloading test.

This paper regards the slope of first unloading as uncracked solid's Young's modulus (*E*), the corresponding ratio of stress versus axial-strain as the Young's modulus (E_{eff}). Based on this, calculate damage factor *(D)* of every point by equation (3).

The results indicate the rocks have the ability to memorize previous damage. Figure 1 shows the typical curve of total AE events versus damage of marble. D_p is previous damage factor and D_k is that factor at Kaiser point.

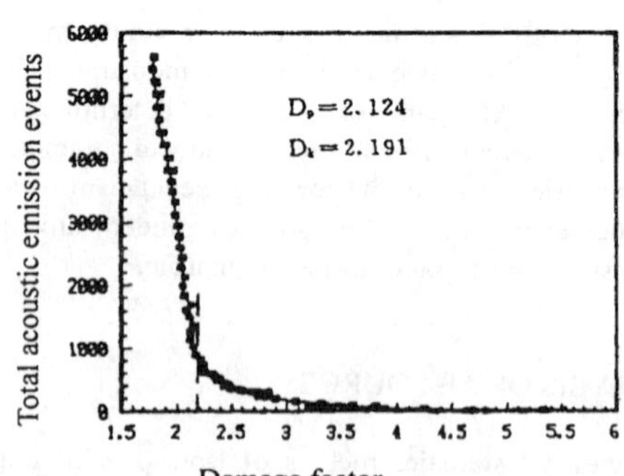

Figure 1 Total AE events versus damage factor

Felicity ratio of KE memorizing previous damage and other parameters for marble sample B0 is shown in Table 1.

Table 1 Felicity ratio of KE memorizing previous damage and other parameters for marble sample B0

	B0-2*	B0-3	B0-4	B0-5	B0-6	B0-7	B0-8
FR(D)	1.032	1.002	1.004	1.025	1.007	1.015	1.025
FR(stress)	0.669	1.022	0.992	0.816	0.756	0.919	0.904
FR(axial-strain)	0.725	1.032	1.000	0.937	0.710	0.953	0.956
FR(transverse-strain)	0.659	1.109	1.038	0.828	0.670	0.938	0.956
FR(bulk-strain)	0.756	0.970	0.950	1.131	0.772	0.978	0.954
FR(Young's modulus)	0.918	0.994	0.990	0.869	1.063	0.967	0.946
FR(Poisson's ratio)	0.909	1.010	1.039	0.885	0.994	0.987	1.000

* B0 means the index of the marble sample and 2 means the reloading time

The nearer to 1 the FR value is, the more accurate the memorize ability is. From Table 1, we can draw conclusion that KE remembers its previous damage factor much better than other parameters.

For all specimens, Felicity ratio of damage factor varied over the range 0.964-1.041 with the average 1.004 to graniodiorite, 0.950-1.046 and 1.008 to marble among more than 80 tests. Their average error is among 2-3%. KE memorizes previous damage factor better than other parameters whose average error is much higher.

Fig 2 illustrates the curve of Kaiser damage versus previous damage factor of all marble samples. The slope of the regression equation is 0.98 that proves the KE can be able to very accurately predict the pre-damage, with the error only 2%. The regression coefficient is 0.991 that proves the marble's memory stability is very good. The slope and regression coefficient of granodiorite is 1.01 and 0.999. Both granodiorite and marble proved the mechanism of Kaiser effect is remembering their previous damage.

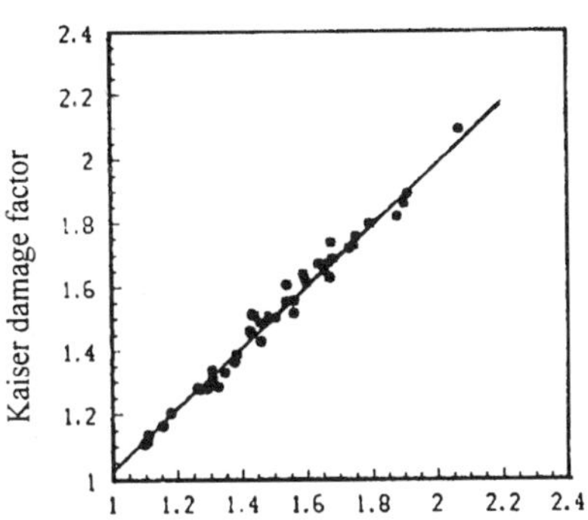

Figure 2 Kaiser damage versus previous damage factor for marble

5 CONCLUSION

Through the analysis of AE source, it is concluded that the real reason of forming AE in uniaxial compression test is the successive damage while reloading, the mechanism of KE in AE activities lies in the memory of previous damage. This paper defines damage-factor (D), the data indicate KE can nicely memorize previous damage, the Felicity ratio is from 0.950 to 1.046 with mean error less than 2-3%, the regression curve of Kaiser damage factor versus previous damage factor illustrates that the correlate coefficient is better than 0.99. This further proves that the mechanism of KE lies in the memory of previous damage.

REFERENCES

Fan, Y. X. 1998. Study on memorial content of Kaiser Effect in rock under uniaxial compression test. *Proceedings of Contemporary Mechanics:* 540~545. Guangzhou: Huanan

Fan, Y. X. 2000. Study on nature of Kaiser Effect in rock under uniaxial compression test. *Geoscience*.14 (1): 95-99

Hardy. H. R. 1992. Recent Kaiser Effect studies on rock. Progress in acoustic emission. The Japanese Society for NDI: 149-157

Li, C and E. Nardlund 1993. Experimental verification of the Kaiser Effect in rock. *Mechanics and rock engineering.* 26(4): 333-351

Walsh. J. B, 1965 The Effect of cracks on the uniaxial elastic compression, *J. G. R,* 70(2): 399-411

Yuan et al.1984 *Technology and Application of Acoustic Emission.* Beijing: Mechanical Industry Press.

Zhang, D. L, Use of the Kaiser effect for estimation of the previous state of stress in rock, *Earthquake geology* 16(1):31-40

Computer Methods and Advances in Geomechanics, Desai et al. (eds) © *2001 Balkema, Rotterdam, ISBN 90 5809 183 X*

Shear characteristics of coarse sand

H.Ghiassian
Civil Engineering Department, Iran University of Science and Technology, Iran

ABSTRACT: The drained stress-strain, volume change, and strength behaviour of coarse Cambria Sand was studied in triaxial compression and extension tests. The specimens were prepared with cross-anisotropic fabric and different height to diameter ratios, and the influence of boundary conditions was examined providing free or fixed top caps. Confining pressures varied from 200 kPa to 4000 kPa. All results were well comparable with those in the previous investigations conducted on sands. In compression tests on specimens with fixed caps, the shear bands were not developed because of end condition effects. The shear bands in tests on specimens with free caps were fully developed with clear inclinations. The modes of failure in extension tests were observed as shearing plane at the early stage of tests and necking at the end of tests. The thickness of the shear band a - peared to be more comparable to the average of those relationships proposed in the literature

1 INTRODUCTION

In studies of the strength and stress-strain behaviour of soils, in which an investigation of the particle crushing is also desired, it is important to use a type of soil with an appropriate particle shape and size distribution so that identification of the broken parti - cles after shearing is possible. In this respect, Carbria Sand which is composed of relatively coarse, rounded particles was employed in the study pre-sented here with the following objectives:

To study the magnitudes of friction angles, onset of shear planes, and volume change behaviour in drained triaxial compression and extension tests over a range of confining pressures up to 4000 kPa.

To examine the influence of boundary conditions on development of shear planes and, consequently, on stress-strain and volume change characteristics.

To compare the characteristics of shear planes (i.e. angle and thickness) to results presented in the literature and to theoretical predictions.

2 SAND COMPOSITION

The soil used in this study was Cambria Sand with subrounded grains and particle sizes between 2.00 mm (Sieve 10) and 0.84 mm (Sieve 20), the same grain size distribution as those used in the previous investigations (Lade & Wasif 1987, Ochiai & Lade 1983). The specific gravity of this sand was measured to be 2.668 and the maximum and minimum void ratios were 0.80 and 0.51 respectively.

3 SPECIMEN PREPARATION AND SATURATION

Based on previous experiments (Ochiai & Lade 1983), in order to prepare dense specimens with a void ratio between 0.54 and 0.53, the sand was poured and shaken under water in the mold in 10 layers with one minute shaking for each layer. De-pending on the range of applied confining pressures, one to three membranes were used to prevent mem brane puncturing. Although all specimens were pro-duced under de-aired water, it was found necessary to perform additional saturation procedures. The d - gree of saturation was checked by measuring the B parameter. This value was in all cases below the ac-ceptable level for full saturation unless a back pres-sure up to 60 psi was applied inside the specimen. B-values measured subsequently were between 0.96 and 1.00 for all specimens, thus indicating that full saturation had been achieved (Ghiassian 1988).

4 TESTING EQUIPMENT AND MESUREMENTS

Two different triaxial apparatuses were used in this study: (a) In the first apparatus a tall specimen with typical height of 19.0 cm and diameter of 7.1cm corresponding to H/D=2.68 was employed. Two different boundary conditions (restrained or free) were provided at the caps of the tall specimens. (b) In the second apparatus a relatively short specimen with a typical height of 13.2 cm and a diameter of 7.I cm corresponding to H/D = 1.86 was employed.

Three readings were recorded in each data set: the vertical load on the specimen either in compression or in tension, the vertical strain, and the volume change showing the amount of compression or dilation. Two corrections were made to the test results. The first and most important correction was made for membrane penetration into the specimen. A co - siderable amount of volume change was measured in the consolidation stage of each test due to membrane penetration. The method applied here was described by Frydman et al. (1973), in which the amount of membrane penetration was basically dependent on the particle size and the effective confining pressure. The second correction was due to the strength of the membrane made to deviator stresses calculated for the extension tests.

5 TESTING PROGRAM AND PROCEDURES

Drained triaxial tests were conducted on specimens consolidated under various effective confining pre - sures (200 - 4000 kPa). At the end of each test, photographs were taken of the specimen to display the mode of failure and any shear banding that might have occurred.

5 .1 *Compression tests*

The first series of tests was conducted as conventional triaxial compression tests in which both tall and short specimens were employed. The tests on tall specimens were carried out at 200 and 500 kPa e - fective confining pressures, because the Lucite cell wall was not strong enough to withstand higher cell pressures. For tests at 200 kPa the two different end conditions were employed in order to investigate their effects on the strength as well as development of shear planes. Similar experiments are presently being conducted for specimens with free caps under much higher confining pressures. For short speci - mens, with restrained caps, the following effective confining pressures were applied: 500 , 1000 , 2000 , 4000 kPa.

5 .1 .1 *Stress-strain, volume change, and strength characteristics*

The results of compression tests are shown on Figure 1. It is seen that all stress- strain curves have smooth curvatures and are gradually dropping after failure. This is due to the type of failure (bulging) caused by the restrained ends used in all specimens in this figure.

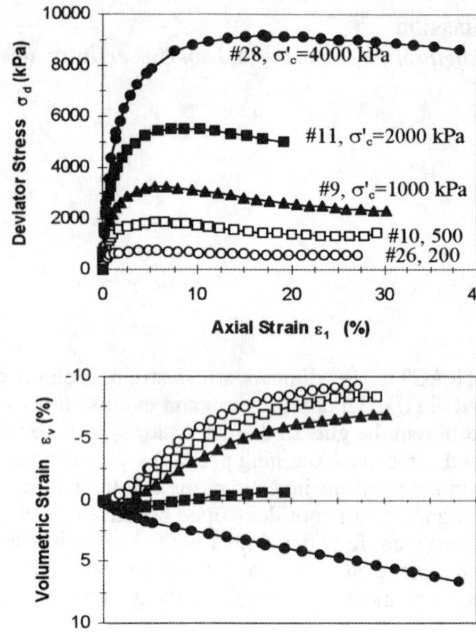

Figure 1. Triaxial compression test results for specimens with fixed caps

As expected from past studies on sand (Lee & Seed 1967), it is seen that the amount of dilation in each test decreases with increasing confining pressure and the volume change becomes compressive in tests at high confining pressures. The rate of dilation is highest at the failure point in each test. In test 28, the specimen is undergoing compression and this continues even at high axial strains (up to 36%). The rate of compression also shows a slight increase as the specimen is strained beyond 20%. The stress-strain curve of this test resembles that of a loose specimen at low confining pressure which has also undergone compression (Holtz & Kovacs 1981). The slope of the curve after failure is small compared to those of other tests in which the specimens dilate. The stress-strain behaviour becomes more brittle as σ_3 reduces, and the strain-to-failure increases as higher confining pressure is applied. This behaviour of sand also has been seen before (Lee & Seed 1967).

Figure 2 shows the result of tests on tall speci - mens with different end conditions. It is seen that the strength obtained for a specimen with regular rough cap (#26) is higher than that with a free cap. This is due to the end restraint which acts as though a slightly higher confining pressure is imposed on the specimen as well as possibly an increased side friction along the loading piston. However, the initial modulus, strain-to-failure, and the amount of volume change up to failure in both tests remain nearly the same.

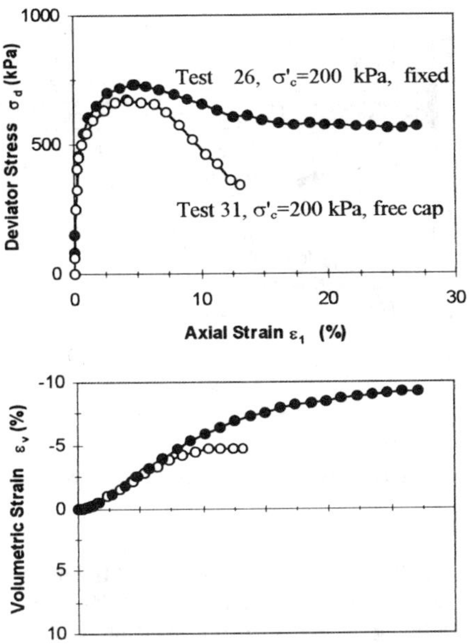

Figure 2. Influence of end conditions in triaxial compression tests on tall specimens (H/D=2.68)

It is interesting also to realize that the stress-strain curves after failure in both tests start with identical initial slopes, and as the tests proceed, the curve of the first test gradually becomes flat due to bulging of the specimen. The curve of the second test becomes more brittle because of shearing until it reaches a constant slope when the shear band is fully developed. The volume change in the second test at the time of shear band formation decreases and has totally terminated at the end of the test. In comparison, the rate of volume change in the first test decreases slowly and the specimen keeps expanding with almost constant rate even at high axial strains (27%).

Figure 3 shows the results of compression tests performed on short and tall specimens at 500 kPa confining pressures. Both specimens were restrained by rough caps, and no shear bands were observed. It is seen again that the strength obtained for the tall specimen is slightly less than those of the short specimen (Bishop 1966).

5.1.2 *Characteristics of shear planes*

Except test 31, in which a free horizontally translating cap was provided for the specimen, all compression tests performed on tall or short specimens and under various minor principal stresses did not exhibit any shear planes. All specimens in these tests failed as they bulged considerably in the middle and exhibited almost zero lateral strain at the ends. In test 31, however, a shear band was allowed to develop clearly as a result of the horizontal cap displacement. The amount of displacement in the horizontal direction at the end of the test was about 1.65 cm, and the thickness of the shear band was about 19 mm. The shear band formed an angle with the major principal stress (vertical) of 45 degrees. It seems that the boundary conditions of different specimens employed in this study have had significant influences on the mode of failures.

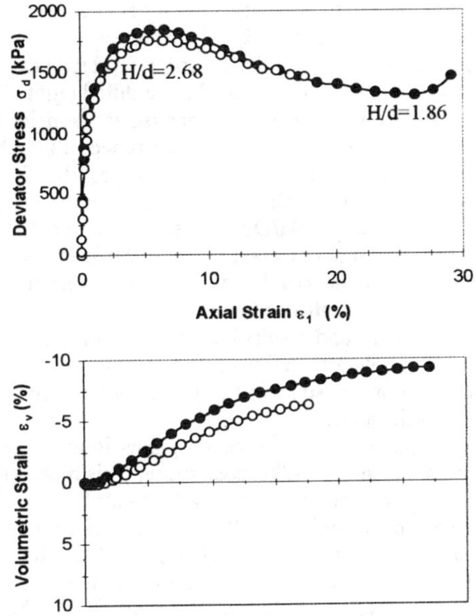

Figure 3. Influence of H/D in triaxial compression tests at σ'_c =500 kPa

Previous studies on granular materials (Muhlhaus & Vardoulakis 1987, Roscoe 1970) show that the thickness of shear band is proportional to the size of particles over a range from l0d to 20d. In a theoretical development (Muhlhaus & Vardoulakis 1987) it has been shown that the thickness of shear bands in compression should be close to 16d in which d is the average diameter of particles. This means that the coarser the grains are, the thicker the shear plane will be.

If the displacement at the cap or base or both were possible, the specimen would deform gradually and adjust itself until the desired displacement would be achieved and a shear band be fully developed. With fixed ends the specimen would start changing its failure path gradually from dilation only in sheaing zone to expansion in the entire specimen until it fails in bulging.

It might be suggested to use a very tall specimen with the same fixed ends as an alternative for the specimen with free ends in order to facilitate the creation of shear bands at the middle height. This may cause another problem, because as the height of the specimen increases, the occurrence of buckling becomes more likely, and this changes the failure mode completely. In the range of appropriate height-to-diameter ratios (H/D=2-3), effects of end conditions on the mode of failure are still present, as can be seen from the results of tests on short and tall specimens with restrained ends.

The shear band results indicate that an average of the two above predictions for thickness of the shear plane is more reasonable and comparable to that observed in the actual tests.

The appearance of a shear plane in test 31 was detected at about 7-8% axial strain. This means that the shear plane became visible after peak failure. This observation is similar to that made before in compression tests on fine sands (Peters et al. 1988). The only difference was that in the test on coarse sand the shear plane appeared closely after peak failure, but in the tests on fine sand the shear planes occurred after considerable straining beyond the peak failure point.

5.2 *Extension tests*

All specimen were isotropically compressed by periodically increasing the cell pressure and measuring the amount of compression at varying time intervals for each increment. This procedure was continued until the final confining pressure had been achieved. As for the compression tests, two different apparatuses were used for the extension tests. Only one test with 200 kPa effective confining pressure was performed on a tall specimen. For tests on short specimens, the range of effective confining pressure was 500 - 1000 - 2000 - 4000 kPa, i.e. the same as those in the compression tests.

5.2.1 *Stress-strain, volume change, and strength characteristics*

Figure 4 represents the stress difference and volume change results in terms of major principal strain (ε_1) in extension tests at different confining pressures. Comparing these curves with those obtained in the compression tests (Fig. 1), it is seen that the range of strain-to-failure for identical specimens (i.e. same minor principal stresses) is much smaller in extension tests than in compression tests. Further, the stress-strain curves after failure are substantially steeper in extension tests than in the corresponding compression tests. The reason is most likely due to the deveopment of shear planes and necking in extension tests, whereas all specimens in Figure 1 failed in bulging. The strength decrease after failure in the extension tests becomes less pronounced as the confining pressure reduces from 4000 to 200 kPa.

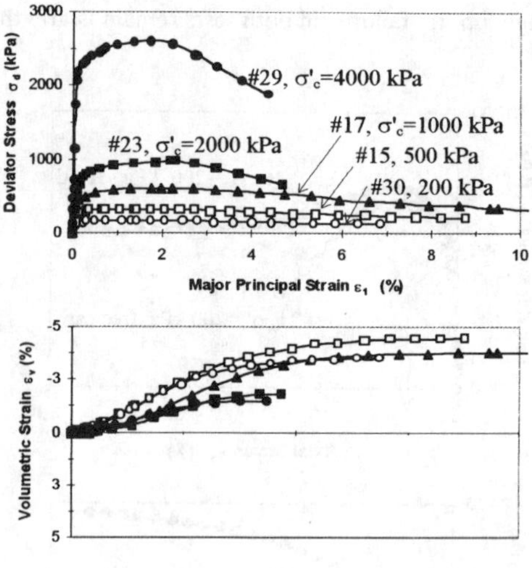

Figure 4. Triaxial extension test results for specimens with fixed caps

It can be also seen in Figure 4 that the major principal strain-to-failure increases from 1.3% to 2.0% as the confining pressure increases from 200 kPa to 1000 kPa, and then it varies irregularly at higher confining pressures. This behavior is also seen for the strain-at-end where the rate of dilation has decreased to zero. For tests up to 1000 kPa, it increases and lasts to high strains (ε_1= 10.4%) but for the tests at 2000 and 4000 kPa, the axial strains-at-end diminish considerably to about 4.3%. The reason for this behavior may be due to development of shear planes in the different tests. Observations showed that the shear planes at 2000 and 4000 kPa confining pressures developed just below the cap with no clear inclinations, but for those at lower pressures the shear planes were observed in the middle sections of the specimens.

The amount of dilation in the extension tests also increases as the confining pressure decreases except in test 30 at 200 kPa. The rate of dilation tapers off earlier in this test than in test 15 at 500 kPa. This may be because a tall specimen was used in test 30 and because a shear plane developed in both tests at about the same axial strain. The tall specimen had a larger volume than the short specimen in test 15 and the volumetric strain (ε_v = dv/v) consequently was calculated to be smaller beyond the point of shear plane formation. The rate of dilation is maximum at failure in all extension tests, and it decreases to zero in response to development of shear planes and necking in the specimens.

In order to examine whether it would be possible to produce shear planes in the mid-section of the specimens instead of beneath the cap, three extension tests were performed on identical specimens at 2000 kPa confining pressures. The results of the tests are shown in Figure 5. All three specimens, however, exhibited necking under the cap. The results of two tests at 500 kPa confining pressures are also presented in this figure. These two sets of data are comparable with the following exceptions:

- The strengths obtained at 2000 kPa confining pressure are functions of void ratio. It means that the higher strength corresponds to the denser specimen. But this relation is not seen for the results at 500 kPa.
- The strain-to-failure for test 23 is a little higher than that for the two other tests (#24, #25).
- The shape of the stress-strain curve for test 23 is different from that of the two other curves (#24, #25).

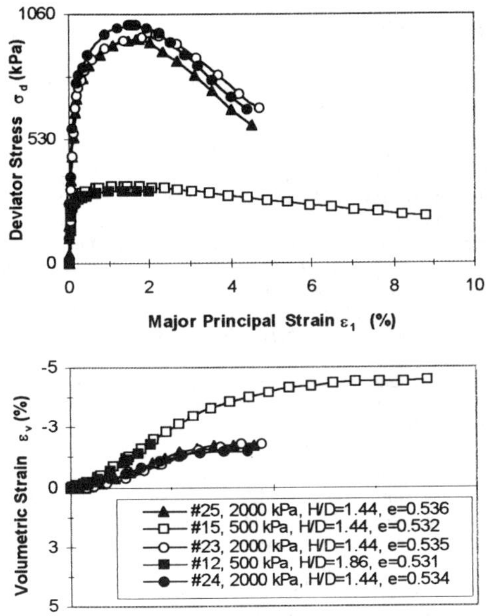

Figure 5. Influence of density or H/D in triaxial extension tests

The strength results obtained in all compression and extension tests are presented in Figure 6 in terms of friction angle versus mean stress at failure. The effects of height-to-diameter and end conditions is exhibited more clearly in this figure. The friction angle values follow a reasonable pattern for all short specimens in the compression tests. They describe a curve which indicates that the magnitude of the fri - tion angle reduces as the mean stress or confining pressure increases. But, the friction angles corresponding to tall specimens in tests 26 and 31 are not located on the curve and represent smaller strengths than obtained from short specimens under equal mean stress conditions. The effect of free cap used in test 31 also results in a smaller friction angle (38.8) than that of test 26 (40.3) in which a fixed cap was employed.

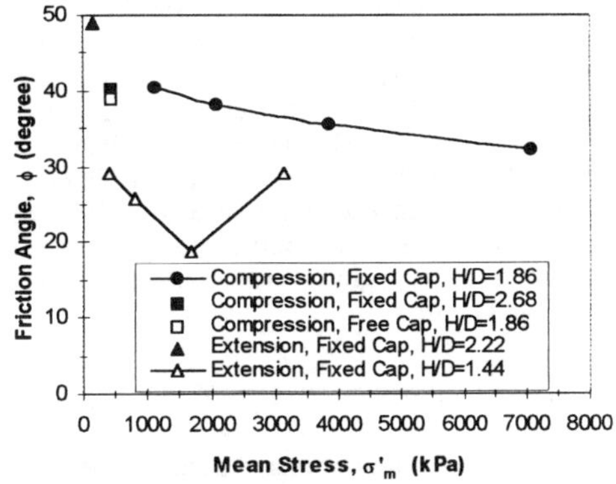

Figure 6. Influence of end conditions or H/D on strengths in triaxial compression or extension tests

The results of the extension tests, however, do not demonstrate a reasonable pattern that can be inter - preted meaningfully as shown in Figure 6. This behaviour is due to development of shear planes, which according to theoretical considerations (Peters et al. 1988) usually occur before peak failure in extension tests. These shear planes initiate and propagate through the specimens in response to local nonun - formities and stress concentrations; consequently, they result in erratic behavior. In comparison, shear planes in compression tests develop after peak fail - ure, and they do not have any influence on the strength in compression. It should be mentioned, however, that in the extension tests performed in this study the shear planes became visible at a point beyond the peak failure.

5.2.2 *Characteristics of shear planes*

The type of failure modes observed at the end of all extension tests was some sort of necking with no clear indications of shear bands. In tests with effective confining pressures higher than 1000 kPa, the failure zones developed below the cap, whereas for tests at lower pressures they developed close to the mid-section of the specimens.

The shear band angle, however, can be measured when the first view of the slip surface is marked. This was possible only in test 30 in which a Lucite cell wall was used. The shear band angle measured in this test with respect to the major principal stress dire - tion was about 15 degrees. The shear band was d - tected at about 8% axial strain.

Calculations for the orientation of the shear band based on theoretical considerations (Frydman 1974, Holtz & Kovacs 1981) in both compression and extension tests do not agree with the measurements and predict higher values. However, the anisotropic fabric may play an important role in shear band orientation, and this is not accounted for in theoretical formulations.

Based on previous studies on fine sands (Peters et al. 1988), the shear bands in extension tests were observed closely after peak failure and accompanied by a sudden and considerable drop in sustained stress difference. But for coarse sand, the results of test #30 in Fig. 4 showed that the shear band became visible at about 7% to 8% axial strain which is rather far beyond the failure point (3.67%). Besides, the shape of the stress-strain curve after failure is almost flat with only a slight decrease in stress difference after peak failure.

6 CONCLUSIONS

The following observations regarding the strength and volume change characteristics of sands under comparable conditions appear to be generally valid independent of the size and type of sand particles employed in the tests :

- The tendency for dilation decreases with increasing confining pressure.
- The friction angle decreases with increasing confining pressure.
- The strain-to-failure increases with increasing confining pressure.
- The strength increases with decreasing height-to-diameter ratio for specimens without free or lubricated ends.
- The maximum rate of dilation occurs at the failure point.
- Specimens with restrained or fixed ends exhibited higher strengths than specimens with free or lubricated ends.

However, other findings in compression tests, such as the mode of failure, appear to have been i - fluenced seriously by the boundary conditions. All specimens with restrained ends failed by bulging with no indications of shear plane developments. In tests on similar specimens but with free cap, shear planes with clear inclinations were fully developed.

The difference between failure modes in tests on fine sand and tests on coarse sand, under comparable boundary conditions, corresponded to the difference in thickness of shear planes mobilized inside the specimens. This thickness, which has been shown theoretically and experimentally to be proportional to particle size, is not pronounced in tests on fine sands. In tests on coarse sands the thickness is relatively large, and development of shear planes requires relative displacements of caps and bases in the horizontal direction.. If the necessary displacements are allowed, the specimens will produce shear planes after peak failure. Otherwise, bulging failures are produced, and these change the stress-strain and volume change characteristics.

The thickness of shear bands in compression tests appeared to be in better agreement with an average of those proposed in the literature.

The inclinations of shear bands were considerably different from theoretical predictions either according to equilibrium considerations or to compatibility considerations. The reason for this discrepancy was not definitively identified, but the anisotropic fabric in the specimen was considered to be one of the factors.

7 ACKNOWLEDEMENT

The financial support of this study was provided by The University of California at Los Angeles under supervision of professor Poul V. Lade, and is gratefully acknowl edged.

REFERENCES

Bishop, A.W., (1966). "The Strength of Soils as Engineering Materials," 6th Rankine Lecture, *Geotechnique*, London, Vol. 16, No. 2, pp. 89 -130.

Frydman, S., Zeitlen, J.G., and Alpan, I., (1973). "The Membrane Effect in Triaxial Testing of Granular Soils," *Journal of Testing and Evaluation*, JTEVA, Vol. 1, No. 1, January, pp.37-41.

Frydman, S., (1974). "Yielding of Sand in Plane Strain," *Journal of the Geotechnical Division*, ASCE, VOL. 100, No. GTS, May, pp. 491- 501.

Ghiassian, H. (1988). "A study of shear characteristics of coarse sand," Master thesis, University of California at Los Angeles.

Holtz, R.D., and Kovacs, W.D., (1981). "*An Introduction to Geotechnical Engineering*," Prentice - Hall, Inc., Englewood Cliffs, NJ, pp. 493 - 519.

Lade, P.V., and Wasif, U., (1987). "Effects of Height-to-Diameter Ratio in Triaxial Specimens on the Behavior of Cross - Anisotropic Sand," *Advanced Triaxial Testing of Soil and Rock*, ASTM STP XXX, R.T. Donaqhe, R.C. Chaney, and M.L. Silver, Eds., American Society for Testing and Materials.

Lee, K.L., and Seed, H.B., (1967). "Drained Strength Characteristics of Sands," *Journal of the Soil Mechanics and Foundation Division*, ASCE, Vol. 93, No. SM6, November, pp. 117 - 142.

Muhlhaus, H.B., and Vardoulakis, I., (1987). "The Thickness of Shear Bands in Granular Materials," *Geotechnique*, London, Vol. 27, No. 3, pp. 271- 283.

Ochiai, H., and Lade, P.V., (1983). Three - Dimensional Behavior of Sand with Anisotropic Fabric," *Journal of Geotechnical Engineering*, ASCE, Vol. 109, No. 10, October, pp. 1313 - 1328.

Peters J.F., Lade, P.V., and Bro, A., (1988). "Shear Band Formation in Triaxial and Plane Strain Tests." *Advanced Triaxial Testing of Soil and Rock*, ASTM STPXXX, ed. By R.T. Donaghe, R.C. Chaney and M.L. Silver, Am. Soc. Test. Mat.

Roscoe, K.H., (1970). "The Influence of Strains in Soil M - chanics," *Geotechnique*, London, Vol. 20, No. 2, pp. 129 - 170.

Computer Methods and Advances in Geomechanics, Desai et al. (eds) © 2001 Balkema, Rotterdam, ISBN 90 5809 183 X

Effect of interface friction on N_c value for plate-penetrometer

Y. Hu

Department of Civil Engineering, Curtin University of Technology, W.A., Australia (Formerly: Centre for Offshore Foundation Systems, The University of Western Australia, Nedlands, W.A., Australia)

P.G. Watson

Ove Arup and Partners, Australia (Formerly: Centre for Offshore Foundation Systems, The University of Western Australia, Nedlands, W.A., Australia)

M.F. Randolph

Centre for Offshore Foundation Systems, The University of Western Australia, Nedlands, W.A., Australia

ABSTRACT: A plate-penetrometer, with a horizontal plate at its tip, has been developed recently for site investigation in offshore geotechnical engineering. It is designed to measure the profile of shear strength, particularly in soft sediments. The effects of plate skin roughness on the bearing capacity factor (N_c) are investigated numerically, physically and analytically. The *H*-adaptive RITSS (Remeshing and Interpolation Technique with Small Strain model) FE method has been used for penetration analysis from the soil surface in normally consolidated soil. Centrifuge testing has been conducted for plates in normal consolidated kaolin clay and calcareous silt. Upper bound analyses have also been undertaken for pre-embedded plates without soil disturbance. The deep penetration analyses suggest N_c values being about 11.7 for both smooth and rough plates with very marginal difference, in spite of about 5% difference in N_c values for pre-embedded plates.

1 INTRODUCTION

The cone penetration test (CPT) has been used widely in geotechnical engineering to obtain continuous soil strength profiles and in situ soil properties. Due to the cone shape on the tip, however, the measured resistance must be corrected for the overburden stress acting on the tip. Especially for CPTs performed in seabed sediments in deep water, the pore pressure acting on the cone tip can be very large, limiting the resolution of the measurement. The overburden pressure itself may be a major part of the measured bearing resistance when the soil deposit is soft, introducing significant uncertainty in the estimated net resistance (Lunne et al. 1997).

After the correction, the net cone resistance, q_{cnet}, can be used for calculating the undrained shear strength of soil, using a cone factor, N_c, as:

$$s_u = q_{cnet}/N_c \quad (1)$$

The N_c value for a standard cone penetrometer is commonly chosen between 8 and 15, though recommended values show considerable scatter among the range of soil types.

In offshore geotechnical engineering, submarine sediments can be very soft over the upper few metres below the seabed, especially for normally consolidated (NC) soil. It is often found that the undrained shear strength of NC soil increases more or less proportionally with depth (Tani & Craig 1995), which can be expressed as:

$$s_u = k\,z \quad (2)$$

in which k is the gradient of shear strength profile and z is the soil depth below seabed.

For foundation design in offshore engineering, the soil strength over the top few metres is crucial. In view of potential errors in correction of CPT data, and uncertainty in the cone factor, N_c, various types of 'full-flow' penetrometer devices have been tested extensively at the University of Wester Australia. These include: (a) cylindrical T-bar; (b) spherical ball; and (c) horizontal flat plate-penetrometer (Watson et al. 1998). The main logic behind these devices is that the flow of soil from front to back of the penetrometer obviates the need to correct the measured penetration resistance for the ambient overburden stress. In addition, plasticity solutions may be derived for the different bearing capacity factors, removing much of the uncertainty associated with the cone penetrometer.

This paper focuses on the plate-penetrometer only. Figure 1 shows a sketch of the penetrometer. The radial skirts beneath the plate are to ensure full interaction with the soil, (and to allow a torsional mode of testing as well (Watson & Randolph 1998), although this is not pursued here.

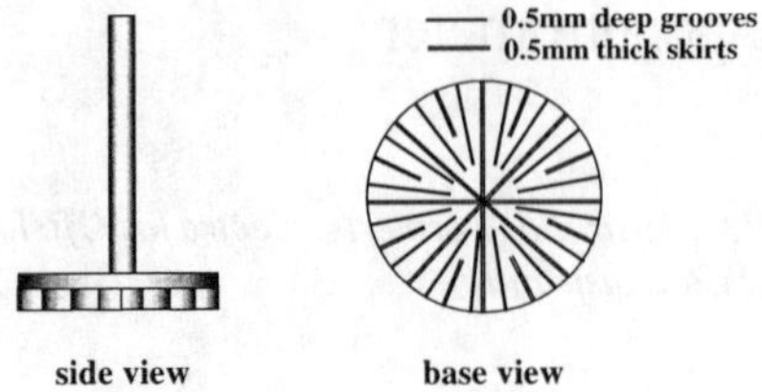

Figure 1. Plate-penetrometer designed at UWA

To investigate the effect of the interface friction between plate and soil, on the bearing capacity factor N_c for the plate penetrometer, finite element analysis, upper bound analysis and centrifuge testing have been performed. As an optimal mesh is essential in the finite element calculations, *h*-adaptive mesh generation has been introduced in FE analysis to ensure high accuracy of the solutions. In *h*-adaptivity, an SPR-strain measure (SPR is Superconvergent Patch Recovery, Zienkiewiecz & Zhu 1993) has been used due to its simple implementation into any existing FE code. The *h*-adaptive FE method has been used in bearing capacity analysis for strip and circular foundation design, and criteria have been proposed to achieve the desired accuracy (Hu & Randolph 1999).

Upper bound analysis is useful for investigating the bearing capacity of foundations on homogeneous and non-homogeneous soil. The method is only valid for 'pre-embedded' foundations, and may not be strictly applicable where foundations undergo large penetration in non-homogeneous soil, due to disturbance of the soil stratigraphy during penetration. The upper bound mechanism used here is an extension of the mechanism for surface circular foundations by Kusakabe et al (1986), allowing for foundation embedment and different foundation shapes, such as spudcans (Randolph, 1998).

It is helpful to conduct lower and upper bound analyses in parallel, in order to gauge the relative accuracy of both. This is addressed in more detail at this conference by Martin & Randolph (2001), where it is shown that the two approaches closely bracket the bearing capacity of plate and shallow caisson foundations. Here, only upper bound results are presented in conjunction with those from finite element analyses.

Two types of investigation have been performed here. The first type is evaluation of the bearing capacity for plate penetrometers that are pre-embedded at different depths. Fully smooth and fully rough plate-soil interfaces are simulated in both FE and upper bound analyses.

The second part is a study of the full penetration behaviour using the *h*-adaptive RITSS FE method and centrifuge testing, for plates penetrating from the soil surface. This method has been used in large penetration problems of offshore foundation systems (Hu & Randolph, 1998[1,2]). Again, fully smooth and fully rough plate-soil interfaces have been modelled.

2 NUMERICAL METHOD

2.1 *H-adaptive finite element method*

In adaptive finite element analysis, the error due to discretization of the computation domain has to be estimated. For an optimal mesh using *h*-adaptivity, the element sizes are adjusted according to the error distribution. Thus smaller elements are generated in regions of high error.

An SPR-strain error estimator has been proposed by the authors (Hu & Randolph 1998[2]). It has been demonstrated that this SPR-strain error estimator works very well in elastic-plastic analysis, with the error in element i defined as

$$e_i^* = \left[\left\{\int_{\Omega_i} (\varepsilon^* - \varepsilon^h)^T (\varepsilon^* - \varepsilon^h) d\Omega\right\} / \Omega_i\right]^{1/2} \qquad (3)$$

where Ω_i is the area of element *i* in two-dimensional analysis (or volume of element in three-dimensional analysis), ε^h are the strains from the FE solution (here using the AFENA finite element package, Carter & Balaam 1992), ε^* are the strains recovered by SPR. This error measure is easy to use due to its non-dimensional character.

Once the discretization error is obtained using Equation 3, the existing mesh can be refined. However, it is not easy to set up a unique criterion for mesh refinement in plastic analysis (Zienkiewiecz et al. 1995). Therefore, a minimum element size (h_{min}) has to be specified. The following procedure has worked well for the refinement:

(1) generate a coarse mesh (using a particular element density function, such as an exponential function (Hu & Randolph 1998)) and compute an initial FE solution ε^h;
(2) calculate the recovered strains ε^* using SPR;
(3) estimate the strain error e_i^* in each element over the mesh using Equation 3;
(4) create a fairly uniform mesh with element size h $= 5.0 * h_{min}$ in the area in which $e_i^* \geq e_p^*$ (see Equation 4 below).
(5) use this new mesh to repeat steps (2) and (3);
(6) assign h_{min} to any elements with $e_i^* \geq 0.9 e_{max}^*$, and halve the element size for elements with $e_i^* \geq e_p^*$ (Eq. 4)
(7) regenerate the mesh using the updated mesh density field.

The permitted error level, e_p^* is given by

$$e_p^* = 0.8 * |e| = 0.8 * \frac{\sum_{i=1}^{n} e_i * \Omega_i}{\sum_{i=1}^{n} |\varepsilon_i^h| * \Omega_i} \tag{4}$$

in which

$$|\varepsilon_i^h| = \left[\left\{ \int_{\Omega_i} (\varepsilon^h)^T * (\varepsilon^h) d\Omega \right\} / \Omega_i \right]^{1/2} \tag{5}$$

There are two cycles involved from step (1) to step (7). The optimal mesh generated using this procedure has worked well.

2.2 *H-adaptive RITSS method*

In large penetration analysis, the *h*-adaptive RITSS FE method is used. As the name implies, incremental small strain analyses are used to solve large deformation problems, by frequent remeshing and interpolation so that the stress field and the soil properties of the old mesh can be interpolated into a new mesh after each remeshing. In *h*-adaptive RITSS analysis, the above 7-step refinement approach is used in the initial mesh generation before remeshing starts. During each remeshing, the mesh density field is only checked once to generate a new mesh according to the updated domain boundary, since the mesh is already close to optimal after the initial refinements. The following procedure is used for large penetration analysis:

(1) generate an optimal mesh for the initial computation domain using the 7-step refinement;
(2) calculate FE results with N steps of small strain analysis using AFENA;
(3) update the domain boundary, check the mesh density field using the error distribution and generate a new mesh;
(4) check if the penetration has reached the required value; if not, go to step (2); otherwise stop.

Steps (2) to (4) are repeated until the desired penetration is obtained. The number of steps of small strain analysis between each remeshing is designed in such a way that the total penetration of N steps remains in a small strain range and is less than half the minimum element size ($0.5h_{\min}$).

3 CENTRIFUGE TESTING

Experimental testing was conducted using the centrifuge facility operated by the Geomechanics Group at the University of Western Australia. The centrifuge, shown in Figure 2, is an Acutronic Model 661 geotechnical fixed beam centrifuge with a radius of 1.8 m and a maximum payload of 40 g-tonnes, and is capable of subjecting 200 kg sample packages to accelerated gravity fields up to 200 g. Control over actuators and other electronic equipment is achieved remotely from the main control room using power slip-rings, with instrumentation data signals and video signals being transferred from the test package to the control room via instrumentation slip-rings. Primary data acquisition and display of results is undertaken in the control room, with additional data being obtained from the in-flight computer which is capable of sampling selected channels at frequencies in excess of 300 Hz. A full description of the centrifuge and associated equipment is available in Randolph et al. (1991), with further detail provided in Watson (1999).

Figure 2. UWA Geotechnical Centrifuge

Testing was performed using the model plate shown in Figure 1. The dimensions of the model are as follows (prototype dimensions at 150g given in brackets):

- diameter = 30 mm (4.5 m);
- thickness of top cap = 2 mm (0.3 m);
- skirt length = 3 mm (0.45 m).

As indicated previously, two different soil types were used as part of the testing program, these being kaolin clay and calcareous silt. Both materials were prepared normally consolidated, and all experimental tests were performed under undrained conditions. A suite of laboratory tests were performed in order to assess the in situ undrained shear strength of the each sample, including CPT, T-bar and vane shear tests. Further discussion of the test methods is provided in Watson (1999).

4 UPPER BOUND MECHANISM

The upper bound mechanism has been used for circular surface foundations by Kusakabe et al (1986). The mechanisms have been extended to allow for embedment of the foundation, either for cylindrical foundation geometries with vertical walls or for geometries that allow complete flow of soil around the foundation (Randolph 1998, Hu & Randolph 1999).

It is straightforward to allow for any given non-homogeneity of the soil strength in such analyses when computing the internal rates of dissipation. Here, the soil strength profile will be restricted to a linear variation with depth from some value, s_{um}, at the soil surface.

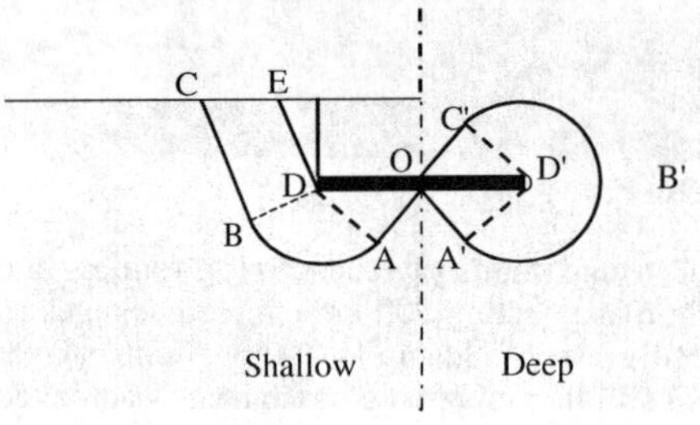

Figure 3. Upper bound mechanism for smooth plate

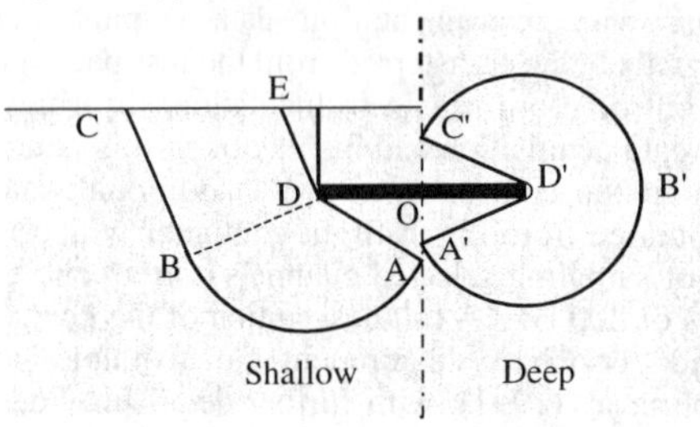

Figure 4. Upper bound mechanism for rough plate

Figures 3 and 4 show the upper bound mechanism for smooth and rough plate with shallow and deep embedment on each side. For shallow embedment, the final exit wedge BCED emerges at an angle that ranges from 45° at very shallow depths, to vertical, with the angle of the fan zone DAB being adjusted accordingly. It should also be noted that the velocity is zero along ABC or A'B'C', so that there is no plastic work done along that surface.

5 RESULTS AND DISCUSSION

The finite element analyses have been conducted using the AFENA finite element package (Carter & Ballam 1990). The soil is modelled as linear elastic, perfectly plastic with Tresca yield criterion. All the analyses simulate undrained conditions with fiction and dilation angles $\phi = \psi = 0$, and a uniform stiffness ratio $E/s_u = 500$ (where E is Young's modulus, s_u is undrained shear strength). The saturated soil self-weight has been taken as $\gamma = 17$ kN/m^3.

Six-noded triangular elements with three internal Gauss points are used in all the analyses. Although this type of element is less suitable than cubic strain element for plastic analysis of axisymmetric problems (Sloan & Randolph 1985), it has been shown that reliable results may be obtained with a refined mesh because of the inexact integration provided by the three Gauss points (Hu & Randolph 1998[1]).

The dimensions of the plate and soil strength profile ($s_u = 0.9z$ kPa, Figure 5) are based on the prototype of centrifuge tests. The plate skirts are ignored. The soil domain is 25 m x 45 m.

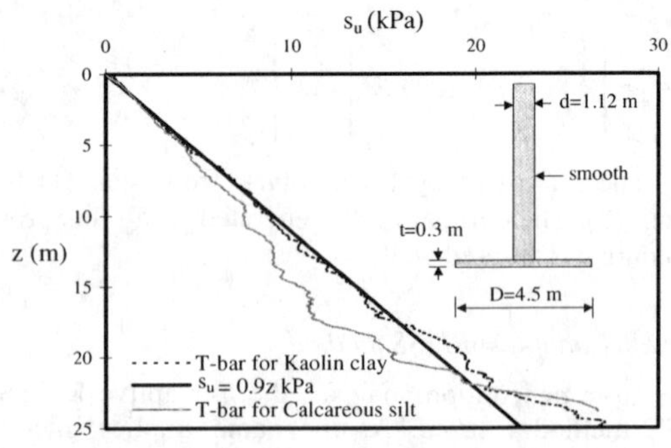

Figure 5. Plate penetrometer and soil strength profile

The penetration analysis starts from the plate base being placed at a depth of 0.3 m, so that the plate upper side is level with the soil surface, avoiding the difficulty of zero surface strength at the plate base.

5.1 *Bearing capacity with pre-embedment*

In bearing capacity analyses, pre-embedment ratios of D_f/D = 0.067, 0.125, 0.25, 0.5, 1, 2, 3 have been analysed, with both fully smooth and fully rough plate-soil interfaces. The embedment D_f is measured from the soil surface to the plate base. The first embedment $D_f/D = 0.067$ is for $D_f = 0.3$ m (the starting depth of penetration analysis).

The minimum element size $h_{min} = 0.1$ m is used. This is larger than the criterion of $h_{min} = 0.005D$ (Hu & Randolph 1998[2]), but mesh dependency analysis has shown that it gave converged results due to the full flow mechanism. The displacement criterion is:

$$\left(\frac{\delta}{D}\frac{E}{s_u}\right)*\left(\frac{kD}{s_{uo}}\right)^{0.8} = 0.03 \qquad (6)$$

but with kD/s_{uo} taken as one when it is less than one.

Soil flow mechanisms from the FE results are shown in Figure 6 for a deeply embedded plate with $D_f/D = 3$. A strong similarity can be seen with the upper bound mechanisms (Figures 3 & 4). However, the zone around the plate outer edge is larger in the upper bound mechanism (distance D'B' in Figure 3 is approximately equal to the plate radius, R) than in the FE results, where D'B' is about 0.6R (smooth) to 0.7R (rough). At shallow embedments, the comparison is similar to the deeply embedded case, with a more extensive deformation field in the upper bound mechanism than found in the FE results.

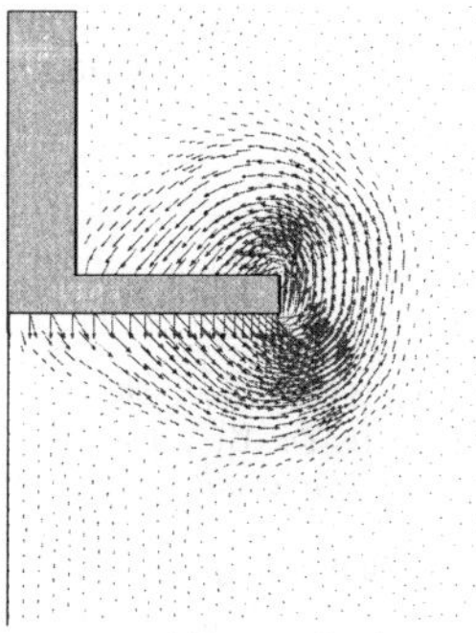

(a) smooth plate

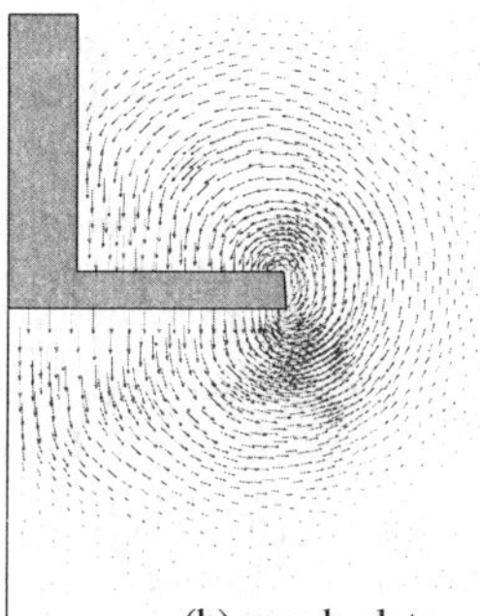

(b) rough plate

Figure 6. Soil flow around plate-penetrometer

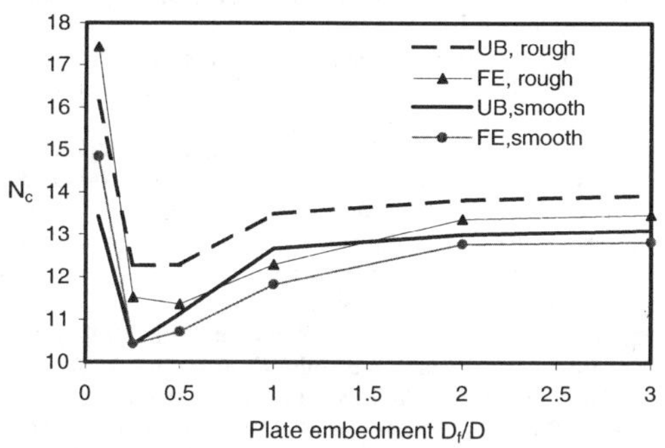

Figure 7. Bearing capacity factor for pre-embedded plate

Figure 7 shows the upper bound (UB) and FE solutions for different embedment. It is apparent that the FE results lie just below the UB solutions, and also lie above the lower bound (LB) solutions for deep embedment, which are 12.4 and 13.1 respectively (for an infinitely thin plate, Martin & Randolph, 2001). The difference in N_c values between fully smooth and fully rough plate is about 5% regardless which method is used.

5.2 *Plate-Penetrometer Penetration Analysis*

In the continuous penetration analysis, the plate-penetrometer is pushed into NC soil to a depth of 3 times the plate diameter (3D), in which D = 4.5m. Soil properties are the same as above. Soil strength profile, $s_u = 0.9z$ kPa (Figure 5) is used and effective soil self weight is $\gamma' = 7$ kN/m^3.

The centrifuge testing has been described in the section above, with the strength profile deduced from T-bar penetration tests. It can be seen in Figure 5 that the soil strength profile of the kaolin clay is very close to the proposed linear variation, although the one for calcareous silt is slightly lower.

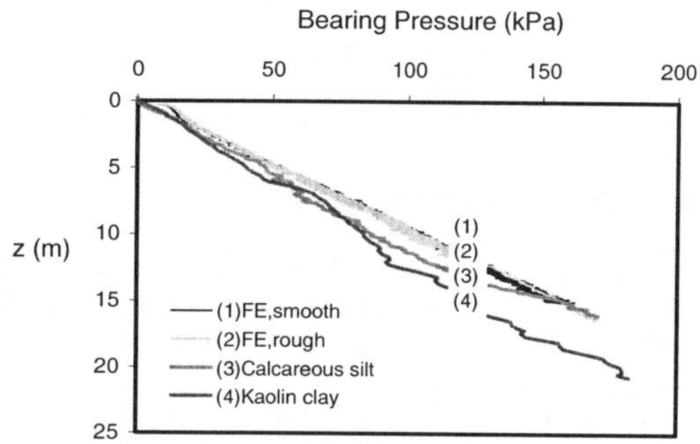

Figure 8. Load-penetration response

The computed and measured penetration-bearing pressure responses are shown in Figure 8. The FE results show little difference between fully smooth and fully rough plate. This is probably due to two compensating effects:

(a) For the rough plate, the plastic zone is larger and the bearing capacity for a pre-embedded plate is larger than for a smooth plate.
(b) For a smooth plate penetrating from the surface, soil can flow around more easily than for the rough plate, hence carrying down less of the low strength soil with the plate.

For calcareous silt (line 3 in Figure 8), friction at the plate-soil interface is expected to be higher than for the kaolin clay (Watson 1999). This may help account for a slightly higher bearing resistance, even though the soil strength profile was slightly lower than for the kaolin clay. Overall though, the FE and experimental results are in excellent agreement.

Bearing capacity factors from penetration tests and FE analyses are shown in Figure 9. Equations 1 and 2 are used for calculating N_c, with no correction for q_{net}. In the centrifuge test results, local s_u values are used directly from the T-bar strength profile.

For the centrifuge data, the plate bearing capacity factor N_c is higher in calcareous silt than in kaolin clay. This may possible be explained by higher plate-soil roughness, or possibly local consolidation occurring during the later stages of the test. The final bearing capacity from the FE results stabilizes at about 11.7, which is lower than for the pre-embedded plate due to carrying down softer soil with the plate.

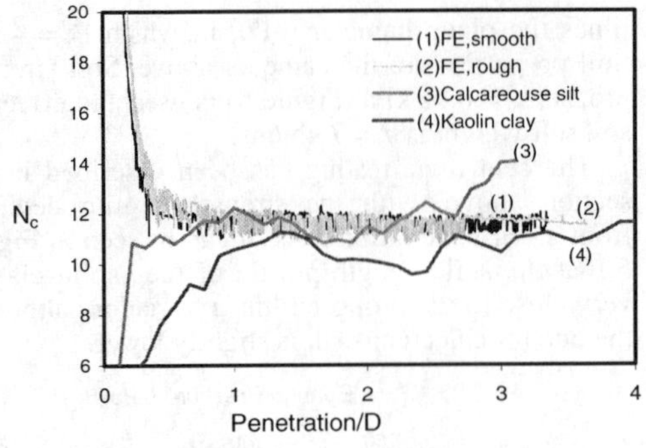

Figure 9. Bearing capacity factor of plate penetrometer

6 CONCLUDING REMARKS

The bearing resistance of plate penetrometers has been investigated for fully smooth and rough plate-soil interfaces, and for pre-embedded and continuously penetrating plates.

The *h*-adaptive RITSS (Remesh and Interpolation Technique with Small Strain model) FE method has provided a powerful technique to investigate the soil behaviour during deep plate penetration. Combined with physical centrifuge modelling, the penetration-load response and soil flow mechanism can be investigated in detail.

The upper bound mechanism for circular plates, modified for the effect of embedment, has shown very similar flow mechanisms and bearing capacity factors to the FE analysis for pre-embedded plates. However, the analytical solutions cannot take account of the disturbance of the soil profile during penetration, and thus overestimates N_c values for deep penetration in normally consolidated soil.

From pre-embedded plates, the difference in N_c values between fully smooth and fully rough plates is about 5 % in NC soil, in all FE, upper bound and lower bound analyses.

For deep penetration, the FE results indicate negligible effect of the interface roughness, and give N_c values that are lower than for the pre-embedded case, with a limiting value at depth of about 11.7 (some 10 % less than for pre-embedded plates).

The N_c values from centrifuge test data are in reasonable agreement with those from the FE analyses, particularly for the test in kaolin clay. The test in calcareous silt shows higher bearing resistance, possibly due to friction (although this would be inconsistent with the FE results). An alternative reason might be the effect of local consolidation, particularly between the skirts of the plate penetrometer.

7 ACKNOWLEDGEMENTS

The research described in this paper forms part of the activities of the Special Research Centre for Offshore Foundation Systems, established and supported under the Australian Research Council's Research Centres Program.

REFERENCES

Carter, J. P. and Balaam, N. 1990. *User's manual, AFENA,* Geotechnical Research Centre, The University of Sydney.

Chen, W. F. 1975, *Limit analysis and soil plasticity,* Elsevier, New York.

Hu, Y. & Randolph, M. F. 1998[1]. A practical approach for large deformation problems in soil, *Int. J. Num. And Anal. Methods in Geomech.,* 22: 327-350.

Hu, Y. & Randolph, M. F. 1998[2]. H-adaptive FE analysis of elasto-plastic non-homogeneous soil with large deformation', *Int. J. Compu. Geomech.,* 23(1/2): 61-84.

Hu, Y., Randolph, M. F. & Watson, P. G. 1999. Circular skirted foundation on non-homogeneous soil: bearing capacity and penetration, *J. Geot. Geoenv. Eng., ASCE,* 125(11): 924-935.

Kusakabe, O., Suzuki, H. & Nakase, A. 1986. An upper bound calculation on bearing capacity of a circular footing on a non-homogeneous clay, *Soils and Foundations,* 114(11): 1277-1291.

Lunne, T., Robertson, P. K. and Powell, J. J. M. 1997. *Cone Penetration Testing in Geotechnical Practice,* Blackie Academic and Professional, London.

Martin, C.M. & Randolph, M.F. 2001. Applications of the lower and upper bound theorems of plasticity to collapse of circular foundations, *Proc. 10th IACMAG Conf.,* Tucson.

Randolph M.F., Jewell R.J., Stone K.J.L. and Brown T.A. 1991. Establishing a new centrifuge facility, *Proc. Int. Conf. on Centrifuge Modelling - Centrifuge 91,* Boulder, Colorado, 3-9.

Randolph, M.F. 1998., Modelling of offshore foundations: Part 1 – Bearing Capacity, EH Davis Memorial Lecture, *Australian Geomechanics,* 33(2): 5-22.

Sloan, S. & Randolph, M. F. 1985. Numerical prediction of collapse loads using finite element methods, *Int. J. Num. And Anal. Methods in Geomech.,* 6: 47-76.

Tani. K. & Craig, W.H. 1995. Bearing capacity of circular foundation on soft clay of strength increasing with depth, *Soils and Foundations,* 35(4): 21-35.

Watson, P. G., Newson, T.A. & Randolph, M.F. 1998. Strength profiling in soft offshore soils, *Proc. 1st Int. Conf. On Sit Characterisation – ISC'98,* Atlanta, 2: 1389-1394.

Watson, P.G. & Randolph, M.F. 1998. Torsional vane and plate load tests, *Proc. Int. Conf. Centrifuge '98,* Tokyo, 1: 167-173.

Watson, P.G. 1999. *Performance of Skirted Foundations for Offshore Structures,* PhD Thesis, The University of Western Australia.

Zienkiewecz, O. Z. & Zhu, J. Z., 1993, The supperconvergent patch recovery and a posteriori error estimates. Part 1: The recovery technique, *Int. J. Numer. Eng.,* 33: 1331-1364.

Zienkiewicz, O.Z., Huang, M. & Pastor, M. 1995. Localization problems in plasticity using finite element solutions with adaptive remeshing, *Int. J. Numer. And Anal. Methods in Geomech.,* 19(2): 127-148.

Computer Methods and Advances in Geomechanics, Desai et al. (eds) © 2001 Balkema, Rotterdam, ISBN 90 5809 183 X

Model discrimination for the dissipation test

E. Imre
Geotechnical Research Group, Hungarian Academy of Sciences and Budapest University of Technology and Economics, Hungary

ABSTRACT: The evaluation of the dissipation test is based on the following assumptions: (i) simple one dimensional model is appropriate if the sensor is situated well above the tip, (ii) the initial condition can be identified with a proper inverse problem solution method. By these, the following difficulties (Baligh & Levadoux (1980)) are resolved: (i) the much more complex (e.g. 2 D, non-linear) nature of the problem, (ii) the uncertainties of the initial pore water pressure distribution. In the paper the pore water pressure response of the existing (the previous and the suggested) 1D linear consolidation models are compared. Simulation, moreover model fittings are made using dissipation test data.

1 INTRODUCTION

A research work was initiated on the evaluation of the rheological cone penetration tests (Table 1). At first, the simplest case (sensor well above the tip) was investigated. Model, non-linear inverse problem solution method and reliability testing method are necessary for the evaluation of the measured rheological cone penetration test data.

It can be shown that the linear consolidation models available are not verified by measured data. The inverse problem solution methods available are not "automatic" in the sense that the global minimum of the merit function is to be found by a trial and error procedure. This is tedious in case when the number of parameters is greater than 3-4.The reliability of the identified parameters is generally not investigated.

Therefore, a model, an inverse problem solution method and, some reliability testing methods were suggested (Imre 1996, Imre & Rózsa 1998). This paper is related to the validation of the pore water pressure response of the suggested consolidation model with the use of the elaborated procedures.

Table1 Rheological cone penetration tests

Type of test	Measured stress variable
Dissipation test	pore water pressure
Piezo-lateral stress cell test	pore water pressure, radial total stress
Simple test	side friction, cone resistance

2 METHODS

The dissipation curve solution of the previous models (Soderberg 1962, Randolph & Wroth 1979) and, the suggested model (Imre & Rózsa 1998) were compared, as follows.

Dissipation curves were simulated for various initial conditions. The resulting theoretical dissipation curves were compared with experimental data known from earlier publications (Poulos et al. 1980, Baligh et al. 1986).

The models were fitted on dissipation test data measured for the Bothkennar soft clay in five filter positions (Lunne et al. 1992). The initial conditions were identified in conjunction with the model fitting and, the reliability of the identified parameters was tested (Imre 1996).

As a further check, the identified initial conditions were related to the strain path prediction (Baligh 1986) and, the identified coefficients of consolidation were compared with the value given by Lehane & Jardine (1994) on the basis of back-computations from in situ tests.

3. SIMULATIONS

3.1 One Dimensional Consolidation Models

The differential equation and, the boundary conditions of the earlier linear 1D consolidation models concerning the pore water pressure are the same.

Therefore, the analytical pore water pressure solutions of the previous models are identical. This solution – the so called Soderberg's model solution - is as follows:

$$u(t,r)=E_{oed}\sum_{k=0}^{\infty}C_k\exp(-\lambda_k^2 c_h t)\,B_{0,k}(r) \quad (1)$$

$$B_{0,k}(r)=J_0(\lambda_k r)+\mu_k Y_0(\lambda_k r) \quad (2)$$

and, the solution of the suggested mode is as follows:

$$u(t,r)=E_{oed}\sum_{k=0}^{\infty}D_k\exp(-\lambda_k^2 c_h t)\left(B_{0,k}(r)-B_{0,k}(r_1)\right) \quad (3)$$

where J_0 and Y_0 are Bessel functions of the first and second kind, zero order; λ_k , μ_k are model constants differing for the two models, depending on the boundary conditions at r_0 and r_1; C_k , D_k (k=1..∞) are coefficient depending on the initial condition, c_h is the coefficient of consolidation.

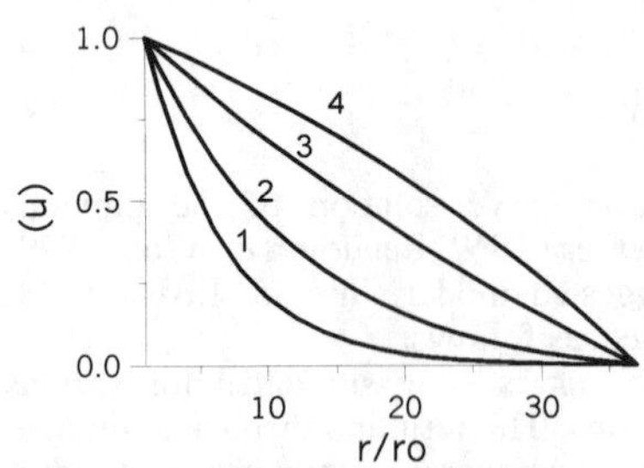

Figure 1 Initial conditions 1 to 4

3.2 Computations, results

The analytical solutions were approximated by 12 terms (Imre & Rózsa 1998). Constants λ_k, μ_k were computed for the ratio of r_1/r_0 = 37 assuming on the basis of the work of Baligh (1986). Parameters C_k , D_k were calculated by linear model fitting for some previously specified closed-form initial condition functions.

The resulting dissipation curves were dependent on the mean initial pore water pressure ($u_m(0)$) the value of which increased from function 1 to 4 as it can be seen in Figure 1.

The elapsed time (t_κ) for any κ was longer the mean of the initial pore water pressure ($u_m(0)$) was greater (i.e. the curves "moved" from left to right) for the Soderberg's model (Figs 1-2).

The same was true for the suggested model in case of convex distributions only (curves 1-3). The reverse was true in case of concave distributions and, as a result, the dissipation curves related to distributions 3 and 4 nearly coincided (Figs 1-2).

For great degrees of dissipation ($\kappa \approx 99\%$) the elapsed times t_κ were about 3 times larger for the Soderberg's model, than for the suggested model,

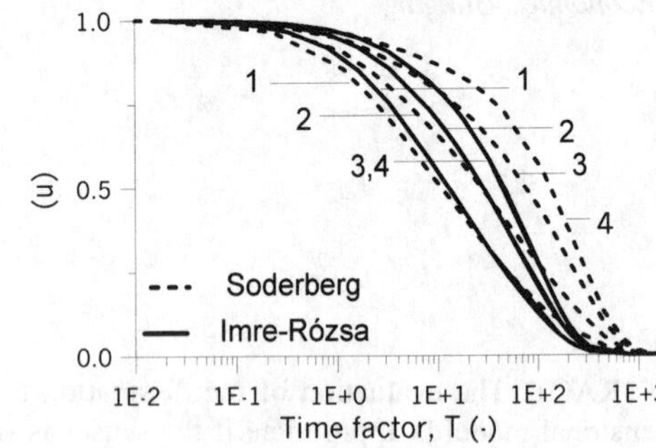

Figure 2 Dissipation curves for initial conditions 1 to 4

4. MODEL FITTING

4.1 Methods

The least squares merit function was used in the following normalised form called fitting error:

$$\left(F(p)\right)=\frac{\sqrt{\sum_i\left[(u_m(t_i)-u(t_i,p)\right]^2}}{u_m(0)\sqrt{n}}\,100[\%] \quad (4)$$

where u, u_m are the computed and the measured pore water pressure at $r=r_0$, $t=t_i$ $(i=1..n)$, n is the number of measured data, $\boldsymbol{p}$ is the parameter vector containing M parameters (M=2..13) as follows.

The parameters were the non-linearly dependent coefficient of consolidation (c_h) and, the linearly dependent initial condition parameters (C_k and D_k, k=1..m). The number of the identified initial condition parameters m was between one to twelve.

The least squares merit function was evaluated in the points of a co-ordinate mesh generated in the subspace of the parameters and, the minimal function value was selected.

The computational work was reduced by a subminimisation (i.e. eliminating the linearly dependent parameters, Imre 1996). 40 SVD solutions (Press et al 1985) and 40 function value evaluations needed for each model fitting.

In the frame of the reliability tests, the uniqueness was tested and, the parameter error was estimated. Results of the latter are presented here only.

The standard deviation of the parameters was determined according to the conventional procedure (Press et al 1985) and, was presented in the form of "variance" as the ratio of the computed standard deviation and the identified parameter.

4.2 Results

Results concerning those cases are presented here where only one initial condition parameter was identified (Table 2, Figs 3-6) since the identified initial condition functions were generally not physically admissible otherwise.

The fitting error and, the variance of c_h decreased from filter position A to E except that filter position D gave worse fit than filter position C (Table 2).

The identified coefficient of consolidation (c_h) decreased by about one order of magnitude. It was always larger for the Soderberg's than for the suggested model (Table 2).

The best solutions - related to filter position E in both cases – can be characterised as follows.

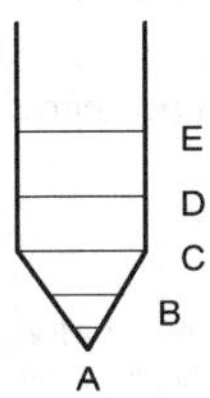

Figure 3. Filter positions for the dissipation test

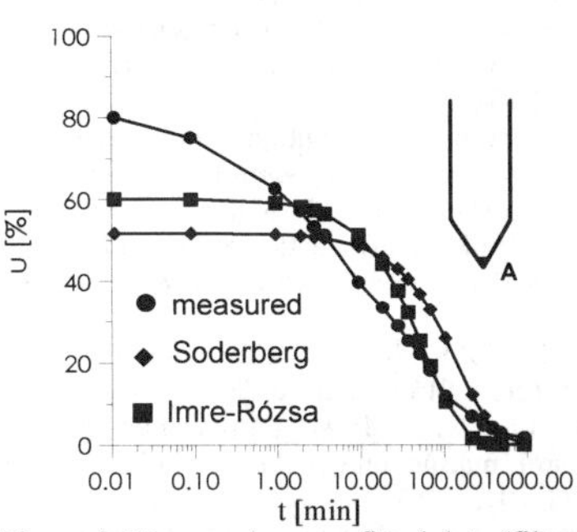

Figure 4. Measured versus fitted data (filter ps. A)

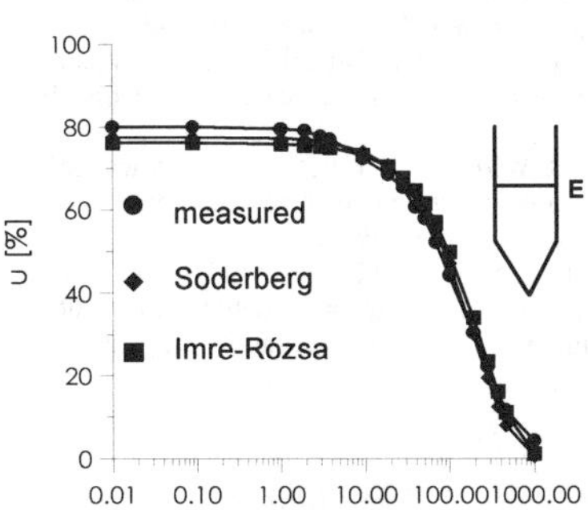

Figure 5. Measured versus fitted data (filter ps. E)

Assuming that the "true" coefficient of consolidation (c_h) is equal to 0.01 cm^2/s (Lehane & Jardine 1994) it can be stated that the suggested model gives better estimation (0.02 cm^2/s) than the Soderberg's model (0.06 cm^2/s).

The identified initial conditions deviate from the expected shape (Baligh 1986) and, the deviation is larger for the Soderberg's model than for the suggested model (Fig. 6).

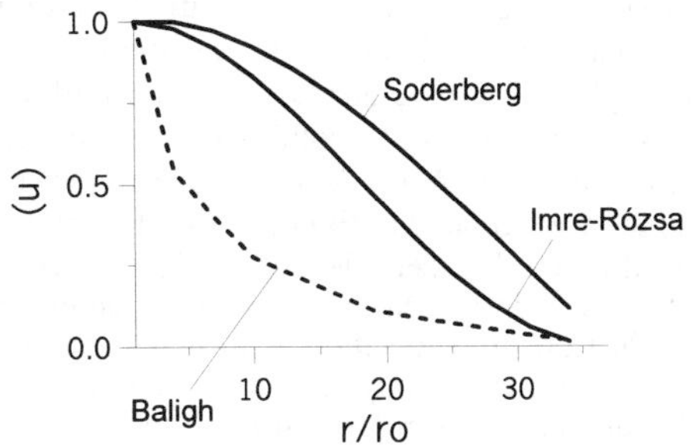

Figure 6. Identified initial conditions

5. DISCUSSION

5.1 Simulation

It is known from earlier works that the rate of pore water pressure dissipation computed by the previous models is greater in the beginning and, smaller later, than the measured one (Figs 7- 8).

According to the results of the simulations (Fig. 2), the dissipation is "slower" at small degrees of consolidation and, "faster" at great degrees of consolidation for the suggested model, than for the Soderberg's model.

These results indicate that the suggested model may explain the differences mentioned by the earlier works (Figs 7, 8).

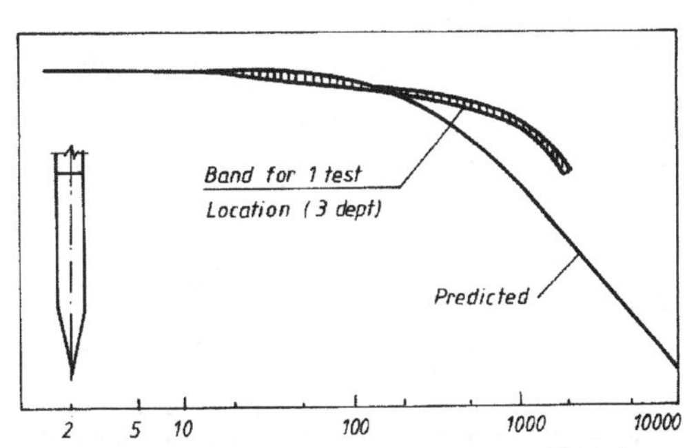

Figure 7. Evaluation of linear 2D uncoupled predictions in Boston Blue Clay (after Baligh et al. 1986)

Table 2. The fitting error, the identified coefficient of consolidation (c_h) and, its variance

	Soderberg			Imre & Rózsa		
	Fitting error [%]	ch [cm2/s]	variance	Fitting error [%]	ch [cm2/s]	variance
A	9	0.31	0.55	9	0.13	0.56
B	8	0.22	0.45	8	0.09	0.45
C	5	0.14	0.29	5	0.06	0.28
D	6	0.07	0.37	6	0.03	0.37
E	3	0.06	0.25	4	0.02	0.23

5.2 Model fitting

The dissipation curves of the two models were fitted on data measured in five filter positions.

The goodness of fit the increased, the error of the identified coefficient of consolidation (c_h) reduced from filter position A to E. The parameter error seemed to be acceptable in filter position E only.

The value of the identified coefficient of consolidation (c_h) decreased by about one order of magnitude from filter position A to E that can be interpreted in such a way that 1D consolidation model is applicable for filter position E only.

These results are in agreement with the observation of Baligh & Levadoux (1986) and Teh & Houlsby (1988) who showed that 1D solution is acceptable in filter position E.

The identified c_h was always larger for the Soderberg's model than, for the suggested model. This can be explained by the difference of the models. The 99% dissipation time (t_{99}) is about 3 times larger for the Soderberg's model, than for the suggested model.

The estimation for c_h of the suggested model was closer to the value given by Lehane & Jardine (1994) than the one of the Soderberg's model. This result needs some further check.

The identified initial conditions deviated from the expected shape. The deviation – indicating some modeling error - is to be checked with a more sophisticated identification strategy.

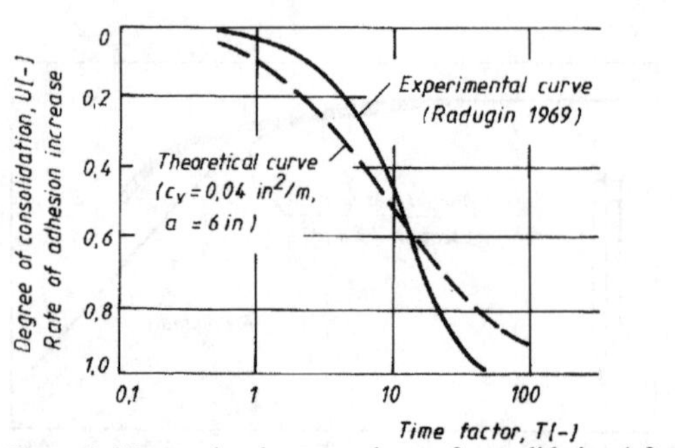

Figure 8. Measured and computed rate of consolidation (after Poulos et al. 1980)

6. CONCLUSION

The evaluation of the dissipation test was investigated in the special case when the sensor is situated well above the tip.

The applied procedure (model fitting, reliability tests) was successfully used implying either the Soderberg's model or, the suggested model.

The results of the simulations and, the results of the model fittings indicate that the suggested model is of better performance than the Soderberg's model. Further research seems to be needed.

REFERENCES

Baligh, M. M.; Martin, R. T.; Azzouz, A. S.; Morisson, M. J. (1985). The piezo-lateral stress cell. *Proc. of the 11th Inter. Conf. on Soil Mech. and Found. Eng.* San Francisco; Vol.2., pp.841-844.

Baligh, M. M. (1986). Undrained deep penetration, II. pore pressures. *Geotechnique*, 36(4): 487-503.

Baligh, M. M.; Levadoux, J. N. (1986). Consolidation after undrained penetration. II. Interpretation. *Jl. of Geot. Eng. ASCE*, 112(7): 727-747.

Imre, E. (1996). Inverse problem solution with a geometrical method. *Proc. of the 2nd Int. Conf. on Inverse Problems in* Engineering. Le Croisic, France.331-338.

Imre, E. & Rózsa, P. 1998. Consolidation around piles. *Proc. of 3rd Seminar on Deep Foundations on Bored and* Auger *Piles.* Ghent 385-391.

Imre, E. & Rózsa, P. (1999). Modelling for radial consolidation. *Proc. of XIIth ECSMGE,* Amsterdam. (1017-1027).

Lehane, B. M.; Jardine, R. J. (1994) Displacement-pile behaviour in a soft marine clay *Canadian Geotechnical Journal*. 31. 181-191.

Lunne, T; Robertson, P.K.; Powell, J.J.M. (1992): *Cone Penetration testing.* Blackie Academic & Professional.

Poulos, H. G.; and Davis, E. H. (1980). *Pile foundation analysis and design.* John Wiley and Sons, New York.

Press, W.H.; Flannery, B.P.; Teukolsky, S.A.; Wetterling, W.T. (1986): *Numerical Recipes*. Cambridge Univ. Press, Cambridge

Randolph, M. F. & Wroth, C. P. (1979). An analytical solution for the consolidation around displacement piles. *I. J. for Num. Anal. Meth. in Geomechanics*, 3:217-229.

Teh, C.I. and Houlsby, G.T. (1988). Analysis of the cone penetration test by the strain path method. Proc. 6th Int. Conf. on Num. Meth. in Geomechanics, Innsbruck.

Computer Methods and Advances in Geomechanics, Desai et al. (eds) © 2001 Balkema, Rotterdam, ISBN 90 5809 183 X

Non-linear finite element analysis of calibration chamber plate load tests for sands

Jun-Hwan Lee & Rodrigo Salgado
Purdue University, West Lafayette, Ind., USA

ABSTRACT: Well designed foundations induce a stress-strain state in the soil that is far from both the elastic and collapse loading ranges. In order to accurately predict working settlements, therefore, analyses more sophisticated than simple elastic analyses are required. Calibration chamber plate load tests are sometimes used to investigate the base load-settlement relationship of non-displacement piles in sand. In this paper, we describe a 3D non-linear elastic-plastic analysis based on a hyperbolic stress-strain relation and the Drucker-Prager failure criterion. Calibration chamber tests are then analyzed using the finite element method based on the 3D non-linear elastic-plastic stress-strain model. Both experimental and numerical results will be presented and compared.

1 INTRODUCTION

Based on the method of installation, piles are classified as either displacement or non-displacement. Calibration chamber plate load tests have been used to investigate the base load-settlement relationship of axially loaded non-displacement piles in sands (Parkin 1991, Ghionna et al. 1994). Although a full-scale pile load test usually provides the best information for design at a specific site, calibration chamber testing is preferable for studying more generically the load-settlement response of piles in sands for various soil states, as calibration chamber tests can be performed at any desired level of relative density and stress state under controlled conditions. In such tests, a cylindrical sand specimen is carefully prepared, consolidated to a desired stress state, and tested (Salgado et al. 1998a). Because calibration chambers have finite sizes, the possibility of size and boundary effects arises. For the results of calibration chamber tests to be effectively applied to pile design, the possible size effect should be properly addressed. If size effects are well understood, calibration chambers can be used to experimentally assess pile base resistance under controlled conditions.

Numerical modeling can also be used to investigate pile load-settlement response. It has the advantages that various boundary and soil conditions can easily be modeled. For successful analyses to be performed, the behavior of soil should be modeled realistically.

In the present paper, a non-linear, elastic-plastic, three-dimensional model based on the modified hyperbolic model proposed by Fahey and Carter (1993) and the Drucker-Prager failure criterion is presented. The three-dimensional elastic-plastic model is then used in the analyses of calibration chamber plate load tests and in the investigation of size effects in such tests. These analyses aim to (1) verify the accuracy of the model predictions of plate resistance for different levels of settlement and (2) assess size effects on the experimentally measured plate resistance. This second goal is important in order to reach conclusions regarding use of calibration chambers to assess pile base resistance. Validation of the numerical method validates the use of numerical analysis for the same purpose.

2 STRESS-STRAIN RELATIONSHIP

2.1 *Non-linear elastic model*

The hyperbolic family of soil models has been often used to describe non-linear soil behavior observed from the early stages of loading (Kondner 1963, Duncan and Chang 1970, Hardin and Drnevich 1972, Tatsuoka et al. 1993). Fahey and Carter (1993) proposed a modification of the original hyperbolic equation as follows:

$$\frac{G}{G_o} = 1 - f\left(\frac{\tau}{\tau_{\max}}\right)^g \qquad (1)$$

where G = secant shear modulus; G_o = initial shear modulus; τ, τ_{max} = current and maximum shear stress; f, g = material parameters. Equation 1 represents the degradation of shear modulus from its initial maximum value G_o according to the stress level τ/τ_{max}.

Sand usually behaves as a linear elastic material with shear modulus G_o for shear strains up to approximately 10^{-5}, after which the stress-strain relationship is strongly non-linear. In this study, the following empirical equation, based on the work of Hardin and Black (1966), was used to estimate the initial maximum shear modulus of sand:

$$G_o = C_g \frac{(e_g - e_o)}{1+e_o} p_a^{(1-n_g)} (\sigma'_m)^{n_g} \tag{2}$$

where C_g, n_g, and e_g = intrinsic material variables; e_o = initial void ratio; p_a = reference pressure = 100 kPa ≈ 1 kgf/cm^2; and σ'_m = initial mean effective stress in the same unit as p_a. Considering an initial anisotropic stress state, Equation 1 may be rewritten as:

$$\frac{G}{G_o} = 1 - f\left(\frac{\tau - \tau_o}{\tau_{max} - \tau_o}\right)^g \tag{3}$$

where τ_o = initial shear stress due to the initial K_o-condition. Using the stress tensor invariants, the non-linear elastic stress-strain relationship given by Equation 3 can be represented for three-dimensional stress conditions as follows:

$$\frac{G}{G_o} = 1 - f\left(\frac{\sqrt{J_2} - \sqrt{J_{2o}}}{\sqrt{J_{2max}} - \sqrt{J_{2o}}}\right)^g \tag{4}$$

where J_2 = second invariant of deviatoric stress tensor. Equation 4 represents the variation of shear modulus in terms of shear stress level for three dimensions. The shear modulus, however, may also increase due to the increase of confinement. In order to account for the influence of confinement on shear modulus during loading, the following relationship is used instead of Equation 4:

$$\frac{G}{G_o} = \left[1 - f\left(\frac{\sqrt{J_2} - \sqrt{J_{2o}}}{\sqrt{J_{2max}} - \sqrt{J_{2o}}}\right)^g\right]\left(\frac{I_1}{I_{1o}}\right)^{n_g} \tag{5}$$

where I_1 and I_{1o} are the first invariants of the stress tensor at the current and initial states, respectively. The parameter n_g is the same as appears in Equation 2: in both equations n_g represents the dependence of the shear modulus on confinement.

The elastic stress-strain relationship may be expressed by two constants; the bulk modulus K and the shear modulus G are often used. The bulk modulus depends mainly on the amount of confining stress (Naylor et al. 1981). Based on the discussion of the K-G model by Naylor et al. (1981), the tangent bulk modulus K_t can be represented by the following equation:

$$K_t = D_s \cdot (\sigma'_m)^{n_k} (p_a)^{(1-n_k)} \tag{6}$$

where D_s = material constant that can be calculated from the initial values of bulk modulus and confining stress; and n_k can be taken as 0.5. The values of initial bulk modulus may be obtained from the initial shear modulus G_o and the initial Poisson's ratio $\nu_o \approx$ 0.1 – 0.15.

2.2 *Plastic model with non-linear failure envelope*

The Drucker-Prager failure criterion was adopted to describe failure and post-failure soil response. The Drucker-Prager failure criterion is given by:

$$F = \sqrt{J_2} - (\alpha I_1 + \kappa) = 0 \tag{7}$$

in which α = frictional shear strength parameter and κ = "cohesion" parameter. The α parameter in the Drucker-Prager failure criterion can be obtained from the peak friction angle as a function of relative density and confining stress. The peak friction angle in sands can be expressed in terms of the friction angle at the critical state and the dilatancy angle. The critical-state friction angle for a given soil is independent of stress state and density. Bolton (1986) proposed the following equation to estimate the peak friction angle in sand:

$$\phi_p = \phi_c + 0.8 \cdot \psi \tag{8}$$

where ϕ_p = peak friction angle, ϕ_c = critical state friction angle and ψ = dilatancy angle given by

$$\psi = \begin{cases} 6.25 I_R & \text{for plane-strain conditions} \\ 3.75 I_R & \text{for triaxial conditions} \end{cases} \tag{9}$$

The dilatancy index I_R in (9) is given by

$$I_R = I_D[Q + \ln(\frac{p_a}{100P'_p})] - 1 \tag{10}$$

where I_D = relative density (as a number between 0 and 1); P'_p = mean effective stress at peak strength in the same unit as p_a; and Q = intrinsic material parameter, approximately equal to 10 for clean silica sands. Equations 8 through 10 were used to define the non-linear Drucker-Prager failure surface.

3 NUMERICAL MODELING OF CALIBRATION CHAMBER TESTS

A total of 30 plate load tests with Ticino sand (see Table 1) were performed within a flexible calibration chamber with a diameter of 1.2 m and height of 1.5 m (Ghionna et al. 1994). Two typical relative densities were used in tests: medium dense ($D_R \approx$

50%) and dense ($D_R \approx 90\%$). The load-settlement curve for each calibration chamber plate load test was predicted numerically using the finite element method with the non-linear elastic-plastic model. The commercial program ABAQUS, with a subroutine specifically written for the three-dimensional elastic-plastic model presented previously, was used in the calculations. The values of f and g in Equation 5 were obtained from triaxial test results by Giuseppe (1991) and Vecchia (1991). The values of f and g used in the analyses were f = 0.97, g = 0.20 for D_R = 50% and f = 0.95, g = 0.26 for D_R = 90%.

Table 1. Properties of Ticino sand (after Ghionna et. 1994).

G_s*	ϕ_c	e_{max}	e_{min}	C_g	n_g	e_g
2.62	34.8°	0.922	0.573	647	0.44	2.27

* Specific gravity

Figure 1 shows a typical finite element mesh used to model the calibration chamber plate load tests. The elements were eight-noded axi-symmetric elements with four internal integration points. The finite element model shown in Figure 1 was plotted with a visual angle equal to 180° for better visualization. The steel casing was modeled as a linear elastic material with very large stiffness. The casing elements were set to be fixed during loading, as they were in the tests.

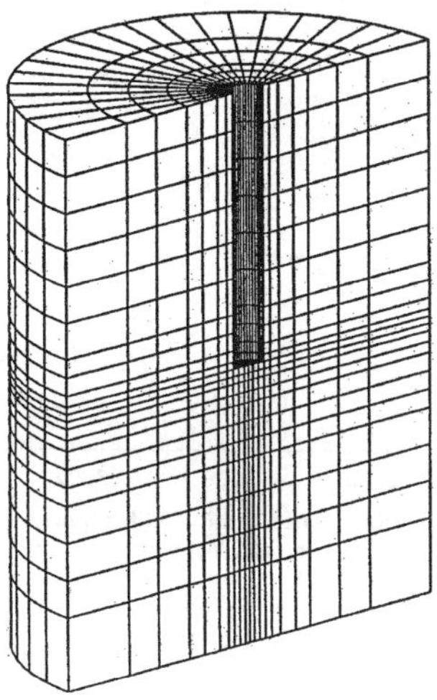

Figure 1. Finite element mesh for modeling calibration chamber tests.

In the calibration chamber tests, four different types of boundary conditions (see Table 2) were used. In the numerical analyses, accordingly, the boundary conditions imposed on the finite element mesh were those used in the actual tests.

Table 2. Boundary conditions in calibration chamber tests.

B.C.	Lateral condition	Top/bottom condition
BC1	σ_h = constant	σ_v = constant
BC2	u = 0*	v = 0**
BC3	σ_h = constant	v = 0
BC4	u = 0	σ_v = 0

* u represents horizontal displacement.

** v represents vertical displacement.

Figure 2 shows typical load-settlement curves for the medium dense and dense sand. The curves extend to a settlement of 15 mm, corresponding to a relative settlement s/B (defined as the ratio of vertical settlement s to pile base or plate diameter B) of 15%. The numerical results compare favorably with the measured responses in both cases. Figure 3 shows measured versus predicted plate unit load at s/B = 10% for all calibration chamber tests. Overall

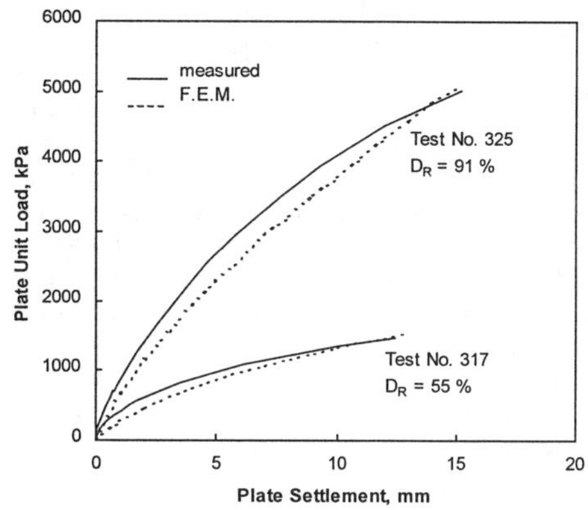

Figure 2. Measured and predicted load-settlement curves for medium dense and dense sands.

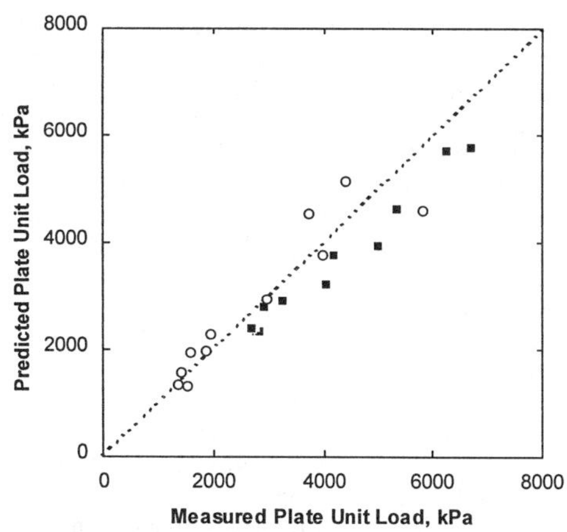

Figure 3. Measured and predicted plate unit load in calibration chamber tests for s/B = 10%.

agreement is very satisfactory, with a maximum error of about 20%. There is a slight underestimation of the plate resistance for the dense sand; this may be due to differences between the dilational response of the sand under triaxial loading conditions (from which f and g were obtained) and plate load test loading conditions.

4 INVESTIGATION OF SIZE EFFECTS

Because of their finite size, calibration chambers do not perfectly reproduce free-field conditions. As a result, measurements made in a calibration chamber sand specimen are different from what would be observed in the field for the same relative density and stress state. These difficulties tend to be more pronounced, the smaller the chamber; hence the term *size effect*. Calibration chamber size effects have been extensively studied in connection with cone penetration testing, where large deformations take place (e.g., Schnaid and Houlsby 1991; Mayne and Kulhawy 1991; Salgado et al. 1998a). It has been determined that size effects in calibration chambers are more significant in dense sands than in loose sands. In plate load testing, the displacements induced in the sand specimen are much less than those caused by cone penetration. On the other hand, the plate diameter used in the study (100 mm) is much larger than typical cone diameters (most commonly 35.7 mm).

In order to investigate size effects in calibration chamber plate load tests, numerical analyses of both full-scale non-displacement piles and calibration chamber plate tests were performed. For studying the loading of non-displacement piles, three 60-cm diameter piles with lengths L = 5 m, 10 m and 20 m were analyzed. The piles were positioned within a granular soil deposit with assumed values of D_R = 30, 50, 70 and 90% in such a way that the initial vertical and lateral effective stresses at the pile base level assumed the following values: (1) σ'_{vo} = 100 kPa, σ'_{ho} = 43 kPa; (2) σ'_{vo} = 200 kPa, σ'_{ho} = 86 kPa and (3) σ'_{vo} = 400 kPa, σ'_{ho} = 172 kPa. All of these states correspond to a normally consolidated condition with K_0 taken equal to 0.43. Plate load tests in a calibration chamber were also simulated numerically for these same values of densities and stresses.

Comparison of the base unit load in the full-scale pile load tests with plate unit load in the calibration chamber tests for the same relative settlement level provides the basis for conclusions regarding size effects. The sand adopted was Ticino sand, whose properties were discussed previously. Both the pile and the soil were modeled with eight-noded axisymmetric elements, and interface elements allowing slippage were used between the pile and the soil.

The comparison of pile load tests and chamber tests were made for all boundary conditions (defined in Table 5). The analysis shows that results for BC4 were similar to those for BC1, and results for BC3 were similar to those for BC2. This illustrates the overwhelming influence of the lateral boundary condition on plate unit load.

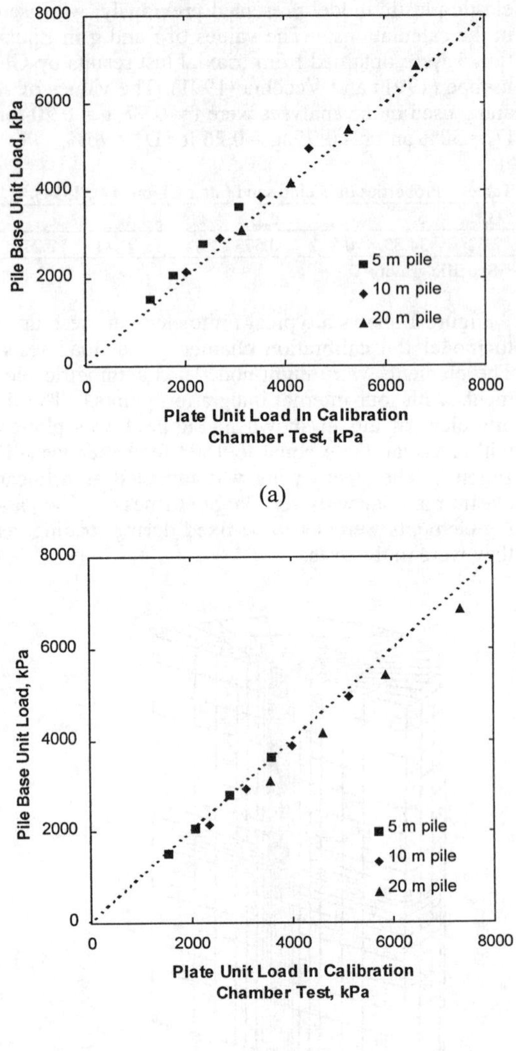

Figure 4. Comparison of pile base resistance with plate resistance in calibration chamber tests: (a) BC1 and (b) BC2.

Figure 4 shows pile base unit loads versus plate unit loads in calibration chamber tests with BC1 and BC2 at s/B = 10%. Differences between pile base unit load and plate unit load are small for all relative density levels. It is observed, however, that the plate unit loads under BC1 conditions were smaller than the pile base unit loads, and more substantially so in the case of short piles, for which confinement is low at the level of the pile base. The differences between pile base unit load and plate unit load were 14 to 16% for L = 5 m, 7 to 9% for L = 10 m, and 1 to 3% for L = 20 m.

The results for BC2 conditions, as shown in Figure 4(b), differ from those for BC1 conditions in that the plate unit loads from the calibration chamber plate load tests were found to be greater than pile base unit load values at corresponding conditions. The differences were 1 to 3% for L = 5 m, 4 to 9% for L = 10 m, and 8 to 15% for L = 20 m. These differences were found to be more significant for the long piles, for which confinement is high at the pile base level.

5 CONCLUSION

Calibration chamber plate load tests have been used to investigate the base load-settlement relationship of non-displacement piles. In the present study, the calibration chamber plate load tests were analyzed through the finite element method using the three-dimensional non-linear elastic-plastic model. The stress-strain model used in this study is based on the hyperbolic stress-strain relationship, and uses intrinsic and state soil variables related to soil shear strength and stiffness as input parameters. A series of calibration chamber tests performed by Ghionna et al. (1994) were modeled and compared with the results of finite element analyses. Predicted plate unit loads were in good agreement with measured results for both dense ($D_R \approx 90\%$) and medium dense ($D_R \approx 50\%$) sand, showing relative errors not larger than about 20% at s/B = 10%.

Calibration chamber size effects, resulting from the finite size of the chamber, were also investigated for different relative densities and boundary conditions using finite element analysis. The comparison was made between pile base unit load for piles loaded under field conditions and plate unit load from the calibration chamber tests. The piles were modeled with three different pile lengths and four different relative densities. Plate unit loads in chamber tests tend to be lower (for BC1, BC4) and higher (for BC2, BC3) than pile base unit loads. The confining stress level at the pile base level also influences size effect. The calibration chamber size effects under BC1 were more pronounced at low confinement, corresponding to shorter piles, while size effects under BC2 were more pronounced at high confinement, corresponding to longer piles. The magnitude of size effect is small, validating the use of chamber tests to simulate pile base loading. In keeping with the recommendation of Salgado et al. (1998b) regarding CPT testing in calibration chambers, it is recommended here that BC1 or BC4 be used in calibration chamber plate load testing. If this is done, our numerical results show that, for practical purposes, no correction is needed to the measured plate unit load in estimating pile base unit load, unless short piles are being simulated.

ACKNOWLEDGEMENT

Access to the calibration chamber data was graciously given to the aurthos by Prof. M. Jamiolkowski, Drs. R. Bellotti, and S. Pedroni. For that, and for inspiring conversation on the topic, the authors are grateful.

REFERENCES

Bolton, M.D. 1986. The Strength and Dilatancy of Sands. *Geotechnique* 36(1): 65-78.

Duncan, J.M., and Chang, C.Y. 1970. Nonlinear Analysis of Stress-Strain in Soils. *Journal of Soil Mechanics and Foundation Engineering Division* 96(SM5): 1629-1653.

Fahey, M., and Carter, J.P. 1993. A Finite Element Study of the Pressuremeter Test in Sand Using a Non-Linear Elastic Plastic Model. *Canadian Geotechnical Journal* 30(2): 348-361.

Ghionna, V.N., Jamiolkowski, M., Pedroni, S., and Salgado, R. 1994. *The Tip Displacement of Drilled Shaft in Sands, Proceedings of Settlement '94 Vol. 2*: 13 – 39. ASCE.

Giuseppe, B. 1991. *Modellazione Numerica Delle Prove Pressiometriche In Camera Di Calibrazione Per La Sabbia Del Ticino*. Graduate School Thesis, Politecnico Di Torino, Italy.

Hardin, B.O., and Black, W.L. 1966. Sand Stiffness under Various Triaxial Stresses. *Journal of Soil Mechanics and Foundation Engineering Division* 92(SM2): 27-42.

Hardin, B.O. and Drnevich, V. P. 1972. Shear Modulus and Damping in Soils; Design Equations and Curves. *Journal of Soil Mechanics and Foundations Division* 98(7): 667 – 692.

Kondner, R. L. 1963. Hyperbolic Stress-Strain Response: Cohesive Soil. *Journal of Soil Mechanics and Foundation Engineering Division* 189(SM1): 115 – 143.

Mayne, P.W. and Kulhawy, F.H. 1991. Calibration Chamber Database and Boundary Effects Correction for CPT Data. *Proceedings of the First International Symposium on Calibration Chamber Testing*: 257 – 264. New York: Elsivier.

Naylor, D.J., Pande, G.N., Simpson, B., and Tabb, R. 1981. *Finite Elements in Geotechnical Engineering*. Pineridge Press, Swansea, Wales, U.K.

Parkin, A.K. 1991. Chamber Testing of Piles in Calcareous Sand and Silt, *Proceedings of the First International Symposium on Calibration Chamber Testing*: 289 – 302. New York: Elsivier.

Salgado, R., Mitchell, J.K., and Jamiolkowski, M. 1998a. Chamber Size Effects on Penetration Resistance Measured in Calibration Chambers. *Journal of Geotechnical and Geoenvironmental Engineering* 124(9): 878 - 888.

Salgado, R., Jamiolkowski, M., and Mitchell, J.K. 1998b. Penetration Resistance in Sand: Analysis and Applications to Liquefaction Potential Assessment and Estimation of Pile

Base Resistance. *Rivista Italiana di Geotecnica* 32(4): 5 – 17.

Schnaid, F., and Houlsby, G.T. 1991. An Assessment of Chamber Size Effects in the Calibration of In Situ Tests in Sand. *Geotechnique* 41(3): 437-445.

Vecchia, G. 1991. Modellazione Con Il Metodo Degli Elementi Finiti Delle Prove Triassiali E Pressiometriche Utilizzando La Legge Constitutiva Di Lade. Graduate School Thesis, Politecnico di Torino, Italy.

Computer Methods and Advances in Geomechanics, Desai et al. (eds) © 2001 Balkema, Rotterdam, ISBN 90 5809 183 X

Numerical modeling of strains induced by suction changes in a compacted soil

A.H.P.Lins, L.M.Costa & J.M.Sobreira
Federal University of Pernambuco, Brazil

ABSTRACT: This paper presents the interpretation of some suction controlled oedometric tests by using the BBM model as constitutive relationship. The soil tested is a clay from the borrowing area of Pirapama Dam, situated in Pernambuco, Brazil. Three series of four samples were statically compacted at a common water content but different dry densities. The oedometric tests are numerically reproduced. A good agreement between experimental results and the simulation of soil behavior is observed.

1 INTRODUCTION

Unsaturated soils are multiphasic porous media, composed by a solid, a liquid and a gas phase. The bases of a comprehensive framework to model the hydromechanical behavior of these soils have been established during the last three decades. The difficulties in finding a single stress concept able to explain the experimental results, and particularly the collapse phenomena, led to the adoption of two stresses fields, set up by combination of total stress, air pressure and water pressure. For experimental convenience, usually it is adopted the net stress (total stress minus gas pressure) and suction (gas pressure minus water pressure).

Among all the constitutive relationships proposed to describe the deformational behavior of unsaturated soils it has been shown that the elastoplastic ones are more suitable to reproduce the complex behavior of these materials, once they can explain the irreversible collapse strain and the stress path dependency (Alonso et al., 1990; Wheeler & Sivakumar, 1995; Cui & Delage, 1995).

This paper presents an application of the BBM elastoplastic model, implemented in a computational code, to back analyze a set of oedometric loading-wetting tests carried out with compacted soil, as part of a research program in progress at the Federal University of Pernambuco.

2 MECHANICAL MODEL

The constitutive model adopted in this paper is the Barcelona Basic Model (BBM), developed by Alonso et al.(1990). Based on the framework of hardening elastoplasticity it's an extension of the modified Cam-clay model. Under isotropic conditions the model uses two independent stress variables, the net stress, $p=(\sigma_{ij}-p_a\delta_{ij})$ and suction, $s=(p_a-p_w)$. In triaxial states the deviator stress, $q=(\sigma_1-\sigma_3)$ is included.

The effects of unsaturation is considered by the introduction of suction on the saturated yield surface expressed by

$$f(p,q,s,p_0^*) \equiv q^2 - M^2(p+p_s)(p_0-p)=0 \quad (1)$$

where

$$\frac{p_0}{p^c} = \left(\frac{p_0^*}{p^c}\right)^{\frac{\lambda(0)-\kappa}{\lambda(s)-\kappa}} \quad (2)$$

$$\lambda(s) = \lambda(0)[(1-r)exp(-\beta s)+r] \quad (3)$$

$$p_s = ks \quad (4)$$

where M is the slope of the critical state line, associated to the friction angle of the soil, p_0 is the yield isotropic net stress for suction s, p_0* is the saturated preconsolidation stress, κ is the elastic stiffness parameter against changes in p, $\lambda(0)$ is the slope of virgin compression line for saturated isotropic loading, $\lambda(s)$ is the slope of virgin compression line for isotropic loading at a constant suction s, β controls the rate of stiffness increase with suction, r is a limiting value of soil stiffness for very high suction, p^c is a reference stress and k describes the increase of the apparent cohesion with suction.

Equation (2) defines the Loading-Collapse yield surface (LC) in the isotropic plane (p,s), (Fig. 1a).

The dependency of the Cam-clay ellipse on suction is represented in Figure 1b. As suction increases the size of the ellipse grows, increasing the elastic domain

A non associated plastic potential is defined by

$$g(p,q,s,p_0^*) = \alpha q^2 - M^2 (p + p_s)(p_0 - p) \tag{5}$$

where α is established in such a way that under K_0 loading lateral strains are zero.

The elastic domain is closed by another yield surface, the Suction Increase (SI) described by

$$f = s - s_0 \tag{6}$$

where s_0 is the maximum suction known by the soil.

The two hardening parameters, p_0* and s_0 depend on the rate of volumetric plastic strain in a coupled way. The hardening laws are:

$$\frac{dp_0^*}{p_0^*} = \frac{(1+e)}{\lambda(0) - \kappa} d\varepsilon^p \tag{7}$$

$$\frac{ds_0}{s_0 + p_{atm}} = \frac{(1+e)}{\lambda_s - \kappa_s} d\varepsilon^p \tag{8}$$

where e is the void ratio, λ_s is the slope of virgin drying line, κ_s is the elastic stiffness parameter against changes in suction and p_{atm} is the atmospheric pressure.

The increment of elastic strain is the sum of two components, one due to stress changes and another due to suction changes

$$d\varepsilon^e = \frac{\kappa}{(1+e)} \frac{dp}{p} + \frac{1}{3G} dq + \frac{\kappa_s}{(1+e)} \frac{ds}{(s + p_{atm})} \tag{9}$$

where G is the shear modulus.

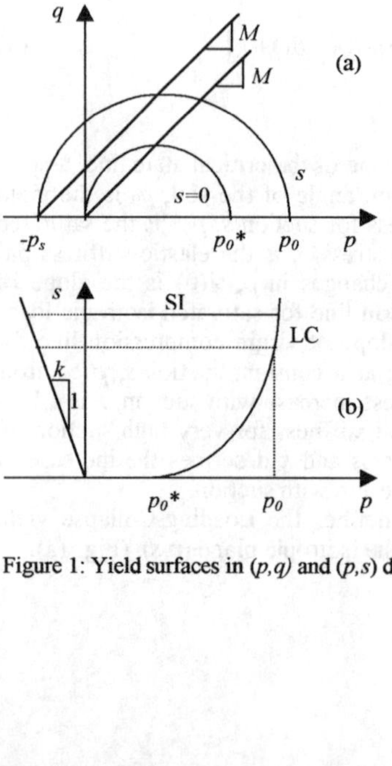

Figure 1: Yield surfaces in (p,q) and (p,s) diagrams.

3 LABORATORY TESTS

Oedometric tests under controlled suction were performed on compacted samples of a gneissic residual soil, obtained from the borrowing area of Pirapama Dam, situated in the State of Pernambuco - Brazil. The soil sample, identified as CL according to the Unified Soil Classification System, presented the following characteristics: w_L = 48%, PI = 32%, Activity = 1.0, w_{op} = 21.2% and γ_{dmax} = 16.25kN/m³.

Several groups of identical specimens were prepared for different initial conditions, as defined by (γ_d, w), presented in Table 1, in order to perform oedometric loading-wetting tests under controlled suction. The specimens for these tests were compacted statically in the oedometer ring.

Table 1. Compaction conditions.

	Test 1a	Test 1b	Test 1c	Test 1d	Test 2a	Test 3a
γ_d (kN/m³)	16.25	15.98	15.75	15.5	16.25	16.25
w (%)	21.2	21.2	21.2	21.2	19.2	17.2

After compaction of the specimen and setting up the apparatus, a suction of 150kPa was imposed to the specimen. When equilibrium under applied suction was reached, the sample was loaded in stages, with a ratio of loading $\Delta\sigma_v/\sigma_v$ =1, up to a maximum vertical pressure of 400kPa, when the sample was soaked under zero suction.

4 NUMERICAL MODELING

The suction controlled oedometric tests are reproduced by numerical simulation. The soil is modeled by BBM model, implemented in the Finite Element code CODE_BRIGHT (Olivella et al., 1995, 1996).

The parameters used in the simulation are listed in Table 2. They are the average value of the tests in a series. A series is made by four samples at same water content but different dry densities (Tests 1a, 1b, 1c and 1d compose a series).

Table 2. Parameters used in simulation.

	Series 1	Series 2	Series 3
κ	0.0085	0.0062	0.0068
κ_s	0.0107	0.0052	0.0012
p^c (kPa)	10	10	10
p_0*(kPa)	70	70	70
$\lambda(0)$	0.045	0.042	0.042
β (MPa⁻¹)	15	15	12.5
r	0.85	0.85	0.76
M	1	1	1

The value of κ is estimated from the slope of stress path AB in Figure 2. $\lambda(s)$ is assessed by the slope of the stress path BC (Fig. 2). β and r control the slope of the loading curve and therefore the amount of collapse. They are estimated by trial and error. The value adopted for β indicates a fast increase in stiffness with suction. Known $\lambda(s)$, r and β, $\lambda(0)$ is assessed by equation (3). According to Gens (1996) different values of r and β (series 3) account for the influence of compaction fabric on soil behavior.

The value of p_0* is estimated by observation of the slopes of the elastic and virgin compression lines in the vertical stress - void ratio diagram. Alonso et al. (1993) suggest that different compaction conditions lead to differences on the initial stress state, p_0* and as compacted suction. According to them each value of compaction dry density corresponds to a value of saturated preconsolidation pressure (p_0*). As dry density variation in the tests analyzed is within a very narrow range (Table 1) we adopted a single value for p_0*.

The parameter κ_s is adjusted in order to reproduce the total collapse strain, which is the sum of the elastic swelling and the plastic collapse. In the simulation carried out it has been adopted a constant Poisson's ratio, ν=0.3 and G is estimated by

$$G = \frac{3(1-2\nu)}{2(1+\nu)} \frac{p(1+e)}{\kappa} \tag{10}$$

Once there is no increase in suction, SI yield surface is not activated and λ_s does not interfere in the tests presented.

Laboratory data and the numerical simulation results are presented in Figures 2 and 3.

From laboratory data shown in Figure 2 we can see that at optimum water content the collapse is almost the same for the different dry densities except for Test 1b, which is larger than the others.

Comparing Test 1a in Figure 2, Test 2a and Test 3a in Figure 3, with the same dry density after compaction but compacted at different water content, we note that for the lower water content the higher is the collapse. The increment value of p_0* at 400kPa before and after flooding indicates the amount of collapse, as expressed by equation (7). The values obtained in the simulation of Test 1a, Test 2a and Test 3a were respectively 15.9kPa, 16.1kPa and 22.9kPa.

The behavior observed indicates that for a same dry density the collapse potential grows with water content reduction and at optimum water content the small differences in dry density values seems to have no relevant influence in soil behavior.

5 CONCLUSIONS

Comparison between laboratory and simulation results shows a very good agreement with experimental data. Loading under controlled suction is very well reproduced for all tests.

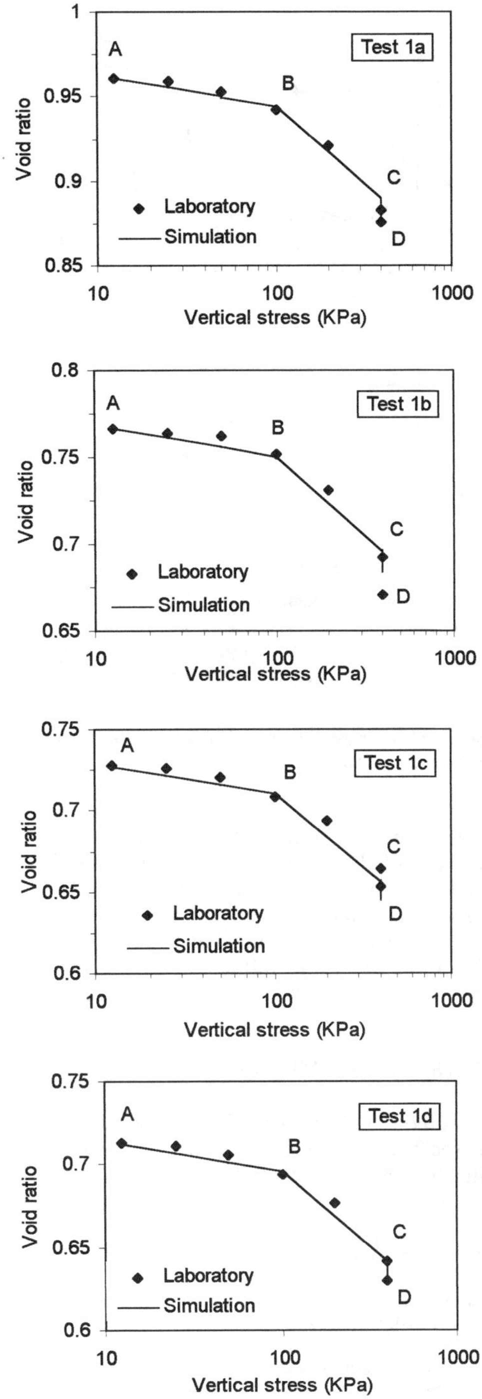

Figure 2: Comparison between laboratory data and simulation results of series 1 oedometric tests.

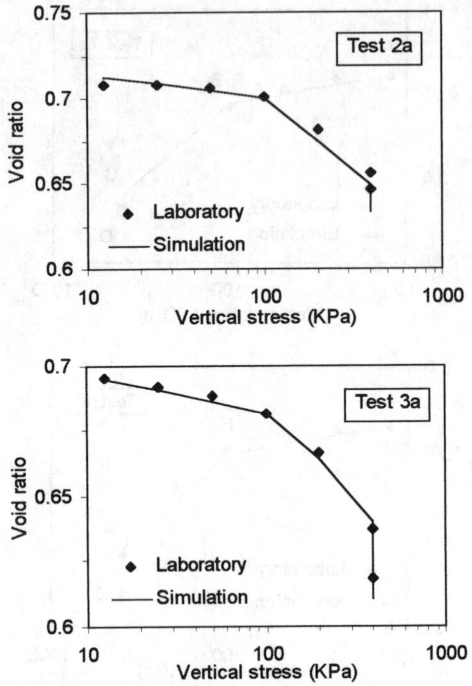

Figure 3: Comparison between laboratory data and simulation results of test 2a and test 3a.

The numerical results obtained demonstrate the capability of the BBM model to reproduce the experimental behavior of compacted soil.

The interpretation of experimental data by BBM provides a better understanding of unsaturated soil behavior as well as the influence of some factors, as compaction water content and dry density.

ACKNOWLEDGEMENTS

The authors wish to thank Prof. Eduardo Alonso of the Technical University of Catalonia (UPC) for the utilization of CODE_BRIGHT.

REFERENCES

Alonso, E. E.,Gens, A. & Josa, A. 1990. A constitutive model for partially saturated soils. *Geotechnique*, vol. 40, nº 3, pp. 405-430.

Alonso, E. E., Josa, A. and Gens, A. 1993. Modelling the behaviour of compacted soils. In *Unsaturated Soils* (eds. S.L. Houston and W. K. Wray), ASCE Geotechnical Special Publication No. 39, pp. 103-114.

Cui, Y. J., Delage, P. & Sultan, N. 1995. An elastoplastic model for compacted soils. In: *Proceedings of the 1st International Conference on Unsaturated Soils*, Paris, Vol. 2, pp. 703-709.

Gens, A., 1996. Constitutive modelling: Application to compacted soils. In: *Proceedings of the 1st International Conference on Unsaturated Soils*, Paris, Vol. 3, pp. 1179-1200.

Olivella, S. 1995. Nonisothermal Multiphase Flow of Brine and Gas Thorough Saline Media. Tesis Doctoral, Universitat Politècnica de Catalunya, Departament D'Enginyeria del Terreny i Cartogràfica.

Olivella, S., Gens, A., Carrera, J. & Alonso, E. E. 1996. Numerical Analysis for a simulator (CODE_BRIGHT) for the coupled analysis of saline media. *Engineering Computations*, 13, pp. 87-112.

Wheeler, S. J. & Sivakumar, V. 1995. An elastoplastic critical state framework for unsaturated soil. *Geotechnique*, vol. 45, nº 1, pp. 35-53.

Computer Methods and Advances in Geomechanics, Desai et al. (eds) © 2001 Balkema, Rotterdam, ISBN 90 5809 183 X

DISO – A computer program for interpreting of in-situ stress measurement by overcoring

M. Lu
SINTEF Civil and Environmental Engineering

ABSTRACT: The DISO (**D**etermination of **I**n-situ **S**tress by **O**vercoring) program has been applied to a great number of projects, since it was first developed at SINTEF Department of Rock and Mineral Engineering in 1990. The current version of DISO can be used for interpreting of 3-D in-situ stress measurements by overcoring in both isotropic and transversely isotropic rock. The doorstopper measurements can also be incorporated. This paper describes the main features, theoretical basis and verification of the code. An application sample is also given.

1 INTRODUCTION

As an effective tool of the 3-D in-situ stress measurement, overcoring technique is widely used in rock engineering. The technique uses the closed form elastic solution in computing the far field rock stress from the strains recorded in stress relief. Theoretically, six independent strain readings are sufficient to determination of the rock stress. However, in practice, at least 9 strain readings are usually obtained, leading to the non-uniqueness of the resulting rock stresses. The accuracy or reliability of the overcoring measurement depends significantly on the geological conditions of the measurement site. The measured strains are inevitably scattered. Therefore, a computerised statistical analysis is needed to estimate the most likely rock stresses.

From the randomly selected strain readings and the elastic constants DISO computes the rock stress based on analytical solutions, in which the rock is assumed to be linearly elastic. Both isotropic rock mass and transversely isotropic rock mass can be handled. Then a Monte-Carlo type statistical analysis is performed to determine the mean and deviations of each principal stress component, both magnitude and orientation. DISO can also incorporate the 2-D doorstopper measurements with the 3-D overcoring, which may be in some cases economically advantageous. The result of DISO is presented in both text and graphic modes.

Generation of input data is extremely simple and execution of DISO program for a common calculation of 20,000 points takes no more than 2 minutes on a Pentium PC. This provides the user with possibility of running the code on the measurement site and decision upon whether it is necessary to carry out additional measurement can be made while the measurement is going on. The elastic constants used in computation can be obtained at field as well by using the so-called biaxial tests that can determine not only the E-modulus and Poisson's ratio for the isotropic rock, but also the five independent elastic parameters for the transversely isotropic rock. A special computer program has been written for this purpose.

The DISO program has been verified with Finite Element modelling and used in tens of locations over the last ten years.

2. THEORETICAL BASIS

2.1 *Overcoring for isotropic rock*

Computation of the far field in-situ stress is based on different analytical solutions for isotropic rock mass and transversely isotropic rocks. For the isotropic rock mass, Lemman's elastic solutions are adopted (Leeman 1966), which are summarised below.

Step 1. Compute the stress components on the borehole surface in the cylindrical coordinate system, from measured strains, ε_x, ε_θ and ε_{45} at three measurement locations, i.e., θ=0, 90 and 225°.

$$\left.\begin{aligned} \sigma_\theta &= \frac{E}{2}\left(\frac{\varepsilon_\theta+\varepsilon_x}{1-\nu}+\frac{\varepsilon_\theta-\varepsilon_x}{1+\nu}\right) \\ \sigma_x &= \frac{E}{2}\left(\frac{\varepsilon_\theta+\varepsilon_x}{1-\nu}-\frac{\varepsilon_\theta-\varepsilon_x}{1+\nu}\right) \\ \tau_{\theta x} &= -\frac{E}{2}\left(\frac{2\varepsilon_{45}-\varepsilon_\theta-\varepsilon_x}{1+\nu}\right) \end{aligned}\right\} \quad (1)$$

The formula is given in Eqn (1), in which E and ν are E-modulus and Poisson's ratio of the rock.

Step 2. By using the Kirsch solution, compute the far field in-situ rock stress components in the local Cartesian system, $\{\sigma_0\}_{xyz}$, from the stress components on the borehole surface in the cylindrical coordinate system, $\{\sigma\}_{\theta x}$:

$$\{\sigma\}_{\theta x} = [A]\{\sigma_0\}_{xyz} \tag{2}$$

where $\{\sigma\}_{\theta x}=[\sigma_{\theta,\theta=0}\ \sigma_{x,\theta=0}\ \tau_{\theta x,\theta=0}\ \sigma_{\theta,\theta=90}\ \sigma_{x,\theta=90}\ \tau_{\theta x,\theta=90}\ \sigma_{\theta,\theta=225}\ \sigma_{x,\theta=225}\ \tau_{\theta x,\theta=225}]^T$, $\{\sigma_0\}_{xyz}=[\sigma_x\ \sigma_y\ \sigma_z\ \tau_{yz}\ \tau_{zx}\ \tau_{xy}]^T$ and [A] is a 9X6 matrix and can be expressed explicitly as:

$$[A]=\begin{bmatrix} 0 & -1 & 3 & 0 & 0 & 0 \\ 1 & -2\nu & 2\nu & 0 & 0 & 0 \\ 0 & 0 & 0 & 0 & 2 & 0 \\ 0 & 3 & -1 & 0 & 0 & 0 \\ 1 & 2\nu & -2\nu & 0 & 0 & 0 \\ 0 & 0 & 0 & 0 & 0 & -2 \\ 0 & 1 & 1 & -4 & 0 & 0 \\ 1 & 0 & 0 & -4\nu & 0 & 0 \\ 0 & 0 & 0 & 0 & -\sqrt{2} & \sqrt{2} \end{bmatrix} \tag{3}$$

It should be noted that Eqn (2) represents 6 unknowns to be solved from 9 equations.

Step 3. A stress transform procedure is then performed to obtain the far field stress components in the global system, $\{\sigma\}_{XYZ}$.

$$\{\sigma_0\}_{XYZ} = [R]^T\{\sigma_0\}_{xyz}[R] \tag{4}$$

where the transformation matrix

$$[R]=\begin{bmatrix} l_1 & m_1 & n_1 \\ l_2 & m_2 & n_2 \\ l_3 & m_3 & n_3 \end{bmatrix} \tag{5}$$

and l_i, m_i and n_i are direction cosines of the three local coordinate axes in the global system, which can be computed from the borehole orientation, say the plunge and elevation angles.

Step 4. The magnitude and orientation of three principal in-situ stresses are subsequently determined by calculating the eigenvalues and corresponding eigenvectors of the stress tensor $[\sigma]_{XYZ}$. Finally, the principal orientations are expressed in terms of trend α and plunge β converted from the enginevector.

Three coordinate systems used are defined in Figure 1. Cylindrical system xrθ: x - borehole axis opposite to drilling direction; r - radial axis in borehole cross section; θ - tangential axis pointing in anti-clockwise direction in borehole cross section counting from horizon. Local Cartesian system xyz: x - borehole axis opposite to drilling direction; y - horizontal in cross section plane of the borehole; z - upwards in cross section plane of the borehole. Global system: X - east; Y - north; Z - upwards.

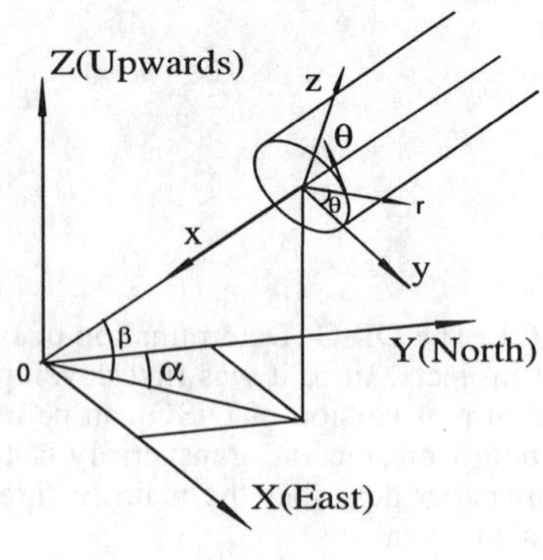

Figure 1. Coordinate systems.

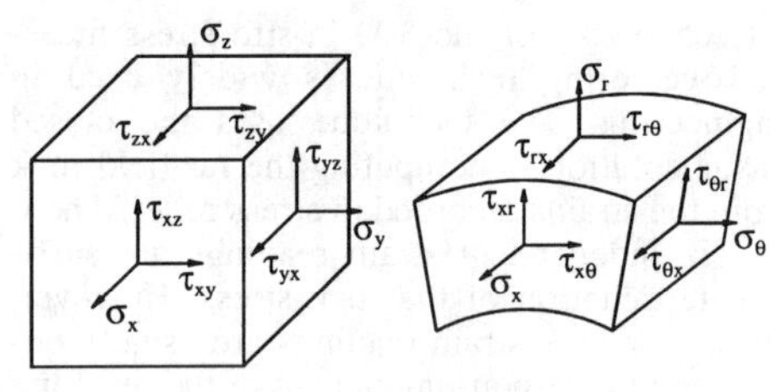

Figure 2. Sign convention of stresses.

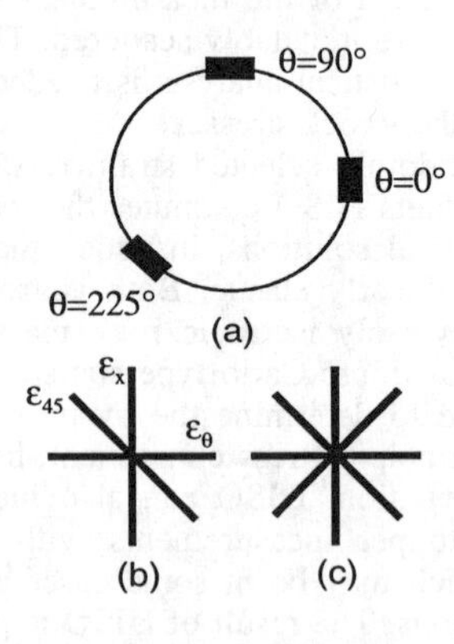

Figure 3. Strain gauge installation (a) layout (b) for overcoring (c) for doorstopper.

Figures 2 and 3 give the sign convention of the stress components and the strain gauge installation, respectively.

2.2 *Overcoring for transversely isotropic rock*

Interpreting overcoring measurement result for transversely isotropic rock mass is much more complicated. The theoretical bases and mathematic formulations used in DISO in handling the anisotropic rock are briefly described here. In addition to the three coordinate systems used for the isotropic rock, one more coordinate system is also used, that is the isotropic system, nst, defined as: n - normal to the isotropic plane; s and t - in the isotropic plane. The strains in the cylindrical system are first transformed to the local Cartesian system:

$$\{\varepsilon\}_{r\theta x} = [T_\varepsilon]_{xyz,r\theta x}\{\varepsilon\}_{xyz} \tag{6}$$

The stress components on the borehole surface in the local Cartesian system is calculated as:

$$\{\varepsilon\}_{xyz} = [C]_{xyz}\{\sigma\}_{xyz} \tag{7}$$

in which the compliance matrix $[C]_{xyz}$ is computed from the compliance matrix in the *nst* system through twice transformation:

$$[C]_{xyz} = [T_\varepsilon]_{XYZ,xyz}[T_\varepsilon]^{-1}_{XYZ,nst}[C]_{nst}[T_\sigma]_{XYZ,nst}[T_\sigma]^{-1}_{XYZ,xyz} \tag{8}$$

where the compliance matrix $[C]_{nst}$ connects the strains and stresses in nst system:

$$\begin{Bmatrix}\varepsilon_n\\ \varepsilon_s\\ \varepsilon_t\\ \gamma_{st}\\ \gamma_{sn}\\ \gamma_{ns}\end{Bmatrix} = \begin{pmatrix} \frac{1}{E_a} & -\frac{v_{sn}}{E_i} & -\frac{v_{tn}}{E_i} & 0 & 0 & 0\\ -\frac{v_{ns}}{E_a} & \frac{1}{E_i} & -\frac{v_{ts}}{E_i} & 0 & 0 & 0\\ -\frac{v_{nt}}{E_a} & -\frac{v_{st}}{E_i} & \frac{1}{E_i} & 0 & 0 & 0\\ 0 & 0 & 0 & \frac{1}{G_i} & 0 & 0\\ 0 & 0 & 0 & 0 & \frac{1}{G_a} & 0\\ 0 & 0 & 0 & 0 & 0 & \frac{1}{G_a}\end{pmatrix}\begin{Bmatrix}\sigma_n\\ \sigma_s\\ \sigma_t\\ \tau_{st}\\ \tau_{sn}\\ \tau_{ns}\end{Bmatrix} \tag{9}$$

The far field rock stresses $\{\sigma_0\}_{xyz}$ are then calculated from $\{\sigma\}_{xyz}$:

$$\{\sigma\}_{xyz} = [F]\{\sigma_0\}_{xyz} \tag{10}$$

The solution for the matrix [F] used in the DISO program is based on the formula derived by Amadei (1982) and revised by Li (1997). The elements of matrix [F] depend on the elastic constants of the rock and location of the point under calculation. When $\{\sigma_0\}_{xyz}$ is obtained the far field stresses in the global system can be calculated from Eqn (4), the same as for the isotropic case. Combining Eqns (6-10) the final form may be written as:

$$\{\varepsilon\}_{r\theta x} = [T_\varepsilon]_{xyz,r\theta x}[T_\varepsilon]_{XYZ,xyz}[T_\varepsilon]^{-1}_{XYZ,nst}[C]_{nst}[T_\sigma]_{XYZ,nst}[T_\sigma]^{-1}_{XYZ,xyz}[F]\{\sigma_0\}_{xyz} \tag{11}$$

The strain and stress transformation matrices $[T_\varepsilon]_i$ and $[T_\sigma]_i$ are calculated from the relative orientation of corresponding coordinate systems. Alternatively, a relation between $\{\sigma\}_{r\theta x}$ and $\{\sigma_0\}_{xyz}$ can also be established through transformation of coordinate systems, such facilitating the same procedures in statistic analysis.

2.3 *Doorstopper*

Sometimes it is economical to combine in a borehole the 3-D overcoring with the 2-D doorstopper measurements, which provide additional information about the three far field stress components in the local Cartesian system, i.e., $\{\sigma\}_y$, $\{\sigma\}_z$ and $\{\tau\}_{yz}$. The complete in-situ stress state is evaluated with the other three remaining stress components, $\{\sigma\}_x$, $\{\tau\}_{xy}$ and $\{\tau\}_{xz}$, provided by the overcoring measurements. A detailed description of the doorstopper measurement can be found in Myrvang and Hassen, 1990.

3. STATISTIC ANALYSIS

As mentioned early the required 6 stress components are evaluated from 9 available equations (Eqn 2). Mathematically, there are C_9^6=84 solutions. Further study shows, however, that only 36 out of the 84 solutions are independent and valid mechanically. Nevertheless, 5-10 measurements are usually undertaken in a single borehole. This provides the possibility of selecting the elements of $\{\sigma\}_{\theta x}$ from different measurement, such greatly enhancing the number of solutions and making a statistic analysis feasible. In the DISO code, a random selection is implemented in two distinct ways:

(1) The stress components in the cylindrical coordinate system for each measurement section, $\{\sigma\}_{\theta x,i}$, are computed first from the measured strain readings. Then, a group of 9 stresses in the cylindrical system are selected randomly across the measurements for computing the far-field stresses in the local Cartesian system (Hassen and Myrvang, 1990). As mentioned above, there are 36 valid combinations out of a total 84 (=C_9^6) for each group of 9 stresses and a total of millions independent solutions can yield.

(2) Alternatively, instead of computing stresses first one can also randomly select the strain readings directly, then compute the stresses. The sole restriction that applies to the strain selection is that only one

axial strain should be included in a single combination, such leaving only 19 independent solutions out of the total 84.

Testing has demonstrated that there is very little difference between the two approaches of random selection. Usually, about 20,000 independent solutions of the far field stresses, $\{\sigma_0\}_{XYZ}$, are computed from measurements in a single borehole. Based on these stresses a Monte Carlo type statistic analysis is performed from which the means, deviations and probability distributions of the magnitude of three principal stresses are obtained. The orientations of the principal stresses are computed from the means of $\{\sigma_0\}_{XYZ}$.

A pre-processing procedure is performed for optimisation. The input strain data are checked first and zero readings are removed. Furthermore, a streamlining procedure is implemented in which the computed stresses, which are extremely deviate the means, are rejected, such leading to more reliable results.

4. RESULT PRESENTATION

The analysis result is presented in both text and graphic modes showing the magnitude and orientation of the resulting principal in-situ stresses. The text output includes the mean, standard and average deviations of the magnitude of three principal stresses as well as their orientation expressed in trend and plunge. An equal area lower hemisphere stereonet plot is also provided presenting the orientation of the mean principal stresses. The probability distributions of the magnitudes of principal stresses are also provided. Given IN Figures 4 and 5 are the text and graphic output of the measurement performed at the Lower Kihansi hydropower project, Tanzania in 1997.

5. ESTIMATION OF ELASTIC PARAMETERS WITH BIAXIAL-CELL TESTS

The elastic parameters used in computing of in-situ rock stress are usually obtained from laboratory tests on core-specimens. Alternatively, they can also be derived from the biaxial cell tests. One of the advantages for the biaxial cell tests is that the test can be performed at site, so that a prompt evaluation of the parameters, and then the in-situ rock stress can be realised, such a decision can be made as to whether further measurements are needed. A numerical procedure has been worked out for evaluation of the parameters for both isotropic rock (E and ν) and transversely isotropic rock (E_1, E_2, ν_1, ν_2 and G_2). Detailed description can be found in Lu, 1997 (b).

```
                 ************************************
                 * In Situ Stress Measurement Report *
                 ************************************

Client Name: Norplan A.S                  Project Name: Niche 1465

Borehole Name: Ch. 1856                   Report Number: Prelim

Borehole orientation:    Trend    94.0        Elevation angle    6.0

Overburden:           725.0

                                 MEASUREMENT RESULT

DEPTH             3-D READINGS                    2-D READINGS     E    nu    rho
 (m)     Theta=0      Theta=90     Theta=225                      (GPa)      (kg/m3)

        a   0   45    a   0   45    a   0   45    0   90  45 135

 8.0  3381405 695  405-250 110  380 945 775    0    0    0    0   29.0 .12   2680
 8.5  2551450 315  485 -90 215  4701055 965    0    0    0    0   29.0 .12   2680
 9.0  2401420 595  380-195 125  300 770 625    0    0    0    0   29.0 .12   2680
 9.5  2051180 455  320-285 -15  205 665 610    0    0    0    0   29.0 .12   2680
10.0  2601165 610  365-195 110  335 755 735    0    0    0    0   29.0 .12   2680
10.5  2101065 410  400-175 210  180 595 525    0    0    0    0   29.0 .12   2680
11.0  1301240 405  325-115 155  220 550 515    0    0    0    0   29.0 .12   2680
11.5  1651125 345  235-230 140  360 815 735    0    0    0    0   29.0 .12   2680

*****************************************************************************
*                STATISTICAL RESULTS OF IN-SITU STRESSES                    *
*           MEAN   AVERAGE DEVIATION   STANDARD DEVIATION   TREND    PLUNGE *
*                                                                           *
* SIGMA1    16.31        2.37                 2.96          100.0     50.4  *
* SIGMA2     8.52        1.87                 2.30          251.6     36.0  *
* SIGMA3     2.34        1.69                 2.20          352.2     14.2  *
*****************************************************************************

*****************************************************************************
*          IN-SITU STRESSES IN VERTICAL AND HORIZONTAL DIRECTIONS           *
*            STRESS                      MAGNITUDE          ORIENTATION     *
*                                                                           *
*          VERTICAL STRESS                 12.77                            *
*     MINIMUM HORIZONTAL STRESS             2.91               178.4        *
*     MAXIMUM HORIZONTAL STRESS            11.49                88.4        *
*****************************************************************************

*****************************************************************************
*                              GRAVITY STRESS                               *
*     VERTICAL STRESS:    19.04                   HORIZONTAL STRESS:   2.55 *
*****************************************************************************
```

Figure 4. Sample text output.

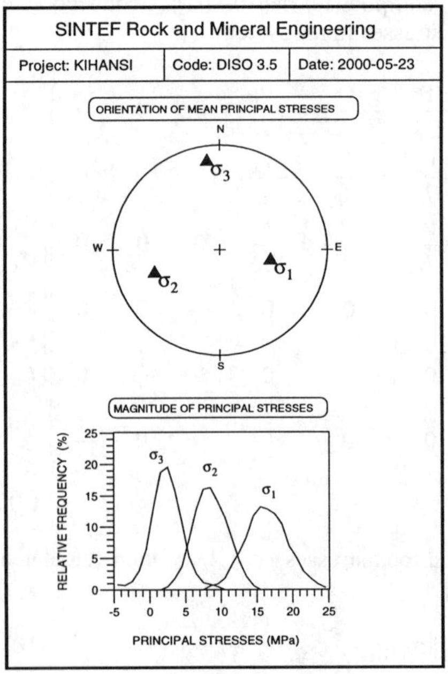

Figure 5. Sample graphic output.

6. PROGRAM VERIFICATION

The DISO programs have been verified with numerical modelling. As shown in Figure 4, the 3-D FE model, which consists of 2560 20-node elements and 11616 nodes, is a 2.0X2.0X2.0 m cube with a cylindrical borehole of 0.1 m in diameter. The commercial code, ABAQUS, is used for the testing. The verification procedure consists of the following steps: (1) Run the ABAQUS model, (2) Compute the strains at the measurement locations from those obtained from numerical simulation; (3) Take the computed strains at measurement as the input data and run DISO program such back calculate the in-situ stresses and (4) Finally compare the back-calculated stresses to those assigned.

Tests have been conducted for both isotropic and transversely isotropic rocks against a variety of parameters, including magnitude and orientation of in-situ stresses, orientation of foliation and elastic constant of the material. The test results have demonstrated the capability of the program. The details of the verification program are addressed in Lu, 1997 (b). So far the DISO code has been used for a great number of projects in many countries.

REFERENCES

Amadei, B. 1982. The Influence of Rock Anisotropy on Measurement of Stresses in-situ, Ph. D. thesis, Univ. of California, Berkley.

Hanssen, T. H. and Myrvang, A.M. 1990. Private communications.

Lu M. 1993. DISO 2.1 A computer program for evaluation of in-situ stress by overcoring and doorstopper technique, SINTEF report STF36 F93026

Lu M. 1997 (a). Evaluation of elastic parameters for transversely isotropic rock from biaxial cell test, SINTEF report STF22 F97004

Lu M. 1997 (b). Verification of new DISO programs, SINTEF report STF22 F97004

Li L.M. 1997. A study of interpretation for in-situ overcoring stress measurements in transversely isotropic rock, SINTEF report STF22 F97004

Leeman E.R. and Hayes D.J. 1966. A technique for determining the complete state of stress in rock using a single borehole, Proc. of the first international congress on rock mechanics, Lisbon, 1966.

Myrvang, A.M. and Hassen, S.E. (1990) Use of the modified doorstoppers for rock stress change measurements, Proc. 31 US Symp. Rock Mech., Balkema, Rotterdam, pp. 999-1004.

Computer Methods and Advances in Geomechanics, Desai et al. (eds) © 2001 Balkema, Rotterdam, ISBN 90 5809 183 X

Effects of end platen roughness on the creep behaviour of clays

G. Manivannan, C.T. Gnanendran & S.-C. R. Lo
School of Civil Engineering, University College, University of New South Wales, Canberra, Australia

ABSTRACT: Non uniformities of stress-strain behaviour due to end restraint in both drained and undrained triaxial tests were investigated numerically with a viscoplastic (creep) model and the Modified Cam clay (MCC) model using the FEM and the results are reported in this paper. The local stress-strain behaviour at selected points predicted by the analysis using the creep model differed slightly from that obtained using the MCC model. The numerical study indicates that soil creep has less impact on local non-uniformities compared to MCC model at large axial strain and the overall difference of non-uniformities predicted by both models are within 10% for drained as well as undrained tests. The secondary compression index (ψ) is the key parameter for predicting the local stress-strain behaviour of the soil. With the increase of ψ, the ultimate local deviator stress values decrease significantly in both drained and undrained tests.

1 INTRODUCTION

Triaxial tests are widely used in the development and verification of constitutive models. In the conventional triaxial testing, frictional forces between the loading caps and the specimen inevitably restrict free lateral movement of the specimen at its ends. As a result, dead zones are formed adjacent to the end platens and the well known "barrelling effect" occurs in the soil. Numerous studies, both experimental and numerical, have been carried out to investigate the stress-strain in-homogeneities within the triaxial specimens to the effects of end restraint and inadequate drainage (e.g., Carter, 1982; Airey, 1991; Schnaz et al., 1994; Sheng et al., 1997). However, none of them considered the effects of soil creep and the study of its influence is the main focus of this paper.

The results of a numerical study using a finite element creep model on the influence of end restraint on the local stress-strain behaviour at three points of triaxial specimen for drained and undrained tests are reported in this paper. The results obtained using the Modified Cam clay model is also presented for comparison. Particular attention is focused on the influence of the secondary compression index on the local stress strain behaviour and effective stress path of selected points in the sample.

2 NUMERICAL DETAILS

The constitutive creep model proposed by Kutter and Sathialingam (1992), based on Perzyna's viscoplastic theory (Perzyna, 1963) was adopted for the current analysis. In the formulation of the creep model, a radial mapping rule was used to represent the stress states of soil similar to that adopted in the bounding surface formulation of Dafalias and Herrmann (1982). In contrast to Perzyna's viscoplastic theory, the viscoplastic strains in the proposed mode are allowed to occur for all stress states, regardless of the position of the stress point in relation to the yield surface. The detailed description of this model can be found in Kutter and Sathialingam (1992). The model was implemented in the author's modified version of AFENA (Carter and Balaam, 1990) for this research program..

3 DETAILS OF ANALYSES AND SELECTION OF PARAMETERS

A 0.05m diameter by 0.1m long conventional size triaxial specimen was adopted for this study. The soil specimen was discretized into 200 six noded linear strain triangles (Fig. 1) for this numerical investigation. The material parameters used for the analyses are summarised in Table 1. Except the shape parameter R and the secondary compression index Ψ, all other material parameters used to simulate the creep behaviour were taken from Westerberg et al. (1995) for the Norrkoping clay. Shape parameter R, governs the shape of elliptical yield surfaces in q-p space. In the Modified Cam clay (MCC) formulation, R was assumed to be equal to 2, irrespective of the clay type. Kaliakin (1985) demonstrated that although R may take any value, it mostly lies in the range of 2-3 in the bounding surface formulation. Methods for determining R are discussed in Kutter and Sathialingam (1992). As the focus was to study only the creep effects on the local stress-strain behaviour, the shape parameter was assumed to be 2.0 as in MCC formulation. Based on the results from a large number of consolidation tests on natural soils, Mesri et al. (1977) concluded that a unique relationship exists between secondary compression index (Ψ) and compression index (C_c) and the values of Ψ/C_c ranges between 0.025 and 0.10. Referring to the values adopted for Ψ for various clay types in the literature, Ψ was assumed to be 0.04 C_c in the present study.

The soil-end platen interaction was modelled using an elastic - perfectly plastic interface with six noded isoparametric slip elements. Yield was determined by Mohr-Coulomb failure criterion with a non associated flow rule.

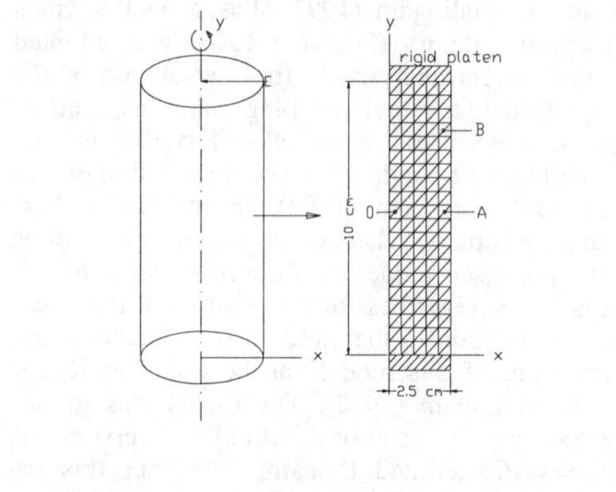

Fig. 1 Finite element discretization of the triaxial specimen

The sample was first subjected to a constant all round pressure of 100 kPa for simulating the initial isotropic consolidation pressure. A displacement controlled analysis was carried out with the overall strain rate of $1\times10^{-5}\ s^{-1}$ to achieve 0.03m total vertical movement, which corresponds to 30% overall compressive axial strain. In the displacement-controlled analysis, the overall axial strain is equal to the total vertical movement of the end platen divided by the initial height of the specimen. In the drained tests, fully drained conditions were assumed. Likewise, fully undrained conditions were assumed in the undrained tests. i.e. every single element in the specimen was undrained.

Two different end conditions of soil-platen contact were considered.

(1) Perfectly smooth contact (SC), i.e. no friction between the specimen and the end platens ($\phi_{interface} = 0$).

(2) Frictional contact (FC), ($\phi_{interface} = 20°$).

Table 1. Material parameters used in simulation

Parameter		Value
Compression index	(λ)	0.2
Recompression index	(κ)	0.02
Critical state slope	(*M*)	1.2
Poisson's ratio	(ν)	0.36
Void ratio on critical state line at p′=1		3.329
Initial preconsolidation pressure (P_0') kPa		120
Shape parameter	(R)	2.0
Secondary compression index	(ψ)	0.0184

4. RESULTS AND DISCUSSION

The end restraint effects for the frictional and smooth contact cases were studied by examining the local stress-strain behaviour at three points, namely O, A and B as marked in Fig. 1. The variation of deviator stress versus axial strain obtained from the numerical study using both MCC and creep models for drained tests are shown in Fig. 2.

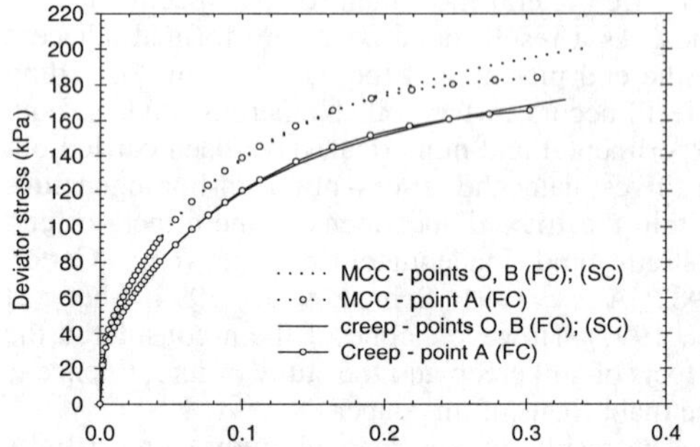

Fig. 2 Variation of Deviator stress with axial strain for drained tests

The local axial and radial strains at the end plate edges were very small compared to that at the mid-height of the specimen due to end plate confinement. Maximum of 11% of local axial strain and 4% of local radial strain were observed for the overall axial strain of 30% in the FC case. The deviator stress at point 'A' was smaller than those at points 'O', 'B' and that of the ideal smooth contact. It is obvious from Fig.2 that both models give the similar local q-ε_a behaviour except that the maximum deviator stresses obtained with the creep model are slightly less due to creep effects. The deviator stress obtained with creep model for point 'O' is 14.6% less than that obtained with MCC model for the overall axial strain of 30%. Similarly, the difference was 9.9% for point 'A'. These differences indicate that the time dependent creep behaviour plays a considerable role on the prediction of local stress-strain behaviour within the triaxial specimen. It is noted that the secondary compression index (Ψ) is the key parameter for predicting the local stress-strain characteristics and its influence will be discussed later.

Table 2. Non-uniformity index (q) of two models for drained tests

Overall axial strain (%)	$Dq/q_{(sc)}$ (%)	
	MCC	Creep
5	3.44	4.93
10	7.41	9.82
15	9.56	11.41
20	11.04	11.63
25	11.69	11.78
30	14.59	12.66

The influence of end restraint on the non-uniformities of local deviator stress, local axial, volumetric strains and local pore pressure due to effect of creep is the main focus of this paper. To examine the soil creep effects on non-uniformities of local deviator stress, the ratio, $Dq/q_{(sc)}$ hereafter referred as non-uniformity index (NUI) was used, where Dq is the maximum difference of deviator stress at any two local points of the three points 'O', 'A', 'B' at a particular overall axial strain. q_{sc} is the corresponding deviator stress for SC case. The calculated results are shown in Table 2. It indicates that for both models non-uniformity increases with overall axial strain and the soil creep has more non-uniformity on local deviator stress compared to MCC predictions until the overall axial strain reaches about 25%. Then the non-uniformity on deviator stress due to the influence of soil creep becomes less compared to MCC predictions. The difference of non-uniformity on local deviator stresses for the two models is always within 3%.

Similar index can be defined in terms of axial strains to compare the non-uniform behaviour on local axial strains. The results are shown in Table 3. For both models, the non-uniformity on local axial strain is very high. Similar observations were reported for MCC model by Sheng et al. (1997). It clearly shows that the local axial strains are consistent at different levels of overall axial strain within the sample for the two models. However, when the overall axial strain exceeds 20%, the non-uniformity on local axial strain from creep model prediction becomes less than that of MCC prediction. Likewise, the non-uniformities on volumetric strains for the two models are presented in Table 4, with NUI defined appropriately in terms of volumetric strains. The non-uniformities on local volumetric strains for the creep model increase until the overall axial strain reaches about 10% and then decrease with the increase of overall axial strain. Whereas, for the MCC model, the non-uniformities increase continuously as the overall axial strain increases. From these findings, it can be concluded that the maximum overall non-uniformity is more in MCC predictions, and the difference in non-uniformity for the two models is within approximately 10% for the overall axial strain of 30%.

Table 3. Non-uniformity index (ε_a) of two models for drained tests

Overall axial strain (%)	$D\varepsilon_a/\varepsilon_{a(sc)}$ (%)	
	MCC	Creep
5	0.00	0.00
10	10.00	10.00
15	29.53	26.80
20	40.00	40.00
25	54.88	47.44
30	66.67	56.67

Table 4. Non-uniformity index (ε_v) of two models for drained tests

Overall axial strain (%)	$D\varepsilon_v/\varepsilon_{v(sc)}$ (%)	
	MCC	Creep
5	6.80	7.32
10	9.51	10.98
15	11.02	10.66
20	11.96	9.16
25	12.00	9.11
30	16.06	8.88

The overall effect of end platen on local deviator stress is always within 15% for MCC model and 13% for creep model. Similarly, the overall effect of end platen on local axial, volumetric strains are always within about 67%, 16% for MCC model and these values for creep model are 57% and 11%. These findings confirm that the soil creep has less impact on local non-uniformities compared to MCC model.

The predicted undrained test results are shown in Fig. 3 for both models. Likewise in drained tests, the two models indicate similar local stress-strain behaviour but reach different ultimate deviator stresses. The ultimate stresses obtained with creep model are slightly less than that with MCC model. The local stress-strain behaviours for the three different points 'O', 'A', 'B', and the ideal

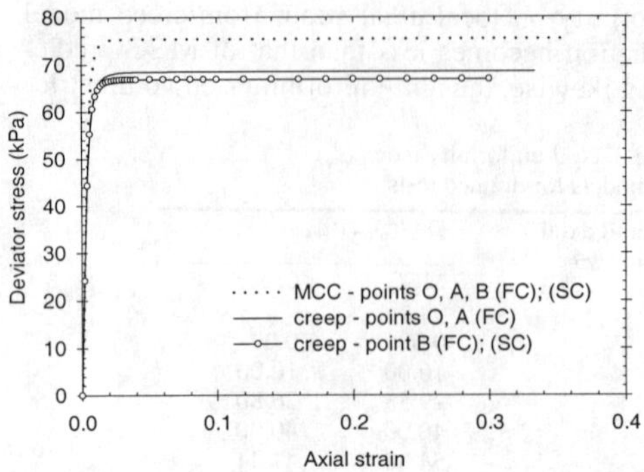

Fig. 3 Variation of deviator stress with axial Strain for undraine dtests

smooth contact case predict the same behaviour for MCC model. But, the creep model predictions for points O, A are slightly higher than that of ideal smooth contact. The similar index calculations were applied to the deviator stresses at different levels of overall axial strains. NUI for q was found to be 0.17% for MCC model and 0.70% for creep model irrespective of overall axial strain levels. These values indicate that the non-uniformities in terms of local deviator stress are consistent within the samples regardless of the type of model. Similar observations were obtained for non-uniformities on local axial strain. However, the non-uniformities on pore pressure, presented in Table 5, show more non-uniform behaviour for MCC model and the maximum difference predicted by the two models is within 10%.

Table 5. Non-uniformity index (u) of two models for undrained tests

Overall axial strain (%)	$Du/u_{(sc)}$ (%)	
	MCC	Creep
5	17.20	7.71
10	18.25	9.33
15	18.48	9.59
20	18.57	9.67
25	18.61	9.72
30	18.66	9.73

The stress paths for drained and undrained tests are shown in Fig. 4 for the two models. In the drained tests, for the MCC model, the local points, 'O', 'A', 'B' and ideal smooth contact follow the same stress path and diverge as the overall axial strain exceeds 20%. For the creep model, the stress paths for the three different points and the ideal smooth contact follow the same stress path and approach the critical state line. For the undrained tests, the two models follow different stress paths and reach

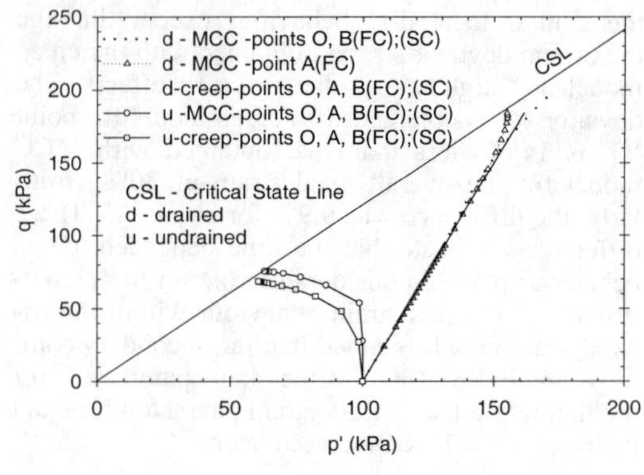

Fig.4 Stress paths for drained and undrained tests in p'-q plane

different critical state positions. However, all the local points within the sample and the ideal smooth contact follow the same stress paths and reach the same critical state to each model.

The deviator stress vs axial strain plots for drained tests are presented in Fig. 5 for different values of secondary compression index (Ψ) for point 'O' (FC). It was observed from the analysis that Ψ has remarkably little influence on the deviator stress-axial strain responses between the

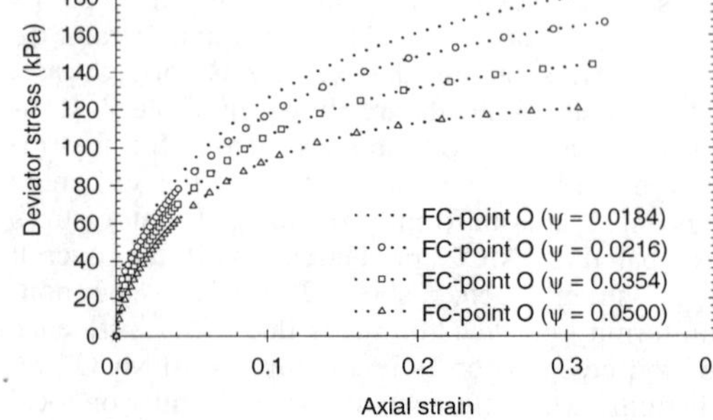

Fig. 5 Variation of deviator stess with axial strain for different Ψ values in drained tests.

different local points 'O', 'A' and 'B'. Therefore, for clarity, the graphs are presented for one local point only. When Ψ increases from 0.0184 to 0.0216 the ultimate deviator stress drops by about 11%'. As Ψ further increases to 0.0354 and to 0.05 the deviator stress values further drop by about 24% and 38% respectively from that of

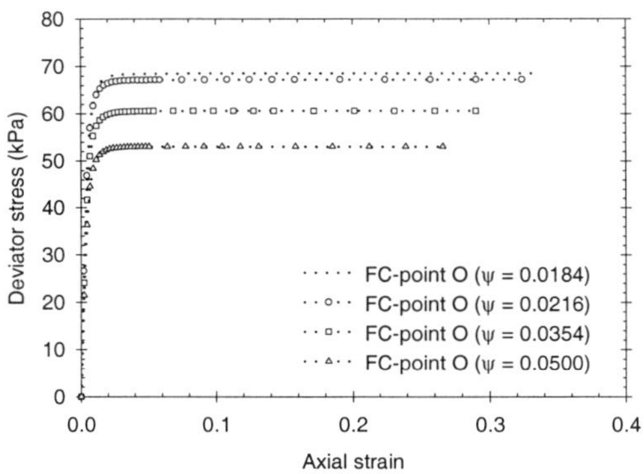

Fig. 6 Variation of deviator stess with axial strain for different Ψ values in undrained tests

Ψ = 0.0184. These plots indicate that the deviator stress decreases as Ψ increases. The undrained test results for different Ψ values shown in Fig. 6 also support this conclusion. Figures 5 and 6 clearly indicate that Ψ has significant influence on local stress-strain behaviour and therefore need to be assessed carefully for practical stress analysis problem.

4 SUMMARY AND CONCLUSIONS

A Numerical investigation was carried out to study the effects of end restraint in both drained and undrained triaxial tests with a creep model using the FEM. The same problem was analysed using the Modified Cam clay model for comparison purposes and the results obtained using both models are reported in this paper. A non-uniformity index was defined to assess the influence of soil creep on the non-uniformities within the triaxial sample. The following conclusions were drawn from the numerical study:

1. The two models predict similar local deviator stress-axial strain behaviour for both drained and undrained tests and the ultimate deviator stress values obtained with creep model are slightly less compared to MCC model due to time dependent creep effects.
2. In large overall axial strain, the soil creep has less impact on non uniformities of local deviator stress, local axial, volumetric strains and local pore pressure compared to MCC model. The maximum difference of overall non-uniformities of local stress-strain and pore pressure predictions obtained using the two models are always within 10% for both drained and undrained tests.
3. In drained tests, for MCC model, local points 'O', 'B', 'A' and ideal smooth case follow the same stress path to certain level of overall axial strain and diverge as overall axial strain increases. Whereas, in the creep model, for a particular Ψ value all three local points and the smooth case follow the same stress path and approach the critical state line.
4. In undrained tests, the two models give different stress paths and reach different critical states. But, all the local points within the sample for each model give the same stress paths and reach the same critical state.
5. With the increase of Ψ, the rate of mobilisation of local deviator stress and the local ultimate deviator stress values decrease for both drained and undrained tests.

REFERENCES

Airey, D.W 1991. Finite element analyses of triaxial tests with different end and drainage conditions. Proc. of the Seventh International Conference on Computer Methods and Advances in Geomechanics. 1:225-230.

Carter, J.P 1982. Predictions of the non-homogeneous behaviour of clay in the triaxial test. Geotechnique. 32:52-58.

Carter, J.P and Balaam, N.P 1990. AFENA user's manual version 5.0.

Kutter, B.L and Sathialingam, N 1992. Elastic-viscoplastic modelling of the rate-dependent behaviour of clays. Geotechnique. 42:427-441..

Mesri, G and Godlewski, P.M 1977. Time and stress compressibility interrelationship. Journal of the Geotechnical Engineering Division. 103, GT5:417-430.Sarma, S.K. 1973. Stability analysis of embankments and slopes. Geotechnique, Vol. 23, No.3, pp. 423-433.

Schnaz, T and Gussman, P 1994. The influence of geometry and end restraint on the strength in triaxial compression in numerical simulations. Numerical Methods in Geotechnical Engineering. Rotterdam, Balkema : 129-133.

Sheng, D; Westerberg, B; Mattsson, H and Axelsson, K. 1997. Effects of End Restraint and Strain Rate in Triaxial Tests. Computers and Geotechnics 21:163-182.

Westerberg, B 1995. Lerors mekaniska egenskaper (Mechanical properties of clays). Licentiate thesis, Lulea University of Technology.

Computer Methods and Advances in Geomechanics, Desai et al. (eds) © 2001 Balkema, Rotterdam, ISBN 90 5809 183 X

A new in-situ direct shear testing method for sands and clays

H. Matsuoka & S. H. Liu
Department of Civil Engineering, Nagoya Institute of Technology, Japan

ABSTRACT: This paper introduces a new in-situ direct shear testing method and presents a number of test results on sands and clays by it. In this new testing method, a latticed shearing frame is directly embedded in a testing ground and then pulled horizontally by a flexible rope or chain with a constant dead load applied on the sample inside the latticed shearing frame. By using this new testing method, a number of tests have been performed on 2 kinds of sands (Toyoura sand and a river sand), Fujinomori clay and 4 kinds of undisturbed clays in fields. The obtained test results are compared with those by such usual tests as standard direct box shear tests and unconfined compression tests on the same samples. Through those comparisons, the validity of this new testing method is demonstrated.

1 INTRODUCTION

The shear strength of soils originates essentially from the frictional forces between soil particles. Based on this consideration, a new in-situ direct shear testing method that is similar to a frictional test on usual materials has been developed and successfully applied to determine the shear strengths of coarse granular materials (Matsuoka et al., 1998, 1999). On the same principle, this new testing method is also applicable to sands and clays. This paper introduces this new testing method and presents the results of tests on sands and clays (Matsuoka et al., 2000).

2 DESCRIPTION OF NEW IN-SITU DIRECT SHEAR TESTING METHOD

Figure1 (a) shows the schematic view of the new in-situ direct shear test and Figure1 (b) its picture. In this new in-situ test, a stainless latticed shearing frame is directly embedded in a testing ground and then pulled horizontally by a flexible rope or chain through the rotation of a wheel manually. For sands, the shearing frame is 14.14cm square (area 200cm^2) by 2cm high. For clays, two kinds of shearing frames are prepared: one is 14.14cm square (area 200cm^2) by 2cm high, the other is 10cm square (area 100cm^2) by 2cm high. The normal (vertical) load is applied by steel weights (dead load). The shear load is measured with a load cell that is connected to the rope or chain. The normal (vertical) displacement is measured with two displacement transducers that are set on the left and right ends of an aluminum bar. The bar is extended from the center of the shearing frame and perpendicular to the shear direction. The shear (horizontal) displacement is measured with a displacement transducer that is set at the middle back of a rigid plate over the shearing frame.

In order to ensure the applied vertical dead load be transmitted uniformly to the shear plane through the specimen inside the latticed shearing frame,

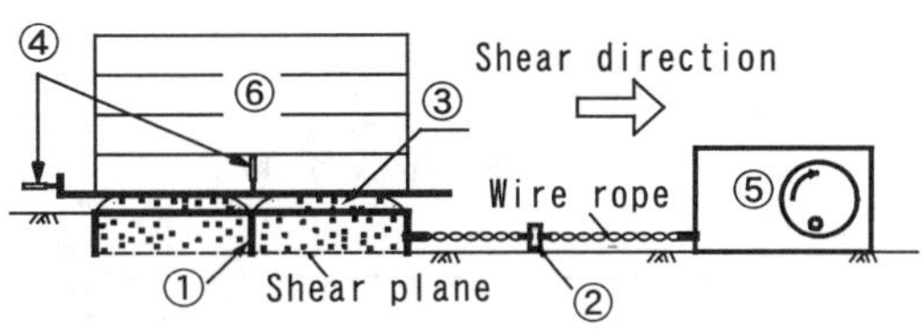

①Latticed shearing frame, ②Load cell to measure shear force, ③Raised specimen, ④Displacement transducers, ⑤Wheel rotated manually, ⑥Steel weight (dead load)

(a)

(b)

Figure1. New in-situ direct shear test

some specimen or other materials are slightly raised among the meshes of the latticed shearing frame, as shown in Figure 1(a). As we know, one of the defects in standard direct box shear test is that the measured shear strengths are affected by the frictional forces between the sample and the internal surfaces of the shear boxes when the sample dilates. In the newly developed in-situ direct shear test, however, this effect is considerably reduced by pulling the shearing frame using a flexible chain.

3 TESTS ON SANDS

The new in-situ direct shear tests have been performed on Toyoura sand and a river sand, the physical properties of which are given in Table 1.

Figure 2 presents the results of tests on these two sands with respect to (a) relationship between peak shear strength and normal stress and (b) relationships among shear-normal stress ratio, horizontal displacement and vertical displacement. In Fig.2 (a), the peak shear strengths are compared with those by standard direct box shear tests in which some improvements were made, as shown in Figure 3, to reduce the effect of frictional forces between the sample and the internal surfaces of the shear boxes

Table 1 Physical properties of sands tested

	Toyoura sand	River sand
G_s	2.65	2.60
D_{max}(mm)	0.3	2.0
D_{50}(mm)	0.2	0.5
U_c	1.3	2.1
e_0	0.7	0.6

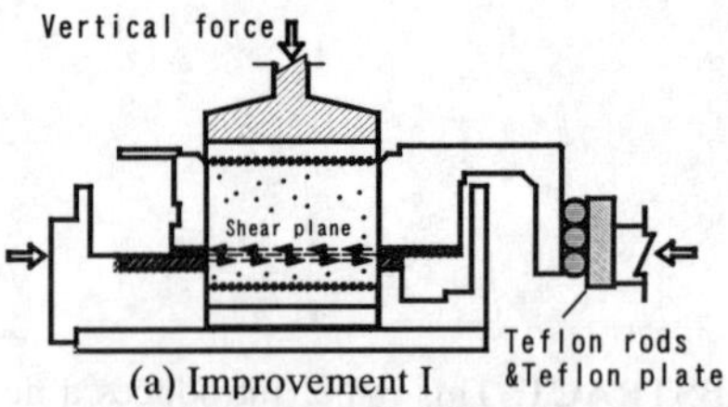

(a) Improvement I

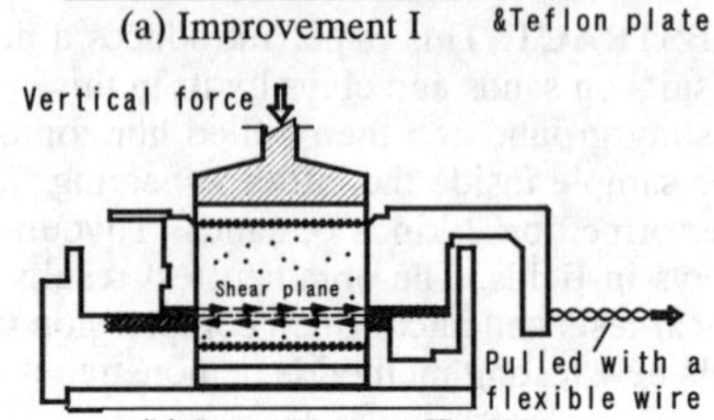

(b) Improvement II

Figure 3. Improvements of standard direct box shear test (Matsuoka et al., 1997)

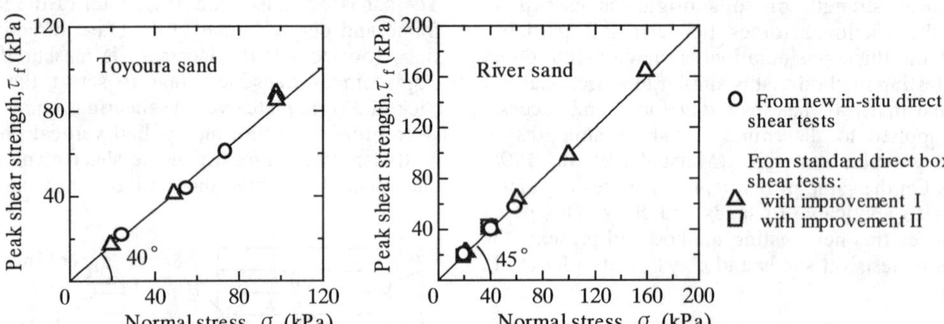

(a)

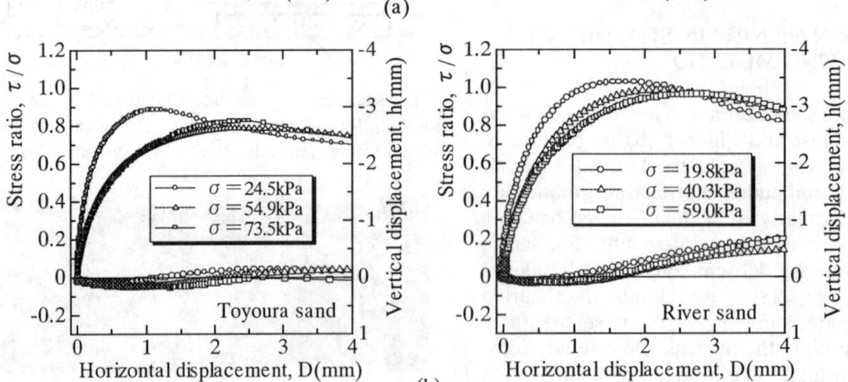

(b)

Figure 2. Results of tests on sands: (a) relation between peak shear strength and normal stress and (b) relationships among stress ratio, horizontal displacement and vertical displacement by new in-situ direct shear tests

(Matsuoka et al., 1997). It is seen from Fig.2(a) that, for both two tested sands, the peak shear strengths measured by the new in-situ direct shear tests are well plotted along a straight line against the normal stresses, and are in good agreement with those by improved standard direct box shear tests. Therefore, it can be considered that the new in-situ direct shear test can measure the shear strengths of sands with a reasonable accuracy. Moreover, the stress ratio-displacement properties can also be reasonably measured by this new in-situ direct shear test, as shown in Fig.2(b).

4 TESTS ON CLAYS

4.1 *Confirmatory tests on Fujinomori clay in laboratory*

When testing on clays, the shear strengths measured by the new in-situ direct shear tests are considered to be unconsolidated-undrained shear strengths, c_u, because clays are of low permeability and unconsolidated during the usual testing period of about ten minutes. In real clay fields, if applying a normal load on the clay specimen, it is too long for clay specimen to complete the consolidation.

Figure 4. Test on Fujinomori clay, showing the necessity of removing any clay around the shearing frame

The application of the new in-situ direct shear test to clays was first confirmed by performing tests on a man-made silty clay called Fujinomori clay. For comparison, standard direct box shear tests with improvement II (cf. Fig.3(b)) and unconfined compression tests were performed simultaneously on the same specimens. Fujinomori clay was made from clay powder taken at Fujinomori, near Kyoto, Japan. The powder was mixed thoroughly with a predetermined amount of water (water content $w \fallingdotseq 80\%$) to obtain a slurry, then about 98kPa of suction was applied to reduce the dissolved air content. Subsequently, the clay slurry was consolidated incrementally at a final vertical stress of 49kPa. The clay cake was removed from the tank and carefully sliced to the dimensions required for the new direct shear, standard direct box shear and unconfined compression tests. The prepared and sampled clay had water content w=50%, liquid limit w_L=41%, plastic limit w_P=23%, and a corresponding plasticity index I_P=18.

Figure 4 shows the new direct shear test on Fujinomori clay. The specimen of Fujinomori clay was carefully put in a box of about 15cm square by 5cm deep. A stainless steel latticed shearing frame, 10cm square by 2cm high, was embedded into it with the intention of causing minimal specimen disturbance. As shown in Fig.4, any specimen around the shearing frame should be removed, otherwise, the cohesive force on the outside surface of the shearing frame will prevent the correct measurement of the shear force. Fig. 5 shows the results of tests on Fujinomori clay by the new direct shear, improved standard direct box shear and unconfined compression tests. The average value of the shear strengths measured by the new direct shear tests is 11.8kPa, Fig. 5(a). It is identical to the average value measured by the improved standard direct box shear tests, Fig. 5(b), and in close agreement with the average value $q_u/2$ of the unconfined compression tests, 12.7kPa, Fig. 5(c). As we know, there is no reason that the undrained shear strength, c_u, measured by direct shear tests, should be identical to that calculated from $q_u/2$ of unconfined compression tests, because they have different modes of failure. The fact in Fig.5 indicates that the new in-situ direct shear test is applicable to determine the unconsolidated-undrained shear strengths of clays.

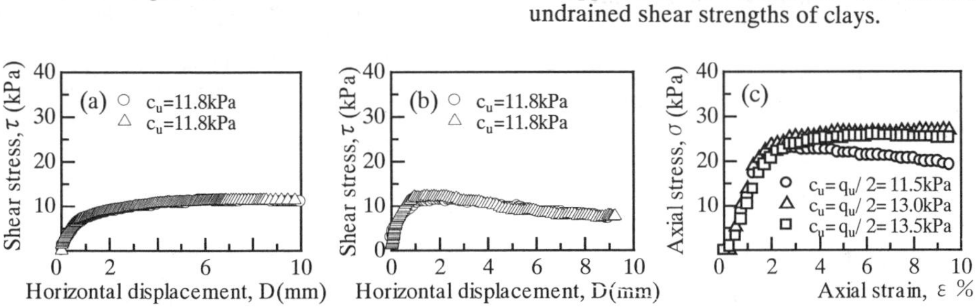

Figure 5. Results of tests on Fujinomori clay:(a) by new in-situ direct shear tests, (b) by standard direct box shear tests with improvementII, (c) by unconfined compression tests

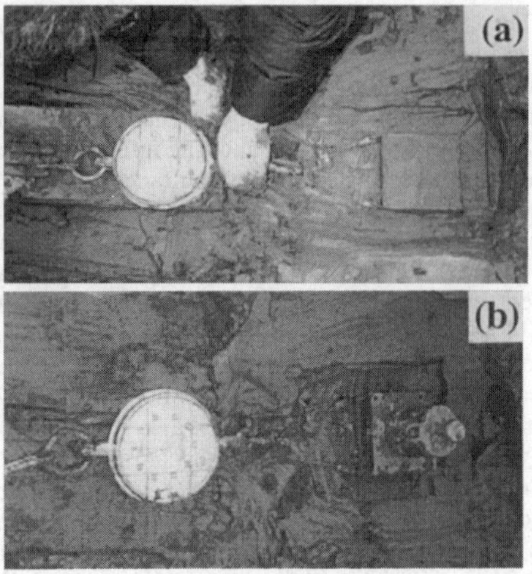

Figure 6. New in-situ direct shear tests on undisturbed clays at HKN

4.2 *Tests on undisturbed clays in various fields*

After verifying the applicability of the new direct shear test to Fujinomori clay, we have performed tests on undisturbed clays at 4 fields. Figure 6 shows an example of this test, performed at HKN. In Fig.6(a), no any external normal load was applied on the sample, while in Fig.6(b), some weight (normal load) was applied on the sample. There were no any difference in the measured shear strengths in these two cases, which gives a more proof that the new in-situ direct shear tests can be applied to measure the unconsolidated-undrained shear strengths of clays. Table 2 summarizes the results of the new in-situ direct shear tests on undisturbed clays at 4 fields, together with the results of three unconfined compression tests. The similarity of the unconsolidated-undrained shear strength c_u by the new in-situ direct shear tests and $q_u/2$ by the unconfined compression tests is also shown in Table 2. Moreover, the measured shear strength does not vary with the area of the latticed shearing frame is shown by the results at location MSM (Table 2). It is pointed out that one such test on clays can be finished within 10 minutes.

Table 2. Comparison of shear strengths of clays by new in-situ direct shear tests and unconfined compression tests

Testing location	Latticed shearing frame used (cm×cm)	c_u by new in-situ direct shear tests (kPa)	$q_u/2$ by unconfined compression tests (kPa)
MBR	14.1×14.1	8	10
SNG	14.1×14.1	37	36
MSM	14.1×14.1	13	—
	10.0×10.0	12	—
HKN	14.1×14.1	4.5	4.4

5 CONCLUDING REMARKS

The new in-situ direct shear testing method described in this paper has the characteristics of simplicity in equipment, easy and quick performance. As the shearing frame is pulled with a flexible wire rope, there are no external forces to affect the accurate measurement of the normal force acting on the shear plane even if the sample dilates. The normal forces acting on the shear plane are merely the sum of the applied dead load and the self-weights of the shearing frame and the sample inside it. The real normal and shear forces acting on the shear plane can be measured. This is the main reason for the correct measurement of shear strength by this new testing method.

ACKNOWLEDGEMENTS

The authors would like to express their sincere gratitude to Dr. D.A. Sun, researcher of Nagoya Institute of Technology, Mr. T. Shinozaki and Ms. H. Aoki, the students of the same institute, for their assistance in this experimental work. The authors would also like to acknowledge the persons in Chubu Branch of Kiso Jiban Consultants Co., Ltd., who have provided us the testing fields of clays.

REFERENCES

Matsuoka, H. and Liu, S.H. 1998. Simplified direct box shear test on granular materials and its application to rockfill materials. *Soils and Foundations*, Vol.38, No.4: 275-284.

Matsuoka, H., Liu, S.H., Satoh, S., Shibahara, T., Kudou, A., Nishikata, U. and Ijiri, K. 1999. Large-sized in-situ direct shear test on coarse-grained materials. *Proc. of 34th Japan National Conf. on Geotechnical Engineering*, 229: 459-460 (in Japanese).

Matsuoka, H., Liu, S.H., Aoki, H. and Shinozaki, T. 2000. Comparison of shear strengths of sands and clays by in-situ direct shear tests and by laboratory tests. *Proc. of 35th Japan National Conf. on Geotechnical Engineering*, 423: 833-834 (in Japanese).

Matsuoka, H., Sun, D. A. and Ueno, Y. 1997. Some improvements of the standard direct box shear test apparatus. *Proc. of 32th Japan National Conf. on Geotechnical Engineering*: 511-512 (in Japanese).

Computer Methods and Advances in Geomechanics, Desai et al. (eds) © 2001 Balkema, Rotterdam, ISBN 90 5809 183 X

Clay-steel pile interface inspection using cylindrical guided waves

W.B. Na & T. Kundu
Department of Civil Engineering and Engineering Mechanics, University of Arizona, Tucson, Ariz., USA
Y.S. Ryu
Department of Ocean Engineering, Pukyong National University, Korea

ABSTRACT: The feasibility of detecting interface delamination or separation of steel piles from the surrounding clay using cylindrical guided waves is investigated in this paper. It is shown that cylindrical guided waves can easily detect these defects. A special coupler between the steel pile and ultrasonic transducer has been used to launch non-axisymmetric guided waves in the steel pile. This investigation shows that the cylindrical guided waves can propagate along the steel pile, embedded in clay and are sensitive to the interface debonding between pile and clay.

1. INTRODUCTION

Delamination at the soil-pile interface, in other words, separation of the pile from the surrounding soil is a crucial problem for geotechnical engineers because it reduces the load carrying capacity of the pile. There have been some investigations on the nondestructive inspection of piles. Rix et al. (1996) used a stress wave technique for evaluating the pile tip elevations by impacting piles with hammer and receiving the reflected waves from the tip of the pile. Lin et al. (1994) used the impact-echo method to identify the presence of delaminations or voids at the concrete/rock interface. They concluded that the presence of the delaminations or voids can be identified only when the acoustic properties of rocks are greater than or equal to those of concrete. In these traditional stress wave techniques, the signal amplitude and the time of flight measurements provide information about the internal or interface defects and the geometries. However, these methods are not very efficient for evaluating the debonding condition at the interface between different materials such as soil and steel.

The purpose of this research is to develop a new inspection technique for the delamination detection using cylindrical guided waves. Several investigators have used cylindrical guided waves for different purposes because it has a number of advantages over other traditional ultrasonic methods. (1) the guided waves give several wave modes; some of those can be very sensitive to target defects (2) they propagate relatively long distance, it reduces the inspection time. Because of these advantages, this technique has been applied to detect defects in pipes and embedded tendons in post-tensioned bridges. Recently, Wang (1999) reviewed the cylindrical guided wave techniques and their potential applications in the petrochemical industry such as pipes, heat tubes, vessels, risers, and heat exchange tubing.

2. METHODOLOGY

Three specimens are fabricated placing steel rod in clay. To make artificial delamination at the interface, the steel rod (model for the pile) is twisted several times. Clay is used as the surrounding medium; the material properties of baseline clay are obtained from ultrasonic testing. The water content and dry density of the clay are used as indicators for saturation and compactness of the clay. These parameters are varied in the three specimens.

To propagate a strong signal along the pile several incident angles are tried out using an innovative transducer holder (Guo and Kundu, 1999). After obtaining the optimum angle, the peak positions in V(f) curves are plotted on the dispersion curves to identify the corresponding mode. Several modes are investigated to study their sensitivity and long distance inspection capability. In this paper, some experimental results are presented with dispersion curves of the specimen.

3. THEORY

Many investigators have studied the problem of elastic wave propagation in cylindrical structures. Gazis (1959) computed the dispersion curves for a hollow cylinder. His derivation has been used by several researchers for wave propagation in traction free cylinders. However, the problem becomes more complex for cylinders embedded in a second medium. In this case, difficulty arises in calculating complex Bessel functions to model leaky guided waves. This problem has been studied by some investigators. These works on the elastic wave propagation in leaky cylindrical systems addressed specific elasticity problems. Recently an expanded work for a general system has been presented by Pavlakovic and Lowe (1999).

In our theoretical model the clay is considered as the surrounding medium around the steel pile. Figure 1 shows the schematic diagram of a two-layered structure. In this figure, layer 1 represents a cylinder of finite diameter embedded in an infinite solid denoted by medium 2. The longitudinal and shear waves (L+-, SV+-, SH+-) combine to form a guided wave. The cylindrical system is axisymmetric and infinitely long. The detailed derivation is well documented in the literature (Pavlakovic and Lowe, 1997).

Using definitions of Silk and Bainton (1979), it can be stated that there are three basic types of cylindrical guided waves: (1) longitudinal axially symmetric modes, L(0,m), (2) torsional axially symmetric modes, T(0,m) and, (3) non-axisymmetric modes, F(n,m). The first index n denotes the circumferential order of the mode; for axisymmetric modes the first index is zero. The second index, m, describes a counter variable.

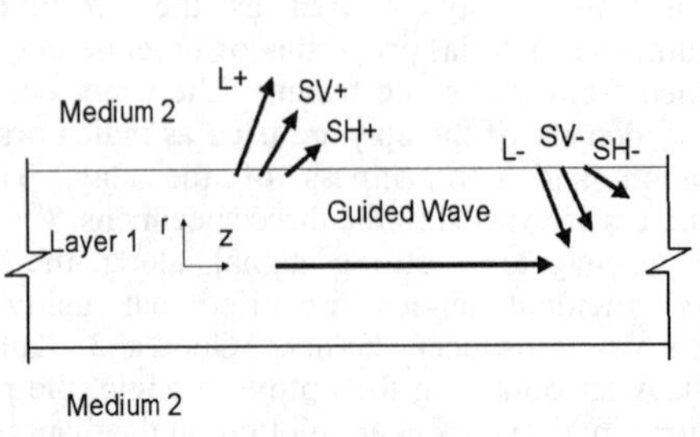

Figure 1. Schematic diagram of a pile (medium 1) embedded in an infinite medium (medium 2).

4. EXPERIMENTAL SETUP

Three types of specimens were fabricated and used in the experiment. These specimens include different amount of water content and dry density as shown in Table 1. To make artificial delamination at the interface, the steel pile was twisted several times. The diameter of steel rods is 7/8 inch (22.225 mm) and the dimension of the surrounding clay is 5×5×12 inch (127×127×304.8 mm) as shown in Figure 2. To propagate strong and sensitive guided wave modes along the steel pile the angle of transmitter was discretely changed using an innovative transmitter holder developed by Guo and Kundu (1999, 2000). The receiver was located at the other end of the steel rod. The transmitter was excited by a tone-burst excitation to generate guided waves. The excitation frequency was continuously varied from a minimum to a maximum value within the bandwidth of the transducer (1 MHz Harrisonic transducer) to find the peak positions that correspond to the guided waves generated in the specimens. The received signal amplitude was then displayed on an oscilloscope screen as a function of frequency. This voltage versus frequency curve is called V(f) curve. The conceptual experimental layout is shown in Figure 2.

Table 1. Water content and dry density of clays used, for computing dispersion curves the clay properties (wave speed, attenuation, etc.) have been obtained from the baseline clay given in the first row.

	Water content (%)	Dry density (kg/m^3)
Baseline clay	21.45	1661
Specimen 1	21.45	1871
Specimen 2	23.32	1555
Specimen 3	13.79	1699

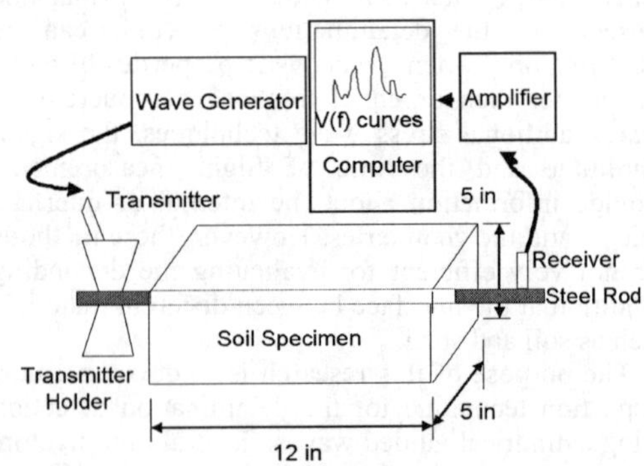

Figure 2. Conceptual layout of experimental setup.

5. MODELING

High attenuation of the guided waves in presence of clay causes the main difficulty for the inspection of steel pile embedded in clay. The attenuation is due to high material damping of the surrounding clay, the energy leaks from the steel pile into the clay. Hence, one should use the modes whose attenuation is small. However, these modes may not be sensitive to the target defects. Thus, ideal situation is to use the modes whose attenuation is low and defect sensitivity is high. From the dispersion curves (attenuation vs. frequency plots for different modes) one can conclude which mode has low attenuation. However, the sensitivity of modes to target defects is not easy to obtain because it requires mathematical modeling considering internal defects – a rather difficult task.

Dispersion curves are computed for the problem geometry shown in Figure 2, without considering any defect (Pavlakovic and Lowe 1997). The problem geometry is idealized as a cylindrical solid steel pile embedded in a medium of clay. The density, velocity and attenuation used in the baseline clay model are shown in Table 2. Here, λ denotes wave number. The properties of clay are measured experimentally using 1 MHz longitudinal and shear wave transducers. The properties of steel are obtained from the handbook and the attenuation in steel is ignored. The phase velocity dispersion curves for the 1st, 2nd and 3rd flexural modes are shown in Figure 3. The solid dots represent the experimental results.

Table 2. Material properties of baseline model.

Property	Clay	Steel pile
Density (kg/m^3)	2018	7932
Longitudinal wave velocity (m/sec)	1519	5960
Shear wave velocity (m/sec)	0.011	3260
Longitudinal wave attenuation (np/λ)	1228	0
Shear wave attenuation (np/λ)	0.155	0

6. EXPERIMENTAL RESULTS

The optimum angle, 27 degree, for transmitting strong signal is found experimentally, see Figure 4. Figures 5, 6, and 7 give the V(f) curves of specimens 1, 2, and 3, respectively. Here, the horizontal line gives frequency (kHz) and vertical line gives amplitude.

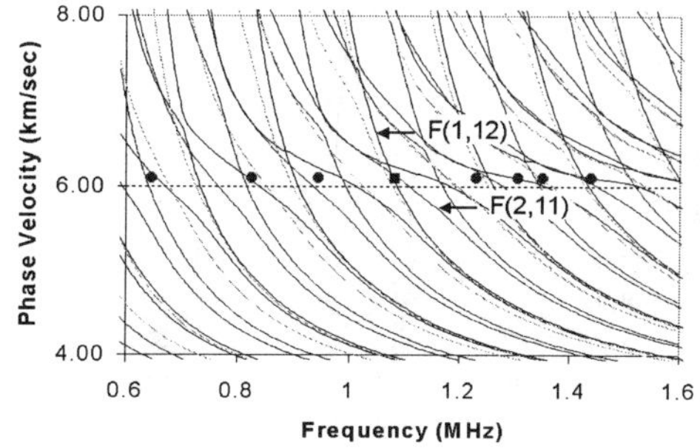

Figure 3. Phase velocity dispersion curves of steel pile embedded in clay.

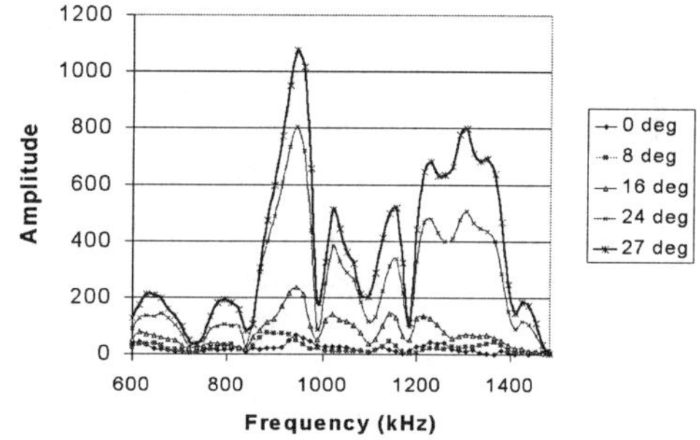

Figure 4. V(f) curves of a specimen for different incident angles.

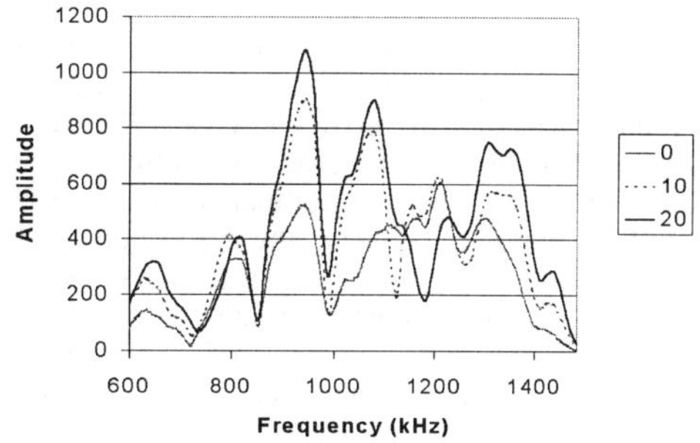

Figure 5. V(f) curves of Specimen 1.

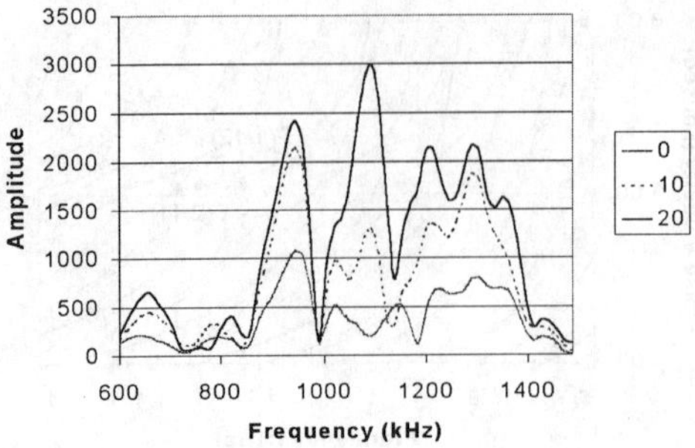

Figure 6. V(f) curves of specimen 2.

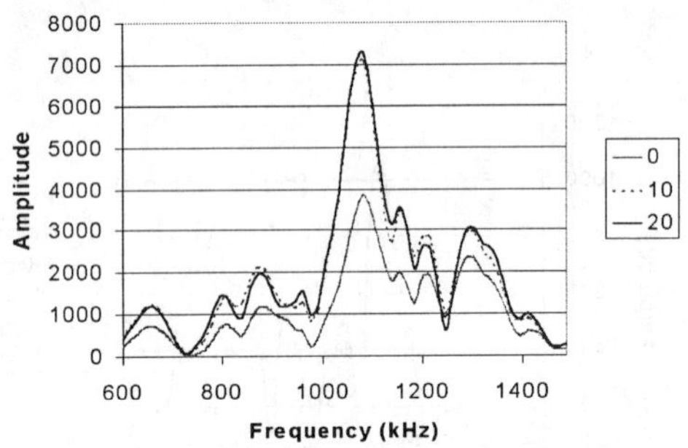

Figure 7. V(f) curves of specimen 3.

The legends of figures 5, 6 and 7 give the number of twists for each case. From these curves, one can see that the number of twists has a strong effect on the V(f) curves at many frequencies. To investigate how the peaks in the V(f) curves match with the theoretical calculation the experimental values are plotted on the theoretical dispersion curves of the steel-baseline clay model in Figure 3. To calculate the phase velocity the Snell's law, $\theta = Sin^{-1}(V_c/V_{ph})$, is used. Where, θ is the angle of inclination of the transducer, V_c is the longitudinal wave speed in the coupling medium between the transducer and the steel pile, and V_{ph} is the phase velocity in the specimen. The experimental points (solid square and circles) represent peak values of V(f) curves for perfectly bonded specimens with 27 degree inclination angle. The solid square corresponds to a strong peak position and matches very well with mode F(1,12) and F(2,11) at 1.08 MHz in Figure 3. The two modes are examined carefully. It is found that for mode F(2,11) the strain energy is mostly confined within the steel pile but have nonzero values at the interface of clay and steel pile. The axial and radial displacements are not vanishingly small at the interface; hence, this mode is sensitive to the delamination at the interface. Thus, mode F(2,11) is recommended for this inspection.

7. CONCLUSIONS

In this paper, it is shown that cylindrical guided waves are sensitive to the delamination at the interface of the steel pile and clay. It is shown that some non-axisymmetric modes are appropriate candidates for this inspection. The modes that have good sensitivity to the interface defects and experience low attenuation as they propagate along the steel pile are considered ideal for this inspection.

REFERENCES

Gazis, D. C. 1959. Three dimensional investigations of the propagation of waves in hollow circular cylinders I analytical foundation. *J. Acoust. Soc. Am.* 31(5): 568-573.

Guo, D. & T. Kundu 1999. Special sensors for generating lamb waves in pipes. In D. O. Thomson and D. E. Chimenti (ed.), *Progress in Quantitative Nondestructive Evaluation*: 18:1155-1162. New York: Plenum Press.

Guo, D., & T. Kundu in press 2000. A new sensor for pipe inspection by lamb waves. *Materials Evaluation.*

Lin, J. & M. Sansalone 1994. Impact-echo response of hollow cylindrical concrete structures surrounded by soil and rock: part II -experimental studies. *Geotechnical Testing J.* 17(2): 220-226.

Pavlakovic, B. & M. Lowe 1997. *DISPERSE user's manual.* Imperial College: University of London: Non-Destructive Testing Laboratory.

Pavlakovic, B. and M. Lowe. 1999. A general purpose approach to calculating the longitudinal and flexural modes of multi-layered, embedded, transversely isotropic cylinders. In D. O. Thomson and D. E. Chimenti (ed.), *Review of Progress in Quantitative Nondestructive Evaluation.* 18:239-246. New York: Plenum Press.

Rix, G. J., L. J. Jacobs, P. B. Rhodes, & R. Q. Raparelli 1996. Nondestructive assessment of pile tip elevation using flexural waves. *Proceedings symposium on the application of geophysics to engineering and environmental problems*: 577-586.

Silk, M. G. & K. F. Bainton 1979. The propagation in metal tubing of ultrasonic wave mode equivalent to lamb waves. *Ultrasonics* 17:11-19.

Wang, W. D., 1999, Applications of Guided Wave Techniques in the Petrochemical Industry. In D. O. Thomson and D. E. Chimenti (ed.), *Review of Progress in Quantitative Nondestructive Evaluation.* 18:277-284. New York: Plenum Press.

Computer Methods and Advances in Geomechanics, Desai et al. (eds) © 2001 Balkema, Rotterdam, ISBN 90 5809 183 X

Parameter calibration of strain hardening/softening of sand from direct shear tests

A. Nobahar & R. Popescu
Faculty of Engineering and Applied Science, Memorial University of Newfoundland, St.John's, Nfld, Canada

I. Konuk
Geological Survey of Canada, Ottawa, Ont., Canada

ABSTRACT: The objective of this study is to obtain a meaningful relationship between the direct shear box test results (recorded shear stress vs. box displacement) and the soil strength parameters and their evolution during plastic shear. A hardening/softening rule expressing the mobilized friction angle and cohesion as a function of the shear strain magnitude in a large deformation analysis is provided for use in routine geotechnical analysis. It is validated for dense sand, sheared under dry condition. The resulting rule is verified based on results of full-scale pipe/soil interaction experiments involving large relative displacements.

1 INTRODUCTION

Soil materials have a pronounced non-linear behaviour, and correct numerical simulation requires sophisticated constitutive models including a relatively large number of parameters and complicated calibration procedures. Occurrence of strain localization at large shear deformation further complicates the numerical modeling. Strain localization is not a continuum phenomenon. It is characterized by a length-scale. Therefore for properly addressing of this phenomenon, a model should include the concept of length-scale either through mesh dimensions (adaptive meshing) or special provision in the constitutive model. Adaptive meshing solutions use elements whose size is the same as or smaller than the shear band (i.e. of the order of millimetres) and therefore they are prohibitive for real boundary value problems. Various methods have been proposed for including the concept of length-scale into the constitutive model and solve the localization problem using finite elements that are much larger than the size of the shear band (see e.g. Borja et al. 2000 among others). However such models have not been implemented in widely available finite element programs.

In this study a relatively straightforward non-associated Mohr-Coulomb soil constitutive model implemented in the finite element program ABAQUS/Standard is customized to account for progressive mobilization (hardening/softening) of soil shear strength parameters, ϕ and c, during shear involving large deformations. The modified model also relates the variation of the dilation angle ψ with mobilized friction angle, according to Rowe's stress dilatancy relation (Rowe, 1962).

The procedure used in this study for estimating hardening/softening behaviour of dense sand is based on the results of direct shear box (DSB) tests as illustrated in Figure 1. First, based on the results of DSB tests performed on dense sand under different confining stresses (Figure 1a), the mobilized shear strength parameters are calculated and expressed as functions of the box displacements (Figures 1b & 1c). Next, finite element simulations of the DSB tests are carried out to find out a correspondence between measured horizontal displacements and state variables defining the material behaviour – the plastic strain magnitude was selected here (Figure 1d & 1e). Finally, the mobilized shear strength parameters are expressed as functions of the plastic strain magnitude (Figure 1f). Based on a good match between numerical predictions and experimental observations, it is assumed that average values of shear strength, as measured in DSB tests, can be used for inferring the hardening/softening rules shown in Figure 1f. The resulting hardening/softening rules are verified against full-scale test results in a soil/structure interaction problem. Based on this approach, hardening/softening curves for a dense sand are provided for use in geotechnical applications.

2 PROCESSING DIRECT SHEAR BOX TEST RESULTS

2.1 *Experimental data*

The proposed method is demonstrated using results of DSB tests on a well-graded coarse sand with mean grain size D_{50}=0.7mm. The specimens were compacted dry up to 95-100% relative density and the normal vertical stress was in the range of 20 to 160 kPa. The samples were sheared to a displacement of 8 to 10 mm. The experimental results are confidential, and therefore they could not be shown here.

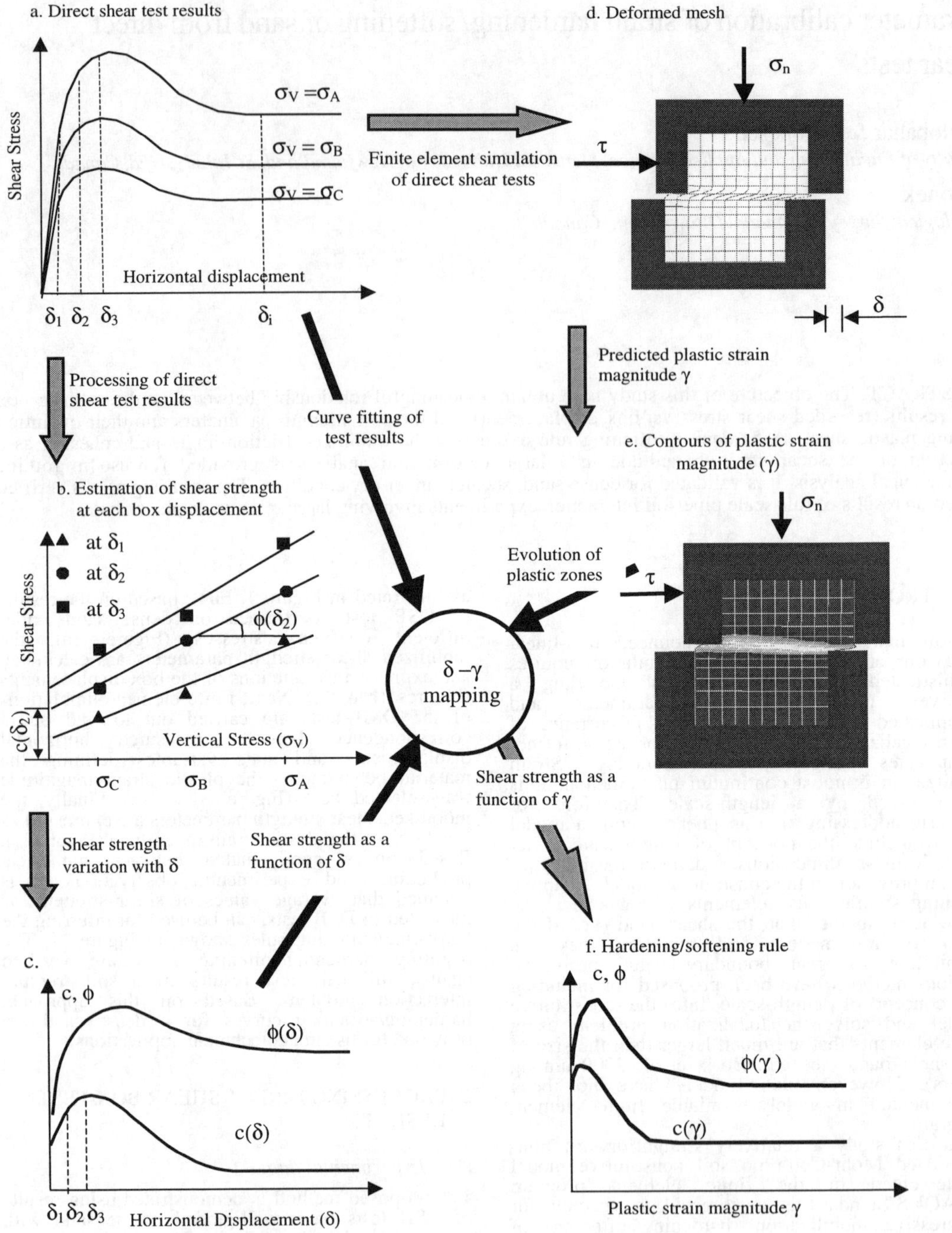

Figure 1. Flow chart illustrating the proposed methodology for estimating progressive mobilization of shear strength parameters based on results of direct shear box soil tests.

2.2 *Progressive mobilization of shear strength*

Usually maximum and residual shear strength measured in the DSB tests are used to obtain peak and residual shear soil strength parameters. However, the DSB test results can provide more information. Each curve shows the variation of mobilized shear strength with box displacements under each normal stress (Figure 1a). A mobilized friction angle and cohesion for each corresponding displacement can be defined from these curves (e.g. Stematiu et al. 1991, Popescu 1993). First, from each τ–δ curve in Figure 1a and for each value δ_i of the box displacement, a point is obtained in the τ–σ space (Figure 1b). Next, for each value of the box displacement, the mobilized friction angle and cohesion are estimated based on the Mohr-Coulomb criterion. Each Mohr Coulomb line can be considered as the instance yield surface in a τ–σ representation during the process of hardening and softening. The friction angle and cohesion obtained from these interpolation processes can then be expressed as discrete functions of the box displacement (Figure 1c) such as:

$$\phi = \phi(\delta) \text{ and } c = c(\delta) \tag{1}$$

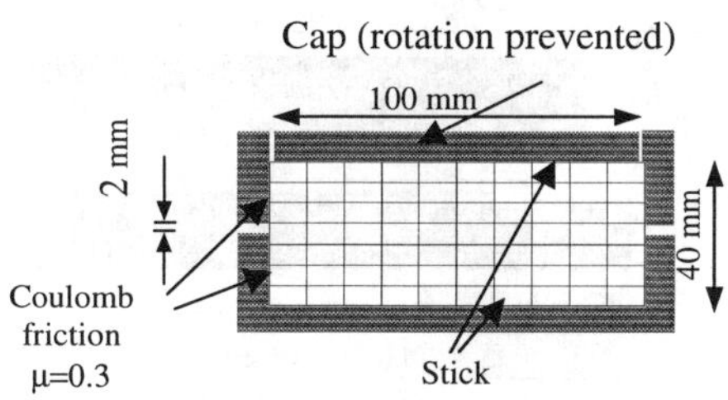

Figure 2. Finite element mesh for simulation of DSB tests

3 NUMERICAL MODEL

3.1 *Mohr Coulomb plasticity*

The non-associated Mohr Coulomb plasticity constitutive model is selected for this study. The Mohr Coulomb constitutive model implemented in ABAQUS/Standard has isotropic hardening, defined in terms of the material cohesion. The plastic flow rule is non-associated, with the flow potential described as a hyperbolic function in the p-q plane, and a smooth elliptic function in the deviatoric plane (Hibbitt et al., 1998). The dilation angle, defining the slope of plastic potential in the p-q plane, can be different from the friction angle, which defines the slope of yield surface.

The model requires the following constitutive parameters: Young's modulus (E), Poisson's ratio (ν), friction angle (ϕ), cohesion (c) and dilation angle (ψ). Also a hardening rule can be introduced by expressing the dependency of one or more soil parameters on a hardening parameter.

It was selected here that the hardening/softening behaviour of dense sand be expressed in terms of cohesion and frictions angle as function of the "plastic strain magnitude", γ, which can be expressed as (Hibbitt et al., 1998):

$$\gamma = \sqrt{\frac{2}{3}\varepsilon^{pl} : \varepsilon^{pl}} \tag{2}$$

Where ε^{pl} is the plastic strain tensor.

To express the dependency of friction angle and cohesion on γ, a correspondence is sought here between shear box displacements, δ, and plastic strain magnitude, γ.

3.2 *Dilation angle*

The variation of the dilation angle, ψ, was assumed related to the mobilized friction angle according to Rowe's stress dilatancy equation (Rowe, 1962) in the form proposed by Vermeer and de Borst (1984) as follows:

$$\sin\psi = \frac{\sin\phi - \sin\phi_{cv}}{1 - \sin\phi \sin\phi_{cv}} \tag{3}$$

where ϕ_{cv} is the friction angle at constant volume and is taken here equal with the residual value of ϕ.

3.3 *Elastic parameters*

"Elastic" behavior is simulated by the constitutive model prior to reaching the yield surface. The parameters describing elastic behavior (Young's modulus and Poisson's ratio) are selected as follows: (1) a value ν=0.30, common for sand materials, has been assumed for the Poisson's ratio, and (2) the Young's modulus has been derived from the secant shear modulus, G, estimated for an assumed "elastic" domain in the DSB tests (i.e. up to $\delta=\delta_1$=0.5mm in Figure 1a).

3.4 *Set-up of the finite element analyses*

2D plane strain finite element analyses of DSB tests are performed with ABAQUS/Standard. The eight-node biquadratic element with reduced integration is used to model the soil material. The mesh size is selected in accordance with the expected shear band width (i.e. 6 to 7 mm, see section 4.1). The soil in a vertical cross-section is discretized into seventy elements as shown in Figure 2. Drained behavior is assumed for the soil. The interface between soil and box walls is modeled using the contact surface capability implemented in ABAQUS/Standard. The lateral walls are simulated by rigid surfaces, and Coulomb friction is assumed at the interface. The steel cap of the shear box is assumed rigid and its rotation is prevented.

4 HARDENING/SOFTENING RULE

In section 2.2, the shear strength parameters are expressed as discrete functions of the DSB displacements. A relation between box displacements, δ, and the soil state variable, γ, used to describe hardening is sought through finite element simulations of DSB tests.

4.1 *Shear band*

Shear band formation in granular soils has been investigated extensively in both experimental and theoretical studies. It is known that during strain localization, shear is concentrated in a narrow band. The width of shear band was studied by several researchers. It was found to be depended on the grain size and confining stress. Most authors defined a ratio between shear band width, h, and mean grain size diameter, D_{50}, e.g. Muhlhaus and Vardoulakis (1987), as shown in Table 1. It is obvious that the ratio of shear band width, h, to D_{50} decreases with increasing grain size. The influence of confining pressure on shear band width has been investigated by Hammad (1991) in biaxial compression tests on sands. Shear band width, h, was found to decrease with increasing confining stress.

Table 1. Experimentally observed shear band width, h, (after Muhlhaus and Vardoulakis, 1987)

	$D_{50\%}$ (mm)	h (mm)	$H/D_{50\%}$
Fine sand	0.2	3.7	18.5
Medium sand	0.33	4.3	13.0

By extrapolating the experimental results of Muhlhaus and Vardoulakis (1987) and accounting for the effects of confining stress as reported by Hammad (1991), a shear band width of about 6 to 7 mm has been assumed for the sand used here, with D_{50}=0.7mm, and for the range of confining stresses considered in this study (20 to 80 kPa).

4.2 *Estimation of plastic strain magnitude*

Due to highly non-uniform shear strain distribution in DSB test, finding a correspondence between box displacement and shear strain magnitude is not straightforward. To solve this problem, the pattern and magnitude of shear strain in different stages are analyzed. At the beginning of the test, the vertical stress (σ_n) is significantly larger than the shear stress (τ), therefore the major principal stress is oriented close to the vertical. A diagonal shear zone develops first (Figure 3a & b), similar to the situation observed in plane stress and triaxial tests. This initial shear zone cannot develop further due to the boundary conditions imposed by the box. With increasing τ, the principal stresses rotate, and the kinematics conditions lead to reorientation of the shear zone (Figure 3c). The predicted shear strain and its evolution during shear are in a close agreement with that obtained by Potts et al. (1987) and Dounias & Potts (1993). The transition between peak and residual strength behavior occurred in this example between δ=2.2mm (Figure 3b) and δ=2.7mm (Figure 3c). For large box displacements, a very well defined shear band was predicted by the numerical model (Figure 3d). It is mentioned that transition to shear localization occurred faster in the numerical model than observed in the experiments. However experimental recorded pre-peak, peak and residual values of shear forces were accurately reproduced by the numerical model for all the values of confining stress between 20 to 80 kPa.

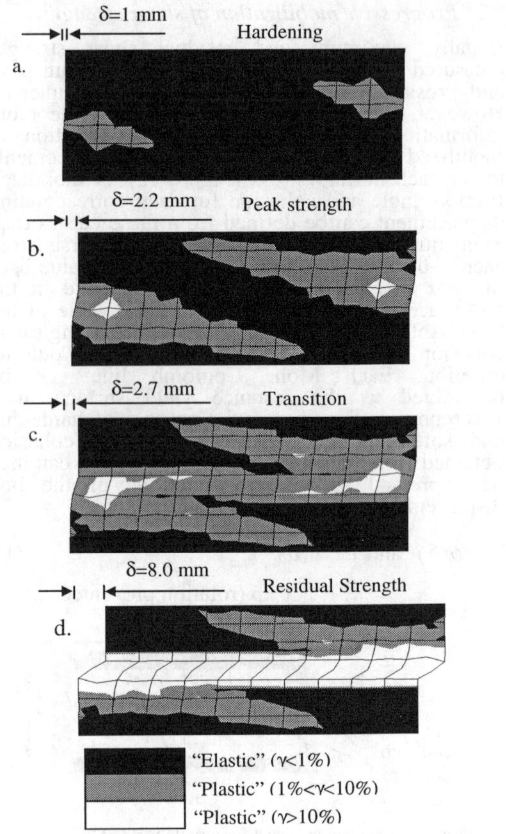

Figure 3. Predicted plastic zone extension: a. pre-peak b. peak strength c. softening (transition) d. residual strength.

To obtain a relation between box displacements, δ, and average shear strain magnitude, γ, a fictional width h of a corresponding uniform shear zone is assumed for each δ. The average γ over this assumed shear zone is calculated for each δ as $\gamma=\delta/h$, and therefore, δ can be expressed as a function of γ in Eqn. 1. These relations (i.e. Eqn.1 and $\delta=\delta(\gamma)$) are used to built the hardening/softening rule:

$$\phi = \phi(\gamma) \text{ and } c = c(\gamma) \quad (4)$$

The values of h at each δ are then adjusted by trial and error and the hardening/softening rules (Eqn. 4) are modified accordingly, until a satisfactory match is obtained between numerical predictions and experimental results in terms of the τ–δ relations. The resulting hardening/softening rule is presented in Figure 4. This rule results in a good match between numerical predictions and DSB test results for a range of confining stresses between 20 and 80kPa. The hardening/softening rule presented in Figure 4 is derived for well-graded Nevada sand with relative density of 95 to 100% and D_{50}=0.7 mm. The rule has been validated for the non-associated Mohr Coulomb constitutive model implemented in ABAQUS/Standard.

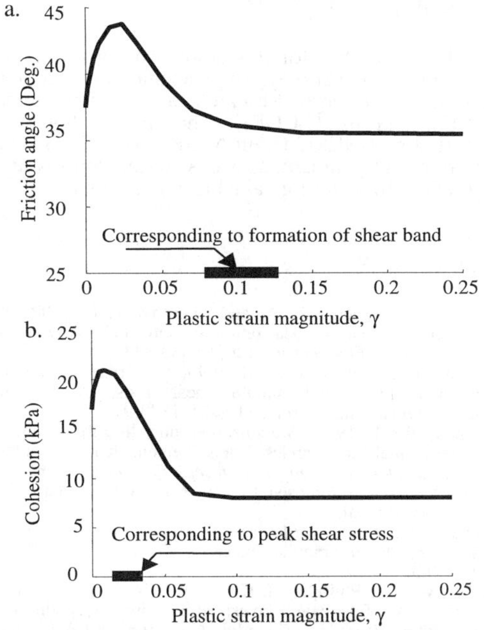

Figure 4. Hardening/softening rule for coarse, well-graded sand, with D_r=95 to 100%: a. friction angle; b. cohesion (γ = 0 corresponds to $\delta = \delta_1$ = 0.5mm) for values at residual state see remark.

Remark: Relatively large values of apparent cohesion at large shear strain did not result from the experimental measurements. They have been imposed by a feature of the Mohr Coulomb model implemented in ABAQUS/Standard that requires non-zero cohesion to accommodate the shape of plastic potential close to the apex (see also Popescu et al. 1999). Therefore, the results presented in Figure 4b can be considered as specific for use with the Mohr Coulomb constitutive model implemented in ABAQUS/Standard.

5 VALIDATION OF PROPOSED PARAMETER CALIBRATION METHOD

An extensive full-scale experimental program for investigating pipe performance in soil was carried out at C-CORE (e.g. Paulin et al. 1998), as part of a large international research project on soil-pipe interaction. Some of the tests consisted in lateral loading of a rigid buried pipe. Results from two such full-scale experiments involving lateral loading of a rigid pipe buried in dense sand (Figure 5a) are used here for validating the proposed methodology for parameter calibration. It is mentioned that the same sand material, at the same relative density has been employed for both the full-scale tests of pipe/soil interaction, and for the DSB tests used here, to infer the hardening/softening rules in Figure 4.

5.1 *Finite element model*

The two-dimensional plane strain simplification is assumed. The finite element mesh (Figure 5b) has 108 eight-node biquadratic continuum elements with reduced integration for the soil material, and 16 three-node quadratic beam structural elements for the cross-section of the pipe. In a drained analysis, the soil is idealized as a one-phase material, and all resulting stresses are effective stresses. The contact surface capability implemented in ABAQUS/Standard, which can handle separation and sliding of finite amplitude, is used in this study. Appropriate surfaces are defined on soil and pipe, and isotropic Coulomb friction is assumed at the interface, with a friction coefficient of μ=0.54 (Popescu et al. 1999).

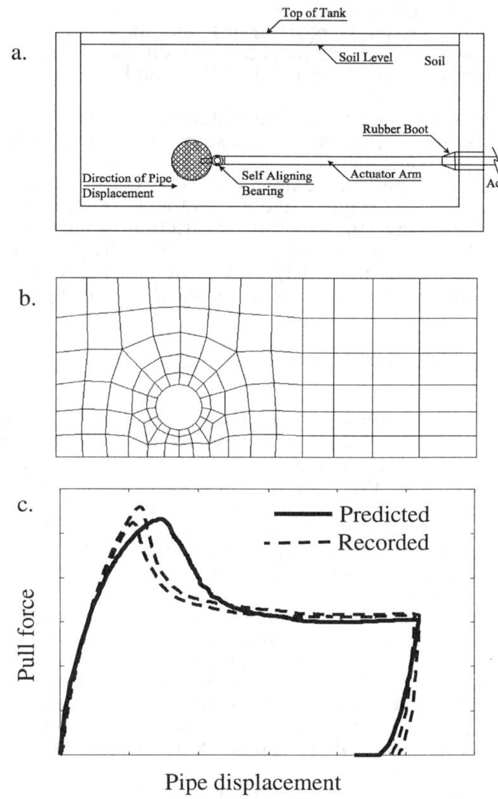

Figure 5. Finite element analysis of lateral loading of a rigid pipe in dense sand: a. full-scale pipe-soil interaction tank (after Paulin et al. 1998); b. finite element mesh; c. comparison between predicted (continuous line) and recorded (dashed lines - after Paulin et al. 1999) force - displacement relations.

5.2 *Numerical predictions*

Predicted and recorded force-displacement curves for the tests performed in dense sand using the hardening/softening rule in Figure 4 are presented in Figure 5c. Good agreement is observed between predicted and recorded values excepting for the location of peak loads.

6 CONCLUSIONS

6.1 *Results*

1 Direct interpretation of DSB test results may induce errors in estimating soil strength parameters, due to highly non-uniform strain and stress distribution in DSB.
2 A practical methodology is presented for obtaining the hardening and softening rule of dense sands from DSB test results, while addressing the above limitation.
3 Based on a good match between numerical model and experimental observations, it is assumed that average values of shear strength recorded in DSB tests can be used for matching the numerical simulation results.
4 The predicted shear strains and their evolution during the test are in a close agreement with those obtained by Potts et al. (1987) and Dounias & Potts (1993).
5 The resulting hardening/softening rule is verified against full-scale test results on a real size structure.
6 A hardening/softening rule for dense, coarse sands, sheared under drained conditions, expressing the mobilized friction angle and cohesion as a function of the shear strain magnitude in a large deformation analysis is provided for use in routine geotechnical analysis.

6.2 *Limitation and comments*

1 The length-scale character of the strain localization phenomenon is overlooked here in favor of computational efficiency. Preliminary studies on the effects of mesh dependency for the cases analyzed in this study (i.e. DSB and lateral loading of rigid pipe) have shown that a drastic change in mesh refinement will induce changes in predicted response in term of shear loads not larger than 10-15 %. As expected, the predicted width of shear band depends on the mesh size.
2 The method is only used for one single type of boundary conditions, and one soil material. The study should be extended to include more sand materials (various fabrics and relative densities) and, pending on availability of reliable experimental data, another type of soil test (e.g. simple shear).
3 The hardening/softening rule (Figure 4) obtained from a confining stress range σ_n=20 to 80 kPa did not work properly for much larger confining stresses (σ_n=160 kPa). This means that there is a dependency on confining stress, which will be addressed in a future study.
4 Average values of shear strength, as recorded from a DSB tests, are used to assess the soil strength in a boundary value application characterized by high stress and strain non-uniformity. Therefore this problem does not have a unique solution. However, successful verification of the proposed hardening rule against full-scale experimental results increased the confidence that the true solution for soil strength had been obtained

ACKNOWLEDGEMENTS

Partial sponsorship for this study was provided by the Geological Survey of Canada through the Program of Energy Research and Development (PERD), and by C-CORE. The first author was supported by NSERC Grant No. RG203795-98. The authors are also indebted to Drs. Ryan Phillips and Peijun Guo for offering valuable suggestions.

REFERENCES

Borja, R. I., Regueiro, R. A. and Timothy, Y. L. 2000. FE modeling of strain localization in soft rock. *Journal of Geotechnical Engineering,* 126(4): 335-343.

Dounias, G. T. and Potts, D. M. 1993. Numerical analysis of drained direct and simple shear tests. *Journal of Geotechnical Engineering,* 119(12): 1870-91.

Hammad, W. I. 1991. Modelisation non lineaire et etude experimentale des bandes de cisaillement dans les sables. *These de Doctorat, Universite Joseph Fourier-Grenoble.*

Hibbitt, Karlsson & Sorensen, Inc. 1998. ABAQUS version 5.8 - Theory Manual.

Muhlhaus, H. B. and Vardoulakis, I. 1987. The thickness of shear bands in granular materials. *Geotechnique,* 37(3): 171-283.

Paulin, M.J., R. Phillips, J.I. Clark, A. Trigg, and I. Konuk, 1998. A full-scale investigation into pipeline/soil interaction. *Proc. 51st Canadian Geotechnical Conference,* pp.241-248.

Popescu, R. 1993. Nonlinear finite element analysis of embankment dams. Ph.D. *Thesis. Dept. Hydr. Struct., Techn. Univ. Civil Eng.,* Bucharest, Romania.

Popescu, R., Philips, R., Konuk, I., and Deacu, D., 1999. Numerical and physical modelling of pipe-soil interaction. *Proc. 52nd Canadian Geotech. Conf.,* Regina. SK, pp 437-444.

Potts, D. M., Dounias, G. T. and Vaughan P. R. 1987. "Finite element analysis of the direct shear box test." *Geotechnique,* 37(1): 11-23.

Rowe, P.W., 1962. The stress-dilatancy relation for static equilibrium of an assembly of particles in contact" *Proc. of Royal. Society of London,* 269(Series A): 500-527.

Stematiu, D., Popescu, R. and Luca, E. 1991. Contact clay problems during the erection of Maru Dam. *Water Power & Dam Construction,* 43(4):28-33.

Vardoulakis, I. and Sulem, J. 1995. Bifurcation analysis in geomechanics. *Blackie Academic & Professional,* Glasgow, UK.

Vermeer, P.A. and de Borst, R. 1984. Non-associated plasticity for soils, concrete and rock. *Heron,* 29(3): 1-64.

Computer Methods and Advances in Geomechanics, Desai et al. (eds) © 2001 Balkema, Rotterdam, ISBN 90 5809 183 X

Evaluation of progressive failure in soils using industrial X-ray CT scanner

J.Otani & T.Mukunoki
Graduate School of Science and Technology, Kumamoto University, Japan

Y.Obara
Faculty of Engineering, Kumamoto University, Japan

ABSTRACT: Progressive failure with strain localization in the soil under triaxial compression is characterized using industrial X-ray CT scanner with new developed triaxial compression test apparatus. First of all, the new triaxial compression apparatus is introduced. Then, this apparatus is examined in the system of X-ray CT scanner. Here, a series of nondestructive tests using X-ray CT scanner is conducted during triaxial compression without moving out the new apparatus from the shield room in the system. Based on those test results, the developed apparatus gives fairly reasonable results as a triaxial compression test apparatus and the progressive failure with progressing strain localization is visualized using CT images. Finally, it is concluded that the proposed test apparatus makes the characterization of the failure possible in the soil without any destructions.

1 INTRODUCTION

It has been recently realized that the failure of soils is caused by progressing strain localization in the soil (Adachi et al. edit.,1997), so that the research on the behavior in the soil is indispensable for the purposes of visualization in the soil and its quantitative discussion as well.

An X-ray computed tomography(CT) method which is one of the nondestructive testing method has been used with the name of an apparatus commonly well-known by the medical diagnostic method which provides the cross sectional images based on the absorption of the x-ray beam through the materials. Recently, an industrial X-ray CT scanner (TOSCANER-23200min:TOSHIBA Corp.) has been installed at the group of Rock Engineering in Kumamoto University, Japan(Sugawara et al. 1997 and Otani et al. 1999).

The purpose of this paper is to develop a new triaxial compression test apparatus for the industrial X-ray CT scanner and the progressive failure with strain localization in the soil under triaxial compression is characterized using this new apparatus. Finally, it is emphasized that the industrial X-ray CT scanner can be used effectively with this new testing apparatus in geotechnical engineering.

2 INDUSTRIAL X-RAY CT SCANNER

In the system of the X-ray CT scanner, the collimated x-ray is moved around the circumference of the specimen by rotating and translating the specimen table. The detected data are assembled and then, the cross-sectional images are reconstructed using engineering workstation. By using all these cross-sectional images around the circumference of the specimen, three-dimensional (3-D) image can be reconstructed. In order to evaluate these nondestructive test results quantitatively, following so-called "CT-value" is used:

$$\text{CT-value} = (\mu_t - \mu_w)\kappa/\mu_w \qquad (1)$$

where μ_t: coefficient of absorption at scanning point; μ_w: coefficient of absorption for water; and κ: material constant. It is noted that the coefficient of absorption for air is zero. When the value of κ is set to be 1000, the CT-value of the air should be −1000. It is known that the CT-value has a linear relationship with the density of the specimen, so that the CT-value is a representative value for the density of the specimen. The CT images are presented with shaded darker or black for lower density region and light gray or white for high density one for all the subsequent black and white images. The total number of the gray level distribution is 256. Precise contents about CT scanner are shown in the reference (Otani et al. 2000).

3 NEW TRIAXIAL COMPRESSION TEST APPARATUS

A new triaxial compression test apparatus for the industrial X-ray CT scanner is developed and is shown in Fig.1. The triaxial compression test under CD

(Consolidated Drained) condition was conducted using Toyoura Standard sand. The test procedure is listed as follows:

1) After consolidating the specimen under the same pressure as the confining pressure, the initial condition of soil specimen was scanned before applying the compression.
2) The specimen was compressed under strain control and is unloaded at the strain levels of 3%. Here, in order to take the effect of stress relaxation, the specimen was unloaded until the deviator stress is zero. Then, the nondestructive test using X-ray CT scanner was conducted. After finish scanning, the specimen was reloaded until the strain level of 6%. Here, the total number of cross sections was 6, which are the locations of the heights of 10, 25, 40, 55, and 70 mm from the bottom of the specimen.
3) The same process of above step 2) was continued until the strain level of 15% in every 3% strain.

4 RESULTS AND DISCUSSION

A series of drained test were conducted using Toyoura Sand under the process of testing described above. Figure 2 shows one the results of stress-strain relationship with volumetric strain, in which the deviator stress is plotted with axial strain and the condition of confining pressure was 49.0 kPa where the density was set to be 1.5 t/m^3. The meanings of step A, step B, step C, step D, and step E in the figure indicate the locations of the strain levels 3, 6, 9, 12 and 15 % respectively where the CT scanning was conducted. There is a peak stress at the strain level of around 7 % and it is realized that the volume of the specimen decreases at the strain level from initial to the stage of 3 % and increases afterwards. As realized easily, the total behavior is the cyclic one and is nothing different with those of conventional apparatus. Thus the usefulness of this proposed apparatus can be confirmed.

Figure 3 shows the cross sectional images for each strain level at six cross sections. The relatively dark area in the image is the area of low density. As realized from those images at the initial condition in Fig.3, there is a dark area around the center of each cross section and it is realized that the specimen is not homogeneous due to the insufficient saturation. Meanwhile, the other type of the low density area appears in the images from step B at the height of 10 mm. This can be considered to be the area of strain localization. It is easily observed from Fig.3 that this local shear banding starts from the bottom of the specimen at the strain level of step B which is the right before the peak stress as realized from Fig.2

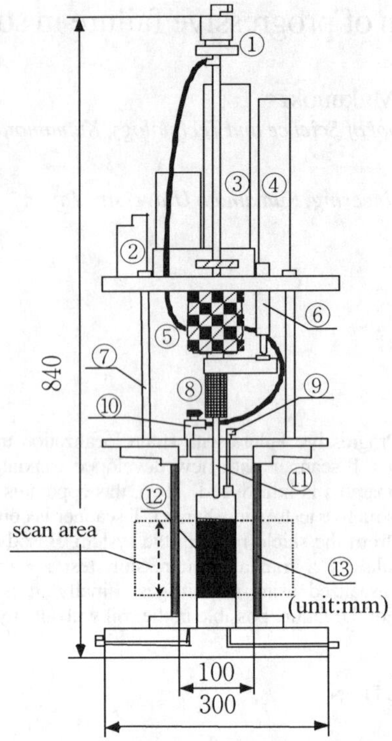

①,⑩pore pressure connection
②motor controller
③loading actuator
④driver ⑤regulator
⑥displacement sensor
⑦frame ⑧load cell
⑨drainage connection
⑪ram ⑫specimen
⑬pressure cell

Figure 1. New triaxial compression test apparatus

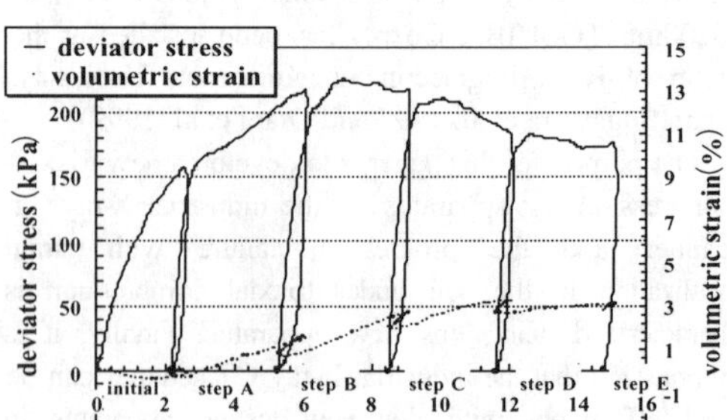

Figure 2. Stress-strain relationship

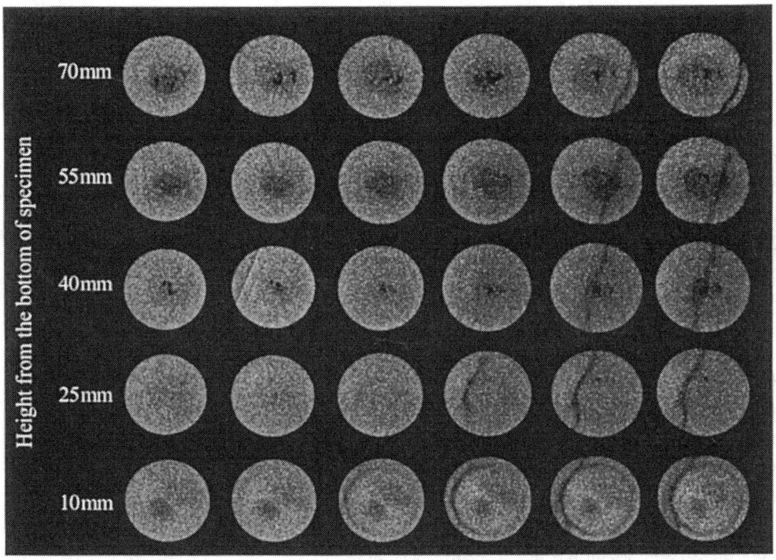

Figure 3. Cross sectional images

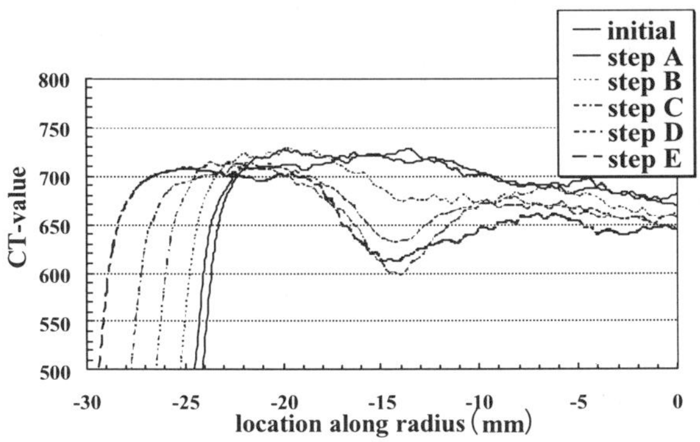

Figure 4. CT-value distributions

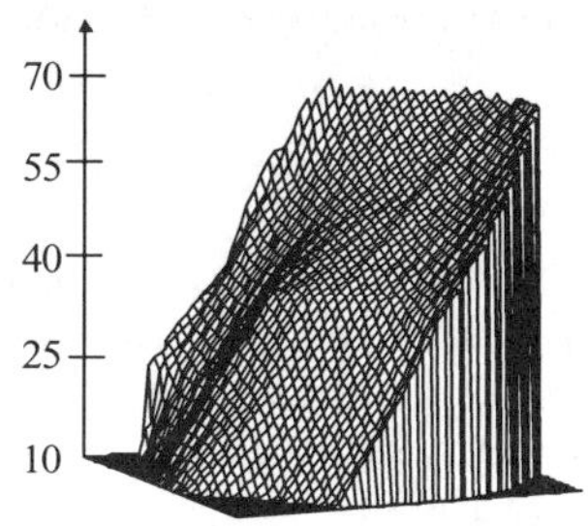

Figure 5. Graphical presentation of failure surface

although the shear band can not be observed from outside of the specimen. The shape of this local shear band in the cross section is the same as that of circumference of the specimen at the bottom of the specimen and this strain localization propagates upwards towards the top of the specimen as the strain increases. It is also observed that the total failure surface as a result of progressing the local ones appears after the peak stress condition, which is the strain level of step D. Thus, the following conclusions are summarized for this specimen from Fig.3:

1) The local shear band appeared before the peak stress in the soil although this was not able to be observed from outside of the specimen by sight.
2) The propagation of the local shear band progressed from the bottom of the specimen to the top.
3) The shape of strain localization is not always plane surface but the curved one.
4) The total failure surface appeared after the peak stress.

Figure 4 shows the distribution of the CT-value for every strain levels (from initial to step E) at the cross section of the height, 25 mm for half cross section in the specimen. As realized from this figure, the density at around the location of 15 mm from the center decreases locally as the strain level increases and this phenomenon shows the progress of the shear band in the cross section. Here, it is interesting to note that the start of the

decrease of the density at that area is observed at even the strain level of step B although the CT image shown in Fig.3 (at 25 mm with step B) did not show any localization in that cross section. Thus, more precise discussion on the progress of the shear band could be possible using the distribution of CT-value in the specimen. Note that the change of the location at the left end of this figure means the volume change of the specimen and the CT-value at the end is going to reach –1000
which is the setting value for the density of the air. Figure 5 shows the graphical image of the total failure surface in the soil at the end of the test, which is the strain level of step E. This result was obtained using the data of the six cross sectional images using computer graphic software. It is easily realized that the shape of the failure surface is clearly reconstructed and this is not plane surface but the complicated curved surface. It is interesting to note that the shapes of total failure surface at the cross section of both top and bottom of the specimen are the same as those of the circumference of the specimen. It can be considered that this is because the shape of the specimen is deeply involved. Thus, the other shape of the specimen should be examined in order to characterize the failure in the soils.

5 CONCLUSIONS

A triaxial compression test apparatus for the industrial X-ray CT scanner was newly developed and the progressive failure with strain localization in the soil was characterized without any destruction using this new apparatus

Finally, it is concluded that the progressive failure in the soil could be precisely discussed without any destruction using industrial X-ray CT scanner and the newly developed triaxial apparatus made it possible. It is also confirmed that the possibility of the other applications of this apparatus will be highly expected.

REFERENCES

Adachi T., Oka,F. and Yashima, A. 1997.Localization and Bifurcation Theory for Soils and Rocks, BALKEMA.

Otani, J., Mukunoki,T. and Kikuchi,Y. 1999. Evaluation of engineering property of light-weight soil with air foam using X-ray CT scanner, Proc. 11th Asian Regional Conference on SMGE, Seoul:57-60.

Otani, J., Mukunoki,T. and Obara,Y., 2000. Application of X-ray CT Method for Characterization of Failure in Soils, Soils and Foundations Vol.40, No.2:111-118.

Sugawara,K., Obara,Y., Kaneko,K., Koike,K., Ohmi,M. and Aoi,T. 1997. Visualization of three-dimensional structure of rocks using X-ray CT method, Proc. Asian Regional Conf. on Rock Mechanics, Seoul:769-774.

Computer Methods and Advances in Geomechanics, Desai et al. (eds) © 2001 Balkema, Rotterdam, ISBN 90 5809 183 X

Determination of plate-loading modulus E_p from E_{50} of unconfined compression test

J.N.Shen & K.Onitsuka
Department of Civil Engineering, Saga University, Honjo, Japan

ABSTRACT: Modulus of elasticity measured by plate-loading test, E_p differs from that measured by unconfined compression test, E_{50}. The former is currently adopted in pavement design method but is costly to measure. The later is popularly used for evaluation of stabilized materials and easily to measure. A process to establish the relationship between E_{50} and E_p was described on the basis of finite element method. The stabilized material was assumed as hyperbolic model. A simplified hyperbolic model was deduced, for which the parameters can be calibrated from unconfined compression test. As an example, the relationship between E_{50} and E_p of lime-stabilized Ariake clay was established.

1. Introduction

Pavement structures are usually identified as elastic multi-layer structures. Pavement materials are thus characterized by elastic modulus. The modulus adopted currently is either measured by plate load test, E_p, or by dynamic triaxial test, called resilient modulus, *Mr*. Both *Mr* and E_p is a very important parameter for both the mechanical-based pavement structure design method and the experimental –based method.

Take a look at the method adopted in Japan for an example. It is called Total Thickness of Asphalt Concrete Method. Pavement layers such as base and subbase of various materials are to be transformed into an equivalent thickness of asphalt concrete layer by means of an factor, the layer equivalent factor, "a_n". In applying the method, the pavement designers determine an total thickness (T_A) using equation (1) by knowledge of factors such as design traffic level (N), subgrade support (representing by CBR), and desired terminal serviceability at first.

$$T_A = 3.84N^{0.16}/CBR^{0.3} \quad (1)$$

where N is the equivalent numbers of the standard axial load during the designed service life. The magnitude of the T_A reflects the degree to which the subgrade must be protected from the effects of traffic. Once the T_A is calculated, it becomes necessary to determine the manner in which the T_A will be achieved, i.e., what the required thickness and quality of each pavement layer should be. This is done by solving the following equation (2):

$$T_A = a_1 T_1 + a_2 T_2 + a_3 T_3 + \quad + a_n T_n \quad (2)$$

where T_A = total thickness of asphalt concrete, a_1, a_2, a_3, a_n = layer equivalent factors of different layers for different materials and T_1, T_2, T_3, T_n = the thickness of different layers.

Examination indicates that the thickness of any particular layer is, to a significant extent, dependent upon the layer equivalent factor. Hence, an accurate determination of layer equivalent factor can have a significant economic impact in regard to the design of the pavement structure. As originally used in the road test study, layer equivalent factors, a_n, were actually regression coefficient, and most commonly determined on the basis of relative layer materials strength or stiffness considerations. As proposed by Institute of Civil Engineering, Japan, as an example, the layer equivalent factor, a_n, can be approximated from the following equation (3):

$$a_n = 0.313\log E_p + 0.616 \quad (3)$$

where E_p is characterized by the modulus of elasticity measured by plate-loading test (10^{-3} MPa).

On the other hand, unconfined compression test is preferably adopted to determine the unconfined compressive strength, q_u, of stabilized materials in pavement and geotechnical engineering. The stiffness of the treated materials is often assessed by a secant modulus from strain–stress curve such as E_{50}, defined by the half maximum strength divided by the corresponding strain, in Japan. It is recognized that E_{50} and E_p are different in value for any pavement material due to its non-linearity of the strain-stress relationship. It is worthwhile to establish the relationship between the two modulus of elasticity of lime-treated material so that the value of E_p can be estimated from E_{50}.

The research is aimed at establishing numerically the relationship between E_{50} and E_p of stabilized pavement materials. A simplified hyperbolic model is deduced on the basis of a general hyperbolic model at first, for which the parameters can be calibrated from the unconfined compression test. Both of the plate–loading test and unconfined compression test were modeled axisymmetrically. As an example, the relationship between E_{50} and E_p of lime-stabilized Ariake clay was established.

2. Simplified hyperbolic material model

Hyperbolic model of the strain-stress relation is as follow (4): Duncan & Chang, (1970).

$$(\sigma_1 - \sigma_3) = \varepsilon / (c + b\varepsilon) \qquad (4)$$

in which σ_1 and σ_3 are the major and minor principal stresses; ε is the axial strain; and c and b are constants which may be determined experimentally. Both of those constants c and b have readily visualized physical meanings: As shown in Figure 1 (a) c is the reciprocal of the initial tangent modulus, Ei, and b is the reciprocal of the asymptotic value of stress difference which the stress-strain curve approaches at infinite strain $(\sigma_1 - \sigma_3)_{ult}$

The values of the coefficient c and b may be determined most readily if the stress-strain data are plotted on transformed axes, as shown in Figure 1 (b), when equation (4) is rewritten in the following form (5)

$$\varepsilon / (\sigma_1 - \sigma_3) = c + b\varepsilon \qquad (5)$$

It may be noted that c and b are respectively the intercept and the slop of the resulting straight line.

When this is done it is commonly found that the asymptotic value of $(\sigma_1\text{-}\sigma_3)$ is larger than the compressive strength of the soil. This would be expected, because the hyperbola remains below the asymptote at all finite values of strain. The asymptotic value may be related to the compressive strength, however, by means of a factor R_f as shown by

$$(\sigma_1\text{-}\sigma_3)_f = R_f\,(\sigma_1\text{-}\sigma_3)_{ult} \qquad (6)$$

in which $(\sigma_1\text{-}\sigma_3)_f$ is the compressive strength, or stress different at failure; $(\sigma_1\text{-}\sigma_3)_{ult}$ is the asymptotic value of stress difference and R_f is the failure ratio.

By expressing the parameters c and b in terms of the initial tangent modulus values and the compressive strength, equation (6) may be rewritten as

$$(\sigma_1\text{-}\sigma_3) = \varepsilon / (1/E_I + \varepsilon\, R_f / (\sigma_1\text{-}\sigma_3)_f) \qquad (7)$$

The hyperbolic model described above is a formation for modeling samples in triaxial test state. In unconfined compression test, the formation and the parameters can be simplified as follows. In unconfined compressive test $\sigma_3=0$, $(\sigma_1\text{-}\sigma_3)_f = q_u$. The tangent modulus, E_t at any point on the stress-strain curve are conveniently expressed as

$$E_t = (1 - R_f\ \sigma_1 / q_u)^2\, E_I \qquad (8)$$

3. Finite element (FE) models

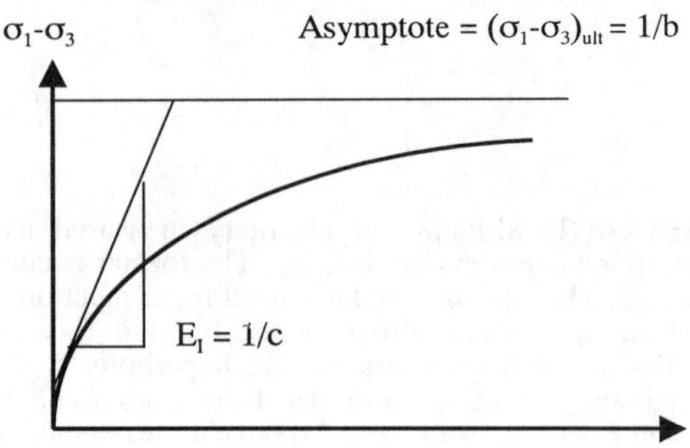

Figure 1 (a) Hyperbolic stress-strain curve

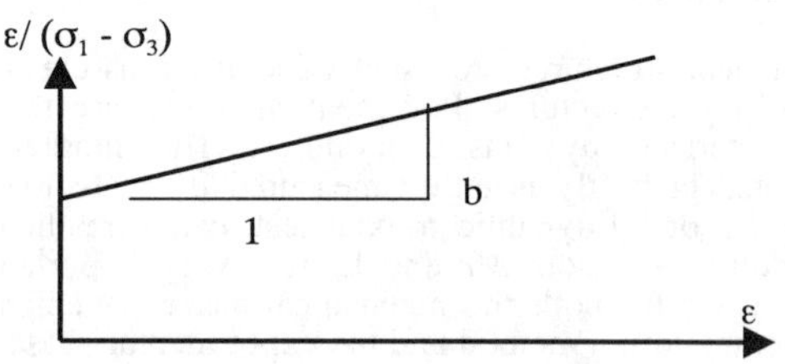

Figure 1 (b) Transformed hyperbolic stress-strain

Rigid plate–loading and unconfined compression tests can be modeled as axisymmetric, because the loads in both of the cases are axisymmetric. Figures 2 and 3 are the FE models of plate-loading test and unconfined compression test used in the study respectively. All of them are axisymmetric.

The real structure of the materials tested by plate–loading is regarded as half infinitive space. When the half infinitive space is modeled for finite element analysis, the size of the geometry is usually determined by a studying of size effect on the deflection at the load center. Various sites of depth with a fixed width were assumed for each case to determine the depth. It is adopted as the minimum value with which the result of calculating is not significantly changed. With the same principle, the width was determined. For plate-loading test, geometry of 1.95 m in depth and 1.5 m in width is used finally. Increasing the size of geometry has no influence on the result but the mesh of element. The width of the model is 20 times as much as the radius of the loading area. The depth is 1.95 m. For the unconfined compression test, the geometry of FE model is determined according to its real size that was the sample size. The axisymmteric geometry is 2.5 cm in width 10 cm in depth for the study discussed.

Mesh of the geometry is another important factor to affect the calculating results. Generally, a finer mesh will get a more accurate result but need more computing time. A common method with changing element size is used to determine how fine the element should be. A finer element is meshed for the loading area and element size becomes more and more away from the loading at both horizontal and vertical directions. Totally, 64 elements are adopted for plate-loading test, and two equal size elements are used for unconfined compression test test.

Boundary conditions for plate-loading test are applied in following ways that no vertical displacement at the bottom and no horizontal displacement at the right side. No vertical displacement is assumed for unconfined compression test test. A linear strain quadrilateral element is used for the two cases.

External load is exerted on the top surface of the model by strain control for unconfined compression test. That means the load is implemented by vertical displacement. The vertical displacement is applied at 50 steps to simulate the real testing speed at strain 1% per minute.

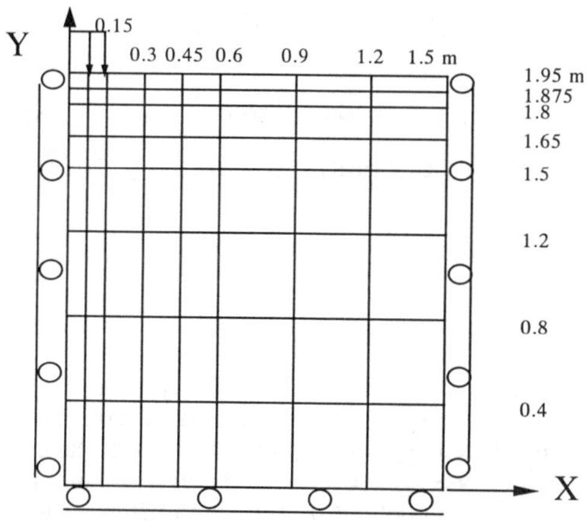

Figure 2 FE model of plate loading test

For the plate-loading test, the load is to apply with pressures on the elements within the plate. The pressures applied on the elements have to in such way that all the displacement on the points within the plate are the same. Displacement is the same on all the points of the plate for rigid plate-loading tests, but the pressure distribution under the plate is not uniform. The pressure distribution under a rigid plate can be expressed as (9), Ullidtz (1987).

$$q(r) = q\, r_0 / 2\, (r_0^{\,2} - r^2)^{0.5} \qquad (9)$$

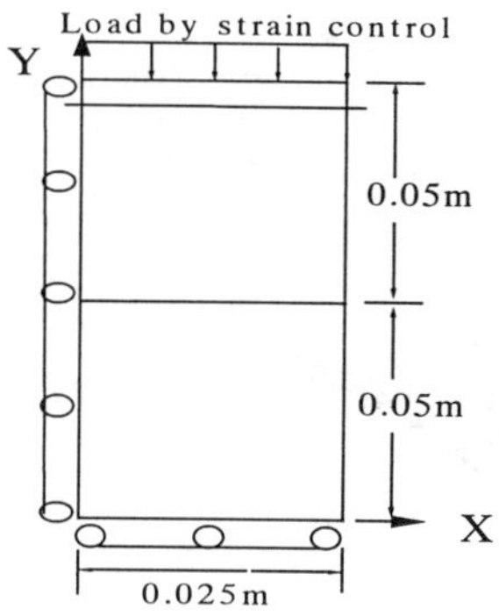

Figure 3 FE model of unconfined compression test

in which r is the distance from center to the point where pressure is to be determined, r_0 is the radius of the plate and q is the average pressure, which is equal to the total load divided by the area.

4. Calibration of the simplified hyperbolic model-from lime stabilized Ariake clay added with fly ash (LAF)

Using the required mathematical formulas, the hyperbolic stress-strain model is developed to predict the stress–strain relation curve. Parameters necessary for the simple hyperbolic model of the LAF are calibrated by means of the unconfined compressive test data and listed in Table 1 for 7 days curing.

Figure 4 presented is an example of the calculated stress-strain curve by the FE model with

Table 1 Hyperbolic model parameters of LAF

Lime (%)	Fly ash (%)	After 7 days curing		
		q_u (MPa)	E_I (MPa)	R_f
10	0	1.10	198	0.50
	10	1.27	275	0.55
	20	1.14	340	0.55
	30	0.95	245	0.60
20	0	1.54	285	0.55
	5	1.56	353	0.54
	10	1.74	305	0.56
	15	1.56	366	0.50
	20	1.62	333	0.55
	30	1.54	330	0.65
30	0	1.82	500	0.58
	10	2.14	625	0.50
	15	2.12	510	0.55
	20	2.03	450	0.50
	30	1.87	400	0.60

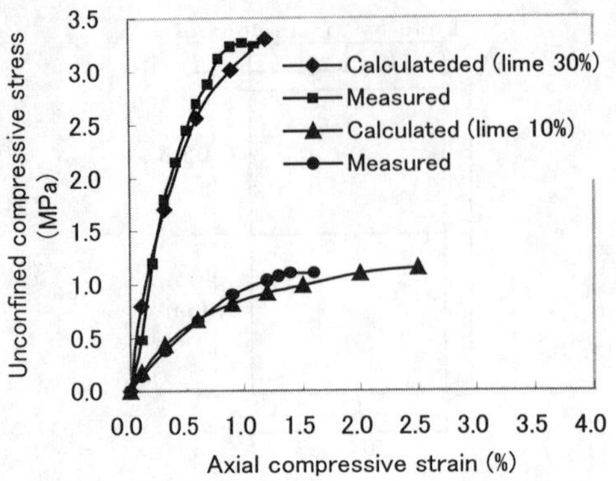

Figure 4 The stress-strain calculated versus measured of LAF mixtures

calibrated parameters together with experimentally determined stress-strain curve for two cases. One is for the case with 10% lime content after 7 days curing. The other is that with 30% lime content after 28 days curing. The stress-strain relationship from tests in both cases can be fitted well by a hyperbolic model, especially when the lime content is lower and within the scope of stress level interested in pavement engineering. For higher content of lime case, it has been noted that the transformed stress-strain diverge somewhat from a linear relation at both low and high values of strain, indicating that the stress-strain curve for this test is not precisely hyperbolic in shape. The LAF mixture was found to fail with brittle, the strength near failure increase sharply. Thus the measured stress-strain curve at the top part a little diverge from the calculated curve in all the cases. Considering the real stress state of the material in a pavement structure, it is reasonable to fit the initial part of the curve well.

5.Relationship between E_{50} and E_p of LAF mixtures

The secant modulus of elasticity adopted in Japan is E_{50}, which is the ratio of half of the maximum unconfined compressive strength q_u to the corresponding vertical axial compressive strain ε.

$$E_{50} = q_u / 2\varepsilon \qquad (10)$$

With regards to the plate-loading test, deformation modulus of tested materials is calculated using the following equation (11) based on elastic theory.

$$E_p = \pi (1-\mu^2)\, q\, r_0 / 2\, w_0 \qquad (11)$$

where q is the average pressure exerted on the plate, r_0, the radius of the plate (0.15 m in the study) and w_0 the displacement corresponding to the pressure μ is Poisson's ratio of the material (0.3 in the study). The ratio of q/ w_0 is not a constant at different pressure of the calculated q-w_0 curve due to the non-linearity of materials. Consequently, the value of E_p varies with the pressure q or the displacement w_0 concerned. Just like the tangent modulus of the material, the value is dependent on the stress state. In the study, it is calculated when the displacement w_0 reaches 1.25 mm, a limit as subbase in pavement structure.

The numerical results were presented in Figure 5 for the LAF mixtures. It is clear that the plate loading measured modulus and the unconfined compression test measured linearly well. From the analysis, a numerical equation to calculate the plate loading modulus from unconfined compression test was obtained as follows:

$$E_p = 1.06E_{50} + 27.2 \qquad (12)$$

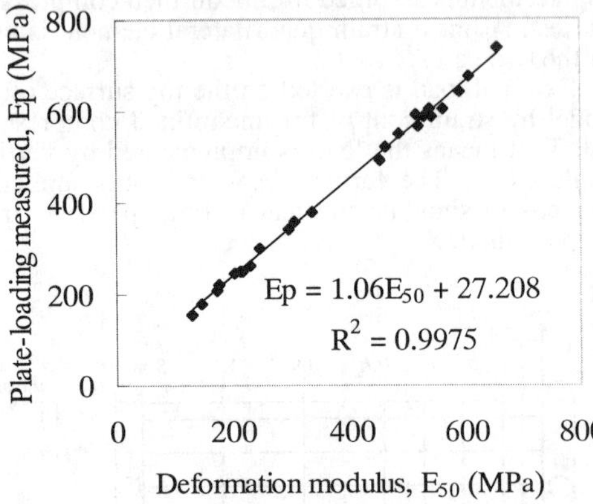

Figure 5 Relationship between E_{50} and E_p

6. Summary and conclusions

A process to establish numerically plate loading measured modulus, E_p and unconfined compression test measured E_{50} was put out. Stabilized pavement material was well modeled by a simplified hyperbolic model. All the parameters can be easilycalibrated from unconfined compression test.

The relationship between plate loading measured modulus, E_p and unconfined compression test measured E_{50} can be established numerically. That is meaningful because the data obtained from laboratory can be used for pavement design by means of the relationship established.

References

Duncan, J. M. and Chang, C.Y., (1970), Nonlinear analysis of stress and strain in soils, J. of Soil mechanics and Foundations Division, Proceeding of The ASCE, pp. 1629-1653.

Ullidtz, P., (1987), Pavement analysis, Elsevier Science, New York.

Computer Methods and Advances in Geomechanics, Desai et al. (eds) © 2001 Balkema, Rotterdam, ISBN 90 5809 183 X

Elasto-plastic finite element analysis of direct shear test on sand

T.Tanaka
Department of Biological and Environmental Engineering, University of Tokyo, Japan

T.Abe
Matsumoto Regional Office, Prefecture of Nagano, Japan

D.Watanabe
Faculty of Agriculture, Meiji University, Japan

ABSTRACT: Although the direct shear box test is commonly used, there have been a few study on applying the finite element method to the test. We analyzed the direct shear box with the finite element method, and compared the results with Rowe's equation and Davis's equation. The constitutive model for this analysis contains non-associated strain hardening-softening elasto-plastic materials. A yielding function of the Mohr-Coulomb type and a plastic potential function of the Drucker-Prager type was employed. An element consisted of a 4-noded iso-parametric element with one point integration. The finite element mesh consisted of various boundary conditions such as whether wall friction exists or not. The results showed that internal friction from FEM in the case of a rough wall side is very similar to Rowe's friction angle and Davis's friction angle.

1 INTRODUCTION

The direct shear box is a test apparatus used to determine soil properties and is commonly used for its simplicity in practice. However the shear strength in direct shear tests does not always agree with those in triaxial compression tests and plane strain tests. The direct shear angle of friction and the plane strain angle of friction do not have the same value.

The relationship of the direct shear angle of friction ϕ_{ds} with the plane strain angle of friction ϕ_{ps} and the angle of dilation ψ is given by the expression

$$\tan\phi_{ds} = \frac{\sin\phi_{ps}\cos\psi}{1-\sin\phi_{ps}\sin\psi} \qquad (1)$$

Jewell & Wroth (1987) derived this equation following Davis. This is the same equation as that presented by Davis for mobilized angle of friction. Various theoretical relationships between the plane strain angle of friction and the angle of dilation in sand have been investigated, and it has been shown that the equation closely fits actual data. Bolton (1986) presented a simple empirical fit for Rowe's stress-dilatancy relationships which is convenient to use :

$$\phi_{ps} = \phi_{ds} + \psi \qquad (2)$$

Equations (1) and (2) provide a relationship between the plane strain angle of friction and direct shear angle of friction measured in the conventional direct shear test.

Rowe (1969) derived a theoretical relationship between the plane strain angle of friction and the direct shear angle of friction, to explain the different strengths measured in biaxial compression tests and direct shear tests. Rowe derived the expression in combination with the full stress dilatancy equation

$$\tan\phi_{ds} = \tan\phi_{ps}\cos\phi_{cv}. \qquad (3)$$

These theories were supported by test results on Leighton Buzzard sand.

On the other hand, there has been few finite element analyses on the direct shear test and their relations. Potts et al (1987) analyzed the influence of volume change, initial stress and strain softening, using the finite element method.

The finite element method has recently been used to analyze various geotechnical problems, and this study of the direct shear box test was done also using the finite element method

The main objectives of this paper are as follows:

a) To interpret the results of finite element analysis of the direct shear test,

b) To clarify the effect of boundary conditions,

c) To compare the theoretical values with the results of finite element analysis.

2 FINITE ELEMENT ANALYSIS

2.1 *Constitutive model*

A constitutive model for a non-associated strain hardening-softening elasto-plastic material is introduced. A yield function of the Mohr-Coulomb type and a plastic potential function of the Drucker-Prager type are employed. The element employed for the analysis is a pseudo-equilibrium model by special reduced integration of a 4-noded Lagrange type element. A dynamic relaxation method with return mapping algorithm is applied to the integration algorithms of the elasto-plastic constitutive relation.

In the return-mapping algorithm (Ortiz & Simo 1986), the elastically predicted stresses are relaxed onto a suitably updated yield surface. A change in stresses can cause an associated change in the elastic strains even by

$$d\varepsilon^e = [D]^{-1}(\sigma_B - \sigma_A) \tag{4}$$

where $[D]^{-1}$ is the elastic matrix. As the total strain does not change during the relaxation process, the plastic strain change is balanced by an equal and opposite change in the elastic strains :

$$sd\varepsilon^p = -d\varepsilon^e = [D]^{-1}(\sigma_B - \sigma_A) \tag{5}$$

s is the ratio of shear band area to finite element area. s is built into the elasto-plastic constitutive model as a characteristic length scale. The plastic strain increment can be obtained by the following equation:

$$d\varepsilon^p = \lambda \frac{\partial \Phi}{\partial \sigma} \tag{6}$$

where λ is a positive scalar multiplier to be determined with the aid of the loading-unloading criterion :

$$\lambda = \frac{f(\sigma_A, \kappa_A)}{s\dfrac{\partial f}{\partial \sigma}[D]^e \dfrac{\partial \Phi}{\partial \sigma} - \dfrac{\partial f}{\partial \kappa}} \tag{7}$$

The yield function (f) and the plastic potential function (Φ) are given by :

$$f = \alpha I_1 + \frac{\bar{\sigma}}{g(\theta)} = 0 \tag{8}$$

$$\Phi = \alpha' I_I + \bar{\sigma} = 0 \tag{9}$$

where I_1 is the primary invariant of deviatoric stress and $\bar{\sigma}$ is the second invariant of deviatoric stress. In the case of Mohr-Coulomb material, $g(\theta)$ takes the following form :

$$g(\theta) = \frac{3 - \sin\phi}{2\sqrt{3}\cos\theta - 2\sin\theta\sin\phi} \tag{10}$$

ϕ is the mobilized friction angle and θ is Lode angle. The frictional hardening-softening functions are expressed by :

$$\alpha(\kappa) = \left(2\frac{\sqrt{\kappa\varepsilon_f}}{\kappa + \varepsilon_f}\right)^m \alpha_p$$

$$\text{hardening regime } (\kappa \le \varepsilon_f) \tag{11}$$

$$\alpha(\kappa) = \alpha_R + (\alpha_p - \alpha_R)\exp\{-(\frac{\kappa - \varepsilon_f}{\varepsilon_r})^2\}$$

$$\text{softening regime } (\kappa > \varepsilon_f) \tag{12}$$

$$\alpha = \frac{2\sin\phi}{\sqrt{3}(3 - \sin\phi)} \tag{13}$$

$$\alpha' = \frac{2\sin\psi}{\sqrt{3}(3 - \sin\psi)} \tag{14}$$

where m, ε_f and ε_r are material constants.

The residual friction angle (ϕ_r), Poisson's ratio (ν) and initial shear modulus (G_o) are chosen based on the data from the test of air-dried dense Toyoura sand. γ=16.1 kN/m^3, G_0=78400 kPa, ϕ_r=33 deg and ν = 0.3. The elastic moduli are estimated using the following equations :

$$G = G_0 \frac{(2.17 - e)^2}{1 + e}\sigma_m^{1/2} \tag{15}$$

$$K = \frac{2(1+\nu)}{3(1-2\nu)}G \tag{16}$$

The peak friction angle (ϕ_P) is estimated from the empirical relations proposed by Bolton (1986) :

$$I_R = D_r\{5 - \ln(\frac{\sigma_m}{150})\} - 1 \quad (\sigma_m \ge 150kN/m^2) \tag{17}$$

$$I_R = 5D_r - 1 \quad (\sigma_m < 150kN/m^2) \tag{18}$$

$$\phi_p - \phi_R = 5I_R \quad (19)$$

The dilatancy angle (ψ) is estimated from Rowe's stress-dilatancy relationship(Rowe 1969).

The material constants are fitted by a trial and error method based on the data set of triaxial compression tests, giving the following values :

$$m = 0.3, \varepsilon_f = 0.1, \varepsilon_r = 0.6 \quad (20)$$

2.2 *Analysis of direct shear box test*

The finite element mesh used for analyzing the direct shear box test is shown Figure 1. It contains 360 elements. The dimensions of the box are 60 mm long by 25 mm deep.

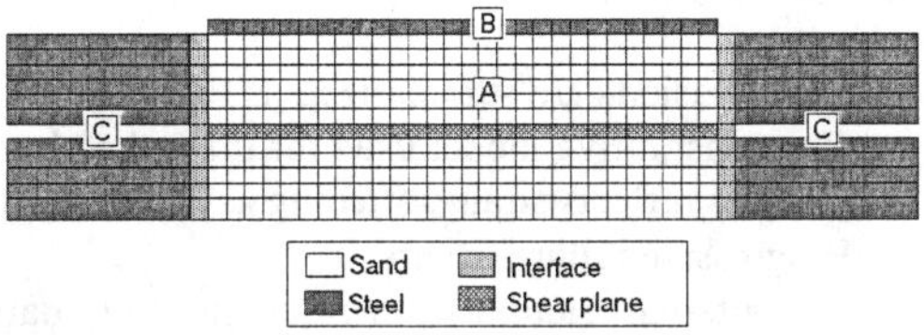

Figure 1. Finite element mesh for analyses

The analysis is performed for plane strain, with the following common boundary conditions. Both the x and y displacements of nodes along the bottom were fixed. Nodes along the upper box side were given a prescribed displacement in the x direction. The center of an element row in the sample represents the gap between the upper box and the lower box. The upper box was free to move relative to the lower box in the vertical direction. Table 1 presents the boundary conditions of the shear box.

Table 1 Boundary conditions in analyses

NO.	I	II	III	IV	V
Constitution	A	A+B	A+C	A+B+C	A+B*2
Bottom	Fix	Fix	Fix	Fix	Fix
Upper side	Smooth	Smooth	Rough	Rough	Smooth
Lower side	Smooth	Smooth	Rough	Rough	Smooth
Top cap	Non	Solid	Non	Solid	Solid *2

Mesh type I is assumed to be a lubricated side and a water-filled rubber bag top, mesh type II a lubricated side and a solid top, mesh type III a rough side and a water-filled rubber bag top and mesh type IV a rough side and a solid top.

The bottom conditions are fixed. The wall side conditions are smooth or rough. 'Smooth' was assumed to be a lubricated wall, using grease on a rubber sheet, with no friction. 'Rough' was assumed to be a sample attached steel wall with angle of wall friction of a maximum of 6 degrees. The top element row was assumed to be a cap for transmitting the normal stress. 'Non' was assumed to be a water-filled rubber bag top. 'Solid' top was assumed to be a free solid material top.

The top row and side box of elements model a solid material, with linear elastic properties, Young's modulus of E=1.96*10^4 MPa and Poisson's ratio of ν =0.3.

In the analysis for mesh type IV, the solid top cap is free to rotate and to move in the vertical direction. In the analysis for mesh type II and V, the solid top cap is not prevented from rotating and is free to move in the vertical direction.

The following approach was adopted to simulate the initial stresses in the sample before shearing. In the approach, the sample was given a very small isotropic initial stress, and the vertical load was applied as a distributed load on the top row. This procedure simulates what happens in the real test.

The shear box test was simulated by giving the upper half of the shear box a displacement increment of d=1.0*10^{-5}cm except for mesh type V which was given a displacement increment of d=1.0*10^{-6}. The corresponding horizontal shear load was obtained by summing the appropriate horizontal stresses in the center layer.

3 RESULTS

3.1 *Stress-displacement behavior*

Figure 2 shows the relations between shear stress ratio and horizontal displacement. In presenting the results of the analyses, the mean shear stress is obtained by dividing the total shear force on the shear plane by the length of the box. The shear stress ratio on the shear plane was obtained as the shear stress by the initial normal stress. All results show that the shear stress decreases after the peak value appeared. There were little differences in the peak shear stresses depending on whether wall friction existed or not and the conditions of the top cap.

3.2 *Contours of strain*

Figure 3 shows the contours of maximum shear strain in the sample. It shows that at sample failure, the maximum shear strain is 0.10. Progressive failure involves the formation of a narrow zone of failure.

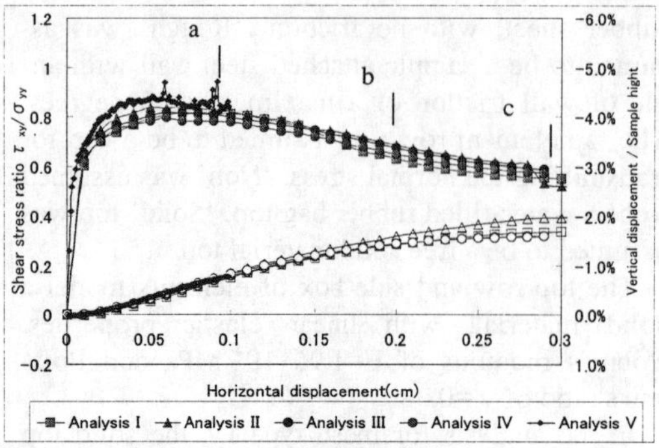

Figure 2 Relations between horizontal displacement and shear stress ratio

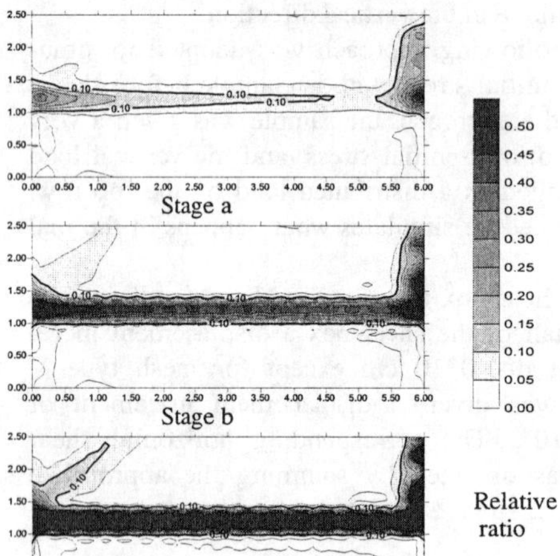

Figure 3 Contours of maximum shear strain within the direct shear test for analysis used Mesh type I

3.3 *Comparison of the angle of internal friction*

Table 2 compares the angle of internal friction. The direct shear angle of internal friction $\phi_{ds\,\mathrm{FEM}}$ is obtained by dividing the mean shear stresses at the peak by the initial vertical stresses. The angle of internal friction obtained for the plane strain condition is expressed by ϕ_{ps} which is estimated from equations (17), (18) and (19) about the initial stress by finite element analyses. $\phi_{ds\,\mathrm{Davis}}$ and $\phi_{ds\,\mathrm{Rowe}}$ could be calculated by using equations (1) and (2) with ϕ_{ps}.

In a results, $\phi_{ds\,\mathrm{Davis}}$ is in good agreement with $\phi_{ds\,\mathrm{Rowe}}$. The values of $\phi_{ds\,\mathrm{FEM}}$ with rough wall side are very close to $\phi_{ds\,\mathrm{Davis}}$ and $\phi_{ds\,\mathrm{Rowe}}$. The difference $\phi_{ds\,\mathrm{FEM}}$ between the maximum value and minimum value is 2.4 degrees. The wall friction influences little to The angle of internal friction on the shear plane.

Table 2 Comparison of the angle of internal friction

NO.	I	II	III	IV	V
ϕ_{ps}	43.4	43.5	44.4	43.4	44.1
$\phi_{ds\,\mathrm{FEM}}$	39.2	40.0	39.4	38.6	41.0
$\phi_{ds\,\mathrm{Davis}}$	38.0	38.1	39.0	38.0	38.7
$\phi_{ds\,\mathrm{Rowe}}$	38.1	38.2	39.1	38.1	38.8

4 CONCLUSION

The conclusions are summarized as follows :

- The constitutive model used for the finite element analysis reflects strain softening as in a real test.
- Progressive failure occurs.
- The internal friction from finite element analysis in the case of a rough wall side is very similar to Rowe's friction angle and Davis's friction angle.

REFERENCES

Arthur, J.R.F., Dalili, A. & Dunstan, T. 1988. 6th Geotechnique symposium in print, Geotechnique 38(1): 140-144.

Bolton, M.D. 1986. The strength and dilatancy of sands. Geotechnique 36(1): 65-78.

Jewell, R.A. & Wroth, C.P. 1987. Direct shear tests on reinforced sand. Geotechnique 37(I): 53-68.

Potts, D.M., Dounias, G.T. & Vaughan, P.R. 1987. Finite element analysis of the direct shear box test, Geotechnique 37(I): 11-23.

Rowe, P.W. 1969. The relations between the shear strength of sand in triaxial compression, plane strain and direct shear. Geotechnique. 19(1): 75-86

Simo, J. C. & Ortiz, M. 1986.An analysis of a new class of integration algorithms for elasto-plastic constitutive relations. *Int. J. Num. Meth. Engng*.23: 353-366

Tanaka, T., Mori, H. & Kikuchi, M. 1997. Finite deformation elasto-plastic analysis of underground structures and retaining wall, Proc. 9th Int. Conf. on Computer Methods and Advances in Geomechanics 2: 1465-1470

Ttsuoka, F., Sakamoto, M., Kawamura, T. & Fukushima, S. 1986. Strength and Deformation Characteristic of Sand in Plane Strain Compression at Extremely Low Pressures, Soils and Foundations 26(1): 65-84.

Computer Methods and Advances in Geomechanics, Desai et al. (eds) © 2001 Balkema, Rotterdam, ISBN 90 5809 183 X

Finite element simulation for dynamic centrifuge test of a simple slope

A.Wakai & K.Ugai
Department of Civil Engineering, Gunma University, Kiryu, Japan
M.Sato
Japan National Research Institute for Earth Science and Disaster Prevention, Japan

ABSTRACT: A dynamic centrifuge test of a slope made of unsaturated silty clay and its numerical simulations based on 2D dynamic elasto-plastic FEM are presented. In the experiment, the residual deformation, accompanied with a clear slip surface, was observed after earthquake. The results observed in the centrifuge test are simulated well by the FE analyses in which a new cyclic loading model of soil is adopted.

1 INTRODUCTION

It has been confirmed that the numerical analysis such as the dynamic elasto-plastic finite element method (FEM) is very effective for evaluating the seismic stability of slopes.

The constitutive model presented in this paper is a modification of the previous Ugai-Wakai's model (Wakai et al., 1999), which was only for $\phi=0$ soils. The new model has been modified so that the strength parameters both c and ϕ are taken into account, in addition, the $G-\gamma$, $h-\gamma$ relationships under cyclic loading are also considered.

2 CONSTITUTIVE MODEL

The details of the newly proposed model are described in a literature (Wakai et al., 2001), and the model formulation will be summarized as follows.

2.1 *Skeleton curve*

The skeleton curve is expressed as a hyperbolic function, such as

$$\overline{\tau} = \frac{G_0\overline{\gamma}}{1+\dfrac{G_0\overline{\gamma}}{\tau_f}} \tag{1}$$

$$\overline{\tau} = \tau(\{\sigma\}-\{\sigma_0\}) \tag{2a}$$

$$\overline{\gamma} = \gamma(\{\varepsilon\}-\{\varepsilon_0\}) \tag{2b}$$

The symbols with "0" denote the initial values. G_0 is assumed to be constant in this study. Young's modulus E is given by $2G_0(1+\nu)$. ν is Poisson's ratio. Here, τ and γ are given by

$$\tau(\{\sigma\}) = \tau_{\max} = \sqrt{J_2}\sin\left(\frac{\pi}{3}+\Theta\right) \tag{3a}$$

$$\gamma(\{\varepsilon\}) = \gamma_{\max} = 2\sqrt{J_2^*}\sin\left(\frac{\pi}{3}+\Theta^*\right) \tag{3b}$$

Θ and Θ^* are Lode angles in the stress and strain space, respectively. The strength τ_f is given by

$$\tau_f = c\cdot\cos\phi + \frac{\sigma_1+\sigma_3}{2}\sin\phi \tag{4}$$

The flow rule is derived from Rowe's stress-dilatancy relationships, and the plastic potential g in plane strain condition (Manassero, 1989) is given by

$$g = \frac{{\sigma_1}^{K_{cv}}}{\sigma_3} \tag{5}$$

K_{cv} is a constant which corresponds to the stress ratio at the phase transformation.

2.2 *Hysteretic loops*

The hysteretic loops are assumed to be

$$\tilde{\tau} = \frac{a\tilde{\gamma}^n + G_0\tilde{\gamma}}{1+b\tilde{\gamma}} \tag{6}$$

$$\tilde{\tau} = \tau(\{\sigma\}-\{\sigma_a\}) \tag{7a}$$

$$\tilde{\gamma} = \gamma(\{\varepsilon\}-\{\varepsilon_a\}) \tag{7b}$$

$(\gamma_a, \tau_a) = (\gamma(\{\varepsilon_a\}), \tau(\{\sigma_a\}))$ is the reference stress point which corresponds to the newest reversal point in the stress-strain history. The value of a always must be adjusted so that the current hysteretic loop should pass through the next target point.

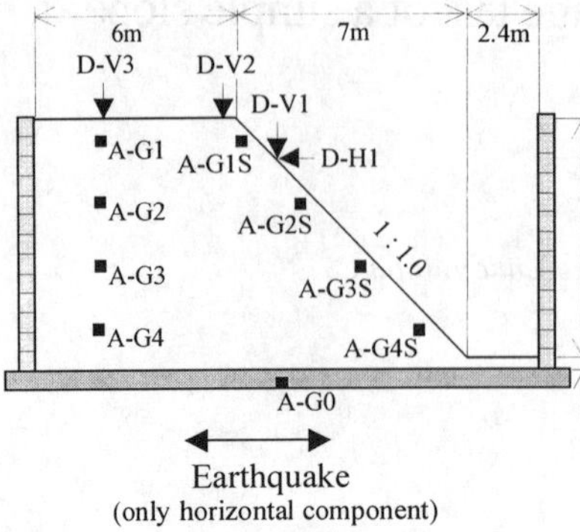

Figure 1. Model slope in the centrifuge test.

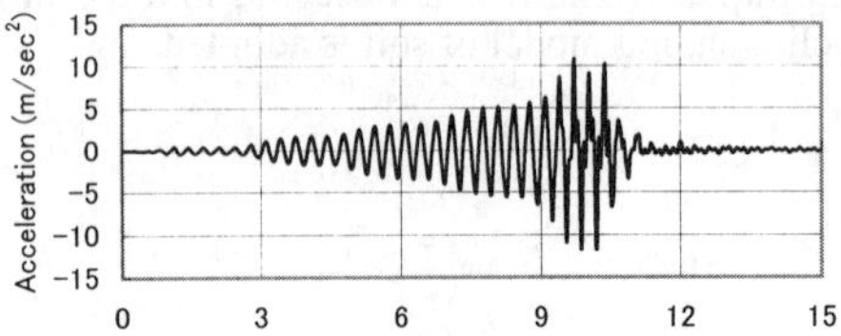

Figure 2. Input horizontal waves (acceleration).

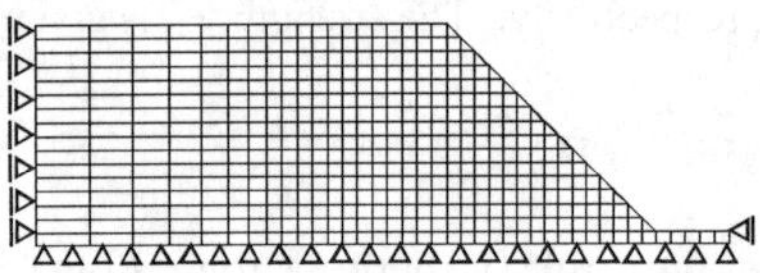

Figure 3. Finite element meshes.

Table 1. Material constants used in the simulation of centrifuge test.

E (kPa)	ν	c (kPa)	ϕ (deg)	K_{cv}	$b \cdot \gamma_{G0}$	n	γ_t (kN/m³)
17000	0.40	11.1	33.3	3.24	8.0	1.5	15.7

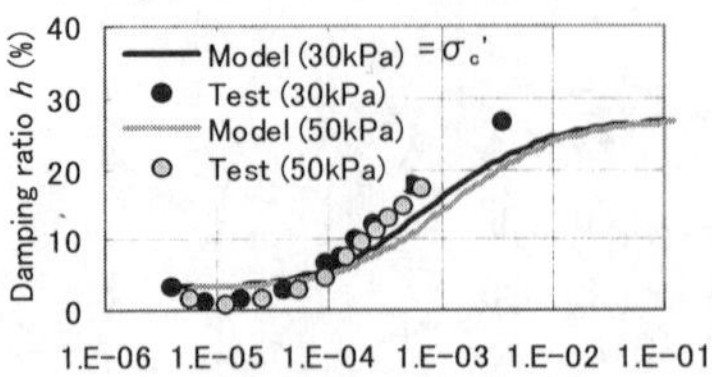

Figure 4. Assumed $h-\gamma$ relationships.

3 CENTRIFUGE TEST AND ITS NUMERICAL SIMULATIONS

3.1 *Testing apparatus*

We performed the shaking table test of a slope under centrifugal acceleration. In the experiment, the slope is made of homogenious and unsaturated silty clay (DL clay). The relative density of the soil is about 75%. The centrifugal acceleration of the test is 20G. Figure 1 shows the sketch of a model slope in the prototype scale, that is, 20 times the size of the model. The points where the history of response is measured are plotted in the figure. The symbols of "D" and "A" denote the measurement of displacement and acceleration, respectively. The history of input waves, which has only horizontal component, is shown in Figure 2.

3.2 *Analytical model for the simulation*

Figure 3 shows the sketch of 2D FE meshes. The finite element used in the analysis is 8-nodes isoparametric element with 4-points reduced integration. Table 1 shows the material constants used in the analysis. Most of these have been determined by the results of drained triaxial compression tests. The strength parameters c and ϕ were evaluated based on the peak strength. E was assumed to be equal to $2 \times E_{50}$, according to the geometrical relationship for a hyperbolic curve. E_{50} is the secant elastic modulus, and was evaluated by the stress-strain curve obtained from a series of triaxial compression tests. In this case, the value of E_{50} was found to be almost constant in the region such that σ_c'= 19.6 to 98 kPa. Shear modulus G_0 was calculated by $0.5E/(1+\nu)$. ν was assumed to be 0.40.

The assumed $h-\gamma$ curves based on the analytical model with parameters in Table 1, when the confining pressure σ_c'=30, 50kPa, are compared with the experimental curves obtained by undrained cyclic triaxial compression tests, as shown in Figure 4. Here, it should be noted that $h \approx 3\%$ Rayleigh damping is also introduced in the analysis. It is found that these $h-\gamma$ curves agree well each other.

3.3 *Comparison of results*

Figure 5 shows the time histories of horizontal acceleration at each point (A-G1, A-G4S), vertical and horizontal displacement at each point (D-H1, D-V1, D-V3). Both the calculated and the measured results are presented in the figures. A good agreement between FEM and the centrifuge test can be seen in these figures. The final values of vertical displacement at D-V1 and D-V3 are almost the same in the analysis, although they are a little different from each other in the experiment. In order to simulate such experimental results precisely, it is desir-

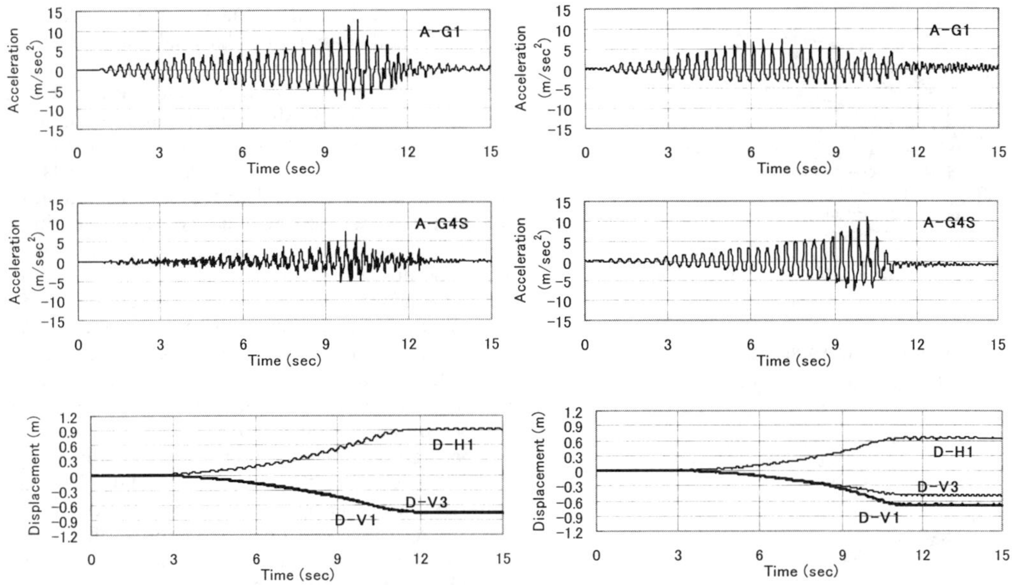

(a) Analytical Results obtained by FEM (b) Experimental Results

Figure 5. Time histories of acceleration and displacement at each point in the slope.

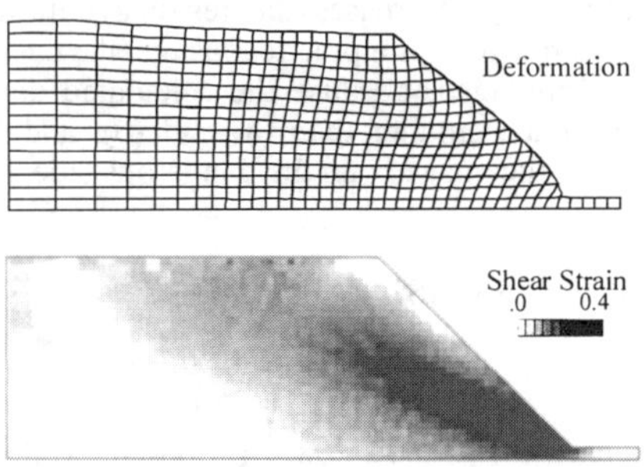

Figure 6. Residual deformation (FEM).

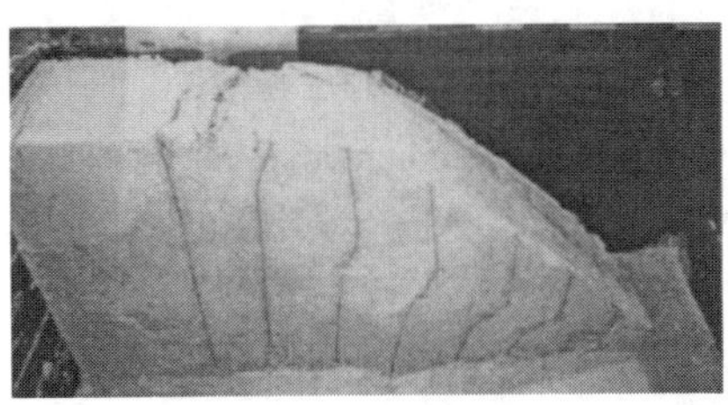

Figure 7. Residual deformation (centrifuge test).

able that the existence of a discontinuous deformation along the slip surface is taken into account in the analysis.

The calculated residual deformation of the slope after earthquake is shown in Figure 6. The slip surface, which is recognized as the area of higher shear strain, is clearly observed in the slope. Large settlement and small upheaving can be seen at the top and the toe of the slope, respectively. These features are almost similar to the experimental result shown in Figure 7.

3.4 *Parametric studies for* G_0

Parametric studies for the value of G_0 are performed, in order to verify the most efficient way for determining the input parameter G_0. Typically, there are 3 ways of determination method for G_0, that is,

(a) Firstly, E_{50} is defined as the secant elastic modulus at the stress that corresponds to half a value of the peak strength in the stress-strain curve. E is assumed to be equal to $2 \times E_{50}$, according to the geometrical relationship for a hyperbolic curve. Then, we obtain the value of G_0, by

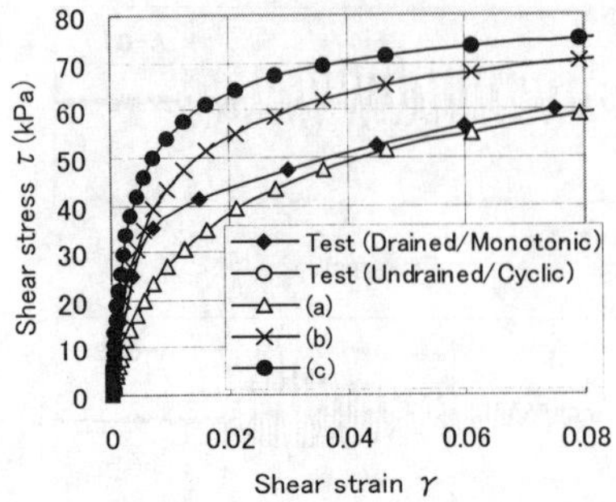

Figure 8. Assumed stress-strain curve ($\tau-\gamma$ relationships) in each case (σ_c'=49kPa).

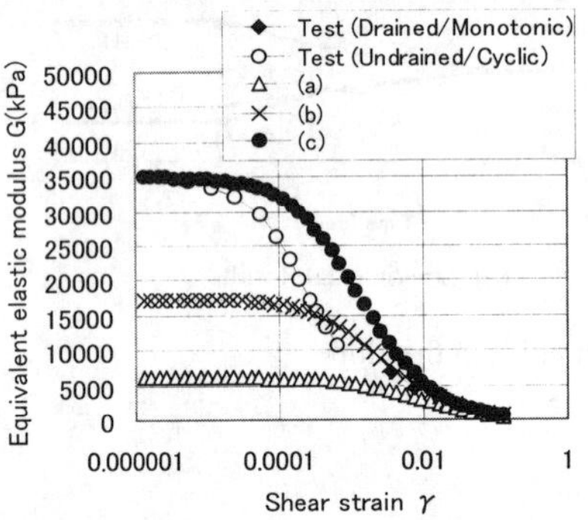

Figure 9. Assumed $G-\gamma$ relationships in each case (σ_c'=49kPa); where $G=\tau/\gamma$.

Table 2. Residual displacement after earthquake in each case.

	Centri-fuge	(a)	(b)	(c)
G_0 (kPa)	-	6071.	17300.	35000.
D-V1(m)	-0.70	-0.77	-0.49	-0.25
D-V3(m)	-0.50	-0.74	-0.47	-0.23
D-H1(m)	0.65	0.91	0.82	0.51

using $G_0 = 0.5E/(1+\nu)$. This method is what we have adopted in the previous paragraph.

(b) According to the reference strain γ_{G_0} in the $G-\gamma$ curve obtained from undrained cyclic triaxial tests, the value of G_0 in the model is determined so that the corresponding $\gamma_{G_0}(=\tau_f/G_0)$ should be consistent with the observed one. By the way, γ_{G_0} is defined as the strain where corresponding equivalent elastic modulus G is equal to $0.5\cdot G_0$.

(c) According to the $G-\gamma$ curve obtained from undrained cyclic triaxial tests, the value of G_0 in the model is defined as the value of G in the region of small strain in the observed $G-\gamma$ curve.

The stress-strain ($\tau-\gamma$) curve and the $G-\gamma$ curve in each case discussed above are plotted in Figures 8 and 9, respectively. Not only curves assumed in the model but also ones obtained from triaxial tests are plotted in these figures. The types of triaxial tests plotted here are such two cases as the 'drained, monotonic' and the 'undrained, cyclic' loading. It should be noted that the points for 'drained, monotonic' appears in large strain area, while the ones for 'undrained, cyclic' appears only in small strain area.

As seen in Figure 8, the stress-strain curve for 'drained, monotonic' test is close to the case (a) especially in large strain area, while it is close to the case (b) in small strain area. On the other hand, as seen in Figure 9, the $G-\gamma$ curve for 'undrained, cyclic' test is close to the case (c) especially in small strain area. Comparing the model curves (a), (b) and (c) with the observed ones, it is found that every model curve has good and bad points. It is difficult to judge the predominance of only one method for determining G_0, immediately.

Finally, we performed the numerical simulations for the centrifuge test. Table 2 shows the observed and calculated residual displacement after earthquake in each case. It is found that the increase of the value of G_0 decreases the residual deformation after earthquake. For a practical usage, the case (c) is not recommended, because the calculated displacement is much smaller than (a) and (b), and it will provide us the answers on the dangerous side.

4 CONCLUSIONS

The results observed in the centrifuge test for a model slope were simulated well by the 2D dynamic elasto-plastic FEM. A newly proposed cyclic loading model was found to be very effective for the actual seismic design of slopes.

REFERENCES

Manassero, M. 1989. Stress-strain relationship from drained self-boring pressuremeter tests in sands, *Geotechnique*, Vol.39, No.2, pp.293-307.

Wakai, A. & Ugai, K. 1999. Evaluation of residual displacement of slopes during earthquake based on a simple cyclic loading model, *Proc. 2nd International Conference on Earthquake Geotechnical Engineering*, pp.685-690, Lisboa, Portugal.

Wakai, A. & Ugai, K. 2001. Numerical analysis for seismic behavior of a slope based on a simple cyclic loading model, *Proc. 4th International Conference on Recent Advances in Geotechnical Earthquake Engineering and Soil Dynamics*, San Diego, USA (*to be published*).

Computer Methods and Advances in Geomechanics, Desai et al. (eds) © 2001 Balkema, Rotterdam, ISBN 90 5809 183 X

A real-time X-ray CT detecting of density damage increment of rock under compression condition

Ding Weihua & Wu Yanqing
State Key Laboratory of Frozen Soil Engineering, Cold and Arid Region Environmental and Engineering Research Institute, Chinese Academy of Sciences, and Institute of Water Resources and Hydro-Electric Engineering, Xi'an University of Technology, People's Republic of China

Pu Yibin & Liao Quanrong
State Key Laboratory of Frozen Soil Engineering, Cold and Arid Region Environmental and Engineering Research Institute, Chinese Academy of Sciences, People's Republic of China

ABSTRACT This paper revises the status of observation on rock meso-damage process under compression. The author proposes a new concept about density damage increment in using CT scanner and deduces a formula between the density damage increment and the radiological density(CT number) . A new method of analyzing rock meso-damage process with CT scanning is developed. On the basis of the new method a meso-damage process of rock can be detected and this gives very important clues about rock damage mechanics and rock constitutive relation.

INTRODUCTION

Detecting rock fracture nucleation and understanding the process of faulting are important keys to earthquake prediction and rock damage mechanics. In rocks subjected to a natural compressive stress-field, little is known about shear fracture genesis, structure, and grown up to macroscopic faulting. Laboratory experiments on rock cores may give deformation models where it is easy to check the stress field and to measure strains with strain gauges stuck on the samples. However, pre-failure phenomena are difficult to analyze without a non-destructive method adapted to the observation of cracks leading to shear fracture during the experiment. Investigations of faulting in rock samples have been carried out since 1960s and fall into three types(H. Kawakata et al 1999). The first is microscopy using an optical microscopy or a scanning electron microscope. The second is observing physical properties, such as P-wave velocity and acoustic emission events .The third is observing sample interiors with an X-ray CT scanning system, which allows one to non-destructively obtain an unlimited number of images of sample cross-sections .The X-ray CT method is more current in recent years.

Since the report on computerized transverse axial tomography was first presented at the April 1972 Annual Congress of the British Institute of Radiology(Ambrose and Hounsfield 1973), CT technology has led to remarkable achievements in the field of medical science for obtaining accurate ,nondestructive, three-dimensional internal representations of human body. A little later the development of industrial CT scanner has enabled observations to be made to internal structure of metals and ice cores.

The internal structure of rocks and characterization of mechanical deformation by X-ray CT method has been analyzed(Suzanne Baynaud et al 1989). It is hypothesized that rock is homogeneous and has the same radiological density (CT number) in every voxel. H. Kawakata et al(1999) observed the spatial fault development in granite undergoing compression. They used a conventional medical X-ray CT scanning system to image the sample interiors. Three-dimensional fault systems was detected with sequential X-ray CT images. It was found that three-dimensional reconstruction by X-ray images yields not only three-dimensional images of the fault system, but also provides fault cross-section images with much less artificial noise (artifacts) than does direct X-ray CT imaging. The common characterization of the two papers is that every CT image is imaged after the sample has been unloaded. Ge Xiurun et al(1999) conducted a real-time CT triaxial testing study of meso-damage evolution law of coal. They used the newest specified loading equipment corresponding to the CT scanner in triaxial and uniaxial compression. That is, when X-ray is scanning the sample do not unload. However the common shortage of these results is that they used the radiological density of the CT image to analyze the structure of rock. Because the radiological density is a function of the gravimetric density and mineralogy, it is difficult for the radiological density to relate to the gravimetric density directly.

For rock material it is impossible to generate a calibration curve plotting the measured CT numbers versus gravimetric density because rock material has complex mine components. In order to overcome this defect that CT number of traditional rock CT image has not definite physics sense, a density damage increment is defined on the basis of CT number and a relation formula between density damage increment and CT number is deduced on the basis of CT physics. In this paper a new method analyzing rock meso-damage process on the basis of CT scanning is developed when a meso-damage process of rock under compression is detected using a medical X-ray computed tomography scanner and the corresponding specified triaxial equipment.

1 A RELATION FORMULA BETWEEN DENSITY DAMAGE INCREMENT AND CT NUMBER

For rock material, a voxel is composed of minerals and voids. If ρ_0, H_0, ρ_i, H_i, ρ_r, H_r, H_g denote the gravimetric density, the CT number of the voxel at initial stress state, the gravimetric density, the CT number of the voxel at arbitrary stress state, the gravimetric density , the CT number of the minerals in the voxel, the CT number of air respectively, a equation can be deduced(Toshiyuki Kawamura et al 1990)

$$\rho_0 = \frac{H_g - H_0}{H_g - H_r} * \rho_r \qquad (1)$$

comparing with void in the voxel of rock, the minerals can be considered not to be compressible, so ρ_r and H_r are constant in the voxel of rock .Other equation can also be deduced

$$\rho_i = \frac{H_g - H_i}{H_g - H_r} * \rho_r \qquad (2)$$

for a medical CT scanner $H_g = 0$. From equation (1)and (2) a formula can be deduced

$$\frac{\rho_i - \rho_0}{\rho_0} = \frac{H_i - H_0}{H_0 + 1000} \qquad (3)$$

We define $\Delta D = \frac{\rho_i - \rho_0}{\rho_0}$ as the density damage increment

2 A EXPERIMENT RESULT AND DISCUSSION

The experimental equipment contain CT scanner and the corresponding triaxial cell . The CT scanner is SIEMENS SOMATOM plus medical X-ray scanner and has a spatial resolution of 0.35mm*0.35mm. The corresponding triaxial cell can be put in the X-ray scanning space with the CT bed. The first scanning is made at the initial state. After this the first step stress is loaded on the sample. When the second scanning is made, the loaded stress state must be steady. After the second scanning is made, the second step increment stress is loaded on the sample. The test is continuous with this method until the sample is destroyed.

The sample of rock-core is of cylindrical shape , 50 mm in diameter , 100mm in length. It comes from Shannxi Province , P.R.China. The density of this sandstone is 2.69g/cm^3.

Fig 1 is the relation curve between the axial

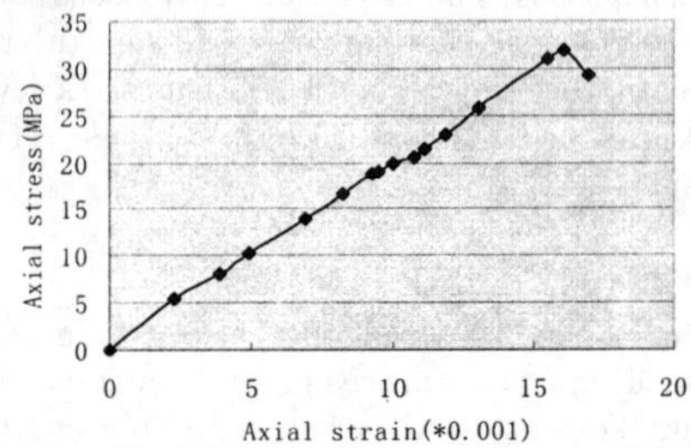

Fig 1 The relation curve between axial stress and axial strain of sample 19

stress and the axial strain of sample 19. When the axial stress is 19.93Mpa the curve is at its yield point.

Fig 2 The enhancement CT image in scanning section –63 of sample 19 at the initial state

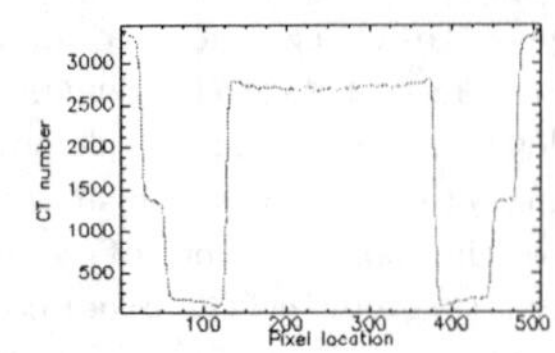

Fig 3 The profile from Fig 2 in its slope line

Fig 2 is the enhancement computerized transverse axial tomography image of the section -63 of sample 19. The region A is the sample of sandstone. The region B is air(in this test the confining pressure is zero). The region C is the plastic crust. The region D is the aluminum cell. The right corner of the upper in this image is the X-ray radiography image of the test.

Fig 3 is the profile from Fig 2 in its slope line of sample 5 when stress is zero .The CT numbers of rock, air , plastic and aluminum are 2700, 0, 1400 and 3800 respectively.

Fig 4 is the sequent CT images of sample 19 in

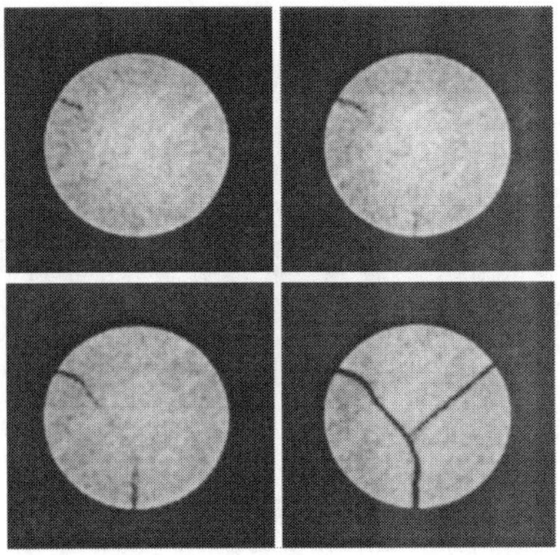

Fig 4 the sequent CT images of sample 19 in CT/-63 cross-section of under axial stress 16.71, 18.85, 19.09, 19.93 MPa respectively

CT/-63 cross-section under axial stress 16.71, 18.85, 19.09, 19.93 MPa respectively .The traditional CT image analyzing method is to analyze the meso-fracture process comparing with the former corresponding CT image. In CT image of rock, the lower grayscale demonstrate the lower radiological density (CT number). The generation of the lower radiography density region can be recognized as meso-fracture aggregate. From Fig 4 the nucleation , joint and run-through of meso-fracture aggregate of sample 19 in CT/-63 cross-section under axial stress 16.71, 18.85, 19.09, 19.93 MPa respectively can be easily seen. But if the variety between the two CT images is little, this method will be inefficient because of the shortage of distinguishability of people's eyes. The other shortage of this method is the radiological density has not a definite physical sense.

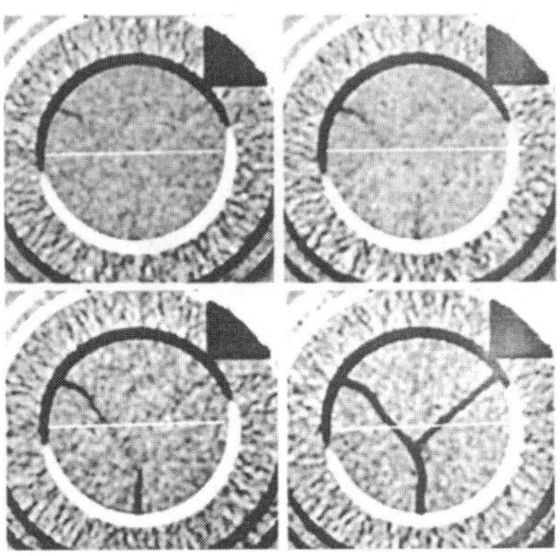

Fig 5 the ΔD sequent images of sample 19 in CT/-63 cross-section of under axial stress 16.71, 18.85, 19.09, 19.93 MPa respectively

Fig 5 is the ΔD sequent images of sample 19 in CT/-63 cross-section of under axial stress 16.71, 18.85, 19.09, 19.93 MPa respectively. In one ΔD image people can see the meso-fracture process. On one hand the ΔD image has a higher distinguish ability than the CT image analyzing method with the help of enhancement method. On the other hand, the ΔD image gives every physical density damage value of voxel of rock in the scanning section. The ΔD value can be easily used in the constitutive relation of rock material. The density damage process of arbitrary region in ΔD image can be analyzed.

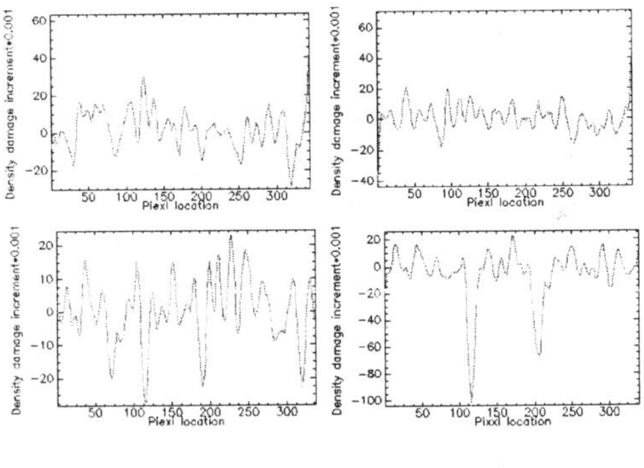

Fig 6 The profiles from Fig 5 in their horizontal lines respectively

Fig 6 is the profiles from Fig 5 in their horizontal lines respectively. It gives every density damage value of the ΔD image in the horizontal lines. The minimum of ΔD value is corresponding to the lowest radiological density region and corresponding meso-fracture aggregate. When the length of this corresponding meso-fracture aggregate reaches a certain extent , the relation curve between the axial stress and the axial strain of sample 19 reaches its yield point. Before the final strength of the sample, the interior has many damage regions or cracks. Because ΔD image can gives the damage at any region and any stress state a correct measure according the ΔD value, this gives very important clues about rock damage mechanics and rock constitutive relation.

3 CONCLUSION

The author proposes a new concept about the density damage increment. In a real-time X-ray CT detecting of rock under compression condition, the relation formula between the ΔD value and the radiological density (CT number) can be deduced . Comparing with the traditional CT image analyzing method , the ΔD image analyzing method is a new progress in using X-ray CT scanning system in rock mechanics test.

Acknowledgements

This work was supported by the research project (No. 9803) of State Key Laboratory of Soil Engineering, Cold and Arid Region Environmental and Engineering Research Institute, Chinese Academy of Sciences, PR China.

References

H. Kawakata, A. Cho, T. Kiyama et al 1999. Three-dimensional observations of faulting process in Westerly granite under uniaxial and triaxial condintions by X-ray CT scan. Tectonophysics. 313:293-305

G. N. Hounsfield 1973. Computerized transverse axial scanning (tomography). British Journal of Radiology. 46:1016-1022

Suzanne Raynaud, Denis Fabre, Frederic Mazerolle et al 1989. Analysis of the internal structure of rocks and characterization of mechanical deformation by a non-destructive method: X-ray tomodensitometry. Tectonophysics. 159:149-159

Ge Xiurun, Ren Jianxi, Pu Yibin ,et al 1999. A real-in-time CT triaxial testing study of meso-damage evolution law of coal. Chinese Journal of Rock Mechanics and Engineering. 18(5):497-502

Toshiyuki Kawamura 1990. Nondestructive, Three-Dimensional Density Measurements of Ice Core Samples by X Ray Computed Tomography. Journal of Geophysical Research. 95(B8): 12407-12412

Computer Methods and Advances in Geomechanics, Desai et al. (eds) © 2001 Balkema, Rotterdam, ISBN 90 5809 183 X

Measuring soil deformation in geotechnical models using digital images and PIV analysis

D.J.White, W.A.Take & M.D.Bolton
Schofield Centrifuge Centre, Engineering Department, Cambridge University, UK

ABSTRACT: Accurate measurement of soil deformation is fundamental to the success of geotechnical modelling. Current practice is to track the movement of a grid of target markers embedded in an exposed plane of soil through a sequence of video images. A new system of deformation measurement is presented which provides improved precision and reliability whilst removing the need for intrusive target markers. Image capture using an inexpensive 2-megapixel digital camera provides a significant increase in resolution and image stability compared to video. Particle Image Velocimetry (PIV) is a velocity-measuring technique in which patches of texture are tracked through an image sequence. Image processing algorithms have been written to apply the PIV principle to images of soil. The resulting software has a precision of 1/15th of a pixel when tracking the movement of natural sand or textured clay. This system allows displacements to be measured to a precision greater than has been previously achieved, without recourse to target markers.

1 INTRODUCTION

Physical modelling of geotechnical events is used to observe soil behaviour, allowing predictive methods to be verified and real constructions to be simulated. To carry out quantitative analysis of the observed behaviour, measurements of deformation must be made. Whilst displacement probes may be placed at the model boundaries to measure the movement of a single point, this paper is concerned with the more valuable measurement of ground movements in the entire model.

The measurement of gross deformations is key to understanding the ultimate failure mechanism of a geotechnical structure. It is through the observation of failure mechanisms that design solutions for ultimate load are verified. Gross, or ultimate, displacements can usually be observed through post-test excavation of a geotechnical model. Such measurements form a 'before' and 'after' snapshot of the model, allowing the failure mode to be estimated.

In contrast to the gross deformations present at the ultimate state, serviceability state design requires an understanding of settlements and ground movements at much lower strains. To achieve this aim, the measurement resolution must be reduced by at least an order of magnitude. Coupled with the reduced scale of geotechnical models, and hence a corresponding reduction in the size of the relevant movements, the measurement of pre-failure deformations remains a difficult task.

Figure 1 presents the strain ranges experienced during various geotechnical processes. The range 0.01% - 1% strain encompasses serviceability and pre-failure deformations. It is the measurement of these movements which is a challenge in small-scale geotechnical modelling. Larger deformations which can be of interest to geotechnical modellers are also shown. Some are associated with catastrophic events (eg landslides, lateral spreading, tunnel collapse) and others with construction processes (eg. CPT/pile penetration).

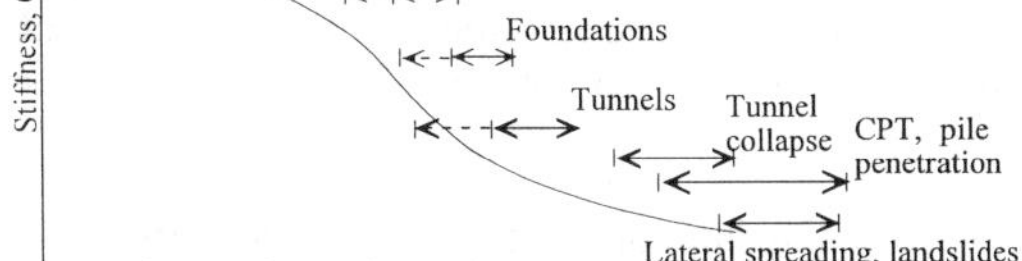

Figure 1. Typical strain ranges experienced in geotechnical engineering (after Mair, 1993).

A successful deformation measurement technique must have sufficient precision to capture the lowest relevant strain throughout the field of interest. A typical centrifuge model might have a 300 mm × 200 mm area of interest. To measure a pre-failure deformation of 0.01% strain in a zone of 10% of the model requires the detection of a movement of 2 μm.

2 BACKGROUND

The first measurements of displacements within a soil mass were carried out by Gerber (1929) using an X-ray method. Lead shots are embedded in the soil model, and successive radiographs are taken to follow the movement of these markers. This system has been widely used to detect incremental strain patterns in large sand models and shear apparatus (Roscoe, Arthur & James, 1963). The technique was successfully used to produce contours of shear and volumetric strain with a precision of 0.1% through large (2.0 m × 0.5 m) models (James, 1965).

Geotechnical centrifuge modelling does not permit the exposure of radiographs in-flight. Instead, modellers adapted the above technique by providing a transparent window exposing a plane of the model into which target markers are placed. Conventional photography is used to record the movement of these markers. The reduced size of centrifuge models compared to the previous large-scale testing coupled with the reduction in scale from model-space to film-space lead to a decrease in precision. Beasley (1973) and Mair (1979) present in-flight strain data to a resolution of 1%, from a measurement precision of 100 μm.

The use of video photography can eliminate the need for time-consuming measurement of exposed film. The image processing technique of centroiding can be used to estimate the location of the array of target markers. Under the conditions of a typical geotechnical model, centroiding is reported to locate target markers to an accuracy of 0.1 pixel (Chen et al, 1996). Taylor et al (1998) report a centroiding-based measurement system which can track movements of 3 mm diameter target markers embedded in clay with a precision of 60-105 μm over a 300 × 200 mm field of view. An example of the resulting displacement measurements are presented by Grant (1998) as contours of strain with a resolution of 1%.

Whilst the technology of the systems described above has evolved considerably in the previous 70 years, the continuing reliance on discrete target markers is a significant drawback. The influence of a dense grid of inclusions in the soil model is not easily quantified. The detail in areas of high strain gradient is unsatisfactory, measurement reliability breaks down when targets become partially obscured by soil, and the points at which observations are made must be determined prior to the modelling event.

This paper presents a new system based on digital photography and PIV image processing, which eliminates the reliance on target markers, and provides displacement measurements to a precision higher than described above. These measurements can be at locations within the model determined by the user after the modelling event has been recorded.

3 IMAGE ACQUISITION

Image frames acquired from a European standard (PAL) video signal have a nominal resolution of 732 × 549 pixels, although the number of photosensitive elements on a typical CCD are considerably less due to the use of interlaced scanning. The precision of any image measurement is inversely proportional to the object-space pixel size. In other words, a twofold increase in pixel resolution corresponds to a doubling of the measurement precision.

The decreasing cost and increasing sophistication of digital still cameras has provided an alternative to video capture in applications where a high frame rate is not required. The images presented in this paper were captured using an inexpensive (≈$400) Kodak DC280 digital still camera which provides a pixel resolution of 1760 × 1168. This represents a fourfold reduction in object space pixel size compared to that of PAL video.

There are other significant differences between a video capture system and a digital still camera. The analogue transfer of video signals through centrifuge slip rings leads to image deterioration, and storage on analogue tape creates line jitter. In contrast, digital still images are stored onboard the camera and transmitted digitally through the sliprings. This prevents the addition of noise during the transfer and storage stages.

Most digital cameras have native control software which can be tailored to suit the experimental application. Remote adjustment of camera settings, image tagging and event logging can all be achieved whilst the camera is operating onboard a centrifuge. The Kodak DC280 and DC215 cameras have both been successfully proof tested on the Cambridge beam centrifuge to 100 g without loss of functionality.

The duration of a typical triaxial or non-dynamic centrifuge test is such that the 10-20 second interval required for image capture and download leads to only a small deformation increment between successive images.

4 PIV IMAGE ANALYSIS

Particle Image Velocimetry (PIV) is a velocity-measuring technique which was originally developed in the field of fluid mechanics. The flow field of a fluid can be examined by seeding the flow with marker particles and tracking the movement of small patches within a larger image (Adrian, 1991).

The deformation of soil can be considered as a low-velocity flow process. In fluid mechanics experiments, polystyrene balls or coloured powder are added to the flow field to provide identifiable texture on which the image processing can operate. In contrast, natural sand has its own texture in the form of different coloured grains, and the light and shadow formed between adjacent grains when illuminating a plane of granular material. Texture can be added to an exposed plane of clay by the addition of coloured 'flock' material or fine sand.

Image processing algorithms have been written to implement the patch-matching PIV principle through a sequence of digitally-captured images of soil. Firstly, a grid of patches is laid out over the area of interest at the locations for which a displacement vector is to be found (Fig. 2). Each patch is then translated at 1-pixel intervals over a pre-defined search zone. At each position, the degree of match between the sample patch and an interrogation patch taken from the same location in the subsequent image is assessed.

This comparison process produces a map of 'degree of match' over the entire search zone. Spline interpolation in the zone of best match allows the patch displacement to be assessed to sub-pixel precision.

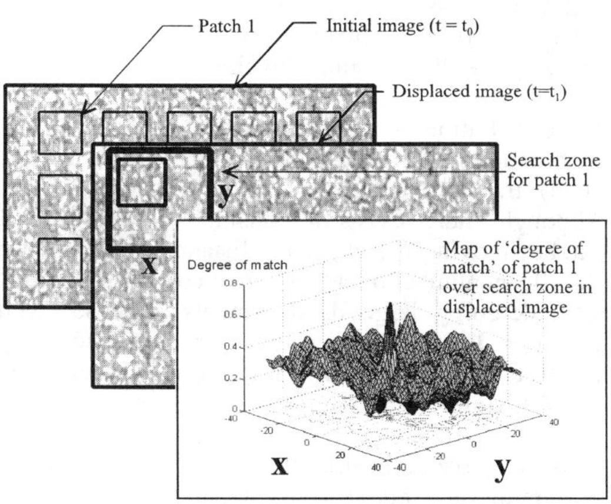

Figure 2. PIV image analysis technique.

A suitable patch size corresponds to a couple of particle diameters. This derives from the requirement that the patch contains sufficient texture to be identifiable in the subsequent image. Figure 3a shows part of a 1760 × 1168 pixel image of Dog's Bay carbonate sand (D_{50} = 0.3 mm) photographed at an image scale of 10 pixels per mm. Patches of size 10 x 10 pixels contain sufficient texture to produce successful PIV analysis, allowing a full image to consist of 176 × 116 measurement points, each of object-space size of 1 mm × 1 mm. This method of obtaining over 20,000 displacement measurements is significantly more practical than embedding a similar number of target markers in a soil model.

To be sure to detect the displaced patch, the search zone should have a width equal to twice the maximum possible soil displacement between each image pair. However, this can increase the computational requirement unnecessarily. In a typical geotechnical model, strain rate, and hence patch velocity, will vary gradually across the model. A more efficient technique of selecting the search zone is to shift the zone from the sample location by the displacement vector of nearby previously-evaluated patches. The size of the search zone can then be reduced to a fraction of the expected displacement.

Having obtained the displacement field using PIV analysis, a calibration procedure for the image-space to object-space transformation is essential when large displacements are under consideration. The non-coplanarity of the CCD and the soil plane, coupled with the radial and tangential lens distortion present in short focal-length optics, leads to a variation in image-object space scale factor across the image.

5 VALIDATION EXPERIMENTS

A series of bench-scale experiments have been conducted to assess the precision and reliability of the PIV technique to measure deformations in soils. The experimental apparatus consists of a rigid soil container which is driven by a micrometer along track in an aluminum base plate with PTFE guides. This apparatus allows a plane of soil to be translated horizontally beneath a rigidly fixed camera. Small known increments of movement are applied to the soil container via a micrometer and the resulting sequence of photos are analysed using PIV. All digital images were acquired using a Kodak DC280 camera. The field of view of the entire 1760 × 1168 pixel image was 176 mm × 117 mm which corresponds to an object-space pixel size of 0.101 mm.

5.1 *Deformations in sand*

A grid of 420 patches of sand was selected from digital images of the translating soil. Each of these patches contains 100 pixels and corresponds to 1 mm^2 of sand (Fig. 3a). To validate the PIV technique, each of these patches was tracked through a series of ten 0.1 mm translation steps. Since the body of sand was translating uniformly, the variation in the detected displacement of each patch indicates the precision of the PIV analysis. In the interests of brevity, only the 0.1 and 1 mm deformations will be examined in this paper. In image space, these displacements correspond to values of 1 and 10 pixels respectively.

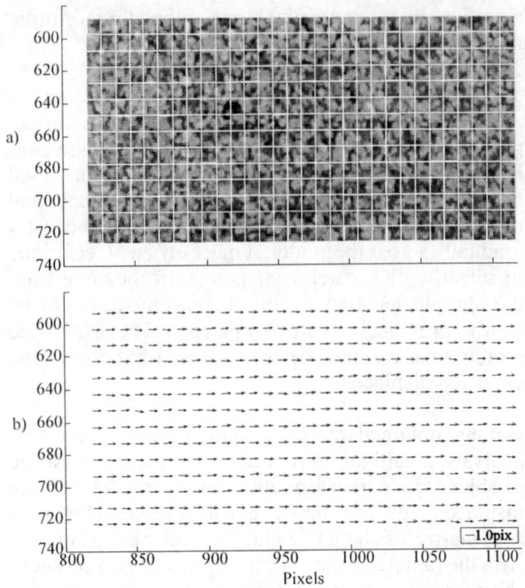

Figure 3. Sand image patches (a), with measured movements during 1 pixel translation (b).

The measured displacement vectors for each patch during the first 0.1 mm (1 pixel) translation step are presented in Figure 3b. The distribution of these vectors is shown as a histogram in Figure 4. The PIV analysis of these patches exhibits excellent internal consistency with the standard deviation of the measured 1 pixel displacements being 0.043 pixels. The mean standard deviation of the incremental patch displacements for the remaining nine image pairs was 0.050 pixels.

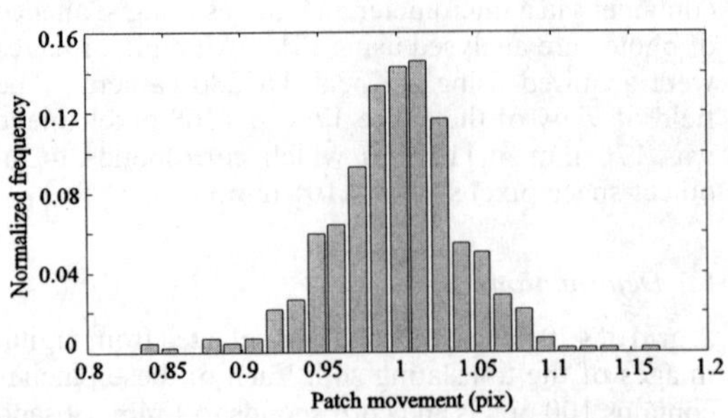

Figure 4. Image-space distribution of measured sand movement at 1 pixel translation.

To further validate the technique against known movement, the sand was translated a full 1 mm using the micrometer. This larger distance was used as a basis for comparison to reduce the significance of operator errors in the micrometer movement. The distribution of measured object-space patch displacements is shown in Figure 5. As can be seen in this figure, the PIV analysis displays an excellent agreement with the applied movement, exhibiting a standard deviation of 6 μm (0.061 pixels in image space).

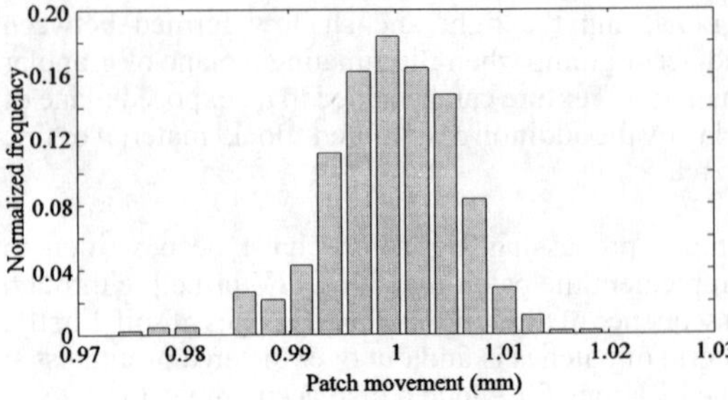

Figure 5. Object-space distribution of measured sand movement at 1 mm translation.

5.2 *Deformations in clay*

Unlike images of sand which have an inherent texture, the particle size distribution of clays dictates that texture must be imparted to the uniformly-coloured clay soil. In the present study, a flock material was lightly dusted onto the clay surface. The density of the flock dusting was purposefully made lower than that required for complete coverage of each of the 420 patches of 10 × 10 pixels to illustrate the issues of applying the PIV technique to clay soils.

The distribution of measured patch movements using each of the 420 soil patches for the first 0.1 mm (1 pixel) translation step is presented in Figure 6. Although there exists a considerable number of patches moving the expected distance of 0.1 mm, a significant number of patches are incorrectly recorded as undergoing large displacements. These 'wild vectors' derive from patches being incorrectly identified in the subsequent image due to insufficient image texture.

As a comparison, the intensity histograms for patches containing sand, pure kaolin clay, and clay with flock are presented in Figures 7a to 7c, respectively. To eliminate the wild vectors, each patch was evaluated for texture content according to the following criteria. If a patch of clay contained at least 10% by area of pixels significantly darker than the background it was accepted. The 124 patches which passed this criterion are shown in Figure 8a.

The measured patch movement for the first 0.1 mm (1 pixel) translation step using the filtered patches is presented in Figure 8b and as a histogram in Figure 9. The PIV analysis of these patches exhibits excellent internal consistency with a standard deviation of 0.076 pixels. This demonstrates that the addition of flock material to uniformly-coloured clay allows successful texture-based PIV analysis.

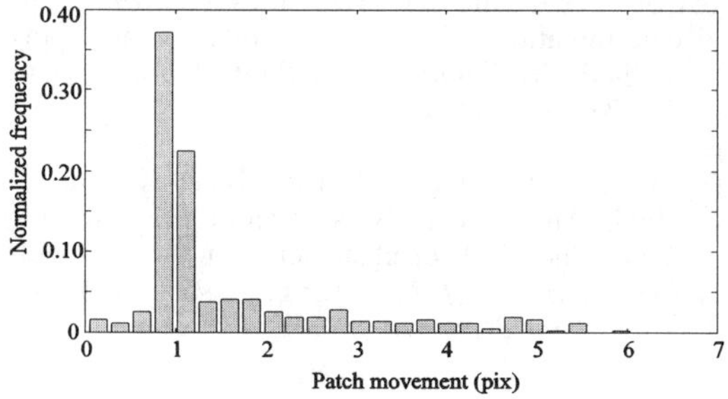

Figure 6. Image-space distribution of measured movement of all clay patches at 1 pixel translation.

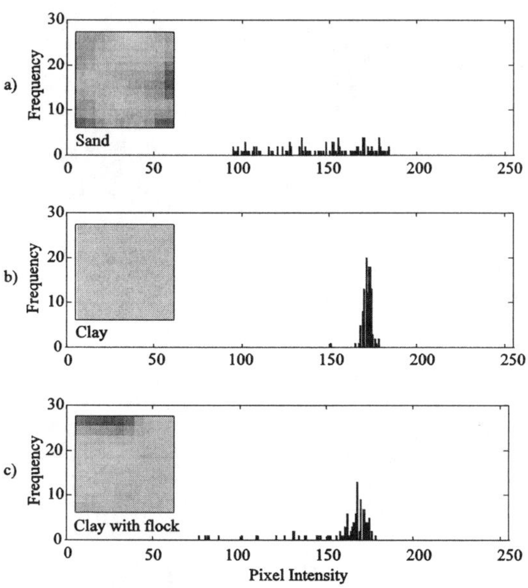

Figure 7. Pixel intensity of patches of sand (a), clay (b) and clay with flock.

6 CONCLUSIONS

Conventional techniques of measuring soil deformation in geotechnical models rely on intrusive target markers and video image capture. The reported precision of such systems when viewing a typical (300 mm × 200 mm) soil model is of the order of 100 μm. Differentiation of the resulting displacement vectors has allowed contours of 1% strain to be produced. The use of target markers has a number of drawbacks, and video image capture is unstable and of low resolution.

A new system of displacement measurement based on digital image capture and PIV image analysis has allowed precision and reliability to be improved. Inexpensive digital still cameras offer over twice the resolution of video capture and provide more stable images. Although these cameras include mechanical operating parts, two models have been successfully tested in the centrifuge environment.

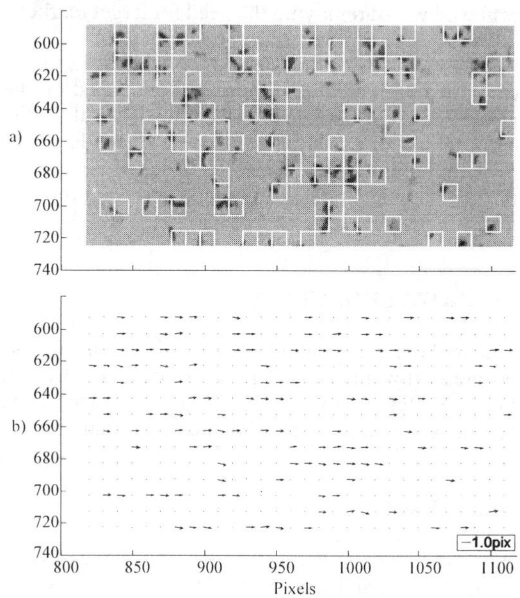

Figure 8. Filtered clay/flock image patches (a), with measured movements during 1 pixel translation (b).

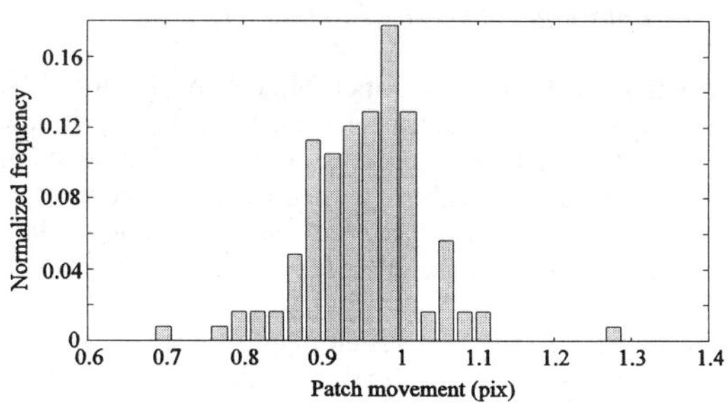

Figure 9. Image-space distribution of measured clay/flock movements during 1 pixel translation.

PIV analysis is a texture-based image processing technique which has been adapted from its fluid mechanics origins to suit geotechnical applications.

Algorithms have been written which allow displacements of small patches of sand and textured clay to be measured to a precision of 1/15th of a pixel.

The combination of digital image capture and PIV analysis allows deformation fields comprising many thousand displacement vectors to be calculated to a precision higher than is possible using centroiding methods. Since the Kodak DC280 digital camera provides 1760 × 1168 pixel images, the resulting precision when viewing a 300 mm × 200 mm soil model is 17 μm. This is a significant increase in precision and measurement density, and has been achieved whilst removing the need for target markers.

The applications of this system are wide. It has been combined with a calibration procedure based on the principles of close-range photogrammetry to allow the deformation of centrifuge and 1-g soil model to be tracked through a large (>50) sequence of images. A further application is non-contact deformation measurement of element tests.

ACKNOWLEDGEMENTS

The authors are grateful for the financial support provided by the UK Engineering and Physical Sciences Research Council, Giken Seisakusho Ltd., and the Commonwealth Scholarship programme.

REFERENCES

Adrian 1991. Particle imaging techniques for experimental fluid mechanics *Annual review of fluid mechanics 23:261-304*

Beasley D.H. 1973. Centrifugal testing of models of embankments on soft clay foundations *Cambridge University PhD thesis*

Chen J., Robson S., Cooper M.A.R. & Taylor R.N. 1996. An evaluation of three different image capture methods for measurement and analysis of deformation within a geotechnical centrifuge *International archives of photogrammetry and remote sensing vol. XXX1, part B5, 70-75, Vienna*

Gerber E. 1929. Untersuchungen uber die Druckverteilung im Oertlick Belasteten Sand. *Dissertation Technische Hochschule, Zurich*

Grant R.J. 1998. Movements around a tunnel in two-layer ground *City University, London, PhD thesis*

James R.G. 1965. Stress and strain fields in sand. *Cambridge University PhD thesis*

Mair R.J. 1979. Centrifugal modelling of tunnel construction in soft clay *Cambridge University PhD thesis*

Mair, R.J. 1993. Developments in geotechnical engineering research: applications to tunnels and deep excavations. *Unwin Memorial Lecture (1992), Proc. ICE, London*, Vol 97 No.1, 27-41.

Roscoe K.H., Arthur J.R.F & James R.G. 1963. The determination of strains in soils by an X-ray method *Civil Engineering & Public Works Review*, 58:873-876,1009-12

Taylor R.N., Grant R.J., Robson S. & Kuwano J. 1998. An image analysis system for determining plane and 3-D displacements in soil models *Proceedings of Centrifuge '98, Balkema, Rotterdam 73-78.*

Computer Methods and Advances in Geomechanics, Desai et al. (eds) © 2001 Balkema, Rotterdam, ISBN 90 5809 183 X

The experimental study on swelling properties of swelling rock

Qing Yang, Jiankui Jiao & Maotian Luan
Department of Civil Engineering, Dalian University of Technology, People's Republic of China

ABSTRACT: In this paper, a new oedometer with laterally-confined conditions and corresponding experimental method are introduced, in order to overcome the drawbacks of Huder-Amberg test method by which the water absorbing content and the radial swelling stress cannot be measured in the whole test process. With use of the new oedometer, the swelling tests for swelling rock sampled from Nanjing Meishan Ferrous Mine have been conducted. Based on the experimental results, the interrelations among the five variables, such as swelling strain, water absorbing content, radial stress, axial pressure and time, are analyzed in the paper. Finally, discussions on the swelling constitutive relationship are made.

1 INTRODUCTION

Swelling rock, composed of a great deal of strong hydrophilic clay mineral, can exert destructive pressure to engineering constructions with volume expansion when it interacts with water. For a long time, the test and description to the mechanism of swelling rock have relied basically on uniaxial test method developed by Hunder-Amberg in 1970 and the linear relationship between swelling strain and the logarithm of axial stress established using this method (Huder & Amberg 1970). Based on this law, a linear semi-logarithmic relationship between volumetric swelling stain and volumetric stress was presented by Einstein (1972) and Wittke (1976). Some valuable results through investigating the lateral stress with the thin cell odeometer were found (e.g., Komornik & Zeitlen 1970). At present, the conventional swelling tests put particular emphases on studies of the stress that affects to the swelling deformation and, however, does not take the condition of water absorbing into consideration. In fact, the swelling deformation obviously depends on water absorbing content. Therefore the swelling test without considering the process of water absorption needs to be improved. A triaxial swelling experimental method has been developed by Yang (1995) and Yang & Jiao (1998). By this method, the whole process of water absorption can be measured. Usually, the triaxial test requires a considerably large specimen, rather long duration for saturation and testing, it is difficult to be applied in engineering practice. In order to overcome these shortcomings, a confined swelling test apparatus and method are developed and presented in this paper. In this new apparatus, systems for measuring radial pressure and water absorption are incorporated. Moreover, the experimental data are used to verify further the results of triaxial swelling test and 3-D swelling constitutive relationship previously obtained.

2 DESIGN OF CONFINED SWELLING TEST

2.1 *The composition of testing apparatus*

The diagram of swelling test apparatus is shown in Figure 1. The test equipment consists of axial loading system, measuring systems respectively for water absorbing content and lateral pressure. The axial loading system is analog to the consolidometer. The main parts of water absorption measurement consist of acidic burette, water filler pipe and graduated cylinder. The water absorbing content can be obtained through the readings of burette. The outlet water can be measured by using the graduated cylinder in order to determine the real quantity of water absorption. The part of confined pressure measurement is composed of transducer, and static resistance strain meter. The correlation between readings of static resistance strain meter and the confined pressure should be calibrated in advance. The calibration is fulfilled by filling water into the container where holds the specimen and sealing it before the test on the pressure measuring board of the triaxial test apparatus. Through several testing calibrations, it is observed that the radial

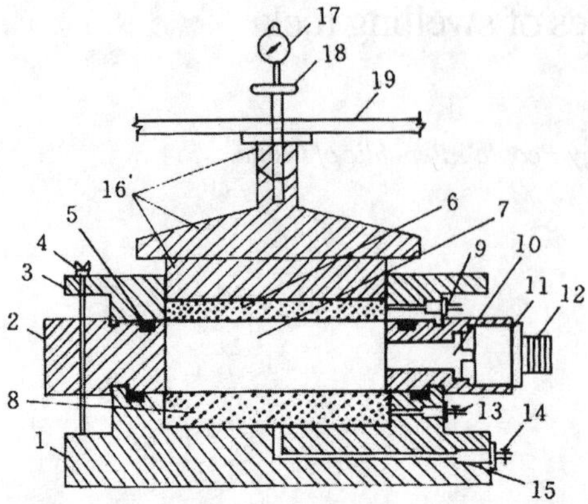

1. Pedestal 2. sample cell 3. cover 4. screw bolts 5. O-type rubber ring 6. upper porous stone 7. specimen 8. lower porous stone 9. water outlet pipe(connected to graduate) 10. pressure transmission bar 11. rubber pad 12. transducer (connected to static resistance strain meter with cable) 13. air exhaust pipe 14. valve 15. Water filler pipe （connected to acidic burette） 16. rigid block 17. percentage scale 18. trestle 19. pressure application level arm of consolidometer

Figure1. The confined swelling test apparatus

pressure is linearly correlated with the strain meter. The size of specimen is 61.8mm in diameter and 20mm in height.

2.2 *Main test procedures*

The test procedures mainly contains the following steps:

1. To install the transmission bar and transducer to the sample cell.

2. To put the porous stone and filter paper onto the base of pedestal and then put on the sample cell and cover (installation of the O-type rubber ring is necessary). Switch on the resistance strain meter and preheat it for 30 minutes and record the initial readings.

3. To put the specimen into the sample cell, and then put the filter paper, porous stone respectively on its top and install the transmission bar.

4. To install the axial percentage scale and impose 1 kPa pressure on the specimen in order to achieve a good contact to each part of test apparatus while recording the readings of percentage scale and strain meter.

5. To increase continuously the required initial pressures by different levels. When the deformation of specimen under the pressure becomes stable (The stabilization standard is that the deformation cannot exceed to 0.01mm per hour), to record the readings of percentage scale and strain meter.

6. To open the water filler pipe and the air exhaust pipe to fill the sample cell with water while opening the valve on the burette. When water drains out of the air exhaust pipe, the valve to air exhaust pipe should be closed immediately while recording the time, the reading of burette and the water drainage quantity of air exhaust pipe.

7. To put the recordings taken from steps 5-6 together as the initial value of status before the specimen absorbing water, to measure and record axial percentage scale, burette and strain meter at a certain interval until the swelling becomes stable.

8. After absorbing-swelling becomes stable under the load, a certain load could be unloaded while recording the instant change of readings from the percentage scale, the burette and the strain meter. Then each reading should be recorded at a certain interval in accordance with the above method. Until the swelling is not developed, the next unloading–swelling test could be carried out.

9. When swelling becomes stable under the final load, the loading can be released and the specimen could be taken out for weighing and drying while calculating the water absorbing content.

3 EXPERIMENTAL RESULTS AND ANALYSES

The samples used in this experimental study are from Nanjing Meishan Ferrous Mine, which is of Jurassic tuff. It is shown by X-ray diffracting analysis that the sample contains 80% of mixed-layer mineral with montmorillonite-illite, 15% of siderite and 5% of impurities. In fact, the undisturbed samples is difficult to achieve and the experimental study is aimed to point out the applicability of the new type lateral confined swelling apparatus and the general rules of swelling during absorbing water of the swelling rock. Therefore remolded samples are used in this experiment. The testing sample with the required size is prepared with the particles with diameter less than 1mm crushed from the sampled rocks. A certain amount of distilled water is added to the particles dried under a constant temperature of 105℃ and the remolded sample is formed by compaction. The Atterberg limit of the remolded sample is 39.5% of liquid limit, 22.2% of plastic limit, 10.32% of shrinkage limit, and the plasticity index is 17.3.

In this study, the swelling tests are conducted on 10 specimens under the different initial states and pressure conditions. The experimental results are similar and only the test results of $1^{\#}$ specimen are to be discussed due to the limited length of the paper. The initial water content and dry density of $1^{\#}$ specimen are respectively at 9.0% and 2.2132g.cm^{-3}. The axial unloading path is 0.3MPa→0.2MPa→ 0.1MPa→ 0.005MPa.

3.1 *Relationship of axial swelling strain, water absorbing content, lateral stress with axial loading*

It is shown in Figures 2-3 that there is a obvious linear relation between the swelling strain ε_u, water absorbing content W_u (alternatively called as swelling limit water absorbing content) at the stable swelling and the logarithm of the axial pressure σ_z.

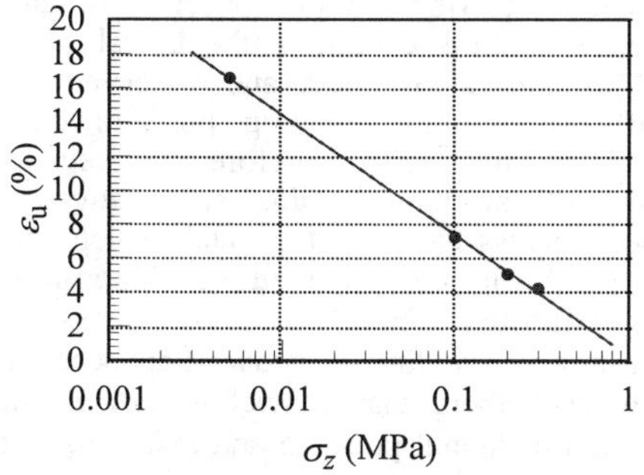

Figure 2. The relation of swelling stable strain and axial pressure

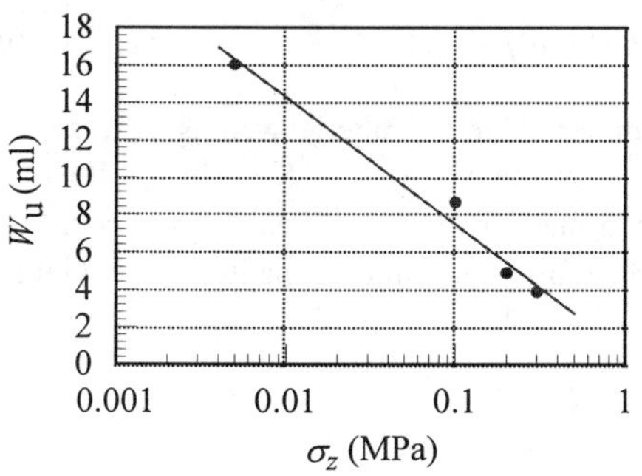

Figure 3. The relation of swelling limited water content and axial pressure

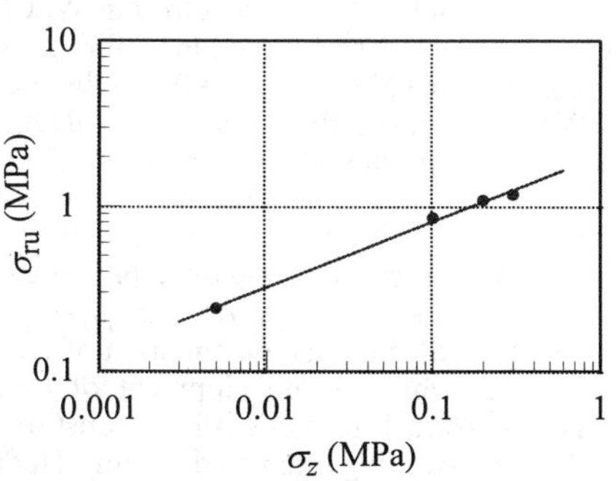

Figure 4. The relation of lateral stable swelling pressure and axial pressure

It is manifested in Figure 4 that there is a linear bilogarithmic relationship between the ultimate lateral pressure σ_{ru} and the axial pressure σ_z. Their regression equations are given respectively as follows

$$\varepsilon_u = K - A\ln\sigma_z = 0.2719 - 3.0662\ln\sigma_z \tag{1}$$

$$W_u = m - n\ln\sigma_z = 0.6734 - 2.9456\ln\sigma_z \tag{2}$$

$$\ln\sigma_{ru} = a + b\ln\sigma_z = 0.7054 + 0.3994\ln\sigma_z \tag{3}$$

The results given by Equation (1) is consistent with the Huder-Amberg relation and the rule represented by Equation (3) is also consistent with the results obtained by Komornik & Zeitlen (1970) among others according to the swelling test with thin cell odeometer. These facts indicate that the experimental results are reliable. Equation (2) implies that the absorbing saturation of the specimen is the function of the state of stresses.

3.2 *Relationship of swelling strain and Water absorbing content*

In this lateral confined swelling test, the water absorbing content and corresponding swelling strains during the whole process of swelling are measured, as shown in Figure 5, in which unloading points are marked with arrows. It can be found that a linear relationship holds between the swelling strain ε and the water absorbing content W, i.e., $\varepsilon = kW = 1.0332W$. This implies that the loading and water content do not affect remarkably the unit water absorption swelling rate k during swelling process of the specimen. Due to the limitation to the absorbing capacity from the state of stresses as illustrated in Figure 3, the swelling strain will not increase unlimited and becomes stable as the water absorbing gradually decreases until ceasing.

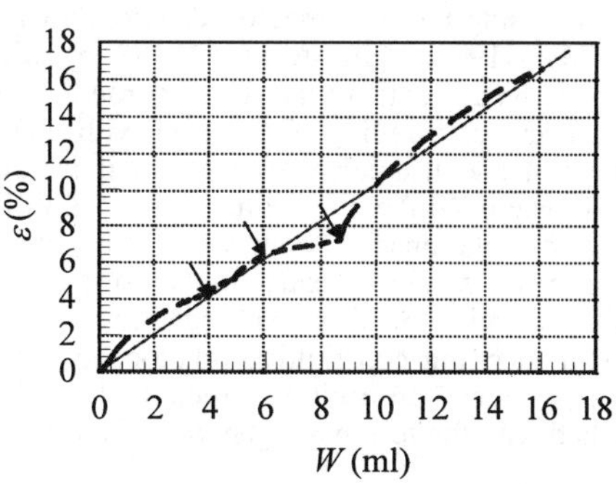

Figure 5. The relation of swelling strain and water absorbing content

3.3 *Development of swelling strain and water absorbing content with time*

The swelling strain increases with the time under the certain axial pressure, however the increasing rate decreases gradually, as shown in Figure 6. And eventually the strain tends to be stable. The whole process can be described in a hyperbola equation as follows

$$\varepsilon = \frac{t}{a + bt} \tag{4}$$

In this equation, coefficient a is the reciprocal of the initial swelling rate when $t \to 0$, and coefficient b is the reciprocal of the terminal swelling strain ε_∞ when $t \to \infty$.

The development of the water absorbing content under the each stage of pressure is of accordance with hyperbola patterns from initially fast rate to slow rate and then finally stopping absorption, and it has a similar tendency as shown in Figure 6.

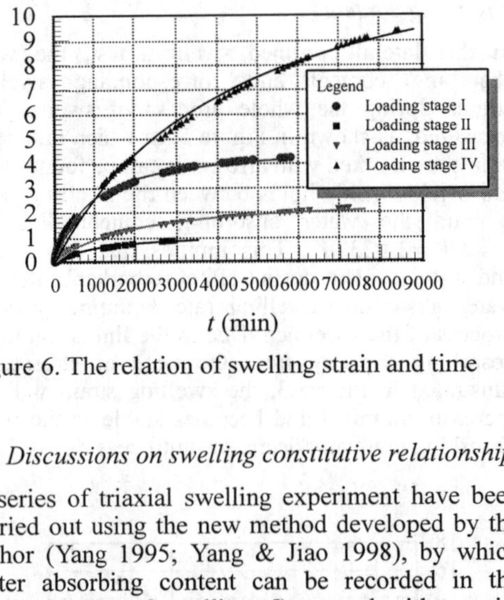

Figure 6. The relation of swelling strain and time

3.4 *Discussions on swelling constitutive relationship*

A series of triaxial swelling experiment have been carried out using the new method developed by the author (Yang 1995; Yang & Jiao 1998), by which water absorbing content can be recorded in the whole process of swelling. Compared with unaxial swelling test and triaxial swelling test, it is found that Einstein-Wittke's volumetric swelling constitutive relationship deduced from unaxial swelling test would greatly underestimate the swelling potential of the swelling rock. A new volumetric swelling constitutive relationship on the basis of water absorbing process has been established with the following general equation

$$\varepsilon_v = A + B\frac{W}{\sigma_v} - C\ln\sigma_v \tag{5}$$

where ε_v, σ_v, W respectively represents the volumetric swelling strain, volumetric stress and unit volume water absorbing content, parameters A, B, C are defined as swelling coefficient relevant to materials.

The above triaxial swelling test is performed under constant confining pressure. However it does not correspond with the actual conditions of lateral pressure increasing with the volume expansion in engineering practice. The lateral confined swelling test method described in this paper is actually a complex 3-D stress condition (the lateral stress is related to both axial stress and water absorbing content). This lateral swelling pressure can be obtained by this test, therefore the regression equation of swelling strain, axial stress and volumetric stress can be established, which can be used to verify Einstein's constitutive relationship.

The Einstein-Wittke's 3-D constitutive relationship of swelling is deduced in accordance with Huder-Amberg unaxial swelling test, in which the relation of lateral pressure and axial pressure is defined by $\sigma_x = \sigma_y = \sigma_z \mu/(1-\mu)$, hence the axial pressure can be given by

$$\begin{aligned}\sigma_z &= (\sigma_x + \sigma_y + \sigma_z)(1-\mu)/(1+\mu) \\ &= I_1(1-\mu)/(1+\mu)\end{aligned} \tag{6}$$

According to Huder-Amberg swelling test, the axial strain is estimated by $\varepsilon_z = K - A\ln\sigma_z$. Based on the confined condition, i.e., $\varepsilon = \varepsilon_v = \varepsilon_z$, the Einstein-Wittk's swelling constitutive relationship can be obtained

$$\varepsilon = \varepsilon_v = K_1 - A\ln I_1 \tag{7}$$

where μ is Poisson's ratio, I_1 is the first stress invariant and $K_1 = K - A\ln((1-\mu)/(1+\mu))$.

Shown in Figure 7 are regression equation of the results of lateral confined test and Einstein-Wittke's equations, e.g., $\varepsilon = 3.3991 - 3.0662\ln I_1$ for μ=0.3 and $\varepsilon = 4.0355 - 3.0662\ln I_1$ for μ=0.5. Obviously, Einstein-Wittke's constitutive relationship deduced from Huder-Amberg unaxial swelling test seriously underestimates actual swelling strain compared with the measured values. It can be concluded that the relationship of lateral pressure and axial pressure can not be simply defined as $\sigma_x = \sigma_y = \sigma_z \mu/(1-\mu)$, as it leads to considerable underestimation of lateral pressure. Hence, while the development of lateral pressure is unknown, Einstein-Wittke's constitutive relationship of swelling deduced from Huder-Amberg test cannot be adopted directly.

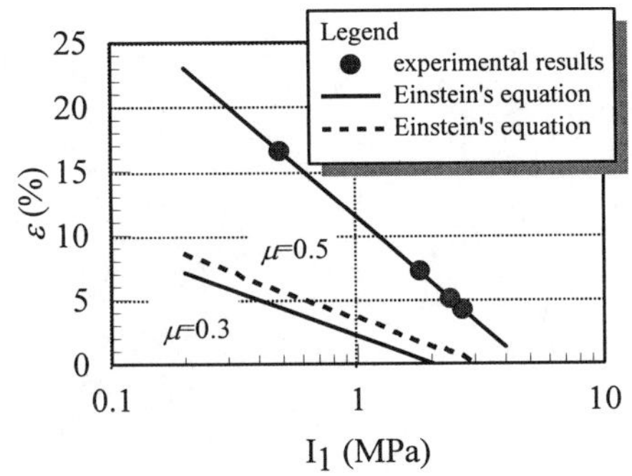

Figure 7. The relation of volumetric strain and logarithm of volumetric stress

Through analyzing of lateral confined swelling test, axial pressure σ_z will be considered as a decisive factor on the lateral confined swelling test. $\ln\sigma_z$ instead of $\ln\sigma_v$ should be used in the original triaxial swelling constitutive relationship represented by Equation (5) and consequently the prediction will be more exact. The regression equation of this experiment is given as follows

$$\varepsilon = 1.8777 + 0.3194\frac{W}{\sigma_v} - 0.9512\ln\sigma_z \tag{8}$$

On the basis of above analysis, the general formula of swelling constitutive relationship can be proposed

$$\varepsilon = A + B\frac{W}{\sigma_v} - C\ln\sigma_z \tag{9}$$

in which σ_Z is the axial stress. Other parameters have been explained in Equation (5).

4 CONCLUSIONS

In this paper, a new lateral confined swelling experimental method is proposed. In this proposed method, the time of water absorption, axial pressure, axial strain, water absorption content and lateral pressure can be simultaneously measured as to establish their relationship. In general, the former swelling experiments cannot measure the last two quantities. The experimental swelling results are mostly accordant with former experimental results. This indicates that the new apparatus and method of lateral confined swelling are reliable. Discussions on 3-D swelling constitutive relationship adopted currently are made. A revised 3-D swelling constitutive relationship is proposed in which volume stress, axial pressure and water absorbing content are included. The volume stress and the water absorbing content are inter-related. However, for actual engineering problems in the semi-infinite space, the vertical stress is the decisive factor in swelling development process of surrounding rock swelling. For example, for tunnel, the supporting strength of invert arch controls the developing process of the tunnel floor heave. In addition, swelling deformation can be predicted according to swelling strain increasing linearly with water absorbing content confined by stress conditions.

ACKNOWLEDGMENTS

The financial supports from National Natural Science Foundation of China through the grant No.59709008 and from the Trans-Century Training Programme Foundation for the Talents offered by the Ministry of Education of China are gratefully acknowledged.

REFERENCES

Einstein H. H., et al. 1972. Verhalten von Stollensohlen in Quellendem Mergel. *Proceedings on Internal Symposium on Underground Works*, Lucerne: 296~319.

Huder J. & G. Amberg 1970. Quellung in Mergei, *Opalinuston und Anhydrit*. Schweiz. Bauzeitung 88:975-980.

Komornik A. & J.G. Zeitlen 1970. Laboratory determination of lateral and vertical stresses in compacted swelling clay. *Journal of Materials, JMLSA*, 5(1):108-128.

Wittke W. & P. Rissler 1976. Bemessung der auskleidung von hohlraumen in quellendem gebrige nach der Finite-Element-Methode. *Veroffentlichungen des instituts fur grundbau*, Bodenmechanik, Felsmechank und Verkehrswasserbau:Heft 6. RWTH Aachen.

Yang Q. 1995. *Swelling rock and tunnel stability*. Beijing: Metallurgical Industry Publishers.

Yang Q. & J.K. Jiao 1998. The Determination of volume expansive properties in swelling clay. *Proceedings of the Second International Conference on Unsaturated Soils*: 450~454. Beijing: International Academic Publishers.

9 Dynamics, earthquake analysis, liquefaction

Computer Methods and Advances in Geomechanics, Desai et al. (eds) © 2001 Balkema, Rotterdam, ISBN 90 5809 183 X

Practical problems in the seismic analysis of bridge pile foundations

W.D.L. Finn
Anabukio Chair of Foundation Geodynamics, Kagawa University, Takamatsu, Japan
T. Thavaraj
Civil Engineering, University of British Columbia, Vancouver, B.C., Canada

ABSTRACT: A simplified method of 3-D continuum analysis, PILE-3D, which can handle nonlinear soil behaviour, kinematic and inertial interaction and seismic porewater pressures, is used to evaluate some current procedures for analysis used in practice, and to present a clearer picture of what happens to a pile foundation during strong shaking. PILE-3D can analyze either single piles or pile groups directly, and because of the continuum nature of the analysis does not require the use of interaction factors. Interaction effects are automatically taken into account.

1 INTRODUCTION

Seismic soil-structure interaction analysis involving pile foundations is one of the more complex problems in geotechnical earthquake engineering. The analysis involves modelling pile-soil-pile interaction, the effects of the pile cap, nonlinear soil response, and inertial and kinematic interactions. Commercial structural analysis programs can not include directly the pile foundation in analysis. Therefore, in the seismic structural analysis of bridges and buildings on pile foundations, discrete springs and dampers associated with the different degrees of freedom of the pile foundations are assigned to the structural analysis model.

Comprehensive methods to establish the properties of the discrete springs and dampers are based on linear elastic behaviour and use either boundary element or finite element methods. Published graphs of the results are limited to relatively small pile groups because of the substantial computing time required. Novak (1991) gave an extensive review of the more widely accepted methods of analysis. The methods are exact for elastic isotropic conditions, but they do not take into account the nonlinear behaviour of soil under strong shaking. The reduction in soil stiffness and the increase in damping associated with strong shaking are sometimes modelled crudely in these analyses by making arbitrary reductions in shear moduli and arbitrary increases in viscous damping ratios.

Dynamic nonlinear finite element analysis in the time domain using the full 3-dimensional wave equations is not feasible for engineering practice at present because of the time needed for the computations. However, by relaxing some of the boundary conditions associated with a full 3D analysis, it is possible to get reliable solutions for nonlinear response of pile foundations with greatly reduced computational effort (Finn & Wu, 1994). The results are very accurate for excitation due to horizontally polarized shear waves propagating vertically. A full description of this method, including numerous validation studies, has been presented by Wu & Finn (1997a,b). The method is incorporated in the computer program PILE-3D. The quasi-3D analysis program, PILE-3D, provides the basis for the direct computation of time-dependent pile impedances, while taking into account nonlinear behaviour of the soil, pile-soil-pile interaction, hysteretic and radiation damping, and kinematic and structural inertia effects. This capability allows a more rational basis for selecting single-valued effective spring and damping constants for use in commercial software.

The most common approach to the analysis of pile foundations is to use nonlinear Winkler springs with dashpots to simulate soil stiffness and damping. Some organizations, such as the American Petroleum Institute (API), give specific guidance for the development of nonlinear soil resistance-deflection p-y curves with depth as a function of soil properties. The API (1993) recommendations are based on static or slow cyclic loading tests. In this case, the stiffness must be evaluated at specified displacements of the pile head which introduces an arbitrary element into the stiffness evaluation. To estimate the stiffness properties of a group of piles, elastic interaction factors are often used. Some organizations use elastic static factors, others use the limited range of elastic dynamic interaction factors that are available and supplement these when necessary by static interaction factors.

In this paper, the API p-y curves will be evaluated using data from the seismic response of a model instrumented pile in centrifuge tests at low and strong levels of shaking. The pile is then analyzed using PILE-3D to show the effects of strong excitation on the distribution of moduli, moments along the pile, and on the time histories of foundation stiffnesses. Comparisons of test data and computed responses shed some light on the limitations of the approximate techniques for the seismic response analysis of foundation-superstructure systems.

The California Department of Transportation (CALTRANS) has adopted a simplified model of a bridge-foundation system that facilitates taking nonlinear soil behaviour and inertial interaction between foundation soils and superstructure into account (Abghari & Chai, 1995). The foundation pile group is represented by a single pile of the group that supports a concentrated mass corresponding to its proportion of the total static force carried by the group. The mass is supported in a single degree of freedom (SDOF) system with a period equal to the first mode period of the superstructure in the direction of interest, assuming fixed supports. The function of the SDOF system is to model approximately the inertial contribution of the superstructure to the response of the pile foundation. The interaction between the soil and the pile is modelled using Winkler springs and dashpots with properties defined by p-y curves. The pile head is maintained fixed against rotation. This model is designated the simplified-pile-superstructure model (SPSM). This model will be evaluated in detail later.

2 SIMPLIFIED 3D SEISMIC ANALYSIS OF PILE FOUNDATIONS

A brief outline is given of the basis of the PILE-3D analysis. For details, the reader is referred to Wu & Finn (1997a,b). The basic assumptions of the simplified 3D analysis are illustrated in Fig. 1. Under vertically propagating shear waves the soil undergoes primarily shearing deformations in the xOy plane except in the area near the pile where extensive compressional deformations may develop in the direction of shaking. The compressional deformations also generate shearing deformations in yOz plane. Therefore, the assumptions are made that dynamic response is governed by the shear waves in the xOy and yOz planes and compressional waves in the direction of shaking, Y. Deformations in the vertical direction and normal to the direction of shaking are neglected. Comparisons with full 3D elastic solutions confirm that these deformations are relatively unimportant for horizontal shaking (Wu & Finn, 1997a,b) Applying dynamic equilibrium in the Y-direction, the dynamic governing equation of the soil continuum in free vibration is written as

$$\rho_s \frac{\partial^2 v}{\partial t^2} = G^* \frac{\partial^2 v}{\partial x^2} + \theta G^* \frac{\partial^2 v}{\partial y^2} + G^* \frac{\partial^2 v}{\partial z^2} \qquad (1)$$

where G^* is the complex modulus, v is the displacement in the direction of shaking, ρ_s is the mass density of soil, and θ is a coefficient related to Poisson's ratio of the soil. Piles are modelled using Eulerian beam theory. Bending of the piles occurs only in the yOz plane. Dynamic soil-pile-structure interaction is maintained by enforcing displacement compatibility between the pile and soils.

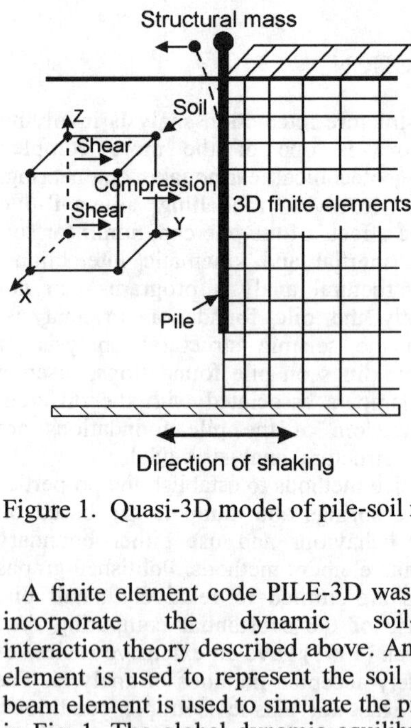

Figure 1. Quasi-3D model of pile-soil response.

A finite element code PILE-3D was developed to incorporate the dynamic soil-pile-structure interaction theory described above. An 8-node brick element is used to represent the soil and a 2-node beam element is used to simulate the piles, as shown in Fig. 1. The global dynamic equilibrium equation in matrix form is written as

$$[M]\{\ddot{v}\} + [C]\{\dot{v}\} + [K]\{v\} = -[M]\{I\} \cdot \ddot{v}_0(t) \qquad (2)$$

in which $\ddot{v}_0(t)$ is the base acceleration, {I} is a unit column vector, and $\{\ddot{v}\}$, $\{\dot{v}\}$ and {v} are the relative nodal acceleration, velocity and displacement, respectively.

The loss of energy due to radiation damping is modelled using the method proposed by Gazetas et al. (1993) for elastic response in which a velocity proportional damping force F_d per unit length is applied along the pile. Direct step-by-step integration using the Wilson-θ method is employed in PILE-3D to solve the equations of motion in Equation (2). The nonlinear hysteretic behaviour of soil is modelled by using an equivalent linear method in which properties are varied continuously as a function of soil strain as originally proposed by Seed & Idriss (1970). Additional features such as tension cut-off and shearing failure are incorporated in the program to simulate the possible gapping between soil and pile near the soil surface and yielding in the near field.

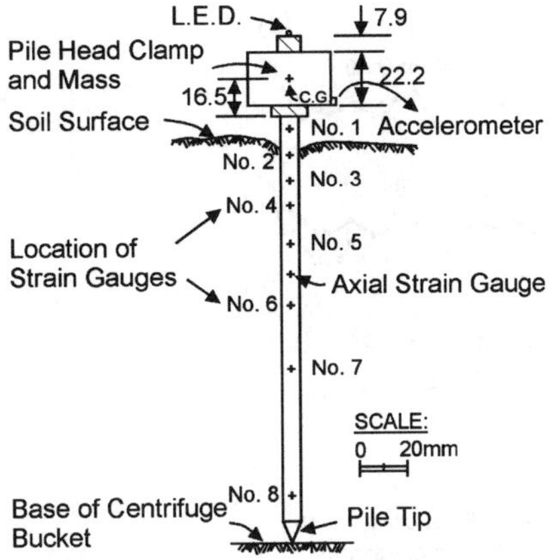

Figure 2. The layout of the centrifuge test for a single pile.

3 SEISMIC RESPONSE ANALYSIS OF A SINGLE PILE

3.1 *PILE-3D Analysis*

PILE-3D was used to analyze the seismic response of a single pile in a centrifuge test which was conducted at the California Institute of Technology (Finn & Gohl, 1987). Fig. 2 shows the instrumented soil-pile-structure system used in the test. The system was subjected to a nominal centrifuge acceleration of 60 g. A horizontal acceleration record with a peak acceleration of 0.158g is input at the base of the system. The distribution of shear moduli was measured prior to shaking using bender elements, while the centrifuge was in flight. Therefore, accurate initial stiffnesses of the pile foundation could be calculated. Fig. 3 shows the finite element mesh used in the PILE-3D analysis.

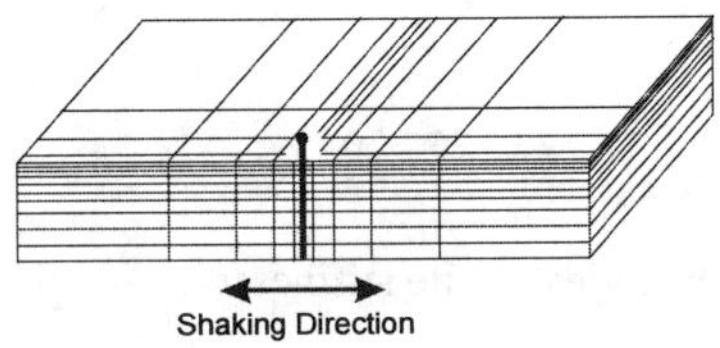

Figure 3. Finite element mesh for analysis of single pile.

The dynamic shear strains generated by strong seismic shaking causes changes in shear moduli and damping ratios. This effect is illustrated in Fig. 4 by the distribution of shear moduli at a depth of 2.1 m in the soil around the pile at a time T = 12.58 secs. The distribution of shear moduli and damping are both time- and space-dependent. This dependence results in a corresponding time dependence of the global stiffnesses of the pile head. Dynamic impedances as a function of time were computed using the time- and space-dependent nonlinear shear moduli. Harmonic loads with an amplitude of unity were applied at the pile head, and the resulting equations were solved to obtain the complex valued pile impedances. The impedances were evaluated at the surface of the sand. The time histories of the stiffnesses are shown in Fig. 5a. The effects of structural inertia on the lateral stiffness, K_{vv}, is shown in Fig. 5b.

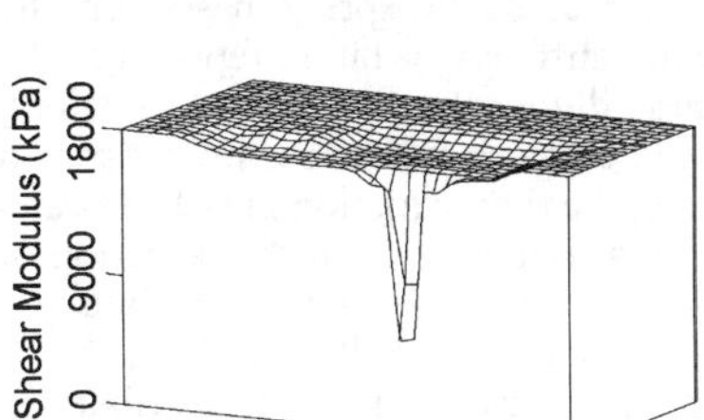

Figure 4. Distribution of shear moduli around pile at a depth of 2.1 m at time 12.58 s into earthquake.

The dynamic stiffnesses (real parts of the impedances) of the single pile during the specified input motions decreased dramatically as the level of shaking increased (Fig. 5). The dynamic stiffnesses experienced their lowest values between about 10 and 14 seconds, when the maximum accelerations occurred at the pile head. It can be seen that the lateral stiffness component K_{vv} decreased more than the rotational stiffness $K_{\theta\theta}$ or the coupled lateral-rotational stiffness $K_{v\theta}$. The equivalent damping coefficients also increased with the level of shaking.

The time histories of stiffnesses in Fig. 5, show clearly the difficulties in selecting a single spring value to represent the lateral or rotational stiffness of a pile foundation. To make a valid selection, one

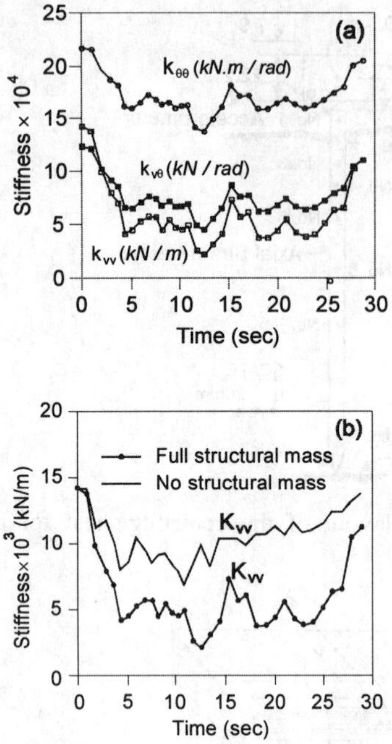

Figure 5. Time histories of pile stiffnesses.

would need to know which segment of the ground motion was most critical in controlling the seismic response of the structure. A spring based on the minimum lateral stiffness would represent the mobilized stiffness during the period of very strong shaking and would be more critical for longer period structures. Clearly, the time variation in stiffness and damping provides a useful guide to the selection of the discrete springs and dashpots required by commercial structural analysis software.

The computed and measured moment distributions along the pile at the instant of peak pile head deflection are shown in Fig. 6. The moments were computed by PILE-3D and agree quite well with the measured moments.

3.2 *Analysis Using API P-Y Curves*

The distribution of bending moments was also determined using the p-y curves recommended for sands by the American Petroleum Institute (API, 1993). These p-y curves are defined by the equation,

$$p = 0.9\,p_u \tanh\left[(kHy)/(0.9\,p_u)\right] \qquad (3)$$

where p_u is the ultimate bearing capacity at depth H, k is the initial modulus of subgrade reaction, y is the lateral deflection, and H is the depth. The relative density of the sand surrounding the pile is $D_r = 38\%$.

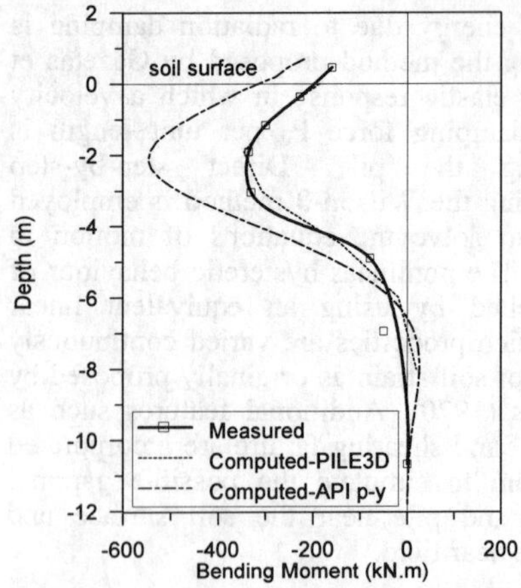

Figure 6. Comparison of measured and computed bending moments.

This corresponds to a k of approximately 15000 kN/m^3 according to the API recommendations. An analysis using this value of k showed that the p-y curves were much too stiff. The distribution of moments for k = 15000 kN/m^3 is shown in Fig. 6. A reasonable approximation to the measured peak moment in the pile is obtained using k = 2500 kN/m^3, which is only 1/6 of the value recommended by API (1993). In another test in the same sand run at very low acceleration levels, with a peak acceleration of 0.04 g, the API stiffness k = 15000 kN/m, gives a very good approximation to the measured bending moments (Fig. 7). The response in this case was almost elastic and the initial stiffness controls the response. These results suggest that the initial stiffness of the API p-y curves is reasonable, but that the curves do not head away fast enough from the initial tangent at the origin. Therefore, stiffness at close to the initial value is being mobilized over too large a displacement range during strong shaking.

The analysis was repeated following the suggestion of Gazetas & Dobry (1984) that the k value should vary with the depth H, with $kH = \delta E_{max}$ (Young's modulus). The shear wave velocity distribution in the centrifuge test was measured during flight and, therefore, accurate measurements of the shear modulus, G, were available. E_{max} was estimated for two values of Poisson's ratio, $\nu = 0.2$ and 0.3. The resulting moment distributions are shown in Fig. 8 for $\delta = 1.0$. This approach to estimating the initial stiffness, gives a better approximation to the measured bending moments, when using the standard API p-y curves.

In view of the widespread use of p-y curves in engineering practice, it would seem necessary to investigate their reliability by more centrifuge tests

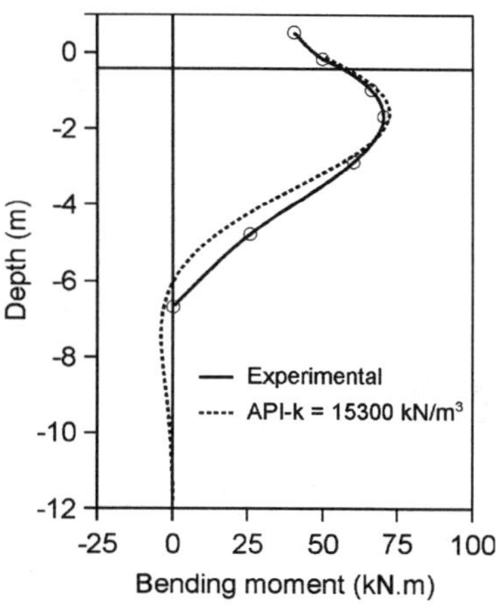

Figure 7. Comparison of measured and computed pile moments for near elastic response using API procedure.

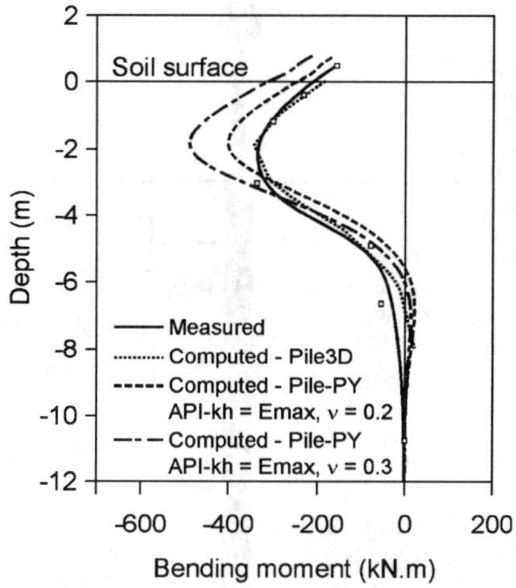

Figure 8. Comparison of computed and measured moments when kH = E_{max}.

and dynamic field studies. It should be noted that the API p-y curves have been shown to be somewhat unreliable also in predicting the response to static and slow cyclic loading tests in the field (Murchison & O'Neill, 1984, Gazioglu & O'Neill, 1984).

4 SPSM AND FULL GROUP ANALYSES

4.1 *Fixed Head Analysis*

The CALTRANS SPSM model, together with soil and pile properties is shown in Fig. 9. The seismic response analysis based on p-y curves has been shown to be potentially unreliable. However, the representative pile concept itself is a very attractive computational feature. Therefore, the concept is evaluated below by treating the foundation soil as a nonlinear continuum in both the representative pile and full pile group analyses using PILE-3D. In this way, the representative pile concept can be tested against full group analysis under identical computational conditions.

Fig. 10 shows the comparison of bending moment profiles for the two different pile groups and the single pile. The piles in the pile groups are spaced at two diameters, centre to centre. The bending moment profiles from the (2×2) and (4×4) analyses do not deviate much from the moment profile in the single pile analysis. The difference in the peak moment is less than 8%. The shear force profiles show similar behaviour. It is evident from Fig. 10 that the group effect does not appear to be a significant factor in the response of the different pile foundation-superstructure systems analyzed here.

CALTRANS adjusts the results of the SPSM for group effects using a group reduction factor.

Table 1 shows the group factor, calculated as a ratio of group stiffness over the single pile stiffness multiplied by the number of piles in the pile group. It is 0.61 for (2×2) pile group and 0.31 for (4×4) pile group under elastic conditions. The group effect corresponding to the minimum stiffnesses for the (2×2) pile group is 0.76 and for the (4×4) pile group is 0.41. The group effect increases with the number of piles in the pile group as expected. Also, there is a reduction in the group effect during the periods of strong shaking at which the pile foundations reach their minimum stiffnesses. This is in keeping with the general perception that the range of pile to pile interaction is reduced as the soil behaviour becomes nonlinear. Despite the group interaction effects, the bending moment and shear force responses were not significantly affected. The reason for this behaviour is explained below.

The fundamental frequencies of the three pile foundation systems were determined corresponding to the initial stiffnesses and the minimum stiffnesses that occur during the time of strongest shaking. The frequencies are shown in Table 2. Although the group effects on foundation stiffnesses of the pile groups were significant, the difference in the global system frequencies of the pile groups were not significantly different from the frequency of the single pile system under either elastic or nonlinear response. The maximum difference is about 15%. This is due to the fact that the global system frequencies result from the combined stiffnesses of the superstructure and pile foundation rather than the stiffness of the pile foundation alone. The similarity in frequencies of the different foundation models is

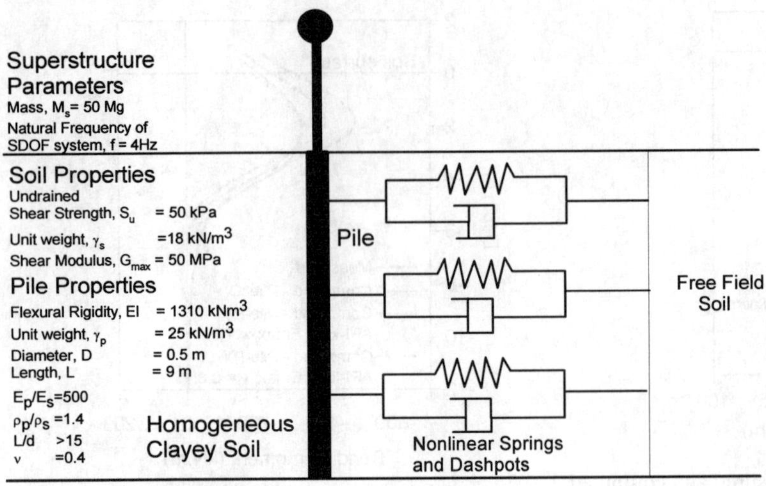

Figure 9. Simplified pile-superstructure model.

responsible for the similarity in outputs. In this study, the superstructure stiffness dominates the system frequency.

The results of the evaluation study suggest that the SPSM model concept for the analysis of pile foundations is only valid if the system frequency of the representative pile-superstructure model and the frequency of the full foundation-superstructure model are approximately the same. This is likely to occur only when the stiffness of the support structures of the bridge dominates the system frequency. For this to happen, the support structures must be much more flexible than the pile foundation in the degree of freedom under consideration. Parametric studies using many earthquakes with different frequency contents would be desirable to

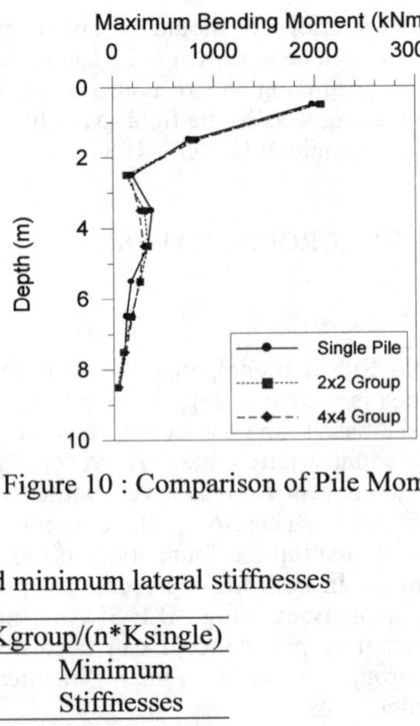

Figure 10 : Comparison of Pile Moment Profiles.

Table 1 : Group effect corresponding to initial elastic and minimum lateral stiffnesses

Type of Pile Foundation	Group Effect,Kgroup/(n*Ksingle) Initial Elastic Stiffnesses	Minimum Stiffnesses
(2×2) Group Pile, $K_{2\times2}$	0.61	~0.76
(4×4) Group Pile, $K_{4\times4}$	0.31	~0.41

*n is the total number of piles in the pile group.

Table 2 : First mode frequency of the superstructure-pile foundation system

Type of Structure	First Mode Frequency (Hz) Initial	Minimum
Single Pile- Superstructure	3.66	1.89
(2×2) Group Pile -Superstructure	3.45	1.82
(4×4) Group Pile -Superstructure	3.07	1.61

explore how full foundation system frequencies compare with representative pile system frequencies for typical bridges and foundation soils under strong shaking. The study would allow the reliability of the representative pile concept to be evaluated properly.

If PILE-3D or any other program is used that can treat the soil as nonlinear continuum and model the whole pile group conveniently enough for engineering practice, the problems associated with system frequencies and p-y curves can be avoided.

4.2 *Effects of Pile Cap Rotation*

Following the CALTRANS procedure, the evaluation analyses were conducted with the assumption that the pile cap was fully restrained against rotation. The analyses were repeated without preventing rotation of the pile cap. In these latter analyses, the only restraint on the pile cap is provided by the moments on the pile cap caused by the axial forces in the piles.

Profiles of maximum bending moments are shown in Fig. 11. In this case there is very poor agreement between results of analyses based on the representative fixed head single pile and the full group analyses, even though the foundation soils are treated identically for all three systems. It would seem that the differences in computed rotations may be primarily responsible for the differences in moments from the representative pile and full pile foundation group analyses. Therefore the

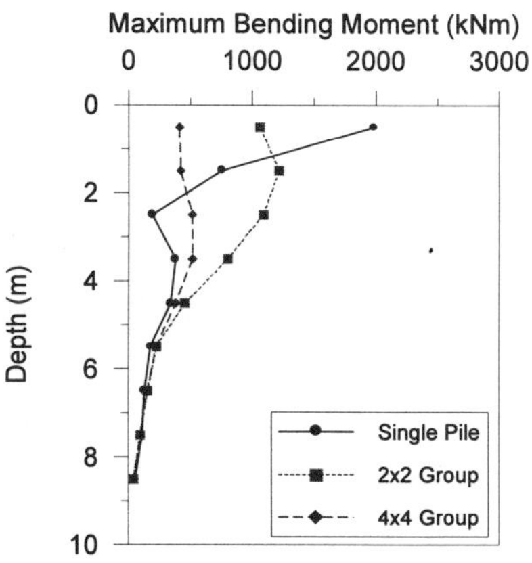

Figure 11. Comparison of pile moment profiles.

representative pile concept is probably not appropriate for small pile groups that cannot effectively restrain the pile cap. However for large pile groups, the restraint of the pile cap by the axial forces in the pile may be sufficient to approximate the fixed condition. Parametric studies to establish a limiting pile group size are necessary.

5 CONCLUSIONS

The p-y method which is often used for analyzing the seismic response of pile foundations was evaluated using data from centrifuge model tests on a pile carrying a superstructure mass. The pile was excited by both low and high levels of shaking. In this study, the p-y curves recommended by the American Petroleum Institute (API, 1993) were used. The evaluation suggests that these p-y curves are much too stiff for the analysis of strong shaking during earthquakes. When these stiffnesses were used to analyze low levels of shaking, the agreement between measured and computed moment distributions in the pile were very good. Therefore, the tentative conclusion is that the initial API stiffness is reasonable, but the p-y curves need to bend away faster from the initial tangent at the origin for seismic analysis of strong shaking.

Structural inertia was shown to have a significant effect on pile stiffness. Inertia can be included directly in a PILE-3D analysis. CALTRANS has proposed a simple computational model for including inertia effects. It consists of a single pile from the pile group carrying its proportion of the structural mass. The dynamic analysis is conducted assuming pile head fixity against rotation. This model was evaluated by single pile and pile group analyses using PILE-3D. It appears that the model is valid only when the system frequencies of the single pile and the pile group-superstructure systems have approximately the same system frequencies. This requires that the frequency of the superstructure system dominate the global system frequency. This occurs when the pile foundation stiffness is much greater than the support system stiffness. Even when this requirement is met, the pile foundation must be able to provide the rotational restraint to approximate the fixed condition. Studies of 2×2 and 4×4 pile groups show that they can not provide the required rotational restraint. When the rotational restraint is inadequate the single pile concept gives very different moments and shears from the pile group analyses.

6 ACKNOWLEDGEMENTS

This work was supported by the National Science and Engineering Research Council of Canada by Grant No. 1498 and by Anabuki Komuten through the Anabuki Chair of Foundation Geodynamics. The assistance of Noboru Fujita in the preparation of the paper is gratefully acknowledged.

7 REFERENCES

Abghari, A. and J. Chai 1995. Modeling of soil-pile-superstructure interaction for bridge foundations. Performance of Deep Foundations Under Seismic Loading, ASCE Geot. Special Publ. No. 51, pp. 45-59.

API 1993. Recommended practice for planning, designing, and constructing fixed offshore platforms. American Petroleum Institute, Washington, D.C.

Finn, W.D. Liam and WB. Gohl 1987. Centrifuge model studies of piles under simulated earthquake loading from dynamic response of pile foundations - experiment, analysis and observation. ASCE Convention, Atlantic City, New Jersey, Geotechnical Special Publication No. 11, pp. 21-38.

Finn, W.D. Liam and G. Wu 1994. Recent developments in dynamic analysis of piles. Proc., 9th Japan Conference on Earthquake Engineering, Tokyo, Japan, Vol. 3, pp. 325-330.

Gazetas, G. and R. Dobry, R 1984. Horizontal response of piles in layered soils. J. of Geotech. Eng., Vol. 110, No. 1, Jan., pp. 20-40.

Gazetas, G., K. Fan and A. Kaynia 1993. Dynamic response of pile groups with different configurations. Soil Dynamics and Earthquake Engineering,12:239-257.

Gazioglu, S. M. and M.W. O'Neill 1984. An evaluation of p-y relationships in cohesive soils. Proc. of the ASCE Symposium on Analysis and Design of Pile Foundations, ASCE National Convention, San Francisco, California, Oct. 1-5, 1984, Edited by J.R. Meyer, pp. 192-213.

ASCE, Vol. 101, GT5:423- 438.

Murchison, J.M. and M.W. O'Neill 1984, An evaluation of p-y relationships in cohesionless soils. Proc. of the ASCE Symposium on Analysis and Design of Pile Foundations, ASCE National Convention, San Francisco, California, Oct 1-5, 1984, Edited by J. R. Meyer, pp. 174-191.

Novak, M. 1991. Piles under dynamic loads. State of the Art Paper, 2nd International Conference Recent Advances in Geotechnical Earthquake Engineering and Soil Dynamics, University of Missouri-Rolla, Rolla, Missouri, Vol. III, pp. 250-273.

Seed, H.B. and I.M. Idriss 1970. Soil moduli and damping factors for dynamic response analysis. Report #EERC70-10, Earthquake Engineering Research Center, Berkeley, CA.

Wu, G. and W.D. Liam Finn 1994. PILE-3D - prototype program for nonlinear dynamic analysis of pile groups. Dept. of Civil Engineering, University of B.C., Vancouver.

Wu., G and W.D. Liam Finn 1997a. Dynamic elastic analysis of pile foundations using finite element method in the frequency domain. Canadian Geotechnical J., Vol. 34, 1:34-43.

Wu., G and W.D. Liam Finn 1997b. Dynamic nonlinear analysis of pile foundations using finite element method in the time domain. Canadian Geotechnical J., Vol. 34, 1:44-52.

Computer Methods and Advances in Geomechanics, Desai et al. (eds) © 2001 Balkema, Rotterdam, ISBN 90 5809 183 X

Nonlinear dynamic analysis of pile foundation: Effect of plasticity of soil

B.K. Maheshwari, K.Z. Truman & P.L. Gould
Washington University, St. Louis, Mo., USA

ABSTRACT: Material nonlinearity of soil such as caused by plasticity and work hardening is rarely considered for the dynamic analysis of pile foundations. In this paper both of these are considered using an advance soil model HiSS. This model is incorporated in a time domain, finite element technique. The algorithm developed is verified with available analytical results. Linear and nonlinear responses of a single end bearing pile are compared.

1 INTRODUCTION

Much of the reported research in the field of dynamic analysis of pile foundation assumes linear behavior of soil media. Much work i.e. Kaynia & Kausel (1982), Gazetas (1984) has been carried out for the analysis of pile groups. But in these analyses as well as by most of the other researchers the nonlinear behavior of the soil media is neglected. This is due to the complexity involved with nonlinear analysis as well as limited computer resources.

To date most of the researchers (in this field) use the frequency domain, which allows linear behavior of soil or at most an equivalent linear technique. Maheshwari & Watanabe (1998) used the equivalent linearization technique to include the material nonlinearity of the soil in the frequency domain. But the level of shear strain in the soil media for which this may be applicable is far lower than those caused by a strong earthquake. Such strong excitation causing severe nonlinearity warrants an analysis to be carried out in real time. Further the possible inhomogeneous nature of soil media dictates use of finite elements. Attempts had been made by Cai et al. (2000) to include the plasticity of soil using a finite element technique in the time domain. But the developed algorithm had not been thoroughly verified using linear elastic analysis.

In this paper plasticity and work hardening of the soil media are considered in the dynamic analysis of the pile foundation. An advanced plasticity based model for soil media, HiSS, as used by Cai et al., is incorporated into the finite element formulation. The soil-pile system is idealized as an assemblage of solid elements with the analysis performed in the time domain.

Due to the rigorous approach using finite elements in the time domain is employed, its verification is imperative. This verification has been executed assuming elastic behavior of soil for harmonic excitation. This approach was used due to the availability of reliable results in the literature. Then linear response (assuming that soil behaves linearly) and nonlinear response (considering plasticity of soil) of a single end bearing pile is compared. Analysis techniques incorporating nonlinear response is being improved and will be extended for use with pile groups.

2 MODELING

A three-dimensional finite element model used for a soil-pile system is shown in Figure 1. The system is idealized as an assemblage of eight-node hexahedral elements. Here pile elements are assumed to be linear but they can also be nonlinear by using an appropriate constitutive relation. The pile is completely embedded in the soil and assumed to be resting on the bedrock.

Since the foundation block is assumed to be resting on bedrock, all the bottom nodes are taken as fixed. Also for simplification, nodes on the side walls of the foundation are assumed fixed and may not be exact for satisfying the radiation condition.

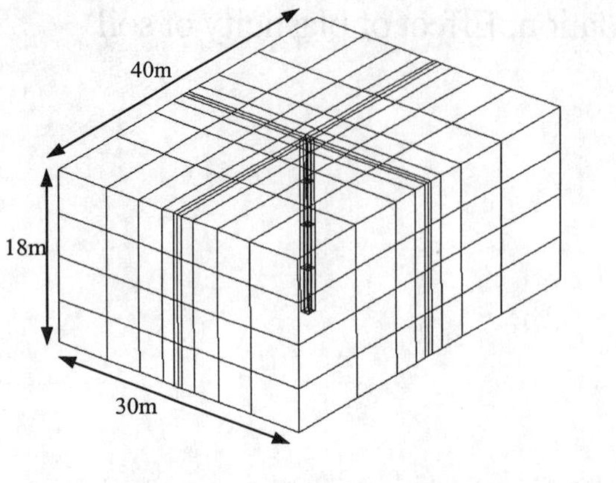

Figure 1 Finite element model for soil-pile system.

Effect of the radiation condition can be diminished by taking a large size block or applying appropriate radiation boundary conditions on the side walls. The latter approach is currently being developed. Seismic excitation is assumed to be acting on the fixed nodes and for simplification assumed to consist of vertically propagating shear waves. Since the analysis is in the time domain, a real three-dimensional excitation can be considered.

3 FORMULATION AND COMPUTERIZATION

3.1 *Governing Equation*

The loading history is digitized at each time increment. Thus the governing equation of motion at time $t+\Delta t$, Bathe (1982), will be:

$$M\,{}^{t+\Delta t}\ddot{U} + C\,{}^{t+\Delta t}\dot{U} + K\,{}^{t+\Delta t}U = {}^{t+\Delta t}R \tag{1}$$

where ${}^{t+\Delta t}R$ is external load at this time step. Employing the Trapezoidal rule (linear acceleration method) of integration, the velocity and displacement at time $t+\Delta t$ can be expressed as:

$${}^{t+\Delta t}\dot{U} = {}^{t}\dot{U} + \frac{\Delta t}{2}({}^{t}\ddot{U} + {}^{t+\Delta t}\ddot{U}) \tag{2a}$$

$${}^{t+\Delta t}U = {}^{t}U + \Delta t\,{}^{t}\dot{U} + \frac{\Delta t^2}{4}({}^{t}\ddot{U} + {}^{t+\Delta t}\ddot{U}) \tag{2b}$$

Using equations 2a and 2b first acceleration and velocity at time $t+\Delta t$ are expressed in terms of displacement ${}^{t+\Delta t}U$ and known quantities at time t, then equation 1 is solved for displacement ${}^{t+\Delta t}U$ as follows:

$$\left(\frac{4M}{\Delta t^2} + \frac{2C}{\Delta t} + K\right){}^{t+\Delta t}U = {}^{t+\Delta t}R + C\left(\frac{2}{\Delta t}{}^{t}U + {}^{t}\dot{U}\right) + M\left(\frac{4}{\Delta t^2}{}^{t}U + \frac{4}{\Delta t}{}^{t}\dot{U} + {}^{t}\ddot{U}\right) \tag{3}$$

Further simplification of equation 3 has been done according to the constitutive relation for the soil.

3.2 *Elastic Model*

Here it is assumed that the constitutive stiffness of the soil is that for the elastic case. Matrix M and K are developed as typical finite element analysis requires while C is taken as Rayleigh damping:

$$C = aM + bK \tag{4a}$$

$$\text{where } a = \frac{2D\omega_i\omega_j}{(\omega_i + \omega_j)} \quad \text{and} \quad b = \frac{2D}{(\omega_i + \omega_j)} \tag{4b}$$

where ω_i and ω_j denotes lowest and highest circular frequency of interest, which is approximately taken here as first and fifth natural frequency (rad/sec) of the soil stratum. Further D denotes the damping ratio. It is taken 5% as for a typical soil used here.

For the elastic case iterative method of solution is not required but analysis is carried out incrementally using the expression:

$${}^{(t+\Delta t)}U = {}^{t}U + \Delta U \tag{5}$$

Using equation 5, 3 is simplified for the displacement increment leading to:

$$\left(\frac{4M}{\Delta t^2} + \frac{2C}{\Delta t} + K\right)\Delta U = {}^{t+\Delta t}R - K\,{}^{t}U + M\left(\frac{4}{\Delta t}{}^{t}\dot{U} + {}^{t}\ddot{U}\right) + C\,{}^{t}\dot{U} \tag{6}$$

At each step of time, displacement is found using equation 6 and then 5.

3.3 *HiSS soil model*

When including soil plasticity the matrix K doesn't remain constant, changing at each step of time requiring an iterative scheme for the solution as:

$${}^{(t+\Delta t)}U^{(i)} = {}^{t+\Delta t}U^{(i-1)} + \Delta U^{(i)} \tag{7}$$

It is assumed that the dissipation of seismic energy through inelastic deformation overshadows the dissipation of energy through viscous damping. Therefore the velocity related damping term is neglected. With this and 7, equation 3 simplifies to:

$$\left(\frac{4M}{\Delta t^2}+{}^{t}K\right)\Delta U^{(i)}={}^{t+\Delta t}R-{}^{t}K^{t+\Delta t}U^{(i-1)}- M\left(\frac{4}{\Delta t^2}\left({}^{t+\Delta t}U^{(i-1)}-{}^{t}U\right)-\frac{4}{\Delta t}{}^{t}\dot{U}-{}^{t}\ddot{U}\right) \tag{8}$$

For determination of the matrix K, δ_0 version of HiSS is considered. This model is based on associative plasticity and assumes nonvirgin loading (loading or unloading), elastic. A brief simplified formulation related to virgin loading in HiSS is described, a detailed formulation can be found in Wathugala and Desai (1993).

Assuming the material parameter $\beta=0$ in this model as is the case for Sabine Clay, the dimensionless yield surface F can be defined as:

$$F=\left(\frac{J_{2D}}{p_a^2}\right)+\alpha_{ps}\left(\frac{J_1}{p_a}\right)^{\eta}-\gamma\left(\frac{J_1}{p_a}\right)^{2} \tag{9}$$

where J_1 is the first invariant of the stress tensor σ_{ij}; J_{2D} is the second invariant of the deviatoric stress tensor; p_a is the atmospheric pressure; γ and η are material parameters; α_{ps} is the hardening function defined in terms of plastic strain trajectory ξ_v, as:

$$\alpha_{ps}=h_1/\xi_v^{h_2} \tag{10a}$$

where h_1 and h_2 are material parameters. The increment of trajectory of volumetric plastic strain $d\xi_v$ is defined as:

$$d\xi_v=\left(\frac{1}{\sqrt{3}}\right)d\varepsilon_v^P \;\; for \;\; d\varepsilon_v^P>0 \tag{10b}$$

$$d\xi_v=0 \;\; for \;\; d\varepsilon_v^P\leq 0 \tag{10c}$$

where $d\varepsilon_v^P$ is the increment in volumetric plastic strain due to virgin loading. Constitutive stiffness tensor for the virgin loading is defined as:

$$C_{ijkl}^{VL}=C_{ijkl}^{e}-\frac{C_{ijnm}^{e}n_{nm}^{F}n_{op}^{F}C_{opkl}^{e}}{H^{VL}+n_{rs}^{F}C_{rstu}^{e}n_{tu}^{F}} \tag{11a}$$

where superscript e denotes elastic quantities and superscript VL denotes quantities associated with virgin loading. Virgin plastic modulus H^{VL} is found from the consistency condition in the theory of plasticity as:

$$H^{VL}=-\frac{\left(\frac{\partial F}{\partial \alpha_{ps}}\right)\frac{\partial \alpha_{ps}}{\partial \xi_v}\frac{n_{kk}^F}{\sqrt{3}}}{\left(\frac{\partial F}{\partial \sigma_{mn}}\frac{\partial F}{\partial \sigma_{mn}}\right)^{1/2}} \tag{11b}$$

and the unit normal tensor n_{ij}^F to the yield surface is determined by:

$$n_{ij}^F=\frac{\left(\frac{\partial F}{\partial \sigma_{ij}}\right)}{\left(\frac{\partial F}{\partial \sigma_{mn}}\frac{\partial F}{\partial \sigma_{mn}}\right)^{1/2}} \tag{11c}$$

3.4 *Seismic Loading*

The control point for seismic loading is assumed at the bedrock and thus the external force in the equation of motion is found by

$${}^{t+\Delta t}R=-MP_F\,{}^{t+\Delta t}\ddot{V}_b \tag{12}$$

where P_F is the pseudostatic response influence coefficient vector and $\ddot{V}_b$ is the bedrock acceleration at time $t+\Delta t$, due to vertically propagating shear waves.

3.5 *Computerization*

A FORTRAN program was developed to perform the analysis. Finite element programming strategies suggested by Zienkiewicz (1977), Bathe (1982), Wathugala (1990) have been incorporated in the development of the program. For nonlinear analysis three types of criteria, namely the displacement criteria, the out of balance load criteria and the internal energy criteria, are simultaneously used to check the convergence of iteration, Bathe (1982). To save space used in the storage of the matrices, a skyline storage scheme (Zienkiewicz, 1977) has been adopted. Special procedures (Wathugala, 1990) have been used to ensure the robustness of the HiSS iterative solution.

4 VERIFICATION OF THE ALGORITHM

Due to the computational effort involved in finite element dynamic analysis of pile foundation using the time domain, it is necessary to carefully check the developed algorithm. This can best be done for the elastic case considering harmonic excitation. This has been done in following two sections with bedrock motion consisting of a sinusoidal wave of amplitude 1 m/s² and frequency of excitation equal to 2 Hz. Following material properties have been used:

For Soil $E = 4147kPa; \upsilon = 0.42; \gamma = 0.047; \eta = 2.4;$
$\rho = 1610Kg/m^3; h_1 = 0.0034; h_2 = 0.78;$

For Pile $E = 25GPa; \upsilon = 0.15; \rho = 2400Kg/m^3;$

4.1 *Free-field Motion*

Here it is observed that the bedrock motion is modified due to the presence of the soil. Figure 2 shows the motion at the ground surface (in the absence of pile) due to bedrock motion. It can be seen that after a few cycles the amplitude of free field motion becomes constant due to damping of the soil. Also it is noted that this constant amplitude is approximately four times the amplitude of the bedrock motion.

Thus due to the soil stratum, the bedrock motion was amplified. It can be seen using the frequency domain analysis for a linear case, Wolf (1985), Gazetas (1984) etc. that, for this frequency and given properties of soil, the motion will be amplified approximately by the same amount. This verification can be done for different frequencies of excitation.

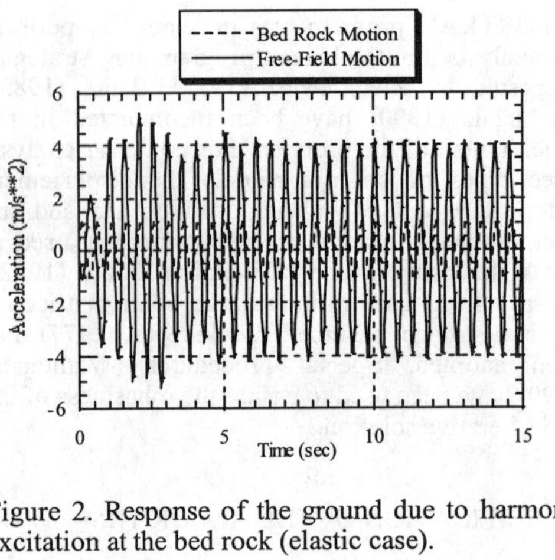

Figure 2. Response of the ground due to harmonic excitation at the bed rock (elastic case).

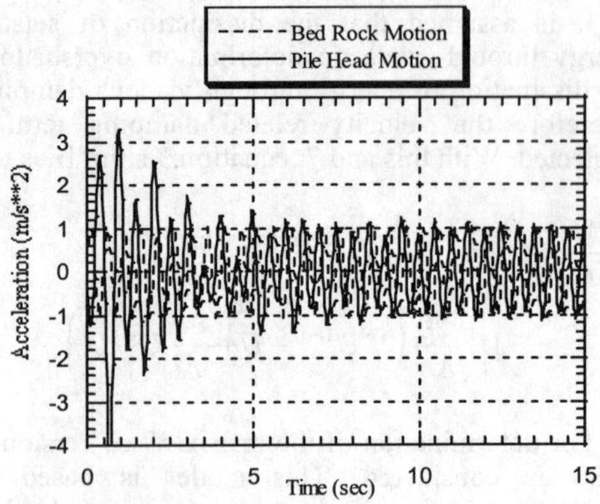

Figure 3. Response of the pile head due to harmonic excitation at the bed rock (elastic case).

4.2 *Pile Head Motion*

The above verification has also been performed in the presence of a single pile. It can be seen from Figure 3 that pile head motion (steady state) for this frequency is about the same as the bedrock motion. An interesting point is that presence of a single pile decreased the amplitude of the motion and it is quite different then the pile less free-field motion. This clearly shows the effect of soil-pile interaction, which beside other factors also depends on relative rigidity between pile and soil. In this case the ratio of rigidity of pile and soil is quite high hence influence of soil-pile interaction is significant. Further if we compare this outcome with Gazetas (1984) we find that it is almost same. Process can be repeated for different frequencies of excitation.

These examples have been used to validate the current algorithm developed for the elastic case. For the HiSS model one has to change the constitutive relation and allow iterations to convergence for the nonlinear case.

5 RESULTS AND DISCUSSION

Using the algorithm developed, linear (elastic) and nonlinear (HiSS) motion at the pile head is compared in Figure 4 for first cycle of loading for harmonic excitation. It can be seen that there is little increase in motion due to plasticity and phase of the motion is same as for the elastic case. A similar comparison is shown in Figure 5 for transient loading that also shows little increase in response due to the plasticity.

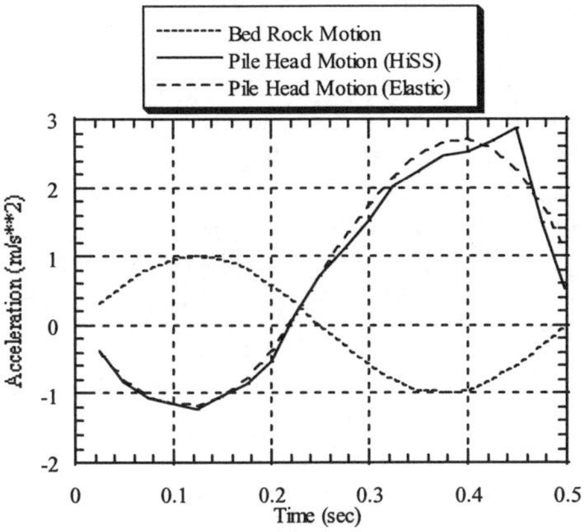

Figure 4. Comparison of linear (elastic) and nonlinear (HiSS) response due to harmonic excitation at the bed rock.

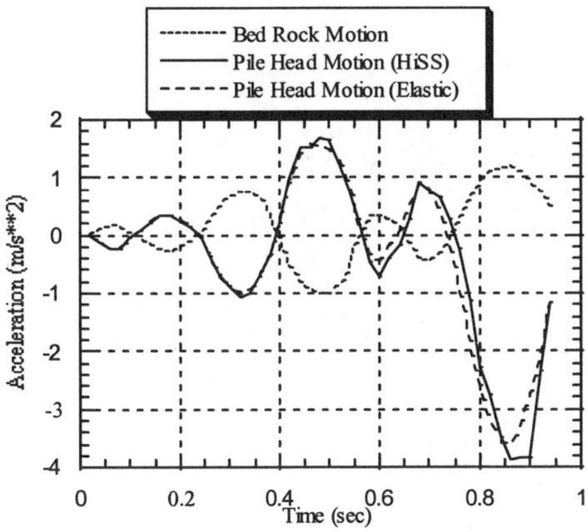

Figure 5. Comparison of linear (elastic) and nonlinear (HiSS) response due to transient loading at the bed rock.

Here comparison of linear and nonlinear motion had been done only for few time steps of loading. This comparison should be for longer time history. Presently the algorithm can have numerical instability when using virgin loading in HiSS. These problems are under consideration and once eliminated, algorithm will be more robust and accurate. Efforts have been performed to improve the robustness by Wathugala and Pal (1996) but additional improvements may required.

For initial time steps elastic and HiSS model are showing same trend is significant for the verification of the algorithm.

6 CONCLUSIONS

This paper presents a methodology for inclusion of soil plasticity and work hardening in the dynamic analysis of pile foundation. It presents verification of the proposed algorithm by computing the free-field response of a soil stratum and pile head response for the linear case.

The algorithm developed is quite versatile, can have different properties of soil and pile for different elements. Therefore, it is easy to represent possible inhomogeneous nature of the soil. Analyses being performed in the time domain it can deal with multiple support, non-uniform excitation. Being based on finite elements, it is easily extendable to pile groups and inclusive of structures as well.

ACKNOWLEDGEMENTS

The research presented here was partially supported by the Mid-America Earthquake Center under National Science Foundation Grant EEC-9701785 and the US Army Corps of Engineers. These supports are gratefully acknowledged. Authors are also thankful to Dr. Y.X. Cai for his cooperation and participation.

REFERCENCES

Bathe, K.J. 1982. *Finite element procedures in engineering analysis.* Englewood Cliffs, NJ: Prentice-Hall, Inc.

Cai, Y.X., Gould P.L. & C.S. Desai 2000. Nonlinear analysis of 3D seismic interaction of soil-pile-structure system and application. *Engineering Structures.* 22(2): 191-199.

Gazetas, G. 1984. Seismic response of end-bearing single piles. *Soil Dynamics and Earthquake Engineering.* 3(2): 82-93.

Kaynia, A.M. & E. Kausel 1982. Dynamic behavior of pile groups. *2nd Int. Conf. on Numerical Methods in Offshore Piling:* 509-532. Texas: Austin.

Maheshwari, B.K. and H. Watanabe 1998. Nonlinear seismic analysis of pile foundation. *11th European conference on Earthquake Engineering.* CD-ROM paper no. 283: 8pp. Paris.

Wathugala, G.W. 1990. Finite element dynamic analysis of nonlinear porous media with applications to piles in saturated clays. *Ph.D. dissertation.* Arizona, Tucson: Univ. of Arizona.

Wathugala, G.W. & C.S. Desai 1993. Constitutive model for cyclic behavior of clays. I: Theory. *J. Geotech. Engrg. ASCE.* 119(4): 714-729.
Wathugala, G.W. & S. Pal 1996. Improving robustness of algorithms to implement HiSS models in FEM. Proc. of the 11th Engineering Mechanics Conference, ASCE: 144-147.
Wolf J.P. 1985. *Dynamic soil-structure interaction*. Englewood Cliffs, NJ: Prentice-Hall, Inc.
Zienkiewicz, O.C. 1977. *The finite element method, 3rd ed.* New York: McGraw-Hill.

Computer Methods and Advances in Geomechanics, Desai et al. (eds) © 2001 Balkema, Rotterdam, ISBN 90 5809 183 X

A cyclic viscoelastic-viscoplastic model for clay and its application to liquefaction analysis of ground

F. Oka, T. Kodaka & Y.-S. Kim
Department of Civil Engineering, Kyoto University, Japan

ABSTRACT: In the present study, a cyclic viscoelastic-viscoplastic constitutive model based on the kinematic hardening rule was proposed, which incorporates three-elements viscoelastic component in the cyclic elastic-viscoplastic constitutive model. The proposed model can well describe both viscoelastic and viscoplastic behaviors of clay in the wide range of strain level. In order to describe actual ground behavior during earthquake, the proposed model was incorporated in the liquefaction analysis code, LIQCA (Oka et al., 1994), and numerical analysis was performed for the man-made islands with clay layers beneath sand fill such as Port Island and Rokko Island in Kobe, Japan. From the numerical analyses, we have found an effect of viscoelastic and viscoplastic components on the dynamic characteristics of sand-clay layered ground, and the proposed model can well simulate phase characteristics of seismic motion compared with the conventional elasto-viscoplastic model for clay.

1 INTRODUCTION

Up to now, many constitutive models for clay have been proposed and studied based on the elasto-plastic or elasto-viscoplastic theory and it has been recognized that the effect of time on the loading process is a salient feature. Since clay behavior is viscoelastic in the infinitesimal and/or small strain levels, a viscoelastic-viscoplastic model for clay is necessary to explain the actual deformation characteristics of clay in the infinitesimal and small strain levels.

After the 1995 Hyogoken-Nambu earthquake, liquefaction phenomena were observed in the wide area, in particular, in reclaimed land such as Port Island, Rokko Island, etc. One of the issues of liquefaction analysis of sand-clay layered ground is the effect of clay layers on liquefaction of sandy layers. As for the dynamic analysis of ground during earthquake, it is necessary to account for the strain dependent characteristics of shear modulus and damping in the small strain level. Hence, in order to perform dynamic analysis accurately it is a better way to consider viscoelastic as well as viscoplastic characteristics of clay.

Outline of the present study is as follows. Firstly, a viscoelastic-viscoplastic constitutive model is proposed which can express viscoelastic and viscoplastic characteristics of clay in the wide range of strain. Secondly, in order to study the strength and deformation characteristics of natural clay cyclic triaxial tests were carried out and then simulated results of cyclic behavior of clay are compared with cyclic triaxial test results. Finally, dynamic response of sand-clay layered ground at Port Island and Rokko Island, Kobe have been numerically analyzed using a proposed model for clay.

2 A VISCOELASTIC-VISCOPLASTIC CONSTITUTIVE MODEL

It is assumed that the deviatric strain rate tensor $\dot{e}_{ij}$ is composed of elastic, viscoelastic and viscoplastic components.

$$\dot{e}_{ij} = \dot{e}^{e}_{ij} + \dot{e}^{ve}_{ij} + \dot{e}^{vp}_{ij} = \dot{e}^{tve}_{ij} + \dot{e}^{vp}_{ij} \tag{1}$$

where $\dot{e}^{e}_{ij}$: elastic component, $\dot{e}^{ve}_{ij}$: viscoelastic component, $\dot{e}^{vp}_{ij}$: viscoplastic component, $\dot{e}^{tve}_{ij}$: total viscoelastic component. As for total viscoelastic component, a three parameter model with Voigt element and elastic spring is adopted. The deviatoric elastic strain rate component tensor is given by

$$\dot{e}^{e}_{ij} = \frac{1}{2G_1}\dot{S}_{ij} \tag{2}$$

where G_1 is the elastic shear modulus, $\dot{S}_{ij}$ is the deviatoric stress rate tensor. Deviatoric viscoelastic strain rate tensor is given by

$$\dot{e}_{ij}^{ve} = \frac{1}{\mu}(S_{ij} - 2G_2 \cdot e_{ij}^{ve}) \quad (3)$$

where μ is the viscosity coefficient, G_2 is the elastic modulus of Voigt element, S_{ij} is the deviatoric stress tensor. Consequently, the total viscoelastic deviatoric strain rate tensor $\dot{e}_{ij}^{tve}$ becomes

$$\dot{e}_{ij}^{tve} = \frac{1}{2G_1}\dot{S}_{ij} + \frac{1}{\mu}(S_{ij} - 2G_2 \cdot e_{ij}^{ve}) \quad (4)$$

For overconsolidated clay, Oka (1982) developed an elasto-viscoplastic constitutive model based on an overstress type viscoplasticity theory and the non-associated flow rule, and the viscoplastic model for overconsolidated clay was extended to the cyclic constitutive model (Oka, 1992). The overconsolidation boundary surface is defined as the boundary in the stress space between the normally consolidated($f_b \geq 0$) and the overconsolidated($f_b \leq 0$) regions.

$$f_b = \bar{\eta}^*_{(0)} + M_m^* \ln \frac{\sigma_m'}{\sigma_{mb}'} = 0 \quad (5)$$

where $\bar{\eta}^*_{(0)}$ is the relative stress ratio defined by

$$\bar{\eta}^*_{(0)} = \left\{(\eta^*_{ij} - \eta^*_{ij(0)})(\eta^*_{ij} - \eta^*_{ij(0)})\right\}^{\frac{1}{2}} \quad (6)$$

$$\sigma_{mb}' = \sigma_{mbi}' \exp(\frac{1+e}{\lambda - \kappa} v^p) \quad (7)$$

The parameter, σ_{mb}' varies with changes in the volumetric viscoplastic strain with the initial condition of $\sigma_{mb}' = \sigma_{mbi}'$. In Eqs.(5)-(7), v^{vp} is the viscoplastic volumetric strain rate, σ_{mbi}' is the initial value of consolidation yield stress σ_{mb}', e is the void ratio, κ is the swelling inex, λ is the compression index, σ_m' is the mean effective stress($\sigma_m' = \frac{1}{3}\sigma_{kk}$), η^*_{ij} is the stress ratio tensor($\eta^*_{ij} = \frac{S_{ij}}{\sigma_m'}$), $\eta^*_{(0)}$ shows the value of η^* at the end of consolidation and M_m^* is the value of $\sqrt{\eta^*_{ij}\eta^*_{ij}}$ at the maximum compression.

Considering the nonlinear kinematic hardening rule, static yield function is given as follows.

For changes in the stress ratio, the following yield function used is

$$f_y = \{(\eta^*_{ij} - \chi^*_{ij})(\eta^*_{ij} - \chi^*_{ij})\}^{1/2} - R_D = 0 \quad (8)$$

where, χ^*_{ij} is the kinematic hardening tensor and R_D is the scalar variable. The evolutional equation for kinematic hardening tensor χ^*_{ij} is given by

$$d\chi^*_{ij} = B^*(A^* de_{ij}^{vp} - \chi^*_{ij} d\gamma^{vp}) \quad (9)$$

where, de_{ij}^{vp} is the deviatoric viscoplastic strain increment tensor, A^* and B^* are material constants ($A^* = M_f^*$, $B^* = \frac{G^p}{M_f^*}$) and $d\gamma^{vp}$ is the second invariant of the viscoplastic deviatoric strain increment tensor. Namely,

$$d\gamma^{vp} = \sqrt{de_{ij}^{vp} de_{ij}^{vp}} \quad (10)$$

B^* follows Eq.(11) in order to express a non-linear characteristics of B^* with a strain level.

$$B^* = B_s + (B_0 - B_s)exp(-B_t \gamma^{vp*}) \quad (11)$$

The plastic potential function is given by

$$f_p = \{(\eta^*_{ij} - \chi^*_{ij})(\eta^*_{ij} - \chi^*_{ij})\}^{\frac{1}{2}} + \tilde{M}^* \ln(\frac{\sigma_m'}{\sigma_{ma}'}) = 0 \quad (12)$$

where,

$$\tilde{M}^* = -\frac{\eta^*}{\ln(\sigma_m'/\sigma_{ma}')} \quad (13)$$

$$\sigma_{mc}' = \sigma_{mb}' \exp\left(\frac{\bar{\eta}_0^*}{M_m^*}\right) \quad (14)$$

in which $\eta^* = \sqrt{\eta^*_{ij}\eta^*_{ij}}$, and $\bar{\eta}^*_0 = \sqrt{\eta^*_{ij(0)}\eta^*_{ij(0)}}$. Also, $\eta^*_{ij(0)}$ and σ_{mb}' are the values of η^*_{ij} and σ_m' at the end of anisotropic consolidation, respectively. $\tilde{M}^*$ can be determined by the current stress and σ_{mc}'. For the change of σ_m', another type of yield and potential functions are used (Oka, 1992). In the present study, however, we neglect second yield function for simplicity.

The viscoplastic strain rate tensor $\dot{\varepsilon}_{ij}^{vp}$ corresponding to $f_y = 0$ is assumed to be given by

$$\dot{\varepsilon}_{ij}^{vp} = \langle \Phi_{1ijkl}(F) \rangle \frac{\partial f_p}{\partial \sigma_{kl}'} \quad (15)$$

where, $\langle \Phi_{1ijkl}(F) \rangle = \Phi_{1ijkl}(F) : F > 0$, and $\langle \Phi_{1ijkl}(F) \rangle = 0 : F \leq 0$.

$$F = f_y = \left\{(\eta^*_{ij} - \chi^*_{ij})(\eta^*_{ij} - \chi^*_{ij})\right\}^{\frac{1}{2}} - R_D \quad (16)$$

in which $F = 0$ denotes the static yield function. In Eq.(15), $\Phi_{1ijkl}(F)$ is the function of F, $F = 0$ denotes the static yield function. In Perzyna's theory(1963), $\Phi_{1ijkl}(F)$ was dealt with as the scalar function. However, it is herein assumed that the first material function is the fourth order isotropic tensor function.

$$\Phi_{1ijkl}(F) = C_{ijkl}\Phi_1'(F) \quad (17)$$

The concrete shape of the material function is determined referring the previous work(Adachi and Oka,1982; Oka,1982)

$$\frac{\Phi_1'(F)}{\sigma_m'} = \exp\left\{m_o' \left\{(\eta_{ij}^* - \chi_{ij}^*)(\eta_{ij}^* - \chi_{ij}^*)\right\}^{1/2}\right\} \tag{18}$$

where, m_o' is the viscoplastic parameter and C_{ijkl} is the fourth order isotropic tensor.

$$C_{ijkl} = a\delta_{ij}\delta_{kl} + b(\delta_{ik}\delta_{jl} + \delta_{il}\delta_{jk}) \tag{19}$$

where, a and b are material constants.($C_{01} = 2b, C_{02} = 3a + 2b$)

Finally, total strain rate is obtained as:

$$\begin{aligned}\dot{\varepsilon}_{ij} &= \tfrac{1}{2G_1}\dot{S}_{ij} + \tfrac{1}{\mu}\left(S_{ij} - 2G_2 \cdot e_{ij}^{ve}\right) + \tfrac{\kappa}{3(1+e)}\tfrac{\dot{\sigma}_m'}{\sigma_m'}\delta_{ij} \\ &+ C_{01}\tfrac{<\Phi_1'(F)>}{\sigma_m'}\tfrac{(\eta_{ij}^* - \chi_{ij}^*)}{\bar{\eta}_x^*} \\ &+ C_{02}\tfrac{<\Phi_1'(F)>}{\sigma_m'}\{\tilde{M}^* - \tfrac{\eta_{mn}^*(\eta_{mn}^* - \chi_{mn}^*)}{\bar{\eta}_x^*}\}\tfrac{1}{3}\delta_{ij}\end{aligned} \tag{20}$$

3 ELEMENT SIMULATION CONSIDERING VISCOELASTIC EFFECT

Clay exhibits non-linear shear stress-strain behavior over a very wide range of shear strain. In the present study, element simulations considering a rate effect were performed by two models such as an elastic-viscoplastic model and a viscoelastic-viscoplastic constitutive model based on the kinematic hardening rule. Four undrained cyclic triaxial test results of clay have been numerically analyzed by the proposed viscoelastic-viscoplastic and the elastic-viscoplastic model. Table 1 shows basic physical properties of natural dilluvial clay used in this experiment.

Table 1. Physical properties of tested natural clay

Parameter / Case	A	B	C	D
Density, $G_s(g/cm^3)$	2.774	2.713	2.725	2.729
Water content, $W(\%)$	42.6	46.9	40.2	35.4
Liquid limit, $W_L(\%)$	44.9	48.5	39.4	41.5
Plastic index, I_p	18.7	22.0	16.7	18.7

From Figs. 1 - 3, it can be seen that stress-strain curve and stress path during cyclic loading are well described by the proposed model. In particular, strain level dependent characteristics in both compression and extension conditions are well reproduced.

Figure 4 shows a relationship between a single amplitude of axial strain and equivalent Young's modulus and a relationship between a single amplitude of axial strain and damping ratio. In Table 2, the parameters used in the calculation are shown.

From Fig. 4, it can be seen that the proposed viscoelastic-viscoplastic model is suitable for reproducing a non-linear deformation characteristics of clay in the wide range of strain. In addtion, it is seen that viscoplastic behavior is significant in the large strain level as approaching a failure state while viscoelastic behavior is notable in the small strain level. From the present study, it is revealed that viscoelastic component of the model is indispensable for explain the dynamic behavior of clay in the small strain level.

Table 2. Parameters used in model simulation

Parameter / Case	A	B	C	D
First elastic shear modulus $G_1(MPa)$	54.1	54.1	54.1	52.7
Second elastic shear modulus $G_2(MPa)$	10.8	10.8	10.8	17.6
Viscous coefficiant $\mu(kPa \cdot sec)$	6.7E+04	6.7E+04	6.7E+04	2.9E+04
Viscoplastic parameter $C_{01}(1/sec)$	1.0E-07	4.0E-07	1.0E-07	4.0E-07
Viscoplastic parameter $C_{02}(1/sec)$	1.0E-10	1.0E-10	1.2E-10	2.5E-10
Viscoplastic parameter m_0'	45	45	45	45
Stress ratio M_f^* (failure state)	1.3	1.11	1.3	1.3
Stress ratio M_m^* (critical state)	1.2	1.05	1.2	1.25
Plastic modulus parameter B_0	60	52	27	40
Plastic modulus parameter B_s	0.5	-7	3	-5
Plastic modulus parameter B_t	1.5	1	2.2	1
Compression index λ	0.1815	0.1948	0.2236	0.196
Swelling index κ	0.0215	0.029	0.0301	0.029
Poisson's ratio ν	0.45	0.45	0.45	0.49
Initial void ratio $e_{(0)}$	1.369	1.331	1.225	1.299
Frequency (Hz)	0.01	0.01	0.01	0.05

4 LIQUEFACTION ANALYSIS OF MULTI-LAYERED GROUND

Port Island and Rokko Island are man-made islands due south of Kobe central business district in Japan. It has been observed that Port Island was extensively liquefied during the 1995 Hyogoken-Nambu earthquake.

Herein, numerical study on the liquefaction phenomenon at Port Island and Rokko Island using the program code referred to as LIQCA (Oka et al., 1994) was performed to reveal that clay layers affect dynamic motion of ground, namely, different types of clay layers result in different extent of liquefaction. In the analysis, the governing equations of the soil skeleton pore water two phase material are obtained through the application of two-phase mixture theory by Biot (1962). u(displacement)-p(pore pressure) formulation is adopted for the liquefaction analysis. The finite element method is used for the spacial discretization of the equilibrium equation, while the finite difference method is used for the spacial discretization of the continuity equation. Newmark's β-method is used for the time discretization of both equations. For sands, a cyclic plastic model for sand (Oka et. al., 1999) was used in the analysis.

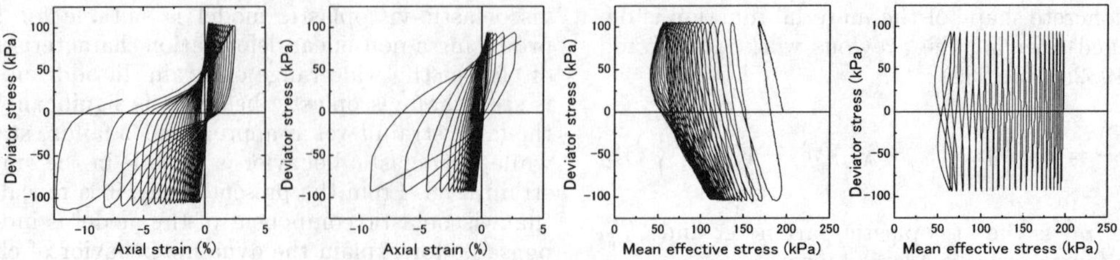

Figure 1. Stress-strain relation and stress path in Case A

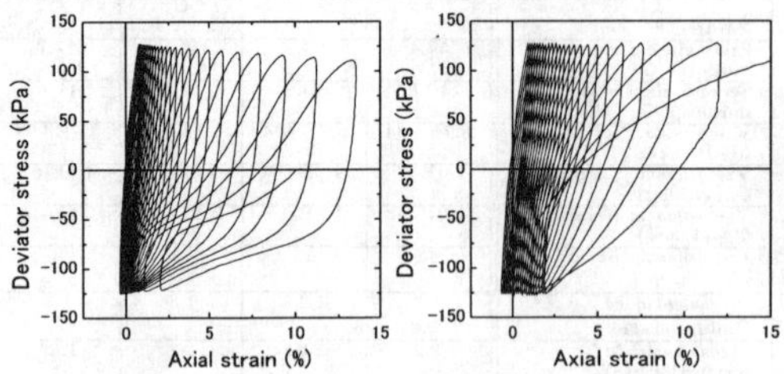

Figure 2. Stress-strain relation in Case B

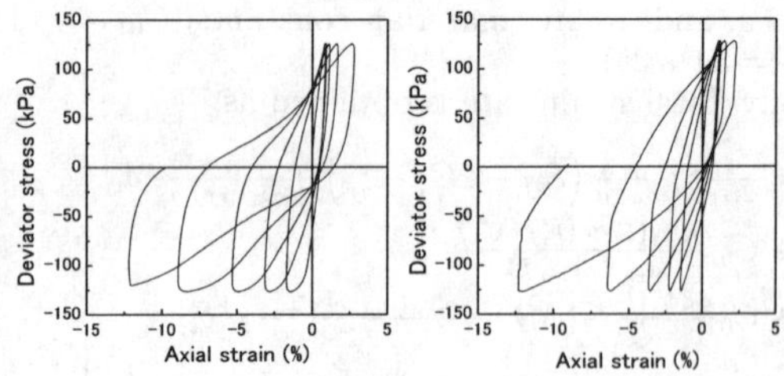

Figure 3. Stress-strain relation in Case C

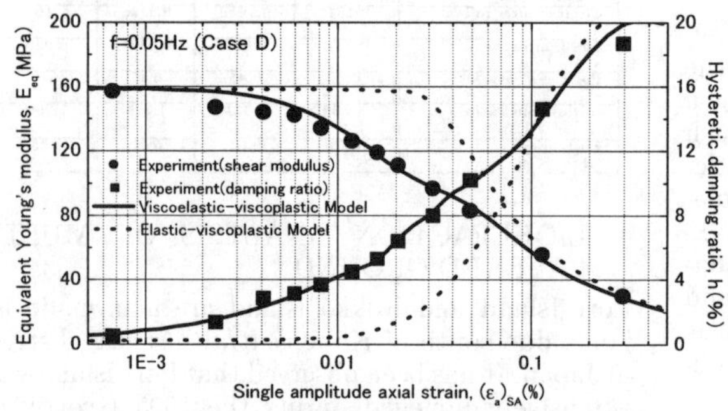

Figure 4. Strain level dependencies of shear modulus and hysteretic damping ratio in Case D

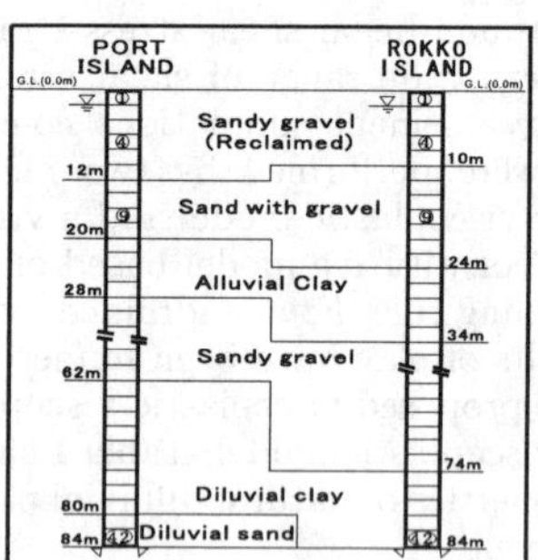

Figure 5. The model ground

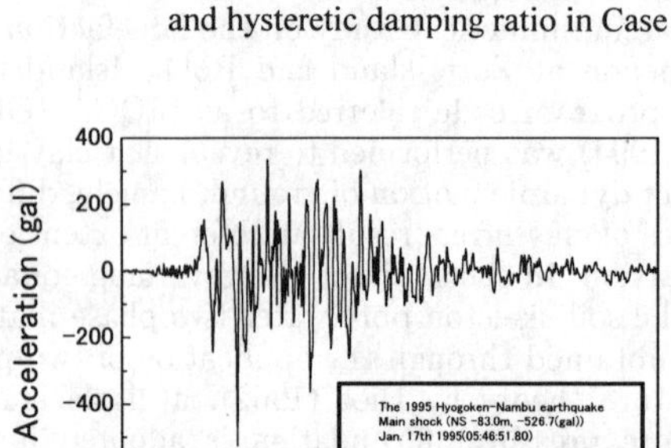

Figure 6. Acceleration record at Port Island (GL.-83.0m)

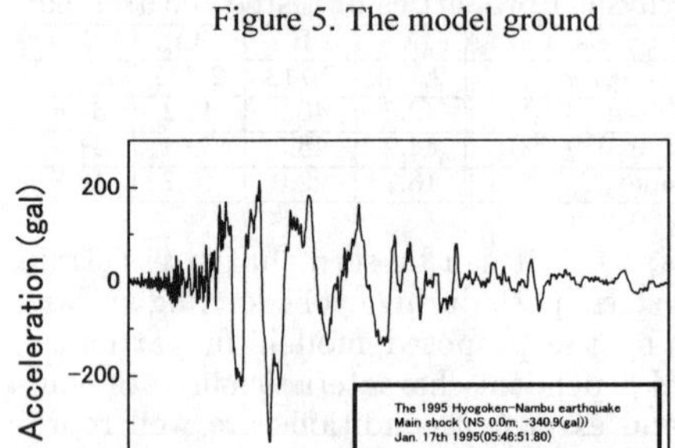

Figure 7. Acceleration record at Port Island (GL.0.0m)

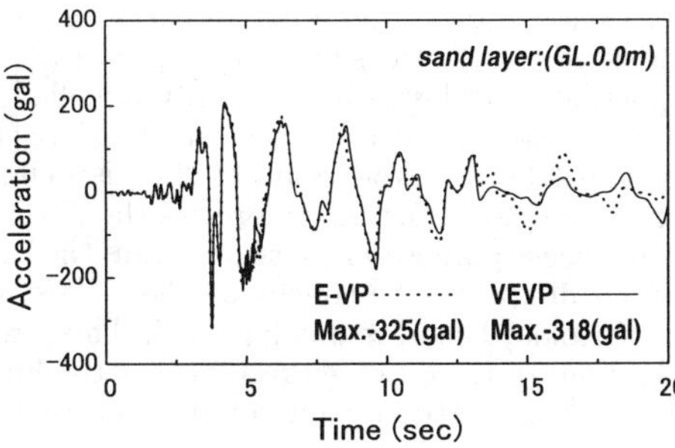

(a) Acceleration response

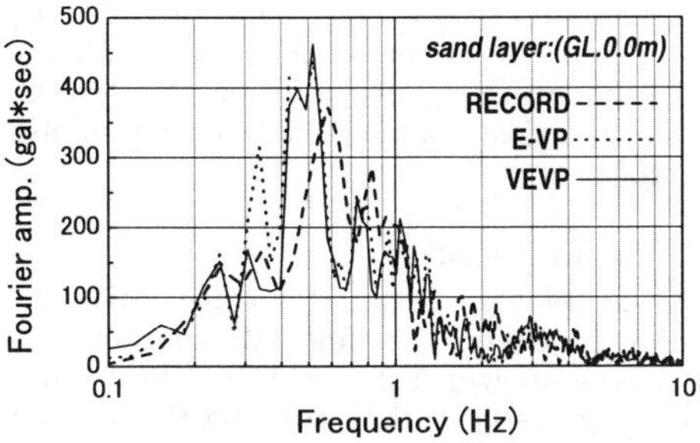

(b) Fourier spectra

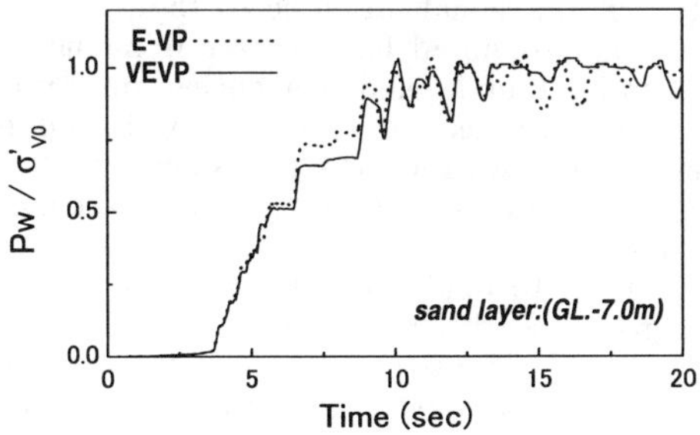

(c) Excess pore water pressure ratio and time profile (GL. -7.0m)

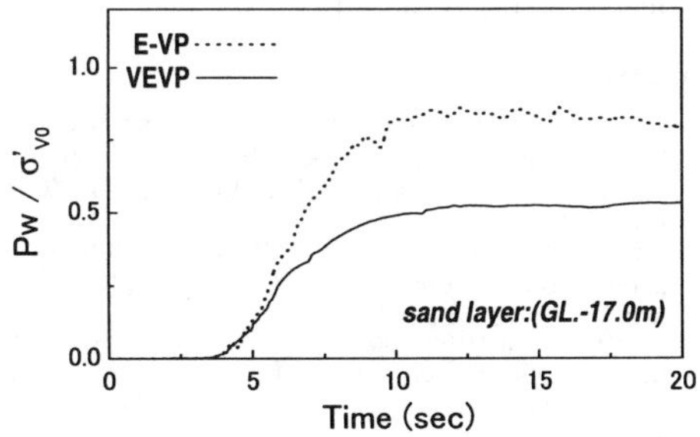

(d) Excess pore water pressure ratio and time profile (GL. -17.0m)

Figure 8. Computed results for Port Island

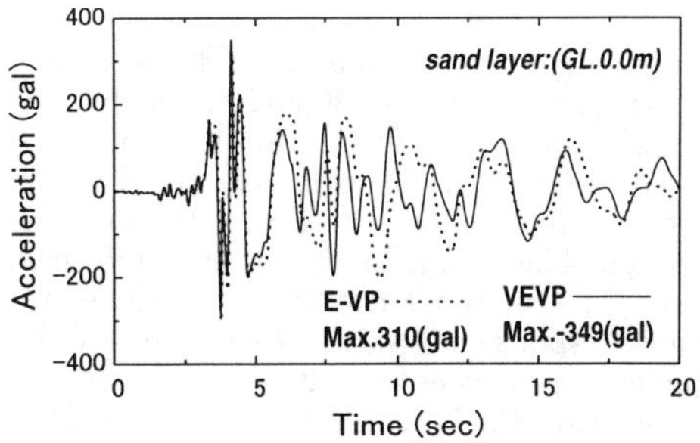

(a) Acceleration response

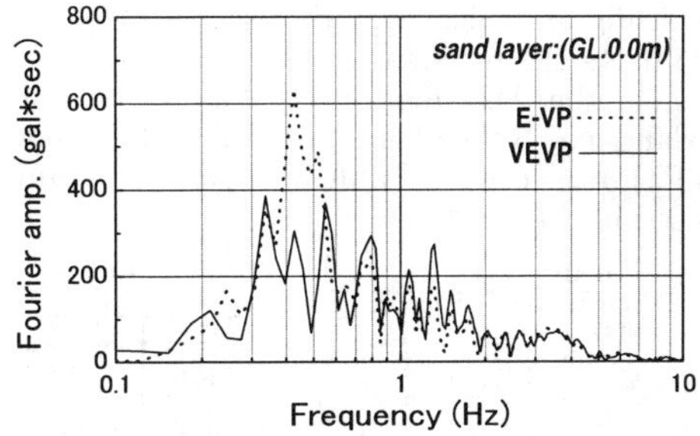

(b) Fourier spectra

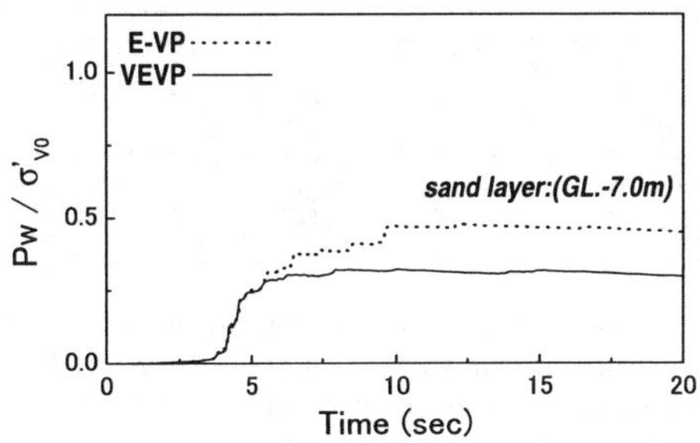

(c) Excess pore water pressure ratio and time profile (GL. -7.0m)

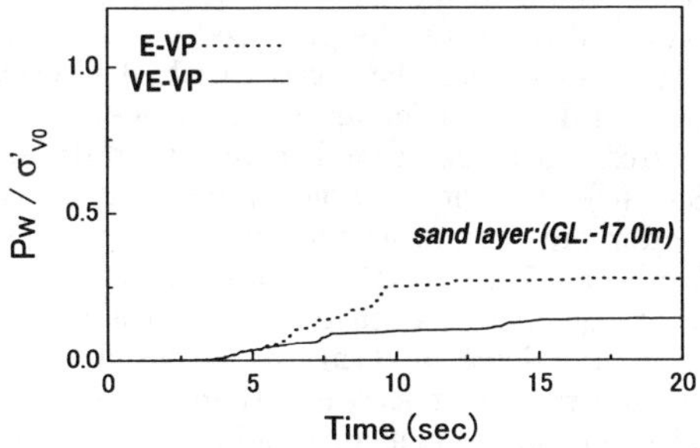

(d) Excess pore water pressure ratio and time profile (GL. -17.0m)

Figure 9. Computed results for Rokko Island

Figure 5 indicates the finite element mesh used in the analysis. This model consists of 42 horizontal layers with 42 elements and 86 nodes. The bottom nodes of soil column are fixed in both vertical and horizontal directions and both sides are fixed in the vertical direction, but are free in the horizontal direction. The ground surface is only a drained boundary. The analyzed results of the element 1 for seismic response analysis and element 4 and 9 for liquefaction analysis are discussed. In the numerical analysis, the borehole array record at Port Island obtained during the 1995 Hyogoken-Nambu earthquake was used for input seismic motion as shown in Fig. 6, which was NS diretion record at G.L.-83m. As the results of PS logging before the earthquake, the shear modulus G_0 of clay layer at Port Island and Rokko Island are 54400(kPa) and 7550(kPa), respectively. Table 3 shows the parameters used in this study.

Table 3. Parameters used in liquefaction analysis

Location	Port Island		Rokko Island	
Soil type	Sand	Clay	Sand	Clay
Parameter / Depth	GL.-7m	GL.-26m	GL.-7m	GL.-26m
$G_1(kPa)$	85907	64281	121049	26080
$G_2(MPa)$	0	21427	0	8693
$\mu(kPa \cdot sec)$	0	1.0E+07	0	1.0E+06
$C_{01}(1/sec)$	-	1.0E-08	-	1.0E-07
$C_{02}(1/sec)$	-	1.0E-10	-	1.0E-09
m'_0	-	30	-	30
B_0	6500	30	8000	35
B_s	100	1	100	1
B_t	1500	1	1500	3
M^*_m	1.20	1.4	1.20	1.4
M^*_f	1.35	1.5	1.35	1.5
λ	0.02	0.3	0.02	0.3
κ	0.00028	0.0135	0.00023	0.029
ν	0.461	0.488	0.343	0.479
e_0	0.6	1.50	0.6	1.50

Figs.8(a)-(b) show the computed accereration time histories on the surface of the ground at Port Island and the Fourier spectra of acceleration and time histories at the surface, respectively. From the comparison of the computed results with the acceleration record shown in Fig.7, it is seen that the computed result by the proposed viscoelastic-viscoplastic model for clay agrees with the acceleration record, in particular, the acceleration response from 15 to 20 second is well reproduced. Figs.8(c)-(d) show the relationship between excess pore water pressure ratio and time histories. At the depth of -17m from the ground surface, it is seen that liquefaction occurred at the time of 10 second. At the depth of -17m, the excess pore water pressure ratio increases less than 1.0 and the ground is not fully liquefied. In addition, Fig.8(d) indicates that the introduction of viscoelasticity into the model for clay depresses the development of pore water pressure. Fig.9 shows the computed results for Rokko Island, from Figs.9(a)-(b), the difference between the phase characteristics by two models for clay can be seen. The relationship between excess pore water pressure ratio and time histories at the depths shown in Figs.9(c)-(d) indicates that sand layers are not fully liquefied for both two cases. Moreover, the computed results for Rokko Island, introduction of the viscoelastic effect on the model for clay depresses the development of excess pore water pressure. From the computed results, it can be concluded that the sandy layer of Rokko Island is not liquefied. This result is consistent of the observed results that the liquefaction was not predominant compared with Port Island.

From the numerical results mentioned above, it has been found that viscous behavior of clay layers plays an important role in the liquefaction analysis. In the future, more study on the determination of viscoelastic and viscoplastic parameters is needed.

5 CONCLUSIONS

1) A viscoelastic-viscoplastic constitutive model for clay based on the kinematic hardening rule was proposed, which incorporates three-elements viscoelastic parts in the cyclic elastic-viscoplastic constitutive model.
2) In the simulation of the cyclic deformation test in infinitesimal strain level the viscoelastic approach is required for accurate dynamic analysis, and strain level dependencies, observed in the laboratory test, of shear strength and hysteretic damping ratio for natural clay can be explained only by the viscoelastic-viscoplatic constitutive model.
3) In order to apply the proposed model to actual ground conditions during earthquake, the numerical analysis was performed at the man-made islands with clay layers beneath sand fill, and it was found that viscoelastic and viscoplastic strain rate components for clay layers affect the seismic response of ground including liquefaction.

REFERENCES

Adachi, T. and F. Oka.,1982. Constitutive equations for normally consolidated clay based on elasto-viscoplasticity, Soils and Foundations, Vol.22, No.4, pp.57-70.

Biot, M.A.,1962. Mechanics of deformation and acoustic propagation in porous media, J. of Applied Physics, Vol.33, No.4, pp.1482-1498.

Oka, F.,1982. Elasto-viscoplastic constitutive equation for overconsolidated clay, Proc. Int. Symp. on Numerical Models in Geomechanics, Zurich, Balkema, pp.147-156.

Oka, F.,1992. A cyclic elasto-viscoplastic constitutive model for clay based on the non-linear hardening rule, Proc. 4th Int. Symp. on Numerical Models in Geomechanics, Balkema, 1, pp.105-114.

Oka, F., A. Yashima, T. Shibata, M. Kato and R. Uzuoka.,1994. FEM-FDM coupled liquefaction analysis of a porous soil using an elasto-plastic model, Applied Scientific Research, Vol.52, pp.209-245.

Oka, F., A. Yashima, Tateishi,A., Taguchi, Y. and Yamashita, S. 1999. A cyclic elasto-plastic constitutive model for sand considering a plastic-strain dependence of the shear modulus, Geotechnique, Vol.49, No.5, Oct., pp.661-680.

Perzyna,P.,1963. The constitutive equations for work-hardening and rate sensitive plastic materials, Proc. Vibrational Problems, Warsaw, 4(3), pp.281-290.

Computer Methods and Advances in Geomechanics, Desai et al. (eds) © 2001 Balkema, Rotterdam, ISBN 90 5809 183 X

Lateral seismic soil pressure – An updated approach

Farhang Ostadan
Bechtel National, San Francisco, Calif., USA

Ignacio Arango
Bechtel Corporation, San Francisco, Calif., USA

ABSTRACT: A new simple method is proposed to compute lateral seismic soil pressure for building walls. The method is based on extensive soil-structure interaction analyses of a wide range building sizes and soil properties using numerous earthquake ground motions. With this method seismic soil pressure profiles can be developed using the dynamic soil properties and the time histories of the design motion incorporating the soil-structure interaction effects. The method was correlated with experimental data and recent observation of seismic soil pressure and eliminates the shortcomings of the Mononobe-Okabe method commonly used for building walls.

1 INTRODUCTION

1.1 *Background*

The effect of ground motion on retaining walls was recognized by Okabe (1924) and Mononobe and Matsuo (1929) following the great Kanto Earthquake of 1923 in Japan. The method proposed by Mononobe and Okabe, currently known as the M-O method, was based on the Coulomb's theory of static soil pressure developed more than 200 years ago. In the last 30 years, a great deal of research work both in the analytical and in experimental areas has been performed to evaluate the adequacy of the M-O method or to extend the method for specific applications. The most notable contributions are by Seed and Whitman (1970), Whitman (1990,1991) and Richards and Elms (1979) all for retaining walls with flexibility to move laterally or rotate at the base. Wood (1973) and Veletsos et al. (1994a, 1994b) developed new methods for non-yielding walls. However, the M-O method continues to be widely used for retaining walls as well as building walls ignoring the basic assumptions made in the development of the method.

1.2 *The proposed Method*

Based on a series of seismic soil-structure interaction parametric studies a simplified method has been developed to compute lateral seismic soil pressure for non-yielding walls. The parametric studies were performed using the Computer Program SASSI (Lysmer et al., 1981). The model and the approach were correlated with the field experimental data from the Lotung SSI experiment (EPRI, 1989, 1991). The parametric study included a wide variation of building dimensions and aspect ratios, soil properties and a large number of ground motions records. Following the development the, it was checked against direct SSI finite element solutions for various building and soil properties to ensures accuracy. The major advantage of the method is in it simplicity recognition of the following points in the solution.

- The walls of the buildings are often of the non-yielding type. The movement of the walls is limited due to the presence of the floor diaphragms, and displacements to allow development of the limit-state conditions are unlikely to develop during the design earthquake. The proposed method considers the non-yielding conditions of the building walls.
- The frequency content of the design motion is fully considered. Use of a single parameter as a measure of design motion such as peak ground acceleration may misrepresent the energy content of the motion, at frequencies important for soil amplifications.
- Appropriate soil properties are included in the analysis. For soil dynamic problems, the most important soil property is the shear wave velocity followed by the material damping, Poisson's ratio, and density of the soil.
- The method is flexible to allow for consideration of soil nonlinear effect where soil nonlinearity is expected to be important.

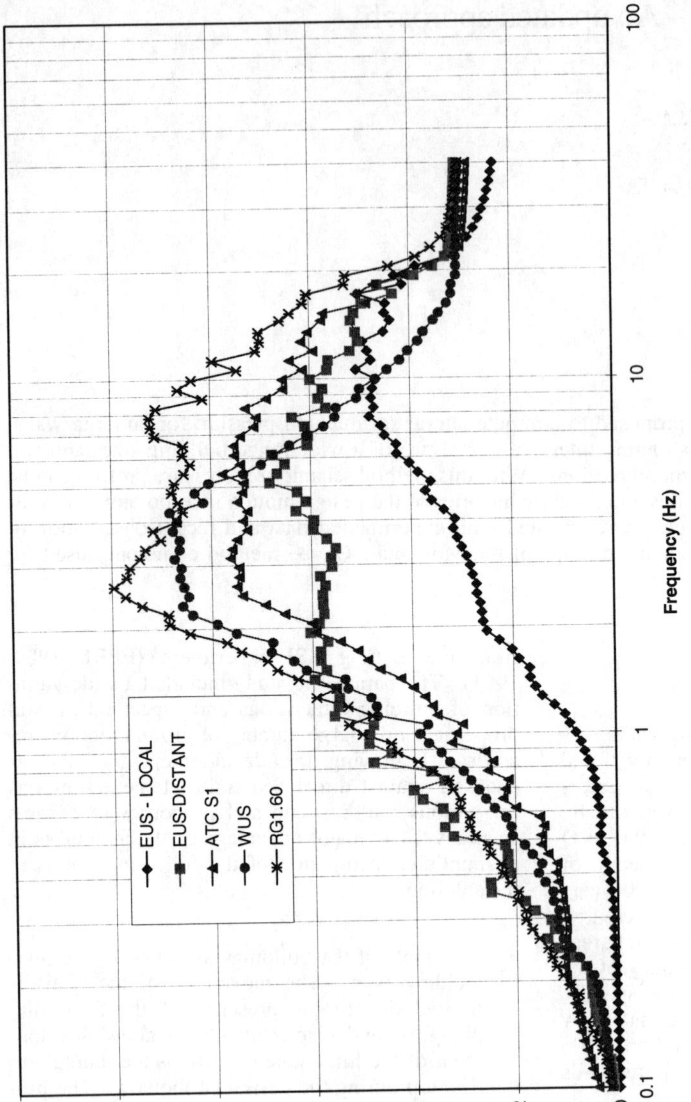

Figure 1
Acceleration Response Spectra (5%)
Comparison of Different Input Earthquake Motions

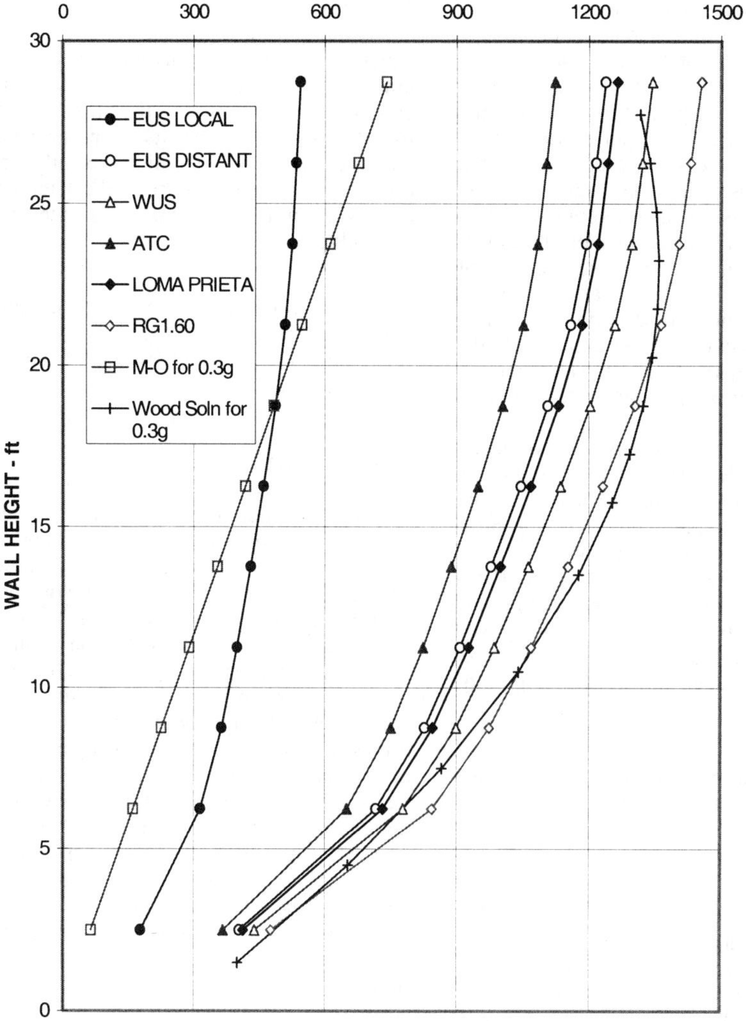

Figure 2
Maximum Seismic Soil Pressure (psf)
WALL 30 ft, Vs=1000 ft/sec, ν=1/3, β=5%
All Motions

The interaction between the soil and the building is represented. This includes consideration for the building rocking motion, amplification and variation of the motion in the soil, geometry, and embedment depth of the building.

2 COMPUTATIONAL STEPS AND COMPARISON WITH OTHER METHODS

2.1 *Computational steps*

To predict the lateral seismic soil pressure for below ground building walls, the following steps are needed:

(a) Perform free-field soil column analysis and obtain the ground response motion at the depth corresponding to the base of the wall in the free-field. The free-field soil column analysis may be performed using the Computer Program SHAKE (Schnabel et al., 1972) with input motion specified either at the ground surface or at the depth of the foundation basemat. The choice for location of control motion is an important decision that needs to be made consistent with the development of the design motion. The location of input motion may significantly affect the dynamic responses of the building and the seismic soil pressure amplitudes. Also use of the equivalent linear method in SHAKE permits consideration of soil nonlinearity effects due to seismic shear strain.
(b) Once the soil column response in terms of acceleration response spectrum has been obtained, number of simple steps involving hand calculations are followed to obtain the seismic soil pressure profile along the height o f the wall.
(c) Application of the methods to many buildings have shown that the effect of earthquake ground motion particularly the energy content at the soil column frequency corresponding to the height of the wall has a significant role in the magnitude of the soil pressure.

Detail discussion of the method and its verification are reported by Ostadan et al. (1997).

2.2 *Comparison to Other commonly used solutions*

Using the proposed method, the seismic soil pressure for a building with 30-ft high wall retaining a soil medium with a shear wave velocity of 1000 ft/sec was computed. In this case the basemat is resting on the rock preventing the building from the rocking motion. Total of six acceleration time histories (two representing an Eastern US site, one representing a Western US site, ATC firm ground motion, RG160 motion, and one record from Loma Prieta Earthquake) all scaled to 0.30g were used. The acceleration response spectra of the 6 records showing different frequency contents are depicted in Figure 1. These records were used in the simplified method and the results are compared with the Wood's solution and the M-O method for 0.3g acceleration in Figure 2. The M-O method results in the smallest pressure values. This is understood since this method relies on the wall movement to relieve the pressure behind the wall. Wood's solution generally results in the maximum soil pressure and is independent of the input motion as long as the peak acceleration is 0.3 g. The proposed method results in a wide range of pressure profiles depending on the frequency contents of the input motion, particularly at the soil column frequency. For those motions for which the ground response motions at the soil column frequency are about the same as the peak ground acceleration of the input motion, e.g., RG1.60 motion, the results of the proposed method are close to Wood's solution. Similar trend in the results is observed if sum of the lateral forces and the overturning moments from the above three methods are.

The simplified method was also extended for application to soil layered system and soil deposits with parabolic distribution of the shear modulus (Ostadan, 1997).

3 SUMMARY

A simple method was developed to predict maximum seismic soil pressures for buildings walls. The method incorporates the dynamic soil properties and the frequency characteristics of the design motion in its formulation. It was found that the controlling frequency that determines the maximum soil pressure is that corresponding to the soil column adjacent to the embedded wall of the building. The proposed method requires the use of conventionally-used simple one-dimensional soil column analysis to obtain the soil column response at the base of the wall. More importantly, this approach allows soil nonlinear effects to be considered in the process. The effect of soil nonlinearity can be important for some applications depending on the intensity of the design motion and the soil properties. Following one-dimensional soil column analysis, the proposed method involves a number of simple hand calculations in order to arrive at the distribution of the seismic soil pressure for design. The accuracy of the method relative to the more elaborate finite element analysis was verified for a wide range of soil properties, earthquake motions, and wall heights.

The method was extended to include buildings on deep soil sites. The complexity of the seismic soil pressure for such cases is compounded by the rocking motion of the structure. The rocking motion is in turn, a function of soil properties, frequency content of the design motion, and embedment ratio of the structure. A wide range of parametric studies was performed that cover many practical cases.

4 REFERENCES

EPRI (1989). "Proceedings: EPRI/NRC/TPC Workshop on Seismic Soil-Structure Interaction Analysis Techniques Using Data From Lotung, Taiwan,"

Electric Power Research Institute Publication No. NP-6154, 2 Volumes, March.

EPRI (1991). "Post-Earthquake Analysis and Data Correlation for the 1/4-Scale Containment Model of the Lotung Experiment," Electric Power Research Institute Publication No. NP-7305SL, October.

Mononobe, N. and Matuo, H. (1929). "On the Determination of Earth Pressures During Earthquakes," Proceeding of World Engineering Congress, Tokyo, Vol. 9, Paper 388.

Okabe, S. (1924). "General Theory of Earth Pressures and Seismic Stability of Retaining Wall and Dam," Journal of Japan Society of Civil Engineers, Vol. 12, No. 1.

Ostadan F., White, W. H. (1997). "Lateral Seismic Soil Pressure- An Updated Approach," Bechtel Corporation, San Francisco, California.

Lysmer et al. (1981). "SASSI - A System for Analysis of Soil-Structure Interaction," Report No. UCB/GT/81-02, University of California, Department of Civil Engineering, Berkeley, California.

Richards Jr., R. and Elms, D. (1979). "Seismic Behavior of Gravity Retaining Walls," ASCE, Journal of Geotechnical Engineering, Vol. 105, No. GT4, April.

Schnabel, P. B., Lysmer, J., Seed, H. B., "SHAKE - A Computer Program for Earthquake Response Analysis of Horizontally Layered Sites," Earthquake Engineering Research Center, University of California, Berkeley, Report No. EERC 72-12, December 1972.

Seed, H.B., and Whitman, R. V. (1970). "Design of Earth Retaining Structures for Seismic Loads," ASCE Specialty Conference on Lateral Stresses in the Ground and Design of Earth Retaining Structures, June.

Veletsos, A, and Younan, A. H. (1994a). "Dynamic Soil Pressure on Rigid Vertical Walls", Earthquake Engineering and Soil Dynamics, Vol. 23, 275-301.

Veletsos, A, and Younan, A. H. (1994b). "Dynamic Modeling and Response of Soil-Wall Systems," ASCE, Journal of Geotechnical Engineering, Vol. 120, No. 12, December.

Whitman, R. V. (1979). "Dynamic Behavior of Soils and Its Application to Civil Engineering Projects," Proceedings of 6th Pan American Conference, Lima, Peru.

Whitman, R. V., (1990). "Seismic Design and Behavior of Gravity Retaining Walls," Proceedings of a Specialty Conference on Design and Performance of Earth Retaining Structures, ASCE, Cornell University, June 18-21.

Whitman, R. V., and Christian, J. T. (1990). "Seismic Response of Retaining Structures," Pola Workshop Meeting, San Pedro, CA.

Whitman, R. V. (1991). "Seismic Design of Earth Retaining Structures," Proceedings of 2nd International Conference on Recent Advances in Geotechnical Earthquake Engineering and Soil Dynamics, March 11-15, St. Louis, Missouri.

Wood, J. H. (1973). "Earthquake Induced Soil Pressures on Structures," Doctoral Dissertation, EERL 73-05, California Institute of Technology, Pasadena, CA.

Computer Methods and Advances in Geomechanics, Desai et al. (eds) © 2001 Balkema, Rotterdam, ISBN 90 5809 183 X

Prediction of reservoir-induced earthquakes by ANN's

M.P. Romo & S.R. García
Instituto de Ingeniería, UNAM

ABSTRACT: Instrumental evidence has shown that reservoir filling of large dams can induce local earthquakes. Although many procedures have been advanced to forecast the severity of reservoir-induced earthquakes (RIE) there is still certain amount of speculation regarding the understanding of the phenomenon. In view of the uncertainties involved in the procedures available, in this paper a knowledge-based procedure is advanced as an alternative to predict the severity of RIE.

1. INTRODUCTION

Man-made lakes such as water reservoirs, may cause deformations and geological readjustments following loading, and also may change the hydrogeological regimes and conditions of the reservoir area resulting in an increased rate of erosion and sedimentation reducing the reservoir capacity, karstification, solution, seepage, etc. The other effect is the seepage of water through the reservoir bottom and consequent changes in the tectonic stress level, which may be responsible for induced earthquakes.

In an extensive sense, the prediction of reservoir-induced earthquakes should include both long-term and short-term prediction. The former refers to the analysis and assessment of the occurrence probability of induced earthquakes in the reservoir area during seismic design of the reservoir to be built and its related structures, while the latter refers to the prediction of earthquakes which may occur after filling and their tendency of becoming larger than before.

Long-term prediction consists of three problems. The first is the probability of induced earthquakes after filling of the reservoir, the second is the evaluation of occurrence location and the third is the maximum magnitude of such earthquakes.

Because of induced earthquakes are less available than ordinary tectonic earthquakes, the prediction and understanding of such earthquakes are greatly limited. The studies on Reservoir Induced Earthquakes (RIE) prediction depend upon whether the information is abundant and complete or not. There is no successful prediction method nowadays but some scientists have made proposals which are worth to be considered (Skipp, 1981; Xia et al, 1981). More recently the theory of fuzzy sets (Baoqi, 1992) and radial basis neural networks (Habibagahi, 1997) have been used to evaluate the severity of RIE. So far, the cause and mechanism of the reservoir induced earthquake is a matter of speculation and it seems that most available evidence of induced seismicity is circunstancial (Meade, 1992). Thus further research is warranted. In this study we explore the applicability of different neural network types. Also, the capabilities of several learning methods and algorithms are evaluated.

2. ARTIFICIAL NEURAL NETWORKS

Neural Networks can be used for a wide variety of learning tasks. They should be considered a mathematical tool, similar to linear regression analysis. The key feature of neural networks over regression analysis is that neural networks use non-linear mathematics and therefore can be used to model highly complex and non-linear functions.

A learning task consists of a set of data from which training examples are formed. Each training example is usually integrated by input data and the desired network response. Use of a neural network to learn this set of data (training the network) requires the user to make a number of decisions related to the use of the available training methods. Selecting the appropriate method consists of configuring a neural network and selecting algorithms (learning rules) for training the network.

A network accepts an input vector and generates a response in the form of an output vector. This network consists of a set of processing elements (nodes) connected by weights. Choosing a neural network includes selecting network architecture and, then, selecting the kind of processing elements comprising the network. The network architecture determines how the nodes are connected; the number and characteristics of processing elements selected for each layer of the network will affect the overall complexity and performance of the network.

The processing element choice is defined as a combination of a processing element function (input function) and a transfer function. The network output response is adapted by modifying the weights on the connections. Using an error function that measures the distance between the desired output vector and the current actual output vector, a learning algorithm adjusts the weights in the network to reduce the error over the set of training examples.

Multilayer Normal Feed Forward (MNFF) is the most commonly used architecture type, and it is generally recommended for most applications. In this paper only the MNFF is examined. The specialized notation for architecture definition used in this study (*m*x*h*x*o*) is interpreted as follows: *m* is the number of input cells, *h* is the number of processing units in the hidden layer and *o* represents the number of output cells.

Input Functions. *Dot Product Input Function (DPF).* The Dot Product is a weighted sum of the inputs plus a bias value. Intuitively, this scales each input according to its relative influence in increasing the net input to the node. *Radial Basis Input Function (RBF).* The Radial Basis Input Function allows nodes to form elliptic regions of activity or inactivity. Results by Poggio and Girosi (1990) and Hartman et al. (1990) prove that RBF networks are universal approximators. *L1 Distance Input Function (L1).* This function calculates the distance between two vectors. Thus the processing element automatically obtains the distance to the input example.

Table 1. Data of reservoir-induced earthquakes used in this study (modified from Baoqi, 1992)

Name of reservoir	Depth H(m)	Capacity V(10^6m)	Surface S(km^2)	Parameter E	Magnitude M	Pattern type
Aswan	84	164000	6200	3.18	3.5	training
Blowering	95	1628	44.5	2.6	3.5	training
Canlles	132	678	16.3	3.17	4.7	training
Eucumbene	106	4761	145	3.23	5	testing
Hoover	191	36703	593	3.09	5	training
Kariba	122	175000	6572	4.58	6.4	training
Keben	182	31000	680	3.99	3.5	training
Kerr	54	1505	60	2.15	3.8	training
Kremasta	120	4750	180	4.55	6.2	testing
Kurubo	186	199	3.5	3.26	4.9	training
Manicouagan	96	10423	207	1.91	4.1	training
Marathon	60.3	41	2.4	3.53	5.7	training
Nurek	285.3	11000	74	1.92	4.5	testing
Oroville	204	4400	63	2.92	5.7	training
Safid Rud	80	1800	56	2.49	4.7	training
Shasta	153	5610	119	3.24	3	training
Talbingo	142	935	19.4	2.95	3.5	training
Uki	63.6	8611	320	2.39	3	testing
Warragmba	104	2053	75	3.8	5.4	training
Xinfengjiang	97	11500	390	3.29	6.1	training
Danjiangkou	92	20896	862	3	4.7	training
Nanshui	81.3	1218	39.4	2.63	2.1	training
Wuxijiang	129	2060	41.9	2.62	2.8	training
Qianjin	50	17	0.8	2.62	3	training
Zhelin	62	7171	387.4	3.35	3.2	training*
Feziling	74.4	488	23	3.51	4.5	training*
Nanchong	45	15.5	1	2.9	2.8	training*
Shenwo	50.3	790	52.6	3.35	4.8	training*
Jocassee	107	1431	30	2.24	3.3	testing*
Jocassee	107	1431	30	2.24	3.2	training*

* testing patterns used by Habibagahi (1997)

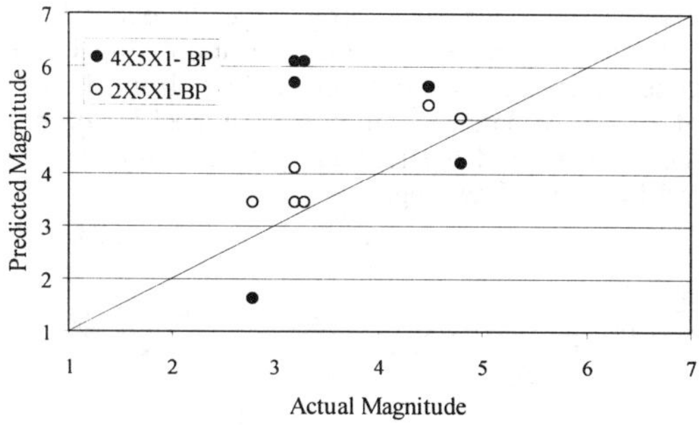

Figure 1. Prediction of the seismic event magnitude with back-propagation network designs.

Learning Rules. There exist a number of methods to carry out supervised learning. The advantages and drawbacks of four of the most commonly used procedures are cursorily listed next: a) Back propagation (BP). It is a steepest descent algorithm that adjusts the weights of the net iteratively until the function error (difference between expected and computed values) is optimized. It has the advantage that it can be easily generalized but for large networks it converges slowly and may get stuck in a local minimum. b) Quick propagation (QP). This algorithm automatically adjusts the size of the search step and detects the best conditions to speed up the training. Thus it has the capability to minimize the time required to define a convenient set of weights. c) Levenberg Marquardt (LM). This algorithm is the transition between the steepest descent and Gauss-Newton methods (Hagan and Menhag, 1994). Its main advantage is its fast convergence to the absolute minimum of the error function. Its main restrictions are that is limited to one output and that it requires important amounts of computer memory. d) General regression (GR). Originally developed in the probabilistic literature, this algorithm was modified by Donald Specht (1991) to approximate functions. It is based on the principle that the contribution an exemplar has on the projection of another one decreases exponentially with the distance between both exemplars. This algorithm has the advantage that its training is fast and can handle both linear and non linear data, however it has trouble with irrelevant inputs (i.e. suffers from the curse of dimensionality) and requires that all the training samples be stored for future use (i.e. prediction). e) Conjugate Gradients (CG). This is a classical method for minimizing arbitrary functions when derivative information is available. When this technique is applied to neural networks becomes an excellent learning procedure which can be used for batch mode training of feedforward networks.

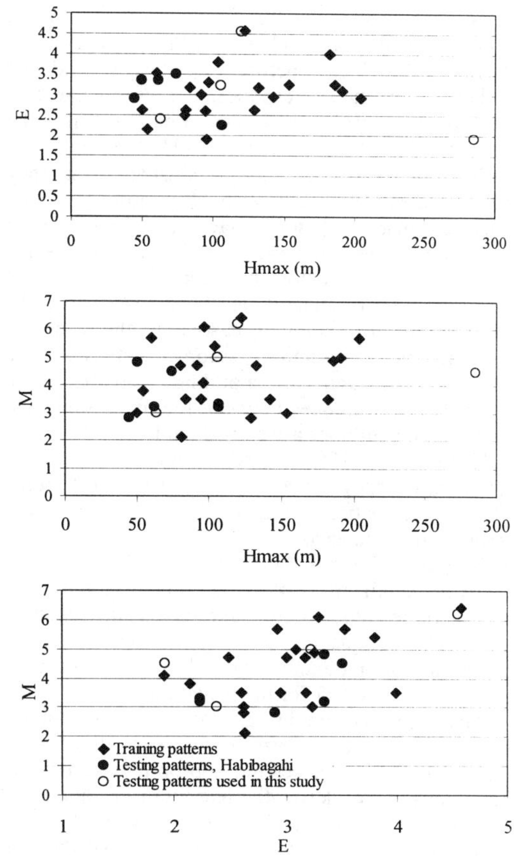

Figure 2. Relations between E-H_{max}, M-H_{max} and M-E

3. MAXIMUM MAGNITUDE OF INDUCED EARTHQUAKES

Excluded the specific geological structure environment in a reservoir area, the "scale" of the reservoir is the dominant factor that influences the intensity of the induced earthquake (Baoqi, 1987, Baoqi, 1992). There are many cases that indicate that the larger the scale of the reservoir, the higher the occurrence probability of induced earthquakes and the greater their magnitudes.

How to select the parameters representing the scale of a reservoir is analyzed in Baoqi (1992). Beacher (Beacher, 1992), in his probabilistic prediction, considered depth and capacity of the reservoir as the main factors to evaluate the severity of the induced earthquakes, while Guha (Guha, 1992) used the reservoir surface area and the maximum reservoir depth as the parameters related to the induced seismicity. Table 1 (Baoqi, 1992) gives three factors that represent the scale for 30 reservoirs which have induced earthquakes: a) depth of reservoir (H_{max}) before the largest induced earthquake, b) the corresponding reservoir capacity (V) and c) the reservoir surface area (S). Since the correlation coefficients of H_{max} $-M$, V-M and S-M are less than 50%, it can not be considered that they are correlated statistically with M. In order to obtain more reasonable relations, Baoqi (1992) suggested to use a characteristic parameter E, given by $S = EV / H_{max}$. The correlation between the E-value of reservoirs and the magnitude, M, is less than 65%. Although the correlation is reasonable, the results of their statistic regression equation are strictly valid for H_{max}>50 m and corresponding V >100 million m^3, and the differences between the predictions and the actual magnitudes are considerable.

Habibagahi (1997) used radial basis neural networks to study the RIE, considering the training and testing patterns indicated in Table 1. His architecture consists of two input neurons representing the parameter E and the maximum reservoir depth H_{max}, five neurons in the hidden layer and the output is the magnitude of the reservoir induced earthquake, M. The coefficient of correlation of predicted magnitudes was 0.69 for the trained data and 0.77 for the tested ones.

These results can be improved performing the following tasks: a) analyze the representativeness of the input data, b) select the appropriate number of input cells and c) define the learning rules and input functions options most adequate to network designs.

To determine the relative importance of the four parameters (E, H_{max} , S and V) on the phenomenon, some multilayer feed forward neural networks (see Table 2) were used to perform sensitivity analysis based the on information of Table 1. The results, shown in the inset of Fig 1, clearly indicate that H_{max} and E are the dominant parameters. This type of information helps to improve significantly network design. Thus, analyzing the input data in terms of E and the output in terms of M, one can more easily select the training and the testing patterns. The results of Fig 1 demonstrate that the overall performance of the network (2x5x1-BP), which includes H_{max} and E as input parameters, is much better than the network that uses H_{max}, E, V and S as input data. This is in agreement with Habibagahi's conclusions.

In order to find patterns and clues that can help on the preprocessing of data to improve the modeling of the phenomenon with neural networks, the data listed in Table 1 was plotted to learn more about the correlation between E and M (Figure 2). From the third plot of Figure 2, it may be seen that there is a significant data dispersion. In these cases it is important to use as much information as possible in the training stage and make sure that extreme cases are left for the testing stage. In general, this will lead to networks that are more capable of generalization. As shown in Table 1, from the total of 30 sets of data, 25 patterns were used for training the network and the remaining five examples were left for network testing.

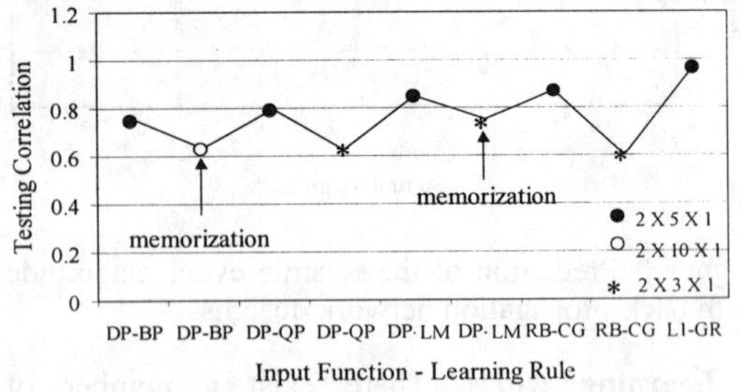

Figure 3. Comparison between different learning rules and input functions

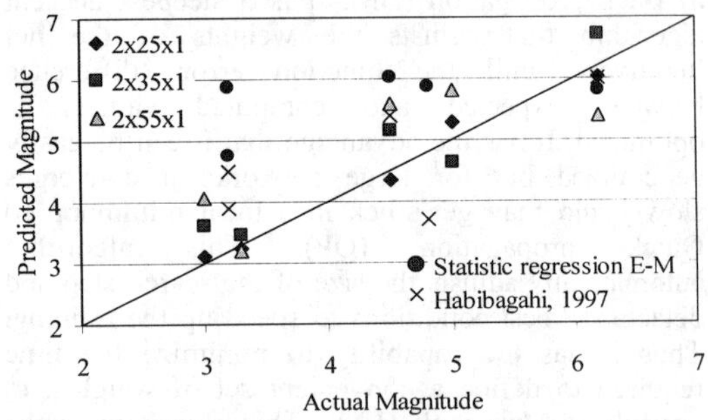

Figure 4. Results of the prediction stage, general regression neural networks

The main problem in the Habibagahi's network designs is the representation of the phenomenon, the examples he used for testing the generalization capabilities of the networks, are more useful in training than in recalling. This seems to lead to contradictions because the adjustments of weights and thresholds do not know where to turn due to the ambiguity brought forth by this kind of patterns. Additionally, the cases he used in testing only cover a small range of the dynamic variations of the input and outputs.

Given the variety of ANN's and the accuracy with which they reproduce specific tasks, a number of possible designs including different learning rules, input and transfer functions were evaluated, with the purpose of optimizing the ANN architecture. Improvement of computing training time and prediction accuracy of the neural network performance can be achieved by testing different designs. The various alternatives studied in this paper are those discussed previously in section 2.

The way they were combined is depicted in Table 2. As in the Habibagahi's design, the optimum number of hidden neurons when using BP, QP, LM –dot product as input function- and CG –radial basis as input function- is five, with correlation coefficients higher than 80% (Figure 3). In this figure it is also shown that the convergence of the general regression neural networks (GRNN) is remarkably superior. The general regression is the fastest converging algorithm for the modeled problem without decreasing its predicting capabilities.

Table 2. General conditions of the ANN designs

	Architecture	Input function	Learning Rule	Transfer Function	Topology
Sensitivity Analysis	MNFF	DP	BP	sigmoid	4x5x1
		DP	BP	sigmoid	4x10x1
		DP	BP	sigmoid	2x5x1
		DP	BP	sigmoid	2x10x1
Optimizing ANN Architecture	MNFF	DP	QP	sigmoid	4x10x1
		DP	QP	sigmoid	4x8x1
		DP	QP	sigmoid	4x5x1
		DP	QP	sigmoid	4x3x1
		DP	QP	sigmoid	2x5x1
		DP	QP	sigmoid	2x3x1
		DP	LM	sigmoid	4x8x1
		DP	LM	sigmoid	4x5x1
		DP	LM	sigmoid	2x3x1
		DP	LM	sigmoid	2x5x1
		RB	CG	sigmoid	4x8x1
		RB	CG	sigmoid	4x5x1
		RB	CG	sigmoid	4x3x1
		RB	CG	sigmoid	2x5x1
		RB	CG	sigmoid	2x3x1
		L1	GR	gaussian	4x55x1
		L1	GR	gaussian	4x50x1
		L1	GR	gaussian	4x35x1
		L1	GR	gaussian	4x25x1
		L1	GR	gaussian	2x55x1
		L1	GR	gaussian	2x35x1
		L1	GR	gaussian	2x25x1

When using GRNN, the number of processing elements (hidden nodes) must equal the training set size, so the design consists of two input neurons (H_{max} and E), 25 processing neurons and one output node (M). The correlation in testing is very high (0.991) and it took just 27 iterations to complete the training stage (in a Pentium II, 450 MHz PC). In order to determine if this was the best design, the training set was modified to include more neurons in the hidden layer. It seems that this does not represent an improvement in the generalization capabilities of the GRNN designs (Fig. 4). The results in this figure show that the (2x25x1) network gives the most accurate predictions. It also represents a significant improvement over the network proposed by Habibaghi (1997).

4. CONCLUSIONS

Cases of reservoir-induced earthquakes are limited, therefore small errors in data acquisition will cause significant dispersion in the information. At the same time, the effects of different factors are not the same. The studies presented here seem to indicate that the GRNN can handle the problems posed by conflicting information and with only 25 patterns it is possible to model the phenomenon. They also show the key role played by the selection of the input patterns and input parameters. It is obvious that this requires a proper understanding of the physics of the problem under study.

The results of the sensitivity analysis corroborate the conclusions of Baoqi (1992) and Gupta (1992) about the importance that depth and scale of the reservoir have on the RIS severity.

The results are promising considering the approximate nature of the parameters employed in this study. Nevertheless, further investigations to incorporate different geological features may be helpful in achieving better predictions.

5. REFERENCES

Baoqi, C. 1987. Correlation between the scale of reservoir and induced magnitude. Journal of South Chinese Seismology, 7, pp. 94-99.

Beacher, G. 1982. Statistical examination of reservoir induced seismicity. Bull. Seism. Soc. Am. 72: 552-569.

Fahlman, S. E. 1988. An empirical study of learning speed in back-propagation networks. CMU Technical Report CMU-CS-88-162.

Guha, S. and Patil, D. 1992. Large water-reservoir-related induced seismicity. In Induced Seismicity, ed. P. Knoll, Balkema, Rotterdam: 243-267.

Gupta, H.K. 1992. Artificial water reservoirs and earthquakes: A world wide status. In Induced Seismicity, ed. P. Knoll, Balkema, Rotterdam, pp. 267-274.

Habibagahi, G. 1997. Reservior induced earthquakes analyzed via radial basis function networks. Soil Dynamics and Earthquake Engineering, 17, pp. 53-56.

Hagan, M.T. and Menhaj, M.B. 1994. Training feedforward networks with the Marquardt algorithm. IEEE Transactions on Neural Networks, Vol. 5. No. 6, pp. 989-993.

Hartman, E.J., Keeler, J.D. and Kowalski, J.M. 1990. Layered neural networks with gaussian hidden units as universal approximators. Neural Computing Journal, Vol. 2, No. 2, pp. 210-215.

Meade, R. 1982. State-of-the-art for assessing earthquake hazards in the United States. Report 19, Miscellaneous paper 5-73-1, Geotechnical Laboratory, U.S. Army Engineer Waterways Experiment Station, June.

Poggio, T. and Girosi, F. 1990. Networks for approximating and learning. Proceedings IEEE, Vol. 78, pp. 1481-497.

Skipp, S. 1981. The potential for induced seismicity - Geological approaches. Dams and Earthquake, London, pp 287-296.

Specht, D. 1991. A general regression neural network. IEEE Transactions on Neural Networks, 2, 6, pp 568-576.

Xia, Q. 1981. On the geological classification of reservoir induced earthquakes. Chinese induced earthquakes, Seismological Press, pp 8-13.

Computer Methods and Advances in Geomechanics, Desai et al. (eds) © 2001 Balkema, Rotterdam, ISBN 90 5809 183 X

Evaluation of a few countermeasures for seismic failure of embankment

K.Suzuki
Saitama University, Urawa, Japan

K.Ogata & N.Mishima
Japan Highway Public Corporation, Tokyo, Japan

N.Ohbo
Kajima Technical Research Institute, Tokyo, Japan

ABSTRACT: This paper investigates the seismic behavior of embankment placed on saturated sands during earthquakes. First, the results of centrifuge test for three kinds of underling saturated sands (D_r=39%, 56%, 93%) were simulated using two-phase nonlinear FEM, and the mechanism of embankment failure due to soil liquefaction was interpreted with aid of numerical model. Secondly, the evaluation of three kind of countermeasures against liquefaction were conducted using the validated numerical model

1 INTRODUCTION

Soil liquefaction induced by earthquakes may cause severe damage to embankment (e.g. JSCE 1986, Matsuo 1996). To prevent the damage of embankment, it is indispensable to comprehend the seismic behavior of embankment under various conditions and develop effective countermeasure against soil liquefaction.

So far, many experimental studies and numerical analysis have been conducted to investigate the seismic behavior of embankment placed on saturated sands (e.g. Abe & Kusano 1992, Koseki et al. 1994, Zheng et al. 1995, Matsuo et al. 2000, Fujii et al. 2000). These previous studies have provided many constructive results, demonstrating the effects of the underlying soil conditions, the embankment conditions and the characteristics of seismic motions. However, to understand the behavior of embankment under various conditions, and especially to comprehend the mechanism of embankment failure due to soil liquefaction, further studies are required because of the limited number of previous studies and the difficulties of the problem.

This paper investigates the seismic behavior of a highway embankment placed on saturated sands and to evaluate the countermeasures against liquefaction. The dynamic centrifuge test results were first simulated and analyzed using two-phase nonlinear FEM, and then the mechanism of embankment failure due to soil liquefaction was studied using validated numerical model. Finally, the evaluation of a few countermeasures against liquefaction was conducted.

2 EFFECTS OF RELATIVE DENSITY

2.1 *Centrifuge tests*

The centrifuge model experiments were performed at 50g centrifugal acceleration. Figure 1 represents the cross section of experiment model together with arrangement of measuring apparatus. The cross section of embankment was selected referring to the standard section of a highway embankment. The embankment was assumed to be placed on saturated sands with relative density of 39%, 56% and 93%. The test models were constructed in following way. The Toyoura sand was first pluviated into a rigid box, and the sands were saturated using silicone oil. The embankment was modeled using Toyoura sand mixed with Kaolin clay, and the ratio between sand and clay was 5:1. Silicone rubber was set between sand and end walls to reduce the boundary effects of end walls. The rigid box was set on the platform of a shaker, and sinusoidal acceleration of 200Hz with amplitude of around 9g was applied at the bottom of rigid box. Pore water pressure transducers, accelerometers and linear variable differential transformers (LVDT) were set up to observe the response of embankment and underlying sands.

2.2 *Numerical Model*

The two-phase nonlinear finite element code DYNAFLOW (Prevost, 1981) was used for numerical analysis. Figure 2 shows the finite element model for simulation analysis in the case without any countermeasure. The multi-yield constitutive model (Prevost, 1985) was adopted to simulate the nonlinear behavior of the soil materials. To find the model parameters, drained triaxial tests and undrained

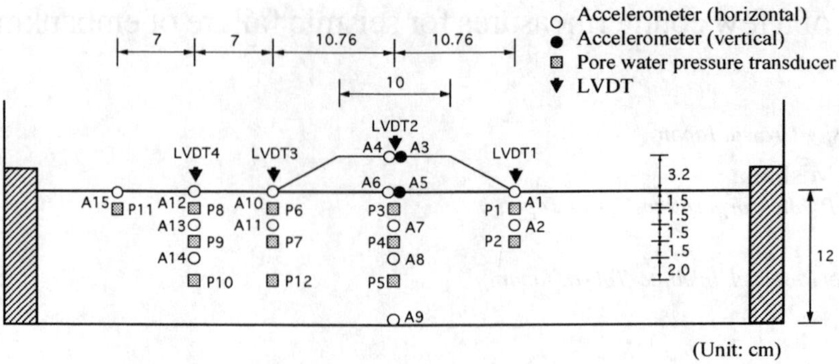

Figure 1. Centrifuge test arrangement.

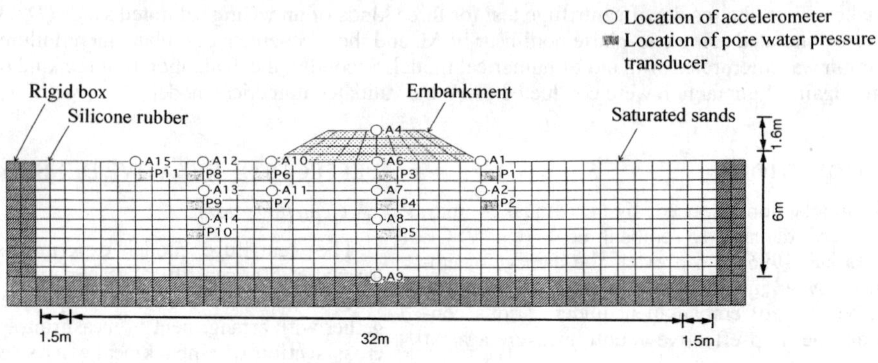

Figure 2. FEM model in prototype term.

Table 1. Material parameters for sands and embankment material.

Property	Sands			Embankment material
	D_r=39%	D_r=56%	D_r=93%	
Solid mass density(kg/m^3)	2648.0	2648.0	2648.0	2683.0
Porosity	0.455	0.436	0.388	0.390
Initial shear modulus(MPa)	85.5	97.1	129.7	86.0
Initial bulk modulus(MPa)	185.2	195.5	281.0	401.1
Reference mean stress(kPa)	98.0	98.0	98.0	98.0
Power exponent	0.4	0.4	0.4	0.3
Fluid bulk modulus(MPa)	2000.0	2000.0	2000.0	-
Friction angle at failure(deg.)	36.8	38.4	44.0	27.0
Cohesion(kPa)	0.0	0.0	0.0	1.0
Maximum deviatoric strain(%)	7.0	6.0	5.0	7.0
Dilation angle(deg.)	30.0	30.0	30.0	27.0
Dilation parameter	0.21	0.15	0.10	-
Permeability(m/s)	6.0E-05	5.0E-05	4.0E-05	-

cyclic triaxial tests were performed. Table. 1 shows the material parameters for sands and embankment material.

2.3 *Verification*

The observed and computed acceleration responses at A15 and A1, and excess pore water pressures at P11 and P1 are presented in Figure 3. Figure 3(a), (b) and (c) show the results of relative density of 39%, 56% and 93%, respectively. All results are presented in prototype terms. It could be seen that the computed results match well with the observed ones both in magnitude of amplitude and wave form.

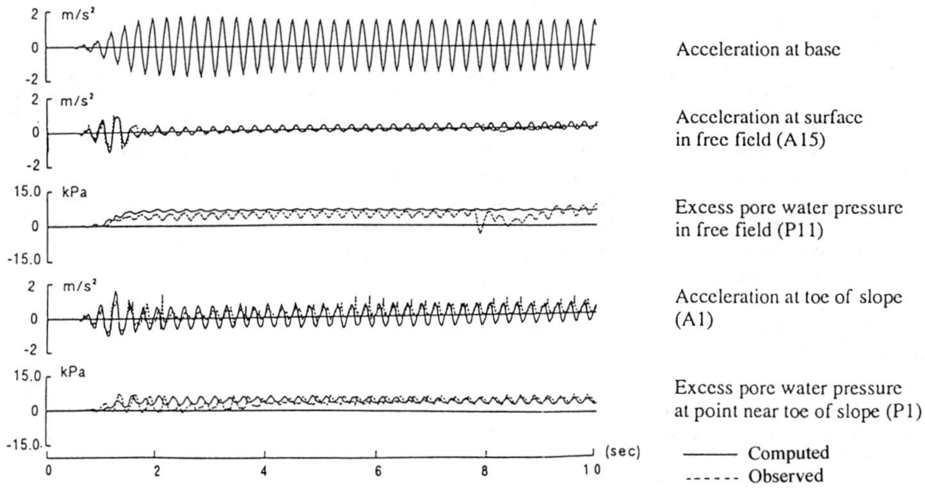

(a) D_r=39%

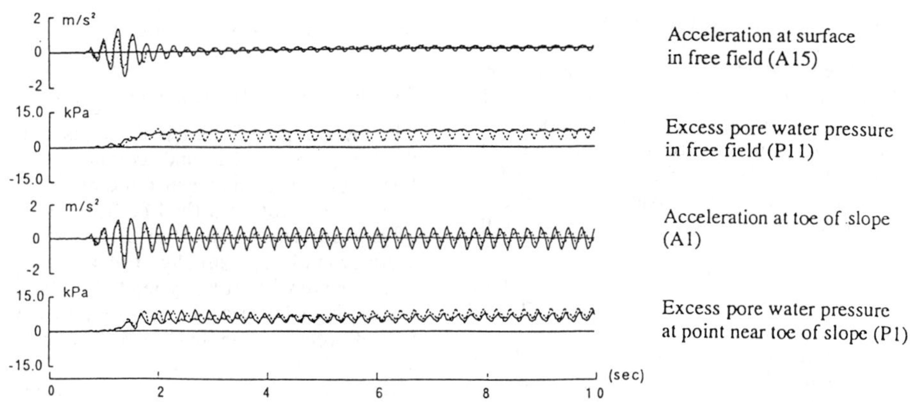

(b) D_r=56%

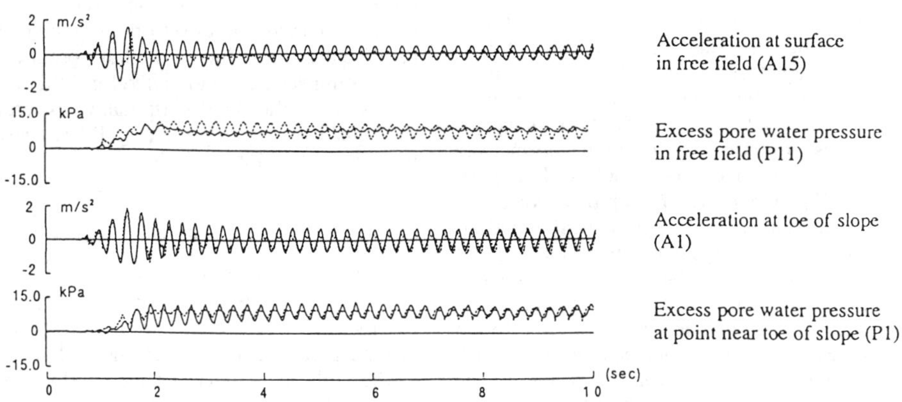

(c) D_r=93%

Figure 3. Computed and observed accelerations and excess pore water pressures.

Good agreement between computed and observed responses is also confirmed at all other observing points, which demonstrates the validity of the presented numerical model.

In the region that is relatively far from the embankment (hereafter called free field), as the amplitude of acceleration increases, the excess pore water pressure begins to rise and soon reaches the maximum value. The amplitude of acceleration quickly decreases as the excess pore water pressure increases, and it becomes very small after the excess pore water pressure arrives its maximum. The maximum excess pore water pressure ratios in all cases are greater than 0.9, indicating that the sand ground liquefies.

At the observing point at the toe of the slope of embankment, the reduction of amplitude of accelerations is not so significant as observed in free field. The maximum excess pore water pressure ratio at this point is less than 0.8, indicating that the sand ground under and near embankment does not liquefy, and thus the reduction of amplitude of acceleration is smaller compared with that in free field. This may result from the fact that the effective vertical stress in sand ground under embankment is higher than that in free field due to the weight of embankment. The wave form of excess pore water pressure is also different from that at observing points in free field, which reflects the effects of embankment on the dynamic response characteristics of its surrounding sands.

As for the effects of relative density of underlying sands, low density leads to fast generation of excess pore water pressure and quick reduction of amplitude of acceleration as shown in Figure 3. The settlement of embankment is shown in Figure 4. It could also be seen that the settlement of embankment in the case with low relative density is significantly large, as reflected in Figure 4. They could be explained as following. Compared with the sand of high relative density, the sand with low relative density has lower liquefaction resistance, and thus is easier to liquefy, and the liquefied zone is also larger.

Figure 5 demonstrates the distribution of excess pore water pressure ratio and horizontal normal effective stress after sand ground liquefies. The negative values in Figure 5 represent compressive effective stress. The sand ground in free field liquefies, while the sand ground beneath and near embankment does not. It can also be seen that low relative density forms large liquefied zone. Regarding the horizontal normal effective stress, tensile stress could be found in the part near the surface of embankment, which implies that the crack may occur in this region. In the centrifuge tests, the cracks occurred at the surface of embankment in the direction perpendicular to the embankment section. Figure 6 shows a typical example of the generated crack for underlying sand with relative density 56%. Therefore, the computed

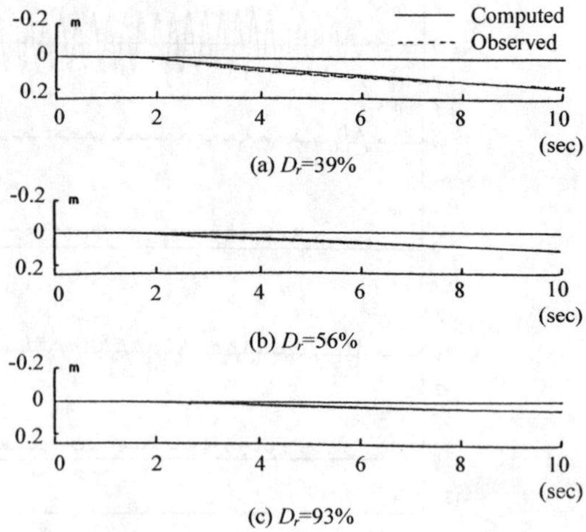

Figure 4. Computed and observed settlement of embankment.

results well explain the test results.

2.4 *Mechanism of embankment failure*

The mechanism of the tensile stress generation is as following. When the sand ground in free field liquefies, the sand ground beneath embankment horizontally extends towards the free field due to the weight of embankment. Thus, the extending of underlying sand ground horizontally pulls the embankment. When the tensile stress generated by the extending of underlying sand exceeds the compressive stress due to self-weight of embankment, the whole effective stress becomes tensile, which could lead to the occurring of cracks if it exceeds the tensile strength of embankment materials.

The pattern of embankment failure due to soil liquefaction is that embankment settles together with cracks occurring in direction perpendicular to embankment section due to the horizontal extension of underlying sand ground. Therefore, this pattern well agrees with the results of many experiments and earthquake damage reports (e.g. JSCE 1986, Nagao & Iriyama 1994, Matsuo 1996).

3 EVALUATION OF COUNTERMEASURES

From the above results, it could be concluded that, to prevent the damage of embankment, one of the effective methods is to constrain the horizontal extending of sand ground under embankment. Thus, to reinforce the soils just under the fee of slope of embankment, or to directly constrain the soil under embankment.

In this study, the following three kinds of countermeasures were chosen for evaluation. The first is a countermeasure using counterweight-fill, which is

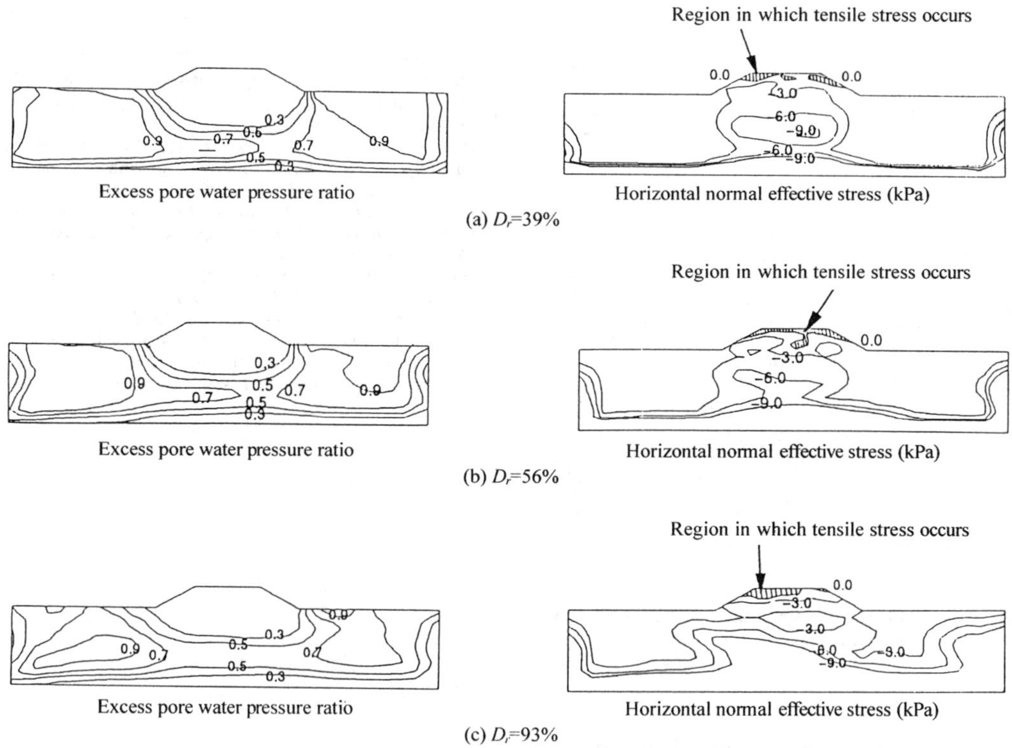

Figure 5. Computed distribution of excess pore water pressure ratio and horizontal normal effective stress.

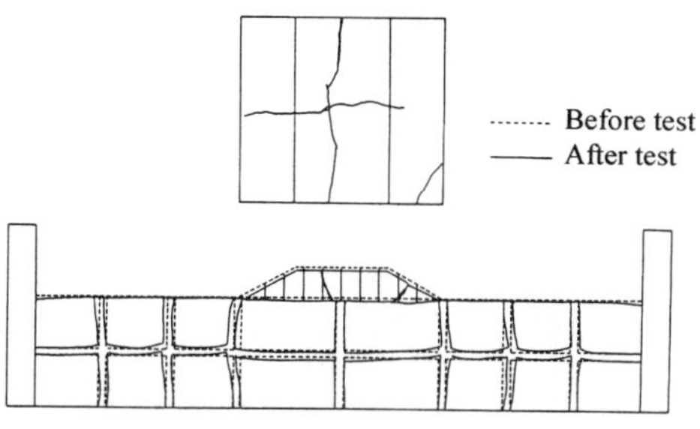

Figure 6. Crack of embankment placed on underlying sand with relative density 56%.

aimed at improving the soil condition around the toe of slope. The second is a countermeasure using steel-wire-net, which is only for constraining the horizontal deformation of embankment. The third is a countermeasure using Boring Hole (BH) pile for constraining the underlying sands. In these cases, the underlying sands (D_r=56%) and the input wave were assumed to be exactly the same as those for the case without countermeasure.

The material parameters of counterweight-fill were assumed as the same as that of embankment. Truss elements for steel-wire-net were applied at the bottom of embankment. Beam elements were used for BH pile.

Figure 7 demonstrates the distribution of horizontal stress after surrounding sands liquefies. Figure 7(a), (b) and (c) show the results used counterweight-fill, steel-wire-net and BH pile, respectively. Compared with the case without countermeasure as shown in Figure 5(b), the depth of tensioned region in embankment becomes shallow. Therefore, the countermeasures used here could reduce the tensioned region in embankment. It needs further consideration for quantitative comparison.

4 CONCLUSIONS

To study the behavior of highway embankment placed on saturated sands, the results of dynamic centrifuge tests were simulated using two-phase nonlinear FEM, and the mechanism of embankment failure due to soil liquefaction was interpreted. Using the validated numerical model, three kinds of countermeasures against soil liquefaction were evaluated. Following are the main conclusions drawn from this

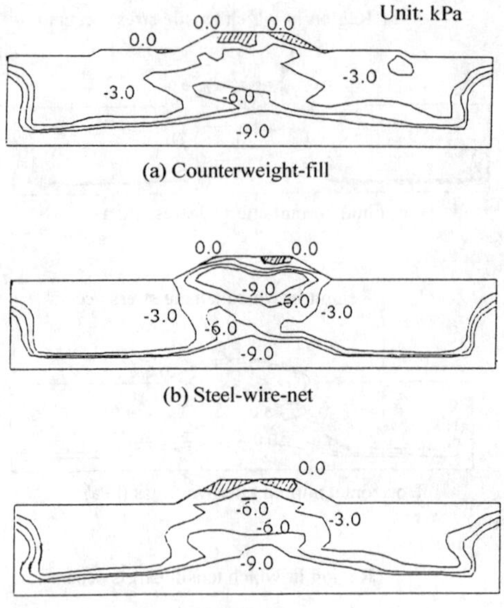

Figure 7. Distribution of horizontal normal effective stress for the cases with countermeasure.

study.

1. The presented numerical model proves to be able of simulating the centrifuge tests, and providing substantial information for comprehending the centrifuge test results.

2. The excess pore water pressure ratio is considerably small in the region close to the bottom of embankment, and the soil under embankment is harder to liquefy compared with the sands in free-field.

3. The pattern of embankment failure is that embankment settles together with cracks occurring due to the horizontal extension of underlying sand ground.

4. The countermeasures used counterweight-fill, sheet-wire-net and BH pile could reduce the tensioned region in embankment.

REFERENCES

Abe, H. & Kusano, Y. 1992. Liquefaction analysis of embankment using 2D FEM. *Proc. 46th Annual Conf., JSCE* 3: 986-987. (in Japanese)

Fujii, T., Hyodo, M., Kusakabe, S. & Fukuda, K. 2000. Online seismic response tests on embankmens foudation. *J. Geotech. Engrg. Div., JSCE* 652(51): 229-243. (in Japanese)

JSCE 1986. *Damage report of the Nihon-kai Central Part Earthquake in 1983*. Tokyo. (in Japanese)

Koseki, J., Koga, Y. & Tateishi, A. 1994. Liquefaction of sandy ground and settlement of embankment. *Proc. Intl. Conf. CENTRIFUGE 1994, Singapore*: 215-220.

Matsuo, O. 1996. Damage to river dikes. *Special Issue Soils and Foudations*: 235-240.

Matsuo, O., Shimazu, T., Uzuoka, R., Mihara, M. & Nishi, K. 2000. Numerical analysis of seismic behavior of embankments founded liquefiable soils. *Soils and Foundations*, 40(2): 21-39.

Nagao, K. & Iriyama, T. 1994. Effects of relative density of underlying sands on the behavior of embankment. *Proc. 50th Annual Conf., JSCE* 3: 96-97. (in Japanese)

Prevost, J.H. 1981. *DYNAFLOW: A nonlinear transient finite element analysis program*. Dept. Civil Engineering Op. Res., Princeton University.

Prevost, J.H. 1985. A simple plasticity theory for frictional cohesionless soils. *J. Soil Dynam. Earthq. Engrg.* 4: 9-17.

Zheng, J., Ohbo, N., Suzuki, K., Suzuki, R., Mishima, N. & Nagao, K. 1995. Analysis of results of centrifuge tests on seismic behavior of embankment. *Proc. 1st Intl. Conf. Earthq. Geotech. Engrg., Tokyo*: 1069-1074.

Computer Methods and Advances in Geomechanics, Desai et al. (eds) © 2001 Balkema, Rotterdam, ISBN 90 5809 183 X

Coupled optimization and finite element dynamic analysis of geotechnical problems

S. Valliappan, A. Hakam & N. Khalili
School of Civil and Environmental Engineering, The University of New South Wales, Sydney, N.S.W., Australia

ABSTRACT. The optimal design of geotechnical structures subjected to dynamic loading, using the finite element method to determine the responses of the system is presented. A three dimensional finite element program for dynamic analysis has been developed using Newmark's integration scheme for the discretization in time-domain. To consider the infinite extent of the domain, unified boundary condition has been adopted. The objective of the optimization problem is to minimize the cost of the structure while the design variables are the dimensions of the structure. The results of the numerical examples show that the geotechnical structures subjected to dynamic loading can be designed both economically and effectively using the proposed method.

1 INTRODUCTION

In practice, it is essential to select an efficient and economical design of a structure from the available alternatives. For this purpose, optimization technique has been used extensively in recent times. The goal of optimization is to optimize an objective function without violating the constraints of the problem. In structural optimization, it is generally the volume or weight of the material which governs the objective function. In many engineering problems, the structural geometry is complex and hence numerical methods such as the finite element method are usually employed to determine the responses under various loading conditions. The combined optimization and finite element technique has been adopted in the past to design structures under static loading. However, this approach to the case where dynamic loading is involved has not been as extensive as for static loading.

Only very little work has been done in the application of optimization technique to geotechnical structures. The combined finite element and optimization technique has been used for the optimal design of raft-pile foundations (Valliappan et al. 1997) and geotechnical structures in Lowland areas (Valliappan et al. 1999) under static loading. The finite difference and semi analytical methods have been used to estimate the sensitivity and the bi-point constraint approximation has been proposed. The optimization of foundations subjected to dynamic loading has not been studied so far. The minimum-weight design of a rectangular foundation due to a harmonic vertical loading in frequency domain has been investigated (Sienkiewicz & Wilczynski 1993). This study was based on the theory of vibration with a single degree of freedom.

In this paper, the optimal design of geotechnical structures subjected to dynamic loading is presented using the combined optimization and finite element method. Numerical examples of foundation subjected to machine vibration and retaining wall due to ground motion are presented in this paper. The design variables are the dimensions of the structures while the constraints are amplitude, velocity and acceleration.

The three main modules required in optimization problems are analysis module, optimization module and interface module. In the analysis module, either analytical or numerical techniques, such as the finite element method, are used to determine the response of the structure due to the applied loads. The optimization module involves the formulation of the optimization problem including the objective function, design variables and constraints. The interface module couples the analysis and the optimization modules. This is an important stage in the optimization process, which involves processes such as sensitivity analysis and constraint approximation. Depending on the constraint formulation chosen, more initial values of each design variable may be required.

Figure 1 shows the flow chart of the proposed method for the optimal design of foundations due to dynamic loading.

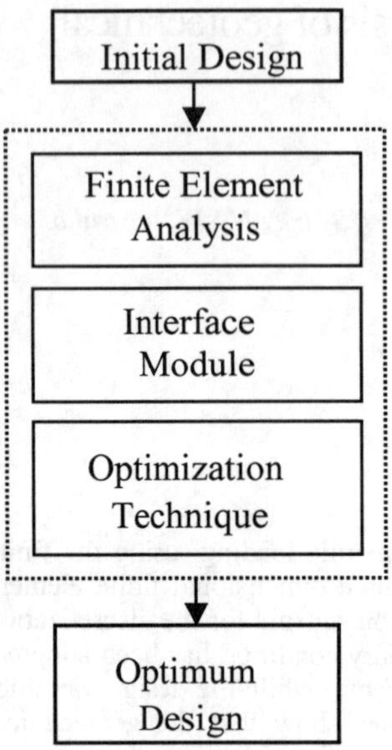

Figure 1. Coupled finite element method and optimization technique.

2 FINITE ELEMENT DYNAMIC ANALYSIS

The finite element method is used as the numerical tool to determine the responses of the structures under dynamic loading conditions. The formulation of finite element method for dynamic analysis is:

$$[M]\{\ddot{u}\} + [C]\{\dot{u}\} + [K]\{u\} = \{f(t)\} \quad (1)$$

where [M], [C], [K] are mass, damping and stiffness matrices respectively, $\{\ddot{u}\}$, $\{\dot{u}\}$, $\{u\}$ are acceleration, velocity and displacement vectors respectively and {f(t)} is external dynamic load vector.

When the wave propagates in the soil, the peak amplitudes will be reduced after each successive cycle and eventually it disappears. This reduction in amplitudes of motion is caused by material damping within the mass of soil. The dissipation of energy for a vibrating foundation in the soil is affected by soil properties. Further, the damping characteristics of soil are determined by the extent of the medium and mass of the system besides the properties of the soil itself. In general, for finite element analysis, the damping of the system is considered to be proportional to the mass and the stiffness for Rayleigh damping and is assumed as:

$$[C] = a[M] + b[K] \quad (2)$$

where a and b are the constants that depend on the natural frequency and the damping ratio.

In dynamic analyses, the size of elements has to be a fraction of the critical wavelength. The reason for this is that in wave propagation analysis within a continuum using the finite element, the mesh acts as a low-pass filter and hence the waves with a frequency larger than the element size cannot propagate through the mesh. The element size of one quarter wavelength was recommended for linear strain element (White 1978). Further, modelling of the infinite medium must be considered in finite element analysis. In this study, the unified boundary condition (White et al. 1977) to model the finite extent of the medium in order to prevent the reflected energy back into the model, is used. This boundary is both frequency independent and simple to incorporate in finite element analysis. Using the unified boundary model, a lumped damping matrix can be developed as:

$$[C_{gm}] = \int [N]^T [D^*] [I]\, dA \quad (3)$$

where $[C_{gm}]$ is a geometric damping matrix for the element at the boundary, [N] is shape function, [I] is identity matrix, A is the area of the boundary and [D*] is energy absorbing matrix of unified boundary which depends on the Poisson's ratio, mass density and wave velocities.

Newmark's integration scheme is used for the discretization in time-domain. This method is a generalisation of linear acceleration method that assumes the acceleration varies linearly in a time interval. With the assumption that the acceleration, velocity and displacement at the pervious time are known, the response at time t(j+1) can be obtained by solving the equation of motion at the time t as:

$$\{u\}\,[\,[M]/(\beta\,\Delta t^2) + [C]\,\gamma/(\beta\,\Delta t) + [K]\,] = \{f_{tot}\} \quad (4)$$

where β and γ are Newmark's parameters to determine stability and accuracy of the solution and Δt is time step. The accuracy can be gained by taking the time step as a small fraction of the period of free vibration of the system.

3 OPTIMIZATION MODULE

The basic formulation of the optimization problem usually includes the objective function, constraints and design variables. It can be written as:

$$\text{Optimize} \quad : F(x)$$
$$\text{subject to} \quad : G_1(x) = 0 \text{ and } G_2(x) \leq 0 \quad (5)$$

where F(x) is the function to be optimized, x is the set of design variables, $G_1(x)$ and $G_2(x)$ are sets of equality and inequality constraint functions respectively.

The objective function is the most important and it can be a collection of functions that are typically expressed in terms of a set of design variables. The solution of the optimization problem can be started from an initial design value to the next step in order to improve the objective function such that the constraints are not violated. The process to obtain the optimum design can be terminated when this process has improved the objective function without violating the constraint.

The selection of design variables is also an important stage that affects the achievement of the optimization solution. In fact the optimization process will search for the optimum values of the design variables to improve the objective function. The design variables should be consistent with the problem to be analyzed. For foundation problems, the design variables can be the dimensions, depth of embedment etc.

The basic procedure to solve the optimization problems usually starts by selecting the active constraints. In many procedures of optimization problems, the choice of active constraint is the important stage that may affect the computational cost. The next step is the direction for searching new design variables based on a set of active constraints and objective function. The direction of search line can be the tangent to these functions. Then, the distance between the new value and the old value provide the perturbed design value. Finally, the convergence criteria will determine the termination of the procedure.

In the past, an optimization technique named general reduced gradient (GRG) method (Lasdon et al. 1978) had been developed which is based on projecting the search direction into sub-space tangent to the active constraints. In this study, this generalized reduced gradient method is used for optimization solver.

4 INTERFACE MODULE

The derivatives of the physical response of a structure with respect to the design variable can be evaluated from sensitivity analyses. The sensitivity analysis is carried out to calculate the gradients of the constraints with respect to each design variable, which are necessary for constructing constraint approximation functions.

Since mass, stiffness and damping matrices are functions of a set of design variables, x, the equation of motion can be expressed as:

$$[M(x)]\{\ddot{u}\} + [C(x)]\{\dot{u}\} + [K(x)]\{u\} = \{f(t,x)\} \quad (6)$$

By solving the equation above, the displacement $\{u\}$, velocity $\{\dot{u}\}$ and acceleration $\{\ddot{u}\}$ can be obtained. It means that u, $\dot{u}$ and $\ddot{u}$ are again functions of x.

The finite difference procedure which is simple and straight forward is adopted to evaluate design sensitivity analyses in this study. The sensitivity value of a constraint function g(x) with respect to the a design variable, x, using forward-difference approach is defined as:

$$\frac{dg(x)}{dx} = \frac{g(x + \Delta x) - g(x)}{\Delta x} \quad (7)$$

where Δx is a perturbation of the design variable, x.

The forward difference approximation is usually preferred because less additional analyses are required than the higher order approximations. For optimization of structure with n design variables, there is an initial analysis and n additional analyses for every perturbed parameter.

There are two sources of error that can happen in the finite difference method due to the step size of perturbation. First it occurs by ignoring higher order in the series of the perturbed function that is commonly known as truncation error. This error will be small if the step size is extremely small. Unfortunately, the relative small perturbation of the design variable can contribute to the round-off error in calculating the response. For optimization of geotechnical structures, to reduce the error in sensitivity analysis, the perturbation of the design variable within range of 5 to 15 % is suggested.

In order to provide correlations between the design variables and the responses of the structure, one or more constraint approximations should be generated. Structural optimization using finite element method involves constraints that are usually complex functions of the design variables. To overcome this problem the approximation of the constraints functions is required to couple the finite element analysis and the optimization technique.

The constraint function can be formed explicitly in terms of the design variables. In the structural optimization the local approximations are popularly adopted to built the constraints. These methods need only one design point for every design cycle. The local approximations usually provide good results in the neighbourhood of a point in the design region. For a linear approximation, the new value of the constraint function can be written as:

$$g(x) = g(x_0) + \sum_{i=1}^{n} \frac{dg(x)}{dx} \Delta x \quad (8)$$

where x is a set of design variables, x_0 initial value of variables, $g(x_0)$ and g(x) are values of the constraints related to the design variables. The linear approximation is based on the first order Taylor series. The derivatives at a single design point are obtained from the sensitivity analysis.

For optimization problems that involve a number of constraints normalization of that functions can be used (Haftka et al. 1990). The normalization of constraint functions is to eliminate such wide variation in the magnitudes and give similar levels of criticality of the constraint values. The typical normalization that can be applied to the constraint such that non-dimensional functions with the values of order one is in the form of:

$$g(x) = 1 - \frac{g(x)}{g(x_c)} \geq 0 \tag{9}$$

where $g(x_c)$ is the limit value of the constraint.

For harmonic input motions, the satisfactory condition can be expressed in terms of displacements related to the frequencies. The limit value for satisfactory vibration condition of foundations caused by operating machines is that can be tolerated by human beings standing and subjected to vertical vibration. There are certain limits for maximum velocity for different sensitivities of response by persons (Richart et al. 1970).

The unsafe condition of a foundation due to dynamic loading occurs if the responses have exceeded a limiting value, which can be expressed in terms of velocity or acceleration. A peak of velocity of 2.54 cm/second and an acceleration of 0.5 g represents the limit for serviceability of machine foundations.

5 RESULTS

Two numerical examples are presented in this paper. Firstly, a 2X2 pile foundation is chosen to be optimised. The schematic illustration of the problem is presented in Figure 2. The piles are driven into a semi-infinite soil medium to support an external vibrating load of 40 kN that is operating with frequency of 200 rpm. The static load resting on the pile cap, including the weight of the machine is 200 kN. The soil medium has shear modulus of 23 MPa, mass density of 1.8 t/m^3 and Poisson's ratio of 0.40. The piles and the cap are made of concrete with the properties: elastic modulus of 21,000 MPa and Poisson's ratio of 0.15. The pile is circular in cross section with the diameter of 50 cm.

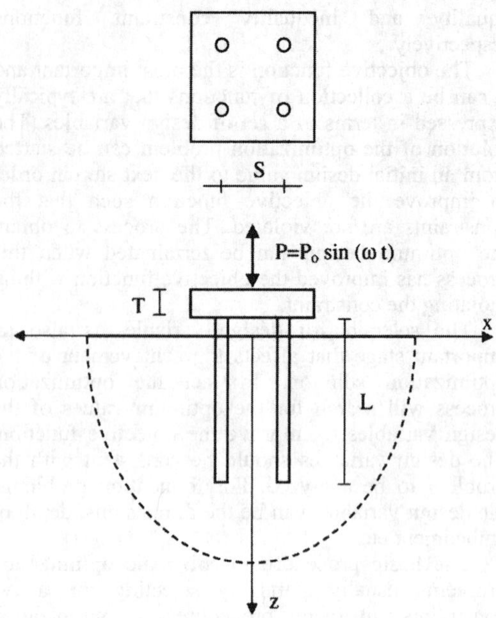

Figure 2 Pile foundation in homogonous soil

The objective of this analysis is to minimize the cost of the foundation that is including the piles as well as the pile-cap. The total cost of the foundation is calculated based on the unit cost from the Commercial/Industrial Building Cost Guide New South Wales in March 1999 (Cordell 1999). The installation cost of the elements of the foundation has been included in the calculation. The design variables are the thickness of the pile cap, the length of the pile and the spacing.

The thickness of the pile cap is limited between 1.0 m to 2.5 m and the pile spacing is between 1 to 6 times of the pile diameter. The length of the pile is essentially not limited to a certain value, however the length of the pile greater than 40 times of the diameter does not have a significant effect on the response of the system as illustrated in Figure 3. However, for economical design of foundation in this analysis, the pile length is limited to 60 times of diameter.

The maximum acceleration of 0.002 g is adopted for this analysis. That value corresponds to the maximum limit to prevent any fault in the machine-foundation system that needs corrections. The very good operation of the machine is desired for this system that implies the maximum velocity of 1.0 x 10^{-3} m/sec. The maximum of the vertical motion is limited to 4.3 x 10^{-5} m.

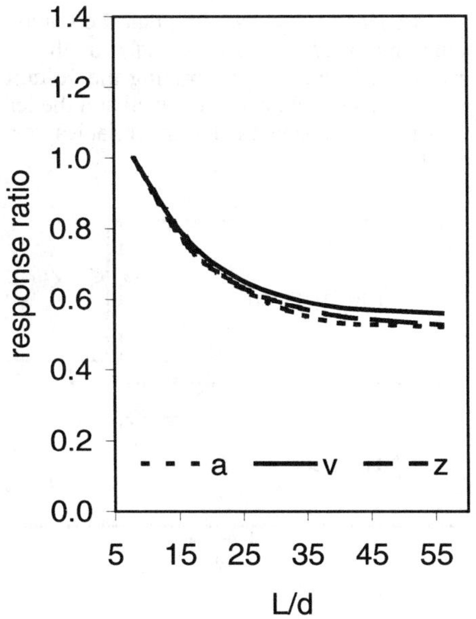

Figure 3 Response ratio versus length-diameter ratio.

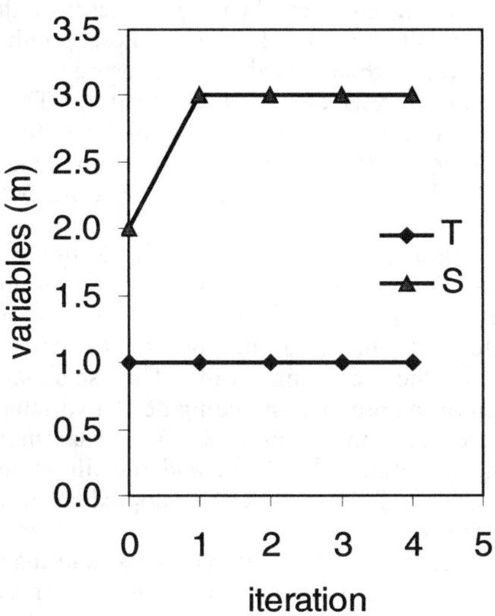

Figure 4. Thickness of the pile cap (T) and pile spacing (S).

During the process of iterations, the design variables of the pile-cap do not change much (Fig. 4). The

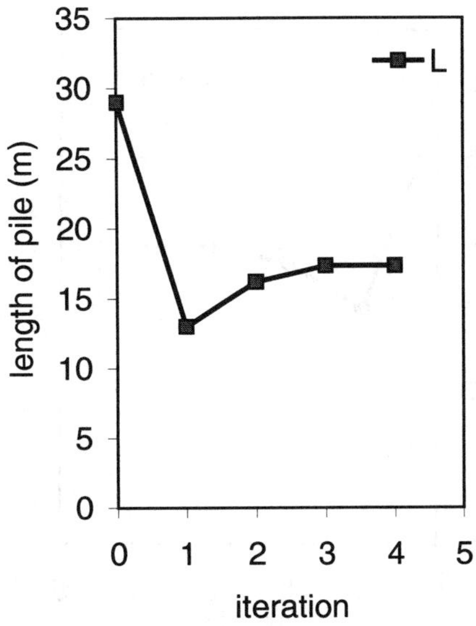

Figure 5. Length of the piles (L) versus iteration.

thickness of the pile cap is still at the initial value during the process, that is 1.0 m. The initial 2.0 m pile spacing is increased to the maximum limit at the first iteration and it does not improve for the further iterations. The initial pile length is 29 m which changes to 13 m at the first iteration and 17 m at the fourth iteration which is the final design (Fig. 5).

The cost of the pile foundation versus iteration number is plotted in Figure 6. The initial cost of the foundation is about \$17,000.00 and it reduced to \$ 11,500.00 at the final design. It shows that the total cost of the foundation can be reduced by two thirds of the initial cost.

The ratio of the constraint values compared to the maximum allowable limits is presented in Figure 7. The acceleration ratio is about one at the initial design. It is violated at the first and second iterations. At the final iteration the acceleration is about the same as the initial value, however at that point the design variables have been altered. The ratio of amplitude of vertical motion is about 0.99 at the initial and about 0.98 at the last iteration. It was violated at the iteration one. During the optimization process, the maximum velocity remains smaller than its upper limit.

In the second numerical example, a cantilever retaining wall with a height of 9 m is designed to retain sand back fill (Fig. 8). The base of the wall is embedded into firm clay soil. The sand back fill has the unit weight of 19.0 kN/m^3 with internal friction

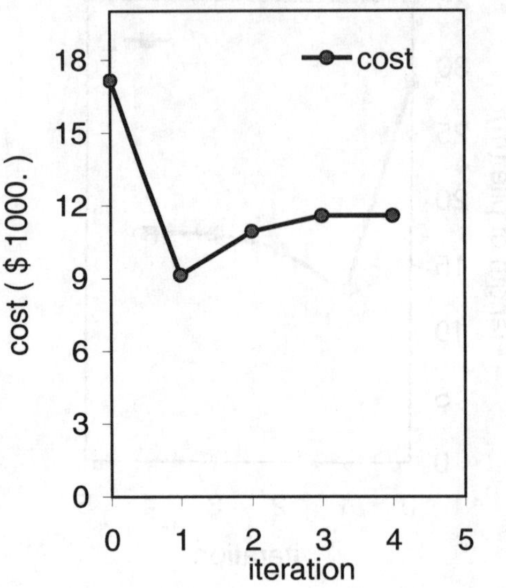

Figure 6: Cost of the foundation

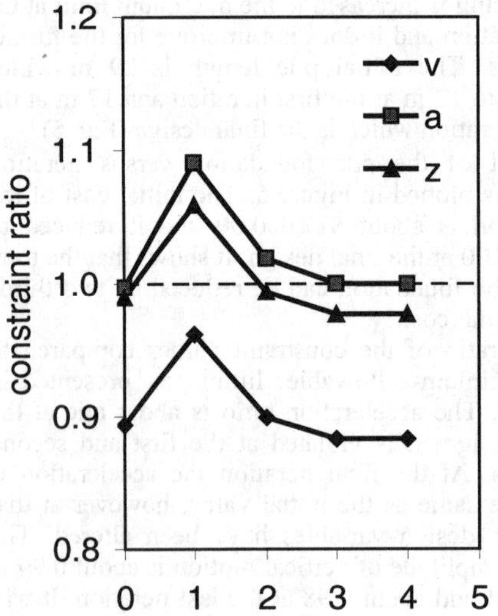

Figure 7: Constraint ratio versus iteration

angle of 30 degrees, modulus of elasticity of 23,500 kPa, and Poisson's ratio of 0.25. The cohesion of the firm clay under the foundation is 167.6 kPa with modulus of elasticity and Poisson's ratio of 34,480 kPa and 0.3 respectively. The wall is made of reinforced concrete. By assuming that the volume of the retaining wall is proportional to the cost, optimization is aimed at minimizing the volume of the material. The thickness of the wall and the length of the base are chosen as design variables for the optimization problem.

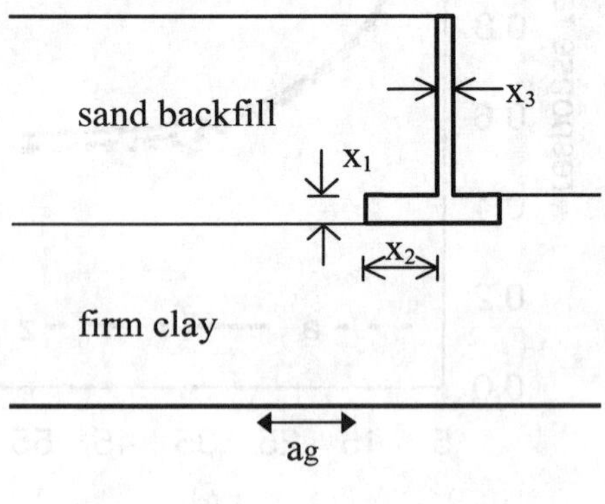

Figure 8. Design variables for the retaining wall (x_1, x_2 and x_3).

The optimum design due to static loading has been done by adopting Rankine's active earth pressure to check the stability against overturning about the toe, sliding along the base and bearing capacity of the foundation (Valliappan et al. 1999). These stability analyses were then used as constraints in optimization analysis due to static loading. Three design cycles were required to obtain the optimal result with the length of the heel of 3.5 m , the length of the toe of 1.0 and the thickness of 0.3 m for both the base and the wall.

Using these values for the initial design, the optimum design of the retaining wall due to ground motion is processed. The El-Centro 1940 earthquake is considered to be input motion for earthquake analysis of the retaining wall. The schematic illustration of the problem including design variables is presented in the figure 8. The maximum horizontal amplitude of 5.2 cm and the tilt of the wall of 0.2 % of the height are adopted as limit values of the constraints.

The result of the optimization process in terms of design variables is presented in Table 1. At the initial design, both the displacement and the tilt of the wall constraints are violated (Fig. 9). Then at the next iterations only the displacement of the wall is critical.

At the third iteration which is the final one, the thickness of the base of 1.0 m and the length of the heel of 5.0 m are reached. Meanwhile, the thickness of the wall has not changed during the process. The final design values give a volume of the wall as 9 m^3/m.

Table 1. Design variables of the retaining wall

Iteration	base thickness (m)	heel length (m)	wall thickness (m)
0	0.3	3.5	0.3
1	0.67	3.5	0.3
2	0.98	3.5	0.3
3	1	5	0.3

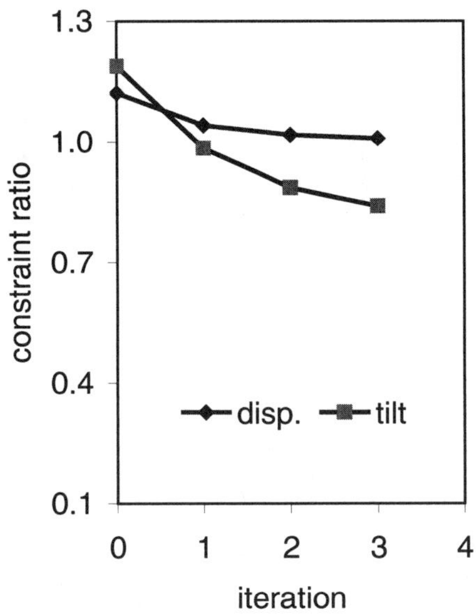

Figure 9. Constraint ratio versus iteration for the retaining wall.

6 CONCLUSIONS

The coupled finite element method and optimization technique to design optimally geotechnical structures subjected to dynamic loading has been presented in this paper. It is found that using the optimization technique, the cost of the structures can be minimised by taking into consideration various design variables and the constraints associated with vibration.

Generally, the optimization process can be accomplished within a relatively few number of design cycles including initial analysis which is not expensive compared to the benefit that can be gained by saving the cost of the structures.

The number of response analyses for every cycle depends on the number of variables used in the optimization. If finite different approach is used, the number of analyses for every cycle is the same as the number of variables with an addition of the initial analysis.

REFERENCES

Cordell Building Information Services. 1999. *Commercial/Industrial Building Cost Guide New South Wales*, Sydney

Haftka R.T., Z. Gurdal & M.P. Kamat 1990. *Element of Structural Optimization.* Dordrecht: Kluer Academic Publishers.

Lasdon, A.D, A.D. Waren, A. Jain & M. Ratner 1978. Design and Testing of a Generalized Reduced Gradient Code for Nonlinear Programming. *ACM Transactions on Mathematical Software*: 4(1):34-50.

Richart Jr., F.E., J.R. Hall Jr. & R.D. Woods 1970. *Vibration of Soils and Foundations*. Englewood Cliffs: Prentice-Hall Inc.

Sienkiewicz Z. & B. Wilczynski 1993. Minimum - Weight Design of Machine Foundation under Vertical Load. *J. Engineering Mechanics ASCE.* 119: 1781-1797.

Valliappan, S., V. Tandjiria, & N. Khalili 1997. Design Sensitivity and Constraint Approximation Methods for Optimization in Non-Linear Analysis. *Communications in Numerical Methods Engineering.* 13: 999-1008.

Valliappan, S., A. Hakam & V. Tanjiria 1999. Optimal Design of Geotechnical Structures for Lowland Areas. *J. Lowland Technology International.* 1/ 1: 67-76

White, W. 1978. Analytical Techniques in Soil Dynamics. in *Finite Element Analysis in Geomechanics* Chap. 10, S. Valliappan, ed., Sydney: UNSW.

White, W., S. Valliappan & I.K. Lee 1977. Unified Boundary Finite Dynamic Models. *J. Engineering Mechanics, ASCE.* 103: 949-964.

Computer Methods and Advances in Geomechanics, Desai et al. (eds) © 2001 Balkema, Rotterdam, ISBN 90 5809 183 X

Spectral analysis of wave propagation in a continuously non-homogeneous soil layer

R. Al-Khoury, A. Scarpas & J. Blaauwendraad
Section of Structural Mechanics, Faculty of Civil Engineering and Geosciences, Delft University of Technology, Netherlands

ABSTRACT: This contribution deals with the use of the computationally efficient spectral element technique as a means of analyzing the dynamic behavior of axisymmetric multi-layer systems consisting of homogenous and/or nonhomogeneous material layers and subjected to a stationary but transient forces. Focus is placed on the mathematical formulation and the numerical simulation of a nonhomogeneous soil material. The soil is assumed isotropic with stress dependent stiffness in the vertical direction and homogenous in the horizontal direction. The numerical results obtained from the simulation of a Falling Weight Deflectometer test for a pavement structure show that the stiffness dependency on confinement can influence the overall behavior and its effects should be taken into account in backcalculation studies.

1 INTRODUCTION

In an earlier publication (Al-Khoury et. al. 2000 (a)), wave propagation in linear elastic homogeneous multi-layer systems was studied by means of spectral analysis. However, because geotechnical materials, especially soils, exhibit stiffness dependency on confinement stresses, the method is extended in this contribution to the case of multi-layer systems consisting of both homogenous and nonhomogeneous material layers.

The soil model proposed by Vrettos (1991) is considered. Also, the Frobenius series solution to the wave equations in a half-space, as suggested by Vrettos (1991), is adopted. The difference here is that the general solution to the wave problem is obtained by use of the spectral element method (Doyle 1997). This method allows the simulation of multi-layer systems consisting of homogenous and/or inhomogeneous layers, which can be subjected to general loading conditions. The spectral element method combines elegantly the exact solution of wave motions in a layer with the finite element organization of the multi-layer system matrices.

In the spectral element, the equations of motion are solved by double summation over the relevant frequencies and wavenumbers. By means of the double summation technique the complications associated with the numerical integration between 0 and infinity are avoided. This contributes positively to the computational efficiency and robustness of the method, which makes it attractive for utilization in solving both forward and inverse problems. This research project involves the development of both forward and inverse models for layered systems. However, in this contribution only the forward model is presented.

After briefly reviewing the formulation of the spectral element for a homogenous half-space, the methodology necessary for the inclusion of the stress dependency of the stiffness in the formulation of the inhomogeneous element is highlighted. As an application, simulation of the response of a pavement structure subjected to the dynamic load pulse of a Falling Weight Deflectometer (FWD) test (Van Gurp 1995) is demonstrated. Results for two analyses are presented; one for a pavement with a homogenous subgrade layer and the other with an inhomogeneous layer.

2 SPECTRAL ANALYSIS OF WAVE MOTION IN A HOMOGENOUS SYSTEM

The equations of motion for a linear elastic homogenous isotropic material can be expressed in terms of displacement by the use of Navier's equations (Achenbach 1973);

$$(\lambda+\mu)\nabla\nabla\cdot\mathbf{u}+\mu\nabla^2\mathbf{u}=\rho\ddot{\mathbf{u}} \tag{1}$$

in which $\mathbf{u}$ is the displacement vector, ρ is the mass density of the material and ∇ is a vector differential operator; $\nabla\cdot\,\mathbf{u}$ is the divergence of $\mathbf{u}$ and $\nabla^2\mathbf{u}$ is the vector *Laplacian* of $\mathbf{u}$. λ and μ are the Lame constants.

Eq. 1 represents a set of coupled differential equations. For a homogenous system, however, this coupling can be separated by use of the Helmholtz decomposition, in which the displacement field of a material is expressed as the sum of the gradient of a scalar potential φ and the curl of a vector potential ψ as

$$\mathbf{u} = \nabla\varphi + \nabla\times\boldsymbol{\psi} \tag{2}$$

For an axially symmetric half-space system (Fig. 1) the vector potential ψ has a component ψ_θ only. This simplifies the solution of the problem to solving only for scalar potentials. For convenience of notation, ψ will be written instead of ψ_θ. Denoting the displacement components in r and z directions by u and w respectively, the relations between the displacement components and the potentials are;

$$u = \frac{\partial\varphi}{\partial r} - \frac{\partial\psi}{\partial z}, \quad w = \frac{\partial\varphi}{\partial z} + \frac{1}{r}\frac{\partial(r\psi)}{\partial r} \tag{3}$$

The two potentials satisfy the following axial symmetric wave equations:

$$\frac{\partial^2\varphi}{\partial r^2} + \frac{1}{r}\frac{\partial\varphi}{\partial r} + \frac{\partial^2\varphi}{\partial z^2} = \frac{1}{c_p^2}\frac{\partial^2\varphi}{\partial t^2} \tag{4}$$

$$\frac{\partial^2\psi}{\partial r^2} + \frac{1}{r}\frac{\partial\psi}{\partial r} + \frac{\partial^2\psi}{\partial z^2} - \frac{\psi}{r^2} = \frac{1}{c_s^2}\frac{\partial^2\psi}{\partial t^2} \tag{5}$$

where c_p and c_s are the compression and shear wave velocities expressed as

$$c_p = \left(\frac{\lambda+2\mu}{\rho}\right)^{1/2}, \qquad c_s = \left(\frac{\mu}{\rho}\right)^{1/2}$$

The corresponding Hooke's law for the stress-displacement relationships is:

$$\begin{aligned}
\sigma_{rr} &= \lambda\left(\frac{\partial u}{\partial r} + \frac{u}{r} + \frac{\partial w}{\partial z}\right) + 2\mu\frac{\partial u}{\partial r},\\
\sigma_{zz} &= \lambda\left(\frac{\partial u}{\partial r} + \frac{u}{r} + \frac{\partial w}{\partial z}\right) + 2\mu\frac{\partial w}{\partial z},\\
\sigma_{\theta\theta} &= \lambda\left(\frac{\partial u}{\partial r} + \frac{u}{r} + \frac{\partial w}{\partial z}\right) + 2\mu\frac{u}{r},\\
\tau_{rz} &= \tau_{zr} = \mu\left(\frac{\partial u}{\partial z} + \frac{\partial w}{\partial r}\right).
\end{aligned} \tag{6}$$

By means of the spectral element technique the general solution to the equations of motion can be obtained by double summation over m wavenumbers and n frequencies, which leads (Al-Khoury et. al 2000 (b)) to

$$u(r,z,t) = \sum_n \sum_m \hat{u}_{mn}(z,k_m,\omega_n)\, J_1(k_m r)\, e^{i\omega_n t} \tag{7}$$

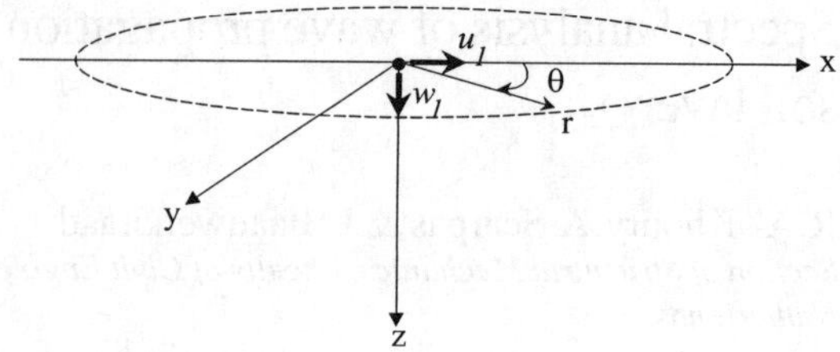

Figure 1. Axi-symmetric half-space.

$$w(r,z,t) = \sum_n \sum_m \hat{w}_{mn}(z,k_m,\omega_n)\, J_0(k_m r)\, e^{i\omega_n t} \tag{8}$$

in which:

$$\hat{u}_{mn}(z,k_m,\omega_n) = -A_{mn} k_m e^{-i\xi_{mn} z} + iB_{mn}\zeta_{mn} e^{-i\zeta_{mn} z}$$

$$\hat{w}_{mn}(z,k_m,\omega_n) = -iA_{mn}\xi_{mn} e^{-i\xi_{mn} z} + B_{mn} k_m e^{-i\zeta_{mn} z}$$

where the "hat" indicates that the solution is in the frequency domain, J_0 and J_1 are the Bessel's functions of the first kind of order 0 and 1, A_{mn} and B_{mn} are wave coefficients determined from the boundary conditions and i is the complex number $\sqrt{-1}$. k_m in Eqs. 7 and 8 is the radial component of the wavenumber and ξ_{mn} and ζ_{mn} are the vertical components of the wavenumber, expressed as

$$\xi_{mn} = \sqrt{\frac{\omega_n^2}{c_p^2} - k_m^2} \quad \text{and} \quad \zeta_{mn} = \sqrt{\frac{\omega_n^2}{c_s^2} - k_m^2}$$

In principle, the vibration shape of the system is obtained by summing over infinitely many solutions over m and n. However, double summation over discrete angular frequencies ω_n with $n = 1 \cdots N$ and wavenumbers k_m with $m = 1 \cdots M$, is adequate to produce the overall vibration shape for a system with a vanishing wave amplitudes at $r = R$ (Al-Khoury et. al 2000 (a)). The summation over N frequencies can be carried out by use of the Fast Fourier Transforms (FFT) and the summation over M wavenumbers can be done by use of the Fourier-Bessel series.

If the system is subjected to an external load pulse $P(r,t)$ with a spatial distribution $S(r)$ and a time variation $F(t)$, the force-displacement relationship can then be determined as

$$u(r,z,t) = \sum_n \sum_m \hat{u}_{mn}(z,k_m,\omega_n)\, \hat{F}_m J_1(k_m r) \hat{F}_n e^{i\omega_n t} \tag{9}$$

$$w(r,z,t) = \sum_n \sum_m \hat{w}_{mn}(z,k_m,\omega_n)\, \hat{F}_m J_0(k_m r) \hat{F}_n e^{i\omega_n t} \tag{10}$$

where $\hat{F}_m$ represents the Fourier-Bessel coefficients of $S(r)$ and $\hat{F}_n$ denotes the Fast Fourier coefficients of $F(t)$. The Fourier coefficient $\hat{F}_n$ can be obtained numerically by means of FFT. The Bessel coefficient $\hat{F}_m$, on the other hand, can be determined analytically. For a cylindrical shape force with a radius a and unit amplitude, $\hat{F}_m$ can be expressed as

$$\hat{F}_m = \frac{2a}{\alpha_m R J_1^2(\alpha_m)} J_1\left(\frac{\alpha_m}{R} r\right) \tag{11}$$

in which α_m denotes the roots of the Bessel function J_0 (Abramowitz & Stegun 1972).

Once the force-displacement relationship for a system is derived the stiffness matrix, similar to that of the finite element method, can be constructed. Detailed mathematical derivation of the stiffness matrix of an axi-symmetric spectral layer element and a spectral half-space element have been presented in Al-Khoury et. al. (2000 (a)). These elements are implemented in the computer program LAMDA (Layered Media Dynamics Analysis) developed at TU-Delft.

LAMDA was utilized for the simulation of a pavement consisting of a top layer of asphalt concrete (E=1000 MPa) 150 mm thick, a granular subbase layer (E=200 MPa) 250 mm thick and a homogenous subgrade layer (E=25 MPa) of infinite thickness. The pavement was subjected to a FWD load pulse (Fig. 2(a)) of 50 ms duration and 0.707 MPa stress amplitude over a loaded area of 150 mm radius. Because waves are described exactly in the spectral element, one element is adequate to describe a whole layer. Consequently, the size of the mesh was 3 elements, each representing a specific material layer; asphalt, subbase and subgrade.

The computed displacements at several radial distances from the center of the load as a result of the FWD load pulse are presented in Figure 2(b).

3 SPECTRAL ANALYSIS OF WAVE MOTION IN A NONHOMOGENEOUS SYSTEM

Having set up the framework for developing the spectral stiffness matrix, this section will focus on generalizing the solution for a homogenous case to that of a nonhomogeneous material layer.

It is a well known fact that the stiffness of soil increases with depth as a consequence of increasing effective overburden pressure. The soil model presented by Vrettos (1991) will be utilized. In this the shear stiffness is postulated to vary as a function of depth according to

$$\mu(z) = \mu_0 + (\mu_\infty - \mu_0)\left(1 - e^{-\alpha z}\right), \quad 0 < \mu_0 \le \mu_\infty \tag{12}$$

in which μ_0 and μ_∞ are shear modulus at the surface and at infinite depth respectively (Fig. 3), and α is a soil constant with the dimension of inverse length. (The mass density ρ and the Poisson's ratio ν are constant.)

Except for few cases of constitutive parameter combinations (Hook 1961 and Alverson et. al. 1963), the Helmholtz decomposition used for uncoupling the equation of motions of the homogenous system is not applicable for the nonhomogeneous systems. Instead, direct treatment of Eq. 1 is necessary. The closed form solution for a half space, suggested by Vrettos (1991), is generalized here to be utilized to multi-layer systems by means of the spectral element technique.

By substituting the stresses of Eq. 6 into Eq. 1, the following partial differential equations, in the time domain, are obtained:

$$\rho \frac{\partial^2 u}{\partial t^2} = \frac{2\mu(1-\nu)}{1-2\nu}\left(\frac{\partial^2 u}{\partial r^2} + \frac{1}{r}\frac{\partial u}{\partial r} - \frac{u}{r^2}\right) + \mu \frac{\partial^2 u}{\partial z^2} + \frac{\mu}{1-\nu}\frac{\partial^2 w}{\partial r \partial z} + \frac{d\mu}{dz}\left(\frac{\partial u}{\partial z} + \frac{\partial w}{\partial r}\right) \tag{13}$$

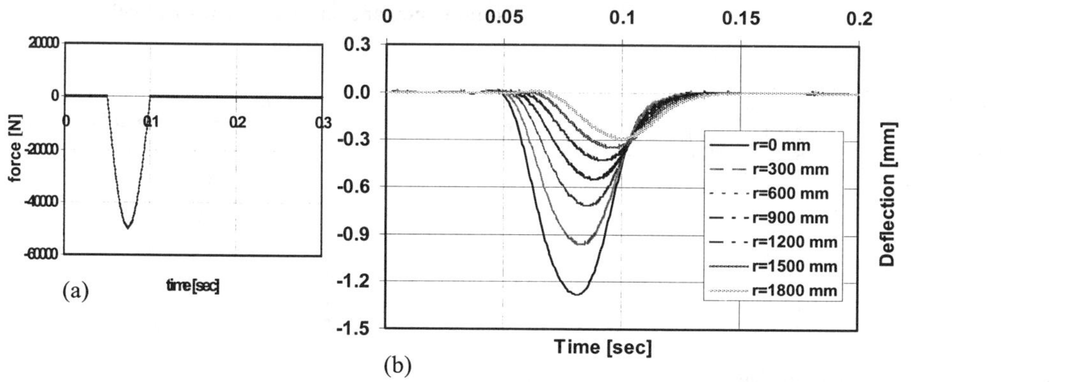

Figure 2. (a) FWD load Pulse. (b) Surface deflection for a pavement with homogenous subgrade.

$$
\begin{aligned}
\rho\frac{\partial^2 w}{\partial t^2} &= \mu\left(\frac{\partial^2 w}{\partial r^2}+\frac{1}{r}\frac{\partial w}{\partial r}\right) \\
&+\frac{2(1-\nu)}{1-2\nu}\left(\mu\frac{\partial^2 w}{\partial z^2}+\frac{d\mu}{dz}\frac{\partial w}{\partial z}\right) \\
&+\frac{2\nu}{1-2\nu}\frac{d\mu}{dz}\left(\frac{\partial u}{\partial r}+\frac{u}{r}\right) \\
&+\frac{\mu}{1-2\nu}\left(\frac{1}{r}\frac{\partial u}{\partial z}+\frac{\partial^2 u}{\partial r\,\partial z}\right)
\end{aligned}
\tag{14}
$$

in which $\mu=\mu(z)$ is defined by Eq. 12. The particular solutions of Eqs. 13 and 14, by employing Fourier transformation, can take the forms:

$$\hat{u}(r,z)=A_{mn}\,\underline{\hat{u}}(z,\omega_n,k_m)\,J_1(k_m r) \tag{15}$$

$$\hat{w}(r,z)=B_{mn}\,\underline{\hat{w}}(z,\omega_n,k_m)\,J_0(k_m r) \tag{16}$$

for each m wavenumber and n frequency, where A_{mn}, B_{mn}, ω_n, k_m, J_0 and J_1 are as defined before and the functions $\underline{\hat{u}}(z,\omega_n,k_m)$ and $\underline{\hat{w}}(z,\omega_n,k_m)$ are need to be determined.

By substituting Eqs. 12, 15 and 16 into Eqs. 13 and 14, two coupled differential equations for the radial and vertical displacements are obtained as

$$
\begin{aligned}
&\mu_\infty(1-H)\alpha^2H^2\frac{d^2\hat{u}}{dH^2}+\mu_\infty(1-H)\alpha^2H\frac{d\hat{u}}{dH} \\
&-\mu_\infty\alpha^2H^2\frac{d\hat{u}}{dH}+\left[\omega_n^2\rho-\frac{k_m^2}{\overline{\nu}}\mu_\infty(1-H)\right]\hat{u} \\
&+\frac{1-\overline{\nu}}{\overline{\nu}}k_m\,\mu_\infty(1-H)\alpha H\frac{d\hat{w}}{dH} \\
&-k_m\,\mu_\infty\,\alpha H\,\hat{w}=0
\end{aligned}
\tag{17}
$$

$$
\begin{aligned}
&\mu_\infty(1-H)\alpha^2H^2\frac{d^2\hat{w}}{dH^2}+\mu_\infty(1-H)\alpha^2H\frac{d\hat{w}}{dH} \\
&-\mu_\infty\alpha^2H^2\frac{d\hat{w}}{dH}+\overline{\nu}\left[\omega_n^2\rho-k_m^2\mu_\infty(1-H)\right]\hat{w} \\
&-(1-\overline{\nu})k_m\,\mu_\infty(1-H)\alpha H\frac{d\hat{u}}{dH} \\
&+(1-2\overline{\nu})k_m\,\mu_\infty\,\alpha H\,\hat{u}=0
\end{aligned}
\tag{18}
$$

in which $H=\left(1-\dfrac{\mu_0}{\mu_\infty}\right)e^{-\alpha z}$ and $\overline{\nu}=\dfrac{1-2\nu}{2(1-\nu)}$.

The analytical solutions to Eqs. 17 and 18, whose coefficients are not constant and exhibit regular singular points, can be attained by Frobenius method (Kreyszig 1999), which leads to

$$
\begin{aligned}
\underline{\hat{u}}_{mn}(z,k_m,\omega_n) &= A_{mn}\sum_{\ell=0}^{\infty}a_\ell^{(1)}H^{i\xi_{mn}+\ell} \\
&+B_{mn}\sum_{\ell=0}^{\infty}a_\ell^{(2)}H^{i\zeta_{mn}+\ell}
\end{aligned}
\tag{19}
$$

$$
\begin{aligned}
\underline{\hat{w}}_{mn}(z,k_m,\omega_n) &= A_{mn}\sum_{\ell=0}^{\infty}b_\ell^{(1)}H^{i\xi_{mn}+\ell}+ \\
&+B_{mn}\sum_{\ell=0}^{\infty}b_\ell^{(2)}H^{i\zeta_{mn}+\ell}
\end{aligned}
\tag{20}
$$

where the a's and b's are the coefficients of the power series, and

$$\xi_{mn}=\sqrt{\frac{\omega_n^2}{c_{p(\infty)}^2}-k_m^2}\quad\text{and}\quad\zeta_{mn}=\sqrt{\frac{\omega_n^2}{c_{s(\infty)}^2}-k_m^2}$$

are the vertical wavenumbers corresponding to the compression and shear waves respectively.

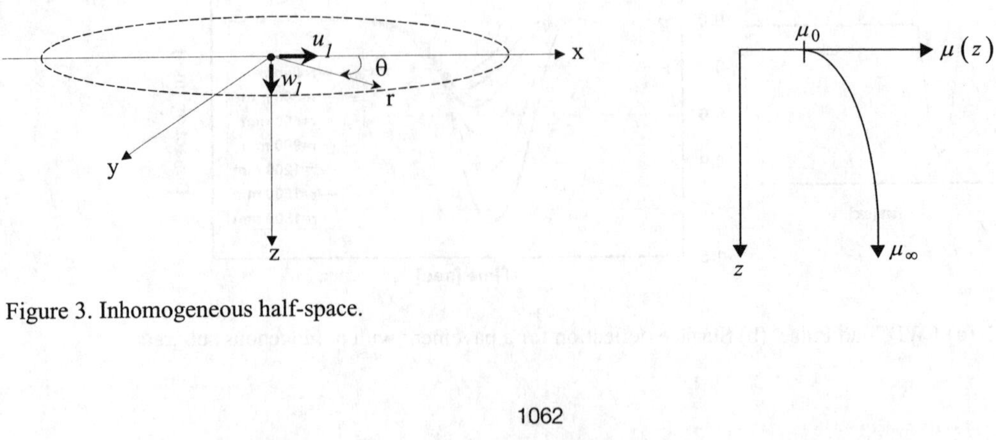

Figure 3. Inhomogeneous half-space.

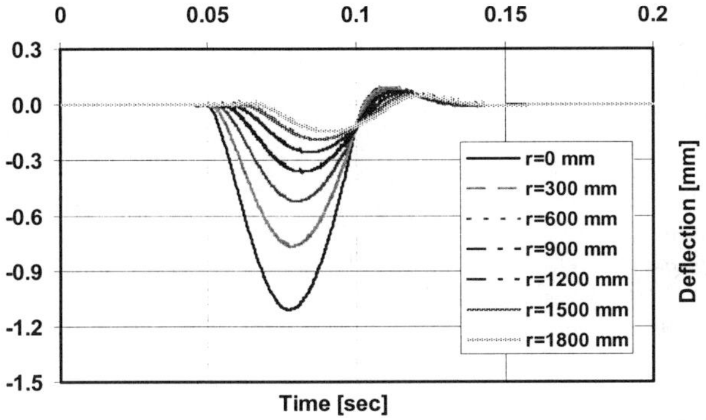

Figure 4. Surface deflection for a pavement with nonhomogeneous subgrade.

By means of the spectral element method, the general solution of wave propagation in an axisymmetric system can then be obtained by the superposition of all particular solutions as

$$u(r,z,t)=\sum_n \sum_m \hat{\underline{u}}_{mn}(z,k_m,\omega_n)J_1(k_m r)e^{i\omega_n t} \tag{21}$$

$$w(r,z,t)=\sum_n \sum_m \hat{\underline{w}}_{mn}(z,k_m,\omega_n)J_0(k_m r)e^{i\omega_n t} \tag{22}$$

The force-displacement relationship can be obtained as given in Eqs. 9 and 10. Once the force-displacement relationship is available, the stiffness matrix for the nonhomogeneous half-space element can be constructed in the same way as for the homogenous system. This element has also been implemented in LAMDA.

LAMDA was utilized for the simulation of a pavement structure similar to that in Section 2 but with nonhomogeneous subgrade layer. For this purpose the elastic stiffness of the soil was assumed to vary from 25 MPa at the top to 100 MPa at infinite depth, and $\alpha = 0.0004\, mm^{-1}$. Same load pulse as that shown in Figure 2(a) was used to simulate the FWD load pulse. The computed vertical displacements at various radial distances from the loading center are presented in Figure 4. Comparison between Figure 3 and Figure 4 reveals that the increase of stiffness with depth influences the displacements' amplitude and shape.

4 CONCLUSIONS

For the type of problems discussed in this paper, the computational advantages of the spectral element method over other methods are:

1. It is able to simulate general cases of layer combinations, material types (homogenous and nonhomogeneous) and load conditions, which is normally not possible in the closed form solutions.
2. It is able to simulate a whole layer with one element without the need for discretization, as typically required for the classical finite element method.

From the results of this contribution it can be concluded that the stiffness dependency on confinement in soil materials can be an important factor in the interpretation of the FWD test results, which may affect erroneously the backcalculation of the pavement layer parameters.

ACKNOWLEDGMENTS

The research project is partially funded by the European Commission under the Standards, Measurements and Testing programme of the 4th Framework under project name "SpecifiQ". The computational support of ir. C. Kasbergen is highly appreciated.

REFERENCES

Abramowitz, M. & I.A. Stegun 1972. *Handbook of Mathematical Functions*. New York: Dover Publication Inc.

Achenbach, J. D. 1973. *Wave Propagation in Elastic Solids*. Netherlands: North-Holland Publishing Company.

Al-Khoury, R., Scarpas, A. Kasbergen, C. Blaauwendraad, J. 2000 (a). Spectral element technique for efficient parameter identification of layered media. I. Forward calculation. *International Journal of Solids and Structures*. Accepted for publication.

Al-Khoury, R., Scarpas, A. Kasbergen, C. Blaauwendraad, J. 2000 (b). Spectral Element Method for the Dynamic Interpretation of FWD test Results. *Proceedings of the Transportation Research Board 79th Annual Meeting*, Washington D.C.

Alverson, R.C., Gair, F.C. & Hook, J.F. 1963. Uncoupled Equations of Motion in Non Homogenous Elastic Media. *Bulletin of the Seismological Society of America.* 53: #5: 1023-1030.

Brigham, E.O. 1988. *The Fast Fourier Transform and Its Applications*. Prentice-Hall.

Doyle, J.F. 1997. *Wave Propagation in Structures: Spectral Analysis Using Fast Discrete Fourier Transforms.* New York : Springer-Verlag.

Hook, J.F. 1961. Separation of the Vector Wave Equation of Elasticity for Certain Types of Inhomogeneous, Isotropic Media. *The Journal of the Acoustical Society of America.* 33: #3: 302-313.

Kreyszig, E. 1999. *Advanced Engineering Mathematics*. John Wiley & Sons.

Van Gurp, C. 1995. *Characterization of Seasonal Influences on Asphalt Pavements with the use of Falling Weight Deflectometers*. Delft: CIP.

Vrettos, C. 1991. Time-Harmonic Boussinesq Problem for a Continuously Non-Homogenous Soil. *Earthquake Engineering and Structural Dynamics.* 20:961-977.

Computer Methods and Advances in Geomechanics, Desai et al. (eds) © 2001 Balkema, Rotterdam, ISBN 90 5809 183 X

Nonlinear analyses of the earthquake behavior of a single pile founded on liquefiable sand

A. Anandarajah
Johns Hopkins University, Baltimore, Md., USA

ABSTRACT: The dynamic behavior of a pile foundation can be highly nonlinear during a strong earthquake shaking, especially when the pile is founded on a liquefiable soil. While fully-coupled elasto-plastic finite element methods are in principle capable of capturing such a nonlinear behavior, they can be rather complicated for use in practice. In the study presented here, the behavior of a single pile, founded on liquefiable sand and subjected to a base shaking in a centrifuge, is modeled using both a fully-coupled finite element method and Beam on Nonlinear Winkler Foundation (BNWF) method, and the results compared. A method of combining the two approaches is explored.

1. INTRODUCTION

While impressive advancements have been made in the area of numerical modeling of geomechanics problems, soil-structure interaction problems involving large pore water pressure changes still remain difficult to model, both from theoretical and application view points. The problem of a single pile founded on saturated sand and subjected to an earthquake loading falls in this class. An accurate modeling of the problem requires the use of a fully-coupled, elasto-plastic analysis method. One such analysis method is implemented into a computer program named HOPDYNE (Anandarajah 1990), which embodies elasto-plastic constitutive models for sands (Anandarajah 1994a, 1994b) and clays (Anandarajah and Dafalias 1986). The validity of computations by using HOPDYNE has been verified in a few recent studies (Anandarajah 1993; Anandarajah et al. 1995). While such approaches are perhaps the only means of analyzing such problems accurately, there is a need for a simpler method for use in routine practice. HOPDYNE also has capability to perform p-y curves based BNWF method of analysis of pile foundations, and equivalent linear method of analysis.

In this paper, HOPDYNE is first used to conduct a two-dimensional analysis of a single pile problem tested in a centrifuge (Wilson, 1998; Boulanger, et al., 1999). It is shown that the results compare reasonably well. A method of combining the fully-coupled approach with the BNWF approach is explored.

2. DESCRIPTION OF THE FINITE ELEMENT METHOD

A two-surface plasticity model was recently developed for describing the stress-strain behavior of liquefying sand (Anandarajah, 1994a, 1994b); a version of this model that involves 9 parameters was used to represent the soil in this study. The coupled finite element analysis is based on the fully-coupled formulation of Zienkiewicz and Shiomi (1984). These procedures are implemented into HOPDYNE.

3. TWO-DIMENSIONAL FINITE ELEMENT ANALYSIS OF THE SINGLE-PILE PROBLEM

Here HOPDYNE is used to analyze one of the problems tested in a centrifuge by Wilson (Wilson, 1998; Boulanger, et al., 1999). The centrifuge model of interest here consisted of two horizontal layers of Nevada sand containing a number of model structures, including single piles and pile groups; here the behavior of a single pile is analysed. The bottom and top layers were 11.4m thick dense Nevada sand (Dr=80%) and 9.3m thick medium-dense (Dr=60%) Nevada sand respectively. The model is referred to as Csp3 and the reader may find the full details in Wilson (1998). Here the structure in prototype dimensions is considered for analysis. The pile has a diameter of 0.667m and a wall thickness of 72cm. The depth of embedment is 16.8m. The free length of the pile above the ground level is 3.81m. A weight of 500 kN was placed at the top of the pile. The cross sectional area, second moment of inertia and Young's modulus of the pile are 0.135 m^2, 0.006 m^4, and 70,000 kPa respectively. An equivalent, one-foot thick domain was used in the 2-D, plane strain finite element analysis. The values of moment of inertia and the cross sectional area were increased by a factor of 1/0.667=1.5; that is, the pile is assumed to extend all the way in the z-direction, and a one-foot thick segment is chosen for the 2-D analysis. Therefore, the influence of that part of the soil in the z-direction that does not contain the pile is ignored.

Full details of analyses and experimental results may be found in Anandarajah (2000). The acceleration time history of the pile head obtained from the finite element analysis is compared in Fig. 1 with the measured acceleration time history. Also presented in Fig. 1 are the response

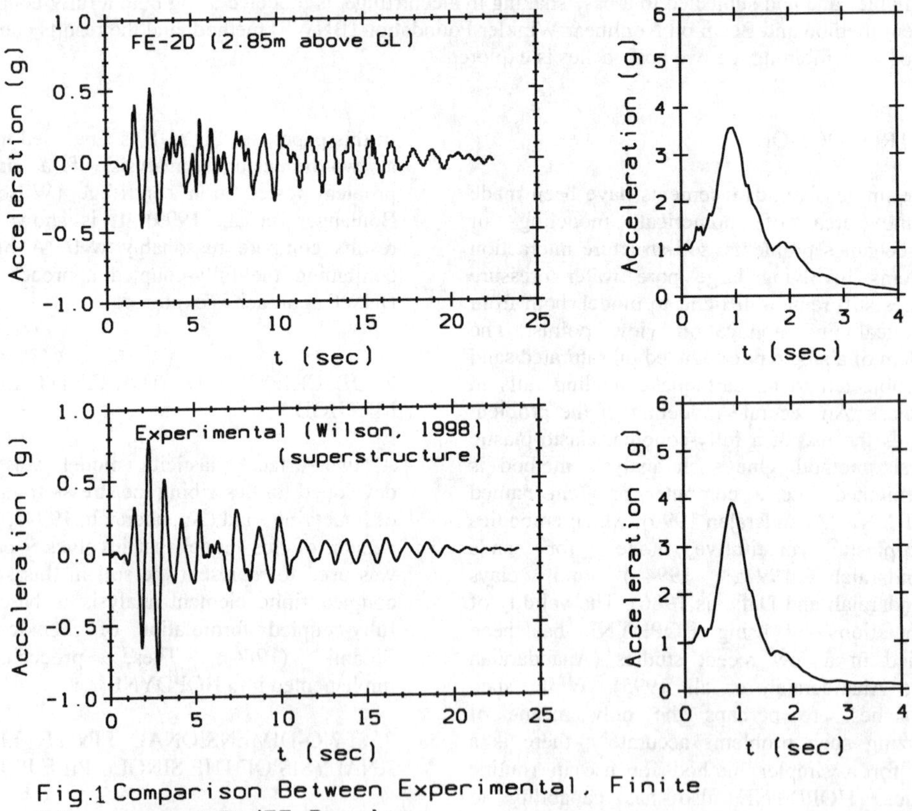

Fig.1 Comparison Between Experimental, Finite Element and Py/FE Results

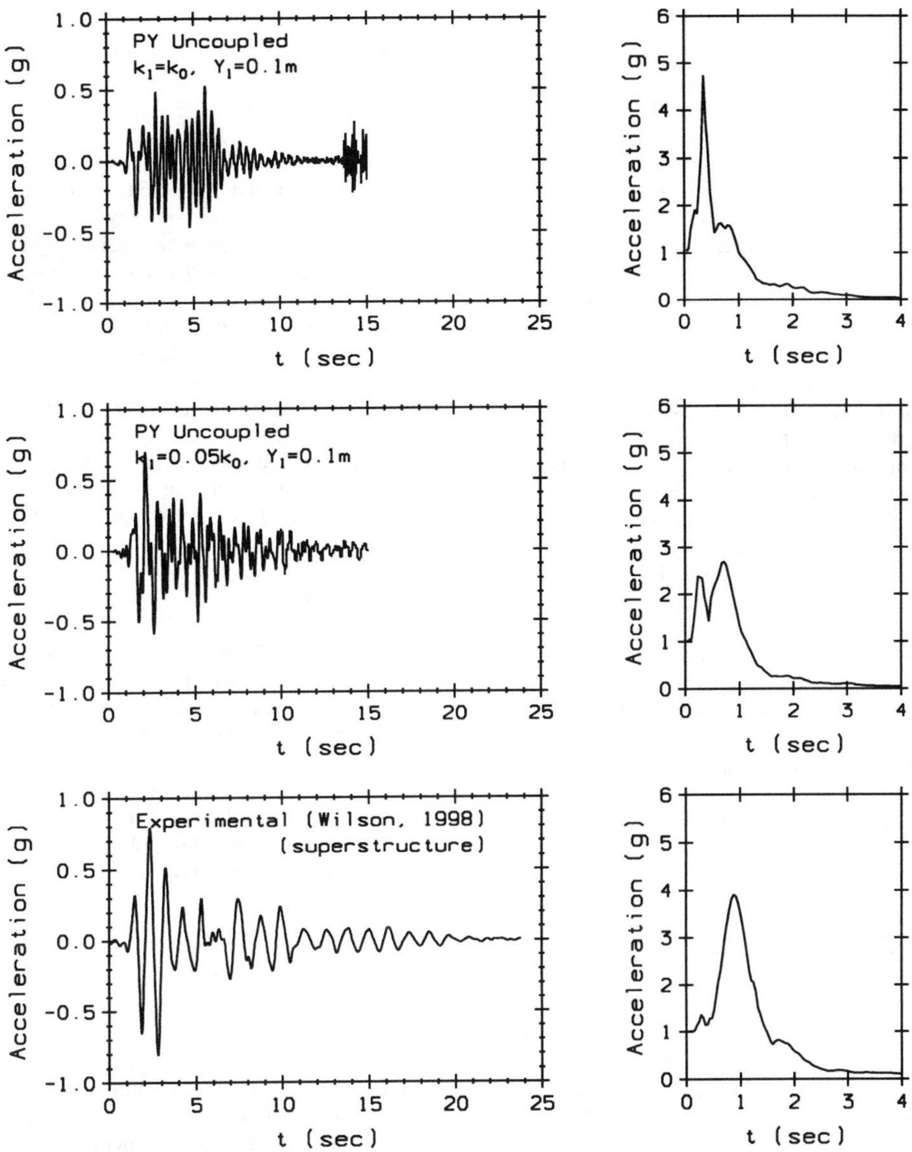

Fig. 2. Comparison of Experimental and Theoretical (p-y Method with Free-Field Respeonse by Fully-Coupled

spectra of the acceleration time histories. It is evident that the results compare reasonably well. The pore water pressure that developed at various points around the pile also compared reasonably well with the finite element results, and the details may be found in Anandarajah (2000).

4. MODIFIED BNWF

Here the influence of the soil surrounding the pile is approximated by a set of springs and dashpots (radiation damping) connected between the free filed and the vertical line representing the axis of the pile. The springs and dashpots are connected in parallel in the study presented here. The pile is subjected to the earthquake acceleration at the base. The free-field motion is fed at the other end of the springs and dashpots. It is, therefore, clear that a method is needed for the prediction of the free-field response of the soil, with the pore water pressure effect properly accounted for. Here, this is done using HOPDYNE (Anandarajah, 1990). The simple model proposed by Berger et al. (1977) was used to approximate the dashpot coefficients, which are to represent the radiation damping of the problem. The force-displacement characteristics of the springs are represented by the nonlinear p-y springs proposed by Reese et al. (1974).

While the magnitude of the predicted accelerations at various points along the length of the pile was reasonably close to that of the measured ones, the frequency content of the predicted accelerations deviated significantly from that of the measured ones. In examining the cause of the discrepancy, it was also discovered that the Berger at al.'s (1977) model overdamped the system. The primary reason for the stiff response obtained from the analysis was, however, found to be due to failure to allow the springs to "soften" as a function of the rising pore water pressure. To delve further into these aspects, The Reese et al.'s p-y relation was replaced by a simple elastic-perfectly plastic relation, with the slope of the elastic part matched with the initial slope of the Reese et al.'s relation. The radiation damping was reduced to a negligible value. The predicted acceleration of the superstructure is shown in Fig. 2a. The predominant period of the motion is about 0.35s, whereas that of the measured acceleration (Fig. 2c) was about 1.0s. An analysis was then repeated with a reduced initial stiffness ($1/20^{th}$); the calculated acceleration and the corresponding frequency response are shown in Fig. 2b. It may now be seen that a new peak begins to develop on the response spectra at a period very close to 1.0s, indicating that the pore water pressure induced softening of the p-y springs must be accounted for in the analysis for accurately capturing the behavior of the pile-soil system; a suitable method of achieving this is currently being developed.

5. CONCLUSIONS

The behavior of a single pile, founded on liquefiable sand and subjected to an earthquake base shaking, is analyzed by two different methods: (1) two-dimensional fully-coupled finite element method, and (2) p-y curve based Beam on Nonlinear Winkler method, with the free-field response computed using the fully-coupled finite element method. The calculated response was compared with that measured in a centrifuge. It is found that the finite element method yields reasonable results. The p-y curves based simplified method has the potential of capturing the behavior provided that pore water pressure induced softening of the p-y spring stiffness can be modeled.

6. REFERENCES

Anandarajah, A. (2000). "Fully-Coupled Analysis of a Single Pile Founded on Liquefiable Sands." *Specilaty Publication: Computer Simulation of Earthquake Effects*, ASCE, GeoDenver 2000.

Anandarajah, A. (1994a). "A Constitutive Model for Granular Materials Based on Associated Flow Rule."*Soils and Foundations*, Japanese Society of Soil Mechanics and Foundation Engineering, 34(3):81-98.

Anandarajah, A. (1994b). "Procedures for Elasto-Plastic Liquefaction Modeling of Sands." *Journal of Engineering Mechanics Division*, ASCE, 120(7):1563-1589.

Anandarajah, A., Rashidi, H. and Arulanandan, K. (1995). "Elasto-Plastic Finite Element Analyses of Earthquake Pile-Soil-Structure Interaction Problems Tested in a Centrifuge." *Computers and Geotechnics*, 17:301-325.

Anandarajah, A. and Dafalias, Y.F. (1986). "Bounding Surface Plasticity, Part 3: Application to Anisotropic Soils." *Journal of Engineering Mechanics Division*, 112(12):1292-1318.

Anandarajah, A. (1993). "VELACS Project: Elasto-Plastic Finite Element Prediction of the LiquefactionBehavior of Centrifuge Models Nos. 1, 3 and 4a," *Proc. Intl. Conf. on Verification of*

Numerical Procedures for the Analysis of Soil Liquefaction Problems, Davis, California, Oct. 17-20, eds. K. Arulanandan and R. F. Scott, 1, 1075-1104.

Anandarajah, A. (1990). "HOPDYNE -- A Finite Element Computer Program for the Analysis of Static, Dynamic and Earthquake Soil and Soil-Structure Systems", The Johns Hopkins University, Baltimore, Maryland.

Berger, E., Mahin, S.A. and Pyke, R. (1977). "Simplified Method for Evaluating Soil-Pile Structure Interaction Effects." *Proc. 9th. Offshore Technology Conference*, OTC Paper No. 2954, Huston, Texas, pp. 589-598.

Boulanger, R.W., Wilson, D.W., Kutter, B.L. and Abghari, A. (1999). "Soil-Pile-Superstructure Interaction in Liquefiable Sand." Transportation Research Record 1569, pp. 55-64.

Reese, L.C., Cox, W.R., and Kooper, F.D. (1974). ``Analysis of Laterally-Loaded Piles in Sands." *Proc. 6th. Annual Offshore Technology Conference*, Huston, Texas, OTC 2080.

Wilson, D. W. (1998). "Soil-Pile Superstructure Interaction in Liquefiable Sand and Soft Clay." Doctoral Dissertation, University of California, Davis, 1998, 173pp.

Zienkiewicz, O. C. and Shiomi, T. (1984). "Dynamic Behavior of Saturated Porous Media: The Generalized Biot Formulation and Its Numerical Solution." International Journal for Numerical and Analytical Methods in Geomechanics, 8:71-96.

Computer Methods and Advances in Geomechanics, Desai et al. (eds) © 2001 Balkema, Rotterdam, ISBN 90 5809 183 X

Wave-induced pore pressure in an anisotropic seabed in the vicinity of a composite caisson breakwater

D.H.Cha & D.S.Jeng
School of Engineering, Griffith University, Gold Coast Campus, Qld, Australia

ABSTRACT: To simplify the complicated problem, all previous investigations for the wave-seabed-caisson interaction have been considered isotropic soil behaviour, although strong evidence of anisotropic soil behaviour has been reported in the literature. In this paper, we consider the wave-seabed-caisson interaction in an anisotropic seabed. Based on Biot's poro-elastic theory, a two-dimensional finite element model is used. The numerical results indicate that the effects of anisotropic soil behaviour, soil types and the degree of saturation on the wave-induced pore pressure cannot always be ignored.

1 INTRODUCTION

Protection of the coastal environment is a vital issue for most countries in the world. Caisson-type breakwaters and foundations are one of the most common-used marine structures. Most of the previous investigations related to caissons have been concerned with the wave forces on the structures (Oumeraci, 1994). Only few studies have attempted to investigate the wave-induced pore pressure and effective stress changes within an isotropic seabed (Mase et al., 1994).

In the field, most marine sediments display certain degree of anisotropic anisotropy with different elastic properties in different directions due to the kind of their deposition, particle shape and stress history. However, Many materials show more limited forms of anisotropy. For example, a cross-anisotropic material possesses an axis of symmetry in the sense that its properties are independent of rotation of a sample about the axis of symmetry. This implies that a cross-anisotropic material has the same properties in all horizontal direction (Pickering, 1970). To simplify the complicated problem, all previous investigations in the area of wave-seabed-caisson interaction have only considered isotropic soil behaviour.

The main intention of this paper is to examine the effects of anisotropic soil behaviour on the wave-induced pore pressure around a caisson. A two-dimensional finite element model proposed by Lin & Jeng (2000) will be adopted here. Based on the numerical model, the effects of cross-anisotropic soil behaviour, soil types and degree of saturation on the wave-induced pore pressure will be examined.

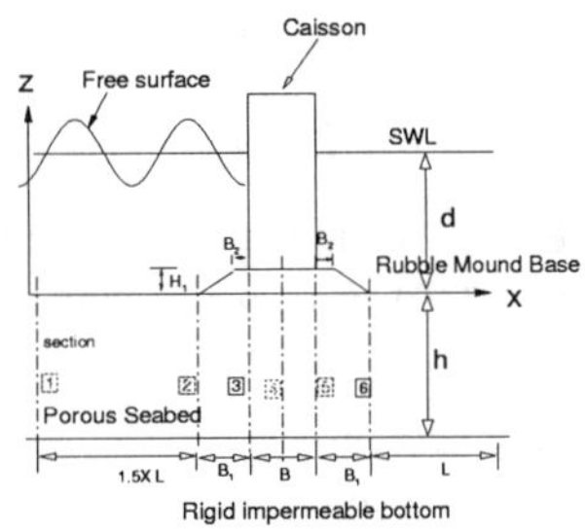

Figure 1: Definition of wave-seabed-caisson interaction problem.

2 THEORETICAL FORMULATIONS

2.1 *Governing equation*

In this study, we consider a fully reflected standing wave in front of a caisson-type breakwater, as shown in Figure 1. Based on conservation of mass, Biot's consolidation equation (Biot's, 1941) is used as the governing equation for the wave-induced seabed response (Jeng, 1997), that is

$$K\nabla^2 p - \gamma_w n_e \beta \frac{\partial p}{\partial t} = \gamma_w \frac{\partial \epsilon}{\partial t}, \tag{1}$$

where, p is the wave- induced pore pressure, K is per-

*Corresponding author. Email: d.jeng@mailbox.gu.edu.au

meability, n_e is soil porosity, γ_w is the unit weight of pore fluid, $\epsilon = \frac{\partial u}{\partial x} + \frac{\partial w}{\partial z}$ is the volumetric strain and $\beta (= \frac{1}{K_w} + \frac{1-S}{P_{wo}})$ is the compressibility of pore fluid (where, K_w is taken as $2 \times 10^9 \mathrm{N/m}^2$. P_{wo} is the absolute water pressure and S is the degree of saturation. Under the condition of plane strain, the equations of equilibrium can be expressed as (Jeng, 1997)

$$C_{11}\frac{\partial^2 u}{\partial x^2} + G_z\frac{\partial^2 u}{\partial z^2} + (C_{13}+G_z)\frac{\partial^2 w}{\partial x \partial z} = \frac{\partial p}{\partial x}, \quad (2)$$

$$G_z\frac{\partial^2 w}{\partial x^2} + C_{33}\frac{\partial^2 w}{\partial z^2} + (C_{13}+G_z)\frac{\partial^2 u}{\partial x \partial z} = \frac{\partial p}{\partial z}, \quad (3)$$

in the x- and z-directions, respectively.

In (2) and (3), the C_{ij} coefficient are given as

$$\begin{aligned} C_{11} &= E_x(1-\mu_{xz}\mu_{zx})/\Delta, \\ C_{13} &= E_x\mu_{xz}(1+\mu_{xx})/\Delta, \\ C_{33} &= E_z(1-\mu_{xx}^2)/\Delta, \\ \Delta &= (1+\mu_{xx})(1-\mu_{xx}-2\mu_{xz}\mu_{zx}), \end{aligned} \quad (4)$$

where E_x, E_z are Young's modulus in the x- and z-directions, respectively, μ_{xx}, μ_{xz} and μ_{zx} are Poisson's ratios.

2.2 *Boundary conditions*

From the governing equations (1)–(3), that involved the soil displacement and pore pressure, can be solved with the boundary conditions at the surface of the seabed and impermeable bottom.

(1) Bottom Boundary Condition (BBC):

$$u = w = \frac{\partial p}{\partial z} = 0, \ \text{at} \ z = -h, \quad (5)$$

(2) Surface Boundary Condition (SBC):

$$n_x\sigma_x' + n_z\tau = -n_x p, \& n_x\tau + n_z\sigma_z' = -n_z p,$$

$$p = \frac{\gamma_w H}{\cos kd}\cos kx \cos \omega t = p_o \cos kx \cos \omega t, \quad (6)$$

for the part subjected to wave motion, and $p = 0$ for the part of no wave action. It is noted that only dynamic wave pressure is considered in the study. The static wave pressure is not included here. In (6), p_o is the amplitude of the wave pressure, H is the wave height, k is the wave number ($k = 2\pi/L$, in which $L=$ the wavelength); and ω is the wave frequency ($\omega = 2\pi/T$; $T=$wave period).

(3) Lateral Boundary Condition (LBC):

Since the influence of structure on the wave-induced seabed response will vanish at points far away from the structure, the results of the wave-seabed interaction without caisson will be used as the lateral boundary of the left-hand side. For the lateral boundary condition of the right-hand side, we assume there is no effect of dynamic wave pressures. Thus, the pore pressure and soil displacements will vanish far away from the caisson, i.e., $p = u = w = 0$.

2.3 *Finite element analysis*

Since the wave induced oscillatory soil response is periodically fluctuating in the temporal domain, the wave-induced pore pressure, effective stresses and soil displacements can be expressed as

$$f(x,z;t) = [f_r(x,z) + i f_c(x,z)]\ e^{-i\omega t}, \quad (7)$$

where subscripts and "r" and "c" represent the real and imaginary parts of the soil response, respectively. f denotes the unknown parameters (i.e. pore pressure, stresses, and displacements).

The first step to solve the wave-soil-caisson interaction is to obtain the lateral boundary conditions on the left-hand side of the composite breakwater. To do so, the wave-seabed interaction can be solved by employing the principle of repeatability (Zienkiewicz & Scot, 1972). This concept is particularly convenient for periodical loading, and has firstly been successfully applied in the problem of wave-seabed interaction (Lin & Jeng, 2000). Once the later boundary conditions are obtained, the whole wave-seabed-pipe problem can be solved with the boundary conditioned described in Section 2.2.

3 RESULTS

To have a basic understanding how cross-anisotropic soil behaviour affects the wave-induced pore pressure around a caisson, a set of input data is tabulated in Table 1.

3.1 *Effects of anisotropic soil behaviour*

To examine the effects of anisotropic soil behaviour on the wave-induced pore pressure, we choose six cross sections as shown in Figure 1. The contours of the wave-induced pore pressure in both isotropic and

Table 1: Input data for case study[1]

Wave characteristics	
Wave period T	10 sec
Water depth d	20 m
Soil characteristics	
Soil depth h	40 m
Degree of saturation S	0.95, 0.975, 1.0
Soil porosity n_e	0.35 (CS[2] and FS) 0.3 (GL)
Permeability K	10^{-1} m/s (GL) 5×10^{-2} m/s (CS) 5×10^{-4} m/s (FS)
Young's modulus E_z	5×10^7 N/m^2 (CS and FS) 5×10^8 N/m^2 (GL)
Anisotropic parameters	
Poisson's ratio μ_{xx}	0.3
Poisson's ratio μ_{xz}	0.35
Anisotropic constant n	0.8
Anisotropic constant m	0.6

[1] For isotropic soil behaviour:
$n = 1, m = 1/2(1+\mu), \mu = \mu_{xx}$
[2] CS=coarse sand, FS=fine sand, GL=gravel

anisotropic seabed are presented in Figure 2. The figure clearly indicates the major differences between isotropic and anisotropic solutions normally occur near the seabed bottom. The maximum difference between two solutions reaches to 5% of p_o in this example. Also, the vertical distributions of the wave-induced pore pressure (p/p_o) versus the soil depth $(z-h)/h$ are illustrated in Figure 3. Figure 3 also presents the different distribution of pore pressure at different sections. In engineering practice, the toe of a composite breakwater (i.e. section 2) and the corner of caisson and rubble mound (i.e., section 3) are commonly damaged under the action of waves. Comparing the distributions of the pore pressure at sections 2 and 3, we found that the gravel rubble mound can reduce the pore pressure to a certain degree. This is the reason why the pore pressure at section 3 is less than that at section 2 beneath the seabed surface, although the wave pressure at section 3 is greater than that at section 2.

3.2 *Effects of degree of saturation*

Soil characteristics are group of parameters, which significant affect the wave-induced seabed response. Herein, we only examine the effects of two most important soil parameters, degree of saturation and soil types. Figure 4 illustrates the effects of the degree of saturation on the wave-induced pore pressure at sections 2 and 3 in a cross-anisotropic seabed. In the figure, both of the coarse sand and fine sand are considered. Generally speaking, the magnitude of pore pressure decreases as the degree of saturation decreases and has similar trend in coarse sand. However, the

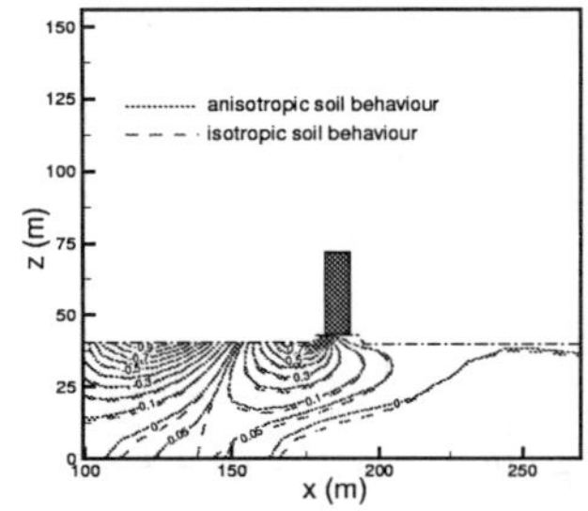

Figure 2: Contours of the wave-induced pore pressures in both anisotropic and isotropic seabeds.

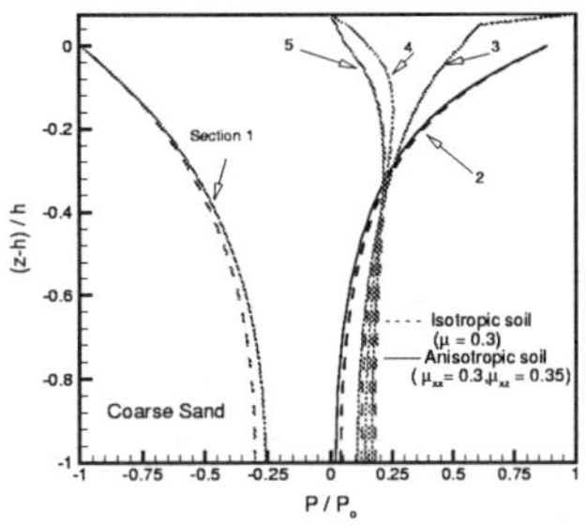

Figure 3: Vertical distributions of the wave-induced pore pressure in both anisotropic and isotropic seabeds.

reduction of pore pressure in saturated fine sand is much rapidly than in unsaturated fine sand, although only 2.5% of gas exists in the pore fluid. This implies that the degree of saturation is more sensitive to the wave-induced seabed response in fine sand, rather than coarse sand.

3.3 *Effects of soil type*

Figure 4 also illustrates the influences of soil type on the wave-induced pore pressure. Basically, the vertical distribution of pore pressure (p/p_o) versus soil depth, $(z-h)/h$, have different trends in coarse sand and fine sand. In fine sand, the pore pressure gradually decreases as soil depth increases. However, the pore pressure decreases to a certain values from $(z-h)/h$ =0 to -0.2, and then increases and approach to a constant near the seabed bottom. This trend is more significant in a fully saturated seabed.

4 CONCLUSIONS

1. Effects of anisotropic soil behaviour on the wave-induced pore pressure may reach 5% of p_o under certain combination of wave and soil conditions.

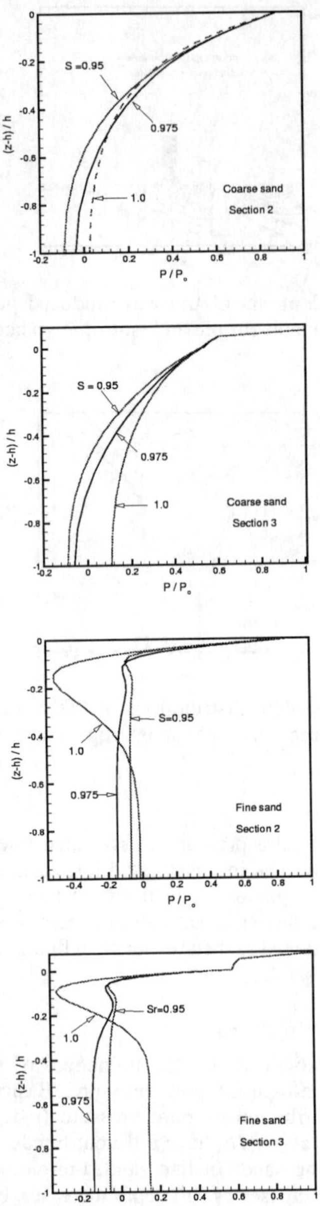

Figure 4: Vertical distributions of the wave-induced pore pressure (p/p_o) versus soil depth ($(z-h)/h$) at sections 2 & 3 for various degrees of saturation in coarse sand and fine sand.

2. Basically, the wave-induced pore pressure is significantly affected by degree of saturation (S), especially in fine sand.

3. The vertical distribution of the wave-induced pore pressure in fine sand is different from that in coarse sand. A dramatically reduction of pore pressure is observed in the region between seabed surface and 20% of the seabed thickness in fine sand.

ACKNOWLEDGMENTS

The second authors is grateful for the financial supports from National Science Council (NSC) of Taiwan and Australian Research Council Small Grant Scheme during 2000.

REFERENCES

Biot. M. A. 1941. General theory of three-dimensional consolidation. *Journal of Applied Physics*, 12:155-164

Jeng, D. S. 1997. Soil response in cross-anisotropic seabed due to standing waves. *Journal of Geotechnical and Geoenviromental Engineering, A.S.C.E.*, 123(1):9-19

Lin, Y. S. & Jeng, D. S. 2000. Effects of variable shear modulus on wave-induced seabed response. *Journal of the Chinese Institute of Engineers*, 24(1):109-115.

Mase, H., Sakai, T. & Sakamoto, M. 1994. Wave-induced pore water pressures and effective stress around breakwater. *Ocean Engineering*, 21(4):361-379.

Oumerachi, H. 1994. Review and analysis of vertical breakwater failure-lessons learned. *Coastal Engineering*, 22:3-29.

Zienkiewicz, O. C. & Scott, F. C. 1972. On the principle of repeatability and its application in analysis of turbine and pump impellers. *International Journal for Numerical Methods in Engineering*, 9:445-452.

Computer Methods and Advances in Geomechanics, Desai et al. (eds) © 2001 Balkema, Rotterdam, ISBN 90 5809 183 X

Numerical simulation of liquefaction-induced lateral spreading

Ahmed Elgamal & Zhaohui Yang
University of California, San Diego, Calif., USA

ABSTRACT: During liquefaction, a large body of experimental research suggests a possible strong influence of soil-skeleton dilation at large cyclic shear strain excursions. Such phases of dilation can result in significant regain in shear stiffness and strength, and are deemed to control the magnitude of observed liquefaction-induced deformation in many actual situations (clean sands and silts). Based on these observations, a constitutive model is developed, and made available (within a fully coupled solid-fluid FE program) over the Internet (**http://casagrande.ucsd.edu**) for on-line execution. A numerical study conducted on this web-site demonstrates the significance of excitation frequency and amplitude in dictating the extent of post-liquefaction lateral spreading in sloping ground.

1 INTRODUCTION

During liquefaction, recent records of seismic site response (Holzer *et al.* 1989, Zeghal and Elgamal 1994) suggest a possible strong influence of soil dilation during cyclic loading excursions. Such phases of dilation may result in significant regain in shear stiffness and strength at large cyclic shear strain excursions, leading to a strong restraining effect on the magnitude of cyclic and accumulated permanent shear strains. Such restraint on shear strain has been referred to as a form of cyclic-mobility in a large number of pioneering liquefaction studies (Seed and Lee 1966, Casagrande 1975, Castro and Poulos 1977, Seed 1979). For the important situations of biased strain accumulation due to an initial locked-in shear stress (e.g., lateral spreading), this pattern of behavior may play a dominant role in dictating the extent of shear deformations.

Currently, the above mentioned effects are thoroughly documented by a large body of experimental research (employing clean sands and clean non-plastic silts), including centrifuge experiments (Taboada and Dobry 1998, Dobry and Abdoun 1998, Balakrishnan and Kutter 1999), shake-table tests, and cyclic laboratory sample tests (Arulmoli et al. 1992). A thorough summary has been compiled (Elgamal *et al.* 1998) of the relevant: i) seismic response case histories, ii) recorded experimental response (centrifuge, shake table, and laboratory), and iii) available constitutive models to simulate this phenomenon.

In the following sections, the main characteristics of the above-described shear stress-strain mechanisms are presented. The framework for a newly developed computational constitutive model and the salient model response characteristics are outlined. This computational framework is integrated into a general two-dimensional (2D) solid-fluid coupled, effective stress Finite Element program. A 1D version of this code (*CYCLIC1D*) is now available for interested researchers to use over the Internet (**http://casagrande.ucsd.edu**). As presented in this paper, a numerical study conducted on this web site reveals the significant influence of input shaking frequency and amplitude on the extent of post-liquefaction lateral spreading.

2 CYCLIC LATERAL SPREADING MECHANISM

An illustration of the dilative-tendency mechanism observed in undrained cyclic laboratory tests is shown in Fig. 1 (Arulmoli *et al.* 1992). This figure depicts the mechanism of accumulation of cycle-by-cycle deformations. This cyclic mobility mechanism can significantly reduce the total accumulated shear strain due to liquefaction. Accuracy in reproducing such response is among the most important goals of a newly developed constitutive model.

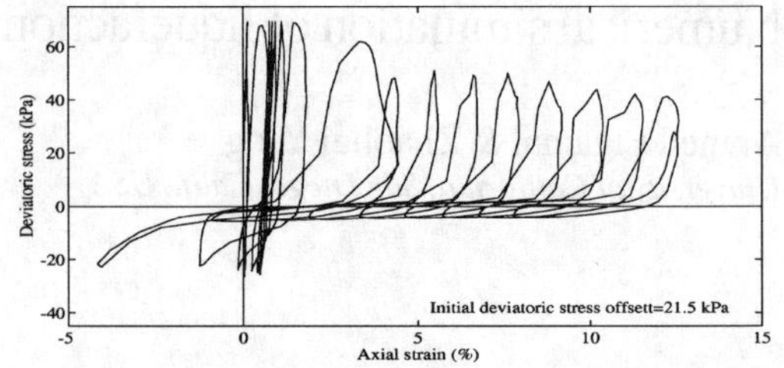

Figure 1. Stress-strain history during an undrained stress-controlled cyclic triaxial test of Nevada sand (Dr=40%) with an imposed static deviatoric stress (Arulmoli et al. 1992).

3 LIQUEFACTION CONSTITUTIVE MODEL

Currently available constitutive models that reproduce important aspects of the above-mentioned dilative shear mechanism include those by Iai (1991), Iai *et al.* (1995), Kramer and Arduino (1999), Tateishi *et al.* (1995), and Cubrinovski and Ishihara (1998). An additional new constitutive model was developed (Parra 1996, Yang 2000), based on the framework of multiple-yield-surface plasticity (Iwan 1967, and Mroz 1967), as proposed and implemented by Prevost (1985) for frictional cohesionless soils. Modified from its original form (Prevost 1985), the new model includes a set of non-associative flow rules (Parra 1996) to reproduce the observed strong dilation/ contraction effects during loading - unloading - reloading (Fig. 2, phases 2-3 and 3-4) near the yield envelope.

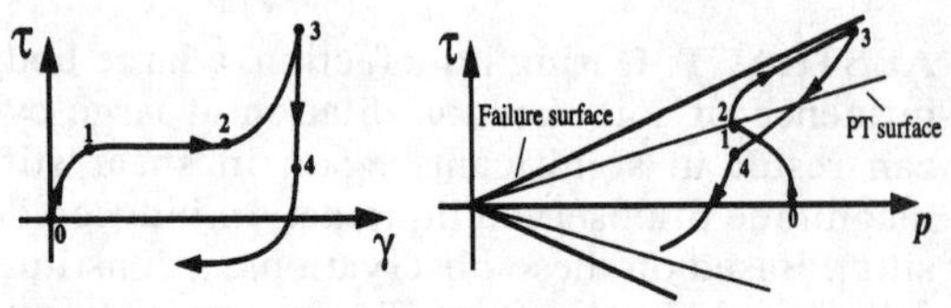

Figure 2. Response characteristics of the new constitutive model: typical shear stress-strain response and effective stress path (Parra 1996, Yang 2000).

The developed model is able to reproduce the shear stress-strain features discussed in Section 2, including the biased accumulation of cyclic shear strains due to the presence of an imposed static shear stress (Fig. 1). In particular, cyclic shear strain accumulation in this model displays a dependence on the degree of load reversal (Parra 1996, Yang 2000), as shown in Fig. 3. Plot (1) in this figure shows that a unsymmetric cyclic shear loading will introduce a net shear strain accumulation (γ_1) in the down-slope direction, followed by a strong dilation phase mimicking the stress-strain response of Fig. 1. On the other hand, plot (2) shows that a symmetric shear loading will cause dilation to take place earlier upon reloading, rendering insignificant accumulation of permanent shear-strain.

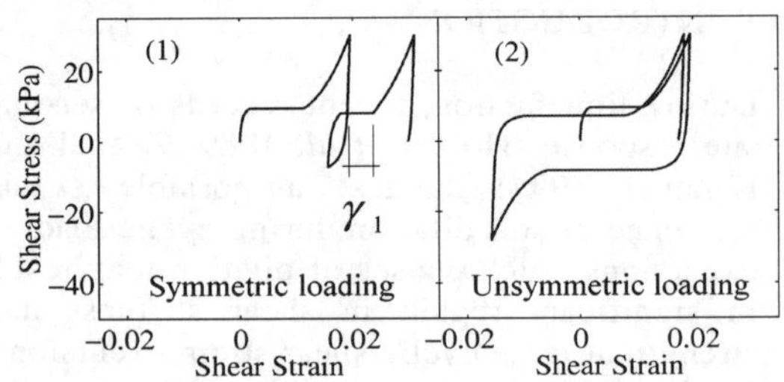

Figure 3. Model shear stress-strain response under symmetric and unsymmetrical (down-slope) shear loading.

The model parameters for medium density (Dr ≈ 40%) Nevada sand were calibrated by extensive laboratory tests (Arulmoli *et al.* 1992) and centrifuge experiments (Parra 1996, Yang 2000). This calibrated set of model parameters was selected in the numerical study to be presented below.

The new constitutive model was integrated into a general-purpose 2D finite element program *CYCLIC* (Parra 1996, Yang 2000). *CYCLIC* implements a fully coupled solid-fluid formulation (known as *u-p* formulation), following the procedures developed by Chan (1988) and Zienkiewicz *et al.* (1990). A 1D version of *CYCLIC, CYCLIC1D,* is now available (**http://casagrande.ucsd.edu)** for on-line execution using commonly available Internet browsers. Currently, the user can: 1) select a sand relative density range, 2) define element height (the soil stratum consists of 10 elements), 3) choose a stratum inclination angle (for mildly inclined infinite-slope situations), and 4) select/scale either a sinusoidal motion (with user specified frequency, amplitude, and number of cycles) or an actual earthquake record. Upon completion, computed results are displayed graphically, and can be downloaded electronically. The numerical study presented in the next section was carried out on this *CYCLIC1D* web site.

4 EFFECTS OF SHAKING FREQUENCY & AMPLITUDE ON POST-LIQUEFACTION LATERAL SPREADING

Shaking frequency and amplitude are two of the major attributes characterizing an input excitation. Therefore, it is important to shed light on how these two factors influence the extent of lateral spreading of a sloping liquefied ground. When these two factors are considered separately, one may easily conjecture that lower shaking frequencies or larger excitation amplitudes will result in more lateral deformations. However, the numerical study of this section shows that more complex mechanisms are involved.

In the *CYCLIC1D* web site, the soil column was chosen to be 10m in height, with 4 degrees mild slope inclination. A sinusoidal base input motion was selected with 6 cycles of full amplitude *A*, followed by 2 cycles of smaller amplitude (Fig. 4). A total of 55 numerical runs were conducted by varying shaking amplitude (*A* = 0.2g, 0.4g, 0.6g and 0.8g) and shaking frequency (from 0.05Hz to 3.0Hz). In most cases, near-surface elements liquefied within 2 or 3 shaking cycles.

Fig. 5 depicts lateral displacement histories (relative to the base) at the ground surface for all shaking amplitudes at a high (3.0Hz) and a low frequency (0.1Hz). In the case of 3Hz, as may be expected, displacements are seen to be proportional to the shaking amplitude. However, in the case of 0.1Hz, the largest permanent displacement was generated by the lowest amplitude (0.2g). At this low frequency, the high-amplitude cases (0.6g and 0.8g) show minimal accumulation of biased permanent deformation throughout.

Fig. 6 summarizes the average displacement accumulation per cycle *d* (after liquefaction), for all simulations. Here, the definition of *d* is such that the difference between the peak displacement at the 3rd and 6th cycles is 3*d* (see Fig. 6 for example). The first two cycles were excluded in the averaging process because during this initial period of shaking, excess pore pressure in the soil profile was relatively low, and soil shear resistance might be considerably higher than in the subsequent cycles. Therefore, *d* may be considered as a measure of post-liquefaction cyclic lateral spreading.

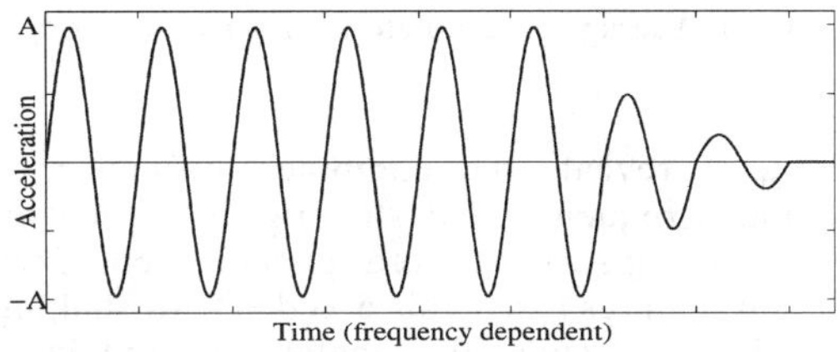

Figure 4. Selected Base input excitation.

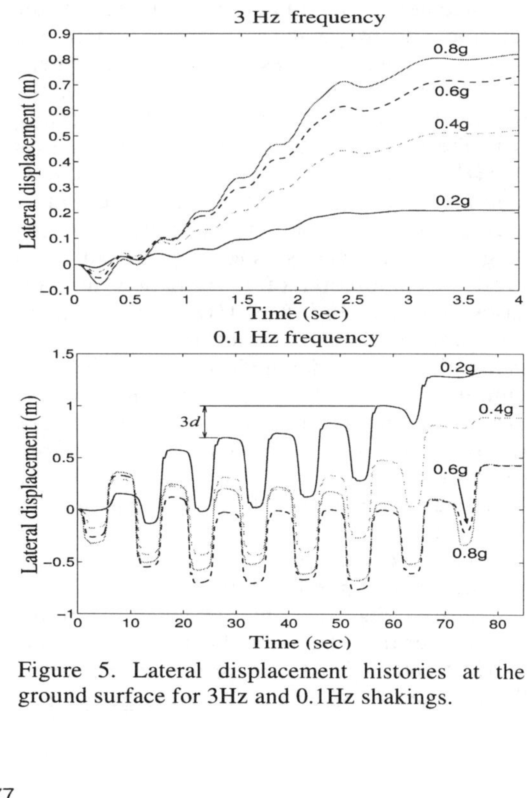

Figure 5. Lateral displacement histories at the ground surface for 3Hz and 0.1Hz shakings.

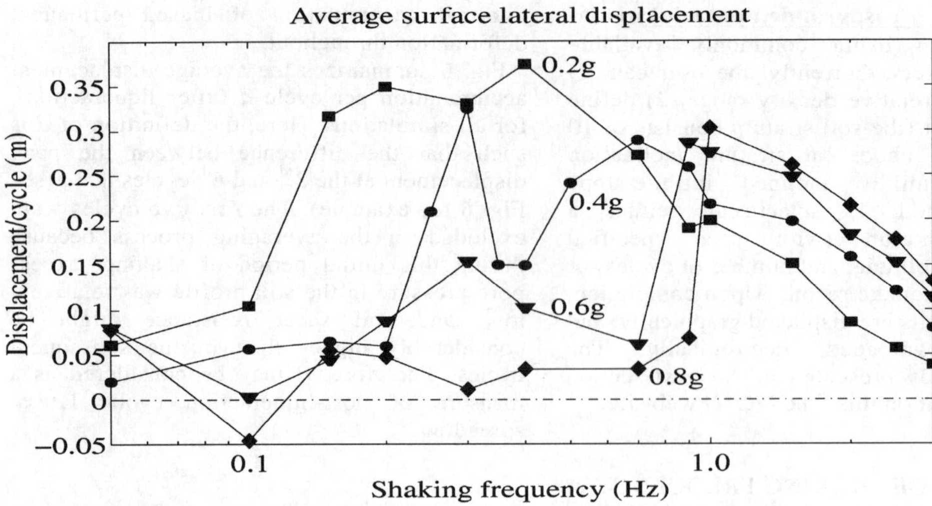

Figure 6. Average accumulated deformation per cycle.

Fig. 6 reveals the combined influence of shaking frequency and shaking amplitude on the resulting surface lateral displacement. Two salient features can be observed: 1) All shaking amplitudes rendered smaller accumulated displacements at the very low (<0.1Hz) and the high frequency (>3Hz) ends, and peak displacement accumulations occurred at the intermediate frequencies (between 0.3Hz and 1.0Hz). The values of these peak accumulations were in the range of 0.3-0.35m per cycle (or 3-3.5% uniform shear strain), with the largest belonging to 0.2g shaking (at 0.4Hz). 2) At lower frequencies (<0.9Hz), low-amplitude shakings in general caused more accumulated deformations, with the 0.6g and 0.8g shaking events showing insignificant cyclic accumulation of deformation in most cases (see also Fig. 5, 0.1Hz case). At these low frequencies, the influence of slope inclination (0.07g down-slope bias) appears to diminish with the increase in shaking excitation amplitude (0.6g and 0.8g cases), resulting in low accumulated down-slope deformations. The situation is reversed at frequencies higher than 1.0Hz (see also Fig. 5, 3Hz case).

In order to explain the above observations, we first investigate the soil shear stress-strain responses (Fig. 7) 3m below the ground surface, for the 0.8g and 0.2g shaking events at 0.1Hz frequency. At this very low frequency, the stronger (0.8g) event induced large cyclic shear stresses and strong dilations of equal magnitude in both up-slope and down-slope directions, a response pattern similar to the symmetric loading case in Fig. 3. Compared to these strong cyclic shear stresses, the effect of the superimposed static shear stress due to inclination became insignificant. The weaker (0.2g) shaking induced up-slope cyclic shear stresses of much smaller magnitudes relative to the stronger (0.8g) event. Therefore, the influence of the superimposed static shear stress became significant even at a low frequency of 0.1Hz for the 0.2g case, resulting in cycle-by-cycle accumulation of permanent shear strain (see unsymmetrical case of Fig. 3). This explains why at the lower frequency range (<1.0Hz), small-amplitude shakings induced more biased shear deformation than large-amplitude events.

Fig. 8 depicts shear stress-strain responses 3m below the ground surface, for the 0.8g and 0.2g shaking events at 1.5Hz frequency. At this frequency, shear stress-strain responses of both events show the same pattern of biased shear strain accumulation. However, since relatively larger biased shear stresses could develop in the stronger (0.8g) event, more shear strain was accumulated within each cycle than in the 0.2g case. This explains why at higher frequency range, large-amplitude shakings induced more biased shear deformation.

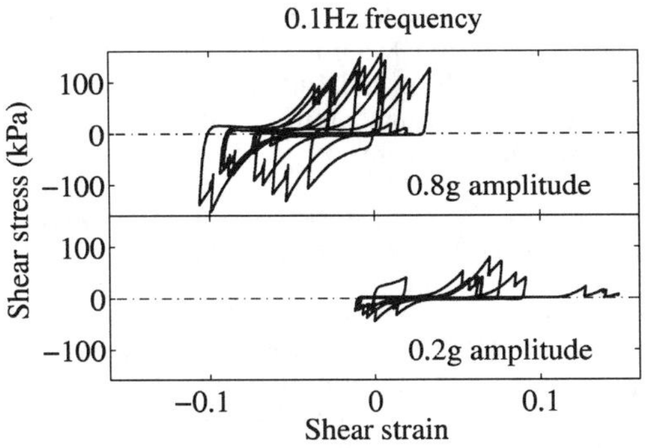

Figure 7. Shear stress-strain responses for 0.8g and 0.2g shaking events at 0.1Hz frequency.

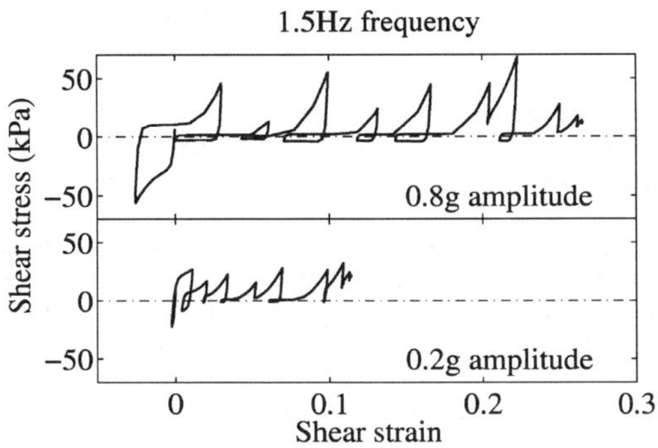

Figure 8. Shear stress-strain responses for 0.8g and 0.2g shaking events at 1.5Hz frequency.

Remarks:

1) It should be noted that all the discussions above concern the behavior of liquefied soil under dynamic excitation. This process is quite complicated, and a number of additional factors may also contribute heavily to the displacement patterns shown in Fig. 6. These factors include for example, the thickness of the liquefied layer, seismic-wave propagation characteristics in liquefied soil, inertia effect, and so forth. Consequently, extensive studies are still needed on this subject.
2) In Fig.6, there are a number of narrow frequency windows within which the average-displacement curves associated with every shaking amplitude display a fast transition from smaller magnitudes to relatively large values. The reason for these fast transitions is not evident at this point. Further investigation is currently underway.
3) The soil response patterns discussed above are largely dictated by the mechanisms of the employed liquefaction constitutive model (Section 3), which in turn was developed based on a limited number of experimental results. Further experimental evidence is needed to verify these numerical predictions.

SUMMARY AND CONCLUSIONS

In clean medium to dense sands, a possible strong influence of soil dilation on liquefaction-induced lateral deformation was discussed. This mechanism is documented in a large body of experimental research (on medium-dense cohesionless soil) including earthquakes records, centrifuge experiments and cyclic laboratory tests. A new constitutive model was developed to reproduce such cyclic shear behavior during liquefaction. The constitutive model was incorporated into a solid-fluid coupled finite element code *CYCLIC*. Currently, a 1D version of *CYCLIC* has been made available for execution over the Internet (**http://casagrande.ucsd.edu**). A numerical study conducted on this web-site demonstrates that the combined effect of excitation frequency with large deformations in the frequency range of 0.1Hz–3.0Hz (for a 10m clean sand stratum with Dr ≈ 40%) may control the extent of resulting lateral spreading.

ACKNOWLEDGEMENTS

Financial support for this research is gratefully acknowledged. This work was supported primarily by the Earthquake Engineering Research Centers Program of the National Science Foundation under Award Number EEC-9701568.

REFERENCES

Arulmoli, K., Muraleetharan, K. K., Hossain, M. M. and Fruth, L. S. (1992). "VELACS: VELACS: Verification of Liquefaction Analysis by Centrifuge Studies, Laboratory Testing Program, Soil Data Report," The Earth Technology Corporation, Project No. 90-0562, Irvine, CA.

Balakrishnan, A. and Kutter, B. L. (1999). "Settlement, sliding and liquefaction remediation of layered soil," *J. of the Geot. and Geoenv. Engrg. Division*, ASCE, 125, No. 11, November.

Casagrande, A. (1975). "Liquefaction and Cyclic Deformation of Sands - A critical Review," Proceedings, 5th Pan-American Conf. on Soil Mech. and Foundation Engrg., Buenos Aires, Argentina; also published as Harvard Soil Mech. Series No. 88, Jan. 1976, Cambridge, MA.

Castro, G. and Poulos, S. J. (1977). "Factors Affecting Liquefaction and Cyclic Mobility," *J. of the Geot. Engrg. Div.*, ASCE, Vol.103, No.GT6, June, 501-516.

Chan, A. H. C. (1988). "A Unified Finite Element Solution to Static and Dyn. Problems in Geomechanics," Ph.D. dissertation, U. College of Swansea, U. K.

Cubrinovski, M. and Ishihara, K. (1998). "State Concept and Modified Elastoplasticity for Sand Modeling," *Soils and Foundations*, Vol. 38, No. 4, 213-225, Dec.

Dobry, R. and Abdoun, T. (1998). "Post-Triggering Response of Liquefied Sand in the Free Field and Near Foundations," Proc. Geot. Eq. Engrg. and Soil Dyn. III, V1, Dakoulas, P., Yegian, M. and Holtz., R. D., Eds., Geot. Special Pub. No. 75, ASCE, Seattle, Washington, Aug 3-6, keynote lecture, 270-300.

Elgamal, A. -W., Dobry, R., Parra, E. and Yang, Z. (1998). "Soil Dilation and Shear Deformations During Liquefaction," Proc. 4th Intl. Conf. on Case Histories in Geot. Engrg., S. Prakash, Ed., St. Louis, MO, March, 1238-1259.

Holzer, T. L., Youd T. L. and Hanks T. C. (1989). "Dynamics Of Liquefaction During the 1987 Superstition Hills, California, Earthquake," *Science*, Vol. 244, 56-59.

Iai, S. (1991). "A Strain Space Multiple Mechanism Model for Cyclic Behavior of Sand and its Application," Earthquake Engineering Research Note No. 43, Port and Harbor Research Institute, Ministry of Transport, Japan.

Iai, S., Morita, T., Kameoka, T., Matsunaga, Y. and Abiko, K. (1995). "Response of a Dense Sand Deposit During 1993 Kushiro-Oki Earthquake," *Soils and Foundations*, 35, 1, March, 115-131.

Iwan, W. D. (1967). "On a class of Models for the Yielding Behavior of Continuous and Composite Systems," *J. of Applied Mechanics*, ASME, Vol. 34, 612-617.

Kramer, S. L. and Arduino, P. (1999). "Constitutive Modeling of Cyclic Mobility and Implications For Site Response," Pro. 2nd Intl. Conf. on EQ Geot. Engrg, Lisboa, Portugal, June, Seco & Pinto (Ed.), Vol.2, 895-900, Balkema.

Mroz, Z. (1967). "On the Description of Anisotropic Work Hardening," *J. of Mech. and Phys. of Solids*, 15, 163-175.

Parra, E. (1996). "Numerical Modeling of Liquefaction and Lateral Ground Deformation including Cyclic Mobility and Dilative Behavior in Soil Systems," Ph.D. dissertation, Dept. of Civil Engineering, Rensselaer Polytechnic Institute.

Prevost, J. H. (1985). "A Simple Plasticity Theory for Frictional Cohesionless Soils," *Soil Dynamics and Earthquake Engineering*, Vol. 4, No. 1, 9-17.

Seed, H. B. and Lee, K. L. (1966). "Liquefaction of Saturated Sands During Cyclic Loading," *J. of the Soil Mech. and Foundations Div.*, ASCE, 92, SM6, Nov., 105-134.

Seed, H. B. (1979). "Soil Liquefaction and Cyclic Mobility Evaluation for Level Ground During Earthquakes," *Journal of the Geotech Engng Division*, ASCE, 105, No. GT2, Feb., 201-255.

Taboada, V. M., and Dobry, R. (1998). "Centrifuge Modeling of Earthquake-Induced Lateral Spreading in Sand," *J. of Geot. and Geoenv. Engrg*, ASCE, 124, No. 12, 1195-1206.

Tateishi, A., Taguchi, Y, Oka, F. and Yashima, A. (1995). "A Cyclic Elasto-Plastic Model For Sand and Its Application Under various Stress Conditions," Proc. 1st Intl. Conf. On EQ Geot. Engng, 1, 399-404, Balkema, Rotterdam.

Yang, Z. (2000). "Numerical Modeling of Earthquake Site Response Including Dilation and Liquefaction," Ph.D. Dissertation, Columbia U., New York, NY.

Zeghal, M. and Elgamal, A. -W. (1994). "Analysis of Site Liquefaction Using Earthquake Records," *J. of Geot. Engineering*, ASCE, 120, No. 6, 996-1017.

Zienkiewicz, O. C., Chan, A. H. C., Pastor, M., Paul, D. K., and Shiomi, T. (1990). "Static and Dynamic Behaviour of Soils: A Rational Approach to Quantitative Solutions: I. Fully Saturated Problems," Proc. R. Soc. London, A429, 285-309.

Computer Methods and Advances in Geomechanics, Desai et al. (eds) © 2001 Balkema, Rotterdam, ISBN 90 5809 183 X

Nonlinear site response simulations at Port Island during the 1995 Kobe earthquake

E. Foerster, H. Modaressi, L. Choppin de Janvry
BRGM, Land Use Planning and Natural Risks, Orléans, France

ABSTRACT: This paper presents an application of the nonlinear dynamic model implemented in the computer-aided design tool called *CyberQuake* (Modaressi & Foerster 2000), to simulate the soil seismic response at Port Island site during the 1995 Hyogoken-Nambu (Kobe) earthquake. In this software, various hydraulic conditions can be defined for the multilayered soil profile, such as totally drained, partly drained (two-phase) or fully undrained conditions. The nonlinear soil behavior is modeled using a cyclic elastoplastic constitutive relationship, which is able to predict irrecoverable displacements or the pore-pressure build-up observed in saturated soils. Results obtained by the nonlinear partly drained and undrained approaches and by the well-known equivalent linear approach are compared. Accelerations computed with the nonlinear model are in good agreement with field data recorded in the Port Island vertical array. In the undrained simulation, liquefaction is obtained between 13 and 16 meters depth, which corresponds to the observations.

1 INTRODUCTION

Observations from many recent strong motion events, such as the 1985 Guerrero-Michoacan, the 1989 Loma Prieta, the 1994 Northridge and the 1995 Hyogoken-Nambu (Kobe, Japan) earthquakes, have shown the importance of local geological site conditions on the seismic ground response. These events have also demonstrated that nonlinear soil behavior strongly affects the seismic motion of near-surface deposits, resulting in shear wave velocity reduction, irreversible settlements and in some cases, the pore-pressure build-up leading to liquefaction. A close analysis of data recorded at Port Island vertical array during Kobe earthquake (Mohammadioun 1997) shows that nonlinearity in soil, evidenced by site liquefaction, resulted in an attenuation of the horizontal components of the acceleration recorded at ground surface and an amplification of the vertical one, compared to the bedrock motion. These observations are in good agreement with theoretical predictions made by Aubry and Modaressi (1992).

Since the early 70s', most studies of strong motion assessment concerning soft soils are performed using the equivalent linear approach (Idriss & Seed 1968, 1970) based on a viscoelastic multilayered soil model. This model can not be considered as a constitutive model able to represent the nonlinear soil behavior and several drawbacks of this approach have already been listed in literature. On the contrary, using an appropriate elastoplastic constitutive model for soil deposits, may permit to reproduce the complex features of soil behavior under seismic loading, such as the variation of the secant shear modulus and damping ratio with respect to distortion, irrecoverable displacement and strain generation and consequently liquefaction potential.

Therefore, the purpose of this paper is first to briefly recall the main features of the nonlinear dynamic model implemented in *CyberQuake* software. Then, results obtained for various simulations (nonlinear and equivalent linear approaches) performed on Port Island site during Kobe earthquake (1995) are presented.

2 THE NONLINEAR DYNAMIC MODEL

The soil profile is modeled as a multilayered dry or saturated porous medium with homogeneous laterally semi-infinite layers lying over a rigid or deformable bedrock. This one-dimensional geometry remains valid when no lateral heterogeneity, either geometric or material, does exist.

The model kinematics may be three-dimensional, with two horizontal and one vertical components of motion.

When a water table is present in the soil profile, layers located above the water level are assumed to be totally drained, whereas a fully undrained or a partly drained condition can be considered for the saturated layers underneath.

2.1 *The transient analysis*

The transient analysis is based on the resolution of the simplified *u-p* model obtained by using Terzaghi's effective stress principle and neglecting the fluid relative acceleration in Biot's macroscopic dynamic model.

The coupled-field equations consist in the dynamic equations of motion coming from the momentum conservation law, and the mass conservation equation for fluid and solid phases, combined with a generalized Darcy's law to describe fluid flow through the porous medium (Modaressi & Foerster 2000).

The boundary conditions are defined using the current engineering language, such as the water table position relative to the ground surface, a rigid or deformable bedrock, etc.

The governing equations are discretized by finite elements and numerically integrated with respect to time by an explicit predictor-corrector Newmark scheme. An adaptive time step is computed by the software, in order to fulfill stability and accuracy requirements for the integration scheme.

2.2 *The constitutive model*

An appropriate constitutive model is essential for evaluating the seismic ground response over a wide shear strain range (from 10^{-6} to 10^{-2}), especially for soft soils subjected to moderate or strong motion. In the Soil Dynamics context, a good constitutive model should be able to reproduce for instance the so-called (G-D, γ) curves, that is the variation of the secant shear modulus and damping ratio with respect to cyclic distortion.

The cyclic elastoplastic constitutive relationship implemented in *CyberQuake* software is successful in reproducing the complex features of nonlinear soil behavior under seismic loading. This model (Modaressi *et al.* 1995, Mellal 1997) is based on the original model developed by Aubry *et al.* (1982). Its main characteristics are as the following:

1. A unique *Coulomb*-type failure criterion is assumed for both monotonous and cyclic loading paths.
2. The plastic yield surface evolves within a volumetric hardening regime, which depends on the consolidation pressure (as in the *Cam-Clay* family models) and the plastic shear strain.
3. To take the dilatancy/contractance behavior of soil into account, that is the evolution of the plastic strain rate, a *Roscoe*-type dilatancy rule is introduced.

3 VALIDATION PROCEDURE

To validate the nonlinear dynamic model implemented in *CyberQuake* software, we simulated the seismic response of Port Island site during Kobe earthquake, using data recorded at various depths down a vertical array on the site. In this strong motion event, nonlinear soil effects did occur, such as site liquefaction, and they clearly appear on the recorded field data. The results obtained by the nonlinear model and the equivalent linear approach were then compared with field records.

3.1 *Description of the performed simulations*

3.1.1 *Soil profile characteristics*

In the Port Island vertical array, accelerations were recorded using four triaxial accelerometers situated at various depths from the ground surface, respectively 0, 16, 32 and 83m. The array was located on the north-west side of the reclaimed island very close to an improved ground area. The array site itself consisted in 19m of unimproved Masado (a decomposed granite fill), 8m of alluvial clays, 34m of sand with inter-layered clays, 18m of diluvial clays and then sand under -79m depth. The water table position is approximately located at 3m depth. The physical characteristics used for the soil profile in our simulations are detailed in Table 1. Most of the data concerning the soil profile, i.e. shear wave velocities, or granulometric features, were provided by the Port and Harbour Research Institute (1995).

Table 1. Soil profile characteristics used in the nonlinear partly drained simulation (water table at –3m).

Material	Depth (m)	V_s (m/s)	V_p (m/s)	ρ (kg/m^3)	k (m/s)
Gravel	-2	170	260	2000	0.007
Gravel	-5	170	330	2000	0.007
Gravel	-12.6	210	400	2200	0.004
Gravelly sand	-19	210	400	2200	0.001
Alluvial clay	-27	180	350	1500	10^{-9}
Diluvial Sand	-32	245	500	2000	0.001
Bedrock	-	245	1330	2000	impervious

3.1.2 *Nonlinear model parameters*

In order to determine the parameters needed for the nonlinear model, a special tool, called "Parameter Wizard" has been developed in *CyberQuake*, which uses various correlations found in literature depending on the material type. The procedure has been proposed by Modaressi and Lopez-Caballero (1999). For granular materials, the determination needs the uniformity coefficient, relative density and grain features as input, whereas for clays, it only requires the overconsolidation ratio and plasticity index.

The (G-D,γ) curves are needed for the equivalent linear approach. Unfortunately, as they were not all available for some layers, they were evaluated directly, using the elastoplastic model simulator integrated in *CyberQuake*.

3.1.3 *Input motion*

The NS (North-South) and EW (East-West) components of the input motion used for the simulations come from the incident wave at -32m. This latter was obtained by an equivalent linear technique performed between -83m and -32m (Ishihara *et al.* 1998). A maximum frequency of 15Hz was assumed in computations for this motion.

3.2 *Results*

The time history for the computed horizontal accelerations obtained at various depths for the nonlinear partly drained simulation are in good agreement with the recorded ones as shown in Figures 1, 2. Peak locations are very well modeled, whereas a slight amplification of the amplitudes can be observed at ground surface. Moreover, the computed accelerations at 16m depth underestimates slightly the motion. The difference may be attributed to our insufficient knowledge of the geomechanical properties for the underlying layers.

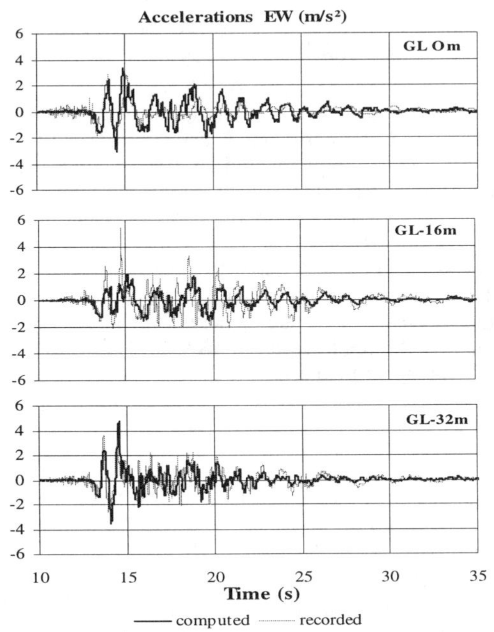

Figure 1. EW components of acceleration recorded and computed with the nonlinear partly drained simulation (0, -16m and -32m).

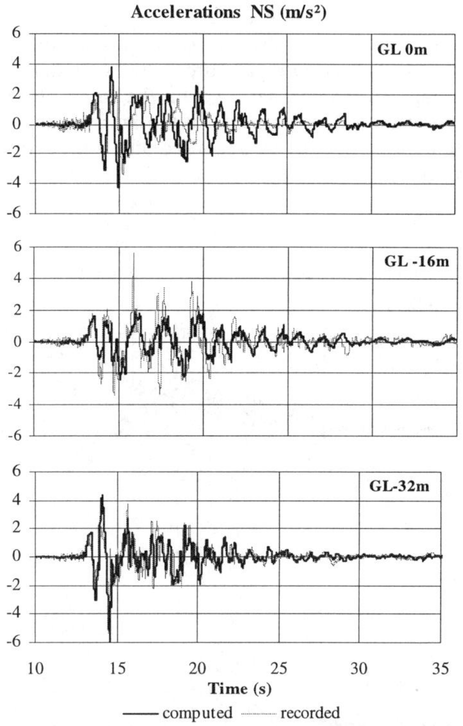

Figure 2. NS components of acceleration recorded and computed with the nonlinear partly drained simulation (0, -16m and -32m).

Figures 3, 4 show normalized response spectra in acceleration computed with a critical damping of 5% and obtained for the various simulations performed with *CyberQuake*, i.e. the equivalent linear, the nonlinear partly drained and fully undrained approaches. They are compared to the spectra obtained for the recorded accelerations.

At ground surface, all methods give response spectra with a little discrepancy in peak location. The amplitude of the response spectrum for the nonlinear partly drained approach is in good agreement with the observed one, whereas it is amplified with the equivalent linear method. In fact, as this technique represents a total stress analysis, it is unable to take into account the irrecoverable settlements and pore-water pressure build-up in the saturated soil. Thus, this approach may overestimate the intensity of the shaking. On the other hand, in the nonlinear fully undrained simulation, as the generated pore-pressure can not dissipate during the shaking, the model is able to reproduce the aforementioned build-up and the annealing of the effective normal stress, as shown in Figures 5, 6. Liquefaction indeed occurs between -14m and –16m, after 15 seconds of shaking, which is in very good agreement with field observations and analyses made by various authors

(Iwasaki & Tai 1996). In Figure 6, it may be seen that, at 15 and 19 seconds, the soil at this depth dilates and the excess pore-pressure reduces suddenly. This strength recovery may explain the large peaks observed at the same moments in the horizontal acceleration components as recorded at 16m depth (Figure 2).

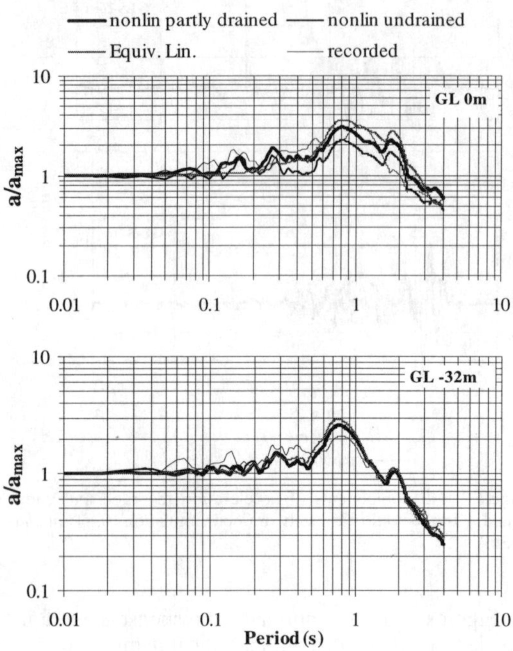

Figure 3. Normalized response spectra in acceleration, computed at ground surface and –32m for EW components (critical damping: 5%).

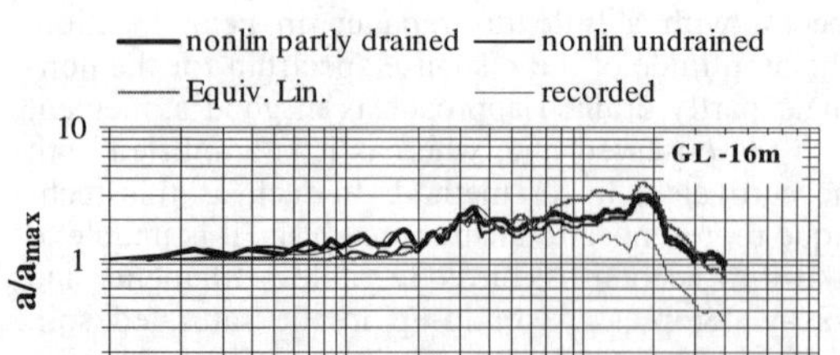

Figure 4. Normalized response spectra in acceleration, computed at –16m for NS components (critical damping: 5%).

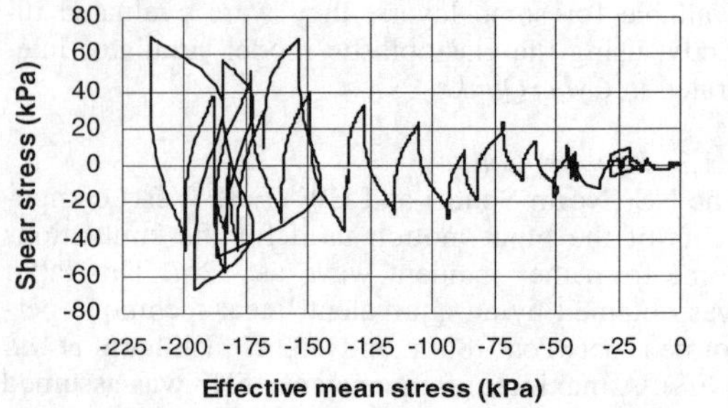

Figure 5. Computed in-plane shear stress vs. effective mean stress at –14.5m.

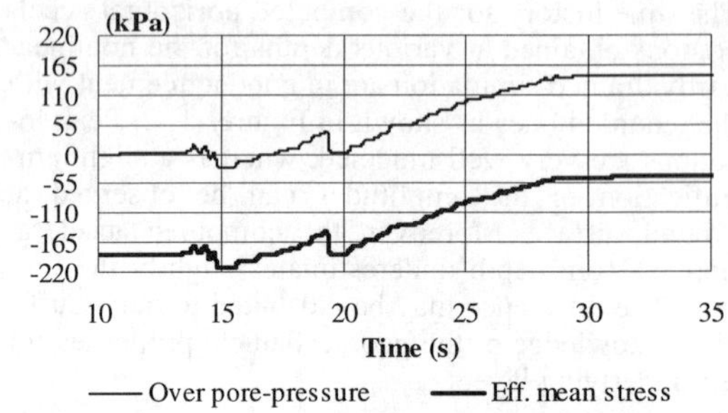

Figure 6. Computed over pore-pressure and effective mean stress vs. time at -14.5m.

4 CONCLUSIONS

Simulations performed for Port Island vertical array site during Kobe earthquake, using the non linear dynamic model implemented in *CyberQuake* software, demonstrate the capability of the model to predict liquefaction potential in soft deposits under moderate or strong ground shaking. Indeed, the simulated responses, either with a partly drained or a fully undrained condition considered for layers underneath water table, are generally in good agreement with the field accelerations recorded at various depths in the vertical array.

REFERENCES

Aubry, D., Hujeux, J.C., Lassoudière, F. & Y. Meimon 1982. A double memory model with multiple mechanisms for cyclic soil behaviour. *Int. Symp. Numer. Models Geomech.*, Zurich, Suisse.

Aubry, D. & H. Modaressi 1992. Seismic Wave Propagation in Soils Including Non-linear and Pore Pressure Effects. In V. Davidovici (ed.), *Recent Advances in Earthq. Eng. Struct. Dynam.*: 209-224.

Idriss, I.M. & H.B. Seed 1968. Seismic response of horizontal soil layers. *J. Soil Mech. Found. Div. ASCE*: 94, 1003-1031.

Ishihara, K., Kokusho, T., Yasuda, S., Goto, Y., Yoshida, N., Hatanaka, M. & K. Ito 1998. Dynamic Properties of Masado Fill in Kobe Port Island Improved through Soil Compaction Method – Summary of Final Report. *Geotech. Research Collaboration Committee on the Hanshin-Awaji Earthquake*.

Iwasaki, Y. & M. Tai 1996. Strong Motion Records at Kobe Port Island. *Special Issue of Soils Found. on Geotech. Aspects of the January 17 1995 Hyogoken-Nambu Earthquake*: 29-40.

Mellal, A. 1997. Analyse des effets du comportement non linéaire des sols sur le mouvement sismique. *Thèse de Doctorat*, Ecole Centrale de Paris.

Modaressi A. and F. Lopez-Caballero 1999. Seismic ground displacement as a tool for town planning, design and mitigation. *Third Progress Report, ECC,* Project ENV4-CT97-0392.

Modaressi, H. & E. Foerster 2000. *CyberQuake* Version 2.0 User's Guide. *BRGM, BP 6009, 45060 Orléans Cedex 2*, France.

Modaressi, H., Foerster, E., Aubry, D., & A. Modaressi 1995. Research and professional computer-aided dynamic analysis of soils. *1st Int. Conf. Earthq. Geotech. Eng*. - IS-TOKYO'95: 1171-1176, Tokyo, Japan.

Mohammadioun, B. 1997. Non-linear response of soils to horizontal and vertical bedrock earthquake motion. *J. Earthq. Eng.*: 1.

Port and Harbour Research Institute 1995. *Technical note No.813*. Ministry of Transport, Japan.

Seed, H.B. & I.M. Idriss 1970. Soil moduli and damping factors for dynamic response analysis of horizontally layered sites. *Earthq. Eng. Research Center,* Report No. UCB/EERC : 70-10. Univ. of California, Berkeley.

Computer Methods and Advances in Geomechanics, Desai et al. (eds) © 2001 Balkema, Rotterdam, ISBN 90 5809 183 X

Designing shallow tunnel linings under seismic effects

N. Fotieva, A. Sammal & N. Bulychev
Tula State University, Russia

J.V. Lemos, A. Vieira & L.R. Sousa
LNEC, Lisbon, Portugal

ABSTRACT: The analytical method of designing shallow circular tunnel linings under seismic effects based on determining the maximal compressive and tensile stresses which may appear in points of the internal lining cross-section outline at any combinations and directions of long longitudinal and shear waves propagating in the plane of the tunnel cross-section due to the earthquake is proposed in the paper presented. A comparison of the analytical solution with a numerical model is also presented.

1 INTRODUCTION

In the design and construction of tunnels in seismic areas it is necessary to take into account that their linings besides usual static loads may be subjected to Earthquakes effects consisting in spreading in the rock (soil) mass the long longitudinal (compressive-tensile) and shear seismic waves of different velocities, the combinations and directions of which respectively to the underground structure are principally unknown.

That is why the original approach to the problem of designing tunnel linings under earthquake effects has been developed at Tula State University (Fotieva 1980). According to that approach the design includes determining the most unfavorable stress state in every normal section of the structure at any combinations and directions of long longitudinal and shear waves propagating in the plane of the tunnel cross-section.

The above approach and methods intended for designing linings of deep tunnels (the depth H>3B, where B is the most linear size of the tunnel cross-section) when the influence of the Earth surface may be neglected have been included in the Russian standards (Instruction 1982, Guide 1996) and widely applied in projects of transport and power-stations tunnels in seismic areas.

A similar method for designing circular shallow tunnel linings taking into account the influence of the Earth surface is proposed in the paper.

2 THE METHOD OF THE DESIGN

According to the basic principles described in the works mentioned above the design of shallow tunnel linings under seismic effects includes the determination of maximal compressive and tensile normal tangential stresses σ_θ which may appear on the internal outline of the lining at any combinations and directions of long longitudinal and shear waves propagating in the plane of the tunnel cross-section.

With this aim the two plane quasi-static contact problems of the elasticity theory for a semi-infinite linearly deformable medium simulating the rock (soil) mass weakened by a circular opening supported by a ring simulating the tunnel lining are considered (Figure 1 a, b). In those figures the semi-infinite homogeneous isotropic linearly deformable medium S_0, restricted by a straight boundary L_0' and a circular curve L_0 of R_0 radius, the mechanical properties of which are characterized by the E_0 deformation modulus and the ν_0 Poisson's ratio, simulates the soil mass. The S_1 ring with the L_1 internal outline of the R_1 radius from the material having E_1 and ν_1 deformation characteristics simulates the tunnel lining constructed at H depth.

In the first problem (Figure 1 a) the S_0 medium is subjected on the infinity to a double-axis compression by stresses $-P$, $-\xi P$, directed under an arbitrary α angle with respect to the horizontal and the vertical, simulating the action of a long longitudinal wave in the compression phase. In the second problem (Figure 1 b) the S_0 medium is subjected on the infinity to a pure shear under the α angle by stresses S simulating the action of a long arbitrary directed shear wave.

The stresses at infinity are expressed by formulae (Guide 1996):

$$P = \frac{1}{2\pi} A K_1 \gamma c_1 T_0, \quad \xi = \frac{\nu_0}{1-\nu_0},$$

$$S = \frac{1}{2\pi} A K_1 \gamma c_2 T_0, \tag{1}$$

where A is the coefficient corresponding to the Earthquake intensity, determined according to the point 2.5 of the Russian standard (SNIP Ï-7-81), K_1 is the coefficient taking admissible damages into account determined by Table 3 of that standard, γ is the soil unit weight, T_0 is the prevailing period of the soil particles oscillation, c_1, c_2 are the longitudinal and shear waves velocities.

The ring and medium undergo deformation together i.e. conditions of the stresses and displacements vectors continuation are satisfied on the L_0 contact line. The L_0 boundary and the L_1 internal outline of the ring are free from loads.

The problems formulated have been solved with the application of the complex variable analytic functions theory using the apparatus of complex series and the analytical continuation of the complex potentials characterizing the lower semi-plane stress-strain state into the upper semi-plane across the straight boundary proposed in the paper by Aramanovich (1955).

From the solution of the first problem the $\sigma^{(P)}$ stresses (here the σ - symbol signifies all components of stress tensor), appearing in the lining due to the action of a long longitudinal wave falling under an α arbitrary angle are determined; from the solution of the second problem the $\sigma^{(S)}$ stresses called forth by a shear wave are obtained.

Then the sum and the difference of general expressions for $\sigma_\theta^{(P)}$ and $\sigma_\theta^{(S)}$ normal tangential stresses characterizing the lining stress state caused by mutual actions of longitudinal and shear waves passing simultaneously (the worst case) are investigated in every point of the internal lining cross-section outline on the extreme relatively the α angle of the waves falling. With this aim the following equations are solved

$$\frac{\partial}{\partial\alpha}[\sigma_\theta^{(P)} \pm \sigma_\theta^{(S)}] = 0 \tag{2}$$

and for every point such a combination of waves and such an angle of their falling at which normal tangential stresses in the points considered are maximal by their absolute are determined. It allows the envelope diagram of normal tangential stresses on the internal lining outline to be obtained analytically.

The stresses along the external lining outline, longitudinal forces and bending moments in every

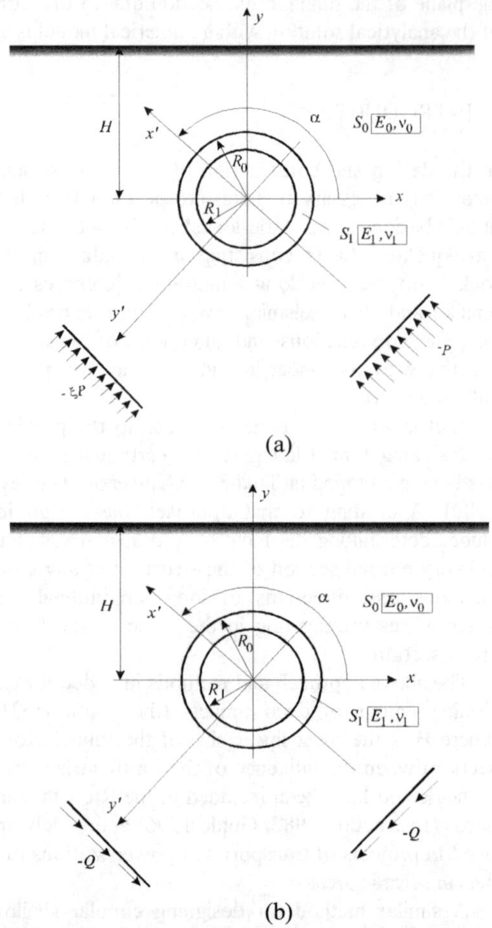

Fig. 1 Design schemes for the determination of the tunnel lining stress state caused by long arbitrary directed seismic waves:
(a) longitudinal, (b) shear.

lining normal section are determined at such a combination and such a direction of waves at which the normal tangential stress in that section has a maximal absolute value.

The stresses and forces obtained that way are assumed to have the signs "plus" and "minus" and summed up in their most unfavourable combinations with stresses and forces caused by the other loads obtained according to the work by Fotieva et al. (1996). After that a sections strength test upon compression and tension is made.

If the lining is not anchored to the massif and is designed with an allowance of the crack formation we assume that the tensile normal loads are not transferred upon the lining. In this case the action of the longitudinal waves in the tension phase is not to be taken into account and the design is made on the base of the two different envelope diagrams of normal tangential stresses, obtained using the maximal absolute values of the compressive (negative) stresses and tensile (positive) ones, called forth by mutual actions of shear waves and longitudinal waves in the compression phase.

3 EXAMPLES OF THE DESIGN

The shallow tunnel lining has been considered for the following input data: $R_0 = 3$ m, $R_1 = 2{,}7$ m, $H = 6$ m, $E_0 = 10000$ MPa, $\nu_0 = 0{,}27$, $E_1 = 23000$ MPa, $\nu_1 = 0{,}2$, $\gamma = 0{,}017$ MN/m^3, $AK_1 = 0{,}1$, $T_0 = 0{,}5$ s.

Diagrams of maximal compressive (negative) and tensile (positive) stresses σ_θ^{in}, in MPa, which may appear in points of the internal outline of the lining cross-section at the Earthquake are given in Figure 2 a by solid and dash lines; the corresponding

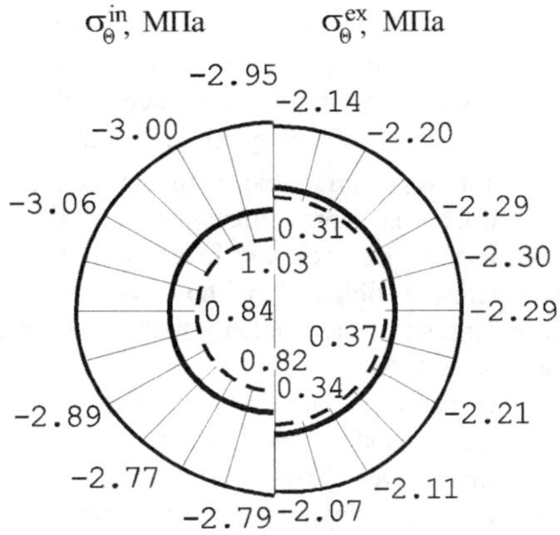

(a) (b)

Fig.2. Diagrams of the extreme stresses σ_θ^{in} on the internal (a) and corresponding stresses σ_θ^{ex} on the external (b) lining face.

them stresses σ_θ^{ex}, in MPa, on the external outline are shown in Figure 2 b.

The calculations show that with increasing the tunnel depth the stresses in the lining decrease approaching to values being determined by the method described in the work [1], i.e. without taking the influence of the Earth surface into account.

Results of comparison of maximal compressive and tensile stresses σ_θ^{in} which may appear along the internal lining cross-section outline at different relative depths of the top of arch $H' = (H - R_0)/R_0$ for the tunnel considered above located in the hard rock with characteristics $E_0 = 10000$ MPa, $\nu_0 = 0{,}27$ and in the soft soil with characteristics $E_0 = 50$ MPa, $\nu_0 = 0{,}37$

Table 1.
Maximal compressive and tensile stresses on the internal outline of the lining

H	I		II	
	compressive	tensile	compressive	tensile
1	-3,061 (18%)	1,029 (47%)	-2,957 (3.5%)	2,308 (3.8%)
2	-2,801 (8.2%)	0,823 (18%)	-2,887 (1,0%)	2,255 (1,4%)
4	-2,666 (3.0%)	0,738 (5,6%)	-2,866 (0,3%)	2,233 (0,4%)
10	-2,604 (0.6%)	0,706 (1,0%)	-2,859 (0,0%)	2,225 (0,0%)
∞	-2,588	0,699	-2,858	2,223

are given in Table 1 (parts I and II respectively). The relative differencies of stresses from the same ones obtained without taking the influence of the surface into account ($H' = \infty$) are given in brackets.

As follows from Table 1 the influence of the Earth surface is more substantial for the tunnel located in the hard rock and inaccuracy of maximal tensile stresses obtained without account of the surface influence is higher then the one for compressive stresses.

We can also notice that the design of tunnel linings only under the action of horizontal longitudinal waves (how was suggested in some works) without taking into account shear waves and different directions of the waves falling may result in inadequate evaluation of a seismic danger. To illustrate this fact the diagrams of stresses σ_θ^{in}, σ_θ^{ex} caused by a single long horizontal longitudinal wave obtained at the same input data as the results shown in Figure 2 are given in Figure 3.

As follows from comparison of results shown in Figures 2 and 3 compressive stresses on the internal lining cross-section outline (solid line in the left paet of Figure 2) may exceed the ones caused by a single horizontal wave in 1.5 times in the top of arch and more then in 20 times in points laying on the horizontal diameter. The stresses in the same points of external outline are higher in 1.4 and 7 times correspondingly. Besides that the tensile stresses in the lining are absent in general at the action of a single longitudinal wave while they may reach $\sigma_\theta^{in} = 1.03$ MPa in the top of arch and $\sigma_\theta^{in} = 0.84$ MPa in points of the horizontal diameter.

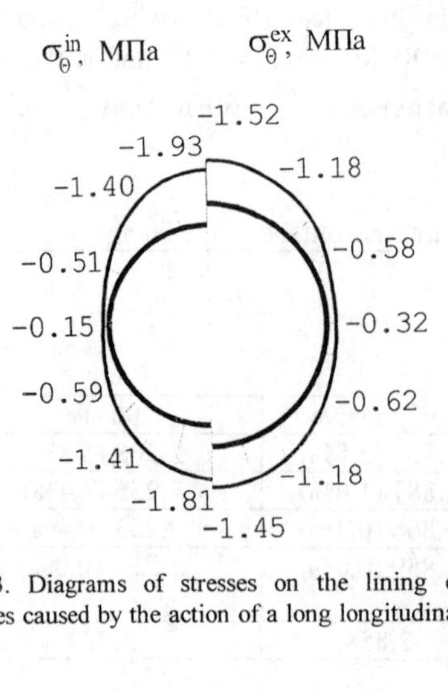

Fig. 3. Diagrams of stresses on the lining cross-section outlines caused by the action of a long longitudinal horizontal wave.

4 COMPARISON WITH NUMERICAL MODELS

Analytical solutions provide a means to evaluate the precision of numerical models, within the range of validity of their assumptions. Numerical models can then be used in more general situations, namely involving non-linear material behavior.

A 2D numerical model of a shallow tunnel in an elastic medium was created with the program FLAC, in a study of comparison of analytical and numerical solutions (Lemos et al. 2000). The tunnel considered had a diameter of 10 m, with the axis at a depth of 15 m. The elastic soil properties were E=50 MPa and ν=0.3; the concrete liner was 0.5 m thick, with E=20 GPa and ν=0.25. The model had absorbing boundaries at the base and lateral free-field boundaries, to simulate a semi-infinite medium. Dynamic analyses were performed in the time domain, considering the cases of vertically propagating P- and S- waves. The earthquake input was based on records generated according to the power spectrum prescribed by the Portuguese seismic code. The analytical solutions

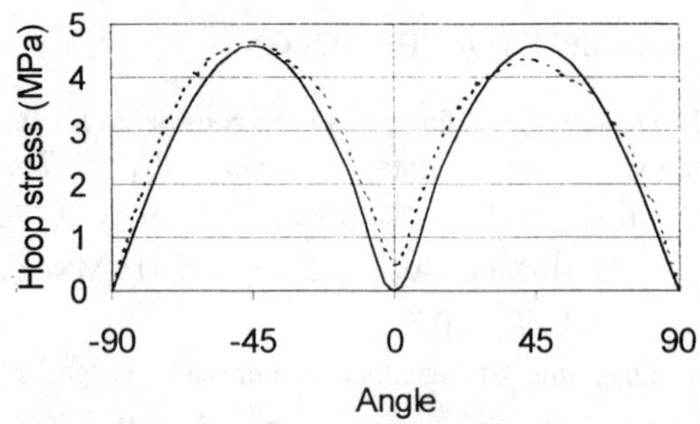

Fig. 4. Comparison of analytical (solid line) and numerical (dashed line) results for internal hoop stress in the lining, for the S-wave case.

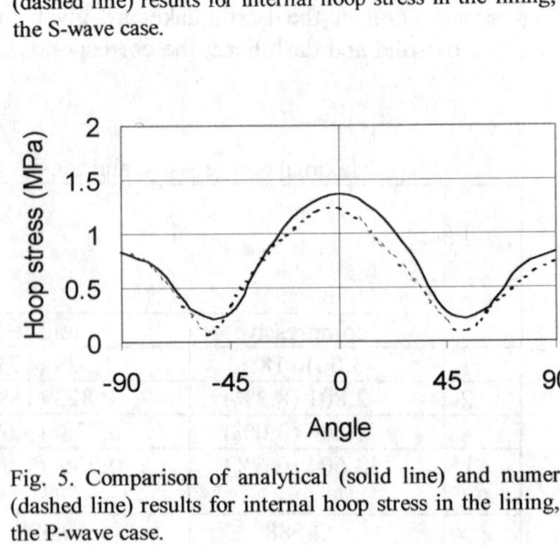

Fig. 5. Comparison of analytical (solid line) and numerical (dashed line) results for internal hoop stress in the lining, for the P-wave case.

were scaled with reference to the numerical free-field stresses obtained at the tunnel axis depth.

The peak absolute values of the hoop stresses along the internal lining face obtained in the numerical model are compared in Fig. 4 with the analytical solution, for the case of a S-wave. The corresponding results for a P-wave are shown in Fig. 5. The match can be considered good, given the differences in the simulation of the seismic action, as the analytical solution corresponds to the quasi-static component of the response, and the numerical results were obtained with a single earthquake record.

5 CONCLUSIONS

1. The design method proposed allows to take into account a variety of seismic effects of the Earthquake and therefore to decrease the risk of the lining failure.
2. The calculations show that for designing the linings of shallow tunnels located in soft soils one can apply the method [1] intended for the design of deep tunnels without substantial errors. Just when the tunnel is located in hard rocks it is necessary to take the influence of the Earth surface into account especially if the lining is designed with the allowance of the crack formation because errors in determining the maximal tensile stresses (the dashed line in the left part of Figure 2) may be unacceptable.

ACKNOWLEDGEMENT

The work presented is supported by NATO Science Programme - Grant EST.CLG 975047.

REFERENCES

Aramanovich, I.G. 1955. Distribution of stresses in an elastic semi-plane weakened by a supported circular opening. Reports of AN USSR, issue 104, No. 3.

Fotieva, N.N. 1980. Design of Underground Structures Support in Seismic Active Regions. Moscow: Nedra.

Fotieva, N.N., Bulychev, N.S. & Sammal, A.S. 1996. Design of shallow tunnel linings. Proc. ISRM Int. Symp. 1996, Torino, Italy. Rotterdam: Balkema.

Guide on designing underground structures in seismic areas 1996. Informative Centre TIMR, Moscow.

Instruction to account seismic effects at the design of mine transport tunnels 1982. Mosow: VPTItransstroy.

Lemos, J.V., Vieira, A., Sousa, L.R. & Câmara, R. 2000. Dynamic analysis of a lined shallow tunnel - Numerical modelling. LNEC Report, Lisbon: LNEC.

Computer Methods and Advances in Geomechanics, Desai et al. (eds) © 2001 Balkema, Rotterdam, ISBN 90 5809 183 X

CIP based liquefaction-induced lateral spreading analysis

S. Hadush, S. Moriguchi, A. Yashima & K. Sawada
Gifu University, Japan

R. Uzuoka
Earthquake Disaster Mitigation Research Center, RIKEN, Hyogo, Japan

ABSTRACT: This paper presents a CIP (Cubic Interpolated Pseudoparticle) based numerical method for liquefaction induced lateral spreading analysis in the framework of fluid dynamics. A Bingham model that takes the residual strength of soil into account is used to represent the behavior of liquefied soil. An old version of a CIP based code is improved by incorporating the Bingham viscosity and an implicit calculation procedure for pressure terms. The numerical method is validated and good agreement is produced in comparison with experimental results. The method is then used to simulate lateral spreading experiments on liquefied subsoil with and without an overlaying unsaturated nonliquefied layer. From the results of simulations, the numerical method is found to reproduce a previously proposed similitude law for liquefied ground flow and the time histories of flow velocity and ground surface displacement.

1 INTRODUCTION

Lateral spreading due to liquefaction is a category of permanent ground movement problems that can cause severe damages on important civil engineering structures such as buried pipelines, foundation piles, etc. In recent years, this problem has been taken into account in earthquake-resistant design standards. The behavior of soil during liquefaction induced lateral spreading has been studied based on case studies of historical earthquakes, shaking table tests and other experimental methods. Based on the results from the above mentioned studies, different empirical methods have been formulated to estimate ground displacements (Hamada et al. 1986, Bartlet and Youd 1992).

Several researchers have also developed numerical methods to study the behavior of liquefied soil, treating liquefied soil as solid by reducing stiffness or as fluid. The point of view that the liquefied soil behaves like a solid states that liquefaction reduces the stiffness of the soil significantly, and the ground deforms as a solid under the influence of gravitational force. In contrast, the point of view that the liquefied soil behaves like a fluid states that liquefaction reduces the stiffness of the soil to a negligible extent, and intrinsically liquefied soil behaves as a fluid during ground flow, but after the earthquake motion ceases, due to dissipation of excess pore water pressure and dialatancy the stiffness of the liquefied soil recovers and it returns to behaving as a solid body.

The numerical methods, which assume the liquefied soil as a solid, have been used to predict the configuration of permanent deformation. For example, Finn et al. (1991) proposed a finite deformation analysis using post-liquefaction stress-strain relations incorporating the residual strength and Aydan (1995) assumed the liquefied subsoil to be a visco-elastic body and calculated the deformation velocity of liquefied soil by an updated Lagrangian method. On the other hand, methods, which assume the liquefied soil behaves as a fluid, have also been used to predict the magnitude of a permanent deformation. Towahata et al. (1992) proposed a closed form solution of permanent displacement by means of the minimum energy principle. Uzuoka et al. (1998) proposed a numerical method that is able to predict the flow process and also lateral spreading load acting on embedded structures. They modeled the liquefied soil as a Bingham fluid and used the VOF (Volume of Fluid) scheme (Hirt & Nicholas 1981), to treat free boundaries in their analyses. They applied the method performing numerical simulations for reported experimental results and a real flow failure.

This paper presents a numerical method for liquefaction induced lateral spreading analysis based on the CIP scheme (Yabe et al. 1991). We used the CIP method because it can correctly define the flow behavior at free boundaries and interfaces between multi fluids and can also be used as a unified scheme for both compressible and incompressible fluids. In this study, we used the Bingham model (Uzuoka et al. 1998) to represent the behavior of liquefied soil. We improved an old version of a CIP based program by incorporating the Bingham viscosity and an implicit calculation procedure for pressure terms. The numerical method is then applied to shaking table tests of a liquefied slope with and without overlaying nonliquefied layer (Hamada et al. 1998, Sakaguchi et al. 1998). The numerical method is found to reproduce the previously proposed similitude law for liquefied ground flow (Hamada et al. 1998, Hadush et al. 1999, 2000) and the time history of flow velocity and ground surface displacement.

2 ANALYTICAL METHOD

The governing equations of fluid motion can be written in a form:

$$\frac{\partial \rho}{\partial t} + (u.\nabla)\rho = -\rho \nabla.u \tag{1}$$

$$\frac{\partial u}{\partial t} + (u.\nabla)u = -\frac{\nabla p}{\rho} \tag{2}$$

$$\frac{\partial p}{\partial t} + (u.\nabla)p = -\gamma p \nabla.u \tag{3}$$

where ρ is the density, u the velocity, p the pressure and γ the specific heat ratio. The key issues of the CIP method are in the representation of advection term and split treatment of other terms. In the CIP method the above equations are split into two phases; nonadvection and advection phases.

2.1 *Nonadvection phase*

In this phase only the right hand side of the above equations is solved with a finite difference method as shown below:

$$\frac{\rho^* - \rho^n}{\Delta t} = -\rho^n \nabla.u^{**} \tag{4}$$

$$\frac{u^{**} - u^n}{\Delta t} = -\frac{\nabla p^{**}}{\rho^n} \tag{5}$$

$$\frac{u^* - u^{**}}{\Delta t} = Q_u \tag{6}$$

$$\frac{p^{**} - p^n}{\Delta t} = -\gamma p^n \nabla.u^{**} \tag{7}$$

$$\frac{p^* - p^{**}}{\Delta t} = Q_p \tag{8}$$

In which Q_u stands for viscous stress and Q_p stands for thermal conduction terms. In the above equations the superscript ^{n} stands for quantities at the present time step and superscript * means one time step after the nonadvection phase, and the quantities having * will be used in the advection phase. For an incompressible fluid, because the speed of sound waves is very large compared to $\Delta x / \Delta t$, the term related to the pressure is solved implicitly. The quantities such as p^{**} and u^{**} are used to make this implicit calculation. In order to determine such implicit equations we take the divergence of eq. (5) and substitute it into eq. (7). Then we obtain an equation:

$$\frac{\nabla^2 p^{**}}{\rho^n} = \frac{p^{**} - p^n}{\gamma p^n \Delta t^2} + \frac{\nabla.u^n}{\Delta t} \tag{9}$$

2.2 *Advection phase*

After ρ^*, u^*, and p^* are obtained in the non-advection phase, the CIP solver devised by Yabe et al. (1991) is used to obtain ρ^{n+1}, u^{n+1}, and p^{n+1}. These two phases complete the numerical procedure and are repeated step by step.

3 METHOD VALIDATION

The numerical method is applied to dam break analysis in which a column of water collapses due to rapid opening of a vertical gate. The column of water is 20 cm in height and 10 cm in length, *L*. Figures 1 (a) and (b) show the numerical model used for this analysis based on the experimental setup and simulated results, respectively. Comparison is made with experimental results shown in Figure 2. The calculated curve matches very well with results from

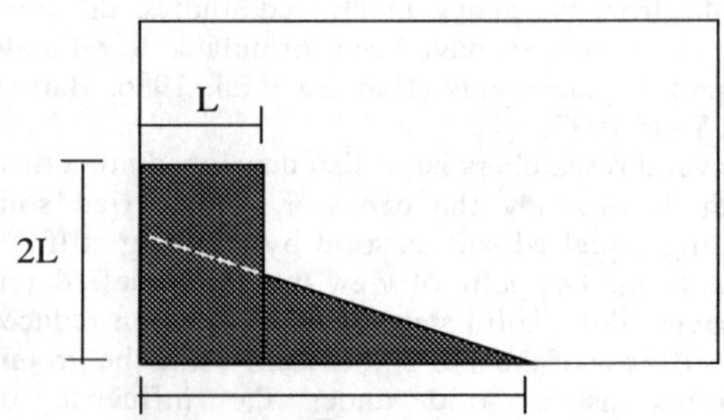

Figure 1 (a). Numerical model, dam break analysis

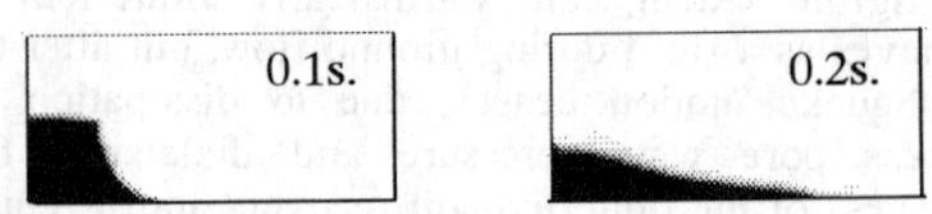

Figure 1 (b). Simulated results at 0.1s. and 0.2s.

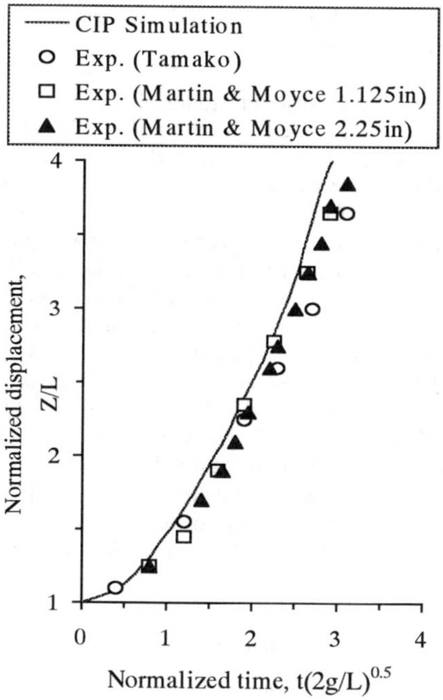

Figure 2. Dam break analysis

experiment, therefore, the numerical method can be used for practical simulations.

4 NUMERICAL SIMULATIONS

The equivalent viscosity η' for the Bingham model (Uzuoka et al. 1998) is given by (Figure 3):

$$\eta' = \eta + \frac{\tau_{\min}}{2\dot{\gamma}} = \eta + \frac{R_{\min} . P}{2\dot{\gamma}} \quad if\ \dot{\gamma} \neq 0$$

$$\eta' = \infty, \quad if\ \dot{\gamma} = 0 \tag{10}$$

in which,

$$\dot{\gamma} = \sqrt{\tfrac{1}{2} \dot{e}_{ij} \dot{e}_{ij}} \tag{11}$$

where $\tau_{\min}$ is the minimum undrained strength (= residual strength), η is the Bingham viscosity after yield, $R_{\min}$ is the minimum undrained strength ratio, p is the hydraulic pressure, $\dot{e}_{ij}$ is the deviatoric strain rate tensor and $\dot{\gamma}$ is the second invariant of the deviatoric strain rate tensor. When $\dot{\gamma}$ is zero at the start of analysis, the equivalent viscosity coefficient is set to be a large value. In this unsteady analysis the equivalent viscosity is treated as a constant in each calculation step and is updated step by step.

The simulations incorporated gravity as the only external force and did not consider the vibration force because its direction was perpendicular to the ground slope. A saturated unit weight of 17.93 kN/m^3 is used to calculate the body force in the model material representing the liquefied soil. In the experiments, Enshunada sand with index properties summarized in Table 1 was used.

Table 1. Index properities of Enshunada sand

D_{50} (mm)	U_c	e_{max}	e_{min}
0.24	1.92	1.251	0.706

D_{50}, mean grain size; U_c, uniformity coefficient; e_{max}, e_{min}, maximum and minimum void ratio.

4.1 *Simulation for similitude law*

Hamada et al. (1998) proposed a law of similitude for liquefied ground flow, which states flow velocity is proportional to the square root of liquefied layer thickness. They used the similitude law as a key basis in formulations of empirical equations for estimating ground displacement during liquefaction induced lateral spreading. The similitude was later verified by Hadush et al. (1999, 2000) using a fluid dynamics based numerical method. In this study we used the CIP based numerical method to simulate the similitude. Table 2 summarizes the analytical parameters used in the simulation.

Table 2. Analytical parameters used for the analysis

θ (%)	η (Pa.s)	$R_{\min}$
3	1.0	0.011
5	1.0	0.022

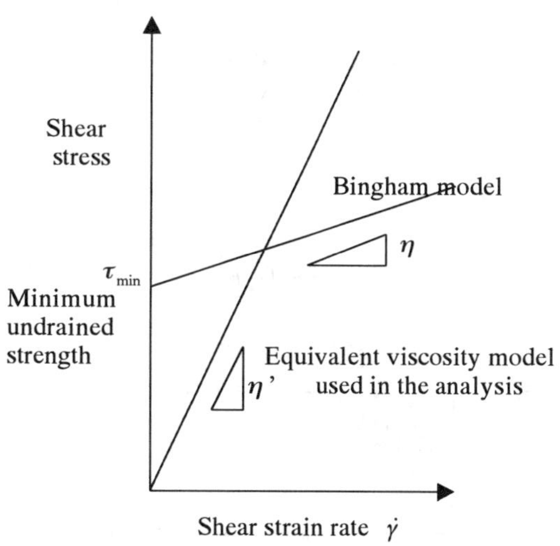

Figure 3. Concept of equivalent viscosity

Figure 4 shows the flow velocity to liquefied layer thickness relations obtained from experiment (Hamada et al. 1998). Figure 5 shows the flow velocity to liquefied layer thickness relations obtained from the CIP based numerical method and experiment, for 3% and 5% ground slopes. From these figures, it can be seen that the CIP based numerical method has well reproduced the proposed similitude law.

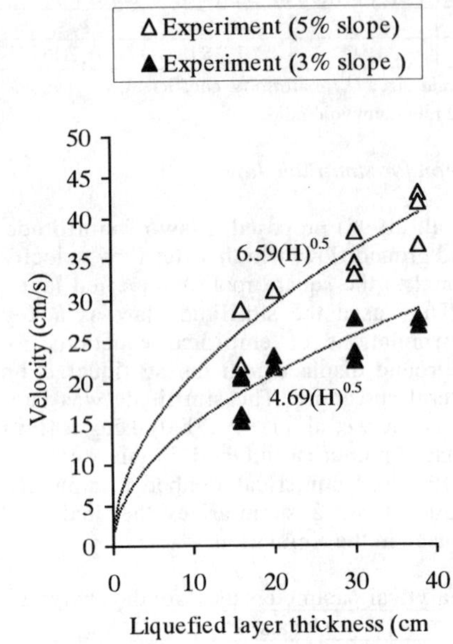

Figure 4. Similitude law from experiment

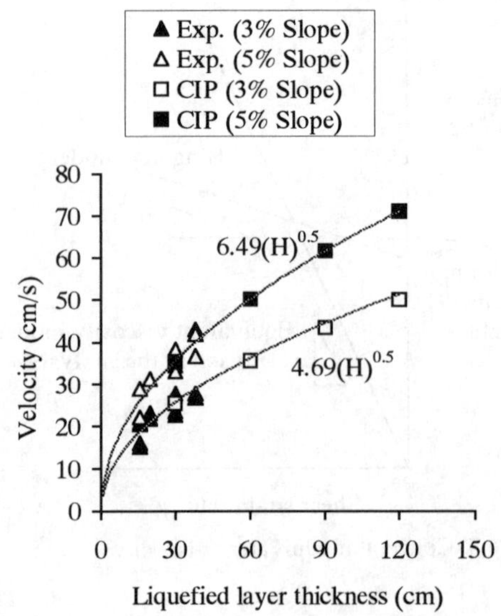

Figure 5. Similitude law from CIP simulation and experiment.

4.2 *Simulation with a nonliquefied layer overlaying a liquefied subsoil*

The shaking table test on liquefied sand with an overlaying unsaturated nonliquefied layer by Sakaguchi et al. (1998) is simulated. The experiment was conducted for two cases of ground conditions: the first with a thin and the second with a thick unsaturated nonliquefied layer overlaying liquefied sand. Enshunada sand was used as a test material.

Figures 6 and 7 show the numerical models used in the analysis, based on the experimental setup, for the thin and thick nonliquefied layer overlaying a liquefied sand, respectively. Table 3 summarizes the analytical parameters used in this simulation. For the nonliquefied layer a large viscosity coefficient (=10000 Pa.s) is used in the analysis.

Figures 8 and 9 show the time histories of flow velocity and displacement at the ground surface obtained from experiment for the thin and thick

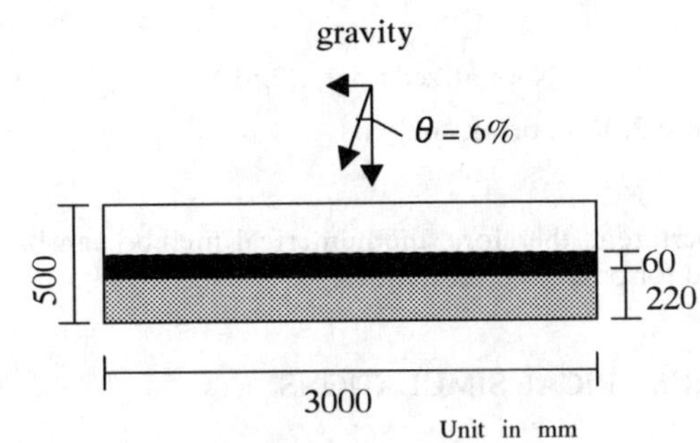

Figure 6. Numerical model, case 1.

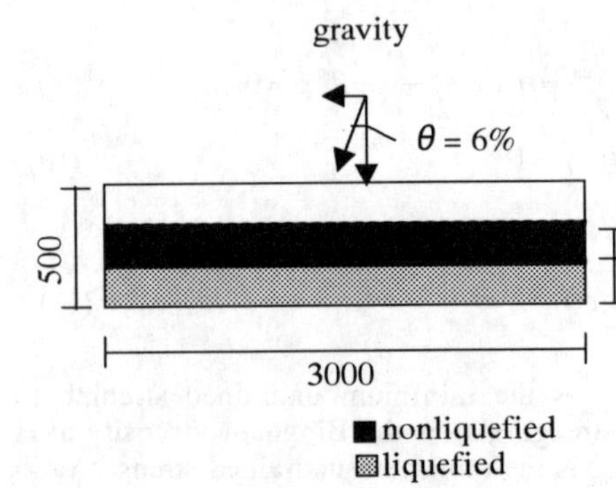

Figure 7. Numerical model, case 2.

Table 3. Analytical parameters

	η (Pa.s)	R_{min}	Notes
Liquefied layer	1.0	0.025	Bingham model
Nonliquefied layer	10000.0		Large viscosity

nonliquefied layer cases, respectively. Figures 10 and 11 show the time histories of flow velocity and surface displacement at 3 cm. depth from ground surface obtained by the simulation together with results from experiment, for the thin and thick nonliquefied layer cases, respectively. From these figures it can be seen that the CIP based numerical method has reasonably reproduced the time histories of flow velocity and ground surface displacement obtained from the experiments.

4.3 *Comparison with an existing numerical code*

The current CIP based numerical method is used to simulate the time history of flow velocity reported by Uzuoka et al. (1998) using an existing fluid

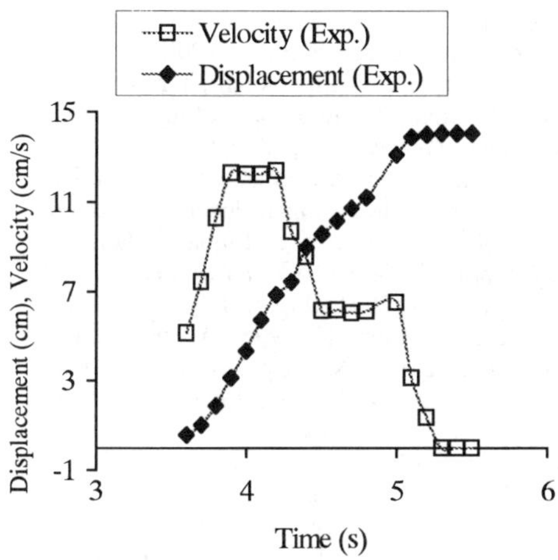

Figure 8. Time history curves for thin nonliquefied layer case, experiment.

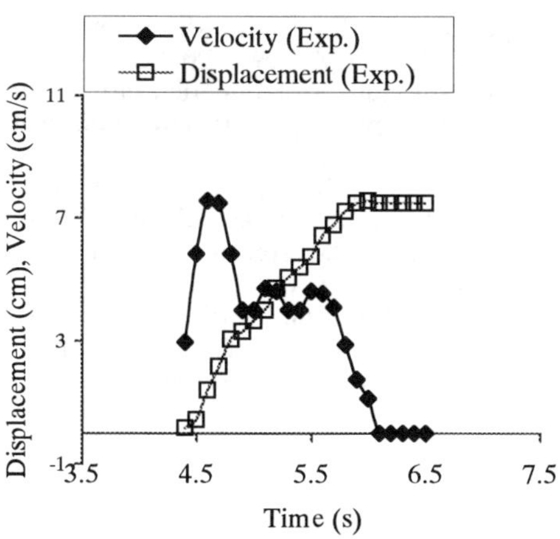

Figure 9. Time history curves for thick nonliquefied layer case, experiment

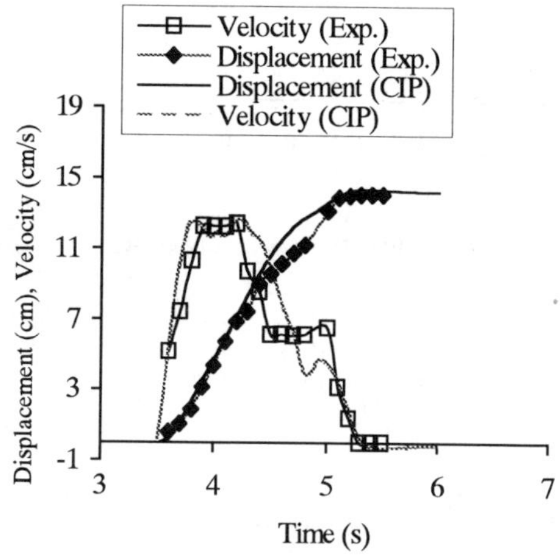

Figure 10. Time history curve for thin nonliquefied layer case, CIP simulation and experiment

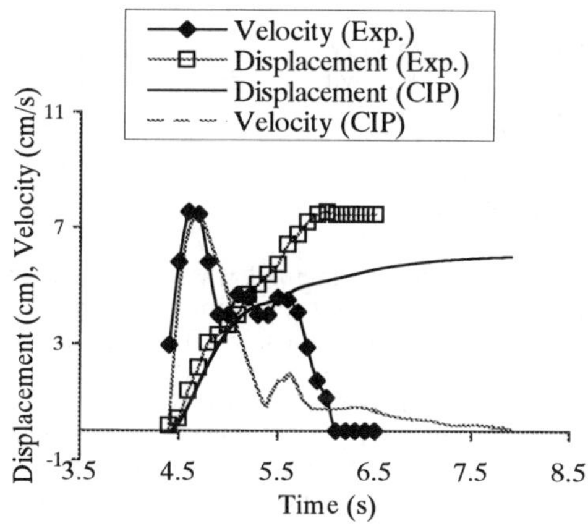

Figure 11. Time history curve for thick nonliquefied layer case, CIP simulation and experiment.

dynamics code, which utilizes the VOF method to treat free boundaries and interfaces between fluids. Table 4 summarizes the analytical parameters used for this analysis.

Figures 12 and 13 show the time history of the flow velocity from an existing fluid dynamics code (Uzuoka et al. 1998) and the CIP based numerical method respectively, for 3% ground slope. From these two figures it can be seen that the CIP based numerical method has reasonably reproduced the

Table 4. Analytical parameters (Uzuoka et al. 1998)

θ (%)	η (Pa.s)	R_{min}
3	1.0	0.018

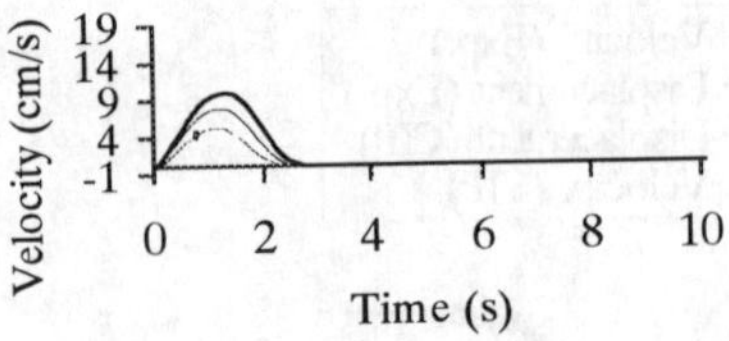

Figure 12. Velocity-time history, existing code.

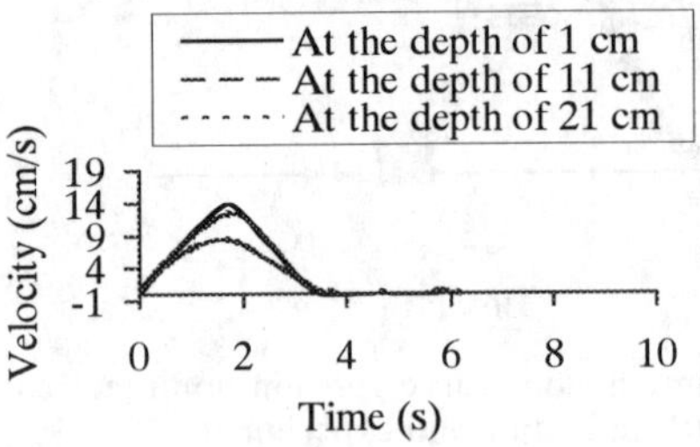

Figure 13. Velocity-time history, CIP simulation

trend of the time history of flow velocity of the liquefied subsoil but with a slightly larger peak velocity.

5 CONCLUSIONS

A CIP based numerical method was developed for liquefaction induced lateral spreading analysis in the framework of fluid dynamics. A Bingham model that takes the residual strength of soil into account was used to represent the behavior of liquefied soil. An old version of a CIP based code was improved by incorporating the Bingham viscosity and an implicit calculation procedure for pressure terms. The numerical method was validated and good agreement was produced in comparison with experimental results. The method was then used to simulate lateral spreading experiments on liquefied subsoil with and without an overlaying unsaturated nonliquefied layer. From the results of simulations, the numerical method was found to reproduce a previously proposed similitude law for liquefied ground flow and the time histories of flow velocity and ground surface displacement.

REFERENCES

Aydan, O. 1995. Mechanical and numerical modeling of lateral spreading of liquefied soil. *Proc. of 1st Int. Conf. On Earthquake Geotechnical Engineering*: 881-886, Tokyo, Japan.

Bartlet, S. & Youd, T. L. 1992. Empirical prediction of lateral spreading displacement. *Proc. from the 4th U.S.-Japan Workshop on Earthquake Resistant Design of Lifeline Facilities and Countermeasures against Soil Liquefaction*, technical report NCEER-92-0019, Vol. 1, 351-365, Honolulu, USA.

Finn, W. D. L., Yogendrakumar, M., Ledbetter, R. & Yoshida, N. 1991. Analysis of liquefaction induced displacements. *Proc. of 7th Int. Conf. on Computer Methods and Advances in Geomechanic*: 913- 921, Cairns, Australia.

Hadush, S., Yashima, A. & Uzuoka, R. 1999. Verification of similitude law for liquefied ground flow. *Proc. from 7th US-Japan Workshop on Earthquake Resistant Design of Lifelines and Countermeasures Against Soil liquefaction*, technical report MCEER-99-0019: 221-236, Seattle, USA.

Hadush, S., Yashima, A. & Uzuoka, R. 2000. Importance of viscous fluid characteristics in liquefaction induced lateral spreading analysis. *Computers and Geotechnics*, in print.

Hamada, M., Yasuda, S., Isoyama, R. and Emoto, K. 1986. Study on liquefaction-induced permanent ground displacements and earthquake damage. *Proc. of Japan Society of Civil Engineers*, No. 376, III-6, pp. 221-229 (In Japanese).

Hamada, M. & Wakamatsu, K. 1998. A study on ground displacement caused by soil liquefaction. *Journal of Geotechnical Eng., JSCE*, , no. 596, Vol. III-43, 189-208 (In Japanese).

Hirt, C. W., and Nicholas, B. D. 1981. Volume of Fluid (VOF) method for the dynamics of free boundaries. *J. Comput. Phys.* 39, 201-225.

Sakaguchi, T. and Yoshida, M. 1998. An experimental and analytical study on characteristics of flow and external force of liquefied ground. *Master thesis*, Waseda University, Japan (in Japanese).

Towahata, I., Sassaki, Y., Tokida, K., Matsumoto, H., Tamira, Y., & Yamada, K. 1992. Prediction of Permanent displacement of liquefied ground by means of minimum energy principle. *Soils and Foundations*, 32(3), 97-116.

Uzuoka, R., Yashima, A., Kawakami, T. & Konrad, J.-M. 1998. Fluid dynamics based prediction of liquefaction induced lateral spreading. *Computers and Geotechnics*, no. 3/4, Vol. 22, 243-282.

Yabe, T. & Aoki, T. 1991. A universal solver for hyperbolic equations by cubic polynomial interpolation, *Comput. Phys. Commun.* 66, 219-232.

Computer Methods and Advances in Geomechanics, Desai et al. (eds) © 2001 Balkema, Rotterdam, ISBN 90 5809 183 X

Effects of facing details on the seismic behavior of segmental walls

S. Helwany
University of Wisconsin-Milwaukee, Wis., USA

D. McCallen
Lawrence Livermore National Laboratory, USA

ABSTRACT: This paper describes a detailed finite element analysis of two 6 m-high segmental walls, subjected to seismic loading, with special attention to connection performance and permanent deformation of the facing. The first segmental wall utilizes concrete facing blocks with pins, and the second wall utilizes the same concrete blocks but without pins. The analysis provides some insight into the behavior of segmental walls particularly the load transfer mechanism between the geosynthetic reinforcement and the segmental facing units during earthquake loading.

1 INTRODUCTION

Segmental retaining walls and slopes have performed well in different seismic events in the USA and Japan. Sandri (1994) conducted a survey of segmental retaining walls greater than 4.5 m in height in the Los Angeles area immediately after the Northridge Earthquake of 17 January 1994 (magnitude = 6.7). The results of the survey showed no evidence of visual damage to 9 of 11 structures located within 23 to 113 km of the earthquake epicenter. Two structures showed tension cracks within and behind the reinforced soil mass that were caused by seismic loading. The facing columns for all walls were intact even though peak horizontal ground accelerations as great as 0.5g were estimated at one site.

The concrete blocks are discrete units that transmit shear through interface friction, concrete keys, mechanical connectors, or a combination of these methods. The modular facing column introduces performance issues not associated with other types of reinforced retaining wall systems (Bathurst and Alfaro, 1996). The interfaces between the dry-stacked facing units provide potential failure planes through the facing column and this requires that stability calculations be carried out to estimate interface shear forces and to compare these forces with available shear capacity. In addition, the connection between the geosynthetic reinforcement layers and the facing column, which is typically constructed by extending the reinforcement layers along the interface between facing units to the front of the wall, must also be evaluated for satisfactory design capacity.

This paper describes a detailed finite element analysis of two 6 m-high segmental walls, subjected to seismic loading, with special attention to connection performance and permanent deformation of the facing. The first segmental wall utilizes segmental facing blocks with pins, and the second wall utilizes the same blocks but without pins. Such analysis is essential for understanding the behavior of segmental walls particularly the load transfer mechanism between the geosynthetic reinforcement and the segmental facing units during earthquake loading. The finite element computer program DYNA3D (developed at Lawrence Livermore National Laboratory) is utilized in the analysis. DYNA3D is capable of rigorously handling the complexity of the segmental wall system, including nonlinear contacts between various wall components, and nonlinear hysteretic energy dissipation soil behavior. This methodology can readily be extended to the general three dimensional analysis situation in which three dimensional acceleration histories can be applied to the structure.

2 THE FINITE ELEMENT COMPUTER CODE DYNA3D

DYNA3D is a finite element code for analyzing the dynamic response of three-dimensional solids and structures. DYNA3D is based on a finite element discretization of the three spatial dimensions and a finite difference discretization of time. The explicit central difference method is used to integrate the equations of motion in time. The element formulations available include one-dimensional truss and beam elements, two-dimensional quadrilateral and triangular shell elements, and three-dimensional continuum elements. Many material models are available to represent a wide range of material behavior, including elasticity, plasticity, thermal effects, and rate dependence. In addition, DYNA3D has a sophisticated contact interface capability, including frictional sliding and single surface contact, to handle a range of mechanical interactions between independent bodies. This capability is essential for the analyses of segmental walls. DYNA3D has been used extensively at Lawrence Livermore National Laboratory and in industry. It has been applied to a wide spectrum of problems, many involving large inelastic deformations and contact.

DYNA3D has been verified (Helwany and McCallen, 2000) by comparing its results with (1) a shake table retaining wall test in which a 1 m-high retaining wall was rotated about its base at a constant angular velocity while being subjected to cyclic loading (Ishibashi and Fang, 1987), (2) a static pullout test on a geosynthetic layer confined in Ottawa sand (Su, 1986), and (3) a typical large-scale connection test. Furthermore, the first author has utilized DYNA3D to solve several soil-structure interaction problems including soil-pavement interaction (Helwany et al 1998, and Helwany, 2000), and dynamic lateral earth pressure on underground structures (Helwany and Chowdhury, 2000).

3 FINITE ELEMENT ANALYSIS OF A 6 m-HIGH SEGMENTAL WALL SUBJECTED TO EARTHQUAKE LOADING

A careful analysis of two 6 m-high segmental retaining walls subject to the El Centro earthquake was conducted utilizing DYNA3D. The segmental retaining wall involved 10 equally-spaced geosynthetic layers (60 cm apart), and the segmental facing consisted of 29 concrete blocks. Each concrete block measured 20 cm high by 53 cm wide, toe to heel, by 45 cm long, and each geosynthetic layer measured 3.65 m in length (60 percent of wall height). The design of this segmental retaining wall was based on current pseudo-static method recommended by NCMA (Bathurst, 1998). The configuration of the segmental wall is shown in Figure 1. The finite element mesh was extended 26 m in the horizontal direction (away from facing) and was bounded by a vertical "nonreflective" surface to simulate an infinite domain and to prevent reflected stress waves from reentering the finite element model.

Two facing types are investigated herein. Both facing types involve the same concrete blocks. Facing Type 1, however, includes two pins between the concrete blocks, whereas, Type 2 includes no pins, i.e., the connection strength (for facing Type 2) relies solely on friction between the geosynthetic layer and the concrete blocks. The pin/geosynthetic rupture strength was assumed 7.0 kN/m for Type 1–a value consistent with that of a typical pin/geogrid system. The geosynthetic reinforcement for both analyses was assumed to behave in an elastic-plastic manner with an assumed ultimate strength of 42 kN/m--a strength consistent with that of a typical geogrid reinforcement.

The soil was discretized into 3,422 solid hexahedron elements. Each concrete block was discretized into 16 solid hexahedron elements, and each geosynthetic layer was discretized into 40 "thin" solid hexahedron elements. For facing Type 1, pin connections were utilized between the concrete blocks. Each pin was simulated by eight beam elements. Eight one-dimensional sliding interface elements were utilized between each pin and the concrete blocks in contact. The number of nodes included in the finite element mesh was 11,822, the total number of brick elements was 4,286, and the total number of beam elements was 448.

The interface between different structural parts of the segmental wall was carefully simulated utilizing the "sliding interface" available in DYNA3D. The sliding interface is a penalty-function based contact surface which allows sliding between disjoint model parts with separation and friction. Large relative motions are permitted, and Coulomb friction is included. Surfaces may separate and come together in a completely arbitrary fashion--a feature essential for the current analysis. A total of 3,058 interface facets were utilized in the current analysis.

The pin/geosynthetic interface (for facing Type 1) was modeled by four discrete elements connecting each geosynthetic layer to the pins. These discrete elements are capable of yielding when the pin/geosynthetic rupture strength is reached. A total of 76 discrete elements were utilized in the analysis. The concrete block/geosynthetic friction coefficient was assumed 0.8. Reasonable friction factors between other materials (i.e., soil/concrete=0.43, concrete/pins=0.6, geosynthetic/soil=0.8, and concrete/concrete=0.8) were assumed.

The soil behavior was simulated by the hysteretic Ramberg-Osgood model. The soil parameters relevant to Ramberg-Osgood model were determined herein following the procedure proposed by Ueng and Chen (1992). These parameters were obtained from the results of numerous laboratory tests, including cyclic simple shear tests and cyclic triaxial tests on sand, compiled by Seed and Idriss (1970). Details of this procedure can be found in Helwany and McCallen (2000).

4 RESULTS OF ANALYSIS

The dynamic relaxation (DR) method in DYNA3D was utilized herein to calculate the static response of the segmental retaining wall at-end-of-construction. This method is based on the observation that the long-time limit of a damped dynamic solution is the quasistatic solution. For facing type 1, the calculated static (t=0 s) lateral displacements at points A, B, and C along the height of the segmental facing, indicated in Figure 1, ranged from approximately 0.5 mm at point C to 2.5 mm at point A as shown in Figure 2. The calculated static lateral displacements at points A, B, and C for facing type 2 are substantially greater than those of facing type 1, and they ranged from approximately 3 mm at point C to 17 mm at point A. This is mainly attributed to the relative displacement between the concrete blocks. Such relative displacement is not possible in facing type 1 because of the presence of pins.

Figure 2 indicates that both facing types have experienced relatively small earthquake-induced lateral

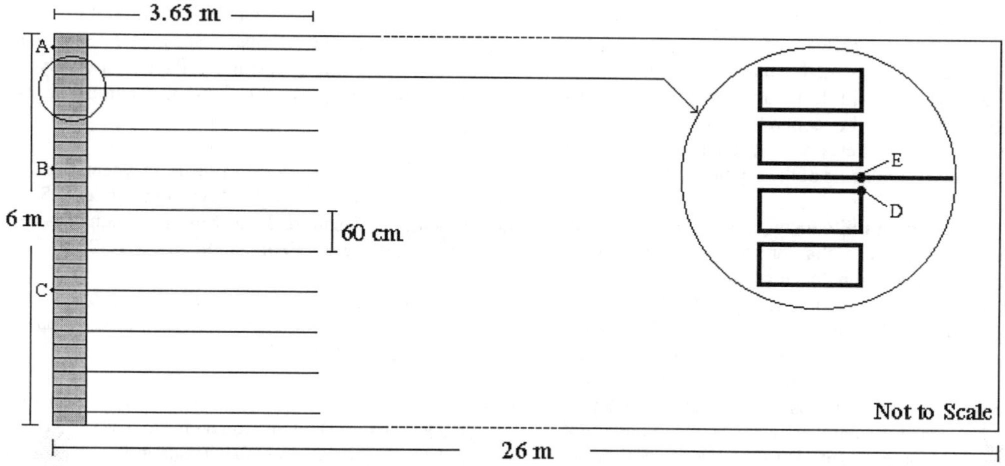

Figure 1: Wall Configuration

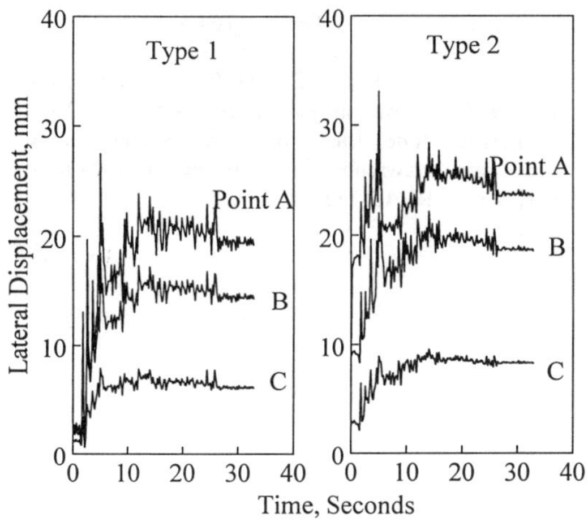

Figure 2: Outward Lateral Displacement of Facing

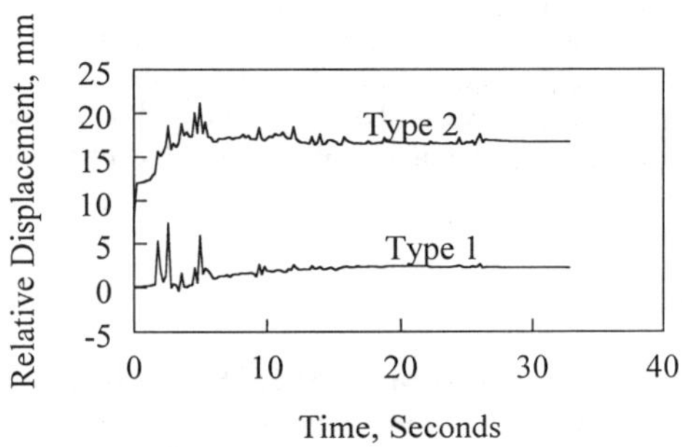

Figure 3: Relative Displacement between Points D and E

deformations. For facing type 1, the maximum total (static+dynamic) lateral "permanent" deformation was approximately 20 mm and located near crest of the segmental facing. It is noted that facing type 2 has experienced smaller dynamic lateral displacements than facing type 1, even though its maximum total lateral deformation was slightly greater than that of type 1. Figure 2 also indicates that most permanent deformations of both types of segmental facing occurred during the first 16 s of earthquake simulation.

Figure 3 presents the calculated relative displacement between point D and point E (see Figure 1). Points D and E initially occupied the same location. Figure 3 indicates that for facing type 1 the relative displacement between points D and E at-end-of-construction was approximately 0. The final relative displacement was approximately 2 mm which can be attributed to the deformation of the geosynthetic layer and to the limited allowance in relative displacement caused by the pin being slightly smaller than the opening in the concrete block. In contrast, the relative displacement between points D and E for facing type 2 was approximately 12 mm at-end-of-construction and 17 mm at the end of earthquake loading.

Nevertheless, the out-of-alignment permanent deformation was approximately 0.5 percent of wall height for both facing types. The analysis indicated that the two types of segmental facing remained intact throughout earthquake simulation with minimal relative deformations and rotations between the concrete blocks.

5 CONCLUSION

The analysis indicated that both facing types experienced relatively small earthquake-induced lateral deformations.

In both cases the out-of-alignment permanent deformation was approximately 0.5 percent of wall height, and the segmental facings remained intact throughout the earthquake simulation with limited relative deformations and rotations between the concrete blocks. This was mainly attributed to the presence of adequate shear strength between various components of the segmental facing (with and without pins).

The preceding finite element analysis is essential for understanding the behavior of segmental walls particularly the load transfer mechanism between the geosynthetic reinforcement and the segmental facing units during earthquake loading. The remarkably small earthquake-induced lateral facing deformations, small earthquake-induced connection forces, and small earthquake-induced axial strains in the geosynthetic layers suggest that the segmental retaining walls have not experienced any significant distress during earthquake simulation. It is to be noted, however, that the analysis conducted herein is limited to one wall configuration and to one ground motion history. Further analysis is required for more conclusive evidence. Moreover, large-scale and/or full-scale seismic tests on segmental walls are exceedingly needed for further verification of DYNA3D and other numerical and analytical seismic design methods.

REFERENCES

Bathurst, R.J. and Alfaro, M.C. (1996). "Review of seismic design, analysis and performance of geosynthetic reinforced walls, slopes and embankments." Invited keynote paper, IS-Kyushu'96, 3'rd International Symposium on Earth Reinforcement, Fukuoka, Kyushu, Japan, 12-14 November 1996, Balkema, Vol. 2, pp. 887-918.

Bathurst, R.J. (1998). NCMA Segmental Retaining Wall Seismic Design Procedure -Supplement to Design Manual for Segmental Retaining Walls (Second Edition 1997) published by the National Concrete Masonry Association, Herdon, VA, 87 p.

Helwany, S., Dyer, J., and Liedy, J. (1998). "Finite Element Analyses of Flexible Pavements." *Transportation Engineering Journal, American Society of Civil Engineers*, Vol. 124, No 5, pp. 491-499.

Helwany, S. and Chowdhury, A. (2000). "Dynamic Lateral Earth Pressure on Underground Structures." *Structural Engineering Journal*, American Society of Civil Engineers (in review).

Helwany, S. and McCallen, D. (2000). "Seismic Stability of Segmental Retaining Walls." *Geotechnical Engineering Journal, American Society of Civil Engineers* (in review).

Helwany, S. (2000). "Finite Element Analyses of Portland Cement Concrete Pavements." Second International Symposium, 3D Finite Element for Pavement Analysis, Design, & Research, October 11-13, Charleston, W. Virginia.

Ishibashi, I. and Fang, Y-S (1987). "Dynamic Earth Pressures with Different Wall Movement Modes." Soils and Foundations, Japanese Society of Soil Mechanics and Foundation Engineering, Vol. 27, No 4, pp. 11-22.

Sandri, D. (1994). "Retaining walls stand up to the Northridge earthquake." Geotechnical Fabrics Report, IFAI, St. Paul, MN, USA, Vol. 12, No. 4, pp. 30-31.

Seed, H.B. and Idriss, I.M. (1970). "Soil Moduli and Damping Factors for Dynamic Response Analysis." Report No. EERC 75-29, Earthquake Engineering Research Center, University of California, Berkeley, California.

Su, C-K (1986). "Investigating Soil-Geotextile Interaction Mechanism." Thesis, department of Civil Engineering, University of Colorado at Denver, 164 p.

Ueng, T.S., and Chen, J.C. (1992). "Computational Procedures for determining Parameters in Ramberg-Osgood Elastoplastic Model Based on Modulus and Damping Versus strain." Lawrence Livermore National Laboratory, Report No. UCRL-ID-111487.

Computer Methods and Advances in Geomechanics, Desai et al. (eds) © 2001 Balkema, Rotterdam, ISBN 90 5809 183 X

Wave-induced oscillatory soil response: Difference between quasi-static and dynamic solutions

D.S.Jeng
School of Engineering, Griffith University, Gold Coast Campus, Qld, Australia

M.S.Rahman
Department of Civil Engineering, North Carolina State University, Raleigh, N.C., USA

ABSTRACT: The behaviour of porous media under dynamic or quasi-static loads was first formulated by Biot (1941, 1960). For the problem of wave-induced seabed response, the quasi-static soil behaviour has been considered as the first approximation since 1970 (Madsen, 1978). This paper attempts to obtain the completed dynamic solution for such a problem, and examine its influences on the wave-induced pore pressure. The numerical results indicate that the dynamic soil behaviour is quite important under certain combination of wave and soil conditions.

1 INTRODUCTION

The subject of wave-induced soil response in a real seabed has attracted the attention of geotechnical and coastal engineers over the last three decades, for which several basic theories have been developed. Since the completely dynamic analysis for the wave-seabed interaction is too complicated, most previous investigations for the wave-seabed interaction have been limited to quasi-static analysis (Madsen, 1978; Jeng, 1997). Recently, the authors (Jeng et al., 1999) derived an analytical solution for such a problem with $u-p$ dynamic form of Biot's poro-elastic theory (Biot, 1960). They concluded that the inertia effects on pore pressure are significant in shallow water. However, the $u-p$ dynamic form they used doesn't include the second time derivatives of the relative fluid displacements.

The influences of dynamic soil behaviour on consolidation problem have been discussed by Zienkiewicz et al. (1980), based on one-dimensional analysis. They proposed two parameters to determine some limits of validity of the various possible assumptions. They are:

$$\Pi_1 = \frac{k_z V_c^2}{g\alpha\omega L^2}, \quad \text{and} \quad \Pi_2 = \frac{\omega^2 L^2}{V_c^2}, \tag{1}$$

where k_z is the soil permeability, ω is the wave frequency, L is the characteristic length (the thickness of the seabed was used in Zienkiewicz et al. (1980) approach), V_c represents a compression wave velocity, and α is the ratio of density of pore fluid and soil.

In Zienkiwicz et al. (1980), they concluded that the dynamic soil behaviour is not important for the case under wave loading. However, the speed of compression wave was fixed as 1000 m/s, which may not reasonable for most cases under wave loading, especially in an unsaturated seabed. Furthermore, only one-dimensional analysis was used in their approach, although the wave loading is a two-dimensional cases.

The objective of this paper is to re-examine the effects of dynamic soil behaviour on the wave-induced pore pressure with a two-dimensional analysis. A set of governing equations will be formulated and represented in non-dimensional parameters. Then, the characteristic of several dynamic soil parameters will be examined, as well as the difference between quasi-static and dynamic solutions.

2 THEORETICAL FORMULATIONS

2.1 *Governing equation*

The problem we considered is illustrated in Figure 1. Biot(1941, 1960) presented a general set of equation governing the behaviour of a linear elastic porous solid under dynamic conditions. They are summarised in the tensor form as bellows:

*Corresponding author.Email: d.jeng@mailbox.gu.edu.au

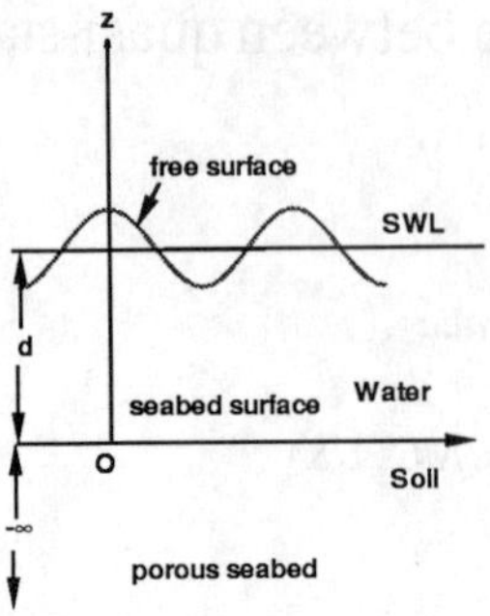

Figure 1: Definition of wave-seabed interaction.

$$\sigma_{ij,j} = \rho \ddot{\mathbf{u}}_i + \rho_f \ddot{\mathbf{w}}_i, \tag{2}$$

$$-p_{,i} = \rho_f \ddot{\mathbf{u}}_i + \frac{\rho_f}{n}\ddot{\mathbf{w}}_i + \frac{\rho_f g}{k_z}\dot{\mathbf{w}}_i, \tag{3}$$

$$\dot{\mathbf{u}}_{ii} + \dot{\mathbf{w}}_{ii} = -n\beta\dot{p} \tag{4}$$

where p is pore pressure, u and w are the displacements of solid and relative displacements of solid and fluid, β is the compressibility of pore fluid.

The definition of effective stresses, σ'_{ij}, which are assumed to control the deformation of the solid skeleton, given for the total stress σ_{ij}

$$\sigma_{ij} = \sigma'_{ij} - \delta_{ij} p \tag{5}$$

Combining (3) and (4), we have

$$\frac{k_z}{\rho_f g} p_{,ii} - n\beta\dot{p} = \dot{\mathbf{u}}_{ii} + \frac{k_z}{g}\ddot{\mathbf{u}}_{ii} + \frac{k_z}{ng}\ddot{\mathbf{w}}_{ii}. \tag{6}$$

If the acceleration terms are neglected in the above equation, it is consolidation equation.

The equations (2)–(4) describe the dynamic soil behaviour of an elastic porous media.

2.2 *Full dynamic solution*

Since the wave-induced oscillatory soil response fluctuates periodically, thus all quantities can be replaced immediately by their complex form

$$f = \bar{f} e^{i(kx-\omega t)} \tag{7}$$

Also, (2)–(4) can be rearranged using new parameters defined as

$$\begin{aligned} K_1 &= G/\left[G/(1-2\mu) + 1/n\beta\right], \\ K_2 &= \left[1/n\beta\right] / \left[G/(1-2\mu) + 1/n\beta\right] \\ V_c^2 &= \left[G/(1-2\mu) + 1/n\beta\right]/\rho, \\ \alpha &= \rho_f/\rho \quad \bar{x} = x/L, \quad \bar{z} = z/L \end{aligned} \tag{8}$$

It is noted that L is the wavelength here.

The equations can be expressed in scalar form as

$$\begin{aligned} & K_1 u_z'' + 2\pi i u_z' + 2\pi i K_2 W_z' \\ &= \left[4\pi^2(K_1+1) - \Pi_2\right] u_x + \left[4\pi^2 K_2 - \alpha\Pi_2\right] w_x, \end{aligned} \tag{9}$$

$$\begin{aligned} & 2\pi i u_x' + (K_1+1)u_z'' + 2\pi i w_x' + K_2 w_z'' \\ &= \left[4\pi^2 K_1 - \Pi_2\right] u_z - \alpha\Pi_2 w_z, \end{aligned} \tag{10}$$

$$\begin{aligned} & 2\pi K_2 u_z' + 2\pi i K_2 w_z' = (4\pi^2 K_2 + \alpha\Pi_2) u_x \\ & + (4\pi^2 K_2 + \frac{\alpha}{n}\Pi_2 + \frac{i}{\Pi_1}) w_x \end{aligned} \tag{11}$$

$$\begin{aligned} & 2\pi i K_2 u_x' + K_2 u_z'' + 2\pi i K_2 w_x' + K_2 w_z'' \\ &= \alpha\Pi_2 u_z + (\alpha\frac{\Pi_2}{n} + \frac{i}{\Pi_1}) w_z \end{aligned} \tag{12}$$

where Π_1 and Π_2 are defined in (1).

These coupled equation are easily solved, the general solution taking the form

$$\begin{aligned} \bar{u}_x &= C_1 e^{\lambda_1 z} + C_2 e^{-\lambda_1 z} + C_3 e^{\lambda_2 z} \\ & + C_4 e^{-\lambda_2 z} + C_5 e^{\lambda_3 z} + C_6 e^{-\lambda_3 z} \end{aligned} \tag{13}$$

with a similar expression for $\bar{u}_z$, $\bar{w}_x$ and $\bar{w}_z$. The λ_i coefficients are the roots of the characteristics equation from the coupled equations:

$$\alpha_3 x^3 + \alpha_2 x^2 + \alpha_1 x + \alpha_o = 0 \tag{14}$$

where α_i are derived from (9)–(12). To complete the solution, boundary conditions must be applied at $\bar{z} = 0$ and $\bar{z} \to -\infty$ to find C_i. These conditions are

$$\begin{aligned} & \sigma'_z = \tau = 0, \ \ \bar{p} = p_o, \ \text{for } \bar{z} = 0 \\ & \bar{u}_x, \bar{u}_z, \bar{p} = 0, \ \text{as } \bar{z} \to -\infty \end{aligned} \tag{15}$$

If all accelerated terms are neglected from (2)–(4), we have the quasi-static case equivalent to the simple consolidation problem, which has been done previously (Madsen, 1978; Jeng, 1997).

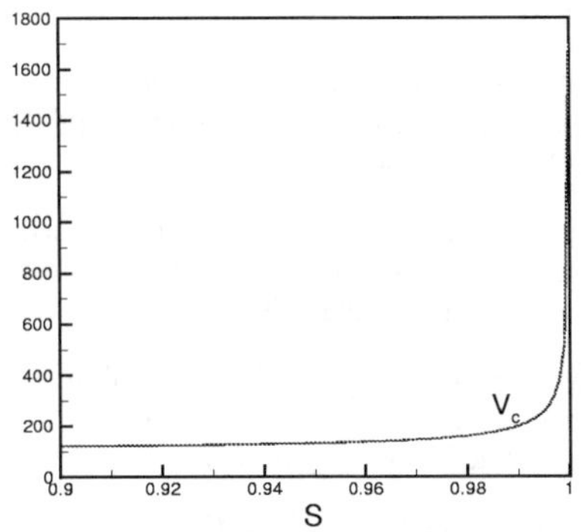

Figure 2: Distributions of the compression wave velocity versa the degree of saturation.

Table 1: Input data for case study.[a]

No	T	D	L	k_z	Π_1	Π_2
1	5	20	38.84	10^{-2}	3.27	8.24E-4
2	5	20	38.84	10^{-4}	3.27E-2	8.24E-4
3	15	20	197.48	10^{-2}	0.378	2.37E-3
4	15	20	197.48	10^{-4}	3.78E-3	2.37E-3
5	5	20	38.84	10^{-2}	1.98E-2	0.136
6	5	20	38.83	10^{-4}	1.98E-4	0.136
7	15	20	197.48	10^{-2}	2.30E-3	0.391
8	15	20	197.48	10^{-4}	2.30E-5	0.391

[a] case 1–4: saturated seabed

[a] case 5–8: unsaturated seabed ($S = 0.98$

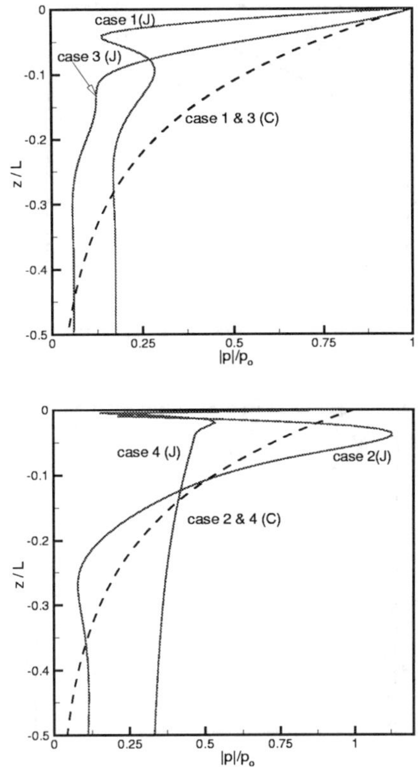

Figure 3: Distribution of the maximum amplitude of pore pressure ($|p|/p_o$) versus soil depth (z/L) for different cases in a saturated seabeds.

3 NUMERICAL EXAMPLES

As mentioned previously, the compression wave velocity (V_c) used in the analysis of Zienkiewicz et al. (1980) was 1000 m/sec. According to the definition of V_c in equation (8), the compression wave velocity depends on the degree of saturation if the soil type is fixed (i.e., G, n and μ are fixed). Figure 2 demonstrates the relationship between the compression wave velocity and the degree of saturation. as shown in the figure, there is a dramatically reduction of V_c from fully saturation ($S = 1$) to nearly saturation ($S = 0.98$). The compression wave velocity reduces from 1800 m/sec to 120 m/sec, for 2 percent of gas existing in the pore between soil particles. This implies that the speed of a compression wave in an unsaturated seabed is much less than in a saturated seabed.

To demonstrate the difference between quasi-static and dynamic solutions on the wave-induced pore pressure, we choose eight (8) cases tabulated in Table 1. Among these cases, cases 1–4 are saturated seabeds, while cases 5–8 are unsaturated seabeds. The odd cases are coarse sandy bed (with high permeability), and the even cases are fine sandy beds (with low permeability). Two different wave periods, 5 sec and 15 sec, are used in the example, which are two extremely cases for most ocean waves. In this paper, the shear modulus (G), soil porosity n, Poisson's ratio (μ) are taken as 10^7 N/m^2, 0.33, 0.3. As shown in Table 1, the values of Π_1 is affected by the degree of saturation (in terms of V_c), soil permeability and wave length, while Π_2 is only affected by the wavelength and degree of saturation.

Figure 3 & 4 illustrate the comparisons between quasi-static and dynamic solutions on the wave-induced pore pressure in both saturated and unsaturated seabeds. In the figure, the solid lines represent the results from the present solution, and dashed lines are quasi-static solution (Jeng, 1997). The figures clearly indicate that the effects of dynamic soil behaviour are significant, especially in a saturated seabed.

According to Figures 3 & 4, it is observed that wave period has greater influence on the the wave-induced pore pressure in saturated seabeds or unsat-

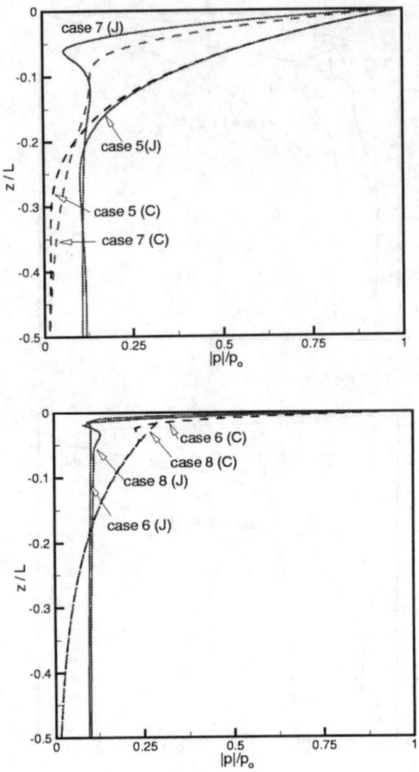

Figure 4: Distribution of the maximum amplitude of pore pressure ($|p|/p_o$) versus soil depth (z/L) for different cases in an unsaturated seabeds.

urated coarse sand, but almost no influence on pore pressure in unsaturated fine sand.

The difference between Figure 3 & 4 indicate the influences of compression wave velocity (V_c) on the pore pressure distribution. As indicated in Figure 2, V_c approach to a constant when S is less than 0.98. This results indicates that V_c also dominates the soil response, when dynamic soil behaviour is considered.

4 CONCLUSIONS

In this paper, the phenomenon of ocean waves propagating over a porous seabed was re-examined by including dynamic soil behaviour. Unlike previous one-dimensional analysis, we directly solve the problem with a two-dimensional analysis. The numerical results indicate that the wave-induced pore pressure are significant affected by dynamic soil behaviour.

Addition to the parameters, Π_1 and Π_2 proposed by Zienkiewicz et al. (1980), the compression wave velocity (V_c, which is denominated by the degree of saturation) is also important in the evaluation of pore pressure with dynamic soil behaviour.

REFERENCES

Biot. M. A. 1941. General theory of three-dimensional consolidation. *Journal of Applied Physics*, 12:155-164

Biot, M. A. 1960. Mechanics of deformation and acoustic propagation in porous media. *Journal of Applied Physics*, 33:1483-1498.

Jeng, D. S. 1997. Jeng, D. S. 1997. Wave-Induced Seabed Response in Front of a Breakwater. PhD Thesis, The University of Western Australia.

Jeng, D. S., Rahman, M. S. & Lee, T. L. 1999. Effects of inertia forces on wave-induced seabed response. *International Journal of Offshore and Polar Engineering*, 9(4):307-313.

Madsen, O. S. 1978. Wave-induced pore pressures and effective stresses in a porous bed. *Géotechnique*, 28:377-393.

Zienkiewicz, O. C., Chang, C. T. & Bettess, P. 1980. Drained, undrained, consolidating and dynamic behaviour assumptions in soils. *Géotechnique*, 30(4): 385–395.

Computer Methods and Advances in Geomechanics, Desai et al. (eds) © *2001 Balkema, Rotterdam, ISBN 90 5809 183 X*

Evaluation of seismic slope stability using a non-linear soil behavior

F. López-Caballero & A. Modaressi
Laboratoire MSS-Mat CNRS UMR 8579, Ecole Centrale Paris, France

ABSTRACT: For evaluating the displacement of natural slopes subjected to earthquakes, different simplified methods for the analysis are proposed in the literature. The most common approach is the rigid block method. This method has been principally used with interface elements having a simple Mohr-Coulomb model. In this paper, to evaluate displacements induced in an infinite saturated clayey slope, a constitutive model, which takes into account complex aspects of soil behavior such as dilatancy or strength degradation, has been used. The effects of the input signal, the state of consolidation and the thickness of the slope have been studied and the obtained results are compared to those from the literature and observed landslides.

1 INTRODUCTION

To study the seismic stability of the slopes several analysis methods are proposed in the literature. The base of these methods is always to make a pseudo-static analysis and to take into account the effect of the earthquake as a horizontal inertial force (Newmark 1965, Makdisi & Seed 1978, Sarma 1979, Moriwaki et al. 1982, Kramer 1996, Stamatopoulos 1996).

Usually, two types of approaches are used. The first produces a factor of safety against the seismic slope failure and the second evaluates the permanent displacements induced by the earthquake. In this work, the second approach will be used.

The goal of this work is to evaluate the seismic displacements induced on a gentle infinite saturated clayey slope.

In the practice the most common approach for the seismic design and stability analysis of soil structures is the rigid block method (Finn et al. 1994, Cai & Bathurst 1996, Al-Homoud & Tahtamoni 2000). Until now, rigid blocks had principally been used with interface elements having a simple Mohr-Coulomb elastoplastic or viscoplastic model (Thika-Vassilikos et al. 1993, Modaressi et al. 1995, Kramer 1996, Sarma 1998). In the framework of this work, a more performant constitutive model which takes into account more complex aspects of soil behavior such as densification and dilatancy or strength degradation, has been used (Aubry et al. 1990).

The importance of these aspects of soil behavior on the magnitude of induced displacements have been observed in *a posteriori* analysis of real landslide, for example in the "Andretta" landslide, Italy 1980 (D'Elia 1992).

The theoretical aspects of this model as well as illustrative examples of its performance will be presented.

2 DEFINITION OF RIGID BLOCK MODEL

The methodology used is a derivation of the Newmark's method. The original method assumes that the slope has a rigid and plastic behavior on an existing slip surface (Newmark 1965, Finn et al. 1994, Kramer 1996, Khasai & Sitar 1998) (Fig. 1). Thus, this behavior can be assimilated with a rigid block that slips on a frictional surface.

In this method, to find the horizontal critical acceleration (k_C), a pseudo-static slope stability analysis

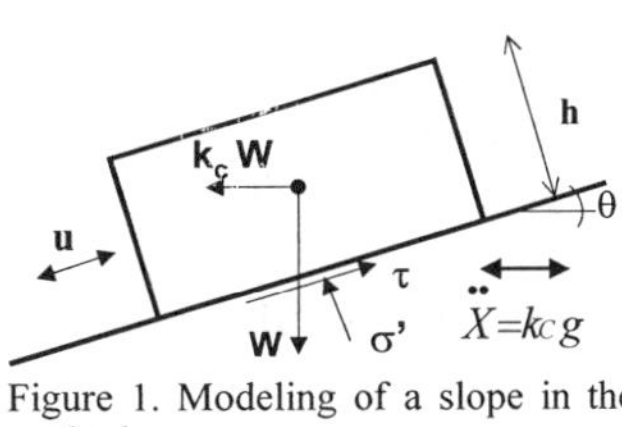

Figure 1. Modeling of a slope in the Newmark's rigid block method.

with the addition of the inertia force given by the earthquake (k_CW) must be made, with W the weight of the block (Fig. 1). Resolving the static equilibrium of the block in terms of total stress, the k_C value is equal to:

$$k_C = (FS_S - 1) Cos\theta \ Sin\theta \tag{1}$$

Where, FS_S is the static factor of safety in terms of total stress and θ is the slope angle. This value of k_C represents the minimum pseudo-static yield acceleration percentage that produces instability in the block and is measured horizontally.

In the case where the stability analysis is carried out in undrained conditions, the reduction of resistance caused by pore-pressure build-up for normally consolidated clays and the dilatancy behavior for overconsolidated clays must be included (Ambraseys & Srbulov 1995, Modaressi et al. 1995). To take into account these effects a factor of correction ξ can be used by introducing the undrained shear strength S_U, which is usually used in geotechnical engineering. The factor ξ is given by:

$$\xi = \frac{S_U}{\sigma'_o \tan \phi} \tag{2}$$

Where σ'_o is the initial effective normal stress on the slope interface and ϕ is the friction angle at critical state of soil interface. Therefore the equation (1) becomes:

$$k_C = \xi (FS_S - 1) Cos\theta \ Sin\theta \tag{3}$$

In the practice the ratio S_U/σ'_o can be obtained by laboratory tests or by using empirical correlations.

3 ELASTOPLASTIC MODEL

In what follows, a brief overview of the type of constitutive model that is used in the GEFDYN programs (Aubry et al. 1990) is given without any detailed description. The model is implemented for seismic analyses of soil interfaces. It should be mentioned that under cyclic loading conditions, soil properties vary still more than under monotonic conditions. So special attention is focused on the shear strain amplitude on which, the shear modulus and damping ratio are strongly dependant.

The plastic yield surface follows an isotropic hardening regime depending on the plastic shear strain γ^P and the influence of volumetric strain is taken into account through the critical stress σ_C as in the Cam-Clay model:

$$f(\sigma',\tau,\varepsilon^P,\gamma^P) = |\tau| + \sigma' \tan\phi \, F \, r(\gamma^P) \tag{4}$$

With

$$F = 1 - b \, Ln(\sigma'/\sigma_c) \tag{5}$$

$$\sigma_c = \sigma_{co} \exp(-\beta \varepsilon^P) \tag{6}$$

Where (σ', τ) and (ε^P, γ^P) are normal and shear stresses and plastic strains on a surface parallel to the natural slope. A kinematical hardening is introduced by updating the shear stress at each load reversal. The parameter b controls the form of the yield surface and varies from b=0 to 1 passing from a Coulomb type surface to a Cam-Clay type one. β is the plasticity compression modulus. The internal variable $r(\gamma^P)$, called degree of mobilized friction, introduces the effect of shear hardening of soil and permits the decomposition of the behavior domain into elastic, hysteretic and mobilized domains, it is given by:

$$r(\gamma^P) = \frac{\gamma^P}{a + \gamma^P} \tag{7}$$

With

$$a = a_1 + (a_2 - a_1)\alpha(\gamma^P) \tag{8}$$

Where

$$\alpha = 0 \qquad \text{if } r^{elas} < r < r^{hys}$$

$$\alpha = \left(\frac{r - r^{hys}}{r^{mob} - r^{hys}} \right)^m \qquad \text{if } r^{hys} < r < r^{mob} \tag{9}$$

$$\alpha = 1 \qquad \text{if } r^{mob} < r < 1$$

4 NON-LINEAR ELASTOPLASTIC MODEL ANALYSIS

The displacement obtained by using the non-linear elastoplastic model for the soil of the interface will be studied.

The infinite slope studied is composed of one layer of clay. Two different slope angles, θ of 10° and 15°, will be used. The water table is assumed to be at the surface and parallel to the slope. The four earthquake records that were selected are: Irpinia NS (Italy) 1980, Kalamata NS (Greece) 1986, Lefkas NS (Greece) 1973 and Tabas EW (Iran) 1978 (Fig. 2) scaled to different values of maximum acceleration (a_{max}).

To define the model's parameters, the behavior of the clay must be studied. The monotonic and cyclic shear tests have been modeled. The monotonic test will define if the soil is normally consolidated or overconsolidated, and the cyclic test gives the curve of the shear modulus (G) reduction and damping (D) increase with the shear strain (γ).

As this is a conceptual example, to identify the soil's characteristics a methodology, which is a combination of empirical relationships and laboratory test data, will be used.

4.1 *Used material*

The material that has been used to model the contact between the rigid block and the sliding surface is a clay with Plasticity Index (I_P) of 30%. Three cases with different initial states have been studied. These conditions correspond to two layer thicknesses (11 and 22m), and two OCR values (1 and 2). The three cases studied are case 1, h=22m and OCR= 1; case 2, h=11m and OCR=2 and case 3, h=11m and OCR=1.

The mechanical characteristics are obtained using the methodology suggested by Modaressi & López-Caballero (1999) and Modaressi & López-Caballero (in press) for the determination of CyberQuake model parameters based on the use of correlations relating parameters describing the nature of the soil (e.g. w_L for clays) to the elastoplastic model parameters. These characteristics are summarized in Table 1.

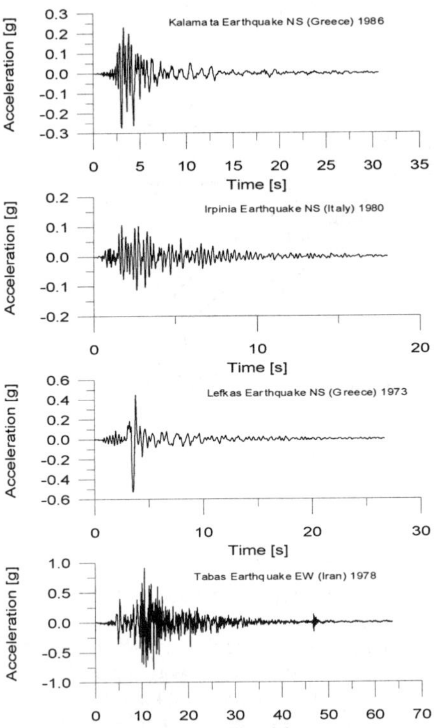

Figure 2. Used earthquake records

4.2 *Definition of elastoplastic model's parameters*

Though the elastoplastic model of the interface in GEFDYN (Aubry & Modaressi 1996) is of the same family as the one in CyberQuake (Modaressi & Foerster 2000), there are some differences concerning the hardening rules. Thus, the principal model parameters have been determined as described above with the same methodology, but the parameters describing the hardening had to be re-calibrated. For this reason, the undrained strain-controlled monotonous and cyclic direct simple shear tests have been simulated. These tests have been made for three initial states corresponding to the three studied cases. The used confining pressures (σ') are those that correspond to the layers' weight applied on the sliding surface level. These values are 91.5, 188 and 93.3 kPa.

Table 1. I_P = 30% clay mechanical characteristics and initial conditions.

	Case 1	Case 2	Case 3
h [m]	22.0	11.0	11.0
ρ [kg/m³]	1880	1866	1830
σ' [kPa]	188	93.3	91.5
φ [°]	26	26	26
C [kPa]	0.0	0.0	0.0
C_C	0.37	0.37	0.37
C_S	0.09	0.09	0.09
e	0.94	0.96	1.05
OCR	1.0	2.0	1.0
S_U [kPa] *	58.0	58.0	28.5
ξ *	0.63	1.28	0.63

* Obtained by the model computation.

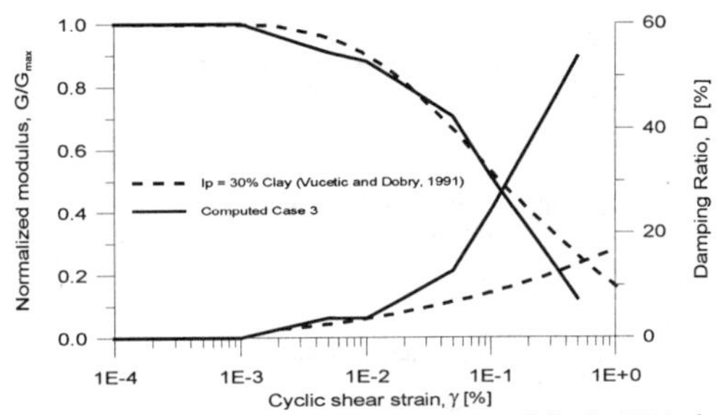

Figure 3. G/G_{max}-γ and D-γ curves computed for I_P=30% clay.

Figure 3 shows the model prediction for the variation of G/G_{max} and the variation of damping ratio (D) with the cyclic shear strain γ for the undrained strain-controlled cyclic direct simple shear test. These curves are compared with the modulus reduction curve and the damping ratio for I_P=30% clay given by Vucetic & Dobry (1991).

The shear modulus (G) obtained by the elastoplastic model is in good accordance with the curve given for I_P=30%, though the model overestimates the damping for high shear strains.

In order to define the different states of the clay (OCR=1 and 2), undrained monotonic simple shear tests are modeled. Figure 4a shows the results of such tests obtained in the τ-γ plane.

In order to verify the validity of the model's parameters proposed, the variation of the shear and normal stress relationship (τ/σ') with the shear strain (γ) has been plotted (Fig. 4b). It has been shown that for the two tests with OCR=1, the soil behavior is always contracting and for the other one with OCR=2, the behavior is dilating and the material exceeds the critical state line before reaching the perfect plasticity. Therefore, for overconsolidated material the resistance degradation is also modeled. In addition the comparison of the S_U values obtained in the numerical simulation are in perfect coherence with the empirical correlations, for example the one given by Jamiolkowski et al. (1985):

$$S_U/\sigma'_o = (0.23\pm 0.04)OCR^{0.8} \quad (10)$$

Where σ'_o is the initial vertical effective stress. In the simulation, S_U/σ'_o values of 0.31 and 0.62 are obtained for OCR of 1.0 and 2.0 respectively. The parameters used in each case are given in the Table 2.

The influence of several parameters on the seismic ground displacements has to be studied. The parameters, which have been studied in the present work, are: the input signal, the state of consolidation (OCR) and the thickness of slope (h). Other parameters such as soil type, damage conditions, etc. will be studied later.

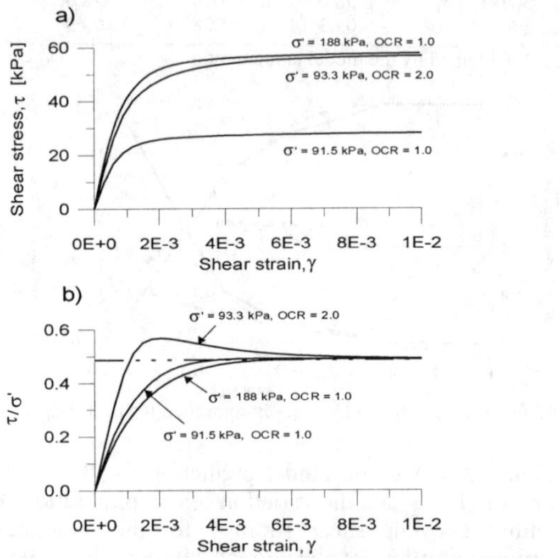

Figure 4. Variation of shear stress τ and τ/σ' ratio with shear strain γ for the 3 undrained simple monotonic shear tests.

5 COMPUTATION OF GROUND DISPLACEMENTS

For this study, the work has been divided into several parts. At first, the effect of the signal on the response of the slope has been studied. Then, the state of consolidation of the soil is changed in order to observe its influence on the response. And finally, the response using two different thicknesses of the block and the same OCR are compared to study its effect.

In order to evaluate the influence of the soil constitutive model, first the displacements using the Newmark integration procedure for the same k_C/a_{max} have been computed and plotted (Fig. 5). These results are compared to those given by Ambraseys & Menu (1988). These displacements are computed using the k_C/a_{max} ratio found for undrained condition. The comparison of all computed displacements with the functions of Ambraseys & Menu (1988) show that displacement curve with 50% probability of exceedance (PE) is a good upper bound to the used acceleration records.

Table 2. Parameters used by the elastoplastic model.

	Case 1	Case 2	Case 3
E_{max} [MPa]	198	160	118
G_{max} [MPa]	76	62	45.4
$\phi = \psi$ [°]	26	26	26
β	16	16	17
σ_{CO} [kPa]	120	120	58.4
b	1.0	1.0	1.0
a_1	0.9E-4	1.4E-4	1.1E-4
a_2	1.8E-4	2.8E-4	2.2E-4
α	0.9	0.9	0.9
γ_{ela}	3.5E-2	3.5E-2	3.5E-2
γ_{hys}	1.0E-1	1.0E-1	1.0E-1
γ_{mob}	5.0E-1	5.0E-1	5.0E-1
m	2.0	2.0	2.0

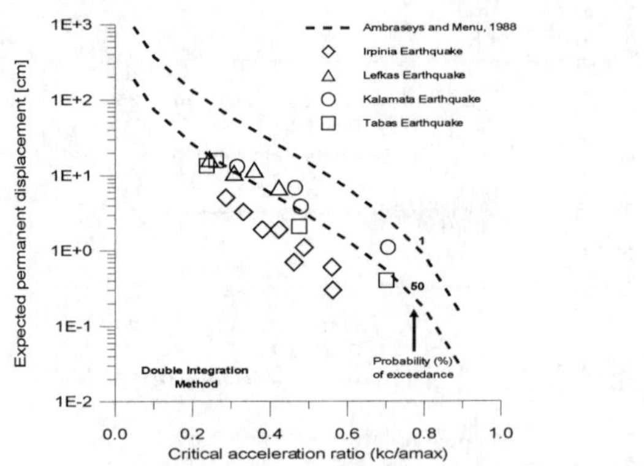

Figure 5. Calculated permanent displacement using the original Newmark integration method.

5.1 *Effect of the input signal*

Initially we worked with the case 1 (h=22m and OCR=1.0). The used slope degrees were 10° and 15°. The signals defined previously were used and the maximum acceleration (a_{max}) was scaled in order to obtain a k_C/a_{max} ratio between 0.24 and 0.8.

Figure 6 shows the variation of displacement with time obtained for the Irpinia earthquake, scaled to 0.34g and a slope of 10°, corresponding to a k_C/a_{max} ratio of 0.56.

According to the computation, the displacement obtained is approximately of 10 cm if the non-linear elastoplastic model is used. Using the equation suggested by Ambraseys & Menu (1988), for the same value of k_C/a_{max} and for 50% PE, the displacement is equal to 1.87 cm.

The comparison between the responses obtained by the non-linear elastoplastic model for each earthquake has been made (Fig. 7), and all responses are grouped near to 1% PE (equation of Ambraseys & Menu 1988). The data indicates that for the particular acceleration records analyzed, the effect of the signal appears not to be important. The use of the non-linear elastoplastic interface results in displacements greater than those obtained by the Newmark integration procedure.

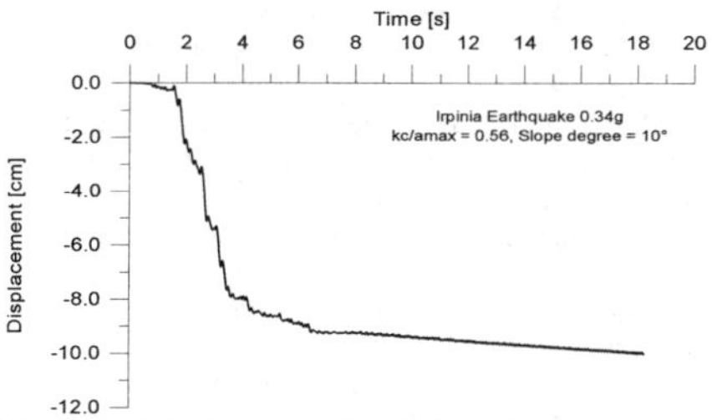

Figure 6. Displacement result from Irpinia earthquake, scaled to 0.34g, $(kc/a_{max}) = 0.56$, Case 1.

5.2 *Effect of OCR*

In order to study the effect of the OCR in the response of the slope, the case 2 has been taken. This model has an OCR value of 2.0 and the thickness of the block is 11m.

The response obtained for OCR=2.0 (symbols) is compared with those obtained by the OCR=1.0 (solid line) (Fig. 8). This figure illustrates that the estimated displacements for OCR=2.0 are much smaller than those obtained for OCR=1.0. This difference is due principally to the dilating behavior of the clay, which exhibits more resistance. Finally, the response is compared to the curves of Ambraseys and Menu (1988) (dashed lines) and the displacements obtained are similar to those of 50% PE (Fig. 8).

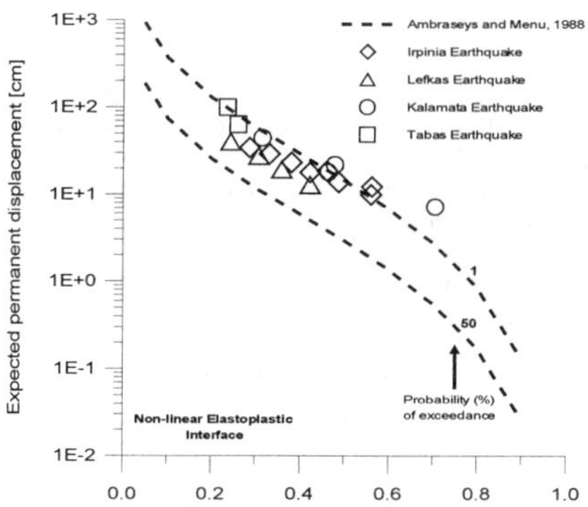

Figure 7. Calculated permanent displacement for the case 1.

5.3 *Effect of the block thickness*

To evaluate the role of the thickness of the block in the response of the slope, the case 1 and 3 have been chosen. The clay is normally consolidated in these two cases. The variation of thickness only changes the confining pressure over the interface and the position of center of the rigid block.

Figure 9 shows the response obtained by the case 3 (symbols) and the comparison with the results obtained by the case 1 (solid line). The comparison shows that the effect of thickness is insignificant because the responses of cases 1 and 3 are grouped.

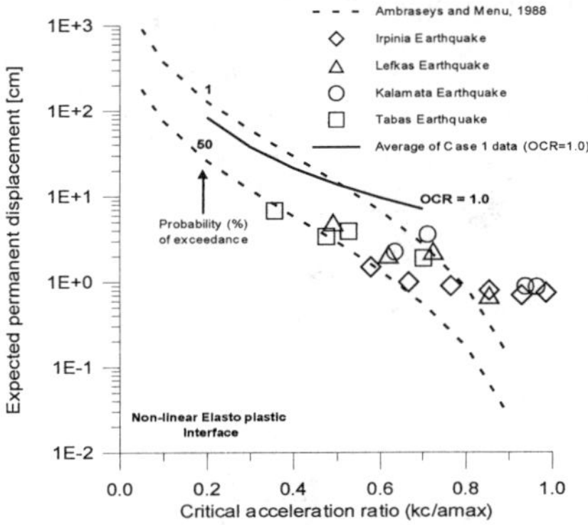

Figure 8. Calculated permanent displacement for the case 2. Comparison with the response of case 1 (Solid line).

Finally, the figure 10 shows the comparison between the responses obtained for OCR=1.0 and 2.0 (solid lines) and both the displacements measured in some cases (symbols) and those of the upper and lower part (areas) of the "Andretta" landslide, Italy, 1980 (D'Elia 1992, Al-Homud & Tahtamoni 2000). In the Andretta case, the gentle slope (≈8°) is formed of saturated overconsolidated hard clay (upper part) and conglomerate deposits (lower part).

Even if the data are not the same, the responses measured have a good accordance with the estimated displacements for both normally and overconsolidated clays.

6 CONCLUSIONS

For a rigid block model with non-linear elastoplastic interface and OCR=1.0, the estimated displacements are greater than those given by the elastic or rigid-plastic approaches due to reduction of resistance caused by undrained condition. For the case of OCR=2.0 the dilatancy behavior of soil increases the resistance and decreases the estimated displacements of the block. This slope behavior has been observed in several historic cases (e.g. the "Andretta" landslide, Italy, 1980).

The obtained responses in the three cases indicate that for the particular acceleration records analyzed, the effect of the signal and the block thickness, appear to be not important in the obtained response. Only the soil behavior has an effect in the estimated displacements of the block.

Numerical models considering the reduction of resistance caused by pore-pressure build-up and the dilatancy behavior of soils must be used in evaluating the slope response to the ground motion records.

The use of factor ξ in the ratio k_C/a_{max} permits to take into account both the behavior of soil and the initial condition for the estimation of ground displacements in an infinite slope.

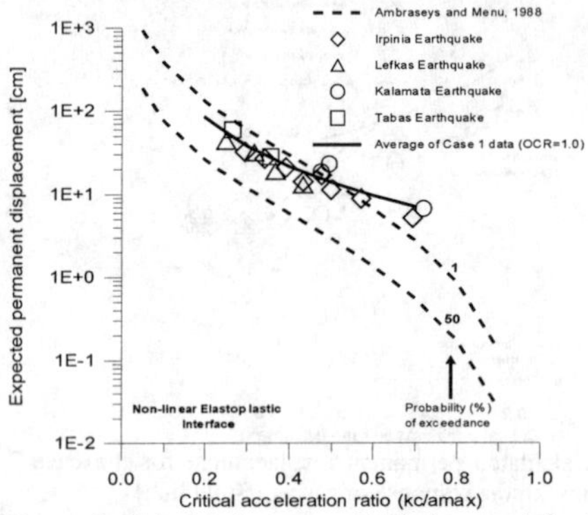

Figure 9. Calculated permanent displacement for the case 3. Comparison with the response of case 1 (Solid line).

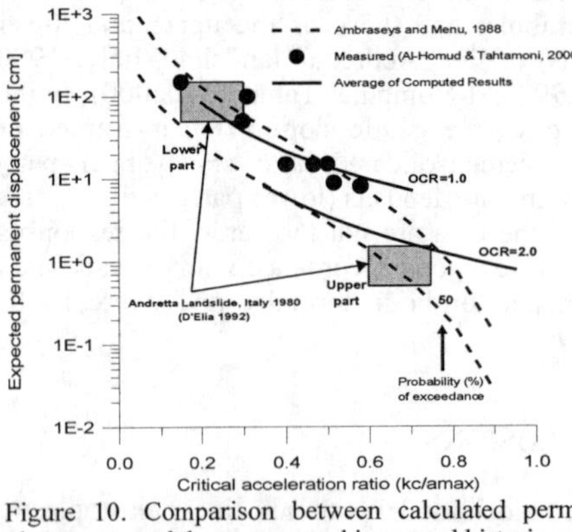

Figure 10. Comparison between calculated permanent displacements and those measured in several historic cases.

ACKNOWLEDGMENTS

This study has been done in the framework of the European Community Contract No ENV-CT97-0392.

REFERENCES

Al-Homoud, A.S. & Tahtamoni, W. 2000. Comparison between predictions using different simplified Newmarks' block-on-plane models and field values of earthquake induced displacements. *Soil Dyn. And Earthq. Engrg.* 19: 73-90.

Ambraseys, N.N. & Menu, J.M. 1988. Earthquake-induced ground displacements. *J. Earthq. Engrg. And Struct. Dyn.* 16: 985-1006.

Ambraseys, N.N. & Srbulov, M. 1995. Earthquake induced displacements of slopes. *Soil Dyn. And Earthq. Engrg.* 14: 59-71.

Aubry, D., Modaressi, A. & Modaressi, H. 1990. A constitutive model for cyclic behaviour of interfaces with variable dilatancy. *Compt. And Geotech.* 9: 47-58.

Aubry, D. & Modaressi, A. 1996. GEFDYN-Manuel Scientifique. Ecole Centrale Paris. France.

Cai, Z. & Bathurst, R.J. 1996. Deterministic sliding block methods for estimating seismic displacements of earth structures. *Soil Dyn. And Earthq. Engrg.* 15: 255-268.

D'Elia, B. 1992. Dynamic aspects of a landslide reactivated by the November 23, 1980 Irpinia earthquake (southern Italy). *Proc. French-Italian Conf. on slope stability in seismic areas*, Bordighera, Italy. pp 25-32

Finn, W.D.L., Ledbetter, R.H. & Marcuson, W.F. 1994. Seismic deformations in embankments and slopes. In Balasubramaniam et al. (eds.) *Symposium on developments in geotechnical engineering*: 233-264. Rotterdam: Balkema.

Jamiolkowski, M., Ladd, C.C., Germaine, J.T. & Lancellotta, R. 1985. New developments in field and laboratory testing of solis. *Proc. 11th Intern. Conf. Soil Mech. And Found. Engrg.* San Francisco, CA. Vol. 1: 57-154.

Khasai, B. & Sitar, N. 1998. Landsliding in native ground: a GIS-based approach to regional seismic slope stability assessment. *Report of Geotech. Engrg. Prog. Dept. Of Civil and Envn. Engrg. Berkeley.* http://www2.ced.berkeley.edu:8002/.

Kramer, S.L. 1996. *Geotechnical earthquake engineering.* Prentice Hall, Upper Saddle River, NJ.

Makdisi, F.I. & Seed, H.B. 1978. Simplified procedure for estimating dam and embankment earthquake-induced deformations. *J. Geotech. Engrg. Div. ASCE.* 104(7): 849-867.

Modaressi, H., Aubry D., Faccioli, E. & Noret, C. 1995. Numerical modelling approaches for the analysis of the earthquake triggered landslides. *Proc.Third Intern. Conf. On Recent Adv. In Geo. Earthq. Engrg. And Soil Dyn.* St. Louis, MO, Vol II: 833-843.

Modaressi, H. & Foerster, E. 2000. *CyberQuake User's manual.* BRGM. France.

Modaressi, A. & López-Caballero, F. 1999. Seismic ground displacements. Progress report No 3. *Seismic ground displacements*. Lab. MSS-Mat Ecole Centrale Paris.

Modaressi, A. & López-Caballero, F. in press. Global methodology for soil behavior identification and its application to the study of site effects. Submitted to *4th Intern. Conf. On Recent Adv. In Geo. Earthq. Engrg. And Soil Dyn.* San Diego, California, March, 2001.

Moriwaki, Y., Idriss, I.M. & Doyle, E.H. 1982. Earthquake-induced deformations of soft clay slopes. *J. Geotech. Engrg. Div. ASCE.* 108(11): 1475-1493.

Newmark, N.M. 1965. Effect of earthquake on dams and embankments. *Géotechnique.* 15(2): 139-160. London.

Sarma, S.K. 1979. Stability analysis of embankments and slopes. *J. Geotech. Engrg. Div. ASCE.* 105(12): 1511-1524.

Sarma, S.K. 1998. Development of new model for sliding block displacements. Progress report No 2. *Seismic ground displacements.* Imperial College of Science, Technology and Medicine, London.

Stamatopoulos, C. 1996. Sliding system predicting large permanent co-seismic movements of slopes. *Earthq. Engrg. And Struct. Dyn.* 25:1075-1093.

Tika-Vassilikos, T., Sarma, S.K. & Ambraseys, N.N. 1993. Seismic displacements on shear surfaces in cohesive soils. *J. Earthq. Engrg. And Struct. Dyn.* 22: 709-722.

Vucetic, M. & Dobry, R. 1991. Effect of soil plasticity on cyclic response. *J. Geotech. Engrg. ASCE.* 117(1): 89-107.

Computer Methods and Advances in Geomechanics, Desai et al. (eds) © 2001 Balkema, Rotterdam, ISBN 90 5809 183 X

An effective stress based liquefaction analysis based on finite deformation theory

Fusao Oka & Takeshi Kodaka
Department of Civil Engineering, Kyoto University, Japan

Tomoyuki Koizumi
Formerly: Kyoto University, Japan

Susumu Sunami
Nikken Sekkei Limited, Osaka, Japan

ABSTRACT: From the field investigation of liquefaction and ground deformation due to Hyogoken Nambu Earthquake, it was found that large deformation occurred near the quay wall. The observed data show that liquefaction analysis method that can describe large deformation is necessary. The aim of the present paper is to propose a liquefaction analysis method using an elasto-viscoplastic constitutive model based on finite deformation theory. From an application of the proposed model to the behaviors of quay wall structures and surrounding ground, we have found that the liquefaction analysis method based on the finite deformation theory is applicable for the liquefaction induced deformation problem.

1 INTRODUCTION

Massive liquefaction occurred at Port Island(man made island) due to the 1995 Hyogoken Nambu Earthquake where weathered granite called Masado was used for reclamation work. Numerical code for liquefaction has been successfully applied to the liquefied ground during Hyogoken Nambu earthquake (Taguchi et al., 1996). Liquefaction analysis using an elasto-plastic constitutive model for sand and Biot's two phase mixture theory has been performed based on the infinitesimal strain theory by many researchers. It, however, has been observed that the large deformation occurred in the ground quaked by the earthquake of high magnitude , in particular, between soil and structures. The development of large deformation is more important in the analysis of soil-structure interaction although slippage may occur between structure and the soil. In the present paper, a liquefaction analysis method based on the finite deformation theory is proposed using a finite strain based constitutive model for sand and updated Lagrangian method. The proposed model was applied to the soil-structure interaction problem.

2 FEM FORMULATION

In the numerical analysis, a cyclic plasticity model for sand(Oka et al., 1999) and a Biot's type two phase mixture theory are used. As for the discretizaion of governing equations for two-phase mixture in space, finite element method was used with the virtual work theorem. In the FEM formulation, u(displacement)-p(pore water pressure) formulation was adopted, in which acceleration of soil skeleton and pore water pressure are taken as independent variables. We consider the equation of motion for water saturated soil and the continuity equation for pore fluid at time $t + \Delta t$. The effective stress concept is used in the analysis as:

$$T_{ij} = T'_{ij} + p\delta_{ij} \tag{1}$$

where T'_{ij} is the effective stress tensor, T_{ij} is the Cauchy's stress tensor and p is the pore water pressure.

Equation of motion of fluid saturated soil is given by

$$\rho a_i = S_{ij,j} + \rho b_i \tag{2}$$

where ρ is the mass density of saturated soil, a_i is the acceleration vector, S_{ij} is the nominal stress and b_i is the body force.

Discretized equation of motion at time $t + \Delta t$ is

$$\begin{aligned}[M]\{a_N\}_{t+\Delta t} + (\Delta t)\left([K]_{t+\Delta t} + [K_L]_{t+\Delta t}\right.\\ \left.+[R]\right)\{v_N\}_{t+\Delta t} + (\Delta t)[K_v]\{\dot{p}_N\}_{t+\Delta t}\\ = \{F\}_{t+\Delta t} - \{S\}_t - (\Delta t)\{T_W\}_t - (\Delta t)\{T_L\}_t \end{aligned} \tag{3}$$

where $[M]$ is the mass matrix, $[K]$ is the stiffness matrix, $\{a_N\}$ is the acceleration vector at node,

$\{v_N\}_{t+\Delta t}$ is the velocity vector, $\{p_N\}$ is the pore water pressure vector, Δt is the increment of time.

Continuity equation is derived by the mass conservation law and the equation of motion for pore fluid as:

$$\rho^F \dot{D}_{ii} - p_{,ii} - \frac{\gamma_w}{k} D_{ii} = 0 \quad (4)$$

The discretized continuity equation becomes

$$\rho^F [K_v]^T \{a_N\}_{t+\Delta t} - \left(\frac{\gamma_w}{k}\right) [K_v]^T \{v_N\}_{t+\Delta t}$$
$$+[K_h]\{p_N\}_{t+\Delta t} = 0 \quad (5)$$

where ρ_f is the mass density of pore fluid, γ_w is the unit weight of pore water, and k is the permeability coefficient. $\{p\}$ is the pore water pressure vector.

In order to discretize equation of motion for water saturated soil and the continuity equation for pore fluid in time, Newmark's β method and finite time difference method are used. In the formulation, acceleration vector $\{a_N\}$ and pore water pressure $\{p_N\}$ are taken as independent variables.

3 CONSTITUTIVE MODEL FOR SAND

In the analysis, we have extended an elasto-plastic model for sand which was derived based on the non-linear kinematic hardening theory by Adachi and Oka (1982), Oka et al.(1999) to a model applicable for the finite deformation theory. In the finite deformation analysis, objective stress rate is necessary. As for the stress rate tensor, we used a Jaumann's stress rate tensor $\hat{T}'_{ij}$ of effective Cauchy stress tensor.

$$\hat{T}'_{ij} = C^{EP}_{ijkl} D_{ij} \quad (6)$$

where C^{EP}_{ijkl} is the elastoplastic stiffness tensor and D_{ij} is the stretching tensor.

Total stretching tensor can be decomposed into elastic(D^E_{ij}) and plastic stretching tensor(D^P_{ij}) as

$$D_{ij} = D^E_{ij} + D^P_{ij} \quad (7)$$

The mode used in the analysis have three main features. The first feature of the model is that the generalized flow rule is used to accurately describe the dilatancy characteristics of sand , and the second one is that cumulative plastic strain dependence of the plastic shear modulus is introduced in order to well simulate a cyclic mobility.

The third one is the introduction of a fading memory of the initial anisotropy. The constitutive model employs an overconsolidation boundary surface which distinguishes an overconsolidated region from a normally consolidated region. The nonlinear kinematic hardening variable is used in the yield function as well as in the plastic potential function.

Overconsolidation boundary surface is introduced to define the N.C and O.C region. The overconsolidation boundary surface, $f_b = 0$, is defined as follows:

$$f_b = \overline{\eta}^*_{(0)} + M^*_m ln \frac{T'_m}{T'_{mb}} = 0 \quad (8)$$

$$\overline{\eta}^*_{(0)} = \left\{ \left(\eta^*_{ij} - \eta^*_{ij(0)}\right) \left(\eta^*_{ij} - \eta^*_{ij(0)}\right) \right\}^{1/2} \quad (9)$$

$$\eta^*_{ij} = T^D_{ij} / T'_m \quad (10)$$

where T'_m is the mean effective stress, T^D_{ij} is the deviatoric stress tensor, M^*_m is the value of the stress ratio expressed by $\sqrt{\eta^*_{ij}\eta^*_{ij}}$ when the maximum volumetric strain during shearing takes place and which could be called the phase transformation stress ratio, and $\eta^*_{ij(0)}$ denotes the value of η^*_{ij} at the end of consolidation. The condition $f_b < 0$ means that the stress state stays in an overconsolidated region (O.C. region), while $f_b \geq 0$ means that the stress state stays in a normally consolidated region (N.C. region). Herein, T'_{mb} in Eq.(8) is given as follows:

$$T'_{mb} = T'_{mbi} \exp\left(\frac{1+e}{\lambda - \kappa} v^P\right) \quad (11)$$

where T'_{mbi} is the initial value of T'_{mb} , κ is the swelling index and λ is the compression index, e is the void ratio, v^P is the plastic volumetric strain.

Such a stress-dilatancy characteristic sometimes gives a rather steeper slope to the liquefaction strength curve than that obtained from laboratory tests. In order to counteract this shortcoming in the original model, the flow rule is generalized using a fourth rank isotropic tensor, H_{ijkl} as

$$D^P_{ij} = H_{ijkl} \frac{\partial g}{\partial T'_{kl}} \quad (12)$$

$$H_{ijkl} = a\delta_{ij}\delta_{kl} + b(\delta_{ik}\delta_{jl} + \delta_{il}\delta_{jk}) \quad (13)$$

One of the problems with the constitutive model for sand is how well the behaviour could be reproduced under cyclic mobility. The original model cannot reproduce the behavior of cyclic mobility so well because shear strain development during cyclic mobility is limited due to the stress path ending up as a steady closed loop at a relatively large mean effective stress. To reproduce the continuous increase in shear strain under cyclic mobility for loose sand, the strain dependence of the plastic shear modulus should be taken into account.

$$G^P = (G^P_{\max} - G^P_{\min}) \exp(-C_f \gamma^{P*}_{(n)}) + G^P_{\min} \quad (14)$$

Table 1. Material parameters used in the analysis

	Soil	Density tf/m^3	Permeability (m/sec)	M^*_m	M^*_f	B_0	B_1	C_f	D_0	n
I	Diluvial sandy gravel	2.00	9.996^{-5}	1.26	1.26	6500	650	2000	1.00	2.00
II	Alluvial sand	1.90	9.996×10^{-5}	1.07	1.45	5000	500	2000	0.90	1.00
III	Alluvial clay	1.60	9.996×10^{-9}	1.26	1.26	8000	180	10000	0.00	1.60
IV	Gravel sand	2.20	3.9984×10^{-4}	1.09	1.33	2200	220	2000	0.80	1.40
V	Sandy gravel(1)reclaimed	2.16	3.9984×10^{-4}	1.06	1.32	2300	230	2000	0.90	1.30
VI	Sandy gravel(2)reclaimed	2.17	3.9984×10^{-4}	1.10	1.40	2500	250	2000	1.00	1.10
VII	Replaced sand	2.20	3.9984×10^{-5}	0.91	1.33	2200	220	2000	0.80	1.40
VIII	Rubble mound	2.20	3.9984×10^{-4}	1.10	1.40	6000	600	2000	0.80	2.00
IX	Sand	2.20	3.9984×10^{-4}	0.91	1.33	2200	220	2000	0.80	1.40
X	Thin element	2.20	9.8×10^{-5}	1.00	1.00	1000	100	2000	0.00	2.00

Swelling index κ=0.0015, Compression index λ=0.02

Caisson: Elastic bulk modulus=3.89$\times 10^6$(kPa), Elastic shear modulus=2.92$\times 10^6$(kPa), density=2.5(tf/m^3)

4 NUMERICAL ANALYSIS

In the present study, we have numerically analyzed a dynamic behavior of caisson type quay wall and surrounding ground at Port Island. Fig.1 shows a cross section of quay wall and ground for numerical analysis. The effective stress analysis mentioned above was performed using methods based on both infinitesimal strain theory and finite deformation theory. As for the finite elements, 20 noded isoparametric element for soil skeleton and 8 noded isoparametric element for pore water pressure are used. 2% Rayleigh damping and $\beta = 0.3025, \gamma = 0.6$ are used in the analysis. The bottom and lateral surfaces are impermeable boundaries, while the ground surface is a drainage boundary.

The 20 second reconstituted acceleration records which direction is perpendicular to the quay wall of west container yard of Port Island, and which was reconstituted from the borehole array records at G.L.-32m were used for input motion in the numerical analysis. In Table 1, material parameters used in the calculation are listed.

In the case of LIQCA-FD, which is a numerical code based on the finite deformation theory expressed in above, very large deformation was occurred in the reclaimed ground just at the back of caisson and computation was stopped due to the large deformation of FE meshes. On the other hand, in the cases of infinitesimal strain theory, LIQCA-208 and LIQCA-3D, such a large deformation was not calculated and computations were carried out for 20 seconds. In order to clarify these differences, we have performed other cases with different permeability of clay. In this case, the permeability of clay was assumed to be equal to that of sandy soil and the computation was carried out for 20 second for three cases. To discuss the comparison between the liquefaction codes with infinitesimal theory and finite deformation theory, the permeability of clay was assumed to be equal to that of alluvial sand layer II. Fig.2 shows the deformed meshes with different methods where displacements are three times magnified. The maximum values of vertical and horizontal displacements are as follows:

Table 2. Maximum displacement

Method (numerical code)	Vertical displacement	Horizontal displacement
LIQCA-FD	1.62m	0.42m
LIQCA-208	0.60m	0.17m
LIQCA-3D	0.49m	0.11m

From Table 2, it is seen that the largest horizontal and vertical displacements are obtained in the case of finite deformation theory. In Fig.3, displacemnt-time profiles at point A are shown. In the case of finite deformation theory, vertical settlement just behind the quay wall is well simulated. Fig.4 shows the acceleration responses at point B. In the case of finite deformation theory, two times larger maximum acceleration, 465*gal* was obtained compared with the cases infinitesimal strain theory. As to the vertical component, it is seen that the phase characteristics of LIQCA-FD is similar to that of LIQCA-208 although the magnitude of finite deformation theory is larger than that of infinitesimal theory.

5 CONCLUSIONS

In the present study, a liquefaction analysis method based on the finite deformation theory is proposed. The proposed theory was derived based on the updated Lagrangian method and the Jaumann stress rate tensor. It is found that the large deformation induced by liquefaction can be simulated by the liquefaction analysis method based on the finite deformation theory. It is, however, necessary to introduce an elaborate adaptive mesh technique to solve the problem of larger gap of integration point in the future.

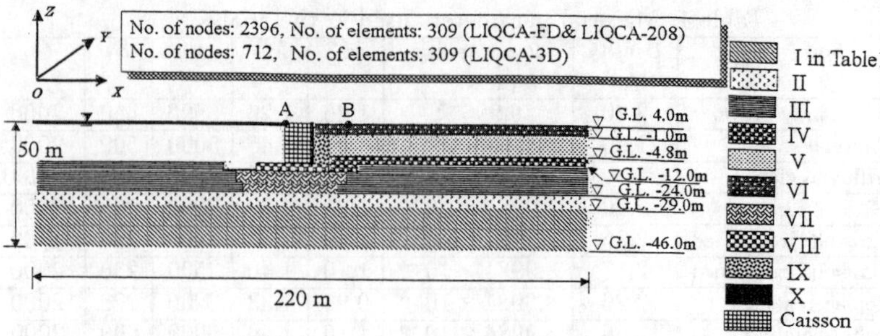

Fig.1. Cross section of quay wall and ground for numerical analysis

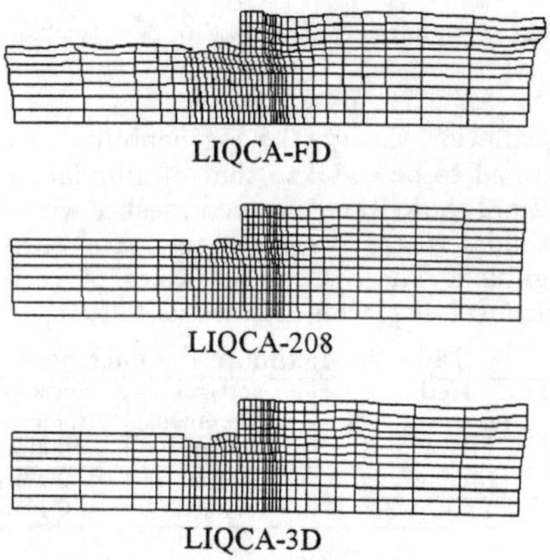

Fig.2. Deformed meshes at t=20.0sec. (×3 displacements)

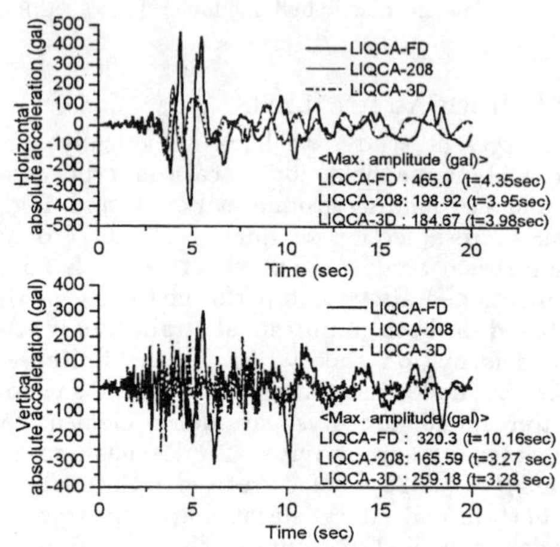

Fig. 4. Acceleration response at point B

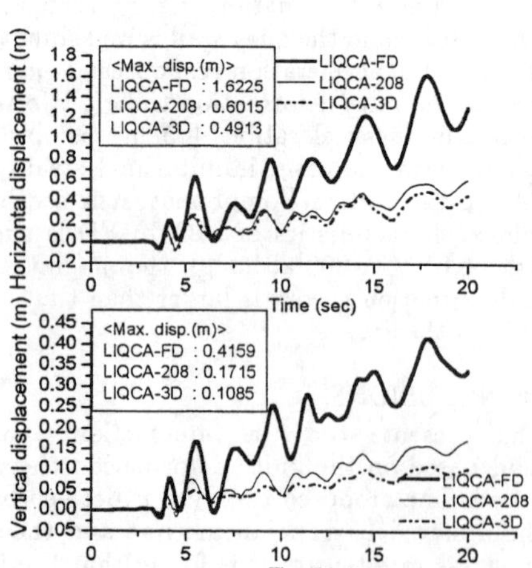

Fig. 3. Displacement – time profile at point A

REFERENCES

Shibata, T., Oka, F. and Ozawa, Y. 1996. Characteristics of ground deformation due to liquefaction, Soils and Foundations, Special issue on the 1995 Jan. 17 Hyogoken-Nambu Earthquake Disaster, Jan., pp.65-79, 1996.

Taguchi, Y., Tateishi, A., Oka, F. and Yashima, A. 1997. Three-dimensional liquefaction analysis method and array record simulation in Great Hanshin earthquake, Proc. 11th World Conference on Earthquake Engineering, Acapulco, Sociedad Mexicana de Ingenieria Sismica, A.C. ed., Balkema(CD-ROM), Paper No.1042, 1997.

Oka,F, A.Yashima, Tateishi,A., Taguchi,Y. and Yamashita, S. 1999. A cyclic elasto-plastic constitutive model for sand considering a plastic-strain dependence of the shear modulus, Geotechnique, Vol.49, No.5, Oct., pp.661-680.

Koizumi, T. 2000. Liquefaction analysis of coastal structures based on finite deformation theory, MS Thesis, Department of Civil Engng., Kyoto University (in Japanese).

Computer Methods and Advances in Geomechanics, Desai et al. (eds) © 2001 Balkema, Rotterdam, ISBN 90 5809 183 X

Effects of a cover layer on the seabed response around a submarine pipeline

P.F. Postma & D.S. Jeng
School of Engineering, Griffith University, Gold Coast Campus, Qld, Australia

ABSTRACT: Recently, considerable efforts have been devoted to the protection of submarine pipelines. To protect the buried pipeline from floatation, it is common to use a cover layer of coarser materials. However, most previous investigations have been limited to experimental approaches, rather than theoretical approaches. This paper presents the effects of a cover layer on the seabed response around a submarine pipeline through finite element analysis. A two-dimensional finite element model established by Jeng & Lin (2000) will be adopted here. The numerical results indicate that by the material of the cover layer significantly affects on the wave-induced excess pore pressure.

1 INTRODUCTION

Submarine pipelines have been extensively used in the transportation of fluids across large stretches of water. They will undoubtably continue to be used in a number of widespread applications, thus seabed protection of an offshore pipeline must therefore be implemented. Pipelines have been damaged from a number of causes ranging from wave and current interaction through to anchor dropping and dragging. To protect the pipeline, it is customary to bury the pipeline in the seabed with a certain range in the cover layer. Since the cost in artificially trenching and refilling is high and the expenditure often covers a large portion of the total budget of a pipeline project. Thus it is necessary to investigate the effects of a cover layer on the wave induced seabed response around the pipe.

It is well known that gravity waves propagating over the ocean exert fluctuations of wave pressure on the sea floor. These fluctuations further induce excess pore pressure and effective stresses, which have been recognised as a dominant factor in analysing the instability of a seabed. Once the wave-induced seabed instability occurs near a pipeline, the phenomenon of pipe floating may occur (Cluket et al., 1989). An inadequate design that leads to flotation of a pipeline will to be a costly exercise as well as an environmental disaster. Thus, the evaluation of wave-induced soil response, including pore pressure, effective stresses and soil displacement, must be taken into careful consideration for marine geotechnical engineers involved in the design of foundations for offshore pipelines.

*Corresponding author. Email: d.jeng@mailbox.gu.edu.au

In engineering practice, a cover layer of a coarser material is often used in the protection of a buried pipeline to minimise the potential of wave-induced seabed instability. This paper will focus on the effects on a cover layer in a porous seabed where the wave induced pore pressure and effective stresses around a buried pipeline and internal stresses will be investigated. To achieve the aims, the two-dimensional finite element model proposed by Jeng & Lin (2000) will be adopted in this study. The pipeline involved will be assumed to be an elastic deformable material in the model. The excess pore pressures will be examined to the extent of showing how they directly influence the stresses placed on and within the pipeline.

2 GOVERNING EQUATIONS

The problem considered in this study is depicted in Figure 1.

2.1 *Porous seabed*

In this study, the consolidation equation (Biot, 1941) is adopted to treat the wave-seabed interaction with variable permeability as

$$K\nabla^2 p - \gamma_w n_e \beta \frac{\partial p}{\partial t} = \gamma_w \frac{\partial}{\partial t}\left(\frac{\partial u_s}{\partial x} + \frac{\partial w_s}{\partial z}\right), \quad (1)$$

where, u_s and w_s are soil displacements in the x- and z-directions, respectively. p is the wave- induced pore pressure, K is permeability, γ_w is the unit weight of pore fluid, $\beta(=\frac{1}{K_w}+\frac{1-S}{P_{wo}})$ is the compressibility of pore fluid (where, K_w is taken as 2×10^9N/m^2. P_{wo}

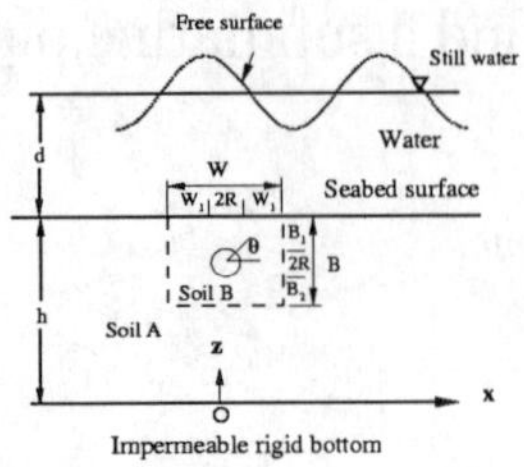

Figure 1: Pipeline buried in a seabed.

is the absolute water pressure and S is the degree of saturation.

Under conditions of plane strain, neglecting the effects of body forces and inertia terms, the equations governing the overall equilibrium of a porous medium can be expressed in terms of the pore pressure and soil displacements as

$$G\nabla^2 u_s + \frac{G}{1-2\mu}\frac{\partial \epsilon}{\partial x} = \frac{\partial p}{\partial x}, \tag{2}$$

$$G\nabla^2 w_s + \frac{G}{1-2\mu}\frac{\partial \epsilon}{\partial z} = \frac{\partial p}{\partial z}, \tag{3}$$

in the x- and z-directions respectively. In (2) and (3) $\epsilon = \frac{\partial u_s}{\partial x} + \frac{\partial w_s}{\partial z}$ is the volumetric strain.

2.2 *Pipeline*

In this study, we assume the pipeline to be an elastic material. Based upon the minimum potential energy theorem, and applying the variation method to the total potential energy of the buried pipeline, we have

$$\delta\Pi = \int_V \sigma_{p,ij}\delta\epsilon_{p,ij}dV - \int_A f_i \delta u_i dA. \tag{4}$$

According to the minimum energy theorem (i.e., $\delta\Pi = 0$), equation (4) can be written as

$$\int_V \sigma_{p,ij}\delta\epsilon_{p,ij}dV = \int_A f_i \delta u_i dA. \tag{5}$$

where f is the external forces acting on the buried pipeline, which consists of two parts: (a) the wave induced pore pressures, and (b) the earth pressure induced by the soil deformations around the buried pipeline. As it is assumed that there is no relative movement between the soil particles and the pipeline, the wave-induced soil deformation can be represnted in terms of the pore pressure.

3 FINITE ELEMENT ANALYSIS

The two-dimensional finite element model that was proposed by the second author (Jeng & Lin 2000) only considered the pipeline as a rigid material. That is the deformation of the pipeline was not considered. In this study the deformation of the pipeline and the cover layer will be included with modifications made to the previous model. Since the wave induced oscillatory soil response is periodically fluctuating in the temporal domain, the wave-induced pore pressure, effective stresses and soil displacements can be expressed as

$$f(x,z;t) = [f_r(x,z) + i f_c(x,z)]\ e^{-i\omega t}, \tag{6}$$

where subscripts and "r" and "c" represent the real and imaginary parts of the soil response, respectively. f denotes the unknown parameters (i.e. pore pressure, stresses, and displacements).

From the governing equations (1)–(3) and (5), that involved the soil displacement and pore pressure, can be solved with the boundary conditions at the surface of the seabed and impermeable bottom. For a porous seabed of finite thickness, as shown in Figure 1, the evaluation of the wave-induced seabed response requires the solutions of (1)–(3) and (5) together with following boundary conditions.

(1) Bottom boundary condition (BBC):

$$u_s = w_s = \frac{\partial p}{\partial z} = 0, \tag{7}$$

(2) Surface boundary condition (SBC):

$$\sigma'_{zs} = \tau_s = 0$$
$$p = p_o(\cos kx + i \sin kx)\ e^{-i\omega t}, \tag{8}$$

where $p_o = \gamma_w H/2\cosh kd$ denotes the amplitude of the wave pressure at the surface of the seabed, d is the water depth, H is the wave height, k is the wave number and w is the wave frequency.

(3) Boundary condition along the surface of the pipe (PBC): It is a reasonable assumption that there won't be any flow through the pipe wall. Thus, the pressure gradient on the surface of the pipe ($r = R$) should vanish, i.e. $\partial p/\partial n = 0$ along the surface of the pipeline

(4) Lateral boundary conditions (LBC): Since the existence of the pipeline only affects the wave-induced soil response in the vicinity of the pipeline, the "disturbed pressure" from the pipeline should vanish at points far away from the pipe.

4 NUMERICAL RESULTS

Based upon the finite element model, the wave induced pore pressures, and effective stresses around the pipeline can be calculated. Two different materials, coarse sand and gravel, are used as the cover layer. To have a basic understanding in the effects of the cover layer on the wave-induced pore pressure and internal stresses of the buried pipe, a case study is performed in the following sections.

In numerical examples, the wave period and water depth are taken as 10 sec and 40 m. The seabed thickness, Poisson's ratio, soil porosity, Young's modulus and degree of saturation are taken as 40 m, 0.4, 7×10^7 N/m^2, and 100%, respectively. For the geometry of the pipe, we have pipe radius $R = 0.5m$, burial depth $b = 2$ m, and both width and thickness of a cover layer are 2 m. In this paper, we consider a concrete pipe with thickness $t_p = 2$ cm, Poisson ratio $\nu = 0.2$, Young's modulus $E_p = 3 \times 10^{10}$ N/m^2 and density of the pipe ρ_p=2400 kg/m^3.

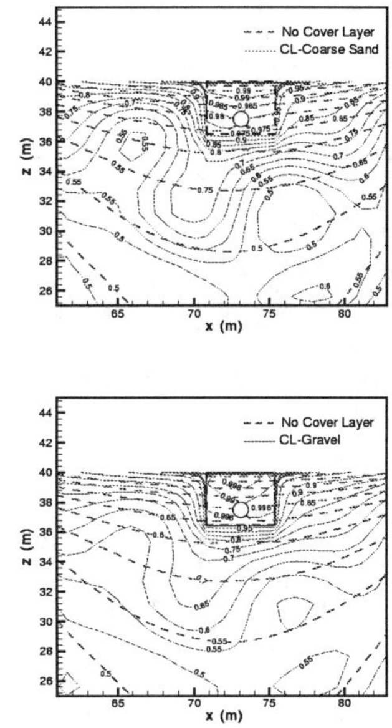

Figure 2: Contours of the wave-induced pore pressures p/p_o in the vicinity of a buried pipeline with a cover layer of (a) coarse sand and (b) gravel.

4.1 *Pore pressure in the vicinity of the cover layer*

Contours of wave induced pore pressure (p/p_o) around the pipe are presented in Figure 2 for the cover layers of coarse sand and gravel respectively. In the figures, solid lines represent the results with a cover layer in place, while the dashed lines are for cases without a cover layer. Figure 2 indicates that the wave-induced pore pressure (p/p_o) varied quite consistently throughout the surrounding soil without a cover layer. The contours were quite regular with very high concentrations in the upper limits. However these values diminished with increasing soil depth. With the presence of a cover layer, a concentration of contours seemed to build up quite significantly around the cover layer. The wave-induced pore pressure was much higher throughout this region and only minor reductions were evident with the increase in depth. This can be contributed to the higher permeability and porosity of the cover materials. The reductions in pore pressure are smallest with the cover layer of gravel, which concludes that the coarser the material is in the cover layer the greater the soil stability in this region.

4.2 *Excess pore pressure along pipeline surface*

The normalised excess pore pressure, $u_e/p_o = 1 - (p/p_o)$, (where p is the pore water pressure and p_o is the amplitude of wave pressure at the seabed surface), around the pipe was conceivably higher with the absence of a cover of a layer. However with the introduction of a coarser and more permeable material the excess pressure was significantly less (Figure 3). This was a direct influence again of the permeability and porosity of the soil in the close proximity of the pipe. Thus the presence of fine sand, coarse sand and gravel yield declining effective stresses respectively. It should be noted that there is only minor variations in excess pore pressure whilst a cover is in place. As higher concentrations of the pore pressure were present in the close proximity of the pipe, this was having a converse effect on the excess pore pressure. It is noted that the excess pore pressure for the cover layer of coarse sand is about one (1) order less than without any cover layer, and that of gravel is about two (2) orders less than without a cover layer. Thus a cover layer significantly reduces the stresses put on the pipeline of concern, which will be of great benefit.

4.3 *Internal stresses within pipeline*

The distributions of radial stress (σ_{rr}), angular stress $(\sigma_{r\theta})$ and shear stresses (τ) are illustrated in Figure 4. In the figures, the results of 3 cases, (a) no cover layer, (b) cover layer of coarse sand and (c) cover layer of gravel, are presented. As seen in the figures, the existence of a cover layer only slightly affects the distribution of internal stresses. In each of the cases presented

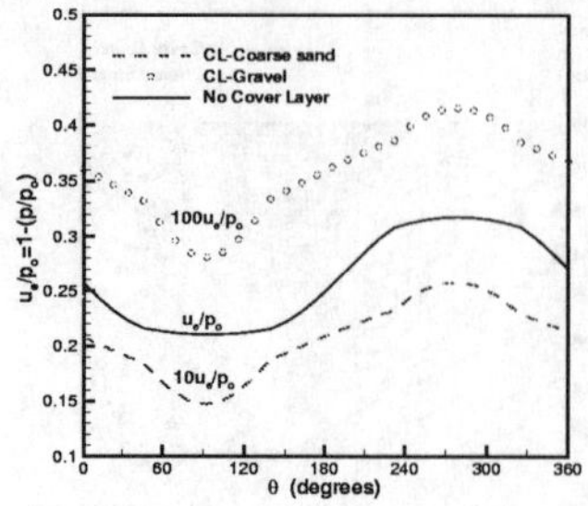

Figure 3: Excess pore water pressure u_e/p_o along pipeline surface.

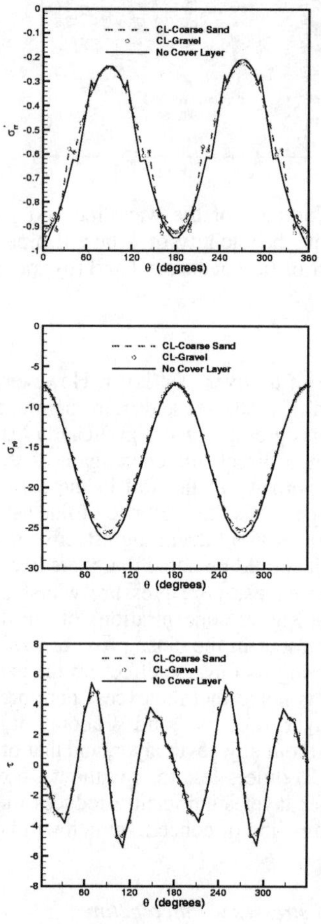

Figure 4: Vertical distributions of the wave-induced internal stresses within pipeline.

the stresses always adhere to similar paths as seen in each of the graphs.

5 CONCLUSIONS

In this study, the two-dimensional finite element model proposed by Jeng and Lin (2000) was adpoted to investigate the effects of a cover layer on the wave-induced pore pressure and internal stresses. Based on the numerical results presented, the follwoing conclusions can be drawn:

1. A cover layer of coarser material (such as coarse sand and gravel) than the surrounding soil significantly increases the soil stability in the vicinity of the pipeline and induces higher pore pressures.

2. The excess pore pressure along the pipeline is reduced two-fold with the introduction of a cover layer.

3. Compared with the pore pressures, the existence of a cover layer insignificantly affects the internal stresses within the pipeline.

ACKNOWLEDGMENTS

The second author is grateful for the financial supports from Australian Academies of Science and National Science Council of Taiwan Exchange Program in 1999.

REFERENCES

Biot, M. A. 1941. General Theory of three-dimensional consolidation. *Journal of Applied Physics*, 12(2):155-164

Clukey, E. C., Vermersch, J. A., Koch, S. P. & Lamb, W. C. 1989. Natural densification by wave action of sand surrounding a buried offshore pipeline. 21st Offshore Technology Conference, 291-300

Jeng, D. S. & Lin, Y. S. 2000. Response of inhomogeneous seabed around a buried pipeline under ocean waves. *Journal of Engineering Mechanics, A.S.C.E.*, 126(4):321-332

Computer Methods and Advances in Geomechanics, Desai et al. (eds) © 2001 Balkema, Rotterdam, ISBN 90 5809 183 X

Modal approaches for the estimation of seismic wave amplification

J.F. Semblat
Laboratoire Central des Ponts et Chaussées, Engineering Modelling Department, Paris, France
A.M. Duval
CETE Méditerranée, Laboratoire de Nice, Seismic Risk Team, France

ABSTRACT : The analysis of seismic wave amplification (or site effects) is possible through many different numerical methods. One can perform either large scale wave propagation computations, local scale dynamic numerical analysis but also investigate, at an intermediate scale, the vibratory response of geological deposits. For that purpose, modal methods are interesting tools to study the modal properties of geological structures. Modal approaches mainly lead to information on such parameters as fundamental frequencies of alluvial basins. It allows the description of both frequencies and eigenmodes of geological structures. In the case of Nice (France), several results are proposed in this paper. Nervertheless, it is difficult to have some quantitative information to discriminate among these frequencies those leading to the largest site effects. To quantify the contribution of the various modes of vibration, this paper also gives some results involving the modal superposition method. Considering several assumptions, this method allows the estimation of some quantitative parameters (modal participation factor, effective modal mass). These parameters show large variations for frequencies corresponding to strong wave amplification. Some comparisons with experimental results are proposed. The agreement between modal numerical results and experimental ones is very satisfactory. Modal methods, involving both eigenmodes determination and modal superposition analysis, then appear as an efficient mean for the qualitative as well as quantitative assessment of seismic site effects.

1. NUMERICAL MODELLING OF SEISMIC WAVE AMPLIFICATION

1.1 *Wave propagation modelling*

The analysis of seismic wave propagation can be achieved by various numerical methods : finite difference, finite element, spectral element, boundary integral approaches... To study local wave amplification, the modelling of the whole propagation process allows accurate investigations on the various wave types and the way they are reflected, transmitted and scattered in the different soil layers. Those approaches are nevertheless costly since one generally has to model a large part of the geological site and make many different computations (time, frequency) to analyse site effects.

1.2 *Vibratory resonance analysis*

Since site effects involve both wave propagation phenomena and vibratory response of surface layers, it is possible to consider modal approaches for the analysis. Modal methods are very interesting because it can easily give some vibratory properties of the alluvial basin (Dobry, 1976, Paolucci, 1999). These methods are generally very efficient but the eigen properties of the site are mainly qualitative results as far as seismic wave amplification is concerned. In this paper, we do not try to directly select some particular, or fundamental, frequencies, but rather compute a complete set of eigenmodes of the basin. We try afterwards to discriminate between these numerous eigenfrequencies those corresponding to strong site effects. This quantitative analysis is made possible using a modal superposition method for a specific type of excitation.

2. SITE EFFECTS ON THE FRENCH RIVIERA

2.1 *Experimental investigations*

The specific site considered in located in the center of Nice (french Riviera). As depicted in figure 1, the alluvial basin is two thousand meters wide and its depth is around sixty meters in the deepest part (West) down to thirty meters in the thinnest part (East). Many experimental investigations have been performed using different techniques : microtremor measurements, weak seismic event records. As shown in previous publications (Duval, 1996, Semblat, 2000b), the main frequencies of the basin are precisely estimated by both experimental methods. Two types of amplification are determined : in the deepest part of the deposit for low frequencies (between 1 and 2 Hz) and in the thinnest part for higher frequencies (above 2.0 Hz). Considering the H/V spectral ratio from microtremor recordings, it is nevertheless difficult to estimate the absolute amplification factor.

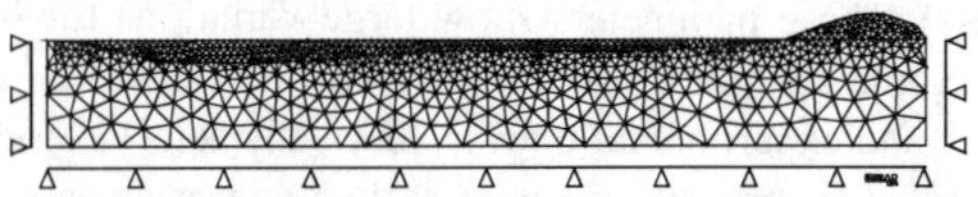

Figure 1 : Finite element model of the geological structure considered for the analysis.

2.2 *Previous numerical analysis*

Previous numerical computations were performed considering the boundary element method thanks to the FEM/BEM code CESAR-LCPC (Humbert, 1989). For sake of simplicity, the surface alluvial layer is assumed homogeneous. To estimate the influence of the mechanical characteristics of the deposit on the seismic motion amplification, the spectral ratios are computed for three different shear moduli of the alluvial layer : μ_1=180 MPa, $\mu_2=2\mu_1/3$ and $\mu_3=\mu_1/2$. As also shown by the experiments (Duval, 1996), it leads to a strong site amplification in the deepest part of the basin (West) for low frequencies and in the thinnest part of the deposit (East) above 2 Hz (Semblat, 2000b). Comparisons with modal approaches are proposed in the following.

3. MODAL APPROACHES

3.1 *Fundamental frequency estimation*

Modal approaches for the vibration analysis of geological structures is increasingly used (Paolucci, 1999, Zhao, 1996). These approaches generally provide the fundamental vibration frequencies of geological structures taking into account their geometry (Dobry, 1976, Paolucci, 1999) or the inhomogeneity of their mechanical features (Zhao, 1996). Considering different types of assumptions, it is then possible to estimate the fundamental frequency of a specific alluvial deposit. It is nevertheless not possible to directly compare various frequencies towards seismic motion amplification.

3.2 *Eigenmodes determination*

In this paper, we propose to directly compute the eigenmodes of any alluvial basin (regular/irregular, homogeneous/non-homogeneous, 2D-3D...) by the finite element method. It only allows the qualitative analysis of the resonance features of the basin. From this qualitative point of view, some comparisons with the results obtained previously by the Boundary Element Method can be made. To lead the analysis further to a quantitative approach, we propose to consider the modal superposition method to discriminate between these (generally numerous) eigenmodes. This quantitative modal approach is detailed in the next section. In the following, we make a qualitative analysis only considering the eigenmodes of the basin.

Considering the finite element method, it is necessary for both modal approach or wave propagation problems to take care of several drawbacks of this numerical method. For eigenvalue problems, there could be some particular issues to address as lower (or upper) bound values or close eigensolutions (Hu, 1998, Lee, 1999). Depending on the integration algorithm and the spatial discretization, the numerical wave dispersion is also difficult to avoid for wave propagation models (Semblat, 2000a).

3.3 *Eigenmodes of the site*

Considering the geological profile chosen in the center of Nice (figure 1), the eigenfrequencies and eigenmodes of the surface deposit and part of the

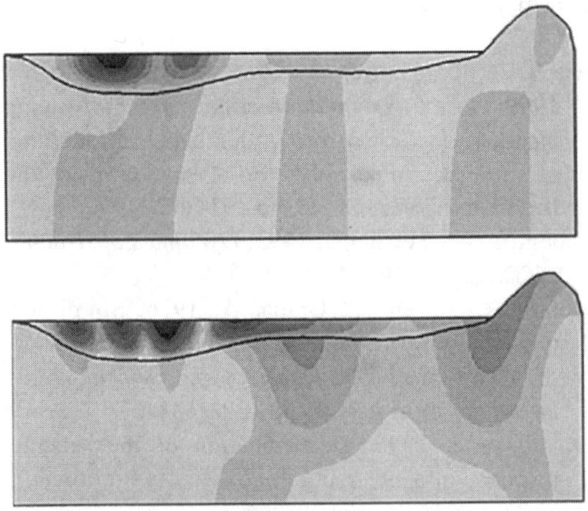

Figure 2 : Eigenmodes of the geological structure at 1.35 Hz (top) and 1.81 Hz (bottom).

surrounding bedrock are estimated by the finite element method. The computations are performed considering a 2D model in plain strain. This model used is depicted in figure 1. Its includes a number of nodes of around 800. For sake of simplicity, the alluvial basin is considered homogeneous : its mass density is ρ=2000 kg/m^3 and its shear wave velocity V_s=300 m.s^{-1} (i.e. shear modulus μ_l=180 MPa). For the elastic bedrock, the values are ρ=2300 kg/m^3 and V_s=1400 m.s^{-1}

The in-plane modes of vibration of the finite element model is computed (plane strain) using the FEM/BEM code CESAR-LCPC. The number of eigenmodes in the frequency range of the problem is 49 different eigenmodes below 2.4 Hz for the model shown in figure 1. Two different eigenmodes (enlarged along the vertical direction) are displayed in figure 2 through the displacement magnitude isovalues at eigenfrequencies 1.35 Hz and 1.81 Hz. These results clearly show that large displacement areas are smaller and numerous for higher frequencies. The larger local motion « amplification » is found in the surface soil deposit for the deepest part of the basin. These results are in agreement with both experimental and previous numerical ones.

The influence of the limit conditions and the model size has been investigated but is not discussed in this paper. The modal superposition method will provide afterwards other reasons to choose the most suitable model.

4. MODAL SUPERPOSITION METHOD

4.1 *Fundamental parameters*

In the modal superposition method, there are two widely used concepts for the analysis of the contribution of the different modes of a structure to its whole dynamic response : the *modal excitation factor* and the *effective modal mass*. Considering a translational excitation at the rigid-base of a structure, the modal response Y_n of the structure in the normal coordinates could be written as follows (Clough, 1993) :

$$M_n\ddot{Y}_n + C_n\dot{Y}_n + K_nY_n = \Xi_n\ddot{v}_g(t) \qquad (1)$$

where M_n, C_n and K_n are the generalized mass, damping and stiffness properties of mode n and Ξ_n is the *modal excitation factor* such as : $\Xi_n={}^T\Phi_n m\{1\}$ with {1} the unit translation applied at the base of the structure.

The modal excitation factor Ξ_n then expresses the contribution of the mode n to the whole response of the structure. To analyze the contribution of each mode in the direction of the dynamic (or seismic) loading, it is possible for this type of translational excitation to estimate the *effective modal mass* of each mode. Considering the kinetic energy of the system, the effective modal mass ζ_n of mode n can be written under the following form :

$$\zeta_n = \frac{\Xi_n^2}{M_n} \qquad (2)$$

The effective modal mass ζ_n corresponds to the part of the total mass of the structure responding to the earthquake in mode n (Bisch, 1999, Clough, 1993). Theoretically, the sum of the effective modal masses of all the modes is equal to the total mass of the structure.

4.2 *Modal excitation factor and effective modal mass*

Considering the finite element model depicted in figure 1, one assumes the seismic loading as a rigid base excitation. Since the basin is very flat and the part of the elastic bedrock not very large, this assumption seems to be acceptable and will be discussed afterwards. The dynamic excitation is then applied as a volumic translational density of force (in the vertical direction) within the whole model. The modal excitation factor and the effective modal mass

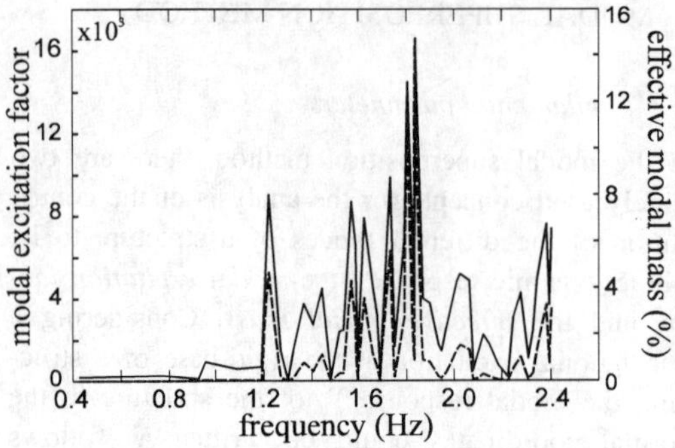

Figure 3 : Modal excitation factor (solid) and effective modal mass (dotted) vs frequency.

for this specific excitation are computed for the fifty first eigenmodes of the geological profile.

The values of modal excitation factor and effective modal mass computed for a vertical translational excitation are given in figure 3. Values of modal excitation factor and effective modal mass are very small for frequencies below 1.2 Hz. They fastly increase between 1.5 and 1.8 Hz. The maximum values are reached for 27th (f_{27}=1.81 Hz) and 29th (f_{29}=1.84 Hz) modes. For these modes, the percentages of effective modal mass are respectively m_{27}^{eff} =12.0% and m_{29}^{eff} =14.7%. The largest part of the mass of the model responding to the vertical earthquake excitation is in a frequency band around 1.8 Hz. These results are in good agreement with experimental ones and previous numerical ones (Duval 1996, Semblat 2000b). The quantification of seismic wave amplification is then shown to be possible through modal approaches.

5. CONCLUSION

A modal approach is considered to analyse the seismic wave amplification in an alluvial basin. To discriminate between these numerous eigenfrequencies towards amplification of seismic motion, the analysis of the corresponding modal excitation factor and effective modal mass for a vertical excitation leads to higher values around highest amplification frequencies. The quantitative comparisons between the numerous eigenmodes of the finite element model lead to very interesting results and the agreement with both experimental spectral ratios and BEM numerical computations is very good (Duval 1996, Semblat 2000b).

REFERENCES :

Bisch, P., Langeoire, A., Prat, M., Semblat, J.F. 1999. *Structures in interaction (in « Finite elements in civil engineering »)*, Chap.7 : Modelling of structures in seismic areas - wave propagation (in French), Hermes ed., pp.467-562.

Clough R.W., Penzien J. 1993. *Dynamics of structures*, McGraw-Hill.

Dobry, R., Oweis, I., Urzua, A. 1976. Simplified procedures for estimating the fundamental period of a soil profile, *Bulletin of the Seismological Society of America*, 66, pp.1293-1321.

Duval, A.M. 1996. Determination of the seismic response of a site using microtremors (in French), Ph.D. thesis (University Paris VI), *Etudes et Recherches des L.P.C*, GT62, L.C.P.C, Paris.

Hu, N., Guo, Z., Katz, I.N. 1998. Bounds for eigenvalues and condition numbers in the p-version of the finite element method, *Mathematics of Computation*, 224(67), pp.1423-1450.

Hughes T.J.R. 1987. *Linear static and dynamic finite element analysis*, Prentice-Hall, Englewood Cliffs, New Jersey.

Humbert P. 1989. CESAR-LCPC : A general finite element code (in french), *Bull. des Laboratoires des Ponts et Chaussées*, 160 : 112-115.

Ihlenburg F., Babuška I. 1995. Dispersion analysis and error estimation of Galerkin finite element methods for the Helmholtz equation, *Int. J. for Numerical Methods in Eng.*, 38 : 3745-3774.

Lee, I., Jung, H., Kim, M., Robinson, A.R. 1999. Eigensolution method for structures with multiple or close eigenvalues, *Jal of Engineering Mechanics (ASCE)*, 125(11): pp.1286-1292.

Paolucci, R. 1999. Fundamental vibration frequencies of 2D geological structures, *2nd Int. Conf. on Earthquake Geotechnical Eng.*, pp.255-260, Lisbon.

Semblat J.F., Brioist J.J. 2000a. Efficiency of higher order finite elements for the analysis of seismic wave propagation, *Jal of Sound and Vibration*, 231(2) : 460-467.

Semblat, J.F., Duval, A.M., Dangla, P. 2000b. Numerical analysis of seismic wave amplification in Nice (France) and comparisons with experiments, *Soil Dynamics and Earthquake Eng.* (to be published).

Zhao, J.X. 1996. Estimating modal parameters for a simple soft-soil site having a linear distribution of shear wave velocity, *Earthquake Eng. and Structural Dynamics*, 25, pp.163-178.

Computer Methods and Advances in Geomechanics, Desai et al. (eds) © 2001 Balkema, Rotterdam, ISBN 90 5809 183 X

Cycle-wise equivalent linear analysis for ground liquefaction

Tadahiko Shiomi
R&D Institute, Takenaka Corporation, Chiba, Japan

Yasushi Nukui, Katsuichiro Hijikata & Takeshi Tsuchida
Construction Department, Tokyo Electric Power Company, Japan

Keisuke Koyama, Fumio Yagishita & Hiroyuki Yoshida
Advanced Engineering Department, Tokyo Electric Power Service Company, Japan

ABSTRACT: Dynamic response analysis of horizontal soil layers subjected to earthquake considering liquefaction has received high attention in the recent seismic engineering. Dynamic effective stress analysis is theoretically best method for this purpose that can solve fully integrated problem such as generation and dissipation of the excess pore pressure based on micro-mechanism. However, there is difficulty to use it in engineering practice. Comprehensive method to determine the coefficients of constitutive laws is not well established. Therefore a new simple method to analysis liquefaction problem for horizontal soil layers is proposed. It is a cycle-wise equivalent linear liquefaction method (CWELL). CWELL is studied with and without liquefaction in this report. CWELL is verified with the dynamic effective stress method and the observed data and provides very good agreement with them. CWELL requires only an additional data, liquefaction strength curve for each soil layer, so the method is easy to use.

1 INTRODUCTION

Dynamic response analysis of horizontal soil layers to earthquake is the most basic and important part in the seismic engineering. It is necessary for prediction of surface acceleration, where degradation of material properties and liquefaction of the soil layer should be considered. In this aspect a procedure SHAKE [Schnabel et. al. 1972] which based on the equivalent linear analysis (ELA) is mostly used. However, there is no well-established procedure that can calculate the response for the liquefied soil layer. SHAKE can be used to calculated the safety factor of liquefaction of soil layer but not for earthquake response with liquefaction.

This paper proposes a new procedure for ground liquefaction, which use is simple as well as SHAKE. Shear stiffness is basically a function of the effective stress. Liquefaction lowers the effective stress of the soil layer so that liquefaction reduces more scant stiffness than that reduced by shear strain in ELA. Very small effective stress is expected for liquefied soil layer. On the other hand it is said that ELA is only valid for the stiffness reduction, which is larger than 40%. So validity of ELA is studied for the large input motion at first.

Since SHAKE is linear analysis with a reduced scant shear stiffness, it theoretically gives error at the large input motion, especially if there is large impact motion during earthquake excitation. Martin et al [Martin et. al. 1982] reported that SHAKE provided good results for the problem subjected to the input motion 0.26g and calculated the double acceleration for the input motion 0.55g. Tanaka et al [Tanaka et al, 1983] showed that maximum acceleration at the surface calculated by the direct integration method in time domain for nonlinear analysis (NLA) is half of the SHAKE's acceleration in some case but it is depend on the shear modulus against shear strain curve (G-γ curve). They used the G-γ curve, which fits the site investigation for ELA and used Ramberg-Osgood model and Heardin-Drnevich model for NLA. Results are depending on the fitted curve. Both studies used the G-γ curve, which is not corresponding to the stress strain curve.

In the above investigation, it should be noticed that they used the inconsistent stress strain relationship with the secant shear modulus and shear strain relationship (G-γ curve). ELA should show little difference from NLA if the G-γ curve, which is obtained from the skeleton curve of stress-strain relationship, is used. Change of the G-γ curve makes much larger difference on the earthquake response. In this context their results are not exact comparison of the numerical methods.

A new procedure is proposed based on the concept of ELA. In this procedure, the equivalent linear method applied to cycle-wise in stead of whole time length of the earthquake. This cycle-wise equivalent linear analysis (CELA) should theoretically have better accuracy. Comparisons are made with nonlinear dynamic effective stress analysis. CELA can be easily applied to liquefaction problem (CWELL).

CWELL only requires the liquefaction strength curve in addition to the data for the ordinary ELA.

At last, liquefaction analyses are compared with the observed data and results obtained by the other program.

2 THEOREY

2.1 *Equivalent linear analysis*

Before explaining the CWELL method, it is good to review ELA, since the concept of CWELL is an extension of ELA.

Basic idea of ELA is to solve the non-linear problem using linear analysis with a reduced scant stiffness, so that ELA can simulate the response at the cycle of the largest amplitude. To obtain the proper stiffness the linear analysis is repeatedly solved until the stiffness and the maximum shear strain response at each layer satisfy the material relationship between stiffness and the shear strain (namely, G-γ curve). Therefore ELA well simulates the response for the cycle, where the maximum magnitude takes place. This is illustrated in Figure 1. Nevertheless, the response of the other cycle theoretically has error, if it is compared with the response of the NLA since stiffness used in ELA is one at the largest amplitude.

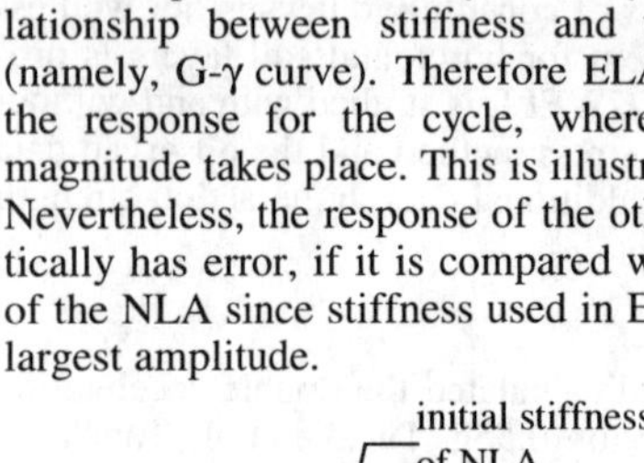

Figure 1. Stiffness corresponds the maximum shear strain, which is used for all response in the ELA.

2.2 *Cycle-wise Equivalent Linear Analysis (CELA)*

As mentioned above, ELA can not change the stiffness and damping during an analysis, so elasto-plastic behaviour of material can not be simulated. To overcome this, cycle-wise equivalent linear analysis (CELA) is introduced. CELA is based on the direct time integration method and it updates stiffness and damping of the dynamic system at each half cycle. The half cycle is defined as period between a zero-crossing of strain time history and the next zero-crossing. This method does not need non-linear stress strain curve and cyclic rule. It uses scant stiffness as well as ELA. There are some efficient and deficient aspects. The efficient aspect is that it can simulate big change of material strength during earthquake such as ground liquefaction. ELA can not analyse two extremely different conditions, i.e. condition before liquefaction and after liquefaction but CELA can. The deficient aspect is that it can not use stiffness of complex number, as SHAKE uses since it is integrated in time domain. This can be overcome by using the general form of normal damping matrix.

$$\mathbf{C} = \mathbf{M}\sum_n a_n \left[\mathbf{M}^{-1}\mathbf{K}\right]^n \tag{1}$$

Stiffness and damping are changed at each cycle. They are calculated as a function of the maximum shear strain of the previous cycle. G = G(coefficient * γ_{max} of the n-1 th half-cycle). In stead of the previous cycle, the results of the previously calculated analysis can be also used. The damping ratio is also obtained by the similar manner.

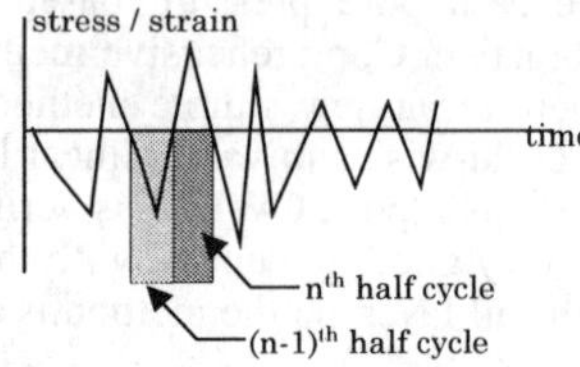

Figure 2. Concept of Cycle-wise Equivalent Linear Method.

Shear modulus can be expressed as a function of the mean effective stress σ_m' and the normalised shear strain γ/γ_{50}. γ_{50} is the shear strain where the shear stuffiness is half of the stiffness at small strain (γ<10^{-5}).

$$G = G_0\left(\frac{\sigma_m'}{\sigma_{m_{ref}}'}\right)\Gamma\left(\frac{\gamma}{\gamma_{50}}\right) \tag{2}$$

σ_m' and γ is the value at (n-1)th half cycle. G_o( ) is a function of normalised confining pressure. Γ( ) is a function represented by the reduction curve of the scant stiffness against the normalised shear strain γ/γ_{50}. This can be obtained by a laboratory test (cyclic triaxial test).

2.3 *Cycle-wise Equivalent Linear Liquefaction Analysis*

Cycle-wise equivalent linear liquefaction analysis (CWELL) [Shiomi et al, 1999] can be easily derived from CELA. In this method, mean effective stress is calculated from excess pore pressure ratio, which is calculated from the accumulated damage ratio. Liquefaction is the phenomenon which reduces the mean effective stress that causes the shear strength and stiffness decrease. The reduction is calculated by equations 4 - 6. The left-hand side of equation includes not only the mean effective stress due to overburden but also liquefaction. γ_{50} also changes as the mean effective stress changes due to liquefaction. The mean effective stress is calculated from the excess pore pressure ratio r_u as follows.

$$G_0(\ )=G_{ref}\times\left(\frac{\sigma'_m}{\sigma_{m_{ref}}}\right)^n\ \&\ \gamma_{50}=\gamma_{50\,ref}\times\left(\frac{\sigma'_m}{\sigma_{m_{ref}}}\right)^n \quad (4)$$

$$\sigma'_m=\sigma'_{mo}(1-r_u) \quad (5)$$

The liquefaction ratio r_u can be obtained from empirical function of the accumulated damage ratio D. Seed et al (Seed et. al. 1976) proposed a function as follows. The various formula are proposed such as by Tatsuoka et al. (Tatsuoka et al. 1980) so that it can simply be proportional, i.e. $r_u = D$.

$$r_u=\frac{1}{2}+\frac{1}{\pi}\sin^{-1}\left(2D^{\frac{1}{\alpha}}-1\right) \quad (6)$$

The accumulated damage ratio D is calculated using the liquefaction strength curve. The curve shows the stress ratio against the number of cycle applied at a triaxial test or a hollow cylinder test. The accumulated damage ratio is calculated by accumulating the incremental damage ratio ΔD for each half cycle. ΔD is obtained by 1/(2N) where N is number of cycle corresponding to the maximum shear stress ratio of the previous half cycle. Figure 3 illustrates whole procedure of CWELL.

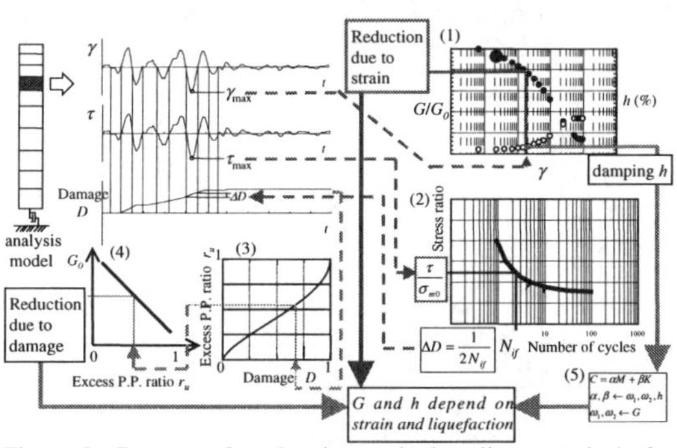

Figure 3. Concept of cycle-wise equivalent linear analysis for liquefaction.

CWELL changes stiffness at the zero-crossing of shear strain so response at the time is almost zero. But velocity at the time is not zero and there also can be some numerical error in displacement. They cause unreasonable change of acceleration when stiffness and damping matrixes are changed. This numerical difficulty can be eliminated applying the unbalance force

$$(\mathbf{C}_{n+1}-\mathbf{C}_n)\mathbf{u}_n-(\mathbf{K}_{n+1}-\mathbf{K}_n)\mathbf{u}_n \quad (7)$$

as a residual force.

2.4 *Iterative method for more accuracy*

As the stiffness of the proposed method was calculated through the previous half cycle of the response, the response can be inaccurate when the small response cycle flows a large response cycle.

Figure 4 shows a outlook of the iterative method. At first an earthquake response is analysed by the ordinary procedure, which uses stiffness depend on the previous response cycle. Then, the response analysis is repeated using the response history of the previous analysis. Figure 5 illustrates the detail of the procedure. The top part shows the ordinary algorithm and the bottom part in the rectangular box shows the algorithm of the iterative method.

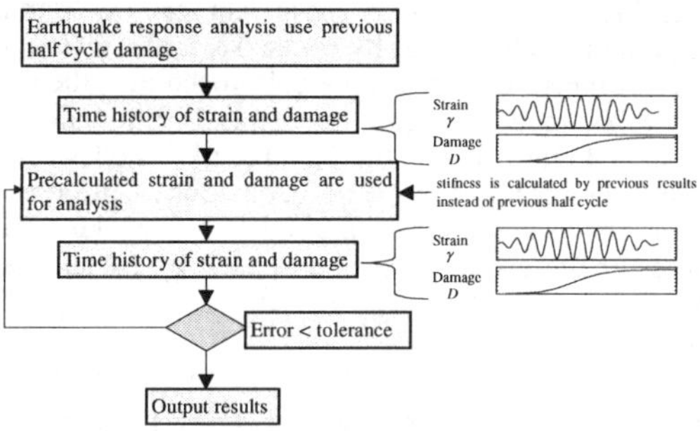

Figure 4. Flow chart of the iterative method.

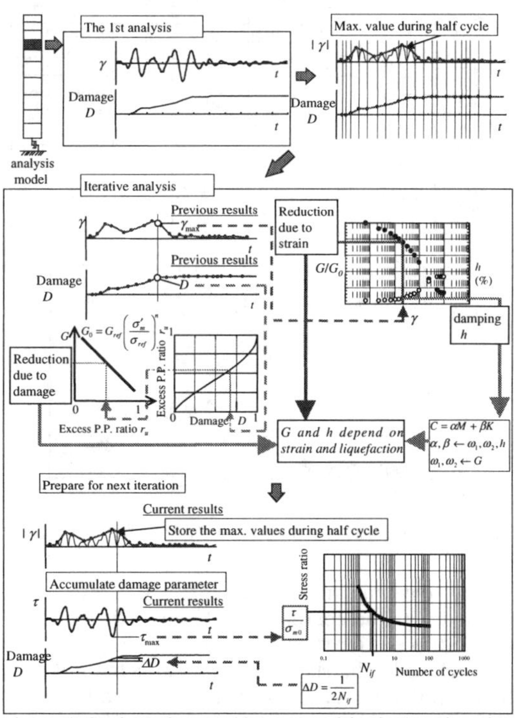

Figure 5. Strain and stress history used in the iterative algorithm of the CWELL.

3 NUMERICAL VERIFICATION

3.1 *Soil layer model and its material properties*

The proposed method CWELL is verified in comparison with ELA and NLA. The analysis is compared in case of "without liquefaction" and "with liquefaction".

For verification study, a soil layer at the Port Island in Kobe City, Japan is used. Material properties are shown in Table 1. G-γ curves listed in Table 1 are shown in Figure 6 for sand (S-1, S-2) and clay (C-1, C-2, C-5). KPU, KPM and KPL in the table are the liquefaction strength curve shown in Figure 7. Surface layers, GL –2.3 m to –18.6 m, are possible to liquefy. In the first study of non-liquefied case, this data are ignored. These data for Port Island are taken from the report by Hatanaka et al, 1997.

To compare the two numerical procedures, the G-γ curve is modelled from the stress strain curves given in Figure 6 into the Ramberg-Osgood stress strain relationship. For cyclic behaviour of the nonlinear analysis, double amplitude of the skeleton curve for the cyclic process proposed by Masing is used.

Table 1. Material properties of soil layer(After Hatanaka et al. 1997)

Layer	Depth (m)	V_s (m/sec)	γ (kN/m^3)	Shear modulus G_0 (kPa)	$G\sim\gamma$ curve	Liquefaction strength curve
sand and gravel	-2.3	170	19.60	57800	S-1	
	-10	210	19.60	88200	S-1	KPU
	-14.7	210	19.60	88200	S-2	KPM
	-18.6	210	19.60	88200	S-2	KPL
alluvial clay	-23	180	16.07	53136	C-1	
	-27	180	16.17	53460	C-2	
sand	-32.8	245	17.15	105044	S-2	
sand and gravel	-50	305	17.93	170236	S-2	
sand	-61	350	18.23	227850	S-2	
dilluvial clay	-67	303	16.66	156075	C-5	
	-79	303	16.66	156075	C-5	
sand and gravel	-83.8	320	19.60	204800		

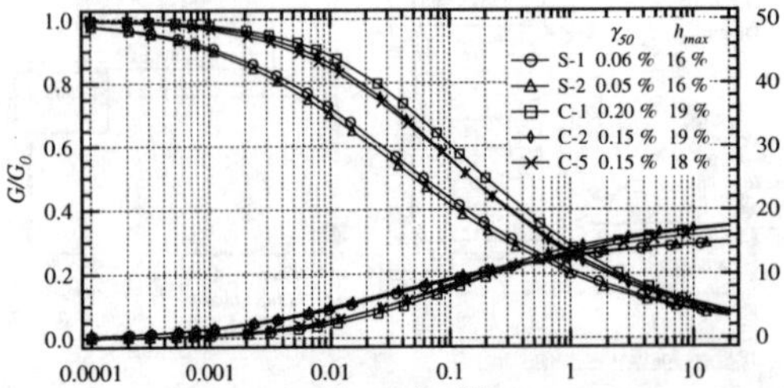

Figure 6. Scant shear modula for soil layers(After Hatanaka et al. 1997).

For dynamic nonlinear analysis, Newmark-β method with the coefficient 0.25 and 0.5, Ramberg-Osgood model for stress strain relationship and K_o = 0.5 are used. The shear failure is affected critically by the initial stress of soil but is not clearly known [Shiomi and Shiogeno 1993]. 0.5 is customary used.

Earthquake data used in this study are three sets of earthquakes recorded at Port Island during 1995 Hyokogen-Nanbu earthquake [Iwasaki and Tai, 1996]. The record at G.L. -83.8m is used as the input motion at the fixed bottom boundary.

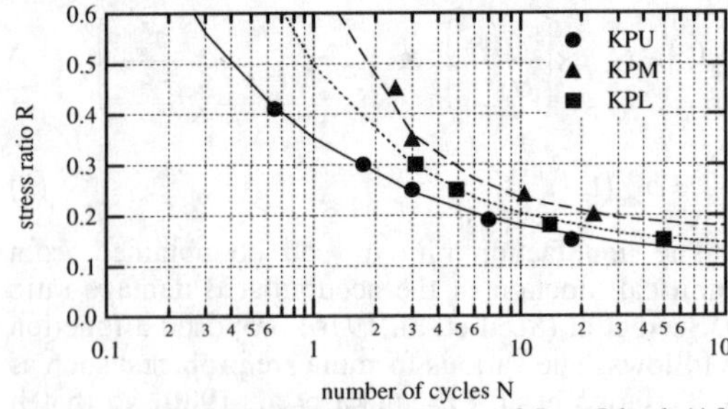

Figure 7. Liquefaction strength curves of Port Island (After Hatanaka et al. 1997).

3.2 *Verification of shear CELA*

CWELL method without liquefaction (is called CELA) is compared with ELA and NLA. The maximum response accelerations at the ground surface by these methods are shown in Figures 8-9. All numerical procedures show the same tendency against the magnitude of the input motions. In these analyses, the larger the input motion is, the smaller the amplification factor is. Amplification factor for the maximum acceleration at the surface for the input motion 50 gal is 1.4, 1.0 for 200 gal, and 0.8 for 723 gal as seen in the profile for depth in Figure 9. In this sense, even for the input motion 50 gal, soil layers may be largely nonlinear.

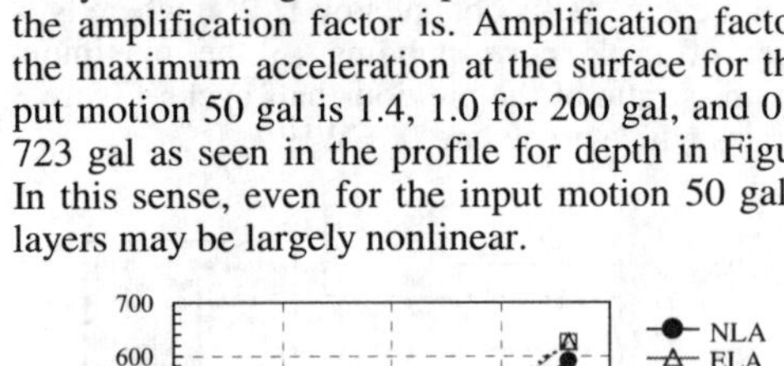

Figure 8. Surface acceleration vs magnitude of earthquake

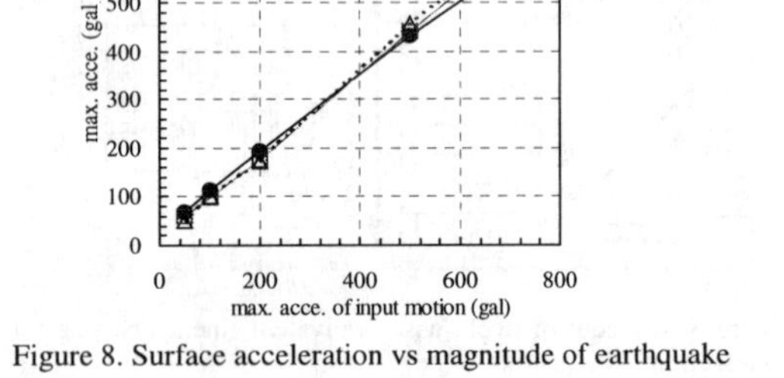

Figure 9. Profile maximum response acceleration by ELA, CELA and NLA

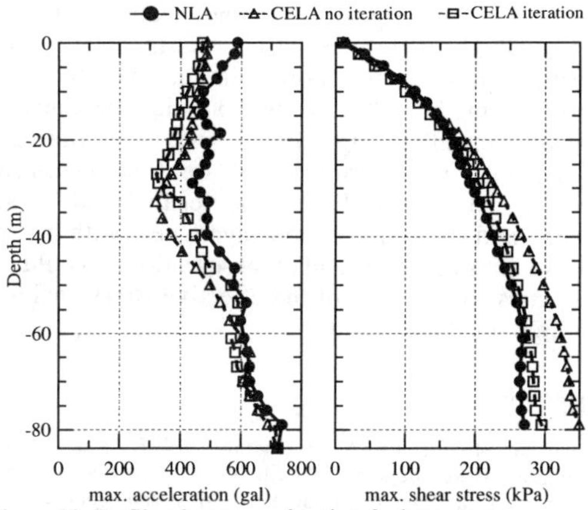

Figure 10. Profile of max. acceleration & shear stress.

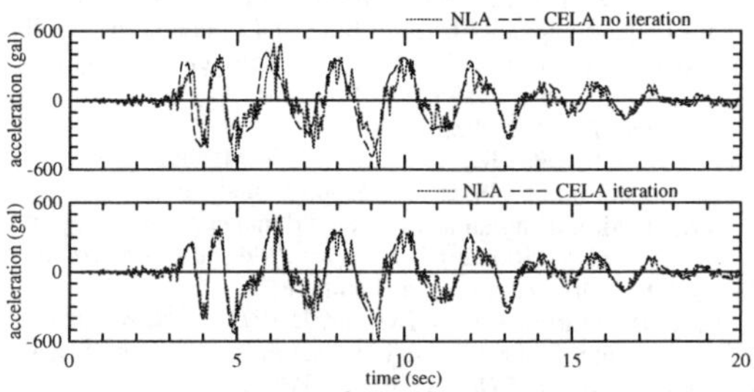

Figure 11.Time history of response acceleration with and without iteration algorithm.

3.3 *Verification of Cycle-wise Equivalent Linear Liquefaction Analysis (CWELL)*

CWELL method (with liquefaciton) is compared with NLA. The program MuDIAN (Shiomi et al, 1993b) is used. In that, constitutive equation is Ramberg-Osgood model for shear behaviour and the densification model for the dilatancy. Similar study is reported by Nukui et al. 1999.

The maximum acceleration at the ground surface of both methods shows similar response for different magnitude of the input motion (Fig. 13). The depth profiles of the maximum accelerations are also show good agreement (Fig. 14). The amplitude of maximum acceleration at the surface is 1.1 - 1.3 times the input motion for 50 gal, 0.8 - 1.0 for 200 gal, 0.6 - 0.7 for 723 gal. There are little differences.

Time history of CWELL and MuDIAN is compared in Figure 14. Both procedures capture the elongation of the period of response wave due to liquefaction but CWELL show less elongation. Liquefaction takes place almost same time of 4 seconds in case of 728 gal (Fig. 15). The transfer functions at the depth G.L. 0.0m, -16.8m and –32.8m against the input motion shows that MuDIAN transfers significantly high frequency components (2 – 5Hz) as well as the observed data but CWELL does not transfer these components.

The iterative method is effective to liquefaction problems as well as the non-liquefaction problem. Two time histories in Figure 17 show its effect. The bottom figure (with iteration) shows the good agreement of iterative CWELL at the first large one cycle (3-5 sec.) with the observation, although non-iterative result (upper figure) shows difference with the observation.

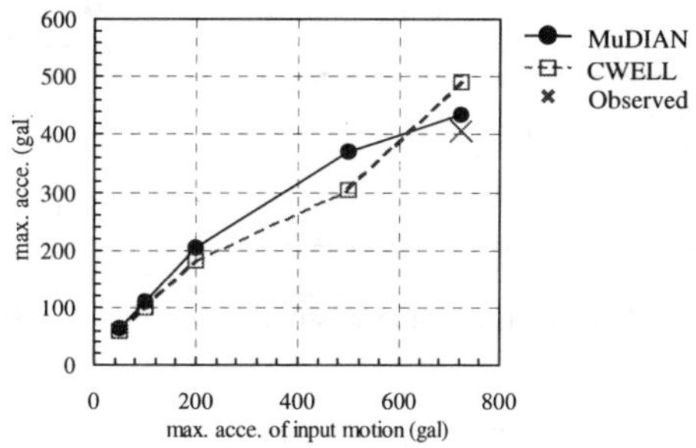

Figure 12. Maximum surface acceleration by CWELL & NLA

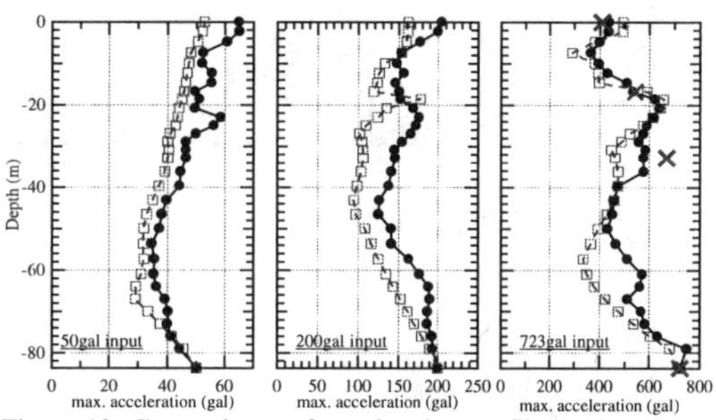

Figure 13. Comparisons of acceleration profile by CWELL & NLA

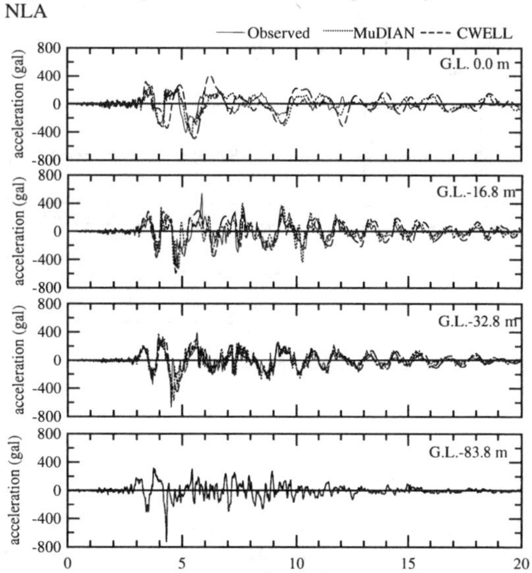

Figure 14. Comparison of two methods on response acc.

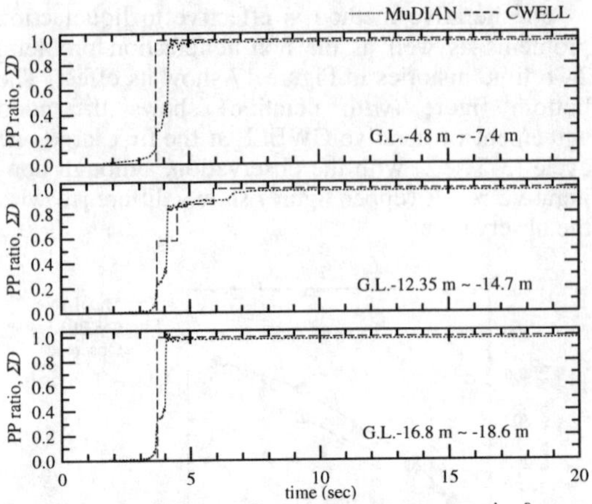

Figure 15. Time history of excess pore pressure ratio & accumulated damage ratio

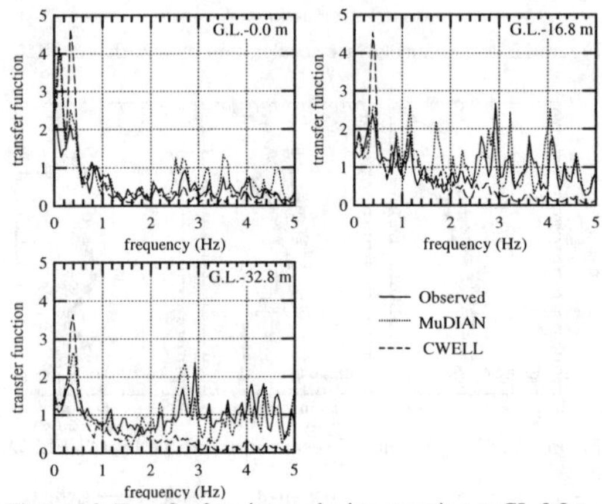

Figure 16. Transfer function to the input motion at GL 0.0m, -16.8m and –32.8m.

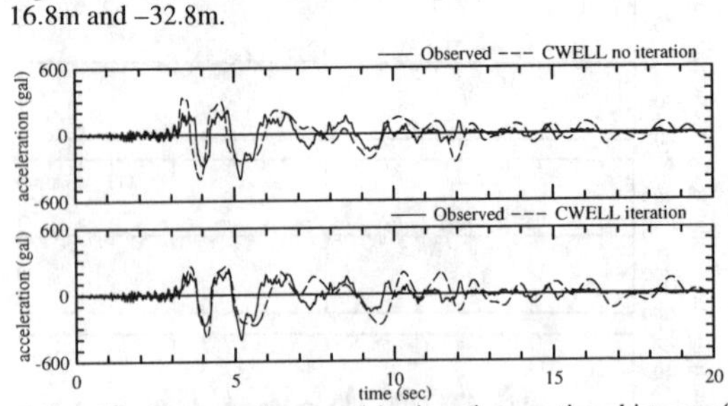

Figure 17. Accuracy improved by iteration on time history of surface acceleration.

4 CONCLUSIONS

A simple procedure for the dynamic response analysis for liquefiable horizontal soil layers subjected to earthquake motion is proposed. The method is the cycle-wise equivalent linear analysis and only uses the liquefaction strength curve as an additional data to the ordinary equivalent linear analysis such as SHAKE. The theory is introduced in details. The verification study is reported in the two steps. At first, the method verified without liquefaction and showed excellent agreement with the ordinary equivalent linear analysis and stepwise nonlinear analysis. At second, the method verified with liquefaction and showed very good agreement with the effective stress analysis and an observation. It could be used as a substitution of the effective stress analysis for undrained condition.

REFERENCES

Hatanaka, M., Uchida, A. & Ohara, J. 1997. "Liquefaction characteristics of a gravely fill liquefied during the 1995 Hyogo-Ken Nanbu earthquake", Soils and Foundations, 37(3):107-115.

Iwasaki, Y. & Tai, M. 1996. "Strong motion records at Kobe Port Island", Special Issue of Soils and Foundations, 29-40.

Martin, P P. & Seed, H B. 1982. "One-dimensional dynamic ground response analyses", *Proc Am Soc Civ Eng, J Geotech Eng Div*, GT7:935-952

Nukui, Y., Hijikata, K., Tsuchida, T., Yagishita, F., Koyama, K. & Shiomi, T. 1999. "A response analysis procedure for liquefaction using an accumulated damage parameter, part 2 Numerical Verification", *Proc. of 1999 conf. on Architecture Institute of Japan*, (in Japanese).

Seed, H B., Martin, P P. & Lysmer, J. 1976. "Pore-water pressure changes during soil liquefaction", *Proc Am Soc Civ Eng, J. of the GT Div.*, GT4:323-346.

Shiomi, T. & Shigeno, M. 1993. "Consideration of initial shear stress on ground liquefaction", *Proc. of Second Asian-Pacific Conf. on Computational Mechanics/ Sydney*, 1077-1082.

Shiomi, T., Shigeno, M. & Zienkiewicz O C. 1993. "Numerical prediction for Model No.1", *Verification of Numerical Procedures for the Analysis of Soil Liquefaction Problems, (eds. K Arulanandan and R. F. Scott)*, 1. 213-219.

Shiomi, T., Hijikata, K., Nukui, Y., Yokoyama, H., Yagishita, F. & Koyama, K. 1999. "A response analysis procedure for liquefaction using an accumulated damage parameter, part 1 Theoretical background", *Proc. of 1999 Conf. on Architecture Institute of Japan*, (in Japanese).

Schnabel, P B., Lysmer, J., & Seed, H B. 1972. "SHAKE: A computer program for earthquake response analysis of horizontally layered sites," *Report No. EERC 72-12,* University of California, Berkeley, Calif., Earthquake Engineering Research Centre.

Tanaka, T., Arisawa, H., Yoshida, N. & Ishihara, K. 1983. " Influence of stress strain model to nonlinear response analysis of ground", *The Proc. of the 18th Japan National Conf. on Soil Mechanics and Foundation Engineering*, pp 631-634 (in Japanese).

Tatsuoka, F. Yasuda, S., Iwasaki, T. & Tokida, K. 1980. "Normalised dynamic undrained strength of sand subjected to cyclic and random loading", *Soil and Foundation*, 20(3):1-16.

Zienkiewicz, O C., Chan, H C A., Pastor, M., Schrefler, B. & Shiomi, T. 1998. *Computational Geomechanics with Special Reference to Earthquake Engineering*, Wiley.

Computer Methods and Advances in Geomechanics, Desai et al. (eds) © 2001 Balkema, Rotterdam, ISBN 90 5809 183 X

Lateral boundary conditions in dynamic analysis under huge earthquake excitation

Fred (Feng) Yi & Weidong Sun
JCD Company Limited

Abstract

As huge earthquake in recent years not only destroyed innumerable houses, buildings and other structures such as bridges but also devastated thousands of people's life, dynamic response analyses under huge earthquake are more and more needed in the design of new structures and reinforcement of existing structures. An issue of great importance is the effect of the boundary conditions on the results of such analysis when deal with a structure in or on the ground. Transmitting boundary is often used in modeling infinite level ground for small or medium scale earthquake. However, for large strain and unrecoverable deformation occurred in large earthquake, transmitting boundary cannot be applied anymore. In this study, the features of several types of lateral boundary conditions are examined and their responses are compared. It is found that soft layer boundary condition is the most efficient among all the type of boundaries.

Keywords: numerical analysis, earthquake, dynamic response, transmitting boundary, spring and damper, soft layer, Fourier spectrum, Strand7

1 Introduction

During 1990's, several huge earthquakes such as Loma Earthquake in 1989, Great Hanshin Earthquake in 1995, Taiwan Earthquake in 1999, occurred all over the world. Only in 1999, there were 12 earthquakes greater than Magnitude 7 (USGS, 2000). These earthquakes not only destroyed innumerable houses, buildings and other structures such as bridges but also devastated thousands of people's life. Now, seismic engineers need to consider not only a small or middle level earthquake but also huge earthquake in the design of a new structure or the reinforcement of existing structures. In these design or strengthening works, dynamic numerical analysis is becoming an increasingly useful tool for predicting the behavior of a structure under huge earthquake excitation. One of the problems in numerical modeling of dynamic problems related to ground is the modeling of boundary conditions in both sides of the model. For small or middle level earthquake, transmitting boundary proposed by Lysmer et al (1972) is commonly used. The application of this theory is the famous soil-structure interaction analysis program -FLUSH (Lysmer et. al. 1975). Unfortunately, this type of boundary conditions can only be applied in frequency domain. As numerical analysis for a huge earthquake that resulting large strain and deformation must be performed in time domain, the set-up of lateral boundary conditions becomes an un-solved problem.

In dynamic model test, shaking table or centrifuge, boundary condition is also a problem (B. Hushmand, et al. 1988; Ging-Song Chang et al. 1989; D. J. Campbell et al. 1991). Campbell et al. (1991) pointed out that an artificial boundary at the sides of a model container could affect the shear stress, normal stress, and vertical and horizontal displacement at the boundary. They summarized the desired boundary conditions for simulation of level ground as:

1) Horizontal normal stress = $K_0\sigma_v'$. The initial lateral strain should be zero.
2) The initial static shear stress on boundary should be zero.
3) The boundary should absorb incident waves.
4) Shear waves traveling parallel to the side of boundaries should be unaffected.
5) If it is desired to model a vertically propagating shear wave, then horizontal traveling P-waves should not be allowed to be transmitted into the sample via side walls of the container.

In numerical dynamic response analysis, the requirements for boundary conditions are the same as in dynamic model tests. The only difference is that there may have exactly free lateral boundary in numerical analysis. In most of the cases, numerical analysis engineers just set the lateral boundary as free or fixed. Although people know that the boundary effect is important and neither free nor fixed boundary is an appropriate boundary condition, few studies are found to deal with this problem. In this study, the features of several types of lateral boundary conditions are examined and their responses are compared. The most efficient boundary conditions are proposed from the viewpoint of engineering application.

2 Possible boundary types and the features

In numerical dynamic response analysis, several types of lateral boundaries can be considered. They can be summarized as:

1) Transmitting boundary
2) Free boundary
3) Fixed boundary
4) Soft layer boundary
5) Spring and damper boundary
6) Spring + damper + earth pressure at rest boundary

Boundary (1) is based on wave propagation theory and can only be used in frequency domain. Although equivalent linear method (Seed and Idress, 1969) is introduced to simulate the nonlinearity of soil (Lysmer et al. 1975), no residual deformation can be obtained after the earthquake excitation. Also, equivalent linear method can only be used when the shear strain occurred is less than around 10^{-3} (Ishihara, 1982). In case of huge earthquake such as Great Hanshin Earthquake, the shear strain is usually larger than several percent. Step-by-step integration method should be used in the analysis.

Boundary (2) appears simple. In fact, this boundary condition is rather complicated. First, it cannot satisfy criterion (1), which suggests a stiff boundary in initial condition. Secondly, even in dynamic condition, theoretically wave propagated to free face reflects completely. However, because free later deformation is arrowed, energy is absorbed partially. The total effect is difficult to evaluate. Boundary (3) is the opposite of (2). It satisfies criterion (1) but does not absorb any incident wave (criterion (3)). In some of the researches, different boundary conditions are used in static and dynamic conditions. In other words, a fixed boundary is used in static condition to calculate initial static stresses. Then, a free boundary is used in dynamic condition (only inertia force is considered at this stage) to calculate dynamic stresses and deformation. This is a good method for earthquake of small scale in that deformation is in an elastic region and superposition theory can be applied. In case of large earthquake in that large plastic deformation occurred, it is believed that the effect of gravity force and inertia force are coupling during earthquake excitation and cannot be simply separated.

Boundary (4) is a general method used in dynamic model tests. This type of boundary satisfies criterion (3) but (1). Also as in model tests, it is difficult to determine the stiffness of a soft layer.

Functionally, boundary (5) should be the same as boundary (4). However, boundary (5) does not have inertia force effect itself. Also, it is expected that the energy transmitted to boundary absorbed by damper clearly. It should work better than boundary (2) and (3).

The above discussion shows that boundary (4) and (5) should give a better simulation of later boundary of ground under dynamic conditions. However, they are still unable to satisfy criterion (1). Because the requirements for an ideal boundary condition are too complicated, it is almost impossible to find a boundary condition to satisfy all of the criteria. Usually, the boundary conditions are contradicted under different criteria. For example, criterion (1) suggests that a stiff boundary should be appeared while criteria (4) and (5) suggest that lateral flexibility should be permitted. Among all the criteria, it is considered the criteria (1) and (3) are the most important. Boundary (6) tries to satisfy these two criteria. Under this boundary condition, earth pressure at rest is applied to the lateral boundary surface to resist the lateral deformation caused by gravity force and spring and damper are used to permit lateral deformation and absorb incident wave during earthquake excitation.

3 Modeling of boundary conditions

A level ground was modeled in this study. The boring log of this ground is shown in Fig. 1. According to the classification of Japanese Road Association, this ground is classified as Type II. The physical properties are shown in Table 1.

In viewpoint of design, gravelly sand layer is used as supporting layer. In this study, the top surface layer was neglected and gravelly sand layer with a thickness of 10.5m was modeled. The numerical model is shown in Fig. 2 that was of 30m in depth and 200m in width.

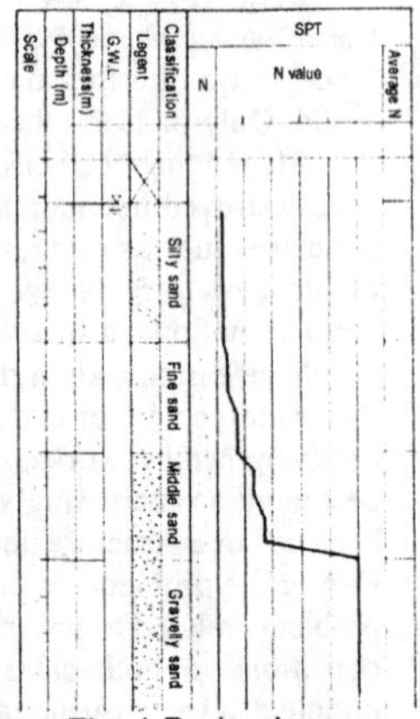

Fig. 1 Boring Log

A real record of earthquake measured in Great Hanshin Earthquake was used in the analysis. The record that has maximum acceleration of 445.9 gals was measured at 33m underground at East Kobe Bridge side. To obtain the input motion at the base of numerical model, one dimensional nonlinear response analysis was carried out. The input motion obtained is shown in Fig. 3 that has a peek acceleration of 385gal.

Six types of boundary conditions, 1) free, 2) fixed, 3) soft layer, 4) spring (XY) + damper, 5) spring + damper (XY) + earth pressure at rest, and 6) spring (X) + damper + earth pressure at rest were adopted in the analysis. Here, (XY) means that shear resistance existed

Table 1 Physical property of soils

Soil	E (kPa)	γ (t/m^3)	ν	ϕ (°)
Silty sand	48250	1.75	0.33	20
Fine sand	108980	1.90	0.33	25
	132020	1.90	0.33	26
Middle sand	211320	2.00	0.33	30
	220610	2.00	0.33	30
Gravelly sand	513000	2.10	0.33	40

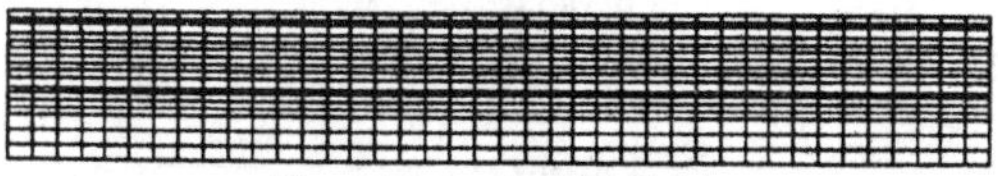

Fig. 2 Numerical Model

while (X) means that only longitude resistance can be obtained. The above boundaries are shown in Fig. 4.

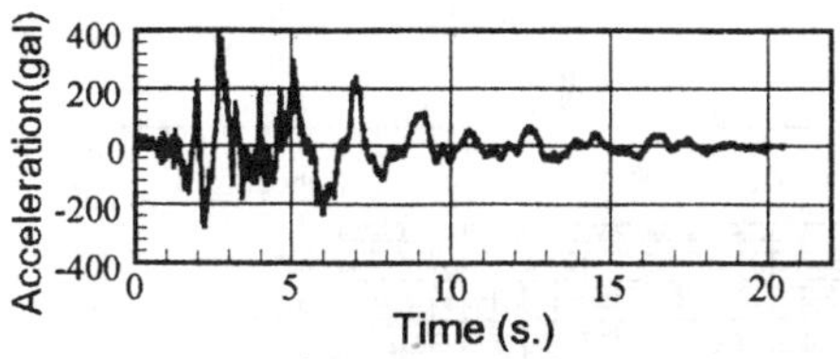

Fig. 3 Input motion

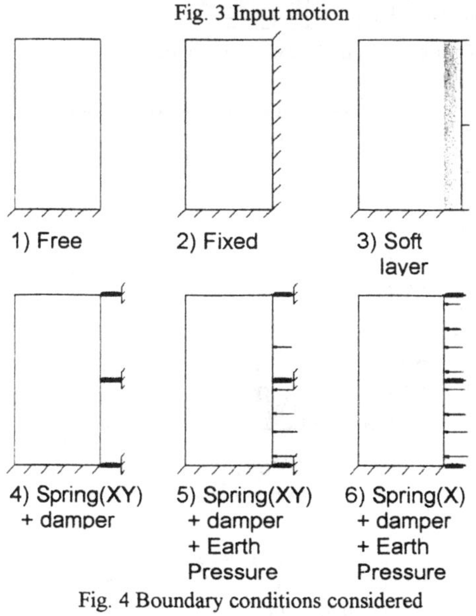

Fig. 4 Boundary conditions considered

In soft layer boundary conditions, the Young's meduli of the layers were taken as 1/20 of that in neighbor layers. Simply considering an elastic condition, this corresponds to another 20m of range modeled.

In case of spring + damper boundary, the value of longitude spring was calculated based on the coefficient of horizontal subgrade reaction. The value of shear spring was taken as 1/3 of that of longitude spring. Assuming the model as a one-degree freedom system supported on the boundary spring, the value of damper can be obtained from

$$c = 2h\sqrt{km}$$

where, h is damping ratio; k corresponds to boundary spring and m is the mass of model. Although this assumption is not accurate enough, it does enable to calculate the value of damper.

Both elastic and elasto-plastic dynamic response analyses were carried out for all the types of boundary conditions to try to obtain the most efficient one. In elastic analysis, the convergent properties in one dimensional equivalent linear response analysis were used to consider the nonlinearity of soils. While in elasto-plastic analysis, Mohr-Coulomb yield criterion was adopted. The nonlinearity of spring was also considered. The passive earth pressure was used as the yield force of spring. Table 2 lists the parameters used.

As for damper, values in Table 3 were used. A damping ratio of 25% was used to calculate damper in elastic analysis. This is extended to consider the nonlinearity under large earthquake excitation even in elastic condition.

Table 2. Springs and yield forces

Layer No.	Thickness (m)	G_{conv}(kPa)	k_H(kN/m)	P_y(kN)	δ_y(m)
1	1.5	214	1008	7.1	0.00314
2	1.0	338	2247	14.1	0.00628
3	2.0	445	3753	35.8	0.00954
4	1.5	552	5118	64.3	0.01256
5	1.5	605	5460	76.2	0.01396
6	1.5	1527	10066	95.2	0.00946
7	1.5	1688	15178	111.0	0.00731
8	1.5	2191	18311	135.0	0.00737
9	1.5	2386	21605	156.0	0.00722
10	2.0	8943	61791	208.0	0.00337
11	1.0	10712	89775	197.0	0.00219
12	1.5	12559	98531	184.0	0.00187
13	1.5	12296	117330	240.0	0.00205
14	1.5	95290	507865	260.0	0.00051
15	3.0	94802	1139985	438.0	0.00038
16	3.0	91030	1352865	660.0	0.00049
17	3.0	87798	1301873	752.0	0.00058

Table 3 Dampers

	Damping ratio h	$c_i(kN \cdot s/mm)$
Elastic	0.25	69.4
Plastic	0.05	13.9

4 Response of different boundaries

A general FEM package-Strand7 (1999), developed by G+D Computing Pty, Australia, was used in this

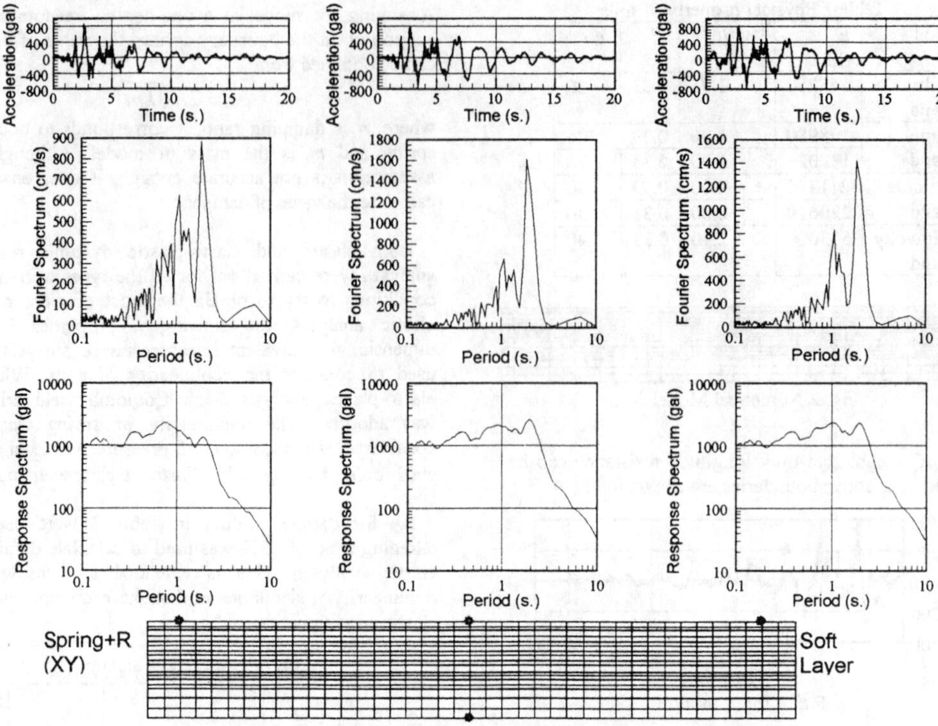

Fig. 5 Time histories, Fourier and Response Spectrum of ground acceleration traces

study. This program has a very powerful pre- and post-processing and is of the ability to perform linear and nonlinear dynamic response analysis.

1) Boundary effects under elastic deformation

The response acceleration, stresses, strains as well as the direction of acceleration were examined in detail. It is found that the response acceleration, especially the Fourier and response spectra of acceleration give better understanding of the influence of boundary conditions. In this paper, discussions are concentrated in the spectra of response accelerations. In the legend of following figures, boundary condition 4) is abbreviated as "Spring (XY)", boundary condition 5) as "Spring + R (XY)" and boundary condition 6) as "Spring + R (X)".

To compare the effects, data were sampled at the points with 10m from the boundary at ground surface to represent the response affected by boundary conditions. Center point at surface was also sampled to represent free field response.

For response acceleration, Fourier spectrum and response spectrum as well as the time history were compared. Fig. 5 shows a sample of the responses at the conditions of spring + earth pressure at rest and soft layer.

As mentioned above, boundary 4) is the common method used in dynamic analysis while boundary condition 5) is applied to eliminate the effect of initial lateral deformation (criterion (1)). Fig. 6 shows the Fourier spectrum under both conditions together with that of free field. The results illustrate that the influence of criterion (1), that is, initial lateral stresses and deformation, is negligible during dynamic condition when spring is considered.

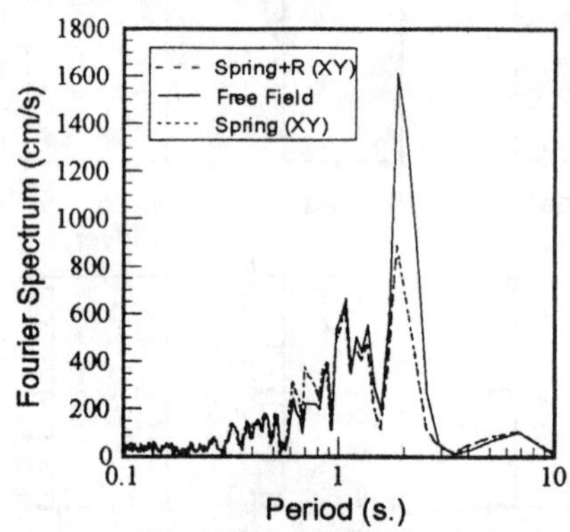

Fig. 6 Effect of initial lateral status

Boundary condition (6) that no shear will occur in lateral boundary during earthquake was used to investigate the effect of shear occurred in lateral boundary. Fig. 7 shows the influence on Fourier spectrum. It is also demonstrated the influence of the lateral boundary shear occurred during earthquake is not significant and can be ignored.

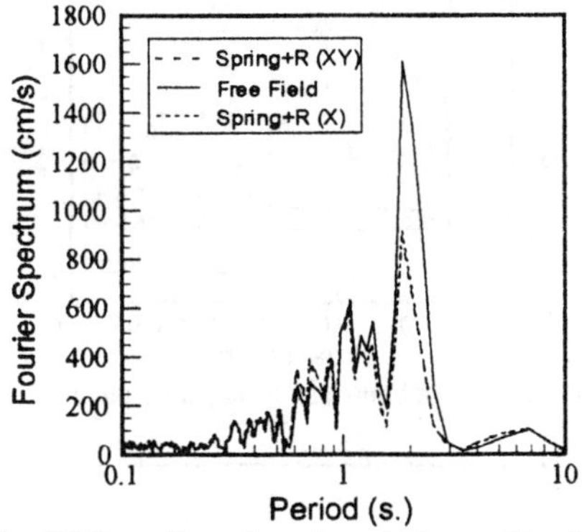

Fig. 7 Effect of boundary shear during earthquake

Fig. 8 shows the response spectrum of the same response acceleration as in Fig. 7. The conclusion obtained from Fig. 7 is confirmed.

Fig. 9 illustrates the Fourier spectra of all the boundary conditions. It can be seen that the effects from boundaries on Fourier spectra are mostly concentrated among a period range of 0.5 to 7 seconds and there is almost no difference in Fourier amplitude when the period is shorter than 0.6 second. In other words, boundary effects are negligible at high frequency region.

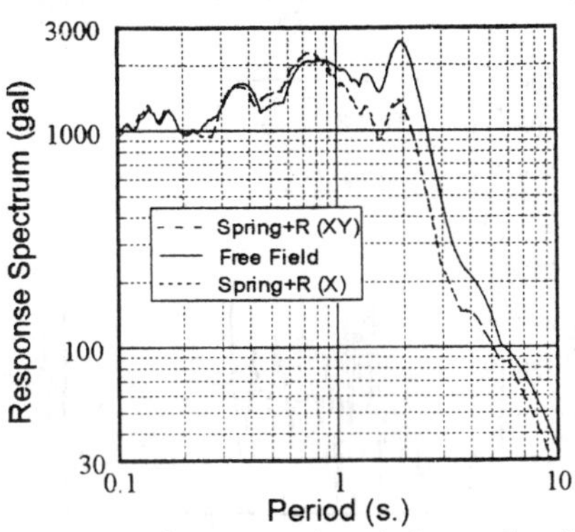

Fig. 8 Effect of boundary shear during earthquake

For different boundary conditions, a free boundary yields the largest response in Fourier spectrum. The influences of boundary types (2), (4), (5) and (6) are similar and the Fourier amplitudes are smaller than that of real one (free field). This can be seen more clearly in Fig. 10 that shows the response spectra of sampled accelerations at different boundary conditions. These results demonstrate that soft layer works the best in simulating boundary conditions.

The Fourier spectra shown in Fig. 9 were normalized using following equation:

$$f_n = \left| \frac{C_{bci}}{C_{freefield}} \right|$$

where C_{bci} is complex Fourier coefficient obtained from *Fast Fourier Transform* of boundary condition *i* and $C_{freefield}$ is that of free field. f_n is smoothed by

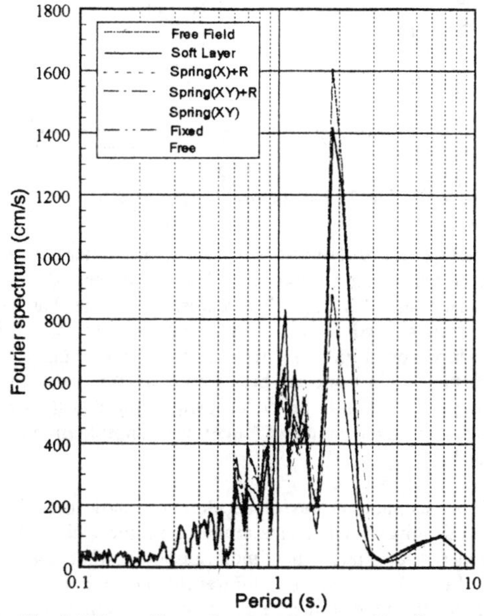

Fig. 9 Effect of boundary conditions on Fourier spectra

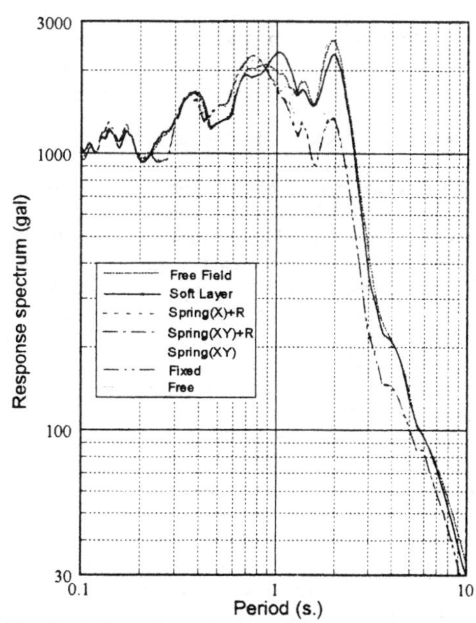

Fig. 10 Effect of boundary conditions on response spectra

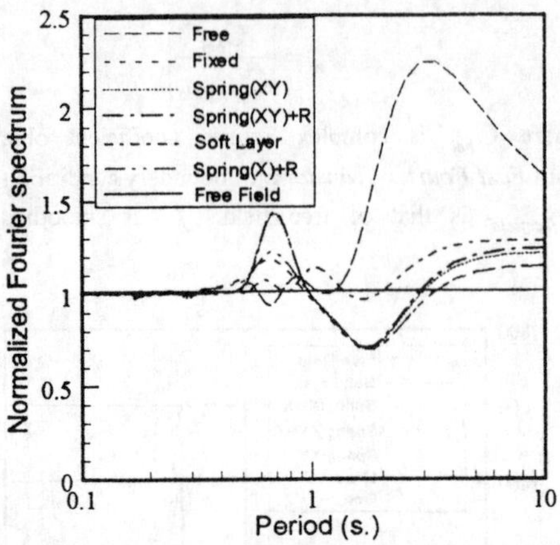

Fig. 11 Normalized Fourier spectra

Parzen's Spectral Window. It is noted that the normalized spectra are within a range of 0.7 to 1.4 except for free boundary (Fig. 11). It is also interesting to note that the results of a fixed boundary are very close to those of boundary condition (6). Also, soft layer boundary is confirmed again to be the best among all the conditions although it gives a larger value than spring and damper boundaries when period is longer than approximately 3 seconds.

The maximum values of acceleration show that free boundary gives smaller maximum amplitude while other boundaries give larger maximum amplitude than that of free field. However, the differences are within 10% percent. When the points are sampled 25m from boundary, the maximum amplitude of acceleration is almost the same as that of free field.

The distribution of stresses, strains as well as the acceleration shows that the affected range is approximately one times the height of the model.

2) Boundary effects under plastic deformation

As mentioned above, Mohr-Coulomb yield criterion was used in elasto-plastic analysis. A bi-linear relationship was adopted in boundary spring. Fig. 12 illustrated the axial force and axial strain relationships of springs in different depth of boundary condition (4). It is known from these results that yielding occurred in boundary springs. It is interesting to note that the boundary springs were always in compression status. This is because that the spring bears initial compression caused by gravity force. Boundary condition (5) tries to eliminate this effect. The results after applying earth pressure at rest are given in Fig. 13. It demonstrates that the applied earth pressure works well in reducing the initial compression in springs.

As discussed on elastic analysis results, the effects of initial lateral deformation and the later shear during earthquake are negligible. To verify whether the same conclusion can be obtained in plastic analysis, Fourier spectra at (4), (5) and (6) are compared in Fig. 14. It can be seen that some influence is demonstrated at period shorter than one second. To see the difference more clearly, compare was made on response spectra (Fig. 15). This figure demonstrates that the effect of initial later status is less than that of lateral shear occurred during earthquake.

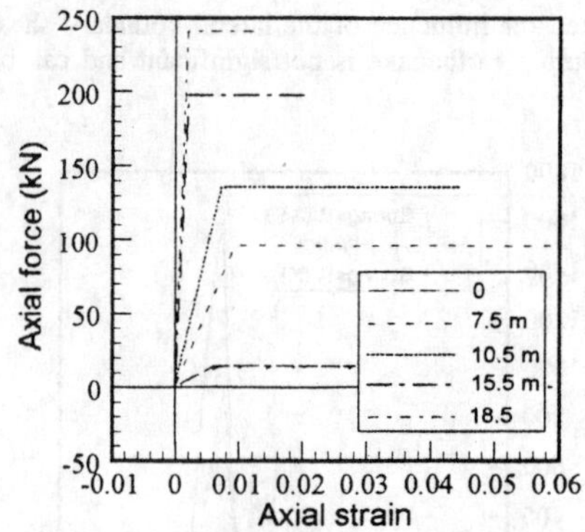

Fig. 12 Yield of boundary spring at boundary condition (4)

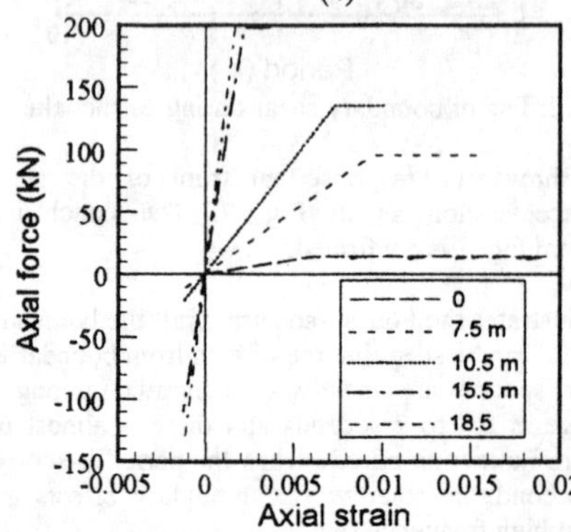

Fig. 13 Yield of boundary spring at boundary condition (5)

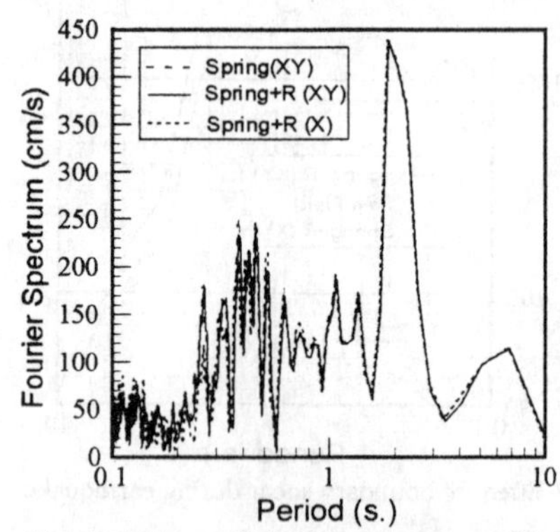

Fig. 14 Effects of initial lateral status and lateral shear during earthquake

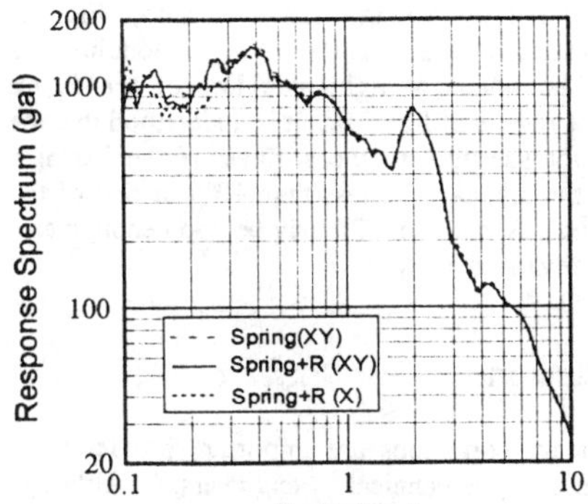

Fig. 15 Effects of initial lateral status and lateral shear during earthquake

The effects of different conditions on Fourier spectra in plastic condition were summarized in Fig. 16. Different from in elastic condition, the effects from boundaries are significant. These effects are specially appeared in short period region except in case of free boundary, which shows strong influence when period is longer than 0.6 seconds. Fig. 17 illustrates the response spectra. Contrary to elastic condition, excepting for free boundary condition, the influences from boundary conditions are negligible when the period is longer than 1.5 seconds. While when period is shorter than 1.5 seconds, the effects are significant and cannot be disregarded. In other words, when the soil comes to plastic status, the weight of high frequency component increases and these high frequency components are affected by boundary condition. The increase of the weight of high frequency components can be confirmed by comparing the Fourier spectrum of free field response both in elastic and plastic conditions (Fig. 9 and 16). In elastic condition, the peek of Fourier amplitude occurred in 1.85 seconds while in plastic condition, it decreases to 0.53 seconds.

Similar to in elastic condition, soft layer boundary seems still to be the most efficient among all the boundary conditions examined in this study.

The normalized Fourier amplitude is shown in Fig. 18. This figure confirms that the boundary effect is concentrated in short period range. The exception is free boundary, which has influence in all range of the period from 0.1 to 10 seconds.

The results of maximum acceleration show that all the boundaries, except free boundary, give smaller maximum vale of acceleration and the difference is as large as 25%. This tendency is the opposite of that in elastic analysis. Even at the points 25m from boundary, the maximum response of acceleration is still approximately 7% smaller than that of free field. This means that although there had reflection from boundary, energy was partially absorbed by plastic deformation.

The distribution of stresses, strains as well as the acceleration also shows that the affected range is wider

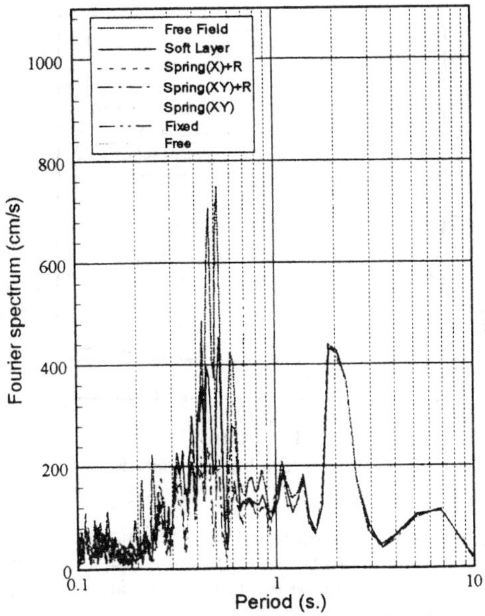

Fig. 16 Effect of boundary conditions on Fourier spectra

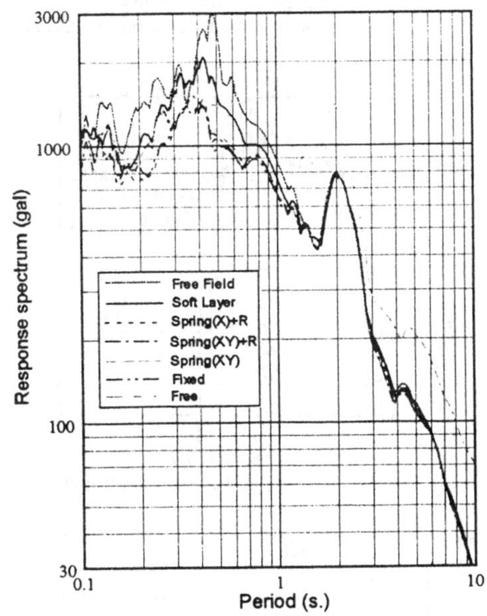

Fig. 17 Effect of boundary conditions on response spectra

than in elastic condition and is approximately 1.5 to 2 times the height of the model. Comparing to shear stress, the affected range to acceleration is much wider.

3) Stiffness of soft layer

The results demonstrated above indicate that soft layer boundary seems to be the most efficient boundary for both elastic and plastic condition. To examine the effect of the stiffness of soft layer, more cases of analysis were

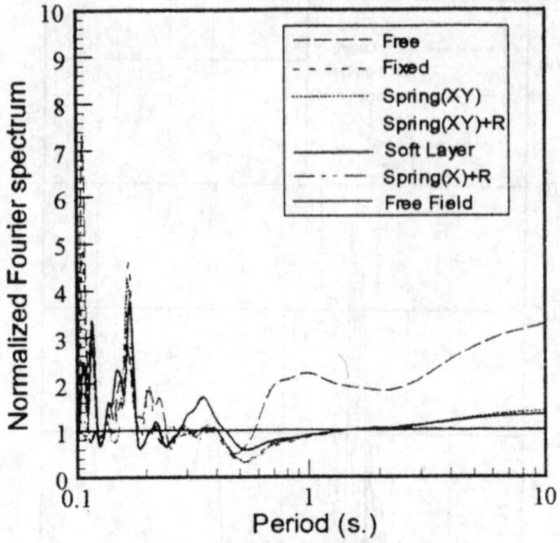

Fig. 18 Normalized Fourier spectra

performed. Fig. 19 illustrated the results in plastic condition. From the response spectrum, it seems that the effect is not sensitive to the stiffness of a soft layer.

To examine the sensitivity of response to the stiffness of boundary soft layer, data were sampled at typical period in Fig. 19 and plotted as a function of stiffness

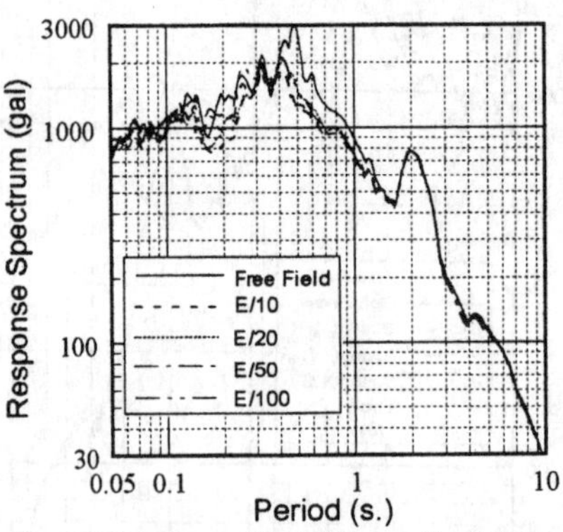

Fig. 19 The effect of the stiffness of soft layer

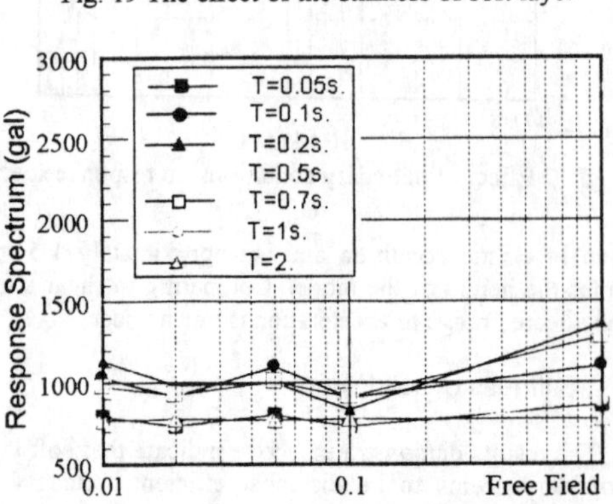

Fig. 20 Sensitivity to boundary stiffness

ratio (represented by Young's modulus ratio). Fig. 20 shows the results, where E' is the Young's modulus of a soft layer. The points at right side line of the figure represents the results of free field. It is understood that the effects are not sensitive to the stiffness if the Young's modulus of the soft layer is less than 10% of that of the neighbor layers. A value of 1/20 may be good enough as a reference in advanced analysis.

5 Conclusions

The boundary conditions are important problems for seismic and geotechnical engineers. Although transmitting boundary is a very useful method in small or medium scale earthquakes, approximate boundary conditions under huge earthquake excitation are yet to be identified. In this study, several possible boundary conditions used in dynamic response analysis were examined in detail. From the discussion above, following conclusions can be made.

1) The effect of a boundary on the response in elasto-plastic analysis is different from that in elastic analysis. In elastic analysis, boundary influence occurred at a period range longer than 0.5 seconds. While in elasto-plastic analysis, boundary effects appeared when the period is shorter than 1.5 seconds (except for free boundary).
2) Among the examined boundary types, free boundary shows the largest influence on response. Especially in elasto-plastic analysis, this effect is demonstrated in all range of the period.
3) The initial lateral deformation gives little influence on the response both in elastic and elasto-plastic analyses.
4) Lateral shear occurred during earthquake almost has no effect on the response in elastic condition. Rather, it shows effect in plastic condition for short period range.
5) Soft layer boundary seems to be the most efficient boundary in the elimination of boundary effects, no matter it is in elastic or plastic condition.
6) The efficiency of soft layer boundary is not sensitive to the stiffness of it. A factor of 1/20 Young's modulus corresponding to that of neighbor layers may be good enough in real application.

REFERENCE

Campell, D. J. Cheney, J. A. and Kutter, B. L. (1991) "Boundary Effects in Dynamic Centrifuge Model Tests," Centrifuge 91, Ko (ed.), 1991 Balkema, Rotterdam.

Chang, G. and Kutter, B. L. (1989) "Centrifugal Modeling of Soil-Pile-Structure Interaction," Engineering Geology and Geotechnical

Engineering, Watters (ed.), 1989 Balkema, Rotterdam.

Hushmand, B., Scott, R. F. and Crouse, C. B. (1988) "Centrifuge Liquefaction Tests in a Laminar Box," Geotechnique 38, No. 2, 253-262

Ishihara, K. (1982) "Evaluation of Soil Properties for Use in Earthquake Response Analysis," International Symposium on Numerical Models in Geomechanics, Zurich, 1982

Lysmer, J. and Drake, L. A. (1972) "A Finite Element Method for Seismology," Methods of Computational Physics, Vol. 11, Chapter 6, Academic Press, 1972

Lysmer, J., Udaka, T., Tsai, C. and Seed, H. B. (1975) "FLUSH – A Computer Program for Approximate 3-D Analysis of Soil-Structure Interaction Problems", Report No. EERC 75-30

Seed, H. B. and Idriss, I. M. (1969) "Influence of Soil Conditions on Ground Motions during Earthquakes," Journal of the Soil Mechanics and Foundation Division, ASCE, Vol. 94, No. SM1, January 1969

Strand7 (1999), http://www.strand.aust.com/

USGS (2000), http://earthquake.usgs.gov/

Computer Methods and Advances in Geomechanics, Desai et al. (eds) © 2001 Balkema, Rotterdam, ISBN 90 5809 183 X

Dynamic effective stress analysis for earthquake subsidence of pile foundations

Huang Yu, Ye Weimin, Tang Yiqun & Chen Zhuchang
Department of Geotechnical Engineering, Tongji University, Shanghai, People's Republic of China

ABSTRACT: A dynamic calculation model is proposed in this paper, including a set of relationships of stress, strain, pore water pressure, earthquake subsidence and pile-soil interface. Based on Biot dynamic consolidation theory, which describes the two-phase porous media system dynamic interaction, the effective stress dynamic analysis with the model is recommended to predict the earthquake subsidence of pile foundations. This method is used to calculate the earthquake subsidence of a typical tall building's pile foundation on liquefied soil. Numerical calculation proves that soil liquefaction could lead to large earthquake subsidence of pile foundations. It is found that the transient subsidence of pile foundations is small during earthquakes and the main earthquake subsidence is developed after the end of earthquakes.

1 INTRODUCTION

Although the earthquake resistance of pile foundations is better than natural ground, buildings with pile foundations can be damaged by large earthquake subsidence during moderate earthquakes. In the Niigata, Japan, earthquake of magnitude 7.5 in 1964, approximately 340 reinforced concrete buildings were damaged as a result of over settlement and incline. Experimental studies also proved that permanent deformation of piles could be caused under cyclic loads (Nauroy et al. 1986).

The current computation methods for pile foundation's settlement under cyclic loads, which have been commonly performed, are the load transfer method (Matlock & Foo 1979), the boundary element method (Poulos 1989; AI-Doury et al 1995) and the finite element method (Boulon et al. 1980; Clausen et al. 1985). All these methods are based on the analysis methods for piles under static loads and have much difficult to consider seismic pore water pressure's influences on soil dynamic behavior, such as liquefaction, softening and so on.

The objective of this paper is to describe a dynamic effective stress analysis for earthquake subsidence of pile foundations and investigate the effects of seismic pore water pressure's increment and dissipation on dynamic pile-soil interaction.

2 CALCULATION MODEL

The equivalent linear approach is used to compute the approximate nonlinear seismic response of pile-soil system approximately in this paper. Firstly, from the study of experimental data, a dynamic computation model is proposed, which is an equivalent viscoelastic model including a set of relationships of stress, strain, pore water pressure, earthquake subsidence and pile-soil interface. Compared with elastoplastic models, the model simulates the soil dynamic behavior in a cycle of load as a whole, not in detail.

2.1 *Modulus*

Initial static stress status has an important effect on soil dynamic behavior. It should be determined before dynamic analysis. The static secant modulus E_s can be determined by Duncan-Chang equation

$$E_s = Kp_a \left(\frac{\sigma_3}{p_a}\right)^n \left[1 - \frac{R_f(1-\sin\phi)(\sigma_1-\sigma_3)}{2c\cos\phi + 2\sigma_3\sin\phi}\right] \tag{1}$$

where σ_1、σ_3 = maximum and minimum principle stress; K and n = dimensionless modulus number and exponent; R_f = failure ratio; c and ϕ = cohesion and angle of internal friction of soil; and P_a = atmospheric pressure. The change of Poisson's ratio of soil is neglected here.

The proposed expression of the dynamic secant shear modulus G at a strain amplitude γ is

$$\frac{G}{G_{\max}} = 1 - H(\gamma) \tag{2}$$

In the model, $H(\gamma)$ and G_{max} are

$$H(\gamma)=\left[\frac{(|\gamma|/\gamma_r)^{2B}}{1+(|\gamma|/\gamma_r)^{2B}}\right]^A \tag{3}$$

$$G_{\max}=6920k_{2\max}\left(\sigma_m^{'}\right)^{\frac{1}{2}} \tag{4}$$

where γ_r = a reference or yield strain; A, B and $k_{2\max}$ = experimental parameters; σ'_m = the effective mean principal stress in Pa.

2.2 *Damping*

The variation of the damping ratio, D, with strain level is

$$\frac{D}{D_{\max}}=\left(1-\frac{G}{G_{\max}}\right)^{\beta} \tag{5}$$

where β = the curve-fitting constant and commonly ranges between 0.2 and 1.2.

2.3 *Pore water pressure generation*

On the basis of the results from undrained cyclic triaxial test data, the seismic pore water pressure buildup of clay may be expressed as

$$\frac{p}{\sigma_m^{'}}=aN^b \tag{6}$$

where p = the pore water pressure; N = the number of uniform stress cycles; and a and b = two experimental parameters which are determined by the ratio of the dynamic shear stress to the effective confining pressure.

For sand, the development of pore water pressure in laboratory cyclic load tests is of the form

$$\frac{p}{\sigma_m^{'}}=(1-ms_l)\frac{2}{\pi}\arcsin\left(\frac{N}{N_f}\right)^{\frac{1}{2\theta}} \tag{7}$$

where s_l = static stress level; m and θ = experimental parameters and N_f = the accumulative number of cycles at the same stress level required to produce a peak cyclic pore water pressure ratio of 100% under undrained conditions.

2.4 *Earthquake subsidence*

The earthquake subsidence can be divided into the volumetric and deviatoric components. In this paper, the analysis only considers the volumetric components. The volume change comes from dissipation of seismic pore water pressure and can be calculated through consolidation equation. In the procedure of finite element method, first the seismic pore water pressure increment Δp can be converted into equivalence nodal force. Then the permanent deformation may be calculated. Equivalence nodal force $\{F\}$ of two-dimensional problem may be expressed as

$$\{F\}=\iint_A [B]^T\{\Delta p \quad \Delta p \quad 0\}^T dA \tag{8}$$

where $[B]$ = strain matrix.

2.5 *Pile-soil interface*

In this paper, the pile-soil interface is simulated by thin-layer element. The constitutive model of interface is the same as those of soil around piles. But their parameters are different from soil.

3 DYNAMIC ANALYSIS METHOD

Biot (1955, 1962) established the basic dynamic equations for porous media. Later, Zienkiewicz et al. (1982, 1984) introduced the numerical solution of these equations. Based on these theories the dynamic effective analysis method is conducted with the dynamic calculation model in this paper.

If the pore water acceleration has been neglected, the formulation of Biot governing equations is represented as the following

$$\sigma_{ij,j}+\rho g_i=\rho\ddot{u}_i \tag{9}$$

$$\left(k_{ij}p_{,j}\right)_{,i}-\dot{\varepsilon}_{ii}-\left(k_{ij}\rho_f g_j\right)_{,i}=0 \tag{10}$$

where σ_{ij} = the total Cauchy stress in the combined solid and fluid mix at any instant; u_i = the displacement; k_{ij} = coefficient of permeability; ρ and ρ_f are the density of the mixed soil and fluid itself respectively. The equations can be solved numerically under given boundary and initial conditions by the finite element method. In this the displacements $\mathbf{u}$ are described in terms of the nodal values $\overline{\mathbf{u}}$ as

$$\mathbf{u}=\mathbf{N}\overline{\mathbf{u}} \tag{11}$$

with a similar discretization for the pressures,

$$\mathbf{p}=\overline{\mathbf{N}}\overline{\mathbf{p}} \tag{12}$$

where $\overline{\mathbf{p}}$ is the vector of nodal pressure values. In the above $\mathbf{N}$ and $\overline{\mathbf{N}}$ are appropriate shape functions. Performing spatial discretization, the finite element formulation are derived as

$$\int_\Omega \mathbf{B}^T\sigma' d\Omega+\mathbf{M}\ddot{\overline{\mathbf{u}}}-\mathbf{Q}\overline{\mathbf{p}}=\mathbf{f} \tag{13}$$

$$\mathbf{H}\overline{\mathbf{p}}+\mathbf{Q}^{\mathbf{T}}\dot{\overline{\mathbf{u}}}=\overline{\mathbf{f}} \tag{14}$$

where **M** = mass matrix; **Q** = couple matrix; **H** = permeability matrix; **f** = nodal earthquake load vector; $\bar{\mathbf{f}}$ = nodal seepage discharge vector.

Many alternative time integration procedures are available for above equations, such as Newmark method and Wilson-θ method.

4 APPLICATION

A 12-storey building showed in Figure 1 is considered to analyze its earthquake subsidence. The pile is 10 m long, with a 500 mm outside diameter. The soil profile consists of saturated sand over rock. The relevant material properties and computation parameters are listed in Table 1. The input base motion is the first 10 s N-S component of the El Centro Earthquake (May 1940) with the maximum acceleration scaled to 0.79m/s^2. The base motion and its power spectrum are shown in Figure 2. In the research two-dimensional analysis is performed by finite element method.

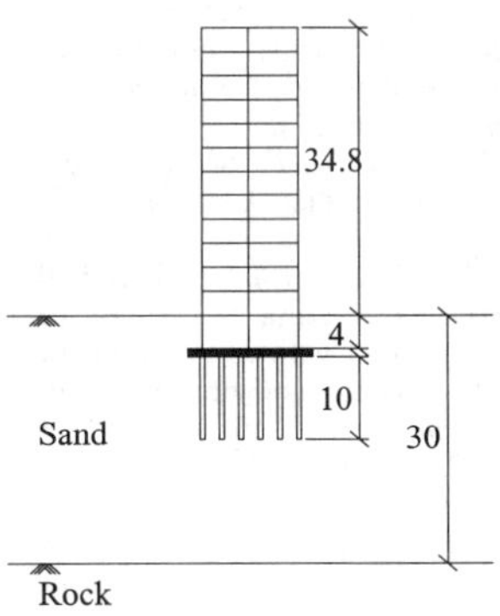

Figure 1. A building with pile foundation

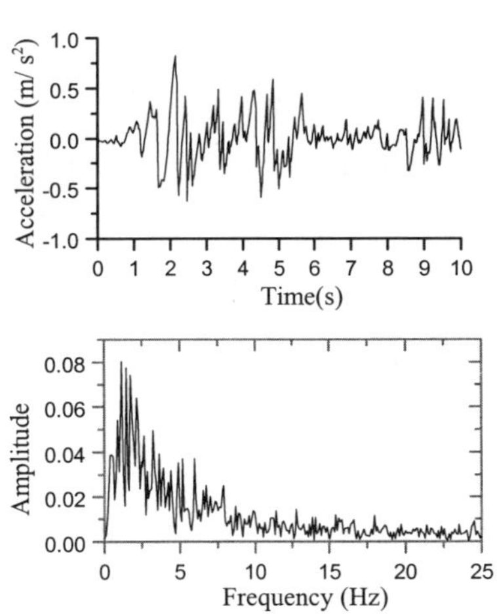

Figure 2. Input earthquake wave and power spectrum.

Table 1. The calculation parameters of soil and interface

	Sand	Interface
ρ (kg/m^3)	1900	1900
c (kPa)	0	0
ϕ	30	20
m	1.1	1.1
K	300	100
n	0.5	0.6
R_f	0.75	0.75
Poisson' s ratio	0.3	0.25
k_{2max}	40	20
D_{max}	0.3	0.3
k (m/s)	5×10^{-5}	5×10^{-5}

4.1 *Acceleration*

The predicted acceleration and power spectrum of foundation center are shown in Figures 3. Its maximum acceleration is 0.83 m/s^2.

The soil deposit acts as a filter when the bedrock earthquake acceleration is transmitted through it. The frequency contents of soil ground acceleration have a tremendous difference in comparison with the bedrock acceleration. The soil deposit filters out a significant portion of the high frequency content of the bedrock acceleration. However, it amplifies the bedrock acceleration about 0.5 s, which is its fundamental frequency.

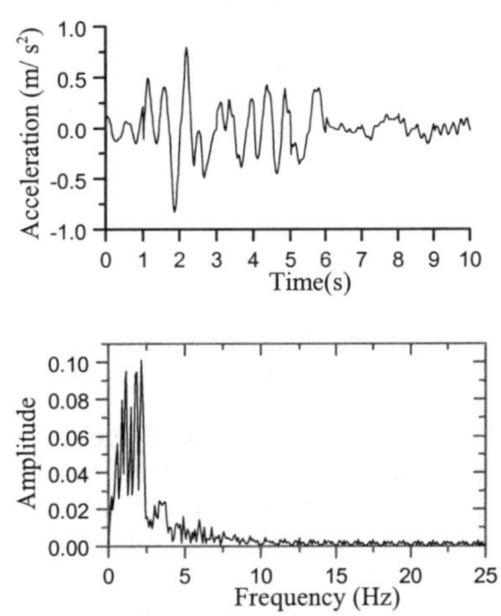

Figure 3. Predicted acceleration time history and power spectrum of foundation center.

4.2 *Seismic pore water pressure*

Computation results indicate that the pore pressure water ratio is substantially lower near the pile foundation than it is in the free field. During the earthquake, while liquefaction of free field occurs, the seismic pore water pressure ratios at similar depths adjacent the pile foundation are less than 0.34. It agrees adequately with other researches. (Yoshimi et al. 1977; Siddharthan et al.1990)

4.3 *Earthquake subsidence*

Figure 4 shows the earthquake subsidence development of pile foundation and soil ground. Their maximum earthquake settlements are 10.06 cm and 25.74 cm respectively. It is to be noted that if soil around pile liquefies, pile foundations can be induced large earthquake settlement. Moreover, at the end of earthquake, the transient settlement of pile foundation is only 0.76 cm and the main earthquake subsidence is developed after the end of earthquake. The duration of earthquake is so short that displacements are small even if pile capacity decreases. Therefore, the seismic stability of pile foundations should be estimated by the step by step numerical computation of whole earthquake.

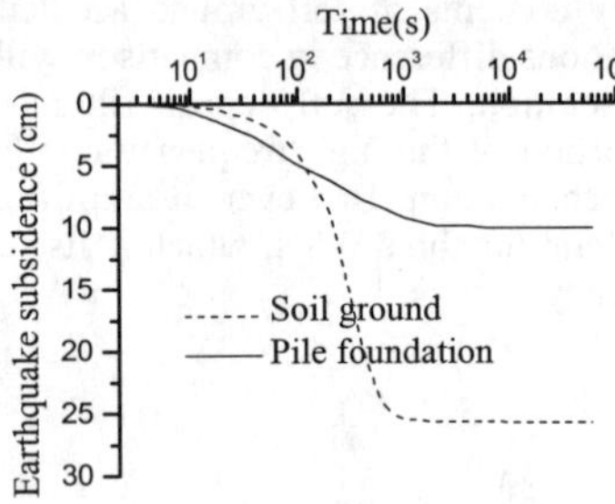

Figure 4. Predicted earthquake subsidence development of pile foundation and soil ground.

5 CONCLUSION

1 An equivalent viscoelastic dynamic model is presented. It contains five formulas for calculating modulus, damping, pore water pressure, earthquake subsidence and pile-soil interface. An approach of dynamic effective stress analysis in the fluid-saturated porous solid is summarized in terms of Biot governing equations and finite element formulation.

2 Numerical calculation proves that the soil deposit filters out the high frequency content and amplifies the low frequency content when the bedrock earthquake acceleration is transmitted through it. The seismic pore pressure water ratio is lower near the pile foundation than it is in the free field.

3 Soil liquefaction could lead to large earthquake subsidence of pile foundations. The transient subsidence of pile foundations is small during earthquakes and the main earthquake subsidence is developed after the end of earthquakes.

REFERENCES

Al-Douri, R.H. & H.G. Poulos 1995. Predicted and observed cyclic performance of piles in calcareous sand. *J. Geotech. Engng., ASCE* 121:1-16.

Biot, M. A. 1955. Theory of elasticity and consolidation for a porous anisotropic solid. *Journal of applied physics* 26: 182-185.

Biot, M. A. 1962. Mechanics of deformation and acoustic propagation in porous media. *Journal of applied physics* 33:1483-1498.

Boulon, M., J. Desrues, P. Foray & M. Forgue 1980. Numerical model for foundation under cyclic loading, application to piles. *Int. Symp. on Soils under Cyclic and Transient Loading*: 681-694. Swansea.

Clausen, C. J. F., O. Kjekstad & O. E. Hansteen 1985. Soil structure interaction study for a piled concrete platform. *Proc. 11th ICSMFE*:1589-1594. San Francisco.

Matlock, H. & S. C. Foo 1979. Axial analysis of piles using a hysteretic and degrading soil model. *Proc. Conf. Num. Methods in Offshore Piling*:165-185. London.

Nauroy, J. F., F. Brucy, P. Le Tirant & J.-P. Kervadec 1986. Design and installation of piles in calcareous formations. *Proc. 3rd Int. Conf. Num. Methods in Offshore Piling*: 461-80. Nantes.

Poulos, H. G. 1989. Cyclic axial loading analysis of piles in sand. *J. Geotech. Engrg., ASCE* 115: 836-852.

Siddharthan, R. & G. M. Norris 1990. Residual porewater pressure and structural response. *Soil Dynamics and Earthquake Engineering* 9: 265-271.

Yoshimi, Y. & K. Tokimatsu 1977. Settlement of buildings on saturated sand during earthquakes. *Soils and Foundations* 17:23-38.

Zienkiewicz, O. C. & P. Bettess 1982. Soils and other saturated media under transient, dynamic conditions; general formulation and the validity of various simplifying assumptions. In G. N.Pande & O. C. Zienkiewicz (eds), *Soil Mechanics-Transient and Cyclic Loads*: 1-16. New York:Wiley.

Zienkiewicz, O. C. & T. Shiomi 1984. Dynamic behaviour of saturated porous media; the generalized Biot formulation and its numerical solution. *International Journal for numerical and analytical methods in geomechanics* 8: 71-96.

Computer Methods and Advances in Geomechanics, Desai et al. (eds) © 2001 Balkema, Rotterdam, ISBN 90 5809 183 X

3-D liquefaction analysis of reclaimed ground with tank and foundation

F.Zhang & A.Yashima
Department of Civil Engineering, Gifu University, Japan

R.Uzuoka
Earthquake Disaster Mitigation, Miki, Japan

ABSTRACTS: In this paper, liquefaction analysis (LIQCA3D) of a petroleum tank-ground-foundation system is conducted using a three-dimensional dynamic finite element-finite difference method. The nonlinearity of the ground is simulated with a kinematic hardening elastoplastic model, which has been verified by a series of hollow cylindrical torsional shear tests and been proved that it can well predict the behaviors of soils such as the liquefaction strength curve, the stress-strain relation as well as the effective stress paths during cyclic loading. In the numerical analysis, an FEM-DEM analytical method is adopted to the soil-water coupled analysis. The petroleum tank is built on a reclaimed ground and is near to seashore. In order to enhance the seismic strength of the tank-soil system, a ring-shaped steel pile wall is designed for the tank. The purpose of the research is to evaluate numerically the effectiveness of the remediation method when a tank is built on a potentially liquefied ground.

1 INTRODUCTION

It is commonly known that liquefaction may lead to big catastrophe to many structures such as embankment, revetment and facilities near seashore. In Japan, due to the shortage of land, large-scale reclaimed ground is constantly made on which many huge structures are built. The geologic materials used for reclaimed ground are usually sandy soils whose liquefaction strength against earthquake should be considered carefully because of the high probability of huge earthquake attack during the life span of these structures. It is very important, therefore, to develop an effective and economical remediation method to protect these seashore structures from earthquake in which liquefaction of ground may probably occur.

Many researches, e.g., model tests including centrifuge model test and numerical analysis, have been made on the liquefaction of the ground related to embankment and revetment. It is also obvious that the ability to evaluate the effect of the remediation method against the liquefaction is absolutely necessary, as the analytical technique can reduce the costs of remediation (Finn et al., 1997).

In this paper, a liquefaction analysis (LIQCA3D) of a petroleum tank-ground-foundation system is conducted using a three-dimensional dynamic finite element-finite difference method. The nonlinearity of the ground is simulated with a kinematic hardening elastoplastic model, which has been verified by a series of hollow cylindrical torsional shear tests and been proved that it can well predict the behaviors of soils. In the numerical analysis, an FEM-DEM analytical method is adapted to the soil-water coupled analysis. The petroleum tank is built on a reclaimed ground and is near to seashore. As a remediation method for the tank, a steel pile ring around the tank is designed. The purpose of the research is to investigate the effectiveness of the ring-shaped steel wall against the liquefaction in huge earthquake when a tank is built on a potentially liquefied ground

2 NUMERICAL PROCEDURE

In the numerical procedure used in this paper, the governing equations for the soil-water coupled problem are described with Biot's two-phase mixture theory and a *u-p* formulation is adopted in the three-dimensional liquefaction analysis, where u stands for the displacement and p is the pore-water pressure. The finite element method (FEM) is used for the spatial discretization of the equilibrium equation while the finite difference method (FDM) is used for the spatial discretization of the pore-water pressure in continuity equation. Equations 1 and 2 show the equilibrium equation and continuity equation for the Biot's two-phase mixture theory.

$$\rho \ddot{u}_i^s = \sigma_{ij,j} + \rho b_i \tag{1}$$

$$\rho^f \ddot{u}^s_{i,i} - p_{,ii} - \frac{\gamma_w}{k}\dot{\varepsilon}^s_{ii} + \frac{n\gamma_w}{kK^f}\dot{p} = 0$$

where ρ is the density of soils, $\ddot{u}_i^s$ is the acceleration of the solid, σ_{ij} is the total stress, b_i is the body force, ρ^f is the density of fluid, p is the pore-water pressure

Table 1 Physical properties and material parameters involved in elastoplastic model of the ground

Soil	e	λ	κ	G_0/σ'_{m0}	γ	k (m/sec)	M_m^*	M_f^*	B_0^*	B_1^*	C_f	D_0	n	γ^P_{ref}	γ^E_{ref}
Reclaimed soil (above water level)	0.600	0.030	0.000133	2654.0	18.62	$5.1*10^{-5}$	0.933	1.051	8000.0	1000.0	3000.0	0.0	0.0	1.000	1.000
Reclaimed soil (under water level)	0.600	0.030	0.000288	1513.0	19.60	$5.1*10^{-5}$	0.802	0.909	6100.0	110.0	2000.0	0.7	2.8	0.003	0.020
Clay	1.300	0.300	0.000830	846.0	16.66	$3.0*10^{-8}$	1.120	1.250	4000.0	960.0	2000.0	0.0	0.0	1.000	1.000
Reclaimed sand	0.600	0.030	0.000823	536.0	19.60	$3.2*10^{-4}$	0.834	0.944	2100.0	910.0	2000.0	0.8	2.5	0.005	0.100
sand	0.600	0.030	0.000823	536.0	19.60	$3.2*10^{-4}$	0.834	0.944	2100.0	610.0	2000.0	0.8	2.5	0.005	0.100

e: Void ratio; λ: compressive index; κ: swelling index; γ: unit weight (kN/m^3); G_0/σ'_{m0}:initial ratio of shear stiffness to stress; k: permeability

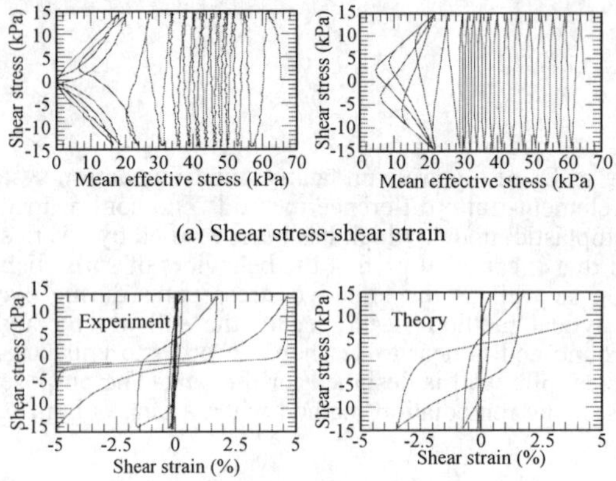

(a) Shear stress-shear strain

(b) Effective stress paths

Figure 1 Comparison of theoretical and experimental results of undrained torsional shear tests (after Matsuo et al., 2000)

Table 2 Physical properties of elastic materials

Material	e	γ	k(m/sec)	E(kN/m^2)	ν
Granite	0.300	20.58	$5.0*10^{-6}$	252000.0	252000.0
Boulder	0.600	19.60	$1.0*10^{-4}$	14900.0	14900.0
Tank and petroleum	-	19.60	-	44423.0	29615.0
Steel pile ring	-	754.87	-	$2.058*10^8$	0.200

E: Young's modulus; ν:Possion's ratio

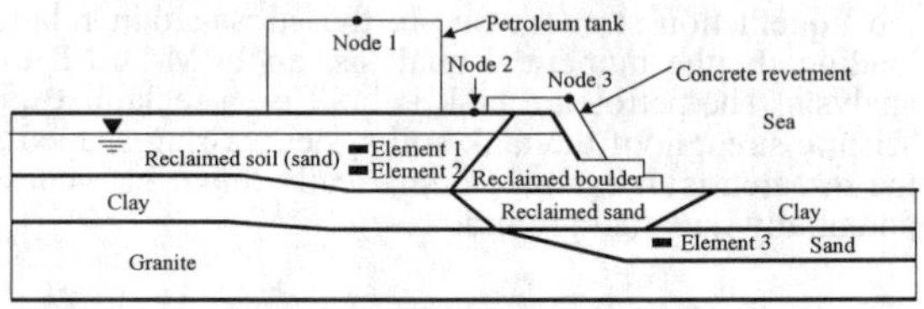

Figure 2 Geologic profile and the position where the results are output

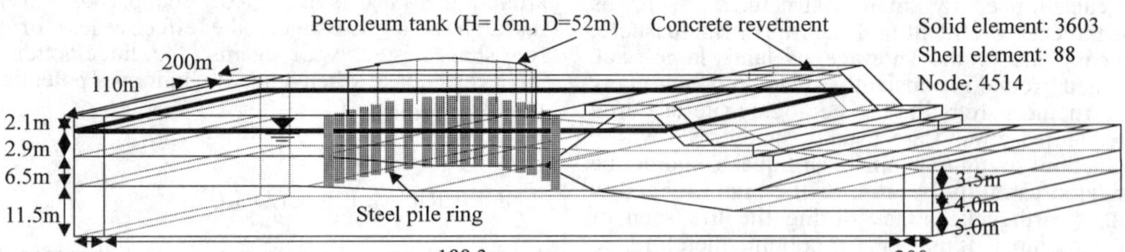

Figure 3 General view of the tank-ground-pile ring system

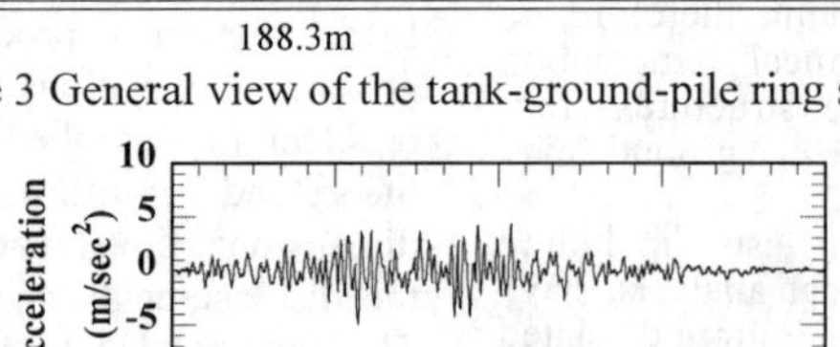

Figure 4 Input earthquake wave

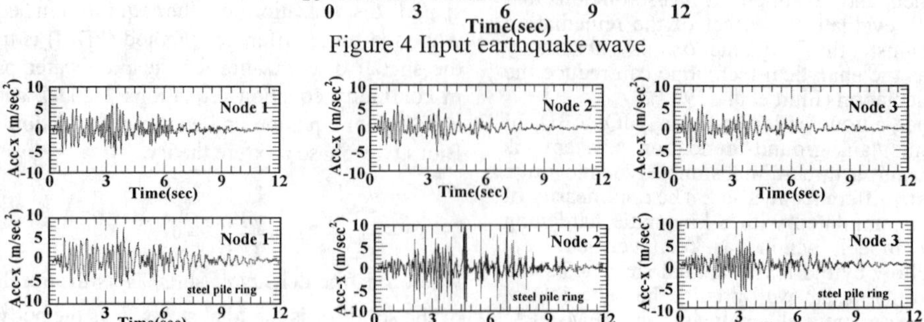

Figure 5 Influence of the steel pile ring on responded horizontal acceleration

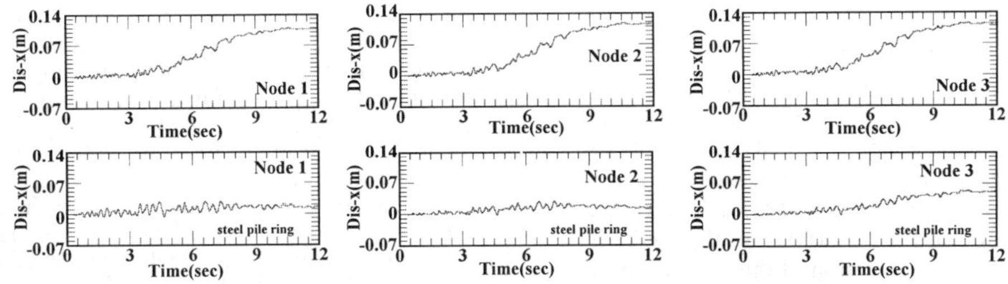

Figure 6 Influence of the steel pile ring on responded horizontal displacement

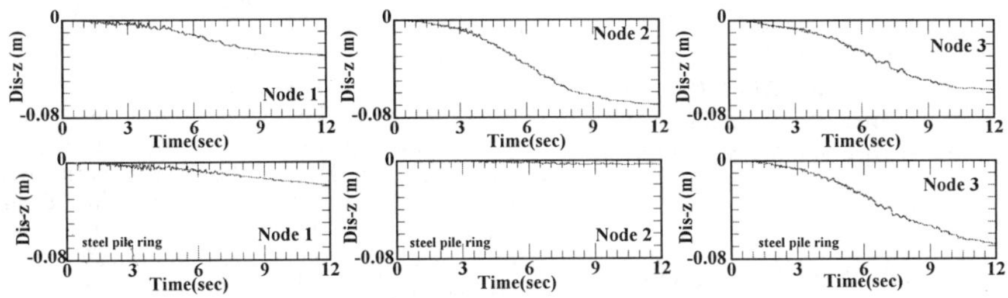

Figure 7 Influence of the steel pile ring on responded settlement

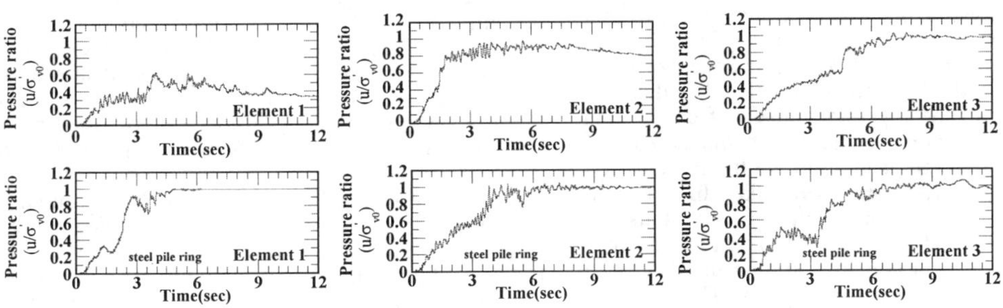

Figure 8 Influence of the steel pile ring on excessive pore-water pressure ratio

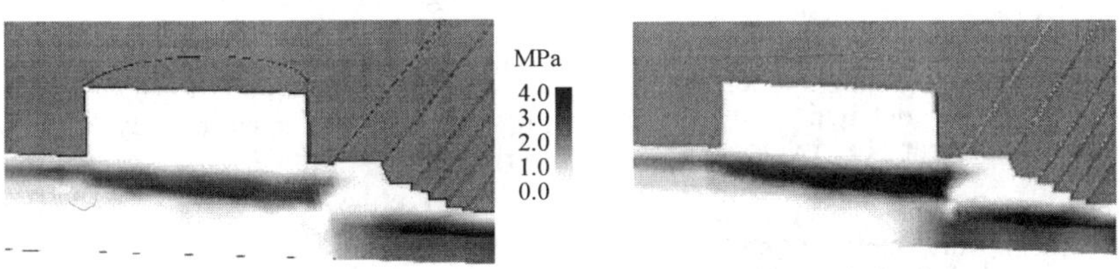

(a) without steel pile ring (b) with steel pile ring

Figure 9 Distribution of excessive pore-water pressure (t=10.0 sec)

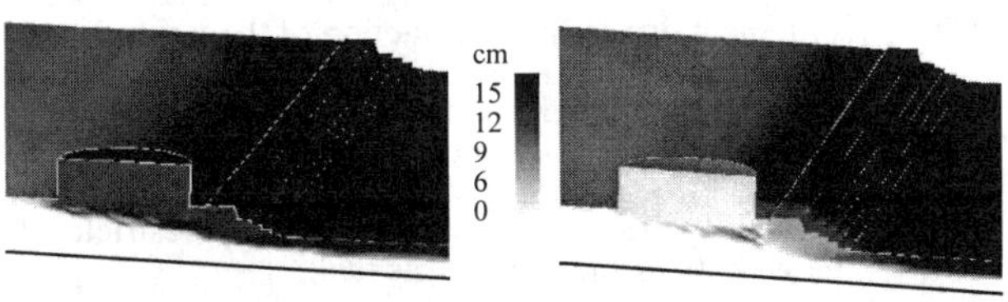

(a) without steel pile ring (b) with steel pile ring

Figure 10 Distribution of horizontal displacement (t=10.0 sec)

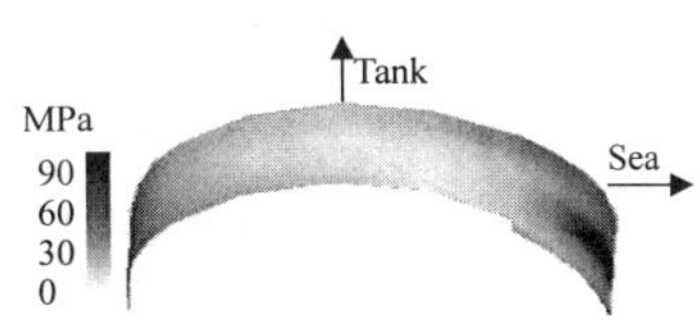

Figure 11 Distribution of stress in steel pile ring (t=10.0 sec)

k is the permeability, K^f is the bulk modulus of the fluid and $\dot{\varepsilon}_{ii}^s$ is the volumetric strain of solid.

The validity of the proposed numerical method was verified by Oka et al. (1994) through a comparison of the numerical results and the analytical solutions for a transient response of saturated porous solids.

As to the constitutive model for sand, a cyclic kinematic elastoplastic model (Oka et al., 1999) is used and the detailed description of the model can be referred to the reference.

Figure 1 shows the performance of the constitutive model for undrained torsional shear tests of Toyoura Standard sand under the condition that the vertical strain is constrained (Matsuo et al., 2000). It is found that the simulated shear stress-shear strain relation and the effective stress path coincide well with the experimental results.

3 NUMERICAL RESULTS AND DISCUSSION

In this paper, the petroleum tank considered in the numerical analysis is located on a reclaimed ground whose geological profile is given in Figure 2. The capacity of the tank is 30,000 KL and its size is 16 meter in high and 52 meter in diameter. The distance from the tank to the revetment made of concrete is only about 20 meters. There exists the danger of horizontal flow of the ground due to the liquefaction of the ground during an earthquake. In order to enhance the seismic strength of the tank, a remediation method is designed in which a ring-shaped steel pile wall is driven around the tank. The thickness of the pile ring is 3.0 cm. The physical properties of the ground and steel pile are listed in Table1 and Table 2. In the numerical analysis, the ground is simulated with 8-node isoparametric solid element while the steel pile ring is simulated with elastic shell element. The nonlinear parameters involved in the kinematic elasto-plastic model are also listed in Table 1. The detailed way of determining these parameters can be referred to references.

Figure 3 shows the general view of the tank-ground-pile ring system. A finite element mesh of 3603 solid elements and 88 shell elements with 4514 nodes are used in the calculation. Because of the symmetric condition, only half of the domain is considered. Figure 4 shows the input earthquake wave with 12 minuets of main vibration time, whose maximum acceleration is 487 gal.

In the dynamic analysis, a stiffness-proportional type of damping is adopted. The direct integration method of Newmark-β is used in the dynamic analysis and the time interval for the integration is 0.01 seconds.

Two cases of the calculations are conducted, one is for normal ground without steel pile ring and another is for the ground with the remediation method. Figure 5 shows the responded acceleration at some points as indicated in Figure 2. It is found that the acceleration of point 2 where the steel pile ring is located is much larger than the one without the ring.

Figure 6 shows the responded horizontal displacements at the points for different cases. It is obvious that due to the existence of the steel pile ring, the horizontal displacement near the ring are much smaller than the ones without the ring. The repression effect of the displacement due to the ring decreases with the distance from the ring. The same phenomenon can be observed for the settlements at these points as shown in Figure 7 in which the vertical displacements near the ring are very small while the settlement a little far from the ring are almost the same as the one without the ring.

Figure 8 shows the change of the excessive pore-water pressure ratios of some elements as shown in Figure 2 for two different cases. It is very interesting to see that the ratios for the case of steel-pile-ring are much larger than those of the case without steel-pile-ring. In element 1, which locates within the ring, the ground is completely liquefied while in the case without steel-pile-ring, the ground does not reach the liquefaction. This phenomenon indicates that the steel pile ring can do nothing for preventing the liquefaction of ground but for preventing the deformation of the ground. The same conclusion can be obtained from the distribution of the excessive pore-water pressure shown in Figure 9. It is easy to see that the pressure within the pile ring is much higher than the ones in the ground without the ring.

Figure 10 shows the distribution of the horizontal displacements obtained from different cases. It is clear that due to liquefaction of the ground, a horizontal flow occurred along the liquefied ground in the case without the pile ring. On the other hand, the ground near the pile ring did not flow due to the resistance of the ring, though the area without flow is only restricted in the vicinity of the ring.

Figure 11 shows the distribution of the stress in the steel pile ring. It is clear that the maximum stress occurred at the boundary where the horizontal flow due to the liquefaction of the ground is retained by the pile ring. The maximum stress is less than the yielding stress of the steel usually used in civil engineering.

4 CONCLUSION

(1) A steel pile ring around the tank which is built on a potentially liquefied ground and is very near to the seashore is useful in reducing the horizontal flow and settlement due to the liquefaction of the ground.
(2) The steel pile ring, however, cannot prevent the ground from liquefaction at all, the only effect that can be expected is the repression of the displacement.
(3) The effect of the steel pile ring as a remediation method against the earthquake is only restricted to the vicinity of the ring, or the tank.

REFERENCES

Finn, W. D. L., Wu, G. and Sasaki, Y. (1997): 'Simulation of response of the Kushiro Rive Dike to the 1993 Kushiro-oki and 1994 Hokkaido Toho-oki earthquake,' Proc. 14th Int. Conf. on SMFE, Hamburg, pp. 99-102.

Oka, F., Yashima, A., Shibata, T., Kato, M. and Uzuoka, R. (1994): 'FEM-FDM coupled liquefaction analysis of a porous soil using an elasto-plastic model,' Applied Scientific Research, Vol. 52, pp. 209-245

Oka, F., Yashima, A., Tateishi, A., Taguchi, Y. and Yamashita, S. (1999): 'A cyclic elasto-plastic constitutive model for sand considering a plastic-strain dependence of the shear modulus,' Geotechnique, 49, No.5, pp. 661-680

Matsuo, O., Shimazu, T., Uzuoka, R., Mihara, M. and Nishi, K., (2000) 'Numerical analysis of seismic behavior of embankments founded on liquefiable soils,' Soils and Foundations Vol. 40, No.2, pp.21-39,

Computer Methods and Advances in Geomechanics, Desai et al. (eds) © 2001 Balkema, Rotterdam, ISBN 90 5809 183 X

Seismic analysis of subway station surrounded by soft soils in Shanghai

Jian Zhou, Yong Chi & Peng Dong
Department of Geotechnical Engineering, Tongji University, Shanghai, People's Republic of China

ABSTRACT:Based on the results of cyclic triaxial test, a seismic analysis of two typical subway stations located in sandy soils and clayey soils in Shanghai is presented in this paper. The method of FE dynamic effective stress analysis and the reliability analysis is used in the analysis. In order to consider the influence of the seismic intensity and different soft soil layers, six different recorded earthquake motions selected from both home and aboard are used in the calculations. The results of dynamic response analysis include the dynamic stresses of subway stations and surrounding soils, pore water pressures of surrounding soils as well as seismic induced settlement. The dynamic reliability of subway stations and surrounding soils is also given in the paper.

1.INTRODUCTION

Shanghai is located in the Yangtze River delta in China. The depth of the overburden of soil is about 280m-300m. The underground water level is about 0.8m-1.0m from the surface of ground and the soil conditions consist essentially of a thin surface layer of mixed silty clay, clay and fill, followed by a 20-40m thick soft clayey soil layer with interbedded seams of silt, fine sand or silty sand. Based on the results of cyclic triaxial test, a seismic analysis of two typical subway stations located in sandy soils and clayey soils in Shanghai is presented in this paper. The method of FE dynamic effective stress analysis and the reliability analysis is used in the analysis. In order to consider the influence of the seismic intensity and different soft soil layers, six different recorded earthquake motions selected from both home and aboard are used in the calculations.

Based on the results of resonant column tests and cyclic triaxial tests, the formulations of pore water pressure and residual strain of the soft soil in Shanghai is given. The six different earthquake motions, Tangshan, Haicheng, El Centro, Niyasakigawa, Lotung earthquake record and Sunan Nuclear Power Station man-made earthquake motion are adopted as different input motions in the analysis. The results of dynamic response analysis include the dynamic stresses of subway stations and surrounding soils, pore water pressures of surrounding soils as well as seismic induced settlement. The dynamic reliability of subway stations and surrounding soils is also given in the paper.

2. TEST RESULTS

Based on the results of cyclic triaxial tests on the soft soil of Shanghai nearby the two subway stations, the following model of cyclic shear strength, model of residual pore-water pressure and model of residual strain of saturated soft soil can be derived (Jian Zhou et al. 1995).

2.1 *Cyclic shear strength and parameter*

If the cyclic stress σ_d is normalized by the mean stress $\sigma'_0(\sigma'_0 = (\sigma_a + 2\sigma_r)/3)$, the relation between the cyclic strength $R_f(\sigma_d/\sigma'_0)$ and the number of load cyclic N_f can be approximated by a straight line in the logarithmic form for both isotropic and anisotropic consolidated conditions:

$$R_f = aN_f^b \tag{1}$$

Where N_f is the number of load cycles required to achieve a 5% double amplitude shear strain for isotropic condition and 5% maximum amplitude shear strain for anisotropic condition respectively, and a and b are experimental constants.

2.2 *Excess pore pressure and parameter*

A uniformed relation can be obtained between the cyclic-induced pore water pressure P_g and the cyclic stress ratio η^* for both isotropic and anisotropic consolidated conditions.

$$P_g / U_f = \eta^* / [c_3 - (c_3 - 1)\eta^*] \tag{2}$$

$$\eta^* = (N_R)^{c_2} / [c_1 - (c_1 - 1)(N_R)^{c_2}] \tag{3}$$

where $\eta^* = (\eta_{p,e} - \eta_s)/(\eta_f - \eta_s)$, $\eta_{p,e}$ is the current effective dynamic stress ratio, η_s is initial effective stress ratio, η_f is effective stress ratio at failure point, U_f is the maximum residual pore water pressure while the 5% double amplitude shear strain for isotropic condition and 5% maximum amplitude shear strain for anisotropic condition is reached, N_R is relative cyclic number, c_1, c_2 and c_3 are the experimental parameters.

2.3 *Residual axial strain and parameter*

The relation between the undrained residual axial strain and the cyclic stress ratio for anisotropic consolidated condition is as follows:

$$\varepsilon_p = \eta^* / [d - (d - 20)\eta^*] \tag{4}$$

Where d is the experimental parameter.

3. DYNAMIC ANALYSIS METHOD

In this paper non-linear elastic constitutive model is applied to subway station structure, Duncan-Chang (1970) model is used for soft soil statistic analysis, and Hardin-Drnevich (1972) model is used for dynamic analysis of soft soil. The related parameters are derived as follows:

$$E_t = K \cdot Pa \cdot (\sigma_3 / Pa)^n \{ \frac{1 - [R(1 - \sin\varphi')(\sigma_1 - \sigma_3)]}{2c' \cos\varphi' + 2\sigma_3 \sin\varphi'} \}^2 \tag{5}$$

$$G = G_{\max} / (1 + \gamma / \gamma_r) \tag{6}$$

$$D = D_{\max} \cdot \gamma / (\gamma + \gamma_r) \tag{7}$$

where K is modular coefficient, n is modular exponent, Pa is atmospheric pressure, R_f is the ratio of the principal stress difference at failure point and the ultimate principal stress difference, G_{max} is initial gradient of stress-strain curve, D_{max} is the maximum of equivalent damping ratio.

Substituting the pore water pressure and residual strain due to dynamic loading into Biot's basic dynamic consolidation equations, and omitting the acceleration of pore water pressure, we got the dynamic differential equation. Then, Galerkin's weighted residual method and 2-D isoparametric element with four nodes is used to formulate the following set of finite equation:

$$[\bar{K}]\{\delta\} + [Q]\{P\} + [M]\{\ddot{\delta}\} = \{F\} \tag{8}$$

$$[Q]^T\{\dot{\delta}\} + [H]\{P\} = \{\bar{F}\} \tag{9}$$

Where $[\bar{K}]$ is stiffness matrix, $[Q]$ is couple matrix, $[M]$ is mass matrix, $[H]$ is permeability matrix, $\{P\}$ is nodal pore water pressure, $\delta, \dot{\delta}, \ddot{\delta}$ is nodal displacement vector, velocity vector and acceleration vector respectively; $\{F\}$ is nodal load vector; $[\bar{F}]$ is nodal seepage discharge vector.

The equation (8) and (9) is solved by front solution method (Jian Zhou & Jian Du 1994).

4. RELIABILITY ANALYSIS

The reliability of the subway stations structure is evaluated using the reliability index value β ,while the damage accumulation is used to indicate the damage level of the soft soil.

If the maximum response of the dynamic shear stress is R, and the cyclic shear strength of subway station structure is R_f, then the structure failure probability is:

$$P_f = P(R > R_f) = 1 - \int_0^\infty (\int_0^N f_R(R)dR) \cdot f_{R_f}(R_f)dR_f \tag{10}$$

where $f_R(x)$、$f_{R_f}(x)$ are the frequency function of R and R_f respectively. when $f_R(x)$ and $f_{R_f}(x)$ are normal distribution, Eq. (10) can be simplified as follow:

$$P_f = \frac{1}{2\pi} \int_{-\infty}^{-\beta} e^{-x^2/2} dx \tag{11}$$

where $\beta = (\mu_N - \mu_R)/\sqrt{\sigma_N^2 + \sigma_R^2}$, μ_N and σ_N are expected value and variance of the structure cyclic shear strength derived from test data, μ_R and σ_R are expected value and variance of the maximum response of the dynamic shear stress derived from seismic analysis.

The damage accumulation of the soft soil is:

$$D = \sum_{i=1}^{n} D_i = \sum_{i=1}^{n} N_i / N_{if} \tag{12}$$

Where N_i is the wave peak number when dynamic shear stress ratio is γ_i, N_{if} is the soil element

Table1. Geologic parameters in Gymnasium Station (layer No.1)

Layer	Elevation of layer bottom (m)	D_{max}	Effective unit weight(KN/m^3)	c' (kPa)	$\sin\varphi'$	$K_{2,max}$	Poisson ratio	Coefficient of permeability (m/s)
Fill	2.80	0.33	9.00	22.30	0.38	8900	0.29	1.5E-08
Very soft clay	6.80	0.30	7.50	10.00	0.21	2470	0.41	1.6E-07
Silty clay	17.50	0.31	8.20	9.15	0.40	3450	0.37	7.1E-07
Silty clay	28.50	0.31	8.60	9.21	0.41	4313.4	0.37	7.7E-07
Sandy silt	38.00	0.30	9.10	9.00	0.31	8300	0.27	5.8E-06
Silty and fine sand	64.60	0.28	9.70	0.00	0.56	10304	0.28	1.1E-05

Table 2. Geologic parameters in Xinkezhan Station (layer No.2)

Layer	Elevation of layer bottom (m)	D_{max}	Effective unit weight(KN/m^3)	c' (kPa)	$\sin\varphi'$	$K_{2,max}$	Poisson ratio	Coefficient of permeability (m/s)
Fill	2.1	0.33	9.0	22.3	0.375	8900	0.29	1.5E-08
silty sand	5.6	0.32	9.0	6.0	0.370	10200	0.29	7.0E-07
silty sand	9.2	0.32	8.6	4.6	0.415	3020	0.26	9.0E-07
Silty sand and very soft silty clay	17.3	0.3	8.0	3.2	0.440	9087	0.30	3.0E-07
silty clay	25.2	0.32	8.2	13.9	0.120	4313	0.37	7.1E-08
silty clay	31.3	0.32	9.8	29.7	0.292	4313	0.37	1.5E-08
Silty fine sand	45.2	0.28	9.7	0.0	0.559	10304	0.28	1.1E-05
Sandy silt	68.3	0.32	9.3	3.7	0.391	8020	0.26	8.6E-06

Table3. Seismic response results of Subway Stations

	Maximum dynamic shear stress ratio			Maximum pore water pressure ratio			Maximum seismic induced settlements (mm)			Maximum shear stress (kPa)		
	GCS*	GLS*	NCS*	GCS*	GLS*	NCS*	GCS*	GLS*	NCS*	GCS*	GLS*	NCS*
Tangshan earthquake	0.12	0.12	0.16	0.10	0.10	0.14	3.0	0.16	6	446.33	853.33	397.74
Haicheng earthquake	0.12	0.11	0.115	0.007	0.006	<10-4	1.3	0.09	<0.1	375.13	1121.1	142.63
El centro earthquake	0.16	0.16	0.16	0.10	0.11	0.04	3.7	0.3	<0.1	431.11	1188.6	295.90
Niyasakigawa earthquake	0.14	0.16	0.13	0.24	0.22	0.25	14	3	<0.1	733.98	1683.0	656.44
Sunan man-made earthquake	0.11	0.13	0.12	0.018	0.16	0.01	1.2	0.8	0.1	653.47	873.35	399.37
Lotung earthquake	0.16	0.17	0.18	0.26	0.27	0.15	17	0.8	6	567.98	1297.6	390.13
Mean value μ	0.14	0.14	0.14	0.12	0.14	0.10	6.70	0.86	2.07	534.67	1169.5	380.37
Coefficient of variation σ / μ	0.16	0.18	0.18	0.89	0.65	1.00	1.04	1.27	1.47	0.26	0.26	0.44

* GCS is Gymnasium Station cross section, GLS is Gymnasium Station longitudinal section, NCS is Xinkezhan Station cross section.

failure cycle number at every γ_i. The soil element assumed damaged when $D=1$.

5. CASE ANALYSIS

The Gymnasium Subway Station and Xinkezhan Subway Station of Shanghai No.1 subway are taken as examples. Limited by the length of this paper, here we just list the soil properties and parameters within 60m shown as Table.1 and Table.2.

As no earthquake has been recorded at bedrock in the city, we use Tangshan, Haicheng, El Centro, Niyasakigawa, Lotung earthquake record and Sunan Nuclear Power Station man-made earthquake motion as input motions. The depth of 280m from the soil layer surface is considered as the input base boundary.

Table 3 shows the seismic response of the subway stations and the surrounding soft soil under the six different input motions. Figure1~Figure 6 show the seismic pore water pressure ratio, the dynamic shear stress ratio and the seismic induced settlement. The results of damage accumulation are shown as Table 4, Figure 7 and Figure 8. The results of the goodness of fit test of the maximum dynamic shear stress probability distribution are presented in Table 5. Known from Table 5 that as $\alpha=0.01$, the critical value of D_n is $D_{6,0.01}=0.364$, therefore we take the structure maximum dynamic shear stress as obeying logarithmic normal distribution. The reliability index value β are shown as Figure 9 and Figure 10.

Table 4. The maximum of damage accumulation

	Layer No.1	Layer No.2
Rayleigh distribution	0.01	0.08
Gaussian distribution	0.014	0.09
Extremal distribution	0.015	0.10
Mean value	0.013	0.09

Table 5. The goodness of fit test of dynamic shear stress distribution

	Gymnasium cross section	Gymnasium longitudinal section	Xinkezha cross section
Normal distribution test	0.2450	0.2790	0.3182
Logarithmic normal test	0.2446	0.3308	0.2535
Exponential distribution test	0.5027	0.4546	0.4692

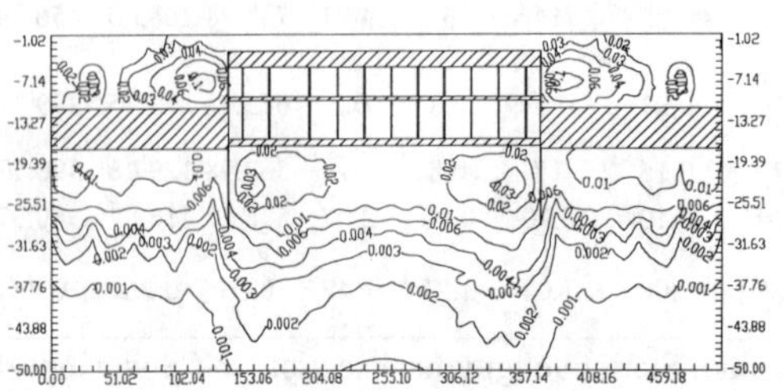

Figure 1. The pore water pressure ratio of soil layer (Gymnasium longitudinal section)

6. CONCLUSIONS

The main conclusions are as follows:

1. The maximum pore water pressure ratio and damage accumulation occur near the top of the subway stations and ground surface under 7-degree earthquake. In this case, there is almost no effect on the subway stations by the generation of pore water pressure and damage accumulation.

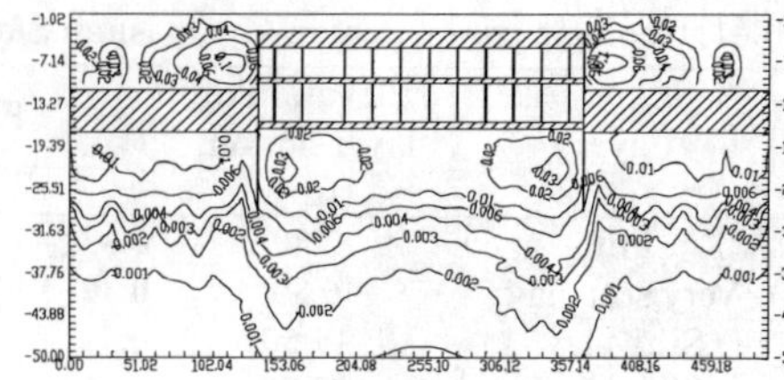

Figure 2. The dynamic shear stress ratio of soil layer (Gymnasium cross section)

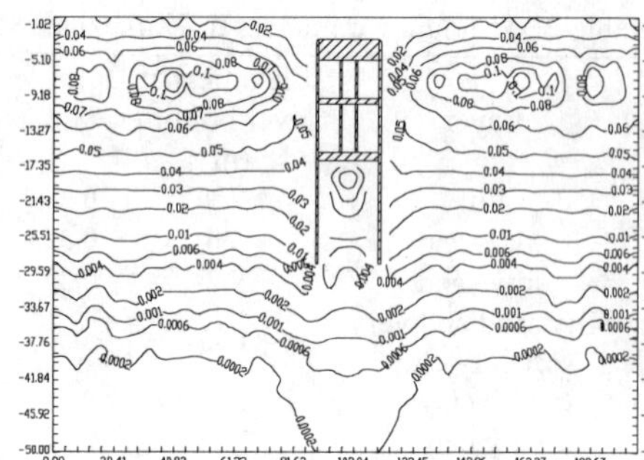

Figure 3. The pore water pressure ratio of soil layer (Gymnasium cross section)

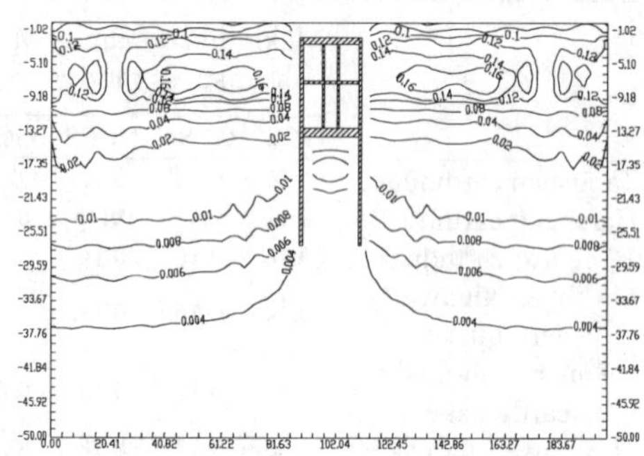

Figure 4. The pore water pressure ratio of soil layer (Xinkezhan Station cross section)

2. The computed results of seismic response are affected by different input motions used in the calculations, such as dynamic stresses, pore water pressure ratio and seismic induced settlement etc. The seismic induced settlements and pore water pressure ratios are relatively sensitive to different input motions, while the different input motions have only little impact on the dynamic shear stress

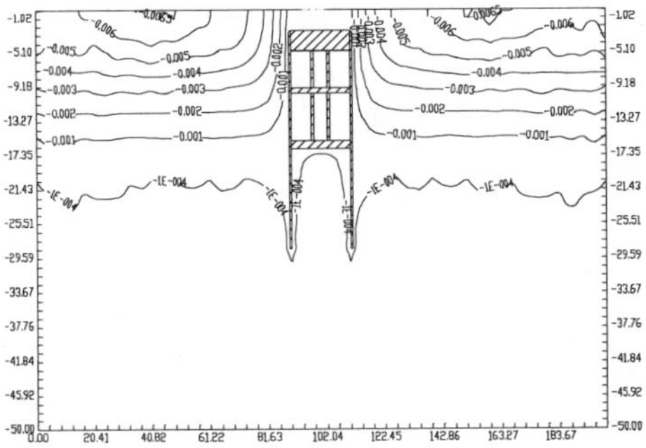

Figure 5. The seismic induced settlement of soil layer in Gymnasium Station

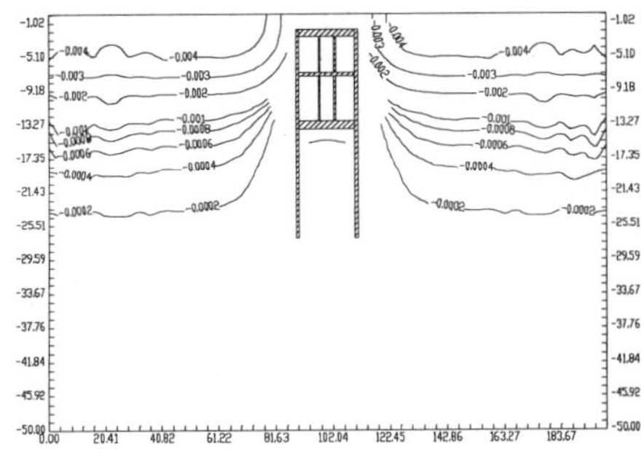

Figure 6. The seismic induced settlement of soil layer in Xinkezhan Station

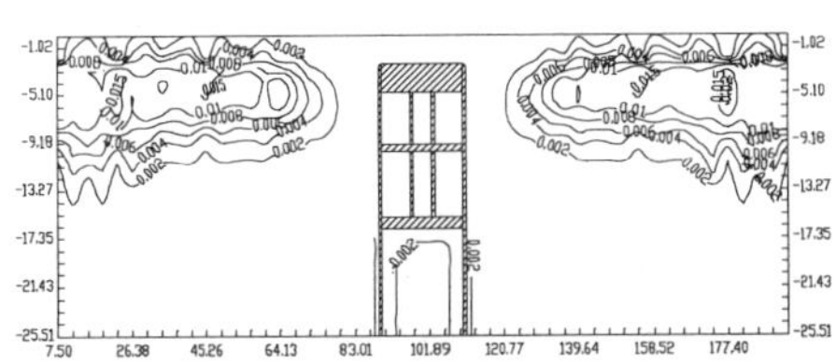

Figure 7. The damage accumulation of layer No.1

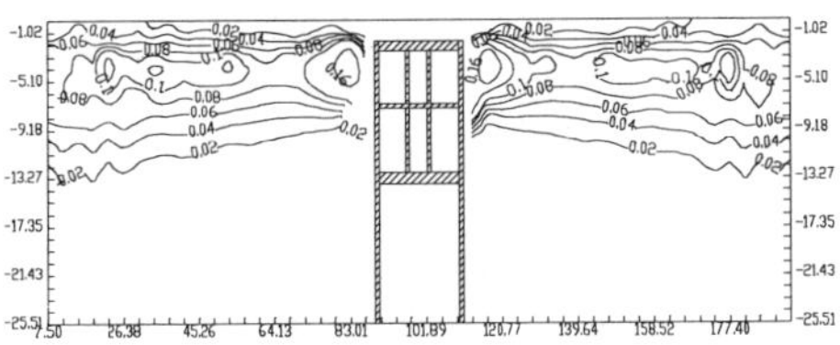

Figure 8. The damage accumulation of layer No.2

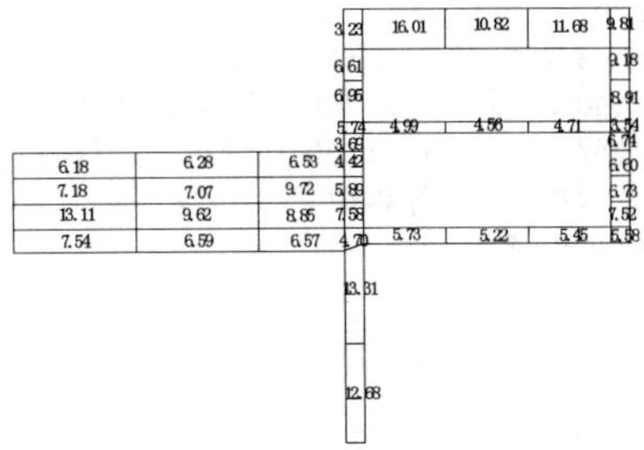

Figure 9. Reliability index β (Gymnasium longitudinal section)

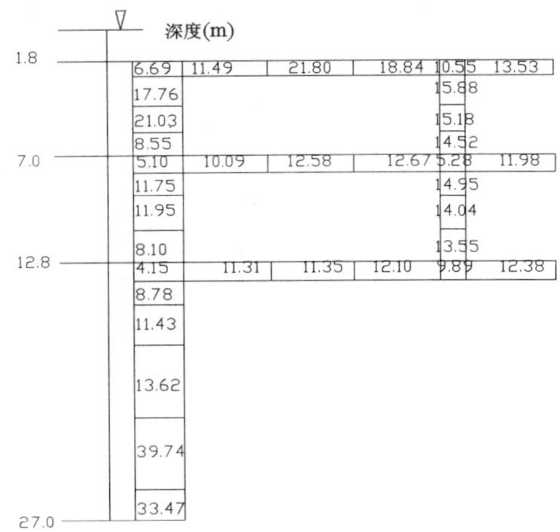

Figure 10. Reliability index β (Xinkezhan Station cross section)

of soils and structure dynamic stresses.

3. The dynamic FE analysis show that the joint internal stress between the subway station and the shield method tunnel increase with the raising of seismic intensity. The maximum compressive stress occurred on the top of the joint, while the maximum tensile stress occurred at the bottom of the joint.

REFERENCES

Duncan, J.M. & Chang, C. Y. 1970. Non-linear analysis of stresses and strain in soil. *Proc.ASCS.JSMFD,* Vol.96, SM5.

Hardin, B.D. & Drnevich, V.P. 1972. Shear modulus and damping in soils: design equations and curves. *Jour.Soil Mechanics and Foundations Division, ASCE* Vol.98,No.7.

Jian Zhou et al.1995. Prediction of settlements of soft clay caused by earthquakes. *Third International Conf. On Recent Adv. In Geotech.*

Earthq. Eng. and Soil Dynamics, St. Louis, Missouri, USA, No. 3.44 (4).

Jian Zhou & Jian Du 1994. Analysis of foundation settlement due to earthque. *ASCE Geotechnical Eng. Division Speciality Conf. Proceedings of Settlements* (6).

10 Soil/Rock-structure interaction

Computer Methods and Advances in Geomechanics, Desai et al. (eds) © 2001 Balkema, Rotterdam, ISBN 90 5809 183 X

Analysis and design of rock-support interaction in underground excavations

C.A.Lauro & A.P.Assis
Department of Civil and Environmental Engineering, University of Brasilia, Brazil

ABSTRACT: Usually in practice, the design of underground excavations in rock is accomplished with empirical or analytical methods, which do not take into account reasonably the rock-support interaction behavior. In order to understand better this phenomenon, firstly a study of the mechanical characteristics of the rock and support has to be done and then an analysis of the behavior of the rock-support interaction be performed. A brief review of empirical, analytical and numerical methods is presented for the rock-support design as well as their evolution and the best use of them all together. It is recommended to follow a direct or indirect methodology for the support design. The indirect approach involves geomechanical classification methods as an evolution of the empirical methods, where geomechanical parameters could be obtained from these modified geomechanical classification methods. These parameters and those related to external loads and to the opening are then input in design tools for predicting support solutions. Based on the interaction rock-support analyses and design, an example based on the case-history of Ipueira mine is shown, with two types of rock masses. The importance of the relative stiffness of the support system to determine its flexibility behavior was verified and hence the stress and strain fields analyzed. Substantial economic support design was verified in comparison to the traditional empirical or analytical methods.

1 INTRODUCTION

The first step for designing an underground structure is to evaluate its functional requirements and environmental conditions before and after the construction. In this terms failure could be defined as the inability of the structure to function as designed (Sinha, 1989).

To study the response of the system (rock mass and support), several types of models may be considered. Empirical, analytical and numerical models are generally less costly and less time consuming than photo-elastic and three-dimensional scaled physical models. Once a model is selected, the governing equations for equilibrium or motion are established and then solved and tested for uniqueness, existence, sufficiency and relevancy. The solution techniques for solving equations for equilibrium or motion could be simple, rigorous, numerical, or empirical, but the solution must exist, be unique and relevant (Sinha, 1989). This physical system (rock mass and support) for analysis can be further divided and classified into different engineering systems according to geometry, loading and material properties. Considering the wide range of loadings, material properties and time effects, the design of underground structures is somewhat more complex than the design of other structures.

The rock load is dependent on several factors such as the rock-support relative stiffness, the elapsed time between excavation and support installation, the rock mass characteristics, the in-situ stresses, the opening size, the water presence, and the construction methods. In case of a flexible support, the rock mass will take the major portion of the load by arching, being the support responsible for a smaller share. A stiffer support attracts more load and a flexible support attracts more displacement (Sinha, 1989).

Then, the tunnel support lining is not an independent structure, acted upon by a well-defined series of loads, but instead, deformations and induced strains are function of the rock mass response to the excavation process. A convenient and often used tool to describe this interaction is the Characteristic Curve (Figure 1), which generally plots internal pressures and radial displacements for a certain rock mass and tunnel support (Schwartz et al., 1980).

In general, rock mass classification methods attempt to formalize an empirical approach to tunnel design, in particular for determining support requirements. While classification schemes are appropriate for their original application, especially if used within the bounds of the case histories from which they were developed, considerable caution must be exercised in applying rock mass classification to other rock engineering problems.

The maximum unsupported span can be estimated from:

$$\text{Maximum span (unsupported)} = 2 \cdot ESR \cdot Q^{0.4} \quad (3)$$

Based upon analyses of case records, Grimstad & Barton (1993) suggest that the relationship between the value of Q and the permanent roof pressure, in kg /cm², P_{roof} is estimated from:

$$P_{roof} = \frac{2\sqrt{Jn} \cdot Q^{-1/3}}{3Jr} \quad (4)$$

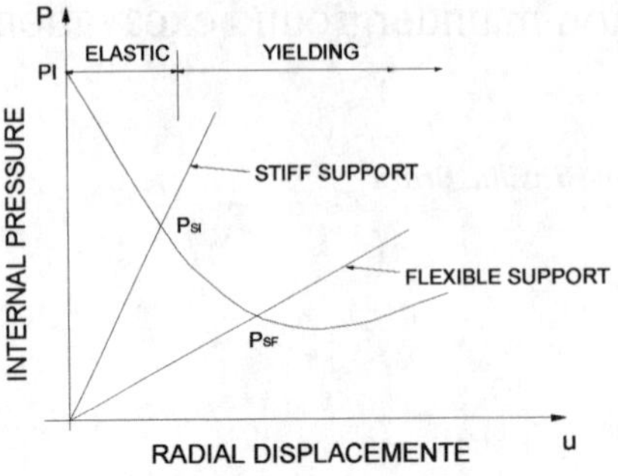

Figure 1 Characteristic curve and the effect of relative stiffness on support loads (Schwartz et al., 1980)

Most of the multi-parameter classification schemes (Wickham et al., 1972, Bieniawski, 1989 and Barton et al., 1974) were developed from civil engineering case histories, in which the most relevant engineering geological components of the rock mass were included. Two widely used classification systems are exposed here, in a sense to understand how they consider the rock support interaction behavior.

Barton et al. (1974) introduced the Tunneling Quality Index "Q" for determining rock mass characteristics and tunnel support requirements. The index Q can be considered a function of three quantities, which are the crude measure of the block size, inter-block shear strength and active stresses. For applying it to a support design, an additional parameter was defined, called Equivalent Dimension (De) of the excavation:

$$D_e = \frac{\text{Excavation span, diameter or height (m)}}{\text{Excavation Support Ratio } ESR} \quad (1)$$

The value of ESR is related to the intended use of the excavation and to the degree of security, which is demanded of the support system installed to maintain the stability of the excavation.

The equivalent dimension, De, plotted against the value of Q, is used to define a number of support categories (9 categories) in a chart published in the original paper by Barton et al. (1974). Grimstad & Barton (1993) have updated this chart to reflect the increasing use of steel fiber reinforced shotcrete in underground excavation support. Barton et al. (1980) provide additional information on rockbolt length, maximum unsupported spans and roof support pressures to supplement the support recommendations published previously. The rockbolt length L can be estimated from the excavation width B and the Excavation Support Ration ESR as:

$$L = \frac{2 + 0{,}15B}{ESR} \quad (2)$$

Bieniawski (1989) published a set of guidelines for the selection of support in rock tunnels for which the value of RMR has been determined. Note that these guidelines have been published for a 10 m span horseshoe shape tunnel, constructed using drill and blast methods, in a rock mass subjected to a vertical stress less than 25 MPa (equivalent to a depth below surface less than 900 m). Entering with the RMR value, one can know the excavation method, the length and lateral spacing of a 20 mm diameter, fully grouted rockbolts, the thickness of shotcrete and the need of steel sets.

The empirical methods are basically for designing the initial support system because they deal only with loosening loads (load of broken rock). Effective use of empirical design approaches require evaluations of both the creator and user's judgments. In order to judge the applicability of empirical design methods, it is important to verify sampling conditions and the biases of the creator on developing the particular empirical methods.

An parameter index call *GSI*, Geological Strength Index, was propose by Hoek (1994) with the idea to link the experience of empirical methods with a rational design method, changing the direct project philosophy. The *GSI* varies from 10 to 100 and is a modified geomechanical classification method, based on the Bieniawski's Rock Mass Rating (*RMR*) and Barton et al.'s *Q* index. It could be calculated as:

For $RMR_{89'} > 23$:
$GSI = RMR_{89'} - 5$, or $GSI = RMR_{76'}$ and

(5)

For $RMR_{89'} < 23$:

$$GSI = 9 Log_e Q' + 44, \text{ Where: } Q' = \left(\frac{RQD}{Jn}\right)\left(\frac{Jr}{Ja}\right)$$

The major advantage of the GSI is that it permits to obtain geomechanical parameters from modified geomechanical classifications by correlation, rather than performing the whole design by empirical methods. Considering the rock mass Hoek & Brown

Table 1 Analytical methods for underground excavation (Lauro et al., 1998)

ANALYTICAL METHOD	GEOMETRY	CONSTITUTIVE LAW	IN-SITU STRESS	SPACE	SUPPORT	INDUCED DEFORM.
Kirsch	circular	elastic	Ko = 1	infinite	absent	Not
Mindlin	circular	elastic	Ko = 1	half infinite	absent	Not
Mohr & Coulomb	circular	elastic-plastic	Ko = 1	infinite	reaction force	Yes
Hoek & Brown	circular	plastic-plastic	Ko = 1	infinite	reaction force	Yes
Burns & Richard	circular	elastic	Ko vary	infinite	structural element	Yes
Einstein & Schwartz	circular	elastic	Ko vary	infinite	structural element	Yes

strength criterion (Hoek, 1994):

$$\sigma_1' = \sigma_3' + \sigma_c \left[m_b \left(\frac{\sigma_3'}{\sigma_c} \right) + s \right]^a \qquad (6)$$

where σ_i' are the principal stresses at failure; σ_c is the unconfined strength of the intact rock; m_b, s, a are rock mass constants.

The Geological Strength Index (GSI) obtains these geomechanical parameters (m_b, a and s) as follows:

$$\frac{m_b}{m_i} = \exp\left[\frac{(GSI - 100)}{28}\right]$$

$$\text{For } GSI > 25\text{:} \quad a = 0{,}5; \quad s = \exp\left[\frac{(GSI - 100)}{9}\right] \qquad (7)$$

$$\text{For } GSI < 25\text{:} \quad a = 0{,}65 - \frac{GSI}{200}; \quad s = 0$$

The variable m_i is the constant m for the intact rock (Hoek, 1994). Another parameter that could be obtained from the geomechanical classification is the deformation modulus. Serafim & Pereira (1983) proposed the following relationship between E_m and RMR (or GSI), obtained by back analyses:

$$E_m = 10^{\frac{(RMR-10)}{40}} \qquad (8)$$

3 ANALYTICAL METHODS

At the beginning, analytical methods tried to define the excavation behavior based on elastic models of the rigid body mechanics, without considering any support type. For example, Kirsch in 1898 (Hoek et al., 1995) presented an analytical solution for a circular geometry, excavated in an infinite space under a biaxial load state. Later, other solutions appeared for different boundary conditions and geometry (Neuber in 1937 and Mindlin in 1940). All these solutions did not consider the rock-support interaction, but only the behavior of the rock mass around the excavation, based on the Elasticity Theory.

With the appearance of the characteristic curve method by Rabcewicz (1969), improvements in the understanding behavior of the rock-support interaction and the stress relaxation phenomenon were made, based on monitoring. This contributed to a better description of the rock yielding process and profit in using the stand-up time for an unsupported span excavated in rock. This characteristic curve method is very important for future analyses, as well as the yielding process.

The problem of total instantaneous loading and strains due to geological in-situ stresses (strains developed along the geological time) was solved, using the superposition principles of Elasticity, and finally obtaining only the strains induced by the excavation. Later, plasticity behavior of the rock was introduced as a linear and non-linear failure criteria (Hoek & Brown, 1980) and support as a distributed reaction force.

Analytical methods for rock-support interaction consider simple geometry and strain deformation parameters implemented in elastic models by Burns & Richard (1964). They did not consider induced deformations, which were later corrected by Einstein & Schwartz (1979). They proposed a method that allows the boundary condition variation between the rock-support interface. The principal characteristics of each method are listed in the Table 1.

4 NUMERICAL METHODS

Many design problems in rock mechanics practice involve more complex geometry and may require more powerful analytical tools, including non-homogeneity of the rock mass and non-linear constitutive behavior. Solutions for more complex excavation design problems may usually be obtained by using computational procedures. The use of these techniques is now firmly embedded in rock mechanics practice (Cundall, 1987; Brady & Brown, 1995). Their application will undoubtedly increase in the future, as computer increases and its cost decreases.

Computational methods of stress analysis fall into two categories: differential and integral methods. In differential methods, the problem domain is divided (discretized) into a set of subdomains or elements. A solution procedure may be based on numerical approximations of the governing equations, i.e. the equilibrium, strain-displacement and stress-strain equations, as in classical finite difference methods.

The characteristic of integral methods of stress analysis is that a problem is specified and solved in terms of surface values of the field variables of traction and displacement. Since the problem boundary is only defined and discretized, the so-called boundary element methods of analysis effectively provide a unit reduction in the dimensional order of a problem. The implication is a significant advantage in computational efficiency, compared with the differential methods.

For the finite element method, nonlinear and heterogeneous material properties may be readily accommodated, but the outer boundary of the problem domain is defined arbitrarily, and discretization errors may occur throughout the domain. On the other hand, boundary element methods model far-field boundary conditions correctly, restrict discretization errors to the problem boundary, and ensure fully continuous variation of stress and displacement throughout the medium. However, these methods are best suited to linear behavior and homogeneous material properties; non-linear behavior and heterogeneity negate the intrinsic simplicity of a boundary element solution procedure. The explicit finite difference method FLAC (Fast Lagrangian Analysis of Continua) is an example of a finite difference program thar can be easily applied to geomechanics, considering the rock-support interaction. This can be done by an interface element that simulates contact between rock and other rock (joints), rock and lining support, and rock and structure. This interface element may be simulated as elastic perfectly plastic (Figure 2).

Figure 3depicts two deformability parameters (horizontal and vertical stiffness, *Ks* and *Kn* respectively) and two strength parameters (shear strength *S* and perpendicular strength *T*). Each node has its own length of influence and when the displacement of the node (i.e. node N moves out of length L_N) goes out of its length of influence, the problem does not converge and then grid had to be corrected, trying to put the total displacement of the node within its length of influence.

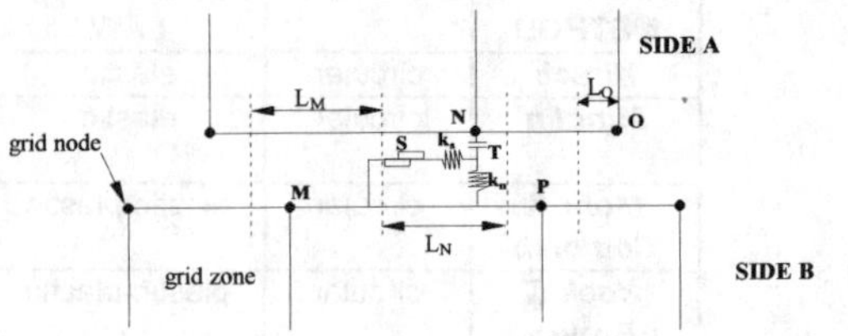

Figure 2 Interface scheme for elastic plastic behavior (ITASCA, 1995)

The other type of interaction is included in the cable element, which permits a full or end node contact of a cable support with the rock. This support element follows almost the same scheme as the interface element (Figure 3). The interaction between the cable and the rock is made with the grout, by a shear stiffness parameter (kbond) and a cohesive strength parameter (sbond), in other words, only uniaxial behavior as in the cable element.

The finite element program PLAXIS (finite element code for soil and rock analysis) applies an interface element to simulate the contact between the rock and lining support, and the output for this interface comprises deformations and stresses. This program simulates the elastic behavior of the rock-support interaction and considers a perfect contact of the rock-support without any losing of stress in the interface (the interface permits a total transfer of the stress and strain behavior of the rock to the support (Brinkgreve et al. 1998).

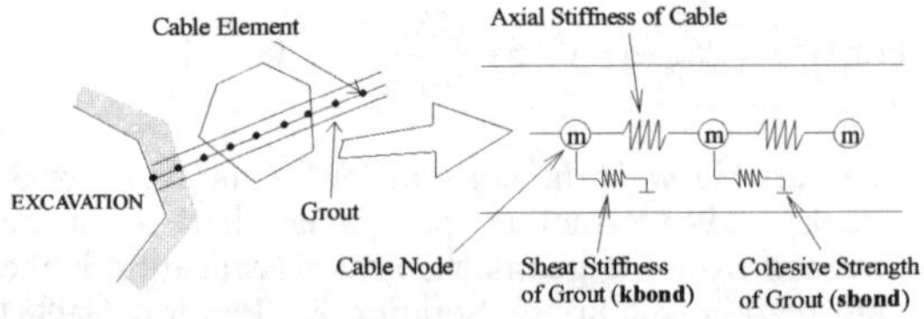

Figure 3 Cable element-grout interface for elastic plastic behavior (ITASCA, 1995)

5 DESIGN APPROACHES FOR UNDERGROUND STRUCTURES IN ROCK MASS

Analyses and designs of underground excavations could be faced by several approaches. Two possible approaches are represented on Figure 4. A preliminary program of geological and geotechnical field investigation helps establish a geomechanical model for the rock mass. Based on this model, geotechnical parameters could be obtained by a direct methodology (laboratory and field tests) or by an indirect methodology (via correlations founded on geomechanical classifications).

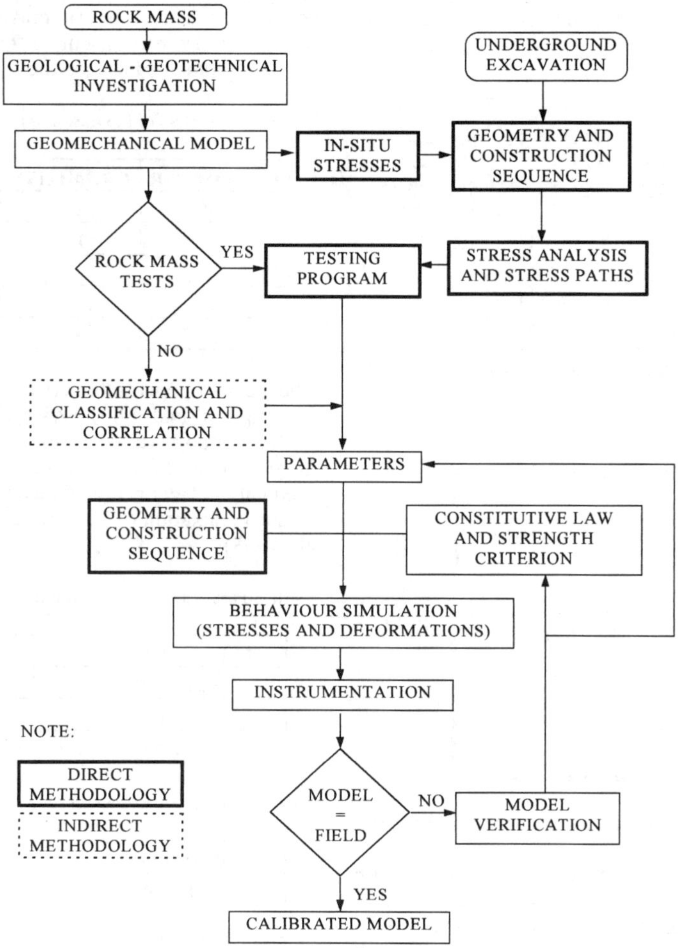

Figure 4 Direct and indirect approaches for designing an underground excavation in rock mass (modified of Vinueza, 1994)

The direct approach combines geometry, construction sequence and in-situ stresses (measured from field tests) for predicting possible stress paths and consequently for defining the testing program to obtain geomechanical parameters. The indirect approach obtains geomechanical parameters from classification methods, especially for cases where a testing program can not be executed. Following these two methodologies, design tools can be involved with analytical or numerical simulation to obtain stresses and strains of the rock and support.

6 CASE-HISTORY OF IPUEIRA MINE, BRAZIL

A cromitite mine, located in Andorinhas, 400 km from Salvador (Brazil), is an underground scheme, 200 m deep, where the representative materials are the cromitite (ore mineral) and the serpentine, as shown in Figure 5. Vinueza (1994) presents in detail the geological data and geomechanical characteristics, which were used for the rock-support interaction analysis performed by Lauro et al. (1998). For the interaction analysis, the indirect approach was followed, using the modified geomechanical classification methods and field data of Vinueza (1994) to obtain the geomechanical parameters. For the numerical analysis, the program FLAC was used. The geometry, initial and boundary conditions of the model are depicted in Figure 6, where four galleries, an interface that represents the contact of the two materials, a distributed load on the top that simulates the material above the modeled region, the location of the liner support (shotcrete) and the boundary conditions are shown. Restrictions of the horizontal displacement on the two walls and restriction of the vertical displacement on the base were also considered.

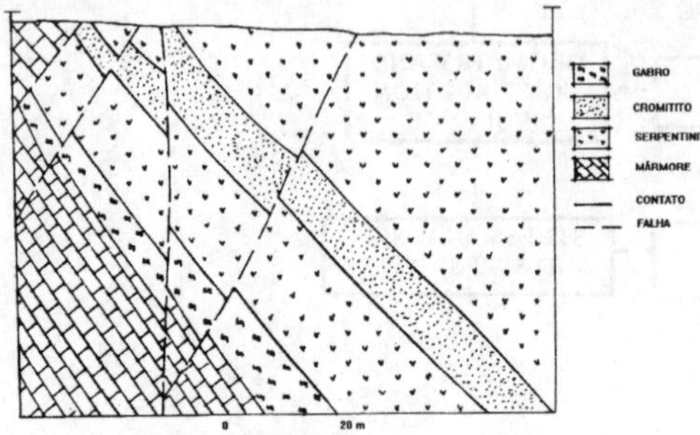

Figure 5 Regional geology

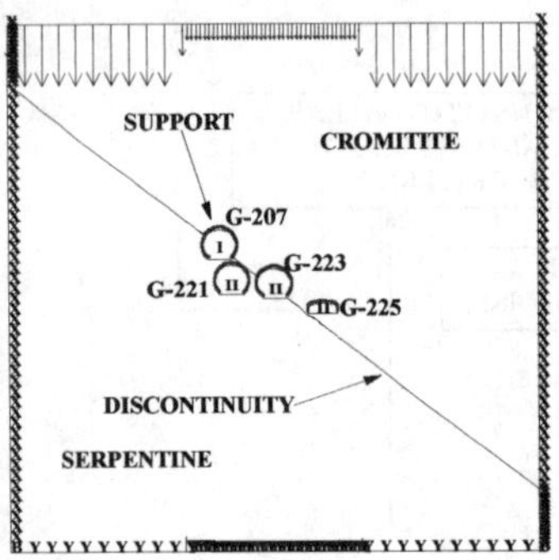

Figure 6 Geometry and initial conditions of the problem

The materials were assumed with a liner elastic perfectly plastic behavior with non-linear Hoek & Brown strength criterion of the rock (Hoek, 1994). The strength parameters were computed through geomechanics classifications supplied by Vinueza (1994) and the adopted values are listed in Table 2. For the discontinuity or contact, it has been adopted parameters of stiffness and strength (Mohr-Coulmb), shown in Table 3. The support applied was shotcrete, assumed with a linear elastic behavior (Table 4).

7 RESULTS

Support flexibility is related directly whit the relative stiffness of the rock-support system (Einstein & Schwartz, 1979), then its observed in the Figure 7, the larger the flexibility of the support (E_{rock} / E_{sup} increases), the smaller the moments and the support begins to dissipate load as deformations increase.

Deformations and support convergences (Figure 8) increase as the flexibility increase too. The maximum displacements are on the tunnel lateral walls, because Ko is greater than one. For the *fs* condition, the displacements are greater than those for the *ns* condition. The support flexibility is advantageous for the support design in rock, to reduce its dimensions, but as the flexibility increases the displacements increase too.

Table 2 Parameters of the rock mass

PARAMETER	CROMITITE	SERPENTINE
E (GPa)	20	30
ν	0,25	0,25
K (s_h/s_v)	1,5	1,5
m_b	0,327	0,838
s	0,00102	0,00386

The rock-support interaction had two conditions: non slip *(ns)* and full slip *(fs)*. The elastic moduli (E) of the support varied from 10 GPa to 20 GPa by increments of 2,5 GPa, aiming to analyze the support flexibility behavior. For these analysis, it was adopted a stress relaxation of 50% of the total value (Lauro et al., 1998).

Table 3 Parameters of discontinuity

PARAMETER	VALUE
K_n e K_s (GPa)	20
c (Pa)	0
ϕ (°)	15

Table 4 Parameters adopted for the support

PARAMETER	VALUE
E (GPa)	10 to 20
ν	0,25
thickness (cm)	5
Inertia (m^4)	$1{,}04 \times 10^{-5}$
Area (m^2)	0,05

8 COMPARISON WITH ANALYTICAL METHODS

A comparison of numerical results with the analytical solution of Einstein & Schwartz is not reasonable, but both follow the same behavior pattern. Then making a rough approximation of the stress relaxation for the analytical method, assuming a reduction of the external load applied to a 50%, the results became closer to the numerical ones (Figure 9 and Figure 10), for moments and axial forces, respectively. Axial force results seem better than the moment ones.

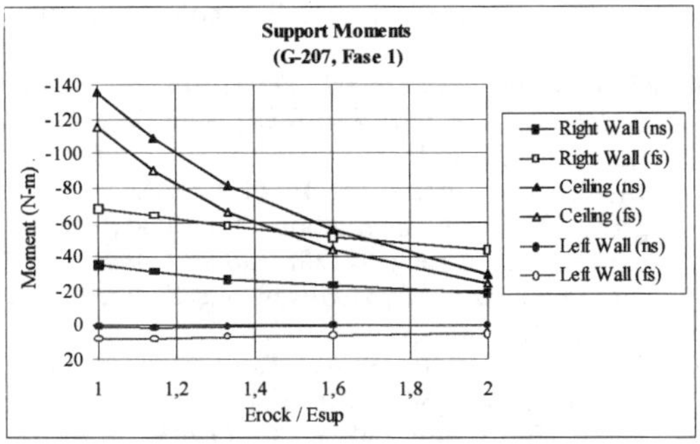

Figure 7 Variation of the support moments with support flexibility

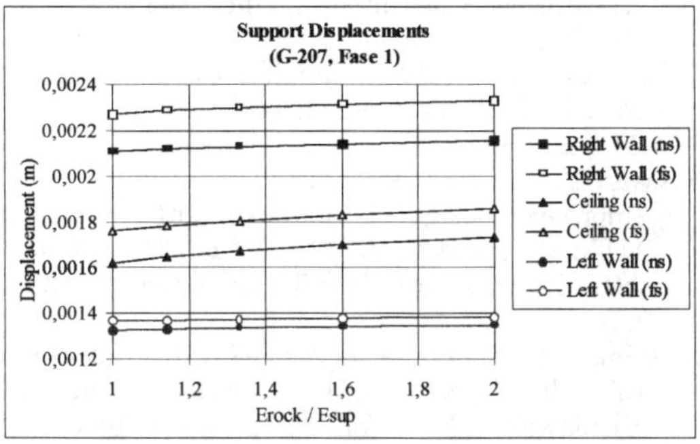

Figure 8 Variation of the support displacement with support flexibility

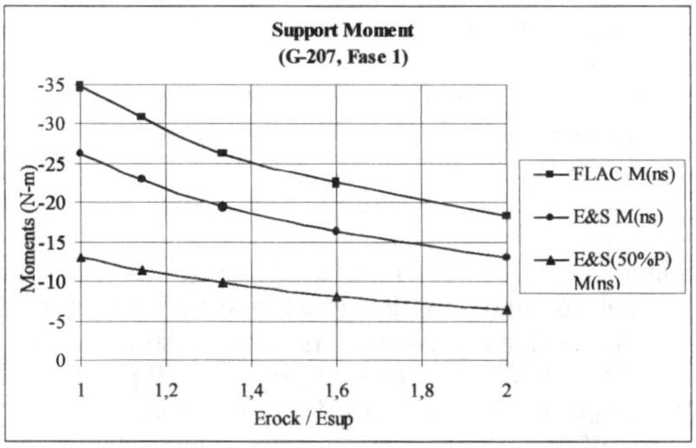

Figure 9 Comparison of support moments

9 COMPARISON WITH EMPIRICAL METHODS

Applied the geomechanical classifications of Bieniawski (1989) with an RMR of 38 (cromitite), category IV (poor rock) was obtained and the support design requires: 20 mm diameter rock bolts, fully grouted, spaced 1 to 1,5 m plus shotcrete with 10 to 15 cm thickness. These values are much greater than those founded in the numerical method, concluding that the empirical method was too conservative.

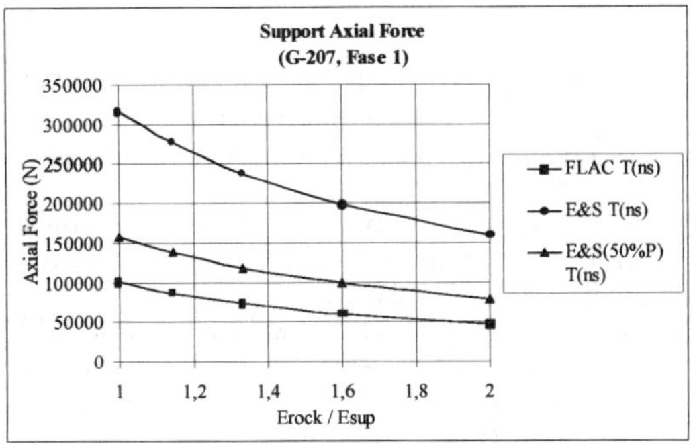

Figure 10 Comparison of the support axial force

10 CONCLUSIONS

The first attempts of support design for underground excavations were using geomechanical classifications, which already consider an idea of rock-support interaction as a global system. After, a clearer way of explaining this physical interaction was presented by the characteristic curve, which was so important for developing next interaction solutions. The evolution continues with analytical methods, which encompassed extremely boundary conditions (full slip or non-slip contact for lining support). Lately, numerical methods appeared to improve this knowledge and incorporated elastic behavior to this interaction as well as failure criteria.

The design approach presents the indirect methodology as an evolution of the empirical design methods (geomechanical classifications) as an attempt to diminish the implicit dependence of the empirical method to many parameters, proposing that the geomechanical classification methods only consider parameters related to rock characteristics. The major advantage of the indirect approach is that the implicit dependence of support solutions would be limited to obtaining geomechanical parameters from classification methods by correlations. This approach is particularly interesting when rock mass parameters cannot be obtained by laboratory or in-situ tests, mainly because scale effects.

The case-history example demonstrated that a more flexible support reduced its absorbed forces and generates greater deformations, leading to find an appropriate equilibrium compromise between both. The analytical method of Einstein & Schwartz shows to be appropriate for a preliminary analysis, regarding the conditions of this model. Also the use of correlations based on geomechanical classifications on the indirect approach works satisfactory for analyzing parameters needed in support design.

REFERENCE

Barton, N. R., Lien, R. & Lunde, J. (1974), Engineering classification of rock masses for the design of tunnel support, *Rock Mechanics.*, 6(4), 189-239.

Barton, N. R. & Choubey, V. (1977), *The shear strength of joints in theory and practice*. Proc. *Rock. Mech.* 10(1-2), 1-54.

Barton, N., Loset, F., Lien, R. and Lunde, J. (1980), Application of the Q-System in design decision. In *Subsurface space*, (ed. M. Bergman) 2, 253-561. New York: Pergamon.

Bieniawski, Z. T. (1976), Rock mass classification in rock engineering. In Exploration for rock engineering, proceedings of de symposia, 97-106.

Bieniawski, Z. T. (1989), *Engineering rock mass classification*, New York, USA., John Wiley & Sons, 251 p.

Brady, B. H. C. & Brown, E. T. (1994). *Rock Mechanics for Underground Mining*, Chapman and Hall, London, 571p.

Brinkgreve, R. B. J. & Vermeer, P. A. (1998), *PLAXIS Finite Element Code for Soil and Rock Analyses*, A. A. Balkema, Rotterdam, Brookfield, Version 7

Burns, J. Q. & Richard, R. M. (1964), Attenuation of stresses for buried cylinders, *Proc. Symp. On Soil-Structure Interaction*, Univ. of Arizona, Tucson, Arizona.

Cundall, P. A. (1987), Distinct element models of rock and soil structure, in *Analytical and Computational Methods in Engineering Rock Mechanics*, Chapter 4, (ed. E. T. Brown), London, George Allen and Unwin, pp 129-163.

Einstein, H. H. & Schwartz, C. (1979), Simplified analysis for tunnel supports, *Journal of the Geotechnical Engineering Division*, J. Geotech. Eng. Div., ASCE, 105(GT4): 499-518.

Grimstad, E. and Barton, N. (1993), Updating the Q-System for NMT. *Proc. Int. symp. On sprayed concrete – modern use of wet mix sprayed concrete for underground support*, Fagernes, (eds. Kompen, Opsahl and Berg). Oslo: Norwegian Concrete Assn.

Hoek, E. & Brown, E. T. (1980), *Underground excavations in rock*, Inst. Min. and Metall., London, 527 p.

Hoek, E. (1994), Strength of Rock and Rock Masses, *International Society for Rock Mechanics (ISRM), News Journal*, Minneapolis, USA, 2(2) 4-16.

Hoek, E., Kaiser, P. K., & Bawden, W. F. (1995). *Support of Underground Excavations in Hard Rock*. A. A. Balkema, Rotterdam, Netherlands, 215 p.

ITASCA, (1995), *FLAC 3.30, User's Manual*, Itasca Consulting Group, Inc., Mineapolis, Minesota, USA.

Kirsch, G. (1898), *Die theirie der elastizitat und die bedürfnisse der festigkeitslehre*, Veit. Ver. Deut. Ing., 42, 797-807.

Lauro, C. A. & Assis, A. P. (1998) Analysis of ground-support interaction for underground excavations in rock, *World Tunnel Congress'98 on Tunnels and Metropolises*, São Paulo, Brazil, pp 321 327.

Mindlin, R. D. (1940), Stress distribution around a tunnel. *Trans, ASCE*, 105, 1117-1140.

Pavoni, R. B. E Portugal, A. C. (1993), *Soluções Analíticas para estudo de tensões e deformações em obras subterrâneas.* Final project (in portuguese), Dept. ENC, Univerisidade de Brasília, UnB, Brasília / DF, Brazil, 140 p.

Rabcewicz, L. V. (1969), Stability of tunnels under rock load. *Water Power* 21(6-8), 225-229, 266-273, 297-304.

Schwartz, C. W. and Einstein, H. A. (1980), Improved Design of tunnel supports , Vol. II, UTMA *Report UTMA-MA-06-0110-80-4-5, MIT*, Department of Civil Engineering, Boston, USA.

Serafim, J. L. & Pereira, J. P. (1983), Consideration of the geomechanical classification of Bieniawski. *Proc. Int. Symp. on Engineering Geology and Underground Construction*, Lisbon 1(II), 33-44.

Sinha, R. C., (1989). *Underground Structures, Design and Instrumentation,* Developments in Geotechnical Engineering, 59A, ELSEVIER, New York, USA.

Terzaghi, K. (1946), Rock defects and loads on tunnel supports, Rock tunneling with steel supports. (eds. R. V. Proctor & T White), Commercial Shearing and Stamping Co., Youngtown, 15-99.

Vinueza, G. (1994) Classificação geomecânica subsidiando a modelagem numérica de uma mina subterrânea. Dissertação de mestrado, Dep. ENC, UnB, Brasília / DF, Brazil, 129 p.

Wickham, G. E., Tiedman, H. R. and Skinner, E. H. (1972), Support Determinations Based on Geological Predictions. *RETC Proceedings*, Vol. 1, June.

Computer Methods and Advances in Geomechanics, Desai et al. (eds) © 2001 Balkema, Rotterdam, ISBN 90 5809 183 X

Numerical analysis of the interaction between groups of piles

F.M. Abdrabbo
Structural Engineering Department, Faculty of Engineering, Alexandria University, Egypt

U.S. El Shamy
College of Engineering and Technology, Arab Academy for Science, Technology and Maritime Transport, Egypt

ABSTRACT: A hybrid approach is utilized for the elastic analysis of the interaction between groups of two piled rafts. This interaction is expected to be significant especially in the case of closely spaced groups of piles found in relatively soft clay. The load distribution amongst piles within each group of piles and the displacement profile of the two rafts are presented for the case of floating pile groups. For the two piled rafts in relatively soft clay, significant differential settlement between the two groups was depicted and relatively high bending moments were developed in the connecting ground beam.

1 INTRODUCTION

The increasing use of piled foundations to support heavily loaded buildings has increased the demand of having reliable, yet easy to apply, numerical approach for the analysis of this complicated problem. The wide spreading of high-speed personal computers encouraged the use of more rigorous analysis methods, such as the finite element method, to be applied in many complicated problems. Many numerical solutions have been developed by many research workers in the past three decades for the analysis of single piles and pile groups. A pioneer boundary integral solution has been presented by Poulos & Davis (1980). A simplified load transfer approach has been developed by Randolph & Wroth (1978) for analyzing axially loaded single piles, and extended later for pile groups (Randolph & Wroth 1979). Chow (1986) used a hybrid approach for the nonlinear analysis of pile groups by using a modified form of Randolph expression, for the single pile response and Mindlin's elastic solution to account for the interaction between piles in a group. Lee (1993) used a nonlinear hybrid approach, which again was based on Randolph expression for single pile response, but the interaction between piles has been taken by an overall expression that evaluates the additional pile settlement as a function of the neighboring pile head load. Abdrabbo & El Shamy (1997) implied an approximate approach to account for the effect of cap flexibility on the load distribution amongst piles of a freestanding pile group. Clancy & Randolph (1993) extended Randolph analysis of pile groups to account for the effect of cap stiffness and the pile-cap-soil interaction. An approximate method of analysis for piled raft of large piles number has been introduced by Clancy & Randolph (1996).

Although such approach has given an effective insight to the response of piled rafts, it still needs to be extended to account to other common situations in the piled foundations practice. One of these situations frequently encountered in practice is the case of closely spaced groups of piles. An important question may arise that whether, or not, the response of each group when exists in a system of closely spaced groups of piles will remain as that in the conventional individual pile group. In order to investigate this effect, an elastic analysis of the response of two closely spaced groups of piles was carried out. Accountancy was also made for the effect of possible presence of a ground beam connecting the two groups.

2 METHOD OF ANALYSIS

Figure 1 presents the analyzed problem, in which a nine square pile group is neighboring to a four square pile group. Modeling of each piled raft was carried out according to Clancy & Randolph (1993). The pile cap was modeled by a 16 degree of freedom four-node rectangular plate bending elements (Smith & Griffiths 1988) as recommended by Griffiths et al. (1991). The piles were idealized as rod elements. The ground beam was simulated by a regular beam element

The pile-soil-pile interaction and the raft-soil-pile interaction are accounted for using Mindlin's solution (Mindlin 1936). The raft-soil interaction is

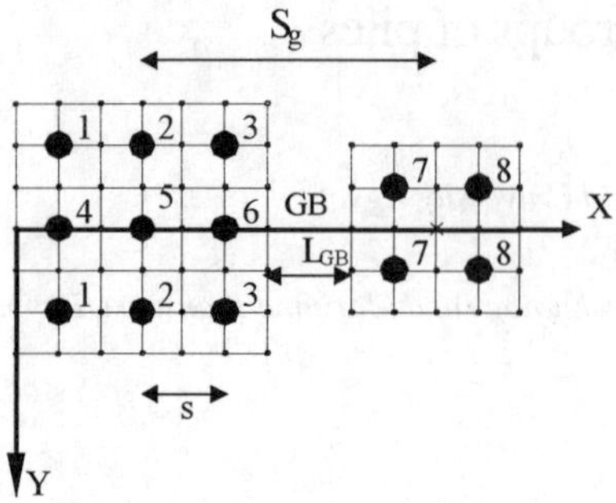

Figure 1. Configuration of the analyzed two piled rafts

considered through an equivalent soil spring at each raft node based on the analytical solution provided by Giroud (1968).

3 MODEL VERIFICATION

The output of the developed computer program was compared with the results presented by Clancy & Randolph (1993) for the analysis of an individual square nine piled raft under uniform load. The discretization scheme of the raft was carried out as indicated in Figure 1. The pile, soil, and raft data are fully defined by Clancy & Randolph (1993). The obtained results are in good agreement with those obtained by Clancy & Randolph (1993), table 1.

Table 1. Verification of computed results.

Item	Computed	Clancy & Randolph (1993)
w_p (mm)	7.966	7.978
w_r (mm)	7.977	8.099
P_p (MN)	29.76	29.66
P_r (MN)	6.24	6.34

w_p = average displacement of pile group
w_r = average displacement of raft
P_p = total load carried by pile group
P_r = total load carried by raft

4 RESULTS

The developed computer program was used to analyze the two groups. Attention was paid to the effect of the spacing (S_g) on the load distribution amongst the piles within the two groups. The pile and raft data used in the analysis are shown in table 2. Two soil profiles were considered and expressed as (E_p/G=1000) which corresponds to relatively stiff clay, and (E_p/G=10,000) which corresponds to relatively soft clay, where (G) is the soil shear modulus; and (E_p) is the pile modulus of elasticity. In both cases the Poisson's ratio of soil was assumed to be 0.5.

Furthermore, to provide more understanding of the effect of the ground beam on the overall response of the system, the analysis was carried out in two extreme cases, that is the groups are connected by a ground beam (GB), and the groups are free of any connection. The beam height (H), considered in the analysis, was as the pile cap height (1.0 m) and its width (B) was taken (0.3 m). The elastic modulus of the beam was taken as that for the raft. A typical stiffness factor for the beam ($E_{PL}I$) will be used, where (I) is the beam moment of inertia.

The analysis was performed applying two concentric loads of (P_{G1}=4.5 MN) on the nine-pile group and (P_{G2}=2.0 MN) on the four-pile group. The own weight of the pile cap and the connecting beam was not considered in the analysis.

Table 2. Pile and raft data used in the analysis.

Pile Data	Raft Data
Length (L_p) = 20 m	Thickness (t_c) = 1.0 m
Radius (r) = 0.25 m	Poisson's ratio (ν_{pl})= 0.15
Pile spacing (s) = 1.5 m	Modulus of elasticity (E_{pl}) = 20 GPa
Modulus of elasticity (E_p) = 20 GPa	

4.1 *Results for two piled rafts in soft clay*

Table 3 introduces the achieved results for the case of (E_p/G=10,000). In addition, the table also includes the results obtained for the load distribution amongst the piles and the raft if each piled raft is analyzed individually. The results show a considerable alteration in the load distribution amongst piles within the two groups than that obtained from individual piled raft analysis for relatively close spacing between the two groups. This difference tends to decrease with the increase of the spacing between the two groups and the load distribution amongst piles within the two groups, with or without a ground beam, approaches that of individual piled raft analysis. It is also noted that some of the applied load on the nine-pile group is transferred to the piles in the four-pile group for close spacing between the two groups because of the interaction provided by the connecting ground beam.

At relatively small spacing between the groups, the presence of the connecting ground beam does not help for obtaining a more uniform pile load distribution but it helps, as expected, in obtaining a more uniform settlement profile of the two groups. This behavior is demonstrated in Figure 2, in which the displacement of the two raft nodes (*w*) along the (X) axis has been normalized, and presented in case of a clear span equal to pile spacing ratio (L_{GB}/s=0.67). This effect becomes less significant when the clear span between the two groups increases to (L_{GB}/s=6.67) as indicated from Figure 3.

Figure 4 presents the effect of the spacing between the two groups on the developed bending moment (M) and shearing force (Q) at the beam end

Table 3. Load distribution amongst piles within the two piled rafts (E_p/G=10,000).

Normalized clear span between groups (L_{GB}/s)		Load distribution in 9-pile group								Load distribution in 4-pile group			
		$\frac{P_1}{P_{av1}}$	$\frac{P_2}{P_{av1}}$	$\frac{P_3}{P_{av1}}$	$\frac{P_4}{P_{av1}}$	$\frac{P_5}{P_{av1}}$	$\frac{P_6}{P_{av1}}$	$\frac{P_r^{9PG}}{P_{G1}}\%$	$\frac{\Sigma P_i^{9PG}}{P_{G1}}\%$	$\frac{P_7}{P_{av2}}$	$\frac{P_8}{P_{av2}}$	$\frac{P_r^{4PG}}{P_{G2}}\%$	$\frac{\Sigma P_i^{4PG}}{P_{G1}}\%$
0.67	With GB	1.395	0.802	1.004	0.83	0.276	0.412	8.67	88.02	0.814	1.136	9.96	97.51
	Without GB	1.226	0.814	1.266	0.694	0.274	0.615	8.93	91.08	0.91	0.902	9.40	90.61
2.67	With GB	1.321	0.772	1.112	0.796	0.282	0.605	8.57	89.9	0.961	0.925	9.14	94.31
	Without GB	1.235	0.777	1.24	0.727	0.281	0.706	8.70	91.30	0.908	0.914	8.90	91.1
6.67	With GB	1.267	0.761	1.183	0.775	0.293	0.704	8.49	91.06	0.972	0.88	8.43	92.6
	Without GB	1.231	0.762	1.231	0.747	0.293	0.743	8.53	91.47	0.916	0.918	8.34	91.66
Pile load in individual group analysis		1.262	0.746	1.262	0.746	0.255	0.746	7.92	92.08	0.926	0.926	7.44	92.56

P_{av1} = Average load per pile in the nine-pile group = $P_{G1}/9$
P_{av2} = Average load per pile in the four-pile group = $P_{G2}/4$
P_r = Load carried by raft of each group

Table 4. Load distribution amongst piles within the two piled rafts (E_p/G=1000).

Normalized clear span between groups (L_{GB}/s)		Load distribution in 9-pile group								Load distribution in 4-pile group			
		$\frac{P_1}{P_{av1}}$	$\frac{P_2}{P_{av1}}$	$\frac{P_3}{P_{av1}}$	$\frac{P_4}{P_{av1}}$	$\frac{P_5}{P_{av1}}$	$\frac{P_6}{P_{av1}}$	$\frac{P_r^{9PG}}{P_{G1}}\%$	$\frac{\Sigma P_i^{9PG}}{P_{G1}}\%$	$\frac{P_7}{P_{av2}}$	$\frac{P_8}{P_{av2}}$	$\frac{P_r^{4PG}}{P_{G2}}\%$	$\frac{\Sigma P_i^{4PG}}{P_{G1}}\%$
0.67	With GB	0.899	0.874	0.892	0.827	0.898	0.774	12.71	86.98	0.845	0.854	15.77	84.93
	Without GB	0.886	0.872	0.913	0.816	0.897	0.795	12.78	87.22	0.853	0.834	15.66	84.34
2.67	With GB	0.898	0.854	0.893	0.834	0.893	0.819	12.80	87.09	0.853	0.839	15.66	84.6
	Without GB	0.896	0.854	0.898	0.832	0.893	0.824	12.82	87.18	0.842	0.845	15.62	84.39
6.67	With GB	0.898	0.845	0.9	0.838	0.888	0.838	12.76	87.23	0.852	0.846	15.09	84.93
	Without GB	0.899	0.846	0.899	0.839	0.888	0.837	12.76	87.24	0.848	0.85	15.08	84.92
Pile load in individual group analysis		0.908	0.846	0.908	0.846	0.891	0.846	12.15	87.85	0.861	0.861	13.93	86.07

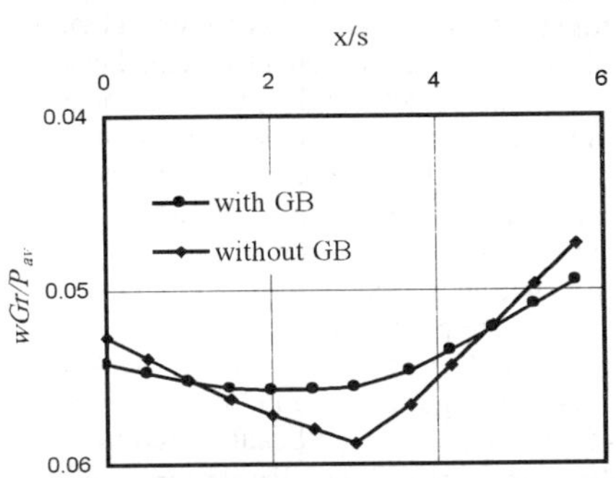

Figure 2. Displacement profile along the (X) axis of the two groups (E_p/G=10,000, L_{GB}/s=0.67)

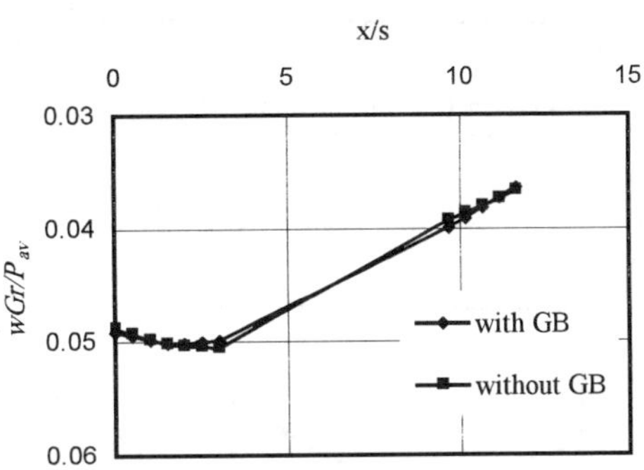

Figure 3. Displacement profile along the (X) axis of the two groups (E_p/G=10,000, L_{GB}/s=6.67)

connected to the nine-pile group. Large values of bending moment and shearing force are developed due to the relatively high differential settlement between the beam-ends in case of relatively close spacing between the two groups. These values tend to decrease significantly with the increase of the spacing between the two groups.

4.2 *Results for two piled rafts in stiff clay*

Results for two piled rafts in relatively stiff clay (E_p/G=1000) are presented in Table 4. The results indicate less alteration in the load distribution amongst piles within the two groups than that in the case of (E_p/G=10,000).

The displacement profile along the (X) axis for this case (E_p/G=1000) is depicted in Figures 5, 6 for a clear span between the two groups to pile spacing ratio of (0.67) and (6.67) respectively. The Figures indicate almost identical profile with or without the existence of the connecting ground beam. It is also noted a more pronounced deflected shape in each group than that in the case of (E_p/G=10,000). This is because the assumed cap thickness resulted in a relatively uniform load distribution amongst piles within the two groups. These almost equally loaded

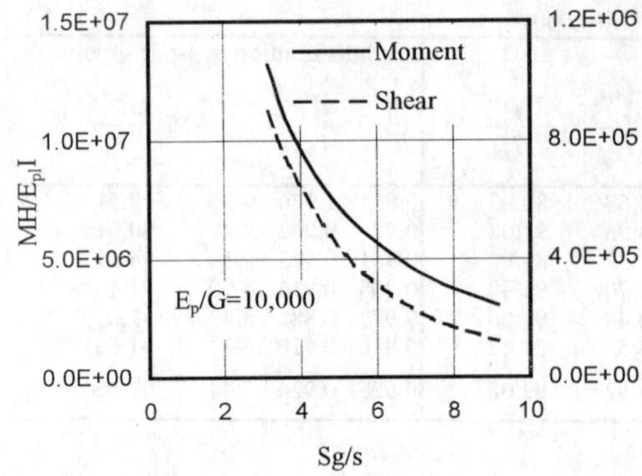

Figure 4. Normalized bending moment and shearing force developed at the beam end connected to 9-pile group

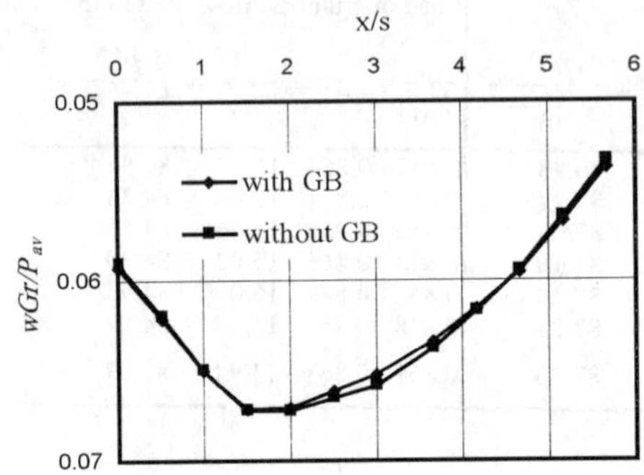

Figure 5. Displacement profile along the (X) axis of the two groups (E_p/G=1000, L_{GB}/s=0.67)

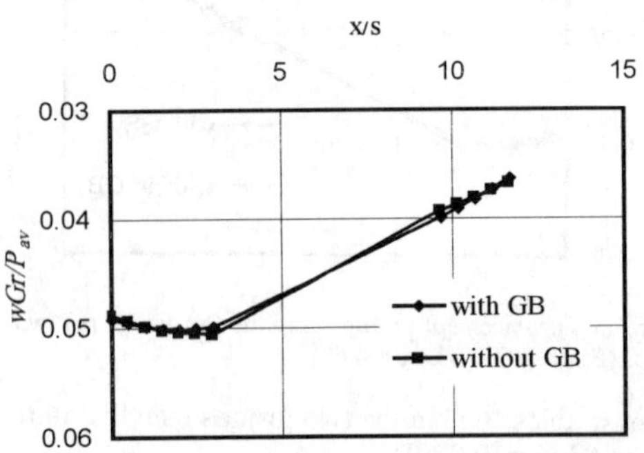

Figure 6. Displacement profile along the (X) axis of the two groups (E_p/G=1000, L_{GB}/s=6.67)

piles resulted in turn in a non-uniform displacement profile.

The developed bending moment and shearing force at the ground beam end connected to the nine-pile group is illustrated in Figure 7. The values of both shear and moment are relatively low because of

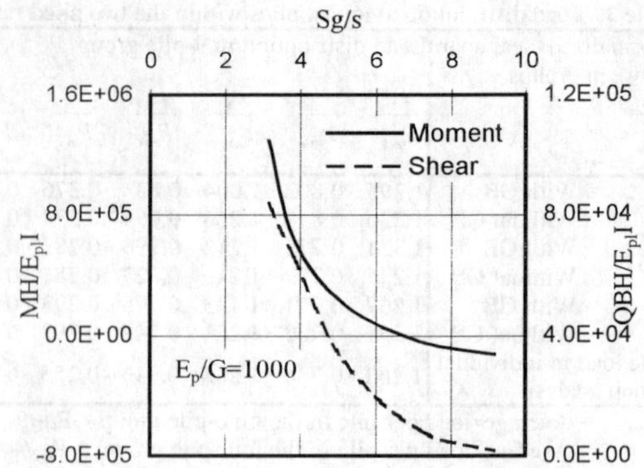

Figure 7. Normalized bending moment and shearing force developed at the beam end connected to 9-pile group

the apparent decrease in the value of the differential settlement between the two ends of the beam.

5 CONCLUSIONS

A hybrid approach was used for the analysis of a more practical situation in the piled foundation. The analysis reveals a considerable alteration in the load distribution amongst piles within the groups when analyzed together than that obtained in individual groups analysis, for relatively small spacing between the groups. Considerably high values of bending moments and shearing forces were developed at ground beam ends, in the case of closely spaced piled rafts in relatively soft clay. Care should be taken when designing the ground beam and the pile cap to provide the stiffness required to minimize the differential settlement between the groups. The analysis also indicates that soil stiffness, cap thickness, load configuration, and spacing between groups of piles are significant factors that considerably affect the load distribution and the displacement profile of the piles and the raft.

REFERENCES

Abdrabbo, F.M. & El Shamy, U.S.1997. Effect of pile cap flexibility on the response of axially loaded pile groups. *Proceedings of Al-Azhar Engineering 5th International Conference*, Al-Azhar University, Cairo, Egypt.

Chow, Y.K.1986. Analysis of vertically loaded pile groups. *International Journal for Numerical and Analytical Methods in Geomechanics*, 10:59-72.

Clancy, P. & Randolph, M.F. 1996. Simple design tools for piled raft foundations. *Geotechnique*, London, England, 46:2, 313-328.

Clancy, P.& Randolph, M.F. 1993. An approximate analysis procedure for piled raft foundations. *International Journal for Numerical and Analytical Methods in Geomechanics*, 17:849-869.

Giroud, J.P. 1968. Settlement of linearly loaded rectangular area. *Journal of Soil Mechanics and Foundation Engineering Division, ASCE,* 94(SM4):813-832, 1968.

Griffiths, D.V., Clancy, P & Randolph, M.F. 1991. Piled raft foundation analysis by finite elements. *Proceedings of the 7th international conference on computer Methods and Advances in geomechanics*, Beer, Booker, and Carter (eds), Balkema, Rotterdam, 1153-1157.

Lee, C.Y. 1993. Pile group settlement analysis by hybrid layer approach. *Journal of the Geotechnical Engineering Division, ASCE*, 119(GT9):1449-1461.

Mindlin, R.D. 1936. Force at a point in the interior of a semi-infinite solid. *Physics*, 7:195-202.

Poulos, H.G. & Davis, E.H. 1980. *Pile Foundation Analysis and Design*. Wiley, New York.

Randolph, M.F. & Wroth, C.P. 1978. Analysis of the deformation of vertically loaded piles. *Journal of the Geotechnical Engineering Division, ASCE*, 104(GT12): 1465-1488.

Randolph, M.F. & Wroth, C.P.1979. An analysis of the vertical deformation of pile groups. *Geotechnique*, London, England, 29:4, 423-439.

Smith, I.M. & Griffiths, D.V. 1988. *Programming the Finite Element Method.* John Wiley and Sons, Chichester, New York, 2nd edition.

Computer Methods and Advances in Geomechanics, Desai et al. (eds) © 2001 Balkema, Rotterdam, ISBN 90 5809 183 X

Differential settlement by soil-pile-mat interaction analysis

K.H.Chang & M.J.Wu
Shannon and Wilson Incorporated, Seattle, Wash., USA

M.Nakhjiri
Boeing Commercial Airplane Group, Renton, Wash., USA

E.D.Loesch
Skilling Ward Magnusson Barkshire Incorporated, Seattle, Wash., USA

ABSTRACT: This paper presents a case history of estimating instantaneous differential settlements for a pile supported concrete mat machine foundation that requires a settlement of less than .05mm and rotation less than 1.5 in 100,000 during operation. A set of in situ, large-strain crosshole shear wave velocity data was utilized as the basis of soil modulus assessment. A Winkler-Boussinesq model was used to simulate the settlements of the pile-soil complex. The foundation mat was modeled as a series of cross-beams connected only at the centers of elements whose contact pressures, settlements, moments, and shears can be obtained by writing and solving a series of finite difference equations based on the basic theory of flexure through a computer program.

1 PROJECT DESCRIPTION

The Boeing FIAT #2 & #3 Machine Foundation System (1994) in Renton, Washington, consists of two similar pile-supported 14 m x 22 m concrete mats that has two settlement criteria for its successful operation. The first criterion is that the transverse and longitudinal rotations should be less than 1.5 in 100,000 during the longitudinal movement of a machine (Aeroflex) weighing 258 kN moving along a central track on the mat. The second criterion is that the settlement of any point within the footprint of the Universal Holding Fixture (UHF) relative to a stationary Aeroflex should be less than 0.05 mm due to a horizontally moving line load in the UHF. This horizontally moving line load, named the "Vertical", weighs 312 kN and is evenly distributed along a distance of 5.3 m, moving longitudinally between the east and west end of the UHF (Fig. 1).

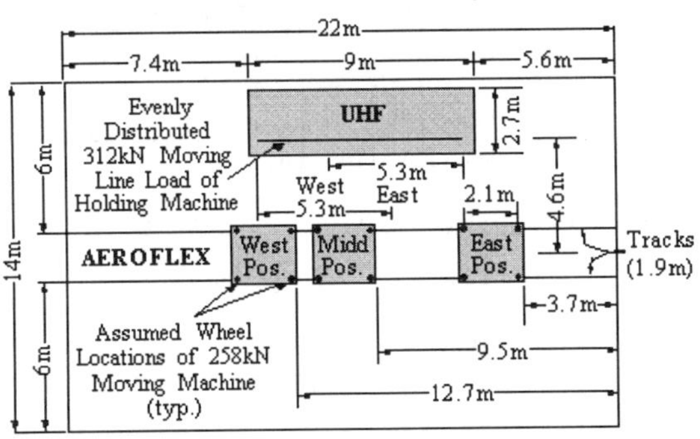

Figure 1. General Plan of Project

Settlement analyses were required to verify whether the settlement criteria could be met for three positions of the Aeroflex and two positions of the moving "Vertical" of the UHF (Fig. 1).

2 SOIL AND IN SITU SHEAR WAVE VELOCITY DATA

The site is underlain by a 2.1 to 3.0 m layer of medium to dense sand fill at the surface overlying soft to medium stiff silt and fibrous peat down to a depth of 10.4 to 11.9 m. Below this is a layer of medium dense to very dense sandy gravel and gravelly sand, about 4.3 to 5.2 m thick, followed by alternating layers of stiff to very stiff fibrous peat, stiff to soft clayey silt, and dense to very dense sand (Table 1).

In situ shear wave velocity data of each soil layer were obtained through regression analyses of large strain cross-hole shear-wave velocity test data obtained from another project in soil with the same geological origin and age. This shear wave velocity data was first grouped into 1.5 m intervals with a corresponding depth (D), standard penetration blowcount per 30 cm (N), measured maximum shear-wave velocity (Vs), and calculated corresponding minimum shear stress (Tv). Regression analyses were then performed between different groups to obtain regression equations for Vs and Tv in terms of D and N. These equations were used to calculate values of Vs and Tv for each soil layer as shown in Table 1. The actual shear stress under the design load for each layer was then checked to see whether they exceed the calculated Tv for that layer and whether the maximum elastic modulus for that layer needs any reduction (Zhang et al. 1981). Using Boussinesq stresses and the octahedral shearing stress as a measure of the actual shearing stress under the design load, it was found, however, that the actual shearing stress under the design load was always less than the shearing stress corresponding to the shear wave velocity (Tv) so that there was no need to reduce modulus obtained from the shear wave velocity data under the design load in this case.

Table 1. Soil and In-Situ Test Data.

Soil Layer Top Depth	Soil Type	SPT Blow-count N	Max. Shear Wave Veloc. Vs*	Max. Elastic Mod. Ev**	Min. Shear Stress Tv*
m		b/30cm	m/sec	MPa	kPa
0	Sand Fill	70	314	552	0.78
2.1	Silt/Peat	2	112	71	0.33
5.2	Silt/Peat	19	175	171	0.56
6.7	Silt/Peat	3	125	88	0.60
11.9	Sand/Gravel	81	314	552	1.62
16.6	Peat	11	166	155	1.14
18.6	Sand	43	279	437	1.54
19.8	Sand	30	237	314	1.49
21.3	Peat	15	188	197	1.43
22.9	Sand	84	314	552	2.31
29.3	Peat	17	209	245	1.94
32.3	Clayey Silt	2	162	147	1.96
35.1	Sand	51	314	552	2.54

* Used regression equations from large-strain, crosshole geophysical tests of SR-509 (Shannon & Wilson, 1992):

Vs(m/s) = 99.4 + 1.66D(m) + 3.44N(b/30cm) (n=27; r=0.79)

Tv(kPa) = 0.047 + 0.056D(m) + 0.00961N(b/30cm) (n=27; r=0.835)

Where: Vs(m/s) = 197(100-314)
D(m) = 20(1.5-40)
N(b/30cm) = 19(3-51)
Tv(kPa) = 1.37(0.08-2.59)

** Take Ev(MPa) = 2.8Gv(MPa) = $0.0056Vs^2$(m/sec)

3 THE WINKLER-BOUSSINESQ MODEL & THE MODIFIED FINITE DIFFERENCE CROSS-BEAM PROGRAM

A Winkler-Boussinesq model of independent springs resting on an elastic continuum was adopted to simulate the settlement behavior of the soil-pile complex under the concrete mat machine foundation. An existing finite difference, cross-beam computer program (Zhang et al. 1984) was modified to accommodate this model. This program assumes that the mat is divided into a series of cross-beams, where all loads, contact pressures, settlements, moments and shears occur at the intersections of the beams, named as nodes. With this simplified assumption, the contact pressure, settlement, and moment/shear can be obtained through a series of finite difference equations based on the basic theory of flexure of beams.

The relation between the contact pressures and the settlements was expressed by following matrix form which takes into account the shortening of piles:

$$\begin{pmatrix} w1 \\ w2 \\ w3 \\ \vdots \\ wn \end{pmatrix} = \begin{pmatrix} d11+k1 & d12 & d13 & \cdots & d1n \\ d21 & d22+k2 & d23 & \cdots & d2n \\ d31 & d32 & d33+k3 & \cdots & d3n \\ \vdots & \vdots & \vdots & \vdots & \vdots \\ dn1 & dn2 & dn3 & \cdots & dnn+kn \end{pmatrix} \begin{pmatrix} p1 \\ p2 \\ p3 \\ \vdots \\ pn \end{pmatrix} + \begin{pmatrix} w01 \\ w02 \\ w03 \\ \vdots \\ w0n \end{pmatrix} \quad (1)$$

where wi = settlement of node i; dij = settlement of soil below pile under node i due to a unit pressure on node j; ki = shortening of pile under node i due to unit pressure on node i; pi = contact pressure under node i; w0i = settlement of node i due to external influences; n = total number of nodes.

The spring constant of each pile was taken as the load to shorten the pile a unit length assuming a full load at the pile top to zero at the bottom (triangular distribution). Boussinesq stresses and elastic assumptions were used in assessing the settlement of soil below the pile, assuming that the foundation is at a depth of 2/3 effective pile length.

4 RESULT OF ESTIMATION

The above method was first used to estimate the settlements of a preliminary design of 1.4-m thick concrete mat supported by 84 augercast concrete piles (35-cm diameter, 20 m deep). It was found that the rotations perpendicular to the motion of the machine cannot meet the required criterion of 1.5 in 100,000. Subsequent estimations were made on a trial and error basis until the settlement criterion was met with a 1.7 m thick concrete mat and 53 augercast piles (35 cm diameter, 25 m deep). The instantaneous settlements estimated for the final design are shown in Tables 2 and 3 and Fig. 2 through Fig.7 for the various positions of the Aeroflex and the "Vertical".

Table 2. Deformations due to Weight of AEROFLEX Machine.

Location	Max. Settlement (0.001mm)	Rotation (radians)* Longitudinal Wheel	Max. Transverse
East End of UHF	105	3.9×10^{-6}	1.1×10^{-5}
Middle of UHF	101	1.4×10^{-6}	1.25×10^{-5}
West End of UHF	101	0.93×10^{-6}	1.3×10^{-5}

* Allowable Rotation is 1.5×10^{-5} radians

Table 3. Deformations due to Weight of "Vertical" of UHF.

Location	Max. Settlement (0.001mm)	Rotation (radians)* Longitudinal UHF	Transverse UHF
West End	200	1.87×10^{-5}	1.63×10^{-5}
East End	198	2.85×10^{-5}	1.63×10^{-5}
Differential	53*	1.69×10^{-5}	0.4×10^{-5}

* Along working surface. Allowable Differential = .051mm

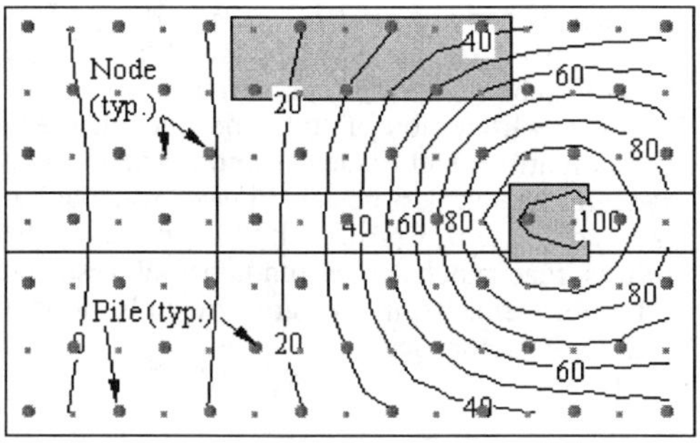

Figure 2. Estimated Settlement due to AEROFLEX at East End of UHF (in units of 0.001mm).

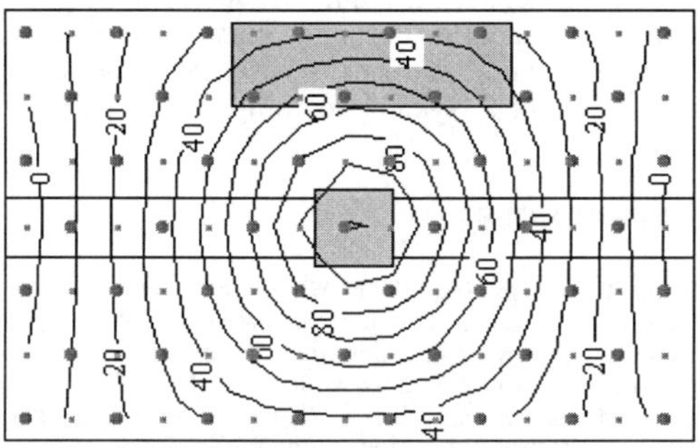

Figure 3. Estimated Settlement due to AEROFLEX at Middle of UHF (in units of 0.001mm).

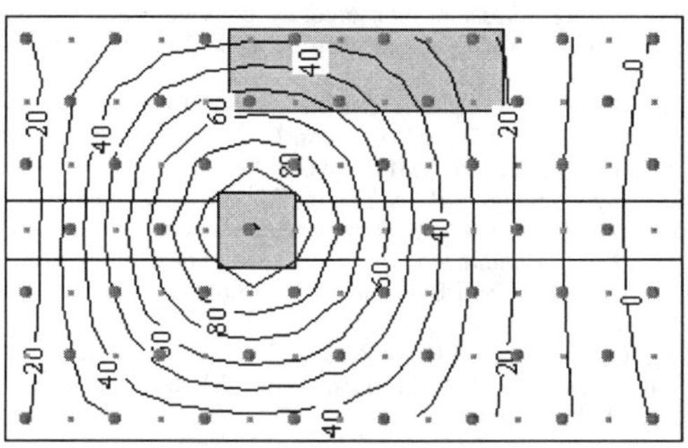

Figure 4. Estimated Settlement due to AEROFLEX at West End of UHF (in units of 0.001mm).

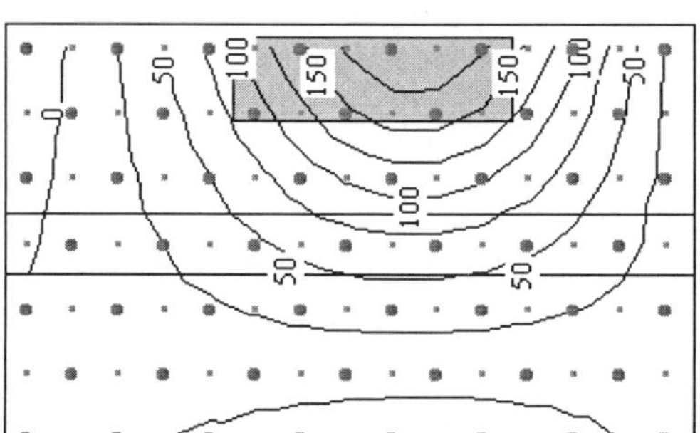

Figure 5. Estimated Settlement due to Vertical at East End of UHF (in units of 0.001mm).

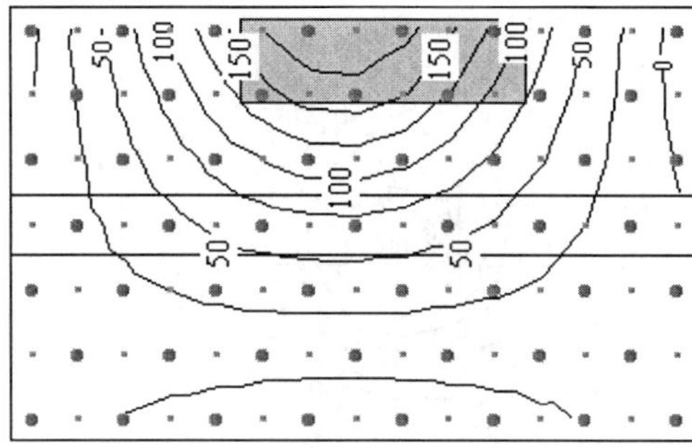

Figure 6. Estimated Settlement due to Vertical at West End of UHF (in units of 0.001mm).

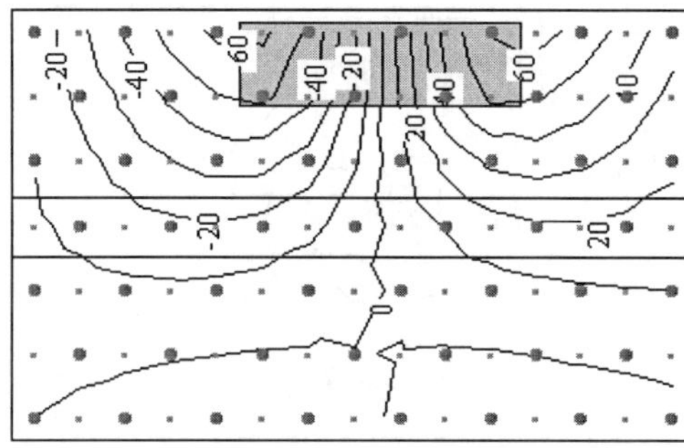

Figure 7. Estimated Differential Settlement due to Moving Vertical of UHF (in units of 0.001mm).

In order to get an idea of how much of the total settlement was contributed respectively by the pile and soil, the case of the Aeroflex located at the middle of UHF was analyzed. First, the load on the piles was calculated as shown on Fig. 8. Then the corresponding shortenings of piles was calculated (Fig. 9) by dividing the load on pile by the spring constant of the pile. The settlements due to soil only were obtained by subtracting the shortening of the pile (Fig. 9) from the total settlements (Fig. 3). It was found that the percentage of settlement contributed by pile varies from about 75 percent under the loads to 0 percent away from the loads.

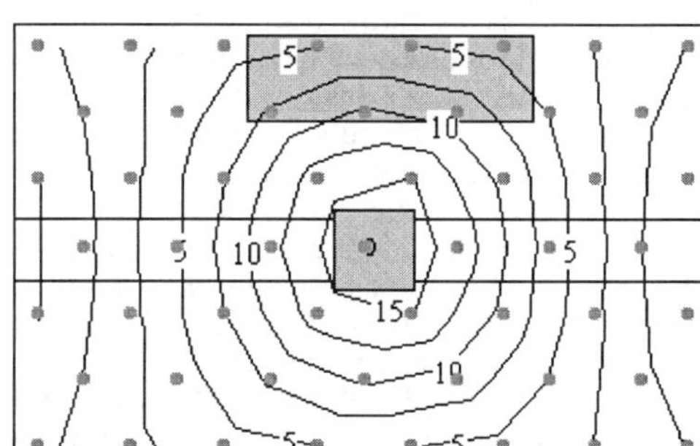

Figure 8. Estimated Load on Piles with AEROFLEX at Middle of UHF (in units of kN).

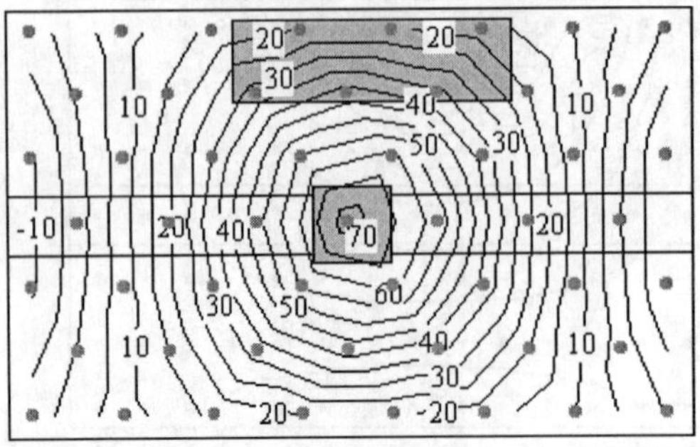

Figure 9. Estimated settlement due to Piles with AEROFLEX at Middle of UHF (in units of 0.001mm).

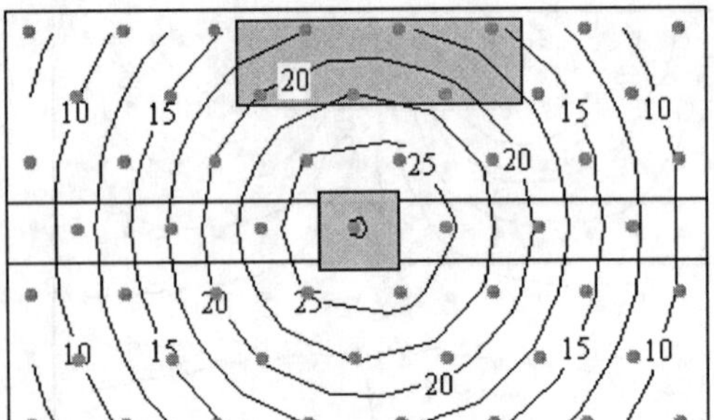

Figure 10. Estimated Settlement due to Soil with AEROFLEX at Middle of UHF (in units of 0.001mm).

The review of the monitored settlement data indicates that the actual settlements are within the range estimated.

5 CONCLUSIONS

The following conclusions were developed from this study:

1) Modern foundations for such structures as earth-satellite stations and high precision automatic assembly line moving machinery sometimes require estimation of settlements accurate to hundreds of millimeters and rotations accurate to a hundred thousands during operation that poise as a challenge to the geotechnical profession. This paper presents a case history of how this was achieved by using in situ shear wave velocity as a basis of soil modulus assessment and a Winkler-Boussinesq finite difference cross-beam model to simulate the soil-pile-mat interaction.
2) The advent of earthquake engineering provides the geotechnical profession with an abundant source of in situ shear wave velocity data of soil (Troncoso et al. 1977, Woods 1985) which can be used in collaboration with other field and laboratory data as a basis of static modulus assessment (Zhang et al. 1981, 1984, 1987). This paper showed another way of using it to estimate soil deformation at very small strain.
3) Due to the limitation of time and budget, many approximations and crude assumptions had to be made on the simplest possible but reasonable basis to complete this project. It is the belief of the authors that much more fundamental research and improvements in the area of soil-pile-mat interaction are needed.

6 ACKNOWLEDGEMENTS

The authors like to express their gratitude to the Boeing Commercial Airplane Group, Skilling Ward Magnusson Barkshire Inc. and Shannon & Wilson, Inc. for their permission to release the information and the help of M. Nykamp to complete this paper.

7 REFERENCES

Shannon & Wilson, Inc. 1992. *SR-509 East-West Corridor Project.* Geotechnical Report prepared for Washington State Department of Transportation, Tacoma, Washington

Troncoso, J.H., Brown, F.R., & Miller, R.P. 1977. *In Situ Impulse Measurements of Shear Modulus of Soils as a Function of Strain.* Proceedings, 6th World Conference on Earthquake Engineering, (3), India.

Woods, R.D., editor 1985. *Measurement and Use of Shear Wave Velocity for Evaluation Dynamic Soil Properties.* Proceedings, Session sponsored by the Geotechnical Eng. Divisions of ASCE in conjunction with ASCE Convention in Denver, Colorado.

Zhang, G.X., Yamane, G., & Wu, M.J. 1987. *A Comprehensive Analysis of SBPT for Ko in Stiff Soils.* Proceedings, 8th Asian Conference of SMFE, Kyoto, Japan.

Zhang, G.X., Zhang, N.R., & Zhang, F.L. 1984. *A Comparison of some Predicted and Observed Settlement Behaviors of Box-Type Foundations in Beijing.* Proc.Third International Conference on Tall Buildings, Hong Kong and Guan Zhou, China.

Zhang, G.X., Zhang, N.R., & Zhang, F.L 1981. *Nonlinear Differential Settlement Analysis.* Proc. XICSMFE, Vol. 2, Stockholm, Sweden.

Computer Methods and Advances in Geomechanics, Desai et al. (eds) © 2001 Balkema, Rotterdam, ISBN 90 5809 183 X

Laboratory testing for study on the interaction between granular column and surrounding soil

Wenfeng Huang & Keizo Ugai
Department of Civil Engineering, Gunma University, Kiryu, Japan

Guangxin Li
Department of Hydraulic Engineering, Tsinghua University, Beijing, People's Republic of China

ABSTRACT: Granular columns as reinforcements have been more and more used to increase the bearing capacity and improve the settlement response of the foundation soil. However, the mechanism of the cooperation of the reinforcement and surrounding soil is still not clear enough for engineering use. This paper presents a method of laboratory testing to study the interaction between reinforcement and surrounding soil. A simple numerical method is also presented to analyze the interaction and is supported by the laboratory testing. The results can provide a deeper understanding of the work of composite ground with granular columns.

Four types of conventional and modified triaxial tests have been performed on gravel-clay specimens under constant confining pressure to get the relationships between the two materials, so that they can describe the interaction between the reinforcement and surrounding soil in details. The results show that the radial stress σ_r of reinforcement is significantly larger than the cell pressure while the foundational clay is weakened a little. According to the numerical analysis, this phenomenon is caused by the dilatancy of gravel. As a result, the gravel column weakens the foundational clay while the foundational clay strengthens the gravel. But for the composite specimen, the strength was increased. The results also show that it is necessary for surrounding soil to provide enough confining pressure for the granular reinforcements so as to increase the strength of the composite ground with gravel columns.

1 INTRODUCTION

Granular column (stone or gravel column) is one of the popular types of foundation treatment all over the world. By model test of composite sample with a single column, Hughes and Withers (1974) found that the settlement and failure of stone column mainly result from distension of upper part of the pile. By testing on composite specimens with cemented stone column in a modified triaxial apparatus, Juran and Riccobono (1991) studied the cementation effect on the performance of the granular column.

Balaam and Poulos (1983) presented a FEM method to calculate the settlement of the composite ground of stone column, in which a section of single column with effective foundation soil was considered as an unit cell. By pressuremeter testing, Hughes and Withers (1975) presented a settlement calculation method in which the volume of stone column is supposed to be constant and the settlement of the column comes from its distension. Cai (1992) suggested that the settlement of single column be calculated from the pressuremeter test results of foundation soil and triaxial test results of stone specimen. By analyzing the relationship between axial strain ε_1 and ratio of axial stress to confining pressure σ_1 / σ_3 together with the relationship between ε_1 and volume strain ε_v, Poorooshasb and Meyerhof (1997) presented an expression of load-settlement relationship of composite ground with stone column.

However, the mechanism of the cooperation of the reinforcement and surrounding soil is still not clear enough for engineering use. This paper presents a method of laboratory testing to study the interaction between reinforcement and surrounding soil. A simple numerical method is also presented in this paper to analyze the interaction. The results can provide a deeper understanding of the work of composite ground with granular columns.

The test results show that the radial stress σ_r of granular reinforcement is significantly larger than the cell pressure; this phenomenon is caused by the dilatancy of gravel. Consequently, the gravel weakened the foundational clay while the foundational clay strengthened the gravel. But for the composite specimen, the global strength increased. The results also show that it is necessary for surrounding soil to provide enough confining pressure for the granular reinforcements so as to increase the strength of the composite ground with gravel columns.

2 LABORATORY TESTING

Test A, B, C and D have been performed on gravel-clay specimens and specimens of the two materials separately. In these tests, the diameter of the specimens (D_0 or $2b$) is 100mm, the diameter of column (if there is, d_0 or $2a$) is 39.1mm, and the height is 200mm. The details of these tests are:

Test A: Conventional triaxial tests to get parameters of gravel and foundational clay respectively. Baihepu Clay and one kind of gravel were adopted in these tests. The parameters are shown in Tabale 1.

Table 1. Parameters of Duncan-Chang *E-v* Model of gravel and foundational Clay

	Gravel	Paihepu Clay
c (kPa)	0	44.5
φ (°)	44.0	28.1
K	460	190
n	0.53	0.23
R_f	0.81	0.89
G	0.756	0.320
F	0.375	0.230
D	0.0264	0.0074

Test B: Modified triaxial tests on composite specimens of gravel and foundational clay to get the volumetric change of the gravel column. The apparatus is shown in Figure 1a and the loading cap and membrane are shown in Figure 1c.

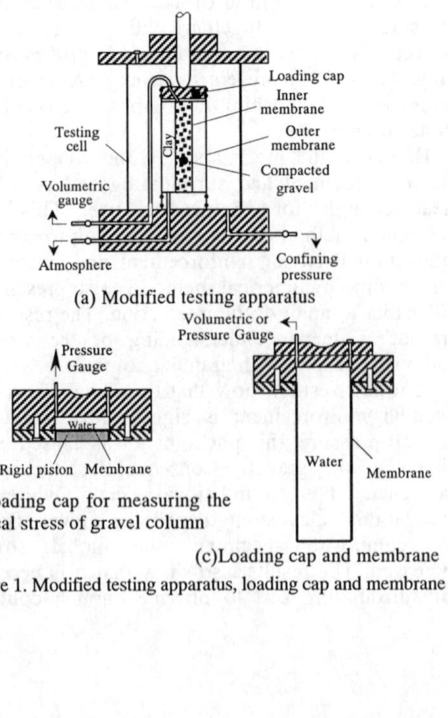

Figure 1. Modified testing apparatus, loading cap and membrane

Test C: Modified triaxial tests on hollow specimens of foundational clay to get the radial stress σ_r on the interface by inputting the same volumetric change got from Test B. The apparatus, the loading cap and the membrane are the same as Test B and the results are shown as the curve 1, 2 and 3 in Figure 2.

Test D: Modified triaxial tests on composite specimens of gravel and foundational Clay to measure directly the vertical load ratio of gravel column to foundational Clay. The loading cap is shown in Figure 1b and the results are shown in Figure 3 and Figure 4. In Figure 4, σ_p, σ_s are axial stress of reinforcement and foundational soil respectively.

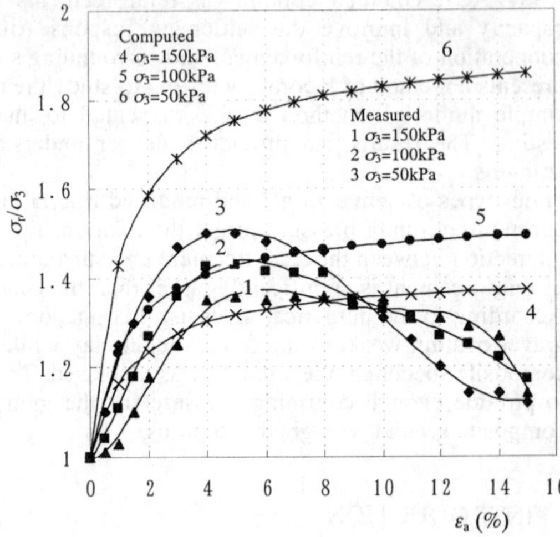

Figure 2. Radial contacting stress on the interface between gravel column and surrounding soil

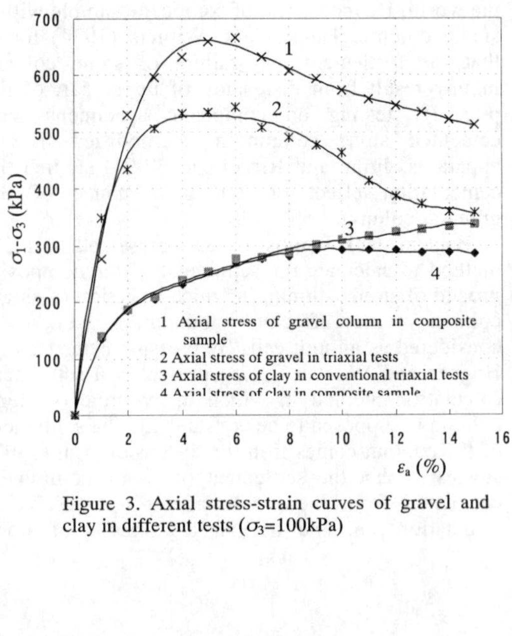

Figure 3. Axial stress-strain curves of gravel and clay in different tests (σ_3=100kPa)

From curve 1, 2, and 3 in Figure 2, it can be seen that the radial contact stress σ_r on interface is from 20% to 50% over the cell pressure σ_3. This means that the confining pressure, which is also the minimum principal stress of the gravel column is increased by the surrounding soil by from 20% to 50%, it also means the reinforcement is strengthened by the surrounding soil. The results were got from Test C and D (see Figure 3). Figure 3 shows that because of the strengthening, the axial stress reaction of gravel column becomes higher than that without surrounding soil. But on the other hand, the surrounding soil is weakened a little by the reinforcement as shown in Figure 3.

Curve 3 in Figure 4 shows that this method to measure the axial stress of reinforcement to surrounding soil is successful. Figure 2 and Figure 4 show that around 4% of axial strain ε_a, the reinforcement is most highly confined and the effectiveness is also the best. This means that the confining is necessary for the granular reinforcement and should be assured in engineering practices.

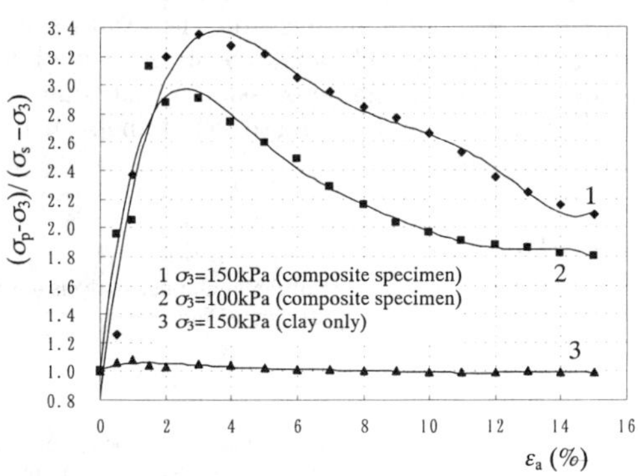

Figure 4. Stress ratio of gravel column to surrounding soil under vertical loading

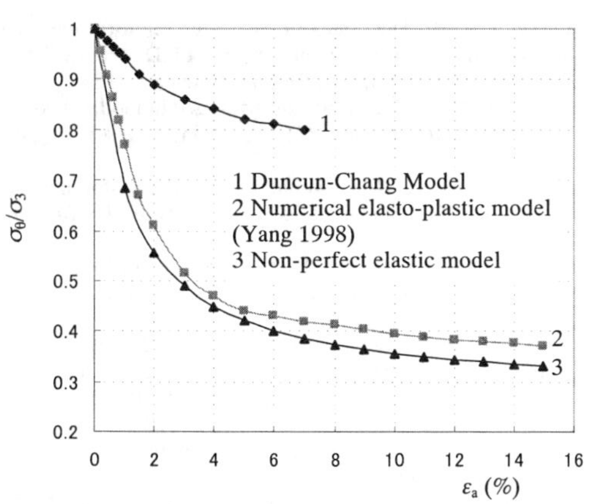

Figure 5. Computed Circumferential stress of the interface between gravel column and surrounding soil (σ_3=100kPa)

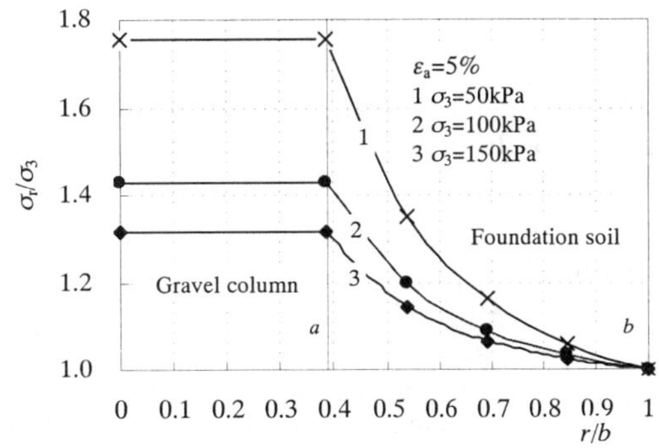

Figure 6. The radial stress distribution along with the radius

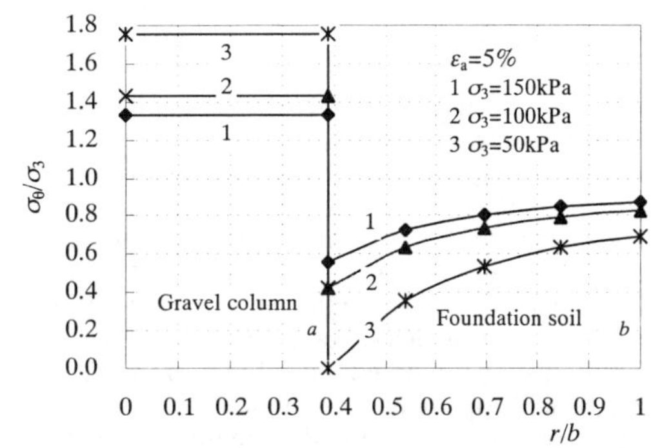

Figure 7 The circumferential stress distribution along with the radius

3 NUMERICAL ANALYSIS

It is well known that elastic model cannot deal with such dilatant materials as gravel and so on. Some elasto-plastic models, for example, Tsinghua Model (Huang 1979) can do so if all the parameters are available. But in most cases, parameters of Duncan-Chang Model are much easier to obtained. Based on Duncan-Chang Model, a simple method called Non-perfect Elastic Model was presented to analyze numerically the stress and strain reaction of specimens with circular symmetry due to axially loading (Huang 1999).

In Non-perfect Elastic Model, lateral deformation coefficient ν_L and tangent Poisson's ratio ν_t (confining pressure being constant) are defined respectively as:

$$\nu_L = -\frac{d\varepsilon_r}{d\varepsilon_a} \tag{1}$$

$$\nu_t = -\frac{\partial\varepsilon_r}{\partial\varepsilon_a}\bigg|_{\sigma_r=\mathrm{const}} \tag{2}$$

According to the definition of complete differentiation, the lateral deformation coefficient ν_L can be written as

$$\nu_L = -\frac{\partial \varepsilon_r}{\partial \varepsilon_a}\Big|_{\sigma_r} - \frac{\partial \varepsilon_r}{\partial \sigma_r}\Big|_{\varepsilon_a} \cdot \frac{d\sigma_r}{d\varepsilon_a} \quad (3)$$

The variation of ε_r adapt to σ_r being assumed as elastic, it can be easily got from elastic theories that

$$\frac{\partial \varepsilon_r}{\partial \sigma_r}\Big|_{\varepsilon_a} = \frac{\nu_0}{\lambda} \quad (4)$$

Then,

$$\nu_L = \nu_t - \frac{\nu_0}{\lambda} \cdot \frac{d\sigma_r}{d\varepsilon_a} \quad (5)$$

where: ε_r, ε_a are radial strain and axial strain separately; σ_r is radial stress, ν_0 is perfect elastic Poisson's ratio (constant, here ν_0 is assumed to be 0.3); ν_t can be worked out with Duncan-Chang *E*-ν Model; λ is *Lamé*'s modulus as:

$$\lambda = \frac{E\nu_0}{(1+\nu_0)(1-2\nu_0)} \quad (6)$$

and E is Young's modulus, which can be worked out from Duncan-Chang Model.

At each stage of strain-controlled loading ($\Delta\varepsilon_a$ is known), assume ν_L a certain value. From equation 1, $\Delta\varepsilon_r$ is available. From Duncan-Chang Model, λ, ν_t, $\Delta\sigma_r$ can be predicted. Then a new value of ν_L can be got from equation 3. Iterate these steps until ν_L is stable, the stress and strain reaction under axial loading can be worked out. Even if $\nu_t > 0.5$, no damage will happen to the computation.

The circumferential stress of the interface between gravel column and foundational soil is show in Figure 5. It can be seen that the circumferential stress is much less than the cell pressure. This means that because of the bulbing of gravel column due to loading, the foundational soil is weakened. This result is quite close to that of a numerical elasto-plastic method (Yang 1998) while the computation using Duncan-Chang Model stopped when ν_t near 0.5.

The radial stress on the interface is shown in Figure 2. Until 5% of ε_a, the maximum error is not beyond 30%. Although this method cannot deal with the softening, which causes that the error becomes larger when $\varepsilon_a > 5\%$, it can help study the mechanism of the interaction between granular reinforcements and surrounding soil.

The radial stress and the circumferential stress along with the radius are shown in Figure 6 and Figure 7. The two figures show clearly the interaction between gravel column and foundation soil. Along with the radius from core to out face, the radial and circumferential stress of column keeps constant and significantly higher than cell pressure, while the radial stress of foundation soil declines from that high value to cell pressure; that of foundation soil, which is also minimum principle stress, is much lower than cell pressure in the interface. It can be also say that foundation soil strengthens the gravel column by weakening itself a little. But for the composite specimen, the strength increases. The figures also show that the interaction between the two materials is sensitive to the confining pressure of the composite specimen. The lower the cell pressure is, the more the strengthening of column and weakening of foundation soil are.

4 CONCLUSIONS

In this paper, a new laboratory method is presented to study the interaction between granular reinforcements and foundational soil. A simple numerical method is also presented to analyze the strain and stress reaction of circularly symmetric specimen due to axial loading.

The test results show that under axial loading, the radial stress σ_r of granular reinforcements is significantly larger than the cell pressure while the foundational clay is weakened a little. The numerical analysis shows that the strengthening of reinforcements and the weakening of foundational clay comes from the dilatancy of gravel column and sensitive to the confining pressure. The results show that the surrounding soil should be strong enough to provide sufficient confining pressure for the granular reinforcements so as to increase the strength of the composite ground.

ACKNOWLEDGEMENT

All the laboratory tests and most of the numerical analysis were finished in Tsinghua University, Beijing, China.

REFFERENCES

Balaam, N.P. & H.G. Poulos 1983. The behavior of foundations supported by clay stabilized by stone columns. *Proc. of 8th ECSMFE*, 1

Cai, F. 1992. The study of granular piles reinforced with geosynthetics, Master Thesis (in Chinese), Tsinghua Univ., Beijing, China

Huang, W.F. 1999. The interaction between reinforcements and surrounding soil in composite ground, Ph.D. dissertation (in Chinese), Tsinghua Univ., Beijing, China

Huang, W.X. 1979. Elasto-plastic stress-strain model theory of soil. *Journal of Tsinghua University* (in Chinese), 19(1): 1-15

Hughes, J.M.O. & N.J. Withers 1974. Reinforcing soft cohesive soils with stone column. *Ground Engineering*. 7(3)

Hughes, J.M.O., N.J. Withers & D.A. Greenwood 1975. A field trial of the reinforcing effect of a stone column in soil. *Geotechnique*. 25(7)

Juran, I. & O. Riccobono 1991. Reinforced soft soil with artificially cemented compacted-sand column. *Journal of Geotechnical Engineering. ASCE*. 117(7): 1042-1060

Poorooshasb, H.B. & G.G. Meyerhof 1997. Analysis of behavior of stone columns and lime columns. *Computer and Geotechnics*. 20(1): 47-70

Yang, G.H. 1998. The mathematical theory for constitutive model of soils and its application. Ph.D. Dissertation (in Chinese), Tsinghua University, Beijing, China.

Computer Methods and Advances in Geomechanics, Desai et al. (eds) © 2001 Balkema, Rotterdam, ISBN 90 5809 183 X

A method for three-dimensional non-linear analysis of pile-soil interaction system in time domain

S. Ben Jamaa & H. Shiojiri
Department of Civil Engineering, Nihon University, Japan

ABSTRACT: Many methods have been developed for the analysis of the dynamic behavior of piles using simplified models where either non-linear behavior of the pile near field was neglected or a full non-linear analysis were conducted with considerable computational efforts. The proposed method has the objective of combining the merits of the above-mentioned methods. It presents a new numerical approach to analyze the pile-soil interaction system taking into account the effects of the behavior of the soil at infinity and at the same time accounting for the non-linearities existing in the soil adjacent to the pile, hence the time domain analysis.

1 INTRODUCTION

The effects of soil-structure interaction on the dynamic seismic response of buildings resting on embedded foundations were subjects of various studies and considerable efforts have been conducted to develop adequate analytical and numerical methods (Waas 1972; Seed and Idriss 1969). However, conceptual and computational difficulties still remain due primarily to the semi-infinite nature of the soil medium and to the embedding of the pile foundation. Indeed, it became a primary concern to develop the appropriate algorithm to express not only the energy generated by the pile-soil interaction but also the propagation of energy toward infinity. Currently there are basically two methods dealing with soil-structure interaction: the *Direct Approach* also referred to as the *complete finite element* approach, and the *Substructure approach* such as the *flexible volume method.*

In the Direct approach, the region of the soil containing the structure is modeled up to the artificial boundary . The latter must be introduced as it is impossible to cover the unbounded soil domain with a finite number of elements with bounded dimensions. Approximate boundary conditions are introduced so as to represent the semi-infinite half-space. Besides modeling the soil's stiffness up to infinity, reflections of the outwardly propagating waves on the artificial boundary should be avoided. Lysmer and Kuhlemeyer (Lysmer and Kuhlemeyer 1969) presented a special frequency independent viscous dashpot boundary on the interface nodes in all directions. This method uses finite element modeling of the core region thus provides a suitable way to study complex structures such as dams or nuclear reactors. However, while it accounts for the non-linearity of near-structure domain, a large number of degrees-of-freedom (d.o.f) arises due to the discretization of the soil region which leads to a significant computational effort even when dealing with moderate or small sized structures such as pile foundations. Moreover, this method does not account correctly for the infinity of the far field.

In the Substructure approach, the soil is modeled as a regular layered homogeneous half-space where a structure is located in the soil's core region. The finite region in the interior of the interaction horizon encompasses beside the structure itself, the geometric irregularities and material inhomogenities of the soil. The infinite half-space is investigated using for example the Thin Layer Element method (TLEM) and is normally formulated in the frequency domain. Although time domain formulation as well as approximate modifications of the TLEM were carried to account for the non-linear dynamic behavior of soil and/or fluids (Kausel 1994) or in analyzing simple-shaped foundations such as footing (Novak and Beredugo 1972; Aboul-Ella and Novak 1980), non-linear response of pile foundations embedded in infinite half-space could not be carried out and the substructure method of analysis remained basically used in linear analysis.

2 GENERAL OUTLINE OF THE PROPOSED METHOD

2.1 *Model Concept and general Assumptions*

The proposed method has the objective of combining the merits of the above mentioned methods and to develop an accurate and efficient approach to analyze the pile-soil system tacking into account the presence of near-pile non-linearities while considering the effects of soil behavior at infinity. As shown in Fig. 1 The whole system is partitioned by an artificial cylindrical energy-transmitting boundary. Displacements are assumed to vary linearly between the layer surfaces of the outer layered soil. The lower boundary of the system is assumed to be rigid and differs from the assumption of support on a homogeneous half space. However, this is of a little consequence since a half space may be simulated by placing the rigid boundary at large depth. The upper boundary of the layered zone is assumed to be the free surface.

A hybrid of Thin Layer Element and Finite Element methods are used to analyze the total soil-substructure system. As the regular unbounded soil on the exterior of the interaction horizon up to the infinity can be assumed reasonably to behave linearly, it can be analyzed in the frequency domain. The irregular zone is analyzed using non-linear FEM.

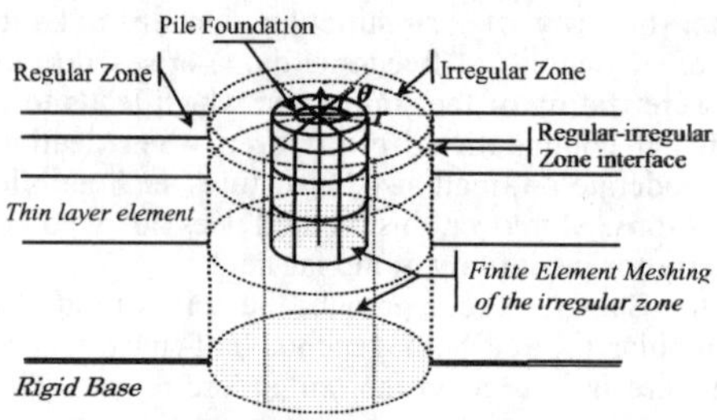

Figure 1: Model Concept of Analyzed System

2.2 *Schematic representation of this method*

One of the basic problems in modeling a semi-infinite domain is to provide boundary conditions that reproduces properly the behavior and interaction of the missing far-field. Here, the unbounded soil is represented by its dynamic stiffness or impedance matrix. The far field equation of motion is solved in frequency domain where the frequency dependency of the impedance coefficients can be readily considered (Wolf and Motosaka 1989a). It is well known, however, that structures are usually designed in such a way that a significant inelastic behavior can be anticipated in the vent of a severe ground motion or external dynamic forces. To compute the dynamic

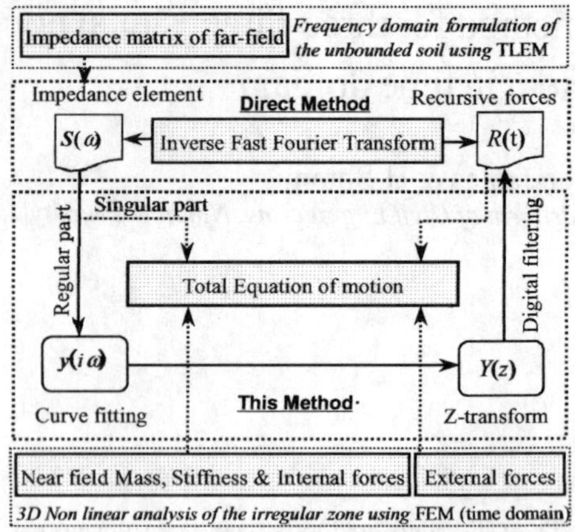

Figure 2: Flowchart of time domain derivation based on element impedance matrix in frequency domain

response in a non-linear range, analysis in the time domain is necessary. Hence the transformation of the impedance matrix to time domain is conducted.

The recursive method (Wolf and Motosaka 1989b; Wolf 1985) and the direct Inverse Fast Fourier Transform (Hayashi and Katakura 1990; Hayashi and Takahashi 1992) were used to formulate the boundary matrix representing the interaction forces representing the far-field's contributions to the total equation of motion in time domain. Fig. 2 illustrates two methods of evaluation of the nodal boundary forces R(t). The direct Inverse Fast Fourier Transform (IFFT) can be used effectively if linear analysis is conducted. However, in the case of non-linear analysis, it becomes necessary to perform IFFT at each time step which proved to be a task of considerable computational effort.

An alternative derivation based on the z-transform is used where each element of the unbounded soil's impedance matrix formulated in the frequency domain is approximated as a ratio of polynomials in $i\omega$. Applying recursive digital filtering, the boundary nodal forces which expresses the contribution of the far-field, are then obtained and added to the total equation of motion with the other components of the infinite domain's constitutive matrices.

3 INTERACTION FORCES IN TIME DOMAIN

3.1 *Impedance matrix*

The TLEM, which consists basically in a partial discretization of the wave equation, became an efficient tool for the dynamic (or static) analysis of systems with material properties that change in one coordinate direction. From the general solution of the wave equation, the equation of motion of the

n-layered far field can be written as (Shimizu et al. 1977):

$$([A]\,k^2 + i\,[B]\,k + [G] - \omega^2\,[M])\,\{\delta\} = 0 \qquad (1)$$

$\{\delta\}$ is a vector of $3n$ components containing the eigenvalues solutions of the above equation. The total force system acting on the regular-irregular boundary ($r = r_0$) can be expressed as:

$$\{P_o\} = [I_{mp}(\omega)]\,\{v_o\} \qquad (2)$$

Where $[I_{mp}(\omega)]$ is the impedance or dynamic stiffness matrix of $3n$ by $3n$ dimension and $\{v_o\}$ is the weighted eigenvectors obtained from equation (1). The detailed form of the matrices of equation (2) and equation (1) can be found in references (Shimizu et al. 1977; Ben-Jamaa 2000).

3.2 *Transformation to time domain*

Considering the fact that each element $s(\omega)$ of $[I_{mp}(\omega)]$ has an asymptotic behavior for high frequencies, a singular part can be determined whose transformation exist only in the sense of a distribution. The transformation from continuous domain to the discrete one would be applied on the remaining regular part:

$$S(\omega) = \underbrace{y(\omega)}_{\text{regular part}} + \underbrace{(k_s + i\omega c_s)}_{\text{singular part}} \qquad (3)$$

Using a curve fitting technique the regular part defined in equation (3) is approximated as ratio of polynomials in $i\omega$:

$$y(\omega) \approx y(i\omega) = \frac{P(i\omega)}{Q(i\omega)} = \frac{\sum_{k=0}^{N-1} p_k (i\omega)^k}{\sum_{k=0}^{N} q_k (i\omega)^k} \qquad (4)$$

The coefficients p_k and q_k are real. Fig. 3 shows the accuracy of the fitting using equally spaced 512 frequency data points. The agreement with the exact impedance coefficient improves as the order of polynomials N increases. Derivation of the regular part of the dynamic stiffness coefficient from frequency to time domain can be obtained using partial fraction form of equation (4). Fig. 4 shows that the proposed approximation is in good agreement with that of direct transformation using IFFT.

The impedance matrix of the energy-transmitting boundary can be transformed to the discrete time domain and thus will be treated as a representation of an ARMA filter. Applying z-transformation to equation (4):

$$y(z) = \frac{b_o + b_1 z^{-1} + \cdots + b_{N-1} z^{-(N-1)}}{1 + a_1 z^{-1} + \cdots + a_N z^{-N}} \qquad (5)$$

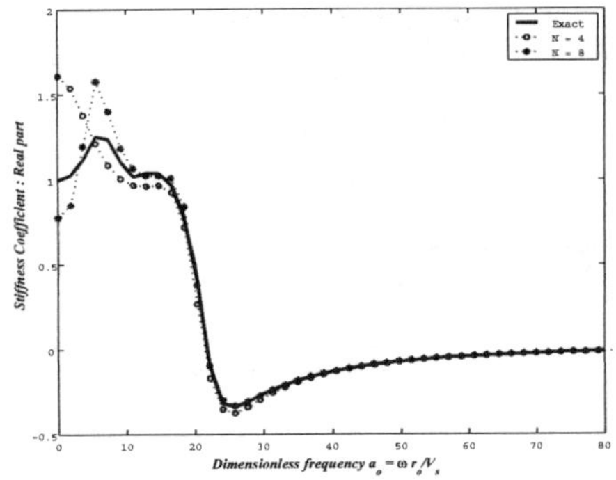

Figure 3: Curve fitting of $y(\omega)$

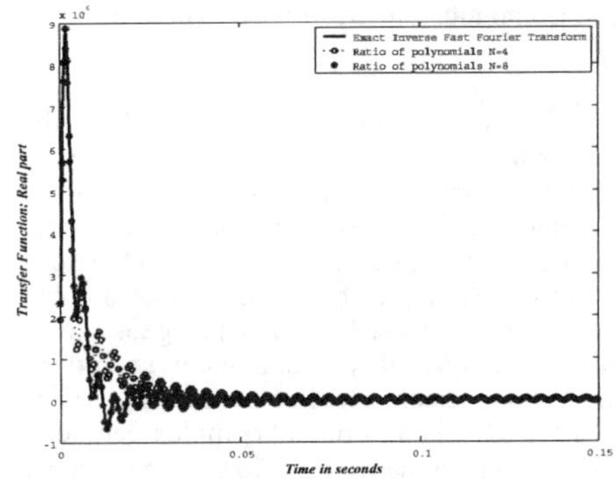

Figure 4: Time domain representation of $y(\omega)$

3.3 *Evaluation of the recursive forces*

The so-called recursive equation of interaction forces acting from the far-field can be written for all degree of freedom system as for time step n:

$$\{R\}_{r,n}^{l} = \sum_{i=1}^{m} R_{r,n}^{li} \qquad (6a)$$

$$\left[R_{r,n}^{lm}\right] = \sum_{i=1}^{N} a_{ik}^{lm} R_{r,n-i}^{lm} + \Delta t \sum_{i=0}^{N-1} a_{ik}^{lm} U_{n-i}^{m} \qquad (6b)$$

where m and l denote the dimension of $[I_{mp}(\omega)]$ and U_{n-i}^{m} is the displacement vector for $(n-i)^{th}$ time step. It is clear from Fig. 5 that using only N previous time steps leads to a great reduction of computational efforts compared to that of a complete IFFT transformation (Bernal and Youssef 1998; Darbre 1990)

4 NON-LINEAR PILE-SOIL INTERACTION

For the non-linear analysis a three dimensional formulation of the finite irregular zone becomes

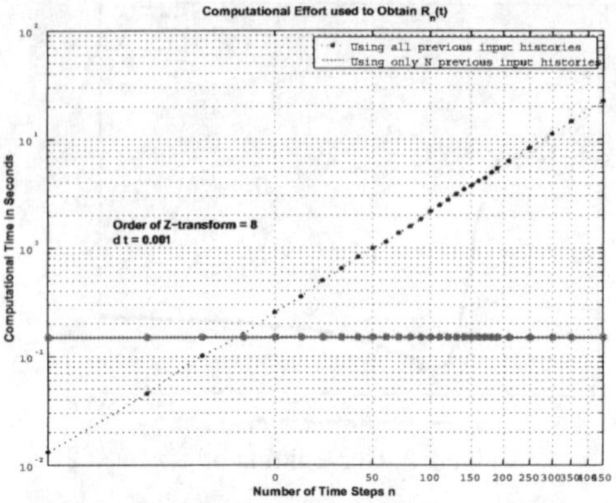

Figure 5: Computational effort used to obtain $R_n(t)$

necessary. The solid elements used in meshing the FE domain which are geometrically compatible with the transmitting element, require a cylindrical surface at one element side. Moreover, since the energy dissipation from the pile region to the unbounded layered soil is assumed to be through finite number of nodes at the interface boundary, geometrical and material compatibility requirement should be satisfied. Although such elements have been derived, they are not well suited to model complex geometry in the discretized domain. Therefore isoparametric brick elements have been adopted and FE analysis is conducted in the cartesian coordinates. Stiffness, damping and mass matrices as well as internal nodal forces should be then transformed to global cylindrical coordinates.

The non-linear solution technique employed here is based on the elimination of residual forces. It requires the maintenance of equilibrium during each load increment:

$$\{\Phi\} = \int_v [B]^T \{\sigma\} dv - \{P\} = 0 \tag{7}$$

In which $\{P\}$ contains all forces due to external loads, initial stresses, initial forces etc. σ is the actual stresses dependent on the strain level reached as governed by the yield criterion. $[B]$ is the strain-stress matrix for the FE domain.

The total incremental equation of motion for the non-linear analysis of pile-soil interaction can be written as:

$$[K^*]^{(i)} \delta u^{(i)} = \{P\}_{n+1} + \{R\}_n + [k_s] u_{n+1}^{(i-1)} - \{F\}_{n+1}^{(i-1)} + [M] \ddot{u}_{n+1}^{(i-1)} + [C^*] \dot{u}_{n+1}^{(i-1)} \tag{8}$$

where the superscript (i) denotes iteration number, $\{F\}_{n+1}^{(i-1)}$ is the internal nodal forces of the FE domain due to the state if the stress at previous iteration. The general equation of motion stated by equation (8), expresses the system response at time $t = (n+1)\Delta t$. The stability of the solution is guaranteed if adequate convergence criteria is used. For dynamic non-linear structural analysis, accumulation of solution errors can be particularly serious and can cause a rapid instability. The incremental effective stiffness matrix $[K^*]^{(i)}$ as well as displacement, velocity and acceleration vectors are obtained using Newmark's integration scheme and can be expressed as:

$$[K^*]^{(i)} = [K]_T^{(i)} + [k_s] - [b_o] + \frac{\gamma}{\beta \Delta t}[C^*] + \frac{1}{\beta \Delta t^2}[M] \tag{9}$$

$$\ddot{u}_{n+1}^{(i)} = \frac{1}{\beta \Delta t^2} \Delta u^{(i-1)} - \frac{1}{\beta \Delta t} \dot{u}_n - \left(\frac{1}{2\beta} - 1\right) \ddot{u}_n \tag{10}$$

$$\dot{u}_{n+1}^{(i)} = \dot{u}_n + \Delta t(1-\gamma)\ddot{u}_n + \gamma \Delta t \ddot{u}_{n+1}^{(i-1)} \tag{11}$$

$$\Delta u^{(i-1)} = \sum_{i=1}^{i-1} \delta u^{(k)} \tag{12}$$

$$u_{n+1}^{(i)} = u_n + \Delta u(i-1) \tag{13}$$

The damping matrix $[C^*]$ is expressed as:

$$[C^*] = [C] + [c_s] \tag{14}$$

$[K]_T^{(i)}$ of equation (9) is the tangential stiffness matrix computed after i^{th} correction of internal stresses of the FE zone. $[k_s]$ and $[c_s]$ contains all singular parts of the impedance matrix as defined by equation (3)and $[b_o]$ contains the z-transform coefficients b_o for each boundary node's d.o.f. All these matrices as well as the vector of recursive nodal forces $\{R\}_n$ are extended to the total d.o.f system and should be assembled over all wave numbers (or Fourier modes).

5 NUMERICAL IMPLEMENTATION

5.1 Analytical model

To demonstrate the capability of handling irregular geometry as well as non-linearity at the pile near-field, a case model is considered from experiments conducted in Funabashi Campus of Nihon university. It consists of a group of piles each having a radius R and depth L embedded in a stratum with depth H. The piles support a structure modeled as a concrete "cap" or platform where the exciter is placed on the middle to simulate the excitation force from the superstructure. Since only a single pile analysis is considered in this study, the model adopted is that of a single pile supporting a portion of the concrete slab and only translational motion in x direction is allowed to simulate the platform remaining part.

In the experiment a sinusoidal force is applied at the top edge of concrete cap and just in between the piles. In this example a force of the same magnitude is directly applied on the top-edge of the pile. Modes of Fourier transform are chosen equal to 0 and 1. Table (1) shows the geometrical and physical material properties of the pile used in this example.

Table 1: Physical and geometrical characteristics of the pile used in analysis

Pile characteristics	Value	Unit
Length	26.600	m
Diameter	0.4060	m
Thickness	0.0095	m
Section Modulus	0.1148	m^2
Mass density ρ	7444.1	kgf/m^3
Young's Modulus	2.3E+04	kgf/ms^2
Poison's ratio	0.3000	–

The parameters defining the material properties of the soil used for the materially non-linear analysis are the soil-cohesion c, angle of friction ϕ and the hardening coefficient H. In the case of perfect plasticity, the later is equal to zero and c and ϕ are constant.Using Drucker-Drager yield criteria, cohesion c and angle of friction ϕ are equal to $20KPa$ and 30° respectively.

5.2 Numerical results

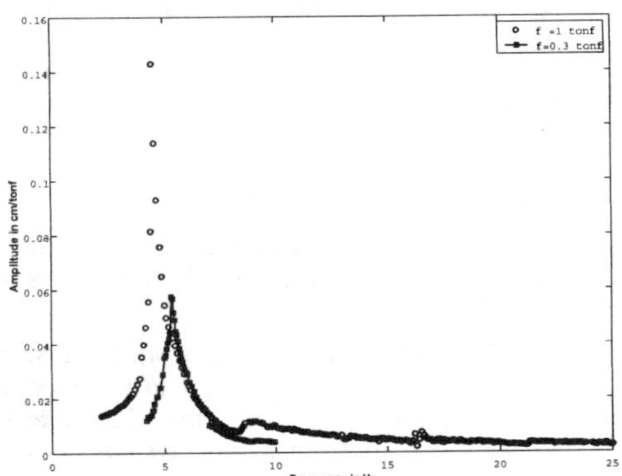

Figure 6: Experimental result with two case loading

The method presented was efficiently programmed and used in the analytical study of dynamic behavior of a single pile embedded in an infinite half-space. Different numerical techniques were also used to

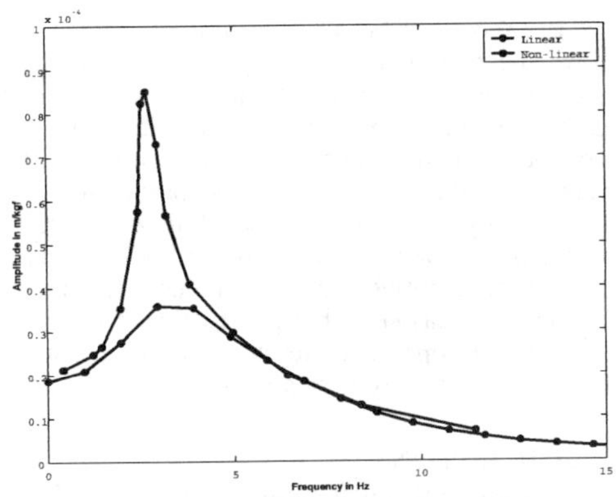

Figure 7: Frequency response computed from the experimental model

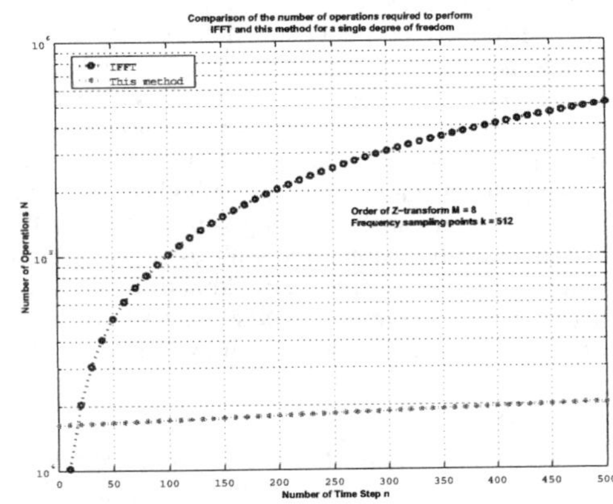

Figure 8: Comparison of the number of operations required to obtain a non-linear response using IFFT and this method

flexibility as well as accuracy to the proposed method. Since the concern of this study is to prove the accuracy of this method in dealing with the non-linear behavior of the near-field, the following results of a case loading are presented. Two different amplitude forces were used and a three dimensional FE non-linear analysis was conducted. The numerical example for non-linear response was computed for different excitation frequencies at time $t = 1s$. The experimental results are shown in Fig. 6 and they are qualitatively comparable to the simulation conducted using the present method (Fig. 7. The quantitative discrepancy is expected since the pile and the soil's material properties can not be exactly simulated. Moreover, the loading condition adding to to the existence of other structure at the site of the experiment makes it difficult -if not impossible- to simulate the exact behavior of a single pile. Fig. 8 shows the total number of operations required to perform IFFT for this non-linear case over a limited number of time steps (not including iterations that might be required to meet convergence criterion). The result shows that the number of operations increases exponentially with the increase of time integration steps. Just for the first few steps the IFFT procedure shows a competitive result with that used in this method. The reason is that this method's algorithm requires more operations to store response histories needed for the next time step.

6 CONCLUSIONS

The efficiency of the combined method of FE and TLEM in three dimensional formulation of a half space with irregularities, can be successfully applied to soil-structure problems for non-linear analysis in time domain. Since the Thin Layer formulation is only performed in frequency domain, an accurate method is needed to transform the dynamic stiffness matrix to the time domain. The recursive procedure requires lesser computer storage and solution time and there us a drastic reduction of the d.o.f compared to using direct transformation methods. That makes the time-domain analysis using substructure method computationally competitive with the corresponding direct non-recursive frequency domain procedure of determining the complex interaction response that is however, applicable only to linear systems.

REFERENCES

Aboul-Ella, F. and M. Novak (1980, December). Dynamic response of pile-supported frame foundations. *Journal Of The Engineering Mechanics Division ASCE 106*, 1215–1232.

Ben-Jamaa, S. (2000, Feburary). *A Method For Linear And Non-Linear Analysis of Pile-Soil interaction in time Domain*. Ph. D. thesis, Nihon University College Of Science And Technology.

Bernal, D. and A. Youssef (1998). A hybrid time frequency domain formulation for non-linear soil-structure interaction. *Earthquake Engineering And Structural Dyanmics 27*, 673–685.

Darbre, G. R. (1990). Seismic analysis of non-linearly base-isolated soil-structure interacting reactor building by way of the hybrid frequency-time domain procedure. *Earthquake Engineering And Structural Dyanmics 19*, 725–738.

Hayashi, Y. and H. Katakura (1990). Effective time-domain soil-structure interaction analysis based on fft algorithm with causality condition. *Earthquake Engineering And Structure Dynamics 19*, 693–708.

Hayashi, Y. and I. Takahashi (1992). An efficient time-domain soil-structure interaction analysis based on the dynamic stiffness of an unbounded soil. *Earthquake Engineering And Structure Dynamics 21*, 787–798.

Kausel, E. (1994). Thin-layer method: Formulation in the time domain. *International Journal For Numerical Methods In Engineering 37*, 927–941.

Lysmer, J. and R. L. Kuhlemeyer (1969, August). Finite dynamic model for infinite media. *Journal Of The Engineering Mechanics Division ASCE 95*, 859–877.

Novak, M. and Y. O. Beredugo (1972, December). Vertical vibrations of embedded footings. *Journal Of The Soil Mechanics And Foundations Division ASCE 98*, 1291–1310.

Seed, H. B. and I. M. Idriss (1969, January). The influence of soil conditions on goround motions during earthquakes. *Journal Of Soil Mechanics And Foundation Division ASCE 94*, 99–137.

Shimizu, N., S. Yamamoto, and Y. Koori (1977, March). Three-dimensional dynamic analysis of soil-structure system by thin layer element method (part 1. outline of analysis method and formulation of layered zone). *Journal Of Structural And Construction Engineering (Trans. of A.I.J.) 253*, 31–41.

Waas, G. (1972). *Linear Two-Dimensional Analysis Of Soil Dynamics Problems In Semi-infinite layered Media*. Thesis presented to the university of california berkeley in partial fulfilment of the requirements for the degree of doctor of philosophy, University of California, Berkeley California USA.

Wolf, J. P. (1985). *Dynamic Soil-Structure Interaction*. Englewood Cliffs New Jersey USA: Prentice-Hall.

Wolf, J. P. and M. Motosaka (1989a). Recursive evaluation of interaction forces of unbounded soil in the time domain. *Earthquake Engineering And Structural Dynamics 18*, 345–363.

Wolf, J. P. and M. Motosaka (1989b). Rercusive evaluation of interaction forces of unbounded soil in the time domain from dynamic-stiffness coefficients in the frequency domain. *Earthquake Engineering And Structural Dynamics 18*, 365–376.

Computer Methods and Advances in Geomechanics, Desai et al. (eds) © 2001 Balkema, Rotterdam, ISBN 90 5809 183 X

The ultimate pullout capacity of anchors in frictional soils

R.S. Merifield, S.W. Sloan, A.J. Abbo & H.S. Yu
University of Newcastle, N.S.W., Australia

ABSTRACT: During the last thirty years various researchers have proposed approximate techniques to estimate the uplift capacity of soil anchors. As the majority of past research has been experimentally based, much current design practice is based on empiricism. Somewhat surprisingly, very few numerical analyses have been performed to determine the ultimate pullout loads of anchors. This paper presents the results of a rigorous numerical study to estimate the ultimate pullout load for vertical and horizontal plate anchors in frictional soils. Rigorous bounds have been obtained using two numerical procedures that are based on finite element formulations of the upper and lower bound theorems of limit analysis. For comparison purposes, numerical estimates of the break-out factor have also been obtained using the more conventional displacement finite element method. Results are presented in the familiar form of break-out factors based on various soil strength profiles and geometries and, are compared with existing numerical and empirical solutions.

1 INTRODUCTION

During the last thirty years various researchers have proposed approximate techniques to estimate the uplift capacity of anchors in sand. A large proportion of this past research has been experimentally based and, as a result, current design practices are largely based on empiricism. Although a number of numerical studies have been undertaken to estimate the capacity of sand anchors, very few of them can be considered rigorous.

Previous numerical studies of anchors in sand have typically utilised simple analytical approaches such as limiting equilibrium, cavity expansion, and limit analysis. These can be found in the works of Meyerhof and Adams (1968,1973), Murray and Geddes (1987,1989), Neely et al(1973), and Yu (2000). More rigorous studies have been performed by Rowe and Davis (1982), Tagaya (1983,1988) and Koutsabeoulis and Griffiths (1989), using the displacement finite element technique. Of these, the most complete numerical study appears to be by Rowe and Davis (1982). In their investigation, results were presented for both horizontal and vertical strip anchors embedded in sand. An elasto-plastic finite element analysis was used and modelled the soil structure interaction at the soil/anchor boundary.

1.1 *Problem Definition*

Soil anchors may be positioned either vertically or horizontally depending on the load orientation or type of structure requiring support. A general layout of the problem to be analysed is shown in Figure 1

For numerical convenience, the ultimate anchor capacity q_u will be presented in a form analogous to

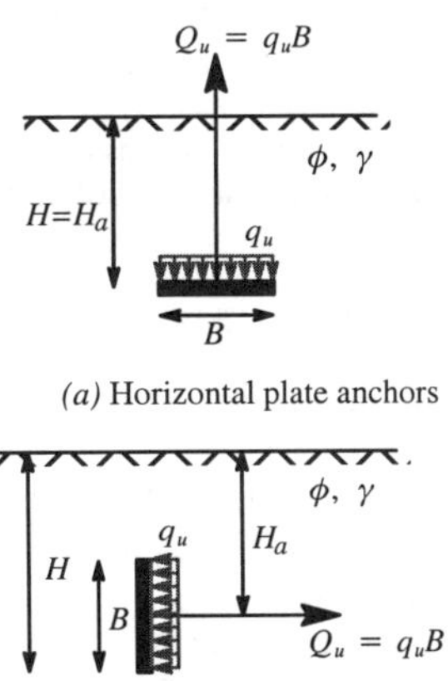

(a) Horizontal plate anchors

(b) Vertical plate anchors

Figure 1 Problem Notation

Terzaghi's equation which is used to analyse surface footings, namely ;

$$q_u = \gamma H N_\gamma \quad (1)$$

where N_γ is referred to as the anchor break-out factor.

2 NUMERICAL ANALYSIS OF ANCHORS

2.1 *Numerical Limit Analysis*

Rigorous bounds on the ultimate pullout load presented in this paper are obtained by using two numerical procedures that are based on finite element formulations of the upper and lower bound theorems of limit analysis. These formulations assume a perfectly plastic soil model with an associated flow rule and generate large linear programming problems. Details of the numerical procedures can be found in Sloan (1988) and Sloan and Kleeman (1995) and will not be repeated here.

2.2 *Displacement Finite Element Analysis (DFEA)*

The use of the finite element method is now widespread amongst researchers and practitioners. Theoretically, the finite element technique can deal with complicated loadings, excavation and deposition sequences, geometries of arbitrary shape, anisotropy, layered deposits and complex stress-strain relationships. However, the method has not been used widely when estimating the capacity of soil anchors.

The finite element formulation used by the authors is that presented by Abbo(1997), and Abbo and Sloan (1998). This formulation was developed with the aim of reducing the complexity of elastoplastic analysis by using advanced solution algorithms with automatic error control. The resulting formulation greatly enhances the ability of the finite element technique to predict collapse loads accurately, and avoids many of the inherent problems discussed by Toh and Sloan(1980) and Sloan and Randolph(1982).

3 RESULTS AND DISCUSSION

3.1 *Horizontal Anchors*

3.1.1 *Numerical Limit Analysis Results*

The computed lower and upper bound estimates of the anchor break-out factorN_γ are shown graphically in Figure 2 for embedment ratios of 1, 3, 5, 7 and 10. It was found that for a given embedment ratio (*H/B*), the break-out factor increases almost linearly with soil friction angle. This linear behaviour is more apparent in the lower bound results than the upper bound results.

The results shown in Figure 2 indicate that the limit analysis is very successful, with only very small error bounds observed between the upper and lower bound solutions. This has enabled the true value of the anchor break-out factor to be bracketed to within ± 5% in all cases. The error bounds were found to increase with increasing friction angle, being less than ± 3% for friction angles less than 34°.

To date, a great deal of uncertainty has surrounded the likely modes of failure for horizontal anchors in sand. The observed upper bound velocity diagrams at collapse, along with the associated zones of plastic shearing, for anchors at several embedment depths are shown in Figure 3

In general, it was found that failure consists of the upward movement of a rigid column of soil immediately above the anchor, accompanied by lateral deformation extending out and upwards from the anchor edge. As the anchor is pulled vertically upwards, the material above the anchor tends to lock up as it attempts to dilate during deformation. As a consequence, in order to accommodate the rigid soil column movement, the observed plastic zone is forced to extend a large distance laterally outwards into the soil mass. This soil locking phenomenon was noticed by Rowe(1978), who found that soil dilatancy had a significant effect on the observed plastic region at failure, which in turn resulted in substantial variations in the predicted anchor break-out factor.

For relatively shallow anchors ($H/B \leq 4$), the bulk of the soil column above the anchor remains non-plastic, while the lateral extent of plastic shearing is extensive. For $\phi = 20°$, the vertical extent of the non-plastic zone extends to the soil surface. However, the vertical extent of the non-plastic zone decreases with increasing friction angle, and no longer reaches the soil surface for friction angles greater than 20°. This is also the case for relatively deep anchors ($H/B \geq 5$).

For relatively deep anchors ($10 \geq H/B \geq 5$) where ϕ is around 20°, the soil column above the anchor no longer "locks up", and the material changes from an elastic to plastic state at ultimate collapse. When $\phi = 20°$, a localised zone of compression and lateral deformation forms above the anchor, which leads to a reduction in the magnitude of the deformations observed throughout the soil mass and at the surface.

Local compression of the soil adjacent to the anchor also produces a velocity jump along the soil/anchor interface. It appears logical to anticipate that when such a velocity jump at the anchor interface exists, the effect of interface roughness will be most significant.

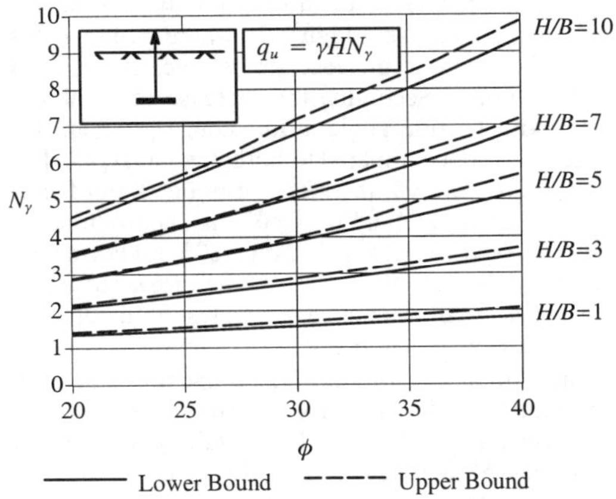

Figure 2 Break-out factors for horizontal anchors in sand.

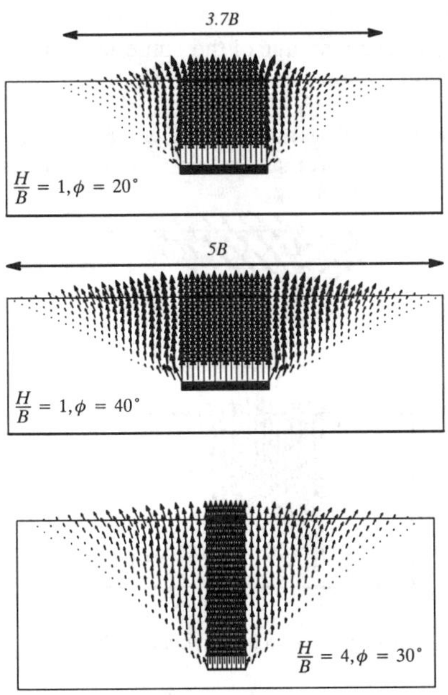

Figure 3 Observed velocity plots from upper bound analyses

The extent of lateral shearing was found to increase with increasing friction angles at a given embedment depth. For example, referring to Figure 3, the lateral extent of vertical deformation observed at the surface for *H/B*=1, increases from approximately 3.7*B* for $\phi = 20°$, up to around 5*B* for $\phi = 40°$. This is also the case for larger embedment depths, where the lateral extent of vertical deformation was found to increase from 2.5*B* to 4.5B, and from 3.5*B* to 6*B* for *H/B*=2 and 4 respectively.

Unlike the case for a horizontal anchor in purely cohesive soil (Merifield et al 1999), no plastic shearing or flow was observed below the anchor for all cases. This is in contrast to the observations of Rowe(1978), who observed quite extensive plastic failure above and below the anchor for $\phi = 15°$. This plastic region was found to shrink, and for $\phi = 45°$, plastic flow below the anchor was non-existent.

3.1.2 Displacement Finite Element Results

The displacement finite element (DFEA) estimates of the break-out factor N_γ are shown graphically in Figure 4 As to be expected the DFEA results plot above the lower bounds estimates, and were found to plot close to the upper bound solutions ($\pm$ 4%).

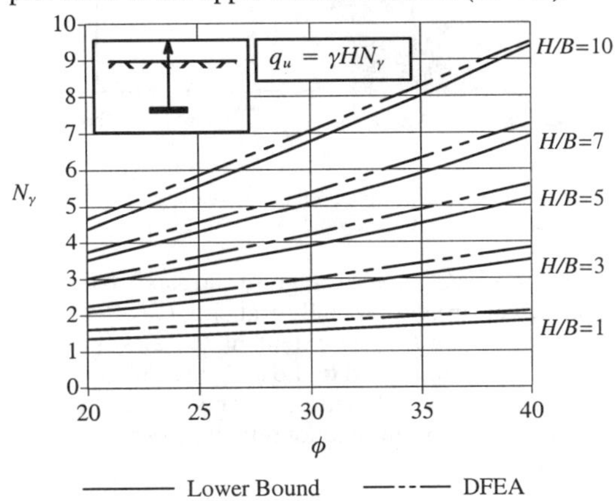

Figure 4 Break-out factors for horizontal anchors in sand.

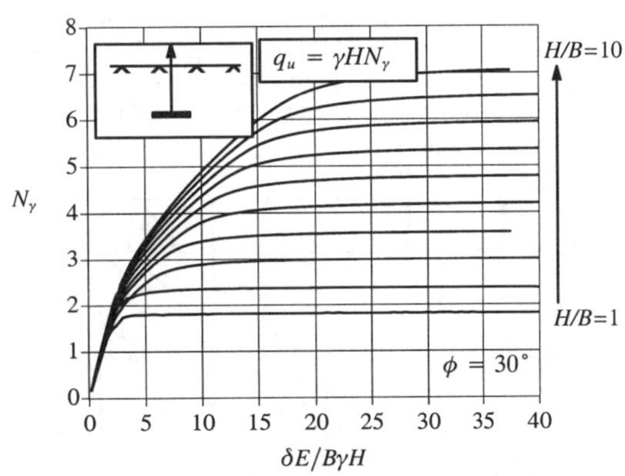

Figure 5 Load displacement behaviour of horizontal anchors

The load displacement curves for a horizontal anchor in a cohesionless soil with $\phi = 30°$ are shown in Figure 5 where δ is the anchor displacement and E is Youngs modulus. These curves indicate the DFEA was able to effectively estimate the anchor break-out factor, and collapse is clearly defined for each embedment ratio(H/B).

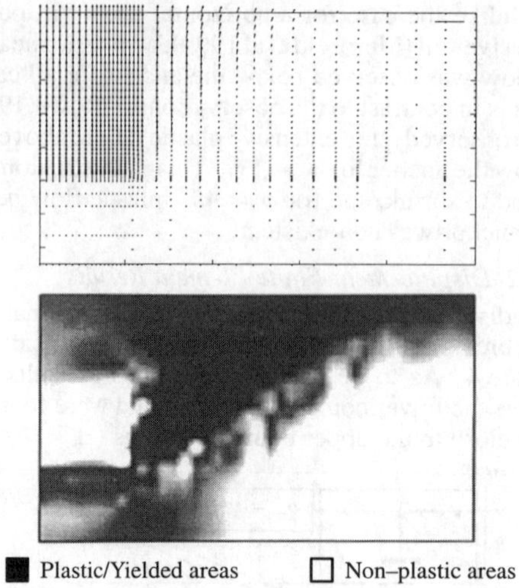

Figure 6 Displacement finite element velocity diagram and associated regions of plastic yielding

The observed velocity diagram and asociated zones of plastic shearing for a shallow anchor ($H/B = 1$) is shown in Figure 6 The extent of the elasto-plastic zones at collapse extend to the soil surface indicating shallow anchor behaviour. The zone of soil immediately above the anchor remains in a non-plastic state for this anchor.

3.2 *Vertical anchors*

3.2.1 Numerical Limit Analysis results

The upper and lower bound estimates of the break-out factor N_γ are shown graphically in Figure 7 The limit analysis procedure was again successful, and the true break–out factor has been estimated to within $\pm 10\%$.

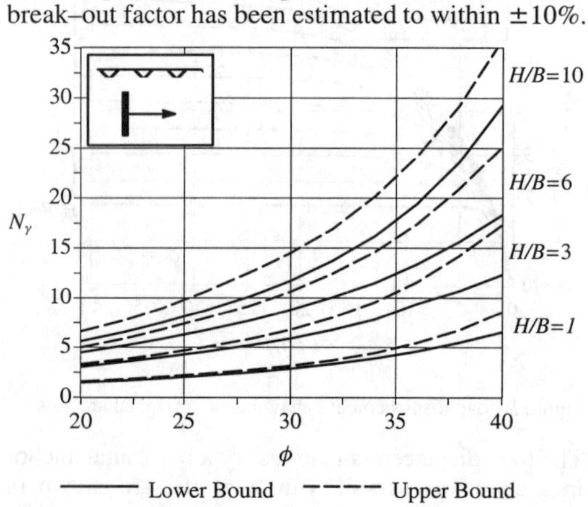

Figure 7 Break-out factors for vertical anchors in sand.

There are two significant features that emerge from the results shown in Figure 7 Firstly, for a given embedment ratio *H/B*, the break-out factor was found to increase in a non-linear manner with soil friction angle ϕ. This is in contrast with the results for horizontal anchors (see Figure 2), where the break-out factor was found to increase almost linearly with soil friction angle. Secondly, the increase in break-out factor with friction angle is far more significant for vertical anchors compared to horizontal anchors. For example, for *H/B*=5, the increase in break-out factor between $\phi = 20^\circ$ and $\phi = 40^\circ$ is approximately 400% for vertical anchors and only 80% for horizontal anchors. The rate of increase in the break-out factor for vertical anchors tends to be greatest for medium to dense sands where $\phi \geq 30^\circ$.

The complex nature of vertical anchor behaviour, due to the lack of symmetry, has led to a great deal of uncertainty regarding the likely failure patterns at collapse. As a result, it is difficult to predict the capacity of vertical anchors using existing approaches that require assumptions regarding the shape of the failure surface. Such approaches include the limit equilibrium method, the method of characteristics, and any analytical upper bound method.

A distinct advantage of the numerical upper bound formulation used, is that the form of the failure mechanism can be obtained automatically without any assumptions being made in advance. The observed modes of failure for several vertical anchor cases are shown in Figure 8

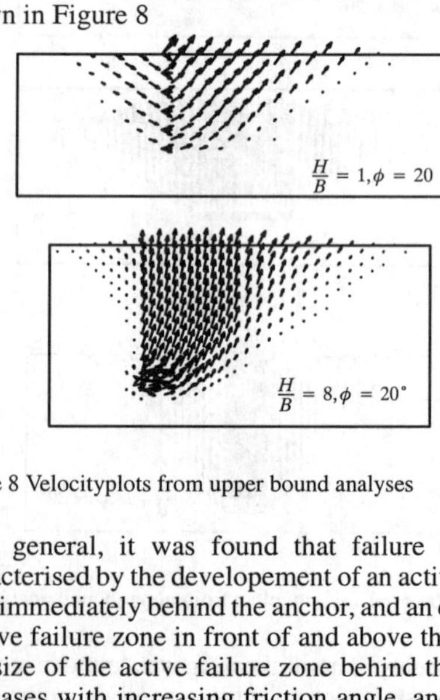

Figure 8 Velocityplots from upper bound analyses

In general, it was found that failure could be characterised by the developement of an active failure zone immediately behind the anchor, and an extensive passive failure zone in front of and above the anchor. The size of the active failure zone behind the anchor decreases with increasing friction angle, and is most extensive for anchors in loose sand ($\phi \leq 20^\circ$).

3.2.2 Displacement Finite Element Results

The displacement finite element (DFEA) estimates of the break-out factor N_γ are shown graphically in Figure 9 The results again compare favourably to the limit analysis estimates and are typically within ±5% of the upper bound.

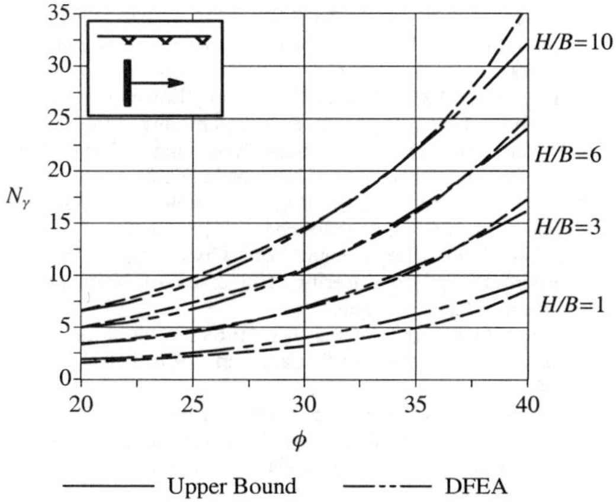

Figure 9 Break-out factors for vertical anchors in sand.

The extent of the elasto-plastic zone at failure confirms the formation of an active zone of collapse behind the anchor, as predicted by the limit analysis results. This is illustrated in Figure 10

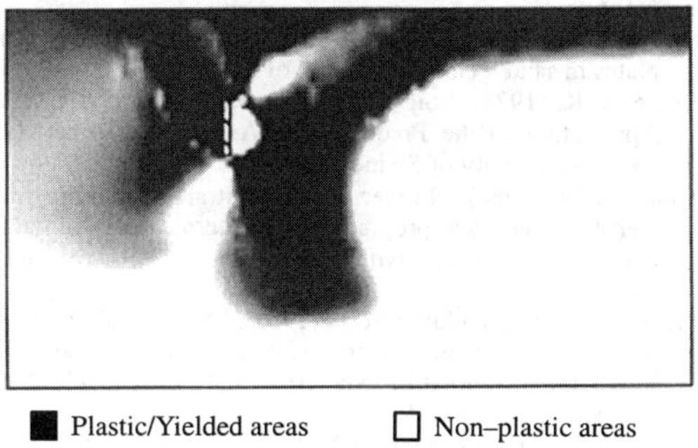

Figure 10 Elasto-plastic zones for vertical anchor with $H/B = 3$, $\phi = 30°$

3.3 *Effect of anchor roughness*

The results discussed in the preceding sections were based on the assumption that the soil-anchor interface could be considered as being perfectly rough. For this case, the angle of friction between the anchor plate and the surrounding soil is taken equal to the soil friction angle ϕ. In reality, the true interface friction angle will vary depending on the type of anchor and surrounding material.

In order to determine the effect interface friction has on the anchor break-out capacity, the specific case of a perfectly smooth interface has been analysed using the limit analysis formulations. For the upper bound analysis the shear strength of the soil-anchor interface is assumed to be zero, whereas for the lower bound analysis the shear stresses at the soil/anchor interface are assumed to be zero. Such analyses will provide an indication as to the likely maximum effect of interface friction.

In order to quantify the effect of anchor roughness, a correction factor is introduced as

$$F_\delta = \frac{q_{u\ rough}}{q_{u\ smooth}} = \frac{N_{\gamma\ rough}}{N_{\gamma\ smooth}} \tag{2}$$

which simply equates the ratio of the pullout capacity for a rough anchor to that of a smooth anchor. The roughness correction factor for $\phi = 30°$ is shown as a function of embedment ratio in Figure 11. The limit analysis, displacement finite element and a line of best fit have been included in this figure.

For horizontal anchors, the interface roughness was found to have little or no effect(<4%) on the calculated pullout capacity for all embedment depths and friction angles analysed. For an anchor with *H/B*=3 and $\phi = 30$, for example, changing the roughness from perfectly rough to perfectly smooth reduces the break-out factor N_γ by just 1%. The effect of anchor roughness is greatest for relatively deep anchors ($10 \geq H/B \geq 5$) where ϕ is around 20°.

Earlier studies on vertical anchors in sand have revealed that anchor interface roughness significantly affects the ultimate anchor capacity (Rowe and Davis 1982).

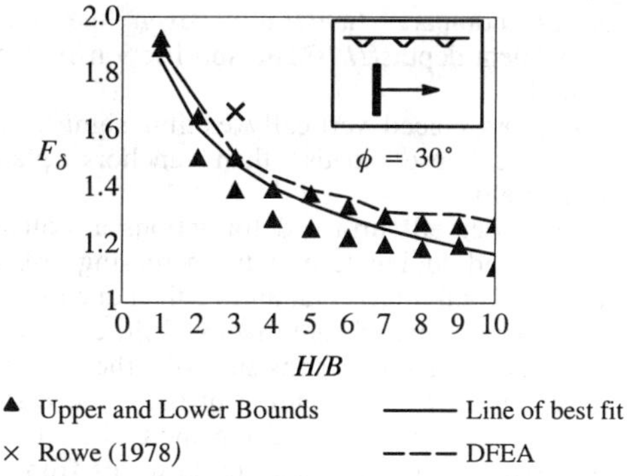

Figure 11 Effect of anchor roughness

The finite element limit analysis results confirm the significance of interface roughness when estimating the capacity of vertical anchors. Changing the interface roughness from perfectly rough to perfectly smooth can lead to a reduction in the anchor capacity by as much as 65%. The effect of anchor roughness was found to decrease with embedment depth for all friction angles. The greatest rate of change in the roughness correction factor occurs between embedment ratios of 1 and 4. For $\phi = 30°$, for example, the anchor correction factor F_δ increases from approximately 1.15 for *H/B*=10 to around 1.9 for *H/B*=1. However, two thirds of the total variation in fact occurs between *H/B*=4 and *H/B*=1.

The reason interface roughness can significantly alter the estimated pullout capacity is because the movement of the soil in front of the anchor is upwards at an angle to the soil/anchor interface. In the case of a rough anchor, significant shear stresses will develop at the anchor interface in response to this upward movement. These shear stresses are resisted by the interface and contribute to the anchor capacity. The shearing resistance provided along a rough anchor interface tends to force the failure zone downwards below the anchor level.

4 CONCLUSIONS

Rigorous lower and upper bound solutions for the ultimate capacity of horizontal and vertical strip anchors in sand have been presented. These have been compared to anchor break-out factor estimates obtained using an advanced displacement finite element formulation. Consideration has been given to the effect of soil friction angle, anchor embedment depth and anchor roughness. The results obtained have been presented in terms of familiar break–out factors in graphical form to facilitate their use in solving practical design problems.

The following main conclusions can be drawn from the results presented in this paper :

- The upper bound, lower bound and displacement finite element estimates for the anchor break-out factors compare favourably over the range embedment depths(*H/B*) and soil friction angles ϕ studied.
- Anchors placed vertically exhibit significantly higher collapse loads than anchors placed horizontally.
- The observed surface deformations at collapse were found to increase with increasing friction angles for both horizontal and vertical anchors.
- In general, it was found that the failure mode for horizontal anchors consists of the upward movement of a rigid column of soil immediately above the anchor, accompanied by lateral deformation extending out and upwards from the anchor edge. As the anchor is pulled vertically upwards, the material above the anchor tends to lock up as it attempts to dilate during deformation
- The failure mode for vertical anchors indicates the development of significant active zones of failure can form behind the anchor.

5 REFERENCES

Davis, E.H. (1968). Theories of plasticity and the failure of soil masses. In: Soil Mechanics, Selected Topics (I.K. Lee, Ed.), Butterworths, London.

Abbo, A.J. (1997). "Finite element algorithms for elastoplasticity and consolidation", PhD Thesis, Department of Civil, Surveying and Environmental Engineering, University of Newcastle, NSW.

Abbo, A.J. & Sloan S. W. (1998). SNAC (Solid Nonlinear Analysis Code), A finite element program for the analysis of elastoplasticity and consolidation. User Manual Version 1.0

Koutsabeloulis, N. C. & Griffiths, D. V. 91989). "Numerical -modelling of the trap door problem". Geotechnique 39:1, 77-89.

Lyamin, A.V. (2000). "Three-dimensionalLower Bound Limit Analysis Using Non Linear programming", PhD Thesis, Department of Civil, Surveying and Environmental Engineering, University of Newcastle, NSW.

Merifield, R. S., Sloan, S. W., Yu, H. S.(1999). "The stability of plate anchors in undrained clay". Research Report 174.02.1999, Department of Civil, Surveying and Environmental Engineering, University of Newcastle. To appear in Geotechnique.

Meyerhof, G. G., and Adams, J. I. (1968). "The ultimate uplift capacity of foundations". Canadian Geotechnical Journal., 5(4), 225-244.

Meyerhof, G. G. (1973). "Uplift resistance of inclined anchors and piles". Proceedings, 8th International Conference on Soil Mechanics and Foundation Engineering, Moscow, 2.1:167-172.

Murray, E. J., and Geddes, J. D. (1987). "Uplift of anchor plates in sand". Journal of Geotechnical Eng., ASCE, 113(3), 202-215.

Murray, E. J., and Geddes, J. D. (1989). "Resistance of passive inclined anchors in cohesionless medium". Geotechnique, 39:3, 417-431.

Neely, W. J., Stuart, J. G. and Graham, J. (1973). "Failure loads of vertical anchor plates in sand". Journal of the Soil Mechanics and Foundations Division, ASCE, 99(9), 669-685.

Rowe, R.K., and Davis, H. (1982). "The behaviour of anchor plates in sand". Geotechnique, Vol 32, No.1, pp. 25-41.

Rowe, R. K. (1978). Soil Structure Interaction Analysis and its Application to the Prediction of Anchor Behaviour. PhD thesis, University of Sydney, Australia.

Sloan, S. W. (1988). "Lower bound limit analysis using finite elements and linear programming". International Journal for Numerical and Analytical Methods in Geomechanics, 12:61-67.

Sloan S. W. and Randolph M. F. (1982). "Numerical prediction of collapse loads using finite element methods." International Journal for Numerical and Analytical Methods in Geomechanics, 6, 47-76.

Sloan, S. W., and Kleeman, P. W. (1995). "Upper bound limit analysis using discontinuous velocity fields". Computer Methods in Applied Mechanics and Engineering, 127:293-314

Toh C. T. and Sloan S. W. (1980). "Finite element analysis of isotropic and anisotropic cohesive soils with a view to correctly predicting impending collapse." International Journal for Numerical and Analytical Methods in Geomechanics, 4, 1-23.

Yu, H. S. (2000). Cavity Expansion Methods in Geomechanics. Kluwer Academic Publishes, London (in press).

Computer Methods and Advances in Geomechanics, Desai et al. (eds) © 2001 Balkema, Rotterdam, ISBN 90 5809 183 X

Interaction between slope and sleeved pile groups

C.W.W. Ng & L.M. Zhang
Department of Civil Engineering, Hong Kong University of Science and Technology, People's Republic of China

K.K.S. Ho
Geotechnical Engineering Office, Hong Kong, People's Republic of China

ABSTRACT: Many large diameter replacement piles in Hong Kong are founded in steep slopes and have to resist severe wind loads during typhoon season. A sleeving technique is sometimes adopted to minimize the transfer of lateral load from the buildings to the shallow depths of the slopes. In this paper, a three-dimensional analysis is carried out to investigate the interaction between a cut slope and a laterally loaded sleeved pile group. The stability of the cut slope with the pile group is evaluated using the strength reduction technique. The sleeving is shown to be capable of decreasing the stresses and displacements in the shallow depths in front of the sleeved pile group and improving marginally the stability of the shallow slope locally. However, the effects of the sleeving are, in general, not noticeable for global stability.

1 INTRODUCTION

Many large diameter replacement piles in Hong Kong are founded in steep slopes and have to resist severe wind loads during typhoon season. The stability of the slope may reduce as a result of the stresses transferred from the piles to the slope in the course of mobilizing lateral soil resistance. In order to minimize the transfer of lateral load into the shallow depths of the slope, an annulus of compressible material, referred to as sleeving, is sometimes constructed between the piles and the adjacent soil (GEO 1984). Ng et al. (2000) analysed the influence of a single sleeved pile on the stability of a cut slope using the strength reduction method and a three-dimensional finite difference code. In practice, piles are likely used in groups. Group interaction may cause changes in the manner of load transfer from the piles to the slope. Thus, the influence of a pile group on slope stability may be different from that of a single pile. This paper is an extension of the earlier work. It aims to evaluate the sleeved pile group-soil interaction and the influence of sleeved and unsleeved pile groups on slope stability.

2 METHODOLOGY OF STABILITY ANALYSIS

Finite element analysis of slope stability does not directly provide a factor of safety (FOS) but it generally utilizes the so-called shear strength reduction technique (Giam and Donald 1988; Griffith and Lane 1999). The factor of safety of a slope is defined as the ratio by which the available shear strength parameters need to be factored down in order to bring the slope to failure. Using this technique, the mobilized shear strength parameters, c'_m and ϕ'_m, are obtained by dividing c' and $\tan\phi'$ by a strength reduction factor (*SRF*), as follows:

$$c'_m = \frac{c'}{SRF} \tag{1}$$

$$\phi'_m = \tan^{-1}\left(\frac{\tan\phi'}{SRF}\right) \tag{2}$$

where c' is the effective cohesion and ϕ' is the effective angle of shearing resistance of the soil. The value of *SRF* is initially assumed to be 1.0. Subsequently, this value is increased step by step until a slope failure occurs. The *SRF* at failure is considered to be the FOS of the slope.

3 NUMERICAL MODEL AND PARAMETERS

In this paper, a pile-slope system in an idealized elastic perfectly plastic soil is investigated using the three-dimensional finite difference program Flac3D (Itasca Consulting Group 1996). Figure 1 shows the plan view and cross-section of the pile group-slope system. A cut slope of 15 m high, with an average slope angle of 32°, is modeled. The pile group consists of three concrete piles in a row at a centre-to-centre spacing of four pile diameters. Each pile is assumed to be 2 m in diameter (*D*), 30 m in length with 22.5 m in penetration (*L*). In the case of the sleeved piles, an annulus of 250 mm thick sleeving material is modeled at the upper 7.5 m portion of each of the embedded piles. The sleeving material is assumed to connect with the pile and the soil perfectly. For simplicity, the ground water table is taken to be below the pile tip in the analysis.

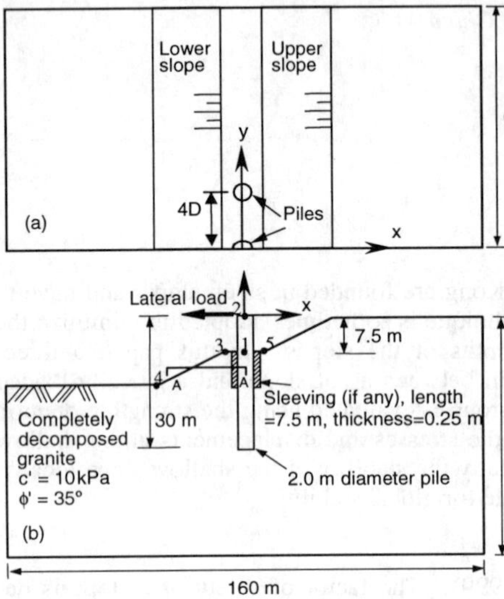

Figure 1 Cross-section of the pile-slope system, (a) plan view, (b) cross section

The pile and the sleeving materials are assumed to be linearly elastic. The soil is simulated with an elastic, perfectly plastic model (Itasca Consulting Group 1996). Soil failure is described by a composite Mohr-Coulomb criterion with a tension cut-off. The parameters adopted are summarized in Table 1. A detailed description of the selection of parameters can be found in Ng et al. (2000).

Table 1. Material parameters for analysis

Properties	Soil	Pile concrete	Sleeving material
(1)	(2)	(3)	(4)
Young's modulus (MPa)	50.0	26000.0	0.3
Poisson's ratio	0.3	0.2	0.3
Bulk modulus (MPa)	41.7	14400.0	0.25
Density (kg/m^3)	1800.0	2400.0	16.5
Internal friction angle (°)	35.0	-	-
Effective cohesion (kPa)	10.0	-	-
K_0 (before excavation)	0.4	-	-

Four load steps are applied for analyses, i.e., 1000 kN, 2000 kN, 3000 kN, and 6000 kN per pile, respectively. In each step, the loads are applied to each of the piles simultaneously. These applied loads per pile correspond to 6%, 12%, 18%, and 36% of the nominal shear resistance V_u of the pile section, respectively.

4 IDENTIFICATION OF SLOPE INSTABILITY

Based on the characteristics of the relationships between *SRF* and slope displacements, slope stability may be described by initiation of instability and global failure. Initiation of instability is said to have occurred when substantial plastic zones in the slope develop and the slope displacements start to accelerate. After the initiation of instability is reached, the slope is at a quasi-stable state. Global failure will subsequently occur when numerical solution loses convergence and uncontrollable large displacements develop in the slope (Zienkiewicz and Taylor 1989).

Figures 2(a) and 2(b) show the relationships between *SRF* and nodal displacement for the slope with sleeved and unsleeved pile groups, respectively, subjected to a lateral load per pile of 12% V_u. In both cases, the increase in displacements with *SRF* is small until a critical *SRF* value of approximately 1.3 is reached. After that, the displacements increase with *SRF* rapidly, and the numerical solution eventually loses convergence at a *SRF* of 2.0. Therefore, the FOSs for the sleeved pile group at initiation of instability and global failure are taken to be 1.3 and 2.0, respectively.

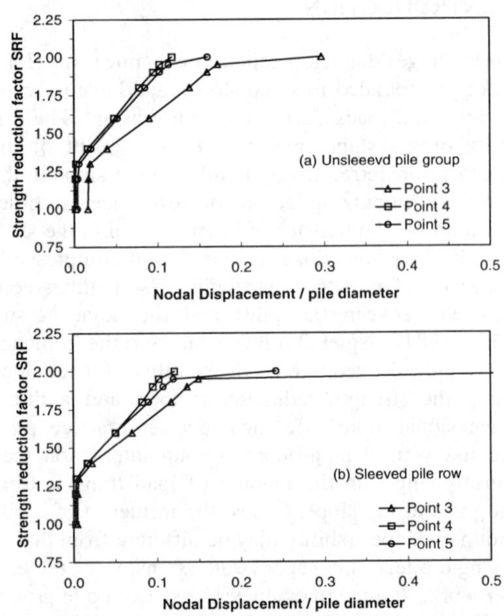

Figure 2 Relationships between *SRF* and nodal displacements for the slope with (a) an unsleeved and (b) a sleeved pile group, lateral load =12% V_u

Figure 3 shows the plastic zones in the slope at *SRF*=1.0, 1.3, and 2.0 when the sleeved pile group is subjected to a lateral load of 12% V_u per pile. As can be seen, there is limited development in the plastic zone in the shallow depth of the slope when *SRF* is increased from 1.0 to 1.3. However, the isolated plastic zones join and develop quickly after *SRF* is greater than 1.3. At *SRF* = 2.0, widespread plastic zone is developed in the slope and no further increase in *SRF* is possible. This clearly illustrates the relation between the slope displacements in Figure 2 and the development of plastic zones in Figure 3.

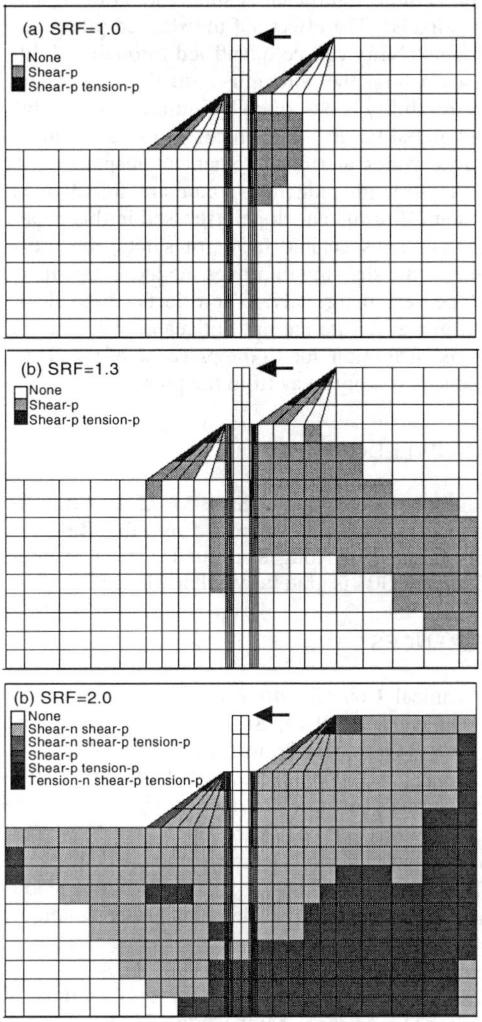

Figure 3 Development of plastic zone in the slope with a sleeved pile group subjected to a lateral load of 12% V_u per pile

5 INFLUENCE OF SLEEVED PILE GROUP ON SLOPE STABILITY

Figure 4 shows the effects of sleeving on the distributions of horizontal stresses σ_{xx} and displacements in the slope immediately in front of the pile group along the y-direction (Figure 1) at a depth of approximately *z/L*=0.2. The laterally loaded piles induce large horizontal stresses and displacements in the shallow depth of the slope. Thus, the stability of the slope will be adversely affected. By the use of the sleeving, both stresses and displacements are substantially reduced. Such reduction is more significant at small lateral loads than at large loads. However, the effects of the sleeving on the stresses and displacements reduce sharply away from the piles, and become negligible at a distance three-diameter outside the outer pile of the pile group.

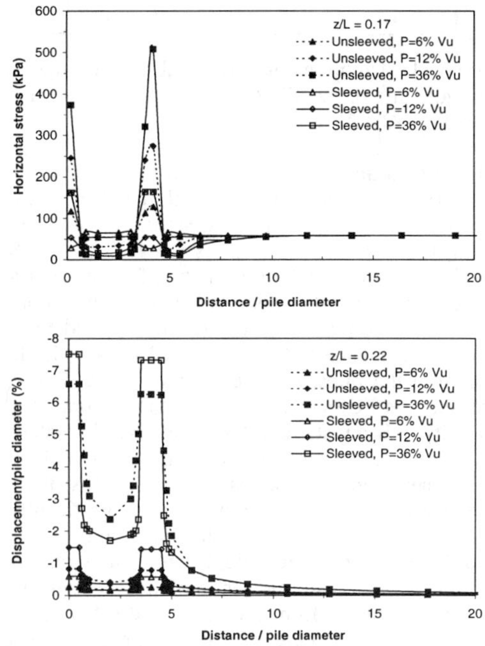

Figure 4. Effects of sleeving on the distributions of horizontal stresses and displacements in front of the pile group along the y-direction

Figure 5 shows the distributions of the horizontal stresses against the middle pile of the pile group along the depth, at a lateral load of 12% V_u. For the unsleeved piles, σ_{xx} increases with depth at shallow depth and the peak stresses occur at approximately *z/L*=0.17. For the sleeved pile, the average σ_{xx} acting on most of the sleeved pile segments is significantly smaller than that on the unsleeved pile. This means that the sleeved pile segment provides much smaller resistance than the corresponding unsleeved pile segment does. For equilibrium of the pile, a large percentage of the applied load has to be transferred to the ground beneath the sleeving material. This results in a significant downward redistribution of the horizontal stresses acting on the sleeved pile and a large peak stress beneath the sleeving. Thus, the pile sleeving functions primarily through a vertical load transfer mechanism (Ng et al. 1995, 2000).

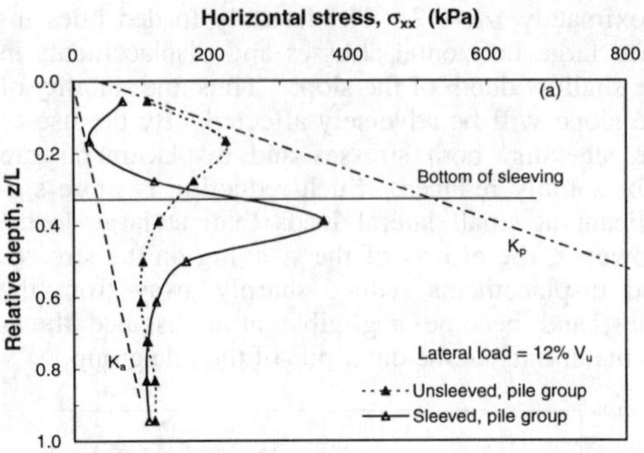

Figure 5. Distributions of horizontal stresses against the middle pile of the pile group along the depth, lateral load of 12% V_u

Figure 6 summarizes the FOSs at the initiation of instability and global failure for both the sleeved and unsleeved pile groups. The FOSs of the cut slope without the construction of piles are 1.90 and 1.28, respectively corresponding to the two failure criteria. According to the earlier discussions in Figures 4 and 5, the FOSs would decrease with the applied loads due to the influence of the transferred stresses in the slope from the piles; but the sleeving is capable of increasing the FOSs of the slope against initiation of instability. By installing the sleeving, the lateral load is transferred into deeper soil layers and the displacements and stresses at shallow depths of the slope in front of the piles are decreased significantly. Nevertheless, the effects of the sleeving on the global FOS are not noticeable, because the stress redistribution due to the presence of the sleeving reduces sharply away from the pile. The influence of the sleeving in the case of the pile group is even smaller than that in the single pile case (Ng et al. 2000). The magnitude of the maximum stresses in the soil in front of the group piles tends to be smaller than that in front of a single pile, because the overlapping of the passive wedges in front of the group piles (group effects) reduces the stress redistribution.

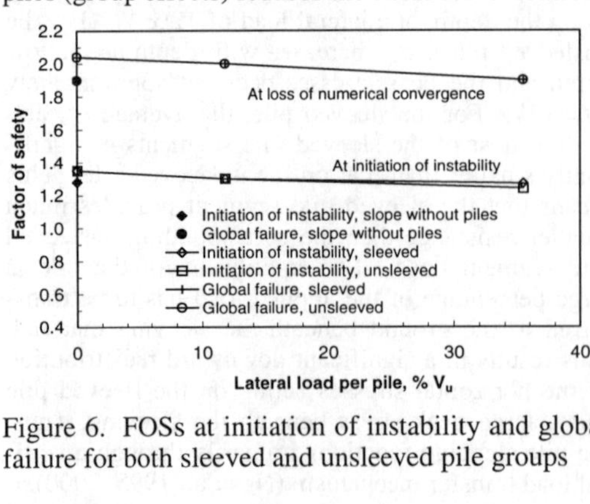

Figure 6. FOSs at initiation of instability and global failure for both sleeved and unsleeved pile groups

CONCLUSIONS

A laterally loaded pile group may induce significant horizontal stresses and displacements in the slope immediately in front of the piles. By the installation of the sleeving, the induced stresses and displacements can be significantly reduced locally. Such reduction is more significant at small lateral loads than at large loads. The effects of the sleeved pile group on slope stability can be quantified through stability analyses using the strength reduction technique. Slope instability is described by initiation of instability and global failure. The processes of development of plastic zones and development of displacements with the strength reduction factor are found to be consistent. For the cut slope discussed in this paper, which has a respectable factor of safety, the introduction of the sleeving improves the local stability of the slope very marginally. The effects of the sleeving on global failure are not noticeable, because the stress redistribution due to the presence of the sleeving reduces sharply away from the piles.

ACKNOWLEDGEMENTS

The authors would like to thank Mr C. K. Choy for assistance. Financial support from the Research Grants Council of Hong Kong under research grant CRC96/99.EG04 is gratefully acknowledged.

REFERENCES

Geotechnical Control Office (GCO). 1984. Geotechnical Manual for Slopes, 2nd edition, Hong Kong Government Printer, 86-88.

Giam, S.K., and Donald, I.B. 1988. Determination of critical surfaces for slopes via stress-strain calculations. Proc. Fifth Australia - New Zealand Conference on Geomechanics, 461-464.

Griffiths, D.V., and Lane, P.A. 1999. Slope stability analysis by finite elements. Geotechnique, 49(3): 387-403.

Itasca Consulting Group, Inc. 1996. FLAC^{3D} Ver. 2.0, Fast Lagrangian Analysis of Continua in Three Dimensions. Minneapolis, USA.

Ng, C.W.W., Zhang, L.M., Ho, K.K.S., and Choy, C.K. 2000. Influence of laterally loaded sleeved piles on slope stability. Proc. GeoEng2000, Australia.

Ng, C. W. W., Lings, M. L., Simpson, B., and Nash, D. F. T. (1995). An approximate analysis of the three-dimensional effects of diaphragm wall installation. *Geotechnique*, 45(3), 497-507.

Zienkiewicz, O.C., and Taylor, R.L. 1989. *The Finite Element Method*, Vol. 1, 4th edition, McGraw-Hill, New York.

Computer Methods and Advances in Geomechanics, Desai et al. (eds) © 2001 Balkema, Rotterdam, ISBN 90 5809 183 X

Numerical analysis of deformations during construction of a wharf built using sheet pile moorings

S.Ohmaki & K.Saeki
National Research Institute of Fisheries Engineering, Fisheries Agency, Japan

S.Shikata
Japanese Institute of Technology on Fishing Ports and Communities, Japan

S.Baba
Toyo Construction Company Limited, Japan

ABSTRACT: In the area behind a wharf built on soft ground using anchor-sheet piles, paper drains are sometimes used to improve the ground for the purpose of building embankments. Numerical analyses were conducted for such a case to evaluate the stress behavior and deformation of the ground behind the wharf. These analyses were conducted under the following conditions: the ground is a two-phase material consisting of soil with an elasto-plastic skeleton structure and incompressible pore water, and sheet piles, anchor piles, tie rods, and foundation layer are elastic undergoing plane strain. The results were then used to examine the effect of differences in construction (extent of ground improvement, width of embankment, etc.) on bending stress and horizontal displacement of pile sheets, axial force of tie rods, and lateral deformation and settlement of ground behind the structure.

1 INTRODUCTION

Fishing ports are usually constructed in sheltered locations, but such locations usually contain cohesive sediments that can easily lead to weak foundations. If gravity-type fishing port structures are built on such soft ground, extensive ground improvement work must be undertaken, thus adding greatly to the construction cost. Therefore, sheet pile structures are used as low cost alternatives for mooring wharves in such situations.

This paper presents a numerical analysis of two-dimensional consolidation deformation of the soft ground and the sheet pile mooring wharf of a fishing port built on top of the soft ground during and after the reclamation of the ground behind the wharf. Since the seabed of the port contains a soft, thick sedimentary layer of cohesive soil, a paper drain was installed in the ground behind the wharf and an embankment was built in an effort to promote consolidation settlement. As it was feared that this work would induce lateral flow of the soft ground and deformation of the sheet piles, numerical analysis was used to investigate the behavior of the ground and structure.

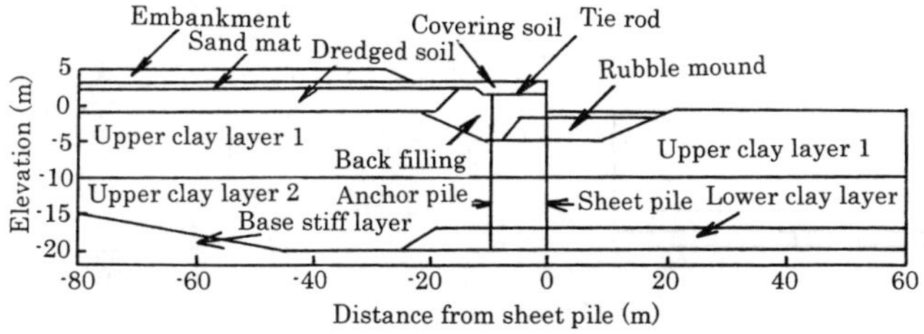

Figure 1. Cross section of the construction site at completion time

2 ANALYTICAL PROFILE AND CONSTRUCTION PROCEDURE

Figure 1 shows the soil model used in the analysis. The soil model was based on the results of an in situ soil survey. As shown in Figure 1, the sheet piles rested on a stiff base layer, corresponding to the original ground, on top of which a 20 m deep clayey layer of sediments had accumulated.

Table 1 shows the schedule of the construction work at the site. Table 2 shows the five cases analyzed which differ from each other in the width of the embankment and extent of paper drainage. The

Table 1. Construction schedule

Process	Period (days)	Cumulative period (days)
Excavation	36	36
Driving sheet piles and anchor piles	3	39
Left untouched	7	46
Rubble mound	12	58
Left untouched	4	62
Back filling	13	75
Left untouched	2,160	2,235
Dredged soil	720	2,955
Left untouched	150	3,105
Laying sand mat and covering soil	240	3,345
Embankment	210	3,555
Diving paper drains	210	3,765
Left untouched	9,235	13,000

Table 2. Cases analyzed

Case No.	Range of embank-ment	Range of paper drain
Case 1	B	B
Case 2	B	None
Case 3	A	A
Case 4	A	None
Case 5	A	B

Range A, B:refer to Fig.2

range of the width of the embankment and the area where the paper drain was installed (indicated by A and B in Table 2) are shown in Figure 2. Case 2 with no ground improvement and a wide embankment was considered to be the worst case of the five. On the other hand, case 5 with paper drain up to B (wider area of ground improvement) and embankment up to A (less wide) was considered to be the best.

3 ANALYTICAL METHOD AND MATERIAL CONSTANTS

In the analysis, the ground was modeled as elasto-plastic following Nakai and Matsuoka (1986) and Nakai (1989, 1990), while the foundation layer, sheet piles, anchor piles and tie rods were modeled as elastic bodies. In addition, the ground was modeled as a two-phase material composed of an elasto-plastic structural skeleton and incompressible pore water. The consolidation deformation was analyzed under plane strain conditions. The time period of the analysis was 13,000 days, as per the construction schedule in Table 1.

The features of the constitutive equation used in the analysis are:

1. The used constitutive equation model is applicable to both cohesive and sandy soils.
2. Through the use of the proposed constitutive equation model it is possible to obtain a thorough explanation of the behavior of soil under various stress states, such as triaxial compression, triaxial extension, plane strain, and three principal stresses, as well as under various deformation conditions.
3. The strain increment in the proposed constitutive model is divided into an elastic component, a plastic component satisfying an associated flow rule in space t_{ij}, and an isotropic plastic component, so that changes in direction of the strain increment due to differences in the stress paths can be expressed.
4. The stress-induced anisotropy of soil can be easily incorporated in the proposed constitutive equation model by using a kinematic hardening equation.

Table 3 shows the soil parameters of each layer. The embankments shown in Figure 1 are made of dredged soil and, therefore, the parameters of the dredged soil are used as the ground constants of the embankments. Additional information about the ground constants can be found in Nakai and Matsuoka (1986) for the t_{ij}-clay model, and in Nakai (1989, 1990) for the t_{ij}-sand model.

The soil constants of the cohesive soil layer were determined from the results of in situ soil surveys and physical experiments, oedometer tests, and triaxial compression tests on samples. The permeability coefficient used in the analysis corresponded to the value for the average effective overburden pressure of each soil layer obtained from oedometer tests.

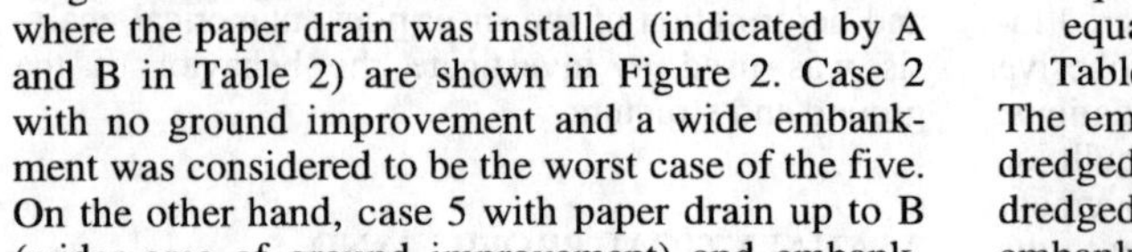

Figure 2. Range of embankment and paper drain

Table 3. Soil parameters used in analysis

Layers	Model	γ_t t/m^3	γ_{sat} t/m^3	λ (clay) or C_t(sand)	κ (clay) or C_e(sand)	m	ϕ deg	D_f	α	ξ	ν	K_0	e_0	k cm/s
Lower clay layer	tij-clay	1.70	1.90	0.109	0.009	–	39.2	–	0.80	0.2	0	0.43	0.9	1×10^{-7}
Upper clay layer 1	tij-clay	1.60	1.75	0.173	0.027	–	36.5	–	0.70	0.2	0	0.42	1.2	5×10^{-7}
Upper clay layer 2	tij-clay	1.60	1.75	0.173	0.027	–	36.5	–	0.70	0.2	0	0.42	1.2	1×10^{-7}
Rubble mound	tij-sand	1.80	2.00	0.0191	0.0055	0.25	42.3	−0.6	0.81	100	0.3	–	–	–
Back filling	tij-sand	1.80	2.00	0.0256	0.013	0.30	43.5	−0.3	1.00	100	0.3	–	–	–
Dredged soil	tij-clay	1.60	1.75	0.173	0.027	–	36.5	–	0.70	0.2	0	0.42	1.2	5×10^{-7}
Sand mat	tij-sand	1.80	2.00	0.084	0.060	0.30	40.5	−0.6	0.85	100	0.3	–	–	–
Covering soil	tij-sand	1.80	2.00	0.0256	0.013	0.30	43.5	−0.3	1.00	100	0.3	–	–	–
		γ_t t/m^3	γ_{sat} t/m^3	E kPa	ν	G kPa								
Base stiff layer	Elastic	1.80	2.00	9.8×10^4	0.333	3.7×10^4								

γ_t: Wet density
γ_{sat}: Saturated density
λ 、κ : Compression and swelling indices for clay
C_t、C_e: Compression and swelling parameters for sand
m : Power of mean effective stress
ϕ : Internal angle of friction
D_f: Slope of $\varepsilon_v \sim \varepsilon_d$ curve at failure
α : Slope of stress ratio and plastic strain increment ratio curve
ξ : Kinematic hardening parameter
ν : Poisson's ratio
K_0: Coefficient of earth pressure at rest
e_0: Initial void ratio
k: Coefficient of permeability
E,G: Young and shear modulus

Table 4. Parameters of steel materials

Member	Type	Unit	A cm^2	I cm^4	W cm^3	E kN/mm^2
Steel sheet pile	FSPⅢA	per 1m of wall width	186	22,800	1,520	206
Anchor pile	H250x250 x9x14	per one H pile	91.43	10,700	860	206
Tie rod	ϕ 28mm	per one tie rod	6.16	–	–	206

A: Cross section W: Modulus of section
I: Geometrical moment of inertia E: Young's modulus

The permeability coefficient of the paper drained, improved ground (see Figure 2) was determined by matching the results of a 1-D compression analysis using finite elements with those of Yoshikuni's (1979) vertical drain analysis which considers well resistance. In the finite element method, the soil layer constituents are the same as the original ground and only the permeability coefficient was varied. Using the method of matching, the permeability coefficient of the improved section of the ground was determined to be 10 times higher than those of the clay layers shown in Table 3.

To express the plastic properties of rubble, rubble was incorporated in the t_{ij}-sand model and its material constants were determined by referring to the results of large-scale tri-axial tests conducted by Ohmaki et al. (1997) on coarse grain materials.

The t_{ij}-sand model was used for the soil used in the back filling behind the sheet piles, the covering soil for the reclaimed area, and the sand mats used for the drains. Since no soil tests were conducted on these materials, values were determined by referring to the published soil test results of Nakai (1989, 1990) and Nakai and Hoshikawa (1991).

Table 4 shows the values of the elasticity constants and other parameters of the sheet piles, anchor piles, and tie rods, which were determined by referring to the values of the respective materials used at the construction site. In the model the sheet piles and anchor piles were considered to be hinged at the bottom.

Joint elements (Nakai 1985) were inserted between the foundation and the steel sheet pile and anchor piles to model slippage of the steel sheet pile. As the joint elements enter the plastic state, slippage can occur if the vertical stress ratio for shear at the contact surface exceeds tan δ (δ = friction angle of the wall). The wall friction angle between the ground and the sheet pile was set at $\delta = 15^{\circ}$ and at $\delta = 5^{\circ}$ for that between the ground and the anchor pile.

As Figure 1 shows, a finite element grid of 37 elements in the lateral direction and 12 in the vertical direction were used in the analysis. In the analysis, the lateral boundary was fixed in the horizontal direction and allowed to vary in the vertical direction, the lower boundary was fixed in both the horizontal and vertical directions, and the upper boundary was allowed to vary in both directions.

The hydrological boundary conditions used in the analysis were as follows:

1. The back filling and rubble placed under the water surface had a higher permeability coefficient than the cohesive soil, so they were modeled as drainage elements.
2. Drainage was not allowed from the lateral and lower boundaries.

3. The drainage at the upper boundary was varied as per the actual condition at the construction site.
4. The sheet piles marked the boundary of no drainage.

4 RESULTS OF ANALYSIS

Figure 3 shows the distribution of the horizontal displacement at each construction stage at sections B to E (see Figure 2) for case 1. From this figure, the following conclusions can be reached:

1. The horizontal displacement increased at all sections during the construction process, but after the construction was completed, some of the displacement was recovered as a result of volume compression and the ground returned back slightly toward its initial position.

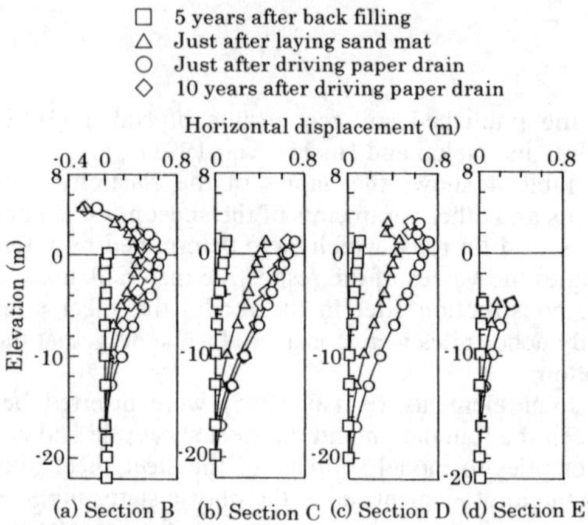

Figure 3. Horizontal displacement at sections B to E (Case 1)

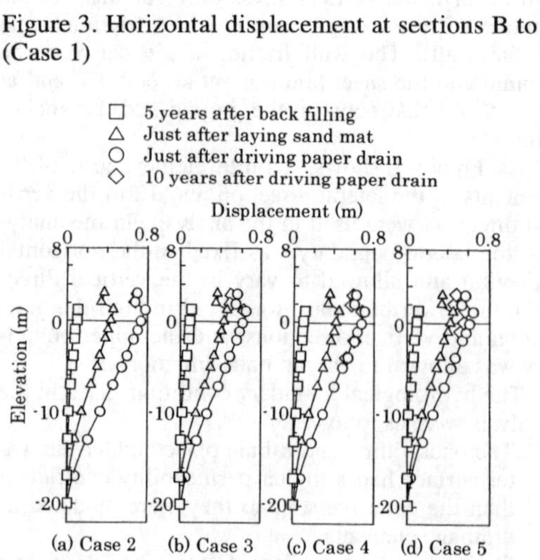

Figure 4. Horizontal displacement of sheet piles at section D (Cases 2 to 5)

2. The sheet pile (section D) and the anchor pile (section C) showed almost the same pattern of deformation due to the constraining action of the tie rods.
3. The lateral displacement of the slope of the embankment toe in section B in ground at elevations below 0 m was similar to that of the sheet and anchor piles under undrained conditions.

Figure 4 shows the distribution of the horizontal displacement at each stage of construction for cases 2 to 5 at section D. Comparing these plots with those for section D of case 1 shown in Figure 3, the following conclusions can be made:

1. In all cases, there was horizontal displacement of the sheet piles from near the bottom of the pile all the way up to the top.
2. In case 2, which corresponds to the worst construction conditions, the horizontal displacement of the sheet piles was the greatest, reaching a value of 80 cm. Also, case 4 with no ground improvement had large displacements.
3. In case 3 and case 5 with narrow embankments and improved ground the maximum horizontal displacement of the sheet piles was a relatively small 70 cm.
4. The recovery after completion of construction of the horizontal displacement near the crown of the sheet pile was greater for the cases with improved ground.
5. In all cases, the horizontal displacement of the sheet piles greatly exceeded the allowable value.

Figure 5 shows the distribution of active soil pressure behind the sheet pile, and the passive soil pressure in front of the sheet pile, at each stage of construction in case 1. In the vicinity of elevation −2 m to -5 m that had been filled in with rubble, a stress concentration was seen in both active and passive soil pressures, indicating a lateral support effect

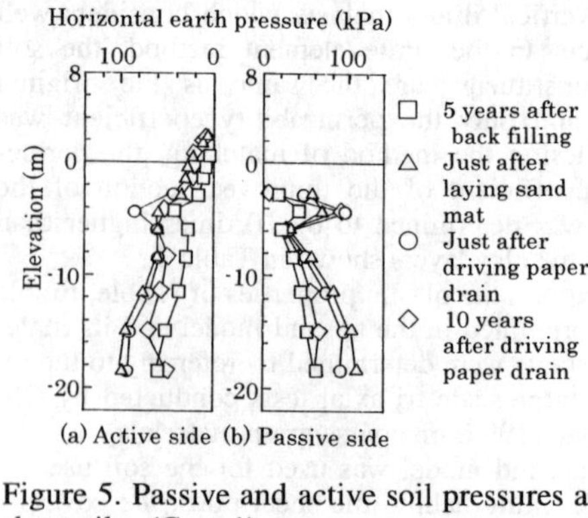

Figure 5. Passive and active soil pressures acting on sheet piles (Case 1)

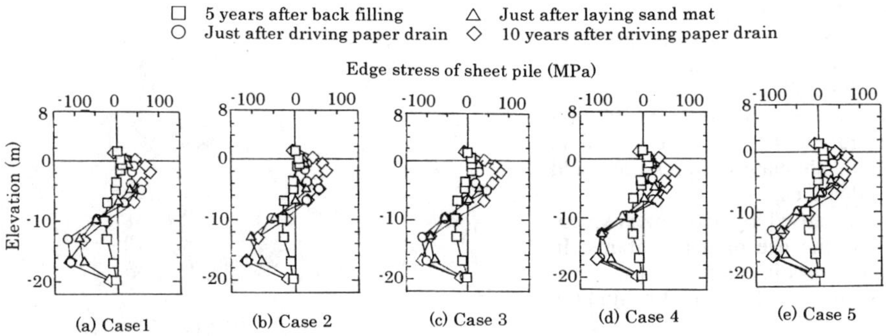

Figure 6. Edge stress on sheet piles

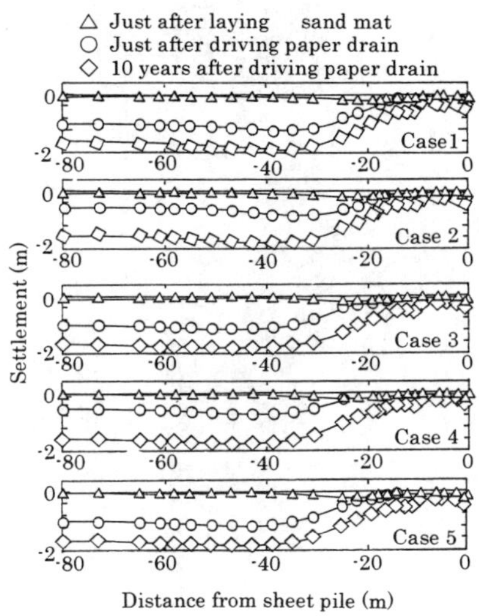

Figure 7. Settlement on ground surface

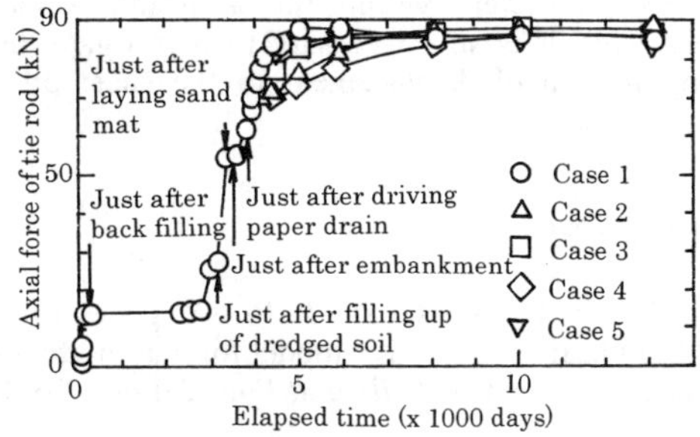

Figure 8. Temporal change in axial force of tie rods (cases 1 to 5)

of the rubble. As the construction progressed soil pressure increased but after completion of the construction it declined slightly as the horizontal displacement of the ground decreased and the ground moved back a little toward its initial position.

Figure 6 shows the distribution of the edge stress (axial stress + bending stress) of the sheet piles at each stage of construction for cases 1 to 5. The edge stress increased as the construction progressed and reached a final maximum of about 120 MPa which is below the allowable maximum of 180 MPa. Only minor differences in edge stress were observed for the different constructions.

Figure 7 shows the distribution of ground settlement on the ground surface (elevation +3 m) for cases 1 to 5 at each stage of construction. The settlement under the embankment section (at a horizontal distance of about −20 m to −80 m from the sheet pile) was slow when paper drains were not installed (cases 2 and 4), but was observed to occur at a much faster pace in cases 1, 3 and 5 with ground improvement.

Figure 8 shows temporal changes in the axial stress of the tie rod for cases 1 to 5. In all five cases, the construction was the same up to the embankment, and the changes observed in the axial force were the same. In cases 1, 3 and 5 with paper drains for improving the ground, the axial force reached its maximum values quickly and stabilized after that, but in unimproved ground (cases 2 and 4), it manifested itself very slowly. These results indicate that the axial force was significantly affected by the consolidation settlement of the ground behind the wharf.

5 CONCLUSIONS

Numerical analysis was used to investigate the behavior of stress and deformation of a wharf built on soft ground using anchor-sheet piles. The results of the analysis indicate the following:

1. As the construction progressed, the sheet piles, anchor piles, and the ground behind the wall showed lateral deformations that can not just be related to drainage effects.
2. The distribution of horizontal displacement of the sheet piles increased in proportion to the

width of the embankment. In addition, the use of paper drains to improve the ground served to suppress, in a small way, the horizontal displacement of the sheet piles

3. By covering the surface of the soft ground with materials such as rubble that are more rigid than the original ground, adequate soil pressure could be maintained in that area.
4. Ground settlement caused by the embankment was accelerated by the use of paper drains. In addition, the axial forces in the tie rods were affected by the settlement of the ground behind the structure.
5. In all sections B to E, the edge stresses of the sheet piles were within the allowable value. However, it is still necessary to investigate the suppression of displacement of the sheet piles and ground.

REFERENCES

Nakai T. & H. Matsuoka 1986. A generalized elasto-plastic constitutive model for clay in three-dimensiomal stress. *Soils and Foundations*, 26-3: 81-98.

Nakai T. 1989. An isotropic hardening elstoplastic model for sand considering the stress path dependency in three-dimensional stresses. *Soil and Foundations*, 29-1: 119-137.

Nakai T. 1990. Relationship between stress and displacement behavior of ground and mechanical properties of ground materials. 38-7: 19-25.

Nakai T. & T. Hoshikawa 1991. Kinematic hardening model for clay in three-dimensional stresses. *Proc. Computer methods and advances in geomechanics*: 655-660

Nakai T. 1985. Finite element computations for active and passive earth pressure problems of retaining wall. *Soil and Foundations,* 25-3:98-112.

Ohmaki S., K. Suzuki, A. Takagi & S. Kageyama 1997. Deformation behavior of coarse materials under axisymmetric stress. *Technical Report of the National Research Institute of Fisheries Engineering, Fisheries Agency*, 18: 111-154.

Yoshikuni H. 1979. *Design and construction management of the vertical drain construction method.* Gihodo: 40-49.

Computer Methods and Advances in Geomechanics, Desai et al. (eds) © 2001 Balkema, Rotterdam, ISBN 90 5809 183 X

3D finite element analysis of rigid pipe interaction with clay

Radu Popescu
Faculty of Engineering and Applied Science and C-CORE Memorial University of Newfoundland, St.John's, Nfld, Canada

Ibrahim Konuk
Geological Survey of Canada, Ottawa, Ont., Canada

ABSTRACT. The current practice in pipeline/soil interaction analysis uses a structural approach where the soil is represented by one dimensional (1D) springs. The shear interaction between different soil zones during complex relative displacements of a buried pipe cannot be captured. Consideration is given here to the effects of: (1) type of pipe displacement, and (2) embedment ratio. The influence of these factors on the failure mechanisms and the magnitude of interaction forces is discussed, and the validity of p-y spring methods used by the current practice is checked for complex pipe/soil displacements in clay soils under undrained loading. Results of two empirical methods for pipe/soil interaction factors are also discussed and compared with the numerical model predictions.

1 INTRODUCTION

Owing to the highly nonlinear behaviour of soil materials, pipe/soil interface phenomena and the possibility of pipe distortion, buried pipe/soil systems have a relatively complex behaviour. Two basic methods are used for numerical modelling of buried pipelines: (1) the "continuum approach", using a continuum finite element representation of both the pipe and the soil, and (2) the "structural approach", using specialized beam-type finite elements for the pipe and Winkler type representation of the surrounding soil.

The current practice in pipe/soil interaction analysis is the structural approach, which is considerably less computationally demanding than continuum modelling methods. The soil continuum is approximated by a series of discrete springs (e.g. *p-y* and *t-s* springs) representing independent, orthogonal soil deformation components (ASCE 1984). Although significant progress has been made regarding the beam type finite elements for simulating pipe behaviour, soil representation by nonlinear springs is rather crude. Besides inability of the structural methods to capture the details of pipe behaviour, such as ovalization and buckling, there are serious limitations related to the modelling of soil plastic, dilatational and hysteretic behaviour, and of shear interaction between different soil zones along the pipe. Some of these limitations can be overcome by use of relatively complicated combinations of nonlinear springs, dashpots and other simple mechanical models. However, lack of physical significance of those soil representations requires extensive experimental data and/or use of questionable empirical formulas for calibration of *p-y* model properties.

As opposed to *p-y* spring representations, plasticity soil constitutive models implemented in continuum finite element codes account for the true three-dimensional (3D) stress state and are based on constitutive equations derived from observed soil behaviour. The parameters used in the soil constitutive models are traditional soil properties that can be estimated from the results of conventional laboratory and/or in-situ soil tests. A credible 3D continuum finite element model could be a real asset in calibration of structural models and development of new simple soil-structure interaction models.

The current practice methods for estimating soil/pipe interaction forces are only considering lateral loading (i.e. translation) of pipes in soil. One objective of this study is therefore to investigate the validity of the structural approach for more complex displacements of a buried pipe. To this end, soil/pipe interaction forces predicted for two types of pipe rotation are compared with the forces predicted for pipe translation. The finite element analysis predictions are also compared with results of two widely used empirical methods for soil/pipe interaction forces, which are subsequently discussed. A finite element model using a full 3D continuum representation of the soil and accounting for the nonlinear behavior of soil materials and the relative slip and separation at the pipe/soil interface is used in a series of large displacement / finite strain analyses. The finite element code ABAQUS/Standard (Hibbitt et al. 1999) is used for the numerical calculations. Only clay materials under undrained loading are considered, and an elastic-perfectly plastic soil constitutive model is used for simulating the soil behaviour.

2 EMPIRICAL RELATIONS

In the current design practice using the structural approach, the force-displacement characteristics of the "soil springs" should relate to the actual nonlinear soil response. Reviews of pipe/soil interaction formulations are presented, among others, by Rizkalla et al. (1992) and Paulin (1998). Exemplified here are the guidelines currently used in North America for estimating lateral pipeline/soil interaction forces in clayey soils under undrained loading. These guidelines (ASCE 1984) suggest a hyperbolic relation between the lateral load per unit length of pipeline, *p*, and the lateral relative displacement between pipeline and soil, *y*:

$$p = \frac{y}{A' + B' y} \tag{1}$$

where *A'* and *B'* depend on the ultimate lateral soil load, P_{ult}, and displacement to ultimate load, Y_{ult}, as follows:

$$A' = 0.15\frac{Y_{ult}}{P_{ult}}, \quad \text{and} \quad B' = \frac{0.85}{P_{ult}} \tag{2}$$

The guidelines state that Y_{ult} occurs at approximately 3 to 5% of *H*, where *H* is the depth to pipeline base. For clays, the ultimate lateral soil load is expressed as:

$$P_{ult} = c_u N_{ch} D \tag{3}$$

where N_{ch} is the bearing capacity factor for vertical strip footings horizontally loaded, *D* is the external pipe diameter, and c_u the undrained shear strength of the soil. Various analytical models used to estimate P_{ult} as well as experimental data show substantial differences, leading to predictions differing by as much as 240% (Rowe and Davis 1982, Trautmann and O'Rourke 1985).

To estimate N_{ch} values for clays, the ASCE (1984) guidelines adapted Hansen's (1961) model for vertical piles subjected to lateral loading. The bearing capacity factor is given as a function of the embedment ratio, *H/D*. Another widely used empirical relation was proposed by Rowe and Davis (1982), based on results of elasto-plastic finite element analyses of vertically oriented anchors. They acknowledged the presence of different failure mechanisms as a function of cover depth, namely general shear failure with plastic flow to the soil surface for shallow cover, and local (punching) failure for deep cover. In their formulation the interaction force is expressed explicitly as a function of both the overburden pressure and a coefficient for the effects of overburden pressure that varies with embedment ratio. These two relations, by Hansen (1961) and by Rowe and Davis (1982), will be discussed in the following and compared with the numerical analysis results.

3 FINITE ELEMENT MODEL

Two embedment ratios are analyzed: deep cover, with H/D = 3.4, and shallow cover, with H/D = 2.2, where H is the depth to pipeline base and D is the pipe diameter (Rowe and Davis, 1982, suggested an embedment ratio of 3 as the transition from shallow to deep cover). The finite element mesh used for the deep cover case is shown in Figure 1. For one of the cases analyzed (Case 2 in the next section), the boundary conditions at the front and back boundaries of the mesh (z=0 and z=3m in Figure 1b) are not compatible with the pipe displacements. Therefore, only the central zone shown in Figure 1b was used for result output.

For the deep cover case (Figure 1), the soil was discretized using quadratic brick finite elements with 20 nodes and reduced integration. These second order elements are known to yield higher accuracy than linear elements at the same computational effort, and are more appropriate for simulating bending and capturing stress concentrations. A shortcoming of the quadratic finite elements consists of nodal forces not being directionally consistent with applied pressures. This induced serious problems when calculating the soil/pipe contact forces. It was therefore decided that linear elements (8-node bricks with reduced integration and hourglass control) be used for the second set of analyses, namely the shallow cover case.

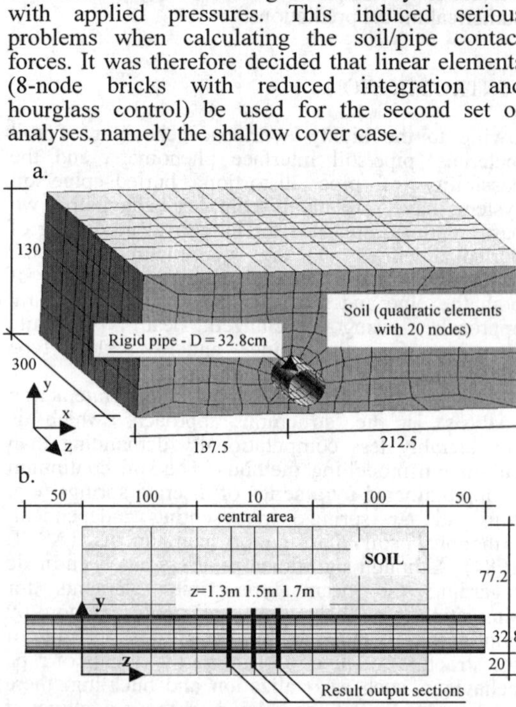

Figure 1. Simulation of a buried rigid pipe subjected to complex loading - deep cover: a. finite element mesh; b. positioning of the pipe and result output sections (dimensions in cm).

The contact surface approach implemented in ABAQUS/Standard is used to simulate the pipe/soil interface. It allows for separation and sliding of finite amplitude and arbitrary relative rotation of the contact surfaces. The model for simulating contact forces includes an equivalent shear stress limit. To allow proper modelling of the pipe/soil interface during large relative rotations, the pipe was assumed to extend out of the soil domain on both sides of the mesh. A clay material under undrained conditions was assumed, with the following properties: undrained shear strength – c_u=50kPa; deformation modulus – E=100*c_u; Poisson's ratio – ν=0.49, to simulate an almost incompressible behaviour; and shear stress limit at soil/pipe interface – $\tau_{max}=c_u/3$.

Finite strain and large displacement analyses were performed. The displacements prescribed to the pipe (in excess of one pipe diameter) induced significant mesh distortions. No remeshing was used, as ABAQUS/Standard does not have rezoning capabilities for three-dimensional analysis. As the ABAQUS linear and quadratic brick elements accommodate mesh distortions, there was little concern related to result accuracy. Moreover, ABAQUS/Standard has been previously used for 2D finite element analyses of pipe/soil interaction involving large relative deformations, and the numerical model has been validated based on comparisons with the results from full scale experiments, for both sand and clay materials (Popescu et al. 1999).

4 SITUATIONS ANALYZED

The pipe motion relative to the surrounding soil was simulated by prescribing nodal displacements in the horizontal direction (direction x in Figure 1), while the pipe was free to move in the vertical (y) direction. The force-displacement relations were obtained in three sections in the central zone of the pipe (Figure 1b). The pipe was assumed rigid in the first two cases addressed in this study, and "partially flexible" in the third case, as explained hereafter.

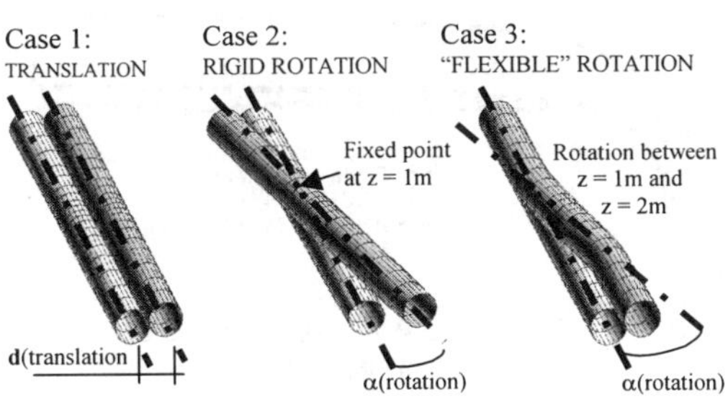

Figure 2. Types of pipe displacement analyzed

Three situations of pipe displacement were analyzed (Figure 2):

Case 1. *Rigid translation of the pipe*. Equal horizontal displacements (in the x-direction) were prescribed at the two ends of a rigid pipe. Ideally, the structural (p-y spring) models should yield results that are similar to this case, involving no shear soil interaction along pipe axis and no deflections of the pipe. Therefore this case was used as a benchmark for comparison with the results of more complex displacements, and for assessing empirical methods.

Case 2. *Rigid rotation of the pipe*. The pipe was "pulled" by prescribing horizontal displacements at two nodes in the section at z=3m, and constrained to rotate with respect to a vertical axis at z=1m. In this way, the central zone of the pipe was subjected to combined translation and rotation. This displacement pattern is considered to simulate the field behavior of a rigid pipe subjected to differential soil displacements.

Case 3. *"Flexible" rotation of the pipe*. The deformation pattern shown in Figure 2 was imposed by prescribing horizontal displacements to all nodes of the pipe. Thus, the pipe was flexible in longitudinal bending, while remaining rigid in cross section (ring deformations). This displacement pattern is believed to better simulate deformations of a flexible pipe in the field.

5 ANALYSIS RESULTS

The predicted normalized force–displacement curves are presented in terms of normalized lateral loads: $N = p / (c_u D)$, where p is the interaction force per 1m of pipe. The interaction forces are calculated along the x-direction.

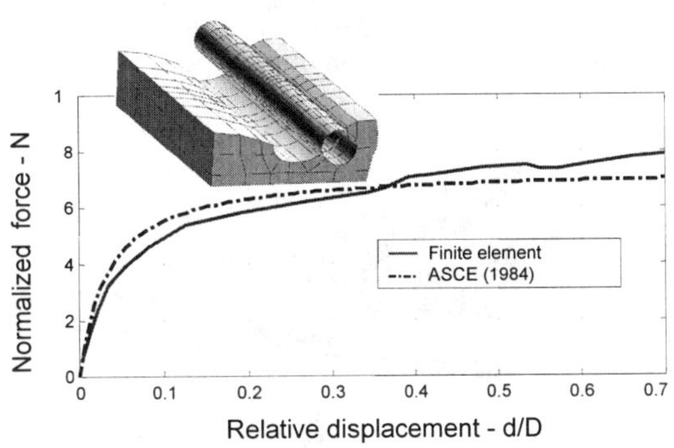

Figure 3. Pipe translation (deep cover): predicted normalized force - displacement curves.

5.1 Deep Cover

Normalized lateral loads obtained for pipe translation (Case 1) are compared in Figure 3 with the results of the empirical procedure recommended by ASCE (1984). The resulted bearing capacity factors at ultimate displacement Y_{ult} = 0.06m (or relative displacement d/D = 0.18) are: $N_u = N(Y_{ult})$ = 5.8 – predicted by the numerical model, and N_{ch} =

6.2 – resulted from both Hansen (1961) and Rowe and Davis (1982) methods. The normalized lateral loads predicted by the numerical model continued to increase with displacement and eventually predicted values exceeded the ones recommended by ASCE guidelines. Within the range of pipe displacements analyzed here, this predicted behavior is confirmed by the results reported by Rowe and Davis (1982) and Paulin et al. (1998). The soil displacement pattern (Figure 7a) indicates a punching failure mechanism. Very large horizontal soil displacements were predicted immediately in front of the pipe. There was some tendency of upward displacement of the soil, but no shear failure surface could be observed. Large lateral extension of plastic zones in the soil predicted by the numerical model (Figure 8a) confirmed the type of failure mechanism indicated in Figure 7a.

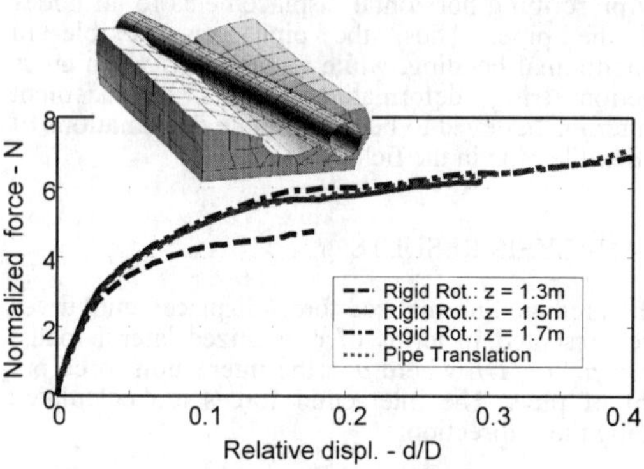

Figure 4. Rigid rotation (deep cover): predicted normalized force - displacement curves for three sections in the central region of the pipe

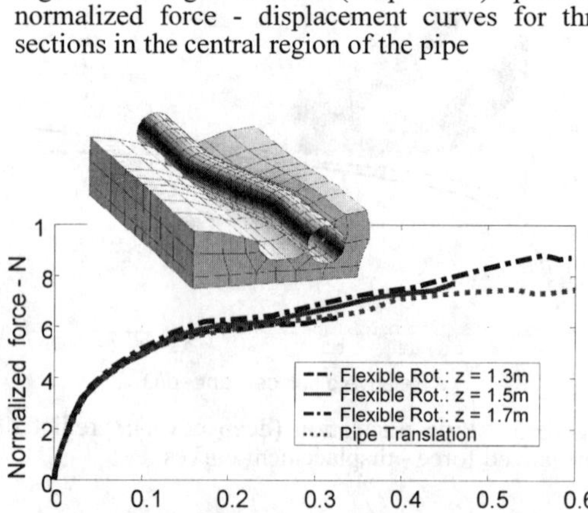

Figure 5. "Flexible" rotation (deep cover): predicted normalized force - displacement curves for three sections in the central region of the pipe

Normalized lateral loads predicted for rigid rotation of the pipe (Case 2) are compared in Figure 4 with the results obtained for pipe translation. Very close results were obtained for sections at z=1.5m and z=1.7m, that are farther away from the rotation axis. The normalized loads predicted for the section at z=1.3m was considerably smaller (N_u = 4.7, as compared to N_u = 5.8, for pipe translation). An explanation can be found in the predicted displacement pattern: the soil situated further in front of the pipe tends to rotate; rotation of the soil closer to the pipe can only be observed at points near the pipe axis of rotation. This indicates that the soil in front of the pipe in this area was partly displaced due to shear interaction with other soil zones along the pipe. This resulted in reduced soil/pipe interaction forces. The resulted bearing capacity factors predicted for Case 3 ("flexible" rotation - Figure 5) are more consistent with the values predicted for pipe translation. No rotation of the soil was predicted in the immediate vicinity of the pipe for this case.

5.2 Shallow Cover

As shown in Figure 6, the bearing capacity factors, N_{ch}, at Y_{ult} = 0.04m (displacement at ultimate load) estimated by the two empirical methods are different in this case: N_{ch} = 5.8 according to ASCE (1984) and using Hansen (1961), and N_{ch} = 4.4 using Rowe and Davis (1982). The finite element results are very close to the empirical results using Rowe and Davis method that better accounts for differences in the failure mechanisms as a function of embedment ratio. The predicted force–displacement curve is flat after reaching the ultimate bearing capacity, in accordance with the results reported by Rowe and Davis (1982) and Paulin et al. (1998). A general shear failure mechanism with large upward displacements of the soil in front of the pipe and a clearly defined failure zone is indicated by the predicted displacement vectors (Figure 7b). The extension of plastic zones is confined to the shear failure area (Figure 8b).

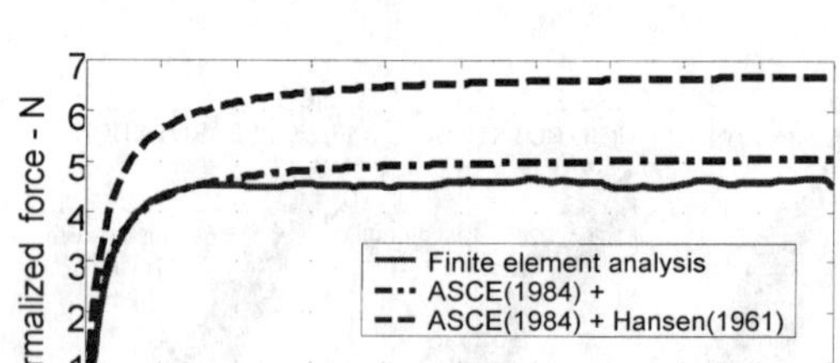

Figure 6. Pipe translation (shallow cover): predicted normalized force - displacement curves.

A summary of normalized force–displacement curves predicted for the three cases analyzed is presented in Figure 9. Bearing capacity factors predicted for rigid rotation of the pipe (Case 2) were more strongly affected by the pipe displacement than in the case of deep cover (compare Figures 9 and 4). Reductions of up to 25% with respect to pipe translation were predicted close to the axis of rotation and overall lower values were predicted along the pipe for shallow cover, as compared with the results obtained for deep cover. An explanation is that soil close to the pipe can rotate easier than in the case of deep cover. The resulted bearing capacity factors predicted for Case 3 ("flexible" rotation) are more consistent with the values predicted for pipe translation. There are however larger differences than predicted for the deep cover case, as follows: bearing capacity factors lower by about 5% close to the rotation axis (z=1.2m), and larger by 5 to 10% in the other two sections.

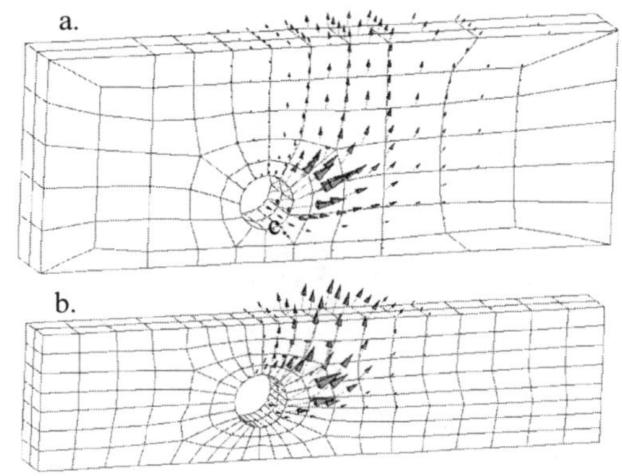

Figure 7. Predicted displacements in soil at the end of analysis: a. deep cover (pipe displacement - d=30cm); b. shallow cover (d = 35cm).

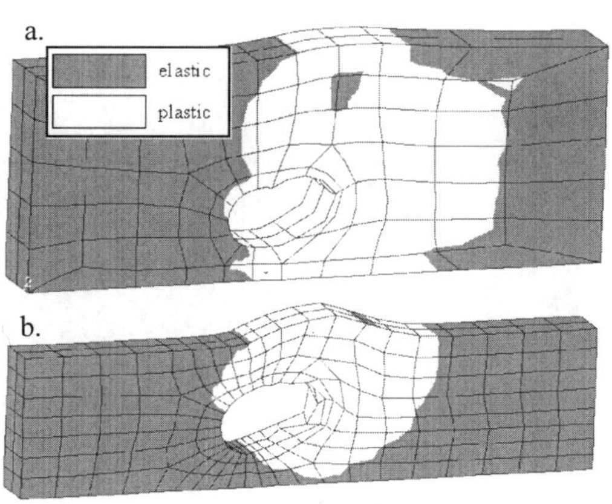

Figure 8. Predicted extension of plastic zones at the end of analysis: a. deep cover; b. shallow cover.

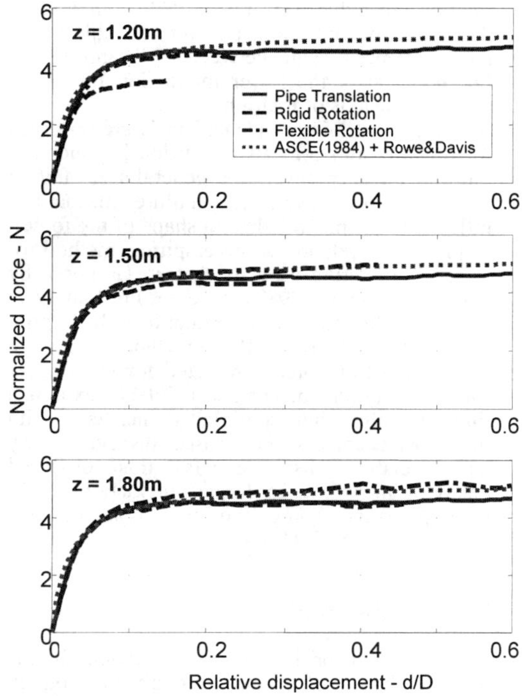

Figure 9. Shallow cover: comparison between normalized force – displacement curves predicted in the central area of the pipe for the three types of pipe displacement.

6 CONCLUSIONS

Three-dimensional continuum finite element analyses of a rigid pipe buried in soil and subjected to complex loading were performed to infer the effects of shear interaction in soil during combined translation and rotation of the pipe. Three types of displacement were imposed to a rigid pipe buried in clay (translation, rigid rotation and "flexible rotation), assuming undrained conditions and elastic perfectly plastic soil. Two embedment ratios were considered: deep cover, with H/D = 3.4, and shallow cover, with H/D = 2.2. The objectives were: (1) to investigate the validity of the structural approach for complex displacements of a buried pipe, and (2) to assess two widely used empirical methods for soil/pipe interaction forces. Based on the situations analyzed in this study, it is concluded that:

1. For large embedment ratios (H/D > 3), the soil/pipe interaction forces are not significantly affected by pipe rotations and by the deflected shape of the pipe. For small embedment ratios (H/D < 3), lateral shear interaction between soil zones along pipe axis becomes significant. Interaction forces for rigid pipes subjected to

rotation are reduced by up to 25%, compared to the forces in translation. For very flexible pipes, the differences range between –5% and +10%. For both cases, the lower interaction forces act closer to the axis of rotation.

2. Different soil failure mechanisms were observed for the two situations analyzed: local (punching) failure - for deep cover, and general shear failure - for shallow cover. The failure mechanism influenced the magnitude and shape of the force - displacement relations. The empirical method for estimating bearing capacity factors proposed by Rowe and Davis (1982) gave results that were closer to the numerical predictions than those resulted from Hansen's (1961) method.
3. The interaction forces predicted for deep cover (punching failure mechanism) did not exhibit a limiting value, but continually increased with pipe displacements. For cases involving very large relative displacements, this observed behavior may lead to larger interaction forces than predicted using the *p-y* hyperbola (1) proposed by ASCE (1984).

ACKNOWLEDGEMENTS

Financial support for this study was provided by the Geological Survey of Canada through the Program of Energy Research and Development (PERD) and by the NSERC Grant No. RG203795-98.

REFERENCES

ASCE (1984), "Guidelines for the seismic design of oil and gas pipeline systems", Committee on Gas and Liquid Fuel Lifelines, ASCE, New York.

Hibbit, Karlsson & Sorensen, Inc., (1999). *ABAQUS Theory Manual*, version 5.8.

Hansen, J.B. (1961). The ultimate resistance of rigid piles against transversal forces. *Bulletin 12*, Danish Geotech. Inst., Copenhagen, pp.5-9.

Paulin, M.J. (1998). *An Investigation into Pipelines Subjected to Lateral Soil Interaction*, Ph.D. Thesis, Fac. Eng. Appl. Science, Memorial Univ. of Nfld., St. John's, NF.

Paulin, M.J., Phillips, R., Clark, J.I. and Boivin, R. (1998). An experimental investigation into lateral pipeline/soil interaction - Phase II. *Proc. Centrifuge '98*, Kimura, Kusakabe & Takemura (eds), Balkema, Rotterdam, pp.699-704.

Popescu, R., Philips, R., Konuk, I., and Deacu, D., (1999). Numerical and physical modeling of pipe-soil interaction. *Proc. 52nd Canadian Geotech. Conf.*, Regina. SK, pp. 437-444.

Rizkalla, M., Poorooshasb, F. and Clark, J.I. (1992). Centrifuge modelling of lateral pipeline/soil interaction. *Proc. 11th Offs. Mech. Arctic Eng. Symp*, 13p. insert.

Rowe, R.K. and Davis, E.H. (1982). The behaviour of anchor plates in clay. *Geotechnique*, 32(1):9-23.

Trautmann, C.H. and O'Rourke, T.D. (1985). Lateral force-displacement response of buried pipe. *Journ. Geotech. Eng.*, 111(9):1077-1092.

Computer Methods and Advances in Geomechanics, Desai et al. (eds) © 2001 Balkema, Rotterdam, ISBN 90 5809 183 X

Scale effect of a shallow anchor in sand masses of different densities

T.Sakai & T.Sano
Faculty of Agriculture, Ehime University, Matsuyama, Japan

T.Tanaka
Graduate School of Agriculture and Life Sciences, Tokyo University, Japan

ABSTRACT: This study was designed to determine the scale effect and progressive failure of a shallow circular anchor in sand masses of different densities by comparing experimental results and the results of FE analysis. In the FE analysis, a shear band effect was introduced into a constitutive equation. In the case of a relatively dense sand mass, the progressive failure differed according to the anchor diameter and a clear scale effect was observed. In a relatively loose sand mass, there was no clear difference in progressive failure according to the anchor diameter and there was no scale effect. The scale effect differed according to the density of the sand mass. Since the scale effect is considered to be progressive failure with shear banding, the difference in the scale effect according to density of the sand mass is attributed to progressive failure.

1 INTRODUCTION

De Beer (1965) discussed scale effect and reported that the bearing capacity factor was presented as a function of the diameter of footing and decreased with an increase in the footing width. The result was subjected to the scale effect. As the scale effect causes trouble in the transformation of model test results to a prototype, it has been studied by many investigators using model tests and analysis (Yamaguchi et. al., 1977, Ovesen, 1980, Tatsuoka et. al., 1991).

The anchor problem provides a useful analogue for soil structure interaction problems. Various rigid plastic analyses of anchor uplift resistance have been carried out by many authors (Ball, 1961, Meyerhof and Adams, 1968, Sutherland, 1988). However, the scale effect can not be evaluated in rigid plastic analysis because this analysis involves arbitrary assumptions regarding the shape of the failure surface observed during the tests and it ignores the effect of progressive failure. In rigid plastic analysis, the dimensionless uplift resistance can be kept at a relatively constant value when the function of depth to anchor width ratio is constant. The scale effect can be caused by progressive failure and crushing of sand particles under high confining pressure. The experimental studies on the scale effect are carried out using a centrifugal testing technique. As the centrifugal test may not be negligible a crushing of sand particle, it is not enough to evaluate the scale effect due to progressive failure. Sakai and Tanaka (1998a, 1998b) discussed the scale effect and progressive failure of an anchor in the case of dense sand and a conventional 1g condition. However, there have been no studies on the scale effect and progressive failure in sand masses of different densities. The present study was therefore designed to determine the scale effect and progressive failure of a circular anchor in sand masses of different densities by comparing experimental results and the results of FE analysis. All tests were conducted in normal gravity conditions, and the shear band effect was introduced into a constitutive equation for FE analysis in order to evaluate the progressive failure with shear banding.

2 TEST APPARATUS AND FINITE ELEMENT ANALYSIS

The test apparatus consisted of a cylindrical container that was 59 cm in diameter and a anchor pullout machine. The model anchors used in the tests were flat circular steel plates of 5mm in thickness and with a diameter (D) of 5 cm, 7 cm or 10 cm. An embedment ratio (h/D, h: depth of sand mass) of two was used in the tests. The sand used for the tests was Toyoura sand. Toyoura sand is a Japanese standard sand with physical properties are G_s=2.64, D_{50}=0.16mm, e_{max}=0.98, e_{min}=0.61 and no fines contents of less than 0.074mm in diameter. The tests were conducted using sands of three different densities to evaluate the scale effect due to progressive failure in sands of different densities. The sand mass was prepared by pluviating air-dried sand. The dry densities of the sand mass were 16.1 kN/m^3 (dense condition), 15.2 kN/m^3 (medium condition) and 14.5 kN/m^3 (loose condition). The anchor was drawn up using a D.C. motor until a shear band had completely passed into the sand mass. The anchor pullout rate was approximately 0.3 mm/min. In order to observe the shear band propagation in the sand mass, an apparatus that was divided in two was used (half-apparatus). Colored sand layers were placed against the front glass wall of the half-apparatus. The shear band propagation observed in the half-apparatus agreed with exposed it at central plane in the full-apparatus.

For the FE analysis, the shear band effect was introduced into a constitutive equation. An elasto-plastic model with a non-associated flow rule and strain hardening-softening yield properties was used for the constitutive model. A yield function of the Mohr-Coulomb type and a plastic potential function of the Drucker-Prager type were employed. The element employed for the analysis was a pseudo-equilibrium model by special reduced integration of a 4-noded Lagrange-type element. A dynamic relaxation method with a return mapping algorithm was applied to the integration algorithm of the elasto-plastic constitutive relation including the shear band effect.

In the return mapping algorithm (Ortiz and Simo, 1986), the elastically predicted stresses (σ_A) are related onto a suitably updated yield surface (σ_B). A change in stresses can cause an associated change in the elastic strain given by

$\varepsilon^e=[D]^{e-1}(\sigma_A-\sigma_B)$,

where $[D]^e$ is the elastic matrix.

Since the total strain does not change during the relaxation process, the plastic strain change is balanced by an equal and opposite change in the elastic strains:

$S\,\varepsilon^p=-\varepsilon^e=-[D]^{e-1}(\sigma_A-\sigma_B)$.

Shear banding was introduced by using the strain localization parameter of S.

The ratio S was incorporated into the elasto-plastic constitutive model as a characteristic length scale. S is the function of the shear band thickness and is the ratio of the shear band area to the finite element area:

$S=F_b/F_e$,

where F_b is the area of the shear band in each finite element and F_e is the area of a finite element.

If the effect orientation of the shear band in each element is ignored, the ratio S is approximately given by

$S=W/\sqrt{F_e}$,

where W is the shear band thickness.

The peak friction angle (ϕ_p) was estimated from the empirical relations proposed by Bolton (1986). The dilatancy angle (ψ) was estimated from Rowe's stress-dilatancy relationship.

Three unit weight (γ) values were used for the analysis: 15.7 kN/m^3, 15.2 kN/m^3 and 1.47 kN/m^3. The values of residual friction angle (ϕ_r), Poisson's ratio (υ) and initial shear modulus (G_0) were based on data obtained from tests conducted on air-pluviated dense Toyoura sand (Tatsuoka et. al., 1986a and 1986b):

$\phi_r=33^\circ$, $\upsilon=0.3$ and $G_0=78400$kPa.

A shear band thickness of 0.3cm was used on the basis of results of studies on shear band thickness (Vardoulakis, 1981, Yoshida et. al., 1994, Sakai, 1997).

Back-prediction of a triaxial compression test was carried out using the material properties, and the calculated results were in good agreement with the experimental result obtained by Tatsuoka et al.(1998b).

3 EXPERIMENTAL AND NUMERICAL RESULTS

Fig. 1 shows the relationships between uplift resistance and displacement curves for sands of different densities and a model anchor with D of 5 cm. In the dense condition, uplift resistance decreased rapidly after peaking and eventually became a constant residual value. As the density of sand mass became looser, peak uplift resistance decreased. In the loose condition, peak uplift resistance was not clearly seen. In all tests, the constant residual values were similar.

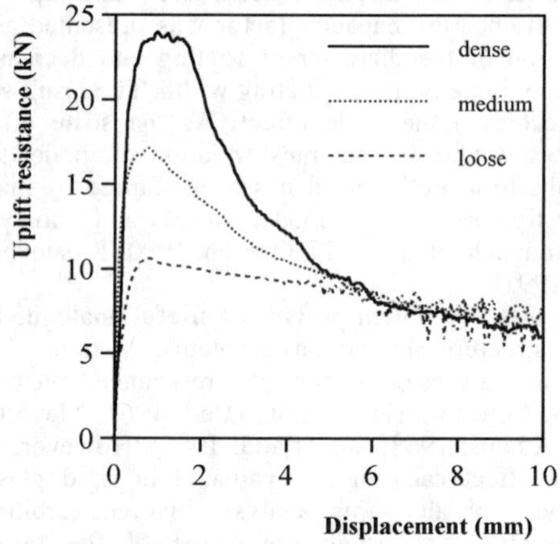

Fig.1 Uplift resistance-displacement curves

Fig.2 shows the calculated and experimentally determined uplift resistance-displacement curves for D=5cm in the dense condition ($\gamma=16.1$ kN/m^3). The maximum resistance obtained by analysis is very close to the experimental value. Fig. 3 shows the relationships between peak dimensionless uplift resistance (Nq) and D obtained by experiments and analysis. In the dense condition, Nq generally decreased with an increase in D and the scale effect was clearly seen. There was no remarkable reduction in Nq with increases in D as the density of the sand mass was decreased, and there was no scale effect in the loose condition.

Fig. 4 shows the shear band propagation at a displacement of 10 mm in dense and medium conditions. An inner secondary shear band developed in the dense condition, but the primary shear band progressed in the medium condition. In the dense condition, shear band propagation was more progressive and progressive failure was remarkable. The interior angles between the direction of primary shear band propagation and the horizontal line is approximately 65 degrees in the dense condition and approximately 70 degrees in the medium condition. The inclination angle depended on the density of the sand mass. This finding corresponds to the result of a trap-door problem (Vermmer, 1985, Tanaka and Sakai, 1987).

Fig. 5 shows the calculated shear strain distribution of an anchor with D of 5 cm at a displacement of 5 mm in sands of different densities. The shear strain shown in this figure is the apparent maximum shear strain averaged at the element level. The direction of the localized narrow zone is calculated to be approximately identical to the direction of shear band propagation observed in the experiment. A steeper concentrated zone of shear strain developed upward as the density became looser.

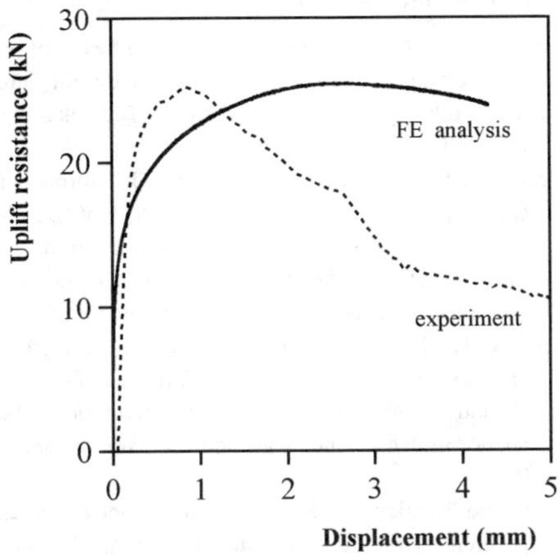

Fig.2 Comparison of the result of FE analysis and experiment

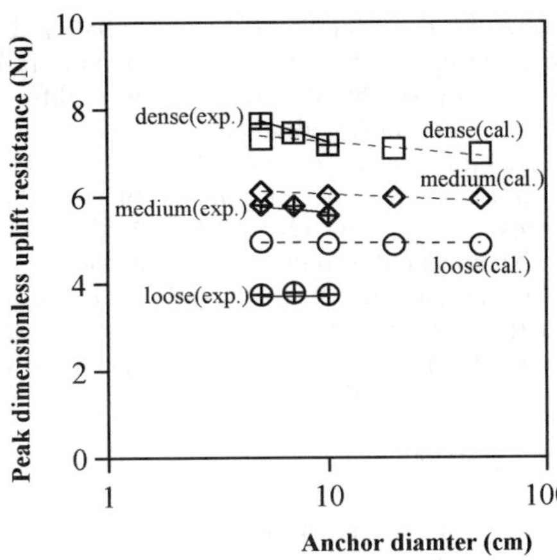

Fig.3 Relationship between peak uplift resistance and anchor diameter

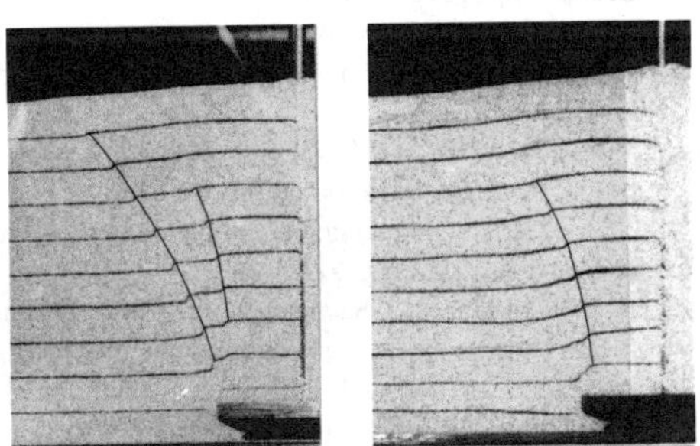

Fig.4 shear band propagation at disp.=10mm

Fig. 6 shows the calculated ϕ_{mob} distribution in dense and loose conditions with anchors of 5 cm and 50 cm in diameter and a displacement of 2mm, where there appeared to be approximately maximum uplift resistance.

ϕ_{mob} was calculated by the following equation:

$\phi_{mob}=\sin^{-1}\{(\sigma_1-\sigma_3)/(\sigma_1+\sigma_3)\}$.

A distribution domain of more than 40 degrees of ϕ_{mob} was not seen in the loose condition, and the uplift resistance in the loose condition was smaller than that in the dense condition. In the dense condition, as the distribution domain of more than

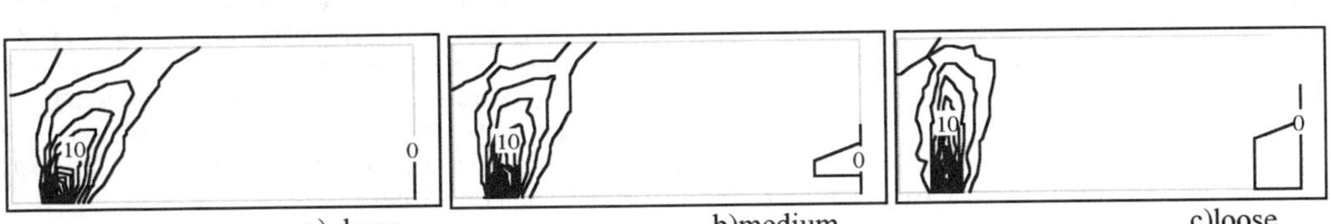

Fig.5 Calculated shear strain distribution at disp.=5mm (unit: %)

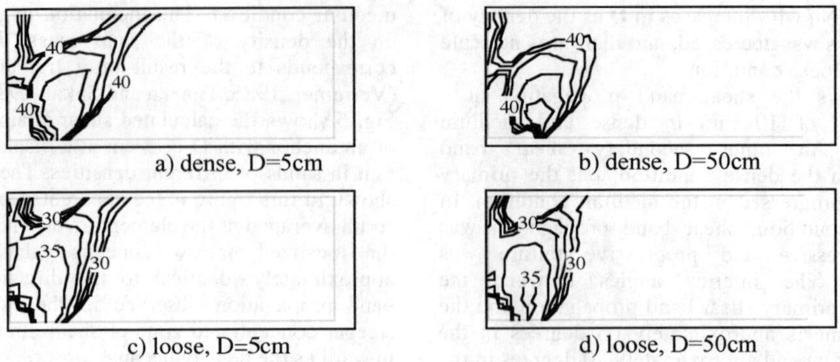

Fig.6 Calculated ϕ_{mob} distribution at disp.=2mm (unit: degree)

40 degrees of ϕ_{mob} for D=50cm was smaller than that for D=5cm, Nq decreased with an increase in D. The results showed that progressive failure differed according to the anchor diameter and there was a clear scale effect in the dense condition. In the loose condition, as the distribution domains of more than 30 degrees of ϕ_{mob} for D=5cm and D=50cm were similar, Nq appeared to remain almost constant even if the anchor diameter was increased. The results showed that there was no difference in progressive failure according to anchor diameter and that there was no scale effect in the loose condition.

4 CONCLUSION

This study was designed to determine the scale effect and progressive failure of a shallow circular anchor in sands of different densities by comparing experimental results and the results of FE analysis. In the FE analysis, the shear band effect was introduced into a constitutive equation. In the case of a relatively dense sand mass, the progressive failure differed according to the anchor diameter and a clear scale effect was observed. In a relatively loose sand mass, there was no clear difference in progressive failure according to the anchor diameter and there was no scale effect. The scale effect differed according to the density of the sand mass. Since the scale effect is considered to be progressive failure with shear banding, the difference in the scale effect according to density of the sand mass is attributed to progressive failure with shear banding.

REFERENCES

Balla, A. 1961. The resistance to breaking-out mushroom foundations for pylons. Proc. 5th Int. Conf. on SMFE: 569-576

De Beer, E. E. 1965. Bearing capacity and settlement of shallow foundations on sand. Proc. of a Symp. held at Duke University.

Bolton, M. D. 1986, The strength and dilatancy of sands. Geotechnique. 36, 65-78, discussion on the strength and dilatancy of sands by Tatsuoka. 1987. Geotechnique. 37: 219-226

Meyerhof, G. G. and Adams, J. I. 1968. The ultimate uplift capacity of foundations. Can. Geotech. J. 5(4): 225-244

Ortiz, M. and Simo, J. C. 1986. An analysis of a new class of integration algorithms for elastoplastic constitutive relations. Int. J. Numer. Meth. in Engrg. 23: 353-366

Ovesen, N. K. 1981. Centrifuge tests of uplift capacity of anchors. Proc. 10th Int. Conf. on SMFE: 717-722

Sakai, T. and Tanaka, T. 1998a. Scale effect of a shallow anchor in dense sand. Soils and foundations. 38(2): 93-99

Sakai, T. and Tanaka, T. 1998b. Particle size effect of anchor problem with granular materials. Proc. 4th European Conf. on Numer. Meth. in Geotech. Engrg.: 191-200

Sutherland, H. B. 1988. Uplift resistance of soils. Geotechnique. 38: 493-516

Tanaka, T. and Sakai, T. A trap-door problem in granular materials. Irrigation Engineering and Rural Planning. 11: 8-23

Tatsuoka, F., Sakamoto, M., Kawamura, T. and Fukushima, S. 1986a. Strength and deformation characteristics of sand in plane strain compression at extremely low pressures. Soils and Foundations. 26: 65-85

Tatsuoka, F., Goto, S. and Sakamoto, M. 1986b. Effects of some factors on strength and deformation characteristics of sand at low pressure. Soils and Foundations. 26: 105-114

Tatsuoka, F., Okahara, M., Tanaka, T., Tani, K., Morimoto, T. and Siddiquee, M. S. A. 1991. Progressive failure and particle size effect in bearing capacity of footing on sand, Proc. ASCE Geotechnical Engineering Congress, Geotechnical Special Publication, 27, Vol. 2: 788-802

Vermeer. P. A. 1985. The uplift resistance of shallow embedded anchors. Proc. 7th Int. Conf. on SMFE: 1-4

Yamaguchi, M., Kimura, T. and Fuhii, N. 1977. On the scale effect of footings in dense sand. Proc. 9th Int. Conf. on SMFE: 795-798

Computer Methods and Advances in Geomechanics, Desai et al. (eds) © 2001 Balkema, Rotterdam, ISBN 90 5809 183 X

An analytical study to estimate horizontal displacement of piled raft subjected to lateral load

T.Tsuchiya
Muroran Institute of Technology, Japan

K.Kida
Taisei Corporation, Japan

ABSTRACT: A hybrid method for analyzing the behavior of piled raft foundations in homogenous soils is developed and presented in this paper. Parametric studies of piled rafts are conducted involving several parameters, such as the number of piles *n*, pile length to diameter ratio *L/d*, pile spacing to diameter ratio *s/d*, stiffness ratio of pile to soil K_{ps}, and stiffness ratio of raft to soil K_{rs}. The behavior of piled raft is significantly affected if the raft is in contact with the soil surface. The horizontal displacement of a piled raft, normalized by $H/(E_s \cdot d)$ where H is the lateral load per unit pile and E_s is the Young's modulus of soil, is found to be strongly related to *n, s/d, L/d* and K_{ps}, while being virtually independent of K_{rs}. Based on these results, a simple method for quickly and easily estimating the horizontal displacement of piled raft is presented.

1. INTRODUCTION

There is an increasing appreciation by foundation designers that the use of piled raft may provide a cost-effective foundation solution when settlement or differential settlements need to be reduced (e.g. Poulos, 1994). However, the application of this type of foundation is a special case in Japan because the stability of the piled raft during earthquakes has not yet been clarified. It is necessary to urgently study the behavior of piled rafts subjected to lateral loads.

In order to analyze the behavior of piled rafts, the interaction between soil, pile and raft must be considered. Analytical techniques such as FEM (Finite element method) or BEM (Boundary element method) generally require the use of a large computer system. Therefore, we developed a hybrid method that combines FEM, Winkler's springs and Mindlin's solutions, and that also takes advantage of symmetry and antisimmetry to reduce the amount of computation that is required. Based on the analytical results, a simple calculation method for estimating the horizontal displacement of a piled raft is presented.

2. ANALYTICAL METHOD AND RESULTS

2.1 *Configuration of piled raft*

Three different piled raft systems are examined in this study, as shown in Figure 1; square groups

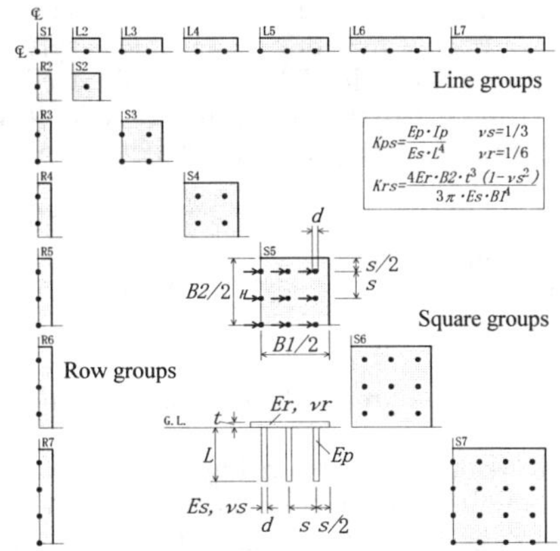

Figure 1. Details of piled rafts

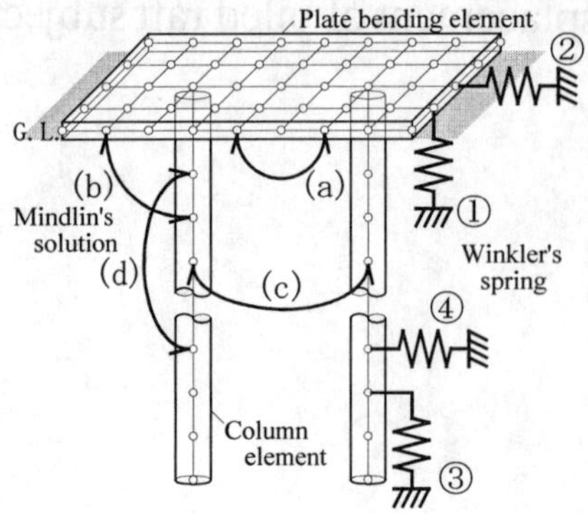

Figure 2. Hybrid model

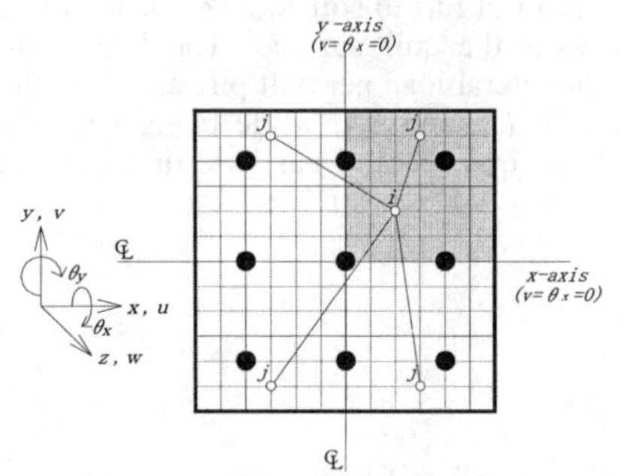

Figure 3. Symmetry and antisymmetry condition

(single pile and 2 × 2 to 7 × 7 piles), line groups (2 to 7 piles), and row groups (2 to 7 piles). The piles have diameter d, length L, and Young's modulus E_p of the equivalent solid pile section arranged at a pile spacing s below the raft. The piled rafts are subjected to a lateral load H at each pile head. The problem can be defined in terms of the following series of dimensionless groups (*; standard model):

pile length	L/d = 12.5, 25*, 50
pile spacing	s/d = 2.5, 5*, 10
pile stiffness ratio	$K_{ps} = 10^{-4}, 10^{-3}*, 10^{-2}$
raft-soil stiffness ratio	K_{rs} = 0.01, 0.1*, 1, 10

2.2 *Hybrid model introduced boundary condition*

A hybrid model is shown in Figure 2. This model combines finite elements (an elastic program with a plate bending element for the raft and a column element for the pile), Winkler's springs (① ~ ④) and Mindlin's solutions ((a) ~ (d)). The analytical model is similar to that used by Chow (1986), Clancy & Randolph (1993) and Matsumoto (1998). However, in order to reduce the amount of computation required, we take advantage of symmetry and antisimmetry with respect to the x-axis (parallel to load; $v = \theta x = 0$) and the y-axis (orthogonal to load; $v = w = \theta x = 0$), respectively. Therefore, we can analyze the behavior of a piled raft subjected to lateral load only in the region of oblique lines. Each pile is composed of 20 column elements and each region of raft adjacent two piles is composed of 4 plate elements, as shown in Figure 3.

2.3 *Analytical results*

The horizontal displacement of rafts for square pile groups, normalized by $H/(E_s \cdot d)$ where E_s is the Young's modulus of the soil into which the piles are driven, decreases as s/d increases, (i.e. area of the raft increases) and the number of piles decreases (i.e. total lateral load decreases), as shown in Figure 4. Although similar tendencies were seen for increases in L/d and K_{ps}, the effect of K_{rs} was very small.

In order to investigate the effect on horizontal displacement when the raft transmits part of the applied lateral load directly to the soil, we also analyzed free standing pile groups in which the raft is not in contact with the soil. The reducing ratio R_r of the horizontal displacement of the raft for square piled rafts to that for free standing square pile groups is shown in Figure 5. R_r decreases as the number of piles, L/d and K_{ps} decrease and s/d increases.

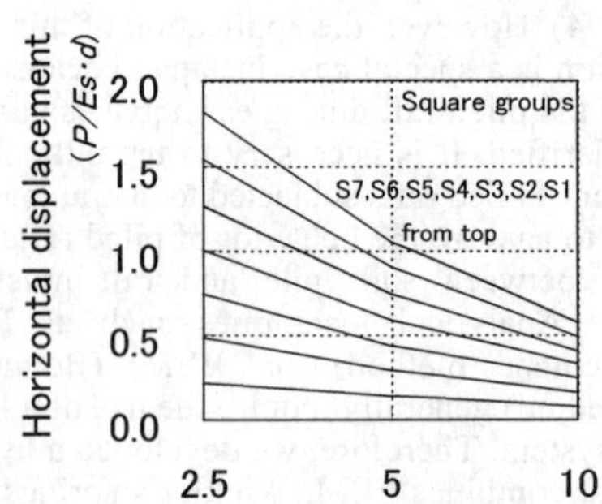

Figure 4. Displacement of piled raft against s/d for square groups

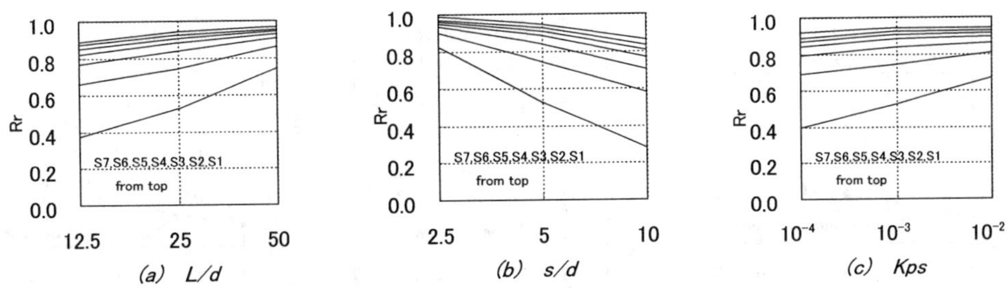

Figure 5. Reducing ratio R_r of horizontal displacement for square groups

3. SIMPLE CALCULATION METHOD FOR HORIZONTAL DISPLACEMENT

Based on these analytical results, a simple method for estimating the horizontal displacement of a piled raft is proposed as follows (see Figure 6):

$$\delta_3 = Fd_{31} \cdot \delta_1 \qquad (1)$$

where,

δ_3; horizontal displacement of piled raft

δ_1; horizontal displacement of single pile, calculated by Chang's solution (fixed at pile head)

Fd_{31}; displacement ratio of piled raft to single pile, calculated by Chang's solution ($= Fd_{21} \cdot Fd_{32}$)

Fd_{21}; displacement ratio of single pile with raft to single pile, calculated by Chang's solution

Fd_{32}; displacement ratio of piled raft to single pile with raft

It has been confirmed that Fd_{21} is affected by many factors of associated with the pile, raft and soil. Therefore, the displacement ratio of the standard model is modified as follows:

$$Fd_{21} = m_L \cdot m_s \cdot m_p \cdot m_r \cdot Fd_{21}^* \qquad (2)$$

where, Fd_{21}^*; standard model (= 0.86), and m_L, m_s, m_p and m_r; correction factor of L/d, s/d, K_{ps} and K_{rs}, as shown in Figure 7.

Fd_{32}, which is not significantly affected by factors other than the number of piles n, can be expressed as

$$Fd_{32} = 1.16n^{1/2} - 0.16 \text{ (for square groups)} \qquad (3)$$

$$Fd_{32} = n^{0.45} \text{ (for line and row groups)} \qquad (4)$$

Fd_{31} calculated by the simple method for various values of K_{ps} is in good agreement to that calculated by the hybrid method, as shown in Figure 8. Similar agreements were found for L/d, s/d and K_{rs}. The differences in the calculations by both methods were within about 20%, even in the cases for which these factors differed significantly from standard values.

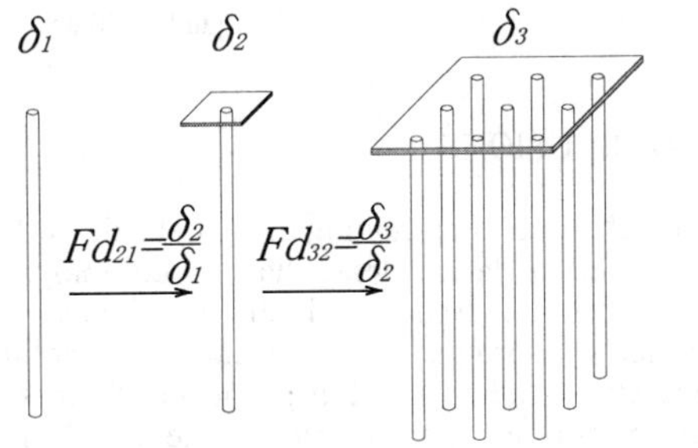

Figure 6. Definition of displacement ratio, Fd_{21} and Fd_{32}

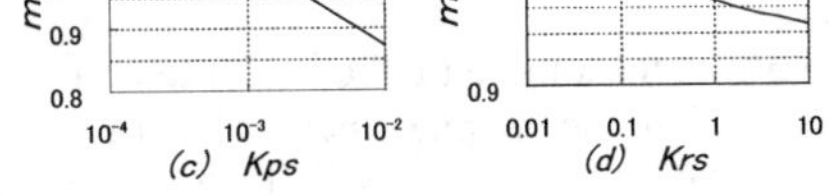

Figure 7. Correction factors; m_L, m_s, K_{ps} and K_{rs}

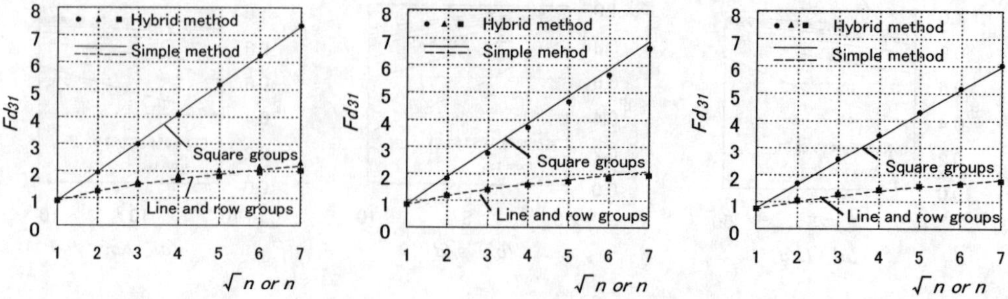

(a) $Kps=10^{-4}$ (b) $Kps=10^{-3}$ (c) $Kps=10^{-2}$

Figure 8. Comparison of solutions for Fd_{31} against K_{ps}

4. CONCLUSIONS

Piled rafts in a deep homogeneous elastic soil mass and subject to lateral loads have been analyzed by the hybrid method, which combines FEM, Winkler's springs and Mindlin's solutions. Parametric studies involving various factors have been performed, and a simple method has been proposed for estimating the horizontal displacement of piled rafts.

The principal results are summarized as follows:

1) For a given number of piles, the horizontal displacement of a piled raft, normalized by $H/(E_s{\cdot}d)$, decreases with increasing pile spacing, pile length and pile stiffness.
2) The reducing ratio R_r of the horizontal displacement of the piled raft to that of the free standing pile group decreases with decreasing number of piles, pile length and pile stiffness, and increasing pile spacing.
3) The difference between the horizontal displacements calculated by the simple method and the hybrid method is less than 20% in all cases.

Therefore, the simple method is suitable for approximating the displacement of a raft in piled raft foundations, which will lead to a clearer understanding of the behavior of this foundation type subjected to lateral loads.

REFERENCES

Poulos, H.G. 1994, Alternative design strategies for piled raft foundations (Keynote paper), 3-rd Int. Conf. on Deep Foundation Practice, Singapore, 239-244.

Tsuchiya, T. 1997, A practical approach for settlement of piled raft, 9-th Int. Conf. of the Association for Computer Methods and Advances in Geomechanics, Wuhan, China, 2117-2122.

Chow, Y.K. 1986, Analysis of vertically loaded pile groups, Int. Jour. Num. and Anal. Methods in Geomechanics, Vol.10, 59-72.

Clancy, P. & Randolph, M.F. 1993, An approximate analysis procedure for piled raft foundation, Int. Jour. Num. and Anal. Methods in Geomechanics, Vol.17, 849-869.

Matsumoto, T. 1998, Analysis of piled raft model provided by ISSMGE TC-18, –Part 4: Analyses of piled raft subjected to vertical loading and lateral loading-, Report to ISSMGE TC-18, Ghent, Belgium.

Computer Methods and Advances in Geomechanics, Desai et al. (eds) © 2001 Balkema, Rotterdam, ISBN 90 5809 183 X

Soil structure interaction analysis transmission line tower foundation

A.Varadarajan & A.Ghosh
Department of Civil Engineering, IIT Delhi, New Delhi, India

ABSTRACT: Soil-Structure interaction analysis of a 66 KV transmission tower foundation is conducted using finite element method. Plate elements and Winkler model are used to characterize the reinforced concrete raft and the sandy silt soil in the linear elastic analysis. A parametric study is conducted by varyiung the relative flexibility of the foundation soil system. Significant differences in contact pressure, shear force and bending moment are observed between the values obtained from the conventional (rigid) and interaction analysis. However, the variation in relative flexibility of the foundation-soil system is found to have relatively less effect on the shear force and bending moment.

1 INTRODUCTION

In recent times tall and heavy towers are used in the transmission of high and extra high voltage. The foundation of the transmission tower is subjected to down thrust, uplift and horizontal thrust which are reversible in nature. The design of tower foundation is often based on the assumption that the foundation is rigid. The interaction analysis of foundation structures assumes the soil medium to consist of linear elastic springs (Winkler model), elastic continuum or more complex elastoplastic continuum for example (Hotenyi, 1946, Cheung and Nag, 1968 Selvadurai, 1979). This paper deals with the soil-structure interaction analysis of a transmission tower foundation using finite element method with Winkler springs to model the foundation medium. The results are compared with those of conventional (rigid) method. A parametric study is also conducted to investigate the relative flexibility of soil-structure system on the behaviour of the foundation structure.

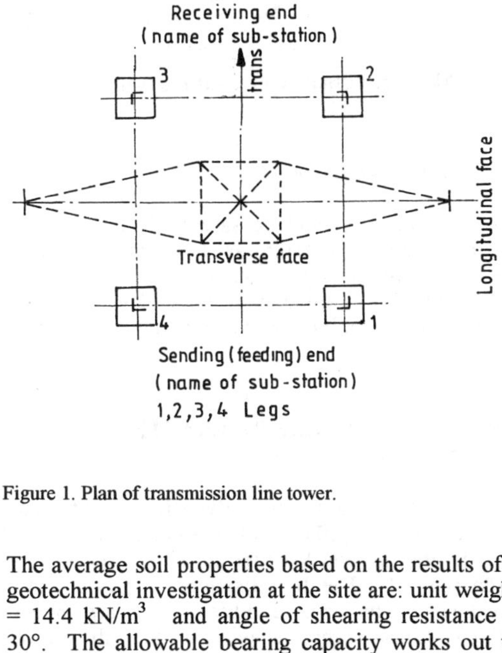

Figure 1. Plan of transmission line tower.

2 TRANSMISSION TOWER FOUNDATION

A narrow base tower for transmission of 66 KV D/C electrical power has been chosen for the present study. The features of the tower are shown in Fig.1. The design loadings for a tower with 10° deviation angle are: down thrust = 467 KN, uplift = 444 KN and horizental shear = 30 KN. The tower is located in Delhi and the soil at the site consists of dry, non-plastic sandy silt with water table at great depth.

The average soil properties based on the results of a geotechnical investigation at the site are: unit weight = 14.4 kN/m^3 and angle of shearing resistance = 30°. The allowable bearing capacity works out to 160-240 kN/m^3 for a 5-6m wide raft foundation at a depth of 3.5m below ground surface.

The raft foundation is designed (i) to resist uplift, (ii) to transmit the downthrust to the soil, without exceeding the allowable bearing pressure and (iii) to resist horizontal load without sliding. Further more,

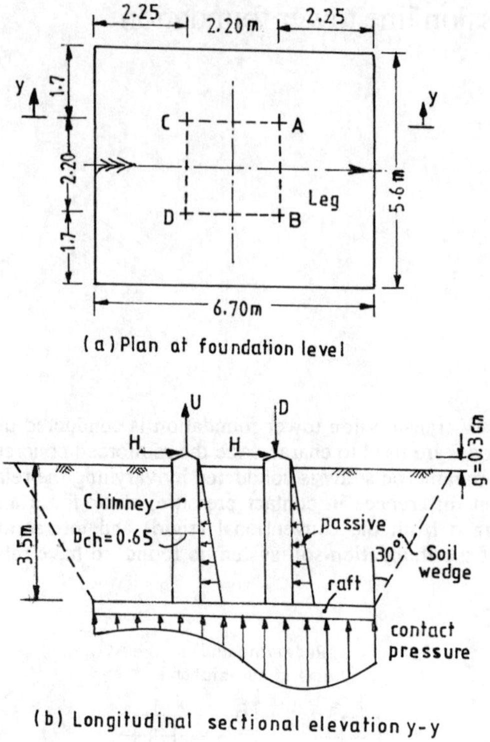

Figure 2. Details of foundation.

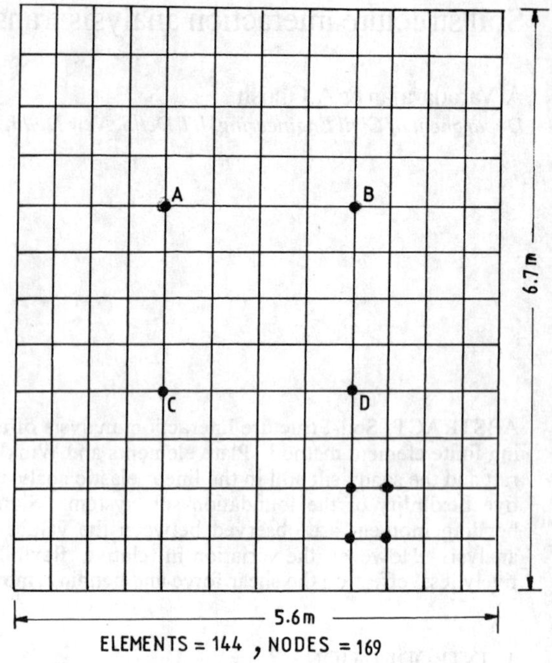

Figure 3. Finite element discretisation.

the foundation should be safe against overturning. Considering all the forces as shown in Fig.2, the dimensions of the raft foundation are decided after several trials and are given in Fig.2, The thickness of the raft is determined based on punching shear consideration.

3 SOIL-STRUCTURE INTERACTION ANALYSIS

Soil-structure interaction analysis is conducted using finite element method. Plate elements are used to model the raft and Winkler spring is used to represent the soil. The discretisations of the raft and the material properties used in the analysis are shown in Fig.3. Linear elastic behaviour of the raft and Winkler spring are considered. The sub grade modulus value has been determined using the relationship given by the Bowles (1982) based on allowable bearing pressure. The contact pressure, shear force and bending moments are obtained from the analysis. For comparison the raft has been analysed by conventional method without considering interaction effect.

The variation of contact pressure as obtained by the two methods along the critical section YY is shown in Fig.4. The interaction predicts low contact pressure where uplift is applied and high contact pressure where down thrust is applied as compared to the contact pressures obtained from conventional method. The difference is as much as 67% and 36% at the two locations.

The variations of shear force and bending moment are markedly different for the two methods of analysis (Fig.5). Although the shearing forces obtained from the two methods are comparable at the center and near the two ends of the critical section, the differences in their magnitudes are high near the locations of the tower legs. The peak value of shearing force near the tower leg as obtained from interaction analysis is found to be about three times the value obtained from the conventional method.

The variations of bending moment obtained from the two analyses along the critical section are shown in Fig.6 The conventional method predicts linear variation of bending moment between the two legs. However, the variation is slightly nonlinear in the remaining portion. The interaction analysis shows highly nonlinear variation of bending moment throughout the section. Although the bending moments obtained from the two methods are comparable at the ends and equal at the mid section, the differences in magnitudes are significantly different at

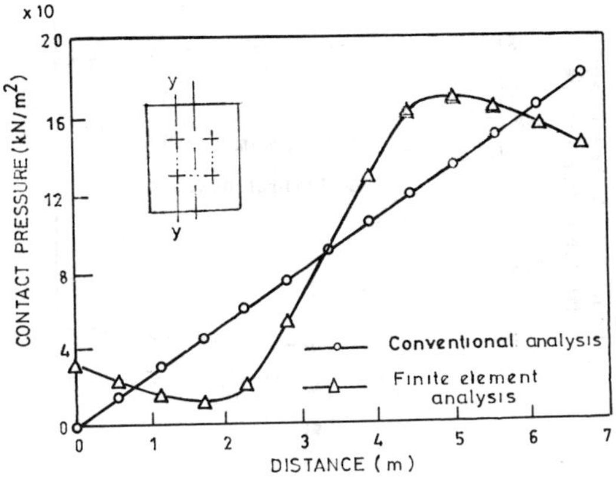

Figure 4. Variation of contact pressure along YY.

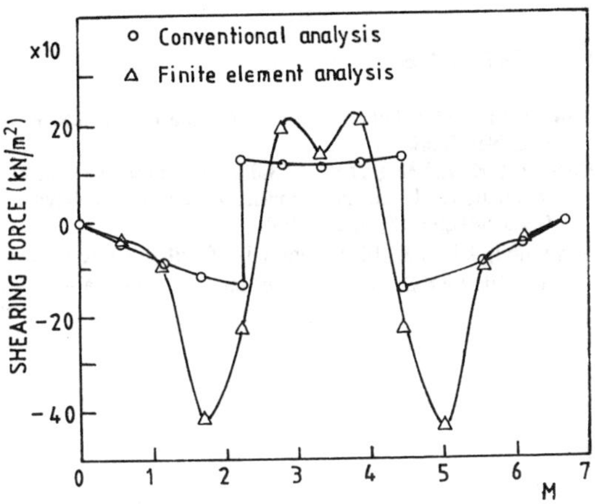

Figure 5. Variation of shear force along YY.

the location of the legs. The peak moments under the legs obtained from interaction analysis are found to be 38 % higher than the values obtained from the conventional method.

4 PARAMETRIC STUDY

The effect of various factors affecting the behaviour of the raft soil system has been investigated by conducting a parametric study. The characteristics of the raft-soil system are represented by relative flexibility of the system. The relative flexibility is affected by the properties of the raft-soil system and is expressed herein in non dimensional form as

f = kb l/b

where

l = length of raft

b= width of raft

k= subs grade modulus of soil

D= flexural rigidity of raft

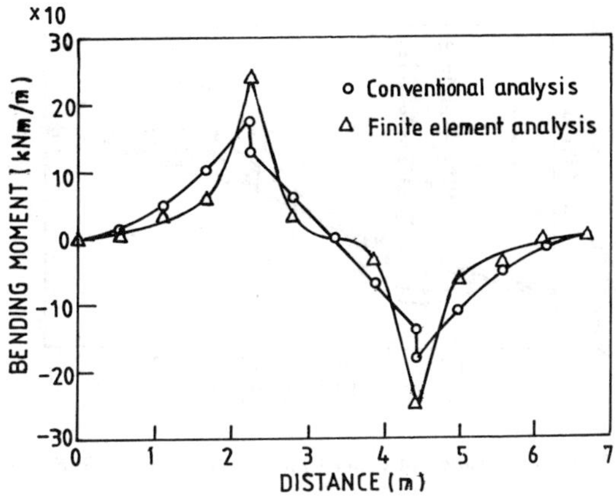

Figure 6. Variation of bending moment along YY.

$$\frac{E_p t_p^3}{12(1-\nu_p^2)}$$

E_p = Young's modulus of raft material

ν_p = Poisson's ratio of raft material

t_p = Thickness of raft

The values of relative flexibility has been varied from 20 to about 800 by varying the thickness of raft and the sub-grade modulus of the soil, keeping the other values constant.

The distribution of contact pressure normalized with respect to average foundation pressure is shown in Fig.7 for various values of flexibility. At low flexibility, the variation of contact pressure is linear and is similar to that predicted by conventional analysis. With the increase in the values of flexibility, the contact pressure variation becomes increasingly nonlinear and the contact pressure tends get concentrated at the location of the legs.

The variation of maximum normalised shear force in the critical section YY is shown in Fig. 8 for various values of flexibility. The maximum difference in the peak values of shearing force for various flexibility values is only about 10%. The maximum normalized bending moment variation in the critical section YY is shown in Fig. 9. The maximum reduction in the bending moment due to flexibility is found to be only 12%. As such the nature of variation of normalized shear force and bending moment along the section YY is not very much affected by relative flexibility.

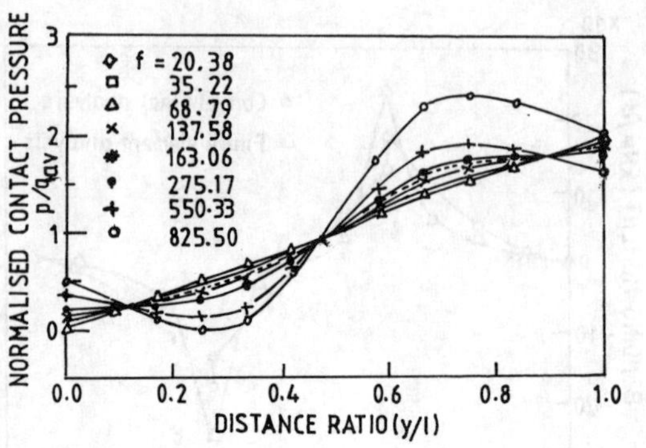

Figure 7. Variation of contact pressure for various flexibility values.

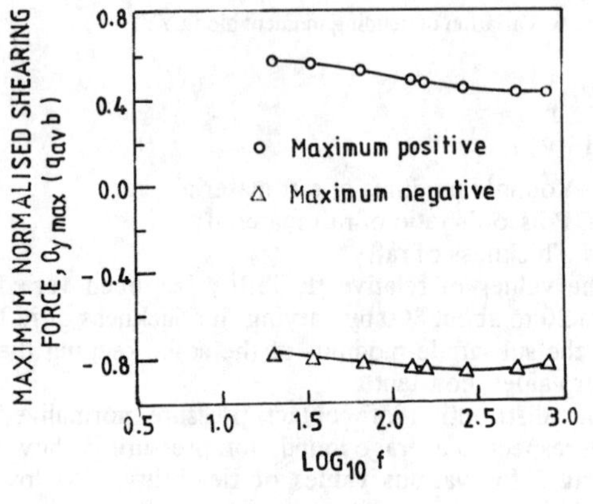

Figure 8. Maximum shear force for various flexibility values.

5 CONCLUSIONS

The predictions by the conventional analysis are significantly different from those by the interaction analysis. The nature of variations of contact pressure is very much affected with the increase in the flexibility of foundation- soil system. However, the nature variation of shear force and bending moment is not very much affect by the change in flexibility. For the range of the values of flexibility studied, the difference in maximum shear force and bending moment values are small.

From this study, it may be concluded that (i) for safe design, it is essential to conduct interaction analysis and (ii) the relative flexibility of the foundation-soil system has relatively less effect on the shear force and bending moment.

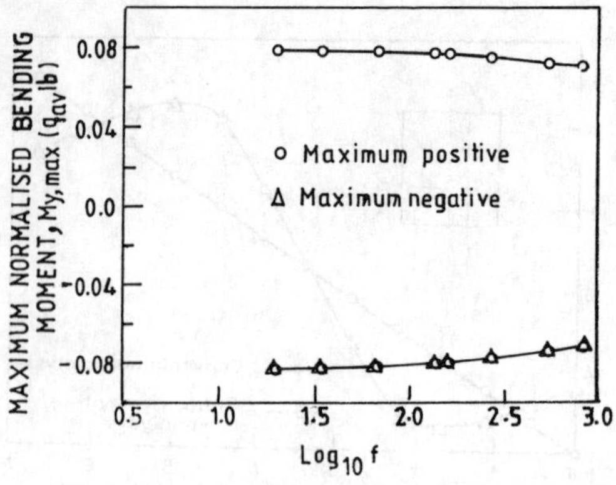

Figure 9. Maximum bending moment for various flexibility values.

6 REFERENCES

Bowles, J.E. 1982, Foundation analysis and design, McGraw Hill, New York.

Cheung, Y.K. and Nag, D.K. 1968, Plates and beams on elastic foundations, linear and non-linear elastic behaviour, Geotechnique, 18(2), 250-260.

Selvadurai, A.P.1979, Elastic analysis of soil-structure interaction, Elsevier Scientific publishing Co. Amsterdam.

11 Ground improvement and geosynthetics

Computer Methods and Advances in Geomechanics, Desai et al. (eds) © 2001 Balkema, Rotterdam, ISBN 90 5809 183 X

Review of numerical modeling of geosynthetic reinforced soil walls

R.J. Bathurst & K. Hatami
Royal Military College of Canada, Kingston, Ont., Canada

ABSTRACT: The use of geosynthetics to construct reinforced soil walls has increased dramatically in recent years. While the economic benefits of modern geosynthetic reinforced soil structures is proven, the mechanical behavior of these structures is less well understood. Correct numerical modeling offers the possibility to investigate the influence of a wide range of geometry and component material properties on structure performance. A challenge faced by modelers has been the treatment of the visco-elastic-plastic properties of polymeric reinforcement materials and the interactions between reinforcement layers, soil and structural facings. The paper reviews numerical modeling efforts of the writers and others to simulate the behavior of reinforced soil walls under both static and dynamic loading conditions. The paper is focused primarily on numerical modeling of geosynthetic reinforced soil walls using the finite element and finite difference techniques. Implications of the results of some numerical modeling studies to current design practice are identified. The few cases in which the results of numerical models have been compared against physical tests are highlighted.

1 INTRODUCTION

Traditionally, conventional Rankine and Coulomb earth pressure theories and modified slope stability methods have been used to carry out analysis and design of geosynthetic reinforced soil walls. These methods have been extended for the case of earthquake loading by simply introducing inertia force components equated to horizontal and vertical ground acceleration coefficients. These methods are limit equilibrium approaches that assume simplified failure mechanisms and employ either global factors of safety (AASHTO 1998, FHWA 1996) or partial factors (i.e. limit states design, BSI 1995) to design against serviceability failure and/or collapse. The performance of structures designed with factors of safety against collapse states (i.e. working stress conditions) are controlled by deformation limits that cannot be accounted for explicitly using limit equilibrium based methods. In addition, reinforced soil walls are complex systems typically involving a facing (such as concrete panels or stacked modular concrete blocks), soil backfill and horizontal layers of polymeric reinforcement. Limit equilibrium and semi-empirical analysis approaches require simplifying assumptions regarding the mechanical properties of polymeric reinforcement products and the interaction between components of reinforced soil wall systems. Finally, these approaches provide very limited insight into the fundamental behavior of reinforced soil walls since failure mechanisms are assumed *a priori*.

Advanced numerical methods hold promise as a design and research tool to investigate the entire response of reinforced soil retaining walls. An advantage of many modern computer codes is that appropriate constitutive models for polymeric materials and soils can also be implemented. Furthermore, numerical models can be used to carry out parametric analyses of the influence of wall geometry, facing type and mechanical properties of the constituent materials on wall behavior. Calibrated numerical models can also be used to extend the database of carefully instrumented field or laboratory-scale structures and hence contribute to the development of rational design methods that are, nevertheless, based on conventional concepts of earth pressure theory.

This paper provides a review of numerical modeling of geosynthetic reinforced walls by the writers and others. The survey is limited to the finite element method (FEM) and the finite difference method (FDM) in numerical codes used to simulate the behavior of geosynthetic reinforced soil walls under static and dynamic loading conditions.

2 STATIC ANALYSIS

2.1 *Numerical modeling studies without calibration*

The studies reviewed in this section are focused on parametric analyses of uncalibrated numerical models (i.e. model predictions are not compared against behavior of physical measurements).

Hermann and Al-Yassin (1978) described two different FEM approaches to model the behavior of reinforced-soil walls: discrete and composite approaches. In the composite approach, the reinforced soil structure comprising the reinforced soil and metallic reinforcement is homogenized and the metallic facing modeled with a structural element with bending capacity. They argued that both the discrete and composite approaches could model the slippage and yielding of reinforcement, buckling of the facing plates and nonlinear material behavior of the backfill material. They concluded that both approaches could be applied with equal accuracy. The advantage of the composite approach is reduced computational effort. However, this advantage is no longer a concern with modern computing platforms. A disadvantage of the composite approach is that there is no explicit information on the interaction of soil and reinforcement including bond stresses, stress concentration, edge effects and localized transfer of stress between the components. Consequently, the composite approach has been abandoned by researchers.

Naylor (1978) studied the effect of metallic reinforcement strip-facing connection rigidity on the behavior of a 10 m high reinforced-soil wall system using a finite element model with a constant linear-elastic law for the soil and a linear elastic stiffness model with a Mohr-Coulomb (M-C) yield criterion for the soil-reinforcement interface. Naylor adopted an equivalent composite approach to model the reinforced soil as a homogeneous mix of soil and metallic strips. The composite approach was justified by the high reinforcement density (i.e. small reinforcement spacing) of the model. Naylor included the flexibility of the reinforcement-facing connection, the density of reinforcement layers and wall foundation condition as variables in his parametric study. He concluded that a non-rigid reinforcement-facing connection allows the soil lateral pressure behind the facing to be reduced. This, in turn, reduces the maximum reinforcement load and shifts the peak reinforcement load location away from the facing and into the soil mass. He observed that increasing the reinforcement stiffness by a factor of 6 increased the maximum reinforcement load by 20% at end of construction. He concluded that Rankine theory underestimates the reinforcement load in heavily reinforced soil walls by a greater amount than for lightly reinforced soil wall systems. Naylor found that a compliant foundation (as opposed to a rigid foundation) results in a modest increase in the maximum reinforcement load and a change in the distribution of vertical stresses at the bottom of the wall. The reinforcement load at the bottom of the wall as well as vertical soil pressures acting at the bottom of the reinforced soil mass were less with a compressible foundation but lateral movements were greater. Shortcomings of the Naylor finite element study were that it did not simulate the staged construction of the wall or soil plasticity.

Ho and Rowe (1993, 1996) and Rowe and Ho (1992, 1996, 1997, 1998) modified the finite element program AFENA to study the response of a hypothetical 6 m high propped-panel wall under self-weight. The soil was modeled using a non-linear elastic model for the soil and linear elastic interface elements with M-C yield criterion. The wall configurations were constructed in 24 lifts.

They noted that three different regions could be defined within the backfill. One region is contained between the Rankine failure plane and the facing (active zone). A second region is a transition zone between the Rankine failure plane and a *zero force line*. The zero force line is a theoretical line beyond which no reinforcement force is required for equilibrium and corresponds to the theoretical stable slope defined by the angle of repose of the backfill frictional soil. The area under the zero force line was defined as the third zone. Numerical results by Ho and Rowe showed that the lateral earth pressures behind the wall were less than predicted values from Rankine theory and oscillated about the pressure line predicted by Coulomb theory. This observation was in agreement with experimental data reported by Andrawes et al. (1990). Ho and Rowe noted that curtailment of reinforcement within the transition zone (i.e. zone 2) results in sharp increase of soil strain within the retained zone. They concluded that the effect of reinforcement is to prevent the development of horizontal strain in the soil. However, they noted that the stiffness range of geosynthetic reinforcement products is not large enough to prevent soil elements from developing into a plastic state. Therefore, geosynthetic reinforced soil walls can be expected to develop a clear active zone within the reinforced region behind the wall. Ho and Rowe found that the locus of maximum reinforcement load along the wall height was closer to the Rankine failure plane than a log-spiral shape. A very important practical observation was that the variation of maximum reinforcement load with depth was closer to a uniform distribution than that predicted by Rankine theory. As a result, the magnitude of reinforcement load was larger than values predicted by Rankine theory close to the top of the wall and considerably less near the bottom. The same conclusion has been reached by Bathurst and Hatami (1998b) who carried out similar parametric analyses using a very different numerical approach (i.e. FDM implemented by program FLAC -Itasca 1998). Ho and Rowe concluded that relatively large reinforcement loads at the top of reinforced soil walls that have been observed in field measurements should not be attributed to compaction effects. Their numerical simulations which did not include compaction of the backfill indicated a significant reinforcement load at the top of the wall which was attributed to interaction between the soil, reinforcement and facing. Ho and Rowe compared the horizontal soil stress with reinforcement load at each reinforcement level. They found that the sum of horizontal soil stresses is not entirely supported by the reinforcement layers but is also carried, in part, by the toe. They further concluded that a rigid foundation has a significant effect on the magnitude of reinforcement load, especially close to the wall base. As a result of their numerical predictions, Ho and Rowe questioned the notion of back-calculating soil lateral pressure from reinforcement loads. They argued that the reinforcement load was not controlled by stress in the soil under active conditions. Ho and Rowe also found that the curtailment of reinforcement in the transition zone (i.e. zone 2) may contribute significantly to wall lateral displacements. They concluded that calculating lateral displacements of the wall from integration of reinforcement strains along the length of the reinforcement will underestimate wall deformations.

Rowe and Ho (1992) found that a rigid toe can increase the vertical soil stress at the bottom of the wall by as much as 25% higher than the nominal expected value. They also found that compaction had little effect on soil horizontal stresses in geosynthetic reinforced soil walls due to wall yielding. Rowe and Ho maintained that the locus of maximum reinforcement load in different layers is not unique in walls with extensible reinforcement under working conditions and may vary depending on construction method, reinforcement stiffness and length, facing type, etc. However, they found that the locus of maximum reinforcement load approaches the Rankine failure plane as the wall approaches a collapse state.

Rowe and Ho (1996) found that continuity of the wall facing panel (i.e. propped versus incremental panel) has a negligible effect on reinforcement load and wall deformations. However, the magnitude of facing rigidity was found to be important, particularly in the case of a rigid foundation. Their numerical results suggested that for the case of a rigid foundation, a more rigid facing would be beneficial to reduce maximum reinforcement loads and wall lateral displacements. Rowe and Ho concluded that the magnitude of elastic modulus of the backfill soil does not have a significant effect on the reinforced zone because this zone typically develops into a plastic state in walls with extensible reinforcement. Rowe and Ho (1996) found negligible influence of foundation flexibility on the magnitude of reinforcement loads and wall deformations. However, the flexible foundation layer in their finite element model was thin (i.e. 0.15H where H is the wall height) and may not apply to thicker foundation layers. Rowe and Ho (1996, 1997, 1998) showed that the soil-reinforcement interface friction angle has essentially no influence on wall response for values larger than $2/3(\phi)$ where ϕ is the friction angle of the cohesionless backfill soil. The reinforcement load in the lower reinforcement layers and wall lateral displacements increased with proportionately lower interface friction angle values.

Ho and Rowe (1996) concluded that the reinforcement length to wall height ratio, L/H, is the most important geometric parameter influencing the wall response. They found little variation in the magnitudes of reinforcement load and soil stress for L/H values larger than 0.7. It is worth noting that the ratio L/H=0.7 is the minimum reinforcement length ratio recommended by FHWA (1996) and AASHTO (1998) design guidelines for static stability of reinforced soil walls. Mitchell and Christopher (1990) concluded from the results of a similar FEM investigation that wall displacements were also largely influenced by the L/H ratio. A design chart to predict wall deformations as a function of L/H and reinforcement type (i.e. extensible or in-extensible) appears in current AASHTO (1998) and FHWA (1996) guidelines.

Ho and Rowe (1996) concluded that placing equally spaced reinforcement layers with L/H=0.7 is an efficient reinforcement distribution and is recommended over other distribution patterns of reinforcement in reinforced soil walls. They expressed caution in extrapolating the response of low-height walls to taller retaining wall systems because of the greater influence of the combined rigid facing and hinged toe boundary condition on low height walls. Ho and Rowe (1999) noted that their conclusions with respect to the optimal L/H ratio and reinforcement distribution arrangement within the backfill were consistent with the results of centrifuge physical modeling (Porbaha 1999).

Rowe and Ho (1997) found that the total force required for internal equilibrium of a reinforced soil wall is essentially independent of material properties other than the backfill friction angle. They noted that the magnitude of this equilibrium force can be determined using Rankine theory. However, a higher total force can be expected for the case where the backfill-facing friction angle is small. In addition, they found that the total force for external equilibrium of the facing is practically independent of material properties adopted in the model and can be calculated using a Coulomb active wedge analysis. The exception to good correspondence between Coulomb theory and numerically predicted results occurs when the facing rigidity or the soil-reinforcement interface friction angle is small.

Rowe and Ho (1998) showed that the magnitude of wall lateral displacement is influenced by the soil friction angle and a reinforcement stiffness factor, Λ, defined as $\Lambda = J/K_a \gamma H S_v$, where J is the reinforcement stiffness, K_a is the Rankine active earth pressure coefficient, γ is the soil unit weight, H is wall height and S_v is the vertical spacing between reinforcement layers. They argued that parameters such as soil-facing friction angle, soil-reinforcement friction angle, soil elastic modulus and facing rigidity are not as important. Rowe and Ho noted that reducing the reinforcement length in the lower half of the wall height may result in a substantial increase of wall lateral displacement.

Hatami et al. (2000) investigated the influence of the simultaneous use of reinforcement layers with different stiffness values (non-uniform reinforcement) in wrapped-face (very flexible facing) reinforced soil walls. The numerical simulation of wall models was carried out using the finite difference based program FLAC. The simulations were carried out using a linear elastic M-C model for the soil and linear elastic (FLAC) cable elements. The simulations included staged construction of the walls. The wall lateral displacements and horizontal earth pressures behind the facing in all non-uniform reinforced wall models showed a clear dependence on relative stiffness values of the individual reinforcement layers. The parametric study showed that wall displacements were less for configurations with alternating primary and secondary reinforcement layers of equal length compared to nominal identical walls with the two different reinforcement stiffness materials grouped separately in the bottom and top halves of the wall structure.

2.2 *Numerical modeling studies with calibration*

This section describes studies in which physical reinforced soil walls were simulated using numerical modeling and the results of numerical and physical models compared.

Karpurapu and Bathurst (1991a,b) modeled the response of the 3 m high *Denver Test Wall* using the non-linear large strain finite element program GEOFEM. They discussed the sources of discrepancies between their prediction of the wall response (Class A prediction) and the observed (measured) performance. They noted that the boundary constraint at the top of the test facility significantly reduced the soil and panel deformations. They reported Class C predictions using the same method of analysis and material properties as used in the Class A prediction analysis but with boundary conditions modeling the actual physical test. Karpurapu and Bathurst (1991b) concluded that proper boundary conditions (including side wall friction) and correct choice of soil properties were essential for successful numerical simulation of the performance of reinforced soil test walls.

Karpurapu et al. (1991), Bathurst et al. (1992), Karpurapu and Bathurst (1992, 1994, 1995) and Rajagopal and Bathurst (1994) used the program GEOFEM to model the behavior of carefully instrumented full-scale walls constructed at the Royal Military College of Canada (RMCC). The material properties used in the numerical simulations were obtained from independent small-scale laboratory experiments. They used the hyperbolic model proposed by Duncan et al. (1980) for the backfill soil and included the soil dilation angle inferred from numerical simulation of the results of direct shear box tests. Karpurapu et al. noted that soil dilation has a significant effect on the normal pressures at the reinforcement-soil interface for well-compacted dense granular soils. For the particular sand used at RMCC, the soil strength was about 20% higher when soil dilation was included compared to the case where soil dilation was neglected. They found it necessary to use appropriate load-strain-time (isochronous) data to model the reinforcement stiffness in order to predict correct strain magnitudes (and hence tensile forces) in the reinforcement layers at soil failure. Karpurapu et al. concluded that their numerical simulation was successful in predicting the peak values and the trend in reinforcement strains at incipient collapse of an incremental panel wall under surcharge loading. However, the numerical model over-predicted the length of the reinforcement in the soil anchorage zone. Nevertheless, this result was viewed as a conservative (i.e. safe) result for the design of reinforced soil walls using the GEOFEM code. The numerical simulation study by Karpurapu et al. was quantitatively consistent with experimentally measured data that showed significant forces generated at the horizontally restrained toe of the wall facing and confirmed that these forces contribute additional stabilizing reactions for low height walls. This additional capacity may account for a large portion of the conservatism inherent in current limit-equilibrium methods of analysis and design. The internal soil failure surface predicted by the finite element simulation was observed to be in good agreement with the excavated failure surface and consistent with failure geometry predicted by conventional Coulomb wedge theory.

Andrawes and Yogarajah (1994) used the finite element program CRISP to study the effect of reinforcement-facing connections on the response of three 2 m high physical reinforced-soil wall models. Equivalent linear reinforcement stiffness values were determined from reinforcement isochronous creep curves. The interface element introduced by Goodman et al. (1968) was used between soil, reinforcement and the facing. Andrawes and Yogarajah found that in the case where the reinforcement is rigidly attached to the facing, the strain distribution along the reinforcement layer is approximately linear with a maximum value at the connection to the facing. If some flexibility was allowed in the reinforcement-facing connection, the maximum reinforcement load was observed to move away from the facing. In this case, the distribution of reinforcement load along its length was nonlinear with negligible values at the connection. Andrawes and Yogarajah concluded that correct modeling of the connection boundary conditions is essential for accurate numerical predictions of reinforced soil wall response.

Ochiai and Fukuda (1996) and Nakane et al. (1996) reported finite element studies of an experimental 6 m high geotextile reinforced retaining wall with a concrete block facing. They concluded that their finite element model satisfactorily replicated experimental measurements of wall lateral displacement, reinforcement strains, soil lateral pressure behind the wall and the facing column toe load. Both groups used a linear elastic-plastic constitutive model with a M-C failure criterion for the soil elements and constant modulus linear elastic bar elements for the reinforcement layers. However, little information was provided on how they determined model input parameter values for backfill soil, reinforcement and joint elements/interfaces.

Nakane et al. (1996) also carried out comparative finite element analyses of geotextile reinforced soil retaining wall models constructed with different facing types. They observed that for the case of a rigid facing, maximum reinforcement forces developed at the connection to the facing panel. In addition, they found a significant foundation reaction at the base of the rigid facing (footing).

Helwany et al. (1999) used the program DACSAR and calibrated their finite element model against the measured response of the *Denver Test Wall*. The non-linear soil model proposed by Duncan et al. (1980) was used to model the soil. They obtained satisfactory agreement between the measured results of facing lateral displacement and reinforcement strain and results from finite element simulation. Helwany et al. carried out parametric analyses to investigate the effects of wall height, backfill type and reinforcement stiffness on wall response. They concluded that backfill type is the most important parameter in the performance of reinforced soil wall systems. They found that the stiffness of the geosynthetic reinforcement has an important influence on wall displacement response when the backfill shear strength and stiffness are low. Helwany et al. developed a series of design charts for selection of backfill type and reinforcement stiffness to meet code requirements for the performance of reinforced soil wall systems.

Walters et al. (2000) reported the monitored behavior of several full-scale reinforced segmental (modular block) retaining walls constructed at RMCC. Program FLAC was used to simulate the wall response during construction and was calibrated against the measured history of wall lateral displacement, reinforcement strain and toe reaction forces. The predicted values from the FLAC program and the measured wall response with respect to all the above response histories were found to be very close. However, the numerical results were very sensitive to the choice of interface properties assigned to the discrete block units during simulated construction.

Ling et al. (2000) modeled the construction of a full-scale, concrete-block, geosynthetic-reinforced soil retaining wall using the finite element method. The soil and geosynthetic reinforcement were both modeled using hyperbolic formulations. Ling et al. compared the numerical and measured experimental results. While the magnitude of recorded wall displacements, lateral earth pressures and bearing pressures were in general agreement, the distribution of values was observed to deviate from measured values.

3 DYNAMIC ANALYSIS

3.1 *Seismic loading*

This section is focussed on seismic modeling of reinforced soil walls. Because of the limited number of studies on geosynthetic reinforced walls under simulated earthquake loading, some early studies of metallic reinforced soil walls are included.

Richardson and Lee (1975) carried out dynamic finite element analyses using the computer program QUAD-4 to interpret shaking table test results of small-scale (i.e. 300 mm high) model walls subjected to sinusoidal base excitation. Their analyses included strain dependent soil modulus and damping. The metallic reinforcement was modeled using elastic bar elements. Richardson and Lee found that using Seed-Idriss nonlinear soil properties resulted in underestimation of measured reinforcement loads. They found that a satisfactory agreement between measured reinforcement load and predicted values from their finite element models could be obtained by using soil properties back-calculated from their small scale shaking table tests.

Segrestin and Bastick (1988) calibrated the finite element program SUPERFLUSH against the dynamic response results of a 3 m high wall reported by Chida et al. (1985). They reported a good agreement between calculated and measured values of reinforcement incremental load (i.e. load in excess of static value) at a given frequency of input harmonic base motion. They then used the calibrated model to investigate the dynamic response of two (6 m and 10.5 m high) metallic reinforced soil wall models to input ground motion. A frequency domain analysis approach was adopted in the program which included an iterative procedure to account for varying soil elastic modulus using the calculated soil deformation. Segrestin and Bastick carried out a set of parametric analyses which included three different foundation soil conditions and two recorded earthquake ground motions, scaled to various peak ground accelerations. The results of Segrestin and Bastick showed that a soft foundation reduces the magnitude of reinforcement load in the wall subjected to ground motion to lower values than the same wall constructed on a rigid foundation. The magnitude of reduction was particularly significant for the reinforcement layers in the bottom half of the wall. Their results showed that the reinforcement incremental load was reasonably uniform along the reinforcement length for most layers and the location of the maximum reinforcement load in each reinforcement layer was the same as that for static loading. Segrestin and Bastick found that the reinforcement incremental load was greater in the lower reinforcement layers. They concluded that lower reinforcement layers would develop higher incremental loads when steel reinforcement strips are wider or there are a greater number in a horizontal plane. This conclusion is consistent with the results of numerical modeling reported by Bathurst and Hatami (1998b, 1999a) who showed that reinforcement incremental loads increased with increasing reinforcement stiffness over a wide range of values that included relatively soft geosynthetic reinforcement materials as well as metallic reinforcement (i.e. $100 \leq J \leq 69{,}000$ kN/m). Segrestin and Bastick found an acceleration amplification factor greater than 2 between the top and the base of their wall models. This result is in agreement with the calculated acceleration amplification by Bathurst and Hatami (1998b) for a comparable model wall height. Segrestin and Bastick proposed a simplified approach to calculate the reinforcement incremental load in reinforced soil walls subjected to ground motion. Their proposed method has been adopted in AASHTO (1998) seismic design guidelines for reinforced earth structures. However, as pointed out by Bathurst and Alfaro (1996) the Segrestin and Bastick simplified procedure is based on a limited analysis that does not include a sufficiently wide range of wall types, material properties and ground motion characteristics. Specifically, the reinforcement stiffness values in their analyses are much greater than values associated with typical geosynthetic reinforcement materials. The subsequent numerical modeling results reported by Bathurst and Hatami (1998b) confirmed that for panel walls with a fixed toe, the incremental reinforcement load was essentially constant with depth for geosynthetic reinforcement materials but increased linearly with depth for metallic reinforcement.

Cai and Bathurst (1995) studied the dynamic response of a modular block geosynthetic reinforced soil retaining wall using the finite element method. The dynamic finite element program TARA-3 was modified to include a hyperbolic stress-strain relationship with Masing hysteretic unload-reload behavior to model the cyclic shear behavior of the backfill soil. The reinforcement material was modeled using a similar hysteretic model which accounted for the measured response of cyclic load-extension tests performed on unconfined geogrid specimens. Interface shear between wall components was simulated using slip elements. Cai and Bathurst found that the relative displacement and interface shear force between modular units increased with the duration and magnitude of base excitation during simulated seismic events. Relative displacements and shear forces were greatest at interfaces where a geosynthetic was present and at some locations the shear capacity between modular units was exceeded. They concluded that an accurate estimate of interface shear properties is particularly important for the seismic design of segmental geosynthetic reinforced soil retaining walls. Cai and Bathurst also observed that dynamic tensile forces in reinforcement layers accumulated during the simulated seismic events and peak tensile forces increased with the magnitude of base peak acceleration. They reported that reinforcement forces calculated using pseudo-static earth pressures (Mononobe-Okabe method) were consistently greater than the forces calculated by the dynamic FEM analyses. Reinforcement forces resulting from static loading conditions prior to base excitation were greater that the forces predicted using conventional Coulomb earth pressure including the topmost layer. The over-prediction of reinforcement forces using the Coulomb approach is likely exacerbated by the low height (i.e. 3.2 m high) concrete wall investigated in combination with the restrained footing condition. The horizontal accelerations predicted at different locations in the unreinforced soil mass, reinforced soil mass and the facing column showed that peak accelerations were time coincident. Cai and Bathurst found little acceleration amplification in their simulation results which, again, can be attributed to the low wall height. They concluded that the conventional practice of using a single acceleration factor in pseudo-static methods of analysis could therefore be justified for the case investigated.

Bathurst and Hatami (1998a,b,c,1999a,b) and Hatami and Bathurst (1999, 2000a,b) used the finite difference based program FLAC to study dynamic response of retaining wall systems to input ground motion. Bathurst and Hatami (1998a,b) investigated the influence of reinforcement stiffness, length and base boundary condition on the seismic response of a geosynthetic reinforced full-height panel wall. They found program FLAC to be a useful research tool that could be used to investigate the response of retaining walls under static and dynamic loading conditions. Several advantages of the program include a fast algorithm capable of tackling large distortions and structural collapse. User-defined constitutive models can be incorporated in the program in addition to the available library of constitutive models. Dynamic loading can be readily applied to the model in terms of acceleration, velocity, stress or force histories. A number of built-in boundary conditions are available to account for truncated modeling of unbounded media (e.g. non-reflecting boundary). The model can also include various structural elements and interfaces within the continuum domain. Bathurst and Hatami found that both reinforcement stiffness and reinforcement length influence the magnitude of wall lateral displacement when subjected to ground motion. However, dynamic response of the simulated walls was most sensitive to the toe boundary condition (i.e. fixed pinned toe viz. sliding pinned toe). The magnitude of permanent wall displacement was observed to decrease with increasing reinforcement stiffness. Permanent wall displacements were relatively insensitive to changes in reinforcement length to wall height ratios in the range L/H = 0.7 to 1.

Bathurst and Hatami (1998b,c) found that significant reinforcement incremental loads can develop when reinforced soil walls are subjected to input ground motions in the vicinity of wall fundamental frequency. The maximum reinforcement load in all simulation runs was consistently observed to occur at the rigid connection of reinforcement layers to the facing. They found that reinforcement load distribution over the wall height is very different for geosynthetic reinforced walls as compared to the walls with metallic reinforcement as noted earlier. A horizontally restrained toe reduced the total and incremental force in the lower layers for cases with reinforcement stiffness values associated with geosynthetic products (e.g. $J \leq 2000$ kN/m). In contrast, the total and incremental reinforcement load increased monotonically with depth for the case of metallic reinforcement. The linear trend of the dynamic load increment with depth below the crest used in the AASHTO (1998) method was not observed in any simulations with reinforcement stiffness values comparable to geosynthetic materials. Bathurst and Hatami (1998c) proposed that a bilinear dynamic load increment distribution be used in empirical models for reinforcement stiffness values typically associated with geosynthetic reinforced soil walls.

Bathurst and Hatami (1998b, 1999b) carried out parametric analyses to investigate the quantitative influences of backfill width, damping ratio and type of far-end truncated boundary on calculated wall response. Bathurst and Hatami (1998b) found that significant acceleration amplification (e.g., more than twice the amplitude of base acceleration) was observed for walls subjected to ground motions in the vicinity of their fundamental frequency. However, they also showed that the magnitude of amplification was influenced by the magnitude of damping ratio used in the numerical models. Bathurst and Hatami showed that the soil in the retained soil zone develops into the plastic state during shaking and the inclination of the failure surface in this region can be predicted satisfactorily using Mononobe-Okabe theory considering the acceleration amplification in the backfill.

Hatami and Bathurst (1999, 2000a) studied the influence of different structural design parameters on the fundamental frequency of reinforced-soil retaining wall models. The design parameters in the study included the wall height, backfill width, reinforcement stiffness, reinforcement length, backfill friction angle and toe restraint condition. The intensity of ground motion, characterized by peak ground acceleration, was included in the study. Hatami and Bathurst concluded that the fundamental frequency of reinforced-soil wall models with sufficiently wide backfill subjected to moderately strong input accelerations can be estimated with reasonable accuracy from linear elastic wave theory using the shear wave speed in the backfill and the wall height. Their numerical analyses showed no significant influence of the reinforcement stiffness, reinforcement length or toe restraint condition on the fundamental frequency of wall models. The strength of the granular backfill, characterized by its friction angle, also did not show any observable effect on the fundamental frequency of reinforced-soil retaining wall models. However, the resonance frequencies of wall models were dependent on the ground motion intensity and to a lesser extent, on the width to height ratio of the backfill.

Hatami and Bathurst (2000b) studied the dynamic response of a segmental (modular block) retaining wall model to recorded ground motions. They compared the magnitude and characteristics of wall response to those obtained under harmonic input base acceleration. They found that the maximum lateral displacement of the facing column and maximum reinforcement load of the segmental retaining wall model subjected to a single frequency, harmonic input acceleration were considerably larger than the corresponding values obtained using a number of earthquake accelerograms with comparable predominant frequencies. They concluded that the random characteristic of actual ground accelerations would partly explain the relatively good performance of reinforced-soil retaining wall systems that were designed without seismic considerations or at best, using simple pseudo-static limit equilibrium methods (Bathurst and Alfaro 1996). Hatami and Bathurst found that low-frequency ground motions with high intensity values can result in significant structural response magnitude of short-period retaining wall systems.

3.2 *Blast loading*

Yogendrakumar et al. (1992) investigated the dynamic response of geosynthetic reinforced soil structures subjected to air blast loading. They adopted a method that was based on an estimate of the pressure-time response at the boundaries of target structures using the 2D gas dynamic code SPLIT2D. The soil-reinforcement response was simulated by modifying the finite element computer program RESBLAST. They compared the response of reinforced soil wall and slope models with and without reinforcement. They used the results of rapid cyclic tensile loading tests to obtain mechanical properties needed for the nonlinear reinforcement model implemented in the finite element code. Yogendrakumar and Bathurst showed that the polymeric reinforcement improved the performance of the structure under air blast loading by reducing the magnitude of dynamic and permanent soil deformations by as much as 30%.

4 CONCLUSIONS

The paper summarizes many examples of the use of computer codes that utilize the finite element and finite difference methods to predict the response of reinforced-soil wall models under static and dynamic loading conditions. The majority of the studies reviewed by the writers investigated the response of idealized reinforced soil walls. Relatively few studies are available in the literature that report direct comparisons of numerical results and experimentally measured results from carefully instrumented walls. In the few cases where direct comparisons are reported there are some significant discrepancies between predicted and measured values. A more serious shortcoming of many studies that have included physical test results is that the physical experiments were only lightly instrumented and hence boundary forces which are an important test of the accuracy of numerical results cannot be evaluated or boundary conditions in the physical tests are over-simplified from typical field cases.

The stress-dependent hyperbolic soil model proposed by Duncan et al. (1980) has been used by many modelers. This model has the advantage that parameters can be determined from independent standard triaxial tests. The model has been used in combination with the M-C failure criterion for granular soils under static conditions. Constitutive models that include soil dilation have been shown to improve the prediction of the response of reinforced soil walls constructed with compacted granular soils. Reinforcement models have been generally linear elastic-type but efforts to incorporate non-linear models are reported by a number of researchers. A practical approach to the problem of visco-elastic (creep) behavior of geosynthetic reinforcement materials is to use modified linear elastic models with parameters interpreted from in-isolation isochronous load-strain data.

For dynamic modeling, most researchers have restricted soil models to linear elastic-type to minimize computational time. This is also true for the reinforcement materials. Nevertheless, hyperbolic stress-strain models with hysteretic unload-reload behavior have been used for the backfill soil and polymeric reinforcement materials. More work is required to investigate whether or not the response of reinforced soil walls under dynamic loading is significantly influenced by the choice of constitutive model for the soil and reinforcement.

A common deficiency of the results of reinforced soil wall modeling is that tensile strains in extensible reinforcement elements propagate deeper into the reinforced soil mass than has been observed in some carefully instrumented physical tests. The modeling of segmental retaining walls constructed with a discrete dry-stacked column of concrete blocks has proven to be a particular challenge. The writers and co-workers have shown that the interface mechanics of the block units are complicated by the presence of reinforcement layer inclusions. Further effort is required to develop more accurate soil-reinforcement interface models that can accurately simulate the construction-induced response of segmental reinforced soil walls.

The response of reinforced soil walls is further complicated by non-quantifiable factors such as quality of construction during building of the structures. Most of the wall deformation that occurs during the in-service life of a modern reinforced soil wall system occurs during construction. Quality of construction may control the behavior of complex multi-component systems such as reinforced segmental retaining walls and render Class A predictions problematic.

Despite the shortcomings in numerical modeling of geosynthetic reinforced soil walls some important lessons have been learned that have practical implications to development of empirical rules for the design of structures in static and earthquake environments. It is clear that the type of connection between the reinforcement and the wall facing plays a major role in determining the location and magnitude of peak loads in the reinforcement. In addition, the boundary condition that exists at the toe of a hard-faced wall has a major influence on wall response. More work is required to investigate the influence of depth of compressible foundation layer on wall response under both static and earthquake loading. Results of numerical modeling have identified or confirmed empirical rules for the optimum vertical arrangement and length of reinforcement layers in typical reinforced soil wall structures.

The numerical simulation work by the writers has shown that the relationship of the fundamental frequency of a reinforced soil wall to the frequency content of base input ground motion may control the seismic response of a wall to a much larger degree than the mechanical properties of the components.

The paper identifies FEM studies that have been used to develop empirical guidelines for static and seismic design of reinforced soil walls. An example is given illustrating how the results of numerical modeling of metallic reinforced soil retaining walls have been used to develop seismic design guidelines. However, a subsequent more exhaustive numerical parametric study has shown that empirical rules for metallic reinforced soil walls may not apply for nominal identical walls constructed with relatively extensible geosynthetic reinforcement materials.

5 REFERENCES

AASHTO 1998. Interims: Standard specifications for highway bridges. *Am. Assoc. Sta. Highw. Trans. Offic.* Washington, DC, USA.

Andrawes, K.Z., K.H. Loke & K.C. Yeo 1990. Application of boundary yielding concept to full scale reinforced and unreinforced soil walls. In A. McGown, K.C. Yeo & K.Z. Andrawes (eds), *Performance of Reinforced Soil Structures - Proc. Int. Reinf. Soil Conf., Glasgow, September 1990:* 79-83. London: Thomas Telford.

Andrawes, K.Z. & I. Yogarajah 1994. Effects of reinforcement connections on the behaviour of reinforced soil retaining walls. In H.J. Siriwardane & M.M. Zaman (eds), *Computer Methods and Advances in Geomechanics - Proc. 8th Int. Conf. Num. Meth. Geomech. Morgantown, WV, USA, May 1994, 2:* 1313-1318. Rotterdam: Balkema.

Bathurst, R.J. & M.C. Alfaro 1996. Review of seismic design, analysis and performance of geosynthetic reinforced walls, slopes and embankments. In H. Ochiai N. Yasufuku and K. Omine (eds), *Earth Reinforcement - Proc. Int. Symp. Earth Reinf., Fukuoka, Kyushu, Japan, November 1996, 2:* 887-918. Rotterdam: Balkema.

Bathurst, R.J. & K. Hatami 1998a. Influence of reinforcement stiffness, length and base condition on seismic response of geosynthetic reinforced retaining walls. *Proc. 6th Int. Conf. Geosyn. Atlanta, GA,USA, March 1998, 2:* 613-616.

Bathurst, R.J. & K. Hatami 1998b. Seismic response analysis of a geosynthetic reinforced soil retaining wall. *Geosyn. Inter. 5(1-2):* 127-166.

Bathurst, R.J. & K. Hatami 1998c. Influence of reinforcement properties on seismic response and design of reinforced soil retaining walls. *Proc. 51st Can. Geotech. Conf. Edmonton, AB, Canada, October 1998, 2:* 479-486.

Bathurst, R.J. & K. Hatami 1999a. Numerical study of the influence of base shaking on geosynthetic reinforced retaining walls. *Proc. Geosynthetics'99, Boston, MA, USA, April 1999, 2:* 963-976.

Bathurst, R.J. & K. Hatami 1999b. Earthquake response analysis of reinforced-soil walls using FLAC. In C. Detournay & R. Hart (eds), *FLAC and Numerical Modeling in Geomechanics - FLAC Symp. Num. Mod. Geomech. Minneapolis, MN, USA, September 1999:* 407-415.

Bathurst, R.J., R.G. Karpurapu & P.M. Jarrett 1992. Finite element analysis of a geogrid reinforced soil wall. In R.H. Borden, R.D. Holtz & I. Juran (eds), *ASCE Geotech. Special Publ. No. 30 – Proc. Conf. Grouting, Soil Improvement and Geosynhetics, New Orleans, LA, USA, February 1992, 2:* 1213-1224. New York: ASCE.

BSI 1995. Code of practice for strengthened/reinforced soils and other fills. *Brit. Stan. Inst. BS 8006: 1995.*

Cai, Z. & R.J. Bathurst 1995. Seismic response analysis of geosynthetic reinforced soil segmental retaining walls by finite element method. *Comp. Geotech. 17(4):* 523-546.

Chida, S., K. Minami & K. Adachi 1985. Tests with regard to the stability of the fill constructed by the reinforced earth technique. *(unpublished report translated from Japanese)*

Duncan, J.M., P. Byrne, K.S. Wong & P. Mabry 1980. Strength, stress-strain and bulk modulus parameters for finite-element analysis of stresses and movements is soil masses. *Rep. UCB/GT/80-01, UC Berkeley, CA, USA.*

FHWA 1996. Mechanically stabilized earth walls and reinforced soil slopes design and construction guidelines. In V. Elias and B.R. Christopher (eds). *Federal Highway Administration (FHWA) Demonstration Project 82*, Washington, DC, USA.

Goodman, R.E., R.L. Taylor & T.L. Brekker 1968. A model for the mechanics of jointed rocks. *ASCE J. Soil Mech. Foun. Div. 94(SM3):* 637-659.

Hatami, K. & R.J. Bathurst 1999. Frequency response analysis of reinforced-soil retaining walls. *Proc. 8th Can. Conf. Erthq. Eng. Vancouver, BC, Canada, June 1999:* 341-346.

Hatami, K. & R.J. Bathurst 2000a. Effect of structural design on fundamental frequency of reinforced-soil retaining walls. *Soil Dyn. Erthq. Eng. 19(3):* 137-157.

Hatami, K. & R.J. Bathurst 2000b. Investigation of seismic response of reinforced-soil retaining walls. *Proc. 4th Int. Conf. Rec. Adv. Geotech. Erthq. Eng. Soil Dyn. San Diego, CA, USA, March 2001. (in press)*

Hatami, K., R.J. Bathurst, P. Di Pietro & P.M. Bianco 2000. Numerical study of retaining walls with non-uniform reinforcement. *Proc. EuroGeo 2. Bologna, Italy, October 2000. (in press)*

Helwany, S.M.B., G. Reardon & J.T.H. Wu 1999. Effects of backfill on the performance of GRS retaining walls. *Geotex. Geomem. 17(1):* 1-16.

Hermann, L.R. & Z. Al-Yassin 1978. Numerical analysis of reinforced soil systems. *ASCE Symp. Earth Reinf. Pittsburgh, PA, USA, April 1978:* 428-457. New York: ASCE.

Ho, S.K. & R.K. Rowe 1993. Finite element analysis of geosynthetic-reinforced soil walls. *Proc. Geosynthetics'93, Vancouver, BC, Canada, March-April 1993:* 203-216.

Ho, S.K. & R.K. Rowe 1996. Effect of wall geometry on the behavior of reinforced soil walls. *Geotex. Geomem. 14(10):* 521-541.

Ho, S.K. & R.K. Rowe 1999. Effect of wall geometry on the behavior of reinforced soil walls - Reply to the discussion by A. Porbaha. *Geotex. Geomem. 17(2):* 125-126.

Itasca Consulting Group 1998. FLAC - Fast Lagrangian Analysis of Continua. Ver. 3.40. *Itasca Cons. Gr., Inc.* Minneapolis, MN, USA.

Karpurapu, R.G. & R.J. Bathurst 1991a. Finite element predictions for the Colorado geosynthetic-reinforced soil retaining walls. In: J.T.H. Wu (ed), *Geosynthetic-Reinforced Soil Retaining Walls - Proc. Int. Symp. Geosynth.-Reinf. Soil Retaining walls, Denver, CO, USA, August 1991:* 329-341. Rotterdam: Balkema.

Karpurapu, R.G. & R.J. Bathurst 1991b. Discussion and class C predictions for the Colorado trial walls. In: J.T.H. Wu (ed), *Geosynthetic-Reinforced Soil Retaining Walls - Proc. Int. Symp. Geosynth.-Reinf. Soil Retaining walls, Denver, CO, USA, August 1991:* 343-346. Rotterdam: Balkema.

Karpurapu, R.G., R.J. Bathurst & P.M. Jarrett 1991. Finite element analysis of an incrementally constructed geogrid reinforced soil wall. *Proc. 44th Can. Geotech. Conf. Calgary, AB, Canada, September-October 1991:* 82/1-82/8.

Karpurapu, R. & R.J. Bathurst 1992. Analysis of geosynthetic reinforced soil walls by finite element methods. In H.J. Siriwardane & M. Zaman (eds), *Computer Methods and Advances in Geomechanics - Proc. 8th Int. Conf. Num. Meth. Geomech. Morgantown, WV, USA, May 1994, 1:* 861-870.

Karpurapu, G. & R.J. Bathurst 1994. Finite element analysis of geosynthetic-reinforced soil walls. *Proc. 13th ISSMFE Conf. New Delhi, India, January, 1994:* 1381-1384.

Karpurapu, R.G. & R.J. Bathurst 1995. Behaviour of geosynthetic reinforced soil retaining walls using the finite element method. *Comp. Geotech.. 17(3):* 279-299.

Ling, H.I., C.P. Cardany, L.X. Sun & H. Hashimoto 2000. Finite element study of a geosynthetic-reinforced soil retaining wall with concrete-block facing. *Geosyn. Int. 7(2):* 137-162.

Mitchell. J.K. & Christopher, B.R. 1990. North American practice in reinforced soil systems. In P.C. Lambe and L.A. Hansen (eds) *ASCE Geotechnical Special Publication No. 25 - Proc. Conf. Design and Performance of Earth Retaining Structures, Cornell University, Ithaca, NY, USA, June 1990:* 322-346. New York: ASCE.

Nakane, A., M. Taki, Y. Yokota & H. Miyatake 1996. FEM comparative analysis of facing rigidity of geotextile-reinforced soil walls. In H. Ochiai, N. Yasufuku & K. Omine (eds), *Earth Reinforcement - Proc. Int. Symp. Earth Reinf., Fukuoka, Kyushu, Japan, November 1996, 1:* 433-438. Rotterdam: Balkema.

Naylor, D.J. 1978. A study of reinforced earth walls allowing strip slip. *ASCE Symp. Earth Reinf. Pittsburgh, PA, USA, April 1978:* 618-643.

Ochiai, Y. & N. Fukuda 1996. Experimental study on geotextile-reinforced soil walls with different facings. *Proc. Rec. Cas. Hist. Devel. Des. Geosyn.-Reinf. Soil Retain. Walls, Tokyo, Japan, November 1996:* 140-147.

Porbaha, A. 1999. Discussion on: Effect of wall geometry on the behavior of reinforced soil walls. *Geotex. Geomem. 17(2):* 121-123.

Rajagopal, K. & R.J. Bathurst 1994. Parametric finite element investigation of reinforced soil retaining walls. *Proc. 5th Int. Conf. Geotex. Geomem. Rel. Prod.: Singapore, September 1994:* 193-196.

Richardson, G.N. & K.L. Lee 1975. Seismic design of reinforced earth walls. *Proc. ASCE J. Geotech. Eng. Div. 101(GT2):* 167-188.

Rowe, R.K. & S.K. Ho 1992. A review of the behavior of reinforced soil walls. In H. Ochiai, S. Hayashi & J. Otani (eds), *Earth Reinforcement Practice - Proc. Int. Symp. Earth Reinf. Prac., Fukuoka, Kyushu, Japan, November 1992, 2:* 801-830. Rotterdam: Balkema.

Rowe, R.K. & S.K. Ho 1996. Some insights into reinforced wall behavior based on finite element analysis. In H. Ochiai N. Yasufuku and K. Omine (eds), *Earth Reinforcement - Proc. Int. Symp. Earth Reinf., Fukuoka, Kyushu, Japan, November 1996, 1:* 485-490. Rotterdam: Balkema.

Rowe, R.K. & S.K. Ho 1997. Continuous panel reinforced soil walls on rigid foundations. *Proc. ASCE J. Geotech. Geoenviron. Eng. 123(10):* 912-920.

Rowe, R.K. & S.K. Ho 1998. Horizontal deformation in reinforced soil walls. *Can Geotech. J. 35(2):* 312-327.

Segrestin, P. & M. Bastick 1988. Seismic design of reinforced earth retaining walls - the Contribution of Finite Element Analysis. In T. Yamanouchi, N. Miura & H. Ochiai (eds), *Theory and Practice of Earth Reinforcement - Proc. Int. Geotech. Symp. Theo. Prac. Earth Reinf. Fukuoka, Kyushu, Japan, October 1988:* 577-582. Rotterdam: Balkema.

Walters, D., K. Hatami, R.J. Bathurst, N. Vlachopoulos, D. Saunders, G.P. Burgess & T.M. Allen 2000. Full-scale testing and modeling of geosynthetic reinforced soil retaining walls. *Proc. 53rd Can. Geotech. Conf. Montreal, Canada, October 2000. (in press)*

Yogendrakumar, M. & R.J. Bathurst 1992. Numerical simulation of reinforced soil structures during blast loadings. *Trans. Res. Rec. 1336, Washington, DC, USA:* 1-8.

Computer Methods and Advances in Geomechanics, Desai et al. (eds) © 2001 Balkema, Rotterdam, ISBN 90 5809 183 X

Numerical simulation of a geosynthetic-reinforced soil wall

K.E.El-Hoseiny
Department of Civil Engineering, Faculty of Engineering, Minufiya University, Shebin El-Kom, Egypt

C.S.Desai
Department of Civil Engineering and Engineering Mechanics, University of Arizona, Tucson, Ariz., USA

ABSTRACT: A Geosynthetic-reinforced soil retaining wall using full-height precast concrete wall facing panel was constructed at Tanque Verde road site for grade separated interchanges in Tucson, Arizona. Numerical simulation of this wall was performed using a finite element code called DSC-SST-2D. The program allows for plane strain, plane stress and axisymmetric idealizations including simulation of construction sequences. The wall was modeled as a plane strain, two-dimensional problem. Material parameters used in the analysis were obtained from experimental results from conventional triaxial compression tests for backfill soils and cyclic multi-degree-of-freedom (CYMDOF) shear tests for interfaces. The soils were modelled using the disturbed state concept (DSC) and the geogrid reinforcement was simulated by a linear elastic model. The interfaces between the reinforcement and soil were modeled using thin layer elements. The results of the finite element analysis were in good agreement with the measured field behavior of the wall.

1 INTRODUCTION

Different approaches were used in designing of reinforced soil structures, two of these approaches are limit equilibrium methods and numerical (finite) element methods. The limit equilibrium based approaches include two groups; force equilibrium analysis and strain compatibility analysis. Numerous simple design methods based on the consideration of limit equilibrium do not provide information concerning deformations or stress distributions in either the soil or the reinforcement (Rowe and Ho, 1992). Also, boundary conditions as well as stress equilibrium at each point within the reinforced mass and along the slippage surface are not involved in the formulation of the conventional limiting equilibrium methods (Leshchinsky, 1992). Furthermore, as reported by Wu (1992), limit equilibrium methods underestimated the failure of two large-scale geosynthetic reinforced soil walls using granular and cohesive backfill. There are two different finite element techniques used in analytical modeling of reinforced soil structures, discrete approach and composite approach. In discrete approach, the reinforcement system is considered as heterogeneous body in which the soil and reinforcing elements are separately represented by different material properties. The important advantage of this type of the model is that detailed information is directly obtained about the behavior of interface between the soil and the reinforcement, and stress concentration in the soil due to reinforcing members. In composite finite element analysis, the reinforced soil is characterized as a "homogeneous" composite structure with the properties of the composite material (Romsted et al., 1976). Superimposing the stiffness of the soil and the reinforcement forms the composite element stiffness. For performing a realistic finite element analysis of a reinforced soil structure, the computer program should have the capability of: modeling the construction sequences, structural elements (i.e. bars and beams) soil elements, and interface elements that allow relative motion between the soil and reinforcement (Collin, 1986).

There are several shortcomings associated with the composite approach, e.g. information about the interaction between the soil and reinforcement such as: bond stresses, stress concentration in the soil due to geometric discontinuities as a result of a the presence of reinforcement and edge effect due to the local transfer of stress between the soil and the reinforcement at the boundaries. Therefore, in a composite finite element analysis the strains and forces in the reinforcements are predicted to be larger than the actual measured values at the edge of the structure, so overestimating the effectiveness of the reinforcement (Romsted et al., 1976). Numerical simulation of reinforced soil structures using finite element analysis can account for the stress-deformation behavior and discontinuous behavior at the interface between soil and reinforcement. The Goodman type zero-thickness interface element (Goodman et al., 1968; Gens et al., 1988) has been used to model the interface problem but does not provide an inaccurate mechanism of deformation (Yi et al., 1995). Different researchers have reported numerical problems when using zero thickness interface element. Desai and Zaman (1984) found that the joint element approach does not provide realistic modeling of the normal stress in soil-structure interaction because it is difficult to reach at an appropriate high value for the normal stiffness. Furthermore, Day and Potts (1994) reported that both ill conditioning of the stiffness matrix and high stress gradients cause numerical instability. In this paper, the thin finite element model with disturbed state concept (DSC) (Desai and Ma1992; Desai, 1999) have been used for modeling soil-geosynthetic behavior at mechanically stabilized earth retaining wall.

2 EXPERIMANTAL GEOGRID-REINFORCED WALL

Between November, 1984 and 1985, forty- three geogrid-reinforced soil walls were constructed at Tanque Verde road site for grade-separated interchanges on the Tanque Verde-Wrightstown-Pantano roads projects in Tucson, Arizona, USA. This project represent the first use of geogrid reinforcement in mechanically stabilized earth (MSE) retaining walls in a major transportation-related application in North America (Tensar, 1989). In this study, wall panel No. 26-32 is simulated using finite element method (FEM) analysis because it is one of the two-instrumented walls for comparing analytical results with the instrumented field data. The wall system was considered as an alternative to a conventional cantilevered retaining wall system or a reinforced earth retaining wall system utilizing steel strips as reinforcement. The Tucson wall height is 4.88 m. The reinforced soil mass was faced with 15.24-cm thick and 3.05 m wide precast reinforced concrete panels. Soil reinforced geogrid were mechanically connected to the concrete facing panels and extended to a length of 3.66 m. On the top of the wall fill, a pavement structure was constructed that consisted of 10.16 cm base course covered by 24.13 cm of Portland cement concrete. Details of the various geometries are reported by Berg et al., 1986; Fishman et al.,1991.

The specified soil reinforcement was Tensar's SR2 structural geogrid. The Tensar SR2 geogrid is a uniaxial product that is manufactured from high density polyethylene (HDPE) stabilized with about 2.5 % carbon black to provide resistant to attack by ultraviolet (UV) light (Tensar, 1986). It is reported to be resistance to chemical substances normally existing in soils (Fishman et al., 1991). The geogrids have maximum tensile strength of 79 kN/m and a secant modulus in tension at 2 % elongation of 1094 kN/m . The allowable long-term tensile strength based on creep considerations is reported to be 29 kN/m at 10 % strain after 120 years. This value was reduced by an overall factor of safety equal to 1.5 to compute a long-term design tensile strength of 19 kN/m for use in design, which is recommended by Netlon (1984).

3 NUMERICAL MODELING

The numerical analysis of the reinforced soil wall was performed using a finite element code called DSC-SST-2D developed by Desai (1998). The program allows for plain strain, plain stress and axisymmetric idealizations including simulation of construction sequences. Various constitutive models, elastic, elasto-plastic (Von Mises, Drucker-Prager, Mohr-Coulomb, Hoek-Brown, Critical State and Cap), hierarchical single surface (HISS), viscoelastic plastic and disturbance (damage) can be chosen for the analysis. The wall was modeled as a plane-strain, two-dimensional problem for the finite element analysis. DSC-SST-2D uses a discrete finite element approach and requires material properties to explicitly model the soil, facing panels, reinforcement layers and all interfaces between components. The DSC-SST-2D program was written to allow incremental fill placement to be simulated (i.e. rows of elements added sequentially as fill placement). The finite element mesh representation of the wall (Case no. 1) and the boundary conditions for the numerical simulation is shown in Figure 1. The finite element mesh consisted of 184 nodes and 167 elements. The nodal points along the bottom boundary were constrained from either horizontal or vertical movement, while the nodal points along the two vertical surfaces at right and the left were restrained in the horizontal direction only.

An additional refined mesh that represents more closely the reinforcement in the retaining wall, was also used. This mesh called Case no. (2) with ten layers of reinforcement. Figure 2 shows the finite element mesh for Case no. (2) for the part near the reinforcement. This mesh consisted of totally 1370 elements and 1188 nodes, which include 605 soil elements, 35 wall facing elements, 250 bar elements (for the reinforcement) and 480 interface elements.

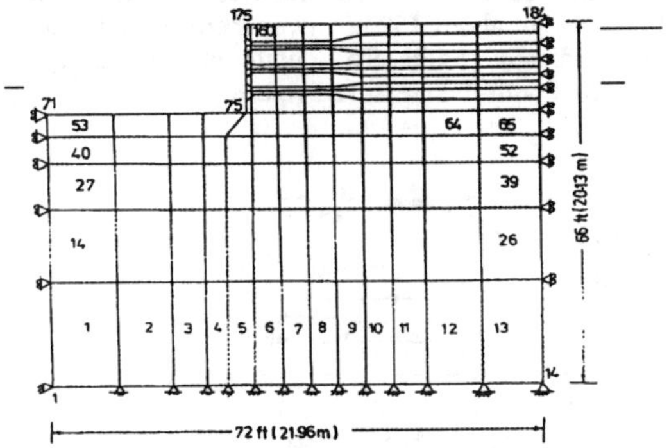

Figure 1. FEM mesh of Tucson wall (Case No. 1)

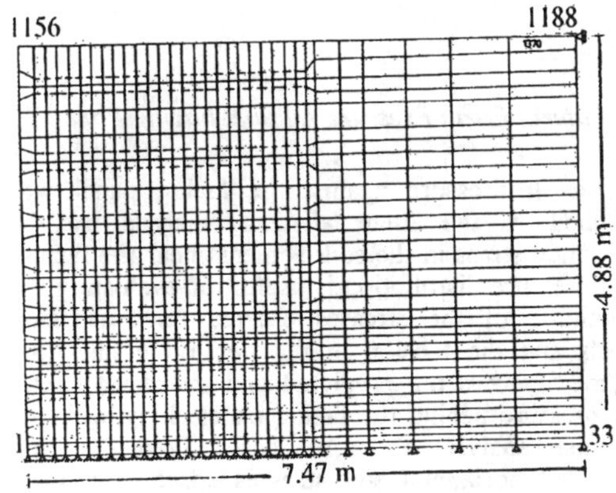

Figure 2. FEM mesh of Tucson wall (case No. 2)

4 MATERIALS PROPERTIES AND MODELS

The properties of the materials used in the analysis were obtained from the experimental results of conventional triaxial compression (CTC) tests for soil backfill and cyclic multi-degree-of-freedom (CYMDOF) shear tests for interfaces. The material models play the major role in finite element analysis. So, it is very important to choose suitable material models to characrize various components of a reinforced soil structure. The backfill, foundation soil, and retained fill were modeled using four-nodded isoparametric elements. The soil was simulated using the disturbed state concept (DSC). The material parameters for the soil used in finite element analysis are given in Table 1.

Table (1) Material parameters used in FEM analysis

Material constants	Symbol	Soil	Interface
Plasticity constants	E or K_n	$f(\sigma_3)$	$f(\sigma 3)$
	ν or K_s	0.3	$f(\sigma 3)$
Plasticity constants	γ	0.12	2.3
	β	0.45	0.0
Phase change paramere	n	2.56	2.8
Groth parameters	a_1	3×E-5	0.03
Nonassociative constant	κ	0.2	0.4
Disturbance parameters	D_u	0.93	
	A	0.368	
	Z	1.6	
Angle of friction	ϕ	40°	
Unit weight	γ	18.84 Kn/m^3	
Coefficient of earth pressure at rest	K_0	0.4	

The concrete facing panels were modeled by four-node isoparametric elements. Young's modulus (E) and Poisson's ratio (ν) for the were concrete E = 2.1 × 10^7 kPa, ν = 0.15, the thickness of 152.5 mm were adopted for the continuous full panel facing. The geogrid reinforcement was modeled using the two noded bar element. The constitutive behavior of reinforcement layers was simulated using the linear elastic model. The stiffness of the reinforcement with thickness of 1.5 mm was described by Young's modulus E = 1.5 × 10^6 kPa. Compressive forces were not permitted to develop within the geogrid element because the geosynthetic reinforcement are extensible and cannot hold compressive forces. The creep of the geogrid was not considered in this analysis.

One of the most important factors in modeling interface zones, relative motion between the soil and reinforcement. The interfaces between the reinforcement layers and soil were modeled using four-node thin layer elements. The disturbed state interface model has been used to simulate sand-geogrid interfaces. The stiffness matrix of these elements is formulated in terms of two independent stiffness values one in the shear direction and the other in the normal direction. The shear strength and stiffness of interfaces were determined using data form direct shear tests performed on sand-geogrid interface. The normal stiffness of the interfaces was calculated from results of normal compression test carried out on sand-geogrid interface. The material parameters used in finite element method simulation of Tucson wall are given in Table 1.

5 PREDICTION DATA AND FIELD MEASURMENTS

To analyze the performance of geogrid-reinforced soil wall, the finite element procedure was used and the predictions are compared with the field data.

5.1 *Wall Face Movement*

Figure 3 shows the measured facing panel displacement and predicted values for Case no. (1) and Case no. (2). Both the finite element analysis and measurements demonstrate that the wall face rotates about the toe of the wall. The predicted deformation agrees qualitatively good but quantitatively only fair.

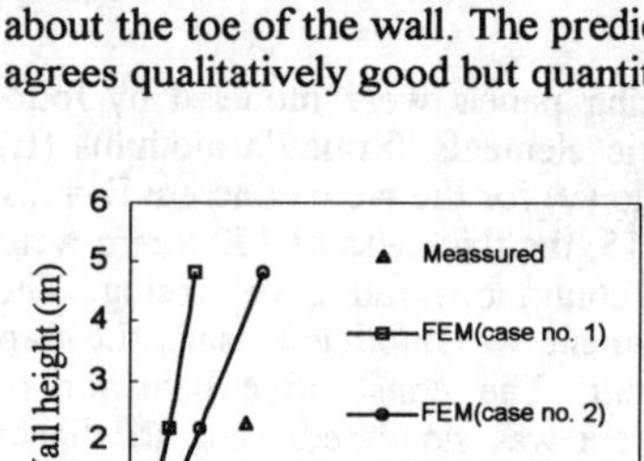

Figure 3. Comparison of predicted & measured wall face movement

5.2 *Vertical Soil Pressure*

The vertical soil stress distributions at elevation 1.53 m from the leveling pad of the wall are shown in Figure 4 for Case nos. (1) ,(2) after opening the road to traffic. The traffic loading was simulated as surcharge with value of 20 kPa. The vertical soil stress near the wall face is less than the overburden value (i.e., $\sigma_v = \gamma h$). This low value may be due to interface effects between backfill and facing panel such as stress relief and arching causing by reaction, and relative motions of the wall. Also, it is observed that the vertical stress distribution along the reinforcement layer is nonlinear. The vertical pressure increases in the zone away from the facing panel until reaching a maximum value, which occurs at distance of about 152 cm away from the wall face. The predicted results from finite element analysis agree well with the measured values.

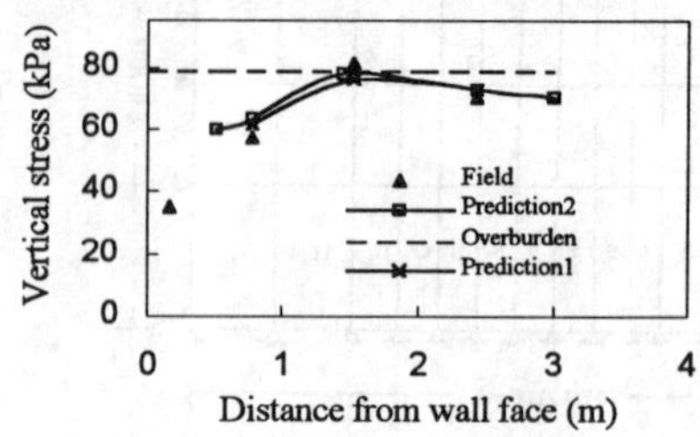

Figure 4. Comparison of field & prediction of vertical stress at ele. 1.53 m after opening to traffic

5.3 *Lateral Earth Pressure Against Facing Panel*

The earth pressure against the facing panel was obtained in the finite element analysis by the horizontal stress in the soil elements near the facing. This pressure distribution is useful for evaluating the magnitude of the stresses exerted on the facing panels and the tension in the geogrid connection. Figure 5 shows the predicted and measured lateral soil pressure behind the facing panel along with the Rankine analysis for one case. Predicted and measured horizontal soil stresses agree very well. The design procedure assumes that no lateral earth pressure is transferred to the wall face, because all of it should transfer to the reinforcement as tension. Except at the bottom of the wall, the horizontal soil stress seems to approximately confirm this assumption. When using full height panels as wall facing, restraint to displacement of the wall facing at the connection between the wall facing at the connection between the wall facing and the footing may inhibit load transfer between the soil and geogrids causing in lateral earth pressure acting on the facing panel.

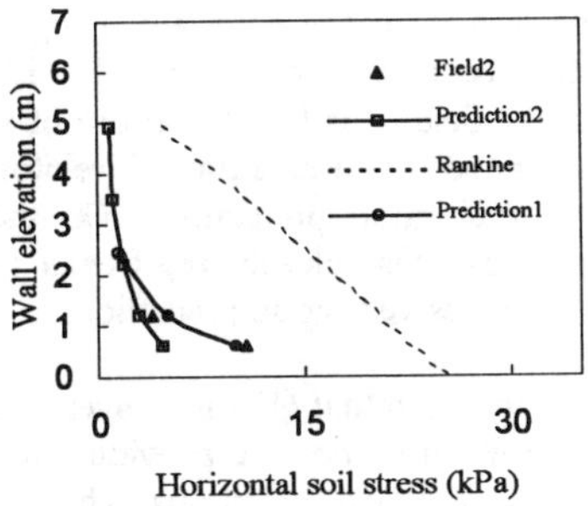

Figure 5. Comparison of field & prediction of horizontal soil stress after opening to traffic

5.4 *Soil Strain*

The predicted values of horizontal soil strains at elevation 2.44m from the finite element analysis are compared with the field data in Figure 6 for Case nos. 1 and 2.

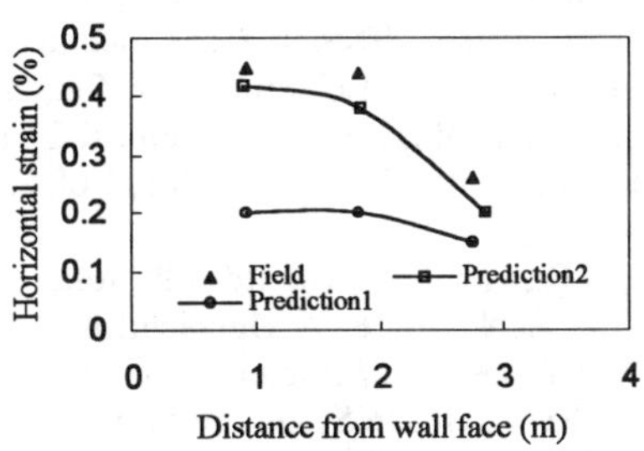

Figure 6. Comparison of field & prediction of horizontal soil strain at elevation 2.44 m after – opening to traffic (Case nos. 1& 2)

It can be seen that there is a good agreement between the predicted and the measured distribution of horizontal soil strain. The horizontal soil strain decrease away from the wall face. Figure 7 displays the predicted and the measured distribution of vertical soil strain at elevation 1.08 m. The measured and the predicted vertical soil strain are qualitatively good.

The measured and predicted valued show an increase in vertical strain with an increase in distance from wall face and this is consistent with the measured and predicted vertical stresses which showed an increase in vertical stress from the wall face to the central part of the reinforced soil mass.

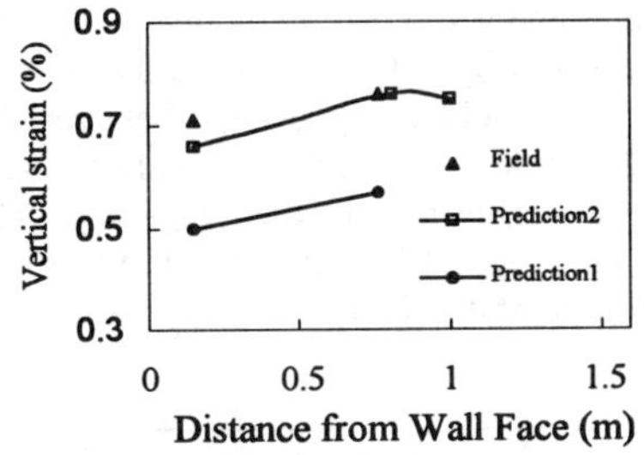

Figure 7. Comparison of field & prediction of vertical soil strain at elevation 1.08 m after opening to traffic

5.5 *Geogrid Strains*

Measured and predicted reinforcement tensile strains at elevations of 1.37 and 4.42 m are shown in Figure 8 The predicted reinforcement strains are not in agreement with the measured response at elevation 1.37 m (Case no. 1). Agreement between the measured and predicted values is considered excellent for Case no.(2). The results demonstrate that tensile strains in the geogrids are less than 0.4 % corresponding to 4.4-kN/m load in the geogrid. Comparison of this load to the maximum tensile strength of the geogrids, which is 79 kN/m (Netlon,1984), indicate that the grids are loaded to 5.5 % of the ultimate strength.

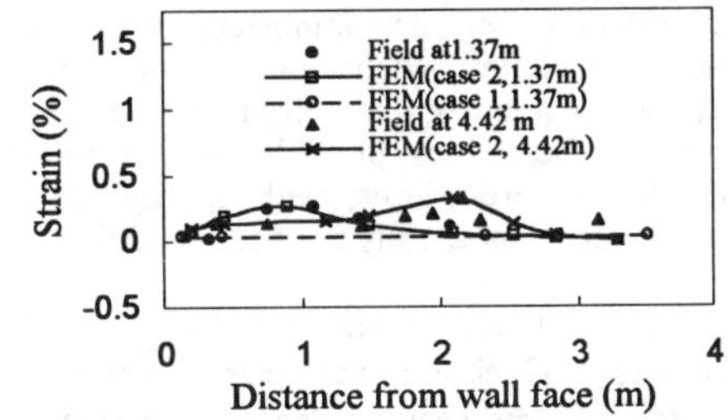

Figure 8. Comparison of field & prediction of geogrid strain

5.6 *Lateral Stress Transferred to Reinforcement*

Figure 9 shows comparison between measured and predicted results at different elevations in terms of horizontal pressure in the geogrids. Also, active earth pressures predicted by Rankine and at-rest theories are presented on the figure for the aim of comparison. There are two sets of measured values from the field. First set measured with strain gages and is refereed to as field 1, and the second set measured by inductance coils and is here after refereed to as field 2. Agreement between the measured and predicted values is considered very good in Case no. 2 as displayed in Figure 9. The predicted results by FEM are close to those determined from K_a state of stress.

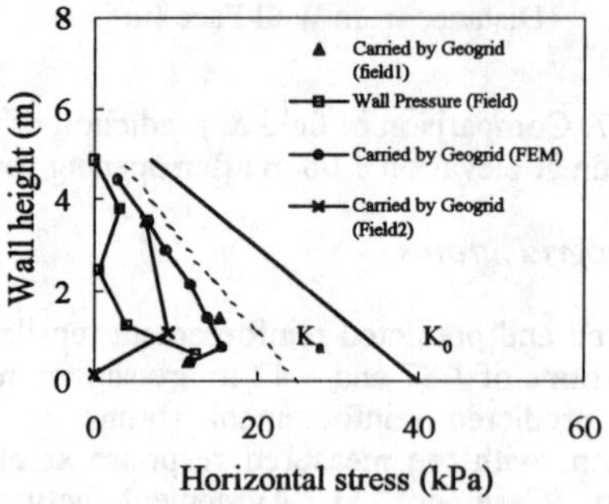

Figure 9. Comparison of field & prediction of horizontal stress carried by geogrid & wall face

6 CONCLUSIONS

For realistic simulation of the behavior of reinforced earth structure, it is necessary to characterize accurately the behavior of backfill soil and interfaces. For consideration of such important factors as the nonlinear material behavior, simulation of construction sequences, geometry and interfaces, and interaction effects, it is possible to use numerical (finite) element procedures, with appropriate mesh layout that represent closely the field conditions.

The disturbed state concept (DSC) that allows consideration of elastic and plastic deformation, stress path, volume change, degradation (softening) in an unified manner can provide accurate representation of the behavior of soils and interfaces. It is used to characterize the behavior of backfill soils and interfaces between the soil and Tensar geogrid reinforcement wall constructed in Tucson, AZ. The parameters for the DSC model are obtained from a series of laboratory triaxial tests for soils and shear tests for the interfaces.

A general finite element procedure (DSC-SST2D) is used to simulate the field behavior of the Tucson wall. The prediction in means of wall movements, stresses and strains in the soils and geogrids, and lateral earth pressures are compared with those measured in the field. It is found that use of a coarse mesh with limited representation of reinforcement dose not provide good predictions. On the other hand, a fine mesh that closely represents the field conditions, provides very good predictions.

It may be concluded that (1) the conventional limit equilibrium procedures do not provide appropriate representation of the field conditions, while the finite element procedures show very good comparisons with the measured data, and (2) the proposed numerical procedures with the DSC constitutive models can provide a realistic tool for analysis and design of reinforced earth and other geotechnical structures.

REFERENCES

Berg, R.R., Bonaparte, R., Anderson, R.P., & Chouery, V.E., (1986), "Design, Construction, and Performance of Two Reinforced Soil Retaining Walls," Proceeding of the 3 rd International Conference on Geotextiles, Vienna, Austria, Vol. 2, pp. 401-406.

Christopher, B.R., Gill, S.A., Giroud, J.P., Juran, I., Mitchell, J.K., Schlosser, F., & Dunnicliff J. , (1989) "Reinforced Soil Structures," Volume II. Summary of Research and System Information, Report No. FHWA-RD-89-043, Federal Highway Administration, Mclean, Virginia, November, 158p.

Collin, J.G., (1986), "Earth Wall Design," Ph.D. Dis sertation, University of California, Berkeley, CA, USA.

Day, R.A., & Pott, D.M., (1994), "Zero Thickness Interface Elements Numerical Stability and Appli cation," Int. J. Numer. Analyt. Meth. Geomech., Vol. 18, pp. 689-708.

Desai, C.S. (1999), "Mechanics of Materials and Interfaces: The Disturbed State Concept,"CRC Press Boca Daten, Florida, Under Publication.

Desai,C.S. (1998), "DSC-SST-2D: Computer Code for Static, Dynamic, Creep and Thermal Analysis-Solid, Structure and Soil-Structure Problems as User's Manuals," Tucson, AZ.

Desai, C.S. & Ma, Y., (1992), "Modeling of Joints and Interfaces Using Disturbed State Concept," International Journal for Numerical and Analytical Methods in Geomechanics, Vol. 16, pp. 623-653.

Desai, C.S., Zaman, M., Lightner, J.G. & Siriwardane, H.J., (1984), "Thin-Layer Element for Inter faces and Joints," Int. J. Numer. Analyt. Meth. Geomech., Vol. 18, pp.19-43.

Fishman, K.l., Desai, C.S., & Berg, R.R., (1991), "Geosynthetic- Reinforced Soil Wall: 4-Year History," in Behavior of Jointed Rock Masses and Reinforced Soil Structures, Transportation Research Record 1330, TRB, Washington, D.C., pp. 30-39.

Gen, A., Carol, I., & Alonso, E.E., (1988), "An Interface Element Formulation for the Analysis of Soil-Reinforcement Interaction," Computers and Geotechnics, Vol. 7, pp. 133-151.

Goodman, R.E., Taylor, R.L., & Brekke, T.L., (1968), " A model for the Mechanics of Jointed Rock," Journal of Soil Mechanics and Foundation Div. ASCE, Vol. 94, pp. 637-659.

Leshchinsky, D. (1992), "Discussion: Strain Compatibility Analysis for Geosynthetics Reinforced Soil Walls," Journal of Geotechnical Engineering, Vol. 188, No. 5, pp.816-819.

Netlon Limited, (1984), "Test Method and Physical Properties of Tensar Geogrids." Technical Guide lines.

Romstad, K.M., Herrmann, L.R., & Shen, C.K., (1976), " Integrated Study of Reinforced Earth II: Behavior and Design," J. Geotech. Eng. Div., ASCE, Vol. 102, No.6, pp. 577-590.

Rowe, R.K., & Ho, S.K., (1992), "A Review of the Behavior of Reinforced Soil Walls," Preprint Special and Keynote Lecture, Int. Symp. On Earth Reinforcement Practice, A.A. Balkam,Rotterdam, pp.47-76.

"Tensar Earth-Reinforced Wall Monitoring at Tanque Verde-Wrightstown-Pantano Roads, Tucson, AZ," (1986), Desert Earth Engineering, Preliminary Report, Pima County Dept. of Transportation and Flood Control District, Tucson, AZ.

"Tensar Geogrid reinforced Soil Wall," (1989), "Experimental Project 1, Ground Modification Techniques," FHWA-EP-90-001-005, US Dept. of Transportation, Washington, D.C.

Wu, J.T.H., (1992), "Predicting Performance of the Denver Walls: General Report," in Geosynthetic-Reinforced Soil Retaining Walls, A.A.Balkema.

Yi,C.T., Chan, D.H., & Scott, J.D., (1995), "A large Slippage Finite Element Model for Geosynthetics Interface Modeling," Geosynthetics' 95, Nashville, Tenneaaee, Vol. 1.

Computer Methods and Advances in Geomechanics, Desai et al. (eds) © 2001 Balkema, Rotterdam, ISBN 90 5809 183 X

Application of polyethylene strip reinforced soils in geotechnical engineering

P. Kuklík & M. Šejnoha
Department of Structural Mechanics, Faculty of Civil Engineering, CTU, Prague, Czech Republic

ABSTRACT: This paper presents several experimental results derived for polyethylene strip reinforced soils. First, standard direct shear tests were performed to asses applicability of scrap bottles as a potential soil reinforcement. The effect of amount, shape and size of the reinforcement was considered. Positive influence of the reinforcement on shear resistance observed for small laboratory specimens was further supported by results obtained during in situ measurements of reinforced gravel piles processed using the Franki technology.

1 INTRODUCTION

The objective of the present contribution is twofold. First, we wish to offer possible solution to a problem associated with ever growing amount of waste and recycled material, such as plastic soft drink bottles, produced every year. Unfortunately, most of the production is destined for the landfill rather than reuse. An application of used polyethylene bottles as a potential candidate for soil reinforcement may provide answer to the above, yet not satisfactorily resolved, environmental problem. Naturally, the principal objective now becomes to confirm applicability of this type of material in such applications. In particular, we attempt to assess the shear strength of soils reinforced by narrow strips made from polyethylene bottles.

In general, the idea of improving the shearing response of soils by introducing various types of reinforcements has been around for more than a decade. To cite a few we mention the work by (Gray and Ohashi 1983) who studied the mechanics of both metallic wires and synthetic fiber reinforced sand mixtures. Our motivation for the proposed type of reinforcement comes from the successful experimentation with chipped automobile tires as general soil reinforcement or lightweight and compressible fill. We refer the reader to (Tatlisozs et al. 1998) for an introductory review. In either case, however, the results indicate an increase of the peak shear strength and limited or no post peak reduction in shear resistance.

Similar potential is expected from the reuse of scrap polyethylene bottles as the soil stabilizer or as a compressible inclusion (a well-defined soil-bottle strips mixture) on earth retaining structures. Such applications not only offer the prospect of an environmentally safe, long-term disposal for plastic ware, but may also result in enhancement of the material properties of ordinary soils.

Section 2 outlines experimental measurements conducted on small laboratory specimens. Application large in situ structures such as gravel piles is discussed in Section 3.

2 LABORATORY MEASUREMENTS

To asses applicability of scrap bottles as a potential soil reinforcement we started with a standard direct shear stress test. A sample of clean quartz sand from Provodin with grain size ranging from 0.1 to 2 mm was selected for the reinforcement test. Standard laboratory direct shear test device was used to carry out experimental measurements. Both reinforced and virgin sands were tested. Reinforced sand was prepared by mixing the sand with polyethylene strips having the width equal to 2 mm and length of 15 and 40 mm. Squared shape of the reinforcement (2x2 mm) was also considered. The degree of reinforcement did not exceed in any test 4 percent of the weight of sand sample.

Results appear in Fig. 1. Fig. 1(a) shows variation of shear stress as a function of sliding displacement derived for virgin soil. Typical decay of shear stress after peak is evident. Fig. 1(b) displays similar results. Here, individual lines are normalized with respect to the applied normal pressure. The figure suggests that with increasing normal pressure the material becomes

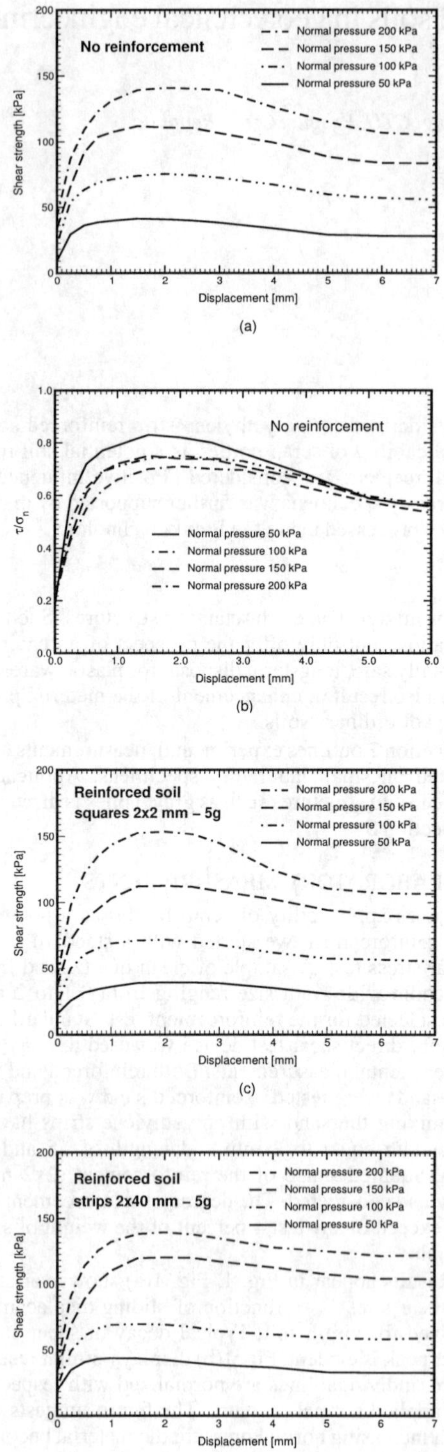

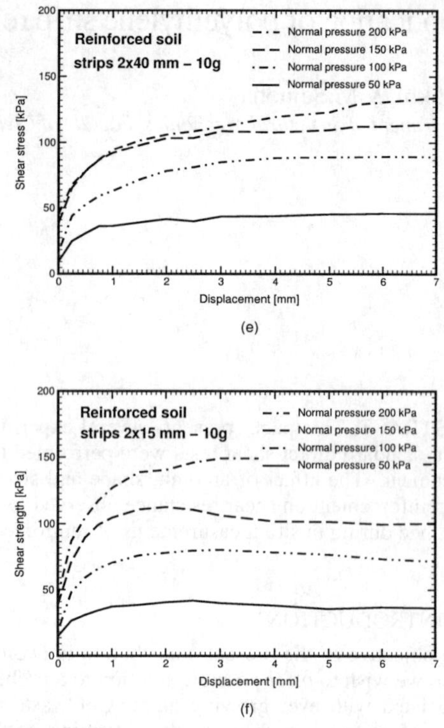

Figure 1: Direct shear stress test

more stable manifested by smaller difference between the attained shear strength and its residual value.

Figs. 1(c)-(f) shows results found for various degree and selected shapes of reinforcement. When inspecting Fig. 1(c) it becomes clear that the small size squared reinforcement does not improve the mechanical behavior of mixture. On the other when mixing the virgin soil with strips of various sizes results in noticeable improvement. Unquestionable is also a positive effect of increased amount of reinforcement in the mixture evident from Figs. 1(e)-(f).

The enhancement of material properties of polyethylene strip-sand mixtures resulting in stable force-displacement curves is further highlighted in Figs. 2(a)-(b) summarizing results for normal pressure equal to 100 kPa and 150 kPa, respectively. It is worthwhile to mention that the soil mixture containing longer strips with higher degree of reinforcement did not experience in any tests undesirable material softening but rather stable behavior throughout the experiment. A slower rise of shear stress in initial stage can be attributed to insufficient compacting resulting in higher porosity of a sample. Strips of smaller size or lower degree of reinforcements , however, did not prevent grains from wedging.

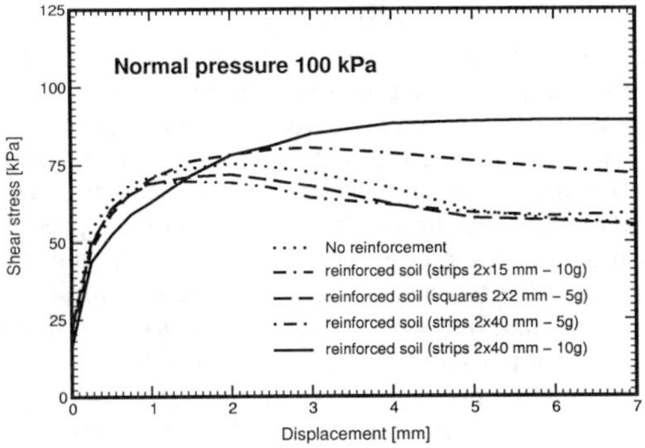

(a)

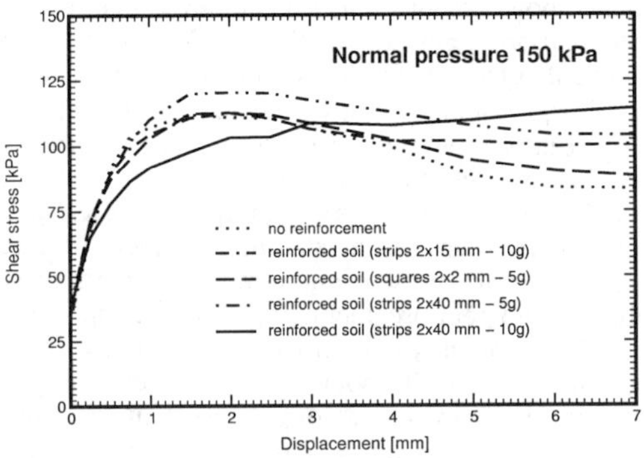

(b)

Figure 2: Direct shear stress test

This conclusion is supported by results displayed in Fig. 2(b), which correspond to distributions already seen in Fig. 1(b). Recall the positive influence of normal pressure on soil behavior.

Although such intuitive explanations are of a certain value, more rigorous prove supported by identified deformation mechanisms is of particular interest. Moreover, experimental measurements alone do not suffice and the back analysis is usually called to substantiate the assumptions used in the design. In this view, successful modeling of the role of grain size, grains and voids distribution and orientation and size and distribution of reinforcement on macroscopic response could suggest interventions on the micro-level. Such efforts involve modeling on different scales starting with the evaluation of shear resistance on grain boundaries to assess shear stress resistance on the meso-scale along certain mesoscopic slip surface. Averaging over a representative volume then results in the shear resistance at a certain macroscopic point. For more details on this subject we refer the reader to (Nemat-Nasser 2000). Some data needed in such a multi-level modeling could be provided through image analyses. An image of the reinforced quartz sand is displayed in Fig. 3. When transformed into binary images they can serve to evaluate e.g. skeleton volume fraction together with volume fraction and distribution of pores, and eventually to determine the fluid permeability of a soil using novel homogenization techniques (Torquato 1991).

As for reinforced soils, the images taken from real material samples can be further used to assess a spatial distribution of reinforcements and to verify a degree of anchoring of reinforcements inside the soil body. In any case, identification of micromechanisms driving the macroscopic deformation plays a vital role in successful design of reinforced soils based on limit equilibrium so popular among civil engineers. This is under current investigation.

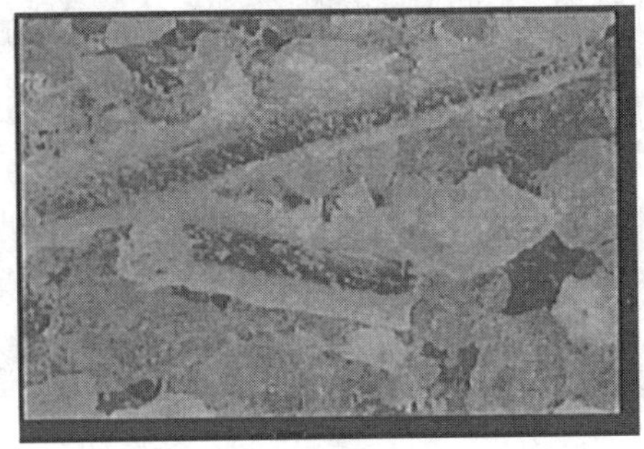

Figure 3: A reinforced sand sample

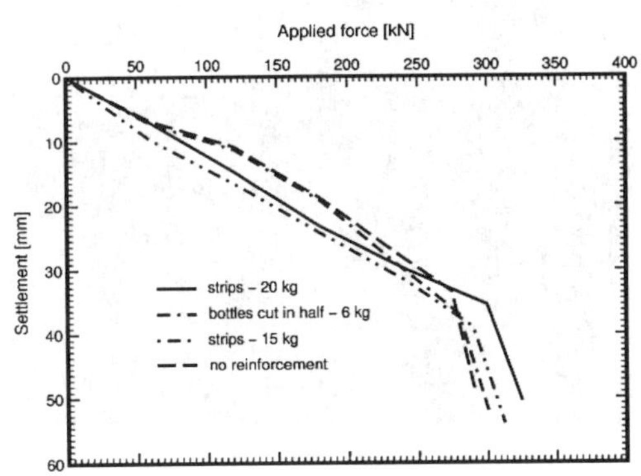

Figure 4: Pile settlement

(a)

(b)

(c)

Figure 5: Gravel pile

3 IN SITU MEASUREMENTS

To further support and also verify encouraging results derived from small laboratory experiments we applied this type of reinforcement to gravel piles processed using the Franki technology. The use of reinforcement made from polyethylene strips in conjunction with Franki piles is promoted by the technological process as it allows for proper soil compaction. In addition, we were able to preserve, to some extent, a similarity between material tested in the laboratory and in situ in terms of particle and reinforcement sizes.

Expected improvement of the soil–reinforcement mixture evident from Fig. 4 was verified. Noticeable is an increase of the normal force applied at the pile head prior to loss of stability. Positive higher degree of reinforcement already proved by laboratory experiments, however, has not been achieved due to technological problems. To conclude we recall one of the principal objectives of the present study, when searching for new application of waste material. The preliminary results show that presence of plastic bottles in soils in any form does not lower the material properties of soils, but rather opposite impact is observed.

To clearly manifest the stabilizing effect of the reinforcement we present photographs of both gravel pile with no reinforcement, Fig. 5(a) and that reinforced by polyethylene strips, Fig. 5(b). Evidently, the unreinforced pile poured out after removing the supporting clayey soil as oppose to reinforced pile, which remained at place. Distribution of reinforcements within the pile are shown in Fig. 5(c).

Although an application of polyethylene strips in piles have not proven its full advantageous as soil reinforcement, it still fulfilled the expected effect as soil stabilizer.

REFERENCES

Gray, D. and H. Ohashi (1983). Mechanics of fiber reinforcement in sand. *J. Geotech. Engr. 109*(3), 335–353.

Nemat-Nasser, S. (2000). A micromechanically-based constitutive model for frictional deformation of granular materials. *J. Mech. Phys. Solids 48*(6–7), 1541–1563.

Tatlisozs, N., T. Edil, and C. Benson (1998). Interaction between reinforcing geosynthetics and soil-tire chip mixtures. *J. Geotech. and Env. Engr. 124*(11), 1109–1119.

Torquato, S. (1991). Random heterogeneous media: Microstructure and improved bounds on effective properties. *Appl. Mech. Rev. 44*(2), 37–76.

Computer Methods and Advances in Geomechanics, Desai et al. (eds) © 2001 Balkema, Rotterdam, ISBN 90 5809 183 X

The application of numerical analysis to innovative design of geosynthetic reinforced soil walls

S.-C.R. Lo
University College, University of New South Wales, Canberra, A.C.T., Australia

ABSTRACT: The use of geosynthetic reinforcement enables the construction of more innovative wall configurations. However, conventional design rules may no longer be adequate. Numerical analyses contribute to the design of these innovative walls even under severe project constraints

1 INTRODUCTION

Design guides for geosynthetic reinforced soil (GRS) walls are often based on a rectangular reinforced zone with a width to height ratio in the range of 0.7 to 1.0. Although GRS walls designed using these guides have been performing satisfactorily, engineers are constrained to the conventional wall configurations. With the range of geosynthetic reinforcement available on the market, and the additional construction flexibility provided by the use of geosynthetic reinforcement, more innovative wall configurations can be constructed. The challenge faced by design engineers is the provision of reliable design of more innovative (hence more economical) wall configurations. This challenge is more obvious when a "Design and Construct" specialist subcontractor contracting system is employed. This paper examines a number of cases where the application of numerical analysis contributes significantly to the effective design of innovative GRS walls in construction projects in Australia.

2 DESIGN CONSTRAINTS

Although it is true that innovative GRS walls can be studied by experimental means, especially with the use of centrifuge testing (Jaber & Mitchell 1991, Zonberg et al 1997), the time constraint imposed by a construction project makes numerical analysis a more attractive option. The successful use of numerical analyses in predicting performance of GRS walls have been reported by a number of researchers (Kapurapu & Bathurst 1995, Rowe & Ho 1997). However, a number of constraints inherent in most projects, particularly in design and construct projects, need to be addressed. Since the fill is yet to be obtained from borrowed areas, it may only be possible to test "similar" soils. This means only simple soil models such as the Mohr-Coulomb elastic-plastic model or the Duncan-Chang non-linear elastic model etc would be appropriate. Since field data clearly suggest that the performance of GRS walls may be significantly affected by the construction procedures, the simulation of construction sequence is considered as essential. A number of wall configurations and wall sections need to be studied numerically. There is virtually no time available for any significant modification of computer code. This means the numerical analysis must be based on well-tested, versatile computer code(s). A comprehensive design specification may also require that the safety margin implied by any innovative design is no less than that of conventional design. This means that the numerical analysis needs to be calibrated against representative conventional walls. The above discussion is largely based on the author's experience in Australia and hence may or may not be equally applicable elsewhere. Despite these severe constraints, numerical analysis contributes significantly to innovative design of GRS walls.

3 SPECIAL NUMERICAL REQUIREMENTS

The behaviour of a GRS wall is highly non-linear even under a working condition. Hence a non-linear soil model is needed in numerical analysis. This can be illustrated by comparing the lateral deflection of a GRS wall obtained by an elastic analysis with that obtained by the Mohr-Coulomb elastic-plastic model. As evident from the comparison presented in Fig. 1 for an 8m high wall, the linear elastic analysis gave unrealistically low wall deflection, whereas the Mohr-Coulomb elastic-plastic model (abbreviated as

MC in Fig. 1) gave considerably larger but more reasonable wall deflection. The wall details are: width of reinforced zone = 4.64m, reinforcement stiffness = 1000kN/m, vertical reinforcement spacing = 0.8m, $\phi = 40^o$ for select fill, $\phi = 30^o$ for general fill. The linear analysis also gave unreasonably low reinforcement tension compared with that of MC analysis.

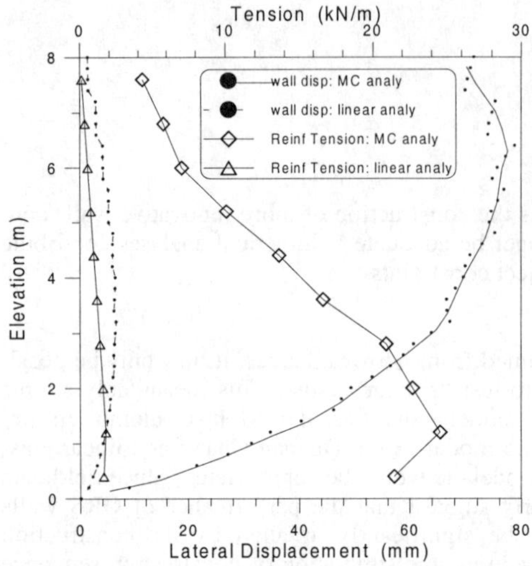

Figure 1. Wall deflection and reinforcement tension

A detailed examination of the stress state in a GRS wall indicates a significant portion of the reinforced block is at a near-failure stress state. This presents numerical problems in the use of simple non-linear elastic soil models such as the Duncan-Chang model. These numerical problems can be overcome by limiting the value of r_f, a Duncan-Chang input parameter, to a certain upper bound value irrespective of test results or setting a lower limit to the Young's modulus (Lo et al 1996). But certain calculated displacements may be dependent on these arbitrarily set values.

Two approaches were used in improving the Duncan-Chang model. The first approach was to have the Poisson's ratio, μ, taken as dependent on stress with the following equation.

$$\mu = \mu_0 + 0.2\sqrt{S} \geq 0.495 \quad (1)$$

where μ_o, is the initial Poisson's ratio and usually has a value of about 0.3; and S is given by

$$S = \frac{r_f (1 - \sin\phi)(\sigma_1 - \sigma_3)}{2c.\cos\phi + 2\phi_3.\sin\phi} \quad (2)$$

where the parameters ϕ, c, σ_1, σ_3 take their usual meanings. Eqns (1) and (2) avoid having unreasonably large volumetric strain.

The second approach is to combine the Duncan-Chang model with the elastic-plastic model. The Young's modulus is given by the Duncan-Chang non-linear elastic equation but with a lower limit, E_{min}, which can be rationally selected based on the stiffness at $S \approx 0.9r_f$. The formulation (in the incremental analysis) will be switched to the Mohr-coulomb elastic-plastic formulation when this lower limit is reached.

Numerical analysis may also be used to develop new design rules or extend currrent design rules, both of which are for the ultimate limit state. This means the numerical analysis must be able to proceed reliably to a near-collaspe state. The difficulties in performing finite element analysis to near-collapse state is well discussed in the literature (de Borst & Vermeer 1984, Fryman & Burd 1997). An alternative is to employ a finite differnce code called FLAC, Fast Lagrangian Continua Analysis (FLAC 1996, Cundall & Board 1988). In FLAC, the equations of equilibrium and stress strain behaviour, being expressed in finite difference form, are solved by an explicit iterative scheme. The analysis can be incrementally taken to collapse along a path prescribed by the user. Such an analysis will be referred to as a FLAC-analysis in this paper. The program includes looping control commands and an internal programming language for implementing user-defined procedures. These features can be used to model construction sequence. FLAC-analyses have been successful used in studying the static and dynamic response of GRS walls (Bathurst & Hatami 1998), and Nγ-bearing capacity factors (Frydman & Burd 1997), and bearing capacity failure of trapezoidal GRS walls (Lo et al 1999).

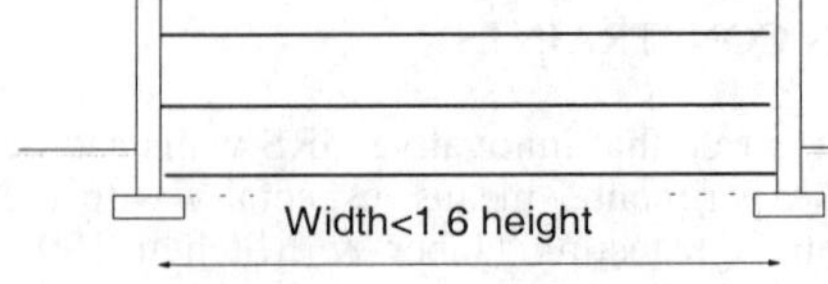

Figure 2. Tied back-to-back wall

4 CASE HISTORIES

A number of case histories on the application of numerical analysis are presented in this section. The wall sections (and data) may only be a simplified version of the actual project in order to avoid addressing unrelated complexities imposed by construction and contractual requirements. In all the cases examined, simple soil models were used; but the influence of soil parameters was always studied. Construction sequences were always modelled.

4.1 *TBB Walls*

Numerical analysis has been applied successfully in the design of tied back-to-back (TBB) GRS walls as shown in Fig. 2. FHWA design guidelines (Christopher et al. 1990, Elias & Christopher 1996) suggested that reinforcement tension in a TBB wall could be considerably higher than that given by conventional design rules. The construction of a TBB wall was also suggested to be problematic. In the Dutton Park to Port of Brisbane Rail Link project, the difficulties in the construction of TBB walls were resolved by the use of high tenacity polyester straps as soil reinforcement. Hence, the caution on higher reinforcement tension was re-examined with a series of non-linear finite element analyses (Lo et al 1996). The results showed that the higher reinforcement tension was related to reinforcement stiffness; and the range of stiffness provided by high tenacity polyester reinforcement led only to a slight increase in reinforcement tension. Both the reinforcement tension and horizontal displacement profile were not sensitive to the choice of soil models and soil parameters. Therefore, simple design rules for the design of TBB walls were established via the finite element results, including the case of heavy set-back surcharge typical of railway loading.

4.2 *Delta values*

In the calculation of active force acting on the reinforced block for overall stability check, the friction angle mobilised at the interface between the general fill and the reinforced zone is needed. This friction angle, referred to as the delta value in this paper, is commonly assumed to be zero in the design of GRS walls. This is supposedly based on the assumption of a highly deformable reinforced block, and hence the necessary relative deformation cannot develop to mobilise a significant delta value. The validity of this assumption depends on the stiffness of the geosynthetics used. For conventional GRS walls, this conservative assumption has little impact on design outcomes. But for a wall with narrow reinforced zone or a wall support sloping backfill, the delta value may have a significant impact on overall stability (sliding and/or bearing capacity), and hence on the reinforced block width. A series of finite element analyses and FLAC-analyses was conducted to examine the delta values of such GRS walls, and this study is partly related to the M2 road project in Sydney. The reinforcement adopted in the analysis was a high stiffness polyester. Eight GRS wall configurations were analysed. Details of this study were reported in Lo & Gopalan (1996). It was concluded that the stiffness of the polyester was adequate to ensure mobilisation of a high delta value.

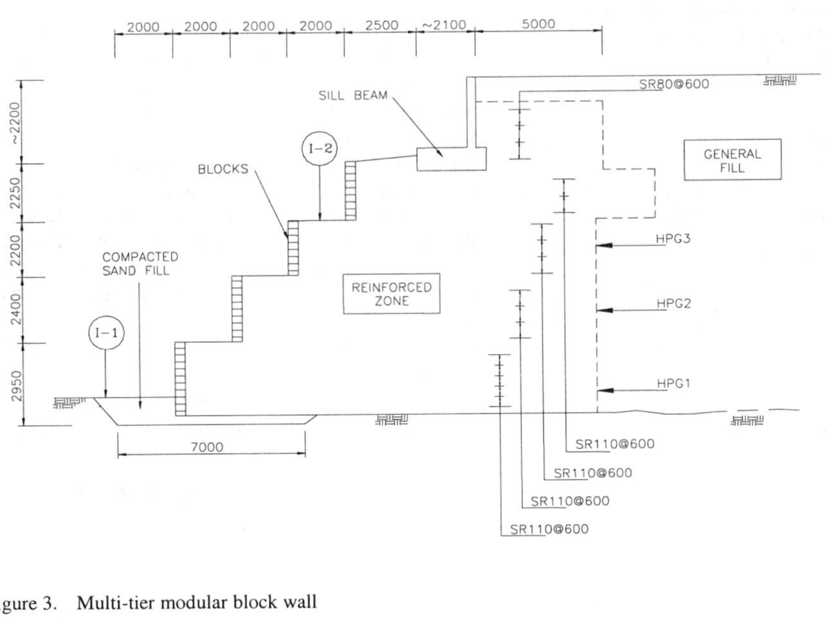

Figure 3. Multi-tier modular block wall

4.3 *Multi-Tier Modular Block Wall*

A 12 m high modular block wall supporting the end span of a major bridge structure as shown in Fig. 3 was constructed using a multi-tier wall arrangement. High density polyethelyne (HDPE) geogrids were used as soil reinforcements. The ground conditions consist of 1 to 3 m of loose silty sand overlying 7 to 10 m of medium dense silty sand. Sandstone bedrock is at approximately 13m depth. The loose sand layer contains pockets and/or lenses of soft silty clay, which had not been located accurately. These silty clay pockets/lenses were considered to have only slight influence on the overall behaviour of the foundation material; but may lead to some differential settlement. A GRS wall was considered to be most suitable in accommodating such differential settlement. In view of the loose and somewhat variable nature of the top layer of the foundation soil, the top 1m was replaced with compacted sand over the front 7 m. The wall was instrumented with horizontal profile gauges (HPG-1 to HPG-3 of Fig. 3), settlement plates, inclinometers (I-1 and I-2 of Fig. 3), load bolts and strain gauges, and earth pressure cells. Both FLAC and limit equilibrium analyses were used in the initial design stage. Details of the initial design, construction, and monitoring of this multi-tier abutment were reported in Won et al. (1996). It is pertinent to note that construction sequence was not modelled due to limitation of computing facility available in the original design stage. The FLAC analysis yielded rather unusual variation of reinforcement tension, with some reinforcement levels having two peaks in the reinforcement tension, with de-tensioning between the two peaks. Such a pattern however, was also reflected in the load bolt readings.

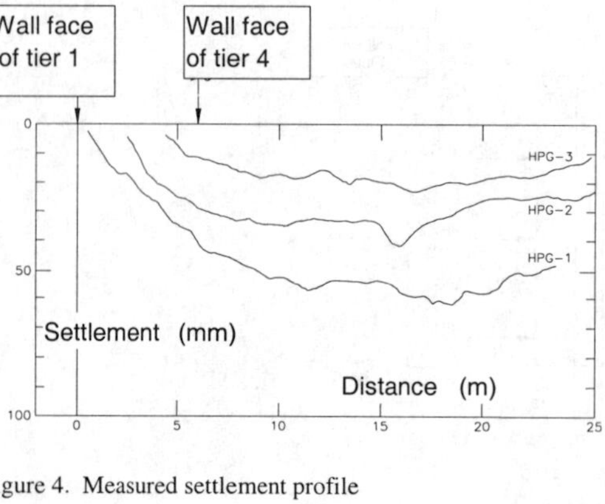

Figure 4. Measured settlement profile

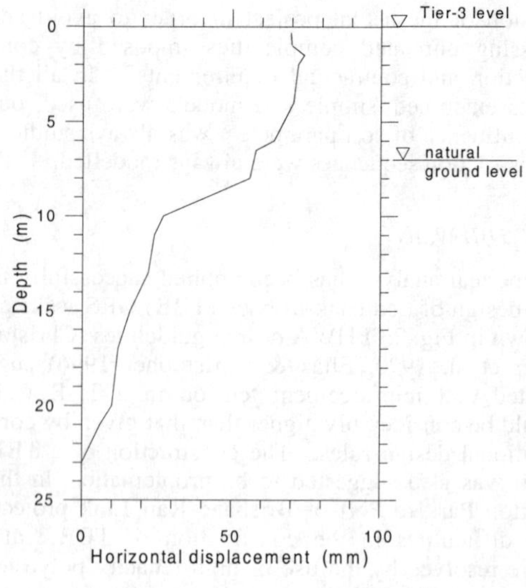

Figure 5. Measured horizontal displacement profile

However, field measurements indicated significant settlement and horizontal displacements. Some of the field measurements taken in mid-December 96 are presented as Fig.4 (for settlement profiles) and as Fig. 5 (for horizontal displacements at I-2). The settlement profiles presented in Fig. 4 were relative to their respective set of initial readings which were taken about a month after filling began. The I-2 inclinometer was installed a few days after the job began and initial readings were taken two weeks after the job began. However, the inclinometer tube was extended with wall construction. As such, some of the displacement occurring during wall construction was not fully registered. The magnitude of this error was considered to be low (10 to 20mm). The initial FLAC analysis can predict the high settlement by assuming low, but tenable, values for the Young's modulus for soils; but the significant horizontal movements were not predicted. Furthermore, the as-measured settlement profile showed a peak near the rear end of the reinforced zone and this was a concern. A series of additional analyses were conducted to study the status of this GRS wall. The final assumptions in the analysis were

- The construction sequence was modelled closely in a layer by layer manner.
- The modular block facing was modelled as 2D elements with horizontal no-tension joint planes.
- Full loading from the bridge deck was applied.

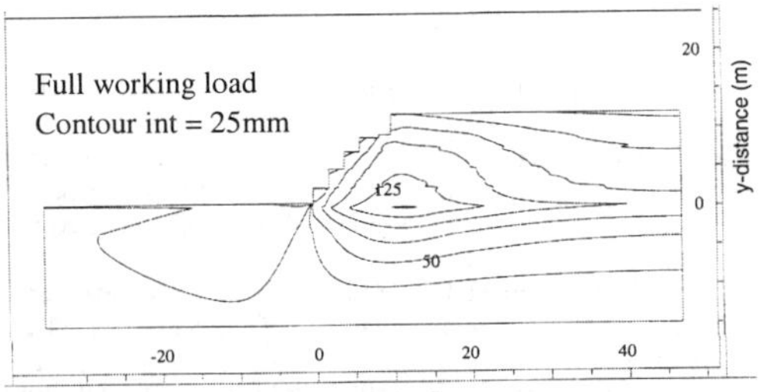

Figure 6. Predicted vertical displacement contours

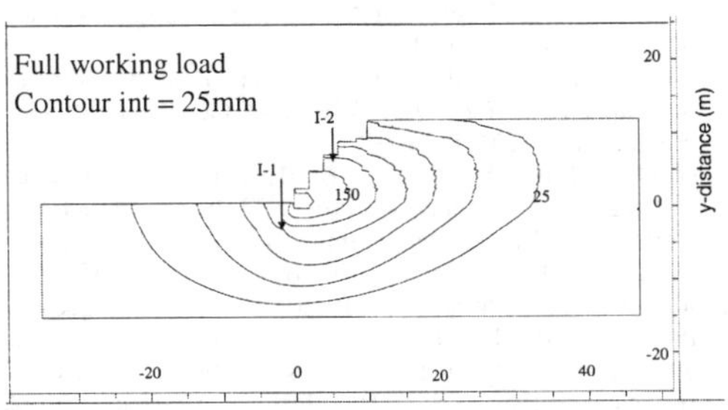

Figure 7. Predicted horizontal displacement contours

The Duncan-Chang model combined with the elastic-plastic formulation as discussed in an earlier section was employed. Eqns (1) and (2) were also included in the formulation. The "best-estimate" parameters adopted in the analysis are listed in Table 1. The sub-surface conditions and the soil parameters adopted in the analysis were considered to be conservative. The predicted vertical and horizontal displacements are presented in Figs. 6 and 7 respectively. The maximum predicted settlement at foundation level was 125 mm but the maximum measured settlement was about 70 mm. The improved analysis was able to predict that both settlement and horizontal displacements were significant. The maximum horizontal deflection at tier-3 was 150 mm but the measured value was 75 to 85 mm. This was somewhat expected, partly because of "worst scenario" assumptions made.

Table 1. Soil parameters

Parameters	Fill	Foundation Soils	
		Loose Sand	Medium Sand
ϕ (deg)	35	30	33
c (kPa)	10	0	0
ν (deg)	0	0	0
K	800	400	600
n	1.0	1.0	1.0
rf	0.9	0.9	0.9

K, n, and r_f are non-dimensional Duncan-Chang parameters
ν is dilatancy angle

The analysis was continued by progressively reducing the strength parameters of the soils. Collapse was detected when the friction angle of the foundation soil was reduced to 21°. Noting that "worst scenario" assumptions were made in the analysis, and that the predicted behaviour at working condition was more severe than that observed, one can conclude that the GRS wall has an adequate safety margin.

4.4 *Fine grained soils*

One possible advantage of GRS walls is that the select fill does not have to comply with tight grading requirements. However, there are still concerns about the use of fine grained soils. Won et al. (1994) reported significant movement of a GRS wall constructed with fine grained soil. The maximum buldge in a facing panel was about 100 mm and continued at a slow rate. The as-measured reinforcement tension was significantly lower than the design value but it manifested a slow increase with time. Although the causes of the higher wall deflection of this particular wall are a matter of debate, compaction of fine grain soil at a moisture content on the dry side of optimum will lead to a high matrix suction that may dissipate with time. The effects of reduction of apparent cohesion can be studied by conducting a FLAC analysis of a "fictitious" GRS wall of 8m height. The foundation soil (7m thick) was assigned constant strength parameters of $\phi = 30°$ and c = 10 kPa, with a Young's modulus increasing from 25 MPa near foundation level to 55 MPa. These wall dimensions and foundation parameters will ensure adequate overall stability. The reinforced zone was 7m wide. The reinforcements were assigned an elastic axial stiffness of 1000 kN/m and high interface strength parameters. This suppressed any form of internal instability. Both the general and select fill was modelled by the Mohr-Coulomb elastic-plastic model with a zero dilatancy angle. For the purpose of this exercise, a friction angle of 30° was assumed for both the general and select fill. An initial apparent cohesion of 60 kPa was assumed in the analysis to represent the initial matrix suction. As such, the analysis is a total stress analysis. Laboratory testing of fine grained soils indicates that these parameters are reasonable. Dissipation of matrix suction after completion of construction was then simulated by progressive reduction of apparent cohesion via time stepping. However, for simplicity, the Young's modulus was maintained constant at 25 MPa. The predicted increase in horizontal wall deflection with reduction in apparent cohesion are presented in Fig. 8. This analysis illustrates that significant delayed wall movement can occur, and such movements can not be taken as indicative of potential instability.

However, the development of delayed movement needs to be considered in the design.

5 CONCLUSION

A number of case histories were examined to highlight the contribution of numerical analysis to the design of GRS walls. The numerical tools used are not advance by research standard, and the emphasis is on selecting the simplest tools that can effectively address the design issues taking into account the project constraints. It is evident that numerical analysis can contribute to more innovative and effective GRS walls even under severe time and resource constraints.

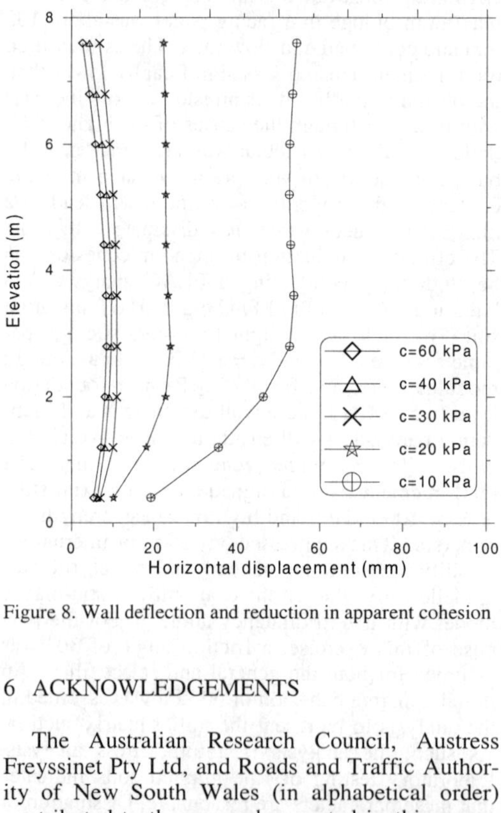

Figure 8. Wall deflection and reduction in apparent cohesion

6 ACKNOWLEDGEMENTS

The Australian Research Council, Austress Freyssinet Pty Ltd, and Roads and Traffic Authority of New South Wales (in alphabetical order) contributed to the research reported in this paper. The opinions expressed in this paper, however, are solely those of the author.

REFERENCES

Bathurst R.J. & Hatami, K. (1998) "Seismic response analysis of a geosynthetic- reinforced soil retaining wall", Geosynthetics International, 5[1], 127-166.

Christopher B.R., Gill, S.A., Juran, I., Schlosser, F., Mitchell, J.K. & Duncliffe, J. (1990) "Reinforced soil structures Vol. 1 Design and construction guidelines" Federal Hwy Administration, publication no., FHWA-RD-89-043.

Cundall, P. & Board, M. (1988) "A microcomputer program for modelling stress-strain plasticity problems" Proc. of 6th Intl Conference in Numerical methods in Geomechanics, Vol. 3, 2101-2108.

De Borst R. & Vermeer R.A. (1984) "Possibilities and limitations of finite element analysis". Geotechnique, 34[2], 199-210.

Elias V. & Christopher B.R. (1996) "Mechanically stabilised earth walls and reinforced soil slopes, design and consruction guidelines". FHWA Demo Project 82, working draft, Jun 96.

FLAC (1996) "Fast Lagragian Analysis of Continua" Version 3.3. User manual, Vol 1-3.

Frydman S. & Burd H.J. (1997) "Numerical studies of bearing capacity factor Nγ" ASCE, J of Geot and Environmental Engr, 123[1], 20-29.

Jaber M. & Mitchell J.K. (1990) "Behaviour of reinforced soil walls at limit states". Proc of Intl Reinforced Soil Conference, British Geotechnical Society, Glasgow, Sept 1990, publisher Thomas Telford of ICE London, 53-58.

Kapurapu R. & Bathurst R.J. (1995) "Prediction of geosynthetic reinforced soil wall performance using the finite element analysis" Proc of Geosynthetic reinforced soil retaining walls, Aug 1991, Denver, Colorado, 1992. Publisher Balkema, 329-342.

Lo S-C.R., Li, S-Q. Gopalan, M., & Gao, Z. (1996) "Analysis and design of a tied back-to-back geosynthetic reinforced soil wall". Geotechnical Engr, SE Asian Geotechnical Society, 27[1], 37-50.

Lo S-C.R. & Gopalan M. (1996) "Detrermination of delta values for reinforced soil walls". 7-th Australian New Zealand conference in Geomechanics. 661-666

Lo S-C.R., Bosler J. & Gopalan M. (1999) "Bearing capacity failure of a trapezoidal geosyhnetic soil wall". Geosynthetic Intl. 6[5], 383-416.

Rowe R.K. & Ho S.K. (1997) "Continuous panel reinforced soil wall on rigid foundation". ASCE, J of Geot and Environmental Engr, 123[10], 912-920.

Won G.W., Haussmann M.R., & Jankulovski, E. (1994) "Performance of Websol soil wall bridge abutment in Sydney, Australia". 5-th Intl Conference on Geotextiles, Geomembranes and Related Products. Singapore, Sept 94, 433-436.

Won G.W., Hull T., & De Ambrosis, L. (1996) "Performance of a geosynthetic segmental block wall structure to support bridge abutments" Proc. of the Intl. Symposium on Earth Reinforcement, Fukuoka, Japan, Nov 96, 543-548.

Zornberg J.G., Mitchell J.K. & N. Sitar (1997) "Testing of a reinforced slope in a geotechnical centrifuge". GTJ, ASTM, 24[4], 470-480.

Computer Methods and Advances in Geomechanics, Desai et al. (eds) © 2001 Balkema, Rotterdam, ISBN 90 5809 183 X

Finite element analysis of embankments supported on geocell layer

K. Rajagopal, G. Madhavi Latha, N. R. Krishnaswamy
Department of Civil Engineering, Indian Institute of Technology Madras, Chennai, India

ABSTRACT: The geocell reinforcement provides 3-dimensional confinement to the soil thus improving both the strength and stiffness of the soil. The improvement in these two parameters is quantified in this paper through standard triaxial compression tests on geocell reinforced sand samples. The results from these experiments have been analyzed to develop simple relations to estimate the shear strength and stiffness of the geocell reinforced soil in terms of the geocell properties. The frame work of hyperbolic model has been employed to develop simple equation to estimate the stiffness of the geocell reinforced soil. This model has been implemented in a finite element program to analyze the performance of geocell reinforced soils. The performance of laboratory-scale embankments has been back-predicted using this model to verify its accuracy. This model was found to be able to simulate the effect of the various parameters, viz. dimensions of the geocell pockets, modulus of geocell material etc.

1 INTRODUCTION

The geocell is a 3-dimensional cellular mattress made of geogrids or geotextiles, which provides all round confinement to the soil. Because of this confinement, the strength and stiffness of the soil increases. This type of soil reinforcement is ideally suited for the construction over extremely soft soils where the use of planar geosynthetics may not be adequate.

The 3-dimensional finite element simulation of problems with geocell reinforcement requires the generation of membrane type elements in a criss-cross manner to model geocell walls. This is rather cumbersome and hence it is preferable to work with a 2-dimensional equivalent model that can very well replicate the behavior of a 3-dimensional system. This paper deals with development of such a model and its verification from back-analysis of laboratory tests on large-scale model embankments.

2 TRIAXIAL COMPRESSION TESTS

The strength and stiffness behavior of geocell encased granular sands was investigated through a series of triaxial compression tests. The tests have used configurations of one geocell, two, three and four interconnected cells. The geocells were made of four different types of geotextiles whose 5% secant stiffness ranged from 0.5 to 70 kN/m. The details of these tests are reported in Rajagopal et al. (1998). Typical test results from the triaxial compression tests on geocell confined soil are shown in Figure 1. The results have shown that the geocell confinement does not change the friction angle of the soil whereas it induces some apparent cohesion even to granular soils.

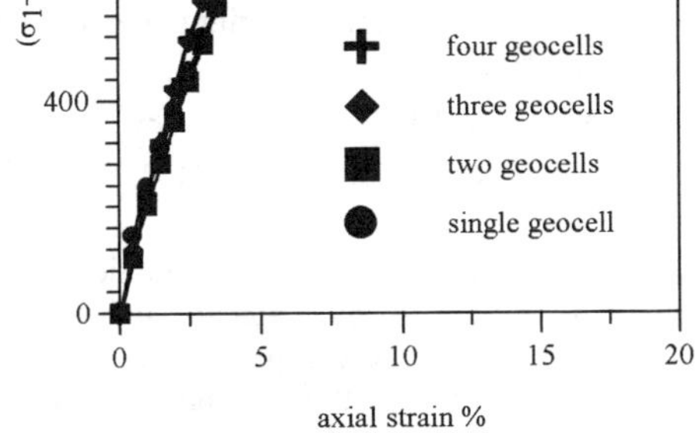

Fig 1. Typical stress-strain results from triaxial compression tests

It could be observed that as the number of geocells is increased, the strength and stiffness of the soil increased. However, when the number of cells was increased beyond three, there was not much further increase in the strength and stiffness. This result shows that the strength and stiffness behavior observed in laboratory tests with three inter-connected cells is representative of geocells in real cases with many inter-connected cells.

3 EQUIVALENT COHESIVE STRENGTH OF GEOCELL CONFINED SOIL

Based on the results from the above triaxial compression tests, an equation has been derived to estimate the apparent cohesion induced to the soil in terms of geometric and material properties of the geocell based on the membrane correction theory. The additional cohesive strength (c_r) due to single geocell encasement was obtained as,

$$c_r = \frac{\Delta\sigma_3}{2}\sqrt{k_p} \tag{1}$$

in which k_p is the Rankine's passive earth pressure coefficient and $\Delta\sigma_3$ is the additional confining pressure due to the membrane stresses given as,

$$\Delta\sigma_3 = \frac{2M\varepsilon_c}{d}\frac{1}{(1-\varepsilon_a)} = \frac{2M}{d_o}\left[\frac{1-\sqrt{1-\varepsilon_a}}{1-\varepsilon_a}\right] \tag{2}$$

The modifications to above equation to consider multiple cells are reported in Rajagopal et al. (1998).

4 STIFFNESS OF GEOCELL CONFINED SOIL

Due to the increase in the confining pressures exerted on the soil by geocell walls, its stiffness will also increase as illustrated in Fig. 1. The increase in the confining pressure depends on the geometric properties of the geocell and the material properties of the geosynthetic used to make the geocell. The following well known hyperbolic equation was used to quantify the effect of the confining pressure on the Young's modulus.

$$E_t = \left[1 - \frac{R_f(1-\sin\phi)(\sigma_1-\sigma_3)}{2c\cos\phi + 2\sigma_3\sin\phi}\right]^2 K\,P_a\left(\frac{\sigma_3}{P_a}\right)^n \tag{3}$$

The different parameters in the above equation were determined as described by Duncan and Chang (1970). It was found that the parameter "n" was nearly equal to 0.70 for all the samples and the modulus parameter "K" was found to depend on the number of geocells and the modulus of the geotextile used to fabricate the geocell. The K_r of the geocell-soil composite was expressed in the proposed model in terms of the K_u of the unreinforced soil and the secant modulus of geocell walls (M in kN/m) as follows,

$$K_r - K_u = \lambda\, 50\, M^{0.16} \tag{4}$$

in which λ is an interaction parameter between the geocell pockets. Its value for the case with three and four cells was found to be equal to 4. The above constants in the equation were determined by regression analysis of the test data. The validity of the above equation was verified by successfully back-predicting the experimental data by implementing the above constitutive model in a finite element program. In these analyses, the geocell confined soil was represented using the equivalent shear strength properties derived from Equations 1 and 2 and the stiffness derived from equations 3 and 4.

5 LARGE-SCALE LABORATORY EMBANKMENT MODEL TESTS

These tests were performed within a steel test tank 1800 mm long, 800 mm wide and 1200 mm height. The tank was fitted with 20 mm thick perspex sheet on one longitudinal side and thick steel plates on all the other sides. All the sides of the tank were lined with two layers of plastic sheets whose inner layers were coated with lubricating oil to create perfect plane strain conditions in the tank.

The clay soil was initially mixed with excessive amounts of water and poked with steel rods to obtain uniform mixing and then consolidated under a surcharge pressure of 10 kPa for one week. After this consolidation, the layer was trimmed to a 600 mm depth for all tests. This procedure was found to produce a soft clay bed with uniform properties. Several undisturbed samples were collected from the bed after consolidation to determine the CBR value and the vane shear strength of the clay soil. The unit weight, moisture content, vane shear strength and CBR value of the clay foundation were maintained almost the same for all the model tests by carefully controlling the amount of water added and using uniform mixing procedures.

On this clay bed a layer of geocells was formed with different pocket sizes and geogrids. After the formation of the geocell layer, pockets of the geocells were filled with soil. The soil within each cell was compacted uniformly by tamping with a steel rod having an enlarged base. The unit weight of infill soil was maintained at 17 kN/m^3 for all the tests. The compaction quality of this layer was verified by testing undisturbed core samples collected from at least six individual cells. Above the geocell layer, a symmetrical half of the embankment was constructed using clayey sand in 50 mm lifts. The number of blows on each layer was adjusted to achieve a uniform average unit weight of 19 kN/m^3 in all the tests.

The embankment was of height 400 mm with crest width of 700 mm. The side slope was 2H:1V. The surcharge pressure on embankment crest was applied through a pre-calibrated proving ring using a hydraulic jack supported against a reaction frame. The uniform distribution of the surcharge pressure was achieved by applying the loads through rigid I-sections running over the full width of the test tank. A thick steel plate and an expanded polystyrene sheet were placed between the I-sections and the embankment crest to ensure uniform load distribution.

The tests were performed with two different pocket sizes with equivalent diameters (d) of 113 and 226 mm. The heights of geocell layers (h) were varied from 100 mm to 250 mm in increments of 50 mm. The aspect ratio (h/d) of geocell pockets in the tests varied from 0.44 to 2.21. In all the tests, the geocell layer was formed over the full length of the test tank except in one test in which it was truncated at the toe of the embankment.

The load was applied in small increments. Each load increment was kept constant until the displacements under that load increment reached steady state. Typically the steady state was obtained within 10 minutes after applying the load. The vertical and lateral deformations in the model embankment were measured using dial gauges placed at different locations of the test tank. The details of the embankment tests are reported by Krishnaswamy et al. (2000).

6 FINITE ELEMENT ANALYSIS OF MODEL EMBANKMENT TESTS

The finite element analysis of the above test embankments were performed using plane-strain idealisation. The geocell layer was represented using hyperbolic model using the properties estimated from the previously described model. The incremental finite element equation was formed with out-of-balance forces on the RHS and the tangent stiffness matrix on the LHS to simulate the full non-linear behavior of soil as closely as possible.

The soft foundation soil and the embankment soil were modelled using elastic-perfectly plastic type models. The geocell layer was simulated using hyperbolic model with equivalent shear strength and stiffness values. The dilation angles for all the soils were assumed to be zero. In the finite element procedure, the excess stresses beyond the yield surface were corrected back along the flow direction and hence it is possible to simulate the dilation behavior of the soil also, Rajagopal and Bathurst (1995). All the interfaces between different materials were simulated using 4-node joint elements. Typical finite element mesh consisted of about 1500 nodal points, 2900 3-node triangular elements and 50 4-node joint elements at the interfaces.

The average Young's modulus (E) of the clayey sand in embankment was obtained as 15,000 kPa from the slopes of the unloading part of the triaxial compression tests performed at different confining pressures. The cohesion and friction angle of this soil were 12 kPa and 34° respectively. The Poisson's ratio (μ) of this soil was assumed as 0.30 which is a reasonable value for this type of soil. Because of the soft consistency of the foundation soil layer, no triaxial tests were possible on that soil. Hence, its Young's modulus (E) value could only be back-analysed from the results of tests on unreinforced embankments. By trial and error the E and μ values were obtained as 200 kPa and 0.45 by matching the experimental data. The predicted surface heave was quite sensitive to the μ value of the foundation soil. A value of 0.49 has given rise to excessive heave predictions and hence its value was reduced to 0.45 by trial and error. The cohesive strength of soft clay bed from vane shear tests was obtained as 10 kPa. Both the embankment soil and the foundation soil were modeled as elastic-perfectly plastic with Mohr-Coulomb flow rule. The dilation angle was assumed as zero.

The geocell layer was modeled using the hyperbolic model as described earlier. The modulus parameter of the unreinforced soil (K_u) was found to be 382. The same for geocell encased soil were estimated from the earlier equations and are reported in Table 1.

Table 1 Young's modulus parameters for different geocell layers

Type of geocell layer	M	K_r	equivalent diameter of pocket (m)	Apparent cohesion (kPa)
UX	267	870	0.226	51
BX	183	842	0.226	42
BX	183	842	0.113	74
NP-1	75	781	0.226	24
NP-2	95	796	0.226	29

The shear strength properties of the interfaces were determined from large-scale direct shear box tests between the different materials. The shear stiffness was reduced to a small value after the shear stress exceeds the shear strength of the interface. The normal stiffness was kept constant at 10^6 kN/m/m/m so as to maintain the continuity at the interface.

The uniform settlement at the crest of the embankment was simulated in the analysis by applying equal vertical displacement to all the nodes on the crest of the embankment. The pressure developed corresponding to the applied displacement was determined as the ratio between the sum of the reaction forces and the embankment crest width.

7 RESULTS

The typical comparison between the measured and predicted lateral deformations in unreinforced and reinforced embankment (height = 100 mm with BX geocells made of BX grid) slopes is shown in Figures 2 and 3. The H1, H2 and H3 are the measurements at three levels of the embankment.

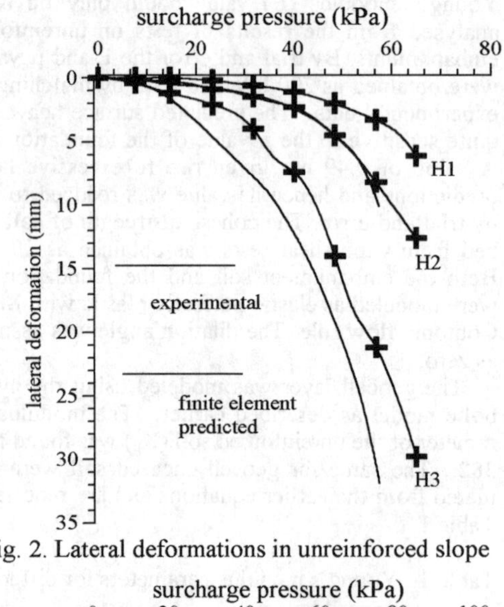

Fig. 2. Lateral deformations in unreinforced slope

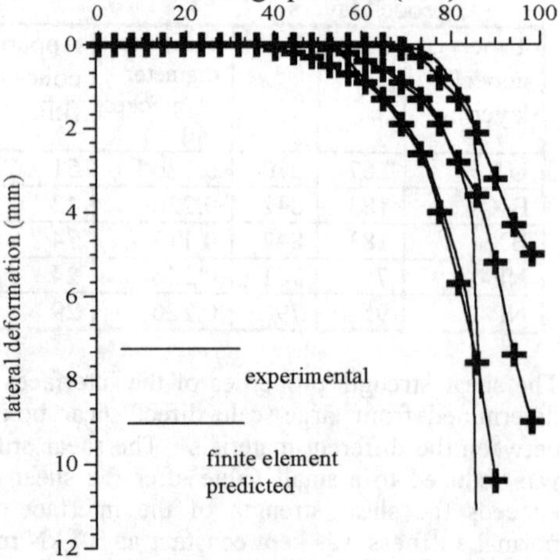

Fig. 3. Lateral deformations in reinforced slope

The different thicknesses of the geocell layers was modeled by assigning the relevant properties to elements within that region of mesh. The different aspect ratios of geocell pockets were modeled by using different apparent cohesion values for the layer as illustrated in Table 2. The excellent finite element predictions for different pocket diameters for BX geocells are shown in Figure 4.

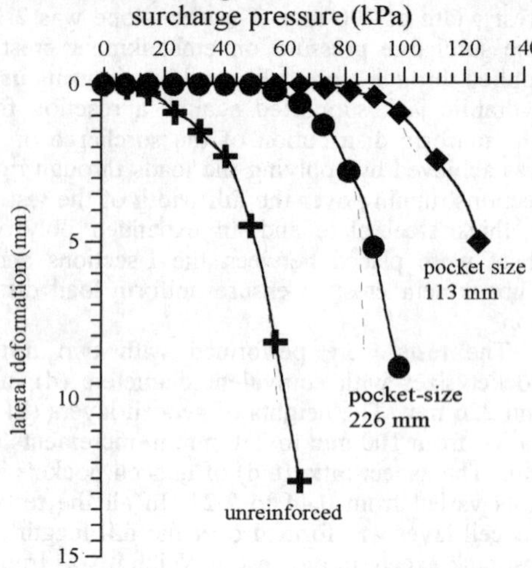

Fig. 4. Effect of pocket size on embankment behavior

It is interesting to note from the above figure that the equivalent model developed is able to account for the effect of the diameter of the geocell pocket on the overall response i.e. the effect of the aspect ratio can also be accounted for using this equivalent model.

8 CONCLUSIONS

The equivalent model developed here is able to replicate the 3-dimensional behavior of geocells using 2-d FE models. Further verification with other test data is being performed.

9 REFERENCES

Duncan, J.M. and Chang, C.Y., 'Non-linear analysis of stresses and strains in soils', J. of Soil Mech. and Foundns. Div., ASCE, 1629-1653, 1970.

Krishnaswamy, N.R., Rajagopal, K. and Madhavi Latha, G., 'Model Studies on Geocell Supported Embankments Constructed Over a Soft Clay Foundation', Geotech. Testing J., ASTM, 23, No. 2, 45-54, 2000.

Rajagopal, K. and Bathurst, R.J., 'Behaviour of geosynthetic reinforced soil retaining walls using the finite element method', Computers and Geotech., 17, 279-299, 1995.

Rajagopal, K., Krishnaswamy, N.R. and Madhavi Latha, G., 'Behaviour of sand confined in single and multiple geocells', Geotextiles and Geomembranes, 17, 171-184, 1999.

Computer Methods and Advances in Geomechanics, Desai et al. (eds) © 2001 Balkema, Rotterdam, ISBN 90 5809 183 X

Pullout resistance of cut/splayed grid reinforcement in MSE walls

Naresh C. Samtani & Michael L. Pegnam
URS, Tucson, Ariz., USA

ABSTRACT: Mechanically Stabilized Earth (MSE) wall system technology has developed rapidly in the last two decades. A common MSE wall system, used by several wall manufacturers, utilizes metallic grid soil reinforcement that mobilizes passive pressure in addition to friction for pullout resistance. This paper provides an opinion regarding the pullout resistance of these reinforcement systems when they are cut and splayed around obstructions that may be placed within the MSE wall backfill. A working hypothesis is developed herein. Several issues are raised in this paper and it is hoped that researchers and practitioners will collaborate on addressing these issues.

1 INTRODUCTION

Advances in the Mechanically Stabilized Earth (MSE) wall system technology have led to a wide range of applications. One of the major applications of this technology has been in the transportation industry. Recently, MSE walls have been used to construct abutments for bridges. In this particular application the bridge may be supported in one of two ways as shown in Figure 1. In the first configuration the bridge girders are supported on a spread footing placed on top of the MSE wall. In the second configuration the bridge girders are supported on a deep foundation system placed within the MSE wall backfill. Design guidance for these configurations is provided by the Federal Highway Administration (FHWA 1997) and the American Association of State Highway and Transportation Officials (AASHTO 1998 Interims).

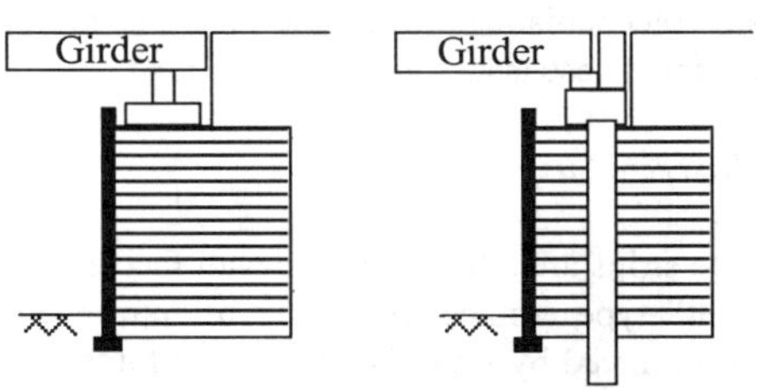

Figure 1. Configurations for supporting bridge abutments on MSE walls.

As shown in Figure 1, the deep foundations are placed through the MSE wall backfill and thereby form "obstructions" to the placement of reinforcements that should be oriented normal to the face of the wall for maximum effectiveness. Similar obstructions also commonly occur due to the placement of other structures such as a catch basin, grate inlet, signal or sign foundation, guardrail post or culvert within the MSE wall backfill. The design of the wall near an obstruction can be modified as follows (FHWA 1997, AASHTO 1998 Interims):

- Option 1: Assume that the reinforcement layers must be partially or fully severed in the location of the obstruction, and design the surrounding reinforcement layers to carry the additional load which would have been carried by the severed reinforcements, or
- Option 2: Place a structural frame around the obstruction (Figure 2) that is capable of carrying the load from the reinforcements in front of the obstruction to reinforcements connected to the frame behind the obstruction, or
- Option 3: If the soil reinforcements consist of discrete strips or bar mats rather than continuous sheets, depending on the size and location of the obstruction, it may be possible to splay the reinforcements around the obstruction.

Guidance on the first two options is provided in FHWA (1997) and AASHTO (1998 Interims). It is the third option with emphasis on metallic grids (or bar mats) that is the focus of this paper. However, before discussing the issue of pullout resistance of cut/splayed grid reinforcement systems a brief review of stress transfer mechanisms is necessary.

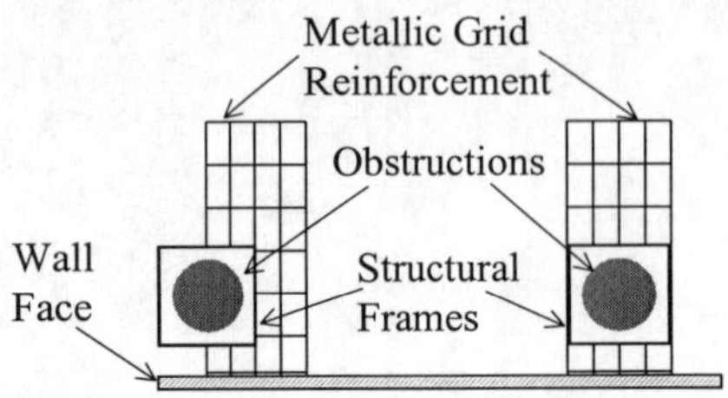

Figure 2. Structural frame concepts.

2 STRESS TRANSFER MECHANISMS

The primary function of soil reinforcement is to restrain soil deformations. By restraining the soil dilation, the reinforcements increase the strength of the soil mass. The maximum increase in shear strength of a reinforced soil mass is obtained when the reinforcement is oriented in the same direction as the principal tensile strain increment that would have occurred in an unreinforced soil mass at failure (Bassett and Last 1978). For retaining walls, the maximum extension or tensile strain occurs normal to the face of the wall. An inextensible reinforcement represents a zero extension line. Thus, maximum benefit can be achieved by placing reinforcement in the direction of maximum tensile strain, i.e. normal to the wall face. When this is done, the reinforcements are primarily placed in tension.

Stresses can be transferred between the soil and the reinforcing element by frictional and/or passive resistance. For sheets or discrete strips, the major stress transfer mechanism is frictional while for grids stress transfer occurs through both passive and frictional mechanisms (see Figure 3). These stress transfer mechanisms are such that the reinforcement is placed primarily in tension.

In a grid reinforced soil system that relies in whole or in part on passive soil resistance, the bearing surface or transverse bars should be normal to the direction of displacement or the pullout force (i.e. earth pressure) to be resisted so that a uniform passive resistance is generated. This is also the basis for such wall systems as the deadman anchor reinforced retaining system.

Figure 4 shows the locus of maximum tensile stresses or potential failure surface for inextensible reinforcement. Only the reinforcement length in the

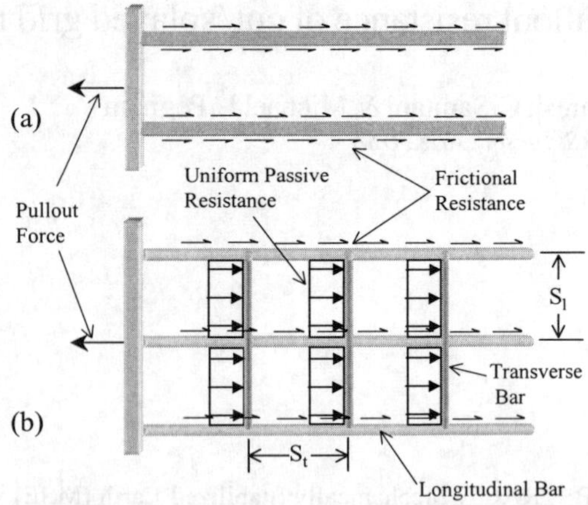

Figure 3: Stress transfer mechanisms in MSE walls (a) Discrete strips, (b) Grids (After FHWA 1997).

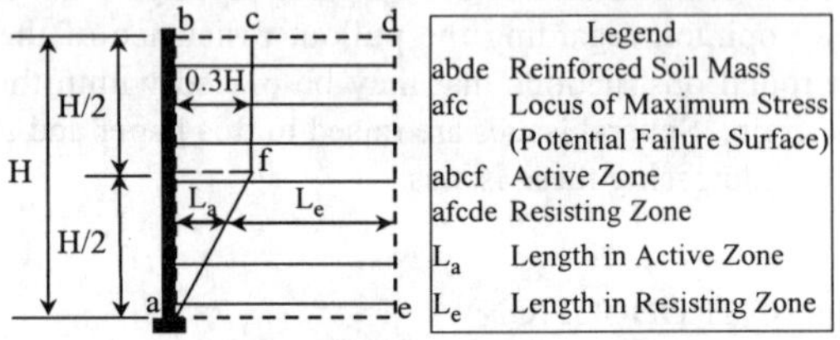

Figure 4. Location of potential failure surface for inextensible reinforcements (FHWA 1997).

resisting zone contributes to the pullout resistance. It is important that the connections of the reinforcements to the wall panels should be tension connections otherwise the stress distribution within the reinforced soil will be altered and the patterns shown in Figure 4 may not be valid.

3 EFFECT OF SPLAY ON STRESS TRANSFER

Splaying reduces the reinforcement length in the resisting zone and thus the pullout resistance. Depending on the reinforcement type and its connection detail with the wall face, splaying can introduce moments in wall connections and/or alter the stress transfer mechanism between the soil and the reinforcement. These issues are discussed next.

3.1 *Discrete Strips*

Discrete strips are attached to the wall face by a "nut and bolt" type connection. Splay of reinforcement can be achieved by first placing the reinforcement at a given splay angle and then tightening the connection; this does not induce additional stresses at the connection. Furthermore, since the principal mechanism of pullout resistance is friction-based the fundamental mechanism of stress transfer is still intact. However, the splay angle should be small enough so that moments are not generated in the reinforcement. The design can then proceed by assuming that the tensile capacity of the splayed reinforcement is reduced by the cosine of the splay angle.

3.2 *Grids (Bar Mats)*

Metallic grids or bar mats usually consist of 4 to 6 longitudinal steel bars spaced at 150 mm welded to transverse steel bars spaced at 150 mm to 600 mm. This configuration results in a rigid grid or bar mat system that is connected to the wall by clevis loops or similar connection. Due to the rigidity of the grid and the nature of the wall connection it may not be possible to splay the grid when an obstruction within the wall backfill is encountered. In this case a structural frame should be used. For wall manufacturers this means additional design and fabrication of connections to the structural frame. Wall manufacturers therefore cut the reinforcements into single longitudinal bars with transverse projections and then force a splay as shown in Figure 5. Forcing a splay introduces moments at the connection between the reinforcement and wall face; thus, the connectors and wall panels should be designed to withstand moments. Furthermore, the entire stress transfer mechanism is altered as shown in Figure 6. For clarity, the obstruction and transverse bars within the active zone are not shown and only one transverse bar in the resisting zone is shown in Figure 6.

In a general case the wall panel may deform unevenly, $\delta_1 \neq \delta_2$, as shown in Figure 6. In the process of wall deformation, point B where the reinforcement is connected to the wall panel will move to point B_1. If the reinforcement is assumed to be inextensible and flexurally rigid then it will be pulled out. In the process of pullout, the reinforcement will change the splay angle from θ_1 to θ_2 with $\theta_2 < \theta_1$. In other words the longitudinal bar will rotate by $\Delta\theta = \theta_1 - \theta_2$. If the following assumptions are satisfied, the transverse bar will experience the same rotation, $\Delta\theta$:

1. The weld at point A and connection at point B have sufficient shear and torsional resistance to prevent breakage, and
2. The transverse bar projections are equal in length ($CA=AD=L_t$) and have sufficient bending resistance to resist deformation.

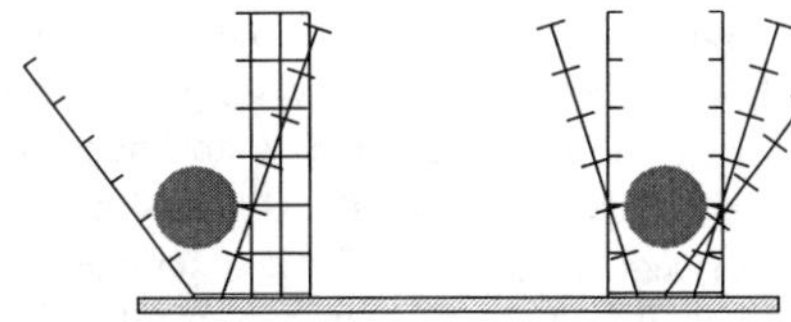

Figure 5: Typical industry schemes for placing metallic grids around obstructions in wall backfill.

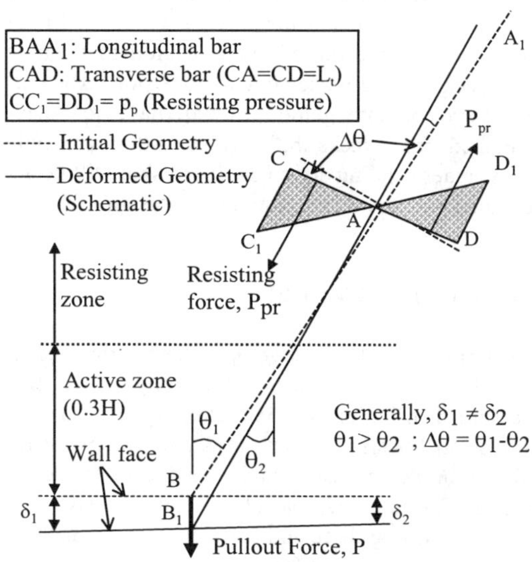

Figure 6: Idealized stress transfer mechanism for splayed metallic grid reinforcement (plan view).

Under the above assumptions, the behavior of the reinforcement will be as follows:

- the length AB will act like a pile that is subjected to a lateral load at its head (top) albeit under a constant overburden pressure corresponding to the depth of reinforcement below the top of wall,
- the transverse bar CAD will pivot around point A and rotate by $\Delta\theta$. Since the bar is rotating into the soil mass, passive resistance will be generated and a resisting couple will be generated subjecting the weld at point A to torsional stress.

It is obvious that the pullout force is no longer transmitted by tension to the reinforcement or shear at the weld connection; instead, the reinforcement is subjected to bending moments and torques (couples). For a given splay angle and overburden pressure, the magnitude of bending moments and torques will vary along the reinforcement. This stress transfer mechanism is different than that described in

Section 2 and thus the stress distribution and failure patterns shown in Figure 4 may no longer be valid.

The current industry practice is to manufacture reinforcement based on the stress transfer mechanisms discussed in Section 2, i.e. tension based stress transfer. It is doubtful that any or all of the assumptions regarding the flexural and torsional rigidity are satisfied under current practice. Violation of any assumption will further change the stress transfer mechanisms and the pullout resistance will be altered. These mechanisms may vary over the life of the structure since the effective section of metallic reinforcement will reduce due to corrosion. The current corrosion rate guidelines will have to be re-evaluated since cracks in the galvanization are more likely under bending action and corrosion will be accelerated with resultant reduction in design life.

4 A SIMPLIFIED MODEL

A simplified model is developed herein for various quantities shown in Figure 6. Assuming the resisting pressure distribution as triangular, F_q to be a non-dimensional bearing factor, σ_v as the effective vertical overburden pressure and t as the effective diameter of a transverse bar at the end of design life, then the resisting force and couple for one transverse bar can be evaluated as follows:

Resisting (bearing) pressure, $p_p = F_q \sigma_v$

Resisting force, $P_{pr} = 0.5(tp_p)(L_t)$

Resisting couple, $M_r=(4/3)(L_t)(P_{pr})=(2/3)(L_t)^2(tp_p)$

Assuming small strain deformations, the total pullout resistance and resisting couple for n transverse bars in the resisting zone may be approximated as nP_{pr} and nM_r, respectively. The safety factor against pullout is assessed by comparing nP_{pr} with pullout force (FHWA, 1997, AASHTO 1998). The quantity nM_r may be used to assess the required torsional weld capacity. Pullout and moment may govern in upper and lower portion of wall, respectively.

The factor F_q may be defined as follows:

F_q =40R at top of the MSE wall,

F_q =20R at a depth of 6 m and below

The term R is a reduction factor when $L_t<0.5S_l$ where S_l is spacing of longitudinal bars. For $L_t=0.5S_l$ use R=1; for $L_t<0.5S_l$, R can be less than 0.5 when L_t is reduced to 10 mm (Schlosser et al. 1985).

For passive resistance to develop the spacing of transverse bars, S_t, should be greater than $(tF_q)/(2\tan\phi)$, where ϕ is the angle of internal friction of the reinforced soil mass (FHWA 1997).

4.1 *Recommendations for Refined Model*

The simplified model should be used with caution, as it does not include the interaction of longitudinal bending with the development of the resisting force/couple on transverse bars. Also, depending on the splay angle and pullout displacement, the total resistance on all transverse bars in the resisting zone might not be approximated by a simple scalar addition such as nP_{pr} or nM_r. For these reasons it is recommended that a comprehensive instrumented laboratory and field investigation program along with analytical studies be undertaken to refine the model.

Assuming standard MSE wall backfill (FHWA 1997) and rigid bars and weld connections, the effects of following parameters should be included in study of pullout resistance of splayed reinforcements: (1) splay angle, (2) length of transverse bars to evaluate R, (3) overburden pressure, (4) interaction between splayed reinforcements within close proximity or on top of others (see Figure 5), (5) the failure surface shape and therefore the resisting length for pullout resistance, and (6) corrosion rates and their effect on design life.

For practical use by the industry, the results should be expressed in terms of conventional earth pressure parameters and in a normalized fashion as presented by FHWA (1997).

5 CONCLUSIONS

Following are the conclusions of this paper:

1. Cutting and splaying of metallic grid reinforcement may alter the stress transfer mechanism and stress distribution in the reinforced soil mass.
2. Research is needed to quantify the pullout resistance of cut/splayed metallic grids in MSE walls.

REFERENCES

AASHTO 1998 Interims. Section 5.8

Bassett, R. H & Last, N. C. 1978. Reinforcing Earth Below Footings and Embankments, *Proc. ASCE Symp. on Earth Reinforcement*, Pittsburg, PA. 202-231.

FHWA 1997. (Authors: Elias V. and Christopher B.) Mechanically Stabilized Earth Walls and Reinforced Soil Slopes - Design and Construction Guidelines, Publication No. FHWA-SA-96-071.

Schlosser, F., Magnan, J.P., and Holtz, R.D., 1985. Geotechnical Engineered Construction, *Proc. 11th Intl. Conf. on Soil Mech. and Foundation. Engineering,* San Francisco, Vol. 5, 2499-2539.

Computer Methods and Advances in Geomechanics, Desai et al. (eds) © 2001 Balkema, Rotterdam, ISBN 90 5809 183 X

Failure modes of soil reinforcements

M. Touahmia & I. Jefferson
Department of Civil and Structural Engineering, Nottingham Trent University, UK

Y. Djerbib
Department of Civil Engineering, Leeds Metropolitan University, UK

ABSTRACT: The mechanisms of failure of soil reinforcements under cyclic loading were examined. Two types of reinforcements namely smooth steel strip and Tensar geogrid strip were tested under a wide range of repeated loading. All the results show that the rate of accumulation of reinforcement movement increases with increase in the number of load repetitions and load amplitude. Failure by pullout resulted with the smooth strip only despite the large number of load applications. The loading-unloading behavior of the reinforcement is non-linear. The effect of repeated loading was to cause significant changes in the distribution of loads along the reinforcement leading to permanent degradation in the frictional resistance at the soil-reinforcement interface.

1 INTRODUCTION

Despite the extensive research on the behavior of reinforced earthworks, many of the fundamental mechanisms, which operate when the reinforcement is loaded, have not been identified. Most of the research has been directed to the static pull-out behavior of a range of element types but little works appears to have been done where the form of loading is non-static. However, there are a number of cases where reinforced earth systems may be subjected to repeated loading. These examples include bridge abutments subjected to traffic loads and retaining structures subjected to significant tidal variations. The development and evaluation of reliable design methods for these systems requires a thorough understanding of the parameters affecting the soil-reinforcement interaction and failure mechanisms of this system under different loading conditions (Raju & Fannin 1998). The available information suggests that the application of an external load causes resistance to be mobilized along the length of the reinforcement according to the laws of bond and bearing, with the mechanics being similar to those of pile loading. Work on the behavior of piles subjected to repeated cycles of loading showed that the frictional resistance between the shaft and the surrounding soil may reduce and the pile may either push in or pull out of the ground (Lin & Liao 1999).

This paper examines the mechanisms of failure of soil reinforcements under slow repeated loading. Large amounts of test data were generated in order to evaluate the effect of cyclic loading on the long-term behavior and life span of the reinforcing elements. Large-scale model was used to avoid errors associated with model testing.

2 EXPERIMENTAL CONFIGURATION

During this study two types of 4 m length reinforcements were tested, namely smooth steel strip and SR2 geogrid plastic strip. The steel reinforcing strip was made from hard rolled steel of 680 N/mm² yield stress and 200 kN/mm² modulus of elasticity. The plastic reinforcement was formed by cutting a row of two ribs in width and 35 bars in length of SR2 geogrid. The width of the strips finally chosen was 32 mm to ensure that there were no edge effects from the test rig walls. The steel reinforcement was provided with special load cells, temperature compensated, at 400 mm center, with the wires from the cells being carried in grooves machined along the length of the strip to an end terminal. The geogrid strip was provided with special axial movement gauges at five locations along the reinforcement. Pullout tests were conducted on a 4 m long, 300 mm square, rigid steel container with the capacity to apply a surcharge pressure up to 200 kN/m² on a sand surface. A medium uniformly graded sand was used, with a bulk density of 1.53 Mg/m^3 and a specific gravity of 2.67. Test strips placed in the middle of the sand mass such that either static or cyclical loading can be used. Complete description of the test apparatus and testing procedures have been presented by Touahmia et al. (1997).

3 CYCLIC LOADING BEHAVIOR

The behavior of the reinforcements was examined under a wide range of cyclic loading. In some cases the tests were taken up to 10^6 load cycles while the confining stress was kept constant at 100 kN/m². The load was changed every 20 seconds to give a square shaped pattern between an upper and a lower load level. The upper and lower load levels were expressed as a percentage of the static pull-out capacity (Pu) of the steel strip and the index load (PI) of the geogrid strip as defined by McGown et al. (1984). The cumulative movements of the reinforcements versus the number of loading cycles are presented in Figure 1.

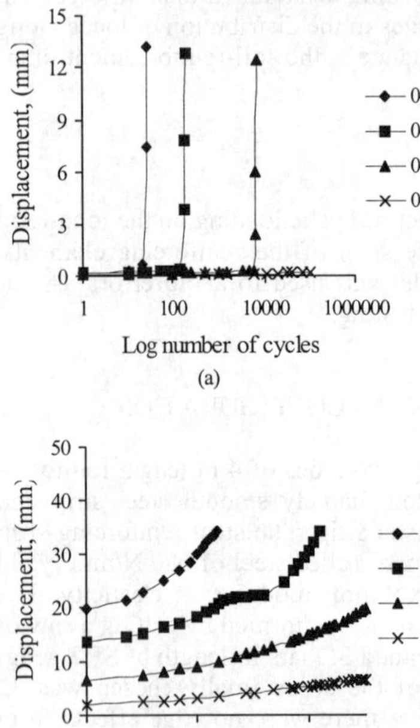

Figure 1. Cyclic pullout failure of the reinforcements: (a) smooth steel strip (b) geogrid strip.

The results demonstrate clearly the significant differences in the manner the two strips behaved. The smooth steel strip was characterized by an initial stable state followed by a short period with an accelerated movement leading to a catastrophic failure. The plastic strip displaced much further under repeated loading, exhibiting a gradual movement and despite the large number of load repetitions this strip did not fail by pulling out. It can be seen that the behavior of the reinforcements was highly dependent upon the magnitude of the repeated loading, with higher amplitude of cyclic load causes a rapid deterioration of the reinforcement performance.

The axial load distribution along the reinforcements and their variation with load reversals were examined and represented in Figure 2 and 3. The steel strip data showed significant changes in the distribution of loads along the reinforcement length with more and more load being transferred towards the distal end of the strip as the number of load repetitions increased. This gradual shift of the axial load toward the back can be attributed to a gradual shift in the mobilization of skin friction from the front half to the rear half of the strip as the load cycles increased. The geogrid strip under axial applied load deformed much further and even at the highest test load the resistance to pullout was mobilized over the front half of the strip only, the remaining half being unstressed. This observation would indicate that unless a very low confining stress is used, it would be impossible to pullout the geogrid strip.

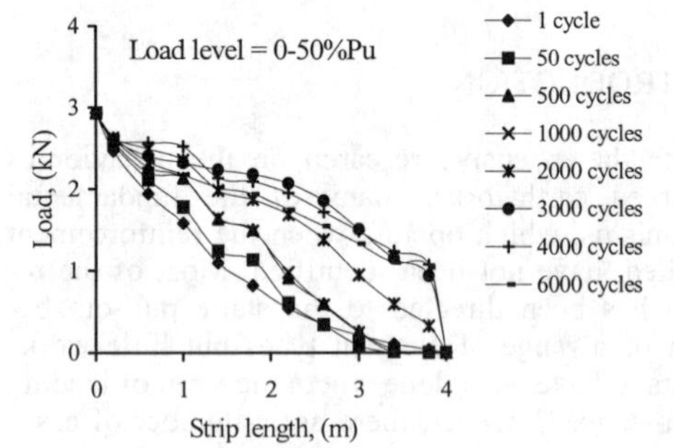

Figure 2. Stress variation along the steel strip.

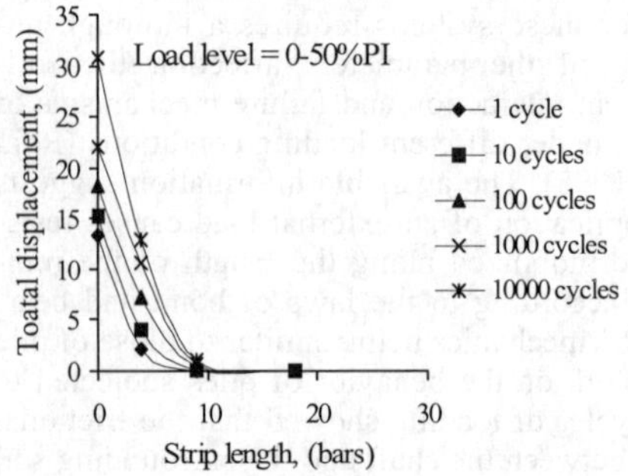

Figure 3. Strain variation along the geogrid strip.

4 THE CYCLIC MODEL

The transfer of stress between soils and various soil reinforcement systems involves two basic mechanisms: friction and/or passive resistance. In systems such as welded wire and anchored earth, passive resistance is the primary mechanism. In others, such as smooth steel strips, smooth rods or sheets, friction dominates. Both mechanisms are active in reinforcement by ribbed strips, and geogrids, but the relative contribution of each has hitherto been indeterminate (Irsyam & Hryciw 1991).

The experimental results showed that as the reinforcement is loaded the resistance available at the soil-reinforcement interface is gradually mobilized along the reinforcement surface. The total pullout resistance of the reinforcement would include then a frictional component as well as passive soil resistance. The transverse members of grid reinforcements can be considered analogous to a series of strip footings in succession which have been rotated through 90° to the horizontal and pulled through the soil.

Consider a reinforcing element which is subjected to incremental loading and unloading. The first increment of load applied to the reinforcement (ΔT_1) develops frictional resistance (ΔF_1) and passive resistance (ΔP_1) along part of the reinforcement length only, the remaining part remains unloaded. At some average adhesion the strength of the interface is mobilised and failure migrates along a distance (Δl_1) of the reinforcement length. Equilibrium of the soil-reinforcement system gives:

$$\Delta T_1 = \Delta F_1 + \Delta P_1 \tag{1}$$

By incrementally increasing the applied load to $T = \sum_{i=1}^{m} \Delta T_i$, where m is the number of increments, the frictional and passive resistance will be mobilised over the full length of the reinforcement. Equilibrium of the soil-reinforcement system would give:

$$T = 2b\sigma_n \int_0^l \mu(x)dx + \sum_{i=1}^{m} \Delta P_i \tag{2}$$

where: σ_n is the applied normal pressure at the soil-reinforcement interface, $\mu(x)$ is the coefficient of friction at the interface, l is the reinforcement length and b is the reinforcement width.

The passive resistance due to the soil bearing in front of the transverse members of the grid can be evaluated using the bearing capacity theory (Palmeira & Milligan 1989; Jewell 1990). The prediction equation for the pullout resistance in this case is based on the Terzaghi–Buisman bearing capacity equation as fellows:

$$\sum_{i=1}^{m} \Delta P_i = [A + B]n \tag{3}$$

$$A = \sigma_n e^{\left(\frac{\pi}{2}+\phi\right)} \tan(45 + \phi/2) \tag{4}$$

$$B = c\left(e^{\left(\frac{\pi}{2}+\phi\right)} \tan(45 + \phi/2) - 1 \right) \tag{5}$$

where: ϕ is the frictional angle of soil, c is the cohesion intercept of the backfill soil and n is the total number of bearing members. The total pullout resistance can be expressed in general form as:

$$T = 2b\sigma_n \int_0^l \mu(x)dx + [A + B]n \tag{6}$$

The increase of the applied load T by ΔT causes an additional displacement $d\Delta$ of the reinforcement. This incremental displacement can be expressed by:

$$d\Delta = \varepsilon(\Delta F + \Delta P).x \tag{7}$$

$$dF = 2b\sigma_n \mu(x)dx \tag{8}$$

where ΔF is the additional frictional load, ΔP is the load taken by the bearing members of the reinforcement and ε is a constant which depends on the reinforcement stiffness and properties. Putting $\Delta F = \alpha \Delta P$ and rearranging equations (7) and (8) gives:

$$d\Delta = \frac{F}{2b\sigma_n \mu(x)} \Delta F \left(1 + \frac{1}{\alpha}\right)\varepsilon \tag{9}$$

which on integration between $F = 0$ and F gives:

$$\Delta = \frac{F^2}{4b\sigma_n \mu(x)} \left(1 + \frac{1}{\alpha}\right)\varepsilon \tag{10}$$

The unloading of the reinforcement may be studied by incrementally adding a reverse or negative loads. Thus, the mechanics of unloading are similar to those described for the loading case, except the friction coefficient is no longer $\mu(x)$ but $\xi(x)$ mobilised in the reversed direction. This will cause a reverse displacement (δ_u) of the reinforcement, which can be expressed by:

$$\delta_u = \frac{F_u^{\,2}}{4b\sigma_n\xi(x)}\left(1+\frac{1}{\alpha}\right)\varepsilon \qquad (11)$$

where F_u is the load transfer change in friction due to unloading of the reinforcement. Figure 4 shows the predicted results of the load-displacement relationship for one cycle of loading. It is clear from this diagram that the loading-unloading behavior of the reinforcement is non-linear.

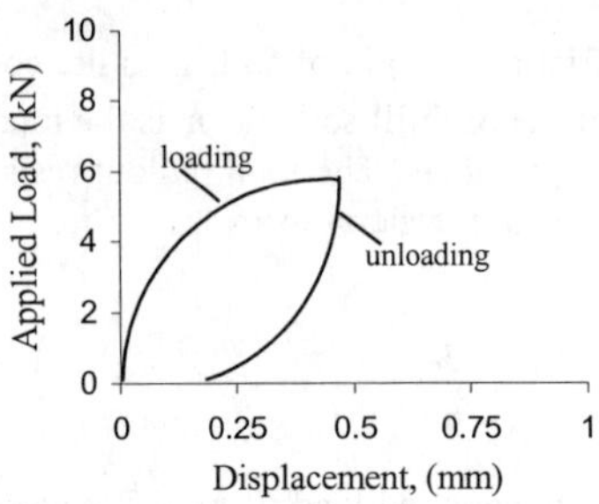

Fig. 4. Loading-unloading-displacement diagram.

5 DISCUSSION

The results showed that under cyclic loading the response of the steel reinforcement is different from that of the geogrid reinforcement. Failure by pullout resulted with the smooth strip only despite the large number of load applications. In no case did any of the applied loads reach the distal end of the geogrid reinforcement. The smooth strip data showed significant changes in the distribution of loads along the reinforcement length with more and more load being transferred towards the distal end of the strip as the number of load repetitions increased.

Based on these observations an analytical model was developed for predicting the pullout resistance of the reinforcement under loading and unloading conditions. The model accounts for both frictional resistance and passive soil resistance. The results showed that the loading-unloading behavior of the reinforcement is non-linear. On reloading, the displacement of the reinforcement may not equal the original displacement, which resulted on first loading. The act of subjecting a reinforcing element to a package of cyclic loading modifies the stress state along the strips and changes the locked-in stress regime. As the number of load cycles increases the longevity of a reinforcing element against pullout or against a particular displacement being reached, decreases and consequently the load levels may have to be reduced to guarantee safety.

6 CONCLUSIONS

Based on the experimental and analytical data obtained during this investigation about the response of earth reinforcing elements to cyclic loading, the following conclusions can be drawn:

1. The smooth steel reinforcement behavior was characterized by an initial stable state, which represented the major part of the cyclic loading period, followed by a sudden pullout failure of the reinforcement.
2. The geogrid reinforcement exhibited a gradual forward movement with increase in the number of load cycles and showed no pullout failure for all levels of applied load.
3. The loading-unloading behavior of the reinforcing elements was non-linear.

7 REFERENCES

Irsyam, M. & Hryciw, R. D. 1991. Friction and passive resistance in soil reinforced by plane ribbed inclusions. *Géotechnique* 41(4): 485-498.

Jewell, R. A. 1990. Reinforcement bond capacity. *Géotechnique* 40(3): 513-518.

Lin, S. S. & Liao, J. C. 1999. Permanent strains of piles in sand due to cyclic lateral loads. *Journal of Geatechnical and Geoenvironmental Engineering* 125(9): 798-802.

McGown, A. Andrawes, K. Z., Yeo, K. C. and Dubois, D. 1984. The load-strain-time behaviour of Tensar geogrids. *Proceeding of the Symposium on Polymer Grid Reinforcement in Civil Engineering*, U.K.: 11-17.

Palmeira, E. M. & Milligan, G. W. E. 1989. Scale and other factors affecting the results of the pullout tests of grids buried in sand. *Géotechnique* 39(3): 551-584..

Raju, D. M. & Fannin, R. J. 1998. Load-strain-displacement response of geosynthetics in monotonic and cyclic pullout. *Canadian Geotechnical Journal* 35: 183-193.

Touahmia, T., Rouili, A. & Hanna, T. H. 1997. Comparative behaviour of earth reinforcing elements under repeated loading. *Proc. XIVth Int. Conf. on Soil Mechanics and Foundation Engineering*, 2, Hamburg, 6-12 September 1997: 1265-1268. Rotterdam: Balkema.

Computer Methods and Advances in Geomechanics, Desai et al. (eds) © 2001 Balkema, Rotterdam, ISBN 90 5809 183 X

Consolidation theory for soft clays reinforced by cement or granular columns

K.H.Xie, P.Gao & X.Y.Xie
Department of Civil Engineering, Zhejiang University, Hangzhou, People's Republic of China

ABSTRACT: In this paper, the consolidation problem of soft clays reinforced by cement or granular columns is studied and an analytical solution is developed. The consolidation theory advanced herein is distinguished from the conventional one of vertical drains for that the load transferred to columns is different from to clay, and the radial flow in columns is considered. A series of diagrams are prepared and numerical examples are presented to show the consolidation behavior of such composite foundation as well as the differences of consolidation rates among the soft clay improved by vertical drains and those reinforced by cement columns or by granular ones. It has been shown that while the rate of consolidation is enhanced owing to the higher stiffness of columns, it may be slowed down due to the poor permeability of cement column or the clog of gravel column with clay particles.

1 INTRODUCTION

Cement columns or granular columns (including sand columns and gravel columns) have been widely used to reinforce soft soils over the world (Navak 1989, Kamon & Bergado 1991, Gong 1992). The soft soil thus reinforced is known as composite foundation. The use of these techniques has three purposes (Mitchell & Katti 1981): (a) to reduce the settlement of highly compressible soils; (b) to accelerate the rate of primary consolidation; and (c) to enhance the bearing capacity. This paper is concerned with the second of them.

Usually, the conventional consolidation theory of vertical drains is adopted to study consolidation behavior of composite foundation, but the true properties of such foundation are seldom taken into account.

As we all known, it is the columns and the soft clay that support jointly the load applied on. The load undertaken by soft clay will result in consolidation that will in turn definitely affect the deformation and the load transfer between columns and soft clay. From such point of view, it is important to distinguish the consolidation theory for composite foundation from the conventional one for vertical drains.

In this paper, the consolidation of composite foundation is studied with its peculiarities considered. An analytical solution for consolidation of such composite foundation is developed, and the consolidation behavior is then analyzed and discussed.

2 MATHEMATICAL MODELLING

2.1 *Analysis sketch and main features of composite foundation*

The analysis sketch for composite foundation is shown in Figure 1, where r_e, r_w, r_s= radius of unit cell, column, and clog or smear zone; d_e, d_w, d_s = diameters of them, respectively; E_w, E_m, E_s = modulus of compressibility of columns, clog or smear zone and soft clay respectively; k_{hw}, k_{vw} = horizontal and vertical permeability of columns; k_h, k_v = horizontal and vertical permeability of soft clay; k_s = permeability in clog or smear zone; H = thickness of soft clay layer.

The main features of composite foundation are as follows:

1. Column is more permeable (except for cement column) and has a higher stiffness than clay.

2. The diameter of column is usually larger than that of vertical drain. So, the radial flow in columns should be taken into consideration, whereas it is neglected in vertical drains (Barron 1948, Zeng & Xie 1989).

2.2 *Basic assumptions*

1. All columns are fully penetrated the soft clay layer.

2. There is no horizontal displacement in the foundation, and the vertical displacements of columns and soft clay are identical at the same depth.

3. Load is instantaneously applied and carried entirely by the pore-water at the beginning of consolidation.

4. The top surface of the clay layer is pervious and the bottom is imperious.

Other assumptions are the same as those of Terzaghi's one-dimension consolidation theory.

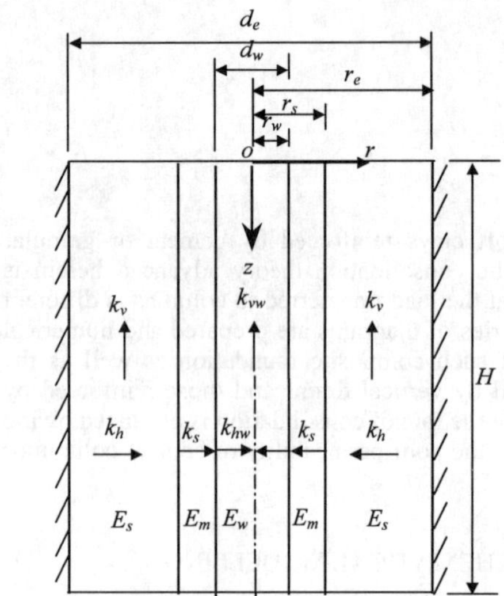

Figure 1. Analysis sketch for composite foundation

2.3 *Governing equations and solution conditions*

The equilibrium at any depth of the unit cell requires:

$$(A-A_m-A_w)\overline{\sigma}_s+A_m\overline{\sigma}_m+A_w\overline{\sigma}_w=AP_0$$

or

$$\frac{n^2-s^2}{n^2}\cdot\overline{\sigma}_s+\frac{s^2-1}{n^2}\cdot\overline{\sigma}_m+\frac{1}{n^2}\cdot\overline{\sigma}_w=P_0 \tag{1}$$

According to the secondary basic assumption, the following equation can be obtained:

$$\frac{\overline{\sigma}_s-\overline{u}_s}{E_s}=\frac{\overline{\sigma}_m-\overline{u}_m}{E_m}=\frac{\overline{\sigma}_w-\overline{u}_w}{E_w}=\varepsilon_z=\varepsilon_v \tag{2}$$

in which $\overline{\sigma}_s$, $\overline{u}_s$ = average total stress and pore-water pressure of soft clay at any depth; $\overline{\sigma}_m$, $\overline{u}_m$ = those in clog or smear zone; $\overline{\sigma}_w$, $\overline{u}_w$ = those for column respectively; ε_z and ε_v = vertical and volume strain of clays at any point; P_0 = the value of uniformly distributed load applied; $A=\pi r_e^2$; $A_m=\pi(r_s^2-r_w^2)$; $A_w=\pi r_w^2$; $s=\frac{r_s}{r_w}$; $n=\frac{d_e}{d_w}=\frac{r_e}{r_w}$.

The combination of Equation (1) and Equation (2) gives

$$\varepsilon_v=\frac{P_0-\overline{u}}{E_c}$$

or

$$\frac{\partial\varepsilon_v}{\partial t}=-\frac{1}{E_c}\cdot\frac{\partial\overline{u}}{\partial t} \tag{3}$$

where E_c is the composite compressibility modulus of foundation and $\overline{u}$ is the average pore-water pressure at any depth, that is:

$$E_c=\frac{n^2-s^2}{n^2}\cdot E_s+\frac{s^2-1}{n^2}E_m+\frac{1}{n^2}\cdot E_w$$

$$\overline{u}=\frac{n^2-s^2}{n^2}\cdot\overline{u}_s+\frac{s^2-1}{n^2}\cdot\overline{u}_m+\frac{1}{n^2}\cdot\overline{u}_w$$

Furthermore, the following equations can be established based on continuity conditions and Barron's equal strain condition (Barron, 1948):

$$\frac{k_{vw}}{\gamma_w}\frac{\partial^2\overline{u}_w}{\partial z^2}+\frac{k_{hw}}{\gamma_w}\left(\frac{1}{r}\frac{\partial u_w}{\partial r}+\frac{\partial^2 u_w}{\partial r^2}\right)=-\frac{\partial\varepsilon_v}{\partial t}$$

$$0\le r\le r_w \tag{4a}$$

$$\frac{k_v}{\gamma_w}\frac{\partial^2\overline{u}_c}{\partial z^2}+\frac{k_s}{\gamma_w}\left(\frac{1}{r}\frac{\partial u_m}{\partial r}+\frac{\partial^2 u_m}{\partial r^2}\right)=-\frac{\partial\varepsilon_v}{\partial t}$$

$$r_w\le r\le r_s \tag{4b}$$

$$\frac{k_v}{\gamma_w}\frac{\partial^2\overline{u}_c}{\partial z^2}+\frac{k_h}{\gamma_w}\left(\frac{1}{r}\frac{\partial u_s}{\partial r}+\frac{\partial^2 u_s}{\partial r^2}\right)=-\frac{\partial\varepsilon_v}{\partial t}$$

$$r_s\le r\le r_e \tag{4c}$$

where u_s, u_m, u_w = pore-water pressure at any point in soft clay, smear zone and column respectively; $\overline{u}_c$ = the average pore-water pressure of clay around columns (including smear zone) at any depth, that is:

$$\overline{u}_c=\frac{1}{\pi(r_e^2-r_w^2)}\left(\int_{r_w}^{r_s}2\pi r\cdot u_m dr+\int_{r_s}^{r_e}2\pi r\cdot u_s dr\right)$$

$$= \frac{s^2-1}{n^2-1}\cdot\bar{u}_m + \frac{n^2-s^2}{n^2-1}\cdot\bar{u}_s$$

Equations (3) and (4) form the governing equations for consolidation of composite foundation.

The solution conditions for governing equations are:

1. $z=0$: $u_s = u_m = u_w = 0$
2. $z=H$: $\frac{\partial u_s}{\partial z} = \frac{\partial u_m}{\partial z} = \frac{\partial u_w}{\partial z} = 0$
3. $r=r_e$: $\frac{\partial u_s}{\partial r} = 0$
4. $r=0$: $\frac{\partial u_w}{\partial r} = 0$
5. $r=r_w$: $u_m = u_w$, $k_s\cdot\frac{\partial u_m}{\partial r} = k_{hw}\cdot\frac{\partial u_w}{\partial r}$
6. $r=r_s$: $u_s = u_m$, $k_h\cdot\frac{\partial u_s}{\partial r} = k_s\cdot\frac{\partial u_m}{\partial r}$
7. $t=0$: $\bar{u} = u_0 = P_0$ (i. e. assumption 3.)

It can be seen from all of the above that the following aspects have been taken into consideration:

1. The difference of stiffness and permeability among columns, smear (or clog) zone and soft clay;
2. The radial flow and vertical flow in columns, smear (or clog) zone and soft clay;
3. The possible difference of permeability of columns in horizontal and vertical directions.

It is therefore obvious that the consolidation theory thus established for composite foundation will be different from the conventional one for vertical drains.

3 ANALYTICAL SOLUTION

The solution to governing equations is given by

$$u_w = u_0\sum_{i=1}^{\infty}\frac{2}{M}\frac{1}{\alpha_v F_p + D_p}[(\frac{k_h}{k_{hw}}\frac{n^2-1}{n^2}\frac{r^2-r_w^2}{2r_w^2} - F)(1-\frac{k_v}{k_{vw}}) + F_p + D_p]\sin\frac{Mz}{H}e^{-\beta t} \quad (5a)$$

$$u_m = u_0\sum_{i=1}^{\infty}\frac{2}{M}\frac{1}{\alpha_v F_p + D_p}[(\frac{k_h}{k_s}(\ln\frac{r}{r_w} - \frac{r^2-r_w^2}{2r_s^2} - F)(1-\frac{k_v}{k_{vw}}) + F_p + D_p]\sin\frac{Mz}{H}e^{-\beta t} \quad (5b)$$

$$u_s = u_0\sum_{i=1}^{\infty}\frac{2}{M}\frac{1}{\alpha_v F_p + D_p}[(\ln\frac{r}{r_s} - \frac{r^2-r_s^2}{2r_s^2} + \frac{k_h}{k_s}(\ln s - \frac{s^2-1}{2n^2}) - F)(1-\frac{k_v}{k_{vw}}) + F_p + D_p]\sin\frac{Mz}{H}e^{-\beta t} \quad (5c)$$

$$\bar{u}_w = \frac{1}{\pi r_w^2}\int_0^{r_w} 2\pi r\cdot u_m dr = u_0\sum_{i=1}^{\infty}\frac{2}{M}\frac{1}{\alpha_v F_p + D_p}\cdot(\frac{k_v}{k_{vw}}\cdot F_p + D_p)\sin\frac{Mz}{H}e^{-\beta t} \quad (5d)$$

$$\bar{u}_c = u_0\sum_{i=1}^{\infty}\frac{2}{M}\frac{1}{\alpha_v F_p + D_p}(F_p + D_p)\sin\frac{Mz}{H}e^{-\beta t} \quad (5e)$$

$$\bar{u} = u_0\sum_{i=1}^{\infty}\frac{2}{M}\cdot\sin\frac{Mz}{H}e^{-\beta t} \quad (5f)$$

where $\beta = \alpha_r\cdot\frac{M^2C_v}{H^2}$; $\alpha_r = \frac{F_p + D_p\alpha_w}{\alpha_v F_p + D_p}\cdot\alpha_e$;

$$\alpha_e = \frac{n^2-s^2}{n^2} + \frac{s^2-1}{n^2}\cdot N_m + \frac{1}{n^2}\cdot N_p;$$

$$\alpha_v = \frac{1}{n^2}\frac{k_v}{k_{vw}} + \frac{n^2-1}{n^2};\quad D_p = \frac{8G}{M^2};$$

$$\alpha_w = \frac{1}{n^2}\frac{k_{vw}}{k_v} + \frac{n^2-1}{n^2};\quad N_m = \frac{E_m}{E_s};$$

$$F_p = F + \frac{k_h}{4k_{hw}}\frac{n^2-1}{n^2};\quad N_p = \frac{E_w}{E_s};\quad G = \frac{k_h}{k_{vw}}(\frac{H}{d_w})^2,$$

factor of well resistance; $C_v = E_s k_v/\gamma_w$, vertical coefficient of consolidation of clay; $C_h = E_s k_h/\gamma_w$, horizontal coefficient of consolidation of clay; $M = (2i-1)\pi/2$, $i = 1, 2, \cdots$;

$$F = (\ln\frac{n}{s} + \frac{k_h}{k_s}\ln s - \frac{3}{4})\frac{n^2}{n^2-1} + \frac{s^2}{n^2-1}(1-\frac{k_h}{k_s})\cdot(1-\frac{s^2}{4n^2}) + \frac{k_h}{k_s}\frac{1}{n^2-1}(1-\frac{1}{4n^2})$$

4 CONSOLIDATION DEGREE AND DISCUSSIONS

According to the solution, the average degree of consolidation under various conditions can be further expressed as:

$$U = 1 - \sum_{i=1}^{\infty} \frac{2}{M^2} e^{-\beta t} \tag{6}$$

The expressions of factors α_r and β in equation (6) are shown in Table 1 for different foundations, where $m = d_w^2/d_e^2 = 1/n^2$, the replacement ratio. It can be seen from Equation (6) and Table 1:

1. When the diameter of column is zero or all the properties of columns are identical with those of soft clays, Equation (6) degenerates into the expression for the case of natural foundation which is Terzaghi's solution.

2. When the horizontal permeability of columns is equal to the vertical one (i. e. $k_{hw} = k_{vw}$), and there is no difference in stiffness between columns and soft clay, Equation (6) changes into that for the case of vertical drains.

3. The difference between expressions of consolidation degree for vertical drains and for sand-column composite foundation is only in α_e. For the latter, the stiffness of sand columns is usually greater than that of soft clay, so $\alpha_e > 1$; while for the former, $\alpha_e \approx 1$. Therefore, even if the permeability, diameter, spacing are identical for sand columns and for vertical drains, the rate of consolidation of sand-column composite foundation will definitely be greater than that of vertical drains, about α_e times.

4. The expression of the degree of consolidation for the gravel-column composite foundation has the same form as that for sand-column one. However, the behavior of clog zone in gravel-column foundation is different from that of smear zone in sand-column one. the cog zone is the mixture of gravel and mud thus both the stiffness and permeability may be higher than soft clay, while the smear zone has a similar property with that of vertical drains and so both the stiffness and permeability are lower than soft clay.

5. For cement-column composite foundation, the column can be regarded as a solid body without permeability ($k_{hw} = 0$) and without pore-water pressure ($k_{vw} = \infty$). So the expression of the degree of consolidation is very simple.

6. When soft clay is replaced by columns completely, Equation (6) turns into the Terzaghi's solution with the coefficient of consolidation of the reinforced body, i. e. $C_{vw} = E_w k_{vw}/\gamma_w$.

It can then be concluded that Equation (6) is the general expression of the average consolidation degree for different foundations.

Table 1. Expressions of α_r and β for different foundations

Foundation classification		Feature parameters	α_r	β
Natural soft clay foundation (i. e. without column)		$r_s = r_w = 0$ $k_v = k_{vw}$ $E_w = E_m = E_s$	1	$\frac{M^2 C_v}{H^2}$
Soft clay foundation with vertical drains	no well resistance	$k_{hw} = k_{vw} = \infty$ $E_w = E_m = E_s$	$\frac{n^2}{n^2-1}(1+\frac{8k_h H^2}{M^2 F k_v d_e^2})$	$(\frac{M^2 C_v}{H^2} + \frac{8C_h}{Fd_e^2})\frac{n^2}{n^2-1}$
	with well resistance	$k_{hw} = k_{vw}$ $E_w = E_m = E_s$	$\frac{F_p + D_p \alpha_w}{\alpha_v F_p + D_p}$	$\alpha_r \cdot \frac{M^2 C_v}{H^2}$
Sand-column composite foundation		$k_h \ge k_s \quad k_{hw} = k_{vw}$ $E_w > E_s > E_m$	$\frac{F_p + D_p \alpha_w}{\alpha_v F_p + D_p} \cdot \alpha_e$	$\alpha_r \cdot \frac{M^2 C_v}{H^2}$
Gravel-column composite foundation		$k_h \le k_s$ $k_{hw} = k_{vw}$ $E_w > E_m > E_s$	$\frac{F_p + D_p \alpha_w}{\alpha_v F_p + D_p} \cdot \alpha_e$	$\alpha_r \cdot \frac{M^2 C_v}{H^2}$
Cement-column composite foundation		$r_s = r_w$ $k_{hw} = 0$ $k_{vw} = \infty$	$1 + \frac{mN_p}{1-m}$	$\frac{M^2[(1-m)E_s + mE_w]k_v}{H^2 \gamma_w (1-m)}$
Foundation replaced by columns completely		$d_e = d_w$	$\frac{C_{vw}}{C_v} \quad (C_{vw} = \frac{E_w k_{vw}}{\gamma_w})$	$\frac{M^2 C_{vw}}{H^2}$

5 CONSOLIDATION BEHAVIOR OF COMPOSITE FOUNDATION

Based on Equation (6), the consolidation behavior is studied by using different numerical examples. It is found that the main factors affecting the rate of consolidation are the replacement ratio m, the ratio of compressibility modulus of columns and that of soft clay (E_w/E_s), also the smear or clog action and the well resistance action for sand column or gravel column composite foundation. The parameters used in the examples and the results are shown in Figures 2-9.

It can be seen from Figures 2-3 that when m or E_w/E_s increases, U versus T_v curve moves left. This shows that the greater the replacement ratio or the higher the stiffness of columns, the greater the rate of consolidation of soft clays reinforced by granular columns (i. e. sand or gravel columns).

Figure 4 shows how the well resistance affects the relationship between U and T_v for granular-column foundation. Just like that for vertical drains, the rate of consolidation is slow down as the factor of well resistance G increases.

The influence of smear action on sand-column foundation can be seen from Figure 5. The rate of consolidation slows down with the increase of k_h/k_s (or s).

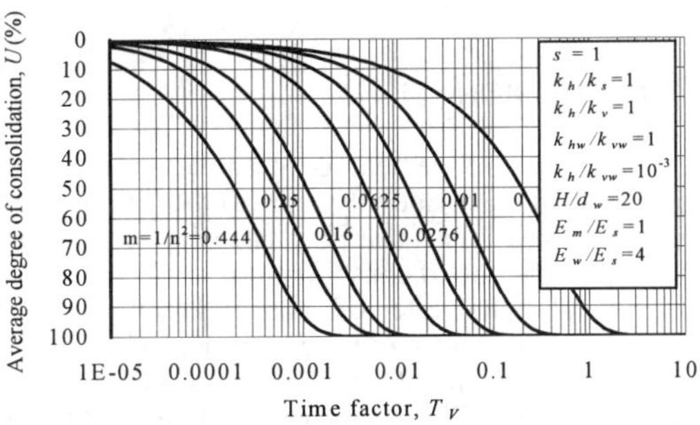

Figure 2. Influence of m on relationship of U and T_v for granular-column foundation

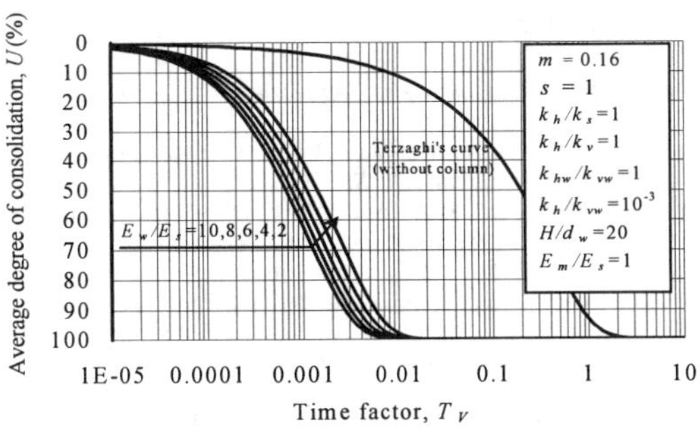

Figure 3. Influence of E_w/E_s on relationship of U and T_v for granular-column foundation

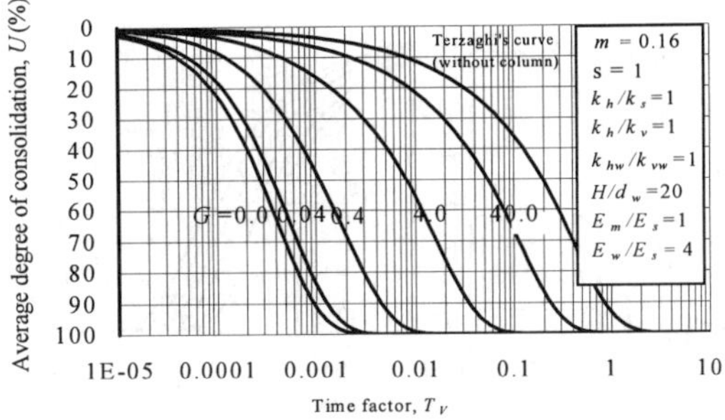

Figure 4. Influence of well resistance on relationship of U and T_v for granular-column foundation

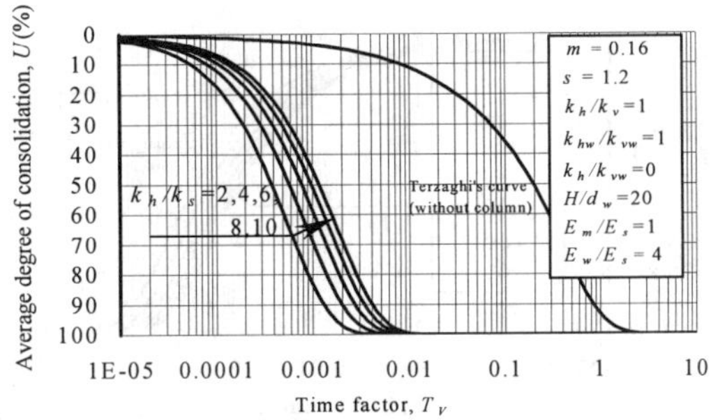

Figure 5. Influence of smear action on relationship of U and T_v for sand-column foundation

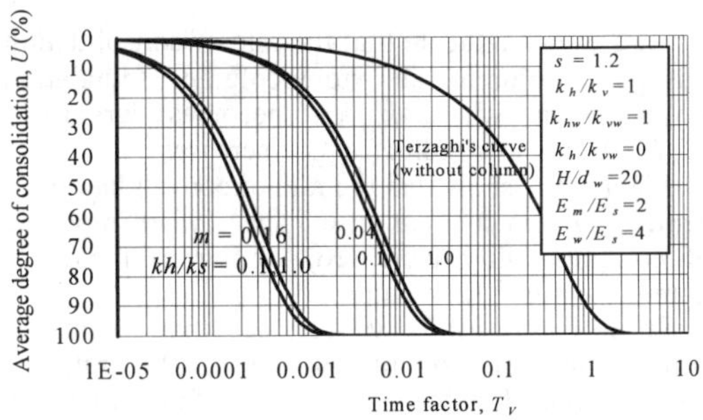

Figure 6. Influence of the permeability of clog zone on relationship of U and T_v for gravel-column foundation

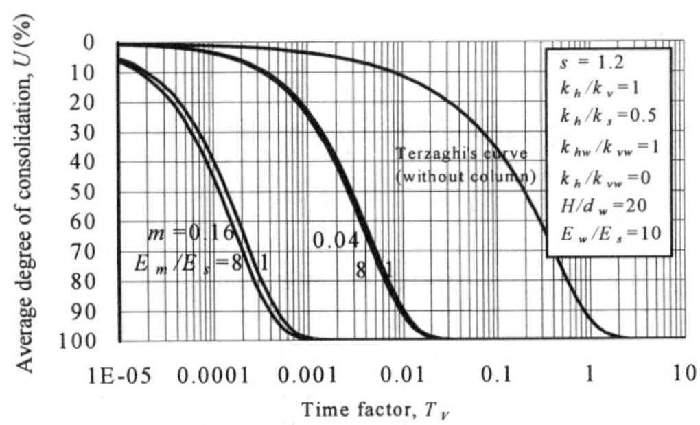

Figure 7. Influence of the compressibility of clog zone on relationship of U and T_v for gravel-column foundation

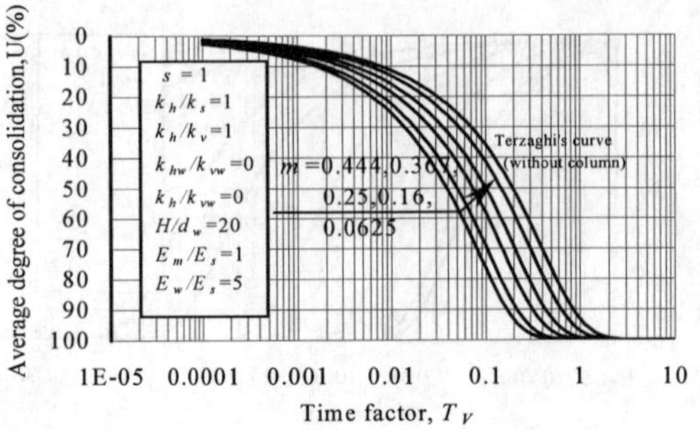

Figure 8. Influence of replacement ratio m on relationship of U and T_v for cement-column foundation

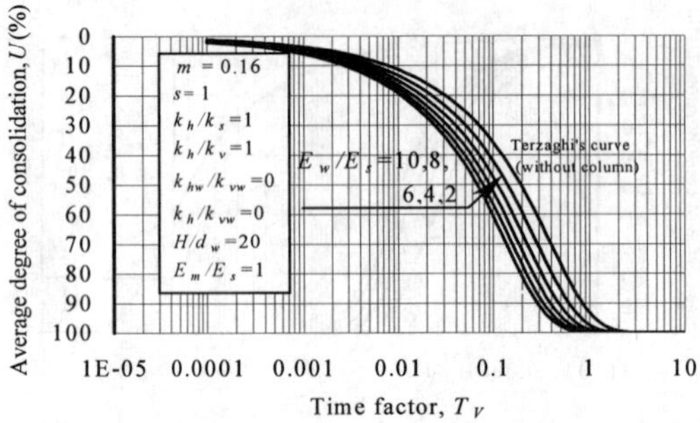

Figure 9. Influence of the stiffness of column on relationship of U and T_v for cement-column foundation

Figures 6-7 indicate that the rate of consolidation for gravel-column (i.e. stone-column) foundation slows down when k_h/k_s becomes larger or E_m/E_s becomes smaller (or s increases).

Figures 8-9 show that the rate of consolidation for cement-column foundation enhances when modulus of compressibility increased, similar to those for granular-column foundation.

From Figures 2-9, we can see that no matter which kind of columns is used, the rate of consolidation for those reinforced foundations is greater than that of natural one.

Table 2 Parameters for different cases

Case	N	$\frac{H}{d_w}$	s	$\frac{k_h}{k_s}$	$\frac{k_{hw}}{k_{vw}}$	k_{vw} (cm/s)	$\frac{E_w}{E_s}$	$\frac{E_m}{E_s}$
1	5	20	1.25	2	1	10^{-3}	1	1
2	5	20	1.25	2	1	10^{-3}	10	1
3a	6.25	25	1.25	0.5	1	10^{-4}	10	2
3b	6.25	25	1.25	0.5	1	5×10^{-3}	10	2
4	5	20	1	1	0	∞	10	--

Case 1=for vetical drains; Case 2=for sand columns; Case 3a and 3b=for gravel columns; Case 4=for cement columns; H=10m; Spacing of drains or culumns = 2.2m(in square pattern); diameter = 0.5m (for gravel columns, the thickness of clog zone is 5cm, so the effective diameter is 0.4m); $k_h = k_v = 10^{-6}$cm/s; $E_s = 0.3$MPa.

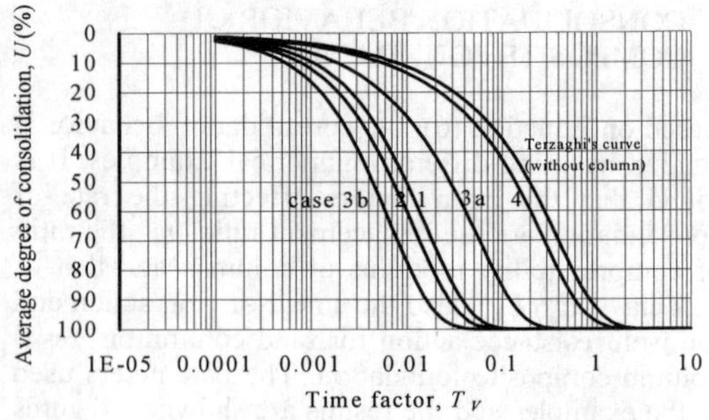

Figure 10. Comparisons of the rate of consolidation for different foundations

Finally, to make a comparison for rate of consolidation of different foundations, 5 cases are studied. Table 2 lists all parameters used. The results are shown in Figure 10. It can be seen that under the particular condition shown in Table2, the consolidation rate of sand-column foundation (case 2) is greater but the one of cement-column foundation (case 4) is smaller than that of vertical drains (case 1). For gravel-column foundation, the consolidation rate is greatly affected by the permeability and stiffness of clog zone. It may be smaller (in case 3a) than that of vertical drains and may also be greater (in case 3b) than that of sand-column foundation.

6 CONCLUSION

1. The use of cement or granular columns to reinforce soft clays not only reduces the deformation and enhances the bearing capacity, but also accelerates the rate of consolidation.

2. While the rate of consolidation can be enhanced by increasing the replacement ratio or the stiffness of columns, it may be slowed down due to the poor permeability of cement column or the clog of gravel column with clay particles.

REFERENCES

Gong, X.N. 1992. *Composite Foundation.* Hangzhou: Zhejiang University Press.

Kamon, M. & D.T. Bergado 1991. Ground improvement techniques. *Proc. 9th ARCSMFE.*

Mitchell, J.K. & R.K.Katti 1981. Soil improvement—state–of-the-art report. *Proc. 10th ICSMFE.*

Navak, N.V. 1989. The recent advances in design and execution of stone column foundations. *Proc.12th ICSMFE.*

Barron, R.A. 1948. Consolidation of fine-grained soils by drain wells. *Trans. ASCE* 113: 718-742.

Zeng, G.X. & K.H. Xie 1989. New development of the vertical drain theories. *Proc.12th ICSMFE.* 2:1435-1438.

Computer Methods and Advances in Geomechanics, Desai et al. (eds) © 2001 Balkema, Rotterdam, ISBN 90 5809 183 X

Finite element analysis of soil reinforced segmental retaining walls in a tiered arrangement

Chungsik Yoo & Joo-Suk Kim
Sungkyunkwan University, Suwon, Korea

ABSTRACT: This paper presents the results of investigation on the behavior of geosynthetics reinforced segmental retaining walls in a tiered arrangement using the finite element method of analysis. A parametric study was performed by varying the location of upper tier wall with respect to the underlying wall. The results of analysis indicate that an equivalent surcharge approach, which is frequently adopted by current design assumptions, may yield inadequate results depending on the location of upper tier wall. Also indicated that the behavior of the upper tier significantly departs from that of a single wall on a rigid foundation. Practical implications of the findings from this study are discussed in a great detail.

1 INTRODUCTION

Segmental retaining wall market in Korea has been growing drastically since late 1990s in both engineered and non-engineered applications. Despite the inherent conservatism in the current design approaches, numerous major and minor structural problems have been reported during and after construction, covering a range of structural damage of local failure to total collapse. Much still need to be investigated to fill the gap between the theory and the practice.

In recent years, there have been a number of investigations into the behavior of reinforced soil walls using small to large-scale laboratory model tests (Juran & Christopher, 1989; Bathurst, 1990; Chou & Wu, 1993; Porbaha & Goodings 1997), field tests (Simac et al., 1990; Bathurst, 1992; Collin & Berg, 1994; Ochiai & Fukuda, 1996), and numerical experiments (Karpurapu & Bathurst, 1995; Ho & Rowe, 1997; Rowe & Ho, 1998; Yoo & Lee, 1999). The majority of the investigations, however, have been focused on independent walls and therefore, the behavior of tiered segmental retaining walls has not been fully understood.

The primary objective of this paper is to provide insight regarding the behavior of geosynthetics reinforced segmental retaining walls in a tiered arrangement and to form a database for use in developing more rational design/analysis method. Attention is focused on a tiered segmental block wall reinforced with extensible reinforcement in a granular backfill resting on a non-yielding foundation. The finite element method of analysis was adopted in the present study since numerous researchers have demonstrated that the technique can successfully capture what might otherwise be difficult when adopting physical modeling techniques. Comparisons are made between the findings of the numerical analyses and current design approaches in order to provide insight concerning the consequences of neglecting these effects in conventional analyses for the type of wall and conditions examined.

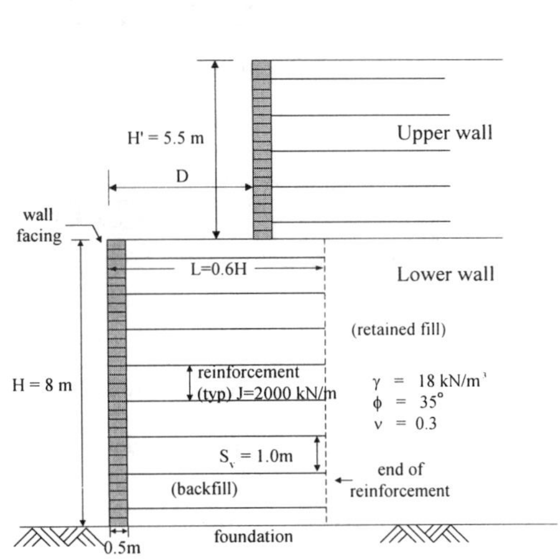

Figure 1. A schematic view of tiered segmental retaining wall system

2 PROBLEM CONSIDERED

2.1 *Wall geometry*

In this study, tiered segmental retaining walls on a rigid foundation with the lower and upper wall heights of 8 and 5 m, respectively, were considered. The geometry of the walls examined in this study is shown in Figure 1 together with the symbols and a set of reference parameters used. It was assumed that the wall facing was constructed with segmental blocks of 0.25 m (height) × 0.5 m (length), and that the backfill soil was cohesionless and free draining. For simplicity, layers of geogrid with a uniform length, L/H=0.6, are placed in the lower and upper walls at a uniform vertical spacing of S_v = 1.0 m. Note that the axial stiffness of the geogrid was assumed to be J=2000 kN/m.

2.2 *Parametric study*

Two cases with different offset distance D of the upper wall were considered (i.e., D=0.5H and 1.0H) in order to investigate the variation of degree of interaction with the location of upper wall. Note that the current design approaches recommend the upper tier wall be replaced by an equivalent surcharge with an intensity depending on the distance between the lower and upper walls. Additional cases with corresponding equivalent surcharge on top of the lower wall were examined to evaluate the validity of current design approaches. Figure 2 summarizes the conditions considered in the present study

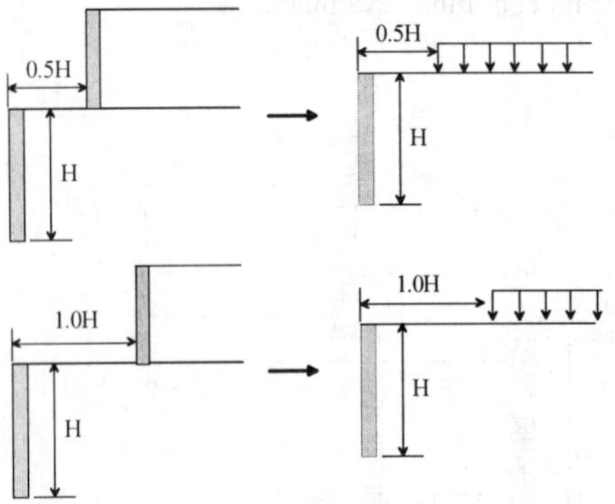

(a) Full model (b) Equivalent surcharge

Figure 2. Conditions considered in the present study

3 FINITE ELEMENT ANAYSIS

3.1 *Finite element model*

A commercial finite element code DIANA (DIANA 1996) developed by TNO Building & Construction Research Co. and Delft Technical University was used for this study. DIANA is a multi-purpose finite element program for use in a range of geotechncial engineering problems including excavation, tunneling, retaining walls, and slopes.

Figure 3 shows the finite element mesh used in the analysis. As can be seen, a very refined mesh, which consists of approximately 980 nodes and 1050 elements, was adopted to minimize the effect of mesh dependency on the results of finite element analysis. In the finite element model, the foundation soil was assumed to extend to one-half times the wall height (H) below the wall base while the lateral boundary at approximately 3.0H away from the wall face.

The backfill soil and the wall facing were discretized using eight-noded isoparamatric plane strain elements (CQ18E), while two-noded truss elements (L4TRU) were used for the reinforcements. In addition, the interface behavior between block/backfill, reinforcement/backfill, and backfill/foundation was modeled using six-noded Goodman type interface elements (CL12I; Goodman et. al. 1968). Figure 4 illustrates the details of the wall/backfill soil/reinforcement interface modeling.

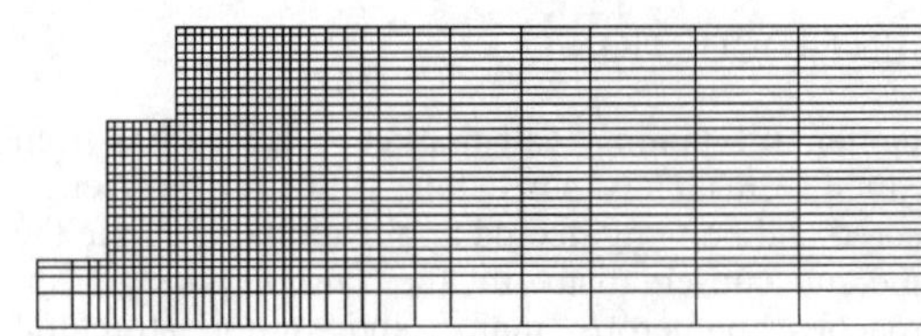

Figure 3. A typical finite element mesh used

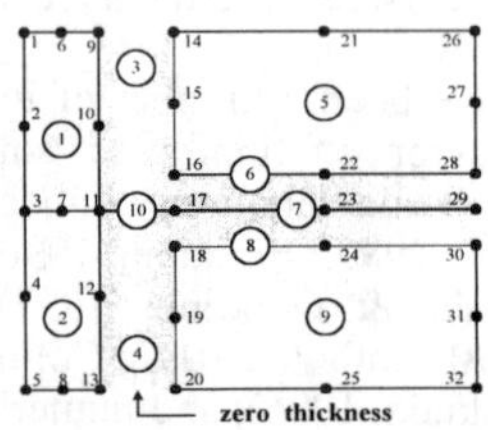

Element	Type	Nodes
1	quadrilateral	9-1-3-11-6-2-7-10
2	quadrilateral	11-3-5-13-7-4-8-12
3	interface	16-14-9-11-15-10
4	interface	20-18-11-13-19-12
5	quadrilateral	26-14-16-28-21-15-22-27
6	interface	17-29-28-16-23-22
7	uniaxial bar	17-29-23
8	interface	18-30-29-17-24-23
9	quadrilateral	30-18-20-32-24-19-25-31
10	nodal link	11-17

Figure 4. Modeling detail for reinforcement/block/backfill soil junction

The non-linear behavior of the backfill soil was modeled using a Mohr-Coulomb failure criterion and the non-associated flow rule proposed by Davis (1968). The dilatancy angle ψ of the soil was related to the friction angle ϕ, using the relationship proposed by Bolton (1986) with a constant critical state friction angle ϕ_{cv}=30°, i.e.

$$\phi = \phi_{cv} + 0.8\psi \quad (1)$$

The wall facing, reinforcement, and interfaces were

assumed to follow a linear elastic behavior. Table 1 summarizes material properties used in the analysis.

Table 1. Material properties used

	Soil/wall properties				
	E_s (kPa)	ϕ (degree)	ψ (degree)	γ (kN/m^3)	EI (kN-m^2)
Soil	3×10^4	35	6	18	-
Wall	-	-	-	21	5,500
	Interface properties				
	K_s (kPa/m)			K_n (kPa/m)	
Soil//Wall	1×10^4			1×10^{10}	
Soil/Reinf.	1×10^5			1×10^{10}	

3.2 *Modeling of construction sequence*

It has been recognized that in the numerical analysis, the final computed results are different for a solution based upon sequential construction and one based upon single stage construction. Therefore, for realistic solutions to segmental retaining wall problems, the construction sequences should be simulated as carefully as possible.

In the present analysis, the step by step wall construction sequence was carefully simulated by activating soil, block, and reinforcement layers until design wall height is reached using the phased analysis option, which is a special feature offered by DIANA. A phased analysis comprises several calculation phases. Between each phase the finite element model changes by addition or removal of elements, constraints and/or loading conditions. In each phase, a separate analysis is performed, in which the results from previous phases are automatically used as initial values. For analysis of the cases with equivalent surcharge, the surcharge was incrementally applied upon completion of the lower wall construction.

4 GENERAL BEHAVIOR

4.1 *Deformation Pattern*

Figure 5 illustrates a displacement vector plot for the case with D=0.5H at the end of the upper wall construction. As can be seen in this figure, significant wall movements occur in the lower and upper walls. Of particular interest to note is the upper wall deformation pattern, which shows significant lateral movements at the wall base. Such a pattern is associated with the lower wall movements and is discussed further later in this paper.

The general deformation pattern shown in the vector plot suggests that the global sliding failure mechanism should be checked with great care for a tiered wall having a relatively short offset distance. Such a trend is better illustrated in a contour plot of maximum shear strain $\varepsilon_{max}=(\varepsilon_1-\varepsilon_2)$ as shown in Figure 6. As illustrated, a well defined shear zone is developed, which initiates at the toe of the lower wall and then propagates into behind the reinforced soil block of the upper wall with an angle equivalent to that of Rankine active failure line. This trend supports the current design approaches, which require to check the global stability of the combined tiered wall system and surrounding soils using a rigorous slope stability analysis.

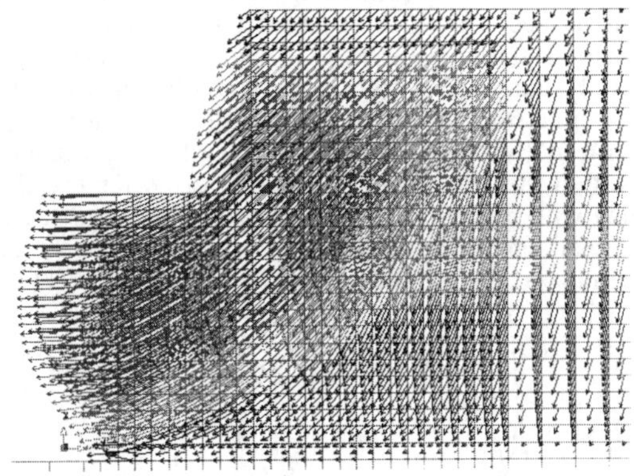

Figure 5. Displacement vector plot for D=0.5H

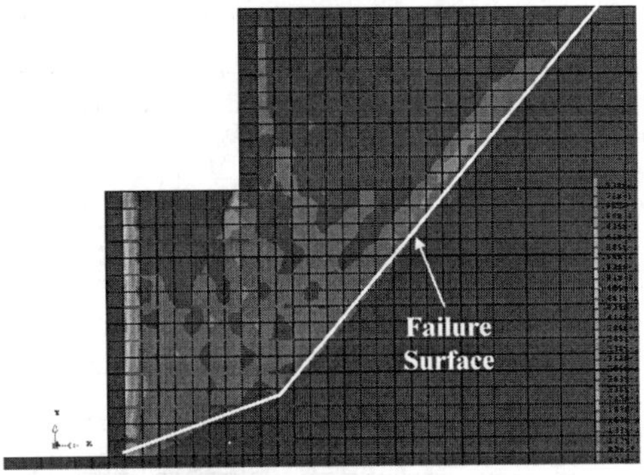

Figure 6. Maximum shear strain contour for D=0.5H

4.2 *Horizontal deformation*

Sources of the horizontal wall deformation include horizontal strains within the reinforced soil block as well as the rigid body movement of the reinforced soil block itself due to lateral thrust acting at the back of the reinforced soil zone. Interaction between the lower and upper walls in a tiered wall system may significantly affect the horizontal wall deformation profile depending on the degree of interaction.

Figures 7 and 8 present horizontal deformation δ_h profiles at the wall face, within and behind the reinforced soil block for D=0.5H and 1.0H, respectively. Also included in these figures are the deformation profiles for walls in a single arrangement for comparison. Note that horizontal deformation within the reinforced soil block (i.e., internal horizontal deformation) is obtained by subtracting the horizontal deformation behind the reinforced soil block from that at the wall face. As shwon Figure 7, the effect of upper wall on the lower wall would be to increase the horizontal deformation within and behind the reinforced soil block and thus the horizontal deformation at the wall face, indicating that the upper wall significantly influences both the internal and external stability of the lower wall. Note that a significant increase in the horizontal deformation within the reinforced soil block occurs for D=0.5H but with negligible increase for D=1.0H. It is therefore assumed that the effect of upper wall on the internal stability can be disregarded when D=1.0H.

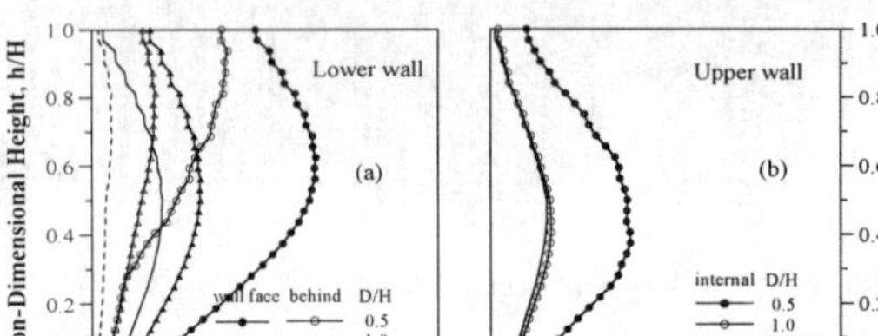

Figure 7. Variation of horizontal deformation profile with D/H for lower wall: (a) At wall face and Behind reinforced soil Block; (b) Within reinforced soil block

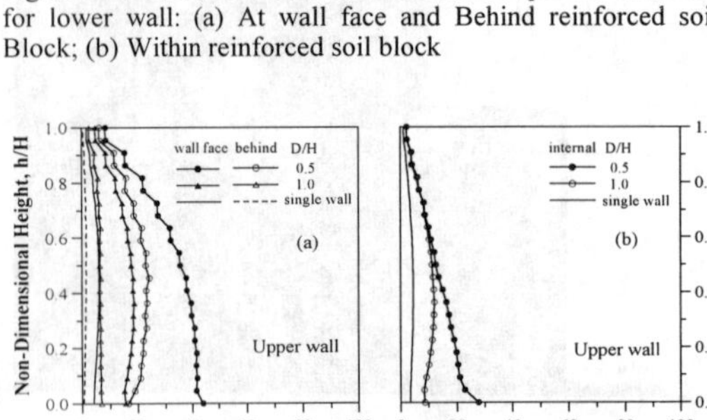

Figure 8. Variation of horizontal deformation profile with D/H for upper wall: (a) At wall face and Behind reinforced soil Block; (b) Within reinforced soil block

Deformation profiles of the upper wall are illustrated in Figure 8. Of key trend shown in this figure is that considerable lateral wall movements are developed in the lower part of the wall, even for the case with D=1.0H. The horizontal deformation at the wall face is accompanied by a considerable increase in the horizontal deformation within and behind the reinforced soil block. The horizontal deformation profiles depart significantly from those of a single wall on a rigid foundation and mainly arise from the lateral movements of the lower wall as observed in Figure 5. This is a direct consequence of interaction between the two walls and not considered in the current design approaches. A possible countermeasure to reduce movements in the lower and upper walls would be to increase the reinforcement density in both walls, respectively. Of practical significance of the findings is that there exists a significant interaction between the walls in a tiered arrangement, and that consideration must be given to both walls when performing a stability calculation. Such a trend is not addressed in the current design approaches and further investigation is required.

5 REINFORCEMENT FORCE

Figure 9 presents maximum reinforcement force distributions for the tiered walls along with those obtained assuming each wall being a single wall on a rigid foundation. Also shown in this figure are the results of the analysis on a model with an equivalent surcharge based on NCMA design approach (NCMA, 1997). Note that an equivalent distributed surcharge of q=27 kPa is suggested for the internal stability calculation for D=0.5H whereas no influence is assumed for D=1.0H. The design earth pressure envelope based on the Rankine active state of stress is also presented in this figure for comparison.

As would be expected for the lower wall shown in Figure 9(a), a considerable increase in reinforcement force is observed for D=0.5H, as much as 100%, when compared to that for the single wall. The increase for D=1.0H, however, appears to be negligible, as can be expected from the internal horizontal deformation profile. A comparison between the full model and the equivalent surcharge model for D=0.5H reveals that the equivalent surcharge model yields somewhat smaller reinforcement force. This is a direct consequence of not being able to consider the soil-structure interaction in the equivalent surcharge model. It is of importance to note that the reinforcement force slightly exceeds that inferred from the Rankine active state of stress in the upper part of the wall for both cases of D=0.5H and 1.0H. Considering the fact that the inherent conservatism in the current design approaches usually yields conservative estimates of the reinforcement force, the current design approach may not be on the safe side. Further investigation is required to draw a general conclusion.

The reinforcement force distributions for the upper walls are illustrated in Figure 9(b) together with that for a single wall on a rigid foundation. As indicated in this figure, for both cases, significantly larger forces are developed for the walls in a tiered arrangement than for walls in a single arrangement, especially at the lower part of the wall. Even for D=1.0H, the force distribution significantly departs from that of the single wall. This phenomenon is a direct consequence of the increased lateral wall movements in the lower tier wall. Larger reinforcement forces should therefore be expected for the upper wall in a tiered arrangement than for a single wall when estimating the force in the reinforcement. Such a trend is consistent with the observation in the horizontal deformation and highlights the need for increasing reinforcement density at the lower part of the upper tier wall to improve the internal stability.

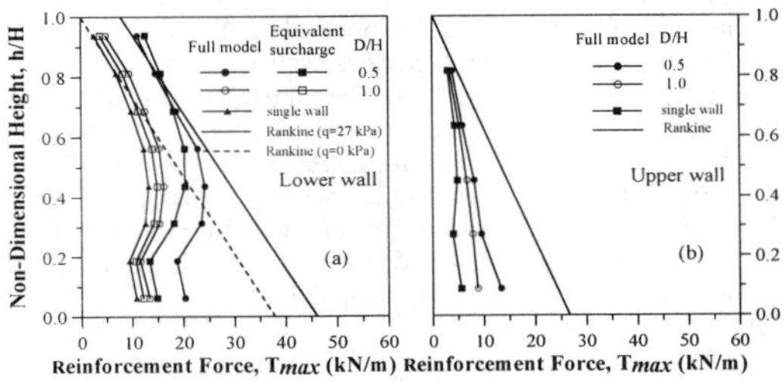

Figure 9. Variation of reinforcement force distribution with D/H: (a) Lower wall; (b) Upper wall

6 STRESS DISTRIBUTION

Design of a reinforced soil wall requires looking into the external stability of a monolithic gravity structure, comprising the facing unit, reinforcement, and backfill soil, against three modes of failure; base sliding, overturning, and bearing capacity. Since the horizontal and vertical stresses at the back and the base of the reinforced soil block play important roles in the external stability analysis, the stress distributions should be carefully examined. In the current design approaches, the vertical and horizontal stress distributions at the back and the base of the reinforced soil block are estimated based on the Meyerhof distribution and the Rankine or Coulomb active stress state, respectively. Vertical and horizontal stress distributions in the tiered walls are examined

6.1 *Vertical stress*

Typical vertical stress distributions at the base of reinforced soil block for the lower wall are shown in Figure 10(a). Also shown in this figure are those inferred from the equivalent surcharge approach with the Meyerhof assumption. The general distribution pattern within the reinforced block is such that the vertical stress is maximum in the vicinity of the wall facing, largely due to the lateral thrust acting at the back of the reinforced soil block and drops to a minimum value then increases with the distance from the facing thereafter. It appears that the equivalent surcharge approach with the Meyerhof assumption yields a reasonable estimate of the maximum vertical stress $\sigma_{v,max}$ for D=0.5H$_1$ but tends to over-estimate for D=1.0H1, implying that the equivalent surcharge approach may give a conservative estimate of $\sigma_{v,max}$ when the offset distance D is larger. For the upper walls shown in Figure 10(b), the distribution pattern is a bit different from that of a single wall on a rigid foundation, showing the maximum vertical stress $\sigma_{v,max}$ near the edge of reinforced soil block. The flexibility of backfill soils under the upper wall and the lateral wall movements may be responsible for such a trend.

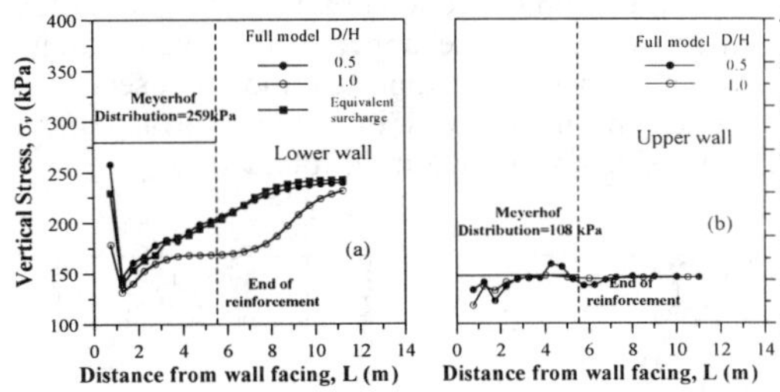

Figure 10. Variation of vertical stress distribution at the base of reinforced soil block with D/H: (a) Lower wall; (b) Upper wall

As observed, although the maximum vertical stress $\sigma_{v,max}$ does not exceed that inferred from the Meyerhof distribution, it appears that the Meyerhof distribution tends to significantly over estimate the average vertical stress within the reinforced soil block. Note that the "average" overburden stress is required when assessing the resistance against base sliding along the foundation or the resistance against slippage of the reinforcement. The use of Meyerhof assumption in calculating the average overburden stress, therefore, may yield inappropriate design, as indicated by Rowe and Ho (1998).

6.2 *Horizontal stress*

Horizontal stress distributions at the back of reinforced soil block are presented in Figure 10. As illustrated in Figure 11(a) for the lower wall, the horizontal stress significantly increases over the entire wall for both cases of D=0.5H and 1.0H, following those inferred the Rankine active state of stress, due to the presence of the upper wall. Note that a considerable increase is observed even for the case of D=1.0H, in which the upper wall is located just edge of the Rankine active zone behind the reinforced soil block, implying that the external stability is affected to some extent. The practical implication of the general observations is that the external stability of reinforced wall should be carefully examined considering the relative distance between the upper wall and the reinforced soil block. Further inspection reveals that the equivalent surcharge model yields similar results as compared to those of the full model, It is therefore concluded that the equivalent surcharge approach can give reasonable estimate of the horizontal stress at the back of the reinforced block.

Figure 11(b) shows the horizontal stress distributions for the upper wall. As seen, the distribution pattern follows those inferred from the Rankine active state of stress for both cases but with somewhat smaller stresses. Note that the horizontal stress distribution remains practically the same regardless of the offset distance D. Furthermore, the horizontal stress at the base level is significantly smaller than that for a single wall, suggesting that the toe resistance is significantly reduced due largely due to the lateral movement of the upper wall. Such a trend highlights the need for adopting a more conservative design for the upper wall in a tiered arrangement than for a single wall on a rigid foundation.

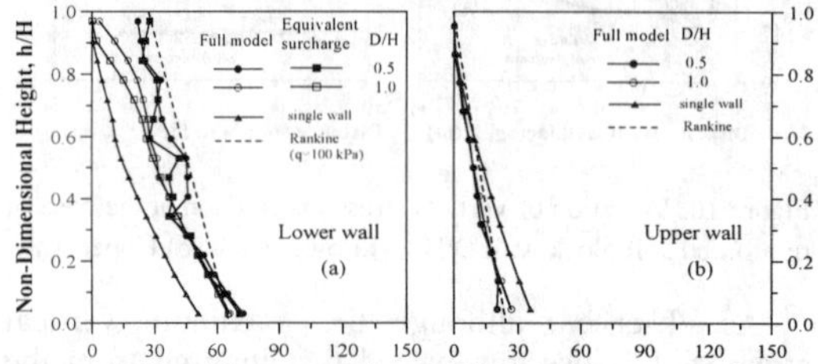

Figure 11. Variation of horizontal stress distribution at the back of reinforced soil block with D/H: (a) Lower wall; (b) Upper wall

7 CONCLUSION

The behavior of tiered soil reinforced segmental retaining walls have been presented. Among the results examined include horizontal deformation, reinforcement force, horizontal and vertical stresses at the back and the base of the reinforced block. The results of analysis indicated the interaction between the upper and lower walls significantly influences both the internal and external stability for both walls. Of significant finding is that the interaction between the two walls results in a significant increase in the horizontal deformation at the wall face and the reinforcement force for the upper wall. Greater reinforcement density is required for the upper tier wall to provide resistance against lateral movements than for a single wall with the same height on a rigid foundation. Also revealed is that the current design approaches based on the equivalent surcharge may yield smaller estimates of the reinforcement force as opposed to the full model. The equivalent surcharge approach however appears to give reasonable estimates of vertical and horizontal stress distributions for use in the external stability analysis.

REFERENCES

Bathurst, R.J. 1990. Instrumentation of Geogrid-Reinforced Soil Walls, *Transportion Research Record, No. 1277, Soils, Geology, and Foundations, Modern Geotechnical Methods: Instrumentation and Vibratory Hammers*, pp. 109-116.

Bathurst, R.J. 1992. Case study of a monitored propped pannel wall, *Proc. 5th Int. Symp. on Geosynthetic-Reinforced Soil Retaining Walls*, Balkema, Rotterndam, Brookfield, pp. 159-166.

Bolton, M.D. 1986. The strength and dilatancy of sands, *Geotechnique*, London, U.K., 36(1), pp. 65-78.

Chou, N. N. S. & Wu, J. T. H. 1993. Investigating performance of geosynthetics reinforced soil walls, *Rep. No. CDOT-UCD-R-93-21*, Colorado Dept. of Transportion.

Collin, J.G. & Berg, R.R. 1994, Permanent geogrid reinforced soil retaining walls – US experience, *Proc. Recent Case Histories of Permanent Geosynthetic-Reinforced Soil Retaining Walls*, Tatusoka and Leshchinsky (eds), pp. 61-68.

Davis, E.H. 1968. Theories of plasticity and the failure of soil masses, *Soil mechanics: Selected topics*, I. K. Lee, ed., Butterworth's, London, U.K., pp. 341-380.

DIANA User's Manual 1996. Release 6.1, *TNO Building and Construction Research*, Netherlands.

Goodman, R. E., Taylor R. L., and Brekke, T. L. 1968. A model for the mechanics of jointed rock, *Journal of Soil Mechanics and Foundation Engineering*, ASCE, Vol. 94, No. SM3, pp. 637-657.

Ho, S.K. & Rowe, R.K. 1996. Effect of Wall Geometry on the Behavior of Reinforced Soil Walls, *Geotextile and Geomembranes*, 14, pp. 521-541.

Juran, I. & Christopher, B. 1989. Laboratory Model Study on Geosynthetic Reinforced Soil Retaining Walls, *Journal of Geotechnical Engineering*, ASCE, Vol. 115, NO. 7, pp. 905-926.

Kapurapu, R. & Bathurst, R.J. 1995. Behavior of Geosynthetic Reinforced Soil Retaining Walls Using the Finite Element Analysis, *Computers and Geotechnics*, Vol. 17, pp. 179-299.

Ochiai, Y. & Fukuda, N. 1996. Experimental Study on Geotextile-Reinforced Soil Walls with Different Facings, *Proc. 1st, European Geosynthetics Conf.*, Netherlands, pp. 113-120.

Porbaha, A. & Goodings, D.J. 1997. Centrifuge Modeling of Geotextile-Reinforced Cohesive Soil Retaining Walls, *Journal of Geotechnical Engineering*, ASCE, Vol. 122, NO. 10, pp. 840-848.

Rowe, R.K. & Ho, S.K. 1997. Continuous Panel Reinforced Soil Walls on Rigid Foundations, *Journal of Geotechnical and Geoenvironmental Engineering*, ASCE, Vol. 123, No. 10, pp. 912-920.

Simac, M.R., Christopher, B. R. & Bonczkiewicz, C. 1990. Instrumented field performance of a 6 m geogrid soil wall, *Proc. 5th Int. Conf. On Geotextiles, Geomembranes and Related Products*", Den Hoedt (ed.), pp. 53-59.

NCMA, 1997. Design Manual for Segmental Retaining Walls, 2nd Ed., NCMA, Virginia.

FHWA, 1997. Mechanically Stabilized Earth Walls and Reinforced Soil Slopes, Design and Construction Guidelines, *FHWA Demonstration Project 82*, FHWA, Washington, DC, FHWA-SA-96-071

Yoo. C. S., & Lee, K. M. 1999. Investigation on Behavior of Reinforced Segmental Retaining Walls, *Journal of Korean Geotechnical Society*, KGS, Vol. 15, NO. 1, pp. 53~62.

12 Petroleum geomechanics, hydraulic fracturing, geological simulation

Computer Methods and Advances in Geomechanics, Desai et al. (eds) © 2001 Balkema, Rotterdam, ISBN 90 5809 183 X

Propagation regimes of fluid-driven fractures in impermeable rocks

Emmanuel Detournay
CSIRO Petroleum, Melbourne, Vic., Australia

ABSTRACT: This paper reviews recent results of a research program aimed at developing a theoretical framework to understand and predict the different modes of propagation of a fluid-driven fracture. The research effort involves constructing detailed solutions of the crack tip region and developing global models of hydraulic fractures. These solutions are then used to identify the limits of applicability of the viscosity- and toughness-dominated regime of crack propagation. They can also be used as benchmarks to check the accuracy of three-dimensional numerical simulators of hydraulic fractures.

1 INTRODUCTION

A particular class of fractures in the earth develops as a result of internal pressurization by a fluid. These fractures are either natural such as kilometers-long volcanic dikes, or man-made like the hydraulic fractures designed to increase production from oil or gas reservoirs. While dikes are driven by magma coming from deep underground chambers, hydraulic fractures are created by injecting a viscous fluid from a borehole.

Mathematical modeling of fluid-driven fractures has attracted numerous contributions, see Khristianovic and Zheltov (1955), Perkins and Kern (1961), Barenblatt (1962), Geertsma and de Klerk (1969), Nordgren (1972), Abe et al. (1976), Spence and Turcotte (1985) for some early contributions. The theoretical research in this area is aimed primarily at establishing models for predicting the evolution of the fluid pressure, opening and size of the fracture given the injection rate, fluid rheology, and rock properties.

Solutions for fluid-driven fractures are notoriously difficult to construct (even for simple geometry), because of the strong non-linear coupling between the lubrication and elasticity equations and the non-local character of the elastic response of the fracture. "Non-locality" simply means that the crack opening w at one point of the fracture depends on the fluid pressure p_f at another point. The "non-linearity" originates from the dependence of the fracture conductivity on the cube of the crack opening. Furthermore, the process is time-dependent with the fracture propagating with unknown speed, which also changes with time. Non-locality and non-linearity conspire to yield a complex solution structure that involves processes at the very small scale near the tip of the fracture. These small-scale processes control, however, the global response of the fluid-driven fracture. The difference of length scales between the tip processes and the overall hydraulic fracture also causes an inherent difficulty in simultaneously modeling numerically both the small and the large scales.

In recent years, the main focus of research in this area has been oriented towards the development of numerical algorithms to simulate the propagation of

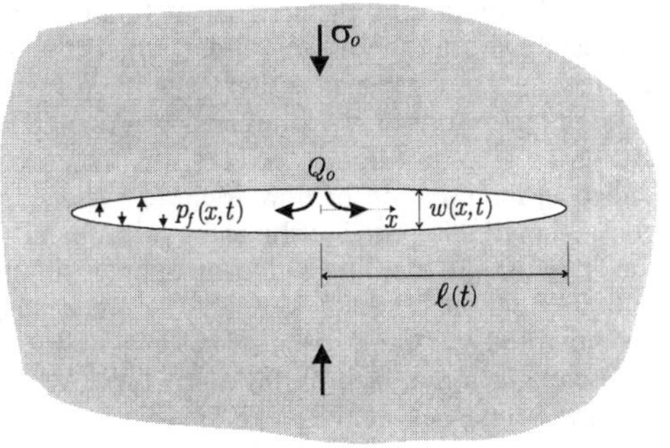

Figure 1: Hydraulic fracture propagating in an impermeable elastic medium.

three-dimensional fractures in non-homogeneous media (e.g., layered strata with different mechanical properties). Nevertheless, there is still interest in obtaining exact solutions for simpler models, giving rigorous consideration to the physical processes taking place. Such solutions can be used not only to benchmark numerical codes, but also to test the influence of various parameters in order to establish the existence of different regimes of propagation.

This paper reviews the main results of a research program, conducted for several years at the University of Minnesota to deal with the seemingly simple problem of a plane strain "KGD" and a penny-shaped hydraulic fracture propagating in an impermeable elastic rock mass. (The plane strain fracture is usually named after Kristianovic, Geersma, and de Klerk who have carried out pioneering research in hydraulic fracturing.) As sketched in Fig. 1, we consider a fracture which is driven by injection of an incompressible, Newtonian fluid at a constant volumetric rate Q_o, and which propagates quasi-statically in a direction normal to the minimum compressive far-field stress σ_o. The main focus of this paper is to review the dependence of the solution (i.e. the fracture half-length ℓ or radius R, opening w, and net pressure $p = p_f - \sigma_o$ as function of position x or radial distance r and time t) on the injection rate Q_o and on the three material material parameters μ', E', K' defined as

$$\mu' = 12\mu,\ E' = \frac{E}{1-\nu^2},\ K' = 4\left(\frac{2}{\pi}\right)^{1/2} K_{Ic} \quad (1)$$

where μ is the fluid viscosity, E and ν the rock Young modulus and Poisson ratio, and K_{Ic} the rock toughness. (While E' is the so-called plane strain modulus, μ' and K' are introduced simply to keep equations uncluttered by numerical factors.) For the sake of convenience, μ', E', and K' will simply be referred to as viscosity, elastic modulus, and toughness, respectively. Actually, an important goal of this research was to establish the conditions under which the solution becomes independent either of K' (viscosity-dominated regime) or of μ' (toughness-dominated regime). In the viscosity-dominated regime, the energy expended in the creation of new fracture surfaces in the rock is small compared to the energy dissipated in viscous flow; in the toughness-dominated regime, the viscous dissipation is small compared to the energy dissipated at the crack tip.

The paper is organized as follows. First, we review some properties of the near tip solution. Under the assumption that the fluid flows up to the tip of a fracture propagating in an impermeable elastic solid with zero toughness, it is shown that the singularity at the fracture tip is weaker than predicted by linear elastic fracture mechanics (LEFM). Next, the existence of a lag of *a priori* unknown length between the fluid front and the fracture tip is taken into account. The presence of a tip cavity filled with inviscid vapors from the fracturing fluid removes the singularity in the fluid pressure and enables the construction of a solution with the "proper" LEFM square root singularity at the fracture tip. Then, we analyze the propagation of a KGD fracture and introduce a viscosity-scaling appropriate for "small" toughness and a toughness-scaling for "large" toughness (or more correctly for "small" viscosity). After reviewing the zero-toughness and large toughness asymptotic solutions, it is shown that the regime of propagation is controlled by a single number $\mathcal{K}$, function only of Q_o, μ', E', K'. Results for the penny-shaped fracture parallels those for a KGD, with the notable exception that $\mathcal{K}$ now depends also on time.

2 NEAR TIP SOLUTION

2.1 *Singular Zero Toughness Solution*

The proper singular behaviour at the tip of a hydraulic fracture, completely filled by an incompressible Newtonian fluid and propagating in a zero toughness elastic solid, was first recognized by Spence and Sharp (1985) and also by Lister (1990), from considerations involving the asymptotic behavior of the singular elastic integral equation and the lubrication equation. The strong coupling between the moving fluid and the zero toughness elastic solid in the tip region actually corresponds to an exact matching singularity between the lubrication and the elasticity equations. This singularity can be deduced using the following simple arguments (Desroches et al. 1994). Consider the stationary propagation of a semi-infinite fluid-driven fracture in an impermeable elastic solid. Noting that any singular behaviour in the net pressure p must be compatible with the elastic behavior near the tip of the fracture, the most general singular solution for p is (Rice 1968)

$$p = A(\alpha)X^{\alpha} \quad (2)$$

where X is a coordinate with origin at the fracture tip (see Fig. 2), and A and α are real-valued constants. The corresponding fracture width w is given by

$$w = B(\alpha)X^{1+\alpha} \quad (3)$$

On the other hand, the simplifying assumptions stated above lead to a particular form of the lubrication equation

$$\frac{dp}{dX} = \frac{\mu' V}{w^2} \quad (4)$$

where V is the crack tip propagation speed. By substituting p and w by their expressions given by (2) and (3) into (4), it can readily be seen that necessarily $\alpha = -1/3$ in apparent contradiction with the classical result $\alpha = -1/2$ of linear fracture mechanics. Using the known elastic expressions for $A(\alpha)$ and $B(\alpha)$, explicit singular expressions for p and w can be found (Desroches et al. 1994)

$$w = 2^{1/3} 3^{5/6} \left(\frac{\mu' V X^2}{E'} \right)^{1/3} \quad (5)$$

$$p = -\frac{1}{6^{2/3}} \left(\frac{E'^2 \mu' V}{X} \right)^{1/3} \quad (6)$$

It is worthwhile pointing out that the far-field stress σ_o does not enter in this solution, except as a reference level for the fluid pressure p_f.

2.2 *Finite Toughness*

If the solid has a toughness $K' > 0$, a lag of *a priori* unknown length λ between the crack tip and the fluid front must necessarily exist to ensure coherence of the mathematical solution (Detournay 1999). In impermeable rocks, the tip cavity is filled with fluid vapors, which can be assumed to be at zero pressure (see Fig. 2). The near tip solution for a hydraulic fracture propagating in rock with finite toughness can be obtained by analyzing the stationary problem of a semi-infinite crack propagating at constant velocity V (Garagash 1998; Garagash and Detournay 2000). The main properties of this solution are summarized next.

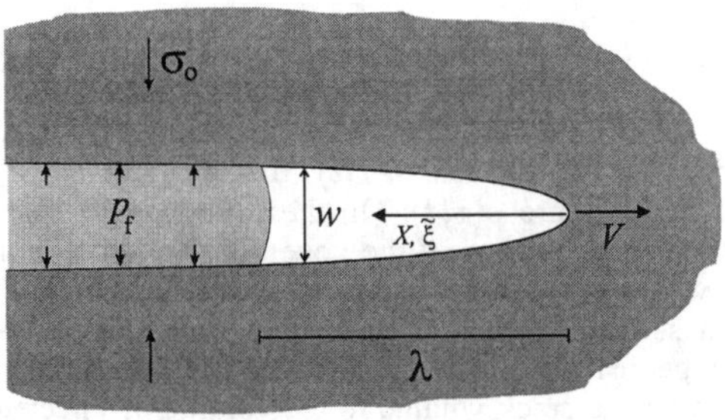

Figure 2: Semi-infinite fluid-driven crack with the lag zone adjacent to the tip.

First we introduce a length scale L_h associated with viscous dissipation and a length scale L_k characterizing the dissipation due to fracturing of the solid, as well as a small parameter ϵ

$$L_h = \frac{\mu' V}{E'} \qquad L_k = \left(\frac{K'}{E'} \right)^2, \qquad \epsilon = \frac{\sigma_o}{E'} \quad (7)$$

It can then be proven that the scaled crack opening $\tilde{\Omega} = \epsilon^2 w / L_h$ and the net pressure $\tilde{\Pi} = (p - \sigma_o)/\sigma_o$ are necessarily of the form $\tilde{\Omega}(\tilde{\xi}; \kappa)$ and $\tilde{\Pi}(\tilde{\xi}; \kappa)$ where $\tilde{\xi} = \epsilon^3 X / L_h$ is the dimensionless moving coordinate and κ a dimensionless toughness defined as

$$\kappa = \left(\frac{\epsilon L_k}{L_h} \right)^{1/2} \quad (8)$$

The near-tip scaled solution depends therefore on *only* one number, the dimensionless toughness κ. This result also implies a universal relationship between the dimensionless lag $\Lambda = \epsilon^3 \lambda / L_h$ and κ. Near the tip, $\tilde{\Omega}$ satisfies the LEFM asymptote (Rice 1968)

$$\tilde{\Omega} \simeq \kappa \tilde{\xi}^{1/2}, \qquad \tilde{\xi} \ll 1 \quad (9)$$

but far away from the tip $\tilde{\Omega}$ behaves according to the zero-toughness singular solution (5)

$$\tilde{\Omega}\left(\tilde{\xi}\right) \simeq 2^{1/3} 3^{5/6} \tilde{\xi}^{2/3}, \qquad \tilde{\xi} \gg 1 \quad (10)$$

Thus the far-field asymptote for the opening $\tilde{\Omega}$ and the pressure $\tilde{\Pi}$ is *independent* of κ. The log-log plot of $\tilde{\Omega}$ along the crack for various values of κ, shown in Fig. 3, provides transparent evidence that the solution behaves in accordance to LEFM in the region immediately adjacent to the tip, and that it converges towards the zero-toughness solution away from the tip.

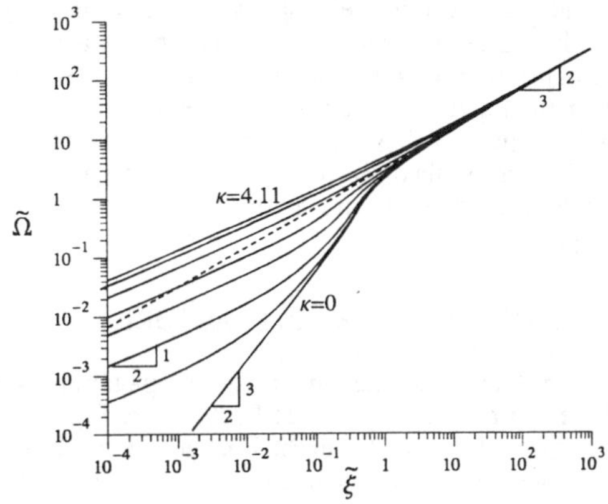

Figure 3: The crack opening $\tilde{\Omega}$ along the crack in log-log scale. The dashed line corresponds to the zero-toughness singular solution (after Garagash and Detournay 2000)

For large enough toughness κ, practically $\kappa \gtrsim 4$, the solution can be further rescaled to become completely independent of κ (Garagash 1998; Garagash and Detournay 2000)

$$\tilde{\Pi} = \kappa^{-2}\,\hat{\Pi}\left(\hat{\xi}\right),\; \tilde{\Omega} = \kappa^{4}\,\hat{\Omega}\left(\hat{\xi}\right),\; \tilde{\xi} = \kappa^{6}\,\hat{\xi} \quad (11)$$

and the asymptotic form of the relation $\Lambda = \Lambda(\kappa)$ is given by

$$\Lambda = \Lambda_{*}\kappa^{6}e^{-\kappa^{2}} \qquad \text{with} \qquad \Lambda_{*} \simeq 4.36\,10^{-3} \quad (12)$$

This rescaled solution behaves according to LEFM in the near-tip region $\hat{\xi} < \hat{\xi}_o$ (with $\hat{\xi}_o \sim 10^{-6}$)

$$\tilde{\Omega}\left(\tilde{\xi}\right) \simeq \tilde{\xi}^{1/2} \quad \text{and} \quad \tilde{\Pi} \simeq \ln\frac{\tilde{\xi}}{\Lambda_{*}} \quad (13)$$

but according the zero-toughness singular solution for $\hat{\xi} > \hat{\xi}_{\infty}$ (with $\hat{\xi}_{\infty} \sim 1$)

$$\tilde{\Omega} \simeq 2^{1/3}3^{5/6}\tilde{\xi}^{2/3} \quad \text{and} \quad \tilde{\Pi} \simeq -\frac{1}{2^{2/3}3^{2/3}\tilde{\xi}^{1/3}} \quad (14)$$

The asymptotes (13) and (14) hold the key for understanding the propagation regime of a fluid-driven fracture. If the characteristic length $\epsilon^{-3}\kappa^{6}L_h$ is small enough compared to the fracture length, then the zero-toughness asymptote (14) develops as an intermediate asymptote. In other words, (14) effectively shields the global solution from the details of the tip solution, including the near-tip behavior (13). Under these conditions, the fracture propagates in the viscosity-dominated regime and solutions can be constructed using the zero-toughness singular solution (5-6) as tip asymptote. Outside the viscosity-dominated regime, the asymptotic behaviour (13) dominates the near tip region at the scale of the overall fracture and $p \sim \ln X$ can be considered as an intermediate asymptote for the pressure.

3 PLANE STRAIN (KGD) FRACTURE

We now turn our attention to the propagation of a plane strain (KGD) hydraulic fracture, see Fig 1. We are especially interested in establishing the conditions under which the solution becomes independent either of K' or of μ'.

3.1 *Governing equations*

The fracture opening $w(x,t)$ and net pressure $p(x,t)$ have to satisfy a non-local elasticity relation (Sneddon and Lowengrub 1969)

$$w = \frac{4\ell}{\pi E'}\int_0^1 p(s\ell,t)G(x/\ell,s)\mathrm{d}s \quad (15)$$

where

$$G(\xi,s) = \ln\left|\frac{\sqrt{1-\xi^2}+\sqrt{1-s^2}}{\sqrt{1-\xi^2}-\sqrt{1-s^2}}\right| \quad (16)$$

and the non-linear Reynolds differential equation from lubrication theory (Batchelor 1967)

$$\frac{\partial w}{\partial t} = \frac{1}{\mu'}\frac{\partial}{\partial x}\left(w^3\frac{\partial p}{\partial x}\right) \quad (17)$$

which is obtained by eliminating the flow rate q between Poiseuille law

$$q = -\frac{w^3}{\mu'}\frac{\partial p}{\partial x} \quad (18)$$

and the continuity equation for an impermeable solid

$$\frac{\partial w}{\partial t} + \frac{\partial q}{\partial x} = 0 \quad (19)$$

The problem is completely formulated with the addition of a propagation criterion and boundary conditions at the fracture inlet $x = 0$ and at the tip $x = 1$. The condition that the fracture is in mobile equilibrium, $K_I = K_{Ic}$, can be expressed as

$$w \simeq \frac{K'}{E'}(\ell - x)^{1/2} \qquad \ell - x \ll \ell \quad (20)$$

Besides the condition $w(\ell,t) = 0$ which is obviously implied by the opening asymptote (20), the boundary conditions consist of

$$q(\ell,t) = 0 \quad \text{and} \quad q(0,t) = Q_o/2 \quad (21)$$

The set of equations (15), (17), (20), and (21) fully define the fracture length $\ell(t)$, the opening $w(x,t)$ and the pressure $p(x,t)$. In the following, we derive particular forms of the governing equations for the viscosity- and the toughness-dominated regime. In these formulations, the inlet flow condition is advantageously replaced by a global continuity equation equating the crack volume to the volume of injected fluid

$$2\int_0^{\ell} w\,\mathrm{d}x = Q_o t \quad (22)$$

3.2 *Scaling*

First, we express the crack opening $w(x,t)$ and the net pressure $p(x,t)$ as

$$w = \varepsilon(t)L(t)\Omega(\xi,t), \qquad p = \varepsilon(t)\,E'\,\Pi(\xi,t) \quad (23)$$

where $\varepsilon(t)$ is a small dimensionless parameter, $\xi = x/\ell(t)$ is the scaled coordinate ($0 \leq \xi \leq 1$), and $L(t)$ a length scale of the same order as the fracture length $\ell(t)$. The two lengths $\ell(t)$ and $L(t)$ are related by

$$\ell(t) = \gamma(t)L(t) \tag{24}$$

The number γ is defined through the mass balance (22)

$$2\gamma \int_0^1 \Omega \mathrm{d}\rho = 1 \tag{25}$$

As a consequence of the scaling (23), the governing equations transform as follows.

- *Elasticity:*

$$\Omega = \frac{4}{\pi}\gamma \int_0^1 G(\rho, s)\,\Pi(s,t)\mathrm{d}s \tag{26}$$

- *Lubrication:*

$$\left(\frac{\dot{\varepsilon}t}{\varepsilon} + \frac{\dot{L}t}{L}\right)\Omega - \frac{\dot{L}t}{L}\xi\frac{\partial\Omega}{\partial\xi} + \left(\dot{\Omega}t - \frac{\dot{\gamma}t}{\gamma}\xi\frac{\partial\Omega}{\partial\xi}\right) = \frac{\varepsilon^3 E' t}{\mu'}\frac{1}{\gamma^2}\frac{\partial}{\partial\xi}\left(\Omega^3\frac{\partial\Pi}{\partial\xi}\right) \tag{27}$$

- *Global fluid volume balance:*

$$\varepsilon L^2 = Q_o t \tag{28}$$

- *Propagation criterion:*

$$\Omega = \frac{K'}{\varepsilon E' L^{1/2}}\gamma^{1/2}(1-\xi)^{1/2}, \quad 1-\xi \ll 1 \tag{29}$$

Consider now the viscosity-scaling, denoted by the subscript μ. The small parameter ε_μ is deduced by imposing that the viscosity does not appear in the lubrication equation (27). The explicit dependence of the lengthscale L_μ on time t then follows from the global mass balance (28). Hence,

$$\varepsilon_\mu = \left(\frac{\mu'}{E't}\right)^{1/3}, \quad L_\mu = \left(\frac{E'Q_o^3 t^4}{\mu'}\right)^{1/6} \tag{30}$$

The dimensionless toughness $\mathcal{K}$ can now be defined as

$$\mathcal{K} = \frac{K'}{\left(E'^3\mu' Q_o\right)^{1/4}} \tag{31}$$

so that the propagation criterion in terms of the opening tip asymptote (29) becomes

$$\Omega_\mu = \mathcal{K}\gamma_\mu^{1/2}(1-\xi)^{1/2}, \quad 1-\xi \ll 1 \tag{32}$$

In the toughness-scaling (denoted by the subscript κ), the small parameter ε_κ is defined by requiring that there are no parameters left in the propagation criterion

$$\Omega_\kappa = \gamma_\kappa^{1/2}(1-\xi)^{1/2}, \quad 1-\xi \ll 1 \tag{33}$$

Taking into account the fluid mass balance (25) yields the explicit power law dependence of both ε_κ and L_κ on time t

$$\varepsilon_\kappa = \left(\frac{K'^4}{E'^4 Q_o t}\right)^{1/3}, \quad L_\kappa = \left(\frac{E'Q_o t}{K'}\right)^{2/3} \tag{34}$$

It is also natural to introduce a dimensionless viscosity $\mathcal{M}$

$$\mathcal{M} = \mu'\frac{E'^3 Q_o}{K'^4} \tag{35}$$

so that $\mathcal{M}$ is the only parameter entering the lubrication equation in the viscosity-scaling.

A power law relation exists, however, between the viscosity $\mathcal{M}$ and the toughness $\mathcal{K}$, and the ratios $\varepsilon_\mu/\varepsilon_\kappa$ and L_μ/L_κ are power laws of $\mathcal{K}$ (or $\mathcal{M}$)

$$\mathcal{M} = \mathcal{K}^{-4}, \quad \frac{\varepsilon_\mu}{\varepsilon_\kappa} = \mathcal{K}^{-4/3}, \quad \frac{L_\mu}{L_\kappa} = \mathcal{K}^{2/3} \tag{36}$$

The transition between the two limiting regimes can thus be understood in terms of one parameter only, which we select to be $\mathcal{K}$. Hence the small viscosity solution can equivalently be referred to as the large toughness solution.

3.3 *Global self-similarity*

In view of the power law dependence of ε and L on t, and since $\mathcal{K}$ (or $\mathcal{M}$) does not depend on time, the set of equations (26-29) accept self-similar solutions in either viscosity- or toughness-scaling. In particular, the lubrication equation (27) can then be rewritten as

$$\bar{\Omega}_\mu - 2\xi\bar{\Omega}'_\mu = 3\left(\bar{\Omega}_\mu^3\bar{\Pi}'_\mu\right)' \tag{37}$$

or

$$\mathcal{M}\left(\bar{\Omega}_\kappa - 2\xi\bar{\Omega}'_\kappa\right) = 3\left(\bar{\Omega}_\kappa^3\bar{\Pi}'_\kappa\right)' \tag{38}$$

where the prime $'$ denotes differentiation with respect to the similarity varable ξ and $\bar{\Omega}_\mu(\xi;\mathcal{K}) = \Omega_\mu/\gamma_\mu$, $\bar{\Pi}_\mu(\xi;\mathcal{K}) = \Pi_\mu$, $\bar{\Omega}_\kappa(\xi;\mathcal{M}) = \Omega_\kappa/\gamma_\kappa$, $\bar{\Pi}_\kappa(\xi;\mathcal{M}) = \Pi_\kappa$. The redefinition of $\bar{\Omega}$ allows to solve first for $\bar{\Omega}$ and $\bar{\Pi}$; then using (25) for γ – which only depends on $\mathcal{K}$ (or $\mathcal{M}$). It is also a simple matter to translate the solution from one scaling to another

$$\frac{\bar{\Omega}_\mu}{\bar{\Omega}_\kappa} = \frac{\bar{\Pi}_\mu}{\bar{\Pi}_\kappa} = \frac{\gamma_\kappa^2}{\gamma_\mu^2} = \mathcal{K}^{4/3} \tag{39}$$

The existence of similarity solutions outside the limit cases $\mathcal{K} = 0$ and $\mathcal{M} = 0$ directly results from the prescription of a *constant* injection rate Q_o. However, the solution is then only self-similar in the global sense, as the tip solution viewed at the spatial resolution of the lag in either viscosity- or toughness-scaling is time-dependent (Garagash 1998; Garagash and Detournay 2001). (The inner and the outer solutions are self-similar in different scalings.) For $\mathcal{K} = 0$, self-similar solutions technically exists for injection rates varying with t according to a power law or an exponential, but they are mere scientific curiosity! For $\mathcal{M} = 0$, there are no restrictions on Q as these zero-viscosity solutions are characterized by a uniform pressure in the fracture (within the assumption of Poiseuille flow).

3.4 *Zero-toughness solution*

An approximate self-similar solution has been derived by Adachi (2001) and Adachi et al. (2001) for the zero-toughness limit case (see also Carbonell (1996) and Carbonell et al. (1999) for an earlier solution to the same problem). Following a methodology originally proposed by Spence and Sharp (1985), this solution $\bar{\mathcal{F}}_{\mu 0}$ is constructed as the sum of a general solution $\bar{\mathcal{F}}^*_{\mu 0} = \{\bar{\Pi}^*_{\mu 0}, \bar{\Omega}^*_{\mu 0}\}$ having the expected singularity (5-6) at the fracture tip and satisfying $\bar{\Pi}^{*\prime}_{\mu 0} = 0$ at the inlet, and a particular solution $\bar{\mathcal{F}}^{**}_{\mu 0} = \{\bar{\Pi}^{**}_{\mu 0}, \bar{\Omega}^{**}_{\mu 0}\}$ characterized by a finite pressure gradient at $\xi = 0$ (to handle the inlet flow condition) and a non-singular behaviour at $\xi = 1$. Hence, $\bar{\mathcal{F}}_{\mu 0}$ is constructed to have not only the correct behavior at the inlet, but also the exact asymptote at the tip

$$\bar{\Omega}_{\mu 0} \simeq 3^{1/2} 2^{2/3} (1-\xi)^{2/3} \tag{40}$$

$$\bar{\Pi}_{\mu 0} \simeq -\frac{1}{2^{1/3} 3 \, (1-\xi)^{1/3}} \qquad \xi \ll 1 \tag{41}$$

In this approach, the elasticity equation is strictly fulfilled by both $\bar{\mathcal{F}}^{**}_{\mu 0}$ and $\bar{\mathcal{F}}^*_{\mu 0}$, with the latter represented in terms of an infinite series of base functions. An n-order approximation $\bar{\mathcal{F}}^{(n)}_{\mu 0}$ is obtained by first truncating the series representation of $\bar{\mathcal{F}}^*_{\mu 0}$, and then minimizing a least-squares functional, constructed using the non-linear lubrication equation, to compute the unknown $n+1$ coefficients of the base functions and the particular solution. The opening $\bar{\Omega}^{(1)}_{\mu 0}$ and pressure $\bar{\Pi}^{(1)}_{\mu 0}$ is illustrated in Fig. 4. The first order solution $\bar{\mathcal{F}}^{(1)}_{\mu 0}$ (see Appendix A) represents already an excellent approximation, as $\bar{\mathcal{F}}^{(1)}_{\mu 0}$ and $\bar{\mathcal{F}}^{(10)}_{\mu 0}$ (not shown here) could hardly be distinguished at the scale of the plot.

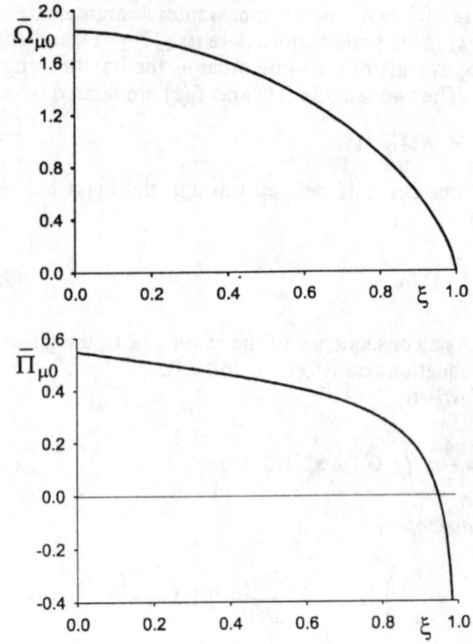

Figure 4: KGD fracture: first order approximation of $\mathcal{K} = 0$ solution (after Adachi 2001).

3.5 *Large toughness solution*

The toughness-scaling is appropriate for cases when the viscosity $\mathcal{M}$ is a small number. The first two terms of the regular expansion of the solution in terms of the small parameter $\mathcal{M}$ have been derived by Garagash (2000), i.e. $\bar{\mathcal{F}}_\kappa \simeq \bar{\mathcal{F}}_{\kappa 0} + \mathcal{M}\bar{\mathcal{F}}_{\kappa 1}$ with $\bar{\mathcal{F}}_{\kappa 0}$ corresponding to the classical solution for an inviscid fluid Huang et al. (1990). The first order similarity solution for opening and pressure is shown in Fig. 5 (see Appendix A for a listing of the solution). As can be seen from Fig. 5, a small viscosity $\mathcal{M} = 10^{-1}$ causes a significant shift of the solution compared to the zero viscosity case.

3.6 *Transition*

Solutions for the intermediate regime of propagation can be calculated using a scheme, originally proposed by Spence and Sharp (1985) and recently improved by Adachi (2001), which relies on expanding the opening Ω in terms of Chebyshev polynomials. The dependence of the propagation regime on the dimensionless toughness $\mathcal{K}$ is best illustrated by Fig. 6, showing how the shape factor γ_μ (in the viscosity-scaling) varies with $\mathcal{K}$. The transition regime corre-

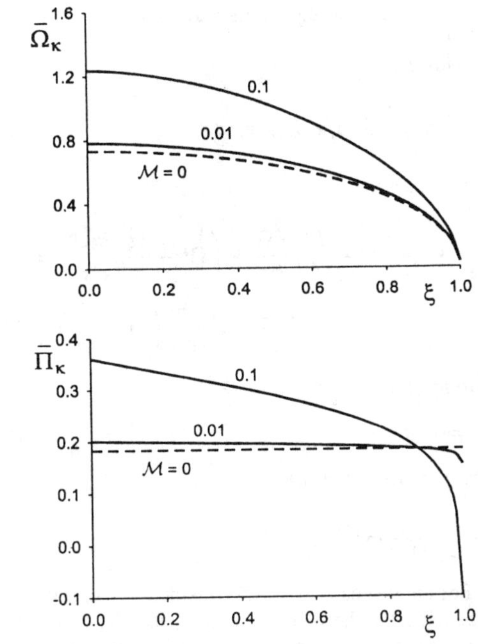

Figure 5: KGD fracture: first order approximation of large $\mathcal{K}$ solution (after Garagash 2000).

sponds to a remarquably small interval $\mathcal{K}_\mu < \mathcal{K} < \mathcal{K}_\kappa$, where the bounds are approximatively given by $\mathcal{K}_\mu \sim 1$ and $\mathcal{K}_\kappa \sim 4$. Thus $\mathcal{F}$ can be approximated by the zero-toughness solution $\mathcal{F}_{\mu 0}$ in the viscosity-dominated regime $\mathcal{K} < \mathcal{K}_\mu$, and by the zero-viscosity solution $\mathcal{F}_{\kappa 0}$ in the toughness-dominated regime $\mathcal{K} > \mathcal{K}_\kappa$.

It is not possible to construct a small tougness solution as a regular expansion in terms of the small parameter $\mathcal{K}$; rather a matched asymptotics approach has to be used (Garagash and Detournay 2001). The root of the difficulty in constructing a complete solution for small $\mathcal{K}$ lies in the different nature of the tip asymptote for $\mathcal{K} = 0$ and for $\mathcal{K} \ll 1$ and the need to explicitly account for the lag. Analysis of the solution in the viscosity-dominated regime shows that the zero-toughness singular solution develops as an intermediate asymptote, which effectively shields the global solution from the tip processes (Detournay 1999). The square root behavior of the opening at the tip is actually restricted to a boundary layer of thickness proportional to $\mathcal{K}^6$ (Garagash 1998).

It is useful to express the relationship between $\mathcal{K}$ and the ratio θ of the energy rate D_s expended in creating new fracture surface over the viscous dissipation D_f, so that the range of each regime of propagation

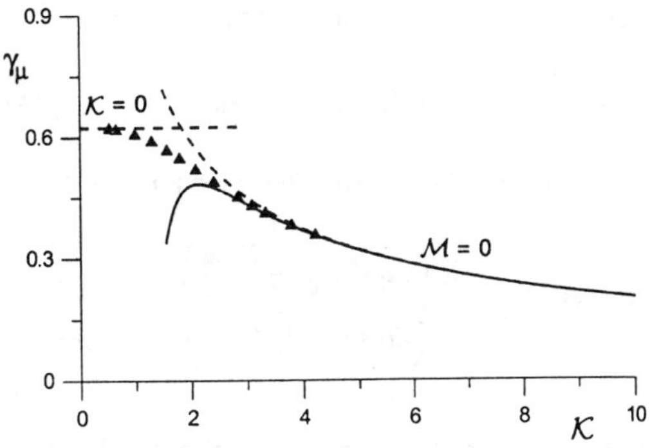

Figure 6: KGD fracture: dependence of γ_μ on $\mathcal{K}$. The dashed horizontal line correspond to $\gamma_{\mu 0}$, the dashed curve to $\gamma_{\mu\infty} = \gamma_{\kappa 0}\mathcal{K}^{-2/3}$, and the continuous curve to the first order small viscosity solution $(\gamma_{\kappa 0} + \mathcal{K}^{-4}\gamma_{\kappa 1})\mathcal{K}^{-2/3}$ (after Garagash 2000).

can be expressed in terms of the more physically intuitive dissipation ratio θ. For example, it can be shown that

$$D_s = \Gamma_\mu \mathcal{D}_s \quad \text{with} \quad \mathcal{D}_s = \frac{\pi}{24}\gamma_\mu \mathcal{K}^2 \tag{42}$$

$$D_f = \Gamma_\mu \mathcal{D}_f \quad \text{with} \quad \mathcal{D}_f = 2\gamma_\mu^2 \int_0^1 \bar{\Omega}_\mu^3 \bar{\Pi}_\mu'^2 d\xi \tag{43}$$

where $\Gamma_\mu = \varepsilon_\mu E' Q_o$ is the characteristic dissipation (function of time) in viscosity-scaling. Hence,

$$\vartheta = \delta_\mu(\mathcal{K})\mathcal{K}^2, \quad \delta_\mu^{-1} = \frac{48}{\pi}\gamma_\mu \int_0^1 \bar{\Omega}_\mu^3 \bar{\Pi}_\mu'^2 d\xi \tag{44}$$

In the viscosity-dominated regime, $\delta_\mu(\mathcal{K}) \simeq \delta_{\mu 0} \simeq 0.197$; hence, this regime of propagation corresponds approximatively to a dissipation ratio $\vartheta \lesssim 0.1$. In the toughness-dominated regime, $\vartheta \sim \mathcal{M}^{-1}$.

4 PENNY-SHAPED FRACTURE

Next we consider the propagation of an axisymmetric hydraulic fracture in an elastic rock, driven by fluid injected at a constant volumetric rate Q_o at the center of the fracture (see Fig. 1, where x should now be interpreted as the radial distance r from the center of the fracture, and $\ell(t)$ as the fracture radius $R(t)$).

4.1 *Governing equations*

The set of equations governing the fracture opening $w(r,t)$ and net pressure $p(r,t)$ consists of a non-local elasticity relation (Sneddon and Lowengrub 1969)

$$w = \frac{8R}{\pi E'} \int_0^1 G(r/R, s)\, p(sR, t) s ds \tag{45}$$

where the kernel G is given in closed form by

$$G(\rho, s) = \begin{cases} \frac{1}{\rho} F\left(\arcsin\sqrt{\frac{1-\rho^2}{1-s^2}}, \frac{s^2}{\rho^2}\right), & \text{if } \rho > s \\ \frac{1}{s} F\left(\arcsin\sqrt{\frac{1-s^2}{1-\rho^2}}, \frac{\rho^2}{s^2}\right), & \text{if } \rho < s \end{cases} \tag{46}$$

with F denoting the elliptic integral of the first kind, and a non-linear differential equation from lubrication theory

$$\frac{\partial w}{\partial t} = \frac{1}{\mu' r} \frac{\partial}{\partial r}\left(r w^3 \frac{\partial p}{\partial r}\right) \tag{47}$$

The fracture tip conditions, i.e., the asymptotic behaviour of the crack opening at the tip $r = R$ with $K_I = K_{Ic}$ as well as the no-flow condition, are identical to those of a plane strain fracture. Since $\partial w/\partial t$ must be finite in the vicinity of the source, the fluid injection condition at the center of the fracture can be expressed as

$$2\pi \lim_{r \to 0} rq = Q_o \tag{48}$$

It follows from the above equation, that $q \sim O(1/r)$ near the source and thus that $p \sim -\ln r$. The source can alternatively be taken into account by the global continuity equation

$$2\pi \int_0^R r\, w dr = Q_o t \tag{49}$$

4.2 *Scaling*

Again we express the crack opening $w(r, t)$ and the net pressure $p(r, t)$ as

$$w = \varepsilon(t) L(t) \Omega(\rho, t) \qquad p = \varepsilon(t) E' \Pi(\rho, t) \tag{50}$$

where $\varepsilon(t)$ is a small dimensionless parameter, $\rho = r/R(t)$ is the scaled radius ($0 \leq \rho \leq 1$), and $L(t)$ a length scale of the same order as the fracture radius $R(t)$. The two lengths $R(t)$ and $L(t)$ are related by

$$R(t) = \gamma(t) L(t) \tag{51}$$

where the number γ is now defined by

$$2\pi\gamma^2 \int_0^1 \rho \Omega d\rho = 1 \tag{52}$$

The governing equations can then be rewritten as follows.

- *Elasticity:*

$$\Omega = \frac{8}{\pi} \gamma \int_0^1 G(\rho, s) \Pi(s, t) s ds \tag{53}$$

- *Lubrication:*

$$\left(\frac{\dot{\varepsilon} t}{\varepsilon} + \frac{\dot{L} t}{L}\right)\Omega - \frac{\dot{L} t}{L} \rho \frac{\partial \Omega}{\partial \rho} + \left(\dot{\Omega} t - \frac{\dot{\gamma} t}{\gamma} \rho \frac{\partial \Omega}{\partial \rho}\right) = \frac{\varepsilon^3 E' t}{\mu'} \frac{1}{\gamma^2 \rho} \frac{\partial}{\partial \rho}\left(\rho \Omega^3 \frac{\partial \Pi}{\partial \rho}\right) \tag{54}$$

- *Global fluid volume balance:*

$$\varepsilon L^3 = Q_o t \tag{55}$$

- *Propagation criterion:*

$$\Omega = \frac{K'}{\varepsilon E' L^{1/2}} \gamma^{1/2} (1 - \rho)^{1/2}, \quad 1 - \rho \ll 1 \tag{56}$$

For the viscosity scaling, considerations similar to those used for the KGD fracture yield the following expressions for the small parameter ε_μ and the length scale L_μ

$$\varepsilon_\mu = \left(\frac{\mu'}{E' t}\right)^{1/3}, \quad L_\mu = \left(\frac{E' Q_o^3 t^4}{\mu'}\right)^{1/9} \tag{57}$$

The dimensionless toughness $\mathcal{K}(t)$ can now be defined as

$$\mathcal{K} = K' \left(\frac{t^2}{\mu'^5 Q_o^3 E'^{13}}\right)^{1/18} \tag{58}$$

so that the propagation criterion in terms of the opening tip asymptote (20) becomes

$$\Omega_\mu = \mathcal{K} \gamma_\mu^{1/2} (1 - \rho)^{1/2}, \quad 1 - \rho \ll 1 \tag{59}$$

In contrast to the KGD model, $\mathcal{K}$ is here a function of time t. The solution is thus not globally self-similar anymore with the implication that transition from the viscosity- to the toughness-dominated regime could in principle take place during fracture propagation.

For the toughness scaling, ε_κ and L_κ are explicitly given by

$$\varepsilon_\kappa = \left(\frac{K'^6}{E'^6 Q_o t}\right)^{1/5}, L_\kappa = \left(\frac{Q_o^2 E'^2 t^2}{K'^2}\right)^{1/5} \tag{60}$$

so that there are no parameters left in the propagation criterion

$$\Omega_\kappa = \gamma_\kappa^{1/2} (1 - \rho)^{1/2}, \quad 1 - \rho \ll 1 \tag{61}$$

It is also natural to introduce a dimensionless viscosity $\mathcal{M}(t)$

$$\mathcal{M} = \mu' \left(\frac{Q_o^3 E'^{13}}{K'^{18} t^2} \right)^{1/5} \tag{62}$$

so that the lubrication equation (54) can now be rewritten as

$$\mathcal{M}\left(\frac{1}{5}\Omega_\kappa - \frac{2}{5}\rho\frac{\partial\Omega_\kappa}{\partial\rho}\right) + \mathcal{M}\left(\dot{\Omega}_\kappa t - \frac{\dot{\gamma}t}{\gamma}\rho\frac{\partial\Omega_\kappa}{\partial\rho}\right) = \frac{1}{\gamma^2\rho}\frac{\partial}{\partial\rho}\left(\rho\Omega_\kappa^3\frac{\partial\Pi_\kappa}{\partial\rho}\right) \tag{63}$$

A power law relation exists between the viscosity $\mathcal{M}$ and the toughness $\mathcal{K}$, and the ratio of ε as well as the ratio of L in the two scalings are power laws of $\mathcal{K}$ (or $\mathcal{M}$)

$$\mathcal{M} = \mathcal{K}^{-18/5}, \quad \frac{\varepsilon_\mu}{\varepsilon_\kappa} = \mathcal{K}^{-6/5}, \quad \frac{L_\mu}{L_\kappa} = \mathcal{K}^{2/5} \tag{64}$$

The transition between the two limiting regimes can thus again be understood in terms of $\mathcal{K}$ only.

4.3 *Zero-toughness solution*

It can readily be shown that the solution for $\mathcal{K} = 0$ is in fact self-similar as none of the governing equations depends on time; in particular, the lubrication equation becomes in this case

$$\rho\bar{\Omega}_{\mu0} - 4\rho^2\bar{\Omega}'_{\mu0} = 9\left(\rho\bar{\Omega}^3_{\mu0}\bar{\Pi}'_{\mu0}\right)' \tag{65}$$

where the prime $'$ denotes here differentiation with respect to the similarity varable ρ and $\bar{\Omega}_{\mu0}(\rho) = \Omega_\mu/\gamma_\mu$, $\bar{\Pi}_{\mu0}(\rho) = \Pi_\mu$, with Ω_μ, Π_μ, γ_μ evaluated at $\mathcal{K} = 0$.

The self-similar zero toughness solution has been constructed by Savitski (2000), see also Savitski and Detournay (2000), following broadly the approach used for the KGD fracture, which takes advantage of the linearity of the elasticity equation. Numerical results show that the first order solution $\bar{\mathcal{F}}^{(1)}_{\mu0}$, shown in Fig. 7, is already a very good approximation. (At the scale of this plot, $\bar{\mathcal{F}}^{(1)}_{\mu0}$ could hardly be distinguished from $\bar{\mathcal{F}}^{(4)}_{\mu0}$.) The first order solution is listed in Appendix B.

4.4 *Large toughness solution*

The first two terms of the regular expansion of the solution in terms of the small parameter $\mathcal{M}$ have been derived by Savitski (2000), using

$$\gamma_\kappa^2\mathcal{M}\left(\int_\rho^1 \xi\Omega_\kappa\, d\xi - \frac{2}{5}\rho\Omega_\kappa\right) = -\rho\Omega_\kappa^3\Pi'_\kappa \tag{66}$$

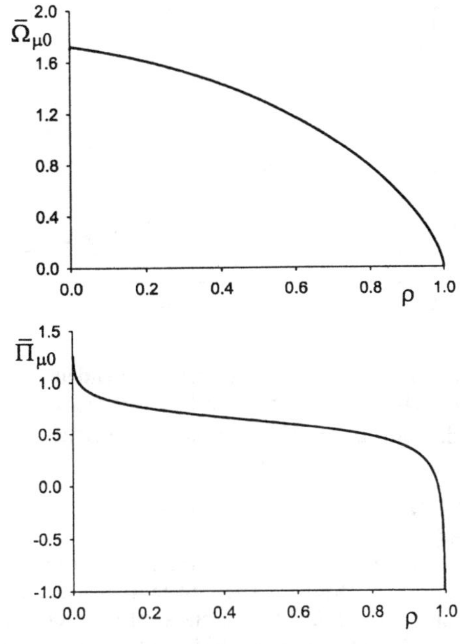

Figure 7: Penny-shaped fracture: first order approximation of $\mathcal{K} = 0$ solution (after Savitski 2000).

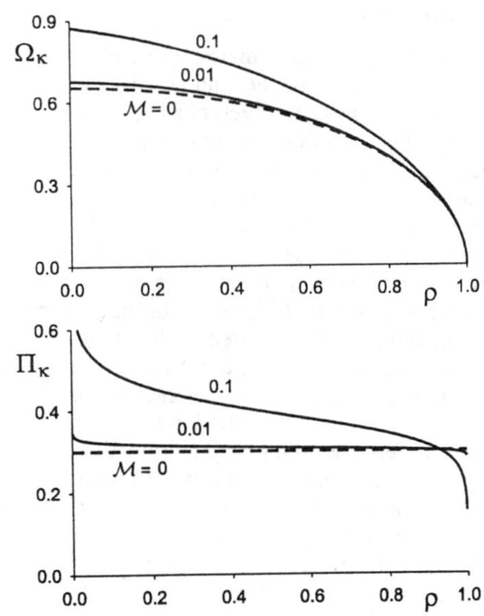

Figure 8: Penny-shaped fracture: first order approximation of large $\mathcal{K}$ solution (after Savitski 2000).

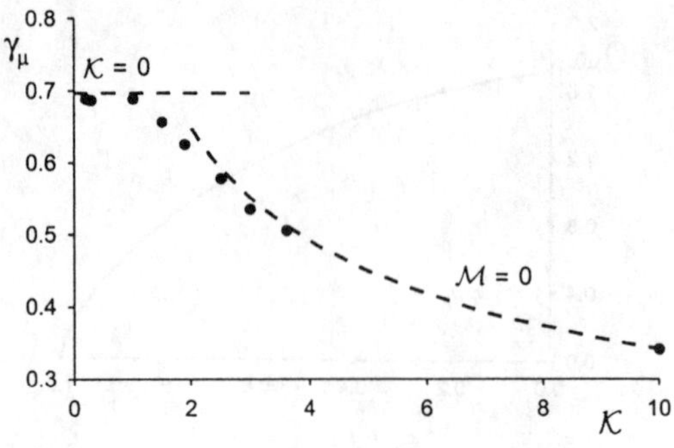

Figure 9: Penny-shaped fracture: dependence of γ_μ on $\mathcal{K}$. The dashed horizontal line correspond to $\gamma_{\mu 0}$, the dashed curve to $\gamma_{\mu\infty} = \gamma_{\kappa 0}\mathcal{K}^{-2/5}$ (Savitski 2000).

which is a simplified lubrication equation, correct to $O(\mathcal{M}^2)$.

Figure 8 shows a plot of the first order solution $\mathcal{F}_\kappa \simeq \mathcal{F}_{\kappa 0} + \mathcal{M}\mathcal{F}_{\kappa 1}$. The first term $\mathcal{F}_{\kappa 0}$ of the expansion corresponds to the classical solution for an inviscid fluid (Abe et al. 1976; Advani et al. 1987). However, as can be seen from Fig. 8, the limit solution $\mathcal{M} = 0$ is not uniformly valid, as even a small viscosity causes the development of a logarithmic singularity of the pressure near the tip and the inlet.

4.5 *Transition*

Solutions for the intermediate regime of propagation were calculated using the numerical code Loramec (refer to Desroches and Thiercelin (1993) or Carbonell et al. (1999) for a description of the code). The dependence of the propagation regime on the dimensionless toughness $\mathcal{K}$ is illustrated in Fig. 9 showing the variation of the shape factor γ_μ with $\mathcal{K}$. The viscosity-dominated regime is bounded by $\mathcal{K} < \mathcal{K}_\mu$, while the toughness-dominated regime starts at $\mathcal{K} > \mathcal{K}_\kappa$, with $\mathcal{K}_\mu \sim 1$ and $\mathcal{K}_\kappa \sim 4$. In other words, the solution can be approximated by the zero-toughness solution $\mathcal{F}_{\mu 0}$ for $\mathcal{K} < \mathcal{K}_\mu$, and by the zero-viscosity solution $\mathcal{F}_{\kappa 0}$ for $\mathcal{K} > \mathcal{K}_\kappa$. As the radial fracture propagates, the curve $\gamma_\mu(\mathcal{K})$ plotted in Figure 9 is in principle travelled from left to right with increasing time.

It should be reminded that these solutions are not applicable at the "early time" of fracture initiation. It is interesting to speculate that a radial fracture actually starts its existence in the toughness-dominated regime, matures in the viscosity-dominated regime, and ages back in the toughness-dominated regime. The most interesting episode of its existence is undoubtedly the one dominated by viscosity!

5 CONCLUSIONS

In this paper, we have reviewed some recent advances in the understanding of the parameters controlling the regimes of propagation of a fluid-driven fracture in an impermeable elastic medium. For both KGD and penny-shaped fractures, the regimes of propagation only depends on a dimensionless toughness $\mathcal{K}$. When $\mathcal{K} \lesssim 1$, the regime is said to be viscosity- dominated as most of the energy dissipation is taking place in the fluid. In this regime, the solution can be approximated by a limit solution constructed on the assumption that the rock has zero toughness. When $\mathcal{K} \gtrsim 4$, the regime is said to be toughness-dominated as most of the energy dissipation is now associated with the creation of new fracture surfaces in the rock. In this regime, the solution can be approximated by a limit solution constructed by assuming that the fluid has zero viscosity.

There is an important difference between the two- and the three-dimensional fracture. For a KGD fracture, $\mathcal{K}$ depends only on the injection rate Q_o, the viscosity μ', the elastic modulus E', and the toughness K', while for a penny-shaped fracture $\mathcal{K}$ depends also on time, although weakly since $\mathcal{K} \sim t^{1/9}$. Thus, a KGD fracture always propagates in the same regime under constant injection rate, while a penny-shaped fracture evolves in principle from a viscosity- to a toughness-dominated regime.

These conclusions have been reached, however, by assuming a perfectly impermeable rock. The next important step will be to extend these results to account for the exchange of fluid between the fracture and the host permeable rock. Such a generalization, although challenging, will provide powerful insights for the design of hydraulic fracturing treatments.

Acknowledgements

This review is dedicated to Anthony Pearson who introduced me to the concept of scaling and with whom I have enjoyed, over the years, numerous stimulating interactions on hydraulic fracturing and many other topics. More importantly, however, I am indebted to Anthony Pearson for his mentoring role, in my early days as a scientist at Schlumberger. His keen interest in the development of junior scientists, his intellectual rigor, his enthusiasm for new results, but also his tempestuous contempt for research not up to his high standard will always be fondly remembered. This paper is essentially based on the outstanding contributions of past and present graduate students: José Adachi, Roberto Carbonell, Dmitry Garagash and Alexei Savitski. This research has been partially funded by grants from the UMN and Schlumberger-Dowell.

REFERENCES

Abe, H., T. Mura, and L. Keer (1976). Growth rate of a penny-shaped crack in hydraulic fracturing of rocks. *J. Geophys. Res. 81*, 5335–5340.

Adachi, J. (2001). *Fluid-Driven Fracture in Impermeable and Permeable Rock.* Ph. D. thesis, University of Minnesota. to be submitted.

Adachi, J., E. Detournay, and R. Carbonell (2001). Plane strain propagation of a hydraulic fracture II: Zero-toughness self-similar solution. *Proc. Roy. Soc. London, Ser. A.* to be submitted.

Advani, S., J. Torok, J. Lee, and S. Choudhry (1987). Explicit time-dependent solutions and numerical evaluations for penny-shaped hydraulic fracture models. *J. Geophys. Res. 92*(B8), 8049–8055.

Barenblatt, G. (1962). The mathematical theory of equilibrium cracks in brittle fracture. *Advances in Applied Mechanics VII*, 55–129.

Batchelor, G. K. (1967). *An Introduction to Fluid Mechanics*. Cambridge University Press.

Carbonell, R., J. Desroches, and E. Detournay (1999). A comparison between a semi-analytical and a numerical solution of a two-dimensional hydraulic fracture. *Int. J. Solids Structures 36*(31-32), 4869–4888.

Carbonell, R. S. (1996, October). *Self-similar solution of a fluid-driven fracture*. Ph. D. thesis, University of Minnesota.

Desroches, J., E. Detournay, B. Lenoach, P. Papanastasiou, J. R. A. Pearson, M. Thiercelin, and A. H.-D. Cheng (1994). The crack tip region in hydraulic fracturing. *Proc. Roy. Soc. London, Ser. A A447*, 39–48.

Desroches, J. and M. Thiercelin (1993). Modeling propagation and closure of micro-hydraulic fractures. *Int. J. Rock Mech. Min. Sci. & Geomech. Abstr. 30*, 1231–1234.

Detournay, E. (1999). Fluid and solid singularities at the tip of a fluid-driven fracture. In D. Durban and J. Pearson (Eds.), *Non-Linear Singularities in Deformation and Flow*, Dordrecht, pp. 27–42. Kluwer Academic Publishers.

Garagash, D. (1998). *Near Tip Processes of Fluid-Driven Fractures*. Ph. D. thesis, University of Minnesota.

Garagash, D. (2000). Hydraulic fracture propagation in elastic rock with large toughness. In *Pacific Rocks 2000, Proc. Of the 4th North American Rock Mechanics Symposium*, Rotterdam, pp. 221–228. Balkema.

Garagash, D. and E. Detournay (2000). The tip region of a fluid-driven fracture in an elastic medium. *ASME J. Appl. Mech. 67*, 183–192.

Garagash, D. and E. Detournay (2001). Plane strain propagation of a hydraulic fracture III: Viscosity-dominated regime. *Proc. Roy. Soc. London, Ser. A*. to be submitted.

Geertsma, J. and F. D. Klerk (1969). A rapid method of predicting width and extent of hydraulic induced fractures. *JPT 246*, 1571–1581.

Huang, N., A. Szewczyk, and Y. Li (1990). Self-similar solution in problems of hydraulic fracturing. *ASME J. Appl. Mech. 57*, 877–881.

Khristianovic, S. and Y. Zheltov (1955). Formation of vertical fractures by means of highly viscous fluids. *Proc. 4th World Petroleum Congress, Vol. II, Rome*, 579–586.

Lister, J. R. (1990). Buoyancy-driven fluid fracture: The effects of material toughness and of low-viscosity precursors. *J. Fluid Mech. 210*, 263–280.

Nordgren, R. P. (1972). Propagation of vertical hydraulic fracture. *Soc. Pet. Eng. J., Trans. AIME 253*, 306–314.

Perkins, T. K. and L. R. Kern (1961). Widths of hydraulic fractures. *J. Pet. Tech., Trans. AIME 222*, 937–949.

Rice, J. (1968). Mathematical analysis in the mechanics of fracture. In *Fracture, an Advanced Treatise*, Volume II, Chapter 3, pp. 191–311. Academic Press.

Savitski, A. (2000). *Propagation of a Penny-Shaped Hydraulic Fracture in an Impermeable Rock.* Ph. D. thesis, University of Minnesota.

Savitski, A. and E. Detournay (2000). Similarity solution of a penny-shaped fluid-driven fracture in a zero-toughness linear elastic solid. *C. R. Acad. Sc. Paris*. submitted.

Sneddon, I. and M. Lowengrub (1969). *Crack Problems in the Classical Theory of Elasticity.* New York: Wiley.

Spence, D. A. and P. Sharp (1985). Self-Similar solution for elastohydrodynamic cavity flow. *Proc. Roy. Soc. London, Ser. A A400*, 289–313.

Spence, D. A. and D. L. Turcotte (1985). Magma-driven propagation crack. *J. Geophys. Res. 90*, 575–580.

A KGD FRACTURE

A.1 *Zero-toughness solution*

The first order approximation $\bar{\mathcal{F}}_{\mu 0}^{(1)}$ of the zero-toughness solution is (Adachi 2001; Adachi et al. 2001)

$$\bar{\Omega}_{\mu 0}^{(1)} = A_0(1-\xi^2)^{2/3} + A_1^{(1)}(1-\xi^2)^{5/3} + B^{(1)}\left[4\sqrt{1-\xi^2} + 2\xi^2 \ln\left|\frac{1-\sqrt{1-\xi^2}}{1+\sqrt{1-\xi^2}}\right|\right] \quad (67)$$

$$\bar{\Pi}_{\mu 0}^{(1)} = \frac{1}{3\pi} \mathrm{B}\left(\tfrac{1}{2},\tfrac{2}{3}\right)\left[A_0\, {}_2\mathrm{F}_1\left(-\tfrac{1}{6},1;\tfrac{1}{2};\xi^2\right) + \tfrac{10}{7}A_1^{(1)}\, {}_2\mathrm{F}_1\left(-\tfrac{7}{6},1;\tfrac{1}{2};\xi^2\right)\right] + B^{(1)}\left(2-\pi|\xi|\right) \quad (68)$$

where $A_0 = 3^{1/2}$, $A_1^{(1)} \simeq -0.156$, $B^{(1)} \simeq 6.63\cdot 10^{-2}$; B is Euler Beta function, and ${}_2\mathrm{F}_1$ the hypergeometric function. Thus, $\bar{\Omega}_{\mu 0}^{(1)}(0) \simeq 1.84$ and $\gamma_{\mu 0}^{(1)} \simeq 0.616$.

A.2 *Large toughness solution*

The zero-viscosity solution $\bar{\mathcal{F}}_{\kappa 0}$ is (Huang et al. 1990)

$$\bar{\Omega}_{\kappa 0} = \frac{\pi^{1/3}}{2}\left(1-\xi^2\right)^{1/2} \quad (69)$$

$$\bar{\Pi}_{\kappa 0} = \frac{\pi^{1/3}}{8}, \quad \gamma_{\kappa 0} = \frac{2}{\pi^{2/3}} \quad (70)$$

The next term of the regular expansion of the solution $\bar{\mathcal{F}}_{\kappa} \simeq \bar{\mathcal{F}}_{\kappa 0} + \mathcal{M}\bar{\mathcal{F}}_{\kappa 1}$ in $\mathcal{M}$ is (Garagash 2000)

$$\bar{\Pi}_{\kappa 1} = \bar{\Pi}_{\kappa 1}^{*} + \Delta\bar{\Pi}_{\kappa 1} \quad \text{with} \quad \bar{\Pi}_{\kappa 1}^{*} \simeq 1.77 \quad (71)$$

$$\Delta\bar{\Pi}_{\kappa 1} = \frac{4}{3\pi^{2/3}}\ln\left(1-\xi^2\right) - \frac{2\xi \arccos\xi}{\pi^{2/3}\left(1-\xi^2\right)^{1/2}} \quad (72)$$

$$\bar{\Omega}_{\kappa 1} = 4\bar{\Pi}_{\kappa 1}^{*}\left(1-\xi^2\right)^{1/2} + \frac{4}{\pi}\int_0^1 G(\xi,\eta)\,\Delta\bar{\Pi}_{\kappa 1}(\eta)\,d\eta \quad (73)$$

where the kernel $G(\xi,\eta)$ is given by (16). Also, $\gamma_{\kappa} \simeq \gamma_{\kappa 0} + \mathcal{M}\gamma_{\kappa 1}$, with $\gamma_{\kappa 1} \simeq -2.72$.

B PENNY-SHAPED FRACTURE

B.1 *Zero-toughness solution*

The first order approximation $\bar{\mathcal{F}}_{\mu 0}^{(1)}$ of the zero toughness solution (Savitski 2000; Savitski and Detournay 2000)

$$\bar{\Pi}_{\mu 0}^{(1)} = A_0^{(1)} - \frac{A_1^{(1)}}{(1-\rho)^{1/3}} - B^{(1)}\ln\frac{\rho}{2} \quad (74)$$

$$\bar{\Omega}_{\mu 0}^{(1)} = \left[C_1^{(1)} - C_2^{(1)}(6-13\rho)\right](1-\rho)^{2/3} + \frac{8}{\pi}B^{(1)}\left[\left(1-\rho^2\right)^{1/2} + \rho\arctan\frac{\rho}{\left(1-\rho^2\right)^{1/2}} - \frac{\pi}{2}\rho\right] \quad (75)$$

$A_0^{(1)} \simeq 0.795$, $A_1^{(1)} \simeq 0.239$, $B^{(1)} \simeq 0.0927$, $C_1^{(1)} \simeq 1.910$, $C_2^{(1)} \simeq 0.071$. Also $\bar{\Omega}_{\mu 0}^{(1)}(0) \simeq 1.72$ and $\gamma_{\mu 0}^{(1)} = 0.695$.

B.2 *Large toughness solution*

The zero-viscosity solution $\mathcal{F}_{\kappa 0}$ is (Abe et al. 1976; Advani et al. 1987)

$$\Omega_{\kappa 0} = \left(\frac{3}{8\pi}\right)^{1/5}\left(1-\rho^2\right)^{1/2} \quad (76)$$

$$\Pi_{\kappa 0} = \frac{\pi}{8}\left(\frac{\pi}{12}\right)^{1/5}, \quad \gamma_{\kappa 0} = \left(\frac{3}{\pi 2^{1/2}}\right)^{2/5} \quad (77)$$

The next term of the regular expansion of the solution $\mathcal{F}_{\kappa} \simeq \mathcal{F}_{\kappa 0} + \mathcal{M}\mathcal{F}_{\kappa 1}$ in $\mathcal{M}$ is (Savitski 2000)

$$\Pi_{\kappa 1} = A_1 - A_2\left[\frac{1}{3}\ln\rho - \frac{1}{5}\ln\left(1-\rho^2\right)\right] \quad (78)$$

$$\Omega_{\kappa 1} = A_3\left(1-\rho^2\right)^{1/2} + A_4\left\{\left[\frac{2(1-\rho)(1-\rho^2)}{\rho}\right]^{1/2}\arctan\frac{\rho}{\left(1-\rho^2\right)^{1/2}} - \frac{5\rho}{6}\arctan\frac{\left(1-\rho^2\right)^{1/2}}{\rho} + \int_\rho^1 \frac{F(\xi,\rho)}{(\xi-\rho)^{1/2}}d\xi\right\} \quad (79)$$

where

$$F(\xi,\rho) = \left(\frac{1-\xi^2}{\xi+\rho}\right)^{1/2}\arctan\frac{\xi}{\left(1-\xi^2\right)^{1/2}} - \left(\frac{1-\rho^2}{2\rho}\right)^{1/2}\arctan\frac{\rho}{\left(1-\rho^2\right)^{1/2}} \quad (80)$$

with $A_1 \simeq 0.638$, $A_2 \simeq 1.709$, $A_3 \simeq 1.091$, $A_4 \simeq 1.240$. Also where the kernel $G(\xi,\eta)$ is given by (16). Also, $\gamma_{\kappa} \simeq \gamma_{\kappa 0} + \mathcal{M}\gamma_{\kappa 1}$, with $\gamma_{\kappa 1} = -0.7349$. Note that $\mathcal{F}_{\kappa 0}$ is self-similar (and can thus be expressed as $\bar{\mathcal{F}}_{\kappa 0}$), but not $\mathcal{F}_{\kappa}$.

Computer Methods and Advances in Geomechanics, Desai et al. (eds) © 2001 Balkema, Rotterdam, ISBN 90 5809 183 X

Localisation of deformation in hollow cylinders based on gradient elastoplasticity

A. Zervos
National Technical University of Athens, Greece & Schlumberger Cambridge Research, UK

P. Papanastasiou
Schlumberger Cambridge Research, UK

I. Vardoulakis
National Technical University of Athens, Greece

ABSTRACT: We present a gradient elastoplastic model for pressure–sensitive, dilatant, strain–softening materials. The theory assumes that the stress increment depends on both the strain increment and its Laplacian. The boundary value problem is solved numerically with the finite element method, using a displacement formulation with C^1–continuity elements. The gradient terms in the constitutive equations introduce an extra material parameter with dimensions of length (*internal length*), regularising the ill–posedness caused by the strain–softening behaviour. The dimensions of localisation zones scale with this length and are no longer indeterminate, allowing robust localisation computations at the post–peak regime. Mesh insensitivity is demonstrated by modelling hollow cylinder tests under external pressure. The displacement field, initially axisymmetric, spontaneously loses symmetry after a loading level is reached. The deformation localises progressively in narrow zones, resembling the failure patterns observed in hollow cylinder experiments.

1 INTRODUCTION

Failure of geomaterials is often characterised by localisation of deformation in shear bands or shear interface layers, which are narrow zones of intense shearing. The material inside these bands deforms and degrades significantly, while the rest of the material remains rather inert. Experimental evidence suggests that the localisation bands have a finite thickness (Mühlhaus & Vardoulakis 1987).

Modelling of progressive localisation of deformation requires the incorporation of material softening to the constitutive equations (strength loss with straining). However, this leads to ill–posedness of the governing equations, since the thickness of the localisation bands remains indeterminate (Vardoulakis & Sulem 1995). Numerical computations in the post–peak part of the stress–strain curve are then pathologically mesh dependent.

The governing equations can be regularised using a higher order continuum theory, which takes into account the material microstructure. Such theories introduce an extra *material length* parameter, typically related to the mean grain size, the micro-crack length or some other characteristic length scale of the material. The dimensions of the localisation zones scale with this parameter and are no longer indeterminate, allowing for robust post–localisation computations to be performed. Higher order theories, presented in the literature, are the Cosserat continuum theory, the Mindlin theory, the non–local continuum, gradient plasticity and its extension gradient elastoplasticity (Zervos et al. 2000a).

The importance of modelling localisation of deformation becomes apparent in the case of the hollow cylinder test. This is a standard laboratory test used in petroleum, geotechnical and mining engineering for assessing the stability of underground openings such as cavities and wellbores. Assessing the ultimate stability of a hollow cylinder requires the reproduction by a model of the experimentally observed localised and failure modes.

Other related work are the bifurcation studies of

the borehole and hollow cylinder test based on classical plasticity (Vardoulakis & Papanastasiou 1988; Papanastasiou & Durban 1999) and higher order theories (Papanastasiou & Vardoulakis 1989; Vardoulakis et al. 1994). The bifurcation solution consisted in determining the lowest load for which an instability could arise. Results predicted warping of the hole surface with infinite (classical theory) or finite (higher order theories) wavenumber. This approach provides only the loading level at which localisation may initiate, without predicting how the localised pattern will evolve. Papanastasiou & Vardoulakis (1992) present a closer related study of progressive localisation, utilising post–bifurcation analysis with a Cosserat finite element model.

In this paper we will use a unified theory called gradient elastoplasticity, which introduces gradient terms of the strain in the evolution equations for state variables (Zervos et al. 2000a; Zervos et al. 2000b). As long as the deformation is relatively homogeneous, like in the elastic and hardening regime, these higher order terms are negligible. In the softening regime, however, where deformation becomes strongly inhomogeneous, the gradient terms dominate and the well-posedness of the governing equations is preserved.

In contrast with other higher order theories, in gradient elastoplasticity there is no need to discretise internal variables, like the plastic multiplier and plastic strains, or to apply boundary conditions on them at the elastoplastic boundary. Only the displacement field needs to be discretised, therefore the usual finite element displacement formulation can be used. Because of the presence of gradient terms in the weak form of the boundary value problem, strains must be continuous so C^1 finite element interpolation is needed.

In the following, we present in section 2 the theory of gradient elastoplasticity and the finite element formulation. In section 3 we describe the geometry of the model used to simulate the hollow cylinder test, along with the necessary boundary conditions and the material parameters. Numerical results are presented, demonstrating computational robustness in terms of mesh insensitivity and calculated failure mechanisms. Final conclusions are drawn in section 4.

2 GRADIENT ELASTOPLASTICITY THEORY

2.1 *Assumptions*

Total strain rates $\dot{\epsilon}_{ij}$ are decomposed into an elastic part $\dot{\epsilon}^e_{ij}$ and a plastic part $\dot{\epsilon}^p_{ij}$,

$$\dot{\epsilon}_{ij} = \dot{\epsilon}^e_{ij} + \dot{\epsilon}^p_{ij} \tag{1}$$

The total (equilibrium) stress rate $\dot{\sigma}_{ij}$ is defined in terms of the elastic strain rate and its Laplacian, as

$$\dot{\sigma}_{ij} = C^e_{ijkl}\left(\dot{\epsilon}^e_{kl} - l_e^2\nabla^2\dot{\epsilon}^e_{kl}\right) \tag{2}$$

where C^e_{ijkl} is the tensor of the elastic constants and l_e is a material parameter with dimensions of length, called the *elastic material length.*

The yield condition can be written as $F(\tau_{ij}, \psi) = 0$. The plastic strain rate is defined through a plastic potential $Q(\tau_{ij}, \psi)$ and a plastic multiplier $\dot{\psi}$

$$\dot{\epsilon}^p_{ij} = \dot{\psi}\frac{\partial Q}{\partial \tau_{ij}} \tag{3}$$

Both F and Q depend on a reduced stress $\tau_{ij} = \sigma_{ij} - \alpha_{ij}$ and on a hardening/softening parameter ψ. α_{ij} is a back stress, evolving according to the law

$$\dot{\alpha}_{ij} = -C^e_{ijkl} l_p^2 \nabla^2 \dot{\epsilon}^p_{kl} \tag{4}$$

where l_p is called the *plastic material length.* Back stresses develop only when the deformation becomes sufficiently inhomogeneous, allowing for a region around a material point to contribute to its strength.

2.2 *Calculation of plastic strain*

The plastic multiplier $\dot{\psi}$, and consequently the plastic strains, can be determined from the consistency condition of plasticity

$$F(\tau_{ij}, \psi) = 0, \;\; \dot{F}(\tau_{ij}, \psi) = \frac{\partial F}{\partial \tau_{ij}}\dot{\tau}_{ij} + \frac{\partial F}{\partial \psi}\dot{\psi} = 0 \tag{5}$$

Neglecting terms of order higher than second and solving the differential equation approximately results in

$$\dot{\psi} = \frac{1}{H}\frac{\partial F}{\partial \tau_{ij}} C^e_{ijkl}\left(\dot{\epsilon}_{kl} + l_c^2\nabla^2\dot{\epsilon}_{kl}\right) \tag{6}$$

with

$$H = H_0 + H_t \quad , \quad H_t = -\frac{\partial F}{\partial \psi} \tag{7}$$

$$H_0 = \frac{\partial F}{\partial \tau_{ij}} C^e_{ijkl}\frac{\partial Q}{\partial \tau_{kl}} \quad , \quad l_c^2 = \frac{H_0}{H}l_p^2 - \frac{H_t}{H}l_e^2 \tag{8}$$

Then both total stress and back stress rates are expressed in terms of the total strain rates as

$$\dot{\sigma}_{ij} \;=\; C^{ep}_{ijkl}\dot{\epsilon}_{kl} - C^m_{ijkl}\nabla^2\dot{\epsilon}_{kl} \tag{9}$$

$$\dot{\alpha}_{ij} \;=\; -l_p^2 C^p_{ijkl}\nabla^2\dot{\epsilon}_{kl} \tag{10}$$

where C^p_{ijkl} is the known *plastic stiffness matrix* of classical plasticity

$$C^p_{ijkl} = \frac{<1>}{H} C^e_{ijmn} \frac{\partial Q}{\partial \tau_{mn}} \frac{\partial F}{\partial \tau_{st}} C^e_{stkl} \quad (11)$$

C^{ep}_{ijkl} is the usual *elastoplastic stiffness matrix* and C^m_{ijkl} is a stiffness matrix for the second gradient terms.

$$C^{ep}_{ijkl} = C^e_{ijkl} - C^p_{ijkl} \quad (12)$$

$$C^m_{ijkl} = l_e^2 C^{ep}_{ijkl} + l_c^2 C^p_{ijkl} \quad (13)$$

If $l_e = 0$ but $l_p \neq 0$, the above equations degenerate to a gradient plasticity model (Vardoulakis & Aifantis 1991; Vardoulakis et al. 1992). For $l_e = l_p = 0$, the classical elastoplasticity theory is recovered.

2.3 *Gradient elastoplasticity as a Mindlin theory*

Following Mindlin (1964) and Vardoulakis & Aifantis (1991), we rewrite equation (9) as

$$\dot\sigma_{ij} = \dot\sigma^{(0)}_{ij} + \dot\sigma^{(2)}_{ij} \quad (14)$$

where

$$\dot\sigma^{(0)}_{ij} = C^{ep}_{ijkl}\dot\epsilon_{kl}, \qquad \dot\sigma^{(2)}_{ij} = -C^m_{ijkl}\nabla^2\dot\epsilon_{kl} \quad (15)$$

$\dot\sigma^{(0)}_{ij}$ is the *Cauchy stress rate*, which is identified as the constitutive stress rate tensor of classical elastoplasticity, and relates to the strain rate. The second term, $\dot\sigma^{(2)}_{ij}$, relates to the Laplacian of the strain rate and is the *relative stress rate*.

We can consider the relative stress rate to be equilibrated by a *double stress rate* $\dot m_{kij}$, energy conjugate to the strain gradient $\dot\epsilon_{ij,k}$. The double stress rate is defined as

$$\dot\sigma^{(2)}_{ij} + \dot m_{kij,k} = 0 \Longrightarrow \dot m_{kij} = C^m_{ijmn}\dot\epsilon_{mn,k} \quad (16)$$

Gradient elastoplasticity corresponds then to a *restricted Mindlin continuum*, i.e. a Mindlin continuum with vanishing relative deformation.

2.4 *Principle of virtual work*

The equations of static equilibrium are obtained through the principle of virtual work. The work of internal forces is written as

$$\delta\dot W_{int} = \int_V \left(\sigma^{(0)}_{ij}\delta\dot\epsilon_{ij} + m_{kij}\delta\dot c_{kij}\right) dV \quad (17)$$

Stress boundary conditions are applied at the S_σ part of the boundary. The external work is written as

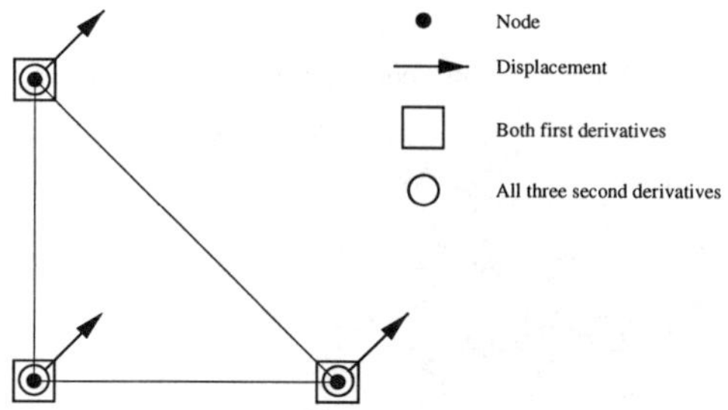

Figure 1: C^1 triangle with 3 nodes

$$\delta\dot W_{ext} = \int_{S_\sigma} \left(t_i\delta\dot v_i + \mu_i n_k \delta\dot v_{i,k}\right) dS \quad (18)$$

where t_i is the applied traction vector, μ_i is the applied double force vector, n_i is the unit normal to S_σ and $\delta\dot v_i$ is the virtual displacement rate vector on S_σ. According to the principle of virtual work, $\delta\dot W_{int} = \delta\dot W_{ext}$

Since strain gradients are included in the formulation, displacement boundary conditions may contain restrictions not only on the displacement, but on its normal derivative at the boundary as well.

2.5 *Finite element formulation*

Since the rates of all constitutive quantities can be expressed in terms of the total strain rate, its gradient and its Laplacian, only the displacement field needs to be discretised. The usual displacement formulation of the finite element method is then quite suitable for solving boundary value problems of gradient elastoplasticity. Nevertheless, the presence of strain gradients in the equations of variational equilibrium dictates continuity of strains, so the employed interpolation scheme should guarantee C^1 continuity.

We use the three-noded C^1 triangle shown in Figure 1. The degrees of freedom at each node are the displacement and all its derivatives of first and second order (Dasgupta & Sengupta 1990).

The displacement field $\mathbf{u}$ inside each element is interpolated using the 2x36 matrix of the shape functions $\mathbf{N}$ and the 36x1 vector of nodal freedoms $\hat{\mathbf{u}}$, as $\mathbf{u} = \mathbf{N}\cdot\hat{\mathbf{u}}$. Strains, strain gradients and strain Laplacians are calculated through derivatives of appropriate order of the shape functions. Stresses, double stresses and back stresses are then calculated through the constitutive matrices (Zervos et al. 2000a). Finally, the virtual work equation gives the following system of equations for the nodal degrees of freedom:

$$\int_V \left(\mathbf{B}_1^T\mathbf{C}^{ep}\mathbf{B}_1 + \mathbf{B}_{2x}^T\mathbf{C}^m\mathbf{B}_{2x} + \mathbf{B}_{2y}^T\mathbf{C}^m\mathbf{B}_{2y}\right) dV \cdot \hat{\mathbf{u}} =$$

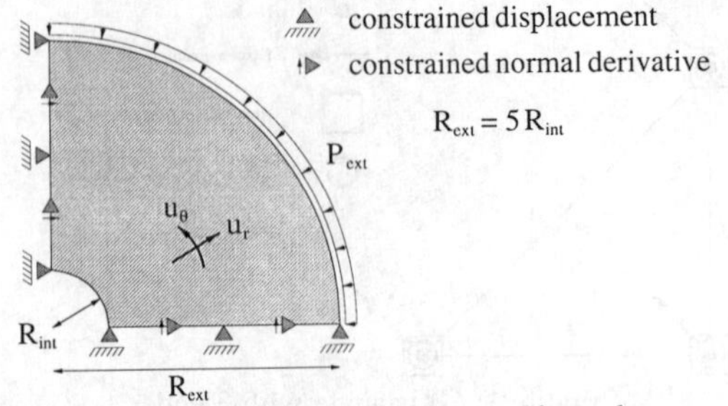

Figure 2: Model geometry, loading and boundary conditions

$$\int_{S_\sigma} \left[\mathbf{N}^{\mathbf{T}} \hat{\mathbf{t}} + \left(n_x \frac{\partial \mathbf{N}}{\partial x} + n_y \frac{\partial \mathbf{N}}{\partial y} \right) \hat{\mathbf{m}} \right] dS \quad (19)$$

where $\hat{\mathbf{t}}$ is the boundary traction vector, $\hat{\mathbf{m}}$ is the boundary double traction vector and $\mathbf{n} = [n_x, n_y]$ is the outward unit normal to the boundary. Matrices $\mathbf{B_1}$ and $\mathbf{B_2}$ contain first and second order derivatives of the shape functions respectively. Equation (19) is solved iteratively with the Newton-Raphson method.

3 NUMERICAL SIMULATION

3.1 *Geometry, loading and boundary conditions*

The geometry of the problem and the loading and boundary conditions are shown in Figure 2. Assuming symmetry along the x and y axis, only one quarter of the specimen is modeled. The cylinder has internal radius $R_{int} = 0.10$ m and external $R_{ext} = 0.50$ m. It is loaded with increasing uniform pressure on the external boundary while the internal pressure is maintened zero.

Special attention must be given to the kinematic conditions needed to establish the two symmetry boundaries along the x- and y-axes, where not only displacements but higher order constraints are also needed to be prescribed. The appropriate boundary conditions turn out to be $u_y = 0$ and $\frac{\partial u_x}{\partial y} = 0$ along the x-axis, and $u_x = 0$ and $\frac{\partial u_y}{\partial x} = 0$ along the y-axis (Zervos et al. 2000b).

3.2 *Material model*

The material behaviour is described by the Mohr-Coulomb failure criterion

$$F = m\tau_1 - \tau_3 - \sigma_c = 0 \quad (20)$$

where $m = (1 + \sin\phi)/(1 - \sin\phi)$ is the friction coefficient and $\sigma_c = (2c\cos\phi)/(1 - \sin\phi)$ is the *equivalent stress*; ϕ is the angle of internal friction and c is the material cohesion. The material parameters were calibrated from triaxial tests on Castlegate sandstone. The elastic constants were found to be $E = 8100$ MPa and $\nu = 0.35$. The friction angle is considered constant, with a value $\phi = 32.54^o$ and an associated flow-rule is assumed, $Q = F$.

Table 1: Finite element meshes

Name	Nodes	Elements	DOFs
coarse	15x41 = 615	1120	7380
medium	20x81 = 1620	3040	19440
fine	25x121= 3025	5760	36300

The hardening/softening behaviour is defined through the equivalent stress $\sigma_c(\epsilon_p)$, which is taken to evolve according to the hyperbolic law

$$\sigma_c(\epsilon_p) = \sigma_{c,0} + \frac{(1 - C_0\epsilon_p)\epsilon_p}{C_1 + C_2\epsilon_p} \quad (21)$$

where $C_1 = 1.323 \cdot 10^{-5}$ and $C_2 = 6.1271 \cdot 10^{-2}$ are calibration constants. $\sigma_{c,0} = 25$ MPa is a conventional threshold value of the equivalent stress defining the state of initial yield. The constant C_0 is an open parameter controlling the rate of softening; it is taken to be $C_0 = 70$. The plastic material length is set to $l_p = 0.2$ mm, equal to the mean grain diameter of the Castlegate sandstone. The elastic material length is taken to be ten times smaller, $l_e = 0.02$ mm, since the existence of microstructure is not as important in the elastic regime as it is in the plastic regime.

3.3 *Numerical results*

We present computations for the three different meshes (coarse, medium and fine) that are shown in Table 1.

Upon loading, the material near the hole yields first following the hardening branch and finally enters the softening regime. Deformation is initially axisymmetric, as can be seen in the vector plot of Figure 3(a) which shows convergence of the hole in terms of the displacement increment. In the same plot the material state is presented. We can see that a zone of softening material forms around the hole wall (light gray), surrounded by a zone of hardening material (black). The rest of the material in the deeper regions is still elastic.

After a loading level is reached the axisymmetry of the deformation breaks spontaneously and the displacement increment assumes a sinusoidal form (Figure 3(b)). Regions of the material which were in the plastic regime unload elastically (dark gray). The unloading areas grow in size with increasing load and, eventually, almost the whole region close to the hole wall unloads elastically, except some narrow bands of plastically softening material (Figure 3(c)). The bands are connected to a familiar breakout form which resembles the failure mechanisms observed in the experiments. It should be noted that axisymmetry breaks while the external load keeps increasing ("tangential bifurcation"). The load eventually reaches a peak and then drops, as the deformation localises.

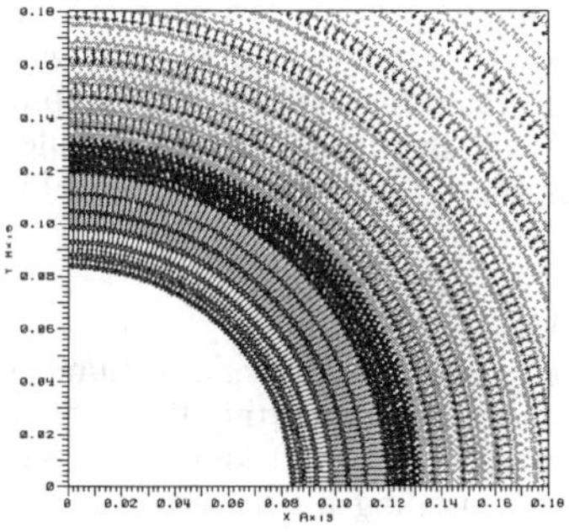

(a) $p_{ext} = 39.22$ MPa

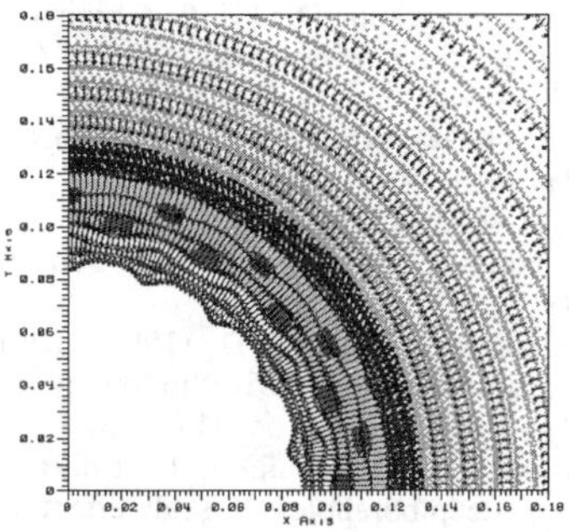

(b) $p_{ext} = 39.89$ MPa

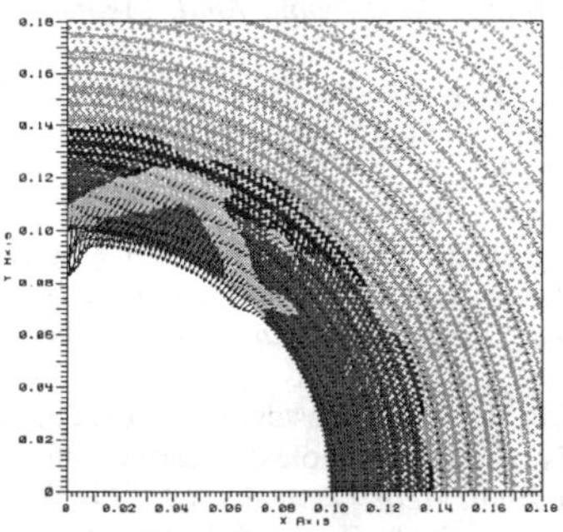

(c) $p_{ext} = 43.09$ MPa

Figure 3: Incremental displacement field and material state (medium mesh)

Loss of symmetry is totally spontaneous. There is no need to perturb the solution with an eigenvector or to introduce imperfect elements. It seems that the round–off error alone is sufficient to push the solution off the trivial path (axisymmetric deformation) to a bifurcated branch that will lead to the localised pattern (breakout). The same observation was made by Papanastasiou & Vardoulakis (1992) for a Cosserat continuum.

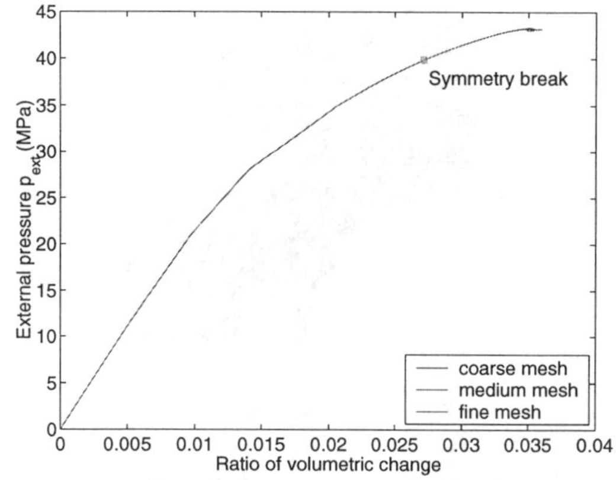

Figure 4: External pressure vs. hole closure for three different meshes

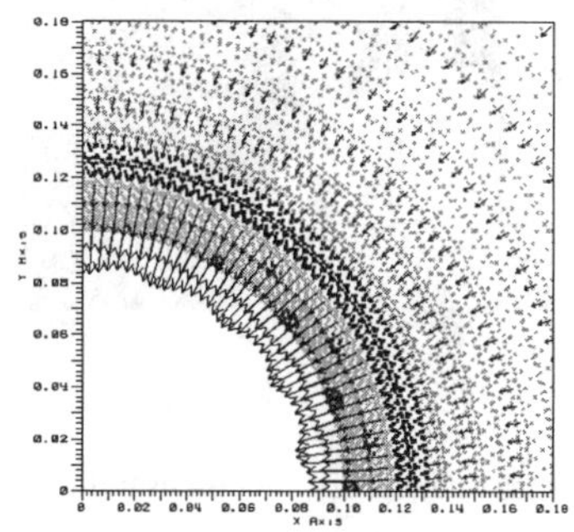

(a) $p_{ext} = 39.87$ MPa

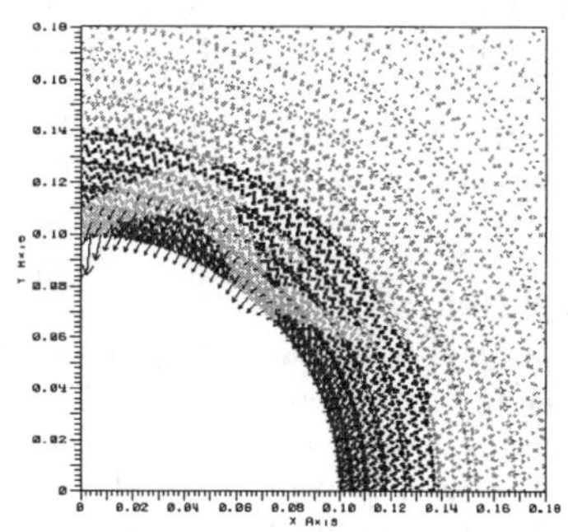

(b) $p_{ext} = 43.20$ MPa

Figure 5: Incremental displacement field and material state, coare mesh

All three meshes predict the same response, as can be seen from the external pressure vs. hole closure

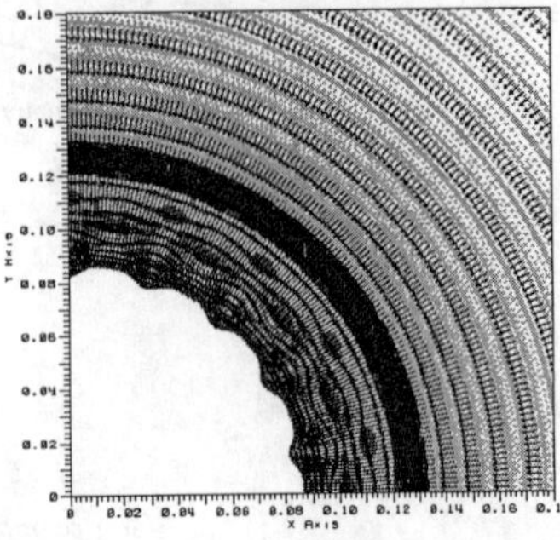

(a) $p_{ext} = 39.87$ MPa

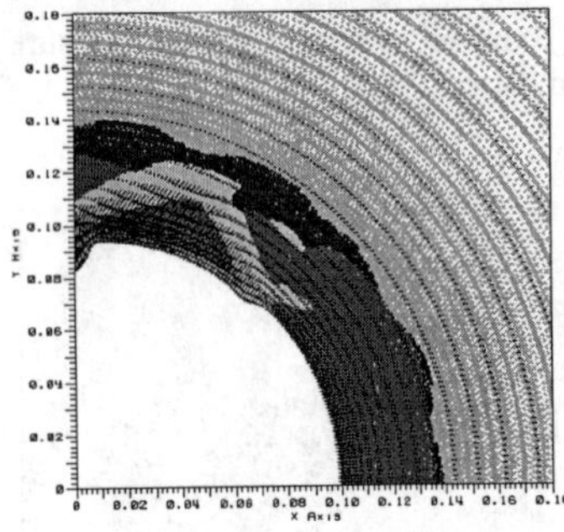

(b) $p_{ext} = 43.05$ MPa

Figure 6: Incremental displacement field and material state, fine mesh

curve of Figure 4. The load corresponding to loss of axisymmetry is the same for all meshes. The warping mode (the number of waves of the sinusoidal instability around the hole) is also the same, as depicted in Figures 3(b), 5(a) and 6(a). The final localised patterns, shown in Figures 3(c), 5(b), and 6(b), are also the same.

4 CONCLUSIONS

Gradient elastoplasticity was used to model the progressive failure of hollow cylinders. The calibrated material corresponds to a weak sandstone with hardening and softening behaviour. The existence of higher order terms in the constitutive relations regularises the material behaviour in the softening regime, allowing for mesh–independent post–peak computations.

The model predicts spontaneous loss of symmetry of the deformation according to a warping mode with finite wavenumber. This takes place in the ascending branch of the load–displacement response. The deformation localises progressively into thin shear bands surrounded by regions of elastic unloading. The final failure patterns are similar to the breakout form observed in experiments. Mesh–insensitivity was demonstrated by showing that the predicted bifurcation mode, the final deformation pattern and the load-displacement curves are the same for different meshes.

ACKNOWLEDGEMENT

The authors would like to thank Schlumberger Cambridge Research for supporting this research. I. Vardoulakis wants also to acknowledge G.S.R.T. of Greece for supporting his research through the program PENED 99 ED 642.

REFERENCES

Dasgupta, S. & Sengupta, D. (1990). A higher-order triangular plate bending element revisited. *Int. J. Num. Meth. Engng 30*, 419–430.

Mindlin, R. (1964). Microstructure in linear elasticity. *Arch. Rat. Mech. Anal. 10*, 51–78.

Mühlhaus, H.-B. & Vardoulakis, I. (1987). The thickness of shear bands in granular materials. *Geotechnique 37*, 271–283.

Papanastasiou, P. & Durban, D. (1999). Bifurcation of elastoplastic pressure-sensitive hollow cylinders. *Journal of Applied Mechanics 66*, 117–123.

Papanastasiou, P. & Vardoulakis, I. (1989). Bifurcation analysis of deep boreholes: Ii. scale effect. *Int. J. Num. Anal. Meth. Geomech. 13*, 183–198.

Papanastasiou, P. & Vardoulakis, I. (1992). Numerical treatment of progressive localization in relation to borehole stability. *Int. J. Num. Anal. Meth. Geomech. 16*, 389–424.

Vardoulakis, I. & Aifantis, E. C. (1991). A gradient flow theory of plasticity for granular materials. *Acta Mechanica 87*, 197–217.

Vardoulakis, I., Papamichos, E., & Sulem, J. (1994). A second-gradient plasticity model for granular rocks. In M. J. Z.P. Bažant, Z. Bittnar & J. Mazars (Eds.), *Fracture and Damage in Quasibrittle Structures*. E&FN Spon.

Vardoulakis, I. & Papanastasiou, P. (1988). Bifurcation analysis of deep boreholes: I. surface instabilities. *Int. J. Num. Anal. Meth. Geomech. 12*, 379–399.

Vardoulakis, I., Shah, K. R., & Papanastasiou, P. (1992). Modelling of tool-rock shear interfaces using gradient-dependent flow theory of plasticity. *Int. J. Rock Mech. Min. Sci. & Geomech. Abstr. 29*(6), 573–582.

Vardoulakis, I. & Sulem, J. (1995). *Bifurcation Analysis in Geomechanics*. Blackie Academic and Professional.

Zervos, A., Papanastasiou, P., & Vardoulakis, I. (2000a). A finite element displacement formulation for gradient elastoplasticity. *Int. J. Num. Meth. Engng (in print)*.

Zervos, A., Papanastasiou, P., & Vardoulakis, I. (2000b). Modelling of localisation and scale effect in thick-walled cylinders with gradient elastoplasticity. *submitted to Int. J. Solids Structures*.

Computer Methods and Advances in Geomechanics, Desai et al. (eds) © 2001 Balkema, Rotterdam, ISBN 90 5809 183 X

Local rock expansion in non-homogeneous productive gas/oil fields

Domenico Baù, Massimiliano Ferronato, Giuseppe Gambolati & Pietro Teatini
DMMMSA, Department of Mathematical Methods and Models for Scientific Applications, University of Padova, Italy

ABSTRACT: The reliability of *in situ* compaction measurements performed by radioactive markers for the evaluation of the mechanical properties of productive gas/oil fields is investigated with a finite element model applied to a typical Plio-Pleistocene sequence of the Northern Adriatic sedimentary basin, Italy. Preliminary results show that with particular layering configurations within the recorded depth interval some low permeable levels may initially expand after the inception of fluid withdrawal thus providing a measurement where the overall compaction may be partially or also totally offset by the local expansion. If such a measurement is used to estimate the *in situ* rock vertical compressibility large errors may arise which are not due to the instrument inaccuracy as at a first glance it might appear. This would suggest that great care must be paid to the detailed lithology of each monitored depth relative to the depleted levels before the marker data may be used to evaluate the actual stiffness of a reservoir.

1 INTRODUCTION

In situ reservoir compaction measurements by radioactive markers represent a promising technique for a most realistic evaluation of the actual mechanical properties of producing gas/oil fields (Mobach & Gussinklo 1994; NAM 1995; Baú et al. 1999a; Baú et al. 1999b). As is at present implemented by Schlumberger and Western Atlas the marker technique is reported to have an accuracy equal to 1 mm for the compaction measurement of a 10.5 m wide depth interval (Mobach & Gussinklo 1994). At such a vertical resolution the reservoir may exhibit significant changes of sediment lithology with fine or very fine grained interbedded layers which may cause a delay of the local rock depletion, and hence a retarded compaction, or even an expansion at the early stage of gas/oil pumping. As an example, Figure 1 shows the typical stratigraphic profile of a vertical wellbore in the Northern Adriatic Sea, Italy, where a number of fine clayey layers alternate with productive sandy units, and the corresponding marker spacing compaction (expansion) measurements. A configuration of this kind may create serious difficulties for the correct interpretation of early data when a prediction of the expected field compaction is of paramount importance for a reliable modeling of the future land subsidence as related to field development.

A numerical study with a coupled linear elastic 3-D finite element (FE) axisymmetric model is performed

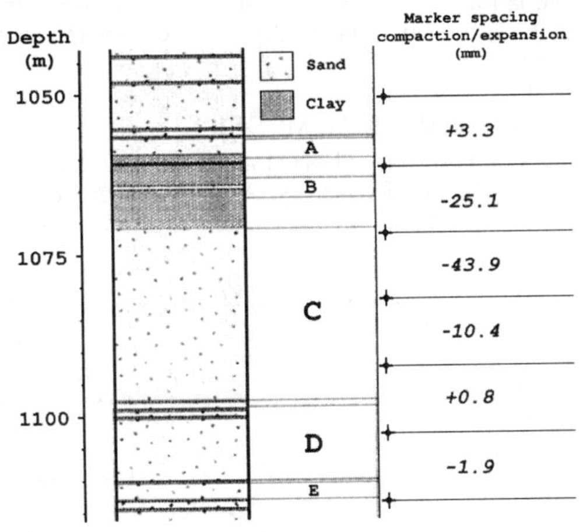

Figure 1: Typical stratigraphic profile of a vertical wellbore in the Northern Adriatic sedimentary basin, Italy. The letters A, B, ..., E indicate producing sandy levels. The marker location and the corresponding compaction/expansion measurements (mm) after 2 years from the inception of gas pumping from level C are reported on the rightmost column.

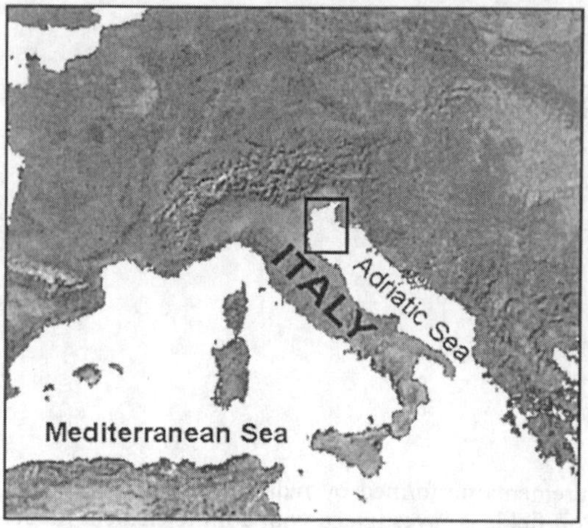

Figure 2: Map of the Mediterranean Sea. The location of the Northern Adriatic sedimentary basin is indicated by the panel.

to address the mechanical behavior of the stressed rock at the marker resolution scale. A test case based on a typical lithologic profile of a Plio-Pleistocene gas reservoir in the Northern Adriatic sedimentary basin (Figure 2) is investigated using a local much detailed FE model able to represent the rock expansion at the marker scale during gas/oil pumping. A few preliminary remarks on the reliability of marker data for the evaluation of the rock vertical compressibility, especially in the early stage of field development, are finally made.

2 MATHEMATICAL MODEL

According to Biot's (1941) formulation, as later modified by van der Knaap (1959) and Geertsma (1966), the coupled model for an isotropic medium reads:

$$(\lambda + G)\frac{\partial \epsilon}{\partial i} + G\nabla^2 u_i = \alpha \frac{\partial p}{\partial i} \qquad i = x, y, z \tag{1}$$

$$\frac{1}{\gamma}\nabla\left(k\nabla p\right) = \left[n\beta + c_{br}\left(\alpha - n\right)\right]\frac{\partial p}{\partial t} + \alpha\frac{\partial \epsilon}{\partial t} \tag{2}$$

where λ and G denote the Lamé constant and the shear modulus of porous medium, α the Biot coefficient, γ the specific weight of water, k the medium hydraulic conductivity, n the porosity, c_{br} and β the volumetric compressibility of solid grains and water, ∇ the gradient operator, ϵ the medium volumetric dilatation and p and u_i the incremental pore pressure and the components of incremental displacement along the i-direction, respectively.

The standard FE element solution to the coupled eqs. (1) and (2) (Zienkiewicz 1991) yields the following system of first-order differential equations:

$$\begin{bmatrix} K & -Q \\ 0 & H \end{bmatrix}\begin{Bmatrix} \boldsymbol{u} \\ \boldsymbol{p} \end{Bmatrix} + \begin{bmatrix} 0 & 0 \\ Q^T & P \end{bmatrix}\begin{Bmatrix} \dot{\boldsymbol{u}} \\ \dot{\boldsymbol{p}} \end{Bmatrix} = \begin{Bmatrix} \boldsymbol{f^u} \\ \boldsymbol{f^p} \end{Bmatrix} \tag{3}$$

with K, H, P and Q the elastic stiffness, flow stiffness, flow capacity and coupling matrices, respectively, $\{\boldsymbol{u},\boldsymbol{p}\}^T$ and $\{\dot{\boldsymbol{u}},\dot{\boldsymbol{p}}\}^T$ the vector of the unknown variables u_i and p and corresponding time derivatives, $\{\boldsymbol{f^u},\boldsymbol{f^p}\}^T$ the vector of the nodal loads ($\boldsymbol{f^u}$) and flow sources ($\boldsymbol{f^p}$). The system (3) is solved using an appropriate time marching scheme (Booker & Small 1975). Detailed expressions for the above terms may be found in Gambolati et al. (2000a).

In our FE implementation we consider a 3-D axisymmetric problem discretized by annular elements with a triangular cross section.

3 PRELIMINARY ANALYSIS

The axisymmetric coupled model is solved over a realistic hydrogeological setting of the Northern Adriatic sedimentary basin (Gambolati et al. 2000b). A typical stratigraphic profile (Figure 1) has been selected with alternating sandy and clayey layers, with minimum thickness of 3 m and 0.5 m, respectively, and a depleted producing level 26.5 m thick located at the average 1084 m depth (level C in Figure 1). A linear pore pressure decline of 50 kg/cm^2 in two years is prescribed within a radius of 200 m from the wellbore axis. For the purpose of the present analysis a high resolution mesh is required in the vicinity of the depleted reservoir (Figure 3) in order to reproduce the detailed lithology of the monitored intervals accurately and to simulate the local rock deformation due to pumping at the marker scale. To reduce the computational burden, a coarser grid is used far from the area of interest. Numerical experiments with different coarser grids show no appreciable discrepancies of the results in the vicinity of the depleted level.

The porous medium is assumed to be fully saturated by groundwater ($\gamma = 1000$ kg/m^3, $\beta = 0.16 \cdot 10^{-9}$ cm^2/kg) with a variable porosity (Gambolati et al. 2000b) and a uniform Poisson ratio $\nu = 0.3$. A compressibility profile derived from *in situ* measurements (Baú et al. 1999b) is used for the definition of the geomechanical properties of the porous medium, with $k = 2.5 \cdot 10^{-6}$ m/s and $k = 10^{-10}$ m/s in the sandy and clayey layers, respectively (AGIP 1996). Standard Dirichlet boundary conditions with zero displacement and zero pore pressure change on the outer and bottom boundaries, zero radial displacement on the symmetry axis, and zero pore pressure decline on the top boundary are prescribed on the overall 10000×27000 m domain. A Crank-Nicolson marching scheme with an appropriate time step $\Delta t_{min} = 10$ min (Ferronato et al. 2000) is adopted.

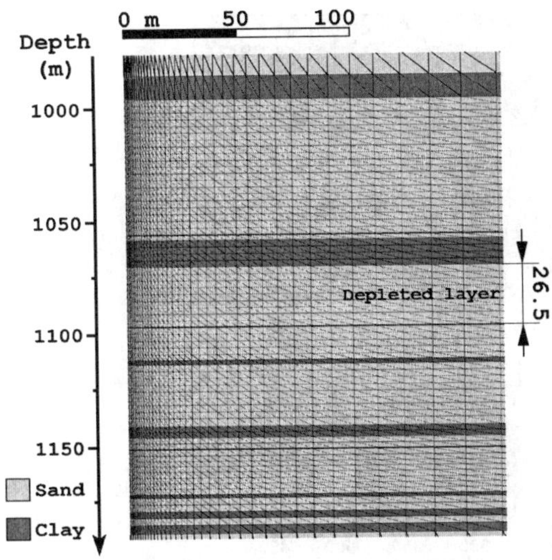

Figure 3: Radial vertical cross section of the 3-D axisymmetric FE mesh used to discretize the porous medium close to the source of stress.

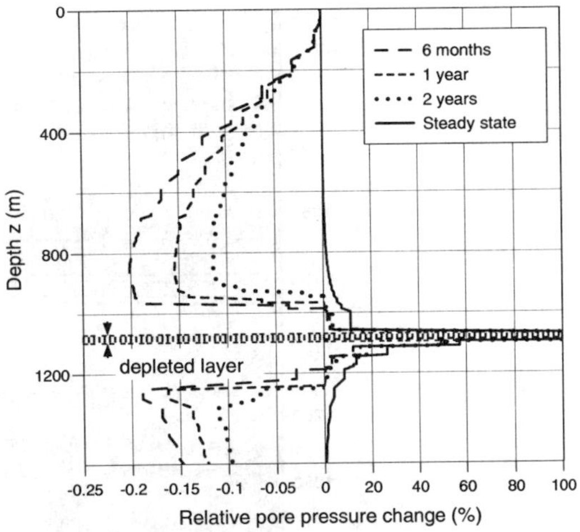

Figure 4: Pore pressure change (%) relative to the pressure decline of the depleted layer at $r = 0$ for different times after the inception of pumping. Negative values indicate overpressure.

As is shown in Figure 4, consistent with findings by other authors (Verruijt 1969; Hsieh 1996), a slight overpressure develops in the strata closer to the depleted layer after the inception of pumping and dissipates with time. The duration and the importance of this effect mainly depend upon the hydraulic conductivity and the thickness of the clayey layers, and increase in low permeable sequences. The overpressure produces a deformation which opposes the overall compaction recorded by early marker spacing measurements. In the present test case, the overpressure still persists after 2 years from the inception of the production (Figure 4), and thus at least partially offsets the compaction measurement of the depth interval which incorporates the expanding layers.

Figure 5 provides a map of the vertical strain ϵ_z at the same time. The areas above and below the depleted layer exhibit a positive ϵ_z, *i.e.* an expansion, that may naturally occur in a 3-D poroelastic medium. This effect is enhanced by the delay of the depletion in the rock closer to the productive unit, which retards the compaction of the confining beds. Moreover, if two radioactive markers span an interval comprising low permeable levels whose compaction is delayed in time, the instrument measurement should not be used for estimating the *in situ* rock vertical compressibility as it would lead to a compressibility pointing to a stiffer sediment.

By our model, the marker spacing measurement has been simulated by computing the vertical deformation Δz of a depth interval 10.5 m wide against depth. This is given at different times in Figure 6 where the numerical Δz values are referred to the depth of the mid-point of each marker interval. It can be observed in Figure 6 that an expansion occurs above and below the depleted layer, and particularly so in the early stage after the inception of pumping.

These few preliminary results show that the permeability contrast at the vertical scale of the measurement (10.5 m overall) may cause a local rock expansion which can actually influence the response of the marker tool if this is not located conveniently. Moreover even though the instrument records a compaction this should not be used to evaluate the rock compressibility until the initial pore pressure of the interbedded fine grained levels is totally dissipated. The expansion in the vicinity of depleted reservoirs may be induced by other factors as well, such as the geometrical configuration of the field (*e.g.* two or more resevoirs each one on top of the other) or the contrast of compressibility (*e.g.* the presence of overconsolidated sediments, plastic deformations in the depleted layers), which are not addressed in the present work.

4 CONCLUSIONS

The local vertical deformation of layering sequences which are typical of gas fields located in the Northern Adriatic sedimentary basin, Italy, is investigated by a coupled axisymmetric FE model. The analysis points to the existence of areas of rock expansion which can be measured by the marker technique depending on the actual installation of the markers. This effect appears to be mainly related to the presence of thin low permeable layers in the basin which can both delay the compaction of the depth interval monitored by the instrument and generate a transient initial

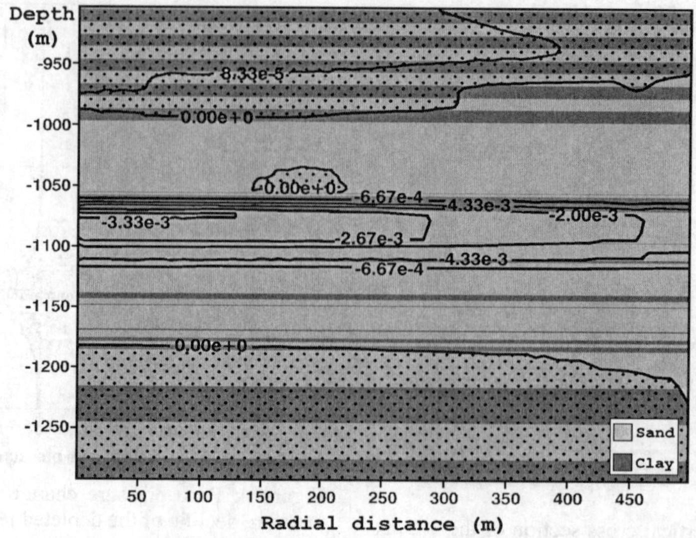

Figure 5: Map of the vertical strain ϵ_z at time $t = 2$ years. Dotted areas exhibit expansion.

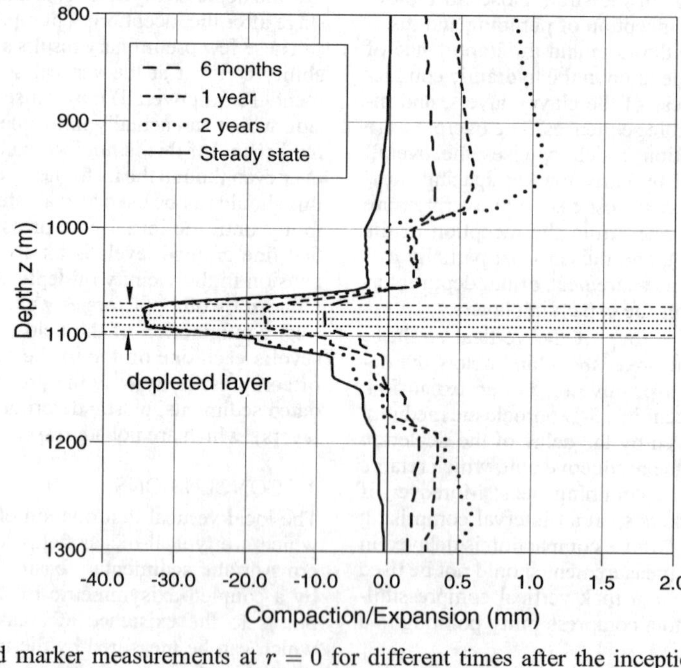

Figure 6: Simulated marker measurements at $r = 0$ for different times after the inception of pumping. The numerical compaction/expansion is referred to the mid-depth of the simulated interval 10.5 m in size.

overpressure with a corresponding layer expansion. If the marker records such anomalous deformations, and these are used to evaluate the vertical rock compressibility, an underestimation may result.

A major conclusion from this study is that the marker measurement is not necessarily inaccurate where an expansion is observed. Therefore a much detailed litho-stratigraphy of the monitored intervals is needed for carefully processing and interpreting the marker measurements. To obtain reliable estimates numerical analyses of the kind developed in the present communication may help.

Other factors than those addressed above may induce local expansions, *e.g.* a contrast in rock compressibility, due for instance to overconsolidation or plasticity, and a special layering configuration. The importance of these factors is currently being investigated at the University of Padova.

ACKNOWLEDGMENT

This study has been partially funded by the Italian MURST project "*Metodologie Innovative per il Monitoraggio, la Gestione ed il Controllo Quali-quantitativo delle Acque Sotterranee*".

REFERENCES

AGIP (1996). Alto Adriatico. Rapporto di sintesi. Technical report, AGIP S.p.A., Milano.

Baú, D., Gambolati, G., & Teatini, P. (1999a). Residual land subsidence over depleted gas fields in the Northern Adriatic basin. *Environmental & Engineering Geoscience V*(4), 389–405.

Baú, D., Gambolati, G., & Teatini, P. (1999b). Statistical analysis of in situ compaction measurements for anthropogenic land subsidence prediction in the Northern Adriatic basin. In A. N. S. J. Lippard & R. Sinding-Larsen (Eds.), *Proc. of 5th Conf. Inter. Assoc. Mathematical Geology*, Volume 1, pp. 299–304. Tapir Publ.

Biot, M. A. (1941). General theory of three-dimensional consolidation. *J. Appl. Phys. 12*(2), 155–164.

Booker, J. R. & Small, J. C. (1975). An investigation of the stability of numerical solutions of Biot's equations of consolidation. *Int. J. Solids Struct. 11*, 907–917.

Ferronato, M., Gambolati, G., & Teatini, P. (2000). Ill-conditioning of finite element poroelasticity equations. *Int. J. Solids Struct.*. Submitted.

Gambolati, G., Ferronato, M., Teatini, P., Deidda, R., & Lecca, G. (2000a). Finite element analysis of land subsidence above depleted reservoirs with pore pressure gradient and total stress formulations. *Int. J. Numer. Analytic. Methods Geomech.*. Submitted.

Gambolati, G., Teatini, P., Baú, D., & Ferronato, M. (2000b). The importance of poro-elastic coupling in dynamically active aquifers of the Po river basin, Italy. *Water Resour. Res.*. In press.

Geertsma, J. (1966). Problems of rock mechanics in petroleum production engineering. In *Proceedings of 1st Cong. Int. Soc. Rock Mechan.*, pp. 585–594. Lisbon.

Hsieh, P. A. (1996). Deformation-induced changes in hydraulic head during ground-water withdrawal. *Ground Water 34*(6), 1082–1089.

Mobach, E. & Gussinklo, H. J. (1994). In-situ reservoir compaction monitoring in the Gröningen field. In *EUROCK'94, Rock Mechanics for Petroleum Engineering*, Rotterdam, pp. 535–547. A. A. Balkema.

NAM (1995). Subsidence caused by natural gas production. Report 27600, 26 pp., Nederlandse Aardolie Maatschappij B.V., Assen, Netherlands.

van der Knaap, W. (1959). Nonlinear behavior of elastic porous media. *Petroleum Transactions of the AIME 216*, 179–187.

Verruijt, A. (1969). Elastic storage of aquifers. In R. De Wiest (Ed.), *Flow Through Porous Media*, New York, pp. 331–376. Academic Press.

Zienkiewicz, O. C. (1991). *The Finite Element Method in Engineering Geoscience* (4th ed.). McGraw Hill, London.

Computer Methods and Advances in Geomechanics, Desai et al. (eds) © 2001 Balkema, Rotterdam, ISBN 90 5809 183 X

Effect of intermediate principal stress on strength and deformation around wellbores in shales

X.Chen, C.P.Tan & B.Wu
CSIRO Division of Petroleum Resources, Melbourne, Vic., Australia

ABSTRACT: Numerical models (under isotropic and anisotropic loading conditions) are developed to investigate the influence of intermediate principal stress on strength and deformation of thick-walled hollow cylinders of a synthetic shale. For each case, different constitutive models and failure criteria (Mohr-Coulomb and Drucker-Prager) are adopted. The numerical predictions obtained are assessed against experimental observations. The investigation shows that there are large differences between predictions by Mohr-Coulomb and Drucker-Prager criteria under both isotropic and anisotropic loading conditions. Generally, anisotropic loading conditions increase borehole strength and reduce borehole deformations. Predictions obtained with Drucker-Prager criterion (intermediate principal stress is considered in this criterion) were found to be in better agreement with the results of the laboratory tests than those from Mohr-Coulomb criterion (intermediate principal stress is not considered in this criterion).

1 INTRODUCTION

The strength and deformation characteristics of geomaterials are important to many geotechnical engineering structures hence, precise prediction of the displacements and change in stress distributions is required. For example, maintaining a stable borehole is one of the major problems encountered in the oil and gas industry since wellbore instability will result in additional high drilling costs. The accuracy of such predictions is dependent on the constitutive model and failure criterion adopted in the analysis (e.g. McLean & Addis 1990). With shales comprising most of the formations penetrated by oil and gas wells (Santarelli et al. 1992), a better understanding of the mechanical behaviour of shales is essential in improving their existing constitutive models and failure criteria which will lead to more effective management of shale instability problems.

Hollow cylinder tests are frequently used for assessing borehole instability and for validating constitutive models and failure criteria (e.g. Addis & Wu 1996, Papamichos et al. 1996, Marsden et al. 1996, Santarelli 1987 & Veeken et al. 1989). For the sake of simplicity in testing equipment and control, the majority of such tests have been conducted under an isotropic loading condition, i.e., a confining fluid is used to apply both external pressure and axial stress, resulting in the axial stress being almost equal to the radial stress on the cylinder external surface. The drilling of the borehole is then simulated by reducing the internal borehole pressure while keeping the external pressure and axial stress constant. The borehole deformation is normally analyzed at the middle height section of the hollow cylinder based on plane strain assumption. However, as most in-situ stress fields in rock masses are anisotropic, the isotropic loading condition adopted in the tests may not be representative of the field situations. Furthermore, interpretation of the test results based on plane strain assumption may result in significant errors associated with the axial deformation generated by the loading.

The aim of this paper is to investigate the influence of intermediate principal stress on strength and deformation of thick-walled hollow cylinders of a synthetic shale. Three numerical models are developed: one for an isotropic loading condition and the other two for an anisotropic loading condition (i.e., axial stress is higher than the radial stress). For each case, different constitutive models and failure criteria (Mohr-Coulomb and Drucker-Prager) are adopted. Due to the low permeability of shales, excess pore pressure will be induced in the material in response to the volume change of the sample during testing. Hence, coupled flow-mechanical interactions were incorporated in the numerical models. The modelling used material parameters obtained from conventional laboratory tests. The numerical predictions obtained are assessed against experimental observations.

2 ASSESSMENT OF BOREHOLE STRENGTH

Failure criteria are normally used in the assessment of borehole stability by predicting the failure initiation. Among them the Mohr-Coulomb failure criterion is probably the most commonly used.

The Mohr-Coulomb criterion (M-C) may be written in terms of principal stresses as:

$$\sigma_1' = \left(\frac{1+\sin\alpha'}{1-\sin\alpha'}\right)\sigma_3' + \frac{2c'\cos\alpha'}{1-\sin\alpha'} \tag{1}$$

where σ_1' and σ_3' are maximum and minimum effective stresses.

It is well known that strength and deformation behaviour of rocks are affected by intermediate principal stress. The Drucker-Prager criterion is therefore often used to include the effect of intermediate principal stress on rock strength.

The Drucker-Prager criterion (D-P) may be written as (Owen & Hinton, 1980):

$$\sqrt{J_2'} - \gamma I_1 = k \tag{2}$$

where

$$I_1 = \sigma_1' + \sigma_2' + \sigma_3'$$

$$J_2' = \left[(\sigma_1' - \sigma_2')^2 + (\sigma_2' - \sigma_3')^2 + (\sigma_3' - \sigma_1')^2\right]/6$$

and γ and k are strength constants. They may be related to internal friction angle α' and cohesion c' by making the Drucker-Prager circle coincides with the outer or inner apices of the Mohr-Coulomb criterion on an octahedral plane:

$$\gamma = \frac{2\sin\alpha'}{\sqrt{3}(3\pm\sin\alpha')}$$

$$k = \frac{6c'\cos\alpha'}{\sqrt{3}(3\pm\sin\alpha')}$$

where "+" corresponds to an inner and "-" to an outer circle.

3 DEVELOPMENT OF NUMERICAL MODELS

This section presents the development of three numerical models to predict the strength and undrained deformation behavior of thick-walled hollow cylinders of a synthetic shale subjected to isotropic and anisotropic loading conditions. The description of the tests can be found in Wu et al. (2000). The models are developed using FLAC, a two-dimensional explicit finite difference program (Itasca 1998).

3.1 *Construction of numerical models*

The basic axisymmetric model of a thick-walled hollow cylinder subjected to both internal and external pressures is shown in Figure 1. The steel platens were included to better model any end effects. Two material constitutive models were used in the analyses: linear elastic-perfectly plastic models with the failure surfaces defined by the Mohr-Coulomb and Drucker-Prager failure criteria respectively, and a non-associated flow rule. The Drucker-Prager criterion may underestimate (by passing through the inner edges of the Mohr-Coulomb pyramid) or overestimate (by passing through the outer edges of the Mohr-Coulomb pyramid) the borehole strength. To avoid the two extreme cases, the mean circle of the Drucker-Prager failure criterion, which lies between the inner and outer circles, was used in the models. The deformation measurement points were at the same locations as the laboratory test sample, and are also shown in Figure 1. The properties of the interface between the platen and the sample were derived from the procedures outlined in the FLAC User Manual (Itasca 1998).

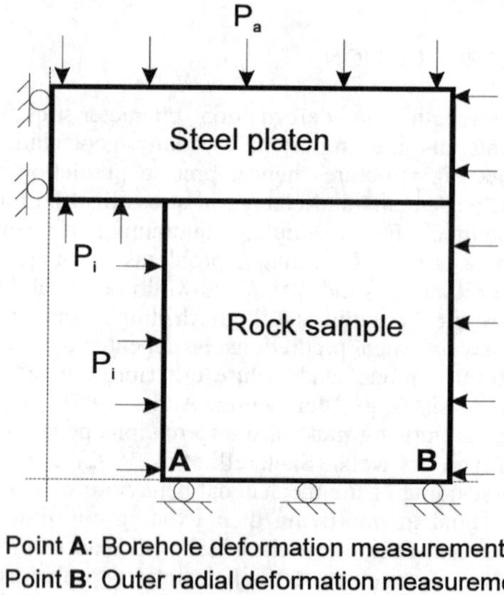

Figure 1. Conceptual model and boundary conditions

The dimensions and material properties used in the model are given in Table 1. The properties of the synthetic shale given in Table 1 were obtained from consolidated undrained triaxial tests (Wu & Tan 1996) and the isotropic loading stage of the borehole collapse tests (Wu et al. 2000).

3.2 *Initial and boundary conditions*

Table 2 summarises the initial pore pressure (P_p), borehole pressure (P_i), axial stress (P_a) and cell pressure (P_o) applied to the models. These values are the same as those adopted in the tests so that direct comparisons between the test and model results can be made. Under isotropic loading, the drilling process was simulated by reducing the borehole pressure to 6 MPa while keeping the cell pressure and axial stress constant. Under anisotropic loading, the drilling process was simulated by reducing the borehole pressure to 0 MPa while keeping the cell pressure constant and axial strain change to zero (i.e., plane strain) during that stage.

Table 1. Model details and input properties

Properties	Model
Inner radius (mm)	12.5*
Outer radius (mm)	40*
Length of model (platen + 1/2 sample) (mm)	120*
Density of sample (kg/m^3)	2278
Density of platen (kg/m^3)	7850
Bulk modulus of sample (GPa)	18.87
Bulk modulus of platen (GPa)	144.9
Shear modulus of sample (GPa)	7.72
Shear modulus of platen (GPa)	78.7
Friction angle of sample (degree)	36.2
Cohesion of sample (MPa)	6.3
Dilation angle of sample (degree)	12
Tensile strength of sample (MPa)	2.07
Porosity of sample (%)	24.23
Permeability of sample ((m/sec)/(Pa/m))	5.14e-17
Bulk modulus of fluid (GPa)	2.0
Density of fluid (kg/m^3)	1000
Normal stiffness of interface (Pa/m)	1.19e15
Shear stiffness of interface (Pa/m)	1.19e15
Cohesion of interface (MPa)	0
Friction angle of interface (degree)	10
Tension limit of interface (MPa)	0

* These values are slightly different in each model since measured values on test samples were used.

Table 2. Summary of initial stress and pore pressure conditions

Loading condition	P_a (MPa)	P_o (MPa)	P_i (MPa)	P_p (MPa)
Isotropic	100	100	60	5
Anisotropic 1	131	100	60	5
Anisotropic 2	149	100	60	5

4 COMPARISON BETWEEN MODEL AND TEST RESULTS

4.1 *Borehole failure initiation*

Table 3 summarises the predicted borehole pressure at failure initiation using the M-C and mean D-P failure criteria.

Table 3. Comparison of predicted and measured borehole failure pressures for different failure criteria and loading conditions

Loading condition	Measured (MPa)	M-C (MPa)	Mean D-P (MPa)
Isotropic	22.5	36	21
Anisotropic 1	19	37	17
Anisotropic 2	10	37	14

As can be seen in Table 3, the borehole strength is enhanced with the increase in intermediate principal stress applied to the samples. However, the predicted borehole failure initiation pressures using the M-C failure criterion are essentially the same for the three cases. It is not surprising since the intermediate principal stress is not considered in the M-C failure criterion. By including the intermediate principal stress in the failure criterion, the predictions based on the D-P failure criterion (mean circle) are significantly improved.

4.2 *Borehole deformation*

Figures 2 & 3 show the comparison between the measured and predicted borehole and external displacements for the isotropic loading condition and Anisotropic Loading Condition 2 respectively.

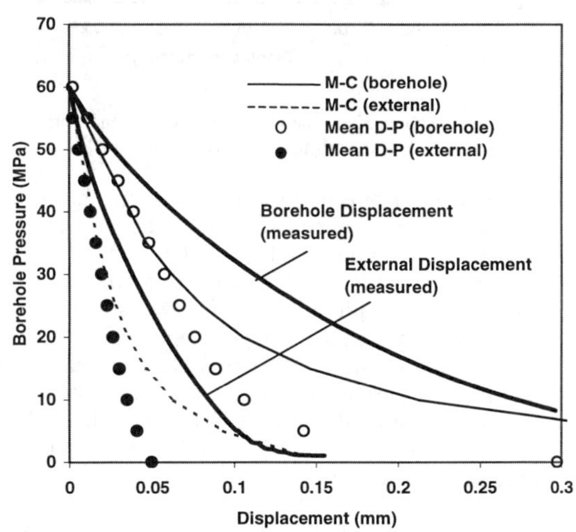

Figure 2. Comparison between measured and predicted displacements for the isotropic loading condition.

It was experimentally observed that anisotropic loading reduced the borehole and external deformations significantly (compare Figures 2 & 3). This was probably due to the stiffening effect of the intermediate principal stress (axial stress) prior to borehole failure initiation and the strengthening effect of the intermediate principal stress on rock strength at and after failure initiation.

For the isotropic loading condition, the models underestimated the measured displacements considerably (Figure 2). One of the possible reasons is that the axial deformation calculated in the numerical simulations was significant different from that generated in the test. It was also found that the displacements calculated based on the D-P failure criterion were considerably smaller than those calculated based on the M-C failure criterion. This result demonstrates the stiffening effect of intermediate principal stress on the borehole.

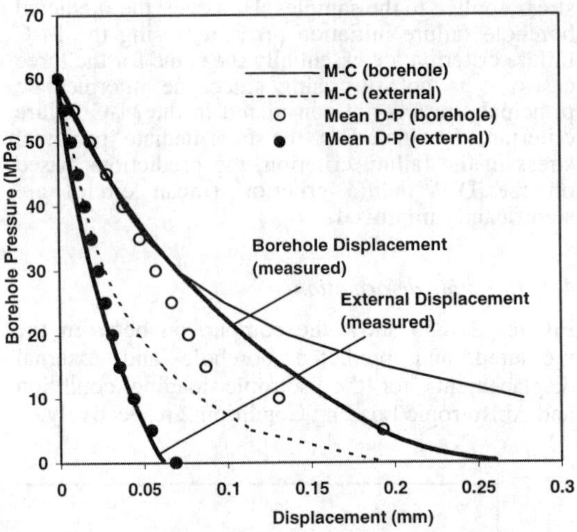

Figure 3. Comparison between measured and predicted displacements for Anisotropic Loading Condition 2.

For the anisotropic loading condition, the model displacements agreed with the measured displacements considerably better than for the isotropic case (Figure 3). However, the model based on the M-C failure criterion overestimated the borehole and external displacements significantly after failure initiation. This is because the intermediate principal stress was not taken into account in the model. By including the intermediate principal stress, the model based on the D-P failure criterion (mean circle) improved the prediction considerably.

As expected, the displacements calculated based on the M-C failure criterion showed no significant difference between the isotropic and anisotropic loading conditions (compare Figures 2 & 3).

5 CONCLUSIONS

In this paper, numerical models are developed to investigate the effect of intermediate principal stress on thick-walled hollow cylinder behaviour. Comparisons are made on the failure initiation pressure and deformations calculated with the Mohr-Coulomb and Drucker-Prager failure criteria. The predicted results are further assessed by comparing them with those determined in the hollow cylinder tests.

The investigation showed that there are large differences between predictions by Mohr-Coulomb and Drucker-Prager criteria under both isotropic and anisotropic loading conditions. In general, anisotropic loading conditions would increase borehole strength and reduce borehole deformations. This trend cannot be predicted by the Mohr-Coulomb criterion. Predictions from Drucker-Prager criterion were found to be in better agreement with the results of the laboratory tests.

In addition, it was found that good agreement cannot be obtained when plane strain assumption is used to predict deformation under isotropic loading conditions in borehole collapse tests.

REFERENCES:

Addis, M. A. & B. Wu 1993. The role of the intermediate principal stress in wellbore stability studies: evidence from hollow cylinder tests, Proc. 34th U.S. Rock Mech. Symp., *Int. J. Rock Mech. & Min. Sci. & Geomech. Abstr.* 30:1027.

Itasca 1998. FLAC user's manual, Itasca Consulting Group Inc., Minneapolis, Minnesota, USA.

Marsden, J. R., B. Wu & J. W. Dennis 1997. Deformation and failure of thick-walled hollow cylinders of mudrock: A study of wellbore instability in weak rock. *Proc. Eurock'96 Prediction and performance in rock mechanics and rock engineering*, 2:759-766.

McLean, M. R. & M. A. Addis 1990. Wellbore stability: the effect of strength criteria on mud weight recommendations, *SPE 20405 presented at the 1990 SPE Ann. Tech. Conf. & Exhib.*

Owen, D. R. J & E. Hinton 1980. *Finite elements in plasticity: theory and practice*. Pineridage Press, Swansea, 594p.

Papamichos, E. & J. Tronvoll, a. Skjarstein & T. E. Unander 1996. The effect of plane-strain and isotropic loading in hollow-cylinder strength. *Proc. Eurock'96 Prediction and performance in rock mechanics and rock engineering*, 1:197-204.

Santarelli, J. F. 1987. Theoretical and Experimental investigation of the stability of the axisymetric wellbore. PhD thesis, Uni. of London.

Santarelli, J. F., M. E. Chenevert & S. O. Osisanya 1992. On the stability of shales and its consequences in terms of swelling and wellbore stability, 1992 IADC/SPE Drilling Conference, New Orleans, USA, 391-400.-

Wu, B. & C. P. Tan 1996. Consolidated undrained and drained triaxial tests of Johnstone, APCRC Restricted Report No. 019.

Wu, B., C. P. Tan & X. Chen 2000. Plane strain effect on strength and deformation of hollow cylinders of shale, *The 4th North American Rock Mechanics Symposium*, Seattle, USA.

Veeken, C. A. M, J. V. Walters, C. J. Kenter, & D. R. Davis 1989. Use of plasticity models for predicting borehole stability. In V. Maury & D. Fourmaintraux (eds), *Proc. Rock at Great Depth, Pau, France,* 2:835-844. Rotterdam: A. A. Balkema.

Computer Methods and Advances in Geomechanics, Desai et al. (eds) © 2001 Balkema, Rotterdam, ISBN 90 5809 183 X

Computer modeling of subsidence and reservoir compaction under field-wide water injection operations

L.Y.Chin & N.B.Nagel
Phillips Petroleum Company, Bartlesville, Okla., USA

ABSTRACT: A full-field, 3-dimensional computer model has been developed to numerically simulate reservoir compaction and surface subsidence for a weak, water-sensitive, hydrocarbon reservoir during field-wide water injection operations. The developed model was used for modeling the compaction and subsidence processes under waterflood operations at the Ekofisk Field in the North Sea. The model was thoroughly validated through the comparison of model results to extensive field measurements with good agreements being achieved. The validated model has been successfully employed as a tool to forecast subsidence and to assist in the development of a subsidence risk assessment.

1 INTRODUCTION

For reservoir rocks with weak mechanical strength (i.e., high compressibility), reservoir compaction and surface subsidence may occur as a result of fluid production and injection operations (Nagel, 2000). At the Ekofisk Field in the North Sea, extensive compaction of the weak reservoir chalk resulted in seafloor subsidence of nearly 8 meters by 2000. Ekofisk subsidence was discovered in late 1984, at a level of nearly 3 meters, leading to a series of engineering efforts to ensure the safety of offshore personnel and the safety and integrity of platforms and structures. These efforts included the raising of platforms in the Ekofisk central complex area by 6 meters in 1987, the installation of a barrier wall around the 2/4 T storage tank in 1989, and the construction and placement of new drilling and processing platforms, the Ekofisk II project, by 1998. These new facilities were designed to withstand approximately 19 meters of subsidence.

Since January 1985, computer modeling of reservoir compaction and seafloor subsidence has played a key role in support of those engineering projects. Over the years, a variety of compaction/subsidence models, with varying degrees of sophistication, were developed to assist in improving the understanding of the physics of the compaction and subsidence processes as well as to provide new forecasts based upon changing field production plans. During the first 15 years of production, the Ekofisk Field was operated primarily under natural depletion with a monotonic decline in reservoir pressure. The compaction/subsidence models developed during this period assumed that the primary mechanism of Ekofisk chalk compaction was pore collapse caused by the increase in effective overburden stress resulting from the decline in reservoir pressure (Boade et al. 1989, Chin & Boade, 1990).

In order to improve field recovery, field-wide waterflood operations were begun by late 1988. By 1993, water injection rate reached 500,000 barrels per day. In this period, subsidence values predicted from the older models were persistently lower than field measurements. Consequently, an improved model was proposed, which took into account the potential effects of overburden behavior, laboratory-to-rock-mass scaling, and shear-induced compaction in order to address the weaker chalk strength observed in the field (Chin et al. 1994). From 1994 to 1998, Ekofisk was operated under a field-average, voidage balancing condition (as much or more fluid was injected as produced per day), though some parts of the reservoir experienced small changes in reservoir pressure. During this period, the subsidence rate for the 2/4 Hotel platform, located near the center of the subsidence bowl, unexpectedly remained nearly unchanged at around 40 cm/year. Under the nearly constant effective stress conditions, the improved model also began to underpredict field subsidence.

With new field observations along with new laboratory test results, it was concluded that under constant effective stress, or even decreasing stress conditions, additional compaction and subsidence could be induced by the weakening effect of injected seawater on the mechanical strength of Ekofisk chalk.

Further, it was determined that the effect of water weakening on reservoir chalk had to be taken into account in order to appropriately model the Ekofisk compaction and subsidence processes.

In this paper, a new computer model for simulating the Ekofisk compaction and subsidence processes under both depletion and waterflood conditions is presented. The primary mechanisms of chalk compaction are assumed to be pore collapse and water weakening, and a constitutive model is developed for properly simulating chalk compaction behavior under these two mechanisms. The detailed description of the model is given in Section 2. The comparison and validation of model results to field measurements are presented in Section 3. Conclusions are provided in Section 4.

2 THE COMPUTER MODEL

The computational domain of the compaction/ subsidence model is schematically shown in Figure 1. The model consists of 4 main regions including overburden, reservoir, sideburden, and underburden. The model numerically simulates the influences that fluid production and injection, from wells located throughout the field, have on reservoir pressure and water saturation, which leads to reservoir compaction and, in-turn, surface subsidence. In the following, the detailed description of the model is presented.

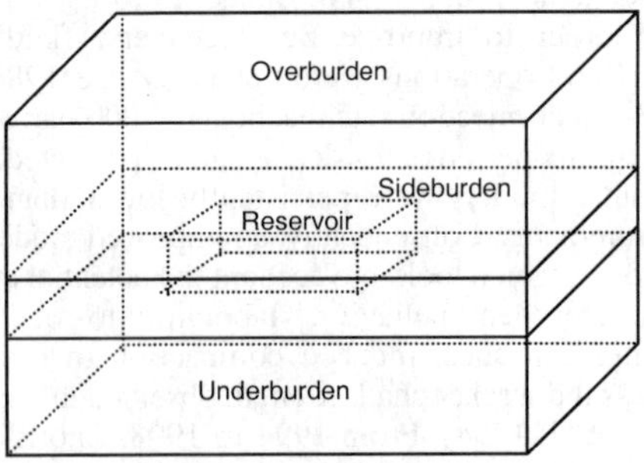

Figure 1. Schematic of the modeled regime.

2.1 *Governing Equations*

The governing equations used in the modeling of the compaction and subsidence processes were formulated based upon the basic principles of the conservation of linear momentum and the effective stress law.

Within the reservoir region:

$$\nabla \cdot \boldsymbol{\sigma} + \rho \mathbf{b} = \mathbf{0} \quad (1)$$

$$\boldsymbol{\sigma} = \boldsymbol{\sigma}' - p\boldsymbol{\delta} \quad (2)$$

And within the overburden, sideburden and underburden regions:

$$\nabla \cdot \boldsymbol{\sigma} + \rho \mathbf{b} = \mathbf{0} \quad (3)$$

where $\boldsymbol{\sigma}$ = solid total stress; ρ = solid bulk density; $\mathbf{b}$ = body force; $\boldsymbol{\sigma}'$ = solid effective stress; p = pore pressure; and $\boldsymbol{\delta}$ = Kronecker delta.

Supplementing Eqs 1 to 3 with pore-pressure and water-saturation distributions in the reservoir and appropriate constitutive equations, the full set of governing equations can be formulated and solved for the primary unknown variable in the compaction and subsidence processes. This primary variable is the solid-displacement vector. Solutions of the displacement field provide simulation results of reservoir compaction, surface subsidence, subsidence bowl contours, stress distribution, etc.

2.2 *Constitutive Models*

2.2.1 *Reservoir Rock*

The main reservoir rock at the Ekofisk Field is high porosity (30-45 %) chalk, which is highly compactable and sensitive to water. Consequently, the compaction behavior of Ekofisk chalk depends on its initial porosity, water saturation, and stress conditions (Rhett, 1998). A constitutive model, that can phenomenologically simulate the elastoplastic behavior of chalk, was developed. The developed model was based on a hypoelastic/hypoplastic 3-D formulation (Prevost & Chin, 1994) that is capable of reproducing uniaxial strain test data for chalk with various initial porosities under a wide range of loading conditions. For water sensitive chalk, the compaction behavior is quantified through the determination of a stress-strain relationship for a given water saturation level, which is interpolated between those of dry and fully water-weakened chalks. Based upon laboratory test results (Rhett, 1998), chalk is considered to be fully water-weakened when its mechanical strength no longer changes with a further increase in water saturation. The saturation level at which this typically occurs in chalk is around 10 to 20 %.

2.2.2 *Overburden, Sideburden, and Underburden Rocks*

It is assumed that no fluid drainage occurs in the overburden, sideburden and underburden regions, and, consequently, the geomaterials in these areas were modeled as single-phase solid materials. An isotropic, linear elastic constitutive model was used for rocks in these regions. However, each region was modeled with the appropriate values for Young's modulus and Poisson's ratio. Also, the overburden was divided into a number of lateral layers, and each layer was assigned values of Young's modulus and Poisson's ratio.

2.3 *Computing Strategy*

An iterative procedure was employed to solve the coupled equations stated in Sections 2.1 and 2.2. The procedure requires the use of two separate models, the compaction/subsidence model and a multi-phase reservoir flow model. For a given iteration pass, the compaction/subsidence model solved for the displacement field of the modeled regime with pressure and water saturation distributions obtained from the numerical results of the previous iteration pass of the multi-phase flow model. A pass is defined as a simulation run with a specified time period. Similarly, for a simulation pass, the reservoir model solved for the pore-pressure distribution and distributions of water, oil, and gas saturations using rock compressibility tables that were updated based upon the calculation results of the compaction/ subsidence model from the previous iteration pass. The iteration process continued until the maximum change of the pore-pressure solution between the current and previous pass was within an acceptable tolerance. Two to three passes were typically required to reach a converged solution for Ekofisk simulation. Also, some additional history matching was performed with the compaction/subsidence model in order to match field data.

The Galerkin finite element method was used to discretize the governing equations of the compaction/subsidence model with the computational domain shown in Figure 1. A typical finite element mesh of the model had about 66,000 elements (67x41x24). The 24 model layers consisted of 8 in the overburden, 14 in the reservoir, and 2 in the underburden. The nonlinear system of equations from the discretization was solved by the quasi-Newton procedure and an ICCG iterative solver was used to solve the resulting linear system of equations. Note also that the multi-phase reservoir flow model simulated only the reservoir section in Figure 1. Further, different reservoir production scenarios were evaluated by repeating the computational process.

3 MODEL VALIDATION

3.1 *Subsidence Versus Time*

Since 1985, significant compaction and subsidence monitoring efforts have been made. Chief among these has been the measurement of platform subsidence through an early GPS system, seabed pressure gauges, and currently by precise GPS. The primary subsidence measurements are taken at the Charlie, Alpha, and Bravo platforms and the total subsidence at these platform locations from 1985 to 2000 are shown in Figure 2. Subsidence calculations for these platform locations over a period about 30 years are also shown in Figure 2. As seen, there is excellent agreement between actual and modeled subsidence, except, perhaps, during the early life of the Bravo platform.

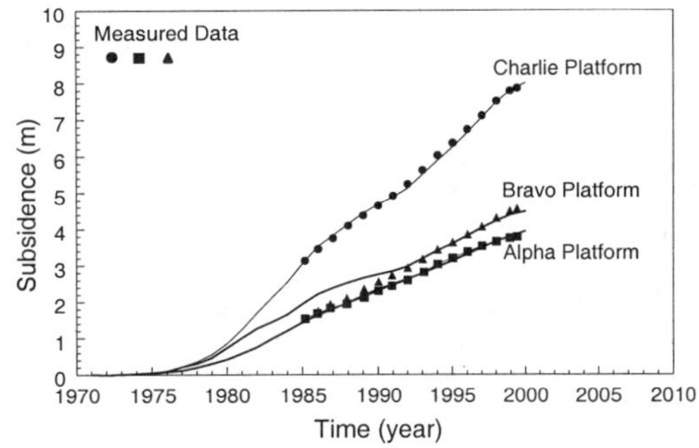

Figure 2. Subsidence versus time at three platform locations.

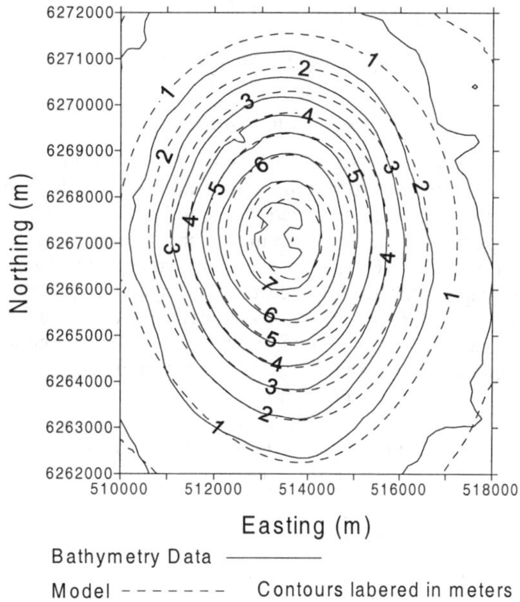

Figure 3. Calculated and measured subsidence contours.

3.2 *Subsidence Contours and Bowl Volume*

Since late 1984, 8 bathymetry surveys have been conducted over the greater Ekofisk area in order to determine the development of subsidence and to estimate subsidence bowl volume. Figure 3 shows a comparison between modeled results and measured subsidence from the 1999 bathymetry survey. As seen, the agreement is reasonably good.

The comparison of modeled and measured subsidence bowl volume as a function of time is shown in Figure

4. Again, a very good agreement between modeled and measured results is obtained.

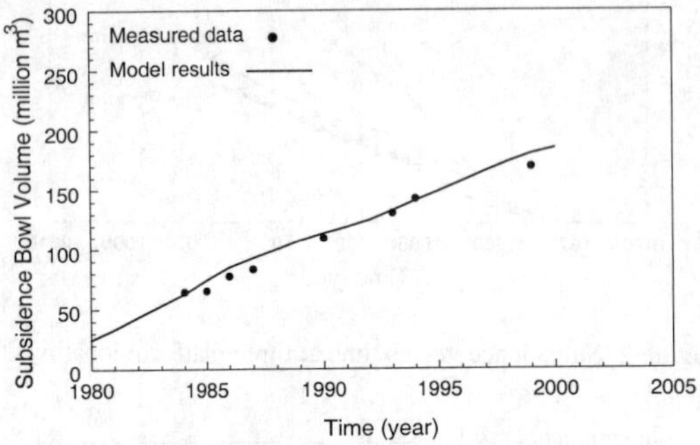

Figure 4. Subsidence bowl volume versus time.

3.3 *Reservoir Compaction Profiles*

In addition to significant subsidence monitoring, extensive reservoir compaction monitoring has also been undertaken. In particular, regular compaction logging in a dedicated compaction monitoring well, well 2/4 C11/C11A has been performed since 1986. Comparisons of log-measured compaction profiles from 2/4 C11/C11A and calculated profiles from the model are presented in Figure 5. As seen, there is good agreement between the measured and computed profiles for the period dominated by depletion (1986 to 1994) and the overall period influenced by both depletion and water weakening (1986 to 1999).

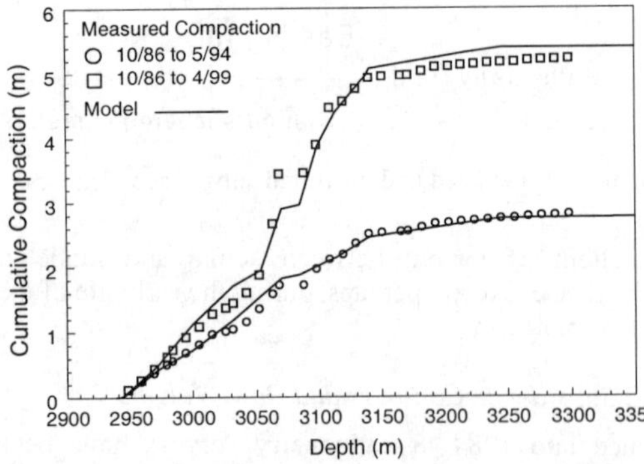

Figure 5. Cumulative compaction versus depth for different time intervals.

4 CONCLUSIONS

A full-field, 3D computer model was developed for numerically simulating the Ekofisk compaction and subsidence processes under both depletion and waterflood conditions. The two primary compaction mechanisms considered in the model are pore collapse and water weakening. Good agreements between modeling results and field measurements were achieved. These results show that in addition to pore collapse compaction caused by reservoir depletion, water-induced compaction is also an important mechanism for water sensitive rock and has a significant impact on predicting compaction and subsidence under waterflood operations at Ekofisk.

ACKNOWLEDGMENTS

The authors acknowledge permission to publish the above paper from Phillips Petroleum Company Norway and Co-venturers, including Fina Exploration Norway S.C.A., Norsk Agip A/S, Elf Petroleum Norge AS, Norsk Hydro Produksjon a.s, TOTAL Norge A.S and Den norsk stats oljeselskap a.s. The opinions expressed in this paper are those of the authors and do not necessarily represent those of the Phillips Norway Group.

REFERENCES

Boade, R.R., Chin, L.Y., and Siemers, W.T. 1989. Forecasting of Ekofisk reservoir compaction and subsidence by numerical simulation. *JPT, July 1989*: 723-728.

Chin, L.Y. and Boade, R.R. 1990. Full-field, 3D, finite-element subsidence model for Ekofisk. *Proc. of the 3rd North Sea Chalk Symposium, Copenhagen, June 11-12, 1990*.

Chin, L., Boade, R.R., Nagel, N.B., and Landa, G.H. 1994. Numerical simulation of Ekofisk reservoir compaction and subsidence: treating the mechanical behavior of the overburden and reservoir. SPE/ISRM 28128, Eurock 1994, 787-794, Delft, Netherlands, 29-31 August, 1994.

Nagel, N.B. 2000. Compaction and subsidence issues within the petroleum industry: from Wilmington to Ekofisk and beyond. Paper presented at *EGS 2000, Nice, France, 25-29 April, 2000*, to appear in *Phy. & Chem. of the Earth*.

Prevost, J.H. and Chin, L. 1994. A constitutive model for simulating reservoir compaction under a constrained path that leads to high shear. *Proc. of The Eighth Intern. Conference on Computer Methods and Advances in Geomechanics, Vol. III, 2139-2144, Morgantown, West Virginia, USA, May 22-28, 1994*.

Rhett, D.W. 1998. Ekofisk revisited: a new model of Ekofisk reservoir geomechanical behavior. *Proc. SPE/ISRM Rock Mechanics in Petroleum Engineering, Eurock 1998, 367-375, Trondheim, Norway, 8-10 July, 1998*.

Computer Methods and Advances in Geomechanics, Desai et al. (eds) © 2001 Balkema, Rotterdam, ISBN 90 5809 183 X

Buckle folding of elastic strata

F.S.Jeng & N.G.Chang
Department of Civil Engineering, National Taiwan University, Taipei, Taiwan

C.Y.Lu
Department of Geology, National Taiwan University, Taipei, Taiwan

ABSTRACT: This paper presents a general solution of an elastic layer embedded in soft material, which has been partly discussed. This solution is then compared to finite-element simulation using a proposed perturbation method, which can exclude the effect of initial geometry imperfection. The numerical simulation reveals both of the waveform and the post-buckle behavior of elastic competent strata. It is observed that, waveform comprised of two frequencies may occur as the exerted force F is greater than a critical force, F_{cr}. When $F < F_{cr}$, the amplitude will decay as the distance to the perturbed end increases. Meanwhile, the mechanism of two-ordered fold and poly-harmonic fold are discussed in this paper.

1 INTRODUCTION

While folding under high confining stress and high temperature condition, the stratum (e.g. metamorphic rocks) exhibits viscous behavior and hence, the folding behavior is frequently analyzed using viscous theories (Biot, 1961; Fletcher, 1977; Mühlhaus *et al.*, 1998; Hunt *et al.*, 1996). However, folding of sedimentary rocks, namely under low confining stress and temperature condition, can still often be observed. Therefore, analyses based on the elastic behavior of stratum were accordingly conducted (Currie *et al.*, 1962; Hunt *et al.*, 1993). Numerical analyses to reveal the mechanism of non-viscous folding have also been conducted (Cobbold, 1977; Lan & Hudleston, 1991).

The simulation of folding using numerical method involves shortening of the strata, a perturbation to induce folding and a subsequent shortening to further deform the folds. The perturbation is induced by the "initial geometric imperfection" of the layer (Zhang *et al.*, 1996), which often is a consequence of natural sedimentary process. However, this initial imperfection can sometimes influence the characteristics (e.g. wavelength) of the yielded fold, which unfortunately interferes the interpretation of the folding mechanism (Mancktelow, 1999). Therefore, it is wished that the effect of the mechanical characteristics of the strata (e.g. stiffness and thickness) upon the folding behavior could be isolated from the effect of the initial perturbation geometry.

Another perturbation method, which excludes the effect of initial geometry imperfection, is therefore proposed in this paper. The layer is first shortened, followed by employing a small rotation at the boundary to apply a moment to the layer to induce folding. The result indicates that this technique yields folds corresponding only to the characteristics of the strata without any influence from the initial perturbation geometry.

To explore the mechanism associated with folding of an elastic layer, this paper presents a complete closed form solution of a layer embedded in soft material, which has been partly discussed (Karman & Biot, 1940; Currie *et al.*, 1962). The solution is then compared to the results of finite-element simulation. The mechanism of polyharmonic fold is also discussed in this paper.

2 STRATUM SURROUNDED BY SOFT MATRIX – ELASTIC THEORY

2.1 *Stratum surround by spring*

First, considering a stratum embedded in soft matrix comprised of elastic springs, the governing equation is (Karman & Biot, 1940):

$$Y''''(X) + \frac{F}{EI}Y''(X) + \frac{K}{EI}Y(X) = 0 \tag{1}$$

where: Y, E and I are the vertical displacement, Young's modulus and moment of inertial of the stratum, respectively; K is the stiffness of the spring; and F is the horizontal force required to buckle the stratum, namely to produce a fold.

Solving Eqn. 1, the characteristic value λ can be determined by the F, E, I and K as:

$$\lambda = \pm\sqrt{\frac{-F \pm \sqrt{F^2 - 4EIK}}{2EI}} \tag{2}$$

The solution of Eqn. 1 depends on the relative magnitude of the exerted force F and the rigidity of the stratum and spring ($\sqrt{4EIK}$). When $F \geq \sqrt{4EIK}$, the required force F for yielding a fold with wavelength (l) can be related as:

$$F = \frac{4\pi^2}{l^2}EI + \frac{Kl^2}{4\pi^2} \tag{3}$$

Fig. 1 illustrated the relation of F with l when $F \geq \sqrt{4EIK}$, based on Eqn. 3.

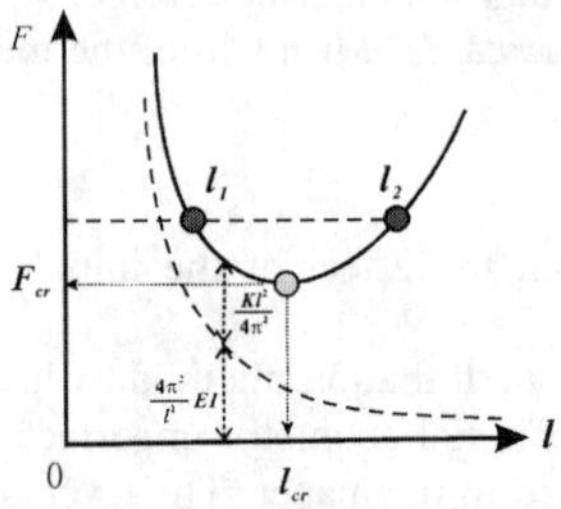

Figure 1 The relation of F with induced wavelength l *when* $F \geq \sqrt{4EIK}$ (after Karman & Biot, 1940).

As illustrated by Fig. 1, when $F \geq \sqrt{4EIK}$, a critical force (F_{cr}) exists with a corresponding wavelength l_{cr}, which are:

$$F_{cr} = \sqrt{4KEI} \tag{4}$$

$$l_{cr} = 2\pi\sqrt[4]{\frac{EI}{K}} \tag{5}$$

For $F > F_{cr}$, the solution of Eqn. 1 is a wave comprised of two frequencies:

$$Y = a\sin mX + b\cos mX + c\sin nX + d\cos nX \tag{6}$$

where m and n are:

$$m = \sqrt{\frac{-F + \sqrt{F^2 - 4EIK}}{2EI}} \tag{7a}$$

$$n = \sqrt{\frac{-F - \sqrt{F^2 - 4EIK}}{2EI}} \tag{7b}$$

The wavelengths corresponding to m and n are (Fig. 1):

$$l_1 = \frac{2\pi}{m}\text{ ; and } l_2 = \frac{2\pi}{n} \tag{8}$$

Yet, when $F < F_{cr}$, folding of the stratum is still possible, which is not well discussed so far. When $F < F_{cr}$, solution of Eqn. 1 has the following form:

$$Y = e^{-mx}(a\cos nx + b\sin nx) \tag{9}$$

where m and n are determined by the exerted force F and the property of material (E, I and K) as:

$$m = \sqrt{\frac{-F + \sqrt{4EIK}}{4EI}} \tag{10a}$$

$$n = \sqrt{\frac{F + \sqrt{4EIK}}{4EI}} \tag{10b}$$

Based on Eqn. 9, the fold has only one frequency and the amplitude tends to decay as x (the distance from the perturbed end) increases.

The relation of F with the wavelength l can be expressed as (Fig. 2):

$$F = \frac{16\pi^2 EI}{l^2} - \sqrt{4EIK} \tag{11}$$

Eqn. 11, indicate that: (1) folding is possible even when the exerted force is close to zero; (2) the less the F, the greater the wavelength l and the severer the decay of amplitude as well; and (3) when F approaches zero, a maximum wavelength (l_3) exists, which can be expressed as:

$$l_3 = \frac{4\pi\sqrt{EI}}{\sqrt[4]{4EIK}} \tag{12}$$

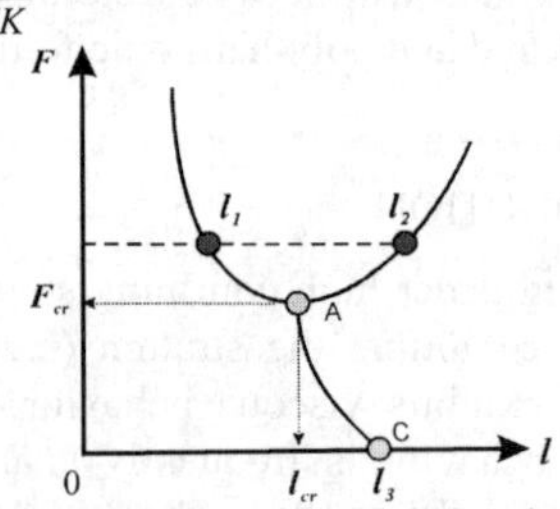

Figure 2 Complete solution of F with induced wavelength l

2.2 Stratum surround by soft matrix

As the stratum is confined by matrix instead of spring, an approximate governing equation can be obtained as (Biot, 1937; Currie *et al.*, 1962):

$$Y''''(X) + \frac{F}{\overline{E}I}Y''(X) + \frac{2\pi\overline{E}_o}{\overline{E}Il}Y(X) = 0 \tag{13}$$

where $\overline{E} = \frac{E}{(1-\nu^2)}$; $\overline{E}_o = \frac{E_o}{(1-\nu^2)}$; E_o = Young's modulus of the matrix; ν = Poisson ratio.

As Eqn. 13 is similar to Eqn. 1, the characteristics of its solution is similar to those described in Section 2.1, except that the term, $\sqrt{4EIK}$, is now replaced by $\frac{3}{2}\sqrt[3]{2\overline{E}I\overline{E}_0^2}$ instead.

The relation of exerted force F and the wavelength can then be determined as (Currie *et al.*, 1962):

$$F = \frac{4\pi^2\overline{E}I}{l^2} + \frac{\overline{E}_0 l}{2\pi} \tag{14}$$

Similarly, F_{cr} and l_{cr} can be obtained as:

$$F_{cr} = \frac{3}{2}\sqrt[3]{2\overline{E}I\overline{E}_0^2} \tag{15}$$

$$l_{cr} = 2\pi h \sqrt[3]{\frac{\bar{E}}{6\bar{E}_0}} \quad (16)$$

where h is the thickness of the layer.

3 NUMERICAL SIMULATION OF BUCKLE FOLDING

To prevent the initial geometric imperfect affecting the yielded buckle fold, the hard layer is set to be flat without initial imperfection. The process to produce buckle fold is as follows: the layer together with the matrix are shortened with an initial shorting strain ε_i, followed by applying moment at one end of the layer to buckle the stratum.

The initial shorten strain ε_i can be expressed in terms of initial shorting length (ΔL) as:

$$\varepsilon_i = \frac{\Delta L}{L} \quad (17)$$

where: L = Length of the layer; A = cross-sectional area of the layer. A dimensionless measure of lateral stress, σ, exerted on the end of strata is defined as:

$$\sigma = \frac{F}{EA} \quad (18)$$

The σ–ε relationship of a layer before and after folding is illustrated by Fig. 3. If the layer is purely shortened without buckling, the σ–ε curve follow a linear segment AB. If the stratum is buckled at C, then it will follow segment CD for the subsequent shortening. This indicates that the buckled stratum (segment ACD) accumulated less strain energy compared to the pure shortening case (segment ACB).

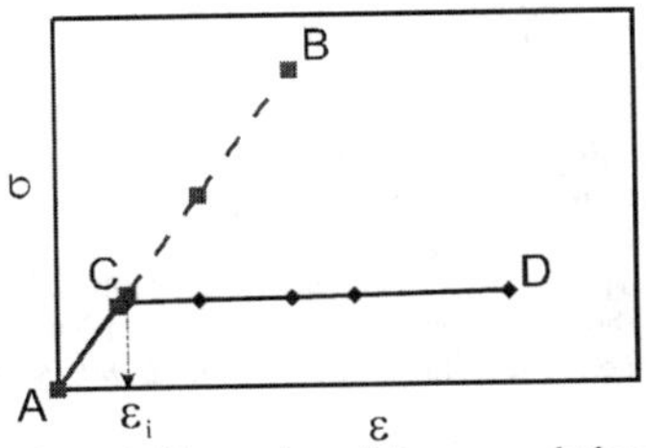

Figure 3 Schematic illustration of the σ–ε relationship of the stratum

When $F < F_{cr}$, the yield fold exhibits decaying nature of its amplitude (Fig. 4) as indicated by Eqn. 8.

Figure 4 Geometry of fold yielded when $F < F_{cr}$

As $F > F_{cr}$, the fold exhibits a "bi-frequency" nature. As shown in Fig. 5, at the commencement of buckling (Fig. 5a), the fold comprises wave of two frequencies. Upon the subsequent shortening of the stratum, the waveform with greater wavelength (l_2) has much greater amplitude than another waveform (l_1) (Fig. 5b), and l_2 hence becomes a dominant wavelength.

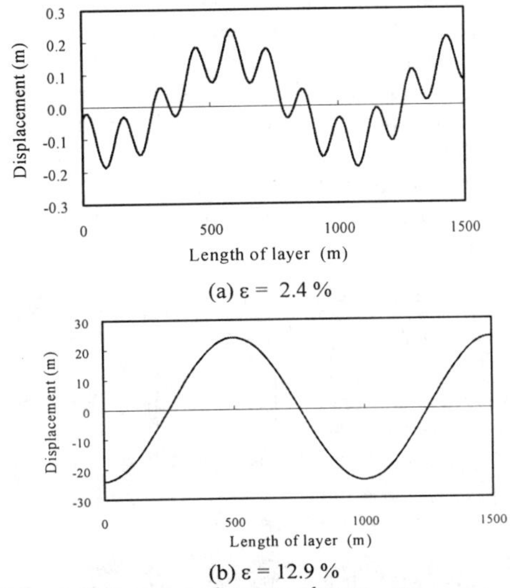

Figure 5 Deformation of stratum when $F > F_{cr}$

The geometry of the fold shown in Fig. 5a is similar to the two-ordered fold observed by geologists. According to Price & Cosgrove (1990), an earlier buckling (with shorter wavelength; second-order fold) increases the effective thickness of the stratum (from a_1 to a_2 shown in Fig.6) and forms a longer wavelength (first-order fold) at a later stage of buckling.

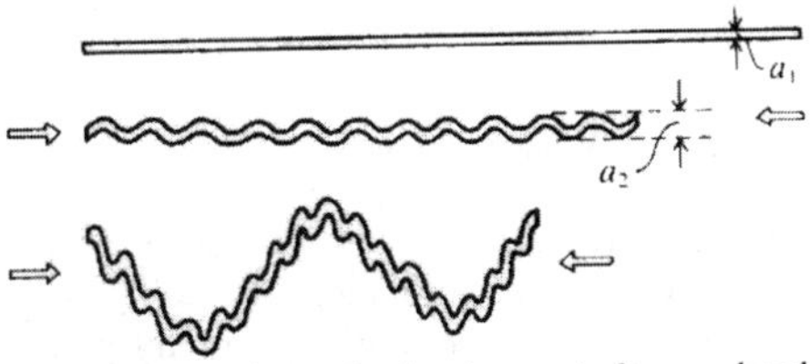

Figure 6 Interpretation to the development of two-ordered fold by Price & Cosgrove (1990)

If the "bi-frequency" nature of fold do exist for $F > F_{cr}$, this simulation result provides another possible mechanism for the underlain mechanism of two-ordered fold. However, the layer may possess viscous behavior, especially under high temperature and high pressure condition; therefore, the impact of viscosity should be accessed before any further interpretation of bi-frequency phenomenon.

The buckle of a multi-layer strata system may induce a poly-harmonic fold. Based on results of numerical analyses, the development of poly-harmonic fold is as follows:

1 When the stratum is compressed, the middle layer is first buckled with a wavelength as indicated in Fig. 7a;
2 Upon subsequent compression, the adjacent layers, hereafter referred as "boundary layer", are then buckled, which possesses a longer wavelength than that of the middle layer (Fig. 7b);
3 As the amplitude of the folded boundary layers is great enough to affect the middle layer, the middle layer is forced to deflect again according to the waveform of the boundary layers.

This process is identical the interpretation by Ramberg (1964).

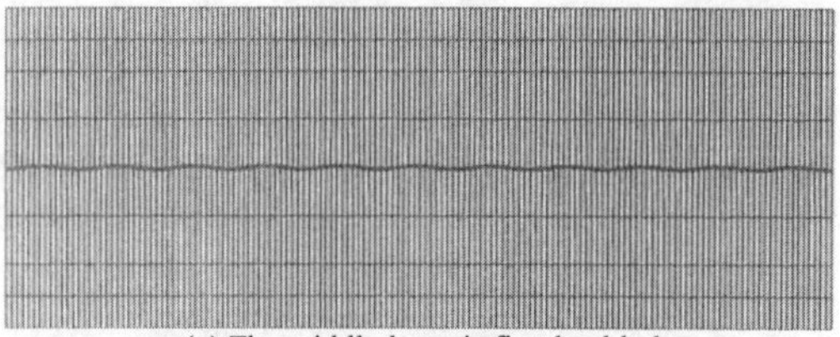

(a) The middle layer is first buckled

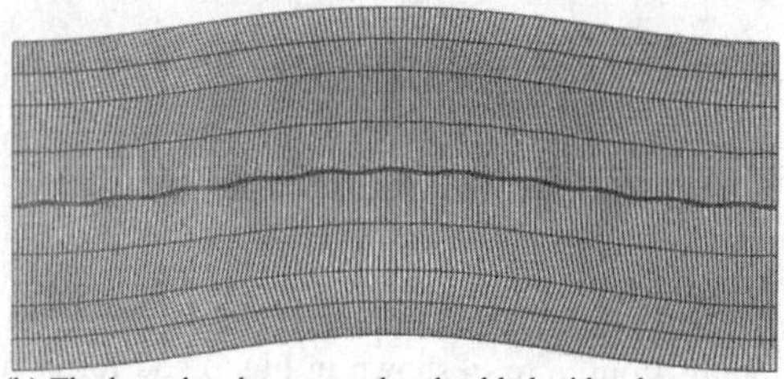

(b) The boundary layers are then buckled with a longer wavelength

Figure 7 Development of poly-harmonic fold of multi-layer strata indicated by numerical analysis.

4 CONCLUSIVE REMARKS

An alternative perturbation method is proposed to achieve numerical simulation of folding without introducing the effect of "initial geometry perturbation". The simulation technique allows studying the underlain folding mechanism related to the characteristics of the layer and matrix only, instead of geometry imperfection.

Meanwhile, a complete general solution for a singe, elastic layer embedded in soft matrix is provided such that the solution and the simulation results can be compared and justified. The solution indicates that folding is possible for both $F \geq F_{cr}$ and $F < F_{cr}$. When $F \geq F_{cr}$, the waveform will may comprises of two frequencies. When $F < F_{cr}$, decay of wave amplitude occurs.

The mechanism associated with the formation of multi-order fold of a single layer as well as the poly-harmonic fold of a multi-layer system is also explored. For a single layer, the "bi-frequency" nature of the waveform, when $F > F_{cr}$, can be a possible cause for multi-order fold.

As revealed by the numerical analysis, the underlain mechanism of poly-harmonic fold of multi-layer can be as follows: high frequency buckle folding along a thinner layer is first initiated at a earlier shortening, followed by buckle of thicker layers at a later shortening. The later folding has a longer wavelength and forces the earlier formed folds to deform with a lower frequency.

REFERENCE

Biot, M.A. 1937. Bending of an infinite beam on an elastic foundation. *J. Applied. Mechanics* **4**: A1-A7.

Biot, M.A. 1961. "Theory of Folding of Stratified Viscoelastic Media and Its Implications in Tectonics and Orogenesis. *Geol. Soc. Am. Bull.* **72**: 1595-1620.

Cobbold, P.R. 1977. Finite-element Analysis of Fold Propagation – a Problematic Application? *Tectonophysics* **38**: 339-353.

Currie, J.B., A.W. Patnode & R.P. Trump 1962. Development of Folds in Sedimentary Strata. *Geol. Soc. Am. Bull.* **73**: 461-472.

Fletcher, R.C. 1977. Folding of a single viscous layer: exact infinitesimal amplitude solution. *Tectonophysics* **39**: 593-606.

Hunt, G.W., H.-B. Mühlhaus & A.I.M. Whiting 1996. Evolution of localized folding for a thin elastic layer in a softening visco-elastic medium. *PAGEOPH* **146(2)**: 229- 252.

Karman, T.V. & M.A. Biot 1940. *Mathematical methods in engineering*, New York :Mcgraw-Hill.

Lan, L. and P. J. Hudleston 1991. Finite-element models of buckle folds in non-linear materials. *Tectonophysics* **199**: 1-12.

Mancktelow, N.S. 1999. Finite-element modeling of single-layer folding in elasto-viscous materials: the effect of initial perturbation geometry. *J. of Structural Geology* **21**: 161-177.

Mühlhaus, H.-B., H. Sakaguchi & B.E. Hobbs 1998. Evolution of three-dimensional folds for a non-Newtonian plate in a viscous medium. *Proc. R. Soc. Lond. A.* **454**: 3121-3143.

Price, N.J. & J.W. Cosgrove 1990. *Analysis of geological structures.* Cambridge: Cambridge Univ. Press.

Ramberg, H. 1964. Selective buckling of composite layers with contrasted rheological properties; a theory for simultaneous formation of several orders of folds. *Tectonophysics* **1**: 307-341.

Zhang, Y., B.E. Hobbs, A. Ord & H.B. Mühlhaus 1996. Computer simulation of single-layer buckling. *J. of Structural Geology* **18 (5)**: 643-655.

Computer Methods and Advances in Geomechanics, Desai et al. (eds) © 2001 Balkema, Rotterdam, ISBN 90 5809 183 X

A numerical method for simulating large displacements of jointed rocks under groundwater pressure

Guangyuan Li
Canada Center for Mineral and Energy Technology, Natural Resources Canada, Ottawa, Ont., Canada

ABSTRACT: This paper presents a numerical technique for analyzing the stability of excavations in jointed rocks. The model uses a discontinuous approach to simulate the large displacements in jointed and faulted rock masses under complex geological and geometrical conditions, such as underground water pressure, in-situ stress, excavation and back fill sequences. A complete modeling package called BSM3D has been developed based on the proposed formulation. An application example is presented to demonstrate the capability of the 3-D model.

1 INTRODUCTION

While planning and designing engineering structures and mining excavations in jointed rocks, it is important to incorporate the influence of geological discontinuities in the stability assessment of the rock mass. The ways of modeling these discontinuities may have a great impact on the analysis results. Basically, there are two types of methods. One uses continuous formulation, such as finite element method (FEM) and boundary element method (BEM). Another type uses discontinuous formulation, such as distinct element method (DEM), displacement discontinuity method (DDM) and displacement deformation analysis (DDA) as well as the proposed block spring model (BSM). In the case of jointed rock mass, the continuous model is obviously not suitable to simulate the progressive failure accompanied with large displacement. Most of the discontinuous models except DDM use a technique of blocky theory (Goodman and Shi 1985), which considers the jointed rock mass as an assembly of separated rigid or deformable blocks. Although a number of block models and programs are available in the literature, very few of them are able to solve the cases of three-dimensional densely jointed rocks imposed with complicated geological conditions.

The proposed 3-D block spring model is developed based on an assumption that the rock structures are formed with isolated rigid blocks separated by joints and the blocks are connected through springs in an equilibrium condition. The blocks are in contact along their interfaces and joint surfaces. The deformation of the whole rock mass is considered as the result of the relative displacements between joint surfaces. The contact forces in normal and shear directions are expressed by the stiffness of the springs and the relative displacements between blocks. A set of equilibrium equations for all the blocks can be formed after introducing boundary conditions and other imposed forces including external load, gravity, in-situ stress, groundwater pressure, unloading caused by the excavation etc. The displacements defined at the centroids of the blocks can be obtained by solving the stiffness equilibrium equations. The contact forces calculated from the relative displacements are then examined by failure criterion of the rock mass. An iterative procedure is used to simulate the large displacement of the blocks if the contact forces do not satisfy the failure conditions. A special technique is used to simulate the unstable blocks, i.e. the blocks that have insufficient supports from other blocks.

2 DISPLACEMENTS AND CONTACT FORCES

As the rock blocks are assumed rigid, there are only six degrees of freedom defined at the centroid of each block. The six components of displacement consist of three translations U_x, U_y, U_z and three rotations θ_x, θ_y, θ_z around coordinate axles as shown in Figure 1. The displacement $\{\boldsymbol{u}\}_i = \{ u_{xi}, u_{yi}, u_{zi} \}^T$ at any point Q inside block i can be determined in terms of the six displacement components of the block.

$$\{\boldsymbol{u}\}_i = [\boldsymbol{B}]_i \{\boldsymbol{U}\}_i \tag{1}$$

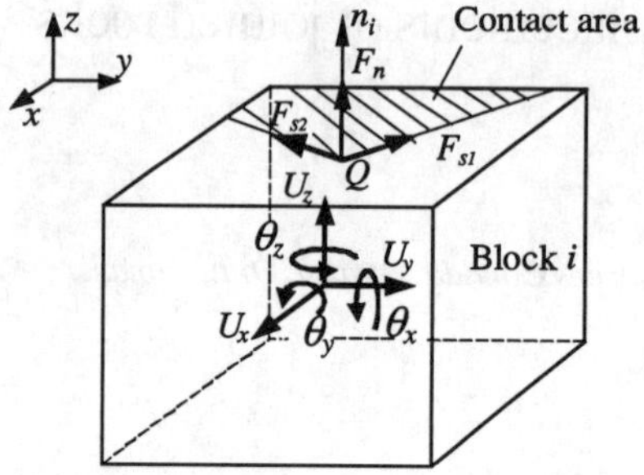

Figure 1. Displacements and contact forces

where

$$[\boldsymbol{B}]_i = \begin{bmatrix} 1 & 0 & 0 & 0 & \bar{z}_i & -\bar{y}_i \\ 0 & 1 & 0 & -\bar{z}_i & 0 & \bar{x}_i \\ 0 & 0 & 1 & \bar{y}_i & -\bar{x}_i & 0 \end{bmatrix} \quad (2)$$

$$\{\boldsymbol{U}\}_i = \{U_{xi} \quad U_{yi} \quad U_{zi} \quad \theta_{xi} \quad \theta_{yi} \quad \theta_{zi}\}^T \quad (3)$$

In which $\bar{x}_i$, $\bar{y}_i$ *and* $\bar{z}_i$ are local coordinates of point Q originated at the centroid of block *i*.

The contact forces between contact points of block *i* and *j* can be calculated from the contact stiffness and the relative displacements.

$$\{\boldsymbol{F}\}_{ij} = [\bar{k}]_{ij}\{\delta\}_{ij} \quad (4)$$

By expressing $\{\delta\}_{ij} = \{u\}_j - \{u\}_i$ in terms of the displacements of block *i* and *j*, we can rewrite Equation (4) as

$$\{\boldsymbol{F}\}_{ij} = [\boldsymbol{k}]_{ij}\{\boldsymbol{U}\}_{ij} \quad (5)$$

$\{\boldsymbol{F}\}_{ij}$ is a vector of forces acting on the centroid of block *i*, which are equivalent to the contact forces; $\{\boldsymbol{U}\}_{ij}$ contains the displacements of block *i* and *j*.

The contact stiffnesses K_n and K_s in normal and shear directions are determined from the stiffnesses of rock joints and those of the intact rocks. The equivalent stiffnesses of joints and intact rocks may be calculated by

$$K_n = \frac{1}{\dfrac{1}{K_n'} + \dfrac{1}{K_n''}}, \quad K_s = \frac{1}{\dfrac{1}{K_s'} + \dfrac{1}{K_s''}} \quad (6)$$

The stiffnesses of intact rocks K_n' and K_s' can be determined using the expression proposed by Kawai *et al.* (1978).

$$K_n' = \frac{E'}{h_i + h_j}, \quad K_s' = \frac{G'}{h_i + h_j} \quad (7)$$

where

$$E' = \frac{(1-\mu)E}{(1+\mu)(1-2\mu)}, \quad G' = \frac{E}{2(1+\mu)} \quad (8)$$

E and μ are Young's modulus and Poisson's ratio of the rock material; h_i and h_j are the distances from the centroids of block *i* and *j* to the contact surface.

The stiffnesses of rock joints may be determined from the rock shear tests along discontinuities.

3 FORCES FROM GROUNDWATER FLOW

The groundwater pressure is considered as the result of underground seepage flow through jointed rocks. The magnitude and direction of the seepage force depend on rock properties, free or confined water table, location of the point in consideration, and boundary conditions of the rock structure. In order to calculate the groundwater pressure or seepage forces on each surface of the block, the water head imposed on the block must be determined. The water head distribution within the rock domain can be obtained by an analysis of steady seepage or fluid potential, which is governed by a set of Laplace's equations:

$$k_x \frac{\partial^2 \phi}{\partial x^2} + k_y \frac{\partial^2 \phi}{\partial y^2} + k_z \frac{\partial^2 \phi}{\partial z^2} = 0 \quad (9)$$

where ϕ is the fluid potential, $\phi = z + h$; z and h are vertical coordinate and water head of the point; k_x, k_y and k_z are permeabilities in x, y and z directions. Fluid potential can be solved using a steady flow FEM program. In the case of modeling excavation sequences, a number of runs may be required to determine the water heads corresponding to different stages of the excavation. The nodal forces induced by the groundwater flow can be calculated by the following expression,

$$\{f\}^e = -\gamma_\omega \iiint_e [N] \left[\frac{\partial \phi}{\partial x} \quad \frac{\partial \phi}{\partial y} \quad \frac{\partial \phi}{\partial z} - 1 \right]^T dxdydz \quad (10)$$

where γ_ω is the unit weight of water; $[N]$ is a interpolation function.

Compared to the method that determines the water head directly from the depth of the point below water table, the results from seepage analysis are more realistic as they reflect the resistance of the rock material to the underground water flow.

The forces at contact points induced by groundwater flow can be calculated from Equation (10). And then the equivalent forces transferred into the centroid of block *i* can be evaluated by

$$\{F_w\}_i = \sum_{b_i} [B]_j^T \{f\}_j^e \tag{11}$$

where $\{f\}^e_j$ is a vector of seepage forces acting at point j of block i; $[B]_j$ is a transformation matrix of block i at point j defined by (2).

4 EQUILIBRIUM EQUATIONS AND SOLUTION OF LARGE DISPLACEMENT

Calculate the internal forces using Equation (5) for all blocks to form a vector of forces from inter-contacts between blocks. Calculate the seepage forces using Equation (11) and other forces from in-situ stress, gravity and boundary displacements to form a vector of external loads. Then a set of simultaneous equations can be established based on the static equilibrium conditions between internal and external forces.

$$[K]\{U\} = \{P\} \tag{12}$$

$[K]$ is the global stiffness matrix of the discrete block system; $\{U\}$ is a column vector of unknown displacements; and $\{P\}$ is a vector of external and initial forces. By solving Equation (12), the unknown displacements and rotations can be obtained. The contact forces are then calculated using Equation (4). The average stresses of the rock mass are evaluated from the contact forces.

Basically, two types of failure modes are considered along the rock joints: slide due to the shear failure and separation due to the tensile failure. A third mode of compression failure may be examined if high stress resulted from the computation. In simulation of the shear failure behavior of the joints, the Mohr-Coulomb failure criterion is used in view of its simplicity, though other failure criteria can be incorporated in the proposed model. Tensile and compression failures happen when normal contact forces exceed the material strength.

The equilibrium state of the forces determined from Equation (12) could be disturbed if failure occurs along joints or block interfaces. As a result, further displacements of the blocks are expected since the forces are now unbalanced. In this case, another cycle of computation is required to simulate the further movement of the blocks and to determine a new equilibrium state of the contact forces. The contact forces on failed surfaces are reduced according to the strength of the rock material. A new stiffness matrix is then formed to reflect the new contact state between blocks and the change of rock properties. In each cycle of iteration, the incremental displacement is accumulated and the new contact forces are checked with failure conditions until no failure is detected or certain blocks do not satisfy the equilibrium conditions. In the case of unstable blocks, a maximum displacement interval and a minimum stiffness are used to damp the system and to prevent the matrix from singularity.

5 EXCAVATION OF AN OPEN-PIT MINE

Figure 2 is a cross section of a simulated open-pit mine near a lakelet. The groundwater flow penetrates through the jointed rocks and comes out from the bottom of the pit. The jointed rock mass around the slopes is cut by two major faults F_1 and F_2 and is divided into four material zones E_1 to E_4. Three sets of joints are simulated in the analysis. The 3rd set of joint (J_3) is parallel to the yz plane. The material properties of the jointed rocks are listed in table 1, in which k ($= k_x = k_y = k_z$) is the permeability of the rock mass. The in-situ stresses are calculated by $\sigma_x = 2.46 + 0.04h$, $\sigma_y = 4.85 + 0.062h$ and $\sigma_z = 0.027h$, where h is the depth below ground surface.

Table 1. Material properties

	E_1	E_2	E_3	E_4	F_1, F_2
E (MPa)	18000	9850	28600	24000	---
μ	0.26	0.28	0.26	0.25	---
γ(MN/m^3)	0.025	0.025	0.03	0.026	---
k (m/day)	0.004	0.007	0.0005	0.0005	---
C (MPa)	0.72	0.5	2.40	1.25	0.06
ϕ (degree)	30	28	32	32	24
K_n (MPa/m)	---	---	---	---	38
K_s (MPa/m)	---	---	---	---	5.2

The fluid potential contours resulting from the seepage analysis are shown in Figure 3 (a) and (b), representing the states before and after the mining excavations. The underground water table is determined by using a trial technique, which requires adjusting the water table several times in order to get a stable level.

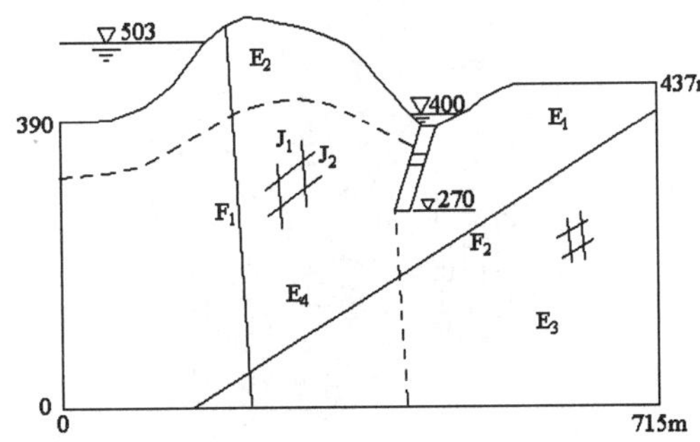

Figure 2. Cross section of an open-pit mine

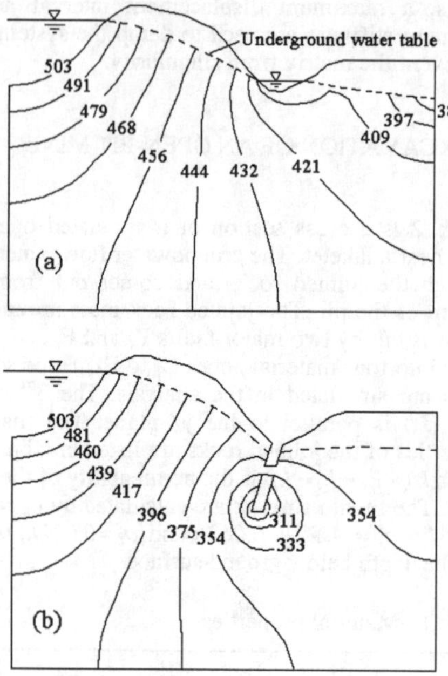

Figure 3. Fluid potentials of the underground flow

Two mining steps are used to simulate the excavations in two levels. The mining stopes are 20m wide, 115m high and 60m long. The sill and rib pillars are 15m thick. Figure 4 is the mesh model used for the analysis.

The displacement vectors induced by in-situ stress and underground water flow are plotted in Figure 5. The jointed rock mass, which is cut by two major faults, is displaced toward the upper right direction due to the groundwater pressure from the left-hand side. In this stage, shear failure occurs along F_2 and some joint surfaces. Although the maximum displacement is 8.4cm on the slope of hanging wall side, the rock blocks are all in stable conditions. The major principal stress prior to the excavation is relatively small, $\sigma_{1max} = 35.7$MPa.

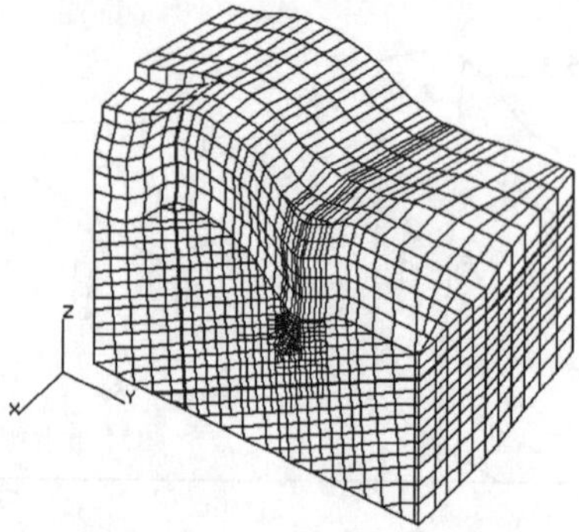

Figure 4. BSM3D mesh model

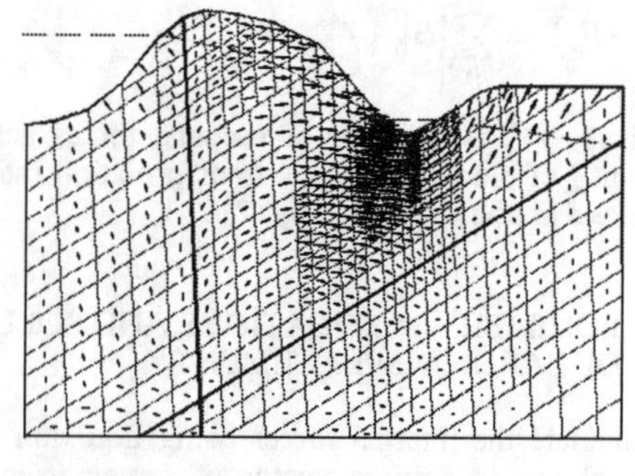

Figure 5. Rock movement induced by in-situ stress and groundwater flow

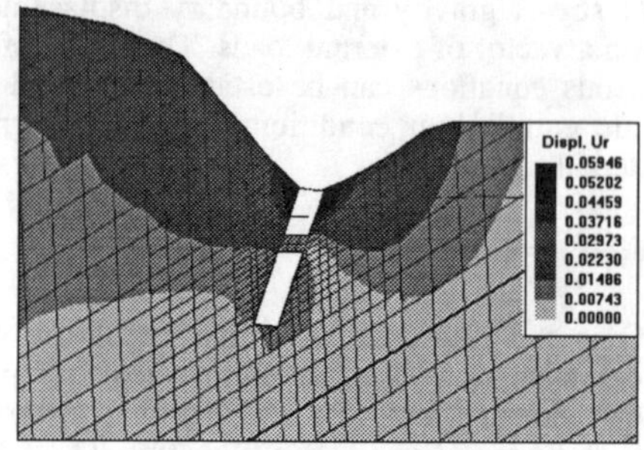

Figure 6. Displacement induced by excavation

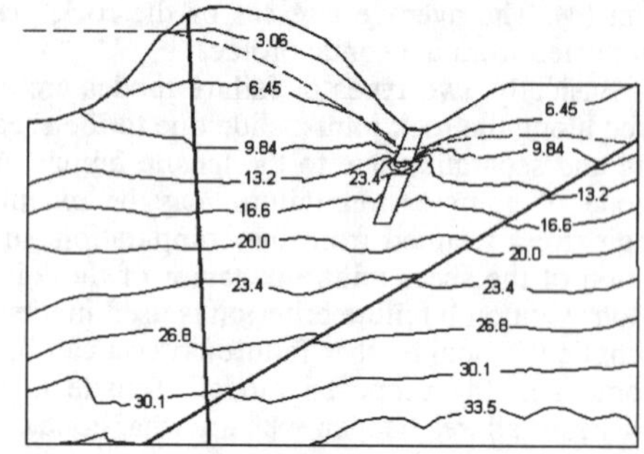

Figure 7. Major principal stress after excavation

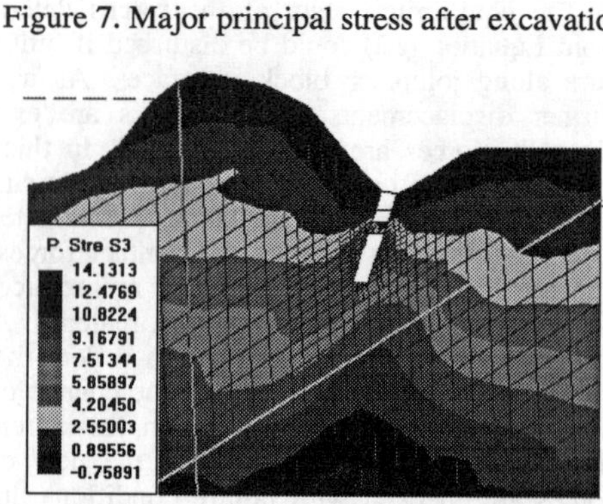

Figure 8. Major principal stress after excavation

After the excavation is completed, the water table is dropped below the ground surface and the groundwater pressure is calculated from the fluid potential in Figure 3 (b). It is seen that high flux is formed around the openings and this will impose a normal pressure on the outer surfaces.

Figure 6 shows the displacement contours induced by the excavation. The deformation of the footwall is larger than that of the hangingwall because the dip angle of J_2 is more critical for the stability of the footwall. The maximum displacement is 5.95cm, occurring at the right-top corner of the opening. In the view of rock mechanics, the sill pillar is very important in supporting the walls of the opening. It can be further predicted that a crown pillar would be more effective in maintaining the stability of the slopes.

The stresses are redistributed after joint failure is detected. Where there is a failure, the stresses are determined by the contact forces resulting from the material strength. Thus the stresses are always smaller than those obtained from the elastic analysis. Figure 7 is the major principal stress contour in mining step 2. There are stress discontinuities across the fault due to the relative displacements between the two sides. Figure 8 is the distribution of the average minor principal stress.

6 CONCLUSIONS

A numerical method, called the 3-D block spring model, has been developed for analysis of jointed rock masses. A discontinuous formulation is used to simulate the large displacements observed in heavily jointed and faulted rocks.

The model uses a static procedure to simulate the stable rock blocks that satisfy the equilibrium conditions. This avoids the damping process as used in the dynamic procedures and saves computer time. The unstable blocks however are simulated by a special technique – an iterative process and a simplified damping procedure.

The 3-D model has been used to analyze the excavation of a simulated open-pit mine. The groundwater pressure obtained from a steady seepage flow was well incorporated in the simulation. The analysis results have shown that the groundwater flow has a major impact on the pattern and magnitude of the rock movement. The declinations and strengths of the rock joints and faults also have greatly affected the deformations of the opening and slopes.

An integrated computer package BSM3D has been developed based on the proposed model. BSM3D includes a solving engine and a Windows based processor to help users preparing models and viewing the results. Complicated geological and boundary conditions can be considered in the computation, such as groundwater flow, in-situ stress, excavation and back fill sequences. The applications of BSM3D in civil and mining cases have demonstrated that it is a useful tool for jointed rock analysis.

7 REFERENCES

Li, G., and Wang, B., 1998. Development of a 3-D block-spring model for jointed rocks. *Proceedings of the 3rd International Conference on Mechanics of Jointed and Faulted Rock*, Vienna, Austria, *Edited by* Hans-Peter Rossmanith. pp. 305-309.

Wang, B. & Garga, V.K., 1993. A numerical method for modeling large displacements of jointed rocks – Part I: Fundamentals. *Canadian Geotechnical Journal:* Vol.30, pp. 96-108.

Goodman, R.E., and Shi G., 1985. Block theory and its application to rock engineering. Englewood Cliffs, NJ: Prentice-Hall.

Farahmand, D. & Li, G., 1986. An application of BEM for underground excavations in jointed rock. Rock Mechanics: Key to Energy Production, *Proceedings of the 27th U.S. Symposium on Rock Mechanics*, pp.299-303.

Kawai, T., Kawabata, Y., Kondou, K., and Kumagai, K., 1978. A new discrete model for analsis of solid mechanics problems. *Proceedings, 1st International Conference on Numerical Methods in Fracture Mechanics*, Swansea, U.K., *Edited by* A.R. Lusmoore and D.R.J. Owen. pp. 26-37.

Computer Methods and Advances in Geomechanics, Desai et al. (eds) © 2001 Balkema, Rotterdam, ISBN 90 5809 183 X

Simulation of the stochastic response of offshore platform to wave forces

Mehedy Mashnad
Department of Civil Engineering, New Mexico State University, N.Mex., USA

Mohammad M. Masud
Department of Computer Science and Engineering, University of Engineering and Technology, Bangladesh

ABSTRACT: This paper presents a computer program that can be used to perform dynamic analysis of single degree of freedom offshore platforms excited by random inertia and drag forces. Stochastic analysis is performed by spectral representation method. The analysis is based on the linear vibration theory, and the assumptions that waves are Gaussian and stationary processes. Sample processes are generated using a cosine series formula. Mean velocity of wind or significant wave height is fed into the program as the main input and response of the platform is obtained. The program can be used to generate non-linear drag force. From the simulated processes such important probabilistic events as threshold crossing, peak distribution can be evaluated, and useful decision can be taken in the preliminary design phase.

1 INTRODUCTION

In a deterministic analysis of offshore platform hydrodynamic forces are commonly estimated from the significant wave height and period. Significant wave height is assumed to be representative of the wave condition in the sea. But real waves involve many wave amplitudes and periods and rarely, if ever, the free surface profile of sea waves is sinusoid or monochromatic. Therefore, wave forces considering a single wave form can not provide desired degree of accuracy. Borgman (1967 & 1969) and Malhotra & Penzien (1970) have described useful techniques for incorporating irregularities of waves into the forces and into the response of structures. Shinozuka (1972), Yang (1973) and Shinozuka & Deodatis (1991) have described useful methods for generating random processes by spectral representation method.

The purpose of developing the computer program, named as SAOP (Stochastic Analysis of Offshore Platform), is to provide the designers with a useful tool for incorporating uncertain properties of waves into the dynamic analysis of offshore platforms. In the program Monte Carlo simulation technique has been adopted in which sample functions are generated by superimposing N number of cosine series. Amplitudes of the cosine functions are estimated from the power spectral density function of the waves and random phase angles from $[0,2\pi]$, uniformly distributed over the interval. Although this concept existed in the past, computation efforts required for such analysis made this method unpopular. However, the advent of fast computers in present days has made engineers and scientists interested in Monte Carlo method. The program can be used to efficiently simulate waves, horizontal velocity and acceleration of water particles, total forces on the pile and deflection of the platform. On the other hand, simulated force can be used in a time domain analysis. Generated waves or other sample processes contain the same statistical properties as the real waves or processes observed at the site (Shinozuka 1972 & Yang 1973).

2 OVERVIEW

The program can be used to analyze the response of an offshore platform that can be modeled as a single degree of freedom structure, and is fixed to the seabed. Inertia force and drag force, obtained by Morison's equation, constitute the total force on the structure. In order to simulate hydrodynamic force, spectrum of force is obtained first. Simulation technique will be described in the next sections. By multiplying the wave spectrum with appropriate transfer function spectrum of force is obtained. In a similar way velocity, acceleration and deflection of platform can be generated.

In the spectrum of drag force, commonly, equivalent linearization of the square of the water particle velocity is used (Borgman 1969). This results in a less peak force than the nonlinear peak force. So non-linearity of the drag force must be taken into account when the drag force has significant contribution to the response of the structure. The program generates velocity of water particles at equal intervals along the depth of water, computes the square of the velocity, and then sum up the squared velocity along the depth of the water. Total nonlinear drag force, thus obtained, can be used to perform dynamic analysis in time domain.

3 STRUCTURAL MODEL AND SPECTRAL REPRESENTATION

We are concerned with the forces due to wave at the location of the platform. According to the linear wave theory the equation of one-dimensional standing wave can be expressed as:

$$\eta(t) = A\cos(-2\pi ft + \varphi) \tag{1}$$

where A = amplitude of wave; f = frequency; t = time; and φ = phase angle. Velocity and acceleration have the following solutions:

$$v(t,z) = A(2\pi f)Z(z)\cos(-2\pi ft + \varphi) \tag{2}$$

$$a(t,z) = A(2\pi f)^2 Z(z)\sin(-2\pi ft + \varphi) \tag{3}$$

$$Z(z) = \frac{\cosh \kappa z}{\sinh \kappa d} \tag{4}$$

where v = velocity; a = acceleration; κ = wave number; z = water elevation measured from the ocean bed; and d = depth of water.

Two different forms of Pierson-Moskowitz spectrum for fully developed sea condition are used in the program (Pierson & Moskowitz 1964). The one-sided power spectral density function of wave as a function of significant wave height is:

$$S_{\eta\eta}(f) = \frac{(H_s/4)^2}{f} 5(f/\bar{f})^4 \exp\left[-5/4(f/\bar{f})^4\right] \tag{5}$$

$$\bar{f} = \frac{1}{2\pi}\left(\frac{5(H_s/4)^2}{\alpha g^2}\right)^{-1/4} \tag{6}$$

where H_s = significant wave height; α = Philip's constant (= 0.0081); and g = gravitational acceleration.

Mean wind velocity dependent one-sided power spectral density function of wave is expressed as:

$$S_{\eta\eta}(f) = \frac{\alpha g^2}{(2\pi)^4 f^5}\exp\left[-\beta(\bar{f}/f)^4\right] \tag{7}$$

$$\bar{f} = g/(2\pi V) \tag{8}$$

where α = Philip's constant (= 0.0081); β = 0.74; g = gravitational acceleration; and V = velocity of wind.

From Equations 2 & 3 the spectral density functions of velocity and acceleration can be expressed respectively as:

$$S_{vv}(f,z) = (2\pi f)^2 Z^2(z) S_\eta(f) \tag{9}$$

$$S_{aa}(f,z) = (2\pi f)^4 Z^2(z) S_\eta(f) \tag{10}$$

The equation of motion of a single degree of freedom system due to total drag and inertia forces can be written as:

$$M\frac{d^2x}{dt^2} + C\frac{dx}{dt} + kx = K_I a + K_D|v|v \tag{11}$$

$$K_I = 1/4\,\rho C_M \pi D^2 \qquad K_D = 1/2\,\rho C_D D \tag{12}$$

where M = total mass due to the mass of platform and the added mass; C = total damping due to structural damping and hydraulic damping; k = spring constant of pile; ρ = density of water; C_D = drag coefficient; C_M = inertia coefficient; D = diameter of pile; v = velocity of water particles along the pile; and a = acceleration. To obtain total forces on the pile the right-hand side of Equation 11 is integrated over the depth of water.

Considering the linearized form of the drag force (Borgman 1969), total hydrodynamic force can be expressed as:

$$F = K_I a + K_D(8/\pi)^{1/2}\sigma_v v \tag{13}$$

in which σ_v = standard deviation of the velocity of water particles. Spectral representation of total hydrodynamic force is:

$$S_{FF}(f) = S_{II}(f) + S_{DD}(f) \tag{14}$$

Spectral density function of total inertia forces is:

$$S_{II}(f) = \left[K_I(2\pi f)^2\frac{1}{\kappa}\right]^2 S_{\eta\eta}(f) \tag{15}$$

Spectral density function of total drag forces is:

$$S_{DD}(f) = \left[K_D(8/\pi)^{1/2}\sigma_\eta(2\pi f)^2 \frac{2\kappa d + \sinh 2\kappa d}{4\kappa\sinh^2\kappa d}\right]^2 S_{\eta\eta}(f) \tag{16}$$

where σ_η = standard deviation of wave heights. It can be noticed that spectral density function of total drag force depends on the depth of water, but the spectral density function of total inertia force does not depend on the depth of water. Wave number κ that appears in the spectra of forces depends on the frequency f of wave and is related by the dispersion relation. Wave number κ can be computed by an iterative method; however, in the program it is computed by the relationship developed by Eckart

(1952), given in Equation 17. The error in this approximation is within 5% and can be considered sufficient for many engineering purposes. Maximum error of 5% occurs when $2\pi d/L \approx 1$.

$$\kappa \approx \frac{(2\pi f)^2}{g\sqrt{\tanh\left((2\pi f)^2 d/g\right)}} \quad (17)$$

Finally the spectrum of response of the structure under total hydrodynamic force is:

$$S_{XX}(f) = |H(f)|^2 S_{FF}(f) \quad (18)$$

$$|H(f)|^2 = \frac{1}{k^2\left[\left(1-(f/f_o)^2\right)^2 + \left(2\xi(f/f_o)\right)^2\right]} \quad (19)$$

where f_o = the natural frequency of the structure; k = spring constant of the pile; and ξ = damping ratio.

4 SIMULATION PROCEDURE

A stationary Gaussian stochastic process with zero mean can be described as (Shinozuka 1972 & Yang 1973):

$$x_o(t) = \sqrt{2}\sum_{k=0}^{\infty}\sqrt{(2S_{xo}(f_k)\Delta f)}\cos(2\pi f_k t + \varphi_k) \quad (20)$$

where S_{xo} (f_k) = two sided power spectral density function of the stochastic process; and φ_k = k^{th} realization of the random phase angles. The stochastic process of Equation 20 can be simulated by considering finite N terms such that $N \to \infty$. Series expression of the simulated process can be expressed as:

$$x(t) = \sqrt{2}\sum_{n=0}^{N-1}\sqrt{(2S_{xo}(f_n)\Delta f)}\cos(2\pi f_n t + \varphi_n) \quad (21)$$

$$\Delta f = f_u/N, \quad f_n = n\Delta f, \quad n = 0,1,..,N-1 \quad (22)$$

where f_u = upper limit of the spectrum. Simulated process is periodic with T_o=1/Δf and time step, $\Delta t \le 1/(2f_u)$.

To generate sea surfaces or forces with similar statistics observed in the sea, theoretically the number of sample N needs to be infinite. It was reported by Yang (1973) that for N = 500 generated process converges to Gaussianness at 2σ range, but has approximately 5% error at 5σ range. If N is as large as 1000 the error is 3% at 5σ range (Shinozuka 1991). The program can consider N well above 1000. For large values of N ensemble expected value $E[x(t)]$ and ensemble autocorrelation function $R_{xx}(\tau)$ of the simulated process are identical to their corresponding target functions. For detailed description of this section readers are referred to Shinozuka & Deodatis (1991).

Adopting Inverse Fast Fourier Transform (FFT) technique can minimize the time required for the numerical calculation of Equation 21. This equation can be expressed in the following form in order to compute in FFT method:

$$x(m\Delta t) = \mathrm{Re}\left[\sqrt{2}\sum_{n=0}^{M-1}\sqrt{(2S_{xo}(f_n)\Delta f)}e^{i\varphi_n}e^{i(n\Delta f)(m\Delta t)}\right] \quad (23)$$

$$n = 0,1,..,M-1, \quad m = 0,1,..,M-1 \quad (24)$$

In Equation 23 Re indicates the real part of the equation. Δf and f_n have similar expressions as given in Equation 22. Time step has following expression:

$$\Delta t = 1/(M\Delta f) \quad (25)$$

Time step, $\Delta t \le 1/(2f_u)$. Here M must be an integer power of 2, and $M \ge 2N$.

5 APPLICTAION OF SIMULATION TO DESIGN

Response of the structure can be determined in frequency domain using Equation 18. Besides, response can be determined in time domain using the simulated force, or the simulated velocity and acceleration.

The ultimate purpose of stochastic process theory is to be able to estimate the reliability of the structure, which has been designed to undergo random excitation (Lin 1967). For knowledge in the first excursion failure or fatigue failure two most important information that should be obtained are (1) threshold crossing, and (2) peak distribution within a time interval. These values can be obtained from the spectrum or from the generated sample process.

6 NUMERICAL EXAMPLE AND DISCUSSIONS

Consider a platform of single pile. Parameters used in this example are as follows: height (h) of the pile from the seabed to the bottom of platform 80 m, Depth (d) of water at the site 72 m, outer diameter (D) of the pile 3.5 m, total mass (M) 1.84e6 kg, spring constant (k) of the pile 2.5e6 N/m, total damping (ξ) ratio 5%, drag coefficient (C_D) 0.7, inertia coefficient (C_M) 2, and density of water (ρ) 1030 kg/m^3.

In this example a significant wave height of 7 m has been considered. Dominant frequency of the waves is 0.076 Hz and corresponding period is 13 s. Natural frequency of the structure is 0.19 Hz. Spectral density functions of the total drag force and total inertia forces were obtained from Equations 15, 16. Total response spectrum was obtained by adding these spectra. Figure 1 shows plots of the simulated processes. This set of stochastic processes was gen-

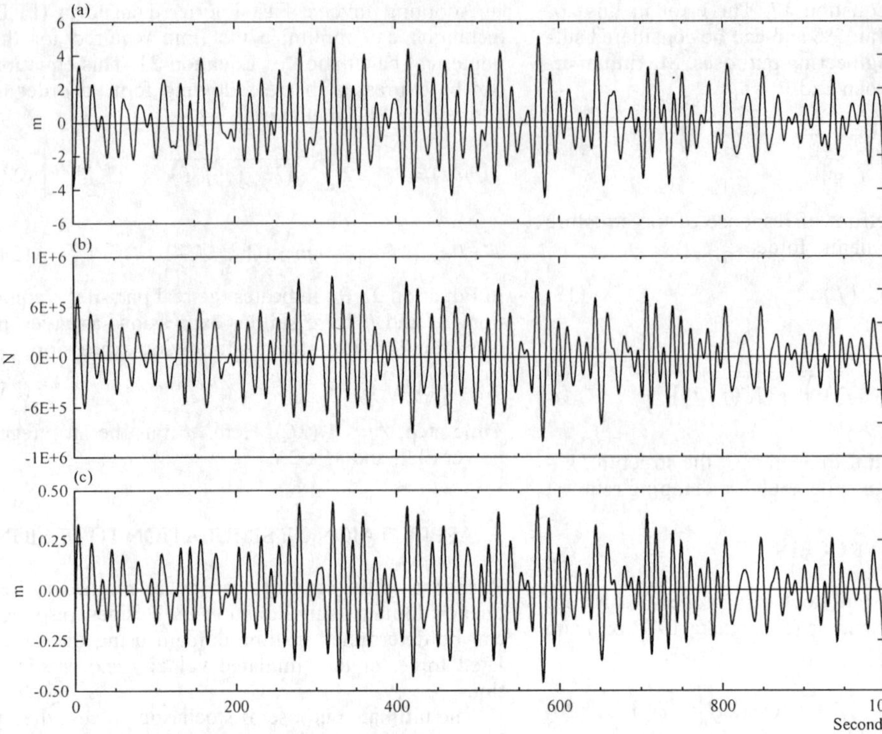

Figure 1. Generated (a) Waves, (b) Total forces on pile, and (c) Response of platform ($H_s = 7$ m, $f_u = 0.5$ Hz, $N = 500$).

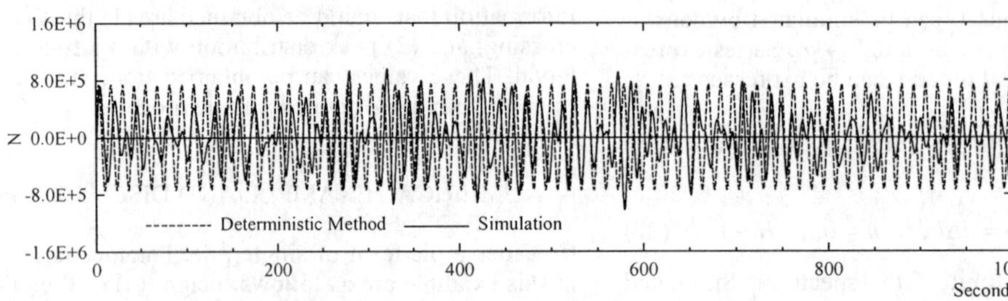

Figure 2. Plots of total forces obtained by deterministic method and simulation ($H_s = 7$ m, $f_u = 0.5$ Hz, $N = 500$).

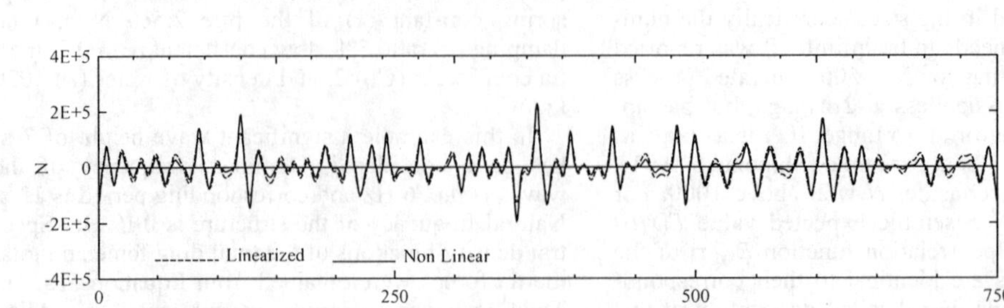

Figure 3. Simulated total linearized and non-linear drag forces on pile ($H_s = 7$ m, $f_u = 1$ Hz, $N = 750$).

erated by means of Equation 23 using the corresponding two-sided power spectral density functions. Upper limits of all the spectra were considered to be 0.5 Hz. This upper limit gives $\Delta f = 0.001$ Hz for $N = 500$, and a period of simulation $T_o = 1000$ s. M was taken equal to 8192 ($= 2^{13}$), which results in a time step of 0.122 seconds $\leq$ upper limit of 1 s. In order to generate random phase angles φ, random numbers were generated from uniform distribution of [0,1] and multiplied by 2π. Simulation can be done by summation of N number of cosine series by the direct application of Equation 21 as well.

Total wave force required for a deterministic response analysis was computed considering the same significant wave height of 7 m and dominant frequency of 0.076 Hz. Corresponding wavelength is 258 m. This force has been plotted along with the simulated force in Figure 2, which shows there are at least 13 occasions when forces obtained in the deterministic method have been exceeded by the simulated forces. Maximum total deterministic force was 735 kN whereas the maximum peak obtained by the simulation was 975 kN (30% higher). So there exist uncertainties in the deterministic analysis. Table 1 shows expected number of times the deflection of the platform obtained in deterministic method would be exceeded within a period of 24 hours if random wave is considered. In Table 1 σ_1 and σ_2 are estimated from the simulated processes. The number of

Table 1. Expected threshold crossings obtained from simulated processes.

H_s	σ_1*	σ_2**	ξ***	Expected crossing
m	m	m	m	Occurrences/day
5	0.1293	0.0736	0.2704	3516
7	0.1591	0.0770	0.3463	1246
9	0.1867	0.0816	0.4152	1014
11	0.2116	0.0861	0.4653	998

* Standard deviation of the deflection of platform.
** Standard deviation of the velocity of platform.
***Threshold value.

threshold crossing decreases as the wave height increases. Because, within a given period the number of peaks decreases as the significant wave height becomes larger.

Figure 3 shows a comparison between the total linearized and nonlinear drag forces. In this example significant wave height is 7 m, upper limit of spectrum is 1.0 Hz, N is 750, and M is 2048 ($= 2^{11}$). Non linear drag force was obtained by the method described in Section 2. Nonlinear drag force has peaks with magnitude greater than that of linearized drag force. Total drag force thus obtained can be used in Equation 11 to perform time domain analysis.

7 CONCLUSION

1. In the preceding example some features of the program have been presented. It is observed that Monte Carlo simulation technique can be a useful mean of incorporating uncertainties of hydrodynamic forces in structural analysis.

2. SAOP can be used to perform probabilistic linear dynamic analysis and to take preliminary decision in selecting the dimensions of the structure under random load environment consisting of drag and inertia forces for most design purposes.

3. The program can be extended to consider multi degree of freedom system, multiple piles and other wave spectra for different sea conditions, and to account for the effects of current, the intermittency of submergence of pile near the free surface, and soil-pile interaction.

8 ACKNOWLEDGEMENT

Professor Derin N. Ural at Istanbul Technical University and Professor Khaled Sobhan at New Mexico State University are acknowledged for their useful suggestions.

REFERENCES

Borgman, L.E. 1967. Spectral analysis of ocean wave forces on piling. *Journal of the Waterways and Harbors Division*. 93(WW2): 129-156.

Borgman, L.E. 1969. Ocean wave simulation for engineering design. *Journal of the Waterways and Harbors Division*. 95(WW4): 557-582.

Eckart, C. 1952. The propagation of gravity waves from deep to shallow water. *Gravity Waves*. Circular no. 521. Washington: National Bureau of Standards.

Lin, Y.K. 1967. *Probabilistic Theory of Structural Dynamics*. United States of America: McGraw-Hill.

Malhotra, A.K. & Penzien, J. 1970. Nondeterministic analysis of offshore structures. *Journal of the Engineering Mechanics Division*. 96(EM6): 985-1003.

Pierson, W.J. & Moskowitz, L. 1964. A proposed spectral form for fully developed wind seas based on the similarity theory of S.A. Kitaigorodskii. *Journal of Geophysical Research*. 69(24): 5181-5190.

Shinozuka, M. 1972. Monte carlo solution of structural dynamics. *Computers & Structures*. 2: 855-874.

Shinozuka, M. & Deodatis, G. 1991. Simulation of stochastic processes by spectral representation. *Applied Mechanics Review*. 44(4): 191-203.

Yang, J.-N. 1973. On the normality and accuracy of simulated random processes. *Journal of Sound and Vibration*. 26(3): 417-428.

Computer Methods and Advances in Geomechanics, Desai et al. (eds) © 2001 Balkema, Rotterdam, ISBN 90 5809 183 X

Modeling borehole stresses in anisotropic stress fields in non-linear rocks

P.A. Nawrocki – *Natural Resources Canada, CANMET Sudbury Lab, Ont., Canada*
S. Mogilevskaya – *Department of Civil Engineering, University of Minnesota, Minneapolis, Minn., USA*
M.B. Dusseault – *Department of Earth Sciences, University of Waterloo, Ont., Canada*
R.K. Bratli – *Applied Rock Mechanics LLC, Abu Dhabi, United Arab Emirates*

ABSTRACT: This paper presents a new approach to model borehole stresses in anisotropic in situ stress fields in non-linear rocks. The numerical approach combines a rapid algorithm based on the Complex Variables Boundary Element Method with a special technique of obtaining input data from a series of triaxial compression tests. Several examples of critical well pressures predicted by applying this approach are given using both Mohr-Coulomb and Lade failure criterion.

1 INTRODUCTION

Anisotropic in situ stress fields and non-linear rock behavior are typically encountered in practice; yet the model most commonly used in the analysis of borehole stresses is the hollow cylinder model (Lamé's solution). It assumes not only linear elastic behavior, but also an isotropic in situ stress field. Numerous solutions exist for this model (elastic, elasto-plastic, poroelastic, linear, non-linear, etc.). On the other hand, the case of anisotropic in situ stress fields has not yet received the attention it deserves. Despite the 100-yr-old linear elastic Kirsch solution and the possibility of addressing this case in complex numerical codes like finite element analysis, other solutions practically do not exist for this case.

The objective of this paper is to present a new numerical approach to model borehole stresses in anisotropic in situ stress fields in non-linear elastic rocks. The approach proposed combines a rapid Complex Variables Boundary Element Method (CVBEM) numerical algorithm with a special technique of obtaining input data from a series of triaxial compression tests; it can be envisioned as an intermediate step between simple linear elastic analysis (Kirsch) and the more general finite element analysis.

2 METHOD AND APPROACH

Consider the problem schematically presented in Figure 1a. A circular opening has been made in non-linear rock in an anisotropic in situ far-field stress state, $\sigma_H > \sigma_h$. If the rock properties are linear, then the Kirsch equations could be used to solve for stresses. In our non-linear analysis, material non-linearities are introduced by dividing the borehole region into zones (Fig. 1b; solution is cyclic, so only one quadrant, $0^o < \theta < 90^o$, has to be analyzed) with elastic properties constant in each zone, but different for different zones. Values of elastic constants are assigned to the zones using triaxial compression test results and the assumption that radial stress σ_r in the field plays the role of confining stress σ_3 in the laboratory (Santarelli et al. 1986). Non-linear rock properties are incorporated using an approach similar to the radius-dependent modulus model (Nawrocki & Dusseault 1995). Analysis consists of extracting elastic and failure parameters from triaxial compression test results, calculation of non-linear borehole stresses, and wellbore integrity analysis.

Continuity of displacements and stresses is imposed on contacting zone boundaries. Then the complex variant of the displacement discontinuity method is used to calculate borehole stresses. It is based on a Complex Hypersingular Equation (CHSIE) for piece-wise homogenous media with a circular opening. The hypersingular equation links displacement discontinuities and tractions at all boundaries. To define the unknown displacement discontinuities or tractions at the boundary, the CHSIE is solved numerically by using the CVBEM (Mogilevskaya 1996, Linkov & Mogilevskaya 1998). Straight segments and circular arcs are used as the boundary elements, and stresses within zones are calculated after defining the unknowns at the boundaries. Several examples of critical well pressures obtained by using this approach are given.

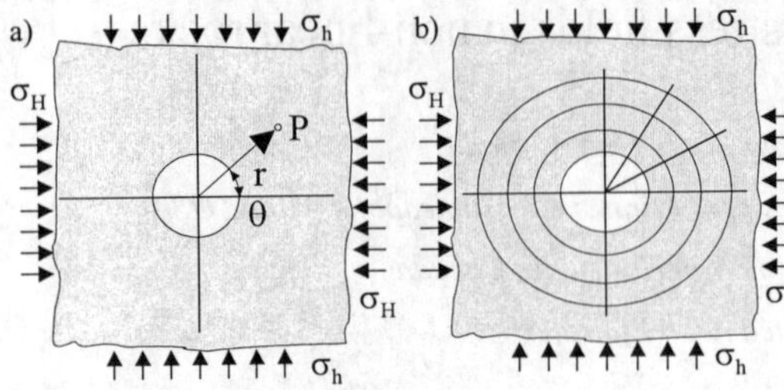

Figure 1: Problem formulation.

3 BASIC EQUATION

The basic hypersingular equation for a piece-wise homogeneous medium with a circular opening was obtained in Mogilevskaya et al. (2000). The resulting CHSIE has the form

$$\frac{1}{2\pi i}\{2\int_S \frac{\Delta u d\tau}{(\tau-t)^2} - \int_S \Delta u \frac{\partial}{\partial t}[d\ln\frac{\tau-t}{\bar\tau-\bar t} + d\ln\left(\bar\tau - r^2/t\right) +$$

$$d\ln\left(\tau - r^2/\bar t\right) - (t - \frac{r^2}{\bar t})$$

$$(\frac{r^2}{\bar t^2} d\frac{\bar\tau - \bar t}{(\tau - r^2/\bar t)^2} - \frac{d\tau}{(\tau - r^2/\bar t)^2} + \frac{d\tau}{\tau^2})] +$$

$$\int_S \overline{\Delta u}\frac{\partial}{\partial t}[-d\frac{\tau-t}{\bar\tau-\bar t} + d\frac{\tau-t}{\bar\tau - r^2/t} - d\frac{\tau-t}{\bar\tau} + \frac{r^2}{\bar t}\frac{d\bar\tau}{\bar\tau^2} -$$

$$r^2\frac{t - r^2/\bar t}{\bar t^2}\frac{d\tau}{(\tau - r^2/\bar t)^2}] + \int_S (2a_1 - a_3)\frac{\sigma d\tau}{\tau - t} +$$

$$\int_S \sigma\frac{\partial}{\partial t}[(a_1 - a_3)\ln\frac{\tau-t}{\bar\tau-\bar t} + (a_1 - a_3)\ln(\bar\tau - \frac{r^2}{t}) +$$

$$a_1\ln(\tau - \frac{r^2}{\bar t}) - a_1 r^2\frac{t - r^2/\bar t}{\bar t^2}\frac{\bar\tau - \bar t}{(\tau - r^2/\bar t)^2} - a_1\frac{t - r^2/\bar t}{\tau - r^2/\bar t} +$$

$$a_1\frac{t - r^2/\bar t}{\tau}]d\tau + \int_S \bar\sigma\frac{\partial}{\partial t}a_1\left[\left(\frac{\tau-t}{\bar\tau-\bar t} - \frac{\tau-t}{\bar\tau - r^2/t} + \frac{\tau-t}{\bar\tau} + \frac{r^2}{\bar t\bar\tau}\right) -\right.$$

$$\left.(a_1 - a_3)\frac{r^2}{\bar t^2}\frac{t - r^2/\bar t}{\tau - r^2/\bar t}\right]d\bar\tau\} = \frac{a_2}{2}\sigma(t) + \sigma^\infty(t) \qquad (1)$$

where S is the totality of external and internal boundaries; r is radius of a circular opening; $\tau, t \in S$; $\Delta u = u^+ - u^-$ is the displacement discontinuity; u^+ and u^- are the limit values of displacements from the left or right sides of the contours respectively; $\sigma = \sigma_n + i\sigma_s$ are tractions (σ_n is normal traction and σ_s is shear traction); and, a bar over a symbol denotes complex conjugation. Constants a_1, a_2, a_3 depend on material properties, defined as follows:

$$a_1 = \frac{1}{2G^+} - \frac{1}{2G^-}; \quad a_2 = \frac{1+\kappa^+}{2G^+} + \frac{1+\kappa^-}{2G^-};$$

$$a_3 = \frac{1+\kappa^+}{2G^+} - \frac{1+\kappa^-}{2G^-} \qquad (2)$$

where $\kappa = 3-4\nu$ for plane strain, $\kappa = (3-\nu)/(1+\nu)$ for plane stress, ν is Poisson's ratio, and G is shear modulus. Plus or minus signs indicate elastic constants for the region which is situated to the left or right of the direction of travel on S. Conditions at infinity are taken into account by the following formulae:

$$\sigma^\infty(t) = -\frac{\kappa^\infty + 1}{2G^\infty}[\frac{\sigma_{xx}^\infty + \sigma_{yy}^\infty}{2}(1 + \frac{r^2}{\bar t^2}\frac{d\bar t}{dt}) +$$

$$\frac{\sigma_{yy}^\infty - \sigma_{xx}^\infty - 2i\sigma_{xy}^\infty}{2}(\frac{r^2}{t^2} + \frac{d\bar t}{dt}) + \frac{\sigma_{yy}^\infty - \sigma_{xx}^\infty + 2i\sigma_{xy}^\infty}{2}$$

$$(\frac{r^2}{\bar t^2} - \frac{2r^2 t}{\bar t^3}\frac{d\bar t}{dt} + \frac{3r^4}{\bar t^4}\frac{d\bar t}{dt})] \qquad (3)$$

where σ^∞_{xx}, σ^∞_{yy}, σ^∞_{xy} are stresses at infinity. These stresses vanish for a finite region.

Additional conditions must be imposed to exclude rigid body movement if tractions are given only on the whole outer contour (Linkov 1983, Linkov & Mogilevskaya 1998):

$$\int_S \frac{\Delta u(\tau)}{\tau - z_o}d\tau = 0; \; \mathrm{Re}\int_S \frac{\Delta u(\tau)}{(\tau - z_o)^2}d\tau = 0 \qquad (4)$$

where z_o is the point belonging to the interior of the region. Tractions (or displacement discontinuities, or prescribed relationships between tractions and displacement discontinuities) may be prescribed on external and internal boundaries.

Equation (1) accounts for cracks ($G^+=G^-=G$; $\kappa^+=\kappa^-=\kappa$, for corresponding parts of S) and for holes ($1/2G^+=0$, $u^+=0$ for a hole). For the case of one and two straight cracks emanating from a borehole, equation (1) coincides with the equation obtained by Atkinson and Thiercelin (1993).

Stresses in the l-th region inner points z can be calculated using the Kolosov-Muskhelishvili formulae (Muskhelishvili 1959) and the following relations for the potentials (integrals with resultant force function f(z) can be reduced to integrals with σ using integration by parts):

$$
\begin{aligned}
\varphi(z) = {} & \frac{2G^l}{2\pi i(\kappa^l+1)}\{\int_S \frac{a_1 f + \Delta u}{\tau - z}d\tau + \int_S [(a_3 - a_1)f - \Delta u] \\
& \frac{d\bar{\tau}}{\bar{\tau} - r^2/z} - \int_S [(a_3 - a_1)f - \Delta u]\frac{d\bar{\tau}}{\bar{\tau}} + \int_S (a_1\bar{f} + \overline{\Delta u})\, d\frac{\tau - z}{\bar{\tau} - r^2/z} \\
& - \int_S (a_1\bar{f} + \overline{\Delta u}) d\frac{\tau - z}{\bar{\tau}} - 2r^2 \int_S (a_1\bar{f} + \overline{\Delta u})\frac{d\bar{\tau}}{\bar{\tau}^3}\} + \\
& \frac{G^l(\kappa^\infty + 1)}{G^\infty(\kappa^l + 1)}\left(\frac{\sigma_{xx}^\infty + \sigma_{yy}^\infty}{4}z - \frac{r^2}{z}\frac{\sigma_{yy}^\infty - \sigma_{xx}^\infty - 2i\sigma_{xy}^\infty}{2}\right)
\end{aligned}
$$

$$
\begin{aligned}
\psi(z) = {} & \frac{2G^l}{2Gi(\kappa^l+1)}\{\int_S [(a_3 - a_1)\bar{f} - \overline{\Delta u}]\frac{d\tau}{\tau - z} + \\
& \int_S (a_1 f + \Delta u) d\frac{\bar{\tau} - \bar{z}}{\tau - z} - \bar{z}\int_S \frac{a_1 f + \Delta u}{(\tau - z)^2}d\tau + \\
& \frac{r^4}{z^3}\int_S [(a_3 - a_1)f - \Delta u]\frac{d\bar{\tau}}{(\bar{\tau} - r^2/z)^2} - \\
& \frac{r^2}{z}\int_S (a_1 f + \Delta u)\frac{d\tau}{\tau^2} + \frac{r^4}{z^3}\int_S (a_1\bar{f} + \overline{\Delta u}) d\frac{\tau - z}{(\bar{\tau} - r^2/z)^2} + \\
& \int_S (a_1\bar{f} + \overline{\Delta u})\frac{d\bar{\tau}}{\bar{\tau} - r^2/z} - \int_S (a_1\bar{f} + \overline{\Delta u})\frac{d\bar{\tau}}{\bar{\tau}} - \\
& \frac{r^2}{z}\int_S (a_1\bar{f} + \overline{\Delta u})\frac{d\bar{\tau}}{(\bar{\tau} - r^2/z)^2} + \frac{r^2}{z}\int_S (a_1\bar{f} + \overline{\Delta u})\frac{d\bar{\tau}}{\bar{\tau}^2} + \\
& \frac{G^l(\kappa^\infty + 1)}{G^\infty(\kappa^l + 1)}\left[\frac{\sigma_{yy}^\infty - \sigma_{xx}^\infty + 2i\sigma_{xy}^\infty}{2}z - \right. \\
& \left.\frac{r^2}{2z}\left(\sigma_{xx}^\infty + \sigma_{yy}^\infty\right) - \frac{r^4}{z^3}\frac{\sigma_{yy}^\infty - \sigma_{xx}^\infty - 2i\sigma_{xy}^\infty}{2}\right]
\end{aligned}
\qquad (5)
$$

where G^l, κ^l are the elastic constants of the l-th region. A numerical algorithm for solving (1) is described in Mogilevskaya et al. (2000).

4 NUMERICAL EXAMPLES

Triaxial test results obtained on weak and strong sandstone (Fig. 2) provided information on material behavior used in this study. Elastic parameters (Young's modulus E, and Poisson's ratio ν) have been extracted from the data separately for each test (for each σ_3). Then shear modulus G has been calculated as $G=E/2/(1+\nu)$. Thus, different values of G, E, and ν correspond to different σ_3 values, Table 1. Shear modulus G and Poisson's ratio ν are next plotted as a function of σ_3 and data are approximated with suitable functions $G(\sigma_3)$ and $\nu(\sigma_3)$. Some physical limitations are imposed: G should be a non-zero, positive number and ν should take values from 0 to 0.5. Also, the $G(\sigma_3)$ function should have a horizontal asymptote at high s tresses. Either stan-

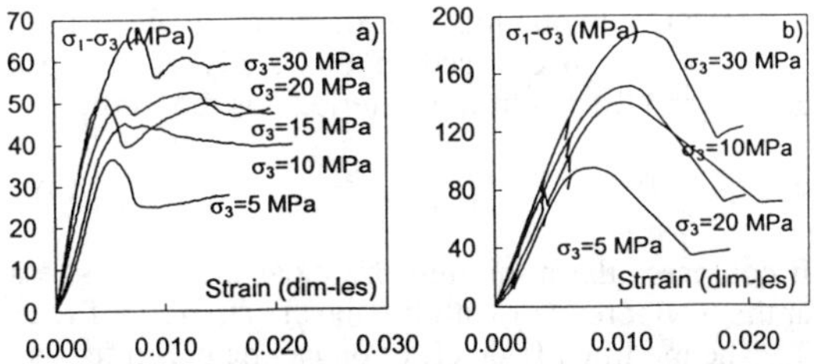

Figure 2: Compression curves for (a) weak and (b) strong sandstone.

dard trendline functions can be used or a function can be postulated and the least-squares procedure can be used to fit experimental data with the postulated function. Simple linear and logarithmic functions have been used for this initial study.

The next step is to use the Kirsch equations to obtain initial linear distributions of both σ_r and σ_θ for different θ'. A linear solution provides a means of comparison with a non-linear solution, and the distribution of linear elastic radial stress σ_r must be used together with the above functions to obtain variations of G and ν around the opening. In doing so, we assume that radial stress plays the role of confining stress σ_3 in the field. Thus, the $G(\sigma_3)$ and $\nu(\sigma_3)$ functions can be transformed into radial variations of shear modulus and Poisson's ratio, and moduli values can be assigned to zones (Fig. 1b). The CVBEM algorithm can then be applied to determine non-linear borehole stresses. Distributions of non-linear borehole stresses are not shown herein as they have been already presented elsewhere both for isotropic and anisotropic field of in situ stresses (Nawrocki et al 1998a; Mogilevskaya & Nawrocki 1998).

Finally, well integrity analysis involves checking the failure criteria for calculated borehole stresses. The goal is to find well fracturing and collapse pressures. Failure criteria in compression (collapse criterion) provide lower bounds for well pressure; failure criteria in tension (fracturing criterion) provide upper bounds for well pressure. The area between these two bounds is called the operational window.It can be used to guide mud weight choices.

Table 1: Elastic parameters for two sandstones

Rock	σ_3(MPa)	E(GPa)	ν	G(GPa)
Weak sandstone	5	9.34	0.017	2.32
	10	10.29	0.033	2.53
	20	11.13	0.143	2.59
	30	11.36	0.127	2.67
Stiff sandstone	5	18.51	0.109	8.34
	15	18.59	0.111	8.36
	20	19.92	0.109	8.98
	30	22.44	0.128	9.94

Mohr-Coulomb and the modified Lade failure criterion have been used in this study as collapse criteria and the fracturing criterion has the form:

$$\sigma_3' + T_0 = 0 \tag{6}$$

It compares the minimum principal effective stress at the wellbore wall with the tensile strength of rock T_o. The modified Lade criterion can be presented as:

$$(I_1'')^3/I_3'' = 27 + \eta \tag{7}$$

where I_1'' and I_3'' are stress invariants:

$$I_1''=(\sigma_1+S_I-p_o)+(\sigma_2+S_I-p_o)+(\sigma_3+S_I-p_o) \tag{8}$$
$$I_3''= (\sigma_1+S_I-p_o)(\sigma_2+S_I-p_o)(\sigma_3+S_I-p_o) \tag{9}$$

σ_1, σ_2, and σ_3 are principal stresses, S_I and η are material constants, and p_o is pore pressure. S_I is related to rock cohesion c_o, while η is a function of the internal friction angle φ:

$$S_I = c_o/\tan\varphi \tag{10}$$
$$\eta = 4\tan^2\varphi(9-7\sin\varphi)/(1-\sin\varphi) \tag{11}$$

The classic Mohr-Coulomb criterion is presented as:

$$(\sigma_1' - \sigma_3') - (\sigma_1' + \sigma_3')\sin\varphi - 2c_o\cos\varphi = 0 \tag{12}$$

where primes stand for effective stresses. Failure parameters extracted from the experimental data are summarised in Table 2.

Table 2: Failure parameters for two sandstones

Rock	UCS (MPa)	φ (deg)	c_o (MPa)	T_o (MPa)	S_I (MPa) η
Weak sandstone	32.7	20.11	11.42	3.27	31.17 5.39
Stiff sandstone	79.4	40.47	18.36	7.94	21.59 36.67

The values of in-situ stresses used in our analysis are taken as σ_H = 103.7 MPa and σ_h=97.2 MPa. These values correspond to in situ stress estimates at the depth of about 4400 m below sea level in the North Sea environment. Pore pressure estimated for this depth is 91 MPa, substantially overpressured.

Following the procedure outlined above, critical well pressure calculations have been carried out. The linear elastic analysis was done first, using the Kirsch equations. Both Lade and Mohr-Coulomb failure criteria were used in the linear elastic analysis. Then, the analysis was repeated assuming non-linear properties for the rocks, as outlined in this paper. Only the modified Lade criterion was used in the non-linear calculations. Results are summarized in Table 3.

Table 3: Collapse and fracturing pressures for two sandstones

Rock	P_{frac} Linear (MPa)	P_{frac} Non-linear (MPa)	P_{coll} Linear MC (MPa)	P_{coll} Linear Lade (MPa)	P_{coll} Non-linear Lade (MPa)
Weak sandstone	191.2	148.7	59.5	51.0	31.5
Stiff sandstone	208.4	178.7	25.1	10.2	0.5

5 CONCLUSION

This paper suggests a new approach to model borehole stresses in anisotropic in situ stress fields in non-linear rocks. Our approach is based on the new CHSIE for piece-wise homogeneous media with a circular opening. This adds substantially to the analytical potential that can be used in well design. To date, the linear elastic Kirsch solution has been the only major tool used in well integrity analysis for wells in anisotropic in situ stress fields (not counting approaches such as the finite element method. However, mesh generation seems to be much simpler in our approach than is required in typical FE calculations, the numerical algorithm used is simple, and the solution is rapid.

Results obtained show that well design based on linear elastic analysis using the Mohr-Coulomb criterion is a safe but not economically viable option. This is because it predicts higher collapse pressures than those obtained from a non-linear model. Also, collapse pressures obtained using the Lade criterion are lower than those using the Mohr-Coulomb criterion (based on linear analysis only, but it is expected that such a statement should also hold for non-linear analysis). Thus, linear elastic design based on the Mohr-Coulomb criterion is a doubly conservative choice: linear theory overestimates collapse pressures, and the Mohr-Coulomb criterion even adds to this effect.

On the fracturing front, the non-linear model predicts lower fracturing pressures than the linear elastic model. Thus, linear elastic analysis seems to suggest use of higher pressures in the fracturing medium (water, steam, etc.) than is really needed to fracture the reservoir. Second, fracturing can initiate earlier than expected, that is, when the well pressure has still not attained the level prescribed by the linear elastic approach. The excessive amount of the fracturing medium is then used to open and propagate fractures, not to initiate fracturing as designed. There are also clear implications on in situ stress measurements using hydraulic fracturing; estimating σ_H is problematic in strongly non-linear systems.

It can be also concluded that the probability of well failure in shear is in fact lower than would be expected based on the results of linear elastic analysis. Note that the collapse pressure calculated for stiff sandstone using non-linear theory is practically zero. Thus, the risk of shear failure seem to be very low for this rock.

Finally, let us state that the approach recommended in this paper is quite general. Here, it is used to calculate borehole stresses around holes in non-linear rocks in anisotropic in situ stress fields. The same approach can also be used to study fracture propagation and the effect of randomly distributed natural fractures (inclusions) on well stresses. Thus, it can be also used in analyzing processes and phenomena commonly encountered in the petroleum industry (cf. hydraulic fracturing, non-linear rock compression, etc.).

REFERENCES

Atkinson, C. & M. Thiercelin 1993. The interaction between the wellbore and pressure induced fractures. *Int. J. of Fracture* 59:23-40.

Linkov, A.M. 1983. Plane problems of the static loading of a piecewise homogeneous linearly elastic medium, *J. Appl. Math. & Mech.* 47:527-532.

Linkov, A.M. & S.G. Mogilevskaya 1998. Complex hypersingular BEM in plane elasticity problems. Ch.9 in V. Sladek & J. Sladek (eds), *Singular Integrals in BEM*, Comp. Mech. Publ., Southampton: 299-364.

Mogilevskaya, S.G. 1996. The universal algorithm based on complex hypersingular integral equation to solve plane elasticity problems. *Comp. Mech.* 18: 127-138.

Mogilevskaya, S.G. & P.A. Nawrocki 1998. Non-linear stress solutions for boreholes in anisotropic in situ stress field. Presentation handouts, PMRI Annual Meeting, Waterloo, Ontario, Nov. 17, 1998.

Mogilevskaya, S.G, Rothenburg, L. & M.B. Dusseault 2000. Interaction between a circular opening and fractures originating from its boundary in a piecewise homogeneous plane. *Int. J. for Num. Anal. Meth. in Geomech* (accepted).

Muskhelishvili, N.I. 1959. *Some Basic Problems of the Mathematical Theory of Elasticity*. Noordhoff, Netherlands.

Nawrocki, P.A. & M.B. Dusseault 1995. Modelling of damaged zones around openings using radius-dependent stiffness, *Rock Mech. & Rock Engng*. 28(4):227-239.

Nawrocki, P.A., Dusseault, M.B. & R. K. Bratli 1998. Use of uniaxial compression test results in modelling stresses around openings in non-linear geomaterials. J. of Petrol. Sci. & Engng, 21(1-2):79-94.

Santarelli F., Brown E.T. & V. Maury 1986. Analysis of borehole stresses using pressure-dependent linear elasticity, *Int. J. Rock Mech. Min. Sci & Geomech. Abstr.* 23: 445-449.

Computer Methods and Advances in Geomechanics, Desai et al. (eds) © 2001 Balkema, Rotterdam, ISBN 90 5809 183 X

Stress-release damage modeling

E. Papamichos
SINTEF Petroleum Research, Trondheim, Norway

ABSTRACT: Stress-release damage in rocks may result in a permanent alteration of the rock properties, which has to be quantified and modeled in order to describe the *in situ* behavior of the rock based on tests from core specimens. A dual yield surface elastoplastic model is introduced to describe stress-release damage in a synthetic sandstone manufactured as an analogue to a reservoir sandstone. The synthetic sandstone, composed by sand and quartz cement, is cemented under *in situ* stress conditions. This process allows the quantification of damage during a stress-release that simulates the coring process. In the model, one yield surface describes the behavior of the sand matrix and the other the behavior of the cement, while the total stresses are given as in mixtures theories. The model is calibrated on synthetic sandstone test results and used for back analysis of the experiments.

1 INTRODUCTION

Core or stress-release damage is a permanent alteration of the rock properties as a result of drill-out and retrieval of the core from the *in situ* environment to the surface. For reservoir rocks, stress release during coring is thought to be a major core damage mechanism. Core damage has been investigated in reservoir sandstones using a synthetic sandstone, analogous to reservoir rocks with respect to rock mechanical and petrophysical properties and texture (Holt et al. 1998, 2000). The synthetic sandstone is created under *in situ* stress conditions to simulate the diagenesis process. Comparative tests are then performed to simulate virgin compaction behavior and the behavior of an unloaded and reloaded simulated core.

The stress paths followed in the virgin compaction test and the simulated core compaction test are shown in Figure 1. In the virgin compaction test, sand was loaded to the *in situ* stress conditions where cementation took place. After cementation, the specimen was compacted under oedometric K_0 conditions. In the simulated core compaction test, after cementation at *in situ* stress conditions, the specimen was unloaded to simulate the coring process. Subsequently the specimen was loaded to the *in situ* stress conditions and oedometric K_0 compaction was performed to simulate a test on a cored specimen. Figure 2 shows a comparison between the virgin compaction and the simulated core compaction for this sandstone. Figure 2a shows only the K_0 part of the tests. The virgin compaction curve is nonlinear with a higher initial stiffness. The stiffness during virgin compaction matches the stiffness of the core compaction above 45 MPa. Thus if core data are used directly for reservoir rock compaction estimation, the initial compaction will be overestimated. In Figure 2b the unloading during the coring simulation is also included. The two specimens behave the same above an axial stress of 45 MPa.

The objective is to model mathematically virgin reservoir rock compaction based on core data. A simple one-dimensional model was developed by Holt et al. (1998, 2000), based on the experience from the synthetic rock experiments. The model was able to describe the axial stress versus axial strain behavior only during uniaxial compaction. This paper describes a novel approach to core damage modeling. An elastoplastic model with two yield surfaces (two-material model) is introduced, with one yield surface for the sand and one for the cement. It is a physically based model that can simulate both virgin compaction and core damage during coring.

2 TWO-MATERIAL ELASTOPLASTIC MODEL

The experimental results show that during coring unrecoverable damage takes place with large plastic strains. A way to describe the core damage process is a model with two yield surfaces. Such models are referred to as overlay or mechanical sublayer models and were first introduced as one-dimensional models

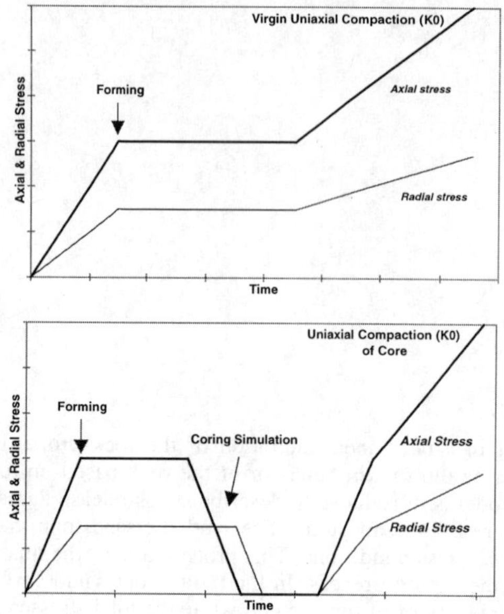

(a)

(b)

Figure 1. Stress paths in compaction tests. (a) Virgin K_0 compaction, and (b) Coring simulation, reloading to *in situ* conditions and K_0 compaction (Holt et al. 1998b).

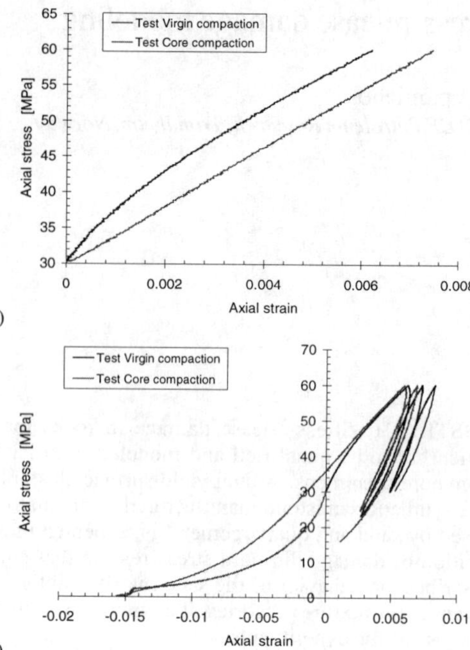

(a)

(b)

Figure 2. Compaction behavior of virgin vs. simulated core of a synthetic reservoir sandstone (Holt et al. 1998). (a) K_0 phase of the tests, and (b) All test segments. Compression is positive.

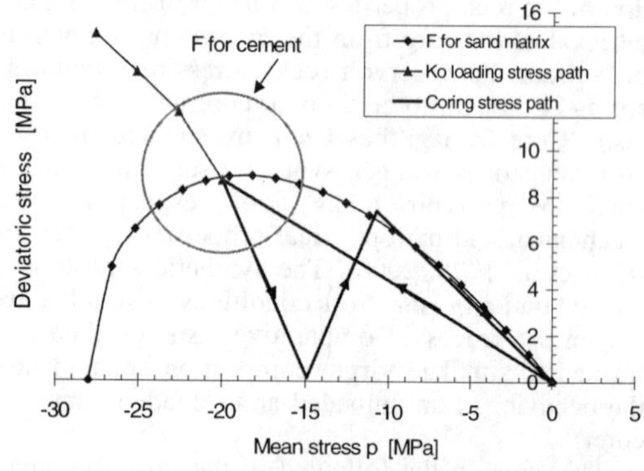

Figure 3. Yield surfaces F for the sand and the cement and stress paths during virgin and core compaction. Compression is negative.

to describe the material behavior in cyclic loading (Duwez 1935, Besseling 1958). Three-dimensional models were later developed and are known as multi-surface or nested yield surface plasticity models (Iwan 1967, Mróz 1967, Prévost 1977, 1978, Chen and Mizuno 1990).

In the proposed model, the one yield surface is for the sand and the other for the cement. The sand at *in situ* stresses is loaded and thus during coring it simply unloads. The yield surface for sand is centered at zero stresses and extends such that the *in situ* stress conditions lie on it. During virgin K_0 loading, the sand is further loaded and the yield surface extends such that the current stress conditions lie always on the yield surface. This is illustrated schematically in Figure 3 where the yield surfaces for the sand and the cement are drawn in the τ-p plane. The cement at *in situ* stresses is not loaded and thus during either coring or virgin K_0 loading, it loads and after an elastic region it yields. In the elastic region of the cement during virgin K_0 loading, the global behavior is not elastic since the sand loads plastically. For coring it is elastic since the sand unloads elastically. The yield surface for the cement is centered at the *in situ* stress conditions and extends such that it gives an initial elastic region upon reduction of the stresses for coring simulation or increase of the stresses for virgin K_0 loading. Yielding leads to the destruction of the cement cohesion, such that the cement yield surface reduces in size during stress changes, which can be either due to coring or virgin K_0 loading. Upon reloading to *in situ* stresses, nonlinear but elastic behavior may be assumed. After that the material continues to yield since the cement has been destructed. Essentially destruction of the cement occurs either by loading or by unloading from *in situ* conditions, such that after complete cement destruction the two materials behave the same.

The proposed two-material model can also be viewed as a model with internal, often referred to also as residual, stresses. The internal stresses result from the fact that the two materials, the sand and the cement act in opposite ways in the case of unloading, that is coring. Internally viewed the compression in the sand is counterbalanced by tension in the cement resulting in zero external stress. It is noted that the sand material is rather the core damaged material and not the sand packing itself prior to cementation. On the other hand, the cement is the material that results in the additional strength of the virgin material, which is destroyed during coring.

An one-dimensional schematic of a sandstone which is viewed as a two component material comprising of sand and cement is shown in Figure 4 together with the corresponding simple sublayer model approximation. During loading, part of the stress is carried by the sand and part by the cement. In the incremental form, which for time-independent behavior is equivalent to the rate form employed in the following and denoted with a dot over a quantity, the total stress rate $\dot{\sigma}_{ij}$ is given as in mixtures theories as

$$\dot{\sigma}_{ij} = \dot{\sigma}_{ij}^{(1)} + \dot{\sigma}_{ij}^{(2)} \tag{1}$$

where $\dot{\sigma}_{ij}^{(1)}$ and $\dot{\sigma}_{ij}^{(2)}$ are the stresses in the cement and the sand, respectively. Both the cement and the sand are described with elastoplastic constitutive stress-strain relations, which can be written as

$$\dot{\sigma}_{ij}^{(a)} = C_{ijkl}^{ep(a)} \dot{\varepsilon}_{kl}\,, \quad a = 1,2 \tag{2}$$

where $\dot{\varepsilon}_{kl}$ is the strain rate, and $C_{ijkl}^{ep(a)}$ is the elastoplastic tangent stiffness matrix of material (a). With Eqs.(1), (2) the global stress-strain relations can be written in the usual form of elastoplasticity, that is

$$\dot{\sigma}_{ij} = C_{ijkl}^{ep} \dot{\varepsilon}_{kl} \tag{3}$$

where

$$C_{ijkl}^{ep} = C_{ijkl}^{ep(1)} + C_{ijkl}^{ep(2)} \tag{4}$$

is the global tangent elastoplastic stiffness matrix.

3 CALIBRATION AND ANALYSIS

For both the sand and the cement the same elastoplastic constitutive model is used, with different material parameters. The selected model was developed to describe the behavior of collapsible rocks (Papamichos et al. 1997). The elasticity of the model is linear isotropic and the plasticity is non-associative and related to modified Mohr-Coulomb yield and plastic potential surfaces, which include a pressure cap to describe the compactive behavior of the sandstone. The yield surface F is written as

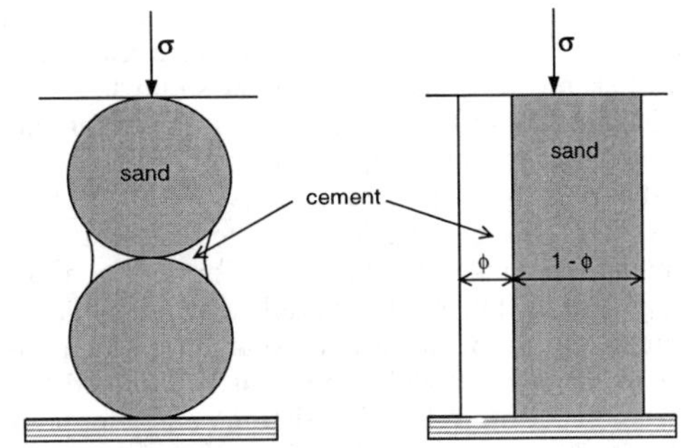

Figure 4. Schematic of a sandstone structure and the corresponding sublayer model approximation.

$$F = \tau\left[\cos\theta - \frac{\mu}{\sqrt{3}}\sin\theta\right] - \mu(q-p)\sqrt{\frac{p+p_c}{q+p_c}} = 0 \tag{5}$$

where is the isotropic (mean) stress, τ the shear stress intensity, and θ the Lode angle with

$$p = I_1/3, \quad \tau = \sqrt{J_2}, \quad \theta = \frac{1}{3}\operatorname{asin}\left(-\frac{3\sqrt{3}}{2}\frac{J_3}{J_2^{3/2}}\right) \tag{6}$$

where I_1 is the first stress invariant, and J_2 and J_3 the second and third deviatoric stress invariants, respectively. The parameters q, p_c and μ represent the size of the yield surface, and can harden or soften with accumulated plastic strain. The parameters q and p_c are identified as the intercepts of the yield surface F with the positive (tension cut-off) and negative (pressure cap) p-axis, respectively, while μ replaces the sinus of the friction angle in the Mohr-Coulomb model. The corresponding plastic potential function Q is

$$Q = \tau\left[\cos\theta - \frac{\mu}{\sqrt{3}}\sin\theta\right] - \beta(q-p)\sqrt{\frac{p+p_c}{q+p_c}} \tag{7}$$

where the parameter β replaces the sinus of the dilatancy angle in the Mohr-Coulomb model.

The calibration of the constitutive model for the sand and the cement is made by considering that the sand corresponds essentially to the core damaged material, as discussed earlier, and the cement to the material that provides the additional strength to the virgin material and which is destroyed during coring.

The calibration for the sand is based on test data from triaxial compression tests at different loading stress paths (Holt 1999). Starting from the in situ conditions of 30 MPa axial stress and 15 MPa radial stress, the stress paths followed in the considered tests were: (i) uniaxial stress path where the radial stress is kept constant, (ii) uniaxial strain path where the radial strain is kept constant, and (iii) isotropic stress path where the radial stress increase is the same as the axial stress increase. The elastic shear modulus G and Poisson's ratio ν were calibrated from the unloading part of uniaxial strain test. The plastic yield surface is described by three parameters, the tension cut-off q, the friction parameter μ and the pressure cap parameter p_c, which can be solved from the test data at the three different loading paths. The unknown parameters are not constant during loading but they generally change with the accumulated plastic shear and volumetric strains. In order to keep the model simpler, it is assumed that the tension cut-off q, and the friction parameter μ are constant. The reason is that the tension cut-off q does not influence substantially the results in the stress regime of interest, while for the friction it may be assumed that at the in situ stresses all friction has already been mobilized. The pressure cap parameter p_c hardens with the accumulated volumetric v^p and shear g^p plastic strains. The dilatancy parameter β can be obtained as a function of the dilatancy function $d = dv^p/dg^p$ (Papamichos et al. 1997).

The stress-strain behavior of the cement can be obtained from test results on virgin and core uniaxial strain path tests. Using Eq.(1) and the virgin and core compaction uniaxial strain tests, the axial and radial stress for the cement is plotted versus the axial strain in Figure 5. The elastic shear modulus G and Poisson's ratio ν were calibrated from the initial part of the curves in Figure 5. The parameters q and μ in the yield surface Eq.(5) are assumed to be constant and reasonable values were taken, as from one data set it is not possible to calibrate the three unknowns parameters, q, μ and p_c. Once q and μ are selected, then p_c can be calculated from Eq. (5). The parameter p_c softens with plastic shear strain and hardens (softens) with compactive (dilatant) plastic volumetric strain. During virgin loading the cement strength does not fall to zero, as seen from the test results, since hardening occurs with compaction. During loading, however, the volumetric strain is dilatant, which together with the softening due to shearing reduces the strength of the cement to zero.

The calibrated model was used for simulation of the virgin and core compaction uniaxial strain path tests. For the simulation of the core compaction test the coring process was simulated before the uniaxial strain test. Figure 6 shows the test results and model predictions for the virgin compaction and Figure 7 for the core compaction. The model predicts well both the axial stress and the radial stress.

4 CONCLUSIONS

An effort has been made to model the behavior of sandstones subjected to virgin and core compaction along various stress paths. A continuum model was developed and calibrated on the test data for a synthetic sandstone. The model is a two material elasto-plastic model, which describes the combined behavior of the sand and the cementing material. It can simulate the unloading behavior during coring and the subsequent loading, in addition to the axial and radial stress development in compaction tests along various stress paths. The model simulations show that the model can predict well the experimental results for both virgin and core compaction. It should

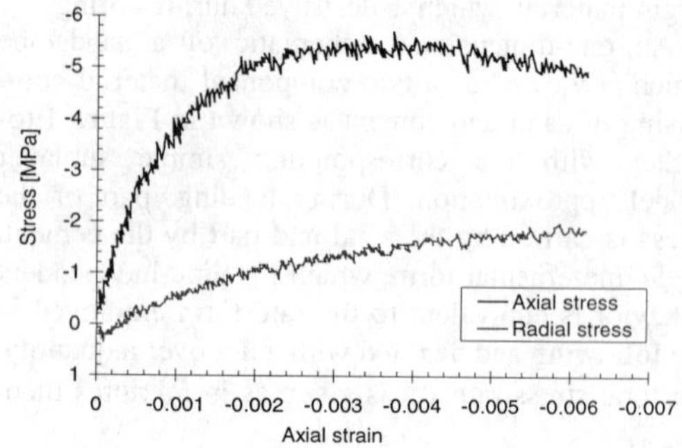

Figure 5. Axial and radial stress at the cement versus axial strain during virgin uniaxial strain path test, as calculated from the virgin and core compaction uniaxial strain tests. Compression is negative.

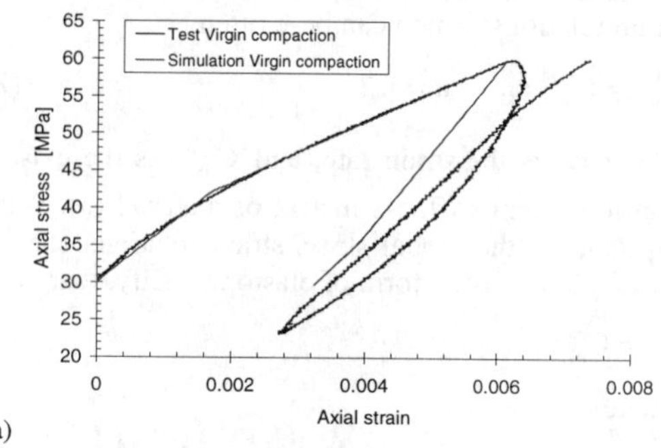

(a)

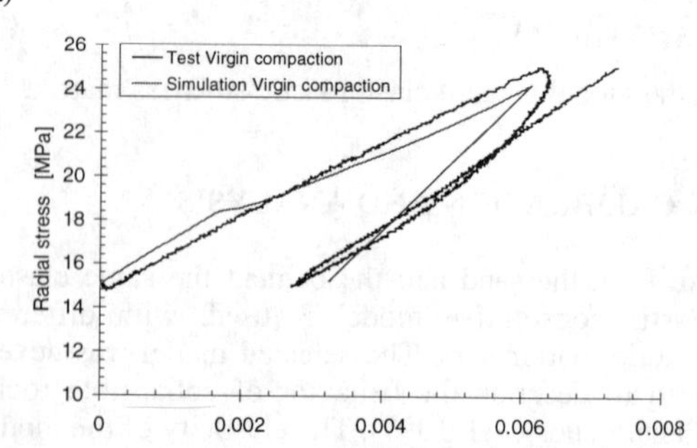

(b)

Figure 6. Uniaxial strain test results and model predictions for virgin compaction starting from in situ stress conditions: (a) Axial and (b) Radial stress versus axial strain. Compression is positive.

be mentioned that the two-material model is a bridge between continuum and micromechanical modeling, as it incorporates two materials, namely the sand and the cement. A similar description is made in discrete element particle models where an assembly of spheres (the sand) held together by bonds (the cement) describes the sandstone.

ACKNOWLEDGEMENTS

The author wishes to thank NAM for supporting this research and granting permission for publication.

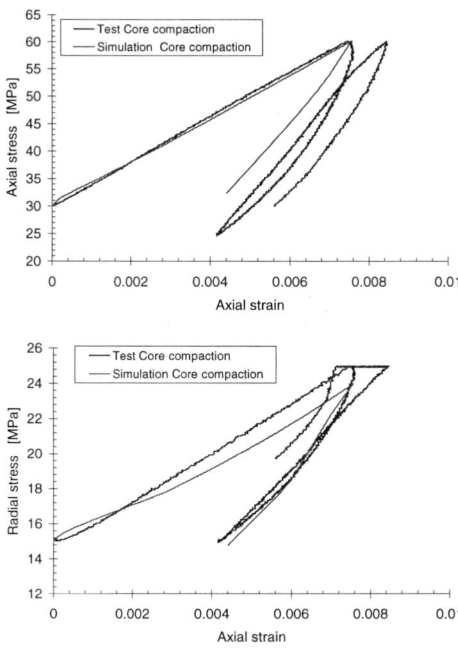

Figure 7. Uniaxial strain test results and model predictions for core compaction starting from in situ stress conditions: (a) Axial and (b) Radial stress versus axial strain. Compression is positive.

REFERENCES

Besseling J.F. (1958). A theory of elastic, plastic and creep deformation of an initially isotropic material showing strain hardening, creep recovery and secondary creep. *J. Appl. Mech., Trans. ASME* 25, 529-536.

Chen W.F. and E. Mizuno (1990). *Nonlinear analysis in soil mechanics. Theory and implementation.* Developments in Geotechnical Engineering Vol. 53, Elsevier, pp.196-216.

Duwez P. (1935). On the plasticity of crystals. *Phys. Rev.* 47, 494-501.

Holt R.M. (1999). Reservoir stress path: Evaluation of core and field data. *Proc. 37th U.S. Rock Mechanics Symposium*, Vail, Rotterdam: Balkema.

Holt R.M., Brignoli M., Kenter C.J. (2000). Core quality: Quantification of coring-induced rock alteration. *Int. J. Rock Mech. Min. Sci.* 37, in press.

Holt R.M., M. Brignoli, C.J. Kenter, R. Meij and P.M.T.M. Schutjens (1998). From core compaction to reservoir compaction: Correction for core damage effects. SPE/ISRM 47263, *Proc. EUROCK '98*; 1, 311-320.

Iwan W.D. (1967). On a class of models for the yielding behavior of continuous and composite systems. *J. Appl. Mech.* 34, 612-617.

Mindlin R.D. and H. Deresiewicz (1953) Elastic spheres in contact under varying oblique forces. J.Appl.Mech. 24, 585-593.

Mróz Z. (1967). On the description of anisotropic hardening. *J. Mech. Phys. Solids* 15, 163-175.

Papamichos E., M. Brignoli and F.J. Santarelli (1997). An experimental and theoretical study of a partially-saturated collapsible rock. *Mech. Cohes.-Frict. Mater.* 2, 3, 251-278.

Prévost J.H. (1977). Mathematical modeling of monotonic and cyclic undrained clay behavior. *Int. J. Numer. Anal. Methods Geomech.* 1, 195-216.

Prévost J.H. (1978). Plasticity theory for soil stress-strain behavior. *J. Eng. Mech. Div. ASCE* 104, EM5, 1177-1194.

Computer Methods and Advances in Geomechanics, Desai et al. (eds) © 2001 Balkema, Rotterdam, ISBN 90 5809 183 X

Elliptical shape perforations: An engineering application based on micro-mechanical modelling

Panos Papanastasiou
Schlumberger Cambridge Research, UK

Stability analysis of circular and elliptical shape perforations, embedded in a compressive stress field of unequal stresses is presented. The study is based on finite element analysis with an advanced material model with Cosserat micro-structure capable of predicting the localization of deformation around the holes which leads to failure. It is shown that an elliptical shape perforation aligned with its major axis parallel to the major principal stress is more stable than a corresponding circular perforation of the same flow capacity.

1 INTRODUCTION

Sand production from wells in weak formations due to high production rates is a growing concern in oil industry. Sand production may lead to a series of problems such as perforation failure, loss of production, increased wear of equipment and sand disposal problems especially offshore. Laboratory sanding experiments showed that rock failure, induced by the applied stresses, occurs before sand production. Rock failure usually takes place in the form of mechanical instabilities and localized deformation in the vicinity of the perforation due to stress concentration. Localized failure is occurs either in the form of shear-banding or axial splitting and spalling. Experimental evidences revealed an extended network of localized deformation bands around the cavity area. The interconnection of the localized zones will result in failure of the surrounded material and the subsequent sand production. Based on these observations, emphasis on perforation design for sand control is given on avoiding rock failure caused by drawdown and reservoir depletion; if the conditions for rock failure are avoided no sanding is expected to occur even at high production rates (Nicholson et al 1998, Papanastasiou et al 1998)

During the last two decades a significant number of papers has been published on failure prediction in geomechanics problems based on higher order continua and bifurcation theories (Vardoulakis and Sulem 1995). Nevertheless, we believe that the bifurcation approach has not been found yet its way to the solution of practical problems. In this paper we present a bifurcation solution to the above mentioned practical problem from petroleum engineering related to the stability of perforation tunnels that are used for the flow of hydrocarbon in the wellbores. We will demonstrate that it is essential to carry out advanced localization analysis in order to reach correct results upon which a new perforation shape design will be based.

It is expected that boreholes drilled parallel to the maximum insitu stress or parallel to the minimum insitu stress are more stable because the difference between the other two insitu stresses acting on a plane perpendicular to the wellbore axis is minimum resulting in less stress concentration near the borehole wall. Another way to distribute uniformly the shear stress avoiding excessive stress concentration is to shoot elliptical rather than circular shape perforations. Elliptical holes can be created by appropriate design of cone explosives in the perforation gun.

In order to investigate the best orientation for optimum stability of elliptical holes we carried out hierarchical modelling based first on classical stress analysis. The problem of an infinite region with an elliptic hole has been solved analytically by Stevenson (1945) and is presented in Jaeger and Cook (1979). The elastic stress analysis shows that an elliptical hole suffers less stress concentration than a circular hole when its major axis is aligned with the direction of the major principal stress and its axis ratio is the same as the insitu stress ratio. This is not convenient for practical applications since it will be necessary to know the magnitude and direction of the horizontal stresses which, in addition, change with the reservoir depletion. Nevertheless, there is no guarantee that an elliptic hole will be more stable than a circular hole since its curvature has also changed compared to the curvature of the circular hole. In previous work on the stability of deep boreholes we found that increase of the tangential stress may cause surface buckling resulting in localization of deformation in shear bands leading to failure in the form of breakouts (Papanastasiou and Vardoulakis 1992, 1994). The conditions for surface buckling near an opening will depend strongly on its curvature. For this reason, in order to assess the ultimate stability of the elliptical shape holes, we carried out finite element computations based on an advanced material model with Cosserat micro-structure which is capable of predicting surface buckling and localization of deformation.

The paper is organized as follows: firstly we outline in section 2 the Cosserat continuum. This model is applied in section 3 to examine the localization of deformation near circular and elliptical holes. The main conclusions are outlined in Section 4. For a more detailed study the reader is referred to Papanastasiou (2000).

2 MODEL

Proper modelling of failure near underground openings requires the incorporation of extreme dilation and extreme strength loss to account for microcracking, weak spots, fissures etc, in rocks. The simplest way to account for these inhomogeneities at macroscopic scale is by smearing them in a dilatant, strain-softening constitutive model. However, a major drawback of strain-softening models is that they lead to mathematically ill-posed boundary value problems, whose numerical analysis is sensitive to mesh refinements. Different methods for regularizing the quasi-static, boundary-value problem with strain softening were proposed in the literature (Vardoulakis and Sulem 1995). In this study, the problem of mesh dependence was resolved by resorting to a polar continuum with Cosserat micro-structure.

In a Cosserat continuum individual grains possess, in addition to their translational degrees of freedom, independent rotational degrees of freedom as well. The displacements are described by the motion of the material points, while the rotation is given independently by the micro-rotation represented by the orthogonal tensor. Grain rotation and its gradient give rise to a non-symmetric stress tensor and to couple stresses, such that the non-symmetric part of the stress-tensor is in equilibrium with the divergence of the couple stresses. From a dimensional analysis point of view, micro-rotation gradients and couple stresses introduce an *internal length* into the problem which is related to the micro-structure of the solid, e.g. grain size. A full description of the Cosserat plasticity theory appears in Vardoulakis and Sulem (1995).

We emphasize that this model is also capable of predicting the existing scale effect in small size holes such as perforations; small holes are more stable than large holes (Papamichos and van den Hoek 1995). We used material parameters derived from triaxial tests on Castlegate sandstone. An extra calibration constant is used to define the material softening required for triggering localization. In addition, the grain size is required in the input parameters. The grain diameter of the Castlegate sandstone is 0.2 mm.

3 RESULTS

We carried out a 2D plane strain analysis by considering the deformation in a plane perpendicular to the perforation axis. We have ignored the stress interaction between wellbore and perforations and the area near the tip of a perforation where the deformation is generally 3D. These assumptions do not affect the outcome of this analysis whose objective is to examine the strength of elliptical perforations relative to circular perforations. First, we performed computations for a circular hole with radius, $r = 0.01$m which have been used as a comparison reference. Computations were performed with elliptical holes of equivalent cross-sectional area in order to compare perforations of the same flowing capacity. In addition, by maintaining the same cross-sectional area we eliminate variation in the cavity strength due to the scale effect which is expected to be important for these small size holes (Papamichos and van den Hoek 1995). As found by the elastic stress analysis the best ellipse orientation in terms of stability is to align its major axis parallel to the major principal axis σ_y. The hole is loaded by increasing the remote compressive stresses stepwise maintaining a constant ratio of $\sigma_y/\sigma_x = \sigma_{max}/\sigma_{min} = 1.5$. The far-field stresses are distributed at the exterior nodal points. The internal pressure in the holes was maintained zero throughout the calculations. The first computations were performed with ellipse axis ratio $a/b = 1.5$. Further computations were performed by increasing the axis ratio to $a/b = 3$.

In contrast with the elastic analysis, the failure analysis based on localization of deformation shows that an elliptical hole will be stronger if its axis ratio is greater than the ratio of the applied stresses. These results are shown in Figure 1 where we plotted the hole closure in terms of a change in cross-sectional area vs applied maximum insitu stress σ_{max} ($\sigma_{max} = 1.5\sigma_{min}$) for the circular hole, $a/b = 1$ and the elliptical holes with axis ratio $a/b = 1.5$ and $a/b = 3$. The initial part of the curve is almost linear although initial yielding at the hole wall takes place quite early. For the circular hole localization of deformation has taken place after the applied stresses have reached the level of $\sigma_x = 33.80$ MPa and $\sigma_y = 50.70$ MPa. The Cosserat model switches to the localized branch

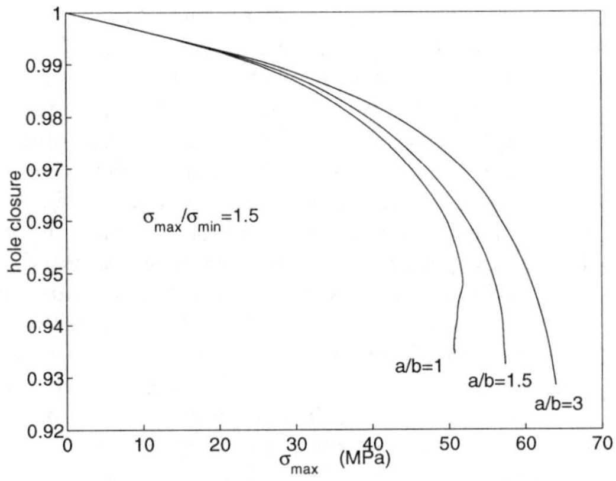

Figure 1: Hole closure vs applied stress for circular and elliptical holes, $\sigma_y/\sigma_x = 1.5$, $R/r = 0.02$

smoothly, without the need to perturb the primary solution with an imperfection. The computations confirmed that in the primary path the Cosserat rotation remains zero everywhere whereas after localization it increases by more than three order of magnitude only on the boundaries of the shear bands. In other words, the micro-rotations are activated only on the boundaries of the shear bands due to the intense shearing. The predicted failure mode is depicted in Figs 2 and 3 where the contours of the plastic strain and the incremental displacement field have been drawn in the vicinity of the hole after localization of deformation has taken place. Once localization of deformation starts, almost all the region close to the borehole wall is unloading elastically, except the narrow zone of the forming shear-band, which continues to plastically soften and progresses towards the interior of the hole. The shear band bends further until it finally meets its symmetric image, leading to a familiar breakout form. Deeper regions remain in the elastoplastic hardening regime or are deformed elastically being below first yield. It appears in these computations that the biaxial loading drives the deformation directly to localization once bifurcation starts. In contrast, in earlier computations with isotropic loading (Papanastasiou and Vardoulakis, 1992; 1994) we found that in the first step after bifurcation the deformation is characterized by a warping mode of finite wave-number, most likely caused by surface buckling. Further continuation of that solution resulted in progressively localized mode similar to the one depicted in Figs 2 and 3.

For the ellipse with axis ratio $a/b = 1.5$ localization of deformation starts after the applied stresses have reached the values $\sigma_x = 38.2$ MPa and $\sigma_y = 57.4$ MPa. Localization of deformation around the ellipse $a/b = 3$ has taken place at higher applied stresses $\sigma_x = 42.6$ MPa and $\sigma_y = 63.9$ MPa. In the last case the

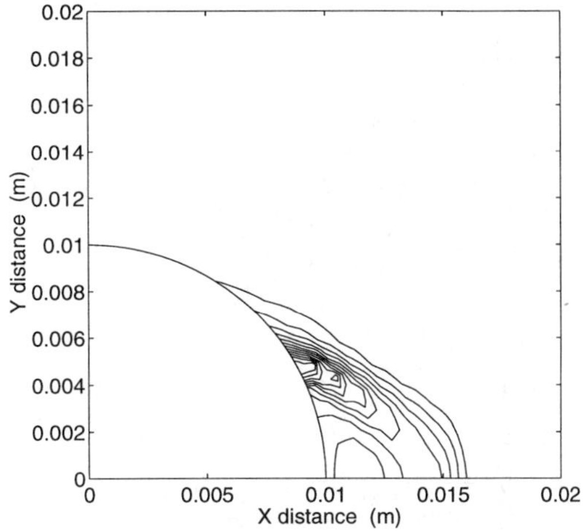

Figure 2: Contours of shear plastic strain after localization of deformation around a circular hole, $\sigma_y = 50.70MPa,\ \sigma_x = 33.80MPa$

localization pattern is shown by the contours of the shear plastic strain in Fig. 4. The shear plastic strain is more uniformly distributed around the elliptic hole compared to the circular hole. The incremental displacement field is shown in Fig. 5. Although the elastic stress analysis has suggested that the more stable ellipse is the one with its axis ratio the same as the ratio of the applied stresses, this localization analysis has shown that an ellipse with axis ratio greater than the ratio of the applied insitu stress is even stronger.

In normal stress fields the vertical insitu stress is

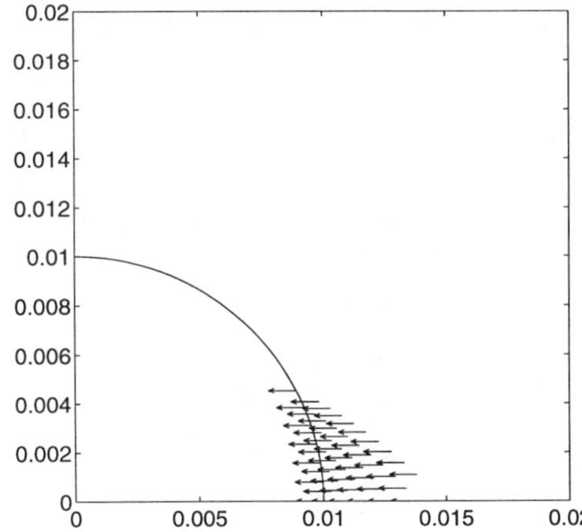

Figure 3: Incremental displacement field after localization of deformation around a circular hole, $\sigma_y = 50.70MPa,\ \sigma_x = 33.80MPa$

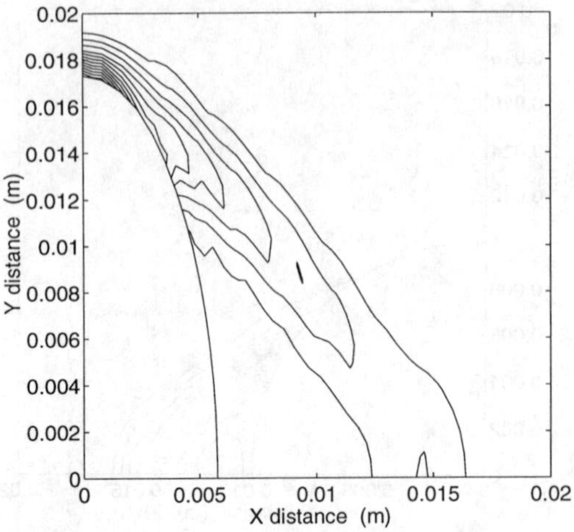

Figure 4: Contours of shear plastic strain after localization of deformation around an elliptical hole, $a/b = 3$, $\sigma_y = 63.94MPa$, $\sigma_x = 42.62MPa$

the major compressive stress. In such a case the elliptical shape perforations will be shot with their major axis vertical. In a tectonic stress environment one of the horizontal stresses becomes the major compressive stress. In such a case the perforations must be aligned with their long axis parallel to the direction of the major horizontal stress.

4 CONCLUSIONS

We have examined the stability of elliptical and circular shape perforations in a compressive stress field.

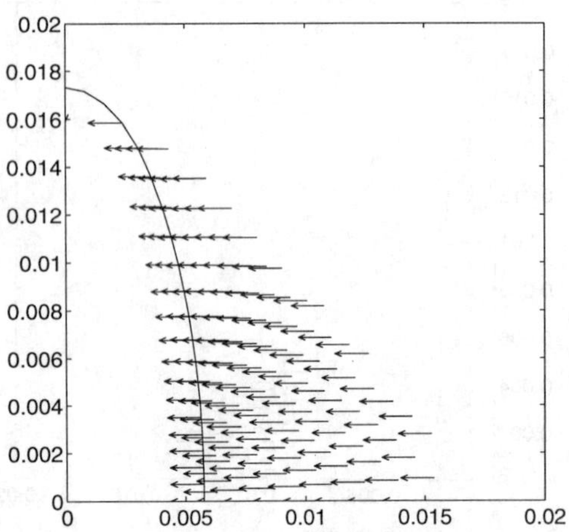

Figure 5: Incremental displacement field after localization of deformation around an elliptical hole, $a/b = 3$, $\sigma_y = 63.94MPa$, $\sigma_x = 42.62MPa$

Objective of the study was to examine the optimum shape design for perforations to avoid rock failure which may cause sand production in petroleum engineering. The study was based on finite element modelling for an extended Mohr-Coulomb elastoplastic model with Cosserat micro-structure. We demonstrated that it is essential to carry out advanced localization analysis in order to reach correct results upon which a new perforation shape design can be based.

The main conclusions for practical applications are summarized as following. Elliptical shape perforations aligned with the major axis parallel to the major principal stress are more stable than the corresponding circular perforations. The larger the ratio of the ellipse axis the stronger the perforation. In perforation design the maximum ellipse ratio can be shot defined by other technical considerations. This is convenient since it will not be necessary to know the magnitude and the direction of the horizontal stresses.

REFERENCES

Jaeger J.C. & N.G.W. Cook 1979. *Fundamentals of rock mechanics*, third edition, Chapman and Hall, London.

Nicholson E.D., Goldsmith G., & J. Cook 1998. Direct observation and modelling of sand production processes in weak sandstone. *SPE 47328 presented at the SPE/SRM Eurock'98 in Trondheim Norway.*

Papamichos E. & P. van den Hoek 1995. Size dependency of Castlegate and Berea sandstone hollow-cylinder strength on the basis of bifurcation theory. *Proc. 35th US Symp. Rock Mechanics.*

Papanastasiou P. & I. Vardoulakis 1992. Numerical treatment of progressive localization in relation to borehole stability. *Int. J. Num. Anal. Meth. Geomech.* Vol.16, 389-324.

Papanastasiou P. & I. Vardoulakis 1994. Numerical analysis of borehole stability problem. In J. Bull (ed) *Numerical Analysis and Modelling of Soil Structure Interaction* Chapter 19, 673-711, E &FN SPON, Chapman & Hall, London.

Papanastasiou P., Nicholson E.D., Goldsmith G. & J. Cook 1998. Sanding prediction: Experimental results and numerical modelling. In Thimus et al. *Poromechanics a tribune to M.Biot*. 457-462, Balkema, Rotterdam.

Papanastasiou P. 2000. Localization of deformation and failure around elliptical perforations based on a polar continuum. *Computational Mechanics*, (in print).

Stevenson A.C. 1945. Complex potentials in two-dimensional elasticity. *Proc. Roy. Soc., London A*, 184, 129-79.

Vardoulakis I. & J. Sulem 1995. *Bifurcation Analysis in Geomechanics*. Blackie A & P, Chapman &Hall.

Computer Methods and Advances in Geomechanics, Desai et al. (eds) © 2001 Balkema, Rotterdam, ISBN 90 5809 183 X

An improved pseudo-3D hydraulic fracturing simulator for multiple layered materials

E. Siebrits, H. Gu & J. Desroches
Schlumberger Cambridge Research, UK

ABSTRACT: We present a new pseudo-3D model for hydraulic fracturing applications in layered materials. The model accurately and efficiently takes into account the different elastic properties in each layer of a reservoir, thereby providing the simulator with an accurate width profile in the fracture height direction. This ensures better control on fracture height and a more accurate distribution of fracture volume, with important consequences for the design of hydraulic fractures in layered reservoirs.

1 INTRODUCTION

Most current pseudo-3D (P3D) hydraulic fracturing simulators employ an algorithm in which the elastic properties of multiple layers are lumped together into a single Young's modulus and Poisson's ratio, which are then used to predict the fracture width given the current net pressure. This is one of the weak areas in P3D hydraulic fracture simulations that has long been recognized in the industry but has not been addressed satisfactorily.

We introduce a new P3D model, which accurately and efficiently takes into account the different Young's moduli and Poisson's ratios of the layered structure of the reservoir and adjacent formations. The new model computes the fracture width as a function of the individual elastic properties in all fractured and near-by layers. This ensures that a more realistic width profile is obtained, because softer and stiffer layers are recognized as separate regions. Fracture height growth is better treated as a result. A more accurate distribution of fluid in the fracture is therefore obtained, which alters fracture volume and fluid efficiency, which in turn impacts fracturing treatment design.

In our approach, we use a displacement discontinuity method (DDM) (Crouch & Starfield 1974) to generate the fracture width profile in each P3D cell. This is based on numerical Green's functions that take into account the modulus variation along the fracture height in an extremely accurate manner, irrespective of the number of layers or their relative thicknesses. A width-height-area-pressure table is then constructed for use by a standard P3D simulator to obtain the coupled hydraulic fracture response.

2 MULTIPLE LAYER SCHEME

We will not present any mathematical details of the multi-layer formulation here, as this is documented elsewhere (Peirce & Siebrits 2000a, b), nor will we present any details of the P3D scheme (Mack *et al.* 1992). We will, however, present details of the implementation of this scheme in a P3D simulator.

We make the following assumptions. All layer interfaces are parallel. The fracture surface is oriented at right angles to the layer interfaces, and can cut any number of interfaces. (Orthogonality is not an intrinsic limitation of the method – any number of fractures can in principle be oriented at arbitrary angles. We place this restriction on the current implementation for simplicity only. The only real restriction imposed by our multi-layer formulation is that all layer interfaces must be parallel.) Layer thicknesses are limited to a large thickness range of 0.2 m to 1,000 m. (Much thinner layers can be handled with the same method, by re-arranging certain scaling factors in the multi-layer algorithm.)

The numerical discretization is performed as follows for each P3D cell. The fracture surface in the height direction is divided into a number of displacement discontinuity elements, each of which has a constant width variation. Piecewise constant displacement discontinuity elements are inaccurate, unless a large number of elements are used. However, we use special fracture tip elements (Ryder & Napier 1985) that ensure a highly accurate width profile is obtained, even if relatively few elements are used in the discretization process.

We employ a graded mesh to ensure that all elements abut interfaces perfectly, and to limit the number of elements in the mesh. A minimum of only one element is required per layer in the thinner layers. In the event that a fracture tip penetrates into a new layer by a distance that is less than the minimum element size, then that element is assumed to be partially fractured (Ryder & Napier 1985), and its influence coefficients are accurately included in the DDM scheme.

3 IMPLEMENTATION IN P3D SCHEME

The general idea for incorporation of the multi-layer solution into a P3D simulator is as follows. The P3D simulator requires a table of data, containing the fracture bottom tip location, the fracture height, the cross-sectional area of the fracture, and the corresponding total fluid pressure (assumed constant over the fracture height). Each row in the table gives the fracture geometry corresponding to the latest choice of pressure loading. This table of data is then used repeatedly by each cell (along the length of the fracture) in the P3D formulation to propagate the solution.

As input to the multi-layer module, we are given the layer elastic properties, toughnesses, confining stresses, and starter fracture location and dimensions. As output to the P3D simulator, we must supply the table data, including the complete width profile along the fracture height for each row of data in the table. We will refer to the combined "table" as the WHAP (width, height, area, pressure) table.

In order to calculate the WHAP table, we proceed as follows. Given the starter fracture dimensions and location, we postulate three different possibilities for the next fracture growth increment (to generate the next row of data in the WHAP table): growth by one element at the bottom tip, growth by one element at the top tip, or growth by one element from both tips simultaneously, as depicted in Figure 1. In order to determine which of the three options is most likely, we invoke the multi-layer module for each case and calculate the pressure required to just satisfy the toughness K_{Ic} (i.e., critical width w_{cr}) requirement at each fracture tip element. The critical width formula is given by

$$w_{cr} = \frac{2(1-\nu)}{G}\sqrt{\frac{2r}{\pi}}K_{Ic} \tag{1}$$

where $r = \Delta y/2$, Δy is the tip element size, and where the shear modulus, G, and Poisson's ratio, ν, are those applicable to the layer in which the fracture top or bottom tip is located.

For each case in Figure 1, we invoke eqn (1) twice – once for each tip. We postulate that the fracture is in a state of propagation for both bottom and top tips. We thus select the maximum pressure from these two calculations, and store this value. The final pressure returned to the table is then taken as the minimum (based on energy minimization arguments) of the three pressures obtained from each case in Figure 1.

It would appear that the above process is inefficient, since six sets of calculations appear to be required per WHAP table entry. However, the process can be streamlined in the following way. The elasticity equations defining fracture width can be written in matrix form as

$$Cw = p - \sigma_c \tag{2}$$

where C is the current (i.e., for the current fracture geometry) multi-layer influence coefficient matrix, p is the fluid pressure, and σ_c is the layer-dependent confining stress vector. Since we assume constant fluid pressure over the fracture height (for each P3D calculation cell along the fracture length), we can write p as $\overline{p}\,i$, where i is a unit vector, and $\overline{p}$ is a scalar constant pressure value. Inversion of (2), and substitution for p leads to the following equations for the bottom and top tip element widths

$$w_b = \overline{p}\left(C^{-1}i\right)_{bb} - \left(C^{-1}\sigma_c\right)_{bb} \tag{3a}$$

$$w_t = \overline{p}\left(C^{-1}i\right)_{tt} - \left(C^{-1}\sigma_c\right)_{tt} \tag{3b}$$

where $(\bullet)_{bb}$ or $(\bullet)_{tt}$ implies the matrix-vector operation utilizing only the bottom or top row of the inverse matrix C^{-1}, respectively. This greatly reduces the number of operations we need to perform. We now choose the left side of (3) to correspond to the critical widths in (1) to give

$$\overline{p}_b = \frac{\left(w_b^{cr} + \left(C^{-1}\sigma_c\right)_{bb}\right)}{\left(C^{-1}i\right)_{bb}} \tag{4a}$$

$$\overline{p}_t = \frac{\left(w_t^{cr} + \left(C^{-1}\sigma_c\right)_{tt}\right)}{\left(C^{-1}i\right)_{tt}} \tag{4b}$$

From considerations based on the fracture being in a state of propagation at both bottom and top tips, we apply

$$\overline{p} = \max(\overline{p}_b, \overline{p}_t) \tag{5}$$

We denote the value of $\overline{p}$ obtained in (5) as $\overline{p}_1$, implying it is obtained from the first possibility in Figure 1. We then repeat the above process for the other two possibilities to obtain $\overline{p}_2$ and $\overline{p}_3$. Finally, based on minimization of energy considerations, we obtain

$$\bar{p} = \min(\bar{p}_1, \bar{p}_2, \bar{p}_3) \quad (6)$$

The new fracture geometry is thus defined by whichever choice satisfies (6).

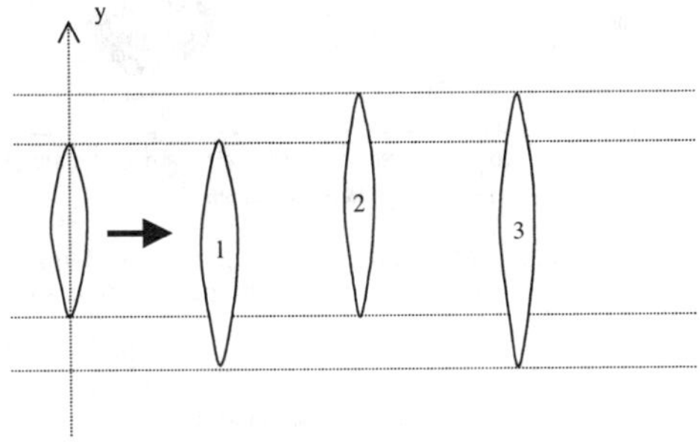

Figure 1. Three fracture growth possibilities from current fracture.

The multi-layer influence coefficient matrix C can be re-calculated for each new fracture geometry. However, this is computationally expensive, and can be avoided by calculating a single parent influence coefficient matrix C for the maximum possible extent of the final fracture. The influence matrices of all smaller fractures are then assembled from the appropriate rows and columns of the parent matrix. Of course, the fracture tips of the sub-fractures will in general not match the parent mesh exactly, and we account for this mismatch by allowing tip elements to be partially fractured, as mentioned earlier. This alteration to the sub-matrices of C can be made rapidly in an on-the-fly fashion, and has been proven accurate through careful testing.

4 APPLICATION TO HYDRAULIC FRACTURING IN A LAMINATED RESERVOIR

Table 1 lists layer elastic properties and thicknesses, obtained from log data (Desroches & Woods 1998), where *TVD* is true vertical depth. In Table 1, the *TVD*s in **bold** symbolize the perforated region of the wellbore, where the hydraulic fracture initiates. The injection schedule consists of 14,000 gal of fluid and 60,000 lb of proppant, injected at a rate of 30 bpm, over a 13.3 minute period.

Figure 2 shows a series of snapshots in time of the fracture width profile, obtained using the layer-averaged (i.e., uniform modulus) standard P3D model with the so-called "equilibrium" option active. (Note that the fracture top and bottom tips are on the left and right sides of the depth axis in the figure, respectively.) An "equilibrium" based P3D model is one in which height growth is calculated based on the assumption of constant fluid pressure over the fracture height in each computational cell in the fracture length direction. However, in this case, even though we use a reduced injection rate of 25 bpm, the model predicts runaway height growth after only 3.8 min. of fluid injection.

We thus resort to a so-called "non-equilibrium" P3D model, where analytic expressions are used to determine the applicable pressure drop due to fluid flow in the vertical direction. Such a model avoids run-away height growth problems. The non-equilibrium uniform modulus model predicts a fracture height of approximately 30 m (Figure 3). In this case, the model "screens out" at the fracture tip after 8.1 min.

In contrast, the new multi-layer equilibrium model screens out after 7.1 min. In this case, the majority of the fracture surface intercepts some higher permeability layers (below a *TVD* of 830 m – see Table 1), thereby allowing more leakoff into the formation, causing earlier screen out compared to the non-equilibrium result. Although Figure 3 also shows fracture growth below a *TVD* of 830 m, the non-equilibrium case screens out later because the majority of the fracture is located above a *TVD* of 830 m – Figures 2 to 5 only show fracture height along the wellbore, and give no indication of fracture coverage away from the wellbore.

Once screen out occurs in the simulator, the fracture tip is frozen, and the remaining injected fluid is accommodated by simply allowing the fracture to balloon. This situation can cause unrealistic fracture pressures and widths.

In order to make an equitable comparison, we thus plot the uniform modulus non-equilibrium result for a 7.1 min injection time in Figure 4, and show the corresponding multi-layer result in Figure 5. We see that the former case has a fracture height of approximately 20 m, but the multi-layer case has a fracture height of approximately 39 m.

In Figure 6, the depth data is measured depth (*MD*) along the wellbore. The top and bottom locations of the perforated interval (of height 20 m in Table 1 and in Figure 6) match the equivalent *TVD* locations shown in bold in Table 1. The discrepancy between the *TVD* and *MD* values is simply due to a shift caused by the wellbore being S-shaped above a depth of 750 m. The wellbore is in fact vertically oriented in the region covered by Table 1.

The multi-layer result in Figure 5 best matches both the magnitude and location of the fracture height inferred from data obtained from radioactive tracer and gamma ray measurements (see Figure 6, where the fracture height location is denoted by the arrow marked $H = 38$ m). We see that the location of the fracture top and bottom tips in relation to the perforated interval (as shown in Figure 5 and Table 1) is in very good agreement with the measured data in Figure 6.

We should point out, however, that our current multi-layer scheme has been implemented for the equilibrium model, and thus is still prone to run-away height growth under some conditions.

Table 1. Layer data for application problem, with perforated interval indicated by **bold** TVD values.

Top TVD (m)	Layer thick (m)	Conf. Stress (kPa)	*E* (GPa)	ν	Perm (mD)	Res. press. (kPa)
792.8	1.19	16,734	5.150	0.39	250	8,536
794.0	2.16	14,693	6.095	0.32	130	8,770
796.1	3.29	17,857	5.206	0.41	130	8,770
799.4	3.81	16,375	5.206	0.20	130	8,770
803.2	1.22	15,982	5.992	0.38	0.1	7,894
804.5	1.86	16,044	5.302	0.39	170	8,515
806.3	1.86	16,830	4.461	0.41	0.1	7,901
808.2	1.25	15,879	5.488	0.37	170	7,922
809.4	3.02	17,982	5.012	0.38	0.1	7,943
812.4	1.92	16,423	6.019	0.35	200	7,943
814.4	2.19	16,017	4.875	0.40	208	9,039
816.6	2.16	18,161	4.309	0.40	0.1	8,005
818.7	6.22	16,396	4.075	0.39	2	9,639
824.9	1.25	14,382	5.861	0.29	0.1	8,067
826.2	0.64	15,658	5.075	0.36	150	8,846
826.8	4.36	16,871	5.012	0.38	0.1	8,115
831.2	3.60	18,140	3.613	0.42	416	8,887
834.8	2.01	15,761	5.123	0.34	400	8,887
836.8	2.16	18,264	5.178	0.36	400	8,736
839.0	30.48	19,236	3.696	0.40	150	8,736

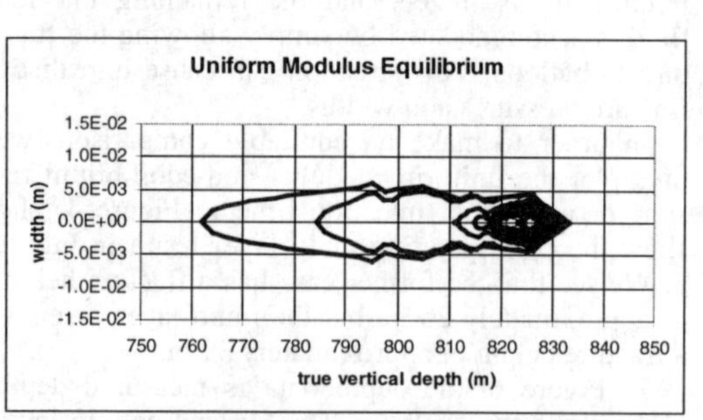

Figure 2. Width profiles along the wellbore at different times from layer averaged equilibrium model, showing run-away height growth (3.8 min. injection time).

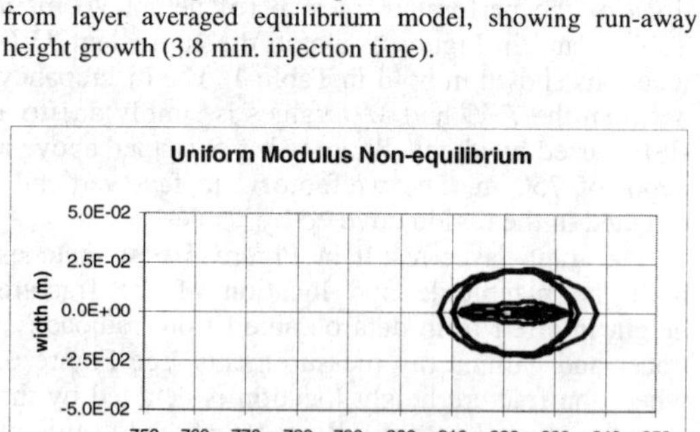

Figure 3. Width profiles along the wellbore at different times from layer-averaged non-equilibrium model (13.3 min. injection time).

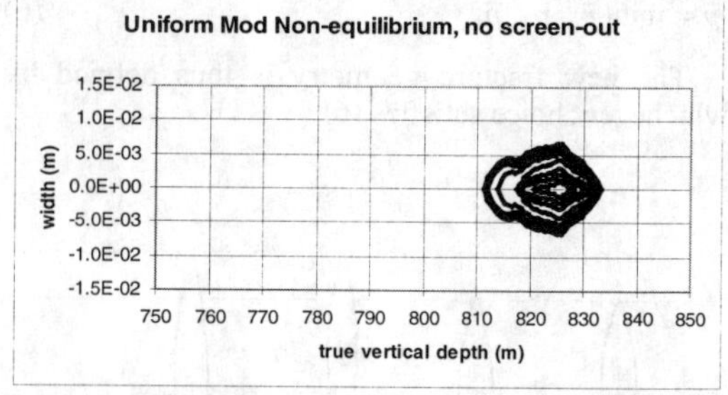

Figure 4. Width profiles along the wellbore at different times from layer-averaged non-equilibrium model (7.1 min. injection time).

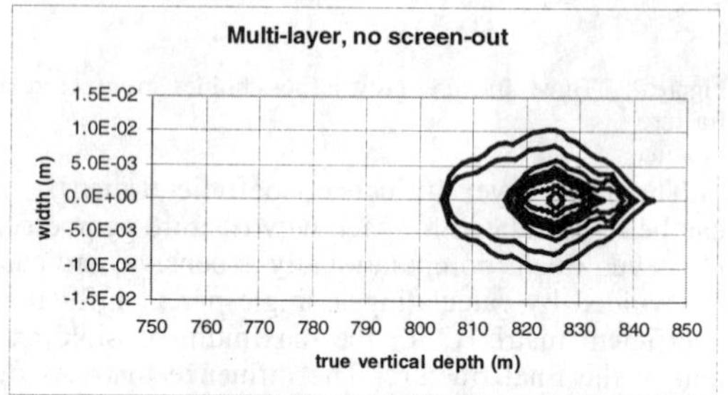

Figure 5. Width profiles along the wellbore at different times from multi-layer equilibrium model (7.1 min. injection time).

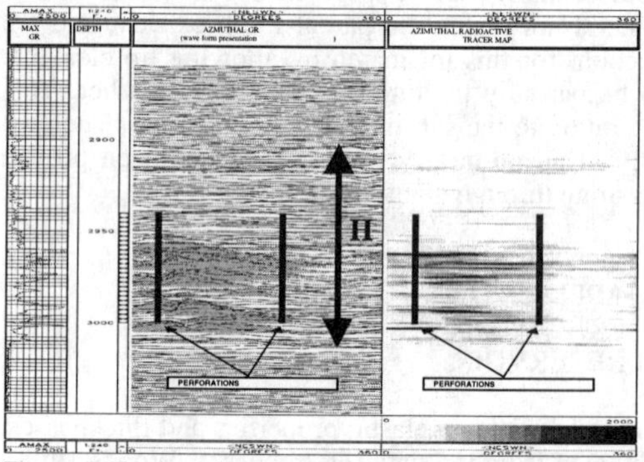

Figure 6. Tracer and gamma ray data showing vertical extent, *H*, of fracture (with measured depth indicated in depth column)

5 CONCLUSIONS

We have shown that a variation in elastic properties along the fracture height can have a significant effect on the final fracture dimensions obtained from a P3D simulator. By accurately modeling the elastic variation along the fracture height, we obtain a better fracture width profile.

We now have a means of designing more aggressive hydraulic fracturing treatments in situations where significant modulus contrasts arise. Contrasts in elastic properties are also linked to contrasts in confining stresses from layer to layer, which can further contribute to variations in the width profile along the fracture height via the coupling of the fluid flow and elasticity equations. Simple averaging of elastic properties, as we have demonstrated, can lead to erroneous fracture height predictions.

Acknowledgments. The authors wish to thank Schlumberger for permission to publish, and the significant contribution of Anthony Peirce, University of British Columbia, who developed the mathematical details of the multi-layered Green's function used in our fracture model.

REFERENCES

Crouch, S.L. & A.M. Starfield 1974. *Boundary Element Methods in Solid Mechanics*. Boston: Allen & Unwin.

Desroches, J. & T.E. Woods 1998. Stress measurements for sand control. SPE paper 47247, presented at the SPE/ISRM Conference *Eurock 98*, Trondheim, Norway, 8-10 July 1998, 219-229.

Mack, M.G., J.L. Elbel & A.R. Piggott 1992. Numerical representation of multilayer hydraulic fracturing. In Tilerson & Wawersik (eds), *Rock Mechanics*: 335-344. Rotterdam: Balkema.

Peirce, A.P. & E. Siebrits 2000. The scaled flexibility matrix method for the efficient solution of boundary value problems in 2D and 3D layered elastic media. *Computer Meth. in Appl. Mech. & Engrg*. To appear.

Peirce, A.P. & E. Siebrits 2000. Uniform asymptotic solutions for the accurate solution of cracks intersecting interfaces in layered elastic media. *Int. J. Fracture*. To appear.

Ryder, J.A. & J.A.L. Napier 1985. Error analysis and design of a large-scale tabular mining stress analyzer. Proceedings of *Fifth International Conference on Numerical Methods in Geomechanics*, Nagoya, 1549-1555.

Computer Methods and Advances in Geomechanics, Desai et al. (eds) © 2001 Balkema, Rotterdam, ISBN 90 5809 183 X

A heterogeneous fracture propagation model for a 3-D hydraulic fracturing simulator

K.Yamamoto, T.Shimamoto & S.Maezumi
Japan National Oil Corporation, Chiba, Japan

S.Sukemura & N.Uryu
Fuji Research Institute Corporation, Tokyo, Japan

ABSTRACT: Multiply generated hydraulically induced fractures are often observed or at least to exist in well stimulation site of the petroleum and geothermal industries. To determine the final geometry of the fractures in consideration of the interaction between each fracture, the authors developed a true 3-D simulation code of the fracturing process. The code can compute the stress field around a fracture tip induced by fracture, and estimate the interaction between fractures of a single well or multiple boreholes. Displacement discontinuity formulation is used to model arbitrary figures of fractures with a fully coupled solution of fracture aperture and injected fluid flow.

1 INTRODUCTION

On the reason of its complex nature involving the behavior of solid materials and fluids under the conditions in deep underground, hydraulic fracturing is a subject of much debate among engineers and researchers. Various numerical models of hydraulic fracturing have been developed, and their numerical application to geomechanics has become a big issue.

A three-dimensional model of the hydraulically induced fractures is necessary for such simulators, especially for massive hydraulic fracturing for well stimulation technique and much greater fracture formation for cuttings reinjection. In such cases, fractures tend to leave the original planes due to the effects of stress and strength anisotropy around the wells, and their geometry and extent are likely to affect the efficiency of hydrocarbon production, capacity of drilling waste disposal, and reliability of environmental safeness.

To consider such three-dimensional and heterogeneous formation of the fractures, interaction between fractures from single or multiple borehole(s) should be taken into account. We often observe such multiple fracture initiation and growth in various conditions, such as; 1) fractures from different intervals of a single open hole completion or multiply perforated extended reach wells, 2) radial growth of fractures from the same interval of a single well in a relatively isotropic stress condition, and 3) fractures from different wells (e.g. from an injection well and a production well of heat dry rock (HDR) power generation site). Such relatively large-scale multiple fracture growth interaction might be governed by stress magnitude and direction induced by the aperture of fractures at each fracture tip.

The authors previously developed a numerical simulator of purely three-dimensional fracture propagation by adopting a displacement discontinuity (DD) formulation, and succeeded in modeling the bending and twisting fracture propagation. In such cases, the fracture tip is driven by the sum of in-situ stress and induced stress by itself. In the former model, two wings of the fracture start advancing on a single plane on which the borehole exists. In this paper, we present the an extended model in which arbitrary numbers of boreholes and fractures can be considered, and the stress field induced by aperture of the fractures can be computed.

2 NUMERICAL MODELLING

2.1 *Displacement-discontinuity (DD) formulation*

The DD formulation is based on the Green's functions which present the relationship between displacement discontinuity and stress tensor. Fracture itself is modeled as a curved or flat plane without thickness. Since only the fracture surface needs to be discretized, the boundary integral method is regarded as suitable for analyzing the fracture propagation problem. The concept of the DD method was presented by Crouch (1976) as a means of discretizing crack opening with finite length segments which have uniform displacement discontinuity throughout each segment. Willes & Curran (1982) and Curran & Vandamme (1992) derived the three-dimensional formulation and Green's functions in uniform, infinite and elastic media for a point source, in order to apply it to the fracture of an arbitrary shape in three-dimensional space.

On the fracture plane, displacement discontinuity is defined by aperture and two kinds of ride (relative share displacement). Assuming arbitrary Cartesian coordinates, the crack opening and rides are modeled as the distribution of the concentrated displacement discontinuities, which is defined by Equation 1 at a point ξ as follows:

$$\phi_{kn}(\xi) = \lim_{\xi_n \to -0} u_k(\xi) - \lim_{\xi_n \to +0} u_k(\xi) = \delta(\xi) \cdot \phi_{kn} \tag{1}$$

where u_k = k-th component of displacement, n = normal component of fracture surface, and $\delta(x)$ is a Dirac's delta function. Stresses at a point x induced by the point source at ξ are obtained by multiplying Green's functions by the intensity vector of the displacement discontinuities as follows:

$$\sigma_{ij}(\mathbf{x}) = T^*_{ijkn}(\mathbf{x}, \xi)\phi_{kn}(\xi) \tag{2}$$

where T^* = the tensor representation of Green's functions.

The stresses induced by the finite extension of a fracture, such as the continuous displacement discontinuity, are obtained by integrating the right-hand side of Equation 2 on the total fracture surface Γ:

$$\sigma_{ij}(\mathbf{x}) = \int_\Gamma T^*_{ijkn}(\mathbf{x}, \xi)\phi_{kn}(\xi) d\Gamma \,. \tag{3}$$

In case of multiple fractures, φ should include the whole fracture surface. The discrete form of Equation 3 is derived by discretizing the surface with triangle elements and appropriate shape functions $N_i(x)$.

$$\sigma_{ij}(\mathbf{x}) = \sum_{e=1}^{E} \left[T^*_{ijkn}(\mathbf{x}, \xi) \int_{\Gamma_E} \left\{ \sum_{i=1}^{3} \phi^{(i)}_{kn} N_i(\xi) \right\} d\Gamma_e \right]. \tag{4}$$

In this model, we choose linear shape functions for the elements inside the fracture and square root shape functions for the distance from the fracture front for front elements in order to represent the fracture opening characteristics predicted as the infinite crack tip stresses by linear fracture mechanics.

The pressure of the fluid should balance the normal component of stresses against the fracture face, hence:

$$\begin{aligned} & p(\mathbf{x})n_i(\mathbf{x}) \\ & = \sigma_{ij}(\mathbf{x})n_j(\mathbf{x}) \\ & = \sum_{e=1}^{E} \left[T^*_{ijkn}(\mathbf{x}, \xi) \int_{\Gamma_E} \left\{ \sum_{i=1}^{3} \phi^{(i)}_{kn} N_i(\xi) \right\} d\Gamma_e \right] \cdot n_j(\mathbf{x}) \end{aligned} \tag{5}$$

where, $p(x)$ = fluid pressure at x; and n_i = normal vector of the surface of fractures.

By applying the numerical integration and coordination transfer, Equation 5 becomes a vector-matrix equation as follows:

$$p^m n_i^m = \sum_{n=1}^{N} T_{ij}^{mn} \phi_j^n \tag{6}$$

where m, n = node numbers; and T_{ij}^{mn} = the influence matrix derived from the tensor T^*. This equation should be formed for the entire system.

2.2 *Coupled model of fluid pressure and aperture distribution*

In this model, the fracturing fluid flow of each fracture wing is solved independently. However, the pressure of each wing of fracture should be same value at the the same point of the borehole. The fracturing fluid is assumed to follow the power law, and the model assumes that the fluid runs through the parallel planes between fracture surfaces as laminar fluid. The numerical method analyzing the fluid pressure distribution of the injected fluid is based on the two-dimensional finite element formulation (Clifton & Abou-Sayed, 1979). Friction loss caused by tortuosity is not considered.

The leak-off rate of any specific point on the fracture is assumed to decrease in proportion to the inverse square root of time after the fracture formation. This assumption is known to be valid for single planar fracture under the condition of sufficiently slow fracture propagation velocity compared to the pressure propagation velocity. In the case of close multiple fractures, pore pressure accumulation by multiple fractures should be taken into account. This issue will be discussed later.

The coupled solution is given by the condition of fluid volume conservation. The fracture volume at step n is given by:

$$\begin{aligned} W &= \sum_e a_e^{(n)} f_{w_e}^{(n)} \\ &= \Delta t \cdot \left(-\sum_e a_e f_{L_e} + \sum_e r_e f_{q_e} \right) + \sum_e a_e^{(n-1)} f_{w_e}^{(n-1)} \end{aligned} \tag{7}$$

where, a_e = area of element; f_w = fracture aperture; r_e = element boundary; and f_q = flow-in through the element boundary. f_L = leak-off rate at each time and is derived as:

$$f_L = \frac{2c_L}{\sqrt{t - t_0}} \tag{8}$$

where, c_L=leak off coefficient defined in each layer.

In the single fracture case, the fracture aperture at each time is obtained so as to satisfy the fracture volume W:

$$\begin{bmatrix} T & -\mathbf{1} \\ \mathbf{a} & 0 \end{bmatrix} \begin{Bmatrix} \phi \\ P \end{Bmatrix} = \begin{Bmatrix} \mathbf{p} \\ W \end{Bmatrix} \tag{9}$$

where, T = the influence matrix given in Equation 6; $\mathbf{-1} = (-1, -1, \ldots -1)^T$; $\boldsymbol{a} = (a^{(1)}n^{(1)}_1, a^{(1)} n^{(1)}_2, a^{(1)} n^{(1)}_3, a^{(2)} n^{(2)}_1, a^{(2)} n^{(2)}_2, a^{(2)} n^{(2)}_3, \ldots a^{(n)} n^{(n)}_3)^T$; $\boldsymbol{\phi} = (\phi^{(1)}_1, \phi^{(1)}_2, \phi^{(1)}_3, \phi^{(2)}_1, \phi^{(2)}_2, \phi^{(2)}_3, \ldots \phi^{(n)}_3)^T$; and $\boldsymbol{p} = p^{(n)}\boldsymbol{n}^{(n)} = (p^{(1)}_1, p^{(1)}_2, p^{(1)}_3, p^{(2)}_1, p^{(2)}_2, p^{(2)}_3, \ldots p^{(n)}_3)^T$.

Optimization of fracture aperture and fluid pressure balance is obtained by the convergence loop as shown in Figure 1.

In the case of multiple fractures from a single borehole, the fracture aperture at each time is obtained so as to satisfy the total fracture volume W (total fracture volume = total injected fluid volume - total leak-off volume):

$$\begin{bmatrix} T_{11} & T_{12} & \cdots & -\mathbf{1} \\ T_{21} & T_{22} & \cdots & -\mathbf{1} \\ \vdots & \vdots & \ddots & \vdots \\ \mathbf{a}_1 & \mathbf{a}_2 & \cdots & 0 \end{bmatrix} \begin{Bmatrix} \phi_1 \\ \phi_2 \\ \vdots \\ P \end{Bmatrix} = \begin{Bmatrix} \mathbf{p}_1 \\ \mathbf{p}_2 \\ \vdots \\ W \end{Bmatrix} \tag{10}$$

On the other hand, the case of multiple boreholes, fluid volume conservation should be satisfied at each borehole if the injection rate is controlled independently by borehole. Therefore, the equation of fracture aperture-fluid pressure relationship and conservation of fluid volume should be written as:

$$\begin{bmatrix} T_{B^{(1)}1B^{(1)}1} & T_{B^{(1)}1B^{(1)}2} & -\mathbf{1} & T_{B^{(1)}1B^{(2)}1} & T_{B^{(1)}1B^{(2)}2} & 0 \\ T_{B^{(1)}2B^{(1)}1} & \ddots & \vdots & T_{B^{(1)}2B^{(2)}1} & \ddots & \vdots \\ \mathbf{a}_{B^{(1)}1} & \cdots & 0 & 0 & \cdots & 0 \\ T_{B^{(2)}1B^{(1)}1} & T_{B^{(2)}1B^{(1)}2} & 0 & T_{B^{(2)}1B^{(2)}1} & T_{B^{(2)}1B^{(2)}2} & -\mathbf{1} \\ T_{B^{(2)}2B^{(1)}1} & \ddots & \vdots & T_{B^{(2)}2B^{(2)}1} & \ddots & \vdots \\ 0 & \cdots & 0 & \mathbf{a}_{B^{(2)}1} & \cdots & 0 \end{bmatrix} \begin{Bmatrix} \phi_{B^{(1)}1} \\ \vdots \\ P_{B^{(1)}} \\ \phi_{B^{(2)}2} \\ \vdots \\ P_{B^{(2)}} \end{Bmatrix} = \begin{Bmatrix} \mathbf{p}_{B^{(1)}1} \\ \vdots \\ W_{B^{(1)}} \\ \mathbf{p}_2 \\ \vdots \\ W_{B^{(2)}} \end{Bmatrix} \tag{11}$$

where, $W_B^{(k)}$ = total fracture volume and $P_B^{(k)}$ = integral constants of borehole k.

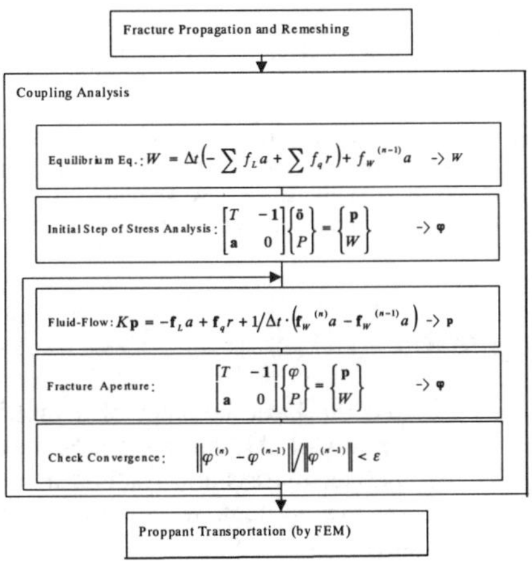

Figure 1. Schematic flow of the coupling analysis

2.3 *Fracture growth model*

The fracture mechanics approach is adopted in the code. Fracture growth velocity is assumed to be determined by comparing the stress intensity factors (K_I) at the fracture front and fracture toughness of the rock mass, which is defined at each stratum, according to Mastrojannis et al. (1980):

$$\Delta d = \Delta t \left(\frac{K_I - K_{IC}}{\left| K_{\mathrm{I}max} - K_{IC} \right|} \right)^{\alpha} \cdot V_{\mathrm{front}} \tag{11}$$

Stress intensity factor K_{I} is derived from the aperture distribution at the fracture tip as follows:

$$\phi = \frac{K_I}{G}\sqrt{\frac{r}{2\pi}} = a\sqrt{r} \tag{12}$$

$$K_I = aG\sqrt{2\pi} \tag{13}$$

This relationship gives only the relative advance of the fracture tip on a fracture, so it is difficult to determine the fracture propagation velocity (V_{front}).

In case of stable propagation, advance of the fracture front should be stopped soon after the shut-in, therefore the propagation velocity should be determined such that the stress intensity factor after the fracture front advanced is equal to the fracture toughness at every point.

$$K_I = K_{IC} \text{ on } \partial\Gamma^{(n)} \tag{14}$$

In the presented model, we determine the fracture propagation velocity at a certain time step to satisfy the equilibrium of fracture volume increase rate (change of fracture aperture and area), leak-off rate and injection rate.

If an inappropriate value is chosen at the initial step, the fracture propagation behavior will oscillate around the numerically stable velocity in order to satisfy the condition of fluid volume conservation.

At present, only the opening mode of the stress intensity factor (K_I) is concerned with the fracture growth velocity. However, the contribution of sliding and tearing mode will be considered in the future.

Fracture propagation direction is determined by the criterion of maximum tensile stress. The stress field around the fracture front is a combination of in-situ stresses and stresses induced by fracture opening and rides, which can be obtained from Equation 3.

In this code, we consider only the mode I crack displacement (fracture opening mode), though we realize that shear modes (mode II and mode III) are also important for fracture propagation in the case of high deviatory stress.

3 SIMULATION RESULTS

3.1 *Multiple fractures from different intervals*

The example of Figure 2 shows the multiple fracture growth from a single borehole in a uniform stress state. In the case, one pair has 30 degree to the other pair.

The result shows that fractures tend to leave the original surface of another one.

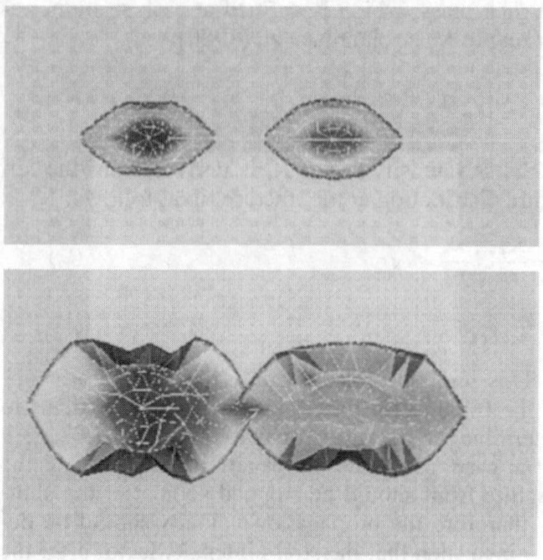

Figure 2. dual fractures from a single horizontal well. The fracture fronts of each fracture tend to leave from each other.

3.2 *Dual boreholes and dual fractures*

Figures 3 and 4 show the cases of interaction between two different fracture systems in slightly deviated stress state ($S_{Hmax} = 1.2S_{hmin}$). In the case of Figure 3, the original surface of the fracture is 15m away from the other, and one wing of one fracture bends to become perpendicular to the other fracture. This is caused by the compressive stress induced by this side of the fracture. The distance is 19m for Figure 4 and the effect seems to be smaller than the former case.

Figure 3. An example the interaction between fractures from different borehole. The second fracture is bent toward the first one, the normal direction to minimum compressional stress axis.

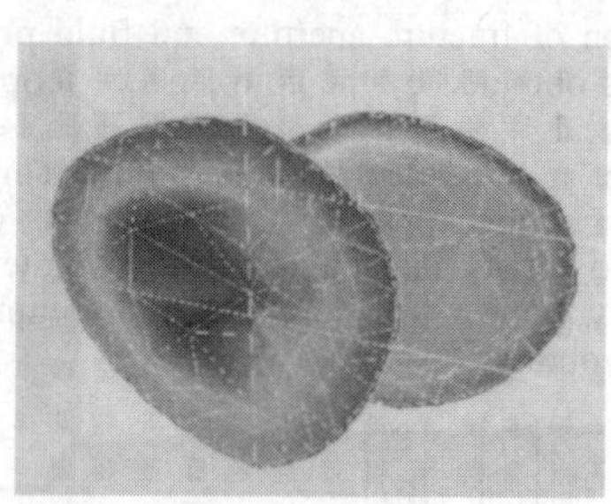

Figure 4. In this case, the distance between the borehole is further than the former case. The wings of each fracture are formed unsymmetrically by the induced stress from the other fracture, but the bending is not occurred.

3.3 *Multiple fractures from single interval*

An example of Multiple fracture formation from a single interval is shown in Figure 5. In the case, one pair has 60 degree to the other one.

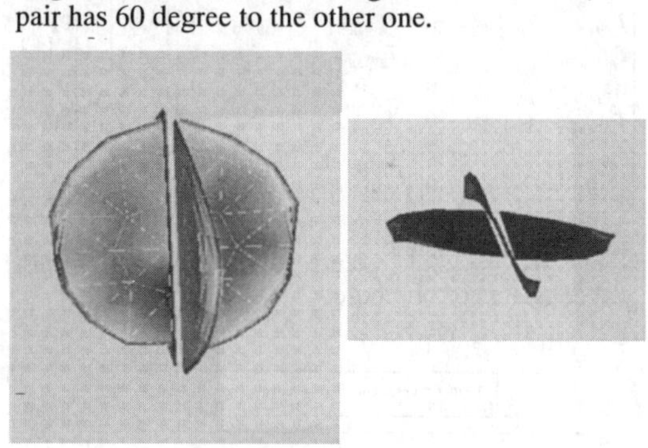

Figure 5. An example the dual fractures from a single interval of the borehole. The smaller fracture is bending to the normal direction of maximum in-situ stress axis, and bigger one leave from the original plane due to the stress induced by the other one.

4 DISCUSSION

We have presented the basic concept of a multiple fracture interaction model. Although the model satisfies the balance between stress and fracture aperture in the multiple fracture system, we should consider the following issues:

4.1 *Fracture propagation model*

In the model, fracture propagation velocity and direction are governed solely by the normal tensile stress against the fracture surface. As a result, the model gives a final fracture geometry in which mode I is predominant.

In real situations, we often observe mode II and mode III shear fractures, and it is anticipated that such shear fracture modes are common in complicated stress states in multiple fracture systems. Such fractures contribute the fracture permeability through the dilation mechanism.

We can derive the shear displacement and stress intensity factors at the fracture front from this numerical model, and if reliable fracture propagation criteria are established, the factors can be applied to the model.

Fracture propagation velocity is also a complicated issue. Although there is much literature about the dynamic behavior of brittle failure for steel, the problem of a rockmass is more difficult with its nature of composite materials. Many micromechanical factors such as inhomogeneousness, accumulation of micro cracks and cavities, and creation of a plastic zone affect the behavior of the fracture tip.

Another factor affecting the fracture propagation velocity is the fluid velocity and pressure at the fracture front. Desroches et al. (1998) pointed out the importance of the pressure distribution of the fracture tip, and regarded the fluid velocity as the key of fracture front advance. It is difficult to derive an accurate value of the fluid pressure distribution of the fracture fronts due to the rough assumption of fracture opening behavior and static model of the fluid flow.

We can say that at least a more accurate model of stress and displacement distribution than the classical square-root model around the fracture front is required, and the code should treat the fluid flow around the fracture tip more carefully.

4.2 Leak-off characteristics of multiple fracture

The leak-off model in the simulator does not consider the accumulation of pore pressure from neighboring wells.

A system of two closed parallel fractures is expected to behave like a single fracture, since the pore pressure between two fractures is saturated soon. However, the result from the simulator gives the leak-off characteristic of two independent fractures, and might give almost double the leak-off volume compared to the model in which pore pressure accumulation is considered, and smaller size for the same injected volume.

To solve the problem, the authors plan to conduct a numerical study of pore pressure change around the advancing multiple fractures in order to establish an accurate relationship between fracture disposition and leak-off rate changes.

5 CONCLUSION

The multiple fracture propagation model presented here can give the final geometry of a multiple fracture system which satisfies the interaction between stress and displacement discontinuity.

This code can be used to predict the fracture formation from a single extended reach well, dual fractures from different wells of HDR, and other situations.

In order to improve the accuracy we have to establish a more reasonable fracture propagation model and pore pressure interaction models.

REFERENCE

Clifton, R. J. & Abou-Sayed, A. S. 1979. On the computation of the three-dimensional geometry of hydraulic fractures, SPE7943: *Proc. 1979 Low Permeability Gas Reservoirs*

Clifton, R. J. & Abou-Sayed, A. S. 1981. A variational approach to the prediction of the three-dimensional geometry of hydraulic fractures. SPE9879: *Proc. 1981 Low Permeability Gas Reservoirs*

Crouch, S. L. 1976, Solution of plane elasticity problems by the displacement discontinuity method, I. Infinite body solution, *Int. J. Numerical Method in Engineering* 10 : 301-343

Mastrojannis, E. N., Keer, L. E. & Mura, T. 1980. Growth of Planar Cracks Induced by Hydraulic Fracturing, *Int. J. Numerical Method in Engineering* : 15 : 41-54

Desroches, J., Lnoach B. & Papanastasiou, P. 1998. On the modeling of near tip process in hydraulic fractures, *Int. J. Mining Science & Geomechanics Abstractl* 30: 7

Simonson, E. R., Abou-Sayed, A. S. & Clifton, R. J. 1978. Containment of Massive Hydraulic Fractures, SPE1089, *Society of Petroleum Engineers Journal*, February 1978

Vandamme, L. & Curran, J.H. 1992. A Three-dimensional hydraulic fracturing simulator, *Int. J. Numerical Method in Engineering* 28 : 99 : 909-927

Willes, T. D. & Curran, J. H. 1982. A general 3-D displacement discontinuity method, *Proc. 4th International Conference for Numerical Method in Geomechanics*

Yew, C. H. 1992. Lecture Notes on Hydraulic Fracturing, presented at Japan National Oil Corporation

Yew. C. H. 1996. Mechanics of Hydraulic Fracturing, presented at Japan National Oil Corporation

Yamamoto, K., Itoh, J., Shimamoto, T., & Ichikawa, M. 1999. Development of a 3-Dl hydraulic fracturing simulator for the fracture propagation analysis in coplex in-situ stress field around oil wells. *Proc. 9 Int. Congress on Rock Mechanics*

Yamamoto, K., Shimamoto, T. & M., Maezumi. 1999. Development of a true 3D hydraulic fracturing simulator. SPE 54265, *Proc. Asia Pacific Oil and Gas Exhibition and Conference*

Yamamoto, K., Gutierrez, M., Shimamoto, T. & Maezumi, M. 2000. Verification of a 3D hydraulic fracturing model against a field case. SPE 59373, *Proc. SPE/DOE Improved Oil Recovery Conference*

13 Pavement geomechanics

Computer Methods and Advances in Geomechanics, Desai et al. (eds) © 2001 Balkema, Rotterdam, ISBN 90 5809 183 X

Interface strength of thin bonded overlay on rigid pavements

M. Zaman & Z. Hossain
School of Civil Engineering and Environmental Science, University of Oklahoma, Norman, Okla., USA

ABSTRACT: The interface strength of thin bonded overlay on rigid pavements obtained from U.S. Highway 69 near Eufaula in McIntosh County, Oklahoma was investigated for different surface conditions. Adverse effects of environmental factors, namely freeze-thaw and wet-dry cycles, on interface strength were evaluated. The increased flexural capacity, failure pattern, and crack geometry were observed by testing beam specimens with different overlay thickness. A series of bond test and Iowa shear test results revealed that the dry and without grout surface contributes to excellent bonding of the interface. The wet-dry cycles were found to be more detrimental than freeze-thaw cycles. From flexure tests it was found that the overlaid pavement significantly increased the stiffness than the original pavement. The presence of a cloud of cracks near the interface in petrographic images of thin sections obtained from specimens failed due to bending indicates the loss of bonding at the interface. More indepth studies are needed to obtain a better understanding of structural and functional performance of an overlaid pavement.

Key Words: Thin bonded overlay, Interface strength, Freeze-thaw, Wet-dry, Flexural strength

1 INTRODUCTION

Proper maintenance can significantly extend the functional and structural capacity of a pavement. However, even a well-maintained pavement can become structurally deficient (e.g., due to increased traffic) after a certain period of service, and rehabilitation of the pavement becomes necessary. The traditional approaches to rehabilitation are either to reconstruct the pavement with new materials or to patch and overlay it with a new wearing surface. Placing an overlay on existing pavement can substantially increase the Serviceability Index (SI) and service life (Easa 1990).

It is generally believed that the effectiveness of bonded overlay is highly dependent on the integrity of the bond between the old pavement and the overlay. If the parent pavement and new concrete is not properly bonded, the pavement is likely to disintegrate and deteriorate within a short span of time. The bond strength can depend on many factors including the condition of the old pavement surface, the overlay thickness and the curing time. Environmental factors such as post construction temperature and humidity may have some influence on the behavior of the interface. The thickness of overlay contributes to the flexural behavior of the rehabilitated pavement. An experimental study was conducted involving bonded overlay on Portland Cement Concrete (PCC) pavement. The main objectives of this study are to characterize the behavior of the interface between an underlying old pavement and the bonded overlay, and to determine the flexural behavior of the bonded overlay. More specific tasks of this study include the following:

- Investigate the condition of the existing pavement and its mechanical properties.
- Conduct appropriate shear tests to determine interface strength and effect of different surface conditions of old pavement on the interface.
- Evaluate the effect of freezing-thawing and wetting-drying cycles on the interface integrity.
- Determine the degree of flexural capacity enhanced due to different overlay thickness.

2 PREVIOUS STUDIES

Figure 1 represents the conceptual structural deficiency and serviceability index (SI). The serviceability index of a pavement decreases with traffic and service time. Placing an effective extra layer on the top of the old pavement can substantially increase its serviceabilty index. It has been reported by ACPA (1990) that 6 to 7 inches (15 to 17.5 cm) of bituminous overlay is required to produce the same edge load stress induced under a 3-inch (7.5 cm) bonded overlay. Construction of PCC overlays can be completed within a short lane closure time, which is very competitive with that of asphaltic concrete overlays (Freeman 1996; Yoo 1997). Another advantage of using PCC pavement is that it reduces the clearance problem below a bridge because of low overlay thickness requirement.

Pavement engineers currently practice three types of PCC overlays on roadway and airport parkway (McGhee 1994). These are bonded, partially bonded and unbonded. A thin layer of concrete of 2-4 inches (5-10 cm) is bonded and directly placed on the pavement surface, when the existing concrete pavement is in good condition. There is a tendency for cracks in the old underlying pavements to appear in the bonded or partially bonded overlay as reflected cracks. For an existing concrete pavement that is in fair-to-poor condition, an unbonded overlay can be used. Before application of the overlay, a separation medium is placed on the old pavement to isolate the existing deteriorations. Asphalt concrete or cold asphalt milling can be used as a separation layer. Generally, if the existing pavement exhibits frequent cracks, a thick (greater than 12.5 mm) separation layer is used. If fewer cracks appear, a thin (less than 12.5 mm) separation layer is applied (ACPA 1990).

Adequate interface shear strength is a major concern for the performance of an overlay (ACPA 1990; McGhee 1994; Yoo 1997). Unclean and improper surface may contribute to debonding rather than enhanced bonding (McGhee 1994). The surface of an existing pavement can be prepared in different ways. It may be kept dry or wet before placing an overlay. Also, grout can be placed at the top surface of the parent pavement to enhance bonding. Due to edge and interior loading conditions, relatively large deflections across the undowelled transverse and longitudinal joints can contribute to flexural stresses at the interface in excess of its capacity (Freeman 1996). Due to excessive bending stress, cracks were found to develop at the bottom of the slab. Cracks propagate upward and may divert their direction along the interface because of its weak surface. The 14-day bond strength of 200 psi (1379 kN/m^2) is considered an acceptable limit by different agencies (Freeman 1996; ACPA 1990; McGhee 1994).

The placement of an overlay can be followed, in regions of rapid weather change, by freezing-thawing or wetting-drying cycles, and the interface characteristics can be greatly influenced by these environmental factors. For example, freeze-thaw damage to Portland cement binder was observed in the Illinois Tollway East-West Extension on I-88 in a study conducted by Rmeli et al. (1992). It was revealed that freeze-thaw damage occurs in concrete that is sufficiently saturated and does not contain an adequate amount of air void. A study by Kazmierowski et al. (1993) concluded that thin overlay, on the other hand, provides a viable and cost-effective surficial rehabilitation alternative for structurally sound pavements in a freeze-thaw environment. Thin-surfacing has exhibited limited success on PCC pavements in Oklahoma (Hixon and Ooten 1993). Zia et. al. (1998) included that one of the important distresses of overlaid pavement is freeze-thaw scaling, and the occurrence of freezing and thawing (frost action) is a major problem for durability. It was also reported that long freeze-thaw cycles are more severe than short freeze-thaw cycles for same number of cycles. Babaei and Hawkins (1990) surveyed the serviceability of overlaid bridge decks in the state of Washington. Freeze-thaw scaling was indicated as one of the most deleterious distresses in their evaluation. On the other hand, tests conducted by Ozyildirim (1993) indicated that the PCC overlay has satisfactory freeze-thaw resistance. Overall, the effect of freeze-thaw and wet-dry cycles on the interface characteristics in bonded overlay remains an important area of research.

3 PROJECT DESCRIPTION AND MATERIAL PROPERTIES

The project site is located on U.S. 69 near Eufaula in McIntosh County, Oklahoma. On December 10, 1996, the ODOT maintenance crews sand blasted the pavement at the site and prepared the surface for sawing operations. As shown in Figure 2, sawing

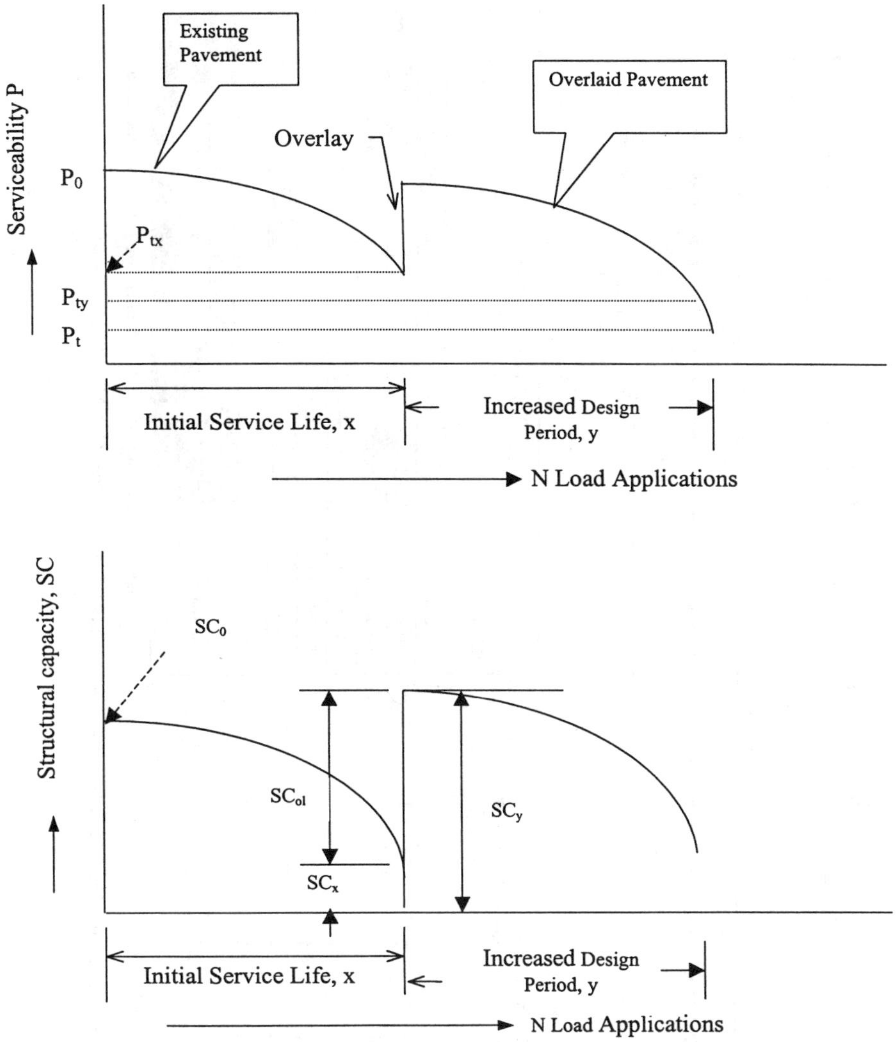

Figure 1 Structural Capacity Loss over Time and with Traffic (Source: Easa 1990)

Table 1. Mechanical Properties of Old Pavement and Concrete

Parameters	Test Method	Old Pavement	New Concrete	Remarks
Compressive strength, f_c'	ASTM C39	6690 psi	4485 psi (7 day) 6018 psi (28 day)	Failure type: both aggregate and mortar. Fracture type: shear, cone and shear.
Split tensile strength, f_t	ASTM C496	645 psi	484 psi (7 days) 624 psi (28 days)	Failure type: both aggregate and mortar.
Static Elastic modulus, E_s	ASTM C78	2.9×10^6 psi	3.63×10^6 psi (28 days)	
Rupture modulus, M_r	ASTM C469	210-440 psi	794 psi (28 days)	Loading type: third point
Poisson's ratio, η	ASTM C469	0.12	0.14	

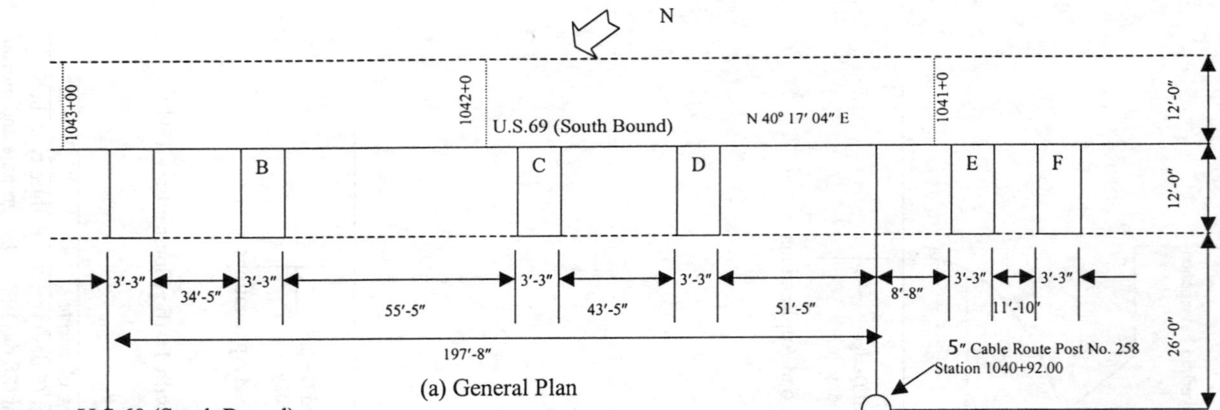

(a) General Plan

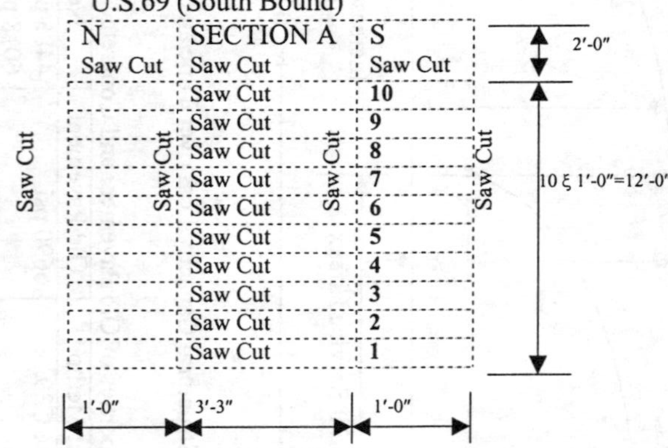

(b) Detail of Sawing Layout

SPECIMEN CHART						
SECTION	A	B	C	D		F
SPECIMEN NO	10	10	10	10	10	10
SPECIMEN NO	9	9	9	9	9	9
SPECIMEN NO	8	8	8	8	8	8
SPECIMEN NO	7	7	7	7	7	7
SPECIMEN NO	6	6	6	6	6	6
SPECIMEN NO	5	5	5	5	5	5
SPECIMEN NO	4	4	4	4	4	4
SPECIMEN NO	3	3	3	3	3	3
SPECIMEN NO	2	2	2	2	2	2
SPECIMEN NO	1	1	1	1	1	1

Note: Shaded sections denote that these specimens were damaged and discarded

Figure 2 General Plan and Details of Sawing Layout Specimens (Source: ODOT 1997)

operations were performed from a grid marked off into 10 one-foot (0.3 m) sections and a single two-foot (0.6 m) section parallel to the centerline of the southbound lanes. The grid was 5.25 ft (1.6 m) wide, perpendicular to the southbound lanes, and saw cuts were made 1 ft (0.3 m) from each end, thus leaving 3.25 ft × 1 ft (1 m × 0.3 m) sections. The thickness of the roadway section was found to be approximately 9 inches (22.5 cm). The sufficiency rating of this site was found to be 71, which places it within the tolerable range (ODOT 1995). Since the condition of the existing pavement is fairly good and the construction of bonded overlay is economical, bonded overlay was selected in this study.

Material Properties

Four-inch (10 cm) diameter cylinders were cored from the old pavement. Using a concrete cutting saw fitted with a diamond blade, the cylinders were sawed to obtain samples of desired height/length (8 inches or 20 cm) for testing. The mechanical properties of the old pavement obtained from laboratory testing are presented in a summarized form in Table 1.

The size distribution of coarse aggregates and fine aggregates is presented in Table 2. The maximum size of aggregates used was 0.75 inches (19 mm). The specific gravity of course aggregates and fine this aggregate was found to be 2.54 and 2.62, respectively. Locally available Portland cement (Type I) cement with a specific gravity of 3.10 was used throughout the testing program. From a systematic exploratory tests, the following mix proportions (by weight) was selected 1.0:1.5:3.0 (cement: sand: aggregate). The optimum water cement ratio (by weight) was found to be 0.42. The mechanical properties of new concrete are presented in Table 1.

Table 2. Sieve Analysis of Coarse and Fine Aggregates Used as Overlay Material

Coarse Aggregates		Fine Aggregates	
Opening (mm)	% Finer	Opening (mm)	% Finer
19.0500	100.0	2.000	99.87
12.7000	62.7	0.850	99.04
9.5250	45.5	0.420	93.25
4.7498	32.8	0.250	61.21
0.8407	21.1	0.149	23.15
0.4191	11.2	0.074	10.04
0.1778	1.8	Pan	0.00
0.1499	1.2		
0.1041	0.5		
0.0737	0.0		

4 INTERFACE PROPERTIES

Surface Preparation and Interface Conditions

A number of previous experimental studies have shown that adequate bond strengths can be achieved without the use of a bonding agent (Freeman 1996; McGhee 1994). Other researchers disagree with this view and contemplate the use of bonding agent (i.e., grout) for enhanced integrity. Some researchers suggest that overlays can be placed successfully on dry or damp surfaces without any bonding agent, although placement of overlays on a damp surface is not recommended by most transportation agencies (McGhee 1994). A thin layer of cement grout can be placed as a bonding agent between the old pavement and the new resurfacing.

To investigate the effect of surface preparation on interface shear strength, four interface conditions were considered: (a) grout placed on dry surface, (b) grout applied on wet surface, (c) without grout and dry, and (d) without grout and wet. A thin layer (0.125-inch [3-mm]) of cement grout (cement: ratio = 1:1) with a water/cement ratio of 0.62 was used to represent the grouted condition.

Sample Preparation

4-inch (10-cm) diameter and full depth (approximately 9 inch [22.5 cm]) cylinders were cored from the old pavement. The height of each cylinder was reduced to 4 inches (10 cm) by trimming the bottom part. For dry and grouted conditions, the grouted surface was dried at room temperature for two hours. Two-hour time span selected for the drying phase because heavily dried grouting leads to debonding instead of enhanced bonding (McGhee 1994). For wet and grouted surface condition, the overlay was placed immediately following the grouting.

Fresh concrete was prepared and placed on top of the old pavement cylinders. The measured slump of the concrete used as an overlay was found to be approximately 0.75 inch (19 mm). The concrete was placed in the mold in two layers of approximately equal thickness. Each layer was compacted with the rounded end of a 3/8 inch (9.5 mm) diameter rod using 12 blows per layer. The compaction was distributed uniformly over the cross section of the mold, and for each upper layer the rod was allowed to penetrate about ½-inch (12.5 mm) into the underlying layer. The mixing and finishing procedures followed the ASTM Designation C 192 recommendations. The cylinders were cast and cured at room temperature (70°F or 21°C), having a relative humidity of approximately 70%.

Effect of Surface Preparation

Interface shear strength varies with surface preparation, as well as with moisture conditions. Figure 3 shows the variation of bond tensile strengths for grout and no grout conditions and measured in accordance with the ASTM C 1245 designation. The interface tensile strength is lower for grouted case (wet and dry) than that of ungrouted case. The dry, without grout surface contributes to excellent bond strength of the interface, and the developed tensile strength is as high as 518 psi (3590 kPa). Reduced interface tensile strength was achieved after 14 days of curing in the case of wet and no grout interface, and it is 80% of the dry but no grouted surface strength. After only two days of curing, the developed shear strength in the case of dry and no grout interface is 210 psi (1450 kPa), this is approximately 30% higher than that in the wet surface condition.

Although the rate of increase in the interface strength in case of grouted interface is lower than that of non-grouted interface, the 5-day strength is approximately 50% and 44% higher than the 2-day strength in dry and wet conditions, respectively. There is a significant reduction in bond strength in the case of dry and grouted interface; only 22% of that in case of dry and ungrouted interface condition. Interestingly, there was no measurable bond strength in the interface in the case of wet and grouted surface after 14 days. Wet and grouted specimens broke during handling; the specimens failed near the interface of old pavement and the grout. This implies that the bond between the old pavement and the grout in the wet condition is not strong enough to hold the grout tightly with the old pavement. In addition, no specimen with a grouted surface achieved the minimum required bond strength. One of the reasons could be the presence of excess pore water near the interface during the hydration of cement. In case of dry specimens without grout, matric suction might have developed as soon as the new concrete was placed over the dry surface, which was unlikely to occur in case of wet specimens. Another possible reason for the reduced bond strength in case of a grouted interface is the presence of particles with lower roughness that leads to lower aggregate interlock at the interface. As noted by Mier (1997), the bond tensile strength between the aggregates and the matrix depends on the mechanical interlock as well as on the physical attraction and/or direct chemical bonding between the cement and the aggregate. The porosity at the interface is another important factor. It is generally accepted that relatively large weak open areas may develop in the grouted interface because of the flow of bleed water in early stages of hardening.

Freezing-Thawing and Wetting-Drying Effect

Freezing-Thawing Effect

The number of freeze-thaw cycles considered in this study is 5, 15, 22, and 29 after a curing period of 2 days. One freezing-thawing cycle consists of alternatively lowering the temperature of the specimen from 40 to 0°F (4.4 to –17.8°C) and raising it from 0 to 40°F(-17.8 to 4.4°C) in 5 hours, according to the ASTM designation C 666 (ASTM 1995). Following the freezing-thawing cycles, Iowa shear tests were conducted on the specimens to determine interface shear strength.

From Figure 4, it is evident that the rate of increase in interface strength in samples subjected to freeze-thaw cycles is slower than that of specimens without subject to freeze-thaw cycles. Samples without any freeze-thaw effect achieved shear strength of 215 psi (1480 kPa), whereas samples including the freeze-thaw effect gained approximately 13% lower shear strength. As number of freeze-thaw cycles increases, the difference in shear strength of the two cases (with and without freeze-thaw) increase, but at a lower rate. At the end of 29 cycles, the interface shear strength is 215 psi (1480 kPa), which is 19% lower than the specimens without any freeze-thaw effect. Note that 29 cycles of freeze thaw (after 2 days curing) take about 14 days. Overall, it is noted that the interface strength decreases when it is subjected to freeze-thaw cycles. However, the 14-day interface strength is greater than 200 psi (1380 kPa), which is the minimum requirement of many transportation agencies (McGhee 1994).

Figure 4 also illustrates the effect of freeze-thaw for the case where this effect is initiated after 4 days of curing. The resulting interface strength of samples subjected to seven freeze-thaw cycles is 210 psi (1450 kPa) that is 9% lower than the case without any freeze-thaw cycles. The reasons behind the reduction in shear strength due to the freezing-thawing effect are manifold. One of the possible reasons is that the hydration reactions of cement compounds are slow at low temperature (ACI 1966). When concrete freezes, ice crystals form, separating the solid constituents, so that when thawing occurs, the mass becomes unconsolidated and spongy, resulting in a hardened concrete that is porous and weak. An adequate temperature level is not maintained during the freezing phase to provide the needed activation energy for reaction. In addition to the hydraulic pressure caused by water freezing in large cavities, the osmotic pressure resulting from the partial

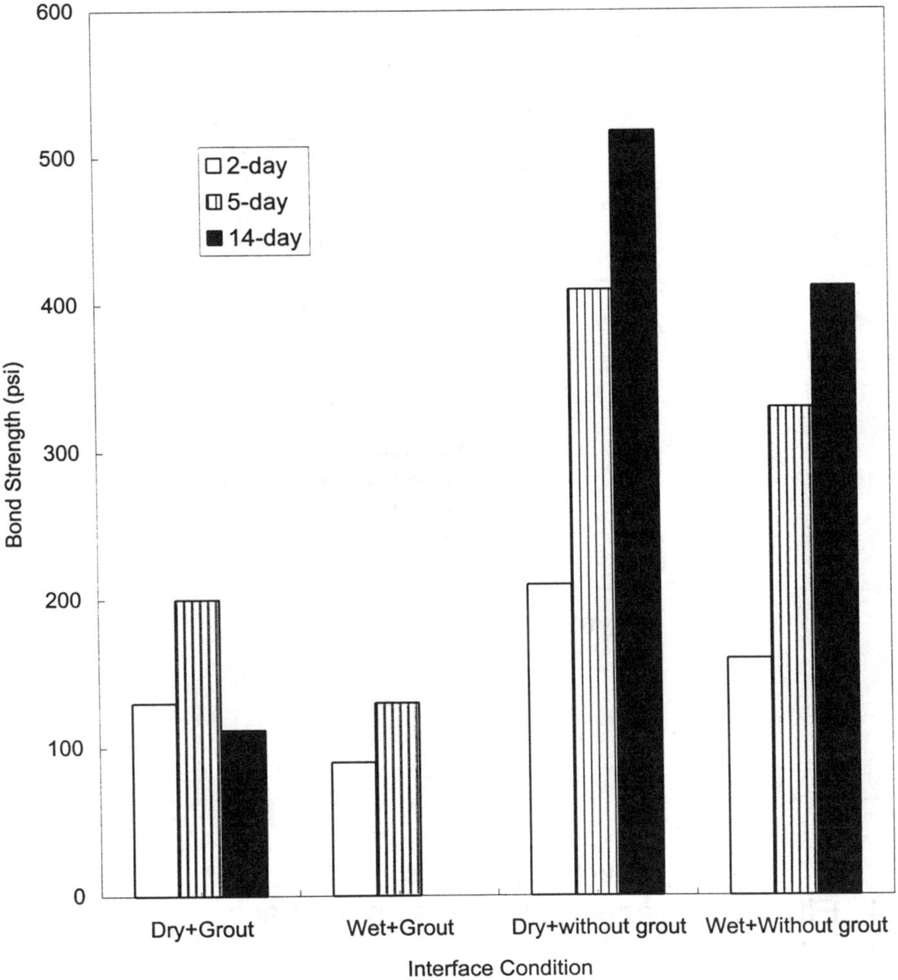

Figure 3 Bond Tensile Strength (ASTM 1245-93) for Different Interface Conditions

Note: 7 days fc' = 5250 psi; 1 psi=6.895 kPa

freezing of solutions in capillaries can be another source of destructive expansions in new concrete. Another possible reason is the progressive expansion of the cement paste matrix due to repeated freeze-thaw cycles.

Wetting-Drying Effect

One drying and wetting cycle used in this study consisted of placing a specimen in an oven at a temperature of 160°F (71°C) for 12 hours and then immersing it in a tap water tank for 12 hours. Since there is no standard approach for wet-dry cycles, the temperature and time span was considered on the basis of practical experience. To evaluate the durability of

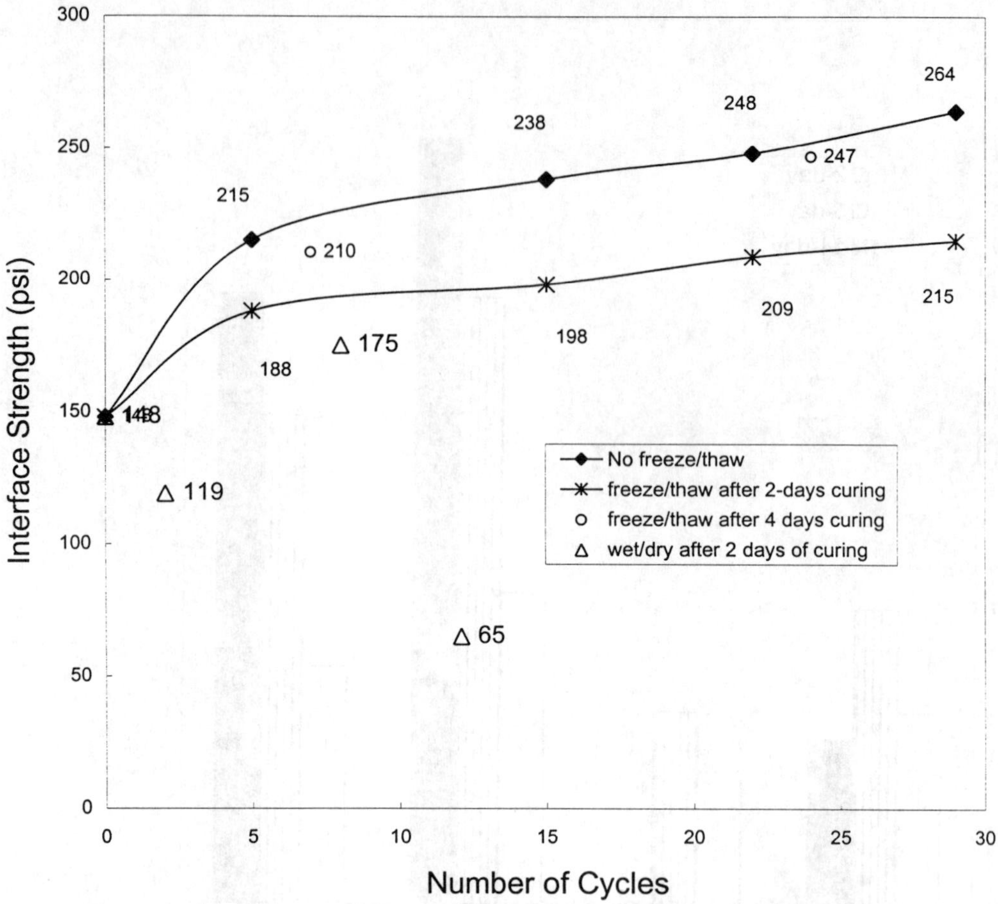

Figure 4 Effect of Freeze-thaw cycles on Interface (Iowa Shear) Strength

Note: 7 days f_c' = 4586 psi, and f_t = 492 psi 1 psi=6.895 kN/m^2
1 Freeze/Thaw cycle = 0 hours 1 Wet/Dry cycle = 24 hours

chemically stabilized aggregates Zhu (1997) used a temperature of 160°F (71°C) for 24 hours in the drying chamber. Following Zhu (1997), three different numbers of cycles, i.e. 2, 8 and 12 cycles, were used in the current testing program to determine the effect of wet-dry cycles on the interface shear strength.

From Figure 4 it is evident that wetting-drying cycles significantly reduce the interface strength. For example, at the end of 2 wetting-drying cycles (4 days after construction; 2 days after curing), the specimens without wetting-drying cycles yield an interface strength of 215 psi (1480 kPa). Comparatively, the samples subjected to two wetting-drying cycles the interface strength is only 119 psi (820 kPa), which is 45% lower than that of the regular sample. Moreover, at the end of 12 cycles (14 days after construction; 12 days after curing), the interface strength reduces to only 65 psi (450 kPa). The reasons behind this deleterious strength reduction could be manifold. For example, drying drives the moisture out of specimens but wetting may not be able to draw enough moisture in them. More moisture is driven out in the next drying cycle. Therefore, the "specimen" becomes very "thirsty," with increasing wet-dry cycles. Another possible reason could be the development of micro-cracks at the transition zone due to high temperature (ACI 1966). Due to repeated drying and wetting, differential movements may occur in the

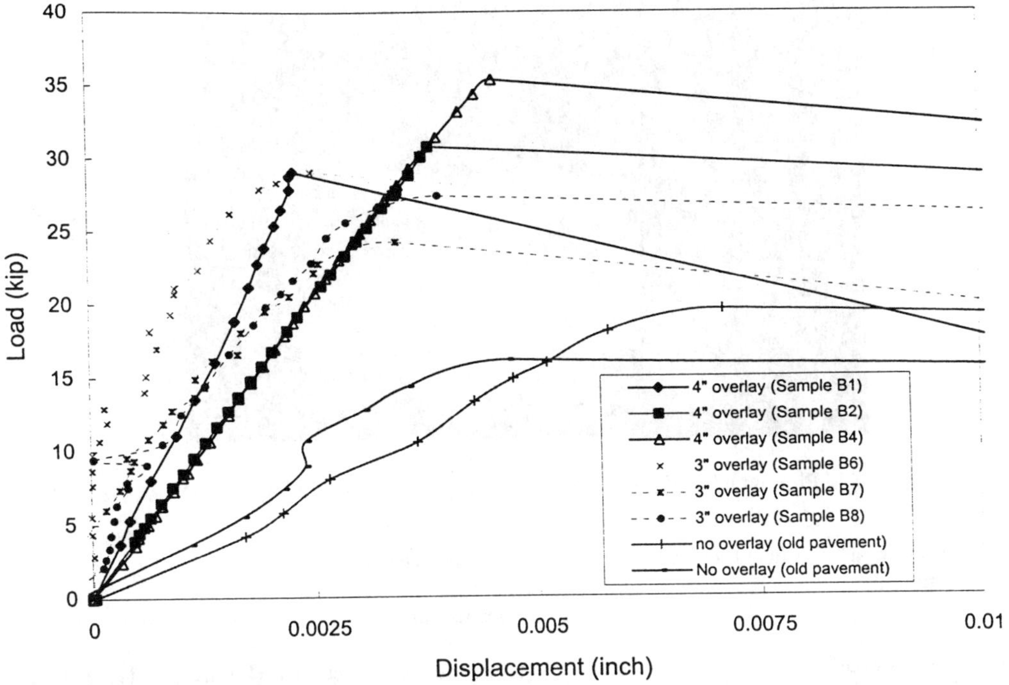

Figure 5 Ultimate Load and Midpoint Displacement of Pavements with Different Overlay Thickness

Note: 1 inch=2.54 cm; 1 kip=1kN

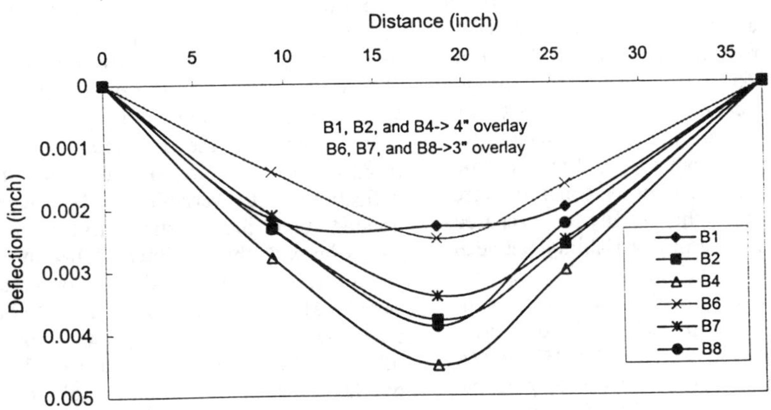

Figure 6 Elastic Curves at Ultimate Load for all Overlaid Beam

Figure 7. Photographic View of Sample B1 (Side A) After Bending Failure

interface resulting in a significant decrease in the interface shear strength.

5 BENDING BEHAVIOR

Sample Preparation and Instrumentation

A total of six beam specimens were prepared maintaining dry and "no grout" interface condition. Three of these specimens had 3-inch (7.5 cm) thick overlays while the other three had a 4-inch (10 cm) thick overlay. The samples were stored and cured at a temperature of approximately 70°F (21°C) with a relative humidity of 65 to 70%. The testing program also included two beam specimens without any overlay for comparison purposes. Mechanical properties, especially compressive and split tensile strength for each batch of overlay concrete were determined to ascertain the quality of prepared specimens and to identify any variability between different mixes.

The instrumentation consists of a series of LVDTs mounted on the top surface. All data were collected, controlled, and stored by an HP 3497A data acquisition unit coupled with a personal computer. A computer code "COMPRS" was used for this purpose. The simply supported beams were tested under a third-point loading in accordance with ASTM Test Method C 78 (ASTM 1995). Deflections along the centerline of each point were obtained by averaging the deflections of LVDTs placed on opposite sides.

Bending Behavior

Flexural Strength

From deflection curves (Figure 5) for various loading conditions it is evident that the ultimate flexural load capacity increases significantly for beams with an overlay. The "straight line plots" beyond this figure is the result of omitting data points to obtain a better resolution at the early stage of loading. The mean third-point modulus of rupture values for beams without and with an overlay are approximately 520 psi (3590 kN/m^2) and 575 psi (3970 kN/m^2), respectively. The ultimate load capacity of beams with 3-inch (7.5 cm) and 4-inch (10 cm) overlay is approximately 50% and 78% higher than the corresponding old pavements, respectively. Figure 6 represents the mid-point deflections with variable overlay thickness. It is noticed that the elastic curve is approximately parabolic with the maximum deflection at the center. However, some differences are observed in case of sample B1 possibly due to nonlinearity of the interface and defects in sample preparation (overlay), among others. The deflection in case of 3-inch (7.5 cm) overlay is 44% less than the old pavement. The deflections of beams with a 4-inch (10 cm) overlay pavement are not always lower than that of a 3-inch (7.5 cm) overlay pavements. This could be attributed to the non-uniform distribution and amount of reinforcing steel in the old pavement. Also, the variation in roughness among specimens could be another factor.

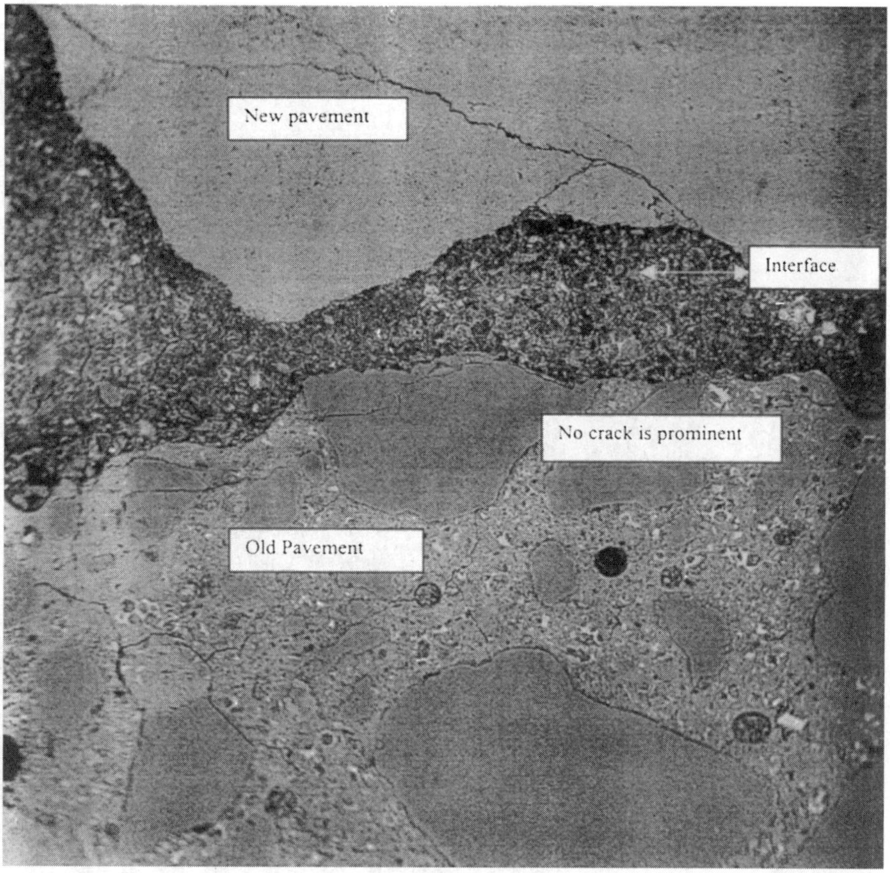

Figure 8 Photographic Image near the Interface of Thin Section Obtained from Specimen before Bending Failure (1500μm×1500μm).

Failure Pattern and Crack Formation

Due to application of load, micro-cracks are likely to form at the interface between old pavement and overlay, particularly in the middle third of the tested beams. From Figure 7, it is evident that the angle of inclination of the cracks with the longitudinal axis is approximately 90° at the bottom surface. These cracks propagated upward until they reached the interface. Afterwards, it propagated horizontally along the interface. As the deflection is increased further, cracks propagated upward, and the specimen failed completely. In some cases, a number of cracks were observed by the naked eye. Although, all specimens appeared to have failed due to bending, it can be argued that some debonding may have occurred at the interface. The deviation of cracks from vertical to horizontal at the interface supports this viewpoint. This phenomenon is more prominent in case of pavements with 3-inch (7.5 cm) overlays. At the time of unloading the overlay separated from the parent beam. Also, it was very difficult to retrieve an intact core from overlaid pavements after bending test.

Prebending and Postbending Interface

Cylinders were cored from overlaid pavement after 28 days of curing, without applying any bending load. A second set of cylinders were cored from tested specimens. 1-inch×2-inch (2.5 cm×5 cm) petrographic thin sections (30-60 μm) were prepared from both cylinders using the epoxy impregnation technique. The thin sections were examined by back scattered electron imaging using a Camera SX50 electron microprobe. The

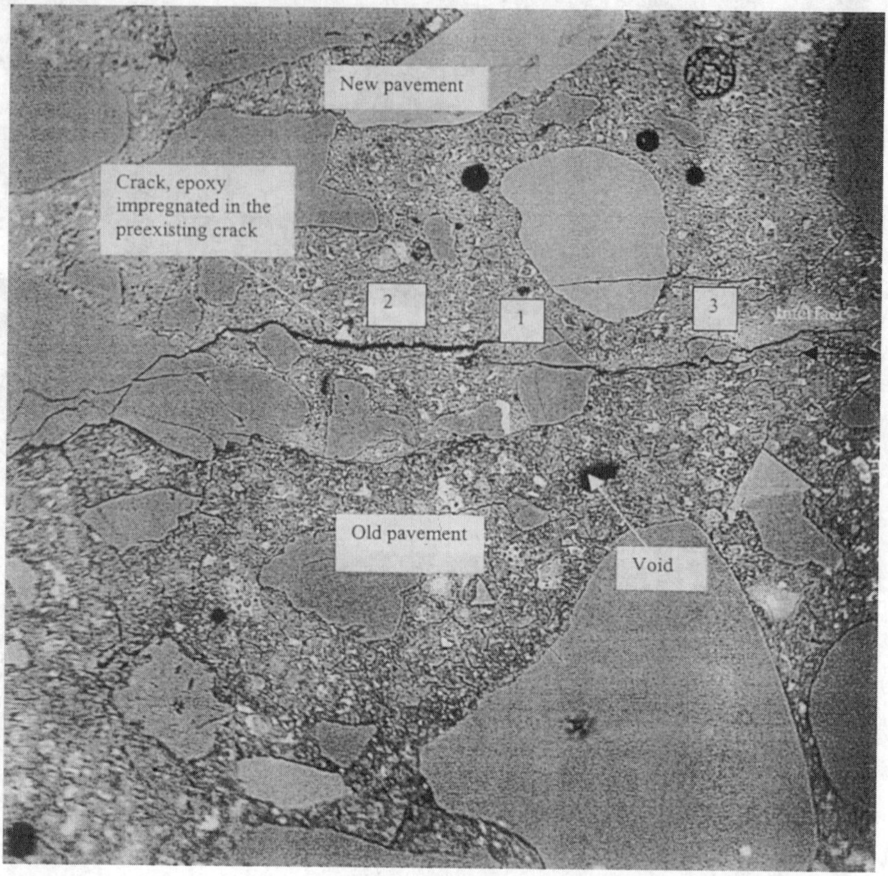

Figure 9 Photographic Image near the Interface of Thin Section Obtained from Specimen after Bending Failure (1500μm×1500μm).

analytical conditions used were 20 kV acceleration and 5 nA beam current. Images were acquired digitally into 1024×1024 pixel arrays and stored in standard TIFF format. Phases were identified by energy-dispersive x-ray analysis using a Kevex Delta class spectrometer.

Figures 8 and 9 represent sample petrographic images of the interface of prebending and postbending specimens, respectively. From Figure 8, it is evident that the zone near the interface is relatively smooth, and no crack due to the influence of bending failure is prominent. Some micro-cracks are evident due to shrinkage after epoxy impregnation. These are shrinkage cracks because they are not filled by fluorescent epoxy. On the other hand, some cracks are observed near the interface in case for specimen subjected to bending (Figure 9). Dark spaces filled with fluorescent epoxy were observed and labeled on the images. These cracks are almost parallel to the interface; no vertical cracks are observed. The probable reason of the formation of these cracks is the propagation of flexural cracks along the interface. Mier (1997) noted similar phenomena in his study. Formation and propagation of cracks and the resulting loss of bond stress are very complex phenomena and need a more detailed study to validate the observations revealed in this study.

6 CONCLUSIONS AND RECOMMENDATIONS

Based on the experimental results and analysis of data, the followings can be concluded

The old pavement with dry and without grout surface contributes to excellent bonding of the interface, and result in a bond strength of as high as 520 psi (3580 kPa). Placing a thin layer of grout on the top of the old pavement before overlaying does not seem to enhance the bond capacity. Application of grout on a wet pavement surface before overlaying leads to a very poor bonding.

Both freeze-thaw and wet-dry cycles have deleterious effect on the quality of bond between the parent pavement and the overlay, but the wet-dry cycles appear to be more harmful than the freeze-thaw cycles.

The ultimate load capacity against flexural failure can be significantly enhanced by placing an overlay. A 3-inch (7.5 cm) overlay can increase the capacity by 50%. The corresponding increase for a 4-inch (10 cm) overlay is about 78%.

Some cracks were observed near the interface in images of thin sections, prepared from specimen failed due to bending, indicating a loss of bond strength due to the development of flexural stress. Additional studies are needed to validate the observations revealed in this study.

REFERENCES

ACI (1966). "Freezing and Thawing of Concrete-Mechanisms and Control." American Concrete Institute Monograph Series, *The Iowa State University Press*, Ames, Iowa.

ACPA (1990). "Guidelines for Bonded Concrete Overlays." Technical Bulletin, Concrete Pavement Technology, *American Concrete Pavement Association*, 3800 North Wilke Road, Arlington Heights, Illinois 6004.

ASTM (1995). "Annual Book of ASTM Standards." Volume 04.02, *American Standard for Testing Materials*, 1916 Race Street, Philadelphia, PA 19103-1187.

Babaei, G. and Hawkins, N.M. (1990). "Performance of Bridge Deck Concrete Overlays: Extending the Life of Bridges." *ASTM Special Technical Publication*, No. 1100, pp.95-108.

Easa, S.M. (1990). "Extension of AASHTO Renaming-Life Methodology of Overlay Design." *Transportation Research Board, No.* 1272, National Research Council, National Academy Press, Washington, D.C. pp.1-14.

Freeman, T.E. (1996). "Evaluation of a Thin-Bonded Portland Cement Concrete Overlay." *Virginia Transportation Research Council*, VTRC 97-R9.

Hixon, C. D. and Oeten, D.A. (1993). "Nine Years of Microsurfacing in Oklahoma." *Transportation Research Board, No.* 1392, National Research Council, National Academy Press, Washington, D.C. pp.13-19.

Kazmierowski, T.J., and Sturm, H. (1994). "Performance of Unbonded Concrete Overlay Project in Canada." *Transportation Research Record*, 1449, National Research Council, Washington, D.C. pp.169-173.

McGhee, K. (1994). "Portland Cement Concrete Resurfacing." Synthesis of Highway Practice 204, *National Cooperative Highway Research Program*, National Academy Press, Washington.

Mier, J. G. M. V. (1997). "Fracture Processes of Concrete," Assessment of Material Parameters for Fracture Models, CRC Press, Inc., New York.

ODOT (1995). "Highway Needs Study and Sufficiency Rating Report." Needs Study Section, Planning Division, Volume I, *Oklahoma Department of Transportation*, 200 N.E. 21st Street, Oklahoma City, Oklahoma 73105.

Ozyildirim, C. (1993). "A Field Investigation of a Concrete Overlay Containing Silica Flume on Route 50 over Opequon Creek." Final Report, *Virginia Transportation Research Council*, Charlottesville, VA. 22 pp. (*FHWA-VA-93/R15; PB94-113412*)

Rmeili, E.H., Johnson, K.D., and Darter, M. I. (1992). "Pavement Evaluation and Development of Maintenance and Rehabilitation Strategies for Illinois Tollway East-West Extension." *Transportation Research Board*, No. 1374, National Research Council, National Academy Press, Washington, D.C. pp.71-80.

Yoo, T.S. (1997). "Thin Bonded Overlay for Concrete Pavement Rehabilitation." http://www.freeway.co.kr/sub05/eng8303.html.

Zhu, J. (1997). "Characterization of Cement-Kiln-Dust Stabilized Base/Subbase Aggegate." A Dissertation submitted to the Graduate Faculty, University of Oklahoma, Norman, OK 73019.

Zia, P., Ahmed, S, and Leming, M. (1998). "High Performance Concretes." A State-of –Art Report, *U.S. Department of Transportation*, Federal Highway Administration, Research and Development, Turner-Fairbank Highway Research Center, McLean, Virginia 22101-2296.

Computer Methods and Advances in Geomechanics, Desai et al. (eds) © 2001 Balkema, Rotterdam, ISBN 90 5809 183 X

Modelling of ballast behaviour in railway track

D.Aubry, A.Baroni, D.Clouteau, A.Fodil & A.Modaressi
MSS-Mat Laboratory, CNRS UMR 8579, Ecole Central Paris, France

ABSTRACT: The aim of this paper is to present a methodology to study permanent displacements of ballast under repeated dynamic loading in high speed railway tracks. The specifications of the problem are that different scales of time and space are present. First, global and local effects due to the 3D geometry of the problem and due to the irreversibility of the ballast behavior, are uncoupled through the use of two softwares with complementary wave propagation approaches. In this manner, the global effects (on several meters of the track) are considered via a linear approach, while non linear effects are considered at a local scale (on a given section of the track). So, applying locally the loading (evaluated with the 3D model) due to the train movement on a railway section, the irreversible behavior of the ballast after almost seven months of traffic is simulated. The non linear behavior of the ballast is considered using an elastoplastic constitutive model in which a damage mechanism based on the plastic energy dissipation during the cycles is introduced.

1 INTRODUCTION

One of the most significant factors in railway track construction is the maintenance cost which is related to the degradation of mechanical characteristics of the rail track specially the ballast which is a key material whose strength plays an important role on the stability and settlements of the track.

The specifications of such a problem are its tridimensional geometry and the non linear and irreversible behavior of the ballast. Unfortunately, the 3D non linear dynamic finite element analysis of such a problem is not yet to be envisaged. So, we suggest to decompose the problem into two parts in which the linear and non linear effects are decoupled.

The overall behavior is considered not to be far from a linear one so that the global effects applying on several meters of the track can be considered to be linear elastic, while the non linearities have only a local effect and can be studied on a given section. We will now explain the weak coupling between a large scale linear model and a small scale local non linear model.

2 COUPLING THE LINEAR AND NON-LINEAR BEHAVIOR

Let Ω be a mechanical unbounded domain on which forces f apply. Under small strain and small displacement assumptions, the displacement field is governed by the Navier equation of dynamics :

$$\underline{Div}\,\underline{\underline{\sigma}}(\underline{u}) + \underline{f} = \rho\partial_{tt}^2\underline{u} \quad \text{in } \Omega \tag{1}$$

To which the initial conditions should be added. $\underline{\underline{\sigma}}(\underline{u})$ is the stress tensor which is supposed to present a small non linearity with respect to $\underline{u}$ such that it can be expressed by:

$$\underline{\underline{\sigma}}(\underline{u}) = \underline{\underline{\sigma}}_0(\underline{u}) + \underline{\underline{\sigma}}_\delta(\underline{u}) \tag{2}$$

where $\underline{\underline{\sigma}}_0(\underline{u})$ is the linear part and $\underline{\underline{\sigma}}_\delta(\underline{u})$ is small with respect to $\underline{\underline{\sigma}}_0(\underline{u})$.

If Ω_l presents a limited zone on which the non linear analysis is to be performed, we can consider that the behavior is linear outside this area and its interaction with Ω can be considered by appropriate boundary conditions. A difficulty is risen as the support of f is defined on a domain much larger than Ω_l.

Lets consider the displacement field u_o which is the solution of the large scale elastic problem :

$$\underline{Div}\,\underline{\underline{\sigma}}_0(\underline{u}_0) + \underline{f} = \rho\partial_{tt}^2\underline{u}_0 \quad \text{in } \Omega \tag{3}$$

The restriction of this equation to Ω_l is used to define the equivalent loading f_l in this domain :

$$\underline{f}_l = -\underline{Div}\,\underline{\underline{\sigma}}_0(\underline{u}_0) + \rho\partial_{tt}^2\underline{u}_0 \quad \text{in } \Omega_l \tag{4}$$

This equivalent loading can be used in the local model instead of f :

$$\underline{Div}\,\underline{\underline{\sigma}}(\underline{u}) + \underline{f}_l = \rho\partial_{tt}^2\underline{u} \quad \text{in } \Omega_l \tag{5}$$

In the particular case, where : $\Omega_l = \Omega$, $\sigma = \sigma_o$ and $f=0$ in Ω-Ω_l , the two formulations are identical. In other cases, f_l is a better approximation of the loading than the restriction of f to Ω_l as long as the non linear response is not far from the linear one. If it is not the case, f_l can not be a good approximation of the loadings as the large scale elastic model is no more applicable.

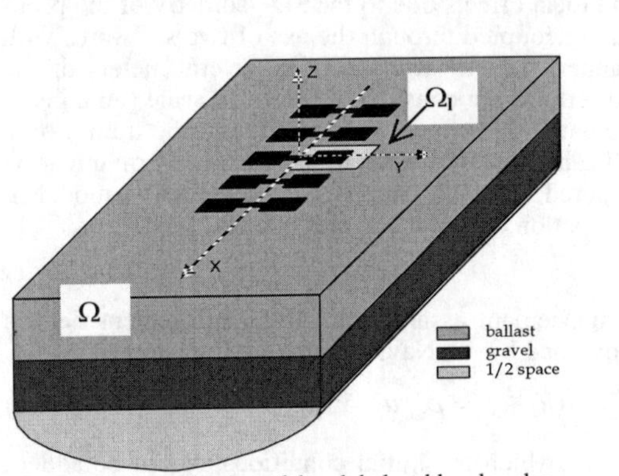

Figure 1: Schematization of the global and local scales

In the case of finite elements - boundary elements coupling the domain Ω_l is meshed, $u_o(t)$ is sampled by the boundary elements method (global scale) on the nodes of the Ω_l mesh (local scale) for a given load f(t). This displacement field is then injected into the finite element code using an elastic behavior to compute $f_l(t)$ which will be injected as unbalanced forces to the finite element code (local scale) with the non linear behavior.

3 SIMULATION MODELS

3.1 *Global Model*

The track is considered to be composed of two 30 cm high horizontal layers placed on a half space. The upper layer is made of ballast while the second one is gravel. The bi-block sleepers representing a mass of 250 Kg, are embedded in another 30 cm of ballast which is not considered geometrically in the model though its weight is taken into account to compute the initial stresses.

The physical parameters used are :

- For the ballast ρ_b =1.323 kg/m³, E_b = 50 MPa, ν=.3
- For the gravel ρ_g= 1.4 kg/m³, E_g = 60 MPa, ν=.3
- For the half space ρ_s =1.8 kg/m³, E_s = 80 MPa, ν=.3

A 5 reasonably rigid sleeper model has been used to model the track. The number 5 results from static computations which show that the stress distribution is negligible from the third sleeper. An analysis with 31 sleepers, did not reveal any significant modification in the dynamic analysis.

Though the Young modulus for the ballast at small strains has been estimated to be 200 MPa, using triaxial results done in the laboratory, a smaller modulus has been used to take into account the non linear variation of the tangent modulus with strains obtained under the traffic load.

3.2 *Influence of the train speed and soil nature*

Based on previous studies (G.Sauvage) a M-shaped signal is considered to represent the forces applied by the moving track to the sleeper. This signal depends on the load of the bogie Q, rigidity of the track (by means of the pulsation ω_0) and it takes into account the existence of the rail which distributes the force of one bogie on the neighboring sleeper. The rail is not modeled in our computations. The traffic is about 300 trains of 13 axles per day. The load signal is thus the superposition of two Gaussians given by:

$$S(t) = Q(e^{-\omega_0(t-t_1)^2} + e^{-\omega_0(t-t_2)^2}) \tag{6}$$

With

Q=55 080 N and ω_0=2πx25 rad/s for a QS1 type soil

Q=68850N and ω_0=2πx30.5 rad/s for a QS2 type soil

t_1=72 ms and t_2=108 ms.

The difference between t_1 and t_2 depends on the distance between the axles and the train speed. In fact they correspond to the arrival of the first and second axle on the central sleeper.

The 31 sleeper model gives the following values for the deflection of the central sleeper in the case where the train speed is 300km/h:

- 2.4 mm for a half space with E=30 MPa
- 1.5 mm for a half space with E=55 MPa
- 0.9 mm for E=80 Mpa

The value of the deflection reduces to 1.3 mm for a speed of 150 km/h on the most deformable half space (E=30 MPa).
The vertical accelerations under the sleeper reach 2g to 3g.

3.3 *Local model*

Once the dynamic unbalanced forces due to the passage of the train are estimated by the linear computation, the induced stresses can be applied to the non linear material to estimate the irreversible strains under hundreds of thousands of cycles. It is noted that due to the large number of cycles to which the model is subjected to, any infinitesimal deformation will be amplified. The stresses obtained at the bottom of the ballast layer are given in Figure 2. As it can be noted the vertical stress is almost multiplied by a factor of 10 and YZ shear stresses are larger than the initial vertical stresses.

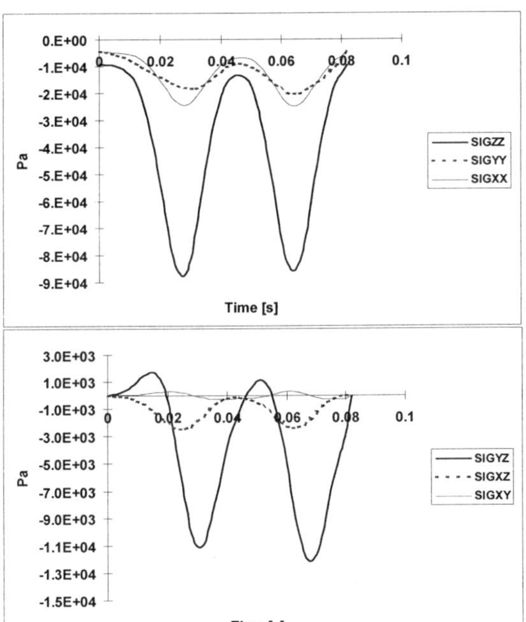

Figure 2: Stress evolution due to the crossing of one axle at the bottom of the ballast layer

4 LABORATORY TESTS

A series of triaxial tests under low confining pressures was performed on used and unused ballast. A large dimension triaxial cell was used to have a representative elementary volume (ϕ=300mm, H=700 mm). The evolution of the deviatoric stresses with axial strain under three confining pressures of 0.02, 0.04 and 0.06 MPa for both ballasts are given in Figure 3. After each test, the grain size distribution was compared to its initial distribution.

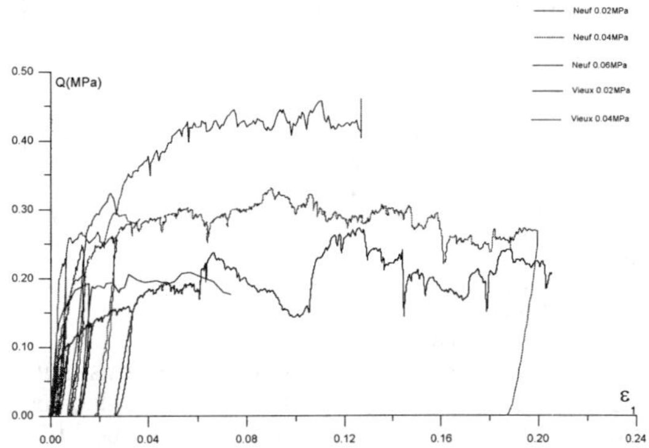

Figure 3: Drained triaxial test results for the used and unused ballast
(σ_3 = 0.02, 0.04 and 0.06 MPa)

5 MODELING THE NON LINEAR BEHAVIOR

The Hujeux's (1985) elastoplastic model has been used to model the behavior. Though this model has been specially developed for simulating cyclic behavior of soil it had never been subjected to such a large number of cycles. The first modification that we had to perform concerned the non linear elasticity. The most common way of introducing non linear elasticity for soil is to take into account the variation of the bulk and shear modulus with respect to mean pressure. Previous analyses, had shown that though the variation of the Young's modulus with only the mean pressure is not thermodynamically correct for an elastic material, the dissipation generated in the analyses in earthquake engineering field are very small. However, these very infinitesimal strains can result in very large irreversibilities even in the elastic domain when cumulated during the cycles.

The second modification concerned the introduction of a damage mechanism into the model. Experimental results and *in situ* observations have resulted in the fact that the characteristics of the ballast decrease with time. The comparison of the grain size distribution of new and old ballast eliminated the hypotheses of grain crushing. We estimated that this loss of properties could be due to the smoothening of the grain angularities. Another observation was that used ballast was much dirtier than the new one and this could also result in a degradation of contact properties of the grains.

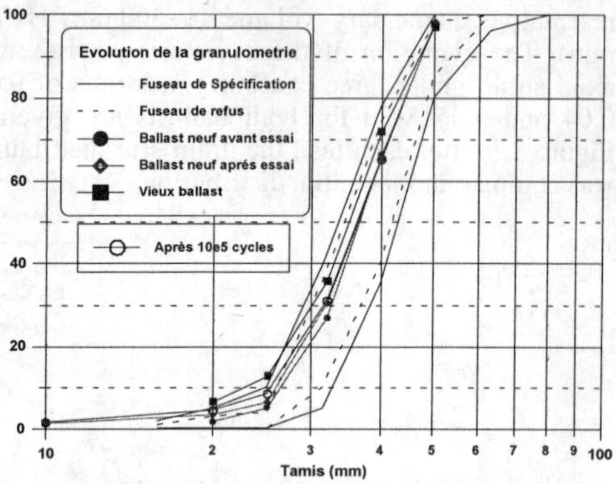

Figure 4: The grain size distribution of used and unused ballsat before and after the triaxial tests

So, it was decided to take into account this material degradation through the friction angle decrease governed by total energy dissipated. This modification resulted into a bilinear variation of volumetric plastic strains with respect to the number of cycles. The evolution of the plastic volumetric strain as a function of the number of cycles is given in Figure 4 for different friction angle evolution model.

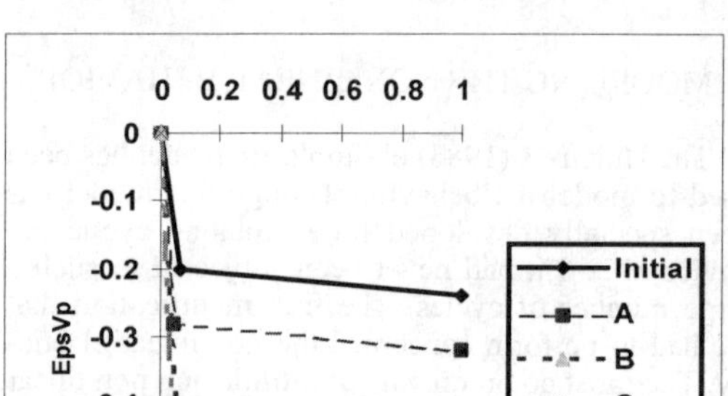

Figure 4: Variation of the volumetric irreversible strains with the number of cycles for different damage parameters

6 CONCLUSION

The objective of this work was to show that the irreversible behavior of ballast in railway track can be modeled numerically. The results are encouraging as to our opinion no similar study had ever been done. Very few elastoplastic models can reproduce one millions of cycles. However, still some modifications could improve the model (such as the consideration of the embankment, more sleepers, elasto-plastic behavior of gravel). A validation of the model by comparison to in situ measurements seems to be absolutely necessary for its future applications.

7 ACKNOWLEDGEMENT

This research has been financed by the French Railway Company (SNCF) under the contract number 0.300.6.412 14.

REFERENCES

Hujeux J.C. 1985. Une loi de comportement pour le chargement cyclique des sols, Génie Parasismique, Ed. Davidovici, Presses Ponts et Chaussées, Paris, 287-302.

Sauvage G. Railway track vertical static behaviour, INRETS-LTN.

Computer Methods and Advances in Geomechanics, Desai et al. (eds) © 2001 Balkema, Rotterdam, ISBN 90 5809 183 X

Asphalt concrete response evaluation: Analytical and experimental aspects

S.M.J.G.Erkens, A.Scarpas & A.A.A.Molenaar
Faculty of Civil Engineering and GeoSciences, Delft University of Technology, Netherlands

ABSTRACT: If test results are to be used for comparison and predictions or as input for the elaborate response models that are needed for reliable pavement performance predictions, they must be accurate, reliable and reproducible. For this reason, tests that can provide such results are not simple and must be designed to minimise the variation by keeping all the important influences constant. In this contribution, a material model for asphalt concrete is presented together with a description of some of the test set-ups and the procedures that had to be developed for determination of the required material characteristics.

1. INTRODUCTION

At Delft University of Technology, in the Netherlands, an extensive project of experimental and analytical investigation on the inelastic response of asphalt concrete has been initiated since 1997. The objectives of the Asphalt Concrete Response (ACRe) project are twofold: the formulation of a 3D, non-linear material model for asphalt concrete and the development of the set-ups and test procedures necessary to obtain the material characteristics that are needed as input for this model. Similar approaches proved very successful in related fields of engineering, such as soil, rock, steel and concrete mechanics. For asphalt concrete this is a rather novel approach, although pioneers like Nijboer (1948) Monismith and Secor (1962) looked into it earlier. Their attempts were never truly incorporated in road engineering design, mainly because of a lack of computational power in those times. With the currently available computers, material and structural modelling of 3D structures is possible and the required material models could well become as important a tool in road engineering as they are in the related fields of rock, soil and concrete mechanics.

Most of the design procedures that are used in road engineering are based on linear elastic multi-layer analysis and simple laboratory tests. Unfortunately, design life and material degradation, which are essential issues in estimating the lifetime of a pavement, can not be addressed by linear elasticity. In the damage accumulation process the complex nature of asphalt concrete is essential and accurate performance predictions are possible only when this nature is modelled. This means that, besides the temperature and strain rate dependency, the state of stress dependency of the material response must be also taken into account. Current design methodologies try to circumvent the inherent problems by means of empirical relations that, however, are not generally applicable. To establish a new design approach that can handle the rapid changes in materials and the number and level of loads, a sound engineering approach of material response is needed.

One of the objectives of the ACRe project is to develop a model for asphalt concrete on the basis of sound mechanics principles (Scarpas et al. 1997 and 1998a and b). Naturally, any model for asphalt concrete should include the strain rate and temperature sensitivity of this class of materials, as well as its sensitivity to confinement. Because the state of stress in a road due to a passing vehicle varies from point to point, the corresponding response will vary as well. In order to get an accurate prediction of the expected damage patterns in a construction, the full 3D state of stress must be considered. To incorporate these influences (temperature, strain rate and state of stress) on the response in a model, their effect on the material behaviour must first be quantified and qualified. For this, experiments that capture the properties of the material as a function of these influences are necessary. For various reasons, standard road engineering tests are usually not suited for this purpose. Therefore, test set-ups and procedures that allow accurate material characterisation have been developed within the framework of the ACRe project. In this contribution, because of space limitations, after a brief description of the developed material model, the issues related to the development of the experimental set up necessary for performing reliable tension tests will be addressed only.

2. THE ACRE MATERIAL MODEL

The ACRe-model is based on the three-dimensional flow surface proposed by Desai for granular materials (1986). This model was selected for a number of reasons, the most prominent one being the fact that the size, shape and position of the flow surface can be easily adapted to available experimental evidence via physically meaningful model parameters. Since it was not known beforehand what a surface for asphalt concrete should look like, this flexibility was of great importance. During the ACRe project the model has been adapted and extended to facilitate the description of asphalt concrete response. The model expression is shown in Equation 1. In broad outlines, the current model works as shown in Figure 1, where the effect of the model parameters on the size and shape of the surface is illustrated.

$$f = \frac{J_2}{p_a^2} - \frac{\left[-\alpha\left(\frac{(I_1 - R)}{p_a^2}\right)^n + \gamma\left(\frac{(I_1 - R)}{p_a^2}\right)^2 \right]}{\sqrt{1-\beta\cos(3\theta)}} = 0 \quad (1)$$

Figure 1, (A) shows the influence of α, the hardening parameter, which controls the surface size. For α=0, the surface becomes a straight line. Figure 1 (B) shows the effect of n, which is related to the onset of dilatation, Figure 1 (C) shows how β controls the shape of the cross-section on the π-plane. Finally in Figure 1 (D) the way in which γ determines the ultimate slope of the surface is illustrated. The fifth model parameter, R, is the triaxial tensile strength. This parameter determines the position of the flow surface, since R is the intercept with the positive I_1-axis. For R=0, the material is cohesionless and the surface passes through the origin, while for higher R values the intercept shifts to the right.

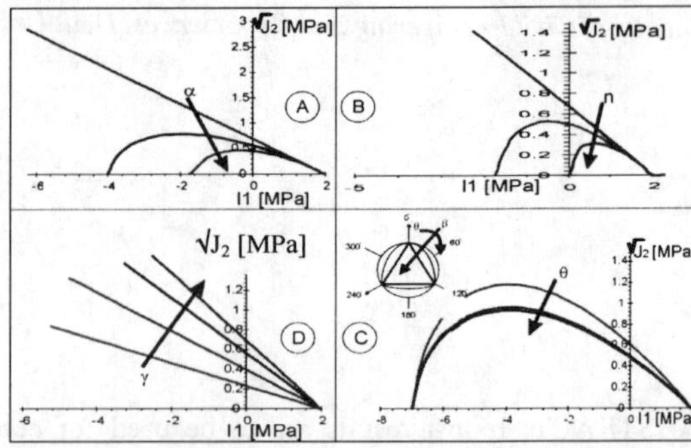

Figure 1: Influence of the model parameters on the size, shape and position of the flow surface in the I_1-$\sqrt{J_2}$ surface

For a given asphalt mix, each set of conditions (temperature and strain rate) corresponds to a set of model parameters. The hardening (an increase of the area within the flow surface, Figure 1A) is described by decreasing α from its original value to zero. From Figure 1a it can be seen that this leads to an open surface (straight line). Softening is modelled by decreasing the slope of this "ultimate response" line via a reduction of the original γ value.

Descriptions of the model at different stages throughout the project and information on how the different types of damage that occur in an asphalt pavement are incorporated in the model can be found in other publications (Scarpas et al. 1997, Scarpas et al. 1998a and b and Erkens et al. 2000).

3. DETERMINING MATERIAL CHARACTERISTICS FROM TESTS

To obtain the full benefits of curren available analysis tools, new test methods that determine material characteristics independent of the set-up and as a function of the (most important) influence parameters are needed. For asphalt concrete these parameters are the temperature, strain rate and state of stress, respectively. These parameters should be kept constant throughout a test and varied systematically between tests. Tests that can fulfil this requirement are, among others, uniaxial compression tests, uniaxial tension tests and triaxial tests.

Triaxial tests are often used for material characterisation. However, it is well known that most of the damage in pavements occurs due to tension-compression states of stress. For this reason it is interesting to characterise the material under conditions other than triaxial compression. On the basis of this consideration, it was decided to use uniaxial compression and uniaxial tension tests as the basis of the test programme. The multiaxial response will be determined using a four point shear test, which allows testing of tension-compression states of stress (Scarpas et al. 1997).

The material that was used for the tests is a down-scaled Dense Asphalt Concrete (DAC), with a maximum aggregate size of 5 mm. DAC is often used in Dutch pavements and it was downscaled to ensure a high ratio between minimum specimen dimension and maximum aggregate size ($D_{min}/D_{max,agg}$=10), without having to resort to very large specimens. This mix composition is shown in Table 1.

Table 1: Composition of the ACRe mix

component	Mass [%]	Volume [%]
Aggregate	77.1	66.3
Filler	14.3	11.8
Asphalt 45/60	8.6	19.3
Air	-	2.6

For reasons of brevity, only aspects related to the uniaxial tension tests will be discussed here. For information about the uniaxial compression tests, the reader is referred to other publications (Erkens et al. 2000a,b and 1998). After presenting the set-up, some results are shown. Since the tests are currently underway, the data do not present a complete overview of the results.

4. UNIAXIAL TENSION TESTS

One of the goals of the ACRe project was to develop a proper tension test set-up. Such a set-up should interfere as little as possible with the state of stress in the specimen. Hinges between the specimen and the set-up are used to ensure that the specimen is subjected to axial forces only. In the ACRe set-up, Erkens (2000c and 2001), three hinges are used Figure 2(right). One underneath the specimen, one between the specimen and the load cell and a third one above the load cell. This combination of three hinges allows the specimen to align itself and ensures that only axial loads can be introduced to the specimen and registered by the load cell. Because failure in a tension test is often brittle, explosive unloading can occur. If the elastic energy that is released by this unloading (i.e. if the elastic deformations in the frame and specimens are large), the test becomes unstable. To prevent this, a rigid loading frame and on-specimen deformation control was utilised.

During testing, displacement transducers give the overall specimen response, Figure 2 (right). After cracking, two different processes occur in the specimen. The crack is opening while the non-cracked parts of the specimen are unloading. However, if only the overall behaviour is registered, it is not possible to differentiate between these two components of response. For this reason a system with strain gages was developed to separately register the unloading. Such a system can only work if cracking at the position of the strain gauges is prevented. Therefore, a specimen with a diameter that varies gradually over the height is used (Figure 2). Failure will occur at the smallest cross section, leaving the remainder of the specimen for the strain gauges. The specimen shape is obtained by using a split-mould inside the gyratory mould, thus compacting the specimen in this shape. This approach was preferred over cutting an originally cylindrical specimen into this shape, since the latter was thought to be a source of surface cracks that might initiate damage. Notches, which would also fix the crack at a given position, are avoided because they lead to initial stress concentrations and, thus, it is not known at which stress the specimen actually fails (tension strength).

Figure 2: Tension set-up (left) and tension specimen (right) used in the ACRe project

The tensile tests are performed under displacement control by applying a monotonically increasing deformation. The test is controlled by the average signal of three displacement transducers that are connected to the specimen at 120° intervals. The complete frame is contained in a temperature-controlled cabinet, which enables testing at specific temperatures. The tests are performed at three temperatures, 0, 15 and 30 °C and at each temperature three different strain rates will be used. Because of the large differences in behaviour that are observed during tension testing of asphalt concrete, the strain rates for each temperature are selected in such a way that they cover the range that is considered of maximum interest for that particular temperature.

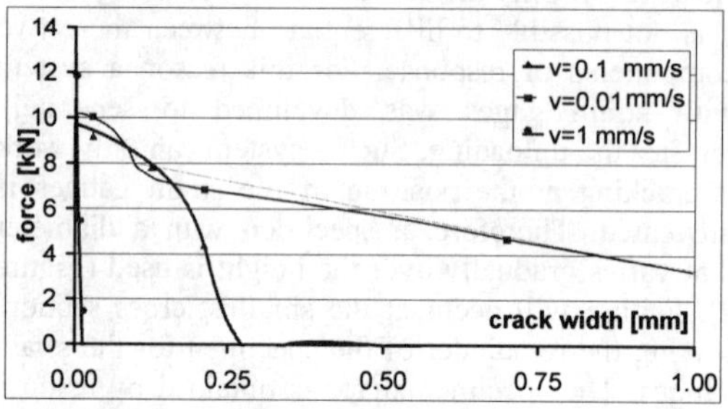

Figure 3: Force- crack opening diagram for some tension tests at 0°C

The results shown in Figure 3 are from three individual tests. The force rather than the stress is plotted because the stress varies over the height of the specimen. It must be noted that failure in tension is very local and, thus it will show more variation that that in compression. Taking this into consideration and looking at the other results (Table 2), it appears that the strength at this temperature (0°C) is independent of the strain rate for the rates tested. The deformation at peak, however, is clearly influenced.

Table 2: Result from tension test at 0°C

V_{test} [mm/s]	$\dot{\varepsilon}$ [s^{-1}]	f_t [N/mm²]	U_{peak} [μm]
0.0067	0.01	5.2	330
0.0067	0.01	5.2	130
0.0067	0.01	6.4	204
0.067	0.1	4.9	45
0.067	0.1	4.5	44
0.067	0.1	4.2	32
0.67	1	5.4	22
0.67	1	6.6	29
0.67	1	4.5	19

Naturally, the graph of the slowest test (0.01 s^{-1} at the centre) eventually also returns to zero force when the two specimen halves have separated completely. However, if the whole branch were shown, the result from the fastest test would not have been visible. The change in shape of the curve for the slowest test (at crack width=0.125 mm) is due to the increased crack growth rate directly prior to failing. Until that point, strength degradation occurs gradually and the overall deformation is still stable and controlled, but at w=0.125 mm the specimen suddenly breaks, causing a rapid increase of the crack width.

5. COMPARISON BETWEEN MODEL AND TEST RESULTS

Since the material characteristics set the requirements for the model, while the model in its turn leads to requirements for the tests, the tests and the model are developed in an interactive approach. This is believed to be essential to arrive at a complete model and material characterisation method. So far, this has led to a series of uniaxial compression and the tension tests, which are currently underway. As a result, not all the tension data is available, nevertheless the available data has been used in combination with the results from the compression tests to arrive at preliminary values for the model parameters. Using these parameters, the compression tests have been simulated. Some of these results are shown in Figure 4. It can be seen that the predictions match the results rather well.

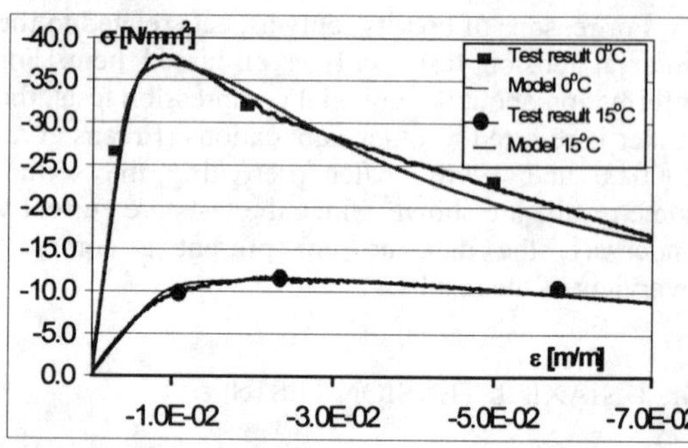

Figure 4: The measured versus the predicted response in uniaxial compression

6. CONCLUSIONS

The rapid changes in traffic and construction materials require a novel approach toward road design, based on structural mechanics rather than empiricism. The combination of the Finite Element Method with plasticity type material models proved successful in developing the insights and establishing the methods necessary for such an approach in related engineering fields, such as concrete mechanics, soil mechanics etc. The ACRe material model is based on the single-surface flow surface proposed by Desai, which has been adapted to describe all components of asphalt concrete response. The parameters that control this model have a clear physical meaning and can, therefore, be determined from laboratory tests. These tests should provide information on the material characteristics, independent of the set-up that is used. This requires a uniform internal state and careful preparation and testing. Most standard road engineering tests result in complicated stress distributions, non-uniform temperatures, a variety of strain rates or combinations thereof. As such, they are not suited for material parameter determination. For this reason, appropriate set-ups and test procedures are developed within the ACRe project. Since the material behaviour, observed in the tests, dictates the requirements for the model and the model parameters in turn set the requirements for the properties that should be recorded during a test, modelling and testing are strongly inter-related. For this reason, the model and the test procedures are developed interactively, to allow maximum flexibility in both aspects throughout the project. It is believed that accurate non-linear material models will result in a better understanding of pavement distress phenomena and the parameters that influence them. This will then lead to more sophisticated design procedures for constructions as well as mixes.

7. ACKNOWLEDGEMENTS

The ACRe project is financed by the Shell Research and Technology Centre and the Technology Foundation STW, applied science division of NWO and the technology programme of the Ministry of Economic Affairs.

8. REFERENCES

Desai, C.S., Somasundaram, S. and Frantziskonis, G., 1986, *"A Hierarchical Approach for Constitutive Modelling of Geologic Materials"*, International Journal of Numerical and Analytical Methods in Geomechanics, 10,3, 225-257

Erkens, S.M.J.G and Poot, M.R., 1998, *"The Uniaxial Compression Test – Asphalt Concrete Response (ACRe)"*, Technische Universiteit Delft, rapportnummer 7-98-117-4, September 1998

Erkens, S.M.J.G and Poot, M.R., 2000 *"Additional Compression Tests – Asphalt Concrete Response (ACRe)"*, Delft University of Technology report 7-00-117-4, (2000a)

Erkens, S.M.J.G., Scarpas, A. and Liu, X., 2000, *"3D Finite Element Model for Asphalt Concrete Response Simulation"*, 2nd International Symposium on 3D Finite Elements in Pavement Engineering, Charleston, West-Virginia, USA, October 11-13, (2000b)

Monismith, C.L. and Secor, K.E., 1962, *"Viscoelastic Behaviour of Asphalt Concrete Pavements"*, Proceedings of the International Conference on the Structural Design of Asphalt Concrete Pavements, University of Michigan, Ann Arbor, USA, Aug. 20-24 1962, pp.476-498.

Nijboer, L.W., 1948, *"Plasticity as a Factor in the Design of Dense Bituminous Road Carpets"*, Elsevier Publishing Company Inc.

Scarpas, A., 1992, *"CAPA –3D Finite Elements System – User's Manual, Parts I, II and III"*, Delft University of Technology, the Netherlands

Scarpas, A., Gurp, C.A.P.M., van, Al-Khoury, R.I.N. and Erkens, S.M.J.G., 1997, *'Finite Elements Simulation of Damage Development in Asphalt Concrete Pavements'*, 8th International Conference on Asphalt Pavements, University of Washington, Seattle, U.S.A., August 10–14 1997

Scarpas, A. and Blaauwendraad, J., 1998 , *"Experimental Calibration of a Constitutive Model for Asphaltic Concrete"*, Proceedings of the Euro-C conferentie on the Computational Modelling of Concrete Structures, Badgastein, Austria, March 31-April 3 1998 (1998a)

Scarpas, A., Erkens, S.M.J.G. and Molenaar, A.A.A., 1998, *"Development of Asphalt Concrete Response Envelopes"*, 12th Engineering Mechanics Conference, ASCE, La Jolla, California, USA, May 17-20 1998 (1998b)

Computer Methods and Advances in Geomechanics, Desai et al. (eds) © 2001 Balkema, Rotterdam, ISBN 90 5809 183 X

Cyclic plasticity characterization of unbound pavement materials

C.W.Schwartz
University of Maryland, College Park, Md., USA

A.Y.Y.Yau
Fugro-BRE, Austin, Tex., USA

ABSTRACT: Plastic deformation behavior of subgrade soils from four field sites was characterized using the Desai *et al.* (1986) δ_0-HISS model with extensions by Bonaquist (1996) for permanent deformations under repeated loading. The model is intended for the mechanistic analysis of permanent deformation distress in unbound pavement materials. Recognizing that many highway agencies are often reluctant to perform the "non-routine" characterization tests required for the model, the study investigated correlations between the Desai-Bonaquist model parameters and other more conventional soil properties.

1 INTRODUCTION

Surface permanent deformations—"rutting"—is a common distress and failure mode in flexible pavement systems, especially in warm climates. Permanent deformations within the surface asphalt layer are often the largest component of the overall rutting, but plastic strains within soft base and subgrade soils may also be significant. Mechanistic pavement distress models such as the Superpave system (Lytton *et al.* 1993) must include all these contributions in order to make accurate predictions of rutting vs. accumulated traffic. This in turn requires realistic material constitutive models, particularly for high cycle repeated loadings.

The plastic deformation behavior of subgrade soils from four field test sites[1] were characterized using an extended version of Desai *et al.*'s (1986) Hierarchical Single Surface (HISS) plasticity model. Descriptions of the four soils are summarized in Table 1. The extended model is based on the Desai δ_0-model for associated flow with extensions prposed by Bonaquist (Bonaquist, 1996; Bonaquist and Witczak, 1996)) for nonlinear resilient response and continuing plastic deformation under repeated loading. The present study follows the general approach used by Bonaquist, but extends it to additional soil types.

The material model is intended for the mechanistic analysis of permanent deformations in unbound materials that is suitable for incorporation into the next-generation flexible pavement performance prediction system (e.g., the final enhanced version of the Superpave system). However, it is recognized that many highway agencies will not be willing or able to perform the "non-routine" material tests required to characterize a subgrade according to the model, even for major projects. Accordingly, a major emphasis of this study was to investigate correlations between the Desai-Bonaquist material parameters and other more conventional and easily determined soil properties.

Table 1. Soils Considered in Study

Soil ID	Description	AASHTO Class	USCS Class
S1	Silty sand from Moulton Pit	A-4(1)	SM
S2	Silty clay from Lyme/Jenks	A-4(8)	CL
S3	Clay from St. Albans	A-6(9)	CL
S4	Stiff clay	A-7-5(11)	ML

2 MATERIAL MODEL

As shown schematically in Figure 1, the basic HISS δ_0 formulation is an isotropic hardening model consisting of a series of yield surfaces that expand with increasing plastic strains (Desai *et al.*, 1986):

[1] The field test sites were part of a larger study being conducted by the Cold Regions Research Engineering Laboratory of the U.S. Army Corps of Engineers.

$$\frac{J_2}{p_a^2}-\left[\gamma\left(\frac{I_1+\frac{k}{\sqrt{\gamma}}}{p_a}\right)^2-\frac{a_1}{\xi^{\eta_1}}\left(\frac{I_1+\frac{k}{\sqrt{\gamma}}}{p_a}\right)^n\right]=0 \qquad (1)$$

in which I_1 is the first stress invariant, J_2 is the second invariant of the deviatoric stresses, p_a is atmospheric pressure, ξ is the plastic strain trajectory, and k, γ, a_1, η_1, n are material parameters to be calibrated from testing. One of the advantages of the HISS model is that the material parameters have physical meaning. The Drucker-Prager failure envelope is characterized by a line in I_1-$J_2^{1/2}$ space with intercept k and slope $\gamma^{1/2}$. Parameters a_1, η_1 describe the strain hardening behavior or the rate at which the yield surfaces expand as a function of accumulated plastic strains. Finally, parameter n relates to the slope of the phase change line (Fig. 1) that separates the volume change behavior into compaction and dilation zones.

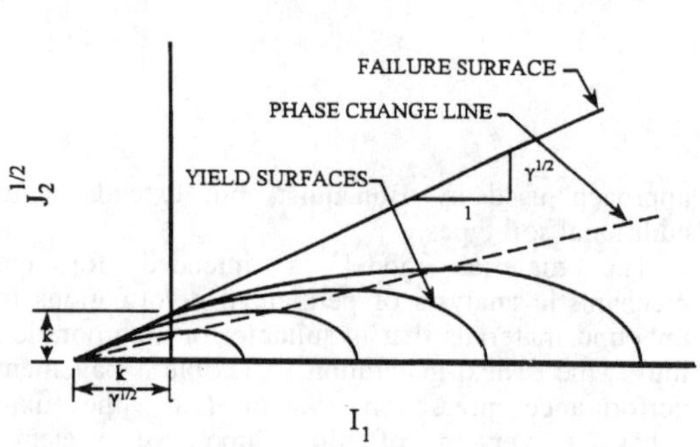

Figure 1. HISS yield surfaces for material with cohesion (from Bonaquist, 1996).

Bonaquist (Bonaquist, 1996; Bonaquist and Witczak, 1996) extended the HISS model to incorporate isotropic cyclic hardening for permanent deformation behavior under the large-cycle repetitive loading typical of pavement systems. The approach is based on two bounding yield surfaces: the first, or inner, surface captures the loading history of the material, while the second, or outer, surface is defined solely by the current stress state. Figure 2 schematically shows the movement of the initial surface toward the stationary outer bounding surface during repeated loading. The outer yield surface is defined by Equation 1, while the geometrically similar inner surface is determined using:

$$\xi_N=\xi_b-\xi_{i(N)}=\frac{1}{N^{a_2}}\xi_b \qquad (2)$$

in which ξ_N is the plastic strain trajectory for load cycle N, ξ_b is the bounding surface for current stress state, N is the number of load repetitions at the current stress level, and a_2 is the cyclic hardening material parameter.

Finally, Bonaquist also proposed a nonlinear elastic model based on the resilient (unloading) behavior of the form:

$$G_S=\sqrt{\frac{2}{3}}\frac{\sqrt{J_2}}{g_1}\left(\frac{I_1}{p_a}\right)^{-g_2}\left(\frac{\sqrt{J_2}}{p_a}\right)^{-g_3} \qquad (3)$$

in which G_S is the secant shear modulus and g_1, g_2, g_3 are material parameters. Equation 3 is similar in form to the so-called "universal" model developed by Witczak and Uzan (1988).

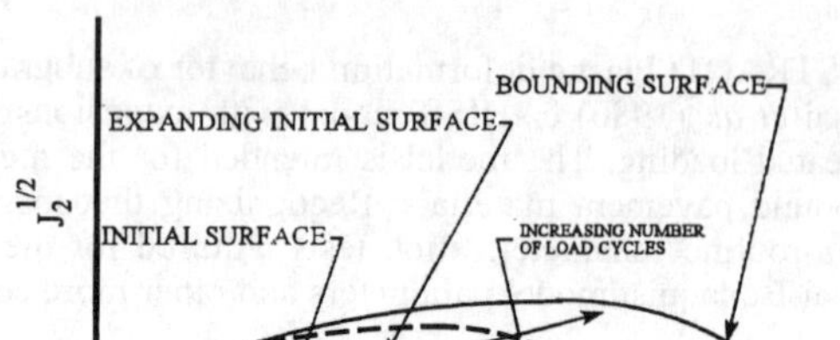

Figure 2. Bounding surface concept for cyclic hardening (from Bonaquist, 1996)

3 MATERIAL TESTING

Cyclic triaxial compression tests were conducted on 4 inch by 8 inch cylindrical specimens of each soil under moisture-density conditions corresponding to the peak of the modified Proctor compaction curve. The stress paths consisted of a series of load-unload-reload haversine deviatoric stress pulses of 0.45 second duration with magnitude increasing by 1 psi per cycle until large deformations (approximately 20% strain) were achieved. A 19.55 second rest period was provided between stress pulses to allow full resilient strain recovery. All tests were performed undrained under constant confining pressures equal to 0, 5 10, and 15 psi. Four radial LVDTs were used to measure radial deformations at midheight, and two specimen-mounted vertical LVDTs were used to measure the vertical deformations over the central portion (gage length = 4 inches) of the specimens. Typical strain results vs. number of cycles (time) are shown in Figure 3. Data from these tests were used to determine material parameters k, γ, a_1, η_1, n for the basic HISS yield surface (Eq. 1) and g_1, g_2, g_3 for the nonlinear resilient response (Eq. 3). The procedures described in Desai (1990) were followed to determine the HISS material parameters. Tables 2 and 3 summarize the material properties determined for the four soils in this study.

The material parameters for Bonaquist's cyclic hardening extension to the HISS model (Eq. 2) were determined from constant amplitude triaxial compression repeated load permanent deformation tests. A haversine stress pulse of 0.45 seconds duration and 1 second rest was applied repeatedly for up to 100,000 cycles per specimen. The applied deviatoric stress amplitude equaled approximately 0.35, 0.5, or 0.7 of the estimated failure stress. Weaker specimens were also tested at a 0.2 stress-to-strength ratio. All tests were performed undrained and without confinement. Radial and axial deformations were measured using the same LVDT arrangement as before. Data from the repeated load permanent deformation tests were used to determine material parameters ξ_b, a_2 for Bonaquist's cyclic hardening model (Eq. 2). A typical fit of incremental plastic strain trajectory to number of cycles is shown in Figure 4. The cyclic hardening parameters as functions of stress-to-strength ratio for each soil are summarized in Table 4.

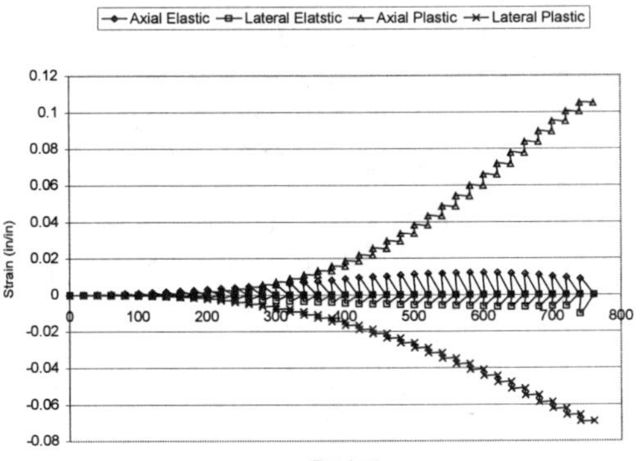

Figure 3. Typical results for strain vs. cycle (time) from triaxial cyclic loading tests (soil S3, unconfined).

Table 2. Summary of HISS material properties (Eq. 1).

Material	k (psi)	$\gamma^{1/2}$	a_1	η_1	n
S1/Silty Sand	4.6	0.268	1.12E-4	0.561	3.50
S2/Silty Clay	9.5	0.228	7.03E-4	0.345	3.14
S3/Clay	14.7	0.0036	3.00E-7	0.259	2.17
S4/Clay	12.2	0.0573	1.03E-4	0.277	2.52

Table 3. Summary of nonlinear resilient properties (Eq. 3).

Material	g_1	g_2	g_3
S1/Silty Sand	0.00566	-0.897	2.083
S2/Silty Clay	0.0160	-0.701	2.121
S3/Clay	0.0275	-0.148	1.919
S4/Clay	0.0261	-0.487	2.360

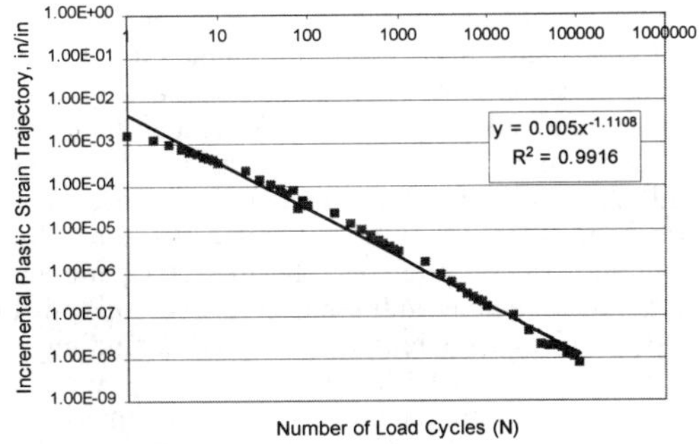

Figure 4. Cyclic hardening behavior for soil S3 at stress-to-strength ration of 0.34.

Table 4. Summary of cyclic hardening properties (Eq. 2).

Soil	r*	ξ_B	a_2
S1 Silty Sand	0.56 (2)	5.00E-5	0.863
	0.68 (2)	4.50E-4	0.974
	0.84	1.30E-3	0.969
S2 Silty Clay	0.32 (2)	2.00E-5	0.878
	0.49	8.00E-4	0.902
	0.64	2.50E-3	0.902
S3	0.26	1.30E-3	1.072
	0.34	5.00E-3	1.111
	0.43	5.00E-3	1.045
S4 Clay	0.25 (2)	7.00E-5	0.880
	0.41 (2)	4.85E-4	0.865
	0.46 (2)	2.45E-3	0.992
	0.55	1.10E-3	0.482
	0.61	2.60E-3	0.626

$$r^* = \text{stress-to-strength ratio} = \frac{\sqrt{J_2}}{k + I_1\sqrt{\gamma}};$$

(n) designates number of replicates

4 CORRELATIONS WITH OTHER PROPERTIES

The intended purpose of the proposed model is for mechanistic analysis of unbound permanent deformation distress in unbound materials in the next-generation flexible pavement performance prediction system (e.g., the revised and enhanced Superpave system). However, it is recognized that many agencies may not be able or willing to perform the "non-routine" material tests required to characterize a subgrade according to this model, even for major pavement projects. Consequently, a major emphasis of this study was placed on investigating correlations between the Desai HISS-Bonaquist material parameters k, γ, a_1, η_1, n, a_2, ξ_B and other more conventional and easily determined soil properties. Bonaquist's earlier results on a limited set of soils (one base material, one subbgase, and one subgrade from the Federal Highway Administration's Accelerated Load Facility at the Turner-Fairbanks Laboratory) indicated that many of the material parameters were strongly correlated with the slope of the ultimate failure surface, $\gamma^{1/2}$, which in turn is directly related to the Mohr-Coulomb friction angle ϕ. The material parameters determined for the additional cohesive soils in this study confirm and strengthen this earlier finding.

Correlations for HISS parameters a_1, η_1, n against the ultimate failure parameter $\gamma^{1/2}$ are summarized in Figure 5. Data for the FHWA ALF soils from Bonaquist's original study are also included in these comparisons. The trendlines labeled "Bonaquist" are from his original study and are based only on the three FHWA ALF soils; the trendlines labeled "Predicted" combine the Bonaquist data with the data from the present study.

Parameters a_1 and η_1 (Figs. 5a and 5b) govern the amount of strain hardening in Equation 1. Parameter a_1 increases with $\gamma^{1/2}$ for all soils. Bonaquist found that η_1 was essentially the same for all of his soils, but the larger data set including the soils from the present study shows an increase in η_1 with $\gamma^{1/2}$. Both of these trends are consistent with physical intuition—i.e., higher quality materials (higher $\gamma^{1/2}$) will exhibit greater degrees of strain hardening.

Bonaquist also found the phase change parameter n to be essentially the same for all of his soils. The augmented data set including the soils from the present study shows an increase in n with $\gamma^{1/2}$ (Fig. 5c). Again, this is consistent with physical reasoning, as higher quality materials with greater friction (larger $\gamma^{1/2}$) would be expected to exhibit more dilation (larger n).

Bonaquist found that the cyclic hardening parameters a_2, ξ_b in Equation 2 were a function of the stress-to-strength ratio rather than to the ultimate failure parameters. The relationships between a_2, ξ_b and the stress-to-strength ratio r (see footnote in Table 4 for definition of r) are summarized in Figure 6. As before, data for the FHWA ALF soils from Bonaquist's original study are also included in these comparisons. Parameter a_2 (Fig. 6a) varies over a relatively narrow range. Although Bonaquist found a_2 decreased with increasing stress-to-strength ratio, the full data set suggests that there is little or no dependence. However, this may in part be because the data from the present study contained very few values at very high stress-to-strength values. Parameter ξ_b similarly varies over a very limited range, and only Bonaquist's data at very high stress-to-strength values show any influence on ξ_b.

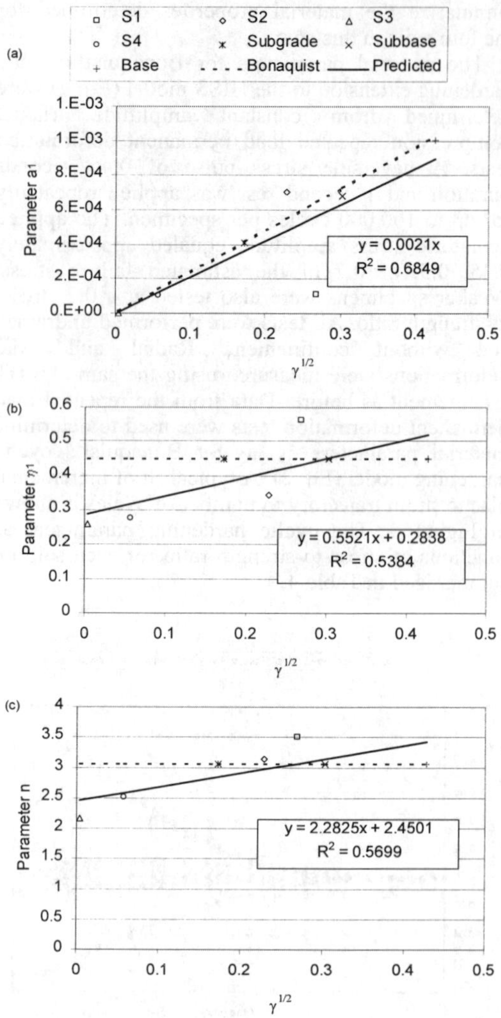

Figure 5. Relationship between HISS yield surface parameters and ultimate failure parameter $\gamma^{1/2}$.

5 CONCLUSIONS

The primary objective of this study was to calibrate Desai's Hierarchical Single Surface (HISS) plasticity model (Desai, *et al.*, 1986) for four subgrade soils taken from field test sites. The HISS model, with extensions by Bonaquist (Bonaquist, 1996; Bonaquist and Witczak, 1996), considers both resilient and permanent deformation responses of unbound pavement materials as well as the effects of repeated loading. The study follows the general approach used by Bonaquist, but extends it to additional soil types.

The intended purpose of the proposed model is for mechanistic analysis of permanent deformation distress in unbound materials in the next-generation flexible pavement performance prediction system. As an aid to practical design, this study investigated possible correlations between the Desai HISS-Bonaquist material parameters k, γ, a_1, η_1, n, a_2, ξ_b and other more conventional and easily determined soil properties. Although the data are still very limited, the strengths of the correlations—at least for the HISS yield surface parameters--are reasonable given the limited number and wide range of soil types considered to date. There may be some promise in the concept of a material property database for use in estimating advanced plasticity model parameters without the need for extensive testing. This database would be an invaluable aid in winning acceptance for such a model in a practical pavement design environment.

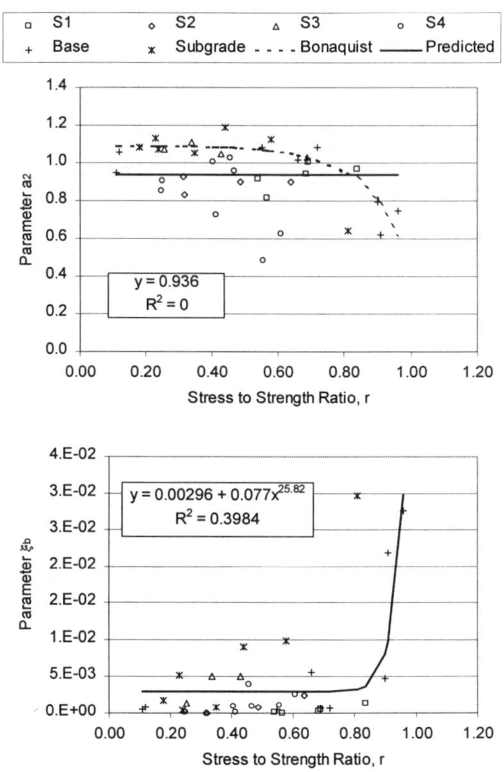

Figure 6. Relationships between cyclic hardening parameters and stress-to-strength ratio.

REFERENCES

Bonaquist, R. 1996. *Development and Application of a Comprehensive Constitutive Model for Granular Materials in Flexible Pavement Structures.* Ph.D. Dissertation, Department of Civil Engineering, University of Maryland, College Park, MD.

Bonaquist, R. and Witczak, M.W. 1996. Plasticity modeling applied to the permanent deformation response of granular materials in flexible pavement systems. *Transportation Research Record 1540*, National Research Council, Washington, DC, pp. 7-14.

Desai, C.S. 1990. Modeling and testing: Implementation of numerical models and their application in practice. In C.S Desai and G. Gioda (eds), *Numerical Methods and Constitutive Modeling in Geomechanics.* Springer-Verlag, Wien, NY.

Desai, C.S., Somasundaram, S., and Frantziskonis, G. 1986. A hierarchical approach for constitutive modeling of geologic materials. *International Journal for Numerical and Analytical Methods in Geomechanics.* Vol. 10, pp. 225-257.

Lytton, R.L., Uzan, J., Fernando, E.G., Roque, R., Hiltunen, D., and Stoffels, S.M. 1993. *Development and Validation of Performance Prediction Models and Specifications for Asphalt Binders and Paving Mixes*, Report SHRP A-357, Strategic Highway Research Program, National Research Council, Washington, DC.

Witczak, M.W. and Uzan, J. 1988. *The Universal Airport Pavement Design System. Report I of V: Granular Material Characterization.* Department of Civil Engineering, University of Maryland, College Park, MD.

Computer Methods and Advances in Geomechanics, Desai et al. (eds) © 2001 Balkema, Rotterdam, ISBN 90 5809 183 X

Implementation of a nonlinear resilient modulus constitutive model for unbound pavement materials

C.W.Schwartz
University of Maryland, College Park, Md., USA

ABSTRACT: An improved model for a stress-dependent resilient modulus has been proposed for unbound materials in pavement structures. Although this model is a convenient functional form for fitting laboratory resilient modulus test data, it is less well-suited to implementation in finite element codes. These problems are discussed in the context of implementation in a conventional, commercially available implicit incremental finite element code. Analysis of a simple example pavement structure illustrates the influence of nonlinear unbound material response on key pavement response quantities such as asphalt concrete tensile stresses and strains and subgrade compressive stresses and strains.

1 INTRODUCTION

The resilient modulus, defined as the unloading modulus after many cycles of repeated loading, is used in pavement engineering as an appropriate measure of stiffness for the unbound (i.e., soil) layers in a pavement structure. The definition of the resilient modulus as measured in the standard resilient modulus cyclic triaxial test is shown in Figure 1, in which σ_a and ε_a are the stress and strain in the axial (i.e., cyclic loading) direction. The sample is initially subjected to a hydrostatic confining pressure σ_c, which induces an initial strain ε_c. This initial strain is unmeasured in the test, but it is assumed the same in all directions for isotropic material behavior. The axial stress is then cycled at a constant magnitude $\Delta\sigma$, which during unloading induces the cyclic resilient axial strain $\Delta\varepsilon$. The resilient modulus M_R is defined simply as the ratio of the cyclic axial stress to resilient axial strain:

$$M_R = \frac{\Delta\sigma}{\Delta\varepsilon} \qquad (1)$$

The resilient modulus for most unbound pavement materials is stress dependent. Many nonlinear models have been proposed over the years for incorporating the effects of stress level on the resilient modulus. A general form for these models can be expressed as (Andrei, 1999):

$$M_R = k_1 p_a \left(\frac{\theta - 3k_6}{p_a}\right)^{k_2} \left(\frac{\tau_{oct}}{p_a} + k_7\right)^{k_3} \qquad (2)$$

in which M_R is the resilient modulus, $\theta = \sigma_1+\sigma_2+\sigma_3$ is the bulk stress, $\tau_{oct} = (1/3)[(\sigma_1-\sigma_2)^2+(\sigma_2-\sigma_3)^2+(\sigma_3-\sigma_1)^2]^{1/2}$ is the octahedral shear stress, σ_1, σ_2, σ_3 are the principal stresses, p_a is atmospheric pressure (to make Equation 2 dimensionless), and k_1 through k_7 are material parameters, subject to the constraints $k_1>0$, $k_2 \geq 0$, $k_3 \leq 0$, $k_6 \leq 0$, and $k_7 \geq 1$. A simplified version of Equation 2 with $k_6=0$ and $k_7=1$ has been adopted for the 2002 Pavement Design Guide currently under development in NCHRP Project 1-37A.

Equation 2 combines both the stiffening effect of the bulk stress (the term under the k_2 exponent) and the softening effect of shear stress (the term under the k_3 exponent). Through appropriate choices of the material parameters k_1-k_7, one can recover the familiar two-parameter bulk stress model for granular materials and its companion two-parameter shear stress model for cohesive soils, the Uzan-Witczak "universal" model (Witczak and Uzan, 1988), and the k_1-k_6 model from the Strategic Highway Research Program's (SHRP) flexible pavement performance models (Lytton *et al.*, 1993).[1]

[1] In fact, the historical precedent of the SHRP performance models, which used material parameters k_4 and k_5 for the stress dependent Poisson's ratio, is the reason that these two material parameters are absent in Equation 2.

Equation 2 and its various specializations are convenient functional forms for fitting laboratory resilient modulus test data. However, they are less convenient—in fact, quite problematic—for implementation in a conventional incremental-implicit (e.g., ABAQUS, 1998) or explicit (e.g., LS-DYNA, 1999) finite element framework for the following reasons:

1. The conventional definition of M_R is neither that of a tangent modulus or a secant modulus, but rather a *chord* modulus. In terms of the axial stress vs. axial strain response for the conventional triaxial resilient modulus (Fig. 1), the resilient modulus is defined as the slope of the chord connecting the confining pressure point ε_c, σ_c to the peak cyclic stress point $\varepsilon_c+\Delta\varepsilon$, $\sigma_c+\Delta\sigma$ on the curve. Note that only the resilient strain increment $\Delta\varepsilon$ is measured during the test; the magnitude of the strain induced by the confining pressure, ε_c, is unknown.
2. The M_R nonlinearity is defined in Equation 2 in terms of *stresses* rather than strains. Contrast this, for example, with the conventional hypoelastic constitutive model defined in terms of strain invariants. Conventional incremental-explicit finite element algorithms take strains (displacements) and strain increments as the fundamental solution quantities.
3. The M_R model in Equation 2 is defined exclusively in terms of stresses at the *peak* of the loading path. For standard incremental-implicit and explicit finite element algorithms, it is usually more convenient to define stiffness at the *beginning* of the load increment.
4. There may be constraints on the material parameters k_1-k_6 in addition to the obvious physical requirements given in the definitions for Equation 2. Specifically, additional constraints (primarily between k_2, k_3, θ, and τ_{oct}) may be required to ensure that the tangent modulus is always positive for pre-peak failure conditions and to ensure solution uniqueness.

Point 3 merits additional discussion. Within the narrow context of the conventional triaxial compression resilient modulus test in which the initial hydrostatic confining stress is held constant and only the axial stress is cycled, the peak stresses are related to the initial stresses as follows:

$$\theta = 3\sigma_c + \Delta\sigma \tag{3}$$

$$\tau_{oct} = \frac{\sqrt{2}}{3}\Delta\sigma \tag{4}$$

Equations 3 and 4 define an implicit relationship among θ, τ_{oct}, and σ_c:

$$\sigma_c = \frac{\theta}{3} - \frac{\tau_{oct}}{\sqrt{2}} \tag{5}$$

Thus, in the standard triaxial resilient modulus test there is a one-to-one correspondence between the stress state θ, τ_{oct} at the peak of the cyclic loading and the cyclic loading stresses σ_c, $\Delta\sigma$. In other words, the standard test provides only one load path to reach each peak θ, τ_{oct} combination. However, Equation 2 is intended for general use in analyzing the more complex stress states in real pavement structures. The relationship in Equation 5 will not hold in this general case, implying that the M_R formulation in Equation 2 can be viewed as independent of the starting stress and strain conditions. The question of whether this conforms to the real behavior of unbound material in the general case requires further study.

Past implementation of stress-dependent resilient moduli models for unbound pavement materials have most commonly employed an iterative secant approach. This approach has been found to exhibit convergence difficulties in many cases (see, e.g., Brown and Pappin, 1981; Tutumluer, 1995). In addition, it is not suitable for incremental finite element formulations. Hjelmstad and Taciroglu (2000) provide an elegant derivation of the simpler Uzan-Witczak stress-dependent resilient modulus model in tangent modulus form. The present paper focuses on the basic incremental formulation for the more general M_R model given in Equation 2. The issues of secondary constraints on the material property values and the potential for nonunique solutions are deferred.

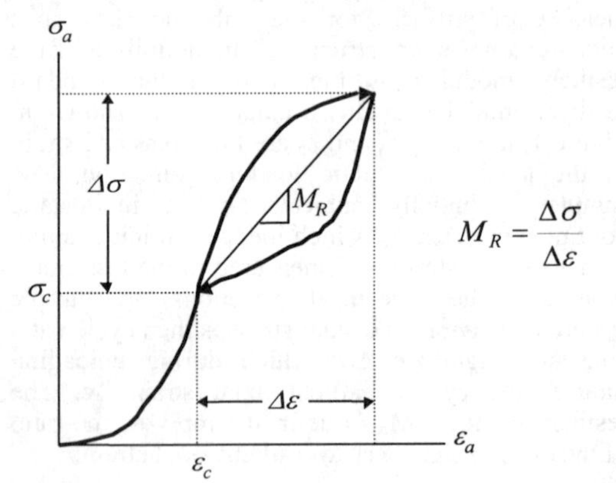

Figure 1. Definition of Resilient Modulus as Chord Modulus.

2 INCREMENTAL M_R FORMULATION

The generalized form of Hooke's law for a nonlinearly elastic material under conventional triaxial compression loading is:

$$\varepsilon_a = \frac{1}{E_s}\left[\sigma_a - 2\nu\sigma_c\right] = \frac{1}{E_s}\left[\Delta\sigma + (1-2\nu_s)\sigma_c\right] \tag{6}$$

in which E_s and ν_s are the secant Young's modulus and Poisson's ratio. Equation 6 can be expressed in terms of the resilient modulus M_R as (Fig. 1):

$$\varepsilon_a = \frac{1}{M_R}\left[\Delta\sigma + (1-2\nu_s)\sigma_c\right] + \varepsilon_c \quad (7)$$

As discussed in the preceding section, Equation 2 implies that M_R is independent of the initial stresses and strains in the general loading case, and thus it is convenient to assume that $\sigma_c=\varepsilon_c=0$ and $M_R=E_S$. Again, the question of whether this conforms to the real behavior of unbound material requires further study. Also note that the standard resilient modulus test provides no insight into the secant Poisson's ratio ν_s in Equation 7; typically, an assumed value is used for this parameter.

The tangent Young's modulus E_T required for conventional incremental finite element formulations can be derived from the secant Young's modulus E_S as[2]:

$$E_T = \lim_{\Delta\varepsilon\to 0}\frac{\Delta\sigma}{\Delta\varepsilon} = \frac{d\sigma_a}{d\varepsilon_a} = \frac{d}{d\varepsilon_a}\left(E_S\varepsilon_a\right) = \frac{dE_S}{d\varepsilon_a}\varepsilon_a + E_S \quad (8)$$

or in terms of M_R as:

$$E_T = \frac{d}{d\varepsilon_a}(M_R\varepsilon_a) = \frac{dM_R}{d\varepsilon_a}\varepsilon_a + M_R \quad (9)$$

Since M_R is defined in terms of stresses rather than strains, it is more convenient to formulate the derivation in terms of the tangent and secant elastic compliances $D_T=(E_T)^{-1}$ and $D_S=(E_S)^{-1}=(M_R)^{-1}$, respectively. Equation 9 can be expressed in terms of compliances as:

$$D_T = \lim_{\Delta\sigma\to 0}\frac{\Delta\varepsilon}{\Delta\sigma} = \frac{d\varepsilon_a}{d\sigma_a} = \frac{d}{d\sigma_a}\left(D_S\sigma_a\right) = \frac{dD_S}{d\sigma_a}\sigma_a + D_S \quad (10)$$

From Equation 2, the secant compliance D_S can be expressed as:

$$D_S = \frac{1}{E_S} = \frac{1}{M_R} = \frac{\left(\dfrac{\theta+\theta_t}{p_a}\right)^{-k_2}\left(\dfrac{\tau_{oct}}{p_a}+1\right)^{-k_3}}{k_1 p_a} \quad (11)$$

[2] An alternative to deriving E_T is the total strain approach employed by Uzan in the SHRP performance models (Lytton *et al.*, 1993), which in compliance terms can be expressed as:

$$\dot{\varepsilon}_x = \frac{\partial\varepsilon_x}{\partial\sigma_x}\dot{\sigma}_x + \frac{\partial\varepsilon_x}{\partial\sigma_y}\dot{\sigma}_y + \frac{\partial\varepsilon_x}{\partial\sigma_z}\dot{\sigma}_z + \frac{\partial\varepsilon_x}{\partial\tau_{xy}}\dot{\tau}_{xy} + \frac{\partial\varepsilon_x}{\partial\tau_{yz}}\dot{\tau}_{yz} + \frac{\partial\varepsilon_x}{\partial\tau_{zx}}\dot{\tau}_{zx}$$

in which $\dot{\varepsilon}_i$, $\dot{\sigma}_j$, $\dot{\tau}_{ij}$ are the strain and stress increments. One consequence of this approach is a nonsymmetric compliance matrix.

and, using $\sigma_a = \sigma_1$, the tangent compliance D_T can be expressed as:

$$D_T = \frac{d}{d\sigma_1}\left[\frac{\left(\dfrac{\theta+\theta_t}{p_a}\right)^{-k_2}\left(\dfrac{\tau_{oct}}{p_a}+1\right)^{-k_3}}{k_1 p_a}\right]\sigma_1 + \frac{\left(\dfrac{\theta+\theta_t}{p_a}\right)^{-k_2}\left(\dfrac{\tau_{oct}}{p_a}+1\right)^{-k_3}}{k_1 p_a} \quad (12)$$

or

$$D_T = \left[\frac{-k_2\left(\dfrac{\theta+\theta_t}{p_a}\right)^{-k_2-1}\left(\dfrac{\tau_{oct}}{p_a}+1\right)^{-k_3}}{k_1 p_a^2} + \frac{-k_3\sqrt{\dfrac{2}{3}}\left(\dfrac{\theta+\theta_t}{p_a}\right)^{-k_2}\left(\dfrac{\tau_{oct}}{p_a}+1\right)^{-k_3-1}}{k_1 p_a^2}\right]\sigma_1 + \frac{\left(\dfrac{\theta+\theta_t}{p_a}\right)^{-k_2}\left(\dfrac{\tau_{oct}}{p_a}+1\right)^{-k_3}}{k_1 p_a} \quad (13)$$

Typical variations of E_S and E_T with strain for various k_2 and k_3 values are illustrated in Figures 2 and 3.

One last modification is required to accommodate tensile stresses within an unbound layer. Tensile stresses may be induced at the bottom of granular base and subbase layers in some pavement structures. A tension cut-off is imposed by setting E_T equal to a very small value when $(\theta-3k_6)/p_a \leq 0$.

3 INFLUENCE OF NONLINEAR M_R ON PAVEMENT RESPONSE

The importance of nonlinear unbound material response on critical stress and strain values in a pavement system has been documented extensively in the literature (e.g., Harichandran *et al.*, 1989; ILLI-PAVE, 1990; DAMA, 1991; Huang, 1993; Zaghoul and White, 1993; Chen *et al.*, 1995; and others). The purpose of this section is to give an example of the implementation of the particular nonlinear M_R formulation given in Equation 2. This implementation is similar to that proposed for use in the new 2002 Pavement Design Guide currently under development in NCHRP Project 1-37A.

The nonlinear M_R formulation has been implemented in the ABAQUS finite element code (ABAQUS, 1998) and used to evaluate the influence of unbound material nonlinearity on the pavement response. The pavement structure analyzed in this study consisted of 150 mm of asphalt concrete over 300 mm of crushed stone base on natural subgrade. The idealized vehicle load consisted of a single dual

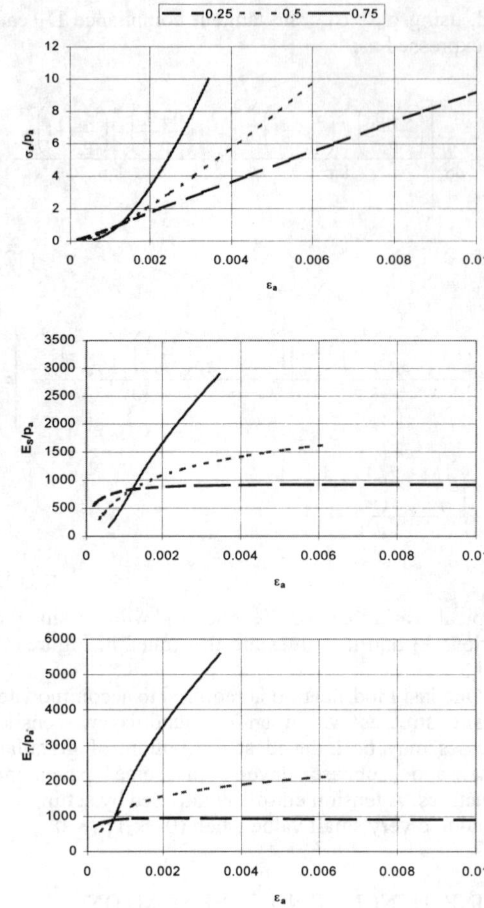

Figure 2. Normalized axial stress, tangent modulus, and secant modulus vs. axial strain (k_1=1000; k_2=0.25, 0.5, 0.75; k_3=-0.3; σ_c=0.

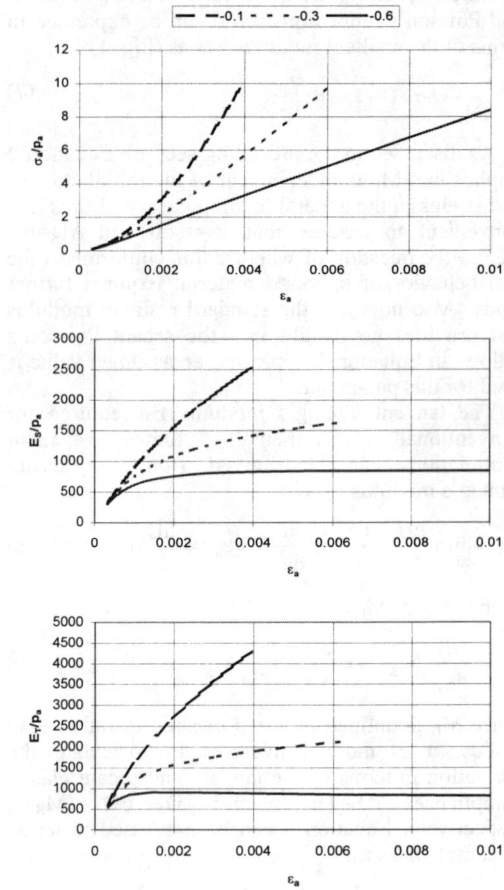

Figure 3. Normalized axial stress, tangent modulus, and secant modulus vs. axial strain (k_1=1000; k_2=0.5; k_3=-0.1, -0.3, -0.5; σ_c=0)

wheel axle with a total load on each wheel of 20 kN. The uniform tire pressure is 500 kPa over a square contact area of 200 mm x 200 mm; centerline spacing between the wheels is 300 mm. Because of symmetry, only one quarter of the problem geometry was discretized with finite elements. The problem geometry was truncated with rigid boundaries at a depth of 900 mm beneath the surface and at a horizontal distance of 1500 mm from the center of the dual wheels. The final finite element mesh based on these idealizations (Figures 1 and 2) consisted of 7524 4-node quadrilateral elements, 8740 nodes, and 26,220 degrees of freedom. The AC layer is divided into 4 element layers while the base is divided into 8 layers. Previous analyses using 4-node quad elements have found this level of discretization sufficient for capturing the bending response of the layers.

The AC and subgrade layers were modeled as conventional isotropic linearly elastic materials and the base layer as a nonlinear elastic material following the relation in Equation 2. For the purposes of the present study, only the bulk stress stiffening term in Equation 2 was considered, which is appropriate for the crushed stone material assumed for the base. All material properties used in the analyses are summarized in Table 1.

Figure 6 illustrates variation of the stress-dependent M_R values with depth in the base layer under the load for the case of a single wheel. The results are computed at the centroid of the finite elements in the vertical column nearest the center of the loading (see Fig. 5). The results in Figure 6 confirm the physical expectation that the k_2=0.8 analysis exhibits the most variation in M_R through the thickness while the k_2=0.2 analysis exhibits the

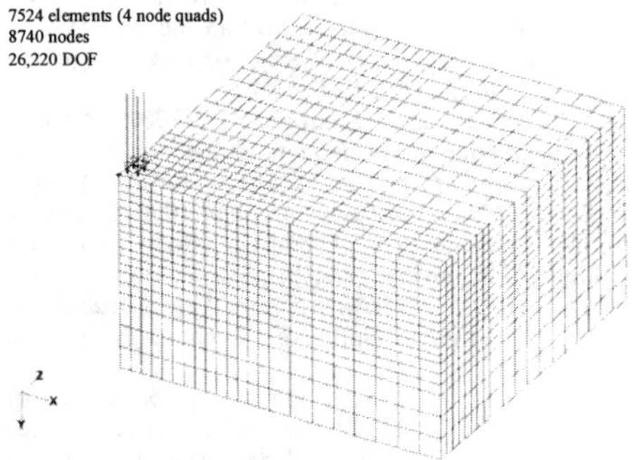

Figure 4. 3D Finite Element Mesh

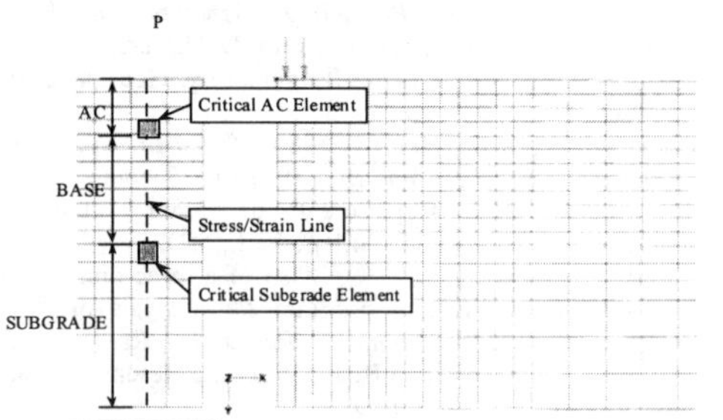

Figure 5. Cross Section of Finite Element Mesh Showing Calculation Locations.

Table 1. Material Properties for Finite Element Analyses

Material	E (MPa)	k_1	k_2	ν	γ (kN/m^3)
Asphalt Concrete	1,400			0.35	24.0
Base					
Low nonlinearity		1278	0.2	0.35	20.0
Med nonlinearity		1305	0.5	0.35	20.0
High nonlinearity		1333	0.8	0.35	20.0
Subgrade	70			0.35	17.5

Note: k_1 and k_2 are based on definition in Equation 2 Combinations of k_1 and k_2 are based on a M_R value of approximately 128 MPa at $\theta \approx 1$ atmosphere.

least. Thickness-averaged M_R values for the base layer in this near-centerline location equaled 93.4, 93.5, and 80.6 MPa for k_2=0.2, 0.5, and 0.8, respectively.

Figures 7, 8, and 9 illustrate the influence of base layer nonlinearity on the surface deflection, horizontal tensile strains at the bottom of the asphalt layer, and vertical compressive strains in the base and subgrade layers beneath the center of one of the dual wheels. The stiffening of the base layer with increasing k_2 is most pronounced for the surface deflections and asphalt tensile strains. It is much less significant for the compressive strains within the base and subgrade layers. This is physically plausible in that changes in the base layer stiffness characteristics would be expected to most strongly influence the overlying layers. The strong effect on the tensile strains in the asphalt concrete has obvious implications for fatigue resistance of the pavement.

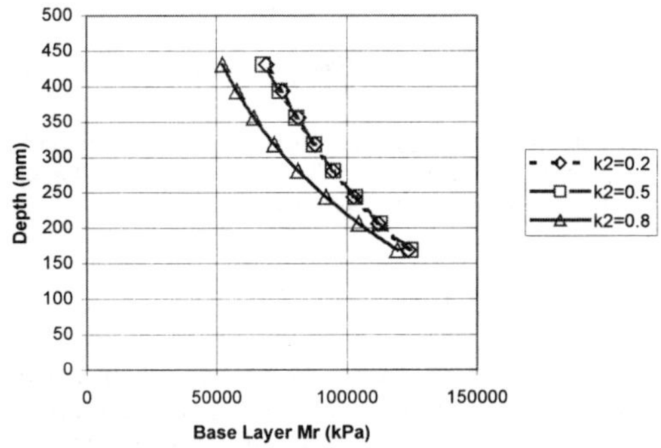

Figure 6. Variation of Base Layer M_R with Depth for Single Wheel Loading.

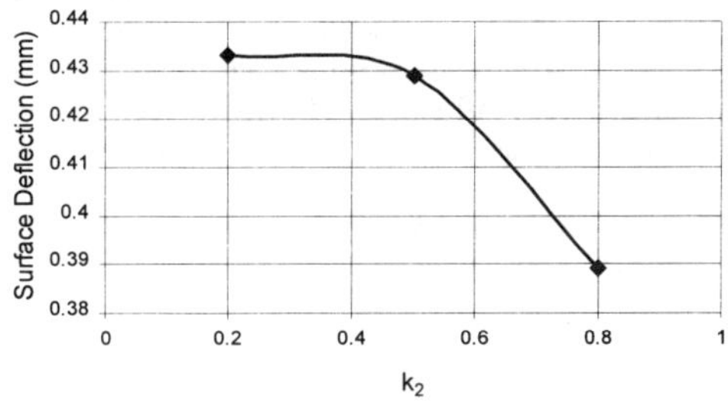

Figure 7. Effect of base layer nonlinearity on surface deflection beneath tire centerline.

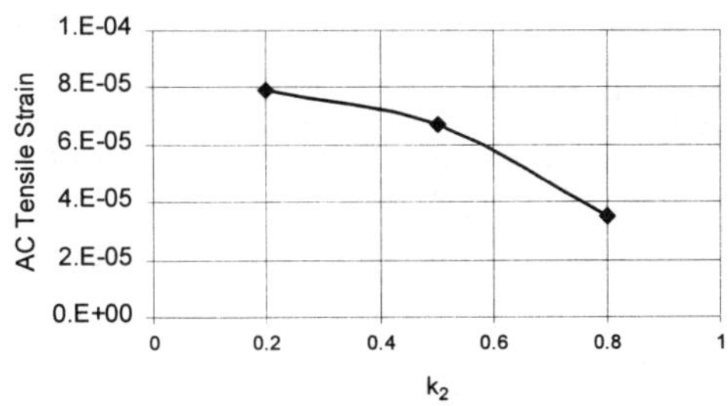

Figure 8. Effect of base layer nonlinearity on horizontal tensile strain at bottom of asphalt layer beneath tire centerline.

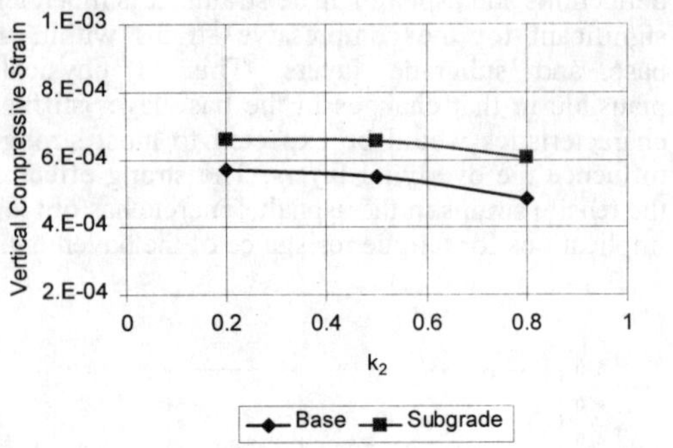

Figure 9. Effect of base layer nonlinearity on compressive strains at the midthickness of the base layer and at the top of the subgrade beneath tire centerline.

4 CONCLUSIONS

An improved model for a stress-dependent resilient modulus has been proposed for unbound materials in pavement structures. Although this model is a convenient functional form for fitting laboratory resilient modulus test data, it is less well-suited to implementation in finite element codes. These problems are discussed in the context of implementation in a conventional, commercially available implicit incremental finite element code. Analysis of a simple example pavement structure illustrates the influence of nonlinear unbound material response on key pavement response quantities such as asphalt concrete tensile stresses and strains and subgrade compressive stresses and strains.

REFERENCES

ABAQUS. 1998. *ABAQUS/Standard User's Manual (Version 5.8),* Hibbitt, Karlsson, and Sorensen, Inc., Pawtucket, RI.

Andrei, D. 1999. *Development of a Harmonized Test Protocol for the Resilient Modulus of Unbound Materials Used in Pavement Design,* Master of Science thesis, University of Maryland, College Park, MD.

Brown, S.F., and Pappin, J.W. 1981. Analysis of pavements with granular bases. *Transportation Research Record,* Transportation Research Board, National Research Council, Washington, DC, pp. 17-23.

Chen, D.H., Zaman, M., Laguros, J. and Soltani, A. 1995. Assessment of computer programs for analysis of flexible pavement structures. *Transportation Research Record 1539,* Transportation Research Board, National Research Council, Washington, DC, pp. 123-133.

Computer Program DAMA—Pavement Structural Analysis Using Multi-Layered Elastic Theory. 1991. Report CP-1 (1991 Revision), The Asphalt Institute, Lexington, KY.

Harichandra, R.S., Yeh, M.S., and Baladi, G.Y. 1989. *MICH-PAVE User's Manual.* Final Report, FHWA-MI-RD-89-023, Department of Civil and Environmental Engineering, Michigan State University, East Lansing, MI.

Huang, Y.H. 1993. *Pavement Analysis and Design.* Prentice-Hall, Inc., Englewood Cliffs, NJ.

Hjelmstad, K.D., and Taciroglu, E. 2000. Analysis and implementation of resilient modulus models for granular soils. *Journal of Engineering Mechanics,* ASCE (in press).

ILLI-PAVE PC Version User's Manual. 1990. NCHRP Project 1-26, Transportation Facilities Group, University of Illinois, Urbana-Champaign, IL.

LS-DYNA. 1999. *LS-DYNA Keyword User's Manual (Version 950),* Livermore Software Technology Corporation, Livermore, CA.

Lytton, R.L., Uzan, J., Fernando, E.G., Roque, R., Hiltunen, D., and Stoffels, S.M. 1993. *Development and Validation of Performance Prediction Models and Specifications for Asphalt Binders and Paving Mixes,* Report SHRP A-357, Strategic Highway Research Program, National Research Council, Washington, DC.

Tutumluer, E. 1995. *Predicting Behavior of Flexible Pavements with Granular Bases,* Ph.D. Dissertation, School of Civil and Environmental Engineering, Georgia Institute of Technology, Atlanta, GA.

Witczak, M.W. and Uzan, J. 1988. *The Universal Airport Pavement Design System. Report I of V: Granular Material Characterization.* Department of Civil Engineering, University of Maryland, College Park, MD.

Zaghoul, S. and White, T. 1993. Use of a three-dimensional., dynamic finite element program for analysis of flexible pavements. *Transportation Research Record 1388.* Transportation Research Board, National Research Council, Washington, DC, pp. 60-69.

Computer Methods and Advances in Geomechanics, Desai et al. (eds) © 2001 Balkema, Rotterdam, ISBN 90 5809 183 X

A technique for reducing settlement difference of road on soft clay

S.L. Shen & N. Miura
Institute of Lowland Technology, Saga University, Honjo, Japan

ABSTRACT: Low embankment roads on soft clay encounter settlement difference between road and crossing-structures supported by piles, such as bridge abutment and box culvert etc. The settlement difference obstructs vehicle running and makes maintenance costs to be very high. This paper proposes a practical design method named Column Approach (CA) method, in which road approach to structure is supported by floating deep mixing (DM) columns with varied lengths. This technique aims to reduce the settlement difference between crossing-structure and road. This paper introduces the basic design concept of CA method and shows the effectiveness of this technique by case histories.

1 INTRODUCTION

Low embankment roads on soft clay ground encounter settlement difference between road and crossing-structures supported by piles, such as bridge abutment and box culvert etc. This kind of settlement difference obstructs vehicle running. General countermeasures to reduce the settlement difference are to cut top pavement on structures and/or to fill the difference by overlaying the pavement after the road is used for a period. However, this repairing method needs much maintenance costs. Miura et al. (2000) reported that for a low embankment road on soft clay, when residual settlement is larger than a certain value, the total construction investment (the sum of initial construction investment and maintenance cost) becomes very high.

This paper introduces a new design concept for approach named Column Approach (CA) method to mitigate the settlement difference of road on soft clay. In this technique, the road approach is supported by a row of soil cement columns in a floating system, of which lengths are varied in order to make settlement along approach reduced gradually. A slab is constructed on the top of columns for evening the stress concentration. Slab is made by mixing the in-situ soil with dry cement. Figure 1 illustrates the design concept of CA method. The proposed design gives the calculation method for the extension (L_a) of approach, the length (H_{col}) and spacing (B) of columns, and the thickness (t_s) of slab. FEM is used to calculate stress in slab. The effectiveness of CA is verified by field investigation on several case histories.

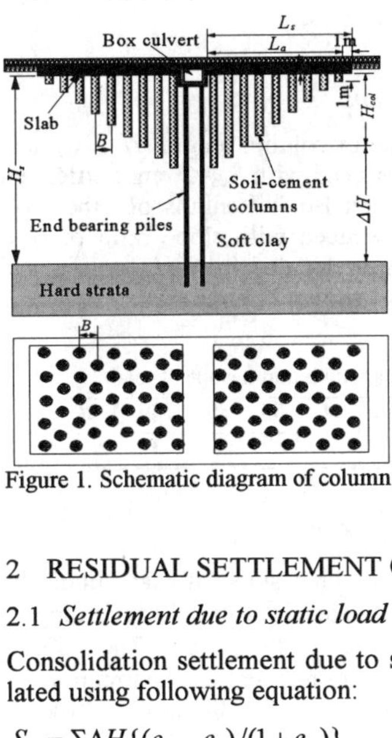

Figure 1. Schematic diagram of column approach method.

2 RESIDUAL SETTLEMENT OF ROAD

2.1 *Settlement due to static load*

Consolidation settlement due to static load is calculated using following equation:

$$S_s = \Sigma\Delta H\{(e_0 - e_1)/(1+e_0)\} \tag{1}$$

where ΔH=thickness of soft clay under column (m); e_0=initial void ratio with respect to in-situ stress p_0; e_1=void ratio with respect to stress level $p_1=p_0+\Delta p_s$; and Δp_s=load of embankment gravity.

2.2 *Settlement due to traffic loading*

Consolidation settlement due to traffic loading is expressed by the equation:

$$S_T = \Sigma \Delta H\{(e_1 - e_2)/(1+e_1)\} \quad (2)$$

where e_2=void ratio with respect to stress level p_2=p_1+Δp_T; Δp_T=equivalent traffic load and its influential depth (Z_T) can be expressed by the following relations (Fujikawa et al. 1996, Miura et al. 2000).

$$\Delta p_T = \exp(4.62 - 0.47 H_{em}) \quad (3)$$

$$Z_T = \exp(2.31 - 0.47 H_{em}) \quad (4)$$

where H_{em}=embankment thickness (m).

Therefore, residual settlement (S_R) is given by:

$$S_R = S_s + S_T \quad (5)$$

3 DETERMINATION OF THE LENGTH OF COLUMNS AND EXTENSION OF SLAB

3.1 *Allowable settlement difference (S_{da}) of CA*

The maximum value for the settlement difference for maintenance is 4 cm according to the guideline of road maintenance (JRA 1978). From practical viewpoint of comfortable vehicle running, 2 cm is adopted as the allowable settlement difference (S_{da}) of CA.

3.2 *Length of columns*

The maximum value of column length (H_{colmax}) next to box culvert is designed with S_{da}=2 cm. Settlement is calculated based on Eq.5. Lengths of other columns are linearly reduced with slope ratio of H_{colmax}/L_a and they can be decided from H_{colmax}, L_a, and column spacing B, as shown in Figure 1.

Table 1. Design speed and maximum slope of road.

Speed (km/h)	30	40	50	60	80	100	120
Slope i (%)	8	7	6	5	4	3	2

3.3 *Maximum slope of CA*

The maximum longitudinal slope of road should be less than the values listed in Table 1, according to guideline of road structure (JRA 1983). Then, 5% of CA slope can be determined as the maximum value because the maximum speed is limited in 60 km/h for the highway of traffic level B, C, D. As illustrated in Figure 2, this value of approach slope (5%) does not obstruct the running of semi-trailer truck, which has the longest axial distance in Japan. Figure 3 illustrates the relationship between clearance under semi-trailer truck and width of box culvert. If setting the minimum width of culvert of 2 m and the maximum CA slope of 5%, the minimum clearance h is 8.4 cm, which is safe enough for semi-trailer truck passing.

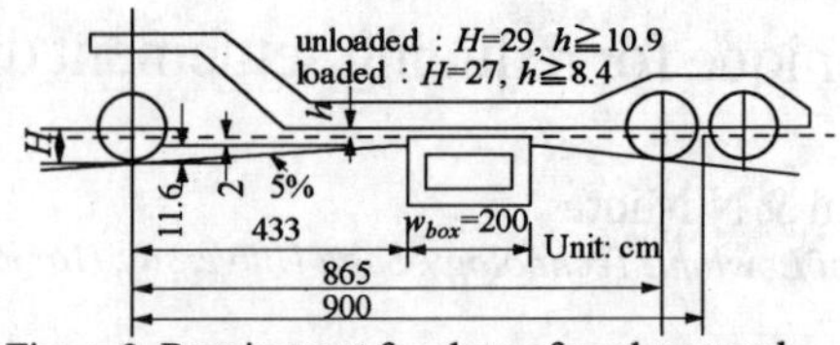

Figure 2. Requirement for slope of road approach.

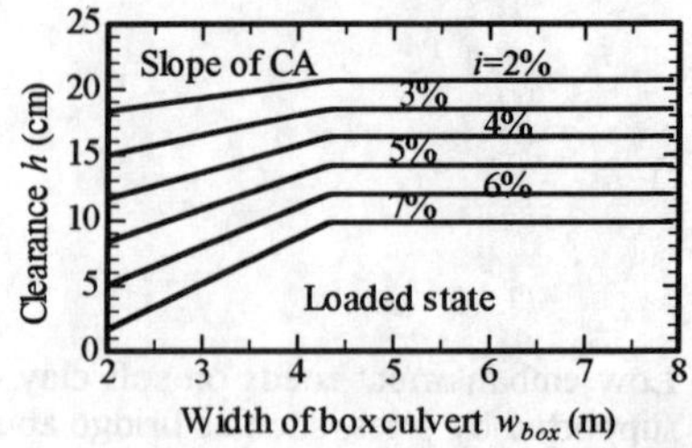

Figure 3. Clearance under semi-trailer bottom.

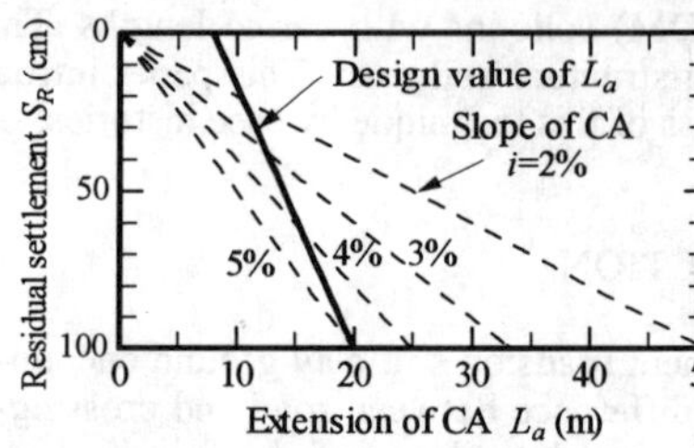

Figure 4. Relationship among CA extension and CA slope.

3.4 *Extension of CA (L_a)*

Since approach is one part of longitudinal curve of road, extension of CA (L_a) can be obtained from CA slope (i) and residual settlement (S_R) as L_a=100S_R/i. If L_a CA is designed as the dotted line in Figure 4, in some cases, L_a becomes very large if residual settlement is larger than 100 cm. This is not economical and CA method cannot be a useful countermeasure. However, in the case residual settlement is less than 10 cm, L_a becomes extremely small. From a practical viewpoint, we adopt a minimum value for L_a of 8 m (JRA 1996) and in case residual settlement is 100cm, the value of L_a is given 20 m. Therefore, the design value of L_a can be expressed by the equation (solid line in Fig. 4):

$$L_a = 8 + (12/100)S_R \quad (6)$$

4 DETERMINATION OF SLAB THICKNESS AND COLUMN SPACING

Road pavement is generally composed of asphalt concrete surface, grain gravel layer and crushed run. The total thickness of pavement is determined by design CBR of subgrade. In CA method, subgrade becomes not uniform due to installation of columns.

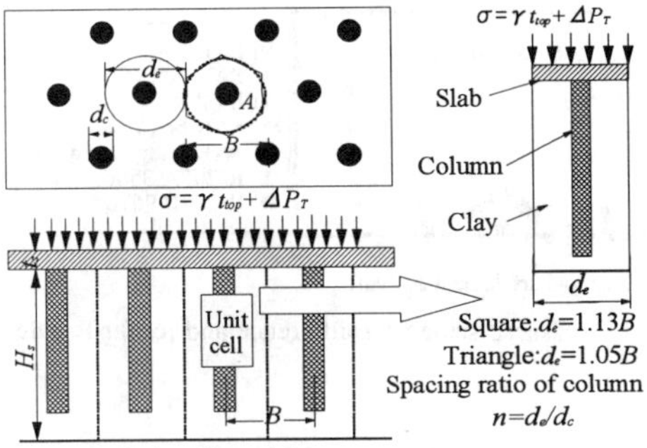

Figure 5. Unit cell model.

In order to reduce stress concentration in pavement and sungrade, slab is placed under the pavement and above columns. Thickness of slab is designed as making the bending stress to be less than bending strength.

4.1 *Unit cell modeling*

In order to obtain bending stresses in slab under pavement gravity and traffic loading, column slab system was simply modeled as a unit cell, as shown in Figure 5. Then, FEM with axi-symmetric stress condition was used to calculate the stresses in slab. Soft clay was modeled as modified Cam-clay and soil-cement column and slab were modeled as elastic material with Mohr-Coulomb criterion. Parameters of clay are listed in Table 2.

Table 2. Parameters used in analysis.

Material	E	γ	ν	λ	κ	M	e_0	p'_c
	MPa	kN/m³						kPa
0~10m		14.0	0.35	0.65	0.065	1.2	3.5	20~50
10~20m		14.0	0.35	0.55	0.055	1.2	2.5	50~90
Slab	200	15.0	0.25					
Column	180	14.0	0.25					

* E=young's modulus; ν=Poisson's ratio; γ=unit weight; λ=virgin loading slope; κ=reloading/ unloading slope; and M=slope of failure line; e_0=initial void ratio; p'_c=yield stress.

Table 3. Load of unit cell.

t_{top} (m)	G_{top}*(kPa)	ΔP_T(kPa)	P_{unit}**(kPa)
0.40	9.6	84.1	93.7
0.80	19.2	69.7	88.9

* G_{top}=gravity of pavement; ** $P_{unit}=G_{top}+\Delta P_T$.

4.2 *Load of unit cell*

The loads on slab surface are gravity of pavement and traffic loading. In FE analysis, equivalent static load is used for traffic loading. The equivalent traffic load Δp_T, which considered the cyclic influence due to several thousands of vehicles passing, is expressed by Eq.3. Table 3 tabulates the equivalent traffic load and the total loads for the thickness of pavement ranged from 0.4 to 0.8 m.

4.3 *Analytical results*

Figure 6 plots the bending stress in slab under the conditions of clay thickness of 10m with end bearing column. If the allowable bending strength of slab σ_b is 120 kPa, thickness of slab can be decided by column spacing. There are following relations between slab thickness and spacing ratio of column.

$$t_{sm}/d_c=\alpha-\exp(\beta-\gamma\cdot n) \tag{7}$$

where t_{sm}=thickness of slab for end bearing columns (m); d_c=diameter of column (m); n=spacing ratio of column; α, β, and γ are the regression constants, as shown in Figure 7, for the given geotechnical and traffic condition, α=1.55, β=1.4, γ=0.72.

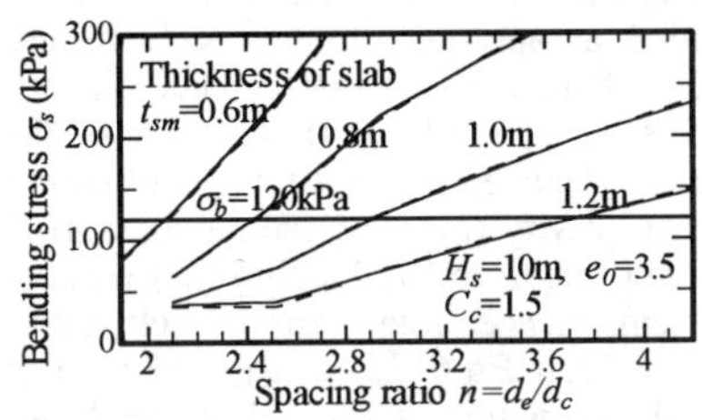

Figure 6. Bending stress in slab.

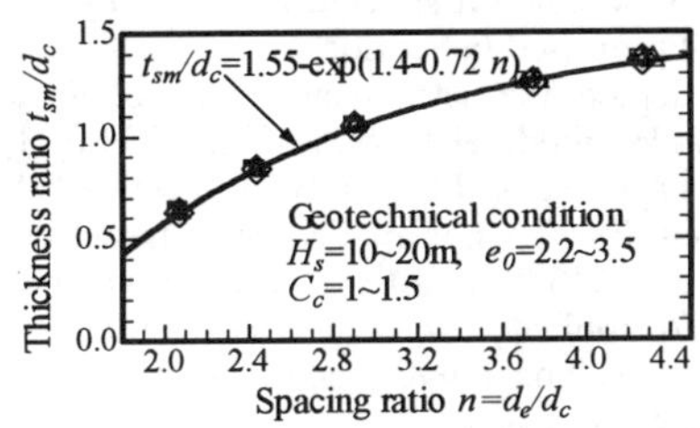

Figure 7. Thickness of slab and spacing ratio of column.

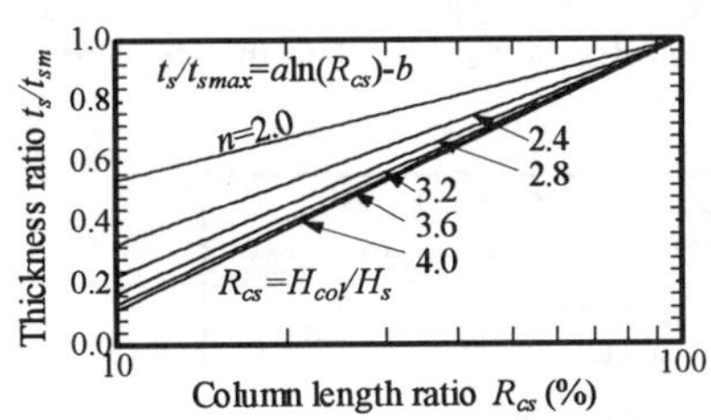

Figure 8. Slab thickness on floating column.

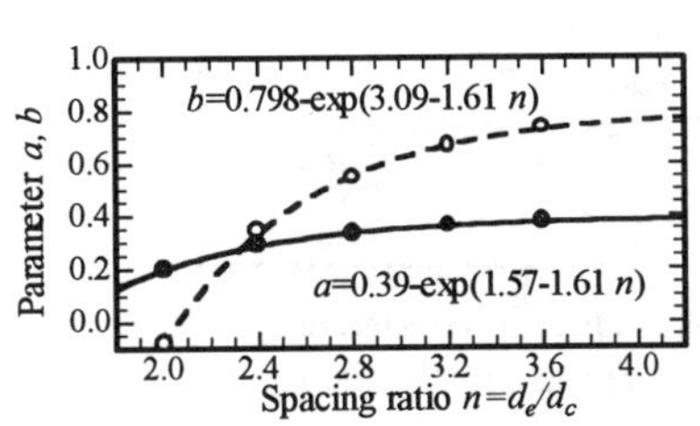

Figure 9. Parameters a and b for slab on floating column.

In case slab is supported by floating columns, stress in slab will become less, and thickness of slab will become thinner. Figure 8 illustrates the relationship between slab thickness ratio (t_s/t_{sm}) and column length ratio ($R_{sc}=H_{col}/H_s$). This relation can be expressed as follows:

$$t_s / t_{sm} = a\ln(R_{cs}) - b \quad (8)$$

where t_s=slab thickness supported by floating columns; t_{sm}= slab thickness supported by end bearing columns; $R_{cs}=H_{col}/H_s$=column depth ratio; a and b are the regression constants, as shown in Figure 9.

5 APPLICATION IN PRACTICE

In order to verify the effectiveness of CA method, it has been applied at approach to box culvert and bridge abutment of the road access to Saga Airport. The design CBR of subgrade of this road was 8% and the thickness of pavement is 40 cm. Settlement due to pavement gravity (S_s) was estimated as 10.3 cm and settlement due to traffic loading (S_T) was estimated as 14 cm. Hence, total residual settlement (S_R) was 24.3 cm. From Eq.6, L_a equals to 10 m. The maximum length of column is about 11m for S_{da}=2 cm. The thickness of slab was estimated as 80 cm with respect to the column spacing B=2.4 m by using the method proposed in Section 4.

Figure 10 depicts the settlement profile of Saga Airport road at longitudinal section of No.279+16.3, where CA was applied, and No.262+15.8, where CA was not applied. For CA not applied case, large settlement difference (13.7 cm after 560 days) was observed. For CA applied case (Fig. 10a), settlement difference between box culvert and road approach was less than 2 cm as expected.

CA method was also applied at more than 30 places of structure and highway conjunction on Saga Plain. Field investigation shows that this technique

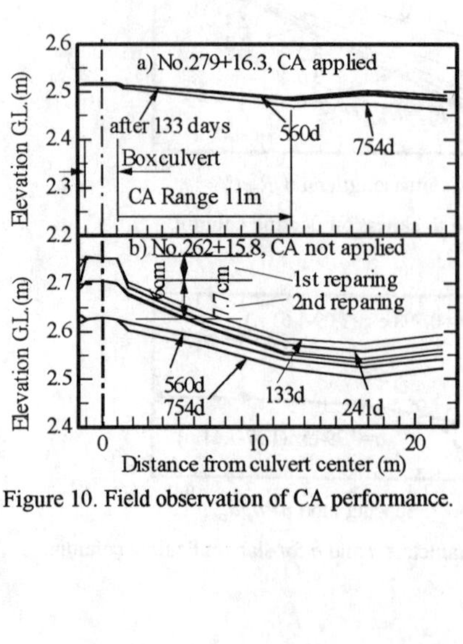

Figure 10. Field observation of CA performance.

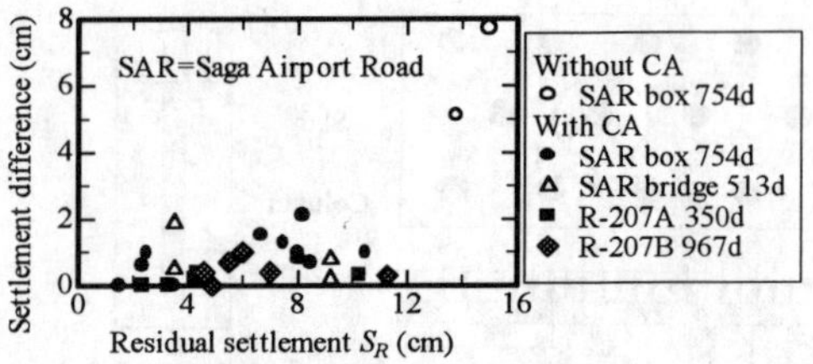

Figure 11. Measured settlement difference and residual settlement.

is very effective. Figure 11 shows the measured settlement difference between structure and road approach. In all cases, if CA was applied, settlement differences were less than 2 cm.

6 CONCLUSIONS

This paper proposed a practical countermeasure with its design method for reducing settlement difference between pile supported structures and road on soft clay. This is a useful technique through applying deep mixing method on engineering practice. It can be concluded as follows.

1. The maximum length of column is decided by controlling settlement difference between structure and road less than 2cm.
2. The extension of CA can be decided by controlling its maximum slope less than 5%, which keeps normal vehicle running.
3. The thickness of slab can be calculated by using FEM. It is decided based on allowable bending stress, traffic condition, column spacing, and ratio between column length and thickness of soft clay.
4. Field investigation results show that this technique is effective on Ariake clay in Saga plain, Japan. The settlement differences of all case histories by this technique were less than 2 cm as expected.

REFERENCES

Fujikawa, K., Miura, N. & Beppu, I. 1996. Investigation of the low embankment road and prediction of the consolidation settlement due to traffic loading. *Soils and Foundations.* 36(4): 147-152.

JRA (Japan Road Association). 1978. *Guideline of Road Maintenance*. Tokyo. (in Japanese).

JRA (Japan Road Association). 1983. *Guideline of Road Structure*. Tokyo. (in Japanese).

JRA (Japan Road Association). 1992. *Guideline of Asphalt Pavement*. Tokyo. (in Japanese).

JRA (Japan Road Association). 1996. *Guidline of Road and Bridge Design*. Tokyo. (in Japanese).

Miura, N., Fujikawa, K., Sakai, A. & Hamatake, A. 2000. Construction investment of the highway on soft ground based on the field full-scale test. *JSCE*. in press.

Computer Methods and Advances in Geomechanics, Desai et al. (eds) © 2001 Balkema, Rotterdam, ISBN 90 5809 183 X

Analysis of piled-pavement system

S.C.Wong & H.G.Poulos
Department of Civil Engineering, University of Sydney, N.S.W., Australia

ABSTRACT: The application of piles provides a convenient means of improving subgrade conditions for road constructions, especially in areas with a soft soil profile. The success of the application of piles in road construction requires understanding of pile behaviour characteristics under traffic loading from the pavement. The three-dimensional nature of a pile-supported road system requires it to be analysed as a three-dimensional problem. In this paper, a three-dimensional finite difference analysis (FLAC^{3D}) has been used to study the characteristics of pile behaviour in pavements and the factors which influence this behaviour. A simplified approach to predict piled pavement response has also been used, and compared with the three-dimensional finite difference analysis. The two approaches show reasonable agreement, and the simplified method generally gives a conservative prediction of piled pavement response. It therefore appears to hold promise as a means of analysing and designing piled pavements.

1 INTRODUCTION

In the design of high performance roads, highways, trunk roads and airport runways over soft soil profiles, it is often essential that the ground is physically or chemically improved. Various methods are available, and the use of piles often provides a cost-effective alternative, particularly in countries such as Thailand, Malaysia, Singapore and Indonesia. One of the advantages is that lesser quantities of imported and expensive granular pavement materials are required, whereas piles are relatively economical.

Ground improvement methods by pile inclusion to physically improve a weak soil profile and strengthen in-situ pavement systems have been recommended by Kjellman (1940) and Eide (1968). Barry et al. (1992a,b) described design considerations for pile-supported road system in East Sumatra, Indonesia. Construction of the A2 London-Rochester-Dover Road in England involved the inclusion of piles as one of the ground improvement schemes (Clarke & Johnson 1971). The rehabilitation of the Thon Buri-Pak Tho Highway in Thailand also involved pile inclusions as a ground improvement method (Ruenkrairergsa 1998).

Pichumani et al. (1974) conducted linear elastic finite element analyses of a piled pavement system. It was concluded that pile inclusions in a road system would decrease surface deflection, decrease the sub-grade stresses, and reduce the horizontal tensile stress in the pavement. However, it has to be noted that the Pichumani et al. (1974) analyses were undertaken primarily to illustrate the advantages of using pile inclusions for pavement construction over a soft soil profile. The application and suitability of assuming the soil block used in the author's analyses with a stress-free boundary was questioned (Poulos et al. 1975). Since then, very little research appears to have been carried out on piled pavement behaviour.

With the increased use of pile inclusions in pavement construction, a more complete understanding of piled pavement behaviour is desirable, and a more reliable method requires to be developed to estimate the response of a piled pavement system to wheel loading. The aim of this paper is to examine in some detail the behaviour of piled pavements via a full three-dimensional analysis, and then to assess the potential for simplified methods of analysis to be used in design.

2 TYPICAL CASE

The typical case examined in this study is shown in Figure 1. The overall system is assumed to behave in a linear-elastic manner. In a piled pavement system, the actual width and length of the pavement structure is large and hence, to simplify the analysis, only a section of piled pavement structure is considered, as illustrated in Figure 2. In the analysis, the pavement structure is idealised as a single layer system although in practice, pavement structures normally consist of two or more layers of different materials. The underlying soil is also assumed to be a single layer.

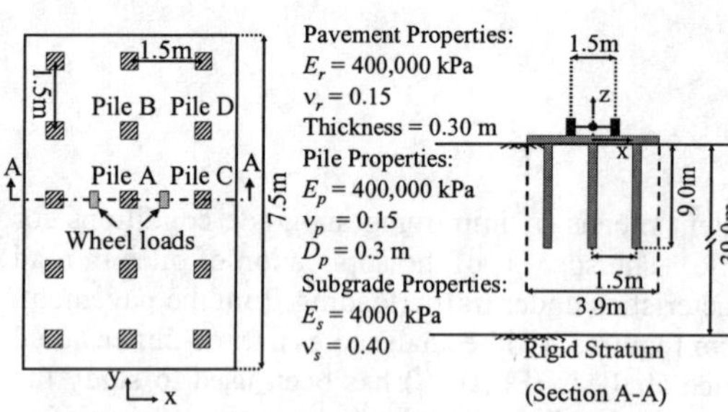

Figure 1. Geometry and property of the typical case.

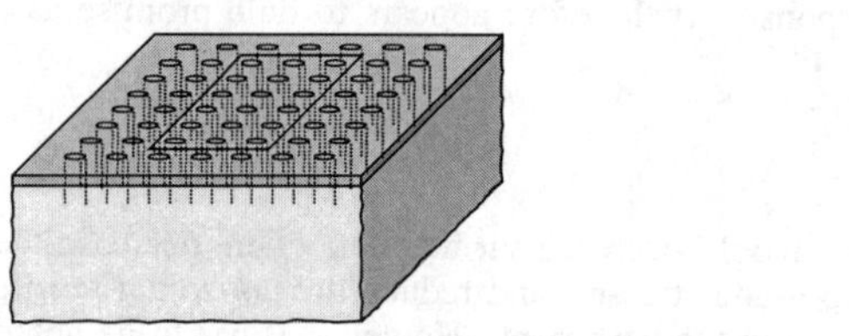

Figure 2. A section of piled pavement system analyzed.

In this study, both vertical and horizontal wheel loads are considered as detailed in Table 1. The application of horizontal wheel load is to simulate cornering manoeuvres. Vertical wheel load has been applied on the piled pavement system in one case, while both vertical and horizontal wheel loads have been applied in a second case.

Table 1. Loading cases analyzed

Case number	Loading conditions	
	Vertical kN	Horizontal kN
1	171.0	0.0
2	171.0	17.1

3 WHEEL LOAD CHARACTERISTICS

In this study, wheel loads are assumed to act as uniformly distributed pressures over a square area. These assumptions are taken to simplify the analysis. Similar assumptions have been adopted in many research works on pavement response, for instance, Chou (1983) and Lau et al. (1994).

A single axle-single wheel was assumed to load the piled pavement structure with the centre-to-centre wheel distance assumed to be 1.5m. The wheel-pavement contact area was taken as 0.09 m^2 with a contact pressure of 950 kPa, which would represent a heavy truck wheel fitted with a newly developed wide base tire.

It is well known that vertical wheel load is the main criterion in any pavement design. However, the importance of horizontal wheel loads in pavement design has also been identified (Lau et al. 1994). The source of horizontal wheel load induced on pavements is mainly due to braking and cornering manoeuvres of a vehicle. The horizontal wheel load due to cornering is given by the product of lateral acceleration and vertical wheel load. AASHTO (1977) guidelines recommended that highways should be designed for lateral acceleration and superelevation with the total coefficient for the two criteria being about 0.1. As the above-mentioned objective of this study does not involves detailed investigation, only the horizontal wheel load contributed by cornering is considered to assess the potential of the simplified approach for design purposes. Wheel loads were assumed to be applied as static loadings on the piled pavement system in order to facilitate the analysis for the present study.

4 MODELING OF PILED PAVEMENT

4.1 *Three-dimensional finite difference (FLAC^{3D}) analysis*

FLAC^{3D} (Fast Lagrangian Analysis of Continua) is a commercially-available three-dimensional explicit finite difference code (Itasca 1997). FLAC^{3D} was used for the benchmarking purpose due to the capability of the code to model three-dimensional geotechnical problems with several built-in constitutive laws. In the present analyses, however, only an elastic model for soil behaviour was considered.

4.1.1 *Model geometry*

In the modeling, polyhedral grids of solid elements were used to represent the soil medium, the piles and the road pavement. The aspect ratio of each element close to the loaded area was maintained below 3.0. However, this condition was not strictly maintained for the elements further away from wheel load area, in order to minimize computer memory requirements and computational time.

The pavement was modeled as a layer on top of the subgrade surface, with a free edge. With such modeling, vertical friction and passive load induced from soil adjacent to the pavement edge was not present. Piles were modeled with a square cross section. In Case 1, the piled pavement system was modeled as a quadrant in FLAC^{3D}, while in Case 2, half of the system was modeled, to allow horizontal loadings to be analyzed.

Boundary conditions for the finite difference (FLAC^{3D}) model have a significant influence on the predicted response, especially for an elastic analysis. In FLAC^{3D} modeling, all boundaries were fixed from moving horizontally on the vertical planes and fixed from moving vertically on the horizontal plane. This was done by fixing the velocity of the boundaries in FLAC^{3D} modeling. In this analysis, the influence of 'truncating' boundaries was assumed to be minimal if they were located at least 20 metres away from the wheel load region. In the analyses carried out, the 'truncating' boundaries were located 30 metres away from the region where detailed analysis was required.

4.1.2 *Preliminary finite difference results*

In any numerical analysis, the nature of spatial discretisation adopted can affect the accuracy of the numerical solution. Hence, different mesh discretisations were investigated to determine a suitably economical mesh which would provide reasonably accurate solutions. In general, a fine mesh was employed for the region where detailed analysis was desired, and a coarser mesh was used for other regions.

To facilitate the choice of appropriate mesh for the study, the geometry of the typical case was maintained, and the Young's Modulus of the piles and pavement was chosen to be 40,000kPa. Results for the two different mesh refinements used are shown in Figure 3 and Figure 4. The coarser mesh was composed of about 45,000 solid elements while the finer mesh consisted of about 125,000 solid elements. Computation time for the coarse mesh was about 5 hours, while the fine mesh took about 50 hours.

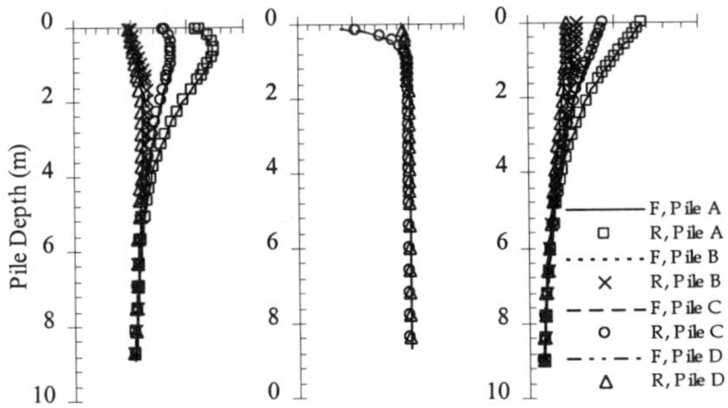

Note: F represents fine mesh and R represents rough mesh.

Figure 3. Distribution of pile responses with depth for different mesh refinements in FLAC^{3D}.

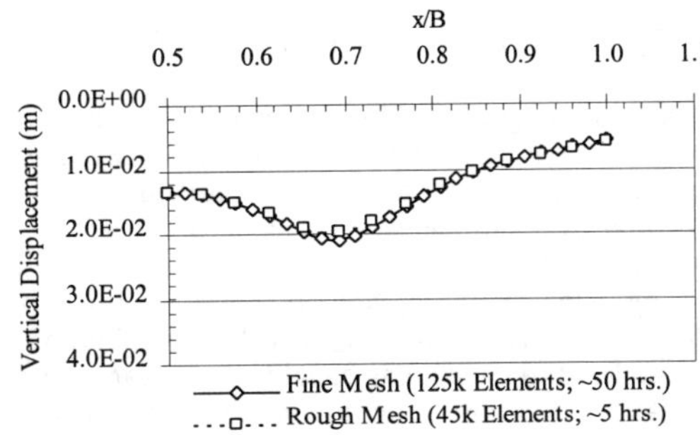

Figure 4. Distribution of pavement displacement for different mesh refinements in FLAC^{3D}.

The main characteristics of piled pavement response are detailed as follows;

- Figure 3(a) shows the distribution of pile axial stress along the length of the piles. It shows that the maximum pile axial stress is at some distance below pile head.
- Figure 3(b) shows the bending moment in the piles in the x-direction. It is obvious that the maximum bending moment induced in the piles is at the pile head.
- Figure 3(c) shows the distribution of pile vertical displacement with pile depth. It is found that the maximum pile vertical deformation is at the pile head.
- Figure 4 shows the transverse distribution of average vertical displacement of pavement, and indicates that the maximum vertical displacement is located under the wheel load.

From the comparisons between the finer and coarser meshes as illustrated in Figure 3 and Figure 4, it is obvious that the coarser mesh provides good agreement with the prediction obtained from the finer mesh. Subsequently, therefore the coarser mesh has been used for further analyses.

In FLAC^{3D} analysis, the bending moments in the piles have been obtained by taking the product of double derivatives of the average lateral pile deformation obtained at nodal points and E_pI_p, where E_p represents Young's Modulus of piles and I_p represents second moment of area of piles. Similarly, the tensile stresses developed at the bottom of the pavement structure were derived from double derivatives of vertical pavement deformation. In obtaining numerical double derivatives for lateral deformation in piles and vertical deformation in pavement, appropriate finite difference expressions were used (Bowles 1974).

4.2 *Simplified approach*

The simplified approach adopted in this study is based on a piled raft analysis via the programs PI-

RAF (Ta 1996) and APRAF (Zhang 2000). PIRAF and APRAF utilise a modified finite layer method to analyse the subgrade. The pavement structure is represented as a thin elastic plate and the analysis is based on a simplified finite element method. Piles are treated as one-dimensional elements and are assumed to be circular in cross section. Since the piles were modeled as a square in cross section in FLAC^{3D}, the circular piles in PIRAF and APRAF were modeled to have areas equal to the square piles in FLAC^{3D}; the corresponding equivalent pile diameter was 0.338m.

In this study, PIRAF was used to predict axial pile response and pavement response while APRAF was used to predict lateral pile response. In the simplified approach, the tensile stress at the bottom of the pavement was obtained from the computed bending moment attained in the plate.

5 COMPARISONS BETWEEN FLAC^{3D} AND SIMPLIFIED METHOD

In the comparisons between FLAC^{3D} and the simplified approach, two cases have been chosen, with variations in loading conditions being considered, while maintaining the other factors as shown in Figure 1. The loading details for each case are tabulated in Table 1.

5.1 *Comparisons for Case 1*

Pavement deformation comparisons across section A-A (see Fig. 1) are illustrated in Figure 5. There is reasonable agreement between the FLAC^{3D} and the simplified approach, with the simplified approach overestimating pavement deformation (by 10% - 20%) compared to the FLAC^{3D} analysis. Both FLAC^{3D} and the simplified approach showed similar trends in the pavement deformation profile.

The comparison of the computed profiles of tensile stress at the bottom of pavement structure is shown in Figure 6. The simplified approach showed good agreement with FLAC^{3D} in predicting tensile stresses, especially the maximum tensile stress.

Figure 7 shows the computed pile vertical deformations and again indicates that the simplified approach over estimates the pile deformation by 10% to 20% compared to FLAC^{3D} analysis.

Generally, comparisons of axial load distribution and bending moment distribution in Pile A and B show relatively good agreement as illustrated respectively in Figure 8 and Figure 9. Figure 8 shows that the maximum axial pile load develops at the top of Pile A, while Pile B has a maximum axial pile load at some distance below pile head. In the simplified approach, the estimated maximum axial pile load location in Piles B is at a slightly greater depth than in the FLAC^{3D} analysis.

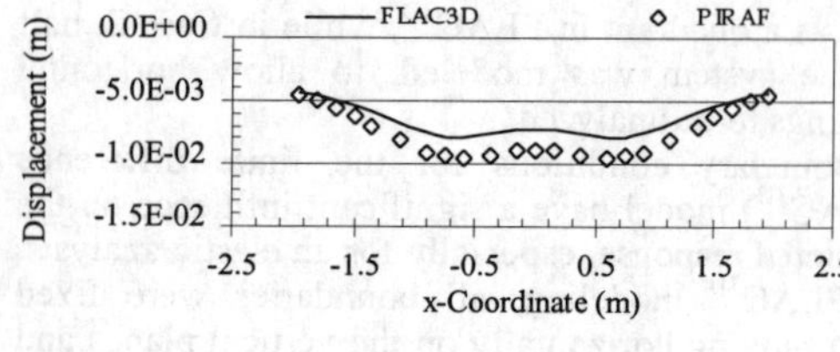

Figure 5. Comparison of pavement vertical displacement.

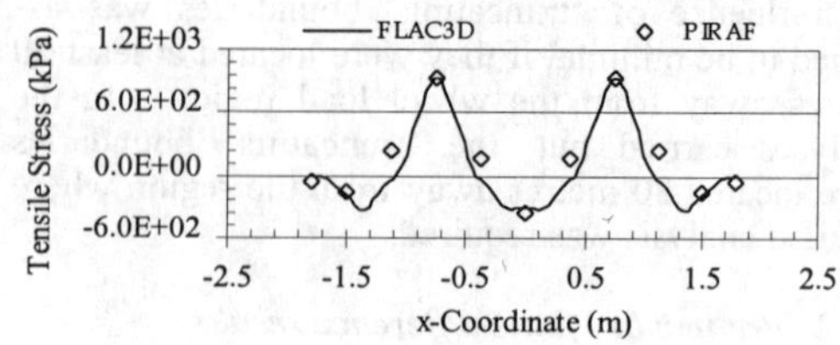

Figure 6. Comparison of transverse tensile stress at the bottom of pavement.

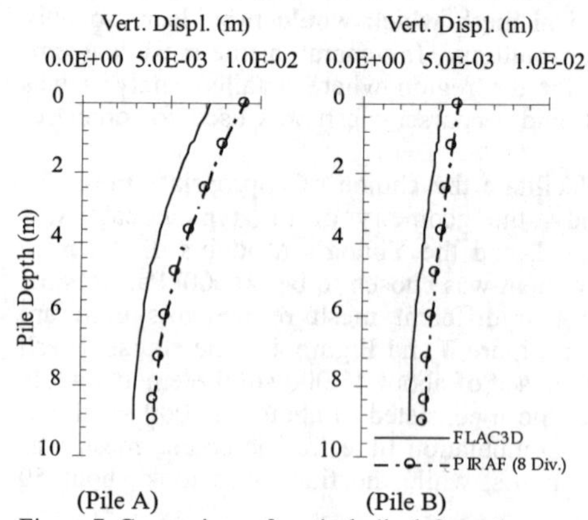

Figure 7. Comparison of vertical pile deformation.

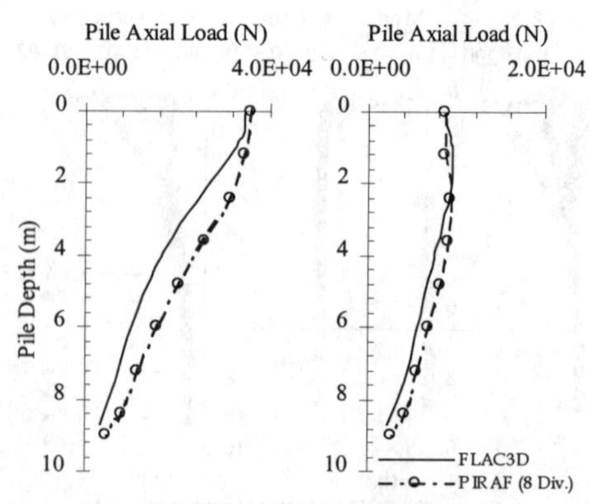

Figure 8. Comparison of axial pile load.

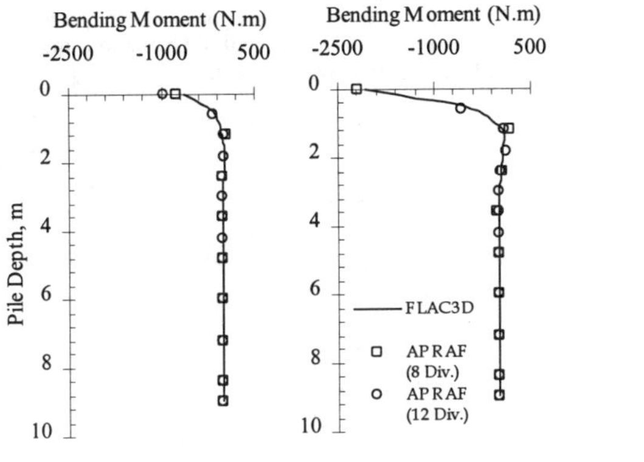

Figure 9. Pile bending moment (x-dir.) comparisons (Case 1).

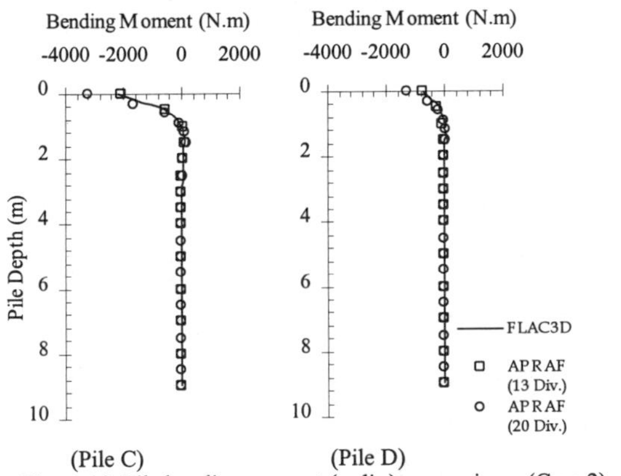

Figure 10. Pile bending moment (x-dir.) comparisons (Case 2).

Figure 9 shows that the maximum bending moment in the piles develops at the pile head. Generally, the simplified approach estimates higher bending moment compared to FLAC^{3D} (by 15% - 25%). This may arise because of the assumptions of thin-plate bending theory in the simplified approach. The computing time taken for Case 1 by the different computer codes is tabulated in Table 2. The simplified analyses were substantially quicker than the FLAC^{3D} analysis.

Table 2. Computing time of different computer codes

Case no.	FLAC^{3D}	PIRAF	APRAF
	hrs.	hrs.	hrs.
1	~5	~0.33	~0.85
2	~50	-	~0.85

5.2 *Comparisons for Case 2*

In this case, the only difference compared to Case 1 is the addition of horizontal load in the x-direction. The introduction of horizontal load into the piled pavement system does not cause a significant difference in axial pile response, but the lateral pile response experiences some changes. Similar trends to Case 1 were found for pavement vertical deformation and pavement tensile stresses developed at the bottom of pavement. However, the shear stress at the pavement-soil interface does experience significant changes. It has been shown that the shear stress effect on bending behaviour in a pavement structure is small for slabs on an elastic foundation (Frederick 1955). Since the scope of this study is focused on determining pavement performance and not on the interface stresses, the interface stress characteristics are not presented herein.

Figure 10 shows the induced bending moment profile in the x-direction in Piles C and D. Generally, Figure 10 shows relatively good agreement between FLAC^{3D} and the simplified approach in estimating bending moment in piles due to application of both vertical and horizontal wheel loads. The consideration of horizontal loading causes an appreciable increase in pile bending moment compared to the case of vertical loading only. It has to be noted that, in Case 1, Piles A and B developed insignificant bending moment, however, in Case 2, the induced bending moment in Piles A and B is significant (about 150 N.m to 200 N.m).

Computing time taken in Case 2 by the different computer codes is tabulated in Table 2. The substantial saving in time by using the simplified methods rather than FLAC^{3D} is clearly demonstrated.

6 CONCLUSIONS

Comparisons have been made between the FLAC^{3D} analysis and the simplified approach based on a piled raft analysis. Both methods show that the maximum pavement vertical deformation and the maximum tensile stress at the bottom of piled pavement are developed directly beneath wheel load. The FLAC^{3D} and the simplified approach analyses show that the maximum pile vertical displacement is at pile head while the maximum pile axial load is at a small distance below ground surface. However, the location of maximum pile axial load may be closer to the pile head in a stiffer pavement configuration. Maximum bending moments induced in the piles are at the pile head but the bending moments diminish rapidly with depth over a short length below the pile top.

The simplified approach shows good agreement with the FLAC^{3D} analysis and generally, the simplified approach provides a conservative prediction of the behaviour of a pile-supported road system. Hence, there are promising indications that the simplified piled raft approach can be used to predict the behaviour of a piled pavement system, and that it can provide a more convenient design tool than the $FLAC^{3D}$ analysis.

ACKNOWLEDGEMENT

The work presented in this paper forms part of a project on 'The use of piles in construction of pavements and embankments' which is supported by a grant from The Australian Research Council. The authors would like to thank Assoc. Prof. John C. Small and H.H. Zhang for fruitful discussions. The first author is grateful to DEETYA for providing financial aids via Overseas Postgraduate Research Scholarships and International Postgraduate Awards.

REFERENCES

American Association of State Highway and Transportation Officials 1977. *Geometric design for resurfacing, restoration and rehabilitation of highways and streets*. Washington D.C.: AASHTO.

Barry, A.J., Brady, M.A. & Younger, J.S. 1992a. Road on peat in East Sumatra. *Geotechnical Engineering* 23(2): 145-161.

Barry, A.J., Trigunarsyah, B., Symes, T. & Younger, J.S. 1992b. Piling for roads over peat in Sumatra. *Intern. Conf. on Geotechnical Engineering, Malaysia, April 1992*: 487-492.

Bowles, J.E. 1974. *Analytical and computer methods in foundation engineering*. McGraw-Hill.

Chou, Y.T. 1983. Subgrade contact pressures under rigid pavements. *Journal of Transportation Engineering* 109(3): 363-379.

Clarke, A.C. & Johnson, B.W. 1971. Comparative designs, performance and costs for a road on peat. *Journal of the Institution of Highway Engineers* 18(3): 25-30.

Eide, O. 1968. *Geotechnical problems with soft Bangkok Clay on the Nakhon Sawan Highway Project*. Publication No. 78. Norwegian Geotechnical Institute.

Frederick, D. 1955. Thick rectangular plates on an elastic foundation. *Trans. ASCE* 122: 1069-1085.

Itasca Consulting Group, Inc. 1997. $FLAC^{3D}$ (Version 2.0) User's Manual.

Kjellman, W. 1940. Calculation of embankment piles (in Swedish). *Tekn. Ts. V.O.V.* 70.

Lau, C.M., Fwa, T.F. & Paramasivam, P. 1994. Interface shear stress in overlaid concrete pavements. *Journal of Transportation Engineering* 120(3): 163-177.

Pichumani, R., Crawford, J. & Triandafilidis, G.E. 1974. Finite element analysis of pile-supported pavement structures. *Journal of Transportation Engineering* 100(TE2): 293-303.

Poulos, H.G., Brown, P.T. & Wiesner, T.J. 1975. Finite element analysis of pile-supported pavement structure (Discussion). *Journal of Transportation Engineering* 101(TE2): 399-401.

Ruenkrairengsa, T. 1998. Recent ground improvement works for highways in Thailand. *13th Southeast Asian Geotechnical Conference, Taipei, Taiwan ROC, 16-20 Nov. 1998*: 17-43.

Ta, L.D. 1996. *A finite layer method for analysis of pile groups, rafts and piled raft foundations in layered soils*. Ph.D. Thesis. School of Civil and Mining Engineering, Univ. of Sydney, Australia.

Zhang, H.H. 2000. *Finite layer method for analysis of piled raft foundations*. Ph.D. Thesis. School of Civil and Mining Engineering, Univ. of Sydney, Australia.

14 Foundations, footings, piles

Computer Methods and Advances in Geomechanics, Desai et al. (eds) © 2001 Balkema, Rotterdam, ISBN 90 5809 183 X

Variational approach for the analysis of pile groups and piled raft foundations

Y.K.Chow & K.Y.Yong
Department of Civil Engineering, National University of Singapore, Singapore

W.Y.Shen
NTU-PWD Geotechnical Research Center, Nanyang Technological University, Singapore

ABSTRACT: A variational approach for the analysis of vertically loaded pile groups and piled raft foundations is described. The raft and pile group is both analyzed by the use of the principle of minimum potential energy. The pile displacements and raft deflections are represented by finite series and the solution procedure is very efficient for this class of three-dimensional problems. The accuracy of the variational approach is validated by comparisons with published solutions of other numerical methods.

1 INTRODUCTION

A number of methods are available for the analysis of vertically loaded pile groups and piled raft foundations. The boundary element method for the analysis of pile groups was developed by Butterfield and Banerjee (1971) and Poulos and Davis (1980). In this method, the soil is modelled as an elastic half-space and its continuity can be taken into account. An approximate analytical method in which the soil was modelled using theoretical load-transfer curves was presented by Randolph and Wroth (1979). The hybrid method that combined the elastic half-space and load-transfer soil models was reported by O'Neill et al. (1977) and Chow (1986) for pile groups and by Clancy and Randolph (1993) for piled raft foundations. The application of the finite element method to pile groups can also be seen among the published literature (Ottaviani, 1975 and Chow, 1987). The finite layer method for the analysis of pile groups and piled raft foundations was reported by Ta and Small (1996).

In recent years, a variational approach has been developed for the analysis of pile groups and rafts by the authors (Shen, Chow and Yong, 1999a, 1999b, 2000). This method makes use of the principle of minimum potential energy. Finite series are introduced to simulate the displacement and shear stress distribution of the piles in a group, and the deflection of a raft. The discretization of the pile shafts into a number of elements is no longer required and the approach is very efficient, especially for large pile groups. A surface stiffness that relates the load-settlement relationship at the pile head-raft-soil interface can be developed and can be easily integrated with the raft, making available an efficient and complete solution of a piled raft foundation.

This paper outlines the main solution procedures of the variational approach for the analysis of pile groups and the extension of the solution procedure to piled raft foundations. Comparisons with other numerical methods are presented to demonstrate the validity of the present method.

2 METHOD OF ANALYSIS

2.1 *Analysis of pile groups*

In a general form, the potential energy of a pile group (Fig.1) can be expressed as

$$\pi_p = U_p + \tfrac{1}{2}\iint\limits_S \{\tau_z\}^T\{w_z\}ds + \tfrac{1}{2}\iint\limits_A \{\sigma_b\}^T\{w_b\}dA - \{w_t\}^T\{P_t\} \tag{1}$$

The first term U_p is the strain energy of the piles in the group. The second and third terms are the work done by shear stresses $\{\tau_z\}$ at the pile shaft and normal stresses $\{\sigma_b\}$ at the pile base, respectively. $\{w_z\}$ and $\{w_b\}$ are the corresponding displacements of the piles in the group, and S and A are the surface area and cross-sectional area of each pile in the group, respectively. The last term is the work done by load $\{P_t\}$ acting on the pile head, and $\{w_t\}$ is the corresponding displacements of the piles at the pile head.

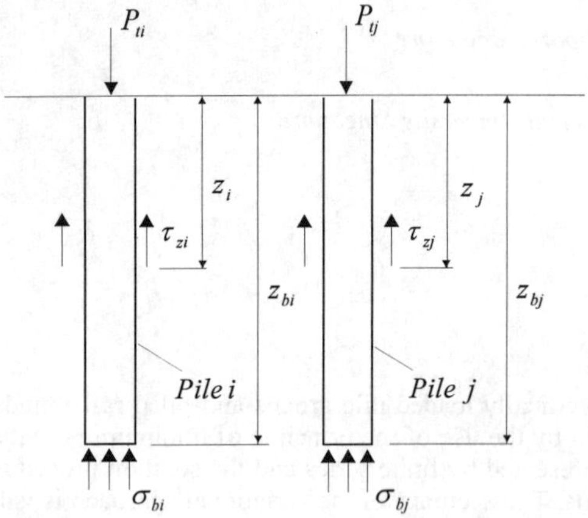

Figure 1 Stresses acting on pile *i* and pile *j* of a pile group

The potential energy expression can be further simplified when Gauss integration is introduced to perform the numerical integration with respect to the piles. In this way, eventually, it can be written as

$$\pi_p = U_p + \tfrac{1}{2}\{P_p\}^{\mathrm{T}}\{w_p\} - \{w_t\}^T\{P_t\} \tag{2}$$

where $\{P_p\}$ are the forces acting at the pile-soil interface but represented in terms of the equivalent forces acting at the Gauss points along the pile shaft and the forces at the pile base, and $\{w_p\}$ are the corresponding displacements. The relationship between $\{P_p\}$ and $\{w_p\}$ that relies on the soil can be expressed as

$$\{P_p\} = [k_s]\{w_p\} \tag{3}$$

where $[k_s]$ is the soil stiffness matrix for group piles. This soil stiffness can be obtained for pile groups embedded in an elastic half-space using Mindlin's equations. Thus, the total potential energy can be expressed as

$$\pi_p = U_p + \tfrac{1}{2}\{w_p\}^{\mathrm{T}}[k_s]\{w_p\} - \{w_t\}^T\{P_t\} \tag{4}$$

A finite series is introduced to represent the displacements of the piles along the pile shaft, which can be written as

$$w_{zi} = \sum_{j=1}^{k_1} \beta_{ij}\left(1 - \frac{z_i}{l}\right)^{j-1}, \; (i = 1,2 \ldots np) \tag{5}$$

where np is the number of the piles in the group. The principle of minimum potential energy then requires that π_p be an extremum with respect to the admissible displacement field characterised by β_{ij}. Hence

$$\frac{\partial \pi_p}{\partial \beta_{ij}} = 0, \quad (i = 1,2 \ldots np), (j = 1,2 \ldots k_1) \tag{6}$$

The potential energy can then be reduced to the following expression

$$\frac{\partial U_p}{\partial \beta_{ij}} + \left\{\frac{\partial w_p}{\partial \beta_{ij}}\right\}^{\mathrm{T}} [k_s]\{w_p\} = \left\{\frac{\partial w_t}{\partial \beta_{ij}}\right\}^T \{P_t\},$$

$$(i = 1,2 \ldots np), (j = 1,2 \ldots k_1) \tag{7}$$

The matrix form of the above equation can eventually be written as

$$[h]\{\beta\} = \{P_T\} \tag{8}$$

where $[h]$ is the coefficient matrix reflecting the pile-soil system, $\{\beta\}$ is a vector made up by a set of the undetermined coefficients of the finite series, and $\{P_T\}$ is a vector related only to the load acting at the pile head. The inverse of the above equation gives

$$[h]^{-1}\{P_T\} = \{\beta\} \tag{9}$$

It is noted that proper addition of rows and columns of the matrix in equation (9) can be made and thereafter the equation is transformed into the load-deformation relationship of group piles at the pile heads and is given by

$$[f_t]\{P_t\} = \{w_t\} \tag{10}$$

in which $[f_t]$ is the flexibility matrix obtained through the proper addition of rows and columns of the matrix $[h]^{-1}$. The stiffness matrix equation is then

$$\{P_t\} = [f_t]^{-1}\{w_t\} \tag{11}$$

The above single stiffness matrix relates the loads and displacements at the pile heads only.

2.2 *Interaction between ground surface and pile groups*

The precursor to the analysis of piled raft foundations is the consideration of the interaction between the load acting on the ground surface and the pile groups (Fig.2). The basic principle is similar to that for pile groups described above. However, in the present case, the work done by the forces acting on the ground surface needs to be taken into account. The potential energy can be expressed as

$$\pi_p = U_p + \tfrac{1}{2}\iint\limits_S \{\tau_z\}^T\{w_z\}ds + \tfrac{1}{2}\iint\limits_A \{\sigma_b\}^T\{w_b\}dA + \tfrac{1}{2}\iint\limits_B \{p_s\}^{\mathrm{T}}\{w_s\}dB - \{w_t\}^{\mathrm{T}}\{P_t\} \qquad (12)$$

The additional term is the work done by the stress $\{p_s\}$ acting on the ground surface at the raft interface. $\{w_s\}$ is the settlement of the ground surface, and B denotes the area of the ground surface where the stress $\{p_s\}$ is applied.

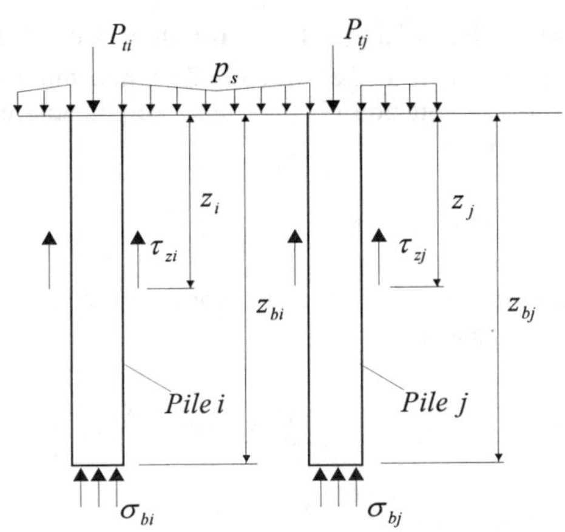

Figure 2 Stresses acting on group piles and ground

The potential energy expression can be simplified as

$$\pi_p = U_p + \tfrac{1}{2}\{w_p\}^{\mathrm{T}}\{P_p\} + \tfrac{1}{2}\{w_s\}^T\{P_s\} - \{w_t\}^T\{P_t\} \qquad (13)$$

where $\{P_p\}$ and $\{w_p\}$ are the forces and displacements at the pile-soil interface, respectively, which are the same as those in equation (2). $\{P_s\}$ and $\{w_s\}$ are the forces and settlements at the ground surface, respectively. The relationships between $\{P_p\}$ and $\{w_p\}$, and $\{P_s\}$ and $\{w_s\}$ in equation (13) can be written as

$$\begin{Bmatrix} P_p \\ P_s \end{Bmatrix} = \begin{bmatrix} k_{spp} & k_{sps} \\ k_{ssp} & k_{sss} \end{bmatrix} \begin{Bmatrix} w_p \\ w_s \end{Bmatrix} \qquad (14)$$

where $[k_{spp}]$, $[k_{sps}]$, $[k_{ssp}]$ and $[k_{sss}]$ are the soil stiffness matrices. These soil stiffness matrices can be solved for pile groups embedded in soil modelled as an elastic half-space using Mindlin's solution (Shen, Chow and Yong, 2000). Thus, the potential energy expression can eventually be obtained as

$$\pi_p = U_p + \tfrac{1}{2}\{w_p\}^{\mathrm{T}}[k_{spp}]\{w_p\} + \{w_p\}^T[k_{sps}]\{w_s\} + \tfrac{1}{2}\{w_s\}^T[k_{sss}]\{w_s\} - \{w_t\}^T\{P_t\} \qquad (15)$$

Applying the principle of minimum potential energy to equation (15), $\dfrac{\partial \pi_p}{\partial \beta_{ij}} = 0$, gives

$$\frac{\partial U_p}{\partial \beta_{ij}} + \left\{\frac{\partial w_p}{\partial \beta_{ij}}\right\}^{\mathrm{T}}[k_{spp}]\{w_p\} + \left\{\frac{\partial w_p}{\partial \beta_{ij}}\right\}^T[k_{sps}]\{w_s\} = \left\{\frac{\partial w_t}{\partial \beta_{ij}}\right\}^T\{P_t\}$$

$$(i = 1,2\ldots np),\ (j = 1,2\ldots k_1) \qquad (16)$$

The matrix form of the above equation can be written as

$$[k_{pp}]\{\beta\} + [k_{ps}]\{w_s\} = \{P_T\} \qquad (17)$$

in which $[k_{pp}]$, $[k_{ps}]$ are the matrices reflecting pile-pile and pile-ground surface interaction. On the other hand, based on equation (14), the following relationship can also be established

$$[k_{sp}]\{\beta\} + [k_{ss}]\{w_s\} = \{P_s\} \qquad (18)$$

in which $[k_{sp}]$, $[k_{ss}]$ are the matrices reflecting ground surface-pile and soil-soil interaction at the ground surface. Equations (17) and (18) can be assembled and then inverted to give

$$\begin{Bmatrix} \beta \\ w_s \end{Bmatrix} = \begin{bmatrix} k_{pp} & k_{ps} \\ k_{sp} & k_{ss} \end{bmatrix}^{-1} \begin{Bmatrix} P_T \\ P_s \end{Bmatrix} \qquad (19)$$

The same addition procedure as that for pile groups previously can be made to the above matrix, and the resulting matrix equation becomes a flexibility relationship at the pile head-ground surface only and is given by

$$\begin{Bmatrix} w_t \\ w_s \end{Bmatrix} = \begin{bmatrix} F_{tpp} & F_{tps} \\ F_{tsp} & F_{tss} \end{bmatrix} \begin{Bmatrix} P_t \\ P_s \end{Bmatrix} \qquad (20)$$

The stiffness matrix equation of a pile group-soil system at the pile head-raft interface is then

$$\begin{Bmatrix} P_t \\ P_s \end{Bmatrix} = \begin{bmatrix} F_{tpp} & F_{tps} \\ F_{tsp} & F_{tss} \end{bmatrix}^{-1} \begin{Bmatrix} w_t \\ w_s \end{Bmatrix} \tag{21}$$

This equation can be rewritten in the form

$$\{P\} = [k_t]\{w\} \tag{22}$$

This stiffness matrix can be combined with that of a raft to give the solution of a piled raft.

2.3 *Piled raft analysis*

The definition of a piled raft foundation is illustrated in Fig.3.

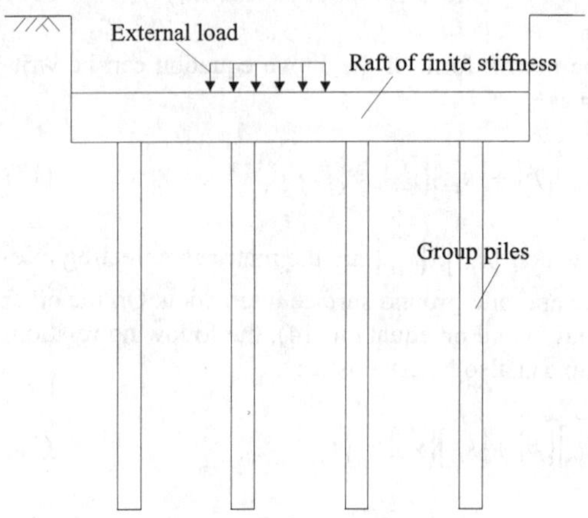

Figure 3 Illustration of piled raft foundation

The potential energy of the raft can, in general, be expressed as

$$\pi_r = U_r + \tfrac{1}{2}\iint_A w_r \sigma_r dA - \iint_A w_r p_r dA \tag{23}$$

in which the first term U_r is the strain energy of the raft, the second and third terms are the work done by the reaction pressure σ_r acting at the bottom of the raft and the external pressure p_r acting at the upper surface of the raft, respectively. $\{w_r\}$ is the deflection of the raft. In order to combine the surface stiffness of a pile group-soil system as obtained previously, the raft is discretised into a number of elements and in this situation, the potential energy can be expressed as

$$\pi_r = U_r + \tfrac{1}{2}\{w_r\}^T\{R\} - \{W_r\}^T\{p_r\} \tag{24}$$

where the vector $\{R\}$ is the reaction force of each element, and $\{W_r\}$ is the integration of the raft deflection over the element area. For a piled raft foundation, by imposing the compatibility condition $\{w_r\} = \{w\}$ and equilibrium condition $\{R\} = \{P\}$ (see equation (22)) at the raft-soil-pile group interface between the raft and the soil-pile group system, the reaction vector $\{R\}$ can be related to the deflection $\{w_r\}$ as

$$\{R\} = [k_t]\{w_r\} \tag{25}$$

in which $[k_t]$ is the surface stiffness matrix of the soil-pile system at the raft-soil-pile group interface obtained in equation (22). Thus, the potential energy becomes

$$\pi_r = U_r + \tfrac{1}{2}\{w_r\}^T[k_t]\{w_r\} - \{W_r\}^T\{p_r\} \tag{26}$$

The deflection of a raft can be represented by a finite series as follows

$$w = w_0 + x\theta_x + y\theta_y + \sum_{m=1}^{k}\sum_{n=1}^{k} A_{mn}\sin\frac{m\pi x}{a}\sin\frac{n\pi y}{b} + \sum_{m=1}^{k} B_m \sin\frac{m\pi x}{a} + \sum_{n=1}^{k} C_n \sin\frac{n\pi y}{b} \tag{27}$$

where x and y are the coordinates, a and b are the length and width of the raft, and w_0, θ_x, θ_y, A_{mn}, B_m and C_n are undetermined coefficients. The principle of minimum potential energy then requires the potential energy to be an extremum with respect to these undetermined coefficients. This gives

$$\frac{\partial U_r}{\partial \delta} + \left\{\frac{\partial w_r}{\partial \delta}\right\}^T [k_t]\{w_r\} = \left\{\frac{\partial W_r}{\partial \delta}\right\}^T \{p_r\} \tag{28}$$

where the parameter δ denotes the representation of the undetermined coefficients. The matrix form of the above equation can be written as

$$\left([k_r] + [Z]^T[k_t][Z]\right)\{\xi\} = [Z_r]^T\{p_r\} \tag{29}$$

where $[k_r]$ reflects the raft stiffness, $[Z]$ and $[Z_r]$ are matrices related only to the coordinate. Equation (29) can then be solved for the unknown constants, i.e. the vector $\{\xi\}$ under the known external load

$\{p_r\}$ applied on the raft. Once these constants have been solved, the deflection and bending moment of the raft, the reaction forces, i.e. the forces at the pile heads and at the ground surface in contact with the raft, all can be determined.

3 RESULTS OF ANALYSIS

Comparisons with published results obtained using other numerical methods are presented in the following to show the validity of the variational approach.

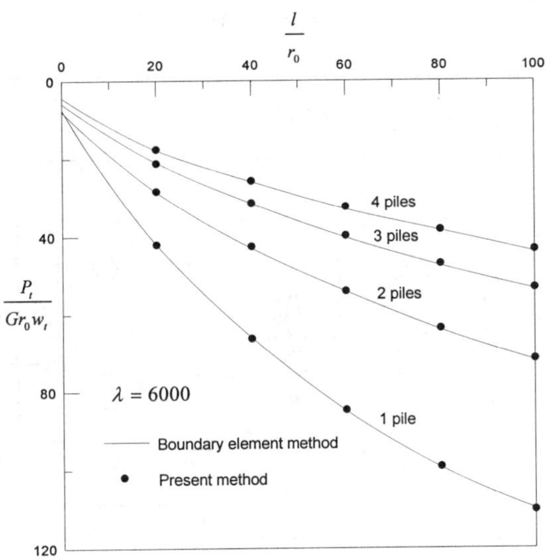

Figure 4 Load-settlement ratio for symmetric pile groups

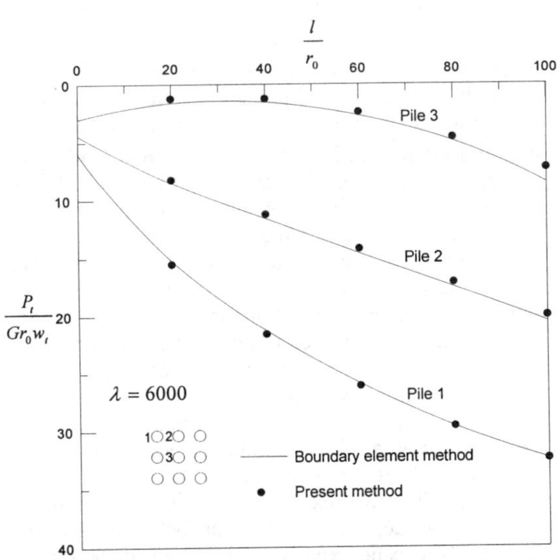

Figure 5 Load-settlement ratio for a 3x3 compressible pile group

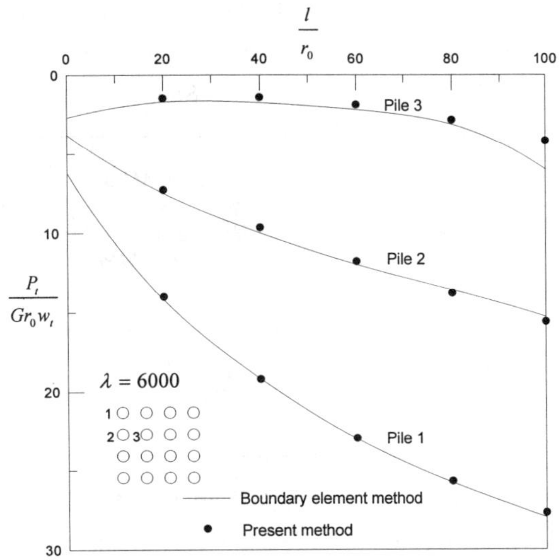

Figure 6 Load settlement ratio for a 4x4 compressible pile group

Comparisons of the load-settlement ratio for pile groups with the solutions reported by Butterfield and Banerjee (1971) using the boundary element method for pile groups in a homogeneous elastic half-space are shown in Figs.4 for symmetric pile groups (including single pile), in Fig.5 for 3×3 compressible pile groups, and in Fig.6 for 4×4 compressible pile groups. Good agreement can be seen between the present method and the boundary element method.

Comparisons with the solutions of Kuwabara (1989) for 3×3 pile groups with rigid caps in contact with the ground are presented in Figs.7 and 8 for the group stiffness, and load taken by the cap and group piles, respectively. The present method shows good agreement with the boundary element analysis of Kuwabara (1989).

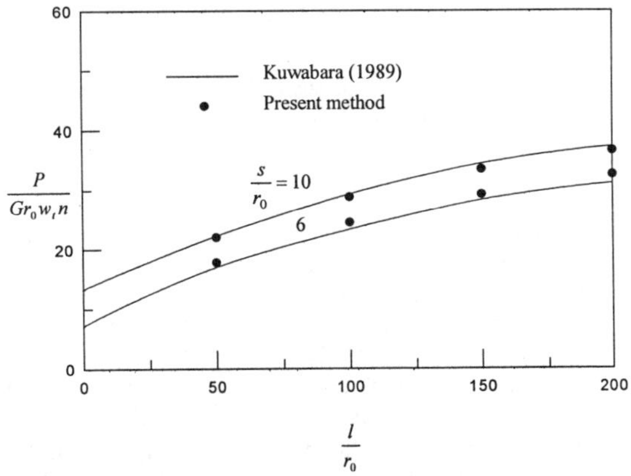

Figure 7 Stiffness of piled rigid rafts with 3x3 pile group ($\lambda = 3000$, $\nu = 0.5$)

Table 1 Properties and cases analysed for a piled raft with concentrated loading

	Soil	Pile	Raft
Properties	E_s 20 MPa ν_s 0.3 H/l 2.0 p_{sult} 0.3 MPa	E_p 30000 MPa l 10 m r_0 0.25 m s 2 m P_{pult} 0.873 MN	E_r 30000 MPa ν_r 0.2 t 0.5 m a 10 m b 6 m
Cases	Case 1 Case 2	P_1 = 1.0 MN P_1 = 1.25 MN	P_2 = 2.0 MN P_2 = 2.5 MN
Note:	p_{sult} Bearing capacity of soil,	P_{pult} Load capacity of pile	

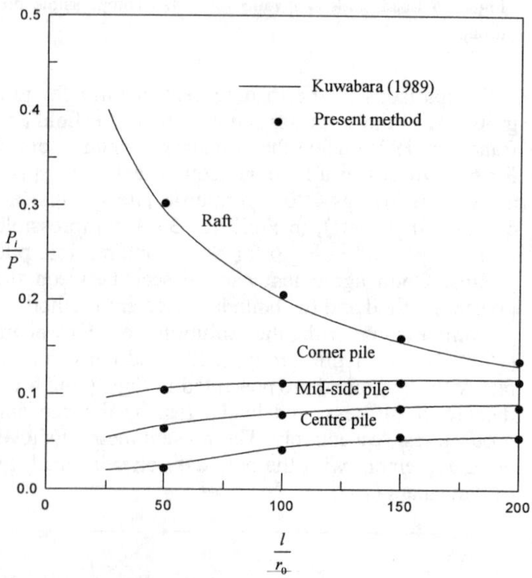

Figure 8 Percentage of load sharing of piled rigid rafts with 3x3 pile group($\lambda = 3000$, $s/r_0 = 10$, $\nu = 0.5$, P_i-load carried by cap and individual piles)

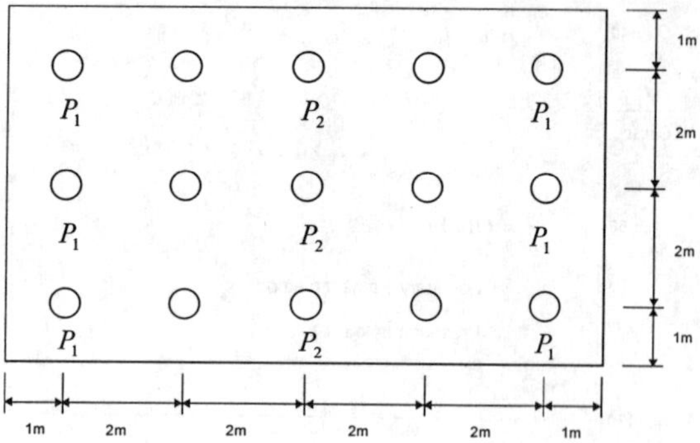

Figure 9 Plan of a piled raft with concentrated loading

Poulos (1997) presented a comparison of some methods for the analysis of piled rafts. A hypotheticcal example consisting of 15 piles with concentrated loads acting on the raft was analyzed by various methods. The plan of the foundation together with the concentrate loads termed P_1 and P_2 acting at the location of a few piles are shown in Fig.9.

The properties and the two cases analyzed for the piled rafts are presented in Table 1. The only difference in these two cases is the magnitude of the applied loads. The results obtained using different methods for the key responses: average settlement, maximum differential settlement, maximum bending, and the proportion of load taken by the piles are presented in Fig.10. The results of Poulos (1994) in Fig.10 are based on an approximate spring approach. The results of Ta and Small (1996) are based on a combined finite element and finite layered method, and the results of Sinha (1997) are based on a combined finite element and boundary element method. All the methods give reasonable results of these key responses. However, the present solutions are in closer agreement with the results of Ta and Small (1996), which involve less assumptions and approximations than the methods of Poulos (1994) and Sinha (1997). It is noted that, in case 2, where the overall applied load exceeds the ultimate capacity of the piles alone, some nonlinear behaviour of the foundation can be expected. The trend of the results for all the methods is similar. The average settle-

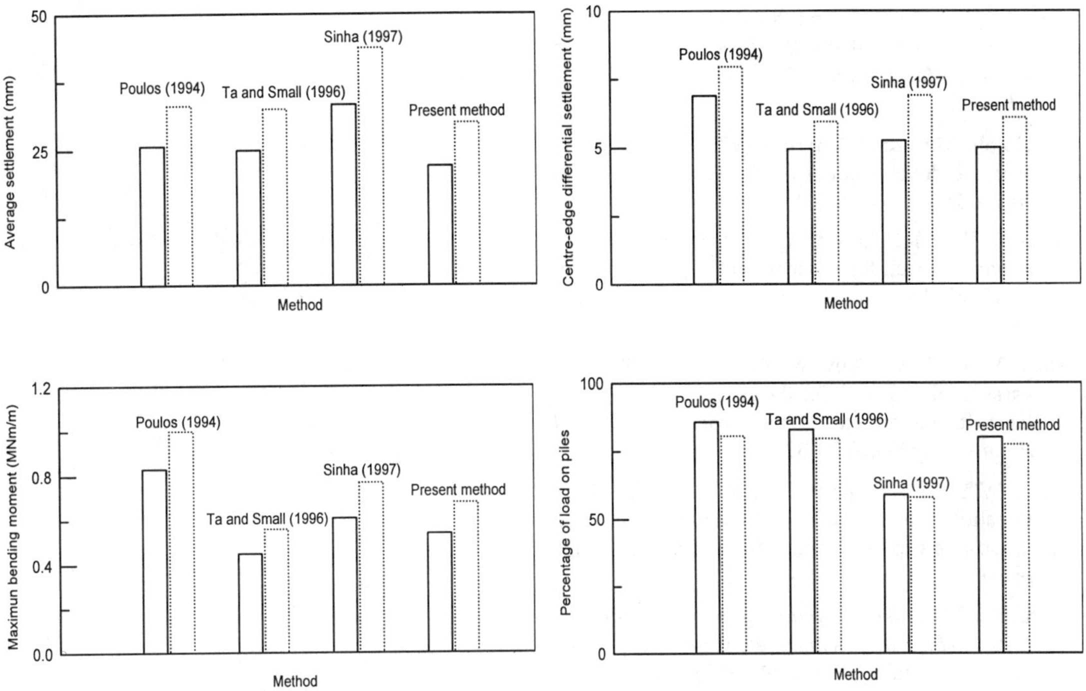

Figure 10 Comparison of methods (Case 1: ——— Case 2: ············)

ment, differential settlement and maximum bending moment will increase and the proportion of load taken by the piles will decrease.

4 CONCLUSIONS

A variational approach for the analysis of pile groups and piled raft foundations is described. This solution procedure is an efficient approach for this class of three-dimensional soil-structure interaction problems. The present solutions have shown to be in good agreement with other numerical methods and can reasonably predict the response of pile groups and piled raft foundations.

REFERENCES

Butterfield, R. & P.K. Banerjee 1971. The elastic analysis of compressible piles and pile groups. *Geotechnique.* 21(1):43-60.

Chow, Y.K. 1986. Analysis of vertically loaded pile groups. *Int. J. Numer. Anal. Methods Geomech.* 10(1):59-72.

Chow, Y.K. 1987. Axial and lateral response of pile groups embedded in non-homogeneous soils. *Int. J. Numer. Anal. Methods Geomech.* 11(6):621-638.

Clancy, P. & M.F. Randolph 1993. Analysis and design of piled raft foundations. *Int. J. Numer. Anal. Methods Geomech.* 17:849-869.

Kuwabara, E. 1989. An elastic analysis for piled raft foundations in a homogeneous soil. *Soils and Foundations.* 29(1):82-92.

O'Neill, M.W., O.I. Ghazzaly & H.B. Ha 1977. Analysis of three-dimensional pile groups with nonlinear soil response and pile-soil-pile interaction. *Proc. 9th. Offshore Technology Conf., Houston.* 245-246.

Ottaviani, M. 1975. Three dimensional finite element analysis of vertically loaded pile groups. *Geotechnique.* 25(2):159-174.

Poulos, H.G. & E.H. Davis 1980. *Pile Foundation Analysis and Design.* Wiley, New York.

Poulos, H.G. 1994. An approximate numerical analysis of pile-raft interaction. *Int. J. Numer. Anal. Methods Geomech.* 18:73-92.

Poulos, H.G. 1997. Comparison of some methods for analysis of piled rafts. *Proc. 14th Int. Conf. on Soil Mech. Found. Eng., Hamburg.* 2:1119-1124.

Randolph, M.F. & C.P. Wroth 1979. An analysis of vertical deformation of pile groups. *Geotechnique.* 29(4):157-166.

Shen, W.Y., Y.K. Chow & K.Y. Yong 1999a. Variational approach for vertical loaded pile groups in an elastic half space. *Geotechnique.* 49(2):199-213.

Shen, W.Y., Y.K. Chow & K.Y. Yong 1999b. A variational approach for the analysis of rectangular rafts on an elastic half-space. *Soils and Foundations.* 39(6):25-32.

Shen, W.Y., Y.K. Chow & K.Y. Yong 2000. A variational approach for the analysis of pile group-pile cap interaction. *Geotechnique (to appear).*

Sinha, J. 1997. *Piled raft foundation on soil subjected to swelling, shrinkage and ground subsidence.* PhD thesis. University of Sydney, Australia.

Ta, L.D. & J.C. Small 1996. Analysis of piled raft systems in layered soils. *Int. J. Numer. Anal. Methods Geomech.* 20:57-72.

Computer Methods and Advances in Geomechanics, Desai et al. (eds) © 2001 Balkema, Rotterdam, ISBN 90 5809 183 X

Bearing capacity of spatially random c – ∅ soils

G.A. Fenton
Dalhousie University, Halifax, N.S., USA

D.V. Griffiths
Colorado School of Mines, Golden, Colo., USA

ABSTRACT: By combining elasto-plastic finite element analysis with random field theory, a preliminary investigation has been performed into the bearing capacity of soils with spatially random shear strength. The main issue is to determine the extent to which spatial variability and cross-correlation in soil properties affects the distribution of the computed bearing capacity. For vanishing coefficients of variation (C.O.V.) in the soil shear strength, the expected value of the bearing capacity tends to the Prandtl solution, N_c. For increasing values of C.O.V., however, the expected value of the bearing capacity falls quite steeply, largely independently of the correlation length and the degree of cross-correlation between c and ϕ. The results of Monte-Carlo simulations on this nonlinear problem are presented in the form of histograms which enable a probabilistic interpretation. In particular, such plots allow the probability of overestimating the bearing capacity to be assessed.

1 INTRODUCTION

The paper presents results obtained using a program developed by the authors which combines nonlinear elasto-plastic finite element analysis (e.g. Smith & Griffiths, 1998) with random field theory (e.g. Fenton, 1990, Vanmarcke, 1984). The program computes the bearing capacity of a smooth rigid strip footing (plane strain) at the surface of a weightless soil with shear strength parameters c and ϕ represented by spatially varying and cross-correlated (point-wise) random fields. These two soil properties were selected to be represented as random fields since they have the greatest impact on soil bearing capacity.

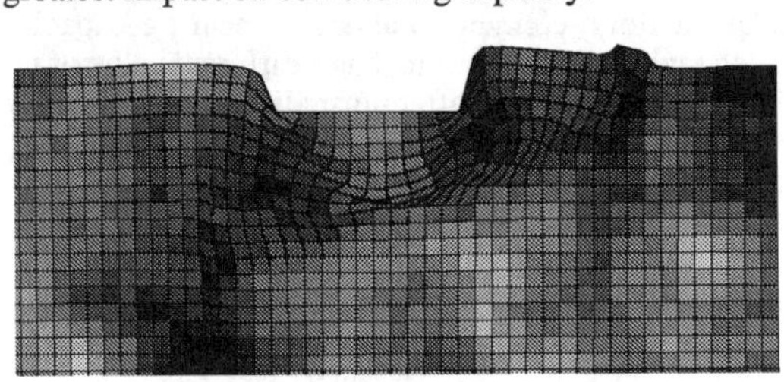

Figure 1. Typical deformed mesh at failure, where the darker regions indicate weaker soil.

Figure 1 shows a typical deformed finite element mesh resulting from a footing's bearing failure on a soil with spatially random properties. Lighter regions in the plot indicate stronger soil and darker regions indicate weaker soil. It is clear, in this case, that the weak (dark) region near the ground surface to the right of the footing has triggered a quite non-symmetric failure mechanism which is often at a lower bearing load than obtained from the traditional 'uniform' and symmetric failure analysis.

The bearing capacity analyses use an elastic-perfectly plastic stress-strain law with a Mohr-Coulomb failure criterion. Plastic stress redistribution is accomplished using a viscoplastic algorithm. The program uses 8-node quadrilateral elements and reduced integration in both the stiffness and stress redistribution parts of the algorithm. The theoretical basis of the method is described more fully in Chapter 6 of the text by Smith & Griffiths (1998). The finite element model incorporates five parameters; Young's modulus (E), Poisson's ratio (ν), dilation angle (ψ), shear strength (c), and friction angle (ϕ). The methodology allows for random distributions of all five parameters, however in the present study, E, ν and ψ are held constant (at 100000, 0.3, and 0, respectively) while c and ϕ are randomized. The finite element mesh consists of 1000 elements, 50 elements wide by 20 elements deep. Each element is a square of side length 0.1m and the strip footing occupies 10 elements, giving it a width of 1m.

Rather than deal with the actual bearing capacity, this study deals with the dimensionless bearing capacity factor, N_c, which is traditionally defined by

$$N_c = \frac{q_f}{c} \quad (1)$$

where q_f is the bearing capacity and c is the cohesion of the soil (traditionally assumed spatially constant). For a soil with spatially constant cohesion and friction angle, the theoretical bearing capacity factor, N_c, is given by Sokolovskii (1965),

$$N_c = (e^{\pi \tan\phi} \tan^2(45 + \phi/2) - 1)/\tan\phi, \quad (2)$$

so that, for example, if $\phi = \mu_\phi = 25$ degrees, then $N_c = 20.7$.

2 THE RANDOM FIELD MODEL

In this study, the soil cohesion is assumed to be lognormally distributed with mean μ_c, standard deviation σ_c, and spatial correlation length $\theta_{\ln c}$. A lognormally distributed random field is easily obtained by first simulating a normally distributed random field, $G_{\ln c}(\underset{\sim}{x})$, having zero mean, unit variance, and spatial correlation length $\theta_{\ln c}$. This 'underlying' normally distributed random field may then be transformed to the desired cohesion field using the relationship

$$c_i = \exp\{\mu_{\ln c} + \sigma_{\ln c} G_{\ln c}(\underset{\sim}{x}_i)\} \quad (3)$$

where $\underset{\sim}{x}_i$ is a vector containing the coordinates of the center of the i'th element, and c_i is the cohesion value assigned to the i'th element.

The friction angle, ϕ, is bounded both above and below, and so neither the normal nor the lognormal distributions are appropriate. In this study, a bounded distribution is selected which arises as a simple transformation of a standard normal random field, $G_\phi(\underset{\sim}{x})$. This approach again allows the generation of a normally random field followed by the transformation

$$\phi_i = \phi_{min} + \tfrac{1}{2}(\phi_{max} - \phi_{min})\left\{1 + \tanh\left(\frac{sG_\phi(\underset{\sim}{x}_i)}{2\pi}\right)\right\} \quad (4)$$

where ϕ_{min} and ϕ_{max} are the minimum and maximum friction angles, respectively, and s is a scale factor which governs the friction angle variability between its two bounds. See Fenton (1990) for more details on the above transformation.

The random fields used in this study are generated using the Local Average Subdivision (LAS) method (Fenton & Vanmarcke, 1990, Fenton, 1994). An isotropic Markovian spatial correlation function is used for both fields, having the form

$$\rho(\tau) = \exp\{-2|\tau|/\theta\} \quad (5)$$

where ρ is the correlation coefficient between the underlying random field values at any two point separated by a distance τ. This correlation function governs the correlation structure of the underlying generated fields $G(\underset{\sim}{x})$.

Cross-correlation between the two soil property fields (c and ϕ) is implemented using a Cholesky decomposition of the cross-correlation matrix between the underlying standard normal two random fields.

3 MONTE CARLO SIMULATION

In the parametric studies that follow, the mean cohesion (μ_c) and mean friction angle (μ_ϕ) have been held constant at 100 kN/m^2 and 25° (with $\phi_{min} = 5^o$ and $\phi_{max} = 45^o$), respectively, while the C.O.V. (= σ_c/μ_c), spatial correlation length (θ), and correlation coefficient, ρ, between $G_{\ln c}$ and G_ϕ are varied systematically according to the following table

Table 1. Random field parameters used in Monte Carlo simulation

θ	=	0.5	1.0	2.0	4.0	8.0	50.
C.O.V.	=	0.1	0.2	0.5	1.0	2.0	5.0
ρ	=	-1.0	0.0	1.0			

In addition, it is assumed that when the variability in the cohesion is large, the variability in the friction angle will also be large. Under this reasoning, the scale factor, s, used in Eq. (4) is set to $s = \sigma_c/\mu_c =$ C.O.V.. This choice is arbitrary, but results in the friction angle varying from quite narrowly (when C.O.V. = 0.1 and $s = 0.1$) to very widely (when C.O.V. = 5.0 and $s = 5$) between its lower and upper bounds, 5° and 45°.

For each set of assumed statistical properties given by Table 1, Monte-Carlo simulations have been performed. These involve 1000 repetitions or "realizations" of the soil property random fields and the subsequent finite element analysis of bearing capacity. Each realization, therefore, has a different value of the bearing capacity and, after normalization by the *mean* cohesion, a different value of the bearing capacity factor,

$$N_{c_i} = \frac{q_{f_i}}{\mu_c}, \quad i = 1, 2, \ldots, 1000, \quad (6)$$

so that the average can be computed as

$$\bar{N}_c = \frac{1}{1000}\sum_{i=1}^{1000} N_{c_i} \quad (7)$$

4 SIMULATION RESULTS

Figure 2(a) shows how the sample mean bearing capacity factor, taken as the average of the N_{c_i} computed over all soil realizations, and refered to as $\bar{N}_c$, varies with the correlation length, soil variability, and cross-correlation between c and ϕ. For small soil variability, $\bar{N}_c$ tends towards the deterministic value of 20.7, which is found when the soil takes on its mean properties everywhere. For increasing soil variability, the mean bearing capacity factor becomes quite significantly reduced from the ideal case. What this implies from a design standpoint is that the bearing capacity of a heterogeneous soil will, on average, be *less* than the Prandtl solution which would be predicted assuming the soil has strength given by mean values.

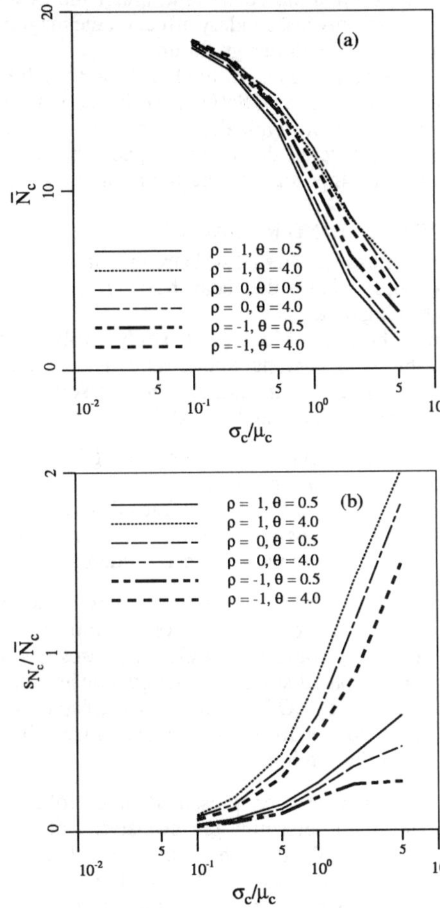

Figure 2. a) Sample mean bearing capacity factor, $\bar{N}_c$, and b) sample coefficient of variation of N_c.

The greatest reduction from the Prandtl solution is observed for low θ values. This can be explained by the fact that as the correlation length decreases, the possibility of a lower strength non-symmetric bearing failure increases because of the more rapidly varying (in space) soil properties. Cross-correlation between c and ϕ is seen in Figure 2(a) to have a only a minor affect on $\bar{N}_c$. Similarly, the variability of $\bar{N}_c$, as seen in Figure 2(b), also is not significantly affected by cross-correlation between c and ϕ. The variability in $\bar{N}_c$, shown in Figure 2(b) increases with the variability in the soil, with the slowest increase occurring at smaller correlation lengths due to the local averaging affect under the footing.

5 PROBABILISTIC INTERPRETATION

Following Monte-Carlo simulations for each parametric combination of input parameters (θ, C.O.V., and ρ), the suite of computed bearing capacity factor values from Eq. (6) can be plotted in the form of a histogram, and a 'best-fit' lognormal distribution superimposed. Figure 3 shows such a plot for the case where $\theta = 2$, C.O.V.$= 1$, and $\rho = 0$.

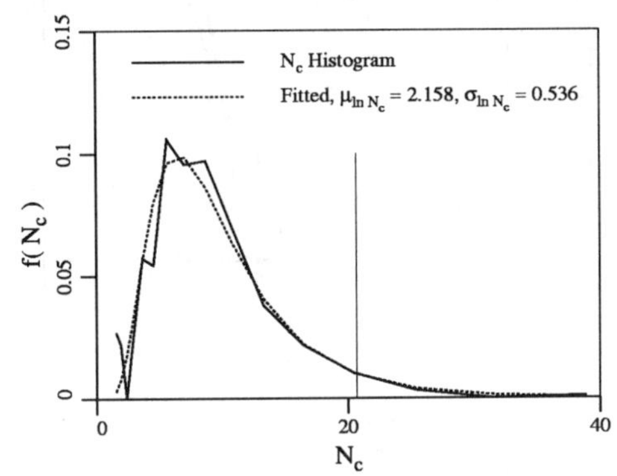

Figure 3. Typical normalized histogram of N_c values with superimposed fitted lognormal distribution.

Since the lognormal fit has been normalized to enclose an area of unity, areas under the curve can be directly related to probabilities. From a practical viewpoint it would be of interest to estimate the probability of 'design failure', defined here as occurring when the computed bearing capacity is less than the Prandtl value based on the mean soil properties, i.e. we have design failure if $N_c < 20.7$, where N_c is computed from Eq. (1).

Assuming that N_c does follow a lognormal distribution, as is roughly indicated by Figure 3, the 'design failure' probability can be computed as

$$\mathrm{P}[N_c < 20.7] = \Phi\left(\frac{\ln 20.7 - \mu_{\ln N_c}}{\sigma_{\ln N_c}}\right) \qquad (8)$$

where Φ is the cumulative normal distribution function. For the particular case shown in Figure 3, the fitted lognormal distribution has parameters $\mu_{\ln N_c} = 2.158$ and $\sigma_{\ln N_c} = 0.536$. Eq. (8) gives $P[N_c < 20.7] = 0.95$, indicating an 95% probability that the actual bearing capacity will be less than the Prandtl value.

Figure 2(a) indicated that the expected bearing capacity of a strip footing on a soil with spatially variable cohesion and friction angle will *always* be lower than the Prandtl value based on the mean soil. However, the design capacity is generally based on the Prandtl solution reduced by a 'Factor of Safety', F. The probability of design failure, in the form of $P[N_c < 20.7/F]$, is considerably reduced, giving a more reassuring result from a design viewpoint.

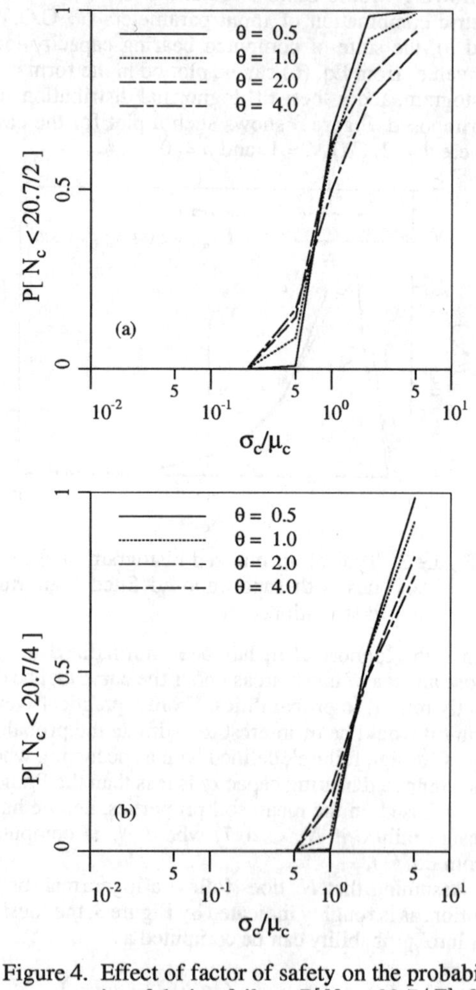

Figure 4. Effect of factor of safety on the probability of design failure, $P[N_c < 20.7/F]$, for a) $F = 2$, and b) $F = 4$.

Figure 4 illustrates the probability of design failure for two different factors of safety, for the case where $\rho = 0$. For example, from Figure 4(a), in which $F = 2$, the probability of design failure for a soil with $\theta = 4$, C.O.V.$= 0.5$ and $\rho = 0$, is about 17% (varying from 14% for $\rho = -1$ to 26% for $\rho = 1$). This probability is reduced to about 0.1% (varying from 0.02% for $\rho = -1$ to 1% for $\rho = 1$) when the factor of safety is increased to $F = 4$, as shown in Figure 4(b).

These plots indicate that quite high factors of safety are required to reduce the probability of 'design failure' to negligible levels. The most important factor affecting the probability of design failure appears to be the soil variability (which includes the cohesion and friction angle variabilities). The correlation length and cross-correlation coefficient, under the assumed model, have only a secondary affect on the magnitude of the probability of design failure.

These results may help explain in a probabilistic context, why Factors of Safety used in bearing capacity calculations are typically higher than those used in other limit state calculations in geotechnical engineering, e.g. slope stability, earth pressures.

6 CONCLUDING REMARKS

The paper has shown that soil strength variability can significantly reduce the mean bearing capacity of a strip footing on a c-ϕ soil.

On the basis of a Monte Carlo study involving 1000 realizations, for each parameter set considered, of the bearing capacity on a spatially random soil, the following observations can be made;

- As the variance of the soil strength increases, the mean bearing capacity decreases. The rate of decrease is only slightly affected by correlation length and cross-correlation coefficient, over the ranges in these parameters considered.
- As the variance of the soil strength increases from zero, the coefficient of variation (C.O.V.) of the bearing capacity also increases. Increasing the spatial correlation length consistently increases the C.O.V. of the bearing capacity due to the reduced local averaging variance reduction under the footing.
- Results have been presented in a probabilistic context to determine the probabitity of 'design failure', defined as the probability that the actual bearing capacity would be lower than a deterministic prediction of factored bearing capacity using Prandtl's formula based on the mean strength of the soil.
- By investigating the role of a Factor of Safety applied to the Prandtl solution, it was observed

that a value of $F = 4$ and greater may be required to reduce the probability of 'design failure' for soils to a negligible amount.

- Cross-correlation between c and ϕ, on a pointwise basis, was found to have only a minor affect on bearing capacity reliability, the negative correlated random fields giving somewhat safer designs.
- The influence of the correlation length on the probabilistic interpretation of the bearing capacity problem was also seen to be not greatly significant, within the range of lengths considered. The major factor influencing the probability of a 'design failure' is the soil C.O.V.

ACKNOWLEDGEMENT

The support of NSF grant number 98-77189 is acknowledged.

REFERENCES

Fenton, G.A. 1990. *Simulation and Analysis of Random Fields,* Ph.D. Thesis, Princeton University , Dept. of Civil Engrg. and Op. Res., Princeton, New Jersey.

Fenton, G.A. 1994. Error evaluation of three random field generators, *ASCE J. Engrg. Mech.*, **120**(12), 2478–2497.

Fenton, G.A. and Vanmarcke, E.H. 1990. Simulation of Random Fields via Local Average Subdivision, *ASCE J. Engrg. Mech.*, **116**(8), 1733–1749.

Smith, I.M. and Griffiths, D.V. 1998. *Programming the Finite Element Method,* (3rd Ed.), John Wiley & Sons, New York, NY.

Sokolovski, V.V. 1965. *Statics of Granular Media*, Pergamon Press, London, UK.

Vanmarcke, E.H. 1984. *Random Fields: Analysis and Synthesis*, The MIT Press, Cambridge, Massachusetts.

Computer Methods and Advances in Geomechanics, Desai et al. (eds) © 2001 Balkema, Rotterdam, ISBN 90 5809 183 X

Applications of the lower and upper bound theorems of plasticity to collapse of circular foundations

C. M. Martin
Centre for Offshore Foundation Systems, The University of Western Australia, Nedlands, W.A., Australia
(Presently: Department of Engineering Science, Oxford University, UK)

M. F. Randolph
Centre for Offshore Foundation Systems, The University of Western Australia, Nedlands, W.A., Australia

ABSTRACT: The bound theorems of plasticity have proved a powerful tool in establishing design methods for the collapse of shallow foundations, buried anchor plates and other geotechnical structures. Closely bracketed upper and lower bound solutions (and in some cases, exact solutions) are now available for a wide range of bearing capacity problems in plane strain. These solutions have been extended to circular geometries in certain cases, often through numerical techniques involving the method of characteristics. With the growth in the use of shallow foundations, suction-emplaced caissons, and vertically loaded anchors for offshore structures, there has been strong practical interest in the capacity of embedded foundations, both under purely vertical loading, and also under combined vertical, moment and horizontal (*V*, *M*, *H*) loading. In turn, this has generated renewed interest in developing plasticity solutions for these problems, taking due account of the variation of shear strength with depth.

1 INTRODUCTION

The trend in the offshore industry towards developments in deep water, with consequential focus on skirted foundations and anchoring systems in generally soft clays and silts, has revived interest in establishing the ultimate capacity of embedded circular foundations. Most practical applications involve combinations of vertical (*V*), moment (*M*) and horizontal (*H*) loading, and the construction of a three-dimensional yield envelope in (*V*, *M*, *H*) space. However, collapse under purely vertical loading is a key aspect of much offshore design, and represents the first step in establishing the full envelope.

While finite element methods offer a general-purpose approach for evaluating the full load–deformation response of foundation systems, much design relies on a framework of bearing capacity calculations, with limiting loads evaluated using plasticity solutions derived from the classical lower and upper bound theorems. In certain cases, exact analytical solutions are possible, but generally it is necessary to resort to numerical approaches, for example using the method of characteristics or finite element implementations of the bound theorems.

The basis of bearing capacity calculations for foundations in 'cohesive' soils (more accurately soils of low permeability where the concept of undrained shear strength may be applied) was established some fifty years ago, with the work of Meherhof (1951), Skempton (1951) and others. Many of the early solutions were based on approximate adjustments of exact solutions, to allow for practical features such as the foundation geometry (circular or square, rather than strip) and depth of embedment.

Caisson foundations such as those used offshore may have skirts that extend up to a diameter below the seabed, penetrating through soil where the strength increases with depth. The bearing capacity of such foundations must therefore consider not only the embedment ratio, d/D (where d is the skirt penetration and D the foundation diameter), but also the degree of strength non-homogeneity and the relative roughness of the skirts.

Referring to Figure 1, the strength profile can be idealised according to

$$s_u = s_{um} + kz \tag{1}$$

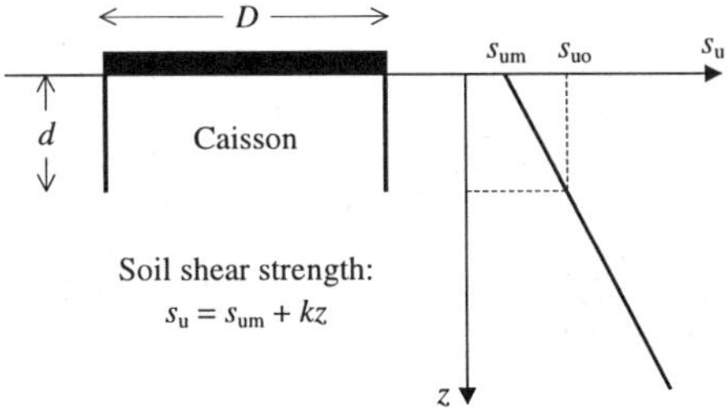

Figure 1. Schematic of caisson and strength profile.

where s_{um} is the strength at the seabed (mudline) and k is the strength gradient with depth, z. It is helpful to normalise the bearing capacity, q_u, by the strength at the effective base of the foundation:

$$q_u = N_c s_{uo} = N_c(s_{um} + kd) \tag{2}$$

where N_c is a bearing capacity factor.

2 THEORY

The bearing capacity analyses in this paper are based on the classical assumptions of rigid–perfectly plastic soil behaviour, with yielding governed by the Tresca criterion and an associated plastic flow rule. All problems considered here involve axially symmetric geometry. Figure 2 defines the (r, z) coordinate system and the notation adopted for the various non-zero components of stress (compression positive) and velocity. The nature of the problems considered is such that the soil self-weight does not affect the solutions, although for completeness it is included in the governing equations below.

2.1 *Lower bound stress fields*

The method of characteristics is a well-known technique for constructing statically admissible stress fields in the solution of plasticity problems. Several authors describe the development of the method for axially symmetric conditions (Shield 1955, Szczepinski 1979, Houlsby & Wroth 1982a) and only a brief summary is appropriate here.

Stresses in the meridional plane are assumed to satisfy the yield criterion ($\sigma_{max} - \sigma_{min} = 2s_u$). This allows the individual stress components σ_{rr}, σ_{zz} and τ_{rz} to be expressed in terms of two auxiliary variables, σ and θ (see Fig. 2b):

$$\begin{aligned} \sigma_{rr} &= \sigma - s_u \cos 2\theta \\ \sigma_{zz} &= \sigma + s_u \cos 2\theta \\ \tau_{rz} &= s_u \sin 2\theta \end{aligned} \tag{3}$$

The hoop stress σ_{hh} is usually equated (by assumption) with one or other of the principal stresses in the meridional plane, i.e.

$$\sigma_{hh} = \sigma + \omega s_u \tag{4}$$

where $\omega = \pm 1$ (this is the Haar/von Karman hypothesis). As discussed by Shield (1955), Houlsby & Wroth (1982a) and others, the choice should be governed by the expected direction of radial plastic flow: $\omega = -1$ if it is outward, $\omega = +1$ if it is inward.

Equations (3) and (4) may be substituted into the equilibrium equations

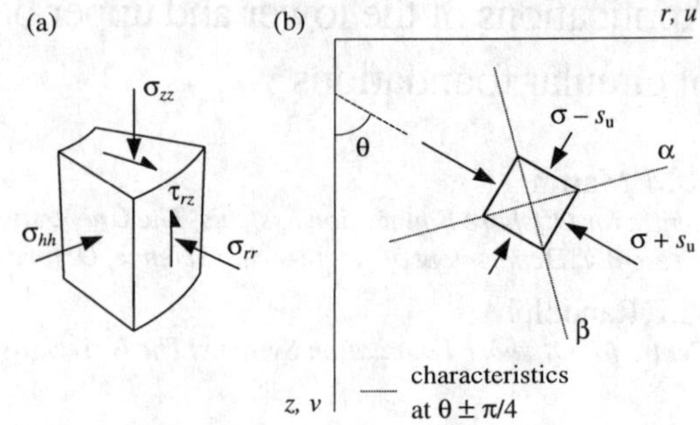

Figure 2. Axisymmetric stresses and characteristic directions.

$$\begin{aligned} \frac{\partial \sigma_{rr}}{\partial r} + \frac{\partial \tau_{rz}}{\partial z} + \frac{\sigma_{rr} - \sigma_{hh}}{r} &= 0 \\ \frac{\partial \tau_{rz}}{\partial r} + \frac{\partial \sigma_{zz}}{\partial z} + \frac{\tau_{rz}}{r} &= \gamma \end{aligned} \tag{5}$$

to yield two partial differential equations in σ and θ. These are hyperbolic in type, with characteristic directions given by

$$\frac{dr}{dz} = \tan(\theta \pm \pi/4) \tag{6}$$

The two families of characteristics are designated α and β (Fig. 2b). In these directions σ and θ vary in accordance with

$$\begin{aligned} d\sigma \pm 2s_u d\theta = &\left(\mp \frac{ds_u}{dz} + \frac{s_u \cos 2\theta + \omega s_u}{r} \right) dr + \\ &\left(\gamma - \frac{s_u \sin 2\theta}{r} \right) dz \end{aligned} \tag{7}$$

Equivalent versions of this equation are given by Houlsby & Wroth (1982a) and Tani & Craig (1995), although the sign of ds_u/dz was misprinted in the former. For the linearly increasing strength profile considered here, ds_u/dz is simply k.

Equations (6) and (7) can readily be implemented in finite difference form and used to solve the various boundary value problems that arise in lower bound bearing capacity analysis. Details of these numerical methods and solution strategies are thoroughly described elsewhere (see e.g. Shield 1955, Booker & Davis 1977, Szczepinski 1979).

2.2 *Upper bound velocity fields*

A stress field constructed using the method of characteristics can also serve as the basis for a 'consistent' velocity field in which the principal strain rate directions coincide with the principal stress directions, and the flow rule (here equivalent to the incompressibility condition) is satisfied. The derivation of the consistent velocity equations for axial symmetry (Shield 1955, Szczepinski 1979, Houlsby & Wroth 1982a) closely follows that of the familiar Geiringer equations in plane strain. A pair of hyperbolic partial differential equations in u and v is obtained, with characteristic directions coincident with those of the stress field. Along the characteristics, the velocity components vary according to

$$du\sin(\theta \pm \pi/4) + dv\cos(\theta \pm \pi/4) = \frac{-u\,dr}{2r\sin(\theta \pm \pi/4)} \quad (8)$$

Numerical finite difference techniques, similar to those employed when constructing the stress field, are used to determine u and v throughout the plastically deforming region. Appropriate velocity boundary conditions for smooth and rough foundations are discussed by Shield (1955) and Eason & Shield (1960) respectively. The extent of the plastically deforming region is identical to that of the 'partial' lower bound stress field (see below).

The form of the consistent velocity field is often quite complex, and furthermore it does not always lead to an optimal upper bound. A family of 'independent' collapse mechanisms can be developed as shown in Figure 3, and these are used extensively in the examples that follow. The velocity fields involved are very simple, consisting of a series of parallel streamlines along which the velocity decays with $1/r$ (thus ensuring satisfaction of the incompressibility requirement). Mechanisms of this type have been used to obtain upper bound collapse loads for a wide range of axially symmetric bearing capacity problems (Levin 1955, Kusakabe et al. 1986, Hu et al. 1999, Randolph et al. 2000).

In the present study, the Hill-type mechanism on the left of Figure 3 is always critical for smooth-based foundations, and is occasionally critical for rough-based foundations (notably when the non-homogeneity ratio kD/s_{uo} is larger than 1 or 2). The Prandtl-type mechanism on the right usually gives a better upper bound for rough-based foundations when kD/s_{uo} is small. The inclination of the (conical) exit region BDEC can be optimised, and varies from about 45° at shallow embedment of the caisson to 90° (annular flow adjacent to the caisson) at deeper embedments. The fan angle, ADB, is adjusted accordingly.

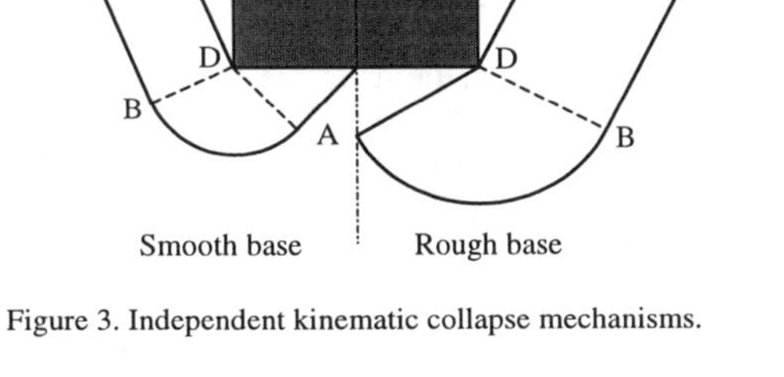

Figure 3. Independent kinematic collapse mechanisms.

3 SURFACE FOUNDATIONS

The indentation of a semi-infinite Tresca medium by a flat-ended circular punch is a fundamental axially symmetric bearing capacity problem, which has been widely investigated using both the method of characteristics and independent upper bound analysis. Studies utilising the former technique have universally adopted the appropriate Haar/von Karman assumption ($\omega = -1$ in the present notation).

3.1 *Homogeneous soil*

For the simplest case of a perfectly smooth punch on homogeneous material, Ishlinskii (1944) used a graphical technique to construct the appropriate mesh of stress characteristics and obtained a lower bound solution of N_c = 5.68 for the average bearing pressure. Shield (1955) obtained an almost identical result of 5.69, using an iterative finite difference approach. More importantly, however, Shield showed that the 'partial' stress field in the vicinity of the punch could be extended throughout the half-space in a statically admissible manner, thus establishing the solution as a rigorous lower bound. He also demonstrated that the consistent velocity field gave principal strain rates $\dot{\varepsilon}$ that were properly associated with the stress field throughout the deforming region, thus proving that the bearing capacity factor of N_c = 5.69 was also an admissible upper bound (and hence the exact solution).

This value has been confirmed subsequently by several authors including Houlsby & Wroth (1983) and Tani & Craig (1995). As part of the present study, analyses with increasingly refined meshes of characteristics have indicated that N_c = 5.689 to three decimal places. The accuracy of Ishlinskii's original 5.68 is certainly remarkable. The value of 5.71 obtained by Meyerhof (1951) from numerical integration of an approximate analytical expression (unrelated to the method of characteristics) also compares favourably with the exact solution. Optimisation of the independent upper bound mechanism of Figure 3 gives N_c = 5.82, some 2% above the method of characteristics solution. Levin (1955) obtained an upper bound of 5.84 using a slightly simpler mechanism.

Indentation by a perfectly rough circular punch was first investigated by Eason & Shield (1960). Using numerical stress and velocity calculations based on those of Shield (1955), a rigorous lower bound and a coincident upper bound were again obtained, the exact bearing capacity factor in this instance being N_c = 6.05. Other computations of this stress field have generally given the same result (Salençon & Matar 1982, Houlsby & Wroth 1983), although Tani & Craig (1995) have recently claimed a surprisingly high lower bound of N_c = 6.34. This analysis appears to have involved a faulty integration procedure, discussed below, and the present work has confirmed that the value of 6.05 is indeed exact (6.048 to three decimal places). Meyerhof's approximate result for a perfectly rough punch was N_c = 6.18, about 2% above the exact solution. The best independent upper bound (Fig. 3) is also 6.18 for this problem.

Because $N_c = 2 + \pi$ for a flat indenter in plane strain, regardless of roughness, exact shape factors for circular foundations on homogeneous, isotropic Tresca soil are 1.11 (smooth) and 1.18 (rough). Clearly there is a sound theoretical basis for the shape factors of about 1.2 that are commonly used in bearing capacity formulae (e.g. Brinch Hansen 1970, Vesic 1975). It should be noted, however, that most real soils exhibit a degree of strength anisotropy, and this can have a surprisingly significant influence on the bearing capacity in both plane strain and axial symmetry (Booker & Davis 1972, Randolph 2000). Further theoretical and numerical work in this area is required.

3.2 *Non-homogeneous soil*

Davis & Booker (1973) studied the bearing capacity of surface strip foundations on cohesive soil where the strength increases linearly with depth. They used the method of characteristics to obtain rigorous lower bound and coincident upper bound solutions for both smooth and rough footings, for all possible values of the non-homogeneity ratio kB/s_{um} (i.e. zero to infinity). The method of characteristics has also been used to investigate the corresponding problem in axial symmetry, but no single study has been as comprehensive. Houlsby & Wroth (1983) listed bearing capacity factors for both smooth and rough circular footings, but only for $0 \leq kD/s_{um} \leq 10$. Salençon & Matar (1982) covered the full range of kD/s_{um}, but only for a fully rough interface. Tani & Craig (1995) considered both smooth and rough foundations on soil with $0 \leq kD/s_{um} \leq 30$, but their rough footing results are erroneous. In all three cases the N_c factors obtained from the method of characteristics have been presented as lower bound collapse loads, but they have not been formally established as rigorous solutions by demonstrating that the stress fields involved are extensible throughout the soil mass. Nor have consistent velocity fields been computed with a view to showing that the lower bound solutions are upper bounds as well (and are therefore exact).

An example of a more thorough approach is given in Figures 4 and 5, for a smooth circular footing on soil with kD/s_{um} = 4. The 'partial' field of stress characteristics (Fig. 4a) gives a lower bound bearing capacity – at this stage non-rigorous – of N_c = 7.56. Having found the consistent velocity field (Fig. 4b), the upper bound is computed by numerical integration of the Tresca dissipation function

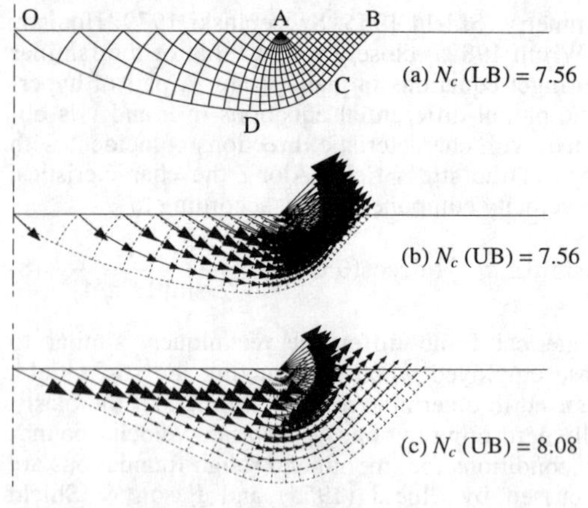

Figure 4. Stress and velocity fields for smooth circular footing on non-homogeneous soil (kD/s_{um} = 4).

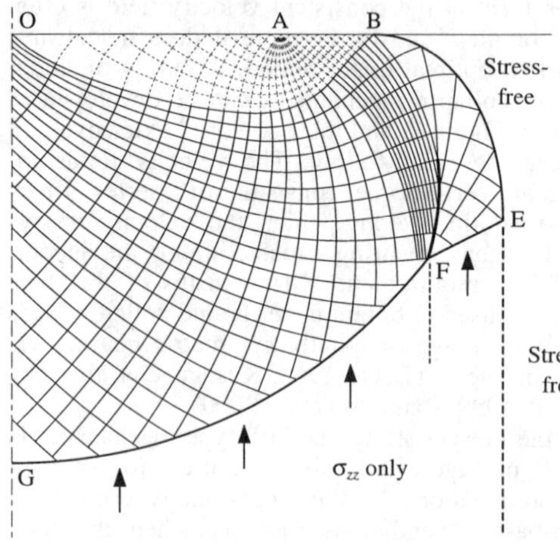

Figure 5. Extension of stress field of Figure 4(a).

$$dW = 2s_u|\dot{\varepsilon}|_{max} \tag{9}$$

over each cell of the mesh. Velocity discontinuities and/or dissipation at the soil–footing interface, although not applicable in this particular example, are handled by introducing degenerate cells. With sufficient mesh refinement, it is found that the consistent upper bound also converges to N_c = 7.56. Presumably the coincidence of lower and upper bounds could also be verified by examining principal strain rates throughout the mesh and checking for proper association with the stress field, as was done by Shield (1955) for homogeneous material.

The best upper bound obtained from the independent collapse mechanism is N_c = 8.08 (Fig. 4c). This is 7% above the consistent upper bound, a somewhat greater differential than the 2% observed for homogeneous soil. Interestingly, however, the accuracy of the independent upper bound eventually improves again with increasing kD/s_{um}, to the point where it gives the exact bearing capacity of $q_u = kD/6$ (analogous to Davis & Booker's $q_u = kB/4$) as s_{um} approaches zero.

The final stage in establishing $N_c = 7.56$ as a rigorous exact solution is to demonstrate the extensibility of the stress field. As shown in Figure 5, the mesh of characteristics can be extended using the strategy of Shield (1955), based on that developed by Bishop (1953) for plane strain conditions. The extension field consists of a region where the stress state is fully plastic (governed by the same equations as the partial stress field), bounded by a stress free region to the right of BE and columns of uniaxial vertical stress below EFG, extending to $z = \infty$. The heavy lines around the plastic region denote statically admissible stress discontinuities. A stress discontinuity also arises within the plastic region, just above F. In this example the uniaxial stresses are found to be acceptable ($|\sigma_{zz}| \leq 2s_u$), but this is not the case when kD/s_{um} becomes very large. It is then necessary to use a different type of extension based on that of Cox et al. (1961), which was also adapted by Davis & Booker (1973) for their work in plane strain. It is always possible to perform a stress field extension of one type or the other.

Figure 6 shows the analysis of a rough circular footing on soil with $kD/s_{um} = 4$. To calculate the lower bound, it is sufficient to extend the mesh of characteristics (Fig. 6a) as far as the final β characteristic FD. Note however that the major principal stress is not vertical along FD, except of course at D on the axis of symmetry. When integrating stresses over AFD to obtain the collapse load, it is therefore essential to use an expression of the form

$$Q_u = \int_0^R \sigma_{zz} 2\pi r dr + \int_0^Z \tau_{rz} 2\pi r dz \qquad (10)$$

where Z is the distance OD. Tani & Craig (1995) appear to have omitted the term involving τ_{rz}, which is negative along FD using the sign convention of Figure 2a. This explains why their N_c values for rough footings are high, most noticeably when kD/s_{um} is small and the 'false head' region OFD is large. When using equation (10) to integrate along AFD it is tacitly assumed that the stress field could be extended into OFD in a statically admissible manner, if desired. In fact an alternative method of calculating Q_u is to perform such an extension and truncate it along the footing boundary, allowing a straightforward integration of σ_{zz} along OFA; this was the method originally adopted by Eason & Shield (1960) for homogeneous soil. Using either technique a lower bound bearing capacity factor of $N_c = 8.74$ is obtained. As above it can be shown that the rough footing stress field is extensible, both for $kD/s_{um} = 4$ and for all possible values.

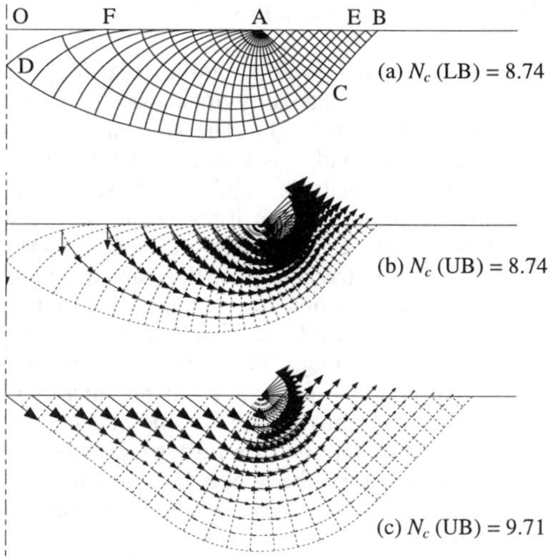

Figure 6. Stress and velocity fields for rough circular footing on non-homogeneous soil ($kD/s_{um} = 4$).

The upper bound obtained from the consistent velocity field (Fig. 6b) coincides with the lower bound at $N_c = 8.74$. By contrast the best independent upper bound (Fig. 6c) is $N_c = 9.71$, 11% above the exact solution. Note that a Hill-type mechanism is critical for this value of kD/s_{um}, even though the foundation is fully rough. As with the smooth footing, it is found that the independent upper bound mechanism gives its best results for large kD/s_{um}, converging to the exact solution (again $q_u = kD/6$) as the mudline strength $s_{um} \rightarrow 0$.

Shape factors for circular foundations on non-homogeneous cohesive soil have been given by various authors, based on either the method of characteristics (Salençon & Matar 1982, Houlsby & Wroth 1983, Tani & Craig 1995) or independent upper bound analysis (Kusakabe et al. 1986). It can now be stated with confidence that shape factors for Tresca soil derived from the method of characteristics are exact, but as discussed above no fully comprehensive numerical study has been performed. An unexpected result is that the shape factors for smooth and rough footings drop below unity for kD/s_{um} values which exceed about 4 and 2 respectively. They continue to drop as kD/s_{um} increases, slowly approaching the theoretical limit of 2/3 as s_{um} approaches zero, regardless of roughness. The traditional shape factor of 1.2 is obviously inappropriate when the non-homogeneity ratio kD/s_{um} is large, and this is a situation frequently encountered offshore.

4 SKIRTED FOUNDATIONS

It is also possible to develop rigorous plasticity solutions for the bearing capacity of caissons or skirted circular foundations (Fig. 1). The soil enclosed by the caisson is normally assumed to be rigid for bearing capacity analysis, and the base of the 'equivalent circular punch' is taken to be fully rough. While these assumptions are clearly appropriate for upper bound analysis, they are questionable as far as a rigorous lower bound solution is concerned – there may be difficulties in extending the stress field for a rough 'base' into the skirt compartment, particularly if the internal boundaries of the caisson are not fully rough. In this paper, however, the usual assumption of a fully rough interface at the skirt tip level will be adopted, subject to a minor caveat regarding the rigour of the lower bounds so obtained.

Only axially symmetric geometry will be considered here, although it is of course possible to develop lower and upper bound solutions for skirted foundations in plane strain. Shape factors can then be derived, but it is found that these are a function of both the embedment ratio (*d*/*D* or *d*/*B*) and the non-homogeneity ratio (kD/s_{um} or kB/s_{um}) such that a compact presentation of results becomes difficult.

4.1 *Homogeneous soil*

Houlsby & Wroth (1982b) used the method of characteristics to obtain lower bound solutions for the bearing capacity of an embedded circular punch ($0 \leq d/D \leq 1$) with smooth sides and base. This work was extended by Martin (1994) to include bases of full and intermediate roughness (but smooth sides), for embedment ratios of up to 2.5. Tani & Craig (1995) developed lower bound solutions for the rough base and rough sides case, but unfortunately they considered only a very limited range of embedment ($0 \leq d/D \leq 0.3$). None of these analyses addressed the question of stress field extensibility. Interestingly, previous studies of the upper bound problem have been focused on offshore applications, and have not considered the case of homogeneous soil.

Figure 7a shows a typical mesh of stress characteristics for a smooth-sided skirted foundation ($d/D = 0.5$) in homogeneous soil. The lower bound bearing capacity factor is $N_c = 7.05$, but the consistent velocity field leads to an upper bound of $N_c = 11.86$ (Fig. 7b). The discrepancy is surprisingly large, and is essentially attributable to a conflict between the direction of shearing stress and the direction of shearing strain along the characteristic XY (which is a velocity discontinuity). In this case, and for embedded foundations generally, the independent upper bound mechanism allows a much improved result, $N_c = 8.90$ (Fig. 7c). Note that, in this example, soil flows vertically upwards past the skirts as the foundation moves down, but there is no dissipation because the interface is smooth.

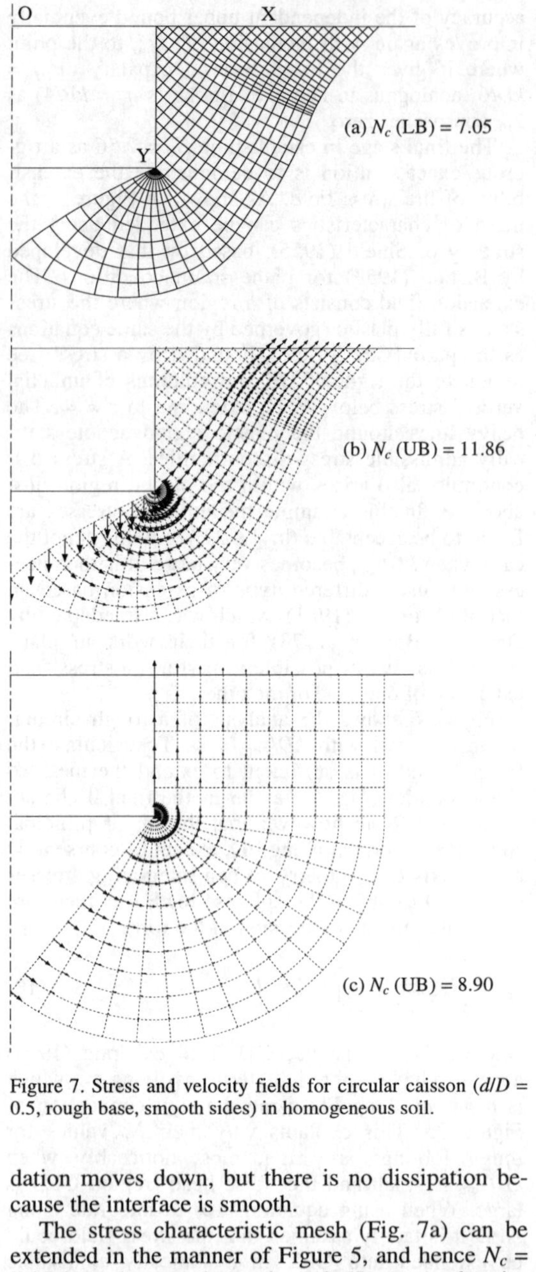

Figure 7. Stress and velocity fields for circular caisson (d/D = 0.5, rough base, smooth sides) in homogeneous soil.

The stress characteristic mesh (Fig. 7a) can be extended in the manner of Figure 5, and hence N_c = 7.05 is a rigorous lower bound. However, the significant separation between lower and upper bounds suggests that it may be possible to improve on the 'standard' lower bound solution. This can indeed be achieved, using an alternative stress field of the type shown in Figure 8. The mesh of characteristics is constructed by working from an artificial (straight) free surface inclined at some angle to the horizontal (cf. Meyerhof 1951). As this inclination is increased, the calculated bearing capacity also increases, since the α characteristics rotate through a larger angle (see equation (7)). There is a limit on the allowable inclination, however, due to the need to extend the partial stress field throughout the soil mass, an aspect overlooked by Meyerhof.

In Figure 8 the artificial free surface is inclined at 47° since this is the maximum that allows an extension of the characteristic mesh to be accommodated beneath the real free surface. In this instance an extension strategy based on that of Cox et al. (1961) is convenient, and this is shown in Figure 9. Heavy lines around the extended mesh denote statically admissible discontinuities. Stresses to the right of EF and FG extend to infinity as indicated, and do not

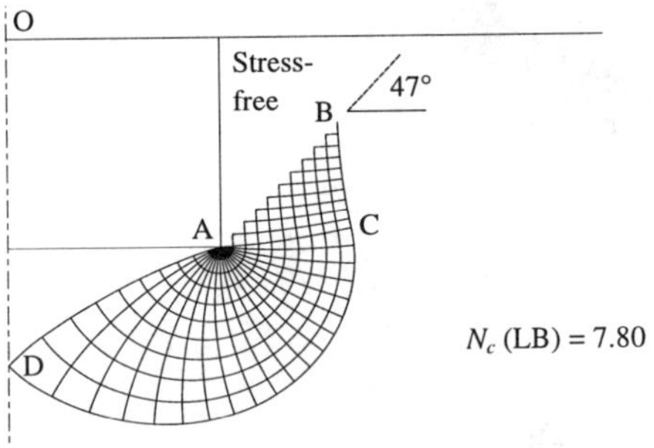

Figure 8. Alternative stress field for caisson of Figure 7.

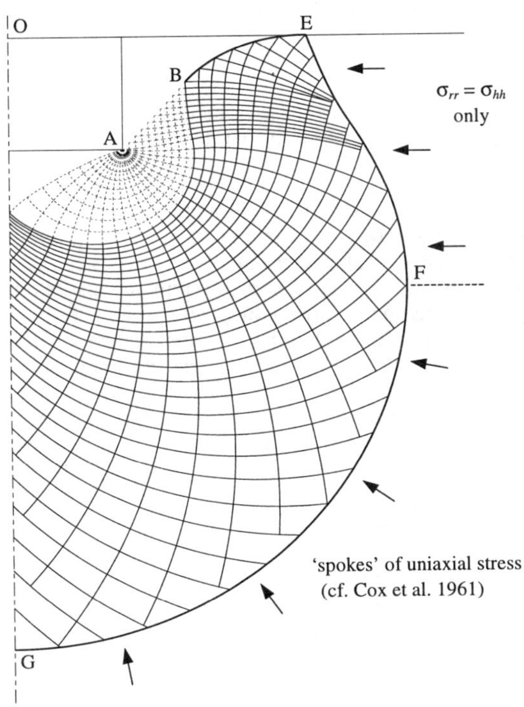

Figure 9. Extension of stress field of Figure 8.

violate yield. This alternative lower bound solution gives N_c = 7.80, a 10% improvement on the 7.05 which was obtained earlier, bracketing the exact solution to within ±7%.

Figure 10a shows upper and lower bound bearing capacity factors calculated in a similar fashion for skirt length ratios $0 \leq d/D \leq 2$. Over most of this range the alternative lower bound gives a significant improvement on the standard lower bound. The consistent upper bound has not been plotted because the independent solution is better (i.e. lower) for all but the shallowest of embedments. There is reasonably close bracketing of the true solution for embedments up to about d/D = 1. The alternative lower bound cuts off at N_c = 9.28 when d/D = 1.27, corresponding to the point where the partial stress field for an artificial free surface inclination of 90° first becomes extensible. This cutoff N_c value can be marginally improved by adopting more complex variations of the Haar/von Karman parameter ω (for example as a function of the principal stress inclination θ) rather than fixing it at ω = −1 when constructing the partial stress field and its extension. Also shown in Figure 10a are the semi-empirical recommendations of Skempton (1951) and Brinch Hansen (1970) for a smooth-sided circular foundation embedded in ho-

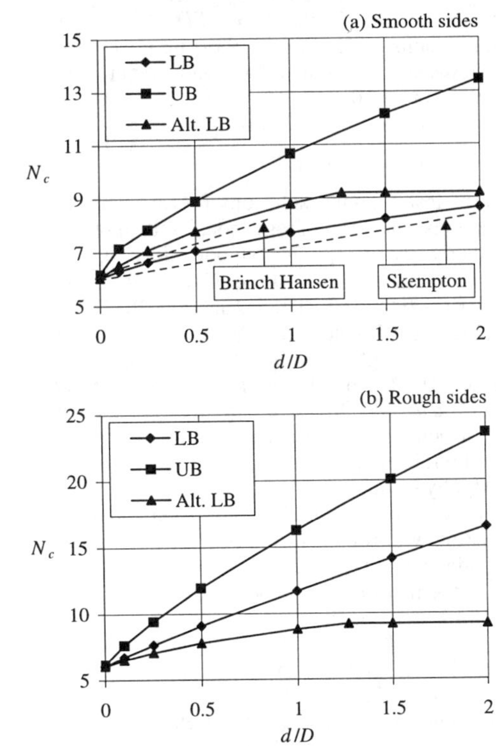

Figure 10. Lower and upper bound solutions for rough-based caissons embedded in homogeneous soil.

mogeneous cohesive soil. Although Brinch Hansen's N_c factors are much less conservative than those of Skempton, they still plot below the alternative lower bound.

Figure 10b shows bound solutions for a skirted foundation with rough sides in homogeneous soil. In the alternative lower bound the soil adjacent to the skirts is stress free (see Fig. 8), hence these N_c values are unchanged from the ones in Figure 10a. By contrast the standard lower bound solution takes full advantage of the rough skirts, mobilising an interface shear stress of s_u over the entire depth d. Bracketing between the lower and upper bounds is again fairly good up to around d/D = 1. Note that the bearing capacity factors given in Figure 10b are inclusive of interface shear stresses on the skirts, whereas Tani & Craig (1995) chose to report only the base resistance component.

Hu et al. (1999) have reported results of finite element analysis of embedded caissons, and noted that an embedment ratio $d/D = 1$ marks the start of a transition from surface failure, to a confined elasto-plastic failure. Their results were for a rigidity index of G/s_u = 167 (where G is the soil shear modulus) and the transition depth will vary with this ratio. For depths greater than this transition depth, the lower and upper bounds are increasingly irrelevant, as the bearing capacity is determined by a cavity expansion limit pressure. At depths up to d/D = 1, the finite element results of Hu et al. (1999) lie close to midway between the best lower and upper bounds shown in Figure 10.

4.2 *Non-homogeneous soil*

Relevant lower bound solutions for circular skirted foundations in non-homogeneous soil have been published by Martin (1994) (smooth sides, $0 \le d/D \le 2.5$, $0 \le kD/s_{um} \le 5$) and Tani & Craig (1995) (rough sides, $0 \le d/D \le 0.3$, $0 \le kD/s_{um} \le 30$). The only available upper bounds are those of Hu et al. (1998) (smooth and rough sides, $0 \le d/D \le 0.5$, $s_u = kz$) and Hu et al. (1999) (smooth and rough sides, $0 \le d/D \le 0.5$, $2 \le kD/s_{um} \le 30$). A comprehensive numerical study of both lower and upper bounds for $0 \le d/D \le 2$ and the full range of kD/s_{um} is currently being undertaken by the authors.

The case chosen for illustration here involves a foundation with rough skirts (d/D = 0.5) in normally consolidated soil (s_{um} = 0, $s_u = kz$). As shown in Figure 11, the method of characteristics yields an (extensible) lower bound of N_c = 9.34 and a consistent upper bound of N_c = 15.95. The 'alternative' lower bound gives N_c = 9.27, marginally below the standard solution. Note that these are bearing capacity factors with respect to s_{uo}, the strength at skirt tip level (see equation (2)). The independent kinematic mechanism gives an improved upper bound of N_c = 12.19 (Fig. 11c). Since the skirts are fully rough, there is no advantage from an 'annular' flow mechanism of the type shown in Figure 7c. The optimum mechanism does, however, involve a fairly steep exit angle, with a wedge of soil that remains stationary with respect to the surrounding soil mass.

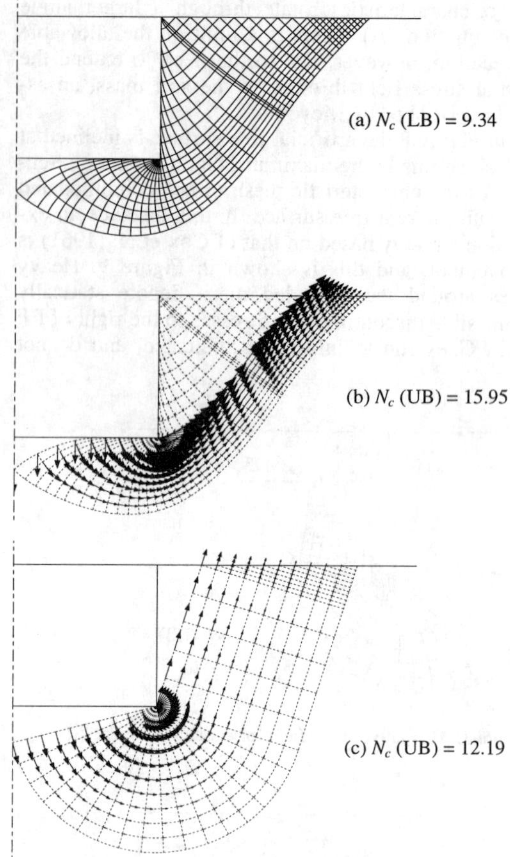

Figure 11. Stress and velocity fields for circular caisson (d/D = 0.5, rough base, rough sides) in soil with $s_u = kz$.

Corresponding solutions for other values of d/D are shown in Figure 12 for smooth- and rough-sided caissons. For the former, the alternative lower bound offers some improvement over the standard lower bound, and provides excellent bracketing up to about d/D = 1.5. For rough-sided caissons, the alternative lower bound is only optimal at very shallow embedments. Comparing Figures 12a and 12b, it is interesting to note that the assumed degree of skirt roughness is not particularly crucial for normally consolidated soil, in contrast to the situation for homogeneous soil (compare Figs 10a and 10b).

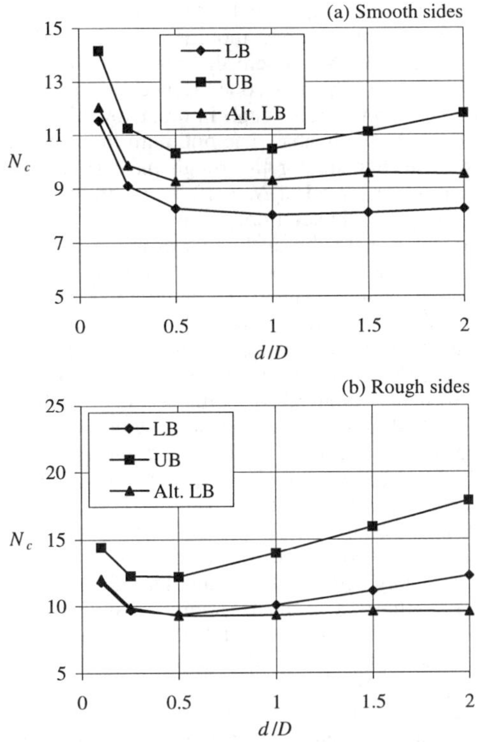

Figure 12. Lower and upper bound solutions for rough-based caissons embedded in non-homogeneous soil ($s_u = kz$).

5 BURIED PLATE FOUNDATIONS

This section presents some new plasticity solutions relating to the collapse of a deeply buried circular plate foundation in cohesive soil. The soil mass is assumed to be homogeneous (uniform s_u) and of infinite extent. It is also assumed that there is no restriction on the development of tensile stress, whether in the soil itself or on the rear face of the plate. While these assumptions allow a relatively simple analysis using the method of characteristics, several complicating factors may be important in a practical design situation (particularly one involving uplift loading), including proximity to the free soil surface, a non-homogeneous profile of s_u, and the possibility of 'breakaway' behind the plate. Indeed, it is an interesting academic question as to whether breakaway affects the bearing capacity of a deeply embedded plate, either in rigid-plastic or elasto-plastic soil.

Consider first the case of a vanishingly thin plate with a fully smooth surface. As shown in Figure 13a, the appropriate mesh of characteristics is symmetrical above and below the plate. The stress field is composed of two Haar/von Karman regions ($\omega = \pm 1$) separated by a buffer zone (Randolph et al. 2000) in which the meridional stress state is defined by

$$\sigma = \sigma_o + \gamma z + s_u z/r \qquad \theta = 3\pi/4 \tag{11}$$

and the hoop stress by

$$\sigma_{hh} = \sigma_o + \gamma z \tag{12}$$

Stresses in the buffer zone satisfy both equilibrium and the yield criterion; there are statically admissible jumps in σ_{hh} at each edge of the zone. The constant σ_o defines an arbitrary reference value of σ on $z = 0$ (cf. Randolph & Houlsby 1984). When building the mesh, new α characteristics are initiated by enforcing continuity of σ and θ (and thus of σ_{rr}, σ_{zz} and τ_{rz}) between the buffer zone and the partially constructed Haar/von Karman regions. The major principal stress rotates from $\theta = 3\pi/4$ in the buffer zone to $\theta = 0$ underneath the plate and $\theta = 3\pi/2$ above it. Integration of the contact stresses leads to a lower bound bearing capacity factor of $N_c = 12.42$. Extension of the stress field to infinity (both below and above the plate) can be achieved using the strategy developed by Randolph et al. (2000) for their rigorous lower bound analysis of a spherical penetrometer.

The consistent velocity field for the smooth plate is computed in the usual fashion (Fig. 13b), and with sufficient mesh refinement gives a converged upper bound of $N_c = 12.42$. This is coincident with the

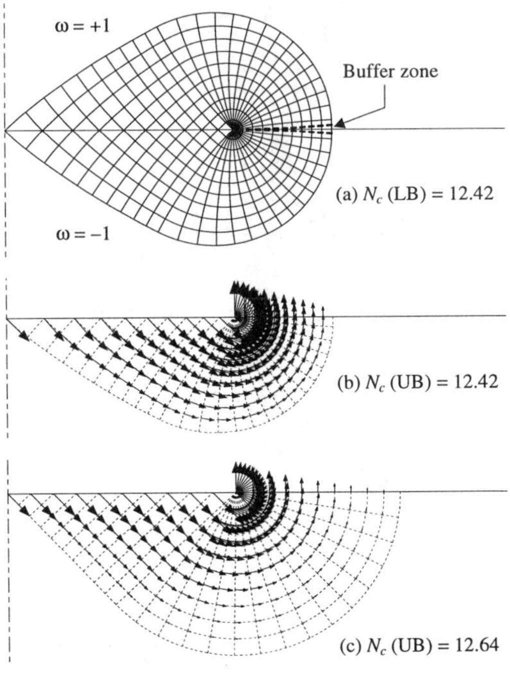

Figure 13. Stress and velocity fields for smooth plate.

lower bound and therefore represents the exact solution. Only half of the field is shown to allow a better appreciation of the velocity profile across $z = 0$. As would be expected, the flow of soil is heavily concentrated in the region close to the edge of the plate. The independent upper bound mechanism (Fig. 13c) yields a good result, $N_c = 12.64$, less than 2% above the exact answer. This is despite the fact that the width of the deforming region is nearly twice that of the consistent velocity field.

Figure 14 shows the corresponding lower and upper bound solutions for a vanishingly thin, perfectly rough circular plate. The bounds obtained from the method of characteristics are again coincident, this time giving an exact bearing capacity factor of $N_c = 13.11$. The lower bound stress field can be extended into the surrounding soil mass using the same technique as for the smooth case. If the stress field is extended into the 'false head' region it is found that the maximum shear stress mobilised on the surface of the plate is $0.76s_u$, where the degenerate α characteristic leaves the fan zone. Hence the interface need not be completely rough in order for this exact solution to be applicable. The optimum independent collapse mechanism for the rough plate (Fig. 14c) gives an upper bound of $N_c = 13.31$, again less than 2% above the exact solution.

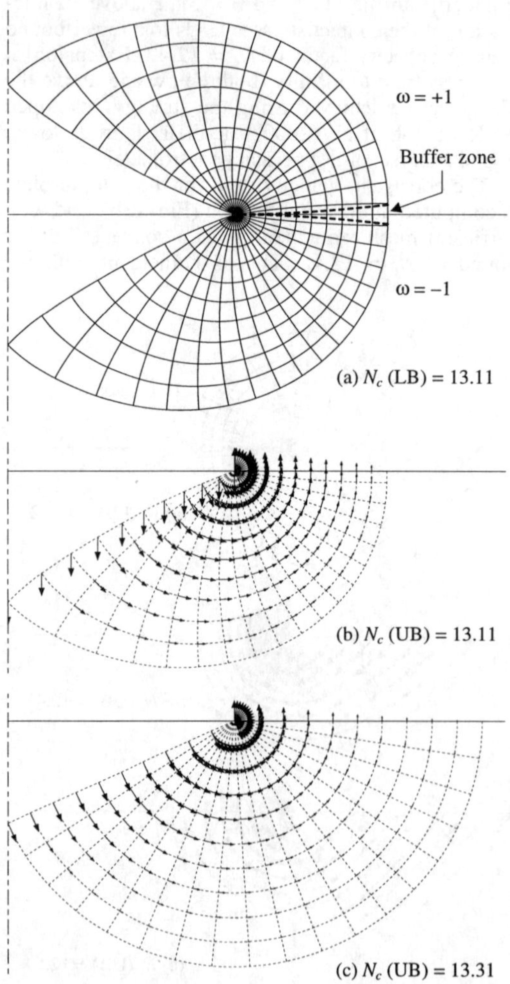

Figure 14. Stress and velocity fields for rough plate.

The solution to the analogous problem in plane strain may be developed analytically using the method of characteristics. As noted by Meyerhof (1951) the collapse load is $N_c = 3\pi + 2 = 11.42$, independent of the roughness of the plate. The fact that this solution is exact does not appear to have been formally established, with workers such as Rowe (1978) and Merifield et al. (1999a) acknowledging $N_c = 11.42$ as an upper bound but highlighting the need for a rigorously extended lower bound stress field (as opposed to Meyerhof's partial stress field). In the course of the present work it has been shown that the requisite extension can indeed be performed, using either the Randolph & Houlsby (1984) method or a plane strain version of the Randolph et al. (2000) method. Details are given in a forthcoming paper by the authors (Martin & Randolph, 2000).

Exact shape factors for infinitely deep circular plates in a homogeneous, isotropic Tresca material are therefore 1.09 (smooth) and 1.15 (rough). It is worth noting that on the basis of his approximate stress field analysis, Meyerhof (1951) concluded that the shape factor should be unity for a deeply buried circular plate, independent of its roughness. While this is clearly an erroneous result, at least within the confines of the highly idealised plasticity problem posed here, certain features of real soils (in particular strength anisotropy) may mean that a shape factor of 1.0 is actually quite appropriate in practice. As noted by Randolph et al. (2000), the observed bearing capacities of cylindrical and spherical penetrometers are in fact very similar in a wide range of soft soils, despite theoretically predicted shape factors (sphere ÷ cylinder) which average around 1.2 for isotropic strength properties. A comparable experimental investigation of plane strain and axisymmetric 'plate penetrometers' would certainly be of interest.

6 COMBINED LOADING

The bearing capacity of caissons under combined (V, M, H) loading is of paramount importance for offshore applications. While rigorous bounds have been developed for plane strain geometries (e.g. Salençon & Pecker 1995, Houlsby & Puzrin 1999), corresponding solutions for circular (or other) geometries remain a significant challenge.

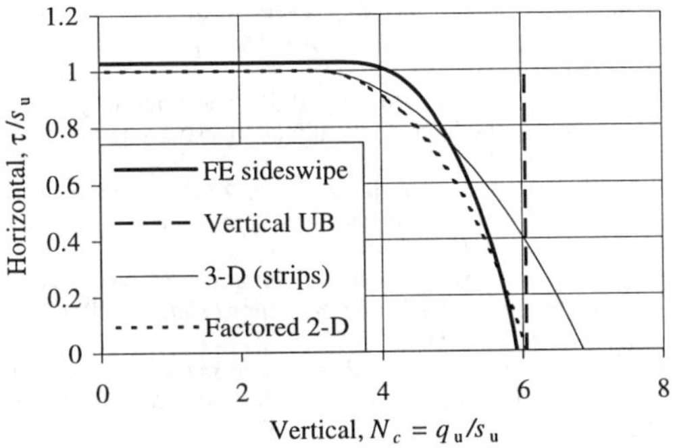

Figure 15. Yield envelopes in (*V, H*) space for rough circular foundation on homogeneous soil.

While there is insufficient space to address this topic in detail in the present paper, it is interesting to consider how best to extend the plane strain solutions to three dimensions, and the form of upper bound collapse mechanisms under general loading. At present, finite element based solutions must be relied on, either through the techniques of Sloan (1988, 1989) for rigid plasticity, which have recently been extended to three-dimensional problems (Merifield et al. 1999b), or using conventional elasto-plastic analysis.

Figure 15 shows an example of a yield envelope in (*V, H*) space for a rough circular foundation resting on the surface of a homogeneous soil. Two rigorous upper bounds are indicated: a vertical cutoff at a bearing capacity of $6.05s_u$; and a three-dimensional solution based on a series of parallel strips, each of which deforms according to the classical Green (1954) solution. The latter is a recent solution developed by Puzrin & Randolph (yet to be published).

Also shown in Figure 15 is a result from an ABAQUS finite element analysis using the Fourier mode for non-symmetric loading. A circular foundation resting on homogeneous soil has been subjected first to bearing failure under purely vertical load, and then a 'sideswipe' horizontal displacement at fixed vertical position. The computed vertical and horizontal limits (normalised by s_u) are 5.91 and 1.03 respectively, compared with exact values of 6.05 and 1. Interestingly, although it would be expected that the sideswipe load path would tend to lie marginally inside the true yield envelope (because of internal plasticity within the soil mass), it actually crosses the upper bound solution.

The final curve on Figure 15 is a scaled version of the plane strain solution (Green 1954), where the vertical load axis has been stretched by the shape factor, 6.05/5.14 = 1.18. Does this represent a true lower bound?

7 CONCLUSIONS

This paper has endeavoured to summarise available bound solutions for axially symmetric foundations in soil where the strength may vary linearly with depth. Classical solutions stemming from the 1950s are put in the context of more recent solutions, where improvements have been made in accuracy (due to modern computer capabilities) and rigour (with new techniques for lower and upper bound solutions).

Many previously published lower bounds have not been rigorously proven by demonstrating extensibility of the partial stress field. Methods of extending lower bound solutions using the method of characteristics have been demonstrated here, for both surface foundations and embedded caissons. The introduction of an artificial free surface has been shown to yield significantly improved lower bounds for smooth-sided caissons.

For surface foundations, an upper bound mechanism that is consistent with the lower bound stress characteristics allows exact bearing capacities to be evaluated. However, for embedded caissons, the (apparently) consistent mechanism includes regions where the shear strain rate is of opposite sign to the lower bound shear stress, leading to very poor upper bound estimates. Instead, a simple independent upper bound mechanism, with only two parameters to be optimised, has been shown to furnish reasonably bracketed solutions, with uncertainty bands generally better than ±10%.

The challenge for the future is to derive rigorous bound solutions for circular and square foundations under combined (*V, M, H*) loading. While the bearing capacity may be ascertained on a case-by-case basis from finite element analysis, an understanding of the collapse mechanisms under combined loading would be a major benefit in terms of permitting simple design calculations, particularly where the soil stratigraphy is non-homogeneous.

ACKNOWLEDGEMENTS

The work described here forms part of the activities of the Special Research Centre for Offshore Foundation Systems, established and supported under the Australian Research Council's Research Centres Program. This support is gratefully acknowledged.

The authors would also like to thank Mr. A. House for performing the ABAQUS analysis depicted in Figure 15.

REFERENCES

Bishop, J.F.W. 1953. On the complete solution to problems of deformation of a plastic-rigid material. *J. Mech. Phys. Solids* 2: 43-53.

Booker, J.R. & E.H. Davis 1972. A general treatment of plastic anisotropy under conditions of plane strain. *J. Mech. Phys. Solids* 20: 239-250.

Booker, J.R. & E.H. Davis 1977. Stability analysis by plasticity theory. In C.S. Desai & J.T. Christian (eds), *Numerical methods in geotechnical engineering*: 719-748. New York: McGraw-Hill.

Brinch Hansen, J. 1970. A revised and extended formula for bearing capacity. *Bulletin No. 28*, Danish Geotechnical Institute, Copenhagen: 5-11.

Cox, A.D., G. Eason & H.G. Hopkins 1961. Axially symmetric plastic deformation in soils. *Proc. R. Soc. London (Ser. A)* 254: 1-45.

Davis, E.H. & J.R. Booker 1973. The effect of increasing strength with depth on the bearing capacity of clays. *Géotechnique* 23(4): 551-563.

Eason, G. & R.T. Shield 1960. The plastic indentation of a semi-infinite solid by a perfectly rough circular punch. *J. Appl. Math. Phys. (ZAMP)* 11: 33-43.

Green, A.P. 1954. The plastic yielding of metal junctions due to combined shear and pressure. *J. Mech. Phys. Solids* 2: 197-211.

Houlsby, G.T. & A.M. Puzrin 1999. The bearing capacity of a strip footing on clay under combined loading. *Proc. R. Soc. London (Ser. A)* 455: 893-916.

Houlsby, G.T. & C.P. Wroth 1982a. Direct solution of plasticity problems in soils by the method of characteristics. *Proc. 4th Int. Conf. on Num. Meth. in Geomech., Edmonton* 3: 1059-1071.

Houlsby, G.T. & C.P. Wroth 1982b. Determination of undrained strengths by cone penetration tests. *Proc. 2nd Eur. Symp. on Penetration Testing, Amsterdam* 2: 585-590.

Houlsby, G.T. & C.P. Wroth 1983. Calculation of stresses on shallow penetrometers and footings. *Proc. IUTAM / IUGG Symp. on Seabed Mech., Newcastle upon Tyne*: 107-112.

Hu, Y. & M.F. Randolph 1998. H-adaptive FE analysis of bearing capacity of skirted foundations. *Proc. Int. Conf. on Offshore and Polar Eng., Montreal* 1: 549-556.

Hu, Y., M.F. Randolph & P.G. Watson 1999. Bearing response of skirted foundations on non-homogeneous soil. *J. Geotech. Eng. Div. ASCE* 125(11): 924-935.

Ishlinskii, A.I. 1944. Osesimmetrichnaia zadacha teorii plastichnosti i proba Brinellia. *Prikl. Mat. i Mekh.* 8(3): 201-208.

Kusakabe, O., H. Suzuki & A. Nakase 1986. An upper-bound calculation on bearing capacity of a circular footing on a non-homogeneous clay. *Soils and Foundations* 26(3): 143-148.

Levin, A. 1955. Indentation pressure of a smooth circular punch. *Quart. Appl. Math.* 13: 381-389.

Martin, C.M. 1994. *Physical and numerical modelling of offshore foundations under combined loads.* D.Phil. thesis, University of Oxford.

Martin, C.M. & M.F. Randolph 2000. Limit analysis of flow-round penetrometers in clay. Submitted for publication.

Merifield, R.S., S.W. Sloan & H.S. Yu 1999a. Stability of plate anchors in undrained clay. *Res. Rept No. 174.02.1999*, Dept of Civil Eng., University of Newcastle, Australia.

Merifield, R.S., A. Lyamin, S.W. Sloan & H.S. Yu 1999b. Three dimensional lower bound solutions for the stability of plate anchors in clay. *Res. Rept No. 178.02.1999*, Dept of Civil Eng., University of Newcastle, Australia.

Meyerhof, G.G. 1951. The ultimate bearing capacity of foundations. *Géotechnique* 2(4): 301-332.

Randolph, M.F. 2000. Effect of strength anisotropy on capacity of foundations. *Proc. John Booker Memorial Symp., Sydney* (in print).

Randolph, M.F. & G.T. Houlsby 1984. The limiting pressure on a circular pile loaded laterally in cohesive soil. *Géotechnique* 34(4): 613-623.

Randolph, M.F., C.M. Martin & Y. Hu 2000. Limiting resistance of a spherical penetrometer in cohesive material. *Géotechnique* 50(5): 573-582.

Rowe, R.K. 1978. *Soil–structure interaction analysis and its application to the prediction of anchor behaviour.* PhD thesis, University of Sydney.

Salençon, J. & M. Matar 1982. Capacité portante des fondations superficielles circulaires. *Journal de Mécanique théorique et appliquée* 1(2): 237-267.

Salençon, J. & A. Pecker 1995. Ultimate bearing capacity of shallow foundations under inclined and eccentric loads. I. Purely cohesive soil. *Eur. J. Mech.* (A) 14: 349-375.

Shield, R.T. 1955. On the plastic flow of metals under conditions of axial symmetry. *Proc. R. Soc. London (Ser. A)* 233: 267-287.

Skempton, A.W. 1951. The bearing capacity of clays. *Proc. Building Research Congress, London* 1: 180-189.

Sloan, S.W. 1988. Lower bound limit analysis using finite elements and linear programming. *Int. J. Num. & Anal. Meth. in Geomech.* 12: 61-77.

Sloan, S.W. 1989. Upper bound limit analysis using finite elements and linear programming. *Int. J. Num. & Anal. Meth. in Geomech.* 13: 263-282.

Szczepinski, W. 1979. *Introduction to the mechanics of plastic forming of metals.* Warsaw: Polish Scientific.

Tani, K. & W.H. Craig 1995. Bearing capacity of circular foundations on soft clay of strength increasing with depth. *Soils and Foundations* 35(4): 21-35.

Vesic, A.S. (1975). Bearing capacity of shallow foundations. In H.F. Winterkorn & H.Y. Fang (eds), *Foundation engineering handbook*: 121-147. New York: Van Nostrand.

Computer Methods and Advances in Geomechanics, Desai et al. (eds) © 2001 Balkema, Rotterdam, ISBN 90 5809 183 X

Bearing capacity of foundations by finite element methods

G.A. Bonner
Dublin Port Tunnel, Ireland

ABSTRACT: This paper describes the use of a computer program which has been developed for the rapid analysis of bearing capacity problems. The finite element grid, loading (including gravity) and boundary conditions may be generated automatically. Post processing of data permits graphical display of computed solutions. The stresses, strains and nodal displacements of any element can be accessed after program execution. Good agreement with published bearing capacity coefficients has been obtained and an efficient method of computing the bearing capacity of single and multi-layer soils is described.

1 INTRODUCTION

This paper describes the use of computer program *MAGNA* (*m*aterially *a*nd *g*eometrically *n*on-linear *a*nalysis) which has been developed by the writer for the rapid analysis of bearing capacity problems. The finite element grid, loading (including gravity) and boundary conditions may be generated automatically, which facilitates the rapid analysis of homogenous single and multi-layer bearing capacity systems. Good agreement with published bearing capacity coefficients has been obtained and an efficient method of computing the bearing capacity of single and multi-layer soils is described. Post processing of data permits graphical display of computed solutions. The stresses, strains and nodal displacements of any element can be found after program execution simply by entering the geographic location of the element at the command prompt. The option to include the effect of small strain large deformation is suppressed in the present analyses.

2 FINITE ELEMENT MODEL

The various computer packages available in the marketplace(e.g. UDEC-Bonner,1999), which may be used for the computation of foundation bearing capacity, are considerably less user friendly and relatively expensive. *MAGNA* utilises a displacement finite element formulation and the viscoelastic technique to simulate the behaviour of elastic-perfectly plastic materials under plane strain conditions using both Von-Mises and Mohr-Coulomb criteria (**Zienkiewicz and Cormeau**, 1974).

The assembled equations for static analysis may be represented by $\mathbf{K} \cdot \Delta\mathbf{u} = \mathbf{R} - \mathbf{F}$, where K represents the non-linear strain stiffness matrix. R is the vector of externally applied incremental nodal point loads and F is the vector of accumulated nodal point forces equivalent to the element stresses, which induce the displacement increments Δu. The global stiffness matrix K is assembled from the stiffness computed for each element by obtaining the volume integral of $\mathbf{B^TCB}\ \mathbf{dV}$, where C is the stress-strain material property matrix and B is the non-linear strain-displacement transformation matrix. The nodal forces accumulated from each load increment are obtained from the incremental visco-plastic strains in a similar manner by adding the result of the volume integral $\mathbf{B^TC}\ \delta\varepsilon_V\ \mathbf{dV}$ for all elements (where soil yielding is computed at the Gauss point) to the F vector (which resulted from the previous iteration), where $\delta\varepsilon_V = \Delta\mathbf{t}$. [rate of change of ε_V] and $\Delta\mathbf{t}$ represents the expressions derived by **Cormeau**(1975) for unconditional numerical stability, depending on the failure criterion adopted for analysis. The rate of change of visco-plastic strains depends on the extent to which the effective stress exceeds the yield criterion.

Stresses are computed at the centre of four-node quadrilateral isoparametric elements from the nodal displacements. Four-node rather than eight node elements were selected in order to reduce conventional memory demand, which maximises the number of elements permitted per model. Dynamic memory allocation is used when available conventional memory is exceeded. Loading may be applied as uniform point loads or as prescribed vertical displacements at nodes beneath the foundation. A constant square mesh in a rectangular grid was used to model most of the problems described herein although fewer elements were required to obtain accurate results for soils with low angles of internal friction. Ten layers of soil may be used in any particular model. In general, the modulus of elasticity used was 1.E5 kN/m^2 and Poisson's ratio 0.3.

3 COMPUTED SOLUTIONS

3.1 Homogenous Cohesive Soils

Figure 1 was plotted by post-processing of computed stresses and displacements using a graphics package. The reader may note that the bearing capacity of a homogeneous soil of cohesion C = 100kN/m^2 approaches the theoretical value $(2+\pi)C$ or $5.14C$ for a surface strip footing in smooth contact with the soil and $6.3C$ for a circular footing. The vertical stress versus displacement under the centre of the foundation was written to file and plotted using Microsoft Excel. The program automatically provides a similar plot on the terminal screen after program execution, which includes full details of the model analysed.

The response of a two layer cohesive soil system is shown in Figure 2. The bearing capacity varies from $5.14C$ for a homogeneous soil with cohesion C_2 to the corresponding value for a soil with cohesion C_1. The ratio C_1/C_2 equals 5, H represents the thickness of layer 1 and B is the foundation width. The finite element results are in agreement with the results of **Reddy and Srinivastan** (1967).

In accordance with **Prandtl** (1920) theory (bearing capacity $q = (2 + \pi)C = 5.14C$, where C is the weightless material cohesion) **Skempton**(1951) demonstrated that the value of the bearing capacity coefficient N_c for a strip foundation on soil of cohesion C increases from $5.14C$ to a limit value of $7.5C$ as the depth of the foundation is increased. Meyerhof has shown that the maximum value where there is no adhesion to the sides of the foundation is equal to $2\pi + 2$ or 8.28. The value corresponding to full adhesion is $2.5\pi + 1$ or 8.85. Figure 3 compares the computed bearing capacity with a plot of the Skempton variation of N_c with depth and the Meyerhof result. Adhesion to the sides of the foundation have not been included in the computations. Excellent agreement has been obtained between the value 8.28 determined by Meyerhof and the maximum value computed using *MAGNA*.

The finite element results obtained for inclined surface loading on an undrained cohesive soil are presented in Figure 4. The foundation was loaded at 10, 20 and 30 degrees to the vertical. Vertical and horizontal loads corresponding to the magnitude of inclined loads were applied to the unrestrained surface nodes beneath the smooth foundation.. The number of elements in contact with the foundation was six, located centrally on an 20 by 10 element rectangular grid. Good agreement was obtained with the values obtained from the Terzaghi bearing capacity correction factor applicable to inclined loading $(1-\alpha/90)^2$, where α is the angle of inclination of the inclined loading to the vertical.

3.2 Cohesive and Frictional (C/Φ) Soils

By employing static analysis of soil rupture surfaces, **Terzaghi** (1943) has shown that the bearing capacity of foundations on C/Φ soils which have weight may be expressed as $q = CN_c + q_0N_q + 0.5B\gamma N_\gamma$, where q_0 is the surcharge due to soil above the foundation depth, B is the foundation width, γ is the soil unit weight. The three terms contain the bearing capacity coefficients representing the contributions from soil cohesion, surcharge and weight respectively. The bearing capacity of foundations on a C/Φ soil can readily be calculated by using the tabulated values of the coefficients N_c, N_q and N_γ. Terzaghi suggested that the effect of depth on the bearing capacity of foundations could be modelled with sufficient accuracy by replacing the soil above foundation level by an equivalent surcharge load q_0. However, this approximation ignores the

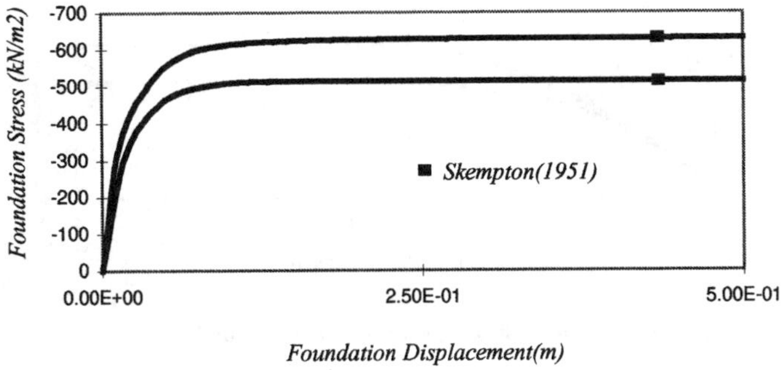

Figure 1 : Computation of Skempton Bearing Capacity Coefficients

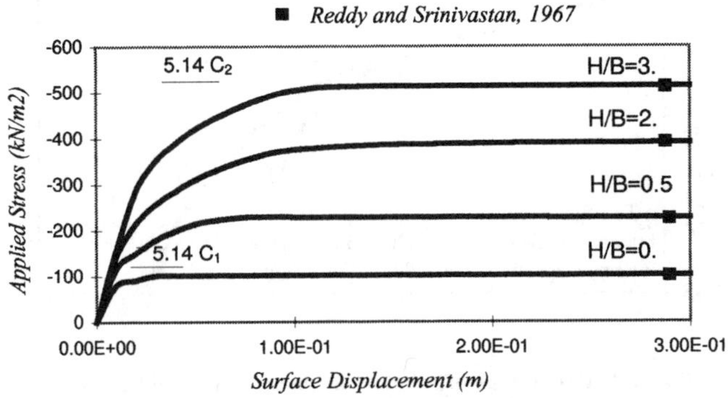

Figure 2 : Computation of Bearing Capacity for Two Layer System

shearing resistance of the soil above the foundation level, which results in an underestimation of bearing capacity. **Meyerhof** (1951) analysed the effect of foundation depth by considering the shear resistance of curved rupture surfaces in the zones of radial shear adjacent to the foundation and provided an equation of the same general form as that of Terzaghi for calculating bearing capacity with revised bearing capacity factors. However, application of the Meyerhof method is somewhat cumbersome for routine use. The equations provided by Terzaghi are still the most widely used because they can be used for all soil types and provide conservative estimates of bearing capacity. These estimates are often close to the allowable bearing pressures when the effect of settlement is considered. The writer has compared the solutions computed using *MAGNA* with the solutions obtained by several investigators. Good agreement with the bearing capacity coefficients N_c by **Prandtl**(1921), N_q by **Reissner**(1924) and N_γ by **Vesic**(1973) has been obtained using the finite element method as shown in Figure 5.

3.3 Multi-layer Soils

There are no general solutions similar to those described above available for calculating the bearing capacity of multi-layer soils with variable C/Φ properties. However, the bearing capacity for such soils may be computed using the finite element method. *MAGNA* may be used to rapidly establish the finite element model for a multi-layer soil system using automatic mesh generation. Input data includes foundation type (circular or strip), the number of soil layers, their cohesion, internal friction angles, unit weights and option for using Von Mises or Mohr-Coulomb failure criteria. The accuracy of the code has been verified by comparing computed solutions with the bearing capacity calculated from assumed rupture surfaces as described above. For

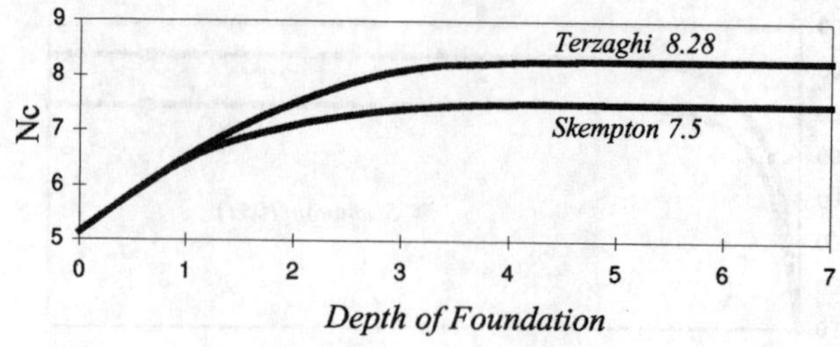

Figure 3 : Increase in N_c with Depth of Foundation

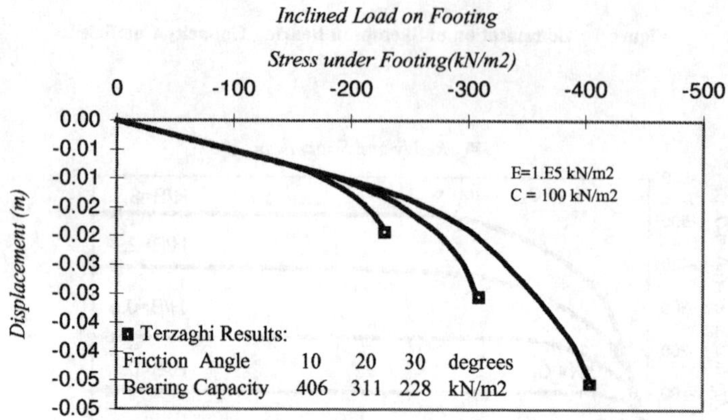

Figure 4 : Comparison of Solutions for Inclined Surface Loading with those of Terzaghi (1943)

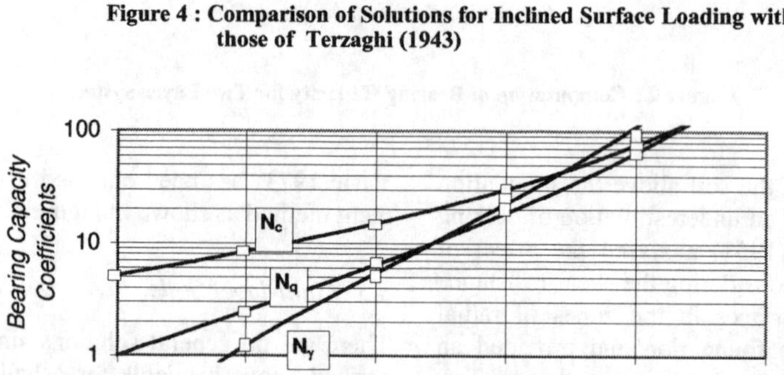

Figure 5 : Bearing Capacity Coefficients (Prandtl N_c, Reissner N_q, Vesic N_γ)

example, the bearing capacity determined from the coefficients plotted in Figure 5 for a foundation of width 0.75m on the surface of a soil with cohesion of 100kN/m^2, angle of internal friction 20 degrees and unit weight 20 kN/m^2 is 1525 kN/m^2. The finite element result obtained for this case after 153 loading steps was 1530kN/m^2. Data input and program execution time amounted to approximately 5 minutes.

4 CONCLUSION

This paper has provided a brief description of computer program *MAGNA* and various computations which were performed to obtain good agreement between calculated and published bearing capacity coefficients. The code permits the automatic generation of the finite element grid, loading and boundary conditions (followed by post-processing of the computed stresses and displacements) and provides a rapid method of analysing bearing capacity problems for homogeneous single and multi-layer soils.

REFERENCES

Bathe, K. J., "Finite Element Procedures in Engineering Analysis", Prentice Hall, Englewood Cliffs, 1982.

Bonner, G. A., "Materially and Geometrically Nonlinear Analysis", *PhD Thesis*, University of Dublin, Trinity College, 1985.

Bonner, G. A., "Norwegian Method of Tunnelling", Engineers Journal, Institution of Engineers of Ireland, May, 1999.

Cormeau, I. C. "Numerical Stability in Quasi-static Elasto-viscoplasticity", *Int. J. Num. Meth. Eng., 9,* 1975.

Meyerhof, G. G., "The Ultimate Bearing Capacity of Foundations", *Geotechnique, 2,* 1951.

Prandtl, L., "Uber die Härte plastischer Körper", *Nachr. Kgl. Ges. Wiss. Göttingen Math. Phys. Kl.*, 1920.

Prandtl, L., "Uber die Eindringungsfestigkeit (Härte) plastischer Baustoffe und die Festigkeit von Schneiden", *Zeitschrift fur Angewandte Mathematik und Mechanik*, Vol. 1, No. 1, Basel, Switzerland, 1921.

Reddy, A. S. and Srinivastan, R. J., "Bearing Capacity of Footings on Layered Clays", *Journal of Soil Mechanics and Foundations Division, ASCE, 93, No. SM2,* 1967.

Reissner, H., "Zum Erddruckproblem", Proc. 1st Int. Conf. Appl. Mech (Delft), 1924.

Skempton, A.W., "The Bearing Capacity of Clays", *Proc. Building Res. Congress*, 1951.

Terzaghi, K., "Theoretical Soil Mechanics", *John Wiley*, New York, 1943.

Vesic, A. S., "Analysis of Ultimate Loads on Shallow Foundations", *Journal of the Soil Mechanics and Foundation Division*, ASCE, Vol. 99, No. SM1, 1973.

Zienkiewicz, O. C. and Cormeau, I. C., "Viscoplasticity, Plasticity and Creep in Elastic Solids, A Unified Solution Approach", *Int. J. Num. Meth. Eng., 8,* 1974.

Computer Methods and Advances in Geomechanics, Desai et al. (eds) © 2001 Balkema, Rotterdam, ISBN 90 5809 183 X

Multiphase model for the 2D and 3D numerical analysis of pile foundations

E. Bourgeois & J.-F. Semblat
Laboratoire Central des Ponts et Chaussées, Paris, France

D. Garnier & B. Sudret
Ecole Nationale des Ponts et Chaussées, CERMMO, Marne-la-Vallée, France

ABSTRACT : The numerical analysis of soil and rock masses reinforced by linear inclusions (pile foundations, bolted tunnels...) is complex and costly when all inclusions are discretized in the computations. This paper presents an elastoplastic model in which the reinforced material is considered as the superposition of a main continuous medium representing the soil and several other continua (or phases) representing the reinforcements. This model has been included in the finite element code CESAR-LCPC. As an example, the settlement of a foundation resting on two groups of symmetrically inclined micropiles is analyzed. The model used makes it possible to discuss the influence of the angle of the piles on the behaviour of the foundation. Then, the behaviour of a square foundation resting on vertical piles (3D model) is discussed. The distribution of the settlements under the foundation and the influence of the inclusions are analysed. The multiphase model allows rather easy and fast numerical computations to investigate the response of pile foundations.

1. MODELLING OF REINFORCED GEOTECHNICAL STRUCTURES

Many geotechnical structures are reinforced by linear inclusions: pile foundations, bolted tunnels, etc. This paper focuses on the influence of piles and micropiles on the behaviour of foundations.

Since it is complex and costly to include a discretization of all the piles in a numerical model, we adopt here a model in which the heterogeneous medium made of the soil and the inclusions is replaced by an equivalent homogeneous material, defined as the superposition of several continuous media representing the initial material (matrix) and the different types of inclusions called the reinforcement phases. This approach is efficient for materials reinforced by a large number of inclusions periodically distributed in space.

2. THE MULTIPHASE MODEL

2.1 *General features*

It is assumed that, at the scale of civil engineering works, a soil or rock mass reinforced by N groups of linear inclusions oriented along N directions, can be represented by a multiphase continuum, i.e. the superposition of a main continuous medium and N additional continua, each having its own velocity field.

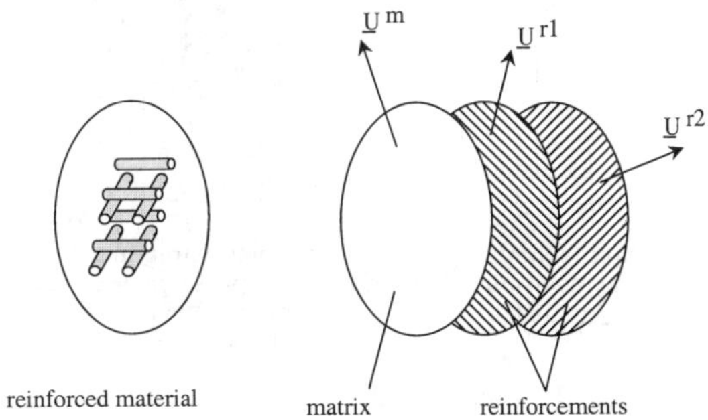

Figure 1 : reinforced material seen as the superposition of several continuous media.

In what follows the different continuous media are referred to as the "phases" and the superposition of the phases is called the "multiphase medium".

The following theoretical formulation has been developed by Sudret and de Buhan (1999). Using the virtual power principle, one is led to defining the internal forces in each phase : we denote by $\underline{\underline{\sigma}}^m$ the *partial stress tensor* in the main phase and $\underline{\underline{\sigma}}^r = \sigma^r \underline{e}_r \otimes \underline{e}_r$ the uniaxial stress tensor for each reinforcement phase where σ^r is the (scalar) partial stress in the inclusions and $\underline{e}_r$ a unit vector in the direction of the inclusions of the corresponding phase. The momentum balance takes the following form for the main phase and the inclusions phases respectively :

$$div \underline{\underline{\sigma}}^m + \rho^m \left(\underline{F}^m - \underline{\gamma}^m \right) + \sum_{r=1}^{N} \underline{I}^r = 0 \qquad (1)$$

$$div \underline{\underline{\sigma}}^r + \rho^r \left(\underline{F}^r - \underline{\gamma}^r \right) - \underline{I}^r = 0 \ ; \ r = 1,...,N \qquad (2)$$

where ρ^m, $\underline{F}^m$ and γ^m (resp. with index *r*) are the mass density, the external volumic force density and the acceleration of the main phase (resp. inclusions phases). $\underline{I}^r$ is the volumic density of interaction forces exerted by the reinforcement phase *r* on the main phase.

The link between σ^r (partial stress in the reinforced phase *r*) and σ_{inc}, the average stress in the inclusions, results from the identification of the force acting on the section of an inclusion in a discrete model and in the multiphase model (figure 2) :

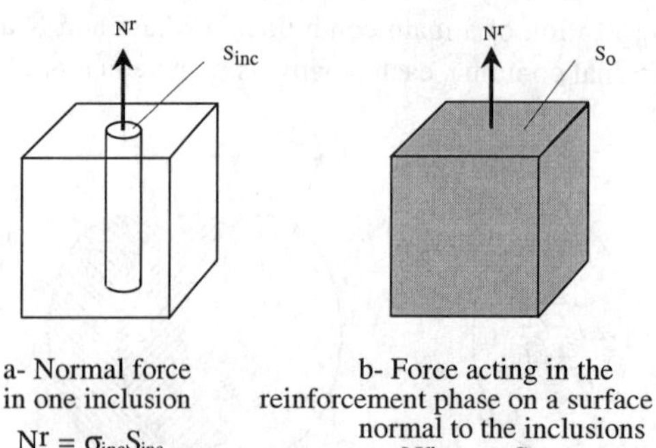

a- Normal force in one inclusion $N^r = \sigma_{inc} S_{inc}$

b- Force acting in the reinforcement phase on a surface S_o normal to the inclusions $N^r = \sigma^r \ S_o$

Figure 2 : Physical meaning of the partial stress σ^r.

Denoting by η^r the volumic ratio of the inclusions in the reinforced medium ($\eta^r = S_{inc}/S_o$), one gets :

$$\sigma^r = \eta^r \ \sigma_{inc} \qquad (3)$$

2.2 Elastoplastic adherent behaviour

From now on, we make the assumption that there is a perfect bond between the different phases. Thus, there is no relative displacement between the main phase and the reinforced phases, and one can define one strain tensor for all phases, denoted by $\underline{\underline{\varepsilon}}$. Besides, the modelling of the interaction forces is no longer needed.

Combining equations 1 and 2, one gets the global momentum balance as follows :

$$div \underline{\underline{\Sigma}} + \rho^m \left(\underline{F}^m - \underline{\gamma}^m \right) + \sum_{r=1}^{N} \rho^r \left(\underline{F}^r - \underline{\gamma}^r \right) = 0 \qquad (4)$$

where $\underline{\underline{\Sigma}}$ is the global stress tensor such as :

$$\underline{\underline{\Sigma}} = \underline{\underline{\sigma}}^m + \sum_{r=1}^{N} \sigma^r \underline{e}_r \otimes \underline{e}_r \qquad (5)$$

Within this framework, the main features of the elastoplastic constitutive model given in Sudret and de Buhan (1999) are the following :

- stresses are related to elastic strains by :

$$\underline{\underline{\sigma}}^m = \underset{\sim}{a}^m : (\underline{\underline{\varepsilon}} - \underline{\underline{\varepsilon}}_p^m) \qquad (6)$$

$$\sigma^r = a^r : (\underline{\underline{\varepsilon}} : (\underline{e}_r \otimes \underline{e}_r) - \varepsilon_p^r) \qquad (7)$$

where $\underset{\sim}{a}^m$ denotes the (fourth-order) elasticity tensor of the continuous medium in which the inclusions are put, a^r the (scalar) stiffness of the reinforcement phase *r*, $\underline{\underline{\varepsilon}}_p^m$ the plastic strain tensor in the main phase, and ε_p^r the scalar plastic strain in the reinforcement phase *r*.

- assuming that all phases follow an associated flow rule, plastic strains are described by :

$$\dot{\underline{\underline{\varepsilon}}}_p^m = \dot{\lambda}^m \frac{\partial f^m}{\partial \underline{\underline{\sigma}}^m} \quad ; \quad \dot{\lambda}^m \geq 0 \qquad (8)$$

$$\dot{\varepsilon}_p^r = \dot{\lambda}^r \frac{\partial f^r}{\partial \sigma^r} \quad ; \quad \dot{\lambda}^r \geq 0 \qquad (9)$$

where $f^m(\underline{\underline{\sigma}}^m)$ and $f^r(\sigma^r)$ are the yield functions of the main phase and of the reinforcements respectively, $\dot{\lambda}^m$ and $\dot{\lambda}^r$ the plastic multipliers associated with the flow rules of the different phases (note that non-associated flow rules could be included in the model).

If no plastic strain occurs, the elasticity tensor takes the following form :

$$\underline{\underline{\underline{\underline{A}}}} = \underline{\underline{\underline{\underline{a}}}}^m + \sum_{r=1}^{N} a^r \underline{e}_r \otimes \underline{e}_r \otimes \underline{e}_r \otimes \underline{e}_r \qquad (10)$$

This expression is similar to the result obtained through a classical elastic homogenization technique (Bernaud, 1995, Greuell, 1993), and shows that the reinforcements modify the moduli relating the strain of the multiphase medium to the global stress tensor $\underline{\Sigma}$. Even with an isotropic main phase, there is some anisotropy in the whole elasticity tensor due to the inclusions.

It is worth noting that the determination of the elastic properties of the multiphase model is quite simple : on the one hand, $\underset{\sim}{a}^m$ is the elasticity tensor of the continuous medium in which the inclusions are put and on the other hand the stiffness a^r of the reinforcement phase r is defined as the product of the Young's modulus of the corresponding inclusions by their volume fraction :

$$a^r = \eta^r E^r_{inc} \tag{11}$$

In a similar way, we adopt for $f^m(\underline{\underline{\sigma}}^m)$ the yield function of the soil with no reinforcement, and for $f^r(\sigma^r)$ the following definition :

$$f^r(\sigma^r) = \max(-\sigma^r - \sigma^r_-, \sigma^r - \sigma^r_+) \tag{12}$$

where σ^r_+ and σ^r_- are the limit values of the traction and compression stress in the reinforcement phase r, defined as the product of the volume fraction of the corresponding inclusions by the limit values of the traction and compression stress of the material they are made of.

3. 2D-FOUNDATION ON MICROPILES

3.1 Description of the problem

We first discuss the settlement of a 2D foundation resting on inclined micropiles. Micropiles are divided into two sets symmetric with respect to the center of the foundation and making opposite angles with the vertical direction. The problem under consideration is the estimation of the displacements induced by a vertical uniform load p applied on the foundation.

Given the symmetry, only half of the problem is represented in the model, for different values of the angles between the micropiles and the vertical : α=0°, ±15°, ±30° and ±45°.

Initial stresses in the ground are assumed to be geostatic, with a volume weight of γ=20kN/m^3 and K_o=1. Elastic strains of the soil are described by a linear elastic law (E = 15 MPa, ν = 0.49), and plastic strains are described by an associated flow rule with a Von Mises yield function (c = 100 kPa, φ = 0°).

The micropiles are hollow tubes of outer diameter 89 mm and inner diameter 79.5 mm, and are made of steel (E^r_{inc} = 200000 MPa). The density of piles is equal in all cases to 1 micropile per square meter in the upper plane of the soil layer.

The foundation width is equal to 2B, the length of the piles is B, the thickness of the soil layer is 4B and the right boundary is located at a distance of 8B from the center of the foundation.

Both components of the displacement field are zero on the lower boundary of the mesh whereas only the horizontal displacement is set to zero on the left and right boundaries.

3.2 Finite element model

Four different models have been used for the computations, including more than 2000 nodes and 1000 6-node triangle elements. Each model is depicted in figure 3 with a dark gray area representing the zone reinforced by two groups of inclusions and a light gray area showing the zone in which there are only micropiles of one group.

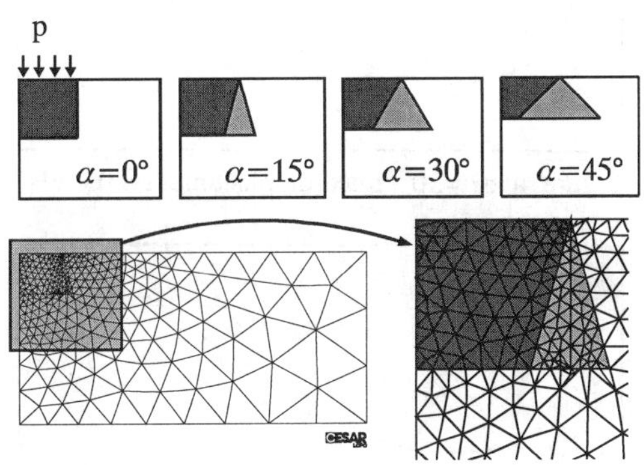

Figure 3 : Schematic of the 2D-foundation and the micro-piles and corresponding finite element model for α=15°.

Note that the micropiles are not discretized, which makes it very simple to discuss the influence of the density of piles, since no modification of the mesh is needed.

3.3 Results and comparisons

Figure 4 shows contour plots of the total displacement divided by the maximum displacement in the ground layer.

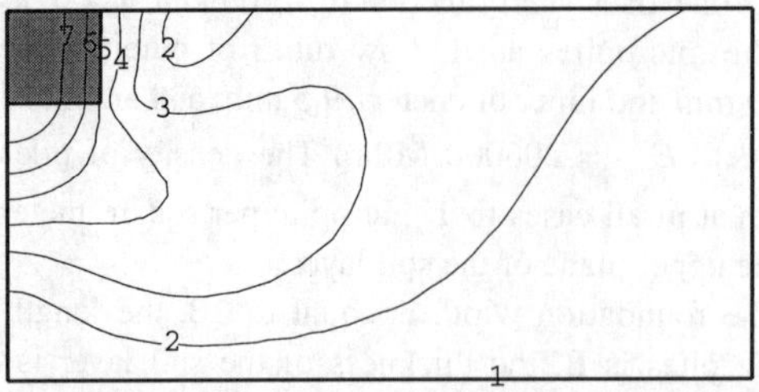

Figure 4a : α=0°. max. displacement 0.0354B

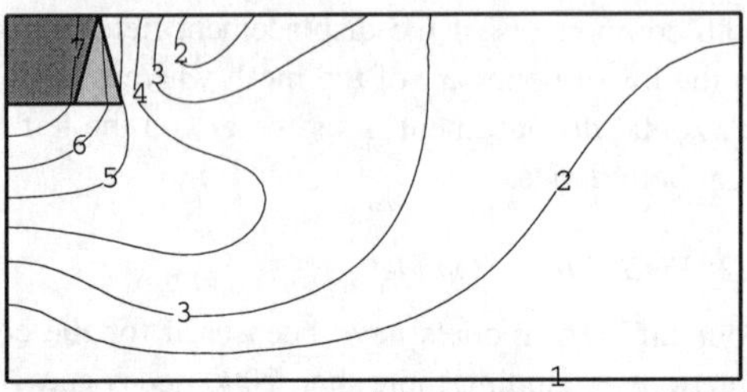

Figure 4b : α=15°. max. displacement 0.0273B

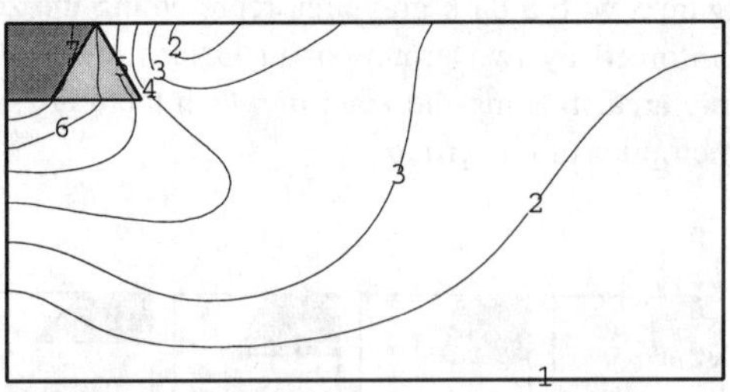

Figure 4c : α=30°. max. displacement 0.0316B

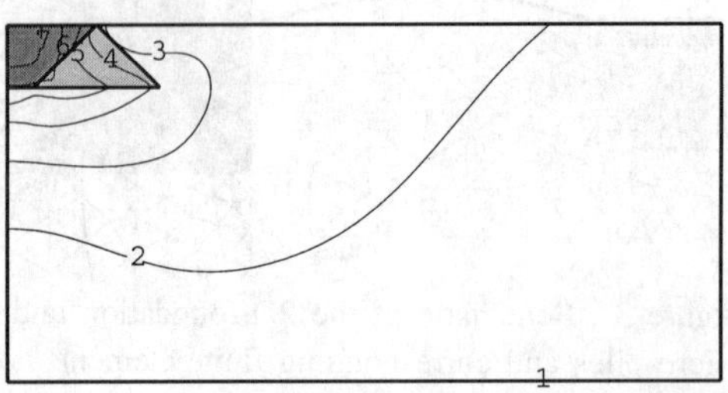

Figure 4d : α=45°. max. displacement 0.0872B

Figure 4 : ratio η of the total displacement divided by the maximum total displacement (for p = 500kPa). Line 1 corresponds to η=0; line 2 : η=12.5%; line 3 : η=25%; etc.

These results make it possible to discuss the influence of the reinforcements on the displacements, from a qualitative point of view.

In the case of vertical piles (α=0°), the largest displacements are mainly located directly under the foundation. For angles of 15° and 30°, the area of large displacements is enlarged to the vicinity of the piles. In other words, when the ground is reinforced by inclined micropiles, the load is distributed over a broader zone, which results in smaller values of the maximum displacement (i.e. of the foundation settlement). For the largest angle, the improvement of the response is located in a much smaller area, and the reinforcement proves less efficient in limiting the settlement.

From a quantitative point of view, the reinforcement of the ground results in smaller vertical displacements of the foundation for a given load p, as can be seen on figure 5. Computations show that the settlement of the foundation strongly depends on the value of α, and is maximum for a value of α close to 15°.

4. SETTLEMENT OF A 3D-SQUARE FOUNDATION

4.1 The 3D problem and model

The second example discussed here is the problem of the settlement of a square foundation resting on vertical piles. Geometrical and material data are taken from Clancy and Randolph (1993). The foun-

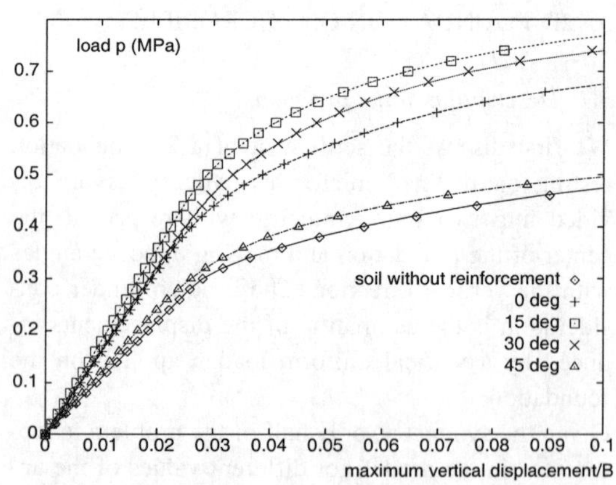

Figure 5 : Maximum vertical displacement under the foundation versus load p for different values of the angle α between the piles and the vertical.

dation is 36 m wide and 5 m thick. It is resting on a 9x9 network of piles of diameter 0.8 m and length 20 m. The distance from center to center between piles is 4 m. The foundation and the piles are made of concrete. The soil is assumed to be linearly elastic and isotropic (E = 280 MPa, ν = 0.4). Taking the symmetry of the problem into account, we built a finite element model with an 1/8th of the foundation (figure 6). Given the rather high value of Poisson's ratio, the size of the model has been set to 30 times the width of the foundation.

The finite element mesh is made of 4200 15-node pentahedra and includes 12500 nodes. There are three different linear elastic laws associated respectively with the foundation, the reinforced zone and the non reinforced zone.

The foundation is submitted to a uniform vertical load of 1 MPa, and computations are carried out to estimate the vertical displacements below the foundation.

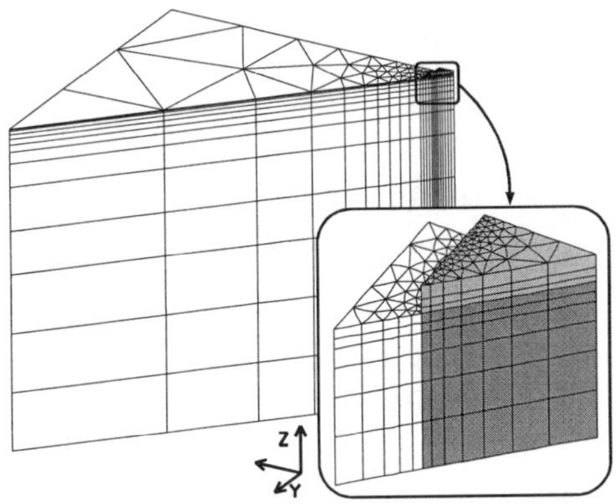

Figure 6 : 3D finite element model and meshing around the foundation.

4.2 Comparisons with former analysis

The results obtained with the multiphase model are compared with simulations carried out by Clancy and Randolph (1993), based on a method developed by Griffiths *et al* (1991), Clancy and Randolph (1993) for pile raft foundations, which can be considered as an extension of the "hybrid" method developed by Chow (1986) for pile groups.

Table 1 gives the values of the average vertical settlement w_{av} under the foundation, and the ratio of the vertical load in the piles to the sum of the loads in the piles and in the soil.

Table 1 – Summary of the results.

	Clancy and Randolph (1993)	Present analysis
average vertical settlement w_{av}	84mm	76mm
ratio of the load in the piles to the total load	67%	68%

The difference between the estimations of the average settlement is less than 10%, which is acceptable. This difference can be explained by the limited size of the mesh (the hybrid method implicitly takes into account an infinite soil mass) and by the fact that the model we have used is based on the assumption that no slippage occurs between the piles and the soil, which results in an overestimation of the stiffness of the pile raft.

Regarding the distribution of the load between the soil and the piles, the agreement between both approaches is very satisfactory.

Eventually, we discuss the distribution of the vertical settlements under the foundation. Clancy and Randolph (1993) introduce a normalized coordinate denoted by λ and defined as shown by figure 7.

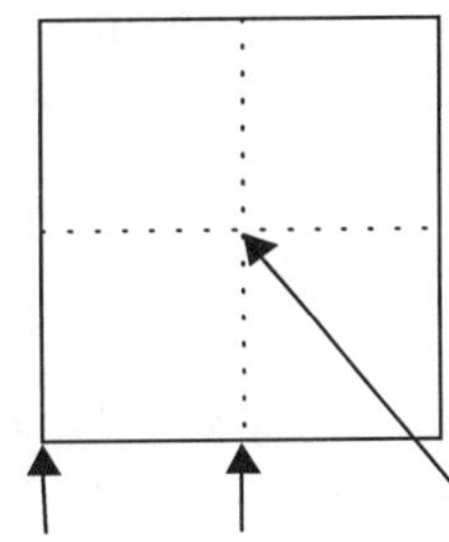

Figure 7 : Definition of the normalized coordinate λ.

The normalized settlement $\overline{w}$ is defined by :

$$\overline{w} = \frac{w - w_{av}}{w_{\max} - w_{\min}} \tag{13}$$

where w_{max} and w_{min} are respectively the maximum and minimum values of the settlement.

The comparison between the multiphase model and the method used by Clancy and Randolph (1993) is shown by figure 8. The agreement between both approaches is very good, which shows that the multiphase approach is well adapted to the problem under

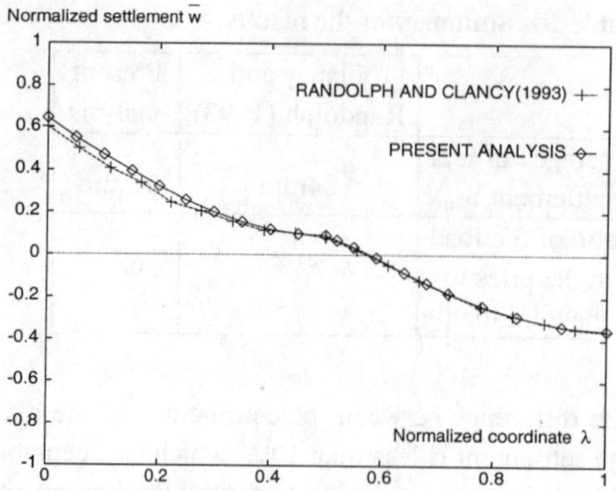

Figure 8 : Comparison between the present analysis with Clancy and Randolph (1993).

consideration (large number of piles periodically distributed in space).

It is worth pointing out that the multiphase model provides the solution of the problem for reasonable computation times, and makes it possible to discuss the influence of the number and diameter of the piles without changing the mesh. The multiphase model takes advantage of the features of the finite element method, and makes it possible to deal with complex geometries. Moreover, since it has been developed in the framework of elastoplasticity, the model makes it possible to account for a non-linear behavior of the soil.

5. CONCLUSION

The multiphase model allows rather easy and fast numerical computations to investigate the response of pile foundations, and results are in good agreement with the hybrid method. However, the method remains to be compared with measures carried out on a real pile foundation. Besides, the model could be improved to deal with the slippage between the piles and the ground (which has been neglected in this paper).

REFERENCES :

Bernaud D., de Buhan P., Maghous S. (1995). Numerical simulation of the convergence of bolt-supported tunnel through a homogenization method, *Int. J. Num. Analyt. Meth. Geomech.*, 199, pp.267-288.

de Buhan, P., Sudret, B. (1999). A two-phase elasto-plastic model for unidirectionally-reinforced materials, *Eur. J. of Mech., A/ Solids*, 18, pp.995-1012.

Chow, Y.K. (1986). Analysis of vertically loaded pile groups, *Int. J. Num. Analyt. Meth. Geomech.*, 23, pp. 61-78.

Clancy P., Randolph M.F. (1993). An approximate analysis procedure for piled raft foundations, *Int. J. Num. Analyt. Meth. Geomech.*, 17, pp. 849-869.

Greuell E., de Buhan P., Panet M., Salençon J. (1993). Comportement des tunnels renforcés par boulons passifs, *XIII Int. Conf. on Soil Mechanics and Foundation Eng.*, New-Delhi, India.

Griffiths D.V., Clancy P., Randolph M.F. (1991). Piled raft foundation analysis by finite elements, *Proc. Computer Methods and Advances in Geomech.*, Beer, Booker and Carter (eds), Balkema, pp. 1153-1157.

Computer Methods and Advances in Geomechanics, Desai et al. (eds) © 2001 Balkema, Rotterdam, ISBN 90 5809 183 X

Finite element analysis of jacket structures with bucket foundations

M.F. Bransby
Department of Civil Engineering, University of Dundee, UK

ABSTRACT: The behaviour of a two-dimensional idealisation of a jacket structure and a monopod with bucket foundations has been studied using finite element analysis, elasto-plastic design methods and lower bound plasticity analysis. Changing mechanisms of system and footing deformation have been observed as loading increases and footings yield. Reasonably good agreement between failure loads suggest the accuracy of failure load prediction. However, the results show the importance of encapsulating footing non-linearity for systems with small numbers of footings if serviceability displacements are to be predicted adequately.

1 INTRODUCTION

When offshore structures are loaded by waves, wind and current, their foundations are subject to combinations of vertical (V), horizontal (H) and moment (M) loading. The stiffness of the footing-soil interaction changes as soil deforms and yields progressively around each foundation under these complex loads. Hence, to predict the behaviour of offshore structures with multiple footings, analysis methods are required which allow for the stiffness changes in the soil-footing interaction.

Recently, both foundation performance and system behaviour have been analysed using plasticity methods. The undrained capacity of individual footings such as spudcans or bucket foundations undergoing combined V-M-H loading is expressed in terms of a yield locus where f(V,M,H) = 0 at yield (e.g. Murff 1994, Tan 1990, Martin 1994, Bransby & Randolph 1998). With the assumption of linear elasticity within the yield surface and the use of a suitable plastic potential and hardening rule, the footing behaviour can be modelled in a work-hardening plasticity framework (e.g. Tan 1990, Martin 1994, Martin and Houlsby 1999).

Analysis of the behaviour of an entire structure-footing system is less well documented. Schotman (1989), Martin (1994) and Martin and Houlsby (1999) analysed jack-up rig behaviour using plastic footing models together with elastic analysis of the super-structure. Murff (1994) calculated the capacity of rigid structures with plastic foundations using bound plasticity theorems. Perol and Meimon (1989) analysed jack-up rigs under cyclic loading conditions using 2D finite element analysis. Bransby and Martin (1999) compared results of a set of centrifuge tests of a jacket structure with analyses of the type used by Shotman (1989) and Murff (1994) and found that methods appeared suitable for the calculation of failure loads. Little work has been carried out to examine the behaviour of the footings before the 'yield' surface is reached and its effect on system behaviour. This is important with bucket foundations because the load state is more likely to remain within the yield locus during environmental loading.

This paper considers the behaviour of both a monotower and an idealised jacket structure, each with bucket foundations founded in undrained clay. The structure is simplified as plane strain and rigid and the environmental loading is simplified by considering the application of a monotonically increasing point load at a fixed height above the seabed. Results from FE analysis are presented and then elasto-plastic and lower bound analysis was carried out to investigate the ability of the different analysis methods to predict the capacity and the load-displacement response of particular offshore structures.

2 BEHAVIOUR OF SINGLE BUCKET FOUNDATIONS

2.1 *Geometry and soil conditions*

Bucket foundations are rigid circular foundations with vertical skirts penetrating into the soil around their circumference, thus appearing like an up-turned bucket. They are used increasingly offshore as an alternative to piles. Bucket foundations will be considered with a diameter or width, D and skirt length, d (Figure 1). Soil within the skirts is usually considered to be rigid as it deflects little during undrained loadings and remains in place during tension

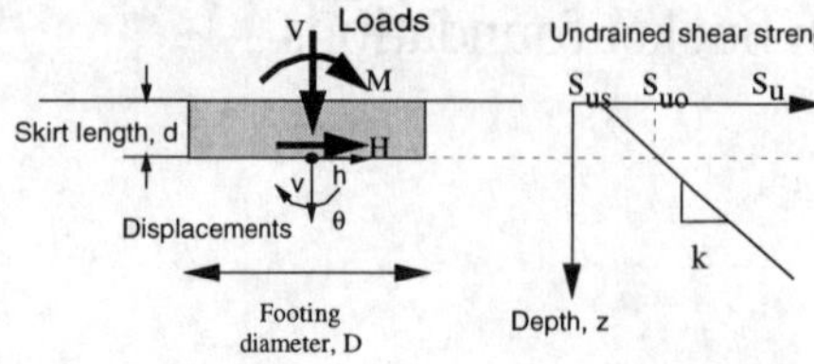

Figure 1. Simplified geometry of a bucket foundation

because of suction (Tani & Craig 1995, Bransby & Randolph 1999). Loads and displacements are defined as shown in Figure 1 and both are measured at the centre of the footing at skirt tip level.

The soil will be idealised with undrained shear strength, s_{us} at the surface and a shear strength gradient, k (Figure 1). The dimensionless group kD/s_{uo} reflects the increase of shear strength with depth. The shear strength at the skirt tip depth will be defined as s_{uo} (i.e. prior to plastic penetration or heave, $s_{uo} = s_{us} + kd$). The base will be considered to be fully rough as it represents soil-soil shear.

2.2 *Elasto-plastic design methods*

2.2.1 *Choice of yield locus*

Many analyses have been performed to investigate the detailed shape of the V-M-H yield locus for bucket foundations (e.g. Ngo-Tran 1996, Ukritchon et al. 1997, Bransby & Randolph 1998, 1999). Additionally, centrifuge model testing of caissons in clay, silt and sand has been carried out to determine the yield locus shape (e.g. Watson et al., 2000). These studies suggest that the shape of the yield locus is irregular but normality holds.

The yield locus equation used in this paper (Figure 2) was used successfully by Bransby and Martin (1999) in their analysis of centrifuge model tests:

$$f = \left(\frac{V}{V_o}\right)^2 - 1 + \left(\left(\frac{M}{M_o}\right)^2 + \left(\frac{H}{H_o}\right)^2\right)^{\frac{1}{1.6}} \quad (1)$$

where V_o, M_o and H_o are the foundation capacities under pure vertical, moment and horizontal loading.

2.2.2 *Load capacities, V_o, M_o and H_o*

The undrained load capacities, V_o, M_o and H_o determine the size of the yield locus. They are calculated with bearing capacity factors N_c, N_m and N_h, which link the bearing pressure to the soil shear strength at skirt tip depth, s_{uo} e.g. $H_o/A = N_h\ s_{uo}$.

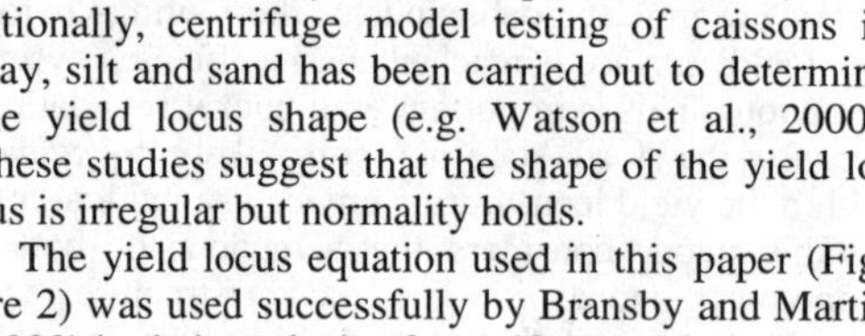

Figure 2. V-M-H yield locus used for the bucket foundations.

Bearing capacity factors were chosen using results presented for strip footings by Bransby and Randolph (1999) and Bransby (2000). These are in good agreement with results from FE analyses.

2.2.3 *Variation of V_o, H_o and M_o with footing penetration (plastic hardening rule)*

Once yield of a footing has occurred, work hardening plasticity is employed using a plastic hardening law and equation 1 as the yield locus and plastic potential. The hardening law is based on the fact that s_{uo} increases with plastic vertical penetration, v^p of the footing. As vertical bearing capacity, $V_o = N_c\ A\ s_{uo}$ then this can be re-expressed as

$$V_o = N_c A\left(s_{us} + kd + kv^p\right) \quad (2)$$

The moment and horizontal capacities, M_o and H_o also work harden with v^p in an analagous manner.

2.2.4 *Stiffness response*

It is assumed that foundation behaviour is linearly elastic inside the yield locus. The elastic stiffness matrix is given by:

$$\begin{pmatrix} \Delta V/_{As_{uo}} \\ \Delta M/_{ADs_{uo}} \\ \Delta H/_{As_{uo}} \end{pmatrix} = \frac{G}{s_{uo}} \begin{pmatrix} C_1 & 0 & 0 \\ 0 & C_2 & C_4 \\ 0 & C_4 & C_3 \end{pmatrix} \begin{bmatrix} \delta v^e/_D \\ \delta\theta^e \\ \delta h^e/_D \end{bmatrix} \quad (3)$$

The values of the stiffness constants (C_1, C_2, C_3 and C_4) were calculated for uniform soil (Martin & Houlsby, 1999) and for non-uniform soil (Ngo-Tran 1996, Bransby & Randolph 1997).

An elasto-plastic model of the footing can be assembled using the above components in the manner of Schotman (1989) and Martin & Houlsby (1999). A 3 by 3 stiffness matrix gives the instantaneous tangential stiffness of the footing for a given footing load and plastic penetration.

2.3 *Finite element analysis*

A single bucket foundation can be analysed using finite element analysis once appropriate constitutive laws and footing idealisations are made (e.g. Bransby and Randolph, 1998). Tani and Craig (1995) suggested that the soil above the skirt tip levels would add little to the capacity of a footing and so it could be ignored. In addition, the soil within the skirts would be rigid and so the simplification shown in Figure 3b was suggested.

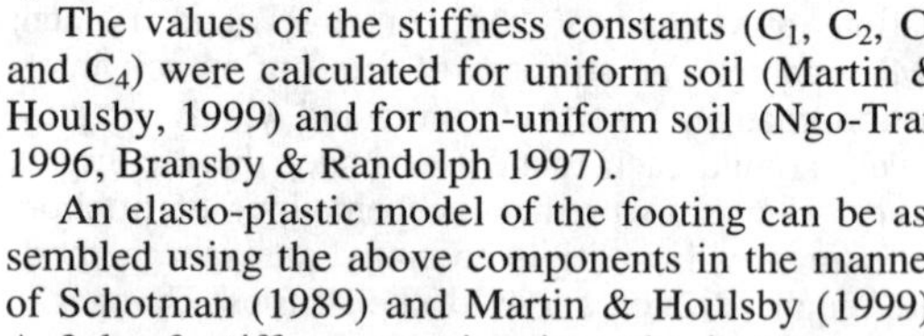

Figure 3. Bucket foundation idealisation.

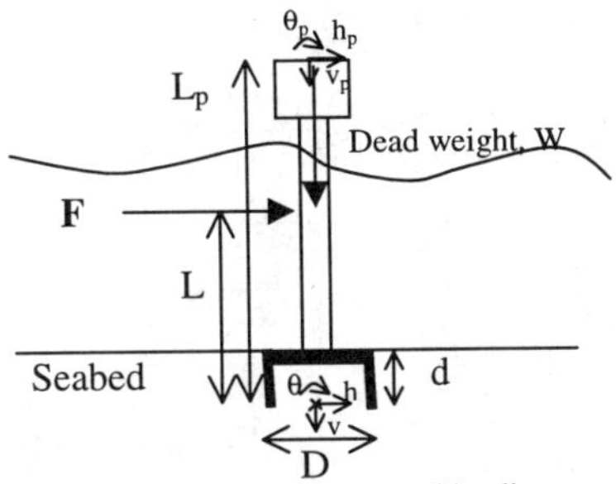

Figure 4. Monotower geometry and loading

Bransby and Randolph (1998) carried out FE analysis to determine the shape of the yield locus and soil deformation mechanisms at failure for buckets on undrained elastic-plastic clay using the surface footing idealisation. They were able to find yield loci, footing capacities and failure mechanisms.

2.4 *Lower bound plasticity*

Murff (1994) presented limit analysis methods for predicting the capacity of rigid offshore structures knowing the geometry of the structure and the yield locus equation of the footings.

Equilibrium is maintained together with the condition of $f = 0$ on each footing. The load distribution on each footing is varied to find the maximum load which can be sustained by the system. When this method is used in this paper, the shape of the yield locus will be given by equation 1.

3 ANALYSIS OF MONOPODS

3.1 *Typical geometry and loading conditions*

A typical monopod geometry is shown in Figure 4. The entire structure of height, L_p with dead weight, W is founded on a single bucket foundation. An idealised environmentally induced force F is applied at a height L above the skirt tips.

The foundation experiences a constant dead weight condition with $V = W$ and $H,M = 0$ followed by an environmental loading with $\Delta V = 0$, $H = F$ and $M = F.L$. Because the system is statically determinate, the load cannot re-distribute as a result of stiffness changes; the foundation will fail (or move excessively) when the load reaches the yield locus.

For the analyses following, a vertical 2D section is studied. The tower is assumed to be rigid and of height, $L_p = 40$ m and weight 193 kN per metre into the plane on Figure 4. The environmental load is applied at height, $L = 28$ m and the buckets are of width, $D = 12$ m. This means that the initial vertical pressure, $V_i/A = 16$ kPa. Undrained soil conditions prevail with $s_{uo} = 10$ kPa, $k = 0$ kPa/m and $G/s_u = 1000$.

3.2 *Finite element analysis*

Load controlled analyses of a single footing will give enough information to calculate the monopod response because the tower is rigid and statically determinate.

Finite element analysis was carried out of a rigid surface footing on soil with shear strength, $s_{uo} = 10$ kPa and $k = 0$ kPa/m. The soil was elastic-perfectly plastic, undrained and had rigidity index, $G/s_{uo} = 1000$.

The analysis was carried out in two parts: the dead weight vertical load was applied ($V = W$) in 200 increments; then the environmental loading was applied ($\Delta V = 0$, $\Delta H = F$ and $\Delta M = F.L$) in a further 200 increments. The vertical displacement v, horizontal displacement, h and rotation, θ of the footing was calculated every 20 increments. The movement of the top of the rigid platform was calculated as: $h_p = h + L_p\theta$; $\theta_p = \theta$; $v_p = v$.

The position of the instantaneous centre of rotation of the rigid tower was calculated and expressed as Cartesian co-ordinates (x, y) which were measured windward and downwards respectively from the footing base and centre.

3.3 *Elasto-plastic modelling*

When the footing loads are within the yield locus, the elastic stiffness matrix (Equation 3) will govern the stiffness response. For the environmental loading stage, this means that $\Delta v = 0$ and $\theta/(h/D) = (C_3/C_2).(L/D)$, where $h = H/(C_3G)$. Hence, the platform displacements, $v_p=0$; $h_p=H/C_3G + L_p.(H.L/C_2GD^2)$ and $\theta_p= H.L/(C_2GD^2)$.

When the footing loading exceeds the initial yield locus, plastic displacements also occur as the yield locus work-hardens.

3.4 *Results*

Figure 5 shows the calculated load-displacement behaviour of the monopod,where displacements are those measured at the top of the platform.

Failure loads are smaller for the elasto-plastic method than for the FE analysis. This is because the yield locus was assumed to be symmetrical in equa-

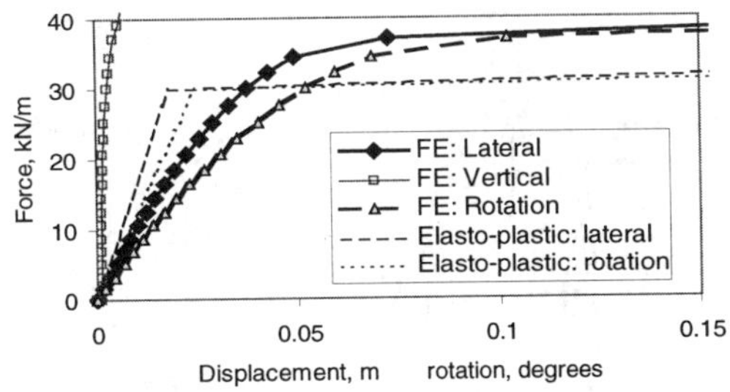

Figure 5. Load-displacement response on the platform.

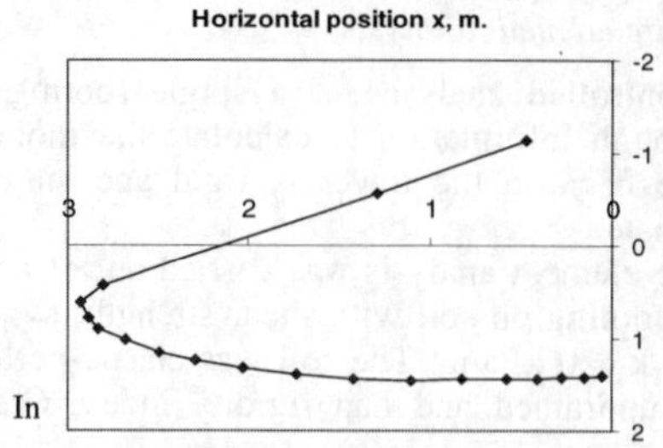

Figure 6. Position of the instantaneous centre of rotation during loading.

tion 1 which is conservative for the positive H and M quadrant (Bransby and Randolph, 1998). Indeed, the soil failure mechanism calculated in the FE analysis was very similar to their upper bound 'scoop' failure mechanism.

The elasto-plastic design method gives a linear elastic system response until it shows a very abrupt failure as the footing yields. In contrast, the FE analysis showed a ductile failure as soil yielded around the footings. Examination of the position of the rotation point during loading (Figure 6) reveals that there is a gradual change of monopod displacement mechanism as load increases.

4 ANALYSIS OF JACKET STRUCTURES

4.1 *Simplified loading and displacement conditions*

A jacket structure with bucket foundations may be loaded as shown in Figure 7. Dead weight, W is applied through the centre of the structure. As with the monopod, the environmentally induced loading is idealised as a single horizontal load, F applied at a height, L above the skirt tips and the structure is considered to be rigid.

The load direction is such that pairs of footings will be subjected to the same loading conditions due to symmetry (Figure 7). The footings are at a spacing w from centre to centre.

The jacket will translate and rotate as a rigid body (Figure 8). The movement can be considered by calculating the equivalent centre of rotation x to the left and y below the centre of the structure at skirt tip level (Figure 8).

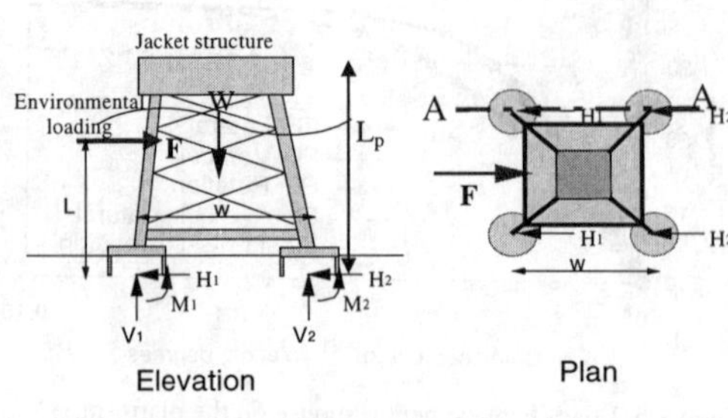

Figure 7. Parallel loading of jacket structure.

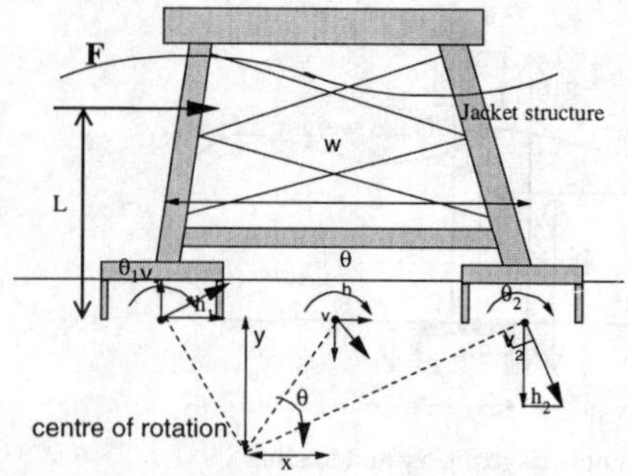

Figure 8. Displacements of a jacket structure.

4.2 *Soil conditions and geometry*

The jacket height, L_p = 55 m with the height of load application, L = 42.5 m. Footings of width, D = 10 m and with spacing, w = 40 m were chosen.

The dead weight loading supplied an initial vertical pressure on each foundations, V_i/A = 25.5 kPa giving W = 510.7 kN per meter width of foundation.

Soil conditions were chosen similar to those for the monotower with s_{uo} = 10, k = 0 kPa/m, G/s_{uo} = 100 and unit weight, γ = 16 kN/m^3. The soil was undrained.

4.3 *Finite element analysis*

Finite element analysis was carried out of the vertical cross-section A-A in Figure 7. The force applied F is given per metre width of the foundation (out of the plane on the elevation on Figure 7).

The FE mesh used 984 cubic strain triangles and its geometry is shown in Figure 9. Soil conditions were modelled using elastic-perfectly plastic soil models and the bucket foundations were modelled as equivalent surface footings. The structure was linear elastic with $E_{structure}$ = 3000 G_{soil} to ensure that the structure remained almost rigid.

Analysis of the system behaviour was carried out in two parts: the dead weight of the jacket structure was applied by increasing the gravity field from 0 to 9.81 m/s^2 in 200 increments; then, environmental loading was applied over another 200 increments.

Interestingly, as dead weight loading was applied, footing-soil-footing interaction caused a small amount of M and H loading on each footing.

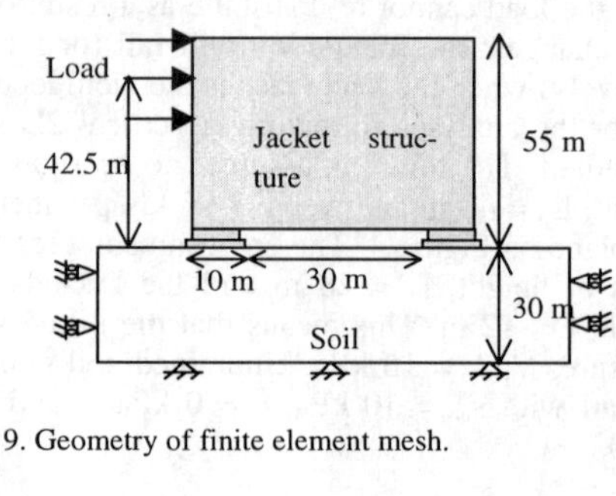

Figure 9. Geometry of finite element mesh.

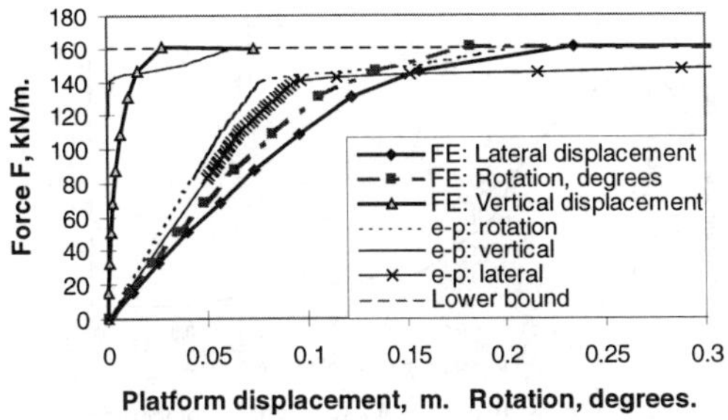

Figure 10. Jacket displacements

4.4 *Analysis by elasto-plastic footing models*

The geometry, loading conditions and footing capacities from the finite element analysis were replicated in the elasto-plastic modelling. Following application of dead load (where the footings remained elastic) the environmental load, F was applied incrementally while the stiffness matrix for the footings changed as loads increased. Typically, 2000 load increments were used to reach a failure state.

4.5 *Lower bound analysis*

Failure loads were calculated using lower bound analysis with the yield locus given in Equation 1.

5 RESULTS FOR THE JACKET STRUCTURE

5.1 *Finite element analysis*

Figure 10 shows the load-displacement response from the FE modelling of the jacket structure for displacements measured at the top of the platform. Failure is observed at F ≈ 160 kN/m although the rate of rotation and lateral movement increased when F > 100 kN/m.

Vertical load reduces on the windward footing (Figure 11) as it initially lifts up (Figure 12) but V increases on the leeward leg as it embeds in to the soil (Figures 11 and 12). Plastic yielding causes continual load re-distribution (Figure 11).

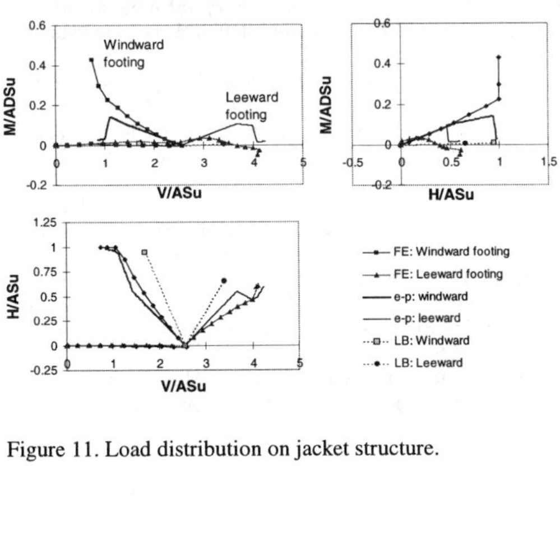

Figure 11. Load distribution on jacket structure.

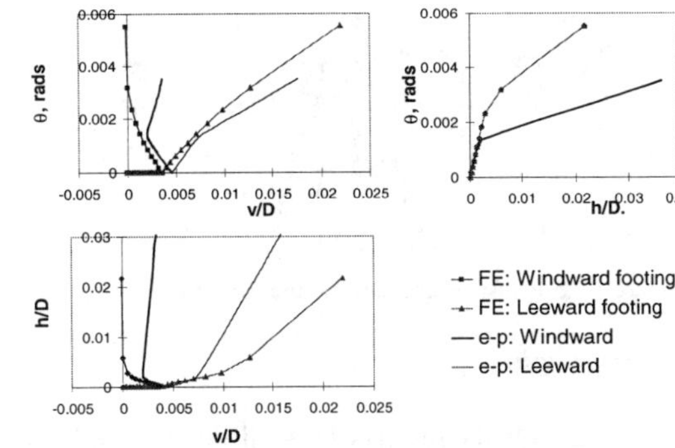

Figure 12. Footing displacements for jacket structure

The footing displacements (Figure 12) show that there is some change of displacement mechanism with increasing loading. This is clarified in the plot the system rotation centre (Figure 13). At failure, the windward footing undergoes pure sliding failure, whilst the soil around the windward footing undergoes the deformation shown in Figure 14 which is similar to the wedge upper bound mechanism of Bransby and Randolph (1998).

5.2 *Elasto-plastic modelling*

The platform displacement for the elasto-plastic modelling is shown on Figure 10. Load distribution is shown in Figure 11 and displacement conditions on Figure 12.

5.3 *Lower bound analysis*

The failure load, F = 160.3 kN/m (Figure 10). Footing loads are shown in Figure 11.

5.4 *Comparison of results*

The failure loads from the three methods are in good agreement although large plastic displacement occur in the elasto-plastic analysis before the other failures. The gradual yielding of the footings in the FE analysis is seen most clearly in Figure 15 which plots system stiffness (dF/dh_p) against h_p. This not reflected in the elasto-plastic analysis where system response is totally elastic (because of elasticity within the yield locus) until the leeward footing yields at F ≈ 120 kN/m. However, the initial and final deformation mechanisms are quite well replicated (Figure 13).

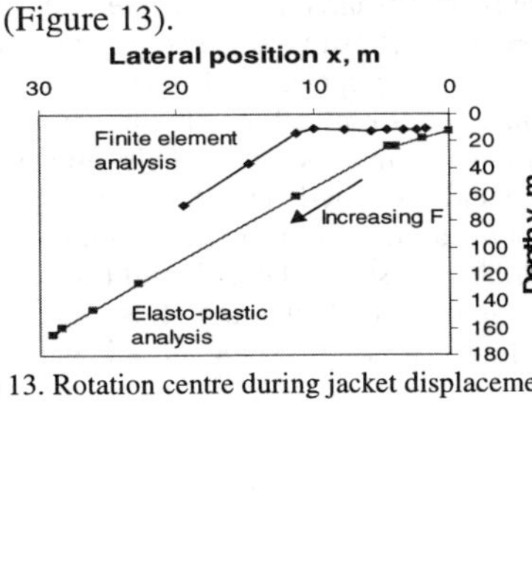

Figure 13. Rotation centre during jacket displacements.

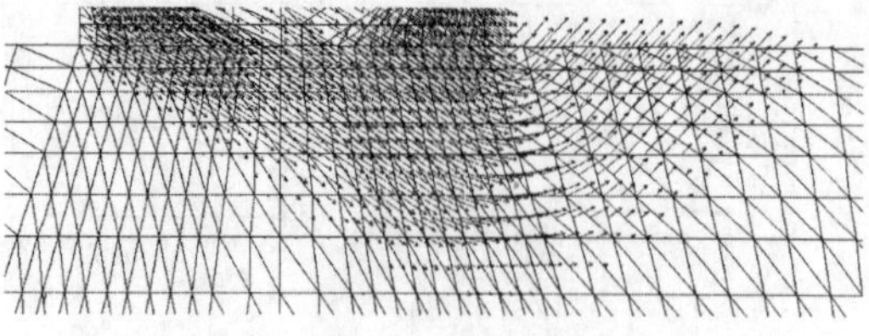

Figure 14. Soil displacements at failure for the leeward footing

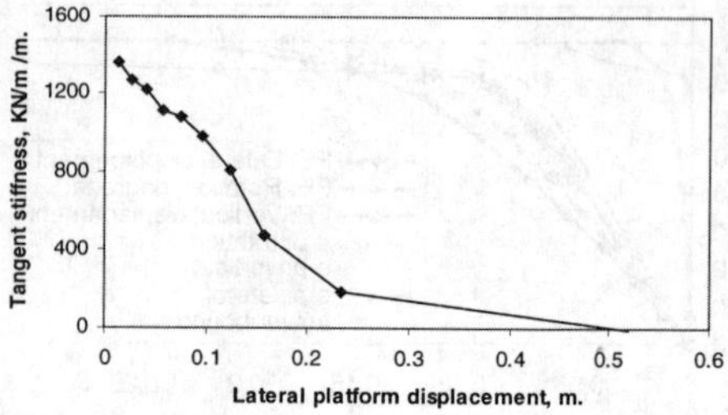

Figure 15. Platform stiffness reduction with displacement.

6 DISCUSSION

Although only the results of two different foundation systems have been presented in this paper, the results have shown some of the inadequacies of using elasto-plastic footing models with linear elasticity within the yield locus to calculate serviceability displacements. This is because the footing loads remain well within the 'yield' envelope until platform loads are almost at failure, especially when there are few footings. In reality, progressive soil non-linearity and yield around the footings prior to failure ensures that the individual footing response and thus the system response is much more non-linear. The FE analyses with simple elastic-perfectly plastic soil models will reflect this behaviour somewhat as soil yields plastically around the footings. This behaviour would be better captured in FE analysis using soil models such as Cam clay which would be ideal for a normally consolidated deposit.

It appears that as the number of independently loaded footings in a system increases, the overall ductility of the system also increases (Bransby and Martin, 1999) and so the assumption of linear elasticity within the yield becomes less unsafe.

7 CONCLUSIONS

Analysis of a jacket structure and monopod have been carried out using finite element analysis, an elasto-plastic design method and a lower bound method. The preliminary results have shown that 2D finite element analysis can be used to investigate offshore foundation system response and it is particularly good at identifying soil and system deformation mechanisms.

Failure loads and mechanisms were found to be in reasonably good agreement between the design methods. However, results suggest that progressive yield of soil around footings will lead to system response which is not replicable using existing elasto-plastic footing design methods. This non-linearity within the yield locus will lead to incorrect prediction of serviceability limit states. Finite element analysis of the type shown here should use soil models which capture this behaviour. Less ductile systems with fewer footings, such as jackets with 2 independently loaded footings and monopods, will be particularly sensitive to small displacement non-linearity.

REFERENCES

Bransby, M.F., 2000. Failure envelopes and plastic potentials for eccentrically loaded surface footings on undrained soil. Submitted to the Int. Journ. of Num. And Analy. Methods in Geomechanics.

Bransby and Randolph, 1997. *Shallow foundations subject to combined loadings*. Proceedings of IACMAG 97, Wuhan, November 1997. Vol. 3, pp. 1947-1952.

Bransby, M.F. &Randolph, M.F. 1998. Combined loading of skirted foundations. *Geotechnique* 48, No 5. pp 637-655.

Bransby, M.F.& Randolph, M.F. 1999. The effects of embedment on the undrained response of caisson foundations to combined loading. *Soils and Foundations*, Vol. 39 - 4, pp. 19-34.

Bransby, M.F.& Martin, C.M. 1999. *Elasto-plastic modelling of bucket foundations*. Proc. of NUMOG VII, Graz, September, 1999, pp. 425-430.

Martin, C.M. 1994. *Physical and numerical modelling of offshore foundations under combined loads*. D.Phil thesis, University of Oxford.

Martin, C.M., Houlsby, G.T. 1999. Jackup units on clay: structural analysis with realistic modelling of spudcan behaviour, *Proc. Offshore Technology Conf.*, OTC 10996.

Murff, J.D. 1994. Limit analysis of multi-footing foundation systems. *Proc of the 8th Int. Conf. On Comp. Methods and Advances in Geomechanics,* Morgantown, 1, pp. 223-244.

Perol, C., Meimon, Y. 1989. Cyclic storm loading soil-structure interaction for a three independent leg jack-up. pp. 223-243.

Ngo-Tran, C.L. 1996. The *analysis of offshore foundations subjected to combined loading*. DPhil thesis, University of Oxford.

Schotmann, G.J.M. 1989. The effects of displacement on the stability of jackup spud-can foundations. *OTC* 6026.

Tan, F. 1990. Centrifuge and theoretical modelling of conical footings on sand. PhD Thesis, The University of Cambridge.

Tani, K. & Craig, W.H. 1995. Bearing capacity of circular foundations on soft clay of strength increasing with depth, *Soils and Foundations*, Vol. 35, No. 4, pp 21-35.

Ukritchon, B., Whittle, A.J., Sloan, S.W. 1997. Undrained limit analysis for combined loading of strip footings on clay. *Jour. Of Geotech. And Geoenvironmental. Eng.*, Vol 124, No. 3, pp.265-276.

Watson, P.G., Randolph, M.F., Bransby, M.F. 2000. Combined lateral and vertical loading of caisson foundations. *OTC* 12195.

Computer Methods and Advances in Geomechanics, Desai et al. (eds) © 2001 Balkema, Rotterdam, ISBN 90 5809 183 X

3D FE analysis on the behavior of laterally loaded rigid piles installed nearby a slope of cohesionless soils

K.S.Chae, K.Ugai & A.Wakai
Department of Civil Engineering, Gunma University, Kiryu, Japan

ABSTRACT: Deep foundations are frequently used to support the transmission towers carrying the overhead power lines. These foundations are usually constructed in the slope, such as mountainous regions, and may be exposed to lateral loads and deformations; and thereby the risk of unserviceable behaviour increases. The aim of this study is to examine the behavior of laterally loaded rigid piles installed nearby a slope of cohesionless soils. Several series of model tests investigating this problem and their 3D elasto-plastic FEM analyses are reported. Their comparisons have shown good agreements.

1 INTRODUCTON

When a short rigid pile is installed nearby a slope, the lateral resistance of the pile is assumed to be lower than when it is placed in horizontal ground, because the horizontal confining pressure acting on the pile in the slope side is smaller.

Wakai et al. (1999) have reported the behaviour of a long single pile and group of piles subjected to lateral loading in a sandy ground by using tests and FE analyses. Prasad et al. (1999) have conducted lateral loading tests on rigid piles in cohesionless horizontal ground, and proposed a method to predict the soil pressure and the ultimate lateral capacity.

This paper deals with FE analyses of the rigid pile installed in horizontal ground and slopes with inclinations of 20 and 30 degrees and the comparison with the measurements of model tests in sand.

2 MODEL TEST

Figure 1 shows schematically the experimental setup. Model tests were conducted using well-graded angular dry sand of unit weight γ=15.3 kN/m^3 with the average relative density D_r =0.75. The pile was a closed-ended 100 mm diameter model. It was made from a 750 mm aluminum alloy tube with a 3 mm wall thickness and instrumented with 10 pairs of strain gauges to measure the bending moments.

The pile, with the free head, was embedded up to depth (L_E) of 500 mm. The lateral loading was applied by a metallic cable connected to the pile at the height (h) of 50 mm above the soil surface. The displacement transducer measured the pile head displacement at the same height. Load was applied on the pile by increments, with each load increment lasting until the rate of the displacement was less than 0.75mm/min.

A data acquisition system consisting of a data logger and a personal computer, which recorded the applied load, the displacement of the pile head and the strain of the pile shaft at each load increment were used.

3 FEM MODELLING

A series of FEM analyses were performed not only to provide comparison with the model test data, but also to obtain more extensive data and to investigate additional effects.

In the 3D FEM, the failure criterion and the plastic potential are based on the Mohr-Coulomb and the Drucker-Prager equations, respectively. The pile material was assumed to be linearly elastic and the soil

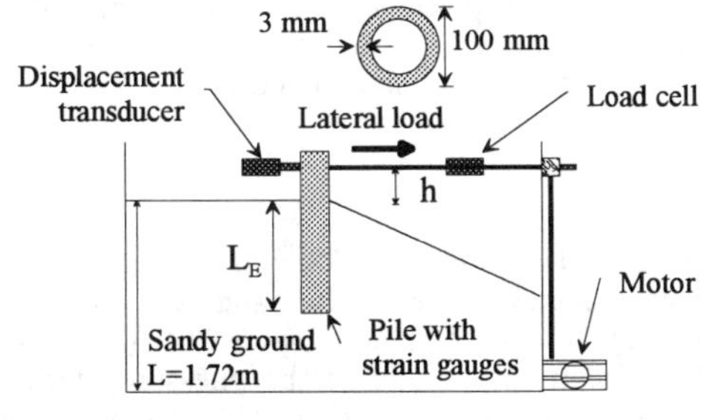

Figure 1. Model test setup

Table 1. Material parameters

	Pile	Sand	Interface
E [MPa]	6.86E+04	Eq. (1)	Eq. (1)
ν	0.345	0.4	0.4
γ [KN/m³]	26.4	15.3	15.3
c [MPa]	Large	0	0
φ	-	42 °	25 °
ψ	-	12 °	0 °

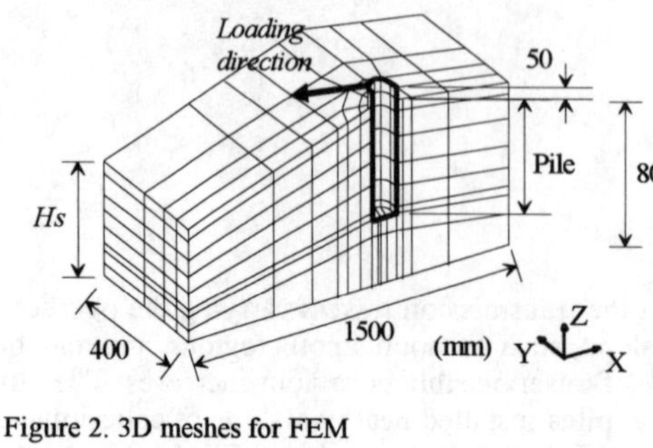

Figure 2. 3D meshes for FEM

was modeled as an elasto-perfectly plastic material with non-associated plasticity, following the MC-DP model (Tanaka, 1992). In addition to these properties, the no-tension analysis (Zienkiewicz et al., 1968) was applied to this sand, in which the minimum principal stress was controlled as positive for each element.

Thin frictional brick elements were inserted between the rigid pile and the soil in order to consider slippage at the rigid pile - soil interface. A typical 3D finite-element mesh is shown in Figure 2, modeled from 20 nodded brick elements with 8 Gaussian points. The mesh consisted of 2479 nodes and 458 elements. Because of the symmetry, the analysis can be conducted on a vertical half of the pile foundation. Bold lines indicate the position of the rigid pile elements. The variable height of the slope side in the mesh, is denoted as *Hs* in figure 2, the slope was 20,30 degrees and zero in the case of horizontal ground.

The material parameters used in the analyses are summarized in Table 1. The soil parameters were determined from the conventional triaxial compression tests. It should be noted that the secant elastic modulus E_{50} of the sand in Table 1 was assumed to be proportional to the square root of the initial mean confining pressure $\sigma=(\sigma_1+\sigma_2+\sigma_3)/3$, as shown in Eq. (1).

$$E_{50} = E_0 \sqrt{\frac{\sigma}{\sigma_0}} \tag{1}$$

where $E_0 = 19.8\ MPa$, $\sigma_0 = 98\ kPa$.

The dilatancy angle of the sand was evaluated according to the equation: $\psi=\phi-30^\circ$ (Tatsuoka, 1993). The friction angle of the pile-soil interface was determined to be equal to 25°, based on the

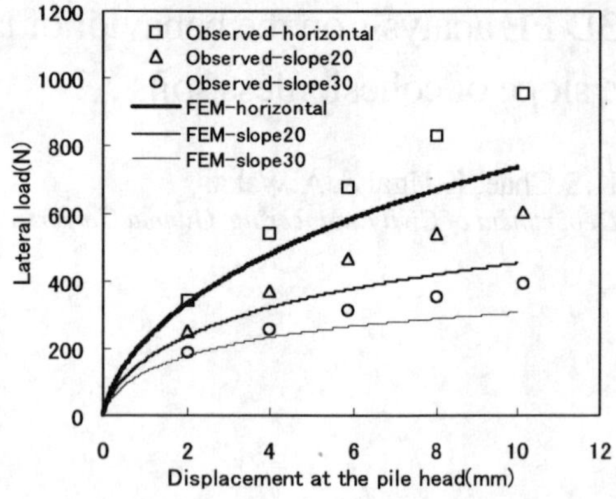

Figure 3. Comparison between observed and calculated load - displacement relationships

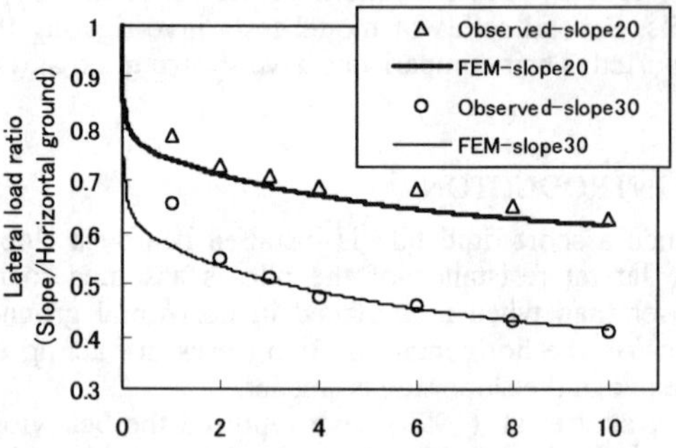

Figure 4. Relation between lateral resistance reduction ratio and displacement

results of the slippage tests between an aluminum plate and the sand.

The calculations were carried out in two steps. In the first step, the initial stress condition was calculated from the dead load of the soil without the pile by using the shear strength reduction method (SSR-FEM, Ugai and Leshchinsky, 1995). The second step was to simulate the soil response with the initial stress condition obtained from the first step, as the displacement was applied incrementally.

4 TEST AND ANALYSIS RESULTS

When a rigid pile with a free head and embedded in the soil is subjected to a lateral load at the pile head, passive earth pressure develops at the front face above the rotation point, which is also the point of zero lateral displacement in the pile, and active earth pressure develops at the corresponding back face of lateral pressure is reversed, and a roughly triangular

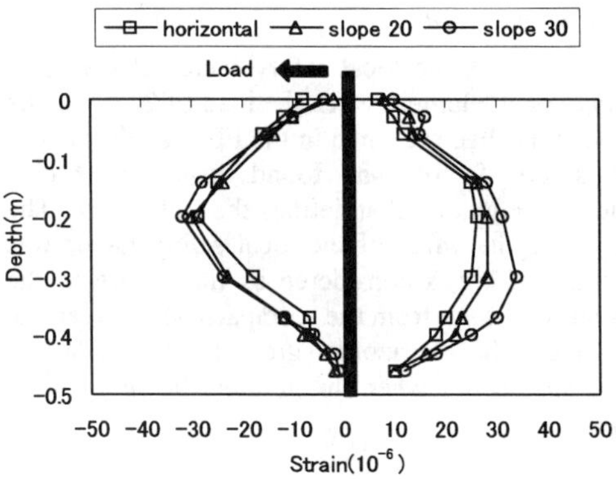

(a) Observed case

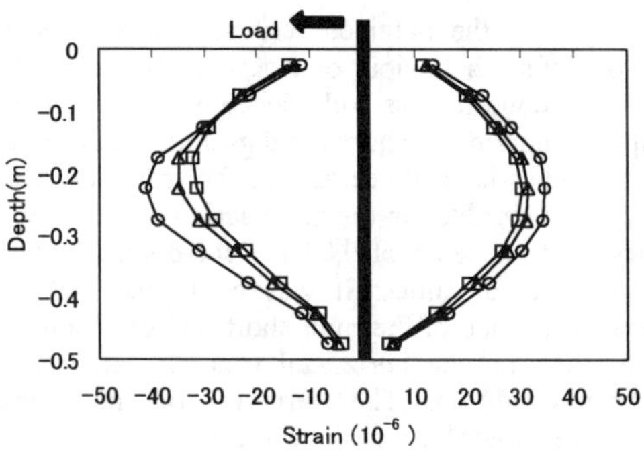

(b) FEM case

Figure 5. Strain distributions at the load 245N

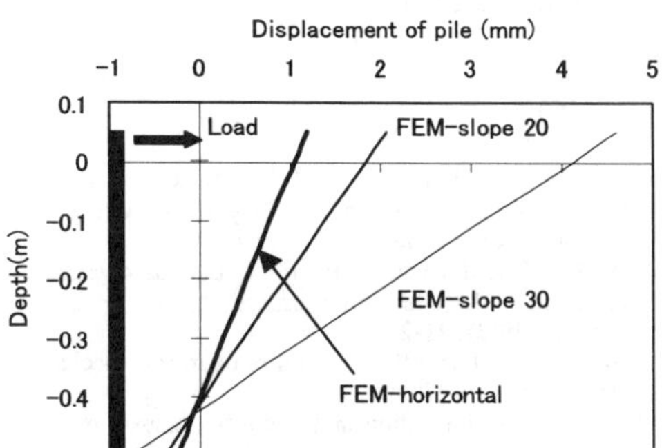

Figure 6.Calculated pile lateral loading and displacement (at center of the pile diameter) at the load 245N.

pressure distribution with a maximum (Roscoe, 1970), can be observed.

4.1 *Pile behaviors*

Figure 3 shows the relationships between the observed and calculated lateral load – displacement of the pile head for the horizontal ground and slopes

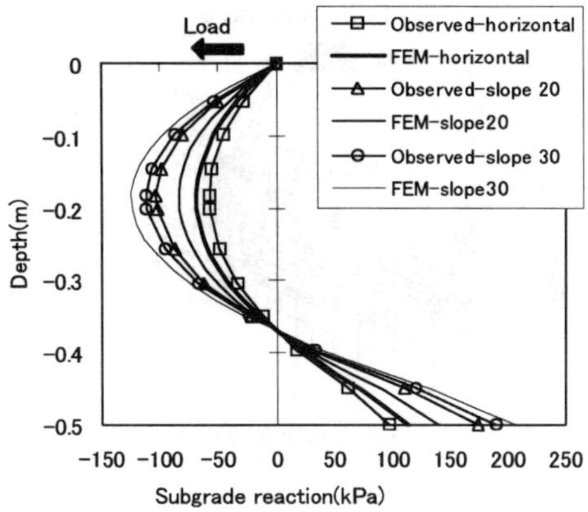

Figure 7. Comparison observed and calculated subgrade reaction at the load 245N.

the pile head for the horizontal ground and slopes with inclination of 20 and 30 degrees. The experimental and calculated curves correlate well with each other.

Figure 4 shows the relationships between the lateral resistance reduction ratio and the displacement at the pile head, for the slope of 20 and 30 degrees. The lateral resistance reduction ratio was defined as the ratio of the lateral load on the pile installed in the slope and in the horizontal ground. For example, the lateral resistance reduction ratio is approximately 0.7 for 20° slope and 0.5 for 30 ° slope at the pile head displacement of 4 mm, as shown in Figure 4. And the ratio decreases slowly as the deformation increases.

Figure 5 (a) and (b) show the observed and calculated strain distributions along piles installed in the horizontal ground and slopes, with 245 N of lateral load. In Figures 5 (a) and (b), the strain along the pile installed in the slope are larger than in the case where it is placed in horizontal ground. It seems that it is caused by a decrease of the confining pressure around the pile. The calculated lateral displacement profiles corresponding to the load of 245 N are shown in Figure 6. The depths of the rotation point on the pile are about 0.4 ~ 0.42 m in the horizontal ground and in the slope, respectively. As the gradient of slope becomes larger, the depth of the rotation point increases.

4.2 *Lateral resistance*

The strain gauges attached on the piles gave the bending moment profiles during lateral loading.

$$p = \frac{1}{d}\frac{\partial^2 M(z)}{\partial z^2} \tag{2}$$

The interpolation of the bending moments $M(z)$ was

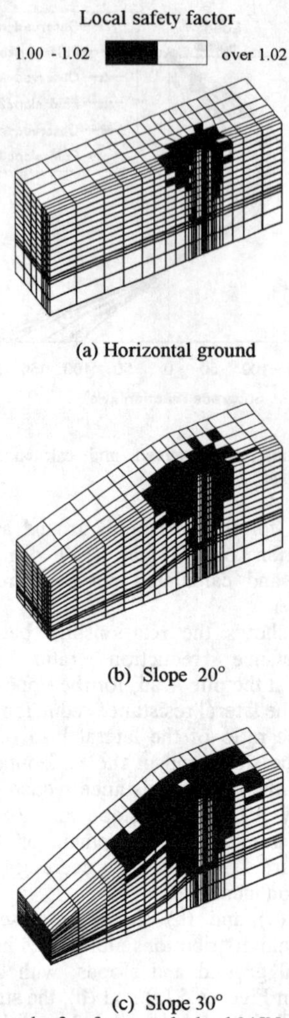

(a) Horizontal ground

(b) Slope 20°

(c) Slope 30°

Figure 8. Local safety factor at the load 245N

fitted by the fifth order polynomial equation and was used to derive the subgrade reaction *P*, and the pile diameter *d* and the depth *z*.

Figure 7 shows observed and calculated subgrade reaction along the pile for the load of 245 N. Based on the results, the subgrade reaction was zero at the ground surface and it increased with depth up to about 0.19 m, that is 0.38 L_E, (L_E is the embedded depth of the pile). Beyond that point the subgrade reaction decreased until it was reduced to zero at the depth of about 0.37 m (0.75 L_E). Below the subgrade reaction zero point, it changed to the opposite direction and increased from zero to the maximum at the bottom of the pile. The subgrade reaction at the bottom of pile was about 1.7 times higher than at 0.38 L_E.

4.3 *Slope behavior*

Figure 8 shows the local safety factor at Gaussian points of each element when the load of 245 N at the pile head has been reached in the FEM analysis. The local safety factor was found from the Mohr-Coulomb's equation that defines the soil failure. The case where the value of the local safety factor was less than 1.02 was considered as the failure of the element. It is seen from the comparison between the slope and the horizontal ground that the failure area is spread out when the gradient becomes more inclined.

5 CONCLUSIONS

Summarizing the obtained results, it was revealed that the lateral behaviour of a rigid short pile nearby slopes of cohesionless soils depends on the ground shape, comparing the horizontal ground with a slope. On the other hand it seems that the front soil area, which was involved in the pile bearing capacity, was restricted to a depth of 0.75 L_E for both, observed and calculated values. It was confirmed that the lateral resistance of the rigid short pile on a slope is lower than on the horizontal ground at the same loading condition. The variation of the lateral resistance distribution along the length of the pile can be assumed as incremental up to about 0.4 L_E. Then it decreased to zero and finally it increased in the opposite direction until it reached the maximum at the bottom of the pile.

REFERENCES

Meyerhof, G.G., Mathur, S.K. and Valsangkar, A.J. 1981. Lateral resistance and deflection of rigid wall and piles in layered soils. *Can. Geotech. J.* 18: 159-170.

Prasad, Y.V.S.N. and Chari, T.R. 1999. Lateral capacity of model rigid piles in cohesionless soils. *Soils and Foundations* 39(2): 21-29.

Roscoe, K.H. 1970. The influence of strains in soil mechanics. *Geotechnique*: 129-170

Tanaka, T. 1992. Deformation and stability analysis by finite element method, in Principle of Soil mechanics (first revision). *The Japanese Geotechnical Society* (in Japanese).

Tatsuoka, F. 1993. Relationships between stress and strain increments, in Introduction to strength of soils and failure of ground (third revision). *The Japanese Geotechnical Society* (in Japanese).

Ugai, K. and Leshchinsky, D. 1990. three-dimensional limit equilibrium and finite element: a comparison of results. *Soils and Foundations* 35(4): 1-7

Wakai, A., Gose, S. and Ugai, K.1999. 3-D elasto-plastic element analyses of pile foundations subjected to lateral loading. *Soils and Foundations* 32: 97-111.

Zienkiewicz, O.C., Valliappan, S. and King, I.P. 1968. Stress analysis of rock as a 'no tension' material. *Geotechnique* 18: 56-66.

Computer Methods and Advances in Geomechanics, Desai et al. (eds) © 2001 Balkema, Rotterdam, ISBN 90 5809 183 X

Direct wave equation analysis on vertically loaded raft-pile

D.W.Chang & C.H.Wen
Department of Civil Engineering, Tamkang University, Tamsui, Taiwan

ABSTRACT: This paper presents the time-domain wave equation analysis on vertically loaded raft-pile. Incorporating the static *t-z* and *Q-z* springs and time-dependent dashpots of the soils with the pile-to-pile interactions for the rafted piles, the foundation response due to various loads are modeled. It is found that the proposed modeling can agree reasonably with the comparative analyses.

1 INTRODUCTION

Direct (time-domain) wave equation analyses of the single piles have been investigated by the author (Chang and Yeh, 1999; Chang and Lin, 1999; Chang *et al.*, 2000[a]) with different stiffness models and time-dependent geometric damping of the soils. In their studies, the frequency-dependent geometric damping ratios of the soils from Novak's impedance functions were first computed. The derived damping parameters were then incorporated with the static *t-z* and *Q-z* soil springs to obtain the equivalent viscous damping coefficients for the soils. Good agreements have been reported in analyzing the theoretic displacement curves (Poulos, 1993) and the records (McCammon and Golder, 1970) from the pile-load-tests. For more realistic modeling, the solutions can be revised taking into account the pile-to-pile interactions and the effects from the top cap. Considering the vertical piles, numerical modeling and validation of the solutions will be discussed by the author in details (Chang *et al.*, 2000[b]). In this paper, the available dynamic solution is simply introduced to the engineers as the recent advance in the pile foundation design practice.

2 PROPOSED MODELING

2.1 *Fundamentals*

Wave equation of a single pile under dynamic vertical load can be derived based on force equilibrium of the pile segment. Applying proper boundaries conditions, the pile response can be obtained from the finite difference representations of these equations (Chang and Yeh, 1999). The soil stiffness derived from the theoretical *t-z* and *Q-z* equations were suggested for the analysis (Chang *et al.*, 2000[a]). A time-dependent viscous damping coefficient, based on frequency-dependent nature of the structure, has been suggested to implement with the stiffness in solving the time-dependent wave equations.

2.2 *Pile-to-pile interactions*

To capture the pile-to-pile interactions, approximate dynamic interaction factor (Dobry and Gazetas, 1988) is adopted herein. The interaction factor α, defined as the ratio of the displacements occurred at i^{th} pile due to a load applied at j^{th} pile, and the displacements at i^{th} pile due to a same load applied on top of it, is suggested as:

$$\alpha = (S/r)^{-1/2} \exp(-\beta \omega S/V_s) \exp(i\omega S/V_s) \qquad (1)$$

where S: pile spacing, r: pile radius, ω: circular frequency, β: material damping ratio of the soil, and V_s: soil shear-wave velocity. This formula is limited to the cases where the number of the piles exceeds 16, and where the pile-to-soil modulus contrast E_p/E_s is less than 300. By assuming that the settlements under the raft are equal, dynamic impedance functions of the soils for arbitrary piles in the pile group can be computed. Therefore, in a 2× 2 square-shape group-pile foundation, dynamic impedance of the soils, K_s^* can be written as:

$$K_s^* = K_S / [1 + 2\alpha(S) + \alpha(\sqrt{2}S)] \qquad (2)$$

where K_s is the original impedance of the soil. To

deriving the impedance functions in the cases where the pile loads are not evenly distributed (e.g., pile orientation of 3× 3 and 4× 4), one needs to compute the pile loads using the matrix analysis. Equations for 3×3 piles that transmitting a total load P to the soils (Figure 1) are summarized as follows.

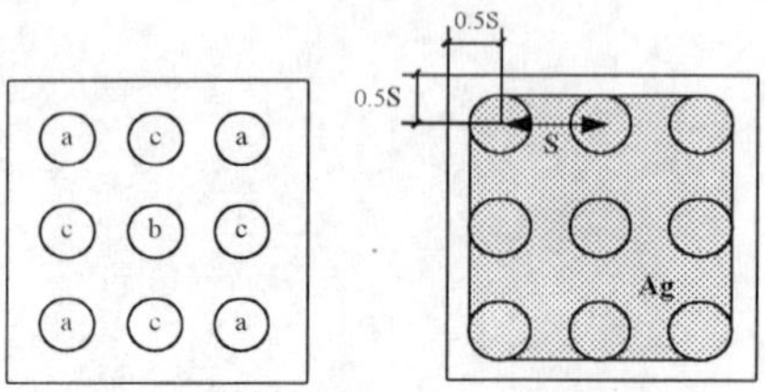

Figure 1. Orientation of the 3×3 square-shape group-pile foundation

At corner piles:

$$K_s{}^* = P_a/W_a = P_a/\{[1+2\alpha(2S)+\alpha(\sqrt{8}S)]\times P_a/K_S + \alpha(\sqrt{2}S)\times P_b/K_S+[2\alpha(S)+2\alpha(\sqrt{5}S)]\times P_c/K_S\}$$

At center pile:

$$K_s{}^*=P_b/W_b=P_b/\{P_b/K_S+4\alpha(\sqrt{2}S)\times P_a/K_S+ 4\alpha(S)\times P_c/K_S\} \qquad (3)$$

At edge piles:

$$K_s{}^*=P_c/W_c=P_c/\{[2\alpha(S)+2\alpha(\sqrt{5}S)]\times P_a/K_S+ \alpha(S)\times P_b/K_S+ [1+2\alpha(\sqrt{2}S)+\alpha(2S)]\times P_c/K_S\}$$

To compute the load distributions of different piles, i.e., P_a, P_b and P_c, the equations $P=4P_a+P_b+4P_c$, $W_a=W_b$, and $W_b=W_c$ are used. The factors including S/d (d: pile diameter), r/V_s and material damping β will affect the load variations. Load distributions of the piles at frequencies of 0, 5, 30 Hz are plotted in Figure 2.

Applying Fourier Transform on the spectra, the corresponding loads on different piles are found distributed equally (Chang *et al.*, 2000[b]). The pile-to-pile interactions will reduce the soil stiffness around the pile. Table 1 lists the reductions at different piles in the pile groups. The corresponding damping ratios can be solved following the procedures suggested by Chang and Yeh. Discrete data of the time-dependent damping are obtained from a FFT routine. Figure 3 illustrates the ratios between the damping coefficients and the stiffness of the soils for the 2× 2 piles, and the corresponding ones around the center pile in the 3×3 and 4×4 pile foundations where S/d=3, respectively. Their ratios are found smaller than 0.05 sec in the studied cases.

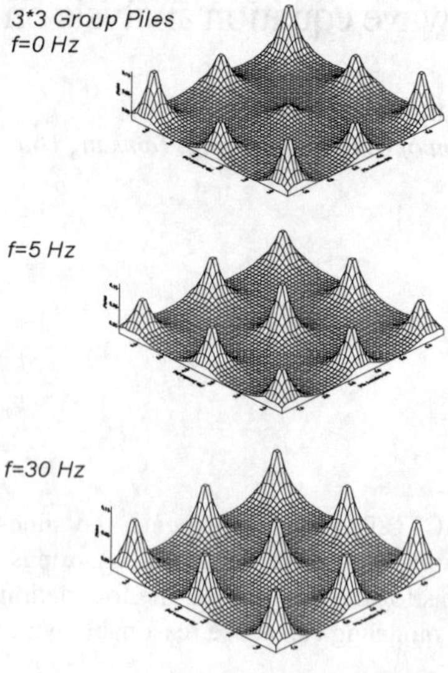

Figure 2. Vertical load distributions on the group-pile foundation at different driving frequencies

Table 1. Ratios between affected stiffness and original stiffness of the soils around the piles at different group-pile foundations where the spacing ratio varies

geometry	4×4			3×3			2×2
S/d	corner	center	edge	corner	center	edge	any
2	24.3%	5.3%	16.8%	32.2%	6.8%	21.9%	41.3%
3	26.3%	9.0%	19.5%	34.3%	13.4%	26.7%	46.3%
4	28.1%	11.8%	21.7%	36.4%	17.8%	30.1%	49.9%
5	29.6%	14.1%	23.5%	38.2%	21.2%	32.8%	52.7%

2.3 *Raft-pile-soil modeling*

In Figure 4, the applicable wave equation is established for arbitrary pile restraint with the cap. The effective cap size can be computed according to the applied load. By assuming that the load distributions are similar to the transmitting forces P, the cross-section area of the effective cap, A_c' can be computed as:

$$A_c' = A_c \times P_i/P \qquad (4)$$

A_c is the original cross-section area of the cap. For the interior piles, the wave equations and the associated boundary conditions are presented as:

For cap: $\dfrac{\partial^2 W}{\partial x^2} = \dfrac{1}{V_{c,c}{}^2}\dfrac{\partial^2 W}{\partial t^2}$

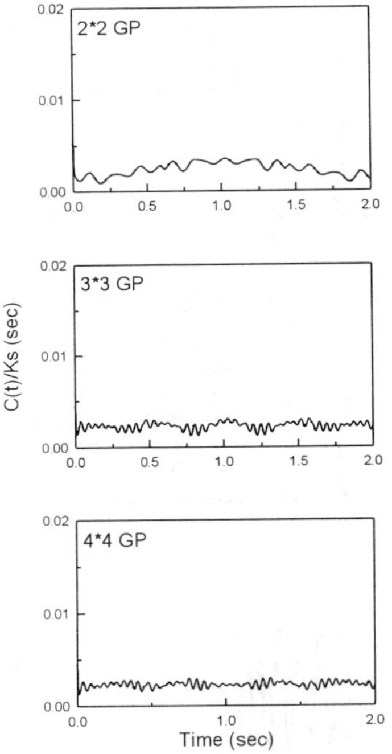

Figure 3. Time-dependent ratios between damping and stiffness of the soils for 2×2, 3×3 and 4×4 group-pile foundations

$$\begin{aligned}
&\textit{For pile: } \frac{\partial^2 W}{\partial x^2} = \frac{1}{V_{c,p}^{\,2}}\frac{\partial^2 W}{\partial t^2} + \frac{C_{s,p}}{E_p A_p}\frac{\partial W}{\partial t} + \frac{K_{s,p}}{E_p A_p} W \\
&\textit{At cap top: } E_c A_c' \frac{\partial W}{\partial x} = -P(t) \qquad (5) \\
&\textit{At cap-pile interface:} \\
&E_c A_c' \frac{\partial W}{\partial x} = K_{b,c} W + C_{b,c}\frac{\partial W}{\partial t} + E_p A_p \frac{\partial W}{\partial x} \\
&\textit{At pile tip: } E_p A_p \frac{\partial W}{\partial x} = K_{t,p} W + C_{t,p}\frac{\partial W}{\partial t}
\end{aligned}$$

where $V_{c,c}$ and $V_{c,p}$: compression wave velocity of the cap and the pile; E_c, E_p: Young's modulus of the cap and the pile; A_c', A_p: cross-section area of effective cap and pile; $C_{b,c}$ and $K_{b,c}$: soil damping and stiffness underneath the cap; $C_{s,p}$, $C_{t,p}$ and $K_{s,p}$, $K_{t,p}$: soil damping and stiffness along the pile shaft and at the pile tip. The side soil resistances of the effective cap are not considered for the interior piles. The cap is regarded as an embedded shallow foundation, whereas the soil resistance underneath the cap can be modeled by Novak's simplified impedance (Chang *et al.*, 2000[b]). The explicit FD formulations are thus reformed. One must know that the displacements resolving at each pile should satisfy the rigid-foundation assumption.

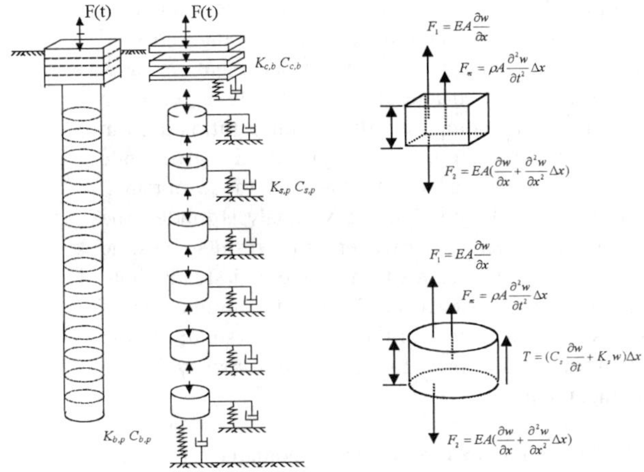

Figure 4. Discrete model of the rafted pile and the equilibrium of the cap and the pile segment

3 COMPARATIVE SOLUTIONS

To validate the proposed solutions, comparative solutions are studied. An axisymmetric lump-mass analysis and the wave equation analysis of the pile foundation represented by a single pier constraint with the raft are made. 2-D and 3-D FEM analyses of the group-pile foundation using the computer program *ABAQUS* are also conducted (Chang *et al.*, 2000[b]). The standard structural geometry and the material parameters used in the numerical examples are summarized in Table 2. The response of the referred pile foundation under monotonic load, impulsive load, and steady-state load are compared with other analyses. Figure 5 shows the results for the 2×2 pile foundation.

The calculations obtained from the wave equation analysis and the lump-mass analysis on the simplified pier model are found compatible. The maximum displacements, obtained from the wave equation analysis for the single pile in the pile group, are however different from the predictions made by the equivalent pier modeling and the FEM analysis in the case of monotonic loading. Response of the wave-equation group-pile modeling seems to be smaller than those obtained from the equivalent pier model. It is reasonable because the material stiffness of the equivalent pier is reduced. The displacement time history also indicates that the equivalent pier solution has large damping effects. Nevertheless, for the impulsive and the steady-state loads where the dynamics becomes apparent, the opposite condition is found. In these cases, the peak responses of the proposed modeling are greater than the equivalent pier analysis. The amount of the damping reflected by the attenuation of the displacement is now governing the structural response. It is easy to understand by comparing the damping ratios used in different modeling. The proposed group-pile solutions are found much less than those obtained from the 2-D and 3-D FEM analysis under the monotonic load. However, the results become agreeable in the cases of impulsive and steady state loads. The influence of the infinite boundaries of the structural system in the FEM analysis under the monotonic load may cause the discrepancy in the comparisons.

Table 2. Parameters of the geometry and the materials used in the analyses

Soil	Density	Young's Modulus	Poisson's ratio	Material Damping
	1890 kg/cm^3	20 MN/m^2	0.5	0.05
Pile	Density	Young's Modulus	Pile Radius	Pile Length
	2399 kg/cm^3	2.46×10^7 kN/m^2	0.45 m	40 m
	S/d	Cap Thickness		
	3	2 m		

4 CONCLUSIONS

The direct wave equation analysis for the piles in vertically loaded group-pile foundations is discussed mainly with the concerns on the effects of pile-to-pile and raft-pile-soil interactions. The proposed solution was compared to similar analyses on equivalent pier approximation. The 2-D and 3-D FEM analyses on the group-pile foundations were also made to show the validity of the modeling. Conclusions are summarized as follows:

1. Considering the pile-to-pile interactions with the assumption of rigid foundation, the loads applied at the piles are different at relatively low frequencies. At the higher frequencies, the transmitting loads are about the same in each pile. The corresponding time-dependent loads are found equal at different piles. The size of the effective cap can be calculated accordingly.
2. The soil resistance around the piles will be reduced by the pile-to-pile interactions. For 3×3 and 4×4 square shaped pile foundations, the reductions are in the order that the center pile> edge pile>corner pile. In general, it will become worse by increasing the number of the piles and decreasing the pile-to-pile spacing.

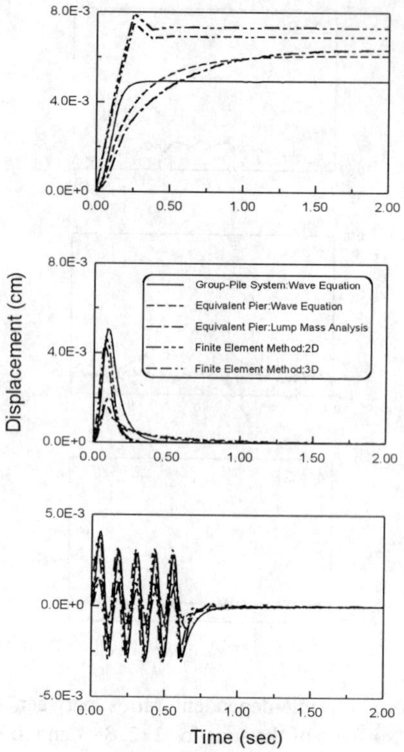

Figure 5. Displacement time-histories obtained from the proposed model and other analyses for 2×2 group pile foundation

3. The ratios between the interaction affected damping coefficient and the soil stiffness are found less than 0.5% sec, which are much smaller than the ratios calculated from a single pile. These damping ratios are approximately constant with the time. Pile-to-pile interactions can reduce the geometric damping of the soils.
4. Comparative solutions have shown the validity of the proposed modeling. It is found that the wave equation analysis for the single piles in a group-pile foundation can provide agreeable solutions with the FEM analysis under the dynamic loads. On the other hand, the FEM analysis at the monotonic loading will provide larger displacements than the proposed ones. Boundary effect of the FEM analysis under such load may be the reason. The modeling based on the equivalent pier approximation will also over-predict the pile displacements at monotonic loading. The corresponding peak displacements are less than the proposed modeling at the impulsive and the steady-state loads. The model

damping used in different analyses is believed to be the main reason of such difference.

ACKNOWLEDGEMENT

This research work is sponsored through the contract NSC-88-2211-E032-009 in Taiwan, ROC. Sincere gratitude is expressed by the authors.

REFERENCES

Chang, D.W., S.H.H. Lee & C.F. Ou Yang 2000[a] Time-domain response of vertically loaded pile with soil modes applied. *Chinese J. of Geotechnical Engineering*. 22(2):162-169

Chang, D.W. & K.C. Lin 1999. Interaction effect on vertical pile response from time-domain wave equation analysis. *Procds.* 2nd Int. Conf. on Earthquake Geotechnical Engineering. 1:407-412. Rotterdam: Balkema.

Chang, D.W., C.H. Wen & C.H. Liu 2000[b]. Wave equation analyses on vertically loaded grouped piles. *Int. J. for Numerical and Analytical Methods in Geomechanics* (in preparation)

Chang, D.W. & S.H. Yeh 1999. Time-domain wave equation analyses of single piles utilizing transformed radiation damping. *Soils and Foundations*. 39 (2):31-44.

Dobry, R. & G. Gazetas 1988. Simple method for dynamic stiffness and damping of floating pile group. *Geotechnique*. 38 (4):557-574.

McCammon, N.R. & H.Q. Golder 1970. Some loading tests on long pipe piles. *Geotechnique*. 20(2):171-184.

Poulos, H.G. 1993. Settlement prediction for driven piles and pile group. *Vertical and Horizontal Deformations of Foundations and Embankments*. Geotechnical Special Publication. 2(40): 1629-1649. ASCE.

Computer Methods and Advances in Geomechanics, Desai et al. (eds) © 2001 Balkema, Rotterdam, ISBN 90 5809 183 X

Geotechnical and structural design of shallow foundations via the Internet

Ashraf M.Ghaly
Civil Engineering Department, Union College, Schenectady, N.Y., USA

ABSTRACT: A *modernistic* approach to the geotechnical and structural design of centrally-loaded shallow foundations resting on homogeneous soil is presented. This approach utilizes the capabilities of the Internet as a platform and the *JavaScript* scripting language as a programming code. In a software-like environment, the designer will be able to interactively enter data and receive results. An Internet site has been developed where the procedure of geotechnical and structural design of shallow foundations is outlined in a logical order leading to the determination of the foundation's cross sectional dimensions, thickness, and reinforcements. This innovative Internet site makes the geotechnical and structural design of shallow foundations an enjoyable experience. It also provides an easily accessible design tool that is globally available to the civil engineering community. The developed program can also be used as a teaching tool to show the effect of different design parameters on the outcome of the design process. An additional feature of flexibility is that the program can be used on any computer platform regardless of the operating system.

1 INTRODUCTION

The design of shallow foundations involves the consideration of geotechnical and structural aspects. The optimization of the design could be a lengthy and tiring task. The geotechnical design of shallow foundations can be conducted using the general bearing capacity theory or Terzaghi's theory. Depending on the shape of the designed foundation, the width, which is the shorter dimension of a strip foundation, or the diameter of a circular foundation are parameters which are required to determine the bearing capacity of the supporting soil. The width or the diameter is usually assumed then the design is verified later. This may require several cycles of assumption and verification. The objective of this paper is to present an interactive method where the user can use the Internet to perform the geotechnical and structural design of shallow foundations in a software-like environment. The *JavaScript* scripting language is used in programming and the designer can enter values of design parameter in designated cells then click buttons to get results of calculations. The advantage of this method is to make the design experience easier and enjoyable. Furthermore, the developed Internet site is freely accessible from anywhere in the world and is independent of the operating system.

2 GEOTECHNICAL DESIGN

Terzaghi (1943) developed the first comprehensive bearing capacity theory for rough shallow foundations. The ultimate bearing capacity q_u is given by the following expressions for strip, square, and circular foundations, respectively. These expressions are developed for the case where general shear is the failure mode of the foundation.

$$q_u = cN_c + qN_q + 0.5\gamma BN_\gamma \quad (1)$$

$$q_u = 1.3cN_c + qN_q + 0.4\gamma BN_\gamma \quad (2)$$

$$q_u = 1.3cN_c + qN_q + 0.3\gamma BN_\gamma \quad (3)$$

Where B is the width of the foundation, c is cohesion of soil, and N_c, N_q, and N_γ are bearing capacity factors that are dimensionless and are only function of the soil friction angle (ϕ). The effective surcharge q = unit weight of soil above the foundation level (FL) times the depth of the foundation (D_f). The factor γ in the third term represents the unit weight of soil located below the FL. Values of both q and γ are affected by the elevation of the ground water table (GWT).

Meyerhof (1963) and Vesic (1973) presented the General Bearing Capacity Theory (GBC) that has taken into consideration the factors that were not accounted for in the Terzaghi's bearing capacity theory. These factors are shape factors (F_{cs}, F_{qs}, and $F_{\gamma s}$), depth factors (F_{cd}, F_{qd}, and $F_{\gamma d}$), and load inclination factors (F_{ci}, F_{qi}, and $F_{\gamma i}$). The general form of the bearing capacity equation is given as:

$$q_u = cN_c F_{cs} F_{cd} F_{ci} + qN_q F_{qs} F_{qd} F_{qi} + 0.5\gamma B N_\gamma F_{\gamma s} F_{\gamma d} F_{\gamma i} \quad (4)$$

Shape factors are dependent on the width (B) and length (L) of the foundation, depth factors are dependent on the depth of the foundation (D_f), and inclination factors are dependent on the angle of inclination of the load acting on the foundation.

A factor of safety (FS) with respect to both bearing capacity and shear failure is usually considered in the design of shallow foundations. The FS with respect to bearing capacity is often in the order of 3-4, while FS with respect to shear failure ranges between 1.4-1.6. The allowable gross bearing capacity (q_{all}) and the allowable net bearing capacity ($q_{all(net)}$) are determined as follows:

$$q_{all} = \frac{q_u}{FS} \quad (5)$$

$$q_{all(net)} = \frac{(q_u - q)}{FS} \quad (6)$$

In equations (1, 2, 3, and 4), the calculation of the bearing capacity is affected by the location of the ground water table (GWT). The term that represents the effective surcharge (q) should be altered based on the elevation of the GWT if located above the foundation level (Case I, Figure 1) or below the same (Case II, Figure 1).

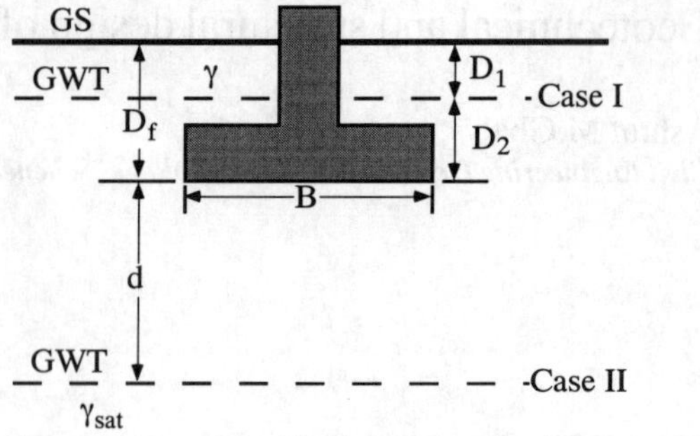

Figure 1. Possible locations of ground water table (GWT).

3 STRUCTURAL DESIGN

Using the specifications of the American Concrete Institute (ACI), the ultimate load carrying capacity (U) of a structural member is:

$$U = 1.4D + 1.7L \quad (7)$$

Where D and L are dead and live loads, respectively. This equation may take different form if the foundation is subjected to different loading conditions. However, in no case should the magnitude of U be less than that given by the equation containing both the effects of dead load and live load.

3.1 *Flexure strength*

The effective depth (d) of the foundation under consideration and the reinforcement bars are designed using the American Concrete Institute's (ACI) concepts of a rectangular section in bending (Figure 2).

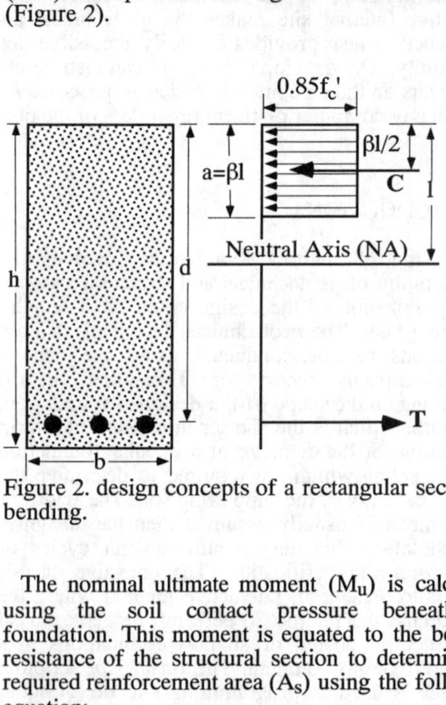

Figure 2. design concepts of a rectangular section in bending.

The nominal ultimate moment (M_u) is calculated using the soil contact pressure beneath the foundation. This moment is equated to the bending resistance of the structural section to determine the required reinforcement area (A_s) using the following equation:

$$M_u = \varphi A_s f_y \left(d - \frac{0.59 A_s f_y}{f_c' b} \right) \quad (8)$$

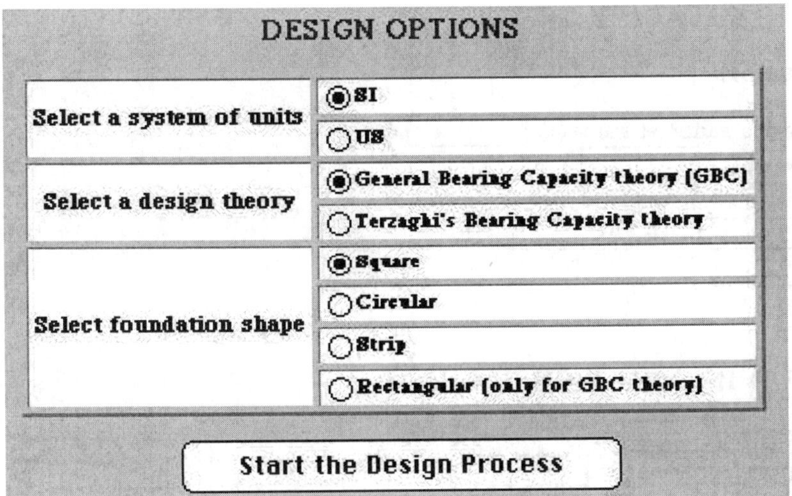

Figure 3. Design options.

Where f_c' = compressive strength of concrete at 28 days, A_s = area of tension steel reinforcement, f_y = yield stress of reinforcement in tension, d = effective depth of section, l = location of neutral axis measured from the top of the compression face, a = βl, β = factor dependent on f_c', b = width of section, and φ = strength reduction factor.

3.2 *Shear strength*

The ultimate shear force (V_u) acting on the foundation is calculated using the soil contact pressure between the foundation and the supporting soil. This shear force is resisted by the concrete and the steel in the section as given in the following equation:

$$V_u = \varphi(V_c + V_s) \tag{9}$$

Where V_c = nominal shear strength of concrete, and V_s = nominal shear strength of reinforcement. The assumed depth of the foundation (d) is verified and adjusted for optimum design.

3.3 *Development length*

Depending on the dimensions of the designed foundation and on the diameter of the bars used as reinforcement, the minimum development length is calculated and checked against the length available for development of sufficient bond between the concrete and steel.

4 INTERNET SITE DEVELOPMENT

4.1 *Geotechnical section*

The developed Internet site is organized in a way that allows the user to select a system of units, the desired design theory, and the shape of the foundation (Figure 3). Both the SI and the US customary systems of units are available. The general bearing capacity theory and Terzaghi's bearing capacity theory can both be used to conduct the geotechnical design of the foundation, however, only the general bearing capacity theory can be used to design a rectangular foundation. The "radio buttons" shown in Figure 3 allow the designer to select only one system of units, one design theory, and one shape of the foundation. Once the desired selections are made, the designer can proceed to the design procedure by clicking the "Start the Design Process" button.

The designer is prompted with a page with cells (boxes) for data entry and buttons to click to perform the calculations. For instance, as shown in Figure 4, the designer is asked to enter the mechanical properties of the soil represented by the angle of shearing resistance (ϕ) and cohesion (c). The bearing capacity factors (N_c, N_q, N_γ) which are function of the angle of shearing resistance of the soil can be calculated by clicking the "Calculate" button. The "Calculate" button activates expressions written in *JavaScript* and programmed to display the calculation results in the corresponding boxes.

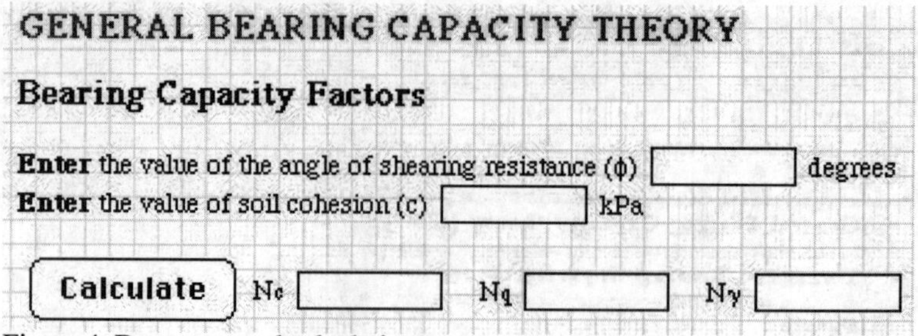

Figure 4. Data entry and calculations.

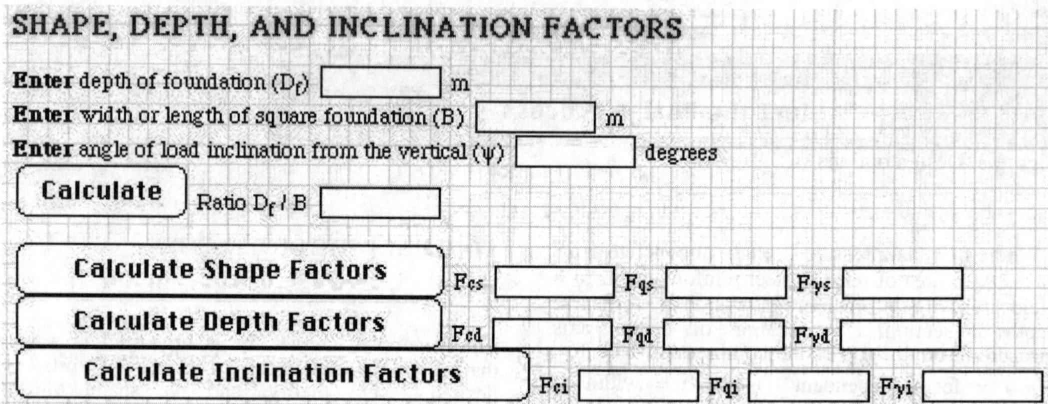

Figure 5. Shape, depth, and inclination factors.

The design proceeds in a similar fashion where the designer is asked to enter values for design parameters then click buttons to calculate results. Figure 5 shows the cells where the designer is asked to enter the depth of the foundation (D_f), width of the foundation (B), and the angle of load inclination (ψ), if any. These values are required to calculate the shape, depth, and inclination factors. The user can always change any value by re-entering that value and the new results will be calculated as the user clicks the buttons following the step where the change is made. Zero is the default value in any cell that is left blank with no value entered.

The designer is to enter the desired factor of safety in the proper box shown in Figure 6. This is used in the determination of the allowable and net allowable bearing capacities. The design can be carried out using either the allowable or the net allowable bearing capacity since the latter is a minor adjustment of the former.

When the designer completes one step of the design, buttons marked "NEXT" can be clicked to move to the next step. The scroll bar can also be used to move the page down. Similarly, the designer can move to the previous step using a button marked "BACK", or by using the scroll bar.

4.2 *Structural section*

A procedure similar to the one used in organizing the geotechnical section of the site, is also used to perform the structural design of shallow foundations. As shown in Figure 7, the designer is required to enter the values of the dead load (DL) and live load (LL) acting on the foundation. The ACI's method of reinforced concrete design outlines several loading scenarios. "Radio buttons" are to be checked by the designer to specify the loading condition acting on the foundation under consideration. The design of the thickness of the foundation, the determination of the reinforcement bars, and the checking of the development length of the reinforcing steel are the main three steps required to conduct the structural design of the foundation. All these steps are organized, and all expressions are programmed in the same way described above. It should be emphasized that the designer can always

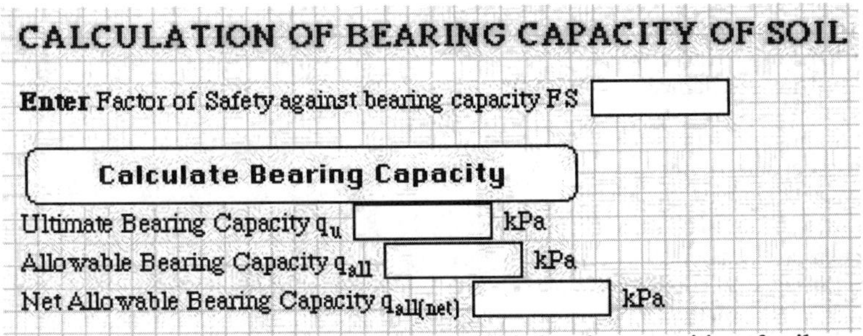

Figure 6. Ultimate, allowable, and net allowable bearing capacities of soil.

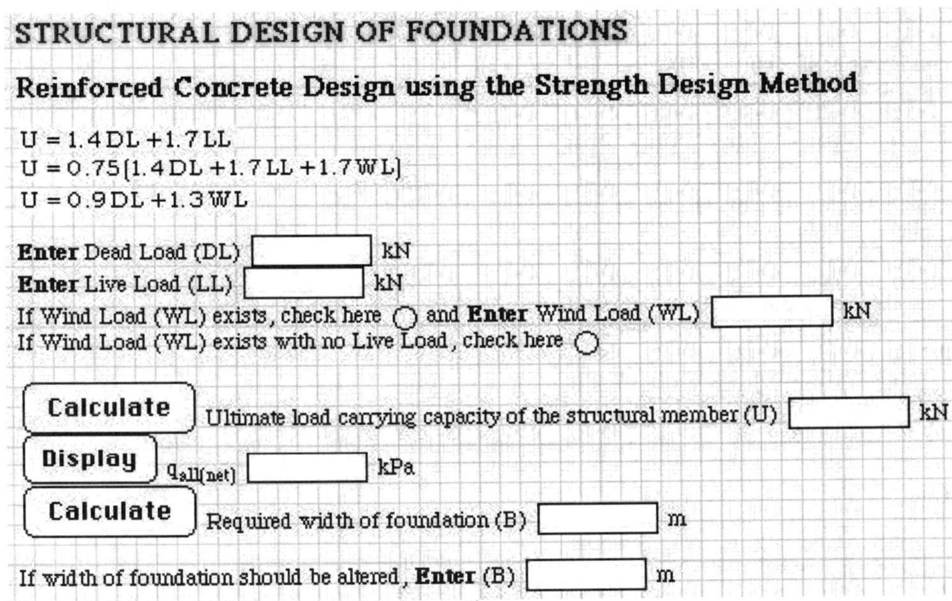

Figure 7. A segment of the structural design section.

alter the input parameters and instantaneously view the new results. This results in great convenience and ability to optimize the design. The design procedure concludes with a "Summary" button that displays all the entered design parameters and the calculated results. As the designer completes both the geotechnical and the structural design of the foundation, a button labeled "HOME" can be clicked to go back to the page where selection of options and design features of a new foundation can be made (Figure 3).

5 TERZAGHI'S BEARING CAPACITY THEORY

The geotechnical design using the general bearing capacity theory or Terzaghi's theory is almost identical. The only difference is that in Terzaghi's method, tables are given for the bearing capacity factors as they vary with the angle of shearing resistance of the soil. Thus, these factors are not calculated as is the case in the general bearing capacity theory, rather they are picked by the designer from given tables. The structural design in both methods is identical.

6 CONCLUSIONS

An innovative approach to the design of centrally-loaded shallow foundation resting on homogeneous soil using the capabilities of the Internet and the *JavaScript* scripting language is presented. The design procedure is outlined in a logical order. The designer proceeds by entering data in special cells designated for data entry, then by clicking specified buttons, the results are displayed in pre-named boxes.

The geotechnical design of the foundation can be conducted using the general bearing capacity theory or Terzaghi's theory. The structural design of the foundation is conducted in accordance with the procedure detailed by the American Concrete Institute, Building Code and Commentary (ACI 318-99). The structural design of shallow foundations detailed in this site addresses all issues related to loading, flexure, shear, and development length of reinforcements. The designer can use the program to conduct the geotechnical analysis and/or the structural analysis.

Available design features include different foundation configurations, soil conditions, and pattern and orientation of the acting external load. The designer can navigate through the procedure in an orderly fashion and may return to review any step at any time. The interactive nature of the technique, coupled with the ease of using it, makes it possible to alter parameters and modify the design at any stage, thus helping the designer optimize the design of the foundation. The designer can easily navigate through the design procedure using special buttons or the scroll bar. Data entry can always be changed to optimize the design. The Internet site is organized in a way that makes it self-explanatory. This, coupled with accessibility from anywhere in the world, makes the site beneficial to the engineering community.

REFERENCES

American Concrete Institute (ACI) 1999. *Building code requirements for reinforced concrete*, ACI 318-99.

http://foundation.union.edu

Meyerhof G.G. 1963. Some recent research on the bearing capacity of foundations, *Canadian Geotechnical Journal*, Vol. 1, No. 1, pp. 16-26.

Terzaghi, K. 1943. *Theoretical soil mechanics*, Wiley, New York.

Vesic, A.S. 1973. Analysis of ultimate loads of shallow foundations, *Journal of Soil mechanics and Foundations Division*, American Society of Civil Engineers, Vol. 99, No. SM1, pp. 45-73.

Computer Methods and Advances in Geomechanics, Desai et al. (eds) © 2001 Balkema, Rotterdam, ISBN 90 5809 183 X

Bending of rectangular rafts subjected to patch loading on an elastic half space

Y.C. How, Y.K. Chow & C.M. Wang
Department of Civil Engineering, National University of Singapore, Singapore

ABSTRACT: This paper presents the bending analysis of thin rectangular rafts on a homogeneous, elastic half-space. The Ritz method was adopted in the bending analysis of raft foundation. The displacement function used is approximated using mathematically complete two-dimensional polynomial. The close agreement of the Ritz bending solutions with those obtained by previous researchers for rectangular rafts subjected to concentrated load and fully distributed load verifies the correctness of the Ritz formulation and software developed and shows that the Ritz method could be an accurate technique in solving this class of raft-soil interaction problems. Due to the dearth of bending results for rafts subjected to patch loads with various sizes and locations and resting on an elastic half space, a series of settlement results for a square raft of arbitrary flexibility, under such a load distribution, are presented. These results should be useful to geotechnical engineers who are designing raft foundations.

1 INTRODUCTION

Raft foundations are widely used to support various types of building structures. To ensure a safe margin against shear failure and excessive local overstressing of the soil, the estimation of settlement and bending moment of the raft foundation at the design stage has to be done. Extensive studies have been reported on the bending analysis of rafts resting on elastic foundation. The most simple foundation model used is that by Winkler who assumed the foundation as a bed of uniform, independent, elastic springs. However, this model is unable to represent the continuous behaviour of the surface soil settlement and leads to a rigid body vertical translation when a raft with free edges is uniformly loaded. To capture the more realistic "dishing" type of raft deformation, researchers have developed various two-parameter foundation models e.g. Filonenko-Borodich model, Hetenyi model and Vlasov model [see Kerr (1964)]. The most widely adopted two-parameter foundation model is that proposed by Pasternak who takes into account the shear interaction between the spring elements. The setback of Pasternak foundation model is the difficulty in determining the modulus of subgrade reaction k and the shear modulus of subgrade G_b. An alternative foundation model that allows for the dishing effect of the raft was proposed by Mindlin (1936) and is referred to as an elastic half-space. A comprehensive review of existing methods of analysis for the bending problem of raft foundation is reported by Selvadurai (1979 a,b), especially for circular rafts resting on an elastic half-space. For other raft shapes, numerical methods have to be employed for the bending analysis. Cheung and Zienkiewicz (1965) presented finite element bending results for flexible rectangular rafts on an elastic half-space. Fraser and Wardle (1976) and Hain and Lee (1978) developed a finite element method for the bending analysis of rafts resting on layered soils. Zaman and Faruque (1985), Rajapakse and Shah (1989) and Shen *et al.* (1999) presented the bending analysis of rectangular rafts under transverse loading on an elastic half space by using a variational approach.

The current study employs the Ritz method for the bending analysis of raft foundations due to its relative simplicity for computer code implementation. The assumed Ritz functions for approximating the raft deflection take the form of a mathematically complete two-dimensional polynomial function which allows exact differentiation and integration to be performed so that the computed solutions may be as accurate as desired. Due to the dearth of bending results for rafts subjected to patch loads with various sizes and locations and resting on an elastic half space, some settlement results for a square raft of arbitrary flexibility, under such a load distribution, are presented. Comparisons of bending results with existing solutions obtained by previous researchers are performed to verify the correctness of the Ritz formulation and software developed. The close agreement of the Ritz results with the previous researchers' solutions shows that the Ritz method could be an accurate technique in solving this class of raft-soil interaction problems.

2 THEORECTICAL FORMULATION

Consider a rectangular raft of uniform thickness h, length L and width B, modulus of elasticity E_r, Poisson's ratio ν_r and shear modulus $G_r = E_r/[2(1+\nu_r)]$ resting on an elastic half-space as shown in Fig. 1.

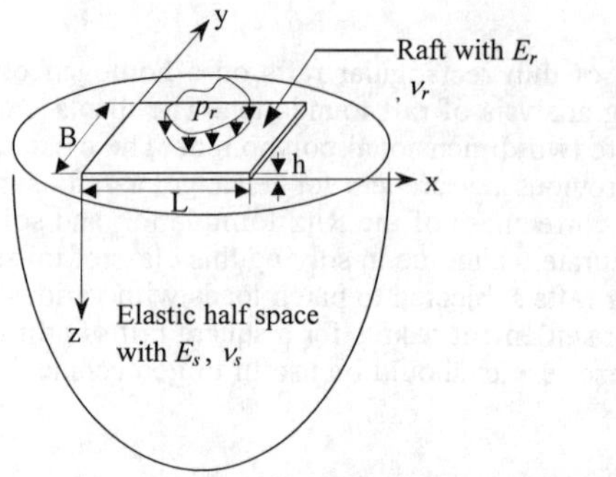

Figure 1 Rectangular raft on elastic half space

The raft is subjected to a transverse load $p(x,y)$ which causes the soil reaction pressure σ_s to develop. According to the classical thin plate theory, the strain energy U_r due to the bending of the raft is given by

$$U_r = \frac{D}{2}\iint_A \left\{ \left(\frac{\partial^2 w}{\partial x^2} + \frac{\partial^2 w}{\partial y^2} \right)^2 - 2(1-\nu_r) \left[\frac{\partial^2 w}{\partial x^2}\frac{\partial^2 w}{\partial y^2} - \left(\frac{\partial^2 w}{\partial x \partial y} \right)^2 \right] \right\} dx\,dy \tag{1}$$

in which w is the transverse deflection of the raft, $D = E_r h^3/[12(1-\nu_r^2)]$ the flexural rigidity of the raft and A the surface area of the raft

The potential energy of the transverse load, V_p and the potential energy of the elastic half-space, V_s are given by

$$V_p = -\iint_A pw\,dxdy \tag{2}$$

$$V_s = \tfrac{1}{2}\sum_{i=1}^{n} R_i w_i \tag{3}$$

where R_i is the resultant force of the i-th soil element due to the distributed reaction pressure σ_s, $w_i = w(x_i, y_i)$ the centre settlement of i-th soil element and x_i, y_i the coordinates of the centre of the i-th soil element. As the raft is resting on an elastic continuum, the transverse displacement of the raft must deform in a compatible manner with the settlement of the soil beneath. Thus, the reaction force vector $\{R\}$ can be related to $\{w\}$ by

$$\{R\} = [F_s]^{-1}\{w\} = [K_s]\{w\} \tag{4}$$

where $[F_s]$ and $[K_s]$ are the flexibility matrix and the stiffness matrix of the elastic half-space, respectively. The elements of the flexibility matrix can be determined by Mindlin (1936)'s point force solution.

The governing static equations for the bending problem at hand may be obtained by minimizing the total potential energy Π with respect to the displacement functions, i.e.

$$\underset{w}{Min}\ \Pi = U_r + V_s + V_p \tag{5}$$

The above problem may be solved by the Ritz method. It is proposed that the Ritz displacement function is approximated by using mathematically complete two-dimensional polynomial, i.e.

$$w(x,y) = \sum_{q=o}^{p}\sum_{i=0}^{q} c_m \phi_m^w(x,y) \tag{6}$$

in which p is the degree set of the complete polynomial space; c_m are the unknown coefficients and $\phi_m = x^i y^{q-i}$. The subscript $m = N - i$ and $N = (q+1)(q+2)/2$. Applying the Ritz method,

$$\frac{\partial \Pi}{\partial c_m} = 0;\ m = 1,2,3,\cdots,N \tag{7}$$

The set of linear equations (7) may be expressed in a matrix form given by

$$([K_r] + [K_s])\{\Phi\} = \{P\} \tag{8}$$

where $[K_r]$ is the stiffness matrix of the raft , $[K_s]$ the stiffness matrix of the elastic half-space, $\{\Phi\} = \{c_1, c_2, c_3, \ldots, c_N\}^T$ and $\{P\}$ is the load vector. The Ritz coefficients are solved from Eq. (8) and back substituted into Eq. (6) for the deflection. The bending moments per unit length, M_{xx}, M_{yy} may then be computed from the following relations

$$M_{xx} = -D\left(\frac{\partial^2 w}{\partial x^2} + \nu_r \frac{\partial^2 w}{\partial y^2} \right) \tag{9a}$$

$$M_{yy} = -D\left(\nu_r \frac{\partial^2 w}{\partial x^2} + \frac{\partial^2 w}{\partial y^2} \right) \tag{9b}$$

3 BENDING RESULTS

3.1 Uniformly loaded rectangular raft

To establish the validity and accuracy of the Ritz method, a comparison of the variation of settlement parameter, $I_w = E_s w / \left[p_r B \left(1 - \nu_s^2\right)\right]$ along raft center-line with the raft-soil stiffness ratio, $K_{rs} = 4E_r\left(1-\nu_s^2\right)h^3 / \left[3E_s\left(1-\nu_r^2\right)B^3\right]$ used by Fraser and Wardle (1976) is shown in Fig. 2. The rectangular raft with aspect ratio $L/B = 2$, $\nu_r = 0.2$ and $\nu_s = 0.5$ is subjected to a uniformly distributed load, p_r. A comparison study for maximum bending moment parameter, $I_M = M/\left(p_r LB\right)$ also is presented in Fig. 3. The Ritz results matched well with Shen *et al.* (1999) solutions and in fair agreement with the finite element results of Fraser and Wardle (1976). The present Ritz results and those obtained by Shen *et al.* (1999) indicate a constant maximum bending moment as $K_{rs} > 10$ while Fraser and Wardle (1976) gives a larger value for rigid rafts ($K_{rs} > 10$). Note that the influence of K_{rs} is more pronounced on bending moments than on settlement.

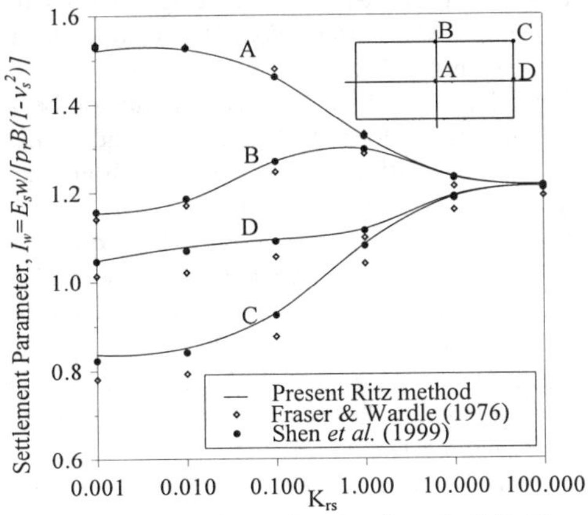

Figure 2 Settlement parameter of rectangular rafts (L/B=2)

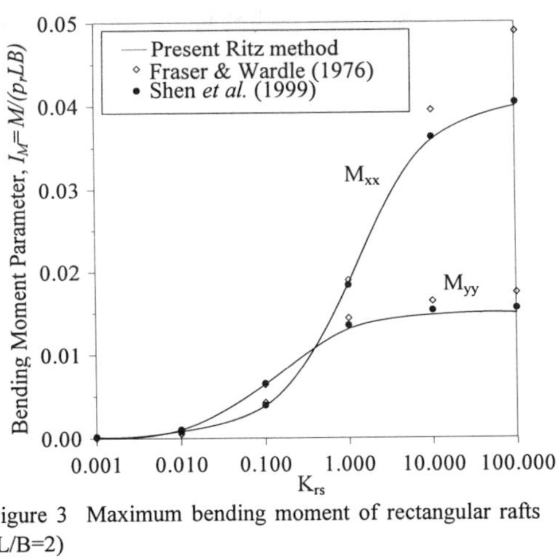

Figure 3 Maximum bending moment of rectangular rafts (L/B=2)

3.2 Centrally loaded square raft

Fig. 4 shows the settlement parameter, $I_{wp} = E_s wL / \left[P\left(1-\nu_s^2\right)\right]$ for a square raft of $\nu_r = 0.15$, $\nu_s = 0.15$ and $K_{rs} = 0.126$ resting on an elastic half space and subjected to a centrally concentrated load, P. The Ritz settlement results are in close agreement with those obtained by Cheung and Nag (1968).

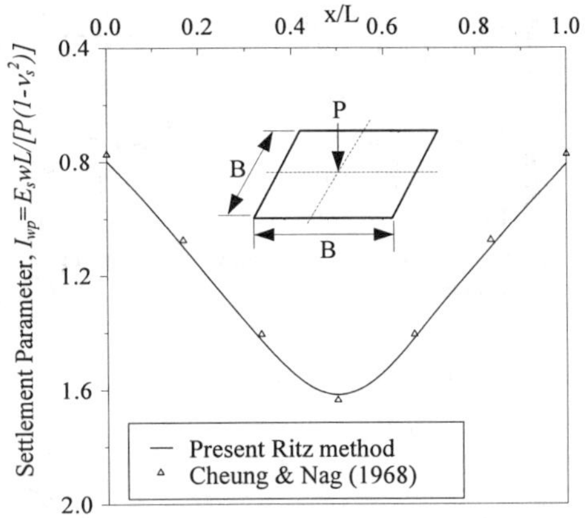

Figure 4 Settlement distribution for centrally concentrated loaded square rafts

3.3 Patch loaded square raft

Some new settlement results for a square raft resting on an elastic half space and subjected to patch loads with ratio of loading area $c/B = 0.25$ at various locations are presented in Figs 5-7. The investigation reveals that the dependence of I_w on K_{rs} is much more apparent on concentrated loaded rafts that that observed for uniformly loaded rafts. It is evident that the differential settlements are much higher in the case of concentrated loaded rafts when compared with rafts subjected to uniformly distributed load.

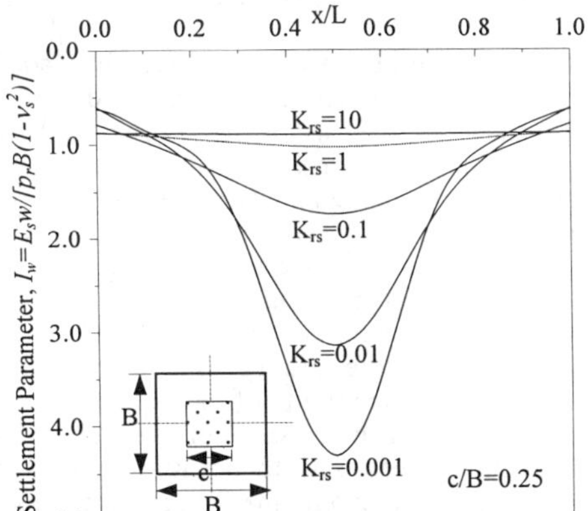

Figure 5 Settlement distribution for centrally patch loaded square rafts

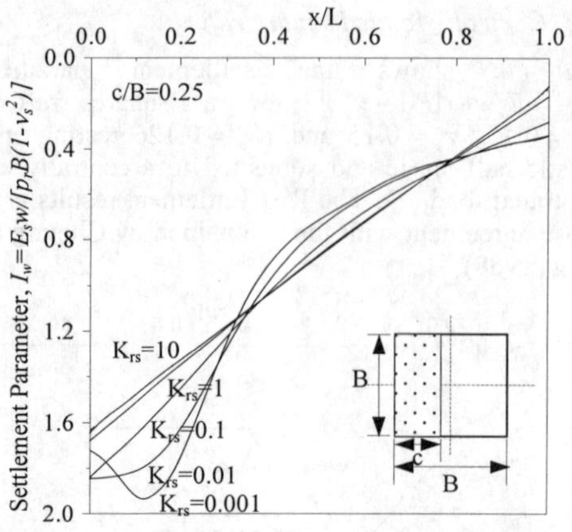

Figure 6 Settlement distribution for sidelong patch loaded square rafts

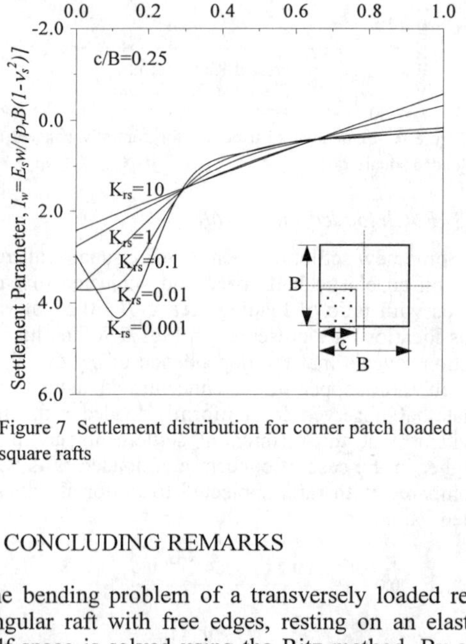

Figure 7 Settlement distribution for corner patch loaded square rafts

4 CONCLUDING REMARKS

The bending problem of a transversely loaded rectangular raft with free edges, resting on an elastic half-space is solved using the Ritz method. By approximating the displacement functions using mathematically complete two-dimensional polynomials of sufficient degree, the Ritz method is shown to be an accurate technique in solving this class of raft-soil interaction problem. An advantage of the present Ritz method is that it does not require the discretization of the raft. This facilitates the task of ensuring the compatibility of the raft deflections and the settlement of the elastic half-space since the latter medium needs only to be discretized. Work is now underway to extend the Ritz method for rafts of arbitrary shape.

REFERENCES

Cheung, Y.K. & D.K. Nag 1968. Plates and beams on elastic foundations – linear and non-linear behaviour. *Geotechnique*. 18:250-260.

Cheung, Y.K. & O.C. Zienkiewicz 1965. Plates and tanks on elastic foundations—an application of the finite element method. *Int. J. Solids Struct.* 1:451-461.

Fraser, R.A. & L.J. Wardle 1976. Numerical analysis of rectangular rafts on layered foundations. *Geotechnique*. 26:613-630.

Hain, S.J. & I.K. Lee 1978. The analysis of flexible raft-pile systems. *Geotechnique* 28:65-83.

Kerr, A.D. 1964. Elastic and viscoelastic foundation models. *J. Appl. Mech.* 31:491-498.

Mindlin, R.D. 1936. Force at a point in the interior of a semi-infinite solid. *Physics*. 7:195-202.

Rajapakse, R.K.N.D. & A.H. Shah 1989. Constraint variational schemes for analysis of rectangular plates on an elastic half space. *Computers & Structures*. 31(6):977-983.

Selvadurai, A.P.S. 1979a. *Elastic Analysis of Soil-Foundation Interaction*. North Holland: Elsevier Scientific Pub. Co.

Selvadurai, A.P.S. 1979b. The interaction between a uniformly loaded circular plate and an isotropic elastic half-space: A variational approach. *J. Struct. Mech.* 7:231-246.

Shen, W.Y., Y.K. Chow & K.Y. Yong 1999. A variational approach for the analysis of rectangular rafts on an elastic half space. *Soils and Foundations*. 39(6):25-32.

Ta, L.D. & J.C. Small 1996. Analysis of piled raft systems in layered soils. *Int. J. Numer. Anal. Methods Geomech.* 20:57-72.

Zaman, M.M. & M.O. Faruque 1985. A variational approach for the analysis of square and rectangular plates resting in smooth contact with an isotropic elastic half space. *Soils and Foundations*. 25:15-26.

Computer Methods and Advances in Geomechanics, Desai et al. (eds) © 2001 Balkema, Rotterdam, ISBN 90 5809 183 X

Axially loaded pile in a group

J. Kos
Department of Geotechnics, Czech Technical University, Prague, Czech Republic

ABSTRACT: The proposed method of an internal group pile analysis is a development of the method of a single pile non-linear analysis. Both methods are based on a column-on-elastic supports model. Winkler's springs, representing a vertical stiffness of a surrounding 'soil pier', are added to the model of a group pile. The pile non-linear load-settlement relationship is not based on predicted (experimental) curves. Only standard soil parameters (γ, φ, c, E, ν) are used. Effects of a pile spacing, piled rafts, edge and corner piles are analysed.

1 INTRODUCTION

The paper deals with the numerical analysis of an axially loaded bored pile in a large group. The large pile group means that the most of piles are internal. These piles can be analysed in the same way as a single pile in a horizontally confined 'soil pier'. No vertical load can leave this 'soil pier'. Edge and corner piles can be analysed in a simplified way as a weighted mean of an internal group pile and the single pile.

The proposed method of a group pile analysis is a development of the method of a single pile non-linear analysis (Kos 1998). Both methods are based on a column-on-elastic supports model. However, the pile non-linear load-settlement relationship is not based on predicted (experimental) p-z or t-z curves. Only standard soil parameters (γ, φ, c, E, ν) are used. The relationship is calculated as the result of the synthesis of three 'partial models'. The first one is a linear elastic column on elastic supports as the basic model of the system. The second one is elastic half-space as the model for stress distribution calculations. The Mohr-Coulomb failure criterion and the bilinear elastic perfectly plastic stress-strain relationship are material models of soil. These three models are periodically used during an iterative process of a non-linear soil stiffness refinement.

A pile loading curve is calculated using the incremental method (usually more than one hundred load increments to minimise computed changes in non-linear soil properties between following loading steps). Nevertheless, the computer time is relatively short, usually less than 15-20 s on a common PC (133 MHz).

2 MODEL DESCRIPTION

The proposed non-linear model of an axially loaded bored pile in a large group is the synthesis of three 'partial models'. Changes in pile/soil contact stresses are computed using a pile on elastic supports model. These stress changes are distributed around a pile using the reduced elastic half-space. New values of pile supports stiffness are computed with respect to the degree of mobilisation of soil shear parameters. These three models are periodically used, so the resultant load-settlement relationship is non-linear.

2.1 *Pile on elastic supports*

A linear elastic column on elastic supports is the basic model of the system, which is used for calculations of pile/soil contact stress changes, pile internal forces and settlement.

There are two types of elastic supports in the model: the Winkler spring under a pile base and 'shear' springs representing a shear stress-settlement relationship of concentric soil cylinders around a pile (Fig. 1- left).

Winkler's springs, representing a vertical stiffness of a surrounding 'soil pier', are added to the model of a group pile (Fig. 1- right).

All supports are linear elastic-perfectly plastic (limited bearing capacity). However, the stiffness of each elastic spring is recalculated as stress dependent during the pile analysis, so the resulting loading curve is non-linear. Usually more than one hundred load increments is used to minimise computed changes in non-linear soil properties between following loading steps.

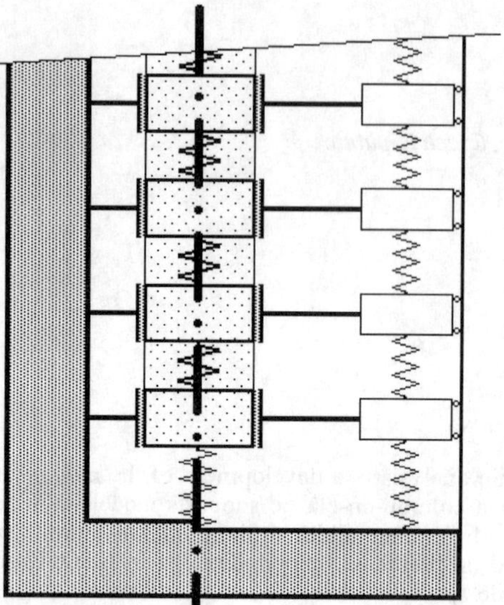

Figure 1. Computer models of single (left) and group (right) piles.

2.2 *Elastic half-space*

The elastic half-space is the model used for stress distribution calculations.

However, stress distribution calculations and continuity conditions are simplified. The Boussinesq half-space under a pile base for the computation of the Winkler spring stiffness, the shear stress distribution in concentric soil cylinders ('shear' springs) and vertical normal stress increments in a 'soil pier' surrounding a group pile (the whole vertical load stays in the 'soil pier'). The influence of horizontal stresses is calculated using oedometer modulus.

The initial state of stresses, the impact of a pile installation process on the state of stresses (drilling the circular cavity in a half-space, pressures of bentonite slurry and fresh concrete - the silo effect) and base and skin bearing capacities are calculated at first (Kos 1998).

2.3 *Soil model*

The Mohr-Coulomb failure criterion and the soil bilinear elastic perfectly plastic stress-strain relationship are used as the third model for the computation of support stiffness.

The breaking point of the bilinear elastic relationship is also calculated using the Mohr-Coulomb criterion. But with reduced (partly mobilised, at rest) strength parameters φ_0, c_0 which are derived from the equivalence of Jáky's coefficient K_0 of earth pressure at rest and Rankin's coefficient K_a of active stress for the partly mobilized internal friction angle φ_0 (Myslivec 1972):

$$K_0 = 1 - \sin\varphi = \frac{1 - \sin\varphi_0}{1 + \sin\varphi_0} \tag{1}$$

$$\varphi_0 = \arcsin\left(\frac{\sin\varphi}{2 - \sin\varphi}\right) \cong \tfrac{2}{3}\varphi \tag{2}$$

Assuming that the soil water suction u (at rest) is constant:

$$u = \frac{c}{\tan\varphi} = \frac{c_0}{\tan\varphi_0} \tag{3}$$

it can be calculated:

$$c_0 = c \cdot \frac{\tan\varphi_0}{\tan\varphi}. \tag{4}$$

Lambe & Whitman (1969) published the equation similar to Eq. 1:

$$K_0 = \frac{(1 - \tan\beta)}{(1 + \tan\beta)} \tag{5}$$

where β = slope of the stress path K_0.

At Jáky's K_0 (and Rankin's active) state of stresses, the minor principal stress is horizontal, the major one is vertical:

$$\frac{\sigma_h}{\sigma_v} = \frac{\sigma_3}{\sigma_1} = K_0(K_a) \Rightarrow \frac{\sigma_1}{\sigma_3} = \frac{1}{K_0}\left(\frac{1}{K_a} = K_p\right) \tag{6}$$

With respect to Eq. 1 and 6, it can be said that limit values of stress at rest can be computed as active and passive (overconsolidated soil) stresses from partly mobilized strength parameters. Generally, stress at rest can be computed using partly mobilized strength parameters from φ_0 (c_0) to zero (because after an isotropic consolidation, the soil is also at rest).

If mobilized values of shear parameters (φ_m, c_m) are less than limit values φ_0 and c_0, the soil behaviour is elastic (several times higher modulus than typical values - the soil is almost at rest). Between φ_0, c_0 limit values and peak values φ_p and c_p, the soil behaves elasto-plastic (typical values of modules), then perfectly plastic. The resultant stress-strain relationship is bilinear elastic-perfectly plastic (Fig. 2).

Standard laboratory tests (oedometer, triaxial) can be analysed using this model. Its parameters can be refined thus.

The capability of the model is illustrated on the analysis of an oedometer test on bentonite powder (Fig. 3). In this case, however, the model was more sophisticated: the stress-strain relationship was not linear but logarithmic (oedometer) and the smooth transition was inserted between elastic and elasto-plastic intervals (Kos et al. 2000).

(The bentonite powder was slightly overconsolidated during the preparation - approx. 40 kPa. Back analyzed parameters of the bentonite powder: $\varphi_u = 30^o$, $c_u = 160$ kPa (due to the suction, w = 7%), C ($\Delta\ln\sigma/\Delta\varepsilon$) = 22, C_0 (elastic, at rest) = 200.)

3 MODEL APPLICATION

Some applications of the group pile model are shown: an analysis of the effect of pile spacing on a bearing capacity, the effect of a piled raft, the effect of pile group subsoil and the possibility of modelling edge and corner piles.

All computations have been made on piles of: d = 1.00 m, L = 20.0 m, E = 23 000 MPa in a soil (stiff ML, MI) of: $\gamma = 20$ kNm^{-3}, $\varphi = 20°$, c = 20 kPa, $E_{def} = 8$ MPa, $\nu = 0{,}4$.

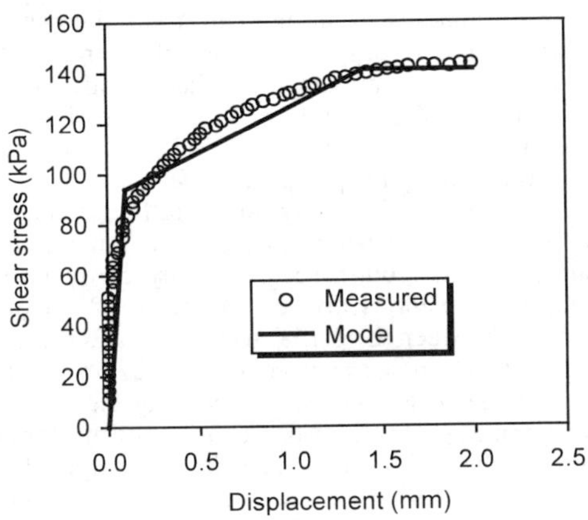

Figure 2. Application of the bilinear elastic-perfectly plastic model on a shear-box test.

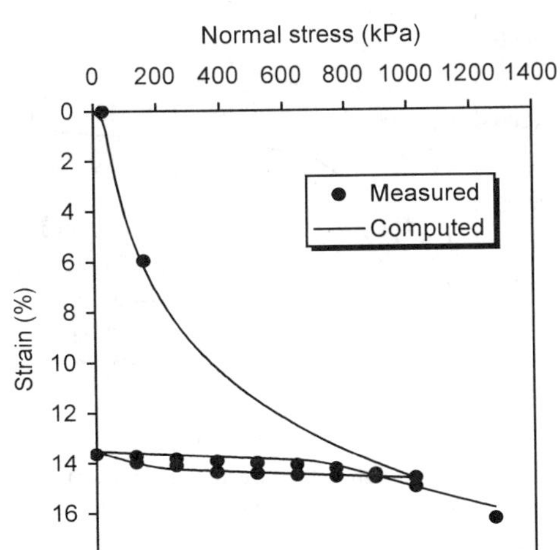

Figure 3. Analysis of an oedometer test on bentonite powder.

3.1 *Effect of pile spacing*

A general assumption of the model of internal group pile is that the whole vertical load (pile head, terrain) stays in the surrounding 'soil pier'. With decreasing pile spacing, the 'soil pier' has to carry rapidly increasing (by square) vertical stress along almost the whole pile length. Its compression and a pile head settlement rapidly increase. The skin resistance becomes concentrated mostly near a pile base (and a pile head as a residuum of initial loading stages - Fig. 4). The pile acts as a column transferring the load to deeper soil layers. The exploitable bearing capacity drops down due to a limited settlement (Fig. 5).

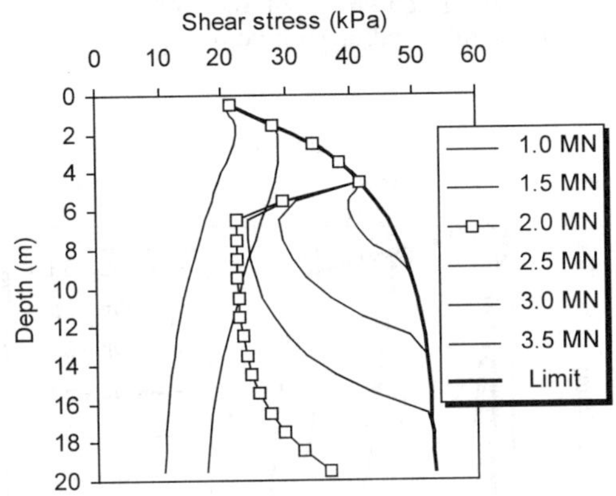

Figure 4. Changes in skin shear stress with increasing load.

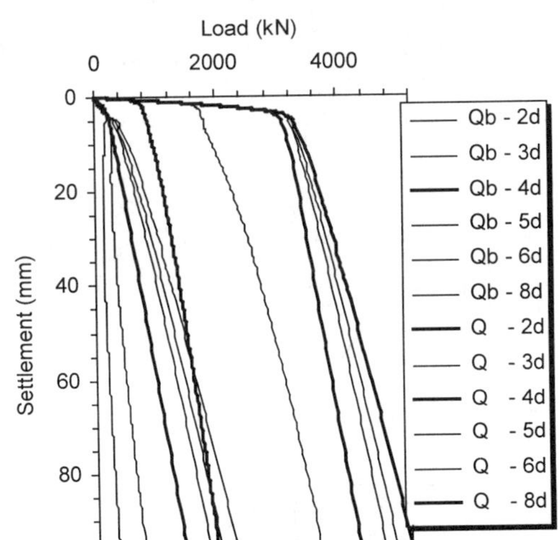

Figure 5. Pile base and total loading curves depending on pile spacing

3.2 *Effect of piled raft*

Quite different experiences with the effect of a piled raft have been published. In some cases, the (experimental) share of the raft on the total bearing capacity is small in other cases it is relatively big.

It can be expected that if the pile spacing is so small that the skin resistance can not be fully exploited (Figs. 4,5), the additional effect of the raft is zero. The certain share of the raft can be measured but at the expense of the pile skin resistance. If the pile spacing is big enough, the positive additional effect of the raft can be observed.

Results of an analysis of the problem are shown in Figures 6, 7. The solution is simplified: the uniformly spread raft/soil contact stress, the constant difference in a settlement of the raft and the pile head (as the raft bending simulation).

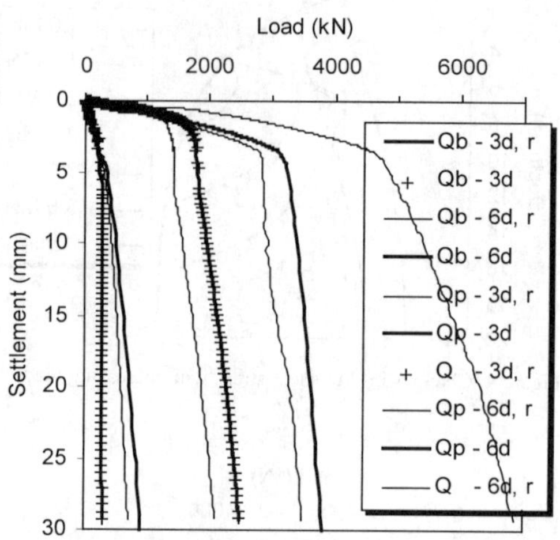

Figure 6. Comparison of loading curves of group piles and a piled raft (r).

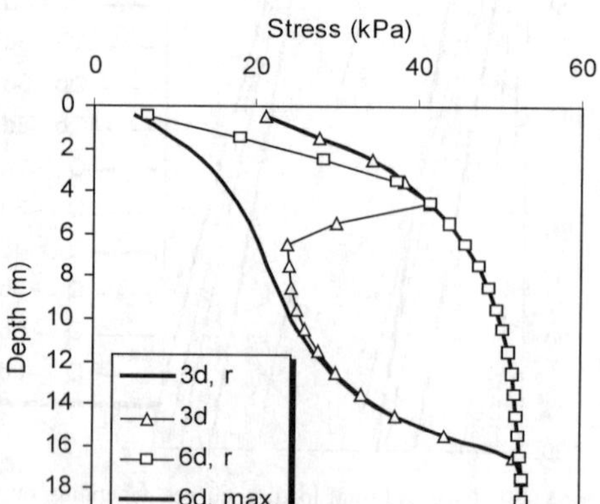

Figure 7. Skin shear stress reduction due to a raft.

3.3 *Effect of pile group subsoil*

The effect of compressible subsoil is calculated using the Boussinesq model of the elastic half-space. The whole vertical load from the 'soil pier' is redistributed in the elastic half-space below a pile base. The resultant range of an active zone and an additional settlement of a pile are dependent on its position in a group. The solution of the effect of compressible subsoil is similar to the solution of a spread foundation settlement.

3.4 *Edge and corner piles*

Bearing capacities of edge and corner piles can be calculated as values from the interval between bearing capacities of an internal group pile and a single pile using following simplified assumptions.

The surrounding of an internal group pile can be divided to four rectangular sectors of a circle with axes oriented to centres of four adjacent piles. (The cross section of the 'soil pier' is square, but the model assumes the circular one.) These sectors are fully influenced by the pile group effect.

The edge pile has only three fully influenced sectors (3/4 of the bearing capacity of an internal group pile). The fourth sector is only partly influenced. This sector is divided to partial 5° sectors. The angle between the partial sector axis and the pile group edge (α) increases from 47.5° to 87.5° and the radius of the sector increases according the expression: r = pile spacing/2/cosα. The share of the partial sector on the pile bearing capacity is computed as 5/360 of the bearing capacity of the internal group pile with the same surrounding 'soil pier' radius. All these computations are based on the condition of the identical settlement of all 'partial piles' (Fig. 8).

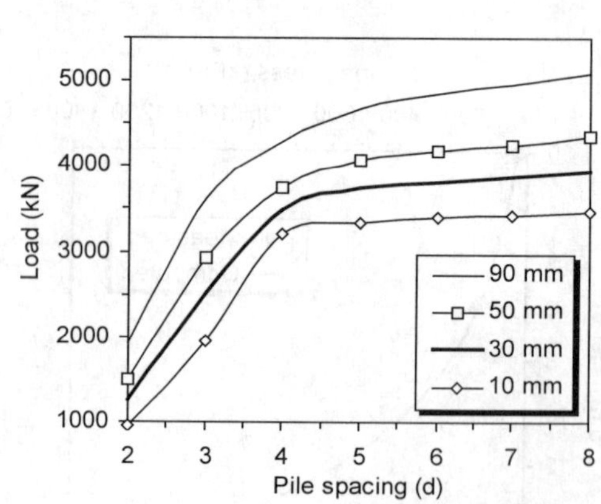

Figure 8. Pile loads capable to induce the identical settlement.

The corner pile has only two fully influenced sectors (2/4 of the bearing capacity of an internal group pile). It has also two 45° partly influenced sectors (described above) and one rectangular uninfluenced sector (1/4 of the bearing capacity of a single pile).

The example of the computation for the (identical) settlement equal to 30 mm (Fig. 8):

Pile spacing	2d	3d
Load: edge/internal	1.40	1.13
corner/internal	1.93	1.27

4 COMPUTATIONS USING FLAC

The problem has been also analyzed using the two-dimensional explicit finite difference program FLAC® (only one pile as the axisymmetric problem).

The material model used has been the simplest (built in the program) linear elastic isotropic model. The model has been modified using the procedures written in the programming language FISH®. After the modification, the program checks values of principal stresses and computes the degree of the mobilisation of shear strength parameters. Then it changes parameters E, ν of the standard linear elastic model in accordance with principles shown in Figure 2. In this way, the behaviour of the soil model is gradually linear elastic, elasto-plastic and almost perfectly plastic with the possibility to model the soil hysteresis.

The computer time is extremely long if compared with the previous model. This is the product of the necessity to stabilise the explicit process using very small movement (load) increments in the soil model with extreme changes of stiffness. The soil medium is very heterogeneous. Some zones may be linear elastic (several times higher modulus of elasticity than typical values). Some zones may be almost perfectly plastic at the same time.

Therefore the FLAC® program has been used for the analysis of a pile installation process to check the analytical solution used in the previous model of an internal group pile.

This model follows the simplified stress path: At the beginning, the soil is at an initial state of stresses (at rest, usually overconsolidated) which have to be verified experimentally or estimated. During drilling the 'circular cavity in a half-space', radial stresses decrease. The soil becomes normally consolidated. The limit values φ_0, c_0 are reached (K_0 state), the soil changes its state from elastic (at rest) to elasto-plastic. The soil cylinders soften from the bore-hole wall and hang themselves on cylinders of bigger radiuses. Vertical stresses in these hanged cylinders decrease. This process can be influenced by an application of bentonite slurry or a steel casing. Then the bore-hole is filled up by fresh concrete and the soil radial stresses increase. The influence of fresh concrete is limited due to the silo effect. In this moment, the skin resistance and the bearing capacity of the pile base (after a similar stress path) are computed.

The scheme of the computation of Winkler's elastic support stiffness is in Figure 9 where: $\sigma_{ol,c}$ = overloading pile/soil contact stress, σ_{ol} = overloading stress distributed in an elastic half-space, σ_s = so called structural strength (the stress interval between the initial stress and the stress when limit values φ_0 and c_0 are mobilised, the soil is at rest, it behaves elastic), t = thickness of the soil with the disturbed structural strength due to a technology (it also influences the maximal computed stiffness of the linear elastic soil at rest), L_a (L_W) length of the active zone (of Winkler's spring).

The scheme for 'shear springs' is similar but with shear stresses.

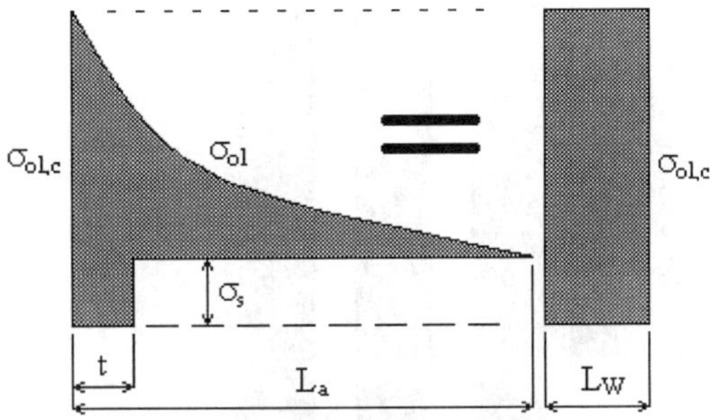

Figure 9. Scheme of the computation of Winkler's elastic support stiffness.

The assumed stress path and the thickness of disturbed soil (t) have been checked using FLAC®. A smaller pile depth has been modelled to reduce the computer time. Different stages of the pile installation process have been modelled changing unit weight, elastic modulus and Poisson's ratio of elements within the pile space.

It can be seen that the thickness of a soil with the disturbed structural strength is relatively small. The most of the soil is at rest (linear elastic).

5 CONCLUSION

The proposed method of a group pile analysis is based on a well known column-on-elastic supports model. Its main advantage is that the pile non-linear load-settlement relationship is not based on predicted (experimental) p-z or t-z curves. Only standard soil parameters (γ, φ, c, E, ν) are used. It has two advantages. The preliminary computations can be easily made on the base of tabulated parameters.

The method is easily adaptable. The effects of group pile spacing, piled rafts, compressible subsoil, edge and corner piles can be analysed.

Pile loading curve is calculated using the incremental method of usually more than one hundred load increments to minimise computed changes in non-linear soil properties between following loading steps. Nevertheless, the computer time of an analysis is short, usually less than 15-20 s on a common PC. So the method can be used for parametric studies and also statistical analyses of an input data variation influence seem to be executable on today's computers.

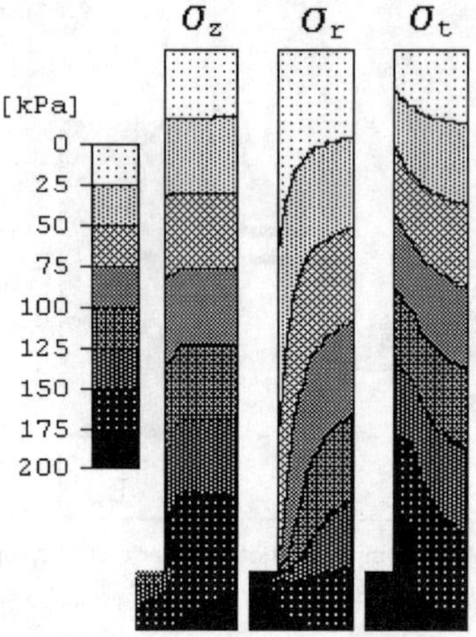

Figure 10. Vertical, radial and tangential stresses around a bentonite slurry filled bore-hole.

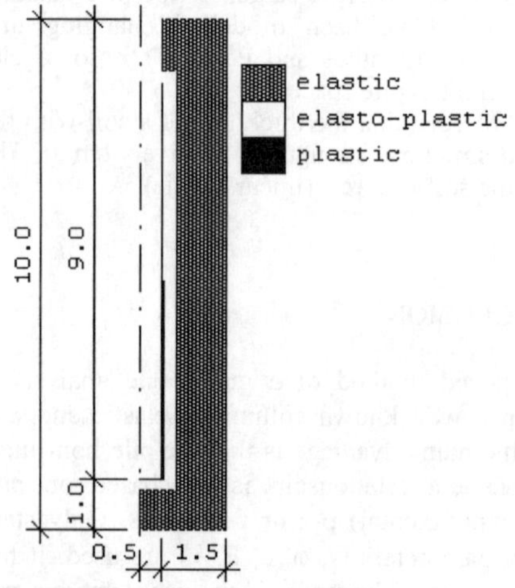

Figure 11. Soil strain state around a bentonite slurry filled borehole

ACKNOWLEDGMENTS

The Czech Grant Agency, through a grant No. 103/98/0632, has supported the research, on which the paper is based.

REFERENCES

Kos, J., Pacovský, J. &Trávníček, R. 2000. Shear strength and swelling of compacted partly saturated bentonite. *Proc. Unsaturated Soils for Asia*, Singapore, pp. 521-526.

Kos, J. 1998. Non-Linear Analyses of Axially and Laterally Loaded Bored Piles. *Proc.4th European Conf. On Numerical Methods in Geotechnical Eng.*, Udine, pp. 121-130.

Lambe, T.W. & Whitman, R.V. 1969. *Soil mechanics.* New York: John Wiley & Sons.

Myslivec, A. 1972. Pressure at rest of cohesive soils. *Proc. 5th EC MSFE.* Madrid, Spain. Vol. 1: 63-67

Computer Methods and Advances in Geomechanics, Desai et al. (eds) © 2001 Balkema, Rotterdam, ISBN 90 5809 183 X

Assessment of the load transfer method for multi-layered soil profiles

K. Lin & M. F. Randolph
The University of Western Australia, Nedlands, W.A., Australia

C.W. Cho
Piletech Consulting Engineers, Korea

ABSTRACT: Although the load transfer method is reasonably straightforward in homogeneous soil, or where the soil modulus varies in a gradual manner, adjustments must be made to the spring stiffnesses and interaction factors when the soil profile shows strong layering. The purpose of this paper is to compare the results of load transfer analysis of pile groups in layered soil profiles with rigorous analysis using a finite layer approach. Suggestions are made for calculating the load transfer spring stiffnesses, and interaction factors, taking account of the layered soil profile and the possible presence of a very stiff underlying layer. The relative accuracy of the load transfer approach using the suggested functions is assessed.

1 INTRODUCTION

The load transfer method is one of the most widely used approaches for analysing piles and pile groups under both axial and lateral loading (Randolph & Wroth 1978, 1979, Chow 1986, Lee 1991, 1993; Mylonakis et al., 1998). In this paper, suggestions are made for calculating the load transfer spring stiffnesses at the pile shaft and the pile base separately, and interaction factors between piles, taking account of the multi-layered soil profile and the possible presence of a very stiff underlying layer. The accuracy of this modified load transfer approach using the suggested formulas is assessed by routine finite element analysis and by the finite layer method described in Lin et al. (2000), which is similar to the method proposed by Small & Booker (1984, 1986) for surface foundations and Zhang & Small (1999) for pile foundations.

2 LOAD TRANSFER METHOD FOR SINGLE PILE RESPONSE

A pile group located in a multi-layered soil is shown in Figure 1. The piles are of embedded length, L, and radius, r_o. The soil is divided into a number of layers with SL and k denoting respectively the total number of layers and a typical soil layer in the zone penetrated by the piles. Similarly, SB and bk are the total number of layers and a typical soil layer below the pile base.

2.1 *Shaft Friction*

In the conventional load transfer approach, the flexibility coefficient F_{ii} denotes the displacement at the pile shaft node i due to a unit load at the same node, defined as in Randolph & Wroth (1978) and Chow (1986) for a homogenous half-space:

$$F_{ii} = \frac{\zeta}{2\pi G_i L_i} \tag{1}$$

In Eq. (1), L_i is the length of pile associated with node i, G_i the local shear modulus and ζ a load transfer parameter given by

$$\zeta = \ln\left(\frac{r_m}{r_o}\right) \tag{2}$$

The radius of influence, r_m, beyond which the soil settlement becomes vanishingly small, has been estimated for piles in a deep stratum as (Randolph & Wroth 1978):

$$r_m = 2.5(1-\nu_s)\rho L \tag{3}$$

where ν_s is Poisson's ratio for the soil and ρ is the soil non-homogeneity factor, taken as $G_{L/2}/G_L$ where $G_{L/2}$ and G_L are the soil shear modulus values at depths of L/2 and L respectively.

For end-bearing piles, r_m is reduced to (Randolph & Wroth 1979)

$$r_m = \left[0.25 + (2.5(1-\nu_s)\rho - 0.25)\frac{G_L}{G_b}\right]L \quad (4)$$

where G_b is the soil shear modulus below the pile base.

In order to consider the influence of both soil non-homogeneity and multiple layering on the shaft response, a general expression for r_m is proposed here as:

$$r_m = \rho_1 \rho_2 (1-\nu_s)L + Br_o \quad (5)$$

Therefore Eq. (2) can be rewritten as

$$\zeta = \ln\left[\rho_1 \rho_2 (1-\nu_s)\frac{L}{r_o} + B\right] \quad (6)$$

where ρ_1 and ρ_2 are the influence factors for the soil multiple layering above and below the pile base respectively. In Eqs. (5) and (6), the parameter B is used to consider the effect of the pile slenderness ratio L/r_o on the pile-soil response. The value of the parameter B may generally be taken as zero. However, for short piles (L/r_o less than 10), the accuracy of the load transfer solution is improved by taking B as 5 (Randolph 1994).

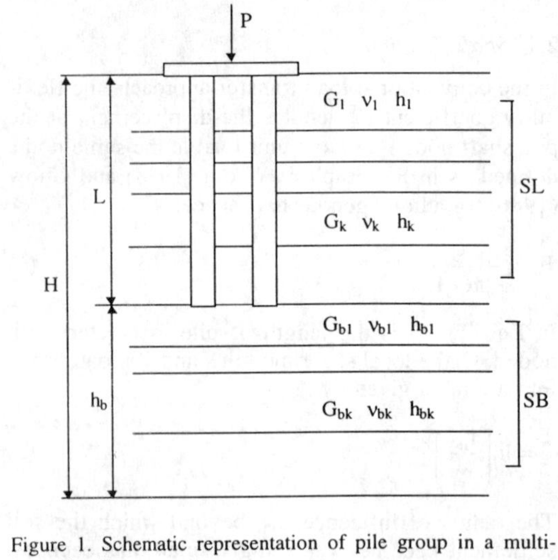

Figure 1. Schematic representation of pile group in a multi-layered soil

It is proposed that the parameter, ρ_1, which represents the effect of multiple soil layers over the depth of penetration of the piles, is retained as representing the ratio of the average modulus to the value at pile tip level. Hence

$$\rho_1 = \frac{\sum_{k=1}^{SL} G_k h_k}{LG_L} \quad (7)$$

where SL is the total number of soil layers above the pile base, and G_k and h_k are the shear modulus and thickness of the kth layer. Thus ρ_1 will equal unity for homogeneous soil, and 0.5 for a soil where the shear modulus varies proportionally with depth (so called Gibson soil).

The parameter ρ_2 in Eq. (6) is used to consider the influence of the soil below the pile base and can be interpreted as

$$\rho_2 = A f_1 f_2 \quad (8)$$

where A is normally taken as 2.5 (Randolph & Wroth 1978, 1979). To describe the effect of an underlying rigid layer on the pile shaft response, the parameter f_1 is defined as

$$f_1 = 1 - e^{-\frac{1.5\,h_b}{L}} \quad (9)$$

where h_b is the total depth of (compressible) soil below the pile base.

Lee (1991) proposed a different form of Eq. (9) as

$$f_1 = 1 - e^{1-\frac{H}{L}} \quad (10)$$

where H is the total depth of the soil layer. This represents a smaller effect of the depth of the soil layer than Eq. (9), especially when the ratio of h_b/L is less than 2.

Figure 2 shows the normalised pile head stiffness of a single pile in a homogenous finite layer soil, computed using the proposed load transfer approach (LT) and a full finite layer analysis (FL). The effect of the ratio h_b/L is shown for two different pile compressibility ratios ($\lambda = E_p/G_L$) of 300 and 1000, and pile slender ratio L/r_o of 40. Good agreement may be seen between the two methods.

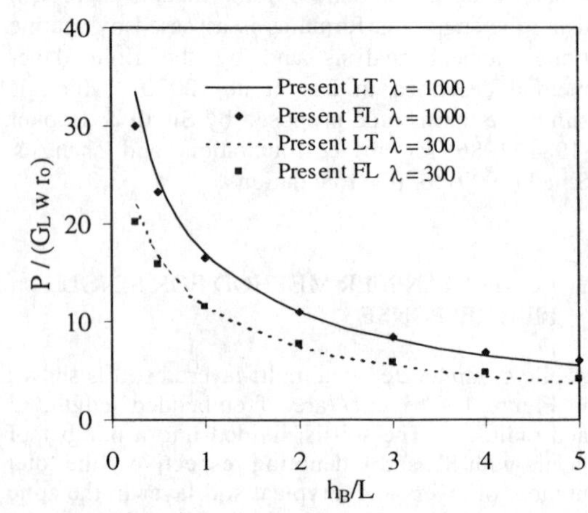

Figure 2. Effect of finite layer on normalised pile head stiffness for $L/r_o = 40$

The parameter f_2 is used to consider the influence of multiple layers below the pile base, with shear modulus G_{b1}, G_{b2}, G_{b3}, etc, and thickness of each layer h_{b1}, h_{b2}, h_{b3} etc. The following expression for f_2 is proposed

$$f_2 = \sqrt{\frac{G_L \sum_{bk}^{SB} h_{bk}\delta}{\sum_{bk}^{SB} G_{bk} h_{bk}\delta}} \tag{11}$$

where SB is the total number of soil layers below the pile base, G_{bk} is the shear modulus and h_{bk} the thickness of the $(bk)^{th}$ soil layer. A weighting factor δ is introduced, decaying with the distance of the layer below the pile base, and defined as

$$\delta = e^{-\frac{h_{bk}^*}{L}} \tag{12}$$

where h_{bk}^* is the depth from the bottom of each layer to the pile base. Equation (11) is consistent with the special case of an end-bearing pile, where

$$f_2 = \sqrt{\frac{G_L}{G_b}} \tag{13}$$

Therefore, the parameter ρ_2 in Eq. (8) can be simplified for particular cases, for example if the soil below the pile base is a homogeneous layer.

To verify the suggested expressions for the parameter f_2 (Eqs. 11 and 12), a single pile in a three-layered soil stratigraphy have been investigated by both the proposed load transfer (LT) and a full finite layer (FL) methods. The soil properties are given in Table 1 for Cases 1 to 3. The pile properties are $L/r_o = 40$ and $E_p/G_L = 1000$.

Table 2 compares the normalised pile head stiffnesses computed by the two methods, and shows reasonably good agreement between them.

Table 1 Soil properties for Cases 1 to 7

Case No.	ν_s	Layer 1 G_1/G_L	h_1/L	Layer 2 G_2/G_L	h_2/L	Layer 3 G_3/G_L	h_3/L
1	0.5	1	1	3	0.25	5	0.25
2	0.5	1	1	3	0.25	10	0.25
3	0.5	1	1	5	0.25	10	0.25
4	0.3	1	0.3	2	0.4	4	1.3
5	0.3	2	0.3	1	0.4	4	1.3
6	0.3	4	0.3	1	0.4	2	1.3
7	0.3	4	0.3	2	0.4	1	1.3

Table 2 Numerical study of parameter f_2

	$\frac{P}{G_L w r_o}$ FL method	LT method	Difference (%)
Case 1	6.81	6.97	2.3
Case 2	6.88	7.38	7.3
Case 3	7.33	7.55	3.0

2.2 *Base Response*

The load-deformation behaviour of soil beneath the pile base is assumed similar to that for a rigid punch resting on the soil surface. The flexibility coefficient for the node i at the pile base is given by (Randolph & Wroth 1978, Chow 1986):

$$F_{ii} = \frac{(1-\nu_b)}{4G_b r_b}\rho_3 \tag{14}$$

where ν_b and G_b are respectively the Poisson's ratio and shear modulus below the pile base, and r_b is the radius of the pile base.

There has been much debate over the parameter ρ_3, which accounts for the effect of the soil finite depth on the base response. Gazetas (1983) suggested that the expression for ρ_3 should be

$$\rho_3 = \frac{1}{1+0.65\frac{d_b}{h_b}} \tag{15}$$

where h_b is the total depth of the soil below the pile base and d_b is the diameter of pile base. Lee (1991) proposed a different formula for considering the effect of the finite layer

$$\rho_3 = 1 - e^{-\frac{H}{d_b}} \tag{16}$$

where H is the total depth of soil. The value of ρ_3 in this equation is close to unity, unless the pile aspect ratio L/d_b is very small, implying that the ratio of H/d_b has almost no effect on the pile base settlement.

In the present method, the parameter ρ_3 is evaluated as

$$\rho_3 = e^{\left(-\frac{d_b}{1.5h_b}\right)} \tag{17}$$

This expression is consistent with Eq. (9), in terms of quantifying the effect of the finite layer on the pile base response and on shaft friction respectively.

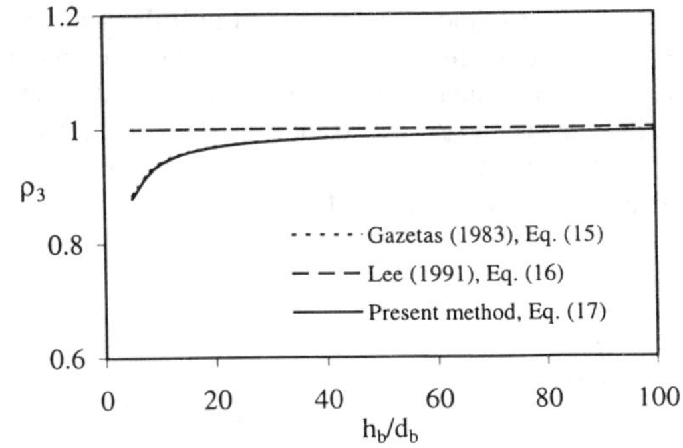

Figure 3. A comparison of parameter ρ_3 ($L/r_o = 40$)

The variation of parameter ρ_3 with h_b/d_b is shown in Figure 3 for the above three methods. The values from the present method are indistinguishable from those using Eq. (15) (Gazetas 1983).

The expression for ρ_3 given by Eq. (17) was used in the load transfer analyses presented earlier (Figure 2 and Table 2), for validating the expressions for f_1 and f_2. Overall, therefore, the method appears to give acceptable accuracy for single piles.

3 LOAD TRANSFER METHOD FOR INTERACTION BETWEEN PILES

Elements of the soil flexibility matrix may be assembled, to give

$$\{w_s\} = [F_s]\{P_s\} \tag{18}$$

The main diagonal elements $[F_{ij}]$ (i = j) of the flexibility matrix $[F_s]$ correspond to the displacement influence coefficients for the individual pile nodes. For linear elastic pile-soil response, these diagonal terms may be evaluated from Eqs. (1) and (14) for the pile shaft and the pile base elements respectively. For non-linear pile-soil response, the equations given by Clancy (1993) may be used.

The off-diagonal terms $[F_{ij}]$ ($i \neq j$) represent the interaction response between node i and node j and are expressed as

$$F_{ij} = F_{ii}\alpha_{ij} \tag{19}$$

The interaction factors, α_{ij}, are evaluated as described in the following sections, for shaft and base respectively. It should be noted that the off-diagonal terms associated with nodal points within the same pile are zero because of the inherent assumption in the load transfer technique for modelling soil behaviour in the individual pile.

3.1 *Shaft Interaction Factor*

The interaction between piles can be evaluated by a variety of methods including Mindlin's solution (often incorporated within a boundary element method), the finite element method or the finite layer method (Lin et al. 2000). Randolph & Wroth (1979) suggested that the interaction factor between shaft elements may be estimated from

$$\alpha_{ij} = \frac{\ln\left(\dfrac{r_{mg}}{s_{ij}}\right)}{\ln\left(\dfrac{r_{mg}}{r_o}\right)} \tag{20}$$

where s_{ij} is the pile spacing between pile i and pile j, and r_{mg} is given as

$$r_{mg} = r_m + r_g \tag{21}$$

Effectively, the maximum radius of influence for a single pile, r_m is increased by r_g, which is taken as the radius of a circle of equivalent area to that covered by the pile group. For square pile groups of n piles at spacing, s, the value of r_g is

$$r_g = \frac{s\left(\sqrt{n}-1\right)}{\sqrt{\pi}} \tag{22}$$

However, this approach overestimates the settlement of pile groups with a large number of piles, particularly when the spacing between distant piles within the group is greater than r_m for a single pile. The evidence of this overestimation will be shown later, and it is suggested here that r_g is taken as zero.

3.2 *Base Interaction Factor*

The base interaction factor derived from the point load solution is given by Randolph & Wroth (1979) as

$$\alpha_{ij} = \frac{2\,r_o}{\pi\,s_{ij}} \tag{23}$$

4 ACCURACY OF LOAD TRANSFER METHOD

In this section, in order to verify the proposed method, the results are compared between those from the present load transfer method (LT) and the finite layer method (FL) (Lin et al. 2000) and other published solutions.

4.1 *Pile Group in Multi-Layered Soil*

The interaction factor for two identical piles embedded in four typical soils was investigated in Lin et al. (2000) using the finite layer method (FL) and Chow (1987) using the finite element method (FE) for different pile spacing ratios.

The results from the three different methods show no significant differences between Case 4, Case 5 and Case 6 in Figures 4 to 6 (see Table 1 for details of shear modulus profiles). This is due to the dominant effect of the relatively stiff underlying stratum. However, for Case 7 in Figure 7, where the base layers are softer than the top, higher values of interaction factor are obtained from the LT method, especially at the large spacing ratio. The reason could be attributed to the LT method being sensitive to the soft underlying base.

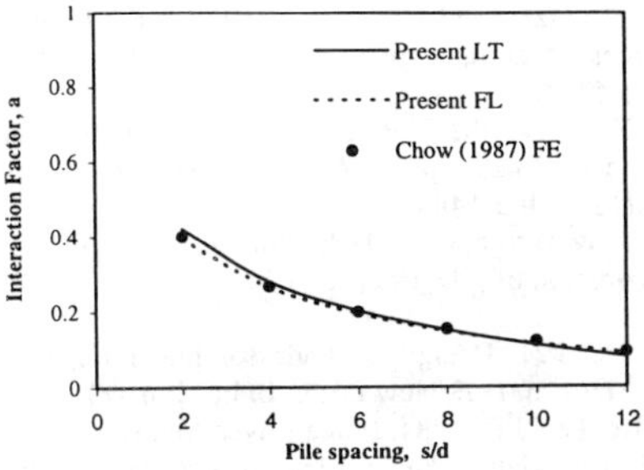

Figure 4 Interaction factors for Case 4

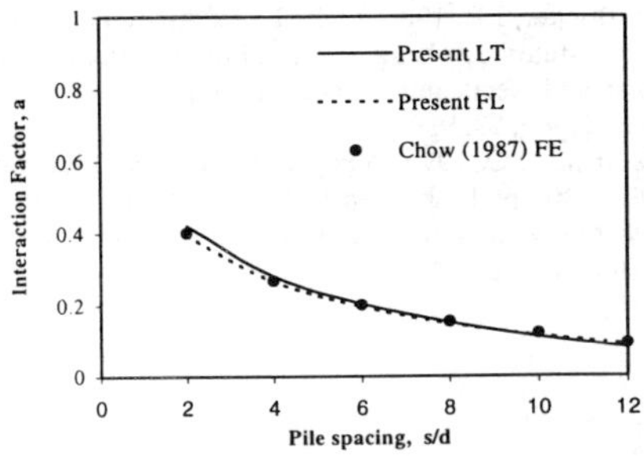

Figure 5 Interaction Factors for Case 5

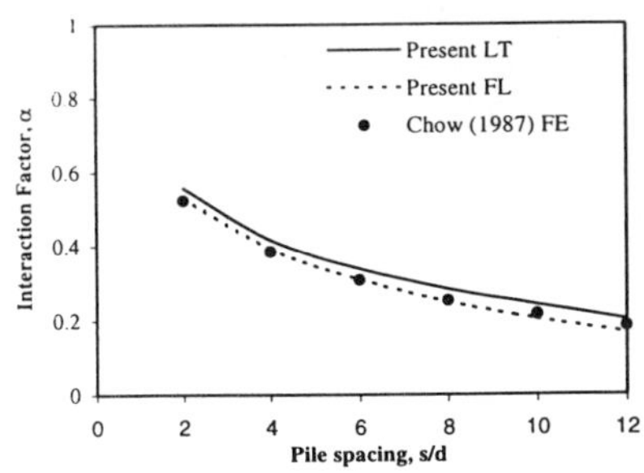

Figure 6 Interaction Factors for Case 6

4.2 *Influence of pile spacing on group settlement ratio R_s of capped pile groups*

Analyses were carried out to explore the effect of pile spacing ratio (s/d = 3, 6, 9, 12, 20) on the settlement ratio R_s for five square groups (n = 2^2, 3^2, 4^2, 5^2, 6^2). A typical soil profile was considered, comprising three separate layers as shown in Figure 8. The following parameters were kept constant in this example: pile slender ratio L/r_o = 40, ratio between the soil depth and pile length H/L = 2, and pile stiffness ratio E_p/G_L=1000.

Figure 9 shows the effect of the number of piles on the settlement ratio R_s. At a given pile spacing ratio, an increase in the number of piles results in an increase in the settlement ratio as expected. Figure 9 also shows that when using r_g = 0 the settlement ratio decreases and converges to 1 as the pile spacing ratio increases. It is interesting to note that, for s/d of 12 or larger, the piles in a group may be treated as individual piles.

In the above analysis, r_g in Eq. (19) was considered to be 0. By contrast, Figure 10 shows the results calculated using the Randolph & Wroth (1979) method, where r_g is taken as the radius of pile group area (Eq. 19). When the pile spacing ratio approaches 20, a significant interaction still exists between piles. This is clearly impractical. Therefore, r_g needs to be considered as 0 in Eq. (19).

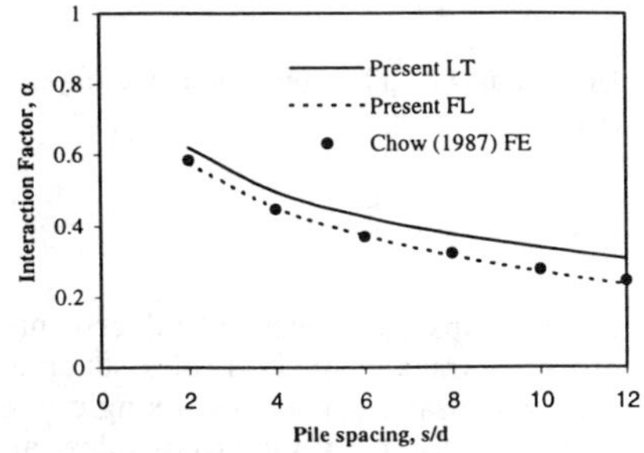

Figure 7 Interaction Factors for Case 7

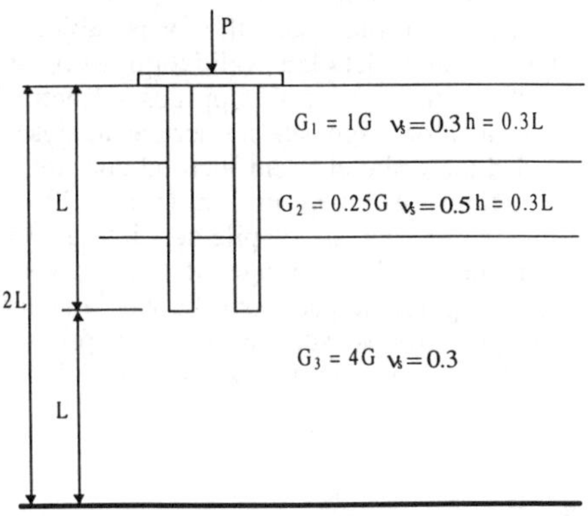

Figure 8 Pile group with a rigid pile cap in a three-layered soil

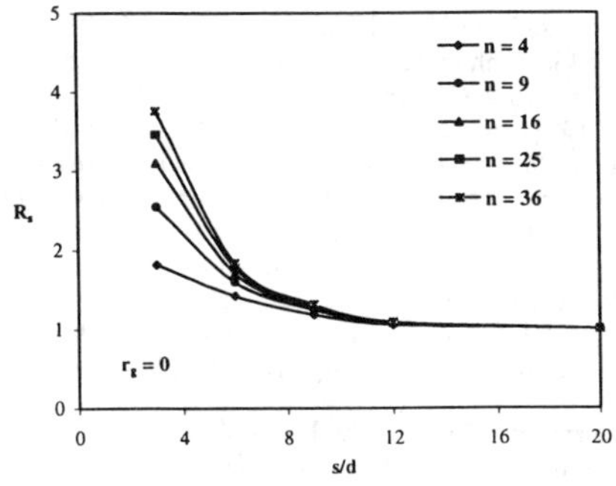

Figure 9 Settlement ratios of pile groups in a three-layered soil

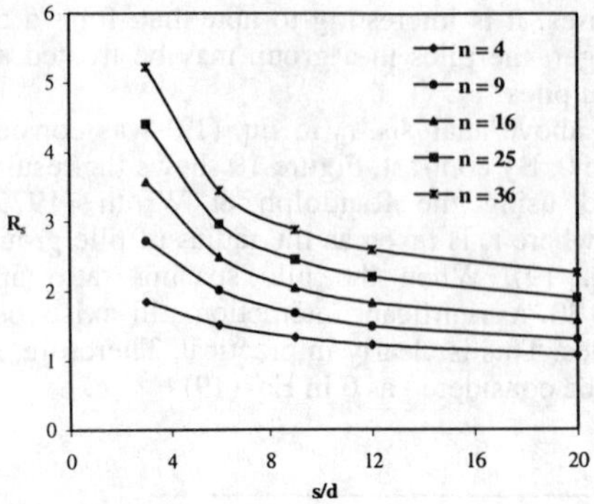

Figure 10 Settlement ratios of pile groups in a three-layered soil using non-zero r_g

5 SUMMARY

A series of relationships have been introduced into the conventional load transfer method for evaluating the shaft friction and base response of a single pile and the interaction factor between group piles, allowing for the effect of finite soil depth and multi-layering within the soil body.

The proposed load transfer method was subjected to an extensive and detailed validation exercise, comparing the results from the proposed method with those from more rigorous numerical analyses. These included the finite element method and the finite layer method described in Lin et al. (2000), which allows a full analysis of pile-soil interaction. An important practical advantage of the proposed method is in being able to carry out the whole analysis for large pile groups, such as with 36 piles or more, in under a few minutes without sacrificing significant accuracy.

REFERENCES

Chow, Y. K. 1986. Analysis of vertically loaded pile groups. *Int. J. for Numer. and Analy. Methods in Geomechanics*, 10 (1): 59-72.

Chow Y. K. 1987. Axial and lateral response of pile groups embedded in nonhomogenous soils, *Int. J. for Numer. and Analy. Methods in Geomechanics*, 11: 621-638

Clancy, P. 1993. *Numerical Analysis of Piled Raft Foundations*. Ph.D. thesis, The University of Western Australia.

Gazetas, G. 1983. Analysis of machine foundation vibration: state of the art, *Soil Dyn. Earthq. Engng* 2(1): 2-41.

Lee, C. Y. 1991. Discrete layer analysis of axially loaded piles and pile groups. *Computers and Geotechnics*, 11: 295-313.

Lee, C. Y. 1993. Pile group settlement analysis by Hybrid layer approach. *J. of Geotech. Engrg. Div.* ASCE, 119(6): 984-997.

Lin, K. & Randolph, M. F. 2000. Pile-soil interaction in multi-layered soil, *TRB 2000*, Washington D.C.

Mindlin, R. D. 1936. Force at a point in the interior of a semi-infinite solid, *Physics* 7: 195-202

Mylonakis, G. & Gazetas, G. 1998. Settlement and additional internal forces of grouped piles in layered soil. *Geotechnique.* 48(1): 55-72.

Randolph, M. F., & Wroth, C. P. 1978. Analysis of deformation of vertically loaded piles. *J. of Geotech. Engrg. Div.* ASCE, 104(12): 1465-1488.

Randolph, M. F., & Wroth, C. P. 1979. An analysis of the vertically deformation of pile groups. *Geotechniqe*, 29(4): 423-439.

Randolph, M. F. 1994. Design methods for pile groups and piled rafts, *XIII ICSMFE*, New Delhi, India, 5: 61-82.

Small, J.C. & Booker, J.R. 1984. Finite Layer Analysis of Layered Elastic Materials Using a Flexibility Approach, Part 1—Strip Loadings. *Int. J. for Numer. Methods in Engng*, 20, 1025-1037.

Small, J.C. & Booker, J.R. 1986. Finite Layer Analysis of Layered Elastic Materials Using a Flexibility Approach, Part 2—Circular and Rectangular Loadings. *Int. J. for Numer. Methods in Engng*, 23, 959-978.

Zhang, H. & Small, J.C. 1999. Analysis Axially and Laterally Loaded Pile Groups Embedded in Layered Soils. *Proc. of the Eighth Australia New Zealand Conference on Geomechanics*, Hobart, 1, 475-483.

Computer Methods and Advances in Geomechanics, Desai et al. (eds) © 2001 Balkema, Rotterdam, ISBN 90 5809 183 X

Effects of spatial variability of soil properties on bearing capacity

A. Nobahar
Faculty of Engineering and Applied Science, Memorial University of Newfoundland, St.John's, Nfld, Canada

R. Popescu
Faculty of Engineering and Applied Science and C-CORE, Memorial University of Newfoundland, St.John's, Nfld, Canada

ABSTRACT: The effects of inherent spatial variability of soil properties on the bearing capacity of strip foundations subjected to vertical loads, and placed on a perfectly elastic-plastic soil deposit are investigated. Two-dimensional finite element analyses are performed assuming undrained loading conditions. Bearing capacity, settlements and foundation rotations predicted by Monte Carlo simulations, accounting for the spatial variability of soil strength, are compared with deterministic analysis results, assuming uniform soil properties.

1 INTRODUCTION

Natural variability of soil properties within geologically distinct and uniform layers has been proven to affect soil behavior. Previous work in this area addressed these effects on seepage through spatially random soils (e.g. Griffiths and Fenton 1993), settlements (Paice et al. 1996), and liquefaction potential (e.g. Popescu et al. 1997). As a general conclusion, it was observed that phenomena governed by highly non-linear constitutive relations are the most affected by spatial variability of soil properties. Bearing capacity of shallow foundations is a function of shear strength of soil, and therefore governed by the strongly non-linear stress-strain behavior of soils.

A Monte Carlo simulation methodology (Popescu 1995), combining digital generation of non-Gaussian stochastic vector fields with non-linear finite element analyses is used to calculate the bearing capacity of a strip foundation on soils with given probability characteristics of their spatial variability. The results of the Monte Carlo simulations, which account for fluctuations in soil properties, are compared to deterministic analysis predictions obtained assuming uniform soil properties. The average value of soil strength used in the Monte Carlo simulations is identical with the soil strength used in the "deterministic" analysis. The resulting random bearing capacities are evaluated statistically based on the average and the 95-percentile values, and compared with the deterministic analysis results. The 95-percentile of a sample is defined here as the value that is lower than 95% of all sample values.

2 MONTE CARLO SIMULATIONS

The Monte Carlo simulation method used here, accounts for the effects of the stochastic spatial variability of soil properties on the system performance. The ensemble of soil properties over the domain of interest is modeled as a bi-dimensional, non-Gaussian stochastic field. The Monte Carlo procedure has three steps (Popescu 1995):

1. Obtain the probabilistic characteristic of the spatial variability of soil properties. For the exercise presented here, these characteristics are assumed.
2. Digitally generate sample functions of a non-Gaussian stochastic field, each sample function representing a possible realization of the relevant soil properties (here the undrained shear strength values) over the domain of interest.
3. Non-linear finite element analyses using stochastic input parameters obtained from the generated sample fields of soil strength.

The simulation methodology in step 2 (Popescu 1995, Popescu et al. 1998a) is based on the spectral representation method. It combines the work by Yamazaki and Shinozuka (1988), Shinozuka and Deodatis (1996), and Deodatis (1996), and extends it to the simulation of multi-variate, multi-dimensional, non-Gaussian stochastic fields. According to the proposed methodology, a Gaussian vector field is first generated taking advantage of the

Fast Fourier Transform technique. Then, this Gaussian vector field is transformed into a non-Gaussian one that is compatible with a prescribed cross-spectral density matrix and with prescribed (non-Gaussian) marginal probability distribution functions. This is achieved through a memoryless non-linear transformation in conjunction with an iterative scheme. For a detailed presentation of the digital simulation algorithm and a discussion on the convergence of the iterative scheme, the reader is referred to Popescu et al. (1998a).

In the third step, for each sample function and at each spatial location (element centeriod), the value of the undrained shear strength determined from step 2 is employed for stochastic input finite element analyses. The analyses are performed for each sample function using the code ABAQUS/Standard (Hibbitt et al. 1998). A series of issues are addressed here, as follows: (1) the data transfer from the discretization used for stochastic field generation to the finite element mesh is done using the mid-point method (Benner 1991), to preserve the prescribed (non-Gaussian) probability distribution function; (2) an appropriate size of the finite element mesh has to be used to capture the essential features of the correlation structure as discussed in section 2; (3) the soil Young's modulus (E) is assumed perfectly correlated to the shear strength, and the Poisson's ratio (ν) is assumed constant over the analysis domain.

3 FINITE ELEMENT MODEL

An appropriate size of the finite element mesh has to be used to capture the essential features of the correlation structure (e.g. Popescu 1995). This mesh size depends on the assumed correlation distance of the soil properties. Der Kiureghian and Ke (1988) suggested that 0.25 to 0.5 of the correlation distance should be taken as maximum mesh size. In this paper, the correlation distances in the horizontal and vertical directions are assumed equal to 5 and 1 meters respectively. Therefore, finite element sizes of 0.25m in vertical direction and 0.5m in horizontal direction are considered acceptable (the mesh size in horizontal direction is also restricted by the necessity of reproducing the stress gradients).

A perfectly elastic-plastic model with a Tresca yield criterion is selected for this study. The Mohr Coulomb model implemented in ABAQUS/Standard is used as a Tresca constitutive model by assuming a zero friction angle, to model the undrained behavior of cohesive soil. The model requires the following constitutive parameters: Young's modulus (E), Poisson's ratio (ν), and cohesion (c_u).

A 4-metre wide strip foundation is placed on the surface of a 12-metre deep layer of over-consolidated clay as shown in Figure 1. The average undrained shear strength is assumed as c_u= 100 kPa. The Young's modulus for undrained loading is computed as E=1500c_u, according to Bowels (1997). A value ν=0.49, appropriate for undrained conditions, is assumed for the Poisson's ratio. A study has been performed to choose the type of finite element to be used for the soil. The soil is discretized using 4-node linear finite elements with reduced integration (CPE4R)-Figure 1. The foundation is assumed rigid. The interface between soil and strip foundation is modeled using the contact surface capability implemented in ABAQUS/Standard. It allows finite relative displacements and rotations of the contact surfaces. The limiting value for the shear resistance at the interface is taken to be about one third of the undrained shear strength of soil. Vertical displacements are gradually imposed to the center of the foundation base, to simulate a vertical central load at the foundation level.

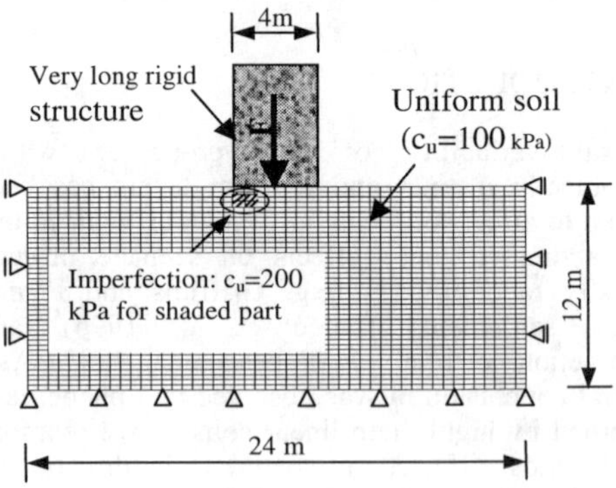

Figure 1. Finite element mesh.

4 DETERMINISTIC ANALYSIS RESULTS

First, a deterministic finite element analysis of the strip foundation, assuming uniform soil with undrained shear strength c_u=100 kPa, is performed. Then the effects of an imperfection in the soil are investigated by considering a small portion of the soil comprised of four finite elements, as shown in Figure 1, with c_u=200 kPa. The results of the two analyses are presented in Figure 2. The imperfection imposed significant changes in the behavior of the soil-structure system. Obviously, there is no rotation in the uniform soil but the soil with imperfection induces a significant rotation of the foundation, which can become a limiting criterion for the allowable bearing capacity. It is concluded that soil spatial variability may cause significant changes in the response of the foundation. The evolution of

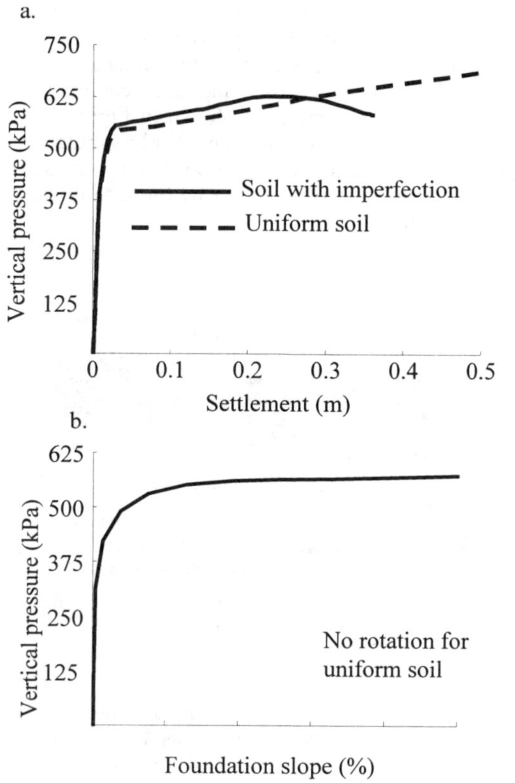

Figure 2. a. Predicted pressure-settlement relationship of the foundation on uniform soil and soil with imperfection; b. Predicted pressure-rotation relationship of the foundation on soil with imperfection.

equivalent plastic strain contours (Figure 3) illustrates the predicted failure mechanism.

Remarks:

1. The displacement-based finite element method tends to overestimate the true limit load (e.g. Nagtegaal et al. 1974, Sloan & Randolph, 1982). For example, the ultimate bearing pressure shown in Figure 2a is about 5% higher than the theoretical value (540 kPa vs 514 kPa). This fact is not of concern here, since only relative bearing capacity values are of interest.
2. Applying the load at the base of the foundation leads to higher predicted ultimate bearing capacity, as compared to the more realistic assumption where the load is applied some where near the centre of the structure. It was selected here to apply the load at the base of foundation to eliminate any effects

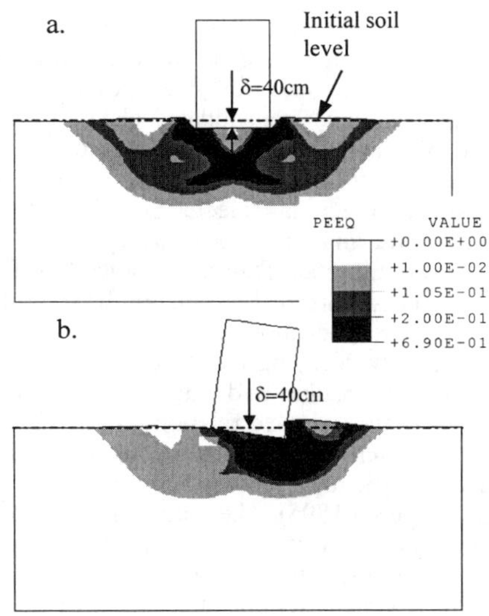

Figure 3. Predicted deformed shape and contours of equivalent plastic strain, γ, at settlement δ=40 cm: a. uniform soil; b. soil with imperfection.

resulting from geometric non-linearity at large deformations.

5 MONTE CARLO SIMULATIONS

5.1 *Selection of probabilistic characteristics for soil variability*

Phoon and Kulhawy (1996) conducted a literature review to estimate the inherent soil variability as observed from some common in-situ soil test measurements. They expressed the observed degree of soil strength variability by means of the coefficient of variation, defined as $C_v = \sigma/\mu$ (where σ is the standard deviation and μ is the average). A coefficient of variation C_v=20 to 40% for clay materials, C_v=20 to 60%, for sand materials, was found from a large number of cone tip resistance measurements. It is mentioned that the scatter in results produced by the testing procedure is much lower than the observed degree of variability (e.g. C_v=5 to 10% for CPT). A coefficient of variation for spatial variability of undrained shear strength, C_v=40% (i.e. toward upper limit from field measurements), is assumed in the present study.

Soil properties follow non-Gaussian probability distributions (e.g. they do not assume negative values). Based on a limited number of field measurement programs, Popescu et al. (1998b) found a general tendency as follows: distributions corresponding to shallow layers are strongly skewed to the left (indicating a stronger influence of a lower bound), while the ones for deep soils (and, by extrapolation for over-consolidated clays) are more symmetrical. In this study, a symmetrical Beta probability distribution function with shape parameters p=q=2.5 is selected to describe the variability of the undrained shear strength.

One of the important features of stochastic fields is the concept of statistical correlation between field values at different locations in space. In this study, an exponentially decaying model, discussed, among others, by Shinozuka and Deodatis (1988), is selected for the auto-correlation function. This model was found to describe relatively well the correlation structure recorded in various soil deposits (Popescu 1995). The main parameter of the auto-correlation function is called scale of fluctuation (or correlation distance) and represents a length over which significant coherence is still manifested. Accounting for the mechanisms of soil deposit formation, leading to different spatial variability characteristics in vertical direction (normal to soil strata) as compared to those in horizontal direction, separable correlation structure models seem appropriate to simulate spatial variability of soil properties. Such a model is assumed in this study, with correlation distances θ_h=5m in horizontal direction (e.g. Przewocki, 2000), and θ_v=1m in vertical direction (e.g. Popescu et al. 1998b).

5.2 *Comparison between Monte Carlo simulations and deterministic results*

Monte Carlo simulations are performed for 100 sample functions representing possible realization of soil strength distribution over the analysis domain. The results are compared with the deterministic analysis predictions in Figure 4. The results of Monte Carlo simulations in terms of pressure-settlement are presented in Figure 4a along with the pressure-settlement curve resulted from the deterministic analysis. The Monte Carlo simulations, accounting for spatial variability of soil strength yielded bearing capacity values that are in general lower than those predicted by the deterministic analysis. It is again mentioned that the average soil strength used in the Monte Carlo simulations is equal to the uniform soil strength used in the deterministic analysis. Non-symmetric soil strength distribution, as accounted for in the Monte Carlo simulations, leads to earlier shear failure than predicted by the deterministic analysis assuming perfect symmetry. In addition, the deterministic analysis does not result in any rotation of the foundation, while the Monte Carlo simulations, accounting for spatial variability of soil properties, result in significant rotations in foundation, which may become the main criterion for the foundation design. Figure 4b shows the pressure versus rotation relations predicted by Monte Carlo simulations. The predicted deformed mesh and the contours of equivalent plastic strains are shown in Figure 5 for one of the samples used in Monte Carlo simulations (pressure-settlement and pressure-rotation curves for that sample function are shown by thick continuous lines in Figure 4).

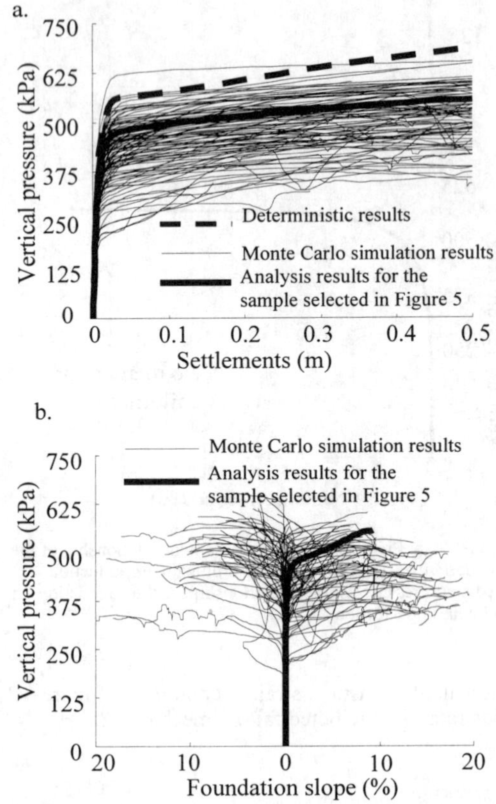

Figure 4. Comparison between Monte Carlo simulations and deterministic analysis results: a. pressure-settlement curves; b. pressure-rotation curves (no rotation is predicted in the deterministic analysis).

5.3 *Result processing*

First, the results of Monte Carlo simulations at a number of displacements and slopes (foundation rotations) are processed and sorted. Next, the average and 95-percentile of these results are calculated at each displacement or slope value (Figure 6). Based on the assumptions in this study and considering a value δ=0.05 m as allowable settlement, the average bearing capacity value resulting from Monte Carlo simulations is 25%

lower than that predicted by the deterministic analysis. As in conventional approaches, a deterministic analysis is performed using a nominal soil strength value considerably lower than the average value (e.g. the 95-percentile of soil strength). A deterministic analysis is performed here based on such a nominal strength value and the results are compared with Monte Carlo simulation results. As it can be seen in Figure 6, the 95-percentile value of bearing capacity resulting from the Monte Carlo simulations exceeds by 38% the bearing capacity resulting from a deterministic analysis using the 95-percentile of soil shear strength.

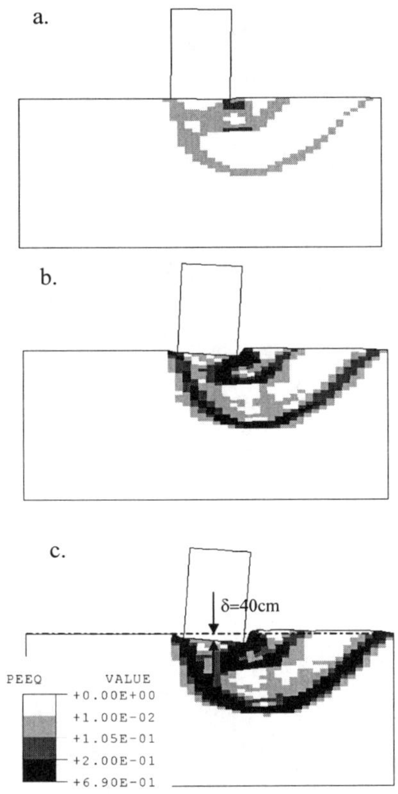

Figure 5. Predicted deformed shape and contours of equivalent plastic shear strain, γ: a. settlement δ=0.05m; b. δ =0.25m c. δ=0.40m.

6 DISCUSSIONS

Based on a comparison between results of Monte Carlo simulations, accounting for the spatial variability of soil strength, and deterministic finite element analyses, assuming uniform soil, it is found that bearing capacity of shallow foundations is strongly affected by the natural spatial variability of soil strength within geologically distinct and uniform layers. For the probabilistic characteristics of soil variability considered in this study, and for the numerical analysis assumptions (undrained loading of purely cohesive soil), the average ultimate bearing capacity predicted by Monte Carlo simulations are 25% lower than that predicted by a deterministic analysis assuming uniform soil properties. However, the 95-percentile of bearing capacity predicted by Monte Carlo simulations is 38% higher than that obtained using the nominal values of soil strength recommended by design codes (based on 95-percentile of soil strength). It is mentioned that the degree of spatial variability assumed in this study is close to the upper limit observed from field measurements.

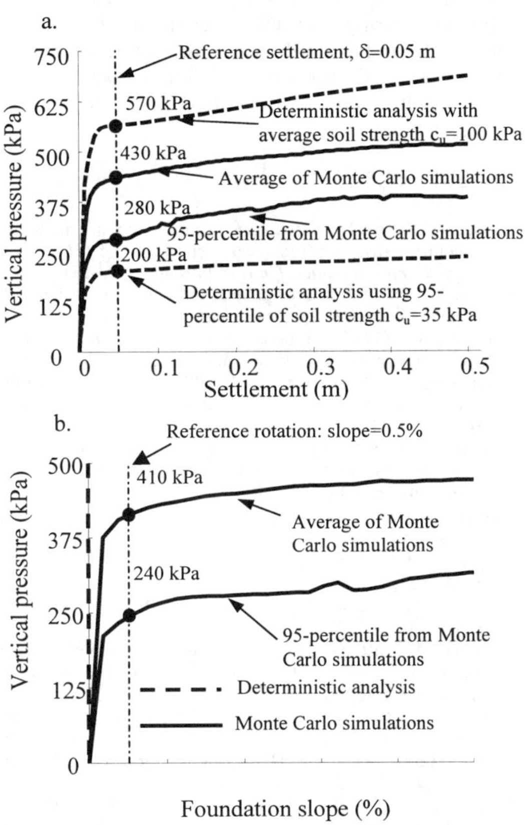

Figure 6. Comparison of deterministic and Monte Carlo simulation results for average and 95 percentile: a. pressure-settlement relationships; b. pressure-rotation relationships.

Design codes account indirectly for variability of soil properties by using a "conservatively assessed nominal value" of soil strength. This value, in general taken as the 95-percentile soil strength, resulted from a number of measurements. This conservative value is not established based on strong theoretical or experimental studies. It is possible to define a characteristic percentile of soil strength that, when used in deterministic analysis, will provide a similar pressure-settlement relationship as was obtained from Monte Carlo simulations for a given confidence interval. For the soil properties considered in this study, a characteristic percentile of 88% of the soil strength used in a deterministic finite element analysis will provide similar results with the 95-percentile of the bearing capacity resulted from Monte Carlo simulations.

In many cases, the rotations resulting from spatial variability of soil, may control the design. For instance, in Figure 6b, limiting the slope of foundation to 0.5% will impose a limit of 240 kPa for bearing capacity, which is 13% lower than the ultimate bearing capacity shown in Figure 6a (namely 280 kPa). In this case, the "nominal value" of soil strength (or characteristic percentile) for deterministic analysis will be the 91-percentile.

REFERENCES

Benner, C.E. (1991). Stochastic finite element methods: Literature review. Technical report 35-91, *Inst. Of Engineering Mechanics, Univ. of Innsburck,* Austria.

Bowels, J. E. (1997). Foundation analysis and design. 5th edition, *McGraw-Hill International Editions*.

Deodatis, G. (1996). Simulation of ergodic multi-variate stochastic processes. *J. of Eng. Mech.,* 122(8).

Der Kiureghian, A. and Ke, J.B. (1988). The stochastic finite element method in structural reliability. *Probabilistic Eng. Mech.,* 3(2): 83-91.

Griffiths, D.V. and Fenton, G.A. (1993). Seepage beneath water retaining structures founded on spatially random soil. Geotechnique, 43(4): 577-587.

Hibbitt, Karlsson & Sorensen, Inc. 1998. ABAQUS version 5.8 - Theory Manual.

Nagtegaal, J.C., Parks, D.M., and Rice, J.R. (1974). On numerically accurate finite element solutions in the fully plastic range. *Computer Methods Appl. Mech. Eng. 4,* pp. 153-177.

Paice, G. M., Griffiths, D. V. and Fenton, G. A. (1996). Finite element modeling of settlement on spatially random soil. *J. of Geotech. Eng.,* 122(9): 777-779.

Phoon, K.K. and Kulhawy, F.H. (1996). On quantifying inherent soil variability. Geotechnical Earthquake Engineering and Soil Dynamics, *Geotechnical Special Publication No. 75, ASCE,* pp. 326-340.

Przewocki, J. (2000). Two-Dimensional Random Field of Mechanical Soil Properties. *J. of Geotech. Eng.,* 126(4): 373-377

Popescu, R. (1995). Stochastic variability of soil properties: data analysis, digital simulation, effects on system behavior. *Ph.D. thesis, Princeton University,* Princeton, NJ.

Popescu, R., Prevost, J.H., and Deodatis, G. (1997). Effects of Spatial Variability on Soil Liquefaction: Some Design Recommendations. *Geotechnique,* 47(5):1019-1036.

Popescu, R., Deodatis, G., and Prevost, J.H. (1998a). Simulation of homogeneous non-Gaussian stochastic vector fields. *Probabilistic Engineering Mechanics,* 13(1):1-13.

Popescu, R., Prevost, J.H., and Deodatis, G. (1998b). Spatial Variability of Soil Properties - Two Case Studies. Geotechnical Earthquake Engineering and Soil Dynamics, *Geotechnical Special Publication No. 75, ASCE,* pp. 568-579.

Sloan, S. W., and Randolph, M.F. (1982). Numerical prediction of collapse loads using finite element methods. *Int. J. Numer. Anal. Methods Geomech.* 6, 47-76.

Soong, T.T. and Grigoriu, M. (1993). Random vibrations of mechanical and structural systems. *Prentice Hall.*

Shinozuka, M. and Deodatis, G. (1996). Simulation of multi-dimensional Gaussian stochastic fields by spectral representation. *Appl. Mech. Reviews, ASME,* 49(1):29-53.

Shinozuka, M. and Deodatis, G. (1988). Response variability of stochastic finite element systems. *Journal of Eng. Mech.,* 114(3): 499-519.

Yamazaki, F. and Shinozuka, M. (1988). Digital generation of non-Gaussian stochastic fields. *Journal of Eng. Mech.,* 114(7):1183-1197.

Computer Methods and Advances in Geomechanics, Desai et al. (eds) © 2001 Balkema, Rotterdam, ISBN 90 5809 183 X

Geometrically non-linear 3-D finite element analysis of a single pile

V.A. Sawant
Mahatma Gandhi Mission's College of Engineering, Kamothe, Navi Mumbai, India
D.M. Dewaikar
Department of Civil Engineering, Indian Institute of Technology, Mumbai, India

ABSTRACT - In the present paper, an attempt has been made to consider the aspect of geometrical non-linearity using 3-D finite element analysis. Pile and soil are discretised into a number of 20 noded isoparametric elements. Interface between pile and soil is modelled using 16 noded surface elements. A Lagrangian approach is adopted throughout, in which, displacements are always referred to original configuration. Incremental approach is used in the analysis, in which, stiffness matrix is updated after every increment of load. Analysis indicates a significant change in response of piles, which have considerable unsupported length and presence of axial load.

1 INTRODUCTION

Piles in the offshore environment are frequently long and subjected to both axial and lateral loads, which may induce large deformations both in the pile and in the surrounding soil mass. As a result, the pile-soil system may experience significant modifications due to changes in the geometry. A situation of this kind may therefore require consideration of both geometric and material non-linearity. Virtually very little work is found in the literature, where, a three-dimensional pile-soil system has been analyzed by considering both material and geometric non-linearity.

In small-strain theory, a linear decomposition of strain into elastic and plastic components is possible. However, for finite strains, a linear decomposition of strain may not be appropriate. One obvious reason, when strains are large, is that elastic recovery on unloading would be definitely affected by the plastic part of the deformation. Procedures are available to include such interdependence of the elastic and plastic strains into the elasto-plastic constitutive laws. However, their implementation into finite element method is rather difficult. Therefore in this paper, only geometrically non-linear 3-D finite element analysis of a pile subjected to combined action of axial and lateral loads is presented.

2 FINITE ELEMENT FORMULATION

The pile and soil are discretised into a number of 20 noded isoparametric continuum elements. The interface between pile and soil is modelled using 16 noded isoparametric surface elements, with zero thickness. These interface elements are useful in simulating the mechanics of stress transfer along the interface.

A Lagrangian approach is adopted throughout, in which, displacements are referred to the original configuration. Equilibrium condition for large displacement analysis can be written as,

$$\{d\psi\} = \left([K_0] + [K_\sigma] + [K_L]\right)\{d\delta\} = [K_T]\{d\delta\}$$

in which,

$\{d\delta\}$ represents incremental displacement vector

$\{d\psi\}$ represents incremental load vector

$[K_T]$ is total tangential stiffness matrix

$[K_0]$ is usual small displacement stiffness matrix

$[K_\sigma]$ is geometric stiffness matrix and,

$[K_L]$ is large displacement matrix

Stiffness matrices, $[K_0]$ for small displacement and $[K_L]$, due to large displacements are given by,

$$[K_0] = \int_v [B_0]^T [D][B_0] dv$$

and,

$$[K_L] = \int_v ([B_0]^T [D][B_L] + [B_L]^T [D][B_L] + [B_L]^T [D][B_0]) dv$$

in which, $[B_0]$, is the same strain-displacement transformation matrix as in linear infinitesimal strain analysis, and only $[B_L]$ depends on displacements.

$[K_\sigma]$, is a symmetric matrix dependent on stress level and given as,

$$[K_\sigma] = \int [G]^T [M][G] dv$$

where,

[G], is the transformation matrix between nodal displacements and derivatives of displacements and,

$$[M] = \begin{bmatrix} \sigma_x I_3 & \tau_{xy} I_3 & \tau_{xz} I_3 \\ & \sigma_y I_3 & \tau_{yz} I_3 \\ & & \sigma_z I_3 \end{bmatrix}$$

in which, I_3, is unit matrix of order 3 x 3, and $\{\sigma_x, \sigma_y, \sigma_z, \tau_{xy}, \tau_{yz}, \tau_{zx}\}$ are stresses in the element. Since, the stiffness matrices, $[K_L]$ and $[K_\sigma]$ are functions of displacement, $\{\delta\}$, they are required to be updated after every load increment. Eq. 1 gives incremental displacements, which are added to get total displacements as,

$$\{\delta\}_i = \{\delta\}_{i-1} + \{d\delta\}$$

where, i and i-1 represent increment numbers.

3 RESULTS

Analysis is carried out on square piles (size 1m), with lengths of 40 m, 45 m and 50 m, with a constant embedded length of 40 m. Material properties of pile, soil and interface are as follows:

Young's Modulus of pile $E_p = 2.0E8 kN/m^2$

Young's Modulus of soil $Es = 4267 kN/m^2$

Poisson's ratio of pile $\nu = 0.15$

Poisson's ratio of soil $\nu_S = 0.45$

Tangential Stiffness of interface $k_s = 1000 kN/m^3$

Normal Stiffness of interface $k_n = 1.0E6 kN/m^3$

In the analysis, following load conditions are considered:

1. Ratio of vertical load P to lateral load H = 0.0
2. P/H = 5.0

Figure 1 shows the comparison of load-deflection behaviour for the case P/H=0.0, as obtained from linear and geometrically non-linear analyses. Both analyses give almost identical results. Results for analysis for the case, P/H=5 are shown in Figure 2. Maximum increase in the pile top displacements from geometrically non-linear analysis for 45 m and 50 m piles, is of the order of 12% and 37%, respectively.

Above results clearly show that geometrically non-linear analysis of laterally loaded piles does not produce any significant change in the response, in the absence of axial loads. With the presence of axial loads, an increase in the pile top displacement is observed with increasing unsupported length.

However, for 40 m pile (i.e., pile without any unsupported length), different behaviour is observed, in which pile top displacements from geometrically non-linear analysis are observed to be lower than those obtained from linear analysis, with a maximum difference of 20%. This is similar to the results reported by Faruque & Desai (1982),who observed that results from geometrically non-linear analysis were 33% lower than those obtained from linear analysis, for pile fully embedded in soil.

This investigation is rather preliminary in identifying effect of geometrical non-linearity in soil-structure interaction of pile foundations. From the results presented, it appears that such effects can be highly significant from viewpoint of analysis and design. Investigations should include verification with respect to experimental observations.

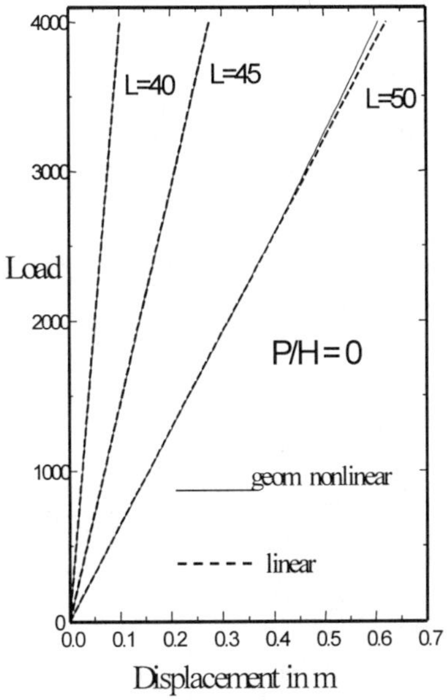

Figure 1 : Load-Pile top displacement relationships from linear & geometrically nonlinear analyses

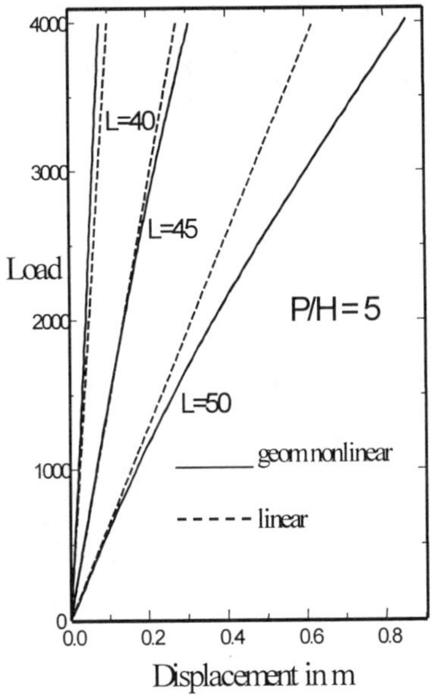

Figure 2 : Load-Pile top displacement relationships from linear & geometrically nonlinear analyses

4 CONCLUSIONS

Following conclusions are drawn from the above study.

1. Geometrically non-linear analysis of laterally loaded piles produces significant change in the response, in the presence of axial loads.
2. An increase in the pile top displacement is observed with increasing unsupported length, with the presence of axial load.
3. For pile without any unsupported length, pile top displacements from geometrically non-linear analysis are observed to be lower than those obtained from linear analysis.
4. Since, offshore piles are subjected to combined action of axial and lateral loads, the geometrically non-linear analysis would tend to become more significant in situations, where, unsupported pile lengths are large enough.

REFERENCES

Faruque, M. O., & Desai, C. S. (1982), '3-D Material and Geometric Non-linear Analysis of Piles', Proc. 2nd International Conference on Numerical Methods in Offshore Pilings, Austin, Texas, 1982.

Lee, E. H. (1959), 'Elasitc-Plastic Deformation at Finite Strains', Journal of Applied Mechanics, Trans., ASME, Vol. 38 , 1959

Computer Methods and Advances in Geomechanics, Desai et al. (eds) © 2001 Balkema, Rotterdam, ISBN 90 5809 183 X

Analysis of piled raft foundations

J.C.Small, H.H.Zhang & A.L.de Ambrosis
Department of Civil Engineering, University of Sydney, N.S.W., Australia

ABSTRACT: The analysis of piled raft foundations subjected to vertical, horizontal or moment loading is complex because the problem is essentially three dimensional in nature. This is because there is interaction among the raft, piles and soil, and because the loadings in general do not lead to conditions of symmetry. In this paper therefore, a method is presented that involves using plate elements for the raft, one dimensional fnite elements for the piles and finite layer theory for the soil. Solutions from the finite layer technique are firstly verified by comparison with 3-dimensional finite element results and it is demonstrated that the present results are in good agreement with those from the full 3-dimensional finite element analysis. Solutions are also presented for problems involving layered elastic soils. It is demonstrated that for piled rafts subjected to either vertical or horizontal loadings, the present technique provides a simple way for obtaining solutions to the piled raft problem.

1 INTRODUCTION

In early methods used for the analysis of a piled raft with the raft or cap on the ground, the raft was generally treated as a completely rigid cap or a perfectly flexible cap such as in the work of Kuwabara (1989), Butterfield and Banerjee (1971), or Davis and Poulos (1972). These methods could not therefore deal with a pile group connected at the heads by a flexible raft of any stiffness. Hain and Lee (1978) considered a piled raft as a flexible elastic plate supported on a group of compressible friction piles subjected to vertical loads only. However, the rotations and horizontal movements of a pile head induced by a vertical load applied to an adjacent pile or the soil surface were ignored. Clancy and Randolph (1993) and Poulos (1994) also developed approximate numerical methods to analyse piled raft foundations. However, their methods may only approximately deal with piled raft foundations subjected to vertical loading or moments and not horizontal loads.

Full 3-dimensional finite element solutions can also be obtained to this problem, but this can be time consuming for fairly large piled raft foundations, even with the fast computers available today. An alternative method of analysis is to use finite layer theory, a semi-analytical technique that allows solutions to be obtained to the piled raft problem with less computational effort. Based on this theory, Ta and Small (1996) developed a method of analysing a piled raft (on or off the ground) subjected to vertical loads only. Zhang and Small (2000) subsequently developed a method of analysing piled rafts subjected to both horizontal and vertical loads where the raft is clear of the ground rather than where the raft was in contact with the ground.

In this paper, a method is presented that involves using plate elements for the raft, one dimensional finite elements for the piles and finite layer theory for the soil. The raft is supported by both the soil and the piles, and it can be subjected to horizontal and vertical loads as well as moments. The movements in three directions (x, y, z) and rotations in two directions (x, y) of the piled raft can be simultaneously solved by using the present program APRAF. Solutions from the finite layer technique are verified by comparison with 3-dimensional finite element results. Solutions are also presented for problems involving layered elastic soils, and for soils that have a stiffness that increases with depth. It is demonstrated that for piled rafts subjected to either vertical or horizontal loadings, the present technique provides a simple way for obtaining solutions to the piled raft problem.

2 METHOD OF ANALYSIS

As shown in Figure 1 (taking a 9-pile raft as an example), the problem of the piled raft foundation can be solved by assuming that the forces between the piles and layered soil can be treated as a series of ring loads applied to 'nodes' along the pile shaft. These loads are both horizontal and vertical, and if enough are used, they closely model the continuous forces that act along the pile shaft.

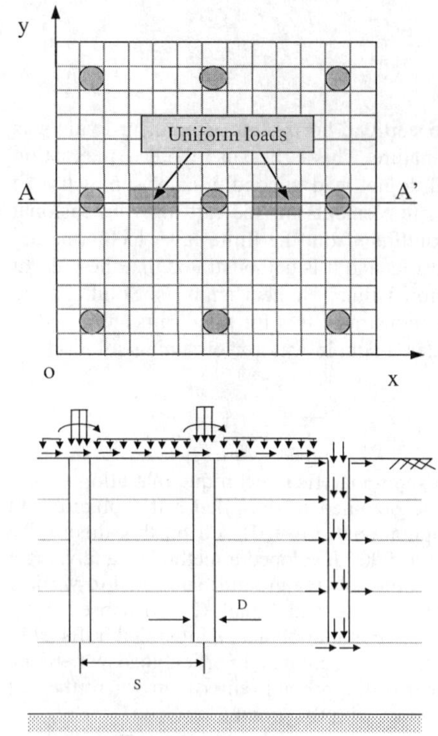

Figure 1. Free body diagram of piled raft

The displacement of the layered soil can then easily be computed as the solution for a layered soil subjected to ring loads at the layer interfaces may be obtained from finite layer theory. The same theory may be used to analyse the deflection of the soil due to vertical and horizontal loads on the soil surface. If the contact stress between the raft and soil can be considered to consist of rectangular blocks of pressure that correspond to each finite element in the raft, then these unknown pressures along with the unknown ring loads along the pile shafts form the unknowns for the problem.

By considering equilibrium of applied forces and moments acting on the piles and raft, and compatibility of displacements of the soil and raft (and of displacement and rotation of the pile head and raft) enough equations may be assembled to obtain the solution under general loading as follows:

$$\begin{bmatrix} ([I_{sp}]+[I_r]) & -\{a\} & -\{b\} & -\{c\} & -\{d\} & -\{e\} & -\{f\} \\ -\{A\} & 0 & 0 & 0 & 0 & 0 & 0 \\ -\{B\} & 0 & 0 & 0 & 0 & 0 & 0 \\ -\{C\} & 0 & 0 & 0 & 0 & 0 & 0 \\ -\{D\} & 0 & 0 & 0 & 0 & 0 & 0 \\ -\{E\} & 0 & 0 & 0 & 0 & 0 & 0 \\ -\{F\} & 0 & 0 & 0 & 0 & 0 & 0 \end{bmatrix} \begin{bmatrix} \{p\} \\ D_x \\ D_y \\ D_z \\ \theta_x \\ \theta_y \\ \theta_z \end{bmatrix} = \begin{bmatrix} \{\delta_0\} \\ -P_x \\ -P_y \\ -P_z \\ -M_x \\ -M_y \\ -M_z \end{bmatrix} \quad (1)$$

where $[I_{sp}]$ and $[I_r]$ are influence matrices for pile-soil and the raft, respectively; D_x, D_y and D_z are the translations of the reference node; θ_x, θ_y and θ_z are rigid body rotations about the reference node (the first node of the raft in this example); P_x, P_y, P_z, M_x, M_y and M_z are applied loads and moments; $\{p\}$ is the vector of interface forces; $\{\delta_0\}$ is the displacement of the pinned raft under applied loadings; and vectors $\{a\}$ to $\{f\}$ and $\{A\}$ to $\{F\}$ are related to the coordinate of each element and area of the raft (See Appendix and Zhang and Small (2000)).

3 COMPARISON WITH FULL THREE DIMENSIONAL FINITE ELEMENT METHOD

The problem to be considered is a 9-pile raft embedded in homogeneous soil subjected to two uniform loadings in the x- or z-direction as shown in Figure 1. The magnitude of the uniform loadings is 100 kPa. The moduli of the soil, raft and piles were chosen to be 20, 20,000 and 20,000 MPa, respectively, and the Poisson's ratios of the soil, raft and piles are taken as 0.4, 0.2 and 0.2, respectively. The raft is 3.9 m square in plan and its thickness is 0.3 m. The pile length was chosen to be 15 m and the thickness of the soil layer was taken as 65 m. For the finite element method, the pile sections are assumed to be 0.3 m square, whereas for the present method, the pile sections are assumed to be circular and the equivalent pile diameter (D) is 0.3385 m. The ratio of the centre-to-centre spacing to the equivalent pile diameter is 4.43. The overhang of the raft (around the perimeter) was one pile diameter.

For both the horizontal and vertical loading cases, normalised lateral or vertical displacement I_u, and normalised moment I_M in the raft are defined as

$$I_u = E_s D u / P \quad (2)$$

$$I_M = M / P \quad (3)$$

where u is the actual horizontal or vertical displacement (u_x or u_z); M is moment per unit length; E_s is the soil modulus; D is the pile diameter; P is the total lateral or vertical load (P_x or P_z).

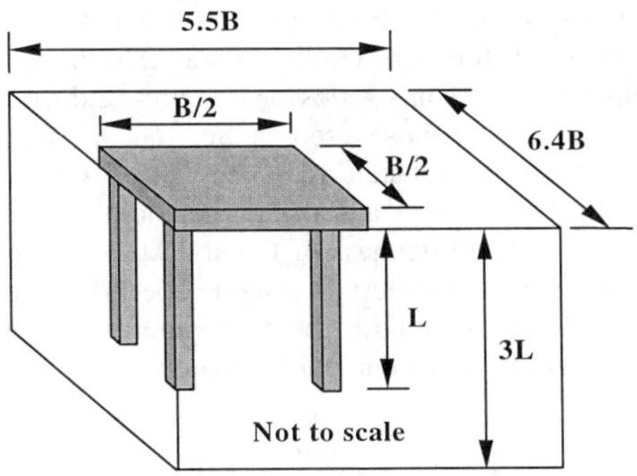

Figure 2. Extent of ¼ of 3-d finite element mesh

The comparisons of the present method with the finite element method are described as follows.

3.1 *Piled raft subjected to vertical loading*

Solutions for the piled raft subjected to two uniform loadings (as shown in Figure 1) are plotted in Figures 3 and 4. The variation of vertical displacement of the raft along section A-A′ is plotted in Figure 3 versus the normalised distance. It may be seen that the difference in vertical displacement between the present method and the finite element method is only about 10%. The figure also shows that the maximum vertical displacement is at the centre of the raft.

The moment per unit length in the raft along section A-A′ in the x and y directions (M_x and M_y) is shown in Figure 4. In this example the moment in the y direction is higher than that in the x direction. The maximum moment occurs at the position of applied loading. The results from the present method are in excellent agreement with those from the finite element method.

3.2 *Piled raft subjected to horizontal loading*

The vertical displacement and moment in the raft along section A-A′ induced by horizontal loading in the x-direction are plotted in Figures 5 and 6. It can be observed from Figure 5 that the vertical displacement calculated from the present method is still higher than that obtained using the finite element method. The maximum difference (about 21%) oc-

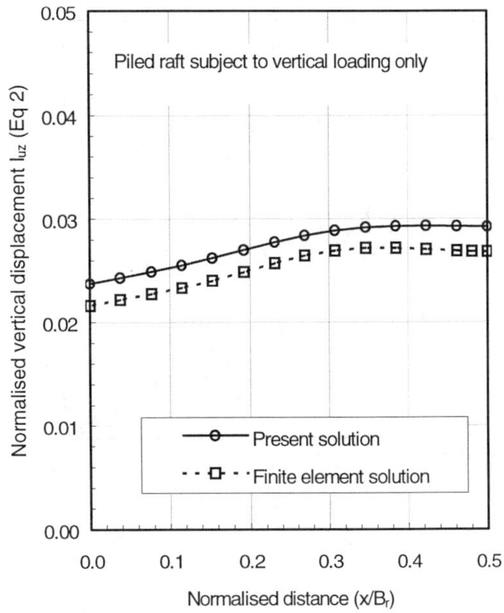

Figure 3. Comparison of normalised vertical displacements along section A-A′

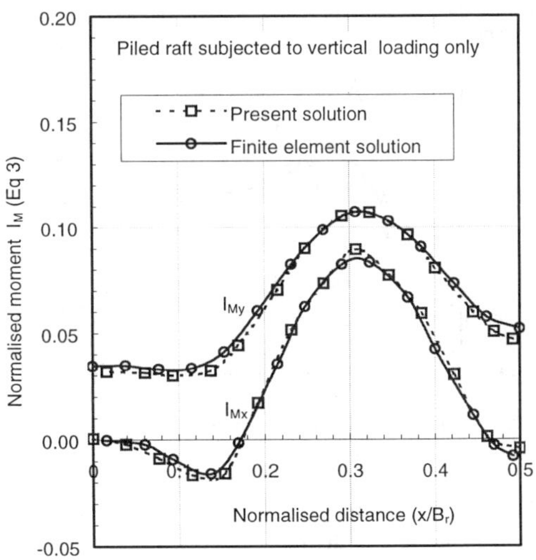

Figure 4. Comparison of moment in raft along section A-A′

curs at the edge of the raft. At the centre of the raft the vertical displacement is equal to zero as the vertical displacement is antisymmetric about the central line of the raft (parallel to the y-axis).

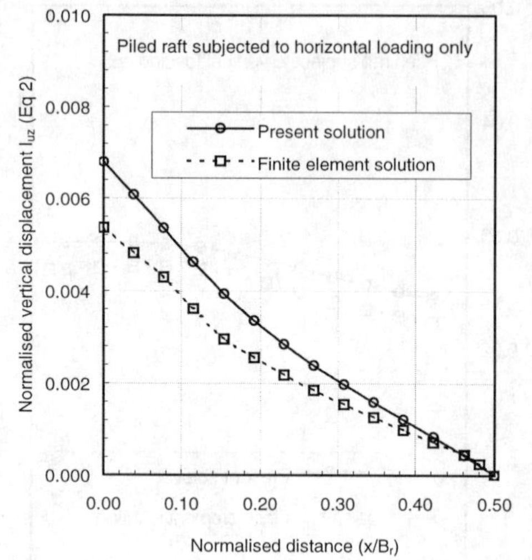

Figure 5. Comparison of vertical displacement along section A-A′

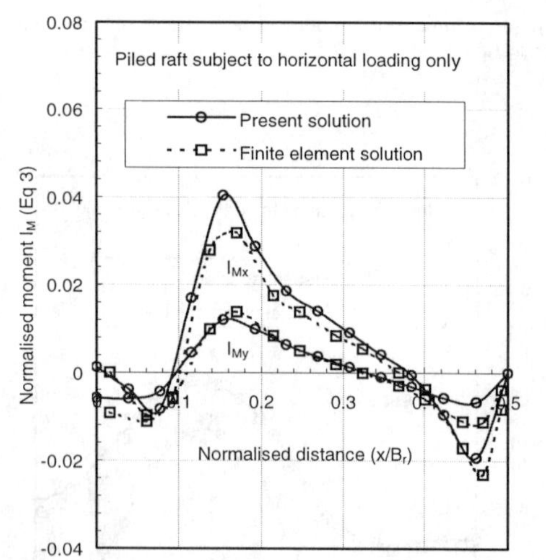

Figure 6. Comparison of moment in raft along section A-A′

The moment in the raft is also antisymmetric as show in Figure 6. It may be seen that the present solution is in close agreement with the finite element solution except for the maximum moment in the x direction where about a 20% difference between the two methods is observed at the pile head, which may be due to the different types of elements used in the two methods. An 8 noded element was used in the finite element solution but a 4 noded one used in the present solution.

4 ANALYSIS OF PILED RAFT EMBEDDED IN LAYERED SOILS

The advantages of using the present method over existing methods is that the present method not only requires less computer memory and computer time to obtain a solution but it can also accurately model layered soils. In this example, a 25-pile raft embedded in two different layered soils was examined. The moduli and Poisson's ratios of the piles and the raft are the same as those used in Section 3. The raft thickness is again taken as 0.3 m. The diameter of the pile, D, is 0.4 m and the pile slenderness ratio, L/D, is 25. The thickness of the soil layer was chosen to be 10 times the pile length. The pile spacing ratio, S/D, is assumed to be 4 and the overhang of the raft was taken as one pile diameter.

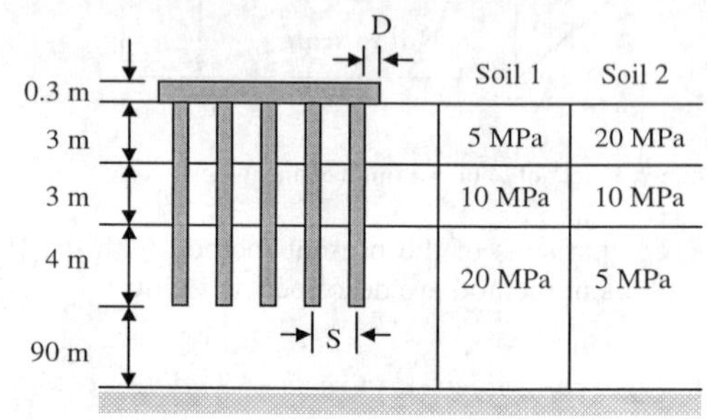

Figure 7. Schematic diagram of 25-pile raft

The profiles of the two layered soils are as shown in Figure 7. For soil 1, the modulus of each layer increases from the top layer to the bottom layer and for soil 2, the modulus of each layer is assumed to reduce from the top layer to the bottom layer.

In this example, the normalised lateral or vertical displacement I_u, and normalised moment I_M in the raft are defined as

$$I_u = E_s Du / qB^2 \tag{4}$$

$$I_M = M / qB^2 \tag{5}$$

where u is the actual horizontal or vertical displacement (u_x or u_z); M is moment per unit length; E_s is the soil modulus; D is the pile diameter; q is the uniform lateral or vertical load (q_x or q_z); and B is the full width of the square raft.

Figure 8 demonstrates the displacement along the pile shaft (pile 1) for the two cases involving horizontal loading. The horizontal displacement along the pile shaft was normalised by dividing by the maximum horizontal displacement at the top of the pile. It may be seen that the horizontal displacement at the top of pile 1 is about 30% higher than that for case 2. But the displacement along the pile shaft reduces more rapidly for case 1 than for case 2. The example indicates that the horizontal displacement of a piled raft induced by horizontal loading, is substantially affected by the stiffness of the upper layer of soil.

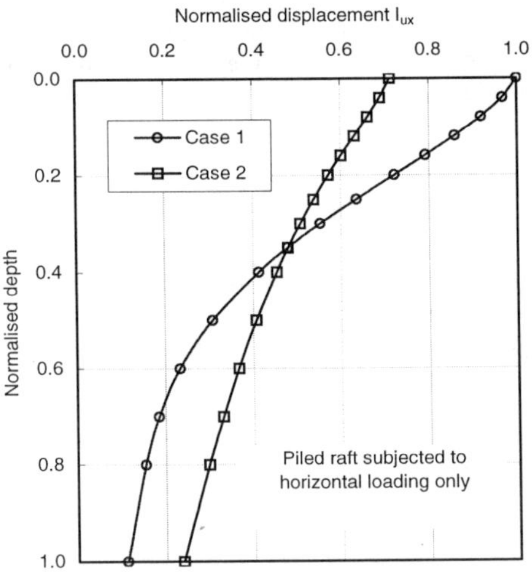

Figure 8. Displacement along shaft of pile 1 in two cases

The moments per unit length in the raft along the centre line in the x direction are firstly examined for a piled raft subjected to uniform horizontal loading and uniform vertical loading. The solutions are plotted in Figures 9 and 10, respectively. For the piled raft subjected to uniform horizontal loading, it may be seen from Figure 9 that the maximum moment occurs on the first pile head (central row of piles). The moment in the raft for case 1 is much higher than that for case 2. This means that the modulus of the top layer of the soil will substantially affect the magnitude of the moment in the raft. However, for a piled raft under uniform vertical loading, the moment in the raft for case 2 is about 30% higher than that for case 1 (Figure 10).

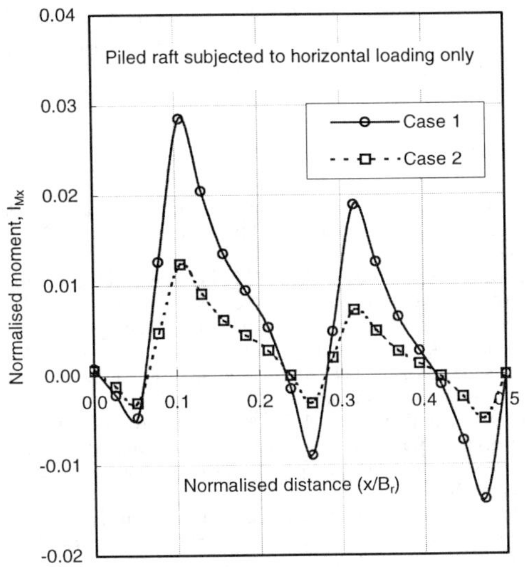

Figure 9. Variation of moment in raft for piled raft embedded in two different layered soil profiles.

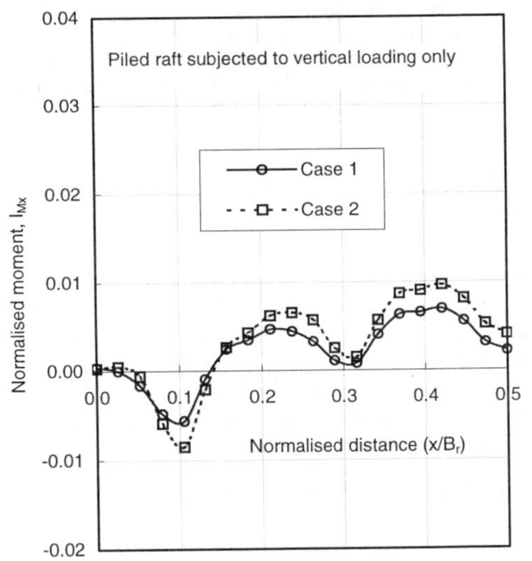

Figure 10. Variation of moment in raft for piled raft embedded in two different layered soil profiles.

5 CONCLUSIONS

From the comparisons of the results of the finite layer method with results from a 3-dimensional finite element analysis and the analysis of examples, the following conclusions can be drawn:

1. The results of the present method are in good agreement with those from 3-d finite element computations for both displacements and moments for piled rafts loaded vertically and laterally and embedded in homogeneous soils.
2. The present program APRAF can not only deal with piled rafts constructed in homogeneous soils, but it can also conveniently be used for analysis of piled rafts embedded in layered soils and soils that have a stiffness that increases with depth.

3. The modulus of the top layer of soil will strongly affect the moment and displacement of a piled raft under horizontal loading.

REFERENCES

Kuwabara, F. 1989. An elastic analysis for piled raft foundations in a homogeneous soil. *Soils and Foundations*, 29 (1): 82-92.

Butterfield, R. and Banerjee, P. K. 1971. The problem of pile group – pile cap interaction. *Géotechnique*, 21 (2): 135-142.

Davis, E. H. and Poulos, H. G. 1972. The analysis of pile raft systems. *Australian Geomechanics Journal*, G2, 1: 21-27.

Hain, S. J. and Lee, I. K. 1978. The analysis of flexible raft-pile systems. *Géotechnique*, 28 (1): 65-83.

Clancy, P. and Randolph, M. F. 1993. An approximate analysis procedure for piled raft foundations. *International Journal for Numerical and Analytical Methods in Geomechanics*, 17: 849-869.

Poulos H. G. 1994. An approximate numerical analysis of pile-raft interaction. *International Journal for Numerical and Analytical Methods in Geomechanics*, 18: 73-92.

Ta, L. D. and Small, J. C. 1996. Analysis of piled raft systems in layered soils. *International Journal for Numerical and Analytical Methods in Geomechanics*, 20: 57-72.

Zhang, H. H. and Small, J. C. 2000. Analysis of capped pile groups subjected to horizontal and vertical loads. *Computers and Geotechnics*, 26 (1): 1-21.

Zhang, H. H. and Small, J. C. 1999. Analysis of axially and laterally loaded pile groups embedded in layered soils. *Proceedings of the 8th Australia - New Zealand Conference on Geomechanics*, 15-17 February, Hobart, Tasmania, Edited by N. Vitharana and R. Coleman,. 1: 475-483.

APPENDIX: EXPRESSIONS FOR AUXILIARY VECTORS

The auxiliary vectors {a} to {f} are dependent on the location of piles under raft elements. For example, assuming the first two elements of the raft are in contact with the soil surface, and the third element of the raft contains a pile, then vectors {a} to {f} corresponding to the first three raft elements may be expressed as follows.

$$\{a\} = (1, 0, 0, 1, 0, 0, 1, 0, 0, 0, 0, \dots \dots)^T$$

$$\{b\} = (0, 1, 0, 0, 1, 0, 0, 1, 0, 0, 0, \dots \dots)^T$$

$$\{c\} = (0, 0, 1, 0, 0, 1, 0, 0, 1, 0, 0, \dots \dots)^T$$

$$\{d\} = (0, 0, x_1, 0, 0, x_2, 0, 0, x_3, 1, 0, \dots \dots)^T$$

$$\{e\} = (0, 0, y_1, 0, 0, y_2, 0, 0, y_3, 0, 1, \dots \dots)^T$$

$$\{f\} = (y_1, -x_1, 0, y_2, -x_2, 0, y_3, -x_3, 0, 0, 0, \dots \dots)^T$$

where x_i and y_i are the coordinates of the centre of element *i* with regard to the first reference point on the raft.

The elements of the auxiliary vectors {A} to {F} are equal to the elements of vectors {a} to {f} multiplied by the corresponding raft element areas. For example, if the third element contains a pile, the vector {A} may be expressed as

$$\{A\} = (A_1, 0, 0, A_2, 0, 0, A_3, 0, 0, 0, 0, \dots \dots)$$

where A_i = the area of element *i* in the raft.

The definitions of positive θ_x, θ_y and θ_z are shown below.

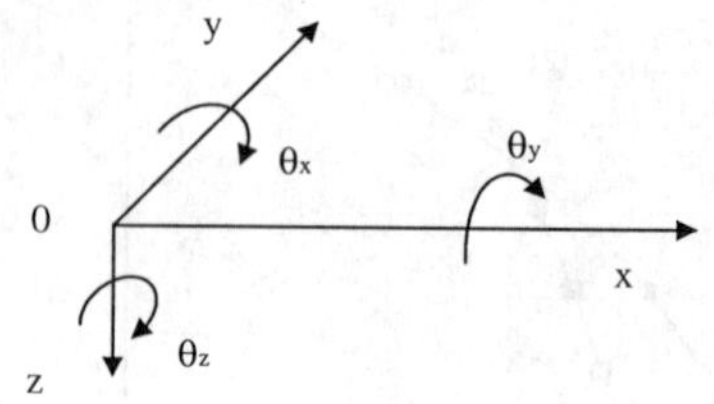

Computer Methods and Advances in Geomechanics, Desai et al. (eds) © 2001 Balkema, Rotterdam, ISBN 90 5809 183 X

Pitfalls in wave equation analysis of pile capacity

M.R.Svinkin
VibraConsult, Cleveland, Ohio, USA

ABSTRACT: The capacity of a driven pile changes with time after installation. Wave equation analysis is a prevalent method for prediction and determination of pile capacity, but on many cases this method yields misleading prediction and results of dynamic testing are misinterpreted. The prediction of pile capacity in pre-driving wave equation analysis can be improved by the use of variable damping as a function of time. Dynamic testing cannot predict pile capacity and any comparison of static and dynamic tests has to be made for tests performed within a short duration.

1 INTRODUCTION

Accurate and reliable determination of pile capacity is extremely important for economic design, construction and reliability of pile foundations. Piles have to withstand their design loads for very long periods of time. Therefore, the consequences of soil modification around the pile are essential with respect to changes of pile capacity.

Soil consolidation and dissipation of excess pore pressure generated during pile driving in the soil-pile interface zone are usually accompanied by an increase in pile capacity. In clayey soils, Seed & Reese (1955) and also Thorburn & Rigden (1980) found an increase in pile capacity (also called set-up factor) of up to 6 times over a period of 30 days after the end of initial driving (EOID). In clays, Svinkin et al. (1994) reported the range of set-up factors between 4.5 and 11.4 for a period of 22-35 days. Schmertman (1991) showed that continued increase of strength and capacity in sand is not a unique phenomenon. Svinkin et al. (1994) revealed upper values of the set-up factors of 3.3 and 1.9 in unsaturated and saturated sandy soils, respectively, over a period of 16-23 days. In saturated sands, York et al. (1994) reported similar results. In glacial deposits, Fellenius et al. (1989) received set-up factors from 2 to 4 for period of 15-30 days. Obviously, there is the necessity of assessment of the actual pile capacity after the completion of driving.

The static loading test is traditionally used to determine ultimate capacity of the pile-soil system or the value of a service load to be supported by a pile. Contemporary dynamic methods have advances in evaluation of the hammer-pile-soil system and in data acquisition during pile driving and restrikes. Hence during the two last decades, dynamic methods have become an integral part of pile capacity prediction and measurement for numerous projects.

Contemporary dynamic methods are founded on the application of the stress wave theory to piles. There are two different techniques of the use of wave equation analysis for determining pile capacity: computation of pile capacity without dynamic measurements on driven piles and a signal matching technique for computed and measured at the pile head force and velocity records.

Dynamic methods have certain advantages and some uncertainties in their application that on many occasion may yield misleading results. The objectives of this paper are illustrations of some pitfalls in wave equation analysis of pile capacity and recommendations how to avoid the pitfalls effect on the accuracy of determining pile capacity by dynamic methods.

2 WAVE EQUATION ANALYSIS

The main goal in using the wave equation method is to provide a better prediction of the pile capacity, as a function of pile penetration resistance, than can be obtained from classical dynamic formulae.

The wave equation analysis program (WEAP) was originally suggested by Smith (1960) to compute the pile capacity at the end of driving. Smith developed a numerical model of the hammer-pile-soil system taking into account accumulated experience of that time of the actual behavior of driven piles and surrounding soil. Although some modifications and improvements of Smith's model were subsequently necessary, results obtained by many researchers confirmed the soundness of the basic approach. Today, the most commonly used wave equation programs are based on either WEAP - Goble & Rausche (1976), TTI - Hirsch et al. (1976) or TNOWAVE - TNO Reports (1985-1996).

Wave equation analysis of driven piles is a prevalent method of pile driving stress calculations. Besides driveability analysis, the wave equation method is used for prediction and determination of pile capacity during both the design stage and for construction control during pile installation, but accuracy of computed pile capacity is unsatisfactory.

2.1 *WEAP Calibration*

Existing programs for wave equation analysis are not the same. Moreover all programs have a number of versions. Each new version gives usually additional beneficial options to users, but it is not clear how each program version ensures the accuracy of pile capacity calculation. There is confusion what program yields more accurate results. The writer has an experience of pile capacity calculation for the same hammer, pile and soil conditions using two versions of the same program. Variation of obtained capacities was about 20 %. This is an evidence of contradictions available between different program versions.

Obviously, to avoid confusion in a choice of the program, it is necessary to calibrate each program version with some standard data of the hammer-pile-soil system.

2.2 *Soil consolidation effect*

The wave equation method is also used for prediction of pile capacity at restrike (RSTR) performed at any time after EOID. By adjusting wave equation analysis input with results of dynamic measurements, some researchers, for example, Hunt and Baker (1988), York et al. (1994) have obtained good correlation between computed and observed pile capacities. However, in most cases, computed pile capacity differs substantially from results of both static and dynamic load tests, Svinkin (1996).

Soil parameters considerably affect WEAP solutions. Dynamic soil resistance parameters (damping and quake) have to be assigned as constant values for wave equation analysis. These parameters do not reflect the changes of soil properties in the pile-soil interface zone after completion of pile driving.

Improvement in WEAP accuracy can be made for taking into account soil consolidation after pile installation.

Existing dynamic models of the pile-soil system mainly use a velocity-dependent approach for calculation of the dynamic resistance as a damping component of the total resistance during pile driving. There are various linear and nonlinear relationships between the damping component and the velocity.

For certain pile capacity, the dynamic resistance depends only on pile velocity and the damping coefficient.

Statistical analysis indicates no correlation between the pile penetration resistance and velocity values. The pile-soil system changes with time after the completion of driving, but the pile velocity is only a pile property and remains in the same range for EOID and RSTRs. The largest values of pile velocity measured at the upper end of the pile and calculated along a pile shaft depend only on pile parameters and energy transferred to the pile and cannot reflect regain in soil strength and pile-soil adhesion after EOID. This is the first cause of unsatisfactory prediction of pile capacity with time after EOID.

There are attempts to determine values of damping and quake from signal-matching solutions for dynamically tested driven piles or from modified SPT. It could be beneficial for some cases, but such approaches cannot resolve the problems. Besides complications with different models in WEAP and signal-matching technique and also with the scale factor effect on the use of SPT results, these approaches yield constant values of soil parameters and cannot be used in prediction of the pile capacity as a function of time after EOID.

The basic disadvantage of many models is the attempt to select the model parameters directly from actual soil properties. This can yield acceptable results for some cases, but in general this approach is not successful in finding good correlation between predicted and actual pile capacity after EOID. The use of the constant damping coefficients for calculation of the dynamic resistance is the second cause of unsatisfactory prediction of pile capacity with time after EOID.

Neither the pile velocity nor the damping constant can reflect time-dependent variation of the pile-soil system after EOID, Svinkin (1996). The existing approach of computing the dynamic resistance does not take into account soil consolidation around the pile after EOID and therefore cannot provide determination of pile capacity as a function of time after pile installation.

For the idealized Smith wave equation model, it is important to find an appropriate combination of parameter values, mainly paying attention to soil variables, in order to achieve the accurate prediction of pile capacity. Probably, there is only one direction to enhance prediction accuracy of the dynamic resistance with the velocity dependent approach. Variation of the pile-soil system after the completion of driving can be taken into account by a *variable damping coefficient* which should be considered as a function of time and other parameters characterizing soil consolidation around the pile, Svinkin (1997). It is assumed that the variable damping coefficient is independent of pile velocity. Inclusion of variable damping is thought to be the next step in the development of Smith's model with the velocity dependent approach for representation of the dynamic resistance.

The damping coefficient as a function of time can be found on the basis of back calculations using the wave equation model of the pile-soil system with known capacity. This procedure is in agreement with Lamb's (1973) equation modified for back analysis by Leroueil & Tavenas (1981)

Observation (pile capacity) + theory (WEAP) → empirical soil parameter (variable damping coefficient)

Since the variable damping coefficient is only one soil parameter reflecting the field soil consolidation after pile installation, this parameter can be successfully back analyzed.

The five soil damping options, available in GRLWEAP program (GRL Manual 1997), were investigated: Standard Smith Damping, Viscous Smith Damping, Case Damping, Coyle-Gibson Damping, and Coyle-Gibson/GRL Damping, Figure 1. A trend of the damping coefficient increase with time after EOID was found for all the considered dynamic soil models and this trend is independent of the damping resistances (Svinkin 1996).

The idea of variable damping has been confirmed by results of statistical analysis performed by Liang and Zhou (1997) who have found that the damping coefficient is affected by the time.

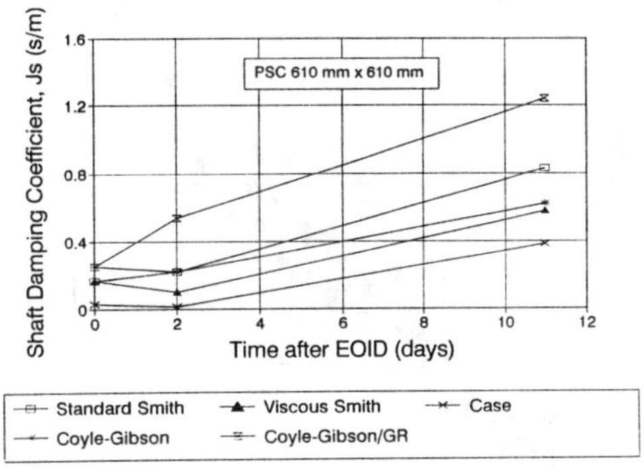

Figure 1. Shaft damping as function of time, after Svinkin (1996)

Soil damping is the key parameter for adjustment of wave equation solutions with time-dependable soil properties in pre-driving analysis. The use of the variable damping coefficient gives an opportunity to compute the time-dependent pile capacity by the wave equation method.

3 DYNAMIC TESTING AND SIGNAL MATCHING

Dynamic testing followed by a signal matching procedure has obvious advantage in determining pile capacity at any time after pile installation. Since dynamic testing is often used to replace the static loading tests, it is important to ascertain the adequacy of both SLT and DT. Design methods predict pile capacity as the long term capacity after soil consolidation around the pile is complete. Independently of the time elapsed between the driving of the test pile and the static loading test, the ratio of the predicted ultimate load to the measured ultimate load from static loading test is used for approximate evaluation of the reliability of design methods, Briaud and Tucker (1988). According to the traditional approach, the main criterion for assessment of the pile capacity prediction based on dynamic measurements is the ratio of capacities obtained by dynamic and static tests or vice versa (Figure 2).

It is necessary to point out that a ratio of DT/SLT or vice versa, taken for arbitrary time between compared tests, is not a verification of dynamic testing results. It is well-known that dynamic testing methods yield the real static capacity of piles *at the time of testing,* Rausche et al. (1985). This is not a

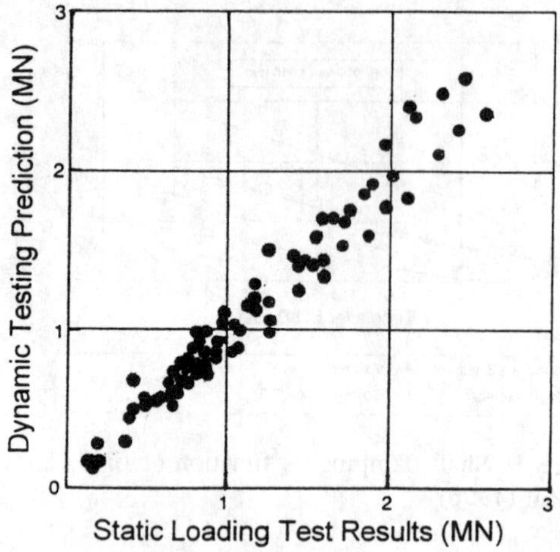

Figure 2. Typical correlation of pile capacity from dynamic and static tests

predicted value. Moreover, the static capacity from SLT is considered as a unique standard for assessment of dynamic testing results. Unfortunately, that is a major error. As a matter of fact, *pile capacity from Static Loading Tests is a function of time* and the so-called actual static capacity from SLT is not a constant value, Svinkin (1998).

For a few separate piles, it is possible to find published information regarding the time between static and dynamic tests. However, for the general case of assessment of reliability of the DT, the ratio of restrikes to SLT results has been considered for various pile types, soil conditions and times of testing lumped together as shown in Figure 2. What is the real meaning of such mixture? Nobody knows. It is not a verification of dynamic testing at restrikes and it is not assessment of real setup factor because everything is lumped together without taking into account the time between different tests. Such a comparison of the pile capacities from SLT and DT is invalid for piles driven in soils with time-dependent properties because the soil properties at the time of DT do not correspond to the soil properties at the time of SLT i.e. soil consolidation is taken into account for restrikes using the DT but is not in the SLT.

Unfortunately, dynamic testing is often used for a capacity prediction without prior knowledge of the static loading test, Goble (2000). This is a misleading interpretation of DT. No in-situ pile test, DT or SLT, can predict pile capacity as a function of time after pile installation. Such a test yields pile capacity only at the time of testing.

Static Loading Tests and Dynamic Testing present different ways of determining pile capacity at various times after pile installation, but for valid correlations of both tests, *static and dynamic capacities must be compared at the same time after pile installation in both SLT and DT methods*, Svinkin (1997) and Svinkin & Woods (1998a).

The adequacy of SLT and DT have to be confirmed by proper correlation of time. Due to the consolidation phenomenon in soils, comparison of SLT and DT can only be made for tests performed immediately one after another. In practice, it is sometimes difficult to make two immediately successive tests, but nonetheless the time difference between both comparable tests should not exceed 1-2 days during which soil setup changes only slightly.

6 CONCLUSIONS

Wave equation analysis is contemporary tool to compute the time-dependent pile capacity. However this method lacks the engineering judgement in predicting and determining pile capacity.

To avoid confusion in a choice of the wave equation analysis program, it is necessary to calibrate each program version with some standard data of the hammer-pile-soil system.

The prediction of pile capacity in pre-driving wave equation analysis can be improved by the use of variable damping as a function of time. Variable damping is the key parameter to enhance accuracy of wave equation solutions because this damping takes into consideration soil consolidation after pile installation.

The main criterion for accurate assessment of pile capacity prediction based on dynamic measurements of force and velocity at the pile head during driving is the ratio of capacities obtained by dynamic and static tests. Such a ratio, taken for arbitrary time between compared tests, in not a verification of dynamic testing results.

Dynamic testing of driven piles followed by a signal matching procedure yield the real, not predicted, static capacity of piles at the time of testing. Both static loading test and dynamic testing yield the pile capacity at the time of testing.

In soils with time-dependent properties, comparison of static loading test and dynamic testing must be made only for tests performed immediately, in short succession.

REFERENCES

Briaud, J.L. and L.M. Tucker 1988. Measured and predicted axial response of 98 piles. *Journal of Geotechnical Engineering*, ASCE, 114(9): 984-1001.

Fellenius, B.H., R.E. Riker, A.J. O'Brien & G.R. Tracy 1989. Dynamic and static testing in soil exhibiting set-up. *Journal of Geotechnical Engineering*, ASCE, 115(7): 984-1001.

Goble, G.G. 2000. "Class A" capacity prediction. *GRL + PDI Newsletter 36*, February.

Goble G.G. & F. Rausche 1976. *Wave Program Documentation*. National Information Service, Washington, D.C.

GRL and Associates, Inc. 1997. *GRLWEAP - Wave Equation Analysis of Pile Driving, Manual*, Cleveland, Ohio.

Hirsch T.J., L. Carr & L.L. Lowery 1976. *Pile driving analysis. Wave equation user manual*. TTI.

Hunt, S.W. & C.N. Baker 1988. Use of stress-wave measurements to evaluate piles in high set-up conditions. In B. Fellenius (ed.), *Proc. Third Inter. Conf. on the Application of Stress-Wave Theory to Piles, Ottawa, 25-27 May*: 689-705. Vancouver: BiTech Publisher.

Lambe, T.W. 1973. Predictions in soil engineering. 13th Rankine Lecture, *Geotechnique*, 23(2): 149-202.

Leroueil, S. & F. Tavenas 1981. Pitfalls of back-analysis. *Proc. 10th Inter. Conf. on Soil Mechanics and Foundation Engrg.*, 1:185-190. Stockholm: ISSMFE.

Liang R.Y. & J. Zhou 1997. Probability Method Applied to Dynamic Pile-Driving Control. *Journal of Geotechnical Engineering*, ASCE, 123(2): 137-144.

Rausche, F., G.G. Goble & G. Likins 1985. Dynamic determination of pile capacity. *Journal of Geotechnical Engineering*, ASCE, 111(3): 367-383.

Schmertmann, J.H. 1991. The mechanical aging od soils. *Journal of Geotechnical Engineering*, ASCE, 117(9): 1288-1330.

Seed, H.B. & L.C. Reese 1955. The action of soft clay along friction piles. *Transactions*, ASCE, 122: 731-754.

Smith E.A.L. 1960. Pile driving analyses by the wave equation. *Journal of the Soil Mechanics and Foundation Division*, ASCE, 1960, 86: 35-61.

Svinkin, M.R., C.M. Morgano & M. Morvant 1994. Pile capacity as a function of time in clayey and sandy soils. *Proc. Fifth Inter. Conf. and Exhibition on Piling and Deep Foundations, Bruges, 13-15 June*: 1.11.1-1.11.8. Rotterdam: Balkema.

Svinkin, M.R. 1996. Soil damping in wave equation analysis of pile capacity. In F. Townsend, M. Hussein & M. McVay (eds.), *Proc. Fifth Inter. Conf. on the Application of Stress-Wave Theory to Piles, Orlando, 11-13 September*: 128-143. Gainesville: University of Florida.

Svinkin, M.R. 1997. Time-Dependent Capacity of Piles in Clayey Soils by Dynamic Methods. *Proc. XIVth Inter. Conf. on Soil Mechanics and Foundation Engineering, Hamburg, 6-12 September*, 2: 1045-1048. Rotterdam: Balkema.

Svinkin, M.R. & R.D. Woods 1998a. Accuracy of determining pile capacity by dynamic methods. *Proc. Seventh Inter. Conf. and Exhibition on Piling and Deep Foundations, Vienna, 15-17 June*: 1.2.1-1.2.8. Rickmansworth: Westrade Group Ltd.

Svinkin, M.R. 1998b. Discussion of 'Probability method applied to dynamic pile-driving control' by Liang & Zhou, *Journal of Geotechnical and Geoenvironmental Engineering*, ASCE, 122(4): 319-321.

Thorburn, S. & W.J. Rigden 1980. A practical study of pile behavior. *Proc. 12th Annual Offshore Technology Conf.*, Houston.

TNO reports 1985-1996. *TNOWAVE - TNODLT Dynamic Load Testing Signal Matching, Users Manual*. TNO.

York, D.L., W.G. Brusey, F.M. Clemente & S.K. Law 1994. Setup and relaxation in glacial sand. *Journal of Geotechnical Engineering*, ASCE, 120(9): 1498-1513.

Computer Methods and Advances in Geomechanics, Desai et al. (eds) © 2001 Balkema, Rotterdam, ISBN 90 5809 183 X

Lateral loads on piles and piers

K. Mohan Vennalaganti & Shailendra N. Endley
PSE, Houston, Tex., USA

Karun Sreerama
Online Constructor Incorporated, Houston, Tex., USA

ABSTRACT: The present study includes the analyses of different pile types with applied lateral loads and moments (i.e., lateral load at an eccentricity from the ground surface) in uniform clay profile. Utilizing the LPILE program, pile head deflections and maximum bending moments were determined for various pile types and pile sizes. About 300 computer runs were made for uniform clay profiles of varying strength. The results of the computer runs were then normalized to obtain simple equations. The equations can be easily used to determine the pile head deflection and the maximum bending moment within the pile. The study is limited to isolated long free-head piles in cohesive soils.

1 INTRODUCTION

Geotechnical Engineers are often asked to provide lateral load capacity of piles or piers. This problem is complex, since it involves bending of a pile along with the movements in the soil. Various methods ranging from chart solutions to finite difference techniques ('p-y' methodology) were developed to solve this problem. Finite difference techniques or 'p-y' methodology involves defining the soil resistance (p) along the length of the pile as a non-linear function of lateral pile deflection (y). The analysis involves generating a set of 'p-y' curves and solving a fourth order differential equation to determine the pile response. The computer program LPILE developed by Lymon C. Reese and Shin-Tower Wang is one of the programs available to perform such an analysis. The program utilizes 'p-y' criteria developed for different kinds of soils by various researchers.

The present study includes the analyses of different pile types with lateral loads and moments (i.e., lateral load at an eccentricity from the ground surface) applied to the pile head at the ground surface. Utilizing the LPILE program, pile head deflections and maximum bending moments were determined for various pile types and pile sizes. The results of the computer runs were then normalized to obtain simple equations. The equations can be easily used to determine the pile head deflection and the maximum bending moment within the pile when a pile is subjected to lateral loads. The equations presented herein should be utilized for preliminary design purposes only. For an accurate determination of pile behavior, LPILE program or other such routines should be used.

2 PRESENT STUDY

The present study was focused on determining deflections and bending moments of laterally loaded piles. These two values often govern the design of a laterally loaded pile. Moreover, in most cases, the ultimate lateral load of a pile is reached at large deflection that may not be acceptable in the design. Utilizing the LPILE data, the goal was to arrive at simple equations that can be used to easily calculate the design values. A similar research was performed by J.M. Duncan et al (1994).

The LPILE program was utilized to perform lateral load analyses for various pile types and pile sizes. The pile types included steel pipe pile with nominal wall thickness (0.25 + d/100 in.) as suggested by the American Petroleum Institute (API) design manual, square concrete precast piles, timber piles and concrete shafts. For each pile type and pile size, lateral load analyses were performed using the following assumptions; 1) cyclic loading conditions, 2) free-head piles, 3) soft clay criteria (Matlock 1970) and, 4) an E_{50} (strain corresponding to one-half the principle stress) value of 0.01. The elastic modulus, E, of the pile material(s), and, the moment of inertia, I, also remained constant. A uniform clay profile was assumed for each pile type and pile size analyzed. The shear strength, Su, of the uniform clay profile was varied between 0.2 ksf (9.58 kPa) and 5.0 ksf (239.4 kPa). From an initial analyses (piles with varying lateral loads), it was determined that the pile head deflections were independent of pile penetration for pile lengths greater than 15 diameters. Therefore, a penetration depth of at least 15d was assumed in this study. It was also determined that for a pile of less than about 8d in length, the pile has only one point of zero deflection and the slope of the pile remains constant along its length. (i.e., pile rotates about a point along its length). Furthermore, for some piles or piers having a length of 10d to15d, some amount of bending is observed and there still remained only one point of zero deflection along its length. In almost all cases, for a pile of length 15d, the pile has a varying slope with two or more points of zero deflection along its length. From the initial analyses, yet another interesting observation was also made with respect to short and long piles. For short piles, observing the soil reaction curves, it was determined that the ultimate soil resistance is mobilized before the bending moment within the pile reaches its yield moment value. Conversely, for a long pile (15d or more in penetration), the maximum bending moment is larger than the yield moment value before the ultimate soil resistance is mobilized.

2.1 *Isolated Loading Conditions*

The results of a typical set of LPILE runs are shown on Figures 1 through 3. Figure 1 is a plot of Su versus H_a for various pile head deflections. As shown, H_a corresponds to that condition where an isolated lateral load (i.e., lateral load only) is applied to the pile head at the ground surface. These lateral loads were determined by fixing the pile head deflection (y) values varying from 0.25 in. (6.4 mm) to 2.0 in. (50.8 mm). Similarly, Figure 2 corresponds to that condition where an isolated moment (i.e., moment only) is applied to the pile head and shows the plot Su versus M_a for various deflections. Figure 3 shows the maximum bending moment (BM_{max}) values for various deflections which correspond to the isolated lateral load values shown on Figure 2. For example, from Figure 3, for a Su of 1.0 ksf (47.9 kPa) and a pile head deflection of 2.0 in. (50.8 mm), the BM_{max} value of 858 kip-ft (1163.2 kN-M) is obtained for the 36-inch (0.91 M) shaft. Similarly, from Figure 1, a lateral load of 109 kips (484.8 kN) can be read for the same deflection and for the same soil shear strength. Therefore, when a lateral load of 109 kips (484.8 kN) is applied to the 36 in (0.91 M) shaft, a shaft head deflection of 2.0 in (50.8 mm) and a bending moment value of 858 kip-ft (1163.2 kN-M) within the shaft will be obtained. For a given pile type and various pile sizes, these three (Figures 1 through 3) plots were generated.

For a particular pile type of various sizes, the lateral loads H_a (such as the ones shown Figure 1) were

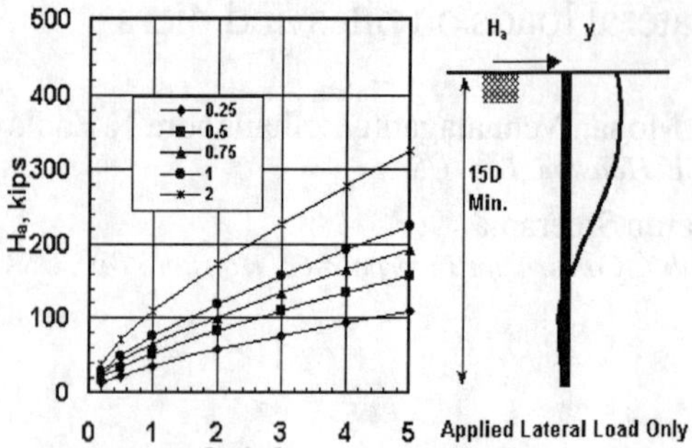

Figure 1. Horizontal load Ha versus Su for various deflections 'y' (36 in. Dia. pier)

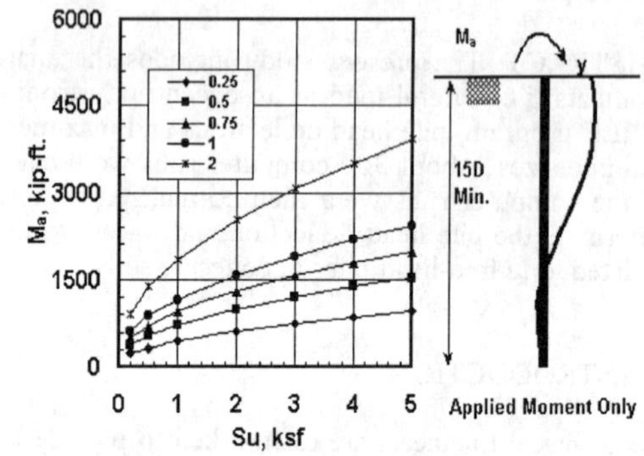

Figure 2. Applied Moment (Ma) versus Su for various deflections 'y' (36 in. Dia. pier)

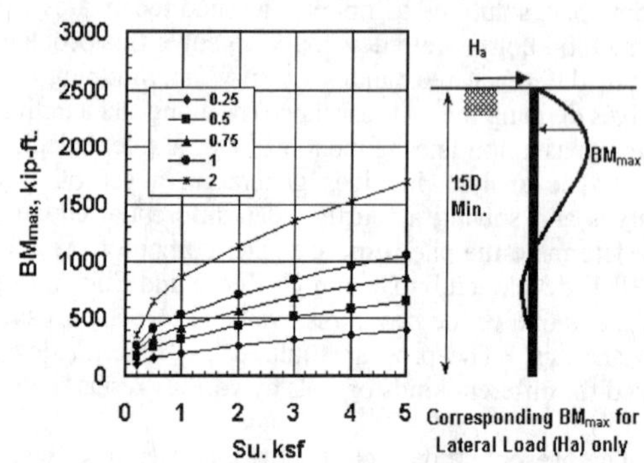

Figure 3. Maximum Bending Moment (BM_{max}) corresponding to Ha versus Su (36 in. Dia. pier)

gathered and were normalized as shown on Figure 4. Since the data is associated with various pile diameters (d) and various deflections (y), the normalization factors y and d were used to arrive at a curve shown on Figure 4. From Figure 4 (a plot of Su versus normalized values of H_a), given the values of d, Su, and H_a, the value of deflection 'y' can be determined. Conversely, given the values of d, Su, and y, the load H_a can be determined. In addition to the normalized curve shown, a regression analyses was performed to generate an equation as shown on Figure 4.

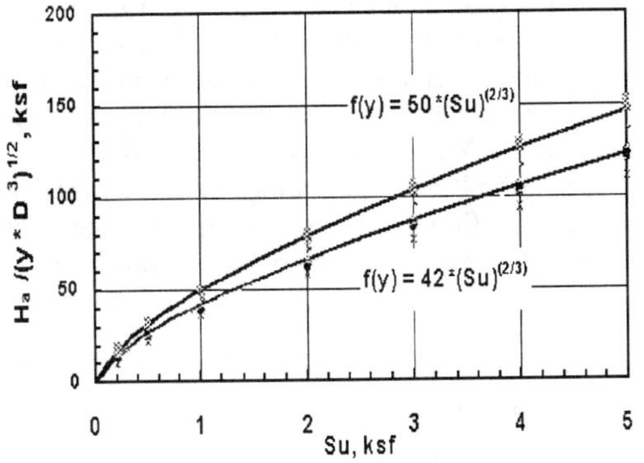

Figure 4 Normalized Plot for Lateral Loads, Ha and Deflection, y (piers).

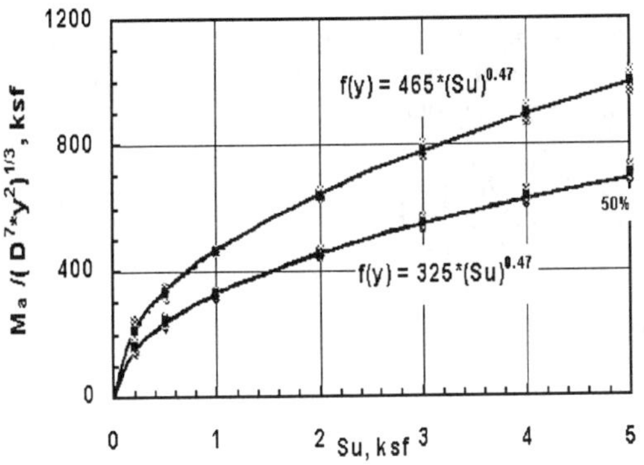

Figure 5 Normalized Plot for Applied Moments, Ma and Deflection, y (piers)

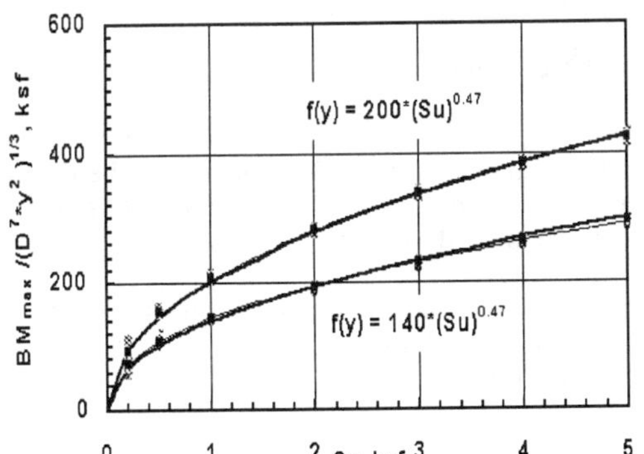

Figure 6. Maximum Bending Moment (BMmax) corresponding to Ha versus Su or various deflections (piers)

In the case of piers, two separate curves were generated assuming both 100%I and 50%I to account for the moment of inertia for a full uncracked section and cracked section, respectively. The equations for these two curves are also shown on Figure 4. In these equations 'f(y)' is the quantity of y-axis.

Similarly, the applied moments (M_a) on the pile head (the data corresponding to Figure 5) and the maximum bending moments (BM_{max}) within the pile (the data corresponding to applied loads H_a) were also normalized. For the normalized quantities, equations were generated as shown on Figure 4 through Figure 6.

Similar curves and equations such as shown in Figures 4 through 6 were also developed for precast square concrete piles, steel pipe piles and timber piles. For precast piles, pile width (d) was used and for timber piles pile butt diameter (d) was used in the respective equations. The generated equations were coincidentally of the similar form with respect to the power of the parameters involved for different pile types. Thus, the equations are reduced to the form as shown herein.

For the case of <u>only lateral loads (H_a)</u> applied to the pile head at the ground surface;

$$\mathbf{H_a = K_H * Su^{2/3} * y^{1/2} * d^{3/2}} \quad (1)$$

For the case of <u>only moments (M_a)</u> applied to the pile head at the ground surface;

$$\mathbf{M_a = K_M * Su^{0.47} * y^{2/3} * d^{7/3}} \quad (2)$$

For maximum <u>bending moments (BM_{max}) corresponding to the lateral loads (H_a) only;</u>

$$\mathbf{BM_{max} = K_{BM} * Su^{0.47} * y^{2/3} * d^{7/3}} \quad (3)$$

In the equations 1 through 3, Su is the average shear strength, d is the width or the diameter of the pile or pier and y is the deflection at ground surface. K_H, K_M, K_{BM} are constants. The constants were derived by regression analyses. These constants have units and are shown in Table 1. The constants are different for different pile types as shown in the table. The constants are appropriately applicable only when the units are kips, kip-ft, ksf and feet; or, kN, kN-m, kN/m^2 (kPa) and meters for the load, moment, shear strength, deflection/diameter, respectively. Table 1 shows the values of these constants for different pile types in both English and SI units. Numbers shown in parentheses correspond to SI units.

Table 1. Constants associated with Equations (1), (2) and (3) for various Pile Type(s)

PILE TYPE(S)	K_H $(F/L^2)^{(1/3)}$	K_M $(F/L^2)^{(0.53)}$	K_{BM} $(F/L^2)^{(0.53)}$
Piers (Shafts- 100%I)	**50 (181.5)**	**465 (3613.5)**	**200 (1554.2)**
Piers (Shafts- 50%I)	**42 (152.5)**	**325 (2525.6)**	**140 (1088.0)**
Concrete Precast Pile(s)	**58 (210.6)**	**625 (4856.9)**	**265 (2059.3)**
Steel Pipe Pile(s)	**53 (192.5)**	**520 (4040.9)**	**225 (1748.5)**
Tapered Timber Pile(s)	**37 (134.4)**	**290 (2253.6)**	**120 (932.5)**

2.2 *Combined Loading Conditions – Lateral loads applied at an eccentricity to the ground surface*

Figure 7 shows the condition when a lateral load, H, is applied to the pile head at an eccentric distance (e) from the ground surface. Loads applied to the pile head at an eccentric distance (e) can be transformed to a load, H, and moment, M (where M = H * e) acting on the pile at the ground surface.

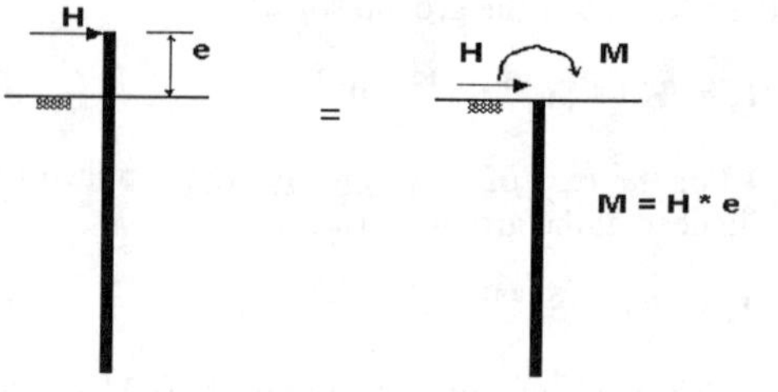

Figure 7 Loads at an eccentricity to the ground surface

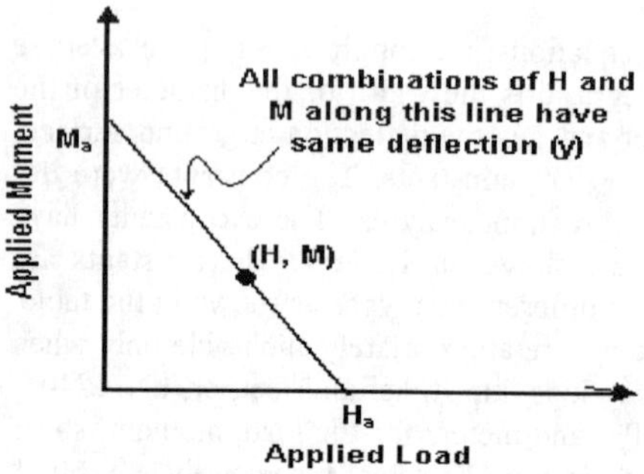

Figure 8 Locus of any given lateral load (H) and moment (M) applied to the pile head.

In this study, a unique feature associated with the LPILE program was discovered. It was determined that, if an isolated value of an applied load (H_a) produces a pile head deflection 'y'; and, if an isolated value of an applied moment (M_a) also produces a pile head deflection 'y'; then, the straight line joining the values H_a and M_a as shown in the Figure 8 represents the locus of all the combinations of those loads (H) and moments (M) which result in same deflection 'y'. This trend seems to work for deflections of up to about 2.0 in. (50.8 mm).

Consequently, the equation of the straight line joining H_a and M_a becomes:

$$\left(\frac{H}{H_a}\right)+\left(\frac{M}{M_a}\right)=1 \tag{4}$$

Where, H_a and M_a are defined by equations (1) and (2), respectively. By substituting equations (1) and (2) in the above equation and rewriting it:

$$\left(\frac{A}{\sqrt{y}}\right)+\left(\frac{B}{\sqrt[3]{y^2}}\right)=1 \tag{5}$$

where,

$$A=\frac{H}{\left(K_H \quad Su^{2/3} \quad d^{3/2}\right)}$$

and

$$B=\frac{M}{\left(K_M \quad Su^{0.47} \quad d^{7/3}\right)}$$

Using Equation 5, pile head deflection can be determined when a load and moment is applied to the pile head at the ground surface.

Similarly, the maximum bending moment $BM_{max(H,M)}$ corresponding to applied H and applied M can be shown as follows.

$$BM_{max(H,M)}=M+K_{BM} \quad Su^{0.47} \quad D_e^{7/3} y^{2/3} \tag{6}$$

Where D_e is defined by

$$D_e = \left(\frac{H}{K_H \, Su^{2/3} \quad y^{1/2}} \right)^{2/3}$$

The value of 'y' in the above equation is determined by equation (5).

3 CONCLUSIONS

Equations (5) and (6) can be easily used to estimate pile head deflections and maximum bending moment within the pile for various pile types subjected to lateral loads and/or moments. The applicability of the equation is limited to pile head deflections of less than 2.0 in. (50.8 mm) and the minimum pile embedment depth of 15 diameters. The equations are applicable to clay soil profiles. In some cases, it may be observed that for Su values of less than 0.5 ksf (23.9 kPa), the calculated deflection could be less than those computed by LPILE program. However, calculated maximum bending moments are within reasonable limits. These equations should be used for preliminary design purposes, since the basic assumption included uniform clay profiles and an E_{50} value of 0.01. Appropriate methodology or programs such as LPILE should be used to refine the lateral load analysis, especially in layered system.

REFERENCES

Broms, Bengt B., "Lateral Resistance of Pile in Cohesive Soils," Journal of soil Mechanics and Foundation Divisions, ASCE, Vol. 90, No. SM2, Part 1, Proc. Paper 3825, March 1964, PP.27-63.

J. Michael Duncan, Leonard T. Evans Jr. and Phillip S.K. Ooi, " Lateral Load Analysis of Single Piles and Drilled Shafts", Journal of Geotechnical Engineering, June, 1994, Vol. 120 No. 5, pp 1018-1033.

Matlock, Hudson and Lymon C. Reese, "Generalized Solutions for Laterally Loaded Piles," Journal of soil Mechanics and Foundation Division, ASCE, Vol. 86, No. SM5, Paper 2626, October 1960, pp 63-91.

Matlock, Hudson, "Correlations for Design of Laterally Loaded Piles in soft Clay," Paper No. OTC 1204, Proceedings, Second Annual Off-Shore Technology Conference, Houston, Texas 1970, Vol. 1, pp. 577-594.

McClelland, Bramlette and John A. Focht, Jr., "Soil Modulus of Laterally Loaded Piles", Transactions, ASCE, Vol. 123, 1958, pp. 1049-1063.

Reese, Lymon C., Discussion of "Soil Modules for Laterally Loaded Piles", by Bramlette McClelland and John A. Focht, Jr., Transactions, ASCE, Vol. 123, 1958, pp. 1071-1074.

Reese, Lymon C., W.R. Cox, "Soil Behavior from analysis of Tests Uninstrumented Piles Under Lateral loading", Performance of Deep Foundations, ASTM special Technical Publications No. 444, ASTM, Philadelphia, PA., 1969, pp. 160-176.

Reese, Lymon C., W.R. Cox, and F. D. Koop, "Analysis of Laterally loaded Piles in Sand," Paper No. OTC 2080 Proceedings, Sixth Annual Offshore Technology Conference, Houston, Texas 1974, Vol. 2, pp. 473-483.

Computer Methods and Advances in Geomechanics, Desai et al. (eds) © 2001 Balkema, Rotterdam, ISBN 90 5809 183 X

Interaction analysis of intact and defective piles

K.J.Xu
Civil Engineering Department, The University of Sydney, N.S.W., Australia

H.G.Poulos
Coffey Geosciences Pty Limited and Civil Engineering Department, The University of Sydney, N.S.W., Australia

ABSTRACT: It is a not uncommon experience to encounter defects in various types of piles. However, very little attention has been paid to the problem of interaction of defective piles, which is necessary to analyze the behaviour of pile group containing defective piles. In this paper, via a general pile analysis program, GEPAN, and based on introducing accurate definitions of interaction factors, the interaction between two piles is analyzed with varying pile and defect parameters. Some simple-but-important relationships between interaction factors for defective piles have been developed. The analysis carried out lead to better understanding of the interaction behaviour of defective piles. As a consequence, programs based on the use of interaction factors for intact piles can be easily extended to analyze defective piles, if the new information on interaction factors is incorporated into these programs.

1 INTRODUCTION

In order to analyse the behaviour of pile group containing defective piles, it is very important to know how the defective piles affect other piles, i.e. to examine pile-soil-pile interaction problems. Elastic continuum-based approaches to the analysis of groups of piles allow for the interaction between piles to be modelled. Although the effects of non-linear soil behaviour are expected to alter the response, the elastic analysis remains a common basis from which to study these phenomena. Interaction factors for two identical piles have been widely used to analyse pile group settlements (Poulos 1968, Randolph & Wroth 1979, Poulos & Davis 1980). Interaction between two dissimilar piles was studied by Hewitt (1988) and Hull (1996). However these available solutions for interaction factors are confined to two-intact piles, and are based on a simplified definition of the interaction factors.

Based on introducing accurate definitions of interaction factors, and using a general pile analysis, GEPAN, the interaction between two piles containing defects is analysed in this paper. The axial interaction of two piles which have one of the more common structural defects, necking, under, is presented. The influence of a number of pile parameters (pile length, diameter, spacing, pile stiffness factor), on the interaction between two piles containing defects is also studied. Some simple-but-important relationships between interaction factors for defective piles are developed from the analysis.

2 BACKGROUND OF ANALYSIS

2.1 *Interaction factor*

The two interacting piles are shown in Figure 1 (assuming pile 1 is intact and pile 2 is defective, i.e. necked). For a "standard" case, these two intact-defective piles are embedded in a layer of soil in which the Young's modulus is assumed as E_s=20*MPa*. Pile 1 is intact with length *L*=20*m*, diameter *d*=1*m*, while pile 2 contains a neck in which z_d/L=0.0, L_d/L=0.3, A_d/A=0.3, *d*/*L*=20, $K=E_p/E_s$=1000 (see Fig. 1 for geometric definition).

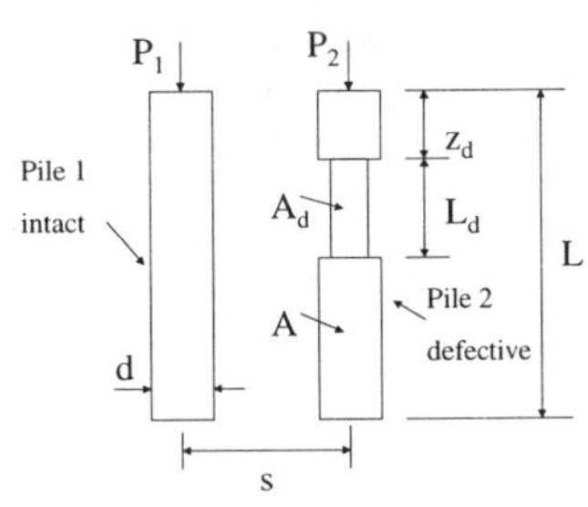

Figure 1. Geometry of two interacting piles

The load-deformation behaviour of these two interacting piles can be accurately expressed in term of the stiffness of corresponding isolated piles and by interaction factors, i.e.

$$S_1 = \alpha_{11}\frac{P_1}{K_1} + \alpha_{12}\frac{P_2}{K_2}$$
$$S_2 = \alpha_{21}\frac{P_1}{K_1} + \alpha_{22}\frac{P_2}{K_2} \quad (1)$$

where
S_i = axial head displacement on pile i
P_i= axial head load on pile i
K_i= axial stiffness of pile i
α_{ij}= interaction factor of pile i on pile j.

The interaction factors for a pile influenced by its own loading are called "auto interaction factors", e.g. α_{11}, α_{22}. The interaction factors due to on the other pile loading are called "cross interaction factors", i.e. α_{12}, α_{21}.

2.2 *Program GEPAN*

To implement the interaction analysis of two piles containing defects, a program called GEPAN has been developed by Xu and Poulos (in press) using 3-D boundary element analysis. It is a general rigorous elastic anlaysis of pile foundations, considering all load/deformation components for each of the piles with arbitrary dimensions, and also incorporates full coupling effects.

3 AUTO/CROSS INTERACTION FACTORS

3.1 *Auto interaction factors*

The auto interaction factors approach 1 when $s/d \geq 5$, but decrease significantly when $s/d<5$ as shown in Figure 2. It should be noted that the values of auto interaction factors are less than 1 due to the existence of the load-free pile. Thus, if the pile spacing is not very close, the auto interaction factors are set as 1, which is the same as the definition by Poulos & Davis (1980).

3.2 *Cross interaction factors*

The cross interaction factors α_{12} & α_{21} are shown in Figure 3. It is found that the value of the cross interaction factors α_{12} (effect of intact pile 1 on defective pile 2) is not reciprocal to the value of α_{21} (effect of defective pile 2 on intact

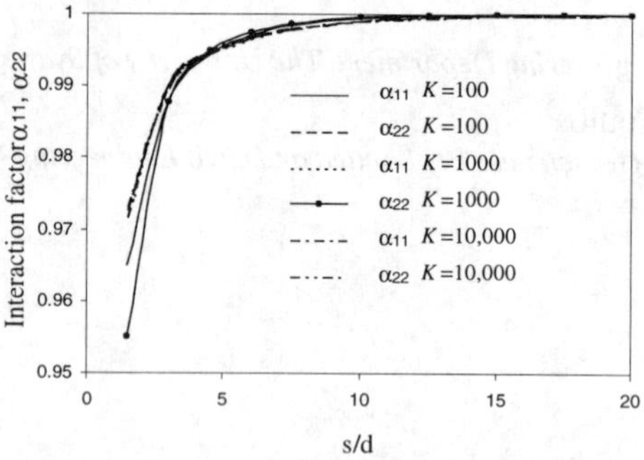

Figure 2. Auto interaction factor of two piles

pile 1). The value of α_{21} is larger than α_{12}, particularly for very compressible piles. α_{12} and α_{21} are plotted in Figure 3, as a function of dimensionless pile spacing *s/d*. As expected, the decreasing interaction with increasing pile spacing is clearly shown in the figure. Also, larger interaction factors are associated with stiffness piles, i.e. larger values of the pile stiffness factor K.

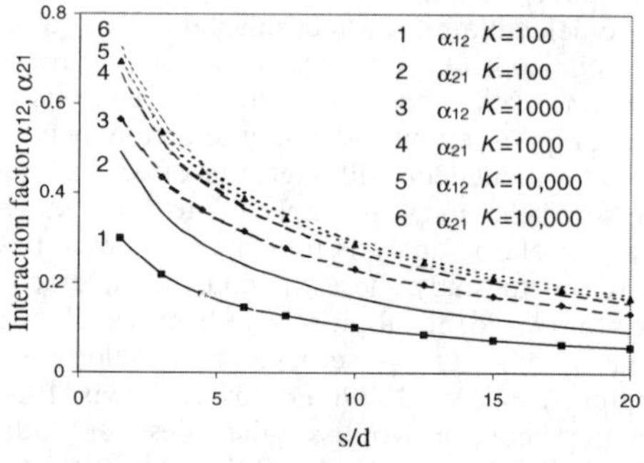

Figure 3. Cross interaction factors of two piles

4 RELATIONSHIPS FOR CROSS INTERACTION FACTORS

From above analysis, it can be seen that the cross interaction factors may be important in an interaction analysis. For convenience in the analysis of the interaction of defective piles, the following symbols for cross interaction factors are defined:

α_{ii} = cross interaction factor of two identical intact piles (α_{12} or α_{21})

α_{dd} = cross interaction factor of two identical defective piles (α_{12} or α_{21})
α_{id} = cross interaction factor for effect of the intact pile 1 on the damaged pile 2 (α_{12})
α_{di} = cross interaction factor for effect of the defective pile 2 on the intact pile 1 (α_{21})

Through a variety of parametric interaction studies on two interacting piles (intact & intact , defective & defective, intact & defective, defective & intact), such as is shown in Figure 4, approximate relationships for the cross interaction factors for defective piles are found to be as follows:

$$\alpha_{di}=\alpha_{ii} \quad (2)$$
$$\alpha_{id}=\alpha_{dd} \quad (3)$$

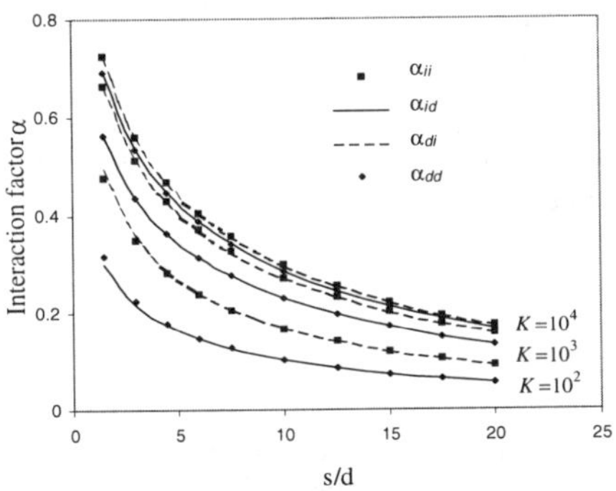

Figure 4. Cross interaction factor of two piles

In other words, the cross interaction factor of the defective pile on the intact pile α_{di} is approximately equal to cross interaction factor of two identical intact-intact piles, α_{ii}. The cross interaction factor of the intact pile on the defective pile α_{id} is approximately equal to the cross interaction factor of two identical defective-defective piles, α_{dd}.
Equations 2 & 3 have considerable significance in the interaction analysis of a group of defective piles. Firstly, the interaction factor α_{id} (or α_{12}) can be easily obtained from the cross interaction factor of two identical defective-defective piles. Secondly, programs based on interaction factors for intact piles, such as DEFPIG (Poulos 1990), PIGLET (Randolph 1989), and DAMPIG (Poulos 1997), can be extended to analyse defective piles, if the interaction factor α_{id} (α_{12}) is incorporated into these programs. Third, the above equations can be used to develop a general "conservation" equation for the interaction factors of two dissimilar piles:

$$\alpha_{di}\alpha_{id}=\alpha_{ii}\,\alpha_{dd} \quad (4)$$

5 CROSS INTERACTION FACTOR FOR EFFECT OF INTACT PILE ON THE DEFECTIVE PILE

As presented above, the cross interaction factor α_{id} is a key factor in the interaction analysis for two piles containing defects. This section will illustrate some parametric effects on interaction factors of the defective piles.

5.1 *Effect of length of defect*

The interaction factors, α_{id}, of two piles for various values of ratios of necked length L_d/L and necked area A_d/A, are shown in Figure 5. For severe necking (such as A_d/A=0.3, 0.1), the value of α_{id} decreases significantly as L_d/d increases for "short necking" (L_d/d <0.5), but α_{id} becomes approximately a constant for "long necking" (L_d/d >=0.5). The effect of the length of defects becomes very apparent if the piles are locally necked severely (*e.g.* A_d/A= 0.1).

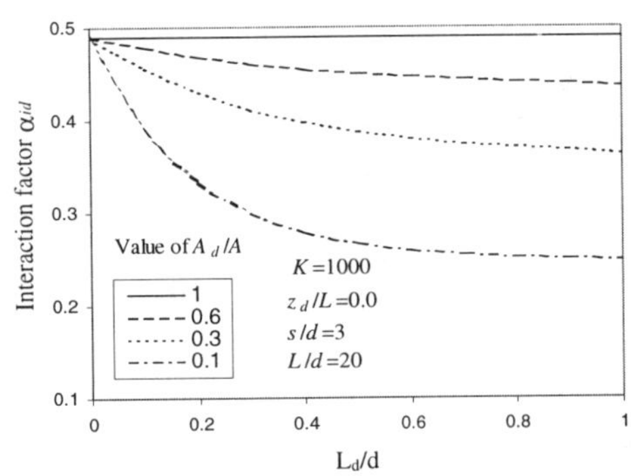

Figure 5. Cross interaction factor vs relative length of defect, L_d/L, for two piles

5.2 *Effect of position of defect*

Figure 6 shows the effect of the location of a defect (neck) along the pile depth. The cross interaction factors, α_{id}, increase as the ratio of necked depth to pile length, z_d/L, increases; this effect is most pronounced for severe necking near the pile top. For necking near the pile tip, the effect of the defect position is small, and all values of α_{id} approach α_{ii}.

5.3 *Effect of pile stiffness factor*

Figure 7 shows the effect of the pile stiffness factor K (Poulos & Davis 1980) on interaction factor α_{id} for a typical case of the interaction of intact-damaged piles. α_{12} increases as the value of K increases. For very stiff piles (*e.g.* $K>10^4$), the effect of the ratio of necked area A_d/A becomes small, and all values of α_{id} approach the value of α_{ii}.

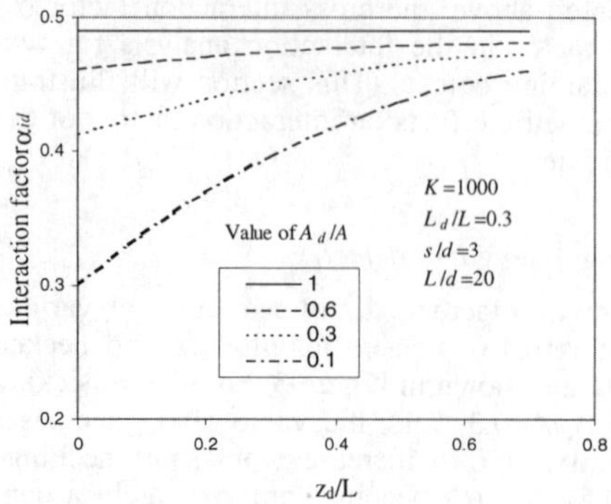

Figure 6. Interaction factor vs relative depth of defect, z_d/L, for two piles

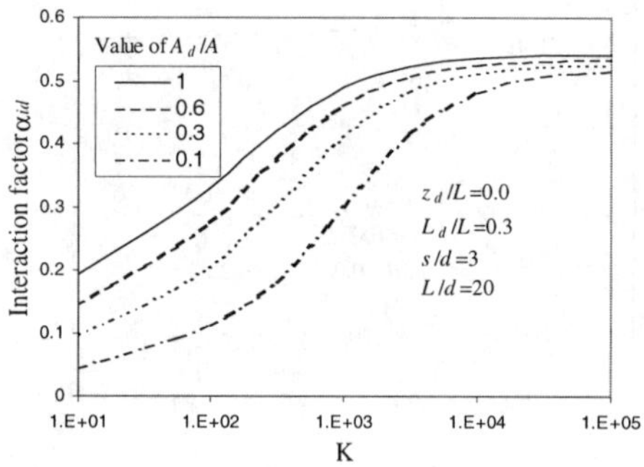

Figure 7. Interaction factor vs pile stiffness factor, K, for two piles

6 INTERACTION OF TWO PILES WITH OTHER DEFECTS UNDER GENERAL LOADING

Due to space limitations, only one kind of defect, i.e. a structural neck, under axial loading, has been considered in this paper. A complete interaction analysis of two interacting piles containing different defects (necking, honey-combing, softened base and their combinations), under general loadings (axial, lateral, moment, torsional), are presented in detail by Xu (2000).

7 CONCLUSIONS

By applying the computer program GEPAN, for the general linear elastic analysis of piles, interaction factors for two piles containing defects have been evaluated. The general interaction factors are defined. The main conclusions drawn from this linear elastic analysis are as follows:

1) The auto interaction factor tends to 1.0 as *s/d* increases, while the cross interaction factors decrease with increasing *s/d*. 2) The cross interaction factor for the effect of an intact pile on defective pile is an important factor in the interaction analysis for two interacting piles. The cross interaction factor for the effect of a defective pile on an intact pile is approximately equal to the cross interaction factor of two identical intact piles. The cross interaction factor for the effect of an intact pile on a defective pile is approximately equal to the cross interaction factor of two identical defective piles.

It is anticipated that soil and structural nonlinear effects may overshadow some of the linear elastic results presented herein, nevertheless, this paper supplies a new and easy-to-apply method to extend the widely used interaction-factor-based programs for intact pile analysis to anlayse pile groups containing defective piles.

REFEERECES

Hewitt, C.M. 1988. Cyclic response of offshore pile groups. *Ph.D. Thesis,* University of Sydney. Australia.

Hull, T.S 1996. Analysis of laterally loaded pile groups with unequal dimensions. *7th Australia New Zealand Conference on Geomechanics*, Adelaide, Australia.521-525.

Poulos, H.G. 1968. Analysis of the settlement of pile groups. *Geotechnique* 18:449-471.

Poulos, H.G. & Davis, E.H. 1980. Pile foundation analysis and design. *John Wiley & Sons*. New York.

Randolph, M.F. & Wroth, C.P. 1979. An anlaysis of the vertical deformation of pile groups. *Geotechnique* 29(4):423-439.

Xu, K.J. 2000. General analysis of pile foundations and application to defective piles. *Ph.D. Thesis,* University of Sydney. Australia

Xu, K.J. & Poulos, H.G. 1999. General elastic analysis of piles and pile groups. *Int. Jour. for numerical and analysis methods in geomechanics* (in press).

Computer Methods and Advances in Geomechanics, Desai et al. (eds) © 2001 Balkema, Rotterdam, ISBN 90 5809 183 X

Simplified method for retaining piles considering effects of ring beam

Gao Yinli
Foundation Institute, China Academy of Building Research, Beijing, People's Republic of China

ABSTRACT: Practice shows that the cooperative action between row of piles and ring beam has a certain influence on the deformation and internal forces of retaining piles. On the basis of the constraint of ring beam on the retaining piles and the deformation coordination between them, a flexural equation method considering the cooperative action between row of piles and ring beam is presented in this paper. This method is very simple and practical, which can be obtained by solving linear equations, avoiding the complicated finite element method calculation. The calculation results are compared with the measured ones and it is shown that they coincide with each other fairly well.

1 INTRODUCTION

During the design and construction of retaining piles, a ring beam is often installed on the pile butts to strengthen the global stiffness of row of piles. Practice shows that the effects of ring beam can improve the state of forces and deformation of piles and reduce the displacement of retaining structure to a certain degree. In actual design of retaining piles, the effects of ring beam are generally neglected and only acted as a safety margin. The corresponding finite element method analysis has been proposed in recent years. But they are fairly complicated. In this paper, the model of beam on elastic foundation is adopted and a simplified method is presented.

2 SPATIAL ANALYSIS OF RETAINING PILES

2.1 *Theoretical model of row piles*

The model of row of piles can be established based on following assumptions:

1. Winkler's assumption is satisfied and the coefficient of foundation horizontal resistance increases linearly with depth, i.e. "m" method;
2. Only horizontal forces impacted on piles by ring beam are considered and the torsion moments are neglected;
3. The spatial effects of earth pressure are not considered and the active earth pressure is equal to Rankine's active earth pressure;
4. The constraint condition of pile tip is free.

The calculated sketch is shown in Fig. 1. The differential equation of pile i can be expressed as follows:

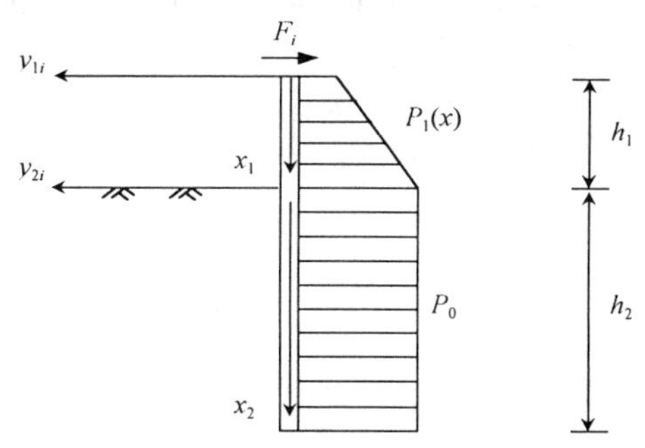

Fig. 1 Calculating sketch of retaining piles

$$\left.\begin{aligned} EI\frac{\mathrm{d}y_{1i}^4}{\mathrm{d}x_1^4} &= bP(x_1)\ (0 \le x_l \le h_1) \\ EI\frac{\mathrm{d}y_{2i}^4}{\mathrm{d}x_2^4} &= -b_0 m x_2 y_{2i} + bP_0\ \ (0 \le x_2 \le h_2) \end{aligned}\right\} \quad (1)$$

where m = proportional coefficient of foundation horizontal resistance; b = pile spacing; b_0 = calculated width of piles; EI = pile rigidity; F_i = horizontal force impacted on pile i by ring beam. The general solution and every order derivatives of formula (1) can be expressed as:

$$
\left.\begin{aligned}
\begin{bmatrix} y_{1i} \\ y'_{1i} \\ y''_{1i} \\ y'''_{1i} \end{bmatrix} &= \begin{bmatrix} 1 & x_1 & x_1^2 & x_1^3 \\ & 1 & 2x_1 & 3x_1^2 \\ & & 2 & 6x_1 \\ & & & 6 \end{bmatrix} \begin{bmatrix} a_{1i} \\ a_{2i} \\ a_{3i} \\ a_{4i} \end{bmatrix} + \frac{b}{EI} \begin{bmatrix} W_1(x_1) \\ S_1(x_1) \\ R_1(x_1) \\ Q_1(x_1) \end{bmatrix} \\
\begin{bmatrix} y_{2i} \\ y'_{2i} \\ y''_{2i} \\ y'''_{2i} \end{bmatrix} &= \frac{1}{EI} \begin{bmatrix} \frac{B_x}{\alpha^2} & \frac{A_x}{\alpha^3} & \frac{E_x}{\alpha^4} \\ \frac{B_\phi}{\alpha} & \frac{A_\phi}{\alpha^2} & \frac{E_\phi}{\alpha^3} \\ \frac{B_m}{1} & \frac{A_m}{\alpha} & \frac{E_m}{\alpha^2} \\ \alpha B_Q & A_Q & \frac{E_Q}{\alpha} \end{bmatrix} \begin{bmatrix} M_{0i} \\ Q_{0i} \\ bP_0 \end{bmatrix}
\end{aligned}\right\} \quad (2)
$$

where M_{0i}, Q_{0i} = moment and shear force of pile i in excavation surface; B_x, $B\phi$, B_m, B_Q, A_x, $A\phi$, A_m, A_Q, E_x, $E\phi$, E_m, E_Q = zero-dimensional coefficients, which can be obtained in relevant references; α = horizontal deformation coefficient of piles calculated by "m" method, $\alpha = \sqrt[5]{\frac{mb_0}{EI}}$; $a_{1i} \sim a_{4i}$ = integral constants; $Q(x_1)$, $R(x_1)$, $S(x_1)$, $W(x_1)$ = one, two, three and four order integral values of earth pressure from the interval of 0 to x_1.

Considering the boundary conditions of pile butts, one can obtain:

$$
\left.\begin{aligned}
Q &= [y'''_{1i}]_{x_1=0} = 6EIa_{4i} = -F_i \\
M &= EI[y''_{1i}]_{x_1=0} = 2EIa_{3i} = 0
\end{aligned}\right\} \quad (3)
$$

therefore $a_{4i} = -\frac{F_i}{6EI}$, $a_{3i} = 0$

Considering the deformation continuity and static equilibrium of excavated surface, one can obtain:

$$
\left.\begin{aligned}
[y_{1i}]_{x_1=h_1} &= [y_{2i}]_{x_2=0} \\
[y'_{1i}]_{x_1=h_1} &= [y'_{2i}]_{x_2=0} \\
EI[y''_{1i}]_{x_1=h_1} &= EI[y''_{2i}]_{x_2=0} = M_{0i} \\
EI[y'''_{1i}]_{x_1=h_1} &= EI[y'''_{2i}]_{x_2=0} = Q_{0i}
\end{aligned}\right\} \quad (4)
$$

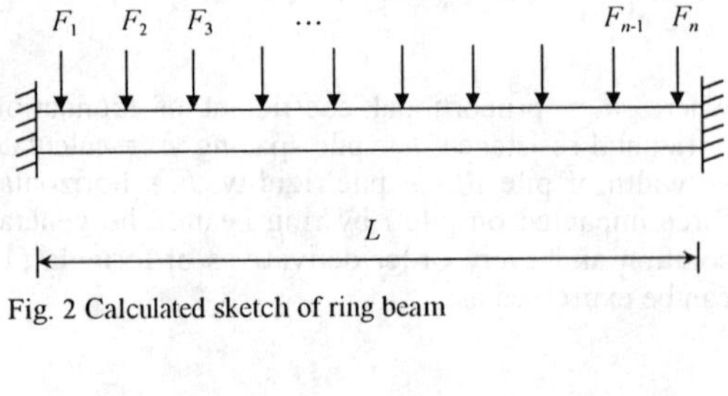

Fig. 2 Calculated sketch of ring beam

Formula (4) can be expressed by form of matrix:

$$
\begin{bmatrix} 1 & h_1 & -\frac{B_{x_0}}{\alpha^2 EI} & -\frac{A_{x_0}}{\alpha^3 EI} & -\frac{h_1^3}{6EI} \\ & 1 & -\frac{B_{\phi_0}}{\alpha EI} & -\frac{A_{\phi_0}}{\alpha^2 EI} & -\frac{h_1}{2EI} \\ & & \frac{1}{EI} & 0 & \frac{h_1}{EI} \\ & & & \frac{1}{EI} & \frac{1}{EI} \end{bmatrix} . \quad (5)
$$

$$
\begin{bmatrix} a_{11,i} \\ a_{12,i} \\ M_{0i} \\ Q_{0i} \\ F_i \end{bmatrix} = \frac{b}{EI} \begin{bmatrix} \frac{E_{x_0}}{\alpha^4} P_0 - W_1(h_1) \\ \frac{E_{x_0}}{\alpha^3} P_0 - S_1(h_1) \\ R_1(h_1) \\ Q_1(h_1) \end{bmatrix}
$$

where B_{x_0}, B_{ϕ_0}, A_{x_0}, A_{ϕ_0}, E_{x_0}, E_{ϕ_0} = zero-dimensional coefficients while $x_2 = 0$ and they can be found in relevant references.

2.2 *Theoretical model of ring beam*

The model of ring beam can be established based on following assumptions:

1. The two end nodes of ring beam have large rigidity and can be simplified as fixed bearings;
2. In the extent of ring beam span, horizontal forces F_i (i=1,2,…n) are impacted on ring beam by piles and the force-displacement relationship complies with the principle of superposition;
3. The torsion moments of piles impacted on ring beam are not considered.

The calculated sketch of ring beam is shown in Fig. 2. For the node i of ring beam, one can obtain:

$$
u_i = \sum_{j=1}^{n} \delta_{ij} F_j \quad (i, j = 1, 2, \ldots n) \quad (6)
$$

where u_i = horizontal displacement of node i of ring beam; δ_{ij} = horizontal displacement of node i of ring beam when the unit horizontal force of node j (F_j = 1) is impacted alone.

According to the deformation compatibility conditions, the horizontal displacement of pile butts is equal to that of ring beam at node i, i.e.

$$
[y_{1i}]_{x_1=0} = u_i \quad (i, j = 1, 2, \ldots n) \quad (7)
$$

therefore

$$
a_{1i} = \sum_{j=1}^{n} \delta_{ij} F_j \quad (i, j = 1, 2, \ldots n) \quad (8)
$$

Substituting Eq. (8) into Eq. (5) and solving $4n$ simultaneous equations by computer, one can obtain a_{1i}, a_{2i}, M_{0i}, Q_{0i} and F_i (i = 1, 2, ... n). Substituting above-mentioned values into Eq. (2), then the displacement, shear forces and moments of every pile in different depth can be obtained.

3 CASE ANALYSIS

An excavation with geological condition of marine sedimentary silt clay. h (excavated depth) = 6.10m, c_{cu} = 15.0kPa, φ_{cu} = 18°, m = 3500kN/m$_4$. ϕ800 row of piles are adopted for the retaining structure. The section dimension of ring beam is 1.0m × 0.6m and the span L = 46.4m. The layout of piles is shown in Fig. 3 (where the numbers denote pile marks).

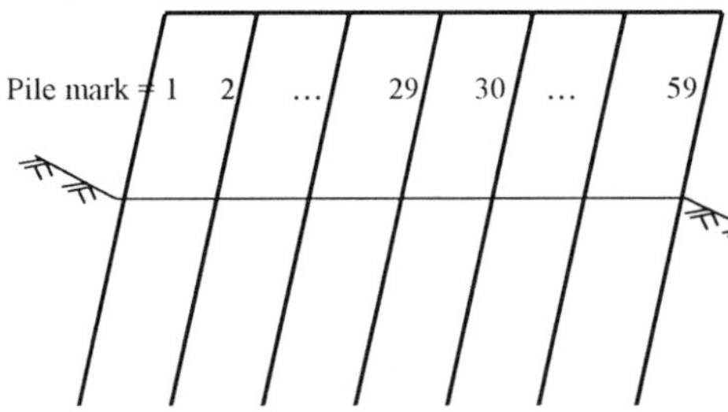

Fig. 3 Arrangement of row of piles

The finite element method and the simplified one in this paper are adopted respectively and the numerical results are shown on Figs 4-5 (where 5, 10, and 15 etc. denote pile marks). The horizontal forces on every pile butts calculated with the method in this paper are shown in Table 1.

Table 1. Calculated values of horizontal forces on pile butts

Pile mark	1	2	3	4	5	6
F_i (kN)	73.99	72.81	69.67	65.14	59.67	53.65
Pile mark	7	8	9	10	11	12
F_i (kN)	47.39	41.13	35.08	29.36	24.07	19.28
Pile mark	13	14	15	16	17	18
F_i (kN)	15.01	11.26	8.04	5.30	3.02	1.15
Pile mark	19	20	21	22	23	24
F_i (kN)	-0.34	-1.52	-2.43	-3.11	-3.60	-3.94
Pile mark	25	26	27	28	29	30
F_i (kN)	-4.08	-4.32	-4.42	-4.47	-4.49	-4.50

* The forces are symmetrically distributed.

As can be seen from above Fig. 4-5, there is a fairly good cooperative action between ring beam and row of piles and the spatial effects can improve the state of the forces and displacement of retaining piles to a certain degree. Especially, the effects can reduce the displacement and moments of the piles near the pit corner and have few influences on the middle piles. The numerical results with the simplified method presented in this paper are well consistent with those with finite element method and the displacement of ring beam coincides with the measured one fairly well.

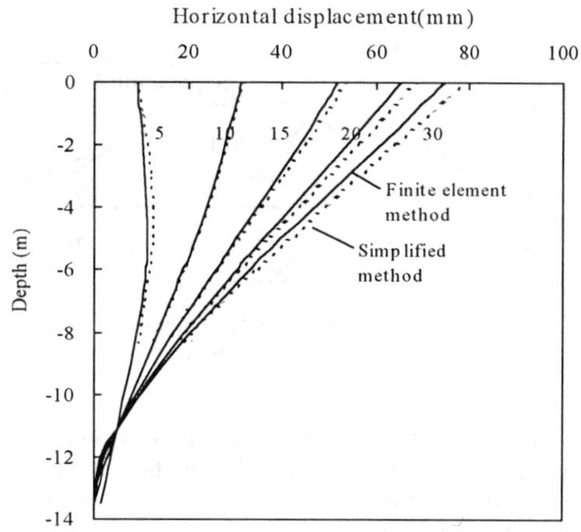

(a) Displacement of piles

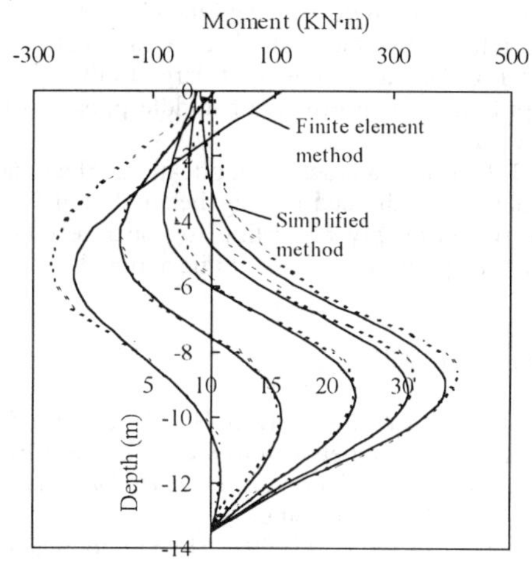

(b) Moment of piles

Fig. 4 Displacement and moment of retaining piles in different position

4 CONCLUSIONS

On the basis of soil resistance method, a simplified method considering the cooperative action between row of piles and ring beam is presented and some conclusions are drawn as follows:

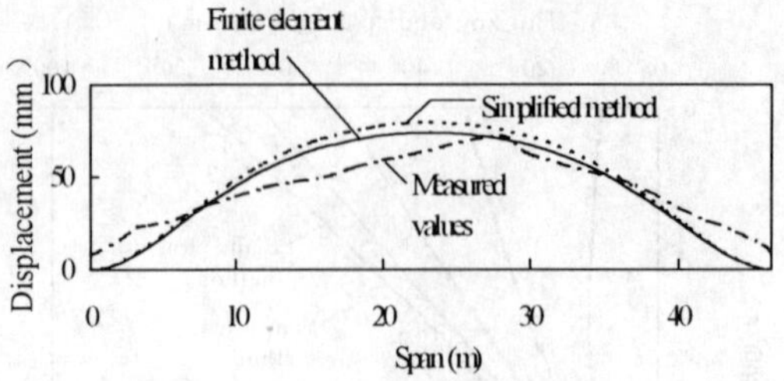

Fig. 5 Comparison between calculated values and measured ones of horizontal displacement of ring beam

1. The simplified method presented in this paper is very simple and practical. The method is used in an actual engineering and the calculated values coincide with the measured ones fairly well;

2. There is a fairly good cooperative action between ring beam and retaining piles. It can improve the state of the forces and deformation of retaining piles to a certain degree. And the closer to the corner the pile is, the more significant the spatial effects are. For the excavations with large ratio of length to depth, the spatial effects of middle piles can be neglected;

3. During the practical applications, the constraint condition of the end of ring beam should be determined reasonably according to concrete conditions and the appropriate δ_{ij} should be adopted.

REFEREENCES

He Jianming & Bai Shiwei (1997). Study of cooperative action between row of piles and ring beam for deep foundation. *Rock and Soil Mechanics*. 18(3):41-46. (in Chinese)

Xiong Juhua et al. (1999). Method for calculating deformation of DMP retaining wall effect of beam at the top. *Rock and Soil Mechanics*. 20(1):84-86. (in Chinese)

Yang Xueqiang & Liu Zude (1998). Research about spatial effect of deep pit supporting. *Chinese Journal of Geotechnical Engineering*. 20(2):74-78. (in Chinese)

Computer Methods and Advances in Geomechanics, Desai et al. (eds) © 2001 Balkema, Rotterdam, ISBN 90 5809 183 X

Extended Mindlin solution and its application to piles

Ai Zhiyong, Yang Min & Pei Jianyong
Department of Geotechnical Engineering, Tongji University, Shanghai, People's Republic of China

ABSTRACT: The extended Mindlin solution for the problem of a concentrated force in the interior of multi-layered elastic solid is derived. The solution is applied to analyze single pile in multi-layered soil. The result and applicability of the present method have been compared with existing solutions.

1 INTRODUCTION

In the analysis of pile foundation, Mindlin solution (Mindlin 1936) is widely used. Most of the traditional methods in this field are based on it. However, natural soils can hardly be regarded as homogeneous elastic body. Usually they present layered structures. Thus considerable error may be introduced if Mindlin solution is applied to calculate the stress and displacement field of pile foundation. Based on the theory for layered elastic solid, Ai Zhiyong (1999) obtained the solution for a point load within a multi-layer elastic body (it is called extended Mindlin solution hereafter), by means of matrix transfer method. Theoretically, it is more reasonable to apply extended Mindlin Solution to the analysis of pile foundations. In this paper, the problem of an axially loaded single pile embedded in a multi-layer soil is analyzed with extended Mindlin solution.

2 EXTENDED MINDLIN SOLUTION

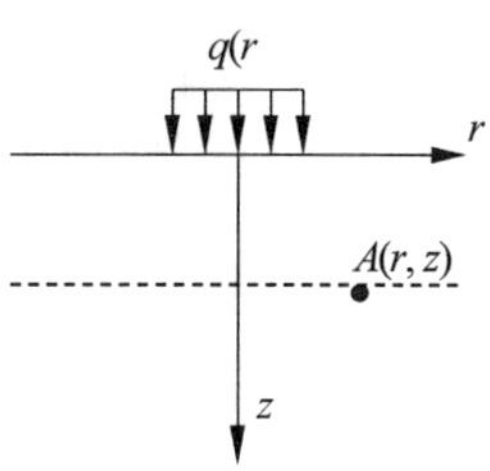

Figure 1. Homogeneous elastic solid acted upon by axisymmetric load

As shown in Figure 1, an axisymmetric load $q(r)$ is applied on the surface of a homogeneous elastic semi-infinite body. For any plane within the elastic body at a certain depth z, which is parallel to the surface, denote the displacements and stresses as $u(r, z)$, $w(r, z)$, $\tau_{rz}(r, z)$, $\sigma_z(r, z)$. Let $\bar{u}(\xi,z)$, $\bar{w}(\xi,z)$, $\bar{\tau}_{rz}(\xi,z)$, $\bar{\sigma}_z(\xi,z)$ be their Hankel transforms with respect to r. That is:

$$\bar{u}(\xi,z) = \int_0^\infty u(r,z) r J_1(\xi r) dr$$

$$\bar{w}(\xi,z) = \int_0^\infty w(r,z) r J_0(\xi r) dr$$

$$\bar{\tau}_{rz}(\xi,z) = \int_0^\infty \tau_{rz}(r,z) r J_1(\xi r) dr$$

$$\bar{\sigma}_z(\xi,z) = \int_0^\infty \sigma_z(r,z) r J_0(\xi r) dr$$

Their inverse transforms can be denoted as:

$$u(r,z) = \int_0^\infty \bar{u}(\xi,z) \xi J_1(\xi r) d\xi$$

$$w(r,z) = \int_0^\infty \bar{w}(\xi,z) \xi J_0(\xi r) d\xi$$

$$\tau_{rz}(r,z) = \int_0^\infty \bar{\tau}_{rz}(\xi,z) \xi J_1(\xi r) d\xi$$

$$\sigma_z(r,z) = \int_0^\infty \bar{\sigma}_z(\xi,z) \xi J_0(\xi r) d\xi$$

in which $J_0(\xi r)$, $J_1(\xi r)$ are Bessel functions of order 0 and order 1 respectively.

Consider an elastic solid with n layers, the m-th layer of which is acted upon by a point load P (Fig. 2). At a certain depth z in the i-th layer, the displacements and stresses can be obtained by the following equation:

$$\bar{g}(\xi,z)=[a_{ij}]_{4\times4}\,\bar{g}(\xi,0) \tag{1}$$

in which:

$$\bar{g}(\xi,z)=\{\bar{u}(\xi,z),\bar{w}(\xi,z),\bar{\tau}_{rz}(\xi,z),\bar{\sigma}_z(\xi,z)\}^T$$

$$[a_{ij}]_{4\times4}=\Phi(\xi,z-H_{i-1})\Phi(\xi,\Delta H_{i-1})\Lambda\ \Phi(\xi,\Delta H_1)$$

where $\Phi(\xi, z)$ is the transfer matrix, H_i and ΔH_i (i = 1, 2, …, n) are the depth and thickness of the i-th layer. Take the inverse Hankel transform of $\bar{g}(\xi,z)$ in Equation (1), the displacements and stresses can be obtained.

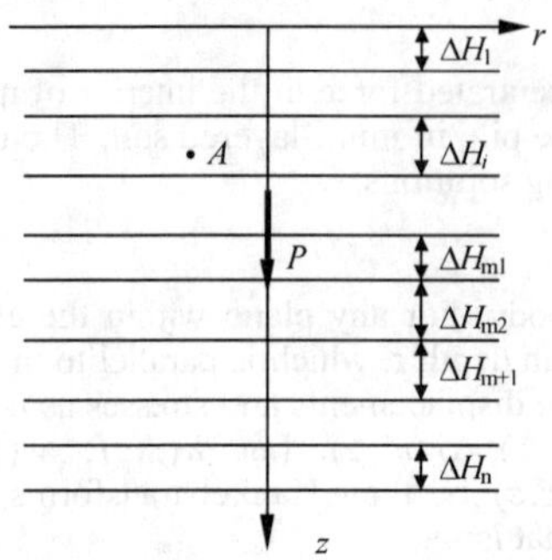

Fig. 2 multi-layer elastic solid acted upon by a point load

Suppose that the deepest layer has a fixed boundary, that is on the boundary $\bar{u}(\xi,H_n)=\bar{w}(\xi,H_n)=0$. It follows that in (1), $\bar{g}(\xi,0)$ can be obtained by:

$$\bar{g}(\xi,H_n^-)=[f_{ij}]_{4\times4}\,\bar{g}(\xi,0)-[s_{ij}]_{4\times4}\{p\}_{4\times1} \tag{2}$$

in which

$$[f_{ij}]_{4\times4}=\Phi(\xi,\Delta H_n)\Phi(\xi,\Delta H_{n-1})\Lambda\ \Phi(\xi,\Delta H_1)$$

$$[s_{ij}]_{4\times4}=\Phi(\xi,\Delta H_n)\Phi(\xi,\Delta H_{n-1})\Lambda\ \Phi(\xi,\Delta H_{m2})$$

$$\{p\}_{4\times1}=\left[0,\ 0,\ 0,\ \frac{P}{2\pi}\right]^T$$

3 THE ANALYSIS OF A SINGLE PILE

Figure 3 shows a single pile embedded in multi-layer elastic subsoil. The flexibility equation for the soil can be written as:

$$\{\rho_s\}=[I_s]\cdot\{p\} \tag{3}$$

in which $\{\rho_s\}$ is the vector of displacements of soil; $\{p\}$ is the vector of vertical stress over the pile shaft and pile base. $[I_s]$ is the soil flexibility matrix, whose elements are readily obtained by integrating extended Mindlin Solution. The stiffness matrix $[K_s]$ of soil can be derived by taking inverse of $[I_s]$, that is $[I_s]^{-1}=[K_s]$.

Taking into account the displacement compatibility of pile and soil, that no slip is assumed to occur along the pile shaft, we obtain:

$$([K_p]+[K_s])\{\rho\}=\{P\} \tag{4}$$

in which $[K_p]$ is the stiffness matrix of pile; $\{P\}=[Q\quad 0\quad \Lambda\quad 0]^T$ is the vector of load. Equation (4) is the displacement equations for solving the problem of a single pile embedded in layered soil. After obtaining the displacements vector {ρ} of pile, the stress vector {p} can be derived from Equation (3).

4 COMPARISON WITH EXISTING SOLUTIONS

Based on the analysis method presented in this paper, computer programs have been developed and some calculation performed. The results are compared with existing results of Poulos and that of FEM.

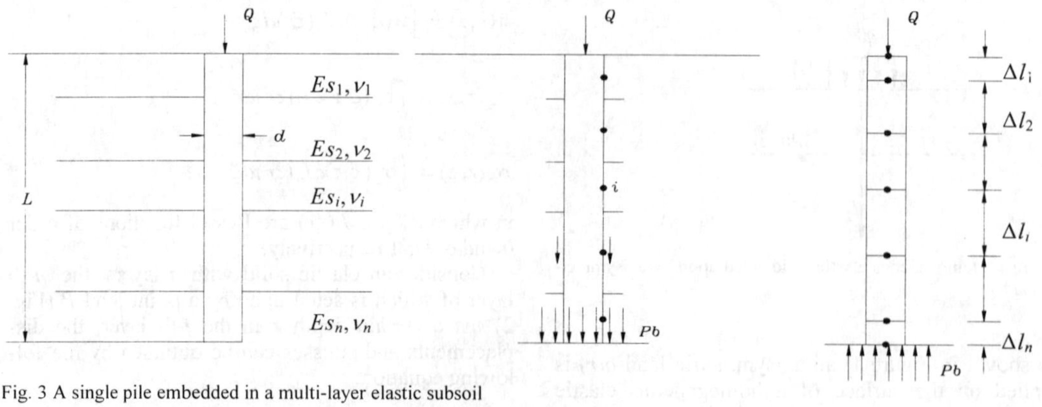

Fig. 3 A single pile embedded in a multi-layer elastic subsoil

4.1 *A single pile in homogeneous soil*

Suppose the homogeneous soil comprises of several soil layers with identical elastic parameters, the extended Mindlin solution can be applied to the analysis. Table 1 compares the dimensionless stiffness ratio $P_0/(G_s r_0 \Delta_0)$ of different l/r_0 with the results presented by Poulos (1980). In Table 1, P_0 is the load acted on pile top, G_s is the shear modulus of soil, r_0 is the radius of pile, Δ_0 is the settlement of pile top, E_p is the Young's modulus of pile, l is the length of pile. It is shown that the results of these two methods agree very well.

Table 1 Comparison of $P_0/(G_s r_0 \Delta_0)$ (single pile in homogeneous soil)

E_p/G_s		10^5	10^4	3×10^3	10^3	3×10^2
$l/r_0 = 40$	Poulos	67.8	65.7	61.3	52.0	36.8
	Present	67.3	65.4	60.8	51.5	36.8
$l/r_0 = 80$	Poulos	112.7	102.2	85.2	61.6	38.0
	Present	111.3	101.7	84.4	60.8	38.2

4.2 *A single pile in a 3-layer soil*

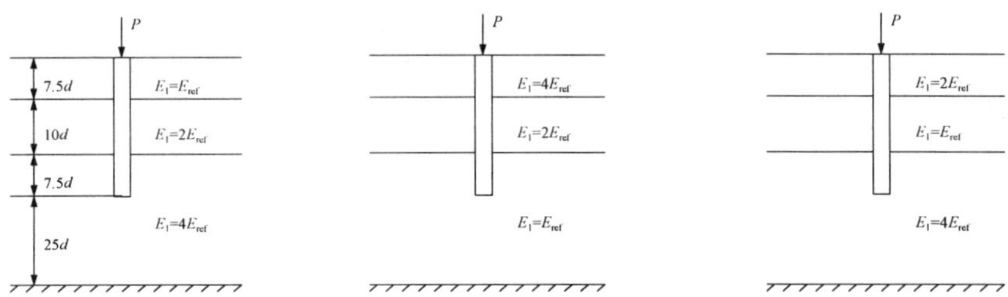

Case 1 Case 2 Case 3

Fig. 4 Single pile in 3-layer soil

Analysis has been made toward the 3-layer soil profile shown in Figure 4. In Table 2, the result is compared with the result presented by Poulos, the result obtained through FEM, and the result of the approximate method in which the soil is supposed to be homogeneous with modulus being the weighed average of all layers. This time, the displacement influence factor $I_\rho = dE_{ref}\rho/P_0$ is compared. In this example, $E_p = 1000E_{ref}$, the Poisson's Ratio of soil is 0.3.

Table 2 Comparison of $I_\rho = dE_{ref}\rho/P_0$ (single pile in 3-layer soil)

Case No.	Present	FEM	Poulos	weighed modulus method
Case 1	0.0376	0.0377	0.0386	0.0381
Case 2	0.0429	0.0430	0.0330	0.0706
Case 3	0.0383	0.0382	0.0366	0.0391

It can be noticed in Table 2, that there's good agreement between the result of extended Mindlin solution and that of FEM. However, FEM consumes much more computer time than present method. This substantiates that the present method is theoretically correct and convenient to use. It can also be noticed from Table 2, that considerable difference exists between the result presented by Poulos and that of FEM and the present method. While the precision of the weighed modulus method is even worse.

5 CONCLUSIONS

Based on extended Mindlin solution, which gives the stresses and displacements due to a point load within a multi-layered elastic body, the analysis methodology of axially loaded single pile in layered soil has been set up. The analysis method can be readily developed into computer program which gives accurate results and consumes less computer time. These features facilitate its application in engineering practice. The method presented in this paper can be readily extended to the analysis of pile groups.

REFERENCES

Ai Zhiyong(1999), Research on Theory of Layered Elastic System and Its Application in Piled Raft Foundation. *Ph. D. Thesis*: Tongji University.

Mindlin, R.D.(1936). Force at a Point in the Interior of a Semi-Infinite Solid. *Physics* 7, 195

Poulos, H.G.(1979). Settlement of Single Pile in Nonhomogeneous Soil. *J. Geotech. Eng. ASCE*, 105(5): 627~642

Poulos, H.G., Davis, E.H(1980). *Pile foundation analysis and Eng., ASCE*, 105(5): 627~ 642.

15 Slopes, dams, walls, landslides

Computer Methods and Advances in Geomechanics, Desai et al. (eds) © 2001 Balkema, Rotterdam, ISBN 90 5809 183 X

An overview on 25 years of numerical modeling of test embankments and tunnels

Ph. Mestat
Laboratoire Central des Ponts et Chaussées, Paris, France

ABSTRACT. The objectives of this paper are to make some conclusions about the ability to predict the behaviour of embankments and tunnels and to provide recommendations for finite element modeling. The Soil and Rock Mechanics division of the Laboratoire Central des Ponts et Chaussées (Paris, France) develops a bibliographical database. Its content is described in this paper and some applications are presented. Approximately 150 well documented case studies are included in the first version of database : 76 for embankments tests on soft soils (without drains) and 78 for tunnel experiments.

1 INTRODUCTION

Finite element methods are now well established as an analysis tool in many engineering problems, and practical applications have been growing at prodigious rate in recent years. With the increase in modeling applications and computing codes, it becomes essential for users that finite element codes be well validated and the range of possible applications assessed in a reliable manner. Unfortunately validating and establishing the capabilities of a complex finite element model are no simple tasks.

In the field of geotechnical engineering, validation of FEM codes is of uppermost importance for three reasons :

- complex models are available and several analyses are possible (monotonous or cyclic loading, static and dynamic loading, seepage, porous mechanics) ;
- complex rheological laws are involved for modeling the behavior of soils and rocks ;
- results are used for the design of structures.

In common cases, we can identify two principal kinds of validation :

- validation of a finite element code by its authors or by experts in terms of computational correctness (verification of routines and reliability of algorithms, comparison with analytical solutions) ;
- validation of finite element models in terms of agreement between numerical results and experimental data (laboratory tests, centrifuge models, in situ measurements).

It is very important to distinguish between the computer code and the numerical model. With a very well validated code, an inexperienced user can make many mistakes.

2 VALIDATION USING MEASUREMENTS

Since 1970, many publications have described comparisons between measurements and finite element results : embankments, tunnels, foundations, dams, retaining structures, etc. Compilations have been presented by Duncan (1994) and by Gens (1995).

Two kinds of comparison can be distinguished (from Lambe, 1973) :

- blind tests or « class A prediction », when the measurements are not known ;
- other comparisons or « class B and C predictions », when some measurements are known.

A very few number of Class A predictions has been achieved and published during last years. In the field of embankment on soft soils, only four International Workshop on predictive behaviour have been organized : Boston (MIT, 1975), Gloucester (1987), Muar Flats (1989) and Haarajoki (1997). Prior to the construction, 4 to 13 experts predicted displacements and pore pressure evolution.

These comparisons have allowed a constant improvement of constitutive soil equations, modeling, finite element codes and may provide further examples for engineers or students.

The objectives of this paper are now to make some conclusions about the ability to predict the behaviour of embankments on soft soils (without drains) and tunnels and to provide recommendations for geotechnical modeling. For this purpose, the Soil and Rock Mechanics division of the Laboratoire

Central des Ponts et Chaussées (Paris, France) develops a bibliographical database which is named MOMIS. A first version has been achieved in April 2000.

3 OBJECTIVES OF MOMIS

The objectives of MOMIS (Modèle numérique d'Ouvrage et Mesures In Situ) are concerned with theoretical and practical aspects :

- by describing case studies, we can extract some lessons on the art of modeling (constitutive laws, mesh, type of elements, boundary conditions, etc.) ;
- by studying the results of finite element modeling, we can estimate the error between a " good model " and in situ measurements.

MOMIS contains the following informations :

- description of embankment or tunnel tests (location, dimensions, construction rate, etc.) ;
- geotechnical data on subsoils (identification tests, in situ tests and laboratory tests) ;
- types of modeling (constitutive laws, mechanical parameters, type of analysis, mesh, finite elements, construction sequences, time steps, etc.) ;
- differences between numerical results and measurements (relative errors on settlements, lateral displacements, excess pore pressure and time evolution) ;
- etc.

An other purpose of this database is to help engineers and searchers in their practical numerical modeling. The informations derived by the database may be useful for modeling an embankment or tunneling problem. The database might provide recommendations about :

- mesh (elements, density), increments of loading, time steps, limit conditions, etc. ;
- how to select appropriate constitutive equations ;
- how to identify properly mechanical and hydraulic parameters ;
- verification of the value of parameters and results from similar case studies ;
- etc.

4 STRUCTURE AND CONTENTS OF MOMIS

4.1 *Structure*

Today, the first version of the database has been constructed with Microsoft Word 97. It contains precise information related to numerical modeling. The data of each numerical model are entered in French, numbers and codes. The data are classified into the following categories :

1) type of analysis (drained static analysis, undrained static analysis, dynamics, consolidation, thermo-poro-mechanics)
2) location
3) soils
4) description of the embankment (or tunnel)
5) constitutive laws of materials and values of parameters
6) numerical model (name of the computer code, descriptions of finite element modeling, number and types of elements, construction sequence, etc.)
7) picture of the mesh (visual information)
8) reference
9) pictures for the comparison between FEM results and in situ measurements
10) estimation of modeling errors.

Each category must be described.

4.2 *Contents : Publications*

Today, about 150 well documented case studies have been captured in the first version of database : 76 for embankments tests (without drains) and 78 for tunnel experiments. These numbers allow representative analyses. As shown on figures 1 and 2, the publication starts from the seventies and seems to be stabilized around eight references each year.

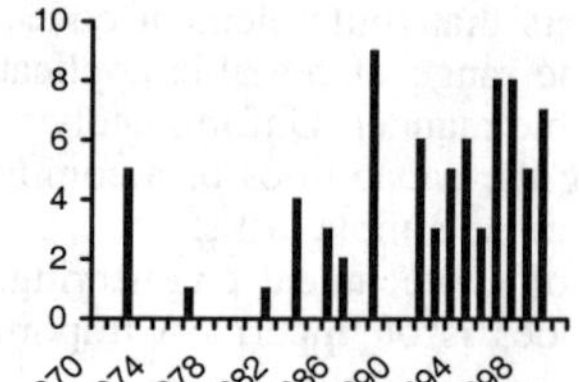

Fig. 1. Number of stored references about embankments

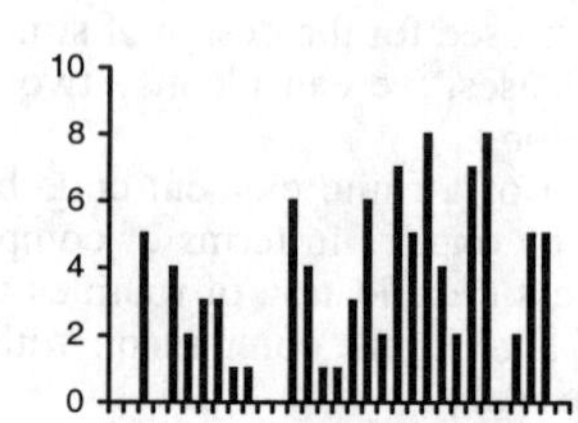

Fig. 2. Number of stored references about tunnels

It must be noticed that the papers published in the proceedings of the International conferences (ISSMGE, IACMAG, NUMOG) represents 65% of all the documents stored in the database. As it could be expected, between those conferences, the production falls down drastically. 25% of the documents are articles published in scientific journals and 10% deals with reports or thesis.

5 FIELDS OF RESEARCH

The stored data in MOMIS aim principally to establish an " art of modelling " based on the comparison between finite element results and in situ measurements. Today the database is a collection of files (one file for one FEM-measurements comparison). All the researches are manual. The next step will be to manage MOMIS with a current software and probably with a Microsoft Access application. This technological evolution could have deep effects on the organisation and the content of each file.

The fields of research are related to the stored informations : location, modeling, mesh, constitutive laws, type of soils, type of results (vertical or horizontal displacements, pore pressure, forces and moments in structural elements and their evolution with time).

6 DATA ANALYSIS FOR EMBANKMENTS

6.1 *Mesh properties*

Analyses of data have revealed some interesting relationships between the dimensions of the mesh and the characteristic length of embankment (height or length of the base) (figure 3). Lm is the maximal length of the mesh ; at this distance, horizontal boundary conditions are imposed. h is the height of natural soils which was considered in the analysis : it is a natural limit (bedrock) or a distance where vertical boundary conditions can be imposed.

The results shown in figure 4 demonstrate that the values of the ratio h/H vary between 1.4 and 15. The mean value is 4.

A similar analysis leads to figure 5. The values of the ratio Lm/Lb vary between 1.8 and 5.8. The mean value is about 3.

Figure 6 shows the increase of node number since 1970. This increase is relatively small because of the consolidation approach which associates a space and time resolution. The cost of computation remains expensive when there are many time steps. The number of nodes is still limited : for recent modeling a mean value seems to be 1500 (figure 6).

It must be noticed that all numerical models entered in the database are performed with the plane strain hypothesis.

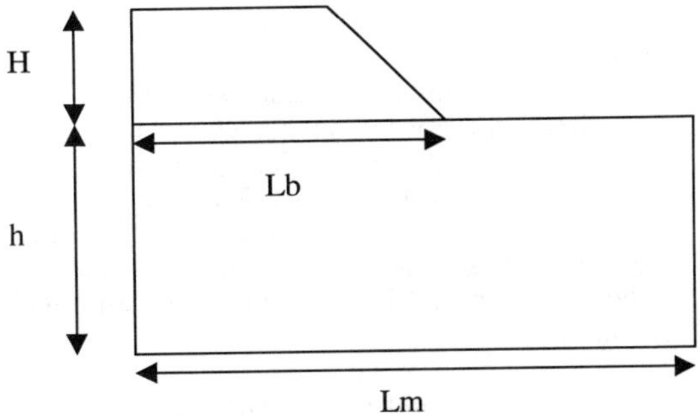

Fig. 3. Dimensions of a typical embankment

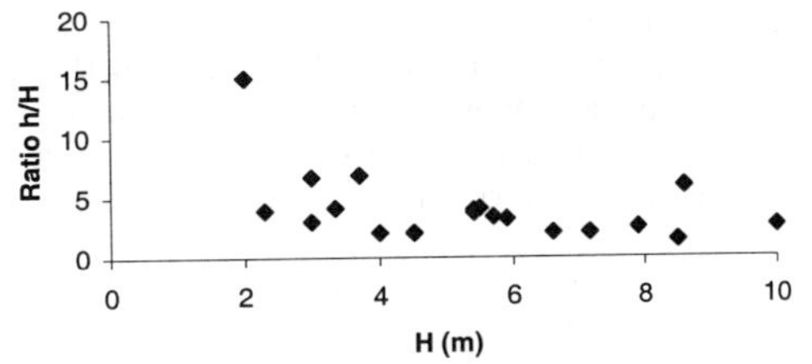

Fig. 4. Relationship between the ratio h/H and the height of embankments H

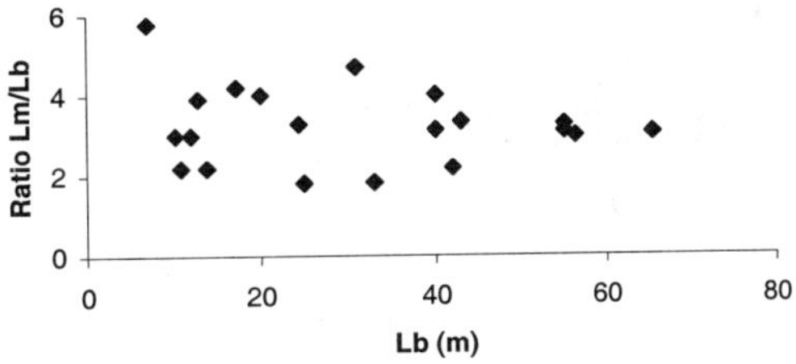

Fig. 5. Relationship between ratio Lm/Lb and width of embankment platforms

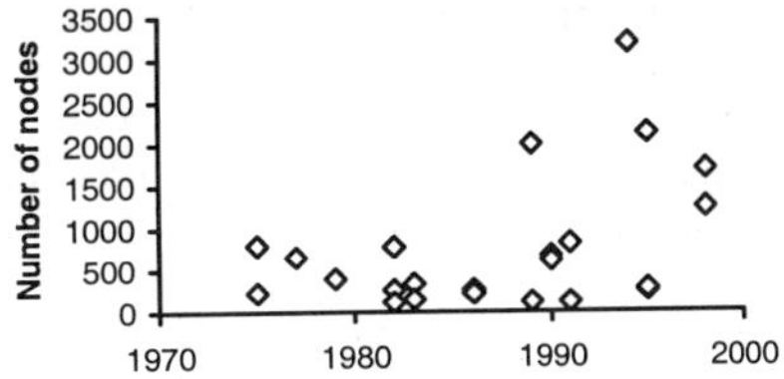

Fig. 6. Evolution of numbers of nodes

6.2 *Constitutive laws for natural soils*

Three kinds of numerical approach are considered in MOMIS : undrained condition (23 %), drained condition (14 %) and consolidation (63 %).

The constitutive laws used for soft soils are essentially :

- linear and nonlinear elasticity (11 %) ;
- elastoplasticity without hardening (14.5 %) ;
- elastoplasticity with hardening (47 %) ;
- elastoviscoplasticity (27.5 %).

Among elastoplastic laws the modified Cam-Clay model is the most popular constitutive law (35 % of all the references and 73 % of all elastoplastic with hardening laws).

6.3 *Mechanical parameters for natural soils*

Mechanical parameters of constitutive laws are also stored in the database. For example, we can deduced the minimal and maximal values for elastic parameters of Cam-Clay model :

$$0.025 < \lambda < 1.9 \text{ and } 0.005 < \kappa < 0.19.$$

Unfortunately, it was impossible to obtain good correlations between these two parameters. Figure 7 shows a large scaterring for the ratio λ / κ.

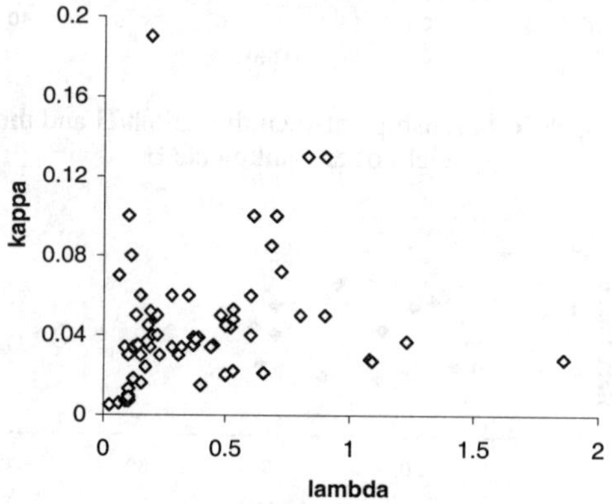

Fig. 7. Parameters λ and κ for the Cam-Clay model (extracted from the database MOMIS)

6.4 *Embankment modeling*

The embankment is represented

- by a trapezoidal pressure applied on the ground level (33 %) ;
- by finite elements with a constitutive law and its weight (67 %). The corresponding mesh can be activated following a construction sequence.

Among constitutive laws, linear elasticity is the most popular (53 %) followed by perfect plasticity (36 %) and non linear elasticity (11 %). Today, an embankment is very often represented by finite elements. For granular fills, the Young's modulus varies between 5 and 50 MPa and the Poisson's ratio between 0.3 and 0.4.

6.5 *Finite element codes*

The most popular finite element code is SAGE-CRISP (15 %), followed by DACSAR (8 %), ROSALIE-LCPC (7 %), PLAXIS (7%), AFENA (7 %) and ABAQUS (7 %).

6.6 *Comparison with in situ measurements*

The finite element class B and C predictions gave relatively good results for the comparison between computed and measured settlements on surface below the embankment axis (figures 8 and 9). Many results are approximately centered on the bisectrix. This fact confirms us in the belief that satisfactory settlement predictions can be achieved nowadays using finite element analyses and hardening elastoplasticity. For some cases, the use of elastoviscoplasticity is necessary for a good representation of time evolution.

For class A predictions (Muar Flats, Gloucester), we can estimate the mean ratio of numerical results/measurements : it varies from 1 to 5 for immediate settlements ; 1 to 1.7 for failure height and 1 to 2.5 for pore pressure.

The quality of pore pressure predictions can be also accepted, even if there is a difficulty to determine hydraulic parameters.

On the other hand, horizontal displacements are frequently poorly predicted in the present state of practice.

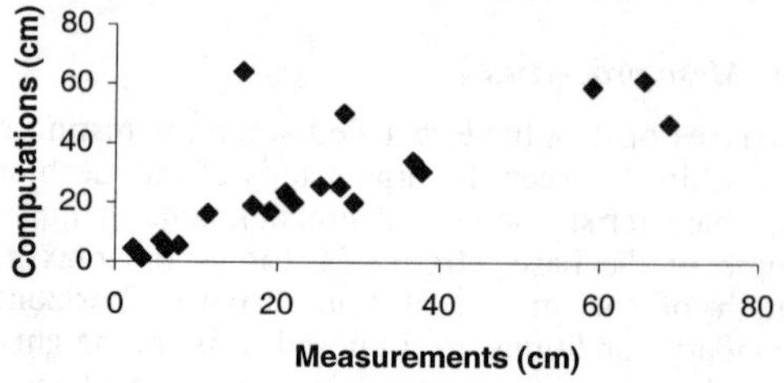

Fig. 8. Comparison between computed settlements and measurements at the end of construction

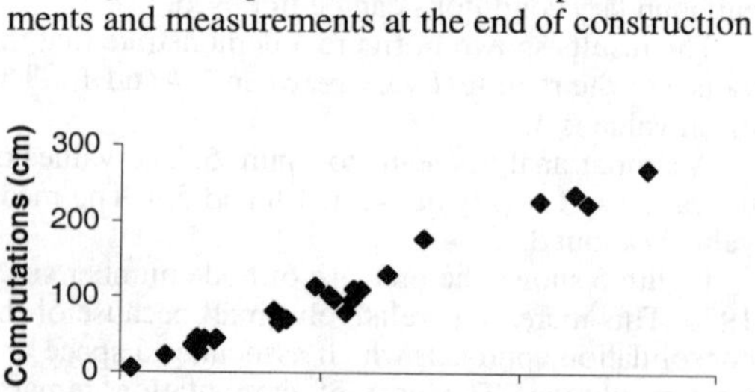

Fig. 9. Comparison between computed settlements and measurements for a long time period

7 DATA ANALYSIS FOR TUNNELS

7.1 *Mesh properties (plane strain model)*

A similar analysis can be made for tunnel modeling. Interesting relationships are obtained between the dimensions of the mesh (figure 10) and the characteristic lengths of tunnel for a plane strain model (diameter D, heigth of recovering H).

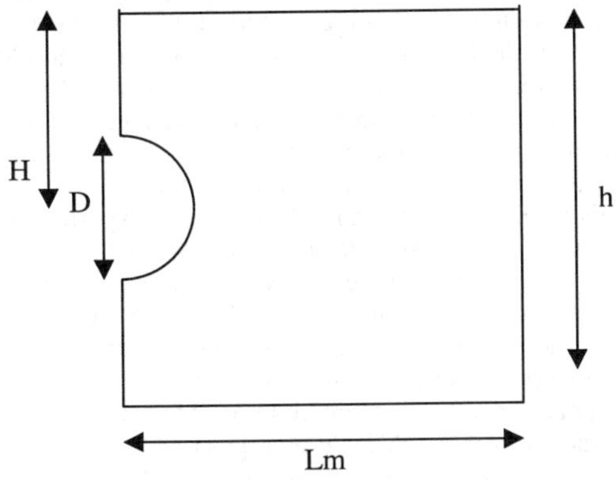

Fig. 10 Dimensions of tunnel model (plane strain)

Figure 11 shows that the values of the ratio Lm/D vary between 2 and 17.7. The mean value is around 5. For the vertical direction, figure 12 indicates that the ratio h/H presents a mean value equal to 2.5.

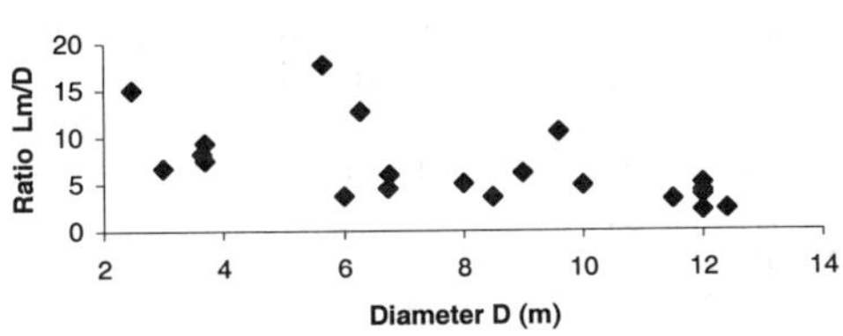

Fig. 11. Relationship between the ratio Lm/D and the diameter of the tunnel D

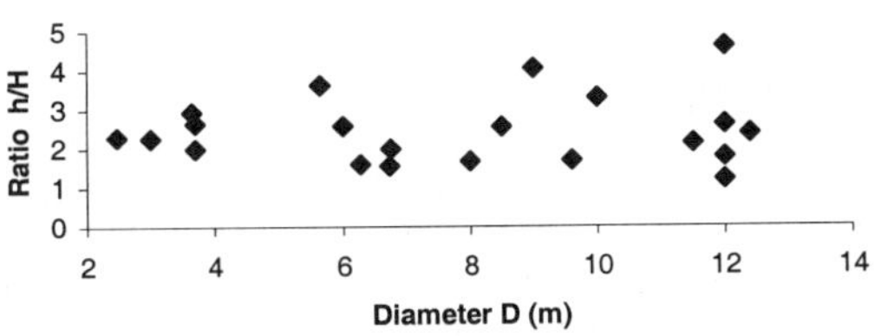

Fig. 12. Relationship between the ratio h/H and the diameter of the tunnel D

These values indicate some basic rules for the construction of tunnel meshes in plane strain. Different kinds of model are entered in the database :

- plane strain model (cross section) : 80 % ;
- plane strain longitudinal model : 17 % ;
- axisymetric model : 1.5 % ;
- 3D model : 1.5 %.

7.2 *Constitutive laws for natural soils*

Three theoretical models are considered : undrained condition (10 %), drained condition (75 %) and consolidation (15 %). The constitutive laws used for natural soils are essentially :

- linear and non linear elasticity (18 %) ;
- elastoplasticity without hardening (37 %) ;
- elastoplasticity with hardening (40 %) ;
- elastoviscoplasticity (5 %).

Unlike embankments on soft soils, it must be emphasized that the part of elastoplasticity without hardening is important. This approach is used for sand and gravel. This demonstrates that satisfactory results can be obtained with the Mohr-Coulomb or Drucker-Prager criterion associated with linear elasticity.

This fact is also readily explained by the weak proportion of clayey soils stored in the case of tunnel modeling.

7.3 *Lining modeling*

The lining is represented :

- by equivalent pressures ;
- by equivalent displacements ;
- by finite elements (solid elements or beams) with a constitutive law and its weight.

The behaviour of concrete lining is frequently isotropic and linear elastic. For shotcrete lining, the Young's modulus varies between 3 500 and 30 000 MPa. This value depends on the age of concrete. The Poisson's ratio is equal to 0.2.

7.4 *Finite element codes*

Only 50 % of our references indicate the name of the code used for computations. It is therefore not surprising that a statistical analysis is not representative. Nevertheless, we can cite the most popular codes : CESAR-LCPC (36 %), EPTUN (10 %), CRISP (8 %), ABAQUS (5 %), ICFEP (5 %), PLAXIS (5 %).

7.5 *Comparison with in situ measurements*

Different numerical results can be compared to measurements :

- surface settlements on the tunnel axis ;
- settlement trough ;
- vertical movements in depth ;
- horizontal displacements ;
- forces and moments in the lining.

Nowadays we have only informations about surface settlements on the tunnel axis and comparison with results of plane strain models (figure 13).

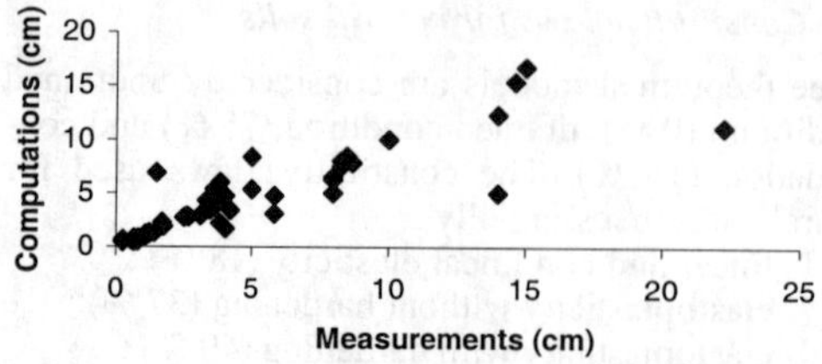

Fig. 13. Comparison between computed maximal surface settlements and measurements (at the end of construction)

Many results are approximately centered on the bisectrix. 50 % of references are optimistic.

The foregoing analysis makes no allowance for the specific geotechnical problems. All references (only class C predictions) are represented in figure 13. A more refined research can be made by type of soils, construction method (NATM for example), type of elements, constitutive laws, plane strain model, 3D model and other points of modeling.

8 ERROR AND MODELING

The analysis of FEM results and in situ measurements can be used to provide a quantitative estimation of modeling error. Class A predictions are the most conclusive but unfortunately there are a few number of cases. It will be useful to collect more informations on this type of benchmarks or to organize new experiments. Class B and C predictions confirm that satisfactory results can be achieved using finite element method.

For embankments on soft soils, settlements at the end of construction are relatively well predicted : 80 % of the case studies stored in MOMIS have a relative error which is less than 40 %. For long term settlements, 88 % of the references lead to a relative error less than 20 % and the maximal relative error is 30%. The long term settlement is more easy to compute than the movement at the end of construction. Similar analyses will be made for pore pressures and horizontal displacements which are generally overpredicted.

For tunnels, the geotechnical problem is more complex because of soil-structure interactions. Class B and C predictions indicate that the surface settlements are reasonably well estimated. 84 % of references provide a relative error less than 50 %.

The estimation of a quantitative modeling error for tunnels needs a more complete study. This paper is only the first step of our project. The next step will be to manage the database with a software. The analysis will be achieve more easily.

9 CONCLUSION

The Soil and Rock Mechanics division of the Laboratoire Central des Ponts et Chaussées (France) has developed a bibliographical database (MOMIS). It is the outcome of a project started two years ago. MOMIS deals with comparisons between finite element predictions and in situ measurements. Nowadays, 25 years of numerical modeling and 150 well documented case studies have been stored in the first version of database : 76 for embankments on soft soils (without drains) and 78 for tunnel experiments. We will update MOMIS and new data will be entered about embankments with drains, retaining structures and reinforced soil. The objectives are to save the practice of numerical modeling and to help engineers and searchers in their study.

This paper has presented some interesting results extracted from MOMIS. It emerges that :

- general rules can be deduced for the construction of embankment and tunnel meshes. Examples have been given on dimensions of mesh. We can also obtain informations about types of element, mesh density, time steps, initial and boundary conditions ;
- for soft soils the Modified Cam-Clay model remains the most popular model ;
- for granular soils the Mohr-coulomb or Drucker-Prager failure criterion associated with linear elasticity is frequently used ;
- comparisons between FEM results and in situ measurements lead to an estimation of error modeling.

Many other points can be investigated from the database. We are in the beginning of the development and the use of MOMIS. The final aim is to offer to the international community the possibility to consult MOMIS and to complete it with well documented case studies.

10 REFERENCES

Duncan J.M. 1994. The role of advanced constitutive relations in practical applications. *14th ICSMFE*, New Delhi, 5:31-48.

Gens A. 1995. General report : prediction, performance and design. *Pre-failure Deformation of Geomaterials*, Shibuya et al. (eds), Balkema, 2:1233-1256.

Lambe T. W. 1973. Prediction in soil engineering. *Géotechnique*, 23:149-202.

Mestat Ph., Riou Y. 1999. À propos des benchmarks en géotechnique. *Revue Française de Génie Civil*, 3:657-672.

Computer Methods and Advances in Geomechanics, Desai et al. (eds) © 2001 Balkema, Rotterdam, ISBN 90 5809 183 X

3D FE analysis on slurry trench stability and active earth pressure

Pavol Oblozinsky & Keizo Ugai
Department of Civil Engineering, Gunma University, Kiryu, Japan

Masaaki Katagiri, Kunio Saitoh & Takeshi Ishii
Nikken Sekkei Nakase Geotechnical Institute, Kawasaki, Japan

Shuji Tomita & Kiyoshi Kuwabara
East Japan Railway, Tokyo, Japan

ABSTRACT: Series of calculations on slurry trench stability in fully saturated sand was carried out, based on the FEM shear strength reduction technique under the three dimensional conditions. The influence of two parameters on the safety factor of 15m deep trench supported by the slurry was investigated. The length of the trench, which was considered 3m, 6m and 9m respectively, and the height of the slurry filling above the ground level, which varied from 0.0 m to +0.75 m. The influence of the guide wall and the shape of the excavation on the safety factor were also investigated.
Active earth pressure acting on the face of the trench excavation was evaluated based on the FE calculation and a simple graphical method for determining the slurry density and/or height of filling is proposed herein. The method consists in a comparison of the calculated active earth pressure and the hydrostatic pressure developed in the slurry.

1 INTRODUCTION

Over the recent decades the diaphragm walling industry has developed rapidly due to the needs of constructing the excavations and underground structures in the highly urbanized areas. Trenches supported by the slurry, often used in foundation engineering, have appeared to be very convenient and cheap mean for constructing the diaphragm walls. The stability of the slurry trench has been analyzed and modelled by many researchers. Conventionally, the safety factor is evaluated by LEM method. Nash and Jones (1963) have proposed a 2D LEM method for analyzing the slurry trench stability problem, which was extended by Morgenstern and Amir-Tahmasseb (1965). Piaskowski and Kowalewski (1965) have introduced an idealized shape of the 3D sliding wedge in their pioneering work on investigation of slurry trench stability under the 3D condition. Washbourne (1984) has suggested the shape of 3D sliding wedge for the deep and shallow slurry trenches and analyzed the stability of the wedge in the same manner as the Coulomb wedge was analyzed in the 2D retaining wall problem. Tsai and Chang (1996) have presented an improved 3D LEM method for slurry trench analysis in the cohesionless soil in which the shell-shaped sliding surface is determined by using the Mohr-Coulomb criterion.

The aim of this paper is to report on the FEM analysis on the slurry trench stability problem. The shear strength reduction technique was employed to evaluate the safety factor of the deep trench supported by the slurry. An active pressure analysis is demonstrated here as a method, which might be used to determine the design parameters of the slurry; the slurry density and/or the height of the slurry filling.

2 CALCULATION METHOD

The shear strength reduction technique was used to evaluate the safety factor of slopes as early as 1975 by Zienkiewicz, later applied by Naylor (1982), Matsui and San (1992). The validity of the shear strength reduction technique was discussed and demonstrated by Ugai and Leshchinsky (1995), who compared the stability of a simple 3D vertical cut analyzed by FEM and LEM and similar results were reported.

The essence of the SSR technique is in reducing the soil strength parameters till the soil fails. The soil strength parameters are defined as follows:

$$c_F = \frac{c}{F}; \qquad \phi_F = \tan^{-1}\left(\frac{\tan\phi}{F}\right) \tag{1}$$

where F is a parameter which reduces the soil strengths. The parameter, F, is a variable, which is increased in the loop in the calculation process. Initially, the value of F is sufficiently small, thus the strength parameters c_F and ϕ_F are of high values and the system is in the elastic conditions. As the value of F is increased within the loop, the soil strength parameters c_F and ϕ_F are reduced. The loop is repeated until the equilibrium of the system cannot be satisfied. The criterion is the number of the iteration needed to outbalance the system. A remarkable increase in the number of iterations needed to satisfy the equilibrium indicates the collapse of the system. At this instant, the value of the parameter, F, comes to be the global minimal safety factor, F_S, with the same meaning as the safety factor defined in LEM.

In FEM analyses, the isoparametric 20 node brick element was employed, with eight Gaussian points for numerical integration. The elasto-perfectly plastic model was adopted in which the failure criterion is governed by the Mohr-Coulomb equation:

$$f = \frac{1}{2}\left[3(1-\sin\phi)\cdot\sin\theta + \sqrt{3}(3+\sin\phi)\cdot\cos\theta\right]\cdot \sqrt{J_2} - I_1\sin\phi - 3c\cos\phi = 0 \tag{2}$$

and the plastic potential obeys the Drucker-Prager equation:

$$\Phi(I_1, J_2) = \sqrt{J_2} - \alpha I_1 \, (\alpha > 0) \tag{3}$$

In Equations 2 and 3,

$$\theta = \frac{1}{3}\cos^{-1}\left(\frac{-3\sqrt{3}}{2}\cdot J_3 / J_2^{3/2}\right)(0 < \pi < \frac{\pi}{3})$$

$$I_1 = \sigma_x + \sigma_y + \sigma_z = 3\sigma_m,$$

$$I_2 = \sigma_x\sigma_y + \sigma_y\sigma_z + \sigma_z\sigma_y - \tau_{xy}^2 - \tau_{yz}^2 - \tau_{zx}^2,$$

$$I_3 = \sigma_x\sigma_y\sigma_z + 2\tau_{xy}\tau_{yz}\tau_{zx} - \sigma_x\tau_{yz}^2 - \sigma_y\tau_{zx}^2 - \sigma_z\tau_{xy}^2,$$

$$J_2 = \frac{1}{2}\left((\sigma_x - \sigma_m)^2 + (\sigma_y - \sigma_m)^2 + (\sigma_z - \sigma_m)^2\right) + \tau_{xy}^2 + \tau_{yz}^2 + \tau_{zx}^2,$$

$$J_3 = I_3 - \frac{1}{3}I_1 I_2 + \frac{2}{27}I_1^3, \qquad \alpha = \frac{\tan\varphi}{\sqrt{9 + 12\tan^2\varphi}}$$

where φ is the dilatancy.

3 FEM ANALYSES

The slurry trench problem is schematically shown in Figure 1. Due to the symmetry of the analyzed problem, only half of the trench length was modelled. As it is indicated in Figure 1, the slurry level was put above the ground surface. The difference between the ground surface and the slurry level is denoted as H_{sl}. The ground surface is divided into two parts, area A1 and area A2, separated by the dash line. Area A1 models the trench wall's face and the sand behind the trench face. This simplified trench model was used in the FEM analyses. The ally of area A1 and area A2 also models the trench width, this was used when the influence of the excavation shape on the safety factor was investigated.

The supporting pressure developed in the slurry was substituted with the equivalent nodal forces. The analyses were carried out in the sense of effective pressure, the underground water table was put at the ground surface. The soil parameters are given in Table 1.

The stability of the slurry trench was investigated using the simplified model trench (area A1 in Fig. 1), in which the trench length was 3, 6 and 9m respectively and 2D case. The depth was 15m in all cases. The height of the slurry filling, H_{sl}, varied from 0.0 to 0.75m above the ground surface. The results are shown in Figure 2 as the plot of the safety factor against the height of filling, H_{sl}. The horizontal line at Fs=1.0 divides the graph into the safe and unsafe

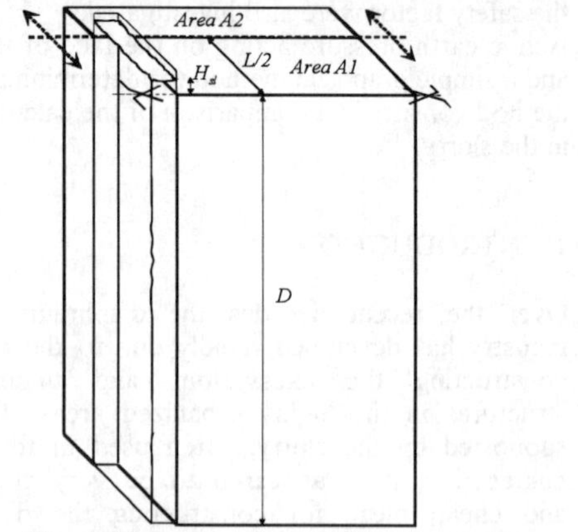

Fig. 1 Slurry trench model

zone. As the guide wall is often essential when constructing the deep trench excavation, the influence of the guide wall on the safety factor was investigated too. The dash lines in Figure 2 show the contribution of the guide wall to the safety factor. As it is seen, the guide wall will enhance the safety factor slightly. The influence is higher when the slurry filling is close to the ground level, up to H_{sl}=0.25m. At the higher height of filling, the influence of the guide wall is almost neglecting.

Bell-Washbourne method (Washbourne, 1984)

Table 1. Soil properties for FEM

	Soil properties:
Submerged unit weight	$\gamma' = 8.7$ kNm^{-3}
Friction angle	$\phi' = 39°$
Dilatancy angle	$\varphi' = 39°$
Cohesion	$c = 0$ kNm^{-2}
Young modulus	$E = 20.000$kNm^{-2}
Poisson ratio	$\nu = 0.3$

was used for evaluating the safety factor of the trenches 6 and 9m long and 15m deep. The results are also shown in Figure 2. As it is seen, the Bell-Washbourne method gives higher values of the safety factor, what may be attributed to the shape of the sliding wedge. The shape of sliding wedge is predefined in the Bell-Washbourne method, having the shape of the prism with the triangular base. This is not necessarily the critical shape of the sliding wedge.

4 SHAPE EFFECT

The arching effect has got a great influence on the stability of any excavation. In order to investigate the influence of the arching effect around the excavation supported by the slurry, a set of analyses was carried out, in which four different excavation shapes were used. A slurry trench of the length of 6m, which was modelled with the area A1 in Figure 1, was compared with the slurry trench 6x1m (length/width) modelled with the area A1+A2. As it was believed that the arching effect around a symmetric excavation would be developed better than around an unsymmetrical one, therefore the square 6x6m and the circular, R=6m, excavations supported by slurry were analyzed too. The depth was 15m in all cases. The results are shown in Figure 3. As it was expected, the arching effect is developed best around the circular excavation, what increases the safety factor and the circular excavation is the most stable one in this comparison. The square excavation, 6x6m, is slightly more stable than the 6x1m trench, what is attributed to more favorable stress distribution around the symmetric excavation due to the arching effect. The less stable is the case in which only the trench face was modelled (area A1 in Fig. 1). However, the difference between the square and rectangular excavation and the case when only trench face was modelled is very small. The author is of the opinion that modelling the slurry trench with area A1 in Figure 1 would be satisfactory enough for design purposes.

5 ACTIVE EARTH PRESSURE ANALYSIS

The FEM analyses were performed in the following fashion: The face of the excavation was subjected to the prescribed displacement directed into the excavation opening. A relatively small horizontal displacement, about 5cm, will cause the development of the active stage in the soil and the reduction in the horizontal stress around the trench, from the initial (K_o) state to the active state. Figure 4 shows the development of active earth pressure acting in the middle of the trench, where the maximal horizontal (bulk) displacement might be expected and thus, the requirement for the support is the highest. The analyses were performed for a trench length of 3, 6 and 9m respectively.

In order to provide the stability of the trench, the horizontal pressure acting on the face of the excavation from the slurry side must be equal to or higher than the active earth pressure acting from the soil side. The dot line in Figure 4 represents the horizontal pressure developed in the slurry reduced by the water pressure, $p_{sl}=H\cdot(\gamma_{sl}-\gamma_w)$, ($\gamma_{sl}$ -unit weight of slurry, γ_w -unit weight of water). Unit weight of the slurry was, $\gamma_{sl}=10.5\ kNm^{-3}$, in this case. If we drew a parallel line with the dot line and tangent to the curve of the active pressure, the intersection of the tangent and the x-axis gives us the magnitude of pressure, p, which needs to be supplied by the slurry at the ground surface. Knowing this value we can determinate the required height of the slurry filling from the equation $p=\gamma_{sl}\cdot H_{sl}$, in which H_{sl} means the height of the slurry above the ground surface and γ_{sl} means the unit weight of the slurry.

If we need to obtain the density of the slurry based on the known or prescribed height of filling, H_{sl}, we can determinate the density of the slurry using the same graph in similar fashion. First we draw the

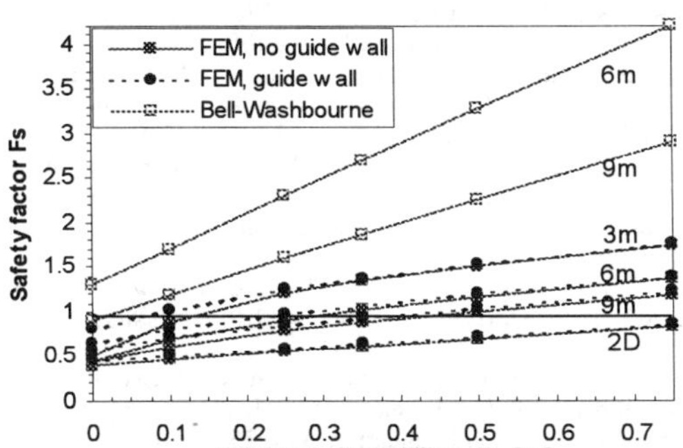

Fig. 2 Safety factor against the height of the slurry filling above the ground

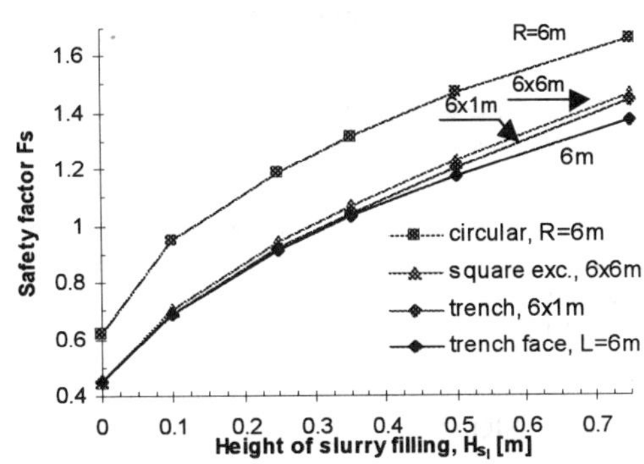

Fig. 3 Stability of the excavations of the different shape, supported with the slurry.

Table 2. Comparison of designed height, H_{sl}, of slurry filling, γ_{sl}=10.5 kNm^{-3}

	Length L=3m		Length L=6m		Length L=9m	
	ΔH	Fs	ΔH	Fs	ΔH	Fs
Prediction by SSR-FEM	15.5 cm	1.00	32.5 cm	1.00	50.0 cm	1.00
Prediction by active earth pressure approach	14.2 cm	0.98	38.1 cm	1.06	77.0 cm	1.10

tangent to the curve of active pressure representing the horizontal pressure in the slurry reduced by the water pressure, $p_{sl}=H\cdot(\gamma_{sl}-\gamma_w)$. The intersection of the tangent and x-axis gives us the required pressure, p, which needs to be supplied by slurry at ground level. Knowing the given height of the filling, H_{sl}, and the pressure in the slurry at the ground level, p, we can obtain such unit weight of the slurry, γ_{sl}, that the equation $p=\gamma_{sl}\cdot H_{sl}$ holds.

Table 2 shows the height of the slurry filling, H_{sl}, predicted by the SSR-FEM when the safety factor, Fs=1.00 for trench length 3, 6 and 9m, respectively. The second row of Table 2 shows the height of the slurry filling, H_{sl}, predicted by the active earth pressure FEM analysis and the safety factor which was evaluated by SSR-FEM for H_{sl} predicted by the active earth pressure analysis. The safety factor of the slurry trench with the height of slurry filling predicted by the active earth pressure analysis is in a reasonably good range, from 0.98 to 1.10. If an engineer chooses this method for determining the preliminary parameters of the slurry, the unit weight and height of filling, it may be expected that the

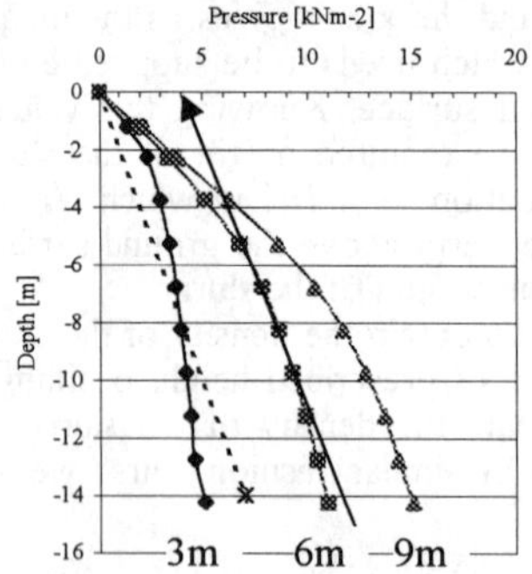

Fig. 4 Active earth pressure acting on the excavation face in the middle of the trench

safety factor would be around the value of 1.00. Then, the slurry parameters might be adjusted in accordance with the required minimal value of the safety factor.

6 CONCLUSIONS

The 3D FEM analyses of the stability of the slurry trench based on the shear strength reduction technique has been performed herein. The calculation has been done for three different lengths of the trench, 3, 6 and 9 m, respectively, and the depth of 15m. The conclusions are as follows:

1. The safety factor F_s of the slurry trench is proportional to the height of the slurry filling, and the safety factor decreased with the increasing length of trench. The SSR-FEM yields lower safety factor compared with the Bell-Washbourne's method.
2. The influence of the guide wall on the safety factor is very small, slightly noticeable up to the height of filling, H_{sl}=0.25, and almost negligible at higher height of the filling.
3. The influence of the excavation shape on the safety factor was investigated. It was found that circular excavation is the most stable and the trench modelled by area A1 is the less stable. This is due to the fact that the arching effect around the symmetric excavation is better developed than around unsymmetrical one.
4. The active earth pressure analysis may be used for the determination of the slurry properties, the density and/or the height of slurry filling. It was shown, that the safety factor from 0.98 to 1.10 might be expected when the slurry properties are designed via active earth pressure analysis.

REFERENCES:

Katagiri, M., Saitoh, K., Masuda, T., Aizawa, F., Ugai, K., 1997, "Centrifuge model tests on stability of slurry trenches", Journal of the JGS, Vol. 45, No. 10:17-19 *(in Japanese)*

Matsui, T., San, K-C., 1992, "Finite element slope stability analysis by shear strength reduction technique", Soils and Foundation, Vol. 32, No. 1:59-70

Piaskowski, A., Kowalewski, Z., 1965, "Application of thixotropic clay suspension for stability of vertical sides of deep trenches without strutting", Proceedings 6th International Conference on Soil Mechanics and Foundation Engineering, Montreal, Vol. 2: 526-529

Tanaka, T., Ugai, K., Kawamura, M., Sakajo, S., Ohtsu, H. "Three-dimensional finite element method analysis of elasto-plastic problems in soil mechanics", Maruzen, 1996 *(in Japanese)*

Tsai, J-S., Chang, J-C, 1996, "Three-dimensional stability analysis for slurry-filled trench wall in cohesionless soil", Can. Geotech. J., 33: 798-808

Ugai, K., Leshchinsky, D., 1995, "Three-dimensional limit equilibrium and finite element analyses: a comparison of results", Soils and Foundation, Vol. 35, No. 4: 1-7

Washbourne, J., 1984, "The three-dimensional stability analysis of diaphragm wall excavation", Ground Engineering, 17(4): 24-29

Computer Methods and Advances in Geomechanics, Desai et al. (eds) © 2001 Balkema, Rotterdam, ISBN 90 5809 183 X

Progressive failure of cut slope in anisotropic ground

T.Adachi, F.Oka, H.Kobayashi & S.Kimoto
Department of Civil Engineering, Kyoto University, Japan

F.Zhang
Department of Civil Engineering, Gifu University, Japan

ABSTRACT: Due to the anisotropic microstructure formed in sedimentary process, a soft sedimentary rock usually shows an anisotropic behavior in its strength, deformation and dilatancy. In present paper, by introducing Boehler's anisotropic theory and generalized Hooke's theory, an isotropic constitutive model with strain softening is extended to an anisotropic model which can not only describe the strain softening behavior, dilatancy but also the anisotropic behavior of soft rock. The validity of the proposed model is verified by a series of laboratory tests. Then, based on the model, a soil-water coupled finite element analysis is conducted to investigate the progressive failure of a cut slope of the soft sedimentary rock under different anisotropic directions. The difference in the long-term stability due to the different anisotropic direction is estimated quantitatively by the coupled finite element analysis proposed in this paper.

1 INTRODUCTION

It is commonly known that soft sedimentary rock can be linked to many geotechnical engineering problems, such as the instability of cut slopes and foundations. Generally speaking, the mechanical behavior of soft sedimentary rock is elasto-plastic, dilatant, strain hardening-strain softening and time dependent. Compared to other geological materials formed in the same epoch, the void ratio is relatively large and a special structure formed during sedimentation. Its mechanical behavior during shearing is largely dependent on the confined stress and the pore-water pressure. The cementation existed in the structure, deteriorates due to the breakdown of the structure. Various processes, such as large shearing deformation, cyclic drying-wetting or stress release cause such a breakdown. The softening behavior of soft sedimentary rock becomes a very important factor in the long-term stability of cut slopes

It is known that progressive failure of cut slopes is usually caused by the following two factors, namely, (a) the deterioration of the structure of geologic materials due to the swelling and the weathering during and after the cut of the slopes, and (b) a reduction in the apparent shear strength due to the dissipation of negative pore-water pressure caused by rapid excavation of the cut slope.

Meanwhile, a soft sedimentary rock usually shows an anisotropic behavior in its strength, deformation and dilatancy due to the anisotropic microstructure formed in sedimentary process.

Adachi et al. (1994) proposed an elasto-viscoplastic model that can not only describe the time dependency but also the strain softening of geologic materials and a finite element analysis was conducted for a cut slope under drained condition to investigate the progressive failure of the cut slope. Adachi et al. (1999) also conducted a soil-water coupled finite analysis based on a strain softening elasto-plastic model to investigate the instability of a cut slope. In this paper, by introducing Boehler's anisotropic theory and generalized Hooke's theory, an isotropic constitutive model with strain softening (Adachi and Oka, 1995) is extended to an anisotropic model which can not only describe the strain softening behavior, dilatancy but also the anisotropic behavior of soft rock. The validity of the proposed model is verified by a series of laboratory tests. Based on the model, a soil-water coupled finite element analysis is conducted for a cut slope under different sedimentary directions. The difference in the long-term stability due to the different sedimentary direction is estimated quantitatively through the analysis proposed in this paper.

2 ANISOTROPIC ELASTOPLASTIC MODEL WITH STRAIN SOFTENING AND FEM PROCEDURE

In this section, a soil-water coupled consolidation analysis by finite element method is conducted to analyze the progressive failure of a cut slope with different sedimentary directions. For a boundary-value problem related to the soil-water coupled analysis, a backward finite difference scheme is adopted for water flow in the analysis based on Biot's theory. The equilibrium equation in a finite element scheme can be given as follow,

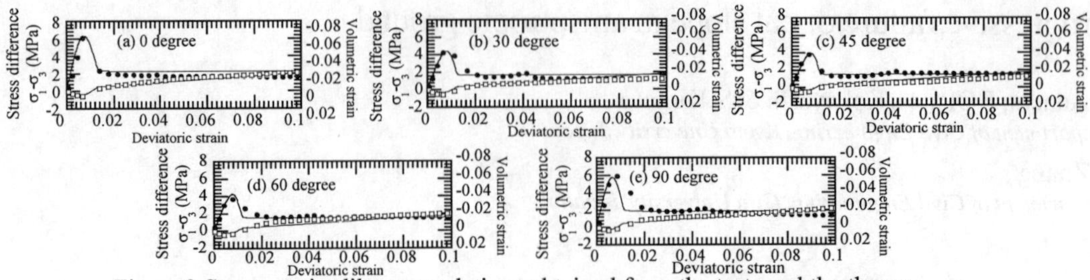

Figure 2 Stress-strain-dilatancy relations obtained from the tests and the theory

$$\begin{bmatrix} K & L \\ L^T & -\alpha \end{bmatrix} \begin{Bmatrix} d\delta \\ u_e|_{t+dt} \end{Bmatrix} + \sum_i^4 \alpha_i u_{ei}|_{t+dt} = \begin{bmatrix} [dF]+[dF_R]+[L]\{u_e|_t\} \\ 0 \end{bmatrix} \quad (1)$$

where

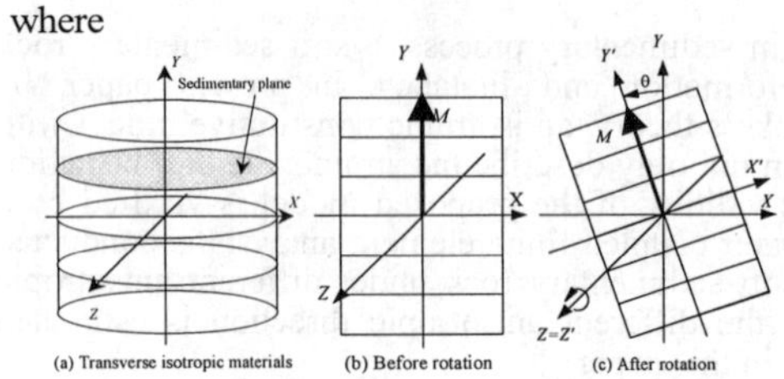

Figure 1 Structural anisotropy and its description

$$[K] = \int_V [B]^T [D][B] dV \quad (2)$$

$$[L] = \int_V [B]^T [M] dV \text{、} [M]^T = [1 \;\; 1 \;\; 0] \quad (3)$$

$$[dF_R] = \int_V [B]^T \{d\sigma_R\} dV \text{、} \{d\sigma_R\} = [D]\{d\varepsilon^p\} \quad (4)$$

$C=[D]^{-1}$ is the flexibility matrix of anisotropic materials in the transverse symmetric plane around y axis, as shown in Figure 1, and can be evaluated with the following equation as,

$$[C] = \begin{bmatrix} 1/E_x & -\nu_{yx}/E_y & -\nu_{xz}/E_x & 0 & 0 & 0 \\ -\nu_{yx}/E_y & 1/E_y & -\nu_{yx}/E_y & 0 & 0 & 0 \\ -\nu_{xz}/E_x & -\nu_{yx}/E_y & 1/E_x & 0 & 0 & 0 \\ 0 & 0 & 0 & 1/2G_{yx} & 0 & 0 \\ 0 & 0 & 0 & 0 & 1/2G_{yx} & 0 \\ 0 & 0 & 0 & 0 & 0 & (1+\nu_{zx})/E_x \end{bmatrix} \quad (5)$$

where, in general, five independent parameters, E_x, E_y, ν_{xz}, ν_{yx}, G_{yx} are needed to be determined. The anisotropic model which can describe the strain softening and the anisotropic behavior of soft rock, extended by introducing Boehler's anisotropic theory and generalized Hooke's theory into original Adachi-Oka's model (Adachi and Oka, 1995), can be expressed as

$$d\varepsilon_{ij}^p = \Lambda \left[\frac{\overline{\eta}_{ij}}{\eta} + (\overline{M} - \overline{\eta}) \frac{\delta_{ij}}{3} \right] \left[\frac{\eta_{kl}^*}{\eta^*} - \eta^* \frac{\delta_{kl}}{3} \right] \frac{d\hat{\sigma}_{kl}^*}{\hat{\sigma}_m^*} \quad (5)$$

$$\Lambda = M_f^{*2} / G' / (M_f^* - \eta^*)^2 \text{、} \overline{\eta}_{ij} = \hat{s}_{ij} / (\hat{\sigma}_m + b) \quad (6)$$

where the stress history tenser is given by

$$\sigma_{ij}^* = \sigma_{ij0} + \int_0^z \frac{1}{\tau} \exp\left(-\frac{z-z'}{\tau}\right)\left(\sigma_{ij}(z') - \sigma_{ij0}\right) dz' \quad (7)$$

$\hat{\sigma}$ is the Boehler's transformed stress tensor based on a simple anisotropic theory and is expressed as

$$\hat{\sigma} = \{(\alpha+\gamma-2\beta) tr(\hat{M}\sigma)\}\hat{M} + \gamma\sigma + (\beta-\gamma)(\hat{M}\sigma + \sigma\hat{M}) \quad (8)$$

where $\hat{M}$ is the structural tensor and can be given by the following matrix as

$$[\hat{M}] = \begin{bmatrix} 0 & 0 & 0 \\ 0 & 1 & 0 \\ 0 & 0 & 0 \end{bmatrix} \quad (9)$$

It is obvious that Equation 8 gives a simplified Boehler's anisotropic model in which three plastic anisotropic parameter α, γ, β are used.

Table 1 Material parameters of Ohya Stone

$E_x(MPa)$	1015.6	$b\ (MPa)$	1.96
$E_y(MPa)$	899.5	$\sigma_{mb}\ (MPa)$	8.82
ν_{yx}	0.061	$\overline{M}_m$	0.50
ν_{xz}	0.27	τ	0.017
$G_{yx}(MPa)$	190.2	α	0.90
G'	6000.0	β	1.00
M^*_f	1.00	γ	1.10

Apart from Young's modulus and Poisson's ratio, 6 parameters are necessary for the original model. For the anisotropic model proposed in this paper, additional three elastic moduli and three plastic parameters are needed to be determined with conventional triaxial compression tests.

Figure 2 shows the comparison of the stress-strain-dilatancy relations between the theoretical and the test results of a soft sedimentary rock 'Ohya Stone' in different direction whose material parameters are listed in Table 1. It is known from the figure that the theoretical simulation agrees well the tested results.

3 FE ANALYSIS FOR ANISOTROPIC CUT SLOPE

The model given by Equation 5 satisfies the uniqueness of the solution for an initial value problem. Therefore, an initial strain method is used in the analysis.

Figure 3 shows a comparison of stress-strain-dilatancy relations of a soft rock with different sedimentary directions in conventional triaxial compression obtained from the theory and the finite element analysis. The material parameters of the rock are listed in Table 2. It is found that the calculated relations by FEM agree well with the theoretical ones, which implies that the finite element analysis is convincible.

Table 2 Material parameters of model ground

$E_x(MPa)$	200.0	$b\ (MPa)$	0.87
$E_y(MPa)$	100.	$\sigma_{mb}\ (MPa)$	16.0
ν_{yx}	0.10	$\overline{M}_m$	1.25
ν_{xz}	0.30	τ	0.025
$G_{yx}(MPa)$	20.	α	0.90
G'	452.0	β	1.00
M^*_f	1.00	γ	1.05
Permeability k (cm/sec)		1.0×10-6	
Unit weight γ' *(kN/m3)*		10.0	
Sedimentary angle θ		0,45,90,135	

Figure 4 shows the finite element mesh used in the analysis of the cut slope. The domain of the ground considered is 1000×360 m with 1053 4-node isoparametric elements and 1120 nodes. The height and the slope gradient of the cut slope are 150 m and 5:1 respectively. In the soil-water coupling analysis, an excessive pore-water pressure is taken as the unknown variable. The initial stress field of the model ground is a gravitational field with a value of K_0=0.43. In the calculation, four different sedimentary directions, θ=0, 45, 90, 135° are considered.

Figure 5 shows the change of the distribution of stress-history ratio. In the present constitutive model, the failure state or the residual state is described by the equation as $\eta^*-M^*_f=0$ where the cohesion or the cementation of the geologic material tends to be zero and only the frictional strength that depends on a confining stress remains. In this case, the stress-history ratio η^* will be the same as the stress ratio and takes the value of M^*_f. In the figure, t=0 means the time immediately after the excavation. At the beginning, the value of η^* is kept as a constant of about 0.80. Afetr1.83 year, it increases abruptly at the toe of the slope and then the phenomenon propagates to other regions. 11 months later, a failure band formed from the toe to surface, in which η^* reaches to the residual value. Finally an unstable block appears in slope, taking the band as its boundary connecting the stable area of ground.

Figure 6 shows the change of the distribution of plastic shear strain. Similar to the stress-history ratio, shear strain develops very quickly in a zone at the time of 2.75 years. The propagation of the shear zone in which a large shear strain occurs takes the same form as the failure zone shown in Figure 5.

Figure 7 shows a change of distribution of excessive pore-water pressure with time. At the time immediately after the excavation, a large excessive pore-water pressure developed in the ground, resulting in an apparent shear strength that keeps the slope stable. After 1.83 years, it dissipated gradually and the failure zone shown in Figure 6 began to develop due to the loss of the apparent shear strength. At the moment, the excessive pore-water pressure reached its minimum value. When the shear zone formed, strain softening occurred and a dilatancy develops in some area, resulting in an increase of excessive pore-water pressure as shown in Figure 7.

From Figures 5-7, it is clear that because of the dissipation of an excessive pore-water pressure due to excavation, the ground of cut slope lost its apparent strength and a strain softening occurs in some area. Then a redistribution of stress leads to a start of the propagation of the softening zone, resulting in the formation of the failure band and the shear zone. The failure band develops gradually and finally a global failure in cut slope happens.

In order to clarify the mechanism of the progressive failure, the time history of field quantities such as stress ratio etc. along the shear zone is studied in detail. Figure 8 shows the change of stress ratio with time and the stress-strain relations. It is found that when the stress ratio reached its peak value, strain softening occurred and the plastic strain developed very quickly, resulting in a sharp reduction of the stress ratio. It is also found that the time when strain softening occurs is different for different elements, starting from the toe to ground surface, showing a clear propagation of the softening zone.

Table 3 Difference of the failure time of slope due to different sedimentary directions

Sedimentary angle θ	0	45	90	135
Failure time (year)	3.50	2.53	∞	0.63

Table 3 shows the failure times of the cut slope with different sedimentary direction. It is found that the relatively stable direction of the cut slope is 90 degree while the relatively unstable direction is 135 degree, showing that the sedimentary direction may greatly affect the progressive failure behavior and long-term stability of a cut slope

Figure 9 shows the stress and stress-history paths of element 423 in isotropic and anisotropic (θ=0°) ground. It is found that the cut slope in anisotropic ground lost its stability more fast than in isotropic ground, which is thought to be the effect of the anisotropic mechanical behaviors of the ground in stress-strain-dilatancy relation and strength.

4 CONCLUSION

(1) A cut slope of soft rock may remain stable for a long time after the completion of a rapid excavation. Sometime, however, a failure band may forms abruptly in the slope and then slope may fail overwhelmingly at a few months.

(2) The propagation of the shear zone in cut slope takes the same form as the propagation of the failure zone.

(3) The progressive failure in a cut slope is caused by the redistribution of a stress due to the stain softening.

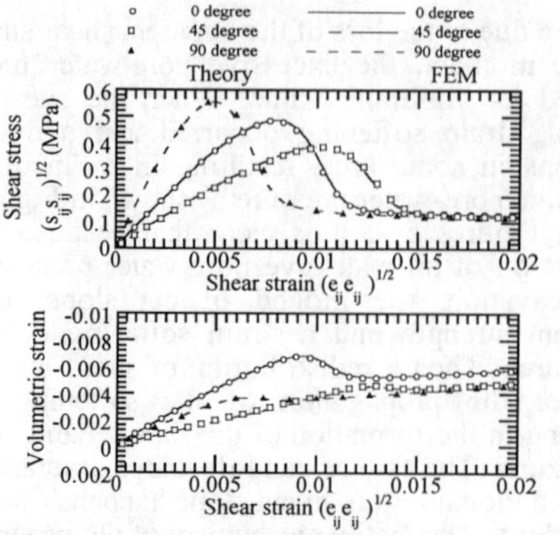

Figure 3 Stress-strain-dilatancy relation from FEM and theory

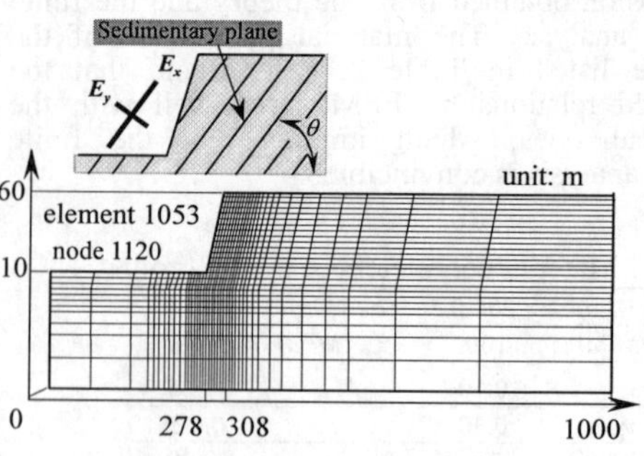

Figure 4 Finite element mesh

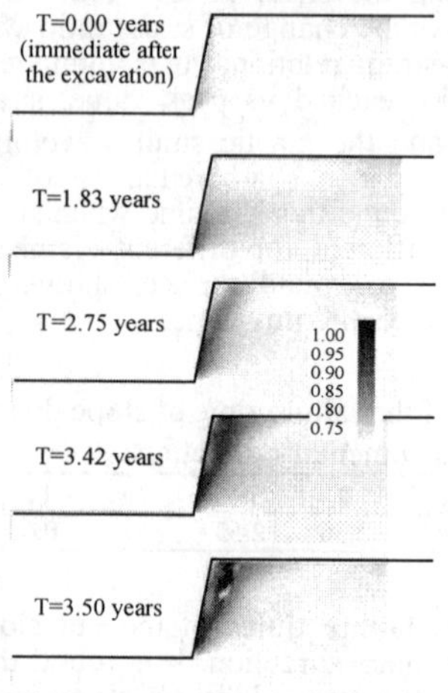

Figure 5 Change of stress-historyratio with time (θ=0°)

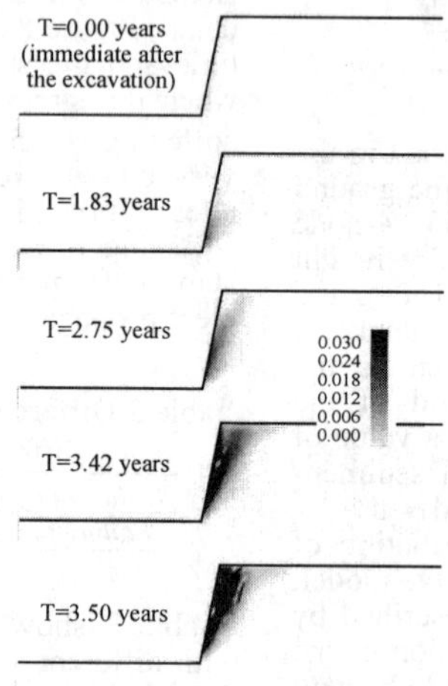

Figure 6 Change of plastic shear strain (θ=0°)

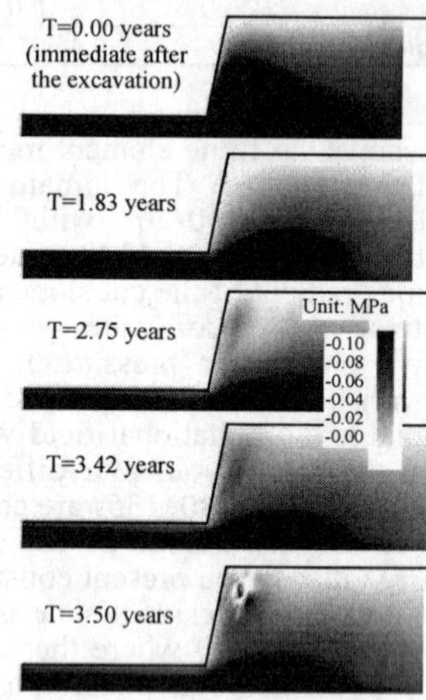

Figure 7 Change of excessive pore-water pressure (θ=0°)

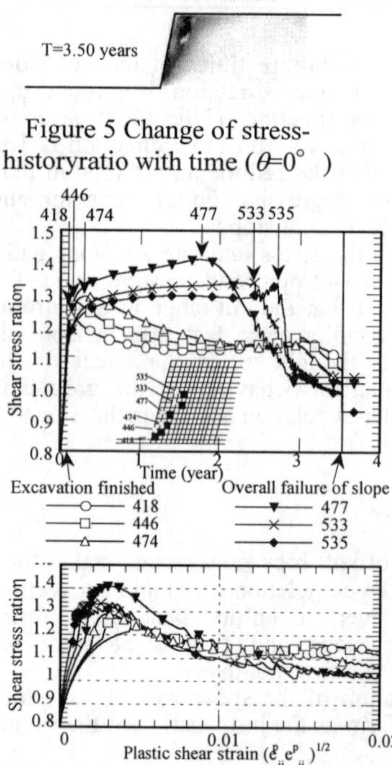

Fig.8 Stress ratio-time profile and stress strain relations (θ=0°)

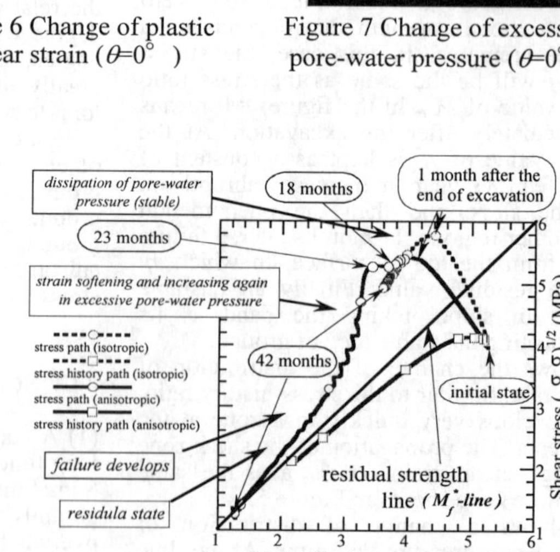

Figure 9 Stress path and stress-history path

(4) The sedimentary direction of a ground may greatly affect the progressive failure behavior and long-term stability of a cut slope.
(5) By conducting a soil-water coupling analysis, it is possible to simulate the time dependent behavior of geologic materials due to pore-pressure dissipation.
(6) The progressive failure of cut slope can be simulated with a soil-water coupling analysis based on an elasto-plastic model with strain softening.

REFERENCES

Adachi, T., Oka, F. and Zhang, F., 1994. An elasto-viscoplastic constitutive model with strain softening and its application to the progressive failure of a cut slope, AMD-Vol.183/MD-Vol.50, Material Instabilities: Theory and Applications, ASME, pp.203~217.

Adachi, T., Oka, F., Osaki, H., Fukui H. and Zhang, F., 1999. Soil-water coupling analysis of progressive failure of cut slope using a strain softening model, Proc. of Int. Conf. on Slope Stability Engineering, Balkema, Vol.1, pp.333-338.

Adachi, T. and Oka, F., 1995. An elasto-plastic constitutive model for soft rock with strain softening, Int. Jour. for Numerical and analytical Method in Geomechanics, Vol. 19, pp.233-247.

Computer Methods and Advances in Geomechanics, Desai et al. (eds) © 2001 Balkema, Rotterdam, ISBN 90 5809 183 X

Seismic lateral earth pressure of cohesive backfill with inclined surface

A.R.Boghrat
Professional Service Industries Incorporated, Geotechnical Department, Columbus, Ohio, USA

M.H.Ganjali
Ferdowsi University, Mashhad, Iran

ABSTRACT: A solution in equation form is presented to determine dynamic normal and tangential stresses against retaining walls. To generalize the solution, a (c-ϕ) soil type and sloping backfill is considered. For verification of the presented dynamic equations, static values have been used and have resulted in equations for the static general condition of (c-ϕ) soil and sloping backfill.

1 INTRODUCTION

One of the main considerations when designing retaining structures in earthquake prone areas is the determination of lateral active and passive soil pressure during seismic activity. In this paper, some equations have been derived to calculate dynamic normal (σ_x) and shear (τ_{xz}) stresses. These are the stresses that act on walls during an earthquake. A general soil type (c-ϕ) and sloping backfill is considered in this analysis.

When the soil is cohesionless and the backfill is horizontal, under a static condition, the stresses in the principal plane are (1990):

$$\sigma_z = \gamma Z \tag{1}$$

$$\sigma_x = k\gamma Z \tag{2}$$

$$\tau_{zx} = 0 \tag{3}$$

In the above equations, (γ) is the unit weight of the soil and (Z) is the depth under consideration. Also, (K) is a coefficient of lateral stress. With respect to the Rankine limiting equilibrium analysis, when a lateral movement is allowed, the value of (K) would be (1976):

$$K_{a,p} = \frac{1 \mp \sin\phi}{1 \pm \sin\phi} \tag{4}$$

The coefficient of lateral stress for active and passive conditions is represented by $k_{a,p}$. Under earthquake conditions, the vertical ($a_z = k_v g$) and horizontal ($a_x = k_h g$) accelerations are considered to be acting on the soil element (Figure 1). Here, "g" is acceleration due to gravity and; k_v and k_h are vertical and horizontal acceleration coefficients. In this case, the equations of equilibrium are (1990):

$$\frac{\partial \sigma_x}{\partial x} + \frac{\partial \tau_{xz}}{\partial z} = k_h \gamma \tag{5}$$

$$\frac{\partial \tau_{xz}}{\partial x} + \frac{\partial \sigma_z}{\partial z} = (1 - k_v)\gamma \tag{6}$$

Equations 5 and 6 depict conditions of no sloping backfill and cohesionless material. The term (τ_{xz}) is a function of (z) only and not (x), and $\sigma_x = K\sigma_z$. Therefore, Equations 5 and 6 could be written as (1990):

$$\sigma_x = K(1 - k_v)\gamma Z \tag{7}$$

$$\sigma_z = (1 - k)\gamma Z \tag{8}$$

$$\tau_{xz} = \tau_{zx} = -k_h \gamma Z \tag{9}$$

The variable "K" is the seismic coefficient of lateral earth pressure. According to Richards, et. al (1990), "K" (for horizontal backfill and cohesionless soil) is defined as:

$$K = \frac{1 + \sin^2\phi}{\cos^2\phi} \pm \frac{2}{\cos\phi}\left(\tan^2\phi - \tan^2\theta\right)^{1/2} \tag{10}$$

Where: $\tan\theta = \dfrac{k_h}{1 - k_v}$

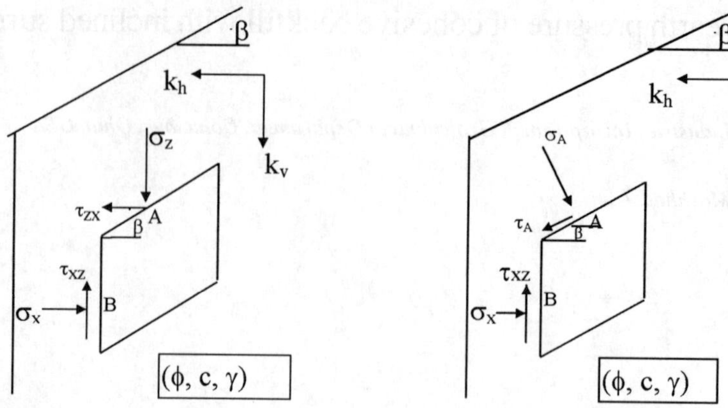

Figure 1: Stresses on an element of soil caused by earthquake forces.

Fang and Chen (1995) presented a procedure to determine active and passive earth pressure for a cohesionless soil with sloping backfill. Their procedures modified the Mononobe (1924) and Okabe (1924) analysis that is based on Coulomb's Method. They studied the effect of direction of vertical acceleration coefficient and concluded that the maximum active force under dynamic condition would result when this coefficient is acting downward.

Furthermore, Budhu, et. al, (1993) for horizontal backfill and (c-ϕ) soil conditions, defines (K) as follows:

$$K = \frac{\sin^2\phi + 1}{\cos^2\phi} + \frac{2D}{1-k_v}\tan\phi \pm \sqrt{\left\{\left[\left(\frac{2D}{1-k_v}\right)\tan\phi + \frac{\sin^2\phi+1}{\cos^2\phi}\right]^2 + \left(\frac{2D}{1-k_v}\right)^2 + \frac{4D}{1-k_v}\tan\phi - \frac{4\tan^2\phi}{\cos^2\phi} - 1\right\}} \tag{11}$$

In Equation 11, the value of $D = \frac{c}{\gamma Z}$ is where (c) is the cohesion of soil. In Equations 10 and 11, the minus and plus signs represent the active and passive cases, respectively.

2 DERIVATION

The technique of analytical geometry is used to determine the essential dynamic equations for a (c-ϕ) soil with sloping backfill and earthquake forces because generalization of other methods would be very complicated. The chosen soil element in this method would have an angle of (β) with a horizontal direction as was used by Terzaghi (1943). The stresses on this element are shown in Figure 1 and can be calculated as follows:

$$\sigma_z = (1-k_v)\gamma\, Z\cos\beta \tag{12}$$

$$\tau_{xz} = -k_h\gamma\, Z \tag{13}$$

The stresses that are perpendicular and tangential to the sloping surface (β) on the element would be:

$$\sigma_A = (1-k_v)\gamma\, Z\cos^2\beta - k_h\gamma\, Z\sin\beta = X_A \tag{14}$$

$$\tau_A = (1-k_v)\gamma\, Z\cos\beta\sin\beta + k_h\gamma\, Z\cos\beta = Y_A \tag{15}$$

(X_A) and (Y_A) are used for convenience and are the coordinates of Point (A) on the Mohr circle of stresses. In Figure 2, (OH) and (AH) are equal to (X_A) and (Y_A), respectively. In this Mohr circle; the circle should pass through Point (A), the center should be on the axis where vertical stress acts (horizontal axis) and should also be tangent to the Mohr-Coulomb failure envelope (O'M) (Figure 2).

If the principles of strength of materials and soil mechanics are used, the coordinates of Point (B) in Figure 2 will result in the necessary dynamic lateral stresses (τ_x) and (σ_x). To simplify the names, the (τ_{xz}) is presented by (τ_x). In the isosceles triangle (BCA), the equal angles (BAC) and (CBA) are equal to (β+θ). Also, the angle (HCA) is an external angle in the triangle (O"CA) and equals (2β+θ). In the triangle (O'CM), if OC=m and CM=R, then:

$$R = (C\cot\text{an}\,\phi + m)\sin\phi = C\cos\phi + m\sin\phi \tag{16}$$

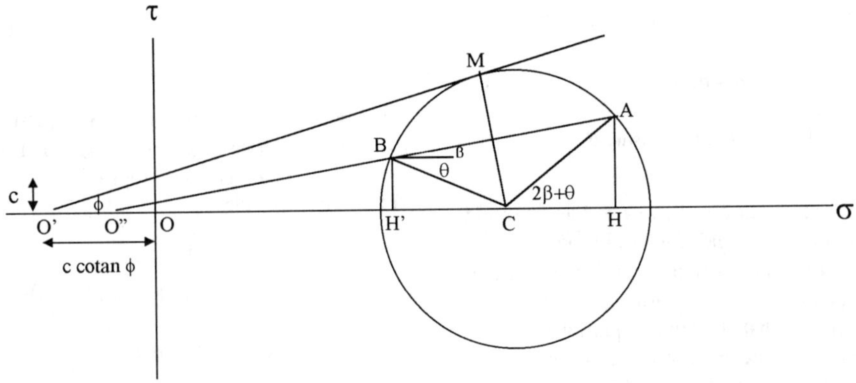

Figure 2: Mohr circle for element of soil in Figure 1.

If the coordinates of (A) are used in the equation of the presented Mohr circle (Figure 2), then (m) can be determined from the following:

$$(X_A - m)^2 + Y_A^2 = R^2 = (m\sin\phi + C\cos\phi)^2 \quad \phi \tag{17}$$

$$\left(1-\sin^2\phi\right)m^2 - 2\left(X_A + C\cos\phi\sin\phi\right)m + \left(X_A^2 + Y_A^2 - C^2\cos^2\phi\right) = 0 \tag{18}$$

After solving Equation 18, (m) will be:

$$m = \frac{1}{\cos^2\phi}\left[(X_A + C\cos\phi\sin\phi) \pm \sqrt{(X_A + C\cos\phi\sin\phi)^2 - \cos^2\phi\left(X_A^2 + Y_A^2 - C^2\cos^2\phi\right)}\right] \tag{19}$$

If the terms under the square root in Equation 19 are equal to (T), then the equation can be simplified as follows:

$$T = X_A^2 + 2C\cos\phi\sin\phi\, X_A + C^2\cos^2\phi\sin^2\phi - \cos^2\phi\, X_A^2 - \cos^2\phi\, Y_A^2 + C^2\cos^4\phi$$

$$T = (X_A\sin\phi + C\cos\phi)^2 - \cos^2\phi\, Y_A^2 \tag{20}$$

Therefore, the equation to determine (m) will be simplified to:

$$m = \frac{1}{\cos^2\phi}\left[(X_A + C\cos\phi\sin\phi) \pm \sqrt{(X_A\sin\phi + C\cos\phi)^2 - \cos^2\phi\, Y_A^2}\right] \tag{21}$$

In Equation 21, the positive and negative signs denote the passive and active cases, respectively. To determine the stresses on the horizontal plane ((σ_x) and (τ_x)), the equations of the Mohr circle and the intersecting line (AB) are used. The equations of the circle and line are as follows, respectively:

$$(X - m)^2 + Y^2 = R^2 \tag{22}$$

$$Y - Y_A = \tan\beta\,(X - X_A) \tag{23}$$

(X) can be determined from the 'line' equation and then used in the equation of the 'circle'.

$$X = \frac{Y - Y_A}{\tan\beta} + X_A \tag{24}$$

After substitutions and simplifications, the value of (Y) would be:

$$Y = -\cos^2\beta\,(X_A\tan\beta - m\tan\beta - Y_A) \pm \cos\beta\sqrt{(X_A\tan\beta - m\tan\beta - Y_A)^2(-\sin^2\beta) + \tan^2\beta\left(X_A^2 - 2mX_A + m^2 + Y_A^2\right)} \tag{25}$$

If the parameters under the square root sign are equal to (H); then, (H) can be simplified as follows:

$$H = (X_A\tan\beta\cos\beta - m\tan\beta\cos\beta + Y_A\sin\beta\tan\beta)^2 \tag{26}$$

Therefore, the equation for (Y) can be found.

$$Y = -\cos^2\beta(X_A\tan\beta - m\tan\beta - Y_A) \pm \cos\beta(X_A\sin\beta - m\sin\beta + Y_A\sin\beta\tan\beta) \tag{27}$$

In Equation 27, (Y_1) would result if the plus sign is used and (Y_2) if the minus sign. (Y_2) would be (τ_x). To determine (σ_x) which is equal to the (X) term of Equation 24, the (Y) term would be replaced by its equivalent (Y_2).

$Y_1 = Y_A$ and $Y_2 = Y_A \cos 2\beta - (X_A - m)\sin 2\beta$

$$\tau_x = Y_A \cos 2\beta - (X_A - m)\sin 2\beta \qquad (28)$$

$$\sigma_x = X_A + \cot\beta\,[Y_A \cos 2\beta - (X_A - m)\sin 2\beta - Y_A] \qquad (29)$$

Therefore for the condition of (ϕ-c) soil and sloping backfill, Equations 28 and 29 could be used to determine the dynamic normal (σ_x) and tangential (τ_x) stresses acting on a wall during an earthquake. It must be verified that dynamic Equations 21, 28, 29 would result in static equations if a seismic condition does not exist. Mazindrani, et. al (1997), presented the static condition and the governing equations for the same soil types and backfill geometry. If k_h and k_v are equal to zero (static condition), then Equations 14 and 15 would change to:

$$k_h = k_v = 0 \Rightarrow \begin{cases} X_A = \gamma Z \cos^2\beta \\ Y_A = \gamma Z \cos\beta \sin\beta \end{cases} \qquad (30)$$

For convenience, let us show the parameter (m) for the dynamic condition as (m_d) and for the static condition as (m_s). Therefore, Equation 21 represents (m) for the dynamic condition or (m_d). If the terms of Equation 30 replace X_A and Y_A in Equation 21, then:

$$m_d = \frac{1}{\cos^2\phi}\left[(X_A + C\cos\phi\sin\phi) \pm \sqrt{(X_A\sin\phi + C\cos\phi)^2 - \cos^2\phi\, Y_A^2}\right]$$

$$m_d = \frac{1}{\cos^2\phi}\left[\gamma Z\cos^2\beta + C\cos\phi\sin\phi\right] \pm \frac{1}{\cos^2\phi} \times$$
$$\sqrt{\gamma^2 Z^2\cos^2\beta\left(\cos^2\beta\sin^2\phi - \sin^2\beta\cos^2\phi\right) + C^2\cos^2\phi + 2C\gamma Z\cos^2\beta\sin\phi\cos\phi} \qquad (31)$$

Equation 31 incorporates the same (m) that is proposed by Mazindrani and Ganjali (1997) for static conditions. For the static condition, the equations for the normal (σ_{xs}) and tangential (τ_{xs}) static stresses can be written as (1997):

$$k_{a,p} = \frac{2}{\gamma Z} m_s - 1 \qquad (32)$$

$$P = k_{a,p}\gamma Z\cos\beta = \gamma Z\cos\beta\left[\frac{2}{\gamma Z} m_s - 1\right]$$
$$= 2m_s\cos\beta - \gamma Z\cos$$
$$\tau_{xs} = P\sin\beta = 2m_s\cos\beta\sin\beta - \gamma Z\cos\beta\sin\beta \qquad (33)$$

$$\sigma_{xs} = P\cos\beta = 2m_s\cos^2\beta - \gamma Z\cos^2\beta = \tau_x\cot\beta \qquad (34)$$

For the same condition as Equation 30, the dynamic tangential stress (τ_{xd}) of Equation 28 can be changed to represent the static condition as follows:

$$\tau_{xd} = Y_A\cos 2\beta - (X_A - m)\sin 2\beta$$
$$= \gamma Z\cos\beta\sin\beta\cos 2\beta - (\gamma Z\cos^2\beta - m)\sin 2\beta$$

This will simply to:

$$= 2m\sin\beta\cos\beta - \gamma Z\cos\beta\sin\beta = \tau_{xs} \qquad (35)$$

It should be noted that in Equation 35, the parameter (m) is equivalent to (m_s) of Equation 31, which represents a static condition.

For the same condition of Equation 30, the dynamic normal stress (σ_{xd}) of Equation 29 can be changed to represent the static condition as presented below:

$$\sigma_{xd} = X_A + \cot\beta(\tau_{xd} - Y_A)$$
$$= \gamma Z\cos^2\beta + \cot\beta(\tau_{xs} - \gamma Z\cos\beta\sin\beta)$$
$$= \gamma Z\cos^2\beta + \tau_{xs}\cot\beta - \gamma Z\cos^2\beta = \tau_{xs}\cot\beta = \sigma_x \qquad (36)$$

Of course, it should be noted that the parameters used in Equation 36 should be static values. The above procedures proved that the presented Equations 21, 28 and 29 could be converted to the static condition.

3 EXAMPLE

Here different wall heights of three, six and nine meters are chosen and the dynamic lateral stresses (σ_x) and (τ_x) are calculated. It is assumed that the backfill has an angle of 20 degrees from the horizontal and the average density of soil backfill is 18 kn./m^3. Different angles of internal friction from 10 to 40 degrees are selected and cohesion of the different backfill soil ranges from 40 to 100 (kpa). The horizontal and vertical acceleration coefficients are chosen to be 0.25 and 0.1, respectively. A simple computer program was written to solve the equations that will result in dynamic lateral stresses and these results are shown in table-1. The results are calculated for three points and each point is a third of the wall height from the top.

TABLE 1: The results of the example. The dynamic stresses are in kpa.

a = active p = passive			$k_h = 0.25,\ k_v = 0.1, \beta = 20^0, \gamma = 18\ kN/m^3$								
			H = 3 m			H = 6 m			H = 9 m		
			H/3	2H/3	H	H/3	2H/3	H	H/3	2H/3	H
$\phi = 10^0$ $c = 100$ kpa	σ_x	a	-144.74	-141.30	-137.85	-141.30	-134.42	-127.55	-137.85	-127.55	-117.27
		p	221.82	233.18	244.55	233.18	255.91	278.65	244.55	278.65	312.78
	τ_x	a	-47.89	-41.85	-35.81	-41.85	-29.77	-17.69	-35.81	-17.69	0.42
		p	85.53	94.45	103.37	94.45	112.30	130.15	103.37	130.15	156.94
$\phi = 20^0$ $c = 75$ kpa	σ_x	a	-91.68	-90.62	-89.58	-90.62	-88.54	-86.50	-89.58	-86.50	-83.46
		p	207.47	225.79	244.12	225.79	262.46	299.17	244.12	299.17	354.26
	τ_x	a	-28.58	-23.41	-18.24	-23.41	-13.07	-2.75	-18.24	-2.75	-12.72
		p	80.30	91.76	103.22	91.76	114.68	137.62	103.22	137.62	172.04
$\phi = 30^0$ $c = 50$ kpa	σ_x	a	-51.63	-52.31	-53.02	-52.31	-53.75	-55.25	-53.02	-55.25	-57.56
		p	182.02	211.14	24.28	211.14	269.44	327.81	240.28	327.81	415.41
	τ_x	a	-14.00	-9.46	-4.93	-9.46	-0.41	8.62	-4.93	8.62	22.15
		p	71.04	86.43	101.82	86.43	117.22	148.04	101.82	148.04	194.30
$\phi = 40^0$ $c = 10$ kpa	σ_x	a	-9.83	-11.78	-13.82	-11.78	-15.89	-20.08	-13.82	-20.08	-26.40
		p	84.61	131.72	178.91	131.72	226.13	320.61	178.91	320.61	462.38
	τ_x	a	1.21	5.29	9.34	5.29	13.37	21.43	9.34	21.43	33.49
		p	35.59	57.52	79.48	57.52	101.46	145.43	79.48	145.43	211.39

4 CONCLUSION

Dynamic equations for determining normal and shear stresses on a wall caused by earthquake forces have been presented in this paper. These equations cover conditions of (c-ϕ) soil and sloping backfill. That is, general conditions are considered. To verify that the presented dynamic equations could be converted to equations representing a static condition when the vertical and horizontal acceleration coefficients are chosen to be zero. This procedure has resulted in the previously presented static equations for (c-ϕ) soil and sloping backfill.

The results of examples of different wall heights and at third points are illustrated here as an example of using the presented equations. Table-1 gives the dynamic lateral stresses for active (a) and passive (p) cases. The results of active cases show that during the earthquake, the soil is completely separated from the wall and this is shown by negative values of horizontal dynamic stress.

The value of tangential stress (τ_x) becomes positive when the soil is less cohesive and the value of friction is high. In passive dynamic cases, the stresses are positive at all times. This means that at critical situations, the passive forces would be much larger.

REFERENCES:

Budhu, M. and Al-Karni, A.V. (1993). "Seismic Bearing Capacity of Soils." *Geotechnique.* 43, No. 1, 181-187

Fang, Y.S. and Chen, T. J. (1995). "Modification of Mononobe Okabe Theory." *Geotechnique* 45, No. 1, 165-167.

Mazindrani, Z. H. and Ganjali, M. H. (1997). "Lateral Earth Pressure Problems of Cohesive Backfill with Inclined Surface." Journal of Geotechnical and *Geo-Environmental Engineering*, Vol. 123, No. 2, 110-112.

Mononobe, N. (1924). "Consideration into Earthquake Vibrations and Vibration Theories." *J. Japan. Soc. Civ. Engg.* 10, No. 5, 1063-1094.

Okabe, S. (1924). "General Theory on Earth Pressure and Seismic Stability of Retaining Wall and Dam." *J. Japan. Soc. Civ. Engg.* 10, No. 5, 1277-1323

Perloff and Baron (1976). *"Soil Mechanics, Principles and Applications."* John Wiley and Sons, Inc. New York, NY.

Richards Jr., R.; Elms, D. G. and Budhu, M. (1990). "Dynamic Fluidization of Soils." *Journal of Geotechnical Engineering*, Vol. 116, No. 5, 740-759.

Terzaghi, K. (1943). *"Theoretical Soil Mechanics."* John Wiley and Sons, Inc. New York, NY.

Computer Methods and Advances in Geomechanics, Desai et al. (eds) © 2001 Balkema, Rotterdam, ISBN 90 5809 183 X

Modeling the effectiveness of groundwater-lowering remediation measures in the Alani-Paty landslide area, Oahu, Hawaii

H.G. Brandes & C.S.L. Tsui
Department of Civil Engineering, University of Hawaii, USA

ABSTRACT: In order to examine slope stability and creep problems in Manoa, one cross-section of the Alani-Paty landslide area was modeled using the computer software package SLOPE/W. The purpose of the study was to determine appropriate water elevations needed to insure adequate stability. The landslide area consists of fine-grained, very soft volcanic colluvium overlying basalt bedrock. The stability analysis suggests that local failures can continue to be expected in the upslope region. To avoid large-scale failure, the water table should be lowered to a depth at least 13 feet below to the ground surface in order to yield a factor of safety of one. This water elevation should be checked against piezometer readings to evaluate the effectiveness of the new drainage system.

1 INTRODUCTION

The Alani-Paty landslide in Manoa valley (Figure 1), on the island of Oahu, Hawaii, is typical of a series of slow-moving slides that occur on most flanks of the deep valleys that cut into the Koolau mountain range in East Honolulu. The volcanic colluvium that is found above the basalt bedrock is mostly fine-grained and very soft. In addition to continuous creep deformations, which are the focus of a companion paper (Tsui et al., 2001), these slopes are highly susceptible to reactivation in response to moderate and extreme rainfall events. Landsliding is a serious problem in these valleys since it typically occurs in high-density residential areas. Accumulated damages in Honolulu are estimated in the hundreds of millions of dollars.

Manoa valley has a relatively flat bottom and steep sides. Through intermittent debris flows and rock falls, weathered residuum, consisting of clasts and soil, has been transported downslope from the sides of Waahila Ridge to form an apron at the base of the ridge on which extensive failure and creep have taken place (Figure 2). This apron is concave upward, with a slope of 2°-6° near the valley floor and 10°-15° near the upslope edge. Average slope of the debris apron is 8°-12°. The affected deposits consist of crudely stratified clayey silt and silty clay containing boulder-, cobble-, and gravel-sized fragments of weathered basalt. The stratification of these deposits is complex and discontinuous, making it difficult to correlate layers even over short distances. The clasts of basalt constitute about 38% of the volume of the bulk soil, not sufficient to affect the stability or creep behavior of the area to a significant extent (Baum and Reid, 1992). The matrix containing the clasts consists of highly plastic silt and clay (Table 1). The head of the landslide has undergone at least 3 to 4 ft of displacement, with some offsets as large as 6 to 10 ft near Lanikaula street. Most of these displacements have occurred during or after intense rainfall events, although creep deformations continue to cause damage.

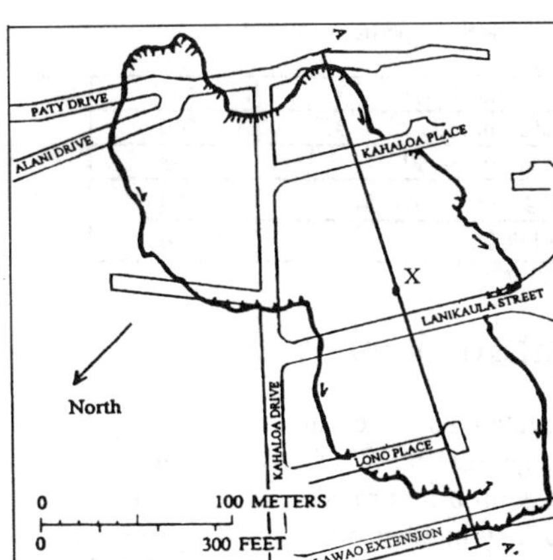

Figure 1. Approximate boundaries of the Alani-Paty landslide

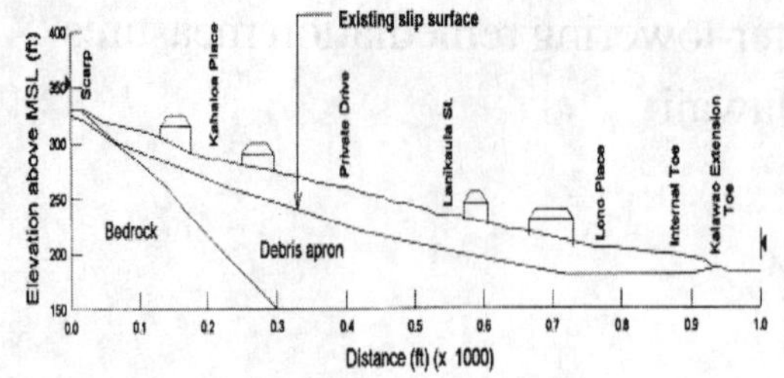

Figure 2. Cross section of landslide

2 REMEDIAL MEASURES

Extensive stabilization work in the area began in May 1993 (Polhemus, 1993). Following the advice of Peck and Wilson (1968) and others, remediation efforts have focused on controlling drainage to lower groundwater levels, grading, and installing tiebacks and various retaining structures. Drainage remedial measures included surface storm drains and repaving, horizontal and vertical subsurface drains, 80-ft deep groundwater pumps, and larger capacity water and wastewater pipes. In all, about 3,700 ft of new sewer lines were installed, along with 5,900 ft of new water lines and 2,600 ft of drains (Figure 3). In addition, an extensive tieback system was installed near the head of the landslide, consisting of side-by-side concrete blocks anchored to the bedrock by 100-ft long anchors (Civil Engineering Magazine, 1998). The purpose of this modeling study was to determine appropriate groundwater elevations that need to be maintained to insure adequate stability (Tsui, 1999).

Table 1. Index properties

Water content, w	47%
Liquid Limit, LL	40
Plasticity Index, PI	43
Average unit weight, γ_{sat}	138 pcf
Sand/Silt/Clay	7/27/66 %

3 SAFE GROUNDWATER TABLE

Computations were carried out using the software package SLOPE/W. This commercial code uses the limit equilibrium method to solve for the safety factor of any number of assumed failure surfaces. The complex soil profile was approximated by the residual strength parameters shown in Figures 4a and 4b. These values were determined from direct shear tests using remolded samples of the sand/silt/clay matrix (Baum et al., 1991). The low effective residual friction angles in Figure 4b have been confirmed by more recent triaxial tests conducted by these authors on remolded specimens. The stability analysis uses residual strength parameters since the soil in the landslide area has undergone substantial shearing. An average unit weight of 138 pcf was assumed for the entire area.

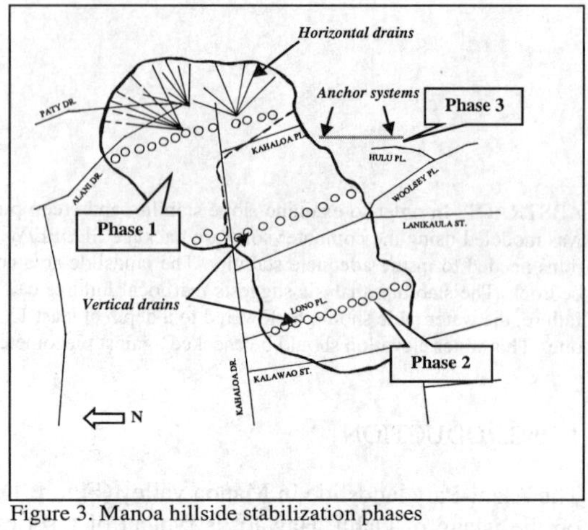

Figure 3. Manoa hillside stabilization phases

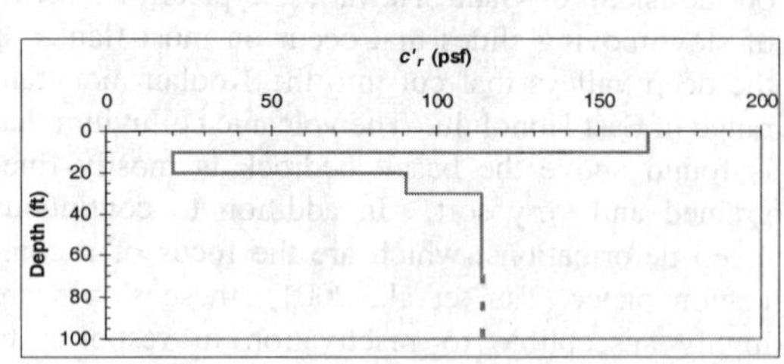

Figure 4a. Variation of residual effective cohesion with depth

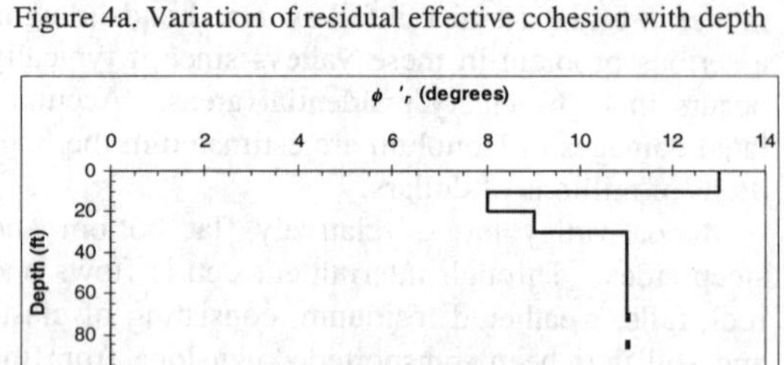

Figure 4b. Variation of residual effective frictional angle with depth

The predominant failure surface in the landslide (Figure 2) has been inferred indirectly, in part from hydraulic conductivity measurements and recorded water elevations. Piezometer data indicates that the freshwater groundwater table is located some 150 ft below the basal slip surface. Nonetheless, piezometer readings also indicate the presence of a prevalent perched water table above the major slip surface. Slug tests reported by Baum et al. (1991) and Baum and Reid (1992) suggest a hydraulic conductivity profile as shown in Figure 5. The location of the predominant failure surface is associated with the interval of low permeability between 20 and 30 ft. Most of the piezometers shallower than 29 ft or deeper than 60 ft responded to rainfall, whereas those between 30 and 60 ft were dry or gave steady readings during the monitoring period.

Hundreds of failure surfaces were investigated. These included numerous circular sections of varying radius, size and location, as well as fully specified surfaces roughly parallel to the presumed major failure surface (Figure 2). In each case, the factor of safety was calculated using the rigorous Morgenstern-Price method, which satisfies both force and moment equilibrium. A constant interslice function equal to 1 was assumed.

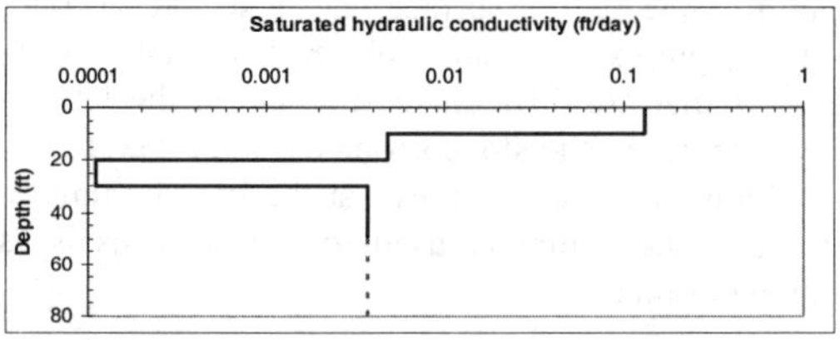

Figure 5. Variation of hydraulic conductivity with depth

Circular slip surfaces near the head of the slide gave minimum safety factors slightly less than one, even with a groundwater table 20 ft below the ground surface (Figure 6). The upslope area is less stable because it has a steeper geometry. However, the fact that the soil in this area is sandier and presumably somewhat stronger than elsewhere may result in true safety factors that are slightly higher. Nonteheless, observations indicate significantly more damage to structures in the upslope area, particularly near Alani Drive and Paty Drive, than anywhere else. The head of the landslide was found to be moving downslope during a number of rainy periods when the United States Geological Survey was monitoring the slope, but the main body of the landslide only moved during two of those rainy periods. A complicating factor, not considered in this study, is that pressure head changes in the upslope region are more rainfall-dependent than in the downslope region.

Twenty fully specified failure surfaces along transect AA' in Figure 1 were analyzed to determine the effect of regional groundwater level on stability throughout the landslide area (Table 2, Figure 7). Slip surface number 12, located 5 ft above the major failure surface and parallel to it, yielded the critical groundwater elevation Figure 8 indicates that a groundwater table of at least 13 ft below the ground surface is required in order to have a safety factor larger than one, and hence avoid large-scale failure. In fact, to allow for uncertainties, the warning elevation should probably be closer to 20 ft below ground. Water levels in the piezometers need to be checked against this value to gauge overall stability, particularly during rain events. If these depths are not maintained, the overall drainage system needs to be reevaluated.

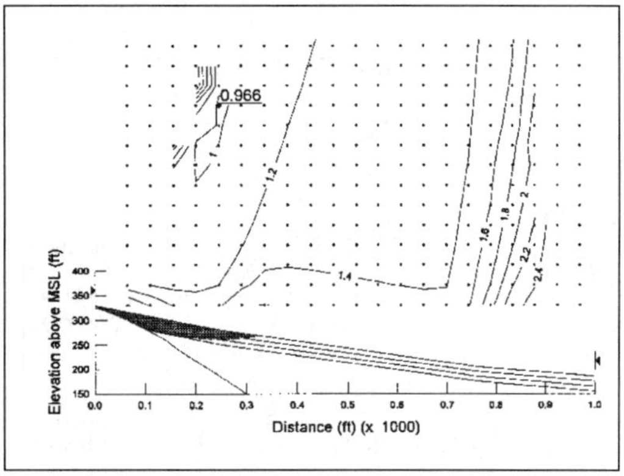

Figure 6. Critical slip surface and contours of safety factors (Water table 20 ft below ground surface)

Table 2. Identification numbers and descriptions for fully specified slip surfaces

No.	Description
Set 1 - Parallel to existing slip surfaces	
1	Existing slip surface
2	5 ft above the existing slip surface
3	10 ft above the existing slip surface
4	15 ft above the existing slip surface
5	5 ft below the existing slip surface
Set 2 - Parallel to ground surface	
6	5 ft below ground surface (about 900 ft long)
7	10 ft below ground surface (about 900 long)
8	15 ft below ground surface (about 900 ft long)
9	20 ft below ground surface (about 900 ft long)
10	25 ft below ground surface (about 900 ft long)
Set 3 - Parallel to existing slip surface and of limited length	
11	Existing slip surface (first 750 ft portion)
12	5 ft above the existing slip surface (first 750 ft portion)
13	10 ft above the existing slip surface (first 750 ft portion)
14	15 ft above the existing slip surface (first 750 ft portion)
15	5 ft below the existing slip surface (first 750 ft portion)
Set 4 - Parallel to ground surface and of limited length	
16	5 ft below ground surface (about 750 ft long)
17	10 ft below ground surface (about 750 long)
18	15 ft below ground surface (about 750 ft long)
19	20 ft below ground surface (about 750 ft long)
20	25 ft below ground surface (about 750 ft long)

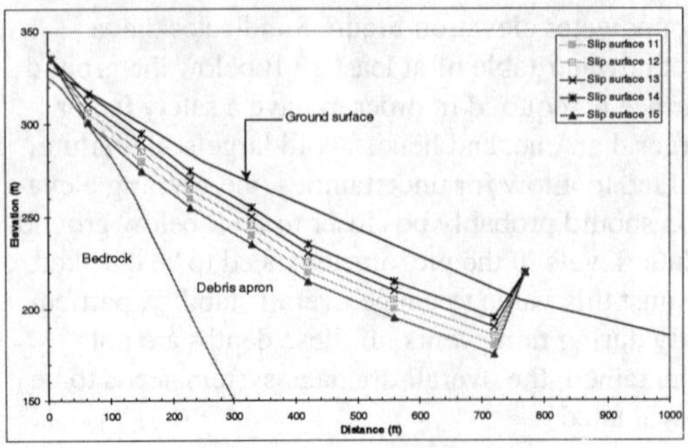

Figure 7. Locations of fully specified slip surfaces

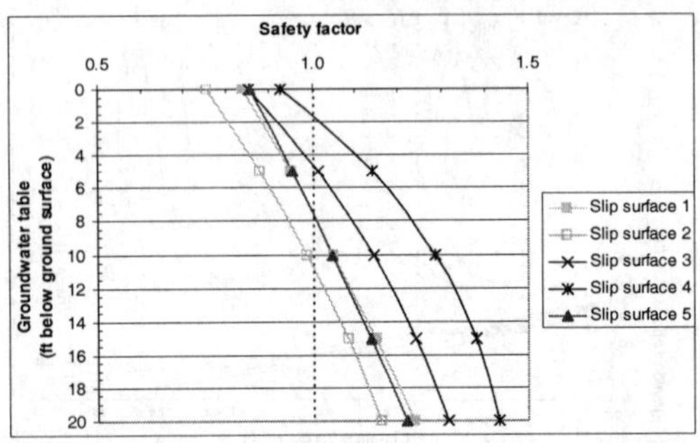

Figure 8. Safety factors for different groundwater conditions

4 ADDITIONAL CONSIDERATIONS

The safety factors determined from this study are conservative in a number of ways. First of all, the new anchor system was not considered in the analysis. The anchor system in the upslope region serves to tie the weak soil to the stronger bedrock underneath, thus improving the safety of the slope. Because presumably only the soil close to the pre-existing slip surface has undergone failure, only this region currently has soil with shear strength parameters reduced to residual values. The soil above and below this region could still be characterized by strength parameters that are closer to peak values. Since residual strength was used for the whole slope, the results for slip surfaces above or below the existing surface are conservative. Furthermore, the residual strength parameters used are based on laboratory tests on remolded samples, which probably underestimate the true residual values. Also, capillary tension above the piezometric line was ignored. Since all the soil below the perched water table was assumed to be submerged, the pressure heads in some regions inside the slope were overestimated, resulting in underestimated safety factors. Finally, no vertical downward movement of water was considered.

Because of all these factors, the field safety factors could actually be higher than those determined from this study. Despite these uncertainties, this study does include realistic cross sections, and to a certain soil properties, and provides meaningful preliminary estimates.

5 BACK-ANALYSYIS

Back-analysis can be used to estimate the peak soil strength parameters at the time of original failure by assuming that such failure occurred when the safety factor reached one. One of the benefits of this type of analysis is that the strength parameters that are obtained represent in situ values, averaged over the length of the failure surface, and take into account fissures, pre-existing shear planes and other imperfections that are difficult to simulate in the laboratory. However, the slope geometry and groundwater conditions need to be known at the time of failure. The assumed geometry used for the analysis is as shown in Figure 9.

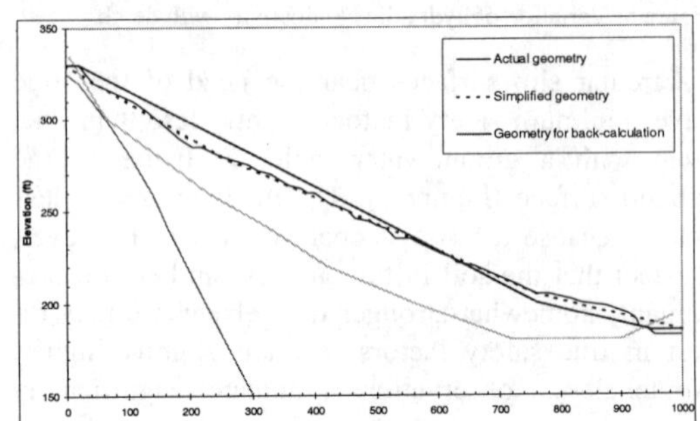

Figure 9. Slope geometry along section AA' used for back-analysis

The results of the back-analysis are summarized in Figure 10, where the strength parameters have been normalized by the baseline values indicated on that figure. Shown in Figure 10 are combinations of ϕ'/ϕ'_r and c/c_r that result in a factor of safety of one for various groundwater locations at the time of failure. Above each straight line the factor of safety is larger than one and below it is less than one. For example, if we assume that the cohesion has not changed due to failure, and we further assume an original groundwater level located at the ground surface (perhaps due to a long rainstorm event), than the peak friction angle would be about 40% higher than the residual value of 10.2^0.

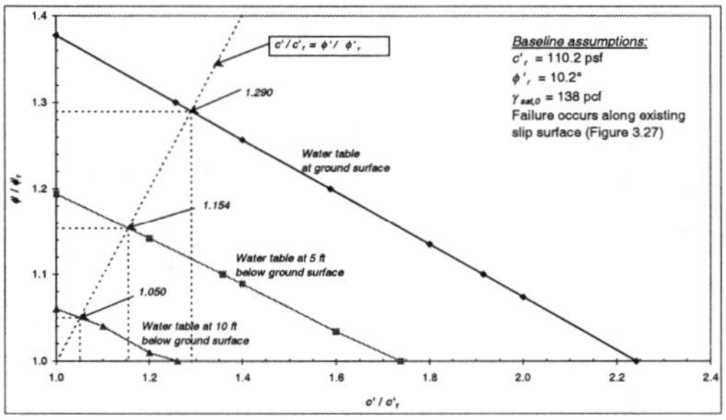

Figure 10. Combinations of c'/c'_r and ϕ'/ϕ'_r for FS = 1 from back-analysis

6 CONCLUSIONS

Morgenstern-Price's slope stability analysis indicates that local circular failures are likely to occur in areas of the upslope region that have not been stabilized, even with the water table as low as 20 ft below the ground surface. These types of failures have indeed been observed in the field, particularly in response to intense rainfall events. Any future large-scale failure will probably develop close to the existing predominant slip surface. To avoid this type of instability, water levels should be kept at least 13 ft below the ground surface, and preferably deeper than that.

7 ACKNOWLEDGEMENTS

This study was supported in part by a grant from the Hawaii Department of Transportation and the Federal Highway Administration. Their support is gratefully acknowledged.

8 REFERENCES

Baum, R.L., M.E. Reid, C.A. Wilburn & J.D. Torikai 1991. Summary of Geotechnical and Hydrologic Data Collected from May 1, 1990 through April 30, 1991, for the Alani-Paty Landslide, Manoa Valley, Honolulu, Hawaii. *U.S. Geologic Survey Open-File Report 91-598*: 102.

Baum, R.L. & M.E. Reid 1992. Geology, Hydrology and Mechanics of the Alani-Paty Landslides, Manoa Valley, Oahu, Hawaii. *U.S. Geological Survey Open-File Report 92-501*: 87.

Civil Engineering, 1998. Soil Anchors, Drains Halt Hawaiian Landslide. *Civil Engineering Magazine*, September 1989: 13-14.

Peck, R.B. & S.D. Wilson 1968. Report on The Hind Iuka Landslide and Similar Movements, *Honolulu, Hawaii*: 21.

Polhemus, D. 1993. "Halting the Manoa slip-slide". *The Honolulu Advertiser*, March 7, A: 18.

Tsui, C.S.L., H.G. Brandes & D.D. Nakayama 2001. Creep Behavior and Modeling of the Slow-Moving Alani-Paty Landslide, Oahu, Hawaii. *Proceedings of the 10th International Conference on Computer Methods and Advances in Geomechanics*, Tucson, AZ.

Tsui, C.S.L. 1999. Modeling the Slope Stability and Creep Behavior of the Alani-Paty Landslide Area, Manoa Valley, Honolulu, Hawaii. *MS Thesis, University of Hawaii*: 159p.

Computer Methods and Advances in Geomechanics, Desai et al. (eds) © 2001 Balkema, Rotterdam, ISBN 90 5809 183 X

Numerical comparison of stability of a slope stabilized with anchors

Fei Cai & Keizo Ugai
Department of Civil Engineering, Gunma University, Kiryu, Japan

ABSTRACT: The effects of anchors on the safety factor of a slope are predicted by 3D elasto-plastic shear strength reduction FEM (SSRFEM). The soil-anchor interaction is simulated with zero-thickness elasto-plastic interface elements. The numerical results are compared with those obtained by Bishop's simplified method, where the anchors are considered only to supply an axial tension. The safety factor of SSRFEM is consistent with that of Bishop's simplified method. SSRFEM shows that the prestressed load has little influence on the safety factor. The better stabilizing effects can be obtained when the anchors are installed with a small angle between the anchor and the horizontal direction.

1 INTRODUCTION

The permanent grouted anchors have been extensively used to provide vertical and lateral support for natural and engineered structures during the past 50 years. The end type of anchorage has been used to stabilize the dangerous slopes, and as a preventive measure in stable slopes because of its significant technical advantages resulting in substantial cost savings and reduced construction period. The basic design concept consists of transferring the resisting tensile forces generated in the anchors into the ground through the friction mobilized at their interfaces. The loads are usually developed by the anchorage of the tendon within the soil mass and tensioning at the surface against a bearing plate or concrete pad. Some of the successful applications of this technique to stabilizing slopes have been reported (Tan et al. 1985, Hashimoto et al. 1986, Corona 1996).

The stability of slopes reinforced with anchors is commonly assessed using LEM, assumed that the anchor only supplies an axial tension to reinforce the slope. The 3D SSRFEM is used to evaluate the reinforcing mechanism of anchors. The soil-anchor interaction is simulated by zero-thickness 3D interface elements. The results of SSRFEM are compared with those of Bishop's simplified method. The effects of the prestressed load, direction angle on the safety factor of slope are numerically studied. Some useful conclusions are obtained.

2 ANALYSIS METHODS

2.1 *Shear strength reduction FEM (SSRFEM)*

SSRFEM has been gradually used to analyze the stability of slope in 2D and 3D situations. The numerical comparison has shown that SSRFEM is a reliable and robust method for assessing the safety factor of slopes and the corresponding critical slip surface. One of the main advantages of SSRFEM is that the safety factor emerges naturally from the analysis without the user having to commit to any particular form of the mechanism *a priori* (Griffiths & Lane 1999). SSRFEM has been successfully used to determine the safety factor of the slopes improved with piles (Cai & Ugai 2000). SSRFEM can evaluate the stability of such slopes under a general frame for slopes, where the soil-structure interaction is considered with zero-thickness interface element.

The process of SSRFEM to predict the safety factor of a slope reinforced with anchors is simply described here. When the shear strength reduction factor, F, is less than unity, the slope completely consists of the soil. When F reaches unity, the materials of the solid elements within the extent of the anchor are changed from the soil into the tendon or the grouted body according to the position of the elements. The materials of the interface elements are simultaneously changed from the soil-soil interface into the soil-tendon interface or the soil-grout interface. The stress in the interface keeps unchanged when the materials are changed. This implies that the grouting pressure is the same as the normal stress in the interface, induced by the self-weight of soil. When F is unity, a pre-stressed load is applied. The prestressed load includes a drag load, applied at the anchor head, and a push load, applied on a circular rigid plate on the slope surface. The value of the two loads is the same, but their directions are opposite. The rigid plate is simulated by the identical deflections of the nodal points on the slope surface under the rigid plate. After the deflection under the prestressed load is finished, the tendon head is fixed to the rigid plate on the slope surface. Therefore, the deflection of the tendon head is the same as that of the rigid plate during the shear strength reduction from unity to the ultimate value. The detailed process for SSRFEM and properties of the interface element can be seen elsewhere (Cai & Ugai 2000).

2.2 *Limit equilibrium method*

Bishop's simplified method of slip circle analysis is employed to determine the safety factor of slopes reinforced with anchors. The reinforced effect of the anchors is considered only as an axial resisting force, P, which is assumed to be the ultimate axial anchor force determined by SSRFEM. Thus the safety factor, F_s, of slopes reinforced with anchors is given by

$$F_s = \frac{\sum\left[cb + (W + P\sin\theta)\tan\phi\right]/m_\alpha}{\sum\left[W\sin\theta - P\cos(\alpha+\theta)\right]} \qquad (1)$$

where θ is the angle between the anchor and the horizontal direction.

The simplex reflection technique is used for locating the critical slip circle that has the lowest safety factor. When the slope is stabilized with anchors, the critical slip surface is found after addition of the reinforcing effects of anchors. Thus a smaller safety factor can be obtained than that considering the effects of anchors with the critical slip surface of the plain slopes.

3 RESULTS AND DISCUSSIONS

3.1 *Model slope*

An idealized slope with a height of 8m, a gradient of 1V:1H, and a ground thickness of 10m is analyzed with a 3D FE mesh, as shown in Figure 1. The diameters of the tendon and grouted body are 32mm and 90mm, respectively, and their lengths are 6m and 6m, respectively. The soil-tendon interface is assumed to be smoothly, and the shear strength to be zero (Table 1). The shear strength of the soil-grout interface is assumed to be the same as the soil. Young's modulus of the grouted body is assumed to be the same as the reinforced concrete. The rigid circular plate on the slope surface is with a diameter of 30cm. The anchors are installed with a horizontal distance between the slope toe and the anchor head, Lx=4m, and a center-to-center spacing, D_1 =1.5m, unless otherwise stated. The material parameters of the soil, soil-tendon interface, soil-grout interface, tendon, and grouted body are shown in Table 1.

When the slope is not reinforced with anchors, SSRFEM gives a safety factor of 1.09, comparing well with 1.084 of Bishop's simplified method. The failure mechanism of SSRFEM, agrees well with the critical slip circle of Bishop's simplified method.

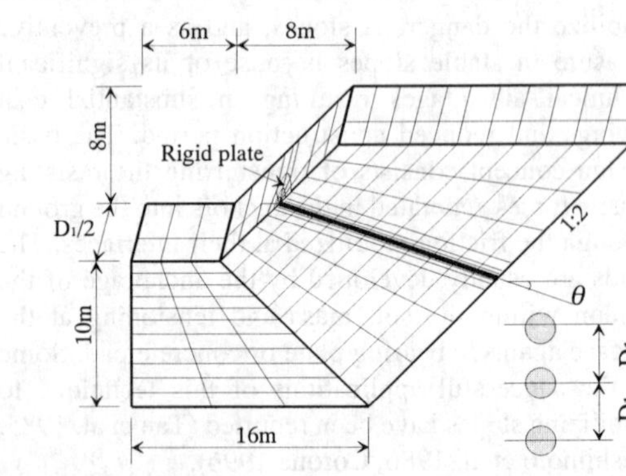

Figure 1. Model slope and schematic FE mesh

Table 1. Material parameters

Parameter	Soil	Soil-tendon interface	Soil-grouted body interface	Tendon	Grouted body
Young's modulus, E (MPa)	200	200	200	210000	24000
Poisson's ratio, ν (-)	0.25	0.25	0.25	0.30	0.20
Unit weight, γ (kN/m^3)	20.0				
Cohesion, c (kPa)	12.0	0.01	12.0		
Friction angle, ϕ (°)	20.0	0.0	20.0		
Dilatancy angle, ψ (°)	0.0	0.0	0.0		

3.2 *Effect of prestressed load*

When the anchors are installed with Lx=4m, θ = 15°, and D_1=1.5m, SSRFEM indicates that the safety factor of the stabilized slope is not influenced by the prestressed load (Table 2). This is because the ultimate axial tension is almost the same for various prestressed loads. Table 2 shows that the safety factor of the stabilized slope of SSRFEM is almost the same as that of Bishop's simplified method. The reinforcing effect of anchors can only come from the increase in the shearing resistance along the slip surface in SSRFEM, and then FEM calculation diverges under a larger safety factor. Thus SSRFEM gives a larger safety factor of the slope reinforced with anchors. Therefore, the reinforcing mechanism in SSRFEM is identical to that in Bishop's simplified method. The failure mechanism in SSRFEM is represented by the nodal displacements difference between just before failure and when the shear strength reduction factor, F, is equal to unity. The failure mechanism in Bishop's simplified method is shown in Figure 2 by the critical slip circle, which has the lowest safety factor of the slope reinforced with anchors. Figure 2 shows that the failure mechanisms, predicted with SSRFEM and LEM, agree well with each other. The prestressed load is assumed to be zero in the following analyses because it has little influence on the safety factor of slope.

Table. 2 Safety factor for various prestressed loads

Prestressed load P_i (kN)	Safety factor		Ultimate axial tension P_u (kN)
	SSRFEM	Bishop	
0	1.21	1.195	59.30
25	1.21	1.196	60.09
35	1.21	1.198	60.75
45	1.21	1.197	60.66

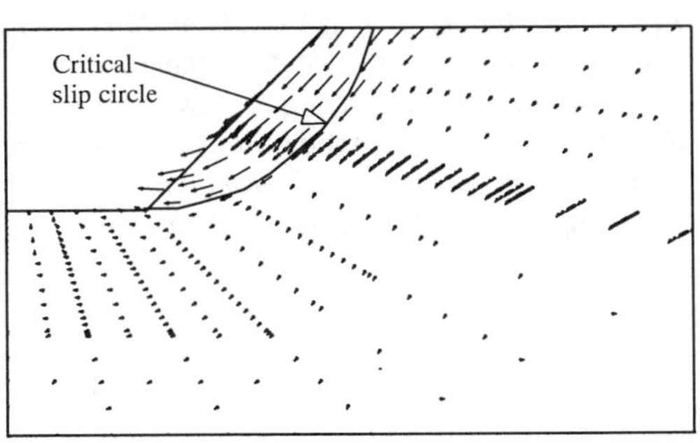

Figure 2. Failure mechanism of slope with anchors

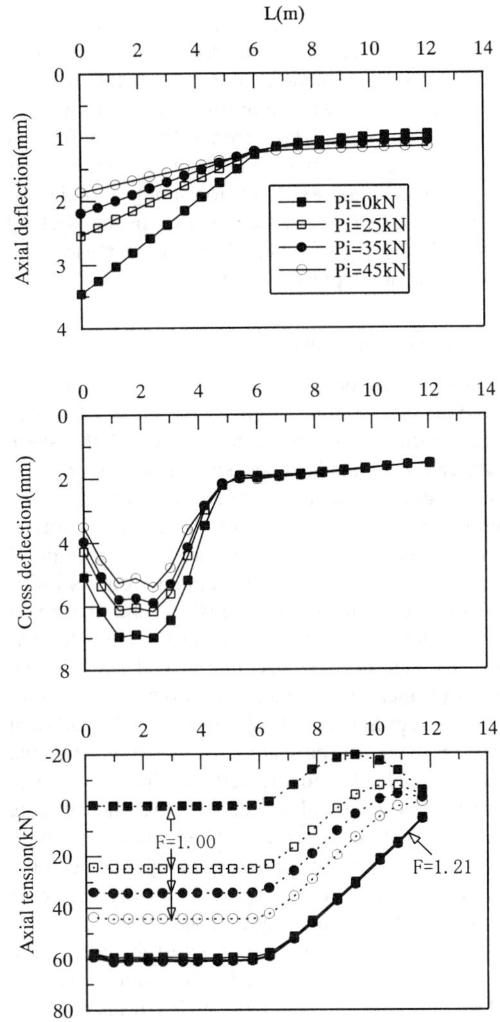

Figure 3. Response of anchor for various prestressted loads

Figure 3 shows the influence of the prestressed load, P_i, on the axial deflection, cross deflection, initial axial tension, and ultimate axial tension of the anchor at the axial distance, L, from the anchor head. When F=1.00, the axial tension is the initial axial tension, which is induced by the prestressed load. When F=1.21, and reaches the ultimate value, the axial tension is the ultimate axial tension, P_u. The ultimate axial tension includes the two parts, induced by the prestressed load, and by the shear strength reduction from unity to the ultimate value, as shown by the solid line in Figure 3. It is indicated that the prestressed load can decrease the axial and cross deflections of the anchor when the slope failure takes place. This implies that the prestressed load can decrease the movement of the slope when the slope failure takes place. The axial tension in the free tendon is the same along its axial direction because of the smooth soil-tendon interfaces. Therefore, the resisting force of the anchor completely comes from the friction between the grouted body and the soil. The distribution of the ultimate axial tension along its axial direction is almost the same for various prestressed loads, although the initial axial tension increases with the increase in the prestressed load.

3.3 *Effect of direction angle*

When the anchors are installed with Lx=4.0m and D_1=1.5m, the direction angle of the anchor has significant influence on the safety factor of the slope reinforced with anchors, as shown in Figure 4. It shows that there is an optimized direction angle of the anchor for the safety factor of the slope reinforced with anchors. The influence of the direction angle of the anchor on the safety factor comes from its influence on the ultimate axial tension, and the angle between the anchor and the slip surface. For the anchors installed horizontally, the normal stress in the soil-grout interface is smaller because of the smaller vertical soil pressure on the grouted body. The normal stress in the soil-grout interface increases with the direction angle due to the self-weight of soil. The experimental results also indicated that the ultimate reinforcing effects are mobilized under a more shearing displacement for a larger angle between the reinforcement and the sliding surface when the angle is less than 90° (Edirisinghe et al. 1996). Figure 4 suggests that the reinforcing effect of anchors is the increase in the shearing resistance along the slip surface under various direction angles of the anchor.

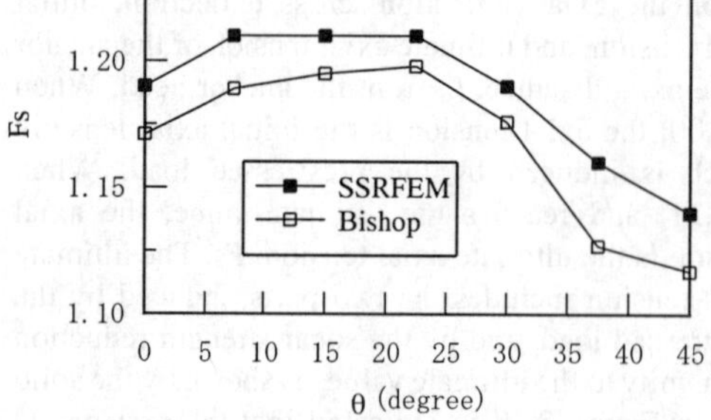

Figure 4. Safety factor versus direction angle

4 CONCLUSIONS

The 3D SSRFEM is used to analyze the stability of a slope reinforced with anchors, and to verify the reinforcing mechanisms of anchors in slopes. The soil-anchor interaction is simulated by the 3D zero-thickness elasto-plastic interface elements. The numerical results are compared with those of Bishop's simplified method. Based on the numerical results, the following conclusions are reached:

(1) The reinforcing mechanisms of anchors in slopes are to increase the shearing resistance on the slip surface due to the axial tension.

(2) SSRFEM shows that the prestressed load has little influence on the safety factor though it can decrease the movement of the slope stabilized with anchors when the failure takes place.

(3) The better stabilizing effects can be obtained when the anchors are installed with the angle between the anchor and the horizontal direction within the range of 7.5° - 22.5° for the analyzed slope.

REFERENCES

Cai, F. & K. Ugai 2000. Numerical analysis of the stability of slope reinforced with piles. *Soils and Foundations*, 40(1): 73-84.

Corona, E.P. 1996. Stabilization of excavated slopes in basalt', *Proc. 7th Int. Symp. on Landslides:* 1771-1776.

Edirisinghe, J., N. Yagi, M. Enoki, R. Yatabe & G. Ohashi 1996. Reinforcing mechanisms and a stability analysis method for reinforced earth structures. *Soils and Foundations*, 36(3): 21-30.

Griffiths D.V. & P.A. Lane 1999. Slope stability analysis by finite elements. *Geotechnique*, 49(3): 387-403.

Hashimoto, I., K. Kawasaki & H. Kodera 1986. The case of landslide control by the dead-anchors. *Proc. 21th Japan National Conf. on Soil Mech. and Found. Engrg.:1483-1486* (in Japanese).

Tan, S.B., S.L. Tan, K.S. Yang & Y.K. Chin 1985. Soil improvement methods in Singapore. *3rd International Geotechnical Seminar:* 249-272, *Nanyang Technological Institute, Singapore*.

Computer Methods and Advances in Geomechanics, Desai et al. (eds) © 2001 Balkema, Rotterdam, ISBN 90 5809 183 X

Numerical analysis of the unloading relaxation effect of rock masses during slope excavation

S.H.Chen & J.Hu
Wuhan University of Hydraulic and Electrical Engineering, People's Republic of China

ABSTRACT: Excavation of slope will cause damage and stress relief of rock masses, as a result, the physical and mechanical parameters will be changed, these effects are usually called "unloading relaxation effects" and will be the dominant factor influencing the local stability of excavated slope. In this paper a new theory and FEM algorithm to simulate the decrease of the deformation and strength parameters of rock masses because of the excavation is proposed. The theory is based on the assumption that if two initially identical samples are tested, of which one is first "unloading relaxation effected" and then loaded whereas the second is loaded and then "unloading relaxation effected", the final stress strain point is the same. A slope engineering example is studied and the correctness of the proposed method is verified.

1 INTRODUCTION

Excavation of rock slope will cause stress redistribution, as a result, from the excavation surface to the inside of slope three bands will be formed. The first is "heavily damaged band" which contains many new cracks and fissures caused by blasting and heavy unloading of initial stress whose depth could vary from 0.5 m to 5 m. The second is "stress relief band" which is mainly induced by tension and shear deformation of the existing discontinuities because of unloading of initial stress whose depth may range from 5m to 30 m. The third band behind the stress relief band is "stress regulation band" and theoretically the depth of the band is unlimited. The physical and mechanical parameters will be changed in these three bands, especially, the Young's modulus(E), the friction angle(ϕ) and the cohesion(c) of the rock masses will be decreased. In the heavily damaged band, the decrease of the above parameters could be over 50%, the decrease of parameters is less pronounced in the stress relief band at the level of 10% to 25%. Inside the stress regulation band the parameters of the rock mass undergo only a slight variation and usually can be neglected. These are usually called "unloading relaxation effects" by practitioners.

The stability of excavated slope is strongly influenced by the unloading relaxation effects, local failure may frequently occur in the heavily damaged and stress relief bands, consequently, rock bolts should be systematically installed depending on the characteristics of the bands. Nevertheless, the conventional numerical analysis of the rock slope or fails to take the unloading relaxation effects into account totally, or simply simulates the effects just by using the decreased parameters in the analysis. Unfortunately these treatments of the unloading relaxation effects can not give out correct solution of the problem.

The interesting of this paper is directed to develop a new theory and FEM algorithm to simulate the decrease of the deformation and strength parameters of rock masses because of the unloading relaxation effects during excavation. The theory is based on the principle that if two initially identical samples are tested, of which one is first "unloading relaxation effected" and then loaded whereas the second is loaded and then "unloading relaxation effected", the final stress strain point is the same. A slope engineering example is studied and the correctness of the proposed method is verified.

2 THE CONCEPT AND THE METHODOLOGY

Generally, the unloading relaxation effects belong to "autogenous strain" phenomenon, i.e. where strains occur due to some cases other than loading. The phenomenon has been wildly observed in the geotechnical engineering structures such as the collapse settlement of embankment dam(Nobari & Duncan 1972). Based on the Nobari and Duncan's experimental finding, Pinto(1983), Maranha et al.(1988) and Naylor et al.(1989) developed a procedure for simulating the effect in finite element analysis using the hyperbolic model and elasto-plastic model. The method is based on the experimental finding that if two initially unsaturated identical samples are tested, of which one is first saturated and then loaded whereas the second is loaded and then saturated, the final stress strain point is the same. This finding suggests that two sets of material properties may be used to describe the soil: the first the soil as placed, i.e. before saturation, the second the saturated soil. The stress-strain curve for the dry soil is followed until saturation occurs. Then a switch is made to the saturated stress-strain curve. This is done in two parts, first a stress change without strain and then a relaxation of the unbalanced stress (Fig. 1).

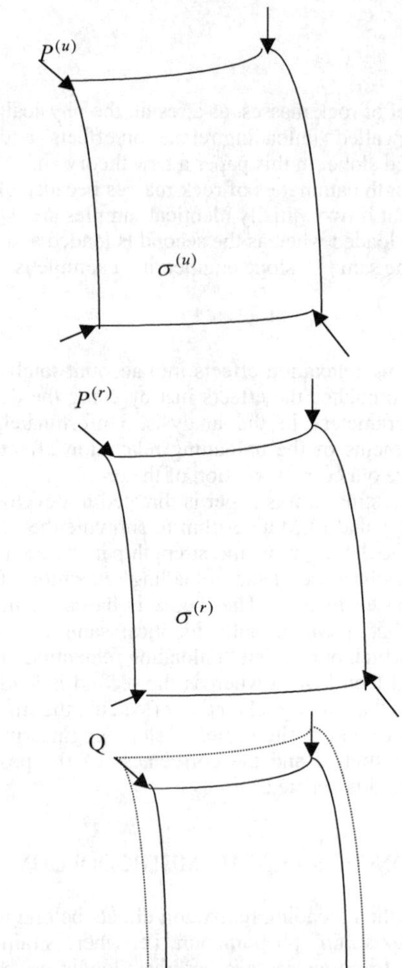

Figure 1. Idealization of effect of saturation(after Naylor et al. 1989)

The technique described is not restricted to the phenomenon of collapse effects but can in principle be applied to any "autogenous strain" situation such as the unloading relaxation effects of rock masses during slope excavation, which is the main topic of this paper. To apply the technique, however, it is necessary to define two sets of material parameters: one relating to the state without unloading relaxation effects and the other to the state with unloading relaxation effects. The feasibility of defining these sets is the main constraint on the range of applications to which the technique can be applied. Fortunately, for some important engineering such as the ship lock slope in the Three Gorges Project, the laboratory and field tests including ultrasonic detection during the excavation have been conducted and the parameters variation has been defined(Chen et al. 2000).

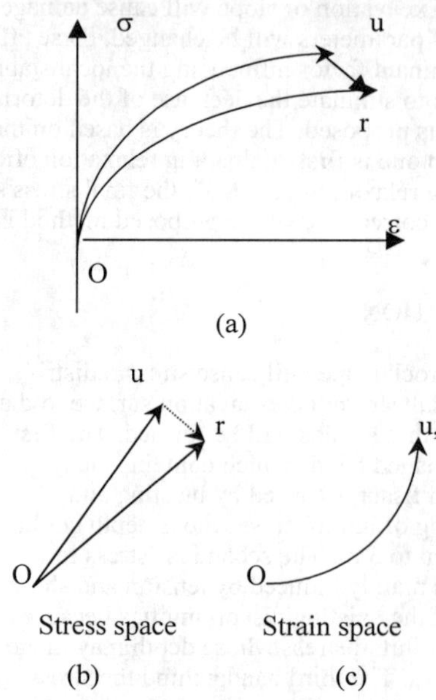

Figure 2. Conceptual stress and strain paths for clamped stress change(after Naylor et al. 1989)

Consider an element of rock under stress $\sigma^{(u)}$ prior to relaxation. It is assumed to be held by imaginary clamps. If the element is separated from the rest of the body the clamps will exert a set of forces $P^{(u)}$ on the element representing the stress transferred across the element boundary. On relaxation the element tries to reduce its volume but the clamps prevent this. Let $P^{(r)}$ represent the set of forces transmitted from the clamps to the element after relaxation. The clamps are then released. This is modelled by the application of nodal forces $Q=P^{(u)}-P^{(r)}$ which will cause deformation in the case of an excavated rock slope being dominantly downwards.

To determine the stress changes in the element due to relaxation in the clamped condition, a further idealization is made. It is imagined that the element in the rock mass was initially unstressed and that after loading it in its original state to the stress $\sigma^{(u)}$ it precisely occupies the boundaries to which the imaginary clamps are applied. A stress path from zero stress to $\sigma^{(u)}$ needs to be assumed. Associated with this stress path is a unique strain path(Fig. 2). The same initially unstressed element is now imagined to be relaxed from the start. It is loaded along this same strain path. Consequently it will precisely fit the clamped boundaries at the end of the path. The final stress, $\sigma^{(r)}$, is then the post- relaxation effective stress prior to release of the clamps. It is in equilibrium with the nodal forces $P^{(r)}$. For lack of a criterion the simplest assumption of a straight stress path has been made. This and the associated strain path is illustrated by the lines OU in Figure2b and OR in Figure 2c.

The nodal forces to be applied to simulate the release of the clamps are obtained from the stress change in conventional manner by the virtual work derived relation:

$$Q = \int_V B^T(\sigma^{(u)} - \sigma^{(r)})dV \tag{1}$$

Where B is a matrix of shape function derivatives and integration is taken over the element volume. Q is assembled for all the relaxation elements to give the total force to be applied. It should followed by the application of the excavation forces.

3 THE INCREMENTAL FORMULATION OF ELASTO-VISCOPLASTIC PROBLEM

The calculation of $\sigma^{(r)}$ will be carried out at each integrating point in each of the relaxed element after the excavation forces have been applied. Let the material observes the elasto-viscoplastic potential theory(Owen & Hinton 1980), at time t, the constitutive equation will take the following form:

$$\{\Delta\sigma\} = \left[D\right]^r(\{\Delta\varepsilon\} - \{\dot{\varepsilon}^{vp}\}\Delta t) \tag{2}$$

in which Δt is the time stepping length, $[D]^r$ is the elastic matrix, Equation (2) suggests that the relaxation occurs at the beginning of the time step. The viscoplastic flow rate is :

$$\{\dot{\varepsilon}^{vp}\} = \gamma\langle F\rangle\left\{\frac{\partial F}{\partial\{\sigma\}}\right\} \tag{3}$$

where

$$\begin{cases} F = aI_1 + \sqrt{J_2} - k = 0 \\ a = \sin\phi^r/[3(3+\sin^2\phi^r)]^{1/2} \\ k = \sqrt{3}c^r\cos\phi^r/\sqrt{3+\sin^2\phi^r} \end{cases} \tag{4}$$

in which ϕ^r, c^r and γ^r are the friction angle, cohesion and fluidity parameter of rock mass after relaxation respectively.

The equilibrium equation at each time step is now:

$$[K]\{\Delta U\} = \{Q\} + \{\Delta F^{vp}\} \tag{5}$$

where $\{\Delta F^{vp}\}$ is the nodal forces relating to the viscoplastic deformation(Owen and Hinton, 1980), {Q}is the nodal forces simulating the release of the clamps which will be discussed as follows.

Let $D^{(u)}$and $D^{(r)}$ be the modulus materices for original and relaxed material respectively, and $\{\Delta\sigma^{(u)}\}$ be the stress increment of the previous time step, then at the beginning of the present time step the first stage is the calculation of the strain path(OU=OR) in the Figure 2c :

$$\{\Delta\varepsilon\} = \left[D^{(u)}\right]^{-1}\{\Delta\sigma^{(u)}\} \tag{6}$$

The stress increment along the path OR is obtained as

$$\{\Delta\sigma^{(r)}\} = [D^{(r)}]\{\Delta\varepsilon\} \tag{7}$$

Introducing the Equation (6) we have:

$$\{\Delta\sigma^{(r)}\} = [D^{(r)}]\left[D^{(u)}\right]^{-1}\{\Delta\sigma^{(u)}\} \tag{8}$$

Then the nodal forces simulating the release of the clamps {Q} can be calculated with the Equation (1).

4 APPLICATION

A rock slope of 70m high excavated by 5 steps is studied(Fig. 3), the depth of the "heavily damaged band" and "stress relief band" are 5m and 15m respectively.

Figures 4-5 are the accumulated displacement after the completion of the slope, it can be seen that main difference is near the top of the slope: the conventional elasto-viscoplastic analysis predicts obvious "rebound" deformation at this part where the vertical displacements are directed upright, while the elasto-viscoplastic analysis proposed in this paper considering the unloading relaxation effects shows

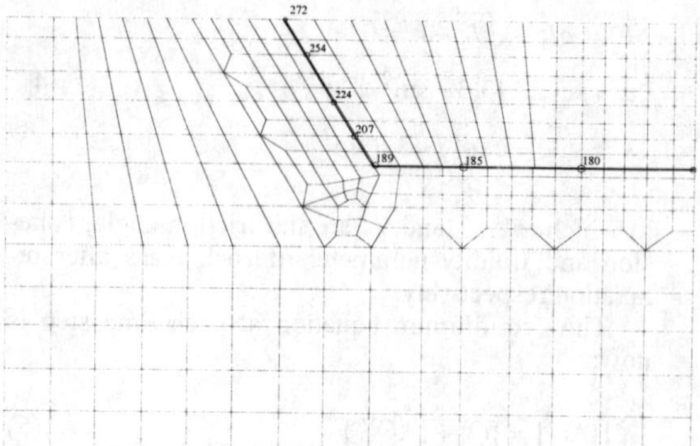

Figure 3. Rock slope and FE mesh

Table 1. The parameters of the rock masses

	E(Mpa)	ν	C(Mpa)	φ(°)
Intact rock	40000	0.22	0.6	25.0
heavily damaged band	20000	0.22	0.1	12.5
Stress relief band	30000	0.22	0.18	22.5

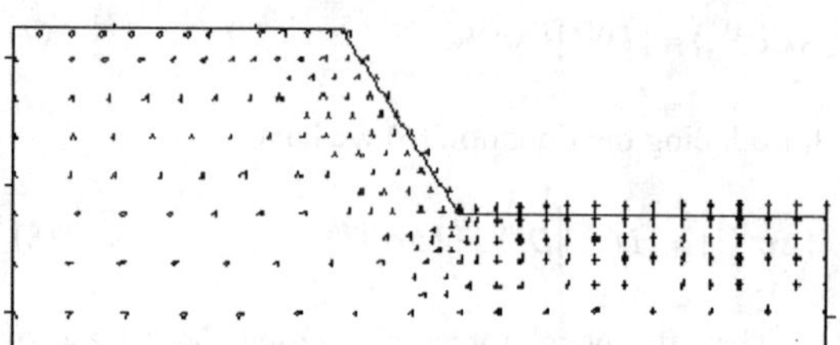

Figure 4. Accumulated displacement after the completion of the slope(conventional analysis)

"collapse" deformation at the top area of the slope, which has been widely observed in the slope excavation practices.

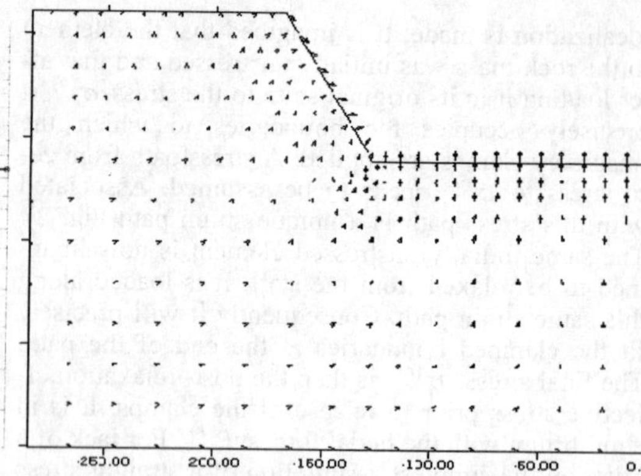

Figure 5. Accumulated displacement after the completion of the slope(relaxation analysis)

5 CONCLUSIONS

The methodology proposed in this paper is based on the fact that usually two sets of parameters before and after excavation could be obtained by the laboratory and field tests. With the help of the assumption similar to the soil collapse, a practical algorithm combined with the elasto-viscoplastic potential theory can be developed successfully. Anyway, it seems quite attractive to generalize the idea of this paper into the area of damage mechanics theory, which can take the decrease of the parameters into account step by step as the evolution of the damage tensor.

REFERENCES

Chen, S. H., S. F. Chen, I. Shahrour & P. Egger 2000. The feedback analysis of excavated rock slope. *Rock Mechanics and Rock Engineering.*

Nobari, E S &J. M. Duncan 1972. Effect of reservoir filling on stresses and movements in earth and rockfill dams, College of engineering. *Office of Research Services , University of California, Berkeley, Report No. TE-72-1.*

Maranha des Neves E & A. A. Veiga Pinto 1988. Modelling of collapse on rockfill dams. *Computer and Geotechnics,* 6:131-153.

Naylor D J, S. L.Tong & A. Shahkarami 1989. Numerical modelling of saturation shrinkage. In *Numerical models in Geomechanics(NUMOGIII):* 636-648. New York:Elsevier.

Owen D R J & E. Hinton 1980. *Finite Elements in Plasticity:Theory and Practice.* Swansea(U.K.):Pineridge Press Ltd.

Veiga Pinto A. A. 1983. *Structural behaviour prediction of rockfill dams.* Lisbon: Doctoral Thesis of L.N.E.C.

Computer Methods and Advances in Geomechanics, Desai et al. (eds) © 2001 Balkema, Rotterdam, ISBN 90 5809 183 X

Engineering characteristics and slope stability of sedimentary rock on weathering grade

Manyoung Choi & Sewook Oh
Measurement Engineering of KRISS, Taejon, Korea

Gwanghee Heo
Department of Civil Engineering, University of Konyang, Nonsan, Korea

Ingoo Lee
Kon A Measurement and Instruments, Taejon, Korea

ABSTRACT : Orientation of discontinuous plane is of paramount importance for rock slope stability studies because they control the unstable conditions and deformation. The failure of the cut slope is chiefly attributed to weathering, which causes change in the shear strength. The weathering is also directly related to slaking and swelling characteristics.

In the paper, the core of diameter 10cm was obtained by digging on a rock slope of Kong-ju in Korea. Then EDX (Energy Dispersive X-ray spectrometer) analysis was carried out to know the elements of chief rock-forming minerals, and uniaxial compression test and point load test were taken to study on the engineering characteristics of conglomerate and red shale.

As a results of test, absorption of conglomerate and shale were estimated 0.329～0.996%, 0.341～2.547% and slaking index was 85.11～99.58, 58.37～99.23, respectively

Finally, a full-scale of rock slope were performed analysis of slope stability with finite difference code direct.

1 INTRODUCTION

Recently, many construction works involving rock slopes or tunnel excavation have caused some problems in terms of time-consuming quality of work, tremendous amount of cost, and planning and construction procedure itself. After regular rainy or thawing season especially, they are exposed to many dangers such as falling rock, land sliding, and collapse of slope, mainly due to water absorption or weathering (Hoek, E., and Brown, 1980).

The characteristics of strength according to weathering is analyzed and correlation between the uniaxial compressive strength and other characteristics were derived with various kind of rock by H. S. Jeong(1997).

However, it is difficult to estimate the stability because mechanical characteristics of rock slope depend on geological condition such as a rock mass with tension crack, joint, fault, and broken zone.

Although all particles of a rock may appear to be of equal quality and size, they are not as a matter of fact. The complex dynamics of each different quality within a rock makes it difficult to define their characteristics and predict the degree of stability that is affected by the geological condition such as a rock mass with tension crack, joint, fault, and broken zone. Therefore, rock slope problems require considering the geometrical relationship between discontinuity planes, the slope and the force vectors involved. In order to properly predict the behavior of discontinuous rock slope is the finite element analysis or finite difference method by general concept useful (Richard E. Goodman, 1989).

This research experiments on a core of diameter 10 cm obtained from a rock slope in Kong-ju, Korea. EDX (Energy Dispersive X-ray spectrometer) analysis was carried out to know its element of chief rock-forming minerals, and a uniaxial compression test and a point load test were taken to study on the engineering characteristics of conglomerate and red shale. Finally, in order to prove the validity of finite element method, a full-scale numerical rock slope model of Kong-ju is analyzed and compared.

2 GEOLOGY AND ROCK MASS

2.1 Geology

Geology is composed of the Metamorphic Rock belonging to Precambrian, and the Sedimentary Rock of Kyongsang System. Also, it is composed of the conglomerate of Metamorphic Rock in contradiction to Shinra in Kyongsang System by rock shape.

Also, most of rock is mainly composed of the conglomerate colored the reddish brown of Metamorphic-Sedimentary Rock. The Sedimentary layer is composed of the unconformity of the sand and gravel that is applicable to the fourth of an alluvium of Metamorphic-Sedimentary Rock.

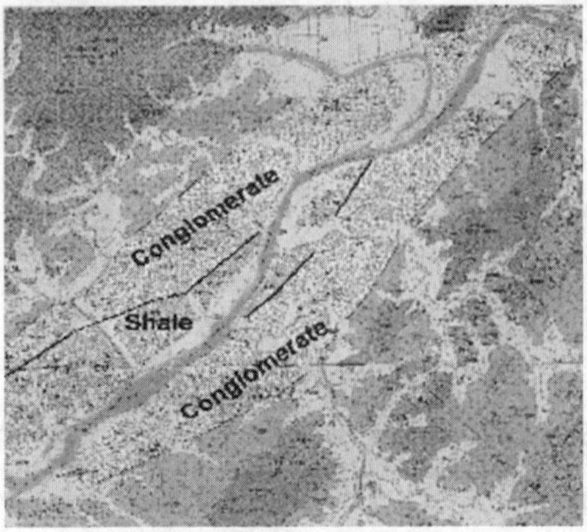

Fig. 1 Plot of geological map

2.2 *State of discontinuity*

The dip of slope was N42E/45NW and most of the section is composed of conglomerate with red and gray and it is growth up fracture zone and a little water has revealed.

Upper part of a slope contained a large quantity of small joint and weathering grade is state of SW(Slightly Weathered)~HW(Highly Weathered) and it is shown rock fall by weathering and bedding plane.

A part of slope is collapsed by joint on toe of slope , fracture and weathering zone

The dip of slope was N40~44E and most of the section intersect conglomerate and sandstone and calcareous red shale distribute.

Strength of rock mass larger than soft rock but most of site II is that shear strength decreased by clayey filling, water, and weathering.

Upper part of a slope occur small sliding on weathering grade is high and most of slope is state of SW(Slightly Weathered)~HW(Highly Weathered) and it is shown rock fall by water and filling.

2.3 *The evaluation by SMR classification*

The slope stability about the section of the subject slope is evaluated by SMR(Slope Mass Ratio) classification method. The slope materials used in evaluation are shown in the table below.

It is proved that the slope stability is very low considering the characteristics of a discontinued surface, and the evaluation also shows that most of the slope is in very unstable condition. In the case of stratified joint whose slope direction and its tendency is very similar, the stability goes remarkably lower. The shape of collapse is assumed to appear as a plane and wedge shape.

3 ENGINEERING CHARATERISTICS OF ROCK

3.1 *EDX(Energy Dispersive X-ray spectrometer) tests*

EDX (Energy Dispersive X-ray spectrometer) tests were performed on mineral of gravel and red shale rock for the analysis of their constituents. The analysis showed that the conglomerate consists of gravel with silica and clay mineral, also that some red shale is made of a number of clay mineral, silica, calcite, mica, and sand plagioclase etc..

An absorption test was also performed to know the way in which rocks were weathered. According to the test, the conglomerate turned out to be denser than red shale, and the range of absorption was also larger than red shale. It is because the conglomerate whose gravel is dense does not permit much empty space due to the absence of bedding surface unlike shale. However, the shale absorbs more water and in larger area than the conglomerate does.

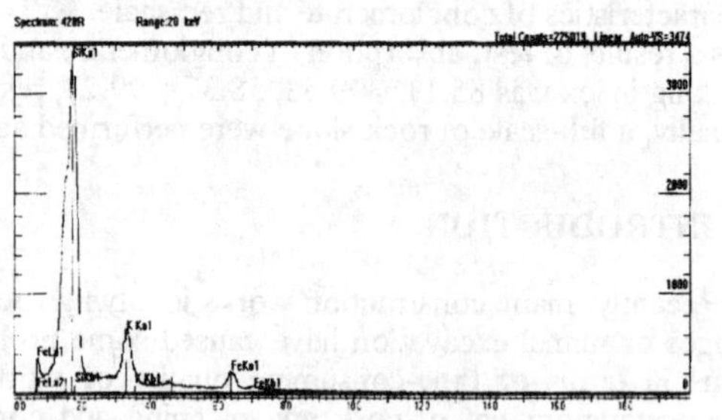

Fig. 1. Plot of results for EDX (Energy Dispersive X-ray spectrometer) tests

3.2 *Absorption test*

Absorption ratio is one of basic physical properties of the rock that indicates the quantity of water permeated in void of rock specimen, and is the ratio that indicates for the mass of the specimen as the maximum absorption quantity that rock specimen can absorb.

Generally, the absorption ratio of strong rock shows below 10%, and that in case of the fresh and dense rock reach not in 1% normally, and it is trend that the absorption ratio increases due to weathering. It indicates a far-reaching range due to the weathering or density of rock from a small number % to 60% in weak rock.

Approximately, the strength of rock is declined without relation with the species of rock if the absorption ratio is over 5%, and it indicates the trend that the strength of rock is very declined if the absorption ratio becomes over 10%.

The method to calculate the absorption ratio is the same as equation (1).

$$\omega = \frac{W_{wn} - D_{wn}}{D_{wn}} \times 100(\%) \qquad (1)$$

Where, ω = the absorption ratio (%)

W_{wn} = The weight of specimen under water of N times

D_{wn} = The weight of specimen after drying of N times

Testing result indicates that the absorption ratio of conglomerate is average 0.731%, and shale is average 1.56%. Therefore, the shale shows that it was a great influence on weathering more than conglomerate.

3.3 Slaking test

Slaking test indicates the measure of resistance of rock for reiterating of the drying and water, and shows the resistance for brittleness or weakness when the reiteration of the drying and water pillow in the rock specimen is performed over 2 times.

To measure slaking durability, the representative rock specimen 10 pieces that the weight is actually estimated by per 40g~60g, is dried in oven after puts in testing device (the cylinder tube of the length 100mm and diameter 140mm made by a wire net of a diameter 2mm) same as Fig.2.

The drum is revolved during about 10 minutes by 200rpm after it is set up the water tank, and then put the water in the water tank to 20mm below the axis of rotation.

The drum is dried in oven by the temperature of 105℃ after it takes out the water tank, and then after it is cooled, the weight of it is estimated, and this stage is reiterated over 2 times.

The slaking durability is calculated by using the equation (2).

$$I_{d2} = \frac{C - D}{A - D} \times 100(\%) \qquad (2)$$

Where, I_{d2} = index of slaking durability

A = the weight of the initial specimen and drum

C = the weight of the final specimen and drum

D = the weight of drum

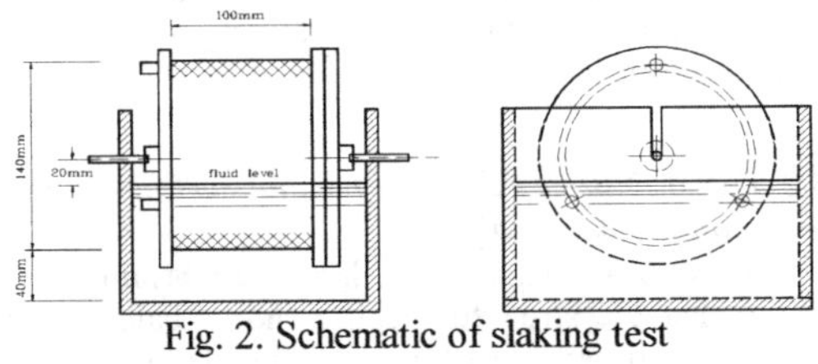

Fig. 2. Schematic of slaking test

Fig. 3, Fig. 4 show the results of slaking test. The conglomerate loses weight by 0.42~14.89% after a couple of slakings while the shale loses by 0.77~41.62%.

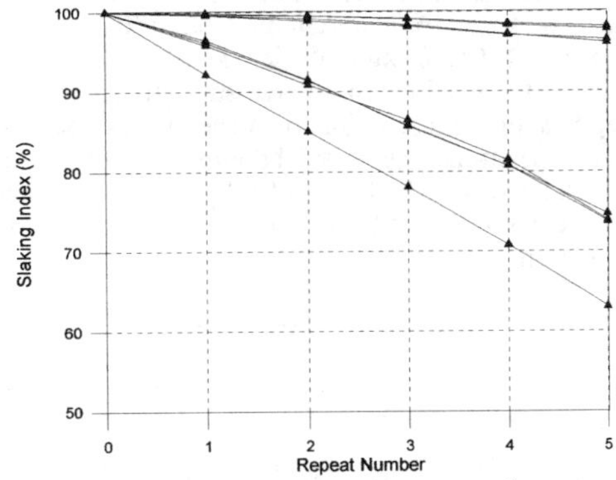

Fig. 3. Slaking index with repeated Number (conglomerate)

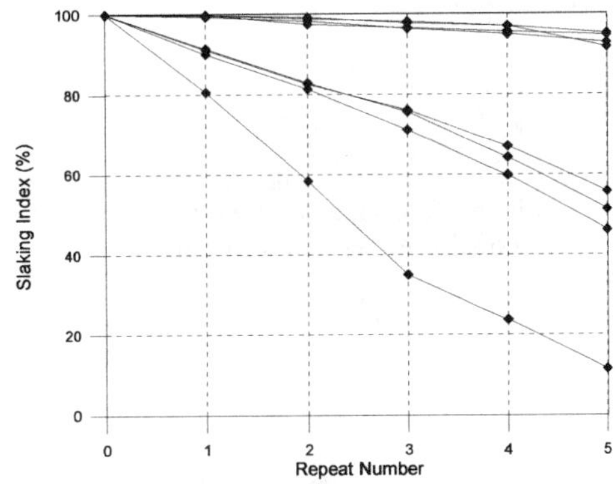

Fig. 4. Slaking index with repeated number (Red shale)

When listing facts use either the style tag List signs or the style tag List numbers.

3.4 Uniaxial Compressive Strength

Suggested techniques for determining the uniaxial compressive strength and deformability of rock material are given by the International Society for Rock Mechanics Commission on Standardization of Laboratory and Field Tests(ISRM Commission, 1979).

The core of diameter 10cm was obtained by sampling on a rock slope of Kong-ju in Korea. Uniaxial compressive strength was carried out to study on the engineering characteristics of conglomerate and red shale.

Specimens were stored, for no longer than 30 days, in such a way as to preserve the natural water content and tested in that condition.

Load were applied to the specimen at a constant stress rate of 5 ~ 10kg/cm^2/sec. The tests were performed along boring depth to know variation of strength with weathering grade of sedimentary rock.

The uniaxial compressive strength of conglomerate was 348.40 ~ 516.08 kg/cm^2 in depth of 0.5 ~ 3m and 444.90 ~ 1101.70 kg/cm^2 in depth of 3 ~ 11.5m. Each difference in compressive strength according to depth is due to discontinuous plane in rock specimen. In the meantime, the uniaxial compressive strength of red shale was 532.41 ~ 733.05 kg/cm^2 in dry condition and 433.15 ~ 517.89 kg/cm^2 in saturated state condition.

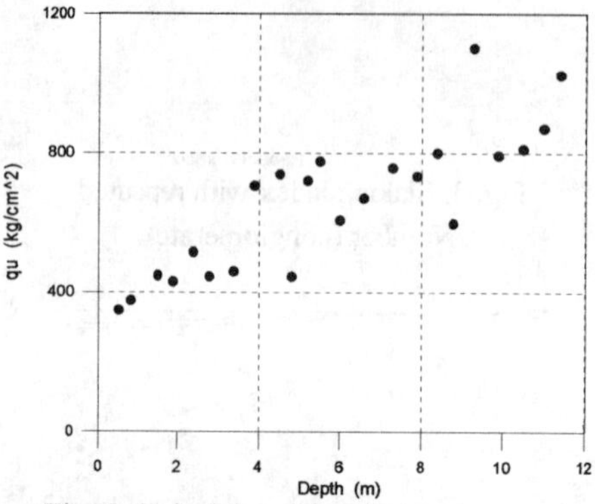

Fig. 5. Uniaxial compression test on conglomerate(natural condition)

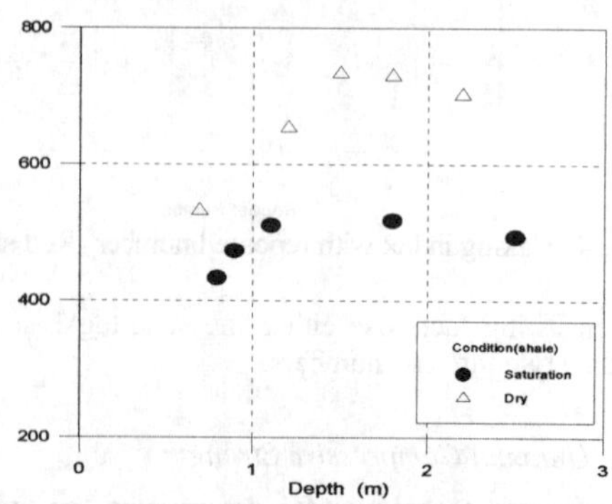

Fig. 6. Uniaxial compression test on red shale.

Table 2. Results of absorption test with rock types

Rock type / Average		Conglomerate	Red Shale
Absorption(%)		0.731	1.558
Slaking Index		0.42 – 14.89	0.77– 41.6
Uniaxial compressive strength	0.5m-3m	384.4-516.08	433.15-733.05
	3m-11.5m	444.9-1101.7	
SMR classification		Ⅳ ~ Ⅴ	

4 THE ANALYSIS FOR SLOPE STABILITY

4.1 The analysis result by stereographic projection

The stereographic projection method is widely being used that it is caculated to consider the direction of cutting surface, the direction of joint surface, and cutting angle of joint surface, and so on that is got by geological survey of the surface and the experiment to analyze the stability of normal surface quickly and approximately. Hoek and Bray(1971) proposed the method to analysis the stability of normal surface by using great circle projection method in rock.

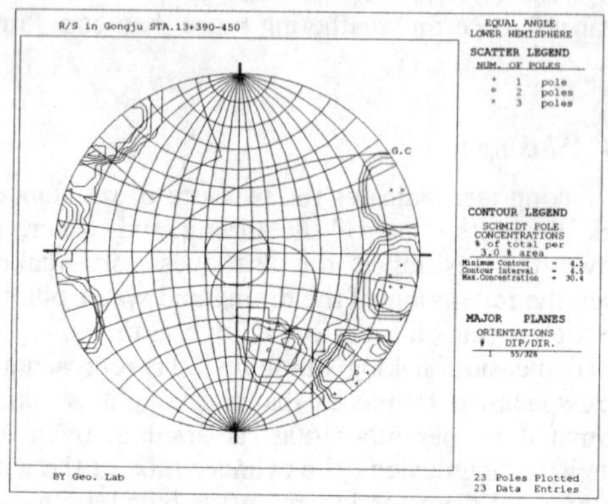

Fig. 7. Plot of analysis result by stereographic projection (Site Ⅰ)

This slope was formed to N42E/45NW, J1 joint group of N55 ~ 72E/35 ~ 60NW concentrated in left and right side of it, J2 joint group of N40 ~ 60W/55 ~ 75NE, and J3 joint group that strike is same, and gradient direction is opposite with this slope.

This section is forecasted plane or circular failure by serious cracking in center part of slope, and plane and partial toppling failure by bedding concentrated in left and right part.

The result executing the stability analysis of slope by a stereographic projection method is forecasted that plane failure is induced by concentrate distribution of J1 joint group into daylight envelope, and that the possibility of a partial toppling envelope is induced by distributing into toppling envelope of J3 joint group.

4.2 The stability analysis of normal surface by limit equilibrium analysis

The condition equation of limit equilibrium is condition that the strength causing sliding balances exactly with strength resisting in sliding. To indicate the stability of normal surface uses the safety factor of

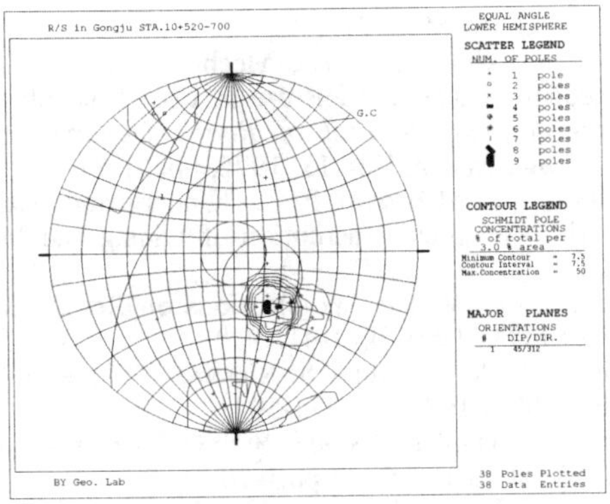

Fig. 8. Plot of analysis result by stereographic projection(Site Ⅱ)

slope, and it is defined to ratio of the sum of strength resisting in sliding for the sum of strength causing sliding.

That is, the coefficient of slope stability(F) is

$$\frac{\text{Resistance}}{\text{Sliding of the rock}} = \frac{\text{Shear strength of the rock}}{\text{Shear Strength causing collapse}}$$

Slope gradient of this subject normal surface is applied overall the gradient of normal surface because is very irregular condition due to falling etc. The tension crack of a small scale is discovered in the part of normal surface, but the analysis was performed by regarding that tension crack don't exist.

Also, the coefficient of shear strength applied in analysis referred testing values of shear strength of joint surface, and assumed to C=2 t/m^2 , $\phi = 20°$ by considering sufficiently field conditions a clayey filling and irrigation and so on.

The another constant except them is applied due to laboratory test, and the normal surface is estimated under the standard of the height to 1 small stair.

Also, to consider the influence on irrigation and storm water, it is analyzed in case of the distribution of water pressure is same with the height of normal surface, 0.5 times of the height of normal surface, and without water pressure respectively.

Analysis result shows the distribution of safety factor, 0.59 in case of being acted buoyancy the most greatly, and is predicted that the stability of normal surface is greatly influenced on storm water and irrigation.

The safety factor after the gradient of slope is relaxed, shows 1.73 in case of being acted buoyancy the most greatly, and 3.49 in case of not considering buoyancy.

Therefore, it is necessary to apply a method of construction of vegetation, and it is applied gradient relaxation of normal surface, and Nail system to a method of construction of foundation in this research.

4.3 The stability analysis of normal surface by FDM

The stability analysis was performed by using FDM(Finite Difference Method) code, FLAC to estimate the application of a method of construction using gradient relaxation together with Nail system applied to this normal surface.

It must utilize to design materials by considering prediction of the strain and settlement around the normal surface, and determine prediction of the behavior and parameters of ground in similar section.

Also, it was investigated that the effect of stress, strain, and displacement after the normal surface is reinforced, give any influence on underground condition after the analysis for effect range is executed.

The objective is that optimizes reinforcement members due to the surrounding condition given, and determines design factor of reinforcement members of normal surface through the understanding and the analysis of stress path due to respective condition of contiguous ground.

The values of physical property of applied ground is applied that density is 2.6t/m^3, the coefficient of displacement is 150×10^3t/m^2, poisson ratio is 0.25, cohesion is 100 t/m^2, friction angle is 20°.

As a results of analysis, maximum slope deformation before reinforced indicate 23cm, 32cm in upper slope and slope surface. Concentrated Stress was shown upper slope and surface and tensile failure is not occurred.

The maximum lateral displacement shows 15cm in case the gradient of slope is 1:1.5, and the stress of the top slope shows to decline 18%.

It is indicates that the maximum displacement of the slope is 1.7cm, and lateral displacement declines to 5.7cm in case of reinforcing the slope by using Nail system, and the stress of the top slope declines about 40%.

5 CONCLUSION

1. Absorption ratio of conglomerate and shale indicate 0.329～0.996% respectively.
Because gravel become stiff, the conglomerate be dense and empty of it small due to never bedding surface as shale. Absorption of shale is showed range larger than conglomerate due to water content be enlarged in weathering.
2. As a results of slaking test, slaking index of conglomerate indicate range of 85.11-99.58 and shale indicate 58.37-99.23. Therefore, we know shale become weak resistance to weathering by shallow depth and it is greatly influenced strength of rock.
3. As a results of analysis, safety factor is 0.59 in case of maximum water level and it is 1.05 in case of

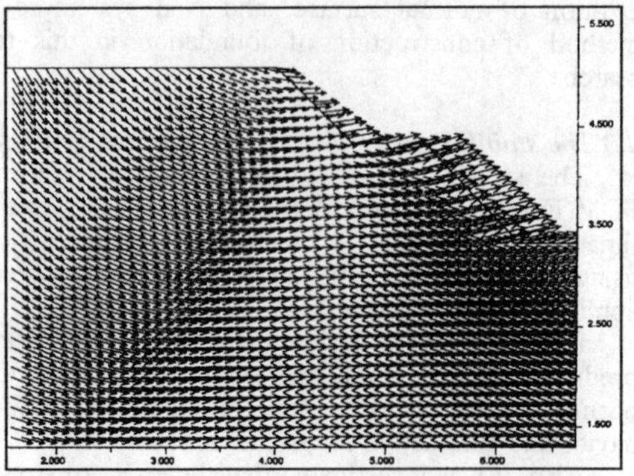

Fig. 9. The analysis result of displacement before reinforcement

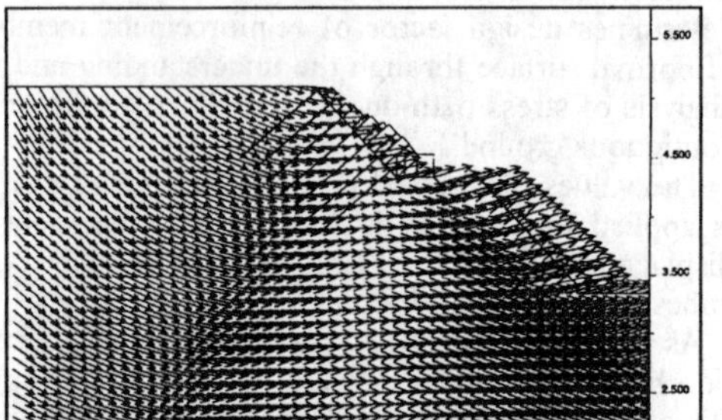

Fig.10. The analysis result of displacement after reinforcement

water level zero. However, after inclination of slope gently down, safety factor indicate 1.73, 3.49 respectively.

4. As a results of finite difference analysis, reinforced slope by nailing is better than inclination of slope gently down 1:1.5.

REFERENCES

1. Brown, E. T., Richards, L. R., and Barr, M. V., "Shear Strength Characteristics of the Delabole Slates", *Proceedings of the Conference on Rock Engineering*, Newcastle University, Vol. 1, 1997, pp. 53 ~ 70.
2. Broch. E. and Franklin, J. A., "The Point Load Strength Test", *International Journal of Rock Mechanics and Mining Science and Geomachanics Abstracts*, Vol. 9., 1972, pp.669-697.
3. Gamble, ISRM, 1971.
4. Van Eeckhout, L., "The Mechanism of Strength Reduction due to Moisture in Coal Mile Shale", *International Journal of Rock Mechanics and Mining Science*, Vol. 13, 1972, pp. 61 ~ 67.
5. Hoek, E., and Brown, E. T., "Underground Excavations in Rock", Institution of Mining and Metallurgy, 1980, pp.153 ~ 154.
6. Duncan C. Wyllie, "Foundations on Rock", E & FN SPON, 1991, pp.1-41,
7. M.G., Anderson and K.S. Richards, "Slope Stability",Wiley, 1987, pp. 475-504.
8. R.J., Chandler, "Slope Stability Engineering", 1991, Thomas Telford pp.1-70.
9. Richard E. Goodman, "Rock Mechanics", Wiley, 1989, pp. 293-339.

Computer Methods and Advances in Geomechanics, Desai et al. (eds) © 2001 Balkema, Rotterdam, ISBN 90 5809 183 X

Numerical analysis and centrifugal modeling of the stability of a natural slope

Jiancheng Du
Sichuan University, Chengdu, People's Republic of China

Limin Zhang
Hong Kong University of Science and Technology, People's Republic of China

ABSTRACT: To evaluate the stability of a natural deposit slope, the nonlinear finite element analysis and centrifugal model tests were carried out. The results of the calculations and the experiments are basically consistent. The slope is stable at the natural state, but at the condition of impounding of the reservoir, the submerged part of the slope will be softened and its stability will decrease. Under the actions of earthquakes with intensity degree Ⅷ, the slope will collapse. At the end of the paper, the authors provide some suggestions to reinforce the slope.

1 INTRODUCTION

The Xiaowan Waterpower Station locates on the Lancangjiang River that is one of the largest branch of the Yangtse River. It is an important project for flood control, water supply, and power production. It has a total installed capacity of 4200MW. An arch dam of 292m height is planed to be built. On the left bank of the reservoir, however, there is a deposit slope of about 4 million cubic meters in volume. The front part of the slope is near the abutment of the arch dam and will be submerged when the reservoir impounds the water. The slope consists mainly of debris rocks, and a soft stratum of 0.15 to 8.05 meters thick in the interface of the bedrock and the deposit slope. Underground water flourishes within the whole slope. The longitudinal cross section of this slope is shown in Figure 1. To evaluate the stability of the slope, the authors conducted a series of centrifugal model tests and calculations.

2 FINITE ELEMENT METHOD ANALYSIS

2.1 *Definition of the factor of safety*

When the slip surface of a slope is certain, the normal stress σ_n and the shear stress τ_s on it can be obtained through finite element method analysis (Pan & Du 1999). To some point, the point factor of safety is defined as the following expression:

$$k_p = \frac{\tau_f}{\tau_s} = \frac{\sigma_n tg\varphi + c}{\tau_s} \quad (1)$$

where τ_f = shear strength, τ_s = actual shear stress, σ_n = normal stress, c= cohesion, and φ = friction angle. If k_p is less than 1.0, the point is supposed to rupture. Furtherly, the following equation can be used to calculate the general factor of safety of the slip surface:

$$k_g = \frac{\sum k_p A_i}{\sum A_i} = \frac{\sum[(\sigma_n tg\varphi + c)/\tau_s]A_i}{\sum A_i} \quad (2)$$

where k_p =point factor of safety, A_i = area of slip surface element.

2.2 *Computing program and constitutive model*

The analysis described herein was conducted using three-dimensional finite element program JOINT3. This program code can carry out the elasto-plastic calculation with the criterion of Drucker-Prager or Mohr-Coulomb (Du & Zhang 1997). Goodman's joint element with thickness of 5cm was used to model the behavior of the interface between the deposit and the bedrock. The entire deposit slope, with the length of 750m and width of 250m, was divided into 3 parts. A total of 998 elements and 1082 nodes were used to perform the analysis, of which 129 were joint elements. The finite element mesh is shown in Figure 2. The values of parameters adopted by the numerical analysis are shown in Table 1. These material properties are provided by Kunming Survey and Designing Institute. They can correctly describe the behaviors of the materials.

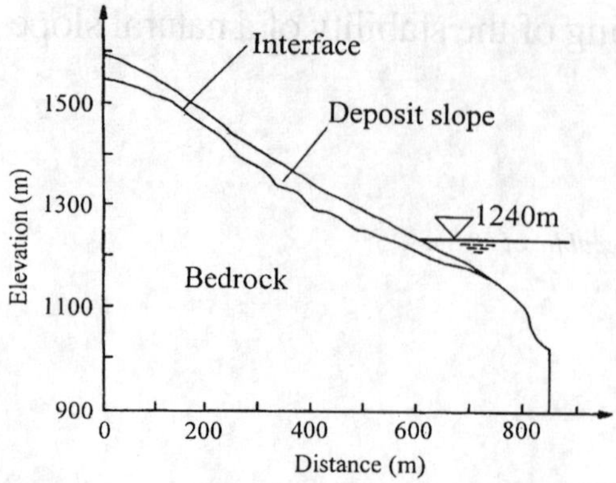

Figure 1. The longitudinal cross-section of the slope

Table 1. The values of material properties in the analysis

Property	Rocks	Deposit	Soft stratum
Unit weight (kN/m^3)	26.3	20.6	20.6
Elastic modulus (MPa)	22000	3000	300
Poisson's ratio	0.2	0.3	0.3
Cohesion (MPa)	1.5	0.1	0.05
Friction angle (degree)	56	45	31
Tension strength (MPa)	0.5	0.1	0.05

3 CENTRIFUGAL MODEL TESTING

3.1 *The centrifuge and the soil gradation*

The centrifugal tests to evaluate the performance of the deposit slope were conducted in the SCU's geotechnical centrifuge (Zhang & Hu 1994). The main characteristics of this machine are:
Capacity: 25gt;
Effective radius: 1.54m;
Maximum acceleration: 250g;
Model container size: 0.6m×0.4m×0.4m;
Signal sliprings: 32 rings;
Driving power: 30 kW;
Rotary joints: 20MPa for oil, 0.6MPa for water.

For the prototype slope, the maximum particle diameter of the deposit is about 2 meter. Therefor, the particle size should be reduced to get the soil accommodated in the model container. According to the results of comparison studies regarding to granular soil (Zhu 1990), performance of the slope can be properly represented when the maximum model material diameter is in the range of 1/40-1/160 that of prototype. In this test series, the maximum diameter is 20mm, the gradation below 2mm diameter is the same as that of the prototype so that their original properties can be guaranteed. The maximum dry density and the optimum water content of the soil are respectively controlled in the range of 19.1-20.5kN/m^3 and 8.5%. The prototype bedrock is modeled by precast concrete in the tests.

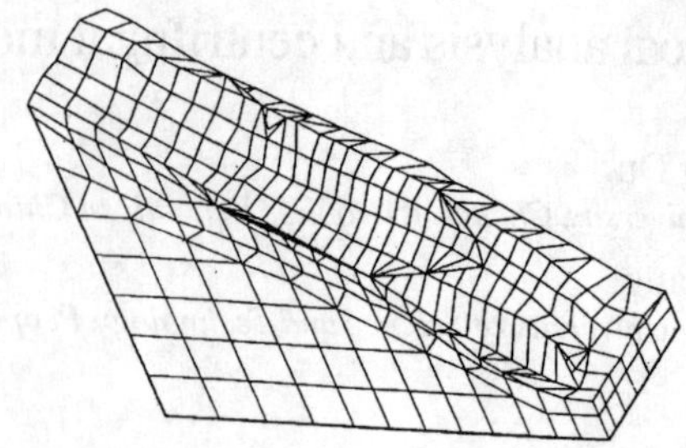

Figure 2. The finite element mesh of the slope

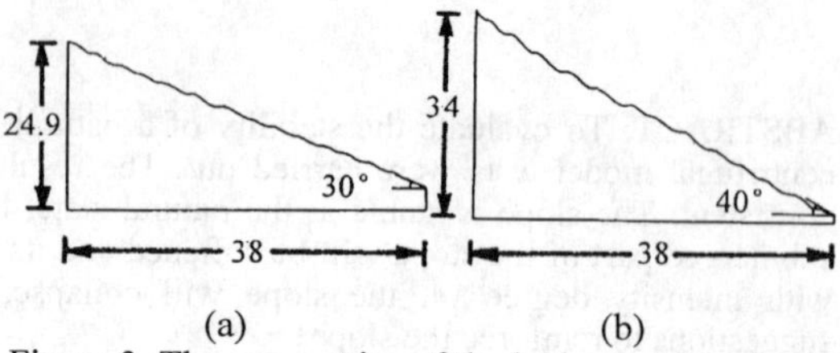

Figure 3. The geometries of the bedrock (a) aseismic condition (b) seismic condition (length unit: cm)

3.2 *The preparation of the model*

Considering the different conditions of operation in the centrifuge tests, two types of bedrock models were employed. The one is used to model the slope bedrock at the condition of natural state, the other is prepared for the condition under the actions of earthquakes. The geometries of the bedrock models are shown in Figure 3. It is noted that because of the limitation of the SCU's centrifuge this paper gives the seismic results of experiments by pseudostatic method.

The centrifuge tests were conducted following the below steps:

1. Place the bedrock model into the container and lubricate the surface with vaseline;
2. Compact the model soil on the surface of the bedrock to form the deposit slope, and place the pressure cells on the concerned positions;
3. Make the deposit slope consolidated at the prescribed acceleration to resume its site stress state;
4. Mark the visible section of the model with a grid;
5. Run the centrifuge at a prescribed acceleration and evaluate the performance of the deposit slope.

4 RESULTS

4.1 *Results of the numerical analysis*

When the finite element method was conducted, underground water in the deposit slope was modeled by exerting water pressure of 500kPa on the slip surface. Moreover, the pseudostatic method was employed to perform the seismic analysis (Zhang & Huang 1995). That is to say, each element was applied an additional horizontal and vertical load to represent the earthquake force. These loads are respectively expressed as follows:

$$P_{hi} = 0.169\, W_i \qquad (3)$$

$$P_{vi} = \frac{2}{3} P_{hi} \qquad (4)$$

where W_i is the weight of the element.

Figure 4 shows the contours of the calculated values of the factor of safety on the slip surface. At the condition of natural state, it is discovered that the factors of safety are largely greater than 1.4, which demonstrates the entire deposit slope is very safe. However, the values distributed near the Drinking Gully are less than 1.2. It shows that the slope has poor stability in the position of the Drinking Gully. When water pressure of 500 kPa is unloaded from the slip surface, the factor of safety will greatly increase. Therefore, it is a practical way to drain underground water for improving the stability of the slope. When the reservoir impounds water, about one tenth of the volume of the deposit body will be submerged. The calculated general factor of safety is still up to 1.47, which is only slightly less than 1.52 at the condition of natural state. It shows that the impounding of the reservoir will not remarkably effect the stability of the slope. When earthquakes with intensity degree Ⅷ attack the slope, the values of the factor of safety (see Figure 4) are largely less than 1.0, those near the Drinking Gully are less than 0.6, the general factor of safety will substancially decrease to the value of 0.92. At the seismic condition, the deposit slope will lose its stability.

4.2 *Results of the centrifuge tests*

The aseismic bedrock was used at the condition of natural state. Firstly, The model experienced 30 minutes running during centrifuge flight at 50g to resume its site stress state. Then the centrifuge was accelerated to 85g and last for 30 minutes. The displacement vectors are shown in Figure 5. At this condition, the displacements of the deposit slope had been obtained. The maximum settlement of the slope occurred at the horizontal distance of about 300m from the toe of the slope. The maximum horizontal displacement occurred near the toe of the slope. These are respectively 17.8cm and 7.6cm for the prototype slope. The observation of the model shows that the slope is very stable. No falling of soil was found beneath the toe of the slope. To model the impounding of the reservoir in the centrifuge test, water was poured into the container with the depth of 12cm before start of the centrifuge. The other procedure is the same as that of the condition of natural state. The maximum settlement and horizontal displacement of the deposit slope occurred approximately in the same positions. The values of the displacements are 17.0cm and 12.6cm. Except that slightly falling of the deposit soil near the toe under the water was found, the deposit slope did not slide along the slip surface. Another bedrock, which slope grade is about 40 degree was used to model the actions of earthquakes with intensity degree Ⅷ. After running of the centrifuge, it was found that the deposit soil markedly moved down along the slip surface. The maximum settlement and horizontal displacement are 24cm and 27cm. Massive falling of the deposit soil near the toe occurred. So the slope will collapse when the earthquakes with intensity degree Ⅷ attack it.

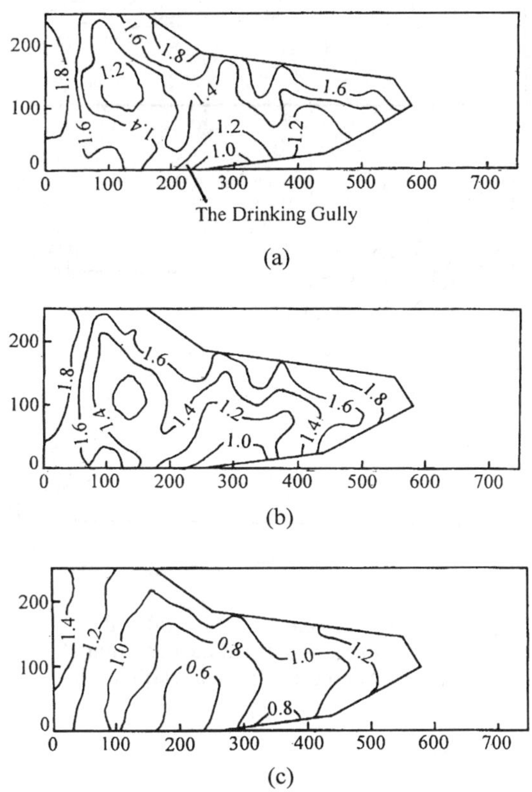

Figure 4. The contours of the calculated factor of safety (a) at the condition of natural state (b) at the condition of impounding (c) at the condition of earthquake (length unit: m)

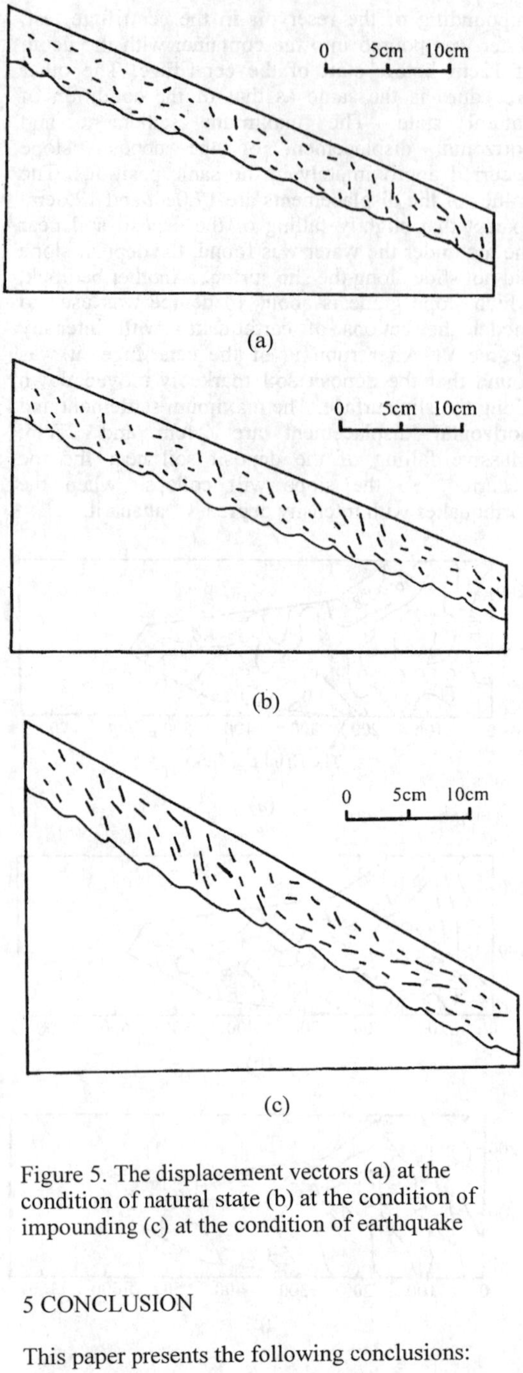

Figure 5. The displacement vectors (a) at the condition of natural state (b) at the condition of impounding (c) at the condition of earthquake

5 CONCLUSION

This paper presents the following conclusions:

1. The results from the three-dimensional finite element analysis show that at the natural state, the general factor of safety of the slope is 1.52. When underground water is completely drained, the general factor of safety will increase. At the condition of impounding, the general factor of safety is 1.47. When earthquakes with intensity degree Ⅷ attack the slope, the general factor of safety will become 0.92. According to the above results, the slope will slide under the actions of earthquakes with intensity degree Ⅷ.

2. Results from the centrifugal model tests show that the slope is safe at the natural state. At the condition of impounding of the reservoir, the submerged part of the slope will be softened and its stability will decrease. Under the actions of earthquakes with degree Ⅷ, the slope will fall.

3. The deposit slope should be reinforced before the arch dam is constructed. Tunnels to drain underground water are proposed to excavate. Site observations should be performed during the period of construction and operation of the reservoir.

ACKNOWLEDGEMENT

Financial support from the National Ninth Five Years Project of China (Grant No. 96-221-05-04-01) is greatly acknowledged.

REFERENCES

Jiancheng Du & Limin Zhang 1997. Three-dimensional stress-strain finite element analysis of high rockfill dam. *Computer Methods and Advances in Geomechanics*: 1789-1792. Rotterdam: A.A.Balkema.

Limin Zhang & Dashou Huang 1995. Effect of soft stratum on the seismic stability of a nature slope. *Earthquake Geotechnical Engineering,* 1109-1114. Rotterdam: A.A.Balkema.

Limin Zhang & Ting Hu 1994. Evaluation of the cut-off structures of a rockfill dam. *Centrifuge 94,* 593-598. Rotterdam: A.A.Balkema.

Pan Hengyong & Du Jiancheng 1999. A study on the stability of the deposit slope on the left bank of the Xiaowan Project. *Research report.* The Hydraulic College of Sichuan University.

Zhu Weixing et at. 1990. Centrifuge model testing of Xibeikou concrete faced rockfill dam. *Research report.* Nanjing Water Conservancy and Hydroelectric Power Institute.

Computer Methods and Advances in Geomechanics, Desai et al. (eds) © 2001 Balkema, Rotterdam, ISBN 90 5809 183 X

Effects of stress path and strength changes on the stability of slopes

C.T.Gnanendran, S.-C.R.Lo & P.Rattana-Ananta
School of Civil Engineering, University College, University of New South Wales, Canberra, A.C.T., Australia

ABSTRACT: Behaviour of a slope under three different stress paths to failure, namely shear strength loss due to heavy downpour of rain, construction near the crest, and undermining near the toe, were investigated using a finite difference analysis approach. The effects of soil dilatancy were also studied and the failure condition indicated by finite difference analysis for each stress-path case was back analysed using Bishop's simplified method.

The analyses indicate that the surcharge load capacity of the slope could increase by about 15% and the failure height of slope due to undermining near the toe could increase by about 25% for *0 to 20°* variation of dilatancy angle. This investigation indicate that soil dilatancy effects require careful consideration in the design of slopes to withstand undermining near the toe as well as surcharging over the crest-width particularly when such designs are based on simplified limit equilibrium approaches with marginal safety factors.

1 INTRODUCTION

Stability of slopes is generally assessed by limit equilibrium type of analyses in terms of a safety factor. Numerous modifications to the basic limit equilibrium method have been proposed in the literature for estimating the factor of safety of slopes under different conditions (e.g., to analyse different geometric shapes for the potential failure surface and to account for the development of cracks, pore pressures etc. in the soil – see for e.g., Morgenstern and Price, 1965; and Sarma, 1973). However, major draw backs of the limit equilibrium approaches such as their inability to estimate the stress-strain state of the soil mass under working conditions, always requiring assumptions for solving the set of equations and implied associative flow rule (hence the collapse is path independent) still remain. Attempts have been made to account for the stress-strain behaviour of the soil mass in limit equilibrium type of analyses for certain cases in the form of design charts on the basis of the results from numerical analyses (e.g., Miller and Hamilton, 1989; and Yu, et al., 1998). However, it is limited to specific cases only and the stress-strain state of the soil prior to failure is mostly not known with sufficient accuracy.

Slope failures, particularly for soils with high friction angle found in tropical and subtropical areas, are commonly associated with heavy downpours of rain and/or flooding and they can occur prior to the build-up of water table. A fundamental explanation for the stability reduction, apparently caused by the arrival of additional water, is the decrease in shear strength of soil associated with the increase in moisture content. Similarly, other common triggering factors of slope failures are the construction near the crest and / or the undermining near the toe. The stability of a slope under such stress path and strength changes is the main focus of this paper. In particular, the stability of a marginally safe slope subjected to these changing conditions is analysed using an explicit nonlinear finite different approach until failure. The predicted failure conditions of the slope from the numerical study are back analysed using a limit equilibrium analysis approach and the results compared. In addition, soil dilatancy may have some influence on the stress-strain behaviour as well as failure of slopes and its effects were also investigated for the three stress paths chosen for study under this research program and the results are reported in this paper.

2 DETAILS OF THE ANALYSES

A finite difference stress analysis program known as FLAC, Fast Lagrangian Analysis of Continua (FLAC 1998), was used for the stress analysis of the slope under different stress path conditions. In this program, the equations of equilibrium and stress strain behaviour are expressed in finite difference form and are solved by an explicit iterative scheme.

The analysis is inherently incremental and can be taken to collapse/failure along a user prescribed path. The numerical principles of FLAC-analysis are detailed in Cundall & Board (1988). For the FLAC analysis of slope reported in this paper, the soil was modelled as an elastic-perfectly-plastic non-associated Mohr-Coulomb type material and a large strain formulation with mesh updating for each step was adopted.

A 10 m high slope having a slope angle of $45°$ (i.e., *1:1* slope) was analysed for three different stress paths representing three common causes of slope failures. In particular, the cases of decreasing shear strength due to heavy rain, surcharging of the crest-width of slope due to construction near the crest and undermining near the toe of slope (or embankment) were investigated. The shear strength parameters of the soil under normal condition was chosen to be $c = 20$ kPa and $\phi = 37°$ and the unit weight (γ) of the soil was assumed as $17.46\ kN/m^3$. The shear modulus, G, and bulk modulus, K, of the soil were assumed as 100 MPa and 200 MPa respectively. The geometric configuration of the slope and the details of finite difference discretisation used for the FLAC analyses for the first two cases are shown in Fig. 1. This slope configuration was considered as reasonable for capturing the behaviour associated with the first two paths to failure but the depth to the rigid base and the distances to the left and right boundaries were increased further for the undermining case to minimise boundary effects on the results.

To account for the effect of shear strength decrease due to heavy rainstorm, the value of the apparent cohesion, c, was decreased gradually until the slope failed. For simulating construction near the crest of embankment, a uniform surcharge pressure was applied over the crest-width of the slope (or embankment) and the corresponding stress and deformational behaviours examined. Again, the surcharge was increased gradually until failure of the slope.

Finite element studies on selected slope configurations of Manzari and Nour (2000) have indicated that soil dilatancy has significant influence on the stability of slopes. Therefore, the influence of soil dilatancy on the effects of shear strength decrease as well as surcharging was also examined by analysing these cases for different dilatancy angles (i.e., $\psi = 0$, $5°$, $10°$ and $20°$).

To study the effects of undermining near the toe of embankments, the same 10 m high, $45°$ (i.e., *1:1*) slope but with the soil shear strength parameters of $c = 5\ kPa$, $\phi = 37°$ was chosen. A lower apparent cohesion of 5 kPa instead of the 20 kPa used for the first two cases was adopted to reduce the amount of undermining (and therefore the computational time) needed to achieve failure. All other properties of the soil used for this case also were the same as the other two cases. The height of the embankment was increased from 10 m by gradual removal of soil layers near the toe of embankment until failure. Again, the influence of soil dilatancy was examined by analysing the stress-strain behaviour of the slope for different values of ψ (i.e., $\psi = 0, 5°, 10°$ and $20°$).

Limit equilibrium type of analyses approaches are generally used for assessing the stability of slopes. To examine the relative merits of such analysis, the minimum factor of safety (FOS) was estimated using Bishop's simplified method for the 3 cases of slope failures indicated by stress analysis with the finite difference approach. The slope stability analysis program GALENA (1999) was used for the back analyses. The results of these analyses are discussed in comparison to those obtained from the FLAC analysis in the next section.

3 RESULTS AND DISCUSSION

The deformed finite difference mesh, indicating the onset of failure, obtained from a typical apparent shear strength loss analysis case prior to failure of the slope is shown in Fig. 1. The variation of settlement at the crest of the embankment (point A in Fig. 1) against the apparent cohesion (hereafter referred as cohesion, c) of the soil obtained from the finite difference stress analysis is shown in Fig. 2. For ease of comparison, the results obtained for different values of dilatancy angle ($\psi = 0, 5°, 10°$ and $20°$) of the soil are superimposed on this figure. The settlement at the crest of the embankment remained constant until the cohesion was decreased from 20 kPa to 10 kPa but increased gradually afterwards under further reduction. The effect of soil dilatancy also didn't have any influence on the settlement at point A until the cohesion was decreased up to 10 kPa but indicated noticeable influence on the settlement behaviour for values of cohesion < 10 kPa.

The settlement increased dramatically when the cohesion was decreased below 5 kPa and the slope failed. However, the value of cohesion at failure didn't have much variation (ranged between 3 and 4 kPa only) for the *0* to $20°$ variation of dilatancy angle investigated in this study. The lower value of $\psi = 0$ represent a loose soil whereas $\psi = 10°$ represent a typical soil that has a friction angle of $\phi = 37°$ and the higher value of $\psi = 20°$ represent a well-compacted soil that has interlocking angular particles. These results suggest that soil dilatancy has only minimal influence on the failure of the slope due to loss of apparent cohesion caused by rain.

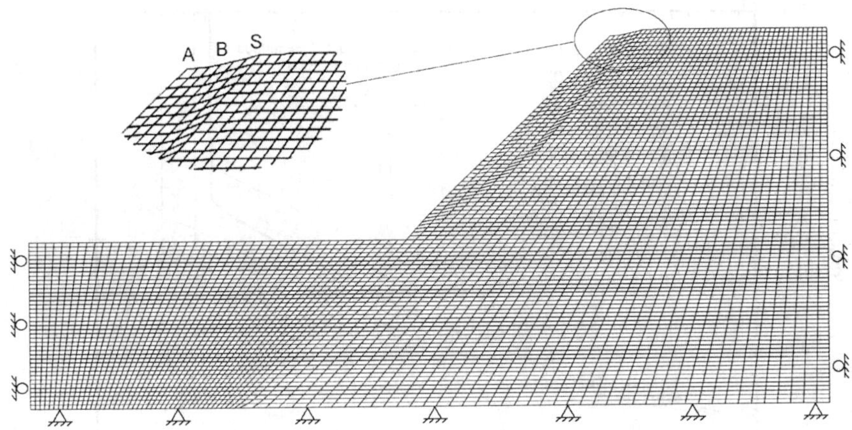

Fig.1 Deformed finite difference mesh and propagation of failure from a typical analysis

For comparison purposes, the number of steps (N) needed for achieving convergence of the solution for each case is also plotted in Fig. 2. As expected, the number of steps needed for convergence increased dramatically when the slope approached failure. The variations of number of steps needed for convergence versus soil cohesion are similar to those of slope-crest settlement and could therefore serve as a verification of failure.

Slope stability analysis using Bishop's simplified method was carried out for determining the minimum factor of safety for the critical values of cohesion corresponding to failure indicated by finite difference analysis. It is noted that the geometric configuration of the slope and other properties of the soil used for this analysis were the same as those used for the finite difference analysis. The slope stability analyses indicated minimum factors of safety of 1.11 and 1.17 for the cases of c = 3 kPa and 4 kPa respectively. This suggest that a slope designed with a marginal factor of safety of about 1.1 using Bishop's simplified approach and design soil parameters of $c = 3\ kPa$ and $\phi = 37°$ may in fact be unsafe and result in failure as indicated by the finite difference analysis. Therefore, considerable caution is needed when selecting the apparent

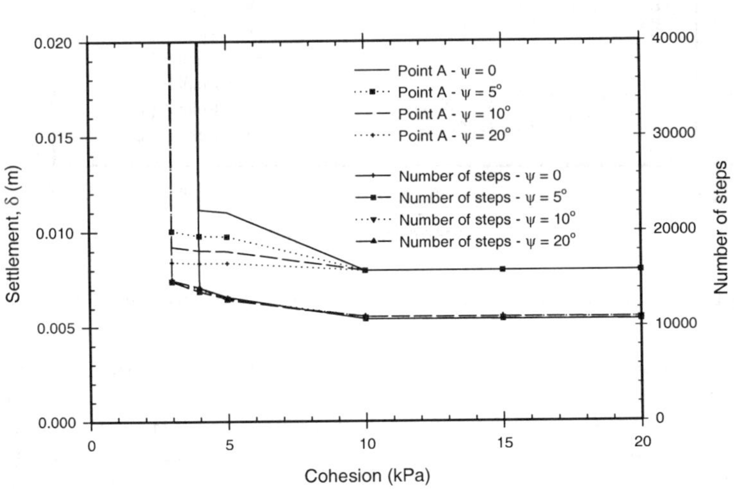

Fig. 2 Variation of settlement and number of steps versus cohesion for different dilatancy angles

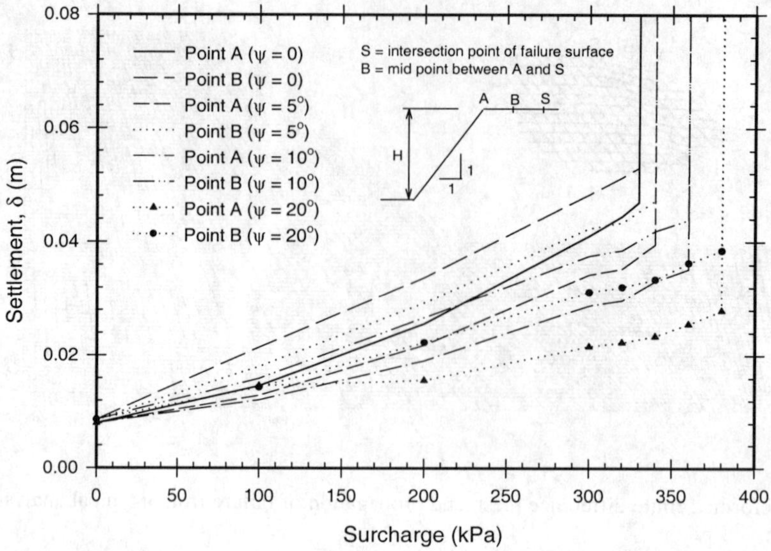

Fig. 3 Variation of settlement at A and B versus surcharge for different dilatancy angles

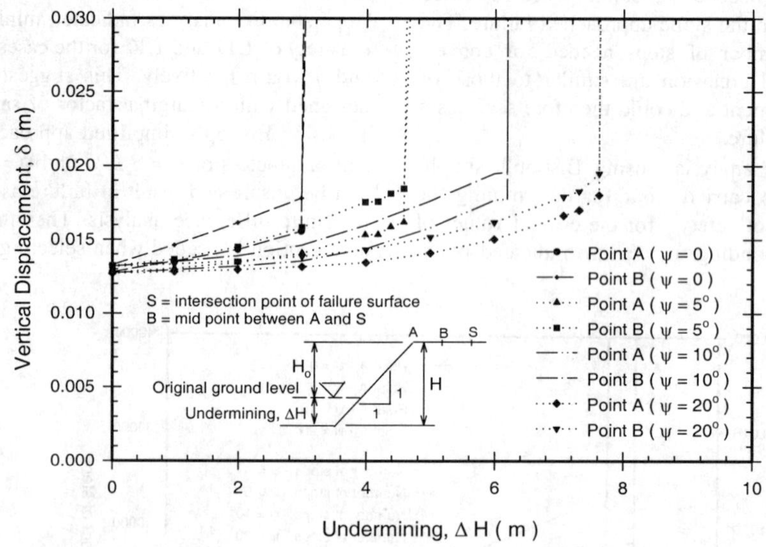

Fig. 4 Variation of settlement at A and B versus undermining for different dilatancy angles

cohesion of the soil for designing the slopes for stress path corresponding to loss of cohesion due to rain, particularly when such designs are based on limit equilibrium type of approaches.

Shown in Fig. 3 are the settlement responses of the slope for the case of surcharging of crest-width due to construction near the crest. In particular, settlements at two points A and B against the magnitude of uniformly distributed surcharge over the entire crest width are plotted. The settlement at both points A and B run off at the same surcharge. loading indicating failure. Here again, the variation of N against surcharge at failure coincided with that of settlement, confirming the earlier observation that it could be used to verify failure. Similar observations were made for the undermining case reported in Fig. 4 (discussed later). However, for clarity, the variations of N against surcharge and undermining are not shown in Figures 3 and 4.

The influence of soil dilatancy was investigated for the range of $\psi = 0$ *to* $20°$ and these results are also shown on Fig. 3. It can be observed that the settlements were less than 0.06 m until the surcharge was increased up to about 330 kPa. The failure surcharge pressure (i.e., surcharge capacity of the slope) increased from 330 kPa to about 380 kPa, indicating an increase of about 15%, when the dilatancy angle was increased from 0 *to* $20°$. Considering this variation, one would expect a surcharge capacity higher than 380 kPa for the associated flow rule case of $\phi = \psi = 37°$, which is the inherent assumption in the slip circle type of limit equilibrium analysis. Therefore, the dilatancy of the soil will have a significant influence on the surcharge load capacity of a slope and require consideration in design.

Back analysis of the slope for the surcharge of 360 kPa (i.e., surcharge corresponding to failure obtained for $\psi = 10°$ from the finite difference analysis) using Bishop's simplified slip circle analysis indicated a factor of safety of 1.07. Consequently, a slope designed for uniform surcharging to account for construction near the crest using Bishop's simplified approach with a marginal factor of safety, say about 1.07, may in fact result in failure.

Fig. 4 shows the settlement versus height of undermining (ΔH) near the toe. As indicated previously, a 10 m high (i.e., H_o = 10 m) and $45°$ angle slope with the shear strength parameters of the soil $c = 5\ kPa$ and $\phi = 37°$ was chosen for this study based on the results presented in Fig. 2 for the effects of changes in cohesion. The influence of soil dilatancy was investigated for $\psi = 0$ *to* $20°$ and the results are summarised in Fig. 4. It can be observed that the settlement at both points A and B increased gradually when the height of slope (H) was increased due to undermining ($H = H_o + \Delta H$). Before failure of the slope, the settlements at A and B decreased with increasing dilatancy angle for a particular height of slope (e.g., the settlements at H = 11 m was higher for $\psi = 0$ compared to that for $\psi = 5°$ which was in turn higher than that for $\psi = 10°$, and so on). Therefore, soil dilatancy has significant influence on the magnitude of settlement that occurs even when the slope is relatively safer (i.e., acceptable factor of safety from limit equilibrium type of analyses) during undermining.

Fig. 4 indicates that the failure height of the slope increased from about 13 m to 16.2 m for an increase of dilatancy angle from 0 *to* $20°$, which represent an increase of about 25% in failure height. Therefore, the influence of soil dilatancy on the failure of the slope due to undermining is significant and requires careful consideration for design. Back analysis of the slope for the case of H = 16.2 m (i.e., failure height indicated by finite difference analysis for $\psi = 10°$) using Bishop's simplified approach indicated a factor of safety of 1.09. Consequently, a slope designed to withstand undermining using Bishop's simplified approach with a marginal factor of factor of safety of 1.09 or less could result in unsafe situation if the actual dilatancy angle of the soil is less than $10°$. Therefore, soil dilatancy effects require careful consideration for the case of undermining near the toe, particularly when marginal safety factors are adopted for design.

Fig. 5 summarises the effects of soil dilatancy on the failure of the slope for the paths of surcharging of crest-width and undermining near the toe. It is obvious from Fig. 5 that the dilatancy angle of the soil had significant influence on both these paths and therefore require consideration in design. Simple limit equilibrium approaches are not capable of accommodating soil dilatancy effects and detailed stress analyses would be beneficial in such cases.

4 SUMMARY AND CONCLUSIONS

Stress-deformation behaviour of a slope under three different stress paths, namely shear strength loss due to heavy downpour of rain, surcharging of crest-width representing construction near the crest, and undermining near the toe of slope, were investigated using a finite difference analysis approach. A typical slope configuration and realistic soil properties were adopted for this numerical investigation. The effects of soil dilatancy on the settlement behaviour of the slope were studied in detail for the three stress paths chosen for study and the results reported in this paper. The failure conditions of the slope indicated by the finite difference analyses for the three stress path cases were back analysed for the minimum factor safety using Bishop's simplified method and the results discussed.

The finite difference analyses indicate that soil dilatancy has only minimal influence on the failure

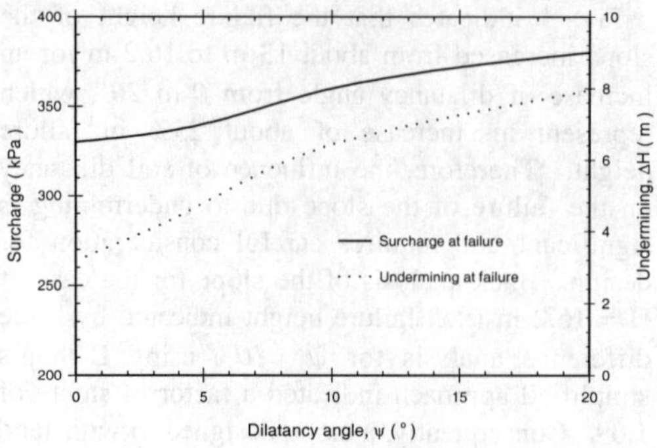

Fig. 5 Variation of surcharge and undermining at failure against dilatancy angle

of the slope for the case of shear strength loss in terms diminishing apparent cohesion due to heavy downpour of rain. However, caution is needed when selecting the apparent cohesion of the soil for design since it has considerable influence on the factor of safety of the slope, particularly when limit equilibrium type of approaches and marginal safety factors are used for the design.

On the other hand, the finite difference analyses indicated that soil dilatancy would have a significant influence (about 15% variation for typical range of dilatancy angle) on the surcharge load capacity of a slope and requires consideration in design. A slope designed for uniform surcharging to represent construction near the crest using Bishop's simplified approach with a marginal safety factor may result in excessive deformation and consequent failure and therefore need considerable care for such design.

Finite difference analyses further indicate that soil dilatancy has significant influence on the magnitude of settlement that occurs even when the slope has acceptable safety factor from limit equilibrium type of analysis for the case of undermining near the toe. This study showed about 25% failure height increase of the slope for a *0 to 20°* variation in dilatancy angle. This investigation further indicate that soil dilatancy effects require careful consideration in the design of slopes to withstand undermining near the toe when such designs are based on simplified limit equilibrium approaches, particularly when marginal safety factors are used.

REFERENCES

Cundall, P. & Board, M. 1988. A microcomputer program for modelling large-strain plasticity problems. Proc. of 6th Intl. Conference in Numerical methods in Geomechanics, Vol. 3, 2101-2108. Published by: A.A. Balkema, Rotterdam, Netherlands. Edited by: H.J. Siriwardane and M.M. Zaman.

FLAC 1998. "Fast Lagrangian Analysis of Continua", Version 3.4. Theory and background. Itasca Consulting Group, Inc., 708 South Third Street, Minneapolis, Minneapolis, USA.

GALENA 1999. Users Guide – Version 3. Slope stability analysis system. Clover Associates Pty Ltd., PO Box 20, Robertson, NSW 2577, Australia.

Miller, T.W. and Hamilton, J.H. 1989. A new analysis procedure to explain a slope failure at the Martin Lake mine. Geotechnique, London, U.K., Vol. 15, pp. 79-93.

Morgenstern, N.R. and Price, V.E. 1965. The analysis of the stability of general slip surfaces. Geotechnique, **15**, 79-93.

Sarma, S.K. 1973. Stability analysis of embankments and slopes. Geotechnique, Vol. 23, No.3, pp. 423-433.

Yu, H.S., Salgado, R., Sloan, S.W. and Kim, J.M. 1998. Limit analysis versus limit equilibrium for slope stability. ASCE J. of Geotechnical and Geoenvironmental Eng. Vol. 124, No. 1, pp. 1-11.

Computer Methods and Advances in Geomechanics, Desai et al. (eds) © 2001 Balkema, Rotterdam, ISBN 90 5809 183 X

Suction variation and infiltration characteristics on slope of highly heterogeneous soils

Low Tian Huat
Jurutera Perunding GEA(M) Sdn. Bhd., Malaysia

Faisal Haji Ali & Saravanan Mariappan
Civil Engineering Department, University of Malaya, Kuala Lumpur, Malaysia

ABSTRACT: Instability is an extremely important consideration in the design and construction of man-made slopes and natural slopes. Slope failures and landslides are influenced by geologic, topographic and climatic factors. In tropical region, most of the slope failure occurs during severe rainfall. Rain water infiltrates into the slope and reduce the soil matric suction and the strength of the soil. Most residual soils especially in slopes are in unsaturated condition and therefore matric suction is an important factor to be considered in the design or analysis of slopes. The suction has an important bearing on water entry, structural stability, and stiffness, shear strength and volume change. The soil matric suction, the water content and the solute content and how they vary with time are often the most important variables in soil engineering design. The field instrumentation of soil matric suction, rainfall, test results from the filed sprinkler system and results from parametric study on perched water table are presented.

1 INTRODUCTION

Construction activities in hilly terrain covered by residual soil frequently confront geotechnical engineers with slope instability problems. Failures in both natural and cut slopes in residual soils of Peninsular Malaysia are usually brought about by rainfall during the monsoon season.

The mechanism of slope failure is that the infiltration of rainwater causes a reduction of matric suction in the unsaturated soil, resulting in a decrease in the effective stress.

Most of the sedimentary residual soils in Malaysia are not homogeneous. The non-homogeneity also caused by the weathering process. They usually overlaid by one or more layer of soils. Due to the different characteristic of the soil layers, ground water condition and seepage flow would be too complicated and their influences to slope stability become unpredictable. Perched water table would probably built up when these non-homogeneous soil strata are exposed to prolong period of infiltration which is common for the climatic condition in this region.

The instrumentation is attempted to study the change of soil matric suction with the rainfall on a cut slope along the link road of The Kuala Lumpur International Airport (KLIA) Malaysia. The cut slope mainly consists of two types of weathered sedimentary residual soil, i.e., weathered sandstone and shale. These residual soils come in alternate bedding which is almost vertical. The weathered sandstone bed basically is the thicker bed and the study is concentrated in one of these beds. The soil consists of very fine sand and silt. Instrumentation was carried out on every berm with respect to different weathering grades of soil.

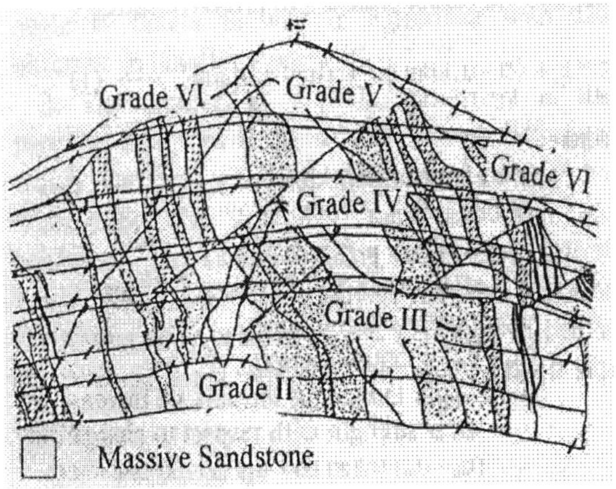

Figure 1.0 Geological Formation of the study site.

2 ROLE OF SUCTION IN SLOPE STABILITY

The principle of effective stress for unsaturated soil was first used by Terzaghi (1923) and proposed by him in the first International Conference on Soil mechanics in 1936. Numerous researchers have carried out work since then in order to confirm. But the validity of the principle for unsaturated soil mechanics has been questioned by Jennings and Burland (1962). Following an extensive research program on unsaturated soil conducted in Imperial College the shear strength of partially saturated soil was hypothesised (Bishop, 1959) to be a function of an effective stress defined as:-

$$\sigma' = (\sigma - u_a) + \chi (u_a - u_w) \quad [1]$$

where σ' and σ are the effective and total stresses respectively, u_a is the pore air pressure and u_w is the pore water pressure. χ is a function that depends on the saturation with value 1 at 100% saturation and 0 for completely dry soil.

Fredlund and Morgentern (1987) showed from a stress analysis that any two combination of the three possible stress variables $(\sigma - u_a)$, $(\sigma - u_w)$ and $(u_a - u_w)$ can be used to define unsaturated soil. The equation for unsaturated shear strength τ is written in terms of the stress state variables for an unsaturated soil and is an extension of the form of equation used for saturated soils

$$\tau = c' + (\sigma - u_a) \tan\phi + (u_a - u_w) \tan\phi^{b} \quad [2]$$

where ,

- c' = effective cohesion
- σ = total stress
- u_a = pore air pressure
- u_w = pore water pressure
- ϕ' = effective angle of friction
- $(u_a - u_w)$ = matric suction
- ϕ^{b} = gradient with respect to changes in $(u_a - u_w)$ when $(\sigma - u_a)$ is held constant.

The factor of safety for slope stability analysis using method of slices can be derived using shear strength equation [2] above. The shear force mobilised at the base of slice can be written as:-

$$S_m = \beta/F \{c' + (\sigma - u_a) \tan\phi + (u_a - u_w) \tan\phi^{b}\} \quad [3]$$

Where

- S_m = the shear force mobilised on the base of each slice.
- F = the safety factor
- β = the sliding surface of the slice.

3 INSTRUMENTATION

20 numbers of tensiometer and 20 numbers of moisture block and a rain gauge were installed on the slope to monitor the changes of matric suction with respect to rainfall. The tensiometers and moisture blocks were installed at different depths. At each berm, 4 numbers of tensiometer and moisture block were installed i.e., with depth of 0.5m, 1m, 2m and 3m. An automatic data acquisition system was designed to record the output from tensiometers and moisture blocks. The data acquisition system was supported by a solar powered set and specifically designed for low power consumption.

An automatic logging tipping bucket rain gauge was installed at the study site. The rain gauge records rainfall events on a real time basis. Figure 2 shows the schematic arrangement of the instrumentation at the study site.

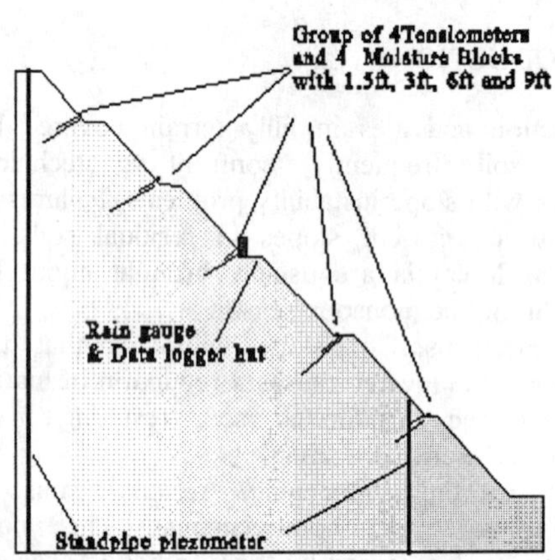

Figure2. Instrumentation Layout of the study Site.

The infiltration characteristics of the soil were deduced by using a sprinkler system installed at the site.A V- Notch fixed with a flow meter was used to measure the surface run-off in a control section in the study (refer to Figure 3). In addition, an infiltrometer P-88 was used to obtain the infiltration capacity at the site with the sprinkler system test results. To investigate the water flow characteristics and suction variation in details, small tip tensiometers were also installed during the sprinkling process.

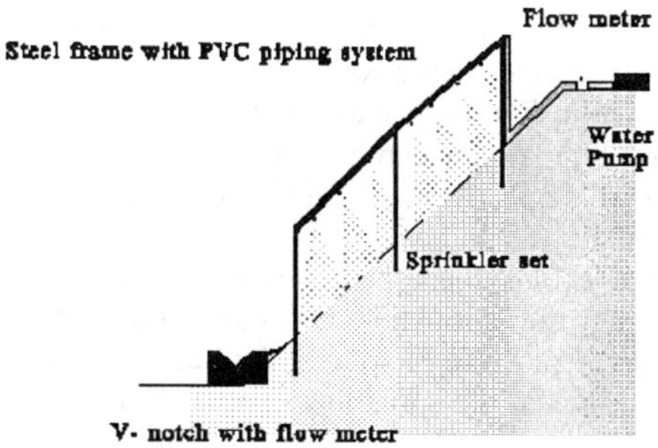

Figure 3. Field Infiltration Sprinkler System

4 MATRIC SUCTION MEASUREMENT

In this study, matric suction of the soil was measured using jet-fill tensiometer and moisture block. Moisture blocks were used because tensiometer can only measure matric suction below 1 bar of negative pressure while moisture block can measure more than 4 bar. The matric suction obtained from the tensiometer was the difference between the gauge reading and the head of the water in the stamp. The longer the tensiometer, the lower the suction it can measure.

In this study the tensiometer was installed perpendicular to the slope surface to reduce the head of water.

During the sprinkling process, extra small tip tensiometers were used to measure the matric suction at the shallower depth, which is 250mm under the slope surface. Figure 4 shows the layout of the tensiometers for the sprinkler model at the study site. S1 to S7 in Figure 4 are representing the position of the small tip tensiometers.

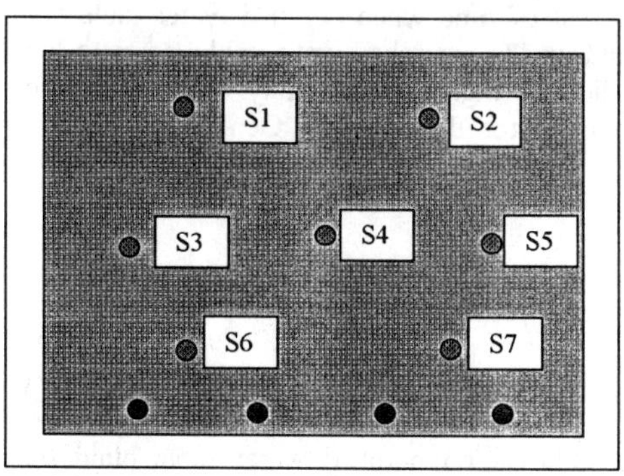

Figure 4. Typical Layout of Tensiometers For Sprinkler Model

5 PARAMETRIC STUDY ON PERCHED WATER TABLE

Due to heterogeneous characteristics of the soils, a simple parametric study was carried out to study the effects of perched water table onto slope stability. In this study, the water flux and seepage on slopes were simulated first by using water seepage software. The seepage output was then exported to slope analysis software to find the safety factor of slope.

Before analyzing slope stability, the moisture condition for transient analysis from seepage output has to be clearly defined. The output at different interval of time was combined with slope analysis to make it possible to determine the factor of safety of each interval of time. For all cases, Bishop method of analysis was chosen for the stability check.

In the analyses, the strength parameters for soils has been defined as follow:

Unit weight (γ)	=	18 kN/m^3
Angle of friction (ϕ)	=	20 °
Cohesion (C)	=	20 kN/m^2

The pattern of rainfall chosen was constant rainfall with intensity 2.1 x 10^{-4} m/s for all cases. The duration of rainfall was 12000 sec (3.333 hrs). Under this type of rainfall intensity perched water table can be generated in most of the cases. Thus the effect of perched water table under these specified parameters can be studied carefully.

The effect of the permeability ratio (ratio of permeability of a permeable stratum to impermeable stratum) was also studied. The critical ratio that causes perched water table was determined.

5.1 Verification of the SEEP/W program

Before the parametric study, verification of the suction values was carried out by comparing the measured suction values in the laboratory hydrological model with the simulated suction values by the seepage analysis software. Figure 5 shows one of the typical variations of with time. The difference of the lowest simulated and measured suction is just 2 kPa. However, there is a time lag of about 100 minutes between the simulated and measured suction values. The rainfall pattern agrees well with the drop in suction. The trend in recovery of the suction values for the simulated and site values are also found to be almost the same.

The initial difference of the suction values might be due to the hysteresis effect of the soil water characteristics of the sample used.

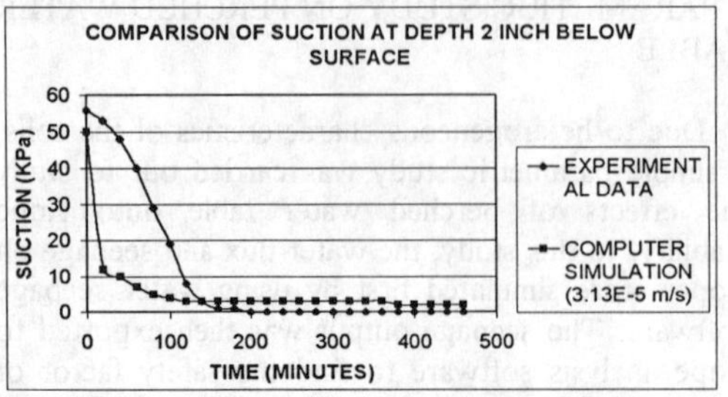

Figure 4 comparison of simulated and measured suction value.

6 DISCUSSIONS

Figure 5 shows a typical suction variation with rainfall (one month duration) for one of the berm at the study site. It is clearly shows that as the depth increase, the matric suction reduces.

During the time interval of 14000 to 24000 minutes, there was no rainfall occurred. All the four tensiometers experiencing increments in matric suction. When the rain started to fall, matric suction did not reduce immediately. Due to the infiltration of rainwater into the ground after rainfall, the matric suction reduced slowly for all depths even without any rainfall after the rainstorm.

From Figure 5, the time interval of 0 to 13000 minutes, suction of the 3.0m depth tensiometers gives very low suction values. This is mainly due to the cumulative rainwater during the raining period.

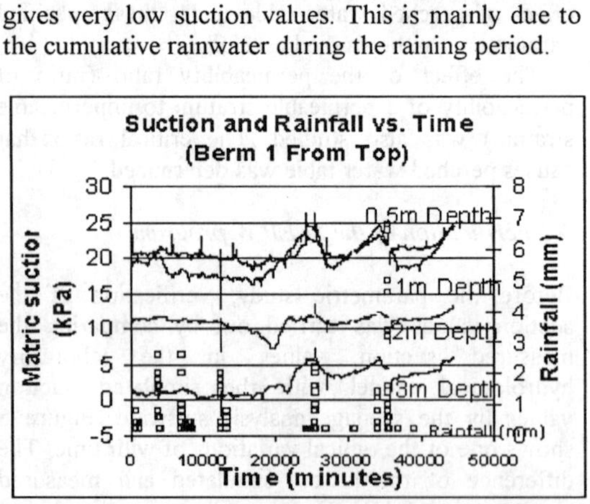

Figure 5. Typical suction variation with rainfall for one-month duration.

From the field infiltration test, the water infiltration rate is 2.31 x10^{-6} m/s. The typical high rainfall intensity at the site can be very high, i.e., 1.13 x 10^{-4} m/s. From Figure 5, it is clearly shown that during heavy rainstorm, the suction drops very fast and also recovers fast. One of the reasons of this behavior is that the rainwater infiltrate into the slope through quartz veins in the soil.

As shown in Figure 5, high rainfall intensity with short duration may not play an important role in the instability of slope. Prolong rainfall even with low intensity could be significant.

Figure 6 shows the typical suction variation on slope during the sprinkling process. The suction reading for small tip tensiometers for S2, S3, S6 and S7 increases when the sprinkling process starts. This might be due to the effects of water flowing on slope that create slight negative pressure at the shallower depth of the slope. The increment is more significant for the above-mentioned small tip tensiometers because the tensiometers are installed at the lower level of the slope and the velocity is much greater than the top slope.

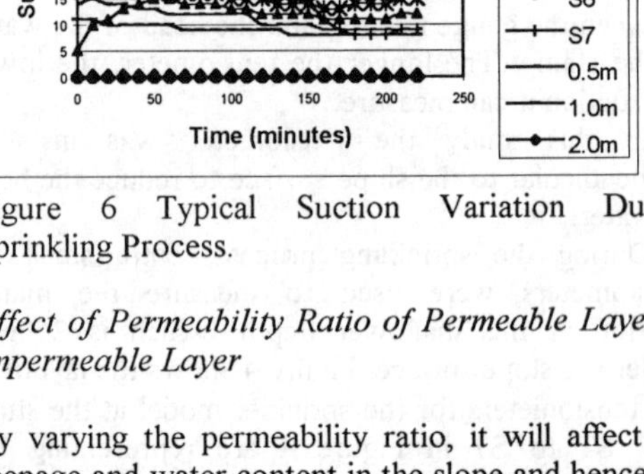

Figure 6 Typical Suction Variation During Sprinkling Process.

Effect of Permeability Ratio of Permeable Layer to Impermeable Layer

By varying the permeability ratio, it will affect the seepage and water content in the slope and hence the stability of the slope. From the results shown in Figure 7, it is clear that safety factors drop at certain interval of time when perched water table starts to build up. The perched water table starts to build up earlier for higher permeability ratio and this causes safety factor drop earlier. Generally perched water table starts to build up from interval time 6000 sec to 8000 sec.

Effect of Rainfall Intensity

Figure 8 clearly shows that for rainfall intensity of 1 x 10^{-5} m/s or less the factor of safety remains constant as no perched water table build up. For rainfall intensity 5 x 10^{-5} m/s and 5.2 x 10^{-5} m/s there are only slight drops i.e., not more than 2%. The drop of safety factor becomes more significant for higher rainfall intensity. Rainfall intensity of 1 x 10^{-4}m/s records a maximum drop of 4.4% while rainfall intensity of 2.1 x 10^{-4} m/s records a maximum drop in factor of safety.

Generally the safety factor starts to drop at time 6000 sec or 7000 sec. The reading is constant for set of data with lower rainfall intensity; as for lower rainfall intensity the effect of perched water table is not significant.

The permeability of the soil also plays an important role in this study because when the rainfall intensity is lower then the permeability of the soil, the infiltrated rainwater can dissipate fast into the slope and does not cause perched water table to build up. But, when the rainfall intensity equal to or more then the soil permeability value, perched water table will start to form.

In the analysis, the flux boundary condition needs to be considered carefully. If the rainfall intensity is more then the permeability of the soil, surface runoff needs to be specified.

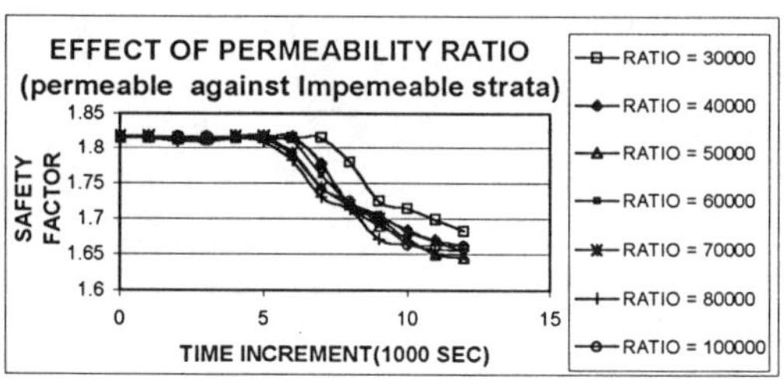

Figure 7. Effect of soil permeability ratio on slope safety factor (Permeability of 1x 10^{-8} m/s for impermeable stratum was used as datum)

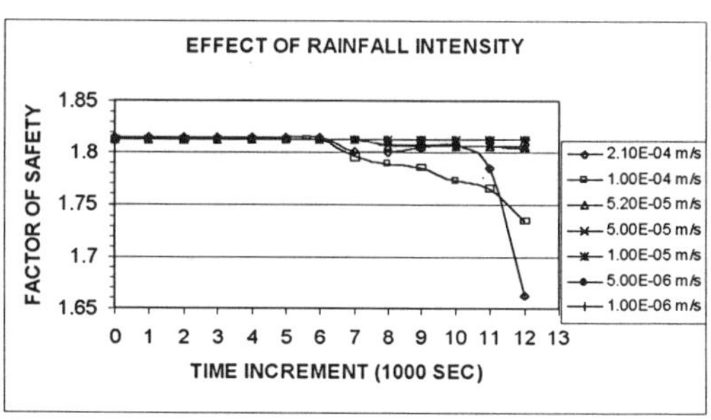

Figure 8 Effect of rainfall intensity on factor of safety

7 CONCLUSIONS

Generally as the depth increases, matric suction reduces. But, in some circumstances, due to the flux condition, this can be the other way round. Heavy rainfall intensity may necessary cause instability to slope but prolong rainfall can be significant. Geological features like quartz vein could be significant in slope instability because it increases the rainwater infiltration. When perched water table is formed, it decreases the factor of safety. Further rainfall does not seem to lower the safety factor.

REFERENCES

Abdullah, Affendi M.L. & Ali Faisal, "Field Instrumentation for Slope Stability in Residual Soil" GEOTROPIKA 92' Johor Bahru, Malaysia.

Anderson , M.G. & Burt, T.P. "Automatic monitoring of soil moisture conditions in a hillside spur and hollow", Journal of hydrology, Vol.33,1977,pp27-36.

Bishop, A.W. and Blight, G.E. (1967) "Effective stress evaluation for unsaturated soils" Jour. Soil Mech. & Found. Engg. Div., ASCE 93 (SM2), pp. 125 – 148.

Fredlund D.G. "Appropriate concepts and tecnology for unsaturated soils". Second Canadian Geotechnical Colloquim, Canadian Geotechnical Journal, No. 16 , 1979, pp 121-139.

Fredlund,D.G., "Slope stability analysis incorporating the effect of soil suction", Slope Stability : Geotechnical Engineering and Geomorphology, Edited by Anderson,M.G., and Richards, K.S., John Wiley and Sons, Chichester,1987, pp113-144.

Low, T.H.. Faisal, H.A, Saravanan, M. "Effects of perched water table on slope stability in unsaturated soils". Proc. Shikoku99, Matsuyama, Japan vol.1 p.p393-397.

Low, T.H.. Faisal, H.A, Saravanan, M. "Field suction variation with rainfall on cut slope in weathered sedimentary residual soils". Proc. Shikoku99, Matsuyama, Japan vol. 1, p.p399-403.

Low, T.H., Faisal, H.A., " Water Infiltration On Slopes". World Engineering Congress, Kuala Lumpur, Malaysia. P.P 211-215.

Low, T.H., Faisal, H.A, Subramaniam, R. "Parametric Studies of Slope Stability in Unsaturated Soils". 13th. South East Asian Geotechnical Conference, Taipei, Taiwan p.p117-122.

Rahardjo.H, Loi, J., and Fredlund, D.G.," Typical matric suction measurements in the laboratory and in the field using thermal conductivity sensors", presented at Indian Geotechnical Conference (IGC-89), Vol.1, Visakhhapatnam, December 1989.

Computer Methods and Advances in Geomechanics, Desai et al. (eds) © 2001 Balkema, Rotterdam, ISBN 90 5809 183 X

Modeling of sandy ground for the stability analysis of slurry trench

Takeshi Ishii, Masaaki Katagiri & Kunio Saitoh
Nikken Sekkei Nakase Geotechnical Institute, Kawasaki, Japan

Shuji Tomita & Kiyoshi Kuwabara
East Japan Railway Company Limited, Tokyo, Japan

Keizo Ugai
Department of Civil Engineering, Gunma University, Kiryu, Japan

ABSTRACT : The modeling of sandy ground for the stability analysis of slurry trenches were studied using the 3D-SSRFEM which was the elasto-plastic 3D-FEM combined with the shear strength reduction method. As the results, the stability condition depends on only the height difference between ground water and slurry levels but also the level of ground water. The ground above the groundwater level can be modeled as the surcharge on the saturated ground.

1. INTRODUCTION

The slurry trench is kept its stability by the effective slurry pressure against the sliding block. Here, the effective slurry pressure is defined as slurry pressure minus groundwater pressure. In practice, however, the stability condition of slurry trenches is defined as only the minimum value of height difference between slurry and groundwater level, ΔH.

Recently, using geotechnical centrifuges that are able to reproduce stress level in field scale, studies on the failure mechanism of slurry trenches in sandy ground have been performed (Higuchi et al.,1994 and Katagiri et al.,1997). They stated that the stability of trenches was dependent on the relative slurry pressure against earth pressure of the ground, an arch effect caused by a trench shape, surcharge, mechanical properties of the ground, and so on.

The stability of the sliding block is dependent on the length/depth of trench, and can not be expressed by the 2-D model. It is therefore necessary to use the 3-D analysis which does not assume the shape of sliding block.

The groundwater level separates saturated and unsaturated parts. If the shear characteristic of the ground is independent on the degree of saturation, the shear resistance parameters of the sliding block are constant regardless of the groundwater level. As the effective self-weight of the sliding block increases with lowering of groundwater level, the shear resistance increases in proportion to the increment of effective normal stress in the sliding block.

In this paper, the effect of groundwater level on the stability of the trenches is investigated using the 3D-FEM combined with the Shear Strength Method (SSRM) proposed by Ugai (1990). And a modeling of the ground above the groundwater level is also proposed.

2. METHOD OF CALCULATING GLOBAL SAFETY FACTOR BY SSR-FEM

The SSR-FEM consists of the elasto-plastic FEM and the SSRM. In the FEM, stresses and strains of soil are calculated by the iteration based on the modified Newton-Raphson method. The stress-strain relation of soil is assumed an elastic-perfectly plastic model. The failure criteria and the plastic potential function of the soil are assumed as the Mohr-Coulomb's equation and Drucker-Prager equation, respectively. 3-D finite elements used in this paper are iso-parametric ones. These consist of 20 nodal points and 8 Gaussian points for the numerical integration.

In the SSRM, the safety factor is defined as the ratio of shear strength parameters at failure to those mobilized in actual conditions. The shear strength of the material are reduced until failure occurs, simultaneously reducing both friction angle ϕ and cohesion c by a shear strength reduction factor F_r. The reduced friction angle ϕ_r and the reduced cohesion, c_r, are defined as follows;

$$\phi_r = \arctan(\tan\phi / F_r) \qquad c_r = c / F_r$$

The initial value of Fr is assumed to be sufficiently small so as to produce a nearly elastic condition. Then the value of Fr is increased step by step. At a point, their shear strength becomes too small to keep the stability of the trench wall, and a divergence occurs during nonlinear iteration. This point is defined as the global failure. The shear strength reduction factor at immediately preceding failure is considered as the minimum global safety factor. The convergence criterion is given by the ratio of the norm of the displacement increment between two successive values of F_r. The convergence is defined to be achieved, when this ratio is less than 1.0×10^{-5} within 500 iterations.

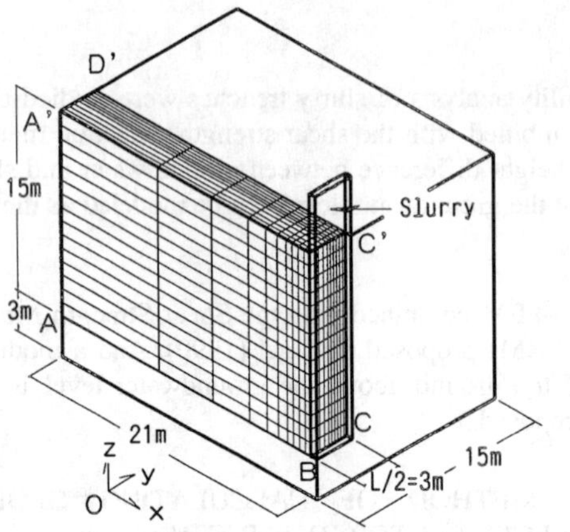

Figure 1. Finite element analysis model for slurry trench of centrifuge model tests.

Figure 1 shows the meshes of the ground with the slurry trench. This finite model was a part of ground in the centrifuge model tests. The centrifuge model was constituted a quarter region because of the symmetry of slurry trench. The size of the slurry trench in Figure 1 is 6 m long, 1 m wide and 15 m deep. The analysis was limited within the part shown in Figure 1, because the safety factor calculated from this limited range was almost consistent with that from the range simulated the centrifuge model tests (Ishii et al., 2000). How to divide the mesh was also followed by Ishii et al. (2000).

The physical and mechanical constants for the model ground and slurry are listed in Table 1. The density of the unsaturated ground is 1.88 g/cm^3. The other hand, the density of the saturated ground is 0.88 g/cm^3

3. RESULTS OF CENTRIFUGE MODEL TESTS

In centrifuge model tests, the model grounds were completely saturated at a designed acceleration, then

Table 1. Properties of ground and slurry.

Properties	Saturated ground	Unsaturated ground
Density, ρ (g/cm^3)	0.88	1.88
Elastic modulus, E (kPa)	23,500	23,500
Poisson's ratio, ν	0.333	0.333
Friction angle, ϕ (degree)	39	39
Dilation angle, φ (degree)	39	39
Cohesion, c (kPa)	0	0
Slurry density, ρs (g/cm^3)	1.05	

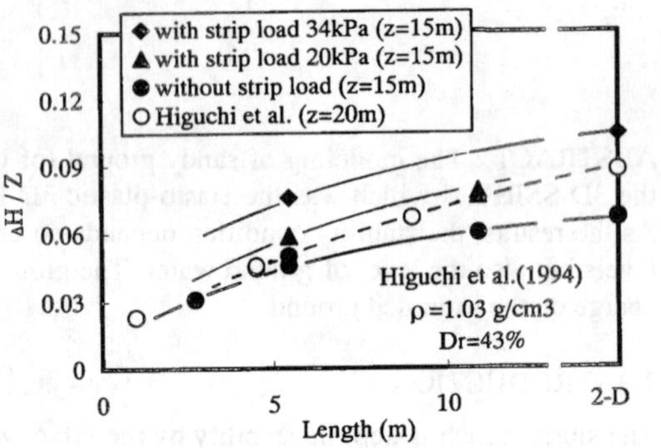

Figure 2. Relationship between length ratio of trench and normalized stability condition.

the yielding condition was determined as shown by Ishii et al.(1998). Here, the definition of the yielding condition was the difference between slurry level and groundwater level at yield point.

Figure 2 shows the relationship between yielding condition and length of slurry trenches with and without the strip load near the slurry trench. The position of this relationship rises with the magnitude of surcharge. Figure 2 also indicates the centrifuge test results by Higuchi et al. (1994). In their tests, the unsaturated model ground was prepared, and the ground water level was increased under the constant slurry level which agrees with the ground surface. The curve obtained by Higuchi et al.(1994) agrees with our curve without the strip load in the range of short length of slurry trench. In the long length of slurry trench and the 2D trench, however, the results by Higuchi et al. (1994) are located above our test results without the strip load and near those with strip load of 20kPa. The test conditions in both the tests were different. That is, the groundwater levels in the tests by Higuchi et al. were located in the ground, and the groundwater levels in our tests were consistent with the ground surface. Weight of soil above the groundwater level may be regarded as the surcharge.

4. MODELING OF GROUND FOR NUMERICAL SIMULATION

From the consideration on the soil above the groundwater level, the effect of surcharge on the safety factor were investigated. Three cases of ground models shown in Table 2 were considered. Case-1 is that the groundwater level is consistent with the ground surface. This model simulates our test. Case-2 is an unsaturated model with the same magnitude of difference between slurry and groundwater levels as the case-1 under the slurry level as same as the ground surface. Case-3 is the situation having safety factor of unity in case-2.

In order to constitute a simple model for a partially saturated ground, the model in which the unsaturated part is replaced with surcharge is investigated as case-4. In this model, as the shear resistance in the unsaturated part is ignored, the stability of the trench is underestimated.

The lengths of slurry trenches are set at 3, 6 and 12m. 2-D trench is also investigated.

Table 2 Outline of calculation cases

[case-1] a simulation model for our centrifuge test in which the groundwater level H_W is consistent with the ground surface H_G. The slurry level H_S at the safety factor of unity is calculated.	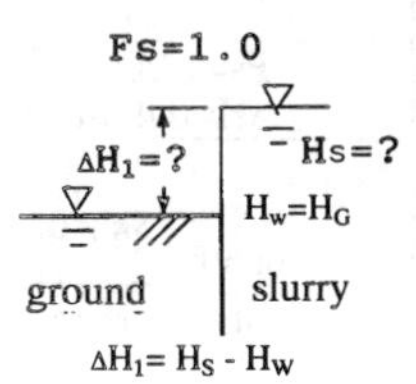
[case-2] an unsaturated model with the same magnitude of difference ΔH_1 between slurry and groundwater level as the case-1 under the same slurry level as the ground surface.	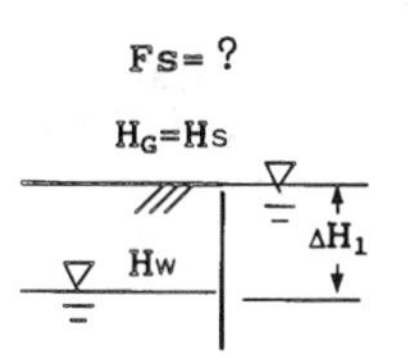
[case-3] a simulation model having safety factor of unity in the case-2.	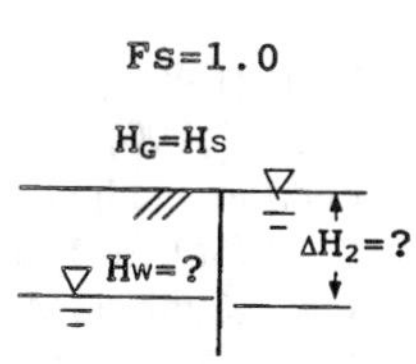
[case-4] the model with surcharge instead of the ground above the groundwater level. The surcharge is equivalent to overburden load in the case-3.	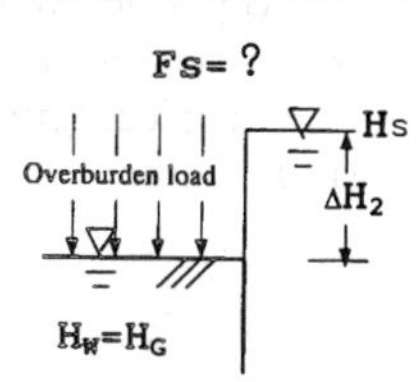

5. NUMERICAL RESULTS AND DISCUSSIONS

Table 3 shows the numerical results of four cases for four types of slurry trenches. In the case-2 where the partially saturated ground has the same slurry level as the ground surface, the safety factors in all the trenches are less than unity. The reduction rate of safety factor increases with the length of slurry trench.

In the case-3 where the partially saturated ground in the condition that the slurry level is consistent with the ground surface has the safety factor of unity, the difference ΔH_2 between slurry and groundwater levels is obtained by trial and error. Each ΔH_2 in the case-3 is larger than ΔH_1 in the case-1. The increment ratios of ΔH_2 are 17 – 37 %, and increase with the length of slurry trench.

In the case-4 where the partially saturated ground is modeled as the surcharge equivalent to overburden load, each safety factor is less than that in the case-3 but is regarded almost the same. In the design of slurry trenches, this model is considered a reasonable one because the safety factors are underestimated.

In addition, comparing the case-1 with case-4, the effect of surcharge on the safety factor of the saturated ground is assessed. When the surcharge is put on the ground, the safety factor of slurry trench decreases.

6. CONCLUDING REMARKS

Using the SSR-FEM, the effect of groundwater level on the safety factor of slurry trenches in sandy ground and a modeling of the unsaturated part above the groundwater level are investigated. The following knowledge can be obtained;

(1) When the difference between slurry and groundwater level is the same, the safety factor decreases with lowering of groundwater level.

(2) The reduction rate of safety factor increases as the length of trench increases.

(3) In the design of slurry trenches, a partially saturated ground can be modeled as a saturated ground with surcharge.

From the above knowledge, the groundwater level as well as the difference between slurry and groundwater level affects the safety factor of the slurry trenches. In the tests controlled by the groundwater level under a constant slurry level, therefore, the failure condition includes the effect of surcharge. When the test results are used, it is necessary to take away that effect.

Table-3 Safety factors of 4 cases

Length	case 1	case 2	case 3	case 4
2D	SWL=1.12m GL=0m GWL=0m Fs=1.00 ΔH=1.12m	SWL=0.0m GL=0m GWL=-1.12m Fs=0.86 ΔH=1.12m	SWL=0.0m GL=0m GWL=-1.53m Fs=1.00 ΔH=1.53m	surcharge SWL=0.0m GL=-1.53m GWL=-1.53m Fs=0.98 ΔH=1.53m
12m	SWL=0.706m GL=0m GWL=0m Fs=1.00 ΔH=0.71m	SWL=0.0m GL=0m GWL=-0.706m Fs=0.92 ΔH=0.71m	SWL=0.0m GL=0m GWL=-0.863m Fs=1.00 ΔH=0.86m	SWL=0.0m surcharge GL=-0.863m GWL=-0.863m Fs=0.98 ΔH=0.86m
6m	SWL=0.35m GL=0m GWL=0m Fs=1.00 ΔH=0.35m	SWL=0.0m GL=0m GWL=-0.35m Fs=0.95 ΔH=0.35m	SWL=0.0m GL=0m GWL=-0.41m Fs=1.00 ΔH=0.41m	SWL=0.0m surcharge GL=-0.41m GWL=-0.41m Fs=1.00 ΔH=0.41m
3m	SWL=0.155m GL=0m GWL=0m Fs=1.00 ΔH=0.155m	SWL=0.0m GL=0m GWL=-0.155m Fs=0.95 ΔH=0.155m	SWL=0.0m GL=0m GWL=-0.185m Fs=1.00 ΔH=0.185m	SWL=0.0m surcharge GL=-0.185m GWL=-0.185m Fs=1.00 ΔH=0.185m

Notes GL : ground level, GWL : groundwater level, SL: slurry level
Datum level is the ground surface.

Reference

Higuchi, Y., Tohda, J., Nakura, K. and Kawasaki, K. (1994), A new stability analysis method for slurry trenches constructed in sandy ground, *TSUCHI-TO-KISO*, **42**-3: 7-12, JSSMFE (in Japanese).

Katagiri, M., Saitoh, K., Masuda, K., Aizawa, F. and Ugai, K. (1997), Centrifuge Model Tests on Stability of Slurry Trenches, *TSUCHI-TO-KISO*, **45**-10: 13-16, JSSMFE (in Japanese).

Ishii, T., Katagiri, M., Saitoh, K., Masuda, T., Kuwabara, K., Aizawa, F. and Ugai, K. (1998), Estimation of Global Safety Factor of Slurry Trenches by 3-D FEM, Proc. 9th IACMAG, vol.3 pp.2101-2108.

Ugai, K. (1990), Availability of shear strength reduction method in stability analysis, *TSUCHI-TO-KISO*, **38**-1: 67-72, JSSMFE (in Japanese).

Ishii, T., Katagiri, M., Saitoh, K., Kuwabara, K. and Ugai, K. (2000), Stability analysis by 3-D elasto-plastic finite element method for slurry trenches constructed in sandy ground, Journal of Geotechnical Engineering, JSCE, (to be printed), (in Japanese)

Computer Methods and Advances in Geomechanics, Desai et al. (eds) © 2001 Balkema, Rotterdam, ISBN 90 5809 183 X

Stability analysis of circular and square excavations

A.V. Lyamin & S.W. Sloan
Geotechnical Research Group, Department of Civil, Surveying and Environmental Engineering, University of Newcastle, N.S.W., Australia

ABSTRACT: This paper uses recently developed upper and lower bound methods to predict the stability of circular and square excavations. Both of these problems are important in practice but have been given little attention in the literature. The only available bounds for a circular excavation differ by a factor of almost two, while the authors are unaware of any rigorous bound solutions for a square excavation. The numerical formulations are based on linear finite elements and quasi-Newton nonlinear programming and are applicable to the solution of one, two and three dimensional stability problems with any type of convex yield criterion. The new methods are much more efficient than the equivalent linear programming techniques developed previously by the second author. The results, obtained for Tresca and Mohr-Coulomb materials, are given in the form of dimensionless stability factors for different height to width ratios of the excavation.

1 INTRODUCTION

The stability of unsupported circular and square excavations (Figure 1) are important practical

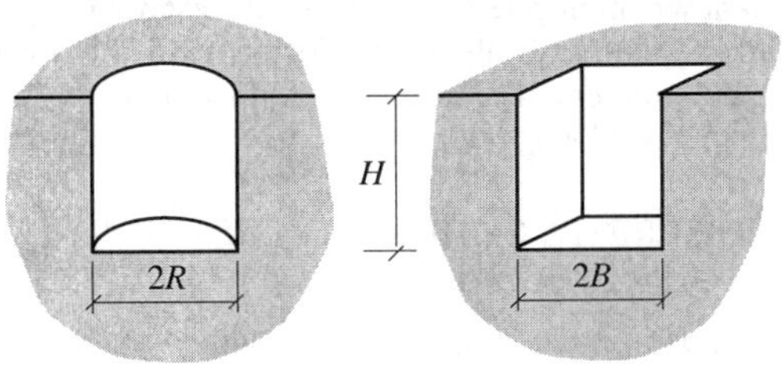

Figure 1. Circular and square excavations.

problems which require accurate solutions for use in the construction industry. In the past, work has been performed on axisymmetric excavations in clays by a number of researchers including Bjerrum & Eide 1956, Prater 1977, Pastor & Turgeman 1979, Pastor 1981, Britto & Kusakabe 1982, and Sloan 1981. All of these studies, however, have focused on soils obeying the Tresca yield criterion and are therefore restricted to undrained loading conditions. No solutions appear to be available in the literature for circular and square excavations in frictional soils.

Most of the previous studies of excavation stability have been based on upper bound techniques involving various failure mechanisms (eg. Britto & Kusakabe 1982). Although useful, upper bound solutions inherently give unsafe estimates on the true collapse loads. In practice, upper bounds are more valuable when they are accompanied by lower bounds, so that the exact solution can be bracketed from above and below.

This study presents upper and lower bounds on the dimensionless stability number $\gamma H/c$ for unsupported circular and square excavations. The bounds bracket the true stability numbers to within a few per cent and are obtained using recently developed 3D finite element formulations of the lower and upper bound theorems.

2 DISCRETE FORMULATION OF LOWER BOUND THEOREM

Consider a body with a volume V and surface area A, as shown in Figure 2. Let **t** and **q** denote, respectively,

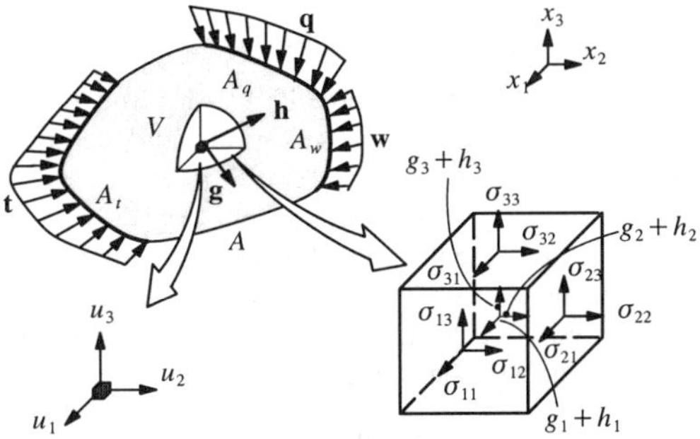

Figure 2. A body subject to a system of surface and body forces.

a set of fixed tractions acting on the surface area A_t and a set of unknown tractions acting on the surface area A_q. Similarly, let **g** and **h** be a system of fixed and unknown body forces which act, respectively, on the volume V. Under these conditions, the objective of a lower bound calculation is to find a stress distribution which satisfies equilibrium throughout V, balances the prescribed tractions **t** on A_t, nowhere violates the yield criterion, and maximises the integral

$$Q = \int_{A_q} \mathbf{q}\, dA + \int_V \mathbf{h}\, dV \tag{1}$$

2.1 *Linear finite elements*

In the present formulation of the lower bound theorem, linear finite elements are used to discretise the continuum and statically admissible stress discontinuities are permitted at all interfaces between adjacent elements (Figure 3). If D is the problem

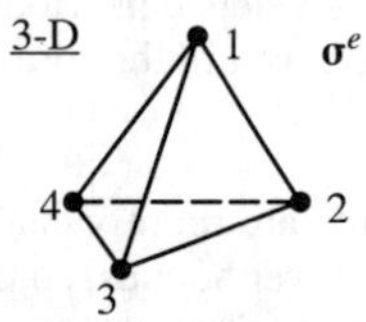

$$\boldsymbol{\sigma}^e = \{\sigma_{11}^e\ \sigma_{22}^e\ \sigma_{33}^e\ \sigma_{12}^e\ \sigma_{23}^e\ \sigma_{13}^e\ h_1^e\ h_2^e\ h_3^e\}^{\mathrm{T}}$$

Figure 3. Simplex finite elements used in lower bound formulation.

dimensionality, then there are D+1 nodes in each element and each node is associated with a $(D^2 + D)/2$-dimensional vector of stress variables σ_{ij}, $i = 1,\ldots,D$; $j = i,\ldots,D$. These stresses, together with the body force components h_i which act on a unit volume of material, are taken as the problem variables. The vector of unknowns for an element e is denoted by $\mathbf{x}^e$ and may be written as

$$\mathbf{x}^e = \left\{ \left\{\sigma_{ij}^1\right\}^{\mathrm{T}}, \ldots, \left\{\sigma_{ij}^{D+1}\right\}^{\mathrm{T}}, \left\{h_i^e\right\}^{\mathrm{T}} \right\}^{\mathrm{T}} ; \quad i,j = 1, \ldots, D \tag{2}$$

2.2 *Optimisation problem statement*

When the stress field is modelled using linear finite elements, the objective function and equality constraints arising from the equilibrium, boundary, discontinuity, and loading conditions are linear in the unknowns, with the only non-linearity arising from the yield inequalities. Thus the problem of finding a statically admissible stress field which maximises the collapse load may be stated as

$$\text{maximise } \mathbf{c}^{\mathrm{T}}\mathbf{x} \tag{3}$$

subject to

$$\mathbf{A}\mathbf{x} = \mathbf{b}$$
$$f_j(\mathbf{x}) \le 0, \quad j \in J_\sigma$$
$$\mathbf{x} \in \mathrm{R}^n$$

where **c** is an n-dimensional vector of objective function coefficients, **A** is an $m \times n$ matrix of equality constraint coefficients, $f_j(x)$ are yield functions, J_σ is the set of constraints on stress variables, **x** is an n-dimensional vector which is to be determined, and n is the number of unknown stresses and element unit weights.

2.3 *Kuhn-Tucker optimality conditions*

Because the objective function and all equality constraints are linear, and the functions f_j are convex, the problem (3) is equivalent to the system of Kuhn-Tucker optimality conditions

$$\left.\begin{aligned} -\mathbf{c} + \mathbf{A}^{\mathrm{T}}\boldsymbol{\mu} + \sum_{j\in J} \lambda_j \nabla f_j(\mathbf{x}^*) &= \mathbf{0} \\ \mathbf{A}\mathbf{x}^* &= \mathbf{b} \\ \lambda_j f_j(\mathbf{x}^*) = 0, &\quad j \in J_\sigma \\ f_j(\mathbf{x}^*) \le 0, &\quad j \in J_\sigma \\ \lambda_j \ge 0, &\quad j \in J_\sigma \\ \mathbf{x}^* &\in \mathrm{R}^n \end{aligned}\right\} \tag{4}$$

where $\mathbf{x}^*$ is the optimal solution and $\boldsymbol{\mu}$ and $\boldsymbol{\lambda}$ are unknown multipliers. A detailed description of a robust nonlinear algorithm for solving the lower bound problem (4) can be found in Lyamin 1999 and Lyamin & Sloan 1997.

3 DISCRETE FORMULATION OF UPPER BOUND THEOREM

Considering the system of forces shown in Figure 2, the objective of an upper bound calculation is to find a velocity distribution **u** which satisfies compatibility, the flow rule, the velocity boundary conditions **w** on the surface area A_w, and minimises the integral

$$W^{internal} = \int_V \boldsymbol{\sigma}\dot{\boldsymbol{\varepsilon}}\, dV \tag{5}$$

An upper bound estimate on the true collapse load can be obtained by equating $W^{internal}$ to the power dissipated by the external loads

$$W^{external} = \int_{A_t} \mathbf{t}^{\mathrm{T}}\mathbf{u}\, dA + \int_{A_q} \mathbf{q}^{\mathrm{T}}\mathbf{u}\, dA + \int_V \mathbf{g}^{\mathrm{T}}\mathbf{u}\, dV + \int_V \mathbf{h}^{\mathrm{T}}\mathbf{u}\, dV \tag{6}$$

3.1 *Linear finite elements*

As we want only linear equality constraints in the final optimisation problem, simplex finite elements are again used to discretise the continuum in the upper

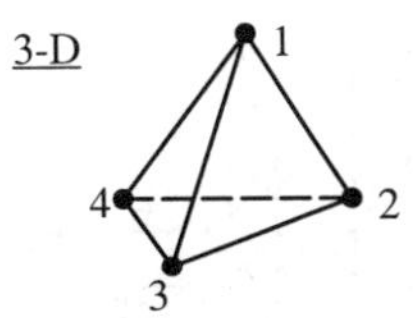

$\mathbf{u}^l = \{u_1^l \, u_2^l \, u_3^l\}^{\mathrm{T}}$
$\boldsymbol{\sigma}^e = \{\sigma_{11}^e \, \sigma_{22}^e \, \sigma_{33}^e \, \sigma_{12}^e \, \sigma_{23}^e \, \sigma_{13}^e\}^{\mathrm{T}}$

Figure 4. Simplex finite elements used in upper bound formulation.

bound formulation (Figure 4). Kinematically admissible velocity discontinuities are permitted at all interfaces between adjacent elements. If D is the dimensionality of the problem, then there are D+1 nodes in the element and each node is associated with a D-dimensional vector of velocity variables $\{u_i\}$, $i = 1,\ldots,D$. These, together with a $D(D+1)/2$-dimensional vector of elemental stresses $\{\sigma_{ij}\}$, $i = 1,\ldots,D; j = i,\ldots,D$ and a $2(D-1)$-dimensional vector of discontinuity variables, $\mathbf{v}^d$, are taken as the problem variables.

3.2 *Optimisation problem statement*

Applying the velocity boundary conditions, the flow rule constraints, and collecting terms for the power dissipated throughout the body, the problem of finding a kinematically admissible velocity field which minimises the internal power dissipation may be stated as

$$\text{minimise} \quad Q = \boldsymbol{\sigma}^{\mathrm{T}}\mathbf{B}\mathbf{u} + \mathbf{c}_{\mathrm{u}}^{\mathrm{T}}\mathbf{u} + \mathbf{c}_{\mathrm{v}}^{\mathrm{T}}\mathbf{v} \quad \text{on } (\mathbf{u},\mathbf{v}) \qquad (7)$$

subject to

$$\begin{aligned} \mathbf{A}_{\mathrm{u}}\mathbf{u} + \mathbf{A}_{\mathrm{v}}\mathbf{v} &= \mathbf{b} \\ \mathbf{B}\mathbf{u} &= \sum_{j \in J_\sigma} \lambda_j \nabla f_j(\boldsymbol{\sigma}) \\ \lambda_j &\geq 0, \quad j \in J_\sigma \\ \lambda_j f_j(\boldsymbol{\sigma}) &= 0, \quad j \in J_\sigma \\ f_j(\boldsymbol{\sigma}) &\leq 0, \quad j \in J_\sigma \\ \mathbf{v} &\geq \mathbf{0} \\ \mathbf{u} \in \mathrm{R}^{n_{\mathrm{u}}}, \ \mathbf{v} \in \mathrm{R}^{n_{\mathrm{v}}}, & \ \boldsymbol{\sigma} \in \mathrm{R}^{n_\sigma}, \ \boldsymbol{\lambda} \in \mathrm{R}^{E} \end{aligned}$$

where $\mathbf{B}$ is a $n_\sigma \times n_{\mathrm{u}}$ global compatibility matrix, $\mathbf{c}_{\mathrm{u}}$ is an n_{u}-dimensional vector of objective function coefficients for the velocities, $\mathbf{c}_{\mathrm{v}}$ is an n_{v}-dimensional vector of objective function coefficients for the discontinuity variables, $n_\sigma = ED(D+1)$, $n_{\mathrm{u}} = DN$ (N is the number of nodes and E is number of elements in the mesh), $n_{\mathrm{v}} = 2D(D-1)Ds$, Ds is the total number of discontinuities, $\mathbf{A}_{\mathrm{u}}$ is an $r \times n_{\mathrm{u}}$ matrix of equality constraint coefficients for velocities, $\mathbf{A}_{\mathrm{v}}$ is $r \times n_{\mathrm{v}}$ matrix of equality constraint coefficients for discontinuity variables, $r = D^2Ds +$ number of boundary conditions and loading constraints, $f_j(\boldsymbol{\sigma})$ are yield functions, λ_j are non-negative multipliers, and $\mathbf{u}$, $\mathbf{v}$ and $\boldsymbol{\sigma}$ are problem unknowns.

3.3 *Kuhn-Tucker optimality conditions*

Since the objective function and the functions f_j are convex, and all equality constraints are linear, the solution to (7), $(\mathbf{u}^*, \mathbf{v}^*, \boldsymbol{\sigma}^*)$, must satisfy the Kuhn-Tucker optimality conditions

$$\begin{aligned} \mathbf{B}\mathbf{u}^* - \sum_{j \in J_\sigma} \lambda_{\sigma j} \nabla f_j(\boldsymbol{\sigma}^*) &= \mathbf{0} \\ \mathbf{c}_{\mathrm{v}} + \mathbf{A}_{\mathrm{v}}^{\mathrm{T}}\boldsymbol{\mu} + \sum_{j \in J_{\mathrm{v}}} \lambda_{\mathrm{v}j} \nabla f_j(\mathbf{v}^*) &= \mathbf{0} \\ \mathbf{B}^{\mathrm{T}}\boldsymbol{\sigma}^* + \mathbf{c}_{\mathrm{u}} + \mathbf{A}_{\mathrm{u}}^{\mathrm{T}}\boldsymbol{\mu} &= \mathbf{0} \\ \mathbf{A}_{\mathrm{u}}\mathbf{u} + \mathbf{A}_{\mathrm{v}}\mathbf{v} &= \mathbf{b} \\ \lambda_{\sigma j} f_j(\boldsymbol{\sigma}^*) = 0, & \quad j \in J_\sigma \\ \lambda_{\mathrm{v}j} f_j(\mathbf{v}^*) = 0, & \quad j \in J_{\mathrm{v}} \\ f_j(\boldsymbol{\sigma}^*) \leq 0, & \quad j \in J_\sigma \\ f_j(\mathbf{v}^*) \leq 0, & \quad j \in J_{\mathrm{v}} \\ \lambda_{\sigma j} \geq 0, & \quad j \in J_\sigma \\ \lambda_{\mathrm{v}j} \geq 0, & \quad j \in J_{\mathrm{v}} \\ \mathbf{u}^* \in \mathrm{R}^{n_{\mathrm{u}}}, \ \mathbf{v}^* \in \mathrm{R}^{n_{\mathrm{v}}}, & \ \boldsymbol{\sigma}^* \in \mathrm{R}^{n_\sigma} \\ \boldsymbol{\lambda}_\sigma \in \mathrm{R}^{E}, \ \boldsymbol{\lambda}_{\mathrm{v}} \in \mathrm{R}^{n_{\mathrm{v}}}, & \ \boldsymbol{\mu} \in \mathrm{R}^{r} \end{aligned} \qquad (8)$$

where the original conditions $\mathbf{v} \geq \mathbf{0}$ are, for the sake of convenience, written as $f_j(\mathbf{v}) \leq 0$, $j \in J_{\mathrm{v}}$.

A detailed description of a two stage quasi-Newton strategy for solving the system (8) can be found in Lyamin & Sloan 2000.

4 NUMERICAL RESULTS

Using the bound methods discussed above, we now consider the stability of circular and square excavations in a soil obeying the Tresca and Mohr-Coulomb yield criteria. The results are presented in terms of the dimensionless stability number $\gamma H/c$, which is a function of both ϕ and H/R (or H/B). To increase the accuracy of the computed stability numbers, the symmetry inherent in both problems is fully exploited. This means that only 15° and 45° sectors are discretised for the circular and square excavations, respectively, as shown in Figure 5 - 7. These plots also show the boundary conditions adopted in the various analyses. For the lower bound meshes, special extension elements are included to extend the stress field over the semi infinite domain (thus guaranteeing the solutions obtained are rigorous lower bounds on the true solutions). Note that the meshes used for the upper bound calculations are designed to suit wall failure modes, as these were found to be critical for H/R ratios not exceeding 7 (Britto & Kusakabe 1982).

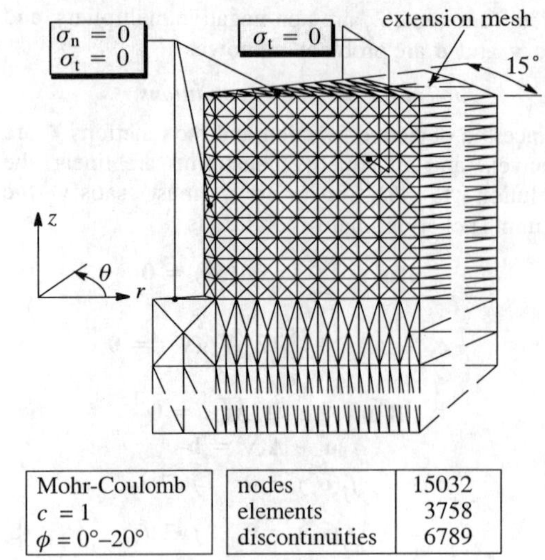

Figure 5. Lower bound mesh for circular excavation.

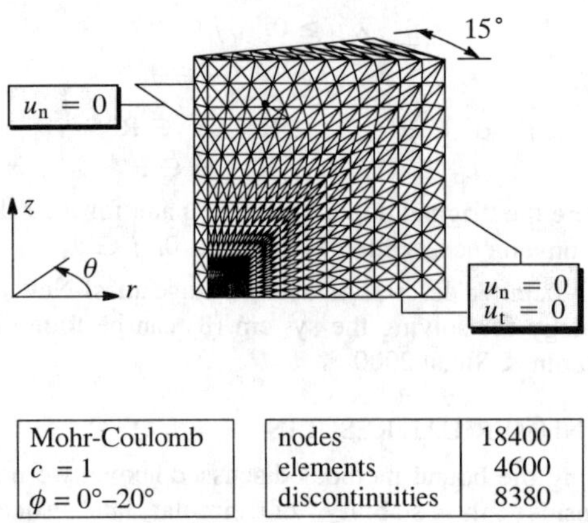

Figure 6. Upper bound mesh for circular excavation.

Table 1. Stability numbers $N_s = \gamma H/c$ for circular excavations.

$\phi°$	Stability number $\gamma H/c$					
	H/R=1		H/R=2		H/R=3	
	LB	UB	LB	UB	LB	UB
0	5.09	5.26	5.95	6.14	6.59	6.83
10	6.48	6.73	7.82	8.20	9.00	9.42
20	8.30	8.85	10.49	11.26	12.54	13.43

The computed stability numbers $\gamma H/c$ for a range of circular excavations are presented in Table 1 and Figure 9. On average, these results bracket the true stability numbers to within 5 per cent. For the undrained case with ϕ=0° and H/R=1, the lower bound of N_s=3.464, obtained by Pastor & Turgeman 1979, has been improved by 47 per cent to N_s=5.09. The corresponding upper bound of N_s=5.26, is a small improvement on the best previous upper bound estimate of N_s=5.298 (Pastor 1981). As no previous results have been found for drained loading, the solutions presented in Table 1 for $\phi > 0$ are new. The accuracy of the upper and lower bounds for these cases decreases as the friction angle increases, with an average discrepancy of 6.8 per cent for ϕ=20° compared to 3.3 per cent for ϕ=0°.

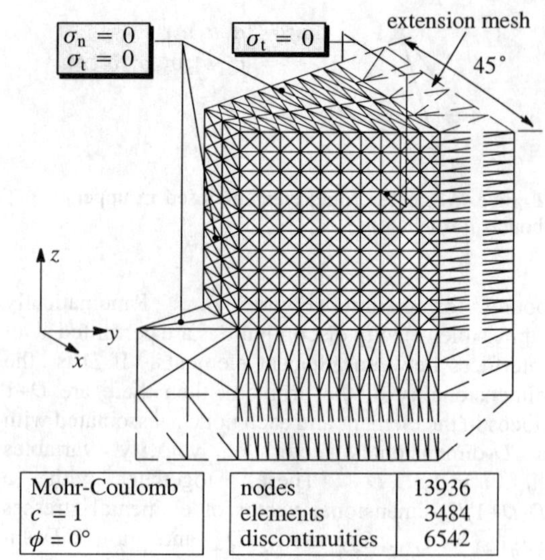

Figure 7. Lower bound mesh for square excavation.

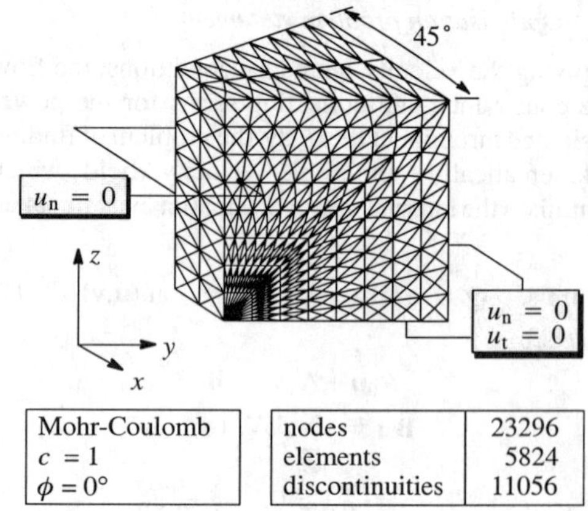

Figure 8. Upper bound mesh for square excavation.

Stability numbers for various undrained square excavations are presented in Table 2 and Figure 10. Because of the coarser meshes used in these cases, the accuracy of the results is less than that reported for the circular excavations, with the maximum discrepancy between the lower and upper bounds reaching 18.5 per

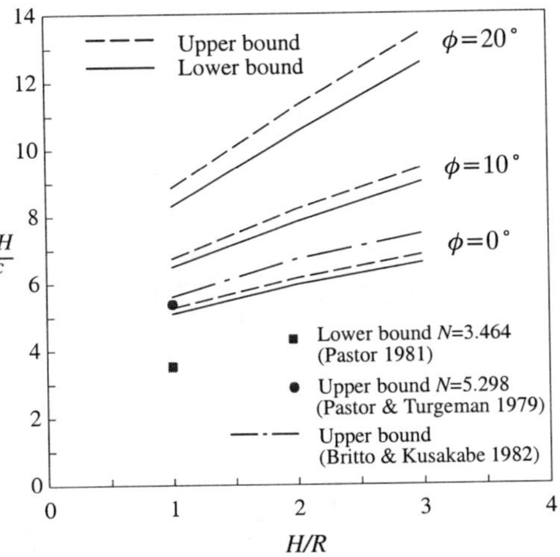

Figure 9. Results for circular excavations.

cent for H/B=1. Comparing the bounds in Table 1 and Table 2, we see that the stability numbers for the square excavations are smaller than their circular analogues. This is expected on physical grounds for cases where $R = B$, and should be taken into account by practicing engineers.

Table 2. Stability numbers $N_s=\gamma H/c$ for square excavations in purely cohesive soil.

ϕ°	Stability number $\gamma H/c$					
	H/B=1		H/B=2		H/B=3	
	LB	UB	LB	UB	LB	UB
0	4.33	5.19	5.12	6.10	5.78	6.84

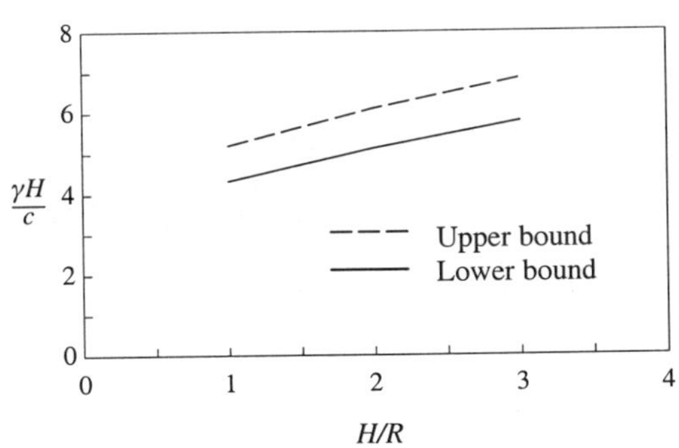

Figure 10. Results for square excavation.

The average number of nodes for the excavation models was around 19000, with the largest mesh having 23296 nodes, 5824 elements and 11056 discontinuities. Bearing in mind that each node has six unknown stresses for the lower bound method, and an even larger number of variables for the upper bound method, these grids generate substantial optimisation problems and it comes as no surprise that the analyses are computationally demanding. The average CPU time for all the analyses was 4500 sec for a Dell Precision 220 workstation with a Pentium III 800EB MHz processor. Although significant, these timings are certainly competitive with those that would be needed for a three-dimensional incremental analysis with the displacement finite element method.

5 CONCLUSIONS

A range of three-dimensional excavations made in Tresca and Mohr-Coulomb soil have been considered and improved lower and upper bounds have been obtained. The exact stability numbers for circular excavations are predicted to within 5 per cent, while those for square excavations are predicted to within 18 per cent. As the power of desktop computers continues to grow, these bounds will be tightened further by the use of finer grids.

REFERENCES

Bjerrum, L. & Eide, O. 1956. Stability of strutted excavations in clay', *Géotechnique,* 6(1): 32-47.

Britto, A.M. & Kusakabe, O. 1982. Stability of unsupported axisymmetric excavations in soft clay, *Géothechnique,* 32(3): 261-270.

Lyamin, A.V. 1999. *Three-dimensional Lower Bound Limit Analysis Using Nonlinear Programming,* PhD Thesis, Department of Civil, Surveying and Environmental Engineering, University of Newcastle, NSW.

Lyamin, A.V. & Sloan, S.W. 1997. A comparison of linear and nonlinear programming formulations for lower bound limit analysis, *Proceedings of the 6th International Symposium on Numerical Models in Geomechanics,* Pietruszczak & Pande (eds), Balkema, Rotterdam: 367-373.

Lyamin, A.V. & Sloan, S.W. 2000. Upper bound limit analysis using linear finite elements and nonlinear programming, *Proceedings of the 5th International Conference on Computational Structures Technology,* Leuven, Belgium, to appear.

Pastor, J. 1981. 'Analyse limite et stabilite des fouilles', *Proceedings, 10th International Conference on Soil Mechanics and Foundation Engineering,* Stockholm, Sweden, 3: 505-508.

Pastor, J. & Turgeman, S. 1979. 'Formulation linéaire des méthodes de l'analys limite en symétrie axiale', *4th Congrès Francais de Méchanique,* Nancy, France.

Prater, E.G. 1977. An examination of some theories of earth pressure on shaft linings, *Canadian Geotechnical Journal,* 14(1): 91-106.

Sloan, S.W. 1981. *Numerical Analysis of Incompressible and Plastic Solids Using Finite Elements,* PhD thesis, University of Cambridge.

Computer Methods and Advances in Geomechanics, Desai et al. (eds) © 2001 Balkema, Rotterdam, ISBN 90 5809 183 X

Integration of Geographic Information System with groundwater models for landslide hazard assessment

Marco Masetti
Applied Geology, D.I.I.A.R., Polytechnic of Milan, Italy

Sterlacchini Simone
Environmental Geology Section, Center for Quaternary and Alpine Geodynamic Studies, National Research Council, Milan, Italy

ABSTRACT: In this study, the integration of a raster-based geographical information system (GIS) with groundwater and deterministic models is proposed in order to perform landslide hazard assessment. Distribution of the factor of safety values for each pixel is calculated using the frequency distributions of the main physical and technical input parameters. Results show that: a) this approach has a good agreement with the morphological evolution of the investigated area; b) the evaluation of unsaturated conditions both for hydrological and geotechnical analysis gives a better understanding and assessment of slope instability processes; c) evaluation of the variance of the safety factors and maps of the probability of failure values can be obtained by the distribution functions of the input parameters; d) real probabilities can be derived considering the calculated probability of failure with the time probability of the considered triggering events.

1 INTRODUCTION

1.1 *GIS and deterministic models*

Deterministic distributed hydrological and slope stability models require maps giving the spatial distribution of input parameters in order to calculate groundwater levels, pore pressures and safety factors. These physically-based models can be implemented in a two-dimensional, raster-based, geographic information system (GIS).

Among a very wide range of deterministic models for slope instability analysis available nowadays, the only one which calculates slope instability on a pixel basis is the infinite slope model. This one-dimensional model considers the spatial distribution of input parameters and describes slope stability in a simple form, resulting very effective for calculation of surficial slip failure with a low depth/length ratio. Neglecting the second dimension the model can be easily applied in a raster based GIS where the slope stability is calculated for the pixels on a map, combining geo-information deriving from several input maps, stored in a relational geo-database (slope angle, soil depth, soil geotechnical characteristic, and so on). Calculation performed on each individual pixel will result in safety factor values or probability values for all pixels, which can be used to create a hazard map. Under this point of view, the pixels can be considered as “homogeneous terrain units” which don’t take into consideration the resulting forces of neighboring areas.

1.2 *General description of the investigated area*

The procedure is applied in order to evaluate the slope stability conditions in a basin in Northern Italy with an approximate area of 40.5 km^2. This basin is characterized by a very high density of landslides, mainly caused by surficial movements of soils on rock substratum, showing low depth/length ratios. For this reason the infinite slope model used in the analysis can be considered a good tool for evaluating the factor of safety.

1.3 *General Structure of the work*

The general scheme of the study is represented in figure 1. The steps that leads to the construction of the geo-database are not described in this work. In the following paragraphs, only the landslide hazard assessment models, validation procedures and the discussion of the results are analyzed (figure 2).

2 HYDROLOGICAL ANALYSES

2.1 *Analytic model*

Broadbridge & al. analytic method (1988) is used to determine the effect of infiltration during rainfall events. The model gives an exact solution to Richard equation for transient single phase flow in a bounded unsaturated soils profile with an impermeable phase.

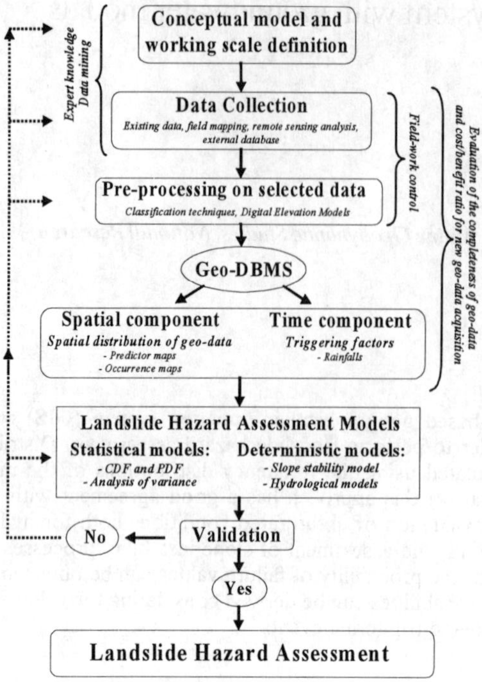

Figure 1: Activity scheme

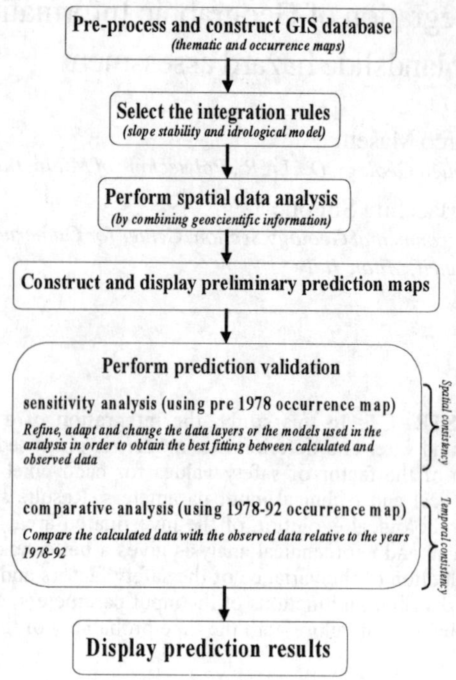

Figure 2: Flow chart of modeling and validation procedures

The equation, solved for a constant rate rainfall infiltration, can give: a) the increasing or the formation of piezometric levels at the impermeable basement b) the evolution of soil water content along a vertical profile.

Given the normalized soil water content as:

$$\Theta = (\theta - \theta_n)/(\theta_s - \theta_n) \quad (1)$$

where θ is actual soil water content, θ_n is the initial water content and θ_s is the saturated water content, hydraulic conductvity K and soil water diffusivity are represented respectively by:

$$K = K_s \Theta^2 (c-1)/(c-\Theta) \quad (2)$$

$$D = h(c) S^2 (\theta_s - \theta_n)^{-2} (c-\Theta)^{-2} \quad (3)$$

where K_s is saturated hydraulic conductivity, S is the sorptivity, c and h(c) are a parameter and a function dependent from the considered soil. Normalized water content and its adimensional relative depth are given by:

$$\Theta = c\left[1 - \left(2\rho + 1 - 2u^{-1}\partial u / \partial \zeta\right)^{-1}\right] \quad (4)$$

$$z_* = c^{-1}\left[\rho(\rho+1)\tau + (2\rho+1)\zeta/2 - \ln u\right] \quad (5)$$

with

$$\tau = 4c(c-1)t_*, \rho = R_* / 4c(c-1) \quad (6)$$

$$t_* = t/t_s, z_* = z/\lambda_s \quad (7)$$

$$t_s = h(c) S^2 / c(c-1) K_s^2 \quad (8)$$

$$\lambda_s = h(c) S^2 / c(c-1)(\theta_s - \theta_n) K_s \quad (9)$$

where z and t are respectively depth and time.

The solution of the function $u(\zeta, \tau)$ and the spatial derivative $\partial u(\zeta,\tau)/\partial \zeta$ are given in explicit form by authors. Soil water content and negative pore water pressure values can be correlated by exponential function using Gupta & Larson (1979) approach.

The hydrologic model is solved outside the GIS but its managing is strictly linked to it, according to the following procedure:

a) crossing of the hydraulic conductivity and the water table depth maps to obtain homogeneous "hydrogeological" zones with the same combination of the two parameters;

b) definition of the triggering rainfall events;

c) solution of the hydrologic model for each homogeneous "hydrogeological" zone with different triggering events;

d) recalculation of new pore-water pressures map with the values computed through a)-c).

Two different situations can result from the application of the model:

1. infiltration front doesn't reach water table: distribution of pore water pressures along vertical profiles in the unsaturated soils;
2. infiltration front reaches water table: change in water table depth is calculated.

2.2 *Numerical model*

A bi-dimensional finite element method, SEEP/W, was applied on some sections, located in areas of particular interest, to better analyze them and to evaluate the approximations resulting from the use of the mono-dimensional model results.

Results derived from numerical analysis show a good agreement with the ones from the analytical model with a general tendency to calculate a lower increase in water table height. This is particularly true with soils with high permeability where differences can reach 20%. This is probably due to the fact that numerical model is bidimensional and can take into account also the slope of soil-rock interface, that means groundwater flow downhill in the saturated zone, while analytical model can analyze a one dimensional flow only.

3 SLOPE STABILITY ANALYSIS

3.1 *Infinite slope model*

Landslides in the investigated area are mainly represented by superficial movement of soils on rock substratum with a low depth/length ratio. For this reason, the infinite slope model is considered a good approximation of reality for the evaluation of the factor of safety.

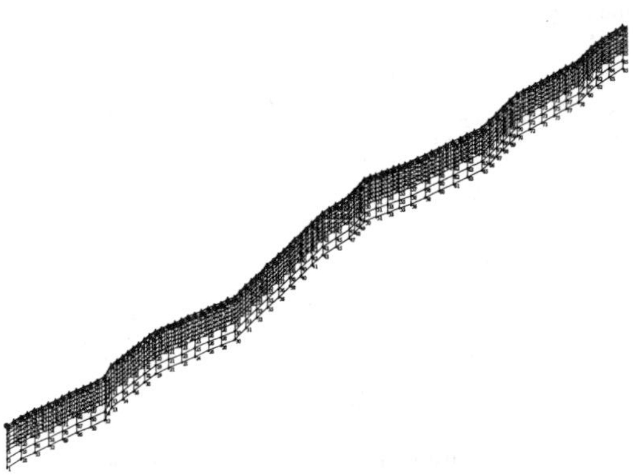

Fig. 3 Schematic representation of numerical model along profile 1

Only for computational reasons, it's more convenient to express the basic formula of the infinite slope model in two parts, as follows:

$$Avg(F) = B_1 * c'_{tot}/z + B_2 * \tan\phi \qquad (10)$$

with:

$$B_1 = 1/\gamma\sin\beta\cos\beta \quad \text{and} \quad B_2 = \frac{1-\frac{m\gamma_w}{\gamma}}{\tan\beta} \qquad (11)$$

where:
F = factor of safety;
c'_{tot} = total cohesion (kPa);
ϕ = effective angle of shearing resistance (°);
β = terrain surface inclination;
γ = unit weight of soil (kN/m^3);
γ = unit weight of water (kN/m^3);
m = z_w/z = groundwater/soil thickness ratio (dimensionless);
z = depth of failure surface below the terrain surface (m);
z_w = height of water table below failure surface (m).

The maps produced in this way resulted in average safety factors (avg(F)) for pre-determined scenarios where average values for c' and tanϕ' are used.

3.2 *Estimating cohesion of soils*

The link between evolution of infiltration and water table and substratum depth is considered to evaluate if formation of slope failure surface was dependent from the triggering event and could also not be coincident with soil-rock contact. This can be considered only with an unsaturated approach also for geotechnical behavior of soils, expressing total cohesion in the infinite slope model as:

$$c_{tot} = c' + (u_a - u_w)\tan\phi^b \qquad (12)$$

where c' is effective cohesion, u_a pore-air pressures, u_w pore-water pressures, ϕ^b shear strength angle due to matrix suction (Fredlund & al 1978).

According to the two different situation of infiltration described in 2.1. paragraph, the following situations were considered:

1) infiltration front reaches water table: factor of safety is calculated at the soil-rock contact surface, total cohesion is given by effective cohesion and parameter *m* is given by height of saturated zone and soil depth ratio;

2) infiltration front doesn't reach water table: factor of safety is calculated at the depth reached by infiltration front, total cohesion is given by (12) and parameter *m* is equal to zero.

3.3 *Statistical analysis*

The recognition of significant degree of uncertainties of many of the input parameters used in the calculation is associated with the real parameter variability, systematic errors, geotechnical and hydrological models. This fact suggests the use of a method which manages the calculation of the variance of the safety factors, based upon the variances of the input variables. Therefore, the result is expressed as the average ± standard deviation.

The following tentative formula is partly based on the equation for error propagation during arithmetical operations; this error propagation is different for summation, subtraction, division and multiplication and is also related to the fact whether the individual parameters are dependent or not (Burrough, 1986). This formula estimates the variance of F based upon the variances of c', tanϕ' and z:

$$VAR(F) = B_1^2\left(\frac{VAR(c')}{z^2} + \frac{c'^2 VAR(z)}{z^4}\right) + B_2^2 VAR(\tan\phi') \quad (13)$$

where:

VAR(F) = variance of safety factor;

VAR(c') = variance of effective cohesion;

VAR(tanϕ') = variance of the tangent of the effective friction angle;

VAR(z) = variance of soil thickness;

B_1 and B_2 are separate terms of equation (10).

In this study, the variances of the cohesion and internal friction angle are calculated from data deriving from shear box and triaxial tests. Simple statistical analysis for cohesion and for ϕ' (or tanϕ') for each soils are performed to obtain the main parameters of distributions. In figure 4, sample and statistical distribution of "argille varicolori" shear strength angles are reported.

The calculation of the variance of the soil thickness is resulted more complex. During data examination, it has been noted a relation between soil thickness and slope angle and, especially, between soil thickness and distance from watershed.

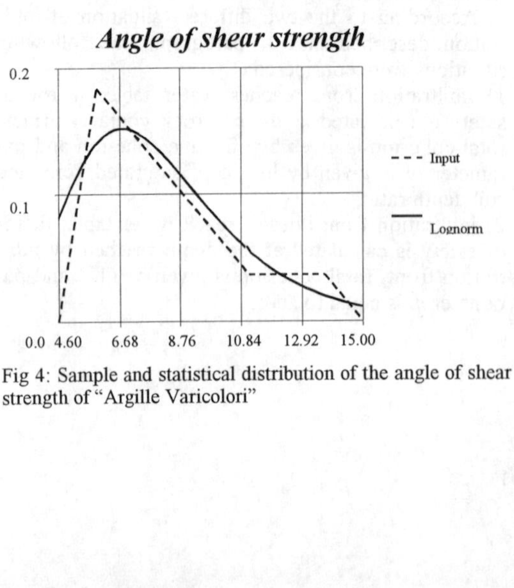

Fig 4: Sample and statistical distribution of the angle of shear strength of "Argille Varicolori"

Therefore, also a relation of the variance of the soil thickness and the watershed distance was to be expected. Using available sampling data, distributed along certain profiles, we have obtained a relation expressed by:

$$VAR(z) = 0.15\ \text{distance} + 0.8 \quad (14)$$

It's also possible to calculate the probability that F≤1 assuming, for any given pixel, a normal distribution of possible F-values (characterized by a mean and a standard deviation).

In general (figure. 6), the F=1 value could be located to the right or to the left of the avg(F); if avg(F)=F=1 then the probability that an F-value is found which is smaller than 1 is 50%. So, in order to calculate the probability that F will be smaller or equal then 1, the values have to be standardized so that we can say that F=1 is so many standard deviation away from avg(F). Using the standardized normal form (Davis, 1986):

$$Z_i = \frac{X_i - avg(X)}{s} \quad (15)$$

it can be obtained, considering X_i=F=1:

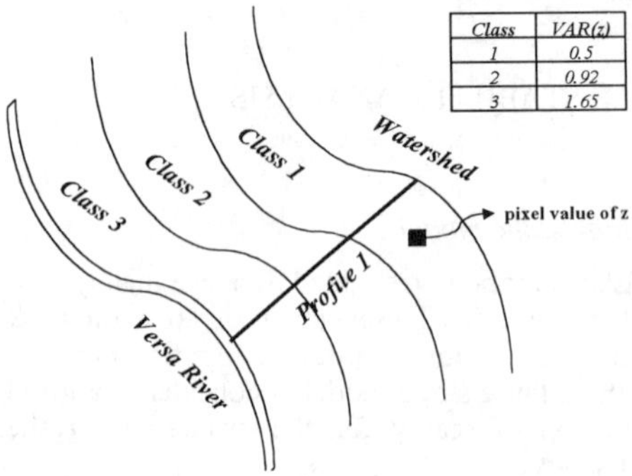

Figure 5: Variance of the soil thickness with the distance from the watershed

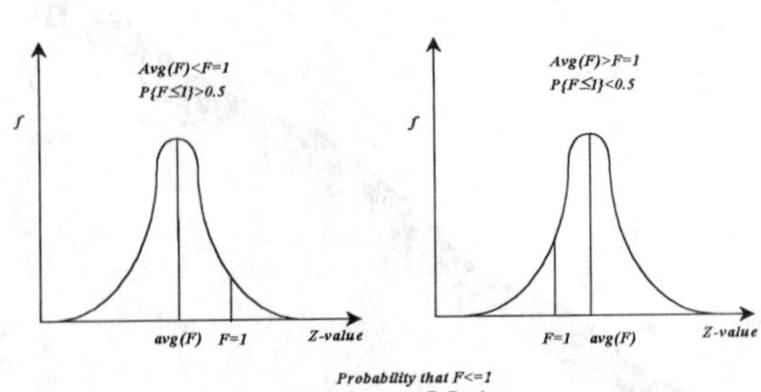

Fig 6: Normal distribution of F-values. The area under the curve to the left of F=1 represents the cumulative probability that F≤1

$$Z = \frac{1 - \text{avg(F)}}{\text{std(F)}} = \frac{1 - \text{avg(F)}}{\sqrt{\text{var(F)}}} \quad (16)$$

In this way, the Z-values can be considered as the distance between F=1 and ordinates which are Z-standard deviation units away from F=1.

So we can consider two cases:

1. Avg(F) < 1. All the area under the normal curve to the left of the F=1 represents the cumulative probability that F≤1, for a given scenario, and will be larger than 0.5.

2. Avg(F) > 1. All the area under the normal curve to the left of the F=1 represents the cumulative probability that F≤1, for a given scenario, and will be smaller than 0.5.

By converting the Z-values in probabilities, the probability that F≤1, only based upon the variances of the input parameters (c', $\tan\phi'$ and z), will be known. To derive the probabilities related to a certain scenario, the obtained values have to be multiplied with the probability that a given event will occur within a given time period (*time probability*). In this case, a design periods of 25 and 50 years are used. Therefore the procedure must follow this steps:

1) select the scenarios for which the calculation has to be performed and define the parameters; 2) calculate the avg(F); 3) calculate the variance of F; 4) transform the values into z-values; 5) calculate the probability that F≤1; 6) define the period for which the hazard (the probability that F≤1) will be expressed; 7) calculate the probability that the event (groundwater peaks) will occur within this period; 8) multiply the results of step 5) and step 7) into the final probability map.

4 MODEL VALIDATION AND RESULTS

4.1 *Model validation*

Two different sets of recorded landslides, 1978 and 1992, are used to respectively determine success and prediction rates (fig. 7,8). In other words, it is assumed that the time of the study was the year 1978 and that all the spatial data available in 1978 are compiled including the distribution of the landslides. These occurrences play an important role in constructing prediction model by establishing relationships between the pre-1978 landslides and the remainder of the input data set. In particular, the estimated probability values of all pixels in the whole area are sorted in descending order. Then the pixels were classified in five classes on the basis of their frequency distribution. The first class consists of the 20% of the pixels having the highest predicted probability values and they are classified as "hazardous areas". The last class consists of the 20% of the pixels having the lowest predicted probability values and they are classified as "non hazardous areas". The five hazard classes are compared with the occurrences of the pre-1978 landslides in order to evaluate the performance of the deterministic models used in the study (success rate). In this way we evaluate the spatial consistency of results. For this reason, for specific rainfall return periods, the pixels in the most hazardous class (with the highest estimated probability values), are compared with the 10% and 30% of the upper part of each landslide.

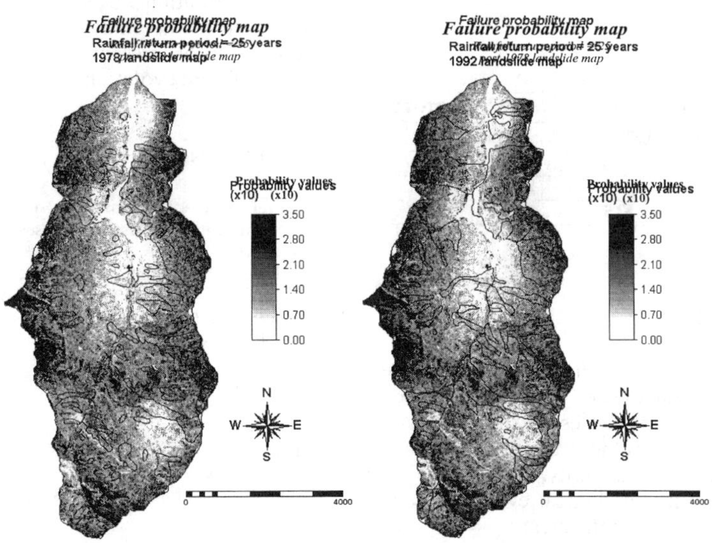

Figure 7: Failure probability maps with a rainfall return period of 25 years.

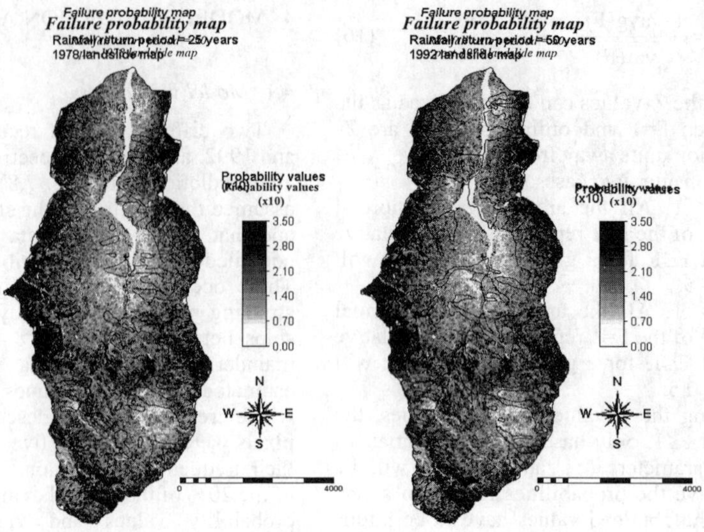

Figure 8: Failure probability maps with a rainfall return period of 50 years.

These upper parts are considered the trigger areas of the landslides. For a return rainfall period of 50 years and considering only the upper 10% of each landslide, 67% of pixels were correctly classified (predicted and observed unstable). Considering the upper 30% of the landslide area, this percentage increased to 76%.

The prediction results are then evaluated by comparing the predicted probability values with the distribution of the landslides that had occurred after 1978, i.e. during the period 1978-1992 (prediction rate). In this way we evaluate the temporal consistency of results.

Tabel 1. Success and prediction rates

	Pre 1978 occurrences		Post 1978 occurrences	
	Success rate		Prediction rate	
	$S_f>1$	$S_f<1$	$S_f>1$	$S_f<1$
Unstable areas	76.2	23.8	81.3	18.7
Stable areas	17.5	82.5	17.1	82.9

Sf = safety factor

5 CONCLUSIONS

In this study we propose the integration of a raster-based GIS with groundwater models to perform landslide hazard assessment. Results show that:

a) this approach allows the determination of areas with different degree of stability, showing a good agreement with the morphological evolution due to slope instability occurred in the area;

b) the evaluation of unsaturated conditions both for the hydrological and geotechnical analysis gives a better understanding and assessment of slope instability processes;

c) in addition to the spatial distribution of safety factors, probability of failure, for a specific triggering event, can also be mapped with the use of distribution functions of the input parameters;

d) real probabilities can be then derived considering the calculated probability of failure with the time probability of the considered triggering events;

e) the failure probability maps and the factor of safety distribution maps for the different scenarios considered in the study, can give synthetical information both on present stability of the area and its possible future evolution.

6 REFERENCES

Burrough, P.A. 1986. Principles of geographical information systems for land resources assessment. *Clarendon Press, Oxford.*

Broadbridge, P., Knight, J.H. & Rogers, C. 1988. Constant rate rainfall infiltration in a bounded profile: solutions of a non-linear model. *Soil Sci. Soc. Am. Jour.* 52: 1526-1533

Davis, P., 1986. Statistics and data analysis in geology. *Wiley, New York.*

Fredlund, D.G., Morgenstern, N.R. & Widger, R.A. 1978. The shear strength of unsaturated soils. *Can. Geot. Jour.* 15: 313-321.

Gupta, S.C. & Larson, W.E. 1979. Estimating soil water retention characteristics from particle size distribution, organic matter percent and bulk density. *Wat. Res. Res.* 15: 1633-1635.

Computer Methods and Advances in Geomechanics, Desai et al. (eds) © 2001 Balkema, Rotterdam, ISBN 90 5809 183 X

Elasto-plastic finite element analysis of retaining wall with passive mode rotated about the bottom – Infinitesimal deformation analysis and finite deformation analysis

H. Mori
Institute of Civil Engineering of Tokyo Metropolitan Government, Japan

T. Tanaka
Department of Biological and Environmental Engineering, University of Tokyo, Japan

ABSTRACT: Retaining walls with a rough wall face were analyzed by elasto-plastic finite element analyses (infinitesimal deformation analysis and finite deformation analysis), and the results were compared with those obtained in model experiments using air-dried dense Toyoura sand. The failure of a sand mass is generally progressive and depends on the development of a shear band of localized deformation. In this analysis, the dynamic relaxation method combined with the generalized return mapping algorithm instead of constituting the stiffness matrix was applied to the integration algorithms of elasto-plastic constitutive relations including the effect of the shear band. The objectivity of the above-mentioned finite element analysis was examined and it was shown that there was little mesh dependency. In the passive mode rotated about the bottom, the coefficient of passive earth pressure obtained by finite deformation analysis was superior to that obtained by infinitesimal deformation analysis. Especially, finite deformation analysis employing the wall friction angle and automatic re-meshing method exhibited good agreement with the experiment. Also, the finite element analysis including the effect of the shear band gave the scale effect of a retaining wall.

1 INTRODUCTION

The classical earth pressure theories of Coulomb in the 18th century and Rankine in the 19th century are still widely used by practicing engineers. As a rigid retaining wall is generally designed based on active earth pressure, little attention has been paid to passive earth pressure. The results of a number of experiments in active modes tended to be identical with those obtained by classical earth pressure theories assuming an isotropic perfectly plastic material in which a peak internal friction angle mobilized along a rupture surface. On the other hand, experimental results in passive modes in which the influence of progressive failure might be remarkable, tended not to agree with classical earth pressure theories. Figure 1 shows the coefficient of passive earth pressure with the relationships between internal friction angle (ϕ) and wall friction angle (δ) by Coulomb, Chen and Sokolovskii's formula. The figure shows that if an internal friction angle or wall friction angle is constant and high, the coefficient of passive earth pressure obtained by the most widely accepted Coulomb's formula will be overestimated.

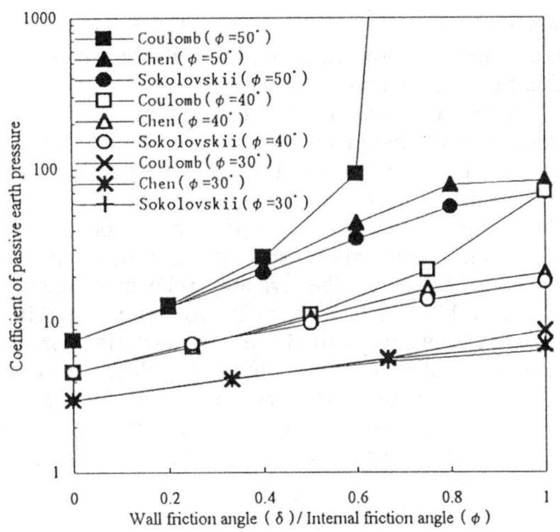

Figure 1. Coefficient of passive earth pressure by classical earth pressure theories

In this study, we attempted to estimate the feasibility of finite element analysis, by comparing the results of experiments by rotation about the bottom in the passive mode using a rigid retaining wall with those of finite element analyses (infinitesimal deformation analysis and finite deformation analysis). The constitutive model for a non-associated strain hardening-softening elasto-plastic material was employed. In order to avoid numerical instability due to the singularity of the non-associated Mohr-Coulomb model, the constitutive model based on the yield function of the Mohr-Coulomb type and the plastic function of the Drucker-Prager type were used. The dynamic relaxation method with return mapping algorithm [Simo and Ortiz, 1986] was applied to the integration algorithm of the elasto-plastic constitutive relation including the effect of the shear band [Tanaka, 1979].

2 TEST APPARATUS

Three earth pressure cells for measuring each earth pressure acting on the central section of the wall face were used, the locations of which are shown in Figure 2. The rough wall face was prepared by attaching sandpaper to the steel plate face. The wall friction angle was about 34° by the direct shear test. For lubrication to reduce the side-wall friction, a thin latex rubber membrane and a thin layer of silicone grease were used. The retaining wall was rotated about the bottom in the passive mode by means of a screw handle until the primary failure surface had completely penetrated into the sand mass. The sand used for the test was Toyoura sand (G_s=2.64, e_{max}=0.977, e_{min}=0.605, D_{50}=0.16 mm). The dry density (γ_d) was about 15.7 kN/m^3.

3 DYNAMIC RELAXATION METHOD

Accurate prediction of the limit load of soil structures is an important problem related to soil mechanics. The failure of a sand mass is generally progressive and depends on the development of a shear band of localized deformation. Though nonlinear geotechnical problems are often solved using the incremental Newton-Raphson algorithm with the stiffness matrix, much of the computational effort is spent in computing displacement increments by solving the stiffness matrix equation. Recently, the dynamic relaxation method has been shown to be more effective than the modified Newton-Raphson method. In this analysis, the dynamic relaxation method combined with the generalized return mapping algorithm instead of the stiffness matrix was applied to the integration algorithms of elasto-plastic constitutive relations including the effect of the shear band.

For elasto-plastic collapse analysis in the static case, the stiffness matrix is often used for redistribution of the unbalanced residual forces. However, it offers computational disadvantages because a long computing time is required. Also, four-node quadrilateral elements show a stiff response generally. To overcome these difficulties, we used the finite element method code with a dynamic relation solver with one point integration (reduced integration). Then, hourglass modes (zero-energy modes) were prevented by the reduced integration in a four-node isoparametric element. The solution is the final (steady state) solution of the transient dynamic problem (the dynamic relaxation method):

$$M\ddot{u} + C\dot{u} + P - P_{init} = F \qquad (1)$$

where M is the mass matrix, C is the damping matrix, P is the vector of internal forces, P_{init} is the vector of nodal forces due to initial stress, F is the vector of external forces, $\ddot{u}$ is the acceleration vector, and $\dot{u}$ is the velocity vector. The central difference expressions are introduced for the derivatives in Eq.(1), and the damping matrix C is determined by:

$$C = \alpha M \qquad (2)$$

where α is the damping ratio. Then, the nodal displacement is expressed as:

$$u^{t+\Delta t} = \frac{M^{-1}}{1+0.5\alpha\Delta t}\Big[(\Delta t)^2\left(F - P - P_{init}\right) + 2Mu^t - \left(1-0.5\alpha\Delta t\right)Mu^{t-\Delta t}\Big] \qquad (3)$$

Solutions are easily obtained by introducing the lumped mass into mass matrix M in Eq.(2). Stability for the time increment Δt is expressed by:

$$\Delta t \le \beta l\sqrt{\frac{\rho(1+\nu)(1-2\nu)}{E(1-\nu)}} \qquad (4)$$

where β<1.0 is constant, l is the minimum distance among nodal points, E is the elastic modulus, ν is Poisson's ratio, and ρ is the density. From Eq.(4) taking Δt=1, the fictitious density for the pseudo mass is obtained:

$$\rho = \frac{E(1-\nu)}{\beta^2 l^2(1+\nu)(1-2\nu)} \qquad (5)$$

For rapid convergence a fictitious density is estimated element wise and updated at each incremental stage especially in the case of variable elastic moduli.

4 FINITE DEFORMATION ANALYSIS

In the case of a buckling collapse for buried pipe or the problem of passive earth pressure for a retaining wall structure, the geometrical nonlinearity can not be ignored [Tanaka et al., 1997]. Finite deformation analysis based on the updated-Lagrangian method was applied [Nagtegaal and De Jong, 1981]. The equilibrium equations for the increment are expressed as:

$$\int_V \left(\Delta S_{ij}\delta_{ki} + S_{ij}\Delta U_{k,i} + \Delta S_{ij}\Delta U_{k,i}\right)\delta U_{k,j}\,dV = \int_S \Delta P_i \delta U_i\,dS \qquad (6)$$

$$\Delta S_{ij} = L_{ijkl}\left(\Delta U_{k,l} + 0.5\Delta U_{m,k}\Delta U_{m,l}\right) \tag{7}$$

where δ_{ki} is the Kronecker delta, S_{ij} is the second Piola-Kirchhoff stress and ΔS_{ij} is the increment in the second Piola-Kirchhoff stress. L_{ijkl} is given by the expression:

$$L_{ijkl} = L_{ijkl}^{(e-p)} - 0.5\left(\delta_{ik}\sigma_{jl} + \delta_{jk}\sigma_{il} + \delta_{il}\sigma_{jk} + \delta_{jl}\sigma_{ik}\right) + \sigma_{ij}\delta_{kl} \tag{8}$$

where $L_{ijkl}^{(e-p)}$ are the elastic-plastic moduli for the usual small-strain analysis. The second Piola-Kirchhoff stress is transformed to the true (Cauchy) stress:

$$\sigma_{ij} = \frac{\left(\delta_{ik} + \Delta_{i,k}\right)\left(S_{kl} + \Delta S_{kl}\right)\left(\delta_{jl} + \Delta U_{j,l}\right)}{J} \tag{9}$$

where J is the Jacobian of the deformation increment. Also, Tanaka [1997] had already pointed out that passive earth pressure obtained by the solution method of Nagtegaal and De Jong [1981] exhibited good agreement with passive earth pressure obtained by the solution method of Nagtegaal and Reblo [1988] using the rotational neutral strain. The automatic re-meshing method was combined with finite deformation analysis for removing the entangled mesh [Tanaka, 1992].

5 EXPERIMENTAL RESULTS (ROTATION ABOUT THE BOTTOM)

Figure 3 shows the relationships among the coefficient of passive earth pressure ($Kp=2Q/[\gamma_d \cdot H^2]$), relative depth (h/H) of action of total passive earth pressure and angle (θ) of wall rotation. Total passive earth pressure (Q) is assumed to be closely fitted to the earth pressure obtained from three earth pressure cells of linear shape. Relative depth (h/H) is the ratio of the height (h) from hinge to active point to the wall height (H). Coefficients of passive earth pressure obtained by Coulomb, Chen and Sokolovskii's formula are indicated in Figure 3, also.

As the peak internal friction angle (ϕ_p) and residual internal friction angle (ϕ_r) obtained by the plane strain compression test at low pressure on dense Toyoura sand [Tatsuoka et al., 1986] were close to 50° and 34°, ϕ was assumed to be 50° and 34°. δ was assumed to be 34°. In the case of ϕ =34°, the experimental result (Kp) existed between Coulomb and Sokolovskii's formula. In the case of ϕ =50°, however, the difference of Kp by those theories was large. Especially, the coefficient of passive earth pressure obtained by the most widely accepted Coulomb's formula was about 20 times as large as the experimental value. The relative depth of action of total passive earth pressure was 0.4 to 0.5 and not in agreement

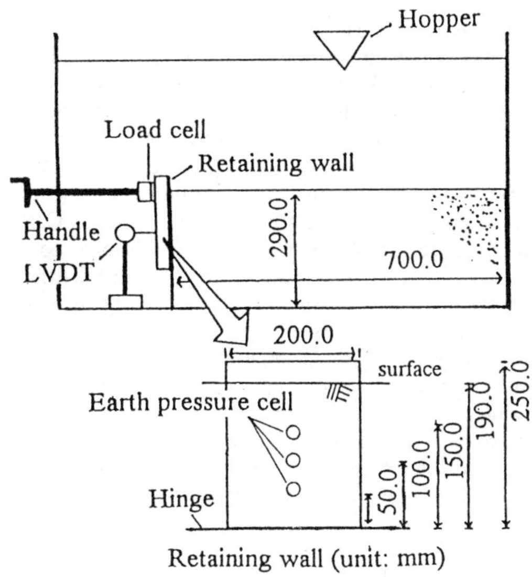

Figure 2. Test apparatus

with Coulomb or Rankine's theories (h/H=1/3) based on general wedge theory.

6 OBJECTIVITY OF FINITE ELEMENT MESH AND THREE PASSIVE MODES

As the finite element analysis for strain softening materials which is closely related to the appearance of the shear band has been shown to depend on the finite element mesh, we investigated the objectivity of the finite element analysis with three passive modes [rotation about the bottom (R.B.), rotation about the top (R.T.) and horizontal translation (H.T.)]. The two types of finite element meshes of retaining wall analyses (Type-1 and Type-2) for examining the objectivity are shown in Figures 4(a) and (b). The parameters used in these analyses were chosen based on the data from the plane strain compression test of air-pluviated dense Toyoura sand [Tatsuoka et al., 1986]. γ_d was 15.7 kN/m^3, ϕ_r 34°, Poisson's ratio 0.2, initial shear modulus 78400 kN/m^2 and thickness of shear band 0.3 cm. For the detailed determination of these parameters, refer to [Mori and Tanaka, 1996].

Figure 5 shows the relationships between coefficient of passive earth pressure (Kp) and angle (θ) of wall rotation and ratio (d/H; d: horizontal distance) of horizontal translation. The wall friction angle employed in these analyses was assumed to be equal to the internal friction angle of sand. Although Kp obtained by the Type-1 mesh at rotation about the bottom was a little smaller than that by the Type-2 mesh, the objectivity of the finite element mesh was assured. Kp obtained by horizontal translation was twice as large as that by rotation about the top, and the peak value was the highest, also. Incidentally, we found that the influence of the finite element mesh was small.

Figures 6(a), (b) and (c) show the calculated contours of maximum shear strain and earth pressure distributions at three passive modes obtained by the Type-1 mesh. Figures 7(a) and (b) show the rupture surfaces and earth pressure distributions observed from the tests on Leighton Buzzard sand at Cambridge University [Roscoe, 1970]. Both the contours of maximum shear strain and earth pressure distributions by analyses were similar to those by the tests. As the experimental results performed at Cambridge University agreed with the calculated results, this analysis can be applied to passive earth pressure problems of retaining walls. The earth pressure distribution by horizontal translation was similar to the pressure distribution defined by Rankine or Coulomb's theories based on general wedge theory.

7 NUMERICAL AND EXPERIMENTAL RESULTS (ROTATION ABOUT THE BOTTOM)

Based on the experiences of finite element analysis for strain softening materials with the passive mode rotated about the bottom, Tanaka and Mori [1996] indicated the necessity to use the finite deformation analysis. We investigated the influences of wall friction angle and finite deformation analysis with the automatic re-meshing method.

Figure 8 shows the coefficient of passive earth pressure obtained by infinitesimal deformation analyses, finite deformation analyses and finite deformation analyses with the automatic re-meshing method. To take the effect of wall friction angle into consideration, the wall friction angle was 34°. Kp obtained by infinitesimal deformation analysis agreed with that by the experiment until 4 degrees of wall rotation and then showed a limit state. The results of finite deformation analysis with re-meshing method were identical with those of the experiment.

8 SCALE EFFECT

The so-called scale effect can be defined as the variation of coefficient of peak passive earth pressure ($Kp_{(peak)}$) with the variation of wall height (H) under normal gravity (1g). Figure 9 shows the relationships between H and $Kp_{(peak)}$ by experiment and by infinitesimal deformation analysis with the assumption $\delta = \phi$. Each experimental data point from the 1g and centrifuge tests performed by variable passive modes and wall friction conditions in dense Toyoura sand is shown [Saho et al., 1985; Terashi et al., 1991; Mori and Tanaka, 1996]. $Kp_{(peak)}$ by this analysis is indicated for the three passive modes. Both the experiment and finite element analysis showed the reduction of $Kp_{(peak)}$ with increasing H. The change of the scale effect be-

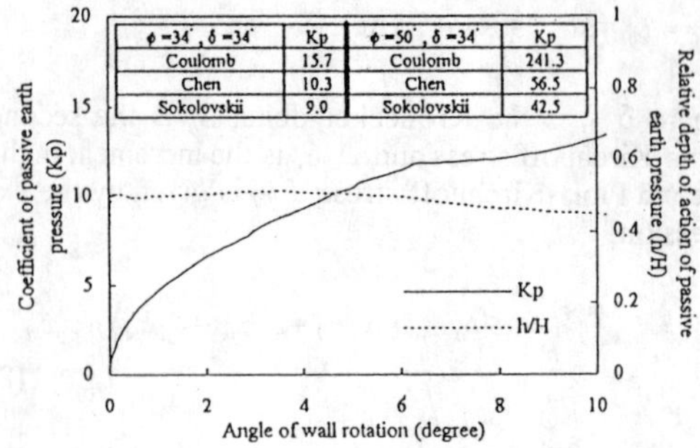

Figure 3. The relationships between Kp and h/H

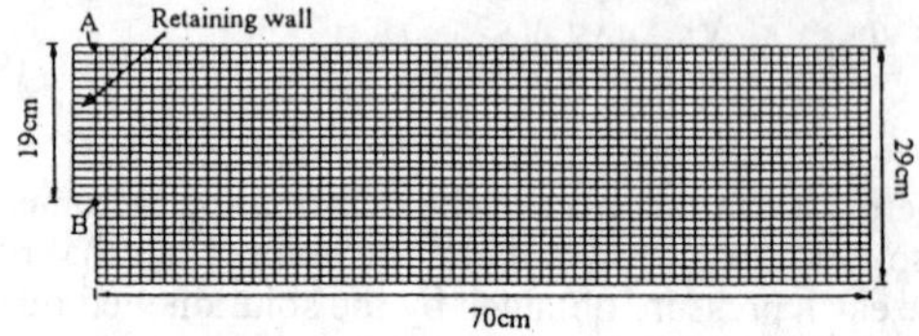

Figure 4(a). Finite element mesh (Type-1 mesh)

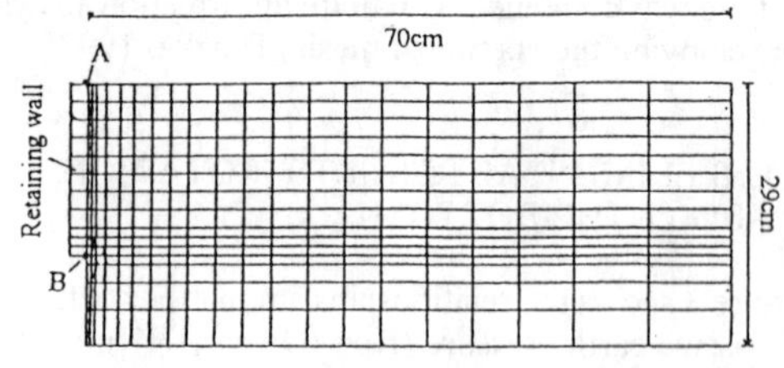

Figure 4(b). Finite element mesh (Type-2 mesh)

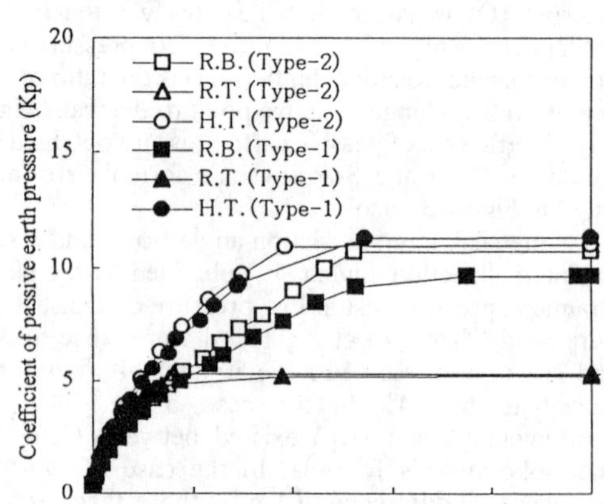

Figure 5. The relationships among Kp, θ and d/H

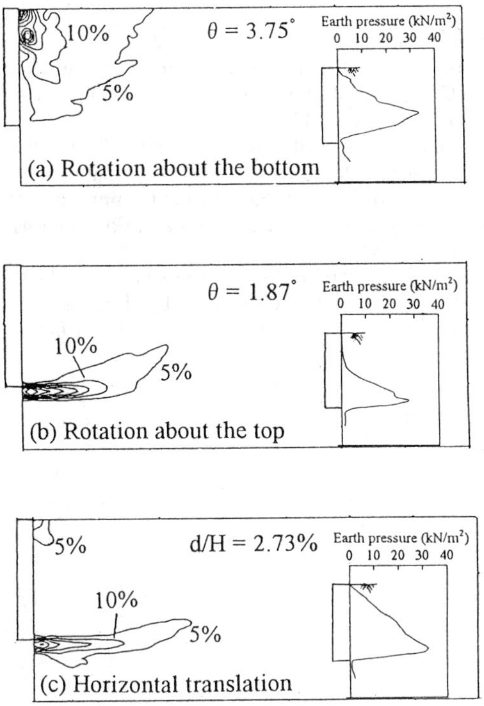

Figure 6(a)(b)(c). The calculated contours of maximum shear strain and earth pressure distributions

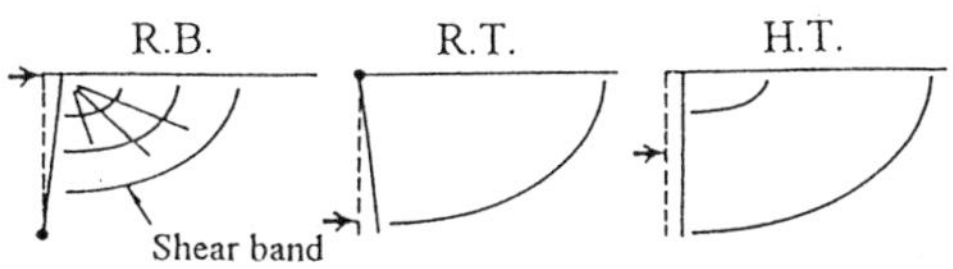

Figure 7(a). Rupture surfaces

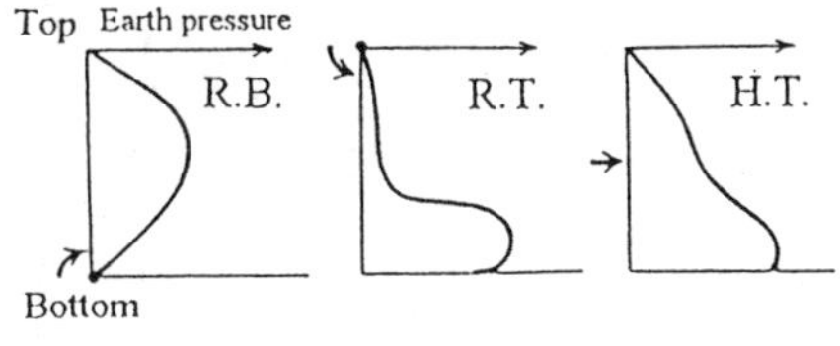

Figure 7(b). Earth pressure distributions

tween H and $Kp_{(peak)}$ was not so large as the scale effect defined as the variation in the bearing capacity characteristics with the variation in the footing width [Siddiquee et al., 1999]. The influence of scale effect by rotation about the bottom was the greatest in other modes.

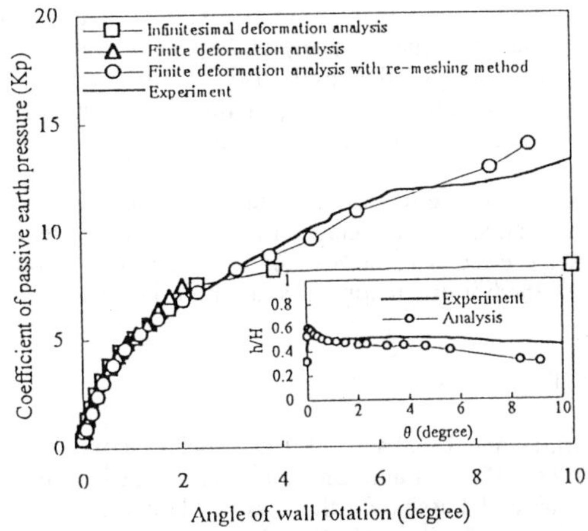

Figure 8. The relationships among the infinitesimal deformation analyses, finite deformation analyses and finite deformation analyses with re-meshing method

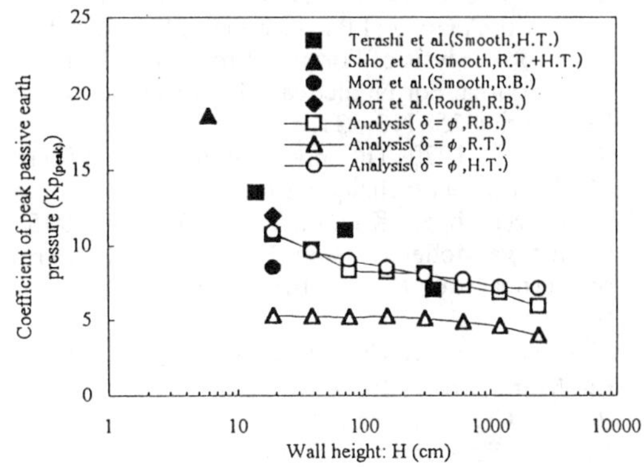

Figure 9. The relationships between H and $Kp_{(peak)}$

9 CONCLUSION

The conclusions are summarized as follows:

1) When the wall friction angle was high, the coefficient of passive earth pressure obtained by classical earth pressure theories introducing a peak internal friction angle overestimated the results.

2) The two types of finite element mesh for retaining wall analyses had little influence. Therefore, the objectivity of the finite element analysis was assured. Both the calculated contours of maximum shear strain and earth pressure distributions for the three passive modes were similar to the tests at Cambridge University.

3) In the passive mode of rotation about the bottom, the coefficient of passive earth pressure obtained by

finite deformation analysis was superior to that by infinitesimal deformation analysis. Especially, finite deformation analysis that took both wall friction angle and an automatic re-meshing method into consideration, exhibited good agreement with the experimental result.

4) The scale effect of retaining wall problems could be evaluated by finite element analysis. The influence of scale effect was largest for the mode of rotation about the bottom in comparison with other modes.

REFERENCES

Mori, H. and Tanaka, T. 1996. Model Test and Elasto-Plastic Finite Element Analysis of Retaining Wall with Passive Earth Pressure. Transactions of Japanese Society Irrigation. Drainage and Reclamation Engineering 183: 57-65 (in Japanese).

Nagtegaal, J. C. and De Jong, J. E. 1981. Some Computational Aspects of Elastic-Plastic Large Strain Analysis. International Journal For Numerical Methods in Engineering 17: 15-41.

Nagtegaal, J. C. and Reblo, N. 1988. On the Development of A General Purpose Finite Element Program for Analysis of Forming Processes. International Journal For Numerical Methods in Engineering 25: 113-131.

Roscoe, K. H. 1970. The Influence of Strain in Soil Mechanics. Geotechnique 20 (2): 129-170.

Saho, H., Saitoh, K., Kimura, T. and Ohta, A. 1985. Centrifuge Model Tests on Passive Earth Pressure on Vertical wall. Proc. 20th. JSSMFE: 1293-1294 (in Japanese).

Siddiquee, M. S. A., Tanaka, T., Tatsuoka, F., Tani, and Morimoto, T. 1999. Numerical Simulation of Bearing Capacity Characteristics of Strip Footing on Sand. Soils and Foundations 39 (4): 93-109

Simo, J. C. and Ortiz, M. 1986. An Analysis of a New Class of Integration Algorithms for Elasto-Plastic Constitutive Relations. Int. J. Num. Meth. Engng 23: 353-366

Tanaka, T. 1979. Generalized Elasto-Plastic Model of Cohesive Soils Including Strain Softening and Finite Element Analyses. The bulletin of the National Research Institute of Agricultural Engineering 18: 101-122 (in Japanese).

Tanaka, T. 1992. Adaptive finite element methods for the elasto-plastic analysis of foundations. Proc. 27th. JSSMFE: 1411-1412 (in Japanese).

Tanaka, T. 1997. Estimation and Effective method for finite deformation analysis of foundations. Proceeding of the Annual congress on JSIDRE: 590-591 (in Japanese).

Tanaka, T. and Mori, H. 1996. Elastoplastic Finite Element Analysis of Retaining Wall with Passive Modes. Third Asian-Pacific Conference on Computational Mechanics 3: 2139-2144.

Tanaka, T., Mori, H. and Kikuchi, M. 1997. Finite deformation elasto-plastic analysis of underground struc-tures and retaining walls. Proc. 9th Int. Conf. on Computer Methods and Advances in Geomechanics 2: 1465-1470.

Tatsuoka, F., Sakamoto, M., Kawamura, T. and Fukushima, S. 1986. Strength and Deformation Characteristics of Sand in Plane Strain Compression at Extremely Low Pressures. Soils and Foundations 26 (1): 65-85

Terashi, M., Kitazume, M. and Matsunaga, Y. 1991. Centrifuge Modeling on Passive Earth Pressure of Sand. Proc. 26th. JSSMFE: 1587-1590 (in Japanese).

Computer Methods and Advances in Geomechanics, Desai et al. (eds) © 2001 Balkema, Rotterdam, ISBN 90 5809 183 X

Slope stability analysis taking into account pre-existing slip lines

S.Ohtsuka & Y.Miyata
Nagaoka University of Technology, Japan

ABSTRACT: This paper presented the analytical procedure for stability assessment of slope including the pre-failed slip lines inside. The pre-failed slip line is generally thin and its strength property is much different from the surrounding soil. It had been difficult to assess the global stability by taking account of the effect of pre-existing slip line with the use of finite element analysis. By developing the rigid plastic constitutive equation for the discontinuous line, it became possible to estimate the global stability of slope including pre-existing slip lines. Through a series of simple case studies, the applicability of the proposed method was clearly exhibited.

1 INTRODUCTION

Landslide is well known to take place repeatedly. The soil mass generally moves along pre-failed slip line, however, the failure mode of slope truly depends on both the shear resistances of soil mass and pre-failed slip lines. When the shear resistance of soil mass is low, the failure mode of slope is uniquely determined by both the shear resistance and the geometry of the slope without any effect of pre-failed slip lines. But, when the shear resistance of pre-failed slip line is comparatively low, the failure mode of slope is much affected by the existence of pre-failed slip lines. The width of slip line is usually thin and it indicates the slope has the discontinuity in strength property from macroscopic viewpoint.

The rigid plastic finite element analysis has been recently introduced into the stability assessment of earth structure. It can assess the feasible stability even though the geometry and the strength distribution of earth structure are complicated. The advantage of the method is that the failure mode of earth structure is determined as a result of analysis and pre-information on feasible failure mode is not necessary different from the conventional methods. *Tamura*(1991) developed the rigid plastic constitutive equation for soil based on the consideration on the limit state achieved by the minimization of upper bound solutions. The advantage of rigid plastic constitutive equation to limit analysis is an expansion of applicability to the non-associated flow rule. In this study the rigid plastic constitutive equation is employed to simulate the soil behavior. However, it is still difficult to assess the global stability of slope including the discontinuity in strength property.

This paper presents a new constitutive equation for contact plane along the pre-failed slip line. With the use of constitutive equation, the stability of slope including pre-failed slip lines is estimated by taking into account the possible failure along the slip lines. The failure mode of slope is finally determined by solving the boundary value problem.

2 RIGID PLASTIC CONSTITUTIVE EQUATION

2.1 *Constitutive equation for soils*

Tamura(1991) developed the constitutive equation for friction material. Following his work, a strength reduction coefficient is newly introduced into the rigid plastic constitutive equation in order to assess the conventional factor of safety employed in slope engineering. The *Drucker-Prager* yield function is expressed with a strength reduction coefficient ω as

$$f(\boldsymbol{\sigma},\omega)=\frac{\alpha}{\omega}I_1+\sqrt{J_2}-\frac{k}{\omega} \tag{1}$$

where $I_1 = tr\boldsymbol{\sigma}$ and $J_2 = \mathbf{s}\cdot\mathbf{s}/2$. The variable of ω gives the factor of safety when the slope attains at the limit state to undergo failure. From the principle of maximum plastic work in plasticity, the stress can be afforded by the strain rate of $\dot{\boldsymbol{\varepsilon}} = \dot{\boldsymbol{\varepsilon}}^p$. However, the strain rate should satisfy the volumetric constraint condition from the associated flow rule such that

$$h(\dot{\boldsymbol{\varepsilon}},\omega)=\dot{\varepsilon}_v-\frac{3\alpha}{\sqrt{3\alpha^2+\omega^2/2}}\dot{e}=0, \tag{2}$$

in which $\dot{e}$ indicates the norm of the strain rate. The rigid plastic constitutive equation was expressed by *Tamura*(1991) such that

$$\boldsymbol{\sigma} = \boldsymbol{\sigma}_1 + \boldsymbol{\sigma}_2 = \gamma \frac{\partial f}{\partial \boldsymbol{\sigma}} + \beta \frac{\partial h}{\partial \dot{\boldsymbol{\varepsilon}}} \tag{3}$$

as shown in Figure 1. The variable of γ is determined by inserting Equation 3 into the yield function of Equation 1. On the other hand, β can not be determined by the constitutive equation itself, however, it can be determined by solving the equilibrium equation with the constraint condition of Equation 2. The rigid plastic constitutive equation is finally given by the following equation:

$$\boldsymbol{\sigma} = \frac{1}{\omega} \frac{k - 3\alpha\beta}{\sqrt{3\left(\alpha/\omega\right)^2 + 1/2}} \frac{\dot{\boldsymbol{\varepsilon}}}{\dot{e}} + \beta \boldsymbol{I} \tag{4}$$

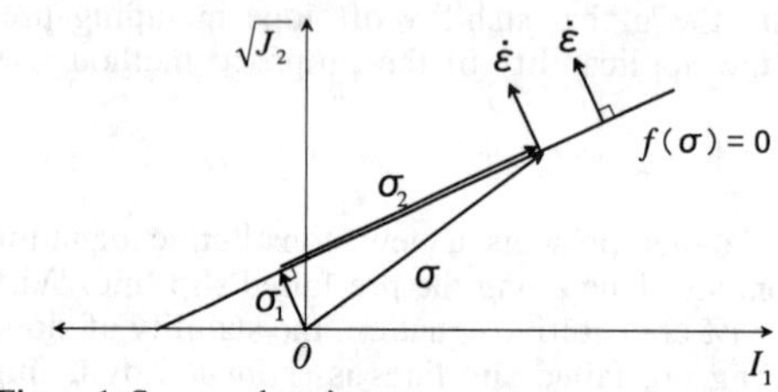

Figure 1. Stress and strain rate.

2.2 *Constitutive equation for contact plane*

The constitutive equation for contact plane can be derived in the same way with soils. Figure 2 illustrates the traction and velocity at the contact plane. The symbols of + and – indicate the corresponding bodies, respectively. The traction $\hat{t}$ at contact plane follows the *Mohr-Coulomb* yield function such that

$$f_c(\hat{\boldsymbol{t}}, \omega) = \left|\hat{t}_s\right| + \hat{t}_n \frac{\tan\phi_c}{\omega} - \frac{c_c}{\omega} = 0 \tag{5}$$

where c_c and ϕ_c denote the cohesion and angle of shear resistance for contact plane. The relative velocity is defined as $\Delta\hat{\boldsymbol{u}} = \Delta\hat{\boldsymbol{u}}^+ - \Delta\hat{\boldsymbol{u}}^-$ by arranging $\Delta\hat{u}_n$ as positive. Considering the associated flow rule, the relative velocity should satisfy the following condition:

$$g_c\left(\Delta\hat{\boldsymbol{u}}, \omega\right) = \Delta\hat{u}_n - \left|\Delta\hat{u}_s\right| \frac{\tan\phi_s}{\omega} = 0 \tag{6}$$

From the consideration on rigid plastic constitutive equation of soils, the traction at contact plane is also afforded by the following equation:

$$\hat{\boldsymbol{t}} = \hat{\boldsymbol{t}}_1 + \hat{\boldsymbol{t}}_2 = \gamma_c \frac{\partial f_c}{\partial \hat{\boldsymbol{t}}} + \beta_c \frac{\partial g_c}{\partial \Delta\hat{\boldsymbol{u}}} \tag{7}$$

The parameter of γ_c is determined by inserting Equation 7 into Equation 5, however, β_c can not determined by the constitutive equation itself as mentioned before for soils. β_c is finally given by solving the equilibrium equation with Equation 6. The rigid plastic constitutive equation is formulated into

$$\hat{\boldsymbol{t}} = c_c \cos\bar{\phi}_c \frac{\Delta\hat{\boldsymbol{u}}}{\Delta\bar{\boldsymbol{u}}} + \beta_c \begin{pmatrix} -\tan\bar{\phi}_c \dfrac{\Delta\hat{u}_s}{\Delta\bar{u}_s} \\ 1 \end{pmatrix} \tag{8}$$

in which $\Delta\bar{u}$ is the norm of $\Delta\hat{u}$ and $\bar{\phi}_s$ is the modified angle of shear resistance such that.

$$\tan\bar{\phi}_c = \frac{\tan\phi_c}{\omega} \tag{9}$$

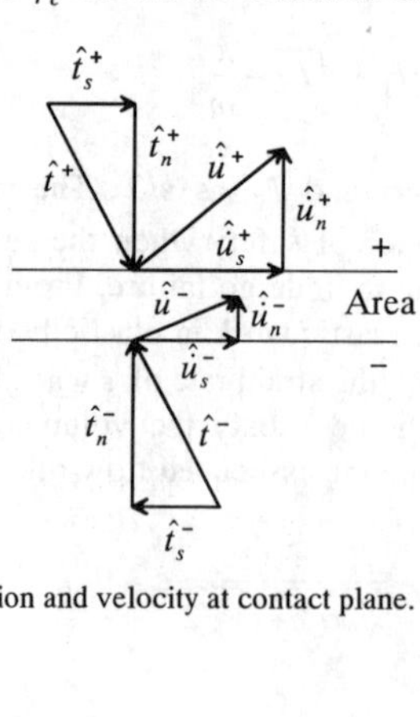

Figure 2. Traction and velocity at contact plane.

3 STABILITY OF SIMPLE SLOPE

Firstly, the stability of simple slope is investigated with the proposed finite element analysis (RPFEM). In order to examine the applicability of proposed method, the obtained factor of safety is compared with those by Taylor and Chen. The result is shown in Table 1. In this computation, the height of slope is *10m* and the slope angle is changed from *30°* to *50°*. The cohesion and unit weight of soil is set as *17.5kPa* and *19.8kN/m³*, respectively.

It is readily seen from the table that the finite element analysis (RPFEM) gives a fairly good estimation on factor of safety for each case. It seems, however, to give closer estimation to Chen's solution based on the upper bound solution in limit analysis. It is noted that the accuracy of the result generally depends on the employed finite element mesh. Finer mesh should be employed to obtain the good estimation.

4 EXAMINATION ON SIMPLE PROBLEM

The performance of the proposed rigid plastic constitutive equation for contact plane is investigated for

Table 1. Comparison of safety factor with conventional methods.

Slope angle	Friction angle	Taylor	Chen	RPFEM
30°	0°	0.701	0.605	0.609
	5°	0.996	0.999	1.013
	10°	1.252	1.274	1.295
	15°	1.476	1.500	1.538
	20°	1.752	1.768	1.747
	25°	2.008	2.017	1.923
40°	0°	0.658	0.605	0.594
	5°	0.877	0.890	0.906
	10°	1.064	1.085	1.122
	15°	1.253	1.276	1.307
	20°	1.426	1.453	1.466
	25°	1.619	1.616	1.600
50°	0°	0.618	0.604	0.581
	5°	0.792	0.800	0.829
	10°	0.940	0.953	1.006
	15°	1.087	1.102	1.156
	20°	1.232	1.247	1.286
	25°	1.376	1.393	1.395

a simple problem as shown in Figure 3. The figure shows a cohesive soil including a discontinuous line inside. By handling the shear resistance of discontinuous line, the factor of safety for gravity force is investigated. The employed soil constants are exhibited in Table 2.

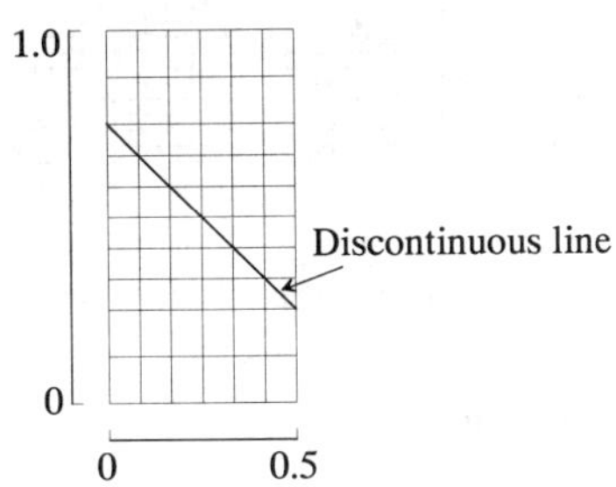

Figure 3. Discontinuous body under plane strain condition.

Table 2. Soil constants for Figure 3.

angle of shear resistance	0°
cohesion	20 *kPa*
unit weight	19.8 kN/m^3

Figure 4 shows the obtained factor of safety for the various shear resistance of discontinuous line. When the shear resistance is low, the factor of safety increases linearly with the increase in shear resistance of discontinuous line. However, after the factor of safety attaining at that for no discontinuous line, it becomes constant regardless of shear resistance of discontinuous line. This result clearly expresses the rigid plastic constitutive equation at contact plane works properly. Figure 5 illustrates the typical velocity fields obtained in case studies. It is readily seen that the failure mode changes from the continuous mode to discontinuous mode depending on shear resistance of discontinuous line.

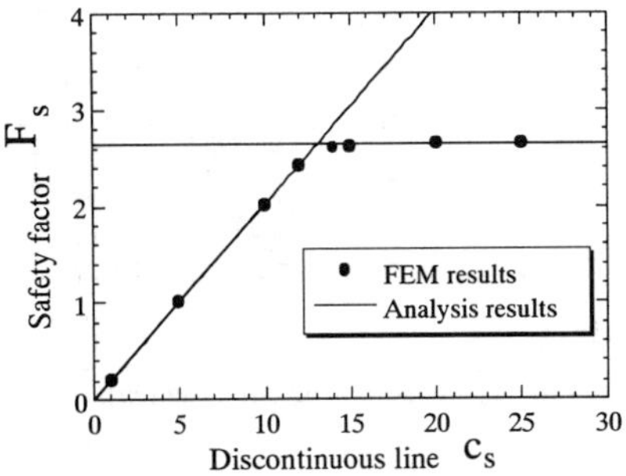

Figure 4. Factor of safety for change in shear resistance at contact plane.

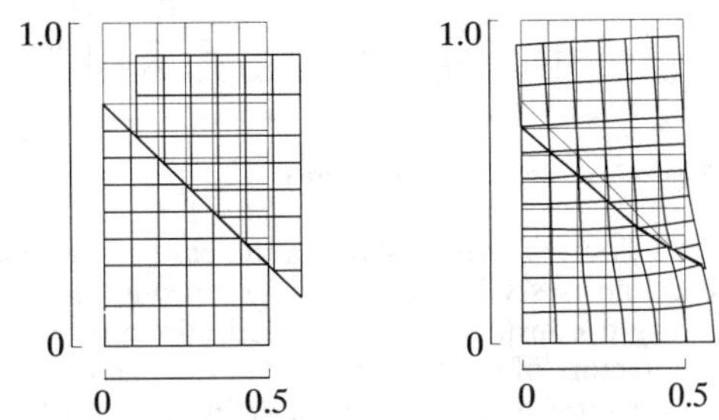

(a) c_s=5 *kPa* (b) c_s=20 *kPa*

Figure 5. Typical velocity fields of Figure 3.

5 STABILITY OF SLOPE WITH DISCONTINUOUS LINE

5.1 Case study: a simple slope

It is well known that landslide generally takes place repeatedly. It is partly because the slope includes a pre-failed slip line inside and its shear resistance is lower to the surrounding soils. The effect of strength property of pre-failed slip line on the global stability of slope is investigated in this section. Figure 6 shows the boundary condition of employed slope. The soil constants are shown in table 3. Figure 7 shows the velocity field obtained for no discontinuous line inside the slope. The obtained velocity field forms the circular slip surface as observed in actual problem. The factor of safety is computed as 1.349.

In order to examine the effect of pre-failed slip line on the global stability, a slip line is set to fit the failure mode observed in Figure 7. Figure 8 shows

Table 3. Soil constants for Figure 6.

slope	angle of shear resistance	20°
	cohesion	9.8 *kPa*
	unit weight	19.8 kN/m^3
pre-failed slip line	cohesion	0.01 *kPa*
	angle of shear resistance	10°

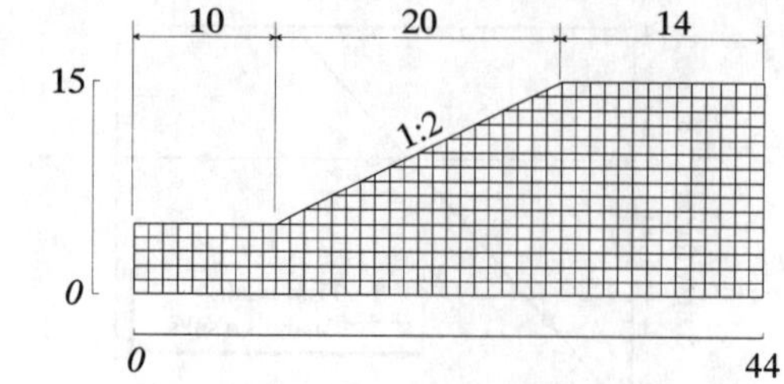

Figure 6. Boundary condition of slope (unit:*m*).

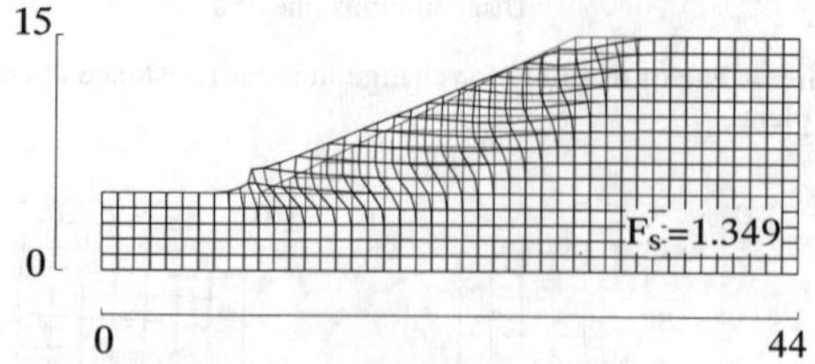

Figure 7. Velocity fields of Figure 6.

the velocity fields of slope in the case of discontinuous line is set in the slope. By considering various strength constants for pre-failed slip line, different two factors of safety and velocity fields are obtained. It is natural that the factor of safety for $c_s = 0.1kPa$ and $\phi_s = 10^o$ is lower than that for no discontinuous line shown in Figure 6, but it is noted that the discontinuity in velocity field is apparently observed and the velocity field forms the rigid block turning as assumed in the circular slip surface method.

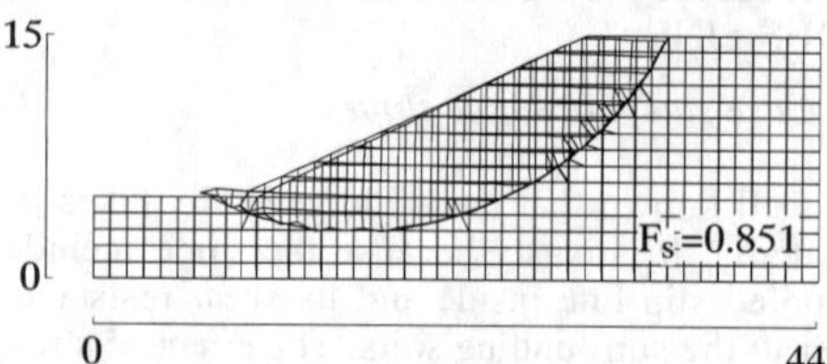

(a) case of $c_c = 0.1kPa$, $\phi_c = 10^o$

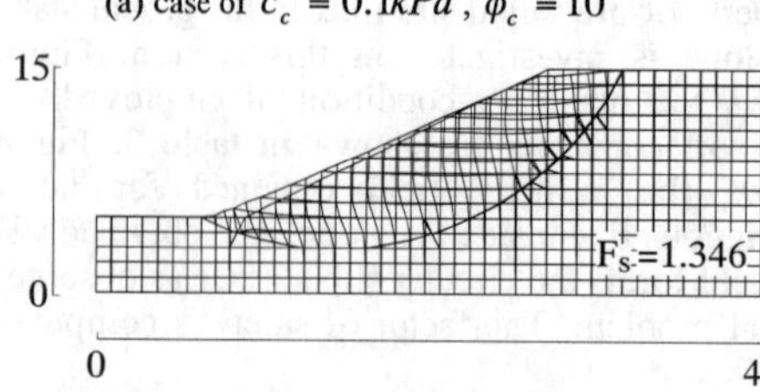

(b) case of $c_c = 100kPa$, $\phi_c = 10^o$

Figure 8. Velocity field of Figure 8 depending on shear resistance along discontinuous line.

5.2 *Case study: cutting*

Stability of the slope in Figure 6 is investigated for cutting of the toe as Figure 9. The soil constants are the same with Table 3. Two cases, with and without a discontinuous line, are investigated. The results are obtained as Figures 10 and 11. It is readily seen that the effect of discontinuous line on the failure mode and global stability of slope.

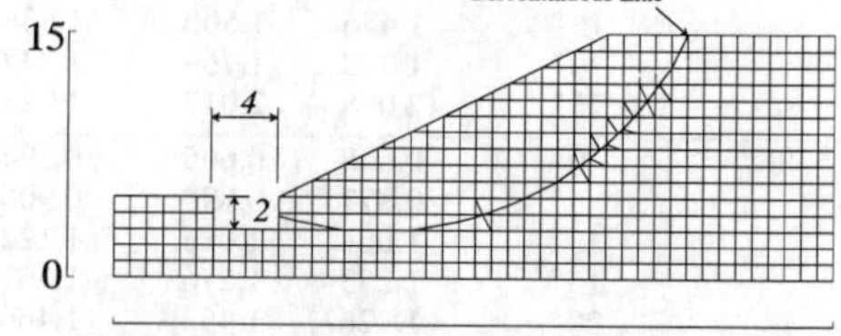

Figure 9. Boundary condition of slope for cutting (unit:*m*).

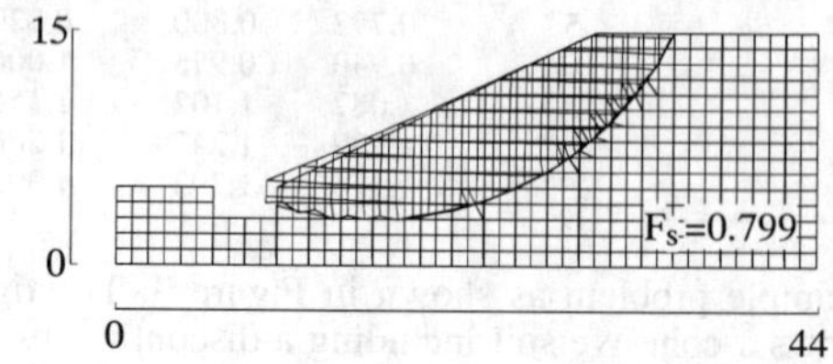

Figure 10. Velocity field for slope including discontinuous line.

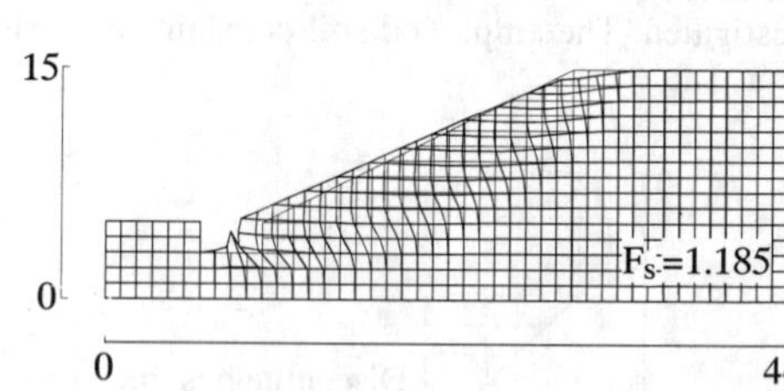

Figure 11. Velocity field for without discontinuous line.

6 CONCLUSIONS

The rigid plastic constitutive equation for contact plane was developed in this study. The strength reduction method was employed for stability assessment in the same way with conventional methods. The comparison with the conventional methods showed a good performance of RPFEM. Through basic numerical examples, the applicability of proposed method to global stability assessment of landslide slope including pre-failed slip lines inside was clearly shown.

REFERENCES

Nash, D.F.T. 1987. A comparative review of limit equilibrium methods of stability analysis. Anderson M.G. & Richards K.S.(eds), *Slope Stability*: 11-75. John Wiley & Sons.

Tamura, T. 1991. Rigid-plastic finite method in geotechnical engineering. *Computational Plasticity*, *Current Japanese Material Research*, 7: 135-164. Elsevier.

Computer Methods and Advances in Geomechanics, Desai et al. (eds) © 2001 Balkema, Rotterdam, ISBN 90 5809 183 X

Method for overloading capacity of dam

Chengxue She
Wuhan University of Hydraulic and Electric Engineering, People's Republic of China

ABSTRACT: The FEM combining embedded softening band in elements for crack propagation with the effective continuum is developed. This method can imitate the crack propagation in intact rock or concrete and sliding or opening of joints and faults in one program. The application in the calculating the overloading capacity of the gravity dam of the Baozhushi Hydraulic Project shows the reasonability of the method.

1 INTRODUCTION

In hydraulic engineering, the overloading capacity and failure pattern of dams are always important in evaluating safety of dams. In overloading state, the dams will experience the following one or two failure patterns:

1). The dam together with rock foundation moves along the joints, faults or the interface of the dam and foundation.

2). Crack propagation in dam or intact rock.

These two problems have been widely studied and many FEM have been developed to imitate the failure process. For example the effective continuum method is used to model the failure process in jointed rock(Zienkiewicz and Pande, 1977), the FEM with embedded softening band is used to calculate the crack propagation process (Klisinski,1991;Ohlsson,1997;.She,2000)

In this paper, these two methods are combined together on the framework of the elasto-viscoplastic theory. Because the faults and joints in rock foundation are determined, the effective continuum theory can be used to imitate sliding or opening along faults and joints The FEM with embedded softening band can model the cracking process in concrete or intact rock. By combining the two methods, the crack propagation in intact rock and sliding or opening in joints or faults can be modeled in one program. The application in the calculating of the overloading capacity of the Baozhushi Hydraulic Project shows the method is reasonable..

2 FEM WITH EMBEDDED SOFTENING BAND

Fig1 is an element, 1,2,3 and 4 are local element node number, ij is an softening band without thickness.

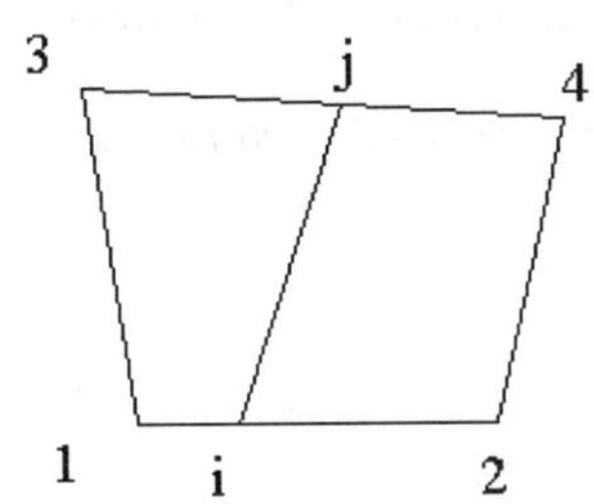

Fig.1 Element with an softening band

The strain in the element can be written as:

$$\boldsymbol{\varepsilon} = \mathbf{B}(\mathbf{d} - \mathbf{H}_l \mathbf{Q}\mathbf{e}) \tag{1}$$

where $\mathbf{d}$ is the vector of node displacement, $\mathbf{e}$ is the relative displacement of the two sides of the softening band. $\mathbf{Q}$ is the transforming matrix of displacement from local to global coordinates system. $\mathbf{H}_l$ is the matrix relating the deformation of the softening band with the node displacements.

The equilibrium equation of FEM and the relationship

between the stress in element and the stress in the softening band can be expressed as

$$\iint_{\Omega} \mathbf{B}^T \boldsymbol{\sigma} dA = \mathbf{f} \tag{2}$$

$$\mathbf{t} = \mathbf{P} \ \boldsymbol{\sigma} \tag{3}$$

in which $\mathbf{P}$ is the matrix transforming the stress from global coordinate system to the local one.

In this paper, we assume the crack passes through the center of the element, $\mathbf{t}$ and $\boldsymbol{\sigma}$ in equation (3) are substituted by the average stress $\bar{\mathbf{t}}$ and $\bar{\boldsymbol{\sigma}}$ of crack and element respectively for convenience.

For crack propagation problem, the shearing component of stress is 0. Thus the simplest crack criterion can be expressed as:

$$F = \overline{\sigma_n} - \sigma_t = 0 \tag{4}$$

where $\overline{\sigma_n}$ is the normal component of stress vector $\bar{\mathbf{t}}$ and σ_t is the tensile strength of rock or concrete, it gets lower as the crack develops.

$$\sigma_t = \varphi(\xi)\sigma_{t0} \tag{5}$$

where σ_{t0} is the initial tensile strength , and the simplest function for $\varphi(\xi)$ can be written as :

$$\begin{cases} \varphi(\xi) = 1-\xi \\ \xi = \omega / G_f^{\mathrm{I}} \\ \omega = \int_0^u \overline{\sigma_n} d u_n \end{cases} \tag{6}$$

in which G_f^{I} is the fracture energy, ω denotes the energy needed for the crack opening from 0 to u, $\xi = 0$ when $\omega = G_f^{\mathrm{I}}$.

The normal displacement rate of crack can be expressed as :

$$\dot{u}_n = \gamma \langle F \rangle \frac{\partial F}{\partial \sigma_n} \tag{7.1}$$

Where γ is fluidity parameter. Thus the increment of vector $\Delta\mathbf{e}$ in time increment Δt is:

$$\Delta\mathbf{e} = \begin{Bmatrix} \dot{u}_n & , & 0 \end{Bmatrix}^T . \Delta t \tag{7.2}$$

Assuming the element except the softening band is in elastic state, the stress increment in the element can be expressed as :

$$\Delta\boldsymbol{\sigma} = \mathbf{DB}(\Delta\mathbf{d} - \mathbf{H}_l \mathbf{Q} \Delta\mathbf{e}) \tag{8}$$

substituting (8) into (2), we can get the increment equation of FEM on the basis of elasto-viscoplastic theory:

$$\mathbf{K}\Delta\mathbf{d} = \Delta\mathbf{f} + \Delta\mathbf{f}_1 \tag{9}$$

where $\mathbf{K}$ is the conventional elastic stiffness matrix , $\Delta\mathbf{f}$ is load vector increment and

$$\Delta\mathbf{f}_1 = \iint_{\Omega e} \mathbf{B}^T \mathbf{DB} \mathbf{H}_l \mathbf{Q} \Delta\mathbf{e} d\Omega e \tag{10}$$

in which Ωe is the element with softening band.

3 EFFECTIVE CONTINUUM THEORY

Zienkiewicz and Pande(1977) researched the multi-laminate model for jointed rock masses. It has been applied in engineering. If there is one set of joints in a rock element and the joint experiences sliding or opening failure then the rate of visco-plastic straining is:

$$\dot{\varepsilon} = \frac{1}{d}\gamma\langle F\rangle\frac{\partial F}{\partial \boldsymbol{\sigma}} \tag{11}$$

where γ is fluidity parameter, d is the spacing of joints and F is the yield function.

The equilibrium equation of FEM for effective continuum medium can be written as,

$$\mathbf{K}\Delta\mathbf{d} = \Delta\mathbf{f} + \Delta\mathbf{f}_2 \tag{12}$$

in which $\mathbf{k}$ and $\Delta\mathbf{f}$ is the same as in equation (9), and

$$\Delta\mathbf{f}_2 = \iint_{\Omega e} \mathbf{BD}\dot{\varepsilon}\,\Delta t d\Omega \tag{13}$$

where Ωe is the element with joints or faults. Obviously, equation (9) and equation (12) are both established on the elatsto-viscoplastic theory, it is not difficult to unite these two equations together on the basis of elasto-viscoplastic theory:

$$\mathbf{K}\Delta\mathbf{d} = \Delta\mathbf{f} + \Delta\mathbf{f}_1 + \Delta\mathbf{f}_2 \tag{14}$$

From the statement above, it is obvious that the expression (14) can be used to calculate crack propagation in intact rock, concrete and sliding or opening in joints or faults in one program.

4 NUMERICAL EXAMPLE

The Baozhushi Hydraulic Project is located in the northwestern China, it is a gravity dam with the elevation of the top of the dam being 132 m(relative to rock foundation) and the maxmum water level being 130.7m.

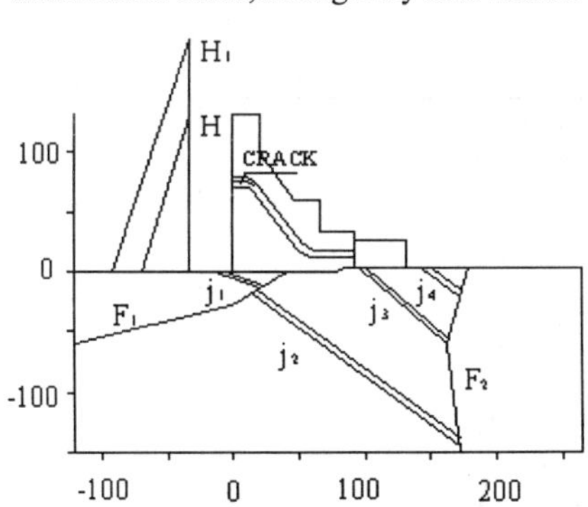

Fig.2 Schematic diagram of dam and foundation

Fig 2 is the simplified section of dam and power plant. The joint between dam and power plant is grouted. There is a water pipe in the dam and it is modeled by decreasing the thickness of the dam in the position. In the rock foundation there are exists joints and faults j1, j2, j3, j4 , F1 and F2. In Fig.2 also shows the overloading method: as the upstream water level increases step by step, the water pressure on the upstream face of the dam increases correspondingly, the procedure lasts until the dam fails. Then the failure pattern and the overloading safety factor can be obtained.

Table 1. Material properties of rock and concrete

	Properties	Value
Rock	Young's modulus	10.0 GN m^{-2}
	Poisson's ratio	0.25
	Frictional angle	45.0
	Cohesion	0.5 Mpa
	Initial tensile strength	0.5 Mpa
	G_f^I	30 N/m
	Properties	Value
Concrete	Young's modulus	18.0 GN m^{-2}
	Poisson's ratio	0.17
	Frictional angle	45.0
	Cohesion	2.5 Mpa
	Initial tensile strength	0.9 Mpa
	G_f^I	50 N/m

Table 2 Material properties of joints and faults

	Properties	Value
j3 j4	Frictional angle	20.0
	Cohesion	0.02 Mpa
	Tensile strength	0 Mpa
	Properties	Value
j1 j2	Frictional angle	22.0
	Cohesion	0.02 Mpa
	Tensile strength	0 Mpa
	Properties	Value
F1 F2	Frictional angle	25.0
	Cohesion	0.03Mpa
	Tensile strength	0 Mpa

Table 1 and Table 2 list the material properties of rock, concrete, joints and faults . The density of rock and

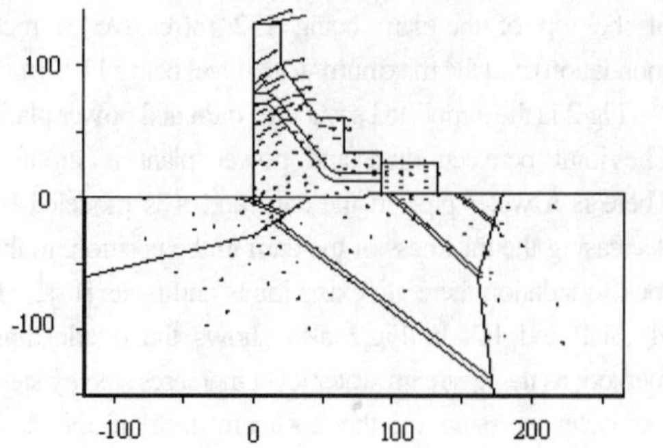

Fig.3 Displacement before failure

concrete are 26500 $N\ m^{-3}$ and 24000 $N\ m^{-3}$ respectively.

The calculation is divided into two steps. The first step calculates the stress field under gravity of the dam and rock. Then the calculated stress field is input as the initial condition for the second step of calculation, in which the water pressure is the load.

Because of the water pipe softening the dam the concrete in the inlet of the water pipe cracks first(see Fig.2). Once the crack occurs, the dam fails rapidly and the calculation is divergent. The water level is 185m when the dam fails.

It can be easily deduced that the overloading safety factor can be expressed as:

$$k_p = \frac{2H_i}{H} - 1 \tag{15}$$

in which H_i = 185m,and H is 130.7m. Substitutes the values into (15), $k_p = 1.83$. This value and the crack position are close to the ones obtained by geomechanical model test(Wu, 1992).

Fig.3 and Fig. 4 are displacement and stress field in dam and foundation.. From Fig.3 it can be seen that the displacement above the crack is much bigger than the lower part of the dam. This means that the propagation of crack will increase the displacement of the dam above the crack. It is reasonable. The tensile stress at inlet of water pipe is big and crack occurs there, the crack propagates alone the trace of principle stress. In the foundation, the joint j1 is drawn to open. Because the influence of joint j1, the tensile stress at the dam heel is not big(see Fig. 4), no crack develops there. Joint j2 , j3 and j4 as well as the upper part of F1 slide alone the joints, but F2 is in elastic state.

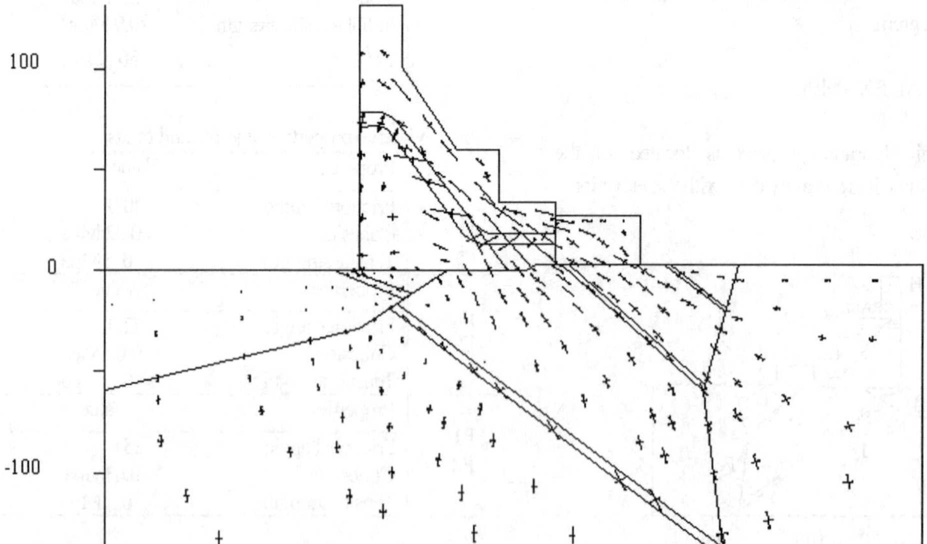

Fig 4 Stress distribution before failure

5 CONCLUSION

In this paper the effective continuum theory and the embedded softening band for cracking are combined together to model the crack propagation in rock or concrete and sliding or opening in joints and faults in one program. Using the proposed FEM to imitate the failure process of the dam of the Baozhushi Hydraulic Project under overloading, reasonable results are obtained. The calculation offers the overloading safety factor, cracking position and propagation direction as well as the sliding or opening sate of joints and faults, it also reflects the influence of joints or faults on the stress field so that the cracking place can be correctly located.

REFERENCES

Zienkiewcz,O.C. & g.n.Pande 1977. Time dependent multi-laminate model of rocks - a numerical study of deformation and failure of rock masses. Int. J. Num. & Anal. Meth. In Geomech., Vol.1: 219-247

Klisinski et al 1991. Finte element with inner softening band. Journal of Engineering Mechanics, Vol. 117: 575-587.

Ohlsson et al 1997. Mixed-mode fracture and anchor bolts in concrete analysis with inner softening bands. Journal of Engineering Mechanics, Vol.123: 1027-1033

Chengxue She 2000. A new numerical method for crack propagation. Rock and Soil Mechanics(in Chinese), Vol.21:130-133.

Wu et al 1992. Geomechanical model test for Baozhushi Hydraulic Project. Report of Wuhan University of Hydraulic and Electric Engineering.

Computer Methods and Advances in Geomechanics, Desai et al. (eds) © 2001 Balkema, Rotterdam, ISBN 90 5809 183 X

Post-seismic displacement of slopes by energy methods

C.A.Stamatopoulos & E.G.Velgaki
Kotzias-Stamatopoulos Consulting Engineers, Athens, Greece

ABSTRACT: A 2-dimensional mass with general shape that slides in translation in "n" slip lines is studied. It is assumed that a line of internal shearing exists through each node where the external slip surface changes inclination. The paper derives an analytical expression that gives the total distance moved by the slide for the case of shear resistance through an energy approach.

1 INTRODUCTION

During major earthquakes in the recent past, several cases of liquefaction - induced slides have been observed. A well-documented case of such a slide is that of the Lower San-Fernando Dam in the 1971 earthquake. During such slides, most of the excessive movement develops as a result of loss of static strength, and not as a result of dynamic loading.

It is of great interest to evaluate the residual strength of liquefaction-induced seismic slides. An approach has been developed to estimate the shear strength of liquefied soils based on in situ test parameters, such as the blow count resistance of the Standard Penetration Test (SPT), by a correlation between SPT blow count and the apparent shear strength back-calculated from observed flow slides (e.g. Kramer, 1996). The above illustrates the need to develop reliable simplified methods that back-estimate the soil strength mobilized during past slides.

Back-analysis for the estimation of the soil strength mobilized during the slide is usually performed using the equilibrium condition, either before or after the slide by performing stability analysis and assuming a factor of safety equal to one. However, the factor of safety exactly before the slide is less than one because motion develops; just after the slide it is greater than one, as a result of inertia.

For soil to move along (external) lines with different inclinations, internal shearing must be allowed along lines intersecting the angles between the external lines. Further, for motion to develop, the stress state along the internal lines must be at failure (Sarma, 1979). According to the above, Ambraseys and Srbulov (1995) proposed two blocks with constant height sliding at two inclinations with internal shearing. The inclination of the line where internal shearing occurs, corresponds to the inclination of least resistance and affects the relative velocity of the two bodies. The equations of motion are solved using force equilibrium, and analytical expressions relating the soil resistance to the distance moved are obtained for some cases. Sarma and Chlimitzas (2000) recently generalized the above model. They considered a slip surface consisting of n linear segments and a sliding mass with general shape. They used a force equilibrium approach which did not produce analytical solutions for the general case.

In the present work the sliding system shown in Fig.1 is considered: A 2-dimensional mass with general shape slides in translation in "n" slip lines. A line of internal shearing passes through each node where the external slip surface changes inclination. All the mass of the slide is either above or below the water table line. During relative motion it is assumed that the mass that is transferred between two adjacent slip lines forms a trapezium. The paper first derives the equation that gives the total distance moved by the sliding system for the case of shear resistance, using an energy approach. The energy approach has the advantage that, in contrast to the force equilibrium approach, it can obtain analytical expressions for the general slope geometry relating soil resistance to the

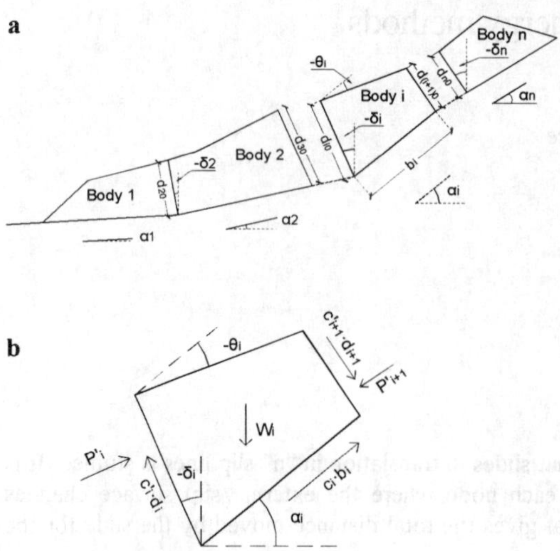

Fig.1 Sliding system considered in the present study (a) initial configuration (b) forces acting on body "i".

distance moved. Then an analytical expression relating the soil strength to the distance moved is obtained.

2 GEOMETRY AND GOVERNING EQUATIONS

2.1 Geometry considered

The sliding mass of Fig.1 is divided in n bodies: body 1 on a sub-plane with inclination α_1, body 2 on a sub-plane with inclination α_2, ... body "i" on a sub-plane with inclination α_i,...., and finally body n on a sub-plane with inclination α_n. The internal slip lines with length d_{i0} and inclination $(90^\circ - \delta_i)$ separate each pair of consecutive bodies "i-1" and "i". The parameters that define the initial geometry of the sliding system are the angles of inclination with the horizontal of the planes, α_1, α_2, ..., α_n the initial cross-sectional areas of the bodies on each of the n sub-planes, A_{10}, A_{20}, ..., A_{n0} their initial contact lengths b_{10}, b_{20}, ..., b_{n0} and the inclination and length of the internal slip lines, $(90^\circ - \delta_i)$ and d_{i0} respectively. Also, the distances moved along each slip sub-plane are denoted as u_i.

The shape of the top ground surface of bodies 2, 3, ..., n can be defined with successive segments and this is necessary because these geometric particularities affect the mass transfer between two adjacent bodies. In particular, in addition to the measures given above, only the geometry of the segments that change inclination need to be defined. Thus, when, for every body "i", the distance moved u_i, is less than the length of the first segment (from left to right) of the body projected along the slip surface, only the angle that the first segment of body "i" makes with the slip surface (θ_i in Fig.1), must be specified.

Shear resistance is taken to act along the external slip surface and the sub-planes of internal shearing. Different values of soil strength are taken in each segment of the external slip line, denoted as c_1,..., c_n and in each interface, denoted as c'_2,..., c'_n.

It is assumed that the angles of the internal slip surfaces, δ_i, do not change as a function of the distance moved. For back analyses, their value can be obtained from the relative distances moved, u_i, along each slip line, using equation (1) below.

2.2 Relations between u_i

As the mass slides, it is assumed that the total mass (or cross - sectional area) lost by body "i" is gained by body "i-1". This compatibility condition gives that the relative displacements of the bodies are connected as:

$$\frac{u_{i-1}}{u_i} = \frac{du_{i-1}}{du_i} = \frac{\cos(\delta_i + \alpha_i)}{\cos(\delta_i + \alpha_{i-1})} = \lambda_{i-1} \quad (1)$$

The quantities λ_i are (n-1), while our sliding system consists of n bodies. Thus, we have a system with only one degree of freedom or only one independent sliding distance. For convenience, this is taken to be equal to the first body's movement, u_1. Also, the maximum distance moved by the blocks can be represented by the quantity u_{1-max}. From u_1 (or u_{1-max}), one can obtain the sliding distance of the body "i", u_i (or u_{i-max}) as:

$$u_i = \frac{u_1}{\prod_{k=1}^{i-1} \lambda_k} \qquad u_{i-max} = \frac{u_{1-max}}{\prod_{k=1}^{i-1} \lambda_k} \quad (2)$$

2.3 Changes in areas and lengths with the distance moved

The contact lengths along the slip surface b_i change as a function of the distance moved as:

$$b_1 = b_{10} + u_1 \tag{3.1}$$

$$b_2 = b_{20} \quad , \ldots, \quad b_{n-1} = b_{(n-1)0}$$

$$b_n = b_{n0} - u_n = b_{n0} - u_1 \cdot \frac{1}{\prod_{k=1}^{n-1} \lambda_k} \tag{3.n}$$

Also, the interslice lengths d_i change as a function of the distance moved as:

$$d_i = d_{i0} + \frac{\sin\theta_i}{\cos(\theta_i + \alpha_i + \delta_i)} \cdot u_i =$$

$$= d_{i0} + \frac{\sin\theta_i}{\cos(\theta_i + \alpha_i + \delta_i)} \frac{u_1}{\prod_{k=1}^{i-1} \lambda_k} \tag{4}$$

Then, the change in the cross sectional area of each body "i" is given as:

$$\Delta A_i = d_{(i+1)0} \cdot \cos(\alpha_{i+1} + \delta_{i+1}) \cdot \frac{u_1}{\prod_{k=1}^{i} \lambda_k} +$$

$$+ \frac{0.5 \cdot \cos(\alpha_{i+1} + \delta_{i+1}) \cdot \sin\theta_{i+1}}{\cos(\theta_{i+1} + \alpha_{i+1} + \delta_{i+1})} \cdot \left(\frac{u_1}{\prod_{k=1}^{i} \lambda_k} \right)^2 -$$

$$- d_{i0} \cdot \cos(\alpha_i + \delta_i) \cdot \frac{u_1}{\prod_{k=1}^{i-1} \lambda_k} -$$

$$- \frac{0.5 \cdot \cos(\alpha_i + \delta_i) \cdot \sin\theta_i}{\cos(\theta_i + \alpha_i + \delta_i)} \cdot \left(\frac{u_1}{\prod_{k=1}^{i-1} \lambda_k} \right)^2 \tag{5}$$

Note that from the change in areas, the change in the weight of each body can be obtained as:

$$\Delta W_i = \Delta A_i \cdot \gamma_i \tag{6}$$

where γ_i equals to the total unit weight γ_t when the slope is dry or above a water table, or to the buoyant unit weight γ_b when the slope is completely submerged.

2.4 Conservation of energy

The change in the kinetic energy of each of the moving bodies should equal the change in the potential energy. In addition, the kinetic energy in both the initial and the final slide configurations is zero. These give:

$$\left\{ T_f - T_{in} = \int_0^{S_f - S_{in}} \sum_{i=1}^{n} \left[F_i \cdot \cos(F_i, v) \right] dS = 0 \right\}_i \tag{7}$$

where T is the kinetic energy, S is the distance moved along the direction of motion, F_i are the external forces that act in the body, (F_i,v) represents the angle that the force F_i makes with the direction of motion and the subscripts "in" and "f" denote the initial and final slide geometry.

The forces acting on body "i" are given in Fig.1b. The distance moved by body "i" is related to the distance moved by body "1" by relation (2). With these, the integral (7) becomes:

$$\int_0^{u_{1\,max}} \left[W_i \sin\alpha_i - c_i\, b_i - c_i' d_i \sin(\delta_i + \alpha_i) - \right.$$

$$- P_i' \cos(\delta_i + \alpha_i) + c_{i+1}' d_{i+1} \sin(\delta_{i+1} + \alpha_i) + \tag{8}$$

$$\left. + P_{i+1}' \cos(\delta_{i+1} + \alpha_i) \right] \cdot \frac{1}{\prod_{k=1}^{i-1} \lambda_k} du_1 = 0$$

3. SOLUTION

When equation (8) is summed from i = 1 to i=n one gets:

$$\int_0^{u_{1\,max}} W_1 \sin\alpha_1 - c_1\, b_1 - c'_2\, d_2 \sin(\delta_2 + \alpha_2) +$$

$$+ P'_2 \cos(\delta_2 + \alpha_1) \; + \; \ldots\ldots\ldots\ldots \; +$$

$$+ \left[W_1 \sin\alpha_i - c_i\, b_i - c'_i\, d_i \sin(\delta_i + \alpha_i) - P'_i \cos(\delta_i + \alpha_i) \right.$$

$$\left. + c'_{i+1}\, d_{i+1} \sin(\delta_{i+1} + \alpha_i) + P'_{i+1} \cos(\delta_{i+1} + \alpha_i) \right] \cdot \frac{1}{\prod_{k=1}^{i-1} \lambda_k} +$$

$$+ \ldots\ldots\ldots\ldots\ldots\ldots + $$

$$+ \left[W_n \sin\alpha_n - c_n\, b_n - c'_n\, d_n \sin(\delta_n + \alpha_n) - \right.$$

$$\left. - P'_n \cos(\delta_n + \alpha_n) \right] \cdot \frac{1}{\prod_{k=1}^{n-1} \lambda_k} du_1 = 0 \tag{9}$$

In the above equation the forces P'_i cancel and we get the following expression:

$$\int_0^{u_{1\max}} \sum_{i=1}^{n} [W_i \sin\alpha_i - c_i b_i - c'_i d_i \sin(\delta_i + \alpha_i) + $$

$$+ c'_{i+1} d_{i+1} \sin(\delta_{i+1} + \alpha_i)] \cdot \frac{1}{\prod_{k=1}^{i-1} \lambda_k} du_1 = 0 \tag{10}$$

After introducing the expressions for the masses and lengths in terms of u_1, [equations (3), (4), (5)], in Eq.(10) we obtain:

$$\int_0^{u_{1\max}} [\, A u_1^2 + B u_1 + C \,] du_1 = 0 \tag{11}$$

where the factors A to C are:

$$A = 0.5\gamma \cdot \sum_{i=1}^{n} \left(\frac{\cos(\delta_{i+1} + \alpha_{i+1}) \sin\alpha_{i+1} \sin\theta_{i+1}}{\cos(\delta_{i+1} + \alpha_{i+1} + \theta_{i+1}) \prod_{j=1}^{i-1} \lambda_j \prod_{j=1}^{i} \lambda_j^2} - \frac{\cos(\delta_i + \alpha_i) \sin\alpha_i \sin\theta_i}{\cos(\delta_i + \alpha_i + \theta_i) \prod_{j=1}^{i-1} \lambda_j^3} \right)$$

$$B = \frac{c_n}{\prod_{k=1}^{n-1} \lambda_k^2} - c_1 + \sum_{i=1}^{n} \left(\frac{\sin(\alpha_i + \delta_{i+1}) \sin\theta_{i+1} c'_{i+1}}{\cos(\delta_{i+1} + \alpha_{i+1} + \theta_{i+1}) \prod_{j=1}^{i-1} \lambda_j^2} - \frac{\sin(\alpha_i + \delta_i) \sin\theta_i c'_i}{\cos(\delta_i + \alpha_i + \theta_i) \prod_{j=1}^{i-1} \lambda_j^2} + \frac{d_{(i+1)0} \gamma \cos(\delta_{i+1} + \alpha_{i+1}) \sin\alpha_i}{\prod_{j=1}^{i-1} \lambda_j \prod_{j=1}^{i} \lambda_j} - \frac{d_{i0} \gamma \cos(\delta_i + \alpha_i) \sin\alpha_i}{\prod_{j=1}^{i-1} \lambda_j^2} \right)$$

$$C = \sum_{i=1}^{n} \frac{1}{\prod_{k=1}^{i-1} \lambda_k} [-c_i b_{i0} + W_{i0} \sin\alpha_i + $$

$$+ c'_{i+1} d_{(i+1)0} \cdot \sin(\alpha_i + \delta_{i+1}) - c'_i d_{i0} \cdot \sin(\alpha_i + \delta_i)]$$

The integral (11) gives:

$$\frac{A}{3} u_{1\text{-max}}^3 + \frac{B}{2} u_{1\text{-max}}^2 + C u_{1\text{-max}} = 0 \tag{12}$$

The above gives that $u_{1\text{-max}}=0$ (which is rejected, since motion develops) and that:

$$u_{1-\max} = \frac{-0.5 \cdot B + \left(0.25 \cdot B^2 - 4/3 \cdot A \cdot C\right)^{0.5}}{2/3 \cdot A} \tag{13}$$

which is the expression that gives the distance moved by the slide, in terms of its strength and initial geometry.

ACKNOWLEDGMENT

This work was supported by the European Commission, DG XII for Science, Research and Development, Climate and Natural risks (project ENV4-CT97-0392).

REFERENCES

- Ambraseys N., Srbulov M. (1995): "Earthquake induced displacements of slopes". *Soil Dyn. and Earthq. Engng*, (14), pp.59-71.

- Kramer S. L. (1996): "Geotechnical earthquake engineering", *Prentice Hall*, New Jersey.

- Sarma S.K. (1979): "Stability analysis of embankments and slopes". *J. Geotech. Engng, ASCE*, 105 (12), pp. 1511-1524.

- Sarma, S. K. and Chlimitzas G. (2000): Edited Report #2 for the project ENV4-CT97-0392, European Commission, DG XII for Science, Research and Development, Climate and Natural risks.

Computer Methods and Advances in Geomechanics, Desai et al. (eds) © 2001 Balkema, Rotterdam, ISBN 90 5809 183 X

Stability charts for cut slope in unsaturated residual soils

R. Subramaniam
KTA Tenaga Sdn Bhd, Malaysia

Faisal Haji Ali
Civil Engineering Department, University of Malaya, Kuala Lumpur, Malaysia

ABSTRACT: In the past, the effects of soil suction, rainfall and conductivity of soil were not considered in the conventional slope stability analysis that was carried out using parameters based on saturated condition. The analysis when linked to shallow failure, deep ground water conditions or climatic fluxes, should however be extended to incorporate unsaturated soil mechanics. This study has been carried out in view of partial saturation with other related parameters involved in the stability of slopes in residual soils using a combination of seepage and slope stability programs.

1 INTRODUCTION

The relatively steep cut slopes in residual soils are initially stable and partly saturated. The partial saturation of the soil allows the negative pore pressure (matric suction) to exist and develops an apparent cohesion, which in majority of the cuts is the predominant stabilizing factor. The rain initiates the absorption of water by the surface layers causing the degree of saturation to increase. The saturation zone advances to greater depths and it alters the hydraulic gradient and the hydraulic conductivity thus changing the flow patterns and the distribution of moisture content in the soil.

This study has been carried out in view of the partial saturation with other related parameters involved in the stability of slopes in residual soils. The following objectives have been outlined.

1. To identify the different hydrological, physical, geometrical and strength parameters that affects the stability of slopes in unsaturated residual soils.

2. To simulate a typical site condition for a Kuala Lumpur – Karak highway slope and to verify the usage of SEEP/W software by studying the changes in suction at the exposed surface due to changes in rainfall intensities for a typical rainfall pattern in a month with selected site condition.

3. To produce a physically based coupled dimensional (2D) dynamic slope hydrological condition (transient) model controlling stability due to various parameters identified above using the available SEEP/W software.

4. To use the predicted hydrological condition (steady state and transient state analysis) as input to a physically based two dimensional (2D) slope stability sub model (SLOPE/W) to study the changes in factor of the safety of slopes based on the various parameters determined above.

5. To apply the combined seepage and slope analysis to arrange a slope form and antecedent condition for the production of stability chart for partially saturated soils, which summarizes the effect of a particular storm event on slope stability.

2 VERIFICATION OF SEEP/W PROGRAM

The SEEP/W program was verified by simulating a particular rainfall pattern for a typical slope at Kuala Lumpur – Karak Highway (Affendi, 1996). The measured suction values at the site were compared with the suction values derived from the SEEP/W analysis.

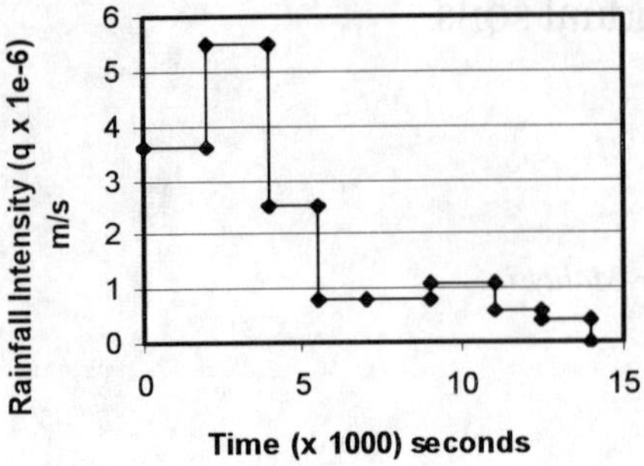

Figure 1 shows the rainfall intensities for 9th April 1992.

Figure 2 shows the variation of suction with respect to elapsed time for a moderately high intensity rain on 9th April 1992 for duration of 4 hours. A total rainfall of 21.50 mm was precipitated in this duration. The simulation gives a lowest value of 11.1 kPa whilst the site value is 12.7 kPa. These lowest values are obtained at the 30th and 60th minute respectively. Both the simulated and site suction values tend to drop gradually and reach approximately the same values with 30 minutes difference. The rainfall pattern agrees well with the drop in suction. The trend in the recovery is found to be the same also.

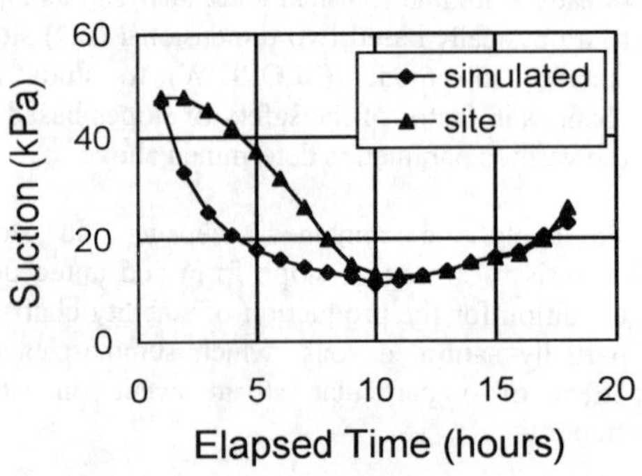

Figure 2 Comparison of simulated and site suction values on April 9, 1992 for a typical slope at KL – Karak Highway

It is observed that the site suction values do not drop immediately as the simulated suction values due to number of reasons. The top surface of the slope is covered by vegetation and it is not considered in this simulation. The small soil particles that exist at this top surface also act as protection by its own cohesive strength. This cohesive strength only decreases gradually as the rainwater infiltrates into the ground whereas in the simulated model the suction drops immediately. These particles also act as colloids and covers the pores that exist the top surface of the slope.

3 DEVELOPMENT OF SLOPE STABILITY CHARTS

Standard approaches to stability analysis usually simplify consideration of the hydrological condition to that of a static, fixed groundwater level. Stability analysis procedures are then used to determine the factor of safety for the slope, given the distribution of positive pressures along the slip surfaces. Soil suctions are generally ignored in such analysis, being assumed zero.

Many researchers have produced chart that enable the stability of simple slopes to be assessed without the necessity of completing detailed calculations. They are also useful for preliminary analysis and enable the designer to quickly assess the sensitivity of a problem to changes in different input parameters.

The main objectives of this exercise are to: -

(1) To find the critical condition where the landslide could occur or the combination of permeability of the soil and the rainfall intensity which will be critical.

(2) To find a linear relationship between the factor of safety and the dimensionless factor tan ϕ^b for unsaturated residual soils.

(3) To establish a stability chart for approximation of preliminary design.

In the computation of the data for the development of stability charts it has been assumed that the slope is homogeneous and constructed of a single material with effective stress strength parameter c' and ϕ'. The critical condition $q_S = k_S$ is chosen when $k_S = 1x\ 10^{-6}$m/s because of its high permeability and the duration of rainfall intensity is fixed at 24 hours as determined from the analysis. The $k_S = 1x\ 10^{-6}$ m/s was found to give the lowest safety factor during the sensitivity analysis for the slope of Kuala Lumpur -Karak Highway.

The range of the parameter is as follows: -

tan β = 2:1 , 1:1, 3:2

ϕ^b/ϕ' = 0.0, 0.25, 0.50, 0.75, 1.00

$c'/\gamma H$ = 0,0.015,0.030

pore water pressure = SEEP/W Heads

An accurate and extensive general solution is made possible by these factors: -

3.1 *Dimensionless number $c'/\gamma H$*

For a given value of the dimensionless number $c'/\gamma H$, the factor of safety depends only on the geometry of the section, expressed as tan β, on the pore-pressure given by SEEP/W and on the angle of shearing resistance, ϕ' and suction resistance, ϕ^b for unsaturated residual soils.

To reduce the amount of computation only three (3) values of $c'/\gamma H$ have been used that is 0.0, 0.015, 0.03. Considering that the cohesion intercept in terms of effective stress is gradually somewhat lower than the cohesion intercept in total stress, these values have been selected as representing the range commonly encountered in effective stress analysis and also a range within which a linear interpolation can be used without significant errors. The intermediate values for $c'/\gamma H$ can be interpolated as shown in Figure 3 for a particular slope angle and strength value. It should however be remembered that for cross sections of natural slopes or wide embankments some errors may be incurred due to the neglect of tension cracks whose effect on stability becomes more pronounced at higher values of $c'/\gamma H$. For these problems, a modified analysis is generally required.

3.2 *Factor of safety (F) varies linearly with the magnitude of the tan ϕ^b*

For a simple soil profile and specified shear strength parameters, it had been found that to a closer approximation, the factor of safety, (F) varies linearly with the magnitude of the tan ϕ^b (dimensionless value of rate of change in shear strength with respect to suction). A typical example of this linear relationship is given in Figure 4.

$$F = f + s.\tan\phi^b \text{ --------(1)}$$

where f and s are termed the stability coefficient for a particular slope and soil properties, tan ϕ^b is the rate of change in shear strength with respect to suction $(u_a - u_w)$.

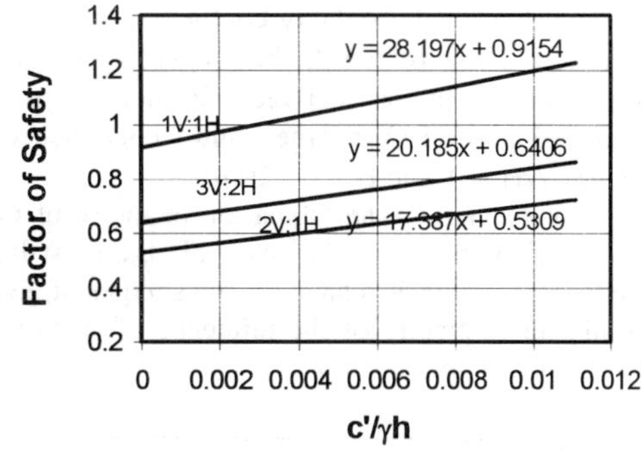

Figure 3 Linear relationship between factor of safety and dimensionless number $c'/\gamma H$ for a particular ϕ' = 30 degrees and ϕ^b = 15 degrees for various slope angle

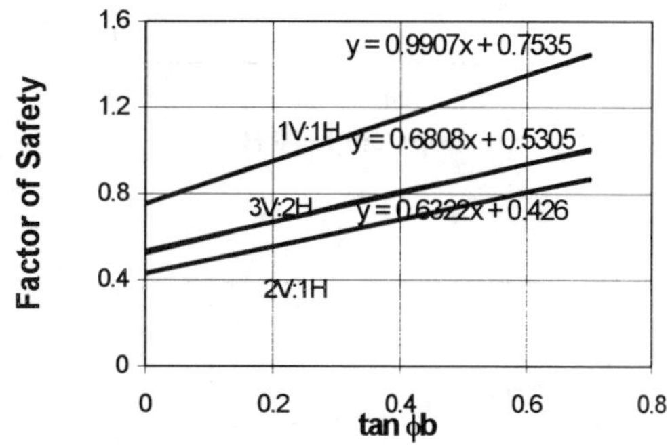

Figure 4 Linear relationship between factor of safety and dimensionless tan ϕb value for ϕ' = 30 degrees and $c'/\gamma H$ = 0.005 for various slope angle

Linear relationship between F and tanϕ^b for a given value of $c'/\gamma H$, ϕ', tan β as shown in Figure 4 can be described in terms of two parameters, f, and s where geometrically, f is the intersection with factor of safety (F) axis of line describing the relationship between F and tan ϕ^b and corresponds to the value of the factor of safety for zero suction value (ϕ^b = 0), and s is the slope of this line. Since the slope of this line is always positive (Fredlund, 1994), the factor of safety increases with increasing ϕ^b value, whilst all other parameters being held constant and may be expressed in the form of equation (1).

These two parameters f and s are determined from the fitting curve passing through all or near all the respective points where ϕ^b/ϕ' = 0.0, 0.25, 0.50, 0.75, 1.00 for various ϕ' values process using the Microsoft EXCEL programme with an average standard error, R^2 value of 0.9 (Fredlund, 1994).

The values of the stability coefficients f and s are then plotted against tan β, the tangent of the slope angle, for varying ϕ' between 20 degrees to 40 degrees. Figures 5 shows the f and s coefficients for values c'/γH = 0, and H = 0 - 20m.

The presentation and tabulation of the results are greatly simplified by the use of these stability coefficients, which also have the advantages of giving an immediate picture of the influence of suction on the factor of safety.

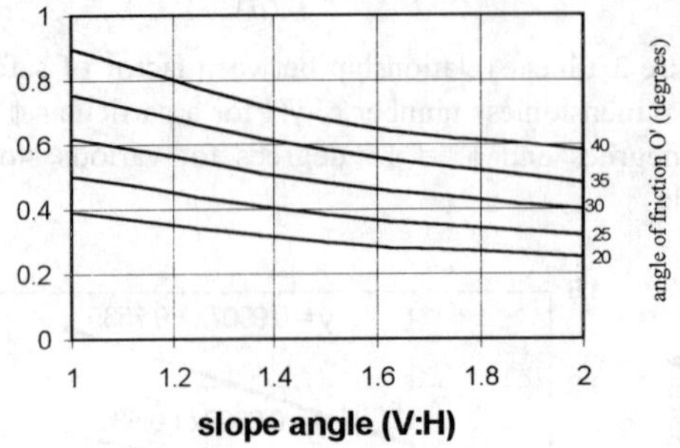

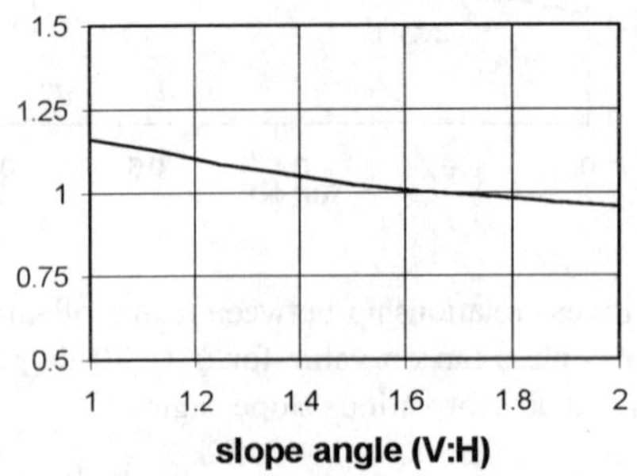

Figure 5 Stability coefficients for c'/γH = 0 and H = 0 – 20 m

To calculate the factor of safety of a section whose c'/γH lies within the range covered by these figures, it is necessary only to apply the equation (1) to determine the factor of safety of the two nearest values of c'/γH and then perform a linear interpolation between these values for the specified c'/γH as the factor of safety (F) is also linearly increasing with dimensionless value c'/γH.

4 CONCLUSION

The following conclusion can be derived from this study: -

1) The simulated suction values from SEEP/W agrees quite well with the site condition values and can be used for further analyis with reasonable limits.
2) Higher suction values give higher factors of safety.
3) The factor of safety for unsaturated soil seems to be increasing with a dimensionless value $\tan\phi^b$.
4) A linear relationship can be obtained for the factor of safety of unsaturated residual soil.

$$F = f + s \tan \phi^b \text{ -------(1)}$$

where

F = factor of safety.

f and s = the stability coefficient

$\tan\phi^b$ = is the rate of change in shear strength with respect to suction (Ua – Uw).

Hence the factor of safety for a particular slope can be determined from the combination of stability coefficients obtained from the stability chart.

REFERENCES

Affendi, A (1996). Field and laboratory study on unsaturated residual soils in relation to slope stability analysis, Ph.D. thesis, University of Malaya. Unpulished.

Fredlund, D.G and Rahardjo, H. (1993). Soil Mechanics for Unsaturated Soils. John Wiley & sons.

Subramaniam, R & Ali, F.H. (1999). Design Chart for cut slope in unsaturated residual soils, Proceedings of the International Symposium on Slope Stability Engineering – IS – Shikoku'99, Matsuyama, pp 375 – 380.

Computer Methods and Advances in Geomechanics, Desai et al. (eds) © 2001 Balkema, Rotterdam, ISBN 90 5809 183 X

Sheet pile extraction analysis by pinball algorithm and excavation method

T.Tanaka
Department of Biology and Environmental Engineering, University of Tokyo, Japan
K. Matsushima & Y. Mohri
National Research Institute of Agricultural Engineering, Ibaragi Prefecture, Japan

ABSTRACT: A series of laboratory model tests with buried pipes and numerical analyses was carried out to study the mechanical behavior of surrounding soil and pipe during sheet pile extraction from sand. The relative density (Dr) and ditch width (B), which is protacted by two sheet piles significantly affect the deflection of a buried pipe and settlement. Looseness of soil around a sheet pile being extracted greatly influences the surrounding ground. As a result, analysis of the extraction behavior, which involves redistribution and transformation of the stresses in ground, is complicated. For contact analysis of the extraction behavior, we developed an extraction method using the pinball algorithm and excavation technique. The concept of the pinball algorithm is to embed pinballs in surface elements, which might contact each other, and to enforce the impenetrability condition only to pinballs. The strain condition of soil around the sheet pile can be represented by numerical analysis, and the relation of extracted length to deflection and to settlement can be simulated.

1. INTRODUCTION

The extraction of buried sheet piles greatly influences the surrounding ground. The void, which is produced by the sheet pile extraction, causes a release and redistribution of in-situ stress of soil around the sheet pile. Accordingly, the walls around both sides of the void displace toward the space, and so the distance between the foundation walls progressively decreases and a contact surface may be generated (Fig. 1). These phenomena of redistribution and transmission of in-situ stress greatly influence underground structures such as a pipeline with sheet piles.

A mechanical contacting system that allows ground contact during sheet pile extraction had been handled numerically to create an element with special functions. Most of the research efforts on ground behavior during sheet pile extraction have focused on underground structure deformation and settlement, and these are certainly the most important problems. However, looseness of soil around a sheet pile during extraction greatly influences the surrounding ground, and visible information such as deformation of an underground structure and settlement appears. Therefore, the mechanism of strain development and change in stress of ground during sheet pile extraction need to be investigated.

Although the behavior of surrounding ground during sheet pile extraction can be studied by laboratory or field tests, the change in stress state and strain condition of soil around a sheet pile cannot be easily evaluated. So, to get a better understanding of the extraction behavior in sand, numerical analyses and laboratory model tests were conducted in this study. The objective of the study was to represent the deflection of a buried pipe and the settlement, and to illustrate the strain condition of soil around a sheet pile by extraction analysis using the pinball algorithm and excavation technique.

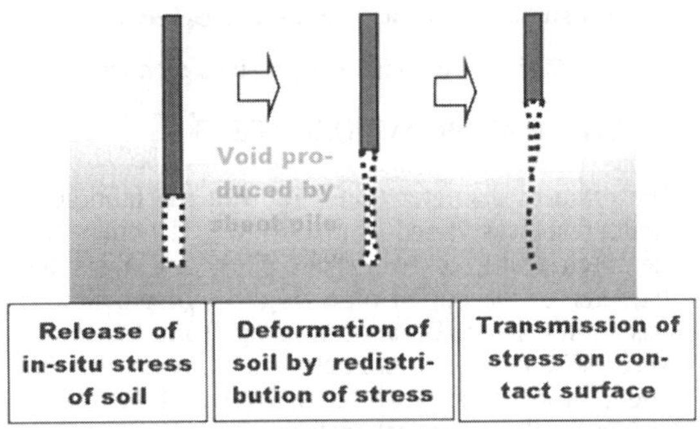

Fig. 1 Concept of extraction behavior

2. FINITE ELEMENT ANALYSIS METHOD

A constitutive model for non-associated strain hardening-softening elasto-plastic material is introduced. The dynamic relaxation method with return mapping algorithm is applied to the integration algorithms of the elasto-plastic constitutive relation.

The contact analysis of sheet pile extraction behavior is performed using excavation techniques which can represent the release of in-situ stress and pinball algorithm which can represent the transmission of stress (Belytschko and Neal 1991). Figure 2 shows a master element and a slave element in which pinballs are embedded. The main concept of the pinball algorithm is to enforce the impenetrability condition between master and slave elements. The contact constraint is embodied in the pinballs rather than the elements themselves. The determination of whether interpenetration has occurred thus becomes a simple check of the distance between two pinballs.

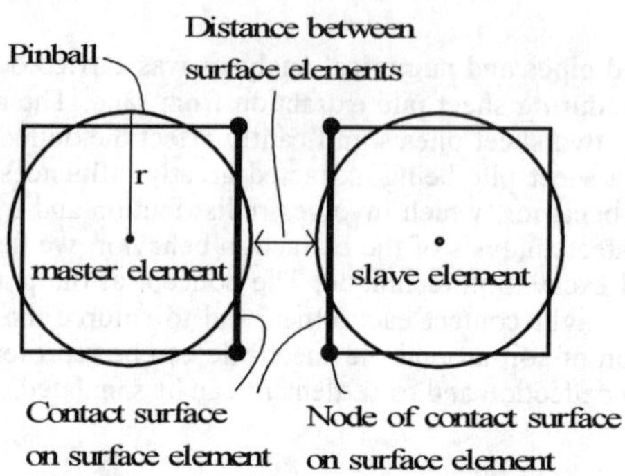

Fig. 2 Pinball embedded in surface element

3. LABORATORY MODEL TESTS

The ditch to diameter (B/D) ratio of the laboratory model tests was varied from 1.74 to 7.71, where B is the ditch width of two sheet piles, and D is the diameter of the buried pipe (Fig. 3). In practice, it was found that B/D ratios ranging from 1.5 to 2.0 were effective for constructing a pipeline. In this study, Tests 8 and 9 with a B/D ratio of 4.96 were chosen for the numerical analysis.

The model sheet pile used for the laboratory test was made of aluminum plate of thickness 1.5 cm and height 75.0 cm. The surface of the sheet piles was smooth to represent the low friction condition. The model pipe was made of aluminum of thickness 2.0 mm and diameter 140 mm ($E=7.28\times10^7$ kN/m^2). The surface of the pipe was covered with sandpaper to represent the full friction condition in sand. Displacement gauges were placed on the middle load plate to measure the settlement and into the buried pipe to measure the vertical and horizontal deflection (Fig. 4). To observe the behavior of a buried pipe, load plates on the ground surface were loaded (62.7 kPa) in the laboratory model tests. While keeping the load condition, two sheet piles were extracted at the same time, at an extraction speed of approximately 3.5 mm/min.

In order to fill the soil foundation, sand particles were allowed to fall freely into the test box by hopper equipment. Then, the test box was laid on its side to assess constant internal friction in ground (Lam and Tatsuoka 1988). The position of the hopper was adjusted during sand "raining" to ensure a constant falling distance of the sand particles. The

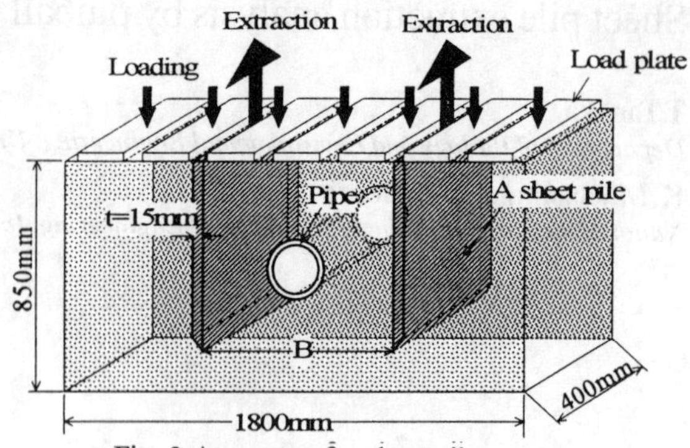

Fig. 3 Apparatus for sheet pile test

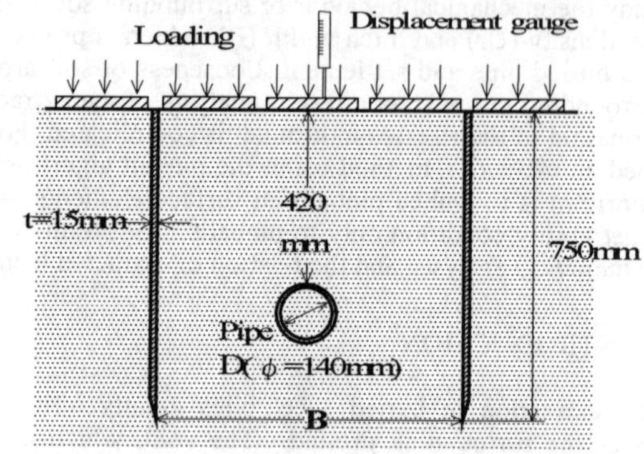

Fig.4 Detail around sheet piles

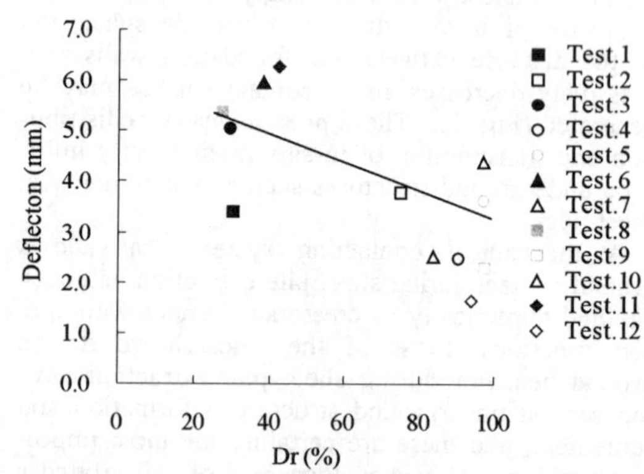

Fig. 5 Relation between relative density and deflection

Table 1 Results of laboratory model sheet pile test

Test No.	Ratio of ditch to diameter (B/D)	Relative Density Dr (%)	Final settlement (mm)	Final deflection (mm)
1	1.74 (loose)	31	45.1	3.42
2	1.76 (a little dense)	75	11.9	3.78
3	2.78 (loose)	29	34.5	5.05
4	2.78 (dense)	90	10.2	2.48
5	2.82 (dense)	97	6.0	3.64
6	3.86 (loose)	38	28.6	5.95
7	3.86 (dense)	97	8.8	4.39
8	4.96 (loose)	27	28.0	5.39
9	4.96 (dense)	97	6.5	2.29
10	4.94 (dense)	84	8.3	2.51
11	7.71 (a little loose)	43	30.0	6.27
12	7.71 (dense)	94	6.4	1.64

*D is the diameter of pipe (140mm)

relative densities of the sand investigated in the laboratory model tests were 27 to 97%.

The results of the laboratory tests for final deflection of the buried pipe and final settlement tested under various conditions are shown in Table 1. The deflection in loose sand is approximately 1.7 times that in dense sand, hence relative density (Dr) greatly affects deflection.

4. RESULTS OF ANALYSIS

Figure 6 shows a finite element mesh with surface elements and extraction elements for analyses (1,198 elements, 1,278 nodes). The surface elements of the sheet pile are embedded pinballs. Table 2 indicates the material constants of dense and loose sands used for calculation. Analyses 8 and 9 in the calculation correspond to Tests 8 and 9 in the laboratory test.

Strain change around sheet pile

To demonstrate the contact condition and the deformation of soil due to the release of in-situ stress and the development of strain around a sheet pile during extraction, the maximum shear strain condition of soil around the sheet pile for extracted lengths of 30.0 cm and 48.0 cm are shown in Fig. 7 (a), (b). Surface elements are displaced toward the void due to the release of in-situ stress and change into the impenetrability condition. In the neighborhood of the sheet pile head, there is a significant shear strain concentration as the surface elements close up on each other. The shear strain development area of Fig. 7 (b) is larger than that of Fig. 7 (a), showing that the shear strain development area was increased by sheet pile extraction.

Deflection and settlement

The deflection and settlement behavior during sheet pile extraction determined from the numerical analysis were compared with the laboratory model tests. Generally, in the case of a flexible pipe, as sheet pile extraction causes looseness of the soil around the buried pipe, the buried pipe may deflect in the horizontal direction as a result of the reduction of passive earth pressure beside the buried pipe. Figure 8 shows the relation between extracted length and deflection. It was found that the buried pipe gradually deflected during sheet pile extraction and almost reached the final deflection at the extracted length of 36.0 cm. Therefore, extracting the sheet pile to the top crown level greatly affects the earth pressure beside the buried pipe. As there is an obvious difference between deflection with loose and dense sands, the influence of stress relaxation range of loose sand is much larger than that of dense sand.

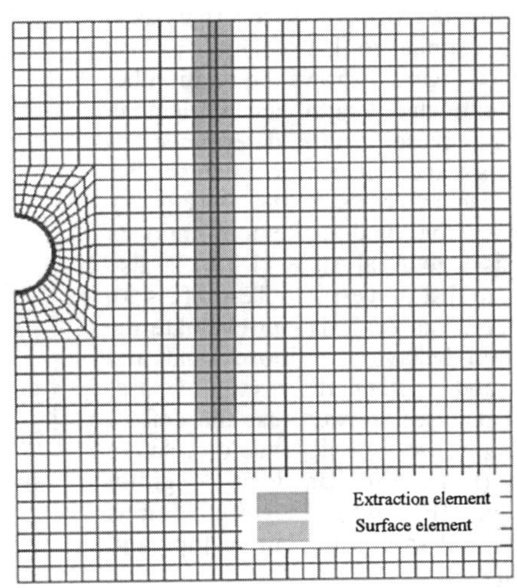

Fig. 6 Finite element mesh for extraction analysis

Table 2 Parameters used for analysis

	Dr(%)	G_0(kN/m^2)	ρ(g/cm^3)	ϕr	S.B(cm)
Analysis 8	97	11,760	1.60	34.0	0.30
Analysis 9	27	5,880	1.60	34.0	0.30

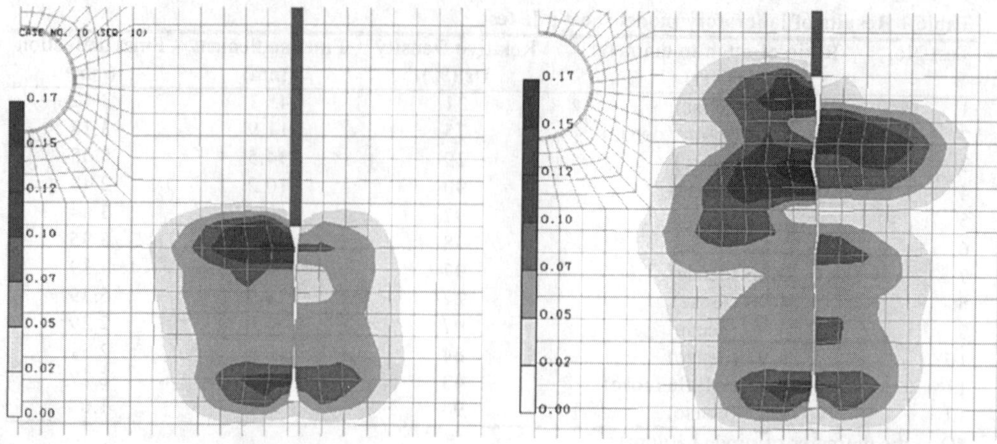

(a) Extracted length=30.0cm (b) Extracted length=48.0cm

Fig. 7 Contour map of maximum strain

In the analysis, it is very important to evaluate the contribution of relative density effect on deflection and settlement. Although the buried pipe gradually deflected due to soil looseness around the sheet pile during extraction, the values determined by Analyses 8 and 9 were much smaller than those by Tests 8 and 9 (Fig. 8). However, to see the final deflection and the final settlement ratio of loose and dense sands, the ratio of final deflection of Analysis 9 to Analysis 8 (1:2.4) is similar to that of Test 9 to Test 8 (1:2.5). Figure 9 shows the relation between extracted length and settlement. Likewise, these results determined by Analyses 8 and 9 also were much smaller than those by Tests 8 and 9 (Fig. 9). However, the final settlement ratio of Analysis 9 to Analysis 8 (1:3.8) is also similar to that of Test 9 to Test 8 (1:4.3). Therefore, this analysis can evaluate the extraction behavior in loose sand and dense sand. Further, the results of parametric analyses showed that the deflection of a buried pipe depends on relative density (Dr) rather than flexural rigidity of the pipe (EI) and soil layer conditions (ϕ, Dr, G_0) around the buried pipe and sheet piles.

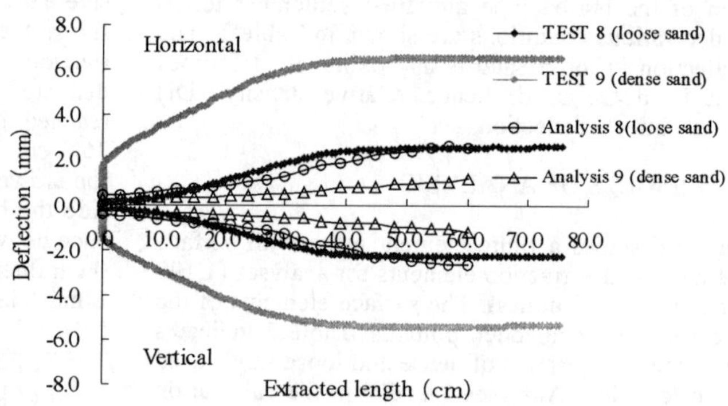

Fig. 8 Relation between extracted length and deflection

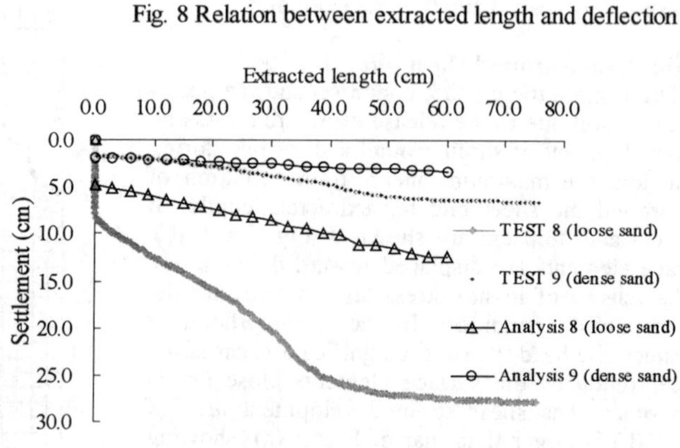

Fig. 9 Relation between extracted length and settlement

5. CONCLUSION

We developed an extraction method using the pinball algorithm and excavation technique in finite element analysis. This method can handle contact problems in ground. Based on the results of laboratory model tests and numerical analyses, we examined the mechanical behavior of surrounding soil of sheet pile and reached the following conclusions.

(1) It was confirmed that surface elements contacted each other and changed into the impenetrabil-

ity condition after sheet pile extraction. The deflection and settlement values simulated by the numerical analysis were much smaller than those by the laboratory test. However, the strain development area was increased by looseness of soil around the sheet pile during extraction, and the deflection and the settlement increased as sheet pile extraction progressed. Therefore, the extraction behavior could be represented by the extraction analysis with pinball algorithm.

(2) Also, the relative density (Dr) significantly contributes to deflection and settlement of soil foundation. It is very important to evaluate the effects of relative density on the interaction between soil and pipe. Regarding relative density, the ratio of final deflection and settlement of dense sand to loose sand calculated by numerical analysis was similar to the laboratory model test results. Therefore, this method could clearly evaluate the difference of relative density.

REFERENCES

1) T. Belytschko and M. O. Neal (1991): "Contact-Impact by the Pinball Algorithm with Penalty and Lagrangian Methods," International Journal for Numerical Methods in Engineering, 31, pp547-572

2) K. Sakai, J. Mitsuo and Kaneto (1992): "The Finite Element Method for Analysis of Ground Movements by Extracting Sheet Piles Considering the Contact Problem," Proc. of 27th JSSMFE, pp.1797-1798 (in Japanese)

3) S. Imamura, *et al.* (1999): "Earth pressures acting on a deep circular shaft and associated with ground deformation," 44th Symposium in Geomechanics, pp261-266 (in Japanese)

4) Y. Mohri, A. Asano, A. Natsuka and Matsuoka. (1991): "Performance of Buried Pipe during Sheet Pile Extraction," Proc. of 27th JSSMFE, pp1797-1798 (in Japanese)

5) W.K. Lam and F. Tatsuoka (1988): "Effects of Initial Anisotropic Fabric and σ_2 on Strength and Deformation Characteristics of Sand," S&F, Vol.28, No.1

Computer Methods and Advances in Geomechanics, Desai et al. (eds) © 2001 Balkema, Rotterdam, ISBN 90 5809 183 X

A stability analysis of masonry walls by discrete finite element method (DFEM)

N.Tokashiki – *Department of Civil Engineering, University of Ryukyus, Japan*
Ö.Aydan – *Department of Marine Civil Engineering, Tokai University, Shimizu, Japan*
Y.Shimizu – *Department of Civil Engineering, Meijo University, Japan*
I.H.P.Mamaghani – *Department of Civil Engineering, Kanazawa University, Japan*

ABSTRACT : This paper presents the results of two dimensional static and pseudo-dynamic stability responses of masonry walls analyzed by using Discrete Finite Element Method (DFEM). The actual masonry walls are analyzed and the applicability of this method to such structures are checked and discussed. The deformation behaviour of masonry walls are analyzed for several geometrical conditions such as height and slope of wall, and material conditions of block and soil mass.

1 INTRODUCTION

Masonry houses presently constitute of almost 70 percent of the residential buildings all over the world and they have a very long building history. In addition, most of historical structures and remains may be categorised as masonry structures (Figure 1). Masonry structures composed of blocks, which are made of natural stones, bricks or both, assembled in different patterns with or without mortar. Compared with other countries, there are a few masonry structures. For example Japanese castle masonry walls and bridges are some of typical examples of masonry structures in Japan. The preservation of Japanese castles as the historical assets of Japan is quite important and it receives a public great attention. However, the mechanical behaviour of masonry walls are less understood inspite of their long use in building history. Furthermore, Japan is a seismically active country and as a result of that the responses and stability of masonry walls during earthquakes are of great concern in engineering community.

In this article, the authors are particularly concerned with the responses and stability of masonry walls under gravitational and seismic loading conditions. Some specific examples of masonry walls are analysed using *Discrete Finite Element Method* (DFEM) and computed results are shown and discussed.

(b) Hagia-Sofia (c) Pizza Tower

(a) Wasser Bridge

Figure 1. Some examples of masonry structures

2 DISCRETE FINITE ELEMENT METHOD

Masonry walls are blocky structure and their mechanical behaviours are less understood. In recent years, many numerical methods are proposed to discontinuum models for blocky media. Although Discrete Element Method (DEM) and Discontinuous Deformation Analysis (DDA) can be used for static and dynamic analysis of masonry structures, the treatment of rate-dependent behaviour of materials in these methods is nothing to do with the actual ones.

Aydan and Mamaghani proposed the discrete finite element method (DFEM) for blocky systems which is based on the principles of the finite element method (Aydan et al., 1996, Managhani et al., 1994). It can be handled deformable blocks and contacts that specify the interaction among them.

Small displacement theory is applied to intact blocks while blocks can take finite displacement, Blocks are polygons with an arbitrary number of sides which are in contact with neighboring blocks, and are idealized as a single or multiple finite elements. Block contacts are represented by a contact element,

Starting from the moment of momentum and employing the conventional finite element discretisation procedure yiels the following set of equations for a typical finite element (Aydan et al., 1996).

$$\boldsymbol{M\ddot{U}} + \boldsymbol{C\dot{U}} + \boldsymbol{KU} = \boldsymbol{F} \qquad (1)$$

Where

$$\boldsymbol{M} = \int_\Omega \rho \boldsymbol{N}^T \boldsymbol{N} d\Omega\ , \quad \boldsymbol{C} = \int_\Omega \boldsymbol{B}^T \boldsymbol{D}_v \boldsymbol{B} d\Omega\ ,$$

$$\boldsymbol{K} = \int_\Omega \boldsymbol{B}^T \boldsymbol{D}_e \boldsymbol{B} d\Omega\ , \quad \boldsymbol{F} = \int_\Gamma \overline{\boldsymbol{N}}^T \hat{\boldsymbol{t}} d\Gamma + \int_\Omega \boldsymbol{N}^T \boldsymbol{b} d\Omega.$$

These are number of procedures to treat the problem of time discretization. If the central difference mehtod is used, Eq.(1) takes the following form:

$$\overline{\boldsymbol{K}}\boldsymbol{U}_{n+1} = \boldsymbol{F}_{n+1} \qquad (2)$$

Where

$$\overline{\boldsymbol{K}} = \frac{1}{\Delta t^2}\boldsymbol{M} + \frac{1}{2\Delta t}\boldsymbol{C}\ ,$$

$$\boldsymbol{F}_{n+1} = \left(\frac{2}{\Delta t^2}\boldsymbol{M} - \boldsymbol{K}\right)\boldsymbol{U}_n - \left(\frac{1}{\Delta t^2}\boldsymbol{M} - \frac{1}{2\Delta t}\boldsymbol{C}\right)\boldsymbol{U}_{n-1} + \boldsymbol{F}_n\ .$$

When Eq.(1) is solved, oscillations may occur depending upon the rate-dependent and rate independent parameters of constitutive models. When the inertia term is considered, contacts and solid blocks are assumed to behave elasto-visco-plasticaly and visco-elasticaly, respectively. On the other hand, if the inertia term is omitted, then the behaviours of contacts and blocks are assumed to be elasto-plastic and elastic, respectively.

In this study, a pseudo-dynamic procedure of this method is employed. So, intact blocks and block contacts can behave elasto-plasticaly. The method of analysis is a pseudo time stepping incremental procedure. To model the elasto-plastic response of materials and contacts in numerical analysis, the initial stiffness method was employed together with the use of *Updated Lagrangian Scheme.*

3 MODELLING THE MASONRY WALL

Masonry walls of Japanese castles usually composed of large stone blocks together with smaller size stones as filling material between larger blocks. Figure 2 shows an idealized masonry wall. Figure 3 shows how a representative element of the masonry wall is modeled in DFEM. Large and small stone blocks are represented as solid elements while contact elements are used between large blocks and small blocks as illustrated in Figure 2. Back filling material and soil mass are modeled through solid elements.

4 RESPONSES OF MASONRY WALLS UNDER GRAVITATIONAL LOADING AND PSEUDO-DYNAMIC LOADING

Two dimensional static and pseudo-dynamic stability respnses of actual and model of masonry walls were analyzed by discrete finite element method.

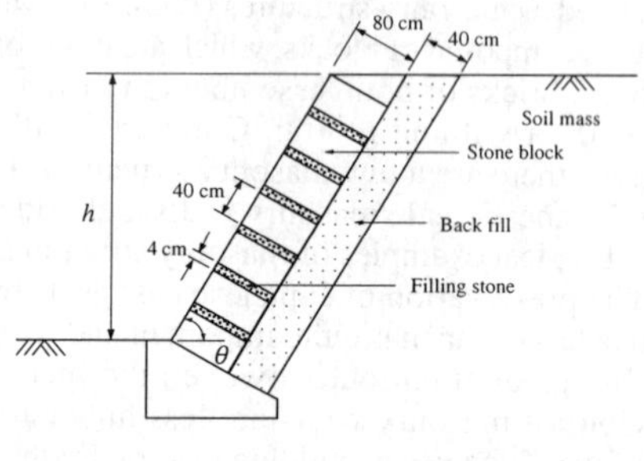

Figure 2. Typical numerical model of masonry wall.

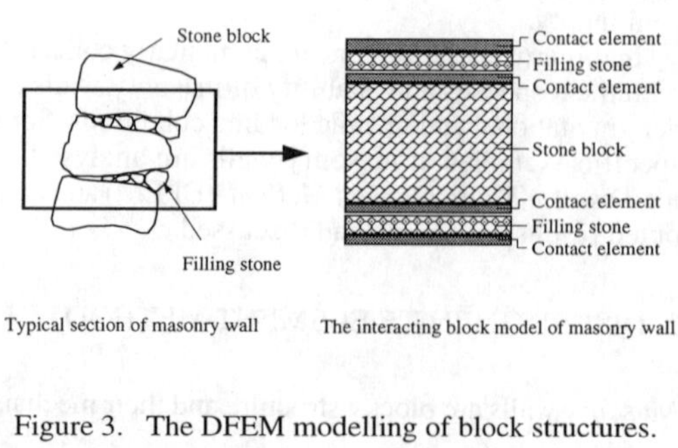

Figure 3. The DFEM modelling of block structures.

Two actual masonry walls were analysed by using physical and mechanical properties given in Tables 1 and 2. The masonry wall shown in Figure 4 is one of historical castles in Japan and it is deformed probably under static and seismic loading. The stability analysis of this masonry wall was first carried out under gravitational loading.

Figure 5 shows the computed deformed configuration of the wall. The computed configuration is barrel-like and it is quite close to the actual one shown in Figure 4.

Figure 4. Deformed wall of Fukui Castle (After Tanaka et al., 1997).

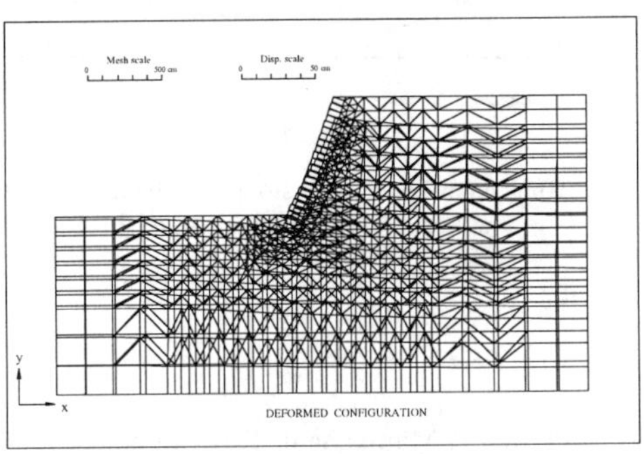

Figure 5. Computed deformed configuration.

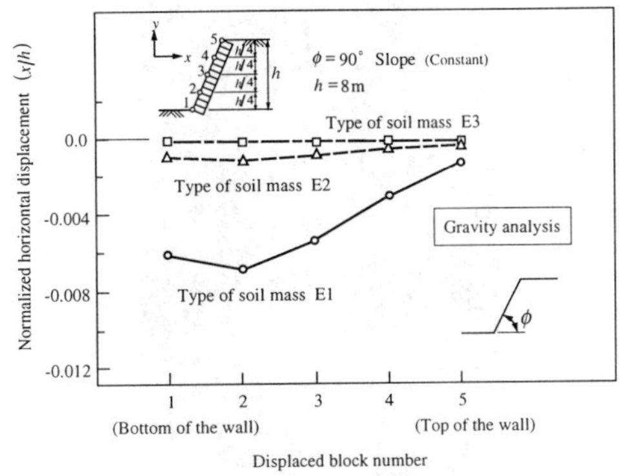

Figure 6. Effect of soil properties on block deformations.

Figure 6 shows the horizontal deformation of the wall under gravitational loading for different soil mass properties. The horizontal displacement of the wall decreases as the soil mass becomes stiffer.

The next example is concerned with the effect of seismic loading on the wall deformation. Figure 7 shows a deformed state of an actual wall in Kobe city, caused by the 1995 Hyogo-ken Nanbu earthquake (M=7.2). The wall composed of granitic blocks built as a dry-masonary wall.

Figure 8 shows the deformed configuration of the wall under a lateral seismic coefficient of 0.2g. As expected, the horizontal displacement of the wall is larger than that for gravitational loading. Figure 9 shows the effect of wall height on its deformation under gravitational and seismic loading. As expected the deformation of wall increases as a result of seismic loading. Furthermore, the increase on the wall height results in larger horizontal displacement of the wall.

Figure 10 shows the effect of wall slope on its deformation under gravitational and seismic loading. As expected the deformation of wall increases as a result of seismic loading. Furthermore, the increase on the wall slope results in larger horizontal displacement of the wall. In all cases the deformed configuration of the wall is barrel-like.

Table 1. Material properties of blocks and contacts.

Type of material	D1		D2		D3	
Properties	Stone block	Filling stone	Stone block	Filling stone	Stone block	Filling stone
E (kgf / cm^2)	1.0×10^4	2.8×10^2	5.0×10^4	7.3×10^2	5.0×10^4	2.0×10^3
ν	0.35	0.37	0.3	0.35	0.3	0.33
ρ (kgf / cm^3)	2.2×10^{-3}	1.3×10^{-3}	2.4×10^{-3}	1.5×10^{-3}	2.6×10^{-3}	1.7×10^{-3}
σ_t (kgf / cm^2)	5.0	4.0	10.0	6.0	20.0	8.0
ϕ (°)	25.0	20.0	35.0	33.0	40.0	35.0
c (kgf / cm^2)	20.0	15.0	30.0	25.0	40.0	35.0
Contact element						
E_n (kgf / cm^2)	2.8×10^2		7.3×10^2		2.0×10^3	
G_s (kgf / cm^2)	2.8		7.3		20.0	
σ_t (kgf / cm^2)	0.1		0.1		0.1	
ϕ (°)	5.0		5.0		5.0	
c (kgf / cm^2)	0		0		0	

Table 2. Material properties of soil mass and contacts.

Type of material	E1		E2		E3	
Properties	Soil mass	Back fill	Soil mass	Back fill	Soil mass	Back fill
E (kgf / cm^2)	1.0×10^2	6.0×10	5.0×10^2	3.0×10^2	5.0×10^3	3.0×10^3
ν	0.4	0.42	0.35	0.40	0.3	0.38
ρ (kgf / cm^3)	1.9×10^{-3}	1.6×10^{-3}	2.0×10^{-3}	1.8×10^{-3}	2.2×10^{-3}	2.0×10^{-3}
σ_t (kgf / cm^2)	0.1	0.06	1.0	0.6	5.0	3.0
ϕ (°)	30.0	18.0	30.0	18.0	30.0	18.0
c (kgf / cm^2)	0.2	0.12	2.0	1.2	2.2	1.3

Type of contact element	a	b	c	a	b	c	a	b	c
E_n (kgf / cm^2)	5.0×10^2	5.0×10^3	5.0×10^4	5.0×10^2	5.0×10^3	5.0×10^4	5.0×10^2	5.0×10^3	5.0×10^4
G_s (kgf / cm^2)	5.0	5.0×10	5.0×10^2	5.0	5.0×10	5.0×10^2	5.0	5.0×10	5.0×10^2
σ_t (kgf / cm^2)	0.1	0.1	0.1	0.1	0.1	0.1	0.1	0.1	0.1
ϕ (°)	25.0	25.0	25.0	25.0	25.0	25.0	25.0	25.0	25.0
c (kgf / cm^2)	0.1	0.1	0.1	0.1	0.1	0.1	0.1	0.1	0.1

Figure 7. Deformed state of a wall in Kobe caused by 1995 Hyogo-ken Nanbu earthquake.

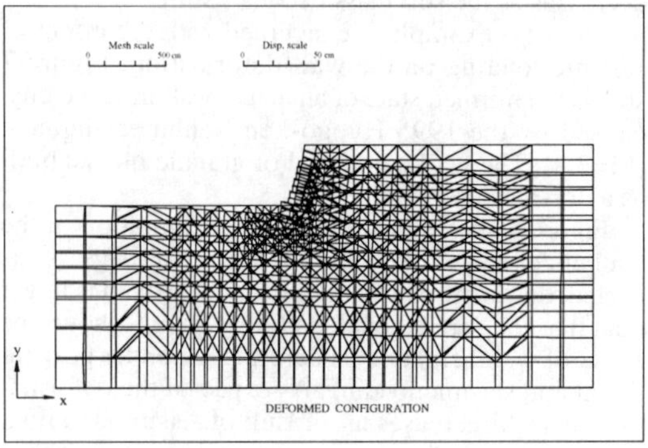

Figure 8. Deformed configuration of wall under seismic loading.

5 CONCLUSIONS

In this paper, the stability of masonry walls was investigated by discrete finite element method. The studies presented are a part of a detailed investigation on the static and seismic stability of masonary structures. The authors have been now performing shaking table tests on the seismic behaviour and stability of the model masonary walls. It seen that the comparison of the computed deformed configurations of the walls with those observed in actual circumtances are quite close to each other and the outcomes are promising. Nevertheless, the dynamic version of discrete finite element method is felt to be necessary for better estimations of the seismic responses and stability of the masonary walls, provided that appropriate viscous characteristics of the walls could be determined for such analyses.

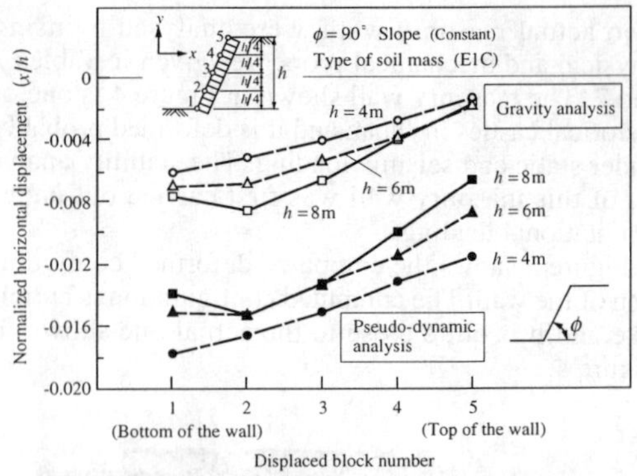

Figure 9. The effect of wall height and loading type on the horizontal displacement of the wall.

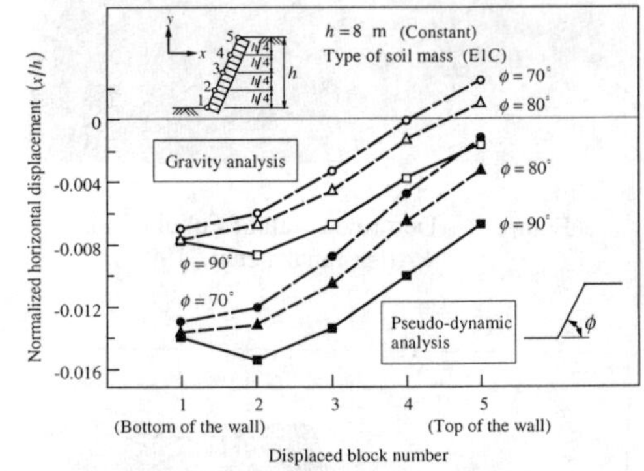

Figure 10. The effect of wall slope and loading type on the horizontal displacement of the wall.

REFERENCES

Aydan, Ö., I.H.P. Mamaghani, T. Kawamoto 1996. Application of discerete finite element method (DFEM) to rock engineering structures, *NARMS '96*, 2039-2046.

Mamaghani, I.H.P., S. Baba, Ö. Aydan, S. Shimizu 1994. Discrete finite element method for blocky systems, *Computer Methods and Advances in Geomechanics*, IACMAG, Morgantown,1, 843-850.

Tanaka, K., Y. Niitani 1997. Transition of Japanese castle's masonry wall and its existing state, *J. of Geotechnical Engineering*, JSCE, IV-37, 101-110.

Computer Methods and Advances in Geomechanics, Desai et al. (eds) © 2001 Balkema, Rotterdam, ISBN 90 5809 183 X

Creep behavior and modeling of the slow-moving Alani-Paty landslide, Oahu, Hawaii

C.S.L.Tsui, H.G.Brandes & D.D.Nakayama
Department of Civil Engineering, University of Hawaii, USA

ABSTRACT: The Alani-Paty landslide area is underlain by soft, high plasticity soils that are susceptible to time-dependant deformation. In order to estimate creep displacements in the field, one cross-section of the landslide area was modeled using the Soft-Soil creep model in the PLAXIS software code. A combination of field measurements over a limited period of time, along with finite element modeling, have allowed for preliminary predictions of future creep displacements.

1 INTRODUCTION

The Alani-Paty landslide in Manoa valley, on the island of Oahu, Hawaii, is typical of a series of slow-moving slides that occur on most flanks of the deep valleys that cut into the Koolau mountain range in East Honolulu. The volcanic colluvium that is found above the basalt bedrock is mostly fine-grained and very soft. Laboratory creep tests show that this type of soil is highly susceptible to time-dependent deformations (Nakayama, 2000), which is also evident in the field by slow but continuous surface deformations that have required millions of dollars in stabilization measures.

This article discusses creep modeling efforts for a representative cross section of the Alani-Paty area, aimed at estimating past and future deformations. A companion paper that focuses on overall stability expands on the geologic setting and geotechnical characteristics of the soils in the landslide area, as well as on recent remediation measures (Brandes and Tsui, 2001). Creep computations were carried out with the commercial code PLAXIS using the SSC (Soft-Soil Creep) model.

2 PROBLEM GEOMETRY AND MODELING APPROACH

Modeling of creep deformations focused on representative cross section AA' in Figure 1. Section AA' can be broadly characterized by inclinations of 5H:1V in the upslope region near the head, 6.8H:1V in the midslope region, and 12H:1V in the downslope region near the toe. In order to simplify the analysis, a single inclination of 6.8H:1V (8.4^0) was used throughout. Initially, a 50-ft long and 5-ft deep section was selected (Figure 2), although the effect of choosing different dimensions is considered as well. The left and right boundaries of the slope section are fixed horizontally, but are free to move vertically. The bottom of the section, on the other hand, is fixed. This means that the height of the slope, *H*, represents the maximum depth to which creep occurs. Remedial measures installed in the area are not considered in this study. Also, the groundwater table is assumed to be at the slope surface to simulate the worst case scenario. The finite element mesh consists of triangular elements with three integration points as shown in Figure 2. Plain strain and drained conditions throughout are assumed.

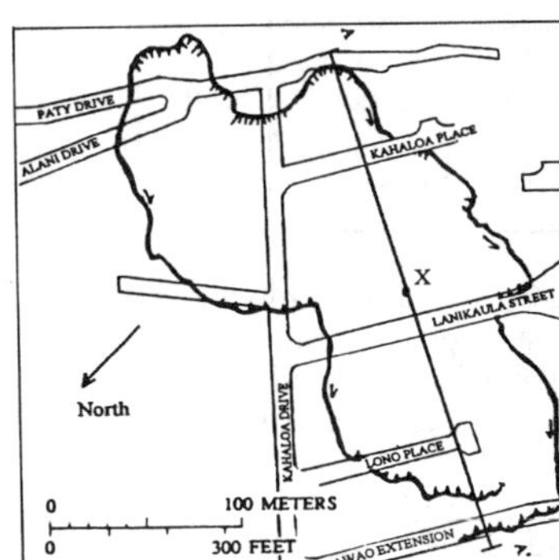

Figure 1. Approximate boundaries of the Alani-Paty landslide

The initial stress due to soil weight is applied to the soil to begin the analysis. An initial time equal to 1 day is input to correspond to the initial stress state. The resulting deformation due to this *gravity turn-on* is reset to zero prior to the second phase, which is the creep phase. During the creep portion of the analysis the soil is allowed to relax and deform with no changes in external loads.

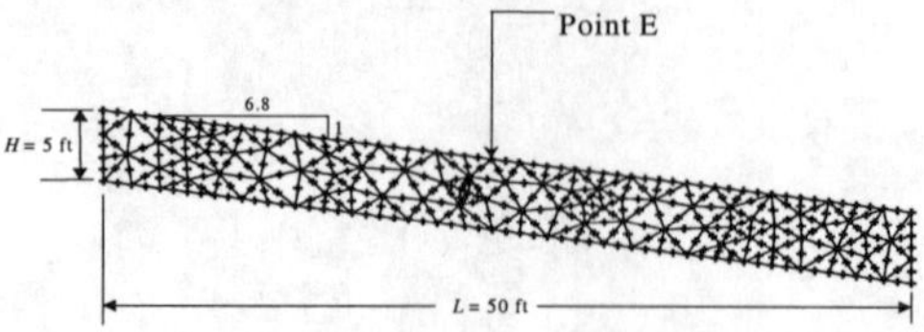

Figure 2. Problem geometry

3 CONSTITUTIVE MODEL AND MATERIAL PARAMETERS

Values of unit weight, cohesion and friction angle are taken from published United States Geological Survey reports (Baum and Reid, 1992). An average saturated unit weight of 138 pcf, an average residual cohesion of 110.2 psf, and an average residual friction angle of 10.2^0 are adopted (Table 1).

Table 1. Index properties

Quantity	Description	Value	Remarks
γ_{sat}	Saturated unit weight	138 lbf/ft^3	
k	Hydraulic conductivity	0.037 ft/d	Same in all direction
c	Cohesion	110.2 psf	$= c'_{residual}$
ϕ	Friction angle	10.2^o	$= \phi'_{residual}$
ψ	Dilatancy angle	0	
λ^*	Modified compression index	0.0703	
κ^*	Modified swelling index	0.0105	= 15% of λ^*
μ^*	Modified creep index	0.0035	= 5% of λ^*

The SSC constitutive model in PLAXIS allows for the elasto-plastic, time-dependent analysis of near-normally consolidated fine-grained soils. The basic volumetric creep equation is given as follows:

$$\dot{\varepsilon}_v{}^t = \frac{\mu^*}{t}\left(\frac{p^{eq}}{p_p^{eq}}\right)^{\frac{\lambda^*-\kappa^*}{\mu^*}} \qquad (1)$$

where:

λ^* = modified compression index (ε_v vs. e)
κ^* = modified swelling index (ε_v vs. e)
μ^* = modified creep index ($\varepsilon_v{}^t$ vs. t')
p^{eq} = final equivalent pressure applied
$p_p{}^{eq}$ = final equivalent isotropic pre-consolidation pressure after creep
$p_{p,0}{}^{eq}$ = initial equivalent isotropic pre-consolidation pressure before loading
t = total time for consolidation and creep
t' = effective creep time
ε_v = total volumetric strain
$\varepsilon_v{}^t$ = volumetric creep strain

Deviatoric creep strains are computed from Equation (1) with the use of the associated flow rule and by assuming p^{eq} as the plastic potential function. Additional details about this model can be found elsewhere (Tsui, 1999; PLAXIS, 1998). Specific parameters selected for the analysis were based on laboratory 1-D consolidation and triaxial. They are listed in Table 1.

4 DISCUSSION OF COMPUTER PREDICTIONS

Figure 3 shows the deformed mesh after 10,000 days. Given that section AA' is approximately 1000 ft long and that the depth of creep is unlikely to reach anything near this length, the geometry of the problem is close to that of a semi-infinite slope. In fact, the deformation pattern in Figure 3 bears this out. After 10,000 days, the mesh reflects shearing caused by the downslope component of gravity and reduction in height due to volumetric compression. The central portion of the slope undergoes nearly uniform changes in stress and displacement, without undue influence from either the left or the right boundaries. Slope lengths of 75 ft and 100 ft were tried as well, but displacements and stresses in the central region, near point E, did not vary much.

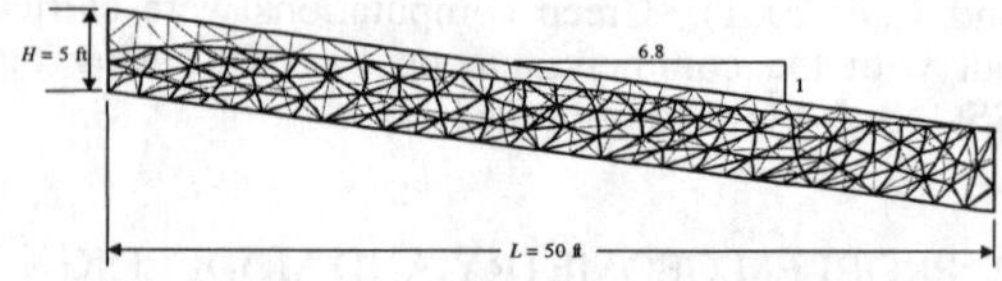

Figure 3. Deformed mesh after t = 10,000 days; displacements are magnified by a factor of 5

However, the same can not be said about the effect of slope height, *H*. A thicker creep zone results in larger surface creep displacements. A number of analyses were conducted using different values of H to evaluate its effect on resulting creep deformations. Because a change in H also affected selection of a representative length, a new appropriate length was calculated for each H so that relative displacements and stresses between points near the center of the slope remained about the same. The variation of surface downslope displacement U_d at point E for slope sections with thicknesses of 2.5 ft, 7.5 ft, 10 ft, 12.5 ft, 15 ft and 17.5 ft are shown in Figure 4. The downslope deformation U_d at point E can be normalized as shown in Figure 5.

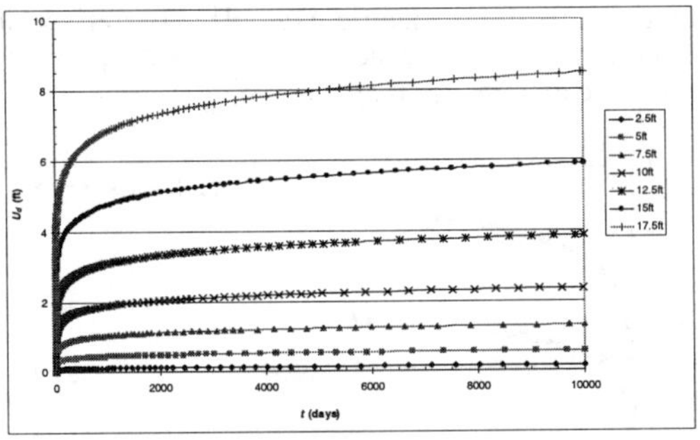

Figure 4. Downslope displacement at point E for different creep depths

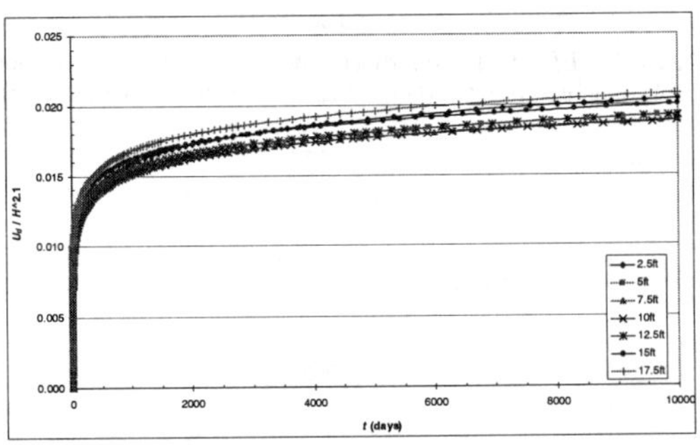

Figure 5. Normalized downslope displacement at point E

Downslope velocity may be a more useful quantity then displacement for estimating past and future deformations due to creep. Normalized velocity as a function of time (Figure 6) can conveniently be expressed by a single relationship:

$$t\, v_d \,/\, H^{2.1} = 0.002 \qquad (2)$$

where:

t = time under sustained loading (in days)
v_d = surface downslope velocity (in ft/day)
H = thickness of slope subjected to creep movement (in ft)

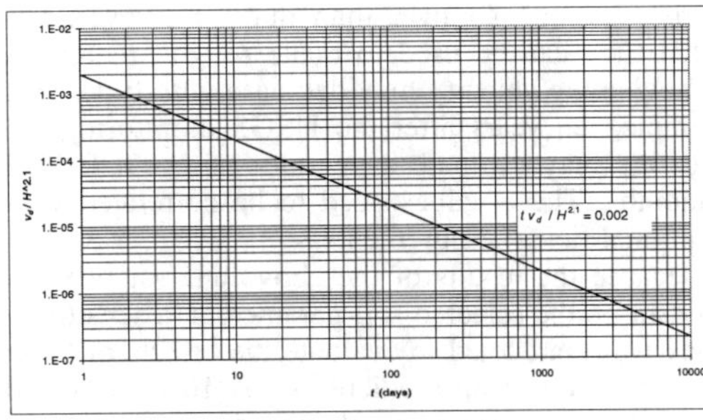

Figure 6. Relationship between surface downslope velocity at point E and effective creep time

Equation (2), in combination with inclinometer measurements, can be used to estimate downslope deformations. However, the start of creep, i.e. $t = 0$, is hard to determine for field slopes. In fact, it is unlikely that such a time even exists for natural slopes that have developed over hundreds or thousands of years. In case of constructed slopes, such an initial time may be more obvious (Brandes et al., 1994). Because the Alani-Paty slope has existed for thousands of years, a small value of t would not be realistic and may lead to overestimating creep deformations. A large value for t is therefore suggested for predictions of future creep movements.

Since the thickness of the Alani-Paty landslide above the inferred failure surface is about 25 ft, a height of 25 ft is a reasonable first choice for the depth to which creep occurs. As an example, downslope velocity v_d can be estimated for two times since the initiation of creep as follows:

Case 1: t = 1,000 days
$v_d = 0.002 \times H^{2.1} / t$
$v_d = 0.002 \times 25^{2.1} / 1000$
v_d = <u>0.0017 ft/day</u> = <u>0.63 ft/year</u> = <u>7.6 in/year</u>

Case 2: t = 100 years
$v_d = 0.002 \times 25^{2.1} / (100 \times 365)$
$v_d = 4.73 \times 10^{-5}$ ft/day
v_d = <u>0.017 ft/year</u> = <u>0.2 in/year</u>

The USGS collected inclinometer measurements during a 20-month period from October 1989 through March of 1991 (Baum and Reid, 1992). During this time, the downslope deformation near the head of the slope attributed to creep was estimated to be about 0.05 ft (Baum et al., 1991). The average creep rate was therefore about 0.0025 ft/month, or 8.33×10^{-5} ft/day. It is interesting to note that by using this value for v_d, along with a height of 25 ft, Equation (2) would give a time from creep initiation of about 57 years. This value of t obviously does not correlate with the actual age of the Alani-Paty slope area. Creep must have been occurring for much longer than that. However, this initial time can be used to make rough predictions on how creep velocity may decrease with time. For example, ten years after the USGS monitoring period one may estimate a velocity of about 0.0022 ft/month. These values need to be correlated with additional field measurements.

Because the results of the above analysis are dependent on the selected soil parameters, a sensitivity study was conducted to evaluate the effects of these parameters on creep predictions. A 10-ft thick and 300-ft long section was selected for this analysis. The results are summarized in Table 2, which indicates what happens if each parameter in turn is increased by 10%. It is apparent that the results are most sensitive to parameters c, ϕ, and μ^*. As with any sensitivity analysis involving soil parameters, the results need to be viewed with caution since it is unreasonable to expect that any of these parameters vary independently of the others.

Table 2. Summary of sensitivity analysis

Parameter in question	Change in parameter	Change in U_d
γ_{sat}	+10%	+4%
k	+10%	0%
c	+10%	-14%
ϕ	+10%	+12%
λ^*	+10%	+1%
κ^*	+10%	0%
μ^*	+10%	+9%

The stability analysis in the companion paper (Brandes and Tsui, 2001) indicated that a groundwater table of at least 15 ft or so below the ground surface is required in order for the slope to be safe. Therefore, creep predictions were carried out using a groundwater table located at an equivalent depth. Because of numerical difficulties encountered when the slope was thicker than about 20 ft, computations were carried out instead on a 10 ft-thick slope with a groundwater elevation at the surface and also at 6 ft below the surface. This depth of 6 ft represents 60% of the slope height, which corresponds to a groundwater depth of 15 ft in the 25-ft slope section used for the stability analysis. The downslope displacement and velocity at point E are shown for both water elevations in Figures 7a and 7b. A 60% drop in the groundwater table results in about a 36% decrease in v_d, independent of the value of H. It is therefore apparent that a lower groundwater table will not only increase slope stability, but it will also reduce creep deformations.

5 CONCLUSIONS

Downslope creep velocity was found to decrease with time according to Equation (2). However, the time for creep initiation is hard to determine for slopes such as the Alani-Paty one which have existed for thousands of years. However, when combined with field measurements, creep modeling can provide estimates of surface displacements as was demonstrated by a specific example. The effect of lowering the groundwater is, as expected, to decrease the amount of creep.

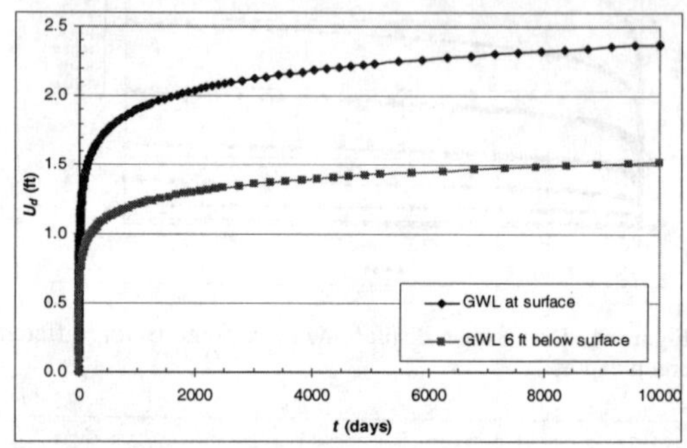

Figure 7a. Effect of groundwater table on surface downslope displacement at point E (H = 10 ft, L = 300 ft, t = 10,000 days)

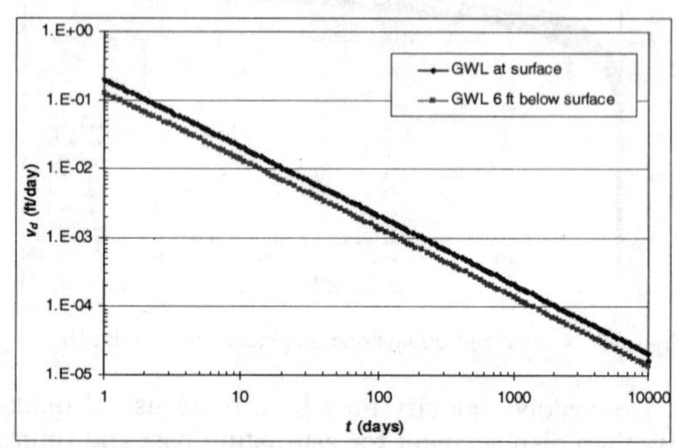

Figure 7b. Effect of groundwater table on surface downslope velocity at point E (H = 10 ft, L = 300 ft, t = 10,000 days)

6 ACKNOWLEDGEMENTS

This study was supported in part by a grant from the Hawaii Department of Transportation and the Federal Highway Administration. Their support is gratefully acknowledged.

7 REFERENCES

Baum, R.L., M.E. Reid, C.A. Wilburn & J.D. Torikai 1991. Summary of Geotechnical and Hydrologic Data Collected from May 1, 1990 through April 30, 1991, for the Alani-Paty Landslide, Manoa Valley, Honolulu, Hawaii. *U.S. Geologic Survey Open-File Report 91-598*: 102.

Baum, R.L. & M.E. Reid 1992. Geology, Hydrology and Mechanics of the Alani-Paty Landslides, Manoa Valley, Oahu, Hawaii. *U.S. Geological Survey Open-File Report 92-501*: 87.

Brandes, H.G. & C.S.L. Tsui 2001. Modeling the Effectiveness of Groundwater-Lowering Remediation Measures in the Alani-Paty Landslide Area, Oahu, Hawaii. *Proceedings of the 10th International Conference on Computer Methods and Advances in Geomechanics*, Tucsion, AZ.

Brandes, H.G., A.J. Silva, M.H. Sadd & G.E. Veyera 1994. Stress-Strain-Time Modeling of Submarine Slopes. *Proceedings of the 8th International Conference on Computer Methods and Advances in Geomechanics*, Morgantown, WV, 3:2435-2445.

PLAXIS, 1998. *PLAXIS Version 7 Users Manual*. R.B.J. Brinkgreve and P.A. Vermeer Eds., Netherlands.

Nakayama, D.D. 2000. Creep Characterization of Tropical Soils. *MS Thesis, University of Hawaii*: 195p.

Tsui, C.S.L. 1999. Modeling the Slope Stability and Creep Behavior of the Alani-Paty Landslide Area, Manoa Valley, Honolulu, Hawaii. *MS Thesis, University of Hawaii*: 159p.

Computer Methods and Advances in Geomechanics, Desai et al. (eds) © 2001 Balkema, Rotterdam, ISBN 90 5809 183 X

3-D adaptive FEM in rock slope stability analysis

H.X.Xia & S.H.Chen
Wuhan University of Hydraulic and Electric Engineering, People's Republic of China

ABSTRACT: In this paper, a 3-D adaptive FEM has been developed for rock slope analysis which includes two essential parts :the mesh generator and the posteriori error estimator. The spline-blended surface interpolation has been used to approximate the irregular surface, and some advises have been pointed out for the error estimation ,too..

1 INTRODUCTION

The adaptive FEM has been widely used in numerical analysis now, because the precision of the calculated results can be improved by the adaptivity, at the same time the cost of pre-processing is reduced greatly. In adaptive FEM, there are two essential ingredients which must be established besides the FEM itself, one is the posteriori error estimator, the other is the mesh generator. Related to error estimator, it has been found that a reliable estimator of the mesh discretization can be achieved by using an interpolation of nodally recovered values, this type of error estimator is referred as Zienkiewicz-Zhu error estimator(O.C.Zienkiewicz & J.Z.Zhu 1992). In particular, it is found that this type of estimator will always be asymptotically exact in the energy norm if the nodally recovery value is superconvergent. Due to the fact that the Zienkiewicz-Zhu error estimator is simple and effective, it is widely used now. Related to the second ingredient, a wide variety of algorithms has been devised for the generation of unstructured grids around complex geometrical shape. The Advancing-front technique(AFT) put forward by Peraire et al is one of these methods(J.Peraire &J.Peiro 1992). Due to the fact that the AFT can resolve the 3D complex domain more successfully than other methods, more attention has been paid to it than to the other algorithm in the adptive FEM.

The 2D adaptive FEM has been implemented in geotechnical engineering(Chen sheng-hong 1998). However, this method is seldom used in the 3D practical rock slope stability analysis because of the complication of the rock slope engineering, this is the research topic which has been investigated in the paper. We use the Spline-blended surface interpolation(William.J.Gordon 1969) to represent the irregular surface, such as the natural ground surface, twisted faults ,which is very common in geotechnical engineering , then the AFT has been used to generate meshes. Considering of the non-homogeneity of material properties ,we propose that the ZZ-error estimation should be proceeded in different material domain, otherwise the new element size can not be determined correctly.

Based on the above improvements, a program system of 3D elasto-viscoplastic adaptive FEM is developed, this program can be used in the stability problem of rock slope with natural irregular ground surface and twisted faults. A engineering example is used to show the robustness of the program.

2 MESH GENERATOR

The AFT requires an initial front in order to start the tetrahedrization of the domain, this means that the surfaces defining the domain to be grided need to be defined, then a modified 2D AFT mesh generator performs triangulation of these surfaces(Peter Moller & Peter Hansbo 1995). It is been known that the quality of the surface discretization has a major impact on the quality of the tetrahedral meshes, which further influence the precision of the calculated results directly.

The definition of curved or irregular surface such as the ground surface of the slope or the large scale faults in the slope is one of the key problem in the generation of 3D FEM meshes of rock slope. There are many methods to represent surfaces in terms of parametric co-ordinates, $X = X(s,t)$, the widely used way is surface patches such as 'coons patches' .For defining these patches, the data of corner point's co-ordinates X, tangents X_s, X_t and the twist X_{st} are needed. As for the twist X_{st} of the irregular surface, it can not be obtained from the survey data directly in geotechnical engineering, so there are some approximated methods, such as 'Adini formular'. Unfortunately, these methods are proved to be not successful, because they frequently lead to some untolerable surface distortion. The other approach is the easiest way, viz $X_{st} = 0$ at all points. But only when sufficiently many points are given or these points are placed along lines of principal curvatures, the results are acceptable, obviously the requirement on survey data can not be satisfied in fact.

In this paper, the spline-blended surface interpolation proposed by W.J.Gordon is used. If the boundary derivation condition of the surface is omitted, in fact, these conditions have no meanings in geotechnical engineering, the surface can be defined only with the co-ordinates of some groups of survey points. There are no special demands on the location of these points, and the number of the points needed is much fewer than the usually' surface patches' methods.

The irregular surface can be represented in terms of parametric co-ordinates,

$$\begin{aligned} x &= x(s,t) \\ y &= y(s,t) \\ z &= z(s,t) \end{aligned} \tag{1}$$

The above three equations can be seen as three surfaces. The Spline-blended surface interpolation will be used in the paper to represent the three surfaces, and then the irregular surface can be modeled.

2.1 The Spline-blended surface interpolation

Let z=z(s,t) be any continuous surface defined over the rectangular domain R:[a,b]*[c,d] in the (s-t) plain.Let $\pi : a = s_1 \le s_2 \le \cdots \le s_n = b,\quad \pi' : c = t_1 \le t_2 \le \cdots t_{n'} = d$ be partitions of the internals I=[a,b] and I'=[c,d]. Consider the interlines of Z and the vertical plains $s = s_i, t = t_j$:

$$\begin{aligned} f_j(s) &= z(s,t_j) \quad (j = 1,n') \\ g_i(t) &= z(s_i,t) \quad (i = 1,n) \end{aligned} \tag{2}$$

These n+n' univariate functions given a net work of curves. There are two sets of splines of odd degree $2p-1 : \{\phi_i(s)\}, \{\varphi_j(t)$ which satisfy the cardinal conditions

$$\phi_i(s_k) = \delta_{ik} = \begin{cases} 0 & if \quad i \ne k \\ 1 & if \quad i = k \end{cases}$$

$$(i,k = 1,m_1 \quad m_1 = n + 2p - 2) \tag{3.1}$$

$$\varphi_j(t_k) = \delta_{jk} = \begin{cases} 0 & if \quad j \ne k \\ 1 & if \quad j = k \end{cases}$$

$$(j,k = 1,m_2 \quad m_2 = n' + 2p - 2) \tag{3.2}$$

The method uses two sets of splines as blending-functions to interpolate the surface z(s,t).

$$z(s,t) = \sum_{i=1}^{m_1} L_i(z)\phi_i(s) + \sum_{j=1}^{m_2} M_j(z)\varphi_j(t)$$

$$- \sum_{i=1}^{m_1}\sum_{j=1}^{m_2} L_i M_j(z)\phi_i(s)\varphi_j(t) \tag{4}$$

For a given $p > 1$, in addition to satisfing the curve network equation(2),It's also satisfied with specified boundary conditions on the first $p-1$ normal derivatives of the surface. Blending with spline functions of degree $2p-1$ results in a surface which is everywhere of class C^{2P-2}.

2.2 The application of spline-blended surface interpolator

Now we can discuss how to represent the irregular surface with spline-blended surface interpolation. If we let $p = 2$ then the spline-blended surface will be class C^2, which will meet the requirement of the mesh generator. If the boundary derivation condition of the surface is omitted,

$$\begin{aligned} z_s(s_1,t) &= 0 \\ z_s(s_n,t) &= 0 \\ z_t(s,t_1) &= 0 \\ z_t(s,t_{n'}) &= 0 \end{aligned} \tag{5}$$

Then the interpolation equation(6) can be derived from equation(4) as follows:

$$z(s,t) = \sum_{i=1}^{n} \phi_i(s) g_i(t) + \sum_{j=1}^{n'} \varphi_j(t) f_j(s)$$

$$- \sum_{i=1}^{n}\sum_{j=1}^{n'} \phi_i(s)\varphi_j(t) z(s_i,t_j) \tag{6}$$

The two sets of splines of odd degree $\{\phi_i(s)\}, \{\varphi_j(t)$ can be computed directly. If we known the co-ordinates of some points, these points is on the networks of the irregular surface, the n+n' univariate functions $f_j(s), g_i(t)$ can be interpolated (the parametric spline curves of 3 degree is used in our programming(P.M.Prenter 1975)).

Using equation(6), the irregular surface which is appeared in geotechnical engineering can be approached. Obviously, the surface will have some distortion near boundary, but compared with the representation of complex geometric shapes in CAD and numerically controlled machine applications, these error can be accepted in geotechnical engineering.

The surface we have now is defined over the rectangular domain R:[a,b]*[c,d]in parametric plain (s-t). In fact, the surface is only an auxiliary surface. The surface we really interested is only a part of the auxiliary surface. For example, if we want to approximate a chamber in rock mass, the crown surface of the chamber can be formed by intersecting between surface (a"b"c"d") and surface F_1, F_2 (In fact, the a"b"c"d" and F_1, F_2 can be any shape,here is only a simple example).In Figuel and Figue2 the real interested part (the crown surface) is shadowed, From Figure2, it can been seen if a proper auxiliary surface is chosen, the distortion near the boundary which is caused by equation(5) will not influence the part which is the real interests in the mesh generator.

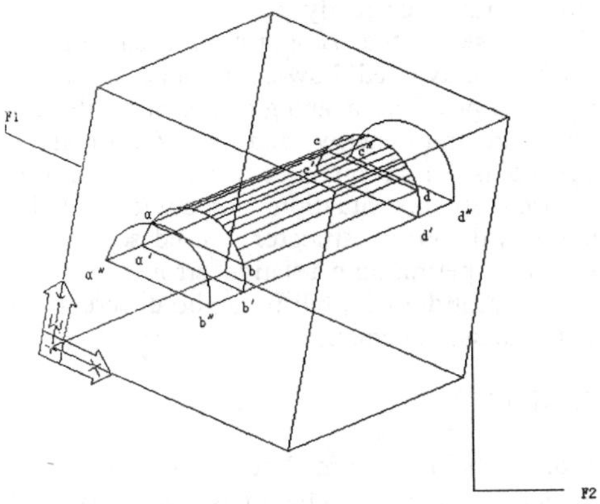

Figure1.The chamber in rock

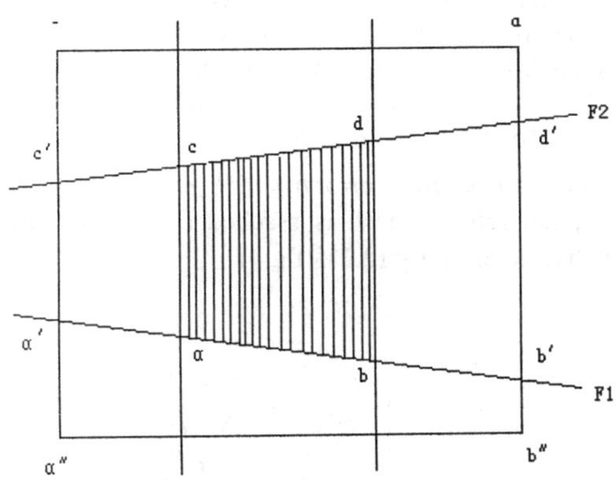

Figure2. The parameter surface s-t

The another question which will be discussed is the parametric surface /surface intersection. From the above example, it can be found that the intersection between faces should be used in the approximation of irregular surface, and if the intersection between faults is in consideration, which is very common in geotechnical engineering, the question becomes very complicated. The surface/surface intersection (SSI) is an essential problem in CAD/CAM, which will not be investigated deeply here. In the programming, a reliable and simple algorithms is adapted, and it is found to be successfully(Yu zheng-sheng & Peng qun-sheng 1999). In the algorithms, one face has the parametric formula, the other is the implicit form, the parametric formula is inserted into the implicit form to get the nonlinear equation, Then the problem for solving the equation can be considered as extracting isocurve in two dimensional scalar field, the intersection curve which we want to represent is the mapping to parametric surface (s-t) of the isocurve, the coordinates of the points on the intersection can be caculated by equation(6)

3 ERROR ESTIMATION

Error is defined as the difference between the exact solution and the caculated solution, the stress error is defined as

$$e_\sigma = \sigma - \sigma^h \tag{7}$$

The point-wise definitions of errors are difficult to specify, so various integral measures are used, There are two most common measures: the so-called L_2-norm and the energy norm.

$$\|e_\sigma\|_E = \left(\int_\Omega e_\sigma{}^T c^{-1} e_\sigma dv\right)^{\frac{1}{2}} \tag{8}$$

$$\|e_\sigma\|_{L_2} = \left(\int_\Omega e_\sigma{}^T e_\sigma dv\right)^{\frac{1}{2}} \tag{9}$$

The equations (8)-(9) are defined on the whole domain Ω, but, the square of each one can be obtained by summing element contributions

$$\|e\|^2 = \sum_{i=1}^{n} \|e\|_i^2 \tag{10}$$

Where $\|e\|_i$ represents the contribution from element i and is obtained by integrating over the element subdomain, n is the total element number. The estimated relative error is

$$\eta = \frac{\|e\|}{\|u\|} \times 100\% \tag{11}$$

Where $\|u\|$ is the corresponding norm of the exact solution. If the error is uniformly distributed in each element and the maximum permissible relative error η is given, then we have

$$\|e\|_i' = \eta \times \frac{\|u\|}{\sqrt{n}} \quad (12)$$

$$\varsigma_i = \frac{\|e\|_i}{\|e\|_i'} = \frac{\|e\|_i}{\eta \times \|u\| / \sqrt{n}} \quad (13)$$

If $\varsigma_i \le 1$, the optimum mesh has been obtained otherwise the new element size should be calculated according to

$$h_i^{new} = h_i^{old} / \varsigma_i^{\frac{1}{p}} \quad (14)$$

Where p is the order of the shape function.

Usually the exact solution which is used in error estimation can't be obtained. In ZZ-estimator, the error (7) is approximated by the difference between the higher order solution and FE solution

$$e_\sigma^* = \sigma^* - \sigma^h \quad (15)$$

Obviously, the validation of the ZZ-estimator depend on the accuracy of the recovered solution σ^*, there are some methods to calculate the recovered solution σ^*, and recently, a local Least-squares fit approach, super convergent patch recovery (SPR) has proven to be more efficient(O.C.Zienkiewicz & J.Z.Zhu 1992), which is been used in our programming.

In geotechnical engineering, the rock mass isn't homogeneous, and material discontinuity is very common. Rock is often dissected by faults and joints and the material properties vary greatly from different subdomains. For example, the intact rock has hundreds times of elastic modulus compared to the weak fault.

According to the above situations in geotechnical engineering, it is recommend that the error estimation should be worked separately in different material domain. This treatment has at least two advantages over the common method..

It's well known that on the material discontinuity interfaces the stress continuity doesn't exist. If the SPR procedure is used in each material domain separately, the recovered stress field σ^* is continuous in each domain but jump across the domain interface ,such as the fault surface(O.C.Zienkiewicz & J.Z.Zhu 1992), obviously the results will be more reasonable and reliable.

The other advantage for the recommendation is related with the error norm. From equation(8) and equation(9) ,the energy norm differs from the L_2 norm only by a weighting maxim c^{-1}. If we adopt equation(8) to calculate the error, the error of each element can be expressed as :

$$\|e\|_i = f(e_{\sigma_i}, c_i^{-1}) \quad (16)$$

According to Eq(10)and (12), the total error is distributed uniformly in each elements, the $\|e\|_i$ will be equal among different material elements. In terms of equation(13)

$$\varsigma_i = \frac{\|e\|_i}{\|e\|_i'} = \frac{f(e_\sigma, c_i^{-1})}{\|e_i\|} \quad (17)$$

Eq(17) means the element size is determined not only by the accuracy of the stress, but by the material quality c_i^{-1} .If the $|c_i^{-1}|$ is greater than $|e_{\sigma_i}|$, the ς_i will be determined mainly by the material quality. In our experience, if the error estimation is worked in the whole domain, the meshes of the fault and other weak domains will be unnecessary density than the domain of complete rock, which can't represent the stress concentration correctly.

Of course, if the L_2-norm is used, all the problem will be avoided. howerver, since the energy norm has a physical meaning (it measures the absolute error in strain energy) and the ZZ-estimator's convergence is based on energy norm , it's frequently used. In this situation, our advice will be useful. When the error estimator is used separately, the maximum permissible relative error η can be same in different domain, but it can be different according to the user's demand.

4 EXAMPLE

The problem being examined here is a tunnel excavated under a rock slope, which has two intersected faults and an irregular ground surface, and the slope's bottom has been restricted in x-y-z. Inputting the initial mesh size, (for approximating the crown surface of the tunnel, the mesh size is smaller than the other domain) the mesh generator gives the initial backgroud mesh shown in Figue3, its estimated relative error η is 36%. Specifying the maximum permissible relative error $\eta = 15\%$, after two iteration, the refined mesh is reached at the estimated relative error of η=12.3%(fig.4)

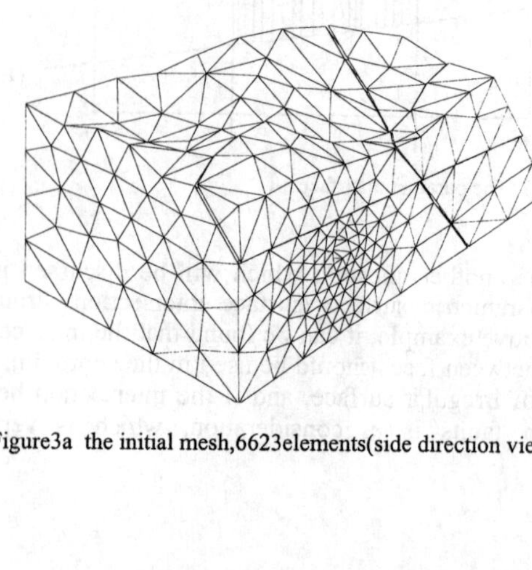

Figure3a the initial mesh,6623elements(side direction view)

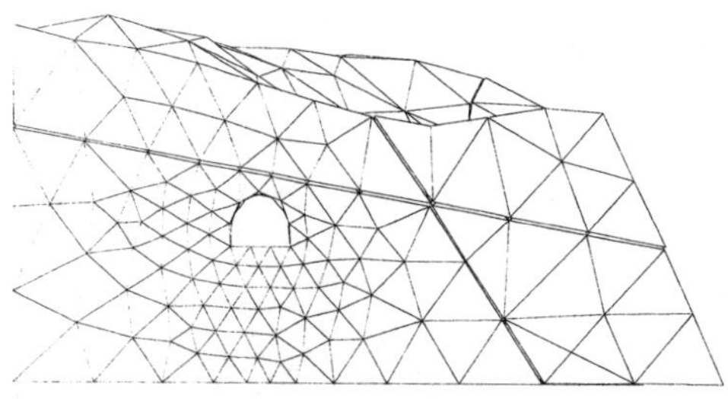

Figure3b the initial mesh,6623elements(front view)

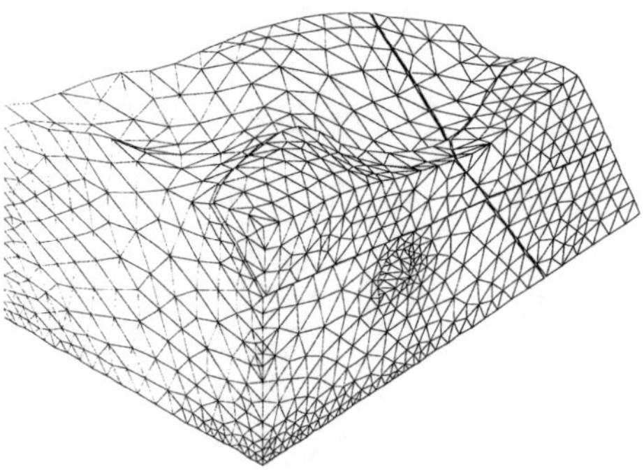

Figure4a the refined mesh,13114elements(side direction view)

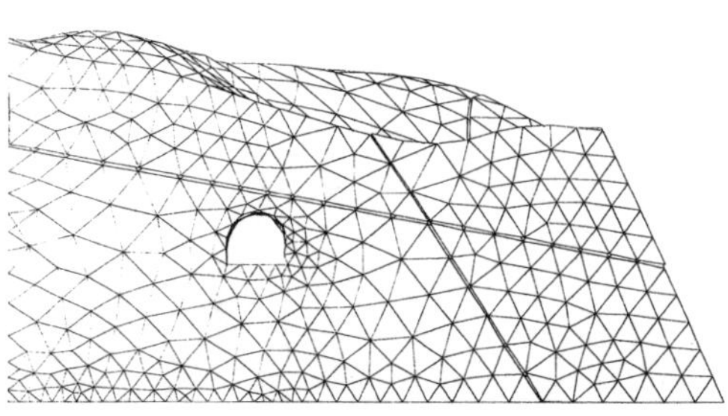

Figure4b the refined mesh,13114elements(front view)

5 CONCLUSIONS

The adaptive FEM illustrated here is only a preliminary results of a series research , the results of the paper showe that the 3-D adaptive FEM could be a powerful tools in the geotechnical engineering. but to achieve the stage where the FEM calculation can be accepted and used more easier by the engineers and the results can be applied in practice more reliably, there are still a lot of works to be done, it is suggested the problem such as the seepage and deformation/stress coupling should be studied in the future.

ACKNOWLEDGMENTS

The work of this paper is mainly supported by the National Science Foundation of China(NSFC) under the grant number 59979021 and the Research Foundation of the Education Ministry of China.

MAIN REFERENCES:

Chen sheng-hong 1998,Elasto-viscoplastic adaptive finite element analysis of the navigation lock in three gorges project, *Rock and soil mechanics*, 19(1):12-19

J.Peraire &J.Peiro 1992, Adaptive remeshing for three-dimensional compressible flow computations, *Journal of computational physics,*103:269-285

O.C.Zienkiewicz & J.Z.Zhu 1992, The superconvergent patch recovery and a posteriori error estimates,Part1 and part2, *Int.j..numer.methods eng.*, 33:1331-1364,1365-1382

O.C.Zienkiewicz & J.Z.Zhu 1992,The superconvergent patch recovery(SPR) and adaptive finite element refinement, *Computer methods in Applied Mechanics and Engineering* 101:207-224

Peter Moller & Peter Hansbo 1995, On advancing front mesh generation in three dimensions, *Int.j..numer.methods eng.*, 38:3551-3569

P.M.Prenter 1975,*Splines and variational methods*, John wiley & Sons

William.J.Gordon 1969,Spline-blended surface interpolation through curve net works , *J.math.mech.*,18,931-952

Yu zheng-sheng, Peng qun-sheng,&Ma li-zhuang 1999,An algorithm for intersection between parametric surface and implicit surface ,*Journal of computer aided design and computer graphics,*11(2):97-99

Computer Methods and Advances in Geomechanics, Desai et al. (eds) © 2001 Balkema, Rotterdam, ISBN 90 5809 183 X

The improved load-distribution method of arch dam considering the interaction between foundation and dam exactly

M.Y.Xu & S.H.Chen
Wuhan University of Hydraulic and Electric Engineering, People's Republic of China

ABSTRACT: This paper presents a combination technique to take advantage of load-distribution method and block analysis theory. The arch dam is considered as arches and cantilevers system and the foundation as block elements system. After supplied with border continuous condition of displacements, the governing equation including arch dam and foundation is established. The coupling analysis is more accurate than the traditional trial load method with F.Vogt assumption and can be used in engineering practices with complicated geology condition.

1 INTRODUCTION

The traditional analysis method of arch dam is trial load method till 1930's that is put forward by the United States Department of the Interior Bureau of Relamation. The method analyzes stress in arch dam according to load distribution of arches and cantilevers. It calculates stress rather exactly in arch dams, but deal with the foundation using F.Vogt assumption that the foundation is homogeneous, half-infinite and elastic. This simplicity is unsatisfied in engineering with complicated geology condition. An improved method is proposed in this paper, in which the foundation will be simulated with finite elements or block elements to adapt various materials and discontinuity system. At the same time, the arch dam will be divided into arches and cantilevers in orthodoxy. To each arch-cantilever element the stiffness equation can be supplied, which relates displacements and forces of nodes. On the contact surface between dam and foundation, the displacements of dam element and foundation element must be the same. With these considerations, the equation is established to solve the displacement and stress of dam as well as the foundation. It requires little computer storage and time in comparison with the conventional trial load method, more important, the method accords with relevant design standard and code nowadays in China. A typical example is studied from which the effectiveness of the proposed method is proved.

2 STIFFNESS OF ARCH AND CANTILEVER ELEMENTS

The whole arch dam will be divided into many arch-cantilever elements by arch and cantilever direction. In every arch-cantilever element, which is showed in Figure 1, all parameter is simplified to middle face. Every angle node has six displacements at global coordinate system. So total degrees of an arch-cantilever element are 24 (4 nodes × 6). Arch direction is expressed as v_a , and cantilever direction is expressed as v_b . So the local coordinates system is obtained with the normal direction defined as $v_r = v_a \times v_b$. At this local coordinate system, every node has three line displacements and two rotate displacements (according to shell theory, normal rotation needn't be considered). The total local degrees are 20.

In every arch-cantilever element many arch and cantilever slices should be carved along arch direction and cantilever direction. Then the equivalent ar-

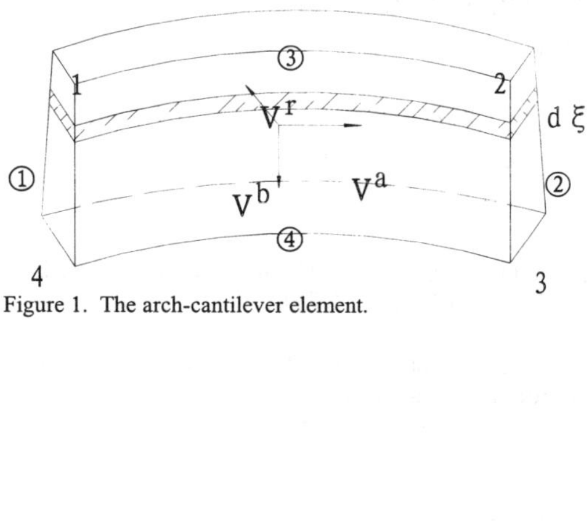

Figure 1. The arch-cantilever element.

ch and cantilever are got after integrating the slices. For example, to an arch slice, the displacements of two end nodes are supposed to be linear interval value of those of neighboring corner nodes:

$$\delta_a'(\xi)=\begin{bmatrix}1-\xi & 0 & 0 & \xi\\ 0 & 1-\xi & \xi & 0\end{bmatrix}\begin{Bmatrix}u_{a1}'\\ u_{a2}'\\ u_{a3}'\\ u_{a4}'\end{Bmatrix}=H_a\delta_a' \tag{1}$$

Where $\delta_a'(\xi)$ is the displacements of two end nodes of the arch slice, δ_a' is the displacement of corner nodes at arch local coordinates system. According to beam stiffness equation:

$$f_a'(\xi)=K_a'(\xi)\delta_a'(\xi) \tag{2}$$

We have

$$H_a^T f_a'(\xi)=H_a^T K_a'(\xi)\delta_a'(\xi)=H_a^T K_a'(\xi)H_a\delta_a' \tag{3}$$

where H_a is a transformation matrix defined as:

$$H_a=\begin{bmatrix}1-\xi & 0 & 0 & \xi\\ 0 & 1-\xi & \xi & 0\end{bmatrix} \tag{4}$$

Equation 3 can be integrated into:

$$\int H_a^T f_a'(\xi)d\xi=\int H_a^T K_a'(\xi)H_a\delta_a' d\xi \tag{5}$$

or

$$f_a'=K_a'\delta_a' \tag{6}$$

in which K_a' is the stiffness of the equivalent arch:

$$K_a'=\int H_a^T K_a'(\xi)H_a d\xi \tag{7}$$

By the same way, cantilever stiffness can be defined as:

$$f_c'=K_c'\delta_c' \tag{8}$$

$$K_c'=\int H_c^T K_c'(\eta)H_c d\eta \tag{9}$$

where

$$H_c=\begin{bmatrix}\eta & 1-\eta & 0 & 0\\ 0 & 0 & 1-\eta & \eta\end{bmatrix} \tag{10}$$

3 EQUILIBRIUM EQUATION OF THE ARCH-CANTILEVER SYSTEM

The stiffness matrix must be transformed from local coordinates system to global coordinates system. The displacements of corner nodes at local coordinates and global coordinates are related as

$$\delta_a'=T_a\delta \tag{11}$$

$$\delta_c'=T_c\delta \tag{12}$$

where T_a and T_b are the transformation matrices from the global coordinates to the local coordinates of the arch and cantilever respectively:

$$T_a=[-v_b \quad v_a \quad v_r \quad -v_b \quad v_a]^T \tag{13}$$

$$T_c=[v_a \quad v_b \quad v_r \quad v_a \quad v_b]^T \tag{14}$$

In equation 6, 8, f_a' and f_c' are forces along the arch and cantilever local coordinates system, they must be transformed to the global coordinate system. Take the arch for example, according to virtual work principle,

$$\delta^T f_a=\delta_a'^T f_a'=(T_a\delta)^T f_a'=\delta^T T_a^T f_a' \tag{15}$$

we have

$$f_a=T_a^T f_a' \tag{16}$$

The same for the cantilever is:

$$f_b=T_b^T f_b' \tag{17}$$

Because of load-distribution theory, the loads distributed to arch direction and cantilever direction should be added up:

$$\begin{aligned}f_{ac}&=f_a+f_c=T_a^T f_a'+T_c^T f_c'\\ &=T_a^T K_a'T_a\delta+T_c^T K_c'T_c\delta\\ &=(T_a^T K_a'T_a+T_c^T K_c'T_c)\delta=K_{ac}\delta\end{aligned} \tag{18}$$

where

$$\begin{aligned}K_{ac}&=T_a^T K_a'T_a+T_c^T K_c'T_c\\ &=\int T_a^T H_a^T K_a'(\xi)H_aT_a d\xi+\int T_c^T H_c^T K_c'(\eta)H_cT_c d\eta\end{aligned} \tag{19}$$

$$\begin{aligned}f_{ac}&=T_a^T f_a'+T_c^T f_c'\\ &=\int T_a^T H_a^T f_a'(\xi)d\xi+\int T_c^T H_c^T f_c'(\eta)d\eta\end{aligned} \tag{20}$$

4 COUPLING OF ARCH-CANTILEVER ELEMENT AND BLOCK ELEMENT

To consider the coupling effect of the arch-cantilever element of dam and the block element of foundation, their displacements must be related in some degree. The displacements of the block of the foundation that contacts to the dam are (Chen 1994):

$$f_b = K_b \delta_b \tag{21}$$

The relation of arch displacements and block displacements is defined as:

$$\delta_a = T\delta_b \tag{22}$$

where δ_a is the displacements of four nodes of the arch-cantilever element,

$$T = \begin{bmatrix} T_1 & & & \\ & T_2 & & \\ & & T_3 & \\ & & & T_4 \end{bmatrix} \tag{23}$$

Since the nodes of the arch-cantilever element are fixed on the block element, every transformation matrix can be written as:

$$\delta_i = \begin{Bmatrix} u_i \\ \theta_i \end{Bmatrix} = T_{ij}\delta_j = \begin{bmatrix} I & A \\ 0 & I \end{bmatrix} \begin{Bmatrix} u_j \\ \theta_j \end{Bmatrix} \tag{24}$$

where:

$$A = \begin{bmatrix} 0 & z_i - z_j & -(y_i - y_j) \\ -(z_i - z_j) & 0 & x_i - x_j \\ y_i - y_j & -(x_i - x_j) & 0 \end{bmatrix} \tag{25}$$

According to virtual work principle, it can be obtained

$$f_b = T^T f_a = T^T K_a \delta_a = T^T K_a T \delta_b \tag{26}$$

Compared with Equation 21, we have

$$K_b = T^T K_a T = \begin{bmatrix} T_1^T K_{11} T_1 & T_1^T K_{12} T_2 & T_1^T K_{13} T_3 & T_1^T K_{14} T_4 \\ T_2^T K_{21} T_1 & T_2^T K_{22} T_2 & T_2^T K_{23} T_3 & T_2^T K_{24} T_4 \\ T_3^T K_{31} T_1 & T_3^T K_{32} T_2 & T_3^T K_{33} T_3 & T_3^T K_{34} T_4 \\ T_4^T K_{41} T_1 & T_4^T K_{42} T_2 & T_4^T K_{43} T_3 & T_4^T K_{44} T_4 \end{bmatrix} \tag{27}$$

and

$$f_b = T^T f_a = \begin{bmatrix} T_1^T f_1 & T_2^T f_2 & T_3^T f_3 & T_4^T f_4 \end{bmatrix}^T \tag{28}$$

If a node not fixed on the block, its transformation matrix is unit matrix. The coupling analysis of dam and basement will be applied when all stiffness matrices are assembled.

5 ENGINEERING APPLICATION

The Xiaowan arch dam is a parabola double curvature arch dam that is 292m high. The elevation of the dam top is 1245m and the normal water level is 1240m. The dam basement is cranny rock. There are several NWW three-rank faults and three sets of discontinuities in the foundation. According to the arch dam's layout, 22 faults and 4 eroded zones that influence the displacement and stability of the abutment obviously are stimulated in the analysis, besides, 10 unloading cranny zones are also considered. With the method mentioned in references (Wang 1998), the foundation of the dam is discretized into block system. Meanwhile the arch-cantilever system of the dam is also formed. The whole system includes 2231 block elements and 88 arch-cantilever elements (9 arch layers and 17 cantilever columns).

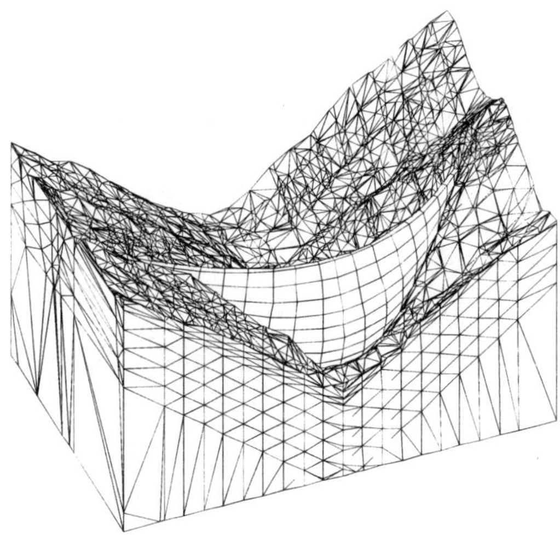

Figure 2. Axonometric perspective of the block (foundation) and arch-cantilever (dam) system

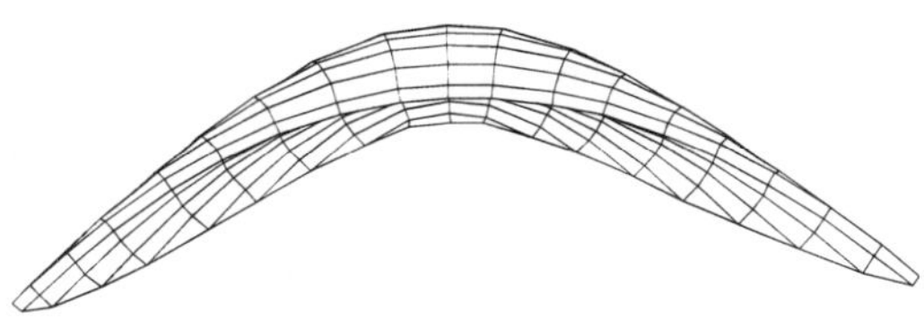

Figure 3. The planform of Xiaowan arch dam

The calculation is carried out for the normal load combination of deadweight + normal water level (▽1240m) + downstream water level (▽1004m) + soil pressure (▽1097m) + temperature drop + seep. Figure 4a, 4b showed the principal stress of upstream and downstream surfaces. The minimum and maximum stresses are -16.07Mpa and 3.84Mpa, which occur at the base of the crown cantilever downstream and upstream respectively. The maximum displacement of left abutment is about 0.027m at the elevation of 1090 ~ 1010m, and that of the right abutment is 0.020m on the same level (Fig. 5). The maximum downstream displacement of arch dam appears on the top of crest cantilever, which is about 0.1544m.

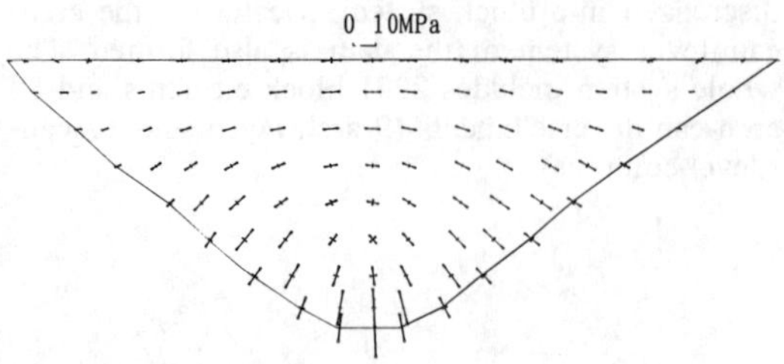

Figure 4a. The principal stress of downstream surface

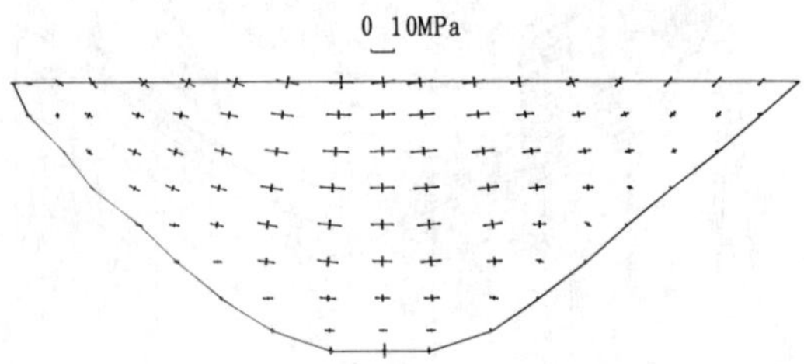

Figure 4b. The principal stress of upstream surface

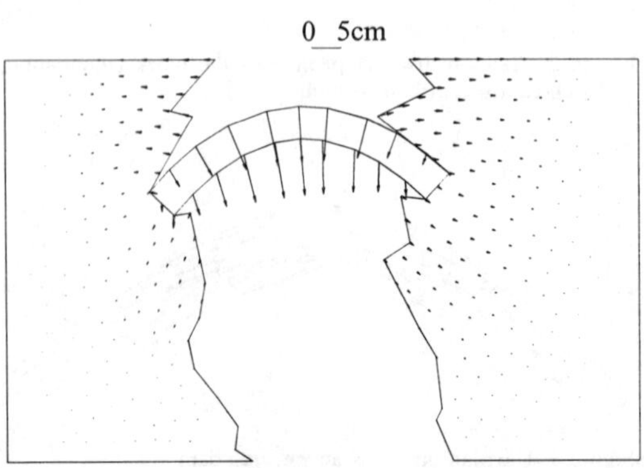

Figure 5. The displacements of the arch dam and the foundation at the elevation of 1090m.

6 CONCLUSIONS

The improved load-distribution method is deduced from the traditional load-distribution assumption, and is convenient to analyze the arch dam and the foundation together using other numerical method at the same time. So the foundation can be simulated by FEM or block theory to consider various rock property and weak discontinuity. The result indicates that the proposed method is reasonable and higher accuracy compared to the conventional method with the assumption of F.Vogt foundation. Besides, it has obvious merit in optimization design of the arch dam because of its little calculation time. This method can be widely used not only in arch dam analysis but also in common thin shell structure design.

ACKNOWLEDGEMENT

The support of the National Science Foundation of China (NSFC) under contract No. 59979021 is gratefully appreciated.

REFERENCES

Chen, S.H. 1994. A new development in elastic-viscoplastic block theory of rock masses. *Computer Methods and Advances in Geomechanics:* 2235-2240. West Virginia (USA): A.A.Balkema.

Huang, W. X. 1984. *The problems of structural mechanics and rock mechanics in hydraulic engineering*. Press of hydraulic and electric engineering of China.

Lin, S. Z. & Z. H. Yang 1987. Load distribution-stiffness method for stress analysis of arch dams. *J. Hydraulic Engineering (in chinese)*. 1: 17-25.

Suresh, P. & R. Natarajan 1981. Analysis of laminated composite shell structures by finite element method. *Computers & Structures*. 14(3-4): 225-230.

Wang, W. M. & S. H. Chen 1998. Automatic identification method for three-dimensional rock block systems. *J. Wuhan University of hydraulic and electric engineering (in chinese)*. 31(5): 51-55.

Zhu, B. F., B. Rao & J. S. Jia 1987, Stress analysis of noncircular arch dam with variable thickness, *J. Hydraulic Engineering (in chinese)*. 11: 17-25.

Computer Methods and Advances in Geomechanics, Desai et al. (eds) © 2001 Balkema, Rotterdam, ISBN 90 5809 183 X

Numerical analysis to earth pressure of retaining walls

Mengxi Zhang
Department of Civil Engineering, Lanzhou Railway University, People's Republic of China
Maoyu Cui
Department of Geotechnical Engineering, Tongji University, Shanghai, People's Republic of China

ABSTRACT: The traditional theories of earth pressure for retaining walls are based on the limit equilibrium method. They incorrectly assume that the earth pressure is independent of the stress state in the soil around the wall. They point out that the earth pressure is closely related to the soil deformation, but not calculate it. In this paper, a numerical analytical method for the calculation of the earth pressure based on stress analysis by finite element method is proposed, which can overcome the weakness of the traditional theories.

1 INTRODUCTION

Whether on fundamental assumptions or in principle, there are some drawbacks in the traditional theories of earth pressure. The primary one is that all the traditional theories neglect the stress state in soil mass around a retaining wall. They also do not give any suggestion about how to calculate the displacement of the wall, though they emphasis the decisive influence of the displacement upon the earth pressure. In addition, they are almost based on the limit equilibrium theory, but they give no evidence that limit state takes place in the fracture surface. It is unreasonable to use the maximum values of c and φ in calculation, corresponding to the peak strength of the soil, for the soil strength will decrease to its residual one as it fails. Finally the traditional theories pay no attention to the difference of earth pressure calculation between the fill retaining wall and the cutting one.

In order to get rid of the above shortcomings of the traditional theories, a new method for the calculation of the earth pressure is proposed in this paper, based on the tracking stress analysis by finite element method (Zhang M.& Chen C. 1991).

2 FORMATION OF EARTH PRESSURE

In study of earth pressure, the immanent cause and the actual process of its formation must be clarified. The relationship between the earth pressure and the stress state as well as that between the earth pressure and the construction sequence also should be made clear.

During construction of a cutting retaining wall, there are two processes, i.e. excavation and wall building. Excavation must cause stress release and shear concentration of the ground. The maximum shear stress usually appears at the bottom corner of the excavation. If the wall is built before excavation, it may be subjected to a deforming earth pressure, because of the wall resistance against the deformation of the ground during excavation. In such case, the wall will play the role of restraining the earth from deformation. On the contrary, if excavation is completed before wall building, fracture surface will inevitably occur because absence of the wall restraint results in to shear overstress in the soil. Thus the wall will be subjected to a sliding earth pressure. Owing to the maximum shear stress at the bottom corner, soil failure takes place there first of all and propagates through the points of shear overstress. The stress release induced by shear overstress promotes further development of soil failure, by which the fracture surface (or plastic band) may be formed. Under such circumstance, the wall is built too late to provide restraint against the soil deformation and will play a role in resisting the landslide.

As for filling retaining wall, it will never be loaded except by sliding earth pressure, because it is difficult to prevent the loose fill from failing. As long as the shear is overstress, there is shear failure. The growth of the shear failure results in failure-sliding surface and then the wall has to be subjected to a sliding earth pressure.

The above discussion only concerns the active earth pressure, which takes places in most cases. In regard to passive earth pressure, it does not easily occur but possibly in case of the fill retaining wall built on soft ground. Under the weight of the fill, the ground surface behind the wall subsides and the wall inclines backward. Therefore, the passive earth pressure will appears at the upper part of the wall back. The passive earth pressure may also be divided into two different types, namely deforming passive earth pressure and sliding passive earth pressure, depended on whether the stress in the fill reaches to its limit state.

Both the active earth pressure and the passive one, it is possible to divide into the deforming pressure and the sliding one. The former represents the pressure produced by soil deformation, while the latter represents the one induced by soil failure. Unfortunately, the traditional theories of earth pressure only study the latter without consideration for the former.

3 THE PRINCIPLE FOR CALCULATION BASED ON STRESS ANALYSIS

The application of finite element method to geotechnics gives a powerful means for the study of a new earth pressure theory based on stress analysis, which will get rid of some old concepts of the traditional theories and makes the calculation of the earth pressure related to stress analysis. On consideration of the close relation between the earth pressure and the stress state as stated above, a new method for calculation of earth pressure is proposed in this paper in accordance with the following fundamental concepts.

(1) The stress state and the deformation in the soil behind the retaining wall are the most important factors that cause the formation of the earth pressure.

(2) As the stress in the soil behind the wall does not reach to the limit state in the case of 'soil-pushing-wall', the retaining wall has to be subjected to an active deforming earth pressure that can be obtained by finite element analysis.

(3) Shear overstress that means the shear stress exceeding its shear strength of soil is the only reason of shear failure. As the shear failure in the soil behind the wall makes a fracture surface (or plastic band) in the case of 'soil-pushing-wall', the wall will be during the action of an active sliding earth pressure.

(4) As there is backward-directed displacement of the wall in the stress analysis, a passive earth pressure will appear on the wall back. If the back displacement makes the soil failure, a passive sliding earth pressure will be induced, otherwise, a passive deforming one will occur, which is proportional to the back displacement.

(5) The earth pressure activity on the wall back is usually the sum of the active sliding one and the passive one. The landslide induces the former and the back displacement conduces the latter. By means of the finite element tracking analysis, the fracture surface and the back displacement of the wall may be obtained. Calculations for the active sliding earth pressure and the passive earth pressure are briefly introduced as follows.

4 ACTIVE SLIDING EARTH PRESSURE

At first, finite element tracking analysis is used to search for the fracture surface. Then, active sliding earth pressure can be calculated by the static equilibrium.

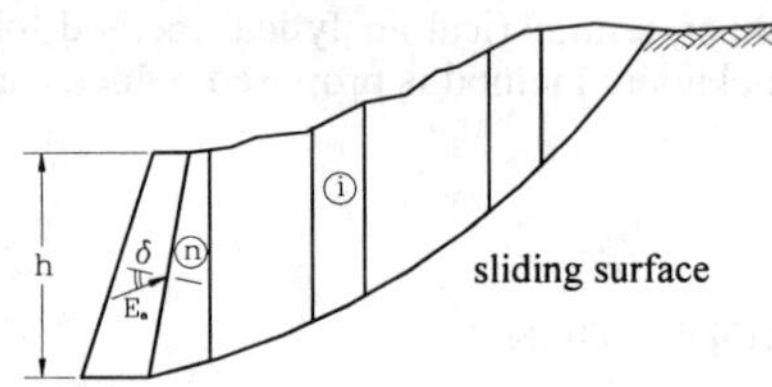

Figure 1. Slice method for irregular sliding surface and sliding body obtained by the tracking analysis.

Figure 1 shows the sliding surface and the sliding body obtained by the tracking analysis. Each slice, such as shown in Figure 2a, is acted on by its weight W_i, by force P_i between any two neighbouring slices and by a set of forces on its base including the shearing force T_i and the normal force N_i. Firstly, force between any two neighbouring slices P_i is calculated, and then the active sliding earth pressure E_a can be obtained. The force between any two neighbouring slices, P_i, whose direction is assumed to parallel the sliding surface of the slice i, can be obtained according to the static equilibrium of slice i.

$$P_i = W_i \sin\alpha_i - W_i \cos\alpha_i \tan\varphi_i - c_i l_i + P_{i-1}\psi_i \quad (1)$$

where the coefficient Ψ_i is given by

$$\psi_i = \cos(\alpha_{i-1} - \alpha_i) - \sin(\alpha_{i-1} - \alpha_i)\tan\varphi_i$$

Where W_i= gravity of soil of slice i; and φ_i=angle of friction of soil. P_i, α_i , W_i and l_i may be shown in Figure 2.

Furthermore, let's consider the static equilibrium of the last slice, as shown in Fig. 2(b), the active sliding earth pressure E_a can be calculated.

$$E_a = \frac{P_n \sin\alpha_n \tan(\varphi_n - \alpha_n) - P_n \cos\alpha_n + W_n \tan(\varphi_n - \alpha_n)}{\sin\delta \tan(\varphi_n - \alpha_n) - \cos\delta} \quad (2)$$

where δ= angle of friction between wall-back and soil.

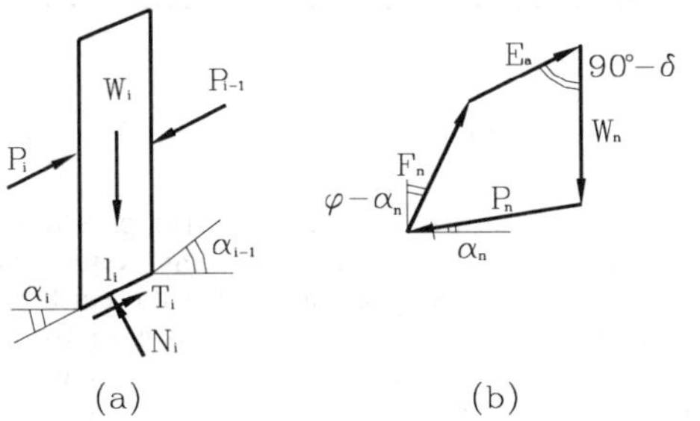

Figure 2. Forces acted on slice *i* and the static equilibrium of slice *n*.

5 PASSIVE EARTH PRESSURE

5.1 *Deformation passive earth pressure*

In the case of 'wall-pushing-soil', the resistance to deformation of the earth depends upon the displacement of the wall. It may be assumed that the passive earth pressure per unit area (σ_p) is proportional to its displacement (u), according to the assumption of Winkler.

$$\sigma_p = k_p \cdot u$$

where k_p = the coefficient of the elastic resistance of the earth, that may be obtained by field experiment. It will reach to its peak value in the limit equilibrium state of the soil, and u may be given by the stress analysis.

The distribution of passive earth pressure is corresponding to displacement situation. Various displacement situations and distributions of passive earth pressure have been presented by authors (Zhang M. et al.1992).

5.2 *Limit passive earth pressure*

Like the calculation for active sliding earth pressure, the limit passive one can be obtained as follows.

The force between any two neighbouring slices P_i can be calculated by the static equilibrium of slice i.

$$P_i = W_i \sin\alpha_i + W_i \cos\alpha_i \tan\varphi_i + c_i l_i - P_{i-1}\psi_i \quad (3)$$

where the coefficient Ψ_i is given by

$$\psi_i = \cos(\alpha_{i-1} - \alpha_i) + \sin(\alpha_{i-1} - \alpha_i)\tan\varphi_i$$

According to the static equilibrium of last slice, as shown in Figure 3, the limit passive earth pressure E_p can be calculated.

$$E_p = \frac{P_n \sin\alpha_n \tan(\varphi_n + \alpha_n) + P_n \cos\alpha_n + W_n \tan(\varphi_n + \alpha_n)}{\cos\delta - \sin\delta \tan(\varphi_n + \alpha_n)} \quad (4)$$

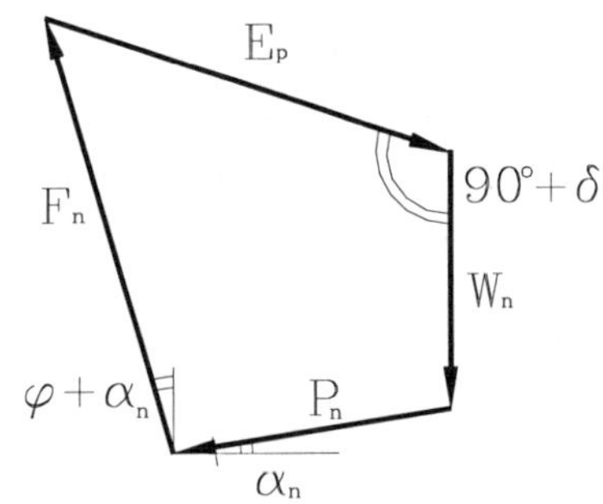

Figure 3. Static equilibrium of slice *n* about limit passive earth pressure.

6 EXAMPLE

Figure 4 shows the results of a fill retaining wall acted by lateral pressure by means of finite element method. The parameters used in analysis are given in Table 1. It presents case of passive earth pressure. The results show that when the lateral displacement of wall is small, there is no failure in soil. If the failure surface appears, there exists limit earth pressure. Meanwhile the failure body and failure surface can be obtained, as shown in figure 4b. The distribution maximum principal stress (figure 4c) shows that the overstress appears at the bottom corner and the soil behind the wall.

According to the location of fracture surface by the finite element tracking analysis as well as above formula, the earth pressure can be easily obtained.

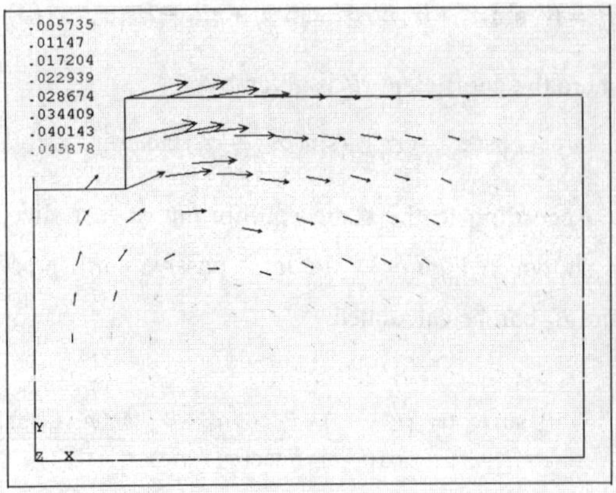

(a) Displacement vectors.

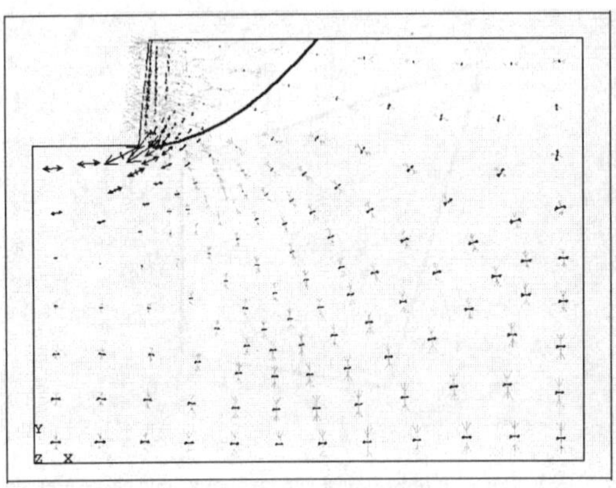

(b) Principal stress vectors.

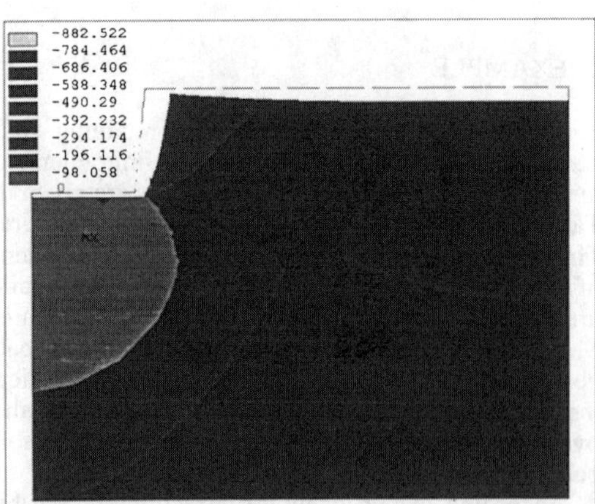

(c) The distribution maximum principal stress σ_1 (kPa).

Figure 4. Results of a fill retaining wall.

Table 1. The parameters of materials.

Property	Value
Unit weight of soil	16.5 kN/m^3
Unit weight of wall	24.0 kN/m^3
Cohesion	15.0 kPa
φ	21.0°
K	160.0
n	0.65
R_f	0.85
Poisson's ratio	0.31

7 CONCLUSIONS

1. The traditional theories of earth pressure for retaining walls, based on the limit equilibrium method, are incorrectly assumed that the earth pressure is independent of the stress state in the soil around the wall. Furthermore, it is not calculated, although the soil deformation is emphasized.

2. The proposed numerical analytical method to calculate the earth pressure, based on stress analysis by finite element method, can overcome the weakness of the traditional theories.

REFERENCE

Clough G. W.& Duncan J.M. 1972. Finite element analysis of retaining wall behavior. *Proc. ASCE 97 (SM12):*1659-1673.

Fang Y.S., Chen T.J. & Wu B.F. 1994. Passive earth pressure with various wall movements. *Journal of Geotechnical Engineering* 21(8): 1307-1323.

Jiang, G.L.& Magnan J.P. 1997. Stability analysis of embankments: comprison of limit analysis with methods of slices. *Geotechnique* 47(4): 857-872.

Pott D.M., Kovacevic N.& Vaughan P.R. 1997. Delayed collaps of cut slopes in stiff clay. *Geotechnique* 47(5): 953-981.

Wang Y.Z. 2000. Distribution of earth pressure on a retaining wall. *Geotechnique* 50(1): 83-88.

Zhang M. & Chen C. 1991. The studies on the stabilization for slopes by tracking analysis with the finite element method. *Chinese Journal of Geotechnical Engineering* 13(6): 35-41.

Zhang M. et al. 1992. Further research on earth pressure of retaining walls. *Chinese Journal of Engineering Mechanics* 9(4): 131-139.

Computer Methods and Advances in Geomechanics, Desai et al. (eds) © 2001 Balkema, Rotterdam, ISBN 90 5809 183 X

Numerical simulations for dam constructions

Thomas Zimmermann & Stéphane Commend
Laboratory of Structural and Continuum Mechanics (LSC), DGC-EPFL, Lausanne, Switzerland

Andrzej Truty
Environmental Engineering Department, Cracow University of Technology, Poland

Jean-Luc Sarf
Stucky Ingénieurs-Conseils SA, Renens, Switzerland

ABSTRACT: In this paper we discuss practical aspects of 3D numerical analyses in relation with dam construction. In addition we propose a novel approach to overcome numerical instabilities in numerical simulation of consolidation problems in saturated two-phase media. Three applications illustrate, the first one a case of consolidation during an earth-dam construction, the second one a safety analysis during construction steps of an arch-dam, and the third one a deformation analysis under thermal loads.

1 INTRODUCTION

Personal computers have reached today a state of performance, which allows three-dimensional analyses to be performed systematically. User-friendly programs, like Z_Soil (Z_Soil 1985-2000) are available, which allow parametric studies to be performed in a matter of hours, while taking into account initial state, construction steps, aging, temperature, transient free surface flow, to mention just a few aspects. While more refined analyses are being undertaken new difficulties appear, which must be taken care of. We dicuss, in this paper, numerical aspects of the problem of two-phase consolidation illustrated by a simulation of an earth-dam construction. Two additional case studies illustrate other aspects of numerical simulations performed with Z_Soil. Recent work performed at LSC-EPFL (Truty et al. 1997) shows that stabilized approaches, already extensively used in fluid mechanics, can be appropriate to overcome locking phenomena in numerical simulation of incompressible or dilatant media. At the same time, numerical pressure instabilities which are known to occur in simulation of consolidation problems when small time steps are necessary can be overcome similarly (Truty 1999).

Finite element solutions of consolidation problems can exhibit oscillating pore pressures, which tend to increase when the time steps are reduced. With classical formulations, a lower limit for the time step has to be enforced (Vermeer et al. 1981), and this can be restrictive depending on the type of analysis which is being carried out. One aim of stabilized approaches described in this report is therefore to eliminate spurious oscillations associated with small time steps in consolidation problems.

2 GOVERNING EQUATIONS

2.1 *Saturated two-phase medium*

The following set of equations applies,

• overall equilibrium in terms of the total stress:

$$\sigma^{tot}_{ij,j} + b_i = 0 \tag{1}$$

where σ^{tot}_{ij} is a component of total stresses and b_i is a component of the body loads vector.

• effective stress concept after Terzaghi (tensile stresses are positive):

$$\sigma^{tot}_{ij} = \sigma_{ij} + \delta_{ij} p \tag{2}$$

where σ_{ij} is a component of the effective stress tensor, p is the fluid pressure, δ_{ij} the Kronecker delta; and $I_1 = \sigma_{kk}$, $J_2 = \frac{1}{2} s_{ij} s_{ij}$, with $s_{ij} = \sigma_{ij} - \frac{\sigma_{kk}}{3}\delta_{ij}$ (with summation over repeated indices), are associated effective stress invariants.

• continuity:

$$\dot{\varepsilon}_{kk} + div\left(v^F\right) - \frac{n}{K^F}\dot{p} = Q^F \tag{3}$$

where $\dot{\varepsilon}_{kk}$ denotes the volumetric strain rate of the solid skeleton, $div(v^F)$ the divergence of the fluid velocity, n the porosity and K^F the fluid bulk modulus.

• an elastoplastic constitutive relation for solid phase expressed in rate form:

$$\dot{\sigma}_{ij} = D^e_{ijkl}\left(\dot{\varepsilon}_{kl} - \dot{\varepsilon}^p_{kl}\right) \text{ with } \varepsilon_{ij} = \frac{1}{2}\left(u_{i,j} + u_{j,i}\right) \tag{4}$$

• Darcy's law :

$$v_i^F = -k_{ij}\left(-\frac{p}{\gamma^F} + z\right)_{,j} \tag{5}$$

• boundary conditions:

tractions on Γ_t: $\sigma_{ij} n_j = \bar{t}$

displacements on Γ_u: $\mathbf{u} = \bar{\mathbf{u}}$, and $\Gamma = \Gamma_t + \Gamma_u$

fluxes on Γ_q: $v_i^F n_i = \bar{q}$

pore pressures on Γ_p: $p = \bar{p}$, and $\Gamma = \Gamma_q + \Gamma_p$

• initial conditions:

$$u_i(t = t_0) = u_{i0}$$
$$p(t = t_0) = p_0$$

Depending on the type of analysis, only equilibrium or equilibrium coupled or uncoupled with continuity will be solved for. Thermal analysis, when applicable, is considered uncoupled and is performed as a preprocessing.

2.2 *Galerkin least-squares (GLS) formulation for two-phase media*

A discretized weak formulation is obtained using a now traditional derivation.

$$\begin{aligned} R_1^f(\mathbf{u}^h, \mathbf{w}^h, p^h) &= \int_\Omega (\mathbf{w}^e)^T \left[\mathbf{B}^T \boldsymbol{\sigma} + \mathbf{B}^T \mathbf{1} \mathbf{N}^p \mathbf{p}^e\right] d\Omega \\ &- \int_\Omega \left(\mathbf{N}^u \mathbf{w}^e\right)^T \mathbf{f} d\Omega - \int_\Gamma \left(\mathbf{N}^u \mathbf{w}^e\right)^T \bar{\mathbf{t}} d\Gamma \\ &= 0 \end{aligned} \tag{6}$$

where $\mathbf{B}$ defines the strain-displacement relation in the usual form:

$$\boldsymbol{\varepsilon}(\mathbf{u}^h) = \mathbf{B}\mathbf{u}^e \tag{7}$$

The discretized weak form of the fluid flow continuity equation can be written:

$$\begin{aligned} R_2^f(\mathbf{u}^h, p^h, q^h) &= \int_\Omega \left(\mathbf{N}^p \mathbf{q}^e\right)^T \mathbf{1}^T \dot{\boldsymbol{\varepsilon}}(\mathbf{u}^h) d\Omega \\ &- \int_\Omega \left(\nabla \mathbf{N}^p \mathbf{q}^e\right)^T \mathbf{v}^F d\Omega \\ &- \int_\Omega \left(\mathbf{N}^p \mathbf{q}^e\right)^T \frac{n}{K^F} \left(\mathbf{N}^p \dot{\mathbf{p}}^e\right)^T d\Omega \\ &+ \int_{\Gamma_q} \left(\mathbf{N}^p \mathbf{q}^e\right)^T \bar{q} d\Gamma - \int_\Omega \left(\mathbf{N}^p \mathbf{q}^e\right)^T Q^F d\Omega \\ &= 0 \end{aligned} \tag{8}$$

The above equations can be written in a combined form as:

$$\begin{aligned} R^f(\mathbf{u}^h, \mathbf{w}^h, p^h, q^h) &= R_1^f(\mathbf{u}^h, \mathbf{w}^h, p^h) \\ &- \theta \Delta t\, R_2^f(\mathbf{u}^h, p^h, q^h) \\ &= 0 \end{aligned} \tag{9}$$

where R_2^f has been premultiplied by a factor $\theta\ \Delta t$, where θ is an algorithmic integration coefficient and Δt the time increment.

2.3 *Stabilization technique*

The discretized variational formulation modified by a least-squares term now takes the form:

$$\begin{aligned} R^f(\mathbf{u}^h, \mathbf{w}^h, p^h, q^h) &= R_1^f(\mathbf{u}^h, \mathbf{w}^h, p^h) \\ &- \theta \Delta t\, R_2^f(\mathbf{u}^h, p^h, q^h) \\ &+ R^{f,GLS}(\mathbf{u}^h, \mathbf{w}^h, p^h, q^h) \\ &= 0 \end{aligned} \tag{10}$$

and it has to be satisfied at each time step. The following least-squares term based only on the residuum of the fluid flow continuity equation is introduced:

$$\begin{aligned} R^{f,GLS}(\mathbf{u}^h, \mathbf{w}^h, p^h, q^h) &= \\ \sum_{e=1}^{N_{el}} \int_{\Omega^e} &\left[\dot{\varepsilon}_{kk}(\mathbf{w}^h) + \frac{1}{\gamma^F} \nabla^T \mathbf{k} \nabla q^h - c\dot{q}^h\right]^T \tau^* \\ &\left[\dot{\varepsilon}_{kk}(\mathbf{u}^h) + \nabla^T\left(\frac{1}{\gamma^F} \mathbf{k} \nabla p^h - \mathbf{k} \nabla z^h\right) - c\dot{p}^h - Q^F\right] d\Omega^e \end{aligned} \tag{11}$$

In the weighting part the term $\mathbf{k}\nabla z^h$ is omitted since the elevation z^h for geometrically linear problems remains unchanged. The scalar factor τ^* has meaning of a stabilization factor. A more detailed development of this stabilization technique can be found in (Truty 1999).

The matrix equations resulting from previous derivations result in a mixed formulation which involves displacements and pore pressures as main variables. These field variables are approximated here using quadrilateral elements with linear interpolation functions.

3 APPLICATIONS

3.1 *Consolidation due to the construction of an earth dam*

A 15 ft depth of fill is placed over a large lateral area having the profile shown in Figure 1. This example is based on an actual field case in Lagunillas, Venezuela, reported in (Lambe et al. 1969).

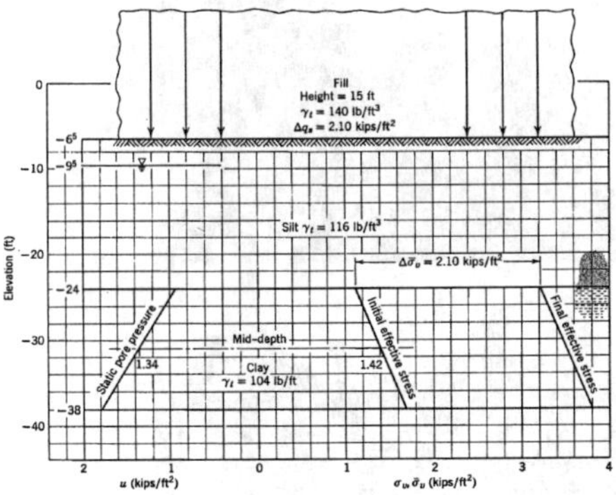

Figure 1. Lagunillas field case geometry (Lambe et al. 1969)

The type of analysis is the following: first, an initial state analysis is performed. Then, the load representing the fill is applied and the consolidation process is initiated ($\Delta t_{init.}$ = 0.01, t_{final} = 10000). Soil is considered linear in this analysis. Material data are given in Table 1.

Table 1. Material properties

	E [kN/m²]	ν [-]	K [m/day]	e₀ [-]
Silt	6900	0	5 10⁻¹	1.00
Clay	690	0	5 10⁻⁵	1.92
	γ_d [kN/m³]	γ_b [kN/m³]	c_v [m²/day]	
Silt	16.34	11.30	2.59	
Clay	18.22	14.80	3.46 10⁻³	

where E is Young's Modulus, ν Poisson's ratio, e_0 the initial void ratio, k the permeability, γ_d the dry body weight, γ_b the buoyant body weight and c_v the coefficient of consolidation.

Remarks:

1) Because the overlying and the underlying soils are much more permeable than clay, there is double drainage (p = 0 at both the top and the bottom of the mesh) and the considered domain for the analysis will be retricted to the clay layer.

2) Poisson ratio ν is taken here equal to zero as the problem is essentially one-dimensional.

3) The whole domain is considered fully saturated, and the water is considered incompressible.

The consolidation process which takes place almost exclusively in the clay layer has the following analytical solution for the evolution of the pore pressure p (Lambe et al. 1969, Z_Soil 1985-2000):

$$\frac{p_w(Z,t)}{q} = \sum_{m=0}^{\infty} \frac{2}{M} \sin(M \cdot Z) \cdot e^{-M^2 T} \tag{12}$$

$$\text{with} \quad M = \frac{\pi}{2}(2m+1) \tag{13}$$

$$Z = \frac{z}{H} \tag{14}$$

$$\text{and} \quad T = \frac{c_v \cdot t}{H^2} \tag{15}$$

The critical time step after (Vermeer et al. 1981) is:

$$\Delta t_{crit} = \frac{1}{4} \frac{\gamma_w \cdot h^2}{E_{oed} \cdot \theta \cdot k} = \frac{1}{4} \frac{h^2}{\theta \cdot c_v} \tag{16}$$

$$E_{oed} = \frac{E \cdot (1-\nu)}{(1+\nu) \cdot (1-2\nu)} \tag{17}$$

h is the element size taken here in vertical direction. θ is an algorithmic coefficient (here θ = 1).

In our case, $\Delta t_{crit} = 13.4$ days.

As the adopted $\Delta t_{init.}$ (= 0.01) is smaller than Δt_{crit}, we expect the Standard Galerkin solution for the pore pressure to oscillate around the exact solution at an early stage of consolidation, and the GLS solution to overcome this problem and give the exact solution (which will be close to q = -96 kN/m²). This is verified in Figure 2.

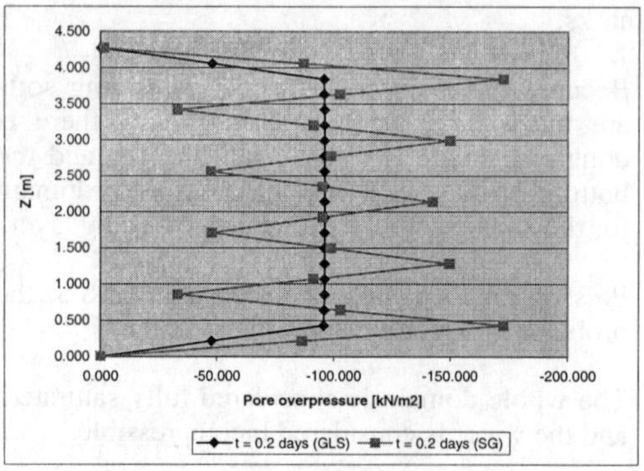

Figure 2. Pressure for t = 0.2 days, comparison between Standard Galerkin and Galerkin Least-Squares (GLS) schemes

3.2 *Shahid Rajaee dam – Iran, case study 1 (by Stucky Consulting Engineers Ltd)*

The Shahid Rajaee dam is a concrete arch structure impounding a reservoir with a capacity of 190 million m^3. The dam foundation is heterogeneous. Whereas the dam toe is located on marly sandstone with low compressive strength (σ_c ~12.5 MPa), the flanks are founded on sound sandstone (σ_c ~7.0 MPa) and the upper arches are abuted against limestone (σ_c ~85 MPa).

The pronounced heterogeneity of the foundation strength and also its deformability necessitated the simulation of the induced displacements in order to compare them with in situ measures. 3D numerical modelling of the dam and the rock foundation was performed using the Z_SOIL 3D (Z_Soil 1985-2000) finite element program to simulate the construction stages (block by block with more than 150 stages). The purpose was also to estimate the safety of the structure against a general shear failure in the marly sandstone.

Figure 3 shows the general layout of the 3D model and the material properties used for numerical simulation.

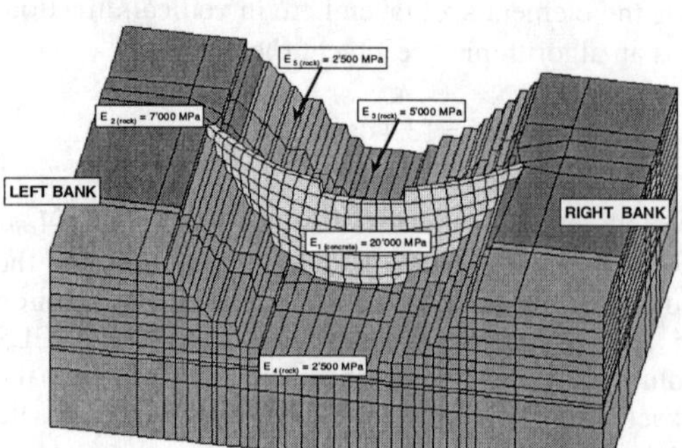

Figure 3. General layout

Figures 4 and 5 represent respectively construction stage after 210 and 485 days.

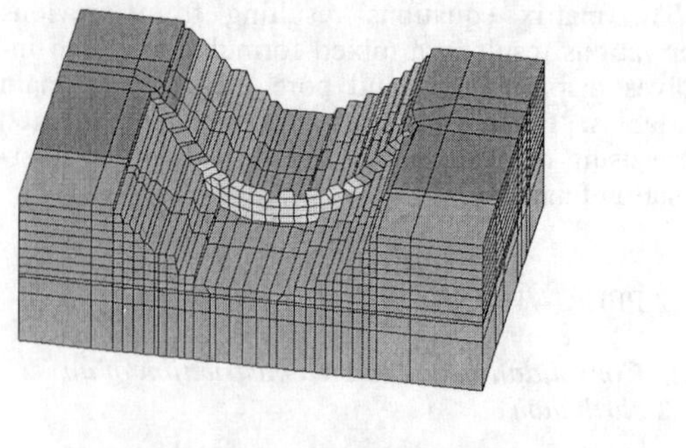

Figure 4. Construction stage after 210 days

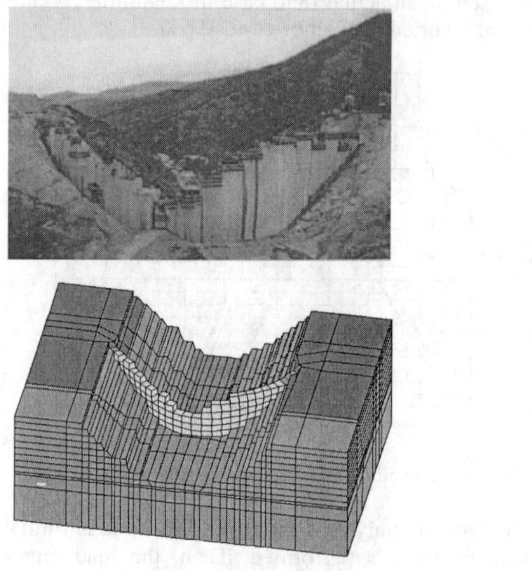

Figure 5. Construction stage after 210 days

Figure 6 shows the comparison of in situ measured displacements with calculated values for the numerical model.

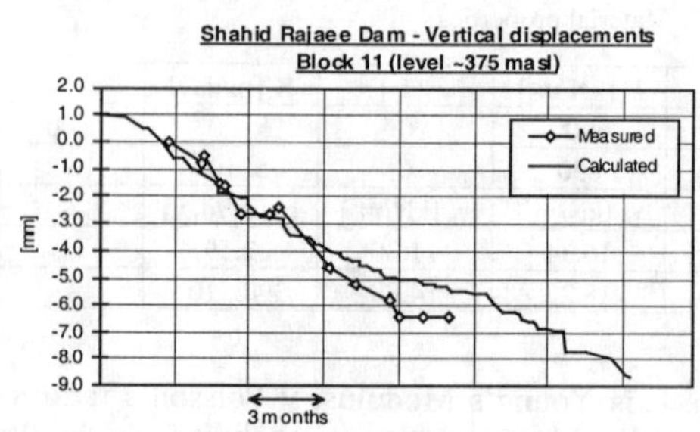

Figure 6. Calculated vs. in situ values

3.3 *Shahid Rajaee dam – Iran, case study 2 (by Stucky Consulting Engineers Ltd)*

In order to perform a rapid control of the behavior of the dam, the measured deflections are compared to the calculated values. These deflections are defined for 2 load types :

1. Water pressure on the upstream face
2. Variation of the concrete temperature with respect to the reference state.

A complete 3D analysis was performed using finite element program Z_SOIL 3D. The first load case studied is the effect of the lake level. The lowest level is chosen at elevation 405. The reservoir level is then raised step by step up to the maximum operating level at 493.00 masl. These steps are shown for the central block 13 in Figure 7.

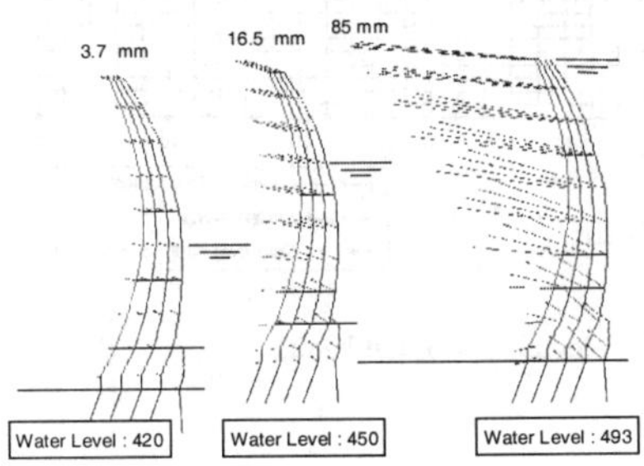

Figure 7. Displacement vectors

Displacements obtained for full lake conditions (level 493) are shown in Figure 8.

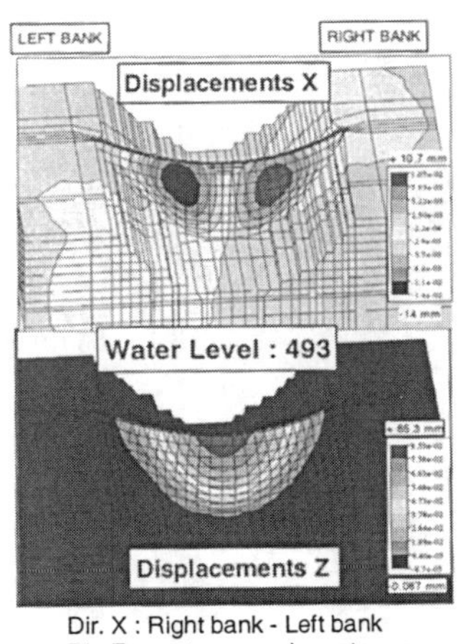

Figure 8. Displacements X and Y

The second load case is considered in order to calculate the displacements due to the effect of the temperature variations of the dam body. A unit temperature increase is applied to each horizontal arch of the dam and its influence at different elevations is calculated. This is shown in Figure 9.

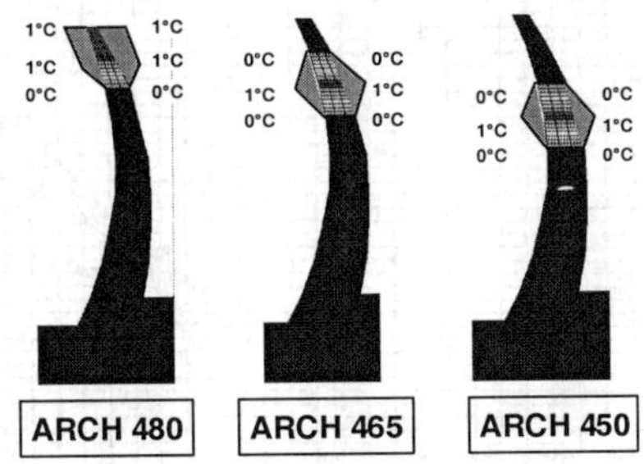

Figure 9. Temperature distribution

Unit temperature increase is applied to each of the 7 arch levels considered (480, 465, 450, 435, 420, 405 and 390). This temperature variation is applied to the whole length of the arch, say i. These 7 temperature loadings enable the radial deflections at the points where plumbline stations are located to be determined. At each level of the different plumbline sections corresponding displacements (a1 to a7) are obtained. In order to compute the actual effect of a temperature difference in the dam body, it is then sufficient to determine the mean temperature difference ΔT_i at levels given previously and to compute displacements at a given point j as:

$$\Delta\delta_j = a_{1j} \bullet \Delta T_1 + ... + a_{7j} \bullet \Delta T_7 \tag{18}$$

(j representing levels 493, 483, 465, 444, 423, 402 and 384)

Calculated displacements have been determined for the upstream / downstream (US / DS) direction.

Measured displacements are then compared to the calculated curves on graphs as functions of the lake level (fig. 10).

4 CONCLUSION

In this paper we illustrate the usefulness of three-dimensional nonlinear finite-element analysis for actual dam structures. In addition, the potential of Galerkin Least-Squares based finite element formulations to overcome numerical oscillations in numerical simulation of consolidation problems is demonstrated.

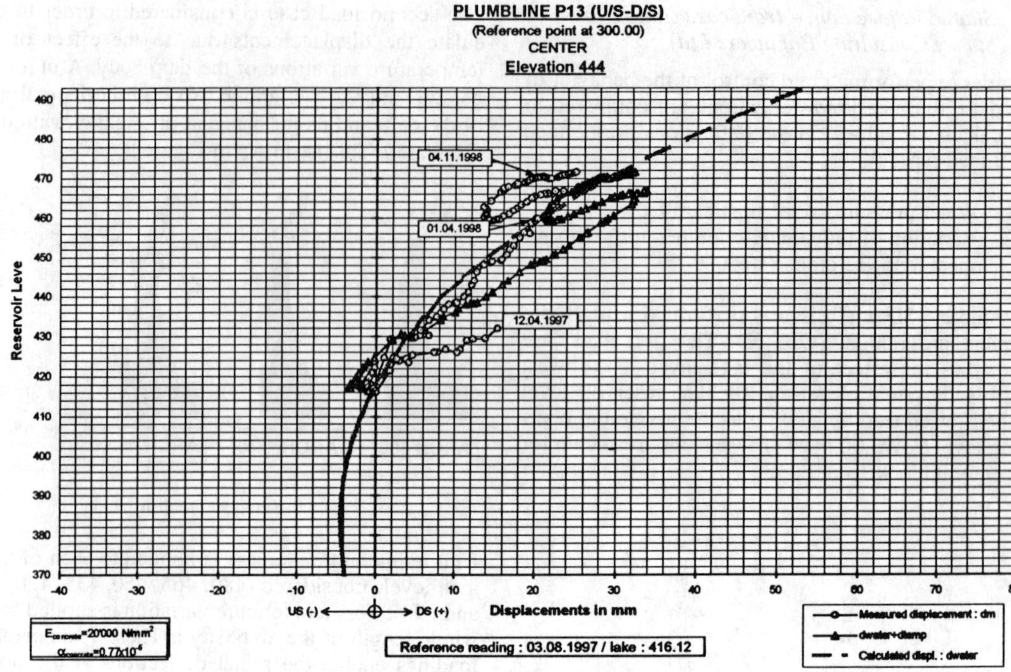

Figure 10. Displacement as function of lake level (plumbline P9, elevation 483, left bank)

ACKNOWLEDGEMENTS

The authors acknowledge the financial support of the Swiss National Science Foundation under grant n° 2100-49404.96 and of the "Fonds du 15ème congrès des Grands Barrages", for the second author.

REFERENCES

Lambe T.W., Whitman R.V., 1969. **Soil Mechanics**. *Wiley & Sons Inc.*

Truty A., Zimmermann Th., 1997. **A robust formulation for FE-analysis of elastoplastic media**. *Numerical Models in Geomechanics (NUMOG VI)*, ed. Pande & Pietruszczak, pp. 381-386

Truty A., 1999. **A stabilized FE-formulation for fully saturated two-phase media**. *LSC Internal report 99/01*.

Vermeer P.A., Verruijt A, 1981. **An accuracy condition for consolidation by finite elements**. *International Journal For Numerical And Analytical Methods in Geomechanics*, 5, pp. 1-14.

Z_Soil.PC User Manual, 1985-2000. Zace Services Ltd., Lausanne, Switzerland.

16 Tunnels, underground works, mining

Computer Methods and Advances in Geomechanics, Desai et al. (eds) © 2001 Balkema, Rotterdam, ISBN 90 5809 183 X

Application of computer methods on tunnelling projects in Korea

I. M. Lee
Department of Civil Engineering, Korea University, Seoul, Korea

ABSTRACT: There are a bunch of tunnelling projects undergoing on in Korea including subway tunnels, railroad tunnels, high speed railroad tunnels, road way tunnels, telecommunication tunnels, electricity tunnels, etc., and more tunnels are being planned to be built in the near future. The author has been trying to apply the numerical and/or analytical tools to the design and safety assessment of tunnels for more than 10 years. In this paper, the author introduced overall view of tunnelling and underground structure projects in Korea and presented some of the author's efforts for applying computer methods to tunnelling projects during the past 10 years.

1 INTRODUCTION

1.1 *General*

Korea is a mountainous country where over 65% of Korean peninsula is occupied with mountain terrain. Since over 65% of the Korean peninsula is mountainous terrain, railroads, roads, and freeways can not avoid the incorporation of tunnels. Construction of intricate subway lines is currently underway across 6 major cities, with even more planned for the future to compensate for the ever increasing traffic. Moreover, several large underground oil and gas storage cavern projects are in progress. In metropolitan cities, power lines, telecommunication lines and water-supply lines are installed within tunnels. The following section will show you an overview of tunnelling projects in Korea (Chung & Lee 2000).

1.2 *Railroad Tunnels*

After about a 30-year lull in construction of new railroads, Korea has become aware of the importance of railroads for the mobilization of material resources. As of such, railroad construction projects have been on the increase since 3 years ago. The projects include construction of new railroads, installation of double tracks for existing single track railroads, the repair of existing railroads, etc. In the construction of new railroads, the incorporation of tunnels can not be avoided. At present, an estimated 13 tunnels with a total length of 37 km are under construction.

1.3 *High Speed Railroad Tunnels*

The first high speed railroad in Korean peninsula is under construction starting from the Seoul metropolitan area to Pusan (see Fig. 1). The total length reaches 411.5 km and a third of the total length consists of tunnels (a total of 60 tunnels).

1.4 *Highway Tunnels*

Highways are constructed and maintained by Korea Highway Corporation. The highways are the backbone of the transportation network in Korea. A total of 151 highway tunnels are under-construction with a total length of 137 km.

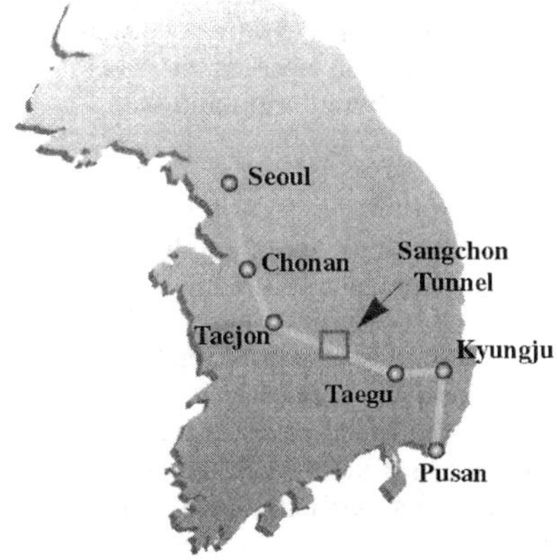

Figure 1. Route of high speed railroad.

1.5 *Subway Tunnels*

In most of the large cities in the Korean peninsula, traffic is one of the most serious social problems. To solve this problem, mass transit transportation systems have been developed in the 6 largest cities in Korea; Seoul, Pusan, Taegu, Inchon, Kwangju, and Taejon.

In the Seoul metropolitan area, the construction of the subway system consists of 3 Phases. Phase I construction, which includes lines 1 through 4, was completed in 1985 with a total length of 116.5 km. Phase II construction, which includes line 5 through 8, will be finished by November of this year (total length of 168.4 km). Phase III (line 9 through 12) is in the planning stage, and among them, line 9, which starts from Kimpo international airport and runs along the southern side of the Hahn river, is expected to undergo design and construction by next year.

The whole track of line 1 was constructed by the cut-and-cover method since this method is the most economical means for construction at shallow depths. However, as the ever-increasing severity of traffic at ground level prohibits extensive use of this method, subway construction is gradually changing from the cut-and-cover method to mined tunnelling. During Phase II construction, more than 60% of the subway tracks have been constructed by tunnlling.

1.6 *Caverns*

At present, 5 underground oil storage caverns and 5 underground LPG gas storage caverns have been completed and under operating. Moreover, additional caverns are undergoing construction, and more are in the planning stages. It goes without saying that hydraulic control is the most important issue regarding storage caverns. Data monitoring analysis system for underground highly pressurized gas storage facilities (MoGas) was developed in Korea (developed by the SK Construction Company). The program capabilities include design and database management for prevention of gas leakage from unlined highly pressurized gas storage caverns. This advanced system has been applied to several LPG caverns that are currently under operation.

1.7 *The Longest Tunnel*

The 'Yongdong-son tunnel', at a length of 16.24 km, will be the longest tunnel in Korea. The design has been completed and construction has just begun. The plan view of the tunnel is shown in Figure 2. As shown in the figure, the tunnel is of a loop shape. The geology at this site is very complex; the tunnel must run through a fractured zone consisting of limestone, large faults and abandoned mines in the immediate vicinity.

2 PARAMETER ESTIMATION IN TUNNELLING PROJECTS

2.1 *General*

In spite of drastic development of underground technology, there are still many uncertainties that exist in the designing and constructing of underground structures. This is mainly due to the discrepancy between the laboratory and in-situ tests and limitations of site investigation techniques during the design stage, etc. The feedback system can be used to estimate the optimum ground properties by minimizing the difference between the predicted and measured ground motions.

Effort was made by the author to formulate a parameter estimation technology by systematically combining the field measurements and the prior information of underground structures. The Extended Bayesian Method (EBM) was adopted for the feedback analysis. The EBM is a technique to give an optimum match between the observed data and prior information. For details on the EBM, see Lee and Kim (1999). Determined in the proposed model were various geotechnical parameters including the elastic modulus (E), the initial horizontal stress coefficient at rest (K_0), the cohesion (c) and the internal friction angle (ϕ).

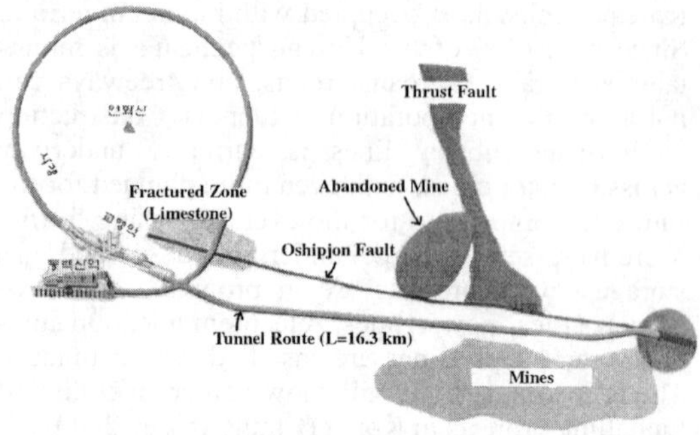

Figure 2. Plan view of Yongdong-son tunnel.

2.2 *Application to a Subway Tunnel*

The proposed parameter estimation model was applied to a subway tunnel in Pusan, Korea. The ground at the site consists of fill layer (cohsionless soil), residual soil and highly weathered rock from the ground surface, as shown in Figure 3. Below the overburden, it is composed of Bulkuksa granite. The tunnel was constructed through the highly weathered rock, which is classified as 'Poor Rock' with a RMR description. The crown subsidence and sidewall convergence measured by tape extensometer are used as the observed data. The location of the measured points is shown in Figure 3. The characteristic lines were obtained from the extrapolation of the measured data, and absolute displacements were obtained. Since the ground motion was almost elastic except at the invert corner zones, the elastic analysis appears appropriate for this site. Since the residual soil with granite origin is mostly cohesionless, it is almost impossible to obtain undisturbed samples with typical sampling techniques. The elastic modulus and the strength parameters can be empirically estimated from correlations with SPT N-value. Estimation of K_0 values is even more difficult. Therefore, some typical values shown in Figure 3 are used as the prior information in this study. The coefficient of variation of SPT N-value is about to be 0.26. The coefficient of variation of the E and K_0 are selected as 0.3 including uncertainty of the SPT N-value.

In order to assess the best model, three models are introduced. The model I-1 is adopted directly from the site investigation (see Fig. 3). Since the elastic modulus of the residual soil selected in the model I-1 seems too small, the model I-2 adopted the larger elastic modulus for the residual soil layer. The estimated elastic modulus of granitic soil for model I-2 was about 98,000 kN/m^2, which was obtained from triaxial test results by Kim (1994). In the model II, residual soils and weathered rocks are considered as a single 'weathered zone'.

The comparison of the proposed model and the optimized parameters were shown in Table 1. As shown in Table 1, the objective functions of the model I-2 and model II were quite small compared with the function of the model I-1.

3 PREDICTION OF DISCONTINUITY ZONE AHEAD OF TUNNEL FACE

3.1 *General*

When fault zones or cavities exist ahead of a tunnel face and the tunnel is excavated without perceiving their existence, it will cause the stress concentration on the region between the face and the discontinuity zone. This is from the influence of the fault zone on the arching effect and may sometimes bring a collapse of underground structures. Recently, abandoned mines or naturally-made underground cavities caused a big problem during tunnel excavation in Korea. Because the underground structure has many unreliable factors in the design stage, the prediction of fault zones ahead of the tunnel face by monitoring displacements of the tunnel and the rapid change of the support systems are required for more economical and safer tunnel construction.

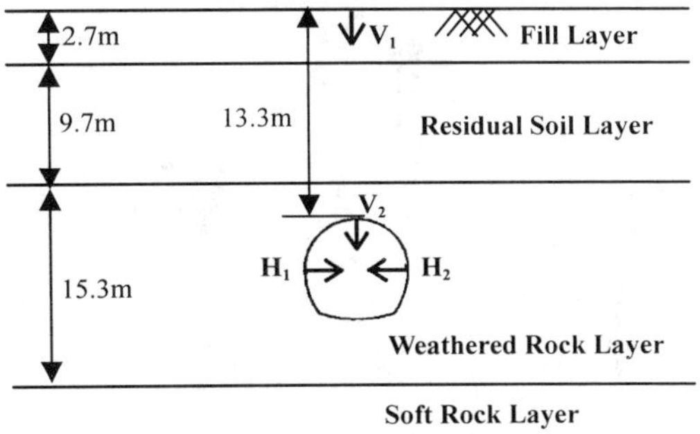

	F.L.	R.S.	W.R.	S.R.
E (kN/m^2)	19,600	29,400	196,000	980,00
K_0	0.5	0.5	0.5	1.0
ν	0.35	0.33	0.23	0.20
γ (kN/m^3)	17.64	18.62	21.56	23.52

Figure 3. Idealized ground condition of the subway tunnel in Pusan.

The conventional method in determining tunnel deformations has many drawbacks. It just shows the relative displacement between two points and shows only two dimensional displacement. By monitoring the 3-dimensional absolute displacement, the more useful information can be obtained without interrupting construction procedure. Components of 3-dimensional absolute displacements are shown in Figure 4. If the longitudinal displacement and settlement measured in the field are considered together in displacement analysis, the prediction of the change in rock mass property is possible. Numerical

Table 1. Results of model identification and parameter estimation

	Ground condition	Elastic modulus (kN/m^2)		Initial horizontal stress coefficient		Objective function
		Prior information	Optimized value	Prior information	Optimized value	
Model I-1	R.S.	29,400	45,178	0.50	0.49	28.60
	W.R.	196,000	221,774	0.50	0.50	
Model I-2	R.S.	98,000	117,502	0.50	0.50	0.26
	W.R.	196,000	192,276	0.50	0.505	
Model II	W.Z.*	196,000	186,102	0.50	0.49	0.76

* W.Z., weathered zone

analysis can play a major role to analyze the monitoring data of the 3-dimensional absolute displacement. The effect of the existence of cavities as well as fault zones on the 3-dimensional displacements can be investigated by performing 3-dimensional numerical analyses.

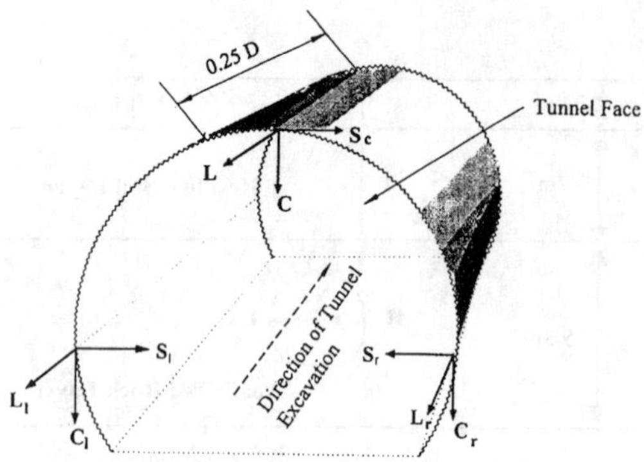

Figure 4. Components of 3-D displacement vector.

3.2 *3-D Analysis Used for Predicting Fault Zones and Cavities*

Circular shape tunnel with a 10m diameter was modeled and the depth from the ground surface to the center of tunnel was assumed to be 5 times tunnel diameter (5D). The thickness of the fault zone was assumed to be 0.25D and the Young's modulus of fault zone was assumed to be 1/10 of that of rock mass.

Three-dimensional FEM analysis was performed for simulating the tunnel excavation in rock mass where fault zones of cavities exist ahead. To compare the influence of fault zones and cavities, it was assumed that the diameter and length of cavity is 0.5D and 1.0D, respectively.

Vertical and horizontal stress in front of the tunnel face and at the crown increase significantly when a fault zone or a cavity exists mainly due to arching effect, compared with the condition of normal rock mass without including any discontinuity zone. As the tunnel approaches to the discontinuity zones, the stress difference between the two cases becomes larger. Because of the larger variation of the longitudinal displacement caused by the concentration of vertical and horizontal stresses, the L/C ratio (the ratio of longitudinal displacement to settlement at crown) also increases as the tunnel advances to that zone.

Figure 5 shows the variation of L/C due to the existence of fault zones or cavities ahead of the tunnel face. The increase in the L/C was significant at some distance ahead of the tunnel face so that the prediction of discontinuity was possible. It begins to increase at about 2.0D ahead of fault zones. The maximum L/C in case a cavity exists was larger than normal case by 6 times, and that in case a fault zone exists by twice (Lee & Kim 1999).

Monitoring program of 3-dimensional absolute displacement was set up in a longest highway tunnel project in Korea and measured displacements were compared with the results of 3-dimensional numerical analysis (for example, Fig. 5) to assess the ground condition ahead of the tunnel face. The tunnel excavation was completed successfully with the help of proposed technique without having any risky situations.

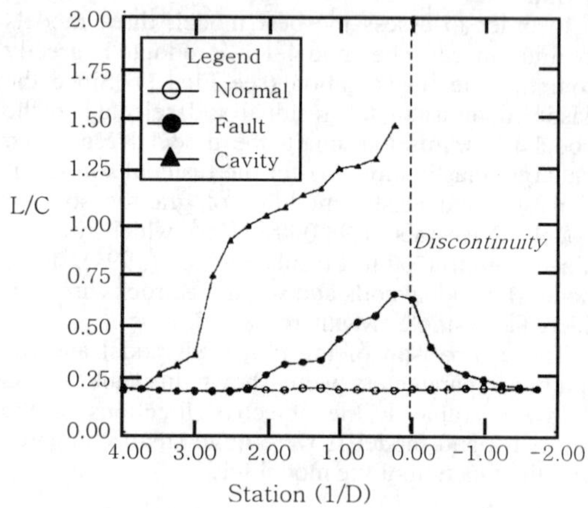

Figure 5. Change of L/C in fault zone and cavities.

4 CONSIDERATION OF SEEPAGE FORCES IN UNDERWATER TUNNELS

4.1 *General*

Subway tunnels in urban area are frequently constructed under the groundwater level, to be called, 'underwater tunnel' (Lee & Nam 2000). The underwater tunnel can be designed either in a waterproof type or a drainage type according to the groundwater control methodology. In the waterproof type, the tunnel lining is designed to support the hydrostatic water pressure under the groundwater level; in the drainage type, the tunnel lining is designed without the consideration of the porewater pressure. In this case, it is presumably supposed that the groundwater will flow into the filter located between the ground and the lining so that the water pressure acted on the lining will be zero. However, if the drainage tunnel is located under the river, the groundwater flow will show steady state flow condition and ground earth pressure will be changed into the state 'effective stress + seepage pressure' by drainage, which will eventually will act on the tunnel lining (Lee et al. 1994).

A numerical analysis of the drainage tunnel with the consideration of the seepage force was performed by the method of superposition of stress change due to excavation with seepage pressure. The effect of the seepage force on the tunnel lining was modeled. A design example of the drainage tunnel under the effect of the seepage force was reviewed through the case study of Seoul subway line 5.

4.2 *Site Condition*

The Seoul subway line 5 underpasses the Han river twice. The one is a waterproof tunnel constructed by NATM; and the other is a box structure constructed by cut-and-cover method.

Here the latter case was studied; the NATM drainage tunnels connected to the box structure. The 700 meter-long box structure constructed by the cut-and-cover method is located underneath the main channel area and both sides of it were connected to the NATM drainage tunnels.

The geological condition of the bedrock around the Han river mostly consists of banded biotite gneiss of the Precambrian. Most of the upper part of bedrock is completely weathered rock and soil of which the thickness varies between 3~10m. The bedrock is covered by a 2~15m thick layer of medium to dense alluvial deposits. Especially in the flood plain area, a 1~8m thick layer of fill material covers the alluvial deposits. The box structure was reinforced by steel bars to withstand the hydrostatic porewater pressure as well as the effective earth pressure.

In the soil stratum under the flood plain area, the drainage tunnel was adopted. The lining of the tunnel was reinforced by steel bars with the consideration of the seepage force and the soil around the tunnel was reinforced by the jet grouting in order to improve the strength of the soil and to reduce the water inflow (see Fig. 6).

4.3 *Numerical Analysis*

The drainage tunnel under the flood plain area was numerically analyzed to check the effect of the seepage force on the tunnel lining . The typical soil profile and properties are shown in Figure 6. The analyses were performed for the three cases-the drainage tunnel without seepage force, with seepage force, and finally the waterproof tunnel.

The results show that the axial stresses were almost zero on the lining in the case of drained condition; in the drainage tunnel under the condition of the steady state flow, there were quite large amount of axial stresses acting on the tunnel lining (experimental study as well as numerical analysis shows that more than 20% of the maximum hydrostatic pressure is acted on the lining due to the seepage effect). The maximum loads which acted on the lining are shown in Table 2.

Table 2. Loads acted on the lining

	Max. moment	Max. axial force	Max. stress
	kN-m	kN	kN/m^2
Drainage tunnel w/o seepage force	0.00	0.00	0.00
Drainage tunnel w/ seepage force	68.992	416.598	5887.252
Waterproof tunnel	229.222	1029.882	18015.634

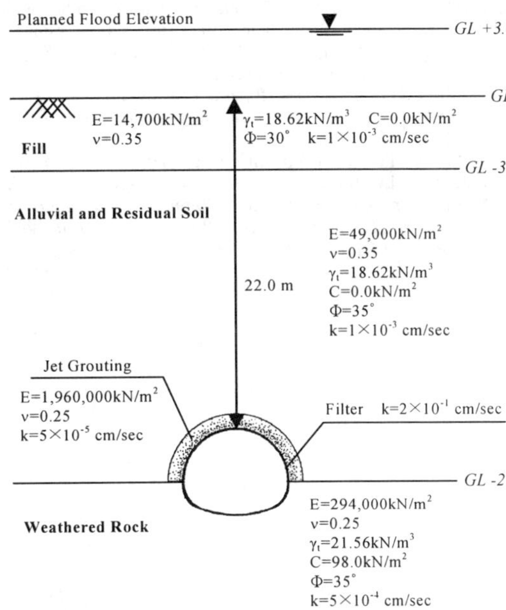

Figure 6. Typical soil profile and tunnel cross section.

5 CONCLUDING REMARKS

The author has tried to apply numerical and analytical tools to the design and safety assessment of tunnelling projects more than 10 years. In this paper, some examples of these efforts are presented. Some geotechnical engineers working in the field have a tendency not to recognize the benefit of numerical analysis and to mostly rely on their engineering judgement. On the other hand, there are some others who are fond of only computer methods. It goes without saying that the two approaches have to be combined for the success of geotechnical engineering projects. Tunnelling projects might be among them. The initial design has to be revised and updated during tunnel excavation; the analytical/numerical tool will play an important role in the tunnelling process.

REFERENCES

Chung, H.S. & I.M.Lee 2000. General report on tunnelling and underground projects in Korea. *Proc. Int. Conf. Tunnels & Underground Struct., Singapore, 26-29 November 2000*, submitted.

Kim, Y.J. 1994. Constitutive characteristics of decomposed Korean granites. Ph.D. Thesis: Korea University.

Lee, I.M. & D.H.Kim 1993. 3-D numerical analysis for the prediction of discontinuity zone ahead of tunnel face. In G.Vouill & P.Gerest (eds), *Proc. Int. Cong. Rock Mech., Paris, 25-28 August 1999*:199-210. Rotterdam: Balkema.

Lee, I.M. & S.W.Nam 2000. Evaluation of face stability with the consideration of seepage forces in shallow tunnels. *Proc. Int. Conf. Tunnels & Underground Struct., Singapore, 26-29 November 2000*, submitted.

Lee, I.M. & D.H.Kim 1999. Parameter estimation using extended Bayesian method in tunnelling. *Comp. Geotech.* 24: 109-124.

Lee, I.M., K.J.Park & S.W.Nam 1998. Analysis of an underwater tunnel with the consideration of seepage forces. In A. Negro Jr. & A.A.Ferreira (eds), *Tunnels & Metropolises-Proc. World Tunnel Cong., Sao Paulo, 25-30 April 1998*: 315-319. Amsterdam: Elseveir.

Computer Methods and Advances in Geomechanics, Desai et al. (eds) © 2001 Balkema, Rotterdam, ISBN 90 5809 183 X

Numerical modeling of deep tunnels including time-dependent behavior

E. Boidy
Coyne et Bellier, Consulting Engineers, France

F. Pellet & M. Boulon
Laboratory 3S, University Joseph Fourier, Grenoble, France

Time-dependent behavior of geomaterials, for instance creep, is of a great importance for the design of deep underground structures such as that for waste disposals. Indeed, in order to ensure safety for several centuries, a good prediction of deformations and damage is needed. For this purpose, a new visco-plastic law has been implemented in a two-dimensional explicit finite-difference program. For such a long time prediction, a special analysis of the time step is required to define an allowable discretisation of the time period of study. This approach is based on experience and theoretical analysis. All these considerations are very important for engineers because for such cases, no back experience is available to modify the models afterwards.

1 INTRODUCTION

The design of underground structures, excavated in rate sensitive materials, like shales, requires the determination of a time-dependent constitutive law able to predict the long term behavior. This study analyses the stability of the numerical solution. The approach connects the modelisation of relaxation test and deep circular tunnel in order to manage an allowable time step discretisation. The time-dependent constitutive law which has been implemented in the finite-difference program FLAC 3.4, corresponds to the model proposed by Lemaitre (1996).

2 LEMAITRE'S VISCO-PLASTIC POWER LAW

This law is based on the overstress visco-plasticity theory formulated by Perzyna's (1966) for rate-sensitive plastic materials and gives incremental constitutive equations for the transient creep phase. The basic assumption is that the strain rate can be decomposed into an instantaneous reversible (elastic) part and irreversible part :

$$\dot{\varepsilon}_{ij} = \dot{\varepsilon}_{ij}^{e} + \dot{\varepsilon}_{ij}^{vp} \tag{1}$$

where irreversible strain rate combines viscous and plastic effects and is given by the following relationship :

$$\dot{\varepsilon}_{ij}^{vp} = \gamma\langle\Phi(F)\rangle\frac{\partial G}{\partial\sigma_{ij}} \tag{2}$$

with the next notations :

γ = viscosity coefficient of the material; G = visco-plastic potential; F = static yield function; Φ = flow rule, controlled by the brackets defined by Macauly : $< \Phi(F) > = ½\ [\Phi(F) + |\Phi(F)|]$, where $|\Phi(F)|$ is the classical absolute value.

The function F includes the overstress concept through the classical plastic function f and the strain hardening parameter κ, which depends only on the update accumulated visco-plastic strains (second invariant of the stress tensor). The assumption of isotropic hardening is also involved. This function F has the following form :

$$F(\sigma_{ij}, \varepsilon_{kl}^{vp}) = \frac{f(\sigma_{ij}, \varepsilon_{kl}^{vp})}{\kappa(\varepsilon_{vp})} - 1 \tag{3}$$

with:

$$\varepsilon_{vp} = \int_0^t \left(\frac{2}{3}\dot{\varepsilon}_{ij}^{vp} \otimes \dot{\varepsilon}_{ij}^{vp}\right)^{\frac{1}{2}} d\tau \tag{4}$$

Practically, all theses expressions are simplified by considering two other assumptions. First, the constitutive law can be defined as associated, for which f correspond to G and can be linked directly to the second invariant of the stress tensor q by assuming a Von Mises model :

$$f = G = q \tag{5}$$

This assumption takes into account that visco-plastic deformations develop without any volume changes, which has been verified experimentally for shales. On the other hand, in order to come out onto a practical and representative tool, power laws for the flow rule Φ and for the strain hardening parameter κ have been introduced like that proposed by Lemaitre. The relationships are summed up below by the equations (6a) and (6b).

$$\Phi(F) = (F+1)^n \tag{6a}$$

$$\kappa(\varepsilon_{vp}) = (\varepsilon_{vp})^{-m/n} \tag{6b}$$

Thus, the irreversible strain rate which combines viscous and plastic effects is given by the following relation :

$$\dot{\varepsilon}_{ij}^{vp} = \frac{3}{2} A.q^{n-1}.(\varepsilon_{vp})^m s_{ij} \tag{7}$$

where s_{ij} is the deviatoric part of the stress tensor; q is the second invariant of the stress tensor. This law needs the identification of the three following parameters :
- A, the viscosity coefficient of the material ($A > 0$),
- n, the stress exponent ($n \geq 1$),
- m^*, the strain hardening exponent ($m^* = -m/n \geq 0$) checking the following restriction : $m^* \leq 1-1/n$.

In order to match to relaxation tests, a strain rate threshold has been introduced into our model. This improvement induces a steady-state behavior as soon as the visco-plastic strain rate goes below a specific threshold detected in laboratory. This law has been developed and implemented in the two-dimensional explicit finite-difference program (FLAC 3.4) with the Fish language. The time stepping is also submitted to stability conditions, especially analyzed and discussed in the following paragraph.

3 STUDY OF CRITICAL TIME STEPPING

3.1 *Numerical explicit stability criterion*

For such a long term prediction, a special analysis of the time step is necessary in order to define the critical discretisation of the time period of study. As the system is required to be always in mechanical equilibrium, the time-dependent stress increment must not be large compared to the strain-dependent stress increment; otherwise, out-of-balance forces will rapidly become large, and inertial effects (which are theoretically absent) may affect the solution. Cormeau (1975) has given the expression of the critical time step for such an explicit mode of resolution in the general case. The stability limit, in the case of Lemaitre's visco-plastic law, becomes :

$$\Delta t_{crit} = \frac{4(1+\nu)}{3E\alpha\beta a.q^{\beta-1}.t^{\alpha-1}} \tag{8}$$

where E is the loading modulus; ν is the Poisson's ratio; α and β are exponents; a represents the viscosity parameter. This three constants are function of the previous parameters m, n and A.

They verify the three next relationships listed in equation (9).

$$\alpha = \frac{1}{1-m};\ \beta = \alpha.n = \frac{n}{1-m};\ a = \left(\frac{1}{\alpha}A\right)^{\alpha} \tag{9}$$

3.2 *Definition of an allowable time step*

For a good evaluation of the time step influence, a theoretical reference must be verified with numerical results. The relaxation test has been chosen (after a quasi-static loading) to avoid the perturbation related to initial conditions. Indeed, the sample represented by an axisymmetric mesh is just jammed between the plates of the frame after loading (at constant strain rate, about 10^{-6} s^{-1}, until 26 MPa of axial compression).

Long term mechanical characteristics for this relaxation modelisation are the following ones :

$$\begin{cases} E = 3620 \text{ MPa} \\ \nu = 0.3 \\ n = 9.25 \\ m = -1.77 \\ A = 10^{-25} \ (\text{MPa, s}^{-1}) \end{cases}$$

with a minimal threshold of steady creep motion state close to 10^{-9} s^{-1}.

Various initial time steps have been tested in order to study how rapidly the initial perturbation is calmed. For this calculation, an allowable time step value would be around 0.1 s. Close to 10 s, it engenders a very net imbalance (Fig. 1a), but not necessarily irreversible at long-term because depending also on the geometrical time progression, defined by the expression :

$$\Delta t_{n+1}\left[t + \Delta t_n(t)\right] = p\Delta t_n(t) \tag{10}$$

where p is the geometrical pace of this progression.

According to Cormeau, the initial critical time step would be around 30 s and also 300 times superior to the allowable value. The safety factor (S.F.) to apply to the critical time step of Cormeau can also be determined. This approach joins Detournay's observations (1998), who recommends practically to take a safety factor between 100 to 1000 on the estimation of the critical time step proposed by Itasca.

At long term, it is the geometrical pace p which is major on the stability of the solution (Fig. 1b). Indeed, regardless of the value of the initial time step, there is stability of the computation, with some reversible oscillations around the analytical solution for a pace equal to p = 1.001. On the other hand, for a not adapted geometrical pace, it is possible to be in conditions of instability or at least with an inadmissible precision of the solution. But even if there is a substantial margin of error for a time period tested, the whole stability is not necessarily challenged in the long term.

The two following figures display successively the influence of the initial time step t_o (Fig. 1a) and the geometrical pace p (Fig. 1b).

Through these different numerical responses, it becomes possible to give the safety factor on the critical time step defined by Cormeau to comply with a good accuracy. This factor reaches 500 to 1000.

In this case, the allowable solution kept here (with t_o = 0.1 s and p = 1.001) induces a factor around 750, as presented on the Figures 2a, b.

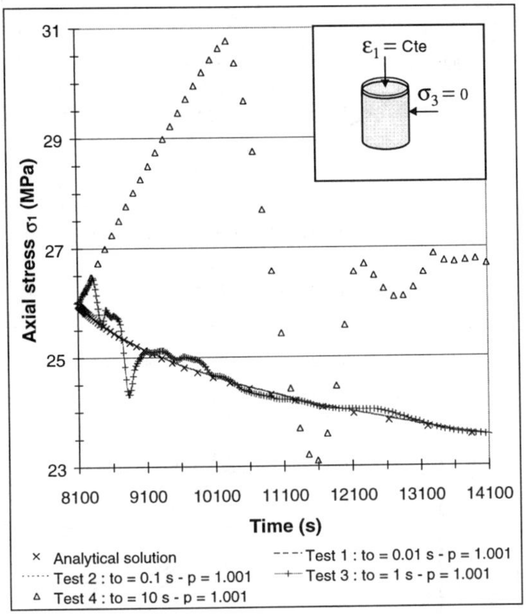

Figure 1a. Influence of the initial time step (t_o) on the numerical stability. Zoom at short term.

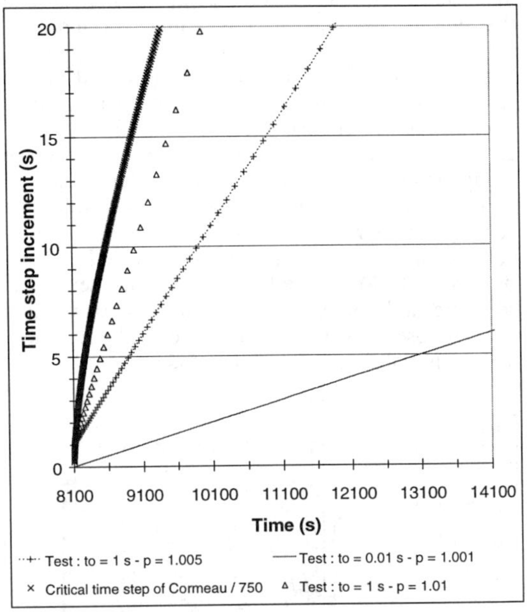

Figure 2a. Definition of an allowable zone of time progression. Zoom at the beginning of the relaxation.

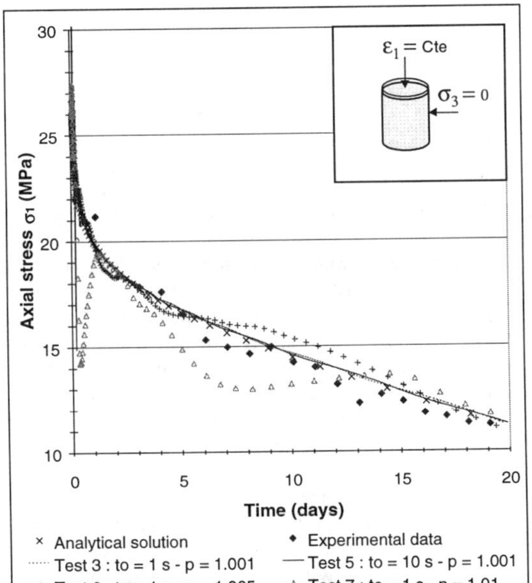

Figure 1b. Influence of the initial time step on the numerical stability. Long term behavior compared to experimental data on shales.

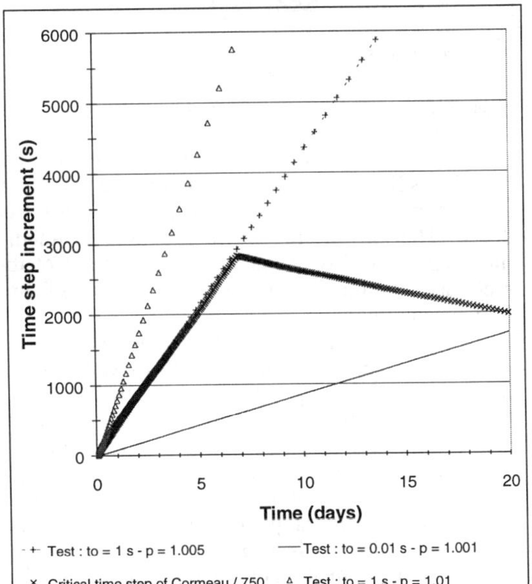

Figure 2b. Long term progression of the allowable time step.

3.3 *Implications for deep circular tunneling*

Practically, for an engineering calculation, it is advisable to choose an allowable time stepping in order to satisfy requirements of precision and computation time. Indeed, the design of a complex underground structure like deep excavations or tunnels can require till 48 hours of calculation time for behavior extrapolation over data for several centuries. Also, precision has a cost and a good management of time stepping allows engineers to optimize the performance of numeral modelisation. Thus, an adequate safety factor on time stepping is very interesting.

To illustrate this aim, we propose to display the behavior of the rock face around a deep circular tunnel. Assumption of stress isotropy is made here with $\sigma_v = \sigma_h = 12$ MPa, and time-dependent phenomena start after a decompression related to the excavation and supposed elastic. Figure 3 shows the 2D-mesh and the initial conditions of the model.

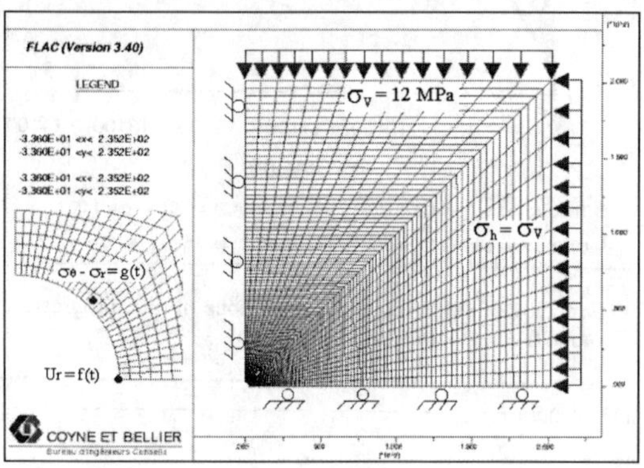

Figure 3. Flac zone geometry and model boundary conditions

The Lemaitre's law uses the same time-dependent proprieties like for the modelisation of the relaxation test but any strain rate threshold hasn't been introduced. The aim of this calculation was to qualify time-dependent displacements and relaxation of the rock face all around the tunnel over a period of time of two centuries.

Thus, the initial time step has been fixed to 10 s in order to reach a good accuracy as soon as one month. Then, the important result demonstrated by Nguyen-Minh and Pouya (1992), concerning the existence of a stationary stress state, can be used for a long time prediction such as this. Considering a safety factor S.F = 750 on the critical time step defined by Cormeau (and justified before on relaxation test modelisation), it is also possible to succeed rapidly in the determination of an allowable geometrical progress for a specific duration of the envisaged study (here 200 years). By introducing this safety factor, the geometric pace p was found around 1.0005.

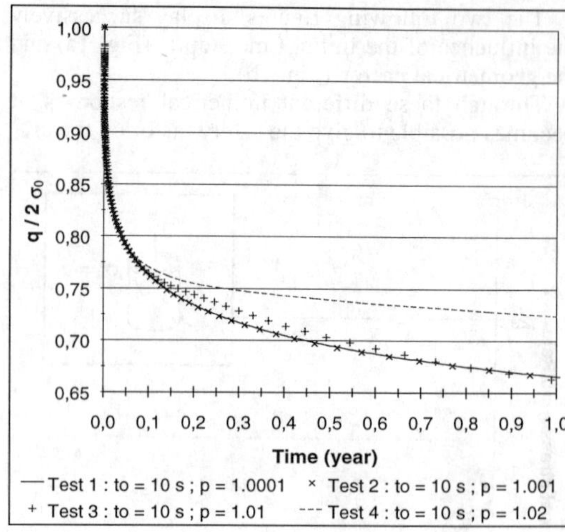

Figure 4a. Deviator variations at the tunnel wall (one year)

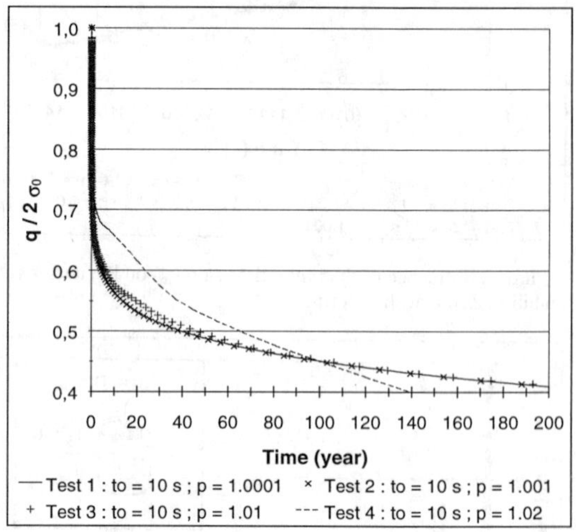

Figure 4b. Deviator variations at the tunnel wall (200 years)

For this example, the reduced deviator near the tunnel wall, $q / 2\sigma_0 = (\sigma_\theta - \sigma_r)_t / (\sigma_\theta - \sigma_r)_{to}$ (Figs 4a, b), and the radial strain at the tunnel wall, U_r / r (Figs 5a, b) can also be indexed dependent on time. The different numerical tests show that the solution is stable with an acceptable precision as soon as the pace p is lower than 1.001. For a pace p higher, there is a real risk of overvaluation of the galleries' closure.

This simple approach is a real investment for engineers because this classic modelisation requires ¼ hour CPU time for a pace p = 1.02, but it rises to more than 2 days with p = 1.0001. Thus with an allowable geometrical progress (p = 1.001), the saving of CPU time is important, about 90% in relation to the smaller progression tested. It is also possible to optimize this computation in 4 hours.

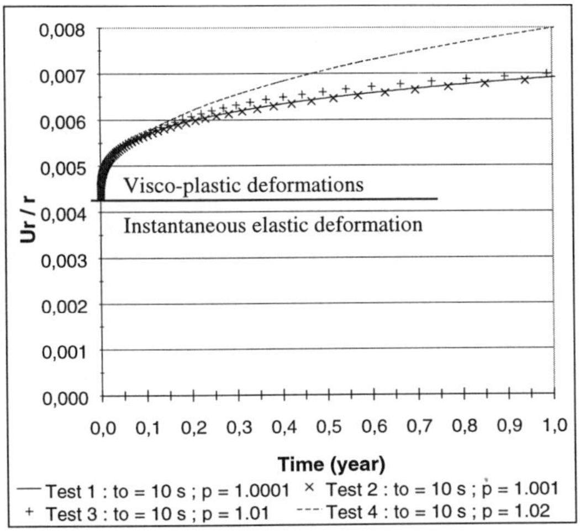

Figure 5a. Gallery's closure variations (one years)

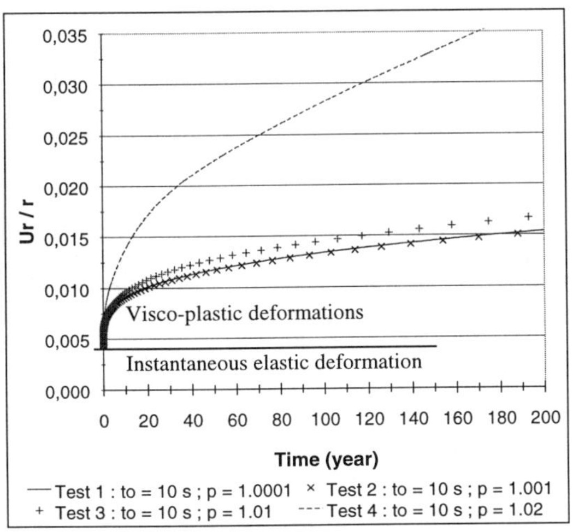

Figure 5b. Gallery's closure variations (200 years)

3.4 *Analysis of unbalanced forces*

In order to challenge the previous results, unbalanced forces variations have been followed during creep, and reported on Figure 7.

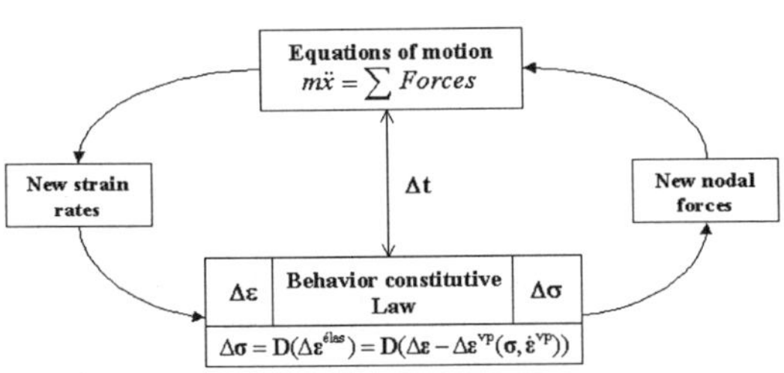

Figure 6. Explicit scheme of resolution during a time step Δt

The equations of motion, which govern the explicit resolution scheme (Fig. 6), must be damped to provide quasi-static solutions (without inertial effects). An alternative damping algorithm is available in FLAC for this kind of situations like creep in which a significant uniform motion drives the model. This damping is called "combined damping" and is more efficient at removing kinetic energy compared to local damping for this special case. The scheme of the algorithm consists of a damping force composed of equal contributions from expressions based on velocity and on unbalanced forces rate. Combined damping scheme was used for the modelisation of relaxation tests where time-dependent motion appeared generalized in the whole model. But for the application on the circular tunnel, local or combined damping induce the same imbalance inasmuch as the damping force is enough strong which is always verified for an allowable time progress.

It is also very interesting to analyze the variations of the maximum unbalanced force during creep and this for each previous geometrical progress. The Figure 6 shows clearly that a pace p = 1.001 is allowable but a pace p = 1.01 becomes critical for this problem.

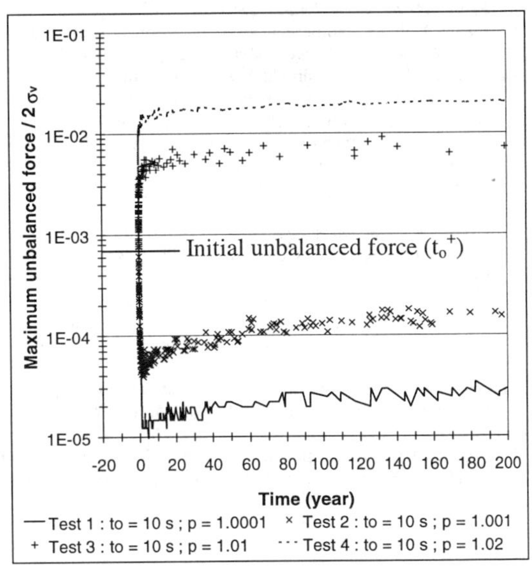

Figure 7. Variations of maximal unbalanced force during creep

4 CONCLUSIONS

The study of time-dependent behavior of geomaterials, requires a specific analyze of the time stepping and a justification for the allowable time step progression chosen. In this way, a special attention has been directed towards the Lemaitre's visco-plastic law which has been implemented in the explicit finite-difference program FLAC 3.4. The theoretical critical time step defined by Cormeau can also constitute a reference on which a safety factor around 100 to 1000 must be apply in order to define an allowable time stepping. For the example presented in this article, unbalanced forces variations during creep confirm that stability was insured in this case.

In the future, we expect to define more precisely some numerical general rules, which could be followed for such time-dependent modelisation. On the other hand, we hope to validate this approach with in-situ measurements to extrapolate this behavior for a very long period of time (several centuries).

5 REFERENCES

Berest, P. 1989. Viscoplasticity in Rock Mechanics. In *Geomaterials. Constitutive equations and modelling*: 239-262. Elsevier Applied Science.

Boidy, E. & Pellet, F. 2000 [to be published]. Identification of mechanical parameters for modeling time-dependent behavior of shales. In ANDRA Workshop, *Behavior of deep argillaceous rocks: Theory and experiment; Proc. Int. Workshop on Geomech., Paris, 11-12 October 2000.*

Billaux, D. & Cundall, P. 1993. Modeling of Geomaterials using the Lagrangian Element method [In French]. *Revue Francaise de Geotechnique* 63: 9-21.

Cormeau, I. 1975. Numerical stability in quasi-static elasto-viscoplasticity. *Int. J. num. Meth. Engng.* 9(1): 109-127.

Detournay, C. 1998. A simple Visco-plastic model in FLAC. *FLAC version 3.4 Users' Manuel Optional Features*. Itasca.

Laigle, F. & Kolmayer, P. 1998. Numerical modelisation for long term behaviour of underground structures [In French]. *Revue Francaise de Geotechnique* 85: 65-78.

Lemaitre, J. & Chaboche, J.L. 1996. *Mecanique des materiaux solides*: Dunod.

Nawrocki, P.A. & Mroz, Z. 1999. A constitutive model for rock accounting for viscosity and yield stress degradation. *Computers and Geotechnics* 25(4): 247-280.

Nguyen-Minh, D. & Pouya, A. 1992. A framework for the analysis of underground excavations in viscoplastic medium, on account of a steady stress state [In French]. *Revue Française de Geotechnique* 59: 5-14.

Malan, D.F. 1999. Time-dependent Behaviour of deep level tabular excavations in hard rock. *Rock Mech. Rock Engng.* 32: 123-155.

Perzyna, P. 1966. Fundamental Problems in Viscoplasticity. In Academic Press (ed.), *Advances in Applied Mechanics* 9: 243-377.

Computer Methods and Advances in Geomechanics, Desai et al. (eds) © 2001 Balkema, Rotterdam, ISBN 90 5809 183 X

Evaluation of reinforced concrete for an underground pressure tunnel

Weizhong Chen, Weishen Zhu & Shucai Li
Institute of Rock and Soil Mechanics, Chinese Academy of Sciences, Wuhan, People's Republic of China

ABSTRACT: In order to verify the feasibility and reasonableness design of reinforced concrete lining of outlet manifold in Shanxi Yellow River Diversion Project (YRDP). Based on field investigations of Young's modulus and in situ stress. 3-D FEM numerical simulation has been done,furthermore, body type and steel ratio of concrete is suggested to satisfy strength requirements as well crack width design criteria. Both owner and designer of YRDP adopt the results in construction.

1. INTRODUCTION

Shanxi Wanjiazhai Yellow River Diversion Project (YRDP) is mainly to solve the problem of shortage in water resources. The project comprises of Wanjiazhai Water Control Project, the general main (GM), the south main (SM), the connection works (CW) and the north main (NM). YRDP will take water at the elevation of 948m from the reservoir area of Wanjiazhai Water Control Project. The main structure on GM, SM and NM are 192Km long tunnel, 101Km long culvert, 39 aqueducts, 6 inverted siphons, 6 pump stations of 5 cascades with the total lift head of 636m, 46.5 MW total installed capacity. The annual diverting amount is $12\times10^8m^3$. The diverting flow is $48m^3/s$. The lining of outlet manifold of GMPS1 & GMPS2 is initially designed to be reinforced concrete, and the internal water head of them is 140m. The rock mass in GMPS1 & GMPS2 is limestone and shale of Cambrian system. The Dip angle of the bedding is $2^0\sim5^0$. The overburden of GMPS1 is 150~180m, and for GMPS2 is 130~160m.The excavation and body type of concrete is shown in Fig. 1.

Some empirical formulae to verify the feasibility of reinforced concrete lining for pressure tunnel are summarized by Norway, U.S.A and Australia. The empirical formulae explain that if the Young's modulus of rock mass is higher than that of concrete, and minimum in situ stress is higher than internal water pressure, then hydraulic jacking failure and a great deal of water leakage phenomena can be prevented. Engineering practices show that if the crack width of concrete were limited in 0.2mm, cracks wouldn't have so much adverse effect on stability of structure and cause much water leakage. In brief, cracked pressure tunnel works doesn't the same as unlined one.

This paper will evaluate the feasibility and reasonableness of reinforced concrete lining for outlet manifold of YRDP. 3-D FEM numerical simulation is based on field investigations of Young's modulus and in situ stress, and body type and steel ratio of concrete is suggested.

2. NUMERICAL MODEL

It is assumed that reinforced concrete is poured after excavation disturbance of surrounding rock mass is stabilized. During numerical simulation, before

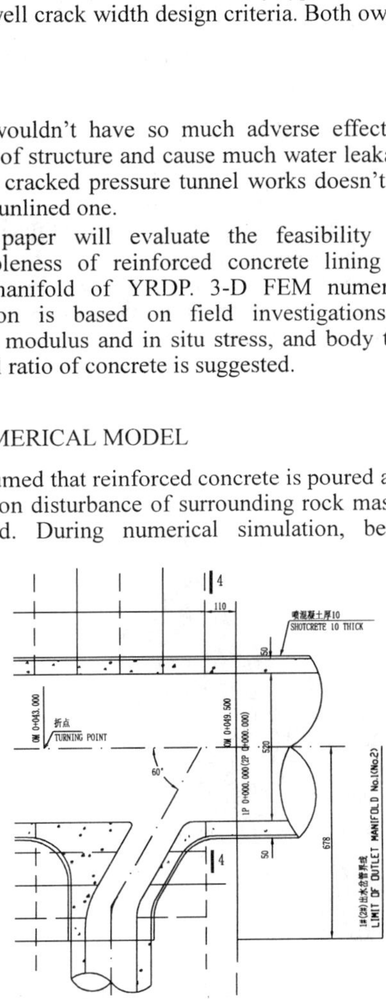

Fig. 1 Excavation shape of manifold

reinforced concrete element cracked, stiffness of which is the superposition of solid concrete element and implicit steel bar element. Zienkiewicz-Pande criterion is applied to simulate non-linear behavior of surrounding rock mass. Yield function F of rock mass can be expressed as (1):

$$F=\sqrt{\alpha}\left(\sigma_m+\frac{\beta}{2\alpha}+\sqrt{(J_2^{+})^2+\gamma-\frac{\beta^2}{4\alpha}}\right) \qquad (1)$$

where, $\sqrt{J_2}^{+}=\sqrt{J_2}/g(\theta_\sigma)$, $\sin(3\theta_\sigma)=\frac{-3\sqrt{3}J_3}{2J_2^{3/2}}$

$$\sqrt{J_2}=\sqrt{\frac{1}{6}[(\sigma_1-\sigma_2)^2+(\sigma_2-\sigma_3)^2+(\sigma_3-\sigma_1)^2]}$$

$$g(\theta_\sigma)=\frac{2k}{(1+k)-(1-k)\sin(3\theta_\sigma)}, \quad tg\overline{\varphi}=\frac{6\sin\varphi}{\sqrt{3}(3-\sin\varphi)},$$

$$\overline{c}=\frac{c\cos\varphi}{\sqrt{3}(3-\sin\varphi)}, \beta=2\overline{c}tg\overline{\varphi}, \alpha=-tg^2\overline{\varphi}, k=-a^2tg^2\overline{\varphi}+\overline{c}^2$$

The elastoplastic constitutive law of rock mass can be expressed as (2):

$$d\{\sigma\}=\left[D-\frac{[D].\{\frac{\partial F}{\partial \sigma}\}.\{\frac{\partial F}{\partial \sigma}\}^T.[D]}{A+\{\frac{\partial F}{\partial \sigma}\}^T.[D].\{\frac{\partial F}{\partial \sigma}\}}\right].d\{\varepsilon\} \qquad (2)$$

where, [D]—elastic matrix, F—yield function, A—strain hardening parameters.

Unloading forces of nodes caused by excavation can be solved from the original stress of element, which can be written as (3),

$$\{F\}_e=\int_V[B]_e^T.\{\sigma\}_e.dV \qquad (3)$$

3. NUMERICAL MODEL AND MECHANICAL PARAMETERS

3.1 FEM mesh and mechanical parameters of rock

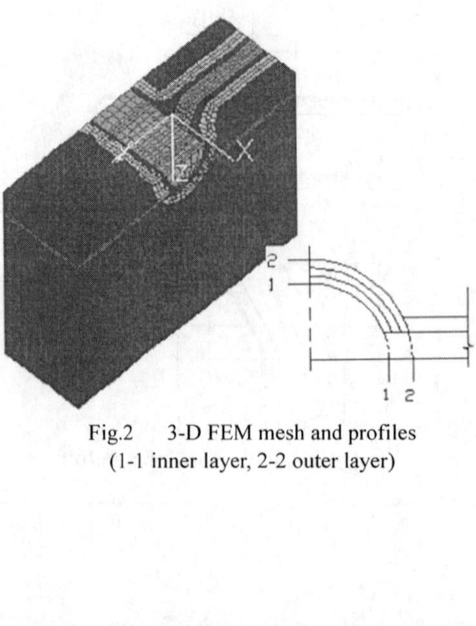

Fig.2 3-D FEM mesh and profiles (1-1 inner layer, 2-2 outer layer)

For outlet manifold of GMPS1 & GMPS2, it is assumed that Y-axis along axis of branch tunnel, X-axis along axis of general tunnel and Z-axis along vertical line. The numerical simulation area is $-10\leqq X\leqq 10$m, $-25\leqq Y\leqq 25$m and $0\leqq Z\leqq 25$m. The model has 6942 nodes and 5936 elements. The reinforced concrete lining is divided into 3 layers (1-1 inner layer, 2-2 outer layer of general tunnel) and shown in Fig.2.

Field investigation of Young's modulus and situ stress of GMPS1 & GMPS2 are shown in table 1, 2.

Original steel ratio of reinforced concrete lining is 1.42%.

Tab.1 Mechanical parameters of rock mass

Rock	E (Gpa)	μ	φ	C (Mpa)
$\in_2Z^5$	12.0	0.27	50°	2.0
$\in_2Z^6$	9.0	0.30	45°	1.0
$\in_3f^1$	9.0	0.26	50°	2.0
$\in_3f^2$	10.5	0.27	50°	2.0
$\in_3f^3$	11.4	0.28	50°	2.0
Concrete	30.0	0.167	58°	2.0

Tab. 2 In situ stress of GMPS1& GMPS2

Location	Item	σ_1	σ_2	σ_3
GMPS1	Stress	-4.97	-3.76	-2.99
	Azimuth	346°	271°	75°
	Dip Angle	-3.4°	77°	13°
GMPS2	Stress	-3.88	-3.11	-2.17
	Azimuth	357°	213°	87°
	Dip Angle	6.5°	82°	4.7°

3.2 Work conditions

Since underground water level is low in GMPS1 & GMPS2, it is not considered into numerical simulation. According to the designer and owner of YRDP, external water head is determined to be half of internal water head. 3 work conditions are as follows:

(1) Elastic 3D-FEM, 190m internal water head,

(2) Elasto-plastic 3D FEM, 190m internal water head,

(3) Elastic 3D-FEM, 95m external water head,

4. NUMERICAL RESULTS

4.1 Stress distribution of concrete

Under internal water pressure, displacement vectors of concrete are pointing outside the tunnel and the maximum tensile stress in concrete is located at the junction of the branch tunnel and general tunnel. While under external water pressure, displacement vectors of concrete are pointing inside the tunnel and most of the concrete lining is in compressive stress state, and the maximum tensile stress in concrete is located at the junction of the branch tunnel and general tunnel.

The principal stress of concrete in different work conditions is shown in table 3, and the contour maps of principal stress of profile 1-1 and 2-2 in GMPS 2 of work condition (2) are shown in Fig. 3,4.

Tab. 3 Principal stress at junction of outlet manifold

Item		Stress (Mpa)	Work conditions		
			(1)	(2)	(3)
GMPS1	1-1	σ_1	2.26	1.97	0.99
		σ_3	-17.3	-16.5	-9.96
	2-2	σ_1	1.80	1.71	0.56
		σ_3	-6.21	-6.43	-6.26
GMPS2	1-1	σ_1	1.98	2.60	0.91
		σ_3	-14.4	-13.90	-14.50
	2-2	σ_1	0.99	1.33	0.58
		σ_3	-5.17	-5.37	-5.25

"-"means compressive stress, "+"means tensile stress.

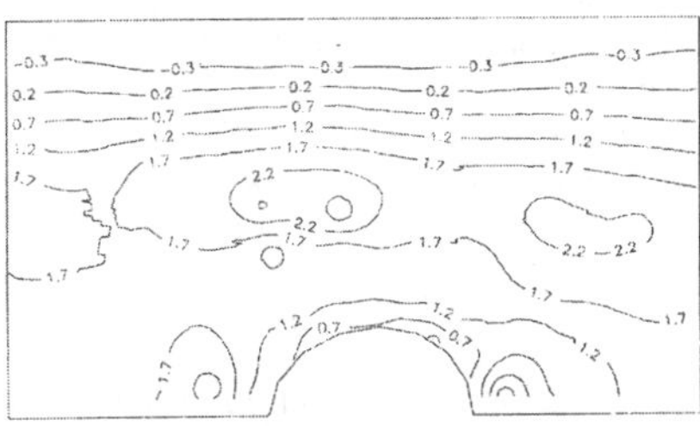

(a) Maximum principal stress

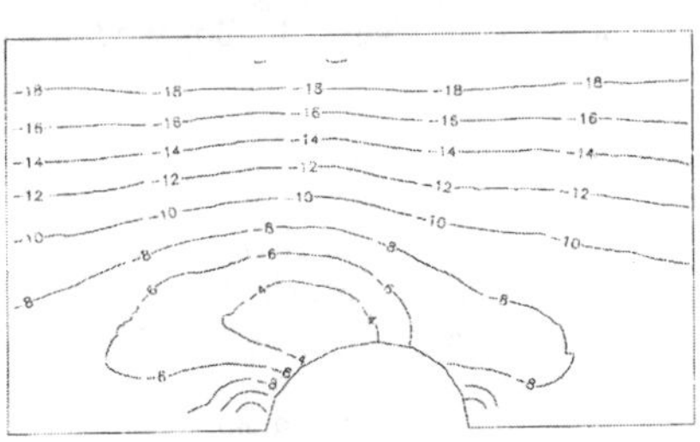

(b) Minimum principal stress

Fig. 3 Principal stress of 1-1,work condition 2 (before optimized)

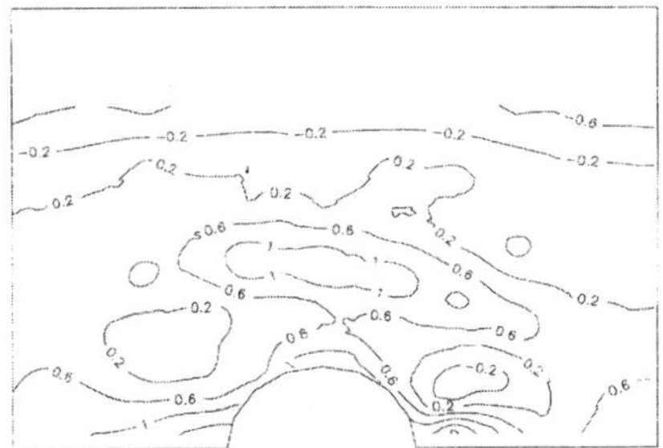

(a) Maximum principal stress

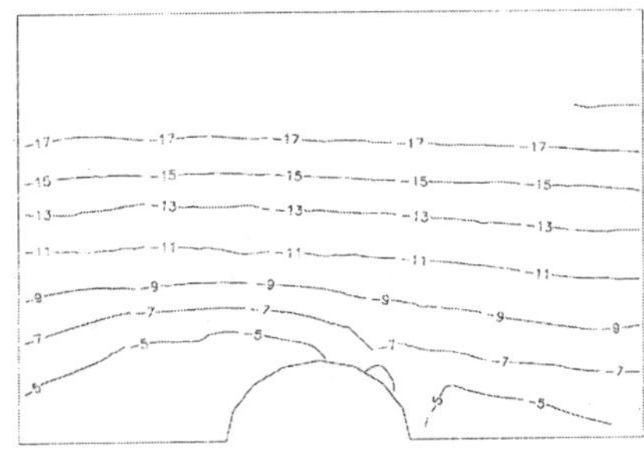

(b) Minimum principal stress

Fig. 4 Principal stress of 2-2, work condition 2 (before optimized)

According to the numerical results, concentration of tensile stress at the junction is still high, so optimize body type to decrease concentration of tensile stress. The best way is to modify junction from tip to circle with radius 0.5m. Based on this new body type, another 3-D FEM mesh with 8036 nodes and 6974 elements is generated. Elastic analysis for plain concrete under internal water head as work condition 4 is calculated. The principal stress in concrete at junction under different work conditions is shown in table 4. Contour maps of principal stress of profile 1-1 and 2-2 in GMPS2 are shown in Fig.5, 6.

Tab. 4 Principal stress at junction of outlet manifold

Item		Stress (Mpa)	Work conditions			
			(1)	(2)	(3)	(4)
GMPS1	1-1	σ_1	2.30	1.50	0.12	2.25
		σ_3	-8.54	-7.62	-4.99	-8.21
	2-2	σ_1	1.40	0.99	-0.07	1.25
		σ_3	-8.30	-8.43	-4.85	-7.82
GMPS2	1-1	σ_1	2.41	1.26	0.17	2.35
		σ_3	-9.12	-8.58	-5.23	-11.9
	2-2	σ_1	1.44	0.80	-0.11	1.20
		σ_3	-9.32	-9.54	-5.20	-8.77

"-"means compressive stress, "+"means tensile stress.

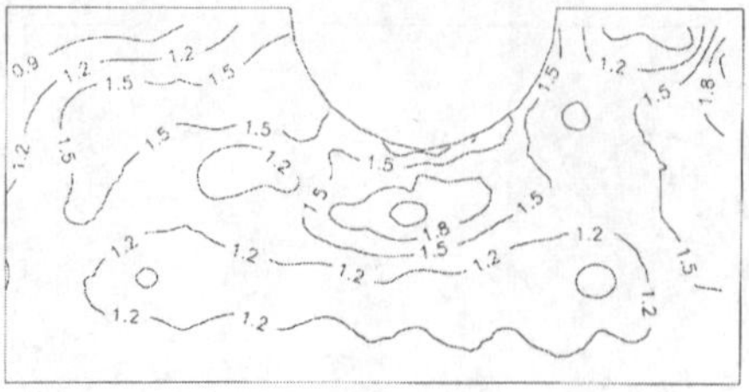

(a) Maximum principal stress

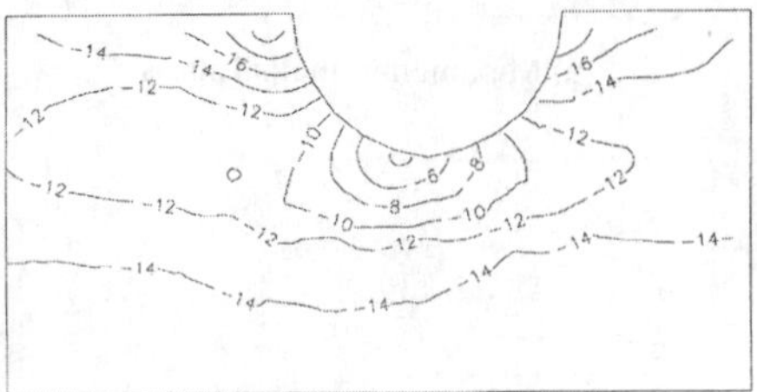

(b) Minimum principal stress

Fig. 5 Principal stress of 1-1 work condition 2 (optimized)

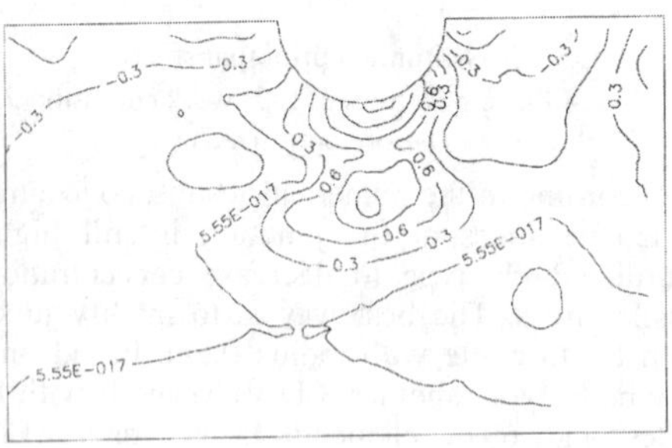

(a) Maximum principal stress

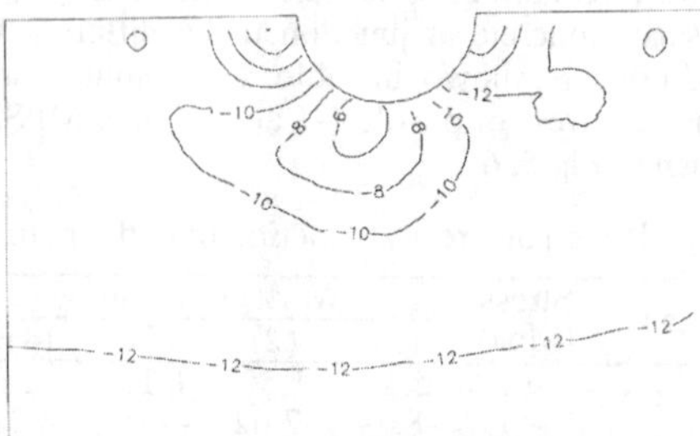

(b) Minimum principal stress

Fig. 6 Principal stress of 2-2 work condition 2 (optimized)

4. 2 Steel ratio of concrete

According to design criteria, in limit state, reinforced concrete will be cracked by internal water pressure, and steel bar will afford all the internal water pressure. Steel ratio of concrete can be determined by numerical results and Lutz theory, and can be expressed as formula (4),

$$\rho = \frac{1}{1 - n + \dfrac{68965[W_{max}]}{\sigma_t \sqrt[3]{2D_e^2 . D_S}}} \quad (4)$$

where, n—ratio of steel Young's modulus to that of concrete, σ_t—tensile stress (Mpa), D_e—thickness of concrete protective layer(mm), D_s—bar space (mm), W_{max}—limit of crack width.

For YRDP, $n=200\times10^3/30\times10^3=6.67$, De = 50mm, Ds=100, 150, 200mm, W_{max}=0.2mm.

The steel ratio with different bar spaces is shown in table 5, which is higher than that of original designed. The steel ratio and body type of reinforced concrete is used in construction.

Tab.5 Steel ratio of concrete

Item	Bar space (mm)	Steel Ratio (%)
	100	1.33
GMPS1	150	1.53
	200	1.70
	100	1.47
GMPS2	150	1.70
	200	1.88

5. CONCLUSIONS

Feasibility and reasonableness of reinforced concrete at outlet manifold of YRDP is discussed, which leads to some useful conclusion:

(1) By increase the steel ratio of reinforced concrete a little at junction, and having a small modification in inner layer shape of concrete, tensile stress concentration in concrete improved a lot, which can fit the requirements of crack width design criteria.

(2) With the help of anchor, 50cm reinforced concrete can bear 90m external water head .

(3) In order to decrease leakage of the tunnel, some anti-seepage measurements should be adopted.

References

1. D.Brox, H.Hagedron. Extreme Deformation and damage during the Construction of large Tunnel. Tunneling and underground space technology, Vol.14, No.1, 23-26.

2. A. Schleiss. Design Criteria applied for the Lower Pressure Tunnel of the North Fork Stanislaus River Project in California. Rock mechanics and Rock engineering, Vol.21,No.3,161-181.

3.J.C.Boot. Elastic Buckling of Cylindrical Pipe Linings with Small Imperfections Subject to External Pressure. Tunneling and underground space technology,Vol.12, Supp.,773-15

Computer Methods and Advances in Geomechanics, Desai et al. (eds) © 2001 Balkema, Rotterdam, ISBN 90 5809 183 X

Numerical modelling of complex slope movements at Savage River Mine, Tasmania

M.A.Coulthard
M.A.Coulthard and Associates, Melbourne, Vic., Australia

K.J.Dugan
BFP Consultants, Melbourne, Vic., Australia

B.J.Hutchison
Australian Bulk Minerals, Burnie, Tas., Australia

ABSTRACT: The Savage River Iron Ore Mine has a history of flexural toppling failures of pit slopes. This mechanism has been studied with the distinct element program *UDEC*, using a rock mass model including both explicit and ubiquitous joints and a Hoek-Brown yield criterion for the rock matrix, and simulation of progressive dewatering of the pit. Two previous failures in the North Pit were back analysed and then the same rock mass parameters were used to examine several options for extended mining of that pit. These options included different ultimate pit depths, dewatering strategies and cases with and without a ramp in the west wall. The models indicated the potential for continued wedge - flexural toppling failures in future west walls but showed that these could be managed by dewatering and monitoring.

1 INTRODUCTION

The Savage River Iron Ore Mine has been operating for over 30 years in the northwest of Tasmania, Australia. A series of flexural toppling failures has occurred in parts of the western walls of Centre Pit and North Pit, limiting overall slope angles to 37° with vertical slope heights of about 150 m.

The stress analysis program *UDEC* (Itasca 1996) has been used to back analyse observed failures in two sections of the North Pit walls and then to examine several options for future mining of that Pit.

A companion paper (Hutchison et al. 2000) provides more details of the geology and flexural toppling history, whilst this paper concentrates upon the numerical modelling strategy and results.

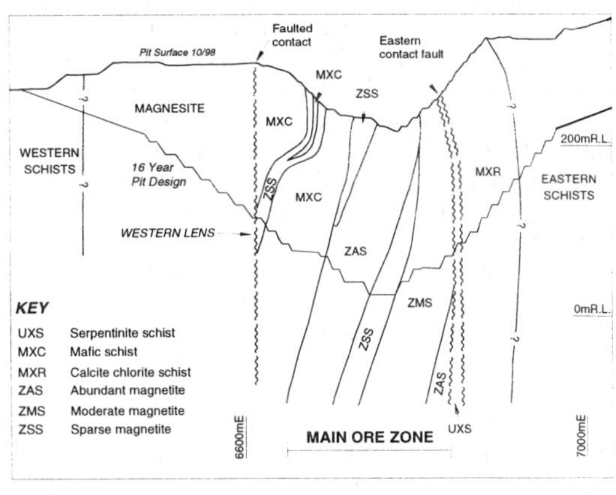

Figure 1. Typical geological section, North Pit.

2 GEOLOGY

A typical geological section within North Pit is shown in Figure 1. Rocks to the west of the orebodies consist of weak mafic schists (designated MXC or MXB) and moderately strong banded mylonites (MYF). The orebody in Figure 1 is shown based on magnetite content but for the modelling it was separated into ore hard (OH) and ore soft (OS) categories, based on perceived deformation characteristics.

The ore is bounded on the east by the near-vertical Eastern Contact Fault comprised of clayey gouge and soft serpentinite schist (UXS). East of the fault is competent and strong calcite chlorite schist (MXR).

Other dominant structural features in the west wall rocks of both pits are steeply dipping, N-S striking foliations and foliation shears, which have a major influence on toppling. Dolerite (in the Centre Pit) and magnesite intrusions also influence toppling failures, but usually by increasing stability. There are two sets of pervasive discontinuities in the west wall of the northern-most area that was modelled. These dip 50-55° to the north-east and south-east. Although they occasionally produce small wedge failures on 60° batter faces, large scale wedge failures have not formed due to the steepness of the potential intersection angle (46°) compared with the overall slope angle of 37°.

3 FLEXURAL TOPPLING HISTORY

Discontinuities in the west walls that strike semi-parallel to the pit walls and dip steeply (typically 60 – 85°) into the slope facilitate the development of toppling failures at Savage River. Most of the initial movement occurs on relatively thin foliation plane shears that are spaced 5 – 15 m apart.

The earliest such failure began in the west wall of Centre Pit in 1989, during excavation of a berm at the base of a slope that had been mined at about 45° over a vertical height of 107 m. Very slow horizontal movements of up to 13 m occurred. The slope was redesigned to an overall angle of 37° but failures became active again on several occasions as mining continued. Depressurisation of the slopes via horizontal drainholes and careful mining and monitoring eventually allowed all the ore to be extracted safely. Mapping of the final slope showed that left lateral vertical slip movement had occurred along discrete foliation shears, leading to development of reverse scarps up to 3-4 m high.

Similar failures at section N9400 in North Pit developed recently and have been managed in the same way. The geological profile was similar to that shown in Figure 1, and accelerated movement began when the 37° slope had been mined to a vertical height of about 90 m. Dewatering led to temporary stabilisation of the slope, but movements increased again when the slope was about 120 m high. In addition to reverse scarps along foliation shears, a large block at the top of the slope dropped about 0.6 m along a fault bounding the magnesite lens to the west. The area of toppling was bounded by two pervasive discontinuities of the previously mentioned SE and NE dipping discontinuity sets.

The influence of similarly oriented but less pervasive discontinuities and the strength of the more competent band of mylonite located at mid slope height was not known. A section further south, at N9250, which did not have these discontinuities, experienced a small planar slip / toppling failure at the top of the slope above the mylonite. This section was therefore analysed first, to obtain the minimum strength of the mylonite that prevented toppling in the modelled upper slope from progressing.

4 ROCK MASS MODEL

Recognition of the critical role of foliation shears in the development of a flexural toppling mechanism at the mine led to a decision to use the 2D distinct element program *UDEC* (Itasca 1996) to model the pit slopes at Savage River. This program allows shear and separation to occur on discrete discontinuities and also permits yield of rock block material, in shear or in tension, via a range of nonlinear constitutive models. In addition, it can perform effective stress analyses based on groundwater pressures acting within joints and within block material.

The 3D equivalent program was to be used if *UDEC* could not emulate potential effects of the two discontinuity sets. However the 2D simulation adequately modelled observed behaviour.

A number of foliation shears were to be represented as explicit joints, then the other joint sets could be simulated with the ubiquitous joint rock material constitutive model and/or by treating rock mass yield by using the modified Hoek-Brown yield criterion (Hoek et al. 1995):

$$\sigma_1 = \sigma_3 + \sigma_c(m_b\,\sigma_3/\sigma_c + s)^a \qquad (1)$$

where σ_1 and σ_3 are the major and minor principal stresses respectively (compressive stresses positive), σ_c is the uniaxial compressive strength of intact rock pieces, and m_b, s and a are rock mass parameters.

The Hoek-Brown parameters m_b, s and a for each material were computed from the equations of Hoek et al. (1995), using values of Geological Strength Index (GSI) estimated by relating field observations of each rock unit to Hoek's table (1997).

Itasca (1995) adapted the Mohr-Coulomb material model to represent the behaviour of uniform Hoek-Brown material with a=0.5. This was done by computing an effective cohesion and friction angle for each zone within the model from the tangent to (1) for the current confining stress, and then updating these parameters as the calculation progressed. Coulthard and Little (1999) generalised their code to treat the modified Hoek-Brown yield criterion and to handle multiple materials in *FLAC* or *UDEC* models. That work was generalised here to function with the ubiquitous joint constitutive model in *UDEC*.

Initial estimates of material parameters for the rock units in the *UDEC* models are given in Table 1. The GSI values and ubiquitous joint dips were obtained from field mapping whilst the density, elastic parameters and UCS values came from laboratory testing for the feasibility study (ABM 1996). Shear strengths were estimated from previous direct shear tests or from back analysis of previous toppling failures on the west wall of Centre Pit using a simple limiting equilibrium method. As shown in the Table, some of the parameters were modified slightly during the back analysis phase of this study.

Active discrete joints in the N9250 model (Figure 2) were those on either side of MXCc and a vertical fault behind the crest. These had cohesion of 50 kPa, friction angle 20° and zero tensile strength. All other block interfaces had high strength, so that any yield would develop through block material, on ubiquitous joints or via rock matrix shear or tensile yield.

Groundwater pressures at each stage were computed in *UDEC* from estimated phreatic surfaces that ware based on standpipe readings from the mine. These pressures were then used in effective stress analysis of the response of joints and block material.

Table 1. Rock material parameters for Savage River Mine *UDEC* models

Rock unit	Density (kg/m^3)	Young's modulus (GPa)	Poisson's ratio	GSI	UCS (MPa)	m_i	Ubiquitous joints			
							Dip (°)	Cohesion (kPa)	Friction angle (°)	Tensile strength (kPa)
MXR	2850	32	0.24	70	125	20	-	-	-	-
Fault1	2500	0.7	0.25	25	10	4	85 [90]	6	26	8
OH	3350	2.9	0.25	30	87	15	-	-	-	-
OS	3350	1.7	0.25	25	53 [20]	15 [25]	-	-	-	-
MXB	2830	2.8	0.25	40	25	8	55 (45)	6 (10)	26	3
MYF	2830	10.3	0.25	55	55	6	80 [75]	20	30	10
MXCa	2830	5.2	0.26	40 (50)	15 (25)	8	70 [80]	6 [50]	26	3
MXCb	2830	5.2	0.26	40 (50)	15 (25)	8	45	6 [50]	26	3
MXCc	2830	5.2	0.26	40 (50)	15 (25)	8	90 [75]	6 [50]	26	3
Magnesite	3000	22	0.25	60	100	20	-	-	-	-
Clay	2000	0.01	0.40	-	-	-	-	50	15	50

Note: Initial estimates for N9250 model given, with adjusted values from back analysis in parentheses. Final values for N9400 models, where different, are shown in square brackets. Strength parameters for clay are for isotropic Mohr-Coulomb model.

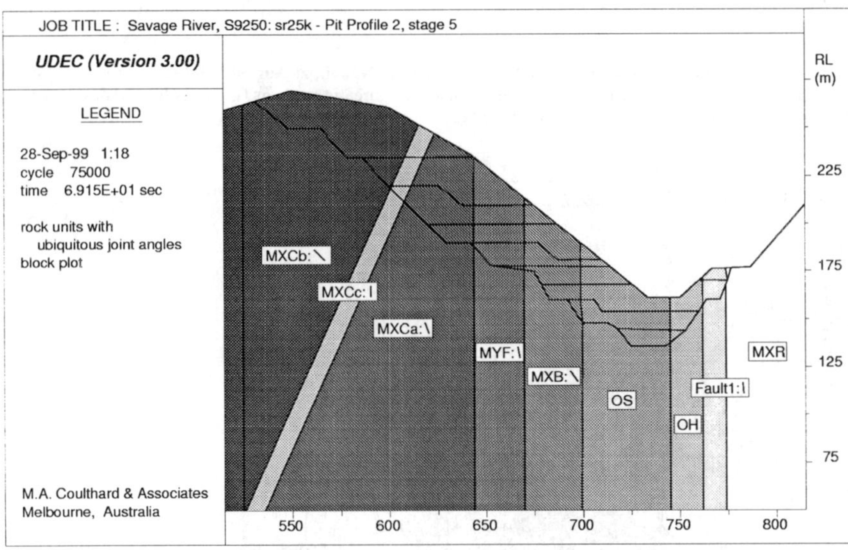

Figure 2. *UDEC* model of North Pit, section N9250. Rock units, with orientation of ubiquitous joints, where modelled, given by slope of lines after rock unit names. Block structure shows future excavation stages from February 1998 to April 1999 pit profiles.

5 NUMERICAL MODELLING

The combined planar slip / toppling failure that had been observed in the upper slopes at Section N9250 of North Pit (see section 3 above) was back analysed by comparison with field observations and trends from surface prism monitoring. The material properties were further adjusted by back analysis of the lower part of that pit slope. A model of the upper parts of the west wall of the pit at N9400 was then compared with observations, to verify the rock parameters before they were used for analyses of future mining options for the extension of North Pit.

5.1 *Back Analyses*

The *UDEC* model of section N9250 was mined in five relatively large lifts from the pre-mining ground profile to the February 1998 pit slope geometry. The rock units represented in that model are shown in Figure 2, which also shows the various excavation stages that were used to simulate mining through to the April 1999 profile. The observed planar failure occurred along 45° outward-dipping joints in the MXCb; these were modelled as ubiquitous joints.

With the rock properties as in Table 1 the upper slope failed even without groundwater pressures, which was clearly contrary to field observations.

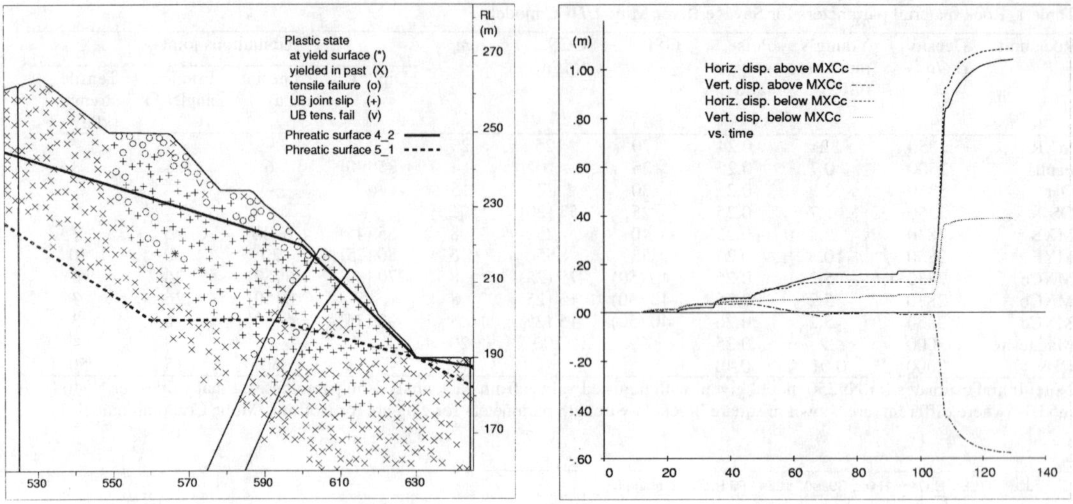

Figure 3. Combined planar slip / toppling failure in upper slope of North Pit, section N9250, at August 1998 stage of mining, stabilises when drainholes lower phreatic surface: left: deformed mesh, plastic yield and phreatic surfaces (heavy solid and dashed lines); right: histories of vertical and horizontal displacement of points on slope face above and below MXCc.

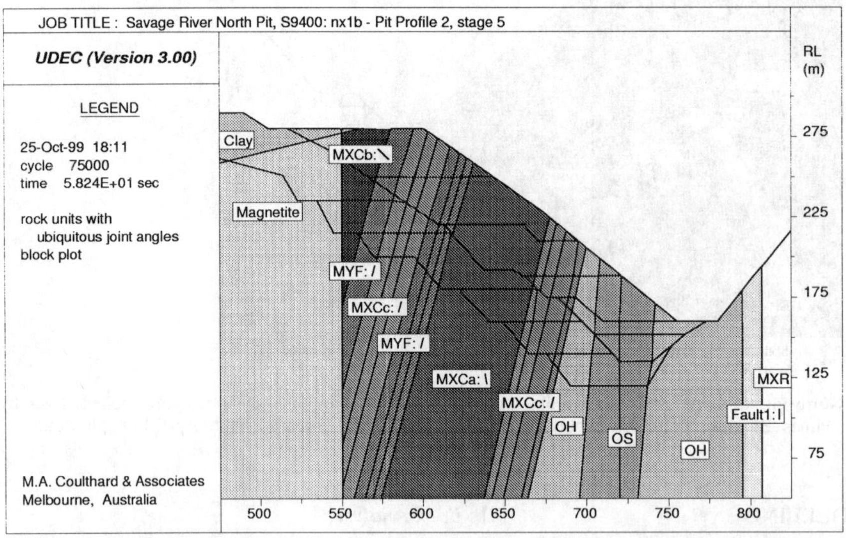

Figure 4. *UDEC* model of North Pit, section N9400. Rock units: ubiquitous joint orientations. Block structure showing future excavation stages from February 1998 to April 1999 pit profile and then on to one possible future pit configuration.

The strength of the mid-slope MXC units was increased until the upper slope was stable with no groundwater (to UCS = 25 MPa and GSI = 50). Next, that model was re-run with estimated phreatic surfaces at each stage. Major instability developed over the range RL 180 – 265 for the pit profile as at August 1998, but this stabilised as soon as a drawn-down phreatic surface was included in the model (Figure 3). Both these aspects were consistent with observations at the mine.

A localised slumping type of failure occurred over RL 150-175 m in the lower west wall of the actual pit. The ubiquitous joint orientation and cohesion of the relevant rock unit (MXB) were adjusted in the numerical model until a similar response was obtained. The final values used in MXB were ubiquitous joint dip of 45° and cohesion (ucoh) of 10 kPa, which produced a somewhat shallower failure than had been observed. With ucoh ≥ 20 kPa that part of the slope was predicted to be stable.

5.2 *Verification Analysis*

The back analysed material parameters were used in a similar model of section N9400 of North Pit (Figure 4). This model contained more discrete joints within the MXC units, to represent extra foliation shears mapped in the field. These joints, the vertical fault at the east boundary of magnetite and the sub-vertical joints in and at the boundaries of the OH and Fault1 units, were all given cohesion of 5 kPa, friction angle 20° and zero tensile strength. As shown in Table 1, some changes were made to the ubiquitous joint orientations in MXC, and the ubiquitous joint cohesion in MXC was increased, to encourage any shearing to localise on the explicit joints rather than being distributed through the rock mass. The clay was modelled as a simple Mohr-Coulomb material, i.e. no ubiquitous joints or Hoek-Brown criterion.

The mechanism here is slightly different in that the upper wedge of MXCb slips downward on the fault at the east of the magnetite unit and so exacerbates the tendency to flexural toppling in the upper foliated units (Figure 5). At later stages of mining slip on foliation shears and toppling begin to develop in the lower slope also, as observed. The response to groundwater drainage again matched well the site behaviour. However, the strength of unit OS had to be adjusted (see Table 1) before the relative movement on the upper and lower shears, originally computed as about 10:1, came closer to the observed ratio of 2:1 (although still computed as 5:1).

This was considered to be a sufficiently good representation of the observed slope response to form the basis for further models of possible future mining of that part of the pit.

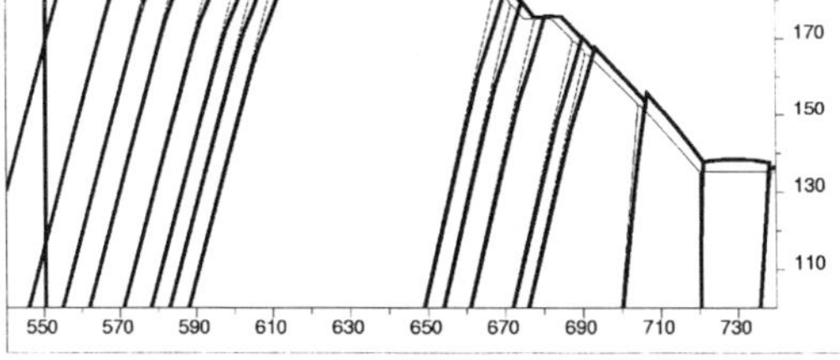

Figure 5. Sliding wedge and toppling mechanism of failure in west wall of North Pit, section N9400, at April 1999 stage of mining. Heavy lines show deformations magnified by factor of 10, with reverse scarps in upper and lower parts of slope.

5.3 *Assessment of Future Mining Options*

Two options for extended mining at section N9400 of North Pit were also analysed in *UDEC* models, using the same material parameters as previously:
Extension 1. incorporating a ramp in the west wall and the pit floor at RL120m; and
Extension 2. without a west wall ramp and with the east wall cut back and the pit floor at RL80m.

Each model was mined to the April 1999 level, then the extension of mining was represented by additional excavation stages. The stability of the west wall was assessed, with and without dewatering.

The Extension 1 model indicated that a wedge-toppling mechanism like that shown in Figure 5 would begin to develop but would probably remain stable without dewatering of the lower west wall. However some further dewatering was recommended to ensure slope stability.

With Extension 2, similar wedge-toppling failures developed at most stages of mining. Dewatering stabilised the slope when the pit floor was at and above RL120, but movements continued slowly as later stages of the computation proceeded, even with extensive dewatering.

6 CONCLUSIONS

UDEC models of the response of the west wall of North Pit at sections N9250 and N9400 reproduced both the observed toppling failures and sensitivity of those failures to groundwater depressurisation. In each case the flexural toppling mechanism was driven by a wedge in the upper slope. At N9250 this "wedge" was formed by slip on the 45° dipping ubiquitous joints in unit MXCb, whilst at N9400 the predominant movement was slip on the vertical fault at the edge of the magnesite unit. The rock mass and joint properties derived from one back analysis were in close agreement with those from the other. This indicated that they provided a consistent model of this complex rock mass, which could then be used to predict the likely response to future mining options at the pit.

Although a similar mechanism begins to develop in the first mining extension option considered at N9400, the pit slope is likely to be stable without dewatering. In contrast, the steeper and deeper Extension 2 was found to be unstable even with extensive dewatering. However, the *UDEC* model and previous experience at the mine both suggest that these failures are likely to progress very slowly, particularly on the relatively flat slopes proposed, and give obvious early warning signs on the surface, such as reverse scarps. Consequently mining to this depth should be achievable with good on-site management, supplemented by monitoring.

ACKNOWLEDGEMENTS

The authors wish to thank Australian Bulk Minerals, the operators of Savage River Mine, for permission to publish this paper.

REFERENCES

Australian Bulk Minerals 1996. Feasibility study for the resumption of operations at Savage River and Port Latta.

Coulthard, M.A. and T.N.Little 1999. Modelling of stability of high westwall at Ok Tedi Copper-Gold Mine. In C.Detournay and R.Hart (eds), *FLAC and numerical modeling in geomechanics:* 39-46. Rotterdam: Balkema.

Hoek, E. 1998. Rock engineering course notes. From Hoek's Corner at www.rocscience.com.

Hoek, E., P.K.Kaiser & W.F.Bawden 1995. *Support of Underground Excavations in Hard Rock.* Rotterdam: Balkema.

Hutchison, B., K.Dugan & M.Coulthard 2000. Analysis of flexural toppling at Australian Bulk Minerals Savage River Mine. To be published in *Proc. GeoEng2000 Conference, Melbourne, 19-24 November 2000.*

Itasca 1996. *UDEC – Universal Distinct Element Code Version 3.0.* Minneapolis: Itasca Consulting Group, Inc.

Itasca 1995. *FLAC – Fast Lagrangian Analysis of Continua Version 3.3.* Minneapolis: Itasca Consulting Group, Inc.

Computer Methods and Advances in Geomechanics, Desai et al. (eds) © 2001 Balkema, Rotterdam, ISBN 90 5809 183 X

LAMODEL – A boundary element program for coal mine design

K.A.Heasley
NIOSH Pittsburgh Research Laboratory, Pa., USA

Z.Agioutantis
Department of Mineral Resources Engineering, Technical University of Crete, Greece

ABSTRACT: LAMODEL 1.0 is a PC-based, boundary-element program for calculating the stresses and displacements in coal mines or other thin, tabular seams or veins. This type of mine modeling software can be used by mine design engineers in the industry to investigate and optimize the pillar sizes and pillar layouts in relation to pillar stress, multi-seam stress and/or bump potential (energy release). This paper introduces the latest version of the LAMODEL program and highlights the new windows-based, pre- and post-processors.

1 INTRODUCTION

Mine planners have a variety of modeling methods, both empirical and numerical, available for analyzing pillar stresses and determining safe pillar sizes for various mine geometries and geologic structures. Empirical methods emphasize the collection and interpretation of case histories of pillar performance. These empirical methods are closely linked to reality, and for many "typical" mining geometries, they work extremely well.

However, it is difficult to apply these empirical methods to mining situations beyond the scope of the original empirical data base. Therefore, when complicated stress conditions arise from complex single or multiple-seam mining geometries, numerical modeling techniques such as finite-element, boundary-element, discrete-element, or finite-difference are usually applied. In general, the numerical, or analytical, design methods are derived from the fundamental laws of force, stress and elasticity. Their primary advantage is that they are very flexible and can quickly analyze the effect of numerous geometric and geologic variables on mine design. Their primary disadvantage is that they require difficult-to-obtain, and/or controversial, information about material properties, failure criteria, and post-failure mechanics.

In order to analyze the displacements and stresses associated with the extraction of large tabular deposits such as coal, potash and other thin vein-type deposits, the displacement-discontinuity (DD) variation of the boundary-element technique is frequently the method of choice. In the displacement-discontinuity approach, the mining horizon is treated mathematically as a discontinuity in the displacement of the surrounding media. Using this technique, only the planar area of the seam needs to be discretized, or gridded, in order to obtain the stress and displacement solution on the seam. Often, this limited analysis is sufficient, since in many practical applications only the distributions of stress and convergence on the seam horizon are of interest. Also, by limiting the detailed analysis to only the seam, the DD method provides considerable computational savings over other techniques, which discretize the entire body (such as finite-element, discrete-element or finite-difference). It is a direct result of this computational efficiency that the DD method is able to handle large areas of tabular excavations as needed in many practical coal mining problems.

A displacement-discontinuity program incorporating a laminated media has recently been developed by the National Institute for Occupational Safety and Health (NIOSH). This new program, called LAMODEL, is designed for calculating the stresses and displacements in coal mines or other thin, tabular seams in layered media. In order to facilitate the transfer of the LAMODEL technology to the mining industry for improving overall mine design, an intuitive, easy-to-use preprocessor, LAMPRE, and postprocessor, LAMPLT, were developed. The LAMPRE program handles all of the numerical parameter input and allows the mine plan to be graphically entered into the program. The new postprocessor, LAMPLT, also uses an intuitive graphical interface, and it allows the user to quickly and easily plot and analyze the output data from the numerical calculation phase of LAMODEL.

2 LAMODEL

Traditional displacement-discontinuity programs use a homogeneous isotropic elastic formulation which simulates the overburden as one solid material. In contrast, the LAMODEL program simulates the geologic overburden stratifications as a stack of layers with frictionless interfaces. Specifically, each layer is homogeneous isotropic elastic and has the identical elastic modulus, Poisson's ratio, and thickness. This "homogeneous layering" formulation does not require specifying the material properties for each individual layer and yet it still provides a realistic suppleness to the mining overburden that is not possible with the classic, homogeneous, isotropic, elastic overburden model. From our experience, this suppleness provides a more accurate strata response than a homogeneous overburden for modeling local deformations, inter-seam interactions and/or surface subsidence (Heasley & Barton, 1999, Heasley & Chekan, 1999, Heasley & Salamon, 1996).

To utilize the LAMODEL program, the user must input: the seam geometry, the seam orientation, the stress field and the geologic material properties. Up to 26 different in-seam materials from 5 different material models (elastic, elastic-plastic, strain-softening, bilinear strain-hardening and exponential strain-hardening) can be used in a model and the grid size can be can be a maximum of 400 X 400. Once the input is developed, the LAMODEL program calculates the stresses and displacements at the seam level and at requested locations in the overburden, or at the surface. The program also has the ability to analyze: 1) the interseam stresses resulting from multiple-seam mining; 2) the effects of topographic relief on pillar stress and gob loading; 3) the stress changes during mining through multiple mining steps and 4) the surface subsidence.

3 LAMPRE - THE PREPROCESSOR

In order to facilitate the transfer of the LAMODEL technology to the mining industry for improving overall mine design, the requirements on the occasional user for data input need to be simplified as much as possible. To this end, an intuitive, easy-to-use preprocessor, LAMPRE, that allows the user to quickly and easily enter both the required material data and the necessary model geometric data was initially developed. LAMPRE runs in the Windows environment using an intuitive graphical user interface with ample context-sensitive on-line help. The preprocessor handles all of the numerical parameter input and also allows graphical input of the material codes for the seam grids. Most recently, a set of "material wizards" for automatically calculating reasonable coal and gob properties from typical values was added to LAMPRE. These wizards greatly reduce the burden on the novice user of specifying the essential material properties for the seam and gob.

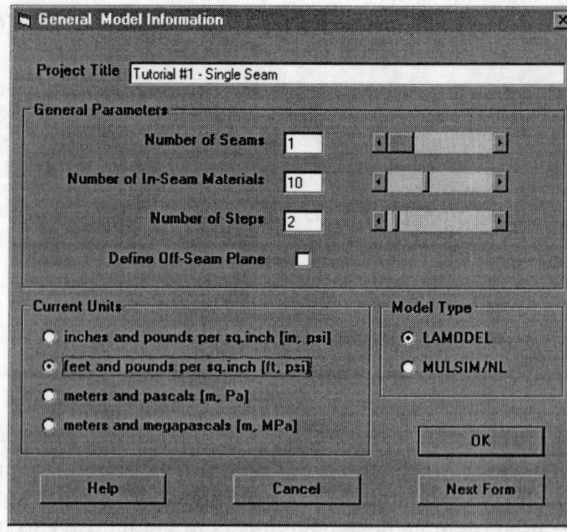

Figure 1: Typical parameter input form in LAMPRE.

A typical LAMPRE input form for parameters values is shown in Figure 1 (in this case the form for the "General Model Information"). In this form, the input parameters are entered using either a text edit, slider or radio button object, whichever is most convenient. If help on the parameter is needed, simply pressing the "F1" function key, while the parameter is active, will open the appropriate help topic. Also, the program checks the value of all of the input parameters before they are saved to make sure that they fall within reasonable prescribed ranges. Using a number of these input forms, all of the geometric and geologic parameters for the LAMODEL program are easily entered.

Once the numerical parameters are entered, the user then needs to define the geometry and material of the pillars and openings in the seam grid. In LAMODEL, the different materials are represented by alphabetical "material codes" (A-Z), and openings are represented by the character "1". LAMPRE provides a graphical editor for creating and modifying the various seam grids as shown in Figure 2. The grid editor essentially works like a spreadsheet. Material codes for individual cells may be set by clicking on a cell and then typing and entering an appropriate letter from A to Z. Blocks of cells may be changed by clicking and dragging the mouse over a range, and then typing and entering the material code. Or, entire rows/columns may be selected and changed using the row/column number. Also, a block of cells may be copied and pasted (greatly simplifying the creation of numerous pillars). Finally, the grid editor allows for automatic yield zone generation based on materials that were automatically defined using the coal property wizard

Once the input file is created, the LAMODEL numerical analysis is run in batch mode for calculat-

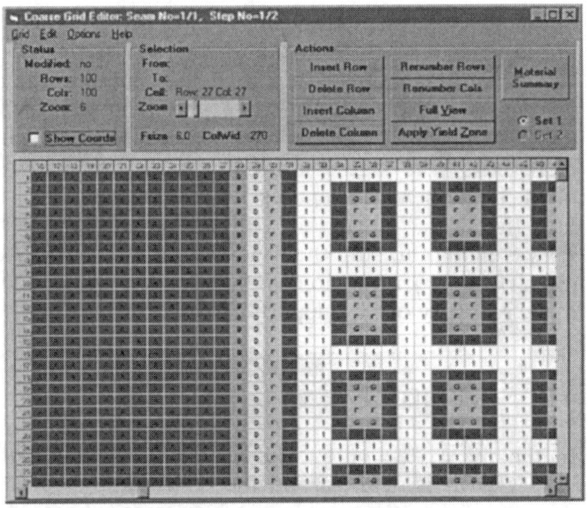

Figure 2: Grid Editor in LAMPRE.

ing the stress and displacements at the seam level. Model runs can take several minutes to several days depending on the computer speed and model complexity (which includes such factors as; number of steps, number of seams, and grid size). The output from the calculation phase is stored in a data file for subsequent analysis by the post-processing program, LAMPLT.

4 LAMPLT - THE POST-PROCESSOR

The post-processing program for LAMODEL, called LAMPLT, allows the user to quickly and easily plot and analyze the tremendous amounts of output data from the calculation phase of LAMODEL. With LAMPLT, the user can interactively examine the seam convergence and the vertical seam stress, and the individual components of the seam stress from: topography, multiple-seams and/or surface effects. These output values can be displayed using either a pseudo 3-D "colored-square" plot (Fig. 3) or a 2-D cross-section. The program also allows interactive

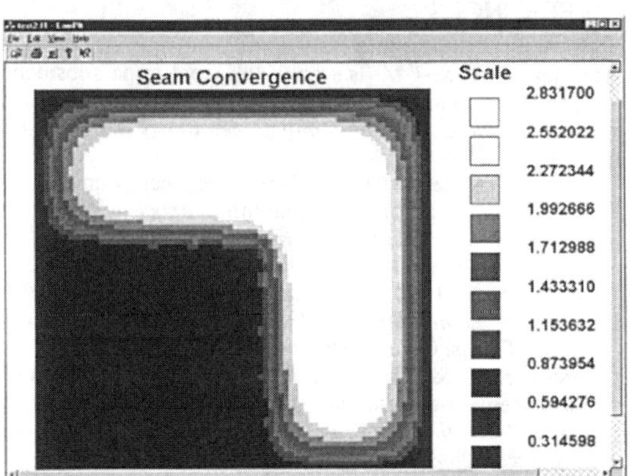

Figure 3: Colored square output in LAMPLT

selection of plot types, step numbers, scaling and color. For additional output, LAMPLT allows cut-and-paste of the graphics for input to other reporting programs.

5 EXAMPLE CASE STUDY

The example case study site is a longwall mine located in Greene County, Pa., USA, and operating in the Sewickley coal seam. This mine is underlain by an abandoned room-and-pillar operation in the Pittsburgh coal seam. The primary problem at this site was the transfer of multiple-seam stress from the lower mine. Yielding of smaller pillars and the subsequent transfer of their load to larger pillars in the lower seam apparently caused increases in vertical stress in the upper seam which were noticed during development of the headgate entries (Fig. 4). Severe pillar spalling and poor roof conditions were experienced when mining the headgate over these large pillars in the lower seam. Mine management was concerned these underlying abutment pillar stresses would continue to be a problem further inby in the headgate and also in longwall panel because there were several areas in the lower seam where similar pillar conditions appeared to exist.

In order to optimally use LAMODEL for accurate stress prediction at a given mine, the program should first be calibrated to the site-specific geo-mechanics based on previously observed stress conditions at that mine. At this site, the model properties were calibrated using underground stress mapping (Heasley & Chekan, 1999). Then, with the calibrated material properties, LAMODEL analyses were created and run in order to predict areas of potential problems within the remaining headgate and the future longwall panel.

Figure 4 shows two areas of the headgate and longwall panel that were modeled using optimized properties from the calibration process. These grey scale plots show the interseam stress, which is the additional stress on the upper mine due to the lower seam mining. In this Figure, Zone 1 covers the upper (inby) part of the headgate panel and the first 365 m (1200 ft) of the longwall panel, and Zone 2 covers the lower part of the headgate (where the stress problems were first noticed) and the last (outby) 330 m (1100 ft) of the longwall panel. In these two zones, the lower mine pillar conditions and the overburden depths appear similar; therefore, the poor pillar conditions encountered in Zone 2 were expected in Zone 1.

However, when comparing the interseam stress between these two zones as shown in Figure 4, it is obvious that the stress is considerably greater in Zone 2 than in Zone 1. Closer investigation reveals two primary reasons for this difference. First, the maximum depth over the gateroads and panel in

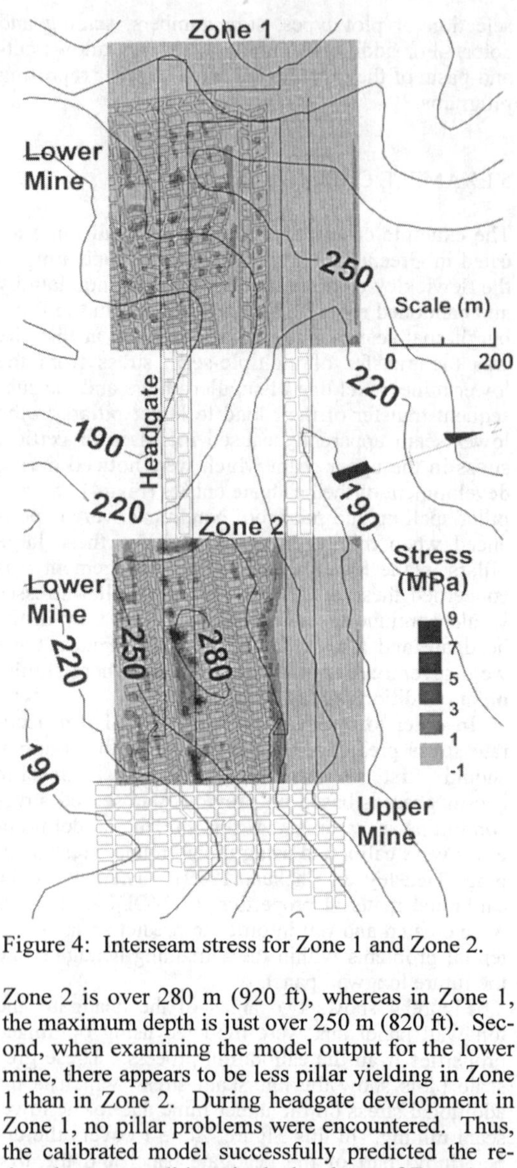

Figure 4: Interseam stress for Zone 1 and Zone 2.

Zone 2 is over 280 m (920 ft), whereas in Zone 1, the maximum depth is just over 250 m (820 ft). Second, when examining the model output for the lower mine, there appears to be less pillar yielding in Zone 1 than in Zone 2. During headgate development in Zone 1, no pillar problems were encountered. Thus, the calibrated model successfully predicted the reduced stress conditions in the headgate of Zone 1.

The mine management was also concerned about the multiple-seam stresses adversely affecting the retreating longwall panel. In particular, a large, irregular shaped barrier pillar in the lower mine is superimposed under the center line of the initial half of the longwall panel in Zone 1 (Fig. 4). However, the interseam stress calculated by the model from this barrier pillar only reaches about 3 MPa (450 psi). When the panel was mined, this slightly increased face stress presented very little problem.

However, in the lower part of the panel near the headgate location where poor ground conditions were first encountered (see Zone 2, Fig. 4), an area of interseam stress up to 9 MPa (1300 psi) is evident in the panel. Because of the underlying barrier pillar, the mine anticipated difficult face conditions in this area. And indeed, when the longwall face reached this area, ground control problems which included severe face spalling and poor roof conditions in the headgate entries were encountered. In fact, the stress interaction with the lower seam was severe enough to stop the longwall face about 15 m (50 ft) short of the longwall recovery chute and make recovery of the supports difficult.

6 CONCLUSIONS

In the paper, the unique laminated formulation of LAMODEL is detailed and the realistic suppleness this provides to the mining overburden is advocated. Also, the program's ability to analyze: interseam stresses, topographic stresses, and surface subsidence is promoted. Next, the preprocessor, LAMPRE, is introduced. This program is described to handle all of the numerical parameter input and to allow graphical input of the material codes for the seam grids in LAMODEL. Then, the postprocessor, LAMPLT, is presented and its two and three dimensional plots are illustrated

Finally, in the case study, the numerous features of LAMODEL (including laminated overburden, multi-seam simulations, strain-softening seam materials, and topographic effects) were used advantageously in simulating an actual mining scenario. Once realistic pillar strengths and load distributions were established by calibration, the mechanics-based overburden behavior in LAMODEL effectively analyzed the complicated stresses and displacements associated with the complex multiple-seam mining scenarios and successfully predicted upcoming high stress conditions in advance of mining for preventive action by mine management.

REFERENCES

Heasley, K.A. & T.M. Barton 1999. Coal mine subsidence prediction using a boundary-element program. *Transactions of the Society of Mining, Metallurgy and Exploration, Inc.*: 306:99-104.

Heasley, K.A. & G. Chekan 1999. Practical boundary-element modeling for mine planning. *Proceedings of the Second International Workshop on Coal Pillar Mechanics and Design,* NIOSH IC 9448: 73-87.

Heasley, K.A. 1998. *Numerical Modeling of Coal Mines With a Laminated Displacement-Discontinuity Code.* Ph.D. Thesis. Colorado School of Mines. 187 pp.

Heasley, K.A. & M.D.G. Salamon 1996. A new laminated displacement-discontinuity program. *Proceedings of the Second North American Rock Mechanics Symposium, Montreal, Quebec, Canada, June 19-21*: 1879-1886.

Computer Methods and Advances in Geomechanics, Desai et al. (eds) © 2001 Balkema, Rotterdam, ISBN 90 5809 183 X

Influence of geometrical distribution of rock joints on deformational behavior of underground opening

Y. Jiang, J. Xiao & Y. Tanabashi
Faculty of Engineering, Nagasaki University, Japan

ABSTRACT: Mechanisms of deformation and failure of underground opening in jointed rock masses are governed by the characteristics of the geometrical distribution of discontinuities. In this paper, the extracting method of geometrical distribution of discontinuities in rock masses by image processing is presented in order to make netmaps of rock joints from the construction field. Next, using those netmaps, fractal characteristics of the discontinuities are described by using the newly proposed model so as to quantitatively evaluate the density of the discontinuous distribution. Finally, a problem of the excavation of underground power plant is analyzed to find the relation between geometrical distribution of rock joints and deformational behavior of underground opening by using DEM.

1 INTRODUCTION

In order to evaluate the deformational behavior and stability mechanism of underground opening appropriately, the key block theory, DDM and DEM have been considering as the useful tools for the support design and stability assessment of underground opening. The analysis modeling of the geometrical distribution of rock joints, however, would influence the results of those analyses, so it's very important and necessary that how to establish a modeling to approach the nature distribution of discontinuities in rock masses.

In nature, almost all rock masses contain discontinuities. The geometrical distribution of rock joints is extremely complicated, and has fractal features. Based on fractal theory, the fractal dimension of the geometrical distribution of rock joints can be calculated by a computer programme. It is thought that the employed fractal dimension can capture the combined influence of joint-size distribution and joints density of rock mass (Kulatilake et al., 1997). As an important parameter, it is useful to support the design of the underground structure.

In this study, the first part of paper shows a new approach to evaluate the fractal feature of jointed rock masses in construction. In the second part, the geometrical distribution of rock joints, with different fractal dimension and orientation distribution are considered in DEM analysis in order to find the influence of those parameters on the deformational behavior of underground opening.

2 AN ANALYTICAL APPROACH TO EVALUATE GEOMATRICAL DISTRIBUTION OF NATURAL JOINTED ROCK MASSES

The DEM and DDM analyses differ from conventional numerical methods in the way that can create the modeling of geometrical distribution of rock joints. This modeling should be oriented toward distribution of nature joints *in-situ* as much as possible. In order to grasp the distribution of rock joints, the following approach is proposed.

2.1 *Extracting methods of the nature joints*

The multiple system given by Figure 1 is for analyzing feature of geometrical distribution of rock joints. Two extracting methods can be constructed; namely the digitizer inputting method and the image processing method. The former is using scanline surveys and borehole survey through the borehole scanner system, invisible discontinuities space data can be estimated so as to obtain a location, orientation, trace,

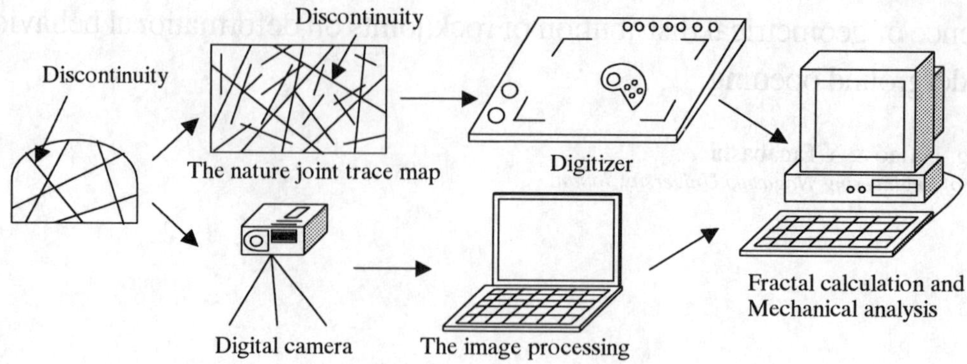

Figure 1. Multiple system for analyzing feature of geometrical distribution of rock joints.

gap etc. of rock joints. Based on the analyzed joint data, the nature joint trace map can be created. Then, using a graphics digitizer instrument and a write computer generator program, the joint trace map can be input to computer through generating a list of x,y-coordinate pairs.

The latter is used to observe the visual area of rock mass *in-situ* such as well-exposed area, the sidewall of underground opening. The photograph of distribution of rock joints around underground opening is recorded by the digital camera with the high resolution. The rock joints are extracted by the image processing. The netmap of rock joints as an analytical object is digitized using the image scanner.

The above two kind of netmaps can be setup as the square cell (360*360 pixels), in which the rock joints are expressed by the fine line of the width of 1 pixel. Then, fractal calculation and mechanical analysis, as indicated in following, are carried out by using those objects.

2.2 *Fractal measure methods*

In a general way, the stochastic method is often used to quantitatively describe the geometrical distribution of rock joints. However, measurement of the density of rock joints is a difficult task, and accuracy of the method is controlled by setting of the probability function. On the other hand, the geometrical distribution of rock joints has the fractal nature. It is considered that the fractal dimension could be higher quality to characterize the geometrical distribution of the joints.

In this paper, the box-counting method is used to determine the fractal dimension of the geometrical distribution of rock joints with the above objects. According to Falconer (1990), fractal dimension D_{Box} by the box-counting method is shown by the following equation.

$$D_{Box} = \lim_{\delta \to 0} \frac{\log N_\delta}{-\log \delta} \quad (1)$$

where D_{Box} = the fractal dimension; N_δ = the of number of boxes, needed to cover the object; and δ = the box size.

Figure 2 shows the graph for calculating fractal dimension of the rock joints. Fractal dimension D_{Box} is defined by the slope of the approximation line on the log-log graph which shows the relationship between cell size and the number of cell filled with the object. The accuracy of the fractal dimension by the box counting is examined by taking account of the lower and upper cut off levels. It is thought that the part of different scale discontinuities have fractal feature in study area. In this paper, optimum cut-off level is obtained by analyzing self-evident fractal object. The lower cut-off level is larger than the resolution of the object figure. The upper cut-off level is required 1/30 to the whole figure size of object. If the following conditions are satisfied, it is thought that the object has fractal feature:(1) The coefficient of determination is over 0.995. (2) The plot points in the relevant are over connected 6 plot points.

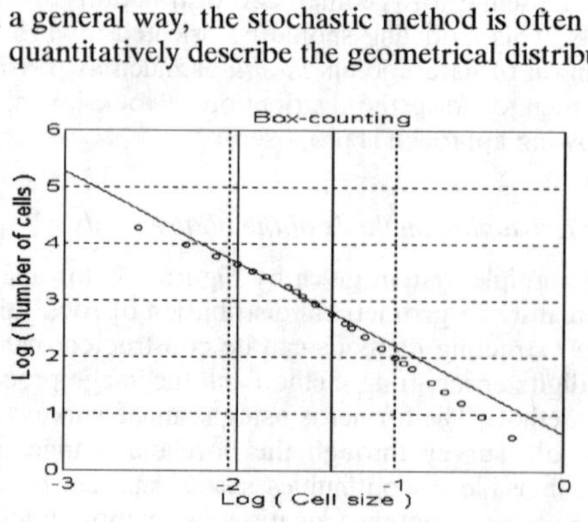

Figure 2. Graph for calculating fractal dimension of modeling.

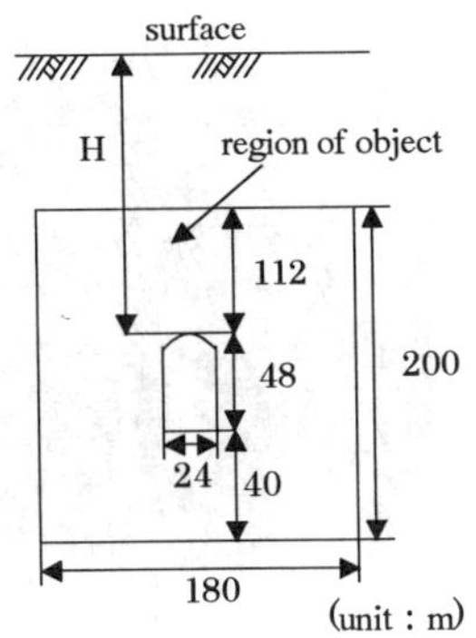

Figure 3. Boundary conditions of model.

Table 1. Properties of the intact rock.

Density	2.50(g/cm^3)
Young's modulus	3.03(GPa)
Poisson's ratios	0.17
Friction angle	40(°)
Cohesion	1.00(MPa)
Tension strength	3.5(MPa)

Table 2. Properties of the joints.

Normal stiffness, K_n	20.0(GPa/m)
Shear stiffness, K_s	0.8(GPa/m)
Cohesion	0.0(MPa)
Friction angle	25.0(°)
Tension strength	0.0(MPa)

Table 3. Joint geometries used for generated joint network.

Case No.	Number of joint sets	Orientation of two sets (deviation)	Fractal dimension	mean values of Joint set (m) Trace	Space	Gap
1	2	0(5) and 90(5)	1.450	60(10)	5(2)	2.5(2)
2	2	0(5) and 90(5)	1.593	30(10)	10(2)	5(2)
3	2	0(5) and 40(5)	1.595	30(10)	10(2)	5(2)

* ( ) is the maximum random deviation from the mean values.

3 THE RELATION BETWEEN FRACTAL DIMENSION AND DEFORMATIONAL BEHAVIOR OF UNDERGROUND OPENING

In order to support the design and construction, it is essential to have a good understand of the influence of the geometrical distribution of rock joints on the deformational behavior of underground opening. In this study, the relationships of the parameter described the joint geometry, such as fractal dimension (density and trace-distribution), and the deformational behavior of underground opening are estimated by the DEM analysis.

3.1 *Description of modeling*

The underground opening, as the underground power plant is the warhead type, height is 50m and width is 25m. Depth of opening is about 160m (see Figure 3).

The ground rock messes are elastic body of the fresh and hard sandstone, the properties are shown in Table 1. The mechanical behavior of rock joints is represented by an elasto-plastic law and the coulomb slip law and the properties are evaluated by the newly designed shear test apparatus under the constant normal stress levels (Jiang et al, 1999). Excavation simulations of all cases are carried out by using UDEC (Cundall, 1971).

3.2 *The relation between fractal dimension and deformation behavior of underground opening*

Figure 4 shows the deformational behavior of rock masses around opening according to fractal dimension D_{Box} and orientation. Table 3 gives a summary of the mean joint set for Cases 1, 2 and 3. The generated square regions have the different joint networks. Using the proposed measure method, the fractal dimension D_{Box} of three cases in total analytical area are calculated as shown in Table 3. Those joint networks are incorporated in DEM analysis. The deformation behavior of underground opening is discussed by comparing all cases. Case 1 and Case 2 are with the different fractal dimension, and Case 2 and 3 are with the different orientation distribution.

Figure 5 shows the maximum displacement according to the fractal dimension, which incorporated the effect of joint orientation distribution. The displacement vectors are significantly influenced by D_{Box} when θ is small than 90°, especially θ = 60°.

As shown in Figure 6, a series of numerical experiments are carried out by changing θ and α. Comparing with those results, followings are clarified: (1) when α = 30° and $\theta >$ 60°, the change of values of maximum displacement is small. (2) there is a peak value of maximum displacement on the curves while θ = 60° and α = 0°, 60°.

4 CONCLUSIONS

In this study, multiple system for analyzing feature of geometrical distribution of rock joints is developed.

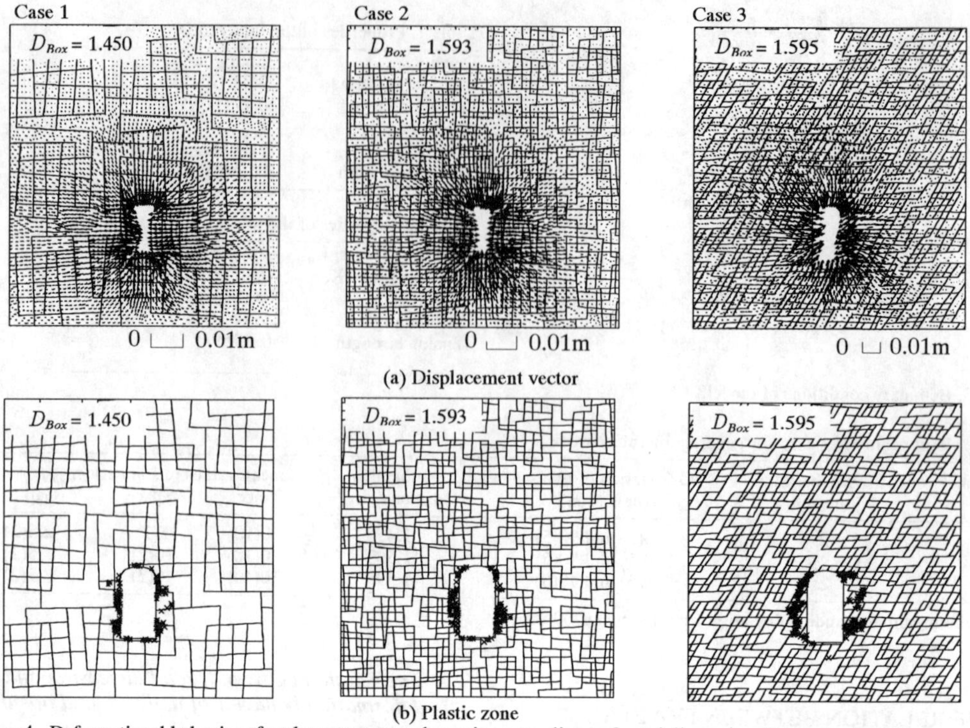

Figure 4. Deformational behavior of rock masses around opening according to fractal dimension D_{Box} and orientation.

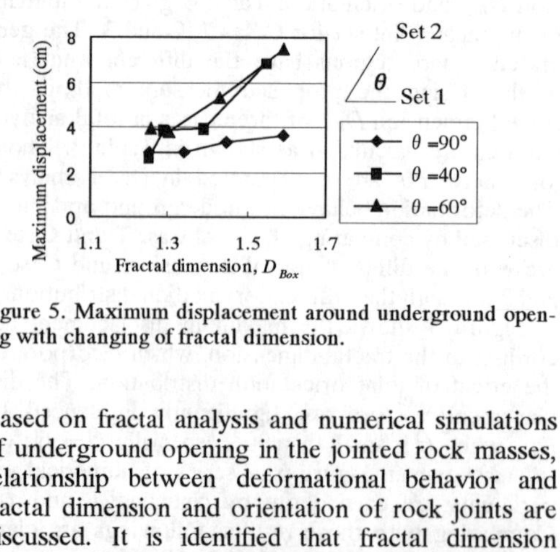

Figure 5. Maximum displacement around underground opening with changing of fractal dimension.

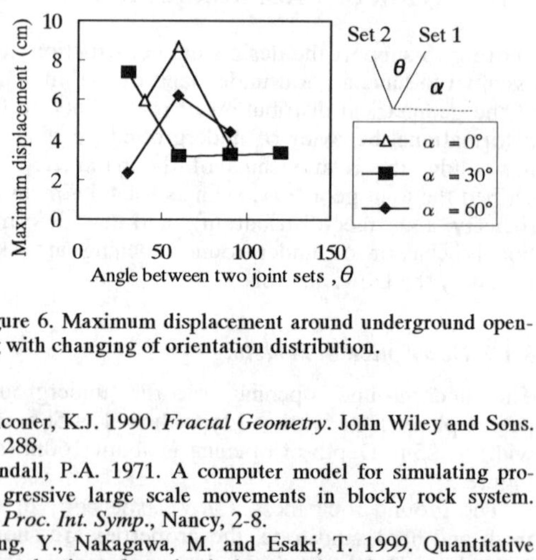

Figure 6. Maximum displacement around underground opening with changing of orientation distribution.

Based on fractal analysis and numerical simulations of underground opening in the jointed rock masses, relationship between deformational behavior and fractal dimension and orientation of rock joints are discussed. It is identified that fractal dimension shows a significant effect on the deformation behavior of opening according to the orientation distributions of rock joints.

5 REFERENCES

Kulatilake, P.H.S.W., Fiedler, R., Panda, B.B. 1997. Box fractal dimension as a measure of statistical homogeneity of jointed rock masses. *Engineering Geology* 48, 217-229.

Falconer, K.J. 1990. *Fractal Geometry*. John Wiley and Sons. 288.

Cundall, P.A. 1971. A computer model for simulating progressive large scale movements in blocky rock system. *Proc. Int. Symp.*, Nancy, 2-8.

Jiang, Y., Nakagawa, M. and Esaki, T. 1999. Quantitative evaluation of mechanical properties of the natural rock joints for analyzing behavior of structures in discontinuous rock masses. *Journal of Geotechnical Engineering*, Japan Society of Civil Engineers, No.624/3-47, 231-243.

Xiao, J., Jiang Y. and Tanabashi Y. 2000. Study on Stability Mechanism of Underground Opening in the Jointed Rock Mass. *Proc. of the 6th Joint Symposium of Cheju National University and Nagasaki University on Science and Technology*, 77-81.

Computer Methods and Advances in Geomechanics, Desai et al. (eds) © 2001 Balkema, Rotterdam, ISBN 90 5809 183 X

Analysis of creep and progressive failure for tunnel in soft rock

P. Jongpradist & H. Horii
Department of Civil Engineering, University of Tokyo, Japan

ABSTRACT: A method of numerical analysis for tunnel design should be able to reproduce the mechanical behaviour of the geological material and effect of tunnel support. In this study, we develop an analytical method based on creep model and progressive failure to reproduce creep failure of tunnel in soft rock. To develop a creep constitutive model, power creep law and creep flow rule are employed with an equivalent stress which is defined for soft rock. Creep analysis of tunnel with no progressive failure is first carried out with the creep model implemented in FEM code. A path of the failure surface is assumed in advance based on the stress fields obtained from the creep analysis. In the creep analysis of tunnel with progressive failure, the shear sliding starts when the failure condition is satisfied. It is shown that the proposed method can reproduce the creep failure in tunnel excavation which is a possible mechanism of troubles in tunnel construction.

1 INTRODUCTION

There are various types of troubles in tunnel construction in soft rock. One of them is an unexpected increase in tunnel deformation after tunnel supports are completed and the tunnelling face moved further forward. The mechanism of such trouble is not clear. Many researches on behavior of soft rock revealed that time dependent deformation property plays an important role (Adachi et al. 1999, Oka et al. 1998). According to the fact investigated from creep tests of soft rock, creep failure may occur in some situation as shown in figure 1. The figure illustrates result of triaxial creep test on mudstone obtained from a tunnel construction site in Japan, with dry density 1.97 g/cm^3, Young`s modulus = 5190 kgf/cm^2, conducted under constant stress state. The creep failure occurred at creep strain of about 1% in this case. Creep failure is one of the candidates of the trouble mechanisms.

To discuss the possibility of creep failure being responsible for this tunnel trouble, the analysis method must be able to predict the creep failure of the geological material. One way to reproduce creep failure is in material model to provide a law for the variation of creep parameters with increasing strain (G. Gioda and A. Cividini 1994). By this procedure, creep deformation proceeds and approaches tertiary creep and ultimately failure. In reality creep strain is distributed. At certain stage, it is localized and form a failure plane. To analyze localization problem, it requires very heavy computations in FEM. It has been known that the results of the analysis in which distributed failure surfaces are localized into one failure surface, are not so different from that of the analysis in which one failure surface is assumes from the beginning. Then, a simplified analysis method was adopted in this study. That is, a path of failure surface in tunnel problem in soft rock is assumed in advance based on stress filed obtained from the creep analysis. Along this predicted path, shear sliding starts when the failure condition is satisfied in the analysis.

In engineering practice , it is well recognized that tunnel supports such as lining and rock bolts are effective to reduce deformation. Therefore the method should also be able to reproduce the effects of tunnel supports. In this work, a method for numerical analysis of tunnel construction by NATM in soft rock which can reproduce creep failure and effects of tunnel supports is presented.

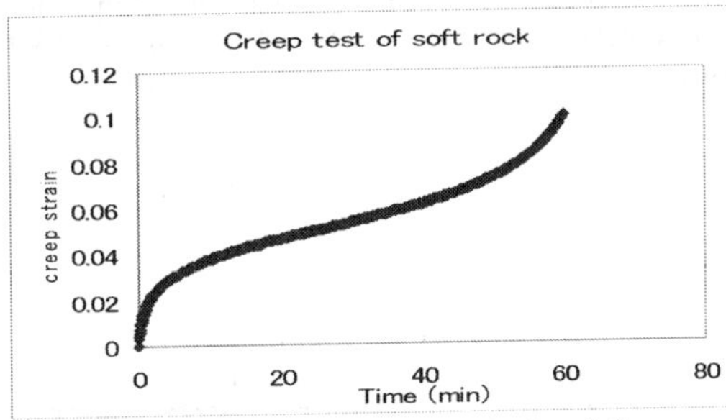

Figure 1. The result of triaxial creep test of soft rock from a tunnel construction site (axial stress = 13.4 Mpa, confining pressure = 2.5 Mpa)

2 CREEP ANALYSIS

2.1 *Creep constitutive model*

In order to reproduce or predict the creep behavior of soft rock, we employ a viscoplasticity constitutive model. Soft rock is said to be c-ϕ material and its behavior depends not only on deviatoric stress but also on mean stress. As the strength of soft rock is characterized by Mohr-Coulomb failure criterion with cohesive stress c and frictional angle ϕ, the dependence of the creep behavior of soft rock on the stress state must be modeled properly. It is treated in the constitutive model with an equivalent stress $\bar{\sigma}$ and in this study we define it as,

$$\bar{\sigma} = \frac{1}{2\cos\phi^*}\left[\sigma_{\max}\left(1+\sin\phi^*\right)-\sigma_{\min}\left(1-\sin\phi^*\right)\right] \qquad (1)$$

where ϕ^* is a material parameter which is different from the internal friction angle. The physical meaning of the equivalent stress is shown in figure 2.

To formulate a creep constitutive equation, an equivalent creep strain must be defined. Here, the equivalent creep strain rate is defined as

$$\dot{\bar{\varepsilon}}^c = \sqrt{\frac{2}{3}\dot{\varepsilon}^c_{ij}\dot{\varepsilon}^c_{ij}} \qquad (2)$$

By adopting the flow rule in theory of plasticity, the creep flow rule is given by

$$\dot{\varepsilon}^C_{ij} = \lambda\frac{\partial g}{\partial \sigma_{ij}} \qquad (3)$$

where λ is a scalar function which represent the dependence of creep strain rate on the stress, strain and time. The mechanism of creep behavior is considered to be similar to that of plastic deformation , the second invariant of deviatoric stress, J_2, is selected for the creep potential g.

$$g = J_2 \qquad (4)$$

A power creep law is introduced as a relationship between equivalent creep strain, equivalent stress and time. The creep law provides the development of creep strain in creep test under constant stress. The following simple relationship is adopted in this study.

Figure 2. Definition of the equivalent stress used in this study.

$$\bar{\varepsilon}^c = A\bar{\sigma}^m t^n \qquad (5)$$

where t stands for creep time which is the time passed after the constant load is applied, A, m and n are material parameters. By this simple relationship, the experimental results of creep test of silt from a tunnel construction site can be reproduced.

Equation (5) is for constant stress. To derive the constitutive relationship under arbitrary loading history, a hardening rule must be employed. In this study, time hardening is adopted for simplicity. It is shown that experimental results in this study can be reproduced by this simple relationship.

$$\dot{\bar{\varepsilon}}^c = nA\bar{\sigma}^m t^{n-1} \qquad (6)$$

By substituting Eq.(3) with Eq.(4) into Eq.(2), we can express the equivalent creep strain rate as:

$$\dot{\bar{\varepsilon}}^c = \lambda\sqrt{\frac{2}{3}s_{ij}s_{ij}} \qquad (7)$$

where s_{ij} stands for the deviatoric stress tensor. From Eq.(7) and Eq.(6), λ is expressed in terms of current stress, current creep strain and time. Substituting it to the flow rule, the following creep constitutive equation is obtained.

$$\dot{\varepsilon}^c_{ij} = \sqrt{\frac{3}{2}}\frac{nA\bar{\sigma}^m t^{n-1}}{\sqrt{s_{kl}s_{kl}}} \qquad (8)$$

The above equation is easily implemented into finite element codes.

2.2 *Creep analysis*

First, creep analysis of tunnel excavation in soft rock are carried out in order to examine whether the proposed model can represent the effects of tunnel supports. The tunnel with a cross section area about 200 m^2 is excavated in soft rock at a depth of 120 m. The geometry, boundary condition and FE mesh used for the analyses are shown in figure 3. Plane strain state and symmetry are assumed. Material parameters used in the analysis are listed in Table 1.

Table 1.Material Parameters for soft rock.

	notation		unit
Elastic Modulus	E	5190	kgf/cm^2
Poisson`s ratio	ν	0.34	
Density of soft rock		1.97	g/cm^3
Creep parameters			
	A	$9.39*10^{-6}$	$(kgf/cm^2)^{-1.748}$ $min^{-0.27}$
	ϕ^*	$\pi/15$	
	m	1.748	
	n	0.27	

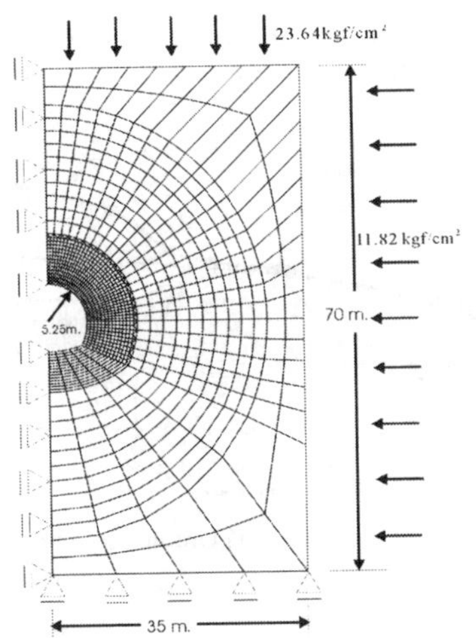

Figure 3. Boundary condition, geometry and FE mesh used for the analysis

2.2.1 *Effect of lining*

The effect of concrete lining on reduction of tunnel wall displacement are investigated in this subsection. The material of concrete lining is assumed to be linear-elastic with Young`s modulus and Poisson`s ratio of 33,000 kgf/cm^2 and 0.1, respectively.

Figure 4 shows the analyzed time history of tunnel wall displacement for various values of lining thickness and result with no lining. It is seen that installation of lining results in significant reduction on wall displacement, compared to the case with no lining. The lining thickness has effect on reduction of the wall displacement.

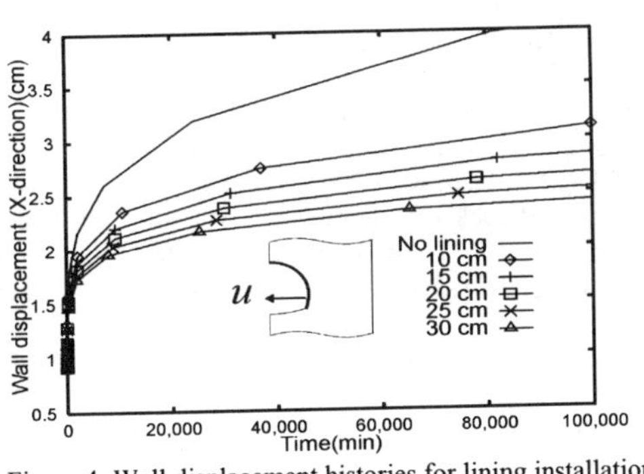

Figure 4. Wall displacement histories for lining installation.

2.2.2 *Effect of rock bolts*

In this subsection, the effect of rock bolts is studied. Steel rock bolts with Young`s modulus of 2.1*10$^6$ kgf/cm$^2$, poisson`s ratio of 0.25 and diameter of 25 mm are used. It is assumed that grouting provides perfect bond between the bolts and rock.

The time history of wall displacement is shown in figure 5 for different length L and spacing d of rock bolts. The figure shows that installation of rock bolts leads to a reduction of wall displacement.

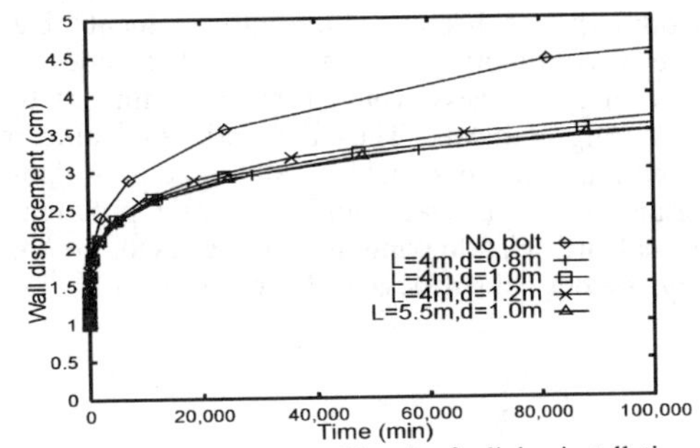

Figure 5. Wall displacement histories for lining installation.

3 PREDICTION OF FAILURE PATH

In this section a method to predict the failure path which will be used in progressive failure analysis in the next section is presented. First the stress state at a certain time is obtained from creep analysis presented in the previous section. The principal stress direction and accordingly the orientation of failure plane at any location is calculated.

The point along tunnel surface which has the maximum shear stress is selected as the starting point for failure path. By solving differential equation with initial value, the failure path is obtained

It should be noted that the predicted failure path obtained from this procedure might not be correct since it is based on a certain stress state obtained from creep analysis without failure and associated stress re-distribution. However, the results of the analysis of progressive failure of soft rock show that there is no significant difference of failure paths predicted from creep analyses with and without progressive failure. Therefore, in this study, the failure path predicted from a certain stress state by creep analysis without progressive failure is used in the analysis.

4 CREEP ANALYSIS WITH PROGRESSIVE FAILURE

4.1 *Finite element mesh with interface element*

The dilatancy of the failure surface is not considered. The interface elements with softening constitutive relation are embedded along the predicted failure path as shown in figure 6. The slip of interface starts and the failure is formed once the criterion of creep failure is satisfied.

The interface element used in this study is zero-thickness four-node element as illustrated in the figure 7. The normal stiffness is chosen so as to make the displacement jump negligibly small compared to the deformations of the surrounding ordinary elements. The failure surface, which is usually a smooth curve, is discretized by linear elements. Due to the discontinuity of the tangent at the node, too high normal stiffness would prevent sliding of the neighboring elements. Thus the optimum value of the normal stiffness, which allows a negligible overlap of the interface surfaces while producing minimal artificial locking has to be used. In this study, the normal stiffness is chosen as 10^6.

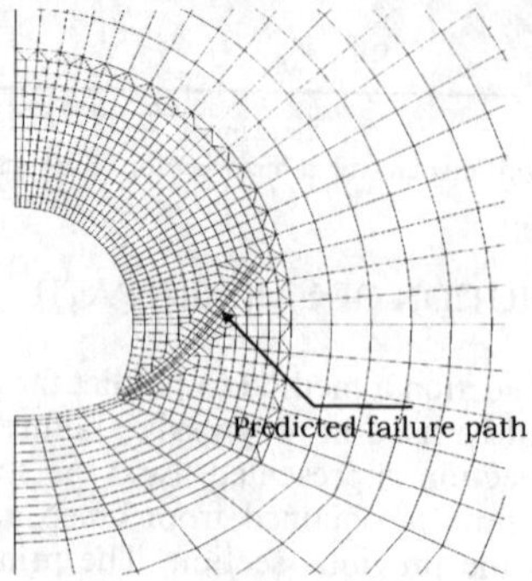

Figure 6. FE mesh included predicted failure path

It is assumed that the creep failure occurs when the equivalent creep strain reaches a critical value. From experimental results of creep test of soft rock obtained from the construction site, the critical equivalent creep strain is selected as 1.0 %. When this creep failure criterion is satisfied, the shear sliding along the failure path starts. After the sliding starts, a relationship between the shear stress and shear displacement jump is satisfied by the interface element. In the present study, the shear stress is assumed to drop to zero just after the failure criterion is satisfied.

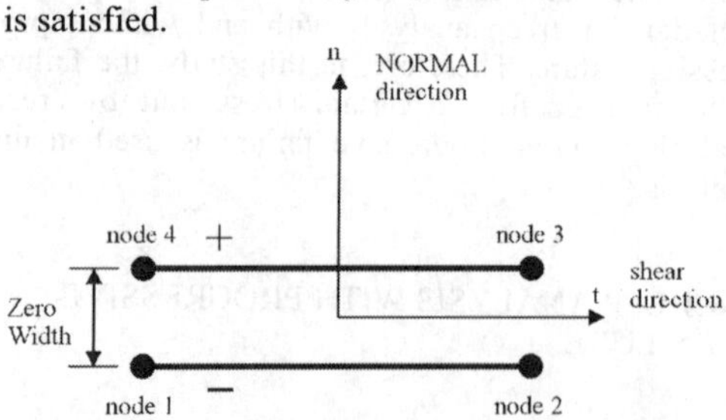

Figure 7. The interface element and used in this study.

4.2 *Result of the analysis*

The creep analysis of tunnel excavation with progressive failure is carried out. The result of this analysis is shown in fig 8. Figure 8 compares the tunnel wall displacements with and without progressive failure. It is seen that the creep failure leads to increase in wall deformation after a long time has passed since the excavation.

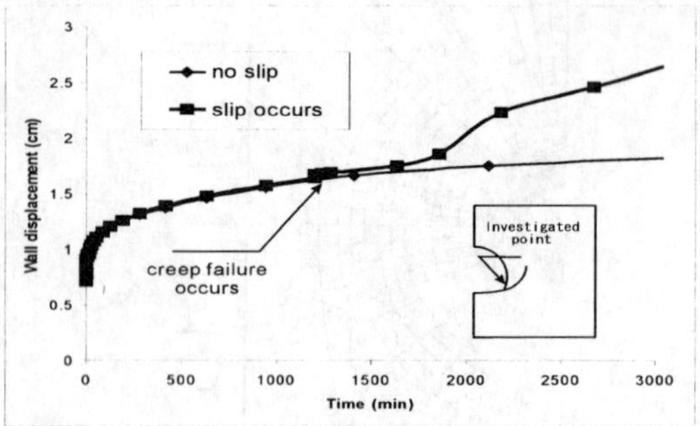

Figure 8. The effect of creep failure on increase of wall displacement

5 SUMMARY

An analysis method was developed to predict creep failure due to tunnel excavation in soft rock. The analysis results show that the proposed method can reproduce the effects of tunnel supports. The creep failure and associated progressive failure leads to increase in the tunnel deformation after some time has passed since the excavation. This seems to explain the mechanism of troubles faced in tunnel construction in soft rock.

REFERENCES

Adachi, T., F.Oka & M. Koike 1999. Time dependent behavior of soft rokc and elasto-plastic constitutive model with strain softening. *Proc. 11th Asian Regional Conf. On SMGE, Seoul*, Vol.1,pp.117-120.

G. Gioda& A. Cividini 1994. Finite Element Analysis of Time Dependent Effects in Tunnels. *Visco-plastic behavior of geomaterials*. New York, Spinger-Verlag.

F. Oka, T Adachi & A Yashima 1998. A constitutive model for soft rock with both strain softening and negative dilatancy. *The Geomechanics of Hard Soils – Soft rocks (Evangelista & Picarelli eds.),* Balkema, pp. 737-744.

Marumi, S.& Y. Yamada 1974. Effects of third invariant of deviatoric stress tensor on transient creep of thick-walled tube. J.*of Engineering Materials and Technology*, July 1974, pp. 207-213.

Computer Methods and Advances in Geomechanics, Desai et al. (eds) © 2001 Balkema, Rotterdam, ISBN 90 5809 183 X

Bored tunnels in soft soil – Fundamentals through three-dimensional numerical applications

W.L.Leendertse & P.S.Jovanovic
Project Organization High Speed Line South, Project Office North Holland, Utrecht, Netherlands

C.B.M.Blom
Delft University of Technology, Netherlands

ABSTRACT: Damages of concrete segments are inherent to shield driving tunneling in soft soils. Almost in all cases they occur during the assembling stage. Experiencing the construction of three large diameter bored tunnels in The Netherlands it becomes clear that the assembling stage is the most critical phase in the construction process. However, due to inability to take into account all aspects which are playing a role in the complex situation that exists during tunnel driving and segment erection, many tunnel designers only consider the serviceability stage ('final situation') of structural behavior. Optimizing lining design and execution procedures should minimize the effects of possible assembling stresses that are leading to damages, leakage and increase of maintenance costs. The main goal of this paper is to present the Dutch experience with three-dimensional Finite Element Method (3D-FEM) modeling of tunnel structure, which seems to be essential for structural analyses of realistic tunnel lining behavior.

1 INTRODUCTION

Design practice show that the structural analyses and calculations of the tunnel lining using engineering's models are usually based on simplified estimation of reality. But it is obvious that the exact parameter values and perfect models are very difficult to obtain. To develop the design process on a proper way it is very important to have knowledge about all the parameters that are influencing safety level.

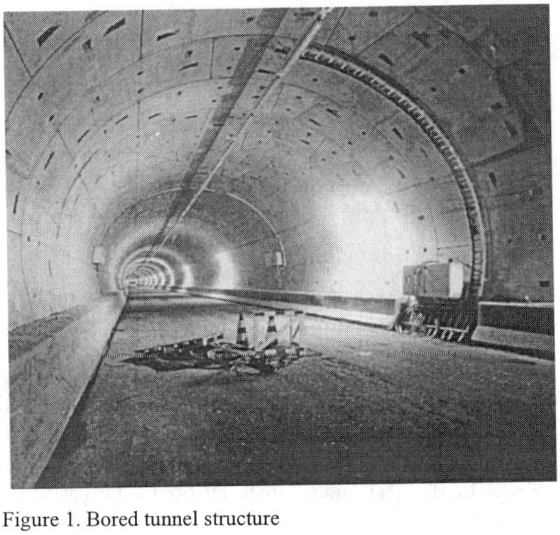

Figure 1. Bored tunnel structure

Knowing these parameter values the estimation of possible risks could decrease or even diminish. On the other hand if the parameters are perfectly matched but the model is not accurate than the results are going to be the same as with non-accurate parameters and perfect model. The statement that the chain is as strong as the weakest link is in this case valid.

2 DESIGN FUNDAMENTALS

Some of the assumptions used by current design practice for the tunnel lining are as follows:

- Serviceability stage is normative. The construction stage is not representative.
- Structure is linear elastic
- Stress Distribution State in the segments is plain-stress
- Load is full overburden
- Supports are the primary ground supports

Two-dimensional models drawn on these assumptions have been used in tunnel lining design for many years. These models are inaccurate because they cannot predict essential aspects of tunnel lining behavior such as damages, leakage, deformations, cracks and deviations, which occur mostly during the construction phase.

3 THREE-DIMENSIONAL MODELING

In order to avoid inaccuracy of two-dimensional models design engineers from Project Office North Holland developed a 3D-FEM model in which essential construction phase aspect of tunneling has been taken into account (see Figure 2.).

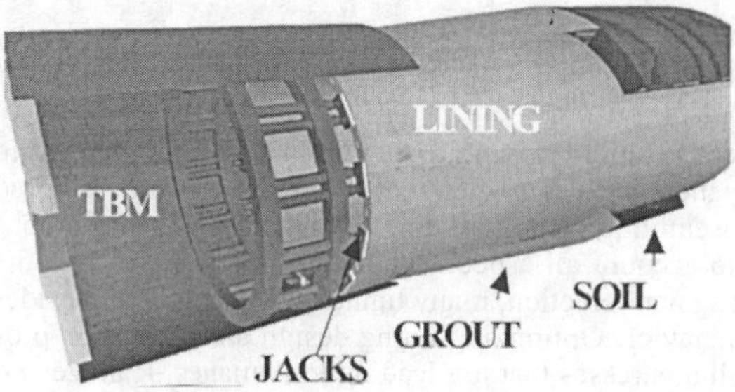

Figure 2. TBM with lining and surrounding during the construction stage

With this model it is possible to analyze a staged simulation of the building process by adding new tunnel rings and altering loading and support after calculating an initial loading situation. The impact of

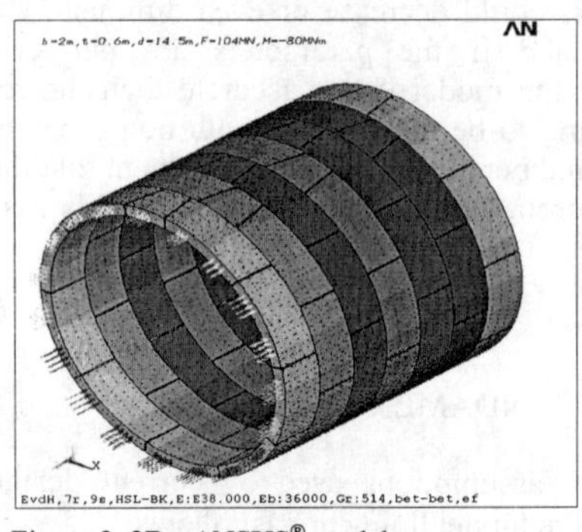

Figure 3. 3D - ANSYS® model

jack forces in combination with grout and soil loads on tunnel lining increase the values of internal stresses causing stress peaks and unwanted segment rotations and sliding in all directions. As a validation of these three-dimensional models several observations and measurements have been carried out on existing tunnels in The Netherlands. Pure model calibration has been done using full-scale test in Stevin Laboratory (TU Delft) of three 9,54 m. diameter rings. The results from the measurement on site, laboratory test and the numerical analysis coincide.

3.1 *Model description*

Tunnel rings with segments are meshed with full-3D linear elastic volume eight-nod elements (ANSYS type Solid45). Contact between segments is modelled with point to point contact elements (see Figure 4), which are capable of supporting only compression in the direction normal to the surfaces and shear (Coulomb friction) in the tangential direction. Surrounding ground as well as the injected grout is modelled as an elastic bedding, using surface elements, where the grout stiffness increases from values close to zero, directly behind the Tunnel Bore Machine, to ground stiffness. Here is assumed that the stiffness of the solid grout is equal to the soil stiffness.

3.2 *Boundary conditions and loads*

In axial direction the rings are supported by a spring system, which represents the stiffness of 10 tunnel rings. Deformation of a ring in its plane is countered by the elastic bedding. The radial loading consists of grouting pressures that change from a certain level directly after being injected to another level, when it became solid, close to the full overburden. It is assumed that the grout stiffness depends on hardening process increase proportionally with an increment to full soil stiffness. The jack forces from TBM are applied in axial direction with their expected values during the standstill.

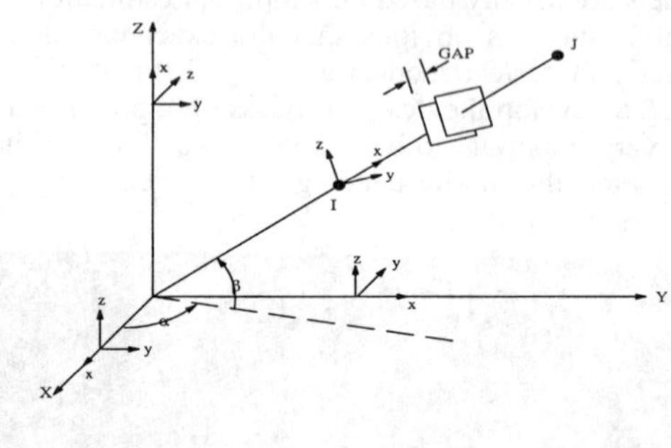

Figure 4. ANSYS® contact element CON52

3.3 *Description of phased calculations*

In the first phase three rings are active, which are surrounded by completely hardened grout. Two rings are fully loaded by radial load; the last third ring is assumed to be half in the TBM and is not fully loaded in radial direction but well in axial direction by TBM jacks. This is the starting point for the phased analysis. After the calculation of the 3 rings loaded 'at once' installation of new rings is started:

- Position of previous ring defines position of the new ring. The data such as deformation is calculated at 19 points in the ring joint. A new ring is going to be placed towards adjunct deformed ring, but without loading.
- Jack forces are moved from previous to new ring
- Radial loads and supports are activated on the new ring
- Repeating the procedure for every new ring

Most of calculations have been carried out with 11 rings but it is possible to add more of them.

4 RESULTS OF CALCULATIONS

Purpose of these calculations was to predict possible damage patterns on the segments as well as the tunnel deformations with joint rotations during the construction stage. Some of the results are given below.

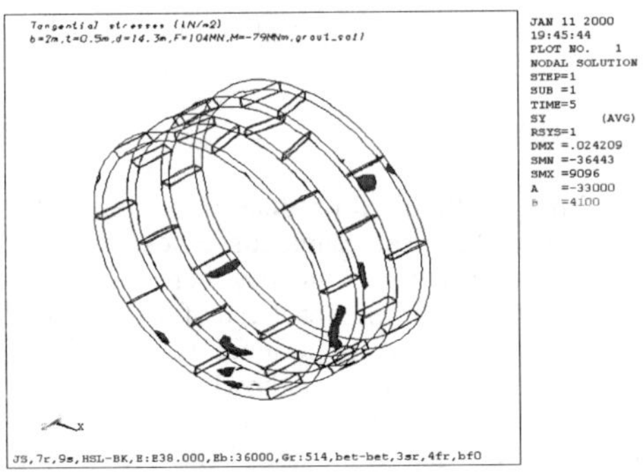

Figure 5. Damage patterns

Figure 5 shows the critical places on the reinforcement concrete segments where the tension stresses are above the allowable values. To avoid these critical zones it is necessary to consider various countermeasures during the construction of the tunnel (see [2]).

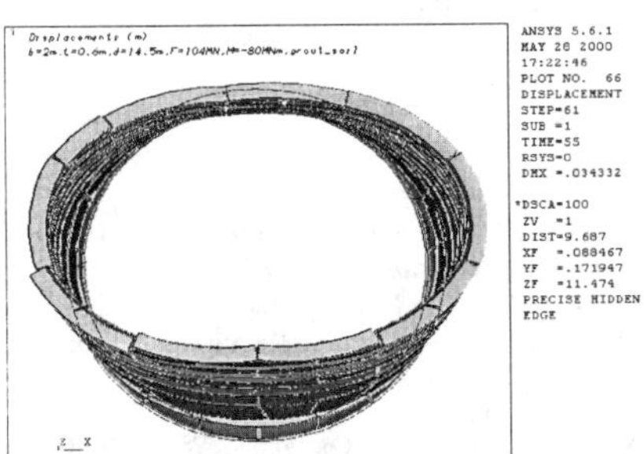

Figure 6. Tunnel deformation

Figure 6 shows the possible global tunnel translation, deformation and joint displacement with rotation. Analyses show that the joint rotation at some places is bigger than predicted in the design. This means that the shape of the joint with clearances between the segments could be adjusted before construction to avoid leakage problems and damages during the tunnel erection. The other problem that may occur is joint sliding. On Figure 6 could be also seen that at some places occur sliding between the segments, which can cause difficulties with assembling of next ring. All relevant tunnel design aspects can be analyzed applying this 3D-FEM model concerning main requirements for the tunnels structure such as: stability, functionality, reliability and water tightness.

5 MODEL CALIBRATION – FULL SCALE LABORATORY TEST

The main goal of the full-scale laboratory test was to calibrate the model, which is used in previous analyses. This is the first test facility in the world where the construction stage of tunneling has been observed.

5.1 *Test facility*

Facility is located at Delft University of Technology, Stevin laboratory, Delft. It consist of: bearing steel frames for longitudinal and radial loads (total steel weight is 200t, beams HE1000M, columns HE800M, axial reaction bars Dywidag FeP1230), outer diameter of 15,0m, 6,5m high; 3 concrete rings (7 segments + key segment) supported by rubber and Teflon plates (total weight 85t, outer diameter of 9,45m, thickness of 0,40m, segment with of 1,5m), possibilities of testing different tunnels from 8m to 11m diameter.

Figure 7. Laboratory test facility

Jacks: 14 hydraulic jack to simulate TBM axial forces (5000kN) and 84 radial jacks for the soil loads (850kN) with enhanced loads steering system. Additional active jacks have solved torque stability problems. Measurement devices are 200 strain gauges with measurement ability in horizontal, vertical and diagonal direction; deformation laser system within the tunnel; 300 deformation measurement devices at joints connected to the data acquisition system.

5.2 *Some of the research topics*

There are so many unknown parameters which should be investigated but the most important one are presented:

- Influence of the segment's complex geometry, joints and packing materials on stresses and deformations under full longitudinal (actual jack forces) and radial loads
- Influence of radial load differences (two rings fully loaded one ring unloaded) on stresses and deformations under the full longitudinal TBM loads.
- Influence of non-accurate assembling and dimension imperfection on stresses and deformations under full longitudinal (actual jack forces) and radial loads. Defining non-accuracy and deviation limits that could be admissible.
- Influence of the available space for the key segment assembling on stresses and deformations under full longitudinal (actual jack forces) and radial loads.
- Influence of the longitudinal jack force variation on stresses and deformations under full radial loads.

5.3 *Test Results*

Among the huge database of the results analyzed by scientists the most important conclusions are stressed below:

- The assembling process causes stresses, which are additional to stresses occurring in later stages as serviceability stage.
- Deformations occurring during assembling mainly maintain. Deformation shapes are still present in later stages.
- Assembling in-tolerances causes stress peaks, which could lead to damage.
- Stresses are non-uniform distributed in a cross section of a segment. Especially not near key segments. Main causes can be ring interaction combined with rotated positions of neighbour rings.
- Stresses near joints drop down because of reduced height in the joint.
- Deformation stiffness is strongly reduced by longitudinal joints. Still it must be mentioned that lower stiffness limits the bending moments in case of tunnels in soft soil.

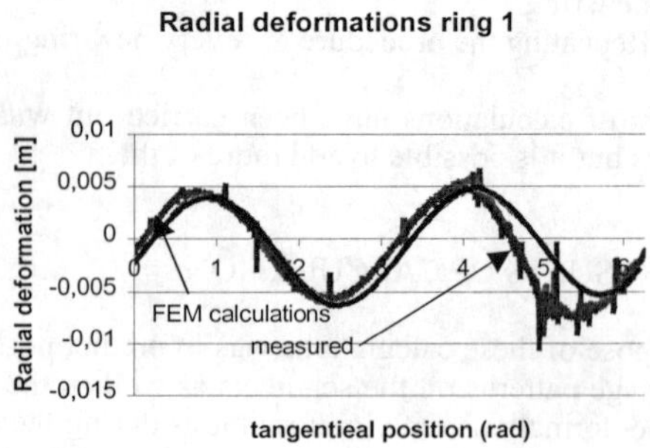

Figure 8. Comparison between measured and calculated values

6 CONCLUSIONS

A tunnel lining is the final product of a complicated process to meet the users demands. Because of the interrelation of so many influences there is an existential need to consider the problems of tunnel lining design with special care. Experience is still of considerable importance but new knowledge gained by the observations, measurements and numerical approach rise to the idea that tunnel behaviour can be predicted. This transit-method is quite expensive and it needs powerful hardware and a lot of process time. On the other hand, it is possible to estimate the real tunnel lining behaviour more accurate defining allowable damage patterns and at the same time to make savings on the tunnel maintenance cost. Comparing these cost it is obvious that using the transit-method of modelling can play an important role in tunnel design process avoiding unnecessary risks during the construction stage.

REFERENCES

Leendertse, W.L. 1997, *Damages of segments on the west tube The Second Heinenoord tunnel*, Bouwdienst Rijkswaterstaat 1933-T-980735, The Netherlands

Blom, C.B.M., Jovanovic, P.S., 1999, Requirements for tunnel design in soft soil, *Proceedings Long Road and Rail tunnels*, Basel, Switzerland

Blom, C.B.M., Jovanovic, P.S., Oudejans, W.L., 2000, Recommendations on guidelines for tunnel lining design in soft soil, *Proceedings of 16th congress IABSE*, Luzern, Switzerland

Computer Methods and Advances in Geomechanics, Desai et al. (eds) © 2001 Balkema, Rotterdam, ISBN 90 5809 183 X

Some problems encountered in numerical analysis of tunnel support

M. Lu
SINTEF Civil and Environmental Engineering

ABSTRACT: Numerical analysis is performed by using one of the most popular commercial codes for checking the design of the temporary support system of a tunnel in soft rock. Problems have been encountered in the analysis and it is finally concluded that the problems are caused by a bug in the code. The even worse thing is that the problem is concealed by another bug in the code for post-processing. This paper describes the problems and discusses the associated issues.

1 INTRODUCTION

In order to check the design of a tunnel temporary support system a numerical analyses is performed by using one of the most popular commercial codes. The 12m wide and 11m high tunnel is excavated and reinforced in two stages. The temporary support includes shotcrete and systematic bolting with fully grouted rock bolts. It is the common knowledge that the bolts in the tunnel roof will generally subject to tension loading due to the tunnel deformation. However, the numerical analysis indicates that the bolts are failed in compression that is obviously wrong. The worse thing is that the mistake is concealed by plotting of the failure mode that shows that the bolts fail in tension. This is caused by another bug in the code. The bugs are then reported to and fixed by the code developer. The reanalysis with the corrected new code indicates that the bolts are failed in tension that is in correspondence with the graphical plots. The problem is described in detail and demonstrated with analysis on two tunnel sections with different rock material parameters. The intention of the paper is to show how a small bug in the code can make a big mistake in analysis; in this case, the tension failure turns to compression failure. The bug itself will not be discussed.

Some parameters in connection with tunnel support analysis are difficult to evaluate, including the stiffness of the interface between shotcrete and rock. Parametric study indicates the analysis is not sensitive to the interface stiffness. However, the analysis performed with the old version of the code shows that the stiffness is really a sensitive parameter.

The conclusion is that the user of a commercial code has to be extremely careful in interpreting the analysis result by using his/her own engineering judgement. Attention should also be paid to those parameters that do not seem important.

2 DESCRIPTION OF PROBLEMS

2.1 *The tunnel and the numerical model*

The road tunnel has a half circle roof and straight walls. Excavation and support are carried out in two steps: the first step is the half-circle roof and the second is the lower part. The support system includes shotcrete and systematical bolting with fully grouted rock bolts. In order to simulate the 3-D and time-effect of the tunnel advancing, the bolts and shotcrete are applied after 50% the tunnel equivalent force is released. As shown in Figure 1, the complete simulation process includes: (1) Generation of initial stress field; (2) Excavation of the upper part to 50%; (3) Application of bolts and shotcrete for the upper part; (4) Excavation of the lower part and (5) Application of bolts and shotcrete for the lower part.

The rock bolt is modeled with one-dimensional element subjected to uniaxial tension or compression, without any bending resistance. The mechanical model is sketched in Figure 2. The mechanical properties of the ϕ25 bolts are: E-modulus, 210GPa; tensile and compressive yield force, 220KN, that is scaled-down based on the bolt spacing in the tunnel axial direction. The rock bolts are rigidly connected to the shotcrete in the model.

The shotcrete is modeled with 1-D beam element. The grout annulus around the bar and the interface between shotcrete and rock are also taken into account. The material parameters for tunnel sections 1 and 2 are given in Tables 1-3.

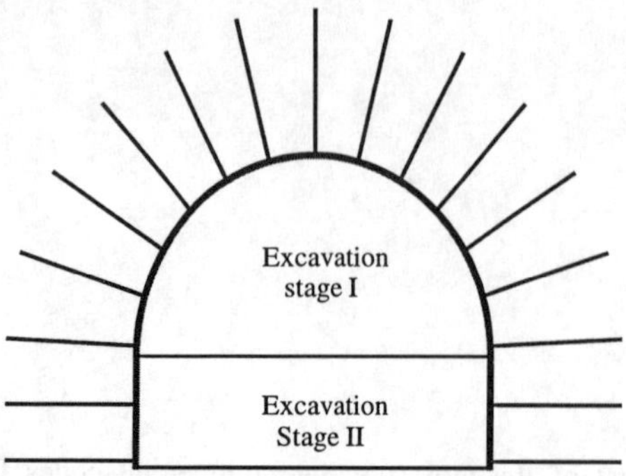

Figure 1. Excavation stages and rock supports.

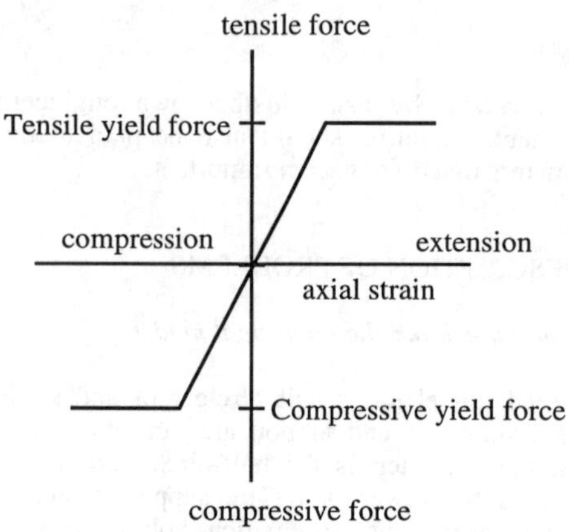

Figure 2. Mechanical model of rock bolts.

Table 1. Mechanical parameters of rock mass.

Section	ρ [kg/m³]	E [GPa]	ν	c [MPa]	ϕ [°]	ψ [°]	σ_T [MPa]
1	2650	10.0	0.30	1.5	32.0	25	0
2	2750	3.75	0.25	0.3	45.0	20	0

Table 2. Mechanical properties of shotcrete.

Section	E [GPa]	ν	ρ [kg/m³]	σ_T [MPa]	σ^r_T [MPa]	σ_c [MPa]
1	14.3	0.17	2550	2.0	0	15.0
2	13.9	0.17	2550	2.4	0	11.3

Table 3. Mechanical properties of interface between shotcrete and rock.

k_n [MPa/m]	k_s [MPa/m]	ϕ [°]	c [MPa]	σ_T [MPa]
1000	1000	45	0.3	0.5

In the tables ρ is density; E is Young's modulus; ν is Poisson's ratio; c is cohesion; ϕ is friction angle; ψ is dilation angle; σ_T, σ^r_T and σ_c are peak and residual tensile strength and compressive strength, respectively. The interface between shotcrete and rock can transfer force in both shear and normal directions. k_n and k_s given in Table 3 are normal and shear stiffness. The vertical in-situ stress is solely caused by overburden, which is 300 and 220m for section 1 and 2, respectively. The corresponding ratio of horizontal over vertical stress is 1.0 and 0.5.

2.2 *Comparison of the analyses with original code and the new code*

Section 1

Figures 3 and 4 show the axial force of rock bolts computed from original and new version of the code, respectively. It can be seen that the axial forces of the bolts are completely different. At the end close to the tunnel surface the axial force of all bolts are in compression from the original code that is apparently wrong. The compressive stress is so high that failure occurs. If the plot of axial failure had showed the compressive failure, it would have been easy to realize that something had been wrong. Unfortunately, this is not the case. The bolt failure plot indicates tension failure instead of compression failure, as shown in Figure 4 (a). Together with the confusion in the sign of the axial force this makes the user difficult to detect the problem. When we started suspect the result, we tried to plot everything which is associated with shotcrete and bolts. It is the plot of axial displacement of the bolts, as shown in Figure 5, made us certain that something must be wrong in the code. This figure shows that the displacement at the connection points between bolts and shotcrete is zero that is absolutely impossible. After we successfully convinced the code owner, they found the bug and corrected it. The plots in Figures 3-5 (b) are from the reanalysis by using the new code. The bolt force now all becomes tension and the yielding zone enlarges to some extent.

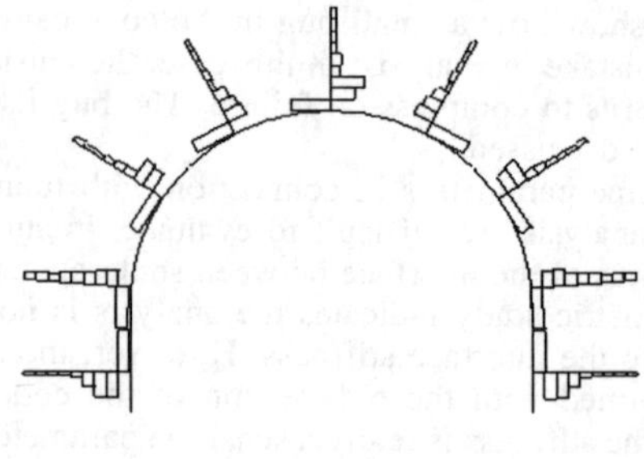

(a) From old code.

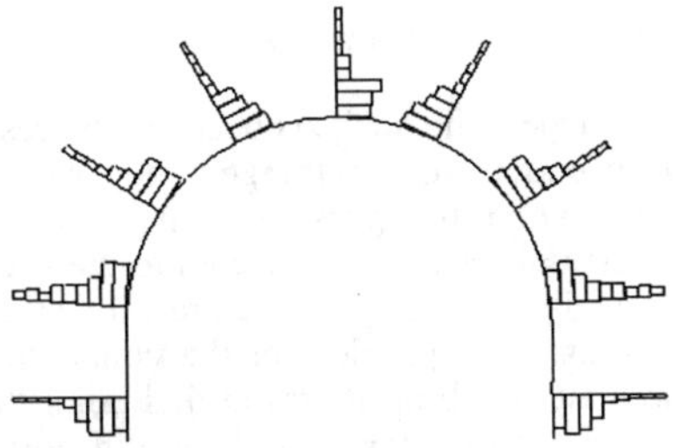

(b) From new code.

Figure 3. Axial force of rock bolts for section 1.

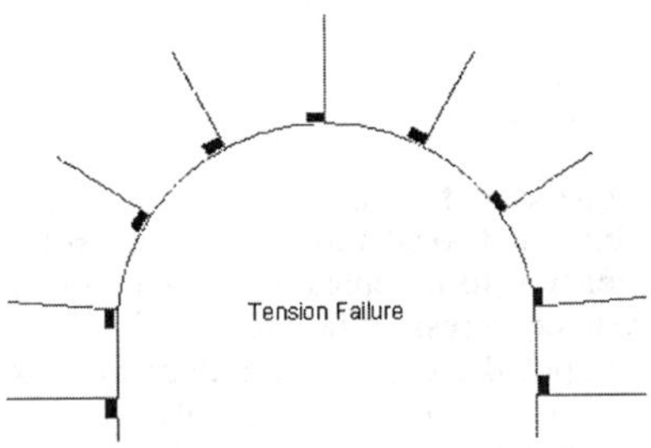

(a) From old code.

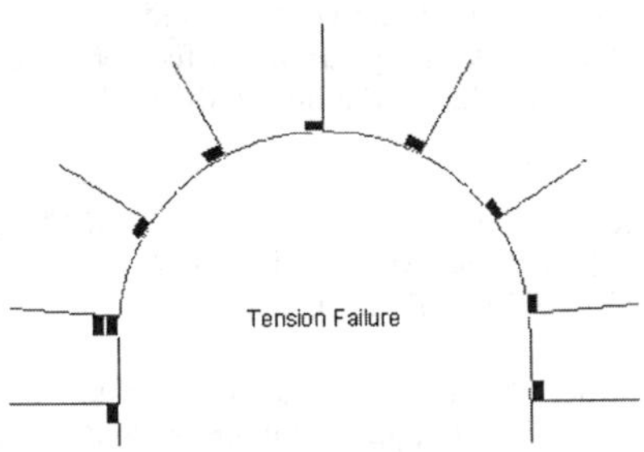

(b) From new code.

Figure 4. Axial failure of rock bolts for section 1.

Section 2

Same as section 1, the axial force of bolts is wrong for the analysis with the old code, see Figure 6 (a). The compression failure is also concealed by the plot of bolt failure where the computed compressive failure is shown as tension failure. Figure 7 illustrates the cause of the error that is the wrong calculation of the displacement at the connection points between shotcrete and bolts. Figure 8 shows distribution of the normal contact force between shotcrete and rock. For the old computation, the compressive contact force is overestimated and a strong concentration exists at the connection points of shotcrete and bolts. The yielding zone is, on the other hand, underestimated. The subsidence of the tunnel crown is 11.1 and 16.2 mm for old and new analysis, respectively, nearly 50% error.

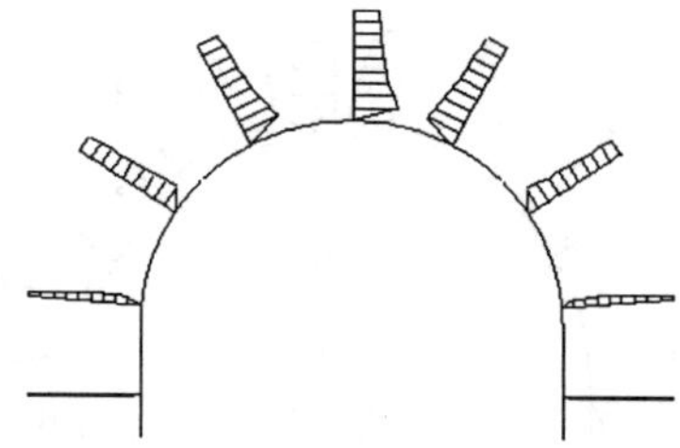

(a) From old code.

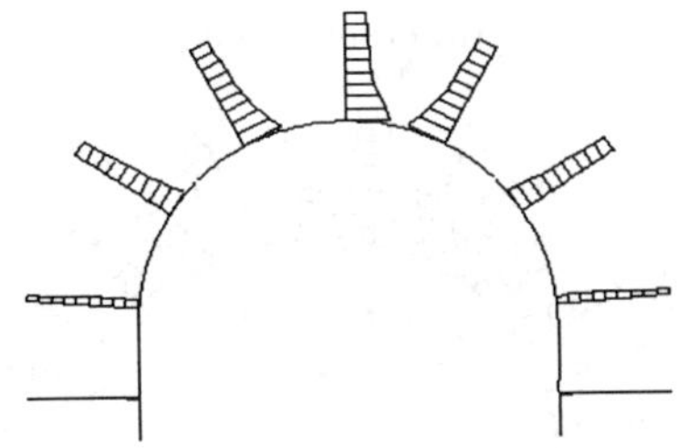

(b) From new code.

Figure 5. Axial displacement of bolts for section 1.

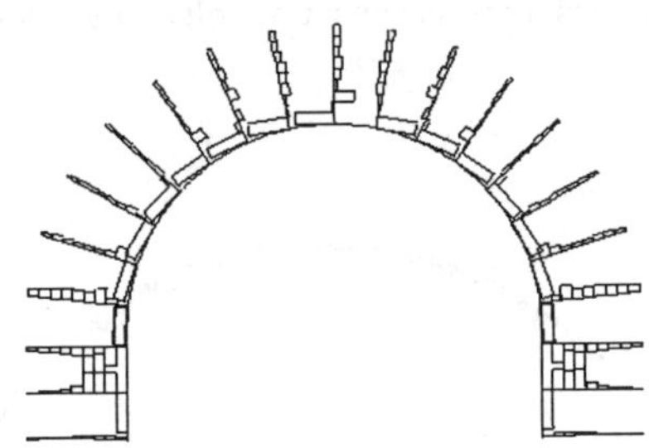

(a) From old code.

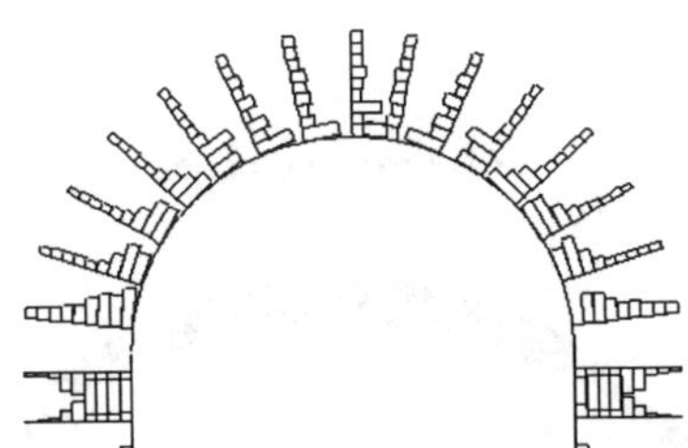

(b) From new code.

Figure 6. Axial force of rock bolts for section 2.

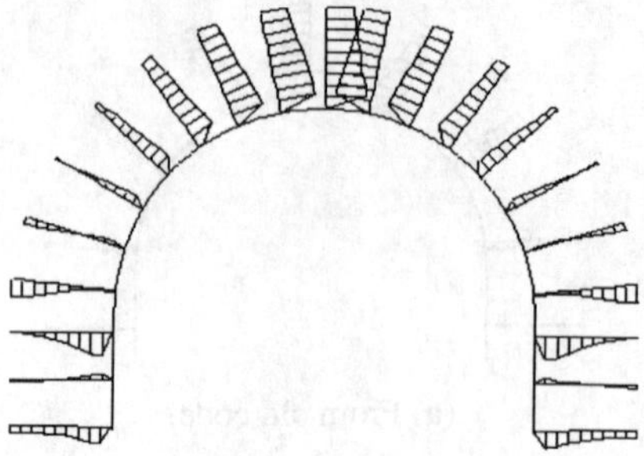

(a) From old code.

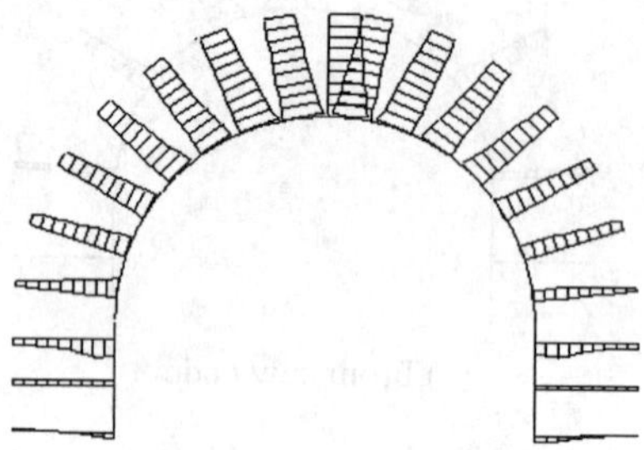

(b) From new code.

Figure 7. Axial displacement of bolts for section 2.

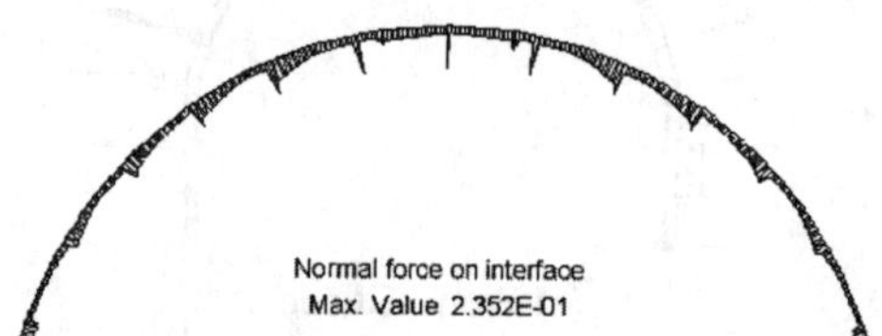

(a) From old code.

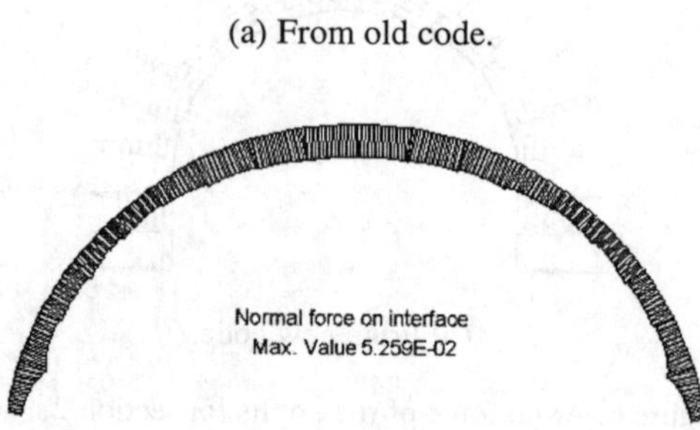

(b) From new code.

Figure 8. Normal contact force between shotcrete and rock.

2.3 *Identification of the problem*

We started suspect the analysis because the resulting deformation pattern looks strange. Then we plotted and printed everything possible. Finally we found that when the bolts are not connected to the shotcrete the problem disappears; so we were almost certain that there must be a problem at the connection between shotcrete and bolt. It was at this point that we contacted the code owner. It took only a few hours for them to correct the bug after they realized it. As later explained by the code developer, the bug, which caused the problem, is an error in an auxiliary operation for speeding-up calculation for quasi-static problems at the connection nodes between the bolt and shotcrete. This is exactly the same as we anticipated.

2.4 *Consequence*

The consequences of the incorrect code include:

- The computed axial force is wrong; sometimes from tension to compression, even from tensile failure to compressive failure.
- The computed compressive failure of rock bolts is shown as tension failure in plots.
- The yielding zone is underestimated.
- The tunnel roof displacement is underestimated.
- The normal contact force (compression) between shotcrete and rock is overestimated.

In one word, we may say the effect of rock bolts is overestimated due to the bug in the code.

3 INFLUENCE OF THE NORMAL STIFFNESS OF THE INTERFACE BETWEEN SHOTCRETE AND ROCK

It is common knowledge that an interface element need be applied between shotcrete and rock for proper simulation of the shotcrete support. However, the normal and shear stiffness of the interface are usually difficult to evaluate. If the effect of the shotcrete is sensitive to the interface stiffness, the model can hardly be justified. In other words, the interface stiffness should not be a sensitive parameter. In order to test whether the parameter is sensitive or not, a testing program on two tunnel sections is conducted in which the normal stiffness of interface ranges from 10^1 to 10^6MPa/m.

3.1. *Sample section 1*

(1) Crown subsidence

Figure 9 shows the dependence of the tunnel crown subsidence upon the interface normal stiffness. The result from the new code indicates that K_n only has a little effect on the displacement, except when the K_n

becomes extremely high, such as 1000GPa/m. However, if the old code is used the influence is dramatically enhanced. For the result from the old code, the roof displacement decreases with increasing normal stiffness and is considerably underestimated.

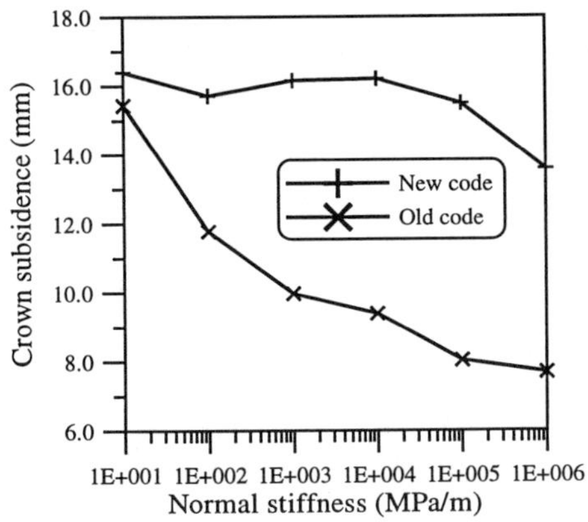

Figure 9. Dependence of tunnel crown subsidence upon normal stiffness K_n, sample section 1.

(2) Yielding zone

The yielding zone computed by using the old code is greatly influenced by the interface normal stiffness, reducing with increasing K_n. When K_n reaches 10^4MPa/m there is virtually no yielding zone in the roof area. On the contrary, there is no significant difference in yielding zone until K_n becomes very large, if the new code is used. Graphs of the yielding zone could not be shown due to space restriction.

3.2 *Sample section 2*

As shown in Figure 10, the tunnel crown subsidence is significantly influenced by K_n for the analysis with the old code; whilst if the new code is used, K_n has little effect. With regard to the yielding zone, K_n has a great influence with the old code and almost no influence with the new code. This is the same as sample section 1.

3.3 *A few points*

- If the computer program is correct, as the new code in this test, the computed rock deformation should be only slightly influenced by the normal stiffness of the interface between shotcrete and rock.
- In the common range of the stiffness, $K_n = 10^3$ - 10^4 MPa/m, the effect of K_n is negligible.
- If the old version of the program is used, the computed tunnel roof displacement is considerably affected by K_n and so is the yielding zone.

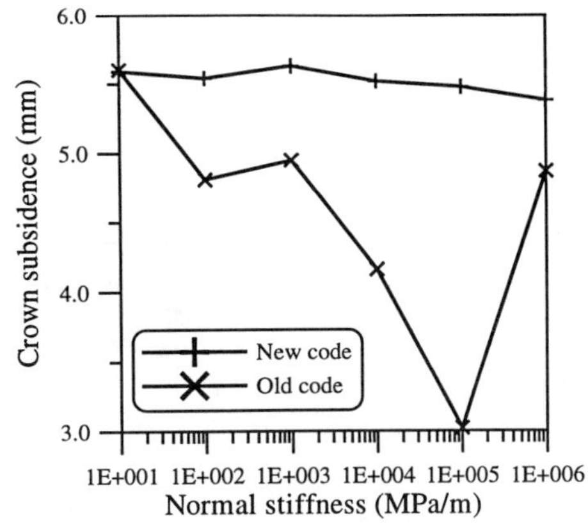

Figure 10. Dependence of tunnel crown subsidence upon normal stiffness K_n, sample section 2.

4 CONCLUSIONS - THE LESSON TO LEARN

The following lessons may be learned from the problem presented here:

- A 'small' bug in the code may result in serious error in the analysis result.
- Bugs may well exist in commercial codes no matter how popular and how mature the code is.
- After decades of development the remaining bugs in some commercial code are deeply hidden and difficult to detect. One should not readily trust the computation result. Instead, one should question the result and make own judgement with engineering experience.
- Probably some users have performed similar analyses with the same code and incorrect results have been presented. One has to admit that this is one of the reasons why some experienced consulting engineers don't believe the numerical analysis, since the result of the numerical analysis is sometimes in contradiction with their experience that has been proved by engineering practice. Of course, many factors may contribute to an incorrect numerical analysis, such as improper numerical model, inaccurate input data and wrong computation technique. However, the errors in the computer program are for sure one of the major sources.
- It is recommended double check be carried out with another independent code, if available. For the project presented here, ABAQUS is used for double check.

Computer Methods and Advances in Geomechanics, Desai et al. (eds) © 2001 Balkema, Rotterdam, ISBN 90 5809 183 X

Numerical simulation of cavity expansion of blast borehole with cylindrical charge

Wenbo Lu
Wuhan University of Hydraulic and Electrical Engineering, People's Republic of China

ABSTRACT: This paper presents the results of numerical simulation of cavity expansion of blast borehole with cylindrical charge. With method proposed by Starfield, after taking into account the differences between real spherical charge and the equivalent element spherical in cylinder, and revising the analytical solution for the waves generated around a spherical explosive filled cavity, process of cavity expansion of borehole is obtained by simulating the cavity expansion of any equivalent spherical charge in cylinder. The simulation results show that the method proposed can improve greatly the simulation accuracy of former method, and some difficulties ever encountered in stimulating stress waves, such as that the main frequency of simulated stress waves was far more higher than that of measured value, etc., are also overcome.

1 INTRODUCTION

Cylindrical explosive charges have become the most important charging method because of their high efficiency of harnessing explosive energy and the better effect of rock fragmentation by blasting. But there are still many theoretical questions kept unsolved, and the expansion process of cavity of blast borehole is just one of the most important.

In order to simulate the propagation of detonation wave along axial direction of cylindrical charge, Starfield & Pugliese (1968), proposed a method by splitting the cylindrical charge into a series of equivalent spheres, and assumed that the total strain of certain point in rock around borehole can be calculated in time domain by the superposition of strain generated by every single spherical charge in accordance with its space coordinate. In their simulation, the shape of strain wave induced in rock by single spherical charge has to be measured experimentally or assumed.

Harries (1983) took some further study on modeling long cylindrical charges of explosive. In his study, the solution of strain wave induced in rock by single spherical charge was obtained by assuming that the adiabatic exponent for the detonation product gases was constant and making a modification to analytical solution for the waves generated around a spherical explosive filled cavity which was first derived by Favreau(1969). Despite of their outstanding works, there are still two important problems unsolved:

1. In the near area around borehole, the calculated value of strain wave deviates seriously with the measured value;
2. The main frequency of simulated strain wave is far higher than that of measured.

In order to coincide the simulated parameters of strain wave well with those of measured, Harries(1983) had to adjust the frequency at which the blast borehole walls vibrate after the detonation: the frequency at which the equivalent spheres oscillate had been reduced to 1/6 of that at which a single isolated sphere would oscillate.

In this paper, differences between the real isolated spherical charge in rock and the equivalent spherical charge in cylindrical charge are discussed first; then, by modifying the analytical solution for stress wave induced by spherical charge, process of cavity expansion of borehole with long cylindrical charge is obtained by simulating the cavity expansion of any equivalent spherical charge in cylinder.

2 EQUIVALENT SPHERICAL CHARGE IN CYLINDER

Practically, there are two important differences between the equivalent spherical charge in cylinder and the real isolated spherical charge in rock with an equivalent diameter (Lu 1995). The first is that they have different expanding processes of blast borehole cavity. As for real isolated spherical charge in rock, its cavity expansion is thoroughly controlled by the factors such as the physical and mechanical parameters of rock mass, diameters of blast hole and explosive charge, characteristic parameters of explosive used and conditions of free surface and stemming, etc. As long as above factors are given, the amplitude, main frequency and some other parameters of strain wave induced in rock can be thoroughly decided. But as for the equivalent element spherical charge in long cylindrical charge, the situation is of great difference: the cavity expansion of every equivalent spherical charge is influenced by both the formerly and latterly detonated series of equivalent spherical charges, this is showed in Figure 1(a). Because of the difference of cavity expanding process between isolated charge in rock and equivalent charge in cylinder, even though they have same initial detonating pressures in wave front at time of detonating, the output stress waves are of different constructs of frequencies and amplitudes.

Another important difference between isolated spherical charge in rock and equivalent one in cylinder is that, they have different initial loading processes of explosion pressure acting on to walls of borehole. For isolated spherical charge in rock, if the initiating point is in the center of charge, the whole wall of borehole is acted simultaneously by explosion pressure on time of explosion gases expanding to wall (Figure 1(b)); as for equivalent spherical charge, because the detonating wave propagates along the axial direction of cylinder with a detonating velocity of D, the wall $AA'BB'$ of borehole can not be acted simultaneously by explosion pressure: if the *ith* equivalent charge is concerned, section AA' is acted by pressure first, and only after detonating wave propagates to section BB', does the loading process of explosion pressure induced by *ith* equivalent charge end (Figure 1(a)).

Difference of aforementioned loading processes between them leads to also the different constructs of frequencies and amplitudes of induced stress waves.

3.CAVITY EXPANSION OF EQUIVALENT CHARGE

Under the condition of effective stemming, compared with the radial movement within borehole, the axial movement of explosion gasses can be ignored. So, as for the *ith* equivalent spherical charge, after detonating, it can be considered approximately that there is no transmission of explosion energy in sections of AA' and BB' (Figure 1(a)). This means that there is a thoroughly stemming effect of

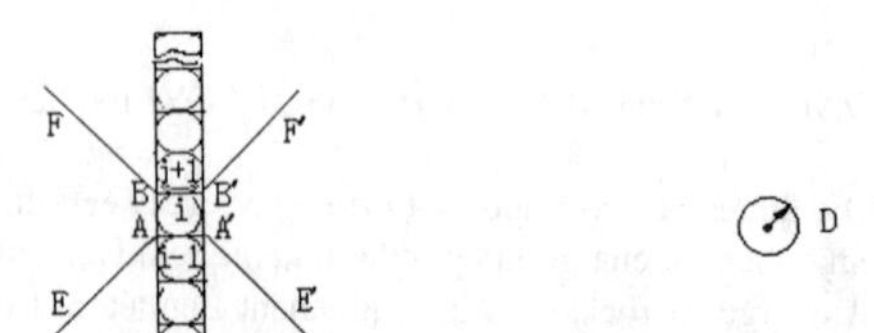

(a) Equivalent spherical charge (b) Isolated spherical charge

Figure1. Equivalent spherical charge in cylinder and isolated spherical charge in rock mass

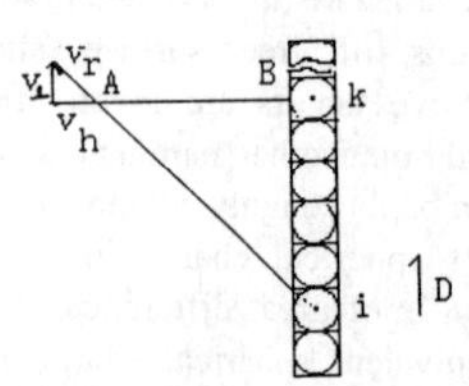

Figure2. Expansion of cavity of equivalent spherical charge

the *(i-1)th* and the *(i+1)th* equivalent charges upon the *ith* equivalent charge. So, the direct acting area of stress wave induced by *ith* equivalent charge is limited to $EABF$ and $E'A'B'F'$ only; and the continuing conditions of stress, displacement and velocity in boundaries BF, AE, $A'E'$ and $B'F'$ of directly arrived stress wave must satisfied. If the movement of any point in boundaries is divided into radial and axial components (Figure 2), given that cylindrical charge has enough length, and take into account the overall effect of formerly and latterly initiated series of equivalent charges, then the condition of plane strain can be used approximately, namely, the axial movement of rock surrounding borehole can be ignored. As a result, as for areas $EAA'E'$ and $FBB'F'$, the radial component of stress wave induced by *ith* equivalent charge acts as a cone-shaped wave which propagates inwards; and by this cone-shaped wave, the stress wave induced by *ith* equivalent charge influences the cavity expansion of every other equivalent charge.

Showed in Fig.4, when the stress wave induced by *ith* equivalent charge arrives to point A, its horizontal component v_h begins to propagate toward center of borehole, and when v_h arrives to the point B of wall of borehole, the *ith* equivalent charge begins to influence the cavity expansion of the *kth* equivalent charge. Parameters of stress wave in point A can be obtained by calculating the stress wave field induced by the *ith* equivalent charge. Now, we need to infer the movement of point B caused by the *ith* equivalent charge from that of point A. As an approximation, we can obtain the horizontal velocity of point B by using relation of geometric divergence in wave front of inward cone-shaped stress wave. If the horizontal component of velocity in point A is given as:

$$v_h(A) = A_0 f(t) \tag{1}$$

Where, A_o = amplitude of horizontal velocity in point A, $f(t)$= function which varies with time. Then, the velocity in point B caused by inward cone-shaped wave can be expressed as:

$$v_h(B) = \left(\frac{R_A}{R_B}\right)^{1/2} A_0 f\left\{t - \frac{R_A - R_B}{C_P}\right\} \tag{2}$$

where, R_A= radial distances between points A and center of borehole; R_B = radial distances between points B and center of borehole; C_p = longitudinal wave velocity of rock mass.

It must be mentioned that, when radial component of stress wave arrives to point B, transmission and reflection of incidence wave happen in point B because there are explosion gases within borehole. So the totally horizontal velocity $v_i(B)$ caused by the *ith* equivalent charge in point B, which is the superposition of both incident and reflective waves, can be expressed as:

$$v_i(B) = Rv_h(B) \quad (3)$$

where, R = coefficient relating to the sonic impedances of rock mass and to that of explosion gases. When method of non-coupling charging is adopted, R is approximately equal to 2.0. The total expansion velocity of point B in wall of borehole shall be the superposition of movements caused both by the formerly and latterly initiated series of equivalent charges and by the *kth* equivalent charge itself, so the combined expansion velocity of point B is:

$$v(B) = \sum_{i=1}^{n} v_i(B) \quad (4)$$

4 SHAPE OF STRESS WAVE INDUCED BY SPHERICAL CHARGE

The analytical solution for stress wave induced by spherical charge in homogeneous isotropic and linear elastic rock mass has first been given by Favreau (1969). Its radial displacement $U(r,t)$ and potential function $\Phi(r,t)$ are related to each other as follows:

$$U(r, t) = \frac{\partial \Phi(r, t)}{\partial r} \quad (5)$$

$$\Phi(r, t) = -\frac{a_0}{\rho \beta r} \int_0^t P(S - \tau) e^{-\alpha \tau} \sin(\beta \tau) d\tau \quad (6)$$

where, r= distance from center of charge; a_0 =initial radius of charge cavity; ρ = density of rock mass; $P(t)$= pressure of explosion gases within borehole; α, β and S are expressed as:

$$\alpha = \frac{1 - 2\mu}{1 - \mu} \frac{C_p}{a_0}, \beta^2 = \frac{1 - 2\mu}{1 - \mu} \frac{C_p^2}{a_0^2},$$

$$s = t - \frac{r - a_0}{C_p} \quad (7)$$

where, μ = Poisson's ratio of rock mass.

So, if $P(t)$ is known, the displacement of any point around explosion source can be obtained , strain and stress in same point can also be inferred by using the relation of stress and strain and the continuing condition, and velocity and acceleration can be obtained by partially differentiating $U(r,t)$ by time. By this way, every parameter of stress wave generated by spherical charge can be thoroughly decided.

Practically, $P(t)$ can not be predetermined and can only be decided numerically step by step by using state equation of explosion gases within borehole. In the mean time, because of the differences between equivalent charge and real spherical charge, only after that the analytical solution of stress wave generated by isolated spherical charge in rock has been modified, can it be used as the basic solution to method proposed by Starfield & Pugliese (1968).

The radial displacement Δr of cavity wall of point B can be calculated by integrating over $v(B)$:

$$\Delta r = \int_0^t v(B) dt \quad (8)$$

Further then, the gases-filled cavity volume $V(k)$ of *kth* equivalent charge can be expressed as:

$$V(k, t) = \pi (a_0 + \Delta r)^2 d_e \quad (9)$$

Where, d_e = diameter of equivalent charge, d_e is related to diameter of charge d_0 as:

$$d_e = \frac{\sqrt{6}}{2} d_0 \quad (10)$$

Assume that the adiabatic exponent for explosion gases is constant, then, the static equation of explosion gases can be expressed as:

$$P(t) [V(k, t)]^{\gamma} = constant \quad (11)$$

Where, γ =adiabatic exponent for explosion gases. As a engineering approximation, when $P(t) \geqslant P_k$, $\gamma = 3.0$; $P(t) < P_k$, $\gamma = 1.4$; P_k= critical pressure of explosion gases.

So, the pressure $P(t)$ of explosion gases within cavity of *kth* equivalent charge can be decided numerically step by step with *eqs.(8)~(11)*; and combining with *eqs.(5)~(6)* and their related discussion, the parameters of stress wave generated by *kth* equivalent spherical charge can be thoroughly calculated, and this numerical result obtained is adopted as the basic solution to superposition method proposed (Starfield& Pugliese,1968).

5. NUMERICAL SIMULATION

In order to test and verify the effectiveness of improvement to method first proposed by Starfield & Pugliese(1968), an example of numerical simulation is present in following paragraphs.

In simulation, assume also that the rock is to be homogeneous, isotropic and linearly elastic, and explosive charge is contained in infinite rock mass. The rock concerned is shale with a density of 2261 kgm^{-3}, Poisson ratio of 0.271, elastic modulus of 17.83 GPa and a longitudinal wave velocity of 2808ms^{-1}; diameters of borehole and explosive charge are 0.1m and 0.05m respectively; the explosive used is $2^{\#}$ rock ammonium nitrate powder with a density of 1000kgm^{-1}, a detonation velocity of 3400ms^{-1} and a critical pressure of explosion gases of 198 MPa.

Stress waves generated by four different conditions are studied in simulation: (1) the isolated spherical charge in rock with a equivalent diameter; (2) cylindrical charge simulated with superposition method, but not taking into account the initial loading process; (3), cylindrical charge simulated with superposition method, and taking into account the initial loading process of equivalent spherical charge (pressure is applied gradually to cavity wall of equivalent spherical charge); (4), cylindrical charge under assumption of instant detonation (pressure is applied instantaneously to the wall of cavity).

The initiated point of isolated spherical charge is assumed to be in center of charge, while that of cylindrical charge is assumed to be in the bottom of charge.

The front pressure P_f of explosion gases is calculated as:

$$P_f = \frac{\rho_e D^2}{2(1+\nu)} \qquad (12)$$

which is 14.45 GPa, where, ρ_e = density of explosive ; D = detonation velocity of explosive.

Then, the initial average pressure of explosion gases which corresponds to time when the expansion of explosion gases begins to act on the wall of charge cavity can be calculated by *eqs.(9)~ (11)*, which is 26.6Mpa.

In simulation, the time of initially linear loading process of explosion load is approximately taken as d_e/D, which is 18 μs.

Because there are many parameters in stress wave generated by explosive charge, only those results that can be compared easily with each other are present in the paper. Those parameters are: the maximum radial displacement (MRD) of wall of charge cavity, the maximum radial expansion velocity (MREV) of cavity wall and its main frequency (MF). The detailed results are showed in Table1.

As for cylindrical charge, parameters present in table 1 are results of cavity expansion of the equivalent charge in the middle of cylinder.

The simulated results show that the maximum radial displacement of cavity of isolated spherical charge with an equivalent diameter is much smaller than that of cylindrical charge, while its maximum radial velocity is much smaller and its main frequency is much lower than those of cylindrical charge. Just as former analysis, this result reveals the combining effect of formerly and following initiated equivalent charges on cavity expansion of the middle one.

The simulated results also show that, as for cylindrical charge, condition (2) has the largest displacement, fastest cavity velocity and lowest main frequency of cavity expansion of equivalent charge, those are 1.1cm, 185m/s and 156KHz respectively. According to theoretical analysis, condition (4) should gain the largest displacement, fastest velocity and lowest main frequency of expansion of charge cavity, and the reasonable results simulated should consist with this kind of tendency. So, only the results of condition (3) are reasonable. Then, we can make a reliable conclusion that, it is necessary to take into account the initial loading process of equivalent spherical charge during simulating the stress wave generated by cylindrical charge.\

Besides, the main frequency of stress wave induced by equivalent charge in cylinder calculated under condition (3) is 1.86KHz, while that of isolated sphere is 11.5KHz,that means the ratio of frequency of equivalent sphere to that of isolated real charge is 1:6.2, this coincides well with the adjusting made by Harries (1983). So, if the numerical

Table 1. Results of numerical simulation

Parameters	cond.(1)	cond.(2)	cond.(3)	cond.(4)
MRD(cm)	0.13	1.1	0.89	0.98
MREV(m/s)	72	185	153	156
MF (KHz)	11.5	1.56	1.86	1.76

result of stress wave induced by equivalent spherical charge in cylinder is used as basic solution in simulating cylindrical charge, it needs no adjusting to its wave shape. This verifies that the improved method put forward in this paper has really overcome the shortcomings of traditional method proposed and so can improve greatly the accuracy of simulated results.

6 CONCLUSIONS

From above analysis and simulation, following conclusions can be made:

1. The equivalent spherical charge in cylinder and isolated real spherical charge with an equivalent diameter in rock are of some differences. They have different cavity expanding and initial explosion pressure loading processes.
2. Because of differences of parameters of stress waves induced, wave shape induced by equivalent charge in cylinder instead of that of isolated real spherical charge shall be used as the basic solution in simulating cylindrical charge with method proposed by Starfield & Pugliese (1968).
3. Results of numerical simulation shows that the improved method proposed in this paper can improve greatly the simulation accuracy and overcome well the shortcomings existed in former method.

It must be added to it that, the presupposition of simulation method proposed is that rock shall be linearly elastic, isotropic and homogenous. When plastic zone is formed or flow of rock happens around borehole during rock blasting, the method proposed is no longer effective.

REFERENCES

Favreau, R.F. 1969. Generation of strain wave in rock by an explosion in a spherical cavity. *J. Geophys. Res.*74: 4276-4280.

Harr]ies, G. 1983.The modeling of long cylindrical charge of explosive. *Proc. First Int. Sym. on rock fragmentation by blasting:* 419~481. Lulea, Sweden.

Lu, Wenbo 1995.*The study of propagation of stress wave and its effect induced by rock blasting* (PH.D thesis). Wuhan Univ. of Hydra. and Electric Eng., P.R. China.

Lu, Wenbo 1996. Simulation of cavity expansion with strip-shaped explosive charge. *Explosion and shock waves.* 16(2):171-177.

Starfield A.M. & J.M. Pugliese 1968. Compression waves generated in rock by cylindrical explosive charges: A comparison between a computer and field measurements. *Int. J. Rock mech. Min Sci.* 5(5): 65-77.

Computer Methods and Advances in Geomechanics, Desai et al. (eds) © 2001 Balkema, Rotterdam, ISBN 90 5809 183 X

Application of computational and statistical techniques to an evaluation of mobile roof supports

Hamid Maleki
Maleki Technologies Incorporated, Spokane, Wash., USA

John Owens
Spokane Research Laboratory, National Institute for Occupational Safety and Health, Wash., USA

ABSTRACT: This study was initiated to evaluate optimum operating conditions for mobile roof supports. Because of the complexity of geologic and mining conditions and the geometry of mine excavations, all of which influence the performance of the machine, the authors combined the strengths of numerical modeling and statistical techniques. Boundary-element techniques were used for stress analyses in a two-seam operation and for the design of field experiments.

To evaluate the influence of machine setting pressure on roof stability, the authors used the results of multivariate regression analysis to analyze data from a field program in which leg pressures on four machines were monitored on a continuous basis. Using the volume of roof falls as the response variable, the authors did not find any strong correlations between machine operating parameters and roof stability. Geometric and mining factors correlated strongly with the volume of roof falls, a finding that emphasizes the importance of prudent mine layout designs.

1 INTRODUCTION

Room-and-pillar mining is one of the oldest methods used for the extraction of tabular ore bodies. In this method, a series of rooms are driven on advance using continuous miners and shuttle cars while the roof is bolted a short distance behind the face. During retreat, the same equipment is used to mine the pillars, which allows roof rocks to cave behind the face. To control the cave line, a series of secondary support systems (posts or mobile roof supports [MRS's]) are installed as mining continues within the pillars.

MRS's consist of a roof canopy, four hydraulic cylinders, a caving shield canopy, and associated electromechanical systems mounted on crawler tracks. The machines are controlled by radio from a remote location and operate on self-contained power units. Typically, an MRS has capacities of 5,340 and 7,120 kN. MRS performance has been monitored both in the laboratory and in the field. Laboratory investigations focused on an evaluation of support stiffness and load-carrying capacity under controlled static loading conditions. The study quantified system stiffness as a function of machine height for both two- and three-stage hydraulic cylinders. Each unit has the load-bearing capacity of six posts and the stiffness of two hardwood posts (Barczak and Gearhart 1997). Three-stage cylinders are needed in thick seams, but reduce support stiffness. Each unit has the load-bearing capacity of six posts and the stiffness of two hardwood posts (Barczak and Gearhart 1997). The advantage of the three-stage cylinder design is greater operating range, providing a lower support profile for transporting and tramming underground. The disadvantage is reduced support stiffness, compared with the two-stage cylinder design.

The mechanics of load transfer from pairs of MRS's to mine strata were analyzed using laboratory results, boundary-element modeling, and analytical solutions (Maleki and Owens 1998). The results showed that MRS's support roof rocks near the machines, but do not have the capacity to control overall roof-floor convergence and overall stress distributions because the MRS's are considerably less stiff than coal-measure rocks. In comparison to posts, however, an MRS is capable of maintaining the yield load after significant amounts of roof-floor deformation. Because the mining cycle is accelerated, MRS's help reduce the potential for time-dependent roof falls (Maleki and others 1999).

This study was initiated to evaluate optimum operating conditions for MRS's. Since the operator has control over both setting pressure and position of the machine, the authors focused on an evaluation of these two parameters. Because of the complexity of geologic and mining conditions and the geometry of mine excavations, all of which influence the performance of the machine, the authors combined the strengths of numerical modeling and statistical analyses. Boundary-element modeling was used to compare the stability of typical pillar extraction methods while controlling geologic factors. In addition, these models were used to design a field program in a cooperating mine that is extracting two-seam coal reserves. The two-seam mining geometry created an opportunity to assess machine performance in the field under different loading and geologic conditions.

The mine is located on the Wasatch Plateau, Utah. The coal seams being mined are the Tank and Blind Canyon, which are located toward the base of the Blackhawk Formation. The Tank Seam is presently being mined in an area partially undermined by the

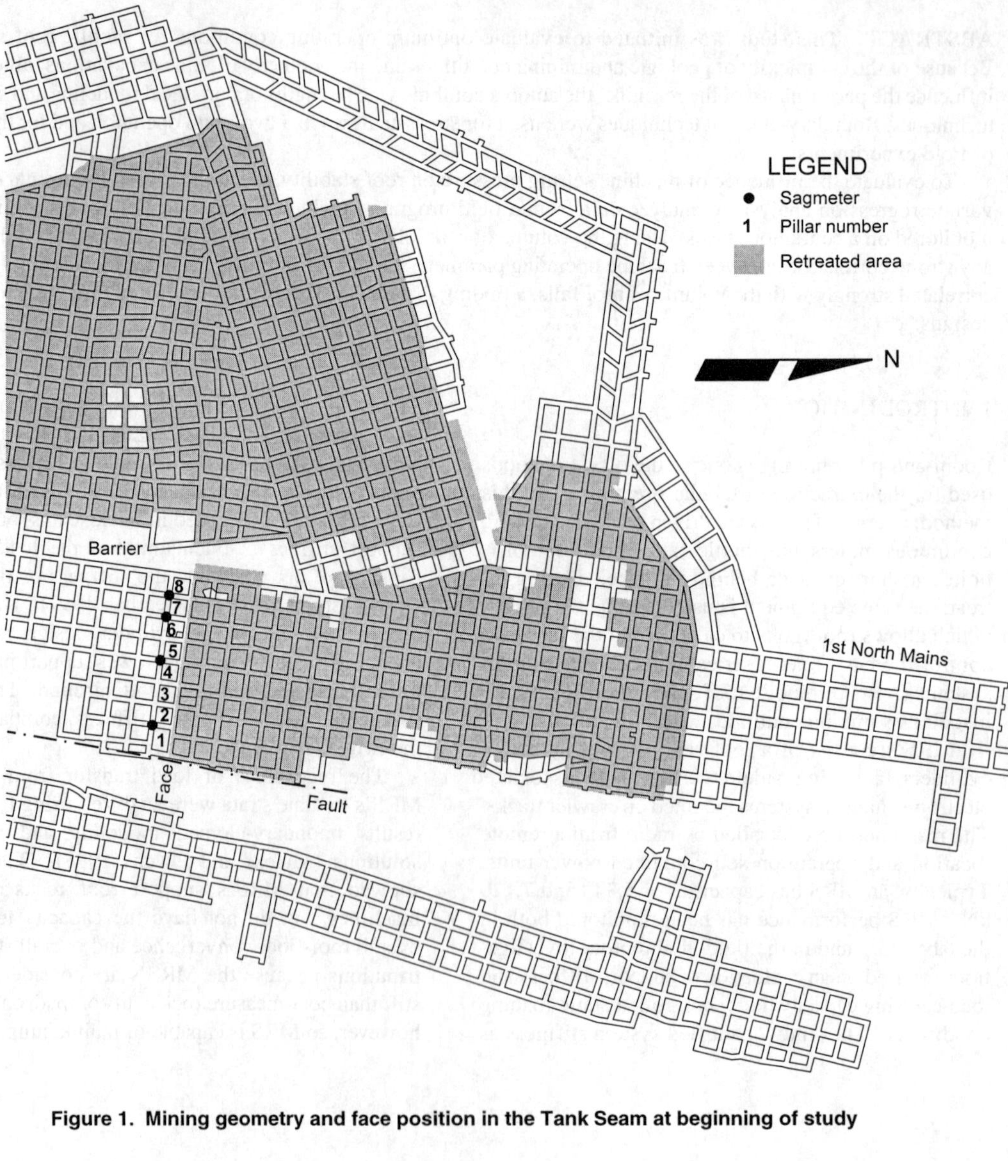

Figure 1. Mining geometry and face position in the Tank Seam at beginning of study

Blind Canyon Seam, approximately 85 m below. The test site is located in a graben bordered to the east and west by two major faults (Fig. 1). Joints are oriented approximately north-south, a common regional structural trend on the Wasatch Plateau. The roof consists of 1 to 2 m of shale overlain by a 9-m-thick sandstone, and rock quality is fair to good, depending on geologic conditions.

2 NUMERICAL MODELING

MRS performance in the field is influenced by geologic conditions and mine layout designs. The presence of retreated areas below the active MRS section influences strata movements and MRS spatial loading conditions. Multistep pseudo-three-dimensional stress analyses (Crouch and Starfield 1990) were completed for the design of the field program and for an assessment of pillar stability because the size of pillars was reduced compared to retreat sections using posts as secondary support. Model input was extracted on the basis of extensive measurements in adjacent mines (Maleki 1981, 1988) and back-analyses of results (Maleki 1990). The coal was modeled elastically. Load transfer through the gob was modeled using a bilinear hardening element. Preclosure of 10.2 cm was allowed before load transfer could take place; beyond this, additional closure was modeled as a linear spring with an equivalent elastic modulus of 1,860 MPa.

Figure 2 presents Tank Seam mining geometry and vertical stress distributions over a portion of the two-seam mining areas during the extraction of pillar 2. A stress profile was also prepared (Fig. 3) along an east-west cross section positioned toward the middle of the modeled area (away from the active face in the 1st North Mains). Results indicate a nonuniform stress distribution over the 1st North Mains with an increase in pillar stresses to the west. Maximum stresses are concentrated over the last row of pillars (row 8) and the barrier pillar. This is in agreement with underground observations and is caused by mining geometry in the Blind Canyon Seam.

Based on this and other multistep stress analyses, it became apparent that pillar stresses exceeded pillar strength (21 MPa) when approximately half a pillar was extracted (Maleki 1992). At this time, the pillar was approaching the postfailure regime. Pillar unloading resulted in increased roof-floor convergence and load transfer to the MRS units nearby. This process was associated with an increase in rate of loading on MRS's. Roof falls may be triggered by additional movement, particularly if smooth joints are present.

3 FIELD PROGRAM

The objectives of the field program were to (1) assess the influence of setting pressure on roof stability and (2) test the possibility of using changes in leg loading rate as a warning of pillar and/or roof failure problems. MRS performance was monitored during extraction of one row of pillars under both moderate and high stress conditions.

A factorial experiment design was initially proposed to evaluate the influence of MRS setting pressure on roof stability. In these designs, MRS performance was to be evaluated by testing the machine at different setting pressures under several geologic and premining stress conditions while changing one variable at a time. Results were inconclusive because the setting pressure could only be varied slightly (from 55 to 110 MPa). The operator did not wish to experiment with higher setting pressures for fear of damaging leg cylinders under dynamic loads (bounces in the roof). For the same reasons, the operator manually reset MRS leg pressures when pressure exceeded approximately 20 MPa.

To monitor MRS performance, leg pressures and convergence were monitored on all four machines at both the gob and plow sides. Thus, in any time window, the distribution of loads on individual machines and over the entire monitored area could be reconstructed. These data were collected using four Campbell Scientific data acquisition systems. By comparing loads on gob and plow jacks, the authors could evaluate differential loading on each machine. Analyses of data from four machines were useful in addressing loading pattern depending on position of machines with respect to pillars and, in some circumstances, inferring ground conditions in the area between machines.

Figure 4 presents mining geometry and roof fall locations during the extraction of pillar 2. Two roof falls occurred while completing approximately 50% extraction in the pillar. The first was outside of areas of active mining during mining cut 17. A few minutes before this roof fall, load (less than 14 MPA) and load rate (100 kPa/sec) increased moderately on MRS 1, which was located closest to the roof fall. The second roof fall buried the continuous miner during mining of cut 19. At this time, the effective area of the pillar was reduced by approximately 50%, and thus the pillar approached the postfailure regime. This assertion is made based on stress analyses (Fig. 2) and loading

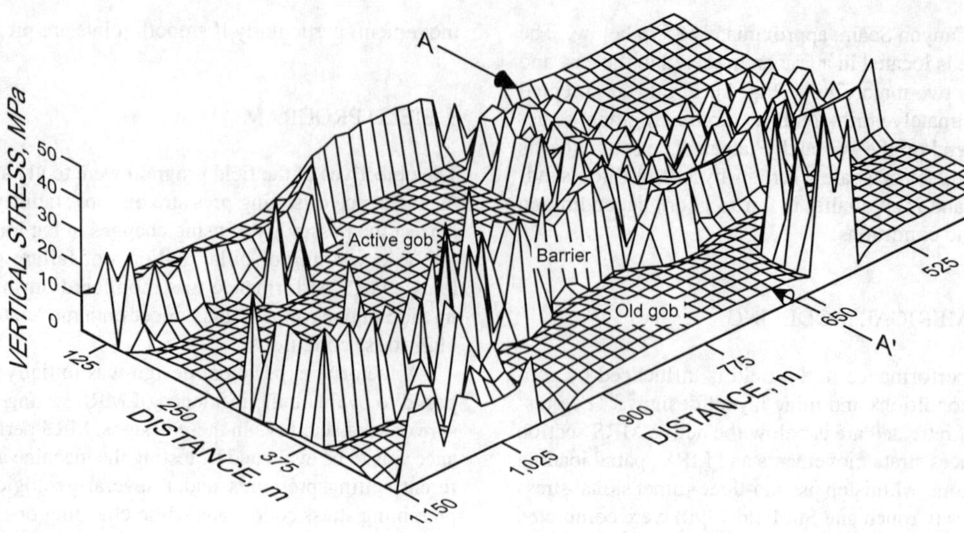

Figure 2. Vertical stress distribution on Tank Seam, step 2

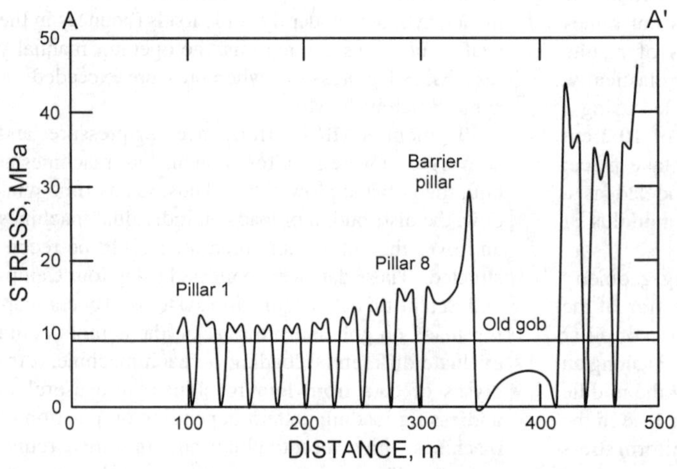

Figure 3. Vertical stress profile at secton A-A'

pattern on MRS's 1 and 4 (Fig. 4). Three minutes prior to this roof fall, the load on both MRS 1 and 4 increased to approximately 20 MPa and load rates to 120 to 200 kPa/sec. We suspect that pillar 2 was in a postfailure load deformation stage because there was a gradual load increase on MRS 4 during equipment recovery operations. At the termination of recovery operations, load was approaching 28 MPa. Pillar failure thus appeared to have triggered movements in roof blocks outlined by the north-south joints.

4 STATISTICAL ANALYSES OF FIELD DATA

To evaluate the influence of MRS operating conditions on roof stability, the authors used multiple regression analysis. This statistical method was suitable because there were a number of geologic, geometric, and support parameters that could have influenced the results. Roof stability was quantified by measuring the volume of roof falls. Six roof falls occurred during the monitoring period, two of which interrupted operations, while the others were located beyond the active limits of mining. Thirteen study locations were selected for statistical evaluation (Fig. 4). The data included information on MRS response at both stable and unstable study locations (approximately 50% each).

The first step of the analyses involved identification of 11 geologic, geometric, and support variables that might have had the potential to contribute to roof falls.

- *Joint frequency index.* Joint frequency index varied between 1 and 3. Index number 1 indicates massive roof and 3 indicates closely spaced jointed roof.

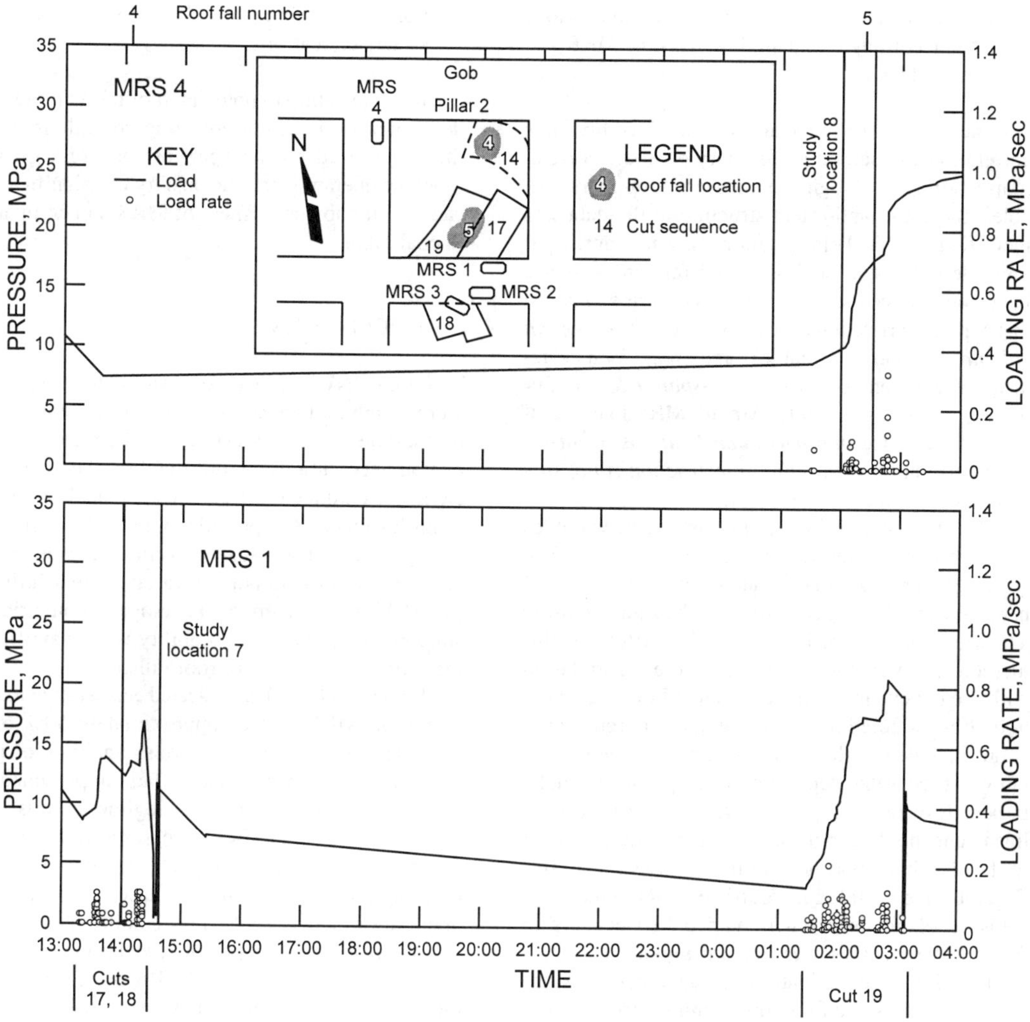

Figure 4. Load history for MRS's 1 and 4 during extraction in pillar 2

- *Stress gradient index.* This index varied between 1 and 1.7, depending on calculated stress distributions over the panel.
- *Cut orientation-span index.* This index varied between 20 and 80 and was calculated by multiplying a cut orientation index by cut span. Cut orientation index assumes a value of 1 where cuts are mined almost perpendicular to joints. Unfavorable cut orientations result when a cut is mined almost parallel to joints (index = 2).
- *MRS distance to unmined coal.* This distance varied from 12 to 25 ft, depending on the distance between the MRS and the elastic coal pillars. The MRS distance to unmined coal pillars (elastic) influences an MRS's response to load because of the high stiffness of elastic coal pillars with respect to MRS stiffness.
- *Sum max load rate index.* This index is equal to the sum of maximum load rates for the two MRS units on the opposite side of a cut. Higher index values are a measure of rapid changes in MRS load rates on the opposite sides and thus may be indicative of a change in overall pillar stability.
- *Sum load rate index.* This index varied between 0 to 35 and is the summation of number of load rate events that exceed an arbitrary critical rate of 0.10 MPa/sec.
- *Sum load rate-distance index.* This index is

similar to sum load rate index, but adjusts load rates according to squared distance of MRS's to unmined coal.

The second step in the analysis involved bivariate correlations. Based on preliminary bivariate correlations among all geologic, geometric, and support variables, the cause-and-effect structure in the data was identified. This helped tailor the procedures for multiple regression analysis using forward stepwise inclusion of dependent variables, as described later in this paper. Correlations between the variable *roof fall volume* (response variable) and *sum load rate-distance index* and *cut orientation-span index* are fair to very good (0.4 to 0.94). Among MRS load variables, for instance, the *maximum load* has a fair-to-good correlation with *set load, extraction ratio, sum load rate index,* and *average load.*

The third step in identifying important variables was to complete multiple regression analyses. The regression procedure consisted of entering the independent variables one at a time into the equation using a forward selection methodology (SPSS 1995). In this method, the variable having the largest correlation with the dependant variable is entered into the equation. If a variable fails to meet entry requirements, it is not included in the equation. The volume of roof fall was used as the dependent or "response" variable. In search of better regression results, we used both linear and nonlinear responses. The square root of roof fall volume was used in the nonlinear analyses. The multiple correlation coefficient (R), which is a measure of goodness-of-fit, varied between 0.9 to 0.96 for these analyses, indicating a good fit.

Based on an examination of standardized regression coefficients, the following mining variables best explain the variations in the volume of roof falls and thus statistically have the most significant influence on roof fall potential.

- *Cut orientation.* Cuts made almost parallel to the joints increase the potential for roof falls. Cut orientation is influenced by panel layout design.
- *Mining span.* The wider the span created during pillar extraction, the higher is the potential for roof falls. Excessive spans can be avoided by leaving small stumps between successive cuts.
- *Pillar extraction ratio.* This factor is negatively related to roof fall volume, indicating that roof falls can occur at moderate extraction ratios (say 50%) during pillar unloading. Depending on stiffness of roof strata, however, the loads can be transferred away from the pillar, allowing additional extraction, particularly if stumps are left at strategic locations.

It is interesting to note that roof fall volume is not dependent on machine operating conditions (within the range used by the operator) or stress gradient. Machine operating pressure history is insignificant because of the lower stiffness of MRS's in comparison to coal pillars.

5 CONCLUSIONS

The strengths of computational and statistical methods were combined in conjunction with detailed field measurements to assess the influence of MRS operating conditions on roof stability. The boundary-element technique was used for stress analyses of two-seam layouts and to provide input to the design of a field program. The statistical techniques were used to identify the most important variables that influence roof stability considering all geologic, geometric, and support variables. Roof stability was quantified by measuring the volume of roof falls.

Review of field data indicated accelerating loading trends on MRS's when approximately 50% of the pillar area was extracted. Stress analyses of pillar geometries indicated that pillars approached the postfailure load-deformation regime at this time. Thus load rate on an MRS increases as pillars unload. There was no unique loading rate or pattern that could be related to pillar failure because of variations in operating practices and stress conditions, including setting pressure, MRS position, and mine seismicity. The loading rates on the MRS's, however, give an indication of the overall stability of the pillars and fenders.

Based on multivariable regression analyses, two variables correlated most directly with roof fall volume. These were (1) mining span and (2) cut orientation with respect to geological discontinuities. The authors did not find any strong correlation between machine operating parameters (within the range used by the operator) and roof stability at this site. Since geometric and mining factors correlated strongly with the volume of roof falls, it is important to develop optimum mine layout designs, including mine orientation, pillar design, and the amount of excavation (to avoid creating very wide spans).

REFERENCES

Barczak, T., and D. Gearhart, 1997. Full-Scale Performance Evaluation of Mobile Roof Supports. Paper in *Proceedings: New Technology for Ground Control in Retreat Mining*, comp. by C. Mark and R. Tuchman (Pittsburgh, PA). NIOSH IC 9446, pp. 99-126.

Crouch, S., and T. Starfield, 1990. *Boundary Element Methods in Solid Mechanics.* Routledge Chapman and Hall.

Maleki, H. 1981. Coal Mine Ground Control. Ph. D. dissertation, Colorado School of Mines, Golden, CO.

Maleki, H., 1990. Development of Modeling Procedures for Coal Mine Stability Evaluation. Paper in *Rock Mechanics Contributions and Challenges: Proceedings of the 31st U.S. Symposium*, ed. by W. A. Hustrulid and G. A. Johnson (CO School of Mines, Golden, CO, June 18-20, 1990). Balkema, pp. 85-92.

Maleki, H., 1992. In Situ Pillar Strength and Failure Mechanisms for U.S. Coal Seams. Paper in *Proceedings of the Workshop on Coal Pillar Mechanics and Design*, comp. by A. T. Iannacchione, C. Mark, R. C. Repshar, R. J. Tuchman, and C. C. Jones (Santa Fe, NM, June 7, 1992). U.S. Bur. Mines IC 9315, pp. 73-77.

Maleki, H., and J. Owens, 1998. Analysis of the Interaction Between Mobile Roof Supports and Mine Strata. Paper in *Design and Construction in Mining, Petroleum, and Civil Engineering: Proceedings of the Fifth South American Congress on Rock Mechanics and Second Brazilian Conference in Rock Mechanics-SAR Rocks 98,* ed. by L. A. da Silva, E. F. de Quadros, and H. H. S. Conçalves (Santos, São Paulo, Brazil, Nov. 22-25, 1998). Univ. de São Paulo, Escola Politécnica, pp. 287-293.

Maleki, H., J. Owens, and W. Howie, 1999. An Overview of Geomechanics Safety Research on Mobile Roof Support. Paper in *Proceedings, 17th International Systems Safety Conference, System Safety at the Dawn of Millennium* (Orlando, FL, Aug. 16-21, 1999). System Safety Society, Unionville, VI, pp. 554-563.

SPSS, 1995. *Statistical Packages for the Social Sciences.* Chicago, IL.

Computer Methods and Advances in Geomechanics, Desai et al. (eds) © 2001 Balkema, Rotterdam, ISBN 90 5809 183 X

Contribution of reinforced shotcrete lining on controlling tunnel failure

Ahmed Moussa
Faculty of Engineering, El-Mataria, Cairo, Egypt

ABSTRACT: Based on the principle of the New Austrian Tunneling Method, shotcrete lining provides such a smooth and flexible support that accept part of ground deformations put guards against failure. Therefore careful design of the lining is very important for critical tunnels. In this study, nonlinear finite element analysis has been employed to predict the influence of lining quality, thickness and reinforcements on the tunnel behavior. . Nonlinearity in terms of strain hardening and softening and cracking of shotcrete lining is considered employing a stress-strain history model. The results of the study can be considered as a guide towards optimum design of tunnel lining which satisfy safety requirements and deformation limits.

1 INTRODUCTION

In most cases, collapse of a tunnel is related to errors in either design or construction or in both. Numerical prediction tunnel behavior may provide a great help towards optimum design and avoiding failure. The employed model in the analysis should yield characteristic quantities as design criteria, which are drawn from a safety concept and different hypotheses for failure possibilities. In-situ monitoring during excavation should check whether the actual deformations and stresses are in accordance with the prognosis of the model. In recent years, great improvement has been achieved in the structural modeling of tunnels. Simple analytical methods were used to describe the behavior of circular tunnels in uniform ground condition. Owing to the great development in computer memory and speed, finite element method has been widely used in analyzing tunnels. Two and three-dimensional models are used successfully to describe the tunnel behavior under different ground conditions and excavation systems. Sophisticated constitutive models are introduced by many authors to describe plasticity, creep and damage in ground materials. Very few researches are concerned with the nonlinear modeling of tunnel lining.

2 EMPLOYED FINITE ELEMENT MODEL

Tunnel excavations are analyzed herein using two dimensional finite element model. Two-dimensional simulation of the three-dimensional effect near the tunnel face has been performed using stiffness reduction model. Excavations and construction works are simulated by loading steps. A special six-node plane frame element is used to model shotcrete in order to simulate the combined action between the tunnel lining and the surrounding soil, which is modeled by a linear strain triangular plane strain element as shown in Fig. 1. The soil is considered as elastoplastic Mohr-Coulomb material. Nonlinearity in terms of strain hardening and softening and cracking of shotcrete lining is considered employing stress-strain history model. This model takes into consideration the effect of damage due to early loading on the long-term development of shotcrete strength, and the plastic deformation due to earlier

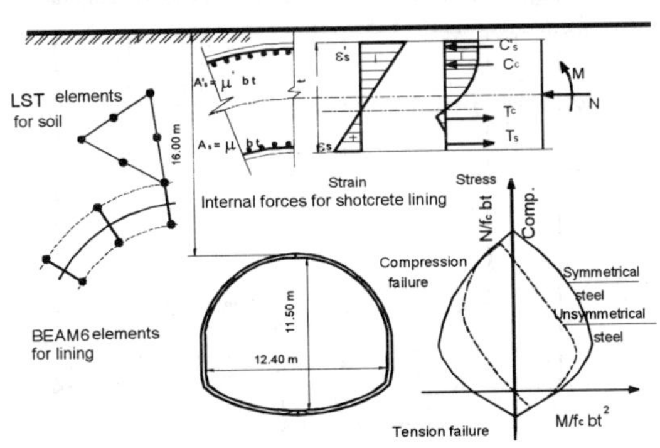

Figure 1: Numerical model

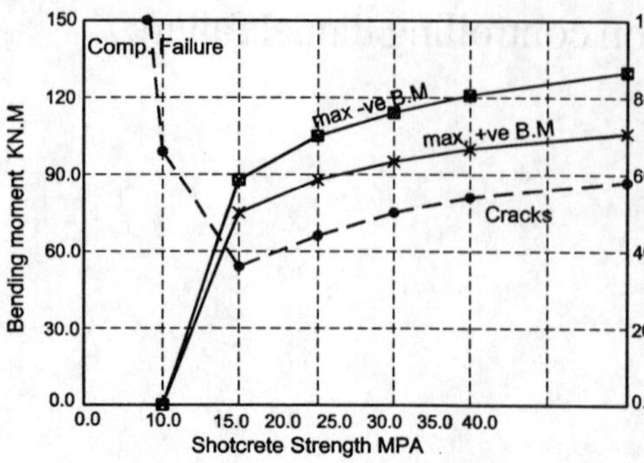

Figure 2: Fcu - B.M & cracks relationships

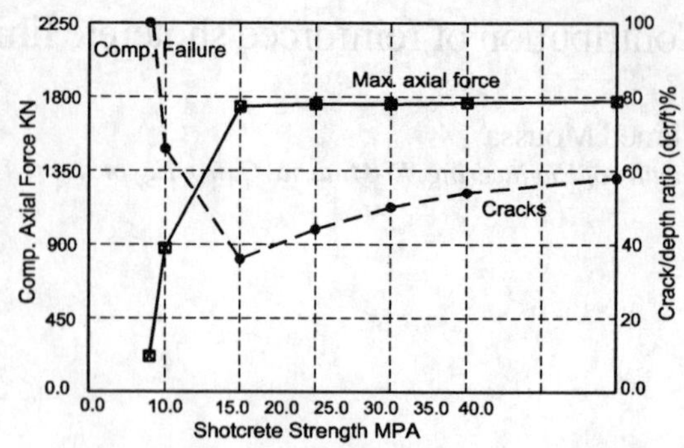

Figure 3: Fcu - N.F & cracks relationships

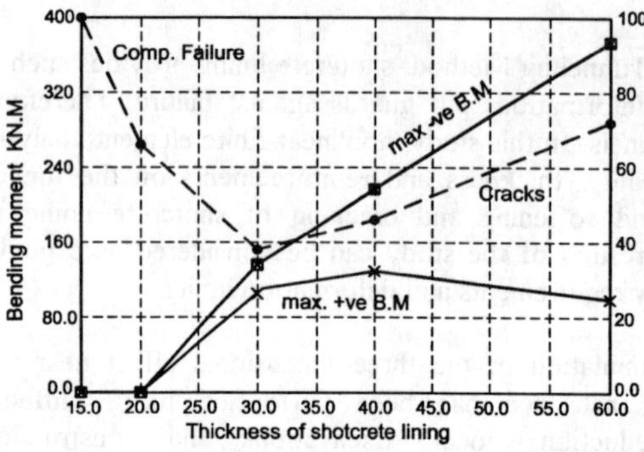

Figure 4: Lining thickness-B.M & cracks relationships

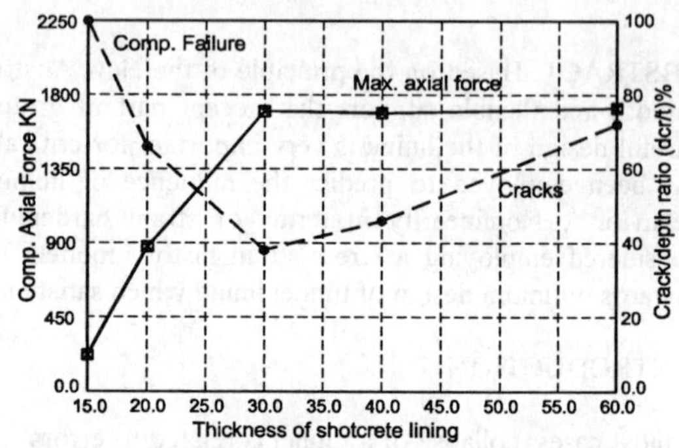

Figure 5: Lining thickness-N.F & cracks relationships

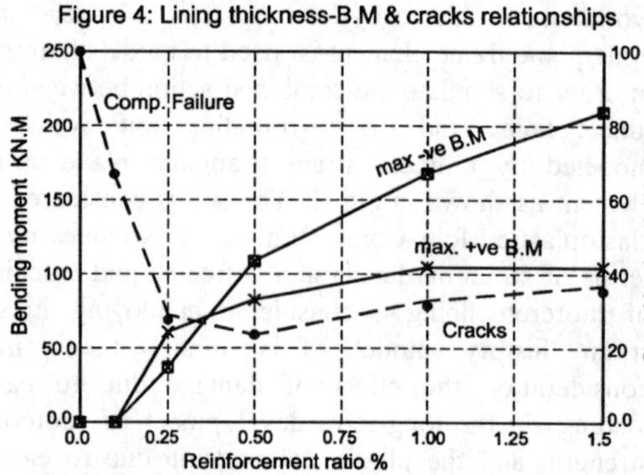

Figure 6: Reinforcements-B.M & cracks relationships

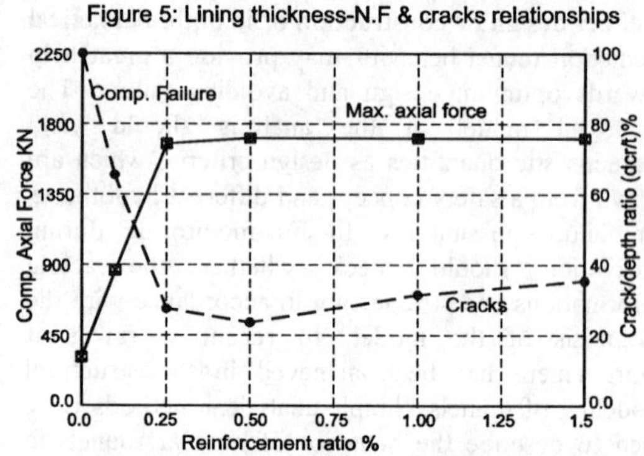

Figure 7: Reinforcements-N.F & cracks relationships

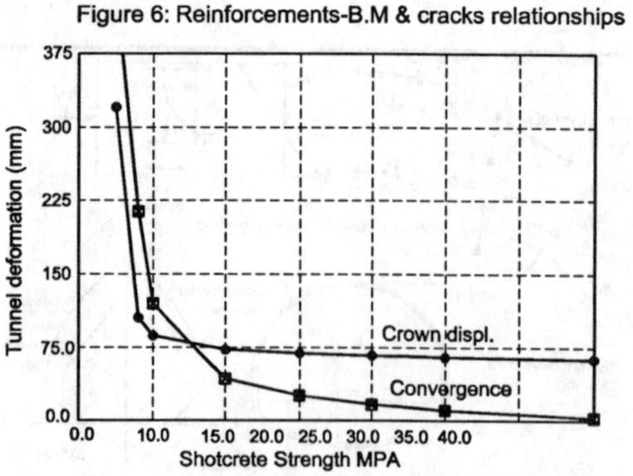

Figure 8: Fcu - lining deformation relationships

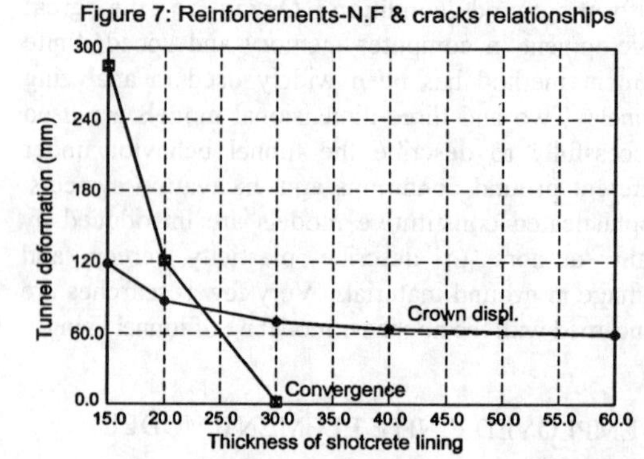

Figure 9: Lining thickness-deformation relationships

Table 1: Main results of tunnel analysis.

Parameter	f_{cu} in MPa	(t) cm	$\mu = As/Ac$ %	Crown displ. $\delta_{surface}$ cm	Crown displ δ_{crown} cm	Vl conv. δv cm	Hz conv. δv cm	Crack depth dcr cm	Crack / thick. ratio dcr/t	Stability condition
Shotcrete quality f_{cu} MPa	5.00	20.0	0.10	15.03	32.19	35.7	43.90	20.0	100.0	Complete collapse
	8.00			7.21	10.54	10.3	21.40	20.0	100.0	Complete collapse
	10.0			5.950	8.690	8.28	12.11	13.20	66.0	2 plastic hinges
	15.0			4.97	7.30	6.67	4.27	7.2	36.0	Stable
	20.0			4.67	6.89	6.20	2.56	8.8	44.0	Stable
	25.0			4.50	6.66	5.93	1.66	10.0	50.0	Stable
	30.0			4.39	6.51	5.76	1.07	10.8	54.0	Stable
	40.0			4.24	6.31	5.53	0.31	11.6	58.0	Stable
Thickness of the lining t cm	10.0	15.0	0.10	8.08	11.83	11.7	28.5	15.0	100.0	Complete collapse
		20.0		5.95	8.69	8.28	12.11	13.2	66.0	2 plastic hinges
		30.0		4.73	6.98	6.37	1.84	11.4	38.0	Stable
		40.0		4.27	6.37	5.66	-0.52	18.4	46.0	Stable
		600		3.85	5.87	4.88	-1.87	28.8	72.0	Stable
Reinforcing steel ratio	10.0	20.0	0.00	6.46	9.45	9.12	16.24	16.8	84.0	Complete collapse
			0.10	5.95	8.69	8.28	12.11	13.2	66.0	2 plastic hinges
			0.25	5.33	7.8	7.27	6.62	7.2	36.0	1 plastic hinge
			0.50	4.91	7.22	6.62	3.48	4.8	24.0	Stable
			1.00	4.50	6.66	5.99	0.904	5.6	28.0	Stable
			1.50	4.31	6.40	5.67	0.07	6.4	32.0	Stable

loadings. For steel reinforcements a tri-linear stress-strain relationship is assumed. Internal forces and stiffness parameters are calculated by numerical integration. The model can represent successfully the post-failure softening behavior of the shotcrete lining.

3 STUDY PARAMETERS

In order to evaluate the contribution of shotcrete lining in controlling tunnel deformation and failure, a collapsed tunnel has been selected as a case study. The geological conditions of the surrounding soil were: Unit weight of 19 KN/m^3, cohesion of 4.0 KN/m^2, angle of internal friction (ϕ) of 20°, Elastic modulus of 80 Mpa, Poisson's ratio (v) of 0.4 and Coeff. of at rest pressure K_O of 0.5. The geometry of the tunnel is as shown in Fig 1. Numerical trials to prevent failure and to limit deformations have been carried assuming different values for lining quality, thickness and reinforcements as shown in Table 1.

4 ANALYSIS OF NUMERICAL RESULTS

Tunnel failure normally starts with formation of plastic hinge at the most critical section in the lining. In case of unstable tunnel, more plastic hinges start to form, while the tunnel deformations increase. In this study, full collapse of tunnel is numerically considered, if the nonlinear solution doesn't converge. In this condition, the more iterations are done, the more deformations and more plastic hinges are obtained. The tunnel is considered safe, if no plastic hinges are formed in the lining. It is considered critical, if partial plastic hinges are formed while the system remains stable.

The relationships between strength of shotcrete and resulted ; lining forces, bending moments, damaged depth of lining, and tunnel deformations are plotted in Figs. 2, 3 and 8. Using tunnel lining with average shotcrete strength of 15 MPa resulted in stable tunnel with reasonable displacements and lining forces, and minimum cracks. The deformations and the lining cracks dramatically increases as the strength decreases, accompanied with softening behavior of the lining forming successive plastic hinges up to complete collapse of the tunnel for a shotcrete strength of 5 MPa. As the average strength of the used shotcrete becomes higher than 15 MPa, slight decrease in tunnel displacements are achieved, bending moments and tension cracks in the lining slightly increase with nearly constant axial forces.

The resulted relationships between the thickness of tunnel lining and the predicted; lining forces, bending moments, damaged depth of lining, and tunnel deformations are plotted in Figs. 4, 5 and 9. It is clear from these results that there is an optimum thickness of lining. As the thickness decreases from the optimum value, the tunnel displacements, horizontal lining convergence, and damaged part of shotcrete lining dramatically increase up to crushing

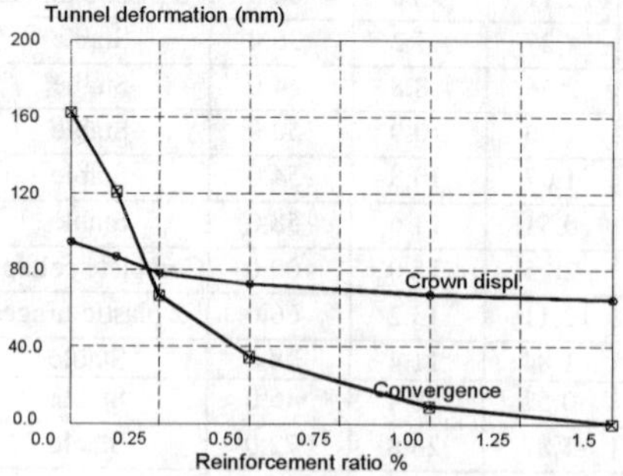

Figure 10: Reinforcements-tunnel deformations relationship

successive plastic hinges in the lining up to complete collapse of tunnel. Increasing the lining thickness than the optimum value results in bending moment concentration accompanied by increase of cracks, especially at the rough corners in lining. For the investigated tunnel, the optimum thickness of the lining was about 30 to 35 cm.

Instability of tunnel associated with excessive horizontal convergence, vertical displacements and plastic hinges were predicted for cases where the lining reinforcement ratio μ was $\leq$ 0.25%. By increasing the steel ratio to 1.0%, complete stability of the tunnel was achieved with reasonable deformations and minimum cracking depth. Increasing steel ratio to 1.5% minimized the horizontal convergence and reduced the vertical crown displacement by about 10% , but the external (-ve) bending moment at the toe of tunnel increased by about 90% with slight increase in crack depth, while the axial forces and bending moments of other parts did not show significant change.

5. CONCLUSIONS

1- The results of the analysis proves the ability of the employed numerical model to describe the behavior of shotcrete tunnels during construction and to simulate successfully the shotcrete nonlinearity and post-failure behavior of shotcrete lining in critical tunnels. Partial and full plastic hinges in the lining are numerically predicted depending on safety level for the lining.

2- Stiffness of the lining should exceed the minimum value required for stability. More stiffening reduces the desired flexibility, attracts more forces and may result in larger cracks. Optimum values for particular tunnel can be obtained based on the numerically predicted tunnel deformations and lining cracks. For the investigated tunnel, stability has been achieved by either increasing the lining thickness, increasing the reinforcing steel, or by increasing shotcrete strength up to certain values.

3- The optimum thickness can be defined as the minimum thickness that satisfy safety and deformation limits with reasonable reinforcements and shotcrete strength. Increasing the lining thickness higher than the optimum value resulted in bending moment concentration accompanied by increase of cracks especially at the sharp corners in the lining.

4- Tension cracks in the lining can be controlled by increasing the lining reinforcements. Very high reinforcement resulted in bending moment concentration at sharp corners of the lining without much improvement in the deformations.

5- Improving shotcrete quality leads to good results concerning safety, deformations and cracks. Concentration of bending moments at corners is much less compared with cases of increasing lining thickness or reinforcing steel.

6. REFERENCES

Eisenstein, Z. and A. Negro. 1985. Comprehensive design method for shallow tunnels. *Proc. Underground Structures in Urban Areas Conf.*, Prague. Vol 1:375-391.

Hisatake, M. and Murakami T. 1992. Non-destructive evaluation of tunnel lining stresses. *Proc. of the 4th Int. Symp. on Num. Models in Geomech. NUMOG IV*, Swansea, UK, 685-695.

Moussa, A. 1996. Prediction of safety level in tunnels. *Proc. of the 7th Int. Colloq. on Structural and Geotechnical Eng.*, Ain shams University, Cairo, Egypt: 281-292.

Sauer, G. 1994. NATM in Soft Ground. In Dr. G Sauer: Design & Consulting, Mining Journal Ltd.: 25-30.

Swoboda, G., Moussa, A., Lukas, W., and Kusterle, W. 1993. On constitutive modelling of shotcrete. *Proc. of the Intr. Sym.on Sprayed Concrete*, Fagernes, Norway: 133-141.

Swoboda G. 1990. Numerical Modeling of Tunnels, in C. S. Desai & G. Gioda. *Numerical Methods and Constitutive Modeling in Geomechanics, Courses and Lectures*, No.311: 277-318 Wien: Springer-Verlag.

Swoboda, G., 1999 *Programmsystem FINAL - finite element analysis program for linear and nonlinear structures*, Version 7.1, Austria: University of Innsbruck.

Computer Methods and Advances in Geomechanics, Desai et al. (eds) © 2001 Balkema, Rotterdam, ISBN 90 5809 183 X

Numerical study on mechanical and hydraulic behavior of concrete plug in underground cavern

Byung-Ki Park, Hi-Keun Lee & Seokwon Jeon
School of Civil, Urban and Geosystem Engineering, Seoul National University, Korea

ABSTRACT: This paper presents the results of the numerical analyses for the performance of a concrete plug in underground storage cavern. In this study, the 2-dimensional finite difference code *FLAC* and the 3-dimensional *FLAC3D* were used. When the plug was installed at a great depth, there was a tendency for the plug to be more stable with increasing in-situ stress ratio (K). The results considering the interface elements at plug-rock contact indicated that the lower asymmetric shaped plug with smooth contact condition was more stable than the symmetric or upper asymmetric plug. The results of the hydraulic analysis showed that the plug slot excavated deeper than the EDZ (Excavated Damaged Zone) was more effective than the slot within the EDZ to prevent leakage.

1 INTRODUCTION

In construction of an underground storage cavern for hydrocarbon, a concrete plug is installed in an access tunnel or a shaft that connects the underground cavern and the ground surface after completion of the excavation and installation of facilities. The plug prevents the stored product (oil, gas, etc) from leaking and the excessive inflow of underground water, so it plays an important role in construction and operation of the storage cavern. Once the plug is constructed, the cavern is isolated from the external access. Therefore, mechanical and hydraulic consideration should be made for the concrete plug to fulfill its function.

But most studies done for the plug in recent years were mainly focused on sealing methods in nuclear waste repository projects, so the studies that deal with mechanical and hydraulic behavior of the plug under high pressure are rare. Therefore, in this study, we conducted numerical analyses on the mechanical behavior of the plug to find optimal shape and thickness of the plug with respect to the installation depth and the condition of rock-plug interface. *FLAC* was used for 2-dimensional analysis and *FLAC3D* for 3-dimensional. Data from the geologic investigation of the area were used for input parameters. We also carried out the same analyses taking account of the effect of Excavation Damaged Zone (EDZ). Finally, we conducted the numerical analyses on hydraulic behavior to examine the flow rate toward cavern through plug and rock mass.

2 2-D MECHANICAL ANALYSIS OF THE PLUG

2.1 *Model setup*

The selected section in this study was Pyong-Taek LPG storage cavern in Korea. The model for the analysis was set up with shaft, plug, and cavern together with axisymmetric model. The boundary was set to 50 m in width and 150 m in height. Mohr-Coulomb failure criterion was adopted for rock mass and the Excavation Damaged Zone (EDZ), and the elastic model for the plug. Three types of plug geometry were used for this analysis as shown in Figure 1.

2.2 *Input data*

Table 1 shows the material properties of rock mass and the concrete plug. The rock mass properties were determined by the results of geologic investigation and laboratory test of the Pyong-Taek LPG storage cavern project [SK Co., 1998]. The calculated stress applied to the top of the plug was about 1.6 MPa.

Table 1. Material properties of the rock mass and concrete plug

Material Properties	Rock mass	Concrete Plug
Density (kg/m^3)	2500	2300
Young's modulus (GPa)	17.8	35
Poisson's ratio	0.2	0.25
Cohesion (MPa)	4.3	-
Friction angle (°)	35	-

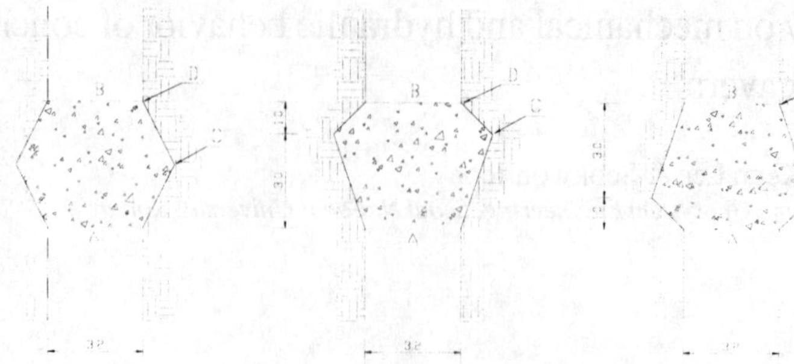

(a) Symmetric (b) Upper asymmetric (c) Lower asymmetric

Figure 1. Plug shape and its analysis point

2.3 Analysis sequence

The analysis procedure followed the real construction sequence as below. The analysis which applied interface element to rock-plug contact was added in STEP 4. In STEP 4, infiltration pressure was also applied to the upper part of the rock-plug interface.

1) STEP 1: Excavate shaft, plug, and cavern.
2) STEP 2: Plug installation.
3) STEP 3: Loading water pressure on the upper surface of the plug.
4) STEP 4: Apply infiltration pressure to rock-plug interface

2.4 Effect of in-situ stress ratio (K value), depth, and plug shape

We have numerically analyzed the plug behaviors with different shapes of the plug, K values, and installation depths. Each analysis applied the various conditions in combination.

- Shapes: upper asymmetric, symmetric, and lower asymmetric
- K values: 0.5, 1.0, 1.5, and 2.0
- Depths: 100, 200, 500, and 1000 m

From the analysis, we obtained several results. The maximum compressive stress occurred in the rock mass at point C (Figure 1). And the maximum value of the compressive stress and displacement within the plug occurred at point B. But, the maximum tensile stress concentration occurred at point A. In case that the plug was installed at a relatively shallow depth (100 m – 200 m), the magnitudes of the principal stress and the maximum displacement were not sensitive to the changes in in-situ stress ratio (K) and the shape of the plug. But, when installed at a great depth (1000 m), there has a tendency for the plug to be more stable with increasing in-situ stress ratio (K).

The criterion to evaluate the stability of the plug was the magnitude and the extent of tensile stress and displacement occurred in the plug. Figure 2 shows the variation of the maximum tensile stress and displacement of each shape with respect to depth and K value. Differences between tensile stress and displacement of each plug were not great. Figure 3 shows the contours of the maximum and the minimum stress around the plug for symmetric shape. Contours for other shaped plug was similar to that of the symmetric shape.

2.5 Effect of rock-concrete interface

The plane of contact between the concrete plug and rock mass was considered by using interface elements, assuming that it follows the Coulomb shear failure criterion. The Coulomb shear strength criterion limits the shear force by the following relation.

$$F_{s\max} = cL + F_n \tan\phi \tag{1}$$

where c is cohesion along the interface, L is effective contact length, and ϕ is friction angle of interface surfaces.

Input parameters used for this analysis are presented in Table 2. In case that the contact plane was rough, the shear stiffness k_s was assumed to be $0.9k_n$. But, for the contact with the intermediate and smooth surface, the shear stiffness of $0.5k_n$ and $0.1k_n$ was considered, respectively [Itasca, 1995].

Table 2. Material properties for the interface

Interface Properties	Rough	Intermediate	Smooth
Normal stiffness (GPa)	20	20	20
Young's modulus (GPa)	20	10	2
Cohesion (MPa)	2	2	2
Friction angle (°)	30	30	30

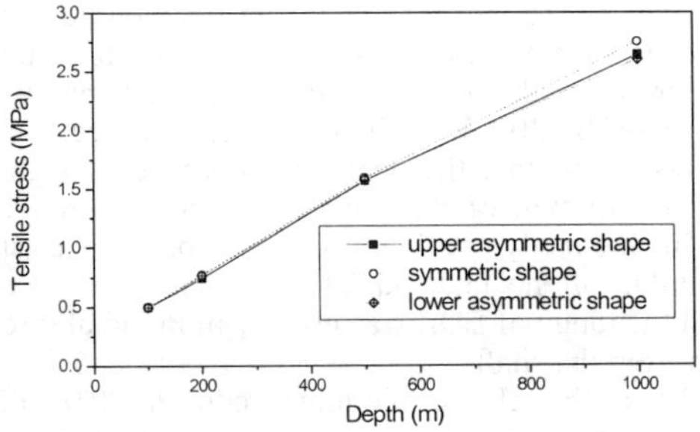

(a) K = 0.5

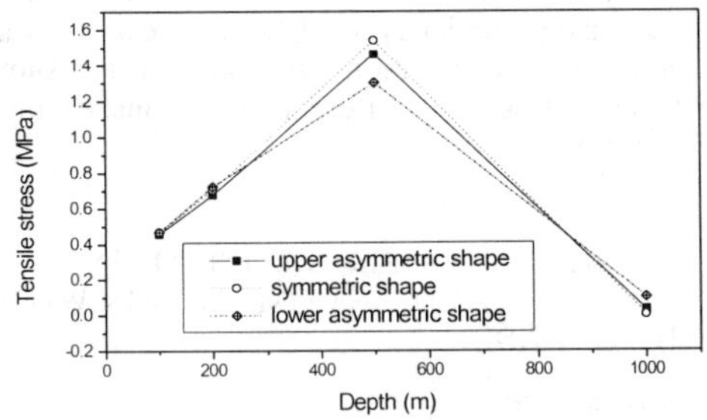

(b) K = 1.0

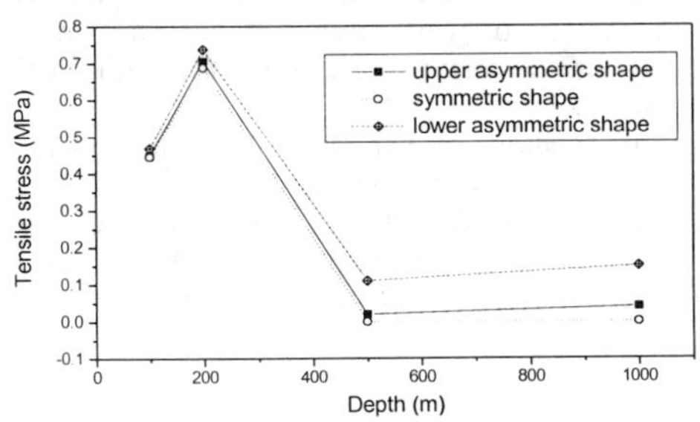

(c) K = 1.5

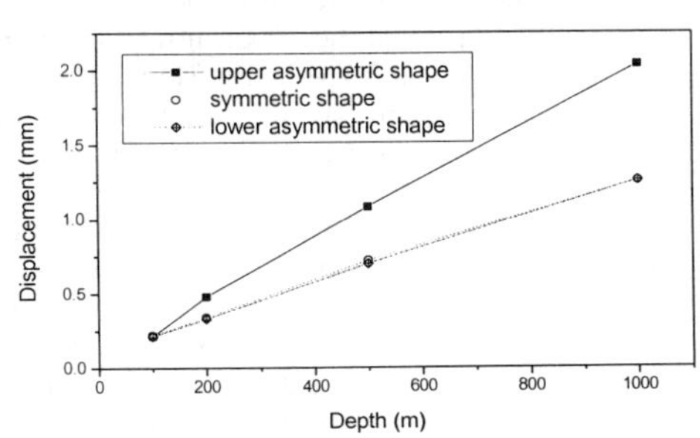

(d) K = 1.0

Figure 2. Variation of the maximum tensile stress and displacement with the changes in depth and K.

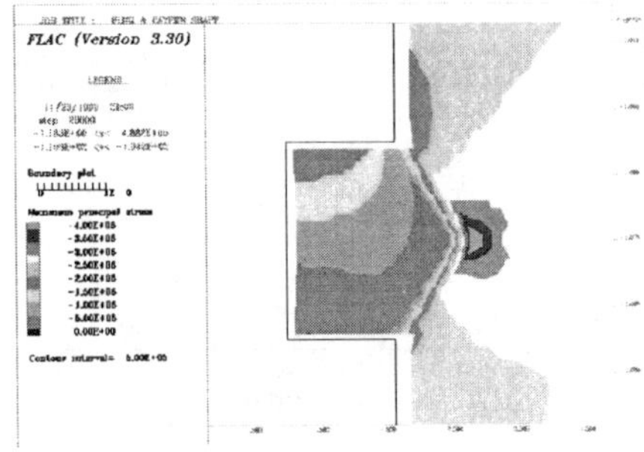

Figure 3. Contour of the maximum principal stresse for symmetric shape (K = 1.0 and Depth = 100 m)

In STEP 3, there was no significant difference compared with the analysis of no interface. The magnitude of the tensile stress occurred at point A lied between the values with the rough and smooth interface analysis approximately. In STEP 4, after the infiltration pressure was applied to the upper part of the interface, the displacement at each point and the tensile stress at point A tended to increase. For the tensile stress, this tendency was more significant in deeper area than in the shallow one, in upper asymmetric or symmetric shape than in lower asymmetric one, and in rough contact condition than in smooth one.

The result of the analysis without the consideration of the infiltration showed that the tensile stress $\sigma_U > \sigma_S > \sigma_L$ at the bottom of the plug, where the subscripts U, S, and L stands for the upper asymmetric, symmetric, and lower asymmetric shape, respectively. But after the infiltration pressure was applied to the upper interface between rock mass and the plug, the results were different. The magnitude and the extent of the tensile stress were smallest in the case of the lower asymmetric shape. This is because that the infiltration pressure applied to the upper interface, which is nearly vertical, plays the role of the lateral stress.

The smoother condition of the interface allowed the more displacement, so that the magnitude and the extent of the tensile stress occurred in the plug were reduced. The displacement was less than 1 mm in all cases satisfying the allowable criterion. Figure 4 shows the tensile stress with respect to interface condition at the installation depth of 100 m.

2.6 *Effect of the EDZ*

When excavating tunnels or underground space, blasting causes additional damages to adjacent rock mass. And as a consequence changes of rock mass strength, density, porosity, internal friction angle, hydraulic conductivity and stress redistribution will result in. Among many criteria for estimating blast

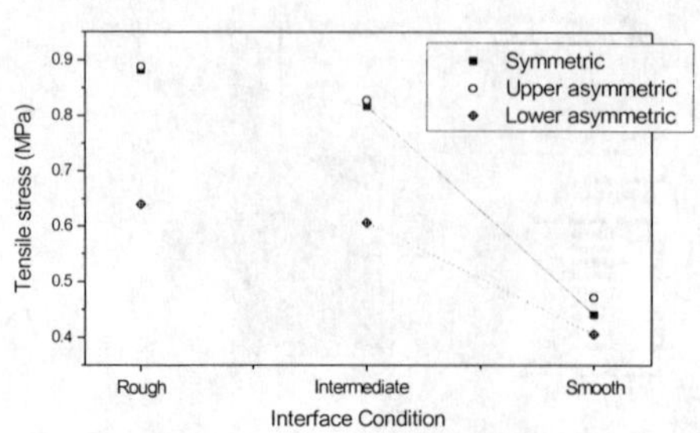

Figure 4. Tensile stress for different interface conditions (K=1.0, Depth = 100 m).

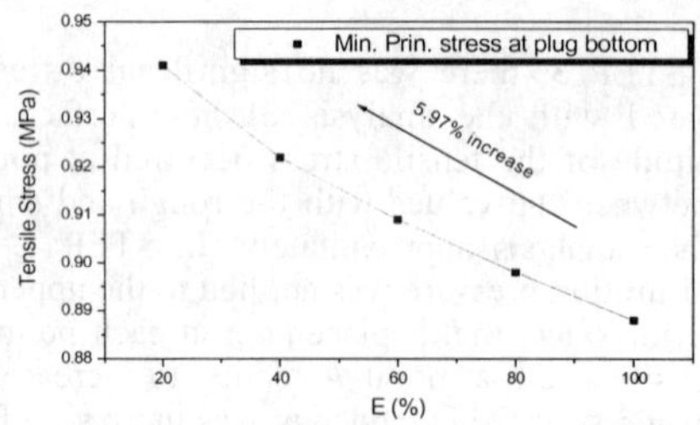

(a) Tensile stress

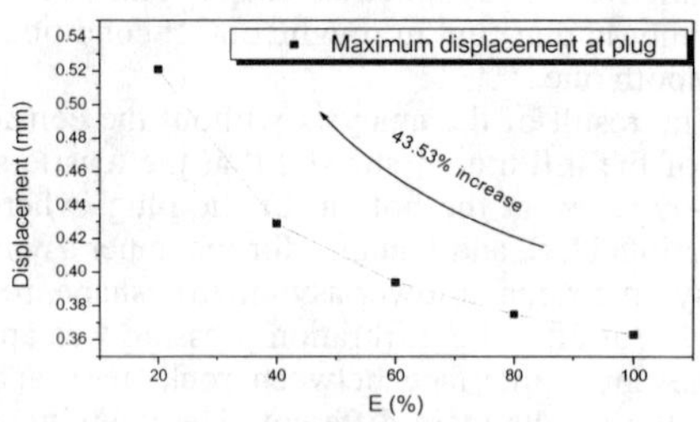

(b) Displacement

Figure 5. Stress and displacement according to the decrease of elastic modulus (for upper asymmetric shape, K=1.0, depth 100m).

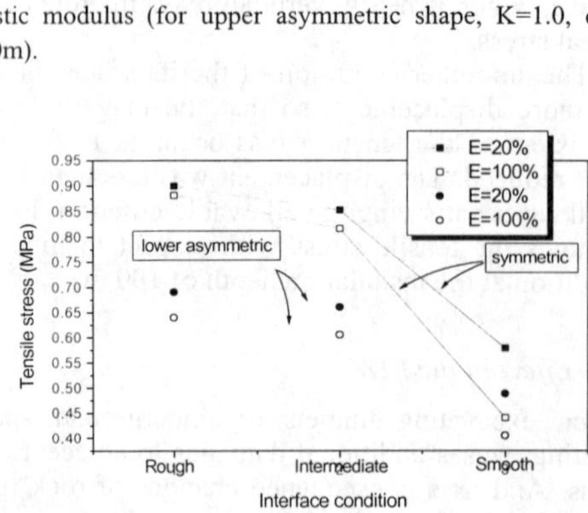

Figure 6. Tensile stress for the different conditions of interface and degree of damage.

induced EDZ, the change in elastic modulus was adopted as indices of EDZ estimation in this study. According to the previous study [3], elastic modulus decreases by 20% near the excavation surface. Also, it was found that the depth of influence, though it varied with type of explosives, was about 0.3 m - 1.0 m. In this study, 80%, 60%, 40%, and 20% elastic modulus of the undamaged rock mass were used. And the range of EDZ was taken 1 m in radial direction from the shaft.

When the elastic modulus reduced 20%, displacement increased 43% and the tensile stress increased by 6%. Therefore, it was observed that the increase of damage has a strong influence on the displacement rather than on stress. Figure 5 shows the increasing tendency of tensile stress and displacement with excavation damage. Figure 6 shows the tensile stress for different conditions of plug-rock interface.

3 3D ANALYSIS OF PLUG-ROCK MASS INTERFACE AND ITS COMPARISON WITH THE 2D CASE

3.1 *Model setup*

For the analysis of an underground storage cavern, which involves many intersections, it is not appropriate to use 2D model. In this study, 3D modeling that consisted of shaft plug and caverns was performed and the results were compared with those of 2D analysis.

Figure 7 shows 3D model. EDZ was considered within radius of 1 m from the shaft. Rough contact was assumed for the plug - rock mass interface. K value was assumed to be 1.0 and the analysis was mainly performed for the case of symmetric shape.

3.2 *Result*

The result of the analysis showed that the maximum tensile stress at the lower part of plug was 0.66 MPa, which was almost the same with the result in 2D analysis (0.67 MPa). And the maximum compressive stress at rock mass was 3.7MPa for 3D analysis and 4.4MPa for 2D, respectively. It was observed that stress concentration was less severe in case of 3D analysis than in 2D.

Figure 8 demonstrated the changes of time-step displacement. Displacement numbers express the points that are indicated in the Figure 7(b). The maximum displacement, 0.45mm, was found to be relatively larger than that of 2D analysis, 0.33mm. In other words, stresses in 3D analysis were smaller than that of 2D, however, displacement showed a little increment. Increment of displacement due to 3D analysis is relatively small compared to that of 2D, so that it is evaluated that it has little influence on plug performance.

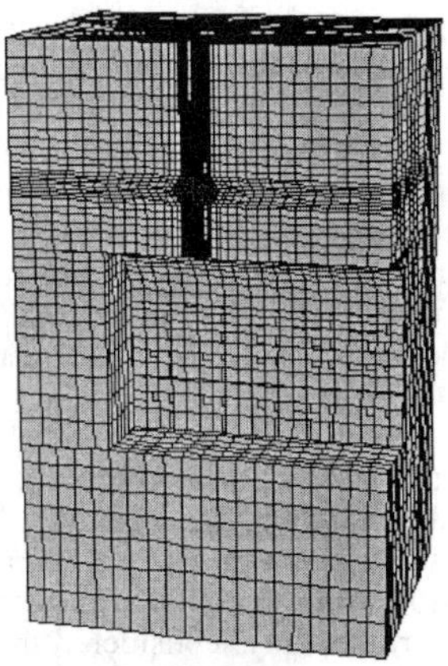

(a) Full grid

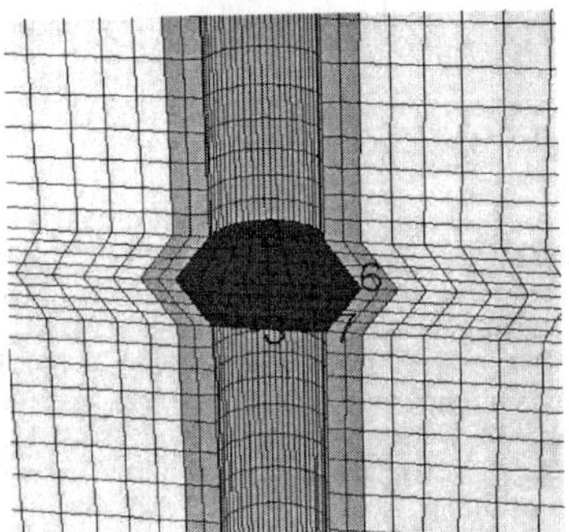

(b) Magnified grid around the plug

Figure 7. 3D finite differential mesh for numerical analysis

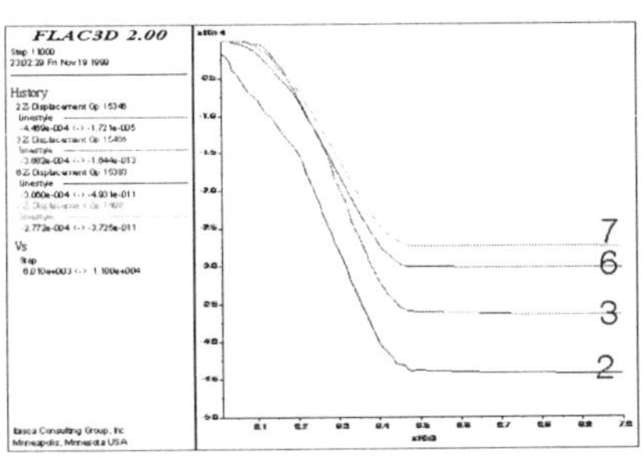

Figure 8. Vertical displacement by excavation step

4 ANALYSIS OF HYDRAULIC BEHAVIOUR OF ROCK MASS ADJACENT TO THE PLUG

4.1 General and scope

It is necessary to conduct a hydraulic study to secure the tightness of storage caverns as well as mechanical stability. Crack initiation and stress redistribution due to cavern excavation increase hydraulic permeability coefficient.

Kelsall (1984) and Pusch (1992) reported that permeability increases up to 10^2 - 10^4 times near the blasted rock mass. Table 3 shows the input parameters used in the hydraulic analysis. Since the area of interest is broad, we assumed the rock mass as an equivalent porous media.

Table 3. Parameters for FDM hydraulic analysis

Water	Density		1000 kg/m^3
	Bulk modulus		2×10^9 Pa
Permeability	Rock mass	Moderate	1×10^{-8} m/sec
		EDZ	1×10^{-6} m/sec
	Concrete		1×10^{-8} m/sec
Porosity	Rock mass	Moderate	1%
		EDZ	10%
	Concrete		10%

Three types of model were used in this analysis; (A) was the case when wedge of the plug was placed within the EDZ (wedge depth = 0.5 m), (B) was when the depth of wedge was identical to the range of EDZ (wedge depth = 1.0 m), and (C) was the case for the wedge depth (1.5 m) to be beyond the EDZ range. In all cases, range of EDZ was assumed to be 1 m. It was also assumed that the upper shaft of plug was filled with water and boundaries of model was fully saturated. In terms of the sequence, after excavation, enough displacement was allowed to produce and converge. Then, the plug was installed and hydraulic behavior was observed when hydraulic pressure acted on. The rest of conditions are identical to those of parameters in mechanical analysis.

4.2 Result

In this study, the relative changes in hydraulic behavior for different depths of plug wedge were analyzed. Water inside the shaft tended to move to EDZ where permeability was relatively higher than concrete and rock mass. There was, however, much difference in discharge between the upper and lower part of wedge according to their wedge depth. In case when the wedge depth was the same as EDZ range, difference in discharge between the upper and lower part of wedge crest was about 1/10. And in case when wedge depth was over the EDZ range, reduction of discharge in the lower part of wedge crest was severe so that it decreased to about 1/100 times. Therefore, it was observed that it is effective to set wedge depth deeper than the range of EDZ to maintain the sealing performance of the plug. Meanwhile, as the depth of wedge increases, discharge in the upper part of wedge crest increases gradually.

Figure 10 (a)~(c) shows the specific discharge vector in three different cases, that is when the wedge is placed within EDZ, contacted with EDZ and beyond EDZ. And it also shows the changes in discharge in unit area according to the increase of wedge depth.

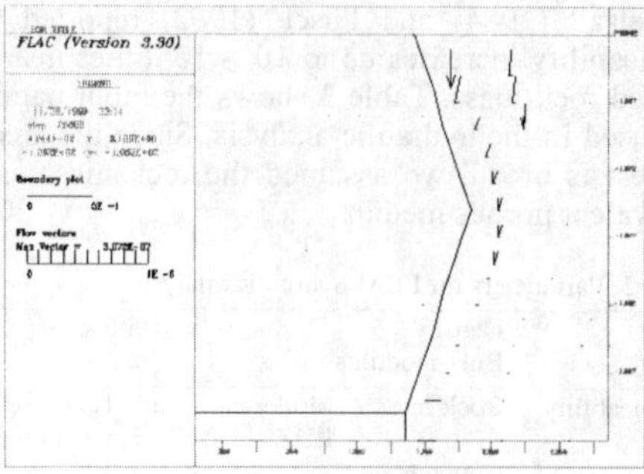

(a) Slot depth = 0.5m

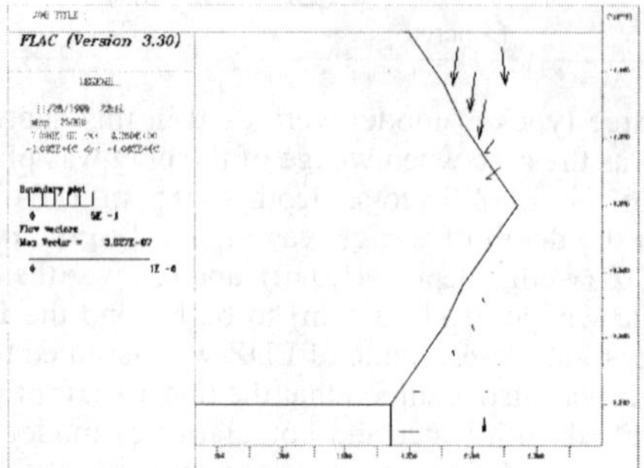

(b) Slot depth = 1.0m

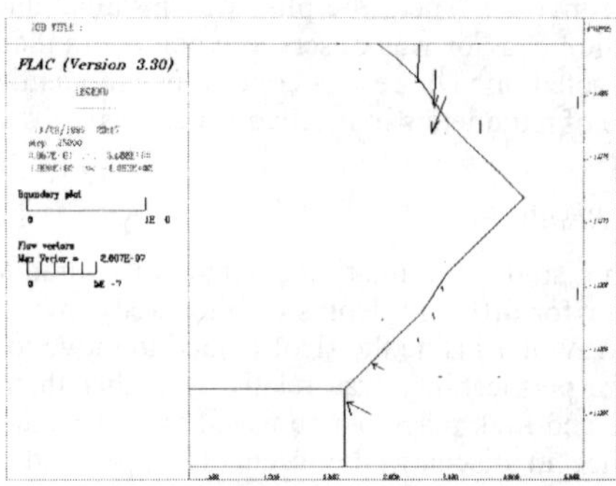

(c) Slot depth = 1.5m

Figure 10. Specific discharge vector around plug (EDZ=1m)

5 CONCLUSIONS

According to the results of the 2D mechanical analysis, as the plug was installed at deeper depth, there was a tendency to be more stable with increasing in-situ stress ratio (K). The result of analysis after applying the infiltration pressure between the plug and rock mass showed the smallest tensile stress in the lower asymmetric shape among three different types in geometry of plug.

The smoother condition of the interface allowed the more displacement, so that the magnitude and the range of the tensile stress in the plug were reduced. The tensile stress and displacement in plug were increased with decreasing deformation modulus of rock mass due to excavation damage and the maximum values for 20% of deformation modulus of undamaged rock mass were increased by 6% and 43%, respectively.

As the result of 3D analysis and its comparison with 2D analysis, it is thought that sufficient safety factor could be achieved with the result of the 2D analysis.

Finally, as the results of hydraulic analysis with respect to the slot depth of the plug, the flow rate when the slot was excavated beyond the EDZ were decreased about 100 times in comparison with that when the slot within the EDZ. Therefore, when the plug is constructed, it is considered that the slot excavated deeper than the EDZ is more effective in order to prevent leakage.

6 REFERENCES

Hökmark, H., 1998, Numerical study of the performance of tunnel plugs, Engineering Geology, 49:327-335.

Itasca Consulting Group Inc., 1995, FLAC2D Manual.

Itasca Consulting Group Inc., 1997, FLAC3D Manual.

Kelsall, P.C., Case, J.B., Chabannes, C.R., 1984, Evaluation of excavation-induced changes in rock permeability, Int. Rock Mech. Min. Sci. & Geomech. Abstr., 21(3):123-135.

Lee, D.H., 1998, Mechanical behavior and cracking characteristics of tunnel lining by model experiment and numerical analysis, Ph.D. Thesis, Seoul Natational Univ., Seoul, Korea.

Pusch, R., 1992, The zone of disturbance around blasted tunnels at depth, Int. Rock Mech. Min. Sci. & Geomech. Abstr., 29(5):447-456.

SK Construction Co., 1998, Report of Pyung-Taek LPG Project.

Computer Methods and Advances in Geomechanics, Desai et al. (eds) © 2001 Balkema, Rotterdam, ISBN 90 5809 183 X

Modeling coal seam damage in cast blasting

Dale S. Preece
Sandia National Laboratories, Albuquerque, N.Mex., USA

Stephen H. Chung
Orica USA Incorporated, Denver, Colo., USA

ABSTRACT: A discrete element computer program named DMC_BLAST (Distinct Motion Code) has been under development since 1987 for modeling rock blasting. This program employs explicit time integration and uses spherical or cylindrical elements that are represented as circles in two dimensions (2-D). DMC_BLAST calculations compare favorably with data from actual bench blasts .

1 Bench Cast Blasting

Cast blasting in surface coal mines employs explosives to not only break the rock but to also move the rock. Cast blasting utilizes significantly more explosives to move the rock than is necessary just for fragmentation. A typical cast blast is shown in Figure 1 several seconds into the blast. This particular blast consisted of 525 blastholes, 320mm in diameter, an average of 90 feet deep and utilized approximately two million lbs. of ANFO. It is now accepted in the industry that muck movement through cast blasting is more economical than using mechanical equipment. Current cast blast designs employ high benches (30 to 50 m), large blastholes (300+ mm) with burdens and spacings on the order of 10 m. Powder factors for cast blast designs are typically 1.0 Kg/m^3 and higher. The object of these designs is to move the maximum amount of muck possible into its' final resting place to eliminate reprocessing with earth moving equipment. These designs work because the high powder factors result in the application of significant explosive energy to the rock. The high benches also allow conversion of significant potential energy into kinetic energy to move the rock further.

A common conflict in cast blasting is between rock movement and maintaining the integrity of the coal seam. Blast designs resulting in very large powder factors (on the order of 1.5 Kg/m^3) can move the rock very well but will also cause significant coal chilling or damage. Coal damage can be controlled to some extent by increasing the buffer between the bottom of the blasthole and the top of the coal seam. However, buffers that are too large will result in unbroken and undiggable rock above the coal. Cast blast design is usually a compromise between rock movement and coal damage. Even with balancing these factors cast blasting has revolutionized surface coal mining over the past 15 years making it more efficient.

2 DMC_BLAST Modeling of Bench Cast Blasting

The cast blast design for the example calculation that will be presented in this paper is given in Table 1. Figure 2 shows the 2-d discrete element model created to represent this design. This model has 3797 spherical discrete elements and five layers of rock represented in gray scale. The model also has eight rows of explosives that are not visible in Figure 2 but the effects of which are visible in Figure 4. Detonation of each row of explosives is modeled with a coupled explosive gas flow and particle motion calculation. Flow of the explosive gases outward from the blasthole through the rock is modeled with a finite difference technique (Preece et al, 1993b). Loading of the rock particles by the explosive gas is done by integrating the gas pressure across each particle (Preece, 1993a).

A material strength capability has been added to DMC_BLAST and has seen several applications (Preece, 1995). Rock strength is incorporated in DMC_BLAST simulations by adding links or bonds between discrete elements as shown in Figure 3. These links are used during the discrete element simulation to hold particles together and thus maintain

Figure 1: Surface coal mine bench cast blast

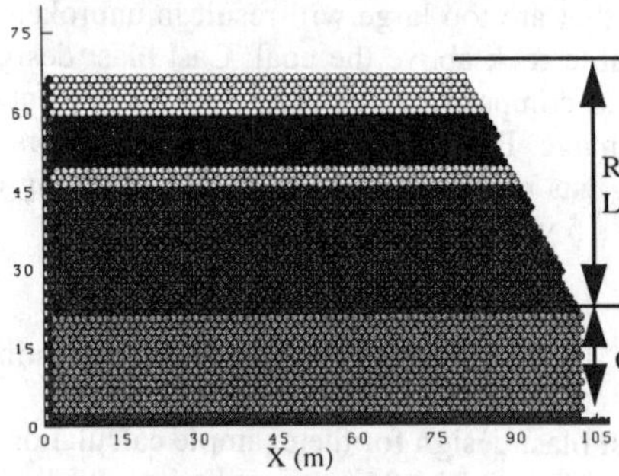

Figure 2: Discrete element model of surface coal mine cast blast.

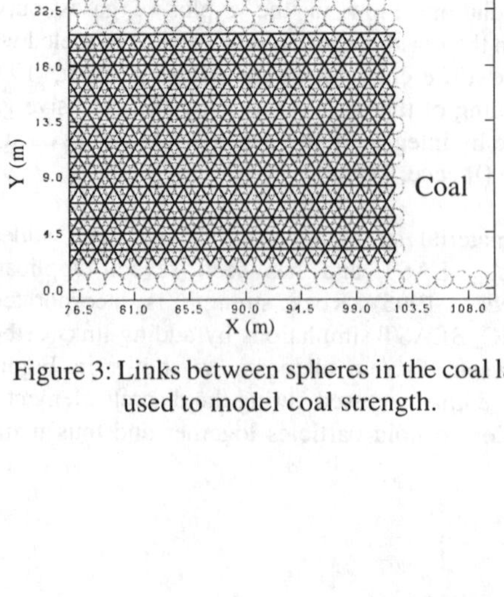

Figure 3: Links between spheres in the coal layer used to model coal strength.

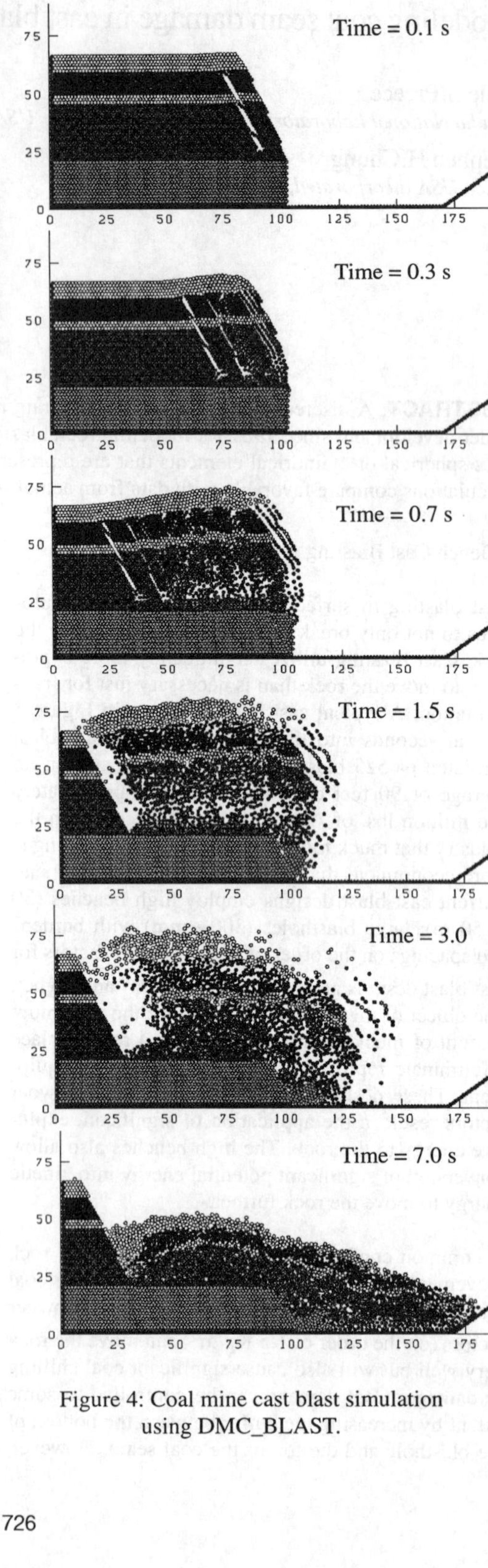

Figure 4: Coal mine cast blast simulation using DMC_BLAST.

the geometrical integrity of the particles and simulate material strength. These links can be broken which allows the discrete elements to move without restriction and thus simulate material dissaggregation. Links can be broken based on simple tension, compression or shear stress in the link exceeding the strength of the material. In this calculation links were only allowed to break in tension since this mode had previously been determined to dominate failure during blasting simulations (Preece, 1995). The tensile strength of the coal layer in this simulation was set at 6.5 MPa (942 psi). Laboratory tensile strength data for coal could not be located. This value is based on an average unconfined compressive strength for coal of 40 MPa (Touloukian et al, 1981). In this calculation only the strength of the coal seam was modeled with interparticle links. It is assumed that the material above the coal seam has a high enough powder factor to totally fragment the rock and thus it is unnecessary to model rock strength in the other layers.

3 Simulating Coal Seam Damage

Predicted rock movement for this blast is shown in Figure 4. Close examination of the top of the coal seam at the bench face shows a shear failure and movement of the top corner of the coal seam. It is proposed that this is the dominant mechanism for coal damage in current cast blasting practice. This region and the interparticle links are shown in more detail in Figures 5 at times 0.3 s and 1.2 s. As seen at 0.3 s in Figure 5, a triangular region is broken at the face and toe of the coal by the detonation of the first row of blastholes. This is caused by compressive transient stress waves emanating from the bottom of the blasthole and reflecting off the surface of the coal at the face producing a tensile wave that breaks a triangular corner at the toe. This phenomenon occurs quite early in the blast and undermines the coal seam to allow a shear failure plane to propagate from the toe of the coal diagonally up to the top surface as shown at 1.2 s This shear failure plane in the coal allows the top corner of the coal at the face to slide into the pit and be lost.

In addition to displaying the intact links between particles, the particles can be colored according to the percentage of links to the particle that are broken. This variable, which is the ratio of the number of links broken to the original number of links is called *damage*. Plots of damage at the same location and times as Figure 5 are shown in Figure 6. The relationship between the link pattern and damage is obvious. It is often more illustrative to plot damage in color or gray shading because the intact links tend to blend together when the plotting scale is larger making it difficult to discern areas of damage. Figure 6 illustrates six frames from the blast simulation with the variable damage being displayed in gray shades. This larger view again shows the front/top corner of the coal seam being damaged and sliding into the pit where it will be lost. Figure 6 also indicates minor coal damage along the top of the coal seam induced by the detonation of each row of blastholes.

4 Simulating Coal Seam Damage

Predicted rock movement for this blast is shown in Figure 4. Close examination of the top of the coal seam at the bench face shows a shear failure and movement of the top corner of the coal seam. It is proposed that this is the dominant mechanism for coal damage in current cast blasting practice. This region and the interparticle links are shown in more detail in Figures 5 at times 0.3 s and 1.2 s. As seen at 0.3 s in Figure 5, a triangular region is broken at the face and toe of the coal by the detonation of the first row of blastholes. This is caused by compressive transient stress waves emanating from the bottom of the blasthole and reflecting off the surface of the coal at the face producing a tensile wave that breaks a triangular corner at the toe. This phenomenon occurs quite early in the blast and undermines the coal seam to allow a shear failure plane to propagate from the toe of the coal diagonally up to the top surface as shown at 1.2 s This shear failure plane in the coal allows the top corner of the coal at the face to slide into the pit and be lost.

In addition to displaying the intact links between particles, the particles can be colored according to the percentage of links to the particle that are broken. This variable, which is the ratio of the number of links broken to the original number of links is called *damage*. Plots of damage at the same location and times as Figure 5 are shown in Figure 6. The relationship between the link pattern and damage is obvious. It is often more illustrative to plot damage in color or gray shading because the intact links tend to blend together when the plotting scale is larger making it difficult to discern areas of damage. Figure 6 illustrates six frames from the blast simulation with the variable damage being displayed in gray shades. This larger view again shows the front/top corner of the coal seam being damaged and sliding into the pit where it will be lost. Figure 6 also indicates minor

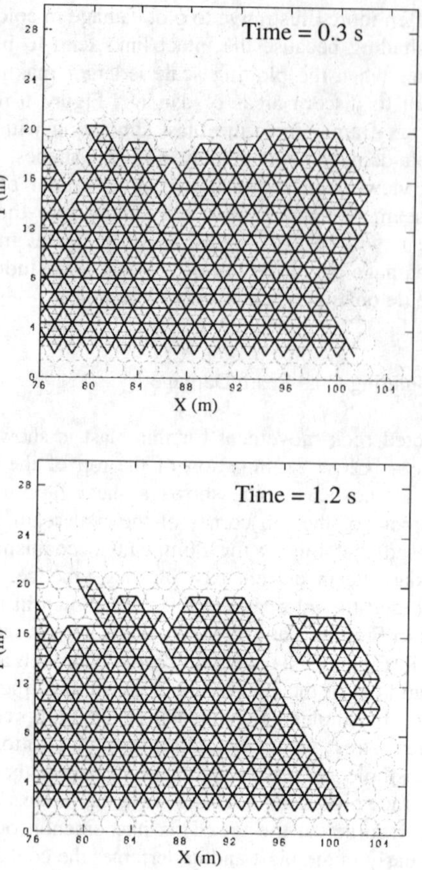

Figure 5: Links broken by tensile stress indicating coal damage in the vicinity of the face of the coal seam. Compare with Figure 3.

coal damage along the top of the coal seam induced by the detonation of each row of blastholes.

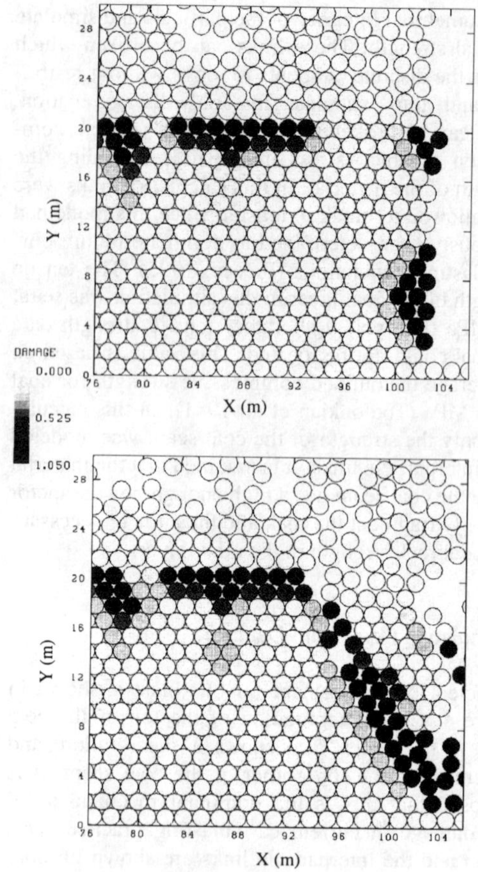

Figure 6: Damage in coal calculated as the percentage of broken links for each sphere. The times and locations for these figures are the same as for Figure 5.

5 Conclusions

A study has been performed to demonstrate the ability of DMC_BLAST to model chilling/damage of the coal seam during surface coal mine cast blasting. With a good cast blast design the dominant damage mechanism is removal of the top corner at the face of the coal seam. The fracture plane allowing removal of this corner extends diagonally upward from a damaged zone at the toe of the coal on the face created by reflected transient stress waves emanating from the first row of blastholes. Some damage of the top surface of the coal by the explosives in blastholes above is also evident. This damage is minimized by the buffer zone left between the bottom of the blastholes and the top of the coal seam.

This capability will be utilized to compare coal seam damage induced by different cast blast designs. Special emphasis will be placed on reducing the coal damage at the face. One obvious solution that is sometimes employed is to leave a buffer of broken rock at the face of the coal instead of a free face. Powder factors and buffer zone thickness can also be varied to learn the effect they have on coal damage.

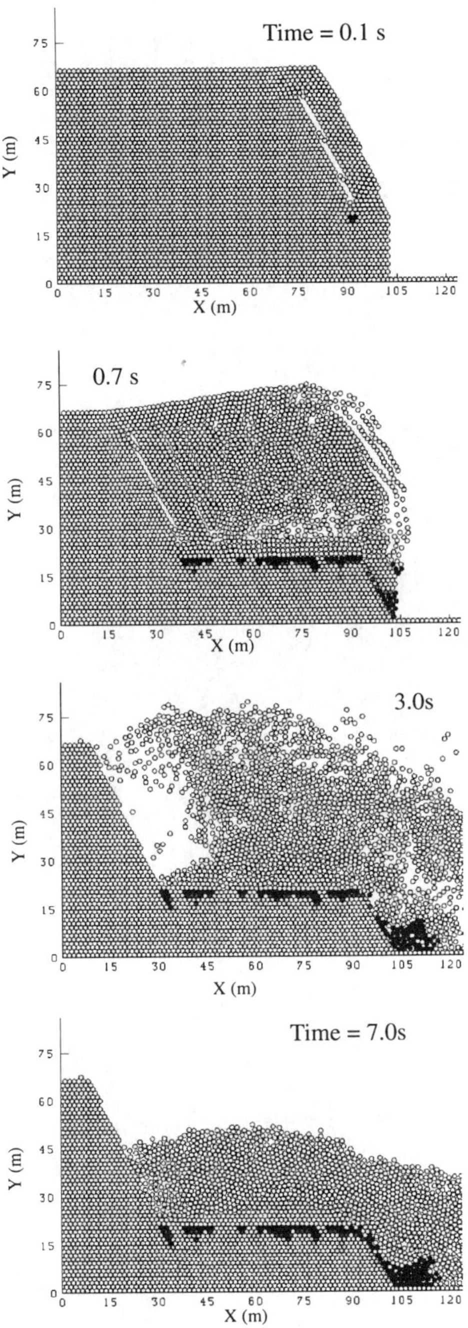

Figure 7: Calculated coal seam damage due to the applied explosive loading.

Table 1: Blast Design Parameters

Parameter	Value
Hole Diameter	311 mm
Hole Depth	45 m
Stemming Length	9 m
Explosive Length	36 m
Explosive Type	Heavy ANFO 50/50
Explosive Density	1.3 g/cm3
No. Rows / Delay	8 / 100 ms
Tensile Strength	6.5 MPa
Buffer Zone between Bottom of Blastholes and Top of Coal Seam	4 m

REFERENCES

Preece, D. S. and Taylor, L. M., 1989, *Complete Computer Simulation of Crater Blasting Including Fragmentation and Rock Motion*, Proceedings of the Fifth Annual Symposium on Explosives and Blasting Research, Society of Explosives Engineers, New Orleans, LA.

Preece, D. S., 1993a, *Momentum Transfer From Flowing Explosive Gases to Spherical Particles During Computer Simulation of Blasting-Induced Rock Motion,* Proceedings of the *Ninth Annual Symposium on Explosives and Blasting Research, Society of Explosives Engineers*, San Diego, California.

Preece, D. S., Burchell, S. L., and Scovira, D. S., 1993b, *Coupled Explosive Gas Flow and Rock Motion Modeling With Comparison to Bench Blast Field Data*, Proceedings of the Fourth International Symposium on Rock Fragmentation by Blasting, Technical University, Vienna, Austria.

Preece, D. S., 1995, *Discrete Element Modeling of Rock Blasting in Benches With Joints and Bedding Planes - Initial Development*, Proceedings of the Eleventh Annual Symposium on Explosives and Blasting Research, Society of Explosives Engineers, Austin, TX.

Preece, D. S., Tidman, J. P., and Chung, S. H., 1997, *Expanded Rock Blast Modeling Capabilities of*

DMC_BLAST, Including Buffer Blasting, Proceedings of the Thirteenth Annual Symposium on Explosives and Blasting Research, Las Vegas, NV.

Touloukian, Y. S., Judd, W. R. and Roy, R. F., 1981, *Physical Properties of Rocks and Minerals,* McGraw Hill / Cindas Data Series on Material Properties Vol II-2.

Computer Methods and Advances in Geomechanics, Desai et al. (eds) © 2001 Balkema, Rotterdam, ISBN 90 5809 183 X

Three-dimensional finite element modelling of compensation grouting

C.Wisser, C.E.Augarde & H.J.Burd
Department of Engineering Science, Oxford University, UK

ABSTRACT: Compensation grouting is a relatively new technique that is used to prevent ground movements caused by tunnel construction damaging nearby surface structures. Compensation grouting schemes are generally designed using a semi-empirical approach. The grouting is controlled on site using measured data on ground and building movements. This paper describes initial work on a numerical model of compensation grouting. Three-dimensional finite elements are used, in conjunction with a complex non-linear material model for overconsolidated clay. Procedures are described for modelling grout injection and for settlement control. A simple test problem with an elastic material model for soil, without tunnelling, is used to illustrate the simulation of grout injection. An example is also given of the analysis of tunnelling under a greenfield site, with surface settlements controlled by grouting.

1 INTRODUCTION

The development of new infrastructure links in urban areas often involves the construction of tunnels, in soft ground, adjacent to existing surface structures. These activities lead to deformations of the surrounding soil and movements at the surface, which can cause damage to existing structures. The prediction of these movements, and the associated damage, is an active area of research in the UK (e.g. Potts and Addenbrooke 1996; Burd et al. 2000). In recent years, the technique of *compensation grouting* has become widely used in the UK to control surface movements. This technique involves the injection of grout into the ground between the tunnel and the ground surface.

While considerable sums have been spent on compensation grouting operations, notably in the construction of the Jubilee Line Extension in London, the method has evolved directly from site practice and there has been little development of techniques to predict in advance the effects of a given scheme of grouting. There is a need for design tools to assess the effectiveness of different schemes of grouting.

This paper describes initial research on the prediction of the effects of compensation grouting for a given tunnelling layout in the presence of surface structures. The research is based on numerical models using finite elements.

2 MODELLING COMPENSATION GROUTING

2.1 *Introduction*

Compensation grouting involves the injection of grout through perforated steel or plastic pipes radiating out from an access shaft, as shown in Figure 1. These pipes, known as *Tubes á manchette* or TAM's, are normally water-jetted into the ground between the tunnel soffit and the surface. As tunnelling proceeds, surface movements are closely monitored at *observation points*, located on the ground surface and on structures that are likely to be affected. Grout is injected once a predetermined value of settlement (a *trigger* level) is reached. Movements are then limited to a given *tolerance* level, usually very close to the trigger level. The purpose of compensation grouting is not to raise buildings once settlements have occurred but rather to nullify any potentially damaging movements as they occur.

Little is known of the behaviour of the injected grout in the ground and there have been few surveys in areas where compensation grouting has been carried out to observe the final grout location. It is therefore difficult, using conventional numerical approaches, to devise anything other than a mechanistic model where the plane of the grout is prescribed. For heavily over-consolidated soils (such as London clay) however, where the coefficient of earth pressure at rest, K_0 is much greater than unity, the formation of horizontal fracture planes seems a reasonable assumption. This is indicated by the observations of Ja-

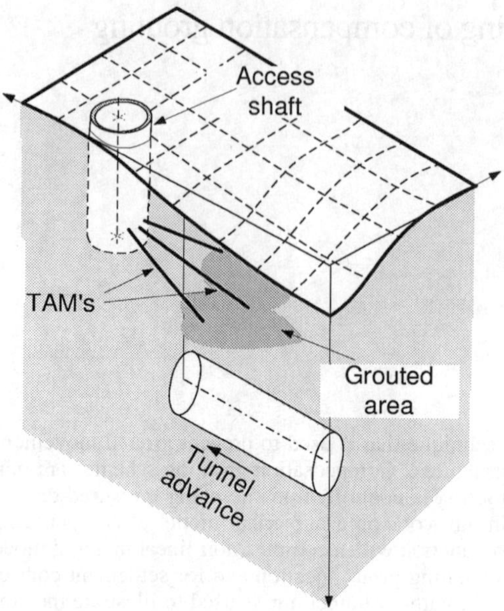

Figure 1: General arrangement of compensation grouting for tunnelling.

worski et al. (1981) and Edmonds (1997).

Examples of previous attempts to use numerical methods to model compensation grouting can be found in Nicholson et al. (1994) and Mayer and Kudella (1996) both of which use conventional two-dimensional continuum finite elements and simulate compensation grouting by increasing the volume of particular elements. Two-dimensional methods are also used by Kovacevic et al. (1996) where interface elements are placed at the anticipated grout plane. Grouting is then modelled by the application of an internal pressure. More recently, Soga et al. (1999) and Buchet and Van Cotthem (1999) have modelled compensation grouting with the three-dimensional finite difference program FLAC. To simulate the grouting process, elements are inflated by applying an internal pressure. There does not, however, appear to be any published research in which a model of compensation grouting is combined with models of tunnel installation and surface structures.

3 DEVELOPMENT OF A NUMERICAL MODEL

3.1 *Introduction*

Work has been underway at Oxford University, UK since 1993 to study the effects of tunnelling settlements on masonry structures. This has led to the development of a three-dimensional finite element model in which incremental construction of a tunnel, as well as a surface structure are included (Augarde 1997; Burd et al. 2000). The model uses tetrahedral elements incorporating a weakly non-linear material model for the soil (Houlsby 1999). This material model simulates the rapid change in stiffness at small strains that is exhibited by undrained overconsolidated soils; this is thought to be important for the deformation behaviour around tunnels. Surface structures are modelled using 2D facades connected to the 3D ground using novel tie elements (Houlsby et al. 2000) and an elastic-no tension model for masonry (Liu 1997). The inclusion of compensation grouting in this three-dimensional numerical model is described below (although no surface structures are included in the particular analyses presented here).

3.2 *Model details*

Following on from previous two-dimensional studies (e.g. Kovacevic et al. 1996) interface elements are embedded in the finite element mesh at the expected final location of the grout. These elements consist of two layers of 6-noded triangles with zero thickness (Ngo-Tran 1997) to match the faces of the 10-noded tetrahedral elements modelling the soil on either side of the grout. The detailed model of grout injection differs in various key aspects from previous studies and is described in detail below.

Mesh generation

The unstructured three-dimensional meshes required for the analysis are generated using the I-DEAS modelling package. In-house software is then used to generate a patch, or patches, of interface elements at the required location(s) in the mesh. The edges of the interface elements coincident with the edge of the patch must be tied together. This can be achieved by altering the formulation of these particular interface elements to include constraints on element node pairs along these edges. An alternative approach is used here where the edges are joined together with separate tie elements that implement the constraints directly by coupling the degrees of freedom of the nodes on the edge. These elements have already been referred to above and are those used to tie the building facade meshes together (Houlsby et al. 2000). This approach

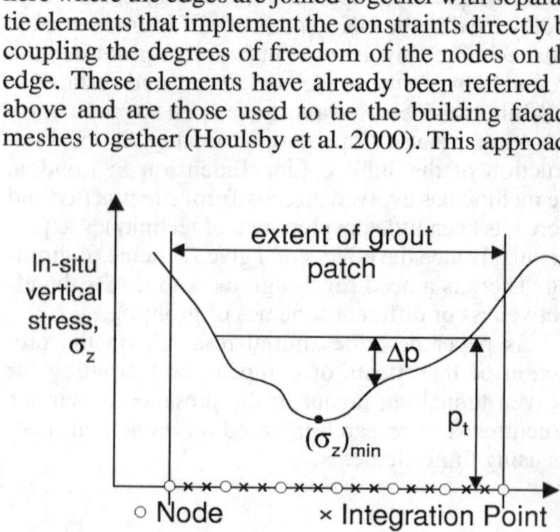

Figure 2: Calculation of grout pressure across a grout patch

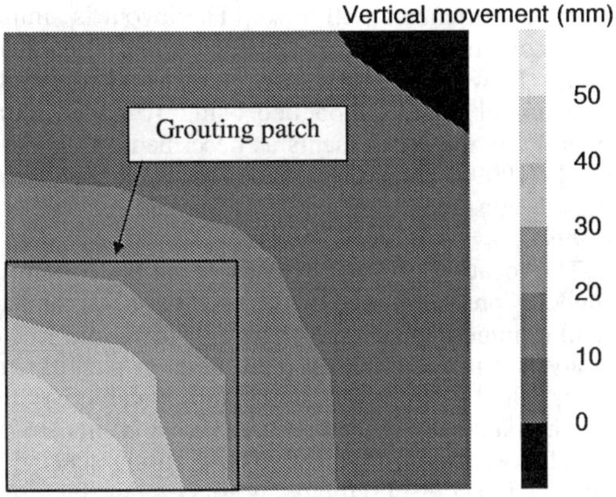

Figure 3: Vertical settlement at surface of simple test block problem.

avoids the need for different types of interface elements, and simplifies mesh generation.

Application of grout pressure
The injection of grout is simulated by applying an internal pressure to the interface elements modelling the grout patches (these are referred to below as grout elements). Clearly, the application of constant incremental pressure in all grout elements, regardless of the spatially varying in-situ vertical stress distribution in the elements, is unrealistic. Instead an approach is used that applies incremental pressure according to existing vertical stress at each stress integration point, in each grout element. This ensures that the resulting total grout pressure is uniform across the grouted areas. To illustrate this procedure, consider a variation in vertical stress σ_z along the example one-dimensional mesh shown in Figure 2. Taking compressive stress as positive, the integration point within the grout elements with the lowest vertical compressive stress $(\sigma_z)_{min}$ is detected. Given a target grout pressure p_t specified by the user, the nodal forces to be applied in this step, to the nodes associated with the grout elements, to model grouting are then calculated as

$$f = \int_A N^T H(\Delta p)\, dA \tag{1}$$

where

$$\Delta p = p_t - \sigma_z \tag{2}$$

and $H(\Delta p)$ is the Heaviside function

$$H(\Delta p) = \begin{cases} 1 & \text{for} \quad \Delta p > 0 \\ 0 & \text{for} \quad \Delta p < 0 \end{cases} \tag{3}$$

A is the area of the element face and N is the vector of shape functions for the six-noded triangular faces of the grout elements. This represents a relatively crude way of modelling the hydrofracturing process, allowing flow to zones only where the vertical stress is less than the target grout pressure.

3.3 *An elastic test problem*

A simple test problem, without tunnelling, is now used to demonstrate the grouting model described above. This is not intended to represent any realistic analysis but it is a useful test of the modelling procedure. The test problem consists of a block of linear elastic soil of dimensions 100m x 100m x 100m and unit weight 20kN/m^3. The block surface is free, the base is fully fixed and the vertical sides are restrained only in the horizontal direction. A horizontal pad of grouting 50m by 50m in plan area, at a depth of 50m, is located at one corner of the block (see plan view in Figure 3). The patch is modelled with automatically generated grout elements. Grout is injected (i.e. the value of p_t is incrementally increased) until the maximum heave at the surface is 50mm.

Figure 3 shows a plot of the vertical displacement at the surface after completion of the grouting process. The plot shows that the maximum heave at the external corner of the patch is limited to the specified value of 50mm. It is noticeable that the heave extends beyond the grouting patch but that the gradients of the surface contours there are much lower. The reason for this sharp change in slope of the surface settlement is that the extent of the grouted area is fixed by the lines of tie elements. While this is an unrealistic feature, the model represents only an initial attempt to model grouting and work is in progress on more realistic modelling procedures. The results of this simple analysis show that this procedure successfully models aspects of grout injection. Its use in a more complex model, including tunnelling and settlement control, is demonstrated below.

4 SETTLEMENT CONTROL IN ANALYSES OF TUNNELLING

The purpose of compensation grouting is to control surface movements. The numerical model allows the comparison of different strategies of movement control during the simulation of grouting. The calculation procedure for the numerical model, including compensation grouting with settlement control is as follows:

- Generation of a horizontal layer of grout elements at appropriate locations.
- Assignment of a suitable stiffness to the grout elements. (The stiffness is selected to ensure that additional flexibility induced by the elements is minimal).
- When a suitable measure of surface settlement reaches the specified trigger level, grout ele-

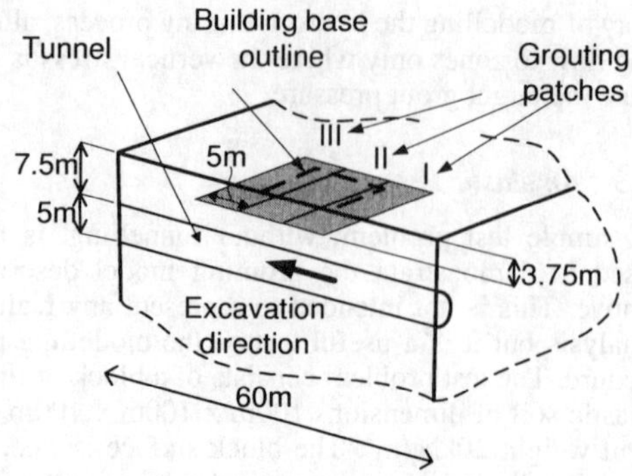

Figure 4: Layout of tunnelling problem

ments are activated by

  - reducing stiffness to zero,
  - applying a suitable grout pressure (based on the procedure described above).

- Following achievement of the required settlement control, as measured at specified positions in the mesh (the observation points), the stiffnesses of the grout elements are returned to a high value to simulate curing.

The simplest case to model is where grouting is started once a specified trigger level of movement (usually settlement) occurs at an observation point. Grouting then continues to limit settlement to less than the specified tolerance. Many other alternative strategies are possible. For instance, it may be important to limit differential rather than total settlements or to limit curvatures along particular building baselines. In the analysis presented here, a simple strategy adopted in which settlements of specified observation points are monitored. When a target level of settlement of 1mm is exceeded at one of these points, grouting is activated to reduce the settlement to 1mm or less (i.e. the tolerance level (1mm) is the same as the trigger level (1mm)). This process is automatic, being built-in to the software. Grout pressure is applied by a process of numerical iteration to ensure that the settlement of all observation points is less than the tolerance.

4.1 *Grouting with incremental tunnel excavation under a greenfield site*

The effectiveness of the modelling procedures outlined above is now demonstrated for a model where installation of a tunnel is included. The installation of a 5m diameter tunnel with its axis at a depth of 10m under a greenfield site is modelled as shown in Figure 4. (One-half of the problem is modelled due to symmetry). The tunnel is excavated in 6 stages. After each increment of excavation the tunnel liner is shrunk to simulate a ground loss of 1% from which surface settlements will arise. (The layout is similar to the one used by Augarde (1997) which used 2576 nodes. However the mesh used here has 8188 nodes). The tunnel liner is modelled using 10-noded tetrahedral continuum elements as described by Augarde et al. (1999). The soil is modelled using the elasto-plastic non-linear soil model described by Houlsby (1999).

The location of grout patches in the finite element model is predetermined by the user, based on the particular tunnelling scheme under consideration. In this analysis it is assumed that a building, of plan dimensions 20m by 10m, is located symmetrically over the tunnel axis (see Figure 4). Observation points are located as shown in Figure 5. The grouting scheme is designed to control settlements affecting the building, although it must be emphasized that no model of the building is included in this analysis, for simplicity.

In practice a grouting zone is defined above a tunnel by lines extended at 45° from the tunnel invert (Haimoni 1997). It is also common practice to avoid pumping grout closer than one tunnel diameter to the tunnel lining soffit (Edmonds 1997) to reduce the possibility of damaging the tunnel lining. Taking this advice, three grout patches (5m x 8.75m each, denoted I, II and III) are included beneath the building base outline, as shown in Figure 4. The patches are located at a depth of 3.75 m, i.e. half way between the tunnel crown and the surface.

Figure 5(a) shows a contour plot of the settlements at the surface after excavation of 25m of tunnel, prior to grouting. In this configuration, the tunnel heading is immediately below grout patch I. Figure 5(b) shows the settlement contours for the same site and tunnel excavation with controlled compensation grouting. To emphasize the difference between the two plots, only an area of 15m x 20m, encompassing the building base outline, is depicted. Grout is injected into patch I only. Figure 6 gives the surface settlements along the tunnel axis for the same calculations.

The settlement plot in Figure 6 shows the expected cumulative probability curve for a greenfield site. The maximum settlement is smaller than predicted by the semi-empirical approach as is common in numerical modelling of this problem (Mair et al. 1996). The plot for the grouted analysis shows the settlements at the observation points to have been reduced to less than, or equal to, the tolerance level of 1mm.

In the direction perpendicular to the tunnel axis the settlements outside the grouting patch before and after the grout injection are quite similar, indicating that this region is not affected by the injection of grout. In the direction along the tunnel axis, however, the slight heave created by the grout injection extends beyond the patch. This can be explained by considering the superposition of the tunnelling settlement profile with a uniform area of heave, as provided by grouting alone. Behind the tunnel face, grouting heave is offset by tunnelling settlement. Ahead of the tunnel

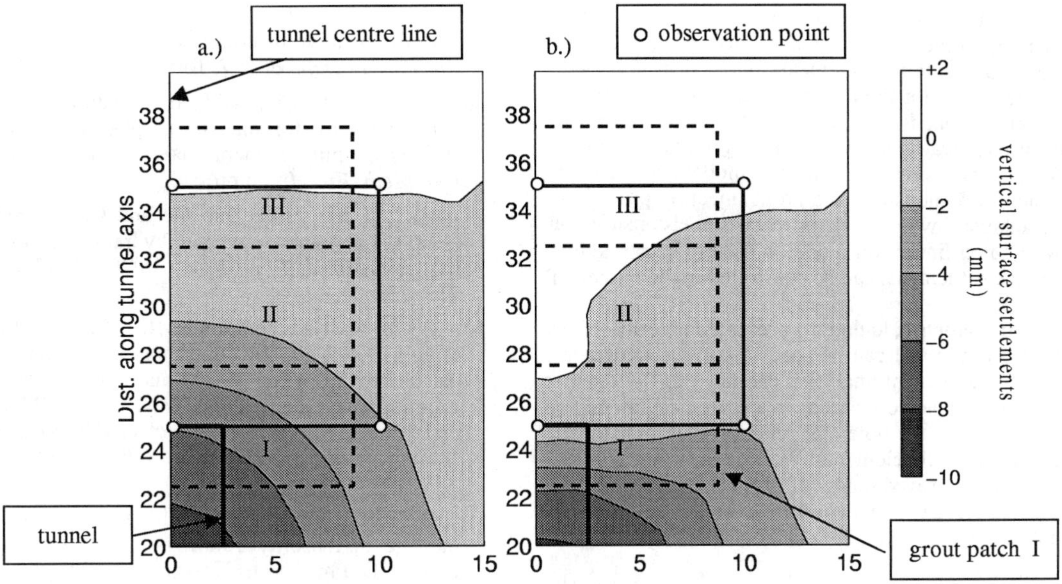

Figure 5: Plan view of vertical settlement at surface of greenfield site (a) without grouting (b) with grouting.

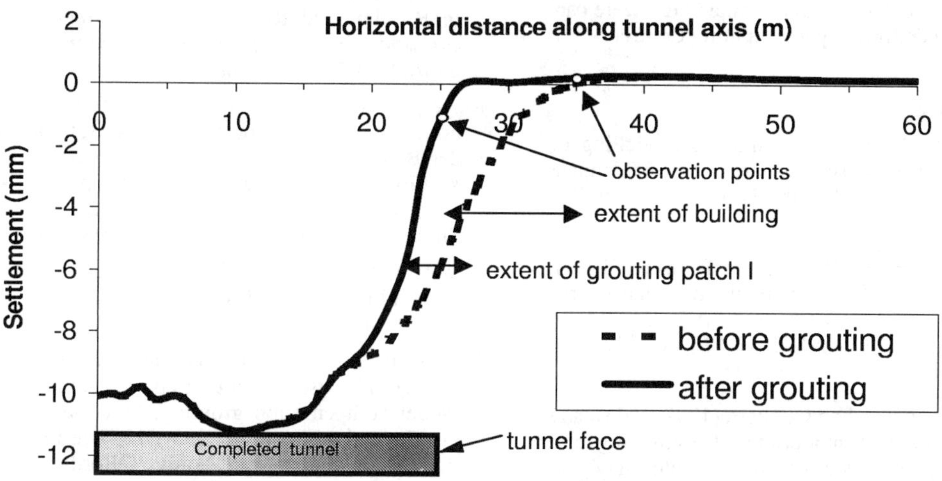

Figure 6: Vertical settlement at surface of greenfield site along tunnel axis.

face there is no settlement and the net effect is a slight heave.

Subsequent stages of this analysis would consist of further increments of tunnel excavation allied with controlled grouting in the patches II and III (see Figure 4). The presence of a building at the surface, adding load and stiffness, would clearly change the effect of grout injection, but the numerical processes used to simulate and control grout injection would be the same.

5 SUMMARY

The use of three-dimensional modelling (rather than two-dimensional analysis) is vital for the study of tunnelling-induced settlement damage to surface structures. Compensation grouting is increasingly being used to control settlements arising from tunnel construction. It is important that modelling procedures are developed to simulate this technique to allow different strategies of grouting for a given tunnelling scheme and building layout to be compared.

Procedures have been described in this paper whereby grout injection is modelled using internally pressurized interface elements inserted in horizontal patches within a three-dimensional mesh of soil elements. Grout flow is controlled by comparison of in-situ vertical stress with the target grout pressure. An analysis using an elastic soil model has been presented to demonstrate the grout injection procedure. An analysis involving incremental tunnel construction with a non-linear soil model is used to illustrate how surface settlements can be controlled in the numerical model.

The example calculations presented here show how a grouting model can be used to control settlements in a finite element analysis. The need to tie together patch edges, however, leads to a somewhat unrealistic situation with the potential generation of large tensile forces in the tie elements. Work is in progress to develop an improved model of grout injection and flow. This will lead to a model where different grouting layouts and control strategies can be studied.

ACKNOWLEDGEMENTS

The first author is supported by an EPSRC studentship. Some calculations presented here were carried out at the Oxford Supercomputing Centre.

REFERENCES

Augarde, C. E. (1997). Numerical modelling of tunnelling processes for assessment of damage to buildings. D. Phil. thesis, University of Oxford.

Augarde, C. E., C. Wisser, and H. Burd (1999). Numerical modelling of tunnel installation procedures. In *Proc. 7th Int. Symp. in Numer. Meth. Geomech. - NUMOG VII*, pp. 329–334. Balkema.

Buchet, G. and A. Van Cotthem (1999). 3d steady state numerical modelling of tunnelling and compensation grouting. In C. Detournay and R. Hart (Eds.), *FLAC and numerical modelling in Geomechanics*, pp. 200–207. Balkema.

Burd, H. J., G. T. Houlsby, C. E. Augarde, and G. Liu (2000). Modelling tunnelling-induced settlement of masonry buildings. *Proc. I.C.E. - Geotechnical Engineering 143*(1), 17–29.

Edmonds, H. (1997). Control and restraint. *Ground Engineering 30*, 38–41.

Haimoni, A. (1997). Tunnelling with compensation grouting, Jubilee Line extension. In *Proc. Going Underground 97*, pp. 20–30. Midlands Geotechnical Soc.

Houlsby, G. T. (1999). A model for the variable stiffness of undrained clay. In *Proc. Int. Symp. Pre-Failure Defm of Soil*, Torino, pp. 443–450.

Houlsby, G. T., G. Liu, and C. E. Augarde (2000). A tying scheme for imposing displacement constraints in finite element analysis. *Comm. Numer. Meth. Eng. 16*. in press.

Jaworski, G. W., J. M. Duncan, and H. B. Seed (1981). Laboratory study of hydraulic fracturing. *J. Geotech. Eng., Proc. ASCE 107*(GT6), 713–732.

Kovacevic, N., H. E. Edmonds, R. J. Mair, K. G. Higgins, and D. M. Potts (1996). Numerical modelling of the NATM and compensation grouting trials at Redcross Way. In R. J. Mair and R. N. Taylor (Eds.), *Geotechnical Aspects of Underground Construction in Soft Ground*, pp. 553–559. Balkema.

Liu, G. (1997). Numerical modelling of settlement damage to masonry buildings caused by tunnelling. D. Phil. thesis, University of Oxford.

Mair, R., R. Taylor, and J. Burland (1996). Prediction of ground movements and assessment of risk of building damage due to bored tunnelling. In R. Mair and R. Taylor (Eds.), *Geotechnical aspects of underground construction in soft ground*, pp. 713–718. Balkema.

Mayer, P.-M. and P. Kudella (1996). Prognose von Bodenverformungen als Bestandteil der Beobachtungsmethode. In *Sonderdruck zur 4. FEM/CAD Tagung, Darmstadt*, pp. 13–19. Marketing RIBTEC, Vaihingerstr. 151, 70567 Stuttgart, Germany: RIB Bausoftware GmbH.

Ngo-Tran, C. L. (1997). The analysis of offshore foundations subjected to combined loading. D. Phil. thesis, University of Oxford.

Nicholson, D. P., C. Gammage, and T. Chapman (1994). The use of finite element methods to model compensation grouting. In *Grouting in the Ground*, London, pp. 297–312. The Institution of Civil Engineers: Thomas Telford.

Potts, D. and T. Addenbrooke (1996). The influence of an existing surface structure on on the ground movements due to tunnelling. In R. Mair and R. Taylor (Eds.), *Geotechnical aspects of underground construction in soft ground*, pp. 573–578. Balkema.

Soga, K., M. D. Bolton, S. K. A. Au, K. Komiya, J. P. Hamelin, A. Van Cotthem, G. Buchet, and J. P. Michel (1999). Development of compensation grouting modelling and control system. In *Proc. Int. Symp. Under. Const. Soft Ground, Tokyo*, pp. 405–410.

Computer Methods and Advances in Geomechanics, Desai et al. (eds) © 2001 Balkema, Rotterdam, ISBN 90 5809 183 X

Tunnel stability assessment based on constitutive modeling

M. Zaki
Civil Engineering Department, Minia University, Egypt

G. Swoboda
University of Innsbruck, Austria

ABSTRACT: This paper presents the classical plasticity approach and the relatively recent damage mechanics approach to describe the behaviour of jointed rock mass. Hoek and Brown's (HB) criterion is considered appropriate for jointed rock mass as it involves the rock quality data as reflected by the constants m and s. The criterion has been discussed comprehensively along with the characteristics of its parameters. On the other hand, continuum damage mechanics, which employs some continuum variables to describe the micro-defects, has been an appealing framework for modeling geomaterials. The geological data regarding the orientation and intensity of joint sets is used along with the suggested damage model which increase the practical importance of the damage model. Finally, a full scale tunneling problem, Schönberg Tunnel (Salzburg, Austria), is analyzed using both Hoek-Brown and damage models. The integrated assessment of field measurements and the output of the numerical models using both approaches reflects the limitations of HB usage and the points of favour of the damage model.

1 INTRODUCTION

The success of the design of a tunnel depends mainly on the ability to predict different ground conditions. Then, the selection of the suitable method of execution including the choice of adequate support measures and final assessment of stability considering long-term performance and final lining design can be performed. Many constitutive laws sufficiently describe the behaviour of soft and medium ground like clay and sand. However, problems always arise when tunneling in rock due to its stochastic and indeterminate nature. Rock masses are seldom found in nature without joints or discontinuities. Discontinuities function as soft and weak links causing relative closure or parting of rock blocks under compression or translational shear. It is the presence of these joints that makes the mechanical behaviour of jointed rock mass significantly complicated. Anisotropy, dilatancy, irreversible strain and strongly path-dependent stress-strain relations, which are generally associated with the existence of a great deal of micro- and meso-cracks and their propagation, are examples of the complex behaviour of jointed rock mass. The mathematical mechanical description of jointed rock is a problem that has not yet been entirely solved.

The numerical analyses supporting tunnels in rock mass, in particular, will in future have a better reliability only if the material laws for jointed rock are considerably improved. Such analyses provide useful aids to the engineering assessment of tunnel stability. This is done by providing guidelines on the potential extent of disturbed ground zones around the tunnel and outlining criteria for rock reinforcement. Numerous construction aspects, such as further development of anchors, use of precast concrete linings, development of new excavation systems, can be supported by numerical models only when constitutive laws are available for jointed rock.

In this area, well-established classical nonlinear theories, for instance the plasticity theory, and the relatively new Continuum Damage Mechanics (Kachanov 1958, Murakami & Ohno 1980, Yazdani & Karnawat 1996) which employs some continuum variables to describe cracks and joints, provide another means of modeling jointed rock mass. Kawamoto et al. (Kawamoto et al. 1988) and Swoboda et al. (Swoboda et al. 1995) introduced the damage theory to establish continuum models of a jointed rock mass. In their models, the anisotropic characteristics of the joints in a rock mass are induced by sets of initial cracks representing the joints where the cracks may also be allowed to propagate. The distributed cracks are characterized by a damage tensor and the behaviour of the rock mass involving such cracks can conveniently be treated by describing the geometry of such cracks, and the properties of the intact rock, separately.

The use of such models for practical purposes demands, however, that the joint system be determined and considered. In order to quickly determine the orientation of a joint system it is imperative that digital models of the joint system be subjected to direct numerical analysis. It is also necessary that the strength characteristic values of rock material be determined through experimental tests or rook mass classifications. This will make it possible to derive rock mass characteristics from the physical properties of the undisturbed rock mass, from the joint properties, joint system orientation and from the rock's failure properties.

2 CONSTITUTIVE MODELING OF JOINTED ROCK

Three main items should be specified to identify the elasto-plastic behaviour of jointed rock :

- the relation between stresses and strains in the elastic stage.
- a failure criterion.
- a flow rule.

Numerous investigations have been performed to obtain a suitable failure criterion and an associated flow rule (Pan & Hudson 1988). It is difficult, however, to find a realistic criterion which considers the behaviour of the massive rock containing few sets of discontinuities or heavily jointed rock (Hoek 1994). Hoek and Brown's (HB) criterion is considered appropriate for jointed rock mass as it involves the rock quality data and the joint information in its constants.

2.1 *Hoek-Brown failure criterion*

To describe the strength of a rock mass, Hoek and Brown (Hoek & Brown 1980) proposed an empirical failure criterion based on their experimental and theoretical experience with the fracture mechanics of rock. The criterion is defined by the equation:

$$\sigma_1 = \sigma_3 + \sqrt{m\sigma_c\sigma_3 + s\sigma_c^2} \tag{1}$$

where σ_1 = major principal effective stress at failure; σ_3 = minor principal effective stress at failure; m and s material constants; and σ_c = uniaxial compressive strength of the intact rock.

A modified criterion was published in 1992 (Hoek et al. 1992) and is given by the equation:

$$\sigma_1 = \sigma_3 + \sigma_c\left(m_b\,\frac{\sigma_3}{\sigma_c} + s\right)^a \tag{2}$$

where m_b = value of the constant m for the rock mass; s and a = constants which depend on the characteristic of the rock mass; and σ_c, σ_1 and σ_3 are defined earlier in equation (1).

The criterion, as described by equation (1), includes three material parameters: s, m and σ_c. Parameters s and m are both dimensionless constants and approximately analogous to the friction angle ϕ and cohesion c of the conventional Mohr-Coulomb (MC) failure criterion. The value for the parameter s ranges between one for intact rock, and zero for heavily jointed or broken rock. The third parameter is the uniaxial compressive strength of intact rock σ_c. It was chosen because it is widely quoted in rock mechanics and is likely to be present in most practical problems. The constants m, s and σ_c can be determined using the results of laboratory triaxial or shear tests (Hoek & Brown 1988). Also, the HB parameters can be derived from the parameters of MC criterion (Baz 1997). Hoek and Brown suggested that the constants of the criterion be related to the most famous rock mass classifications (Hoek 1994), namely the Bieniawski's RMR classification and Barton's Q system, which facilitates the practical applicability of the criterion.

HB criterion can be applied perfectly in case of homogeneous and isotropic rock mass. For heavily jointed rock masses contain four or more closely spaced discontinuity sets, and none of these discontinuities is significantly weak as compared to the others to which HB criterion can be applied. However, when one of the continuity sets is very weak such that it causes the behaviour of the rock mass to be anisotropic, HB criterion should not be used unless allowance is made for this anisotropy (Hoek 1994).

To check for the validity of HB criterion, an excavation problem was analyzed using both HB criterion and the well known MC criterion. A set of relationships was derived by Baz (Baz 1997) to express the corresponding MC strength parameters (c and ϕ) in terms of HB parameters (m, s and σ) and vice versa in three different cases:

- *condition 1 : at specified normal stress* σ_n
- *condition 2 : at specified minor principal stress* σ_3
- *condition 3 : at equivalent uniaxial compressive strength of rock mass* $\sigma_c(mass)$

Figure (1) geometrically shows the relationship between Mohr-Coulomb (MC) and Hoek-Brown (HB) envelopes. When the uniaxial compressive strength of the rock mass based on MC criterion, $\sigma_{c(mass)MC}$, is equivalent to that based on HB criterion, $\sigma_{c(mass)HB}$ (condition 3), the following procedure can be followed to estimate HB parameters from MC parameters:

$$\sigma_{c(mass)HB} = \sigma_{c(mass)MC}$$

$$s = \left[\frac{2\,c\,\cos\phi}{(1-\sin\phi)\sigma_c}\right]^2$$

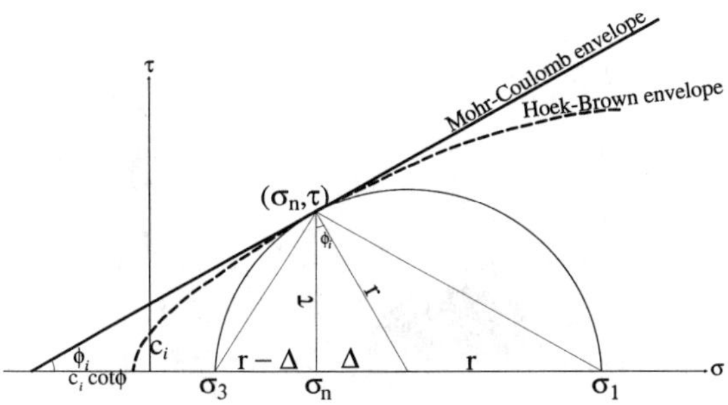

Figure 1: Geometrical relations between Hoek-Brown and Mohr-Coulomb criteria

$$m = 4\sqrt{s}\,\tan\phi\,\frac{1+\sin\phi}{\cos\phi}$$

The material constants for HB criterion were calculated based on the previous relations and with a cohesion value of 10.0 kN/m^2 and ϕ of 20°. Figure (2) illustrates the plastic zones formed due to the circular excavation by using both MC and HB criterion. Good agreement can be found in the results of both criteria, which indicates the validity of HB criterion for more practical applications.

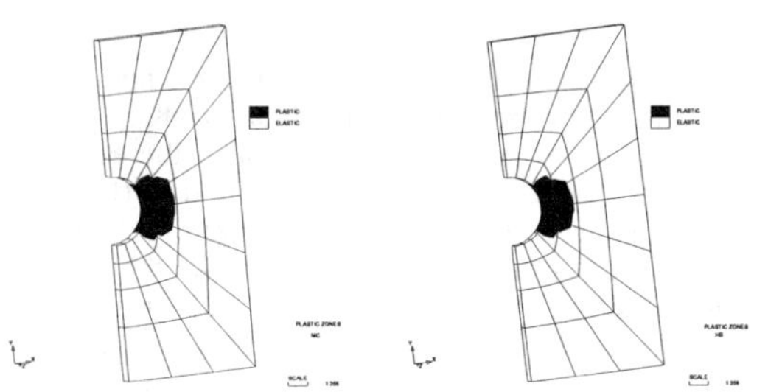

Figure 2: Distribution of plastic zones using both MC and HB criteria

2.2 *Damage model*

The paper utilizes an anisotropic nonlinear damage model in strain space to describe the behavior of jointed rock and applies it to the mechanical analysis of tunneling. This work focuses on rate-independent and small-deformation behavior during static isothermal processes. The prime features of this model include:

- the properties of damage-dependent elasticity tensors based on geological information on the jointed rock mass;
- the damage evolution law presented on the basis of thermodynamics combined with the endochronic theory;
- the symmetric form of the effective stress concept which held between the elasticity tensor of the damaged material and that of the fictitious undamaged material; and
- The global damage tensor based on the work-equivalence principle and the local geological data of multi-joint sets.

The damage model adopted here is a second-order tensor whose definition for one set of parallel joints is given by:

$$\Omega_{ij} = \omega n_i n_j, \quad i,j = 1,3 \tag{3}$$

where $\mathbf{n} = [l,m,n]^T$ is the direction vector of damage tensor Ω of the crack set; and ω is the separation factor of one set of cracks, which is taken as the crack density. The separation factor is part of the geological information of the jointed rock mass.

A one-parameter damage elasticity tensor is introduced as:

$$\mathbf{D} = \phi \cdot \mathbf{D^0} \cdot \phi \tag{4}$$

$$\tilde{D}_{ijkl} = \lambda\phi_{ij}\phi_{kl} + \mu(\phi_{ik}\phi_{jl} + \phi_{il}\phi_{jk}) \tag{5}$$

$$\phi_{ij} = \delta_{ij} - m\tilde{\Omega}_{ij} - (1-m)\tilde{\Omega}_{im}\tilde{\Omega}_{mj}, \qquad 0 \le m \le 1 \tag{6}$$

where m is assumed to be the only damage-related material constant and λ and μ are the Lame's constants of the undisturbed rock mass. For jointed rock m is related to the crack system and is a function of the crack layer thickness. The condition $0 \le m \le 1$ is needed to satisfy the positive definite requirement.

Damage evolution law is one of the most important aspects of the constitutive relationship of rock. Here it is assumed that the damage evolution is completely determined by its conjugate force, which is also known as *the strain energy density release rate* (Krajcinovic *et al.* 1991).

To account for further calculations for the elasto-plastic damage, the effective stress concept and the energy equivalence principle should be held between the elasticity tensor of the damaged material and that of the fictitious undamaged material. Calculation of the symmetric effective stress of fictitious undamaged state is given as (Zheng & Betten 1996):

$$\boldsymbol{\sigma}^{effective} = \frac{\boldsymbol{\sigma}^{Cauchy}}{(\mathbf{I}-\boldsymbol{\Omega})} \tag{7}$$

Next, in order to obtain a symmetric stress tensor, the following equation is used:

$$\tilde{\sigma_{ij}} = \sqrt{\sigma_{ij}^{effective}\sigma_{ji}^{effective}} \tag{8}$$

where there is no summation for i and j. Therefore we can find the square root of the damage tensor in the principal space of damage tensor as follows:

$$\tilde{\boldsymbol{\sigma}} = (\mathbf{I}-\boldsymbol{\Omega})^{-\frac{1}{2}}\boldsymbol{\sigma}^{Cauchy}(\mathbf{I}-\boldsymbol{\Omega})^{-\frac{1}{2}} \tag{9}$$

For the stress σ^{Cauchy} which is calculated in the global coordinate system, it is necessary to make a go and back transfer with the eigen-matrix $\mathbf{T}$ of the damage tensor Ω. Then we find the effective stress tensor $\tilde{\sigma}$ in the global coordinate system:

$$\tilde{\sigma} = \mathbf{T}(\mathbf{I} - \Omega)^{-\frac{1}{2}}(\mathbf{T}^T \sigma^{Cauchy}\mathbf{T})(\mathbf{I} - \Omega)^{-\frac{1}{2}}\mathbf{T}^T \quad (10)$$

The main advantage of the damage model is the relatively small number of parameters needed to establish the model. In addition, the model can be implemented easily in the existing Finite Element codes.

3 PRACTICAL USE OF GEOLOGICAL DATA

The orientation of the joint set is usually identified in the field by sampling along a line, as in a borehole, or over an area, as on an outcropping surface. Statistical descriptions of the orientation data are always expressed as the orientation of the normal line of the joint plane. Spherical coordinates (α, β, r) or the direction cosines (l, m, n) are generally used for this purpose.

The direction vector of the damage tensor Ω according to equation (3) is defined on the basis of the direction vector $\mathbf{n} = [l, m, n]^T$. Calculation of the global damage tensor for multiple sets of joints is based on the principle of energy equivalence: The energy dissipated by the global damage tensor is equivalent to the sum of the energy dissipated by the damage tensor related to each set of joints.

The equation for calculating the global damage tensor is given in the following:

$$\Omega_g = \mathbf{I} - \left[\sum_{n=1}^{N}(\mathbf{I} - \Omega_n)^{-1} - (N-1)\mathbf{I}\right]^{-1} \quad (11)$$

where Ω_g is the global damage tensor; Ω_n is the damage tensor for the nth joint; N is the total number of joint sets; $\mathbf{I}$ is a second-order unit tensor.

4 CASE STUDY

The three kilometers long Schönberg Tunnel was newly built for the by-pass around the city of Schwarzach (Austria). The tunnel had to be constructed in heavily jointed phyllite rock. The construction was a combination of a TBM pilot tunnel and NATM for final construction. A 3D model was used to simulate the impact of the jointed rock on the construction sequences and to investigate the stability of the tunnel vicinity.

In the numerical model, Figure (3), it was very important to very carefully simulate all construction steps from the pilot tunnel to the crown excavation,

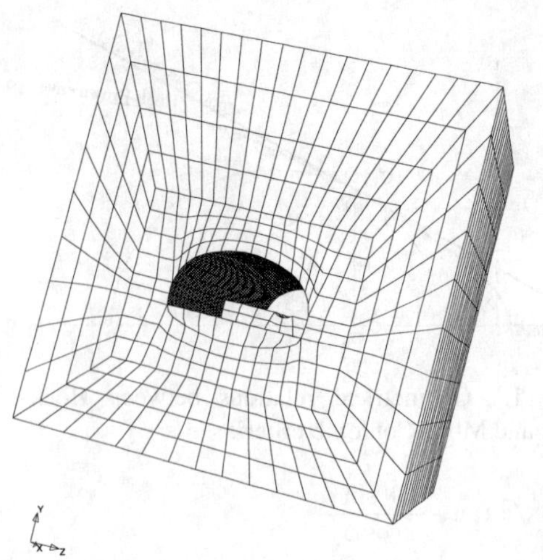

Figure 3: Finite Element Mesh of Schönberg Tunnel

Table 1: Orientation of the joint system (Km 00.782)

Set	α	β	ω
S	40°	50°	0.6
K	110°	65°	0.3

to the middle part and bench excavation and the final lining. The measurements at station (Km 00.782) were chosen and the geological situation there was investigated carefully for the damage model. Based on the geological information, the principle joint system has two main sets of joints, K and S, with the orientation and intensity shown in Table (1) and Figure (4). The orientation of the model is described by the angle $\gamma = 105^o$, which is the angle between the north direction and the Z-axis of the 3D model. Based on this data the initial global damage tensor is calculated and used in the damage model.

To investigate the impact of the jointed rock on the stress field and the displacement field, two calculations were performed using HB criterion and the dam-

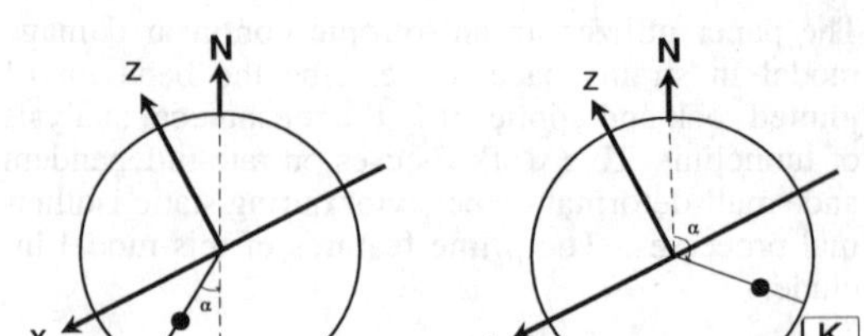

Figure 4: Orientation of joint system based on the geological data

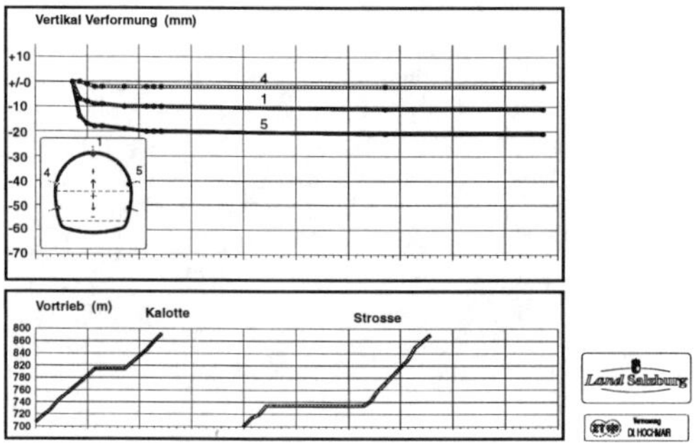

Figure 5: Field measurements at (Km 00.782), Schönberg Tunnel

age model. Field measurements indicated an unsymmetrical deformation profile, Figure (5). The damage model also resulted in an unsymmetrical deformation profile, whereas HB model indicated a symmetrical one, Figure (6). The displacement values are in good agreement with field data and can be further improved by readjusting the initial stress field and the mechanical rock parameters. By comparing the displacement

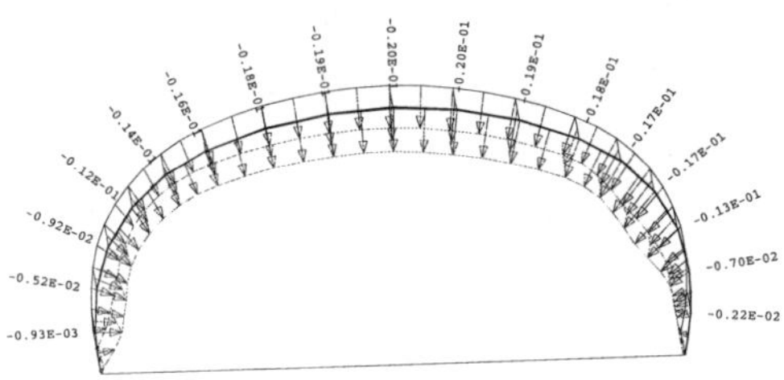

a. Damage model

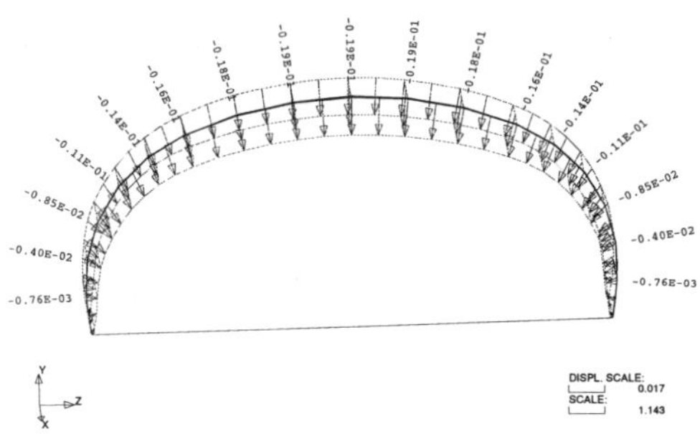

b. HB model

Figure 6: Deformed shape of lining for both damage model and HB model

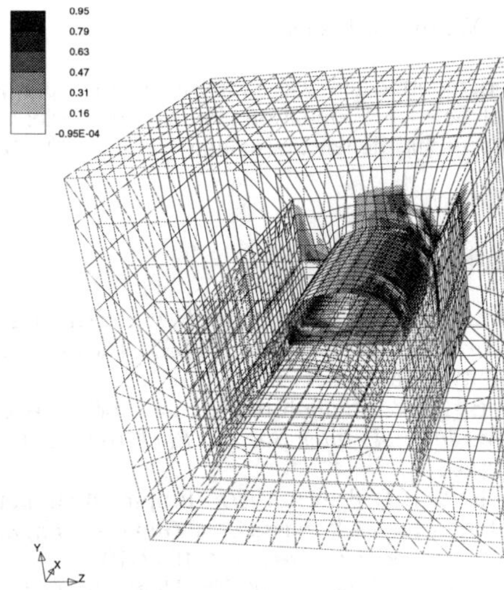

Figure 7: Distribution of maximum principle active damage tensor

profiles with field data one can rely more on the rest of results of the damage model as it better reflects the rock mass anisotropy.

Figure (7) illustrates the distribution of the damaged due to tunneling. The damage tensor (Ω_{max}) is the information about the opening of the joint system. From the point of view of rock stability assessment, the damage model shows that rock loosening appears in a great area around the tunnel with different values. This dictates the use of some anchor bolts in the tunnel vicinity to ensure tunnel stability. On the other hand, the conventional elastoplastic calculation using HB criterion did not show any plastic region at all.

5 CONCLUSIONS

HB criterion is introduced as one of the most suitable criteria for jointed rock mass. However, attention should be paid to some limitations on the criterion applicability with the existence of rock mass anisotropy. The damage theory and HB criterion presented in this paper were applied in the mechanical analysis of the Schönberg Tunnel (Salzburg, Austria) where one set of joints is dominant. The numerical results of damage distribution are reasonable and coincide well with *in situ* measurements. The distribution of the damaged area which is much larger than the plastic zones reflects the idea that the resulting carrying capacity is reduced due to the consideration of material defects in the damage model. Hence, the choice of the suitable constitutive model is essential to provide useful aids to the engineering assessment of tunnel stability.

ACKNOWLEDGMENTS

The work reported here was supported by the Austrian National Science Foundation, Fonds zur Förderung der wissenschaftlichen Forschung, project S08004-TEC.

REFERENCES

Baz, M. 1997. *Deep tunnel analysis Practical and theoretical considerations*. Ph.D. Thesis. University of Innsbruck.

Hoek, E. 1994. Strength of rock and rock masses. *News Journal - Inter. Soc. for Rock Mech. ISRM*. 2(2):4-16.

Hoek, E. & E.T. Brown 1980. Empirical strength criterion for rock masses. *J. Geotech. Engng. Div, Am. Soc. Civ. Eng.*. 106:1013-1035.

Hoek, E. & E.T. Brown 1988. The Hoek-Brown failure criterion- a 1988 update. *In Rock engineering for underground excavation, Proc. 15th Canadian Rock Mech. Symp.*:31-38.

Hoek, E., D. Wood & S. Shah 1992. A modified Hoek-Brown criterion for jointed rock masses. In Hudson, J.A., editor, *Proc. of Rock Characterization Symp. ISRM*:209-214.

Kachanov, L.M. 1958. Time of the rupture process under creep condition, *IVZ Akad Nauk, S.S.R., Otd Tech Nauk*. 8:26-31.

Kawamoto, T., Y. Ichikawa & T. Kyoya 1988. Deformation and fracturing behavior of discontinuous rock mass and damage mechanics theory. *Int. J. Numer. Anal. Meth. Geomech.* 12:1-30.

Krajcinovic, D., M. Basista & D. Sumarac 1991. Micromechanically inspired phenomenological damage model. *J. Appl. Mech.* 58:305-316.

Murakami, S. & N. Ohno 1980. A continuum theory of creep and creep damage. *Proc. 3rd IUTAM Symp. on Creep in Structures*: 422-443. Ponter, A. R. S. and Hayhust, D. R. Ed., Springer-Verlag, Berlin.

Pan, X.D. & J.A. Hudson 1988. A simplified three dimensional Hoek-Brown yield criterion. *Rock Mechanics and Power Plants*, Roman(ed.):95-103.

Swoboda, G., Q. Yang & F. Ito 1995. Damage propagation model and its application to rock engineering. *Proc. 8th Int. Congress on Rock Engineering*: 159-163. Tokyo.

Yazdani, S. & S. Karnawat 1996. A constitutive theory for brittle solids with application to concrete. *Int. J. Dama. Mech.* 5:93-110.

Zheng, Q.S. & J. Betten 1996. On damage effective stress and equivalence hypothesis. *Int. J. Dama. Mech.* 5:219-239.

Computer Methods and Advances in Geomechanics, Desai et al. (eds) © 2001 Balkema, Rotterdam, ISBN 90 5809 183 X

Technological opportunities for managing the highwall stability in opencast mines

P.Zlatanov, G.Trapov & P.Stoeva
University of Mining and Geology 'St. Ivan Rilski', Sofia, Bulgaria

ABSTRACT: Opencast mining at the Maritsa-east coal basin is often accompanied with origination of landslides. They bring to significant damages and risj for people and machines. This requires continuous observation of highwall condition with an aim of immediate impact on the deformation process by changing the highwall configuration during the operation of existing equipment. Effect is achieved when time for occurring of irreversible deformations (according to threshold speed of deformation and total deformation) is estimated correctly as well as real time for excavator heading to the dangerous zone and its excavation. A simulation model is composed under conditions of indetermination of technological parameters for estimation of the time needed for technological impact. This model is able to grow into a member of an overall system for monitoring and control of mining works that guarantees mine opening stability.

The Maritsa-east opencast mines in Bulgaria exploit lignite coal. Highwalls are composed of Pliocene clays. The Pliocene profile is presented of black over-coal clays, coal and over-coal suite – black, gray-green and yellow-brown oxidized clays, reaching the earth surface. Exploitation is accompanied by rheological deformation process that almost always completes with landslide phenomena. These phenomena are a danger for highwall stability and safety of mining machines and people working in the opencast mine. Origination of landslide phenomena depends not only on specific geomechanical properties of loose, plastic clays but also on mining technology and respective parameters of benches and highwalls. Stability calculations are related to a deformation mechanism as well as to a scheme for formation of zones with different resistance of existing low angles of internal friction and clay cohesion and relatively low density of Pliocene rocks. Long "in situ" observations and laboratory testing proved the following scheme of deformation: a prism of active rock stress in the upper part of the highwall, central block of horizontal displacement over existing mirror surface and a prism of passive stress (Figure 1). The scheme proved both by calculations and "in situ" provides an opportunity for monitoring the deformation processes and respective technological control of exploitation factors and conditions (Georgiev, 1971). Stability of highwalls is determined by two main groups of factors and conditions: technological and natural.

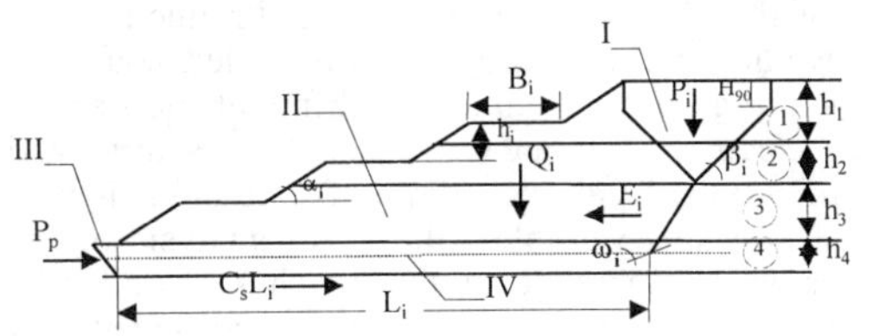

Figure 1. Principle scheme for calculating the highwall stability
1-blue clay; 2-black clay; 3-lignite coal; 4-thin-layered clay; I-active prism; II-central block; III-passive prism; IV–landsliding surface.

The first can be managed, and the second can only be estimated. On this basis the working highwall is an object of management. The highwall has changeable characteristics – height of bench and width of working sites. Impact on the deformation process of working highwalls is reduced to non-mining of coal splits and unloading the massif in the vicinity of profiles, where movement is observed. In practice one or several excavators direct to dangerous areas. The most often case is when an excavator goes to a preliminary defined profile on its own and then cuts into the massif. The main difficulty is real estimation of time (way passed, respectively) for reaching the area to be subjected to mining and time for its excavation. It this total time, presented in shifts and days is shorter than time for reaching the threshold total deformations (Δl near 500 mm) at a maximum speed of deformation of the massif (v=2 mm/day) then moving of excavator in the process of mining is the most advisable. Not completing this condition a possible acceleration can be achieved by mining of not complete split. In case of critical situation (active deformation process) the excavator can go back to the dangerous zone and unload the active prism by excavating the clays below the level of staying. In this case the landsliding force E, caused by the prism of active pressure reduces, as $E=f(H^2)$. When lower excavation is applied height H of the active prism is reduced. In situaions like that one the willingness for maximum loading of excavation-transportation mechanization is evident. Practice shows that shift schedules for excavator productivity and measures for its achieving are not always observed. As a result the excavator is not at the provided location in the required moment of time. This fact is due to the contradiction between determining approach in the compilation of time schedules and indirect modeling of excavator movement and random character of changes in its productivity. Movement of excavator during one shift is a function of three variable values – productivity Q, bench height h and split width A. All of the above values is random. Productivity of excavator is influenced by rock properties, condition of working head, reliability of different mechanical and electrical aggregates, climatic factors and qualification of staff etc. Width of split and height of bench are also variable values. An analysis of the upper and lower edge of the bench highwall of a random level from the plan is sufficient for achieving an idea. The above reveals that the shift advance $l=f(Q, h, A)$, which is a function of these random values is also a random value. Under these conditions, the most suitable approach for prediction of the way passed by excavator for certain period of time is the simulation modeling. On the basis of algorithm for comparing the time for moving the excavator along a fixed way and time needed for achieving the total threshold deformations a MOVEL software programme for personal computer is compiled. The movement of bucket wheel excavator Rs-1200 No8 on fourth level of "Troyanovo-2" (Maritsa-east mines) is modeled with the aim of testing the method for managing the stability of working highwall by change of technological parameters. Empirical laws of distribution of used random values are constructed. Empirical distribution of excavator productivity is determined on the basis of 220 observations during first and second shift in 1999. Excavator productivity was varied within 250m^3/shift to 7700m^3/shift during the tested period. The average productivity for the sample period is 4171 m^3/shift.

Empirical distribution of split width in the fourth level is obtained on the basis of 170 measurements on the month situational plans of the mine for the period 1996-1999. Width of split varied within 36-65 m and the arithmetic mean value was 48.8 m. Empirical distribution of the height of fourth level was obtained on the basis of 168 measurements on the situational plans for the period 1996 – 99. Height of level varied within 3.8 – 20.4 m. input values for the model are the number of shifts for which movement of excavator will be predicted (critical time for reaching the dangerous deformations) and number of needed predictions. Random values of shift excavator productivity, height of bench within the split and split width are obtained using the empirical functions of distribution after starting the programme. The obtained values are used for finding the length of excavator advance for the first shift modeled etc. Results generated by the programme are compared to the real way passed by the excavator for one month (from the end of March 1999 to the end of April 1999). At the end of March the excavator was at a distance of 56 m after profile 62 and was moving to the south edge. After one month of operation it reached a point located at a distance of 16 m from profile 42. Excavator passed a distance of near 360 m for one month. The result obtained by the MOVEL programme for the used empirical distributions and given number of shifts for one month (60 shifts) is 351 m. Error admitted in the modeling is about 2.5%. As any task related to probabilities this one also admits different modifications in its algorithm. Insignificant correction is required for solving the inverse problem – finding out the most probable time for passing a given way. The last parameter is important for the work of excavator moving one against another and using railway transport and one and the same track.

The probability model of excavator movement can be used as a background for compilation of a dynamic probability model of the entire mine, for predicting the development of mine works and estimation of rate of correlation with designed decisions.

REFERENCES

Georgiev, G.K. & T. Hristov. 1971. Scheme for calculation of highwall stability. *National Conference, Varna:* 173-184.

17 Flow, seepage, consolidation

Computer Methods and Advances in Geomechanics, Desai et al. (eds) © 2001 Balkema, Rotterdam, ISBN 90 5809 183 X

Numerical modeling of a grouting treatment

Annamaria Cividini
Department of Structural Engineering, Politecnico di Milano, Italy

ABSTRACT: A numerical study is presented of the grouting of granular soils based on the use of diluted silicate solutions. This represents one of the possible techniques to drastically reduce the "natural" permeability of soils and to limit the seepage flow towards underground facilities. First, the equations governing the grouting process, seen as an advective-diffusive-dispersive flow of a liquid containing a solute, are outlined and their finite element formulation is presented. Then, an empirical law is discussed that allows introducing in the calculations the variation of permeability produced by grouting. Finally, the results of a numerical simulation are presented and some conclusions are drawn on the effectiveness of the treatment.

1 INTRODUCTION

A notable increase of the water table level is taking place in the city of Milan (northern Italy), since about 1975, which is caused by a series of industrial and agricultural changes.

Due to this process, the phreatic surface gradually reached an elevation higher than that of various underground facilities, like underground parking, subway tunnels, etc. (Colombo, 1997).

These structures, which have been completed between 1960 and 1980, were not properly waterproofed. In fact, at that time they were well above the phreatic level and the possible increase of the water table level was not accounted for in their design. As a consequence, some extensive leakage of water took place at various locations within these structures that, in the worst cases, led to their partial flooding.

A series of provisions could be adopted to eliminate, or reduce, this negative effect. Some of them concern essentially the reinforced concrete members interacting with the soil. These are based, for instance, on the sealing of their joints or on the installation of waterproofing membranes. Other provisions, of geotechnical nature, could involve the installation of pumping wells in the vicinity of the underground structures, to locally lower the phreatic level. Alternately, the granular soil surrounding them could be grouted through silicate solutions, to decrease its permeability and to reduce the seepage flow toward the structures.

In the following, the main features of a numerical study concerning the mentioned second provision is outlined, that was carried out in the framework of a research project supported by the Milan Metro (Metropolitana Milanese).

First the equations governing the seepage flow of a liquid containing a solute, and their finite element formulation, are recalled.

Then, on the basis of the results of a series of laboratory tests, an empirical law is proposed which governs the variation with time of the permeability of a granular soil subjected to the seepage flow of silicate solutions. Finally, this law is adopted in a numerical example that aims to get an insight into the effectiveness of the proposed provision for reducing the seepage flow toward underground structures.

2 GOVERNING EQUATIONS

The grouting through diluted silicate solutions is modeled here as the seepage flow of a liquid, containing a given substance as a solute, through a saturated porous medium. The mass of substance per unit volume of the liquid is expressed by its concentration c.

The macroscopic flux of the substance is customarily subdivided into three parts, referred to as advective, dispersive and diffusive fluxes (see e.g. Bear & Verruijt, 1987, Peters & Smith, 2000), that depend on different physical causes. The advective flux is related to the macroscopic velocity (in Darcy's sense) of the liquid flow. The dispersive flux depends on local fluctuations of the velocity field, and can be related to the gradient of the concentration. The diffusive flux is due to the molecular diffusion of the substance within the liquid caused by the random movements of the fluid molecules.

This last part of the flux could be incorporated into that related to the mechanical dispersion (see e.g. Rowe & Booker, 1985; 1997). It should be considered, however, that the diffusive flux depends on a coefficient of molecular diffusion, which has a very small order of magnitude for most cases of interest in geomechanics. Consequently, its contribution is often negligible with respect to those of the first two components and it will not be considered in the following.

Let consider an infinitesimal volume of the porous medium, in a Cartesian reference system x,y,z, fully saturated by an incompressible liquid. If a seepage flow takes place, with macroscopic velocity components v_x, v_y and v_z in the Cartesian directions, the rate of accumulation of the substance mass within a unit volume of the porous medium is,

$$m_a = -\left(\frac{\partial c v_x}{\partial x} + \frac{\partial c v_y}{\partial y} + \frac{\partial c v_z}{\partial z}\right) \quad . \tag{1}$$

The components of the seepage flow velocity in eq.(1) are related to the gradient of the hydraulic head through Darcy's law,

$$\underline{v} = -\underline{K} \cdot grad(h) \quad . \tag{2}$$

Here, $\underline{v}$ is the discharge velocity vector, $\underline{K}$ is the matrix of the coefficients of permeability and h represents the hydraulic head

Substitution of eq.(2) into eq.(1) leads to,

$$m_a = div\left(c \cdot \underline{K} \cdot grad(h)\right) \quad . \tag{3}$$

Let consider now the dispersive part of the substance flux. The relationship between the components of this flux, collected in vector $\underline{q}_d$, and the gradient of the concentration can be written in the following compact form, where $\underline{D}$ is the matrix of the dispersivity coefficients,

$$\underline{q}_d = -\underline{D} \cdot grad(c) \quad . \tag{4}$$

For isotropic granular media, and assuming a unit value of the Peclet number, the coefficients of matrix $\underline{D}$ (having dimension of square length divided by time) can be expressed in the following compact form (see e.g. Bear, 1961),

$$D_{ij} = \frac{1}{n}\left(a_T \cdot v \cdot \delta_{ij} + (a_L - a_T) \cdot \frac{v_i \cdot v_j}{v}\right) \quad \text{(i,j=x,y,z)} . \tag{5}$$

In the above equation v is the modulus of the discharge velocity of the liquid, v_i and v_j are its components, δ_{ij} is the Kronecker symbol ($\delta_{ij}=1$ for i=j and $\delta_{ij}=0$ for $i \neq j$) and a_L and a_T are the longitudinal and transversal dispersivity (having dimension of a length).

If a dispersive flux in the Cartesian directions is present, the rate of accumulation of the mass of solute within a unit volume of the porous medium can be expressed as,

$$m_d = -\left(\frac{\partial(n_x q_{dx})}{\partial x} + \frac{\partial(n_y q_{dy})}{\partial y} + \frac{\partial(n_z q_{dz})}{\partial z}\right) . \tag{6}$$

By taking into account eq.(4), and by assuming that for an isotropic porous medium the surface porosities n_x, n_y and n_z coincide with the volume porosity n, eq.(6) reduces to

$$m_d = div\left(n \cdot \underline{D} \cdot grad(c)\right) \quad . \tag{7}$$

Finally, the continuity equation for the advective-diffusive flux is obtained by imposing that the mass rate $m_a + m_d$ compensates the variation of the mass of substance contained within the porous element. This, in turn, is due to the variation m_{cn} of the concentration and/or porosity during the time increment, and to the contribution m_{qc} of possible internal sources having an imposed flux q_c,

$$m_{cn} = \frac{\partial cn}{\partial t}, \quad m_{qc} = q_c \cdot n \quad . \tag{8a,b}$$

Hence, the continuity equation reads,

$$m_a + m_d = m_{cn} + m_{qc} \quad . \tag{9}$$

Substitution of the relevant relationships into eq.(9) leads to the following continuity equation,

$$div\left(c \cdot \underline{K} \cdot grad(h) + n \cdot \underline{D} \cdot grad(c)\right) = \frac{\partial cn}{\partial t} + q_c \cdot n \quad . \tag{10}$$

Note that, when the concentration c is constant in time and space, eq.(10) reduces to the well-known continuity equation of the seepage flow of an incompressible liquid into a saturated porous medium,

$$div\left(\underline{K} \cdot grad(h)\right) = \frac{\partial n}{\partial t} \quad . \tag{11}$$

3 FINITE ELEMENT FORMULATION

The finite element formulation of the problem governed by eq.(10) can be based on a procedure quite similar to that adopted for seepage or diffusion problems. The details of this procedure, which are omitted here for briefness, can be found e.g. in (Cividini, 2000).

The medium is dicretized into isoparametric finite elements within which the variation of the hydraulic head h(x,y,z,t) and of the concentration c(x,y,z,t) depends on the corresponding nodal values, collected in vectors $\underline{h}(t)$ and $\underline{c}(t)$, and on the vector of "shape functions" $\underline{f}(x,y,z)$,

$$h(x,y,z,t) = \underline{f}(x,y,z)^T \, \underline{h}(t) = \underline{h}^T \underline{f} \quad , \tag{12a}$$

$$c(x,y,z,t) = \underline{f}(x,y,z)^T \, \underline{c}(t) = \underline{c}^T \underline{f} \quad . \tag{12b}$$

Following the mentioned procedure, eq.(10) can be reduced to the following matrix form,

$$\underline{A}_1 \cdot \underline{c} + \underline{A}_2 \cdot \underline{c} + \underline{A}_3 \cdot \dot{\underline{c}} = \underline{a}_1 \quad , \tag{13}$$

where

$$\underline{A}_1 = \int_V \underline{B}^T \underline{K} \cdot \underline{B} \cdot \underline{h} \cdot \underline{f}^T dV \quad , \tag{14a}$$

$$\underline{A}_2 = n \cdot \int_V \underline{B}^T \underline{D} \cdot \underline{B} dV \quad , \tag{14b}$$

$$\underline{A}_3 = n \cdot \int_V \underline{f} \cdot \underline{f}^T dV \quad \text{and} \quad \underline{a}_1 = -n \cdot \int_V \underline{f} \cdot q_c dV \ . \tag{14c,d}$$

In eqs.(14) $\underline{K}$ and $\underline{D}$ are, respectively, the permeability and diffusivity matrices and $\underline{B}$ is the matrix of the derivatives of the shape functions,

$$\underline{B}^T = \left[\frac{\partial \underline{f}}{\partial x} \ \ \frac{\partial \underline{f}}{\partial y} \ \ \frac{\partial \underline{f}}{\partial z} \right] \ . \tag{14e}$$

It is worthwhile observing that the finite element evaluation of the nodal values of the hydraulic head is governed by the following equation derived from eq.(11),

$$\underline{A}_4 \cdot \underline{h} = \underline{a}_2 \quad , \tag{15}$$

where

$$\underline{A}_4 = \int_V \underline{B}^T \underline{K} \cdot \underline{B} dV \quad \text{and} \quad \underline{a}_2 = \int_V \underline{f} \cdot q_w dV \ . \tag{16a,b}$$

The compressibility of the soil skeleton is neglected in the above equations, i.e. the porosity does not change with time, and q_w represents the imposed flow due to possible internal sources.

4 TIME INTEGRATION PROCEDURE

Various numerical schemes have been proposed in the literature for the time integration of advective-dispersive flow problems. Recently Wendland & Schmid (2000) presented an overview of these schemes and a novel integration procedure, showing that some of them may lead to spurious oscillations, and even to negative values, of the concentration.

Here, a two-step time integration is adopted. The first step consists in determining the hydraulic head vector $\underline{h}$ by means of eqs.(15). If the coefficients of permeability are constant, and if the boundary conditions do not vary with time, the assumptions of incompressibility of the soil skeleton and of the liquid phase lead to a time independent distribution of the hydraulic head. Hence, under the above assumptions, only the time dependent change of the concentration vector $\underline{c}$ has to be evaluated during the second step of the analysis through an iterative procedure.

This step is subdivided into a series of time increments Δt during which the following expressions of vector $\underline{c}$ and of its time derivative are adopted,

$$\underline{c}(t) = \vartheta \underline{c}_{t+\Delta t} + (1-\vartheta)\underline{c}_t \quad , \quad \text{where } 0 \le \vartheta \le 1 \quad , \tag{17a}$$

$$\dot{\underline{c}}(t) = (\underline{c}_{t+\Delta t} - \underline{c}_t) / \Delta t \quad . \tag{17b}$$

Substitution of eqs.(17) into eq.(13) leads to the following matrix relationship,

$$\left[\vartheta \Delta t \cdot (\underline{A}_1 + \underline{A}_2) + \underline{A}_3\right] \cdot \underline{c}_{t+\Delta t} =$$

$$= \Delta t \cdot \underline{a}_1 - \left[(1-\vartheta)\Delta t \cdot (\underline{A}_1 + \underline{A}_2) - \underline{A}_3\right] \cdot \underline{c}_t \quad . \tag{18}$$

Since $\underline{A}_1$ is in general non-symmetric, its contribution can be moved to the right hand side of eq.(18) to preserve the symmetric nature of the system matrix.

This leads to the following recursive solution equation for $\underline{c}_{t+\Delta t}$,

$$\left[\vartheta \Delta t \cdot \underline{A}_2 + \underline{A}_3\right] \cdot (\underline{c}_{t+\Delta t})_i =$$

$$= \Delta t \cdot \underline{a}_1 - \left[(1-\vartheta)\Delta t \cdot \underline{A}_2 - \underline{A}_3\right] \cdot \underline{c}_t +$$

$$- \Delta t \cdot \underline{A}_1 \cdot \left[\vartheta (\underline{c}_{t+\Delta t})_{i-1} + (1-\vartheta)\underline{c}_t\right] \quad , \tag{19}$$

where i denotes the iteration number. Note that the evaluation of the nodal pore pressures and concentrations are "uncoupled", with a consequent saving in the size of the relevant arrays of the finite element code.

It turned out that the mentioned spurious oscillations of the computed nodal concentrations can be drastically reduced if values of ϑ greater than 0.5 are adopted in the calculations.

The extension of this integration technique to the case of non-constant permeability (due e.g. to the time dependent variation of the viscosity of the liquid-solute mixture) is rather straightforward. To this purpose it is necessary to define a law expressing the variation of the solution viscosity, and hence of the permeability of the granular soil, with time. The characteristics of this law are discussed in the subsequent Section.

Since the variation with time of the coefficient of permeability leads to a time dependent distribution of the hydraulic head, an iterative procedure should be employed to evaluate both vectors $\underline{h}$ and $\underline{c}$ at the end of each time increment Δt.

For the first iteration the nodal concentrations at time t+Δt are set equal to those at time t and the permeability of the soil elements reached by the grouting mixture is determined through the adopted empirical law. Then, a first approximation of the nodal hydraulic heads and concentrations at time t+Δt is evaluated following the described two-step scheme. The time increment is repeated, introducing the computed values of the concentration at time t+Δt in eq.(17a) and evaluating updated values of the coefficient of permeability. The iterations are re-

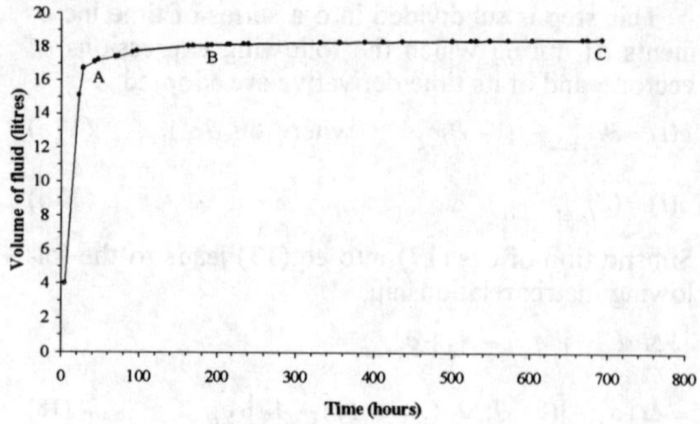

Figure 1. Volume of fluid leaving the sample during time for test performed with 0.3 moles per liter of solution and 0.03 moles per liter of phosphoric acid (pH=11.09). Points A, B and C denote the beginning and the end of grouting, and the end of the test.

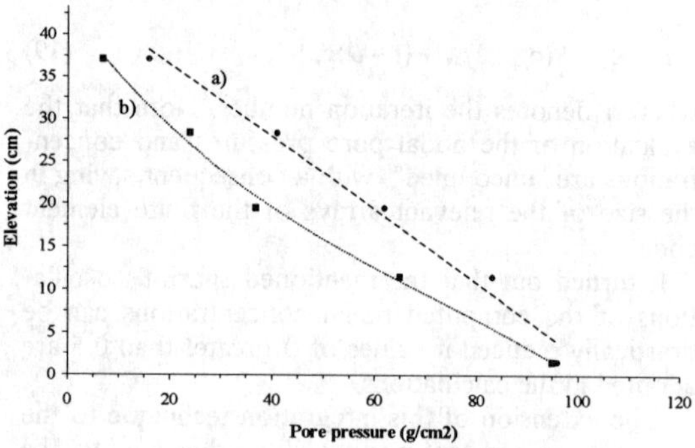

Figure 2. Experimental distribution of the pore pressure along the sample height: a) before grouting, b) during the test, before the complete formation of the silica gel.

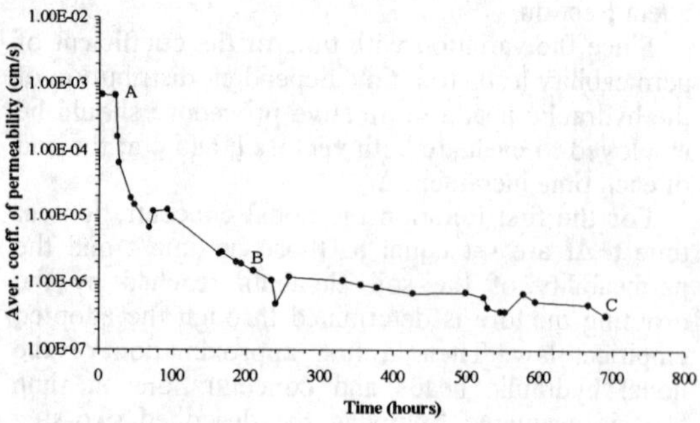

Figure 3. Variation of the average coefficient of permeability of the sample during time for test referred to in Figure 1.

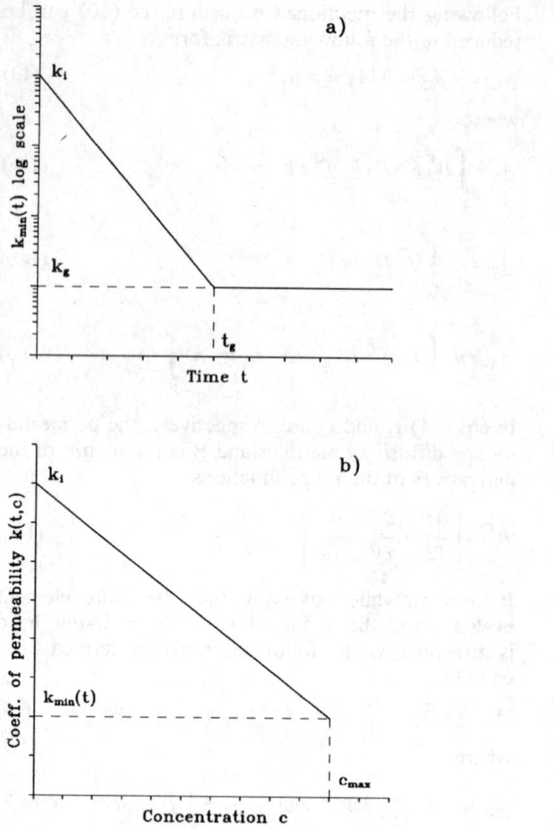

Figure 4. Adopted variation of the coefficient of permeability of the soil with time a) and concentration b).

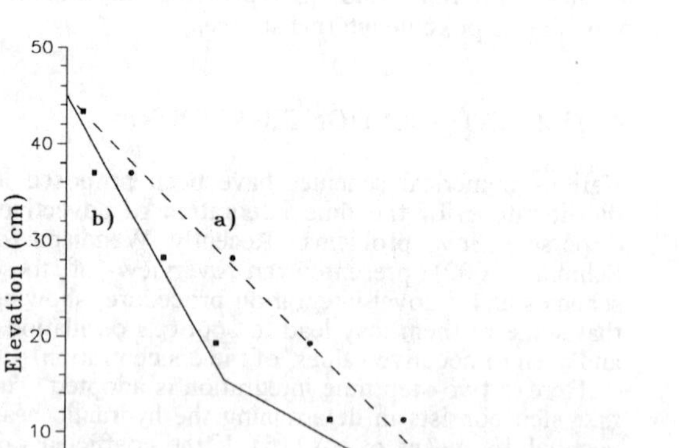

Figure 5. Numerical interpretation (dashed and solid lines) of the experimental results shown in Figure 2.

peated until further changes of the final nodal variables become negligible.

When convergence has been reached, the iterative procedure is applied to the solution for the subsequent time increment.

As to the time required for solution, a rapid convergence was observed for all examined cases, leading to a computational effort comparable to that necessary to solve non-linear or unconfined seepage flow problems.

5 VARIATION OF THE COEFFICIENT OF PERMEABILITY WITH CONCENTRATION AND TIME

As previously mentioned, a law governing the variation of the grouted soil with time is required to carry out the numerical analyses. To this purpose, a laboratory investigation was carried out by subjecting samples of granular soil to the seepage flow of silicate solutions.

The soil was placed within a "permeameter", consisting of a Plexiglas pipe with internal diameter of 10cm, within which the fluid flow takes place in the vertical direction. Five pressure gauges were placed along the pipe to detect the local variation of the pore pressure during the tests.

The soil was compacted in layers of small thickness within the permeameter, using the "moist tamping" technique. The height of the samples after preparation ranged between 42 and 45 cm.

The hydraulic load (of the order of 50 cm) was generated by means of a container filled with water, or with the silicate solution, located at an elevation higher than that of the sample base.

The tests were subdivided into 5 stages:

1) sample saturation,
2) determination of the initial, or natural, coefficient of permeability,
3) grouting,
4) determination of the final coefficient of permeability,
5) constant head water flow through the grouted samples.

The details of the testing procedure have been presented elsewhere (Massone, 1998; Cividini, 2000).

The average value of the initial coefficient of permeability, determined through a variable head test, was close to $4.0 \cdot 10^{-3}$ cm/s, with a minor dispersion of data.

The grouting stage of the tests was performed by connecting the base of the sample to the tank containing the silicate solution and recording during time the volume of fluid leaving the sample and the pore pressure values at different elevations within it.

The final coefficient of permeability was determined at the end of this stage, through the same procedure adopted at the beginning of the test. Then the samples were subjected to a water flow, with a constant hydraulic head, to evaluate possible changes of the seepage velocity during time. Depending on the adopted grouting solution, the total duration of tests ranged between 20 and 30 days.

It is well known that the chemical properties of the adopted solutions play a major role in the effectiveness of the grouting treatment (see e.g. Karol, 1965; Nonveiller, 1989). It seems sufficient to mention here that three different solutions of sodium silicate have been used during the laboratory tests. They consist of solutions of SiO_2 and of Na_2O with different ratios between the two components (namely, 1:1, 1:2 and 1:3). Phosphoric acid was added to them, since the time required for the formation of the silicate gel depends, in addition to its concentration, on the pH of the solution. Further details on these aspects of the tests can be found in (Grimoldi, 1998).

Some typical experimental results are reported in Figures 1 to 3. They refer to the seepage tests performed with ratio of 1:3 between silicate and sodium solutions.

Figure 1 reports the increase during time of the total volume of fluid leaving the sample. Figure 2 shows the pore pressure distributions recorded along the sample height at the beginning of the test and during the grouting process, before the formation of the silica gel has been completed. Finally, figure 3 presents the variation of the average coefficient of permeability for the entire sample computed on the basis of the data reported in Figure 1.

These results show that, for the tested granular soil, the grouting through diluted solutions of sodium silicate may lead to a reduction of the coefficient of permeability of two to three orders of magnitude.

It was observed that a main scope of this experimental investigation was to derive an empirical law expressing the variation of the coefficient of permeability with time, to be adopted in the numerical analyses.

To this purpose, the linear laws depicted in Figure 4 were adopted as an engineering approximation of the test results. Here k_i and k_g denote, respectively, the initial (natural) coefficient of permeability and that of the grouted soil, t_g represents the time necessary for the formation of the silicate gel and $k_{min}(t)$ is the coefficient of permeability corresponding to the maximum concentration of the solution.

On these bases the time dependent value of the $k_{min}(t)$ can be expressed as follows

$$k(t) = k_i \cdot (k_g / k_i)^{t/t_g} \quad , \quad 0 \le t \le t_g \quad . \tag{20}$$

The assessment of the validity of this empirical law was based on the numerical simulation of the experimental results already shown in Figure 2.

The experimental data (dots) and the numerical results (dashed and solid lines) are presented in Figure 5. This comparison shows that the adopted relationship between permeability, time and concentration leads to an acceptable approximation of the experimental data (from the engineering point of view) and, hence, that it can be adopted for the numerical analysis of in situ treatments.

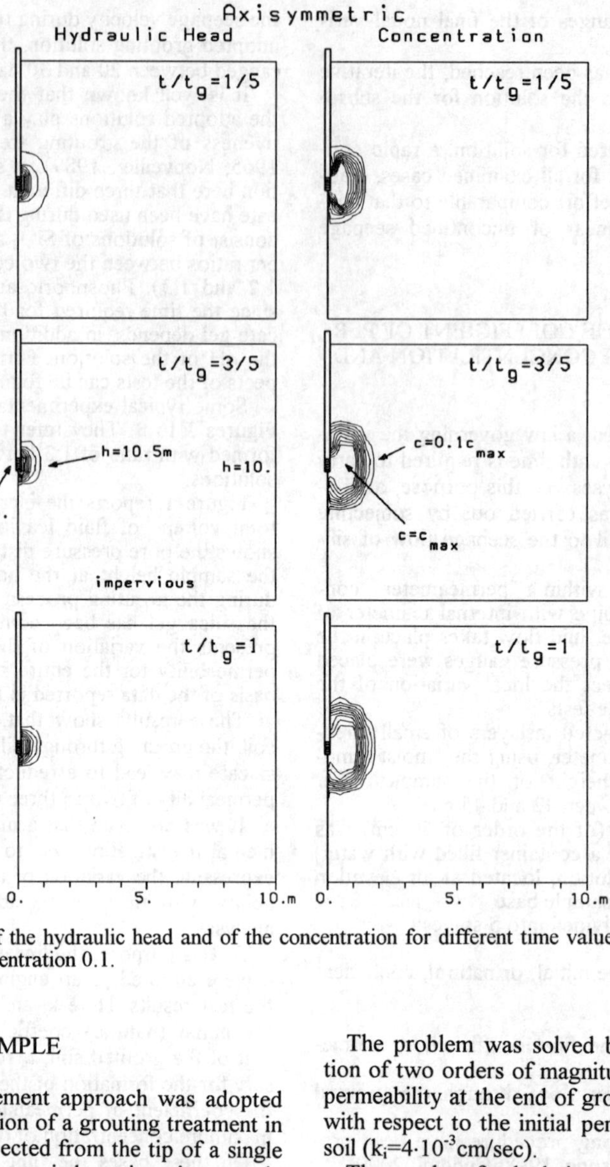

Figure 6. Contour lines of the hydraulic head and of the concentration for different time values. Contour line intervals: hydraulic head 0.5m, concentration 0.1.

6 NUMERICAL EXAMPLE

The described finite element approach was adopted in the numerical simulation of a grouting treatment in which the solution is injected from the tip of a single pipe. The analyses were carried out in axisymmetric regime, adopting the mentioned empirical law for expressing the time dependent variation of the permeability of the grouted soil.

A soil portion of 10x10m was discretized into square, 4 node finite elements, 0.5m in size. At the beginning of calculation this zone is subjected to a constant hydraulic head distribution of 10m. The hydraulic head at the pipe tip is kept equal to 15m. The bottom side of the mesh and its left vertical side (representing a line of symmetry of the problem) are impervious, while the hydraulic head is imposed on the remaining sides.

The problem was solved by considering a reduction of two orders of magnitude of the coefficient of permeability at the end of grouting (k_g=4·10^{-5}cm/sec) with respect to the initial permeability of the natural soil (k_i=4·10^{-3}cm/sec).

The results of calculations are summarized in Figure 6 that shows the contour lines of the hydraulic head and of the concentration for different time values. They indicate that the formation of the silica gel, and the parallel reduction of the coefficient of permeability, induces a concentration of the hydraulic gradient in the vicinity of the pipe tip.

Another consequence of this reduction is the limited spreading of the solution within the granular soil. In fact, in spite of the provision adopted to delay this effect (based on the use of phosphoric acid to modify the solution pH), the formation of the silica gel initiates since the early stages of the injection. This prevents an extensive spreading of the solution within the soil surrounding the pipe.

It should be observed that some additional calculations, performed by doubling the increment between in situ and injection fluid pressures did not lead to an appreciable increment of the size of the grouted zone.

On this basis it could be concluded that a further delay between the beginning of grouting and the initiation of the silica gel formation seems necessary to increase the spreading of the solution and, hence, to improve the effectiveness of the treatment.

7 CONCLUSIONS

A numerical study has been presented concerning the grouting of granular soils by means of silicate solutions.

This study originates from the increase of the water table level within the urban area of Milan (northern Italy), which led to the leakage of ground water within underground structures and subway tunnels.

To limit this negative effect, the Milan Metro (Metropolitana Milanese) undertook a technical investigation to evaluate the performance of some provisions aimed at reducing the mentioned water leakage. One of them is based on the grouting of the granular soil surrounding the structures, to reduce its "natural" coefficient of permeability.

The numerical treatment of this problem was approached by considering it as equivalent to the advective, dispersive and diffusive flow, through a saturated porous medium, of a liquid containing a solute.

A finite element approach was developed for the time integration of this problem. It turned out that a simple two-step iterative procedure, and a suitable choice of the integration parameter ϑ, permits to follow both the spreading of the solution within the granular soil and the time dependent reduction of its permeability, drastically reducing the oscillations of results.

A series of laboratory grouting tests permitted to derive an empirical law, expressing the reduction of the coefficient of permeability with time and concentration, that was adopted in the numerical simulation of a grouting injection.

The results of calculations permit to evaluate the size of the zone where the spreading of the solution and, hence, the reduction of permeability, takes place. This information could be useful in the design of actual in situ treatments and for evaluating their overall costs.

It was observed that, for the examined problem, the rapid decrease of the coefficient of permeability during time leads to a limited extension of the grouted zone.

To limit this drawback it seems necessary to further delay the initiation of the silica gel formation with respect to the beginning of the in situ treatment.

ACKNOWLEDGEMENTS

The geotechnical investigation was carried out at the Laboratory of the Department of Structural Engineering of the Politecnico di Milano. The assistance of E. Iscandri is gratefully acknowledged.

REFERENCES

Bear J. 1961. On the tensor form of dispersion. *Journal of Geophysics Research* 66(4): 1185-1197.

Bear J. & Verruijt A. 1987. *Modeling groundwater flow and pollution*. Dordrecht: D.Reidel Publishing.Co.

Cividini A. 2000. A study of the low pressure grouting of granular soils. *Proc. .of the John Booker Memorial Symposium, Sydney;* Rotterdam: Balkema.

Colombo A. 1997. The increase of the phreatic level in Milan (in Italian), *Il Giornale dell'Ingegnere* 19.

Grimoldi A. 1998. *Reduction of the permeability of the granular soil in the vicinity of underground structures through the injection of silicate solutions (in Italian)*. Laurea Thesis, Department of Applied Chemical Physics, Politecnico di Milano.

Karol R.H. 1965. Chemical grouting technology. *Journal of Soil Mechanics and Foundation Engineering Division, ASCE* 91: 41-65.

Massone G. 1998. *A technique for reducing the permeability of granular soil (in Italian)*. Laurea Thesis, Department of Structural Engineering, Politecnico di Milano.

Nonveiller E. 1989. *Grouting theory and practice*. Elsevier.

Peters G.P. & Smith D.W. 2000. *Numerical study of boundary conditions for solute transport through a porous medium.* Research Report No. 188.02.2000, Dept. of Civil, Surveying & Environmental Engineering, The University of NewCastle, Australia.

Rowe R.K. & Booker J.R. 1985. 1-D pollutant migration in soils of finite depth. *Journal of Geotechnical Engineering, ASCE* 111(4): 479-499.

Rowe R.K. & Booker J.R. 1997. Recent advances in modelling of contaminant impact due to clogging. *Proc.9th Int.Conf. of International Association for Computer Methods and Advances in Geomechanics; Wuhan, People Republic of China*: 43-56, Rotterdam: Balkema.

Wendland E. & Schmid G. 2000. A symmetrical streamline stabilization scheme for high advective transport. *International Journal for Numerical & Analytical Methods in Geomechanics* 24: 29-45.

Computer Methods and Advances in Geomechanics, Desai et al. (eds) © 2001 Balkema, Rotterdam, ISBN 90 5809 183 X

Coupled ground-water processes in porous fractured media

Günther Schmid & Britta Rosen
Department of Civil Engineering, Ruhr University Bochum, Germany

ABSTRACT: In this paper partly saturated flow and solute transport in porous fractured media is addressed and the relevant physical properties are discussed. Under the assumption of a linear rock mass and nonlinear mechanical behavior or fractures interaction between goundwater flow and solid deformation is considered. As a numerical application a sandstone block with and without a fracture is analyzed.

1 INTRODUCTION

Soil and rock can be modelled as porous and fractured media where pores and fractures are filled with fluids, in this case water and air. Acting forces will produce a motion in the system resulting in a deformation of the solid part and a flow of the fluid parts. In general these physical processes are coupled. In this presentation the solid part is assumed to consist of porous fractured rock. Loose soil is not considered here.

2 FLOW IN POROUS FRACTURED MEDIA

Many types of natural rock, as for example sandstone, are porous and contain as well macro–fractures. The diameter of the pores is in the order of 10^{-6} m to 10^{-5} m; the aperture of the fractures is in the order of 10^{-4} m to 10^{-3} m. As the permeability of a rock depends on these geometric quantities, the transmissivity of the fracture is in general several orders of magnitude larger than the transmissivity of the porous rock matrix. The network of fractures form dominant flow paths or pronounced channels within the fracture, through which a large part of the water is transported.

The volumetric part of voids formed by the pores is essentially higher as that arising from the fractures. This correlation remains even if the rock is strongly fractured. Therefor the storage capacity of the porous rock matrix is much larger than the storage capacity of the fractures.

The combination of the dominant storage capacity of the rock matrix with the larger permeability of the fractures creates a hydraulic system with short time delay in its response and with large storage capability.

Assuming laminar flow, the flow can be described by Darcy's law:

$$q_i = -k_{ij}\frac{\varrho g}{\mu}\frac{\partial}{\partial x_j}\left(\frac{p}{\varrho g}+x_3\right) \quad . \qquad (1)$$

The Darcy velocity q_i is proportional to the gradient of the hydraulic head, consisting of a part due to water pressure p and a part due to its geodetic height x_3.

The permeability tensor k_{ij}, the water density ϱ and the dynamic viscosity μ are material properties. The gravity force is denoted by g.

With the assumption of laminar flow between two parallel planar smooth plates the flow velocity in the fracture can also be described by Darcy's law in a generalized form:

$$q_i = -\frac{b^2}{12}\frac{\varrho g}{\mu}\frac{\partial}{\partial y_i}\left(\frac{p}{\varrho g}+x_3\right) \quad . \qquad (2)$$

Here the transmissivity is determined by the square of the hydraulic aperture b. The hydraulic aperture is an equivalent aperture determined empirically from the physical aperture m. Barton (1982) gives the following relation:

$$b = \frac{m^2}{JRC^{2,5}} \quad . \qquad (3)$$

In equation (2) the coordinates y_i are oriented parallel (y_1, y_2) and normal (y_3) with respect to the plane of the fracture.

The flow across the fracture and the flow in the porous rock are connected by the law of balance of mass. The water pressure across the fracture is continous. The tangential velocity component is discontinous.

The equation of balance of mass leads to:

$$\frac{\partial S}{\partial t} + \frac{\partial}{\partial x_i}(q_i \varrho) = Q \varrho^Q \quad , \tag{4}$$

where S is the storage capacity,
q_i the Darcy velocity,
ϱ the water density,
Q the source term (m^3/s) and
ϱ^Q the density of the source.

The time rate of the storage term consists of a part arising from change of matrix volume and a part arising from change of water volume due to change of pressure. In the fully saturated case one obtains for porous media:

$$\frac{\partial S}{\partial t} = (\alpha n + \beta(1 - n))\frac{\partial p}{\partial t} \tag{5}$$

For the porous rock α is the the bulk modulus. For the fracture α corresponds to change of fracture aperture due to pressure change. β is the modulus of compression of water.

3 SOLUTE TRANSPORT

We consider solute transport in ground water. Its concentration c [kg/m^3] is described by the transport equation

$$\begin{aligned} &(nS^w \varrho)\frac{\partial c}{\partial t} + nS^w \varrho \frac{\partial}{\partial x_i}(v_i c) \\ &- \frac{\partial}{\partial x_i}\left(nS^w \varrho\left(D_{ij} + D_{mol}\delta_{ij}\right)\frac{\partial}{\partial x_i}c\right) \\ &= q(c^* - c) \quad , \end{aligned} \tag{6}$$

where non–conservative terms as sorption, radioactive decay, biological degradation and hydro–chemical relations are not considered.

The first term in equation (6) repesents the mass storage, followed by advective and diffusive–dispersive flux terms with dispersion tensor D_{ij}, diffusion coefficient D_{mol}, porosity n, water saturation S^w and distance ground–water velocity v_i. On the right of equation (6) q is the source term with concentration c^*.

Within the fractures the mass transport is dominated by advective processes due to the high flow velocities.

In the porous rock the diffusive transport dominates due to the low flow velocities in the pores. Solutes are mainly transported in the dominant flow paths in the fractures resulting in a high concentration gradient between the rock matrix and the fracture. Along this gradient the solute diffuses from the fracture into the matrix.

After the solute in the fracture is transported away, the concentration gradient is reversed. Diffusion from the porous rock into the fracture occurs. This results in tailing effects in the mass transport accompanied with a reduction of the maximum concentration at the transport front and in a retardation effect after the source has ended (Himmelsbach & Wendland (1999); see Fig. 1). As the sorption in the porous rock is essentially higher than in the fracture, the tailing effects are even more pronounced.

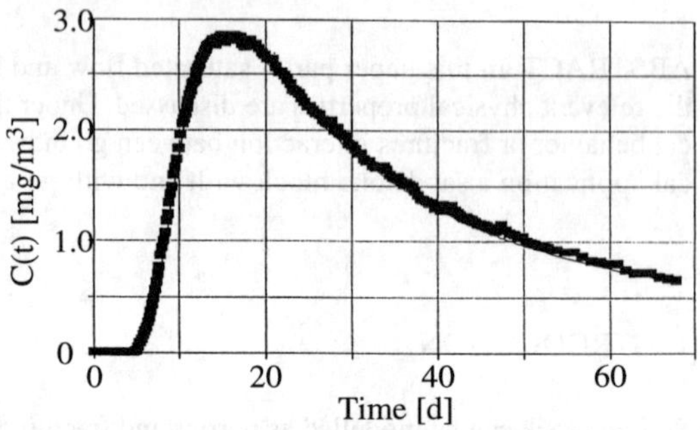

Figure 1. Results of tracer test in a fractured sandstone; Himmelsbach & Wendland (1999).

4 UNSATURATED FLOW

Capillary forces pull the water into the pores above ground–water table which is defined as the free surface with atmospheric pressure. The different pore diameters pull the water to different heights above ground–water table. The region partly filled with water, partly filled with air is the partially saturated zone, where saturation is defined as the fraction of water volume of the total void volume.

The dependency of the saturation of capillary pressure is a characteristic soil property which has to be obtained experimentally. Examples of such functions are e.g. given by Brooks and Corey (1964) or van Genuchten (1980). If the voids are filled with more than one fluid phase the permeability of both phases decrease. The phases impede each other in their flow. The reduction of the permeability of one phase with decreasing saturation is represented by multiplying the permeability with the relative permeability. If one assumes that air can move unimpeded through the pores, the pore pressure will be equal to the atmospheric pressure everywhere. In this case the ground–water flow can be analyzed independent of the air flow and the negative capillary pressure corresponds to the water pressure. The stated assumption can be considered as a sufficient accurate approximation to analyze ground–water flow.

The generalized Darcy's law given in (1) leads with the relative permeability k^{rel} to:

$$q_i = -k_{ij}\,k^{rel}\,\frac{\varrho g}{\mu}\frac{\partial}{\partial x_j}\left(\frac{p}{\varrho g} + x_3\right) \qquad (7)$$

The storage term is extended to:

$$\frac{\partial S}{\partial t} = \varrho n\frac{\partial S^w}{\partial p} + S^w(\alpha n + \beta(1-n))\frac{\partial p}{\partial t} \qquad (8)$$

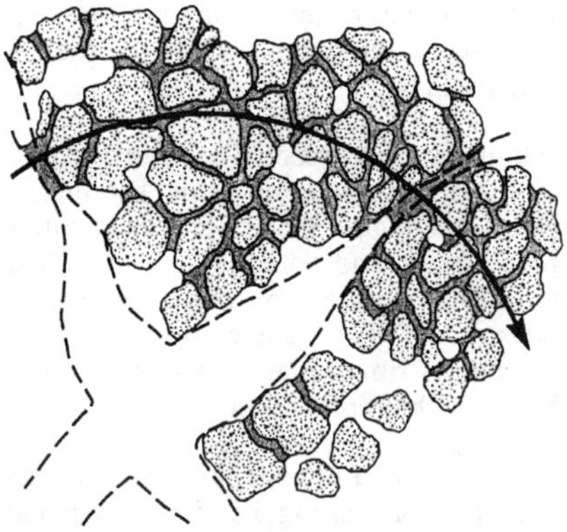

Figure 2. Schematic representation of flow in partially saturated porous fractured media; Wang & Narasimhan (1985).

The capillary pressure in the fractures is much smaller than in the pores, as the cavity of a fracture compared with the diameter of the pores is much larger. In addition water flow in a fracture is only possible, if the wetted regions are interconnected. If the water connection is interrupted the fracture acts as flow barrier. The hydraulic properties of the fractured porous rock in a partially saturated region are determined by the different capillary pressures in the pores and in the fractures. In contrast to the fully saturated flow, where the fractures form the main flow paths, the fractures act as flow barrier in the partly saturated regions. Water is retained only in zones of small aperture. The air in between forms in the direction of the fracture a complete flow resistance. Normal to the fracture water flows through bridges of pore material and water at zones of contact in the fracture.

5 MULTIPHASE FLOW

Flow of several non miscible fluids in pores and fractures are described by multiphase flow. The fluids may be gaseous or liquid. The essential difference of the gaseous and liquid phase is its compressibility. The character of a multiphase flow consisting of two liquid phases, as e.g. water and mineral oil, is essentially different from that of flow consisting of a liquid and a gaseous fluid, as e.g. water and air. Simulation strategies for two phase flows are given by Riedel (1998) and Flensberg (1999).

The flow for each phase α of n non–miscible fluids is given by the equation

$$\frac{\partial n S^\alpha \varrho^\alpha}{\partial t} - \frac{\partial}{\partial x_i}\left(\frac{\varrho^\alpha}{\nu^\alpha}k^{rel,\alpha}K_{ij}\frac{\partial}{\partial x_j}(p^\alpha g x_3)\right) \qquad (9)$$

$$-\varrho^\alpha q^\alpha = 0 \quad , \quad \alpha = 1,..,n$$

where the flow phases are coupled through the condition

$$\sum S^\alpha = 1$$

6 FLUID – SOLID INTERACTION

The ground–water flow is strongly influenced by the deformation state of the rock. In the porous matrix the deformation state above all influences the storage capacity but also its permeability which decreases with larger rock pressure. This effect, however, is only remarkable for large pressure changes. Therefore the deformation state influences above all – through the storage capacity – the time dependent behavior of flow in porous rock.

The interaction of flow and stress state is especially strong, where time–depending loads and flow boundary conditions lead to time–depending solutions. Already small changes in the state of deformation of the rock mass results in an essential change of permeability in the fracture, as those are approximatively dependent upon the square of its aperture. The storage capacity of the fracture is also directly influenced by stress changes. But this has only a minor influence upon the total system response.

On the other hand the ground–water flow influences the stress and deformation state of the rock mass. The loads of the system, i.e. dead loads, surface loads and tectonic stresses, are not only carried by the rock matrix but also partially by the water in the pores and fractures. In the fractures the water pressure acts against the fracture closing rock pressure. If the water pressure becomes larger than the rock pressure normal to the fracture the tension strength of the fracture may be exceeded leading to a separation and enlargement of the fracture or to the formation of new fractures.

The differential equations of the coupled hydro–mechanical processes are for the saturated porous media:

$$-\frac{\partial}{\partial x_i}\left(k_{ij}\frac{\varrho^w g}{\mu}\frac{\partial}{\partial x_j}\left(\frac{p}{\varrho^w g} + x_3\right)\right)$$
$$+\alpha\frac{\partial}{\partial t}\frac{\partial u_i}{\partial x_i} + \left(\frac{(\alpha - n)}{K^s} + \frac{n}{K^w}\right)\frac{\partial p}{\partial t} = Q\,, \qquad (10a)$$

$$G\frac{\partial^2 u_i}{\partial x_j \partial x_j} + (G+\lambda)\frac{\partial^2 u_j}{\partial x_i \partial x_j} - (1 - \frac{K^d}{K^s})\frac{\partial p}{\partial x_i} + F_i = 0 \quad , \qquad (10b)$$

and for the saturated fracture:

$$-\frac{\partial}{\partial y_i}\left(\frac{b^2}{12}\frac{\varrho^w g}{\mu}\frac{\partial}{\partial y_i}\left(\frac{p}{\varrho^w g} + x_3\right)\right) + \frac{1}{K^w}\frac{\partial p}{\partial t} + \frac{\partial}{\partial t}\frac{\partial u_i}{\partial x_i} = Q \quad , \qquad (11a)$$

$$\frac{\partial}{\partial x_j}[D_{ijkl}\frac{1}{2}\left(\frac{\partial u_k}{\partial x_l} + \frac{\partial u_l}{\partial x_k}\right) - (1 - \frac{K^d}{K^s})\delta_{ij}p] + F_i = 0 \quad , \qquad (11b)$$

where u_k are the displacements, G and λ are Lamé's constants, and K^d and K^s are the bulk moduli of the rock structure and rock material, respectively. D_{ijkl} represents the material properties of the fractures and F_i are body forces. The complete description of the interaction of the hydro–mechanical system in partially saturated porous fractured rock are given in Rosen (2000).

7 NUMERICAL MODEL

The discretization of the differential equations may be performed with different numerical procedures. The selection of a method depends on the physical processes, the size of the region, available computer techniques and the accuracy aimed for. Different models may be used depending upon the physical properties of the rock, the number and geometry of the fractures, the model size and scale lengths.

In case of numerous equally distributed fractures with similar aperture and with a model region being large compared to fracture length, the system can be described by an analogous continuum model. Such a model is numerically simple and stable. The model creation is straight forward as discrete fractures have not to be represented. A disadvantage arises in the analysis of transport processes as the concentration fronts are smeared and do not follow the physical flow paths.

Under similar geometrical conditions, as stated above, the double–porosity model gives more refined results, as the fractures and the porous matrix are represented in two continuum models, respectively and connected through exchange terms. The advancing concentration front, arising from the high advective part of the fractures can be modelled in an average way through this procedure. A problem doing this is the determination of the exchange term.

If the physical processes are determined almost exclusively through dominant fractures a discrete fracture model should be chosen. Here the transmissivity of the fractures has to be determined according to equations (2) and (3) and the ground–water flow is modelled as a flow system of interacting fractures. This is appropriate in cristaline rock, where the rock matrix permeability can be neglected.

If the rock matrix also essentially influences the flow processes the discrete fracture model should be combined with a rock matrix model. The rock matrix with its micro–fractures and pores is then modelled as a continuum. The complete system is known as fracture–matrix model. Its advantage is that potential and concentration distribution are not smeared over and local effects can be representend.

For the discrete fracture model and the fracture–matrix model the preprocessing, i.e. the mesh generation, is of importance as the geometry of the fractures and its intersections make the mesh geometry very complicated. Roehrig (1999) gives an effective data structure with an algorithm capable of generating finite element models of complicated geological structures.

Due to its irregulare geometry the fractures act in many cases not as two–dimensional flow planes but form one dimensional flow channels where most part of the flow and transport occurs. This is particularly the case in fracture intersection.

8 APPLICATION

A block of sandstone rigidly supported along its lower side through an impermeable bearing is subjected to its dead weight and pressure P. Boundary conditions and material data are given in Figure 3. The flow boundary conditions are chosen in such a way that without the influence of the deformation state a constant potential will be produced. The block is fully saturated. Initially at all nodal points the displacements are zero and the potential is specified as h = 2 m. For the deformation law of the fracture a logarithmic relation between fracture opening Δu and normal stress σ_n across the fracture is assumed resulting in a fracture normal stiffness of

$$K_n = 2,8087 \cdot 10^{6,987 \cdot \Delta u}$$

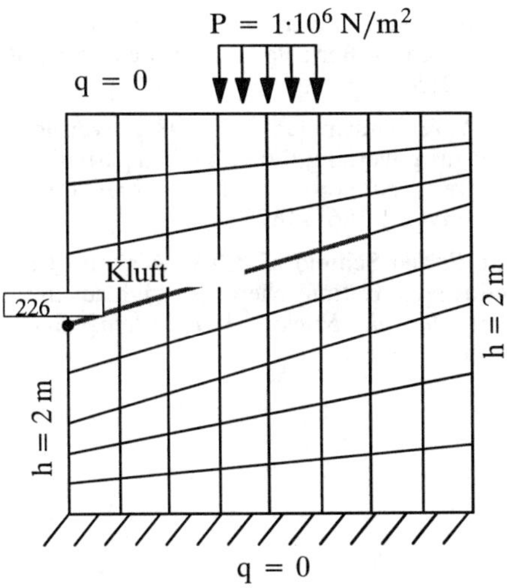

Permeability	$k = 10^{-8}$ m/s
Porosity	$n = 0,06$
Bulk modulus	$K_d = 8,3 * 10^7$ N/m^2
Poisson's ratio	$\nu = 0,30$
Density sandstone	$\gamma = 25$ kN/m^3

Initial fracture aperture	$\Delta u_0 = 500$ µm
Initial deformation	$u_{01} = u_{02} = 0$ m
Initial potential	$h_0 = 2$ m

Figure 3. Model of sandstone block with fracture.

The shear stiffness of the fracture is obtained according to an empiric relation as a function of the Joint Roughness Coefficient (JRC) and the mean Joint Compressive Strength (JCS) (Bandis et al. 1983):

$$K_s = 100\sigma_n \tan\left[\mathrm{JRC}\log\left(\frac{\mathrm{JCS}}{\sigma_n}\right) + 30\right] \quad (12)$$

In a first analysis the block is assumed without fracture. In a second analysis an oblique fracture through the block is assumed.

The mechanical load of the rock is carried by the matrix skeleton and the water. The initially large pore pressure due to the load decreases with time, since the water can flow of. In the same way as the water pressure decreases, the load is taken by the rock matrix. Simultaneously the matrix displacements increase and approach asymptotically the final state. Figure 4 shows the time history of the corresponding fluid potential in the block, without and with fracture.

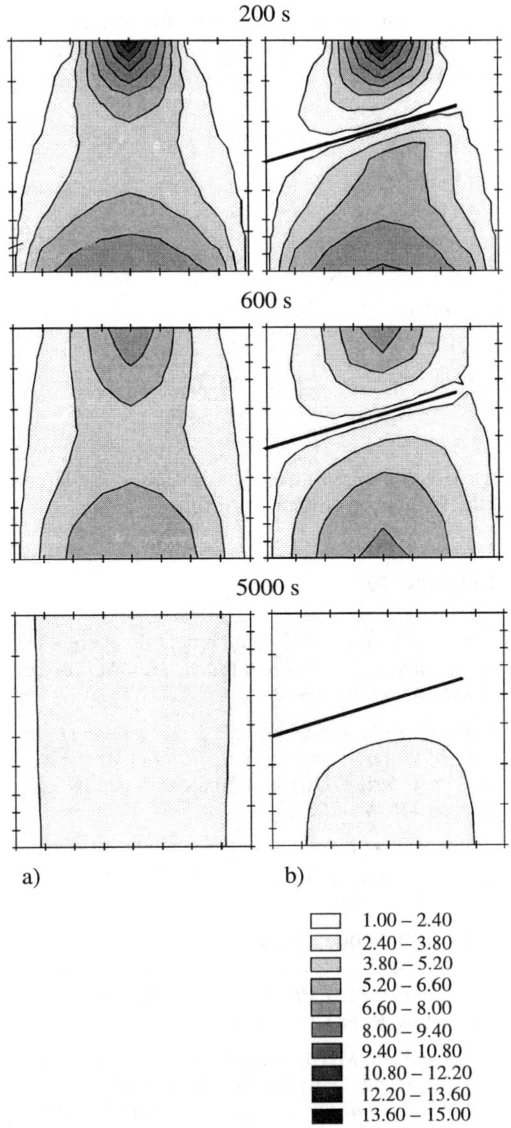

Figure 4. Development of the potential [m] in a) the porous sandstone block, b) the porous and fractured sandstone block after different elapsed times.

Figure 4 also recognizes the compensating effects of the fracture. Due to its large transmissivity the potential and pressure differences are immediately cancelled out. The fracture acts as drainage. The rise of potential and pore water pressure below the fracture is due to the supporting forces on the lower part of the block.

Figure 5 shows the aperture of the fracture at its most left end as a function of time. The fracture deformation asymptotically approaches its final value.

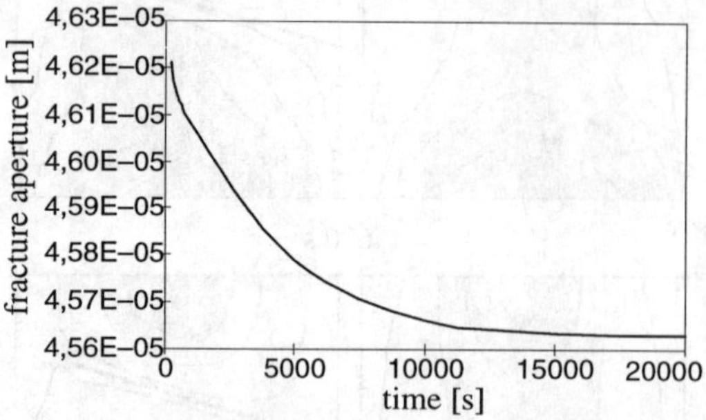

Figure 5. Aperture of the fracture at the most left end of the fracture as a function of time.

REFERENCES

Bandis, S.C. et al. 1983. Fundamentals of rock joint deformation. *Int. J. Rock Mech. Min. Sci. & Geomech. Abstr.* 20:249–268.

Barton, N. 1982. *Modeling rock joint behavior from in situ block tests; implications for nuclear waste repository design.* Office of Nuclear Waste Isolation. Report ONWI–208.

Brooks, R.H. & Corey, A.T. 1964. *Hydraulic properties of porous media.* Hydrology Papers. Colorado State University.

Flensberg, D. 1997. *Numerical treatment of a two–phase flow in a convection–dominated nonlinear case* (in German). Ph.D. thesis. Department of Mathematics. Ruhr University Bochum.

van Genuchten, M.Th. 1980. A closed–form equation for predicting the hydraulic conductivity of unsaturated soils, *Soil Sci. Soc. of Am. J.* 44:892–898.

Himmelsbach, Th. & Wendland, E. 1999. *Heavy metal transport in sandstone under high saline pore water conditions – experiment and modelling* (in German). Grundwasser. 03/99.

Riedel, D. 1998. *Analysis of the propagation of non–miscible fluids in heterogenous porous media* (in German). Ph.D. thesis. Department of Civil Engineering, Ruhr University Bochum.

Roehrig, J. 1999. *Data structures for solid modelling and mesh modelling* (in German). Ph.D. thesis. TWM Nr. 99–2. Department of Civil Engineering. Ruhr University Bochum.

Rosen, B. 2000. *Numerical modelling of interaction of saturated/unsaturated ground–water flow and deformations in fractured rock* (in German). Ph.D. thesis. Fortschritt–Berichte VDI. Dusseldorf. ISBN 3–18–322215–9.

Wang, J.S.Y. & Narasimhan, T.N. 1985. Hydrologic mechanisms governing fluid flow in a partially saturated, fractured, porous medium. *Water Resources Research.* 21:1861–1874.

Wendland, E. and Schmid, G. 2000. A symmetrical streamline stabilization scheme for high adventive transport. *Int. J. Numer. Anal. Meth. Geom.* 24:29–45.

Computer Methods and Advances in Geomechanics, Desai et al. (eds) © 2001 Balkema, Rotterdam, ISBN 90 5809 183 X

On some recent developments in poroelasticity

A.P.S. Selvadurai
Department of Civil Engineering and Applied Mechanics, McGill University, Montreal, Que., Canada

ABSTRACT: This paper deals with the application of Biot's classical three-dimensional theory of poroelasticity to some novel problems with applications to geomechanics. In particular we consider the computational treatment of quasi-static crack extension in fluid saturated media and mechanics of continuum damage evolution and the attendant alterations in both the fluid transport characteristics and elastic stiffness on the resulting response of poroelastic materials. Examples are provided to illustrate the application of the various developments to specific problems of interest to geomechanics.

1 INTRODUCTION

The classical problem in geomechanics dealing with the stress-induced isothermal flow of a fluid through a deformable porous medium, referred to, as the process of soil consolidation, is a cornerstone in the development of "soil mechanics". Although the development of the one-dimensional theory of soil consolidation is generally attributed to Terzaghi (1923), the important contributions of others and Fillunger (1913) in particular, are now being recognized. The formal theory of soil consolidation presented by Biot (1941) is a generalization of the one-dimensional theory to three dimensions. This is a complete theory, which is rigorous in the sense of a continuum formulation applicable to a medium with voids. It is also an elegantly simple theory that continues to flourish after nearly six decades. The theory employs linear elasticity to characterize the mechanical behaviour of the porous soil skeleton and Darcy's law governs fluid flow through the porous space. Extensive expositions of both fundamental aspects of Biot's theory of poroelasticity and its applications to the analytical solution of problems in geomechanics are documented Mandel (1953), de Josselin de Jong (1953), McNamee and Gibson (1960), Cryer (1963), deWiest (1969), Agbezuge and Deresiewicz (1974), Szefer and Gaszynski (1975), Chiarella and Booker (1975), Atkin and Craine (1976) Rice and Cleary (1976), Booker and Small (1975,1984), Bedford and Drumheller (1878,1983), Booker and Randolph (1984), Bear and Corapcioglu (1985),Whitaker (1986), Kassir and Xu (1988),Ehlers (1989), de Boer and Ehlers (1990), Bear and Bachmat (1990), Tsang (1991) and Detournay and Cheng (1993), Selvadurai and Yue (1994), Yue and Selvadurai (1994,1995) and Lan and Selvadurai (1996). The volumes by Coussy (1995), Selvadurai (1996), Cheng et al. (1998), Thimus et al. (1998), Lewis and Schrefler (1998) Drew and Passman (1991) and de Boer (2000) also give recent advances in the applications of Biot's theory. These references, along with the review articles by Scheidegger (1960), Paria (1963) and Schiffman (1985), inter alia, contain a significant number of references to further work concerning the application of Biot' classical theory of three-dimensional soil consolidation, via both analytical and computational modelling, to a wide range of problems including geomechanics of energy exploration, environmental geomechanics, biomechanics, materials science and materials engineering.

In this paper we apply Biot's theory of poroelasticity to certain non-classical problems in geomechanics. These include the study of the quasi-static extension of a plane crack in a poroelastic medium the examination of the influence of micro-mechanical damage and the attendant alterations in the fluid transport characteristics of the poroelastic medium on the deformational response of a half-space region. These problems are investigated by appeal to computational modelling of the poroelastic response. Numerical results are presented in the paper to illustrate the influence of the non-classical effects.

2 GOVERNING EQUATIONS

The basic equations governing Biot's formulation of the isothermal theory of poroelasticity are well known and are summarized here for completeness. The constitutive equations governing the quasi-static response of a poroelastic medium, which consists of a porous isotropic soil skeleton saturated with a compressible pore fluid, take the forms

$$\boldsymbol{\sigma} = 2\mu\,\boldsymbol{\varepsilon} + \frac{2\mu\nu}{(1-2\nu)}(\nabla.\mathbf{u})\mathbf{I} + \alpha\, p\,\mathbf{I} \tag{1}$$

$$p = \beta\zeta_v + \alpha\,\beta(\nabla.\mathbf{u}) \tag{2}$$

where $\boldsymbol{\sigma}$ is the total stress dyadic, p is the pore fluid pressure, ζ_v is the volumetric strain in the compressible pore fluid; ν and μ are the "drained values of Poisson's ratio and the linear elastic shear modulus applicable to the porous fabric and $\mathbf{I}(=\mathbf{ii}+\mathbf{jj}+\mathbf{kk})$ is the unit dyadic. In (1), $\boldsymbol{\varepsilon}$ is the soil skeletal strain dyadic, which is defined by

$$\boldsymbol{\varepsilon} = \frac{1}{2}(\nabla\mathbf{u} + \mathbf{u}\nabla) \tag{3}$$

and $\mathbf{u}$ is the displacement vector. The material parameters α and β which define, respectively, the compressibility of the pore fluid and the compressibility of the soil fabric are given by

$$\alpha = \frac{3(\nu_u - \nu)}{\tilde{B}(1-2\nu)(1+\nu_u)} \tag{4}$$

$$\beta = \frac{2\mu(1-2\nu)(1+\nu_u)^2}{9(\nu_u-\nu)(1-2\nu_u)} \tag{5}$$

where ν_u is the undrained Poisson's ratio and $\tilde{B}$ is the pore pressure parameter introduced by Skempton (1954). The effective stress dyadic $\boldsymbol{\sigma}'$ of the porous skeleton is given by

$$\boldsymbol{\sigma}' = \boldsymbol{\sigma} - \alpha\, p\,\mathbf{I} \tag{6}$$

In the absence of body forces, the quasi-static equations of equilibrium for the complete fluid saturated porous medium take the form

$$\nabla.\boldsymbol{\sigma} = \mathbf{0} \tag{7}$$

The velocity of fluid transport within the pores of the medium is governed by Darcy's law,

$$\mathbf{v} = -\kappa\nabla p \tag{8}$$

where $\mathbf{v}$ is the vector of fluid velocity and κ $(= k/\gamma_w)$ is a permeability parameter which is related to the hydraulic conductivity k and the unit weight of the pore fluid γ_w as indicated above.

The mass conservation equation for the pore fluid can be stated as an equation of continuity, which, for quasi-static flows can be stated in the form

$$\frac{\partial\zeta_v}{\partial t} + \nabla.\mathbf{v} = 0 \tag{9}$$

Considering the thermodynamic requirements for a positive definite strain energy potential (see e.g. Rice and Cleary, 1976) it can be shown that the material parameters should satisfy the following constraints:

$$\mu > 0\,;\, 0 \le \tilde{B} \le 1;\, -1 < \nu < \nu_u \le 0.5;\, \kappa > 0 \tag{10}$$

The governing equations for a poroelastic medium can be reduced to the following:

$$\mu\nabla^2\mathbf{u} + \frac{\mu}{(1-2\nu)}\nabla(\nabla.\mathbf{u}) + \alpha\nabla p = \mathbf{0} \tag{11}$$

and

$$\kappa\,\beta\nabla^2 p - \frac{\partial p}{\partial t} + \alpha\beta\frac{\partial}{\partial t}(\nabla.\mathbf{u}) = 0 \tag{12}$$

for the displacement vector $\mathbf{u}$ and the scalar pore fluid pressure p and

$$\nabla.\nabla = div(\,grad\,) = \nabla^2. \tag{13}$$

To complete the mathematical formulation of the problem it is necessary to prescribe boundary conditions and initial conditions applicable to the dependent variables $\mathbf{u}$ and p. The boundary conditions can be posed in terms of the conventional Dirichlet, Neumann or Robin type boundary conditions applicable to $\mathbf{u}$ and p.

3 FINITE ELEMENT FORMULATIONS

Finite element methods have been widely applied for the analysis of problems in poroelasticity. The earliest of these studies is due to Sandhu and Wilson (1969) with other noteworthy studies by Ghaboussi and Wilson (1973), Booker and Small (1975), Verruijt (1979), Huyakorn and Pinder (1983), Valliappan et al. (1984) and Simon et al. (1986). Complete discussions of the computational aspects of poroelasticity are also given by Desai and Christian (1977), Smith and Griffiths (1988), Britto (1992), Selvadurai (1996), Cheng et al. (1998), Thimus et al. (1998), Lewis and Schrefler (1998) and de Boer (2000). In the ensuing we shall present a brief overview of the basic aspects of the finite element approach to the modelling of classical poroelasticity and supplement these developments as and when extensions are made to accommodate special aspects of problem areas of interest. The Galerkin approximation technique is applied to transform the partial differential equations into a discretized matrix form. The approximations used for the displacements **u** and the pore pressure p can be obtained through

$$\mathbf{u} = N^u\{\mathbf{u}\} \quad ; \quad p = N^p\{p\} \tag{14}$$

where $\{\mathbf{u}\}$ and $\{p\}$ are the nodal displacements and pore pressure vectors and N^u and N^p correspond to the nodal shape functions for the displacement and pore pressure fields respectively. In general N^u and N^p can be different but both must satisfy C^0 continuity.

The application of the Galerkin procedure to the governing equations gives rise to the following incremental forms for the equations governing poroelastic media:

$$\begin{bmatrix} \mathbf{K} & \mathbf{C} \\ \mathbf{C}^T & \{-\gamma\Delta t\mathbf{H} + \mathbf{E}\} \end{bmatrix} \begin{Bmatrix} \mathbf{u}_{t+\Delta t} \\ p_{t+\Delta t} \end{Bmatrix} = \begin{bmatrix} \mathbf{K} & \mathbf{C} \\ \mathbf{C}^T & \{(1-\gamma)\Delta t\mathbf{H} + \mathbf{E}\} \end{bmatrix} \begin{Bmatrix} \mathbf{u}_t \\ p_t \end{Bmatrix} + \{\mathbf{F}\} \tag{15}$$

where **K** is the stiffness matrix of the soil skeleton; **C** is the stiffness matrix due to interaction between the soil skeleton and the pore fluid; **E** is the compressibility matrix of the pore fluid; **H** is the permeability matrix; **F** are force vectors due to external tractions, body forces and flows; $\mathbf{u}_t$ and p_t are, respectively, the nodal displacements and pore pressure at time t; and Δt is the time increment. The time integration constant γ varies between 0 and 1. The criteria governing stability of the integration scheme given by Booker and Small (1975) requires that $\gamma \geq 1/2$. According to Selvadurai and Nguyen (1995) and Lewis and Schrefler (1998), the stability of the solution can be achieved by selecting values of γ close to unity (see also Wang et al.(2000)). In the ensuing we shall apply the classical theory of poroelasticity to the specific problems identified in the introduction.

4 STEADY CRACK EXTENSION IN A POROELASTIC BODY

In poroelastic and other materials, which exhibit dissipative phenomena, the extension of any defect can be dynamic and unsteady at the initial stages. If the geometry of the medium permits we can consider situation where for long durations (relative to some intrinsic time) the crack extension can be identified with an idealized steady state. For poroelastic materials, the flow of energy into the pore fluid can stabilize the crack growth to achieve a near steady state. Also to preserve conditions for steady state crack extension, the traction or the displacement distribution initiating the quasi-static crack growth should be time-invariant in a reference coordinate system, which moves with the crack tip. This is invariably associated with poroelastic media of infinite extent. Examples of developments along these lines are given by Palmer and Rice (1973), Rice and Cleary (1976), Rice and Simons (1976), Rudnicki and Koutsibelas (1991), Atkinson and Craster (1991), Craster and Atkinson (1996) and Selvadurai and Mahyari (1998). Although most applications involving steady crack growth deal with plane problems, Selvadurai and Mahyari (1998) have identified an axisymmetric problem, which gives rise to a steady growth situation.

Considering the two-dimensional case, we consider the situation where the tip of an open crack moves in the x-direction with velocity V. We consider the transformation

$$x = X - Vt \quad ; \quad z = Z \tag{16}$$

where x-z is a coordinate system, which moves with the crack tip. The resulting time-independent equations of poroelasticity take the form

$$\mu\hat{\nabla}^2\hat{\mathbf{u}} + \frac{\mu}{(1-2\nu)}\hat{\nabla}(\hat{\nabla}.\hat{\mathbf{u}}) + \alpha\hat{\nabla}p = \hat{\mathbf{0}} \tag{17}$$

$$\kappa\beta\hat{\nabla}^2 p + V\mathbf{i}\hat{\nabla}p - \alpha\beta V\,\mathbf{i}\,\hat{\nabla}(\hat{\nabla}.\hat{\mathbf{u}}) = 0 \tag{18}$$

where the operations involving $\hat{\nabla}, \hat{\nabla}^2, \hat{\mathbf{u}}$, apply only to the two-dimensional case. Since the time dependency is now eliminated through the transformation (16), for a well-posed boundary value problem we require only Dirichlet, Neumann or Robin-type boundary conditions to be prescribed through the dependent variables $\hat{\mathbf{u}}$ and p. The finite element procedures applicable to the study of this problem are described by Selvadurai and Mahyari (1998). The procedure accommodates the singular behaviour of the stress field at the crack tip by provision of special elements. The accuracy of the method has also been verified by comparison with certain analytical solutions given by Rice and Simons (1976) and boundary integral equation solutions given by Cheng and Liggett (1984). The computational procedure has been applied to examine the variation in the Mode I stress intensity

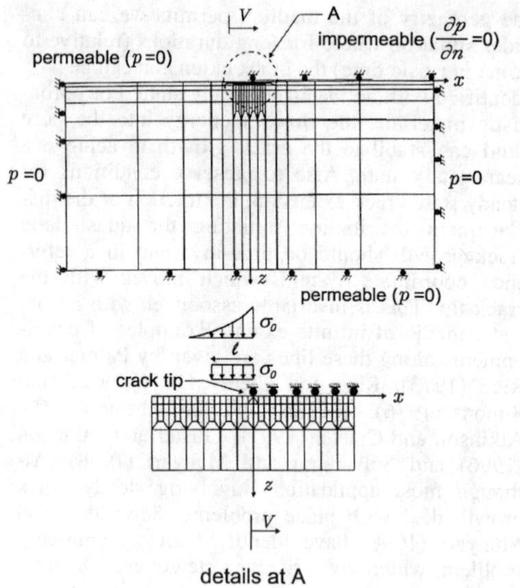

Figure 1. Finite element discretization for the steady crack extension problem

factor at the tip of a semi-infinite crack the faces of which are subjected to an exponentially decaying total normal stress

$$\sigma_{zz} = \sigma_0 e^{x/a} \quad ; x \in (-\infty, 0) \tag{19}$$

where a is a length parameter.

Figure 1 illustrates the typical mesh configuration employed in the computational modelling. The elements are plane and quarter-point isoparametric elements with the conventional specification of the pore pressures only at the corners. The computational procedure could be improved by a choice of an adaptive mesh configuration around the crack tip element; these refinements are best left for further studies. Figure 2 illustrates the normalized stress intensity factor at the crack tip as a function of the normalized propagation velocity aV/C where

$$C = \frac{2\kappa\mu(1-\nu)(1+\nu_u)^2}{9(1-\nu_u)(\nu_u-\nu)} \tag{20}$$

and for the computations we have set $\nu = 0.3$ and $\nu_u = 0.4$. The stress intensity factor for the poroelastic body is normalized with respect to the stress intensity factor for the plane elastic body under the same surface tractions (18). In this case it may be noted that the stress intensity factor is independent of the elastic constants of the material. i.e.

$$K_c = \sqrt{2a\sigma_0} \tag{21}$$

The second part of Figure 2 illustrates the variation of the pore fluid pressure ahead of the crack tip, for $\nu = 0.2$; $\nu_u = 0.3$ and $aV/C = 1$. It may be noted that at high values of aV/C, the pore pressure field at the crack tip exhibits oscillatory behaviour. As aV/C becomes small, the computational scheme yields results consistent with the analytical results of Atkinson and Craster (1991). In interpreting these results, we note that

$$\lambda = \frac{2\tilde{B}(1+\nu_u)(1-\nu)\sigma_0(K_I/K_c)}{3(1-\nu_u)}; \; X = \frac{Vx}{C} \tag{22}$$

In general the results indicate that significant pore fluid suction can develop ahead of the crack tip leading to alterations in the apparent strength of the material in that zone.

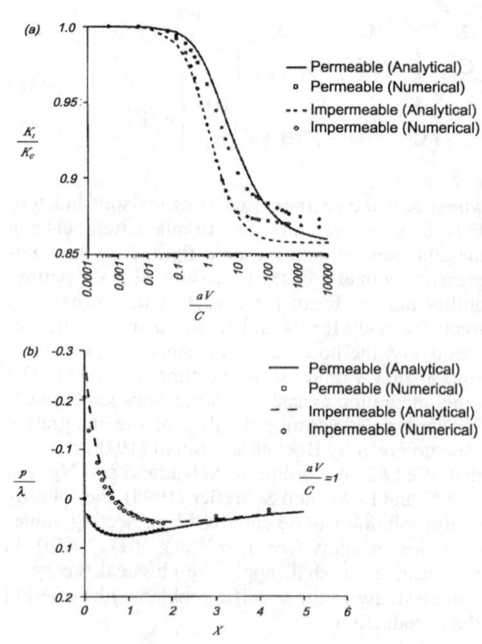

Figure 2. (a) Variation of the stress intensity factor K_I with $a_1 = aV/C$; and (b) pore pressure distribution ahead of the crack tip for normalized velocity $a_1 = 1$. (Analytical results due to Craster and Atkinson (1996))

5 POROELASTIC MEDIA SUSCEPTIBLE TO DAMAGE

In the classical theory of poroelasticity it is implicitly assumed that the porous structure remains unaltered during either loading or deformation of the poroelastic medium. In many instances this assumption is a satisfactory approximation, provided there are no appreciable alterations in the poroelastic constitutive response. Extreme examples of alterations of poroelastic behaviour can range from the development of irreversible plastic deformations in the porous soil skeleton (see e.g. Prevost (1984), Borja (1994)) to development of discrete fractures, which can introduce discrete discontinuities where the fluid transport characteristics are altered. Here, we are primarily concerned with geomaterials that maintain their predominantly elastic behaviour but suffer micro-mechanical damage. A result of such damage can be the inhomogeneous alteration of the elastic behaviour of the porous skeleton and the attendant alteration in the permeability characteristics of the medium. Such material characteristics alterations can be introduced in zones of poroelastic media, which experience continuum damage according to prescribed damage evolution laws. The subject of continuum damage mechanics is a well-researched area and the coupling of elasticity and damage modelling has been investigated by a number of researchers including Kachanov (1958), Sidoroff (1980), Mazars (1982), Krajcinovic (1984) and Lemaitre and Chaboche (1990). Attention is focused on the type of behaviour where the geomaterial experiences isotropic damage defined by a scalar damage variable D, which can be related to the initial area A_0 and the reduced area $\bar{A}$ through the relationship

$$D = \frac{A_0 - \bar{A}}{A_0} \tag{23}$$

The damage variable varies between 0 and D_c which is the critical value corresponding to the development of fracture of the material. (The critical damage parameter can be viewed as a normalizing parameter against which damage evolution can be estimated. For isotropic damage, the net stress dyadic $\boldsymbol{\sigma}^n$ is related to the stress dyadic in the undamaged state $\boldsymbol{\sigma}$ according to

$$\boldsymbol{\sigma}^n = \frac{\boldsymbol{\sigma}}{(1-D)} \tag{24}$$

Considering the 'strain equivalence hypothesis' proposed by Lemaitre (1984), the constitutive equation for the damaged skeleton of the poroelastic medium can be written as

$$\boldsymbol{\sigma} = 2(1-D)\mu\,\boldsymbol{\varepsilon} + \frac{2(1-D)\mu\nu}{(1-2\nu)}(\nabla.\mathbf{u})\mathbf{I} + \alpha\, p\,\mathbf{I} \tag{25}$$

which implies that Poisson's ratio remains constant, which is an added constraint when considering the material behaviour in three-dimensions. In addition to the specification of the constitutive relations for the damaged geomaterial skeleton, it is also necessary to prescribe damage evolution criteria which can be postulated either by appeal to micro-mechanical considerations or determined by experiment. For example, based on a review of the results of experiments conducted on rocks it has been shown (Cheng and Dusseault, 1993) that

$$\frac{\partial D}{\partial \xi_d} = \eta \frac{\gamma \xi_d}{(1+\xi_d)}\left(1 - \frac{D}{D_c}\right) \tag{26}$$

where ξ_d is related to the second invariant of the deviator strain dyadic and η and γ are positive material constants. The evolution of the damage variable can be obtained by the integration of (26) between limits D_0 and D where D_0 is the initial value of the damage variable corresponding to the intact state. (e.g. D_0 is zero for materials in a virgin state.) The deformability parameters applicable to an initially isotropic elastic material which experiences isotropic damage can be updated by adjusting the elastic constants in the elasticity matrix $\mathbf{C}$ by its equivalent applicable to the damaged state $\mathbf{C}^d$ but maintaining Poisson's ratio constant. Using the result (26), the evolution of D can be prescribed as follows:

$$D = D_c - (D_c - D_0)(1+\gamma\xi_d)^{\eta/\gamma D_c} .exp(-\eta\xi_d / D_c) \tag{27}$$

If damage in poroelastic media results in the creation defects such as micro-cracks and micro-voids, it is foreseeable that such defects will lead to alterations of the permeability characteristics of poroelastic media. The limited literature on permeability evolution in porous media is more focused on experimental evaluation of alteration in permeability of geomaterials, which are subjected to triaxial stress states. Zoback and Byerlee (1975) have documented results of experiments conducted on granite and Shiping et al. (1994) give similar results for tests conducted on sandstone Figure 3). Kiyama et al. (1996) also documents results of triaxial tests conducted on granite. While the general trends indicate alterations in the permeability characteristics with increase in stress/strain, it is not clear whether such alterations are restricted to zones where strain localization is present.

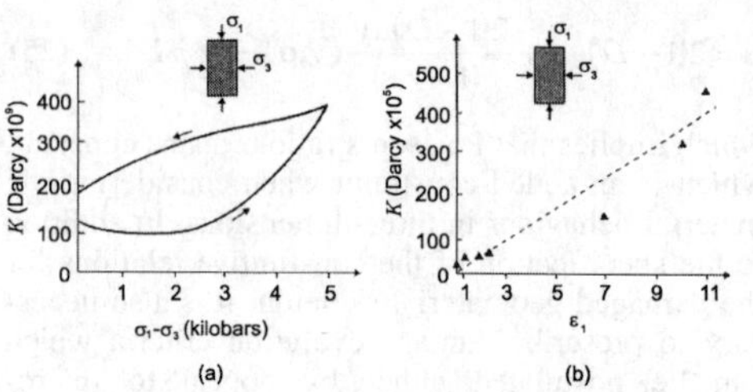

Figure 3. Permeability evolution in saturated geomaterials (a) After Zoback and Byerlee (1975) and (b) after Shiping et al. (1994)

The studies by Mahyari and Selvadurai (1998) suggest the following postulates for the evolution of permeability κ (expressed in m^2) as a function of the parameter ξ_d, which, is related to the second invariant of the stress deviator dyadic: i.e.

$$\kappa^d = (c_1 + c_2\xi_d)\kappa^0 \quad ; \quad \kappa^d = (c_3 + c_4\xi_d^2)\kappa^0 \qquad (28)$$

where c_1, c_2, c_3 and c_4 are constants and κ^0 is the permeability of the undamaged material. The linear- and quadratic-dependency of the permeability on ξ_d are plausible estimates, which are adopted strictly for purposes of the computational modelling. An incremental finite element procedure is developed using the damage and permeability evolution criteria described previously. The scalar damage variables are first obtained at the nine Gauss points within the finite elements. The constitutive matrix $\mathbf{C}^d$ and the permeability κ^d are updated at these locations to account for evolution of damage. The discretized governing equations are then solved to obtain the state of strain at each integration point using the updated values of $\mathbf{C}^d$ and κ^d. An iterative process using a standard Newton-Raphson technique solves the coupling between the state of strain and the state of damage in each step. The convergence criterion adopted in the analysis is based on the norm of the evolution of the damage variable in relation to a prescribed tolerance (Simo and Ju, 1987). The details of the iterative procedure are further summarized by Mahyari and Selvadurai (1998). As an illustration of the application of the theory of poroelastic media susceptible to damage, we consider the problem of the axisymmetric indentation of a poroelastic halfspace by a rigid indentor with a smooth, permeable interface (Figure 4). The poroelastic medium is subjected to a total load P with a time variation in the form of a Heaviside step function.

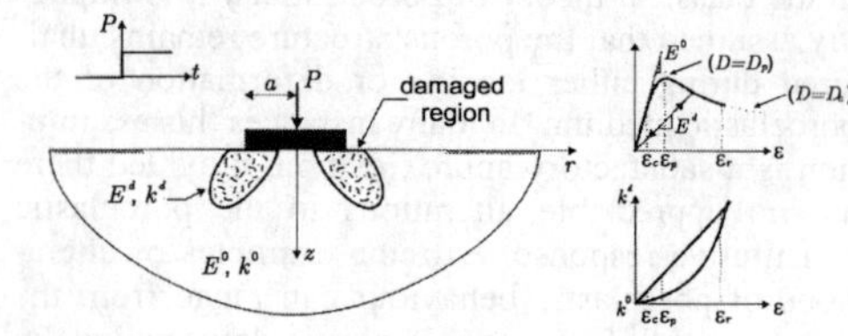

Figure 4. Indentation of a brittle geomaterial susceptible to damage.

The porous skeletal material has a Poisson's ratio of ν and the pore fluid is assumed to be nearly incompressible, giving $\nu_u = 0.499$. Again we note here that damage evolution precedes strain softening and the theoretical basis is applicable to elastic responses prior to strain softening. The material parameters employed in the numerical simulations are as follows:

$$E = 8300\,MPa; \nu = 0.195$$

$$\gamma = \eta = 130; D_c = 0.75$$

and the peak stress in uni-axial compression is 30 *MPa*. Other parameters c_1, c_2, etc., pertaining to the evolution of permeability are deduced from the experimental results cited previously. Estimates for these parameters are

$$c_1 = c_3 \cong 1\,;\; c_2 \cong 2\times10^{-3}\;;\; c_4 \cong 0.46\times10^{-6}$$

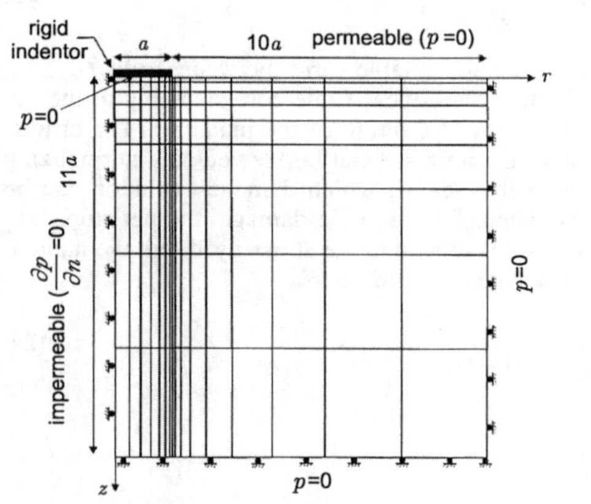

Figure 5. Indentation of a poroelastic geomaterial susceptible to damage: the finite element discretization

The finite element discretization of the indentation problem and the associated boundary conditions are shown in Figure 5. No provision is made for to account for singular stress fields that can exist at the boundary of the indenting region. Some mesh refinement is provided to allow for high strain gradients at this boundary region. If singular fields are to be accounted for, the entire treatment has to be changed to accommodate the attainment of peak stresses and possibly brittle fracture at the boundary of the indentation region (Selvadurai, 2000). In this study the objective is to establish in a global sense the influence of the strain induced damage in the brittle matrix and the attendant alterations in the stiffness characteristics of the porous skeleton and the permeability characteristics of the geomaterial on the overall poroelastic response.

Figure 6 illustrates the time-dependent behaviour of pore fluid pressure in the damage susceptible poroelastic medium, at two different locations of the poroelastic medium, as a function of the damage evolution characteristics of the medium. It is observed that the damage in the poroelastic medium enhances pore pressure development and accelerates the consolidation process.

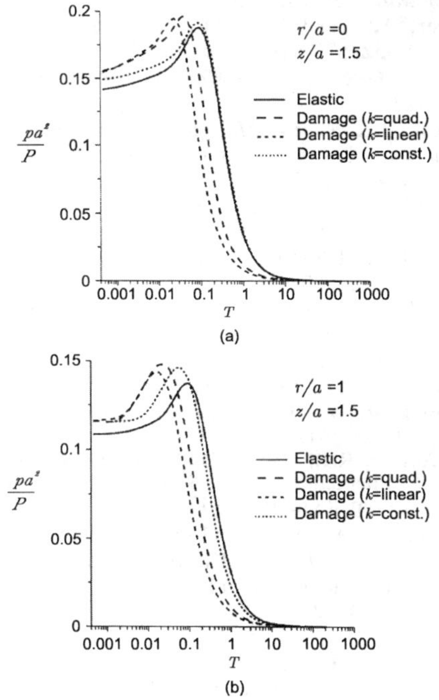

Figure 6. Pore pressure evolution at various depths of the poroelastic medium

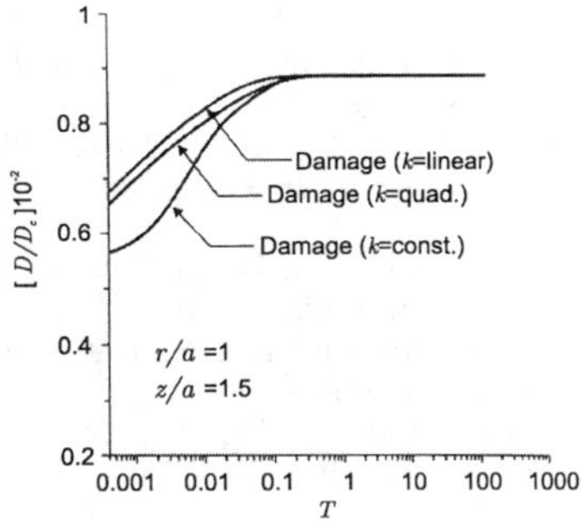

Figure 7. Damage evolution at the edge region of the indentor

Similarly, Figure 7 illustrates the time dependent evolution of damage in the poroelastic medium as a function of the permeability evolution characteristics of the poroelastic medium. The time-dependent evolution of damage is observed with all forms of permeability evolution laws.

6 CONCLUDING REMARKS

The classical theory of poroelasticity proposed by Biot forms a cornerstone in the development of modern continuum geomechanics. Despite its highly linearized nature, involving linear elastic response of the porous skeleton and linear response of the fluid transport characteristics, the theory has found applications not only in the field of geomechanics but also in a variety of other areas in materials engineering. The objective of this paper is to highlight further areas of application of the basic theory of poroelasticity where the essence of the original theory is retained but the scope of the problem area is enhanced to take advantage of computational modelling of poroelasticity via finite element methods. The two categories of extensions to the classical theory dealing with quasi-static crack extension and damage evolution in poroelastic media are examined using finite element techniques. It is shown that these modifications can be implemented relatively conveniently into computational schemes and results of both fundamental and engineering interest can be obtained from such extensions.

7 REFERENCES

Agbezuge, L.K. and Deresiewicz, H. (1974) *Israel J. Tech.* **12**: 322-338.

Atkinson, C. & Craster, R.V. (1991) *Proc. Roy. Soc.*, Ser. A. **434**: 605-633.

Atkin, R.J. and Craine, R.E. (1976) *J.Inst.Math. Applics.*, **17**: 153-207.

Bear, J. & Corapcioglu, M.Y. (1985) Fundamentals of Transport Phenomena in Porous Media, Nijhoff, The Netherlands.

Bear, J. & Bachmat, Y. (1990) *Introduction to Modelling of Transport in Porous Media*, Acad. Press, New York.

Bedford, A. & Drumheller, D.S. (1978) *Arch. Rational Mech. Analysis,* **68**: 37-51.

Bedford, A. & Drumheller, D.S. (1983) *Int. J. Engng. Sci.,* **21**:863-960.

Biot, M.A. (1941) J. *Appl. Phys.*, **12**, 155-164.

Booker, J.R. and Small, J.C. (1975) *Int.J.Solids Struct.*, **11**: 907-917.

Booker, J.R. and Small, J.C. (1984) *Int.J.Num. Analyt. Meth. Geomech.*, **8**: 343-357.

Booker, J.R. and Randolph, M.F. (1984) *Q.J.Mech.Appl.Math.*, **37**:479-495.

Borja, R.I. (1994) *Computer Methods and Advances in Geomechanics* (H.J.Siriwardane and M.M.Zaman, Eds.), A.A.Balkema, **1**:753-758.

Bowen, R.M. (1982) *Int. J. Engng. Sci.*, **20**: 697-735.

Britto, A.M. et al. (1992) *Soils and Foundations*, **32**: 13-25.

Cheng, A.H.-D. and Liggett, J.A. (1984) *Int. J.Num. Meth. Engng.*, **20**: 255-278.

Cheng, A.H.-D. et al. (1998) Poroelasticity: M.A.Biot Memorial Issue, *Int. J. Solids Structures,* **35**: 4513-5031.

Cheng, H. and Dusseault, M.B. (1993) *Int.J.Rock mech. Min. Sci.*, **30**:1323-1331.

Chiarella, C. and Booker, J.R. (1975) *Q.J.Mech. Appl. Math.*, **28**:317-328.

Coussy, O. (1995) *Mechanics of Porous Continua*, John Wiley, New York.

Craster, R.V. and Atkinson, C. (1996) *Mechanics of Poroelastic Media* (A.P.S.Selvadurai, Ed.) Kluwer Academic Publ. 23-45.

Cryer, C.W. (1963) *Q.J. Mech. Appl. Math.*, **16**:401-412.

de Boer, R. (2000) *Theory of Porous Media*, Springer-Verlag, Berlin.

de Boer, R. and Ehlers, W. (1990) *Int. J.Solids Struct.*, **26**,43-51.

de Josselin de Jong, G. (1953) *J. Appl. Phys.*: 922-928.

de Wiest, R.M.J. (Ed)(1969) *Flow Through Porous Media*, Academic Press, New York.

Desai, C.S. and Christian, J.T. (1977) *Numerical Methods in Geomech.*, McGraw-Hill, New York.

Detournay, E. & Cheng, A. H. -D. (1993) Comprehensive *Rock Engineering*, Pergamon Press, **II**: 113-171,

Drew, D.A. & Passman, S.L. (1999) *Theory of Multi-Component Fluids*, Appl. Math. Sci. **135**, Springer, Berlin

Drumheller, D.S. & Bedford, A. (1980) *Arch. Rational Mech. Analysis*, **73**: 257-284.

Ehlers, W. (1989) Arch. Mech. Stosow., 41, 73-93.

Fillunger, P. (1913) *Osterr. Wochenschr. Fur den offentlich. Baudienst,* **19**, 532-516, 567-570.

Ghaboussi, J. and Wilson, J.L. (1973) *Int. J.Numerical Meth. in Engng.*, **5**, 419-442.

Huyakorn, P.S. and Pinder, G.F. (1983) *Computational Methods for Subsurface Flow,* Academic Press, New York.

Kachanov, L.M. (1958) *Isv. Akad. Nauk SSr Odt.Tek,* **8**:26-31.

Kassir, M.K. and Xu, J. (1988) *Int.J.Solids Struct.*, **24**: 915-936.

Kiyama, T. et al. (1996) *Proc. 2nd North Amer. Rock Mech. Symp.,* A.A.Balkema, 1643-1650.

Krajcinovic, D. (1984) *Appl. Mech. Rev.*,**37**: 1-6.

Lan, Q. and Selvadurai, A.P.S. (1996) *J.Appl. Math. Phys.(ZAMP),***47**:695-716.

Lemaitre, J. (1984) Nucl. Engng. Design, 80: 233-245.

Lemaitre, J. and Chaboche, J.L. (1990) *Mechanics of Solid Materials*, Cambridge University Press.

Lewis, R.W. and Schrefler, B.A. (1998) *The Finite Element Method in the Deformation and Consolidation of Porous Media,* John Wiley, New York.

Mandel, J. (1953) *Geotechnique*, **3**: 287-299.

Mazars, J. (1982) *Advances in Fracture Research,* Pergamon Press, 1499-1506.

McNamee, J. and Gibson, R.E. (1960) *Q. J. Mech. Appl. Math.***13**: 98-111,210-227.

Palmer, A.C. and Rice, J.R. (1976) *Proc. Roy. Soc., A* **332**: 527-548

Paria, G. (1963) *Applied Mechanics Reviews,* **16**: 421-423.

Prevost, J.H. (1984) *Mechanics of Engineering Materials* (C.S. Desai and R.H.Gallagher Eds.) John Wiley, 515-533.

Rice, J.R. and Cleary, M.P. (1976) *Rev. Geophys. Space Phys.*, **14**: 227-241.

Rice, J.R. and Simons, D.A. (1976) *J. Geophys. Res.*, **81**: 5322-5334.

Rudnicki, J.W. and Koutsibelas, D.A. (1991) *Int. J. Solids Struct.*, **27**: 205-225.

Sandhu, R.S. and Wilson, E.L. (1969) *J. Eng. Mech. Div. Proc ASCE,* **95**:621-652.

Scheidegger, A. (1960) *Applied Mechanics Reviews,* **13**:313-318.

Schiffmann, R.L. (1985) Fundamentals of Transport Phenomena in Soils (J.Bear and M.Y.Corapcioglu, Eds), 617-669.

Selvadurai, A.P.S. (Ed)(1996) *Mechanics of Poroelastic Media,* Kluwer Acad. Publ, Dordrecht, The Netherlands.

Selvadurai, A.P.S. (2000) *Mechanics of Cohesive Frictional Materials,* **5**: 325-339.

Selvadurai, A.P.S. & Mahyari, A.T. (1998) *Int. J. Solids Structures*, **35**: 4869-4885.

Selvadurai, A.P.S. & Nguyen, T.S. (1995) *Computers and Geotechnics,* **17**: 39-73.

Selvadurai, A.P.S. & Yue, Z.Q. (1994) *Int.J.Num. Analyt. Meth. Geomech.*, **18**:161-175.

Shiping, L., et al. (1994) *Int.J.Rock Mech.Min.Sci.*, **31**: 383-391.

Sidoroff, F. (1980) *Int. Theo. Appl. Mech. Symp.*, Senelis (J.Hult and J.Lemaitre, Eds) 237-244.

Simo, J.C. and Ju, J.W. (1987) *Int. J. Solids Struct.*, **23**:841-869.

Simon, B.R. et al. (1986) *Int.J.Num. Analyt. Meth. Geomech.*, **10:461**-482.

Skempton, A.W. (1954) *Geotechnique*, **4**:143-147.

Smith, I.M. and Griffiths, D.V. (1988) *Programming The Finite Element Method*, John Wiley, New York.

Szefer, G. and Gaszynski, J. (1975) *Arch. Mech. Stos.* **27**: 497-515.

Terzaghi, K. (1923) *Akad.dr Wissen. Wien, Math.nature.Klasse,II,* **132**, 125-138.

Teufel, L.W. (1987) *Proc. 28th U.S. Symp. Rock Mech.*, Tucson, AZ., 473-480.

Thimus, J. -F. et al. (Eds.)(1998*) Poromechanics: A Tribute to M.A.Biot,* A.A.Balkema,The Netherlands.

Tsang, C.-F. (1991) *Reviews in Geophysics*, **29:** 537-551.

Valliappan, S., Lee, I.K. and Boonlualohr, P. (1974) *Finite Element Methods in Flow Problems* (J.T.Oden et al. Eds) U. Alabama Press, 741-755.

Verruijt, A. (1977) *Finite Elements in Geomechanics* (G.Gudehus, Ed.), John Wiley,293-317.

Yue, Z.Q. and Selvadurai, A.P.S. (1994) *Appl. Math. Modelling*, **18**:170-185.

Yue, Z.Q. and Selvadurai, A.P.S. (1995) *Int. J. Engng. Sci.*, **33**:1633-1662.

Wang, X., et al. (2000) *Modelling in Geomechanics* (M.M.Zaman, G.Gioda and J.R.Booker, Eds.)John Wiley, New York, 51-64.

Whitaker, S. (1986) *Transport in Porous Media*, 1:3-25, 105-125, 127-154.

Zoback, M.D. and Byerlee, J.D. (1975) *J.Geophys. Res.*, **80**: 752-755.

Computer Methods and Advances in Geomechanics, Desai et al. (eds) © 2001 Balkema, Rotterdam, ISBN 90 5809 183 X

Conceptual model for finite element analysis of artificial recharge

M.J.Johnson, S.L.Houston & W.N.Houston
Arizona State University, Tempe, Ariz., USA

ABSTRACT: The NWWRP recharge facility, operated by the City of Mesa, Arizona since 1990, has averaged a recharge rate less than one-half the design value due to being underlain by a geologically heterogeneous alluvial formation with a higher fraction of fine-grained material than has been generally observed at other areas immediately adjacent to the Salt River. Possible remedial activities to increase the hydraulic loading capacity will be studied and recommendations will be made regarding possible manners of hydraulic improvement. The conceptual model for the NWWRP recharge facility research has been developed based on information collected from varied sources including both 'hard' and 'soft' data. Computer visualization and manipulation of three-dimensional data is necessary for construction of a sound conceptual model for flow modeling in a complex geologic formation. Stratigraphic data from visual classifications during soil borings in difficult to drill alluvial deposits may differ significantly from data obtained from piezometric monitoring.

1 INTRODUCTION

Countries with Western European or North American standards of living require more than 2000 cubic meters of water per person per year (Bouwer, 1997). Groundwater is vulnerable to overdraft and depletion, especially in arid to semi-arid regions where natural rates of recharge are low. Artificial recharge can restore the sustainability of these groundwater resources.

The feasibility of an artificial recharge project is tied to the hydraulic loading capacity. The hydraulic loading capacity is determined by the intake capacity at the soil-water interface, the conveyance capacity of the unsaturated zone, and the transmission capacity of the recharging aquifer. Any one of these may act as the hydraulic control on recharge.

The City of Mesa, Arizona operates the Northwest Water Reclamation Plant (NWWRP) and recharge facility. The NWWRP recharge facility, operational since 1990, consists of four recharge basins with area totaling approximately 27 acres. Figure 1 shows the arrangement of the basins. The design capacity of the NWWRP site was for 8 million gallons per day (MGD) based on recharge rates measured at other sites along the ephemeral Salt River. The volume of water recharged has averaged less than one-half of the design capacity (i.e. about 3.5 MGD). Previous studies have concluded that the site is underlain by a geologically heterogeneous alluvial formation with a higher fraction of fine-grained material than has been generally observed at other areas immediately adjacent to the Salt River channel.

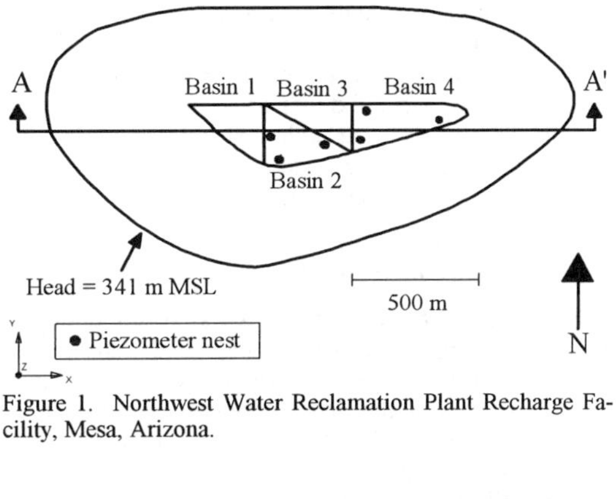

Figure 1. Northwest Water Reclamation Plant Recharge Facility, Mesa, Arizona.

Research has started for the purpose of studying the hydraulics of groundwater flow beneath the spreading basins. Possible remedial activities to increase the hydraulic loading capacity will be studied and recommendations will be made regarding possible manners of hydraulic improvement.

The conceptual model is the first and most important step in obtaining meaningful results from numerical modeling of groundwater flow, especially so for formations exhibiting significant material heterogeneity. The conceptual model for the NWWRP recharge facility has been constructed based on information collected from varied sources. These sources include 'hard data', such as observational field data, material properties derived from laboratory and field testing, and soil borings as well as 'soft data', such as geological interpretation and computer-aided interpolation.

The graphical software package GMS v2.1 (Owen et al. 1996) developed by researchers at the Engineering Computer Graphics Laboratory at Brigham Young University was used to organize, perform operations on, and visualize the collected data.

2 FORMATION STRATIGRAPHY AND HYDROGEOLOGIC PROPERTIES

2.1 *Information from collected boring logs and reports*

The elevation of the ground surface within the basins is 353.3 m above mean sea level (m MSL). The thickness of the upper alluvial unit (UAU) has been reported as 36 to 46 m. The UAU is a well-graded alluvium consisting of boulders, cobbles, gravel, sand, silt, and clay deposited by the ancestral Salt River. Drilling and sampling at depth is difficult and possible only by percussion hammer techniques.

Material may be grouped into 3 types; (1) 'clean' materials classified as either well or poorly graded sands or gravels, (2) 'silty' materials classified as silty-sands or gravels and, and (3) 'clayey' materials classified as clayey-sands or gravels and clays. Several geotechnical investigations have reported the presence of low permeability layers. The depths, thickness, and lateral extents of these layers differed among the reports. The collected borings within the footprint of the basins extend to a maximum depth of about 18 m beneath the basin floors.

2.2 *Hydraulic conductivities*

Laboratory saturated vertical hydraulic conductivity testing on samples of clayey sand yielded results of the order of 10^{-7} m/sec. Borehole testing for saturated horizontal hydraulic conductivity produced results of the order of 10^{-5} m/sec for clayey sand and 10^{-3} m/sec for clean alluvium (AGRA, 1994). Corell & Corkhill (1994) have reported a regional anisotropy ratio of 20 for the UAU.

Schoneinz & Drewes (2000) presented results from a two-dimensional finite difference simulation of the impact of recharge from the NWWRP facility on local flow patterns. The calibrated value of horizontal hydraulic conductivity of the aquifer underlying the basins was of the order of 10^{-3} m/sec.

3 FIELD DATA COLLECTION

3.1 *Recharge Basin Infiltration Rates*

Table 1 provides the average wetting cycle infiltration rates for the four basins. Data from December 1990 to July 1999 were compiled for basins 1 and 2. Recharge commenced in basins 3 and 4 in 1996.

Table 1. Average wetting-cycle infiltration rates.

Basin	Rate m/day	Cycles
1	0.23	106
2	0.08	68
3	0.08	31
4	0.12	51

3.2 *Piezometer Water Levels*

Six piezometer nests were constructed in basins 2 and 4. Each nest consists of up to six piezometers installed at depths ranging from 1.5 to 9.1 m. Intensive piezometric data was collected during two inundation cycles. The first data set, collected during July 1999, followed a drying cycle ranging from 5 to 12 days (10 days for basin 2 and 12 days for basin 4). The second data set, collected during March 2000, followed two to three months of basin activity. The piezometer data indicate saturated conditions established beneath the basin floor in the initial stages of wetting cycles and free water recession to a depth of 1.5 to 8.2 m depending upon the length of the drying cycle. Regions of relatively large vertical head-loss were observed at several sampling locations indicating the presence of impeding layers. Table 2 provides a summary of the observations of significant vertical headloss. The larger vertical headlosses observed in basin 2 are consistent with the lower infiltration rates.

Table 2. Observations of significant vertical head-loss.

Nest	Vertical Region m MSL	Headloss m-H_2O
2 West	347.2 – 344.1	4.0
2 South	350.2 – 348.7	4.6
	345.6 – 344.1	1.5
2 East	345.6 – 344.1	3.4
4 West	351.7 – 350.2	1.2
4 North	347.2 – 345.6	2.1
4 East	351.7 – 348.7	2.1
	347.2 – 345.6	3.0

3.3 *Water Quality Data*

Stainless steel six-inch screened intervals were installed alongside of the piezometers for collecting water samples. Measurements of dissolved oxygen, sulfate, nitrate, manganese, specific conductivity, pH, ion concentration, and dissolved organic carbon were obtained by fellow Arizona State University researchers (Rengaraj, 2000). Sulfate and nitrate concentrations are strong indicators of redox potential. As oxygen is removed from the system, the reductive environment becomes anoxic. In this stage nitrogen is still available as an electron acceptor, therefore nitrate reduction occurs and concentrations fall. As the available nitrogen is removed from the system, the reductive environment becomes anaerobic. As the system becomes anaerobic, sulfate is still available as an electron acceptor, therefore sulfate reduction occurs and concentrations fall. As sulfur is present as MnS or FeS in its reduced state, manganese concentrations provide additional evidence of very strong reducing conditions. Strong reducing conditions under recharge basins indicate regions of nearly stagnant water, in other words, regions through which most of the applied water is not flowing. Specific conductivity and ion concentrations provide water quality signatures useful in discerning flowpaths.

Table 2 provides a summary of the water level and quality data obtained from the piezometer nests. The values within the table indicate depths in meters at which each indicator was observed

Table 3. Summary of water quality data.

Nest	Anaerobic Conditions	Different Signature
2 West	344.1*, maybe 348.7	344.1, maybe 348.7
2 South	347.2	347.2
2 East	344.1	344.1
4 West	347.2	348.7 and 347.2
4 North	348.7	348.7
4 East	348.7, maybe 347.2	348.7

* m MSL

4 CONCEPTUAL MODEL

4.1 *Material distributions*

The collected field data and boring logs were combined with the knowledge that stratigraphic units in alluvial deposits tend to be elongated parallel to the direction of flow for construction of a likely material distribution. The thickness and lateral extent of each unit will be calibrated to the measured infiltration rates and the piezometer nest water levels during the numerical modeling stage of research. Figure 2 shows in plan view the initial estimate of the distribution of low conductivity sediments. The set of numbers associated with each unit are the elevation of the top of the layer in m MSL and the thickness in m.

4.2 *Boundary Conditions*

For this and many other modeling studies, determining an appropriate set of lateral boundary conditions is difficult. Classical modeling boundaries such as perennial rivers, rock outcroppings, and groundwater divides are located at a great distance from the site of interest. In cases such as this, it is usually the practice to establish the location of the lateral boundaries in the ad hoc manner of simply moving the boundaries farther and farther away until the result of interest is no longer affected. This approach is not feasible for the present study due to practical

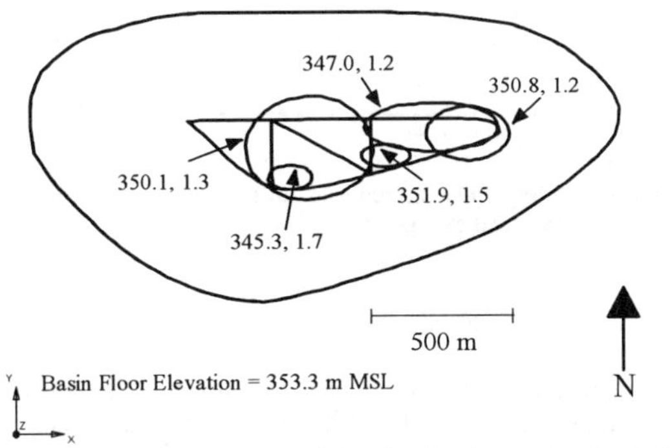

Figure 2. Initial estimate of the distribution of interbedded low conductivity sediments.

considerations regarding computer resources. The flow modeling will be performed using FEMWATER (Lin et al. 1997) a three-dimensional finite element model for variably saturated single-phase flow through porous media. The high degree of nonlinearity involved in the solution of the unsaturated flow equation requires fine spatial and temporal discretizations. Couple this with the general requirement of domain sizes 10 to 20 times the radius of the recharge source for accurate modeling of the aquifer response and the memory requirements for three-dimensional modeling soon exceeds the capabilities of most systems. In order to save computer resources, the technique of telescopic mesh refinement is employed. The results of a larger scale two-dimensional simulation of the impact of the NWWRP recharge facility (Schoneinz & Drewes, 2000) are applied as boundary conditions. More specifically, the lateral boundary is defined as the contour line of hydraulic head equaling 341 m MSL shown on Figure 1. The basal boundary condition is defined as no-flow at the base of the UAU unit at a depth of 36 m. The boundary condition at the recharge basins is defined as a dynamic head boundary equal to recorded basin levels.

4.3 *Initial hydraulic condition*

Figure 3 shows the initial pressure head distribution along cross-section A-A′ (see Figure 1) as defined by three-dimensional interpolation of piezometer water levels collected from July 1999 coupled with the results of the Schoneinz & Drewes (2000) simulation.

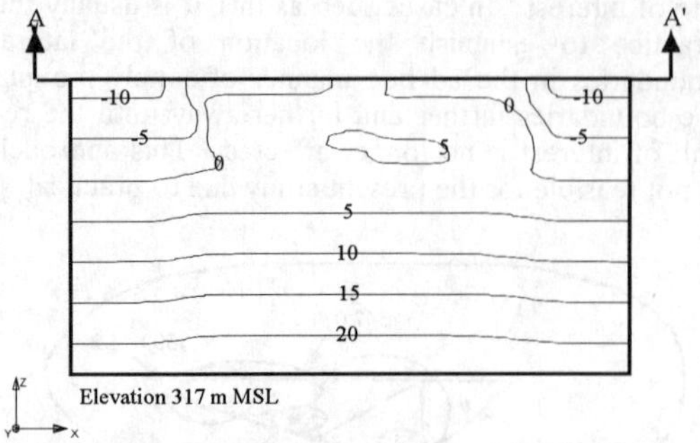

Figure 3. July 1999 interpolated initial pressure head, in m-H_2O. 30x vertical exaggeration.

5 CONCLUSIONS

Construction of a sound conceptual model is vital for obtaining meaningful results from a modeling effort. This is especially true when modeling flow in a geologically heterogeneous formation.

Computer visualization and manipulation of three-dimensional data is necessary for construction of a sound conceptual model for flow modeling in a complex geologic formation.

Stratigraphic data from visual classifications during soil borings in difficult to drill alluvial deposits may differ significantly from data obtained from piezometric monitoring.

6 ACKNOWLEDGEMENTS

This work was supported in part by the cities of Phoenix, Mesa, Tempe, Glendale, and Youngstown, Arizona, Tucson Water, the American Water Works Association Research Foundation (AWWARF), the Water Environmental Research Foundation (WERF), the National Water Resources Institute (NWRI), the US Environmental Protection Agency (EPA), and the Arizona Department of Water Resources (ADWR).

7 REFERENCES

AGRA Earth and Environmental. 1994. *City of Mesa percolation ponds.* AGRA Job No. E94-124 Phase I Report No. 2. prepared for the City of Mesa, Arizona.

Black & Veatch. 1991. *Soil boring results - percolation pond cells 2 and 3.* Report prepared for the City of Mesa, Arizona.

Bouwer, H. 1997. Role of groundwater recharge and water reuse in integrated water management. *The Arabian Journal for Science and Engineering*. 22, 1C, 123 – 131.

Corell, S., & Corkhill, E. F. 1994. A regional groundwater flow model of the Salt River Valley - Phase II – Numerical model, calibration, and recommendations. *Arizona Department of Water Resources Modeling Report No. 8.*

Corkhill, E. F., Corell, S., Hill, B. M., & Carr, D. A. 1993. A regional groundwater flow model of the Salt River Valley - Phase I – Hydrologic framework and basic data report. *Arizona Department of Water Resources Modeling Report No. 6.*

Dames and Moore. 1994. *City of Mesa Northwest Water Reclamation Plant percolation ponds performance and alternatives evaluation.* Report prepared for the City of Mesa, Arizona.

Hydrosystems Inc. 1995. *NWWRP recharge impacts determination report.* Report prepared for the City of Mesa, Arizona.

Kenneth, D. Schimdt and Associates. 1987. *Hydrological conditions at the Northwest Water Reclamation Plant Recharge Site.* Report prepared for Black and Veatch, Phoenix, Arizona.

Kenneth, D. Schimdt and Associates. 1990. *Hydrologic impact area for underground storage at City of Mesa water reclamation plants.* Report prepared for the City of Mesa, Arizona.

Lin, H. J., Richards, D. R., Talbot, C. A., Yeh, G. T., Cheng, J. R., Cheng, H. P., & Jones, N. L. 1997. FEMWATER: a three-dimensional finite element computer model for simulating density-dependent flow and transport in variably saturated media. *Technical Report CHL-97-12. U.S. Army Engineers Waterways Experiment Station.*

Owen, S. J., Jones, N. L., & Holland, J.P. 1996. A comprehensive modeling environment for the simulation of groundwater flow and transport. *Engineering with Computers* 12: 235-242.

Rengaraj, S. 2000. *Water quality transformations below the percolation basins at the Northwest Water Reclamation Plant, Mesa.* MS Thesis, Arizona State University, Tempe, Arizona.

Schonheinz, D. & Drewes, J. E. 2000. Groundwater flow modeling for the NWWRP Mesa. *Arizona State University Department of Civil and Environmental Engineering Technical Report.*

Computer Methods and Advances in Geomechanics, Desai et al. (eds) © 2001 Balkema, Rotterdam, ISBN 90 5809 183 X

Field and theoretical study on the settlement of soils with geometric and material nonlinearity under short term increasing loads

Awad M. Kassem
Department of Civil Engineering, Assiut University, Egypt
Ayman A. El-Maghraby
Department of Mathematics and Engineering Physics, Cairo University, Fayum Branch, Egypt

ABSTRACT: Some structures subjected to a short term increasing loads related to their way of construction or the nature of loading during their lifetime. The soil layer situated just below them settled according to the type of this load. This paper presents a completed mathematical formulation and solution of the fully saturated soil with geometric and material nonlinearity subjected to short term increasing loads. The soil layer is assumed homogeneous, nonlinear and resting on an impervious base. The used model is based on Biot's formulation of fully saturated soil in one dimension including the nonlinear terms. Finite difference method is used to solve the equations. Field measurement of soil settlement and a determination of their properties in the Laboratory were carried out to verify the numerical solution. The used model can handle complex surface loading and time varying loads. Also, the model express the behavior of the developed pore water pressure and its associated effective stress along the depth of the consolidating soil layer. The effect of nonlinearity for both geometry and material is highly improve the calculated settlement values and considerably express the interactions between the soil constituents.

1. INTRODUCTION

Some structures, such as tanks and silos is constructed by means of slip forms subjected to gradually increasing loads. The soil layer situated just below them are thus subjected to short term raising loads which is the subject of the present study.

Most practical settlement calculation are based on the small-strain version of Terzaghi's one dimensional consolidation theory. A group of years later, the consolidation problem was treated as a purely continuum mechanics composite problem. The first rigorous general three dimensional theory was established by Biot (1940). Zienkiewicz (1985) has formulated the basic equations for porous media established by Biot. Siddharthan (1993) was presented the dynamic response of a layered nonlinear soil deposit subjected to moving load. A non linear elasto-viscous model to explain compression process of clay in a unified manner is proposed by Yoshikuni et al (1994).

This paper presents a complete mathematical formulation and solution of the fully saturated soil with geometric and material nonlinearly subjected to short term increasing load.

2. MATHEMATICAL FORMULATION

2.1 General formulation

The equation of motion for the solid-fluid system for a unit volume is presented by Zienkiewicz, as

$$\sigma_{ij,j} + \rho b_i - \rho \ddot{u}_{si} - \rho_f (\ddot{u}_{fi} + \dot{u}_{fk}\, \dot{u}_{fi,k}) = 0 \qquad (1)$$

Also the equation of motion of the fluid of a unit volume, is written as,

$$-P_{,i} + \rho_f b_i - \rho_f \left[\ddot{u}_{si} + \frac{1}{n}\left(\ddot{u}_{fi} + \dot{u}_{fk}\, \dot{u}_{fi,k} \right) \right] - \frac{\rho_f g}{k} \dot{u}_{fi} = 0 \qquad (2)$$

While the constitutive relations for solid skeleton are given as,

$$\sigma_{ij} = \sigma''_{ij} - \alpha\delta_{ij}P = D_{ijkl}\left(\overset{\circ}{\varepsilon}_{kl} - \overset{\circ 0}{\varepsilon}_{kl}\right) + \overset{\circ}{\omega}_{ik}\sigma_{kj} + \overset{\circ}{\omega}_{jk}\sigma_{ki} - \alpha\delta_{ij}P \tag{3}$$

The final equation is the mass conservation equation applied to the fluid flow

$$u_{fi,j} + \alpha\varepsilon_{ii} = -\frac{P}{Q} \tag{4}$$

Where σ_{ij} is the total stress in the combined solid and fluid mix u_{si} and u_{sf} are the displacements of solid skeleton and fluid relative to the skeleton, ρ and ρ_f are the total density of mix and fluid respectively; b_i is the body force acceleration, P the pore pressure in the fluid, g is the gravity acceleration, k is the permeability coefficient of soil and n is the porosity.

$\varepsilon_{ij} = \frac{1}{2}\left(u_{si,j} + u_{sj,i}\right)$ is the rate of deformation tensor and (^{o}) refer to the total derivative .

$\overset{\circ}{\omega}_{ij} = \frac{1}{2}\left(u_{si,j} - u_{sj,i}\right)$ is a spain tensor, $\overset{o0}{\varepsilon}_{kl}$ is the rate of thermal or other initial strain, σ''_{ij} is identical to effective stress providing the coefficient α is unity and D_{ijkl} is the tangent modulus tensor.

$\frac{1}{Q} = \frac{n}{k_f} + \frac{\alpha - n}{k_s}$ where k_s and k_f are the bulk modulus for skeleton and fluid.

2.2 Nonlinear one dimensional problem

The prescribed three dimensional consolidation Equations (1), (2), (3), and (4) can be reduced to contain the compressibility of porous media in the vertical direction only including the nonlinear terms as,

$$\frac{\partial\sigma_z}{\partial z} = \rho\ddot{u}_s + \rho_f(\ddot{u}_f + \dot{u}_f\frac{\partial u_f}{\partial z}) \tag{5}$$

$$-\frac{\partial P}{\partial z} = \rho_f\ddot{u}_s + \frac{\rho_f}{n}(\ddot{u}_f + \dot{u}_f\frac{\partial u_f}{\partial z}) + \frac{\rho_f g}{k}\dot{u}_f \tag{6}$$

For i = j = k = l = 1

1) The tangent modulus $D_{ijkl} = 2\tau G$

where $\tau = \frac{1-\nu}{1-2\nu}$, ν is Poisson's ratio which is assumed constant and G is the nonlinear shear modulus. The relation between the shear modulus (G) and Young's modulus (E_t) is

$$G(z,t) = \frac{E_t(z,t)}{2(1+\nu)} \tag{7}$$

2) $\overset{\circ}{\varepsilon}_{ij} = \frac{1}{2}(u_{si,j} + u_{sj,i}) = \frac{\partial u_s}{\partial z}$ for i = j = 1

3) $\overset{\circ}{\omega}_{ij} = \frac{1}{2}(u_{si,j} - u_{sj,i}) = 0$ for i = j = 1

4) δ ij = 1

$$\sigma_z = 2\tau G\frac{\partial u_s}{\partial z} - \alpha P \tag{8}$$

$$-\frac{\partial u_f}{\partial z} = \alpha\frac{\partial u_s}{\partial z} + \frac{1}{Q}P \tag{9}$$

Equations 5, 6, 8 and 9, are the nonlinear governing equations in one dimension. The soil considered as a column supporting a load $P_o = -\sigma_z$ that is confined laterally and resting on impervious layer.

2.3 Hyperbolic stress – strain relationship

Kondoner et al. (1963) had shown that the nonlinear stress–strain curves for number of soils could be approximated by a hyperbola represented by the equation.

$$\sigma_z - \sigma_3 = \frac{\varepsilon_a}{\frac{1}{E_i} + \frac{\varepsilon_a}{(\sigma_z - \sigma_3)_{ult.}}} \tag{10}$$

where,

ε_a is the axial strain, E_i is the initial tangent modulus, σ_z is the vertical total stress and σ_3 is the confined stress.

Experimental work by Jambu (1967) explain the relationship between the initial tangent modulus and confined pressure as:

$$E_i = K_h P_a\left(\frac{\sigma_3}{P_a}\right)^{n^*} \tag{11}$$

where,

K_h is an experimental parameter called the modulus number, P_a is the atmospheric pressure and n^* is the modulus exponent. The parameters K_h and n^* are dimensionless numbers experimentally determined.

Based on Mohr-Coulomb failure criteria , the relationship between compressive strength and confining pressure is expressed as:

$$(\sigma_z - \omega_3)_f = \frac{2c.\cos\phi + 2\sigma_3 \sin\phi}{1 - \sin\phi} \tag{12}$$

in which c and ϕ are the cohesion intercept and the internal friction angle of the soil, respectively. The compressive strength or the stress difference at failure $(\sigma z - \sigma_3)_f$ can be related to the asymptotic stress difference as,

$$(\sigma z - \sigma_3)_f = R_f\, (\sigma z - \sigma_3)_{ult.} \tag{13}$$

in which R_f is the failure ratio which is always smaller than unity and varies from 0.5 to 0.9 for most soils.

The slope of the tangent to the stress-strain curve can be defined as,

$$E_t = \frac{\partial}{\partial \varepsilon_a}(\sigma_z - \sigma_3) \tag{14}$$

By differentiating equation (10) with respect to ε_a and substituting equations (11), (12) and (13) into the resulting expression, the tangent modulus can be expresses as:

$$E_t = K_h P_a \left(\frac{\sigma_3}{P_a}\right)^{n^*} \left[1 - \frac{R_f(\sigma_z - \sigma_3)(1 - \sin\phi)}{2c.\cos\phi + 2\sigma_3 \sin\phi}\right]^2 \tag{15}$$

This equation can be used to calculate the appropriate value of tangent modulus for any stress conditions.

Substituting equation (15) into equation (7), the nonlinear shear modulus can be expressed as:

$$G = \frac{K_h P_a}{2(1+v)} \left(\frac{\sigma_3}{P_a}\right)^{n^*} \left[1 - \frac{R_f(\sigma_z - \sigma_3)(1 - \sin\phi)}{2c.\cos\phi + 2\sigma_3 \sin\phi}\right]^2 \tag{16}$$

A numerical scheme based on the finite difference method is designed to solve the problem. The domain is divided into uniform cells and the governing equations are discritized over it. All variables are defined and are computed at boundaries of the cells.

3.FILED AND LABORATORY MEASURMENT

To verify the proposed mathematical model, a field tests were carried out to measure the settlement of both fully saturated clay and sand strata. The clay layer of about 6 m thick. underneath by a stiff clay of about 4.5 m thick which considered practically impervious. The clay is normally consolidated and classified as medium clay with cohesion c of 62 kPa, K = 10^{-7} m/sec, LL = 64, PL = 29, Gs = 2.71 and dry density is 15.5 kN/m^3. After over consolidation (OCR=1.5) the properties of the clay are k $\cong 10^{-8}$ m/sec, c=120 kPa and the dry density is 16.3 kN/m^3. The sand layer is formed in the same site as column 0.4 m diameter and 3m deep surrounded by a medium clay layer. The used sand is classified as well graded sand with some fines with Gs = 2.67. For loose sand, c = 20 kPa, $\phi = 24^o$, and dry density = 17 kN/m^3.For medium dense sand and medium dense sand, c = 20 kPa, $\phi = 30^o$ and dry density = 19.3 · kN/m^3 .

Both sand and clay layers were loaded by a short term increasing load acting on a plate 0.30X0.30 m with different rates. The maximum rate occurs when the stress increased gradually from zero to 320 kPa within 600 sec. . The settlement is recorded every one minute. The obtained field settlement values and their corresponding numerical values are compared and discussed in the next section.

4. NUMERICAL AND FIELD RESULTS

The results of the prescribed numerical scheme is compared with the field results. Figures 1 and 2 show the field settlement of both normally consolidated clay and loose sand respectively as compared with its corresponding numerical results. The comparison indicate that both field and numerical results are quit close specially for normally consolidated clay. Figure 3 shows a similar comparison for saturated loose sand where the loads increased by different rates (R) of 9, 22 and 32 kPa/min. The field and numerical results seems to be closer as the rate of loading decreases. The nonlinear terms for both material and geometry seems to be significant and effective in the calculated settlement to be closer with its corresponding field values. The effect of nonlinearity for both geometry and material is highly improve the calculated settlement values and

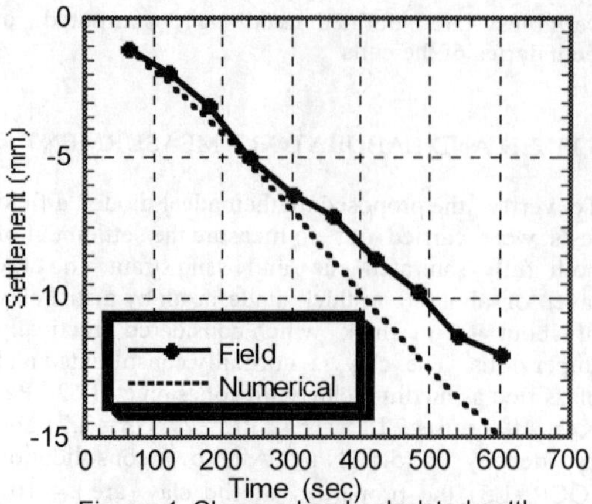

Figure 1. Observed and calculated settlement for normally consolidated clay.

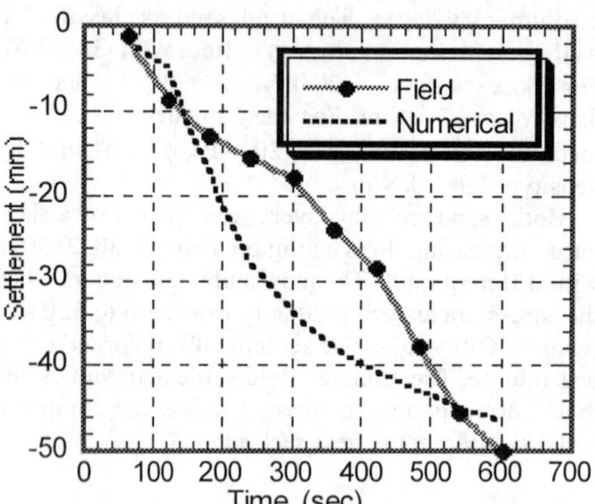

Figure 2. Observed and calculated settlement for loose sand.

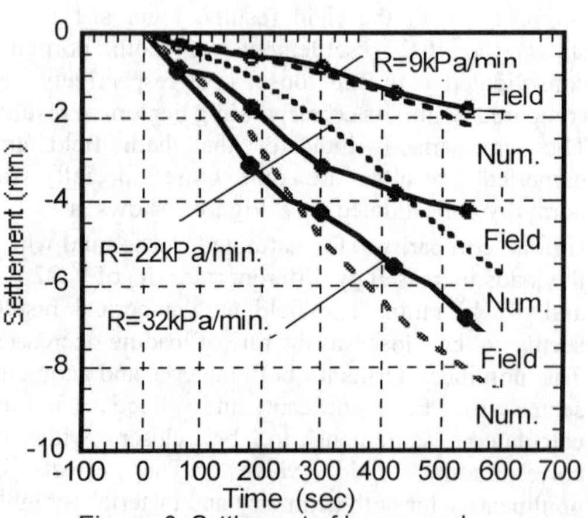

Figure 3. Settlement of loose sand with different rates of loading.

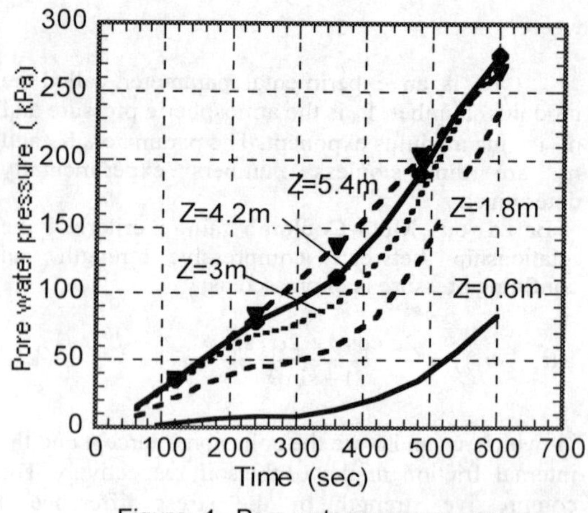

Figure 4. Pore water pressure versus time for normally consolidated clay.

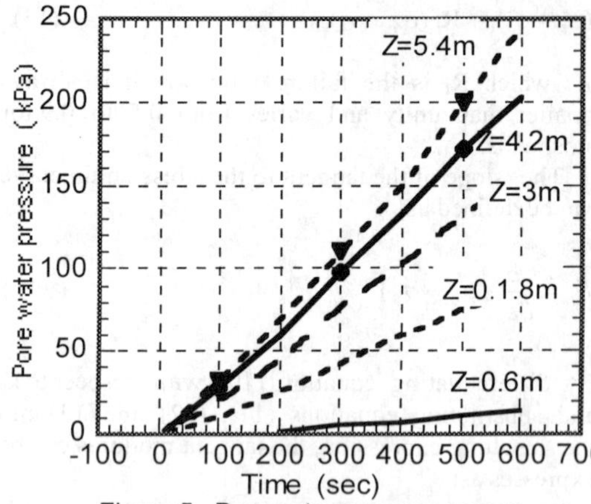

Figure 5. Pore water pressure versus time for over consolidated clay.

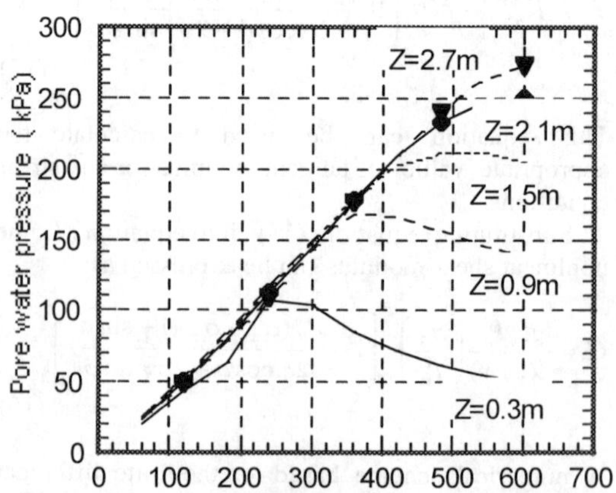

Figure 6. Pore water pressure versus time for loose sand

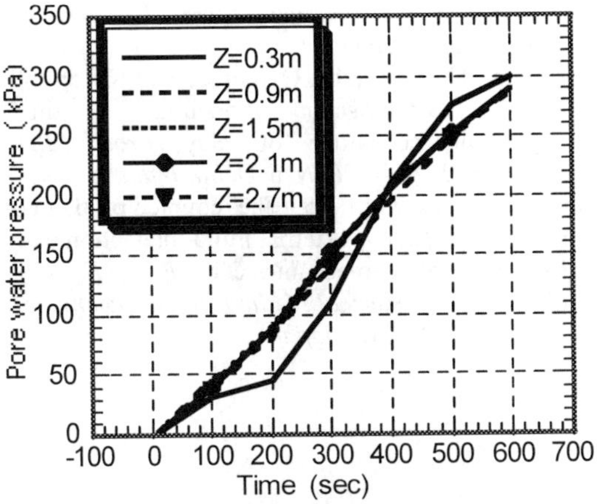

Figure 7. Pore water pressure versus time for medium dense sand.

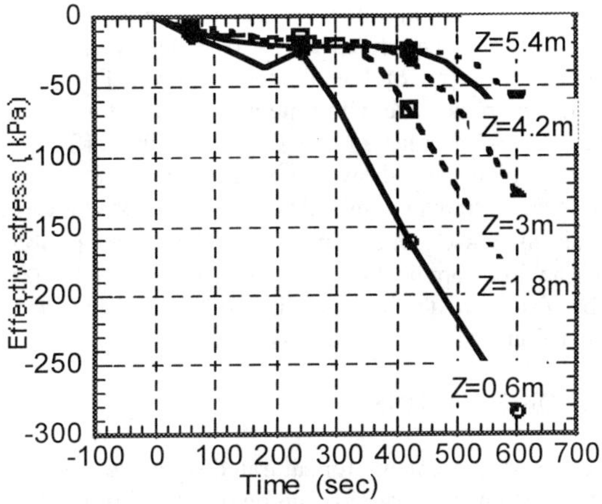

Figure 10. Effective stress versus time for loose sand. .

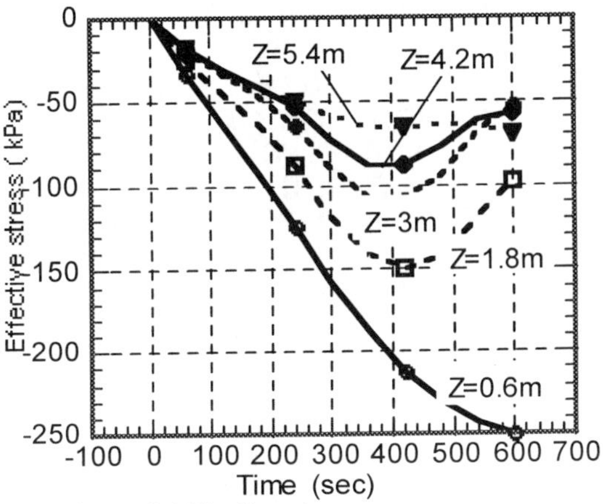

Figure 8. Effective stress versus time for normally consolidated clay.

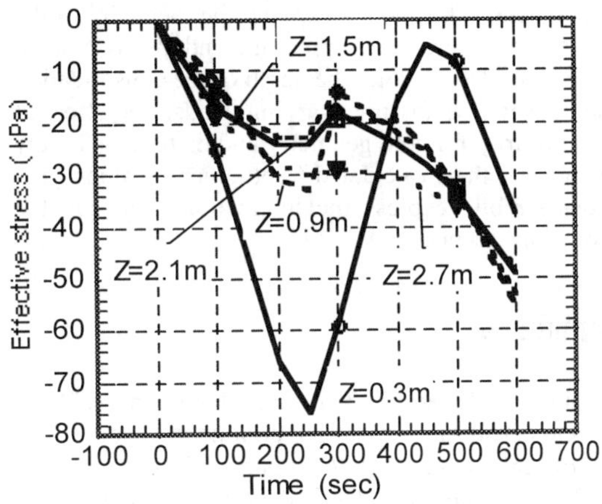

Figure 11. Effective stress versus time for medium dense sand.

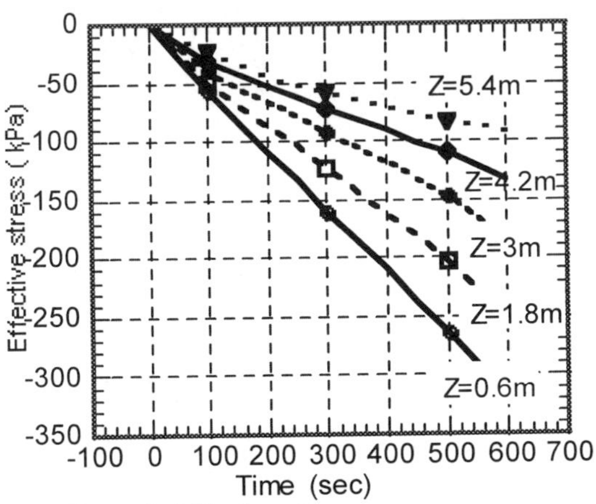

Figure 9. Effective stress versus time for over consolidated clay.

considerably express the interactions between the soil constituents.

Figures 4 and 5 show the development of pore water pressure (pwp) for over and normally consolidated clay with time along different levels of the consolidating soil layer. During the linear increase of the applied load, pwp values increases with respect to time by different rates. These rates also increase as the depth from the ground surface increases for both normally and over consolidated clay. This behavior could be explained based on the nature of the applied short term increasing load and the volume of voids for both clay types. Similarly, Figures 6 and 7 show the development of pwp values for loose and medium dense sand respectively. In case of medium dense sand, the pwp values generated and continue to increase with time by different rates within the period of load application. These rates seems to be nearly the same along the depth of the loaded soil layer. For loose sand, pwp increase with time up to cretin limit, then its values decrease. The corresponding effective stress values are shown in Figures 8, 9, 10, 11 for normally and over consolidated clay and loose and medium dense sand respectively. Within the period of the load application, the effective stress values changes with depth depending on the soil type and the volume of voids for the same soil type.

5. CONCLUISIONS

The field settlement for saturated soils due to short term increasing loads could be presented through a mathematical model based on Biot's theory of consolidation. The comparison between field and calculated settlement values indicate good agreement. The model explain the behavior of the developed pore water pressure and its associated effective stress along the depth of the consolidating soil layer due to time varying loads. The effect of nonlinearity for both geometry and material is highly improve the calculated settlement values and considerably express the interactions between the soil constituents.

REFERENCES

Bichir, S., Elmaghraby, A. and Mahmoud, A. 2000- Gometric and material nonlinearity soil under dynamic load (I). *Bulletin of Faculty of* Engineering, Cairo University, Cairo, Egypt.

Biot, M. A. 1940. General theory of three dimensional consolidation. Journal of Applied Physics. 12 : 155-164.

Janbu, N. 1967. Soil Compressibility as determined by odometer and triaxial tests. *Proc. European Conf. On Soil Mechanics and Foundation Engineering : 1 Weisbaden, Germany.*

Kondner, H. Y. and Zelasko, J. S. 1963. A Hyperbolic stress-strain formulation for sands. *Proc. 2nd Pan-American Conf. On soil Mechanics and Foundation Engineering : 1. Barzil.*

Siddhartthan R., Zafir Z., Gary M. N. 1993. Moving load response of layered soil, II : Verification and application. *Journal of Engineering Mechanics. 119 (10) : 2072-2089.*

Wong. K. S., and Duncan, J. M. 1974. Hyperbolic stress-strain parameters for nonlinear finite element analysis of stresses and movement of soil masses. *Report No. TE – 73 - 3 to National Foundation Geatechnical Engg. University of California, Berkeley.*

Yoshikuni, H., Kusakabe, O., Hirao, T., I Kegami. S. 1994. Elasto–viscous modeling of time-dependent hehaviour of clay. *Proc. 13 th ICSMFE. 1: 417-420. New Delhi, India.*

Zienkiewicz, O. C. 1985. The coupled problem of soil –Pore fluid – external Fluid interaction basis for a general geomechanics code. *Proc. Fifth Int. Conf . on Numerical Methods in Geomechanics: 1731-1740. Nagoyg Japan.*

Computer Methods and Advances in Geomechanics, Desai et al. (eds) © 2001 Balkema, Rotterdam, ISBN 90 5809 183 X

Stochastic modelling of regional groundwater flow in karstic areas

G. Klubertanz, O. Jaquet & P. Siegel
Colenco Power Engineering Limited, Switzerland

H. Benabderrahmane
Agence Nationale Pour la Gestion des Déchets Radioactifs (ANDRA), France

ABSTRACT: Modeling transient groundwater flow in karstic regions is a challenge both in terms of hydrogelogical conceptualization and grid generation. Based on a real case, this article presents new techniques first for generating stochastically in the previously conceptualized limestone layers karstic networks and then for building a 3D global finite element mesh that incorporate the continuous hydrogeological features 3D and the discrete patterns 1D for karstic network 2D for the fractures, required for a computation of this type.

The model area consists of 7 stratigraphic layers, three of which are karstic. Using a lattice gas simulator, karstic networks are generated stochastically at three levels independent of each other. Conditioning is achieved by imposing directional preferences by means of a potential field that includes spring locations. The density of the network is conditioned by surficial karstic manifestations and represented via a contribution to the potential field, which also includes main fault locations in the area. Finally, a layered 3D finite element mesh is built that incorporates the three karstic networks and the hydrostratigraphic units.

Based on field data, the regional groundwater flow model in this karstic area is established using observations.

1 INTRODUCTION

Karstic formations constitute aquifers with important water reserves all over the world. It is estimated that 25% of the global population is supplied with drinking water largely or entirely from karst aquifers. Responses of such, mainly limestone, aquifers to recharge by rain or inflow are very rapid, complex, and can be rather localised. These characteristics make karstic aquifers particularly vulnerable to pollution and overexploitation.

The improvement of groundwater management and protection in karstic areas require adequate numerical modelling. Due to their heterogeneity these aquifers are difficult to describe from a modelling point of view. The heterogeneous structure is characterised by highly permeable conduits (karstic networks) embedded in a relatively low permeability matrix that includes geological formations with spatially variable hydraulic properties. This system complexity has led to the selection of a stochastic approach in order to model karstic geometry.

The modelling approach is applied to a regional karstic limestone aquifer situated in the north-eastern part of France. Its vulnerability with respect to potential surface-pollution related to an industrial project needs to be assessed. The region to be modelled presents the following characteristics: (a) the matrix includes seven hydrostratigraphic units, (b) three of these units incorporate karstic networks connected via vertical conduits, (c) the stratigraphic layers are partially eroded and (d) faults and rivers are represented.

In this paper, an approach is developed to condition the karstic network simulation to observations; i.e. the simulated networks account for the location of karstic conduits and springs, as well as main flow directions as indicated by field data. Conditioning information includes observations of the karstic manifestation density, the location of springs and river water losses, conduit locations from borehole and piezometer data as well as fault locations and tracer data.

Karstic networks are modelled with a specific random walk approach. Conduit generation is achieved using a lattice gas simulator based on a stochastic differential equation accounting for a simplified representation of physico-chemical processes which lead to the formation of karstic networks (Jaquet. 1998). Within the numerical model, conduits are represented by 1D elements in a 3D finite element mesh.

2 GEOLOGY OF THE MODEL AREA

2.1 *Conceptual model*

The development of a hydrogeological model requires the conceptualisation of the model area's geometry and hydrogeology. Therefore, it is necessary to define the hydrostratigraphic units to be modelled as well as their respective properties. This information can be provided by geological and hydrogeological studies, including the interpretation of borehole data, seismic surveys and speleologic investigations. In the given case, a multitude of information of varying quality from various sources is available (see e.g. Barthélemy et al., 1998; Fauvel, 1997; Fillion et al., 1999; Normand et al., 1997; Normand & Stollsteiner, 1998; Paris et al., 1998; Paris & Robert, 1997; Vinsot et al., 1998; Wendling et al., 1996).

As shown in Table 1, seven main hydrostratigraphic units are observed in the region (units 1 through 7), three karstic networks are incorporated into units 3, 5 and 7 (k1 through k3) as well as some important faults (f). The karstic networks are interconnected vertically. Numerous springs are present all throughout the model area. Material properties and geometries are subject to uncertainties as indicated in Table 1.

2.2 *Hydrogeologic units*

The épikarst (unit 2) is a thin, fractured and highly permeable layer on top of the limestone formations. It outcrops at the surface if there is no Crétacé inférieur (unit 1). These two units concentrate the water flow close to the surface and direct it towards the karstic networks.

The units "calcaires cariés", "calcaires de Dommartin" and "calcaires lithographiques" are highly karstified aquifers and each supposedly contains a karstic network. These 3 units are separated by semi-permeable layers of Oolithe de Bure and Pierre Châline, i.e. non-karstic units which may be crossed by the vertical karstic connections.

3 GENERATION OF KARST NETWORKS

Due to the complexity of the karstic geometry and the small amount of available information (e. g. the location of most network conduits remains unknown), a stochastic approach was chosen for the modelling of the network geometry.

The karstic geometry is modelled by solving the stochastic differential equation of Langevin which is a simplified description of the coupled physico-chemical processes governing the karstification (Jaquet, 1995). The non-linearity of this equation requires a numerical resolution obtained with a lattice gas simulator (Doolen et al., 1990).

The applied simulator, GARST (**GA**z sur **R**éseau pour la **S**imulation de la **T**ubulure karstique; Jaquet, 1998) is based initially on a regular network representing a fractured medium within which particles might move according to a random walk process

Table 1: Hydrogeologic units

Lithology	Thickness[m]	Karstification	Hydrogeologic unit		Permeability [m/s]
Crétacé inférieur (limestone, clay and sand)	0 – 20		1		1E-06 to 1E-02
Epikarst (shallow and covered karst) (limestone)	10 (if Crét.=0)	yes	2		1E-06 to 1E-02
Calcaires cariés (limestone aquifer)	Variable	yes	3	karst network k1	matrix: 1E-08 to 1E-03 network > 1E-01
Oolithe de Bure (compact limestone)	1.8 – 2.2	-	4		1E-09 to 1E-06
Calcaires de Dommartin (limestone aquifer)	34	yes	5	karst network k2	matrix: 1E-08 to 1E-03 network > 1E-01
Pierre Châline (marl + limestone)	8 – 10	-	6		1E-09 à 1E-06
Calcaires lithographiques (limestoneaquifer)	37	yes	7	karst network k3	matrix: 1E-08 to 3E-03 network > 1E-01
Base of the Calcaires du Barrois					
Vertical connections	1D	-		k4	network > 1E-01
Faults	0 (2D)	-		f	1E-04 à 1E-03

with memory. The trajectories of the particles displaced within the lattice gas constitute the network of simulated conduits. Furthermore, these karstic simulations are conditional, i.e. the generated karstic networks respect the available hydrogeological information. This conditioning is achieved during the deterministic and stochastic component of the particle motion. Regarding the deterministic component, the conditioning information is integrated using a potential field that comprises:

- General flow direction.
- Karstic observations at the surface: variability in the potential according to the observations.
- Karstic springs (by hydrologic unit): karstic conduits are forced to pass the spring locations because of an attractor in the potential field. The force of the attractor is proportional to the spring outflow.
- Major faults: variability in the potential according to fault location.

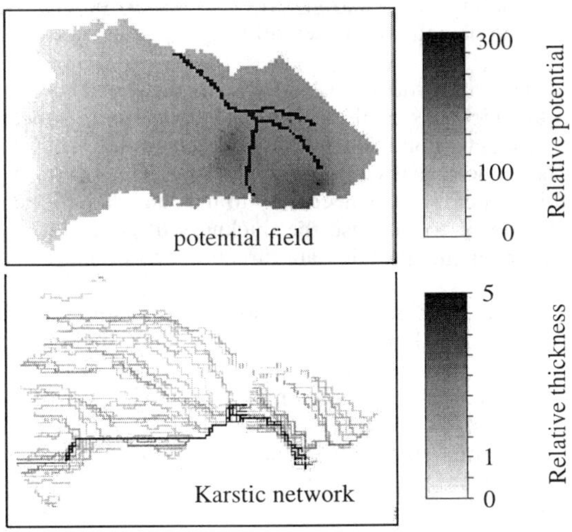

Figure 1: Potential field and conditioned karstic network

Concerning the stochastic component, the conditioning is achieved by:
- Respecting the geological medium heterogeneities: first, the particles tend to follow major faults present in the region and second, karstic observations.
- Describing coupled (memory) effects: the larger the conduit, the more the particles tend to cross it.

In addition, the following observations were also included in the karstic network simulation:

- Known conduit: if a borehole or piezometer is connected to the karstic network, this location was chosen as a starting point for at least one particle to assure a connection to the simulated network.
- Erosion: some hydrostratgraphic units have been eroded. The limits of the generated network were adapted either to the observed or the supposed limits of their corresponding unit.
- Water loss of rivers: some rivers in the region show a loss of water due to karstic conduits. In order to create a connection to the simulated network, these locations were selected as particle starting point.

A consequence of this approach is that the stochastic component becomes less important with time as more preferential pathways are created and it has less influence on the particle motion, the closer the particle is to an attractor (e.g. to a spring). The pathways followed by the particles injected randomly at the upstream limit of the modelled region reaches, after some time, steady state conditions, i.e. no new conduits are created; the karstic simulation is in equilibrium with the conditioning information.

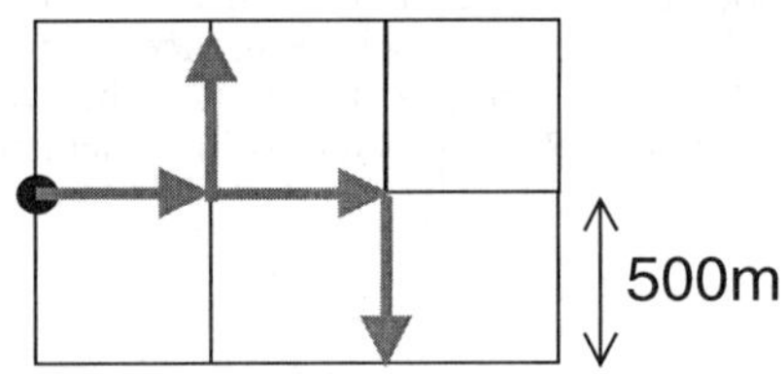

Figure 2: Principle of the lattice gas generator

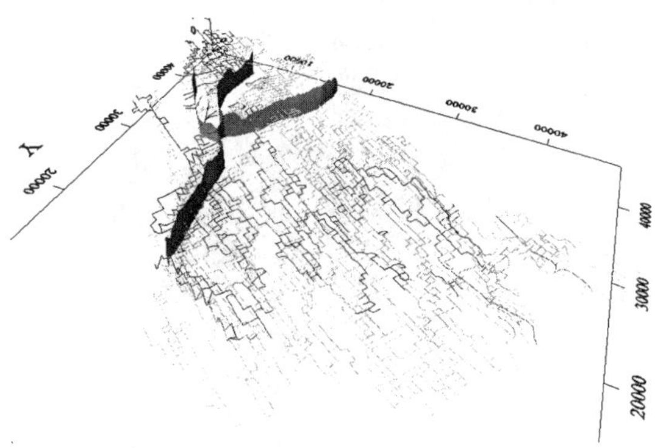

Figure 3: Discretisation of the shear zone and the 3 karstic networks at different levels

This generation process is applied independently to the three karstic layers, considering the individual information available for each unit. At the end, vertical connections from each of the networks towards the surface are generated according to the density of surficial karstic observations. Sources and piezometers that are connected to one of the networks are also linked via vertical tubes to their respective layers.

This conditioning procedure was developed and integrated in the GARST simulator within the framework of this project (Klubertanz, 1999).

4 GRID GENERATION

4.1 *2D base mesh*

The layered grid to be built requires the generation of a 2D base mesh integrating all information and limits to be meshed in the third dimension, i.e geologic and hydrogeologic constraints as well as the karstic networks and all emplacements of sources, piezometers and rivers.

To obtain an appropriate 2D discretisation, the Delaunay mesh generator MAGICS (**M**esh **A**utomatic **G**enerator for **I**ntegrated **C**omplex **S**tructures; Siegel, 2000) generates a finite elements mesh according to the severe restrictions concerning geometry and element size (see Figure 4). The final 2D mesh contains 8'815 nodes and 10'621 elements.

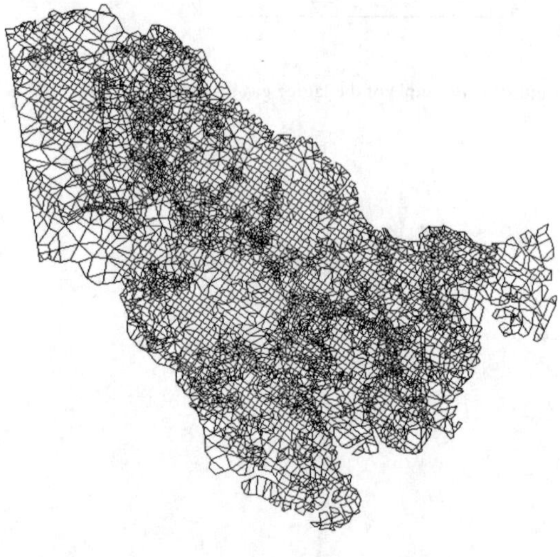

Figure 4: 2D base mesh

4.2 *3D grid*

The 3D mesh is generated using a multilayer technique, i.e. the 2D base mesh, which contains all necessary information, is projected into the third dimension, based on the thickness information of the respective layers. As some layers are partially eroded, discontinuous or even missing over large parts of the model area, the expansion to 3D requires particular attention. Thickness information of the layers is first obtained from boreholes and seismic surveys. Then kriging is applied as an interpolation technique where necessary. In particular, the following information has been considered:

- Topography, based on a numerical elevation model; where necessary, corrected for known elevations at the sources and also to assure the monotonic decrease of altitude in the river valleys.
- Lower limit of the Crétacé.
- Thickness information for each layer, provided by seismic surveys or field observations; if no information was available, a constant thickness was assumed
- Lower limit of the calcaires lithographiques, including information provided from boreholes and piezometric observations.

This information is sufficient to determine the total thickness of the considered formations, which is used to control the 3D grid generation and to check the consistency of the layer-thickness data. In this context, it was necessary to introduce additional erosion or to increase the thickness of some layers wherever the altitude data turned out to be not consistent.

Finite element generation was done with RASTELM (Rivera et al., 1998). This multilayer mesh generator uses the 2D mesh and the obtained layer thickness information at every node of the 2D base mesh to generate the 3D finite element grid.

The final network contains more than 270'000 1D, 2D and 3D elements and more than 309'000 nodes.

4.3 *Incorporation of karstic networks*

The generated karst networks are incorporated into the grid by including them as 1D finite elements in the horizontal plan at the middle of their respective limestone layer. The varying transmissivity of the karstic network is modelled by attributing different values of permeability to the elements, depending on the number of particles that cross the corresponding conduit section during the lattice gas simulation process. The karstic networks are interconnected or directly connected to the surface through vertical conduits according to the density of observed surface karstic manifestations. Surface connections are forced where springs and piezometers are located.

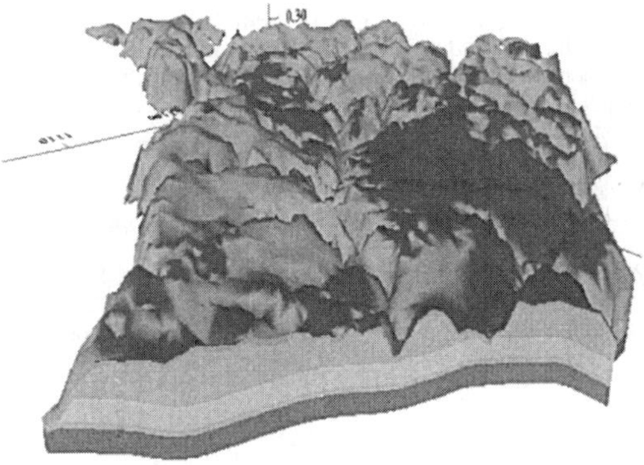

Figure 5: Cross-section through the model with colors corresponding to the hydrostratigraphic layers

4.4 *Boundary conditions*

The conceptual model extensions and features were defined in order to be able to apply the appropriate boundary conditions. The outer boundaries and the bottom of the model are no-flow boundaries, with the exception of the north-eastern part, where a fixed potential is imposed. On the land surface and based on the available meteorological data, four zones with different infiltration rates were defined. Particular attention was devoted to the main rivers in the region where hydraulic interactions between these rivers and the karstic networks are generally characterised by springs and/or water losses. The rivers are subject to significant volume losses at some places, providing an important water supply to the karstic networks. This supply has been accounted for by imposing a variable volume flux at the locations of the observed losses. Elsewhere rivers are modelled imposing a fixed potential corresponding to 0 pressure at the surface.

5 MODEL CALIBRATION

The model calibration is performed by adjusting the computed head and flux until they match the measured values. First, a steady state flow model has to be calibrated in order to adjust the model parameters and to produce an initial condition for the next transient flow model. A transient flow model simulates the hydraulic behaviour of the aquifers system during one year with time step of one month. A second transient model is extracted from the first one in order to simulate the fast groundwater flow which characterise the karstic aquifers. The preliminary computed results seem to be in good agreement with the field observations.

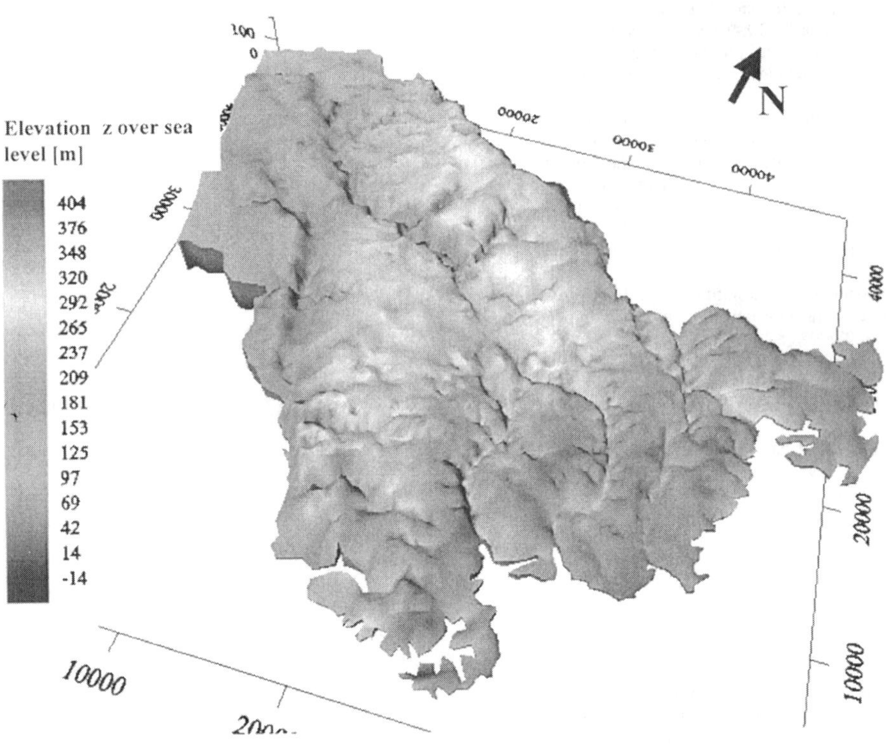

Figure 6: View of the final model with colors corresponding to absolute altitude

6 CONCLUSIONS

A numerical hydrogeological model of a complex karstic region in northern France has been built. The chosen approach, based on a stochastic generation of the karstic network, includes a new procedure to account for "hard" geological and "soft" statistical information of both geology and karstic networks. It has been shown that the proposed approach yields good results and that the generated hydrogeological model describes well the main features in the region. Grid generation is demonstrated to be feasible by combining the very high number of constraints while at the same time respecting severe criteria on the element size. Both of these are important conditions in order to obtain reliable numerical results using e.g. the finite element technique.

REFERENCES

Barthélemy Y., Eberschweiler A. & Gutierriez A., 1998. Site de l'Est synthèse des reconnaissances hydrogéologiques, D RP 0ANT97-058/A

Doolen G.D., Frisch U., Hasslacher B., Orszag S. & Wolfram S., 1990. Lattice gas methods for partial differential equations, Addison-Wesley publishing company, New York, 555p.

Fauvel P-J., 1997. Données géologiques, hydrogéologiques et spéléologiques concernant les karsts, Journées Scientifiques, Atlas des posters, Bar-le-Duc, E.H.2, p.43

Fillion E., Vinsot A. & Altman S., 1999. Note de synthèse sur la modélisation hydrologique du système aquifère des calcaires du Barrois: Note Technique Andra ; D NT AHYG 99-004

Jaquet O., 1998. Modèle stochastique de la géométrie des réseaux karstiques, Thèse, Université de Lausanne, 131p.

Jaquet O., 1995 Modèle probabiliste de réseaux karstiques: équation de Langevin et gaz sur réseau, Cahiers de Géostatistique, fascicule 5, Ecole des Mines de Paris, pp.69-80

Kinzelbach, W., Marburger M. & Chiang W.-H., 1992. Determination of groundwater catchment areas in two and three spatial dimensions. J. of Hydrol., 34, pp.221-246.

Klubertanz G., 1999. Erstellen eines konditionierten FE Netzes mit eingebetteten, eindimensionalen Elementen, Colenco Memorandum, No 4627/002

Normand M., Stollsteiner P. & Allemmoz M., 1997. Modélisation hydrologique du système aquifère des calcaires du Barrois, Phase 1 – Collecte des données et études, D RP 0ANT 97-061/A

Paris B., Cavarroc J., Beach J., Chalot J.C., 1998. Modélisation hydrogéologique de secteur – site est, rapport final, Volumes 1, 2 & 3 , D RP O INA 98-001

Normand M. & Stollsteiner P., 1998. Modélisation hydrologique du système aquifère des calcaires du Barrois, Phase 2 – Synthèse des données disponibles et proposition d'un schéma hydrologique conceptuel , D RP 0ANT 98-007/A

Paris B., Cavarroc J., Beach J. & Chalot J.C., 1998. Modélisation hydrogéologique de secteur – site est, rapport final, Volumes 1, 2 & 3 , D RP O INA 98-001

Paris B. & Roberts R., 1997. Mesures de transmissivités par essais de puits et analyse de sensibilité, Journées Scientifiques, Atlas des posters, Bar-le-Duc, E.H.6, p.51

Rivera A., Jaquet O. & Genter M. 1998. Laboratoire souterrain Andra – site du Gard – Modélisation hydrogéologique à l'échelle du secteur, Volumes 1 & 2, BRP 0CPE 98-001

Siegel P. 2000, MAGICS: mailleur automatique contraint pour structures complexes à 2 et 3 dimensions, Colenco Memorandum, No 4627/003

Vinsot A., Fillion E. & Altmann A., 1998. Note de synthèse sur les modélisations régionale et de secteur du site de l'Est, DNT AHYG 98-004

Wendling J., Barthélemy Y. & Vandenbeusch M., 1996. Modélisation hydrogéologique régionale – Etape 2, Modèle régional – numérisation des données – tests de cohérence site Est, B RP O. ANT 96.122, Rapport ANDRA

Computer Methods and Advances in Geomechanics, Desai et al. (eds) © 2001 Balkema, Rotterdam, ISBN 90 5809 183 X

A one-dimensional model of EPS fill on soft soils

C.J.Leo & Y.Zou
Department of Civil Engineering, University of Western Sydney, N.S.W., Australia

ABSTRACT: Time-dependent long-term deformation of EPS fill overlaying soft soil is of practical interest to engineers and designers. A model of EPS fill resting on consolidating soft soil presented in this paper can enable designers to assess the long-term performance of the EPS fill solution used for overcoming settlement problems on soft soils. The model considers two distinct materials: the viscoelastic material which is the EPS fill and the poro-viscoelastic material which is the soft soil. The deformations from the two materials are superimposed to obtain the total settlement with time.

1 INTRODUCTION

Expanded Polystyrene (EPS) geofoam is essentially used in ground fill applications where the light-weight material is required to reduce the stresses on underlying soils. The long-term behaviour of the foam has always been of practical interest to engineer and designers needing to make a critical assessment of its performance for the duration of its design life.

Long term behaviour of a material is defined by various aspects of the material response under stress and strain loading and includes the time-dependent phenomena of creep and stress relaxation. In fill applications, it is the creep response in the geofoam that will be a major design consideration. The authors in the present paper collected data from their laboratory creep experiments so as to develop a viscoelastic model of the EPS geofoam which include creep effects. This is then combined with the time-dependent consolidation of soft soil to produce a model that is capable of simulating the long term settlement behaviour of a typical EPS fill application.

It is worth noting from the onset that EPS fill and the underlying soft soils have to be regarded quite differently since the earlier does not deform in the same manner of consolidating soil where excess pore water are being expelled under loading. Thus the EPS geofoam is treated as a viscoelastic material whereas the consolidating soil is considered a poro-viscoelastic material. An illustrative example is given showing the relative merits of using the EPS fill where the foundation soil is very soft.

2 DEVELOPMENT OF A MODEL

The model in the present paper considers the deformation of two material types: EPS fill and the underlying consolidating soft soil. A schematic of this is illustrated in Figure 1.

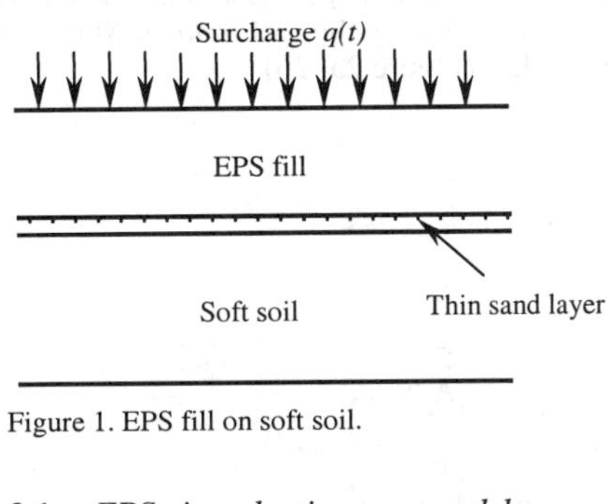

Figure 1. EPS fill on soft soil.

2.1 *EPS viscoelastic creep models*

With few exceptions, the fundamental equation for creep models has the following qualitative form (Findley, 1976):

$$\varepsilon_z(t) = \varepsilon_0 + \varepsilon_c(t) \tag{1}$$

where: $\varepsilon_z(t)$ = the total strain at time t after the stress application; ε_0 = the instantaneous strain upon a stress application and $\varepsilon_c(t)$ = the time-dependent creep component of strain at some time t after the stress application. For the small stress ranges considered in the present paper and which occur under representative conditions in pavement and embankments, the instantaneous strain gives,

$$\varepsilon_0 = \frac{\sigma_z}{E_1} \tag{2}$$

where σ_z is the applied stress, E_1 is the linear spring constant (or the Young's modulus) of the geofoam.

If the stress history is piecewise differentiable, then the creep component of strain under any given stress history may be expressed as a convolution integral,

$$\varepsilon_c(t) = \int_0^t J(t-\tau)\frac{\partial \sigma_z(\tau)}{\partial \tau}d\tau \tag{3}$$

where the kernel $J(t)$ is the compliance for the creep component of strain whose form depends on the model adopted. Equation (3) forms the basis of hereditary creep in that the material "remembers" the stress history it has been subjected to.

Laboratory creep tests on EPS results in typically parabolic creep curves presented in Figure 3. The present authors have made an evaluation of the viscoelastic mechanical models to obtain a suitable compliance relationship that is considered applicable to EPS geofoam. Mechanical models of various forms are already widely used to model a number of material and has been described as being suitable for polystyrene material. These models are made up of combinations of (Hookean) springs and (Newtonian) dashpots as exemplified by the well known Maxwell and Kelvin models (Figure 2a, 2b).

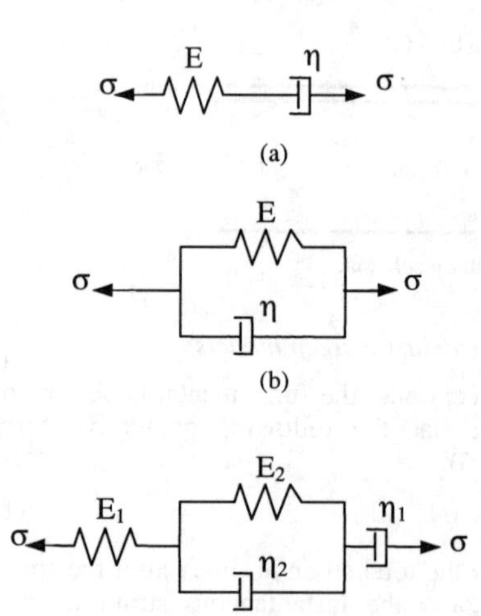

Figure 2. (a) Maxwell model (b) Kelvin model (c) Burgers model.

Initially, a four-element or Burgers model (Figure 2c, where a Maxwell model and a Kelvin model are connected in series), was thought to be suitable for capturing the creep characteristics of EPS geofoam since it could model both primary and secondary creep. In a Burger's model, the strain response when a constant stress σ_0 is applied at $t = 0$ is,

$$\varepsilon_z(t) = \frac{\sigma_0}{E_1} + \frac{\sigma_0}{\eta_1}t + \frac{\sigma_0}{E_2}\left(1 - e^{-\frac{E_2}{\eta_2}t}\right) \tag{4}$$

so that the compliance equation gives,

$$J(t) = \frac{1}{\eta_1}t + \frac{1}{E_2}\left(1 - e^{-\frac{E_2}{\eta_2}t}\right) \tag{5}$$

It is observed that the strain behavior described in Equation (4) allows for an instantaneous strain as well as the strain rate to approach asymptotically the value of σ_0/η as t tends to infinity. The fact that this model captures both primary and secondary creep characteristics is useful, but as experimental data yields typically parabolic creep curves a modification was introduced to enable the creep strain rate asymptote to a parabola as t tends to infinity. The strain equation for the modified four-element model for a constant stress σ_0 is applied at $t = 0$ is,

$$\varepsilon_z(t) = \frac{\sigma_0}{E_1} + \frac{\sigma_0}{\eta_1}t^n + \frac{\sigma_0}{E_2}\left(1 - e^{-\frac{E_2}{\eta_2}t}\right) \tag{6}$$

where n is a creep parameter and the compliance equation for creep gives,

$$J(t) = \frac{1}{\eta_1}t^n + \frac{1}{E_2}\left(1 - e^{-\frac{E_2}{\eta_2}t}\right) \tag{7}$$

In Figure 3a where the experimental data and the model results of both the 4-element model and modified 4-element model are plotted, it quite clearly shows a superior goodness of fit for the later model. Similar results were also presented by Taylor et al. (1997), in applying mechanical models to flexural creep defection of structural insulated panel.

The present research is done for a geofoam material supplied by an Australian firm, with a small stress range (0-40 kPa) in mind. This stress range has practical interest since it typically occurs in representative conditions in road sub-base and embankments where EPS fills are being used. Creep data from the laboratory creep tests were collated and normalised with respect to the applied stress i.e. ε_c/σ_0, then the creep parameters were determined using a Levenberg-Marquardt nonlinear least squares regression procedure. The parameters of the modified 4-element model for EPS foam of specified 20 kg/m^3 density adopted in the current paper are tabulated in Table 1.

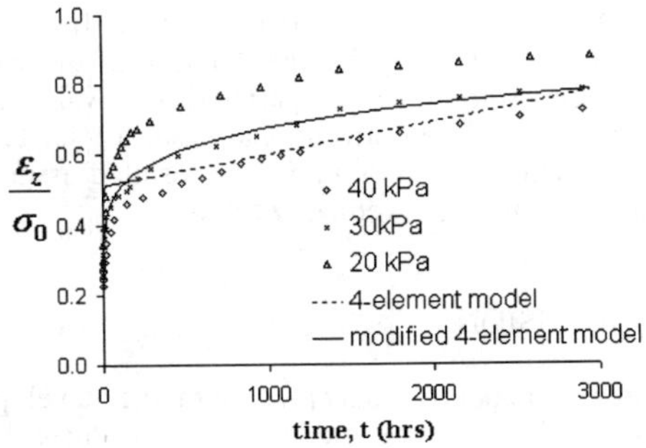

(a)

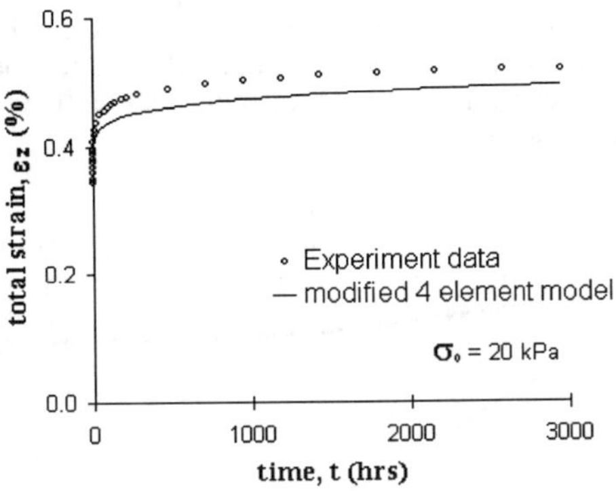

(b)

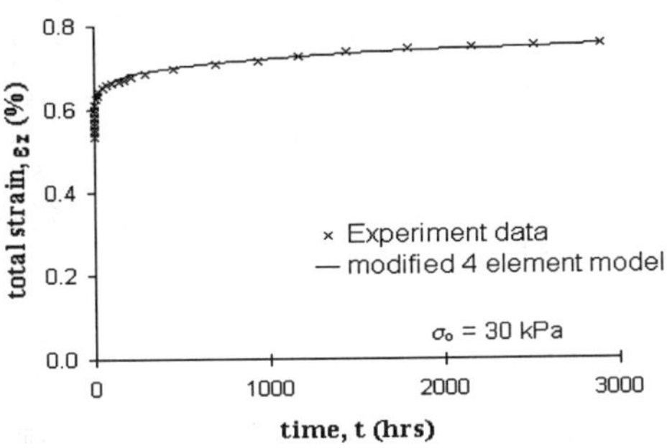

(c)

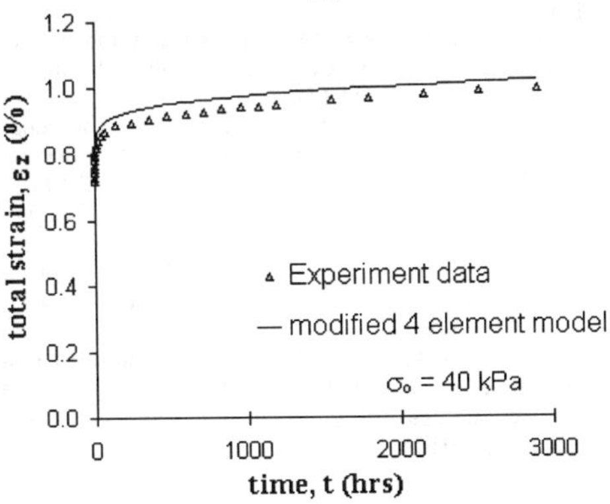

(d)

Figure 3. Experimental and model results of creep tests

Table 1. Parameter values of modified 4-element model for EPS

n	E_1 (kPa)	η_1 (kPa-yr)	E_2 (kPa)	η_2 (kPa-yr)
0.165	5966.2	13553.	77520.	0.131

Figure 3b,c,d shows good fit for the strains produced at various stress levels using the modified 4-element model.

2.2 *Soft Soil*

In field applications, the EPS blocks rest on a soft soil foundation, commonly separated by a thin layer of sand used both for leveling and as a drainage layer (Figure 1). The consolidation of the soil with time effects can be described by the constitutive equation:

$$\varepsilon_c(t)=\frac{\sigma_z'(t)}{E_1}+\int_0^t J(t-\tau)\frac{\partial\sigma_z'(\tau)}{\partial\tau}d\tau \tag{8}$$

where σ_z' = effective vertical stress. If a 4-element (Burgers) model had been use for the consolidating soil as is the case in the present paper then the constitutive equation gives,

$$\varepsilon_z=\frac{\sigma_z'}{E_1}+\int_0^t\frac{\sigma_z'(\tau)}{\eta_1}d\tau+\int_0^t\frac{\sigma_z'(\tau)}{\eta_2}e^{-\frac{E_2}{\eta_2}(t-\tau)}d\tau \tag{9}$$

where: E_1, E_2, η_1, η_2 = model parameters. The constitutive equation has to be solved in combination with the continuity equation,

$$\frac{\partial\varepsilon_v}{\partial t}=\frac{\partial\varepsilon_z}{\partial t}=-\frac{k_v}{\gamma_w}\frac{\partial^2 u}{\partial z^2} \tag{10}$$

in which: ε_v = volumetric strain; k_v = coefficient of vertical permeability; γ_w = unit weight of fluid; u = excess pore pressure. Interested readers may refer to the companion paper (Leo and Xie, 2001) for a discussion on the details of the consolidation of a layered system.

3 SOLUTION OF EPS GEOFOAM ON SOFT SOILS

The solution of the Equation (3) containing a convolution integral is significantly facilitated by introduction of a Laplace transform defined as,

$$\bar{\varepsilon}_c=\int_0^t\varepsilon_c(\tau)d\tau \tag{11}$$

which from Equations (1),(2) and (3) gives,

$$\bar{\varepsilon}_z = \left(\frac{1}{E_1} + s\bar{J} \right) \bar{\sigma} \quad (12)$$

where $\sigma(0) = 0$. It is noted here that in the present model $\sigma(t) = q(t)$, the surcharge load which is simulated by a ramp loading representing the construction sequences from beginning until completion when a steady load is imposed, although other loading functions may also be applied. The compliance Equation (7) yields,

$$\bar{J} = \frac{1}{\eta_1} \frac{\Gamma(n+1)}{s^{n+1}} + \frac{1}{s(s\eta_2 + E_2)} \quad (13)$$

where Γ is the Gamma function. The strain in real time is found by numerically inverting Equation (12) using Talbot's (1979) inversion algorithm. The settlement arising from the deformation of the EPS layer is obtained from the product of the strain and the depth of EPS fill layer. This result is superimposed on to the settlement from the consolidating soft soil (Leo and Xie, 2001) to obtain the total settlement of the system.

4 ILLUSTRATIVE EXAMPLE

The present model is being developed as an analytical tool to assess the long-term settlement of EPS fill resting on very soft consolidating soil. The same model can also be used to gauge these results against the 'no EPS' case where conventional fill material are used and considerably higher self-weight would be imposed on the consolidating soil.

The illustrative case considered in this paper has a 5 m high 20 kg/m^3 density EPS embankment situated above 5 m of very soft clay, this in turn overlies fairly firm stratum. The values of the model parameters for the EPS fill are provided in Table 1 while those for the very soft soil are presented in Table 2 below.

Table 2. Parameter values of 4-element model for soft soil

E_1 (kPa)	η_1 (kPa-yr)	E_2 (kPa)	η_2 (kPa-yr)	c_v (m^2/yr)
1000	0.5×10^6	1000	25	5

A pavement was constructed directly above the EPS fill during a 6 months period, which imposes an ultimate surcharge of 20 kN/m^3. The present model is applied to evaluate the expected surface settlement over the next 100 years. In the 'no EPS' case where a conventional fill material has been used in the construction of the embankment, it is assumed that this would cause an ultimate surcharge of 100 kN/m^3 (instead of the 20 kN/m^3) based on a unit weight of 16 kN/m^3 for the fill material. The expected settlement results for both these cases are plotted in Figure 4. The results indicate that the EPS fill performed quite effectively for the duration of the 100 years in that settlement is restricted to a value of about 0.25 m. In the conventional case, a total settlement of about 1.1 m would have resulted assuming no other soil improvement was done.

5 CONCLUSION

The present paper is concerned with the development of an EPS-soft soil model used to evaluate the long-term effectiveness of EPS fill solution for overcoming settlement problems in soft soil.

The developed model is simple to use and can be extended to simulate other aspects of construction and geotechnical design considerations currently not provided for.

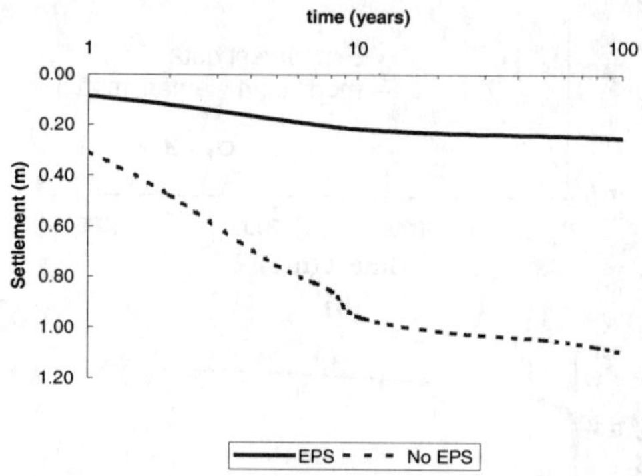

Figure 4. Settlement plot for use and non-use of EPS foam in illustrative example.

REFERENCES

Findley W.N., Lai J.S and Onaran K. (1976), 'Creep relaxation of nonlinear viscoelastic materials', *Dover Publications, Inc.*, New York, N.Y.

Leo C.J. and Xie K.H. (2001) 'An efficient solution of one-dimensional consolidation of layered soft soils with time effects', submitted to 10th IACMAG.

Zou Y. and Leo C.J. (1998) 'Laboratory studies on the engineering properties of expanded polystyrene (EPS) material for geotechnical applications', *Proceedings of the 2nd International Conference on Ground Improvement Techniques*, pp 581-588, 7-9 Oct, Singapore.

Talbot A. (1979), 'The accurate numerical integration of Laplace transforms', *Journal Inst. Mathematical Applications*, Vol. 23, pp97-120.

Taylor S.B, Manbeck H.B., Janowiak J.J. and Hiltunen D.R. (1997), 'Modeling structural insulated panel (SIP) flexural creep deflection', *Journal of Structural Engineering, ASCE*, Vol. 123 (12), pp1658-1665.

Computer Methods and Advances in Geomechanics, Desai et al. (eds) © 2001 Balkema, Rotterdam, ISBN 90 5809 183 X

An efficient solution of one-dimensional consolidation of layered soft soils with time effects

C.J.Leo
Department of Civil Engineering, University of Western Sydney, N.S.W., Australia

K.H.Xie
Department of Civil Engineering, Zhejiang University, Hangzhou, People's Republic of China

ABSTRACT: A transfer matrix method is developed to determine the consolidation of soft soils with time effects caused by the construction of an embankment. The soil is represented as poro-viscoelastic material incorporating a 4-element (Burgers) viscoelastic model, but the methodology presented is also generally applicable to other viscoelastic model of choice. A brief discussion of the parameter identification methodology is given as well as an illustrative example to complete the treatment of this subject.

1 INTRODUCTION

When Terzaghi laid the theoretical foundations for the consolidation problem, he considered the soil as a two-phase system consisting of solid particles and water. The theory predicts that as water gets expelled from the pore spaces in a soil sample under a given loading, the resulting consolidation conforms to a unique settlement curve when suitably scaled. The consolidating soil sample only experience primary consolidation according to the theory in that the settlement asymptotes to an ultimate value as time approaches infinity and as the excess pore pressure dissipates.

In practice, however, the theoretical results do not always correspond closely with the measured data indicating that with certain samples, a significant secondary consolidation caused by creep effects is present with increasing elapsed time, after the pore pressure dissipation is complete and effective stresses become constant. As well, both instant compression and creep occur simultaneously during primary consolidation as is normally observed in the deformation in other common non-porous materials. The above shows that conventional consolidation analyses that ignore creep effects will tend to underestimate the magnitude and the rate of settlement and over-estimate the rate of pore pressure dissipation in the field. These effects can be important in many practical applications.

Until recently, time-dependent effects like creep are commonly neglected in the geotechnical engineer's calculations because of a lack of appropriate analytical tools as well as a perceived difficulty with obtaining the time effects parameters that are needed to go into the analysis. Modern computers and the development of efficient techniques of analysis (e.g. Booker and Small, 1985) have changed all this and made it rather easy for the geotechnical engineer to solve the consolidation problems where time effects are significant.

The present paper shows how time effects like primary and secondary creep can quite easily be included in the analysis of consolidating soft soil. Here, the consolidating soil is treated as a poro-viscoelastic material where the viscoelasticity is modelled as a combination of springs and viscous dashpots. An efficient transfer matrix method is then developed to solve the problem of settlement caused by the placement of surcharge loading on a soft soil foundation. To obtain the values of the model's parameters, a brief discussion of a proposed parameter identification methodology is also provided.

2 GOVERNING EQUATIONS

Assuming a 4-element viscoelastic Burgers model, the consolidation of the soil with time effects can be described by the constitutive equation:

$$\varepsilon_z = \frac{\sigma'_z}{E_1} + \int_0^t \frac{\sigma'_z(\tau)}{\eta_1} d\tau + \int_0^t \frac{\sigma'_z(\tau)}{\eta_2} e^{-\frac{E_2}{\eta_2}(t-\tau)} d\tau \quad (1)$$

where ε_z = vertical strain; σ'_z = effective vertical stress; E_1, E_2, η_1, η_2 = model parameters. Although a 4-element mechanical model consisting of a Maxwell and a Kelvin model connected in series (Figure 1) is discussed in the present paper, the methodology presented allows for other viscoelastic model of choice to be adopted.

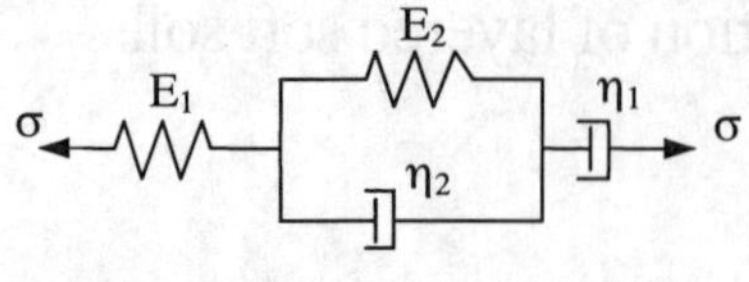

Figure 1. The 4-element Burgers model

Equation (1) has to be substituted into the continuity equation,

$$\frac{\partial \varepsilon_v}{\partial t} = \frac{\partial \varepsilon_z}{\partial t} = -\frac{k_v}{\gamma_w}\frac{\partial^2 u(z,t)}{\partial z^2} \tag{2}$$

where: ε_v = volumetric strain, k_v = coefficient of vertical permeability, γ_w = unit weight of fluid, $u(z,t)$ = excess pore pressure, to give the 1D equation consolidation with time effects,

$$c_v \frac{\partial^2 u(z,t)}{\partial z^2} = \frac{\partial u}{\partial t} + \frac{E_1}{\eta_1} u(z,t) + \frac{E_1}{\eta_2}\int_0^t \frac{\partial u(z,t)}{\partial \tau} e^{-\frac{E_2}{\eta_2}(t-\tau)} d\tau + f_q(t) \tag{3}$$

in which $c_v = k_v E_1/\gamma_w$,

$$f_q(t) = -\frac{\partial q(t)}{\partial t} - \frac{E_1}{\eta_1} q(t) - q(0) e^{-\frac{E_2}{\eta_2}t} - \frac{E_1}{\eta_2}\int_0^t \frac{\partial q}{\partial \tau} e^{-\frac{E_2}{\eta_2}(t-\tau)} d\tau$$

and $q(t)$ = applied loading. Readers may also immediately observe the relationship $E_1 = 1/m_v$ where m_v is the coefficient of volume compressibility.

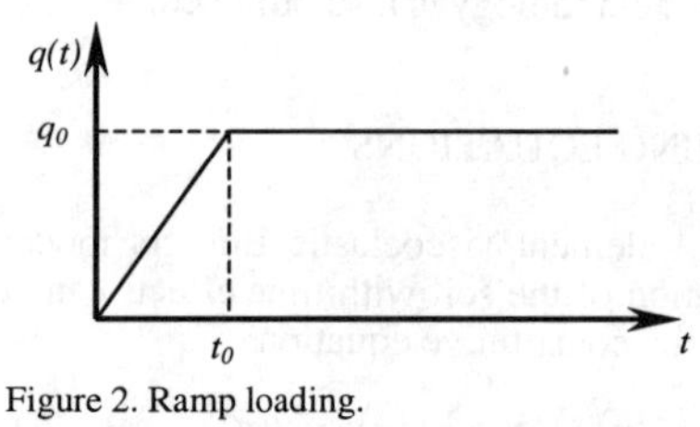

Figure 2. Ramp loading.

The time loading from commencement of construction to completion of an embankment may commonly be represented as a ramp shown in Figure 2 in which case,

$$q(t) = \begin{cases} \frac{q_0}{t_0} t & 0 < t \le t_0 \\ q_0 & t > t_0 \end{cases} \tag{4}$$

where q_0 = the ultimate loading; t = elapsed time; t_0 = the time taken from commencement to completion of construction of the embankment.

3 DEVELOPMENT OF PROPOSED TRANSFER MATRIX METHOD

The consolidating soil is often layered as a result of natural deposition process and so can be represented approximately as a finite number of distinct homogenous layers 1,2,…J (Figure 3). It is possible each soil layer may have its distinct consolidating as well as creep characteristics governed by Equation (3). The boundary conditions for the system are: $z = 0$; $u(0,t) = 0$; $z = H$; $u(H,t) = 0$ or $\partial u(H,t)/\partial z = 0$.

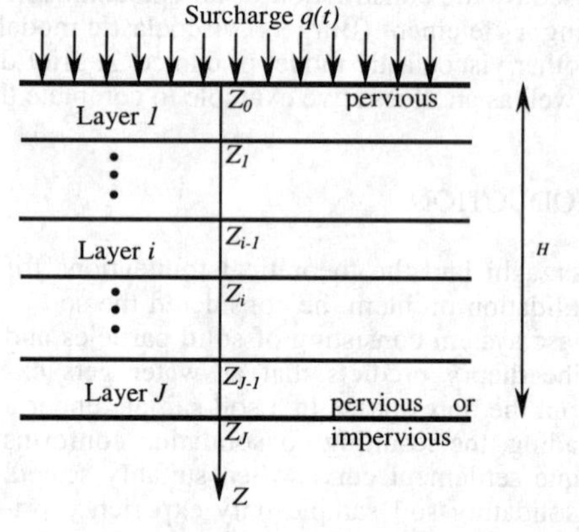

Figure 3. Layout of layered system.

A Laplace transformation is now introduced to eliminate the time variable in the governing equations and so facilitate the solution of the consolidating problem with time effects. Defining the Laplace transformation as,

$$\bar{u}(z) = \int_0^t u(z,\tau) d\tau \tag{5}$$

it follows from the consolidation Equation (3) that,

$$c_{vi}\frac{\partial^2 \bar{u}}{\partial z^2} = \bar{\phi}_i \bar{u} - u(z,0) - \frac{u(z,0)E_{1i}}{s\eta_{2i} + E_{2i}} + \bar{f}_{qi} \tag{6}$$

where the subscript 'i' is introduced for sake of clarity to denote that the parameters belong to a typical i^{th} soil layer and,

$$\bar{\phi}_i = s + \frac{sE_{1i}}{s\eta_{2i} + E_{2i}} + \frac{E_{1i}}{\eta_{1i}} \tag{7a}$$

$$\bar{f}_{qi} = \begin{cases} \Xi_i & 0 < t \le t_0 \\ \left(1 - e^{-st_0}\right)\Xi_i & t > t_0 \end{cases} \tag{7b}$$

$$\overline{\Xi}_i = -\frac{q_0}{t_0}\left[\frac{1}{s} + \frac{E_{1i}}{s^2\eta_{1i}} + \frac{E_{1i}}{s(s\eta_{2i} + E_{2i})}\right] \tag{7c}$$

In the case of the ramp loading represented by Equation (4), $u(z,0)$ vanishes immediately. It is worth noting here that other complex time loading may also be considered where necessary provided it is possible to implement the Laplace transform procedures for the desired loading. It is not difficult to show that the solution to Equation (6) is given by,

$$\overline{u} = Ae^{\alpha_i z} + Be^{-\alpha_i z} - \frac{\overline{f}_{qi}}{\overline{\phi}_i} \tag{8a}$$

$$\alpha_i = \sqrt{\overline{\phi}_i / c_{vi}} \tag{8b}$$

where the coefficients A, B are to be determined. Using Equation (8), it is found that for the layer i the flow ($v = -k_{vi}/\gamma_w \partial u/\partial z$) is given by,

$$v = -\frac{\alpha_i k_{vi}}{\gamma_w}\left[Ae^{\alpha_i z} - Be^{-\alpha_i z}\right] \tag{9}$$

so that at the interlayer boundary $z = z_{i-1}$,

$$\overline{u}_{i-1} + \frac{\overline{f}_{qi}}{\overline{\phi}_i} = Ae^{\alpha_i z_{i-1}} + Be^{-\alpha_i z_{i-1}}$$

$$\overline{v}_{i-1} = -\frac{\alpha_i k_{vi}}{\gamma_w}\left[Ae^{\alpha_i z_{i-1}} - Be^{-\alpha_i z_{i-1}}\right]$$

yielding,

$$\begin{Bmatrix} \overline{u}_{i-1} + \dfrac{\overline{f}_{qi}}{\overline{\phi}_i} \\ \overline{v}_{i-1} \end{Bmatrix} = \begin{bmatrix} e^{\alpha_i z_{i-1}} & e^{-\alpha_i z_{i-1}} \\ -\dfrac{\alpha_i k_{vi}}{\gamma_w} e^{\alpha_i z_{i-1}} & \dfrac{\alpha_i k_{vi}}{\gamma_w} e^{-\alpha_i z_{i-1}} \end{bmatrix} \begin{Bmatrix} A \\ B \end{Bmatrix} \tag{10}$$

Similarly at the interlayer boundary $z = z_i$,

$$\begin{Bmatrix} \overline{u}_{i} + \dfrac{\overline{f}_{qi}}{\overline{\phi}_i} \\ \overline{v}_{i} \end{Bmatrix} = \begin{bmatrix} e^{\alpha_i z_{i}} & e^{-\alpha_i z_{i}} \\ -\dfrac{\alpha_i k_{vi}}{\gamma_w} e^{\alpha_i z_{i}} & \dfrac{\alpha_i k_{vi}}{\gamma_w} e^{-\alpha_i z_{i}} \end{bmatrix} \begin{Bmatrix} A \\ B \end{Bmatrix} \tag{11}$$

Rearranging Equation (11) gives,

$$\begin{Bmatrix} A \\ B \end{Bmatrix} = \frac{\gamma_w}{2\alpha_i k_{vi}} \begin{bmatrix} \dfrac{\alpha_i k_{vi}}{\gamma_w} e^{-\alpha_i z_i} & -e^{-\alpha_i z_i} \\ \dfrac{\alpha_i k_{vi}}{\gamma_w} e^{\alpha_i z_i} & e^{\alpha_i z_i} \end{bmatrix} \begin{Bmatrix} \overline{u}_{i} + \dfrac{\overline{f}_{qi}}{\overline{\phi}_i} \\ \overline{v}_{i} \end{Bmatrix} \tag{12}$$

Now, substituting Equation (12) into (10) yields,

$$\mathbf{w}_{i-1} = \mathbf{H}_i \mathbf{w}_i + \mathbf{\Lambda}_i \tag{13}$$

in which,

$$\mathbf{w}_{i-1} = \begin{Bmatrix} \overline{u}_{i-1} \\ \overline{v}_{i-1} \end{Bmatrix} \quad \mathbf{w}_i = \begin{Bmatrix} \overline{u}_{i} \\ \overline{v}_{i} \end{Bmatrix} \tag{14a}$$

$$\mathbf{H}_i = \begin{bmatrix} a_i & b_i \\ c_i & d_i \end{bmatrix} \qquad \mathbf{\Lambda}_i = \begin{Bmatrix} \overline{\Lambda}_{i1} \\ \overline{\Lambda}_{i2} \end{Bmatrix} \tag{14b}$$

where,

$$a_i = d_i = \frac{1}{2}\left(e^{\alpha_i \Delta z_i} + e^{-\alpha_i \Delta z_i}\right) \tag{15a}$$

$$b_i = \frac{1}{2}\left(\frac{\gamma_w}{\alpha_i k_{vi}}\right)\left(e^{\alpha_i \Delta z_i} - e^{-\alpha_i \Delta z_i}\right) \tag{15b}$$

$$c_i = \frac{1}{2}\left(\frac{\alpha_i k_{vi}}{\gamma_w}\right)\left(e^{\alpha_i \Delta z_i} - e^{-\alpha_i \Delta z_i}\right) \tag{15c}$$

$$\Lambda_{i1} = \left(\frac{\overline{f}_{qi}}{\overline{\phi}_i}\right)\left\{\frac{1}{2}\left(e^{\alpha_i \Delta z_i} + e^{-\alpha_i \Delta z_i}\right) - 1\right\} \tag{15d}$$

$$\Lambda_{i2} = \frac{1}{2}\left(\frac{\alpha_i k_{vi}}{\gamma_w}\right)\left(\frac{\overline{f}_{qi}}{\overline{\phi}_i}\right)\left(e^{\alpha_i \Delta z_i} - e^{-\alpha_i \Delta z_i}\right) \tag{15e}$$

$$\Delta z_i = z_i - z_{i-1} \tag{15f}$$

If Equation (13) were to be applied recursively, it leads to the equation,

$$\begin{aligned} \mathbf{w}_0 = \mathbf{H}_1\mathbf{H}_2 \cdots \mathbf{H}_J \mathbf{w}_J + \mathbf{H}_1\mathbf{H}_2 \cdots \mathbf{H}_{J-1}\mathbf{\Lambda}_J + \\ \mathbf{H}_1\mathbf{H}_2 \cdots \mathbf{H}_{J-2}\mathbf{\Lambda}_{J-1} + \cdots \mathbf{H}_1\mathbf{\Lambda}_2 + \mathbf{\Lambda}_1 \end{aligned} \tag{16}$$

Equation (16) immediately allows the determination of the unknown boundary pore pressure or flow values of the layered system. For example if $u_0 = 0$ and $u_J = 0$, it enables v_0 and v_J to be obtained. Other boundary conditions may be dealt with in the same manner, if required. Then using Equation (13) recursively, the pore pressure and flow values at the interfaces of the layers can be determined accordingly.

To obtain the strains, use is made of Equation (1) where its Laplace transform version is:

$$\overline{\varepsilon}_z = \overline{\sigma}'_z\left[\frac{1}{E_1} + \frac{1}{s\eta_1} + \frac{1}{(s\eta_2 + E_2)}\right] \tag{17}$$

in which $\overline{\sigma}'_z = \overline{q} - \overline{u}$. Commencing from the base ($z = z_J$), the strains are integrated over each layer and summed cumulatively to give the settlements at the various depths from the surface. Finally, the Laplace transform solutions are inverted numerically to produce the real-time solutions by way of the inversion algorithm by Talbot (1979).

4 IDENTIFICATION OF MODEL PARAMETERS

Before the model can be applied to problems of practical interest, the values of the model parameters need to be established. The identification process is based on consolidation and creep results obtained from the oedometer test. A Laplace transform solution has been found (Leo, 2000) for the consolidation of the soil sample with time ($\overline{w}$) in an oedometer test, this is given by:

$$\overline{w} = \frac{2q_0}{s^2 E_1}\sqrt{c_v \overline{\phi}}\,\frac{\left(e^{\alpha H}-1\right)}{e^{\alpha H}\left(e^{\alpha H}+1\right)} \qquad (18)$$

where: q_O = the applied constant surcharge; H = twice the length of the drainage path in the soil sample. The model parameters are determined by regressing the model results given in Equation (18) against the experimental data using a Levenberg-Marquardt (Levenberg 1944, Marquardt 1963) non-linear least squares regression procedure. As an example, Figure 4 shows the results of the proposed methodology where experimental data from a typical oedometer test are presented along with the solutions yielded by a 4-element model determined from Equation (18). In the given example, a 2 cm thick clay sample was consolidated under a surcharge pressure of 20 kPa and the regression yields the following results: c_v = 909 m^2/year, E_1 = 10.3 MPa, E_2 = 11.5 MPa, η_1 = 15.e20 MPa-year, η_2 = 14.8e3 MPa-year.

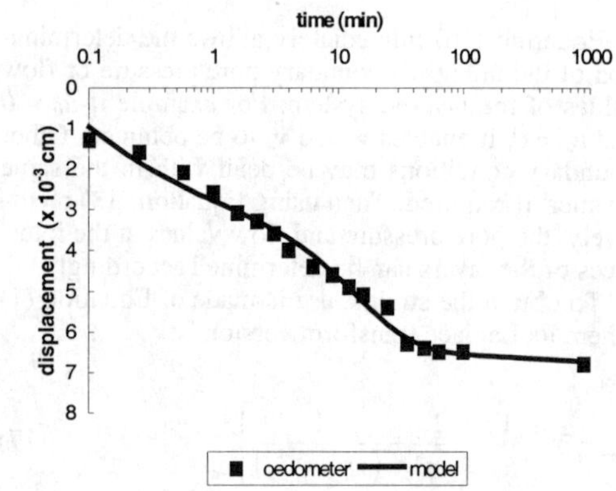

Figure 4. Oedometer test data vs. 4-element model results.

5 ILLUSTRATIVE EXAMPLE

The predicted solutions of the dimensionless vertical displacement and degree of consolidation in respect of pore pressure, Up, of a consolidation problem where a uniformly constant surcharge q_O is applied at $t = 0$ on a very soft soil layer of thickness h are shown in Figure 5. The results apply for the case where $E_1/E_2 = 1$ and $\eta_1/\eta 2 = 2 \times 10^4$. The model shows that the time effects for the consolidating soil are quite apparent and significant particularly at the end of primary consolidation (i.e. as $Up \rightarrow 100\%$).

6 CONCLUSION

A transfer matrix method is proposed in the present paper to solve the one-dimensional consolidation problem with time effects. A discussion of a methodolgy to estimate the parameters that go into the model is provided to show that the proposed method can be applied to solve consolidation problems of practical interest.

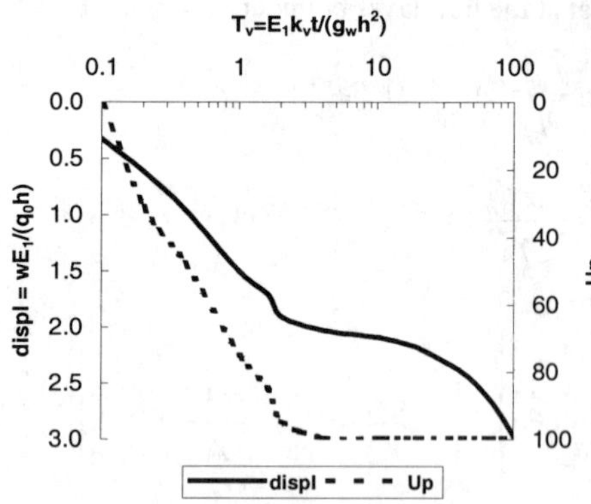

Figure 5. Predicted vertical displacement and degree of consolidation (U_p) for the illustrative example.

REFERENCES

Booker J.R. and Small J.C. (1985), 'Finite layer analysis of settlement, creep and consolidation using micro-computers', *Proceedings 5th International Conference of Num Meth in Geomechanics*, Nagoya, Japan, V(1), pp3-18.

Levenberg K. (1944), 'A method for the solution of certain problems in least squares', *Quarterly of Applied Mathematics*, 2, pp164-168.

Leo C.J. (2000), 'A methodology for parameter identification of poro-viscoelastic models of soft soil', *Research Report CE16*, UWS Nepean.

Marquardt D. (1963), 'An algorithm for least-squares estimation of nonlinear parameters', SIAM Journal of Applied Mathematics, 11, pp431-441.

Talbot A. (1979), 'The accurate numerical integration of Laplace transforms', Journal Inst. Mathematical Applications, Vol. 23, pp97-120.

Computer Methods and Advances in Geomechanics, Desai et al. (eds) © 2001 Balkema, Rotterdam, ISBN 90 5809 183 X

Modeling and parameter estimation of non-linear flow through rockfill structures

T. Morii
Faculty of Agriculture, Niigata University, Japan

K. Shimada
Faculty of Agriculture, Tokyo University of Agriculture and Technology, Japan

T. Tateishi
Nippon Giken Incorporated, Nagoya, Japan

ABSTRACT: Rock has been advantageously employed in the hydraulic structures such as flowthrough dams, cofferdams, gabion weirs and drain works. To predict flow transmissibility of the structures, an accurate evaluation of hydraulics of flow through rockfill is needed. The analysis of flow through rockfill is complicated because of non-linear relationship between the discharge velocity through rockfill and the hydraulic gradient applied. In this study, a procedure to evaluate the transmissibility of the rockfill is proposed. A head loss equation of flow through rockfill is investigated based on laboratory permeability tests. An accuracy of the head loss equation is examined by laboratory water flume tests and non-linear finite element method (FEM). Then a parameter estimation procedure combining FEM with a genetic algorithm to determine non-linear coefficients included in the head loss equation is developed. Effectiveness of the procedure is examined by the large water flume tests.

1 INTRODUCTION

There has been a gradual increase in the use of rockfill for hydraulic structures. To predict the flow transmissibility of such structures as flowthrough dams, cofferdams, gabion weirs, and drain works, an accurate evaluation of hydraulics of flow through rockfill is needed. The analysis of flow through rockfill is complicated because of non-linear relationship between discharge velocity through rockfill and hydraulic gradient applied.

In the present study, a procedure to predict the flow transmissibility of the rockfill is developed based on laboratory tests and numerical calculations. Firstly, a head loss equation of flow through rockfill is determined from one-dimensional permeability tests using river gravel 5-25mm in diameter. A hydraulic mean radius of voids is effectively introduced into the head loss equation of flow to describe the effects of the shape and size of rock particles and the voids within the rockfill on the flow hydraulics. The effect of water temperature is also included in the head loss equation of flow. Secondary, the numerical code based on the finite element method (FEM) is developed to evaluate the flow transmissibility of the rockfill structures. Numerical difficulty due to the non-linearity in the head loss equation of flow is solved by a method of successive approximation. Laboratory water flume tests are conducted to examine practical accuracy of the proposed head loss equation of flow and the FEM calculation code developed. Applicability of the head loss equation of flow should be examined for the lager rock particles. For this, lastly, a parameter estimation procedure combining the nonlinear FEM with a simple genetic algorithm (GA) is developed. The applicability of the head loss equation of flow as well as the effectiveness of the procedure, FEM-GA, is examined by the large water flume tests using river gravels 75-100 mm in diameter.

2 HEAD LOSS EQUATION OF FLOW

2.1 *One-dimensional permeability tests and material*

One-dimensional permeability tests of rock particle column were conducted to study the head loss property of the flow through rockfill. Rock particles were filled and compacted lightly by hand into the cylindrical column 10 cm diameter and about 100 cm length. Mass and volume of the rockfill specimen were measured to calculate porosity, n, and void ratio, e, of the rockfill specimen. The rockfill specimen being placed vertically, water was supplied into the bottom of the specimen from the water reservoir. A flow rate through the rockfill specimen was measured by mass, and divided by the cross-sectional area of the specimen to determine a discharge velocity of flow through the rockfill specimen, V. The total head along the rockfill specimen was measured by eight piezometers connected 10cm interval to the

Table 1 Size, shape and physical properties of the rock particles used in the one-dimensional permeability tests.

Class of Particle	Particle size, mm	Arithmetic mean of particle size d, mm	Rock particle shape [a]					Specific gravity of dried particle G	Water absorption of particle, %
			Blades	Disks	Spheroids	Rods	Mean shape coefficient r [b]		
(1)	(2)	(3)	(4)	(5)	(6)	(7)	(8)	(9)	(10)
D1	20-25	22.5	0.32	0.22	0.14	0.32	1.9314	2.563	1.4
D2	15-20	17.5	0.22	0.18	0.26	0.34	1.8600	2.550	1.5
D3	10-15	12.5	0.22	0.24	0.20	0.34	1.8774	2.557	1.4
D4	5-10	7.5	0.32	0.16	0.06	0.46	1.8944	2.575	1.6

a) Rock particle shapes such as blades, disks, spheroids and rods are classified according to the *Zingg* classification (Garga et al. 1991). Numerals in the cells show frequency of rock particle shapes determined by measuring three orthogonal axes of rock particle. Fifty rock particles are selected from each class of rock particle size to determine the frequency of rock particle shapes.

b) r is a mean shape coefficient of rock particles, and is calculated as a mean of r_e weighted by the frequency given in the rows (4)-(7). r_e is the shape coefficient for each shape of rock particles, and is given by Sabin & Hansen (1994).

wall of the specimen column. A hydraulic gradient, i, applied to the rockfill specimen was calculated from the slope of the linear regression line through a plot of the total head along the rockfill specimen. i was increased successively from about 0.05 to 0.8 by seven to eight steps in each test of the rockfill specimen. Water temperature of flow through rockfill specimen, T, was also measured at the beginning and the completion of the test.

River gravel was used in the tests. Being washed and air-dried, the rock particles were sieved into four classes of size, D1, D2, D3 and D4. Classes of particle, and their representative diameter, rock particle shape and physical properties of the rock particles in each class of size are given in Table 1. The representative diameter of rock particles, d, is defined as an arithmetic mean of particle size. The rock particles were classified according to the *Zingg* diagram (Garga et al. 1991) into four shapes of blade, disk, spheroid and rod based on the measurements of three orthogonal lengths of selected rock particles. Their frequencies are given in the rows (4)-(7) of Table 1. The mean shape coefficient of rock particles in each class of size, r, in the row (8) is calculated as a mean of r_e weighted by the frequencies measured in the rows (4)-(7), in which r_e is the shape coefficient for each shape of rock particles measured by Sabin & Hansen (1994).

Sixteen rockfill specimens, that is four classes of rock particles given in Table 1 repeated twice plus four mixed classes of rock particles repeated twice, were tested. T was 15-23 degrees centigrade. The *Reynolds* numbers, R_e, which was calculated by the following equation, ranged between 9-323:

$$R_e = \frac{V \cdot m}{n \cdot \nu} \quad (1)$$

where m is the hydraulic mean radius of voids within the rockfill which will be defined in Section 2.2, and ν is the kinematic viscosity of water. Typical results of the one-dimensional permeability test of rockfill are shown in Figure 1, where solid lines are quadratic regression curves. It is shown that the discharge velocity of flow through the rockfill specimen has a non-linear relationship with the hydraulic gradient applied to it, and that much lager rate of water flows through the rockfill as compared with the *Darcy* flow through soil.

2.2 *Head loss equation of flow*

It has been well known that there are two types of equation describing the non-linear head loss of flow through coarse materials such as rockfill: The *Forchheimer* equation and the *Missbach* equation (Basak 1977, Parkin 1990, Kazda 1990). A superior

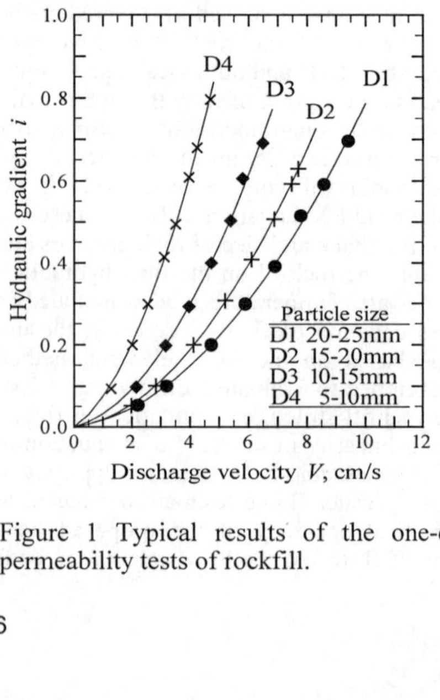

Figure 1 Typical results of the one-dimensional permeability tests of rockfill.

feature of the former is that it can be theoretically derived from the *Navier-Stokes* equation of fluid, while the latter has been empirically determined.

The *Forchheimer* equation is given by

$$i = A \cdot V + B \cdot V^2 \qquad (2)$$

where A and B are coefficients which depend on the void structures of the rockfill as well as the water viscosity of flow, and have their dimensions $[T^1L^{-1}]$ and $[T^2L^{-2}]$, respectively. According to the theoretical considerations done by Irmay (1958), Scheidegger (1974), and Bear & Verruijt (1987), the coefficients A and B in Equation (2) can be described as

$$A = \frac{A_0 \cdot \nu}{g}, \qquad B = \frac{B_0}{g} \qquad (3)$$

in which g is the acceleration of gravity, and both A_0 and B_0 are the coefficients depending only on the void structures of rockfill. A_0 and B_0 have the dimensions $[L^{-2}]$ and $[L^{-1}]$, respectively. A reciprocal of A_0 is corresponding to a physical or intrinsic permeability of porous media that has been well established in the theory of *Darcy* flow through soil.

Effects of the void structures on the head loss property of flow through the rockfill should be evaluated by taking the size and shape of the rock particles as well as the size and distribution of voids within the rockfill into account. Martins (1990) shows that all the effects above mentioned could be well described by the hydraulic mean radius of voids, m, which is defined as (Sabin & Hansen 1994)

$$m = \frac{e \cdot b}{6 \cdot r_e} \qquad (4)$$

in which b is a representative diameter of rock particles. It may be right to think that the coefficients A_0 and B_0 in Equation (3) are closely related to m. Figure 2 shows such relationships of A_0 and B_0 with m determined from the one-dimensional permeability tests of rockfill. The values of A and B in Equation (2) being firstly determined by the regression analysis as shown in Figure 1, then A_0 and B_0 were calculated by using Equation (3) with the known value of ν at the water temperature T. In calculating m, b and r_e in Equation (4) were effectively replaced by d and r in Table 1, respectively. Close relationships of A_0 and B_0 with m can be seen in Figure 2. According to Scheidegger (1974) and Bear & Verruijt (1987), a power type equation may be accepted to describe the functional relationships of A_0 and B_0 with m shown in Figure 2:

$$A_0 = 31.815 \cdot m^{-1.5011}, \quad B_0 = 0.6282 \cdot m^{-1.1547} \qquad (5)$$

in which A_0, B_0 and m are in cm^{-2}, cm^{-1} and cm, respectively. The hydraulic mean radius of voids being determined from the size and shape of rock particles and the void ratio of the rockfill deposition, and the water temperature of the flow being measured or supposed, then the head loss property of the flow through rockfill will be able to estimated by using Equations (2)-(5) with the known values of A and B.

3 EXAMINATION OF HEAD LOSS EQUATION OF FLOW

3.1 *Water flume tests*

To examine the accuracy of the head loss equation of the rockfill proposed in the preceding chapter, a series of the model embankment tests was conducted in the acrylic-walled water flume 20 cm wide and 40 cm deep in the laboratory. The same rock particles as used in the one-dimensional permeability tests were filled into the water flume in layers and compacted lightly with a steel rod to construct the model embankment. The mass and the volume of the model embankment were measured during and after the

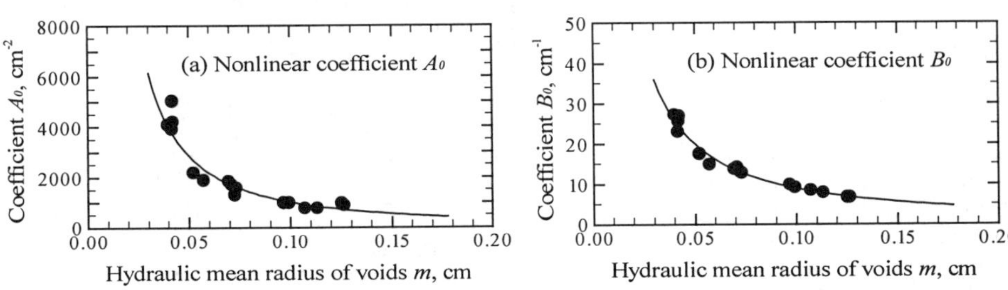

Figure 2 Relationships of the non-linear coefficients A_0 and B_0 with the hydraulic mean radius of voids determined from the one-dimensional permeability tests of river gravel.

construction, respectively, to calculate n and e. Both the upstream and downstream slopes of the model embankment were retained by the wire mesh with the squared openings of 2 mm to ensure the stability of the slopes and to prevent any erosion of rock particles from the downstream slope. The total head ϕ within the model embankment were measured from the tapping points installed along the base and on the wall of the water flume. The free surface of flow within the model embankment and its exit point on the downstream slope were observed through the acrylic wall of the water flume. The upstream water level, h_u, was raised successively from about 10 cm to 30 cm by four to six steps. After the steady state of flow was attained at each step in the water flume test, the bulk discharge through the model embankment was measured by mass at the outlet of the water flume. The flow discharge per unit width of the water flume, q, was calculated by dividing the measured bulk discharge by the width of the water flume. No regulation was given on the downstream water level in any water flume tests.

3.2 *Non-linear FEM*

To estimate the flow transmissibility of the rockfill structures, a numerical calculation code was formulated with the help of the FEM. A numerical difficulty due to the non-linearity in the head loss equation of flow, Equation (2), was successfully solved by the method of successive approximation (Kazda 1990). In each iterative calculation, the flow is assumed to be *Darcy* one, and a fictitious hydraulic conductivity, K_f, is introduced to calculate the flow velocity (Kazda 1990, Kells 1993):

$$K_f = \frac{-A + \sqrt{A^2 + 4B \cdot i}}{2B \cdot i} \tag{6}$$

In the non-linear FEM, the iterative calculation to determine the position of the free surface of flow was performed until the difference between the new height of a point of the free surface and its previous value is less than 0.1 % of the upstream water level.

3.3 *Accuracy of head loss equation*

Fourteen model embankments with different combination of the rock particle size, the downstream slope and the top width of the model embankment were tested in the water flume. The values of T and R_e measured in the tests were 8-24 degrees centigrade and 15-145, respectively. In calculating R_e by Equation (1), the average velocity through the vertical section within the embankment under the upstream water level was used as V. R_e of the water flume tests were sufficiently included within the range of the one-dimensional permeability tests shown in Section 2.1.

Figure 3 shows a comparison of q between the water flume tests and the FEM calculations. The first, second and third terms of the test number represent the rock particle size given in Table 1 (D13 is the case of D1 mixed with D3), the specified downstream slope (11=1V:1H, 12=1V:2H and V= Vertical), and the specified top width of the embankment in cm, respectively. The model embankment with the rectangular section, the upstream and downstream halves of which were made of the rock particle size D1 and D3, respectively, was also tested and indicated as D1(u)+D3(d)-V-60 in Figure 3. The height of the model embankments was about 30 cm above the base of the water flume in all tests, and the upstream slope was 1V:1H except for the vertical D1-V-60 and D3-V-60. In the FEM calculations, the non-linear coefficient A and B were determined by using Equations (3)-(5) with the known values of e, d, r and T. It's can be seen in Figure 3 that the FEM calculations which incorporate the head loss equation proposed in Section 2.2 predict well the measurements in the water flume tests. In Figure 4, the free surface of flow and the total head of water within the model embankment are compared between the FEM calculations and the water flume tests at the step of the highest position in the upstream water level for the tests D1-11-10 and D1(u)+D3(d)-V-60. There can be see again good predictions by the FEM in Figure 4.

3.4 *Parameter estimation*

There is no evidence that proves the head loss equation of flow determined from the permeability tests using small size rock particles can be effectively ap-

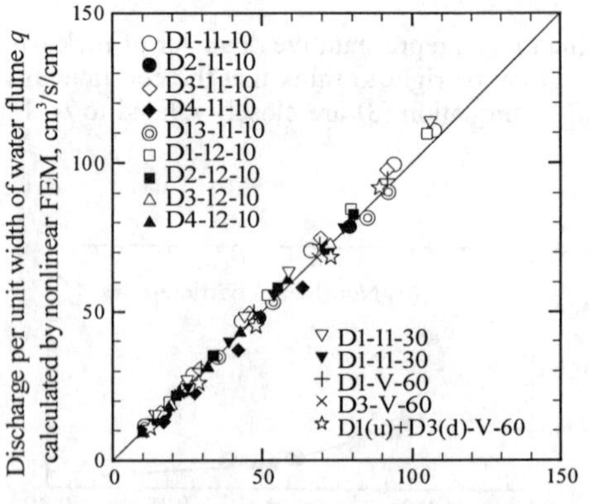

Figure 3 Comparison of flow discharge between the measurements in the water flume tests and the FEM calculations.

plied to the larger rock particles. Meanwhile it is difficult to measure i-V relationship of larger rock particle deposit by using the one-dimensional permeability test because the laborious and huge rockfill specimen should be constructed in the laboratory. To solve these problems, a parameter estimation procedure combining the nonlinear FEM with the GA was developed. The GA (Goldberg 1989) is a search algorithm formulated on the basis of simulation of reproduction, crossover and mutation of living individuals. A special feature of the GA is that it does not require any mathematical derivatives of object functions that should be minimized or maximized. The FEM-GA was applied to the water flume tests to search a set of the non-linear parameters A_0 and B_0 which describe optimally a stage-discharge rating curve, h_u-q, observed in the model embankment.

The model embankment was constructed by using the river gravel 75-100 mm in diameter in the large water flume 50 cm wide and 60 cm depth. The mass and volume of the model embankment were measured to calculate n and e, and the value of m was calculated by Equation (4). Three model embankments with the different downstream slope were tested. In the test, h_u was raised from about 15 to 45 cm by 4 to 5 steps, and, after the steady state of flow was attained in each step of h_u, q through the model embankment was measured. Then the non-linear coefficients A and B were determined by the FEM-GA so that the deviation of q at given h_u between the observation and the calculation become minimum. Crossover rate and mutation rate were 0.3 and 0.01, respectively in the FEM-GA analysis. The coefficient A_0 and B_0 were then calculated by using Equation (3) with the known value of T measured in the test. Figure 5 shows the plots of A_0 and B_0 with the

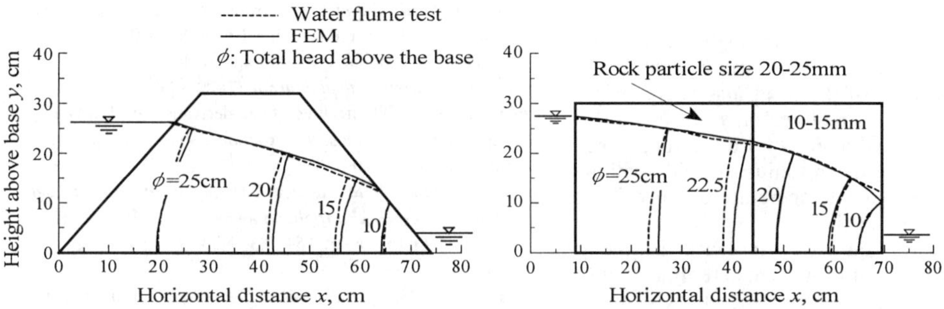

(a) Test D1-11-10 (b) Test D1(u)+D3(d)-V-60

Figure 4 Comparison of the free surface of flow and the total head within the model embankment between the water flume tests and the non-linear FEM calculations.

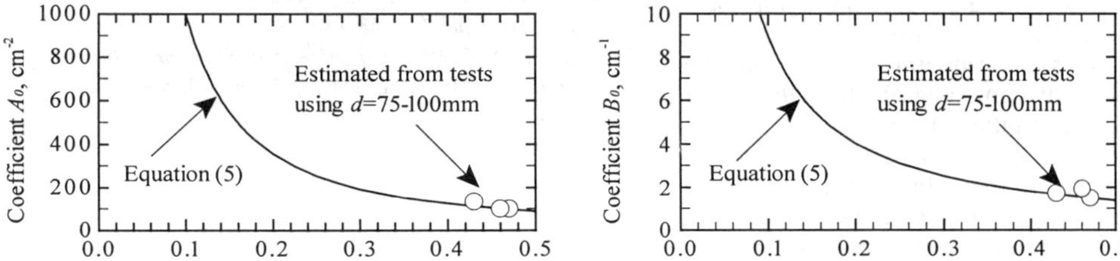

(a) Coefficient A_0 (b) Coefficient B_0

Figure 5 Non-linear coefficients A_0 and B_0 estimated by the FEM-GA from the large water flume tests using river gravel 75-100 mm in diameter. The solid lines show the relationship determined by the one-dimensional permeability tests, Equation (5).

measured values of m on the line determined from the one-dimensional permeability tests shown in Chapter 2. Good correspondence of the estimated parameters with the line show that Equation (5) can be validly applied to the wide range of rock particle size, at least 5-100 mm in this study.

Further trials of the parameter estimation in the laboratory as well as in the field will be needed to obtain the more accurate and practical relationship of A_0 and B_0 with the wide range of m. Once the relationship is determined, it may be possible to calculate the value of m by inserting the estimated A_0 and B_0 into this relationship. The value of m may provide us the practically important information about the rock particle properties, the deposition of the rockfill, or the degree of clogging due to the soil particles within the rockfill.

4 CONCLUSIONS

The head loss property of flow through the rockfill was successfully modeled by the *Forchheimer* equation. A special feature of the head loss equation of flow developed is that it requires only two parameters to estimate the hydraulics of flow through the rockfill: The hydraulic mean radius of voids within the rockfill and the water temperature of flow. The followings are summarized:

1) The non-linear parameters included in the head loss equation of flow were determined by the one-dimensional permeability tests using the river gravel, and were effectively correlated to the hydraulic mean radius of voids.
2) By the comparison between the measurements in the water flume tests and the non-linear FEM calculations, the accuracy of the head loss equation of flow developed was examined.
3) From the FEM-GA analysis of the large water flume tests, it was shown that the head loss equation of flow developed is valid for the wide range of rock particle size.

Rockfill structures are economical ones from the geotechnical point of view, and are friendly to our environment. Understanding the rockfill hydraulics may be helpful to design safely such structures as the flowthrough dams, the gabion weirs, the drain works and the river control works, and as well may give us a new idea of mitigation works for our environment.

ACKNOWLEDGMENTS

The present study was supported by the Grant-in-Aid for Scientific Research (C), No. 12556036, made by the Ministry of Education, Science, Sports and Culture of Japan. The authors are grateful to Messrs. T. Sada, T. Kobayashi and S. Soyama for their help to conduct the laboratory tests.

REFERENCES

Basak, P. 1977. Non-Darcy flow and its implications to seepage problems. *Journal of the Irrigation and Drainage Division, Proceedings of the American Society of Civil Engineers* 103(4): 459-473.

Bear, J. & A. Verruijt 1987. *Modeling Groundwater Flow and Pollution*: 41-43. Holland: D. Reidel Publishing.

Garga, V. K., R. Townsend & D. Hansen 1991. A method for determining the surface area of quarried rocks. *Geotechnical Testing Journal* 14(1): 35-45.

Goldbergh, D. E. 1989. *Genetic Algorithms in Search, Optimization, and Machine Learning*: 412 Addison-Wesley Pub. Co.

Hansen, D., V. K. Garga & R. Townsend 1995. Selection and application of a one-dimensional non-Darcy flow equation for two-dimensional flow through rockfill embankments. *Canadian Geotechnical Journal* 32: 223-232.

Irmay, S. 1958. On the theoretical derivation of Darcy and Forchheimer formulas. *Transactions, American Geophysical Union* 39(4): 702-707.

Kazda, I. 1990. *Finite Element Techniques in Groundwater Flow Studies: With Applications in Hydraulic and Geotechnical Engineering*: 164-183. New York: Elsevier Science.

Kells, J. A. 1993. Spatially varied flow over rockfill embankments. *Canadian Journal of Civil Engineering* 20: 820-827.

Martins, M. 1990. Principles of rockfill hydraulics. In Maranha das Neves, E. (ed.), *Advances in Rockfill Structures*: 523-570. Netherlands: Kluwer Academic Publishers.

Parkin, A. K. 1990. Through and overflow rockfill dams. In Maranha das Neves, E. (ed.), *Advances in Rockfill Structures*: 571-592. Netherlands: Kluwer Academic Publishers.

Sabin, C. W. & D. Hansen 1994. The effects of particle shape and surface roughness on the hydraulic mean radius of a porous medium consisting of quarried rock. *Geotechnical Testing Journal* 17(1): 43-49.

Scheidegger, A. E. 1974. *The Physics of Flow through Porous Media*: 152-170. Third edition. Tronto: University of Toronto Press.

Computer Methods and Advances in Geomechanics, Desai et al. (eds) © 2001 Balkema, Rotterdam, ISBN 90 5809 183 X

An estimation of flow channels in a natural rock using fractal of fractures

Shinji Nakaya, Hidenori Kondoh & Minoru Nishikuni
Shinshu University, Nagano, Japan

Makoto Nishigaki
Okayama University, Japan

ABSTRACT: The geometry data of fracture was extracted from the fracture trace in scan plane (10mx10m) on the sandstone as large number of line elements. From data for fractal, length distribution according to power law and porosity distribution on fractures, the method (FFN model) for reproducing the natural fracture pattern as flow channel was examined. As a result of applying the FFN model to natural fracture pattern of a sandstone, it was able to reproduce on the part over 70% of the fracture pattern by introducing priority from porosity distribution to order of fracture generation.

1. INTRODUCTION

Concerning resources and energy-problem and civil engineering problem, the method for estimating spatial distribution of fracture, the linkage and flow-channel on the subsurface rock mass is desired. Because it is generally known that distribution and shape of fracture have the self-similarity(e.g. Hirata, 1989), some researchs on network modeling of the natural fracture using the fractal has already been proposed(e.g. Ohno & Kojima, 1987; Wtanabe & Takahashi, 1995). In this study, in order to estimate fracture network as flow channel of fractured rock mass, fracture characterization was extracted from natural fracture pattern observed from the rock mass exposure, and the method for reproducing natural fracture pattern was studied. By applying this technique, the reproducibility of original natural fracture pattern and the problem of method are described.

2. METHOD

2.1 Digitization and fractal of natural fracture pattern

In this study, the coordinate of double end of fracture was obtained, when each fracture of the 10mx10m region was linearized using natural fracture (Figure 1) which Odling & Webman (1991) measured on scan-plane of the sandstone. When fractal dimension D as a distribution density of fracture was measured by the box counting method, D=1.95 was obtained in order to observe the fractal nature (self-similarity) of this natural fracture pattern. It is not limited to 10m x 10m region, when we judge that used natural fracture pattern has the self-similarity, and in addition, the result of this study is also expanded on the region of large moderate size.

2.2 The reproduction technique of fracture

In this technique, the position of fracture is caught with spatial distribution of fracture center point on two dimensional plane, and the center point position is being generated using cascade model. The procedure of this technique(here, Fractal Fracture Network model, FFN model) is shown in the following.

First, fracture is separated from the orientation to some groups. Here, fractures are divided into fracture group of the 4 directions shown in Figure 2. On the way of division of the group, it is different by distribution of orientation of fracture in individual object region.

Second, the boxes of the $(2^k)^D$ piece are chosen from fractal dimension D and square division box of 2^k x 2^k (k=1,2,..., 6 steps) in every fracture group shown in Figure 2. In the time, the boxes of the $(2^k)^D$ piece are chosen with the priority decided from the porosity of the box. This procedure is repeated from k=1 to 6 steps. However, the box which was not chosen at the pre-step is not chosen even in the step. The spatial distribution of the box chosen according to such procedure should agree with fractal

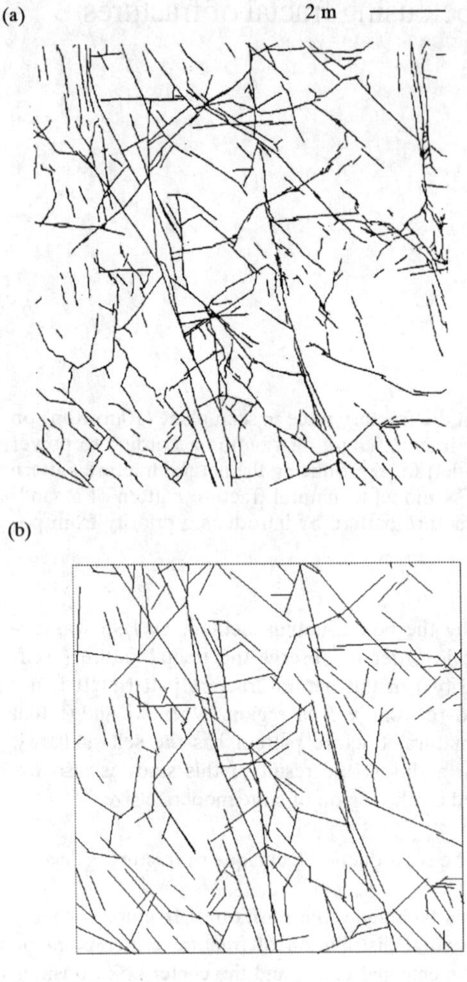

Figure 1. Trace map of natural fracture measured on scan-plane of sandstone. (a) Original natural fracture pattern (after Odling and Webman(1991)), (b) Fracture pattern linearized (10mx10m) using original natural fracture pattern.

dimension measured by box counting method of fracture center location within the rounding error of original calculation of box number. Next, fracture is generated for center location of blocks chosen at the k=6 step as center location of fracture. In the time, fractures with the fracture length according to power law(e.g. Balberg et al., 1991; Bour & Davy, 1997) are generated with the order of longer fracture length. After the porosity which generated fracture contributes is deducted from the porosity of each block, the porosity is recalculated, and the priority is re-determined. This operation is repeated on the box of the $(2^6)^D$ piece. The characteristic values of natural fracture pattern for this reproduction technique are

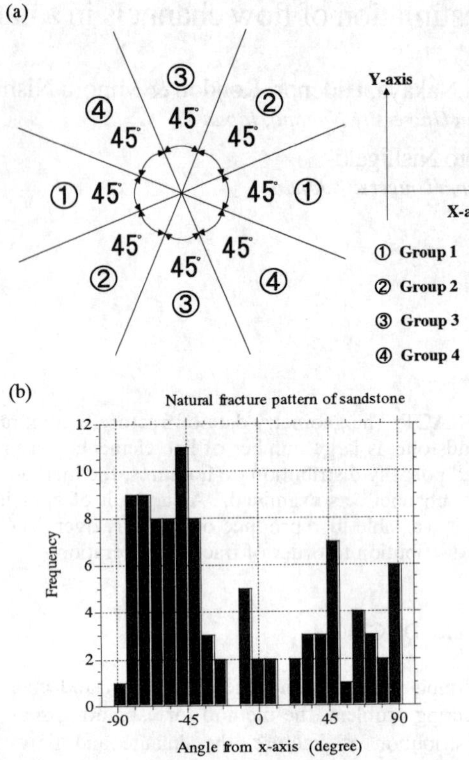

Figure 2. Fracture groups divided from orientation data. (a) Fracture groups, (b) Histogram of fracture orientation.

shown in Table 1 and Figure 3. And, porosity, n of each box is calculated from the following equation, when fracture aperture, t is assumed being proportional for crack length, L(Ohno & Kojima, 1987).

$$n = L' \cdot t / S ,$$

where,

L' : Length of fracture cut by the box,
S : Area of the box,
$t = C \cdot L$, $(C = 1 \times 10^{-5})$

3. REPRODUCTION RESULTS OF THE NATURAL FRACTURE

3.1 On the reproducibility of fracture pattern

Fracture pattern was reproduced by applying fractal dimension, fracture length distribution and porosity distribution to the FFN model (Figure 4). Reproduced fracture pattern was compared with natural fracture pattern (Figure 1) on ① fracture spatial distribution pattern, ② fractal dimension for 4 fracture

Table 1. Characteristic values of natural fracture pattern.

	Fracture groups divided from orientation			
	1	2	3	4
D*	1.584	1.556	1.838	1.848
a**	5.881	3.535	21.14	8.910
b**	-1.327	-1.338	-1.679	-1.872

*: N(r) r^{-D}. **: log(N(L)) = a – b · log(L).

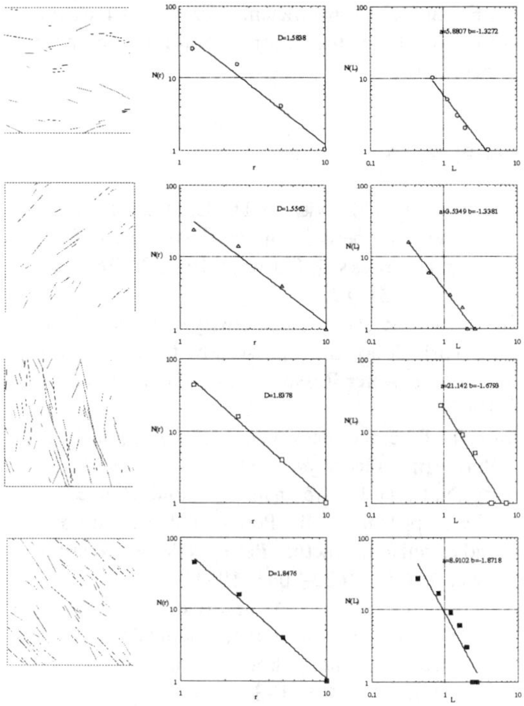

(a) (b) (c)

Figure 3. Fractal dimensions, D and length L distribution for natural fracture pattern. (a) Fracture pattern, (b) N(r) r^{-D}, (c) log(N(L)) = a – b · log(L) for each groups.

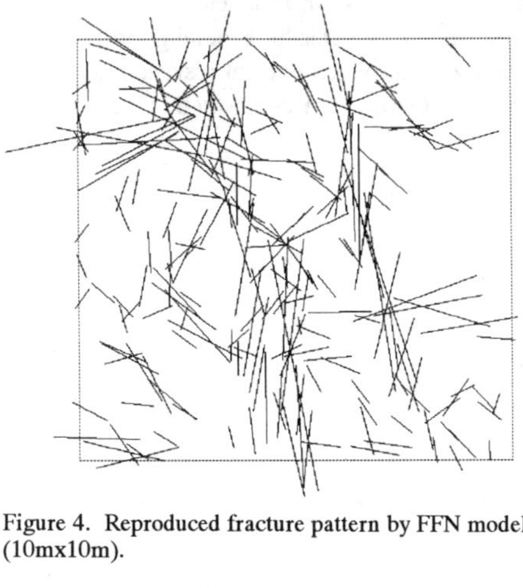

Figure 4. Reproduced fracture pattern by FFN model (10mx10m).

Table 2. Comparison of fracture patterns.

Number of 0-block* for natural fracture pattern , Nn	411
Number of 0-block* for reproduced fracture pattern , Nr	352
Number of duplicated 0-block* for Nn and Nr, No	295
Nn - No	116
Nr - No	57
No / Nn	0.718
No / Nr	0.838
(Nn-No) / Nn	0.282
(Nr-No) / Nn	0.139

*: block in which fracture exists for 25x25 square blocks.

Table 3. Comparison of fractal dimension.

D*	Fracture groups divided from orientation			
	1	2	3	4
Natural fracture pattern	1.584	1.556	1.838	1.848
Reproduced fracture pattern	1.585	1.525	1.827	1.818

*: N(r) r^{-D}.

Table 4. Comparison for distribution of fracture length.

	Fracture groups divided from orientation			
	1	2	3	4
Natural fracture pattern				
a*	5.881	3.535	21.142	8.910
b*	-1.327	-1.338	-1.679	-1.872
Reproduced fracture pattern				
a*	5.391	3.194	20.842	8.272
b*	-1.373	-1.365	-1.727	-1.970

*: log(N(L)) = a – b · log(L).

Table 5. Comparison of flux through flow channel(Q).

Flow direction	Qn* (m^3/s)	Qr** (m^3/s)	Qr/Qn
Up-Down	1.290×10^{-8}	8.083×10^{-9}	0.626
Left-right	5.121×10^{-10}	1.319×10^{-9}	2.576

*: Q in natural fracture pattern.

**: Q in reproduced fracture pattern.

Table 6. Comparison of fractal dimension for backbones.

Flow direction	Dn*	Dr**
Up-Down	1.744	1.786
Left-right	1.694	1.797

*: D for backbone of natural fracture pattern. N(r) r^{-D}.

**: D for backbone of reproduced fracture pattern. N(r) r^{-D}.

orientation groups, ③ characteristics of the fracture length distribution in order to observe the reproduci bility. Fracture spatial distribution pattern was com pared on reproduced fracture pattern and natural fracture pattern by dividing the block in which fracture is included and block in which fracture is not included into 2 kinds of blocks 0 and 1, respectively. For 10m x 10m region, the block division for the comparison is 25 x 25. Comparison results of ①, ② and ③ are shown in Table 2, Table 3 and Table 4, respectively. From Table 2, it is evaluated that 71.8% part piled up and 28.2% part did not agree on both fracture patterns. From Table 3 and Table 4, errors of both patterns on crack length distribution according to the direction and fractal dimension according to the direction were respectively 2% and 11% at most. From them, basic data were obtained on the reproducibility in reproducing natural fracture pattern from the characteristic values in present FFN model.

3.2 On the reproducibility of flux through flow-channels

On 10m x 10m region, the finite element model which makes fracture to be line element is made, after the fracture network as flow-channel is extracted from the fracture pattern. Each fracture assumes the laminar flow between parallel plates of fracture aperture, t. Steady state flux, Q which flows flow-channel from the upper surface to the lower surface can be calculated by FEM, under the condition for making the upper-and-lower surface in the region to be constant head boundary, and for making the right and left plane to be the impermeable boundary. Steady state flux, Q of crosswise direction is also calculated in the same way. On reproduced fracture pattern and natural fracture pattern, calculated steady state flux through flow-channels are shown in Table 5. It remained the reproducibility of about 40～60% on steady state flux through flow-channels. And, there were about 6% differences of fractal dimension, D (of center point distribution) between both fracture patterns for fractures of flow-channe in top-bottom direction and right-left direction, respectively(Table 6).

4. CONCLUSIONS

We studied the method for reproducing the natural fracture of rock mass, and a method of fracture reproduction (FFN model) was obtained. And using natural fracture pattern, the reproducibility of fracture pattern which reprodued according to FFN model was examined. The following conclusions can be drawn from the present study.

1) The algorithm of FFN model which deterministically reproduced the spatial distribution of original fracture was explained, when fractal dimension of fracture, crack length frequency distribution characteristics data according to power law and porosity distribution data for each groups clustered from fracture orientations were measured.

2) As a result of applying the FFN model to natural crack pattern of a sandstone, it was able to reproduce on the part over 70% of the fracture pattern.

3) On the reproducibility of steady state flux, Q which passes through flow-channel, it was 40%～60% of Q for original natural fracture.

In the future, it is necessary to advance the study which reproduces natural fracture pattern by applying the FFN model to insufficient servey data measured in situ, while case studies for natural fracture pattern are increased.

REFERENCES

Balberg, I., B. Berkowitz, and G. E. Drachsler, Ap plication of a percolation model to flow in frac tured hard rocks, J. Geophys. Res., 96(B6), 10015-10021, 1991.

Bour, O. And P. Davy, Connectivity of random fault networks following a power low fault length dis tribution, Water Ressour. Res., 33(7), 1567-1583, 1997.

Hirata, T., Fractal dimension of fault system in Japan, Pure Appl. Geophys., 131, 157-170, 1989.

Odling, N. E. And I. Webman, A "Conductance" Mesh Approach to the Permeability of Natural and Simulated Fracture Patterns, Water Ressour. Res., 27(10), 2633-2643, 1991.

Ohno, H. and K. Kojima, An approach of site char acterization based on statistical similality of rocks, Proceedings of the 19th Symposium on Rock Mechanics, 416-419, 1987(in Japanese).

Yamamoto, H. and K. Kojima, Spatial consentration characteristics of fracture system in rocks –an ex pression using cascade-, Proceedings of second workshop on characterization of subsurface cracks, 75-79, 1992(in Japanese).

Wtanabe, K. and H. Takahashi, Fractal geometry characterization of geothermal reservoir fracture networks, J. Geophys. Res., 100(B1), 521-528, 1995.

Computer Methods and Advances in Geomechanics, Desai et al. (eds) © 2001 Balkema, Rotterdam, ISBN 90 5809 183 X

Conventional finite element analysis and infinite element techniques in unbounded seepage analysis

G. Pokharel
Public Works Research Institute, Tsukuba, Japan

Y. Honjo
Department of Civil Engineering, Gifu University, Japan

ABSTRACT: In this paper, a set of finite element computations for a typical time dependent seepage flow through an unbounded confined aquifer are presented to illustrate how a given conventional finite element mesh with intermediately placed constant head boundary can produce sufficiently accurate results up to a certain limit of time depending on the material parameters. Once it exceeds the time limit, the computed drawdown starts deviating away and produces very large discrepancy that may go beyond the limit of acceptable accuracy. The infinite element technique proposed by Honjo and Pokharel (1993) is shown to be one of the most efficient methods to solve such unbounded problem in reduced costs and efforts. In this technique, the infinite extent of the constant head boundary is included in FE analysis by modeling the far-field using special elements called infinite elements while minimizing the size of near-field FE Mesh without compromising the accuracy of the analysis.

1 INTRODUCTION

In Geotechnical Engineering or Geomechanics, unbounded problems are frequently encountered or considered them as unbounded problems especially when the area is either really very big or the area of interest (here-in-after called as near field) under the study is very small compared to the total influence area. In conventional finite element methods, the infinitely spaced zero displacement or zero drawdown (*constant head boundary*, e.g. Figure 1) is assumed to be at a finite (*at an intermediate distance*) distance (*on the outer boundary of the Finite Element Mesh*). When the FE mesh size is increased, the time for data generation to the time of computation increases tremendously. This ultimately increases the total cost of computations. Because of such trade-off relation, the size of FE mesh involving unbounded media has to be compromised with the accuracy and the total cost of analysis.

In this paper, a time dependent seepage flow through an unbounded confined aquifer is taken as a representative problem and a set of finite element computations were carried out to illustrate how a given conventional finite element mesh with intermediately placed constant head boundary can produce sufficiently accurate results up to a certain limit of time depending on the material parameters. Once it exceeds the time limit, the computed drawdown starts deviating away and produces very large discrepancy that may go beyond the limit of acceptable accuracy. To solve such a problem, various methods were proposed and the infinite element technique(e.g. Honjo and Pokharel, 1993) is one of the most efficient methods to solve unbounded problem in reduced costs and efforts (Bettess, 1977). In this technique, the infinite extent of the constant head boundary is included in FE analysis by modeling the far-field using specialized elements called infinite elements while minimizing the size of near-field FE Mesh without compromising the accuracy of the analysis.

In this paper, an infinite element technique is briefly summarized in the perspective of seepage analysis and some typical results are presented. Meanwhile, a dimensionless size (*that consist of material parameter and time*) for use in conventional FEM with constant head boundary on the periphery of FE Mesh, is proposed such that it can produce fairly accurate results up to a certain dimensionless time limit. The coupling of the infinite element technique is recommended as early as the dimensionless time approaches or exceeds the limit.

Overall, the proposed size of conventional FE Mesh is believed to help the practicing engineers to easily decide the size of conventional FE Mesh and it may further help them in the selection of an appropriate type of boundary to suit the accuracy requirements under a given limited pre- and post-processing costs.

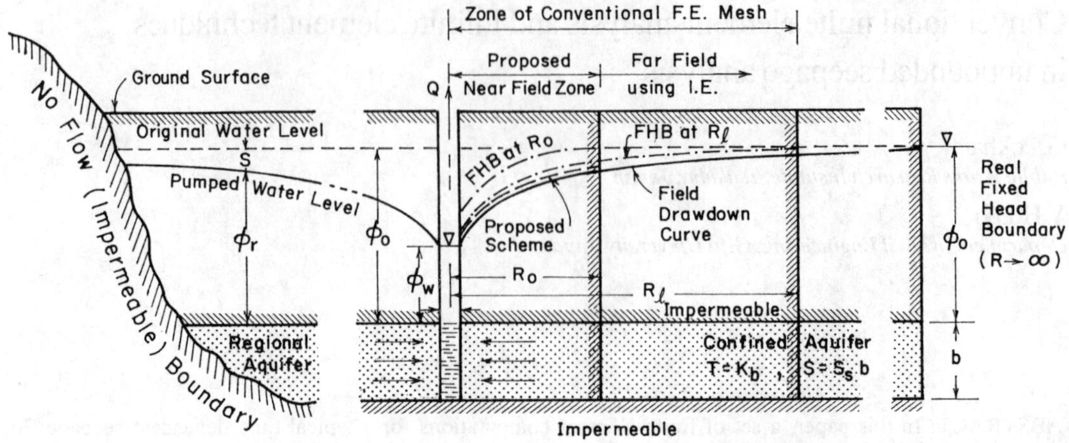

Figure 1 Illustration of an unbounded confined aquifer with various boundary conditions.

2 FORMULATION OF SEEPAGE FLOW

Throughout this paper, the unbounded aquifer media refers to a relatively large confined aquifer compared to the influence area under study e.g. construction site, a tube well pumping. This is practically appropriate because the real aquifer properties outside the periphery of the influence area, e.g. construction site, is assumed to be same as on the periphery of the influence area. Theis (1935) solution for a central point sink (well) can be considered a very well known unbounded seepage flow problem. The time dependent 2-D flow through a confined aquifer may be explained through the following governing differential equation:

$$\frac{\partial}{\partial x}\left(T\frac{\partial\phi}{\partial x}\right)+\frac{\partial}{\partial x}\left(T\frac{\partial\phi}{\partial x}\right)+f=S\frac{\partial\phi}{\partial t} \qquad ...(1)$$

where ϕ : piezometric head (≅drawdown),
T : Transmissivity of aquifer ($T=K_{aquifer}\times b_{\text{thickness of aquifer}}$),
S : storage coefficient of aquifer,
x,y : orthogonal co-ordinates,
t : time,
f : source / sink (e.g. pumping well)
r : radial distance from the well

The governing equation for 2-D axi-symmetric flow (e.g. single point well) in radial (r) coordinates system under the transient conditions applied to a confined aquifer is as follows:

$$T\left(\frac{\partial^2\phi}{\partial r^2}+\frac{1}{r}\frac{\partial\phi}{\partial r}\right)=S\frac{\partial\phi}{\partial t} \qquad ...(2)$$

The selection of an appropriate governing equation is not sufficient to get a complete and useful solution unless the initial and boundary conditions are chosen properly. The initial condition (e.g. spatial distribution piezometric head at the beginning, t=0, where t= time) in time dependent flow not only works as a datum point, but, it also controls the direction and duration of flow to approach a practically steady state condition. There are two additional surface conditions sometime referred to as the boundary conditions. First one of them is the head boundary condition that refers to the specified piezometric head (constant or function of time, but, directly independent of flow equation) and the other one is the flow boundary condition, e.g. a sink or a well (constant or function of time, but, directly independent of flow equation). Mathematically, these conditions may be expressed as follows:

***Initial Condition*:**

$$\phi(x,y,t=0)=\bar{\phi}_o(x,y,0) \qquad On\,(x,y)\in\Omega \qquad ...(3)$$

***Boundary or Surface Conditions*:**

(a) *Head Boundary Condition*

$$\phi(x,y,t)=\hat{\phi}(x,y,t) \qquad On\,(x,y)\in\partial\Omega_\phi \qquad ...(4)$$

(b) *Flow Boundary Condition*

$$-K\nabla\phi\cdot n=\bar{q}(x,y,t) \qquad On\,(x,y)\in\partial\Omega q \qquad ...(5)$$

where ∇ is grade operator, while $\partial\Omega\phi$, $\partial\Omega q$ *and* Ω are Dirichilet boundary, Neumann boundary and Flow Domain, respectively.

3 CONVENTIONAL FE FORMULATION FOR SEEPAGE FLOW ANALYSIS

The Eq. (1) can be simplified to the following form after some mathematical steps (details may be referred to *Smith and Griffiths*, Programming the FEM,1992):

$$M\frac{d\underline{\phi}(t)}{dt}+K\,\underline{\phi}(t)=\underline{Q}(t) \qquad \ldots(6)$$

where,

$$K=\sum_{\alpha=1}^{N}\left[K^{e}_{\alpha\beta}\right]=\sum_{\alpha=1}^{N}\left[\int_{\Omega^e}\left(T_x\frac{\partial N_\alpha}{\partial x}\frac{\partial N_\beta}{\partial x}+T_y\frac{\partial N_\alpha}{\partial y}\frac{\partial N_\beta}{\partial y}\right)dx\,dy\right],$$

$$M=\sum_{\alpha=1}^{N}\left[M^{e}_{\alpha\beta}\right]=\sum_{\alpha=1}^{N}\left[\int_{\Omega^e}S\,N_\alpha N_\beta\,dx\,dy\right] \quad \textit{and}$$

$$\underline{Q}=\sum_{\alpha=1}^{N}\left[Q^{e}_{\alpha}\right]=\sum_{\alpha=1}^{N}\left[\int_{\Omega^e}f_2 N_\alpha\,dx\,dy+\int_{\partial\Omega^e_q}\delta\alpha^{*}\bar{q}\,N_\alpha\,ds\right]$$

while N_α and N_β are shape and interpolation functions in local coordinates. The Eq. (6) is a basic equation in the analysis of seepage flow problems via a conventional FE Method .

4 FORMULATION OF A TYPICAL INFINITE ELEMENT TECHNIQUE

Fundamental concept behind an infinite element technique is to *finitely map* the infinite extent of infinite elements starting from the near-far field interface of conventional isoparametric FE Mesh to infinite (Figure 3). The mapping functions corresponding to infinite element nodes at interface are defined in such a way that the infinitely placed nodes of the I.E. can be mapped with the same mapping function, and is in contrast with the conventional FE shape functions, where the shapes functions have value 1 at self node, while the value is 0 at all other remaining nodes. While the head interpolation functions are defined same as in the conventional finite element analysis, i.e. same as in near field discretization. Tables 1 and 2 below illustrate the shape and head interpolation functions for conventional and infinite element techniques.

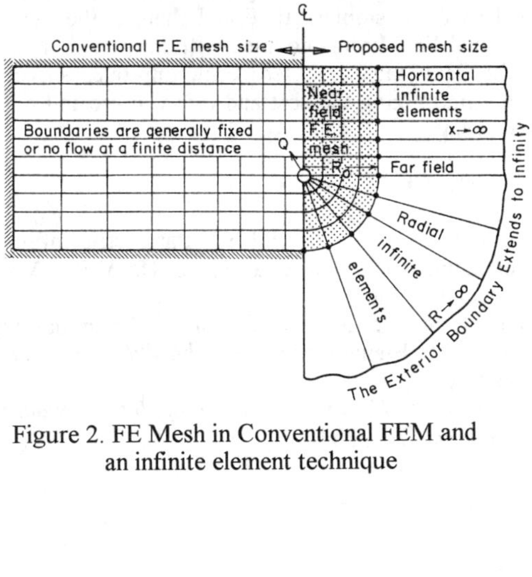

Figure 2. FE Mesh in Conventional FEM and an infinite element technique

Table 1: Coordinate mapping and head interpolation functions for isoparametric finite elements

Node Num.	Shape Functions for Coordinate Mapping and Head Interpolations	
1	$N_1^{cmhi}=\frac{(1-\xi)(1-\eta)}{4}$	Near Field coordinate mapping (x,y)) functions, and Near & Far Field head (φ)interpolation functions have same functions as in Isoparametric finite elements.
2	$N_2^{cmhi}=\frac{(1+\xi)(1-\eta)}{4}$	
3	$N_3^{cmhi}=\frac{(1+\xi)(1+\eta)}{4}$	
4	$N_4^{cmhi}=\frac{(1-\xi)(1+\eta)}{4}$	

Table 2: Coordinate mapping functions in infinite element techniques (*Honjo and Pokharel*, 1993)

Node Num.	Coordinate Mapping Functions Far Field (Infinite elements) (x,y)		
1	$N_1^c=\frac{(1-\eta)}{2}\sum_{m=1}^{n}C_m\left(\frac{2}{1-\xi}\right)^m$		N^c_3 and N^c_4 do not appear, because the explicit summation (i.e. numerical integration) is carried out over the near-far field interface only.
2	$N_2^c=\frac{(1+\eta)}{2}\sum_{m=1}^{n}C_m\left(\frac{2}{1-\xi}\right)^m$		
$C_m=$ $C_{\xi=-1}\frac{(-1)^{m+1}}{m!m^m2^m}$		$C_{\xi=-1}=\frac{1}{\sum_{m=1}^{n}\frac{(-1)^{m+1}}{m!m^m2^m}}$	

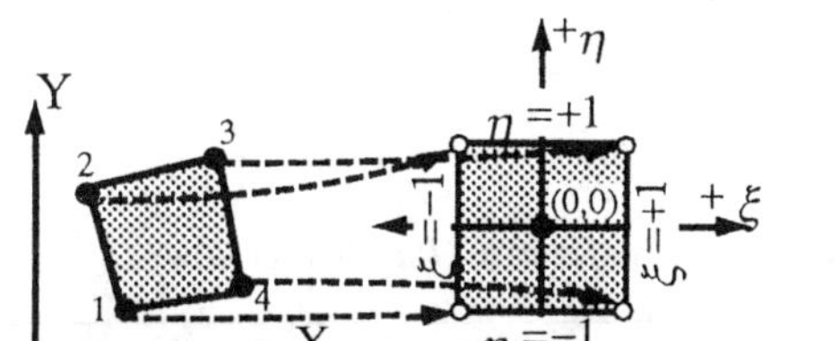

real world finite elements *after mapping*

(a) Conventional FE Mapping

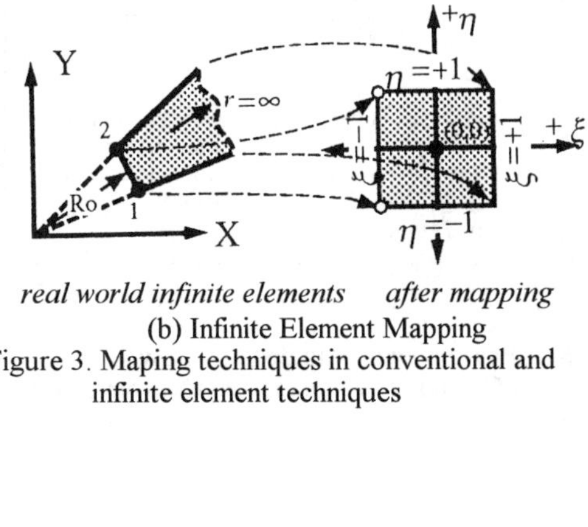

real world infinite elements *after mapping*

(b) Infinite Element Mapping

Figure 3. Maping techniques in conventional and infinite element techniques

5 PROPOSED MODEL AND ITS VALIDATION

The main objective of this paper is to propose a new model for an estimation of a FE mesh size in the perspective of conventional finite element analysis for an unbounded seepage flow problem assuming fixed head boundary at an itnermediate distance (e.g.the periphery of the FE mesh). In this study a pumping well at the center of an unbounded confined aquifer media with radial flow is analyzed through conventional FEM (i.e. fixed head boundary around the periphery of mesh) and the numerical analysis was carried employing the infinite element technique discussed in earlier sections. Several analyses were carried out, but, only a few typical results are shown in Figures 4.a, 4.b and 5. Based on all the computations, the Figure 6 is proposed. The Figure 6 shows that the conventional FEM with fixed head boundary on outer boundary gives fairly acceptable results until certain degree of dimensionless time, and once it exceeds the limit, the result starts deviating, after that either an infinite element technique needs to be coupled to include the effect of far field behavior or the lateral size of conventional FE mesh should be expanded.

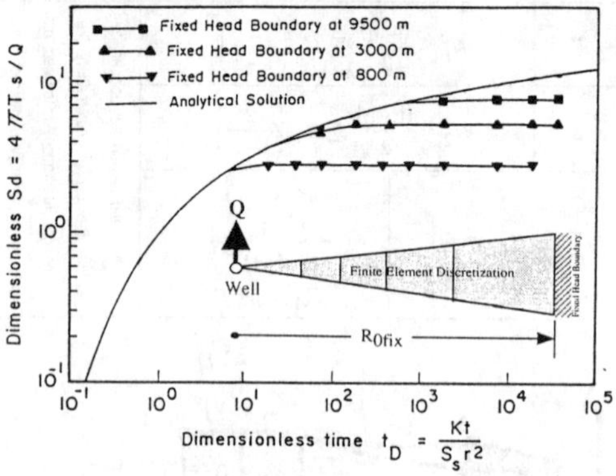

(a) Confined aquifer case (*Purely 2-D flow*)

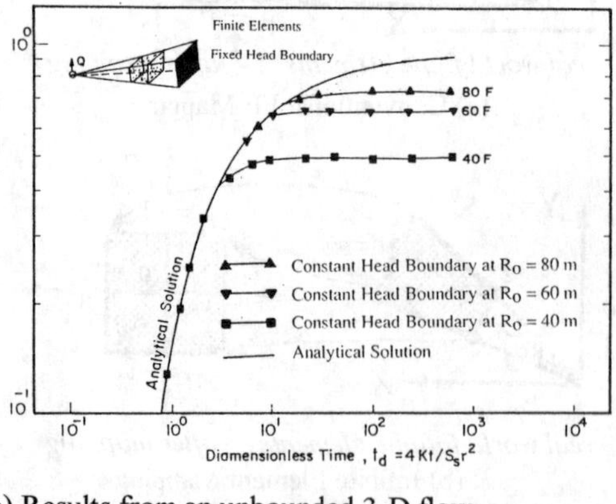

(b) Results from an unbounded 3-D flow

Figure 4. Computated results from Conventional FEM

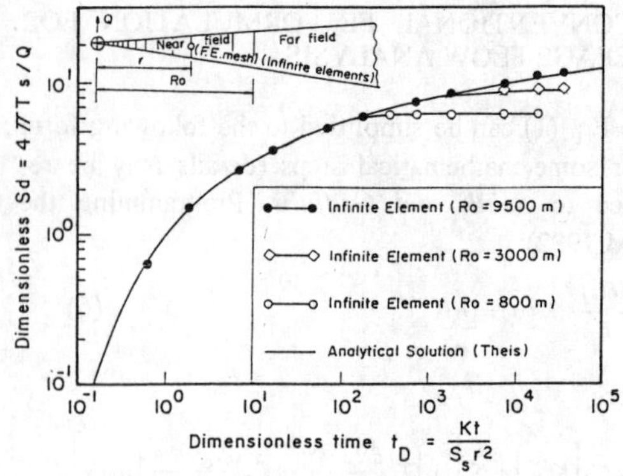

Figure 5 Typical results from infinite elements coupled finite element analysis (*confined aquifer*).

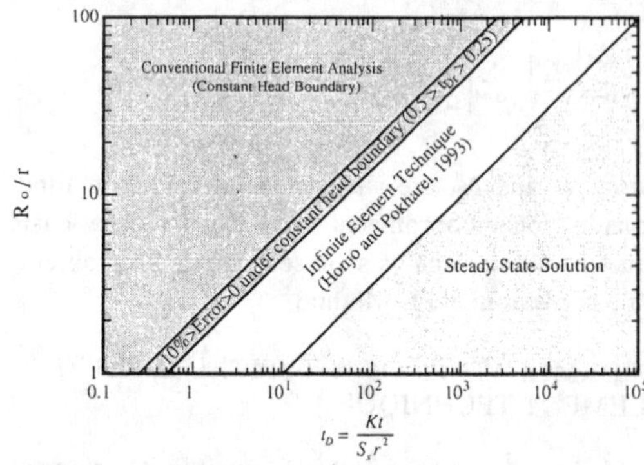

Figure 6 Proposd FE Mesh size for conventional FE analysis along with the one proposed by Honjo and Pokharel, 1993.

6 CONCLUSIONS

Infinite element technique is used by researchers in modeling various unbounded geotechnical problems. The conventional FEM in unbounded seepage flow analysis fairly produces good results to certain level of dimensionless time and then, if the conventional FE Mesh is coupled with infinite elements in far field, the computed result improves sharply while minimizing the cost and effort in pre and post processing.

REFERENCES

Honjo, Y. & G. Pokharel 1993: Parametric infinite elements for seepage analysis, IJNAMG, Vol. 17(1) pp. 45-66.

Smith, I. M. and D.V. Griffiths 1988. Programming the Finite Element Method, *John Wiley and Sons*, 2nd Ed.

Bettess, P. 1977. Infinite elements. *Int. J. Numer. Methods Eng.*, Vol 11, p.53-64.

Computer Methods and Advances in Geomechanics, Desai et al. (eds) © 2001 Balkema, Rotterdam, ISBN 90 5809 183 X

A model for the erosion of fine particles due to groundwater pumping and induced surface settlements

D. Sterpi
Department of Structural Engineering, Politecnico di Milano, Italy

ABSTRACT: The pumping of groundwater from drainage wells is often used in urban environment as a short-term provision to temporarily lower the water table level. The possible consequences of the erosion of fine particles from the soil surrounding the well are discussed, first in terms of local variations of volume and mechanical characteristics of the soil, subsequently in terms of the induced surface settlements. To this purpose, some laboratory tests are first presented, aimed at investigating the erosion from soil samples subjected to a seepage flow. The modelling of the processes of erosion and transport of the fine particles is then attempted through a numerical approach, based on a suitable law of erosion and on the conservation of mass of the moving particles. The procedure, calibrated on the basis of the laboratory tests, is subsequently adopted for the evaluation of settlements in an actual case.

1. INTRODUCTION

The progressive rising of the groundwater table, taking place in the urban area of Milano (northern Italy), is presently causing serious problems of water inflow within deep underground structures, such as parking lots and subway tunnels. Moreover, the general increase of the pore pressure, and the consequent reduction of the effective stress distribution, might lead to differential settlements of the foundations of major buildings.

Among the possible solutions currently under investigation, a temporary provision is represented by the local lowering of the water table through pumping wells located in the vicinity of the damaged underground structures.

The characteristics of the soil (mainly consisting of sand and gravel) and the need to keep high values of well discharge (to reach a sufficient lowering of the water table) have drawn the attention to the possible erosion of fine particles and to the consequent settlements of nearby buildings. In fact, the erosion could lead to a local decrease in volume, due to the gradual loss of fine material and to the possible rearrangement of the coarser fraction of soil. In addition, the mechanical characteristics of the soil may be affected by the reduction of the fine fraction and by possible changes of the void ratio.

Here the term "erosion" refers to the detachment of the fine particles from the soil primary fabric having a coarser grading. The "fine particles" are the soil grains passing through a chosen ASTM standard sieve. The term "transport" refers to the mechanism of macroscopic migration of the eroded particles due to the water flow.

To get some insight into the mentioned effects, a laboratory investigation was carried out on samples of silty sand subjected to the flow of water under controlled hydraulic gradients. During these tests the amount of eroded particles was measured at different time intervals.

The experimental program included also a series of triaxial compression tests carried out on reconstituted samples, having different values of the relative density and of the fine particle content. They intend to evaluate the effects of erosion on the deformability and shear strength of the tested soil.

The seepage tests provided the experimental basis of a law governing the phenomenon of particle erosion from an element of soil. In particular, it expresses the variation of the quantity of removed particles with time and hydraulic gradient.

The law of erosion, associated with the condition of mass conservation for the particles moving within the soil element, leads to an equation governing the problem at hand. This equation has been adopted in the finite element analysis of the process of groundwater pumping from a vertical well. The calculations permit to estimate the development of the surface settlements during time caused by the erosion and transport processes. On these bases, some conclusions are drawn on the potential use of the developed numerical approach for the analysis of actual field problems.

2. LABORATORY TESTS

The internal erosion of soils is defined as the process of removal of the fine particles from their primary fabric, caused by the seepage of fluids (Kenney & Lau, 1985). To investigate this process, various experimental studies have been presented in literature that mainly concern the evaluation of stability criteria for filter design (e.g. Hadj-Hamou et al., 1990; Fisher & Holtz, 1996). They are based on the application of a seepage flow through the soil samples, under controlled hydraulic gradient, and on the measurement of the amount of removed particles.

The laboratory tests here presented were based on a similar setup. The 14 cm high soil samples were reconstituted by the moist tamping procedure (Ladd, 1978) within a cylindrical mould, 7 cm in diameter. The initial relative density of 70% corresponds to a void ratio e_0 of 0.49.

The soil, recovered from borings in the urban area of Milano, mainly consists of compacted sand and gravel, with a non negligible percentage of fine material (silt μ_0=20%) and with a relatively high average permeability (k=10^{-2} cm/s).

The experimental equipment is depicted in fig.1. The upward seepage flow is induced by the assigned value of the hydraulic head, which depends on the difference in elevation between the upper reservoir and the overflow valve. A manometer is connected to the base of the sample to measure the inflow pressure. This allows for the direct evaluation of the hydraulic gradient, thus avoiding possible errors due to hydraulic head losses taking place at the nozzles or within the inlet tube.

A porous stone ensures a uniform flow at the sample base. The mould is subjected to slight vibrations to prevent the formation of preferential seepage channels. A gentle flow of air is applied at the top of the sample, through a thin tube, to avoid the deposition of the eroded particles on the sample head.

A standard ASTM 200 sieve is placed above the lower reservoir, which collects the water outflow, to separate the fine particle (with diameter of 0.074 mm) from coarser grains possibly eroded from the soil primary fabric. The amount of coarser grains collected in the sieve was anyway negligible, even for the highest applied gradients.

The lower reservoir is removed and substituted at constant time intervals and the quantity of water, contained within it, is measured. This permits to evaluate the changes of the soil permeability during the test. The weight of the eroded particles is also measured, after sedimentation and oven drying.

The results of 5 tests, carried out with various hydraulic gradients, are shown in fig.2. The diagrams report the increase with time of the percentage in weight of eroded fine particles μ. The continuous lines in the same figure represent the interpretation of the experimental data through the law that will be discussed in the next section.

The coefficient of permeability did not substantially change during the tests. Only for the case with gradient equal to 0.75 a sudden increase of water outflow was observed after 5 hours from the beginning of the seepage flow, which involves an appreciable increase of permeability.

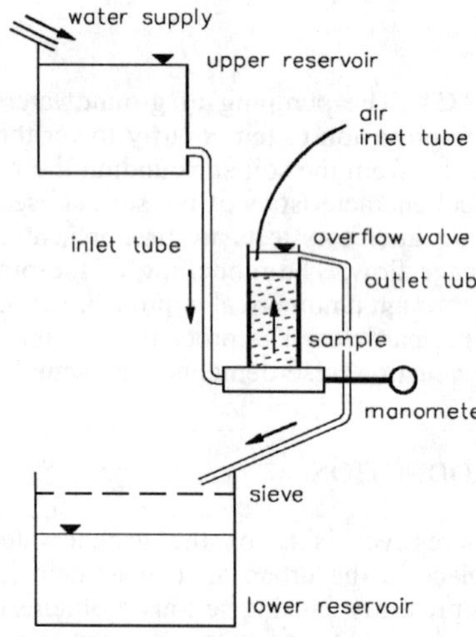

Fig.1. Experimental setup for seepage tests.

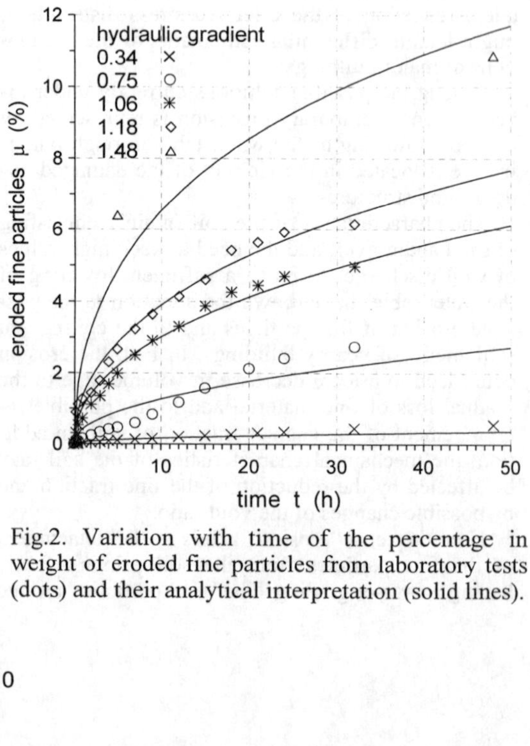

Fig.2. Variation with time of the percentage in weight of eroded fine particles from laboratory tests (dots) and their analytical interpretation (solid lines).

The erosion process could produce a variation of the soil microstructure and, hence, a change of its volume. With this respect, three alternative hypotheses seem possible:

a) the erosion is not associated with a new arrangement of the soil primary fabric (i.e. the total volume V_t is constant);
b) the volume change is only due to a loss of solid material (i.e. the volume of voids V_v is constant);
c) the volume change is also due to an induced compaction of voids such that no changes of the void ratio occur.

Fig.3 shows the variation of the void ratio (dashed lines) and volume strain (solid lines) derived, for the soil at hand, from the three mentioned hypotheses. A possible fourth case, in which the soil undergoes a greater compaction, leading to a further decrease of the void ratio, has not been considered. For the examined soil, in fact, this would lead to a degree of compaction greater than the one the primary fabric could permit.

A series of drained triaxial compression tests has been also carried out in order to investigate the variation of stiffness and shear strength (Peruzzi, 1999). The samples were prepared without fine content or with a fine content equal to half the original one (i.e. 10% in weight). In both cases, they were compacted up to relative densities of 70% and 30%. The former represents the initial density, while the latter is the final value of density after the removal of the fine particles, according to assumption a).

The results, which are not shown here for briefness, indicate that the partial or total removal of the fine particles produce an increase of stiffness and shear strength, for both dense and loose samples. Further investigation on this important aspect (see e.g. Salgado et al., 2000; Thevanayagam & Mohan, 2000) is presently ongoing.

3. A MODEL FOR EROSION AND TRANSPORT OF FINE PARTICLES

Based on the results of the seepage tests, a law has been derived that describes the process of erosion from a unit volume of soil.

The considerable length of the flow path, compared to the grain mean size, permits to consider the phenomena of re-deposition and void clogging as already accounted for in the evaluation of the amount of eroded particles (Khilar et al., 1985; Reddi et al., 2000).

A more rigorous approach would require the analysis of the soil microstructure (seen as a 3D net of channels representing the interconnected voids) and the evaluation of the probability that the erosion of a particle is prevented if the channels have diameter smaller than the one of the particle (Atmatzidis, 1989).

For interpreting the laboratory seepage tests the following relationship is proposed here between the quantity of eroded fine particles μ, the hydraulic gradient i and time t

$$\mu = \mu_0 \left[1 - \exp\left(-\left(t/t_0\right)^b \cdot i^c / a\right)\right] \quad . \tag{1}$$

Here μ_0 represents the initial fine content in weight, t_0=1 h, and a, b, c are non-dimensional parameters (Borelli, 2000).

These parameters have been back calculated by minimising an error function that represents a measure of the discrepancy between experimental and numerical results. The "optimal" solution for the examined samples (a=22.6, b=0.5, c=2.64) led to the solid lines shown in fig.2.

It has to be pointed out that eq.(1) involves two main consequences. The first one is that erosion occurs for any non-vanishing value of the hydraulic gradient (i.e. no lower limit exists below which the process is absent). The second consequence is that the erosion process will completely remove the fine particles, if sufficient time is allowed (i.e. $\mu \rightarrow \mu_0$ if $t \rightarrow \infty$). However, this second drawback should not be crucial, since the theoretical time necessary to complete the erosion becomes exceedingly large when the gradient value is within the range met in practical applications.

If microscopic mechanisms, such as clogging, have already been accounted for, the transport phenomenon can be related solely to the water flow velocity and direction (den Adel et al., 1994; Indraratna & Vafai, 1997).

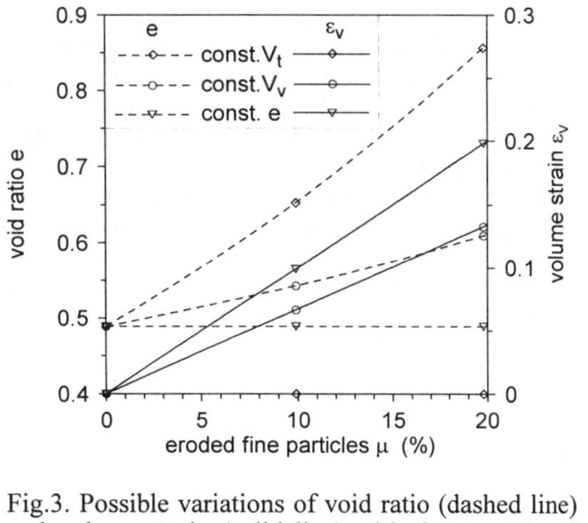

Fig.3. Possible variations of void ratio (dashed line) and volume strain (solid line) with the quantity of eroded fine particles.

Considering a unit volume of soil, the conservation of mass of the fine particles transported by a one-dimensional water flow in direction s can be written as

$$\frac{\partial \rho}{\partial t} = -\frac{\partial}{\partial s}\left[v_s \cdot (\rho - \rho_f)\right] \quad . \qquad (2)$$

Here, ρ, ρ_f and v_s are functions of t and s. They represent, respectively, the density of fine particles within a unit volume of soil, the density of non-moving fine particles in the same volume and the velocity of flow. The left-hand side term in eq.(2) represents the rate of accumulation of the mass of fine particles, while the right-hand side term is the balance between inflow and outflow masses of moving particles.

The density of non-moving particles ρ_f can be defined on the basis of the law of erosion (1)

$$\rho_f = \mu_0 - \mu = \mu_0 \exp\left(-(t/t_0)^b \cdot i^c / a\right) \quad . \qquad (3)$$

The rate of erosion can be worked out, on the basis of eq.(3), as a function of the hydraulic gradient or, introducing Darcy law, as a function of the water flow velocity and soil permeability:

$$\frac{\partial \rho_f}{\partial t} = -\frac{\partial \mu}{\partial t} = g(t, v_s, k) \quad . \qquad (4)$$

The governing equation relating the unknown functions $(\rho-\rho_f)$ and v_s is finally arrived at by combining eqs.(2) and (4)

$$\frac{\partial(\rho - \rho_f)}{\partial t} + \frac{\partial}{\partial s}\left[v_s \cdot (\rho - \rho_f)\right] + g(t, v_s, k) = 0 \quad . \qquad (5)$$

Equation (5) can be easily extended to 3D problems, in a co-ordinate system $\underline{x}$, by expressing the flow velocity through vector $\underline{v}(\underline{x},t)$ and by replacing the space derivative with the divergence operator.

The analysis of boundary value problems, where both the flow velocity and the density of moving fine particles are unknown, is obtained by coupling eq.(5) with the one governing the seepage flow, in the framework of a suitable method for numerical solution. For instance, in terms of the finite difference method, eq. (5) can be re-written as

$$(\rho-\rho_f)_x^{t+\Delta t} = (\rho-\rho_f)_x^t - \Delta t\left[(\rho-\rho_f)_s^t \cdot \frac{v_{sx}^t - v_{s(x-1)}^t}{\Delta x} + v_{sx}^t \frac{(\rho-\rho_f)_x^t - (\rho-\rho_f)_{x-1}^t}{\Delta x} + g(t, v_{sx}, k)\right] \quad . \qquad (6)$$

Once the density of moving particles $(\rho-\rho_f)$ has been determined, the total particle density ρ within a unit volume can be evaluated through eq.(3).

If the erosion does not appreciably modify the coefficient of permeability, the problems of seepage flow and particle erosion and transport become uncoupled and can be separately solved.

In particular, in the case of steady state flow, a preliminary seepage analysis permits to evaluate the velocity field, constant with time. Then, the step-by-step time integration of the erosion and transport problem can be based on eq.(6), where only the function $(\rho-\rho_f)$ is unknown. The case of transient flow would just require the separate evaluation, within each time increment, of the velocity and density functions.

4. EFFECTS ON THE SURFACE SETTLEMENTS

Having evaluated the quantity of eroded and transported particles, a stress analysis is required to estimate the effects of the loss of fine material on the distribution of stresses and strains within the soil mass.

To this purpose, one of the three assumptions discussed in Section 2, and relating the amount of erosion to the consequent volume strain (fig.3), should be adopted. Then, the fine particle density ρ, evaluated for each time increment, is "converted" into volume strains, through the chosen relationship, and the corresponding components of the "initial" strain vector $\underline{\varepsilon}_o$ are evaluated.

If the stress analysis is based on the finite element method, the nodal force vector $\underline{f}_o$ equivalent to the "initial" strains $\underline{\varepsilon}_o$ is determined according to the following relationship

$$\underline{f}_0 = \int_V \underline{B}^T \cdot \underline{D} \cdot \underline{\varepsilon}_0 \, dV \quad , \qquad (7)$$

where $\underline{B}$ represents the matrix of shape function derivatives, $\underline{D}$ the elastic constitutive matrix and the integral is extended over the element volume V.

These forces are then applied as external loads and, depending on the problem at hand, the consequent stress/strain changes can be evaluated in elastic or elasto-plastic regime.

5. APPLICATION

The described solution procedure for the erosion and transport of soil particles has been applied to the analysis of the surface settlement induced by a drainage trench.

The water table is initially located 14.m below the ground surface. The level of water is then lowered within the vertical trench, 0.8 m large and 30.m deep, up to -26.m from the ground surface.

The problem was treated in plane strain conditions and, due to its symmetry, only a portion of the soil mass on the right hand side of the trench was subdivided into finite elements.

The mesh bottom, located -40 m below the ground surface, is seen as an impervious and rigid boundary the displacements of which are completely constrained.

The soil is assumed to be homogeneous and isotropic, with an initial fine content equal to 20% in weight. The erosion of fine particles can freely take place through the walls of the trench (i.e. the possible presence of a filter is neglected), while no changes of the initial fine content takes place on the right vertical boundary. The initial stress state depends on the soil own weight (γ_{sat}=22.kN/m^3) and on the coefficient of earth pressure at rest (K_0= 0.5).

A linear elastic material behaviour was assumed in the calculations (E=10^5 kPa, ν=0.13), since the stress state induced by the lowering of the water table and by the erosion-transport phenomenon is well within the elastic range.

The seepage problem was treated in steady state conditions, neglecting possible changes of the coefficient of permeability. Hence, a steady state unconfined seepage problem was carried out first (Cividini & Gioda, 2000) the solution of which led to the position of the phreatic surface and to the flow velocity field. Then the time integration of the transient erosion-transport problem was initiated.

The results of calculations are summarised in figs. 4 to 6. Fig. 4 presents the contour lines of the percentage of non-moving ρ_f and total ρ fine particles after 1 month from the initiation of pumping.

Note that the quantity of non-moving particles (fig.4a) is relatively low in the vicinity of the trench (min.value=7%). In the same zone, however, the total amount of particles (fig.4b) is close to its initial value (20%), due to the large quantity of transported material.

On the contrary, after 12 months, the steady decrease of the quantity of non-moving particles (min.value =3% in fig.5a) involves also a reduction of the total amount of particles, up to a minimum of 11% (cf. fig.5b). A limited increase of the particle erosion was calculated for considerably larger time intervals beyond 12 months.

Figure 6 reports the surface settlements calculated in three different conditions. In the first case (dashed line) only the effects of seepage, due to the water flow and to the lowering of the water table, were accounted for.

The effects of erosion, and of the consequent volume change, were considered in the second and third cases, after 1 and 12 months from the initiation of pumping. In both cases the settlements were evaluated assuming a constant void ratio (i.e. by adopting the largest volume deformation shown in fig.3) and by neglecting the increase of soil stiffness caused by the loss of fine particles.

These assumptions are likely to lead to overestimated values of the surface settlements caused by erosion.

6. CONCLUSIONS

A model for the erosion and transport of the fine particles of granular soils, caused by the action of a seepage flow, has been discussed. This model is based on an erosion-transport law that has been developed and calibrated considering the results of laboratory seepage tests on reconstituted samples of granular soil.

By coupling the erosion-transport model with a finite element code for unconfined seepage analysis, and with a stress analysis program, it is possible to estimate the surface settlements induced by the erosion phenomenon.

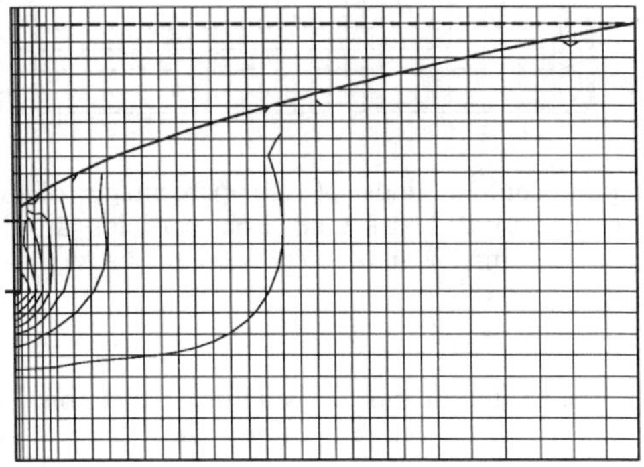

(a)

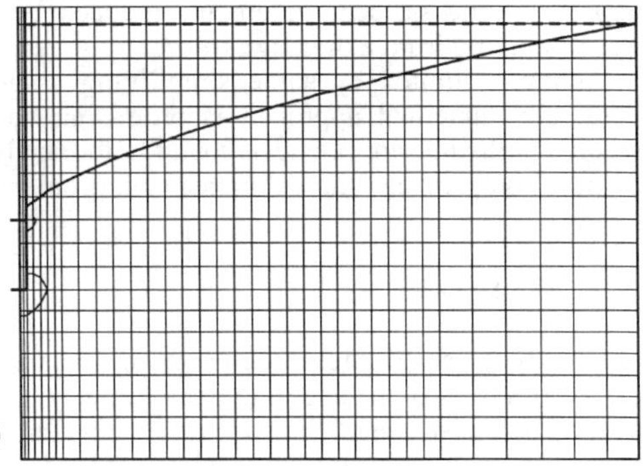

(b)

Fig.4. Contour lines of the percentage of non-moving (a) and total (b) fine particles after 1 month (min=7%, max=19%, increment=2%).

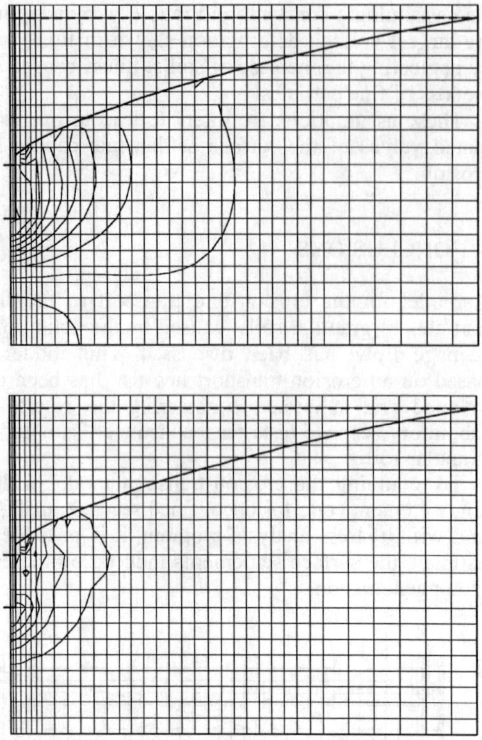

Fig.5. Contour lines of the percentage of non-moving (a) and (b) total fine particles after 12 months (min=3%, max=19%, increment=2%).

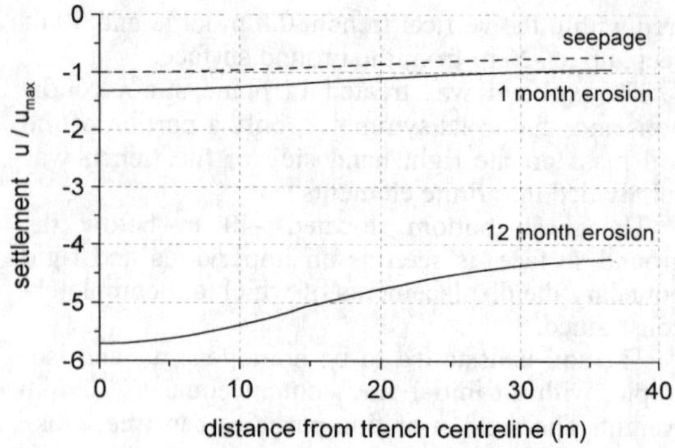

Fig.6. Surface settlements due to seepage (dashed line) and erosion (solid lines) after 1 and 12 months.

The results of this study indicate that some further experimental investigation is necessary to get a better insight into the various aspects of the problem.

The proposed model has been applied to the analysis of the surface settlements induced by the water pumping from a drainage trench.

Although limited by some simplifying assumptions, the numerical approach is promising and it seems worthwhile to extend it towards the analysis of more complex, e.g. three dimensional, problems.

ACKNOWLEDGEMENTS

This study is part of a research program financed by the Metropolitana Milanese (Milan Subway Company). The assistance of Enzo Iscandri in performing the laboratory tests is gratefully acknowledged.

REFERENCES

Atmatzidis D.K. (1989). A study of sand migration in gravel. *Proc. 12th Int. Conf. Soil Mech. Found. Eng.*, Vol.1, 683-686.

Borelli, S. (2000). Sull'asportazione di parti fini dal terreno indotta dal pompaggio da pozzo (in italian). MS Thesis, Politecnico di Milano.

Cividini A. & G.Gioda (2000). Finite element analysis of free surface seepage flow. In Zaman-Gioda-Booker (eds), *Modeling in geomechanics*, Chap.20, Wiley&Sons.

den Adel H., Koenders M.A. & K.J. Bakker (1994). The analysis of relaxed criteria for erosion-control filters. *Can. Geotech. J.*, 31, 829-840.

Fischer G.R. & R.D.Holtz (1996). A critical review of granular soil filter retention criteria. *Proc. 2nd Int. Conf. Geofilters*, eds. Lafleur-Rollin, 409-418.

Hadj-Hamou T., Tavassoli M.R. & W.C.Sherman (1990). Laboratory testing of filters and slot sizes for relief wells. *J. Geotech. Eng. ASCE*, 116 (9), 1325-1346.

Indraratna B. & F.Vafai (1997). Analytical model for particle migration within base soil-filter system. *J. Geotech. Geoenv. Eng. ASCE*, 123 (2), 100-109.

Kenney T.C. & D.Lau (1985). Internal stability of granular filters. *Can. Geotech. J.*, 22, 215-225.

Khilar K.C., Fogler H.S. & D.H.Gray (1985). Model for piping-plugging in earthen structures. *J. Geotech. Eng. ASCE*, 111 (7), 833-846.

Ladd R.S. (1978). Preparing test specimen using under compaction. *Geotech. Testing J.*, 1, 16-23.

Peruzzi C. (1999). Avulsione di parti fini dal sottosuolo dovuta al pompaggio da pozzi (in italian). MS Thesis, Politecnico di Milano.

Reddi L.N., Lee I.-M. & M.V.S.Bonala (2000). Comparison of internal and surface erosion using flow pump tests on a sand-kaolinite mixture. *Geotech.Testing J.*, 23 (1), 116-122.

Salgado R., Bandini P. & A.Karim (2000). Shear strength and stiffness of silty sand. *J. Geotech. Geoenv. Eng. ASCE*, 126 (5), 451-462.

Thevanayagam S. & S.Mohan (2000). Intergranular state variables and stress-strain behaviour of silty sands. *Géotechnique*, 50 (1), 1-23.

Computer Methods and Advances in Geomechanics, Desai et al. (eds) © 2001 Balkema, Rotterdam, ISBN 90 5809 183 X

Numerical analysis of fluid flow in naturally fractured media

K.Xia & M.Zaman
School of Civil Engineering and Environmental Science, University of Oklahoma, Norman, Okla., USA
M.Bai
Rock Mechanics Institute, University of Oklahoma, Norman, Okla., USA

ABSTRACT: Fluid flow in naturally fractured porous media can be considered as an unbounded domain problem in which assumed boundary conditions can significantly influence results. In this paper, a three-dimensional finite-infinite element model, based on the poroelasticity dual porosity approach, is presented. The advantage and efficiency of the three-dimensional finite-infinite element model is demonstrated by comparing solutions with those of conventional finite element results under two extreme boundary conditions. It is observed that the mapped infinite element is a useful approach for modeling fluid flow in porous media.

Key Words: Unbounded domain, Finite-infinite element, Poroelasticity

1 INTRODUCTION

Fluid flow through porous media is an important phenomenon in oil reservoir engineering, and also in environmental and geotechnical engineering. In the past few decades, finite element method (FEM) has been widely used to analyze fluid flow and solid deformation in porous media (see e.g., Bai & Meng 1994, Bibby 1981, Duguid & Adbel 1974, Elsworth & Bai 1992, Ghafouri & Lewis 1996, Khaled et al. 1984, Lewis & Schrefler 1987, Young 1981).

While using FEM, a difficulty is often encountered in numerical analysis of fluid flow problems dealing with an unbounded exterior domain. In conventional FEM analyses, an unbounded domain is represented approximately by a finite domain by placing the desired boundaries sufficiently away. Usually large numbers of elements are needed in such idealizations and certain boundary conditions could not be simulated rationally. To overcome these weaknesses, the infinite element concept was developed and adopted by researchers (Bettess 1977, Damjanic & Owen 1984, Lynn & Hadid 1981, Marques & Owen 1984, Ungless 1973, Zienkiewicz et al. 1983). The infinite element concept has also been applied for numerical analysis of soil consolidation (Simoni & Schrefler 1987, Zaman et al. 1991). Most of the formulations for infinite elements in the literature are generally based on two methods. One deals with elements with decay functions in the infinite direction that multiply the usual shape functions. The other type is the mapped infinite element involving mapping of the coordinates.

In this paper, three-dimensional mapped infinite elements are employed and coupled with a three-dimensional finite element algorithm based on the dual porosity poroelasticity approach. The numerical results are compared with those of conventional three-dimensional FEM model without any infinite element. Accuracy and efficiency of the finite-infinite element method (FIEM) are demonstrated through a parametric study of a boundary value problem.

2 THREE-DIMENSIONAL MAPPED INFINITE ELEMENTS

To idealize a three-dimensional unbounded domain in a rational manner, one-directional, two-directional and three-directional infinite elements are very useful (Marques & Owen 1984). These three types of three-dimensional infinite elements are coupled here with conventional three-dimensional finite elements. The procedure involves evaluating the products of shape functions based on the mapping techniques reported by Marques & Owen (1984) and by Zienkiewicz et al. (1983).

Based on the aforementioned mapping technique, an 8-noded one directional (Figure 1(a)) and two-directional (Figure 1(b)) three-dimensional infinite elements are developed that can be conveniently used in discretizing an unbounded domain extending to infinity in one and two directions, respectively. A two-directional infinite element can be used in the corner area extending to infinity in the η and ς directions, as shown in Figure 1(b). For all the infinite elements employed here, the mapping relationship between the global co-ordinate system (ξ, η, ς) and the local co-ordinate system can be written as follows:

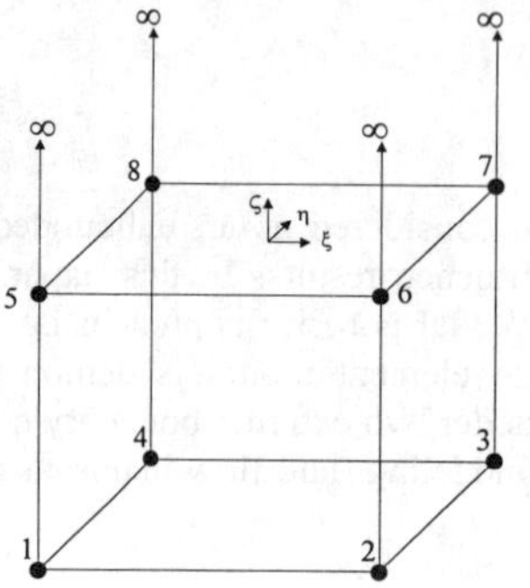

(a) One-directional infinite element.

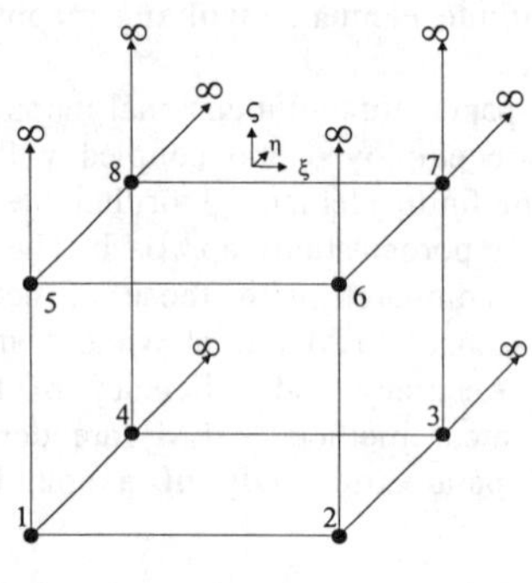

(b) Two-directional infinite element.

Figure 1. Three-dimensional mapped infinite elements.

$$x = \sum_{i=1}^{N} M_i(\xi,\eta,\varsigma)x_i$$
$$y = \sum_{i=1}^{N} M_i(\xi,\eta,\varsigma)y_i \qquad (1)$$
$$z = \sum_{i=1}^{N} M_i(\xi,\eta,\varsigma)z_i$$

It may be noted that the integration procedures to evaluate the element matrices (such as K_s, L_1, L_2, etc., discussed later in this paper) for the mapped infinite elements need to be adjusted for the Gauss abscissas and the weight coefficients W_i in the infinite directions. The integration parameters for infinite elements and the corresponding finite element are shown in Table 1.

Table 1. Gauss-Legendre integration parameters for finite-infinite elements.

Finite element		Mapped infinite element	
Abscissa	Weight	Abscissa	Weight
-0.5773503	1.0	-0.7320508	0.8038475
0.5773503	1.0	2.73250511	11.1961541

Because of the use of infinite elements, the final global matrix generally is a full rank matrix that can be solved directly without introducing the corresponding boundary conditions at infinity, as required for conventional finite elements.

3 OVERVIEW OF POROELASTICITY DUAL POROSITY FORMULATION

For fluid flow in a dual-porosity porous medium, external loads will create a pressure gradient between the fluid within the matrix pores and the fluid in the adjacent fractures. The fluid within the matrix is squeezed out and injected into the mutually-connected fissured continuum due to the generated pressure gradient. Consequently, fluid begins to flow towards and through the fissured network. In this model, the derivation of the governing equations is presented by viewing the system as an elastic skeleton infiltrated by a two-state fluid, one flowing through the fissures and the other flowing through the matrix. In conformity with the classical Biot's theory (1941), the equilibrium equations for the rock mass are written in terms of total effective stress and a two-state Darcy's law is used to specify the fluid flow in the two distinct continua as given by Equation (2).

$$v_i^{\lambda} = \frac{-k_{\lambda}}{\mu} p_{\lambda,i} \qquad (2)$$

where k_λ is the permeability, p_λ is the fluid pressure, μ is the fluid dynamic viscosity, and the subscripts $\lambda = 1$ and 2 represent the fracture and matrix, respectively.

For a dual porosity system, the continuity equation of fluid flow in fracture and matrix may be expressed as follows (Elsworth & Bai 1992, Khaled et al. 1984).

$$\frac{-k_f}{\mu}p_{1,ii} = \alpha_f\delta_{ij}C^1_{ijkl}D^{12}_{klmn}\dot{\varepsilon}_{mn} - \beta_1\dot{p}_1 + \Gamma_f(p_1 - p_2) \quad (3)$$

$$\frac{-k_m}{\mu}p_{2,ii} = \alpha_m\delta_{ij}C^2_{ijkl}D^{12}_{klmn}\dot{\varepsilon}_{mn} - \beta_2\dot{p}_2 - \Gamma_f(p_1 - p_2) \quad (4)$$

Here Γ_f is fluid transfer coefficient denoting the rate of inter-porosity flow due to pressure difference between fracture and matrix, and can be expressed as $\Gamma_f = \rho_f \frac{4n(n+2)k_1}{S^2\mu}$ based on the concept proposed by Warren and Root (1963). In this paper, S is considered as the statistical mean value of fissure spacing. C^1_{ijkl} and C^2_{ijkl} are compliance matrices for fracture and matrix, respectively (Bai & Meng 1994), and are defined in terms of E, υ, k_n and S. k_f and k_m are the permeability of the fracture and matrix, respectively. n_m and n_f denote the porosity of matrix and fracture, respectively. K_f represents the fluid compressibility modulus, and α_f and α_m are the pore pressure ratio factors of the fracture and matrix. Also, K_t and K_s are the bulk modulus of the skeleton and solid grain, respectively, S represents fracture spacing, K_n represents fracture stiffness.

Within each continuum of the dual porosity system, the relationship between total stress, effective stress, and pore pressure can be expressed as follows:

$$\begin{aligned}\sigma^1_{ij} &= {}_f\sigma'_{ij} + \alpha_f p_1 \delta_{ij} \\ \sigma^2_{ij} &= {}_m\sigma'_{ij} + \alpha_m p_2 \delta_{ij}\end{aligned} \quad (5)$$

$$\begin{aligned}{}_f\sigma'_{ij} &= D^1_{ijkl}\varepsilon^1_{kl} \\ {}_m\sigma'_{ij} &= D^2_{ijkl}\varepsilon^2_{kl}\end{aligned} \quad (6)$$

where ${}_f\sigma'_{ij}$, and ${}_m\sigma'_{ij}$ refer to the effective stresses within the fracture and matrix, respectively.

Based on the assumption that the matrix and fracture are lumped together, total strain (ε_{ij}) for the dual porosity medium can be expressed as

$$\varepsilon_{ij} = \varepsilon^1_{ij} + \varepsilon^2_{ij} \quad (7)$$

Based on Equations (5) and (6), Equation (7) can be expressed in the following form

$$\varepsilon_{ij} = (C^1_{ijkl} + C^2_{ijkl})\sigma_{kl} - C^1_{ijkl}\delta_{kl}\alpha_f p_1 - C^2_{ijkl}\delta_{kl}\alpha_m p_2 \quad (8)$$

or

$$\sigma_{ij} = D^{12}_{ijkl}(\varepsilon_{kl} + C^1_{klmn}\delta_{mn}\alpha_f p_1 + C^2_{klmn}\delta_{mn}\alpha_m p_2) \quad (9)$$

Following the work of Bai & Meng (1994), for any volume of the matrix and fracture of a dual porosity system having a surface area S and volume Ω, the governing equations can be expressed as

$$\sigma_{ij,j} + F_i = 0 \text{ in } \Omega. \quad (10)$$

$$\sigma_{ij}n_j = \overline{T}_i \text{ on } S. \quad (11)$$

Substituting Equation (9) into Equation (10), the equilibrium equation can be written as

$$\begin{aligned}&(D^{12}_{ijkl}\varepsilon_{kl} + D^{12}_{ijkl}C^1_{klmn}\delta_{mn}\alpha_f P_1 \\ &+ D^{12}_{ijkl}C^2_{klmn}\delta_{mn}\alpha_m P_2)_{,j} + F_i = 0\end{aligned} \quad (12)$$

The governing Equations (3), (4) and (12) can be discretized using the finite-infinite element procedure outlined earlier. For simplification, the shape functions for each of the variables are assumed to be identical. Thus, spatial approximations of the displacement (U), pore pressures (P_i) and strain (ε) are written by

$$\begin{aligned}U(t) &= N\overline{U}(t) \\ P_i(t) &= N\overline{P}_i(t) \\ \varepsilon &= BU(t)\end{aligned} \quad (13)$$

where B is associated with the strain-displacement transformation.

By applying the weighted residual method (Galerkin procedure) and the Green's theorem, and substituting appropriate spatial approximations of the unknowns into the equilibrium equation of solid grain, the governing Equations (13), can be transformed into Equation (14).

$$\overline{K_s}\frac{dU}{dt} + L_1\frac{dP_1}{dt} + L_2\frac{dP_2}{dt} = \frac{d\overline{f_s}}{dt} \quad (14)$$

where

$$\overline{K_s} = \int_\Omega B^T D^{12}_{ijkl} N_j d\Omega \quad (15)$$

$$L_1 = \int_\Omega \alpha_f B^T m D^{12}_{ijkl} C^1_{klmn} N d\Omega \quad (16)$$

$$L_2 = \int_\Omega \alpha_m B^T m D^{12}_{ijkl} C^2_{klmn} N d\Omega \quad (17)$$

$$\overline{f_s} = \int_\Omega N^T \rho b_i d\Omega + \int N^T \overline{T_i} ds \qquad (18)$$

Using the same procedures, continuity Equations (3) and (4) corresponding to the fracture and matrix, respectively, can be rewritten as Equations (19) and (24).

$$L_1 \frac{dU}{dt} - F_1 \frac{dP_1}{dt} + Z_2(P_1 - P_2) + Z_3 P_1 = \overline{q_1}$$

for fissured continuum (19)

$$F_1 = \int_\Omega N^T \frac{n_f}{K_f} N d\Omega \qquad (20)$$

$$Z_2 = \int_\Omega N^T \Gamma N d\Omega \qquad (21)$$

$$Z_3 = \int_\Omega \nabla N^T \frac{k_f}{\mu} \nabla N d\Omega \qquad (22)$$

$$\overline{q_1} = \int_s N^T q_1^s ds \qquad (23)$$

$$L_2 \frac{dU}{dt} - F_2 \frac{dP_2}{dt} - Z_2(P_1 - P_2) + R_3 P_2 = \overline{q_2}$$

for matrix continuum (24)

$$F_2 = \int_\Omega N^T \frac{n_m}{K_f} N d\Omega \qquad (25)$$

$$R_3 = \int_\Omega \nabla N^T \frac{k_m}{\mu} \nabla N d\Omega \qquad (26)$$

$$\overline{q_2} = \int_\Omega N^T q_2^s ds \qquad (27)$$

Equations (14), (19), and (24) can be rewritten in a matrix form as:

$$\begin{bmatrix} K_s & L_1 & L_2 \\ L_1^T & -F_1 & 0 \\ L_2^T & 0 & F_2 \end{bmatrix} \frac{d}{dt} \begin{bmatrix} U \\ P_1 \\ P_2 \end{bmatrix} + \begin{bmatrix} 0 & 0 & 0 \\ 0 & Z_2 + Z_3 & -Z_2 \\ 0 & -Z_2^T & Z_2 + R_3 \end{bmatrix} \begin{bmatrix} U \\ P_1 \\ P_2 \end{bmatrix} = \begin{bmatrix} \frac{d\overline{f_s}}{dt} \\ q_1 \\ q_2 \end{bmatrix} \qquad (28)$$

These coupled equations should be rearranged prior to solving in the time domain. A fully implicit finite difference scheme in the time discretization domain, as depicted by Equation (29), is employed because the coefficients matrix are time dependent functions.

$$\dot{U}^{t+\Delta t} = \frac{1}{\Delta t^{t+\Delta t}} (U^{t+\Delta t} - U^t)$$

$$\dot{P}_1^{t+\Delta t} = \frac{1}{\Delta t^{t+\Delta t}} (P_1^{t+\Delta t} - P_1^t) \qquad (29)$$

$$\dot{P}_2^{t+\Delta t} = \frac{1}{\Delta t^{t+\Delta t}} (P_2^{t+\Delta t} - P_2^t)$$

Substituting Equation (30) into Equation (28), thus yields:

At each time step, the displacement and pore pressure are calculated as follows:

$$\frac{1}{\Delta t^{t+\Delta t}} \begin{bmatrix} K_s & L_1 & L_2 \\ L_1^T & -F_1 + (Z_2 + Z_3)\Delta t^{t+\Delta t} & -Z_2 \Delta t^{t+\Delta t} \\ L_2^T & -Z_2 \Delta t^{t+\Delta t} & F_2 + (Z_2 + R_3)\Delta t^{t+\Delta t} \end{bmatrix} = \frac{1}{\Delta t^{t+\Delta t}} \begin{bmatrix} K_s & L_1 & L_2 \\ L_1^T & -F_1 & 0 \\ L_2^T & 0 & F_2 \end{bmatrix} \begin{bmatrix} U \\ P_1 \\ P_2 \end{bmatrix}^t + \begin{bmatrix} \frac{d\overline{f_s}}{dt} \\ q_1 \\ q_2 \end{bmatrix} \qquad (30)$$

$$U^{n+1} = U^n + \frac{\Delta t^{n+1}}{\Delta t^n} (U^n - U^{n-1}) \qquad (31a)$$

$$P^{n+1} = P^n + \frac{\Delta t^{n+1}}{\Delta t^n} (P^n - P^{n-1}) \qquad (31b)$$

$$P^{n+1} = P^n + \frac{\Delta t^{n+1}}{\Delta t^n} (P^n - P^{n-1}) \qquad (31c)$$

4 NUMERICAL EXAMPLES

A computer program was developed based on the three-dimensional finite-infinite algorithm outlined above. A numerical example was solved for the illustration of the model. The example involves fluid flow (due to consolidation) in a fully saturated unbounded domain consisting of a dual-porosity porous medium and subjected to a step uniform normal loading $q = 1.0$MPa with a ramp between $t = 0.0$ and 0.001 second. The finite-infinite element mesh used to discretize the problem is shown in Figure 2. Here both one-directional and two-directional infinite elements are used. For the purpose of comparison, two FEM analyses are conducted based on the same domain as above, but

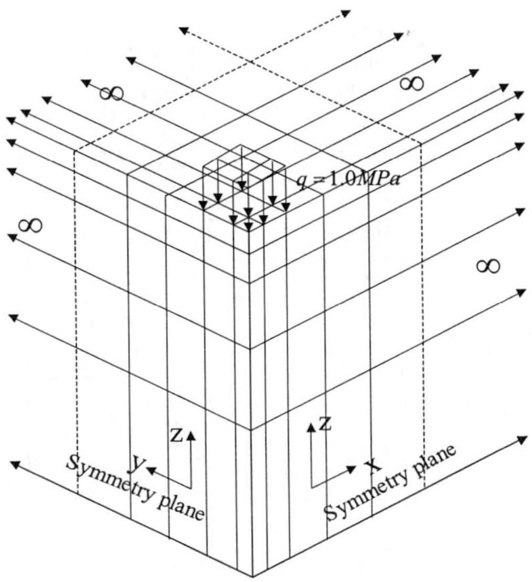

(a) Finite-infinite element mesh.

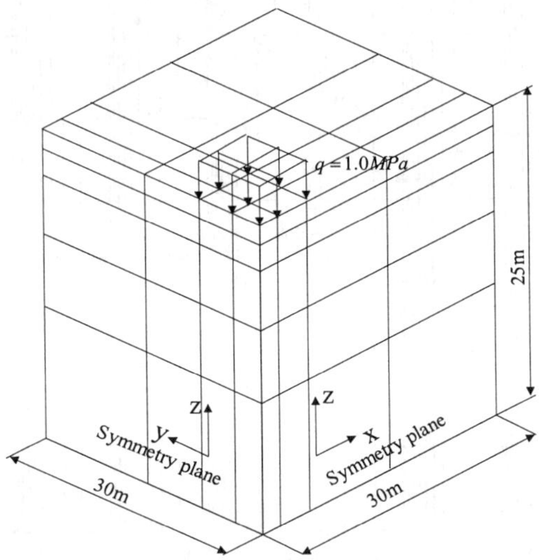

(b) Finite element mesh.

Figure 2. Finite-infinite element mesh used for numerical example.

different boundary conditions are used. The material properties used in both cases are given in Table 2.

For the analyses involving the finite-infinite element idealization, the pore pressure is assumed to be zero at the top surface, and the bottom is assumed to be impervious. Due to spatial symmetry, we select only a quarter of the infinite domain for numerical analysis and the symmetric faces (see Figure 2) are assumed to be impervious/fixed. While using infinite elements to simulate the far field domain surrounding the finite elements, the boundary conditions along the infinity directions will be automatically satisfied (e.g., zero displacement). Therefore, it is not necessary to specify the conditions at the outer nodes of the infinite elements.

For analysis using the conventional FEM (i.e., without using infinite elements) (Figure 2), the boundary conditions for all the boundaries are to be

Table 2. Parameters used for numerical examples.

Parameter	Definition	Magnitude	Units
E	Modulus of elasticity	1000.0	MN/m^3
υ	Poisson ratio	0.2	-
K_n	Fracture stiffness	600.0	$MN/m^2/m$
K_w	Fluid bulk modulus	5.0	MN/m^2
n_f	Fracture porosity	0.05	-
n_m	Matrix porosity	0.2	-
k_1/μ	Fracture permeability	1×10^{-3}	m^4 /(MNs)
k_2/μ	Matrix permeability	1×10^{-5}	m^4 /(MNs)
S	Fracture spacing	0.1	m
K_s	Modulus of solid grain	5400.0	MN/m^2
α	Biot's coefficient	0.90	-

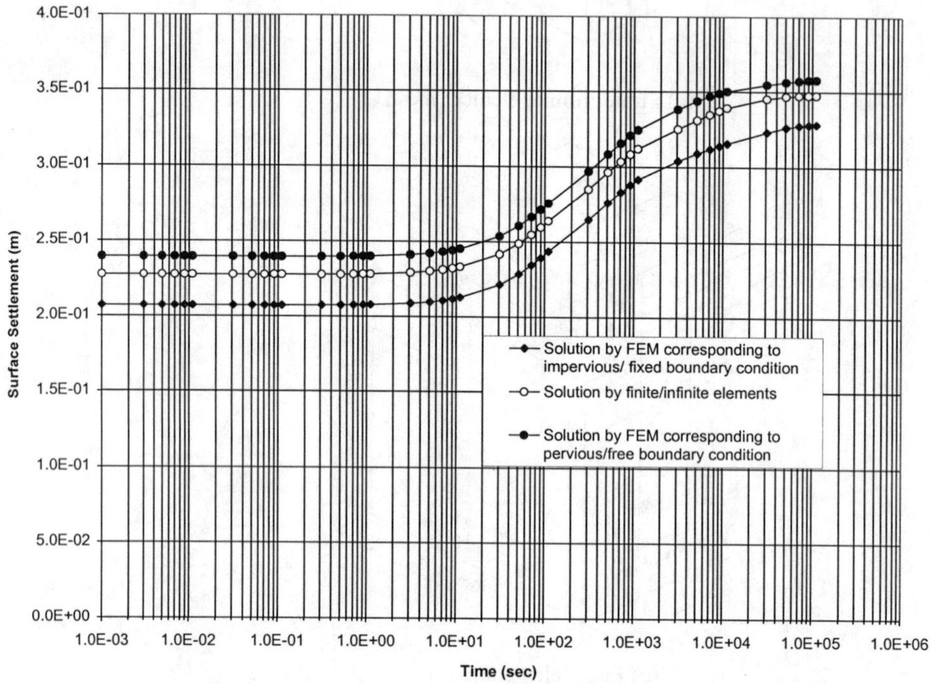

Figure 3. Surface settlement versus time at node 1 by FEM and FIEM.

specified, besides the two symmetrical surfaces. Two extreme boundary conditions are employed: in one case it is assumed that the two faces corresponding to the interfaces of finite-infinite elements are fixed horizontally and are impervious (called Case I), while for the other case these boundaries are assumed to be free and pervious (Case II), even though their finite element meshes are the same. The main purpose of these boundary conditions is to demonstrate the advantage of the infinite element in automatically simulating the boundary conditions.

The consolidation settlements of node 1 (0,0,0) versus the time history obtained from the FEM analyses for Case I and Case II and finite-infinite element model are compared in Figure 3. It is observed that the surface settlement obtained from the FEM model corresponding to free boundary conditions (Case II) is the largest, while that of the

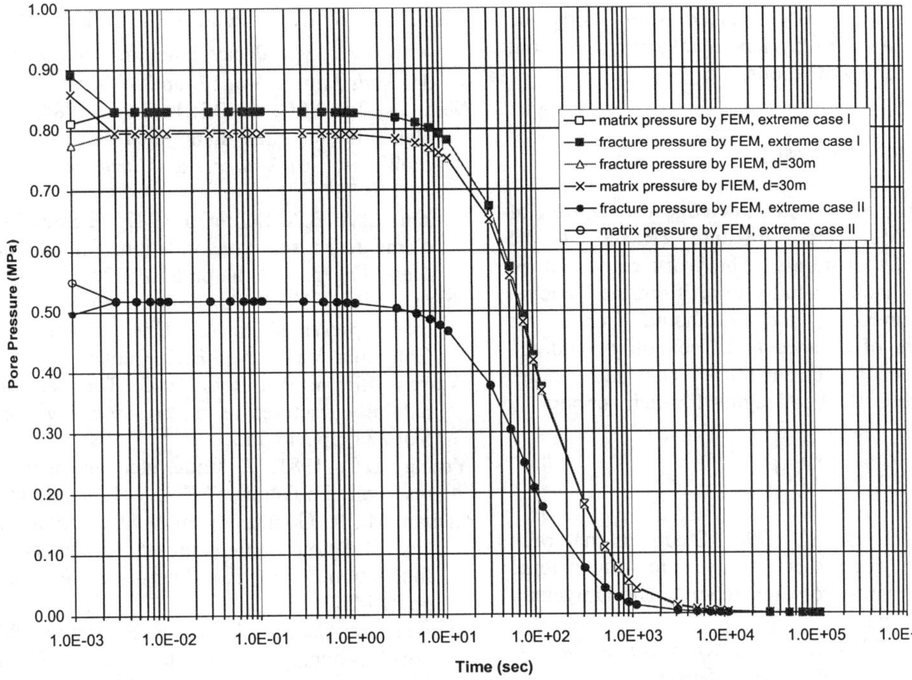

Figure 4. Pore pressure versus time at node 25 by FEM and FIEM.

fixed boundary conditions (Case I) is the smallest. The surface settlement obtained from the finite-infinite element algorithm is between the two bounded FEM solutions. This is attributed to the different boundary conditions used for the FEM analyses. The boundary conditions simulated by the finite-infinite element algorithm are closer to the actual boundary conditions. As noted earlier, when using the mapped infinite elements, it will automatically satisfy the boundary condition requirements at the infinity that is close to the field situation.

Figure 4 shows the transient pore pressure history at node 25 (0,0,–2.0). One can see that fluid pressures within fracture and matrix continua are very close for the single-phase flow. Pore pressure difference between the fracture and matrix only exists at the beginning of the consolidation process and this difference may be associated with the parameters (e.g., time integration) used in the numerical analysis. Once the fluid within the fractured network and matrix are fully connected, the fluid pressures within these two different continua will compatibly maintain the same value and simultaneously decrease due to dissipation.

For Case II FEM idealization, fluid pressure depletes at the fastest rate, among the three cases, due to its special boundary condition (pervious boundary) feature adopted. Fluid pressure dissipates most slowly for Case I because the fluid can only depart vertically through the top surface due to the impervious condition imposed at the other boundaries. The finite-infinite element model also presents an intermediate solution for excess pore pressure, as expected.

Based on the above results, one can conclude that for unbounded domain problems, because of inherent difficulties in positioning and defining realistic boundary conditions for finite element analyses, a finite-infinite element algorithm provides a suitable alternative.

5 CONCLUDING REMARKS

A three-dimensional finite-infinite element algorithm, based on the poroelasticity theory of dual porosity, was implemented and applied to examine an unbounded fluid flow problem associated with consolidation. Results are compared with the solutions from finite element analyses corresponding to different boundary conditions. The results show that the mapped infinite element is a very useful approach. The main advantage of the infinite element lies in automatically modeling the boundary conditions in a rational way. Also, the number of elements can be greatly reduced when using the combined finite-infinite element algorithm without losing the accuracy of results.

ACKNOWLEDGEMENTS

The authors gratefully acknowledge the support of the Rock Mechanics Institute (RMI) at the University of Oklahoma. The work could not be done without the financial support from the National Science Foundation, the Oklahoma Center for Advancement of Science and Technology and the Rock Mechanics Consortium. The authors are indebted to their RMI colleagues for their support.

REFERENCES

Bai, M. & F. Meng 1994. Study of naturally fractured reservoirs using three-dimensional finite elements (research report), The University of Oklahoma.

Bettess, P. 1977. Infinite elements, *Int. J. Numer. Methods Engrg.*, Vol.11, pp.53-64.

Bibby, R. 1981. Mass transport in dual-porosity media, *Water Res.*, 17(4), pp.1075-1081.

Biot, M.A. 1941. General theory of three-dimensional consolidation, *J. Appl. Phys.*, Vol.12, pp.155-164.

Damjanic, F. & D.R.J. Owen 1984. Mapped infinite element in transient thermal analysis, *Computer & Structure*, Vol.19, No. 4, pp.673-687.

Duguid, J.O., & J.F. Adbel 1974. Finite element galerkin method for analysis of flow in fractured porous media, In *Finite element methods in flow problem*, Oden, J.T., UAH press, Huntsville, Alabama.

Elsworth, D. & M. Bai 1992. Flow-deformation response of dual-porosity media, *J. Geotech. Engrg.*, ASCE, Vol.118, pp.107-124.

Ghafouri, H.R. & R.W. Lewis 1996. A finite element double porosity model for heterogeneous deformable porous media, *Int. J. Numer. Analy. Meth. Geomech.*, Vol.20, pp.831-844.

Khaled, M.Y., D.E. Beskos & E.C. Aifantis 1984. On the theory of consolidation with double porosity-III A finite element formulation, *Int. J. Numer. Analytic. Meth. Geomech.*, Vol.8, pp.101-123.

Lewis, R.W. & B.A. Schrefler 1987. *The finite element method in the deformation and consolidation of porous media*, Wiley & Sons, New York.

Lynn, P.P. & H.A. Hadid 1981. Infinite elements with $1/r^n$ type decay functions. *Int. J. Numer. Methods Engrg.*, Vol.17, pp.347-355.

Marques, J.M.M.C. & D.R.J. Owen 1984. Infinite elements in quasi-static materially nonlinear problems, *Computer & Structure*, Vol.18, pp.739-751.

Simoni, L. & B.A. Schrefler 1987. Mapped infinite elements in soil consolidation, *Int. J. Numer. Meth. Engrg.*, Vol.24, pp.513-527.

Ungless, R.L. 1973. An infinite element, master thesis presented to the University of British Columbia, Vancouver, B.C., Canada.

Warren, J.E. & P.J. Root 1963. The behavior of naturally fractured reservoirs, *Soc. Pet. Eng. J.*, Vol.228, pp.245-255.

Young, L.C. 1981. A finite element method for reservoir simulation, *SPE J.*, Vol.21, pp.115-128.

Zaman, M., A. Gopalasingam & J.G. Laguros 1991. Consolidation settlement of bridge approach foundation, *J. Geotech. Engrg.* ASCE, Vol. 117, pp.219-240.

Zienkiewicz, O.C., C. Emson & P. Bettess 1983. A novel boundary infinite element, *Int. J. Numer. Meth. Engrg.*, Vol.19, pp.393-404.

Computer Methods and Advances in Geomechanics, Desai et al. (eds) © 2001 Balkema, Rotterdam, ISBN 90 5809 183 X

Viscous elastic plastic consolidation model and finite element analysis for soft foundation

Chen Xiaoping
Institute of Rock and Soil Mechanics, Chinese Academy of Sciences & Wuhan University of Hydraulic and Electric Engineering, People's Republic of China

Bai Shiwei & Yang Chunhe
Institute of Rock and Soil Mechanics, Chinese Academy of Sciences, Wuhan, People's Republic of China

ABSTRACT: The one of important mechanical behaviors of soft clay is time-dependent. It affects the stress-strain behavior of soil. A viscous elastic plastic consolidation model is presented and is implemented in a finite element program in order to describe a time dependent behavior of soft foundation.. The viscous elastic plastic consolidation behavior for soft foundation is studied and some valuable conclusions are drawn in this paper.

1 INTRODUCTION

Many research results show that the stress-strain behavior of soft clays exhibits pronounced time-dependent. Besides the nature of consolidation, viscosity is another main source. The coupling effect of viscous and consolidation will determine the long-term stability of structure in soft foundation. Based on of viscous elastic plastic constitutional model and Biot's consolidated theory, a model of viscous elastic plastic consolidation (VEPC) has been established, which can describe elastic plastic nature as well as creep and consolidation characteristic of soft soil.

2 VISCO ELASTIC PLASTIC MODEL AND BIOT'S CONSOLIDATION EQUATION

2.1 *Viscous elastic plastic constitutional model*

When elastic-plastic nature and viscous nature are took into account comprehensively, the increment of total strain $d\varepsilon_{ij}$, at a certain time, can be divided in an elastic-plastic part $d\varepsilon_{ij}^{ep}$,a visco-elastic part $d\varepsilon_{ij}^{ve}$, and a visco-plastic part $d\varepsilon_{ij}^{vp}$:

$$d\varepsilon_{ij} = d\varepsilon_{ij}^{ep} + d\varepsilon_{ij}^{ve} + d\varepsilon_{ij}^{vp} \tag{1}$$

where, $d\varepsilon_{ij}^{ep}$ is determined by Yin's double yield surface model(Yin,1988); $d\varepsilon_{ij}^{ve}$ is described by Kelvin model; and $d\varepsilon_{ij}^{vp}$ is defined by Bingham model.

Substituting the exact expression of every $d\varepsilon_{ij}$ for equ.1. When the associative flow rule $Q \equiv F$, is adopted, the constitutional equation of viscous elastic plastic can be expressed as (Zhao&Shi,1996):

$$\{d\varepsilon\} = [C]\{d\sigma'\} + \frac{[A]\{d\sigma'\}}{E_1}(1-e^{\eta_e dt}) + \left[\left\langle\Phi(\frac{F_1}{F_{10}})\right\rangle\left\{\frac{\partial F_1}{\partial\sigma}\right\}^T\left\{\frac{\partial F_1}{\partial\sigma}\right\} + \left\langle\Phi(\frac{F_2}{F_{20}})\right\rangle\left\{\frac{\partial F_2}{\partial\sigma}\right\}^T\left\{\frac{\partial F_2}{\partial\sigma}\right\}\right]\frac{dt}{K_p} \tag{2}$$

where, $\{d\varepsilon\} = [d\varepsilon_x \ d\varepsilon_z \ d\gamma_{xz}]$; $[C]$ = elastic plastic flexible matrix; $\{d\sigma'\} = [d\sigma'_x \ d\sigma'_z \ d\tau_{xz}]$, denotes effective stress; $[A]$ = stress matrix; $\eta_e = E_1 / K_e$, E_1 = elastic modulus of Kelvin model, Ke = visco-elastic coefficient; Kp = visco- plastic coefficient; F_0 is a reference value that makes F/F_0 dimensionless; $\left\langle\Phi(\frac{F}{F_0})\right\rangle$ = switch function; F_1 and F_2 are the first and the second yield surface of Yin's model, respectively, which have relation to compression and expansion of soil mass:

$$F_1 = p + \frac{q^2}{M_1^2(p+p_r)-q} = \frac{h\varepsilon_v^p}{1-m\varepsilon_v^p}p_a$$
$$F_2 = \frac{aq}{G}\sqrt{\frac{q}{M_2(p+p_r)-q}} = \varepsilon_s^p \tag{3}$$

in which, ε_v^p = hardening parameter of plastic volumetric strain; ε_s^p = hardening parameter of plastic deviator strain; G= shearing modulus of elasticity: $G = K_G p_a (p/p_a)^n$; and a is the parameter reflecting dilatancy of soil.

There are 11 parameters in the modeling: a、K_G、n、M_1、M_2、h、m、K_1、n_1、η_e、K_p.

2.2 *Biot's consolidation equation*

For plane question, the Biot's consolidation differential equation of satureted soil can be given by:

$$\text{Equilibrium equation}\quad \left.\begin{aligned} &\frac{\partial \sigma'_x}{\partial x} + \frac{\partial \tau_{xz}}{\partial z} + \frac{\partial u}{\partial x} = 0 \\ &\frac{\partial \tau_{xz}}{\partial x} + \frac{\partial \sigma'_z}{\partial z} + \frac{\partial u}{\partial z} + \gamma = 0 \end{aligned}\right\} \tag{4}$$

$$\text{Continuity equation}\quad \frac{\partial \varepsilon_v}{\partial t} = -\frac{1}{\gamma_w}\left(k_x \frac{\partial^2 u}{\partial x^2} + k_z \frac{\partial^2 u}{\partial z^2}\right) \tag{5}$$

where, u= pore water pressure; σ'= effective stress; τ= shear stress; γ= unit weight of soil; k_x and k_z= the seepage coefficient in horizontal direction and in vertical direction; $\partial\varepsilon_v/\partial t$ = the rate of unit volume compressibility, for saturation soil, it is equal to the sum of net discharge flowing unit bulk specific. Under the hypothesis of small deformation, ε_v can be expressed by displacement:

$$\varepsilon_v = \frac{\partial w_x}{\partial x} + \frac{\partial w_z}{\partial z} \tag{6}$$

where, w_x and w_z =deformation in x direction and z direction separately.

3 COUPLING VEPC MODELING AS WELL AS ITS FINIT ELEMENT SOLVE

According to the principle of virtual displacement and the condition of continuity, the finite element formula of Biot's consolidated differential equation is as follows:

$$\begin{bmatrix} \bar{K} & K' \\ K'^T & \tilde{K} \end{bmatrix} \begin{Bmatrix} \Delta w \\ u \end{Bmatrix} = \begin{Bmatrix} R - R_t \\ 0 \end{Bmatrix} \tag{7}$$

where, $\{\Delta w\}$ = node displacement increment; $\{u\}$ = node neutral pressure gross; $\{R\}$ = node load vector; $\{R_t\}$ = node load balanced by effective stress at the time of day:

$$\{R_t\} = \sum_{\Delta t} \iint_e [B]^T \{\Delta\sigma'\} dxdz \tag{8}$$

and $[K]$ = coefficient matrix of the equation, obtained by assembled cell stiffness matrix $[k]$; $[k]$ is determined by formula(9)-(11), respectively:

$$[\bar{k}] = \iint_e [B]^T [D][B] dxdz \tag{9}$$

$$[k'] = \iint_e [B]^T \{M\}[N] dxdz \tag{10}$$

$$[\tilde{k}] = \frac{\Delta t}{\gamma_w} \iint_e \left(k_x \{\frac{\partial N}{\partial x}\}\{\frac{\partial N}{\partial x}\}^T + k_z \{\frac{\partial N}{\partial z}\}\{\frac{\partial N}{\partial z}\}^T\right) dxdz \tag{11}$$

where, $[B]$ = strain matrix, $[D]$ = the matrix of elastic-plastic, $\{M\} = [1\ \ 1\ \ 0]^T$; $[N]$= form function. Coupling viscous elastic plastic constitutive model and Biot's consolidation equation, the finite element formula of viscous elastic plastic consolidation can be gained:

$$\begin{bmatrix} \bar{K} & K' \\ K'^T & \tilde{K} \end{bmatrix} \begin{Bmatrix} \Delta w \\ u \end{Bmatrix} = \begin{Bmatrix} R - R_t + R_{Vep} \\ 0 \end{Bmatrix} \tag{12}$$

Comparing equ.12 with equ.7, a annexation item $\{R_{Vep}\}$ is obtained, which is node annexation load of soil body cell and results from the fluxion of visco-elastic and visco-plastic:

$$\{R_{Vep}\} = \iint_e [B]^T [D](\{\varepsilon_{Ve}\} + \{\varepsilon_{Vp}\}) dxdz \tag{13}$$

Formula (12) is a control equation of finite element analysis in VEPC model. The dissipation of pore water pressure and the creep of soil skeleton are took into account. If the initial stage condition and boundary condition are given, the numerical value solution of the model can be obtained.

4 EXAMPLE ANALYSIS

A soft foundation in GuangDong Province is given in figure 1, the accumulation preloading method was put into action for improving its bearing capacity. The VEPC modeling introduced above is adopted to calculate changes of settlement with time.

The model parameters determined by laboratory soil test are listed in table 1 and table 2.

Paving and unpaving plastic plate drain are calculated respectively in this analysis. Figure 2 and figure3 are W_z-t and W_x-t curve for surface settlement of soft foundation and the representative node in foundation. Where, W_z denots vertical settlement and W_x denots horizontal displacement.

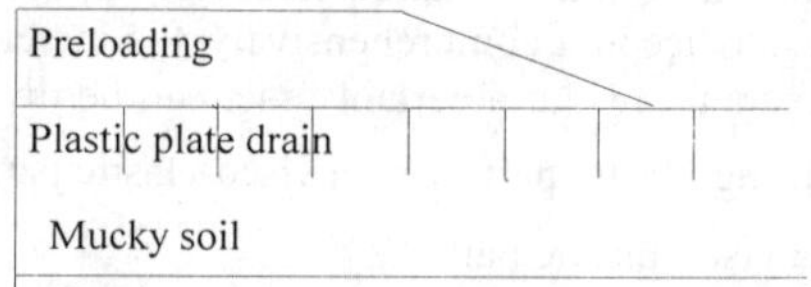

Figure 2. Soft foundation and preloading

Table 1. Physical and mechanical index of soil property

Soil strata	Layer thick(m)	γ kN/m^3	w %	φ' °	c' kPa	k_X 10^{-7}cm/s	k_Z 10^{-7}cm/s	E_{1-2} Mpa^{-1}	ν	e_0
Mucky soil	9~11	16.6	56.5	12.0	15.3	1.48	3.44	1.58	0.37	1.648
Clay	Undrill through	19.2	30.4	19.6	21.4	4.8	24.8	5.60	0.42	0.866

Table 2. Viscous elastic plastic modeling parameters

Soil strata	a	K_G	N	M_1	M_2	h	M	K_1	n_1	K_e Mpa.d	K_P Mpa.d
Mucky soil	0.19	30	0.4	1.34	1.37	10.4	4.1	110	1.07	46	15
Clay	0.25	90	0.5	1.34	1.37	11.5	7.1	210	1.20	48	20

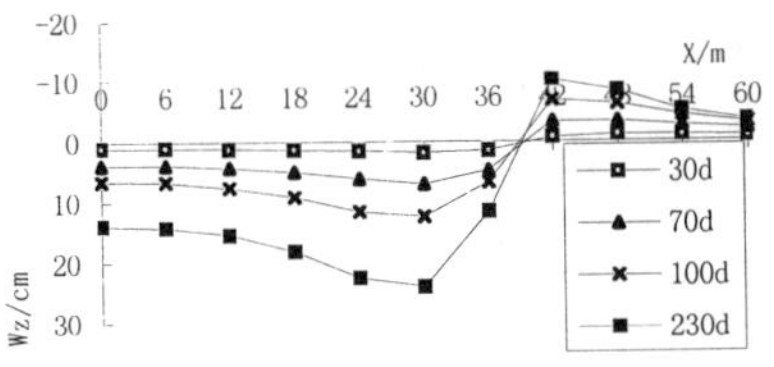

(a) Unpave plastic plate drain

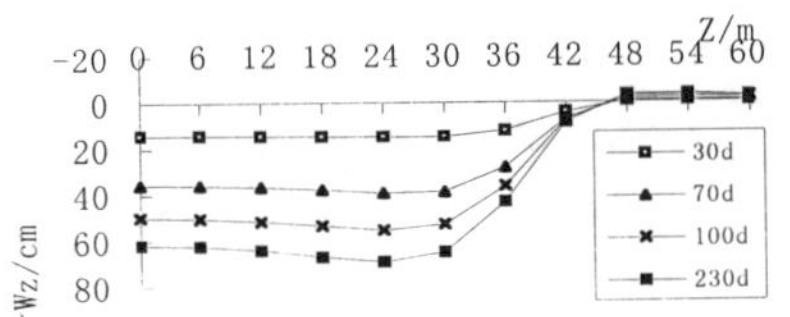

(b) Pave plastic plate drain

Figure 2. W_z-t and W_x-t curve for surface settlement of soft foundation

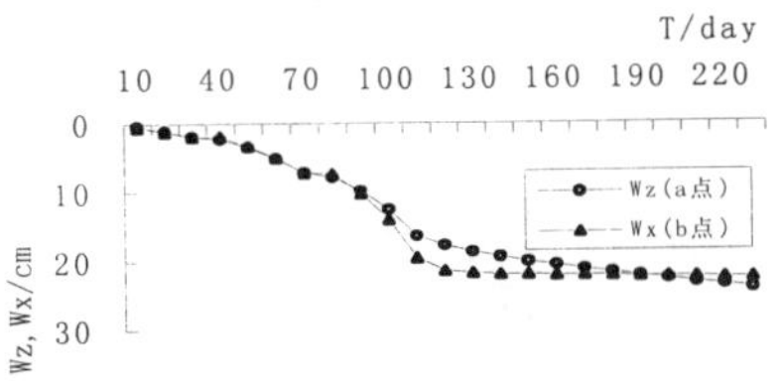

(a) Unpave plastic plate drain

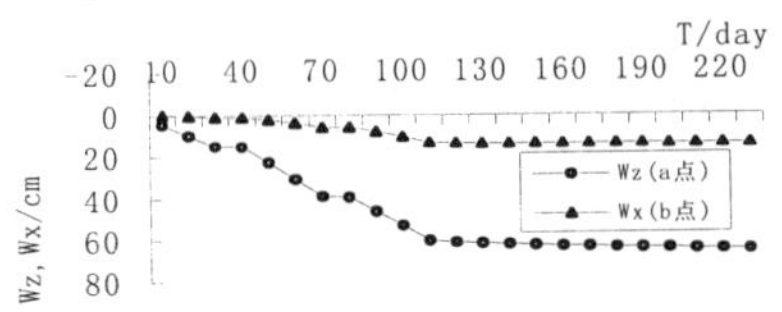

(b) Pave plastic plate drain

Figure 3. W_z-t and W_x-t curve of the representative node in foundation

The results of finite element analysis indicates that the deformation is very small in the loading time, when unpaving plastic plate drain, because the permeability is very low and the degree of consolidation only about 30% be reached. Due to distinct creep nature, after loading is complete, the deformation continual increases and it may lead to loss of stability finaly(Fig.2 a and Fig.3 a).

When the plastic plate drains are paved , the perviousness of soil is improved. A part of consolidation finishes when loading accomplished. the viscous effect on deformation is reduced evidently because of the full consolidation of soil. The surface settlement of foundation accretes and horizontal movement recedes. The subgrade is stable (Fig.2 b and Fig.3 b).

5 CONCLUSIONS

Through the VEPC model of soft foundation, the coupling effect of viscous and consolidation on deformation behavior has been investigated. The results indicate that the viscous nature and consolidation action are reciprocal and interactional. The lower the permeability of soft soil, the more obvious the viscous nature. In addition, the viscous effect on the vertical displacement is lager than lateral displacement.

REFERENCES

Yin zongze 1988. A couple yield surface stress-strain modeling of soil. *Chinese Journal of Geotechnical Engineering* 10(4):64-71

Zhao weibings & Shi jianyong 1996. Consolidation and rheology of soft soil. NanJing: *River and Ocean University Publishers:*177-243

Computer Methods and Advances in Geomechanics, Desai et al. (eds) © 2001 Balkema, Rotterdam, ISBN 90 5809 183 X

A study on the overburden pressure in layered soils considering sedimentation

K.H.Xie & A.F.Hu
Department of Civil Engineering, Zhejiang University, Hangzhou, People's Republic of China

C.J.Leo
Department of Civil Engineering, UWS Nepean, Kingswood, N.S.W., Australia

ABSTRACT: In this paper, the equations are derived for predicting overburden pressure in layered soils based on the classical e~logσ' relationship. In these equations, the sedimentation is considered and as a result, specific gravity, instead of unit weight of soil is used in calculating overburden pressure. Several numerical examples are presented and the results are compared with those obtained from the conventional method. It shows that the equations developed herein give more reasonable results and can be used in one-dimensional large strain consolidation analyses for layered soils.

1 INTRODUCTION

In nonlinear or large strain consolidation analyses, the self-weight of soil usually needs taking into consideration (Gibson et al. 1981, Xie & Leo 1999), and the overburden pressure therefore should be calculated. Conventionally, the overburden pressure of a soil layer is calculated based on the assumption that the unit weight of soil is constant over the layer. The overburden pressure obtained thus increases linearly with depth for a given layer. In actual case, however, the unit weight of soil is usually increased with the increase of depth due to sedimentation. This implies that the overburden pressure will increase with depth in a nonlinear way. Although it may be not important to take consideration of sedimentation in the calculation of overburden pressure for small strain analyses, it is necessary for large strain analyses, otherwise this will lead to unreasonable results, especially for the case where clay deposit is very thick and highly compressible.

Many studies have been made on the overburden pressure considering sedimentation. Furgaroli & Prager (1969) made the first study on the effective unit weight of a normally consolidated clay considering sedimentation. They established a relevant formula based on the classical void ratio versus logarithm of the effective stress (e~logσ') relationship. Simon & Viggiani(1975) extended the work of Furgaroli & Prager and proposed equations for calculating the overburden pressure of normally consolidated clay. In 1994, Li presented equations for predicting the overburden pressure and settlement of a submerged homogeneous normally consolidated clay. In addition, results for clays which do not obey the linear e~logσ' relationship are also presented in his study.

However, the all studies so far made on the overburden pressure considering sedimentation are only limited in homogeneous soil and nothing has been done in layered soils. In this study, equations are derived for calculating overburden pressure in layered soils which obey the linear void ratio versus logarithm of effective stress (e~logσ') relationship. The specific gravity, instead of unit weight of soil is used in calculating overburden pressure considering sedimentation. Numerical examples for several conditions are presented and the results are compared with those obtained from conventional method. It can be shown that the theory presented herein gives more reasonable results and is of fundamental significance to one-dimensional large strain consolidation analysis for layered soils.

2 THEORY

Layered normally consolidated clay deposit is considered herein (Fig.1). The total thickness and layer numbers of the stratum is H and n repectively. The thickness of the ith layer is denoted by h_i (i=1,2,⋯n). z represents the spatial coordinate which takes its origin at the mud-line and z_i is the vertical distance from the origin to the bottom of layer i. Thus, z_0=0 and z_n= H.

The effective unit weight (denoted by γ') of a normally consolidated clay at depth z can be given by:

$$\gamma' = \frac{G_s - 1}{1+e}\gamma_w \qquad (1)$$

where γ_w, G_s and e is respectively the unit weight of water, the specific gravity of the clay and the void ratio at depth z.

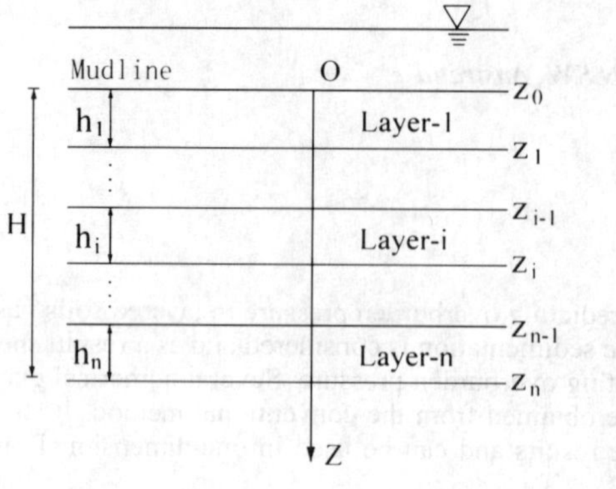

Figure 1. Layered clay deposit

The unit weight of water γ_w and the specific gravity G_s of a given clay can be taken as constants because the pore water and the soil particle can be assumed to be incompressible. However, due to sedimentation e is no longer a constant but decreases with the increase of z. For a normally consolidated soil with a linear e~$\log\sigma'$ relationship, e can be obtained from the expression:

$$e = e(z) = e_1 - C_c \log\left(\frac{\sigma'(z)}{\sigma'_1}\right) \qquad (2)$$

where C_c = compression index; $\sigma'(z)$ = overburden pressure at depth z; e_1 = the void ratio corresponding to the overburden pressure σ'_1 which is a reference value.

From equations (1) and (2) the overburden pressure $\sigma'(z)$ can be given by:

$$\sigma'(z) = \int_0^z \gamma' dz = \int_0^z \frac{\gamma_w (G_s - 1)}{1 + e_1 - C_c \log\left(\frac{\sigma'(z)}{\sigma'_1}\right)} dz \qquad (3)$$

As can be seen from equations (1) to (3), when sedimentation is considered, effective unit weight γ' is no longer a constant but increases with the increase of depth z, and overburden pressure $\sigma'(z)$ is a nonlinear function of z.

To show the variation of effective unit weight γ' with depth z, Figure 2 is prepared based on equations (1) to (3), in which three curves are included corresponding to the case of γ' = const (in this case void ratio e in equation (1) is replaced by e_1) and these of C_c =0.65 and C_c=0.89. The other soil parameters used in the computation are: G_s=2.75, e_1=1.571, σ'_1=100kPa.

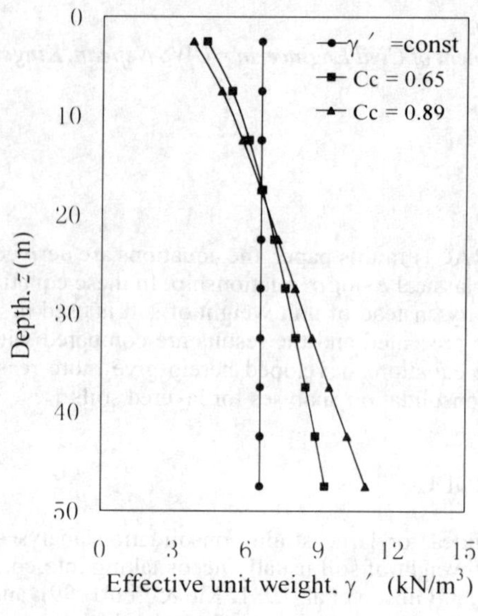

Figure 2. Variation of γ' with depth

Equations (1) to (3) are appropriate for the homogeneous soil layer. For the layered soils shown in Figure 1, the overburden pressure $\sigma'(z)$ at a depth z in the ith layer can be expressed as

$$\sigma'(z) = \sigma'(z_{i-1}) + \int_{z_{i-1}}^{z} \frac{\gamma_w (G_{si} - 1)}{1 + e_{1i} - C_{ci} \log\left(\frac{\sigma'(z)}{\sigma'_{1i}}\right)} dz$$

$$z_{i-1} \le z \le z_i \qquad (4)$$

where the subscript i denotes layer i and $\sigma'(z_{i-1})$ is the overburden pressure at the top of the ith layer (i.e. at z=z_{i-1}).

Equation (4) can be re-arranged to give:

$$\int_{\sigma'(z_{i-1})}^{\sigma'(z)} [1 + e_{1i} - C_{ci} \log\left(\frac{\sigma'(z)}{\sigma'_{1i}}\right)] d\sigma'(z)$$

$$= \int_{z_{i-1}}^{z} \gamma_w (G_{si} - 1) dz \qquad (5)$$

Integrating both sides of Equation (5) yields:

$$\lambda_i\sigma'(z)[\log\frac{\sigma'(z)}{\sigma'_{1i}}-\beta_i]+z=$$

$$\lambda_i\sigma'(z_{i-1})[\log\frac{\sigma'(z_{i-1})}{\sigma'_{1i}}-\beta_i]+z_{i-1} \quad z_{i-1}\le z\le z_i \qquad (6)$$

where

$$\beta_i=0.434+\frac{1+e_{1i}}{C_{ci}} \qquad (7)$$

$$\lambda_i=\frac{C_{ci}}{\gamma_w(G_{si}-1)} \qquad (8)$$

For i=1, $z_{i-1}=z_0=0$, and $\gamma'(z_0)\to 0$, $\sigma'(z_0)\to 0$ (Li 1994), Equation (6) becomes:

$$\lambda_1\sigma'(z)\{\log[\frac{\sigma'(z)}{\sigma'_{11}}]-\beta_1\}+z=0 \qquad 0\le z\le z_1 \quad (9)$$

or

$$z=\frac{\sigma'(z)}{\gamma_w(G_{s1}-1)}\{1+e_{11}+\frac{C_{c1}}{\ln 10}[1-\ln(\frac{\sigma'(z)}{\sigma'_{11}})]\} \qquad (10)$$

Equation (10) is the same as that proposed by Li (1994). This shows that the equation derived herein for layered soils can be reduced to the one for homogeneous layer when n = 1.

Equations (6) to (9) forms the basic formulas for predicting overburden pressure $\sigma'(z)$ in layered soils considering sedimentation. When these equations are used for calculating $\sigma'(z)$, the following procedure is suggested:

1. Calculate first the overburden pressure of layer 1 by using Equation (9).
2. Take the overburden pressure at the bottom of the first layer, i.e. $\sigma'(z_1)$, as the one at the top of the second layer and introduce it into the right side of Equation (6). Thus, the overburden pressure in the layer 2 can be obtained from the equation.
3. Follow the same way to find the overburden pressure in next layer until $i = n$.

If the effective unit weight γ' is required, the overburden pressure $\sigma'(z)$ obtained above at the corresponding depth can be introduced into Equation (2) to find the corresponding void ratio e, and γ' can then be obtained by introducing e into Equation (1).

3 NUMERICAL EXAMPLES AND DISCUSSION

In order to study the effect of sedimentation, the overburden pressure in several cases are calculated using the equations developed herein. The results of these numerical examples are presented and compared with those given by conventional method without considering sedimentation.

In conventional method, the effective unit weight γ' is a constant over a soil layer, i.e.

$$\gamma'=\frac{(G_s-1)}{1+e}\gamma_w=const \qquad (11)$$

Thus the overburden pressure varies linearly with depth, i.e.

$$\sigma'(z)=\gamma' z=\frac{G_s-1}{1+e}\gamma_w z \qquad (12)$$

Whereas when sedimentation is considered, γ' is no longer a constant, and likewise the overburden pressure will not vary linearly with depth, as shown by Equation (6).

A soil stratum of three clay layers is studied. The relevant geotechnical data of the layered system used in the computation are shown in Table 1. The results are shown in Figure 3.

Table 1. geotechnical data used in computation

Layer i	h_i(m)	G_{si}	e_{1i}	σ'_{1i}(kPa)
1	20	2.75	2.071	50
2	30	2.68	1.578	50
3	30	2.72	1.715	50

In Figure 3, the curve marked with hollow dots (case 1) is obtained based on Equation (6) considering sedimentation. It reasonably shows the nonlinear variation of overburden pressure with depth. The compression index used in the analyses for the three layers is respectively 1.25, 0.95 and 1.02. The curve marked with solid squares (case 2) represents the results obtained from conventional method, i.e. in Equation (11) e_{1i} is used instead of e. As can be seen from Figure 3, there is significant difference between the two results, and the difference is increased with the increase of depth.

Although the formulas developed in this paper

gives more reasonable overburden pressure, it is not so convenient to use as conventional method. In order to find a way to predict overburden pressure not only reasonable but also convenient, we made two numerical tests. One is finding the effective unit weight of soil at the middle point of each soil layer by using the equations given by the authors and then calculate overburden pressure by using conventional method employing the obtained effective unit weight of each soil layer (Case 3). Another one is similar to above but it is the effective unit weight of soil at one-fourth depth (instead of middle point) of each layer that is calculated and employed (Case 4). The results of the two numerical tests are also shown in Figure 3. It can be seen from this figure that there seems no difference between the two curves of Case 1 and Case 3. This indicates that the overburden pressure concerning sedimentation can be approximately predicted by using conventional method employing the effective unit weight at the middle point of each soil layer calculated by the formulas presented.

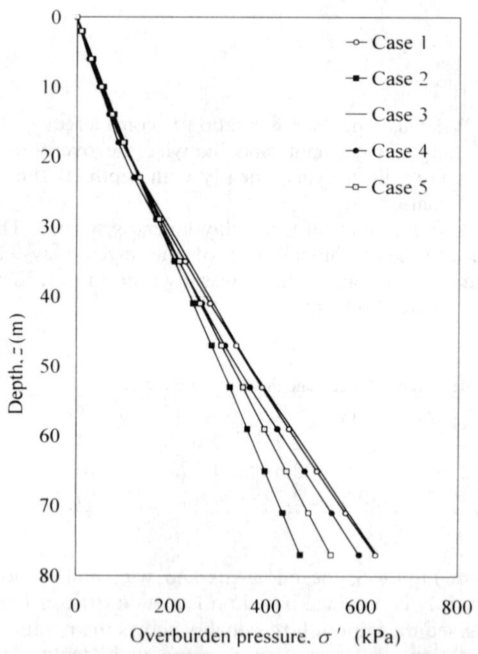

Figure 3. Variation of overburden pressure along depth

Finally, Case 5 is studied to investigate the influence of compressibility of soil on overburden pressure considering sedimentation, in which the compression index of the three layers is respectively 0.68, 0.43 and 0.57, a little smaller than those assumed in Case 1. It can be seen from Figure 3 that the influence is noticeable.

4 CONCLUSIONS

1. The formulas are developed based on the classical $e \sim \log\sigma'$ relationship for predicting overburden pressure in layered soils considering sedimentation, in which, specific gravity, instead of unit weight of soil is used in the prediction. The overburden pressure thus obtained increases reasonably with depth in a nonlinear way. Thus these formulas can be used in one-dimensional large strain consolidation analyses for layered soils to take account of self-weight of soil.

2. There exists significant difference between the overburden pressure predicted by the theory presented herein and that calculated by conventional method. The influence of sedimentation and that of compressibility of soil on overburden pressure are also noticeable.

3 When sedimentation is considered, the overburden pressure can be approximately predicted by conventional method using the effective unit weight at the middle point of each soil layer calculated from the formulas presented.

ACKNOWLEDGEMENT

The financial support from China National Nature Science Foundation (No.59895410) is gratefully acknowledged.

REFERENCES

De Simon, P. & C. Viggiani 1975. Settlement of a thick bed of normally consolidated uniform clay. *Geothchnique*, 25: 390-393.

Fungaroli, A.A. & S.R. Pranger 1969. Unit weights of a normally consolidated soil. *Proc 7th International Conference on Soils Mechanics and Foundation Engrg., Mexico*, 1: 131-134.

Gibson, R. E., R.L.Schiffman & K.W.Cargill 1981. The theory of one-dimensional consolidation of saturated clays: II. Finite nonlinear consolidation of thick homogeneous layers. *Canadian Geotechnical Journal*, 18(2): 280-293.

Xie, K.H. & C.J. Leo 1999. An investigation into the nonlinear consolidation of layered soft soils. *Research Report CE11*. School of Civic Engineering and Environment, UWS Nepean, Australia.

Li, K.S.1994. Predicting the density profile and settlement of marine deposits. *International conference on computational methods in structural and geotechnical engineering, Hong Kong*, 4:1562-1567.

Computer Methods and Advances in Geomechanics, Desai et al. (eds) © *2001 Balkema, Rotterdam, ISBN 90 5809 183 X*

3D FEM coupling analysis of seepage and stress field in deep excavation

Ping Yang, Bai Shiwei & Zheng Shaohe
Institute of Rock and Soil Mechanics, Chinese Academy of Sciences, Wuhan, People's Republic of China

ABSTRACT: This paper describes a three-dimensional finite element model used to predict the soil deformation characteristic of deep foundation pit. The model considers the elastic-plastic property of clay, the coupling effect between stress and seepage, the seepage in unsaturated zone and so on. Finally, a case of deep excavation with field measurement of support piles had been calculated by the use of the model and computed results had close agreement to field data.

1. INTRODUCTION

Design and construction of a deep excavation is an old but still challenging undertaking of both geotechnical and structural engineers. In a typical project, three major aspects are commonly considered: stability, deformation, and groundwater control. Prediction of excavation wall deformation or ground surface settlement has been studied using plane strain finite element analysis and 3D FEM simulation. Plane strain analysis might give conservative results for sections near the corner of deep excavation because excavations are in nature three-dimensional problems. On the other hand, the accuracy of predicted values are affected by the existence of ground flow. The prediction of 3D FEM without seepage analysis is not fitted for the practical engineering too. Therefore, the objectives of this paper are to establish a three-dimensional finite element analysis procedure considering interaction of seepage field and stress field based on Biot's consolidation theory and to develop a computer program to model deep excavation. Furthermore, the finite element formulation for this work was designed to simulate various phrases of excavation such as dewatering, soil excavation process, strut preloading and so on. Due to space limitations, only a brief summary will be presented here.

2. ANALYTICAL DEVELOPMENT

2.1 Basic Theory of Biot's Consolidation Theory

On the basis of some hypotheses such as linear elastic property of soil skeleton, seepage flow following Darcy's theorem and small deformation analysis the increasement format of Biot's consolidation equation can be expressed as:

$$\begin{bmatrix} [K] & -[L] \\ -[L]^T & \left(-\frac{\Delta t_i}{2}\right)[T] \end{bmatrix} \begin{Bmatrix} \Delta\delta_i \\ \Delta P_i \end{Bmatrix} = \begin{Bmatrix} -\Delta F_i \\ \Delta t\{Q_i\} + \Delta t_i\{P_{i-1}\} \end{Bmatrix} \quad (1)$$

Where: *[K]* stiffness matrix; *[T]* seepage matrix; *[L]* coupling matrix which represents transformation relationship from pore water pressure to nodal force; $\Delta\delta_i$ displacement increasement with time variable; Δp_i increasement of pore water pressure with time variable; ΔF_i increasement of load acting on the node; $\{Q_i\}$ flux through node; Δt_i time interval.

2.2 modeling of structure and soil

In this paper, soil was assumed to an elastic-plastic material. Its stress-strain relationship was considered hyperbolic model which was put forward by Duncan and change in 1977. Young's modulus E is equal to tangent modulus E_t and Poission's ratio ν is equal to tangent modulus ν_t, E_t and ν_t were defined as fellow:

$$E_t = [1 - \frac{R_f(1-\sin\varphi)(\sigma_1 - \sigma_3)}{2C\cos\varphi + 2\sigma_3\sin\varphi}]^2 KPa(\frac{\sigma_3}{Pa})^n \quad (2)$$

$$\nu_t = \frac{G - F\log(\frac{\sigma_3}{Pa})}{1-A} \quad (3)$$

Where $A = \frac{(\sigma_1 - \sigma_3)D}{KPa(\frac{\sigma_3}{Pa})^n[1 - \frac{R_f(1-\sin\varphi)(\sigma_1-\sigma_3)}{2C\cos\varphi + 2\sigma_3\sin\varphi}]}$

K=stiffness modulus number for primary loading; n=stiffness modulus exponent, R_f =failure ratio; *F,D*=oblique indifferent conditions; *G*=section length; *C*=soil cohesive force; φ=friction angle. All these parameters are obtained by triaxial test. Additionally, Pa is standard atmosphere pressure. In finite element analysis, if calculated of ν_t was greater than 0.49, ν_t was equal to 0.49, near or at failure condition of soil elements. If the non-linear deformation exists in the simulation, the matrix *[K]* will calculate by the use of the model mentioned above.

2.3 Simulation of Excavation Process

Simulation of excavation by the finite element method was done with the removal of elements representing the excavated soil. In the present study, the degrees of freedom of excavated elements were excluded from the formulation of global stiffness matrix in order to conserve computer storage. The equivalent nodal forces generated by the removal of elements and by the dewatering were calculated as fellow:

$$\{f\} = \int[B]^T[\sigma]dV - \int[N]^T\gamma dV + \int[B]\{U\}dV \quad (4)$$

Where *{f}* = equivalent nodal forces; *[B]* = strain-displacement transformation matrix; $[\sigma]$ = stress vector; $[\gamma]$ = body force; *{U}* = magnitude of variable in pore water pressure.

2.4 Initial Conditions

The magnitude of initial stresses has a major effect on the results of analysis during nonlinear analysis. In the present paper, the effective horizontal stress is equal to the effective vertical stress multiplying the coefficient of the at-rest lateral earth pressure (K_0). The total stresses are equal to the sum of the stresses and pore water pressure.

2.5 Variable of coefficient of permeability after dewatering

In general, Variable of water level or effective stress has important effects on the permeability of aquifer. Though many in situ observations, coefficient of permeability after dewatering can be calculated as fellow:

$$K = K_0^{'}\exp(\beta\Delta H)$$

Where K = seepage coefficient after dewatering; $K_0^{'}$ = seepage coefficient before dewatering; ΔH=variable of water level β= experimental constant

2.6 Coefficient of permeability in unsaturated zone

After dewatering, an unsaturated seepage zone is formed near the region of deep excavation. The prediction of seepage in an unsaturated soil has been of increasing concern in recent years. Theoretically, the coefficient of permeability for an unsaturated soil is dependent on the degree of saturated and void ratio. Generally, coefficient of permeability in unsaturated zone can be derived from the soil-water characteristic curve or evaluated by the formula postulated by van Genuchten (1980):

$$k_r = \frac{\{1-[\alpha'(u_a-u_w)]^{n'-1}\{1+[\alpha'(u_a-u_w)]^{n'}\}^{-m'}\}^2}{\{1+[\alpha'(u_a-u_w)]^{n'}\}^{-m'/2}} \quad (5)$$

Where: α' = constant representing the air entry pressure; m' n' = experiment indices; (u_a-u_w) = matric suction

3.SELECTION OF ELEMENT TYPE

The tetrahedron element and hexahedron element are two types of elements commonly used in three-dimensional finite element analysis. Clough (1969) and Zienkiewicz (1989) commented that for a regular boundary shape, the hexahedron element used individually is superior to either the tetrahedron element or the hexahedron element formed by the combination of tetrahedron elements.

A hexahedron element can also be classified as a high order element or a linear order element depending on the order of the displacement shape function. Theoretically, the use of higher order elements in three-dimensional deep excavation problems can reduce the total number of elements required in the analysis. However, the finite element mesh for excavation must take into account the depth of excavation, location of struts the dewatering scheme, soil layering and so on. These considerations usually result in a higher mesh density on the excavation side. Consequently, the use of higher order elements may not be able to reduce the total number of elements as expected. For this reason and for simplicity, the linear-order hexahedron element was used to model the behavior of soil and excavation wall.

The temporary strut during the excavation is often formed by steel members or a concrete slab. The strut is normally subjected axial force, and therefore the bar element was used to simulate the behavior of struts in the study.

4.CASE STUDY

4.1 Soil profile and properties

A building, located in first-degree ladder area in Wuhan city, is in the center of Hankow city near Yangzte River in Hubei province, china. The chief computational parameters are listed in Table1.

Table 1. Soil Properties at the Building Site

Type of Soil	Depth (M)	E (Mpa)	ν	C (Kpa)	φ(°)	K_h	K_v
Fill	1.0						
Soft Clay	9.5	87.5	0.32	30	27	5.16×10^{-6}	5.16×10^{-6}
Silty Clay	35.0	153.4	0.28	16	22	2.53×10^{-2}	3.17×10^{-3}

The soft clay has high water content (above 95%), low shear strength; On the other hand, the silty clay is heavily overconsolidated, high plastic and contains microfissures. It is hard plastic material which is typical in the region. The soil layer has high shear strength, but its strength would sharply drop to much lower level with the failure of the microstructure. Prior to construction, groundwater was near ground surface and approximately hydrostatic.

4.2 Numerical analysis

Due to uncertain material properties of top layer and relatively small thickness (about 1.0m), it was assumed to possess the same material properties as the soft clay layer that simplified numerical simulation. The deep excavation in the building site was in depth of 12.3 and large excavated scope. Support system was sheet piles (Diameter equal to 1.0m, embedded depth to 13.0 m). During construction, Two brace was installed on these piles and measure devices were install on the piles to monitor the stresses and deformation of pile and ground heave behind the piles. The detail excavation steps were completed according to the following procedure:

1. Excavate down to 4.7 m below the ground surface
2. Construct first level of strut at 4.1 m below the ground surface
3. Excavate down to 8.8 m below the ground surface
4. Construct second level of support strut at 8.2 m below the ground surface
5. Excavate down to 12.3 m below the ground surface

In order to minimize the effects of the finite element boundaries on predicated ground displacement and pile deflection, the mesh extended from excavation to a total depth of 45.0m and laterally to a distance of 100.0m from the centerline. The overall mesh includes 1060 elements and 922 nodal freedom degrees. The horizontal boundaries were restrained with rollers and the base was supported with hinges. Goodman interface elements were used at the interface between the soil elements and pile elements. Namely, soil elements and the unbraced piles elements were separated by interface elements. All elements were in form of eight-nodal hexahedron elements (Fig1). No cracking of these piles was simulated, that may not be strictly correct for small parts of the piles at final stage of excavation.

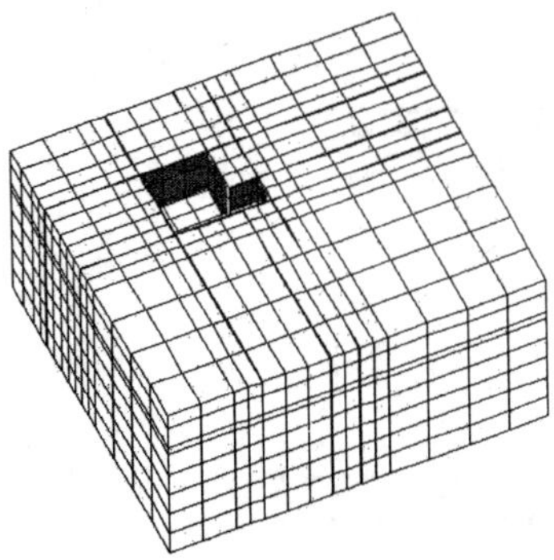

Fig1. Mesh of the Case

4.3 Computational Results

Fig 2 shows a comparison of measured and calculated soil displacement occurred near the corner of the excavation region. Maximum lateral displacement is 65mm in final stage by computed method. Compared to measured method, it is less; but compared to the computational results without considering the coupling effects, it is more accuracy. During the three stages of excavation level, there is no significant difference between measured and computed results. Practicing engineers are very concerned with ground movements around deep excavation and ground heave at the bottom of deep excavation in urban area, especially in residence zone. The consequence of unpredicted magnitude and extend of ground movements could be very serious, particularly when there are high buildings close to the edge of deep excavation. Fig2, Fig3 shows the ground movement behind the sheet piles and the value of ground heave at the bottom of excavation pit. While measured movements are compared with computed results, the deformation pattern is not significant altered, maximum movement of soil nearly equal both surveying and calculating results, about 70mm. It should be noted that ground maximum movement behind sheet piles happened within 10m to piles.

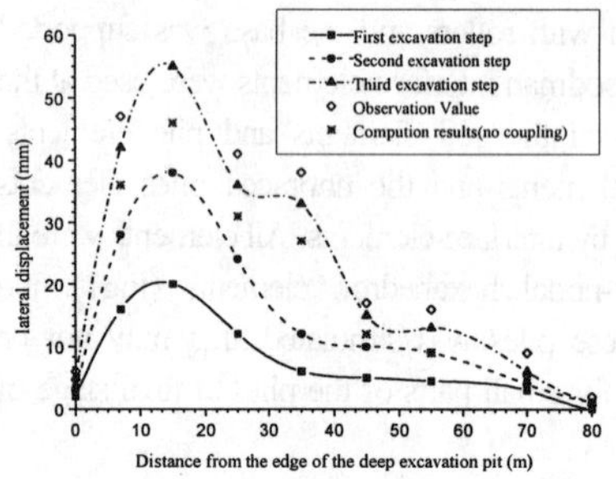

Fig2. Lateral displacement

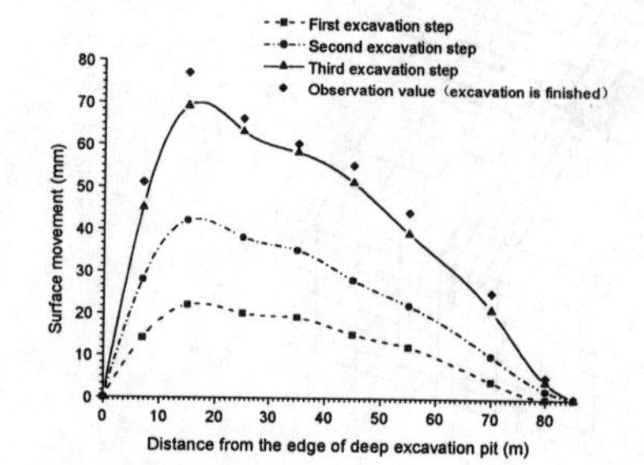

Fig3. Ground Movement

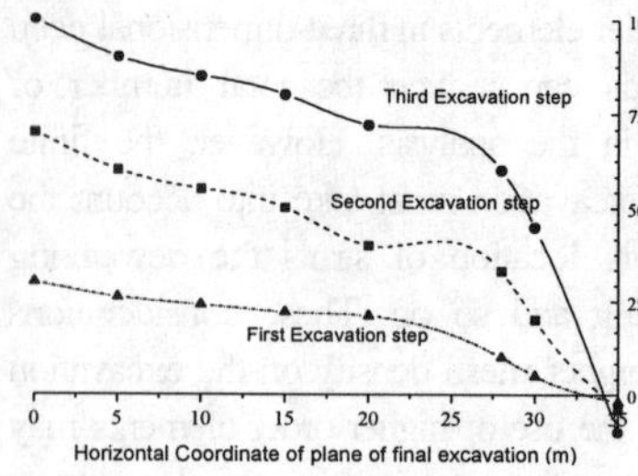

Fig4. Value of Heaving at the bottom of the deep excavation pit

5.CONCLUSION

The paper has presented 3D nonlinear finite element method considering the coupling interaction between deformation and seepage and resulted with numerical simulation. Compared with measured data, Analysis results have shown good agreement between measured and calculated data. Furthermore, Numerical results show that model and algorithm adopted in consideration of factors stated above have great advantages in the analysis of braced excavation process.

REFERENCES

Clough, G.W., and Denby, G. M. (1977) “stabilizing beam design for temporary walls in clay.” *J. Geotech. Engrg. Div.*, ASCE, 75-90

Zienkiewicz, O, C. (1989). *The finite element method,* fourth ed., McGraw-Hill Book Co., Inc., New York, N.Y.

Duncan, J. M. and chang. C. Y., 1970, No-linear Analysis of stress and strain in soils, *proc, ASCE, JSMFD,* Vol.96, No. SM5, 1629-1653

Frantziskonis, G. & Desai, C. S., 1987, Constitutive Model with strain softening, *Int. J. Solids&. Structures,* Vol.L3, No.6

Computer Methods and Advances in Geomechanics, Desai et al. (eds) © 2001 Balkema, Rotterdam, ISBN 90 5809 183 X

Planar-channel coupled model for fluid flow in discrete fracture network

Q. Yu, Y. Ohnishi, H. Ohtsu & S. Nishiyama
Geofront Laboratory, Department of Civil Engineering, Kyoto University, Japan

ABSTRACT: This paper presents a new approach, the planar-channel coupled model, to the analysis of fluid flow in discrete fracture networks. The new model allows the simultaneous existence of both channel and planar flows in one individual fracture network. Through representing the many small stochastic fractures with channels, and large-dimension discontinuities with deterministic planar fractures, it can possess the advantages of both a channel model and a planar model.

1 INTRODUCTION

The presented models simulating fluid flow in discrete fracture networks of fractured rocks may be roughly divided into two kinds: channel model (e.g., Cacas et al, 1990) and planar model (e.g., Andersson, 1984). In the former, first a planar fracture (e.g., disc-shaped) network is generated, and then it is simplified to a network of one-dimensional channels in three-dimensional space; in the latter, after the generation of a planar network, a 2D meshing grid in 3D space is constructed for the analysis of fluid flow and solute transport. The strength of a channel model lies in its high efficiency: for a specific computational effort, it can include much more fractures than a planar model; the weak point of it is its roughness when representing the geometry of flow path. The advantage of a planar model is its flexibility in the description of pathway geometry; the weakness of it is its high computational requirement. Each of them can reflect only one flow form: either channel flow or planar flow.

This paper presents a coupled planar-channel model because:

1. It is reasonable to suggest that in a complex fracture network both channel flow and planar flow exist. Thus, a planar-channel coupled model is likely to be a more practical conceptual model for the actual regime of fluid flow in fracture networks.

2. A fracture network is usually composed of a few large-dimension discontinuities and a great number of small fractures; for the large discontinuities, a channel representation is difficult; and for the many small fractures a planar method is too computational demanding.

3. Such a method can make use of some physical measurements, for instance some observed data on the shapes and apertures of some large discontinuities. This is especially clear when there are some artificial facilities whose location, shape, size, and aperture (or transmissivity) may be well defined.

2 SIMULATING STOCHASTIC FRACTURE

Fracture networks provide pathways for ground water flow in fractured rocks. In a discrete network method, mathematical description of the geometry of fracture network makes a foundation for the analysis of fluid flow. To describe a fracture network is to define all the geometric parameters of all fractures in the fracture network, including the location, orientation (dipping direction and dip angle), 3D size and shape of each fracture. Usually, even in a very small scale, the number of fractures in an actual rock mass is enormous. And, among the great number of fractures only a very little part can be measured or observed through outcrops, excavations or boreholes. Thus, the geometric parameters of fractures are generally treated as random variables; almost all presented models for discrete fracture network are stochastic models.

2.1 *Geometric parameters of stochastic fractures*

In a stochastic discrete fracture network model, every geometric parameter of each set of fracture is represented with a probabilistic distribution, which may vary with fracture set and the type of geometric parameter.

In this paper, the geometric parameters (location, orientation, shape, size and transmissivity) of large-dimension discontinuities are treated as input data. The locations of stochastic fractures are defined with Poisson process. The distributions of fracture poles are fitted to one of the following distributions: Fisher, Bingham, Bivariate Fisher, Bivariate Bingham, Bivariate Normal, and Empiric distribution. In 3D space, individual fractures are assumed to be disc-shaped, and their diameters have the same distribution forms as the traces (e.g., lognormal or exponential distribution). The mean and standard deviation of the diameters are determined through inverse method. The following equation is used to initialize the estimate on the number of fractures existing in a given volume of rock mass: $N = \sum C_i \cdot d_2 \cdot V \cdot f_1$. Here N is the number of the fractures that will be generated for the rock of volume V; C_i is a correcting parameter, which is determined through inverse method; d_2 is the number of the fractures within unit area of sampling excavation; f_1 is the fracture frequency along the boreholes that are perpendicular or sub-perpendicular to the sampling excavation.

2.2 *Simulating process of fracture network*

An Inverse approach to the building stochastic fracture network model are used because that,

1 It is difficult to define the three-dimensional density of fractures with foreword methods.

2 Fracture sizes in 3D space cannot be measured directly. Based on the assumption that fractures are disc-shaped, the probability distribution of fracture diameter may be inferred from the distribution of trace length, which can be measured through excavation surfaces or natural outcrops. But the direct estimation of trace length are often affected by varied biases, and the distribution inferred from trace length involves an integral that can only be evaluated numerically, and this also becomes a difficulty when building a statistical model with Monte Carlo simulation.

The method to build a statistical model of fracture network is an inverse method, including the following major steps:

1 Construct an initial variable generator for each geometric parameter of each fracture group (including location, orientation, and diameter). The mean and standard deviation of fracture diameters may be initialized with that of trace lengths.

2 Generate fractures with Monte Carlo simulation. The parameter C_i may be initialized with 1.0. To avoid boundary effect, fracture generation region should be little larger than actual studied region.

3 Sample the simulated fractures in the same way as that the in situ actual fractures are sampled; investigate the trace length and density of sampled fractures. This implies that the sampling excavations and the boreholes should be identical with the actual ones regarding dimension, location, shape and direction.

4 Compare the simulated fractures with the measured with regarding to trace length and one-dimensional density.

5 If the simulated and measured fractures do not make a satisfying match, improve the mean and standard deviation of fracture diameters, and the parameter C_i with inverse method. This paper uses Downhill Simplex method (Press et al. 1989).

6 Repeat Step 2 through Step 5 until the simulated and the measured fractures make a satisfying match

Through above steps, the final variable generators for all the geometric parameters are determined which will be used to generate stochastic fractures for the analysis of water flow.

3 CONSTRUCTING FLUID FLOW MODEL

3.1 *Channel-representation for stochastic fracture*

As shown in Figure 1, in this study the channels and nodes are designed as: For each fracture in the network, nodes are designed at the centers of its intersections and the center of the fracture-disc; each node at intersection-center is connected to the node of disc-center by a channel. A channel has a shape of round pipe.

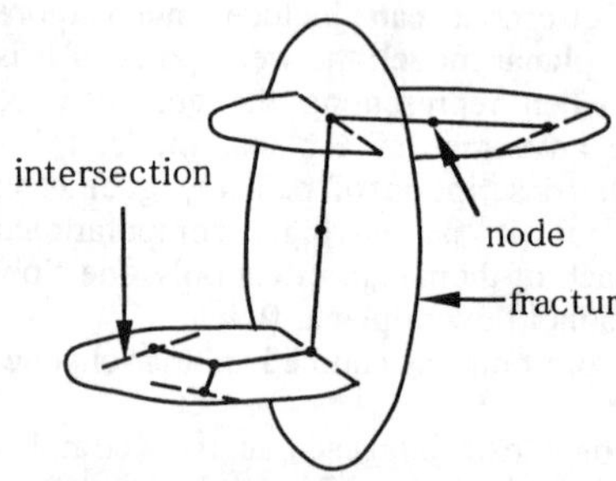

Figure 1. From fracture-disc to channel network

After simplifying a fracture-disc network to a channel network, the next issue is to determine the transmissivity (or the diameter) of the channels. Yu (2000) discussed the methods to determine the channel diameters through model calibration by using the data of observed flow rates and specific flow rates.

3.2 *Coupling planar and channel*

In the coupled planar-channel model, for the normal nodes, i.e. the nodes where there is no fluid exchange between channel element and planar element, the equations will be constructed in the same way as in a planar model or in a channel model. Some special treatments are needed only when two varied kinds of elements are tied.

As shown in Figure 2, if a channel element mn crosses planar element ijk at node i (when a channel crosses a triangular element, the crossing point will be shifted to the node that is nearest to the crossing point among the three nodes of the element. This may simplify calculation essentially). Node i is a node in both planar and channel grids. Conservation equation for node i is:

$$\sum_e [T\frac{\overline{pb}}{\overline{ij}}(h_j - h_i) + T\frac{\overline{qb}}{\overline{ik}}(h_k - h_i)] + Q_i = 0 \qquad (1)$$

Where Q_i is the recharge from the channel, and

$$Q_i = \frac{\pi D^4 \rho g}{128\mu}\left(\frac{h_m - h_i}{\overline{mi}} + \frac{h_n - h_i}{\overline{ni}}\right) \qquad (2)$$

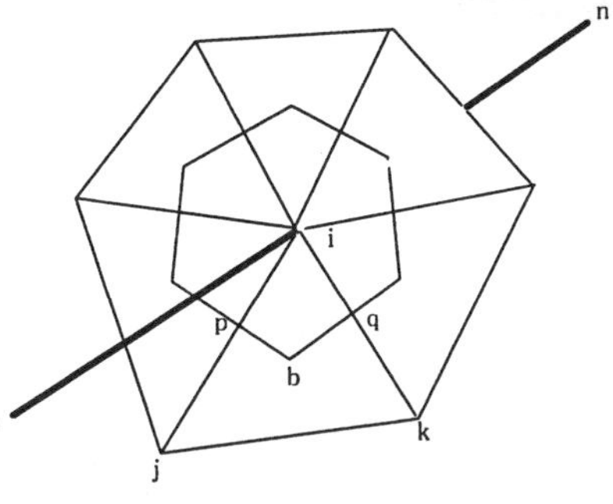

Figure 2. Tied planar and channel elements

When the case that the channel cross the planar element at a point within the element being considered, the crossing point forms a new node, node c, for the channel network, and the recharge from channel would be distributed among the three nodes i, j, and k. If the channel crossing planar element at (x_c, y_c), the recharges distributed to the three nodes are:

$$Q_i = QW_i; \quad Q_j = QW_j; \quad Q_k = QW_k \qquad (3)$$

where:

$$W_i = \frac{1}{2A_e}\{(x_j - x_c)(y_k - y_c) - (x_k - x_c)(y_j - y_c)\}$$

$$W_j = \frac{1}{2A_e}\{(x_k - x_c)(y_i - y_c) - (x_i - x_c)(y_k - y_c)\}$$

$$W_k = \frac{1}{2A_e}\{(x_i - x_c)(y_j - y_c) - (x_j - x_c)(y_i - y_c)\}$$

$$Q = \frac{\pi D^4 \rho g}{128\mu}\left(\frac{h_m - h_c}{\overline{mc}} + \frac{h_n - h_c}{\overline{nc}}\right)$$

$$W_i + W_j + W_k = 1$$

4 EXAMPLE

Figure 3a shows the model region for this example. It is a 20 by 20 by 200m cubic rock mass block; the average flow is along the x-axis; the recharge and discharge boundaries are hydraulic head specified, and the average hydraulic gradient is 8/1000. Along y-axis there is a gallery with diameter of 2.2m; around it is an excavation disturbed zone 0.5m thick. When simulating water flow, the gallery (it is imagined that this gallery will be used to store canisters of hazardous waste) is specified as impermeable. The disturbed zone is represented with 4 deterministic planar fractures (this is adopted from an international comparison project, Yu, 2000). The output is the flow rate across the disturbed zone, because it is likely to be of one of the major concerns for the performance-assessment of a repository of hazardous waste.

Stochastic fracture model is constructed based on the observed fracture data of the rock around the Test Pit at Kamaishi mine, Japan. The rock has 3 groups of fractures. The fracture orientations are Fisher distributed; the diameters of fracture-disc and the pipe-diameters are lognormal distributed. The means of the geometric parameters are shown in Table 1 (Yu, 2000).

Table 1 Parameters of stochastic fractures

	Orientation	Diameter	Density	Pipe-diameter
Set1	$337\angle 78''$	1.22m	$3.5/m^3$	0.497mm
Set2	$44\angle 63''$	1.68m	$0.4/m^3$	0.598mm
Set3	$240\angle 24''$	1.89m	$0.2/m^3$	0.400mm

Figure 3b shows the 2D traces of the simulated fractures with the model boundaries, and Figure 3c the channels for the stochastic fractures. Six realizations are analyzed, and the flow rates across the disturbed zone are between 1.29E-9m^3/s and 3.44E-9m3/s, the mean flow rate is 2.32E-9m^3/s.

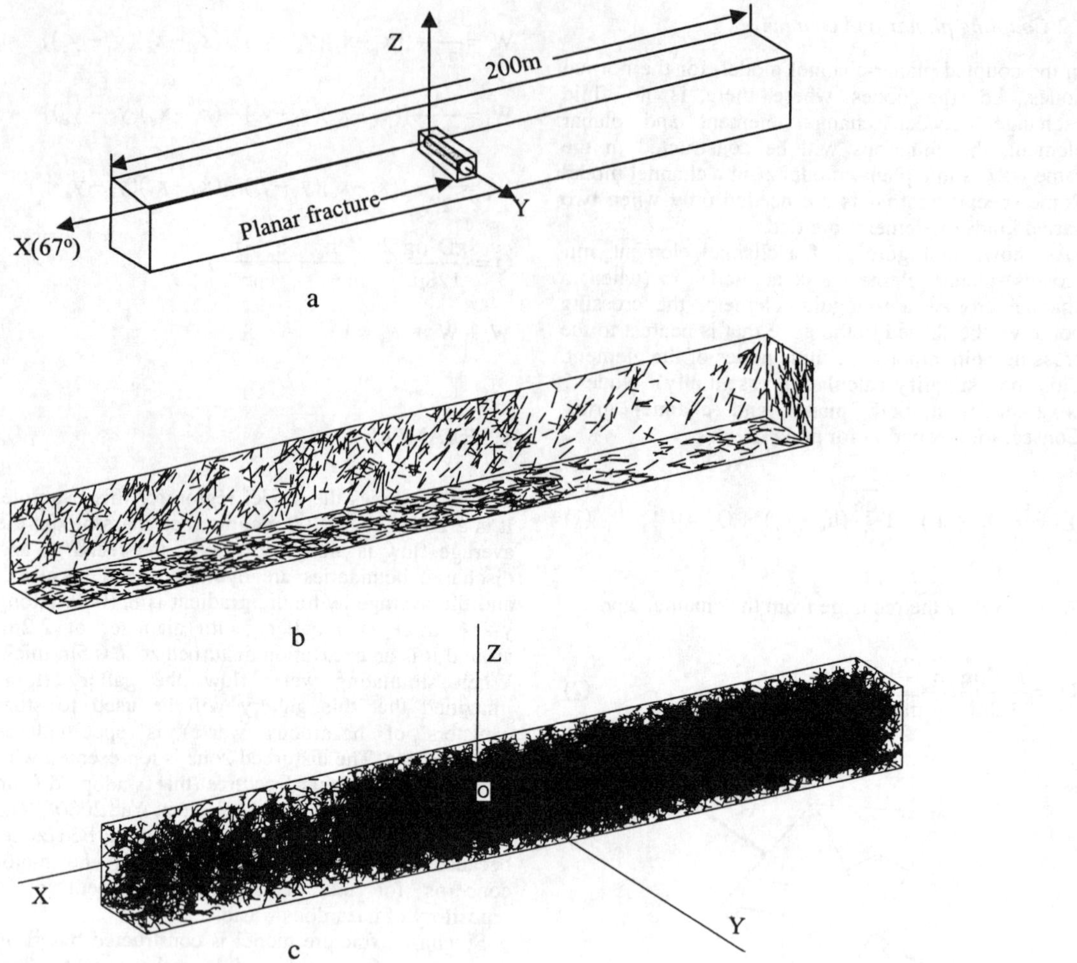

Figure 3. An example of planar-channel coupled model

5 CONCLUSIONS

For the complex fracture networks in actual rock masses, both channelized and planar flows may exist, the planar-channel coupled model is a more practical conceptual model. Planar-channel model possesses the strong point of a channel model of high computational proficiency and the advantage of a planar model of pathway-fine-description, which is of special importance when there are some artificial facilities whose shapes are not negligible.

REFERENCES

Andersson, J. 1984. A stochastic model of a fractured rock conditioned by measured information, Water Resources Res. 20:79-88.

Cacas M. C., E. Ledoux, G. Marsily, B. Tillie, A. Barbreau, E. Durand, B. Feuga, and P. Peaudecerf 1990. Modeling fracture flow with a stochastic discrete fracture network: calibration and validation, 1 the flow model. Water Resources Res. 26:479-489.

Press, W. H., B. P. Flannery, S. A. Teukolsky, and W. T. Vetterling 1989. Numerical Recipes, the Art of Scientific Computing (FORTRAN version), Cambridge University Press.

Yu, Q. 2000. Analyses for Fluid Flow and Solute Transport in Discrete Fracture Network. Ph.D. thesis. Kyoto University, Kyoto.

18 Education

Computer Methods and Advances in Geomechanics, Desai et al. (eds) © 2001 Balkema, Rotterdam, ISBN 90 5809 183 X

Enhancing geomechanics education using modern communications technologies

Muniram Budhu
Department of Civil Engineering and Engineering Mechanics, University of Arizona, Tucson, Ariz., USA

ABSTRACT: The purpose of this paper is to present an interactive multimedia geotechnical engineering courseware that is suitable for web-based instructions and/or delivery by a CD-ROM. The courseware consists of text-based teaching materials, interactive animations of the various concepts of geomechanics, images, a glossary, notation, a notepad, a virtual geotechnical laboratory, interactive problem solving, electronic quizzes and computer program utilities. The courseware is intended to enhance Geomechanics education at the undergraduate level. Two modules of the courseware are discussed. One deals with interactive learning of critical state soil mechanics while the other deals with a virtual triaxial test.

1 INRODUCTION

The Internet or Web is no stranger to the academic community. We have used the Internet in its infancy mainly to exchange scientific information. The recent phenomenal growth of the Internet has created demands for products and services at the click of a mouse 24 hours daily. Education is no exception to the demands. Indeed, education is now touted to be the new wave of the Internet. The question that arises for educators in geomechanics is "How can we enhance geomechanics education within the distributed environment of the Internet?" We are surrounded with a host of visual media – TV, computer games, video games, etc. – and obtain much of our information from them. Research has shown (Carpenter, 1991) that hearing and seeing (lecture and demonstration) lead to 50% knowledge retention compared with 20% for hearing only. The question that arises is " How can we improve visual contents to capture students' interests and improve knowledge retention?" The answer to the two questions posed lie in the selective use of modern communication technologies that include multimedia (streaming digital videos, animations, sound, texts and graphics), chat room, web pages, email, conferencing and listservers. The intention of this paper is to discuss and illustrate how selective use of modern communication technologies can enhance geomechanics education.

Geomechanics is a civil engineering course that involves the integration of fundamental principles and laboratory exercises. It has become relatively easy to develop web-based text material describing the fundamental principles. However, to capture today's students' interests and to increase knowledge retention we need to include dynamic visualization techniques and integrate laboratory exercises with text-based materials.

2. eLEARNING

There is a number of buzzwords, eLearning is one of them, that are now commonly used to describe electronically delivered instructional materials that provide media-rich content, self-paced instructions, interactivity, evaluation, feed-back and help. Prophecies abound about the values of eLearning and some foretold even complete replacement of classroom instructions. Dede et. al. (1982) wrote " As society moves to a knowledge–based economy, increasing numbers of people will use communications technologies as a major tool of learning, whether educators choose to encourage this process or not. If this shift takes place in an unplanned, reactive fashion, a variety of undesirable side effects may occur. On the other hand, a deliberate transformation that maximizes the strengths and minimizes the weaknesses of these tools could greatly improve current educational practice. Hopefully, society will realize the importance of encouraging such a planned transition." eLearning, like classroom instructions, has limitations but by carefully integrating the advantages of each, we can enhance the transmission and retention of knowledge.

In this paper, two teaching modules that take advantages of modern communications technologies to enhance learning and retention in geomechanics are described. One teaching module is on the concept of critical state soil mechanics while the other is on virtual laboratory tests. These modules are part of a geomechanics interactive CD (Budhu, 2000a) that contains text, interactive animations of the various concepts of geomechanics, images, a glossary, notation, a notepad, a virtual geotechnical laboratory, interactive problem solving, electronic quizzes and computer program utilities.

3. INTERACTIVE CRITICAL STATE CONCEPT

The concept of critical state – a state of constant stress and constant volume – has been recognized (Casagrande, 1936; Taylor, 1948; Schofield and Wroth, 1968) as a characteristic of soil. Roscoe and Schofield (1963) and Schofield and Wroth (1968) further developed the ideas of critical state into a framework or model that provided an understanding of the mechanical behavior of soils that are of the greatest importance to geotechnical engineers. However, the critical state model is not widely taught to undergraduates. The author has not done any comprehensive survey for the reason or reasons for not introducing the critical state model to undergraduates. One reason that was expressed by some faculty is the complexity of the various mechanics elements such as yielding and plasticity that are embodied in the critical state model. We may ask, can modern communication technologies be used to enhance the understanding of these mechanics elements to enable students to understand the critical state model? The author developed a multimedia interactive courseware on the critical state model with the following learning goals.

- To describe soil yielding and simple plasticity concepts necessary for the understanding of the critical state model
- To estimate failure stresses under drained and undrained conditions
- To estimate the volume changes under drained condition and the changes in excess pore water pressure under undrained condition.
- To estimate stress-strain behavior for loading conditions below failure.
- To evaluate possible soil stress and strain states if the loadings on a geotechnical system were to change from those anticipated.

It is assumed in the courseware that the students are familiar with stress and strain invariants. The author (Budhu, 2000 a,b) has developed another courseware that teaches students these invariants. The critical state soil mechanics courseware is structured as follows.

- Gains student's attention by presenting a practical situation where the critical state model can be used to provide a solution.
- Informs students of the learning outcomes
- Presents the course material in a way that connects the learner with the material to be learned. Students interact with the courseware and learn at their own pace. Interactive animations and mathematical solutions are used in concert to improve learning and retention. Two screen shots (Figures 1 and 2) illustrate the kind of self-paced interactivity developed. As the soil is gradually stressed, in this case isotropically, the yield surface progressively expands until the maximum stress is attained (Figure 1). The soil is then unloaded to a desired value. If after unloading, the soil is then sheared under drained triaxial conditions, the effective stress path (q/p') will have a slope of 3; q is the deviatoric stress and p' is the mean effective stress. Projection geometry is used to demonstrate the initial yield point and the failure point (Figure 2).

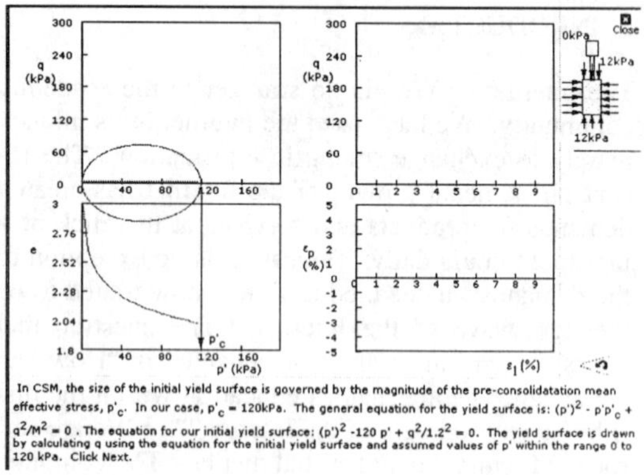

Figure 1 Initial yield surface of the critical state model

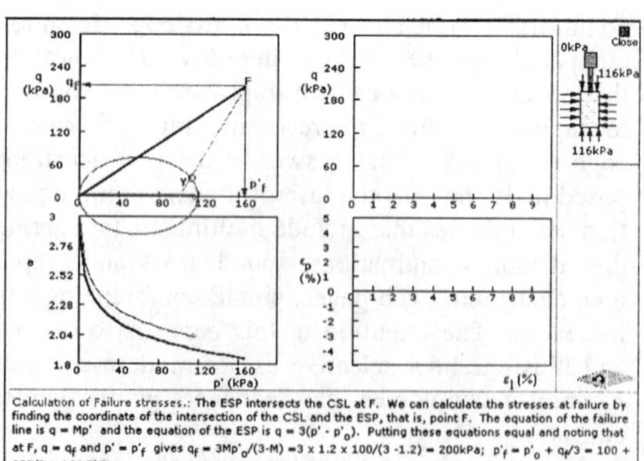

Figure 2 Failure stresses of a drained triaxial test from the critical state model

The method of calculating the initial yield and failure stresses, and the void ratios at initial yield and at failure are presented (Figure 3). If a small increment of stress is applied beyond initial yield, the yield surface expands and plastic strains develop.

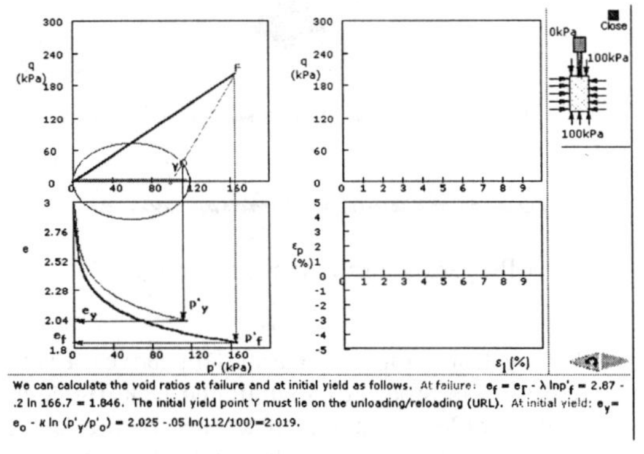

Figure 3 Illustration and calculation of yield and failure void ratios in a drained triaxial test from the critical state model

The elastic and plastic volumetric strains (shown as void ratio changes) resulting from a small increment of stress is depicted in Figure 4. The module continues by showing how to determine the shear strains and how failure is achieved. The stress-strain diagrams following each loading increment are animated.

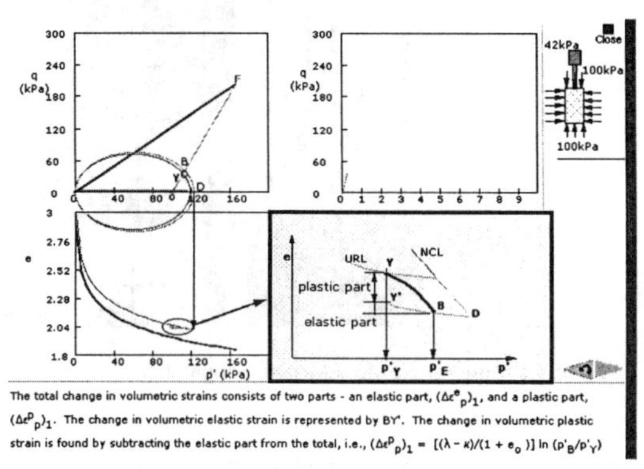

Figure 4 Illustration of elastic and plastic volumetric (void ratio) changes in a drained triaxial test from the critical state model

- Provides a summary of key points at strategic locations in the courseware to ensure that essential knowledge is grasped.
- Allows a student to explore "what-if" situations. For example, a student can use a set of soil parameters and let the critical state model predicts the stress-strain behavior under drained conditions. The student can use the same or different parameters to investigate undrained behavior. If experimental results are available, the student can compare the predictions with the measured results.

4. VIRTUAL LABORATORY TESTS

Undergraduate courses in Geomechanics usually have laboratory components. Two of the tests that are often run by students are the direct shear test and the triaxial undrained compression test. The triaxial test apparatus is a versatile device in which a variety of axisymmetric tests can be conducted with and without drainage of pore water. However, it is practically impossible for a student to make full use of the versatility of the apparatus because of time constraints. For example, students do not usually conduct drained tests on fine-grained soils or back saturate a fine-grained soil prior to testing because of the lengthy time required for such tests. Even in industry, drained triaxial tests are not frequently conducted. At some institutions, students do not prepare a test sample. Rather, a teaching assistant or a technician will prepare the sample, set it up in the apparatus and then the students, in groups, would conduct the consolidation and shearing of the soil sample.

Virtual geomechanics laboratory tests (Budhu, 2000a) have been developed to allow students to conduct tests independent of location and time. All the procedures that would normally be conducted in a real (3D environment) laboratory test are simulated. For example, in performing a triaxial test in the virtual laboratory environment, a student prepares a soil sample and can conduct a variety of drained and undrained tests with or without back pressures, with or without pore water pressure measurements and apply loads to follow different stress paths.

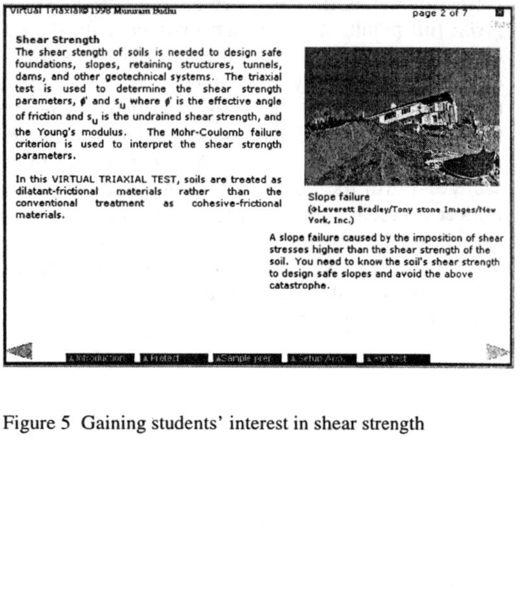

Figure 5 Gaining students' interest in shear strength

The course structure of the virtual laboratory tests is as follows.

- Gain student's attention to instill in him/her the reason for understanding shear strength of soils. A house in a precarious position when the slope adjacent to it failed is used as an attention grabber for shear testing (Figure 5).
- Present clear objectives of the test.
- Present learning outcomes. The student is informed what knowledge he/she is expected to know after completing the virtual laboratory. These outcomes are assessed by a quiz at the end of the laboratory exercise.

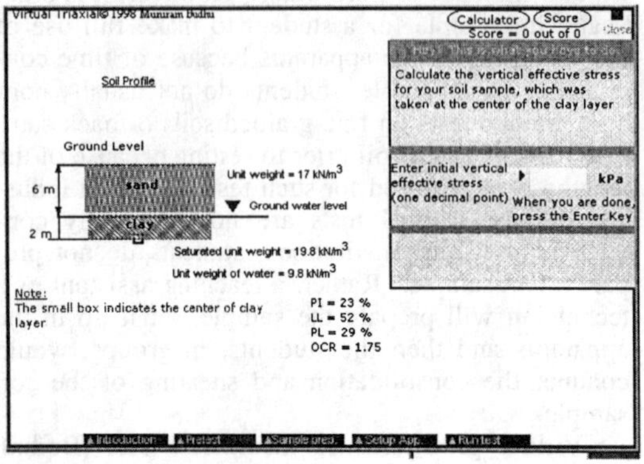

Figure 6 Testing pre-requisite knowledge

- Test pre-requisite knowledge. For example, a student must know how to calculate, at least, effective stresses and initial void ratio to be able to interpret the test results. A soil profile is randomly generated and presented with basic information on the physical properties of the soils. The student uses this information to determine the vertical effective stress on his/her virtual soil sample (Figure 6). If the student calculates the vertical effective stress correctly in the first or second attempt, he/she is given full points and can then proceed with the rest of the courseware. If the vertical effective stress is calculated incorrectly, the student is given up to three attempts to get the correct answer. After the third attempt, the solution is presented and the student is given a score of zero. A record is kept of the score and also the question that the student did not answer correctly. An instructor can use this information to take remedial actions in the course. To relate the virtual soil sample to real life, the student can access a digital video clip of a real soil boring and soil sampling operation.

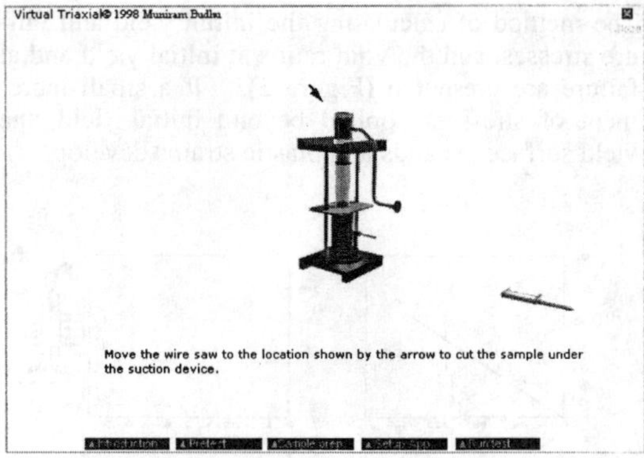

Figure 7 Preparing a sample for testing

- Prepare test sample. The student moves a simulated sample tube into a simulated jack, tightens a screw to hold the tube in place and then clicks on the jack handle to extract the top portion of the soil sample. Using a simulated wire saw, the student cuts out the top portion of the soil sample and removes it. The student then clicks on the jack handle once more to extract the sample into its sample preparation holder. The sample is then cut below the holder using the wire saw (Figure 7).

Figure 8 Setting up the virtual triaxial apparatus.

- Apparatus setup. The student sets up the sample and assembles the apparatus just as if the real apparatus were to be set up (Figures 8 and 9). Drag and drop guided self-paced interactions are used.
- Testing. All procedures required by ASTM are followed in conducting a virtual test. The application of cell pressure to consolidate the virtual sample is illustrated in Figure 10.

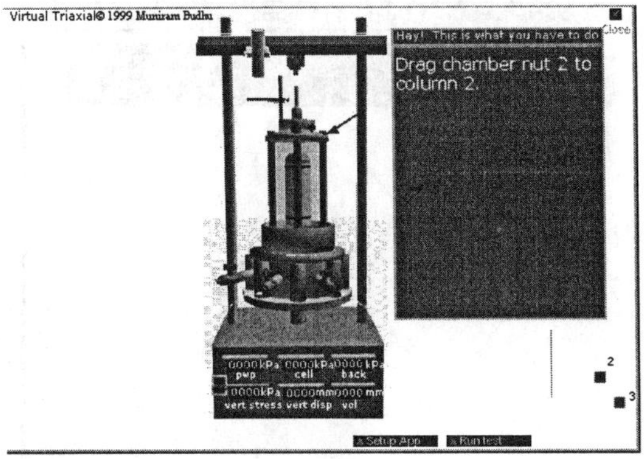

Figure 9 Instruction to drag nut to seal top plate to the triaxial chamber

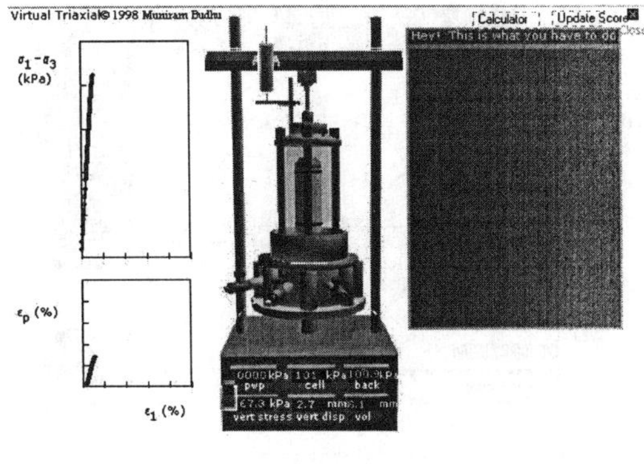

Figure 11 A drained test in progress in the virtual triaxial laboratory module

The cam-clay model (Roscoe and Burland) is used to predict the response of the soil to the user loading conditions. Soil parameters for the model are randomly generated within limits of real values reported in the literature. In the case of the cam-clay model, water content and index properties are randomly generated and the required model parameters, e.g. compression index, λ, are generated using empirical relationships (Wood and Wroth, 1978).

A virtual triaxial test in progress and a complete set of results for three tests on the same soil at different cell pressure are shown in Figures 11 and 12.

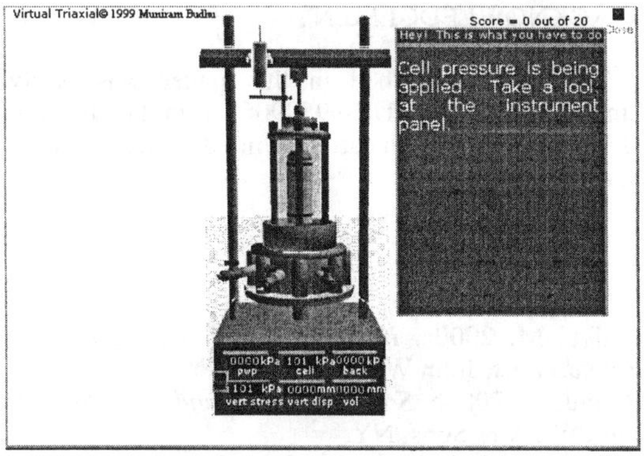

Figure 10 Application of the desired cell pressure.

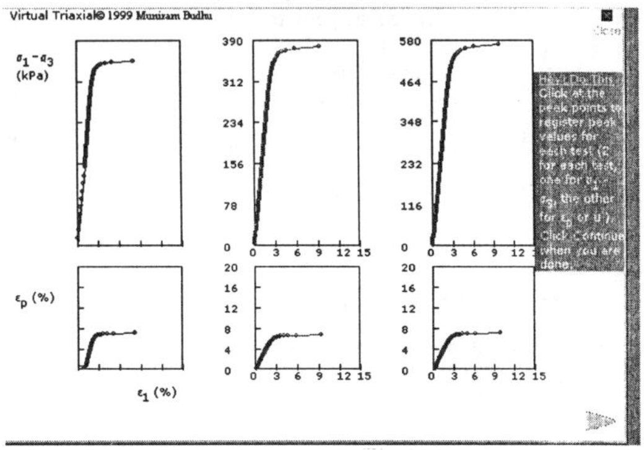

Figure 12 Three virtual triaxial test results

Since actual test results do not always show continuously smooth variations, perturbation functions are randomly introduced to modify the model predictions to produce some minor jumps in the results. In addition, because of sample preparation and the vagaries of soil samples, a test sample may fail prematurely. Algorithms in the program randomly select a sample for premature failure. If a selection is made, the student is asked to give possible reasons for the premature failure assuming that the test is real.

- Interpretation of the test results. The student is guided to interpret the results and is scored on the correctness of the interpretation or calculation. A screen shot of the guided interpretation of a drained triaxial test using Mohr's circle is depicted in Figure 13.
- Exploration. Students can explore various loading or initial soil conditions to address "what-if" situations. For example in the virtual triaxial test, if the student conducted a drained test, he/she will be prompted to conduct an undrained test on the same soil (Figure 14) or test a soil at a different over-consolidation ratio.

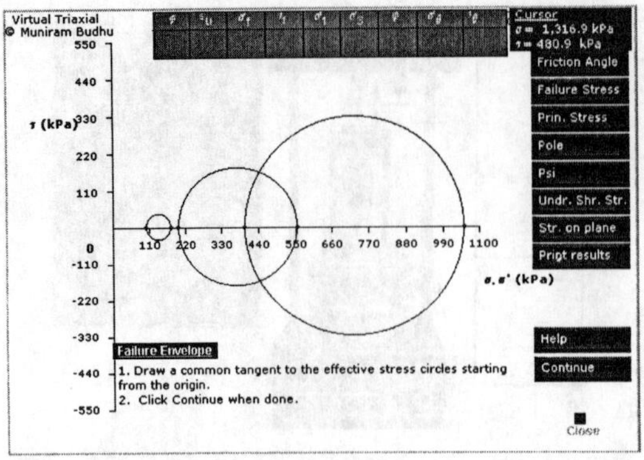

Figure 13 Interactive Mohr's circle to interpret virtual triaxial test results.

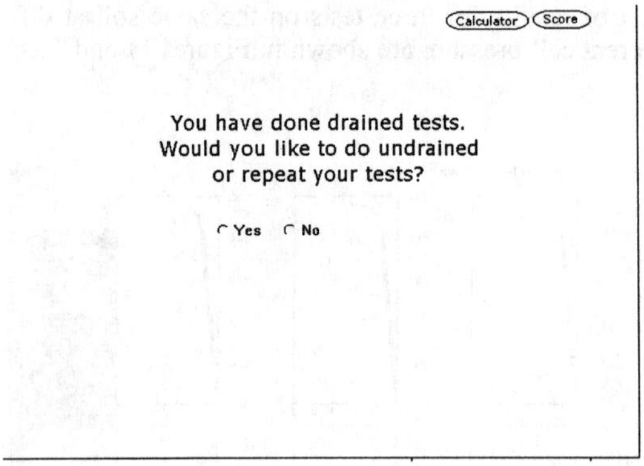

Figure 14 Student is prompted to explore a different test condition on the same soil

- Electronic quiz. At the end of a virtual test, the student must answer a quiz to evaluate knowledge retention. An example of a question on the quiz is shown in Figure 15.

4. EVALUATION

Preliminary evaluation at the University of Arizona showed that students who use these modules were able to perform better in tests to assess understanding and retention than students who did not use these modules. In a questionnaire to evaluate the effectiveness of the modules to transmit information, 98% of the students agree that they understood the material better because of the use of interactive animation. The same percentage of students claimed in the questionnaire that they think that they are visual learners.

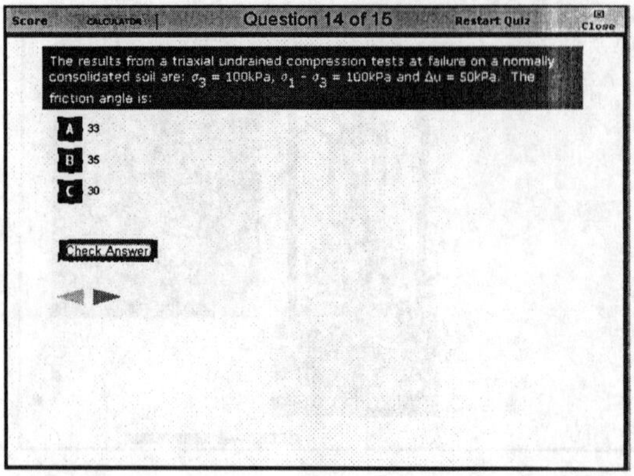

Figure 15 A sample question in the electronic quiz

5. CONCLUSIONS

In appears from limited evaluations that modern communication technologies when used judiciously offer the potential to enhance the understanding of difficult concepts and improve knowledge retention. The two modules presented in this paper appear to be effective but more assessment is required.

6 ACKNOWLEDGEMENT

The work described in this paper was partly funded by a grant (DUE-9950906) from the Division of Undergraduate Education, United States National Science Foundation.

7 REFERENCES

Budhu, M. 2000a. *Interactive Soil Mechanics and Foundations*, John Wiley & Sons, NY.

Budhu, M. 2000b. *Soil Mechanics and Foundations*, John Wiley & Sons, NY.

Carpenter, Mark 1999. *Oh, Now I get it!," Journal of Engineering Education*, ASCEE, Vol. 88 (4), pp. 387-383.

Casagrande, A. 1936. *Characteristics of cohesionless soils affecting the stability of slopes and earth fills.* J. Boston Soc. Civ. Eng., pp. 257-276.

Dede, Christopher; Bowman, Jim & Kierstead, Fred 1982. *Communications technologies and education: The coming transformation*, Communications and the Future, ed. Howard Didsbury (Jr), World Future Society, pp. 181.

Roscoe & Burland 1968. *On the generalized stress-strain behaviour of 'wet' clay*, in J. Heyman and F. A. Leckie (eds), Engineering plasticity, Cambridge University Press, pp. 535-609.

Roscoe, K. H. $ Schofield, A. 1963. *Mechanical behaviour of an idealized wet clay,* Proc. 2nd European Conf. Soil Mech., pp. 47-54.
Schofield, A. & Wroth, P. 1968. *Critical state soil mechanics*, McGraw-Hill, London.
Taylor, D. W. 1948. *Fundamentals of soil mechanics,* Wiley, New York.
Wood, D. M. & Wroth, C. P. 1978. *The correlation of index properties with some basic engineering properties of soils*, Canadian geotechnical Journal, Vol. 15, No. 2 pp. 137-145.

Computer Methods and Advances in Geomechanics, Desai et al. (eds) © 2001 Balkema, Rotterdam, ISBN 90 5809 183 X

An undergraduate experience in subsurface exploration and data interpretation

Eric C. Drumm & N. Randy Rainwater
Department of Civil and Environmental Engineering, University of Tennessee, USA

ABSTRACT: A "sub-contractor" program was developed, in which teams of students from the junior level Introductory Soil Behavior course act as sub-contractors to provide geotechnical support to student teams enrolled in the Capstone Senior Design Course. The sub-contractor program exposes the seniors to the process of preparing a request for proposals to secure geotechnical services, identification of appropriate lab tests, and the use actual lab data (and all the inherent problems) for the project design. The introductory students are exposed to the process of preparing a competitive proposal/statement of qualifications, and the successful teams are exposed to the senior design experience and provided an opportunity to work on a "real" project. The interaction has generally benefitted both groups of students.

1 INTRODUCTION

The Department of Civil and Environmental Engineering at the University of Tennessee includes two required geotechnical courses (introductory soils mechanics/behavior and foundations design) and a Capstone Senior Design Course. The Capstone design The Senior Design course is usually a "real" project incorporating most of the sub-disciplines within Civil Engineering is a project oriented class, in which the students are divided in teams for the various aspects of the project (e.g. geotechnical, structural, environmental). The Capstone Senior Design course offerec at the University of Tennessee is intended to combine many of the sub-disciplines of civil and environmental engineering in a single project, often with a real client.

This is a major part of the geotechnical evaluation (which is quite distinct from the foundation design aspects which are really structural considerations). In my view, most of the 330 folks are not well prepared to be doing much engineering, but can serve as technicians to provide data to support the work of the CE 400 folks.

A "sub-contractor" program was developed, in which teams of students from the Introductory Soil Behavior course act as sub-contractors to provide geotechnical support (subsurface investigation, determination of index properties, and consolidation & strength testing) to the senoior students enrolled in the Capstone Senior Design Course. . The sub-contractor program exposes the seniors to the process of preparing a request for proposals to secure geotechnical services, the specification of boring locations, and the identification of lab tests necessary for the project. The senior students then use actual lab data (and all the inherent problems) for the project design. The introductory students are exposed to the process of preparing a competitive proposal/statement of qualifications, and the successful teams are exposed to the senior design experience and provided an opportunity to work on a "real" project.

This paper will provide a description of haowe the students in these two classes have been drawn together to benefit both.

2 SENIOR CAPSTONE DESIGN CLASS

Teams are formed for various tasks, and faculty often serve as consultants to the various teams. As a consult to the geotechnical team, I decided to introduce to the senior students several aspects of geotechnical practice that were not covered in the course materials

The process usually begins with an offer to the instructors of the Capstone course to have the undergraduate students supply real material properties for the design the project. This is really the deliverable items that most of the students engineers want. If the project is suitable and the instructor is interested, a meeting is scheduled with the geotechnical team. The geotechnical team should determine what kind of information/data various aspects of the project (e.g. geotechnical, structural, environmental). To improve this experience for the students in the geotechnical groups, a subcontractor program was developed in conjunction with the junior level soil mechanics class.

The process usually begins with an offer to the instructors of the capstone course to have the junior soil mechanics students supply real material properties obtained from the actual project site for use in design. If the project is suitable and the instructor is interested, a meeting is scheduled with the seniors identified as the geotechnical team.

The geotechnical team must determine what kind of information/data is required and/or desired, and from where it should be obtained. This is a major part of the geotechnical evaluation (which is quite distinct from the foundation design, many aspects of which are really structural considerations).

I usually meet with the geotechnical team to discuss the following tasks, based on the anticipated needs of the project:

1) discussion and preliminary decision making about the number and location of borings, the number and types of samples, and the anticipated types of laboratory tests
2) identification of any existing subsurface information about the site
3) preparation of a request for proposals and statement of qualifications to provide drilling, sampling and testing services,
4) scheduling of a preliminary site visit for the layout of borings, notification of utility location agencies

After the proposals and statements of qualifications are received from the junior soil mechanics teams "bidding" to provide the investigation and testing services, the senior geotechnical group usually participates in the selection of the successful team. In some instances, the work has been divided into parts with two of more subcontracts for different aspects of the site investigation or laboratory testing.

After the subsurface report has been received, we usually meet again to review the provided boring logs and test reports, and to discuss the selection of appropriate material properties to be used in design.

3 JUNIOR SOIL MECHANICS CLASS

Every student in the junior level introductory soil mechanics course must complete a term project. The objective of this exercise is to combine the concepts learned in the lecture and the laboratory, and apply them to a "practice" type problem. This includes site investigation and sampling, testing for basic soil classification and strength, simple slope stability analysis, and the preparation of a final report summarizing the findings/conclusions. The projects are completed in small teams.

While the typical problem involves the stability analysis of a slope, other problems may be selected, one of which is to provide the subsurface and lab testing service for the senior design class. The competition to participate in the senior capstone design class is usually intense, with more teams "bidding" than can be accommodated. One of the benefits of providing these services is that the report must be completed near the mid-point of the semester in order that the results can be utilized by the seniors and meet design deadlines. This eliminates the term project at the end of the semester which is often considered attractive.

To support the senior capstone design class, the following services are typically included:

1) subsurface investigation and sampling
2) laboratory testing
3) report preparation

Each of the above tasks is described below.

3.1 *Subsurface investigation and sampling*

The junior team will drill a series of exploratory borings at the project site, based on the boring program developed by the senior geotechnical team. The senior geotechnical team is usually present during the drilling process. The borings are usually advanced only to a depth of about 15 feet (5 m) or until refusal is encountered. Although these depths may not be representative of many projects in practice, this simulates a typical exploration program considering the limits of the drilling equipment, and is is usually sufficient for the projects encountered by the capstone design projects.

The Department of Civil and Environmental Engineering and the Department of Geological Sciences at UT share a J.K.SMIT GW-15 Winkie drill, Figure 1, which is a lightweight portable gasoline powered drill (JKS Boyles 2000). The drill was designed to be carried into remote areas by two people, but to facilitate its use as a "hands on" subsurface investigation experience in the junior level class, it has been mounted on a custom trailer, Figures 2 - 4.

The manufacturer suggests that the Winkie was designed to recover 1 inch (25mm) rock cores to a depth of 400' (120 m), and that in some formations, cores up to 2 1/8 inch (54 mm) in diameter can be recovered to lesser depths. When drilling through soil with continuous flight augers, a reduction gear is added between the motor and auger, Figure 1, to reduce the rotation rate and increase the torque.

At the University of Tennessee, 4 foot (1.2 m) long continuous flight solid stem augers, Figure 4, of 2.75 inch (70 mm) diameter are used for subsurface exploration and sampling of soils. Auger cuttings are used for visual identification, and for finer than #200 by-wash sieve analysis.

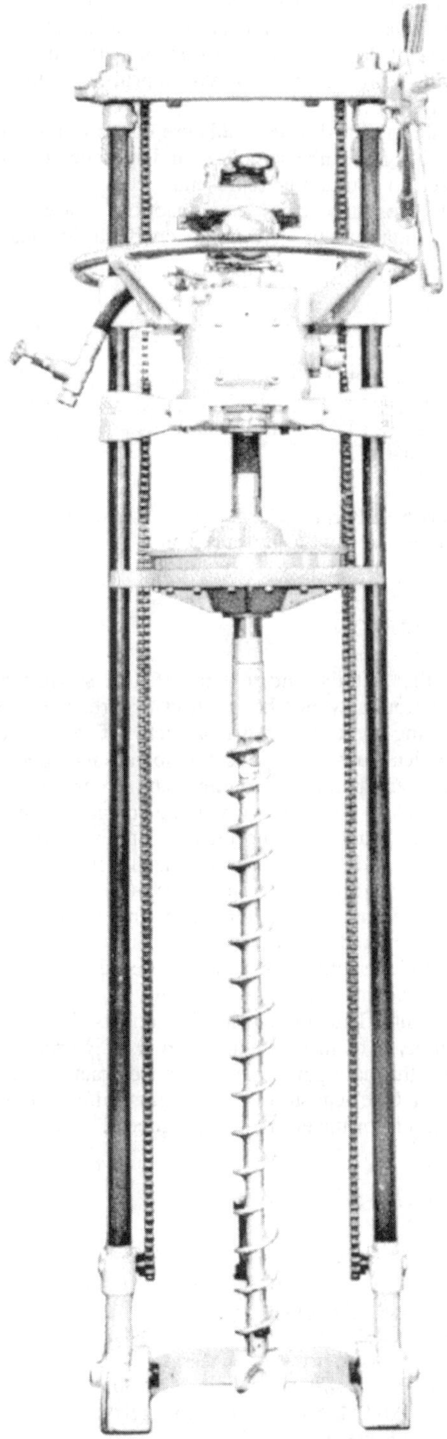

Figure 1 J.K.Smit model GW-15 winkie drill

Figure 2 Trailer mounted "Winkie" drill rig drilling on campus

Figure 3 Winkie drilling at a remote site

Figure 4 Solid stem auger and roller track mount

For sampling, the auger is withdrawn and the Winkie moved back out of the way on a rolling carriage incorporated into the trailer, Figure 4.

Samples 2 inch (50 mm) diameter by 6 inch (150 mm) long are obtained with an AMS slide hammer and core sampler, and are used for unit weight determination and strength testing. Instu penetration resistance is measured with a Dynamic Cone Penetrometer (DCP), the results of which can be correlated with SPT blow resistance (Sowers & Hedges 1966).

3.2 *Laboratory testing*

Laboratory testing on the auger cuttings generally consists of natural water content and Atterberg limit determination. A finer that #200 sieve by wash grain size determination is usually sufficient to classify the soils. For most projects, undisturbed tube samples are tested in unconfined compression to determine the undrained shear strength. Occasionally, unconsolidated undrained (UU) triaxial, direct shear, or consolidation tests are performed.

Projects which contain a significant amount of cut and fill may have the soils evaluated for potential borrow materials. This will include Standard Proctor compaction tests on candidate soils, and perhaps unconfined compression tests on compacted borrow materials. In one instance, a project was encountered in which the borrow materials were to be used as a hydraulic barrier beneath wood composting facility to prevent leachate from reaching the groundwater. Compaction mold hydraulic conductivity tests (falling head) were performed to evaluate the suitability of these materials.

3.3 *Report Preparation*

A typical report from the subsurface investigation team is expected to include the following components:

a) A cover/transmittal letter similar to one which a consultant would prepare for their client. The report should summarize the results of any analysis or tests conducted, and provide recommendations to the client. The cover letter should include signatures of all members of the project team.
b) A simple map showing the locations of the borings relative to any local landmarks or structures. Each boring should be identified with a number or letter.
c) A simple boring log for each hole identified in the map. The boring log should include:
   1) a description of each soil encountered and approximate range of depths. This description should include color, a visual classification, and origin (residual soil, alluvial soil, fill, etc.)
   2) a plot of the natural water content and Atterberg limits versus depth
   3) a plot of the measured penetration value versus depth
   4) symbols or notation to indicate the depths at which tube samples were obtained
d) An idealized soil profile drawing obtained from the individual borings indicating the location and extent of the strata or layers encountered in each boring. An example of a typical profile is shown in Figure 5.
e) A table indicating the moisture content and unit weight of the tube sample(s) and an indication of which soil strata this represents
f) The results from compaction, strength, consolidation or hydraulic conductivity tests performed.

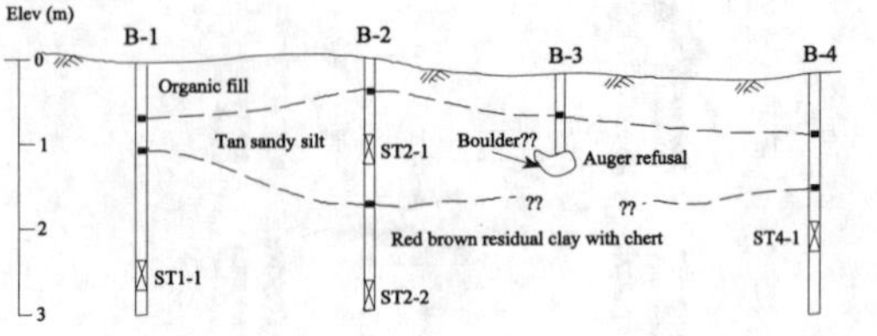

Figure 5 Soil Profile along Section A-A' (STx-y = indicates location of Shelby tube number y for boring x)

4 CONCLUSIONS

While the details and extent of the subsurface investigation may not be in strict accordance with engineering practice, most aspects of a typical geotechnical subsurface investigation are included in the above subcontractor program. The senior students are exposed to concepts such as the consideration of the number and depth of borings, the number and type of lab tests, and routine tasks such as the clearance of utilities prior to drilling. These are tasks that many geotechnical engineers encounter routinely, yet are not typically covered in undergraduate course work. The senior students must make design decisions on the basis of actual data, which may be flawed or inconsistent. The junior level students have an opportunity to contribute data to a real design project, and those that have participated as subcontractors enter the senior level capstone design class with a better appreciation for the entire design process.

REFERENCES

JKS Boyles International Inc. 2000. North Bay, Ontario, Canada. Internet address: http://www.jksboyles.com/

Sowers, G. F. & C. S. Hedges 1966. Dynamic cone for shallow in-situ penetration testing, *Vane Shear and Cone Penetration Resistance Testing of In-Situ Soils*, ASTM STP 399, American Society of Testing Materials, pp 29-37.

Computer Methods and Advances in Geomechanics, Desai et al. (eds) © 2001 Balkema, Rotterdam, ISBN 90 5809 183 X

Thoughts on the future of geotechnical and geosynthetics education

D.J. Elton
Auburn University, Ala., USA

J.J. Bowders Jr
University of Missouri-Columbia, Mo., USA

ABSTRACT: Geotechnical engineering knowledge and practice has expanded considerably from the days when Terzaghi and Peck wrote the landmark Soil Mechanics in Engineering Practice in 1948. Many new developments, especially in practice and materials used for geotechnical engineering practice now exist.

University geotechnical educational programs have also changed to somewhat accommodate the ever-increasing knowledge base. The changing face of higher education and the increasing amount of knowledge make further changes inevitable. Many more dramatic changes are expected.

This paper considers some of the changes that are needed/expected in geotechnical and geosynthetic engineering education.

INTRODUCTION

The engineering profession is changing, and education must change to keep up. Furthermore, educational practices, due to previously unenforced market restraints are requiring universities to change the way they do business. These will have a significant effect on geotechnical and geosynthetic engineering education. Geotechnical education is considered first, followed by a few thoughts on an important specific area of geotechnical education, geosynthetics.

THE CHANGING FACE OF HIGHER EDUCATION

The world has become affected by a global economy. Conventional businesses, including geotechnical engineering consulting and construction, now routinely face global competition, whereas only a few short years ago they did not.

Geotechnical engineering education now faces the same challenges. While not as competitive as the 'practice' side of geotechnical engineering, education faces an increasing pressure to:

1. be accountable to the 'customer', usually the student or those funding the student or the university, and
2. compete for students in a global arena, rather than just local. This is somewhat an outgrowth of (1).

These problems are compounded by a continuously changing field, filled with new technical challenges and knowledge. Moreover, the public who hires the graduates wants the new graduate to understand the legal, economic, political, and societal considerations of major projects (Mitchell 1999). The increasingly broader scope of geotechnical engineering projects requires even the new engineer to interface effectively with regulators, other disciplines (engineering and non-engineering), financial institutions, and attorneys to name a few. Improved written and spoken communication skills are needed to be effective.

Universities that have traditionally operated in a 'social vacuum' are striving mightily to change the face of their new graduate, who now wants (expects) to graduate in an ever-shortening time period. The competition for students (and their tuition dollars) has led university administrators (usually at the higher levels) to seek ways to make their university more attractive to potential students. A shorter path to graduation can be an effective measure.

The effect of these pressures has been startling to many administrators and funders of higher education.

CURRENT RESPONSES OF TRADITIONAL UNIVERSITY PROGRAMS

Universities have greater competition for students than ever before. The modern university student has more educational choices then ever before. To attract the new student, the university has to have more of an 'edge' than simply education to offer the student. Many students that, ten years ago, would not have considered college are in the market. This is due to expanded world wealth and increased amount of knowledge.

Many universities have responded to the increased competition for students. The primary responses seem to include:

1. reducing the credit hours for graduation,
2. seeking new markets off-campus (mostly through 'distance learning'),
3. reducing the amount of full-time instructors (increase in part-time instructors), and
4. reducing overhead by reduction in staff and outsourcing.

Reducing (or freezing) tuition rates has occasionally occurred along with increased scholarship funds.

Of these measures, reducing the credit hours for graduation has the most impact on geotechnical and geosynthetic engineering education. Fewer courses mean less learning. This is compounded by increasing 'core curricula' requirements that expose the student to non-engineering information. While these requirements have value, and contribute to the students functioning in society, they displace domain-specific courses at a time when the size of the domain is getting larger very rapidly! The lack of geosynthetics in geotechnical engineering education is a good example of new knowledge largely missing from university geotechnical education (Bowders and Elton 1999).

The reduction in the credit hours to graduate (which is not likely to be reversed, for economic reasons), and the competition for students through increased 'distance learning' opportunities leaves little time for the geotechnical engineering student to gain appreciation for the many types of geotechnical problems, the interfaces of his future work with others or the larger impacts of his work on society.

'Distance learning', a euphemism for learning away from campus, has had a pronounced effect on learning. This student market has exploded with the pressure from licensing agencies for continuing education for licensure, and the need from industry for updating skills (here 'updating' may be too kind a word. Many graduates were not taught the skills they need before graduation. New geotechnical knowledge is generated too fast for many universities to keep up). Issa (1997) gives a review of many distance learning technologies and the attendant advantages and disadvantages. All are viable.

Distance learning has many forms, mostly focused on individuals that need to learn on their own, rather than in a classroom setting with peers. Self-help books, videotapes, and closed satellite broadcasts were the traditional methods used. They are still in use. Web-based courses are beginning to dominate this area. These courses seem to be either traditional lecture courses broadcast on the Web, or more recently, interactive software for learning provided on the Web. A live instructor is not used in the latter. This venue extends beyond geotechnical engineering.

Distance learning is usually inexpensive. Web-based learning opens markets heretofore unknown to the university. Suddenly, anyone in the world near a PC is a potential student. The economic benefits of this expanded market are beginning to be embraced by geotechnical educators, albeit slowly.

Universities have embraced outsourcing as they seek to cut costs. Custodial and food services, once completely university-run are now done by specialists hired from outside. Other functions (machining and fabrication, telecommunications, printing, financial) are following suit.

A more important 'outsourcing' is the hiring of non-tenure track instructors in large numbers. Liberal arts staffs have traditionally had large percentages of non-tenure track instructors, due in part to the glut of qualified personnel. In the more recent decade, the push to reduce costs (coupled with an attractive non-academic job market for well-educated engineers) has led to widespread hiring of personnel who have no vested interest in either the university or the department they work in. These instructors (often retirees, post-docs and graduate students) do not contribute to university governance or the development of curricula. The burden of program guidance is left to a smaller amount of faculty, already pressed by existing teaching and research loads.

Non-tenure track personnel are often not accountable to the university as are tenure track faculty. While the university may evaluate their teaching skills, their compensation (and pay raises) may not be as closely tied to it. In fact, the universities may be more likely to ignore their teaching efforts, since the reason they were hired was to inexpensively fill classroom assignments. They are less likely to be discouraged from leaving.

A more optimistic response of US universities is their response to the demands of accreditation bodies for more practice-oriented graduates. A final year 'capstone' design course is required in many US curricula. The course frequently requires students to integrate technical information from many civil engineering subdisciplines, including geotechnical. Students form teams, hold meetings, do analyses and make presentations. The oral and written aspects of these courses are usually a new experience, and most beneficial! In the best of cases, experienced practitioners oversee these courses. Their perspective is often very different from the academic experience of the students' professors. Here, the importance of teamwork, communication and economic solutions are emphasized. This experience has been good for geotechnical and geosynthetic engineering education.

The net result from all but the last of these changes is a 'leaner, meaner' university operation, frequently resulting in less education for the student, but at a lower financial cost. The cost to the geotechnical engineering community is higher.

FUTURE RESPONSES

Geotechnical Engineering - Anticipated University Responses

Geotechnical engineering education will change. The economic forces driving higher education in general are much different than those a decade ago, to say nothing of when Terzaghi was an instructor.

In keeping with the change to more competition for end-user's and capital, geotechnical education will change. Geotechnical programs struggle to cover the "basic core" subjects and include more temporal topics as they arise. Many universities will continue to increase "basic core" knowledge courses to take hours from curriculum majors. Increasing competition for *numbers* of students is driving universities to offer 'discount degrees' – ones having fewer hours for completion than their competition. The reduction in domain–specific courses (like geotechnical) will continue as more core courses are added. This phenomenon is compounded by the changing roles of engineers in business. Bordogna (1998) predicts that engineers will continue to have even greater input in shaping society, and must prepare for this role with 'holistic baccalaureate education' beyond the scope of many current university curricula.

Universities must continue to find ways to attract and produce their product (graduates) at lower costs. Web-based instruction offers the best of both worlds at the moment – access to the broadest market, and lowest cost of delivery.

Effective Web-based instruction, however, requires major revision of delivery methods. Most of the pedagogical principles are in place, from years of educational research. Most have not been applied. Most geotechnical engineering faculty have not been trained in educational principles, but have received only on-the-job training. Fewer yet understand the concepts needed to create an effective Web-based course.

Universities will have to form closer collaborations with between geotechnical faculty and education faculty / consultants to form effective 'distance learning' media.

The demise of graduate programs, with the advent of a few strong graduate programs is expected. The cost of research equipment, space and faculty is prohibitive. There have been occasional professional discussions on the need for so many geotechnical graduate programs. While it was once desirable for many faculty to be engaged in research, it will likely be viewed as a privilege, rather than an obligation soon. More faculty will likely take up more of their time with distance learning media, rather than research media. The financial returns to the university will likely be greater.

A potential solution to this dilemma will be a major change in the way engineering education is done. Pressure from employers for better skilled graduates will be balanced with the university need to produce a graduate with the lowest education possible. The middle ground between these competing market forces is to enhance the fundamentals for the bachelor's degree and encourage students to stay for further training. This extra year can be devoted to electives of the student's choice. Many civil engineering university students find at the end of their four-years that they are just beginning to find an interesting subarea. A fifth year of education could be attractive.

The American Society of Civil Engineers has recognized the Master's degree as the first professional degree in the US (ASCE 1998). This may become a trend. The university gets the student enrollment for another year or so, and the employer gets a better-educated student. This proposal would work best with the cooperation of industry, which would give openly preferential treatment to these new graduates.

The nature of graduate programs is likely to change. More, specialized courses will become available to students than ever before because of the Web (and, perhaps, other unforeseen media). Universities will no longer be able to constrain student credit simply to their own courses. Rather, an integration of courses from several universities (or, at least, sources) will form the students' plan of study. The Western Governor's School in the US is an example of how this is beginning to take hold. A brief search (Oct 1999) for 'virtual' engineering degrees (ones where the student, or, end-user, does not have to set foot on campus) yielded 42 universities with a multitude of degrees. More conventional universities will have to compete to remain viable.

This will be good for geotechnical engineering, as students will be better able to study, for example, geological engineering. Terzaghi (cited in Goodman 1999) was a keen proponent of geology, many times saying that to be a good geotechnical engineer, one must understand geology. Courses in niche areas, newly developing areas and specialty areas will contribute to the degree. In situ testing, geosynthetics, ground improvement and others will become more readily available. Clough (1998), cited in Mitchell (1999) noted that "technological advances are likely in the areas of geo-construction technology, information and computer technology applied to geotechnical problems, site characterization technology, and in the understanding of soil and rock behavior fundamentals". Legal aspects, environmentally acceptable construction and risk management might be added.

This newfound flexibility will attract more students, but perhaps not to any particular university. Universities that do not accept this flexibility will be left behind as the crop of newer students, seeking specialty courses, comes forward.

An interesting sidelight of the survey conducted by Bowders and Elton (1999) was the wide range of credit hours required for an advanced degree in civil engineering. It is hoped that universities will make this number more uniform, similar to undergraduate degree requirements, which, while creeping downwards, at least are quite similar.

It is hoped that undergraduate curricula will focus more on fundamentals of engineering – particularly more math, physics, chemistry, geology, and engineering sciences, including solid mechanics, fluid mechanics, dynamics, and thermodynamics. Mitchell (1999) makes a strong case for undergraduate emphasizing fundamentals. These courses are the foundation for future geotechnical engineers.

Others have addressed various aspects of these issues (Burland 1987; Poulos 1997; Manassero 1998; Shackelford 1998 and Bowders and Elton 1999; Mitchell 1999).

Geosynthetic Engineering - Anticipated University Responses

Geosynthetics is a small part of geotechnical engineering used as an example of the ways that geotechnical engineering is expanding. The future of geosynthetic engineering is examined below as a case study of how education in other geotechnical specialties might evolve.

The role of geosynthetics in traditional geotechnical areas (slopes, foundations, earth retaining, drainage, and filtration) is increasing rapidly. However, it is underrepresented in the formal academic setting. The informal survey by Bowders and Elton (1999) showed that many US universities include some component of geosynthetic education in some aspect of the civil engineering curriculum. However, few include a course devoted to geosynthetics. It does not appear, for the reasons noted above, that many more courses devoted to geosynthetics will appear. A decrease in 'specialty' courses is expected, including geosynthetics.

Currently, university geosynthetic education forms parts of other courses, rather than a stand-alone course. This trend is expected to expand. Foundation engineering, retaining walls, erosion control, geoenvironmental and other areas where geosynthetics are appropriate will continue to see the integration of geosynthetics in the curriculum. The Industrial Fabrics Association International / National Science Foundation Professor Training Course for Geosynthetics presented geosynthetics information in a modular format. Modules could be integrated into existing courses (likely), or combined to form a stand-alone course (unlikely). McCarthy's (1998) textbook is a good example of integration of geosynthetics into conventional geotechnical designs, rather than forming a separate chapter. This model is expected to dominate future geosynthetic instruction. Moreover, this course likely contains the model of industry involvement in education that will become more prominent. Seven organizations sponsored the course over its five year run: the US National Science Foundation, the Industrial Fabrics Association International, the North American Geosynthetics Society, the International Erosion Control Association, the Geosynthetics Institute, the Federal Highway Administration, and the PVC Geomembrane Institute.

Geosynthetic education will become more 'appropriate' in the future. As the darkness of geosynthetic ignorance is dispelled, geotechnical engineers will no longer think that generic geosynthetics are the answer to all their problems. Rather, the need for geosynthetic *design* will become recognized, including an awareness of when geosynthetics are inappropriate.

The cost of geosynthetics is expected to continue to decrease, encouraging the 'commodity' mentality that makes design costs unreasonable, resulting in less design being done. The low product cost doesn't justify the engineering expense. That is, less engineering is justified when the cost savings will be minimal compared to the increased engineering time. However, less design may lead to increased number of failures!

Universities will not be the main conveyors of geosynthetic training. Graduates will obtain knowledge of geosynthetics from outside the conventional classroom such as industry-sponsored literature, short courses, books, workshops, symposia, and the Web. Geosynthetics education will be primarily post-BS and transmitted via Web-based and other distance learning mechanisms. Software (Web-based or otherwise) will be prominent. Poulos (1999) lists software for general geotechnical education.

Professional organizations, such as the International Geosynthetics Society, American Society of Testing and Materials, North American Geosynthetics Society, Geo-Institute and American Society of Civil Engineers), and trade organizations (e.g. the Industrial Fabrics Association International, Erosion Control Technology Council, International Erosion Control Association, Geosynthetic Institute , International Association of Geomembrane Installers and the PVC Geomembrane Institute), and government bodies such as the Transportation Research Board, Federal Highway Administration, companies or universities and individuals (Geosynthetic Institute) and others with a vested interest in geosynthetics will take more responsibility for education.

Finally, geosynthetics will remain heavily tied and associated with geoenvironmental work. We expect that landfills and site remediation will be primary areas for geosynthetics usage and, thus, the demand for education in these areas will be the highest.

CONCLUSIONS

Geotechnical and geosynthetic education will continue to evolve more rapidly than ever before. Universities will change their processes. If universities adapt to new demands of their clientele, their role will not diminish. Great increases in flexibility of delivery and content will take place. The market share (%) of universities will decrease with time, as other educational venues expand. However, the absolute amount of educational business they do will likely expand as the market expands.

Other forms of education will grow, eclipsing some conventional university education. Videotaped courses (reusable), satellite broadcast, CD ROMs and the Web will see exponential growth and cost reductions. Low production and distribution costs, particularly for Web-based materials, will drive down the cost (and eventually the availability!) of other instructional media. These modes of instruction have attractive pricing and allow great flexibility to the student when and what to learn. This will apply to both geotechnical and geosynthetics education.

Manufacturers/vendors/contractors in many areas of geotechnical practice will continue to encourage market growth through education of users. However, the continued consolidation of manufacturers/vendors/contractors in the global economy will affect the design procedures and education expenditures. Most industry-sponsored education is expected to focus on applications, not understanding.

University geotechnical education will likely become less specific, as other 'core' courses encroach on discipline-specific courses. It is hoped that a trend toward more general science/math courses will take place.

In spite of these changes, however, the current demand for entry level geotechnical engineers is very great, at least in the U.S. Thus the need for well-educated geotechnical engineers remains strong.

Others have addressed various aspects of these issues (Filz 1994; Bordongna 1998; Koerner 1999; Poulos 1999; Guimaraes 1999). The interested reader is referred to these timely articles.

REFERENCES

ASCE (1998) "ASCE Board Supports Master's as First Professional Degree," *ASCE News*, 23(11).

Bordongna, J. (1998), "Tomorrow's Civil Systems Engineer – The Master Integrator," Journal of Professional Issues in Engineering Education & Practice, Vol. 124, No. 2, April, pp. 48-50.

Bowders, J., Elton, D.J. (1999) Future Trends In Geosynthetics Education, in Proceedings, GRI-13, Future Trends in Geosynthetics, R. Koerner, Ed., Geosynthetics Institute, Folsom, PA, p400-415..

Burland, J.B. (1987). The Teaching of Soil Mechanics – A Personal View - The Nash Lecture, Proc. 9th European Conference on Soil Mechanics and Foundation Engineering, Dublin, Ireland, Vol. 3, pp. 1427-1441.

Clough, G.W. (1998) A View of the Future of Geotechnical Engineering, prepared for the ASCE Geo Congress, Boston, MA, October 19, 1998.

Filz, G. (1994) "Environmental Geotechnics Curricula," *Geotechnical News*, 12(1): 33-36.

Goodman, R.E. (1999). Karl Terzaghi – The Engineer as Artist, ASCE Press, Reston, VA, 340 pp.

Guimaraes, R.B. (1999) "The role of didactical model in teaching geotechnical engineering," *Proceedings* of the 11th Pan-American conference on Soil Mechanics and Geotechnical Engineering, Foz do Iguassu, Brazil, 08-12 August, Vol 2, pp731-737.

Issa, R.A., (1997)," Critical Issues in the Development of Distance Education Capabilities," *Computing in Civil Engineering*, ASCE, New York, NY, pp.81-88.

Koerner, R.M. (1999) "Vehicles for geosynthetic education," *Geotechnical Fabrics Report*, 17(6):16-17.

Manassero, M. (1998). Education in Environmental Geotechnics – European Practice, Environmental Geotechnics, P. S. Seco e Pinto, Editor, Balkema, Vol. 4, pp. 1335-1342.

McCarthy, D.F. (1998) Essentials of Soil Mechanics and Foundations – Basic Geotechnics, 5th ed., Prentice Hall, Upper Saddle River, N.J.

Mitchell, J.K. (1999) "EDUCATION IN GEOTECHNICAL ENGINEERING – Its Evolution, Current Status, and Challenges for the 21st Century" in Proceedings of the XI PanAmerican Conference on Soil Mechanics and Geotechnical Engineering, Iguassu Falls, Brazil

Poulos, H. (1999) "Geotechnical Education for 2000 and Beyond". Keynote Lecture, Proceedings, 11th Asian Regional Conference on Soil Mechanics and Geotechnical Engineering, Seoul, Korea, Balkema, Rotterdam, Vol.2.

Shackelford, C.D. (1998) "Educational Programs in Environmental Geotechnics / Geoenvironmental Engineering at USUCGER Universities," *Proceedings* 3rd Intl. Congress on Environmental Geotechnics, Lisbon, Portugal, A.A. Balkema Publ., Rotterdam, 1342-1351.

Terzaghi, K., Peck, R.B. (1948) Soil Mechanics in Engineering Practice, Wiley, N.Y.

Computer Methods and Advances in Geomechanics, Desai et al. (eds) © 2001 Balkema, Rotterdam, ISBN 90 5809 183 X

Teaching mechanics to freshman

Christopher D. Pionke & J. Roger Parsons
Engineering Fundamentals Division and Department of Mechanical, Aerospace and Engineering Science, University of Tennessee, USA
Richard M. Bennett
Engineering Fundamentals Division and Department of Civil and Environmental Engineering, University of Tennessee, USA

ABSTRACT: Advanced topics in numerical analysis typically used in geomechanics problems are introduced to freshman in simple terms. The introduction of the concepts is accomplished through the use of team projects, which enable larger projects to be assigned than the typical homework problem. Several projects that have been used in the *engage* freshman engineering program at The University of Tennessee are described.

1 INTRODUCTION

Numerical analysis of geomechanics problems is often considered to be a topic for upper level undergraduate courses or graduate courses. This is due to the need to understand matrix analysis, energy methods, numerical integration, nonlinear behavior, and other advanced topics. However, it is our contention that many of these topics can be introduced to freshman in a simple way. The University of Tennessee accomplishes this through the use of an integrated freshman program. A key aspect to this freshman program is a team project component that allows the introduction of applications of material outside the normal classroom homework problems. The integration of projects and numerical analysis has the added advantage of students learning the need to validate computer results.

This paper will provide an introduction to the Freshman program at The University of Tennessee. Some of the projects that are used to show applications of mechanics are briefly discussed. One project that incorporates many of the features required in geomechanics analysis is presented in greater detail.

2 THE *ENGAGE* PROGRAM

The University of Tennessee College of Engineering has recently revamped its freshman engineering program. Five existing courses were replaced by two new six-hour semester courses (Gillliam et al. 1998). The first course, Engineering Approaches to Physical Phenomena, emphasizes problem-solving, teamwork, design, and computer tools (engineering graphics using AutoCAD's Mechanical Desktop and computer programming using MATLAB). The course is based on the study and use of introductory mechanics, such as position-velocity-acceleration relationships, vectors, forces, moments, work, and energy. The second course, Fundamentals of Engineering Mechanics, integrates an entire semester of statics and an entire semester of particle dynamics into one course (Pionke et al. 1999a). This course assumes mastery of the techniques and tools from the first course, and forces students to make extensive use of those tools. The statics and dynamics are taught in an integrated manner with the use of lectures, problem solving sessions, a hands-on laboratory, design projects, and computer modeling. Both courses are taught by the newly created Engineering Fundamentals Division and both courses are taken by all freshman engineering students. The entire new integrated curriculum is entitled the *engage* program.

Both courses are divided into four basic components: lecture, hands-on lab, analysis and skills sessions, and team project time. The lecture format is fairly "traditional" with a professor presenting and outlining new technical material. The purpose of the hands-on lab is to help the students develop a physical "feeling" for the concept most recently presented in the lecture. Each lab briefly restates and reinforces the background and concept, gives students a physical experience with the new concept, and requires students to analyze the physical results in light of the covered concept (Yoder et al. 1998). Analysis and skills sessions are small sections conducted by graduate teaching assistants (GTAs). In the first course the majority of the time is used to teach basic computer skills (Pionke et al. 1999b). In the second course the majority of the small section time is used for statics and dynamics "problem sessions." The team project time provides the opportunity for students to be exposed to real world applications that are beyond the scope of normal homework problems. The team project sessions are conducted by a professor with the assistance of GTAs and undergraduate "team facilitators." The GTAs help in technical guidance for the students, and schedule and supervise non-class time openings of the project room. The undergraduate facilitators assist the students in dealing with issues such as teamwork and conflict resolution. All facilitators are undergraduate engineering students who must enroll in a special class where they are trained in group dynamics, conflict resolution, and team facilitation (Knight et al. 1998). The use of the team project time to introduce concepts related to computer analysis and geomechanics is the focus of this paper.

For the first half of the first semester, a series of short design/analysis projects (e.g., a mechanical dissection of a common kitchen appliance) are assigned to the students. The purpose of these small projects is to teach teamwork, engineering communications, and the basic approach to engineering design and analysis. A tie-in with the basic mechanics material is not intended. In the second half of the first semester, two projects are assigned that require the integration of basic mechanics material developed in the lecture portion of the course. Since the students are still learning the basics of computer-programming at this point, numerical techniques are not utilized in these two projects. Applied numerical techniques are reserved for the second semester projects.

3 THE DESIGN PROJECTS AND NUMERICAL MODELING

3.1 *The "Robusto" chair*

One project that has been used to teach basic statics is the "Robusto" chair. This project is used after the first introduction to free body diagrams and equilibrium. Students are required to build a chair out of foamboard that meets certain dimensional requirements. The design criteria is that the chair be able to hold a light person (approximately 600N), a heavy person (approximately 1000N), and then be able to support a 220N horizontal pulling load on the back of the chair with the heavy person sitting in the chair. As part of the report, students must draw free body diagrams and perform simple static equilibrium calculations on the chair.

3.2 *Tractor pull*

A favorite project to end the first semester is the tractor pull. Students build small tractors out of foamboard, 25mm PVC pipe, duct tape, paper clips, and rubber bands. For testing, the tractors are hooked to a string which through a pulley system is attached to a weight. The students run their tractor down a track and see how high their tractor can pull the weight. Through this project students begin to learn the concepts of energy, friction, and momentum. A common occurrence is for students to store significant energy in the rubber band, but forget about the mechanism needed to transfer the energy to lifting the weight. The tractor will simply spin out without lifting the weight at all. Although a bit embarrassing to the student, it provides a valuable lesson in design.

3.3 *Bridge over "Trouble Gorge"*

A more rigorous statics project used in the second semester is to build a bridge to cross "Trouble Gorge." The bridge is built out of wood strips (approximately 20mm x 20mm), twine, and cotter pins. Each piece of material is assigned a price, and the total cost of the bridge is calculated. The bridge has to span 2 meters, going over a "Trouble Gorge," which is a small swimming pool filled with apple sauce and fish crackers. The bridge is first tested by having one team member walk across the bridge. The second test is to have as many team members as possible stand on the bridge before it fails, and the team members end up in Trouble Gorge. The teams are scored on a weight/cost ratio. Some teams decide to build a very low cost bridge, and only attempt to hold two light people. Other teams build a very stout bridge. These bridges will often easily support the whole team. However, since the judging basis is weight/cost and the teams have run out of team members to continue to load the bridge, the stout bridges tend not to do as well in weight/cost. Students learn the importance of identifying what is the actual design criteria.

As part "Trouble Gorge" project, all teams are required to perform a complete truss analysis on their bridge. Doing these calculations by hand would be a very time consuming endeavor. Therefore, after the students are taught how to do truss analysis by the method of joints and the method of sections on very simple 2-D trusses, they are shown how to automate the process using MATLAB. In the analysis and skills portion of the program, the GTA's work with the students individually to develop a small program that will analyze medium sized 2-D trusses (10-30 members). Then each team works together to extend and combine their programs to a level that will analyze their 3-D "Trouble Gorge" bridge. The objective of this exercise is to show the students that the basic concepts they learned to analyze very simple linear 2-D trusses can be extended to analyze very large linear 2-D and 3-D trusses. The concepts are the same for both systems, only the number of required calculations has changed. The students realize that they can extend the concepts they learn in class to "real" problems, however, they may have to rely on the use of the computer to get accurate results in a reasonable amount of time.

3.4 *The bungee cord project*

An example of a second semester dynamics project is a variation of a project that has been used at other universities. The project was entitled "The Bungee Omelette - NOT" project. This project illustrated many of the concepts that are needed in geomechanics analysis. Difficult concepts were presented in a simple way to freshman through this team project.

The students were told that their instructors were going to expand their investment holdings by developing amusement park rides. The first venture was a multi-passenger bungee jump. The goal is to design a bungee jump that will safely drop a specified number of passengers from a given height. To test the validity of the design, a small scale test model will be designed, built, and tested. Eggs will be used to model the passengers since they come in various sizes and weights and are reasonably fragile, just like people. A sandwich bag will be used for the passenger compartment. A single bungee chord will be used, but the length and number of strands of the chords are free variables. When dropped from a given height, the lowest portion of the passenger compartment should just barely touch down in a sand pit. The students were given the drop distance and the number of eggs on the morning of the test.

The students had to resolve at least two problems related to their "real" system. The first was the variability in mass of each individual egg. Most teams resolved this problem by creating ballast weights to be added at the time of the drop. The mass of the eggs was determined just prior to testing and enough ballast weights were added to bring the total mass up to a "pre-determined" value.

The second problem was the non-linearity of the bungee cord. All teams were required to resolve the second problem by writing a MATLAB program of the work-energy equation of the entire system. The difficulty is that bungee cords are non-linear springs. The non-linear characteristics of the bungee cord were determined experimentally by the students. Although not directly a part of the project, the students learned the phenomenon of creep, or time dependent deformation, in the experimental portion. The longer they left the weight hanging, the more the chord would stretch. The students were also required to develop a "normalized" plot of the force-deformation curve. A percent stretch was calculated which was really the strain in the bungee chord, although the term strain was not used. The force was "normalized" by the number of strands, so the force/strand was plotted. This, of course, is analogous to stress, although again the word stress was not used. The students begin to see the value of "normalized" terms for communicating results. A typical "normalized" plot is shown in Figure 1.

The basis of the solution was energy methods. At the bottom of the drop, the potential energy of the

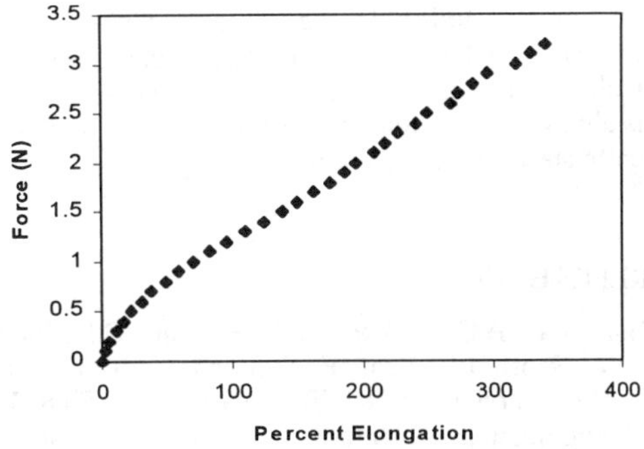

Figure 1. Normalized force-deformation curve for bungee cord.

bungee cord equals the potential energy of the total mass at the drop height, i.e., conservation of energy is achieved. To account for the nonlinear force-deformation curve of the bungee cord, the students used various polynomial curve fits of the data until they were confident in the level of precision of the model. In the MATLAB program, the "best fit curve" was numerically integrated to calculate the potential energy at full extension. Due to the nonlinearity, an iterative solution was required. The input parameters were the drop mass and distance. The MATLAB program output the length and number of cords required to safely drop the sandwich bag containing the eggs.

The objective of this exercise is to demonstrate to the students that the basic concepts that apply for linear systems can be extended to non-linear systems with the use of the computer. In addition, the students gain an appreciation of the ease of use of approximating a system as linear in order to understand the concept. They also understand that many "real" systems are non-linear but that the concepts and techniques used for linear systems can be extended and modified to solve these problems.

4 CONCLUSIONS

Advanced concepts in numerical analysis typically used in geomechanics can be introduced to freshman in simple terms. One means of doing this is the use of team projects, which enable larger projects to be assigned. Several projects that have been used in the *engage* freshman engineering program at The University of Tennessee are described. The projects range from teaching the fundamentals of mechanics, such as free body diagrams, to advanced concepts such as nonlinear behavior, energy methods, and iterative solutions. By integrating both analytical and experimental aspects of the projects, students see the relationship between the two. The experimental portion also adds a degree of fun to the project. The test day, whether it be the "Robusto" chair, the tractor pull, the "Trouble Gorge" bridge, or the bungee drop, is always an exciting day for the students, which adds enthusiasm to the program.

REFERENCES

Gilliam, F., P.G. Klukken,, J.R. Parsons, C.D. Pionke, T. Scott, J.E. Seat, F. Symonds, F.E. Weber, & D.C. Yoder, 1998. "Renovating the First Year Engineering Program at the University of Tennessee," *Proceedings of the Frontiers in Education Conference*, Phoenix, AZ, November, 1998, 814-819.

Knight, D., W. Poppen, P.G. Klukken, J.R. Parsons, & J.E. Seat 1998. "Training Upperclassmen to Facilitate Freshman Design Teams," *Proceedings of the ASEE Annual Meeting*, Seattle, WA.

Pionke, C.D., Parson, J.R., Seat, J.E., Weber, F.E., and Yoder, D.C. 1999a. "Integration of Statics and Dynamics in a Hands-On Project-Oriented Environment," *Proceedings of the ASEE Annual Meeting*, Charlotte, NC.

Pionke, C.D., Weber, F.E., Parsons, J.R., Seat, J.E., Yoder, D.C. 1999b. "A Strategy for Ensuring Minimum Competency in the Use of Engineering Computer Skills." *Proceedings of the Frontiers in Education Conference*, San Juan, PR.

Yoder, D.C., Parsons, J.R., Pionke, C.D., Weber, F., "Hands-On Teaching of Engineering Fundamentals," *Proceedings of the ASEE Annual Meeting*, Seattle, WA, June, 1998.

Computer Methods and Advances in Geomechanics, Desai et al. (eds) © 2001 Balkema, Rotterdam, ISBN 90 5809 183 X

A geomechanics-based educational software environment

Timothy Robert Wyatt
Applied Research Associates, Raleigh, N.C., USA

Emir Jose Macari
Department of Civil and Environmental Engineering, Louisiana State University, Baton Rouge, La., USA

J. David Frost
School of Civil and Environmental Engineering, Georgia Institute of Technology, Atlanta, Ga., USA

ABSTRACT

A virtual reality soil laboratory environment, the Visual Immersive Geotechnical Interactive Laboratory (VIGIL), has been developed and used at Georgia Tech as an authentic, interactive environment for soil mechanics education. The VIGIL tool, shown in Figure 1, is intended to present advanced computational models for soil behavior in a concrete, virtually real context, in order to permit students to obtain "hands-on" interaction with theoretical concepts that would normally be accessible only in equation form. It is believed that this will enhance the development of intuitive understanding of soil behavior that is such a desired quality for the practicing engineer, and will aid in the transfer of this conceptual understanding of theoretical principles to practical applications such as the prediction of soil response to actual engineering activities. While the intent is not (necessarily) to convey the intricacies of the underlying constitutive model, it is supposed that the ability to interact with the model in goal-directed activity will reinforce classroom instruction in soil behavior. A scaffolding framework is provided to introduce the student gradually to soil mechanics concepts in the context of soil laboratory test procedures via "training" modules for isotropic consolidation, consolidated drained, and consolidated undrained triaxial shear testing. These three training modules corresponded with the learning objectives shown in Table 1. This paper describes only the consolidated drained (CD) training module and presents assessment data obtained for that module.

Table 1. Training Modules Developed for VIGIL, and their Associated Goals and Objectives.

Lab Test	Capability Introduced	Educational Goal	Learning Objective
Isotropic Consolidation	isotropic pressure	understanding of consolidation	interpretation of consolidation parameters
Consolidated Drained Shear	deviatoric stress	understanding of shear behavior	interpretation of shear strength parameters
Consolidated Undrained Shear	drainage	understanding of effective stress	interpretation of effective strength parameters

During the Fall 1999 academic semester, VIGIL was assessed in conjunction with CE 4400, the undergraduate *Geosystems Engineering* course. Seventy students were enrolled in this course. On three separate occasions in lieu of the regularly scheduled class lecture, CE 4400 students met in a computer laboratory to undergo training sessions on VIGIL. The sessions opened with a short review lecture of material previously covered in the class, followed by a five-minute anonymous pre-test quiz to gauge overall student understanding with respect to the corresponding learning objective. Students were then asked to complete the VIGIL training module for the associated laboratory test. Successful completion of the training module required the student to answer quizzes, administered by the software program, to demonstrate proficiency with the same concepts assessed on the pre-test quiz.

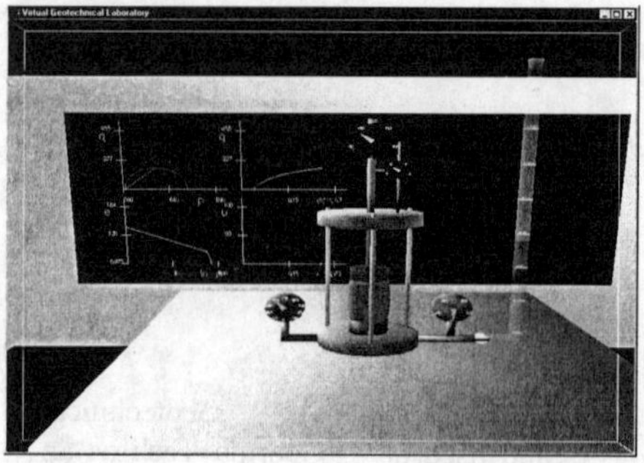

Figure 1. Screen View of VIGIL.

Following the training session, students were given an assignment to be completed using the VIGIL software in non-training mode over a one- to two-week period. Each assignment required the student to generate laboratory data using VIGIL, interpret the simulated results, and compare to actual laboratory data. These assignments also assessed student understanding of some of the same concepts from the pre-test, after undergoing training and after active generation of simulated data. Traditional assessment methods were employed in the form of written examinations before and after each training session. These are the conventional measures of proficiency for students in soil mechanics, and therefore cover a wide range of topics to assess student understanding of soil behavior.

Log files recorded all student activity on the software, both during training and during completion of the assignment. Log file results were used to determine the level of individual student usage of VIGIL, in both the training modules and project performance. These results were used to group students according to level of completion of each training module.

CONSOLIDATED DRAINED LABORATORY ASSIGNMENT

Of the 49 students who attended the CD training session, 46 students were observed to have undergone CD training. These 46 students spent 26 minutes on average using the training module. The training session followed standard lab test procedures, encouraging the student to perform three drained shear strength tests at various levels of confinement, interpret the shear strength at failure for each test, and use the results to generate a failure envelope.

Twenty-five of these students successfully completed CD training, meaning that they completed all three tests in the CD test series (normally consolidated, lightly overconsolidated, and highly overconsolidated) and successfully answered the quiz at the end of each test which required them to interpret the shear strength at failure.

At the end of the training session, students were given a "laboratory" assignment to be completed using VIGIL. There were five parts to the assignment. In Part 1, students were given real laboratory consolidation data, from which they were to interpret soil parameters. These parameters were then provided to the VIGIL software for modeling purposes. In Parts 2 – 4, students were advised to perform three CD triaxial shear tests within VIGIL, at different degrees of consolidation (NC, lightly OC, and highly OC), and to report the shear strength at failure.

Part 1 of this assignment dealt with course material (consolidation parameters) that had been introduced in the previous VIGIL training session (isotropic consolidation testing), so students were expected to do well on this part. The curve in Figure 2 was obtained from VIGIL using soil parameters of e_0 = 1.46, p'_0 = 380 kPa, C_C = 0.41, and C_R = 0.11. The actual data values are included on the plot to demonstrate the close fit between actual data and simulation.

Parts 2 – 4 each required the student to perform a CD shear test at a different degree of consolidation. Scoring for these questions was equally allocated to submission of the τ-ε output plot from VIGIL for each test, and to correct interpretation of the shear strength at failure for each test type. Partial credit could be deducted from the plot grade if the plot demonstrated poor testing procedures, such as not permitting excess pore water pressure to dissipate between axial load increments. Partial credit could be deducted from the numerical responses if the student ran the test at a cell pressure that was not commensurate with the type of test (*e.g.*, running a "normally consolidated" test with an OCR > 1), or for improper interpretation of the strength at failure. Students were advised in both the pre-training review lecture and during the training module itself to interpret failure as either the point of maximum shear resistance, or the point of 20% axial strain.

Finally, in Part 5, students were asked to report the failure envelope parameters obtained from the three tests.

Twenty-six students were observed to have used VIGIL to perform the assignment. These 26 students spent an average of 34 minutes using VIGIL for the assignment. Because each student was requested to submit a plot generated from the software, and 55 assignments were submitted with plots, this indicates that students either worked in groups to generate the output, with only one student logged into the program, or some students submitted copies of work performed by others.

A possible solution for the NC test, generated using VIGIL for the estimated parameters shown above, is shown in Figure 3. Only seven students performed the test at a cell pressure that was significantly less than their estimate of preconsolidation pressure. The mean cell pressure used for this test was σ_c = 594 kPa. The mean value of τ_f reported was 261 kPa. This value is somewhat larger than the value computed in the example solution, partially due to the larger average test cell pressure used by students, and partially due to misinterpretation of the point of failure. Many students interpreted failure as the point of ultimate resistance, even if that corresponded to an axial strain of much greater than 20%.

NC Test:

σ_3 = 400 kPa (OCR = 1)

$\tau_{fail} = q_{fail}$ = 164 kPa (Figure 3)

$\sigma_1 = 2\ \tau_{fail} + \sigma_3$ = 728 kPa

$p_{fail} = (\sigma_1 + \sigma_3)/2$ = 564 kPa

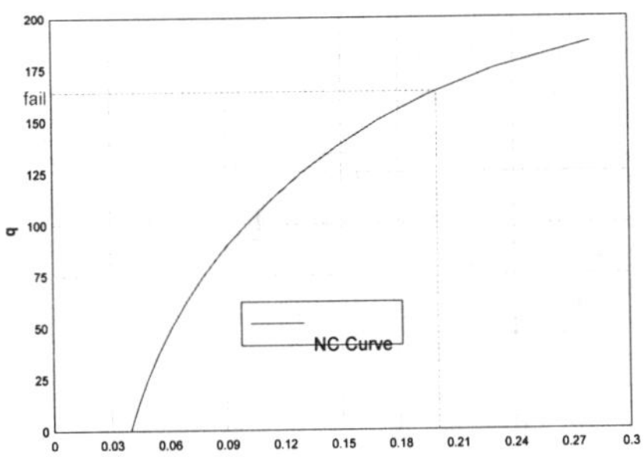

Figure 3. NC Test Output Generated by VIGIL, with Failure Interpreted at 20% Axial Strain.

A possible solution for the lightly OC test, generated using VIGIL for the estimated parameters shown above, is shown in Figure 4. Four students performed the test at a cell pressure that was greater than their estimate of preconsolidation pressure, and eight students performed the test at a cell pressure less than half of their estimate of preconsolidation pressure (corresponding to a highly overconsolidated OCR > 2). The mean cell pressure used for this test was σ_c = 269 kPa. The mean value of τ_f reported was 123 kPa. These values are certainly comparable to those in the example solution.

Lightly OC Test:

σ_3 = 250 kPa (OCR = 1.5)

$\tau_{fail} = q_{fail}$ = 112 kPa (Figure 4)

$\sigma_1 = 2\ \tau_{fail} + \sigma_3$ = 474 kPa

$p_{fail} = (\sigma_1 + \sigma_3)/2$ = 362 kPa

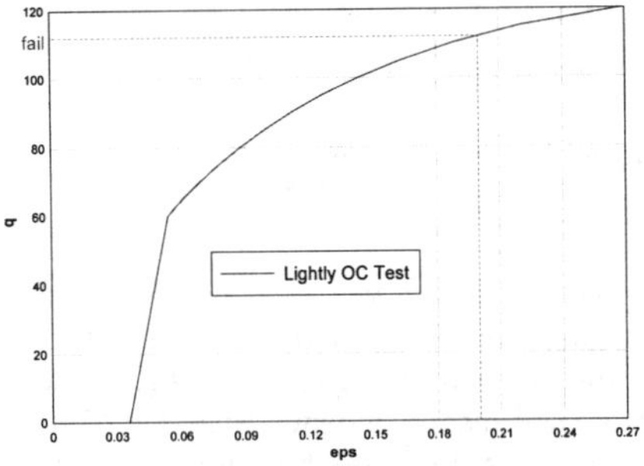

Figure 4. Lightly OC Output from VIGIL, with Failure Interpreted at 20% Axial Strain.

A possible solution for the highly OC test, generated using VIGIL for the estimated parameters shown above, is shown in Figure 5. Eight students performed the test at a cell pressure greater than half of their estimate of p'_0 (corresponding to a lightly overconsolidated OCR < 2). The mean cell pressure used for this test was σ_c = 161 kPa. The mean value of τ_f reported was 84 kPa. This value is somewhat larger than the value computed in the example solution, partially due to the larger average test cell pressure used by students, and partially due to misinterpretation of the point of failure. Some students interpreted failure as the point of 20% strain, rather than the point of peak shear stress.

The average score of the final question (interpret strength parameter) was 1.28/2, or 64%. This compares to 43% success immediately prior to CD training and 51% success at the conclusion of CD training. One possible solution for the strength parameters, obtained using the simulated test results shown above, is shown in Figure 6. Values reported by students could vary widely, depending on their estimated values of soil parameters and the cell pressures at which the three tests were performed.

Heavily OC Test:

$\sigma_3 = 100$ kPa (OCR = 3.8)

$\tau_{fail} = q_{fail} = 67.5$ kPa (see Figure 5)

$\sigma_1 = 2\,\tau_{fail} + \sigma_3 = 235$ kPa

$p_{fail} = (\sigma_1 + \sigma_3)/2 = 167.5$ kPa

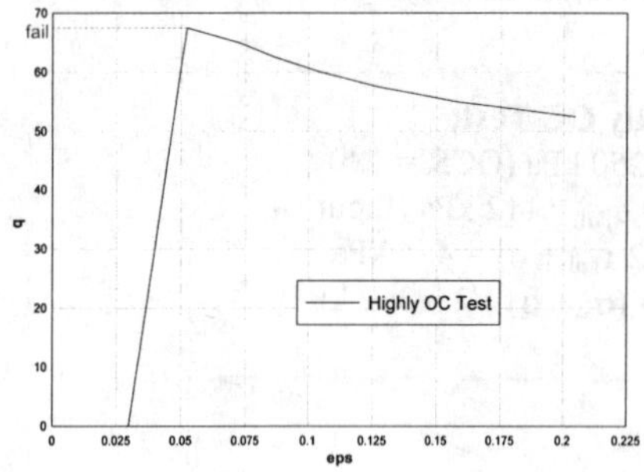

Figure 5. Highly OC Output from VIGIL, with Failure Interpreted as Peak Shear Stress.

The average values reported were $c = 18.5$ kPa and $\phi = 24.5°$. The main reason for the differences between the example solution values and the average student responses was that many students tried to "force" their failure envelope to cross through a cohesion intercept of $c = 0$ kPa. Thirty-one of the 59 students (53%) reported $c = 0$ kPa.

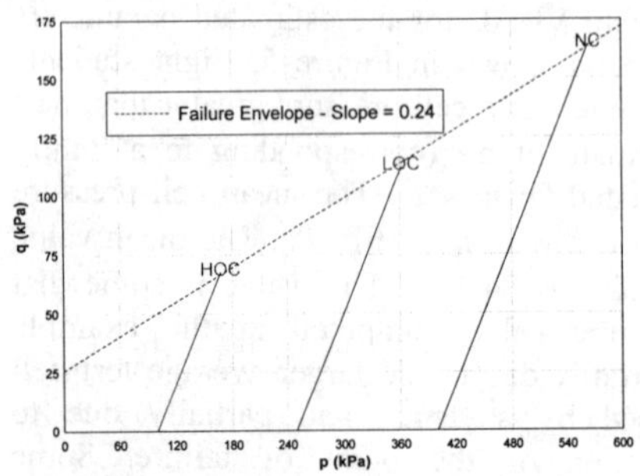

$a = 26$ kPa (q-intercept)

$\phi = \sin^{-1}(0.24) = 14°$

$c = a / \cos\phi = 27$ kPa

Figure 6. Interpretation of Failure Envelope for the Simulated CD Triaxial Test Series.

TRADITIONAL ASSESSMENT

A traditional written exam was given to the students after the CD training session. This was the third course exam, and covered topics including elastic (low strain) and plastic (large strain and failure) response of soils, stress paths, and strength properties of soils. As shown in Table 2, there was a definite correlation between level of completion of CD training in VIGIL and performance on the exam, which dealt with strength-related issues. Students who successfully completed CD training received an average grade of 74.8 on the exam. Students who attempted CD training but did not successfully complete the entire test series received an average grade of 70.0 on the exam. Students who did not attempt the training module at all received an average grade of 67.3. The average score for all students on the exam was 70.8.

Course exam scores were employed in order to gauge the influence of VIGIL with respect to the traditional means of measuring student understanding of soil mechanics. Exams assess student knowledge over a broad range of soil behavior concepts, and thus these scores represent a more "holistic" understanding of soil mechanics than do the specific concepts assessed on the pre-test quizzes and project submissions. These scores were employed in the evaluation of VIGIL in recognition of the fact that specific soil mechanics concepts do not exist in isolation independently of one another (*i.e.*, the role of isotropic confinement, shear stress, and excess pore pressure all have an interrelated influence on overall soil behavior). The degree of student participation and level of completion of each VIGIL training module was shown to correspond to improved average performance on course exams, even after removing the statistical correlation with prior exam scores in order to account for student performance that would have been "expected" in absence of the VIGIL learning opportunities.

Table 2. Grades after VIGIL Session for Various Levels of CD Training

Level of CD Training	# Student	*After CD Training Session*	
		Grade	Exam Grade
Successful Completion	25	9.2	74.8
Incomplete Attempt	21	8.0	70.0
No Attempt	24	7.6	67.3
All Students	70	8.3	70.8

REFERENCES

1. Arduino, Pedro, Augusto Op den Bosch, and Emir Jose Macari, "Geotechnical triaxial soil testing within virtual environment," *Journal of*

Computing in Civil Engineering, American Society of Civil Engineers, v. 11, n. 1, January 1997.

2. Whitaker, William, "Teaching soil mechanics laboratories with computer assistance," *Computing in Civil Engineering*, ASCE, Washington DC, 20-22 June 1994.
3. Soloway, Elliott, Mark Guzdial, and Kenneth E. Hay, "Learner-centered design: the challenge for HCI in the 21st century," *Interactions*, April 1994.
4. Jackson, Shari L., Joseph Krajcik, Elliot Soloway, "The design of guided learner-adaptable scaffolding in interactive learning environments," ACM Computer-Human Interactions Conference (CHI98), Los Angeles CA, April 1998.

Author index